DICTIONARY OF
ENGINEERING
MATERIALS

About the Authors

Harald Keller is a technical writer and translator with a broad background in mechanical engineering and materials testing. He is author of several dictionaries on abbreviations and acronyms and materials-related databases.

Uwe Erb is a professor in the Department of Materials Science and Engineering at the University of Toronto, Canada. He has published widely in the fields of materials science, nanomaterials, metallurgy and applied physics, and is author of over 200 scientific and technical articles and several books and book chapters on grain boundaries and nanocrystalline materials.

DICTIONARY OF ENGINEERING MATERIALS

Harald Keller

Uwe Erb

⟨W⟩WILEY-INTERSCIENCE

A JOHN WILEY & SONS, INC., PUBLICATION

This text is printed on acid-free paper. ∞

Copyright © 2004 by John Wiley & Sons, Inc. All rights reserved.

Published by John Wiley & Sons, Inc., Hoboken, New Jersey.
Published simultaneously in Canada.

For general information on our other products and services please contact our Customer Care Department within the U.S. at 877-762-2974, outside the U.S. at 317-572-3993 or fax 317-572-4002.

Wiley also publishes its books in a variety of electronic formats. Some content that appears in print, however, may not be available in electronic format.

Library of Congress Cataloging-in-Publication is available.

ISBN 0-471-44436-7

Printed in the United States of America.

10 9 8 7 6 5 4 3 2 1

Contents

Major Groups of Materials Covered

Preface

This *Dictionary of Engineering Materials* is a one-stop reference book with close to 40000 entries for generic and proprietary engineering materials providing answers to questions, such as: "What type of material is it?" "What is its composition?" "What are its general properties?" and "What are its applications?"

One of the main attributes that distinguishes man from the animal is his ability to make tools, parts, complex objects and whole engineering systems. The substances available to him for this purpose are referred to as "engineering materials." The role of materials throughout the long history of mankind was so significant that anthropologists and archeologists identified the times early in the development of human cultures by the most important materials used, e.g., the Stone Age, Bronze Age and Iron Age. While early cultures were mostly limited to a precious few materials, throughout the ages man has learned how to mine, extract and process materials and develop technologies for synthesizing a wide range of different engineering materials: metals, polymers, ceramics, semiconductors and composites.

In the broadest sense, an "engineering material" is a material used in the manufacture of technical products including machinery, equipment, installations and systems, structural members, machine parts, instruments, tools, buildings, bridges, highways, land and space vehicles, ships, and consumer products. Important types of engineering materials include alloys, ceramics, concrete, glass, plastics, elastomers, composites, semiconductors, superconductors, and natural substances, such as cotton, wood, stone, and animal skin or hide. They are characterized by properties, such as strength, hardness, toughness, dimensional stability, formability and machinability, chemical, electrical and thermal properties, heat and flame resistance, and resistance to corrosion, acids, and solvents. Engineering materials are often identified on the basis of the particular engineering discipline in which they are used, e.g., semiconductors as engineering materials for electrical engineering and electronics, and cement and concrete as engineering materials for architecture, building construction and civil engineering.

To compile a comprehensive Dictionary on engineering materials the authors had to put aside their professional biases for one or the other material groups. They felt that the audience would be best served by an objective approach based on a balanced treatment of the full spectrum of engineering materials. Consequently, this Dictionary does not only cover traditional materials, but also includes more recently developed engineering materials, such as bioengineered materials, diamond-like carbon, dendrimers, fullerenes, giant and colossal magnetoresistance materials, grain-boundary-engineered materials, nanomaterials, optical and optoelectronic materials, and photonic materials. In addition to listing the final engineering materials or end products, some of the constituent materials used in their preparation are also included. However, since this Dictionary is intended to primarily deal with engineering materials and not with chemicals, biochemicals, and precursor materials, the listing of such ingredients is not intended to be comprehensive. With the exception of a few cases, gases, liquids, solvents, organic and inorganic synthesis chemicals, monomers, crosslinking agents, organometallics, intermediates, oils, greases, and laser dyes are not included in this Dictionary. There are already many excellent dictionaries on the market that cover these substances. However, because of their relatively large importance as raw materials, minerals and the different species of wood are extensively covered in this Dictionary. For a comprehensive listing of the types of materials included in this Dictionary the reader is referred to page vii (*Major Groups of Materials Covered*).

To achieve their goal of a balanced coverage, the authors have tried their best to treat engineering materials used in the construction of highways, bridges, buildings, plant, machinery, aircraft, automobiles, ships or consumer goods as being equal in importance to engineering materials used in the realization and perfection of the information superhighway. In the fascinating and diverse mosaic developed over decades and centuries, each engineering material has its rightful place. Consequently, in this Dictionary, the reader can find engineering materials that are or have been of interest in science, engineering, industry and commerce over the past several decades including the "advanced" engineering materials developed during the mid-20th Century and the 1960s, 1970s and 1980s, and are now deemed "traditional," as well as the novel engineering and engineered materials introduced throughout the 1990s and up to and including the year 2003. Nevertheless, some readers may ask the question why both older and newer materials are covered in the same Dictionary. Aren't "older" materials obsolete by today's standards? Absolutely not! Although some of them may no longer be made today, many of the infrastructures, buildings, plants, machines and devices built several decades ago by civil, electrical, mechanical, structural and other engineers in this country and abroad are still in existence today, and the materials used in their construction were the "advanced" materials of that time. Who would have thought in 1907, when L.H. Baekeland invented a synthetic resin by reacting phenol with formaldehyde, later known as *Bakelite*, that phenol-formaldehyde plastics would still be used a century later? Who could have even imagined early in the 20th Century that plastics, considered "curiosities" then, could replace metals and alloys in many engineering applications just a few decades later? Conversely, will the novel materials of today still be around half a century from today? In all probability they will, but they may then be known as "traditional materials" and complemented by the novel materials of that time.

This Dictionary is a multidisciplinary one providing information on a wide range of different engineering materials used in many different fields of science, engineering, industry and commerce. From alloys to zeolites, abrasives to *Zytel* polyamides, and ABS plastics to zirconia-toughened ceramics, the reader can find engineering materials used in (i) architecture, building construction and civil engineering, (ii) biomedicine, biotechnology and dental engineering, (iii) aeronautical and aerospace engineering; (iv) chemical and nuclear engineering; (v) electrical engineering, electronics, telecommunications and information technology; (vi) automotive and mechanical engineering; (vii) materials science and engineering; (viii) environmental engineering, and many other fields and branches. As such this Dictionary is designed to serve a very broad audience ranging from academics and students to consulting and practicing engineers, technologist, technicians, purchasing agents, science writers and journalists, technical and scientific translators, and the general public interested in understanding scientific and technical terminology. In order to best serve the different needs of such a large and diversified audience, the authors had to find a compromise between the language of the academic and that of the nonacademic. Consequently, as is the case with any specialized dictionary, readers with limited technical background will find some of the entries far too technical, while the expert in a particular area of materials may have expected much more depth and breadth of coverage.

In order to make the most of this book, the reader is reminded that it is an engineering materials dictionary, i.e., an alphabetical compilation of definitions and explanations of engineering materials, and not an encyclopedia or handbook keeping the reader's attention fixed on one engineering material or material group. A multitude of excellent encyclopedias, handbooks, property data books, databases and websites dealing with materials classes, groups, families or grades, such as metals and alloys, polymers, ceramics, composites, liquid crystals, or photonic materials, are available (see *Appendix*), and readers who are particularly interested in such specific data are encouraged to consult these for detailed information. By definition, the first and foremost purpose of a dictionary is to record usage of a term, not to give lengthy descriptions on virtually every aspect of it, and this Dictionary has been compiled accordingly. However, for some generic and proprietary engineering materials it was deemed necessary to include particular types or grades, because they are the commercially most important members.

An important feature of this Dictionary are the definitions for thousands of proprietary engineering materials. These have been selected based on their nature, properties, applications and overall importance, and the authors do not favor one supplier's products over those of another. Entries for materials found in the literature as trademarks or trade names are so identified. Wherever possible, the particular trademark or trade name is followed by a company name and, in many cases, the company specified is also the original and/or current proprietor or holder of the trademark or trade name. However, it is important to note that in certain cases the company specified may not be the holder of the trademark or trade name, but the licensee in a particular country. Although every care has been taken to ensure the correctness of proprietary information, owing to the facts that companies go out of business, or merge with or are acquired by other companies, *the authors do not assume any legal liability or responsibility for the accuracy, completeness or usefulness of any trademark or trade name listed in this Dictionary. No entry shall be regarded as affecting the scope, validity or ownership of any trademark or trade name.* In this respect, the authors found that even the most up-to-date handbooks, databooks, directories and company websites cannot always keep up with the pace of such proprietary changes.

The compilation of the many exciting engineering materials and their presentation in a comprehensive and useful Dictionary has been a monumental, yet challenging and exciting task. However, as with so many other reference books in science and engineering, some members of the audience may not be in total agreement with some of the definitions given in this Dictionary. The authors welcome and encourage any feedback from the audience. Suggestions that are deemed constructive in making reprints or future editions of this reference book even more useful will be taken into consideration.

HARALD KELLER
UWE ERB

Ontario, Canada
January 2004

Guide to This Dictionary

This guide deals with the different kinds of entries and their arrangement and presentation in this dictionary. To use this book most effectively, it is recommended to read this section first.

1. Arrangement of Entries

The terms in this dictionary are given in alphabetical order, letter by letter, with each key entry printed in boldface type and the definition that follows it in lightface type. Uppercase spellings precede lowercase forms, and entries containing spaces, ampersands, slashes, dashes or numerals follow those without such characters.

Italic type is often used in the following instances: indication of scientific (Latin or Greek) names of species and genera of plants and animals (e.g., as sources of natural fibers, leather, or wood), repetition of a key entry within a definition, listing of special grades of a material as part of a key entry, and specification of synonyms for a key entry (See 3.1 Directional Cross-References).

In general, entries containing Arabic or Roman numerals are alphabetized as if the numerals were spelled out: *IV-VI compounds* ("four-six compounds") comes after *Four-Ply* and before *Four Star,* and *1-2-3 superconductors* ("one-two-three superconductors") follows *One Ton Brass* and precedes *One-Up Bond F.* An exception to this rule is sometimes made for entries consisting of number/letter combinations: *3D photonic crystal* (three-dimensional photonic crystal) is found at the beginning of letter section D preceding all other entries.

In general, entries containing Greek letters are alphabetized and spelled out: *beta brass* follows *Betabrace* and precedes *Beta-C.* This also applies to Greek prefixes for chemical elements, inorganic compounds and metallurgical phases: α-*titanium alloys* is spelled out as *alpha titanium alloys* and comes after *alpha titanium* and before *Alphatized steel.* However, in alphabetizing, many of the prefixes for organic and organometallic compounds are disregarded, since they are not considered an integral part of the key term. Examples include *alpha (α), beta (β), gamma (γ), meta- (m-), ortho- (o-), N-, tert-, sec-, sym-, cis-* and *trans-* as well as all numerals denoting structure.

Wherever possible throughout this dictionary, a basic format for listing the various engineering materials has been adopted: each key entry is followed by a definition which in most cases is followed by important material properties and applications and, if applicable, relevant synonyms, symbols and abbreviations. When there are more than one definition for a key entry, each definition is introduced by a numeral in parentheses, e.g., (1), (2), or (3).

Materials properties typically given include density (in g/cm^3 unless otherwise stated), melting and boiling point (in °C and °F) and refractive index. Other structural or property data given for many entries include: crystal system, crystal structure, dimensions (usually in SI and imperial units), service (or use) temperature (in °C and °F), flash point (in °C and

°F), superconductivity critical temperature (in K), tensile and yield strength, hardness, dielectric strength, and band gap. Chemical formulas are usually given as molecular (or atomic) formulas. In some cases, however, it was deemed more appropriate to use structural formulas. For wood species and other natural products (e.g., natural resins and gums) the geographical origin is introduced by the word "Source" and for minerals by the word "Occurrence".

2. Guide Words

A set of guide words, separated by a bullet, appears at the top of each page indicating the alphabetically first and last key entries on that page. For example, on page 16 the guide words are *admiralty metal* and *advanced materials* indicating that these two entries and all those falling alphabetically between them are found on that page.

3. Cross-References

The following cross-references are used:

3.1. Directional cross-references: These begin with "See also..." and direct the reader to look elsewhere in this dictionary for additional information. For example:

carbon fibers. ...See also graphite fibers.

Also, in some cases italic type refers to terms under which more information can be found. For example, in the definition of "nanointermetallics" the term "intermetallic compounds" is given in italic type to direct the reader to this term for further information.

3.2. Synonymous cross-references: These are introduced by "See..." and direct the reader to the definition of an entry. For example:

aluminum oxide. See alumina.

4. Synonyms, Symbols, and Abbreviations

The definition of an engineering material is often followed by relevant synonyms, symbols and/or abbreviations. Synonyms are introduced by "Also known as..." and followed by the corresponding term or terms in italic type:

condensation polymer. ... Also known as *step-reaction polymer; step-growth polymer.*

In general, the synonym and its definition can be substituted for the term looked up, i.e., the term "condensation polymer" and its definition can be substituted for the term "step-reaction polymer." If applicable, symbols and abbreviations for a particular term appear after the definition and are introduced by the words "Symbol" and "Abbreviation", respectively. For example:

iron. .... Symbol: Fe.

magnesia-stabilized zirconia. .... Abbreviation: MSZ.

A

A-Alloy. (1) Trade name of the National Physical Laboratory (UK) for wrought aluminum alloys developed by the Alloys Research Committee of the British Institution of Mechanical Engineers (BIME). *A-Alloy 1* contains 20% zinc and 3% copper, and is non-age-hardenable, and *A-Alloy 2* is age-hardenable and contains 4.7% copper, 1.85% nickel, 1.5% lead, 1.34% magnesium, and 0.5% iron.

    (2) Trade name of Alloy Engineering & Casting Company (USA) for a casting alloy containing 66-68% nickel, 19-21% chromium, 0.4-0.8% carbon, and the balance iron. Used for heat- and corrosion-resistant parts.

ABA. Trademark of Wesgo Metals, Division of Morgan Advanced Ceramics, Inc. (USA) for a series of active brazing alloy and filler-metal systems.

ABAC. Trademark of Wesgo Metals, Division of Morgan Advanced Ceramics, Inc. (USA) for active brazing alloys that are specially conditioned ceramic bodies used for sealing metals to ceramics by single-step brazing. Usually used with *ABA* active brazing alloys. *Note:* ABAC stands for "Active Brazing Alloy Compatible Ceramics."

abaca. See Manila fibers.

abaca fibers. See Manila fibers.

ABA copolymer. A block copolymer that has three sequences (A, B, A), but only two domains (A and B). See also block copolymer; copolymer.

Abassi cotton. A white, pure variety of *Egyptian cotton* available as lustrous staple fibers with a length of approximately 30 mm (1.2 in.).

Abbcut. Trade name of Abbott Ball Company (USA) for steel media used in vibratory mass-finishing operations.

abbot. See monk's cloth.

Abcite. Trademark of E.I. DuPont de Nemours & Company (USA) for powder coating resins.

Abco. (1) Trade name of Abbott Ball Company (USA) for a range of steel burnishing balls and shapes used in metal finishing.

    (2) Trade name of A & B Plastics, Inc. (USA) for flame-retardant polyurethanes.

abele. See white poplar.

abelsonite. A pink-purple or dark reddish brown mineral composed of nickel porphyrin, $C_{31}H_{32}N_4Ni$. Crystal system, triclinic. Density, 1.3-1.4 g/cm³. Occurrence: USA (Utah). Used for electronic applications

abernathyite. A yellow mineral of the meta-autunite group composed of potassium uranyl arsenate trihydrate, $KUO_2AsO_4 \cdot 3H_2O$. Crystal system, tetragonal. Density, 3.3-3.6 g/cm³; refractive index, 1.608. Occurrence: USA (Utah).

abestrine. A term sometimes used for fabrics made from asbestos.

Abestron. Trade name for asbestos fibers used for making fire- and heat-resistant fabrics.

Abex. Trade name of Abex Corporation (USA) for a series of copper casting alloys, leaded and unleaded bronzes, brasses (containing varying amounts of iron, aluminum, nickel, zinc, lead, tin, and/or chromium), electrode and blast-furnace coppers, etc. See also copper casting alloys; bronze; brass.

Abfab. Trademark of Abco Industries Limited (USA) for non-woven fabrics based on polyester. Used in the manufacture of boats, skis, shower stalls, etc., and as reinforcing layers in polyester-resin composites.

abherent. A substance in the form of a dry powder, solution, suspension, or soft solid which prevents or greatly reduces the adhesion of a material to itself, or to another material. Examples of such substances include mica, diatomaceous earth, bentonite-water and soap-water solutions, tallow waxes, fluorocarbons, etc. Used as dusting agents and mold washes in the adhesives, plastics and rubber industries. Fluorocarbon resin coatings are used on cooking utensils.

abhesive. A material, usually in the form of a coating or film, which prevents or greatly reduces adhesion to other surfaces. Examples include mold release agents and parting agents used to minimize sticking and facilitate heat sealing.

ablation material. A protective material, such as an all-carbon composite or a glass- or ceramic-reinforced plastic, that is capable of rapidly dissipating high amounts of heat from a substrate by allowing its surface region to be removed by a pyrolytic process involving melting or sublimation. It has low thermal conductivity, high thermal capacity, high heat of fusion and evaporation, high heat of dissociation of its vapors, and low material loss. Used on spacecraft nose-cones for maximum-temperature exposure up to 1650°C (3000°F). Also known as *ablative agent; ablative material; ablator.*

ablation plastics. Plastics, usually glass- or ceramic-reinforced, which are capable of rapidly dissipating high amounts of heat with low material loss and char rate from a substrate by allowing its surface region to be removed by a pyrolytic process. For example, fluorocarbon polymers were used on early spaceships. Also known as *ablative plastics.*

ablative agent. See ablation material.

ablative material. See ablation material.

ablative plastics. See ablation plastics.

ablator. See ablation material.

Ablebond. Trade name of Ablestik Laboratories (USA) for a series of epoxy and polyimide compounds used as metal-glass adhesives.

Ablefilm. Trademark of National Starch & Chemical Corporation (USA) for a series of epoxy and polyimide compounds supplied in film form, and used as adhesives.

Ablex. Trademark of Dart Industries, Inc. (USA) for special polyolefin extrusion blends.

AboCast. Trade name of Abatron, Inc. (USA) for an extensive series of solventless epoxy resins and compounds for casting, patching and moldmaking applications.

AboCure. Trade name of Abatron, Inc. (USA) for an extensive series of hardeners for epoxy resins.

Aboglas. Trade name for sheeting made of asbestos and glass fibers, and used in thermal insulation.

AboWeld. Trade name of Abatron, Inc. (USA) for an extensive series of epoxy resins.

Abracorr. Trade name of Société Nouvelle des Acieries de Pompey (France) for a series of abrasion- and oxidation-resistant steels containing 0.1% carbon, 4% chromium, 0.75% nickel, 0.65% aluminum, 0.5% manganese, 0.2% molybdenum, and the balance iron. They possess good weldability and formability, and are used for structural applications.

abradables. A group of soft coatings of aluminum, polyester, graphite, etc., which are applied to workpieces to provide clearance control and allow rotating parts to "machine in" their own tolerances during operation.

Abradur. Trade name of Société Nouvelle des Acieries de Pompey (France) for a series of plain- and medium-carbon steels and wear-resistant manganese steels. Used for machinery parts, fixtures, tools, etc.

Abrasalloy. Trade name of Atlantic Steel Corporation (USA) for a tough, wear-resistant steel containing 0.35% carbon, 1% chromium, 0.75% manganese, 0.35% silicon, 0.25% molybdenum, and the balance iron. Used for coal screens, dredge pumps, and hammers for crushers.

abrasion-resistant cast irons. A group of cast irons including (i) *Pearlitic white irons* with 2.8-3.6% carbon, 0.2-2.0% manganese, 0.3% phosphorus, 0.15% sulfur, 1.6% silicon, 2.5% nickel, 1.0-3.0% chromium and 1% molybdenum, and the balance iron; and (ii) *Martensitic white irons* with 2.3-3.7% carbon, 0.5-1.5% manganese, 0.3% phosphorus, 0.1% sulfur, 1% silicon, 1-28% chromium, 0.5-5.0% nickel and 1% molybdenum, and the balance iron. Used for strong, tough, abrasion-resistant castings, malleable-iron castings, chilled-iron rolls, and grinding balls. Also known as *abrasion-resistant irons*.

abrasive ceramics. A group of ceramic materials including alumina (corundum), silica, and tungsten carbide which possess high hardness, toughness and wear resistance, and are used to cut, grind or wear away substances softer than themselves. See also abrasives.

abrasive fabrics. Textile fabrics, usually cotton, that are used as backings for grinding and polishing abrasives. The abrasives are usually bonded to the fabrics by means of an adhesive resin.

abrasive garnets. Crushed, ground, separated and graded *garnets* derived from minerals, such as *almandite, andradite, rhodolite* or *uvarovite*. Used as abrasives in garnet paper and wheels for woodworking operations.

abrasive grit. Abrasive particles used in the manufacture of grinding wheels. They are sorted into various sizes by passing through screens whose mesh sizes determine the grit number.

abrasive paper. A coated abrasive product in the form of a tough, strong paper or cloth (e.g., kraft paper or cotton) to which abrasive grains (e.g., silica, emery, flint or garnet) have been bonded by means of an adhesive, such as glue or resin. Used for grinding and polishing operations. Examples are sandpaper and emery paper. See also emery paper; garnet paper; flint paper; sandpaper.

abrasive powder. (1) An abrasive material, such as alumina, silica, or garnet, made into a powder of varying mesh size for grinding, polishing or similar applications. See also abrasives.

(2) An impure corundum, or aluminum oxide powder used as an abrasive in grinding and polishing operations. Usually available in mesh sizes from "Coarse" (8 to 24 mesh) to "Fine" (24 to 240 mesh). See also emery powder.

abrasives. A class of finely divided, hard refractory materials (hardnesses from 6 to over 10 Mohs) which cut substances softer than themselves (e.g., glass, plastics, stone, or wood). They may be natural materials, such as sand (silica), tripoli, garnet, corundum, emery, pumice, rouge (iron oxide), feldspar or diamond dust, or synthetic materials, such as silicon carbide, boron carbide, cerium oxide or fused alumina. The synthetic materials are commonly sold under trade names or trademarks, such as *Aloxite, Carborundum* or *Crystolon. Abrasives* are used in various forms (powder or block, bonded to paper or cloth, suspended in liquid or paste, on grinding wheels, etc.) for cleaning, reducing, grinding, honing, lapping, superfinishing, polishing, pressure blasting, or barrel finishing.

abrasive sand. Any sharp-grained sand, usually graded to a mesh size, which is used for abrasive and grinding operations, such as sandblasting, glass grinding, or stone cutting.

abrasive slurry. A fine abrasive powder (800-1000 mesh) of boron carbide, silicon carbide, aluminum oxide, diamond, etc., mixed with a liquid vehicle, such as water. Used in ultrasonic machining, and in metallography for polishing, and with wire saws for sectioning.

Abrasoweld. Trade name of Lincoln Electric Company (USA) for an abrasion- and impact-resisting cast iron containing 2.1% carbon, 1.1% manganese, 0.75% silicon, 0.4% molybdenum, 6.5% chromium, and the balance iron. Used for hard-surfacing and arc-welding electrodes.

Abrazo. Trade name of British Steel plc (UK) for abrasion-resisting carbon steel plate containing up to 0.6% carbon, 0-0.25% silicon, 0.6-1.5% manganese, up to 0.5% chromium, up to 0.2% molybdenum, and the balance iron. Used for liner plates, construction equipment, storage bins, etc.

Abros. British trade name for a stainless, corrosion-resistant alloy containing 88% nickel, 10% chromium and 2% manganese. Used for resistance components and heating elements.

Absalyn. Trade name of Hercules Inc. (USA) for modified natural resins used in adhesives.

Absafil. Trademark of Wilson-Fiberfil International (USA) for a glass-fiber-reinforced acrylonitrile-butadiene-styrene (ABS) copolymer.

Absaglas. Trademark of Wilson-Fiberfil International (USA) for glass-fiber-reinforced acrylonitrile-butadiene-styrenes.

Absco Metal. Trade name of Abex Corporation (USA) for a series of heat- and wear-resistant high-carbon cast irons containing varying amounts of iron, carbon, silicon, manganese and sulfur. Used for pulleys, gears, cylinders, crankshafts, tool shanks, housings, molds, press parts, forming dies, and water valves.

ABS copolymer. See acrylonitrile-butadiene-styrene copolymer.

Abselex. Trademark of British Celanese (UK) for acrylonitrile-butadiene-styrene (ABS) resins.

Absinol. Trademark of Allied Resinous Products (USA) for acrylonitrile-butadiene-styrene (ABS) resins and products.

Absolute. Trade name for a silicone elastomer used for dental impressions.

Abson. Trademark of Mobay Chemical Corporation (USA) for acrylonitrile-butadiene-styrene plastics and blends with outstanding strength and toughness, good physical and chemical properties, good resistance to heat distortion, poor resistance to flames and some organic solvents, and a maximum service temperature of 120°C (250°F). Used for highway safety devices, lawn and garden equipment, refrigerator linings, appliance housings, automotive trim, cases, piping, impellers, helmets, knobs, handles, and toys.

Absorbacote. Trademark of Sternson Limited (Canada) for a series of interior wall coatings.

absorbent. (1) Any substance or compound that takes up liquid or vapor, and changes physically or chemically during the process.

(2) A material that does not transmit certain wavelengths of incident radiation.

absorbent cotton. Cotton that has been made absorbent by a chemical treatment that removes natural surface waxes and fats. Used especially for surgical dressings.

absorbent paper. A paper, such as blotting paper, that can take up and hold liquids in the pores between or within the cellulosic-fiber mat. It is usually thick and loosely formed, and used chiefly in chemistry, medicine and the graphic arts.

absorber. (1) A material that absorbs or stops radiated energy and dissipates it.

(2) A substance that absorbs neutrons without propagating them. Used in nuclear reactors.

absorber steel. A steel, often an austenitic grade, whose neutron absorption coefficient has been increased by the addition of up to 5wt% boron, usually as a dispersion of boron carbide (B_4C) or iron chromium boride [$(Fe,Cr)_2B$]. Used as absorber or shielding material in nuclear reactors.

Absorbit. Trade name of Absorbit Inc. (USA) for absorptive rayon-based textile fibers.

ABS-PA alloys. Thermoplastic polymer blends of amorphous acrylonitrile-butadiene-styrene (ABS) and crystalline polyamide (PA) usually supplied in molding grades. ABS-PA alloys containing about 50% polyamide (nylon) exhibit high toughness. They have high impact resistance, good abrasion resistance, outstanding processibility, excellent flow characteristics, good chemical and heat resistance, good surface finish, and reduced moisture sensitivity. Used for lawn and garden equipment, power-tool housings, sporting goods, and automotive components.

ABS-PC alloys. Thermoplastic polymer blends composed of more than 50% acrylonitrile-butadiene-styrene (ABS), and the balance polycarbonate (PC). Commonly supplied in injection molding and plating grades, they have good mechanical and thermal properties, good impact resistance, good interfacial adhesion, low surface tension, and relatively low viscosity for easier processing. Used for business machine and appliance housings, automotive interior and exterior applications, e.g., instrument panels, grilles and wheel covers, power tools, electrical components, such as terminal blocks and switches, and food trays.

ABS plastics. See acrylonitrile-butadiene-styrene plastics.

ABS-PUR alloys. Polymer blends which uniquely combine the outstanding toughness and abrasion resistance of thermosetting polyurethane (PUR) with the lower cost and rigidity of thermoplastic acrylonitrile-butadiene-styrene (ABS).

ABS-PVC alloys. Thermoplastic polymer blends of acrylonitrile-butadiene-styrene (ABS) and polyvinyl chloride (PVC) that are available in several grades including self-extinguishing, high- and medium-impact, and high- and low-gloss. Plasticized blends of ABS and PVC are used in automobile dashboard assemblies. ABS-PVC blends are usually processed by extrusion or injection molding.

ABS resin. See acrylonitrile-butadiene-styrene resin.

ABS-SAN alloys. Thermoplastic polymer blends of amorphous acrylonitrile-butadiene-styrene (ABS) and amorphous styrene acrylonitrile (SAN) exhibiting excellent impact properties.

ABS-SMA alloys. Thermoplastic polymer blends of acrylonitrile-butadiene-styrene (ABS) and styrene-maleic anhydride (SMA) resins with improved heat resistance supplied in injection molding, extrusion, and plating grades.

Abstracto. Trade name of Vereinigte Glaswerke (Germany) for a patterned glass with irregular geometrical designs.

Abstrex. Trademark of Rexall Drug & Chemical Company (USA) for several polstyrenes.

abura. The uniformly light pinkish-brown wood of the tree *Mitragyna ciliata*. It has a moderately straight grain and a fine and very even texture, seasons well, has a moderate blunting effect on tools, low durability, and takes preservatives well. Average weight, 560 kg/m³ (35 lb/ft³). Source: West Africa. Used as general utility timber, and for furniture and light constructional work.

Abyssinian gold. (1) A rich low brass containing 8-11.5% zinc, and the balance copper. It has a thin facing of gold applied by rolling, and is used for costume jewelry and ornaments. Also known as *talmi gold*.

(2) A yellow or golden-colored aluminum bronze containing 90-95% copper and 5-10% aluminum.

acacia. (1) The coarse bast fibers obtained from various trees of the species *Acacia*. Used for textile fabrics.

(2) See black locust.

acacia gum. A dried, water-soluble gum obtained from the stems of various trees of the species *Acacia* including *A. senegal* found in the Sudan, West Africa and Nigeria. It is available as white or yellowish powder, flakes or granules, and used for cold-water paints, inks, leather, show card colors, adhesives (mucilages), polishes, textile printing, emulsifiers, and as a binder for improving the green strength of porcelain enamels and glazes. Also known as *gum arabic; Senegal gum*.

Acadal. Trade name of Alumetal (Italy) for an aluminum alloy used for color anodizing applications.

Academy Gold. Trademark of Ivoclar Vivadent AG (Liechtenstein) for a palladium-free dental casting alloy (ADA type II/III) containing 77.2% gold, 12.7% silver, 8.5% copper, and less than 1.0% of zinc, indium, iridium, tantalum and platinum, respectively. It has a rich yellow color, a density of 15.9 g/cm³ (0.57 lb/in.³), a melting range of 940-980°C (1730-1795°F), an as-cast hardness of 125 Vickers, high elongation, excellent castability and processibility, and good biocompatibility. Used for crowns, inlays, onlays, and short-span bridges.

acanthite. A black opaque, or lead-gray mineral composed of silver sulfide, Ag_2S. Crystal system, monoclinic. Density, 7.20-7.24 g/cm³; hardness, 2-2.5 Mohs.

acapau. The heavy, blackish, insect-resistant wood of the large ebony-type tree *Voucapapoua americana*. Source: Brazil (Amazon). Used for fine inlay work and furniture. Also spelled acapu. Also known as *partridgewood*.

acaroid resin. A red, or yellow gum resin derived from aloe-like trees of the genus *Xanthorrhoea* found in Australia and Tasmania. It is available in powder and lump form for use in sealing waxes, varnishes, inks, and wood stains. Also known as *gum accroides; yacca gum*.

Accelagold. Trade name of Atofina Chemicals, Inc. (USA) for yellow chromium conversion coatings used for automotive and aerospace applications.

accelerated cement. A cement produced from Portland cement and small additions of quartz or iron ore by controlling the proportions of the ingredients, and heating in a rotary kiln with gradual temperature increase to about 1455°C (2651°F). After burning, the clinker is ground with some gypsum to control the set. It sets hard in about 3-4 days. See also Portland cement; clinker.

accelerating agent. (1) A substance which when added to another substance increases the rate of chemical reaction between them. Also known as *accelerator; accelerator catalyst.*

(2) A substance which when added to a catalyst or a resin, greatly reduces the time required for the polymerization of resins, or the vulcanization of natural and synthetic rubbers while, at the same time, improving the aging and other physical properties. Also known as *accelerator; accelerator catalyst; promoter.* See also curing agent.

(3) A substance, such as potash alum or an alkali carbonate, which when introduced into a batch of concrete, mortar, plaster or grout, acts as a catalyst and hastens hydration, thereby shortening the time of set. Also known as *accelerator; accelerator catalyst.*

accelerator. (1) See accelerating agent (1), (2), (3).

(2) A substance, such as copper, manganese or silver, which when added in minute quantities to crystalline substances, such as cadmium sulfide, zinc sulfide or certain phosphors, will induce or increase luminescence. Used in the manufacture of television screens. Also known as *activator.*

accelerator catalyst. See accelerating agent (1), (2), (3).

Accello. Trademark of Anti-Hydro Canada Inc. for a concrete antifreeze and accelerator.

Accepta. Trademark of KoSa (USA) for a series of heat settable polyester fibers that have low dyeing temperatures (less than 104°C or 220°F), good colorfastness, good dimensional stability, and blend well with a wide range of other fibers including nylon, acetate, spandex, and wool. Used for wearing apparel (garments) including velours and velvets, upholstery, and home furnishings. Also known as *Accepta Easy Dye; Easy Dye.*

Access. Trade name for an addition-curing elastomer used for dental impressions.

Access Crown. Trade name for a dental resin for temporary bridge and crown restorations.

Acclaim. (1) Trademark of Bayer Corporation (USA) for polyurethane elastomers used for automotive components, sporting equipment, machine parts, and construction and transportation applications.

(2) Trademark of Degussa-Ney Dental (USA) for a dental alloy containing 49% palladium and 41% silver. It has a light oxide layer, high yield strength, good injection castability, a hardness of 250 Vickers, and is compatible with most non-greening porcelains. Used in restorative dentistry especially for long-span bridges.

Acclaim Lite. Trademark of Degussa-Ney Dental (USA) for a low-density dental alloy containing 50% palladium and 40% silver. It has a light oxide layer, very high yield strength, exceptional sag resistance, good castability and injection castability, good laser weldability, and a hardness of 300 Vickers. Used for dental restoration work.

Acclear. Trademark of BP Amoco (USA) for a series of high-clarity homopolymer and random copolymer polypropylenes.

Accoe. Trade name of GC America (USA) for a biocompatible condensation-curing silicone-based dental impression material.

Accoloy. Trade name of Alloy Engineering Casting Company (USA) for an extensive series of corrosion- and/or heat-resistant austenitic and ferritic steels used for furnace and heat-treating parts and equipment, and chemical and mining equipment.

Accotile. Trade name for asphalt tiles supplied in various standard shapes and sizes and produced by mixing asphalt with various fillers and pigments.

account-book paper. A high-quality book paper, often containing cotton or linen rags.

Accpro. Trademark of BP Amoco (USA) for a series of polypropylenes which are available in enhanced, extrusion and injection grades.

Acctuf. Trademark of BP Amoco (USA) for a series of impact-modified copolymer polypropylenes.

AccuBead. Trademark for polystyrene beads available in various sizes from 0.5 to 100 μm (0.00002 to 0.004 in.), and manufactured to very close tolerances. Used for the determination of particle retention capabilities of industrial liquid and air filters, and membrane separators.

Accubrade. Trade name of S.S. White Technologies Inc. (USA) for alumina abrasive blends for deburring operations.

Accucoat 917. Trade name of Accurate Plastics, Inc. (USA) for epoxy-coated copper foil used for the lamination of printed circuit boards.

Accuflex. Trademark of Bayer Corporation (USA) for flexible polyurethane integral skin foams and reaction-injection molding elastomers used for automotive components, furniture, sports equipment, and packaging applications.

AccuForm. Trademark for a sheet steel with good mechanical properties and an electrolytically-cleaned, smut-free surface contributing to superior paint adhesion.

Acculam. Trademark of Accurate Plastics, Inc. (USA) for an extensive series of durable, lightweight, uniformly dense glass-, fabric-, or paper-reinforced thermoset plastic laminates supplied in several grades. The laminating resins used include epoxy, melamine, phenolic, polyester and silicone. *Acculam* laminates have high mechanical strength, stability, good thermal and moisture resistance, and good electrical insulating properties.

Acculam Epoxyglas. Trademark of Accurate Plastics, Inc. (USA) for strong plastic laminates with high mechanical strength and good electrical and dielectric properties, made by impregnating woven glass fabrics with epoxy resin binder.

Acculam Melaglas. Trademark of Accurate Plastics, Inc. (USA) for hard plastic laminates with good dimensional stability and chemical and arc resistance, made by impregnating continuous woven glass fabrics with melamine resin binders.

Acculam Phenolfab. Trademark of Accurate Plastics, Inc. (USA) for light tan- to brown-colored laminates with good mechanical properties composed of continuous woven cotton fabrics impregnated with phenolic resin binders. They are supplied in medium weave canvas (e.g., *Acculam Phenolfab Canvas C/CE*), and fine weave linen (e.g., *Acculam Phenolfab Linen L/LE*) grades.

Acculam Phenolkraft. Trademark of Accurate Plastics, Inc. (USA) for tan-colored laminates including *Acculam Phenolkraft Paper X/XX/XXX* composed of kraft paper impregnated with a phenolic resin binder. They have moderate mechanical properties, and good electrical insulating properties.

Acculam Polymat. Trademark of Accurate Plastics, Inc. (USA) for plastic laminates made by impregnating nonwoven fiberglass mats with polyester resin binders. They have good electrical insulating properties.

Acculam Siliglas. Trademark of Accurate Plastics, Inc. (USA) for plastic laminates with good heat and arc resistance, and good dielectric properties, made by impregnating woven glass fabrics with silicone resin binders.

Acculoy. Trade name of Fugi Iron & Steel Company (USA) for an alloy of gold and tin with high electrical conductivity, and high corrosion and etch resistance. Used as a brazing alloy for electronic components.

accumulator metal. A lead alloy that contains 9.2-9.3% tin and 0.7-0.8% antimony, and is used for condenser foils, battery plates, and as an antifriction metal for bearings.

Accunamel. Trade name of Bethlehem Steel Corporation (USA) for a cold-rolled, low-carbon sheet steel containing 0.03% carbon, 0.15% manganese, 0.004% boron, and the balance iron. Used for porcelain-enameling applications, dishwashers, bathtubs, sinks, gas and electric ranges, and washing machines.

Accupeen. Trade name of S.S. White Technologies, Inc. (USA) for glass beads used for surface finishing operations.

Accurate Metal. Trade name of Connecticut Metals Corporation (USA) for a corrosion-resistant, nonmagnetic alloy of 68% copper, 29-32% nickel, and 0.4-0.7% iron. Used for pump valves, marine hardware, ferrules, and electronic components.

Accuricon. Trademark of Anti-Hydro Canada, Inc. for a concrete hardener and sealer.

Accurloy. Trade name of Baldwin Steel Company (USA) for a heat-treated, shock-resisting steel containing 0.51% carbon, 1.05% chromium, 0.97% manganese, 0.53% nickel, 0.25% molybdenum, 0.21% vanadium, and the balance iron. Used for piston rods, shafts, and pins.

AccuShield. Trade name of Mathison's (USA) for a vinyl laminating product with an ultrathin, thermally applied, scratch- and UV-resistant coating. Supplied with mat or glossy surfaces, it is used for banners, signage and posters.

Accuspin 418. Trade name of AlliedSignal Inc. (USA) for a porous, silica-based, low-dielectric-constant material produced by spin-on coating. Used as ultra-large-scale integrated interlayer dielectrics for ultra-large-scale-integrated circuit applications.

Accuspin T18. Trademark of AlliedSignal Inc. (USA) for methylsilsesquioxane that upon baking and curing develops a crosslinked network of ladder structure. It has very high thermal stability (typically above 500°C or 930°F), high cracking resistance and a low and stable dielectric constant (k = 2.7 at 1 MHz). It is used as an interlayer dielectric for integrated circuit devices, and deposited in the form of a thin film by spin-on coating.

Accu-Star. Trade name of J.F. Jelenko & Company (USA) for a high-palladium ceramic dental alloy containing 6% gold. It has high strength, good finishability and polishability, and is used for fusing porcelain to metal.

A.C.E. (1) Trademark of Honeywell Performance Fibers (USA) for high-tenacity industrial polyester yarns having excellent mechanical properties, excellent dye pick-up, and low creep and elongation. Used for cordage, narrow- and broad-woven fabrics including seat belts, tow-straps, truck tiedowns and other automotive safety products, and outdoor furniture.

(2) Trademark of AlliedSignal, Inc. (USA) for polyamide (nylon) filament yarns.

Ace. Trade name of American Hard Rubber Company (USA) for a hard, but resilient rubber. It has high tensile and dielectric strength, a distortion temperature of 78°C (172°F), and low water absorption (0.04%). Used for electrical components.

Acelan. (1) Trade name of Taekwang Industrial Company (South Korea) for acrylic fibers and spandex filament yarns used for textile fabrics.

(2) Trade name of Daehan Synthetic Fiber Company (South Korea) for polyester fibers and yarns.

Acelba. Trade name for rayon-based textile fibers.

Acele. Trade name of E.I. Du Pont de Nemours & Company (USA) for a series of rayon and cellulose acetate fibers and yarns.

Acella. Trademark of J.H. Benecke GmbH (Germany) for film materials based on flexible polyvinyl chloride (PVC) or PVC copolymers.

Acelon. Trademark of Acelon Chemicals & Fibers Corporation (Taiwan) for polyester fibers and yarns.

Acelose. Trade name of American Cellulose Company (USA) for cellulose acetate plastics.

acenaphthene. A saturated hydrocarbon obtained from coal tar. It is available in the form of colorless crystals or white needles in technical and high-purity grades with a density of 1.024 g/cm³, a melting point of 93-95°C (199-203°F), a boiling point of 279°C (534°F) and a refractive index of 1.605 (100°C or 212°F). Used chiefly in plastics, as a dye intermediate, in the preparation of acenaphthenequinone, and in genetics. The high-purity, zone-refined grade is also used as an aromatic reference standard. Formula: $C_{12}H_{10}$.

Acenor. Trade name of Acenor SA (Spain) for a series of carbon and alloy machinery steels, case-hardening and nitriding steels, quenched-and-tempered alloy steels, free-cutting leaded steels, chromium bearing steels, spring steels, and various die and mold steels.

ACEramic. Trademark for a series of ultrahigh-temperature mold and tool materials including pourables, blocks, glues, paints, and thinners. They have high service temperatures up to 1650°C (3000°F), high thermal-shock capabilities, and low coefficients of thermal expansion. Used for various types of composite materials, glasses, and metal alloys.

Aceroid. Trade name of Central Foundry Company (USA) for an alloy of copper and zinc used for castings.

Acesil. Trademark of Industria Tessile di Vercelli SpA (Italy) for cellulose acetate fibers and yarns.

Acetabel. Trade name of Bellignite for spun cellulose acetate textile fibers that contain casein, and have enhanced strength, crimp and surface roughness.

acetal copolymer. One of the two basic types of engineering thermoplastics of the acetal resin family (the other being acetal homopolymer). It is manufactured by the copolymerization of trioxane with small amounts of a co-monomer which randomly distributes carbon-carbon bonds in the polymer chain. It has a high degree of thermal stability, good resistance to strong alkalies, good resistance to hot water and good stability in long-term high-temperature service. Used in the manufacture of acetal resins.

acetal-elastomer alloys. Blends of crystalline polyacetal polymers and elastomers, such as butadiene rubber or ethylene-propylene terpolymer (EPDM). They combine the high resiliency, elongation, and impact resistance of elastomers with the strength, stiffness, and chemical resistance of polyacetals. Used for sporting goods, automotive and industrial components, and shoe soles.

acetal homopolymer. One of the two basic types of engineering thermoplastics of the acetal resin family (the other being acetal copolymer). It is manufactured by the polymerization of formaldehyde and subsequent capping with acetate end groups. It is a highly crystalline linear polymer with high hardness and rigidity, high tensile, flexural and fatigue strengths, and relatively low elongation. Used in the manufacture of acetal resins.

acetal plastics. A group of exceptionally hard and tough plastics based on acetal resins. They are manufactured either by polymerization of formaldehyde and subsequent capping with acetate end groups (homopolymer), or by copolymerization of trioxane with small amounts of a comonomer which randomly distributes carbon-carbon bonds in the polymer chain (copolymer). See also acetal copolymer; acetal homopolymer.

acetal resins. A family of hard, rigid, strong, tough, resilient, and highly crystalline engineering thermoplastics (polyformaldehyde and polyoxymethylene resins) made by polymerizing formaldehyde or formaldehyde with trioxane. They may be homopolymers or copolymers. Special glass-filled grades with higher strength and stiffness, and polytetrafluoroethylene-filled grades with exceptional frictional and wear properties are available. Unmodified acetal resins are translucent white and can be readily colored. They have high strength and tensile modulus, good fatigue-life and resilience, excellent creep resistance, excellent dimensional stability, high-gloss and low-friction surfaces, high abrasion resistance, good impact resistance over a wide temperature range, low moisture absorption (less than 0.4%), low moisture sensitivity, high solvent and chemical resistance, good electrical properties, a service temperature range of -30 to +120°C (-22 to +248°F), fair UV (ultraviolet) resistance, excellent resistance to all common solvents (but are attacked by strong acids and alkalies), and are combustible, but slow burning. They may be processed by conventional injection molding, blow and rotational molding, and extrusion techniques. Used as substitutes for metals (platable), for automotive and appliance parts, automobile instrument clusters, industrial parts, pump impellers, conveyor links, drive sprockets, spinning reel housings, gear valve components, bearings, machine parts, oil and gas pipes, hardware, communication equipment, and aerosol containers for cosmetics. Abbreviation: POM. Also known as *acetals; polyacetals; polyacetal resins; polyoxymethylenes.*

acetals. See acetal resins.

acetamide. A usually colorless mineral, that is sometimes gray from admixed carbonaceous matter. It is a functional carboxylic acid derivative. Crystal system, rhombohedral (hexagonal). Density, 1.15-1.17 g/cm³; melting point, 81°C (178°F); boiling point, 222°C (432°F). Occurrence: Coal wastes. Formula: CH_3CONH_2. Used as a soldering flux, in explosives, and as a general solvent. Also known as *ethanamide.*

acetate. A salt or ester of acetic acid. Copper acetate, $Cu(CH_3COO)_2$ is an example of a salt and ethyl acetate, $CH_3COOC_2H_5$, of an ester.

acetate fabrics. Woven or knitted fabrics made from acetate fibers. They have good drapability, good mildew and moth resistance, low flammability, and can be blended with other fibers, such as cotton, nylon and viscose rayon. Used for dresses, ribbons, linings and furnishings.

acetate fibers. Combustible, thermoplastic fibers produced from cellulose acetate in which 74-92% of the hydroxyl (OH) groups of the original cellulose are acetylated (ethanoylated). They weaken above 80°C (176°F), become tacky at 176°C (350°F), and have a moisture absorption of 5-7%, relatively high strength and flexibility, good resistance to microorganisms, and poor resistance to acetone, concentrated solutions of strong acids, and alkalies. Used as textile fibers for wearing apparel, and as industrial fibers. These fibers were formerly referred to as "acetate rayon," or "acetate silk". Also known as *acetates; cellulose acetate fibers.*

acetate film. A durable and transparent film made from cellulose acetate polymer and available in thicknesses from about 0.035 to 0.3 mm (0.001 to 0.01 in.). It has a heat-sealing temperature of 176-232°C (349-450°F), high permeability to carbon dioxide and oxygen, excellent resistance to grease, oil, dust and air, high tear strength, good nondeforming characteristics, good hygienic properties, and relatively poor flame resistance. Used as photographic film, pressure-sensitive and magnetic-recording tape, window cartons and envelopes, in document preservation, as a packaging material, and in laminates. Also known as *cellulose acetate film.*

acetate green. A yellowish-blue pigment consisting of *chrome green* and lead acetate.

acetate of lead See lead acetate.

acetate rayon. See acetate fibers.

acetates. See acetate fibers.

acetate silk. See acetate fibers.

Acetech XLS. Trade name of A.L. Hyde Company (USA) for a low-stress acetal copolymer supplied in the form of sheets with a thickness of 10-40 mm (0.375-1.500 in.), and a width of 1.2 m (4 ft.). It has good machinability and physical properties, and is used for food-processing applications.

Acetite. Trademark of Anchor Packing Division, Robco Inc. (Canada) for blue asbestos braided packing.

acetone-soluble cellulose acetate. Cellulose acetate that contains approximately 54% combined acetic acid and is soluble in acetone. It is obtained by the hydrolysis of primary cellulose acetate. See also cellulose acetate (1); primary cellulose acetate; secondary cellulose acetate.

Acetovis. Trade name for rayon-based textile fibers.

Acetron. Trademark of DSM Engineered Plastics (USA) for engineering thermoplastics based on homopolymer or copolymer acetal resins incorporating solid lubricants. Available in extrusion and injection molding grades, they have superior bearing characteristics, very low wear rates, and low coefficients of friction. Used for thrust washers, valve seats, seals, wear plates, slides, and bushings. *Acetron GP* is a porosity-free general-purpose acetal copolymer used for electrical components, gears, and guide rollers, and *Acetron NS* an internally lubricated grade with excellent wear resistance, and a low coefficient of friction, used for bearings and bushings.

acetylated cellulose. See cellulose acetate (1).

acetylated cotton. Cotton, usually in the form of fibers or fabrics, with improved heat, rot and mildew resistance obtained by treating with acetic acid, acetic anhydride or perchloric acid in the presence of a suitable catalyst. This treatment converts the raw cotton fiber to cellulose acetate. Also known as fully acetylated cotton. See also cellulose acetate; cotton; partially acetylated cotton.

acetylene. A colorless, illuminating, unsaturated gaseous hydrocarbon in which the carbon atom has a triple bond ($-C\equiv C-$). It has a sweetish odor, is generated industrially by a controlled reaction of calcium carbide in water, and can also be manufactured from methane, heavy gas oil, and naphtha. *Acetylene* produces high flame temperatures (over 2700°C, or 4892°F) when burned in an atmosphere of oxygen, and is used in preparing compounds to make synthetic fibers and vinyl plastics, in the manufacture of carbon black, and in oxyacetylene welding and cutting operations. Formula: C_2H_2. Also known as *acetylene gas; ethyne.*

acetylene black. A very graphitic *carbon black* resulting from thermal decomposition or incomplete combustion of acetylene. It has a high liquid adsorption capacity, good retention of high bulk volume, high electrical conductivity, very low oiliness, low ash content, and an apparent density of 0.21 g/cm³. Its particle size is intermediate between that of *furnace black* and *channel black*. Used for dry cell batteries, as a gloss suppressor in paints, as a surface carburizing agent in steel hardening, as a pigment in printing inks, and as a filler in conductive polymers.

acetylene polymer. A linear polymer of acetylene with alternate single and double bonds. It is electrically conductive, but can be altered by doping to have a wide range of conductivities from insulators and n- or p-type semiconductors to strongly conductive forms. It is available in film and fiber form for use in batteries. The semiconductor grade has a typical electron mobility of about 10^{-4} cm^2 V^{-1} s^{-1}, and is used in solid-state devices, e.g., Schottky devices. Also known as *polyacetylene*.

acetylides. See metal acetylides.

Acetyloid. Japanese trade name for cellulose acetate plastics.

Achatit-Biochromatic. Trade name for a dental silicate cement.

achavalite. A dark gray mineral of the nickeline group composed of iron selenide, FeSe. Crystal system, hexagonal. Density, 6.58 g/cm^3. Occurrence: Argentina.

Acheson graphite. A synthetic graphite produced from coke in a resistance-type heating furnace.

Achrolyte. Trade name of Enthone-OMI Inc. (USA) for a tin-cobalt alloy electroplate and plating process.

achromatic glass. A glass that is capable of transmitting light without decomposing it into its constituent colors.

Acibar. Trade name of British Steel Corporation (UK) for a free-cutting, hot-rolled steel containing 0.1% carbon, 1% manganese, 0.06% silicon, 0.04% phosphorus, 0.3% sulfur, and the balance iron. Used for machine-tool parts, gears, and shafts.

Acibel. Trade name of British Steel Corporation (UK) for a free-cutting, case-hardening steel containing 0.07-0.13% carbon, 0.85-1.15% manganese, 0.15-0.25% lead, 0.23-0.33% sulfur, and the balance iron. Used for machine-tool parts, gears, and shafts.

Acibrade. Trade name of British Steel Corporation (UK) for a steel containing 0.6-0.75% carbon, 1.25-1.55% manganese, 0.05% silicon, and the balance iron. It has excellent abrasion resistance, and high toughness in the as-rolled condition. Used for rod mills, coke crushers, mineral dressing equipment, chutes, conveyors, and buckets.

ACI casting alloys. (1) A group of highly corrosion-resistant cast nickel-base alloys as specified by the Alloy Casting Institute (ACI) designations. This designation system consists of a letter-number combination, e.g., CZ-100 specifies the standard grade of cast nickel, and CY-40 refers to an Inconel-type cast nickel alloy typically containing 0.2% carbon, 1.5% silicon, 1.0% manganese, 15.5% chromium and 8.0% iron.

(2) A group of cast corrosion-resistant steels as specified by the Alloy Casting Institute (ACI) designations. This designation system consists of a letter-number combination, e.g., CA-15 specifies a cast steel containing 0.15% carbon, up to 1.5% silicon, up to 1.0% manganese and nickel, respectively, 11.5-14.0% chromium, up to 0.5% molybdenum, and the balance being iron. CF-20 refers to a cast steel containing 0.20% carbon, up to 2.0% silicon, up to 1.5% manganese, 8.0-11.0% nickel, 18.0-21.0% chromium, and the balance iron.

acicular cast iron. Alloy cast iron with a bainitic microstructure of a needle-like (acicular) appearance.

acicular ferrite steel. Low-carbon steel with a microstructure containing either acicular ferrite, or a mixture of acicular and equiaxed ferrite.

acicular powder. A term used in powder metallurgy to identify a powder composed of needle- or sliver-like particles.

acid-bath tinplate. Sheet iron or steel that has been passed through an acid bath to produce a protective tin coating on its surface. Used chiefly for food-preservation containers. See also tinplate.

acid Bessemer steel. A steel produced from a low-phosphorus pig iron in a Bessemer converter with an acid refractory lining, i.e., a lining consisting of materials, such as silica brick, siliceous rock, sand, etc. Used for structural purposes, boiler plate, and other low-stress applications. See also basic Bessemer steel; Bessemer steel.

acid brick. Firebrick made from acid refractory materials, such as silica or sand. Also known as *acid firebrick*. See also basic brick; firebrick.

acid bronzes. A series of acid- and corrosion-resistant bronzes containing 73-88% copper, 2-17% lead, 8-10% tin, and up to 2% zinc and nickel, respectively. Used for bearings, bushings, fittings, valves, and chemical and pump equipment.

acid casein. Casein that has been precipitated from skimmed milk by dilute acid, and is used as a raw material in the manufacture of adhesives, paper, coatings, textiles, food products, etc. See also casein.

acid copper. Copper that has been electrolytically deposited from a solution of a copper salt, such as copper sulfate.

acid core solder. A solder in wire or bar form that has a corrosive or acid flux, usually ammonium chloride or zinc chloride, as a core.

acid-cured rubber. Rubber that has been cured in a sulfur chloride bath.

acid firebrick. See acid brick.

acid glaucine blue. A ceramic color containing approximately 45% china stone, 33% cobalt oxide, 7% carbon black, and 15% flint.

acid gold. A gold decoration on a glaze whose surface has been etched with hydrofluoric acid to enhance adherence.

acid gold bromide. See hydrogen tetrabromoaurate.

acid gold chloride. See hydrogen tetrachloroaurate.

acid gold trichloride. See chloroauric acid.

acidic rock. See acid rock.

acid lead. A fully refined lead (99.90% pure) that has 0.10% copper added to improve its elevated-temperature and fatigue properties.

acid metal. A leaded tin bronze containing about 88% copper, 10% tin, and 2% lead. It has excellent resistance to corrosion by acids, and is used for chemical equipment, bearings, and bushings.

acid open-hearth steel. A pig iron with low phosphorus content that has been treated in an open-hearth furnace lined with acid or siliceous refractories (e.g., silica or sand). See also basic open-hearth steel; open-hearth steel.

acidproof brick. A low-porosity, low-permeability brick with excellent resistance to attack or penetration by most commercial acids and corrosives.

acidproof cement. A type of cement made by mixing the powder with a silicate solution (e.g., sodium silicate). Used for lining of chemical and laboratory equipment (e.g., tanks, sinks, drains, etc.).

acid refractory materials. See acid refractories.

acid refractories. Refractory materials, composed chiefly of silica, which may undergo reactions with alkalies, lime, and basic fluxes and oxides at elevated temperatures. Examples include silica refractories (92% or more SiO$_2$), and siliceous refractories (about 78-92% SiO$_2$). Used for lining furnaces, converters, etc. Also known as *acid refractory materials.* See also basic refractories; neutral refractories; silica refractories; siliceous refractories.

acid-resistant alloys. Nickel-base alloys containing significant additions of chromium, molybdenum and copper. Sold under trade names or trademarks, such as *Illium* or *Parr Metal,* they have excellent general corrosion resistance, and are used for machinery parts, and chemical apparatus and equipment.

acid-resistant cast irons. A group of alloy cast irons containing about 0.4-3.8% total carbon, 0.4-4.5% manganese 2.5-17% silicon, 0.1-0.3% phosphorus, 0.1% sulfur, up to 5% chromium, 2% nickel and 1% molybdenum, respectively, and the balance iron. *Medium-silicon gray* and *ductile irons* (about 2.5-7% silicon) have good acid resistance and improved heat resistance, and *high-silicon irons* (about 7-17% silicon) have excellent resistance to corrosive acids, but are hard, difficult to cast and do not machine well. *Acid-resistant cast irons* have good high-temperature properties, good corrosion resistance, good shock resistance, and low magnetic hysteresis. Often sold under trade names or trademarks, such as *Duriron, Durichlor, Silal* or *Superchlor,* they are used for transformer sheets, cathodic protection anodes, pump parts and chemical equipment. Also known as *silicon cast irons; silicon irons.* See also alloy cast irons.

acid-resistant concrete. Concrete that has been treated with gaseous silicon tetrafluoride (SiF_4) to transform any free lime (CaO) to calcium fluoride (CaF_2). The resulting concrete products have increased density, and are more acid- and wear-resistant than ordinary concrete. See also concrete.

acid-resistant brick. A fired clay brick with high resistance to acids and chemicals, and low-water absorption. It is normally used with acid-resisting mortars.

acid-resisting enamel. A porcelain enamel with a high resistance to acids especially household, fruit and cooking acids. Also known as *acid-resisting porcelain enamel.* See also porcelain enamel.

acid-resisting porcelain enamel. See acid-resisting enamel.

acid rock. An igneous rock that contains 65% or more silica, and is used in civil engineering and building construction. See also basic rock.

acid salt. A salt derived from an acid whose hydrogen has been replaced only in part by a metal or radical. When dissolved, it yields a solution in which there are more H^+ than OH^- ions. Examples include sodium bicarbonate ($NaHCO_3$) and sodium bisulfate ($NaHSO_4$).

acid slag. A slag in which the content of the acid ingredients (chiefly silica) is substantially greater than that of the basic ingredients (chiefly lime and magnesia). See also basic slag.

acid spar. A grade of *fluorite* (fluorspar) containing more than 98% calcium fluoride (CaF_2) and 1% or less silica (SiO_2). Used in the manufacture of hydrofluoric acid, refrigerants, plastics and chemicals, and in aluminum reduction.

acid steel. A grade of steel melted in a furnace with acid (siliceous-type) refractory bottom and lining. It may be an acid Bessemer steel, or an acid open-hearth steel. See also basic steel.

Acidur. Trade name of Maschinenfabrik AG (Germany) for a corrosion-resistant alloy containing 83-84% iron and 16-17% silicon. Used for chemical apparatus.

Acier. Trade name of Creusot-Loire (France) for a series of plain-carbon machinery and tool steels, several alloy machinery steels (chromium-molybdenum, nickel, nickel-chromium, and nickel-chromium-molybdenum types including several case-hardening grades), cold- and hot-work tool steels, and austenitic manganese steels.

Acieral. (1) Trade name of Acieral Company of America (USA) for a silvery-white aluminum alloy containing 2.3-6% copper, 0.1-1.4% iron, and up to 1.5% manganese, 0.9% magnesium, and 0.4% silicon, respectively. It has high strength, low weight, and good corrosion resistance. Used in the manufacture of aircraft, automotive components, and railway equipment.

(2) Trade name of Acieral Company of America (USA) for an aluminum casting alloy containing about 6% copper, 0.9% nickel, 0.4% zinc, 0.4% silicon, and 0.1% iron. Used for automotive engine parts.

Acipco. Trade name of ACIPCO Steel Products Division (USA) for a series of cast ferrous alloys including several low-, medium- and high-carbon machinery steels, alloy steels (chromium-molybdenum and nickel-chromium-molybdenum types), gray cast irons, corrosion-, heat- and/or wear-resistant alloy cast irons, acicular cast irons, austenitic, ferritic and martensitic stainless steels, heat-resistant steels, etc.

Ack Die Steel. Trade name of Ackerlind Steel Company, Inc. (USA) for a water-hardening steel containing 1% carbon, 0.4% chromium, 0.2% vanadium, 1.6% tungsten, and the balance iron. Used for tools and dies.

Aclacell. Trade name of Acla-Werke GmbH (Germany) for polyurethane foams and foam products.

Aclar. Trademark of AlliedSignal, Inc. (USA) for a series of clear, nonflammable fluoropolymer films based on chlorotrifluoroethylene (CTFE). They have useful properties from about -200 to +200°C (-330 to +390°F), exceptional resistance to oils and chemicals, outstanding moisture resistance, excellent electrical properties (including low dielectric constants over a wide frequency range and high dielectric strengths), low melt temperature, and good physical properties. Used for food packaging, as moisture barrier films for electronic components, as bonding films for microwave circuitry applications, in electroluminescent encapsulation, holographics, tank linings, and as wire coverings.

Aclathan S. Trade name of Acla-Werke GmbH (Germany) for polyurethane resins and products.

Acliner. Trade name of Teijin Fibers Limited (Japan) for polyester filaments used for the manufacture of textile fabrics.

Aclon. Trade name of Allied Signal Inc. (USA) for fluoropolymers.

Acme. Trade name of Acme Chemicals & Insulation (USA) for epoxy casting and potting compounds for electric applications.

Acmite. Trade name of Columbia Tool Steel Company (USA) for a tungsten-type high-speed tool steel (AISI type T4) containing 0.75% carbon, 18% tungsten, 5% cobalt, 4% chromium, 1% vanadium, and the balance iron. It has a great depth of hardening, excellent wear resistance, high hot hardness, and good machinability. Used for lathe and planer tools, milling cutters, boring tools, and shaper tools.

acmite. A brown to dark green mineral of the pyroxene group composed of sodium iron silicate, $NaFeSi_2O_6$. Crystal system, monoclinic. Density, 3.50-3.58 g/cm³; refractive index, 1.806; hardness, 6-6.5 Mohs. Occurrence: Greenland, Canada (Labrador), USA (New Jersey). Also known as *aegrine; aegrite.*

Acoplan. Trade name of Aco Plastics (USA) for glass-fiber-reinforced unsaturated polyesters.

A-Copper. Trade name of Isabellenhütte Heusler GmbH (Germany) for a copper-base resistance alloy containing 0.5% nickel. It has a maximum service temperature of 200°C (390°F), and is used for electrical instruments and equipment.

Acorn. (1) Trade name of Acorn Aluminum Products Company

(USA) for a sealed insulating glass.

    (2) Trade name of Swedish Iron & Steel Corporation (USA) for an extensive series of water-hardening plain-carbon tool steels (AISI type W), molybdenum- and tungsten-type high-speed tool steels (AISI types M and T, respectively), chromium, tungsten and molybdenum-type hot-work tool steels (AISI type H), high-carbon high-chromium cold-work tool steels (AISI type D), medium-alloy, air-hardening type cold-work tool steels (AISI type A), oil-hardening cold-work tool steels (AISI type O), low-alloy special-purpose tool steels (AISI type L), carbon-tungsten-type special-purpose tool steels (AISI type F), mold steels (AISI type P), and shock-resisting tool steels (AISI type S).

    (3) Trade name of Aluminium Industrie AG (Germany) for an age-hardenable, high-strength aluminum alloy containing 4% copper and 1% silicon, used for automotive engine components.

Acousta-Pane. Trade name of Amerada Glass Company (USA) for a laminated glass used in sound-transmission control.

Acousta Vu. Trade name of Multiplate Glass Corporation (USA) for a laminated glass with high acoustic absorption.

acoustical board. A lightweight board, made from a pulp of wood, cane or other cellulosic fibers, which combines good strength with good acoustical and thermal insulating properties. It is supplied in a wide range of sizes in tile form with 203 mm (8 in.) square, and sheet form with 1.2 m (4 ft.) in width, and 3.2 m (10 ft.) or more in length. The standard thickness range is 13 to 25 mm (0.5 to 1.0 in.). It usually has a factory-applied finish and a textured surface pattern (e.g., fissures, slots, perforations, etc.) to absorb sound waves. Also known as *acoustical insulation board; acoustical slab.*

acoustical insulation board. See acoustical board.

acoustical materials. Materials designed to absorb sound waves, and thus reduce reflection or reverberation. Examples include tile, plaster, perforated fiberboard, fiber sheathing board, mineral wool, hair felt, foamed plastics, or simple sheathing paper for application to the interior walls of buildings. Also known as *sound barrier materials; sound insulators.*

acoustical plaster. A low-density, finishing-type of plaster with a chemically or mechanically textured surface, which absorbs or prevents the transmission of sound.

acoustical sealants. A group of durable, usually latex-, or elastomer-based (e.g., butyl rubber) construction sealants designed for the reduction of sound transmission in ceiling, wall and floor systems by application to joints, cutouts and other penetrations. They are usually non-hardening (flexible), and bond well to most building materials including concrete, drywall, metal, and wood

acoustical slab. See acoustical board.

acoustical tile. A thin, square tile, usually about 610 × 610 mm (24 × 24 in.) in size, made of a sound-absorbing material, such as ceramic or plaster. Used as decorative covering for ceilings and walls.

Acousti-Celotex. Trade name of Owens-Corning Fiberglas (USA) for a series of acoustic products, such as wallboard, paneling and tile, manufactured from bagasse fibers. They have perforated surfaces to improve the acoustic absorption efficiency.

Acoustifibre. Trade name of Produits D'Isolation Cellulose PIC, Inc. (USA) for cellulose-based acoustic insulation materials.

Acousti-Shell. Trade name of Manville Corporation (USA) for acoustic ceiling panels of the fiberglass lay-in type.

acoustooptical material. A material in which an acoustic wave results in a change of optical properties, e.g., the refractive index.

Acowell. Trade name of Aco Plastics (USA) for glass-fiber-reinforced unsaturated polyesters.

Acpol. Trademark of Freeman Chemical Corporation (USA) for thermosetting polyester resins.

Acquatic. Trade name of Agate Lacquer Manufacturing Company, Inc. (USA) for a waterborne lacquer coating.

Acra-Flow. Trade name for a self-cure dental acrylic.

Acraglas. Trade name of Acraglas Company (USA) for plexiglass-type polymethyl methacrylate resins.

Acralen. Trademark of Bayer Corporation (USA) for aqueous polymeric dispersions for bonding synthetic and/or natural fibers, reinforcing nonwovens, textiles and paper, and priming nonwovens, textiles and paper substrates for adhesive application.

Acralite 88. Trade name for a vinyl acrylate copolymer formerly used for denture bases.

Acribel. Trademark of Sté Fabelta (Belgium) for acrylic fibers used for the manufacture of clothing, mats, drapes, and coverings.

Acridur. Trademark of Sarvetnick Industries Inc. (USA) for biocompatible, rigid acrylic sheeting for orthopedic, prosthetic and podiatric devices.

Acrifix. Trademark of Roehm GmbH (Germany) for polymethyl methacrylate (PMMA) resins and products.

Acrifluor. Trademark of Acrilex, Inc. (USA) for translucent, fluorescent acrylic sheeting used in the manufacture of exterior and interior fluorescent signs and displays.

Acriglas. Trademark of Sarvetnick Industries, Inc. (USA) for cast acrylic sheeting used in the manufacture of exterior and interior store fixtures, purchase displays, and signs.

Acrilan. Trademark of Solutia Inc. (USA) for acrylic staple and filament fibers with woolly textures based on an acrylonitrile-vinyl acetate copolymer. They are supplied in a wide range of grades including abrasion-resistant, pigmented UV-resistant, producer-colored, bicomponent, and technical. *Acrilan* fibers have medium tenacity, low water absorption properties, good to excellent acid resistance, fair to good alkali resistance, and good resistance to bleaching agents, organic solvents, mildew and moths. Used as textile fibers for the manufacture of apparel, blankets, carpets, draperies, upholstery and other home furnishings, pile fabrics, craft yarns, brake fibers, friction applications, etc.

Acriplex. Trademark of Roehm GmbH (Germany) for polymethyl methacrylate (PMMA) resins and products.

Acri-Shield. Trade name of Porter Paints (USA) for exterior acrylic paints.

Acrive. Trade name of Swedlow (USA) for a series of acrylic plastics supplied in cast sheet, casting, general-purpose, and high-impact grades.

Acrolite. Trade name of Consolidated Molded Products Corporation (UK) for phenol-formaldehyde plastics.

Acrolon. (1) Trade name of Sherwin-Williams (USA) for acrylic enamel primers and acrylic/polyurethane finishes.

    (2) Trade name for acrylic fibers used for the manufacture of textile fabrics.

Acron. (1) Trade name of Alusuisse (Switzerland) for an age-hardenable, high-strength aluminum alloy containing 4% copper and 1% silicon. Used for automotive engine parts.

    (2) Trade name for a range of acrylic dental products including *Acron Duo* and *Acron Trio* heat-cure denture acrylics,

Acron Hi high-impact denture acrylics, *Acron Rapid* fast-setting, heat-cure denture acrylics, *Acron Standard* standard-grade heat-cure denture acrylics, and *Acron MC* microwave-curing acrylics and *Acron Self-Cure* self-curing acrylics for dental restorations.

Acronal. Trademark of BASF AG (Germany) for synthetic vinyl compounds supplied in the form of emulsions, solutions or solids for various industrial applications. Also included under this trademark are various acrylic resins.

Acrybel. Trade name for acrylic fibers used for the manufacture of textile fabrics.

Acrycal. Trademark of Continental Polymers, Inc. (USA) for a series of modified acrylics usually produced by injection molding or extrusion. They are supplied in the form of pellets and sheeting for use in medical applications, and for the manufacture of appliances and recreational vehicles.

Acry-Cote. Trademark of Phillips Paint Products Limited (Canada) for acrylic latex paint.

Acrygel. Trade name for a high-impact denture acrylic.

Acry-Kote. Trade name of West Point Pepperell, Inc.–Industrial Fabrics Division (USA) for acrylic coatings for cotton and synthetic industrial fabrics.

Acryl-AC. Trade name of Pavco Inc. (USA) for a clear acrylic lacquer applied over plated surfaces by dipping.

Acrylador. Trade name of Rohm & Haas Company (USA) for acrylic-based baking enamels for automotive applications. See also baking enamel.

Acrylafil. Trade name of Wilson-Fiberfill International (USA) for flame-retardant, chemical-resistant, injection moldable styrene-acrylonitrile (SAN) reinforced with glass fibers. It is supplied in the form of granules or pellets for use in electronics, and for automotive applications, and appliances.

acrylamate polymers. A class of highly crosslinked polymers formed by the reaction of two liquid components–an acrylesterol (i.e., a hybrid of an urethane and an acrylic), and a liquid-modified diphenylmethane-4,4'-diisocyanate (MDI). They have good high-temperature properties, and are used for strong, high-modulus carbon- or glass fiber-reinforced composites for the automotive and aerospace industries, and for recreational products, agricultural equipment, etc. Also known as *urethane acrylic polymers; urethane hybrids.*

acrylamide. Colorless crystals that are available in technical grades (97+% pure), usually stabilized with 25-30 ppm cupric ion, and high-purity grades (99+%). It has a density of 1.122 g/cm^3 (30°C/86°F), a melting point of 84-86°C (183-187°F), a boiling point of 125°C (257°F)/25 mm, and is used as a monomer (polymerizes violently on melting), in the form of gels in electrophoresis and isoelectric focusing, as a crosslinker, as a size for paper and textiles, in adhesives, in permanent-press fabrics, in dye synthesis, in ore processing, and in biochemistry and biotechnology. Formula: C_3H_5NO.

acrylamide copolymer. A thermosetting engineering resin formed from acrylamide and an acrylic resin.

acrylamide gel. See polyacrylamide gel.

Acrylast. (1) Trademark of Chemtron Manufacturing Limited (Canada) for a sealant based on acrylic resin.
  (2) Trade name for acrylic fibers used for the manufacture of fabric, e.g., medical bandages.

ACRYLAT D. Trademark of Hanno-Werk GmbH & Co KG (Austria) for acrylate-based adhesives and sealants suitable for a wide range of applications.

acrylate-butadiene rubber. An elastomeric copolymer of an es-

ter of acrylic acid and butadiene, used for the manufacture of rubber goods. Abbreviation: ABR.

acrylate polymer liquid crystal. A liquid crystal material made by inserting liquid crystals into a polymer that contains submicron-sized spherical acrylate particles. It is used to produce flexible liquid-crystal displays. See also liquid crystals.

acrylate resins. See acrylates.

acrylate rubber. A wear-resistant elastomer made from acrylate esters with or without the addition of acrylonitrile. It has good resistance to hot oils and lubricants, and is used for gaskets, wire insulation, hose, and tubing.

acrylates. The polymerization products of acrylic and methacrylic acids or their esters (e.g., methyl or ethyl acrylates). They have great optical clarity, high degrees of light transmission, and are used in surface coatings, emulsion paints, paper and leather finishes, etc. Also known as *acrylate resins.*

Acryl Black. Trade name of Pavco Inc. (USA) for temporary and permanent, black acrylic lacquers for plated surfaces.

Acryli-Bond. Trademark of U.S.E. Chemicals (Canada) for an acrylic bonding agent.

acrylic adhesives. A class of fast-setting adhesives that are available in the form of two-part liquids or pastes, and derived from solutions of rubber-base polymers in methacrylate monomers or one-part solvent-free methacrylate. They are supplied in a wide range of viscosities ranging from flowable to thixotropic. *Acrylic adhesives* are usually hardened by light curing, although they can be formulated to cure using activators and heat as well. They have excellent impact resistance, high peel and shear strength, good moisture resistance, good chemical and heat resistance, low shrinkage during curing, good elevated-temperature properties, and a maximum service temperature of 105°C (221°F). They are easy to mix, can be used to dissolve grease, and bond well to many substrates, even oily or poorly prepared surfaces. Used for bonding of various materials, such as wood, glass, aluminum, brass, copper, steel, dissimilar metals, most plastics, and many composites.

acrylic cement. An acrylic adhesive used to form clear joints in cast acrylic sheets.

acrylic-coated fabrics. Fabrics made waterproof by the application of a coating, usually to one side only, based on an acrylic resin. See also waterproof fabrics.

acrylic coatings. Water-white coatings based on acrylic resins. They have good chemical and heat resistance, and good outdoor durability, and are used for automotive finishes (topcoats), appliances, coil coating, on aluminum siding, and for waterproofing textiles.

acrylic fibers. A generic term for strong, lightweight, combustible textile fibers obtained by the polymerization of acrylonitrile (C_3H_3N). They have low water absorption, and are used for the manufacture of modacrylic fibers, knitwear, pile fabrics, blankets and carpets. Also known as *acrylics.*

acrylic fabrics. Soft, light, bulky textile fabrics woven or spun from acrylic fibers. They have good resistance to mildew and moths, good absorbency, and a wooly feel. Used especially for clothing, e.g., skirts, pants, dresses, etc.

acrylic latex sealants. Durable, weather- and mildew-resistant, waterborne emulsion sealants based on acrylic rubber latex. They are available in a wide range of colors for gun application, and may or may not contain silicone. Acrylic sealants bond well to porous and nonporous surfaces including wood, aluminum, glass and ceramics, and are used in building and construction for interior and exterior sealing applications. See also latex; latex sealant.

acrylic plastics. A family of engineering thermoplastics based on polymers or copolymers of acrylic acid, methacrylic acid, esters of these acids, or acrylonitrile. Depending on the monomer and method of polymerization, they can be hard, stiff, brittle solids, fibrous, elastomeric substances, or viscous liquids. *Acrylic plastics* are available as sheets, blocks, rods and tubes, and can be processed by casting, extrusion, injection molding, compression molding, and other molding techniques. The grades supplied include: regular with natural, water-white color; transparent with excellent optical properties (light transmission factor about 92%); translucent and opaque; colored; high-impact blended with rubber; and crack-resistant. *Acrylic plastics* have excellent resistance to shattering, excellent weathering characteristics, relatively high thermal expansion, good resistance to ultraviolet radiation, an upper service temperature of 90°C (194°F), good resistance to dilute acids and alkalies, fair resistance to concentrated acids, poor resistance to aromatic hydrocarbons, greases, oils, halogens and ketones, are either slow burning, or self-extinguishing. Used as glass substitutes, e.g., for regular windows, aircraft windows and canopies, and for decorative parts, lighting fixtures, decorative illuminated signs, contact lenses, prosthetic fixation, furniture components, and in nitrile rubber manufacture. Acrylic solution polymers are used in the manufacture of coatings and paints for paper, textiles, wood, metal, etc., and aqueous emulsions are employed in the production of adhesives, laminated structures, and woven and nonwoven fabrics. Also known as *acrylic polymers; acrylics.*

acrylic polymer concrete. A concrete in which an acrylic polymer, such as polymethyl methacrylate, is used as the binder. It has high freeze-thaw resistance, outstanding weathering resistance, good waterproofing and chemical properties, low water absorption, and low setting shrinkage. Used for curbstones, stairs, facade plates, and as a patching material for civil engineering applications (e.g., bridge deck repairs). See also polymer concrete.

acrylic polymers. See acrylic plastics.

acrylic resins. Polymers or copolymers of acrylic acid, methacrylic acid, esters of these acids, or acrylonitrile. See also acrylic plastics.

acrylic rubber. A synthetic rubber based on acrylonitrile (C_3H_3N). It has excellent weathering and high-temperature resistance, good long-term resistance to abnormal temperatures, oxygen and ozone environments and compounded oils, and poor water and low-temperature resistance. Used for gaskets, seals, O-rings, hoses, tubing, etc. Also known as *alkyl acrylate copolymer.*

acrylics. See acrylic fibers; acrylic plastics.

acrylic sheer. A heavy, open-weave, often two-color net fabric, usually in a plain weave. It may be made from 100% acrylic fiber, or acrylic-flax or acrylic-polyester blends. Used for curtains.

Acrylic Solder. Trade name for an acrylic resin/resin bonding system for dental applications.

acrylic strippable coatings. Transparent, high-gloss, high-strength acrylic-base coatings that are used as temporary protection for metal parts, and can be readily removed.

acrylic-styrene-acrylonitriles. Rubber-modified styrene-acrylonitriles in the form of random amorphous terpolymers made by grafting or polymerization techniques. They are commercially available in unmodified and modified types, and as blends. *Acrylic-styrene-acrylonitriles* have excellent outdoor weather-ability and color stability and good ultraviolet resistance, and are used for car and truck body moldings, exterior and interior automotive trim, outdoor furniture, swimming pool and pump components, boat hulls, window frames, siding, downspouts, eavestroughs, etc. Abbreviation: ASAs. See also styrene-acrylonitriles.

acrylic-vinyl. See acrylic-vinyl fiber.

acrylic-vinyl fiber. A staple fiber made from a copolymer of 60% vinyl chloride and 40% acrylonitrile. Used especially for industrial fabrics, and blankets. Also known as *acrylic-vinyl.*

Acrylite. Trademark of Cyro Industries (USA) for acrylic resins based on polymethyl methacrylate (PMMA), and supplied in cast sheet, casting, general-purpose, and high-impact grades. They have good optical properties from clear to colored, good electrical resistance, good environmental resistance, moderate strength, low heat resistance, and a maximum service temperature of 60-93°C (140-200°F). Used for injection molding and extrusion compounds, automotive lenses, instrument covers, diffusers, signs, nameplates, decorations, display items, dials, glazing, bottles, and sheeting products.

Acrylite Plus. Trademark of Cyro Industries (USA) for a series of acrylic molding and extrusion compounds.

Acryloid. Trademark of Rohm & Haas Company (USA) for a series of acrylic polymers including various coating resins, modifiers, and oil additives.

Acrylon. Trademark of Borden Chemical (USA) for a series of acrylic rubbers with exceptional resistance to ozone, oxidation, oils, greases, and diester lubricants. Used for gaskets and rubber parts.

acrylonitrile. A colorless, mobile liquid (99+% pure) usually inhibited with monomethyl ether hydroquinone to prevent polymerization. It has a density of 0.806, a melting point of -83°C (-117°F), a boiling point of 77°C (171°F), a flash point of 32°F (0°C), and a refractive index of 1.391. Used as a monomer in the manufacture of acrylonitrile-butadiene-styrene (ABS) resins, acrylonitrile-styrene copolymers (e.g., its copolymer with butadiene is nitrile rubber), acrylic and modacrylic synthetic textile fibers, semiconductive polymers, and high-strength whiskers. It is also used in organic synthesis, in the cyanoethylation of cotton, and in the manufacture of adiponitrile. Formula: $CH_2=CHCN$, or C_3H_3N. Also known as *propenenitrile; vinyl cyanide.*

acrylonitrile-butadiene copolymer. A random copolymer of acrylonitrile and butadiene made with an organometallic catalyst. The repeating structure may be represented as $-CH_2CH=CH-CH_2CH_2CH(CN)-$. The acrylonitrile content, which may vary from 18-50%, increases strength, hardness, abrasion and heat resistance, oil and fuel resistance, but decreases resilience and low-temperature flexibility. It has high elasticity (elongations of 400-600%), high flexibility even at low temperatures, excellent resistance to swelling in organic solvents, excellent resistance to vegetable, animal and petroleum oils, exceptional wear resistance, poor tear strength, unless specially formulated, and a service temperature range of -50 to +150°C (-58 to +302°F). Often sold under trade names and trademarks, such as *Butaprene, Chemigum, Hycar, Perbunan, Paracil* or *Buna-N*, it is used for oil well parts, fuel tank liners, gasoline, chemical and oil hoses, O-rings, gaskets, packing oil seals, grommets, pump parts, hydraulic equipment, adhesives, shoe heels and soles, kitchen mats, sink topping, printing rolls, and jet aircraft tires. Also known as *acrylonitrile-butadiene rubber; acrylonitrile copolymer; acrylonitrile rubber; butadiene-acrylonitrile* *copolymer; NBR rubber; nitrile-butadiene rubber; nitrile rubber; poly(acrylonitrile-co-butadiene).*

acrylonitrile-butadiene plastics. See acrylonitriole-butadiene-styrene resins.

acrylonitrile-butadiene resins. See acrylonitrile-butadiene-styrene resins.

acrylonitrile-butadiene rubber. See acrylonitrile-butadiene copolymer.

acrylonitrile-butadiene-styrene copolymer. A copolymer made from the three monomers acrylonitrile, butadiene and styrene. Its repeating structure can be given as $-CH_2CHCNCH_2CHCH-CH_2CH_2CH(C_6H_5)_n-$. Used in the manufacture of ABS resins. Abbreviation: ABS copolymer. Also known as *acrylonitrile-butadiene-styrene terpolymer; poly(acrylonitrile-co-butadiene-co-styrene).* See also acrylonitrile-butadiene-styrene resins.

acrylonitrile-butadiene-styrene plastics. See acrylonitrile-butadiene-styrene resins.

acrylonitrile-butadiene-styrene/polyamide alloys. See ABS-PA alloys.

acrylonitrile-butadiene-styrene/polycarbonate alloys. See ABS-PC alloys.

acrylonitrile-butadiene-styrene/polyurethane alloys. See ABS-PUR alloys.

acrylonitrile-butadiene-styrene/polyvinyl chloride alloys. See ABS-PVC alloys.

acrylonitrile-butadiene-styrene/styrene-acrylonitrile alloys. See ABS-SAN alloys.

acrylonitrile-butadiene-styrene/styrene maleic-anhydride alloys. See ABS-SMA alloys.

acrylonitrile-butadiene-styrene resins. A class of engineering thermoplastics made from acrylonitrile, butadiene and styrene by polymerization, grafting, physical mixing, or combinations thereof. They can be processed by extrusion, injection molding, blow molding, calendering, vacuum forming and other processes, and are platable and metallizable. *ABS resins* have outstanding strength and toughness, good impact strength, good dimensional stability from -40 to +71°C (-40 to +160°F), high resistance to heat distortion, good electrical and low-temperature properties, good chemical resistance, low water absorption, poor resistance to nitric and sulfuric acids and aldehydes, ketones, esters and chlorinated hydrocarbons, good resistance to alcohols, aliphatic hydrocarbons, mineral and vegetable oils, and slow-burning properties. Special grades including high-impact, low-temperature impact and heat-resistant high-strength are also available. Used for autobody parts and fittings, boats, radiator grilles, machinery and appliance housings, grilles for hot air systems and pump impellers, business machines, refrigerator door liners, garden equipment, plastic pipe and building panels, helmets, shower stalls, telephones, bottles, heels, luggage, and packaging materials. Abbreviation: ABS resins. Also known as *acrylonitrile-butadiene plastics; acrylonitrile-butadiene-styrene plastics; polyacrylonitrile-butadiene-styrene plastics.*

acrylonitrile-butadiene-styrene terpolymer. See acrylonitrile-butadiene-styrene copolymer.

acrylonitrile copolymer. See acrylonitrile-butadiene copolymer.

acrylonitrile fibers. Synthetic textile fibers based on acrylonitrile (C_3H_3N). They have high dielectric strength, good dimensional stability, and good to excellent resistance to water and solvents. Used as staple fibers and for monofilaments, screens and weaving.

acrylonitrile-methyl acrylate copolymer. A copolymer of about 94 wt% acrylonitrile and 6 wt% methyl acrylate with a typical molecular weight of 100000. Used in the manufacture of synthetic resins. Also known as *poly(acrylonitrile-co-methyl acrylate).*

acrylonitrile rubber. See acrylonitrile-butadiene copolymer.

acrylonitrile-styrene copolymer. A polymer alloy of acrylonitrile and styrene monomers that produces rigid thermoplastic parts when processed by injection molding or extrusion processes. It has good dimensional stability and scratch resistance, and is suitable for use in contact with foods. Used for containers, food service trays, bottles, etc.

acrylonitrile-vinyl acrylate copolymer. A copolymer of 65 wt% acrylonitrile, 33 wt% vinyl chloride and 2 wt% acrylamide methylpropane sulfonate. Used in the manufacture of resins. Also known as *poly(acrylonitrile-co-vinyl chloride).*

Acrylux. Trademark of Westlake Plastics Company (USA) for polymethyl methacrylate (PMMA) resins and products.

Acrynar. Trademark of PPG Industries, Inc. (USA) for durable spray-applied acrylic extrusion coatings with excellent weatherability and mar resistance, used for high-traffic areas, such as storefronts, condominium buildings, hospitals, schools, etc.

Acrypol. Trademark of CIL Paints (Canada) for a series of acrylic rubbing and polishing compounds.

Acryseal. Trademark of Collins & Aikman Corporation (USA) for pile fabrics based on acrylic fibers.

Acrysol. Trade name of Rohm & Haas Company (USA) for acrylic resins, and artificial leather made from polyacrylic acid.

Acrysteel. Trademark of Aristech Acrylics LLC (USA) for impact-resistant continuously cast thermoplastic acrylic sheeting for thermoforming applications. Also known as *Aristech Acrysteel.*

Acrytex. Trademark of Roehm GmbH (Germany) for polymethyl methacrylate (PMMA) resins and products.

Acrytite. Trademark of Sternson Limited (Canada) for a water repellent based on acrylic resins.

ACS SAN. Trade name of Biddle-Sawyer (USA) for a series of styrene-acrylonitrile (SAN) resins supplied in regular, high-heat, fire-retardant, high-impact, UV-stabilized, and 30% glass fiber-reinforced grades.

Act Carbide. Trade name of Atlantic Steel Corporation (USA) for a sintered tungsten carbide with a cobalt binder. Used for tools and cutters.

Acticoat. Trade name for a wound dressing material consisting of a rayon/polyester nonwoven core sandwiched between layers of high-density polyethylene mesh coated with silver. When applied to a wound, it forms an antimicrobial barrier and prevents wound adhesion.

Actimer. Trade name for a series of reactive fiber-reinforcing monomers.

Actinic. Trade name of Pennsylvania Wire Glass Company (USA) for amber-colored, heat-absorbing glass.

actinic glass. A glass with yellow tint that transmits more of the visible components of light, and less of the infrared and ultraviolet components. Used for skylights and factory windows.

actinides. See actinide metals.

actinide metals. The series of radioactive metals starting with atomic number 89 and including actinium (89), thorium (90), protoactinium (91), uranium (92), neptunium (93), plutonium (94), americium (95), curium (96), berkelium (97), californium (98), einsteinium (99), fermium (100), mendelevium (101), nobelium (102), lawrencium (103) as well as all elements beyond 103. All of them are radioactive, fill the 5*f* electron sublevel and, with the exception of uranium and thorium, have been artificially produced. Also known as *actinides.*

actinium. A rare, radioactive silvery-white metallic element that exists as the isotope ^{227}Ac in uranium ore, and may also be prepared by neutron bombardment of radium. It is the first member of the actinide series of the Periodic Table. Crystal system, cubic. Crystal structure, face-centered cubic. Density, 10.07 g/cm^3; melting point, 1050°C (1922°F); boiling point, approximately 3200°C (5790°F); atomic number, 89; atomic weight, 227.028 (longest-lived isotope); trivalent. ^{227}Ac has a half-life of 21.8 years. Used as a neutron source. Symbol: Ac.

actinium-uranium. See uranium-235.

actinouranium. See uranium-235.

actinolite. A dark green variety of asbestos that belongs to the amphibole group of minerals, and is composed of calcium iron magnesium silicate hydroxide, $Ca_2(Mg,Fe)_5Si_8O_{22}(OH)_2$. Crystal system, monoclinic. Density, 2.9-3.2 g/cm^3; hardness, 5-6 Mohs; refractive index, 1.61-1.64. Occurrence: Canada, Europe, USA. Used as a minor source of asbestos, as a building material, and for acid-resistant and high-temperature applications.

Actionwear. Trade name of Actionwear Inc. (USA) for nylon fibers used for the manufacture of sports and outdoor clothing.

Actisorb Silver 220. Trade name for an activated carbon cloth impregnated with silver and used as a dressing material for wound care applications. When applied to a wound the release of silver provides antimicrobial action and bacterial toxin management.

activated alumina. A granular, highly porous form of alumina (Al_2O_3) that has a strong affinity for moisture, gases and vapors. It will not soften, swell or disintegrate in water, and is extremely resistant to thermal and mechanical shock and abrasion. Used as a desiccant, catalyst, catalyst carrier and absorbent, in water purification (fluoride removal), and in the dehydration of organic solvents. See also alumina.

activated bauxite. Bauxite that has been ground, screened and calcined in sizes from 20-60 and 30-60 mesh. Used as a filter medium in oil refineries. Also known as *filter bauxite.* See also bauxite.

activated carbon. An amorphous, porous, highly carbonaceous substance that is usually obtained by the destructive distillation of carbonaceous materials, such as wood, nut shells, animal bones, etc., and subsequently heated with steam or carbon dioxide to develop high adsorptive properties. It has a very large surface area per unit volume, and is used as an adsorbent for gas masks and cigarette filters, in water, air and acid purification, in the recovery of solvents, in the decolorization of liquids, in chromium electroplating, in air conditioning, and in the purification of stack gases. Also known as *activated charcoal; active carbon; active charcoal; filter carbon.*

activated carbon fiber. A carbon fiber that has been specially processed to increase its porosity and absorptive properties. It is prepared by first spinning the precursor material, usually a phenolic, polyacetate, cellulose, polyacrylonitrile or an isotropic pitch, followed by an anti-flammable process at temperatures between 200 and 400°C (390 and 750°F), and a final activation process in which the spun material is heated in the temperature range between 800 and 1200°C (1470 and 2190°F) in the presence of water vapor or carbon dioxide. Used in air, water and blood filtration. Abbreviation: ACF.

activated charcoal. See activated carbon.

activated clay. A clay, such as bentonite or fuller's earth, that is treated with an acid to enhance its absorptive properties or

bleaching action. Used as an absorbent, and in oil bleaching. Also known as *active earth; bleaching clay; bleaching earth.*

activated lime mortar. A lime mortar to which an activator, such as potash alum or an alkali carbonate, has been added to shorten the time of set. Used for external plastering, e.g., on building walls. See also lime mortar.

activated rosin fluxes. A group of soldering fluxes composed of water-white rosin dissolved in an organic solvent, such as alcohol, turpentine or petroleum. The additive enhances activity, and thus wetting by the solder, and the flux residue is noncorrosive and nonconductive. Used especially in electric and electronic soldering, and for other critical soldering applications. Also known as *noncorrosive fluxes.* See also rosin.

activating agent. (1) A substance which when added to another substance will promote or speed up a physical or chemical change. Also known as *activator.*

(2) See accelerator (5).

activator. (1) See accelerating agent (1).

(2) See dopant (2).

active carbon. See activated carbon.

active charcoal. See activated carbon.

active earth. See activated clay.

actively smart material. See intelligent material.

active material. (1) The chemical material, e.g., lead oxide, contained in the plates that provides the electrical action of a storage cell.

(2) A radioactive substance.

(3) The phosphor coating used in cathode-ray tube screens.

(4) A fissionable material that releases significant amounts of atomic energy.

Active Ni. Trade name of British Driver Harris Company Limited (UK) for a commercially pure nickel containing 0.2% silicon, and traces of magnesium. Used for thermionic valves.

Actobond. Trademark of Actol Chemicals Limited (Canada) for polyvinyl acetate emulsions.

Actolene. Trademark of Actol Chemicals Limited (Canada) for thermoplastic starch resins.

Acton Tungsten Carbide-Cobalt. Trade name of Acton Materials, Inc. (USA) for cobalt-bonded tungsten carbide.

Actores. Trademark of Actol Chemicals Limited (Canada) for thermoplastic starch resins.

Actoresin. Trademark of Actol Chemicals Limited (Canada) for polyvinyl acetate emulsions.

Actorez. Trademark of Actol Chemicals Limited (Canada) for polyvinyl acetate emulsions.

Actose. Trademark of Actol Chemicals Limited (Canada) for modified starches.

Actoseal. Trademark of Actol Chemicals Limited (Canada) for starch adhesives.

Actox. Trademark of The New Jersey Zinc Company, Inc. (USA) for a series of lead-free zinc oxides produced either from zinc ore or from the metal. They are supplied as pellets and free-flowing powders, and used as reinforcing agents and rubber accelerators.

Acupren. Trade name for a silicone elastomer used for dental impressions.

Acusil. Trade name for a silicone elastomer used for dental impressions.

ACuZinc. Trademark of General Motors Corporation (USA) for strong, wear-resistant zinc die-casting alloys with copper contents between 5 and 11%, suitable for structural applications at elevated temperatures. *ACuZinc5* is a zinc alloy containing 5%

copper and 3% aluminum, optimized for hot-chamber die casting, and *ACuZinc 10* with 10% copper and 3.5% aluminum is designed for cold-chamber die casting. *ACuZinc* alloys have enhanced resistance to corrosion, damage and creep.

acyclic compound. An organic compound, such as methane, in which the atoms are arranged in an open (straight or branched) chain. See also aliphatic compound.

Adabraze. Trade name of Adamas Carbide Corporation (USA) for a sintered carbide coated with a thin layer of pure cobalt. Used to improve brazing to base-metal tool shanks or supports.

Adair. Trade name of Arriscraft Corporation (Canada) for marble and marble products.

Ad-Aluminum. Trade name for an aluminum brass containing 82% copper, 15% zinc, 2% aluminum and 1% tin.

Adalloy. Trade name of Adamas Hardfacing Company (USA) for iron-chromium-base hardfacing electrodes for build-up and repair on industrial equipment, e.g., shovels, pulverizers and plows.

Adamant. (1) Trade name of Firth Brown Limited (UK) for normalized medium- or high-carbon steel castings used for tools and fixtures.

(2) Trade name of Richard W. Carr & Company Limited (UK) for a water-hardening tool steel containing 1.45% carbon, 4.75% tungsten, 0.5% chromium, 0.25% vanadium, and the balance iron. It keeps a keen cutting edge, and is used for rifling and burnishing tools.

(3) Trade name of Magnolia Metal Company, Inc. (USA) for tough babbitts, and bearing and antifriction alloys and metals of tin, copper and antimony. Used for marine, airplane and internal-combustion engine bearings.

(4) Trademark of Resco Products, Inc. (Canada) for refractory bricks and wet refractory cement.

(5) Trademark of United States Gypsum Company (USA) for plasterboard and asphalt roofing.

adamantane. A white, crystalline compound (99+% pure) whose molecular structure consists of four fused cyclohexane rings. It has a melting point of 209-212°C (408-414°F) (sublimes), and is used in the form of derivatives to enhance the chemical, thermal or solvent resistance of certain plastics, in pharmaceuticals and synthetic lubricants, and in the life sciences. Formula: $C_{10}H_{16}$.

Adamantine. Trade name of Babcock & Wilcox Company (USA) for a wear-resistant steel containing 0.7% carbon, 0.7% manganese, 0.7% chromium, and the balance iron. Used for steel balls for grinding mills.

Adamas. Trade name of Adamas Carbide Corporation (USA) for a series of straight and complex cemented carbides. The straight grades contain tungsten carbide in cobalt binders, and the complex grades contain either tungsten carbide and tantalum carbide in cobalt binders, tungsten carbide and titanium carbide in cobalt binders, tungsten carbide, tantalum carbide and titanium carbide in cobalt binders, or titanium carbide and molybdenum carbide in nickel binders. Used for cutting tools for ferrous and nonferrous alloys, drawing dies, and wear parts.

Adamite. Trade name of Gulf & Western Manufacturing Company (USA) for a wear-resistant high-carbon cast iron containing 3-3.5% carbon, 0.5-2% silicon, 1-1.5% chromium and 0.75% nickel. It has good tensile strength, and is used for castings, drawing and forming dies, mill guides, rolls, and furnace parts.

adamite. A greenish-yellow mineral composed of zinc arsenate hydroxide, $Zn_2(AsO_4)(OH)$. Crystal system, orthorhombic.

Density, 4.34-4.46 g/cm³; refractive index, 1.742; hardness, 3.5 Mohs. Occurrence: Europe, Mexico.

Adams' catalyst. A crystalline platinum dioxide (PtO_2) with a purity of 99.5%, a density of 11.8 g/cm³, a melting point of 450°C (842°F), and a surface area of 75 m²/g or more. It is prepared by the Adams' nitrate fusion method, contains 80-85% platinum, and is low in iron. Used in the hydrogenation of platinum.

adamsite. A greenish-black variety of *muscovite* mica from Vermont (USA).

Adams Lake. Trade name of Holding Lumber Company Limited (Canada) for spruce lumber.

Adamull. Trademark of Resco Products, Inc. (Canada) for wet refractory cements and firebricks.

Adaptaloy. Trade name of American Smelting & Refining Company (USA) for a zinc casting alloy containing 1.8% copper, 1% silicon and 1% manganese. It has high elongation and impact strength, and is used for structural castings, valve handles, and ornamental parts.

Adaptic. Trade name for a conventional self-cure composite formerly used in restorative dentistry. *Adaptic II* is a light-cure hybrid composite restorative, *Adaptic LCM* a light-cure microfine composite restorative and *Adaptic Universal* a self-cure hybrid composite restorative.

Adaset. Trademark of Resco Products, Inc. (Canada) for a wet refractory cement.

Adastral A. Trade name of Stone Manganese–J. Stone & Company Limited (UK) for a tin-base bearing alloy with good self-lubricating properties containing 4.5% antimony and 3% copper.

Adatosil. Trade name for medical-grade liquid injectable silicone for skin reconstruction and plastic surgery.

Adcote. Trademark Morton Thiokol, Inc. (USA) for a series of polymer adhesives, and decorative and protective organisols and plastisols. The adhesives include heat-seal, pressure-sensitive, bonding and laminating grades that bond well to film, foil and paper substrates.

Addcrete. Trade name of Pacific Industries Limited (Canada) for concrete additives.

Addiment. Trademark of Heidelberger Zement (Germany) for concrete additives.

addition-cured silicone. See A-silicone.

addition polyimides. Crosslinked polyimides that are the products of addition reactions between unsaturated groups of imide monomers or oligomers. Examples include bismaleimides, reverse Diels-Alder polyimides, and acetylene-terminating polyimides.

addition polymer. A polymer, such as polyethylene, polypropylene or polystyrene, formed by a chemical reaction in which bifunctional unsaturated monomer units are added to each other to form linear (long-chain) polymer macromolecules. Also known as *addition resin; chain-reaction polymer.*

addition resin. See addition polymer.

additive. (1) In general, a substance added in relatively small quantities to another substance to add or enhance certain properties.

(2) In polymer engineering, a substance, such as a plasticizer, initiator, curative, blowing agent, heat or light stabilizer, antistatic agent or flame retardant, added to a plastic or rubber to enhance its properties.

(3) In casting, any carbonaceous (e.g., seacoal, pitch, or graphite) or cellulosic (e.g., cereal, or wood flour) material

incorporated in molding sand mixtures, but not involved in the bonding process.

Additivato. Trade name of Agriplast Srl (Italy) for durable UV-stabilized polymer films with excellent mechanical properties, and good photo-oxidation and weathering resistance. Used for protecting flowers and crops.

Adekit. Trade name of Axson North America (USA) for a series of adhesives including epoxies for structural bonding, polyurethanes for flexible bonding, and methacrylates for flexible bonds with high mechanical resistance.

adelite. A colorless, gray, or bluish to yellowish-gray mineral of the descloizite group composed of calcium magnesium arsenate hydroxide, $CaMgAsO_4(OH)$. Crystal system, orthorhombic. Density, 3.71-3.76 g/cm^3; refractive index, 1.721; hardness, 5 Mohs. Occurrence: Sweden.

Adell. Trade name of Adell Plastics Corporation (USA) for a series of thermoplastic resins. *Adell A* are polyamide 6,6 resins supplied in regular, 10 and 30% carbon fiber-reinforced, 10, 33 and 60% glass fiber-reinforced, 20% polytetrafluoroethylene- or molybdenum disulfide-lubricated, 25 or 50% glass bead-filled, mineral-filled, glass fiber- and bead-reinforced, fire-retardant, high-impact, supertough, and UV-stabilized grades. *Adell B* refers to polyamide 6 resins supplied in regular, 30% carbon fiber-reinforced, 10, 30 and 60% glass fiber-reinforced, polytetrafluoroethylene- or molybdenum disulfide-lubricated, 40% glass bead- or mineral-filled, glass fiber- and bead-reinforced, casting, elastomer copolymer, stampable sheet, fire-retardant, high-impact, supertough, and UV-stabilized grades. *Adell DR* are polycarbonate resins supplied in regular, 30% carbon fiber-reinforced, 20 and 30% glass fiber-reinforced, 15% polytetrafluoroethylene-lubricated, fire-retardant, high-flow, UV-stabilized, and structural foam grades. *Adell DS* refers to polycarbonate resins supplied in regular, 30% carbon fiber-reinforced, 20 and 30% glass fiber-reinforced, 15% polytetrafluoroethylene-lubricated, fire-retardant, high-flow, UV-stabilized, and structural foam grades.

Adetex. Trade name of Adetex SL (Spain) for latex foams.

Adherence M5. Trade name for a dual-cure dental resin cement.

Adherite. Trade name of MacDermid Inc. (USA) for a series of iron conversion coatings.

adhesion agent. See adhesion promoter (1), and (2).

adhesion promoter. (1) A special coating, usually a low-viscosity fluid, applied to a substrate prior to the application of an adhesive to improve the adhesion and flowing properties of the latter. Also known as *adhesion agent; adhesion-promoting agent; primer (3)*.

   (2) A substance (admixture) that is introduced in small amounts to a batch of concrete to improve its adhesion characteristics. Also known as *adhesion agent; adhesion-promoting agent; antistripping additive; antistripping agent; bonding additive; dope; nonstripping agent*.

adhesive film. A dry, thin film of thermosetting resin adhesive used in the manufacture of bonded or laminated products, such as plywood, densified wood, etc.

adhesively bonded nonwovens. Nonwoven fabrics that in contrast to thermally pressed or mechanically interlocked types have been bonded by the application of a suitable adhesive.

adhesives. A group of substances in film, liquid, powder or paste form capable of bonding two or more similar or dissimilar materials together by surface attachment. Commercial adhesives can be grouped into the following categories: (i) *Inorganic adhesives*, such as water glass, phosphate cements, Port-

land and hydraulic cements; (ii) *Natural organic adhesives*, such as animal glues (from fishes, hides, or bones), vegetable glues (e.g., tapioca paste, casein and rubber latex, gums, mastics and mucilages), and mineral-base adhesives (e.g., asphalt and pitch); and (iii) *Synthetic organic adhesives*, such as thermoplastic and thermosetting resins (e.g., polyethylenes, polyamides, epoxies, and phenol-formaldehydes), elastomer-solvent cements, chlorinated rubber, and silicone polymers and cements. See also natural adhesives; synthetic adhesives; glues.

adhesive tape. See tape (1).

Adic. Trade name of Eagle & Globe Steel Limited (Australia) for an air-hardening hot-work tool steel containing 0.4% carbon, 1% silicon, 5% chromium, 1.3% molybdenum, 1.0% vanadium, and the balance iron. Used for aluminum, zinc and magnesium die-casting tools, hot-work tools and dies, extrusion dies and mandrels, die inserts, piercing and blanking tools, and shear blades.

ADI irons. See austempered ductile irons.

adipocellulose. A cellulose found in corky tissue and composed of normal cellulose and cutin.

adipocerite. See hatchettine.

Adiprene. Trademark of E.I. DuPont de Nemours & Company (USA) for a series of wear- and chemical-resistant polyurethane rubbers made by reacting diisocyanate with polyalkylene ether glycol. They are viscous liquids that can be converted into solid products by conventional processing techniques. Used for belts, hoses, rolls, wheels, bearings, bushings, seals, pads, linings, and sporting equipment.

Adlerstahl. Trade name of Zweigbetrieb der Carp & Hones Deutsche Edelstahlwerke GmbH (Germany) for a series of plain-carbon, cold-work and high-speed tool steels, and several alloy machinery and chromium steels.

Admer. Trademark of Mitsui Chemicals, Inc. (Japan) for a modified thermoplastic polyolefin resin with functional groups, supplied in the form of pellets and powders, and used as adhesive layer in co-extrusion applications (e.g., bottles, tubes, sheets, and films), and for coatings. It has good mechanical strength and excellent heat and chemical resistance, and is designed to bond, by thermal reaction, to a variety of ionomers, polycarbonates, polyamides, polystyrenes, ethylene vinyl alcohols (EVOHs), polyethylenes, polyethylene terephthalates, and polypropylenes.

Admira Bond. Trade name for a dental resin used for dentine/enamel bonding.

Admira Flow. Trade name for an ormocer of flowable consistency used in restorative dentistry. See also ormocer.

admiralty alloy. An alpha brass containing 70-73% copper, 0.75-1.20% tin, and the balance zinc. Available in the as-cast, as-rolled and annealed condition, it has excellent cold workability for forming and bending, improved hardness and strength, good corrosion resistance, good resistance to dilute acids and alkalies, seawater and moist sulfurous atmospheres, and a melting point of 935°C (1715°F). Used for condensers, evaporators, distiller and heat exchanger tubing, ferrules, pump rods, fixtures, and marine equipment. Also known as *admiralty brass; admiralty bronze; admiralty metal*.

admiralty brass. See admiralty alloy.

admiralty bronze. (1) See admiralty alloy.

   (2) A copper-tin alloy containing 88% copper, 10% tin, and 2% zinc. Used for gears, trolley wheels, and worm wheels.

admiralty gunmetal. A strong, corrosion-resistant golden-colored casting alloy containing 88% copper, 10% tin, and 2%

zinc. It has a density of 8.7 g/cm³, and good castability and bearing properties. Used for valves, pump parts, steampipe fittings, bearings, hydraulic castings, and formerly also for cannons, hence its name. Also known as *G bronze; gunmetal.*

admiralty metal. See admiralty alloy.

admiralty nickel. A corrosion-resistant alloy composed of 70% copper, 29% nickel and 1% tin, used for condenser and heat exchanger tubes, valve diaphragms, and chemical equipment.

admiralty white metal. A bearing alloy containing 84-90% tin, 8-9% antimony and 2-7% copper.

Admiral Wood. Trademark of Georgia-Pacific Corporation (USA) for pressure-treated lumber used for paneling.

Admiro. Trade name of Allgemeines Deutsches Metallwerk GmbH (Germany) for a corrosion-resistant copper-zinc alloy containing 43-48% copper, 35% zinc, 10-15% nickel, 3% manganese, 2% aluminum, and 2% iron. It is available in the as-cast and as-rolled condition, and used for tubes and turbine bearings.

admixture. A material, other than aggregates, cement and water, that is introduced in small amounts to a batch of concrete to chance its performance characteristics in the desired manner.

admontite. A colorless mineral composed of magnesium borate hydrate, $Mg_2B_{12}O_{20} \cdot 15H_2O$. Crystal system, monoclinic. Density, 1.88 g/cm³; refractive index, 1.5. Occurrence: Europe (Austria).

Admos. Trade name of Allgemeines Deutsches Metallwerk GmbH (Germany) for an alloy containing 40-55% copper, 22-40% zinc, 3-15% nickel, 1-3% manganese, 1-2% iron, and 0.5-3% aluminum. It has excellent resistance to corrosion and erosion due to superheated steam. Used for bearings, bushings, valves, gears, worm wheels, turbine blades, and condenser and heat exchanger tubing. Also included under this trade name are several cast leaded coppers used for bearings and bushings.

Adnic. Trade name of Century Brass Products Inc. (USA) for admiralty nickel. See also admiralty nickel.

adobe. (1) A large brick composed of earth or clay reinforced with straw, and molded to standard shape by baking in the sun. Also known as *adobe brick; sun-dried brick.*

(2) A term used for calcareous sandy silts and sandy, silty clay deposits found in the semiarid regions of the southwestern United States and in Mexico. It corresponds to loess, a similar deposit found in Europe and Asia. Used for making adobe brick (1). Also known as *adobe clay.* See also loess.

adobe brick. See adobe (1).

adobe clay. See adobe (2).

Adorbond G. Trade name for a high-gold dental bonding alloy.

Adpo. Trade name of Nova Chemicals, Inc. (USA) for polypropylene resins.

Adr. Trade name of Creusot-Loire (France) for an iron alloy containing 40% nickel. It has a very low coefficient of expansion at room temperature, and is used for instrument components, lead wires, etc.

Adretta. Trademark of Adretta for polyvinyl chloride (PVC) resins and film materials.

Adriatical. Trade name of Alfa Romeo (Italy) for an aluminum alloy, similar to duralumin, containing 4-5% copper. Used for light-alloy parts. See also duralumin.

ADS3. Trade name of Sumitomo Metal Industries Limited (Japan) for a high-tensile-strength steel containing 0.6% molybdenum and 0.3% vanadium, and small, grain-refining additions of niobium for high-temperature tempering resulting in a microstructure of tempered martensite with finely dispersed carbides. Its low contents of phosphorus, sulfur and manganese

reduce grain boundary embrittlement. *ADS3* steel has excellent resistance to hydrogen embrittlement and cracking, and is used for high-tensile bolts.

adsorbent. A solid or liquid substance that has the ability to cause a liquid, vapor or gas (called the adsorbate) to adhere to its surface without a chemical or physical change during the process. Examples include activated carbon, activated alumina, activated clay, magnesium silicate, silica gel, mercury and water.

Adspec. Trade name of N.C. Ashton Limited (UK) for a wrought aluminum bronze containing 89% copper, 8.5% aluminum, and 2.5% iron. Used for hardware, fasteners, and decorative trim.

adularia. A colorless, translucent to transparent variety of orthoclase feldspar with a bluish opalescence, found in pseudo-orthorhombic crystals. See also feldspar; orthoclase.

Advagum. Trade name for a highly-plasticized vinyl copolymer used as a rubber extender.

Advalite. Trademark Carlisle Chemical Works, Inc., Advance Division (USA) for various organic and organometallic compounds used as heat and light stabilizers in vinyl coatings.

Advance. (1) Trade name of US Reduction Company (USA) for a babbitt metal containing varying amounts of tin, copper, and antimony. Used for bearings.

(2) Trade name of Driver Harris Company (USA) for a resistance alloy containing 55% copper, 43.5-45% nickel, and 0-1.5% manganese. It has a low temperature coefficient of resistance, and is used for pyrometers.

3) Trademark of Dentsply/LD Caulk (USA) for a resin-modified dental ionomer cement used for luting restorations, and as a cavity liner and base. See also ionomer.

advanced aluminide. A compound based on nickel aluminide (Ni₃Al), and characterized by high hardness, excellent oxidation and thermal-shock resistance, and a low coefficient of thermal expansion. Used for turbine blades, combustion chambers, glass-processing equipment, and flame-sprayed coatings. See also nickel aluminide.

advanced ceramics. A group of ceramics with unique chemical, electrical, magnetic, mechanical and/or optical properties, produced from powders by any of various processing technologies (e.g., hot pressing, reaction bonding, or sintering). Examples include aluminum nitride, aluminum oxide, boron carbide, boron nitride, silicon carbide, titanium diboride, and zirconium oxide. Used for heat, internal-combustion and turbine engines, cutting tools, military armor systems, in energy generation, conversion and storage, and in electronic packaging. Also known as *high-tech ceramics.*

advanced composites. Carbon-, ceramic-, polymer- or metal-matrix composites that are reinforced with continuous fibers whose elastic modulus exceeds that of fiberglass.

advanced composite thermoplastics. High-performance composite materials with a relatively high proportion of reinforcing fibers (typically 60 vol%), usually in continuous unidirectional, or woven form, contained in thermoplastic resin matrices.

Advanced Duty Oilite Bronze. Trade name of Chrysler Corporation (USA) for an Oilite-type bearing material containing 87.5-90.0% copper, 0-1% iron, 0-1.75% carbon, and 9.5-10.5% tin. See also Oilite.

advanced materials. Materials, such as metals, alloys, ceramics, polymers and composites, that are employed in advanced modern technology, i.e., for high-tech applications. They can be improved conventional materials, or novel, specially engineered materials. Examples include materials used in the manufacture of electronic, microelectronic and optoelectronic components (e.g., diodes, transistors, integrated circuits, or liquid crystal displays), computers and magnetic information storage systems, video recorders, camcorders, compact disk players, laser systems, and materials for fiber-optic and microwave communication systems and waveguides, satellite and rocket systems, aerospace systems (e.g., USAF Stealth fighter airplane, NASA Space Shuttle, Mars Pathfinder, Hubble telescope and Space Station). Also known as *high tech materials.*

advanced polymers. Novel polymeric materials, such as ultra-high molecular-weight polyethylenes, liquid crystal polymers and thermoplastic elastomers, which due to their unique combination of properties are suitable for advanced modern technology, i.e., high-tech applications. Also known as *high tech polymers.*

Advancer. Trademark of Norton Pakco (USA) for silicon nitride-bonded silicon carbide.

Advanta. Trademark of DuPont Dow Elastomers (USA) for specialty thermosetting elastomers supplied in several grades. They have good fluid resistance, fair compatibility with fuel, lubricating and hydraulic oils, vegetable oils, aliphatic hydrocarbons, alcohols and water, good high-temperature performance, a hardness of 75 Shore A, a service temperature range in air of -12 to +175°C (-10 to +347°F), a glass transition temperature of -16°C (3°F). Used for O-rings, hoses, tubing, valve stem seals, valve lifter seals, transmission and radiator seals, shaft seals, engine gaskets, vibration damping systems, wire insulation and jacketing, and business machine rollers.

Advantage. Trade name for a dental bonding resin used in orthodontics.

Advantex. Trade name of Owens Corning (USA) for glass-fiber reinforcements supplied in continuous and chopped strand grades. Their properties are intermediate between those of *E-glass* and *E-CR glass* combining good electrical and mechanical properties with high acid and heat resistance. Used in the form of rovings, mats and surfacing veils in composite materials.

Advantra. Trademark of H.B. Fuller Company (USA) for packaging adhesives.

Advastone. Trade name for a die stone for dental applications.

Advawax. Trade name for a series paraffin waxes used for paper coatings.

Advitrol. Trademark of Süd-Chemie, Inc. (USA) for a range of thixotropic materials based on castor oil, or castor oil derivatives.

ADX Alloys. Trade name of General Motors Corporation (USA) for a family of creep-resistant magnesium-aluminum-calcium-strontium casting alloys with good corrosion resistance, high-temperature properties and castability. Used for powertrain components, such as transmissions and engine blocks.

Aegis. Trade name of Republic Steel Corporation (USA) for a steel containing 0.15-0.45% carbon, 0.1-0.2% silicon, 0.5-0.8% manganese, 3.25-3.75% nickel, and the balance iron. Used for machinery parts, gears, shafts, and axles.

Aegisglass. Trademark of Diamonex (USA) for a superhard diamond-like carbon coating used on glass substrates to provide sapphire-like wear resistance.

aegrine. See acmite.

aegrite. See acmite.

Aelener Zinc. An alloy containing 51.4% zinc, 21.6% lead, 20.5% tin, 3.5% copper and 3% antimony, used for solders and bearings.

Aelitefil. Trademark of BISCO Inc. (USA) for a non-sticky, light-cure bisphenol glycidylmethacrylate (Bis-GMA) hybrid composite that is filled with particles of specially milled barium glass with a controlled size distribution of 0.7 μm (28 μin.), and also contains submicron silicon. It has high strength and polishability, controlled polymerization shrinkage, and outstanding optical and physical properties. Supplied in a wide range of matching shades, it is used in dentistry for all types of restorations.

Aeliteflo. Trademark of BISCO Inc. (USA) for a light-cure, radiopaque bisphenol glycidylmethacrylate (Bis-GMA) micro-hybrid composite filled with 60% barium glass particles with a controlled size distribution of 0.7 μm (28 μin.). It is flowable, yet thixotropic to prevent slumping, and has a low modulus of elasticity, high strength, good stain and wear resistance, and excellent polishability. Used in dentistry for restorations, porcelain repair, porcelain veneer cementation, and as a restorative liner and pit and fissure sealant.

Aeliteflo LV. Trademark of BISCO Inc. (USA) for a light-cure, radiopaque, low-modulus microhybrid filled with barium glass particles with an average size of 0.7-0.8 μm (26-32 μin.). It has low viscosity and high flowability, and is used in dentistry as a pit and fissure sealant.

aenigmatite. A black, or brown mineral of the amphibole group composed of sodium iron titanium silicate, $Na_2Fe_5TiSi_6O_{20}$. It can also be made synthetically. Crystal system, triclinic. Density, 3.14-3.87 g/cm³. Also known as *enigmatite.*

Aeonite. Trade name of Shanks & Company Limited (UK) for a white metal containing varying amount of copper and nickel. Used for sanitary appliances. See also white metal.

Aeralloy. (1) Trade name of Engelhard Corporation (USA) for a hard platinum-ruthenium contact alloy with high ruthenium content, and long wear life. Used for aircraft magneto contacts.

(2) Trade name of Engelhard Corporation (USA) for a heat- and arc erosion-resistant platinum-iridium alloy used for electrical contacts, thermostats, and vibrators.

aerated cement. See aerated concrete.

aerated concrete. A lightweight ceramic product consisting of Portland cement, cement-pozzolan, cement-silica, lime-pozzolan or lime-silica pastes, and containing a high proportion of entrapped air or gases introduced by chemical means (e.g., gas-forming chemicals, foaming agents, etc.), or mechanical means (e.g., whipping air into the mix). It has improved insulating properties, and is used mainly for making precast building units (blocks, slabs, etc.). Also known as *aerated cement; cellular concrete; foamed concrete; gas concrete; gas-formed concrete.*

Aeress. Trade name for modacrylic fibers used especially for industrial fabrics, and carpets.

Aerex 350. Trademark of SPS Technologies Inc. (USA) for a high-strength, high-temperature multiphase nickel superalloy containing 25% cobalt, 17% chromium, 4% tantalum, 3% molybdenum, 2% titanium, 2% tungsten, 1.1% niobium, 1% aluminum, 0.015% carbon, and 0.015% boron. It has outstanding tensile and impact strength, excellent thermal stability and stress relaxation, good mechanical properties up to about 730°C (1350°F), good ductility, high resistance to creep and stress rupture, high toughness, good corrosion resistance, relatively high cryogenic impact strength, excellent forgeability, and a low coefficient of thermal expansion. Used for fasteners (bolts, etc.) in advanced gas turbine engines, engine blades and disks.

AerMet 100. Trademark of Carpenter Technology Corporation

(USA) for a corrosion-resistant nickel-cobalt steel strengthened by carbon, chromium and molybdenum additions. The nominal composition is: 0.23% carbon, 13.4% cobalt, 11.1% nickel, 3.1% chromium, 1.2% molybdenum, and the balance iron. It has high strength and toughness, good fracture toughness, optimum combination of strength and toughness, and good stress-corrosion-cracking resistance. Used for aerospace applications, e.g., landing gears, arresting hooks, structural members, fasteners, armor, jet engine shafts, driveshafts, helicopter masts, automotive driveshafts, high-strength gun breeches and bolts, high-impact tooling, racing bike frames, etc.

AerMet 310. Trademark of Carpenter Technology Corporation (USA) for a corrosion-resistant steel composed of 0.245% carbon, 2.4% chromium, 11.0% nickel, 15% cobalt, 1.4% molybdenum, and the balance iron. It is similar to *AerMet 100*, but provides higher strength and toughness. Used for aircraft and aerospace applications, e.g. landing gears.

Aero. Trade name of Magnolia Anti-Friction Metal Company (USA) for a tin-base alloy for high-speed and high-load bearings.

Aero-Board. Trademark of Canadian Forest Products Limited for hardboard.

aeroclay. A fine clay, especially china clay, made by drying and air separation.

Aerocapsule. Trade name of Teijin Fibers Limited (Japan) for a series of polyester filaments used for the manufacture of textile fabrics.

Aerocor. (1) Trade name of PPG Industries Inc. (USA) for glass fibers.

(2) Trade name of Fiberglas Canada Inc. for superfine glass wool used for thermal and acoustical insulation applications.

Aerocote. Trade name of Aerocote Corporation (USA) for iron, manganese and zinc phosphate coatings.

Aerocrete. Trade name for an aerated concrete made by the addition of aluminum powder or flakes to cement to effect a chemical reaction. It sets hydrogen-free making the concrete porous, and has improved insulating properties.

Aerodux. (1) Trademark of Dyno Industrier AS Company (Norway) for a series of aminoplast and phenoplast resins.

(2) Trademark of Ciba-Geigy Limited (UK) for a series of resorcinol-formaldehyde adhesives.

Aeroflex. (1) Trade name for a series of textile glass-fiber base mats for ducts, automotive insulation, and sound-deadening materials.

(2) Trade name for polyethylene extrusions.

Aerofluid. Trade name of Metalor Technologies SA (Switzerland) for a series of dental casting alloys including *Aerofluid 2PF*, a hard alloy containing 78% gold and 1% platinum, *Aerofluid 3*, an extra-hard alloy containing 71% gold, 2% platinum and 2% palladium, and *Aerofluid M*, an extra-hard alloy containing 71% gold, 4% platinum and 2% palladium.

Aerofoam. Trademark of Borden Chemical Company Limited (USA) for a series of foamed polystyrene products used for thermal insulation applications, and aircraft seat cushions.

aerogel. (1) A highly porous, low-density solid made from a gel using a gaseous dispersing medium. Examples include flexible and rigid plastic foams. It is the opposite of an aerosol. See also aerosol; gel.

(2) A porous, low-density, three-dimensional silica or resorcinol-formaldehyde particulate system with a fractal microstructure obtained by sol-gel polymerization.

Aerogun. Trade name of North American Refractories Corpora-

tion (USA) for refractory gunning mixes.

Aerolam. Trademark of Hexcel Corporation (USA) for glass- or carbon fiber-reinforced epoxy tooling prepregs.

Aerolastic. Trademark for a fuel-proof joint-sealing compound used in jet engines.

Aerolite. (1) British trade name for an aluminum alloy containing 1.2% copper, 1% iron, 0.5% silicon, and 0.4% manganese. Used for automotive components, pistons, etc.

(2) Trade name of British Aluminium Company Limited (UK) for an aluminum alloy containing 12% copper and 2% manganese. It has good machinability, and is used for automotive parts and engine pistons.

(3) Trade name of PPG Industries, Inc. (USA) for extremely thin laminated sheet glass used primarily for aircraft fuselage windows.

(4) South African trade name for glass-fiber insulation bonded with an inert thermosetting resin. Used for building insulation applications.

(5) Trade name of Aero Research Limited (UK) for urea-formaldehydes including several adhesives.

(6) Trade name of Smith-Alsop Paint & Varnish Company (USA) for synthetic exterior enamel paints.

(7) Trade name of AEROSTAR International, Inc. (USA) for a durable, lightweight fabric for hot air balloons.

Aerolor. Trade name of Carbone de America for a composite material consisting of vapor-infiltrated carbon matrix reinforced with polyacrylonitril-based carbon fibers. Used for aerospace applications.

Aerometal. Trade name for an aluminum alloy containing 0.2-4.0% copper, 0.3-1.3% iron, 0-1.2% manganese, 0-3.0% manganese, 0-3.0% zinc, and 0.5-1.0% silicon. Available in the as-cast and as-rolled condition, it is used for automotive engine parts.

Aeromin. German trade name for a wrought aluminum alloy containing 6.2% manganese, 0.8% iron, and 0.3% silicon. Used for light-alloy parts.

Aeron. (1) Trade name for an aluminum alloy containing 1.5-20% copper, 1.0% silicon, and 0.75% manganese.

(2) Trade name of Flexfirm Products, Inc. (USA) for vinyl-coated nylon.

Aerophenal. Trademark of Ciba-Geigy Limited (UK) for a series of phenol-formaldehyde resins.

aeroplane fabrics. Closely woven, tear-resistant fabrics used as outer coverings of light and superlight aircraft, gliders and hang-gliders. They are usually made of polyamide or polyester, and may be laminated with polymer films. Also known as *airplane fabrics*.

Aeroplastic. Trade name of Aeroplastics Limited (UK) for phenol-formaldehyde plastics.

Aeroplex. Trade name Aeroplex Company Limited (UK) for a laminated safety glass.

AeroRove. Trade name Glass Fibers Products Inc. (USA) for glass fiber rovings, chopped strands, and reinforced mats and yarns.

Aero-Seal. Trademark for a vulcanized natural rubber that is introduced into asphalt in hot, liquid form, and used as a sealing compound.

Aerosil. (1) Trademark of Degussa AG (Germany) for ultrafine, white fumed silica powder made from silicon tetrachloride. Used as a flatting agent for paints, in rubber and polymer compounding, and as a grease thickener.

(2) Trade name of Georgsmarienwerke Selesiastahl GmbH (Germany) for a series of austenitic, ferritic and martensitic

stainless steels.

aerosol. A relatively stable suspension of ultramicroscopic solid or liquid particles dispersed in a gaseous medium, such as air, dimethyl ether, isobutane or dimethyl ether. Common examples of natural aerosols include smoke, dust, fog, clouds, mist, haze, and fumes. Fine sprays (e.g., paints, perfumes, insecticides, pesticides, inhalants and deodorants) are examples of manufactured aerosols. See also aerogel.

AeroSorb. Trade name of National Nonwovens (USA) for a lightweight, anti-microbial nonwoven composite that can absorb 20 times its weight of fluids and water vapors. Used in the aerospace industries for containing condensation and/or residual moisture.

Aerostretch. Trademark of Rubafilm (France) for macroperforated polyethylene stretch films for packaging food and flowers, and for palletization applications.

Aerotuf. Trademark of Anchor Plastics Company, Inc. (USA) for stiff, resilient extruded polypropylene rods, shapes and tubing.

Aeroweb. Trademark of Ciba-Geigy Corporation (USA) for laminates of paper, plastics, woven and nonwoven fabrics and wood, supplied in unexpanded, cellular, and honeycomb form. The honeycomb composites consist of fiberglass cells in phenolic-resin matrices, and are used for aircraft, missile and space vehicle components. Also included under this trademark are carbon and glass fiber and whisker reinforcements, and cellular or honeycomb-based sandwich structures.

Aertex. Trade name for an absorbent, cellular cotton fabric with open texture, used for shirts, underwear and sportswear.

aerugite. A green mineral composed of nickel arsenate, $Ni_9As_3O_{16}$. Crystal system, monoclinic. Density, 5.9 g/cm³. Occurrence: Germany.

aeschynite. (1) A dark brown opaque mineral of the columbite group composed of cerium titanium niobium oxide, $(Ce,Ca,Fe,Th)(Ti,Nb)_2O_6$. It can also be made synthetically. Crystal system, orthorhombic. Density, 5.2-5.6 g/cm³. Also known as *eschynite*.

(2) A mineral of the columbite group composed of yttrium titanium niobium oxide, $(Y,Er,Ca,Fe,Th)(Ti,Nb)_2O_6$. It can also be made synthetically. Crystal system, orthorhombic. Density, 5.13 g/cm³; hardness, 5-6 Mohs. Also known as *eschynite*.

Aeternamid. Trademark of CMG (Germany) for a series of polyamides (nylons).

afara. The light, blond wood from the tree *Terminalia superba*, often sold under the name of "korina." It has an open grain with a hardness and texture comparable to mahogany, sometimes with a dark heartwood and grayish-brown or nearly black markings. Darker, figured wood looks like walnut. It works easily and finishes well, has a tendency to split in nailing, and low resistance to decay, insects and termites. Average weight, 550 kg/m³ (34 lb/ft³). Source: West Central Africa and Congo. Used as general utility timber for light constructional work (paneling, blond furniture and fixtures), and in plywood manufacture. Also known as *limba*.

Afax. Trademark Himont, Inc. (USA) for amorphous polypropylenes.

(2) Trademark for a series of high-quality continuous casting mold fluxes. *Afax GR* are low-density spherical granular fluxes with good flowability. *Afax P* refers to specially processed powder blends of synthetic materials.

AFC. Trademark of Asia Fiber Company (Thailand) for nylon 6 fibers and yarns.

AFC media layers. See antiferromagnetically coupled media layers.

Afco. Trade name of Alloy Foundry Company Limited (Canada) for nonferrous and gray-iron castings.

Afcodur. Trademark of Armosig/Pechiney SA (France) for polyvinyl chlorides.

Afcolene. Trademark of Pechiney SA (France) for polyvinyl chloride (PVC) supplied in plasticized (flexible) and copolymer forms.

Afcoloy. Trade name of Atlas Foundry Company (USA) for a series of cast ferrous alloys including various low- and medium-carbon and low-alloy steel castings, cast corrosion- and/or heat-resistant chromium and chromium-nickel stainless steels, cast heat-resistant iron-chromium and iron-chromium-nickel steels, and various cast corrosion- and heat-resistant nickel-iron-chromium alloys.

Afcomet. Trade name of Atlas Foundry Company (USA) for a series of gray, ductile and alloy cast irons.

Afcoplast. Trademark of Pechiney SA (France) for plasticized (flexible) polyvinyl chloride (PVC) supplied in paste and other forms.

Afcoryl. Trademark of Pechiney SA (France) for acrylonitrile-butadiene-styrenes.

Afcovyls. Trademark of Pechiney SA (France) for a series of polyvinyl chlorides.

Affimet. Trade name of Aluminium Français (France) for a series of cast aluminum-copper, aluminum-magnesium, aluminum-silicon and aluminum-zinc alloys.

Affinity. Trademark of Dow Chemical Company (USA) for transparent polyolefin plastomers made by a proprietary technology (known as INSITE). Supplied in blown and cast film, profile extrusion and coating/lamination grades, they have a density range of 0.868-0.910 g/cm³ (0.031-0.033 lb/in.³), high toughness, tear resistance and durability, good hot-tack strength and low seal initiation temperature, and are used for packaging applications, e.g., for personal care products (shampoos, detergents, powdered soaps, etc.), health and hygiene products (e.g., baby diapers), and durable goods, such as flooring.

afghanite. A blue mineral of the cancrinite group composed of sodium calcium aluminum silicate chloride sulfate, $Na_8Ca_4Si_6Al_6O_{25}(SO_4,Cl)_3$. Crystal system, hexagonal. Density, 2.6 g/cm³; refractive index, 1.523. Occurrence: Afghanistan, Canada.

Aflas. Trademark of Dyneon LLC (USA) for a series polytetrafluoroethylene elastomers with carbon black fillers. They have excellent resistance to aggressive oils, coolants and bases, a glass-transition temperature of 2°C (36°F), and a hardness of 70 Shore A. Used for automotive parts, tank linings, and as a viscosity modifiers for tetrafluoroethylenes.

A-Flex. Trade name of Action Technology (USA) for polyvinyl chloride supplied in the form of tubing.

Aflon. (1) Trade name of Asahi America (USA) for a series of fluoropolymers.

(2) Trademark of Finnaren & Haley, Inc. (USA) for thermally curing enamels.

African bass. A coarse, durable, water-resistant, reddish-brown vegetable fiber obtained from West African palm trees, and used for sweeps and brooms.

African boxwood. The wood of the hardwood tree *Buxus macowani*. It is fine-textured, durable and very similar to common *boxwood*, but somewhat softer. Source: Southern Africa. Also known as *cape boxwood*. See also boxwood.

African cherry. The strong wood from the hardwood tree *Tieghemella heckelii*. It is usually pale red with darker red lines, but may be dark reddish brown. Important properties include a fine, even texture, good strength properties, stability, durability and overall appearance, and fairly good workability. Average weight, 630 kg/m^3 (39 lb/ft^3). Used for furniture and cabinetwork. Source: West Africa (Ivory Coast, Ghana and Nigeria). Also known as *cherry mahogany; makore; makori.*

African ebony. The heavy, hard, close-textured wood of the ebony tree *Diospyrus denta.* The heartwood is black and the sapwood brownish-white. Average weight, 1250 kg/m^3 (78 lb/ft^3). Source: West Africa. Used for piano keys, turnery, and cabinetwork. Also known as *black ebony.* See also ebony.

African fiber. A fiber obtained from the leaves of the Algerian palmetto, and used for stuffening mattresses.

African greenheart. See greenheart (2).

African ironwood. The hard wood of the tree *Lophira alata,* that is usually chocolate-brown in color, and occasionally dark red with a speckled surface due to yellowish deposits in its pores. It is very heavy (will not float on water), has high strength and durability, high distortion in use, and is very difficult to work. Average weight, 1030 kg/m^3 (64 lb/ft^3). Source: West Africa. Used for outside work including wharf and dock construction, piling, planking for bridges, etc. Also known as *azobe; ekki.*

African mahogany. The commercial name of a class of highly figured light pinkish-brown to dark reddish-brown hardwoods from trees of the genus *Khaya.* Most *African mahogany* is from red mahogany *(K. ivorensis)* and closely resembles *American mahogany,* but has a slightly coarser texture, more pronounced grain patterns, and greater shrinkage. Important properties include good machinability and finishability, good seasoning properties, and moderate decay resistance. Average weight, 530 kg/m^3 (33 lb/ft^3). Source: Tropical West Africa (Ivory Coast, Ghana and Nigeria). Used for fine furniture, interior paneling, store fixtures, art objects, boat construction, veneer, and plywood. See also khaya; mahogany; red khaya.

African pencil cedar. The relatively hard and heavy wood from the East African tree *Juniperus procera,* used for furniture.

African satinwood. The strong, light wood of the tree *Turraeanthus africanus.* It is dull white to pale yellow, but darkens to golden yellow, has a beautiful clean grain, works and finishes well, and is available as lumber and veneer. Source: African rainforest. Used for fine cabinetwork and decorative purposes. It is very popular for marquetry. Also known as *avodire.*

African teak. The strong wood of the tree *Chlorophora excelsa* which is brown with yellow bands, has a coarse, open grain, high durability and stability, good resistance to decay and termite attack, and is rather hard to work. Source: Tropical Africa. Used for building construction work, ship construction, high-class joinery, and carving. Also known as *iroko.*

African walnut. The golden-brown wood of the tree *Lovoa trichilioides* belonging to the mahogany family. It has a fine texture, and an interlocking grain. Average weight, 640 kg/m^3 (40 lb/ft^3). Source: West Africa. Used for chairmaking, cabinetwork, flooring, paneling veneers, and high-class joinery. Also known as *amonilla.*

African wool. A term usually referring to the fine, soft, white wool obtained from merino sheep raised in southern Africa.

afrormosia. The stable and durable hardwood of the large tree *Pericopsis elata* of the teak family. It does not resemble the latter in general appearance, is generally brownish with yellowish-brown streaks, fairly easy to work, tends to split when nailed, and has quite high strength and good resistance to fungi and termites. Average weight, 690 kg/m^3 (43 lb/ft^3). Source:

West Africa. Used as a substitute for teak, and for boat construction, ship decking, furniture, veneer, cabinets, paneling, and furniture. Also known as *kokrodua.*

afwillite. A colorless mineral composed of tricalcium disilicate trihydrate, $3CaO \cdot 2SiO_2 \cdot 3H_2O$. Crystal system, monoclinic. Density, 2.64 g/cm^3; refractive index, 1.6204. Occurrence: South Africa, Northern Ireland, USA (California). It also occurs artificially in some hydrated Portland cement mixtures. Used in ceramics and the construction trades.

afzelia. A durable and stable wood of a genus of trees *(Afzelia)* found in Tropical Africa. It has high strength, a pleasing appearance, is somewhat hard to work, and tends to split when nailed. Average weight, 820 kg/m^3 (51 lb/ft^3). Used for docks, harbors and gantries, moderately heavy constructional work, and high-class exterior and interior joinery.

Ag-ABA. Trade name Wesgo Division of GTE Products Corporation (USA) for a silver-based active brazing alloy. See also ABA.

Agalain. Trade name for cellulose nitrate formerly used for denture bases.

Agalit. Trade name for a series of glass-ceramics.

Agalite. Trade name of Agalite Bronson Company (USA) for a toughened glass.

Agaloy. Trade name for a tubular perforated metal.

Agalyn. Trade name for a cellulose nitrate formerly used as denture base.

agar. A gelatinous substance that is obtained from seaweed and marine algae, and can also be made synthetically. It is a polysaccharide mixture composed of agarose and agaropectin, and is available in the form of a pale buff powder or thin shredded pieces that are strongly hydrophilic and swell enormously in water. Used as a gelling agent in foods, as a culture medium in the growing of microorganisms, in dental impressions, in photographic emulsions, in laboratory reagents, and in the preparation of agarose. Also known as *agar-agar; gum agar.*

agarose. A biocompatible polysaccharide with a gel matrix that is a purified linear galactan hydrocolloid obtained from agar or agar-bearing marine algae. It is available as a powder in various grades with melting temperatures, gel points and gel strengths typically ranging from about 50 to over 95°C (122 to over 203°F), 8 to over 42°C (46 to over 108°F) and less than 75 to over 2000 g/cm^2, respectively. Used in immunology, virology, microbiology, electrophoresis, immunodiffusion and immunoelectrophoresis, and ion-exchange chromatography, and in biotechnology as a precursor or component in the manufacture of tissues and implants.

agardite. A green-blue mineral of the mixite group composed of copper yttrium arsenate hydroxide hydrate, $Cu_6(Y,Ca)(AsO_4)_3 \cdot (OH)_6 \cdot 3H_2O$. Crystal system, hexagonal. Density, 3.72 g/cm^3; refractive index, 1.782. Occurrence: Morocco.

agate. A fine-grained variety of the mineral *chalcedony* with its colors arranged in stripes, blended in clouds, or displaying dendritic, or fern-like patterns. Occurrence: Brazil, Uruguay, USA. Used for mortars and pestles, instrument bearings, grinding balls, burnishers and polishers, textile rollers, ornaments, in the manufacture of knife edges, and as a gemstone.

Agateen. Trademark of Agate Lacquer Manufacturing Company, Inc. (USA) for clear air-dry and baking cellulose lacquers and water-base coatings.

agate glass. A multicolored glass made by blending glasses of different colors while in the molten state, or by rolling a clear glass into other glasses of different color. It resembles the mineral variety *agate* in appearance.

agatized wood. See silicified wood.

agave fibers. The hard vegetable fibers obtained from the leaves of any of a genus (*Agave*) of tropical American plants including henequen (*A. fourcroydes*), sisal (*A. sisalana*), maguey (*A. lurida* and *A. tequilana*), and the century plant (*A. americana*). Used especially for binder twine, cordage, etc. See also cantala fibers; henequen; sisal fibers; pita; maguey fibers.

agba. The plain, light-colored wood of the tree *Gossweilerodendron balsamiferum* with pale brown tone, and rather close, even texture. It has high resistance to decay, works to a good finish, has good nailing, screwing and gluing qualities, and takes stains and polishes. Average weight, 520 kg/m³ (32 lb/ft³). Source: Tropical Africa. Used as a lightweight general utility hardwood, and as utility joinery timber.

AGC. Trademark of Wieland Dental + Technik GmbH & Co. KG (Germany) for a special, yellow-colored, extra-hard gold-platinum dental ceramic alloy.

Ag/Cu Eutectic. Trade name of Johnson Matthey plc (UK) for a silver brazing alloy containing 28-29% copper. It has a melting point of 778°C (1430°F).

Agedroc. Trade name of Atlas Turner Inc. (Canada) for asbestos cement panels and boards.

Agee. Trade name of Australian Window Glass Proprietary Limited for hollow glass blocks.

age-hardenable alloys. Metal alloys that can be hardened through the formation of precipitates by aging at room temperature or above. Typical age-hardenable alloys include aluminum-copper, aluminum-silicon, aluminum-nickel-copper, copper-beryllium, copper-silicon, copper-tin and magnesium-aluminum alloys and certain ferrous alloys. Also known as *precipitation-hardenable alloys*.

Agepan. Trade name of Agepan Holzwerkstoffe GmbH (Germany) for wood and plywood shuttering boards for concrete.

Agfenit. Trade name of former Reiher Brothers KG (Germany) for phenol-formaldehyde resins and products.

Aggerlit. Trade name of Dörrenberg Edelstahl GmbH (Germany) for a series of high-temperature steels containing 6-22% chromium, and 1.5-2.2% silicon. Used for furnace equipment and fixtures, and heat-treating equipment.

Aggerstahl. Trade name of Dörrenberg Edelstahl GmbH (Germany) for a water-hardening, wear-resistant steel containing 1.45% carbon, 1.4% chromium, 0.6% manganese, 0.25% silicon, and the balance iron. Used for sleeves, bearings, and liners.

agglomerated material. A material composed of small particles which are bonded together into an integrated mass. Examples include concrete, brick, asphalt pavements, grinding wheels, and powdered metals.

aggregate. (1) In general, any granular mineral material.

    (2) In powder metallurgy, a mass of powder particles.

    (3) In the construction trades, coarse particles, such as sand, gravel, crushed stone, or rock, lightweight materials, such as blast-furnace slag, shale, clay, or carbon, or synthetic inorganic materials, used with a cementing medium to make concrete or mortar. It is also used in base courses, road-building, paving compositions, and as railroad ballast.

    (4) Any hard, coarse, fragmented mineral material held together by a suitable resin (e.g., an epoxy binder). Used for plastic tools, flooring, and as a surfacing medium.

aggregate composites. Composite materials, such as concrete, dispersion-strengthened metals or particulate composites, in which the matrix phase is reinforced with a dispersed particulate phase. See also concrete; dispersion-strengthened metals; particulate composites.

Agilene. Trade name for crosslinked polyethylene that has high impact and tensile strength, high toughness and excellent high-temperature resistance.

Agilon. Trade name for nylon fibers used for the manufacture of clothing.

AgION. Trade name for an antimicrobial compound with slow-release encapsulated silver ions supplied in epoxy and polyester film, and as coating for application to stainless and carbon products to inhibit the growth of microbes, bacteria, mold, mildew and other organisms on surgical and sterilization equipment, hospital beds, instrument trays, carts, countertops, heating and ventilation equipment, food processing equipment, etc.

A-glass. A high-alkali soda-lime-silica glass with a typical composition of 72% silica (SiO_2), 10-15% soda (Na_2O), 8% lime (CaO), and a total of 5-10% magnesia (MgO) and alumina (Al_2O_3), respectively. It has good chemical resistance and reinforcing properties, and is used as glass-fiber reinforcement in composites, and for windows and containers. Also known as *high-alkali glass*.

A-glass fibers. Fibers of high-alkali glass (A-glass) that have a density of about 2.5 g/cm³ (0.09 lb/in.³), good reinforcing properties and chemical resistance, and low thermal expansion. Used chiefly as reinforcing fibers in composites. Also known as *high-alkali glass fibers*.

Agno. Trade name for 100% rayon fibers used for the manufacture of light textile fabrics for clothing, e.g., jerseys.

aglime. See agricultural lime.

Agmolite. Trade name of Engelhard Corporation (USA) for a sintered alloy of silver and molybdenum, used for electrical contacts in circuit breakers.

Agnilite. Trade name of Engelhard Corporation (USA) for a sintered alloy of silver and nickel, used for electrical contacts in circuit breakers.

Agomet M. Trade name for a low-temperature curing, quick-setting methacrylate adhesive with good processibility, and good low-temperature and moderate room-temperature lap-shear strengths. Used for joining metals.

agra gauze. An open-sett gauze-like silk fabric in a plain weave and with a stiff finish.

agrellite. A white, grayish or greenish mineral composed of sodium calcium fluoride silicate, $NaCa_2Si_4O_{10}F$. It may also contain lanthanum. Crystal system, triclinic. Density, 2.87 g/cm³; refractive index, 1.579. Occurrence: Canada.

Agraphite. Trade name of Engelhard Corporation (USA) for a sintered alloy of silver and graphite, used for electrical contacts in circuit breakers.

Agricola. Trade name of Manco Products, Inc. (USA) for an alloy containing 70% copper and 30% lead, used for heavy-duty engine bearings, bushings, seals, etc.

Agricola Bronze. Trade name of British Metal Corporation Limited (UK) for a copper alloy containing 29% lead and 1% silver. Used for bearings and bushings.

agricolite. See eulytite.

agricultural lime. Ground calcium oxide (unslaked lime) or calcium hydroxide (slaked lime) used in agriculture to neutralize soil acidity. Also known as *aglime; agricultural limestone; agstone*.

agricultural limestone. See agricultural lime.

agricultural mass-finishing media. Materials, such as sawdust, corncobs and walnuts shells, used in mass finishing operations, such as barrel, disk or vibratory finishing. See also mass-finishing media.

agricultural tile. A round, hollow, unglazed tile used to drain agricultural lands.

agricultural varnishes. Varnishes used to protect or improve agricultural machinery and implements.

Agrilite. Trade name of American Injector Company (USA) for a self-lubricating bearing metal containing 70% copper, 25% lead, and 5% antimony. Used for thrust bearings in boats and ships, and for rolling-mill bearings and bearings for appliances, such as ironing and washing machines.

agrinierite. An orange mineral of the becquerelite group composed of potassium calcium strontium uranyl oxide hydrate, $(K_2,Ca,Sr)(UO_2)_3O_4 \cdot 4H_2O$. Crystal system, orthorhombic. Density, 5.70 g/cm^3; refractive index, 2.0. Occurrence: France.

Agrilux. Trade name of Agriplast Srl (Italy) for diffused-light thermal films composed of mineral-filled ethylene-vinyl acetate copolymers. They provide high and uniform visible light transmission, and are antidrop treated to minimize the risk of fungal diseases. Used to protect greenhouse crops against external influences of the climate (e.g., frost, or unfavorable light conditions).

Agrisilo. Trade name of Agriplast Srl (Italy) for plastic films supplied in white or black mono- or co-extruded grades. They have high opacity, excellent tear and puncture resistance and good aging resistance, and are used for ensilage covering applications.

Agro. Trade name for rayon fibers used for the manufacture of clothing.

Agrolam. Trade name for rayon fibers used for the manufacture of clothing.

agstone. See agricultural lime.

Agua-Rock. Trademark of Rock-Tred Corporation (USA) for a fast-drying, 52% solids water-based epoxy coating with excellent abrasion and adhesion resistance. Used on industrial and commercial floors.

aguilarite. An iron black mineral composed of silver selenide sulfide, Ag_4SeS. Crystal system, orthorhombic. Density, 7.59 g/cm^3. Occurrence: Mexico.

ahlfeldite. A green to yellow mineral composed of nickel selenite hydrate, $NiSeO_3 \cdot 2H_2O$. It can also be made synthetically. Crystal system, monoclinic. Density, 3.51 g/cm^3; refractive index, 1.744. Occurrence: Bolivia.

Aich's metal. An oxidation-resistant brass containing 56-60% copper, 38-42% zinc, 0.9-1.8% iron, and traces of other elements. Commercially available in cast and forged form, it has exceptional malleability at red heat, and is used for hydraulic cylinders and forgings.

Aidion. Trade name of John Finn Metal Works (USA) for an antifriction alloy containing tin, antimony and lead. Used for bearings.

Aiken metal. An alloy of 50% copper and 50% lead, used for nonelastically deformable metal parts.

aikinite. A grayish-white mineral of the stibnite group composed of bismuth copper lead sulfide, $CuPbBiS_3$. It can also be made synthetically. Crystal system, orthorhombic. Density, 6.95-7.19 g/cm^3; hardness, 2 Mohs. Occurrence: Russia, Turkey, UK. Also known as *needle ore.*

Aim. Trademark of Dow Chemical Company (USA) for advanced styrenic resins based on polystyrene. They combine the high gloss and impact resistance of acrylonitrile-butadiene styrene resins with the ease of processing and handling of high-impact

polystyrenes. Sold as natural color resins in several grades, they are usually processed by extrusion, injection molding or thermoforming. Used for kitchen appliances, electrical lawn and garden equipment, toys, medical test kits, etc.

Air-4. Trade name of Bethlehem Steel Corporation (USA) for a free-machining air-hardening medium-alloy cold-work tool steel (AISI type A4) containing 0.95% carbon, 2% manganese, 2.2% chromium, 1.1% molybdenum, 0.15-0.35% lead, and the balance iron. Used for screw-machine parts, fasteners, etc.

Airaloy. Trade name of Republic Steel Corporation (USA) for a medium-alloy air-hardening tool steel (AISI type A4) containing 1% carbon, 2% manganese, 0.9% chromium, 0.9% molybdenum, and the balance iron. It has excellent deep-hardening properties, good wear resistance, and is used for dies, punches, shear blades, and mandrels.

air barrier. A material used to protect the structure and insulation of a building from air damage. An effective air barrier material must be durable, resistant to air movement, and strong enough to withstand wind pressures. Examples of such materials are drywall, plywood, polyethylene and spun-bonded olefin sheeting, glass, wood, poured concrete, and rigid or flexible sealants. Also known as *air barrier material.*

air barrier material. See air barrier.

Airblanket. Trade name for a cured rubber blanket material.

air-blown asphalt. A black, solid substance produced by blowing air at an elevated temperature (usually 204-316°C or 400-600°F) through molten, petroleum-derived asphalt with subsequent cooling. This treatment increases the softening point to 93-121°C (200-250°F). Used primarily for roofing applications. See also blown asphalt.

air-blown mortar. A mixture of cement, sand and water transported through a hose and applied to a surface using high-velocity compressed air. It will adhere firmly to the surface. Also known as *shotcrete; sprayed mortar.*

air brick. A fired brick, essentially of standard size, that has several holes along its lateral axis to allow the circulation of air in structures.

Air Cap. Trademark of Sealed Air of Canada Limited for a plastic packaging material.

Aircast. Trade name for castable room-temperature-vulcanizing (RTV) silicone rubbers.

Aircell. Trade name for a rubber foam made from natural rubber latex.

Air-Chrom. Trademark of Kloster Steel Corporation (USA) for a medium-alloy air-hardening cold-work tool steel (AISI type A2) containing 1% carbon, 0.6% manganese, 5.25% chromium, 0.25% vanadium, 1.1% molybdenum, and the balance iron. It has excellent deep-hardening properties, and good hot hardness and wear resistance. Used for tools and dies.

air-classified powder. A ceramic or metal powder that has been separated into particle size and/or weight fractions in an air stream of appropriate velocity.

Airco. Trade name of Airco Vacuum Metals (USA) for an extensive series of welding and hardfacing electrodes, wires and rods of plain-carbon and low-alloy steels, mild steels, high-strength steels, welding-grade austenitic, ferritic and martensitic stainless steels, heat-resistant stainless steels, nickel and nickel-iron superalloys, gray and alloy cast irons, aluminum, manganese, phosphor and silicon bronzes, nickel silvers, commercially pure aluminums, various aluminum-magnesium and aluminum-silicon alloys, cemented tungsten carbides, etc.

Aircoflex. Trademark of Air Reduction Company, Inc. (USA) for a series of ethylene vinyl acetate emulsion copolymers used in pigment binders, paper and textile coatings, and adhesives. They are also available in latex form.

Aircomatic. Trade name of Airco Vacuum Metals (USA) for a series of brazing and braze welding rods based on copper, aluminum, manganese and phosphor bronzes, and several aluminum-magnesium and aluminum-silicon-magnesium alloys. Also included under this trade name are welding wires of mild and austenitic stainless steels.

Aircon. (1) Trade name of Duplate Canada Limited for double-glazing units.

(2) Trade name of PPG Industries Inc. (USA) for electrically heated windshields for non-pressurized aircraft.

air-cooled blast-furnace slag. A product made by cooling molten blast-furnace slag under normal atmospheric conditions. See also blast-furnace slag.

Aircosil. Trade name of Airco Vacuum Metals (USA) for an extensive series of silver-base brazing filler metals.

Airco Vinal. Trade name for a vinal (polyvinyl alcohol) fiber used for textile fabrics.

Aircraft. British trade name for a corrosion-resistant alloy of 82-88% copper, 10-11.5% aluminum, 4-6% nickel and 4-6% iron. Used for aircraft parts, and corrosion-resistant parts.

aircraft steels. A group of steels suitable for use in aircraft manufacture. They have been made by rigidly controlled manufacturing processes to ensure the highest level of quality and meet the most demanding requirements [e.g., MIL-SPEC (Military Specifications) and AMS (Aerospace Material Specification) specifications] regarding cleanliness, and ultrasonic and magnaflux magnetic testing. Also known as *aircraft-quality steels.*

Aircrat. Trade name of Marshall Steel Company (USA) for an air-hardening tool steel containing 1% carbon, 1.1% molybdenum, 0.5% manganese, 0.2% vanadium, 5% chromium, and the balance iron. Used for tools, dies and punches.

Airdi. Trade name of Crucible Materials Corporation (USA) for a series of high-carbon, high-chromium type cold-work tool steels (AISI type D). They have high abrasion and wear resistance, high hardness, and good nondeforming properties. Used for dies and die blocks, shear blades, punches, and bending and seaming tools.

air-dried enamel. An enamel cured by exposure to air, making use of a drying process involving both solvent evaporation and oxidation.

air-dried lumber. Lumber dried in stacks in storage yards by exposure to outside air to reduce their moisture content to less than 24%. Also known as *air-seasoned lumber; natural-seasoned lumber.* See also partially air-dried lumber.

Air-Eez. Trade name for a free-machining low-temperature tool steel containing 0.95% carbon, 2% manganese, 0.35% silicon, 2.2% chromium, 1.1% molybdenum, 0.3% lead, and the balance iron.

air-entrained cement. A cement into which minute air bubbles or spheroids have been introduced during grinding or mixing to improve its properties. Used for civil engineering and construction applications.

air-entrained concrete. A concrete into which minute air bubbles or spheroids have been introduced during grinding or mixing to improve its durability and other properties. Its density and strength is lower than that of ordinary concrete, and it has exceptionally high resistance to frost. Used for civil engineering and road construction applications. Also known as *air-entraining concrete.*

air-entraining agent. A chemical substance, such as a grease, resin or soap, that is added to a hydraulic cement, mortar or concrete to reduce the surface tension of water, facilitate the entrapment of very fine air bubbles, and improve properties, such as workability and sub-zero durability. It may be added during mixing (mortars and concrete), or grinding (cement). Abbreviation: AEA. Also known as *air-entrainment agent; air-entrainment compound.*

air-entraining hydraulic cement. A *hydraulic cement* containing an amount of air-entraining agents adequate to cause the product to entrap air in the mortar.

air-entraining Portland cement. Portland cement containing an amount of air-entraining agents adequate to cause the product to entrap very small air bubbles in the concrete. The concrete is characterized by numerous pores, and is lighter than that obtained with ordinary *Portland cement.*

air-entrainment agent. See air-entraining agent.

air-entrainment compound. See air-entraining agent.

air-equivalent material. A material, such as impregnated paper or a light plastic with conductive graphite, that has essentially the same effective atomic number as air. Used for walls of ionization chambers. Also known as *air-wall material.*

AiResist. Trade name of Cannon-Muskegon Corporation (USA) for a series of cobalt-base heat-resistant casting alloys containing about 0.20-0.45% carbon, 62.0-64.0% cobalt, 19.0-21.0% chromium, 4.5-11.0% tungsten, 4.3-3.4% aluminum, 7.5-2.0% tantalum, and 0.1% yttrium. Used for metallurgical and industrial furnaces, turbochargers, gas turbines, power plant equipment, and plant and equipment for the manufacture of cement, glass, rubber, etc.

Airex. (1) Trade name of Columbia Tool Steel Company (USA) for a tough, shock-resistant tool steel containing 0.55% carbon, 2% silicon, 0.75% manganese, and the balance iron. Used for chisels, pneumatic tools, etc.

(2) Trade name of Airex AG (Germany) for plastics, foams and rubbers.

(3) Trademark of Alcan Airex AG (Germany) for a range of core materials for sandwich and composite reinforcement applications.

Airfill. Trade name for polyester filler pastes available in high- and low-temperature grades.

air-float clay. A clay that has been coarsely ground and subjected to an air separation process, usually carried out in an air classifier. This treatment results in a product with very fine particles.

Air Form. Trade name of Guardian Industries Corporation (USA) for an automotive safety glass that combines the best features of laminated and tempered glass. The manufacturing process is also referred to as "Air Form."

Air-Hard. Trade name of Teledyne Vasco (USA) for an air-hardening medium-alloy cold-work tool steel (AISI type A2) containing 0.95-1.05% carbon, 0.2-0.4% silicon, 0.5-0.7% manganese, 5-5.5% chromium, 0.95-1.25% molybdenum, 0.2-0.3% vanadium, and the balance iron. Free-machining and precision-ground grades are also available. It has good deep-hardening properties, high strength, good nonshrinking and nonwarping properties, good abrasion and wear resistance, and high strength and toughness. Used for press tools, blanking, forming and trimming dies, drawing and thread-rolling dies, lathe centers, gages, shear blades, punches, punch plates, rolls, knurls, and fixtures.

air-hardening refractory cement. A finely ground *refractory cement* containing chemical agents that cause hardening at temperatures between room and vitrification temperature. Used for setting mortars, and as a jointing cement for furnace brickwork.

air-hardening steel. (1) An alloy steel that contains enough carbon and other alloying elements to harden fully during cooling in air or gas from a temperature above its transformation point. It must be subjected to isothermal annealing or spheroidizing prior to machining. Also known as *self-hardening steel*.

(2) A tungsten-type tool steel containing 1.85% carbon, 5-9% tungsten, 2.5% manganese, and 1.5% silicon, formerly used for hard cutting tools.

Airhold. Trade name for pressure-sensitive tapes and fabrics.

air-jet-milled quartz. A fine, dry *quartz* powder produced in a mill employing air-jet nozzles. Abbreviation: AJM SiO_2.

Airkap. Trade name for a pressure-sensitive tape/*Kapton* polyimide film carrier.

Airkool. Trade name of Crucible Specialty Steel (USA) for an air-hardening medium-alloy cold-work tool steel (AISI type A2) containing 0.95-1.00% carbon, 5.00-5.25% chromium, 1.00-1.15% molybdenum, and 0.50% vanadium. It has a great depth of hardening, good nondeforming properties, and good wear resistance and toughness. Used for press tools, punches, gages, blanking, forming and trimming dies, thread-rolling dies, and plastic molding dies.

Airkool V. Trade name of Crucible Specialty Steel (USA) for an air-hardening medium-alloy cold-work tool steel (AISI type A7) containing 2.25% carbon, 5.25% chromium, 4.25% vanadium, and 1.00% tungsten. It has a great depth of hardening, excellent abrasion and wear resistance, and high hot hardness. Used for forming and drawing dies, burnishing tools, liners for sandblasting equipment, brick molds, and gages.

air-laid composites. Composites produced from fibers of wood or recycled paper by bonding with thermosetting resins.

air-laid fabrics. Fabrics formed by dispersing staple textile fibers into a stream of air and condensing onto a permeable conveyor or cage.

Air-Loc. Trademark of General Latex & Chemical Corporation (USA) for a series of rubber-based materials for vibration absorption applications.

Airloy. Trade name of Allegheny Ludlum Steel (USA) for an air-hardening medium-alloy cold-work tool steel (AISI type A5) containing 1% carbon, 3% manganese, 1% molybdenum, 1% chromium, and the balance iron. Used for punches, gages, blanking and forming dies, and thread-rolling dies.

air-mail paper. A thin, wood-free writing paper that may contain cotton or linen rags. It is white or colored, fully-sized, smoothened, and normally opaque.

air-melted steel. A steel made by conventional melting methods, such as open-hearth, basic-oxygen, or electric-furnace processes.

AirMo. Trade name of Teledyne Firth-Sterling (USA) for an air-hardening cold-work tool steel (AISI type A4) containing 1% carbon, 2% manganese, 1% molybdenum, 1% chromium, and the balance iron. Used for cold-work dies, and cold chisels and punches.

Airmold. Trade name of Crucible Materials Corporation (USA) for a deep-hardening mold steel (AISI type P4) containing 0.1% carbon, 5% chromium, 0.5% molybdenum, and the balance iron. Used for cold-hubbed plastic molds.

air-oxidizing coatings. A group of coatings including oleoresinous, alkyd, silicone-alkyd and phenolic types cured by air oxidation of the drying oils.

Airpad. Trade name for uncured, non-silicone tooling rubber.

air-placed concrete. See dry-mix shotcrete.

airplane fabrics. See aeroplane fabrics.

Airplane Babbitt. Trade name of Southern Metals Corporation (USA) for a lead babbitt containing 16% tin, 10% antimony, 2% copper, and a trace of bismuth. It has a melting range of 315-370°C (600-700°F), and is used for bearings.

Airpro. Trade name of Bethlehem Steel Corporation (USA) for a cold-work tool steel containing 1% carbon, 0.6% manganese, 5.25% chromium, 1.1% molybdenum, 0.25% vanadium, and the balance iron. Used for thread-rolling dies, trimming and blanking dies, broaches, and reamers.

Airque. Trade name of CCS Braeburn Alloy Steel (USA) for an air-hardening medium-alloy cold-work tool steel (AISI type A2) containing 1% carbon, 0.7% manganese, 5.25% chromium, 1.2% molybdenum, and the balance iron. It has a great depth of hardening, good nondeforming properties, and good wear resistance and toughness. Used for press tools, punches, blanking, forming and trimming dies, thread-rolling dies, and plastic molding dies.

Airque Special. Trade name of CCS Braeburn Alloy Steel (USA) for an air-hardening nondeforming cold-work tool steel (AISI type A2) containing 0.8% carbon, 5.25% chromium, 1.1% molybdenum, 0.25% vanadium, and the balance iron. It has a great depth of hardening, good nondeforming properties, and good wear resistance and toughness. Used for press tools, punches and shears, and blanking, forming and trimming dies.

Airque V. Trade name of CCS Braeburn Alloy Steel (USA) for an air-hardening medium-alloy cold-work tool steel (AISI type A3) containing 1.25% carbon, 0.5% manganese, 0.3% silicon, 5.25% chromium, 1.15% molybdenum, 1% vanadium, and the balance iron. It has a great depth of hardening, high wear and abrasion resistance, and good toughness and machinability. Used for swaging, forming, blanking and thread-rolling dies.

Air-Seal. Trade name of Air-Seal Windows Limited (UK) for a double glazing unit in which the glass is sealed into a soft polyvinyl chloride gasket, backed by an extruded aluminum section.

air-seasoned lumber. See air-dried lumber.

air-setting cement. A cement that will take permanent set in air under normal conditions of temperature and atmospheric pressure.

air-setting mortar. A mortar that will take permanent set in air under normal conditions of temperature and atmospheric pressure.

air-setting refractories. Ground refractory materials, such as refractory mortars, plastic refractories, ramming mixes, gunning mixes, cements, etc., that when tempered with water produce a strong bond upon drying.

air-setting refractory cement. A finely ground refractory material containing chemical agents, such as sodium silicate, that facilitate hardening at room temperature. Used as jointing cement for furnace brickwork.

Air Shock. Trade name of Peninsular Steel Company (USA) for an air- and/or oil-hardening, shock-resisting tool steel (AISI type S7) containing 0.5% carbon, 0.7% manganese, 3.25% chromium, 1.4% molybdenum, and the balance iron. It has great depth of hardening, good toughness, and relatively high hot hardness. Used for rivet sets, punches, chisels, plastic molds, hot gripper dies, and die-casting dies.

Airsil. Trade name for high-temperature silicone rubber.

air-slaked lime. The product resulting from the exposure of quicklime to air for gradual absorption of carbon dioxide and moisture from the atmosphere. It is a powder containing various proportions of the oxides, hydroxides and carbonates of calcium. Also known as *lime powder; powdered lime.*

Air Space. Trade name of Air Space Inc. (USA) for insulating glass units.

Airtac. Trade name for sprayable rubber adhesives.

Airtem. Trade name of Lehigh Steel Corporation (USA) for an air-hardening tool steel containing 1.3% carbon, 0.5% manganese, 1% silicon, 5% chromium, 1.25% molybdenum, 0.3% vanadium, and the balance iron. It has excellent wear and shock resistance, and good nondeforming properties. Used for punches, knives, and blanking and forming dies.

Airthane. Trademark of Air Products & Chemicals, Inc. (USA) for polyurethane prepolymers used in the manufacture of polyurethane coatings, plastics, and elastomers.

Air-Tite. Trade name of Venetian Mirror & Glass Company (USA) for an insulating glass.

Air-Tough. Trade name of Carpenter Technology Corporation (USA) for a tough, air-hardening tool steel containing 0.7% carbon, 2% manganese, 1% chromium, 1.25% molybdenum, and the balance iron. Used for tools and dies.

Air-True. (1) Trade name of Simonds Worden White Company (USA) for an air-hardening tool steel containing 1% carbon, 3% manganese, 1% molybdenum, 1% chromium, and the balance iron. It has a good nondeforming properties, and is used for punch and die facings, templates, and gages.

(2) Trade name of Precision-Kidd Steel Company (USA) for an air-hardening, medium-alloy cold-work tool steel (AISI type A2) containing about 1% carbon, 0.7% manganese, 0.3% silicon, 5.25% chromium, 0.2% vanadium, 1.1% molybdenum, and the balance iron. It has a great depth of hardening, good wear resistance and hot hardness, and good nondeforming properties. Used for drill rod.

Airtrue. Trade name of Simonds Worden White Company (USA) for an air-hardening medium-alloy cold-work tool steel (AISI type A2) containing 1% carbon, 5% chromium, 1% molybdenum, 0.25% vanadium, and the balance iron. It has a great depth of hardening, good nondeforming properties, and good wear resistance and toughness. Used for blanking, forming and trimming dies, knives, and gages.

Airtrue LC. Trade name of Simonds Worden White Company (USA) for an air-hardening medium-alloy cold work tool steel (AISI type A8) containing 0.55% carbon, 1% silicon, 5% chromium, 1.5% molybdenum, 1.25% tungsten, and the balance iron. It has a great depth of hardening, high hot hardness, good wear resistance and toughness, and good to moderate machinability. Used for trimming, blanking, forming, coining and notching dies, and punches.

Airtuff. Trademark of Ludlow Composites Corporation (USA) for a composite consisting of firm open-cell styrene-butadiene rubber latex foam with sheet or fabric facing.

Airvan. Trade name of Teledyne Firth-Sterling (USA) for an air-hardening, nondeforming tool steel (AISI type A2) containing 1% carbon, 5.25% chromium, 1.15% molybdenum, 0.25% vanadium, and the balance iron. It has excellent wear resistance, and is used for tools, heavy-duty blanking and forming dies, cold-work dies, cutters, shear blades, punches, and gages.

air-wall material. See air-equivalent material.

Air-Wear. Trade name of Carpenter Technology Corporation (USA) for an air-hardening, wear-resistant tool steel containing 1.5% carbon, 12% chromium, 1% vanadium, and the bal-

ance iron. Used for dies and tools.

Airweave. Tradename of Airtech Europe SA for breather cloths available in a wide range of sizes. Used in vacuum-bag molding.

AISI-SAE steels. Steels whose compositions have been standardized by the American Iron and Steel Institute (AISI) and the Society of Automotive Engineers (SAE). They are chiefly plain-carbon and/or low- to medium-alloy steels. The AISI-SAE designation is a four-digit number system in which the first two numbers indicate the alloy content, and the last two the nominal carbon content, e.g. AISI-SAE 1060 is a plain-carbon steel containing 0.60 wt% carbon. The following numbers and/or digits are used: 10xx, 11xx, 12xx, and 15xx (carbon steels); 13xx (manganese steels); 23xx and 25xx (nickel steels); 31xx, 32xx, 33xx, and 34xx (nickel-chromium steels); 40xx and 44xx (molybdenum steels with the latter having a higher alloy content); 41xx (chromium-molybdenum steels); 43xx, 47xx, 81xx, 86xx, 87xx, 88xx, 93xx, 94xx, 97xx, and 98xx (nickel-chromium-molybdenum steels in the order of increasing alloy contents); 46xx and 48xx (nickel-molybdenum steels with the latter having higher nickel contents); 50xx, 51xx, and 52xx (chromium steels in the order of increasing alloy contents); 61xx (chromium-vanadium steels); 72xx (tungsten-chromium steels); 92xx (silicon and silicon-chromium steels). The letters B or L between the first and last two numbers indicate the presence of minute quantities of boron or lead, respectively. Also known as *SAE steels.*

AISI tool steels. The most commonly used tool steels which have been arranged into the following seven categories by the American Iron and Steel Institute (AISI) and the Society of Automotive Engineers (SAE): (i) *High-speed tool steels* [subdivided into molybdenum types (group M), and tungsten types (group T)]; (ii) *Hot-work tool steels* [subdivided into chromium types (groups H1-H19), tungsten types (groups H20-H39), and molybdenum types (groups H40-H59)]; (iii) *Cold-work tool steels* [subdivided into high-carbon, high-chromium types (group D), medium-alloy, air-hardening types (group A), and oil-hardening types (group O)]; (iv) *Shock-resisting tool steels* (group S); (v) *Mold steels* (group P); (vi) *Special-purpose tool steels* [subdivided into low-alloy types (group L) and carbon-tungsten types (group F)]; and (vii) *Water-hardening tool steels* (group W).

Aitch metal. A high-strength copper alloy containing 38-40% zinc and 1-1.5% iron. Used for hydraulic cylinders and forgings.

Ajax. British trade name for a bearing alloy containing 30-70% iron, 25-50% nickel, and 5-20% copper.

Ajax Bull. Trade name H. Kramer & Company (USA) for a babbitt containing 76% lead, 17% antimony, 7% tin, and traces of other elements. Used for bearings and bushings.

Ajax metal. A bearing alloy containing 30-70% iron, 25-50% nickel, and 5-20% copper.

Ajax P. Trademark for a white-firing air-floated kaolin with fine particle size. Used for whitewares and ceramic tiles.

ajoite. A bluish green mineral of the cancrinite group composed of potassium sodium copper aluminum silicate hydroxide trihydrate, $(K,Na)Cu_7AlSi_9O_{24}(OH)_6 \cdot 3H_2O$. Crystal system, triclinic. Density, 2.96 g/cm^3; refractive index, 1.583. Occurrence: USA.

akaganeite. A mineral composed of iron oxide hydroxide, β-FeO(OH). It can also be made synthetically. Crystal system, tetragonal. Density, 3.51 g/cm^3. Occurrence: Japan.

Akalit. German trade name for casein plastics.

Akatherm. Trademark of AKATHERM GmbH (Germany) for rigid polyethylenes.

akatoreite. An orange-brown mineral composed of manganese aluminum silicate hydroxide, $Mn_9(Si,Al)_{10}O_{23}(OH)_9$. Crystal system, triclinic. Density, 3.48 g/cm^3; refractive index, 1.704. Occurrence: New Zealand.

akdalaite. A white mineral composed of aluminum oxide hydrate, $(Al_2O_3)_4 \cdot H_2O$. Crystal system, hexagonal. Density, 3.68 g/cm^3; refractive index, 1.747. Occurrence: Kazakhstan.

akermanite. A colorless mineral of the melilite group composed of calcium magnesium silicate, $Ca_2MgSi_2O_7$, and which may contain small additions of iron and manganese. It has a simple silicate structure, and can also be prepared synthetically from basic calcium carbonate and silicon dioxide by a prescribed calcining regime. Crystal system, tetragonal. Density, 2.94-2.95 g/cm^3. Used in ceramics.

akhtenskite. A black mineral composed of manganese dioxide, ϵ-MnO_2. Crystal system, hexagonal. Density, 4.0 g/cm^3.

Akilen. Trademark of Kolon Industries, Inc. (South bKorea) for polyester fibers and yarns.

Aklo. Trade name of Corning Glass (USA) for a blue-green, rolled glass with excellent heat-absorbing properties.

Akrit. Trade name of Thyssen Edelstahlwerke AG (Germany) for a series of coated cobalt-chromium hardfacing electrodes for overlaying valves, valve parts, wear rings, etc.

akrochordite. A brownish yellow to reddish brown mineral composed of magnesium manganese arsenate hydroxide tetrahydrate, $Mn_4Mg(AsO_4)_2(OH)_4 \cdot 4H_2O$. Crystal system, monoclinic. Density, 3.26 g/cm^3; refractive index, 1.676; hardness, 3 Mohs. Occurrence: Sweden.

Akron. British trade name for a corrosion-resistant alloy containing 63% copper, 36% zinc, and 1% tin. Used for hardware and fittings.

Akroplast. Trade name for thermoplastic color concentrates.

Akryl. Trade name for acrylic fibers used for the manufacture of textile fabrics.

Aksacryl. Trademark of AKSA AS (Turkey) for acrylic fibers.

aksaite. A colorless mineral composed of magnesium borate tetrahydrate, $MgB_6O_{10} \cdot 5H_2O$. It can also be made synthetically. Crystal system, orthorhombic. Density, 1.99 g/cm^3; refractive index, 1.503. Occurrence: Kazakhstan.

AKS tungsten. Tungsten to which minute quantities of aluminum (Al), potassium (K) and silicon (Si) have been added. Used for nonsagging light-bulb filaments.

aktashite. A gray-black mineral of the nowackiite group composed of copper mercury arsenic sulfide, $Cu_6Hg_3As_4S_{12}$. Crystal system, rhombohedral (hexagonal). Density, 5.5 g/cm^3. Occurrence: Russia.

Aku CMC. Trademark of DSM Engineering Plastics, Inc. (USA) for carboxymethylcellulose polymers.

Akulene. Trademark of DSM Engineering Plastics, Inc. (USA) for polyester fibers and plastics.

Akulon. Trademark of DSM Engineering Plastics, Inc. (USA) for engineering thermoplastics based on nylon 6 or nylon 6,6, and available in the form of resins, fibers, and cast and blown films. The resins are supplied in various grades including neat, fire-retardant, high-impact, UV-stabilized, glass fiber-reinforced, mineral-filled and lubricated, and used for injection-molded electrical connectors, relays, switches, wire and cable, and various other industrial and consumer applications. The fibers are used especially for carpets, and the tough, puncture-resistant films are used food and medical packaging applications, vacuum bags for plastic and composite molding, chemical bag liners, and cooking bags and pouches. *Akulon* nylon has a density of

1.13 g/cm^3 (0.041 $lb/in.^3$), good processibility and physical properties, a service temperature range of -40 to +160°C (-40 to +320°F), good dielectric properties, and good resistance to alcohols, alkalies, hydrocarbons, greases, oils and ketones.

Akulon F. Trademark of DSM Engineering Plastics, Inc. (USA) for nylon 6 supplied in the form of film, sheet, rods, powder, granules, and monofilaments. It has good processibility and physical properties, a service temperature range of -40 to 160°C (-40 to 320°F), good dielectric properties, and good resistance to alcohols, alkalies, hydrocarbons, greases, oils and ketones. Used for electrical components, and wire and cable.

Akulon K. Trademark of DSM Engineering Plastics, Inc. (USA) for nylon 6 supplied in the form of film, sheet, rods, powder, granules and monofilaments. It has good processibility and physical properties, a service temperature range of -40 to +160°C (-40 to +320°F), good dielectric properties, and good resistance to alcohols, alkalies, hydrocarbons, greases, oils and ketones. Used for electrical components, and wire and cable.

Akulon S. Trademark of DSM Engineering Plastics, Inc. (USA) for strong, stiff nylon 6,6 supplied in various grades including regular, fire-retardant, high-impact, supertough, UV-stabilized, 30% glass fiber-reinforced, and 40% mineral-filled. It has good abrasion resistance, dimensional stability and heat distortion properties, an upper service temperature of 80 to 160°C (176 to 320°F), and very good resistance to alcohols, alkalies, hydrocarbons and ketones. Used for bearings, gears, fasteners, automotive under-hood parts, powder tool housings, and sports equipment.

Akulon Ultraflow. Trade name of DSM Engineering Plastics, Inc. (USA) for nylon 6 with good flowing and processing characteristics, supplied in unreinforced and 30 and 40% glass fiber-reinforced grades for injection molding applications.

Akuloy. Trade name of DSM Engineering Plastics, Inc. (USA) for a series of modified polyamide (nylon) alloys with good impact strength, high heat resistance and low moisture absorption. *Akuloy RM* is available as an unfilled or glass fiber-reinforced polyamide alloy that ranges in its properties between nylon 6 and polypropylene. *Akuloy* polyamides have good impact strength and heat resistance, and low moisture absorption, and are used for handles, appliance housings, consumer electronics and appliances, climate-control components, automotive under-the-hood components, etc.

Akvaflex. Trademark of Norfil-Norpack A/S (Norway) for polyethylene and polypropylene fibers and monofilaments.

Alabama marble. A white, pure marble with low porosity from Alabama, USA, used for statuaries.

alabandite. An iron-black mineral of the halite group composed of manganese sulfide, MnS. It can also be made synthetically. Crystal system, cubic. Density, 4.00-4.05 g/cm^3; refractive index, 2.70; hardness, 3.5 Mohs. Occurrence: Rumania. Also known as *manganblende*.

alabaster. A compact, marble-like, fine-grained, smooth, white, translucent variety of gypsum, $CaSO_4 \cdot 2H_2O$. It has high softness, good workability, and good carving and polishing qualities. Occurrence: Algeria, Egypt. Used for ornamental building work, vases, lamps, and novelty items.

alabaster glass. A milky-white variety of glass that contains inclusions of materials with different refractive indices.

Alabond E. Trade name for a silver-free palladium dental bonding alloy.

Aladar. British trade name for an alloy containing 12.5% silicon, 0.4% iron, 0.13% copper, 0.09% manganese, and the balance aluminum. Used for castings and light alloys.

Aladdin. Trade name of Aladdin Welding Products, Inc. (USA) for an alloy containing varying amounts of zinc, aluminum, and copper. Used for welding rods and white-metal die castings.

Aladdinite. Trade name for a casein plastic used for ornamental items, buttons, etc.

Alais. Trade name of Compagnie des Mines (France) for a series of chromium, nickel, nickel-molybdenum and nickel-chromium-molybdenum case-hardening steels, and several plain-carbon tool steels.

Alakron. Trade name of William Mills & Company Limited (UK) for a corrosion-resistant alloy containing 95% aluminum and 5% silicon. Used for light-alloy castings.

alalite. See diopside.

Alamo. (1) Trade name of Alamo Iron Works (USA) for a nitriding steel containing 0.3% carbon, 0.6% manganese, 1-1.5% aluminum, 0.6-1% molybdenum, and the balance iron. Used for gears, pinions, cams, shafts, etc.

(2) Trade name for a high-duty firebrick used for linings of furnaces, and ladles.

alamosite. A white, or colorless mineral of the pyroxenoid group composed of lead metasilicate, $PbSiO_3$. It can also be made synthetically by fusing litharge and silica. Crystal system, monoclinic. Density, 6.49 g/cm^3; melting point, 766°C (1411°F); hardness, 4.5 Mohs; refractive index, 1.955. Occurrence: Mexico, South West Africa. Used as secondary lead ore, for glazing pottery, in glassmaking and ceramics, and for fireproofing textiles.

Alane. Trade name for polyolefin fibers and products.

alapat. A fine two-ply spun *coir* (coconut) fiber yarn. See also plied yarn.

Alar. Trade name of Alar Limited (UK) for a series of cast aluminum-silicon alloys.

alargan. A German alloy composed of aluminum and silver over whose surfaces platinum black has been powdered and worked in by hammering or pressing. Used as a platinum substitute in jewelry.

Alarm Kinonglas. Trade name of former Glas- und Spiegelmanufaktur N. Kinon GmbH (Germany) for a glass product consisting of two, three or four sheets of glass laminated together and connected with an alarm system by a thin wire running through the interlayer.

Alaska cedar. The lightweight, moderately heavy, strong and stiff softwood obtained from any of several medium-large evergreen trees of the cypress family, especially *Chamaecyparis nootkaensis* and *Cupressus sitkaensis*. It has a fine texture and an uniform, straight grain. The heartwood is bright, clear yellow and very decay-resistant, and the sapwood white to yellowish and hardly distinguishable from the heartwood. *Alaska cedar* has moderate hardness, high durability and shock resistance, low shrinkage, high stability after seasoning, good workability, and takes a high polish. Average weight, 495 kg/m^3 (31 lb/ft^3). Source: Pacific Coast of North America from Alaska to Oregon. Used for boatbuilding, furniture, joinery, and interior finish. Also known as *Sitka cypress; yellow cedar; yellow cypress.*

Alaskan pine. The lightweight, strong, moderately hard, nonresinous wood of the tall hemlock *Tsuga heterophylla*. It has a straight grain and fairly even texture. The heartwood and sapwood are almost white with a purplish cast. *Alaskan pine* is easy to work, and has an average weight of 480 kg/m^3 (30 lb/ft^3). Source: Pacific Coast from Alaska to Northern California.

Used for construction lumber, plywood corestock, containers, boxes, pallets, crates, flooring and woodenware, as a substitute for Douglas fir, and as a pulpwood. Also known as *gray fir; hemlock spruce; Prince Albert fir; West Coast hemlock; western hemlock; western hemlock fir.*

Alaska White Brass. Trade name for an alloy of copper, zinc and nickel, used for heavy-duty bearings and bushings.

Alastin. Trademark of Fabelta Ninove NV (Belgium) for viscose rayon fibers.

Alastra. Trade name for viscose rayon used for clothing, and blended with other fibers for suitings, furnishings, carpets, linings, etc.

Alathon. Trademark of E.I. DuPont de Nemours & Company (USA) for a series of polyethylene resins supplied in low-density, high-density, glass fiber-reinforced and UV-stabilized grades. They have good toughness from -57 to +93°C (-70 to +200°F), good chemical, moisture and electrical resistance, low coefficients of friction, and good processibility. Used for piping, ducts, containers, housings, insulation, houseware, toys, packaging (also food items), coatings, and films.

Alba. Trade name of Vacuumschmelze GmbH (Germany) for a heat-treatable, corrosion-resistant alloy of 60% silver, 30% palladium, and 10% gold. Used as a dental casting alloy, and for fountain-pen points.

Albabond. Trade name of J.F. Jelenko & Company (USA) for a palladium-silver dental bonding alloy.

Albacast. Trade name of J.F. Jelenko & Company (USA) for a platinum-colored, hard dental casting alloy (ADA type III) containing 70% silver, 25% palladium, and 5% gold. It has a density of 10.6, a melting range of 840-920°C (1545-1690°F), and provides high strength, and good burnishability, grindability, and polishability. Used for crowns, hard inlays, and fixed bridgework.

Albalith. Trade name for a white pigment (lithopone) containing 70% barium sulfate and 30% zinc sulfide. Used in paper products, inks, rubber articles, etc.

Albaloy. Trade name of Hanson-Van Winkle-Munning Company (USA) for a corrosion-resistant alloy containing 55% copper, 30% tin, and 15% zinc. Used for bright electrodeposits.

Albamit. Trademark of Albert/Hoechst AG (Germany) for melamine-formaldehyde resins and plastics.

Albanene. Trade name of Keuffel & Esser Company (USA) for tracing papers.

Albanoid. Trade name of Henry Wiggin & Company Limited (UK) for a copper casting alloy of 62% copper, 20% nickel, and 18% zinc. Used for architectural and sanitary fittings.

Albany. (1) Trade name of Allegheny Ludlum Steel (USA) for a low-alloy special-purpose tool steel (AISI type L2) containing 0.75% carbon, 0.9% chromium, 0.6% manganese, 0.2% vanadium, and the balance iron. It has excellent impact resistance and toughness, and is used for general dies, cutlery dies, punches, knives, shear blades, and machine-tool components, e.g., collets.

(2) Trade name for a die steel containing up to 1% carbon, 1% chromium, 0.45% tungsten, 0.45% vanadium, and the balance iron. It has high hardness, and excellent toughness and wear and fatigue resistance.

Albany Caroga. Trade name AL Tech Specialty Steel Corporation (USA) for a special-purpose tool steel (AISI type L2) containing 0.5-1.1% carbon, 1% chromium, 0.2% vanadium, and the balance iron. It has a great depth of hardening, excellent toughness, good to moderate machinability, and moderate wear resistance. Used for forming tools, forming and trimming dies, and structural parts. Also known as *Caroga.*

Albany clay. A clay of high flux content and fine particle size that fuses at relatively low temperature. It is found in the vicinity of Albany, New York, and also in Michigan. Used for dark to greenish brown glazes on electrical porcelain, stoneware, etc., and as a bond in the manufacture of grinding wheels. Also known as *Albany slip.*

Albany slip. See Albany clay.

Albatra metal. A copper alloy containing 20-22.5% zinc, 18.75-20% nickel, and up to 1.25% lead. Used for hardware, fittings, and automotive trim.

albatross. A fine, soft, lightweight, plain-woven wool or wool-blend fabric with pronounced napped texture resembling the breast of an albatross. Used for blouses, dresses, scarves, sleepwear, and children's wear.

AlBeCast. Trademark of Brush Wellman Inc. (USA) for investment cast aluminum-beryllium composite alloys.

AlBeMet. Trademark of Brush Wellman Inc. (USA) for lightweight composite alloys composed of 62% beryllium and 38% aluminum. They are available as hot-isostatically-pressed powder-metallurgy shapes, and extruded and rolled bars, and have a high modulus of elasticity, high thermal conductivity and stability, and good isotropic properties. Used for structural parts and heat sinks for aerospace and satellite applications.

Albene. Trade name for cellulose acetate fibers used for textile fabrics.

Alberene. Trademark for Alberene Stone Company (USA) for a dense bluish-gray soapstone from Virginia, USA. Available in medium-hard and hard grades, it is used for structural and architectural applications. The medium-hard grades are used for trim, and laboratory tables and sinks, and the hard grades for stair treads and flooring.

Alberni. Trademark of MacMillan Bloedel Limited (Canada) for kraft linerboard.

Albertit. Trademark of Albert/Hoechst AG (Germany) for urea-formaldehyde resins and plastics.

Albertson Special. Trade name of Sioux Tools Inc. (USA) for an alloy cast iron containing 3.2-3.4% carbon, 0.35-0.4% molybdenum, 0.3-0.4% chromium, 1.75-2% silicon, 0.6-0.7% manganese, 1.25-1.5% nickel, and the balance iron. Used for valve-seat inserts.

Albi Cote. Trade name of StanChem, Inc., Albi Manufacturing Division (USA) for fire-retardant coatings and paints.

Albidur. (1) British trade name for a nonhardenable aluminum alloy used for light-alloy parts.

(2) Trademark of Hanse Chemie GmbH (Germany) for thermosetting resins supplied in standard, casting and silicone elastomer-modified grades. Used in the manufacture of plastic goods, and for adhesives and coatings.

Albin. British trade name for a white brass containing varying amounts of copper, zinc and nickel. Used for fittings, fixtures, and hardware.

albino asphalt. A light colored asphaltic resin.

Albis. Trade name of Albis Corporation (USA) for polyamide (nylon) resins.

albite. A colorless or grayish-white plagioclase mineral of the feldspar group composed of sodium aluminum silicate, $NaAlSi_3O_8$. Crystal system, triclinic. Density, 2.62 g/cm³; hardness, 6-6.5 Mohs; refractive index, 1.53. Occurrence: Canada, Germany, Switzerland, USA (California, New York, North Carolina, Pennsylvania, Texas, Virginia). Used for glazes in ceram-

ics, and as a refractory raw material in the manufacture of porcelain enamels. Also known as *soda feldspar; sodium feldspar; white feldspar; white schorl.*

Albolit. German trade name for a phenolic resin.

albolite. A plastic cementitious material made chiefly of magnesia (MgO) and silica (SiO_2).

Albond. Trademark of Brent America Inc. (USA) for a chromate conversion coating that imparts paint bonding characteristics and corrosion protection to aluminum.

Albondur. (1) German trade name for an age-hardenable, high-strength aluminum alloy containing 3.5-5.5% copper, 0.3-0.5% silicon, 0.3-1% manganese, and 0.2-0.7% magnesium. Used for general-purpose applications.

(2) Trade name of Vereinigte Leichtmetallwerke GmbH (Germany) for an aluminum alloy containing 0-1.5% manganese, 0-1.2% silicon, and 0-1% magnesium. It has good weldability and formability, and is used for containers, tanks, furniture, and marine parts.

Albor. Trade name of Jessop-Saville Limited (UK) for an oil-hardening chromium-molybdenum die steel containing 0.90% carbon, 0.90-1.20% chromium, 0.30% molybdenum, and the balance iron. It has great toughness and good deep-hardening properties, and is used for dies for stamping hard metals.

Alborium. Trade name of J.F. Jelenko & Company (USA) for an extra-hard dental casting alloy (ADA type IV) containing 45% silver, 25% palladium, and 15% gold. Used for hard inlays, thin crowns, partial dentures, and fixed bridgework.

Albral. Trade name of Foseco Minsep NV (Netherlands) for a flux used for the oxide removal from, and purification of aluminum- or silicon-containing copper alloy melts. It achieves a significant reduction of dross formation, and improves the fluidity of the metal.

Albrifin. Trade name of Rhodia (USA) for a clear acrylic resin coating that is dyed prior to the curing process.

Albristar. Trade name of Rhodia (USA) for a bright nickel electrodeposit and plating process.

Albro. Trade name of Bronze Die Casting Company (USA) for a corrosion-resistant aluminum casting bronze containing 89% copper, 10% aluminum, and 1% iron. Used for acid-resisting parts.

Al Bronze. Trade name of Manganese Bronze Limited (UK) for wrought aluminum bronzes containing up to 10% aluminum, and up to 2% nickel and iron, respectively. It has good strength, and excellent corrosion and wear properties. Used for pump parts, fasteners, etc.

Albula. Trade name for rayon fibers and yarns used for textile fabrics.

albumin. A water-soluble simple protein found in plant and animal tissue, and milk, eggs and blood serum. It can also be made synthetically. It is available in powder and crystalline form, readily coagulated by heat, and converted to amino acids or amino acid derivatives on hydrolysis. Used in biology, biochemistry, medicine, affinity chromatography, and as a bioabsorbable biomedical polymer coating in thromboresistant devices.

albumin glue. A glue containing chiefly soluble dried blood. It has a bonding strength inferior to that of most animal glues, and is used chiefly for plywood joints.

albuminoid. Any of the water-insoluble simple proteins found as keratin in hair, horn and feathers, as elastin in tendons, and as fibroin in silk. They are dissolved by hydrolysis and boiling with concentrated acid solutions. Also known as *scleroprotein.*

Albus 190. Trade name for a high-gold dental bonding alloy.

Alcaloy. Trade name of Alcaloy Inc. (USA) for a corrosion-resisting, nonsparking copper casting alloy containing 87-91% copper, 9-11% aluminum, 0-2% iron, and 0-0.5% beryllium. Used for safety tools, chains, chain fittings, and hand tools.

Alcan. Trademark of Aluminum Company of Canada Limited for an extensive series of fabricated and semifabricated aluminum products including commercially-pure aluminums, cast and wrought aluminum-silicon alloys, wrought aluminum-magnesium, aluminum-magnesium-silicon, aluminum-copper and aluminum-zinc alloys, cast aluminum-magnesium and aluminum-magnesium-silicon alloys, cast-aluminum copper, etc. They are commercially available in bar, sheet, plate, tube and wire form, and also as forgings, castings and extrusions. Used for architectural applications, structural parts, decorative purposes, automotive components, costume jewelry, toys, furniture, electrical parts, machine parts, aluminum-clad aluminum alloys, etc.

Alcantara. Trademark of Alcantara SpA (Italy) for a high-quality artificial leather that imitates suede and is used for clothing (coats, jackets, etc.), suitcases, bags, briefcases, wallets, etc.

Alcar. Trade name for a general-purpose acetal prime resin.

Alcast. Trademark of Clayburn Refractories Limited (Canada) for a series refractory specialty products.

Alceru. Trademark of lyocell textile fibers.

Alchrogold. Trade name of American Chemical & Equipment (USA) for a yellow chromate conversion coating for aluminum and aluminum alloys.

Alchrome. (1) Trade name of Heatbath Corporation (USA) for a chromate conversion coating for aluminum and aluminum alloys.

(2) Trade name of Driver Harris Company (USA) for magnetic iron-chromium alloys containing up to 15% chromium, and 5.5% aluminum. Used for heating elements.

Alclad. Trademark of Alcan Rolled Products Company (Canada) for a wrought aluminum product supplied in sheet or tube form. It is composed of a high-strength aluminum-copper base metal, such as duralumin, coated or clad with a corrosion-resistant high-purity aluminum (99.7+%). Since the cladding is anodic to the base metal, the product is electrolytically protected against general and seawater corrosion.

Alco. (1) Trade name of Cytemp Specialty Steel Division (USA) for a shock-resisting tool steel (AISI type S1) containing 0.5% carbon, 2.2% tungsten 1.7% chromium, 1% silicon, 0.5% molybdenum, 0.2% vanadium, and the balance iron. Used for hot- and cold-work tools, chisels, punches, and pneumatic drills.

(2) Trade name of Dirigold Corporation (USA) for an alloy of copper and zinc.

(3) Trade name of Skandinaviska Armaturfabriken (Sweden) for a tough, hard, corrosion-resistant bronze containing 85% copper, 10% tin, and 5% nickel. Used for surgical instruments, propellers, ship forgings, and skates.

(4) Trade name of Alco-NVC, Inc. (USA) for asphalt-base construction products.

Alcoa. Trade name of Alcoa–Aluminum Company of America (USA) for an extensive series of commercially pure aluminums and wrought and cast aluminum alloys containing one or more alloying elements, such as silicon, magnesium, manganese, copper, zinc, etc. Used for equipment for the chemical, petrochemical, petroleum, pharmaceutical, brewery and food processing industries, domestic appliances, aircraft and automotive applications, shipbuilding, cables, solders, cladding, etc.

Alcodie. Trade name of Columbia Tool Steel Company (USA)

for an oil- or air-hardening chromium-type hot-work tool steel (AISI type H12) containing 0.35% carbon, 5% chromium, 1.5% molybdenum, 1.3-1.5% tungsten, 0.3-0.4% vanadium, and the balance iron. It has a great depth of hardening, high toughness, good machinability, high hot hardness, and good abrasion resistance. Used for die-casting dies, extrusion dies, hot blanking and forging dies, bolt header dies, trimmer dies, shear blades, punches, and piercing tools.

alcogel. An aquagel that has been submerged in alcohol. See also aquagel; aerogel.

alcoholate. See alkoxide.

Alcolite. Trade name for a cellulose nitrate formerly used for denture bases.

Alcoloy. Trade name of Olin Corporation (USA) for an alloy of 73.5% copper, 22.7% zinc, 3.4% aluminum, and 0.4% cobalt. It has good deep-drawing properties, and good resistance to corrosion and stress-corrosion. Used for connectors, diaphragms, and springs.

Alcomax. British trade name for a permanent magnet alloy containing 21-25.5% cobalt, 11.9-13.5% nickel, 7.4-8.1% aluminum, 3-4% copper, and the balance iron. Small amounts of niobium (columbium) and silicon may also be present. It has a high coercive force and magnetic energy product, and is anisotropic with maximum flux along preferred axes. Used for electrical and magnetic equipment, e.g., loudspeakers, and magnets for motors and dynamos.

Alcon. Trade name of Wieland Werke AG (Germany) for a corrosion-resistant, age-hardenable aluminum alloy containing 0.4-0.8% magnesium, and 0.3-0.7% silicon. Used for structural and decorative parts.

Alco Ni-Iron. Trade name for a nickel cast iron.

Alconit. German trade name for a series of cast permanent magnet alloys containing varying amounts of iron, aluminum, cobalt, nickel, and titanium.

Alcop. Trade name of Janney Cylinder Company (USA) for a bronze containing 62% copper, 35.5% zinc, 2.5% aluminum, and a trace of lead. Used for centrifugal castings, bushings, liners, and bearing cages.

Alcor. Trade name of Iso-Cor Inc. (USA) for nickel-based sealing compounds used on anodized aluminum.

alcosol. A colloidal solution in which the dispersion medium is alcohol and the dispersed phase a solid, liquid or gas.

Alcress. British trade name for an electrical resistance alloy containing 70-75% iron, 20-25% chromium, and 5-7.5% aluminum. It has excellent heat resistance, and is used for resistors.

Alcrome. Trademark of Heatbath Corporation (USA) for a line of corrosion-resistant chromate coating products for aluminum and aluminum alloys including *Alcrome 2*, which yields a light yellow to dark golden chromate conversion coating, and *Alcrome 16*, which produces a light iridescent golden to tan chromate coating on aluminum alloys as well as zinc and cadmium surfaces. They can be used as primary coatings, or paint bases.

AlcroPlex. Trade name of Alon Surface Technologies, Inc. (USA) for a dual-layer diffusion coating system produced by first diffusing chromium or chromium/silicon into the base material as a barrier, and then diffusing aluminum or aluminum/silicon which forms a stable aluminum oxide layer. It is resistant to high-temperature corrosion (up to 1120°C or 2050°F) as well as carburization, coking, erosion and metal dusting. Used on steam cracker furnace tubes, reformer tubes, direct iron reduction equipment, furnace tubes and fittings, etc.

Alcryn. Trademark of E.I. DuPont de Nemours & Company (USA) for a series of environmentally resistant, melt-processible, thermoplastic elastomers based on partially crosslinked, chlorinated olefin interpolymer alloys. They have a service temperature range of -40 to +121°C (-40 to +250°F), very good resistance to oil, ozone, heat, ultraviolet light and weather conditions, and require neither compounding nor vulcanizing. Used for hoses, tubing, seals, gaskets, weatherstripping, conveyor belts, and fabrics, and in the high-productivity manufacture of rubber parts.

Alcufont. Trade name of Società Alluminio Veneto per Azioni (Italy) for a series of wrought aluminum-copper alloys for light-alloy parts, pistons, etc.

Alculoy. Trade name of NL Industries (USA) for a series of die-cast aluminum-copper-silicon alloys.

Alcuman. Trade name for an age-hardenable, high-strength aluminum alloy containing 4.8-6.0% copper, 1.5% magnesium, 0.5% manganese, and 0-0.2% silicon. Used for screw-machine products, and construction and transportation equipment.

Alcumet. Trade name for a white-colored nickel bronze containing 10-40% nickel. It has excellent corrosion resistance, and is used for hardware, plumbing fixtures, fasteners, etc.

Alcumite. Trade name of Duriron Company Inc. (USA) for a corrosion-resistant alloy of 88-92% copper, 7.5-7.7% aluminum, 2-3.5% iron, and 1-1.5% nickel. Used for blowers, pumps, ventilators and tanks to resist weak acids and other corrosives.

Alcunic. Trade name of Century Brass Products, Inc. (USA) for a corrosion-resistant aluminum brass containing 70% copper, 27% zinc, 2% aluminum, and 1% nickel. Used for belt buckles, and condenser and heat-exchanger tubing.

Alcuplate. Trade name of Texas Instruments, Inc. (USA) for a wrought high-aluminum alloy with copper bonded to both sides. It has good electrical properties, good solderability, and is used for formed and stamped parts, and electrical contacts.

Alda. Trade name of British Oxygen Company Limited (UK) for a carbon steel containing 0.1-0.2% carbon, and the balance iron. Used for copper-coated welding rods.

Aldal. Trade name of Société du Duralumin (France) for a dur-alumin-type alloy containing 3.5-4.5% copper, 0.4-1% manganese, 0.3-0.75% magnesium, and the balance aluminum. Used for light-alloy parts, and aircraft components.

Aldecor. Trade name of Republic Steel Corporation (USA) for a steel containing up to 0.15% carbon, 0.35-1% copper, 0.16-0.28% molybdenum, 0.25-1.5% chromium, 0.25-2% nickel, and the balance iron. It has good forming and welding properties, and is used for transportation equipment, and bus and truck bodies.

aldehyde. Any of a group of aliphatic or aromatic compounds that contain the aldehyde group (–CHO) and are obtained from alcohols by either dehydrogenation or partial oxidation. They are volatile, transparent, colorless liquids with suffocating odor, and used as a chemical building blocks in organic synthesis (synthetic resins, etc.). The generic formula is RCHO. Examples of *aldehydes* include acetaldehyde (CH_3CHO), benzaldehyde (C_6H_5CHO), and formaldehyde (HCHO).

aldehyde condensation polymer. A polymer made by the condensation of a substance, such as an aromatic alcohol, ethylene oxide, melamine, etc., with an aldehyde, such as formaldehyde. Also known as *aldehyde polymer.*

aldehyde polymer. See aldehyde condensation polymer.

aldehyde resins. A class of resins made by the interaction of an aldehyde, such as acetaldehyde, butyraldehyde or formaldehyde, with another substance. Examples include urea and mela-

mine resins, polyacetal resins, and phenolic resins.

Aldek. Trade name Reynolds Extrusion Company (Canada) for aluminum extrusions, extruded pipes and tubes, etc.

alder. The wood of any of a genus of trees (*Alnus*) of the birch family. The black alder (*A. glutinosa*) and the red alder (*A. rubra*) are of commercial importance. See also black alder; red alder.

aldermanite. A colorless mineral composed of magnesium aluminum phosphate hydroxide hydrate, $Mg_5Al_{12}(PO_4)_8(OH)_{22}$·$32H_2O$. Crystal system, orthorhombic. Density, 2.0 g/cm³; refractive index, 1.50. Occurrence: Australia.

Aldivan. Trade name of Latrobe Steel Company (USA) for a hot-work tool steel containing 0.4% carbon, 2.5% chromium, 0.3% vanadium, and the balance iron. Used for aluminum casting dies.

Aldo-Kote. Trade name of Aldoa Company (USA) for a line of conversion coatings.

Aldox. Trademark of Pennwalt Corporation (USA) for an acidic powdered aluminum deoxidizer and desmutter used before spot welding, or after etching operations.

Aldrey. Trade name of Aluminum Walzwerke Singen GmbH (Switzerland) for a heat-treatable, corrosion-resistant aluminum alloy containing about 0.4-0.6% magnesium, 0.2-1.0% silicon, and up to 0.3% iron and titanium, respectively. It has good electrical conductivity, and is used for transmission-line wire, marine hardware and structures, automobile engine components, and aerospace and architectural parts.

Aldur. Trade name of Voest International, Inc. (Austria) for a series of steels containing 0.09-0.2% carbon, 0.25-0.5% silicon, 0.6-1.7% manganese, and the balance iron. Used for steel plate.

Aldural. Trade name of Alcan-Booth Industries Limited (UK) for wrought duraluminum-type aluminum-copper alloys used for structural parts, fasteners, rivets, and aircraft skins and parts. See also duralumin.

Aldurbra. Trade name of IMI Kynoch Limited (UK) for corrosion-resistant wrought alloys containing 76% copper, 22% zinc, and 2% aluminum. Used for condenser and heat-exchanger tubes.

Aldurol. Trademark of Hoechst AG (Germany) for urea-formaldehyde resins and plastics.

Alecra. Trade name of Atofina Chemicals Inc. (USA) for a chromium electroplate and plating process.

aleksite. A pale gray mineral of the tetradymite group composed of lead bismuth tellurium sulfide, $PbBi_2Te_2Si_2$. Crystal system, hexagonal. Density, 7.59 g/cm³. Occurrence: Russia.

Alemite. Trade name of Stewart Warner Corporation (USA) for a corrosion-resistant alloy containing 90% aluminum and 10% silicon. Used for die castings and light-alloy parts.

Alençon lace. A type of *needlepoint lace*, usually machine-made, having a fine net ground and a floral or other design outlined with heavy (cordonnet) thread. Named after Alençon, a city in northwestern France where this lace was originally made.

Alert. Trade name of Jeneric-Pentron Corporation (USA) for light-cure, packable, high-viscosity dental resin composite for posterior restorations.

Alessandra. Trademark of Basofil Fibers, LLC (USA) for fire-resistant yarns and barrier fabrics based on a proprietary core-spun technology. The yarns consist of a dual core comprising a fine-denier glass and nylon filament, covered with a first sheath of fire-resistant *Basofil* melamine fibers and modacrylic fibers, and a second sheath of polyester or other fibers.

Alex. (1) Trade name of DAP, Inc. (USA) for acrylic latex caulking used for exterior and interior applications.

(2) Trademark of Argonide Corporation (USA) for nanosized metal powders.

alexandria. A lightweight fabric made of cotton and wool and having small woven floral or other patterns. Used for blouses, dresses and children's wear.

alexandrite. (1) A chromium-bearing variety of the mineral *crysoberyl* (beryllium aluminate) that changes from emerald-green to columbine red or violet in artificial light. Occurrence: Russia, Sri Lanka. Used as a gemstone.

(2) A crystalline synthetic mineral composed of chromium-doped *crysoberyl* (beryllium aluminate) with a broad wavelength tuning range of 710-800 nm, and the capability to store and multi-joule energy pulses. Used especially for tunable solid-state lasers.

Alex Plus. Trade name of DAP, Inc. (USA) for weather- and mildew-resistant acrylic latex caulking containing silicone, used for exterior and interior applications.

Alfa. Trade name of Allegheny Ludlum Steel (USA) for a stainless steel containing 0.025% carbon, 0.035% manganese, 0.3% silicon, 13% chromium, 3-4% aluminum, 0.4% titanium, and the balance iron. Used for heat exchangers, recuperator parts, kiln linings, furnace parts, automotive exhaust system components, and refractory reinforcement fibers.

alfa grass. See esparto fiber.

alfalfa fiber. A cellulose fiber obtained from a herbaceous plant (*Medicago sativa*) of the pea family, native to Southeast Asia but also cultivated in southern Europe, Central and South America, and the southern and western United States. Used for twine and cordage.

Alfater. Trade name of Lavergne Performance Compound Division (USA) for a series of thermoplastic elastomers.

alfenide. A nickel silver containing 59-60% copper, 30% zinc, 9.5-10% nickel, and 0-1% iron. It may or may not have an electroplate of silver. Used for fittings, hardware, and ornamental parts.

Alfenol. Trade name of Carpenter Technology Corporation (USA) for a wear-resistant, magnetically soft material containing 84-88% iron and 12-16% aluminum. It is brittle at room temperature, and processed by rolling at about 300°C (570°F). It has high permeability, and is used for electrical and magnetic equipment, heads for tape recorders, and transformer cores.

Alfer. (1) Trade name for a series of high-permeability soft magnetic iron alloys containing about 12-14% aluminum.

(2) Trade name of Texas Instruments, Inc. (USA) for a material composed of a thin strip of steel coated or clad with aluminum. The cladding on each side is about 10% of the total thickness.

Alferium. Trade name of Creusot-Loire (France) for a heat-treatable aluminum alloy containing 2.5-4.5% copper, 0.4-1% manganese, 0.3-0.75% magnesium, and up to 0.3% silicon. Used for aerospace and automotive parts.

Alferon. Trade name of Harrison Alloys Inc. (USA) for a heat-resistant alloy cast iron.

Alfesil. Trademark of CC Chemicals Limited (Canada) for low-carbon fly ash.

alfin. A catalyst obtained from an alkali alkoxide, such as sodium isopropoxide, and an olefin halide, such as allyl chloride. Used in the polymerization of olefins.

alfin rubber. An elastomer prepared using an alfin catalyst.

Alfo. Trade name of Ato Findley (France) for vinyls including copolymers, emulsion and adhesives for wood, wood products, and furniture.

Alfol. Trademark of Alfol, Inc. (USA) for aluminum foil supplied in crumpled form, and on nonmetallic substrates, such as paper. Used as thermal insulation in buildings.

Alfor. German trade name for a plaster with a hardness of 44 MPa (6.4 ksi) that requires a plaster-to-water ratio of 1.61:1. Used in the ceramic industries for casting and jiggering molds for making round hollowware, such as cups, bowls, etc.

Alform. Austrian trade name for a series of special-grade structural steels including cold-rolled types in sheet form with a mat finish, and hot-rolled types in plate and sheet form in the unpickled condition. They have good cold workability, high yield strength of 165-500 MPa (24-72.5 ksi). Used for drawn parts, pressings, gas cylinders, motor vehicle frames, special profiles, etc.

alforsite. A colorless mineral of the apatite group composed of barium chloride phosphate, $Ba_5(PO_4)_3Cl$. It can also be made synthetically. Crystal system, hexagonal. Density, 4.81 g/cm³; refractive index, 1.696. Occurrence: USA (California).

Alfrax. Trademark of Harbison-Carborundum Corporation (USA) for electrically fused aluminum oxide supplied in the form of refractory cements and bonded refractories. Used for furnace linings, and as catalyst carriers.

Alftalat. Trademark of Hoechst AG (Germany) for a series of alkyd resins.

Alganol. Trade name for a zinc oxide/eugenol dental cement.

Algerian onyx. See onyx marble.

Alger metal. A white alloy that contains about 90% tin and 10% antimony, and may also have up to 0.3% copper. Used for jewelry and antifriction bearings.

Algier's metal. A corrosion-resistant bearing alloy containing 75-90% tin and 10-25% antimony.

Algil. Trade name for a modacrylic fiber used especially for clothing and industrial fabrics.

algin. A polysaccharide obtained from brown algae (seaweed), and usually supplied in the form of sodium alginate. Used in ointments, toothpastes and cosmetics, and in biology, biochemistry and biotechnology. See also alginic acid; sodium alginate.

Alginate. Trade name for a regenerated protein (azlon) fiber.

alginate fiber. A generic term for a nonflammable regenerated polysaccharide fiber based on metallic salts of alginic acid (e.g., calcium alginate or sodium alginate). It has high specific gravity, and is soluble in soap solution. Used in the manufacture of clothing and camouflage netting, and as a carrier or scaffolding yarn for weaving certain fabrics. See also alginates; alginic acid.

alginates. A group of hydrophilic, colloidal salts of alginic acid, such as ammonium, calcium, potassium and sodium alginate. Used as binders, emulsifiers, stabilizers, thickeners and suspension agents in ceramic materials, as waterproofing agents in concrete, as flotation agents for enamels, in the manufacture of yarns and fibers, and in biochemistry, biotechnology and medicine.

alginic acid. A colloidal chemical product that is obtained from algae and seaweed. It is a straight-chain polyuronic acid composed primarily of anhydro-β-D-mannuronic acid residues linked in the 1:4 position, and commercially available as a white to yellow powder with a melting point above 300°C (570°F). It is insoluble, but swells in water, possessing exceptional hydrophilic properties. It is also available as the ammonium-calcium salt and the sodium salt (i.e., sodium alginate). Used in the manufacture of alginates, as an emulsifier, thickener, stabilizer, suspension agent, as a waterproofing agent for concrete, in boiler water treatment, and in biochemistry, biotechnology and medicine. Formula: $(C_6H_8O_6)_n$. See also algin; sodium alginate.

algodonite. A steel-gray to silver-white mineral of the zinc group composed of copper arsenide, Cu_6As. Crystal system, hexagonal. Density, 8.38 g/cm³; refractive index, 1.523. Occurrence: Chile, USA (Michigan).

Algoclad. Trademark of Enthone-OMI Inc. (USA) for a decorative gold electroplate.

Algoflon. Trademark of Ausimont USA, Inc. for an extensive series of fluoropolymers based on polytetrafluoroethylene (PTFE). They are available in various grades including unfilled and filled with molybdenum disulfide, calcium fluoride, carbon, petroleum coke, graphite, ceramics, glass fibers, stainless steel, bronze, etc.

Algoform. Trade name of Algoma Steel Corporation Limited (Canada) for a series of high-strength low-alloy (HSLA) steels containing up to 0.1% carbon, 0.35-1.45% manganese, 0.04% aluminum, 0.05-0.11% niobium (columbium), and the balance iron. Used for steel sheet and plate.

Algogold. Trade name of Enthone-OMI Inc. (USA) for a gold electroplate and process.

Algoma. Trade name of Algoma Steel Corporation Limited (Canada) for steel billets, blooms and slabs, seamless line, oil-well-casing and standard steel pipe, mechanical tubing, cold- and hot-rolled steel sheet and strip, hot-rolled steel plate, normalized and quenched-and-tempered steel plate and floor plate, standard and wide-flange structural steel shapes, special hot-rolled steel sections, etc. Also included under this trade name are sinters, and siderite ores.

Algonquin. Trade name of Atlas Specialty Steels (Canada) for a water-hardening tool steel (AISI type W1) containing 1% carbon, 0.25% manganese, 0.2% silicon, and the balance iron. It has good resistance to decarburization, high to moderate shock resistance and toughness, moderate wear resistance and machinability, and low red hardness. Used for lathe centers, mandrels, arbors, cones, blanking, drawing and forming dies, coining, embossing and trimming dies, die plates, knife blades, knurls, springs, etc.

Algotuf. Trade name of Algoma Steel Corporation Limited (Canada) for a series of high-strength low-alloy (HSLA) structural steels.

Algrip. Trademark of ALGRIP 2000 (USA) for rigid steel plate having abrasive grains rolled into one surface, and used for nonskid flooring, loading ramps, docks, platforms, catwalks, etc.

Algrital. Trade name of Aluminum Press- und Walzwerk (Switzerland) for an aluminum casting alloy containing 4.5-5.5% silicon. Used for aircraft and automobile parts.

Alhead. Trade name of Allhegeny Ludlum Steel (USA) for a tough die steel containing 1% carbon, 1.5% tungsten, 1.5% cobalt, 0-0.15% molybdenum, and the balance iron. It has excellent abrasion and wear resistance, and is used for cold-heading dies.

Aliaf. Trademark of Sherkat Sahami Aliaf (Iran) for nylon 6 fibers and yarns.

Alicia. Trade name of Magnolia Anti-Friction Metal Company (UK) for a tin-base alloy used for medium-duty bearings.

alicyclic compound. (1) Any aliphatic cyclic (or ring) compound.

(2) A saturated organic compound, such as benzene, in which the carbon atoms are arranged in the form of a ring. The generic formula is C_nH_{2n}. Also known as *homocyclic compound*. See also cycloparaffins; cycloolefins.

aliettite. A pink mineral of the mixed-layer group composed of magnesium silicate hydroxide, $Ca_xMg_6Si_8O_{20}(OH)_4 \cdot xH_2O$. Crystal system, orthorhombic. Density, 2.15 g/cm³; refractive index, 1.50. Occurrence: USA (New Jersey).

aligned discontinuous fibers. Discontinuous fibers of glass, carbon or other materials that have been aligned by the controlled flow of a fiber-liquid slurry.

Alike. Trade name for a dental resin used in restorative dentistry for crowns and bridgework.

aliphatic amine. Any of a group of compounds, such as diethylenetriamine or triethylentetramine, containing one or more amine groups ($-NH_2$), and one or more aliphatic groups (e.g., an olefin, such as ethylene) in its molecular structure; other chemical groups may or may not be present. Frequently used as room-temperature and moderate temperature curing agents for epoxy resins.

aliphatic compound. Any of a group of organic compounds of carbon and hydrogen whose basic structure consists of carbon atoms arranged in straight or branched (but not closed) chains. They are the major ingredient of all fats, greases and waxes. Examples include paraffin, olefin, acetylene hydrocarbon and their derivatives (e.g., the fatty acids). See also acyclic compound.

aliphatic hydrocarbon. A saturated organic compound having hydrogen and carbon in its straight open-chain structure. For example, gasoline, methyl alcohol, propane, trichloroethylene, etc.

aliphatic polyol epoxies/epoxy esters. A group of modified vinyl ester resins providing good resiliency, thermal and mechanical shock resistance and impact strength. Used for corrosion-resistant reinforced plastics.

Aliron. (1) Trade name of Texas Instruments Inc. (USA) for a low-carbon steel clad on one or both sides with aluminum alloy. Used for electron tubes.

(2) Trade name for a material consisting of a core of copper clad on both sides with a layer of iron and a layer of aluminum. The core amounts to about 40% of the thickness, and each layer amounts to 30%.

Alistran. Trade name for metallic fibers.

Alite. Trademark of US Stoneware Company for a series of sintered oxides.

Alithalite. Trademark of Alcoa–Aluminum Company of America (USA) for a series of low-density aluminum-lithium alloys with good mechanical properties even at cryogenic temperatures, high strength-to-weight ratios, high elastic moduli, improved durability and damage tolerance, and good manufacturability and processibility. Used for aerospace and cryogenic tank applications.

Aljep. Trade name for an aluminum alloy containing 3.8% copper, 1.5% nickel, 1.3% manganese, and 0.5% vanadium. Used for structural parts.

alkadiene. See diene.

Alkadur. Trade name of Kreidler Werke GmbH (Germany) for a series of heat-treatable, high-strength, fatigue-resistant aluminum alloys containing 3.5-4.5% copper, 0.2-1% silicon, 0.3-1.2% manganese, and 0.4-1.4% magnesium. Used for fasteners, fittings, rivets, and aircraft parts.

alkali. A carbonate, hydroxide or oxide of the alkaline or alkaline-earth metals. It dissolves in water forming an alkaline solution, neutralizes acids, and forms salts with them. It turns red litmus blue, and has a pH value greater than 7.0.

alkali alcoholate. See alkali alkoxide.

alkali alkoxide. An organic compound in which the hydrogen of the hydroxyl (–OH) group is replaced by an alkali metal, e.g., sodium ethylate, C_2H_5ONa. Also known as *alkali alcoholate.* See also alkoxide.

alkali blue. A class of dry, blue pigment powders with very high tinting strength used especially in printing inks and interior paints. They are usually made from *p*-rosaniline or fuchsin by phenylation, immersion in hydrochloric acid and washing and subsequent treatment with sulfuric acid.

alkali cellulose. Cellulose, usually cotton linters or woodpulp, that has been steeped in 18-20 wt% sodium hydroxide (NaOH). It is the first product formed in the manufacture of viscose rayon and other cellulosics, and upon pressing contains approximately 30% cellulose, 15% sodium hydroxide, and the balance water.

alkali-doped fullerenes. A class of carbon cluster compounds composed of 60- or 70-carbon-atom (C_{60} or C_{70}) molecular crystals that become conductive when doped with alkali metals, such as cesium, potassium, or rubidium. They can be synthesized by reacting solvent-free, chromatographically purified C_{60} or C_{70} with alkali metal vapors in sealed tubes either under partial helium pressure or in high vacuum. In general, *alkali-doped fullerenes* have face-centered-cubic (fcc) structures in which the fullerene molecules are located at the corners of the cube, and the alkali atoms in the center of each face. Depending on the particular alkali-metal dopant and fullerene used, the superconductivity critical temperatures may range from as low as 8K to over 40K. Owing to their unique electronic properties and states, which are quite similar to those of organic superconductors, they are usually classed with the latter. See also fullerene superconductors; organic superconductors.

alkali feldspars. Alkali-rich feldspars composed of potassium feldspar and sodium feldspar, e.g., albite ($NaAlSi_3O_8$), microcline ($KAlSi_3O_8$), orthoclase ($KAlSi_3O_8$) and anorthoclase ($KAlSi_3O_8$–$NaAlSi_3O_8$). See also feldspar.

alkali halides. Compounds of a halogen and an alkali metal. Examples include lithium fluoride (LiF), sodium chloride (NaCl), and cesium iodide (CsI). *Note:* Alkali halide crystals, such as cesium iodide thallide, CsI(Tl) and sodium iodide thallide NaI(Tl) are used in the discriminator circuits of scintillators for spectroscopy. See also alkali metals; cesium iodide thallide; sodium iodide thallide.

alkali hydroxides. A group of compounds consisting of an alkaline metal (e.g., lithium, potassium, or sodium) and a hydroxyl (–OH) group. Examples include sodium hydroxide (NaOH), potassium hydroxide (KOH), etc. Used chiefly as electrolytes for electropolishing tungsten, lead, zinc, tin, etc., and in metallographic etching, electroplating, and alkaline cleaning.

alkali lead. A lead alloy hardened with small amounts of alkali metals, such as calcium, lithium and/or sodium. It can be made by electrolysis of the fused alkali salts using a cathode of molten lead, and has good bearing strength at elevated temperatures, a melting point of 370°C (700°F), and relative high susceptibility to corrosion. Examples of such metals are Bahnmetall, Frary metal, Ferry metal, Mathesius metal, etc. Used for bearings. Also known as *lead-alkali metal; tempered lead.*

alkali metals. The six metals in Group IA (Group 1) of the Periodic Table, namely, lithium, sodium, potassium, rubidium, cesium and francium. They are all monovalent, strongly electropositive, have low densities (ranging from 0.53 g/cm³ for lithium to 1.87 g/cm³ for cesium) and melting points (ranging from 27°C or 81°F for francium to 180.5°C or 357°F for lithium) and, except for francium, are all silver-white and ductile, and react strongly with water. Also known as *alkaline metals.*

alkaline aluminosilicate. A compound of aluminum silicate with an alkali metal, such as lithium, sodium or potassium.

alkaline-earth metals. See alkaline earths (1).

alkaline-earth oxides. See alkaline earths (2).

alkaline earths. (1) The six metals in Group IIA (Group 2) of the Periodic Table, namely, beryllium, magnesium, calcium, strontium, barium and radium. They are predominantly divalent and electropositive, and have densities ranging from 1.53 g/cm³ for calcium to 5.0 g/cm³ for radium, melting points ranging from 650°C (1,202°F) for magnesium to 1280°C (2336°F) for beryllium and, in general, are malleable, extrudable, and machinable. Also known as *alkaline-earth metals.*

(2) The oxides of the alkaline-earth metals, especially barium, calcium, magnesium and strontium, corresponding to the general formula MO. Used chiefly as fluxes in porcelain enamels and glazes, and also in refractories, and in metallurgy. Also known as *alkaline-earth oxides.*

alkaline electrolytes. A group of electrolytes based on alkali carbonates (e.g., K_2CO_3), chromates (e.g., K_2CrO_7), cyanides (e.g., NaCN or KCN), hydroxides (e.g., KOH or NaOH), or phosphates (e.g., $Na_3PO_4 \cdot 12H_2O$), and used in metallography for electropolishing of gold, silver, tungsten, lead, zinc, and/or tin.

alkaline glaze. A glaze containing substantial amounts of alkaline-earth oxides, such as barium oxide (BaO), calcium oxide (CaO) and magnesium oxide (MgO).

alkaline metals. See alkali metals.

alkaline oxides. The oxides of the alkali metals, especially lithium, sodium, potassium, rubidium and cesium, corresponding to the general formula M_2O.

alkali-reactive aggregate. See reactive aggregate.

alkali-resistant glass. A highly alkali-resistant glass essentially composed of silicon, sodium and oxygen with small additions of zirconium, lithium, potassium and boron. It has high tensile strength, and is used in fiberglass for composite reinforcement, and as reinforcement in Portland cement concrete.

alkali-resistant oxynitride glass. An alkali-resistant glass usually based on the calcium-aluminum-silicon-oxygen-nitrogen, magnesium-silicon-aluminum-oxygen-nitrogen, or sodium-calcium-silicon-oxygen-nitrogen system, and typically containing 1-2 wt% nitrogen. It has improved chemical stability over ordinary AR-glasses.

alkali-resisting porcelain enamel. A porcelain enamel that is highly resistant to soaps, detergents, cleaning fluids and household alkalies. Also known as *alkali-resisting enamel.*

alkali silicate coating. A relatively soft, low-melting, highly fluxed ceramic coating prepared from powdered alkali-alumina borosilicate glass powders (frits), and used for long-term elevated temperature applications, e.g., on aircraft turbines and heat exchangers.

alkane. Any of a class of saturated aliphatic hydrocarbons whose carbon atoms are arranged either in a straight or branched chain. The generic formula is C_nH_{2n+2}. Examples include ethane (C_2H_6), methane (CH_4) and propane (C_3H_8).

Alka-Star. Trade name of Engelhard Corporation (USA) for a bright nickel for replating bumpers.

(2) Trade name of Atotech USA Inc. for an alkaline zinc electroplate and process.

Alkathene. Trademark of ICI Limited (UK) for tough, electrically insulating, chemically resistant low-density polyethylenes supplied in the form of sheets, rods, tubes, granules, powders and fibers. UV-stabilized grades are also available. Used for packaging film, fabrics, and molded articles.

alkene. Any of a class of unsaturated aliphatic hydrocarbons containing one carbon-carbon double bond ($-C=C-$). The generic formula is C_nH_{2n}. Examples include ethylene ($H_2C=CH_2$) and propylene ($CH_3CH=CH_2$). See also olefins.

alkine. See alkyne.

Alkophos. Trade name for a white aluminum orthophoshate ($AlPO_4$) powder used in ceramics as an alkaline flux, and as a bonding agent for high-temperature cement.

Alkor. (1) Trademark of Atlas Minerals & Chemicals, Inc. (USA) for an anticorrosive thermosetting furan-type resin cement with excellent resistance to acids, alkalies and chemicals. Used as a mortar cement for temperatures below 193°C (380°F).

    (2) Trademark of Alkor GmbH (Germany) for plastic sheeting, film, panels, etc., used in the manufacture of curtains, wall covers and other decorative items, and for belts, portfolios, handbags, etc.

Alkorerom. Trademark of Alkor GmbH (Germany) for flexible polyvinyl chloride (PVC) supplied in film and other forms.

Alkorfol. Trademark of Alkor GmbH (Germany) for polyvinyl chloride (PVC) and acrylonitrile-butadiene-styrene film materials.

Alkorit. Trademark of Alkor GmbH (Germany) for flexible polyvinyl chloride (PVC) supplied in film and other forms.

Alkoron. Trademark of Alkor GmbH (Germany) for polyethylene and polyamide (nylon) resins and film products.

Alkorthylen. Trademark of Alkor GmbH (Germany) for polyethylene film materials.

alkoxide. The product of a reaction in which the hydroxyl hydrogen of an alcohol is substituted by an alkali metal. Lithium methoxide (CH_3OLi) and potassium ethoxide (C_2H_5OK) are examples. Also known as *alcoholate*. See also metal alkoxide.

alkoxyaluminum hydrides. A class of organometallic compounds obtained by reacting aluminum hydride with an alcohol, such as ethanol or methanol, in a tetrahydrofuran (THF) solvent. They correspond to the general formula H_nAlOR_{3n}, and are used as reducing agents in organic and organometallic synthesis.

Alkronid. Trade name of SWB Stahlformguss Gesellschaft mbH (Germany) for a creep-resistant steel containing 0.34% carbon, 1.4% chromium, 1.1% aluminum, and the balance iron. Used for oil-refinery equipment.

Alkrothal. Trade name of Kanthal Corporation (USA) for a resistance alloy containing 80.5% iron, 15% chromium, and 4.5% aluminum. Available in wire, ribbon, strip and foil form, it is used for electrical resistance components for service up to 1050°C (1920°F).

Alkumag. Trade name of Kreidler Werke GmbH (Germany) for a series of age-hardenable aluminum alloys containing 2.5-5% copper, 0.2-1.8% magnesium, and 0.3-1.5% manganese. Used for aircraft structures and parts.

Alkvalon. Trade name for a series of nylon fibers.

Alkydal. Trademark of Bayer AG (Germany) for a series of alkyd resins supplied in the form of emulsions or solutions.

alkyd coatings. A class of high-gloss coatings based on alkyd resins. They have good to excellent color retention, durability, versatility and flexibility, moderate drying speeds and heat resistance, fair chemical and salt-spray resistance, poor alkali resistance, and are relatively low in cost. Used as wall paints and baking enamels for automobiles and appliances, and as

trim paints, exterior enamel and house paints, marine paints, and general metal finishes.

Alkyd Molding Compounds. Trademark of Allied Chemical Corporation (USA) for thermosetting polymers based on unsaturated polyesters modified with inorganic mineral fillers, etc. Used for electrical and electronic components, electrical insulation, and potting and encapsulation.

Alkydon. Trademark of Bayer AG (Germany) for a series of alkyd resins.

alkyd plastics. A class of thermosetting plastics based on alkyd resins. They possess high stiffness, excellent heat resistance up to 150°C (300°F), good dielectric strength, moderate tensile and impact strength, low moisture absorption, short curing time, and good moldability. Used for electronic and electrical components and hardware including insulators, motor-controller components, switchgear, encapsulation, and components for automotive ignition systems, such as distributor caps, etc. Also known as *alkyds*.

alkyd resins. A group of thermosetting synthetic resins which are condensation products of a dihydric or polyhydric alcohol, such as glycerol or ethylene glycol, and a polybasic acid, such as adipic, azelaic, maleic, phthalic, or succinic acid, usually with a modifying agent, such as a drying oil. They are chemically similar to polyester resins, and used in paints, varnishes and lacquers, as caulking compounds, in adhesives for glass fibers, and in the manufacture of alkyd plastics. Also known as *alkyds*.

alkyds. See alkyd plastics; alkyd resins.

alkyl. A monovalent aliphatic hydrocarbon radical with the generic formula C_nH_{2n+1} whose combination, e.g., with a hydroxyl ($-OH$) group yields a simple alcohol. It can be obtained from an alkane by losing one hydrogen. Examples include ethyl (C_2H_5), methyl (CH_3) and propyl ($CH_3CH_2CH_2$). Usually represented in formulas by the letter R. Also known as *alkyl radical*.

alkyl acrylate copolymer. See acrylic rubber.

alkylate. The product resulting from the introduction of an alkyl group into a chemical compound.

alkylated polymer. A polymer made by the introduction of an alkyl radical into an organic molecule.

alkyl halide. A hydrocarbon derivative in which one or more hydrogen atoms have been replaced by halogen atoms. For example, ethyl chloride (C_2H_5Cl) or methyl fluoride (CH_3F).

alkyl lithium. Any of a class of organometallic compounds prepared by reacting an alkyl halide with metallic lithium in an inert solvent, such as ether or tetrahydrofuran (THF). The general formula is RLi. Examples include ethyllithium (C_2H_5Li) and methyllithium (CH_3Li).

alkyl magnesium halide. An organometallic compound formed by reacting an alkyl halide with metallic magnesium in an inert solvent, such as as ether or tetrahydrofuran. Examples include ethyl magnesium iodide (C_2H_5MgI) and phenyl magnesium chloride (C_6H_5MgCl). The general formula is RMgX. See also Grignard reagent.

alkyl radical. See alkyl.

alkyne. Any of a class of unsaturated aliphatic organic compounds characterized by a carbon-carbon triple bond ($-C≡C-$). The general formula is C_nH_{2n-2}. Examples include acetylene ($HC≡CH$) and allylene ($CH_3C≡CH$). Also known as *alkine*.

Allabond. Trade name for fast-cure epoxies.

allactite. A light to dark red, or brownish-red translucent mineral composed of manganese arsenate hydroxide, $Mn_7(AsO_4)_2(OH)_8$. Crystal system, monoclinic. Density, 3.83 g/cm^3; refractive index, 1.772; hardness, 4.5 Mohs. Occurrence: Sweden.

Alladin. Trade name for a zinc alloy used for white-metal die castings and welding rods.

allanite. A black mineral of the epidote group composed of calcium rare-earth aluminum iron silicate hydroxide, $(Ca,Ce,La,Y)_2(Al,Fe)_3Si_3O_{12}OH$. Crystal system, monoclinic. Density, 3.5-4.2 g/cm^3; refractive index, 1.783; hardness, 5.5 Mohs. Occurrence: Africa. Also known as *bucklandite*.

Allan metal. See Allan red metal (1).

Allan red bronze. A tough bronze containing 62.5-66.0% copper, 25-30% lead, and 7.5-9.0% tin. Used for hardware and bearings.

Allan red metal. (1) A mechanical mixture of about 50% copper and 50% lead, used for bearings, crankpins, etc. Also known as *Allan metal.*

(2) An alloy of 60% copper and 40% lead, used for high-speed bearings and piston wearing rings.

Allantal. Trade name for an age-hardenable aluminum alloy containing 4.5-5.5% copper and 0.2-0.5% silicon. It has high tensile and fatigue strength, and is used for aircraft parts, fittings, fasteners, and rivets.

allargentum. A silvery mineral of the zinc group composed of antimony silver, $Ag_{1-x}Sb_x$ (x = 0.09-0.16). It can also be made from the elements under prescribed heating conditions, and is the epsilon phase in the silver-antimony system. Crystal system, hexagonal. Density, 10.00-10.12 g/cm^3. Occurrence: Canada (Ontario).

Allautal. Trade name of VAW Vereinigte Aluminum-Werke AG (Germany) for a hard, high-strength *Lautal*-type aluminum-copper alloy with commercially pure aluminum rolled on.

All-Bond. Trademark of BISCO, Inc. (USA) for a dental adhesive that bonds dentin to precious and nonprecious casting alloys.

All-Bond 2. Trademark of BISCO, Inc. (USA) for a light and dual-cure methacrylate-based universal dental adhesive that bonds well to dentin, enamel, composites, precious and nonprecious casting alloys, amalgams, and silane-treated porcelain.

All-Bond C&B Luting. Trademark of BISCO, Inc.(USA) for a self-cure dental composite cement for luting crowns and bridges.

Allcast. Trade name of Apex International Alloys, Inc. (USA) for an extensive series of corrosion-resistant die-, sand- and permanent mold-cast aluminum-copper, aluminum-copper-silicon, aluminum-magnesium, aluminum-silicon-copper and aluminum-zinc alloys. Used for architectural and marine castings, chemical equipment, instrument casings, hardware, casings, housings, machine parts, aircraft fittings, brake shoes, wheel flanges, valve components, bearings and bushings, and automotive and aircraft engine components, e.g., pistons, crankcases, gear cases, oil pans and cylinder heads.

AllCeram. Trade name of Degussa-Ney Dental (USA) for a biocompatible, translucent dental veneering porcelain with high corrosion and wear resistance, high strength, smooth surface finish, and low heat conductivity. Used for brilliant-color ceramic restorations.

Allcorr. Trademark of Teledyne Industries, Inc. (USA) for a premium, corrosion-resistant high-performance alloy containing 31.0% chromium, 10.0% molybdenum, 0.04% carbon, 2.0% tungsten, and the balance nickel. Supplied in the form of sheets, plates, bars, rods, billets, ingot, wire, tubes, pipes and shapes, it is a single-phase, non-age-hardenable alloy with high resistance to general, pitting, crevice and intergranular corrosion and stress-corrosion cracking, high strength and ductility, and

good weldability, machinability and workability. Used for applications in severe corrosion conditions as found in deep gas wells containing hydrogen sulfide, carbon dioxide and brine, such as chemical and petrochemical processing equipment, equipment for the pulp and paper industries, flue-gas desulfurization equipment, etc.

alleghanyite. A brown to pink mineral of the humite group composed of manganese silicate hydroxide, $Mn_5(SiO_4)_2(OH)_2$. Crystal system, monoclinic. Density, 3.8-4.1 g/cm^3; refractive index, 1.782. Occurrence: USA (Colorado, North Carolina).

Allegheny. Trade name of Allegheny Ludlum Steel (USA) for an extensive series of corrosion- and/or heat-resistant steels including various austenitic, ferritic and martensitic stainless grades as well as several electrical and tool steels, and various cobalt-base superalloys, iron-nickel low-expansion, and soft-magnetic alloys.

Allegheny Metal. Trade name of Allegheny Ludlum Steel (USA) for an extensive series of austenitic stainless steels containing about 0.02% carbon, 15-20% chromium, 5-10% nickel, varying amounts of other elements, and the balance iron.

allemontite. A gray or reddish native antimony arsenide, AsSb. Crystal system, rhombohedral. Density, 5.8-6.2 g/cm^3; hardness, 3-4 Mohs. Also known as *arsenical antimony.*

Allen Alumi-Solder. Trade name of L.B. Allen Corporation (USA) for an aluminum solder.

allene. (1) Any of a class of unsaturated aliphatic hydrocarbons with two carbon-carbon double bonds (–C=C–), e.g., 1,2-butadiene, 1,2-propadiene and 1,4-pentadiene.

(2) A colorless, unstable gas that can be easily liquefied and boils at 34.5°C (94°F). Also known as *dimethylenemethane; propadiene.*

Allenite. Trade name of Edgar Allen Balfour Limited (UK) for a series of sintered tungsten carbides used for cutting tools, drawing dies, and woodworking tools.

Allenoy. Trade name of Allen Manufacturing Company (USA) for a series of molybdenum and chromium-molybdenum steels containing 0.35-0.53% carbon, 0.6-1% manganese, 0-1.1% chromium, 0.4-0.7% nickel, 0.15-0.3% molybdenum, and the balance iron. They have high strength and toughness, and good machinability. Used for cap screws, set screws, nuts, and pipe plugs.

Allen red bronze. A tough red bronze containing 55-70% copper, 20-40% lead, 5-10% tin, and 1% sulfur. Used for bearings and hardware. Also known as *Allen red metal; Allen's metal.*

Allen red metal. See Allen red bronze.

Allen's Imperial Special. Trade name of Allen Manufacturing Company (USA) for a high-speed tool steel containing 0.68% carbon, 0.3% manganese, 1.6% chromium, 19% tungsten, 1.4%, and the balance iron. Used for tools, cutters, dies, and punches.

Allen's metal. See Allen red bronze.

Allen's Oil Hard. Trade name for a nondeforming tool steel containing 0.8% carbon, 1.64% carbon, and the balance iron. Used for tools and dies.

Allevyn. Trade name for a polyurethane foam material used as a dressing for wound care applications.

Allfen. Trade name of Allfen-Dämmglas W. Voss KG (Germany) for insulating glass.

Allglas. Trade name of Porzer Doppelglas Gesellschaft mbH (Germany) for a double glazing unit with glass-to-glass seal.

Alliage. Trade name of Gilby-Fodor SA (France) for a series of resistance alloys containing 77-98% copper and 2-23% nickel. Depending on the nickel content, their service temperatures range from 300 to 510°C (570 to 950°F). Used for heating elements.

alligator. The light, tough leather made from the skin of alligators, crocodiles and other large lizards. It has characteristic platelike scales on the surface, and is used for luggage, shoes, pocketbooks, wallets, etc. Also known as *alligator leather.*

alligator-grained leather. Alligator leather imitations made from the hides of sheep, calves and cattle by embossing. The grain resembles that of alligator leather.

alligator leather. See alligator.

all-in aggregate. A natural mixture of concrete aggregates of various grain sizes including both ballast and crusher-run materials.

Allis-Chalmers. Trade name of Allis-Chalmers Manufacturing Company (USA) for a series of abrasion-, erosion- and impact-resistant iron-base alloys containing varying amounts of carbon, chromium and molybdenum. Used for hardfacing and arc-welding electrodes.

Allisite. Trade name of Allis-Chalmers Manufacturing Company (USA) for a high-strength cast iron containing 3% carbon, and the balance iron. Used for machinery castings.

Allite. (1) Trade name of Allied Products Corporation (USA) for a zinc alloy used to make forming and stamping dies.

(2) Trade name for allyl plastics with a low density (approximately 1.37 g/cm³), high clarity, good colorability, a refractive index of 1.57, low moisture absorption, high hardness, high compressive strength, high dielectric strength, and good chemical resistance. Used as glazing for automobiles and aircraft, and for lenses, prisms, reflectors, consumer electronics, and electrical components.

Allmul. Trademark of Babcock & Wilcox Company (USA) for a firebrick composed chiefly of mullite, and available in various shapes. Used for lining glass furnaces.

alloclasite. A silver-white mineral of the marcasite group composed of cobalt iron arsenic sulfide, (Co,Fe,Ni)AsS, in which nickel can replace all of the iron. Crystal system, orthorhombic. Density, 5.91-6.16 g/cm³. Occurrence: Rumania, Russia

Alloderm. Trade name for an artificial skin grafting material consisting essentially of healthy human skin with all cellular components removed, and then preserved by freeze drying.

AlloGro. Trademark of AlloSource (USA) for a demineralized bone matrix (DBM) that consists of an osteoinductive, demineralized cortical bone powder which can stimulate growth of new bone cells.

allomelamins. A group of *melanins* (black pigments) produced by the polymerization of polyphenols. They contain graphitic cores, and yield graphitic acids (e.g., mellitic acid) on oxidative cleavage.

allophane. A colorless, blue, brown, yellow, or pale green clay mineral composed of hydrated aluminosilicate gel, $Al_2O_3 \cdot SiO_2 \cdot xH_2O$, which may also contain considerable amounts of phosphorus pentoxide (P_2O_5).

Alloprene. Trademark of ICI Limited (UK) for chlorinated rubber supplied in powder form, and used to make film coatings, printing inks, and textileproofing compounds.

allover lace. A wide lace with a pattern that is repeated over the whole width of the fabric.

alloy. (1) In metallurgy, a metallic substance composed of two or more elements. In most cases one of the elements is a metal to which metallic or nonmetallic elements are added during melting. The final product has enhanced properties, and may be a solid solution of one or more elements, or a mixture of several phases.

(2) In polymer engineering, a blend of two or more immiscible polymers or copolymers. The end product is a plastic resin with improved performance properties. Alloying permits blending of polymers that cannot be polymerized. Also known as *blend; plastic alloy; polymer alloy; polymer blend.*

alloy adhesives. Adhesives made by mixing thermosetting and thermoplastic, or thermosetting and elastomeric polymers. They combine the properties of both polymers resulting in improved adhesive substances.

Alloybond. Trade name of Southern Dental Industries Limited (Australia) for a fluoride-releasing dental amalgam bond based on a dimethyl methacrylate (DMA) resin with high dentin bond strength.

alloy cast irons. A group of cast irons usually containing a total of about 3-5% alloying elements, such as aluminum, chromium, copper, manganese, molybdenum, nickel, silicon, silicon and vanadium, either singly or in combination, added to improve their strength, corrosion, heat and/or wear resistance, facilitate heat treatment, or obtain certain microstructures (e.g., martensitic, austenitic or ferritic irons). Used for automotive parts, such as cylinders, pistons and crankcases, and for machine tools, crushing and grinding machinery, dies, etc.

alloy cast steels. See alloy steel castings.

alloy constructional steels. See low-alloy steels.

alloyed malleable irons. Special malleable irons chiefly alloyed with copper (0.25-1.25%), or copper and molybdenum. Other alloys may be added to effect either an increase (aluminum, titanium), or a retardation (vanadium, chromium) of graphitization. Compared to unalloyed grades, copper-alloyed grades have higher ultimate strength, yield points, fatigue limits and greatly enhanced corrosion resistance even to sulfurous gas atmospheres, and grades alloyed with both copper and molybdenum have added high tensile and yield strengths and good elongation and ductility. Used for housings, railroad castings, flasks and flanges, industrial machinery, and equipment exposed to atmospheric corrosion. See also special malleable irons.

alloyed powder. See alloy powder.

alloyed tungsten carbide. A cemented carbide based on cobalt-bonded tungsten carbide, but also containing considerable additions of other refractory carbides, such as tantalum carbide and titanium carbide. It has excellent shock resistance, excellent to good resistance to thermal cracking, deformation and cratering, good to moderate wear resistance, and better hot hardness than unalloyed grades. Used for indexable inserts for cutting tools. See also cemented carbides.

Alloy Finishing. Trade name of CCS Braeburn Alloy Steel (USA) for a nondeforming alloy tool steel (AISI type F2) containing 1.4% carbon, and 3% tungsten, and the balance iron. Used for cutting, finishing and shaping tools.

alloying agent. See alloying element.

alloying element. A chemical element added to a molten metal, and remaining in the final product, to effect changes in microstructure and properties. For example, elements, such as silicon, manganese, chromium, nickel, molybdenum, tungsten, vanadium and aluminum are added to molten steel in order to enhance its properties. Also known as *alloying agent.*

alloy mechanical coating. A thin coating consisting of a combination of metals, such as zinc, cadmium and tin, deposited onto workpieces by a kinetic energy method.

Alloymet. Trade name of Alloy Metal Products Inc. (USA) for an extensive series of master alloys including several copper-base alloys (copper-chromium, copper-iron, copper-manganese, copper-nickel, copper-nickel-iron and copper-silicon), iron-base alloys (iron-nickel, iron-nickel-molybdenum, iron-nickel-chromium and iron-nickel-copper), and nickel-base alloys (nickel-chromium-manganese, nickel-copper, nickel-copper-iron, nickel-copper-iron-chromium and nickel-iron-chromium). Used as alloying additions in the manufacture of ferrous and nonferrous alloys.

alloy nuclear fuel. A nuclear fuel composed of an alloy of one or more nonfissionable metals, and a fissionable material, such uranium-235 (^{235}U) or plutonium-239 (^{239}Pu).

alloy plate. A coating produced by electrodeposition, and composed of two or more elements of which at least one is a metal.

alloy powder. A metal powder composed of at least two components which are completely or partially alloyed with each other. Also known as *alloyed powder.*

alloy steel castings. Cast steels in which special alloying elements, such as chromium, manganese, molybdenum, nickel or vanadium, in combination with suitable heat-treatments result in the enhancement of certain desirable properties. They can be grouped into two categories: (i) *Low-alloy steel castings* with alloy contents of up to 8%; and (ii) *High-alloy steel castings* with alloy contents exceeding 8%. *Alloy steel castings* have high tensile and shear strengths, excellent impact and shock resistance, and high toughness, and are used for marine equipment, engine casings, mining machinery, pump parts, valve bodies, railroad car frames, forging presses, turbine wheels, and machine elements, such as gears. Also known as *alloy cast steels.*

alloy steels. Steels that contains appreciable concentrations of alloying elements, such as chromium, nickel, molybdenum, manganese, silicon and/or tungsten, added to effect changes in hardenability, toughness, wear resistance, mechanical strength, corrosion resistance, etc. See also low-alloy steels; high-alloy steels.

All-Po. Trade name of Grand Northern Products Limited (USA) for a coated hardfacing electrode containing 0.9% carbon, 0.4% manganese, 9.5% chromium, 0.6% molybdenum, and the balance iron. It has good abrasion resistance, and is used for overlay or buildup on scrapers, loaders, dozer buckets, shovels, and blades.

Allround 4. Trade name of J.F. Jelenko & Company (USA) for a deep yellow-colored gold-platinum porcelain-to-metal dental alloy for use with low-fusing dental porcelains.

Allround 55LF. Trade name of J.F. Jelenko & Company (USA) for a rich yellow-colored microfine high-gold ceramic dental alloy for use with low-fusing dental porcelains.

All-State. Trade name of Chemetron Corporation (USA) for a series of welding products including several steel and cast-iron arc-welding and arc-cutting electrodes, copper-base brazing and braze welding rods, silver soldering and brazing alloys, magnesium alloy solders, and zinc welding rods.

Allstep. Trademark of Ludlow Composites Corporation (USA) for closed-cell polyvinyl chloride foam.

Allstop. Trade name of Flachglas AG (Germany) for bulletproof glass.

Allstop Privat. Trade name of Flachglas AG (Germany) for an insulating safety glass.

All Strike. Trade name of MacDermid Inc. (USA) for an alkaline electroless nickel strike.

alluaudite. A yellow to brownish yellow mineral of the varulite group composed of sodium iron manganese phosphate, NaFe$(Mn,Fe)_2(PO_4)_3$. Crystal system, monoclinic. Density, 3.58 g/cm^3; melting point, 1000°C (1830°F); refractive index, 1.802; hardness, 5-5.5 Mohs. Occurrence: Europe (France).

Alluloy. Trade name of Anti-Attrition Company, Limited (UK) for a lead-base white metal with excellent anti-attrition properties. Used for heavy-duty and eccentric bearings, and capping metal.

Allvac. Trademark of Teledyne Allvac (USA) for a series of vacuum-cast or double-vacuum-cast nickel-chromium aging alloys, nickel-base superalloys, titanium alloys and austenitic stainless steels.

All Weather Thermo. Trade name of General Glass Corporation (USA) for an insulating glass.

allylics. Thermosetting polyester resins made by polymerization of chemical compounds containing the allyl group (CH_2=CH-CH_2-), e.g., esters of allyl alcohol, and an organic acid or anhydride. They possess good resistance to moisture and chemicals, good abrasion resistance and thermal properties, good electrical resistivity, and low shrinkage. Used for laminating adhesives and coatings (prepregs), varnishes, potting and encapsulation of electronic components, etc. Also known as *allyl resins; allyls.*

allyl plastics. Thermosetting plastics based on allyl resins, e.g., diallyl phthalate (DAP), diallyl isophthalate (DAIP), diallyl maleate (DAM), etc.

allyl resins. See allylics.

allyls. See allylics.

Allymer. Trademark of PPG Industries, Inc. (USA) for a series of polymerizable allyl resins used in the manufacture of cast and molded products, and for laminating and impregnating applications.

Almac. Trade name of MacDermid Inc. (USA) for zinc-aluminum coatings applied by mechanical deposition.

Almadur. German trade name for a series of aluminum-base bearing alloys.

Almag. (1) Trade name of Acme Aluminum Alloys, Inc. (USA) for a series of strong, corrosion-resistant nonheat-treatable cast aluminum-magnesium alloys used for architectural and marine castings, machinery parts, and ornamental purposes.

(2) Trademark of Refratechnik GmbH (Germany) for refractory masses and stones used for fireproof linings.

Almalek. German trade name for an aluminum-magnesium alloy used for electrical applications.

Almanite. Trade name of Meehanite Metal Corporation (USA) for a series of wear- and abrasion-resistant cast irons containing varying amounts of carbon, manganese, aluminum and iron. Used for mill liners, crushing rolls, grinding balls, and mixer blades.

almandine. A fine deep red to brownish red mineral of the garnet group composed of iron aluminum silicate, $Fe_3Al_2(SiO_4)_3$. Crystal system, cubic. Density, 4.25-4.31 g/cm^3; hardness, 7-7.5 Mohs; refractive index, 1.815; resembles ruby. Occurrence: Canada (Ontario), USA (New York, South Dakota). Used as an abrasive on paper and cloth. Fine varieties are also used as gemstones. Also known as *almandite.*

almandite. See almandine.

Almar. (1) Trade name of AL Tech Specialty Steel Corporation (USA) for a standard precipitation-hardening stainless steel containing 0.05% carbon, 14-14.5% chromium, 6.2-7% nickel, 0.5-0.9 titanium, 0.5% manganese, 0.3% silicon, and about 0.03-0.04% phosphorus and sulfur respectively. It has exceptional high-temperature resistance, and is used for corrosion-resistant parts, such as cutlery, razor blades, valves, etc.

(2) Trade name of AL Tech Specialty Steel Corporation (USA) for a series of high-nickel, solution-treated and aged maraging steels containing 0.03% carbon, up to about 25% nickel, varying amounts of chromium, cobalt and molybdenum, small amounts of titanium, aluminum and niobium (columbium), and the balance iron. Used for machine parts, such as shafts, compressor and turbine disks, etc.

Almasilium. Trade name of Société du Duralumin (France) for a corrosion-resistant, heat-treatable aluminum alloy containing 1% manganese and 2% silicon. Used for light-alloy parts.

Almatex. Trademark of Mitsui Toatsu Chemicals, Inc. (Japan) for a series of acrylic resins used in the manufacture of paints.

Almax. Trade name of Mitsui Company Limited (Japan) for a continuous fiber composed of 99.5wt% alpha aluminum oxide (α-Al_2O_3). It has an average diameter of 10 μm (394 μin.), a density of 3.6 g/cm^3 (0.13 lb/in.^{3}), a tensile strength of 1765 MPa (256 ksi), a tensile modulus of 325 GPa (47 msi), and excellent high-temperature stability and corrosion resistance. Used especially for the reinforcement of composites, and for thermal insulating applications.

Almec. Trade name of Thermoset Plastics Inc. (USA) for liquid, single-component epoxy hybrid resins for electrical and electronic encapsulation applications.

Almeco. Trade name of Pechiney/Société Vente de l'Aluminium (France) for a plated aluminum-alloy wire used as insulated building wire.

Almeg. Trademark of US Granules Corporation (USA) for aluminum granules produced from recovered aluminum foil products.

Almelec. Trade name of Pechiney/Société Vente de l'Aluminium (France) for an aluminum alloy containing 0.5-0.7% magnesium, and 0.5-0.6% silicon. Used for overhead line conductors.

Almen. Trade name for zinc-aluminum-copper castings used for bearings.

almendro. The yellowish brown, heavy wood of a species of tonka bean trees *(Coumarouna panamensis)* with a high resistance to marine borers. Source: Panama, Costa Rica. Used for marine construction, and as a substitute for greenheart. See also greenheart.

almerite. See natroalunite.

Almet. (1) Trade name of H.K. Porter Company, Inc. (USA) for an extensive series of corrosion- and/or heat-resistant austenitic, ferritic and martensitic stainless steels including several precipitation-hardening grades.

(2) Trademark of Ball Chemical Company (USA) for an aluminum paint.

Alminal. Trade name of Renfrew Foundries Limited (UK) for a series of wrought and cast aluminum alloys containing varying percentages of other alloying elements, such as silicon, magnesium, manganese, copper, etc. Used for structural parts, architectural parts, and aircraft engine parts.

Almo. Trade name of Colt Industries (UK) for a series of high-strength chromium-molybdenum alloy steels used for automotive components, such as axles, crankshafts, connecting rods, and gears.

Almod. German trade name for a plaster supplied in several grades (e.g., *Almod 60* and *Almod 70*) with different hardnesses and plaster-to-water ratios. Used in the ceramic industries for molds employed in the manufacture of tableware articles, such as

beaks, bowls, figurines, jugs and handles.

Almold-20. Trade name of AL Tech Specialty Steel Corporation (USA) for a mold steel (AISI type P20) containing 0.35% carbon, 0.7% manganese, 0.5% silicon, 1.25-1.70% chromium, 0.40% molybdenum, and the balance iron. It has high core hardness, excellent resistance to decarburization, a great depth of hardening, and good toughness and machinability. Used for zinc die-casting dies, and plastic molds for injection, compression and transfer molding.

almon. The hard, strong, coarse-textured wood of the lauan tree *Shorea eximia*. The heartwood is reddish-brown, and the sapwood light red. Source: Philippines, Malaya. Used for general construction purposes, and furniture.

Almos. Trade name of Compagnie des Mines (France) for a case-hardening steel containing 0.08-0.16% carbon, 0.8-1.2% chromium, 0.15-0.3% molybdenum, 0.6-0.9% manganese, and the balance iron. Used for machine parts, such as shafts, gears, etc.

Almute. Trade name for a lightweight aluminum-alloy powder product developed by NDC Company Limited, Nissan Automotive Group (Japan), and made by sintering prealloyed powders into panels of up to 0.6 × 1.2 m (2 × 4 ft.) in size, and 2.5 mm (0.1 in.) in thickness. It has a pore volume of about 40%, a pore diameter of approximately 0.1 mm (0.04 in.), outstanding sound-absorption characteristics, excellent weatherability, good thermal and corrosion resistance, and is supplied in micro- and mactrotextured surface finishes. Used for architectural, automotive and other acoustic insulation applications.

AlNel. Trademark of Advanced Refractory Technologies, Inc. (USA) for several grades of aluminum nitride powder for use in the electronic, structural and refractory industries. *AlNel Grade A-100* has high purity, high sinterability, high surface area; high thermal conductivity, a low oxygen content, and is used for electronic applications, e.g., package and substrate fabrication. *AlNel Grade A-200* is a high-purity, low-cost powder used for structural and refractory applications, e.g., wear resistant parts, crucibles for molten metals, heat radiation plates, nitriding furnace parts, as substitute for boron nitride powders, in SiAlON production, etc.

Alneon. Trade name of Strasser Co. (Switzerland) for an age-hardenable alloy containing 80-90% aluminum, 7-22% zinc, 2-3% copper, and traces of copper and nickel. It has a high endurance limit, and is used for light-alloy parts.

Alnesium. American trade name for a corrosion-resistant alloy containing 97% aluminum and 3% magnesium. Used for cases and containers.

Alni. Trade name of Harsco Corporation (USA) for a permanent magnet alloy containing 11-13% aluminum, 24-28% nickel, 0-6% copper, 0-0.5% cobalt, 0-0.1% carbon, 0-0.5% titanium, and the balance iron. It has high permeability, and is used for magnetic and electrical equipment.

Alnic. Trade name for a permanent magnet alloy containing 63% iron, 25% nickel, and 12% aluminum. It has high coercive force, and is used for electrical and magnetic equipment.

alnico. A magnetic alloy containing 34-63% iron, 10-28% aluminum, 3-35% nickel, 3-12% cobalt, and small amounts of copper and/or titanium. It is available in cast and sintered form. The cast alloys tend to have better magnetic properties, while the sintered alloys have fine-grained microstructures, and higher strength. The magnetic orientation of the grain structure is developed by precipitation hardening. *Alnico* has strong magnetic properties in all directions, high magnetic remanence, high coercive force, and a high magnetic energy product. Used in the manufacture of permanent magnets. Also known as *alnico alloy; aluminum-nickel-cobalt alloy.* See also alnico magnet.

alnico magnet. A permanent magnet made of *alnico* (aluminum-nickel-cobalt-iron alloy). A heat-treating regime, consisting of heating, cooling and reheating, provides the magnet with a high degree of stability and strength, and increased resistance to demagnetization. *Alnico magnets* are available in many shapes and sizes including bars, disks, cylinders, horseshoes, rings, etc.

Alnicus. Trade name of US Magnet & Alloy Corporation (USA) for a cast alnico-type alloy containing about 51% iron, 24% cobalt, 14% nickel, 8% aluminum, and 3% copper. The magnetic orientation of the grain structure is developed by directional solidification. *Alnicus* has a high maximum energy product, and high magnetic retention. Used for magnetic chucks, meters, electrocardiographs, starting devices, and oriented permanent magnets.

Alnide. Trademark of Boride Products, Inc. (USA) for aluminum nitride supplied in several grades. It has a density of about 3.25-3.29 g/cm^3, high thermal conductivity, excellent electrical insulating properties, and a maximum service temperature of about 400-1900°C (750-3450°F). Used for structural components, heat sinks, crucibles, insulators and fixtures.

Alnifer. Trade name of Texas Instruments Inc. (USA) for a thin-gage product composed of a low-carbon steel coated or clad with aluminum on one side, and nickel on the other. Used for electronic components.

Alniloy. Trade name of NL Industries (USA) for an aluminum casting alloy containing 4% copper, 4% nickel, 2% silicon, and up to 2% iron. Used for ornaments.

Alnovol. Trademark of Hoechst AG (Germany) for synthetic resins based on phenolic novolacs.

Alomite. Trade name for a blue *sodalite* ($Na_8Al_6Si_6O_{24}Cl_2$) found at Bankcroft, Ontario, Canada, and used as an ornamental stone.

Alon. (1) Trademark of Surmet Corporation (USA) for a transparent, lightweight polycrystalline aluminum oxynitride made by powder technology. It has high strength, toughness and durability, and is used for scanner windows, missile domes, underwater sensors, watch crystals, scratchproof lenses, etc. See also aluminum oxynitride ceramics.

(2) Trade name for rayon and cellulose acetate fibers used for the manufacture of lingerie, pyjamas, shirts, ties and swimwear, and in staple form in blends for suitings, sportswear, knitting yarns, household textiles, carpets, cable insulation, etc.

AlON ceramics. See aluminum oxynitride ceramics.

Alonized steel. A wrought or cast steel, including several plain-carbon and low-alloy, ferritic and austenitic stainless and highly alloyed nickel-chromium grades, into whose surface aluminum has been vapor-diffused by a proprietary processing technique known as "Alon Processing" (developed by Alon Processing, Inc., USA) to provide it with excellent heat and corrosion resistance.

Alowalt. Trade name for fused alumina used as a refractory.

Aloxite. Trademark of Harbison-Carborundum Corporation (USA) for aluminum oxide products made by fusing high-alumina materials, such as bauxite. Used for abrasive powders and grains, coated abrasives, grinding wheels, hones, filters, and refractory products, such as cements.

Aloyco. Trade name of Walworth Company (USA) for a series of corrosion-resistant cast nickel-chromium and chromium-nickel stainless steels, and numerous corrosion-resistant cast nickels

and cast nickel-copper, nickel-chromium-iron, nickel-molybdenum-iron and nickel-molybdenum-chromium alloys. Used for chemical equipment, valves, pumps, fittings, etc.

Aloy-Num. Trade name of E.N. Egge Company (USA) for an age-hardenable aluminum alloy containing 4% silicon and 2% copper. Used for pistons and cylinders.

alpaca. (1) A nickel silver containing about 55-66% copper, 19-31% zinc, 16-26% nickel, and 0-2% silver. It has a silvery-white color, and excellent corrosion resistance. Used as a base metal for silver-plated tableware. Also spelled alpacca, alpakka or alpaka.

(2) The long, soft, silky wool of the alpaca, a domesticated grazing animal *(Lama pacos)* of the mountainous regions of Peru and Bolivia. Used for yarn. See also alpaca (3).

(3) The soft, warm cloth made from alpaca yarn. See also alpaca (2).

(4) A glossy, wiry cotton or wool cloth, usually black in color.

alpaca crepe. A soft fabric with a dull surface, usually made from acetate or polyester fibers.

Alpase. Trademark of Pase Aluminum, Inc. (USA) for a series of wrought aluminum-copper plates supplied in various thicknesses and lengths. They have good weldability, anodizability and machinability, good dimensional stability, consistent hardness (95 Brinell), and good nickel platability. Used for molds for blow, structural-foam, reaction-injection, rotational-transfer and rubber molding.

Alpaste. Trade name of Alcan Aluminum Company (Canada) for paint pigments and aluminum paint.

Alpax. British trade name for a series aluminum-silicon and aluminum-silicon-magnesium casting alloys containing up to 13% silicon, and small additions of magnesium, manganese, titanium, and/or iron. They have a density of 2.66, good castability, good corrosion resistance, and relatively good mechanical properties. Used for lightweight castings, water pumps, hydraulic components, engine parts, gear boxes, etc.

Alpearl. Trademark of Asahi Chemical Industry Company (Japan) for acrylic fibers.

Alperm. Japanese trade name for a soft-magnetic alloy containing 84-86% iron and 14-16% aluminum. It has high magnetic permeability, and is used for magnetic laminations.

Alpex. (1) Trademark of Reichhold Chemical Company (USA) for cyclicized rubber used for moistureproof barrier coatings on food wraps, and as a vehicle for printing inks.

(2) Trademark of Centrala Handlu Zagranicznego Company (Poland) for chipboard and flaxboard.

3) Trademark of Hoechst AG (Germany) for granular and liquid resins used as lacquer and paint binders.

Alpha. (1) Trade name of Atlas Specialty Steels (Canada) for water-hardening tool steels (AISI type W1). *Alpha-8* contains 0.8% carbon, 0.25% manganese, 0.2% silicon, and the balance iron. *Alpha* tool steels have good toughness and shock resistance, low wear resistance, low red hardness, and moderate machinability. Used for auger bits, well drills, stone channellers, cold chisels, cold sets, vise jaws, blacksmithing tools, mauls, sledges, fullers, machinery parts, etc.

(2) Trade name of Dexter Corporation (USA) for a series of polyvinyl chlorides (PVCs) including flexible PVC foams for bottle cap liners, flexible unreinforced and nylon-reinforced PVC compounds for gaskets and cable clips, and elastomeric and flexible extrusion-and injection-molding-grade PVCs for automotive components and weatherstripping.

(3) Trade name of Atotech USA Inc. (USA) for electroless nickel plates and plating processes.

(4) Trade name of Samsung/OmegaDent (South Korea) for dental casting alloys for crown and bridge restorations.

(5) Trade name of Amoco Fabrics & Fibers Company (USA) for polypropylene fibers and yarns.

alpha alumina. White, ultrafine alumina powder or pellets with hexagonal crystal structure. It has a density of 4.0, a hardness of 9 Mohs, and occurs in nature as the mineral *corundum*. Used as an abrasive for lapping and grinding, in dentifrices, in the treatment of clay and bauxite, for refractories, as a catalyst base and bed support, and in the manufacture of fibers and bioceramics. Abbreviation: α-Al_2O_3. Also known as *alpha aluminum oxide*. See also alumina; corundum; high-density alumina.

alpha alumina trihydrate. See alumina hydrate.

alpha aluminum oxide. See alpha alumina.

alpha americium. See americium.

alpha amino acid. Any of the amino acids that have an amino ($-NH_2$) group attached to the alpha-carbon atom, and are building blocks of proteins. They have important functions in the metabolism, growth and repair of plant and animal tissue. Also known as *α-amino acid*. See also amino acid.

alpha arsenic. See arsenic.

Alpha-Base. Trade name for a glass-ionomer dental cement used as a base lining material.

Alpha BCF. Trade name of American Fibers & Yarns Company (USA) for polyolefin fibers and yarns used for textile fabrics.

alpha-beta alloy. An alloy whose microstructure contains two principal phases–alpha (α) and beta (β)–at specified temperatures.

alpha-beta brass. An alloy composed of 57-63% copper and 37-43% zinc, sometimes with a small addition of lead. It is a two-phase brass consisting of a heterogeneous mixture of face-centered cubic alpha (α) solid solution, and body-centered cubic beta (β) solid solution. *Alpha-beta brass* has good hot and cold workability, and good machinability. Used for wire, fittings, hardware, fasteners, structural parts, and profiles. Abbreviation: α-β brass.

alpha-beta titanium alloys. A group of heat-treatable titanium alloys containing a mixture of alpha (hexagonal close-packed crystal structure) and beta (body-centered cubic crystal structure) phases at room temperature. Typical alloying elements include aluminum, vanadium, tin, zirconium, molybdenum, chromium and manganese. The most common of these alloys are: Ti-6Al-4V, Ti-6Al-6V-2Sn, Ti-8Mn, and Ti-7Al-4Mo. They have an upper service temperature of 427°C (800°F), good formability, medium toughness, and moderate weldability. *Alpha-beta titanium alloys* with less than 2-3% beta (β) titanium, are usually referred to as *near-alpha titanium alloys*. Abbreviation: α-β titanium alloys. See also alpha-titanium alloys; beta-titanium alloys; near-alpha titanium alloys.

Alpha-Bond. (1) Trade name for a glass-ionomer dental cement.

(2) Trade name of Sumsung/OmegaDent (South Korea) for dental porcelain-bonding alloys.

alpha berkelium. See berkelium.

alpha beryllium. See beryllium.

alpha boron. See boron.

alpha brass. An alloy consisting of a solid solution of up to 36% zinc in copper, sometimes with a small addition of lead. It is very ductile, easily cold-worked, and has face-centered cubic crystal structure. Other properties include good mechanical properties, excellent corrosion resistance, high brittleness at 500°C (932°F), and fair to poor machinability. Used for deep-drawn parts, pipes, sheets, springs, fasteners, fixtures, and hardware. Abbreviation: α brass.

Alpha Bronze. Trade name of Makin Metal Powders Limited (USA) for a water-atomized prealloyed bronze powder containing 91% copper and 9% tin. Sintered parts produced from this powder possess good surface finishes, have finely distributed pores, and improved hydrodynamic lubrication properties. Used for bearings.

alpha cadmium iodide. See cadmium iodide (i).

alpha calcium. See calcium.

Alphacan. Trade name of Atofina Alphacan (France) for polyethylene, polypropylene and polyvinyl chloride tubing.

Alpha Cel. Trade name of International Filler Corporation (USA) for alpha cellulose used as a filler in rubber and plastics.

alpha cellulose. A very pure cellulose which is that fraction of the total carbohydrate component of a wood substance which is empirically defined as insoluble in 17.5% aqueous sodium hydroxide (NaOH) at 20°C (68°F). It is the major component of wood and paper pulp. Abbreviation: α cellulose. Also known as *chemical cellulose*.

Alpha-Cem. Trade name for a glass-ionomer dental luting cement.

alpha cerium. See cerium.

alpha cobalt. See cobalt.

Alpha-Core. Trade name for a creamy composite paste used in restorative dentistry for core build-ups.

alpha curium. See curium.

Alphadie MF. Trade name of Schuetz Dental (Germany) for a urethane-containing dental die stone.

alpha dysprosium. See dysprosium.

alpha fiber. A bleached fiber based on esters or ethers of cellulose and containing at least 94% alpha-cellulose.

Alpha-Fil. Trade name for a glass-ionomer luting cement.

alpha ferrite. See ferrite (1)

alpha flock. Finely divided alpha cellulose used as a filler in rubber. See also alpha cellulose.

alpha gadolinium. See gadolinium.

alpha gypsum. Gypsum of low consistency with very high compressive strength, usually in excess of 34.5 MPa (5 ksi).

alpha hafnium. See hafnium.

alpha iron. See iron.

Alphalac. Trade name of LG Chemicals (USA) for a series polystyrenes available in several grades.

alpha lanthanum. See lanthanum.

alpha manganese. See manganese.

alpha molybdenum monoboride. See molybdenum boride (1).

alpha neodymium. See neodymium.

alpha neptunium. See neptunium.

alpha-phase alloy. An alloy in which only one phase, namely the alpha phase, is present, e.g., alpha brass.

Alphaplast. Trademark of DMG Hamburg (Germany) for a self-cured macro-filled quartz composite with high mechanical strength used for dental restorations.

alpha plutonium. See plutonium.

alpha polonium. See polonium.

alpha praseodymium. See praseodymium.

alpha promethium. See promethium.

alpha quartz. See low quartz.

alpha samarium. See samarium.

alpha scandium. See scandium.

Alpha-Seal. Trade name for a light-cure dental resin used as a fissure sealant.

Alphaseal. (1) Trademark of Alphagary Corporation (USA) for elastomeric flexible compounds based on polyvinyl chloride, styrene-butadiene or styrene-butadiene-styrene, and used for extruded and molded goods.

(2) Trade name for a light-cure dental resin used for cavity lining.

alpha selenium. See selenium.

alpha-SiAlON. An engineering ceramic obtained by combining alpha silicon nitride (α-Si_3N_4), aluminum nitride (AlN) and small additions of yttria, ytterbia and neodymia by hot pressing at a prescribed temperature and pressure in a nitrogen atmosphere. It has high strength, hardness and toughness, and is used for bearings, cutting tools and structural applications. Abbreviation: α-SiAlON. See also Sialon; silicon aluminum oxynitride ceramics.

Alpha-Silver. Trade name for a silver-filled glass-ionomer dental cement.

alpha sodium. See sodium.

alpha stabilizer. In alpha-titanium alloys, an alloying element, such as aluminum, tin or zirconium, that dissolves chiefly in the alpha phase, and increases the alpha-beta transformation temperature. Abbreviation: α stabilizer.

Alpha Star. Trademark of C-E Minerals (USA) for a dense, homogeneous calcined aggregate composed of 89.5-90.3% alumina (Al_2O_3), 3.9-4.0% titania (TiO_2), 3.9% silica (SiO_2), 1.24-1.75% ferric oxide (Fe_2O_3) and up to 0.4% alkalies (calcium, sodium, magnesium, and potassium oxides). Supplied in several grain sizes, it has a bulk density of 3.4-3.56 g/cm^3 (0.12-0.13 $lb/in.^3$), low apparent porosity (2.7%) and is used in the manufacture of blast furnace ladles and troughs, and for low-cement castables.

alpha strontium. See strontium.

alpha sulfur. See sulfur.

Alphatech. Trade name for PVC-free elastomeric flexible compounds.

(2) Trade name of Ferro Corporation (USA) for organic powder coatings.

alpha terbium. See terbium.

alpha thallium. See thallium.

alpha thorium. See thorium.

alpha tin. A gray powder modification of *tin* obtained, due to a crystallographic transformation (tin pest), from solid, white beta tin by cooling to a temperature between +13 and -40°C (+56 and -40°F). It has semiconductive properties. Abbreviation: α-Sn. Also known as *gray tin*. See also tin.

Alpha Titanium. Trade name for a titanium-aluminum dental alloy.

alpha titanium. The allotropic modification of *titanium* which is stable below 885°C (1625°F), and has a close-packed hexagonal crystal structure. Abbreviation: α-Ti.

alpha-titanium alloys. A group of non-heat-treatable titanium alloys containing varying amounts aluminum, tin, zirconium, molybdenum, niobium (columbium), vanadium, tantalum and/or nickel. The most common of these alloys are: Ti-0.3Mo-0.8Ni, Ti-5Al-2.5Sn, and Ti-2.25Al-11Sn-5Zr-1Mo. They have good creep resistance at high temperatures, good oxidation resistance, good stability from -250 to +540°C (-420 to +1000°F), high strength and toughness, good weldability, and moderate forgeability. Used for high-temperature and cryogenic applications, and aerospace applications. Abbreviation: α-Ti alloys. See also alpha-beta titanium alloys; beta titanium alloys; near-alpha titanium alloys.

alphatized steel. Steel on whose surface chromium has been deposited by a diffusion process.

alpha uranium. One of the three allotropic forms of *uranium* (the other two being beta and gamma uranium) which is stable below 688°C (1270°F). It has an orthorhombic crystal structure, and its ductility increases with temperature. Abbreviation: α-U.

Alpha-Weld. Trade name of Alpha Associates, Inc. (USA) for coated fiberglass fabrics used for welding curtains and blankets.

alpha ytterbium. See ytterbium.

alpha yttrium. See yttrium.

alpha zirconium. The allotropic modification of *zirconium* that is stable below 870°C (1598°F). It has a close-packed hexagonal crystal structure. Abbreviation: α-Zr.

Alphos. Trade name of Alchem Corporation (USA) for a phosphate coating and process.

Alpine. Trade name of Vereinigte Edelstahlwerke (Austria) for an extensive series of steels including various alloy machinery steels (chromium, molybdenum, chromium-manganese, chromium-vanadium, nickel-chromium types), some of which are case-hardening grades, as well as several chromium-aluminum and chromium-aluminum nitriding steels, and numerous plain-carbon, cold-work and hot-work tool steels.

Alplate. Trade name of Reynolds Metals Company (USA) for aluminum-clad steel wire.

Alplatin. Trade name of W. Seibel AG (Germany) for a strong, heat-treatable wrought aluminum alloy containing 4-5% zinc and 2-4% magnesium. Used for aircraft parts.

Alpolit. Trademark of Albert/Hoechst AG (Germany) for an unsaturated polyester-based gel coat.

Al Polymer. Trademark of Amoco Chemical Corporation (USA) for an aromatic polymer based on trimellitic anhydride ($C_9H_4O_5$) which produces a polyamide-imide when cured at medium temperatures. Used for wire enamel insulation, glass and asbestos laminates, for cloth, film and sleeving impregnation, and in dipping varnishes.

Alpro. Trade name of Belmont Metals Inc. (USA) for an extensive series of aluminum-, copper- and nickel-base master alloys including several aluminum-copper, aluminum-copper-nickel, aluminum-iron, aluminum-iron-manganese, aluminum-manganese, aluminum-nickel, aluminum-nickel-tin, aluminum-nickel-manganese and aluminum-silicon alloys, a wide range of copper-aluminum, copper-aluminum-iron, copper-aluminum-nickel, copper-manganese, copper-nickel and copper-phosphorus alloys, and various nickel-aluminum-copper, nickel-chromium, nickel-manganese, nickel-iron, and iron-nickel-chromium alloys. Used as alloying additions, deoxidizers and hardeners in the manufacture of ferrous and nonferrous alloys.

Alpropat. Trade name of Belmont Metals Inc. (USA) for a low-shrinkage aluminum alloy containing small quantities of copper and silicon. Used for match plates and patterns.

Alpulp. Trademark of Anchor Packing Division of Robco Inc. (Canada) for *Teflon*- and asbestos-braided packing materials.

Alrac Nylon. Trade name for nylon-4, i.e., a 2-pyrrolidone-based polyamide, used for synthetic textile fibers.

Alraman. German trade name for a non-heat-treatable, corrosion-resistant aluminum alloy containing 0.9-1.4% manganese, and 0-0.3% magnesium. Used for trim and roofing.

Alsex. Trade name of Allgemeine Elektrizitäts-Gesellschaft (Germany) for aluminum alloy containing 0.5% silicon, and 0.5%

magnesium. It has an electrical conductivity (about 50% IACS), and is used for light-alloy parts and forgings.

Alsia. Trade name of Société Alsia (France) for an aluminum casting alloy containing 20% silicon, 1% copper, and 0.7% iron. Used for pistons.

Alsi ABF. Trade name of AGS Minéraux SA (France) for a hard clay obtained from a kaolinitic sedimentary ball clay mined in southwestern France. It is supplied as a powder with a particle size ranging from submicron to 100 µm (less than 40 to over 4000 µin.). A typical chemical analysis (in wt%) is: 49% silica (SiO_2), 34% alumina (Al_2O_3), 1.7% ferric oxide (Fe_2O_3), and 1.3% titania (TiO_2). The loss on ignition is about 12%, the density 2.6 g/cm^3, the pH-value 5.0, and the specific surface area 22 m^2/g. Used in the manufacture of ceramic products.

Alsibronz. Trademark of Franklin Mineral Products Company (USA) for wet-ground *muscovite* mica used as a filler in engineering polymers and paper, and as a paint pigment.

AlSiC. Trademark of Ceramics Process Systems (USA) for a lightweight metal-matrix composite consisting of an aluminum matrix reinforced with silicon carbide (SiC) particles. It has high strength and thermal conductivity, and a low coefficient of thermal expansion, and is manufactured into near-net shape components by the company's patented "Pressure Infiltration Casting" (PIC) process. Used for lightweight applications, e.g., in aerospace, thermal management, for microwave packages, ceramic and plastic flip-chip ball-grid array (BGA) lids, and as thermal substrate.

Alsichrom. Trade name of J.C. Sodbach & Halberg (Germany) for a series of heat- and oxidation-resistant steels containing 0.05-0.1% carbon, 0.6-0.8% silicon, 0.3-0.4% manganese, 21-22% chromium, 4.8-5.2% aluminum, 0.5% cobalt, and the balance iron. Used for electric-furnace heating elements.

Alsifer. Trade name for a master alloy composed of 20% aluminum, 40% silicon and 40% iron. Used as a deoxidizer in steelmaking.

Alsifilm. Trade name for a heat- and oil-resistant bentonite gel supplied in the form of thin, transparent, hard sheets, and used essentially for electrical insulation applications, and as a substitute for mica.

ALSIL-Q. Trade name of BPI Inc. (USA) for a free-flowing aluminum silicate powder with uniform particle size. A typical chemical composition is 63-68% silica (SiO_2), 30-35% alumina (Al_2O_3), 1.1% rare-earth oxides, 0.5% iron, 0.5% sodium, and 0.15% carbon.

Alsilit. Trade name of Metallwerk Olsberg GmbH (Germany) for an aluminum alloy containing 5-7% silicon, 2-3% copper, 0.3-0.8% manganese, and 0.1-0.4% magnesium. Used for machinery and vehicles.

Alsilox. Trademark of Eagle-Picher Industries, Inc. (USA) for a fusion product of 65% lead oxide (PbO), 34% silicon dioxide (SiO_2), and 1% aluminum oxide (Al_2O_3). Used in ceramic glazing.

Alsiloy. Trade name of NL Industries (USA) for a series of cast aluminum-silicon and aluminum-silicon-copper alloys.

AlSiMag. Trademark of CeramTec North America (USA) for custom-engineered ceramic components based on alumina, silica and magnesia.

Alsimag. (1) Trade name for a series of ceramic materials produced from ground *talc* and sodium silicate, or phosphate-bonded *steatite*, and commonly known as "lavas." Used for electrical insulating parts, gas-burner tips, nozzles, and thyratron tubes.

(2) Trade name for fine-grained silicon carbide used for brazing fixtures, kiln furniture, and refractories for sintering furnaces.

(3) Trademark of Alsimag Technical Ceramics, Inc. (USA) for extruded, molded, pressed or machined fired ceramic shapes supplied in the form of sheets, rods, block and tubes. They possess excellent corrosion and heat resistance, and are used as electrical insulation and for various other industrial applications.

Alsimica. Trade name for wet-ground muscovite mica used as a filler in engineering polymers.

Alsimin. Trademark for a ferrosilicon aluminum containing 50% aluminum with the balance being silicon and iron. Used as a steel deoxidizer, and as an addition to silicon to aluminum casting alloys.

Alsint. Trade name of W. Haldenwanger Technische Keramik GmbH & Co. KG (Germany) for sintered aluminum and aluminum alloys.

Alsiplate. Trade name of Texas Instruments Inc. (USA) for an aluminum sheet product clad on both sides with silver, and used for electrical contacts.

Alsoco. Trade name of Aluminum Solder Corporation (USA) for tin-lead and tin-lead-zinc solders used for soldering aluminum and its alloys.

Alsteel. Trade name of H. Boker & Company (USA) for a carbon steel containing 0.2-0.7% carbon, and the balance iron. Used for tools.

alstonite. A colorless mineral of the aragonite group composed of barium calcium carbonate, $BaCa(CO_3)_2$. Crystal system, triclinic. Density, 3.63 g/cm^3; refractive index, 1.671. Occurrence: UK. Also known as *bromlite*.

Altac 10. Trade name of Ato Findley (France) for adhesives used for bonding carpets, and other coverings.

Altair. (1) Trade name for opaque, continuously cast acrylic or acrylic/ABS composite sheeting.

(2) Trademark of Altair Technologies, Inc. (USA) for titanium dioxide (TiO_2) powders and slurries including nanoparticle *anatase*- and *rutile*-based crystal products. See also antase; rutile.

(3) Trade name of Monarch Tile Inc. (USA) for ceramic floor tile.

altaite. A tin-white mineral of the halite group composed of lead telluride, PbTe. Crystal system, cubic. Density, 8.19 g/cm^3; hardness, 47-51 Vickers; melting point, 917°C (1683°F). Occurrence: Asia (Altai Mountains), Rumania, Canada (Ontario).

Altam. Trade name of NL Industries (USA) for an alloy containing 68-88% titanium, 11% aluminum, 1.25% iron, 1.2% silicon, and the balance titanium dioxide (TiO_2). Used in the production of ductile titanium.

Altdeutsch. Trade name of former Vereinigte Glaswerke (Germany) for a series of antique glass products with patterns in lozenge-shaped or rectangular panels.

Altek. Trademark of AOC Limited (USA) for unsaturated polyester resins used in the manufacture of watercraft and for other marine applications.

Altemp. Trade name of Allegheny Ludlum Steel (USA) for a series of cobalt- and nickel-type superalloys.

Alten. Trade name of Alten Foundry & Machine Works (USA) for a series of heat- and wear-resistant cast irons.

alternating copolymer. A copolymer in which two different monomer units, A and B, alternate positions along the molecular chain, –A–B–A–B–A–B–A–. It is usually referred to as poly(A-

alt-B), e.g., a copolymer of ethylene and chlorotrifluoroethylene may be referred to as poly(ethylene-*alt*-chlorotrifluoroethylene). Also known as *alternate copolymer; alternating polymer.* See also copolymer.

alternating polymer. See alternating copolymer.

Altex. Trade name of Altex Gronauer Filz GmbH & Co. KG (Germany) for nonwovens and geotextiles.

Alt Blitz. Trade name of Fakirstahl Hoffmanns GmbH & Co. KG (Germany) for a high-speed steel containing 0.79% carbon, 18% tungsten, 4.7% cobalt, 4.3% chromium, 1.5% vanadium, 0.7% molybdenum, and the balance iron. Used for lathe and planer tools, taps, reamers, and hobs.

althausite. A light gray mineral of the triplite group composed of magnesium phosphate hydroxide, $Mg_2PO_4(OH)$. Crystal system, orthorhombic. Density, 2.97 g/cm^3; refractive index, 1.592. Occurrence: Norway.

Altic. Trade name of Nottingham Aluminum Company Limited (UK) for a castable composite consisting of an aluminum matrix reinforced with fine titanium carbide particles.

Altior. Trade name of Birkett, Billington & Newton Limited (UK) for a series of tin-lead solders for aluminum, and antifriction metals of tin, lead, and antimony, and for bearings.

Altmag. Russian trade name for an age-hardenable aluminum alloy containing 5-8% magnesium, 0.5-1% manganese, and 0.1-0.5% titanium. Used for light-alloy parts.

Alton. Trade name for cellulose acetate fibers used for wearing apparel and as industrial fabrics.

Altowhite. Trademark of Dry Branch Kaolin Company (USA) for calcined, fractionated *kaolin*-based pigment extenders used in adhesives, paper, paints, inks, and reinforced plastics.

ALTRA. Trademark of Rath Performance Fibers, Inc. (USA) for high-alumina fibers manufactured by a proprietary sol-gel process, and supplied needled into blankets with alumina fiber contents of 72%, 80% or 97% for applications at temperatures exceeding 1600°C (2910°F).

Altraseal. Trade name of Chemtech Finishing Systems Inc. (USA) for a series of corrosion-inhibiting water-based and high-solids coatings.

Altuglas. Trade name of Atoglas, Atofina Chemicals Inc. (USA) for polymethyl methacrylates supplied in granules, beads, resins, cast and extruded sheets, and general-purpose and high-impact grades. Also included under this trade name are acrylic adhesives for polymethyl methacrylates.

Alual. Trade name of American Steel Company (USA) for a commercially pure aluminum (99.9%).

Alu-Alftalat. Trademark of Hoechst AG (Germany) for a series of alkyd resins.

Alubond. Trademark of Swiss Aluminum Limited (Switzerland) for semifinished products based on aluminum and aluminum alloys, and aluminum-based composites. The semifinished products are available as bars, blocks, panels, sheets, strips, tubing, and sections, and are used in the manufacture of industrial goods. The composites are supplied in the form of panels, and are used for wall panels for commercial vehicles, facing exterior and interior building walls, partitions, balcony railings, license plates, and indicating and warning panels.

Alucable. Trade name of Société Electro-Cables (France) for a nonhardenable aluminum alloy containing 0.55% silicon, and 0.4% magnesium. Used for light-alloy parts.

Aluchrom. Trade name of ThyssenKrupp VDM GmbH (Germany) for an extensive series of strong, heat-resistant iron-chromium-aluminum alloys used for electrical resistances and heating ele-

ments.

Aluchrom O. Trade name of ThyssenKrupp VDM GmbH (Germany) for a resistance alloy composed of 70% iron, 25% chromium, and 5% aluminum. It is commercially available in wire and powder form with a density of 7.1 g/cm^3 (0.26 $lb/in.^3$), a melting point of 1520°C (2770°F), high tensile strength, high electrical resistivity, a low temperature coefficient, a low coefficient of expansion, and a maximum service temperature (in air) of 1250°C (2280°F). Used for electrical resistances and heating elements.

Aluco. Trade name of Aluco AG (Switzerland) for a double glazing unit that has silica gel in the hollow chamber of the extruded aluminum frame section, sealed with cotton wool plugs to prevent leakage.

Alucoat. (1) Trademark for a golden-yellow, corrosion-resistant, self-healing chemical film produced by a chromate-conversion process. It is available as a liquid concentrate, contains no cyanide, and is applied by immersion or spraying. It has good paint-adhesion properties, very low electrical contact resistance, and does not inhibit spot, seam, or resistance welding. Used for coating aluminum and aluminum alloys, for repairing damaged coatings, etc.

   (2) Trademark of Altena Cleaning BV (USA) for an epoxy coating for air conditioning, refrigeration and heating systems.

Alucopan. Trademark of Alcan Airex AG (Germany) for aluminum composite sandwich panels for architectural and building applications.

Alucore. Trademark of Alcan Service Center BV (Netherlands) for an aluminum honeycomb sandwich panel with strong, decorative cover sheets, used in interior decoration, e.g., for ceiling and wall cladding applications.

Aludecor. Trade name of Alumet GmbH (Germany) for aluminum sheet and strip covered with polyvinyl chloride film. Used for decorative applications.

Aludur. Trade name of Aluminum-Werke Wutöschingen GmbH (Germany) for a series of wrought aluminum-copper, aluminum-magnesium, and aluminum-silicon alloys. Used for structural parts, hardware, fasteners, aircraft parts, transmission lines, etc.

Alufer. German trade name for a non-heat-treatable aluminum-iron alloy used for light-alloy parts.

Alufix. Trade name for high-luster buffing and polishing compounds for aluminum and other light-metal alloys, chromium, and stainless steel.

Aluflex. Trade name of Pechiney (France) for an alloy containing 99.25% aluminum and 0.75% magnesium. It has high electrical conductivity, and is used for electrical conductors, electrical wires, and cables for braiding.

Alu-Flex. Trademark of Twinpak Inc. (Canada) for paper and plastic film vacuum-metallized with aluminum.

Alufont. Trade name of Swiss Aluminum Limited (Switzerland) for a series of corrosion-resistant cast aluminum alloys containing 2-5% copper, 0-2% silicon, 0-0.6% titanium, 0-0.2% magnesium, and 0-0.8% manganese. It may also contain small amounts of iron and zinc. Used for valve and pump components, machine parts, aircraft parts, etc.

Alufran. Trade name of Pechiney/Société Vente de l'Aluminium (France) for various grades of aluminum available as ingots, billets, and casting alloys.

Alugard. Trade name of Premier Refractories and Chemicals, Inc. (USA) for silicon carbide plastic refractory mixes.

Alugir. Trade name of Ateliers de la Gironde (France) for a heat-treatable aluminum alloy containing 3% copper, 1.2% nickel, and 0.8% magnesium. Used for light-alloy parts.

alum. (1) *Ammonia alum.* A colorless or white crystalline substance obtained from a mixture of ammonium and aluminum sulfates. Density, 1.64 g/cm^3; melting point, 93.5°C (199°F); boiling point, loses 10 H_2O at 120°C (248°F). Used in the manufacture of lakes and pigments, as a setting-up agent for porcelain enamel ground coats and acid-resistant cover coats, in sizing of paper and retanning of leather, and in water and sewage purification. Formula: $AlNH_4(SO_4)_2 \cdot 12H_2O$. Also known as *aluminum-ammonium sulfate; aluminum-ammonium sulfate dodecahydrate; ammonia alum; ammonium alum; ammonium-aluminum sulfate; ammonium-aluminum sulfate dodecahydrate.*

(2) *Potash alum.* A colorless or white crystalline substance that is soluble in water. Density, 1.757 g/cm^3; melting point, 92°C (197.6°F); boiling point, loses $9H_2O$ at 64.5°C (148°F); hardness, 2-2.5 Mohs. Used as a cement hardener, as a dyeing mordant, in papermaking, matches, paints, tanning agents, waterproofing agents, and for water purification applications. Formula: $AlK(SO_4)_2 \cdot 12H_2O$. Also known as *aluminum-potassium sulfate; aluminum potassium sulfate dodecahydrate; potassium alum; potassium-aluminum sulfate; potassium aluminum sulfate dodecahydrate.*

(3) *Soda alum.* A colorless crystalline substance obtained by adding sodium chloride to a hot solution of aluminum sulfate. Density, 1.675 g/cm^3; melting point, 61°C (142°F). Used in waterproofing of fabrics, in dry colors and ceramics, in tanning and paper sizing, in matches, in engraving, and in water purification. Formula: $AlNa(SO_4)_2 \cdot 12H_2O$. Also known as *aluminum-sodium sulfate; aluminum-sodium sulfate dodecahydrate; sodium-aluminum sulfate; sodium-aluminum sulfate dodecahydrate.*

(4) *Cesium alum.* A colorless, water-soluble crystalline substance obtained by the addition of a solution of cesium sulfate to a solution of potassium aluminum followed by concentration and crystallization. Density, 1.202 g/cm^3; melting point, decomposes at 110°C (230°F). Used in water purification, in the manufacture of cesium salts, in the purification of cesium by fractional crystallization, in mineral waters, and in materials research. Formula: $AlCs(SO_4)_2 \cdot 12H_2O$. Also known as *aluminum-cesium sulfate; aluminum-cesium sulfate dodecahydrate; cesium-aluminum sulfate; cesium-aluminum sulfate dodecahydrate.*

Aluma. Trade name of Aluminium Belge SA (Belgium) for a corrosion-resistant alloy containing 98.5% aluminum and 1.5% manganese. Used for light-alloy parts.

Alumabond. Trade name for a series of alumina-based ceramic adhesives that have been specially formulated for adhesion to metals, ceramics, glass and quartz, and operating temperatures up to 1760°C (3200°F). They have high dielectric strength, and good thermal conductivity, and mechanical strength even at high temperatures. Used for electronic components, instrumentation and appliances, and in many metallurgical applications especially in high-temperature environments.

Alumabrite. (1) Trade name of Sun-Fin Chemicals Corporation (USA) for a bright-dip for aluminum used in metal finishing.

(2) Trademark of Micro Abrasives Corporation (USA) for abrasive grains and powders of alumina, silicon carbide, etc., supplied in fused and unfused grades. Used for refractory applications, and for buffing, polishing and grinding.

Alumafilm. Trademark of Aluminum Company of Canada Limited for vacuum-aluminized plastic film.

Aluma-Foil. Trade name of Inland Steel Company (USA) for aluminum-coated carbon and high-strength low-alloy (HSLA) steel foil.

Aluma-Fuse. Trade name of Inland Steel Company (USA) for a highly formable, heat-resistant aluminum-coated steel for high-temperature applications.

Alumag. Trade name of Trefileries & Laminoirs du Havre (France) for a series of corrosion-resistant wrought aluminum-magnesium and aluminum-magnesium-manganese alloys. Used for marine and aircraft parts, automotive trim, chemical equipment, etc.

Alumagnese. Trade name of Pechiney World Trade (USA) Inc. for a series of corrosion-resistant aluminum alloys containing 1-6% magnesium, 0.6% silicon, 0.75% iron, 0-0.15% copper, 0-1% manganese, and 0.2% titanium. Used for light-alloy parts.

Alumal. Trade name for wrought aluminum-manganese alloys.

Alumaloy. Trade name of Alumaloy Castings Limited (Canada) for cast aluminum products.

Alumalun. Trade name of American Abrasive Metals Company (USA) for a wear-resistant aluminum-silicon casting alloy with abrasive grains embedded in its wearing surface. It has excellent antislip properties, and is used for floor plates, stair treads, and car steps.

Alumalux. Trademark of Alcoa-Aluminum Company of America (USA) for a series of ultrapure ceramic aluminas available in unground and superground form in purities from 99.9% to 99.99%. They have high strength, high wear resistance and toughness, and a smooth surface finish. Used for cutting tools, electronic substrates, sodium vapor-lamp tubes, in the manufacture of single-crystal sapphires, yttrium-aluminum garnets, and optical ceramics.

Aluman. Trade name of Swiss Aluminum Limited (Switzerland) for corrosion-resistant wrought aluminum-manganese alloys. Used for containers, roofing, paneling, siding, sheetmetal work, heat exchangers, cooking utensils, chemical equipment, storage tanks, etc.

Alumangan. Trade name of Versevorder Metallwerk (Germany) for a non-heat-treatable, corrosion-resistant aluminum alloy containing 0.9-1.4% manganese and up to 0.3% magnesium. Used for roofing, automotive trim, etc.

Alumaningot. Trade name of Specialloy Inc. (USA) for a foundry alloy of 60% manganese and 40% aluminum. Used in the manufacture of high-purity aluminum, and as manganese additions to iron alloys.

Alumanodic. Trademark of Alumicor Limited (Canada) for anodized aluminum products.

Aluma-Plate. Trademark of Ferro Corporation (USA) for trowelable wear coatings containing 70% sapphire-hard *Aluma-Sand* ceramic spheres suspended in epoxy resin. Used in the construction trades.

Alumar. Trade name of Arcos Alloys (USA) for a series of aluminum and aluminum-alloy gas welding and brazing rods for pure aluminum, and various aluminum-silicon, aluminum-manganese, aluminum-silicon-manganese, and aluminum-magnesium-chromium alloys.

Aluma-Sand. Trademark of Ferro Corporation (USA) for small sapphire-hard ceramic spheres used as grinding media in high-speed mills, and in the manufacture of *Aluma-Plate* coatings.

Alumasod. Trade name of Continental Industries Corporation (USA) for a soldering alloy composed of 76% tin, 23% zinc and 1% cadmium. Used for soldering aluminum.

Aluma-Tek. Trade name of RPM Inc. (USA) for anti-corrosion aluminum paints for metals used for roofing, flooring and similar applications.

Aluma-Ti. Trade name of Inland Steel Company (USA) for heat-resistant aluminum-coated steel for high-temperature applications.

Aluma-Tube. Trademark of Ferro Corporation (USA) for a series of monolithic ceramic tubes pressed from alumina (Al_2O_3). They have sapphire-like hardness, exceptionally high abrasion, wear and chemical resistance, and a high service life. Used for lining small pipes, and wear control of tubes and pipes.

Alumaweld. Trade name of Johnson Manufacturing Company (USA) for a series of tin-zinc-lead solders for joining aluminum, magnesium, aluminum bronze, cast iron, etc.

Alumazite "Z". Trade name of Tiodize Company, Inc. (USA) for a family of aluminum-pigmented coatings used to inhibit galvanic corrosion on metallic surfaces. Used for aerospace, industrial and marine applications.

Alumbro. Trade name of IMI Kynoch Limited (UK) for an aluminum bronze containing 76% copper, 22% zinc and 2% aluminum. Used for plates, and condenser and heat-exchanger tubing.

Alumec. Trade name of British Alcan Aluminum Limited (UK) for a heat-treatable, high-strength aluminum alloy containing 3.5-4.5% zinc, 2.5-3.5% magnesium, 0-1% copper, and 0-0.5% chromium. Used for aircraft and automotive components, and for general mechanical engineering applications.

Alumec 99. Trademark of Uddeholm Corporation (USA) for high-strength aluminum plate with high thermal conductivity, structural stability and machinability, and excellent resistance to stress-corrosion and exfoliation corrosion. Used for plastic molding parts.

Alumel. Trademark of Hoskins Manufacturing Company (USA) for a thermocouple alloy containing 94-95% nickel, 1.25-2.00% aluminum, 2.00-2.50% manganese, 1.00-1.75% silicon, and up to 0.50% iron and/or cobalt. Commercial available in wire and insulated wire form, it has a density of 8.5-8.7 g/cm³ (0.30-0.31 lb/in.³), a melting point of 1315-1390°C (2400-2535°F), a negative electromotive force, high tensile strength, good corrosion and heat resistance, high resistivity, low coefficient of expansion, a maximum service temperature (in air) of 1100°C (2010°F). Used as a pyrometric thermocouple wire, and for lead wires.

Alumend. Trade name of Arcos Alloys (USA) for a series of coated aluminum-base arc-welding rods for welding commercially pure aluminum, aluminum-silicon and aluminum-manganese alloys.

Alumet. (1) Trade name for aluminum and aluminum-alloy die castings.

(2) Trademark of Teich AG (Austria) for aluminum foil used in building insulation.

(3) Trademark of Alumet Manufacturing, Inc,. (USA) for roll-formed aluminum and steel, and extruded aluminum for use in awnings, canopies, screen rooms, enclosures, etc.

Alumetal. Trade name of Braun-Steeples Company Limited (USA) for a corrosion-resistant alloy of 95% aluminum and 5% silicon. Used as an aluminum solder.

alumetized steel. Iron or steel whose surface has been made resistant to scaling and oxidation by dip coating with aluminum, or an aluminum-silicon alloy, and heating at 800-1000°C (1470-1830°F). The aluminum forms an aluminum-iron or aluminum-iron-silicon alloy with the iron or steel surface, and produces a thin, hard, tightly adhering diffusion coating. Used for machinery parts, tubes, plates, pots, refinery tubes, etc. Also known as *aluminized steel; calorized steel.*

Alumex. Trademark of J.M. Huber Corporation (USA) for a high-grade white kaolin.

Alumibond. Trade name of ACI Chemicals Inc. (USA) for a conversion coating for aluminum and aluminum alloys.

Alum-I-Flex. Trade name of National Paint & Oil Company (USA) for flexible asbestos roof coatings containing aluminum.

Alumifoil. Trademark of Alcan Canada Foils Division of Alcan Limited (Canada) for aluminum foil used as wrapping material for tobacco products.

Alumigold. Trade name of Atofina Chemicals Inc. (USA) for a chromium conversion coating for aluminum used for automotive and aerospace applications.

Alumi-Guard. Trademark of J.H. & McNairn Limited (Canada) for a series of aluminum building papers.

Alumilastic. Trademark of Parr Industries Limited (Canada) for aluminum-containing elastomeric waterproofing compounds.

Alumilit. Trade name of Aluminiumwerke Rorschach AG (Switzerland) for aluminum-copper alloys.

Alumina. Trade name of A.P. Green Industries, Inc. (USA) for an extensive series of aluminum oxide ceramics with varying composition and properties. Used for electrical insulators, high-temperature electronics, and refractory products.

alumina. A white, crystalline substance available as powder, balls, or lumps of varying mesh size in standard and high-purity grades (99+%). Crystal structure, cubic (gamma phase), or hexagonal (alpha phase); density, 3.4-4.0 g/cm³, melting point, 2030°C (3686°F); boiling point, 2980°C (5396°F); upper continuous-use temperature, 1800°C (3270°F); hardness, 9 Mohs. The mineral *corundum* is a natural aluminum oxide, and *emery, ruby* and *sapphire* are impure varieties. It can also be made synthetically. Used in the manufacture of aluminum, as an abrasive for grinding and polishing, in the production of refractories and strong, tough, heat and chemical-resistant ceramics, in refractory coatings and surface finishes, for protective surfaces of transistors, in glass, cermets and whitewares, in crucibles and laboratory wares, in electrical insulators, spark plugs, light bulbs and electronic and optoelectronic products, in heat-resistant fibers, for clutch and brake linings, mill linings, nuclear fuel elements, welding-rod coatings, jewel bearings, gas-turbine parts, radomes and rocket equipment, as a catalyst, in fluxes, in papermaking, and as artificial gems. Formula: Al_2O_3. Also known as *aluminum oxide.*

alumina abrasives. A group of abrasives based on alumina (Al_2O_3) and including natural products, such as corundum and emery, as well as synthetic materials, such as the various forms of fused alumina. Used especially for grinding applications. Also known as *aluminum oxide abrasives.*

alumina aerogel. A gel that can be prepared at room temperature by mixing equal proportions of aluminum *sec*-butoxide and ethyl acetoacetate, adding an ethanol solvent and then gradually adding ethanol-diluted water. The product of this hydrolysis is an alumina (Al_2O_3) sol that can be converted to a gel by maintaining at elevated temperature (typically 60°C, or 140°F) for several days. *Alumina aerogels* have high heat resistance, and can contain additives, such as silica, zirconia, magnesia, silicon carbide, and/or barium, lanthanum or phosphoric oxide. Also known as *aluminum oxide aerogel.* See also alumina gel.

alumina balls. (1) Spheres made of 99% alumina (Al_2O_3) and typically ranging from 6.4 to 19 mm (0.250 to 0.750 in.) in diameter. They have high thermal and chemical resistance, and are used in reactor and catalytic beds. Also known as *alumi-*

num oxide balls.

(2) Spheres made of high-density, abrasion-resisting alumina (Al_2O_3) and used as grinding media in ball mills. Also known as *aluminum oxide balls.*

alumina brick. Refractory brick containing more than 50% alumina (Al_2O_3), used for high-temperature applications, such as furnace and kiln linings. Also known as *alumina firebrick.* See also basic brick.

alumina bubble brick. A lightweight ceramic product made by pouring molten alumina (Al_2O_3) through a blast of air, and pressing and baking. The resulting product is a lightweight refractory brick with innumerable small air bubbles.

alumina cement. A hydraulic cement made by sintering mixtures of bauxite with limestone. It contains a high quantity of alumina (Al_2O_3), is resistant to elevated temperatures, and will set to maximum strength in about 24 hours. See also aluminate cement.

alumina ceramic coatings. White, hard ceramic coatings based on alumina (Al_2O_3) usually applied by flame or plasma-arc spraying. They have excellent abrasion resistance, good corrosion and oxidation resistance, a melting range of 2040-2100°C (3705-3810°F), a porosity of 8-12%, and bulk density of 3.3 g/cm³ (0.12 lb/in³). Used to protect metals against oxidation and corrosion at room and elevated temperatures.

alumina ceramics. A group of hard, strong ceramics composed principally of alpha aluminum oxide (α-Al_2O_3), or *corundum* with compositions ranging from 50-100% α-Al_2O_3. They have melting points of up to 2038°C (3700°F), exceptional electrical resistivity, high dielectric strength, good insulating and high-temperature properties, good resistance to most chemicals and industrial atmospheres, good room-temperature strength, high abrasion resistance, and close dimensional tolerances. Used for electrical insulators, noble-metal thermocouple protection tubes, vacuum and heat-treating furnace parts and tubes, radiant and laser tubes, chemical equipment, aerospace parts, bearings, chute linings, dies, discharge orifices, pump plungers, wear-resistant applications, prostheses and bone replacements, etc.

alumina-diaspore fireclay brick. Fireclay brick containing 50-70% alumina (Al_2O_3).

alumina fibers. Strong ceramic fibers composed chiefly of alumina (up to 99% pure). Commercially available in the form of whiskers, continuous and discontinuous filament, yarns, and fabrics, they have high moduli of elasticity, a high melting point, exceptional corrosion resistance, and an upper service temperature of 1300-1320°C (2370-2410°F). Used for composite reinforcement applications, as insulating fibers, and as dielectrics. Also known as *aluminum oxide fibers.*

alumina-fiber metal-matrix composites. Advanced composite materials that consist of matrix materials, such as aluminum, magnesium, lead or copper reinforced with alumina fibers. They have good creep and fatigue resistance, good wear properties, and good strength and stiffness. Used for weight-sensitive applications in aerospace industry, reinforced automotive pistons, engine components, and bearing materials.

alumina firebrick. See alumina brick.

alumina gel. A white gelatinous precipitate usually obtained by reacting an aluminum chloride or sulfate solution with sodium hydroxide, sodium carbonate or ammonia. Density, approximately 2.4 g/cm³ (0.86 lb/in³). Used in the manufacture of glassware, ceramic glazes, and vitreous or near-vitreous ware, in waterproofing of fabrics, as a dyeing mordant, in the manufacture of lakes, as a filtering medium, and in paper sizing. Formula: $Al_2O_3{\cdot}xH_2O$. Also known as *aluminum hydroxide gel; gelatinous aluminum hydroxide; hydrous aluminum oxide.* See also alumina aerogel.

alumina hydrate. A white crystalline substance derived from *bauxite,* and supplied as powder, balls or granules. It occurs in nature as the mineral *gibbsite.* Density, 2.423 g/cm³; melting point, loses H_2O at 300°C (572°F). Used as a source of aluminum, in iron-free aluminum and aluminum salts, in the manufacture of activated alumina, glass, vitreous enamels, china glazes and other ceramics, as a base for organic lake pigments, in flame retardants, as mattress batting, as a water repellent in textiles, and as a rubber reinforcing agent, paper coating and filler. Formula: $Al_2O_3{\cdot}3H_2O$, or $Al(OH)_3$. Also known as *alpha alumina trihydrate; alumina trihydrate; aluminum hydrate; aluminum hydroxide; hydrated alumina; hydrated aluminum oxide.*

alumina porcelain. A dense, strong porcelain of superior quality composed chiefly of alumina (Al_2O_3). Used in the manufacture of spark plugs and electric insulators.

alumina-reinforced aluminum composites. Lightweight composites consisting of aluminum-alloy matrices (e.g., aluminum-magnesium or aluminum-lithium) reinforced with alumina fibers or particulate. Used for automotive engines and components, and aerospace structural applications.

alumina-reinforced epoxy composites. Lightweight composites that have thermosetting epoxy matrices reinforced with alumina fibers and exhibit good mechanical properties.

alumina-reinforced magnesium composites. Lightweight composites consisting of magnesium-alloy matrices reinforced with alumina particulate. Used for automotive parts, and aerospace applications.

Aluminark. Trade name of Westinghouse Electric Corporation (USA) for coated aluminum welding electrodes containing 5% silicon.

Alumina-seal. Trademark of Bren Cor Chemicals Limited (Canada) for a series of roof coatings.

alumina-silica-boria ceramics. Ceramic compounds containing typically 60-70% alumina (Al_2O_3), 20-30% silica (SiO_2) and 2-15% boria (B_2O_3). Commercially available in fiber and fabric form, they have a density of 2.7 g/cm³ (0.097 lb/in³), zero apparent porosity and nil water absorption, a refractive index above 1.56, high tensile strength, an upper continuous use temperature of 1400°C (2550°F), good resistance to dilute acids and metals, and fair resistance to concentrated acids and alkalies. Used as reinforcements for advanced composites, and for thermal insulation applications.

alumina-silica-boria fibers. A series of polycrystalline oxide fibers with a typical composition of 60-70% alumina (Al_2O_3), 20-30% silica (SiO_2) and 2-15% boria (B_2O_3). Commercially available in continuous and discontinuous fiber, roving, yarn, fabric, bulk, mat and blanket form, they have low dielectric constants, high melting points, exceptional handling qualities, and a glossy appearance. Used for reinforcement and thermal insulation applications.

alumina-silica cement. A highly refractory cement based on calcined kaolin, diaspore, mullite, sillimanite, or combinations thereof. It has good electrical insulating properties, good thermal properties, and good resistance to attack by most metals and slags.

alumina-silica fibers. Noncombustible ceramic fibers that typically consist of 45-96% alumina (Al_2O_3) and 4-55% silica (SiO_2). Commercially available in various forms including continuous and discontinuous filaments, yarns, fabrics, bulk, mats, blankets and sheets, they have low heat conductivity, high thermal-shock resistance, high tensile strengths and elastic moduli, a service temperature range of 1200-1600°C (2190-2910°F), an upper-use temperature in oxidizing atmospheres of 800°C (1490°F), excellent chemical resistance, and fair to poor resistance hydrochloric acid, phosphoric acid and concentrated alkalies. Used as reinforcements for resins, ceramics and light metals, in woven and nonwoven fabrics, and for cordage, thermal insulation, furnace linings, high-temperature piping, welding insulation, and various insulating applications in the aerospace industries.

alumina-silica hybrid multidirectional composites. Advanced composite materials consisting of three-directional orthogonal preforms of polycrystalline alumina (Al_2O_3) fibers densified with colloidal silica-matrix phases. Used for radomes.

alumina-silica refractories. A group of refractories, such as high-alumina fireclay and kaolin refractories, composed principally of alumina (Al_2O_3) and silica (SiO_2). Usually supplied in brick form.

alumina silicate. An engineering ceramic composed of 53% silica (SiO_2) and 47% alumina (Al_2O_3), and supplied in fiber form. It has a density of 2.7 g/cm^3 (0.097 $lb/in.^3$), a melting point of 1790°C (3255°F), an upper service temperature of 1260°C (2300°F), good high-temperature stability, excellent thermal-shock resistance, excellent sound absorption properties, low thermal conductivity, and good resistance to dilute acids. Used for thermal insulation applications, and as an asbestos substitute.

alumina-silicate fibers. Ceramic fibers based on about 42% silica (SiO_2) and 38% alumina (Al_2O_3) with 20% organic binder. They have a density of 0.92 g/cm^3 (0.03 $lb/in.^3$), and are used in ceramics, and as reinforcements for advanced composites.

alumina sol. The initial hydrolyzate obtained in the preparation of an alumina aerogel by the sol-gel technique. See also alumina aerogel.

aluminate cement. A slow-setting, rapid-hardening cement consisting of up to 40% alumina (Al_2O_3), 30-40% lime (CaO), 5-10% silica (SiO_2), 0-15% ferric oxide (Fe_2O_3) and small amounts of magnesia (MgO). It is usually made by burning chalk with bauxite, has a density of about 3.0 g/cm^3 (0.11 $lb/in.^3$), and will set in about 24 hours. Used for bank walls, in road laying, and as a heat-resisting cement. Also known as *aluminous cement; calcium-aluminate cement; high-alumina cement; high-speed cement.*

aluminates. A group of compounds consisting of any of various metal oxides and aluminum oxide, and having the general formula $M_xO_y·xAl_2O_3$. They possess high tensile strength and oxidation resistance, and have melting points in the range of approximately 1400-2140°C (2550-3885°F). Used for structural applications.

alumina trihydrate. See alumina hydrate.

alumina whiskers. Very fine, axially-oriented, single-crystal alumina (Al_2O_3) fibers with exceptionally high tensile strengths and elastic moduli, excellent inertness at high temperatures, high heat capacity, and an upper continuous-use temperature in oxi-dizing atmospheres of 1700°C (3090°F). Used as reinforcements in the manufacture of advanced composites with metal, plastic or ceramic matrices.

alumina whiteware. A ceramic whiteware containing substantial amounts of alumina (Al_2O_3), used for spark plugs, wall tile, sanitary ware, artware, dinnerware, etc.

aluminian chromite. A brown to brownish-black variety of the mineral *chromite* composed of iron chromium aluminum oxide, $Fe(Cr,Al)_2O_4$. It usually contains about 8-31% aluminum oxide (Al_2O_3). Crystal system, cubic. Density, 5.09 g/cm^3; melting point, 1850°C (3362°F); hardness, 5.5-7.0 Mohs. Occurrence: Cuba, Greece, India, Morocco, New Caledonia, Philippines, Russia, South Africa, Turkey, USA (California, Maryland, Montana, Oregon, Pennsylvania), Zimbabwe. Used as a source of chromium, in chromium chemicals, pigments, in foundry sand, and in the manufacture of refractories. See also chromite.

aluminian goethite. A yellowish aluminum-bearing variety of the mineral *goethite* containing iron aluminum oxide hydrate, $(Fe,Al)_2O_3·H_2O$. It is a principal constituent of most bauxites. See also goethite.

aluminide coating. A diffusion coating, usually based on cobalt or nickel aluminide, used to protect superalloys against high-temperature oxidation.

aluminides. A group of intermetallic compounds of aluminum and any of the transition metals, such as chromium, nickel, tantalum, titanium, vanadium and zirconium. They have good strength and oxidation resistance at temperatures up to 1095°C (2000°F). An example is nickel aluminide (NiAl or Ni_3Al).

Aluminite. (1) Trade name for a non-heat-treatable aluminum alloy containing 23% zinc, 2.7% copper, 0.4% iron, and 0.2% silicon. Used for light-alloy castings.

(2) Trade name for high-alumina refractories available in block form for temperatures up to 1093°C (2000°F). Used for furnace linings.

(3) Trade name of Wolke Paint Company (USA) for an aluminum paint.

aluminite. A white mineral composed of aluminum sulfate hydroxide heptahydrate, $Al_2(SO_4)(OH)_4·7H_2O$. Crystal system, monoclinic. Density, 1.66 g/cm^3. Occurrence: Hungary.

Aluminized Poly Bond. Trade name of Kleen-Flo Tumbler Industries Limited (Canada) for aluminized autobody filler.

aluminized steel. See alumetized steel.

aluminocopiapite. A lemon-yellow mineral of the copiapite group composed of magnesium aluminum iron sulfate hydroxide hydrate, $(Mg,Al)(Fe,Al)_4(SO_4)_6(OH)_2·20H_2O$. Crystal system, triclinic. Density, 2.16 g/cm^3; refractive index, 1.535. Occurrence: Alaska.

aluminosilicate. A compound of aluminum silicate and a metal oxide, or radical.

aluminosilicate gel. A homogeneous, transparent gel that can be prepared by a sol-gel process either by hydrolyzing aluminum isoproxide with water using nitric acid as a catalyst, or by first mixing appropriate quantities of ethanol and tetraethyl orthosilicate (TEOS), and then adding appropriate quantities of water and aluminum nitrate. The resulting solid is then allowed to gel in an air oven at elevated temperatures.

aluminosilicate glass. A glass containing typically about 51% silica (SiO_2), 22% alumina (Al_2O_3), 9% lime (CaO), 5% magnesia (MgO), 5% barium oxide (BaO), the balance being other oxides. It has a density of 2.65 g/cm^3 (0.095 $lb/in.^3$), a refractive index of 1.54, an upper continuous-use temperature (annealed condition) of approximately 650°C (1200°F), high thermal-shock resistance, an exceptional resistance to corrosion due to water and weathering, and fair acid resistance. Used for high-temperature thermometers, traveling-wave tubes, power tubes, combustion tubes, and advanced ceramics. Abbreviation: ASG.

aluminosilicate refractories. Refractories, such as diaspore, fireclay, mullite, sillimanite and bauxite, in which alumina and silica are the predominant compounds.

aluminous abrasives. Abrasive products made by fusing alumina.

aluminous cement. See aluminate cement.

aluminous fireclay refractory. A fireclay refractory of the aluminosilicate type composed of about 38-45% alumina (Al_2O_3).

aluminous refractory products. Refractory products which contain more than 45% alumina (Al_2O_3).

aluminum. A silvery-white, lightweight, malleable, ductile metallic element belonging to Group IIIA (Group 13) of the Periodic Table. It is commercially available in the form of structural shapes of all types, ingots, plates, sheets, foil, microfoil, microleaves, flakes, powders, granules, rods, wires, tubes and single crystals. The crystals are usually grown by the Bridgeman technique. *Aluminum* is obtained from bauxite chiefly by the Bayer process. Crystal system, cubic. Crystal structure, face-centered cubic. Density, 2.708 g/cm^3; melting point, 660°C (1220°F); boiling point, 2450°C (4440°F); atomic number, 13; atomic weight, 26.982; trivalent. It has high electrical conductivity (65% IACS), excellent thermal conductivity, high light reflectivity and softness, is strongly electropositive, and develops a white, protective coating of aluminum oxide making it highly resistant to ordinary corrosion. Used as the metal or in alloys for interior and exterior parts in the automotive and aerospace industries and for chemical and food-processing equipment, machine elements, permanent magnets, building products, window frames, structural parts, hardware, power transmission lines, photoengraving plates, parts for cryogenic equipment, tubes for toothpastes, etc., as rocket fuel, as an ingredient in thermite, in pyrotechnics devices, as a catalyst, in metallizing and coating (e.g., mirrors), as foil for packaging, and as a matrix material for composites. Particulate, fibers and whiskers are used as reinforcements for advanced composites, and powders in paints and protective coatings. Symbol: Al.

aluminum acetylacetonate. A white or pale yellow powder (99% pure) that is soluble in water, toluene, ethanol and pentanedione. It has a density of 1.213 g/cm^3, a melting point of 184-189°C (363-372°F), a boiling point of 320°C (608°F) (decomposes). Used as an organometallic precursor for superconductor research, as a catalyst, and in the deposition of aluminum. Formula: $Al(O_2C_5H_7)_2$.

aluminum alloys. A group of alloys in which aluminum is the predominant metal. Commercial aluminum alloys contain variable amounts of bismuth, chromium, copper, lead, magnesium, manganese, nickel, silicon, tin, and/or zinc. They can be grouped into the following categories: (i) *Wrought alloys;* (ii) *Clad alloys;* and (iii) *Casting alloys.* Although a wide range of properties is possible, the following are common to all aluminum alloys: low densities and melting points, high reflectivity to heat and light, good machinability, weldability, and workability, good corrosion resistance, good electrical and thermal conductivities, good low-temperature characteristics, and good fabricability. See also wrought aluminum alloys; clad aluminum alloys; aluminum casting alloys.

aluminum–alumina metal-matrix composites. Strong, lightweight engineering composites consisting of cast aluminum or aluminum-alloy matrices reinforced with long or short alumina fibers. They can be made by vacuum infiltration of the ceramic preform, and are used for lightweight applications, e.g., aerospace. Abbreviation: Al/Al_2O_3 MMC. See also aluminum-matrix composites.

aluminum amalgam. An amalgam obtained by adding fine aluminum filings to a 5% mercury chloride solution, and washing the resulting product with alcohol.

aluminum-ammonium sulfate. See ammonium-aluminum sulfate.

aluminum-ammonium sulfate dodecahydrate. See ammonium-aluminum sulfate.

aluminum antimonide. A hard, brittle stoichiometric intermetallic compound with semiconductive properties made by the fusion of pure aluminum and pure antimony, and subsequent zone refining. Crystal system, cubic. Crystal structure, sphalerite. Density, 4.218 g/cm^3; melting point, 1057°C (1935°F); upper continuous-use temperature, 538°C (1000°F); hardness, 4000 Knoop; dielectric constant, 11; band gap, 1.6 eV; refractive index, 3.2. Used in the manufacture of transistors, rectifier, and electronic and optoelectronic products. Formula: AlSb.

aluminum arsenide. A compound of aluminum and arsenic that possesses semiconducting properties. Crystal system, cubic. Crystal structure, sphalerite. Density, 3.81 g/cm^3; melting point, 1740°C (3164°F); hardness, 5000 Knoop. Used in the manufacture of distributed Bragg reflectors, and semiconductor devices. Formula: AlAs.

aluminum borate. (1) A white, granular powder obtained by the interaction of aluminum hydroxide and boric oxide. Melting point, dissociates at about 1035° C (1895°F). Used in the manufacture of glass and other vitreous and semivitreous products, and as a polymerization catalyst. Formula: $2Al_2O_3 \cdot B_2O_3 \cdot 3H_2O$. Also known as *boroaluminate.*

(2) A white compound of aluminum oxide and boron oxide. Melting point, about 1950°C (3542°F). Used in ceramic bodies requiring good thermal-shock resistance and refractoriness under load. Formula: $9Al_2O_3 \cdot 2B_2O_3$. Also known as *boroaluminate.*

aluminum boride. Any of the following compounds of aluminum and boron: (i) *Aluminum diboride.* Density, 3.16 g/cm^3; melting point, 1654°C (3009°F); hardness, 2600 Knoop. Used in ceramics and composites. Formula: AlB_2; (ii) *Aluminum decaboride.* Density, 2.54 g/cm^3; melting point about 2420°C (4388°F); hardness, 2650 Knoop. Used in ceramics. Formula: AlB_{10}; and (iii) *Aluminum dodecaboride.* A powder formed when AlB_2 dissociates. Density, 2.54-2.60 g/cm^3; melting point, 2165-2215°C (3929-4019°F); hardness, 2400-2600 Knoop; high neutron absorption. Used in nuclear shielding. Formula: AlB_{12}.

aluminum borosilicide. A fine powder of aluminum, boron and silicon used in ceramics and coatings. Formula: AlBSi.

aluminum brass. A corrosion-resistant brass containing about 20-42% zinc, 1-6% aluminum, 0-0.6% arsenic, and the balance copper. It has excellent cold workability, fair to poor machinability, and is used for condensers, evaporators, heat-exchanger and distiller tubing, and ferrules.

aluminum brazing alloy. An alloy with a brazing range of about 571-621°C (1060-1150°F) used for brazing aluminum and its alloys. Typically, it is an aluminum alloy containing 6.8-13.% silicon, 0.25-4.7% copper, 0.05-3.0% magnesium, and small additions of zinc, manganese and iron.

aluminum bromide. A moisture-sensitive, white to pale yellow powder (98+% pure). Density, 3.01 g/cm^3; melting point, 97.5°C (207°F); boiling point, 268°C (514°F). It is also available as a 1.0M solution in dibromomethane. Used in organic synthesis as a catalyst for alkylation, bromination and isomerization reactions. Formula: $AlBr_3$.

aluminum bromide hexahydrate. A moisture-sensitive, crystalline, yellowish powder (98+% pure). Density, 2.54 g/cm³; melting point, decomposes at 93°C (199°F). Formula: $AlBr_3 \cdot 6H_2O$.

aluminum bronze. A bronze containing about 2-12% aluminum, and sometimes additions of nickel, iron, silicon and tin. It has a density range of 7.50-8.19 g/cm³ (0.27-0.29 lb/in.³), excellent corrosion resistance, high strength, ductility, hardness, and resistance to shock and fatigue, good hot and cold workability, and relatively poor machinability. Used for condenser and evaporator tubes, ferrules, nuts and bolts, bushings, hardware, structural and wear-resistant applications, nonsparking tools, etc.

aluminum bronze powder. See gold bronze (1).

aluminum carbide. A compound obtained by heating coke and aluminum oxide. It is available as yellow or yellowish crystals, a brown powder, or a gray, massive substance. Crystal system, hexagonal. Density, 2.36 g/cm³; melting point, 1400°C (2550°F); boiling point, 2200°C (3990°F). Used in metallurgy, as a drying and reducing agent, in methane generation, as a catalyst, in ceramics, in the synthesis of nanowires and ribbons, and in composite reinforcement. Formula: Al_4C_3.

aluminum carbide nanostructures. Nanoribbons and nanowires of aluminum carbide (Al_4C_3), tens of micrometer long, synthesized in a solid-state reaction. The Al_4C_3 nanowires synthesized so far have typical diameters of 5-70 nm (0.2-2.8 μin.), and the nanoribbons have thicknesses of 5-70 nm (0.2-2.8 μin.) and widths of 20-500 nm (0.8-19.7 μin.). See also nanoribbon; nanowires.

aluminum-cesium sulfate. See cesium-aluminum sulfate.

aluminum-cesium sulfate dodecahydrate. See cesium-aluminum sulfate.

aluminum casting alloys. A group of aluminum alloys including principally the aluminum-silicon, aluminum-magnesium and aluminum-copper alloys. They are available as heat-treatable and non-heat-treatable alloys for sand, permanent-mold, and die casting. Aluminum-silicon alloys cast exceptionally well, and produce strong castings with dense, fine-grained microstructures. Other properties include good machinability, weldability and polishability, and good corrosion resistance (in absence of copper). Aluminum-magnesium alloys are not quite as strong as aluminum-silicon alloys, and their castability decreases with increasing magnesium content. However, they have higher corrosion resistance, and good machinability, weldability and polishability. Aluminum-copper alloys are age-hardenable high-strength casting alloys. Their strength increases with increasing copper and magnesium content. However, since they are only moderately castable, they are usually used for sand castings only. Uses of aluminum casting alloys include aircraft and automotive parts, such as gear boxes, engine blocks, cylinders, pistons, supports, and brackets.

aluminum chloride. White or yellowish-gray, moisture-sensitive crystals, granules or powder (99.9+% pure). Density, 2.44 g/cm³; melting point, 190°C (374°F); boiling point, sublimes at 180°C (356°F). It is also available as a 1.0M solution in nitrobenzene. Used as an intermediate in butyl rubber and hydrocarbon resins, and as a nucleating agent for titanium dioxide pigments. Formula: $AlCl_3$.

aluminum chloride hexahydrate. White or yellowish, crystalline, hygroscopic powder (99% pure). Density, 2.398 g/cm³; melting point, decomposes at 100°C (212°F). It is also available as a 1.0M solution in nitrobenzene. Used for roofing granules, special papers, pharmaceuticals, pigments, cosmetics, textiles and photography, and as a reagent in organic and organome-

tallic synthesis. Formula: $AlCl_3 \cdot 6H_2O$.

aluminum chops. Aluminum wire chopped into small pieces and supplied in spherical, irregular, flake, shot, needle or crystal form in high-purity grades (up to 99.999%), and low-purity recycled aluminum-alloy scrap grades. It has a Mohs hardness of approximately 2-3, and is used chiefly for abrasive blasting applications. See also aluminum shot.

aluminum-chromium. A master alloy of either 90% aluminum and 10% chromium, or 80% aluminum and 20% chromium.

aluminum-clad steel. A steel, usually in strip form, coated or clad with aluminum.

aluminum-clad wire. A wire with a copper core to which a thin layer of aluminum has been bonded by a metallurgical process.

aluminum-coated paper. Paper on whose surface a thin layer of aluminum has been deposited by a metallizing process. It combines the flexibility and lightness of paper with the brightness and reflectivity of aluminum. Used for packaging applications.

aluminum-coated steel. A steel with an aluminum surface coating that has been applied by batch or continuous hot dipping, electroplating, electrophoresis, ion vapor deposition, chemical or vacuum vapor deposition, cladding, thermal spraying or pack diffusion. It has a bright metallic appearance, high reflectivity, good electrical conductivity, good corrosion and oxidation resistance, and good formability and weldability. Used for corrugated roofing and siding, ductwork, outdoor signs, panels and housings, weather shields, clothes driers, agricultural implements, prefabricated steel buildings, heat-treating equipment, chemical equipment, heat-exchanger parts, recuperator and refinery tubing, turbine vanes and blades, chimney caps, fasteners including nails, studs and anchor bolts, pole-line hardware, aluminum railing, lighting posts, and wire products.

aluminum coating. A layer of aluminum deposited on the surface of a metal or nonmetal by any of various processes, e.g., spraying, immersion or electrodeposition.

aluminum columbate. See aluminum niobate.

aluminum copper. A double metal alkoxide supplied as a liquid with aluminum and copper contents of 3.6-3.9% and 4.2-4.5% respectively. The aluminum-to-copper ratio is 2:1. Used for sol-gel and coating applications. Formula: $Al_2Cu(OR)_x$.

aluminum-copper alloys. A group of high-strength aluminum alloys supplied in wrought and cast forms. They contain 4.0-10.0% copper, 0.2-2.0% magnesium, and small additions of manganese, nickel and silicon. Important properties include good age-hardening properties, good ductility and hardness, and relatively high susceptibility to intercrystalline corrosion. Used for truck wheels and frames, screw-machine products, and autobody parts.

aluminum-copper catalyst. A powder containing 50% aluminum and 50% copper, used as a catalyst.

aluminum-copper-magnesium alloys. A group of lightweight, hard high-strength aluminum-base alloy containing 3.5-5.5% copper, 0.5-0.8% magnesium, 0.5-0.7% manganese, a total of up to 0.7% silicon and iron. Available in cast and wrought form, they have good resistance to corrosion by acids and seawater, and high hardness resulting from the precipitation of copper on aging. Used for aircraft parts, railroad cars, boats, and machinery. See also duralumin.

aluminum-copper metal-matrix composites. Lightweight composite materials consisting of aluminum-copper matrices reinforced with continuous silicon carbide fibers. Commercially available in sheet, tube and rod form, they have high tensile and yield strengths, and are used in the aerospace industry for aircraft, missiles and engines, and for lightweight pressure vessels. Abbreviation: Al-Cu MMC.

aluminum-copper-silicon alloys. A group of aluminum casting alloys with good machinability, and high hardness and strength. A typical composition is 4-10% copper, 2-4% silicon, up to 0.3% magnesium, and the balance aluminum. Used for sand and permanent-mold castings.

aluminum-deoxidized steel. A steel whose oxygen content has been reduced by the addition of aluminum to the ladle. Deoxidizing is carried to the point where no reaction takes place between carbon and oxygen as the metal solidifies. Also known as *aluminum-killed steel.*

aluminum decaboride. See aluminum boride (ii).

aluminum diacetate. See basic aluminum acetate (1).

aluminum diboride. See aluminum boride (i).

aluminum disilicate. See aluminum silicate (iii).

aluminum dodecaboride. See aluminum boride (iii).

aluminum enamel. A rather low-firing porcelain enamel designed for application to aluminum and its alloys. It has excellent resistance to chemicals, weathering, corrosion and thermal and mechanical shocks, and good color rendition and dielectric properties. Also known as *aluminum porcelain enamel.*

aluminum flake. A lustrous, flaky aluminum powder made by stamping, and used as a paint pigment, and for silvering rubber, plastics, etc.

aluminum fluoride. A white, moisture-sensitive, crystalline powder (98+% pure) obtained by a reaction of fluosilicic acid on aluminum hydrate, or by a reaction of hydrogen fluoride on alumina trihydrate. Density, 3.1 g/cm^3; sublimes at about 1260°C (2300°F). Used as a source of alumina, in the manufacture of aluminum, as a flux and opacifier in ceramic glazes and enamels, in the preparation of doped fluoride glasses, in mixed-metal fluoride catalysts, and as a versatile catalyst. Formula: AlF_3.

aluminum fluoride trihydrate. A white, hygroscopic, crystalline powder made by interaction of hydrogen fluoride on alumina trihydrate. Density, 2.88 g/cm^3; melting point, 250°C (482°F). Used in the manufacture of white porcelain enamels. Formula: $AlF_3 \cdot 3H_2O$.

aluminum fluosilicate. A white powder with a density of 3.58 g/cm^3, used in porcelain enamels, glass, and artificial gems. Formula: $Al_2(SiF_6)_3$. Also known as *aluminum silicofluoride.*

aluminum foam. An aluminum or aluminum-alloy structure made by adding titanium or zirconium hydride which releases hydrogen to produce a uniform, cellular material. It has a density about that of water (1.0 g/cm^3, or 0.036 $lb/in.^3$), good strength properties, and good absorption of shock impact without elastic rebound. Used as a core material for lightweight sandwich composites, and fiber-reinforced light-metal foams for automobile bodies, and roofing and building panels. Also known as *foamed aluminum.*

aluminum foil. A flat-rolled aluminum product usually made from commercially-pure aluminum in thicknesses ranging from 0.006-0.200 mm (0.002-0.008 in.). High-purity aluminum foil for electrical applications is also available as is microfoil with a thickness of 0.001-1.0 μm (0.04-40 μin.) and a purity of 99.999%. *Aluminum foil* has high luster, but poor tear resistance. Used for household wrap, honeycomb cores for sandwich panels, and aluminum-faced plywood.

aluminum gallium arsenide. A semiconductor obtained either by doping gallium arsenide with controlled amounts of alumi-num atoms (p-type dopant), or by an epitaxial process in which a layer of aluminum arsenide (AlAs) is grown on a gallium arsenide (GaAs) substrate. Used for photodetectors, diodes and other electronic components, and for thin-film applications. Abbreviation: AlGaAs.

aluminum gold. A deep-red gold alloy containing 78% gold and 22% aluminum. It has a melting point of 1060°C (1940°F), and is used for jewelry and ornamental purposes.

aluminum hydrate. See alumina hydrate.

aluminum hydride. A white or grayish powder that decomposes at 160°C (320°F). Used for electroless coatings on metals, plastics, textiles, and fibers, as a polymerization catalyst, and as a reducing agent. Formula: AlH_3.

aluminum hydroxide. See alumina hydrate.

aluminum hydroxide gel. See alumina gel.

aluminum hypophosphite. A solid crystalline substance obtained by heating a solution of an aluminum salt and sodium hypophosphite. Melting point, decomposes at 218°C (424°F). Used as a finishing agent for polyacrylonitrile fibers. Formula: $AlH_6O_6P_3$.

aluminum iodide. A moisture-sensitive, off-white powder (95+% pure) with a density of 3.98 g/cm^3, a melting point of 191°C (375°F), and a boiling point of 360°C (680°F). Used as a catalyst in organic and organometallic synthesis. Formula: AlI_3.

aluminum-iron bronze. A strong, corrosion-resistant aluminum bronze containing 85-89.5% copper, 6-9.5% aluminum, 3.5-7.5% tin, and up to 0.4% lead, 0.5% manganese and 0.09% phosphorus, respectively. Used for structural parts, propeller blades, pump parts, and bearings.

aluminum isopropoxide. A white solid (98+% pure) obtained by reacting isopropanol with aluminum. Density, 1.035 g/cm^3 (at 20°C, or 68°F); melting point, 138-142°C (280-288°F); soluble in ethanol, isopropanol and toluene. Used as a dehydrating agent, catalyst, waterproofing compound for fabrics, in paints, and in organometallic research. Formula: $Al(OC_3H_7)_3$. Also known as *aluminum isopropylate.*

aluminum isopropylate. See aluminum isopropoxide.

aluminum-killed steel. See aluminum-deoxidized steel.

aluminum lactate. The aluminum salt of lactic acid available as a white to yellowish, water-soluble powder (95+% pure). Melting point, above 300°C (570°F). Used as fire foam, in biochemistry, and in organic and organometallic synthesis. Formula: $Al(O_3C_3H_5)_3$.

aluminum leaf. Aluminum sheets hammered or rolled very thin. Microleaf with a thickness of 0.15-1.0 μm (6-40 μin.), and a purity of 99.999% is also available.

aluminum-lithium alloys. A group of lightweight, wrought aluminum alloys containing 1-5% lithium, 1-5% copper, up to 1% magnesium, up to 0.6% zirconium, and sometimes small quantities of silicon, silver and iron. Available in the form of sheets, plates and profiles, they have a density of 2.50-2.60 g/cm^3 (0.090-0.094 $lb/in.^3$), high tensile strength and moduli, high fatigue durability at elevated and low temperatures, and good elasticity and weldability. Used for metal-matrix composites, structural applications, weight-sensitive applications, e.g., in the aircraft and aerospace industries. Also known as *lithium-aluminum alloys.*

aluminum-lithium metal-matrix composites. Lightweight composite materials consisting of aluminum-lithium alloy matrices reinforced with continuous silicon-carbide fibers. Commercially available as sheet, tubes and rods, they have high tensile and yield strengths, and are used in the aerospace industry for aircraft, missiles and engines, and for lightweight pressure vessels.

aluminum lithium oxide. See lithium aluminate.

aluminum magnesium. See magnesium aluminum.

aluminum-magnesium alloys. A group of wrought aluminum alloys containing about 3.0-10.0% magnesium, 0.5-2.0% silicon, and 0.2-1.0% zinc. They have low density, good to medium strength, excellent corrosion resistance, excellent resistance to seawater and alkaline solutions, good weldability, solderability and brazeability, and poor castability. Used for aircraft, automotive and marine equipment and fittings, marine hardware, fasteners, building hardware, levers, brackets, appliances, welded structures and vessels, storage tanks, cryogenic equipment, reflectors, and costume jewelry.

aluminum-magnesium silicide alloys. A group of heat-treatable, medium-strength aluminum alloys containing sufficient silicon and magnesium to form magnesium silicide (Mg_2Si). A typical composition is 98% aluminum, and 1% magnesium and silicon, respectively. Supplied in wire and tube form, they have low density, good resistance to industrial and seawater corrosion, and good formability, machinability and weldability. Used for marine hardware, screw machine parts, architectural parts, bridge railings, bicycle frames, furniture, and busbars. Also known as *aluminum-silicon-magnesium alloys.*

aluminum-manganese. An alloy of 70-75% aluminum and 25-30% manganese used in the manufacture of nonferrous alloys (e.g., duralumin-type alloys).

aluminum-manganese alloys. A group of wrought aluminum alloys containing about 0.2-1.2% manganese, 0.1-0.2% copper and 0.4-1.0% magnesium. They have high corrosion resistance, and good weldability, brazeability and solderability, moderate mechanical properties, and fair machinability and cold workability. Used for aircraft engines, chemical equipment and apparatus, storage tanks, sheetmetal work, hardware, etc.

aluminum metal-matrix composites. Strong, lightweight engineering composites consisting of aluminum or aluminum-alloy matrices reinforced with fibers, whiskers or particulate of aluminum oxide, silicon carbide, titanium carbide, titanium nitride, etc. Abbreviation: Al-MMC. See also aluminum-lithium metal-matrix composites; aluminum-magnesium metal-matrix composites; aluminum-silicon metal-matrix composites.

aluminum metaphosphate. A white crystalline powder with a melting point of approximately 1530°C (2785°F). Used in porcelain enamels, glazes and glasses, and in high-temperature insulating cement. Formula: $Al(PO_3)_3$.

aluminum-molybdenum. An alloy of 75% aluminum and 25% molybdenum, used in the manufacture of nonferrous alloys.

aluminum monohydrate. A white or yellowish powder with a density of 3.4 g/cm³. Used as a source of alpha alumina or corundum in ceramic bodies formed by hot pressing, and as a coating material, binder, high-temperature adhesive, inorganic thickener, and suspension agent. Formula: $Al_2O_3 \cdot H_2O$.

aluminum monopalmitate. The aluminum salt of palmitic acid available as a yellow or yellowish-white amorphous solid, or a fine white powder obtained by heating a solution of aluminum hydroxide and palmitic acid. Density, 1.072 g/cm³; melting point, 200°C (392°F). Used in waterproofing leather, paper, and textiles, as a viscosity improver for lubricating oils, as a varnish and paint drier, in the production of high-gloss leather and paper, and as a lubricant for plastics. Formula: $Al(C_{16}H_{31}O_2)_3 \cdot H_2O$. Also known as *aluminum palmitate.*

aluminum monostearate. A yellow or yellowish-white powder with a density of 1.020 g/cm³, a melting point of 155°C (311°F).

Used in waterproofing, as a viscosity improver for lubricating oils, as a varnish and paint drier, in the production of inks, greases, waxes and high-gloss leather and paper, and as a stabilizer for plastics. Formula: $Al(C_{18}H_{35}O_2)_3 \cdot H_2O$. Also known as *monobasic aluminum stearate.*

aluminum naphthenate. A brownish yellow rubbery substance obtained by reacting an aqueous solution of an aluminum salt and an alkali naphthenate. It has high thickening power, and is used as a drier and bodying agent for paints and varnishes.

aluminum-nickel catalyst. A Raney-type alloy composed of 50% aluminum powder and 50% nickel powder. Used in the manufacture of Raney nickel catalysts. See also Raney nickel.

aluminum-nickel-cobalt alloy. See alnico.

aluminum nickelide. (1) A compound of aluminum and nickel usually supplied as a fine powder and used in ceramics, and as a composite reinforcement. Formula: Al_3Ni.

(2) A compound of aluminum and nickel, usually supplied as a fine powder, and used in ceramics. Formula: Al_3Ni_2.

aluminum niobate. A compound of aluminum oxide (Al_2O_3) and niobium (columbium) pentoxide (Nb_2O_5) with a melting point of 1549°C (2820°F). Used in ceramics. Formula: $Al_2Nb_2O_8$ ($Al_2Cb_2O_8$). Also known as *aluminum columbate.*

aluminum nitrate. A white, hygroscopic, crystalline substance made by the interaction of nitric acid on aluminum. Melting point, 73.5°C (163°F). Used in the production of incandescent filaments, in nucleonics, as a textile mordant, in leather tanning, as a catalyst in petroleum refining, and as an anti-corrosion agent. Formula: $Al(NO_3)_3 \cdot 9H_2O$. Also known as *aluminum nitrate nonahydrate.*

aluminum nitrate nonahydrate. See aluminum nitrate.

aluminum nitride. A clear white or blue crystalline solid, or gray powder (98+% pure). Crystal system, hexagonal. Crystal structure, wurtzite (zincite). Density, 3.05-3.26 g/cm³; melting point, 2227°C (4041°F); upper continuous-use temperature, 1200°C (2192°F); hardness, 6-7 Mohs; translucent to ultraviolet radiation. Used for melting crucibles for iron and aluminum, as a semiconductor in electronics, and in nitriding of steels. Formula: AlN.

Aluminum Oilite. Trade name of Chrysler Corporation (USA) for an aluminum-base powder metal containing up to 4.5% copper, 1% magnesium and 1% silicon, respectively. It has a density of 2.4-2.6 g/cm³ (0.087-0.094 lb/in.³) and is used for bearings.

aluminum oleate. A yellowish-white, viscous metallic soap obtained by heating a solution of aluminum hydroxide and oleic acid. Used as a varnish and paint drier, as a lacquer for metals, as a waterproofing compound, as a thickener for lubricating oils, and as a lubricant for plastics. Formula: $Al(O_2C_{18}H_{33})_3$.

aluminum orthophosphate. A white crystalline substance obtained by interacting solutions of aluminum sulfate and sodium phosphate. Density, 2.566 g/cm³; melting point, above 1500°C (2730°F). Used as a binder in refractory products and high-temperature cements, as an alkaline flux for ceramics, in dental cements, in paints and varnishes, and in the pulp and paper industry. Formula: $AlPO_4$. Also known as *aluminum phosphate.*

aluminum oxide. See alumina.

aluminum oxide abrasives. See alumina abrasives.

aluminum oxide aerogel. See alumina aerogel.

aluminum oxide balls. See alumina balls (1), (2).

aluminum oxide fibers. See alumina fibers.

aluminum oxide trihydrate. See alumina hydrate.

aluminum oxynitride. A compound of aluminum nitride (AlN) and aluminum oxide (Al_2O_3) with good high-temperature properties used in the manufacture of *SiAlON* advanced *silicon aluminum oxynitride ceramics*, and for various high-temperature applications. Abbreviation: AlON.

aluminum oxynitride ceramics. A relatively recent class of high-temperature ceramics based on aluminum oxynitride (AlON) and usually synthesized by powder technologies, e.g. hot-press sintering. They have a density of about 3.6 g/cm³ (0.13 lb/in.³), excellent chemical, mechanical and optical properties including high strength, fracture toughness and durability, and good oxidation resistance at temperatures above 1500°C (2730°F). Used for structural, high-temperature, electronic and optical applications. Abbreviation: AlON ceramics.

aluminum paint. A mixture of finely divided, flaky aluminum particles and oil varnish. It reflects up to 75% of the sun's rays, and is used on tanks, water pipes, roofing, recreational vehicles, etc.

aluminum palmitate. See aluminum monopalmitate.

aluminum paste. An intimately ground aluminum powder suspended in oil and used in the manufacture of aluminum paints.

aluminum phosphate. See aluminum orthophosphate.

aluminum phosphide. A dark gray, or dark yellow crystalline substance. Crystal system, cubic. Crystal structure, sphalerite. Density, 2.42-2.87 g/cm³; hardness, 5.5 Mohs; melting point, approximately 1827°C (3320°F). Used in semiconductor technology. Formula: AlP.

aluminum porcelain enamel. See aluminum enamel.

aluminum-potassium borate. A white crystalline substance. Density, 3.42 g/cm³; melting point, below 1800°C (3270°F). Used in ceramics. Formula: $(AlO)_2K(BO_2)_3$.

aluminum-potassium sulfate. See potassium-aluminum sulfate.

aluminum-potassium sulfate dodecahydrate. See potassium-aluminum sulfate.

aluminum powder. Aluminum in the form of granules, or flaky or spherical particles with sizes ranging from less than 25 to over 850 μm (0.001 to over 0.033 μin.). Used in powder metallurgy, and as a paint pigment.

aluminum resinate. A brown solid substance obtained by heating *rosin* and aluminum hydroxide. Used as a varnish drier. Formula: $Al(O_5C_{44}H_{63})_3$.

aluminum selenide. A compound of aluminum and selenium usually supplied as moisture-sensitive, black pieces or powder (99.9+% pure). Crystal system, hexagonal. Crystal structure, wurtzite (zincite). Density, 3.44-3.91 g/cm³; hardness, 5.5 Mohs; melting point, 977°C (1791°F). Used in semiconductor technology. Formula: Al_2Se_3.

aluminum shot. Spheroidal particles of aluminum up to 12.7 mm (0.500 in.) in size, used in the deoxidation of steels. See also aluminum chops.

aluminum silicate. (1) Any of the following compounds of aluminum and silicon: (i) *Aluminum trisilicate.* A white solid substance available in lump and powder form. Density, 3.15 g/cm³; melting point, decomposes at 1810°C (3290°F). Used as a refractory component in glass and ceramic compositions, and as an extender pigment in coatings. Formula: $Al_2Si_3O_9$; (ii) *Trialuminum disilicate.* A white powder (mullite) used in ceramics. Formula: $Al_6Si_2O_{13}$; and (iii) *Aluminum disilicate.* A white powder usually supplied as the dihydrate. Used in ceramics, electronics, and organometallic research. Formula: $Al_2Si_2O_7 \cdot 2H_2O$.
(2) See calcined aluminum silicate.

aluminum silicate clay. Any of several types of clay containing varying amounts of alumina (Al_2O_3) and silica (SiO_2). See also kaolin.

aluminum silicate fabrics. Fabrics woven from aluminum silicate fibers and having good insulating properties and filtering characteristics. Used for insulating and protective clothing, filters, gaskets, and belting.

aluminum silicate fibers. Lightweight, resilient ceramic fibers of superior fineness manufactured by passing molten alumina and silica through a stream of high-pressure air. They have a melting point of approximately 1650°C (3000°F), good chemical stability, good heat and flame resistance, good electrical insulating properties, and an upper continuous-use temperature of approximately 1260°C (2300°F). Used for insulating paper, panelboard, as an extender for plastics and composites, for chemical-resistant rope, and for fabrics (e.g., blankets, tapes and broad goods), and paper.

aluminum silicate paper. A paper made from aluminum silicate fibers. It has good high-temperatures resistance, good dielectric properties, low thermal conductivity, and is used as insulating paper.

aluminum silicate refractories. Refractory materials made from blends of alumina (Al_2O_3) and silica (SiO_2) with natural materials, such as andalusite, bauxite, diaspore, gibbsite, kyanite or sillimanite.

aluminum silicide. A compound of aluminum and silicon, available as a fine powder, and used for refractories and coatings. Formula: Al_4Si_3.

aluminum silicofluoride. See aluminum fluosilicate.

aluminum-silicon alloys. A group of lightweight aluminum casting alloys containing about 5-20% silicon, and a total of about 5% iron, copper, magnesium, zinc and titanium. They have dense, fine-grained microstructures, good castability and corrosion resistance, good thermal conductivity, and good machinability and weldability. Used for pistons, cylinders, housings, pipe and marine fittings, thin-walled castings, etc.

aluminum-silicon bronze. A bronze containing 91% copper, 7% aluminum and 2% silicon. It has good corrosion resistance, and moderate to good strength and hardness. Used for marine hardware, bolts, connectors, sleeves, gears, studs, and pump parts.

aluminum–silicon carbide metal-matrix composites. Strong, lightweight engineering composites consisting of cast aluminum or aluminum-alloy matrices discontinuously reinforced with silicon carbide particulate. They can be made by vacuum infiltration of the ceramic preform, and are used for lightweight applications, e.g., aerospace components. Abbreviation: Al/SiC MMC. See also aluminum-matrix composites.

aluminum-silicon-magnesium alloys. See aluminum-magnesium-silicide alloys.

aluminum-silicon metal-matrix composites. Lightweight composite materials composed of aluminum-silicon-alloy matrices reinforced with continuous ceramic fibers, usually of the silicon carbide type. They have high strength and good thermal properties.

aluminum soap. A water-insoluble soap of aluminum and oleic, palmitic, or stearic acid, used in heavy lubricating greases, and as a drier in paints and varnishes.

aluminum-sodium sulfate. See sodium-aluminum sulfate.

aluminum-sodium sulfate dodecahydrate. See sodium-aluminum sulfate.

aluminum solder. An alloy with a soldering range of about 287 to 410°C (550 to 770°F) used for soldering aluminum and its alloys. Typically, it is an alloy of about 50-75% tin and 25-

50% zinc, 20-90% zinc and 10-80% cadmium, or 95% zinc and 5% aluminum.

aluminum stearate. A white powder obtained by reacting aluminum salts with stearic acid. Density, 1.01 g/cm³; melting point, 115°C (239°F). Used as a cement additive, as a paint and varnish drier, as a waterproofing agent for textiles and concrete, in lubricating greases, cutting compounds and flatting agents, and as a flux in soldering compounds. Formula: $Al(O_2C_{18}H_{35})_3$. Also known as *aluminum tristearate.*

aluminum steel. A low-alloy steel that contains about 0.35-0.42% carbon and has about 1.0% aluminum added to control the case depth during gas nitriding.

aluminum subacetate. See basic aluminum acetate (1).

aluminum sulfate. A white crystalline powder obtained by treating pure kaolin or aluminum hydroxide, or bauxite with sulfuric acid. Density, 2.71 g/cm³; melting point, decomposes at 770°C (1480°F). Formula: $Al_2(SO_4)_3$.

aluminum sulfate octadecahydrate. A colorless crystalline substance. Density, 1.69 g/cm³; melting point, decomposes at 86.5°C (187°F); refractive index, 1.474. Used as a size and pH control in papermaking, in white leather tannage, as a dyeing mordant and foaming agent, in fireproofing of textiles, as a waterproofing agent for concrete, in petroleum refinery processes, and in sewage precipitation and water purification. Formula: $Al_2(SO_4)_3 \cdot 18H_2O$. Also known as *aluminum sulfate octadecahydrate; cake alum; filter alum; papermakers' alum; patent alum; pearl alum; pickle alum.*

aluminum sulfide. A compound of aluminum and sulfur supplied as moisture-sensitive, yellow crystals or powder, or gray pieces (99+% pure). Crystal system, hexagonal. Crystal structure, wurtzite (zincite). Density, 2.55 g/cm³; melting point, 1127°C (2061°F). Used in semiconductor technology. Formula: Al_2S_3.

aluminum sulfocyanate. A yellowish powder used in pottery and ceramics. Formula: $Al(SCN)_3$. Also known as *aluminum thiocyanate.*

aluminum telluride. A compound of aluminum and tellurium supplied as moisture-sensitive black crystals (99+%). Crystal system, hexagonal. Crystal structure, wurtzite (zincite). Used in semiconductor technology. Formula: Al_2Te_3.

aluminum thiocyanate. See aluminum sulfocyanate.

aluminum-tin alloys. Alloys containing varying amounts of aluminum and tin as well as small amounts of copper, magnesium, and nickel. Usually supplied as sand or permanent-mold castings, they have high load-carrying capacities and fatigue strength. Used for tool housings, and automotive components, such as connecting-rod and crankcase bearings.

aluminum titanate. A compound made by heating a mixture of aluminum oxide and titanium dioxide. Density, 3.68 g/cm³; melting point, 1865°C (3390°F); stable from 1260°C (2300°F). Used in the manufacture of thermal-shock-resistant ceramics. Formula: Al_2TiO_5.

aluminum titanium bronze. A strong, tough alloy containing 88-90% copper, 9-10% aluminum, 1% iron, and 0-1% titanium. Used for propellers.

aluminum triricinoleate. A yellowish or yellowish-brown plastic mass obtained by reacting aluminum salts with castor oil. It has a melting point of 95°C (203°F), and is used as a solvent-resistant lubricant, as a gelling agent, and as a waterproofing agent. Formula: $Al(C_{18}H_{33}O_3)_3$.

aluminum trisilicate. See aluminum silicate (i).

aluminum tristearate. See aluminum stearate.

aluminum wire. A wire of aluminum, or steel-reinforced aluminum. Hard-drawn aluminum wire is usually commercially pure aluminum with small amounts of alloying elements, such as magnesium. It has a high tensile strength, a conductivity of about 61% IACS, a resistivity of 28.28 nΩm (17.01 Ω circmil/ft.), and is used for electric equipment. Steel-reinforced aluminum wire is composed of an aluminum wire with steel core, and is used for power transmission especially for long spans.

aluminum-zinc alloy coatings. Protective coatings with a nominal composition of 55% aluminum, 43.4% zinc and 1.6% silicon, often applied to steel sheets and wires by continuous hot-dip processes. They have good durability, good high temperature resistance, superior atmospheric corrosion resistance, cut-edge protection, and/or heat oxidation resistance.

aluminum-zinc-magnesium alloys. A group of strong, corrosion-resistant, heat-treatable aluminum alloys containing varying amounts of zinc, magnesium, copper and chromium. They have good machinability and poor elevated temperature characteristics. Used for structural parts, highly stressed components, and airframe structures.

aluminum yarn. A glossy textile yarn, such as *Lurex,* that has been combined with aluminum fibers. Used for clothing.

Alumite. (1) Trade name of Kapp Alloy & Wire Inc. (USA) for a lead-free solder that forms permanent, noncorrosive joints on aluminum and zinc alloys, pot metal and white metal. It is also used as a general-purpose high-strength solder. Supplied in rod form, it has a density of 6.95 g/cm³ (0.25 lb/in.³) , a melting range of 380-390°C (715-735°F), a hardness of 100 Brinell, and good ductility.

(2) Trademark of US Granules Corporation (USA) for aluminum granules produced from recovered aluminum foil products. Used for exothermic applications including thermit and exposives manufacture.

alum leather. Leather made by tanning with alum, usually in combination with salt, egg yolk, or other substances.

alumohydrocalcite. A white, yellow, light blue, gray or violet mineral composed of calcium aluminum carbonate hydroxide hydrate, $CaAl_2(CO_3)_2(OH)_4 \cdot 3H_2O$. Crystal system, triclinic. Density, 2.31 g/cm³; refractive index, 1.553. Occurrence: Europe (Germany), Siberia.

Alumolite. Trade name of MSC Laminates & Composites Inc. (USA) for a laminate with mirror-like appearance and high reflectance composed of a steel or aluminum substrate coated with adhesive, metallized, and then covered with a thin polyester film and a scratch-resistant coating.

alumopharmacosiderite. A white mineral of the pharmacosiderite group composed of potassium aluminum arsenate hydroxide hydrate, $KAl_4(AsO_4)_3(OH)_4 \cdot 6.5H_2O$. Crystal system, cubic. Density, 2.68 g/cm³; refractive index, 1.565. Occurrence: Chile.

Alum-Seal. Trade name of Plating Resources Inc. (USA) for an anodizing sealant.

Alum Sol. Trademark of Cramco Solder Alloys, Division of Cramco Alloy Sales Limited (Canada) for a series of aluminum solders.

alumstone. See alunite.

alum-tanned leather. Leather, such as kidskin, that has been tanned or retanned with aluminum sulfate, or a double salt composed of the sulfate of a monovalent metal (e.g., potassium) and a trivalent metal (e.g., aluminum or chromium). For example, aluminum sulfate is used in white leather tanning, and aluminum-ammonium sulfate in retanning.

Alundum. Trademark of Norton Company (USA) for pure, crystalline, granular fused alumina supplied as grains, cements, and refractory shapes. Used for grinding wheels, as abrasives for polishing and tumbling, as electrical insulation for radio and television tubes, as a reagent for the determination of carbon in steel, as a filtering material, for embedding electrical resistors and setting refractory bricks, in the production of refractory materials, such as bricks, plates, muffles, tunnel kiln parts, tubes, laboratory ware, etc., and for chemical apparatus.

alunite. A white, grayish or reddish brown mineral composed of potassium aluminum sulfate hydroxide, $KAl_3(SO_4)_2(OH)_6$. Crystal system, rhombohedral (hexagonal). Density, 2.6-2.82 g/cm^3; hardness, 3.5-4.0 Mohs; refractive index, 1.57-1.59. Occurrence: Australia, Italy, USA (Arizona, California, Colorado, Nevada, Utah, Washington). Used in the manufacture of high-alumina refractories, aluminum-potassium compounds, millstones and potash alum, and as a substitute for bauxite in aluminum manufacture. Also known as *alumstone.*

alunogen. A white mineral composed of an aluminum sulfate hydrate, $Al_2(SO_4)_3 \cdot 18H_2O$. Crystal system, triclinic. Density, 1.77 g/cm^3. Occurrence: Canada, Europe, USA.

Alupur. Trade name of Sollac/Usinor (France) for mild steel sheets that have a 40 μm (1.6 mils) pure aluminum coating on to each side, applied by continuous immersion in a molten bath. They have excellent resistance to industrial, marine and urban atmospheres, aggressive hydrocarbon combustion products and high-temperature corrosion, good sunlight refraction, and good fire and heat resistance up to about 450°C (840°F). The aluminized sheets are supplied in coils, sheets and slit strips in two qualities: (i) *Commercial* (C) with a maximum ultimate tensile strength of 500 MPa (72 ksi), and (ii) *Bending & Roll-Forming* (C320) with a minimum yield strength of 320 MPa (46 ksi), a minimum tensile strength of 390 MPa (56 ksi), and an elongation of 15-17%. Used for roofing applications, oil-tank cladding, and as insulation in refineries and thermal power stations.

Alurite. Trademark of Alcan Rolled Products Company (Canada) for coatings used on aluminum products.

Alusad. Trade name of Metalltechnik Schmidt GmbH & Co. (Germany) for an aluminum alloy containing 6-6.5% copper, 0.6-1.2% silicon, 1.1-1.5% iron, 1-1.5% zinc, 0.3-0.6% manganese, and up to 0.1 lead and 0.2% magnesium, respectively. Used as shot for abrasive-blast cleaning applications.

Alusi. Trade name of Sollac-Usinor (France) for sheet steel coated by continuous immersion in a molten bath containing an alloy of 90% aluminum and 10% silicon. Standard coating weights range from 40-200 g/m^2 (0.13-0.66 oz/ft^2). Supplied in several grades (e.g., *Alusi BEA* and *Alusi BEC*) with widths from 0.5-1.5 m (1.6-4.9 ft.) and thicknesses from 0.5 to 3.0 mm (0.02 to 0.12 in.), it has high mechanical strength, good corrosion and high-temperature oxidation resistance, good (cold) formability, and good compatibility to foodstuffs. Used for heat exchangers, industrial furnaces, ovens and dryers, flues and chimney liners, fire doors and domestic appliances, such as water heaters, cookers, barbecues, toasters, etc.

Alusil. German trade name for an aluminum alloy containing 20-21% silicon, 1-2% copper, 0.7% nickel and 0.7% iron, respectively. It has low thermal expansion, and is used for engine pistons, cylinder liners, pumps, and welding wire.

(2) Trade name of Alusil-Preßstoffwerk (Germany) for phenol-formaldehyde resins and products.

Alusite. Trademark of Harbison-Walker Refractories Company (USA) for a low-porosity *alumina brick* with good abrasion resistance and good resistance to penetration by molten slags. Used for reverberatory furnaces, rotary kilns for dolomite, lime, lime sludge, walls of soaking pits, lead furnaces, nonferrous melting and refining furnaces.

Alustar. Trade name of Hoogovens Aluminum Walzprodukte (Germany) for wrought aluminum-magnesium alloys containing small additions of iron, zinc, copper, chromium, titanium, and zirconium. Supplied in the form of sheets of varying thickness, they have a density of 2.66 g/cm^3 (0.096 $lb/in.^3$), high strength before and after welding, good weldability and bendability, excellent corrosion resistance, and high ductility. Used for marine parts and in shipbuilding.

Alustre. Trademark of Alcan Limited (Canada) for a coating used on aluminum products.

Alu-Tel. Trade name of Glasfaser GmbH (Germany) for glass felt coated on one side with aluminum film and backed with tar paper on the other.

Aluvac. Trade name of Fonderie de Precision SA (France) for a heat-treatable alloy containing 96% aluminum and 4% copper. Used for light-alloy parts.

Aluwangan. Trade name of Versevorder Metallwerke GmbH (Germany) for an aluminum alloy containing 0.5-1.5% manganese, and up to 0.3% chromium. It has good welding and forming properties, and is used for cooking utensils, heat exchangers, tanks and furniture.

Aluwe. Trade name of Westfälische Kupfer- und Messingwerke AG (Germany) for a series of aluminum-base alloys supplied in sheet, strip, rod and wire form.

Alva. German trade name for a wrought aluminum alloy containing 3% antimony, 3% lead, 2% copper, and up to 1% manganese. Used for bearings.

Alva Extra. Trademark of Crucible Materials Corporation (USA) for a water-hardening tool steel (AISI type W2) containing 1% carbon, 0.25% vanadium, and the balance iron. It has excellent machinability, good to moderate depth of hardening, wear resistance and toughness, and low hot hardness. Used for drawing, forming, heading and threading dies.

alvanite. A bluish green to bluish black mineral composed of aluminum vanadium oxide hydroxide hydrate, $Al_6(VO_4)_2(OH)_{12} \cdot 5H_2O$. Crystal system, monoclinic. Density, 2.41 g/cm^3. Occurrence: Kazakhstan, Russia.

Alvar. Trade name of Uddeholm Corporation (USA) for a hot-work tool steel.

Alvars. Trade name for a series of polyvinyl acetals used in lacquers, adhesives, and formerly for phonographic records.

Alzen. Trademark of Voest Alpine AG (Austria) for aluminum-alloy products containing 30% zinc and 5% copper, and zinc-alloy products containing 30% aluminum and 5% copper. Supplied as ingots, forgings, castings, and cored and solid bars, they are used for extrusions, light-alloy parts, and bearings.

Alzinc. French trade name for a casting alloy containing 80% aluminum and 20% zinc.

amakinite. A pale green mineral of the brucite group composed of iron magnesium hydoxide $(Fe,Mg)(OH)_2$. Crystal system, rhombohedral; (hexagonal). Density, 2.98 g/cm^3; refractive index, 1.707. Occurrence: Russia.

Amalcap Plus. Trade name for an amalgam alloy.

Amalcoden. Trade name for a glass-ionomer/cermet dental composite used as an amalgam bonding agent.

Amalfine. Trade name for a low-copper dental amalgam alloy.

amalgam. A solution of a metal, such as cesium, gold, lithium, potassium, silver, sodium, tin, or zinc, in mercury, or an alloy of mercury with one or more of these metals. An example is *dental amalgam* which is a mixture of mercury with a silver-tin alloy. Used for dental fillings, mirrors, in titanium production, in metal separation, and in organic synthesis.

Amalgambond. Trade name of Parkell Inc. (USA) for a resin-based adhesive dental cement.

Amalgambond Plus. Trade name of Parkell Inc. (USA) for a self-curing resin-based dental bonding agent with a thin to medium film thickness that strongly bonds the tooth structure (dentin and enamel) to composite-resin and amalgam restorations.

Amalite. Trade name for an alkyd resin employed for surface finishes.

AmAlOx 68. Trademark of Astro-Met Inc. (USA) for a high-density alumina-based ceramic engineering material (99.8% pure) It has excellent mechanical stability to about 1650°C (3000°F), exceptional wear resistance and hardness, high resistance to corrosion and oxidation, high compressive strength, a low coefficient of thermal expansion, and high dielectric strength. Used for piston caps and rings, cylinder, exhaust-port and manifold liners, and blades, vanes, shrouds, rotors, water pump seals, valve trains components, spray nozzles, sandblast nozzles, wear plates, cutting tool inserts, seals and bushings for pumps, extrusion dies, and fixtures.

AmAlOx 87. Trademark of Astro-Met Inc. (USA) for nonporous, inert alumina ceramics (99.95% pure). It has excellent high-temperature, wear and corrosion resistance, good dimensional stability, and good biocompatibility. Used for extrusion dies, nozzles, orifices, valve assemblies, pump pistons and cylinders, biomedical implants, etc.

Amaloy. (1) Trade name of Armstrong Brothers Tool Company (USA) for a water-hardening tool steel containing 0.6-0.7% carbon, 0.7-0.9% manganese, up to 0.5% molybdenum, and the balance iron. Used for tools and dies.

(2) Trade name of AMF Inc. (USA) for a corrosion-resistant alloy of 90-97.5% lead and 2.5-10.0% tin. Used for protective coatings on ferrous and copper alloys.

(3) Trade name for a heat- and corrosion-resistant alloy of nickel, chromium and tungsten.

Amalloy. Trade name of American Alloys Corporation (USA) for a corrosion- and impact-resistant aluminum casting alloy containing 7% magnesium, 0.15% iron, 0.14% titanium, 0.1% manganese, and 0.01% copper. Used for sand, permanent-mold and die castings, agricultural and dairy equipment, transportation and oilfield equipment, aircraft parts, hardware, and cookware.

Amalreco. Trade name for a series of low-copper dental amalgam alloys.

Amanimphy. Trade name of Creusot-Loire (France) for a non-magnetic nickel alloy used for motors and electrical equipment.

Amanox. Trade name of Krupp Thyssen Nirosta GmbH (Germany) for a series of strong, corrosion-resistant, nonmagnetic austenitic chromium-nickel and chromium-manganese-nickel stainless steels supplied in cold- and hot-rolled strip, sheet, and plate form.

amaranth. A strong, durable hardwood from trees belonging to the amaranth family (*Amaranthaceae*). Its heartwood changes from brown to purple on exposure, and has a fine, open grain, and high hardness. It is commercially available in boards and veneer. Average weight, 850 kg/m³ (53 lb/ft³). Source: Central and South America. Used for furniture, inlays, and turnery. Also known as *purpleheart*.

amarantite. A red to brownish red mineral composed of iron sulfate hydroxide hydrate, $FeSO_4(OH) \cdot 3H_2O$. Crystal system, triclinic. Density, 2.2 g/cm³; refractive index, 1.598. Occurrence: Chile.

amarillite. A light yellowish-green mineral composed of sodium iron sulfate hexahydrate, $NaFe(SO_4)_2 \cdot 6H_2O$. Crystal system, monoclinic. Occurrence: Chile.

Amasteel. Trade name of Ervin Industries, Inc. (USA) for cast steel grit and shot for abrasive blasting applications requiring fast cleaning speeds.

Amavar. Trade name for an alkyd resin used for surface finishes.

Amax. Trade name of Climax Performance Materials Corporation (USA) for an extensive series of high-purity and commercially-pure molybdenum and molybdenum-tungsten alloy plate, sheet, foil and powder products, and several high-purity and commercially-pure coppers, and nickel-cobalt-chromium-molybdenum alloys.

amazonite. A bright green to bluish green variety of the mineral *microcline*. It is a potash feldspar composed of potassium aluminosilicate $(KAlSi_3O_8)$. Occurrence: Russia, United States (Colorado, Virginia). Used as a gemstone, and for ornamental purposes. Also known as *amazon stone*.

amazon stone. See amazonite.

ambary. See kenaf fibers.

amber. A yellow to brownish translucent fossil resin derived from the extinct pine *Pinus succinifera*. It is amorphous, brittle, takes a fine polish, has excellent electrical insulating properties, and accumulates static charge by rubbing. Density, 1.05-1.11 g/cm³; hardness, 2-2.5 Mohs. Occurrence: Europe, Burma. Used for gems, ornaments, and in varnishes and lacquers.

amber glass. A special light-reflecting glass tinted to colors ranging from pale yellow to brown and brownish-red by the addition of controlled mixtures of iron oxide and sulfur compounds to the batch. Used for medicine bottles.

AmberGuard. Trade name of Eastman Chemical Company (USA) for amber-colored polyethylene terephthalate providing UV/visible light shielding for beer and medicine bottles.

Amberlac. Trademark of Rohm & Haas Company (USA) for fast-drying oil- or acrylic-modified alkyd resins with good color and gloss retention. They are available in various formulations, and are used in varnishes.

Amberlite. (1) Trademark of Rohm & Haas Company (USA) for an extensive series of ion-exchange resins available in various grades. Used in anion and cation exchange, as monobed and selective ion-exchange resins, and for many other applications.

(2) Trade name of former Resinous Products & Chemical Company (USA) for phenol-formaldehyde plastics.

amber mica. Brownish-yellow mineral composed of potassium magnesium aluminum silicate hydroxide, $KMg_3(Si_3AlO_{10})$-$(OH)_2$. It can also be made synthetically. Crystal structure, monoclinic, or hexagonal. Density, 2.78-2.86 g/cm³; hardness, 2.5-3.0 Mohs; refractive index, 1.54-1.59; excellent heat resistance; good insulating properties. Occurrence: Argentina, Brazil, Canada (Ontario, Quebec), India, USA (New Jersey). Used as an electrical and thermal insulator. Also known as *bronze mica; brown mica; magnesium mica; phlogopite*.

Amberoid. Trade name of Barker & Allen Limited (UK) for a nickel silver containing 62% copper, 23% zinc, and 15% nickel. Used for optical equipment, hollowware and costume jewelry.

Amberol. Trademark of Resinous Products & Chemical Company (USA) for low-cost phenolics with excellent thermal stability to above 150°C (300°F) which can be compounded with a variety of filler and resins. Used for motor housings, electrical fixtures, and automotive under-the-hood components.

Ambetti. Trade name of Hartley Wood & Company Limited (UK) for a type of translucent antique glass with small opaque specks of crystallized particles.

Amblerite. Trade name for an asbestos product that is available in the form of a thin sheet composed of asbestos fibers held together with a resilient binder. Used as packing for superheated steam and chemical fittings.

Ambloid. Japanese trade name for casein plastics.

amblygonite. A white to grayish-white natural mineral composed of lithium aluminum fluorophosphate, $LiAlPO_4F$, containing on average 8-10% lithia (Li_2O). Crystal system, triclinic. Density, 3.10 g/cm^3; melting point, 1170°C (2138°F); hardness, 6 Mohs; refractive index, 1.578-1.598. Occurrence: Africa, Brazil, Germany, USA (Maine, California, South Dakota). Used as a lithium ore, a source of alumina-phosphate, a flux in low-temperature porcelain enamels, in glass dinnerware to promote opacity, and in ceramic glazes and coatings.

Ambo. Trade name of Ambo Stahl-Gesellschaft (Germany) for an extensive series of cast and wrought carbon machinery and structural steels, alloy machinery and structural steels (e.g., silicon-manganese, chromium-molybdenum, chromium-vanadium, nickel-chromium, nickel-chromium-molybdenum and nickel-molybdenum types), austenitic, ferritic and martensitic stainless steels, high-temperature steels, austenitic manganese steels, plain-carbon tool steels, and cold-work, hot-work and high-speed tool steels.

amboyna. A species of wood that is usually rosy red with tiny knots resembling bird's eyes. Source: East Indies. Used for inlay work.

Ambra. Trade name of American Brass & Aluminum Foundry Limited (Canada) for nonferrous metals (e.g., aluminum, copper, magnesium, tin, zinc, lead, etc.) and nonferrous metal-alloy castings.

Ambrac. Trade name of Anaconda Company (USA) for a series of nickel silvers containing 20-30% nickel, 4-7.5% zinc, up to 3% tin, 5% lead, 1% manganese, respectively, and the balance copper. Used for hardware, fixtures, laundry and paper machinery, and plumbing.

Ambralloy. Trade name of A.M. Byers Company (USA) for a steel containing 0.4% carbon, 0.7% manganese, 1.8% nickel, 0.8% chromium, 0.2% molybdenum, and the balance iron. Used for structural members, cranes, derricks, bridges, and high-strength plate.

Ambraloy. Trade name of Anaconda Company (USA) for a series of wrought aluminum bronzes and brasses containing 76-95% copper, 20.96-22% zinc, 2-10% aluminum, 0.04%-0.05% arsenic, and up to 2.5% iron. They have high to medium strength, good to excellent resistance of seawater and impingement corrosion. Used for condenser and heat-exchanger tubes, ferrules, hardware, and marine parts.

Ambraze. Trade name of American Brazing Alloys Company (USA) for an extensive series of brazing filler metals including numerous nickel silvers, phosphor coppers, bronzes, silver-base alloys, and magnesium-aluminum alloys.

Ambro. Trade name of Ambro Steel (USA) for a corrosion-resistant aluminum bronze containing 76% copper, 22% zinc and 2% aluminum. Used for condenser and heat exchanger tubing.

Ambroid. Japanese trade name for casein plastics.

Ambroin. Trademark for Vereinigte Isolatorenwerke (Germany) phenol-formaldehyde and other plastics and molding com-

pounds.

Ambronze. Trade name of Anaconda Company (USA) for a series of wrought, corrosion-resistant leaded and lead-free copper-zinc alloys. Used for springs, weatherstripping, heat-exchanger tubing, and electrical contacts and clips.

Amcarb. Trade name of American Carbide Corporation (USA) for a series of straight and complex cemented carbides. The straight grades contain tungsten carbide in cobalt or nickel binders, and the complex grades contain tungsten carbide and titanium or tantalum carbide in cobalt or nickel binders. Used for cutting tools for ferrous and nonferrous alloys, drawing and extrusion dies, and wear parts.

Amco. Trademark of American Manufacturing Company (USA) for polyethylene and polypropylene fibers and monofilaments.

Amco Copper. Trade name of Climax Performance Materials Corporation (USA) for an electrolytic tough-pitch copper (99.90% pure). It has good corrosion resistance, and excellent cold and hot workability. Used for electrical applications and for general applications in the form of sheet, strip and rod.

Amcoh. Trade name of A. Milne & Company (USA) for a series of oil-hardening tool steels containing 0.80-0.95% carbon, 1.00-1.25% manganese, 0.50-0.90% chromium, 0-0.50 tungsten, and the balance iron. Used for hollow dies, such as ring and drawing dies, tools, and punches.

Amcoloy. (1) Trade name of A. Milne & Company (USA) for a series of oil-hardening low-alloy-type special purpose tool steel (AISI type L6) used for lathe tools, shafts, cams, and dies.

(2) Trade name of Ampco Pittsburgh Corporation (USA) for beryllium and manganese bronzes.

Amco OFHC. Trade name of AMAX Corporation (USA) for an oxygen-free high-conductivity copper (99.99% pure). It has good corrosion resistance, excellent hot and cold workability, and good forgeability. Used for electrical and electronic equipment, magnetrons, and triodes.

Amcosite. Trade name of American Hard Rubber Company (USA) for hard rubber.

Amcrom. Trade name of Climax Performance Materials Corporation (USA) for an age-hardenable alloy containing 98.8-99.6% copper and 0.4-1.2% chromium. It has good resistance to softening at high temperatures and high electrical conductivity (80-90% IACS). Used for electrical contacts, current-carrying components, resistance-welding tips, and rocket nozzles.

Amdry. Trademark of Sulzer Metco (USA) for an extensive series of coating, hardfacing and diffusion-alloy products including: (i) thermal abradable coatings composed of blends of an aluminum-silicon alloy and thermoplastic polyimide powders or yttria-stabilized zirconia (YSZ) powders for use as thermal-barrier coatings on turbofan engine parts, turbine blades, gas turbine engines, and diesel-engine components; (ii) nickel and nickel-chromium-base hardfacing powders for flame spraying applications, and brazing materials including nickel and cobalt materials as well as brazing cements and stop-offs; (iii) bond coat and one-coat tungsten carbide and ceramic coatings for thermal spray application to metallic and nonmetallic substrates for abrasion, wear, corrosion, heat- and/or oxidation-resistant applications in the aerospace industry; (iv) nickel-chromium-cobalt-base activated diffusion alloys supplied in powder, paste, tape, preform, sheet and ribbon form for brazing and surface-restoration applications; (v) nickel-chromium-molybdenum, nickel-chromium-cobalt, nickel-chromium-iron and cobalt-chromium superalloy powders for surface restoration and wear coatings; (vi) chromium, aluminum-cobalt, and alumina powders for pack-diffusion applications; (vii) various high-velocity oxyfuel (HVOF) powders; and (viii) various brazing powders, pastes, tapes, sheets, ribbons and preforms.

Amdry Braz-Bond. Trademark of Sulzer Metco (USA) for powdered metals for the production of protective coatings used for aircraft, automotive, hydroelectric, glass-industry, machinery, and watercraft applications.

ameghinite. A colorless mineral composed of sodium borate dihydrate, $NaB_3O_5 \cdot 2H_2O$. Crystal system, monoclinic. Density, 2.03 g/cm^3; refractive index, 1.528. Occurrence: Argentina.

Amelogen Universal. Trademark of Ultradent Products, Inc. (USA) for a radiopaque, submicron-particle bisphenol glycidyl-dimethacrylate (BisGMA) posterior/anterior composite resin used for dental restorations.

Amera-Mag. Trade name of American Tank & Fabricating Company (USA) for a high-tensile steel containing 0.08-0.15% carbon 1.2% manganese, 0.12% phosphorus, 0.05% aluminum, up to 0.1% nickel, up to 0.1% molybdenum, and the balance iron. Available in the form of bars, billets, ingots, plates, and sheets, it is used for autobodies, tanks, etc.

Amercoat. Trademark of Ameron International Corporation (US) for an extensive series of volatile-organic-compound (VOC) compliant epoxy-base primers and topcoats.

Amercut. Trade name of US Steel Corporation for various free-machining grades of carbon steels (chiefly of the AISI-SAE 11xx series). Used for screw-machine parts, bolts, fasteners, and gears.

Amer-Glo. Trade name of Cellanese Corporation (USA) for cellulose acetate and cellulose nitrate plastics.

Amerhead. Trade name of US Steel Corporation for various alloy machinery steels including several carburizing grades (manganese, molybdenum, nickel, nickel-chromium, chromium-molybdenum, nickel-molybdenum and nickel-chromium-molybdenum types). Used for screw-machine parts, fasteners, wire, bolts, shafts, cold-headed parts, etc.

American. Trademark of American Manufacturing Company (USA) for polyethylene and polypropylene fibers and monofilaments.

American alloy. A heat-treatable alloy of 95% aluminum, 3% copper, 1% manganese and 1% magnesium, used for light-alloy parts.

American ash. The heavy, strong, stiff, and durable wood of the tree *Fraxinus americana.* The heartwood is brownish with a texture similar to oak, and the sapwood nearly white. *American ash* has a straight grain, high hardness, high elasticity, good bending properties, high shock resistance, and works fairly well with hand tools. Average weight, 660 kg/m^3 (41 lb/ft^3). Source: Eastern USA; Canada (Southern Ontario, Quebec, New Brunswick and Nova Scotia). Used for sporting goods, such as tennis rackets, baseball bats, cricket stumps, snowshoes and skis, for millwork, furniture, cabinets, upholstered frames, boxes, crates, interior decorating, agricultural implements, tool handles, oars and musical instruments, and as a fuelwood. Also known as *Biltmore ash; Canadian ash; cane ash; eastern ash; white ash.*

American aspen. The lightweight wood of the quaking or trembling aspen (*Populus tremuloides*) and the bigtooth aspen (*P. grandidentata*). It is sometimes mixed with cottonwoods and sold as poplar or cottonwood. The heartwood is grayish white to light grayish brown, and merges gradually into lighter-colored sapwood. *American aspen* is straight-grained with a fine, uniform texture, has low strength and hardness, and works very

well. Average weight, 420 kg/m^3 (26 lb/ft^3). Source: USA (Great Lake states, Northeastern states, Rocky Mountains). Used as lumber for light structural work, and also for pallets, boxes, crating, pulpwood, particleboard, matches, excelsior, and as a pulpwood. Also known as *American poplar.* See also cottonwood; poplar.

American basswood. The light-brown, porous wood of the linden tree *Tilia americana* which is the softest and lightest commercial hardwood. The heartwood is yellowish-brown, and merges gradually into the wide, whitish sapwood. It has a fine, even texture, a straight grain, works very well, and has low strength and durability, large shrinkage, high resistance to warpage, and high gluing and nailing qualities. Average weight, 420 kg/m^3 (26 lb/ft^3). Source: Eastern United States and Canada. Used for cabinetwork, finish work and paneling, sash and door frames, moldings, boxes, Venetian blinds, drawing boards, food containers, woodenware, domestic utensils, as a corestock for plywood, also for veneer and cooperage, and as a pulpwood. It is one of the best woods for carving. Also known as *American lime; American linden.*

American beech. The hard, heavy, and durable wood of the tree *Fagus grandfolia.* It has a fine grain, a close uniform texture, light to reddish brown heartwood, and nearly white sapwood. It has medium to high strength, high shock resistance, large shrinkage, and turns well. Average weight, 753 kg/m^3 (47 lb/ft^3). Source: USA (Central and middle Atlantic states) and southern Canada. Used for flooring, furniture, tool and brush handles, handles, woodenware, shoe lasts, clothespin, containers, boxes, grates, cooperage, railway ties, gunpowder, charcoal, veneer, as a pulpwood and fuelwood, and as a substitute for sugar maple.

American Bemberg. Trade name of American Bemberg Corporation (USA) for rayon fibers and yarns used for textile fabrics.

American blackheart malleable iron. A term formerly used in North America for blackheart malleable cast iron containing 2.8-3.5% carbon, 0.6-0.8% silicon, a trace of manganese, and the balance iron. Used for pipes, fittings, plumbing, and hardware. See also blackheart malleable iron.

American black walnut. The heavy, hard, durable, brownish wood of the medium-sized walnut tree *Juglans nigra.* The heartwood is light to chocolate brown, and the sapwood nearly white. *American black walnut* is normally straight-grained with beautiful grain patterns, has a fine texture, high to medium strength, very high stiffness and shock resistance, and excellent machining and finishing properties. Average weight, 640 kg/m^3 (40 lb/ft^3). Source: USA (from Vermont to Great Plains and southward into Louisiana and Texas). Used for quality furniture, fine cabinetwork, interior paneling, gunstock, veneers, plywood, bowls, and other turned items.

American brass. A wrought, free-cutting copper alloy containing 35% zinc and 3% lead, available in plate and bar form. See also brass.

American cherry. The hard, strong, heavy wood of the tree *Prunus serotina.* The heartwood is light to dark reddish-brown with satin luster, and the sapwood narrow and white. *American cherry* has an uniform texture, a close grain with beautiful grain pattern, high stiffness and shock resistance, good dimensional stability after seasoning, and good machinability. Average weight, 530 kg/m^3 (33 lb/ft^3). Source: Southeastern Canada and eastern United States. Used for lumber, cabinets, fine furniture, fine veneer for paneling, caskets, etc., and medium and small-sized foundry patterns. Also known as *black cherry; wild black cherry; wild cherry.*

American chestnut. The light brown to yellowish wood of the tree *Castanea dentata* that is very scarce, and is available only in wormy grades due to a fungus disease known as "chestnut blight." It has a reddish brown heartwood, a coarse texture and an open grain, high durability, moderate strength, and is easily worked with hand or machine tools. Source: New England to Northern Georgia. Formerly widely used for poles, railway ties, furniture, boxes, fences, veneers, musical instruments, and caskets. Some "wormy chestnut" is still used in paneling, trim, and picture frames.

American cloth. A lightweight, glossy, air- and waterproof fabric, usually cotton, coated on one side with a linseed oil-base mixture.

American cord. See rattail cord.

American cotton. A term referring to cotton cultivated in America, and consisting of fibers with an average length of 23-32 mm (0.9-1.3 in.).

American elm. The hard, tough, durable wood of the tall tree *Ulmus americana*. The stand is severely threatened by a fungus disease known as "Dutch elm disease." The heartwood is light brown, and the sapwood nearly white. *American elm* is fine-grained, and has a fairly coarse texture with open pores, good bending and machining properties, and high strength and durability. Average weight, 641 kg/m³ (40 lb/ft³). Source: Eastern United States and Canada. Used for barrel staves, bent handles for axes, baskets, and specialty furniture. Also known as *white elm*.

American gold. See coinage gold (1).

American holly. The hard, heavy wood of the evergreen holly tree *Ilex opaca*. It has ivory-white heartwood and sapwood, an uniform, compact texture, a fine, close grain, high shock resistance, can be dyed (e.g., black for use as a replacement for ebony), works well, and cuts smoothly. Average weight, 750 kg/m³ (47 lb/ft³). Source: Southeastern United States (Atlantic Coast, Gulf Coast, Mississippi Valley). Used for furniture inlays, veneers, piano keys, scientific and musical instruments, ship models, and sporting goods. Also known as *white holly*.

American hornbeam. The heavy, tough wood of the small hardwood tree *Carpinus caroliniana* with poor decay resistance. Source: Eastern United States and southeastern Canada. It is not of great commercial importance, but some is used for small woodenware, and as fuelwood. Also known as *ironwood*.

American horse chestnut. The soft, dense wood of the tree *Aesculus glabra* which is similar to sweet or yellow buckeye. The heartwood is yellowish-white and the sapwood white. *American horse chestnut* has an uniform texture, a straight grain, low shock resistance, and is difficult to machine. Source: USA (from Ohio westward to Kansas and southward to Texas, and in the Appalachian Mountains from Pennsylvania to Northern Alabama). Used for lumber, boxes, furniture, artificial limbs, and planing-mill products, and as a pulpwood. Also known as *Ohio buckeye*. See yellow buckeye.

American Hylastic. Trade name of American Steel Foundries (USA) for a tempered steel containing 0.26-0.36% carbon, 1.4-1.7% manganese, 0.1% vanadium, and the balance iron. Used for railroad and structural castings.

American larch. The wood of a small or medium-sized tree *Larix laricina* belonging to the pine family. The heartwood is yellowish to reddish-brown, and the sapwood narrow and whitish. It has a coarse texture, and moderate weight and strength. Source: USA (from Maine to Minnesota, Lake states, and Alaska); Canada. Used for lumber, railroad ties, fenceposts, poles,

sills, boats, and as a fuelwood and pulpwood. Also known as *eastern larch; hackmarack; tamarack*.

American lime. See American basswood.

American linden. See American basswood.

American malleable iron. See blackheart malleable iron.

American Marine. Trade name of United American Metals Corporation (USA) for a babbitt containing varying amounts of lead, tin, and antimony. Used for bearings.

American Marine Genuine. Trade name of United American Metals Corporation (USA) for a shock-resistant babbitt containing varying amounts of lead, tin, and antimony. Used for bearings for locomotives and marine engines.

American mahogany. The tough, hard yellowish-brown to deep reddish-brown wood of any of various trees of the genus *Swietenia* especially *S. mahagoni* and *S. macrophylla* with pale to dark reddish-brown heartwood. It has a straight, beautifully figured grain, high strength and dimensional stability, high stability, low shrinkage and warpage, and works and finishes well. Average weight, 550 kg/m³ (34 lb/ft³). Source: Central America (Mexico), South America. Used for cabinetwork, fine furniture, interior decorative work, models, foundry patterns, boat construction, radio and television cabinets, caskets, interior trim, paneling, and precision instruments. Also known as *caoba*.

American poplar. See American aspen.

American red oak. The reddish brown wood of a group of broad-leaved trees including the northern red oak (*Quercus rubra*), the southern red oak (*Q. falcata*), the scarlet oak (*Q. coccinea*), and the black oak (*Q. velutina*). The *northern red oak* is found throughout southeastern Canada and the northeastern and central United States, and the *southern red oak* throughout the southeastern and southern United States. The *scarlet oak* grows throughout the eastern United States and southeastern Canada, and the *black oak* throughout the eastern and central United States and southeastern Canada. The heartwood of *American red oak* is reddish-brown, and the sapwood white. It has a coarse texture, a beautiful grain, high hardness, moderate strength and durability, moderate suitability for exterior use, is difficult to work, and is inferior to white oak. Average weight, 770 kg/m³ (48 lb/ft³). Used for lumber, veneer, flooring, furniture, boxes, pallets, crates, boxes and interior trim, agricultural implements, caskets, and handles. When preservative-treated, it is also used for fencing, posts, mine timbers, and railroad ties.

American silver. A corrosion-resistant nickel silver containing 49-59% copper, 21-24% zinc, 11-24% nickel, 0-4% manganese, 0-1.3% iron, 0.5%-5 tin, 0-3% lead; 0-1.5% aluminum, and a trace of sulfur. Used for ornaments and hardware.

American standard lumber. Lumber that meets the minimum requirements of the American softwood lumber standard–a product standard published by the US Department of Commerce–which sets down the basic principles of grading lumber according to appearance, dimensions, etc.

American sycamore. The wood of a large hardwood tree *Platanus occidentalis*. The heartwood is reddish-brown, and the sapwood light brown. *American sycamore* has a coarse, interlocking grain, a fine texture, medium density, hardness, stiffness and strength, good shock resistance, and saws well, but has a slight tendency to bind. Average weight, 641 kg/m³ (40 lb/ft³). Source: USA (Eastern United States from Maine to Nebraska, southward to Texas and eastward to Florida). Used for railroad ties, barrels, boxes, fenceposts, as a fuelwood, and as lumber for pallets, cabinetwork, furniture, flooring, handles and butchers' blocks, and as veneer for fruit and vegetable baskets and boxes. Also known as *buttonball; buttonwood; eastern sycamore.* See also *planewood; sycamore.*

American Tripoli. Trade name of American Tripoli, Inc. (USA) for a water-insoluble, free-flowing silica (SiO_2) powder with a density of about 2.1-2.6 g/cm³, and a melting point of about 1640-1710°C (2985-3110°F). It is supplied in several particle size ranges in two grades: (i) Rose tripoli, a rose-colored powder containing about 94% silica, with the balance being other oxides including alumina, and ferric oxide; and (ii) Cream tripoli, a cream-colored powder containing about 98.3% silica, with the balance being other oxides, such as alumina, calcia and ferric oxide. *American Tripoli* is used an abrasive in metal finishing, in polymers, and in automotive friction materials.

American vermilion. See Chinese scarlet (1).

American walnut. The heavy, hard, strong wood of the medium-sized walnut tree *Juglans nigra.* The heartwood is light to chocolate brown, and the sapwood nearly white. *American walnut* has a fine texture, a straight grain with beautiful grain pattern, excellent machining and finishing properties, high durability and stability in use, and high stiffness and shock resistance. Average weight, 640 kg/m³ (40 lb/ft³). Source: USA (Eastern States from Vermont to Great Plains, and southward into Louisiana and Texas). Used for veneers, quality furniture, fine cabinetwork, interior finishing, gunstocks, plywood, decorative purposes, and turned items (bowls; plates; handles, etc.). Also known as *black walnut.*

American white birch. The nearly white wood of the large tree *Betula papyrifera.* The tree is named for its papery white bark which was once used by North American Indians for making canoes. *American white birch* has medium weight, a fine texture, and is softer and weaker than sweet and yellow birch. Source: Canada, United States (mainly from northeastern and Lake states). Used as lumber, and for woodenware, pulp, fuel, turned products (e.g., spools, handles and toys), and plywood. Also known as *canoe birch; paper birch; white birch.* See also sweet birch; yellow birch.

American white oak. The strong, durable, light brown wood of a group of broad-leaved trees including the white oak (*Quercus alba),* the chestnut oak (*Q. prinus),* the post oak (*Q. stellata),* the overcup oak (*Q. lyrata),* the bur oak (*Q. macrocarpa),* and the live oak (*Q. virginiana).* Most commercial production is from the southern, south Atlantic, and central United States. The heartwood is grayish-brown and has good decay resistance, and the sapwood nearly white. *American white oak* has a close, coarse grain, outstanding hardness, works well with power tools, and has good gluing and nailing quality. Average weight, 770 kg/m³ (48 lb/ft³). Used for lumber, high-quality millwork, interior finish, furniture, carvings, railroad car and truck flooring, ship and boat structures and planking, bent parts, tight barrels and kegs, railroad ties, fencing, posts, and veneer.

American whitewood. The light, soft wood of the large hardwood tree *Liriodendron tulipifera* of the magnolia family. The heartwood is pale olive-brown often streaked with black, blue, green or red, and the sapwood is white to grayish-white. *American whitewood* has an uniform texture, a straight grain, good resistance to warpage, and works very well with hand and machine tools. Average weight, 480 kg/m³ (30 lb/ft³). Source: USA (from Connecticut and New York southward to Florida and westward to Missouri), Canada (Southern Ontario). Used as lumber, and for furniture, interior finish, siding, veneer, corestock for plywood, boxes, trunks, toys, cooperage staves, as a pulpwood, and for excelsior. Also known as *Canary whitewood;*

hickory poplar; tulip poplar; tulipwood; yellow poplar.

American willow. The hard, tough wood of the willow tree *Salix nigra.* The heartwood is light gray to light reddish brown, and the sapwood whitish to whitish-yellow. *American willow* has a fine, open grain, excellent resistance to shrinkage, and moderate strength. Average weight, 481 kg/m³ (30 lb/ft³). Source: Eastern and Central United States. Most commercial production is from the Mississippi Valley (Louisiana to southern Missouri) and Illinois. Used for lumber, furniture, corestock for plywood, wall paneling, toys and novelty items, artificial limbs, barrels and kegs, veneer, excelsior, charcoal, as a pulpwood, and for fenceposts. Also known as *black willow.*

americium. A synthetic, radioactive, metallic element of the actinide series of the Periodic Table. It was first produced by bombardment of uranium-238 with high-energy alpha particles in a cyclotron. Silver-white, ductile metallic americium is obtained from americium trifluoride. Density, 13.6 g/cm³; melting point, 1176°C (2149°F); atomic number, 95; atomic weight, 241; divalent, trivalent, tetravalent, pentavalent, hexavalent; half-life, 458 years; forms compounds with halides, oxygen and lithium; superconductive properties. Two allotropic forms are known: (i) *Alpha americium* (close-packed double hexagonal) that is stable below 1074°C (1965°F); and (ii) *Beta americium* (face-centered cubic) that exists between 1074 and 1176°C (1965 and 2149°F). Used in gamma radiography, electronics and crystal research, in aircraft fuel gages, fluid-sensing gages and distance-sensing devices, for smoke detectors and glass thickness meters, and as a neutron source. Symbol: Am.

Amerikagrün. Trade name of Schott DESAG AG (Germany) for a green-tinted spectacle glass.

Ameripol. Trademark of Ameripol Corporation (USA) for butadiene and isoprene rubber products.

Amerith. Trade name of Celanese Corporation (USA) for cellulose nitrate plastics.

Amer-Led. Trade name of US Steel Corporation for a series of free-cutting steels containing up to 0.15% carbon, 0.75-1.35% manganese, 0.25-0.36% sulfur, 0.15-0.35% lead, and the balance iron. Used for screw-machine products, screws, nuts, and bolts.

Amerlock. Trademark of Ameron International Corporation (US) for a series of high-solids, surface-tolerant maintenance epoxy resins for the protection of steel structures including bridges, offshore installations, marine equipment, industrial facilities, tank exteriors, water towers, roofing, pipes, etc.

Amerlock Sealer. Trademark of Ameron International Corporation (US) for a 100% solids epoxy sealer with excellent resistance to corrosive environments for use on steel and old coatings.

Ameroid. Trade name of American Plastics Corporation (USA) for a thermoplastic casein resin available in translucent and opaque grades. It has a density of 1.30-1.40 g/cm³ (0.047-0.050 lb/in.³), and good machining properties. Used for ornamental items, buttons, beads, etc.

Amershield. Trademark of Ameron International Corporation (USA) for a series of high-solids, aliphatic polyurethane coatings supplied in various grades, usually applied by standard or airless spray equipment, brush or roller, and providing single-coat protection from corrosion. They have excellent abrasion resistance, good impact resistance, high resistance to flexing, and may be applied over various substrates including clean, galvanized or phosphatized steel, aluminum, concrete, and masonry. Used as protective coatings on concrete walls and floors, railcar exteriors, hopper linings, and as refresher coats over old paint.

Amersil. Trade name for fused silica.

Amerspring. Trade name of US Steel Corporation for a spring steel containing 0.7-1% carbon, 0.2-0.6% manganese, 0.12-0.3% silicon, and the balance iron. Used in the form of music wire for springs.

Amerstitch. Trade name of US Steel Corporation for a steel containing 0.4-0.95% carbon, 0.5-1.2% manganese, 0.1-0.3% silicon, and the balance iron. Used for metal stitching wire.

Amerstrip. Trade name of US Steel Corporation for water-hardening steels containing 0.05-1.35% carbon, and the balance iron. Used for springs.

Amerthane. Trademark of Ameron International Corporation (US) for elastomeric polyurethane/polyurea coatings.

Amervan. Trade name for a foundry alloy containing 63.5% iron, 35% vanadium, and 1.5% silicon. Used as a vanadium addition in alloy steel manufacture.

Ames Copper. Trade name for a black copper dental cement.

amesite. A pale to apple-green mineral of the kaolinite-serpentine group composed of magnesium aluminum silicate hydroxide, $(Mg,Fe)_4Al_4Si_2O_{10}(OH)_8$. Crystal system, hexagonal. Density, 2.77 g/cm^3; refractive index, 1.597. Occurrence: USA, Antarctica. It also occurs as a variety without iron having a triclinic crystal structure, and a density of 2.69 g/cm^3. Occurrence: USA (Massachusetts); Antarctica.

A-metal. An alloy composed of 49-51% iron, 44% nickel, and 5-7% copper. It has high magnetic permeability, and provides non-distortion characteristics in the magnetized condition. Used for loudspeakers and transformers in sound transmission.

amethyst. A clear purple or bluish-violet variety of *quartz* (SiO_2) occurring in Russia, Uruguay, Brazil, USA (Arizona), and used as a gemstone, and for ornamental purposes.

Amfab. Trade name of Amfab (USA) for woven glass-fiber fabrics integrated with microscopic glass fibers between the threads and impregnated with synthetic resin. *Nonporous Amfab* is used as an insulating material supplied in tape form for high-voltage cables, motor windings, etc. *Porous Amfab* is used as a filtering medium.

Amfkote. Trade name for a series of high-temperature oxidation-resistant modified aluminide and silicide coatings deposited by pack-cementation diffusion on refractory metals, such as niobium.

Amherst sandstone. A dense, hard, gray to dull yellow sandstone quarried in Ohio, USA. It is composed of up to 95% silica (SiO_2), 4% alumina (Al_2O_3), and some iron oxides. Used in civil engineering and building construction.

Ami. Trade name of Alloy Metals Inc. (USA) for a low-carbon alloy steel containing 0.35% carbon, 5% chromium, 1.5% molybdenum, 1.35% tungsten, 0.5% vanadium, and the balance iron. Used for brass forging dies and inserts, hot-working tools and dies. Also known as *Ami Special*.

amianthus. Any fine, silky, long-fibered grass-green or greenish variety of asbestos, such as antigorite or chrysotile. See also antigorite; asbestos; chrysotile.

Amibrite. Trade name of Sur-Fin Chemicals Company (USA) for a bright dip for stainless steel.

amicite. A colorless mineral of the zeolite group composed of sodium potassium aluminum silicate pentahydrate, $K_2Na_2Al_4Si_4O_{16}\cdot5H_2O$. Crystal system, monoclinic. Density, 2.06 g/cm^3; refractive index, 1.498. Occurrence: Germany.

Amicor. Trade name of Acordis UK Limited for an antibacterial acrylic fiber that contains an organic chemical additive (Triclosan) and is used for sportswear, leisure wear, workwear and underwear. *Amicor Plus* is an antibacterial and antifungal fiber blend used for towels and odor-free socks and shoe linings, and *Amicor Pure* a blend of antibacterial and antifungal fibers used for bedding and hosehold furnishings to eliminate microbes and dust mites.

amide. (1) An ammonia compound in which one hydrogen atom has been replaced by a metal. For example, the replacement of one hydrogen atom (H) of ammonia (NH_3) with sodium (Na) yields sodium amide $(NaNH_2)$.

   (2) A carboxylic acid derivative in which the hydroxyl (OH) of the carboxyl group (–COOH) has been replaced by an amino group $(-NH_2)$, thus containing the $CONH_2$ radical. Examples include acetamide (CH_3CONH_2) and formamide $(HCONH_2)$.

Amidel TN. Trade name of Amoco Performance Products (USA) for polyamide resins supplied in transparent and amorphous 30% glass fiber-reinforced grades.

Amiesite. Trade name for asphalt mixed with synthetic or natural rubber.

Ami-Flex. Trade name of Auburn Manufacturing, Inc. (USA) for aramid-based textiles used for high-temperature applications.

Ami-Glas. Trade name of Auburn Manufacturing, Inc. (USA) for fiberglass textiles used for high-temperature applications.

Amilan. (1) Trademark of Toray Industries, Inc. (Japan) for a series of thermal engineering plastics supplied as nylon 6, nylon 6,6 and nylon 6,10 resins. *Amilan Nylon 6* and *Amilan Nylon 66* offer excellent chemical, oil, wear and abrasion resistance, excellent toughness, excellent long-term heat-resistance up to 80-150°C (175-300°F). Reinforced grades also have outstanding elastic modulus and strength. They are supplied in wide range of grades including unreinforced, carbon fiber-reinforced, glass fiber-reinforced, glass bead-filled, silicone-, polytetrafluoro-ethylene- and molybdenum disulfide-lubricated, mineral-filled, fire-retardant, high-impact, UV-stabilized, casting, elastomer, copolymer, monofilament and stampable sheet. *Amilan Nylon 6/10* refers to copolymer resins with excellent impact strength and flex fatigue properties and, depending on the particular grade, good transparency, flexibility, and adhesiveness. They are supplied in various grades including unreinforced, carbon fiber-reinforced, glass fiber-reinforced, silicone- and polytetra-fluoroethylene-lubricated and fire-retardant.

   (2) Trademark of Roray Industries Inc. (Japan) for nylon 6 fibers, monofilaments and yarns.

Amiloy. Trade name of Toray Industries, Inc. (Japan) for a series of crystalline engineering thermoplastics based on acetal copolymer resins. Supplied in various grades including general-purpose, high-flow and high-viscosity, they have high fatigue and creep resistance, and good elasticity and elongation. Used for automotive components (e.g., fuel-pump parts), consumer electronics and office equipment (keyboards, floppy disks, etc.), gears, and industrial fasteners.

Amilus. Trademark of Toray Industries, Inc. (Japan) for a series of engineering plastics based on polyacetal resins.

amine. Any of a group of organic compounds derived from ammonia (NH_3) by replacing one or more hydrogen atoms by alkyl or aryl groups, e.g., methylamine, CH_3NH_2 and phenylamine, $C_6H_5NH_2$. Depending on the number of hydrogen atoms replaced (one, two or three), the amine is known as "primary," "secondary," or "tertiary." The general formula is RNH_2 (primary), R_2NH (secondary), or R_3N (tertiary), where R is any group of atoms.

amine resins. Synthetic resins derived by reacting urea, thiourea, melamine or any related compound with aldehydes, particularly formaldehyde.

aminimide. Any of a wide variety of short- or long-chain aliphatic and aromatic compounds of nitrogen obtained by the reaction of an epoxide, such as epichlorohydrin or ethylene oxide, with 1,1-dimethylhydrazine in the presence of a carboxylic acid ester. The molecular chain type (short- or long-chain) and hydrocarbon class (aliphatic or aromatic) depends on the particular epoxide and ester used. Used in the manufacture of elastomers, adhesives, sealants, coating formulations, as a rubber adhesion improver in tire-cords dips, etc.

aminized cotton. Cotton, usually in the form of fibers, yarns or fabrics, obtained by treating with 2-aminoethylsulfuric acid ($C_2H_7NO_4S$) in a strong alkaline solution. This treatment improves its lightfastness, chemical reactivity and receptiveness to dyes (especially acid wool dyes) and subsequent treatments for rot resistance and water repellency.

amino acid. Any of a class of organic acids that contain an amino ($-NH_2$) group and are fundamental units in protein molecules. They are broken down from proteins in the digestive system and then used by the body to form its own protein. The following eight of the twenty-two known amino acids are considered "essential," and must be obtained from dietary sources because they cannot be synthesized in the body: isoleucine, leucine, lysine, methionine, phenylalanine, threonine, tryptophan, and valine. The general formula is $R-CH(NH_2)-COOH$ where R is an alkyl, aryl or heterocyclic radical. *Amino acids* can also be made synthetically, and are used in biotechnology, biochemistry and medicine, and also in the manufacture of biopolymers and biocomposites, and certain plastics (e.g., nylon 11). See also poly(amino acid).

α-amino acid. See alpha amino acid.

aminoaldehyde resins. A group of amino resins which are made by condensation polymerization of an amine and an aldehyde. The most common of these resins is urea-formaldehyde.

aminobenzene. See aniline.

aminoffite. A colorless mineral composed of beryllium calcium silicate hydroxide, $Ca_3(BeOH)_2Si_3O_{10}$. Crystal system, tetragonal. Density, 2.94 g/cm³; refractive index, 1.647. Occurrence: Sweden.

aminohexylamide gel. A hygroscopic, gel-like polymer of acrylic acid and 6-aminohexylamide used chiefly in the biosciences. Formula: $[-CH_2CH[CONH(CH_2)_6NH_2]-]_n$.

Aminolac. Trade name of Etablissements Kuhlmann (UK) for urea-formaldehyde resins.

Aminoplast. Trade name for urea-formaldehyde resins.

aminoplastics. A family of thermosetting plastics based on amino resins, and having high surface hardness and abrasion resistance, good machinability, excellent compressive strength, good tensile and flexural properties, good thermal and electrical insulating properties, and high hardness. Used for electrical and electronic components including circuit breakers, receptacles, switches, etc., and for closures, buttons, stove hardware, small machinery housings, household appliances, and dinnerware. Also known as *aminos*.

aminoplast resins. A class of thermosetting resins made by the reaction of an amine, such as urea or melamine, with an aldehyde, such as formaldehyde. Used for molding compounds, protective coatings, adhesives, laminates and permanent-press fabrics, and in papermaking, leather treatment, foundry sands, graphite resistors, structural foams, and ion exchange. Also

known as *amino resins; aminos*.

aminopolystyrene. See aminopolystyrene resin.

aminopolystyrene resin. A synthetic resin usually made from 200-400 mesh polystyrene, and used in protein analysis for diisothiocyanate (DITC) and C-terminal coupling of peptides. Abbreviation: APS resin. Also known as *aminopolystyrene; poly(4-aminostyrene); poly(p-aminostyrene)*.

aminos. See aminoplastics; aminoplast resins.

Amiran. Trademark of Schott DESAG AG (Germany) for a nonreflective glass for single glazings and laminated windows. It provides high light transmission (98%) and reduced UV transmission. Used for windows, display cases, show cases, etc. Also supplied as toughened safety glass.

Ami Special. See Ami.

Ami-Therm. Trade name of Auburn Manufacturing, Inc. (USA) for aramid-based textiles used for high-temperature applications.

ammine. A complex inorganic compound in which ammonia molecules (NH_3) are linked to metal ions by coordinate bonds. Examples include $[Ag(NH_3)_2]^{+1}$, $[Cu(NH_3)_4]^{+2}$, $[Ni(NH_3)_6]^{+2}$, and $[Zn(NH_3)_4]^{+2}$.

ammonia. (1) A colorless pungent gas composed of nitrogen and hydrogen. It is very soluble in water forming ammonium hydroxide, lighter than air, and can be easily liquefied into a colorless liquid by cooling or compression. Used as a refrigerant, fertilizer, in explosives, as a nitriding compound for steel, in the manufacture of certain plastics, and as a general reagent in the chemical industry. Formula: NH_3.
    (2) An aqueous solution of ammonia gas. Formula: NH_4OH. Also known as *ammonia water*. See also ammonia solution.

ammonia alum. See alum (1); ammonium-aluminum sulfate.

ammonia amalgam. An amalgam composed of ammonium and mercury.

ammonia solution. A colorless solution of up to 30% ammonia in water. Density range, 0.900-0.995 g/cm³. Used as an electrolytic reagent and etchant in metallography, and in ceramics, as a fireproofing compound, in the manufacture of rayon and rubber, in refrigeration, and in lubricants. Formula: NH_4OH. Also known as *ammonium hydrate; ammonium hydroxide; aqua ammonia*.

ammoniated nickel nitrate. Green, water-soluble crystals used in nickel plating. Formula: $Ni(NO_3)_2\cdot4NH_3\cdot2H_2O$. Also known as *nickel-ammonium nitrate; nickel nitrate tetramine*.

ammonioborite. A white mineral composed of ammonium borate pentahydrate, $(NH_4)_2B_{10}O_{16}\cdot5H_2O$. Crystal system, monoclinic. Density, 1.76 g/cm³; refractive index, 1.487. Occurrence: Italy.

ammoniojarosite. A light yellow mineral of the alunite group composed of ammonium iron sulfate hydroxide, $NH_4Fe_3(SO_4)_2(OH)_6$. Crystal system, rhombohedral (hexagonal). Density, 2.94 g/cm³; refractive index, 1.80. Occurrence: USA (Utah).

ammonium acid phosphate. See ammonium dihydrogen phosphate.

ammonium acid phthalate. The ammonium salt of phthalic acid. It is an alkaline metal biphthalate available in the form of single crystals with high plasticity and fissionability. Used as analyzing crystals in X-ray spectral analysis. Abbreviation: NH_4AP.

ammonium alum. See ammonium-aluminum sulfate.

ammonium-aluminum sulfate. A colorless or white crystalline substance obtained from a mixture of ammonium and aluminum sulfates. Density, 1.64 g/cm³; melting point, 93.5°C (199°F); boiling point, loses 10 H_2O at 120°C (248°F). Used in the manufacture of lakes and pigments, as setting-up agent for porcelain enamel ground coats and acid-resistant cover coats; in paper sizing, in retanning of leather, and in water and sewage purification. Formula: $AlNH_4(SO_4)_2 \cdot 12H_2O$. Also known as *aluminum-ammonium sulfate; ammonia alum; ammonium alum; ammonium-aluminum sulfate dodecahydrate.*

ammonium-aluminum sulfate dodecahydrate. See ammonium-aluminum sulfate.

ammonium amalgam. A pasty mass obtained by pouring sodium amalgam in an ammonium chloride solution.

ammonium biborate. A colorless crystalline substance obtained by the interaction of ammonium hydroxide and boric acid. Density, 2.33-2.95 g/cm^3; melting point, decomposes at 198°C (388°F). Used as a fireproofing compound for textiles, and in capacitors. Formula: $NH_4HB_4O_7 \cdot 3H_2O$. Also known as *ammonium borate.*

ammonium biphthalate. See ammonium acid phthalate.

ammonium borate. See ammonium biborate.

ammonium carbonate. A mixture of variable quantities of ammonium carbonate and ammonium carbamate supplied in the form of colorless, crystalline plates, or a white powder. Decomposition temperature, 58°C (136°F). Used in fire-extinguishing compounds, as a mordant for fabrics, and in ceramics. Formula: $(NH_4)(NH_2)CO_2$. Also known as *ammonium sesquicarbonate; hartshorn.*

ammonium chloride. A white crystalline substance (99.5+%) obtained from the interaction of sodium chloride solutions and ammonium sulfate. It is found in nature as the mineral *sal ammoniac.* Density, 1.527-1.530 g/cm^3; melting point, sublimes at 340°C (644°F); boiling point, 520°C (968°F); refractive index, 1.642. Used as an electrolyte in dry cells, as a soldering flux, as a pickling agent in zinc coating and tinning, in electroplating, in textile printing, in organic and organometallic synthesis, and in urea-formaldehyde resins. Formula: NH_4Cl.

ammonium cobaltous phosphate. See cobaltous ammonium phosphate.

ammonium dihydrogen phosphate. White crystals (98+%) with a density of 1.803 g/cm^3 and a melting point of 190°C (374°F). It is antiferroelectric below 148K, has good piezoelectric and electro-optical properties, a high nonlinear optical coefficient, a high optical damage threshold, and a wide range of transmission wavelengths. Used as wavelength-dispersive spectrometer crystals, in the manufacture of fourth-harmonic wave generators, as lasing and frequency multiplier for dye lasers, in flameproofing, and in analytical and food chemistry. Formula: $(NH_4)H_2PO_4$. Abbreviation: ADP. Also known as *ammonium acid phosphate; monobasic ammonium phosphate.*

ammonium diuranate. See ammonium uranate.

ammonium fluosilicate. A white, crystalline powder with a density of 2.01 g/cm^3, used in the etching of glass, in light-metal castings, and in electroplating. Formula: $(NH_4F)_2SiF_6$. Also known as *ammonium silicofluoride.*

ammonium heptamolybdate. The ammonium salt of molybdic acid available in the form of white crystals, or powder, and obtained by dissolving molybdenum trioxide in aqueous ammonia. Density, 2.498 g/cm^3; melting point, decomposes. Used as a metallographic etchant, in pigments, in the manufacture of molybdenum metal, as an adherence-promoting agent in some porcelain-enamel ground coats, in petroleum and coal refining, and in electron microscopy. Formula: $(NH_4)_6Mo_7O_{24} \cdot 4H_2O$. Also known as *ammonium molybdate tetrahydrate; ammonium paramolybdate; commercial ammonium molybdate.*

ammonium hydrate. See ammonia solution.

ammonium hydroxide. See ammonia solution.

ammonium metatungstate. A white powder (99.9+% pure) with a melting point above 100°C (210°F). Used in electroplating and microscopy. Formula: $(NH_4)_6W_{12}O_{39} \cdot xH_2O$.

ammonium metavanadate. A substance available in the form of white crystals, or a white to light yellow powder (99+% pure), and obtained by reacting vanadium pentoxide with ammonium chloride. Density, 2.326 g/cm^3; melting point, decomposes at 200°C (392°F). Used as a colorant to produce yellow, green and turquoise porcelain enamels and glazes, in organic and organometallic synthesis, in paint, varnish and ink driers, and in microscopy and photography. Formula: NH_4VO_3. Also known as *ammonium vanadate.*

ammonium molybdate. (1) A colorless crystalline substance with a density of 2.27 g/cm^3, used in pigments. Formula: $(NH4)_2MoO_4$.

(2) The ammonium molybdate of commerce, i.e. *ammonium molybdate tetrahydrate*, $(NH_4)_6Mo_7O_{24} \cdot 4H_2O$.

ammonium molybdate tetrahydrate. See ammonium heptamolybdate.

ammonium molybdophosphate. A yellow, crystalline powder obtained from the interaction of ammonium molybdate and nitric and phosphoric acids. Used in ion exchange, as a fixing and oxidizing agent in photography, in microscopy, and in the manufacture of certain water-resistant adhesives, cements and plastics. Formula: $(NH_4)_3PMo_{12}O_{40} \cdot xH_2O$. Also known as *ammonium phosphomolybdate.*

ammonium molybdosilicate. Yellow, crystalline granules obtained from the interaction of ammonium molybdate and nitric and silicic acids. Used as an additive in electroplating, as a catalyst and ion exchanger, as a fixing and oxidizing agent in photography, and in the manufacture of certain water-resistant adhesives, cements and plastics. Formula: $(NH_4)_4SiMo_{12}O_{40} \cdot xH_2O$. Also known as *ammonium silicomolybdate.*

ammonium-nickel chloride hexahydrate. Green crystals with a density of 1.65 g/cm^3, used in electroplating, and as a mordant. Formula: $NiCl_2 \cdot NH_4Cl \cdot 6H_2O$. Also known as *nickel-ammonium chloride hexahydrate.*

ammonium-nickel sulfate. Dark blue-green crystals with a density of 1.93 g/cm^3, used in the preparation of nickel dip solutions, and in nickel electroplating. Formula: $NiSO_4 \cdot (NH_4)_2SO_4 \cdot 6H_2O$. Also known as *ammonium-nickel sulfate hexahydrate; double nickel salt; nickel-ammonium sulfate; nickel-ammonium sulfate hexahydrate.*

ammonium-nickel sulfate hexahydrate. See ammonium-nickel sulfate.

ammonium nitrate. Colorless or white crystals (98+% pure) obtained by a reaction of ammonia vapor with nitric acid. Density, 1.725 g/cm^3; melting point, 169.6°C (337°F); boiling point, 210°C (410°F) (decomposes). Used as a reagent and catalyst in inorganic, organic and organometallic synthesis, in the manufacture of nitrous oxide, as an absorbent, as an oxidizer in solid rocket propellants, as a fertilizer, herbicide and insecticide, in explosives and pyrotechnics, and in biochemistry, medicine and nutrition. Formula: NH_4NO_3.

ammonium paramolybdate. See ammonium heptamolybdate.

ammonium paratungstate. A white crystalline substance usually obtained by the interaction of tungstic acid and ammonium hydroxide. Used in the preparation of ammonium phosphotungstate, tungsten trioxide, and tungsten alloys. Formula: $(NH_4)_6W_7O_{24} \cdot 6H_2O$. Also known as *ammonium tungstate; am-*

monium wolframate.

ammonium perrhenate. A white powder (99+% pure) that can be precipitated from the flue dust of molybdenum-bearing copper ores. Density, 3.97 g/cm³; melting point, decomposes at 365°C (689°F). Used as a source of rhenium. Formula: NH_4ReO_4.

ammonium peroxydisulfate. A white, corrosive, crystalline substance obtained electrolytically from concentrated ammonium sulfate solutions. Density, 1.982 g/cm³; melting point, decomposes. Used as a metallographic etchant, as an etchant for printed circuit boards, in copper electroplating, as a battery depolarizer, as a bleaching and oxidizing agent, and as a polymerization catalyst for gel formation in polacrylamide gel electrophoresis (PAGE). Formula: $(NH_4)_2S_2O_8$. Also known as *ammonium persulfate.*

ammonium persulfate See ammonium peroxydisulfate.

ammonium phosphate. White powder or crystals obtained from the interaction of phosphoric acid and ammonium hydroxide. It has a density of 1.619 g/cm³, and is used as a soldering flux for nonferrous metals, such as brass, copper, tin and zinc, and as a flame retardant for textiles, wood and paper. Formula: $(NH_4)_2HPO_4$. Also known as *dibasic ammonium phosphate.*

ammonium phosphomolybdate. See ammonium molybdophosphate.

ammonium salt. A salt in which the molecular structure includes a central nitrogen atom bonded to four hydrogen atoms (the cation) and a negatively charged acid radical (the anion), e.g., ammonium chloride (NH_4Cl) or ammonium nitate (NH_4NO_3).

ammonium selenite. Colorless or slightly reddish crystals used as a colorant in the production of some red glasses. Formula: $(NH_4)_2SeO_3 \cdot H_2O$.

ammonium sesquicarbonate. See ammonium carbonate.

ammonium silicofluoride. See ammonium fluosilicate.

ammonium silicomolybdate. See ammonium molybdosilicate.

ammonium silicotungstate. See ammonium 12-tungstosilicate.

ammonium stearate. A waxy, tan-colored solid substance. Density, 0.87-0.91 g/cm³; melting point, 74°C (165°F). Used as a waterproofing additive in hydraulic cements, concrete, stucco, paper and textiles. Formula: $C_{17}H_{35}COONH_4$.

ammonium sulfamate. A white solid substance obtained by hydrolysis. It has a melting range of 132-135°C (269-275°F), and is used as a flameproofing agent for paper and textiles, and in electroplating. Formula: $H_2NSO_3NH_4$.

ammonium sulfate. A white to brownish-gray crystalline substance. Density, 1.769 g/cm³; melting point, decomposes at 235°C (455°F). Used as an ingredient in glass batches to improve melting, in fireproofing compounds, and in biochemistry and molecular biology. Formula: $(NH_4)_2SO_4$.

ammonium sulfocyanate. A colorless or white crystalline substance. Density, 1.305 g/cm³; melting point, 149°C (300°F). Used in pickling iron and steel, as a separator of zirconium and hafnium, and gold and iron, in electroplating, in zinc and black-nickel coatings, as a curing agent for plastics, as a polymerization catalyst, in adhesives, and in liquid rocket propellants. Formula: NH_4SCN. Also known as *ammonium thiocyanate.*

ammonium tetrachlorozincate. White hygroscopic crystals or powder. Density, 1.8 g/cm³; decomposes without melting at 150°C (300°F). Used in welding and soldering fluxes, in glass-to-metal and ceramic-to-metal seals, in dry batteries, and in galvanizing. Formula: $(NH_4)_2ZnCl_4$. Also known as *ammonium-zinc chloride; zinc-ammonium chloride.*

ammonium tetrathiotungstate. A yellow-orange, crystalline powder. Density, 2.710 g/cm³; melting point, decomposes. Used in lubricants and semiconductors, as a source of tungsten disulfide, and in organometallic research. Formula: $(NH_4)_2WS_4$.

ammonium thiocyanate. See ammonium sulfocyanate.

ammonium thiosulfate. A white crystalline substance (99% pure). Density, 1.679 g/cm³; melting point, decomposes at 150°C (300°F). Used in cleaning compounds for zinc die-casting alloys, as a brightener in silver plating baths, and as a fixing agent in photography. Formula: $(NH_4)_2S_2O_3$.

ammonium tungstate. See ammonium paratungstate.

ammonium 12-tungstosilicate. The ammonium salt of 12-tungstosilicic acid available as granules or powder for use in electroplating, and in microscopy as a metal stain. Formula: $(NH_4)_4SiW_{12}O_{40} \cdot xH_2O$. Also known as *ammonium silicotung-state.*

ammonium uranate. A reddish or orange powder used in painting of porcelain. Formula: $(NH_4)_2U_2O_7$. Also known as *ammonium diuranate.*

ammonium uranyl carbonate. Yellow crystals that decompose at 100°C (212°F) in air. Used in uranium-yellow glazes. Formula: $UO_2CO_3 \cdot 2(NH_4)_2CO_3 \cdot 2H_2O$. Also known as *uranium-ammonium carbonate; uranyl ammonium carbonate.*

ammonium vanadate. See ammonium metavanadate.

ammonium wolframate. See ammonium paratungstate.

ammonium-zinc chloride. See ammonium tetrachlorozincate.

Amnic. Trade name of Climax Performance Materials Corporation (USA) for a series of high-purity alloys of 70-90% copper and 10-30% nickel. Used for electronic and cryogenic applications.

Amoco. Trade name of Amoco Chemical Company (USA) for a series of thermoplastic resins. *Amoco HIPS* refers to high-impact polystyrene resins supplied in regular, fire-retardant, and UV-stabilized grades, *Amoco PP* to polypropylene homopolymers supplied in regular and UV-stabilized, and *Amoco PS* to polystyrene resins supplied in various grades: regular, medium-impact, UV-stabilized, silicone-lubricated, 30% glass fiber-reinforced, and structural foam.

Amodel. Trademark of Amoco Performance Products (USA) for crystalline thermoplastic engineering resins based on polyphthalamide (PPA), and available in glass-, carbon- or mineral fiber-reinforced and polytetrafluoroethylene-lubricated grades. They have good mechanical and chemical properties including high tensile strength, stiffness, and fatigue and creep resistance over a broad temperature range, high wear resistance, and high static conductivity. Used for automotive valves and transmission stub tubes, bearings, gears, sucker-rod guides for oilfields, and electronic components, e.g., DIP switches.

Amol. Trade name of Atlas Metal & Alloys Company Limited (USA) for a water-hardening tool steel containing 0.8-0.9% carbon, and the balance iron.

A-Monel. Trademark of Inco Alloys International, Inc. (USA) for a series of cast alloys containing 62-68% nickel, 26-33% copper, 2.5% iron, 1.5-2.0% silicon, 1.5% manganese, and 0.35% carbon. Used for pumps, valves, and fittings. See also Monel.

amonilla. See African walnut.

Amorphic Carbon. Trade name for fluorinated diamond-like carbon coatings composed of an amorphous carbon network with densely and uniformly packed diamond-structured carbon nodules with a size of 100-200 nm (4-8 μin.). They have good chemical inertness, good resistance to atomic oxygen erosion, low coefficients of friction and improved lubricity, yet are almost as hard as diamond. They adhere well to many substrates including ceramics, polymers and metals.

amorphous alloys. Noncrystalline, metastable alloys prepared under non-equilibrium processing conditions by various methods including rapid solidification (cooling rates typically 10^5-10^8 °C/s), vapor deposition, electrodeposition, electroless deposition and mechanical attrition. As a result of their unique structures, which lack long-range order, these materials exhibit many unusual physical, chemical and mechanical properties not found in their crystalline counterparts of the same composition. Applications range from soft magnets to wear- and corrosion-resistant coatings. Also known as *amorphous metals; glassy alloys; metallic glasses.* See also amorphous materials.

amorphous boron. Boron in the form of a highly flammable, water-insoluble, brown to black, amorphous powder with a density of 2.25 g/cm³, and a melting point of approximately 2200°C (3990°F). Used in the manufacture of amorphous alloys. Abbreviation: a-B.

amorphous carbon. Pure carbon obtained by thermal decomposition or partial combustion of coal, crude oil, wood or a three-dimensionally crosslinked polymer. Available in foil, rod, fiber, powder and foam form, it is a black, amorphous substance with a density of 1.88 g/cm³, a sublimation point of 3500°C (6330°F), high flexural strength, low permeability to gases, and outstanding chemical resistance. Used for electrodes, laboratory equipment, as bioceramic coatings on stainless steel, and as fibers for orthopedic composites. Abbreviation: a-C. Also known as *glassy carbon; vitreous carbon.*

amorphous graphite. A type of natural graphite that contains more than 80% graphitic carbon. Used for foundry facings. Abbreviation: a-Gr.

amorphous iron. Any of various amorphous (glassy) ferrous alloys often containing boron and silicon as their primary alloying elements. They lack grain boundaries, are easily magnetized, and can be used for soft magnetic applications (transformer cores, etc.). All have high tensile strengths and moduli, hardnesses of 700-900 Vickers, maximum service temperatures (in air) usually below 150°C (300°F), and a crystallization temperature range between 450 and 550°C (840 and 1020°F). Abbreviation: a-Fe.

amorphous magnetic alloys. A category of amorphous binary, ternary or multi-component alloys with soft magnetic properties, usually produced as continuous tapes or ribbons of varying width by melt spinning or planar-flow casting. They are characterized by high magnetization and magnetic permeability, low to nil magnetostriction, and low magnetic exciting power and losses, which make them useful for recording heads, transformer coils, etc. Examples include alloys of iron, cobalt and nickel with one or more metalloids, such as boron, carbon, phosphorus, germanium and/or silicon, and multi-component alloys, such as iron-boron-carbon-silicon, iron-nickel-phosphorus-boron, and iron-cobalt-boron. See also amorphous alloys.

amorphous materials. Noncrystalline materials, i.e., materials, such as liquids, glasses, most plastics and amorphous alloys, that have no determinable crystal structure and lack long-range atomic order. See also amorphous alloys.

amorphous metals. See amorphous alloys.

amorphous phosphorus. A violet-red, amorphous powder with a density of 2.34 g/cm³, an autoignition temperature of 260°C (500°F), and high electrical resistivity. Used in phosphors, electroluminescent coatings, phosphor bronzes, metallic phosphides, compound semiconductors, safety matches, and in the manufacture of phosphorus compounds. Abbreviation: a-P. Also known as *red phosphorus.*

amorphous plastics. Plastics with irregular structures that lack long-range atomic ordering, and do not have exact melting points, but change from brittle, glassy materials to flexible, rubbery materials at a well-defined temperature (the so-called glass-transition temperature, T_g). They have good dimensional stability and impact properties and, generally, poor chemical resistance. Examples include acrylics (e.g., ABS), polycarbonates, polyarylates, polyphenylene oxides, polysulfones, polyether sulfones, polyetherimides, and polystyrenes. Also known as *amorphous polymers; glassy plastics; glassy polymers; polymer glasses.*

amorphous polymers. See amorphous plastics.

amorphous powder. A powder that is composed predominately of noncrystalline particles.

amorphous ribbon. An amorphous metal alloy in the form of a strip, typically 25-50 μm (1-2 mils) thick and 1-150 mm (0.04-5.9 in.) wide, made by ejecting a melt through an orifice onto a rotating copper drum on which the material rapidly solidifies.

amorphous selenium. A finely divided, amorphous, red powder obtained by reduction from selenous acid. It has a density of 4.26-4.28 g/cm³, softens at 40°C (104°F), and becomes black on standing and crystalline on heating. Used as a semiconductor. Abbreviation: a-Se. Also known as *vitreous selenium.*

amorphous semiconductor. A noncrystalline semiconductor, i.e., a semiconductor that lacks long-range atomic ordering. It is capable of conducting electricity when chemical, optical or thermal energy is supplied. There are elemental amorphous semiconductors, such as silicon, germanium, sulfur, selenium and tellurium, and compound amorphous semiconductors, such as gallium arsenide (GaAs), germanium selenide (GeSe), germanium telluride (GeTe) and arsenic selenide (As_2Se_3). Used in photovoltaic cells, photoconductive coatings, electronic and optoelectronic devices, etc.

amorphous silica. A natural form of silica, such as the mineral *opal*, that does not exhibit long-range atomic ordering. Abbreviation: $a-SiO_2$.

amorphous silicon. An organic semiconductor with good electrical and optical properties. It is a noncrystalline allotrope of silicon available in the form of a medium- to dark-brown powder with a density of 2.35 g/cm³, or as a thin film on various substrates. Used in the manufacture of solar cells, thin-film transistors, pin diodes, photovoltaic/electrochromic window coatings, flat-panel displays, and electroluminescent devices. Abbreviation: a-Si. See also hydrogenated amorphous silicon.

amorphous silicon carbide. A hydrogenated noncrystalline form of silicon carbide used for solar cells. Abbreviation: a-SiC:H.

amorphous silicon germanide. A hydrogenated noncrystalline form of silicon germanide used for solar cells. Abbreviation: a-SiGe:H.

amorphous solid. A solid substance in which the atoms in their equilibrium positions do not exhibit any long-range periodic pattern, i.e., lack crystalline periodicity.

amorphous sulfur. A pale yellow amorphous allotropic form of sulfur. Density, 1.92 g/cm³; melting point, approximately 115°C (240°F). Used as a semiconductor. Abbreviation: a-S.

amorphous tellurium. One of the two allotropic forms of tellurium. It is available as a brownish black powder. Density, 6.00 g/cm³; melting point, 452°C (845°F); boiling point, 1390°C (2535°F). Used as a semiconductor. Abbreviation: a-Te.

Amorton. Trade name of Sanyo Corporation (Japan) for an amorphous-silicon film material used for solar-cell applications.

amosite. See grunerite.

Amotun. Trade name of Atlantic Steel Corporation (USA) for a high-speed tool steel containing 0.85% carbon, 8% molybdenum, 6% cobalt, 4% chromium, 1.75% vanadium, 1.5% tungsten, and the balance iron. Used for cutting tools, dies, and taps.

Amoutun. Trade name of Atlantic Steel Corporation (USA) for a high-speed tool steel (AISI type M1) containing 0.8% carbon, 8% molybdenum, 4% chromium, 1.5% tungsten, 1% vanadium, and the balance iron. It has high red-hardness, and is used for lathe and planer tools, drills, taps, reamers, broaches, hobs, form cutters, end mills, etc.

Ampal. Trade name of Ampal Inc. (USA) for a series of aluminum powders (99.7+% pure) containing 0.1% silicon and 0.16% iron. They are available in various grades including atomized, free-flowing, medium-coarse and ultrafine. Used as a fuel material in explosives, as a reducing agent in chemical processing, in master alloys for ferrous alloys, in aluminothermic reactions including thermit welding, for cored welding filler wire, in pyrotechnics, as plastic compound fillers, in refractory cutting systems, in the production of metallic pigments, in aluminized coatings, and in aluminum powder metallurgy.

Ampco. (1) Trade name of Ampco Metal, Inc. (USA) for a series of aluminum bronzes containing 6-15% aluminum, 1.5-5.25% iron, traces of tin, manganese, nickel and/or silver, and the balance copper. They have good resistance to cavitation-pitting, corrosion, erosion, fatigue, and wear. Used for bearings, bushings, gears, etc.

(2) Trade name of Ampco Metal, Inc. (USA) for a commercially pure cast nickel.

Ampcoloy. (1) Trade name of Ampco Pittsburgh Corporation (USA) for a series of wrought and cast coppers and copper alloys including high-conductivity coppers, iron-aluminum and nickel-aluminum bronzes, beryllium, lead, tin and manganese bronzes, copper nickels, and high-conductivity copper alloys.

(2) Trade name of Ampco Pittsburgh Corporation (USA) for a series of nickels and nickel alloys including commercially pure nickel, aging and non-aging nickel-chromium alloys, and cast and wrought nickel-copper alloys.

Amperit. Trademark of H.C. Starck Inc. (USA) for a series of thermal-spray powders for flame, plasma, high-velocity-oxyfuel (HVOF) and laser spraying applications. They are supplied in a variety of compositions and particle sizes including pure metals, alloys, borides, carbides, nitrides, oxides and silicides. Used for aerospace and general industrial applications.

Amphenol. Trade name of American Phenolic Corporation (USA) for polystyrene plastics.

amphibole asbestos. A group of asbestos minerals which includes the following silicates of magnesium, iron, calcium and sodium: (i) *grunerite asbestos*, or *amosite;* (ii) *riebeckite asbestos*, or *crocidolite;* (iii) *anthophyllite asbestos;* (iv) *tremolite asbestos;* and (v) *actinolite asbestos.* Amphibole asbestos fibers are normally brittle and cannot be spun, but have better chemical and high-temperature resistance than serpentine asbestos. Also known as *amphibole.* See also asbestos; serpentine asbestos.

ampholyte. A substance, such as water that, at the same time, can react as an acid or a base since it can both accept or donate a proton. In solution, it yields either hydrogen (H^+) ions or hydroxyl (OH^-) ions depending on whether the solution is acidic or alkaline. Also known as *amphoteric electrolyte.*

ampholytoid. A particle in suspension that, depending on the pH value, has the capacity of adsorbing hydrogen (H^+) ions or hy-droxyl (OH^-) ions. Also known as *amphoteric colloid.*

Amphos. (1) Trade name of Climax Performance Materials Corporation (USA) for a 99.97% pure oxygen-free copper containing 0.03% phosphorus. It has good hot and cold workability, and is used for condenser and heat-exchanger tubing, piping, high-strength wrought metals, and for bright acid plating anodes.

amphoteric compound. A substance, such as aluminum oxide, chromic oxide, aluminium hydroxide or zinc hydroxide, that has the capacity of reacting either as an acid or a base. During electrolytic dissociation it may produce either hydrogen (H^+) or hydroxyl (OH^-) ions depending on the environmental conditions.

amphoteric colloid. See ampholytoid.

amphoteric electrolyte. See ampholyte.

amphoteric hydroxide. A metal hydroxide, such as aluminum or zinc hydroxide, that dissociates to hydrogen (H^+) ions or hydroxyl (OH^-) ions. It is insoluble in water, but soluble in acids or alkalies.

amphoteric metal. A metal, such as lead, tin or zinc, that is soluble in acids and alkalies.

Amplate. Trade name of Fidelity Chemical Products Corporation (USA) for amorphous/nanocrystalline nickel-base alloy coatings electrodeposited from mildly alkaline solutions. The alloy electrodeposits consist of 60% nickel, 39% tungsten and 1% boron, and have an as-plated hardness of 650 Vickers which can be increased to 1000 Vickers by heat treating at 300°C (570°F) for a prescribed period of time. Used to provide corrosion resistance to ferrous and nonferrous substrates. The electrodeposition process is also known as "Amplate."

Amplify. Trade name of Dow Plastics (USA) for engineering plastics that are alloys of thermoplastic polyurethane and acrylonitrile-butadiene-styrene resins. They have good mechanical properties, chemical resistance and high flow-rate processibility, and are used for automotive bumpers, power-tool housings, and electronic equipment.

Amplum. Trade name for rayon fibers and yarns used for textile fabrics.

AmPorOx. Trade name of Astro Met Inc. (USA) for a series of refractory, open-cell, porous ceramic foam materials composed of 90% alumina (Al_2O_3) toughened with about 10% zirconia (ZrO_2). Available in various shapes, they have superior strength, high crushing strength, low nominal density, a maximum temperature capability of 1732°C (3150°F), good thermal shock resistance, and high thermal transfer. Used for in-mold tundish, pouring-cup or baffle-type filters, as filtering weirs, in the filtration of molten metals, and for kiln and sintering furniture, such as cones, wedges, disks, plates, tubes, boxes, cups, and saggers.

AmPorMat. Trade name of Astro Met Inc. (USA) for sintered, porous ceramic and metal structures available in high-purity iron, titanium, tungsten, noble metals, and all commercially sinterable materials. Used for furnace hearth, flame arrestors, and molten metal filters.

Amrel. Trademark of Rynel Limited, Inc. (USA) for partially processed hydrophilic polyurethanes supplied as cast foam sheets, and die-cut and laminated foam. Used in the manufacture of household and industrial goods.

Amsco. (1) Trade name of Abex Corporation (USA) for an extensive series of plain-carbon, manganese and stainless steels used for hardfacing and welding rods and electrodes.

(2) Trade name of Angel Manufacturing & Supply Company, Limited (Canada) for cast iron and nonferrous castings.

Amsil. Trademark of Climax Performance Materials Corporation (USA) for a commercially-pure copper containing 0.05% silver. It has high electrical conductivity, high creep resistance, and very high resistance to embrittlement, and is used for commutators, and rotor and stator windings.

Amsulf. Trademark of Climax Performance Materials Corporation (USA) for a free-cutting copper containing 0.3% sulfur. It has high electrical conductivity, and is used for clips, connectors, clamps, and screw-machine products.

Amtel. Trademark of Climax Performance Materials Corporation (USA) for a free-machining copper containing 0.5% tellurium. It has very high resistance to hydrogen embrittlement, and is used for screw-machine products, welding nozzles and tips, hardware, and fasteners.

Am-Tuf. Trade name of Coyne Tempered Glass Products (USA) for toughened glass.

Amviloy. Trade name of CMW Inc. (USA) for a series of sintered alloys of tungsten, nickel and copper, used for electrical parts.

amyloid. A compound obtained from normal cellulose by treatment with sulfuric acid.

amylopectin. One of the two polysaccharides in starch (the other being *amylose*) that is a water-insoluble branched chain polymer of glucose with both α-1,4 and α-1,6 glucosidic linkages.

amylose. One of the two polysaccharides in starch (the other being *amylopectin*) that is water-soluble and composed of a linear chain of glucose molecules connected by α-1,4-glucosidic linkages.

Amzirc. Trademark of Climax Performance Materials Corporation (USA) for a zirconium copper consisting of 99.8% pure oxygen-free copper, and 0.15% zirconium. It has good high temperature strength, good corrosion resistance, high conductivity, good softening resistance, and a softening temperature of 580°C (1076°F). Used for electrical components, such as rectifier bases and rotor wedges, and for resistance-welding wheels and tips.

AmZirOx 86. Trademark of Astro-Met Inc. (USA) for yttria-stabilized zirconia ceramics with excellent wear and corrosion resistance, and high strength and toughness. Used as reinforcements for metals and ceramics, and in high-temperature applications. See also yttria-stabilized zirconia.

Anaconda. Trade name of Anaconda Inc. (USA) for a series of brasses and bronzes, high-copper alloys, and beryllium coppers.

Anacos. Trade name for commercially-pure wrought copper that contains small additions of cadmium and silver, and traces of phosphorus. Used for hard-drawn and annealed wire.

anaerobic adhesive. An adhesive that does not cure in air or oxygen, but in absence of air when subjected to metal ions (copper, iron, etc.). See also aerobic adhesive.

analcime. A colorless, white, or slightly colored mineral with vitreous luster composed of sodium aluminum silicate hydrate, $NaAlSi_2O_6 \cdot H_2O$. Crystal system, cubic. Density, 2.22-2.29 g/cm³; hardness, 5-5.5 Mohs. Occurrence: Canada (Nova Scotia), Europe (Italy), USA (California, Colorado, New Jersey, Michigan, Wisconsin). Used as a natural zeolite, and as an ion exchanger in water softening. Also known as *analcite*.

analcite. See analcime.

Anancio. Trade name of Teijin Fibers Limited (Japan) for polyester filaments used for the manufacture of textile fabrics.

anandite. A black mineral of the mica group composed of barium iron silicate hydroxide, $BaFe_3(Si,Fe)_4(O,OH)_{10}(OH,S,Cl)$. Crystal system, monoclinic. Density, 3.94 g/cm³; refractive index, 1.855. Occurrence: Sri Lanka.

Ana-Norm. Trade name for a light-cure hybrid composite used in restorative dentistry.

anapaite. A pale green mineral composed of calcium iron phosphate tetrahydrate, $Ca_2Fe(PO_4)_2 \cdot 4H_2O$. Crystal system, triclinic. Density, 2.80 g/cm³; refractive index, 1.614; hardness, 3-4 Mohs. Occurrence: Italy, Ukraine.

anaphe silk. Wild silk that is produced by the larvae of the Anaphe moth, and is quite similar to tussah silk. See also tussah; wild silk.

anatase. A colorless, white to brown, dark-blue or black mineral that is one of the three polymorphic forms of titanium dioxide (TiO_2), with the other two being *brookite* and *rutile*. Crystal system, tetragonal. Density, 3.8-4.2 g/cm³; melting point, above 1560°C (2840°F); hardness, 5.5-6 Mohs; refractive index, 2.5. Occurrence: Brazil, Canada, France, Switzerland, USA (North Carolina, Rhode Island). Used as an opacifying agent in porcelain enamels, glazes and glass, in the manufacture of white pigments for ceramics, paints, plastics, etc., and as a secondary titanium ore. Also known as *octahedrite*.

anatomical alloy. A fusible alloy for anatomical impressions and casts that has a nominal composition of 54% bismuth, 19% tin, 17% lead, 10% mercury, and traces of cadmium. Used in the manufacture of bone replicas.

anauxite. A clay-type mixture of the minerals *kaolinite* and *quartz* with the formula $Al_2(SiO_7)(OH)_4$. Also known as *ionite*.

Anavor. Trade name for polyester fibers and yarns used for textile fabrics.

Ancarez. Trademark of Air Products & Chemicals, Inc. (USA) for waterborne epoxy resins with low volatile organic compound (VOC) content.

Anchor. (1) Trademark of Anchor Wall Systems (USA) for mortarless retaining wall stones used for commercial applications.

(2) Trade name of Anchor Packing Company for an extensive series of asbestos and non-asbestos packing and gasket materials supplied in sheet, tape, rope, wick and other product forms.

Anchor Chaser Die. Trade name of Teledyne Vasco (USA) for a series of high-speed, alloy and carbon steel grades that are true to shape and size, and uniform in analysis and surface smoothness. They have uniform microstructures, good nondeforming properties, and excellent resistance to decarburization. Used for tools, dies, chasers, etc.

anchorite. A zinc-iron phosphate coating for ferrous materials.

Anchor-Tite. Trade name of Tungsten Widia Tool Corporation (USA) for cemented tungsten carbide used for tools and dies.

Anchorweld. Trademark of Roberts Company Canada Limited for a series of adhesives.

Ancoloy. Trade name of Hoeganaes Corporation (USA) for sponge iron powder containing about 1.75% nickel, 1.55% copper, 0.6% molybdenum, 0.02% carbon, and the balance iron. Used for powder-metallurgy parts.

Ancor. Trade name of Hoeganaes Corporation (USA) for a series of austenitic, ferritic and martensitic stainless steel powders, soft-magnetic nickel-iron powders, and sponge iron powders. Used for powder-metallurgy as well as electronic, electric and magnetic applications.

Ancorloy. Trade name of Hoeganaes Corporation (USA) for a series of high-performance powder-metallurgy steels supplied as binder-treated powder premixes in several grades: (i) *Ancorloy 2* and *Ancorloy 4* based on 0.5 wt% a molybdenum steel powder (*Ancorsteel 50 HP*); (ii) *Ancorloy DH-1* and *Ancorloy HP-1* based on highly alloyed steel powders; (iii) *Ancorloy MDA* based on straight carbon steel powder containing 0.7 wt% silicon; and (iv) *Ancorloy MDB* and *Ancorloy MDC* based on nickel-molybdenum alloy steel powders containing 0.7 wt% silicon.

Ancormet. Trade name of Hoeganaes Corporation (USA) for sponge iron powder containing about 0.18% carbon, and the balance iron. It provides fast carbon pickup during sintering and small dimensional change, and is used for medium-density, high-strength powder-metallurgy parts, e.g., electric motors.

Ancorspray. Trade name of Hoeganaes Corporation (USA) for a series of atomized nickel-base powders used for thermal-spray applications.

Ancorsteel. Trade name of Hoeganaes Corporation (USA) for a series of ferrous powder-metallurgy materials based on low-alloy steel powders. Used for parts that require strong, tough cores, and hard, wear-resistant surfaces especially machine elements, such as gears, cams, etc.

Ancrolyt. Trade name of Rasselstein AG (Germany) for chrome-plated steel sheet and strip.

ancylite. A pale yellow to yellowish brown mineral composed of calcium strontium cerium lanthanum carbonate hydroxide monohydrate, $(Sr,Ca)(La,Ce)(CO_3)_2(OH)\cdot H_2O$. Crystal system, orthorhombic. Density, 3.95 g/cm^3; refractive index, 1.700. Occurrence: Greenland.

andalusite. A mineral composed of aluminum silicate, Al_2SiO_5, that on firing at about 1350°C (2460°F) dissociates to yield mainly *mullite*. It is usually pink, reddish brown, rose red, or whitish, but sometimes grayish, yellowish, violet, or greenish. Crystal system, orthorhombic. Density, 3.0-3.5 g/cm^3; hardness, 7-7.5 Mohs; refractive index, 1.64. Occurrence: Australia, Brazil, Europe, South Africa, Sri Lanka, USA (California, Connecticut, Maine, Massachusetts, New Hampshire, Nevada, New Mexico, Pennsylvania). Used as a component in *sillimanite* refractories, for spark plug insulators and laboratory ware, in superrefractories, and as a gemstone.

Andaman marblewood. See marblewood.

Andard. Trade name of Duke Steel Company Inc. (USA) for a water-hardening tool steel containing 1.1% carbon, and the balance iron.

andersonite. A yellow-green mineral composed of sodium calcium uranyl carbonate hexahydrate, $Na_2Ca(UO_2)(CO_3)_3\cdot 6H_2O$. Crystal system, rhombohedral (hexagonal). Density, 2.80 g/cm^3; refractive index, 1.530. Occurrence: USA (Utah).

andesine. A soda-lime feldspar composed of 50-70% *albite* (NaAlSi$_3$O$_2$) and 30-50% *anorthite* (CaAl$_2$Si$_2$O$_8$). Crystal system, triclinic. Density, 2.6-2.7 g/cm^3; hardness, 6 Mohs.

andesite. A crystalline volcanic rock with fine-grained texture composed chiefly of soda-lime feldspar (*oligoclase* or *andesine*), and one or several ferromagnesian minerals, such as *pyroxene*, *hornblende*, and/or *biotite*. Used as concrete aggregate.

andiroba. The reddish-brown wood of the carapa tree *Carapa guianensis* with a coarse grain, good durability, good resistance to decay and insects. It works well and has good gluing and painting qualities. Average weight, 641 kg/m³ (40 lb/ft³). Source: Tropical America (Brazil and Guianas). Used for flooring, furniture, cabinetmaking, decorative veneer and plywood, and as a substitute for mahogany. Also known as *carapa; cedro macho; crabwood; tangare.*

Andisil. Trademark of Anderson & Associates Limited (USA) for stable, low-viscosity dimethyl silanol fluids. They are supplied as clear, water-white liquids with a flash point (closed cup)

above 101°C (214°F) for the treatment of fillers and anti-structuring additives for high-consistency silicone rubber and room-temperature-vulcanizing (RTV) silicones.

Andoran. Trademark of Bayer AG (Germany) for polycarbonate resins including glass-reinforced grades. Formerly used for denture bases.

andorite. A light lead-gray to black mineral of the lillianite group composed of silver lead antimony sulfide, $AgPbSb_3S_6$. Crystal system, orthorhombic. Density, 5.31 g/cm^3. Occurrence: Czech Republic, Rumania, Bolivia, Japan.

andradite. A yellowish, greenish, or brownish-red mineral of the garnet group composed of calcium iron silicate, $Ca_3Fe_2(SiO_4)_3$. It can also be made synthetically. Crystal system, cubic. Density, 3.75-3.85 g/cm^3; refractive index, 1.88.

Andralyt. Trade name of Rasselstein AG (Germany) for electrolytic tinplate supplied in sheet, coil and strip form.

andremeyerite. A pale emerald-green mineral composed of barium iron silicate, $BaFe_2Si_2O_7$. Crystal system, monoclinic. Density, 4.15 g/cm^3; refractive index, 1.740. Occurrence: Zaire.

andrewsite. A bluish-green mineral composed of copper iron phosphate hydroxide, $(Cu,Fe)Fe_3(PO_4)_3(OH)_2$.

Andrez. Trade name of Anderson Development Company (USA) for styrene-butadiene rubber resins.

anduoite. A mineral of the marcasite group composed of ruthenium arsenide, $RuAs_2$. Crystal system, orthorhombic. Density, 8.25 g/cm^3. Occurrence: China.

Andur. Trademark of Anderson Development Company (USA) for polyether, polyester and polyurethane prepolymers.

Anemone. Trade name of Chance Brothers Limited (UK) for decorative table glassware with anemone pattern.

Anfriloy. Trade name of Wellman Dynamics Corporation (USA) for an alloy composed of varying amounts of copper, tin, zinc and lead. Used for heavy-duty bearings.

Angelbrick. Trade name of Arriscraft Corporation (Canada) for building brick used for veneer walls.

angelellite. A blackish brown mineral composed of iron arsenate, $Fe_4As_2O_{11}$. Crystal system, triclinic. Density, 4.87 g/cm^3; refractive index, 2.2. Occurrence: Argentina.

angelique. The strong, durable wood of the tree *Dicorynia guianensis*. It is olive-brown with reddish spots when freshly cut, but becomes either dull brown (gris angelique), or reddish-brown (rouge angelique). It has high hardness and excellent resistance to decay and marine borers. In the dried condition, it can be worked only with carbide-tipped tools. Average weight, 801 kg/m³ (50 lb/ft³). Source: Guianas, Surinam, and Lower Amazon. Used for heavy marine construction and boats, and as a substitute for greenwood. Also known as *basra locus*.

Angelstone. Trade name of Arriscraft Corporation (Canada) for a building stone with one naturally textured surface, used for veneer walls.

angico. The very hard, reddish-brown wood of the curupay tree *Angico rigada*. The light-brown variety of this wood is known as *angico vermelho*, or *yellow angico*. Angico has a close grain, and an average weight of 1121 kg/m³ (70 lb/ft³). Source: Brazil. Used for furniture and cabinetwork. Also known as *queenwood*.

Angioseal. Trademark of AHP (USA) for a biodegradable polymer based on poly(DL-lactide-*co*-glycolide) used for angioplastic plugs.

angle iron. A rolled structural shape of steel with equal, or unequal legs used for joining two or more members at an angle.

angle-ply laminate. A laminate constructed of two or more plies or layers with the fiber direction of adjacent layers being oriented at alternating angles.

anglesite. A colorless, white, gray, or pale yellow mineral of the barite group composed of lead sulfate, $PbSO_4$, and containing as much as 68% lead. It can also be made synthetically. Crystal system, orthorhombic. Density, 6.10-6.40 g/cm^3; hardness, 2.5-3 Mohs. Occurrence: Australia (New South Wales), Italy, Southwest Africa, UK, USA (Idaho, Pennsylvania, Utah). Used as a valuable lead ore, and as a source of lead oxide in ceramics. Also known as *lead spar; lead vitriol.*

Anglobond. British trade name for a silver-free palladium dental bonding alloy.

Angola. (1) A yarn spun from a blend of wool and cotton or other natural or synthetic fibers.

(2) A fabric made with Angola yarn (1) weft (filling) threads and cotton warp threads, usually in a plain or twill weave.

Angora. (1) Trademark of Anchor Packing Division, Robco Inc. (Canada) for cotton and wool-braided packing.

(2) Trade name of Vertex NP (Czech Republic) for glass yarn, cloth, or tape made from staple fibers.

angora. Coarse to fine natural protein fibers obtained from the long, soft, usually white hair of the Angora rabbit. Used for the manufacture of yarns, especially for knitting. Also known as *angora rabbitt hair.*

angora goat hair. See mohair.

Angsburg. A brass containing 72% copper and 28% zinc used for tubing, fittings, fixtures, etc.

angular powder. A powder consisting of sharp-edged particles.

angular sand. A sand consisting of sharp-edged grains.

anhydride. (1) A chemical compound formed from another compound (e.g., an acid) by removal of one or more molecules of water. For example, carbon dioxide (CO_2) is the anhydride of carbonic acid (H_2CO_3).

(2) An oxide of a nonmetal (acid anhydride) or metal (basic anhydride) that reacts with water to form an acid or a base, respectively.

anhydrite. A colorless or grayish white mineral composed of calcium sulfate, $CaSO_4$. Essentially, it is anhydrous gypsum, and can also be made synthetically. Crystal system, orthorhombic. Density, 2.93-2.96 g/cm^3; melting point, 1450°C (2640°F); hardness, 3-3.5 Mohs; refractive index, 1.575. Used as a drying agent, as a gypsum substitute for cement, as a source of sulfuric acid, and in plaster. Also known as *cube spar.* See also artificial anhydrite.

anhydrous borax. Sodium borate from which all water has been removed. It is available as a white, hygroscopic, free-flowing powder (99+% pure). Density, 2.367 g/cm^3, melting point, 741°C (1366°F); boiling point, decomposes at 1575°C (2867°F); refractive index, 1.501. Used in ceramics as a flux, as a glass former in glass, glazes and porcelain enamels, as a metal flux, e.g., in soldering, as a chemical reagent, and in biology, biochemistry and medicine. Formula: $Na_2B_4O_7$. Also known as *anhydrous sodium borate; anhydrous sodium tetraborate; sodium tetraborate; burnt borax; calcined borax; dehydrated borax; fused borax.* See also borax glass.

anhydrous calcium sulfate. See calcium sulfate.

anhydrous compound. A compound resulting when the water of crystallization is removed from a hydrate, e.g., anhydrite ($CaSO_4$) is the anhydrous compound resulting when the water of crystallization is removed from gypsum ($CaSO_4 \cdot 2H_2O$).

anhydrous gypsum plaster. See calcium sulfate.

anhydrous magnesium silicate. See magnesium silicate (ii).

anhydrous plumbic acid. See lead dioxide.

anhydrous lime. See calcium oxide.

anhydrous phosphoric acid. See phosporus pentoxide.

anhydrous sodium borate. See anhydrous borax.

anhydrous sodium phosphate. See monosodium phosphate.

anhydrous sodium tetraborate. See anhydrous borax.

anhydrous strontium bromide. See strontium bromide.

anhydrous strontium chloride. See strontium chloride.

A Nickel. Trade name of Inco Alloys International Inc. (USA) for a commercially-pure nickel (99+%) containing 0.18-0.35% manganese, 0.1-0.2% iron, 0.05-0.15% silicon, 0.05-0.15% copper, 0.01-0.05% titanium, 0.01-0.08% magnesium, 0.06-0.15% carbon, 0.005-0.008% sulfur, and traces of cobalt. Commercially available in the form of wire, bar, rod, sheet and strip, it has an annealed electrical conductivity of 20% IACS, and is used for grids and anodes of electronic valves, and for other electrical and electronic applications, e.g., lead wires, base pins, cathode shields, etc.

Anid. Trade name for a series of nylon fibers.

anidex fiber. A man-made fiber in which the fiber-forming substance is any long-chain polymer consisting of 50 wt% or more of at least one ester of a monohydric alcohol and acrylic acid. Also known as *anidex.*

Anil. Trade name for a series of nylon fibers.

Anilana. Trademark of Chemitex-Anilana (Poland) for acrylic fibers used for the manufacture of textile fabrics.

aniline. A simple aromatic amine derived from chlorobenzene or nitrobenzene. It is a pale brown liquid at room temperature that darkens with age, slightly soluble in water, and has a melting point of -6°C (21°F), and a boiling point of 184°C (363°F). Used in the manufacture of aniline formaldehyde and other synthetic resins, in the production of certain rubber accelerators and antioxidants, in varnishes and leather finishes, isocyanates for urethane foams, dyes, explosives, pharmaceuticals, etc. Formula: $C_6H_5NH_2$. Also known as *aminobenzene; phenylamine.*

aniline dyes. A large group of dyes obtained from benzene and aniline, and used in the manufacture of leather, textile fabrics, pharmaceuticals and various other products.

aniline-formaldehyde resins. Thermoplastic aminoplastic resins made by the condensation polymerization of aniline and formaldehyde. Used for molded and laminated insulating products with high dielectric strength and good chemical resistance.

aniline leather. A leather made from calfskins, kidskins, or heavier hides finished with aniline dye to provide a texture-revealing transparent effect, and used chiefly for shoe uppers.

Anilite. Trade name for a phenolic impregnate supplied in the form of thin sheets. It has high dielectric strength, and is used as an insulating material.

anilite. A dark-colored mineral composed of copper sulfide, Cu_7S_4. It can also be synthesized from high-purity copper and sulfur. Crystal system, orthorhombic. Density, 5.56-5.59 g/cm^3. Occurrence: Japan.

Anim/8. Trade name for anidex fibers.

animal black. A finely divided, high-grade black substance made from ivory (*ivory black*) or calcined animal bones (*bone black*). It has a carbon content of about 10%, and a density of 2.7 g/cm^3 (0.10 $lb/in.^3$). Used in the manufacture of activated carbon, as a cementation reagent, as a absorptive medium in gas masks, as a filtering medium, in water purification, as a paint and varnish pigment, in the clarification of shellac, and as a decolorizing and deodorizing agents. Also known as *bone black; bone* *char; bone charcoal.*

animal charcoal. Charcoal made by the destructive high-temperature distillation of bones and other animal matter. Used in the adsorption of organic coloring matter, and in filtration.

animal fibers. A group of natural protein fibers of animal origin including various silks, wools, furs and hairs. Used especially for the manufacture of textile fabrics.

animal glue. A semitransparent hot-melt glue obtained from animal bones and hides. See also bone glue; hide glue.

animi gum. A hard, yellowish-red, brittle, usually fossil-type resin belonging to the East African copals. It is found chiefly on the island of Zanzibar and the adjacent African mainland. Density, 1.06-1.07 g/cm^3; melting point, 245°C (473°F). Used in varnishes and lacquers. Also known as *Zanzibar copal; Zanzibar gum.*

anion exchanger. An *ion exchanger*, usually an organic compound, whose positively charged functional groups are attached to a support by covalent bonds. It it has the capacity to enter into a reversible process involving the exchange of negatively charged atoms. Also known as *anion-exchange material.* See also cation exchanger.

anionic polymer. A polymer produced by a chain-reaction polymerization initiated by a base, such as lithium amide, or an organometallic compound, such as *n*-butyllithium. See also ionic polymer; cationic polymer.

anisotropic laminate. A laminate whose strength properties are different in different directions along the laminate plane. See also isotropic laminate; laminate.

anisotropic material. A material whose physical properties differ in various directions. Also known as *anisotropic substance.* See also isotropic material.

anisotropic metal. Any of a group of metals including beryllium, magnesium, tin, titanium, uranium, zinc and zirconium that have noncubic crystal structures, and different optical characteristics in different crystallographic directions. See also isotropic metal.

Aniversario Angosto. Trade name of Idrieria Argentina SA (Argentina) for patterned glass with narrow horizontal ribs.

Anka. Trade name of Dunford Hadfields Limited (UK) for an austenitic stainless steel containing 0.08-0.2% carbon, 0.6% silicon, 0.6% manganese, 12-20% chromium, 6-12% nickel, and the balance iron. Used for chemical plant equipment, tanks, vessels, and fittings.

ankerite. A gray, white, or orange-brown mineral of the calcite group composed of calcium iron magnesium carbonate, Ca(Fe,-Mg,Mn)(CO$_3$)$_2$. Crystal system, hexagonal. Density, 2.95-3.04 g/cm^3; refractive index, 1.726. Occurrence: USA (Pennsylvania).

Anko. Russian trade name for a series of permanent magnet alloys containing 8-10% aluminum, 14-20% nickel, 3-4% copper, 15-24% cobalt, and the balance iron. Used for electrical and magnetic equipment.

Ankorite. (1) Trademark of Harbison-Walker Company (USA) for hot-setting refractory cements and mortars containing high amounts of alumina (Al$_2$O$_3$). Used for laying fireclay brick.

(2) Trade name of Anchor Packing Company (Canada) for asbestos and non-asbestos packing, gaskets and sheets.

Ankorlok. Trade name of Anchor Packing Company (Canada) for asbestos and non-asbestos packing materials.

annabergite. An apple-green to light yellow-green mineral composed of nickel arsenate octahydrate, Ni$_3$(AsO$_4$)$_2$·8H$_2$O. It can also be made synthetically. Crystal system, monoclinic. Den-

sity, 3.07-3.22 g/cm^3; refractive index, 1.658. Occurrence: Europe.

AnnaCarbid. Trade name of St. Gobain Industrie Keramik Roedental GmbH/Anna-Werk (Germany) for a series of silicon carbide-base products of varying purity with a density of 2.5 g/cm^3 (0.09 lb/in.^{3}), a porosity ranging from about 15-22%, a service temperature range of 1400-1500°C (2550-2730°F), and excellent thermal-shock resistance. Used for saggers for the porcelain industry, electrical and sanitaryware ceramics, kiln furniture, and abrasion- and oxidation-resistant furnace linings.

Annahütte. Trade name Max Aicher GmbH & Co. KG Stahlwerk Annahütte Hammerau (Germany) for a series of specialty steel products including prestressing and reinforcing steels, special reinforcing steels, welded wire fabrics, wire rod, and anchor steel.

Annalin. Trade name of Anna-Werk (Germany) for a finely divided gypsum.

AnnaMullit. Trade name of St. Gobain Industrie Keramik Roedental GmbH/Anna-Werk (Germany) for a series of mullite-base ceramics with varying purity having a density of 2.5-2.6 g/cm^3, a porosity of about 20%, a service temperature range of 1550-1750°C (2820-3180°F), good mechanical properties, high refractoriness under load, and good to excellent thermal-shock resistance. Used for industrial furnaces, kilns, firing chambers, and kiln furniture.

AnnaSicon. Trade name of St. Gobain Industrie Keramik Roedental GmbH/Anna-Werk (Germany) for a series of silicon nitride-bonded silicon carbide products of varying purity with a density of 2.6-2.8 g/cm^3 (0.01-0.10 lb/in.^{3}), a porosity ranging from about 10-18%, a service temperature of about 1650°C (3000°F), good refractoriness under load, and excellent thermal-shock resistance. Used for metallurgical furnace linings, nonferrous foundry parts, kiln furniture, burner nozzles, and abrasion-resistant linings.

annealed glass. Glass which after being formed has been reheated in an oven to relieve internal stresses and then slowly cooled to room temperature to avoid introducing new stresses.

annealed powder. A soft, compactible, heat-treated metal powder.

annealed steel. A steel that has been heated to a specific temperature lower than its melting point, held at this temperature for a specified period of time, and then gradually cooled, usually at a slow rate. This heat treatment removes stresses and induces softness, and thus makes the steel more tenacious and less brittle.

annealed wire. A wire that has been softened by annealing.

annealing carbon. See temper carbon.

annite. A black, green, or brown mineral of the mica group composed of potassium aluminum iron silicate hydroxide, KFe$_3$-AlSi$_3$O$_{10}$(OH)$_2$. Crystal system, monoclinic. Density, 3-3.2 g/cm^3; refractive index, 1.690; hardness, 3 Mohs.

Anobond. Trade name of Lorin Industries (USA) for a transparent anodized aluminum bonding film with open-pore structure. It can be applied to coil 1-2.5 mm (0.04-0.1 in.) in thickness. Used to enhance the adhesiveness of coatings, laminates, veneers, finishes or adhesives to aluminum sheet products.

Anodal. Trade name of Clariant Corporation (USA) for a low-temperature anodizing sealant.

anode copper. A shaped slab of copper obtained in blister copper refining, and used for anodes for the manufacture of electrolytic copper.

anode metals. Cast or rolled, shaped or unshaped metals or metal alloys of commercial purity used as the positive terminals in electroplating.

anodic coatings. Heavy, stable oxide films produced on the surface of aluminum, copper, nickel, cadmium, iron, silver, zinc, beryllium, magnesium, titanium and their alloys by electrolytic oxidation carried out in a bath using sulfuric, oxalic or chromic acid electrolytes, and with the particular metal being the anode. The artificially produced films are mineral in nature, have high chemical resistance and hardness, good electrical insulating properties, and are much thicker than the oxide films that naturally protect these metals. Used chiefly for protective and decorative applications, and as wear resistance and paint adhesion improvement. Also known as *anodized coatings*.

anodic material. A material that acts as the anode in an electrolytic cell.

anodic metal. A metal that, in reference to other metals, has the tendency to corrode, dissolve or oxidize preferentially.

Anodisal. Trade name of Metalloxyd GmbH (Germany) for anodized aluminum sheeting.

anodized aluminum. A natural- or surface-colored aluminum or aluminum alloy on whose surface a hard protective coating of aluminum oxide (Al_2O_3) has been produced by anodic oxidation, i.e., by converting the aluminum surface to Al_2O_3 by making the aluminum or alloy the anode in an electrolytic cell, and using an oxalic, sulfuric or chromic acid electrolyte. These coatings are usually from 3 to 60 μm (0.1 to 2.4 mils) thick, as compared to the natural oxide film with less than 0.1 μm (0.004 mil). It has excellent chemical resistance, high hardness, good electrical insulating properties, good abrasion and wear resistance, and good corrosion and oxidation resistance.

anodized coatings. See anodic coatings.

anodized magnesium. Magnesium with an anodic surface coating of magnesium oxide (typically 3 μm or 0.1 mil thick) produced by immersion in one of various electrolytes, mainly composed of fluorides, phosphates or chromates, and application of an electric current. Anodized magnesium possesses enhanced corrosion resistance and improved paint adhesion.

Ano-Fol. Trade name for anodized aluminum strip.

Anolok. Trademark of Alcan Aluminum Limited (Canada) for anodized aluminum and aluminum alloys.

Anorefract. Trade name of Acieries et Forges d'Anor (France) for a series of heat-resisting steels containing 0.25-0.35% carbon, 24-26% chromium, 12-14% nickel, 0-1.5% niobium (columbium), and the balance iron.

Anoresist. Trade name of Acieries et Forges d'Anor (France) for a series of heat-resisting steels containing 0.4-0.6% carbon, 17-19% chromium, 37-39% nickel, and the balance iron.

Anorexact. Trade name of Acieries et Forges d'Anor (France) for a series of tool steels including several low- and high-alloy, shock-resisting, and cold- and hot-work types.

Anorinox. Trade name of Acieries et Forges d'Anor (France) for a series of austenitic, martensitic and precipitation hardening stainless steels.

Anormat. Trade name of Malory Metallurgical Products Limited (UK) for a hardened dental alloy containing a total of 78.3% gold and platinum. It has a melting range of 915-917°C (1680-1685°F).

anorthite. A mineral of the plagioclase subgroup of the feldspar group. *Low anorthite* is colorless, white, grayish, reddish or black, and composed of calcium aluminum silicate, $CaAl_2Si_2O_8$, while *high anorthite* is usually colorless, white, gray or pale golden yellow, and composed of sodium calcium aluminum silicate, $(Ca,Na)(Si,Al)_4O_8$. Mineralogically, *anorthite* is considered to be composed of 90-100% low anorthite and 0-10% *albite*. Crystal system, triclinic. Density, 2.68-2.76 g/cm³; hardness, 6.0-6.5 Mohs; refractive index, 1.56-1.57. Occurrence: Italy, Norway, USA (Minnesota, Oregon). Used in glass, glazes, concrete, porcelain enamel, abrasives, insulating compounds, etc. Also known as *calciclase; calcium feldspar; lime feldspar.*

anorthoclase. A soda-potash mineral composed of 60-90% *albite* ($NaAlSi_3O_8$), 10-40% *orthoclase* ($KAlSi_3O_8$), and a small quantity (typically up to 20 mol%) *anorthite* ($CaAl_2Si_2O_8$). Crystal system, triclinic. Density, 2.56-2.65 g/cm³; hardness, 6-6.5 Mohs. Also known as *anorthose; soda microcline.*

anorthose. See anorthoclase.

Anoseal. Trade name of Novamex Technologies Inc. (USA) for a low-temperature light-resistant sealer for anodized aluminum.

Anoxin. Trade name of Röchling Burbach GmbH (Germany) for an extensive series of austenitic stainless steels containing about 0.03-0.15% carbon, 16-20% chromium, 8-14% nickel, 0.3% molybdenum, and the balance iron. Small additions of copper, titanium, and niobium (columbium) may also be present. They have high strength and corrosion resistance, good high-temperature properties, good weldability, and moderate machinability. Used for structural applications, such as aircraft and transport equipment, cutlery, chemical equipment, food-processing equipment, and surgical instruments.

Anscol. Trade name of Acme Steel Company (USA) for a series of structural steels containing 0.15-0.2% carbon, 0.7% manganese, up to 0.04% silicon, 0.015% or more niobium (columbium), and the balance iron. Used for railroad and mine cars, bridges, booms, pressure vessels, and derricks.

Ansala. Trade name of Situkku Oy (Finland) for double glazing units in which a rigid plastic material acts as an adhesive and spacer.

Ansemax. Italian trade name for a cobalt magnet steel containing 0.5-0.8% carbon, 24% cobalt, 14% nickel, 8% aluminum, 3% copper, and the balance iron.

Anso. Trademark of Allied Signal/Honeywell Performance Fibers (USA) for a family of tough, resilient, durable, abrasion-, soil- and stain-resistant nylon 6 fibers including *Anso Caress, Anso CrushRegister, Anso Soft* and *Anso Premium*. Used for textile fabrics including clothing, blankets, carpets and rugs, upholstery, industrial fabrics, synthetic turf, etc.

Anso Caress. Trademark of Honeywell Performance Fibers (USA) for a nylon carpet fiber that produces a naturally tough, yet silk-like soft, comfortable walking surface.

Anso CHOICE!. Trademark of Honeywell Performance Fibers (USA) for a durable, abrasion-resistant nylon carpet fiber with exceptional natural toughness and improved soil and stain resistance.

Anso CrushResister. Trademark of Honeywell Performance Fibers (USA) for a family of durable, abrasion- and crush-resistant nylon fibers. *Anso CrushRegister III* is produced by a proprietary cross-bonding technology (CrossBond) that ensures better and more permanent fiber twist, and is used for carpets with a high level of appearance retention. *Anso CrushResister III ACT* and *Anso CrushResister III TLC* are modified versions of Anso CrushRegister III. The former is produced by a proprietary crossbonding and fusion technology [Advanced Cross-Bond Technology (ACT) and Fusion™] and the latter by a proprietary fiber twist-lock and fusion technology [Twist-Lock Control (TLC) and Fusion™]. Both types are used for carpets with improved texture and loop styles, and higher levels of appearance retention.

Anso f(x). Trademark of Honeywell Performance Fibers (USA) for a nylon carpet fiber with outstanding durability and lightfastness, excellent crush, matting and wear resistance, excellent soil and stain resistance, and a beautiful texture. Supplied in a wide range of colors, it is particularly suited for commercial and institutional applications.

Anso Premium. Trademark of Honeywell Performance Fibers (USA) for a nylon carpet fiber with outstanding toughness and improved durability and abrasion resistance.

Anso Soft. Trademark of Honeywell Performance Fibers (USA) for a soft, durable nylon fiber that provides a soft, warm feel, and excellent washability and colorfastness. Supplied in a wide range of bright, rich colors and unique lusters, it is used for the manufacture of bath rugs.

Anso-tex. Trademark of Allied Signal/Honeywell Performance Fibers (USA) for a tough textured nylon fiber with excellent tear and puncture resistance used for consumer, industrial and military appli-cations, e.g., for the manufacture of automotive safety products, leisure and recreation fabrics, engineered reinforcements, tractor hose covers, and backpacks.

Anso Total Comfort. Trademark of Honeywell Performance Fibers (USA) for nylon fibers with improved durability and abrasion resistance used for high-quality carpets providing exceptionally comfortable walking surfaces.

Antaciron. Trade name of Antaciron, Inc. (USA) for an iron-silicon alloy with excellent acid resistance, used for chemical parts and equipment.

Antarctic beech. The fine, even-textured wood of the tree *Nothofagus procera* which is not a true beech, but resembles the latter in general appearance. It has good decay resistance, and satisfactory nailing qualities. Average weight, 540 kg/m³ (34 lb/ft³). Source: Chile. Used for interior and exterior joinery. Also known as *rauli*.

antarcticite. A colorless mineral composed of calcium chloride hexahydrate, $CaCl_2 \cdot 6H_2O$. It can be made from a solution of calcium chloride by slow evaporation at room temperature. Crystal system, hexagonal. Density, 1.71 g/cm³; refractive index, 1.550.

Antarctic pine. Those species of araucarian pine which grow in southern Chile and Argentina. Its wood is soft, whitish-yellow with rose-colored veins, and used for poles, handles, and interior trim. See also araucarian pine.

Antelio. Trademark of Saint-Gobain (France) for solar control glass.

antelope finish suede. Calfskin, goatskin or lambskin that has been sueded and specially finished to imitate genuine antelope leather.

antelope leather. A fine, soft, scarce leather obtained from the skin of antelopes. It is sueded on the flesh side, and has a velvety texture and luster.

Anthella. Trade name for rayon fibers and yarns used for textile fabrics.

Antherta. Trade name for rayon fibers and yarns used for textile fabrics.

anthoinite. A white mineral composed of aluminum tungstate trihydrate, $Al_2W_2O_9 \cdot 3H_2O$ Crystal system, monoclinic. Density, 4.60-5.20 g/cm³; refractive index, 1.81-1.82. Occurrence: Zaire

anthonyite. A pale reddish-violet, pleochroic mineral composed of copper chloride hydroxide trihydrate, $Cu(OH,Cl)_2 \cdot 3H_2O$. Crystal system, monoclinic. Refractive index, 1.602. Occurrence: USA (Michigan).

anthophyllite. A brownish white, yellowish, or greenish long-fiber asbestos mineral of the amphibole group composed of magnesium iron silicate hydroxide, $(Mg,Fe)_7Si_8O_{22}(OH)_2$. Crystal system, orthorhombic. Density, 3.09-3.38 g/cm³; hardness, 5.5-6.0 Mohs; refractive index, 1.630. Occurrence: South Africa (Transvaal), USA (Alabama, Georgia, Kentucky, New Hampshire, New York, Ohio, Tennessee, Vermont, Virginia, West Virginia). Used for insulation, fireproof clothing, and acid-resistant fabrics. Also known as *bidalotite*.

anthracene. A crystalline tricyclic aromatic hydrocarbon with electroluminescent, semiconducting and photoconductive properties, usually obtained by the distillation of coal tar. It exhibits blue-violet fluorescence, and is soluble in alcohol and ether, but insoluble in water. Density, 1.27 g/cm³; melting point, 217°C (423°F); boiling point, 340°C (644°F); flash point, 250°F (121°C). Used in the manufacture of dyes (alizarin) and anthraquinone, in coating applications and calico printing, as scintillation counting crystals, and as an organic semiconductor and photoconductor in electroluminescent devices. Formula: $C_{14}H_{10}$.

anthracite. A hard, black lustrous coal whose ratio of fixed carbon to volatile matter (e.g., hydrogen, oxygen and nitrogen) is 10:1. It has a density of approximately 1.7 g/cm³ (0.06 lb/in.³), a calorific value of 34 MJ/kg (14600 Btu/lb), is difficult to ignite, burns with a short intense flame, and produces no smoke. Used as a fuel for household and industrial furnaces, and for carbon refractories. Also known as *anthracite coal; hard coal*.

anthracite-base refractory. A carbon refractory manufactured from calcined anthracite coal.

anthracite powder. Finely divided anthracite coal used as a filler in plastics and rubber.

anthraquinoid dyes. A large group of dyes including alizarin and inanthrone which are derivatives of anthraquinone.

anthraquinone. A polynuclear aromatic compound prepared by first reacting phthalic anhydride and benzene in the presence of aluminum trichloride, and then reacting the resulting product (*o*-benzoylbenzoic acid) with sulfuric acid at elevated temperatures. It is available in the form of yellow needles (97+% pure) with a melting point of 286°C (547°F), a boiling point of 380°C (716°F) and a flash point of 365°F (185°C), and used in the manufacture of dyes and printing inks, and in organic synthesis. Formula: $C_6H_4(CO)_2C_6H_4$.

antiacid bronze. An acid-resistant, high-leaded casting bronze used for chemical parts.

anti-adhesive paper. A kraft paper treated with synthetic resin and used as an interleaf for adhesive materials. Also known as *casting paper; release paper*.

Antichoc. Trade name for wear-resistant medium-carbon steels used for machine components, fasteners, and cutting and hand tools.

Anticorodal. Trade name of Swiss Aluminium Limited (Switzerland) for a series of strong, highly corrosion-resistant aluminum alloys including cast aluminum-silicon and aluminum-silicon-magnesium alloys, and wrought aluminum-magnesium-silicon alloys. Used for architectural purposes, and structural, machine and pump parts.

anticorrosive paint. A metal paint that is designed to inhibit corrosion. It usually contains a corrosion-resistant pigment, such as lead or zinc chromate or red lead, and a chemical- and moisture-resistant binder, and can be applied directly to metals.

antiferroelectric liquid crystal. A chiral liquid crystal that exhibits spontaneous polarization in which an antiferroelectric liquid crystal (AFLC) phase appears at a temperature below that of an ordinary ferroelectric liquid crystal. It differs from the latter in the arrangement of the molecules in adjacent layers. In the AFLC phase the director (i.e., the direction of perferred molecular orientation) is tilted in opposite direction in each subsequent molecular layer and lies on a theoretical cone with the spontaneous polarization facing in the other direction. Abbreviation: AFLC. See also chiral liquid crystal; liquid crystal.

antiferroelectric materials. Dielectric materials of high permittivity in which there are spontaneous electrical polarizations with equal numbers of dipoles in opposite directions. Examples of such materials include barium stannate titanate, cadmium niobate, lead zirconate, and lead lanthanum zirconate titanate. Also known as *antiferroelectrics*.

antiferromagnetically coupled media layers. Multi-layered structures that consist of two ferromagnetic layers and a nonmagnetic ruthenium interlayer which is only three atoms thick. The magnetization of the magnetic top and bottom layer is antiparallelly coupled (i.e., coupled in opposite directions). This antiferromagnetically coupled multilayer permits increased areal data densities at decreased magnetic media thicknesses. *Antiferromagnetically coupled media layers* are thermally stable and used for magnetic data-storage media, such as magnetic hard-disk drives and random-access memory devices with significantly increased data densities. Abbreviation: AFC media layers. See also magnetic medium; magnetic recording materials.

antiferromagnetic materials. Weakly magnetic materials, such as chromic chloride, ferrous oxide, ferrous bromide, manganous fluoride, manganous oxide, etc., in which there exists a state similar to that of ferromagnetic materials, but with the adjacent spins antiparallel (parallel, but of opposite direction) instead of parallel. Their susceptibility increases with temperature up to a critical temperature (Neél temperature), above which they exhibit paramagnetism. Also known as *antiferromagnets*. See also canted antiferromagnetic materials; ferromagnetic materials.

antiflux. A compound, such as aluminum oxide, calcium oxide, cerium oxide, magnesium oxide, tin dioxide, titanium dioxide, zinc oxide or zirconium oxide, that acts as a flux at high temperatures, but hinders fusion at lower temperatures.

Anti-Fly. Trade name of Pilkington Brothers Limited (UK) for amber-colored glass that transmits light of a color which is repellent to flies.

antifoamer. A substance, such as dimethylpolysiloxane, organic phosphates, sulfonated oils, silicone fluids and octyl alcohols, that prevents or reduces foaming by greatly decreasing the surface tension. Also known as *antifoaming agent; defoamer; defoaming agent*.

antifouling coating. A special organic coating or paint that usually contains copper, arsenic or mercury, and is used to prevent or inhibit fouling of underwater structures, such as ship or boat hulls and bottoms, by the attachment and growth of marine organisms, such as barnacles, algae or teredos. Also known as *antifouling composition; antifouling marine paint; antifouling paint*.

antifouling composition. See antifouling coating.

antifouling paint. See antifouling coating.

antifriction babbitt. An alloy of 80% lead, 15% antimony, and 5% tin. It has a low coefficient of friction, a melting point of 272°C (522°F), and is used for low-speed bearings.

antifriction materials. Soft, pliable materials, such as babbitt,

rubber, lignum vitae or overlay metals, used on hard, wear-resistant machine elements.

antifriction metals. Metals, such as babbitt or white-metal alloys, that have low coefficients of friction, and are thus used for bearing surfaces. See also babbitts.

antigorite. A brownish-green or greenish-black asbestos mineral of the kaolinite-serpentine group composed of magnesium silicate hydroxide, $Mg_3Si_2O_5(OH)_4$. Crystal system, monoclinic; lamellar structure. Density, 2.52-2.65 g/cm³; refractive index, 1.56. Occurrence: New Zealand, USA (California). Used for ornaments. See also chrysotile; serpentine.

Antikorro. Trade name of Westa-Westdeutsche Edelstahlhandelsgesellschaft (Germany) for an extensive series of austenitic, ferritic and martensitic stainless steels.

Antikorrosiv. German trade name for specialty protective synthetic resin coatings based on chlorinated rubber. They have good weatherability, good resistance to impact, marring and abrasion, and good resistance to alkalies and acids. Used to prevent local cell formation in light-metal to heavy-metal assemblies.

antimonate of lead. See lead antimonate.

antimonial admiralty brass. A brass containing 71% copper, 27.96% zinc, 1% tin and 0.04% antimony. Commercially available in the form of flat products, tube and wire, it has good to excellent corrosion resistance, excellent cold workability, and high elongation. Used for distiller, evaporator and heat-exchanger tubing, condensers, ferrules, and fixtures on marine equipment. Also known as *inhibited admiralty brass*.

antimonial glass. A clear, deep red, vitrified mass obtained from antimony trisulfide by partial roasting followed by melting. Used for coloring glass and porcelain. Also known as *antimony glass; vitreous antimony; vitreous antimony sulfide*.

antimonial lead. A lead alloy containing about 4-30% antimony, and traces of other elements, such as tin, copper and arsenic. The antimony has a hardening and strengthening effect on the lead. *Antimonial lead* has a density of approximately 11 g/cm³ (0.4 lb/in.³), and a melting point of 245-300°C (473-572°F). Used for pipes, collapsible tubes, bullets, cable sheaths, roofing, type metal, chemical plants, storage-battery plates, grids and terminals, tank linings, bearing metals, and small cast objects. Also known as *antimonial lead alloy; hard lead*.

antimonial solder. Tin or tin-lead solders with an addition of approximately 4-6% antimony to increase the tensile and high-temperature strength.

antimonial tin solder. A tin solder containing 5% antimony that has excellent soldering and strength characteristics as well as good creep strength and fatigue resistance. Used in plumbing and refrigeration. Also known as *tin-antimony solder*.

antimonic acid. See antimony pentoxide.

antimonic anhydride. See antimony pentoxide.

antimonides. A group of binary compounds of antimony with a more electropositive metallic element, such as aluminum, bismuth, gallium, indium, lanthanum, thallium or zinc. Many *antimonides* are semiconductive, and used as semiconductors, phosphors, and for other electronic applications.

antimonite. See stibnite.

antimonous chloride. Colorless, or off-white hygroscopic crystals (99% pure). Density, 3.140 g/cm³; melting point, 73.4°C (164°F); boiling point, 283°C (541°F). An electrical grade with 99.999% purity is also available. Used in bronzing iron, in antimony plating, in the manufacture of lakes, as a cotton mordant, as a catalyst in petroleum refining, in organometallic synthesis, in electronics, in fireproofing fabrics, and in antimony salts. Formula: $SbCl_3$. Also known as *antimony chloride; antimony trichloride; caustic antimony.*

antimonous oxide. A colorless or white crystalline substance (99% pure) that is also available in high-purity (99.999%) electrical grades. It occurs in nature as the minerals *valentinite* and *senarmonite.* Density, 5.2-5.67 g/cm³; melting point, 656°C (1213°F); boiling point, sublimes at 1550°C (2820°F). Used as a paint pigment, as a fining agent and decolorizer for glass, as a ceramic opacifier, as a colorant in lead glazes, as a mordant, in phosphors and infrared transparent glass, in staining iron and copper, in flameproofing paper, textiles and plastics, and as a catalyst; Formula: Sb_2O_3. Also known as *antimony oxide; antimony trioxide; antimony white.*

antimonous oxysulfide. A brown powder that is a double salt of antimony oxide and antimony sulfide. Used as a rubber vulcanizer. Formula: $Sb_2S_3 \cdot Sb_2O_3$. Also known as *antimony flowers.*

antimonous sulfide. See antimony sulfide (i).

antimony. A bluish- or silvery-white, brittle, crystalline, metallic element belonging to Group VA (Group 15) of the Periodic Table. It is commercially available in various forms, such as ingots, shot, powder, granules, lumps, pieces, foils, microfoils, and single crystals. Common antimony ores include *stibnite* (antimonite), *kermasite, tetrahedrite, livingstonite* and *jamisonite.* Crystal system, orthorhombic. Crystal structure, rhombohedral. Density, 6.7 g/cm³; melting point, 630.5°C (1167°F); boiling point, 1380°C (2516°F); atomic number, 51; atomic weight, 121.760; trivalent, tetravalent, and pentavalent; semiconductive; low thermal conductivity. Used in lead alloys (as a hardener), tin alloys, solders, type metals, bearings, cable sheaths, lead batteries and infrared detectors, as a semiconductor, and in pyrotechnics. Symbol: Sb.

antimony-124. A radioactive isotope of antimony having a mass number of 124. It has a half-life of 60 days and emits beta and gamma radiation. Used as a tracer in solid-state studies, as a marker of interfaces between products in pipelines, and as a portable neutron source. Abbreviation: ^{124}Sb.

antimony-125. Radioactive antimony with mass number of 125. It has a half-life of 2.4 years and emits beta and gamma rays. Used in nuclear research. Abbreviation: ^{125}Sb.

antimony black. See antimony sulfide (i).

antimony blende. See kermesite.

antimony bronze. A tin-free copper alloy containing 7.5% antimony and 2% nickel.

antimony chloride. See antimonous chloride.

antimony flowers. See antimonous oxysulfide.

antimony fluoride. See antimony trifluoride; antimony pentafluoride.

antimony glance. See stibnite.

antimony glass. See antimonial glass.

antimony oxide. See antimonous oxide.

antimony pentachloride. A reddish-yellow, moisture-sensitive, oily liquid (99+% pure). Density, 2.336 g/cm³; melting point, 2.8°C (37°F); boiling point, 79°C (174°F)/22 mm; refractive index, 1.601. Used in testing for cesium and alkaloids, in organometallic synthesis, as a dyeing intermediate, and as a chlorine barrier in organic chlorination reactions. Formula: $SbCl_5$. Also known as *antimony perchloride.*

antimony pentafluoride. A colorless, moisture-sensitive, viscous liquid (99+% pure). Density, 2.99 g/cm³; melting point, 7°C (44°F); boiling point, 149.5°C (301°F). Used as a catalyst and source of fluorine in fluorinations, and in organometallic synthe-

sis. Formula: SbF_5. Also known as *antimony fluoride.*

antimony pentasulfide. See antimony sulfide (ii).

antimony pentoxide. A white, or yellowish powder (99+% pure). Density, 3.80 g/cm³; melting point, 450°C (842°F), loses oxygen above 300°C (570°F). A high-purity electrical grade (99.999%) is also available. Used for electrical and electronic applications, as a flame retardant for textiles, in organometallic synthesis, and in the manufacture of antimony compounds. Formula: Sb_2O_5. Also known as *antimonic acid; antimonic anhydride; stibic anhydride.*

antimony perchloride. See antimony pentachloride.

antimony persulfide. See antimony sulfide (ii).

antimony red. See antimony sulfide (ii).

antimony selenide. A gray substance with a melting point of 611°C (1132°F), used as a semiconductor. Formula: Sb_2Se_3.

antimony sodiate. A white, granular powder (97+% pure) with a melting point of 1427°C (2601°F), used as an opacifier in enamels for glass and cast iron, as an ingredient of acid-resisting sheet-steel enamels, and as a high-temperature oxidizing agent. Formula: $NaSbO_3$. Other forms include sodium metaantimonate, $2NaSbO_3 \cdot 7H_2O$ and sodium pyroantimonate, $Na_2H_2Sb_2O_7 \cdot H_2O$. Also known as *sodium antimonate.*

antimony sulfide. Any of the following compounds of antimony and sulfur: (i) *Antimony trisulfide.* A black or orange-red crystalline substance, or gray or black lustrous amorphous powder (98+% pure) that occurs in nature as the mineral *stibnite.* The orange-red variety is only produced synthetically. Density, 4.12-4.64 g/cm³; melting point, 546-550°C (1015-1022°F); boiling point, approximately 1150°C (2100°F). Used as a vermilion pigment, in camouflage paints, in the manufacture of ruby and amber glass, as an opacity promoter in opal glass, as an adherence promoter in certain porcelain enamels, in organometallic research, in antimony salts, and in pyrotechnics. Formula: Sb_2S_3. Also known as *antimonous sulfide;* and (ii) *Antimony pentasulfide.* An orange-yellow powder with a density of 4.12 g/cm³. It decomposes at 75°C (167°F), and is used as a red pigment, and as a rubber accelerator. Formula: Sb_2S_5. Also known as *antimony persulfide; antimony red; golden antimony sulfide.*

antimony telluride. A gray substance usually supplied in lump form. Crystal system, hexagonal. Crystal structure, tetradymite. Density, 6.44-6.50 g/cm³; melting point, 622°C (1152°F). Used in electronics, and as a semiconductor. Formula: Sb_2Te_3.

antimony trichloride. See antimonous chloride.

antimony trifluoride. A substance (98+% pure) available in the form of white to gray, hygroscopic crystals, or off-white to beige powder. Density: 4.38 g/cm³; melting point, 292°C (558°F); boiling point, sublimes at 319°C (606°F). Used in organic and organometallic synthesis, in the manufacture of porcelain and pottery, and in dyeing. Formula: SbF_3. Also known as *antimony fluoride.*

antimony trioxide. See antimonous oxide.

antimony trisulfide. See antimony sulfide (i).

antimony vermilion. An inorganic red pigment composed of antimony sulfide.

antimony white. See antimonous oxide.

antimony yellow. See lead antimonate.

Antinit. Trade name of Vereinigte Edelstahlwerke (Austria) for an extensive series of austenitic and martensitic stainless steels.

Antiox. Trade name of Chiers-Chatillon (France) for a series of hardenable martensitic stainless steel containing up to 0.15% carbon, 11.5-24% chromium, and the balance iron. Used for flat springs, cutlery, tableware, surgical instruments, gears, shafts, gages, and needle valves.

antipitting agent. An agent added to an electroplating bath to prevent the formation of large pores and pits in the deposit.

Antique Cathedral Glass. Trademark of Schott DESAG AG (Germany) for a colored antique glass with finely structured surface. Used in the restoration of colored historic windows.

antique paper. A laid or wove paper with a surface finish somewhat rougher than eggshell, produced in wet presses and calender stacks operated at lower pressures. Used principally for books. See also laid paper; wove paper.

antique satin. A textile fabric, usually with a dark, texture-enhancing warp, whose two sides imitate different fabrics, one side resembling *shantung* and the other *satin*. It is a popular fabric for draperies.

antique taffeta. A stiff textile fabric, usually of silk or synthetics, in a plain weave and with a slubbed weft. See also taffeta.

antirad. A substance that can be added to a plastic or rubber during processing to increase its resistance to deterioration by radiation.

antireflection coating. (1) A thin layer of dielectric material applied to a substrate to decrease its reflection, and improve its transmission of light or other electromagnetic radiation.

(2) A coating applied to a glass surface to reduce the amount of light reflected at that surface.

anti-rubbers. See negative Poisson's ratio materials.

antirust paint. A paint that is designed to inhibit rusting of iron and steel surfaces. It usually contains a corrosion-resistant pigment (e.g., lead chromate or red lead), and a chemical- and moisture-resistant binder, and can be applied directly to ferrous surfaces. Also known as *rust-inhibiting paint.*

antisagging agent. A thixotropic agent, such as clay, used in paints.

antisettling agent. A surface-active agent, such as lecithin, used in paints to control settling.

antiskinning agent. A synthetic organic product, usually a liquid antioxidant, that prevents the formation of skin on the surface of a liquid paint or varnish exposed to the atmosphere.

Antislip. Trade name for synthetic fibers that have received a special slip resistance treatment.

antislip metal. A metal, such as aluminum, bronze or iron, with abrasive grains of silica sand or aluminum oxide cast or rolled into it. Used for floor plates, stair treads, car steps, loading ramps, catwalks, etc.

antislip paint. A paint to which sand, aluminum oxide or carbide granules or wood flour has been added to increase its frictional coefficient. It bonds to metals, concrete, wood, brick, tile, etc., and provides long-lasting protection against slipping and skidding on floors, ramps, stairs, catwalks, loading docks, etc.

Antisol-Profilit. Polish trade name for heat-absorbing glass.

antistatic nylon. A term used for nylon fibers that have been chemically treated, or have antistatic agents incorporated to reduce their tendency to accumulate static charges. Used for fabrics with reduced clinging and sparkling properties.

antistatic tile. A floor tile containing a material, such as carbon, that will dissipate or disperse static electricity, and minimize sparking. Used in areas where combustible and explosive atmospheres may be present.

antistripping additive. See adhesion promoter (2).

antistripping agent. See adhesion promoter (2)

Antisun. Trade name of Pilkington Brothers Limited (UK) for softly tinted blue-green glass that combines high light transmission with relief from solar radiation.

anti-tarnish paper. An impregnated or coated paper that protects other materials against corrosion either by giving off vaporized substances, or by direct contact.

Antitherm. Trade name of Vereinigte Edelstahlwerke (Austria) for a series of corrosion- and heat-resistant austenitic, ferritic and martensitic stainless steels.

Antix. Trade name of Sovirel (France) for a glass giving protection against X-rays.

antlerite. A green mineral composed of copper sulfate hydroxide, $Cu_3(SO_4)(OH)_4$. It can also be made synthetically. Crystal system, orthorhombic. Density, 3.88 g/cm³; refractive index, 1.738; hardness, 3.5 Mohs. Occurrence: Chile. Used as an ore of copper.

Antron. Trademark of E.I. DuPont de Nemours & Company (USA) for strong, lightweight nylon 6,6 fibers and yarns available in several grades including *Antron, Antron II, Antron Advantage, Antron Legacy* and *Antron Lumena*. Used for the manufacture of textile fabrics, especially clothing with excellent resistance to abrasion, wrinkling, mildew and moths.

Antwerp blue. A blue pigment similar to *Prussian blue,* but of lower grade and made by mixing iron and zinc ferrocyanides.

Anval. Trade name of Anval (USA) for high-performance gas-atomized powders based on nickel and cobalt alloys, stainless steels, or high-speed and other tool steels. They are suitable for hot-isostatic pressing, plasma-transferred-arc spraying, and metal-injection molding applications.

Anvil brass. A yellow brass containing 62.5% copper and 37.5% zinc, used for tubes and hardware.

Anviloy. Trade name of CMW Inc. (USA) for a series of alloys containing 90% tungsten, 4% nickel, 4% molybdenum, and 2% iron. Used for high-temperature tools, hot-extrusion dies, high-strength applications, and welding rods for welding tungsten.

Anvyl. Trade name for vinyl extrusions.

Anywear. Trade name for acrylic fibers used for the manufacture of clothing.

AOD steel. See argon-oxygen decarburized steel.

Apac. Trademark of Atlas Turner Inc. (Canada) for asbestos-cement insulating and building materials supplied in the form of panels and boards. Used for siding and partitions in buildings.

Apache. Trade name AL Tech Specialty Steel Corporation (USA) for a wear-resistant, air-hardening medium-alloy cold-work tool steel (AISI type A6) containing approximately 0.70% carbon, 2.00% manganese, 0.30% silicon, 1.25-1.35% molybdenum, 1.00% chromium, and the balance iron. It has outstanding size stability in heat treatment, good nondeforming properties, and a great depth of hardening. Used for forming, blanking and trimming dies, forming tools, punches, shear blades, bending tools, stripper plates, master hubs, and gages.

apachite. A blue mineral composed of copper silicate hydrate, $Cu_9Si_{10}O_{29} \cdot 11H_2O$. Crystal system, monoclinic. Density, 2.80 g/cm³; refractive index, 1.650. Occurrence: USA (Arizona).

apamate. The pale to dark brown wood of the tree *Tabebuia rosea.* It works and finishes well, has high resistance to fungi and high durability, and is similar to *American white oak* in bending and compression strength. Source: Southern Mexico, Central America, Venezuela, Colombia, Ecuador. Used for general construction, paneling, furniture, flooring, trim, and boats. Also known as *mayflower; roble.*

apatite. A common phosphate mineral, usually composed of calcium fluorophosphate, $Ca_5F(PO_4)_3$ or calcium chlorophosphate, $Ca_5Cl(PO_4)_3$, and containing up to 20% phosphorus pentoxide (P_2O_5). It is commonly green or brown with vitreous luster, but violet, blue, yellow, and colorless varieties have also been found.

Crystal structure, hexagonal. Density, 3.1-3.2 g/cm^3; hardness, 5 Mohs. Occurrence: Africa, Brazil, Canada, Europe, French Oceania, North Africa, Russia, USA (California, Florida, Idaho, Kentucky, Maine, New York, Pennsylvania, Tennessee). Used as an opacifier in the manufacture of opal glass, as a substitute for bone ash in whiteware bodies, in laser crystals, as a source of phosphorus, in biomaterials, and for ornamental purposes. See also hydroxyapatite.

APEC. Trademark of Bayer Corporation (USA) for high-heat polycarbonate resins available in general-purpose, UV-stabilized, flame-retardant and specialty grades. Used for automotive electrical components, lighting components, electrical and electronic equipment, medical devices, and laboratory equipment.

Apex. Trade name of Apex International Alloys, Inc. (USA) for an extensive series of die-, sand-, and permanent mold-cast aluminum-copper, aluminum-magnesium, aluminum-silicon, aluminum-zinc, magnesium-aluminum and zinc-aluminum alloys.

aphthitalite. A colorless mineral composed of potassium sodium sulfate, $K_3Na(SO_4)_2$. Crystal system, hexagonal. Density, 2.70 g/cm^3; refractive index, 1.494.

Aphtit. British trade name for an alloy of 70-75% copper, 20-21% nickel, 2.4-5.5% zinc and 1.8-4.5% cadmium, used for corrosion-resistant parts.

Apis. Trade name of Saarstahl AG (Germany) for an air-hardening, nondeforming tool steel containing 1.65% carbon, 12% chromium, some cobalt, and the balance iron. Used for punches, cutters, blanking and forming dies, etc.

apitong. The reddish-brown, heavy wood of the tree *Dipterocarpus grandiflorus*. It has high strength and hardness, and machines well. Average weight, 705 kg/m^3 (44 lb/ft^3). Source: Philippines, Malaya, Borneo. Used as structural timber for truck floors, chutes, flumes, pallets, and boardwalks.

apjohnite. A colorless, white, yellow, or pale rose-green mineral of the halotrichite group composed of manganese aluminum sulfate hydrate, $MnAl_2(SO_4)_4 \cdot 22H_2O$. Crystal system, monoclinic. Density, 1.78 g/cm^3; refractive index, 1.482; hardness, 1.5 Mohs. Occurrence: South Africa, Italy.

Apligraf. Trade name for an artificial skin grafting material based on a natural biological substrate, comprising collagen and glycosaminoglycans, suitable for seeding with human dermal cells.

aplite. A fine-grained granitic rock consisting principally of *quartz* and *feldspar*. Used as a source of alumina in glass, porcelain and whitewares, and as a flux for ceramics.

aplowite. A pink mineral of the starkeyite group composed of cobalt sulfate tetrahydrate, $CoSO_4 \cdot 4H_2O$. It can also be made synthetically. Crystal system, monoclinic. Density, 2.33 g/cm^3. Occurrence: Canada.

Apolam. Trade name of Axson (France) for epoxy laminating resins.

Apollo. (1) Trade name of Cytemp Specialty Steel Division (USA) for an oil-hardening tool steel containing 0.45% carbon, 2% chromium, 0.2% vanadium, and the balance iron. Used for die-casting dies, flying shears, etc.

(2) Trade name of Time Steel Service Inc. (USA) for an oil- or air-hardening high-carbon, high-chromium tool and die steel (AISI type D7) containing 2.5% carbon, 12% chromium, 4% vanadium, 0.8% molybdenum, and the balance iron. It has a great depth of hardening, excellent wear resistance, and relatively high hot hardness. Used for dies, and drawing tools.

(3) Trade name of Cyberbond LLC (USA) for a series of one-component rapid-setting cyanoacrylate adhesives. They are available in a wide range of viscosities and service temperatures (from -54 to +121°C, or -65 to +250°F), and bond well to metals, plastics, elastomers, and wood. Examples include popular products, such as *Krazy Glue* and *Super Glue*.

(4) Trade name for a light-cure composite cement used in restorative dentistry.

(5) Trade name for high-gold dental casting alloys with rich yellow color.

Apollo Chromsteel. Trade name of Apollo Metals, Inc. (USA) for a nickel-chromium-plated steel with good heat resistance up to 427°C (800°F) used for heat-resistant parts.

Apollo Crom. British trade name for stainless chromium-plated sheet zinc used for corrosion-resistant parts.

Apolloy. Trade name of Apollo Steel Company (USA) for a weather- and corrosion-resisting steel containing 0.08% carbon, 0.25% copper, and the balance iron. Used for various applications in building construction.

appearance lumber. A category of softwood construction lumber including the following grades: select, finish, paneling and siding lumber. It is a strong lumber of good appearance.

Appeel. Trademark of E.I. DuPont de Nemours & Company (USA) for sealant resins with controllable heat-seal strength and a broad sealing temperature range. They adhere well to many substrates including polypropylene, polyethylene, polyvinyl chloride, metal foil, and paper. Used as seal layer in flexible lidding and labels.

Applause. Trade name of Degussa-Ney Dental (USA) for a white gold-color dental casting alloy containing 54.9% palladium and 35% silver. It has a light oxide layer, a melting range of 1170-1225°C (2140-2240°F), a casting temperature of 1370°C (2500°F), exceptional sag resistance, high yield strength, good induction castability, and a hardness of 240 Vickers. Used for porcelain-to-metal restorations subject to stress.

apple. The hard, strong and tough wood of the tree *Malus pumila*. It has medium weight, high stability when dried, poor decay resistance, and saws, machines and polishes well. Source: Chiefly Europe and western Asia, but also grown in the USA and Canada. Although not of great commercial importance, it is suitable for crafts and small-shop production, and for carving.

Applika. Trade name of former Gerresheimer Glas AG (Germany) for hollow glass blocks to which colored antique glass has been applied.

Appretan. Trademark of Hoechst AG (Germany) for polyvinyl acetate (PVAC) resins and products used in the leather and textile industries.

Appryl. Trade name of Atofina Appryl (USA) for polypropylene homopolymers and copolymers. The copolymers are supplied in regular and UV-stabilized grades, and have better impact strength and low-temperature properties than the homopolymers. They have a density of 0.9 g/cm^3 (0.03 $lb/in.^3$), low water absorption, and good resistance to acids, alcohols, alkalies and halogens. The homopolymers are used for appliance housings, packaging applications, housewares and fibers and monofilaments, and the copolymers for containers, pipes, automotive components, boat hulls, etc.

Apro. Trade name for expanded polypropylene resins with excellent resilience and impact resistance used for automotive components, e.g., bumper cores.

apron leather. Any of several leathers used in the manufacture of workman or blacksmith aprons, tool belts, etc.

apuanite. A black mineral composed of iron antimony oxide sulfide, $FeFe_4Sb_4O_{12}S$. Crystal system, tetragonal. Density, 5.33 g/cm^3. Occurrence: Italy.

Apyral. Trade name of VAW-Vereingte Aluminium-Werke AG (Germany) for aluminum oxide used as a flame-resistant filler for plastics and elastomers.

AQ Bond. Trade name of Sun Medical Company Limited (Japan) for a light-curing, self-etching, self-priming, methacrylate-based dental bonding agent containing the adhesive monomer 4-methacryloxy-ethyl trimellitate anhydride (4-META). It forms a strong, durable bond between the composite restoration and the tooth structure (dentin and enamel).

aqua ammonia. See ammonia solution.

Aquablak. Trade name for water dispersions of various grades of *carbon black* including *bone black*, *channel black* and *furnace black*. Used as pigments in paints, inks and leather finishes.

AquaBlok. Trademark of Owens-Corning Fiberglass (USA) for a stiff, flexible high-strength glass-fiber reinforcement with an uniformly applied water-resistant polymer coating. Used as a nonflammable, protective central member of optical cables for telecommunication applications.

AquaBoxyl. Trade name for a zinc polycarboxylate dental cement for the cementation of bridges, crowns and inlays, and for lining and filling tooth cavities.

Aquacem. Trade name for a glass-ionomer dental cement.

Aqua Cenit. Trade name for a resin-modified glass-ionomer dental cement.

Aquacidox. Trade name for steel tubing containing 0.11% carbon, 0.4% manganese, 0.3% copper, up to 0.3% silicon, traces of sulfur and phosphorus, and the balance iron. Used for boiler feed tubes and other tubing applications.

Aqua-Cote. Trade name of Chemco Manufacturing Company Inc. (USA) for a water-based spray-booth coating.

Aquacrete. British trademark for a water-repellent cement.

Aquadag. Trademark of Acheson Industries Inc. (USA) for a colloidal suspension of up to about 22% graphite in water. It is used as a lubricant for dies, tools and molds, as an electrically conductive coating, as a coating in cathode-ray tubes, in the collection of secondary electrons, and as a post-deflection acceleration anode.

Aqua-Fil. Trademark for a glass-ionomer dental filling cement.

Aquagel. Trademark of National Lead Company (USA) for a gel-forming *bentonite* used in drilling muds, and for waterproofing concrete.

aquagel. (1) A low-density solid made from a gel by washing with water. See also aerogel; alcogel.

(2) A porous, low-density water-silica body prepared by mixing a *sodium silicate* (*water glass*) solution with hydrochloric acid and, after gelling during a period of approximately 24 hours, washing the acid out by the addition of water.

Aquagem. Trade name for a light-blue synthetic aquamarine spinel. See also spinel.

Aquaguard. Trademark of Westroc Industries Limited (Canada) for water-resistant *gypsum wallboard* for building interiors.

Aqua Ionobond. Trade name for a glass-ionomer cement used in restorative dentistry as a lining material.

Aqua Ionofil. Trade name for a glass-ionomer cement used in restorative dentistry as a filling material.

Aquakent. Trade name of Kent Dental/Hejco (UK) for a glass-ionomer dental luting cement.

Aqualac. (1) Trade name of RAE Products & Chemicals Corporation (USA) for air-drying water-based lacquer coatings.

(2) Trade name of MacDermid Inc. (USA) for a water-based lacquer.

Aqualat. Trade name of Inter-Africa Dental (Zaire) for a highly biocompatible, anhydrous polycarboxylate dental cement with low solubility and, when mixed with water, excellent adhesion to dentin. Used in restorative dentistry for cavity lining, temporary fillings, and for the cementation of crowns, bridges, onlays, inlays, orthodontic brackets, etc.

Aqua-Link. Trademark of AT Plastics Inc. (USA) for water cross-linkable ethylene-vinyl silane copolymers and catalyst concentrates. Used in the manufacture of adhesives, sealants, packaging films, pipe and pipe fittings, electrical wire insulation, tubing, sheeting, and foams.

Aqualite. Trade name of National Vulcanized Fiber Company (USA) for phenol-formaldehyde plastics.

Aqualock. Trademark of Sovereign Engineered Adhesives LLC (USA) for water-based adhesives for industrial bonding applications.

Aqualoid. Trade name of Sico Inc. (Canada) for water-base coatings.

Aqualon. (1) Trademark of Kanebo Limited (Japan) for acrylic fibers.

(2) Trademark of Aquafil SpA (Italy) for nylon 6 fibers and yarns.

Aqualox. Trade name for zinc polycarboxylate dental cements for the cementation of bridges, crowns and inlays, and for lining and filling tooth cavities.

Aqualoy. Trade name of ComAlloy International Corporation (USA) for a series of reinforced engineering thermoplastics including chemical-, impact- and high-temperature-resistant polypropylenes reinforced with glass or other fibers, and chemical- and wear-resistant, glass-, hybrid-, or otherwise -reinforced polyamides (nylon 6,6). Used for bearings, wear parts, etc.

Aqualure. Trade name Glidden Company Division of SCM (Canada) Limited for water-soluble coatings.

Aqualux. Trademark of Sico Inc. (Canada) for water-base industrial enamel.

aquamarine. A blue, or sea-green variety of the mineral *beryl*, $Be_3Al_2(SiO_3)_6$. It can also be made synthetically. Crystal system, hexagonal. Density, 2.7 g/cm^3; hardness, 7.5-8 Mohs; refractive index, 1.58. Occurrence: Brazil, Madagascar, Russia, USA (Connecticut, Pennsylvania, Maine). Used as a gemstone.

Aqua Meron. Trade name for a glass-ionomer cement used in restorative dentistry as a lining material.

Aquamid. Trade name of Schenectady Chemicals Inc. (USA) for insulating and protective coatings for foils, core plates, and magnet wires. Also included under this trade name are several insulating varnishes.

Aquanel. Trademark of Schenectady Chemicals Inc. (USA) for insulating varnishes.

Aquapearl. Trade name of Aquapearl Catalin Corporation (USA) for phenol-formaldehyde plastics.

Aquaphos. Trade name of Texo Corporation (USA) for patented organic overcoatings used on zinc phosphates.

Aquaplate. Trade name Semon Bache & Company (USA) for a plate glass.

Aquaplex. Trade name for a synthetic alkyd resin dispersed in aqueous medium for use on porous surfaces.

Aquapol. Trade name of Monopol AG (Switzerland) for anticorrosive and dip primers, and several finishing coats.

Aqua-Poly. Trade name of Polysciences, Inc. (USA) for water-soluble, nonfluorescing media used in biochemistry and medicine for mounting sections for immunofluorescent analysis techniques.

Aqua-Quench. Trademark for a series of polyethyloxazoline-based quenching media with slow cooling rates that eliminate many of the disadvantages (oil smoke, fire hazards, etc.) of conventional quenching oils. Used for quenching ferrous alloys especially forgings, castings, and high-hardenability alloy steels.

Aquarius. Trademark of Ivoclar Vivadent AG (Liechtenstein) for a palladium- and copper-free ceramic dental alloy containing 86.0% gold, 11.0% platinum, 2.5% indium, and less than 1.0% tin, iridium, tantalum and lithium, respectively. It has a rich yellow color, a density of 18.5 g/cm^3 (0.67 $lb/in.^3$), a melting range of 1010-1135°C (1850-2075°F), an as-cast hardness of 160 Vickers, moderate elongation, and excellent biocompatibility. Used for crowns, inlays, onlays, and bridges.

Aquarius Hard. Trademark of Ivoclar Vivadent AG (Liechtenstein) for a ceramic dental alloy containing 86.1% gold, 8.5% platinum, 2.6% palladium, 1.4% indium, and less than 1.0% ruthenium, tantalum, iron and lithium, respectively. It has a rich yellow color, a density of 18.5 g/cm^3 (0.67 $lb/in.^3$), a melting range of 1050-1145°C (1920-2095°F), an as-cast hardness of 205 Vickers, low elongation, and excellent biocompatibility. Used for crowns, inlays, onlays, posts, and bridges.

Aquarius HPF. Trademark of Ivoclar Vivadent AG (Liechtenstein) for a palladium-, silver- and copper-free ceramic dental alloy containing 85.9% gold, 12.1% platinum, 1.5% zinc, and less than 1.0% iridium, indium, tantalum, iron and manganese, respectively. It has a rich yellow color, a density of 18.9 g/cm^3 (0.69 $lb/in.^3$), a melting range of 1055-1170°C (1930-2140°F), an as-cast hardness of 235 Vickers, low elongation, and excellent biocompatibility. Used for crowns, inlays, onlays, posts, and bridges.

Aqua-Satin. Trade name of Pratt & Lambert Paints (USA) for satin-finish latex enamel paints.

Aquasil. Trade name of Dentsply/Caulk (USA) for an addition-polymerizing silicone elastomer used for dental impressions.

Aquastik. Trademark of DuPont Dow Elastomers (USA) for polychloroprene latex supplied in several grades for use in adhesives, coatings, etc.

Aquatex. Trade name of ASG Industries Inc. (USA) for a glass with intermediate patterns.

Aquathene. Trade name of Quantum Chemical Company (USA) for polyethylene resins.

Aquator. Trade name for a *Tactel*-type nylon 6,6 fiber used for sportswear.

Aquatough. Trade name of British Steel Corporation (UK) for a series of water-hardening tool steels containing 0.7-1% carbon, 0.3% manganese, and the balance iron. Used for chisels, caulking tools, cold-heading dies, rivet snaps, arbors, axes, forming dies, and springs.

Aquazimo. Trade name of Monopol AG (Switzerland) for a water-dilutable zinc-rich primers.

Aqu-Bar. Trade name of Copperweld Steel Company (USA) for hot-rolled carbon and alloy steel bars specially developed for applications involving machining and cold, warm and hot forging to close tolerances.

Aquila. Trade name of Westa-Westdeutsche Edelstahlhandelsgesellschaft (Germany) for a tough, case-hardening steel used for machine elements, such as cams, camshafts, axles, and gears.

Arabesco. Trade name of Vetreria Milanese Luccini Perego SA (Italy) for a glass with Mauresque pattern.

Arabesk. Trade name for a light-cure dental hybrid composite.

Arabian. Trade name of Central Glass Company Limited (Japan) for a glass with curvilinear pattern.

Aracast. Trademark of Ciba-Geigy Corporation (USA) for epoxy resins obtained by the interaction of epichlorohydrin and hydantoin. They have excellent mechanical properties, high dielectric strength, good resistance to ultraviolet light, and good light transmission properties. Used for electronic and electrical applications.

Aracon. Trademark of E. I. DuPont de Nemours & Company (USA) for lightweight, flexible aramid fibers with copper, nickel or silver coatings of varying thickness. They possess high strength and good conductivity, and are used for EMI shielding and specialized conducting applications.

aragonite. A colorless, white, gray, or pale yellow mineral composed of calcium carbonate, $CaCO_3$. Crystal system, orthorhombic. Density, 2.95 g/cm^3; melting point, decomposes at 825°C (1517°F); hardness, 3.5-4.0 Mohs; refractive index, 1.68. Occurrence: Austria, Canada, Italy, Spain, UK, USA (New Mexico, Arizona). Used in refractories, whitewares, glass, electronic ceramics, etc. Also known as *Aragon spar.*

Aragon spar. See aragonite.

Arakote. Trademark of Ciba-Geigy Corporation (USA) for a series of polyester resins designed for curing with blocked isocyanates to produce high-gloss polyester/urethane coatings. They have good weather resistance, good flow properties, and smooth finishes. Used on architectural, automotive and outdoor equipment, on appliances and furniture, in the general metal finishing industries, and also for powder coatings.

Aralac. Trade name of National Dairy Products Corporation (USA) for a white, translucent, silky casein fiber obtained from milk protein and used in blends with rabbit hair for making felt hats and with wool, mohair, rayon and cotton for making fabrics and garments.

Araldico. Trade name of Vetreria di Vernante SpA (Italy) for glass with checkerboard pattern.

Araldite. Trademark of Ciba-Geigy Corporation (USA) for epoxy resins available in solid and liquid form for casting, foaming, reinforced molding, laminating and coating. Bisphenol, cresol novolac, phenol novolac and several water-based types are available. The molding compounds are supplied as glass fiber-reinforced, mineral filled and/or high-heat grades, and the resins in general-purpose, high heat, flexible, and aluminum-, glass-, mineral-, or silica-filled casting grades. Heat-curing and room-temperature-curing grades for adhesives are also available. *Araldite* provides an excellent combination of mechanical properties and corrosion resistance, and has good dimensional stability, good chemical and electrical resistance, good resistance to most organic solvents, low shrinkage, excellent adherence to metal, glass, ceramics, etc. Used for precision castings, high-strength laminates, advanced composites, adhesives, protective coatings and paints, as embedding media in microscopy, and for electrical moldings, tools and dies.

aralkyl. An arylated alkyl, i.e., a radical in which one hydrogen atom of an alkyl is replaced with an aryl group. Also known as *arylalkyl.*

aralkyl halide. An organic halide derived from an *arene* in which the halogen is not attached directly to the aromatic ring.

Arall. Trademark of Alcoa–Aluminum Company of America (USA) for a family of structural hybrid materials combining the advantages of high-strength aluminum with those of strong aramid fibers. They consist of alternating layers of thin aluminum sheet bonded by an adhesive (usually an epoxy resin) impregnated with strong aramid fibers. These advanced composites have high strength and excellent fatigue resistance, and provide ease of machinability and formability. Used for advanced composites.

Aramado. Trade name of Vidrobrás (Brazil) for patterned glass.

aramayoite. An iron-black mineral composed of silver antimony bismuth sulfide, $Ag(Sb,Bi)S_2$. Crystal system, triclinic. Density, 5.60 g/cm³. Occurrence: South America (Bolivia).

aramid. Generic name for a distinctive class of highly aromatic long-chain polyamide fibers derived by condensation polymerization from *p*-phenylene diamine and terephthaloyl chloride. They have high thermal stability, outstanding flame-retardant properties, high tensile strength, high modulus of elasticity, good fatigue and creep resistance, high toughness, excellent damage tolerance characteristics, and a low dielectric constant. Usually sold under trade names or trademarks, such as *Conex*, *Kevlar*, *Nomex* or *Spectra*, they are used as reinforcing fibers in advanced composites, and in protective clothing, tire cord, bullet-resistant structures, and dust-filter bags. Also known as *aramid fibers; para-aramid fibers; p-aramid fibers; polyaramid.*

aramid composites. Engineering composites consisting of aramid fibers embedded in organic or inorganic matrices.

aramid-epoxy composites. Advanced composite materials consisting of unidirectional aramid fibers in elevated-temperature-curing epoxy resin matrices. They have excellent strength properties at elevated temperatures, and are used for boat hulls, aircraft surface panels, rocket motor cases, circuit boards, etc.

aramid fibers. See aramid.

p-aramid fibers. See aramid.

aramid-fiber-reinforced plastics. Advanced engineering materials with polymer matrices (e.g., epoxy or polyester) and aramid fiber reinforcements.

aramid paper. A thin, lightweight paper based on discontinuous aramid fibers, and used in sandwich constructions, and in the form of wet-laid papers, composed of chopped aramid fibers and pulp, as an asbestos substitute in gaskets, and in composites for printed-circuit boards, aerospace parts, etc.

aramid-polyphenylene sulfide composites. Composite materials consisting of polyphenylene sulfide matrices reinforced with aramid fibers. Commercially available in the form of rods and sheets, they have excellent high-temperature strength and high toughness, and are used chiefly for aerospace applications.

aramid pulp. A discontinuous filament form of aramid composed of very short fibers (2-4 mm, or 0.08-0.16 in.) with numerous attached fibrils. It has a large surface area, high aspect ratio (greater than 100%), and is used as an asbestos substitute in gaskets, sealants, caulks, friction products, and coatings.

aramid-vinylester composites. Composite materials consisting of vinylester matrices reinforced with aramid fibers. They are commercially available in rod, tube, sheet form, and have very low density, very high tensile strength, and excellent impact resistance and fracture toughness. Used chiefly for aerospace and other weight-critical high-strength applications.

aramina fiber. See urena fiber.

Aramith. Trade name of Saluc SA (Belgium) for phenol-formaldehyde plastics.

Arapaho. Trade name of AL Tech Specialty Steel Corporation (USA) for a general-purpose air-hardening, shock-resisting steel (AISI type S7) containing 0.5% carbon, 0.8% manganese, 0.25% silicon, 3.25% chromium, 1.45% molybdenum, and the balance iron. It has a great depth of hardening and excellent toughness. Used for blanking and forming dies, and plastic molds requiring exceptional toughness.

Arasox. Trademark of A&P Technology (USA) for sleevings braided with aramid fibers, and supplied as light, medium and heavy fabric grades. Used as reinforcements in composites.

Arathane. Trademark of Ciba-Geigy Corporation (USA) for polyurethane adhesives used for automotive and industrial bonding applications.

Araton. Trademark of Owens-Corning Fiberglas Corporation (USA) for glass mats, rovings, and chopped fibers and strands used for reinforcing synthetic resins.

Arazole. Trade name for a polybenzimidazole (PBI) fiber with high strength, excellent resistance to most chemicals and solvents (except concentrated acids), and excellent flame and high-temperature resistance. Used as a reinforcement fiber.

araucarian pine. The soft, yellowish-white wood of several trees of the genus *Araucaria* especially *A. brasiliensis*, and *A. angustifolia*. Although called a "pine," this tree is not a true pine. *Araucarian pine* has a fine, uniform texture, a beautiful grain pattern, a heartwood that is often streaked red, high to medium strength, and low decay resistance, unless suitably treated. Average weight, 540 kg/m³ (34 lb/ft³). Source: Southern Brazil, Paraguay, Argentina. Used for framing lumber, interior trim, veneer, and furniture. Also known as *Brazilian pine; Parana pine; pinheiro do Parana; pinho do Parana.*

Aravite. Trademark of Ciba Specialty Chemicals Corporation (USA) for a series of cyanoacrylate and acrylic adhesives with setting times ranging from 15 to 60 seconds. Used for bonding of metals and plastics.

Arbocel. Trademark of J. Rettenmaier & Söhne GmbH & Co. (Germany) for fibrous materials based on asphalt, bitumen, cellulose polymers, etc., and used for chemical and building applications.

Arboga. British trade name for a sintered tungsten carbide-base alloy used for dies and hard cutting tools.

arborescent powder. A powder whose particles have the typical dendritic, or fir tree-like structure. It is usually an electrolytic powder. Also known as *dendritic powder.*

Arborite. Trade name of Arborite Limited (UK) for melamine-formaldehyde plastics.

arborvitae. The wood of any of several Asian and North American evergreen trees of the genus *Thuja* belonging to the cypress family. North American species include eastern white cedar (*T. occidentalis*), and western red cedar (*T. plicata*), the most familiar Asiatic species is the Chinese arborvitae (*T. orientalis*). See also eastern white cedar; western red cedar.

Arbosol. Trademark of Arbonite Corporation (USA) for polyvinyl chloride coatings and lining materials used on metal structures, tanks, and in industrial processing.

Arc. Trade name of Creusot-Loire (France) for a series of austenitic stainless steels used for chemical processing equipment, tank cars, pressure vessels, oil refinery equipment, digesters, evaporators, mixers, valve parts, kettles, exhaust systems, and aircraft parts.

Arcalloy. Trade name for a corrosion-resistant cobalt-chromium dental bonding alloy.

Arcaloy. Trademark of Chemetron Corporation (USA) for an extensive series of covered austenitic, ferritic and martensitic stainless steel welding electrodes. *Arcaloy Shield-Bright* refers to a series of all-position, gas-shielded, flux-cored stainless steel welding wires providing excellent slag removal, extremely smooth arcs, outstanding out-of-position welding performance, low spatter, attractive bead profiles, and high deposition rates.

arcanite. A colorless to white mineral of the olivine group composed of potassium sulfate, K_2SO_4. It can also be made synthetically. Crystal system, orthorhombic. Density, 2.66 g/cm³; refractive index, 1.495. Occurrence: South America (Peru). Also known as *glaserite*.

Arcast. (1) Trade name of C.E. Philips & Company (USA) for a cast iron welding electrode containing 3% carbon, 2% silicon, and the balance iron.

(2) Trade name of Anwood Corporation (USA) for a heat-treatable casting alloy containing 5-9% zinc, 0.9-1.4% magnesium, 0.1-0.4% titanium, 0.025-0.15% chromium, and the balance aluminum. Used for general-purpose castings, casings, housings, and hardware.

arc carbon. High-purity carbon in rod form for use in weatherometers, searchlights, and high-intensity electric discharge lamps for motion-picture projection, live theater lighting, outdoor movie production lighting, photography, blueprinting, etc.

Arcel. Trademark of Arco Chemical Company (USA) for moldable polyolefin copolymers (e.g., polyethylene). They have high strength, toughness, durability and water resistance, and excellent moldability. An expandable polyethylene grade for packaging applications is also available. Used for sporting goods, recreational equipment, materials handling systems, flotation devices, and marine parts.

Arc Flux. Trade name for a series of fluxes for submerged-arc welding applications.

archerite. A colorless mineral composed of potassium hydrogen phosphate, KH_2PO_4. It can also be made synthetically. Crystal system, tetragonal. Density, 2.33 g/cm³; refractive index, 1.511. Occurrence: Peru, USA.

architectural bronze. A corrosion-resistant, free-cutting copper alloy containing 38-44% zinc, 2.5-3.2% lead, up to 0.25% tin, and traces of manganese and iron. It has excellent machinability and hot workability, and excellent forging and extruding characteristics. Used for architectural and ornamental parts, fittings, hardware, trims, hinges, locks, store fronts, showcases, handrails, casements, and forgings. Also known as *art bronze*.

architectural coating. A high-quality coating that has been applied on-site to the exterior or interior surfaces of a building or other structure.

architectural concrete. A high-quality, blemish-free concrete used for ornamentation or finish on the exterior or interior surfaces of residential, commercial, institutional or industrial buildings, and for wall and basement construction.

architectural terra-cotta. A hard-burnt, machine-extruded or hand-molded clay unit, glazed or unglazed, and usually larger than brick or ordinary facing tile. Used in the form of blocks, slabs, and special shapes for decorative applications, such as wall facings.

Arco. (1) Trade name of Armco International (USA) for a series of quenched-and-tempered steels containing up to 0.2% carbon, 0.4-0.7% manganese, 0.2-0.35% silicon, 1.15-1.65% chromium, 0.25-0.4% molybdenum, 0.2-0.4% copper, 0.0015-0.005% boron, 0.04-0.10% titanium or vanadium, and the balance iron. Used for structural applications, such as buildings, bridges, ordnance, ships, earth-moving equipment, and crane booms.

(2) Trade name for refractory brick containing 60-80% alumina. It has excellent heat resistance up to 1835°C (3335°F), and is used for kiln and furnace linings.

Arcol. Trademark of Bayer Corporation (USA) for flexible polyurethane foams and polyurethane elastomers used for automotive components, furniture, packaging, machine parts, electrical equipment, construction and transportation applications, etc.

Arcolite. Trade name of Consolidated Molded Products Corporation (USA) for phenol-formaldehyde plastics.

Arcoloy. Trade name of American Radiator Company (USA) for a corrosion-resistant copper casting alloy containing 2.6-5.0% silicon, 0-0.12% iron, and 0.01% phosphorus. Used for storage tanks, and range boilers.

Arcometal. Trademark of Aluminum Reduction Company (Canada) for aluminum- and zinc-alloy ingots.

Arcos. Trade name of Arcos Alloys (USA) for a series of coated arc-welding electrodes based on corrosion- and heat-resistant austenitic, ferritic and martensitic stainless steels, chromium-molybdenum alloy steel, or nickel-chromium-iron, copper-nickel or nickel-copper alloys. Formerly known under trade names, such as Chlorend, Chromar, Chromend, and Stainlend.

Arcosarc. Trade name of Arcos Alloys (USA) for flux-cored welding wires based on carbon, low-alloy, or stainless steel.

Arcplus. Trade name of Arcweld Products Limited (Canada) for arc-welding electrodes.

arc silica. Microcrystalline silica made by heating silica sand in an arc furnace at high temperatures. See also microcrystalline silica.

Arctic. (1) Trade name of Pilkington Brothers Limited (UK) for a patterned glass.

(2) Trade name of British Steel plc (UK) for a low-carbon structural steel.

(3) Trade name of Crane Company (USA) for a ferritic alloy steel containing up to 0.15% carbon, 0.5-0.8% manganese, 3-4% nickel, and the balance iron. Used for low-temperature applications.

(4) Trade name of Mathison's (USA) for a range of mounting products for the graphic arts and allied trades, supplied in several sizes, and including (i) *Arctic Dura-Mount* consisting of a polyester substrate coated on both sides with a pressure-sensitive adhesive and a silicone-treated release liner; (ii) *Arctic Front* consisting of an optically clear polyester substrate coated on both sides with a clear, pressure-sensitive adhesive; (iii) *Arctic Lo-Tack* consisting of an optically clear polyester film coated on one side with a removable adhesive and on the other with a permanent adhesive; (iv) *Arctic Premium* consisting of a polyester substrate coated on both sides with a pressure-sensitive acrylic adhesive; (v) *Arctic Removable* consisting of a white, opaque, rigid polyvinyl chloride substrate coated on one side with a removable adhesive with silicone-treated paper release liner and on the other with a permanent adhesive; and (vi) *Arctic White* consisting of a white polyester substrate coated on both side with a permanent, pressure-sensitive acrylic adhesive with silicone-treated paper release liner.

(5) Trade name of Mathison's (USA) for a range of thin polymer films coated on one or both sides with pressure-sensitive adhesives, and used in the graphic arts and allied trades for overlamination applications. They are supplied in several grades including (i) *Arctic Gloss* glossy vinyl films; (ii) *Arctic Graffiti* graffiti-proof, UV-resistant *Tedlar* polyvinyl fluoride films; (iii) *Arctic Grain* textured vinyl films; (iv) *Arctic Lustre* semi-gloss vinyl films; (v) *Arctic Matte* mat-finished vinyl films; (vi) *Arctic Photo Mount* high-gloss polyester films; (vii) *Arctic Polycarbonate* velvet-mat polycarbonate films; and (viii) *Arctic Write Erase* polypropylene films for dry erase maker applications.

arctite. A colorless mineral composed of sodium calcium fluorophosphate, $Na_2Ca_4(PO_4)_3F$. Crystal system, hexagonal. Density,

3.13 g/cm³; refractive index, 1.578. Occurrence: Russia, Siberia.

Arctic White. Trade name of Aardvark Clay & Supplies (USA) for medium-smooth, white stoneware clay (cone 5).

arc-welding electrode. A consumable or nonconsumable rod, usually of metal or carbon, and either bare or covered with a flux coating, used to strike and maintain an electric arc, and thus form part of the welding circuit and, if consumable, to provide filler metal to the joint. Also known as *electrode; welding electrode*.

Ardajoint. Trade name of Ato Findley (France) for epoxy-based joint fillers.

Ardal. (1) Trade name of Ardal Limited (UK) for a non-heat-treatable aluminum alloy containing 2% copper, 1.5% iron, and 0.6% nickel. Used for pistons, bearings, wire, and tubes.

(2) Trade name of Ato Findley (France) for a wide range of adhesive products including hot melts for wood and furniture, acrylics for ceramic and floor tiles, and cements for various tiling applications.

ardealite. A white, or light yellow mineral composed of calcium hydrogen phosphate sulfate tetrahydrate, $Ca_2H(PO_4)(SO_4) \cdot 4H_2O$. It can also be made synthetically. Crystal system, monoclinic. Density, 2.30 g/cm³. Occurrence: Rumania.

ardein fiber. A regenerated fiber obtained from ardein, a purified protein found in chili pepper (*Capsicum*) nuts.

Ardel. Trademark of Amoco Performance Products, Inc. (USA) for tough, dimensionally stable polyarylate resins. They have good resistance to elevated temperature, a heat deflection temperature of 174°C (345°F), good weatherability, excellent surface finish, and good coatability and metallizability. Used for automotive components and mine safety devices.

ardennite. A yellow to yellowish-brown mineral composed of manganese aluminum vanadium arsenate silicate hydroxide, $Mn_5Al_5(Si,As,V)_6O_{24}(OH)_2$. Crystal system, orthorhombic. Density, 3.60 g/cm³; refractive index, 1.74. Occurrence: Belgium.

Ardent. Trade name of Osborn Steels Limited (UK) for an abrasion-resistant cold-work tool steel containing 1.5% carbon, 4% tungsten, 0.5% chromium, and the balance iron. Used for forging dies, upsetters, and cold drawing dies.

Ardho. Trade name of Spencer Clark Metal Industries Limited (UK) for a series of chromium and nickel-chromium tool steels. Used for blacksmith tools, cold chisels, punches, and tools.

Ardil. Trademark of ICI Limited (UK) for an azlon textile fiber made from peanut protein obtained from groundnuts. Used for carpets and clothing.

Ardolon. Trade name of Ato Findley (France) for a waterproofing coating.

Ardoloy. Trade name of Herbert Cutanit Limited (UK) for a series of cemented carbides used for cutting tools.

Ardonax. Trade name of Vereinigte Zwieseler & Pirnaer Farbenglaswerke AG (Germany) for a bluish-green, heat-absorbing glass.

Ardopox. Trade name of Ato Findley (France) for an epoxy adhesive.

Ardorit. Trade name of Hoffmann & Co. KG (Germany) for a series of heat- and corrosion-resistant austenitic, ferritic and martensitic stainless steels.

Ardux. Trade name for a heat-curing epoxy-phenolic adhesive with medium peel strength, and good lap shear strength at temperatures from -73 to +260°C (-99 to +500°F). Used as a structural adhesive for joining metals.

Aremco-Bond. Trademark of Aremco Products Inc. (USA) for a series of regular and conductive epoxy adhesives.

Aremcolox. Trademark of Aremco Products Inc. (USA) for a series of pure and mixed ceramic oxides and nitrides available in medium, high-density, unfired and fired grades. Pure grades are based on alumina, beryllia, silica, boron nitride, etc. Supplied in the form of plates and rods, they have densities ranging from 2.8 to 3.0 g/cm³ (0.10 to 0.11 lb/in.³), a hardness of 5-5.5 Mohs, zero porosity, high dielectric and mechanical strength, and good machinability. Used for high-temperature coil forms and lamp housings, high-voltage insulators, thermal switches, radiation components, soldering fixtures, and arc barriers.

Aremco-Seal. Trademark of Aremco Products Inc. (USA) for a series of ceramic coatings.

Aremite. Trade name of Robbins & Myers Inc. (USA) for a synthetic cast iron used for hardware, fittings, and pipes.

Arena. Trade name of Fabbrica Pisana SpA (Italy) for patterned glass.

arenaceous clay. A sandy clay or soil, such as *loam*.

arene. Any of a class of hydrocarbons that contain a benzene ring and one or more aliphatic groups, e.g., ethylbenzene, toluene, xylene, styrene, etc. Also known as *aromatic-aliphatic compound*.

Arenka. See Twaron.

arfvedsonite. A black mineral of the amphibole group composed of sodium iron aluminum silicate hydroxide, $(Na,K)_{2.6}Fe_5(Si,Al)_8O_{22}(OH)_2$. Crystal system, monoclinic. Density, 3.37 g/cm³; refractive index, 1.694. Occurrence: Greenland.

Argal. Trade name of Gerhardi & Co. (Germany) for an aluminum alloy containing 2-4% manganese, and up to 0.4% magnesium and 0.3% chromium, respectively. It has excellent resistance to seawater corrosion, and is used for aircraft tanks and fittings, marine parts and hardware, and fuel lines.

Argalium. Trade name Gebrüder Rieger (Germany) for an aluminum alloy containing 1.5-3% magnesium, 0.6-1.3% manganese, up to 1.3% silicon, 0.3% chromium and 0.2% titanium. It has good formability and weldability, and is used for roofing, hydraulic tubing, and architectural trim.

Argelite. (1) British trade name for a non-hardenable aluminum alloy containing 6% copper, 2% silicon, and 2% bismuth. Used for light-alloy parts.

(2) Trademark of The Argen Corporation (USA) for precious metal alloys used for dental applications.

Argen. Trade name of The Argen Corporation (USA) for gold-based dental alloys and solders.

Argenco. Trade name of The Argen Corporation (USA) for a series of high-, medium- and low-gold dental casting alloys, and silver-free palladium dental bonding alloys.

Argent. Trade name of Plouff Metallographic Institute (USA) for a solder containing 98% tin and 2% silver, used for soldering aluminum and aluminum alloys.

Argental. (1) British trade name for a corrosion-resistant alloy containing 60-75% aluminum, 15-16% silver, 7-20% zinc, and 3-5% copper. Used for jewelry.

(2) Trade name for a corrosion-resistant copper-base jewelry alloy containing 10% tin and 5% cobalt.

Argentalium. British trade name for a non-heat-treatable aluminum alloy containing 5% silver, and 0.1-1% manganese, used for light-alloy parts.

argentan. See nickel silver.

Argentan lace. A type of *needlepoint lace* that features a fine net ground larger than that of *Alençon lace* and does not have the heavy (cordonnet) thread outlining of the latter.

Argentea. Trade name for rayon fibers and yarns used for textile fabrics.

argent français. An alloy of copper, nickel and zinc used for jewelry and ornaments. Also known as *French silver.*

argentic oxide. See silver peroxide.

argentiferous lead. A silver-bearing lead or lead alloy.

Argentin. British trade name for an antifriction alloy containing 85% tin, 14.5% antimony, and 0.5% copper. Used for bearings.

argentina. A unglazed porcelain that has been covered chemically with copper, gold, or silver.

argentine. (1) A white metal that has been coated with silver.

(2) A variety of *calcite* ($CaCO_3$) with a white pearlescent luster.

Argentine metal. A silvery casting alloy that contains 85-85.5% tin and 14.5-15% antimony, and expands on cooling. Used for statuettes, small ornaments, and toys.

Argentine wool. A good- to medium-quality wool obtained from Argentinian hybrid merino sheep. Used especially for clothing and carpets.

argentite. A lead-gray to black, or grayish-black mineral with metallic luster composed of silver sulfide, Ag_2S, and containing theoretically 87.1% silver. Crystal system, monoclinic. Density, 7.2-7.36 g/cm^3; hardness, 2-2.5 Mohs. Occurrence: Bolivia, Canada, Chile, Europe, Mexico, Peru, USA (Arizona, Colorado, Montana, Nevada, Virginia). Used as an important silver ore. Also known as *argyrite; silver glance; vitreous silver.*

argentojarosite. A yellow or brownish mineral of the alunite group composed of silver iron sulfate hydroxide, $AgFe_3(SO_4)_2(OH)_6$. It can also be made synthetically. Crystal system, rhombohedral (hexagonal). Density, 3.62 g/cm^3; refractive index, 1.8895. Occurrence: USA (Utah).

Argentomerse. Trade name of Technic Inc. (USA) for bright immersion silver deposits, up to 20 µin. (0.5 µm) in thickness, applied over copper or copper alloys.

argentopentlandite. A reddish mineral of the pentlandite group composed of iron nickel silver sulfide, $(Fe,Ni)_8Ag_{1-x}S_8$. Crystal system, cubic. Density, 4.66 g/cm^3. Occurrence: Finland.

argentopyrite. A steel-gray mineral of the cubanite group composed of silver iron sulfide, $AgFe_2S_3$. Crystal system, orthorhombic. Density, 4.25 g/cm^3. Occurrence: Germany.

argentous oxide. See silver oxide.

Argeste. Trade name of Ergst Edelstahlwerke (Germany) for an extensive series of austenitic, ferritic and martensitic stainless steels.

Argical. Trade name of AGS Mineraux SA (France) for thermally processed clays obtained from a kaolinitic sedimentary *ball clay* mined in southwestern France. It is supplied as a powder with a particle size ranging from submicron to 10 µm (less than 40 to over 400 µin.). A typical chemical analysis (in wt%) is: 55% silica (SiO_2), 39% alumina (Al_2O_3), 1.6% ferric oxide (Fe_2O_3), and 1.7% titania (TiO_2). The loss on ignition is about 1%, the density 2.2 g/cm^3 (0.08 lb/in.³), the pH-value 5.7, and the specific surface area 17 m^2/g. Used in the manufacture of ceramic products.

Argicast B. Trade name for a medium-gold dental casting alloy.

Argicraft 1. Trade name for a palladium-silver dental bonding alloy.

Argident. Trade name for a range of high-gold dental bonding alloys.

Argiflex. Trade name of AGS Mineraux SA (France) for a surface-treated hard clay obtained from a kaolinitic sedimentary ball clay mined in southwestern France. It is supplied as a powder in particle sizes ranging from submicron to 10 µm (less than 40 to over 400 µin.). A typical chemical analysis (in wt%) is: 49% silica (SiO_2), 34% alumina (Al_2O_3), 1.7% ferric oxide (Fe_2O_3), and 1.3% titania (TiO_2). The loss on ignition is about 13%, the density 2.6 g/cm^3 (0.09 lb/in.³), the pH-value 9.4, and the specific surface area 22 m^2/g. Used in the manufacture of ceramic products.

argil. See potter's clay.

Argilite. (1) French aluminum alloy containing 6% copper, 2% silicon, and 2% bismuth. Used for automotive engine parts.

(2) Trade name for medium-gold dental bonding alloys.

argilite. A compact rock formed from *siltstone, shale,* or *claystone* and having a higher degree of induration than the latter. Used as a concrete aggregate.

Argilite 50. Trade name for a palladium-silver dental bonding alloy.

argillaceous hematite. A brown or red mineral composed chiefly of ferric oxide (Fe_2O_3) with a substantial amount of clay, or sand. Also known as *clay ironstone; iron clay; red ironstone clay.*

argillaceous limestone. A limestone that contains a substantial amount of clay, and is used in the manufacture of certain cements.

argillaceous material. A material made, or composed of clay or clay minerals.

argillaceous rock. See clay rock.

argillaceous sandstone. A sandstone, such as *bluestone,* that contains a substantial amount of clay or clay minerals.

argillaceous slate. A slate containing a considerable quantity of clay.

Argion. Trade name for a glass ionomer/cermet dental cement.

Argipal. Trade name for a silver-free palladium dental bonding alloy.

Argirec 96. Trade name of AGS Minéraux SA (France) for a kaolinitic sedimentary *ball clay* mined in southwestern France, and supplied as a powder with a particle size ranging from submicron to over 10 µm (less than 40 to over 400 µin.). A typical chemical analysis (in wt%) is 52% silica (SiO_2), 36.5% alumina (Al_2O_3), 1.6% ferric oxide (Fe_2O_3), and 1.7% titania (TiO_2). The loss on ignition is about 6.5%, the pH-value 5.5, the density 2.4 g/cm^3 (0.09 lb/in.³), and the specific surface area 20 m^2/g. Used as a mineral filler and extender in the manufacture of paper, composites, elastomers, etc.

Argistar 45. Trade name for dental bonding alloy of light yellow color containing 45% gold.

AR-Glas. Trademark of Schott Rohrglas AG (Germany) for alkali-resistant glass tubing used in the manufacture of pharmaceutical primary packaging, medical pipettes, test tubes and disposables, and for cosmetic and food-processing applications.

AR-glass. See alkali-resistant glass.

Argo-Bond. Trade name of Johnson Matthey plc (UK) for a cadmium-bearing silver brazing alloy containing 35% copper, 27% zinc, 23% silver, and 15% cadmium. It has a melting range of 616-735°C (1141-1355°F).

Argo-Braze. Trade name of Johnson Matthey plc (UK) for a series of silver brazing alloys containing varying amounts copper, zinc and nickel, and sometimes cadmium or manganese. Depending on the alloy combination, the melting range is from 600 to 830°C (1310 to 1525°F). Used for brazing various alloys and carbides.

Argofil. Trade name of Imperial Metal Industries (UK) for an alloy wire containing 0.25% silicon, 0.25% manganese, and the balance copper. It has a melting point of 1063°C (1945°F), and is used as filler rods for argon-arc and inert-gas shielded-metal-arc welding of copper.

Argo-Flo. Trade name of Johnson Matthey plc (UK) for a cadmium-bearing silver brazing alloy containing 40% silver, 21% zinc, 20% cadmium, and 19% copper. It has a melting range of 595-630°C (1103-1165°F). Used for brazing iron, copper, copper alloys, and nickel.

Argo High Speed. British trade name for a high-speed tool steel containing 0.8% carbon, 0.2% manganese, 1.7% chromium, 4.9% tungsten, 0.1% vanadium, and the balance iron. Used for tools, cutters, dies, reamers, gages, and punches.

Argoid metal. A British nickel silver containing 52.6% copper, 25.8% zinc, and 21.6% nickel. Used for ornamental parts.

Argon. Trade name for a kaolin brick containing a considerable quantity of calcined grog. See also kaolin brick.

argon. A colorless, odorless, tasteless, nonmetallic element belonging to the noble gas group (Group VIIIA, or Group 18) of the Periodic Table. It is a monatomic gas that forms a very small part (0.94%) of the air, and does not combine chemically with any known element. Density, 1.38; melting point, -189.3°C (-309°F); boiling point, -185.8°C (-302°F); atomic number, 18; atomic weight, 39.948; zerovalent. Used as a shielding medium or protective atmosphere in arc welding, furnace brazing, plasma-jet cutting and heat treating, in the decarburization of stainless steel, in titanium and zirconium refining, in gas-filled electric lamps (e.g., incandescent, neon, fluorescent and sodium-vapor), and in gas-discharge tubes and rectifiers, Geiger counters, and lasers. Symbol: Ar.

argon-oxygen decarburized steel. A highly refined stainless steel produced by first transfering molten, unrefined electric-furnace steel to a pear-shaped, refractory-lined vessel, and then blowing oxygen, gradually replaced by argon, through the melt. Abbreviation: *AOD steel.*

Argos. Trade name of Solutions Globales (France for an antistatic, flame- and heat-resistant textile fabric made from viscose fibers. Used for commercial and industrial applications.

Argo-Swift. Trade name of Johnson Matthey plc (UK) for a silver brazing alloy containing 30% silver, 28% copper, 21% zinc, and 21% cadmium. It has a melting range of 607-685°C (1124-1265°F).

Argozoil. Trade name for a corrosion-resistant nickel silver containing 54% copper, 20-28% zinc, 14% nickel, 2-10% lead, and 2% tin. Used for ornamental parts.

Argus. (1) Trade name of Nassau Smelting & Refining Company Inc. (USA) for a bearing alloy containing varying amounts of tin, lead and antimony.

(2) Trade name of Pittsburgh Corning Corporation (USA) for non-light-directing glass blocks with smooth exterior surfaces, and rounded flutes on the interior surfaces at right angles to each other.

argutite. A colorless to light gray mineral of the rutile group composed of germanium oxide, GeO_2. It can also be made synthetically. Crystal system, tetragonal. Density, 6.28 g/cm³. Occurrence: France.

Arguzoid. German trade name for a nickel silver containing 48-56% copper, 13-21% nickel, 23-31% zinc, and up to 4% tin and lead, respectively. Used for cutlery, ornaments, and utensils.

Argyroid. British trade name for a nickel silver containing varying amounts of copper, zinc and nickel. Used for ornaments

and resistances.

argyrite. See argentite.

argyrodite. A black mineral of the cancrinite group composed of silver germanium sulfide, $Ag_8(Ge,Sn)S_6$. Crystal system, orthorhombic. Density, 6.20 g/cm³. Occurrence: Bolivia, Germany. It can also be made synthetically, but the synthetic material does not contain tin, and has a density of 6.25 g/cm³. Used as a source of germanium.

Argyrolith. British trade name for a nickel silver containing 50-70% copper, 10-20% nickel, and 5-30% zinc. Used for ornaments and resistances.

Argyrophan. British trade name for a nickel silver containing varying amounts of copper, zinc and nickel, used for ornaments, and utensils.

arhbarite. A blue mineral composed of copper arsenate hydroxide hexahydrate, $Cu_2(AsO_4)(OH)·6H_2O$. Occurrence: Morocco.

Aridall. (1) Trademark of Chemdal International Corporation (USA) for a crosslinked copolymer of acrylamide and acrylic acid that can absorb large amounts of water (about 400 times its weight). Supplied in granule form, it is used for hygiene products, such as disposable diapers, for laboratory applications, and in horticulture.

(2) Trade name of American Colloid Company (USA) for bentonite clay products.

aridized plaster. A plaster that has been treated at elevated temperatures with calcium chloride to improve its uniformity and strength.

Ariel bronze. A cast bronze containing 10% tin, 0.25% lead and 0.5% phosphorus, used for bearings and bushings.

Aries. Trademark of Ivoclar Vivadent AG (Liechtenstein) for a ceramic dental alloy containing 63.7% palladium, 26.0% silver, 7.5% gallium, 7.0% tin, 1.5% indium, and less than 1.0% ruthenium and rhenium respectively. It has a white color, a density of 10.8 g/cm³ (0.39 lb/in.³), a melting range of 1165-1290°C (2130-2355°F), a hardness after ceramic firing of 185 Vickers, high elongation, and excellent biocompatibility and processibility. Used for crowns, onlays, posts and bridges.

Ariloft. Trademark for cellulose acetate fibers and yarns used for clothing.

Arimax. Trade name of Ashland Chemical Company (USA) for a series of high-strength, low-viscosity composites composed of acrylamate resin matrices reinforced with glass fibers. They are usually processed by reaction-injection molding, and used for automotive and aerospace components, electronic components, and agricultural parts.

aristarainite. A colorless mineral composed of sodium magnesium borate octahydrate, $Na_2MgB_{12}O_{20}·8H_2O$. Crystal system, monoclinic. Density, 2.03 g/cm³; refractive index, 1.498. Occurrence: Argentina.

Arista. Trademark of Schott DESAG AG (Germany) for a fusible glass of varying color and shape made by fusing in an oven. Supplied in thicknesses from 2.5-8.5 mm (0.10-0.34 in.), it is used for windows, doors, etc., and in the manufacture of laminated safety glass and insulating glass.

Aristaloy. Trade name of Goldsmith & Revere (USA) for dental amalgam alloys including *Aristaloy 21* dispersed phase amalgam alloy capsules and *Aristaloy CR* high-copper spherical amalgam alloy capsules.

Aristech. Trademark of Aristech Acrylics LLC (USA) for an acrylic resin supplied in the form of optical or colored sheets for signs, and other industrial applications. Also known as *Aristech Acrylics.*

Aristech Acrysteel. See Acrysteel.

Aristocore. Trade name for a creamy, composite paste used in restorative dentistry for core and crown build-ups.

Aristocrat. (1) Trade name of Lumen Bearing Company (USA) for a tin babbitt containing 85% tin, 7.5% antimony, and 7.5% copper. Used for bearings.

(2) Trade name for rayon fibers and yarns used for textile fabrics.

Aristofil DP. Trade name for light-cure microfine composite resin used in restorative dentistry.

Aristoloy. (1) Trade name of Engelhard Corporation (USA) for an amalgam composed of 67-70% silver, 25-29% tin, 3-5% copper, up to 1% zinc, and the balance mercury. Used for dental fillings.

(2) Trade name of American Radiator Company (USA) for a series of heat- and corrosion-resistant austenitic, ferritic and martensitic stainless steels, and leaded and lead-free chromium-molybdenum alloy steels.

Aristonol. Trade name for zinc-oxide/eugenol dental cement.

Aristopane. Trade name of Aristocrat Division of Pacific Coast Company (USA) for insulating glass.

Arizona ruby. A dark crimson to ruby-red *pyrope* of the garnet group from the southwestern United States particularly Arizona and New Mexico. Used as a gemstone.

Arjalloy. Trade name for a dental amalgam alloy.

Ark. Trade name of Jessop-Saville Limited (UK) for a high-speed tool steel containing 0.7% carbon, 3.7% chromium, 0.6% molybdenum, 14% tungsten, 1% vanadium, and the balance iron. Used for cutting tools of all types, punches, etc.

Arkansas pine. A collective term used in the US lumber trade for mixtures of loblolly, longleaf, pond, shortleaf and slash pines.

Arkansas stone. A variety of *novaculite* found in Arkansas, USA, and used as an abrasive, and for whetstones.

Arkaywall. Trade name of Robinson King & Company Limited (UK) for a glass having a ceramic color applied and fired in. Used for infill panels in curtain wall constructions.

Arkit. French trade name for an alloy of 38% cobalt, 30% chromium, 16% tungsten, 10% nickel, 4% molybdenum, and 2.5% carbon. It has high heat resistance, and is used for high-speed cutting tips of lathe tools.

Arko. Trade name for drawn or spun brass containing 80% copper and 20% zinc. Used for tubes, fittings, and hardware.

Ark Superior. Trade name of Jessop-Saville Limited (UK) for a high-speed tool steel containing 0.8% carbon, 4.25% chromium, 0.6% molybdenum, 18% tungsten, 1.3% vanadium, and the balance iron. Used for cutting tools of all types, drills, broaches, reamers, and gear cutters.

Ark Superlative. Trade name of Jessop-Saville Limited (UK) for a high-speed tool steel containing 0.8% carbon, 4.7% chromium, 0.6% molybdenum, 18.5% tungsten, 1.6% vanadium, 5.7% cobalt, and the balance iron. Used for cutting tools of all types, milling cutters, and other various other tools.

Ark Supreme. Trade name of Jessop-Saville Limited (UK) for a high-speed tool steel containing 0.8% carbon, 5% chromium, 22% tungsten, 0.6% molybdenum, 1.7% vanadium, 17% cobalt, and the balance iron. Used for cutting tools of all types, and for boring and planing tools.

Ark Triumph. Trade name of Jessop-Saville Limited (UK) for a high-speed tool steel containing 1.2% carbon, 4.4% chromium, 14% tungsten, 4.5% vanadium, and the balance iron. Used for cutting tools of all types, and for form tools.

Ark Triumphant. Trade name of Jessop-Saville Limited (UK) for a high-speed tool steel containing 1.23% carbon, 4.5% chromium, 13% tungsten, 3.7% vanadium, and the balance iron. Used for cutting tools of all types.

Arkturus. Trade name of Gebrüder Hover Edelstahlwerk (Germany) for a tough, oil-hardening cold-work tool steel used for tools, headers and upsetters.

Arlcite. Trademark of Ferro Corporation (USA) for a high-density aluminum oxide (3.4 g/cm³, or 0.12 lb/in.³) having a hardness of 9.0 Mohs, and a very consistent wear rate. Commercially available in the form of linings and grinding balls in 85% and 90+% purities, it is used as grinding medium in high-solids, high-viscosity milling systems, and for lining grinding mills.

Arleccino. Trade name of Vetreria di Vernante SpA (Italy) for a glass with mosaic-type of pattern.

Arlen. Trademark of Mitsui Chemicals, Inc. (Japan) for a modified polyamide 6T with high rigidity and dimensional stability, high melting point, excellent chemical and heat resistance, and very low water absorption. It is available in three product series: (i) *Arlen A/G* for mechanical and structural components; (ii) *Arlen AE* for tribological applications; and (iii) *Arlen C* for electrical and electronic components.

Arlene. Trademark of Aquafil SpA (Italy) for polypropylene filament yarns.

Arlequin. Trade name of Boussois Souchon Neuvesel SA (France) for patterned glass featuring a broken-line design.

Arley. Trade name of Hall & Pickles Limited (UK) for a high-carbon die steel containing 1.4% carbon, a total of 3% chromium and tungsten, the balance being iron. Used for extrusion dies.

Arlex. (1) Trade name for American Falcon, Inc. (USA) for a range yarns and fibers.

(2) Trade name for converted polyester film.

Arlon. Trademark of Greene Tweed & Company (USA) for a series of ductile, high-temperature polyetheretherketone thermoplastics, either unreinforced or reinforced with glass or carbon fibers. Used for bearings, bushings, insulators, valves and pumps.

Arloy. Trade name of Arco Chemical Company (USA) for a series of polycarbonate/styrene-maleic anhydride (PC-SMA) blends. They have good ductility and toughness, good impact and shock resistance, excellent moldability and processibility, good melt-flow characteristics, and good chemical and thermal resistance. Grades with high resistance, and/or enhanced low-temperature ductility are also available. Used for automotive-trim applications, hand-held power tools, tool housings, camera components, dishware, and food containers.

Arma. (1) Trade name of Phosphor Bronze Company Limited (UK) for a tin-base bearing metal containing varying amounts of antimony and copper. Used for high-duty bearings.

(2) Trade name for rayon fibers and yarns available in several types including *Arma Flisco and Arma Lamo*, and used for textile fabrics.

Armacast. Trade name of Duraloy Blaw-Knox Corporation (USA) for a tough, cast steel containing 0.25-0.3% carbon, 2.65-3.15% chromium, 0.45-0.55% molybdenum, and the balance iron. Used for road construction and railway equipment.

Armacor. Trademark of Amorphous Technologies International (USA) for a series of hardfacing materials which transform under abrasive wear or surface grinding from two-phase alloys to hard, amorphous materials. They have high toughness, excellent abrasion and corrosion resistance, very good oxidation resistance up to 925°C (1700°F), good hot hardness, and low coefficients of friction. *Armacor M* is a hard, wear-resistant high-temperature iron-base coating material that partially transforms to an amorphous state when abraded or dry ground.

armalcolite. A gray mineral of the pseudobrookite group composed of iron magnesium titanium oxide, $FeMgTi_4O_{10}$. It can also be made synthetically. Crystal system, orthorhombic. Density, 3.92 g/cm^3. Occurrence: Found in lunar rock samples.

Armalon. Trade name of Saint-Gobain (France) for *Teflon* fluorocarbon impregnated fiberglass.

armangite. A brown, or blackish mineral composed of manganese arsenite, $Mn_3(AsO_3)_2$. Crystal system, rhombohedral (hexagonal). Density, 4.23-4.41 g/cm^3; refractive index, 2.01. Occurrence: Sweden.

Armaspray. Trade name of Armstrong World Industries (USA) for sprayable asbestos and non-asbestos plasters and fireproofing compounds.

Armasteel. Trade name of General Motors Corporation (USA) for a series of pearlitic malleable cast irons containing 2.45-2.75% carbon, 1.25-1.55% silicon, 0.3-0.5% manganese, 0.025-0.12% sulfur, up to 0.05% phosphorus, and the balance iron. They have high ultimate and yield strengths, and are used for automotive components, such as camshafts, universal joints, gears, rocker arms, connecting rods, etc.

Armat. Trade name for a chromium permanent-magnet steel with high coercive force and large magnetic hysteresis.

Armater. Trademark of Colbond Geosynthetics/Acordis Group for a honeycomb-type geocomposite used in civil engineering to confine and stabilize soil top layers on slippery surfaces.

Armator. Trade name for a high-gold dental bonding alloy.

Armature Electric. Trade name of Follansbee Steel Company (USA) for a silicon steel containing 0.04% carbon, 0.5% silicon, and the balance iron. It has high magnetic permeability, and is used for electrical applications, e.g., rotating equipment and field poles.

armature varnish. An *insulating varnish* applied the windings of electric generators and motors to protect them against moisture, and hold them in place.

Armaver. Trade name of Saint-Gobain (France) for reinforced glass cloth used on pipelines as anti-corrosive sheathing.

Armco. Trade name of Armco International (USA) for an extensive series of wrought and ingot irons, stainless steels, silicon steels, carbon steels, etc.

Armco ingot iron. See Armco iron.

Armco iron. Trade name of Armco International (USA) for a ductile, high-purity iron (99.85+%) made by the open-hearth process. It typically contains about 0.012% carbon, 0.017% manganese, 0.025% sulfur, 0.005% phosphorus, and the balance iron. It has a density of 7.858, a melting point of 1530°C (2786°F), excellent corrosion resistance, high elongation, and good soft magnetic properties. Used for electromagnetic cores, steelmaking, porcelain enameling, pipes, culverts, tanks, roofing, etc. Also known as *Armco ingot iron.*

Armco SMC. Trade name of BP Chemical Company (USA) for polyester sheet-molding compounds supplied in regular, low-profile, high-impact and fire-retardant grades.

Armelec. Trade name of Empire Sheet & Tin Plate Company (USA) for an alloy of 99.5% iron and 0.5% silicon. It has high magnetic permeability, and is used for armatures of motors and generators.

armenite. A grayish green mineral composed of the osumilite group composed of barium calcium aluminum silicate dihydrate, $BaCa_2Al_6Si_8O_{28}\cdot2H_2O$. Crystal system, hexagonal. Density, 2.76 g/cm^3; refractive index, 1.559. Occurrence: Norway.

Armet. Trademark of Armet Industries Limited (Canada) for plastics, fluorocarbon, fluorosilicone and silicone rubber supplied in the form of blocks, rods, sheeting, tubing, coatings, and calendered, extruded and molded products including washers, gaskets, O-rings, hoses, shrouds, insulating tape, and for adhesives and gummed tape.

Armex. (1) British trade name for a series of austenitic stainless steels containing 0.07-0.08% carbon, 17-20% chromium, 8-9% nickel, 2-4% molybdenum, and the balance iron. Used for welding electrodes.

(2) Trademark of Church & Dwight Company Inc. (USA) for corrosion- and rust-inhibiting coating compositions for ferrous and nonferrous metals and alloys.

(3) Trade name of Church & Dwight Company Inc. (USA) for soft blasting media based on sodium carbonate. Used for cleaning and restoring brick, stone, concrete, wood, metals, and glass.

(4) Trade name of ABCO Manufacturing Company (USA) for epoxy patching compounds.

Armico. Trade name for strips made of commercially pure iron and containing 0.2% copper, 0.15% nickel, 0.03% carbon and 0.005% silicon.

Armide. Trade name of Armstrong Bothers Tool Company (USA) for a series of bonded tungsten carbides used for cutting tools, drawing and forming dies, gages, wear parts, etc.

Armite. (1) Trade name for a *synthetic cast iron* with high tensile strength.

(2) Trademark of Spaulding Fibre Company, Inc. (USA) for vulcanized fiber products and laminates available in the form of rolls, sheets, strips, blocks, and boards. They have high dielectric strength, and high mechanical strength. Used for electrical insulation (e.g., impregnated fish paper), and as backing for sheet abrasives.

Armon. Trade name for rayon fibers and yarns used for textile fabrics.

Armonia. Trade name of Fabbrica Pisana SpA (Italy) for a glass with trellis or lattice patterns.

Armor. Trade name of McGean (USA) for an electroless nickel plate and electroplating process.

armored fabrics. A class of *protective fabrics* made of asbestos, cotton, glass, rayon or other fibers and coated or impregnated with synthetic resin (e.g., polyvinyl chloride), natural or synthetic rubber or certain chemicals (e.g., cellulose nitrate). Used especially for industrial applications involving harsh and severe environments.

Armored Poly-Thermaleze 2000. Trade name of Phelps Dodge Magnet Wire Company (USA) for polymer-coated magnet wire.

armored wood. Wood that has been reinforced with a sheetmetal facing on one or both sides. See also wood laminates.

Armor-Gard. Trade name of ASG Industries Inc. (USA) for burglar-resistant glass consisting of two or more clear, bronze, gray, or green plates or sheets of glass laminated with an extra-tough, 15 mm (0.06 in.) thick plastic interlayer.

Armorlite. Trade name of ASG Industries Inc. (USA) for bullet-resistant glass consisting of two or more layers of plate glass laminated with 0.38 mm (0.015 in.) thick plastic interlayers.

armorplate. An extremely heavy, surface-hardened low-carbon steel plate produced in special rolling mills. A typical composition is 0.3% carbon, 2.75% chromium, 2.5% nickel, 0.25% molybdenum, and the balance iron. Used for the protection of warships and military vehicles.

Armorthane. Trademark of ArmorThane Coatings Inc. (Canada) for wear-resistant polyurethane coatings for application on tanks, truck beds, and metal, wood, fiberglass and concrete deck surfaces.

Armor Weld. Trade name of Missouri Paint & Varnish Company (USA) for epoxy adhesives and floor cements.

Armour. Trade name of Iko Industries Limited (Canada) for asphalt shingles.

Armourbex. Trade name of BX Plastics (UK) for cellulose acetate plastics.

Armourcast. Trade name of Pilkington Brothers Limited (UK) for a toughened rough-cast glass.

Armourclad. Trade name of Pilkington Brothers Limited (UK) for a ceramic-enameled glass used for cladding applications.

Armourfloat. Trade name of Pilkington Brothers Limited (UK) for a toughened float glass.

Armourglass. Trade name of Pilkington Brothers Limited (UK) for a toughened patterned glass.

Armourlight. Trade name of Pilkington Brothers Limited (UK) for toughened glass insulators, flameproof well-glasses and toughened-glass lenses used for glass-concrete construction.

Armourlite. Trade name of Duplate Canada Limited for a toughened sheet glass.

Armourplate. Trade name of Pilkington Brothers Limited (UK) for toughened float and plate glass.

Armourplate Tuyere. Trade name of Pilkington Brothers Limited (UK) for a blue, toughened plate glass used for furnace peepholes.

Armoursheet. Trade name of Pilkington Brothers Limited (UK) for a toughened sheet glass.

Arm-R-Brite. Trade name of C-E Glass (USA) for an insulated glass building panels consisting of ceramic-enameled toughened glass, an insulating core of polyurethane foam or fiberglass, and a back-up panel.

Arm-R-Clad. Trade name of C-E Glass (USA) for a toughened glass.

Armstrong. (1) Trade name for a heat-resisting, stainless steel containing 0.5% carbon, 12% chromium, 5% silicon, and the balance iron. Used for heat- and/or corrosion-resistant parts.

(2) Trade name of Valspar Corporation (USA) for an extensive line of paints, enamels, and varnishes.

(3) Trade name of Armstrong World Industries (USA) for asbestos- or magnesia-based high-temperature insulation blocks and pipe covering products.

armstrongite. A brown mineral composed of calcium zirconium silicate hydrate, $CaZrSi_6O_{15} \cdot 2.5H_2O$. Crystal system, monoclinic. Density, 2.58 g/cm^3; refractive index, 1.569. Occurrence: Mongolia.

Armstrong metal. (1) A corrosion-resistant steel containing of 0.1% carbon, 17% chromium, 8% nickel, 4-6% manganese, 3% copper, and the balance iron. Used for heat-resistant parts.

(2) An alloy containing 70-80% nickel, 17.5% chromium, 4-6% manganese, 2.9% copper, and 0.1% carbon. Used for corrosion-resistant drawn or pressed shapes.

Arndt alloy. An alloy containing 60% magnesium and 40% copper.

Arngrim. Trade name of Uddeholm Corporation (USA) for a series of magnet alloys containing 0.65-0.75% carbon, 0.25-0.65% chromium, 6% tungsten, and the balance iron. Used for permanent magnets.

Arne. Trade name of Uddeholm Corporation (USA) for an oil-hardening cold-work tool steel (AISI type O1) containing 0.9% carbon, 1.2% manganese, 0.5% chromium, 0.5% tungsten, 0.1% vanadium, and the balance iron. It has a great depth of hardening, good machinability, and low tendency to shrinking and warping. Used for tools and dies.

Arnel. Trademark of Celanese Chemical Company (USA) for thermoplastic cellulose triacetate fibers and yarns that have a melting point of approximately 300°C (570°F), good resistance to dilute solutions of weak acids and most common solvents, poor resistance to strong alkalies and oxidizers, and excellent resistance to mildew and sunlight. Used to make dimensionally stable woven and knitted fabrics that are resistant to shrinkage, creases, and color fading. Also used for protective coating applications.

Arnew. Trade name of Uddeholm Corporation (USA) for an oil-hardening cold-work tool steel (AISI type O1).

Arnite. Trade name of DSM Engineering Plastics (USA) for thermoplastic polyesters, especially polybutylene terephthalate and polyethylene terephthalate, supplied in various grades. *Arnite A, Arnite B & Arnite C* are strong, stiff polyethylene terephthalate (PET) resins supplied in various grades: crystalline, unreinforced, UV-stabilized, fire-retardant, and glass fiber-reinforced. They possess good dimensional stability, low water absorption, good chemical resistance, and are used for automotive and electrical components. *Arnite D* includes several grades of amorphous polyeth-ylene terephthalate (PET) resins, *Arnite G* several polyethylene terephthalate (PET) resins and products, and *Arnite T* refers to flame-retardant polybutylene terephthalate (PBT) supplied in various grades including unreinforced and reinforced with glass contents of 10-40%. It has excellent mechanical strength, heat resistance, weatherability, UV-resistance and surface appearance, and is used for automotive, electrical and electronic applications.

Arnitel. Trademark of DSM Engineering Plastics (USA) for high-performance thermoplastic elastomers based on copolyester-esters and supplied in three grades (*Arnitel E, P* and *U*).

Arno. Trade name of Schering Corporation (USA) for adhesive tape.

Arnochrome. Trademark of Arnold Engineering Company (USA) for a ductile, isotropic permanent-magnet alloy composed of 26-30% chromium, 7-10% cobalt, and the balance iron. Commercially available in the heat-treated form as bar, rod and wire, it can be drawn, cold-headed, formed and machined prior to magnetic heat treatment. It has a density of 7.6 g/cm^3 (0.275 $lb/in.^3$), a magnetic coercivity of 50-300 Oe, a magnetic remanence of 9-12 kG, and a Curie temperature of 625°C (1157°F). Used for sensor systems and low-coercivity applications.

Arnox. (1) Trademark of General Electric Company (USA) for a series of one-part liquid and solid epoxy resins available in various grades including compression and transfer molding, injection molding, filament winding and pultrusion.

(2) Trademark of Arnold Engineering Company (USA) for a series of hard sintered ceramic magnet materials of the barium hexaferrite ($BaO \cdot 6Fe_2O_3$) or strontium hexaferrite ($SrO \cdot 6Fe_2O_3$) type. They have high coercivities and energy products, and are used for permanent magnets for motors, generators, etc.

Aro. Trade name of Time Steel Service Inc. (USA) for an oil- or water-hardening, shock-resistant tool steel (AISI type S2) containing 0.5% carbon, 0.7% manganese, 0.7% silicon, 0.2% vanadium, 0.45% manganese, and the balance iron. It has outstanding toughness, a great depth of hardening, moderate wear resistance, and moderate machinability. Used for pneumatic tools, impact tools, chisels, and punches.

Arobond. Trade name for a nickel-chromium dental bonding alloy.

AroCy. Trademark of Ciba-Geigy Corporation (USA) for high-performance polymer resins including various monomeric and prepolymer bisphenol A dicyanates and cyanate esters.

Arodure. Trademark of Ashland Inc. (USA) for a series of urea-formaldehyde resins.

Arofene. Trademark of Ashland Inc. (USA) for low-cost phenolics with excellent thermal stability to above 150°C (300°F) that can be compounded with a variety of fillers and resins. Used for motor housings, electrical fixtures, and automotive under-the-hood components.

Aroflat. Trademark of Reichhold Chemicals, Inc. (USA) for rosin-modified alkyd resins used for flat interior paints.

Aroflint. Trademark of Reichhold Chemicals, Inc. (USA) for hard, durable polyester resins with good adhesion properties.

Arolon. Trademark of Reichhold Chemicals, Inc. (USA) for acrylic and styrenated acrylic emulsions used for thermosetting coatings.

aromatic-aliphatic compound. See arene.

aromatic amine. Any of a group of organic compounds, such as m-phenylenediamine or 4,4'-methylenedianiline, containing one or more amine groups ($-NH_2$) and one or more benzene groups ($-C_6H_4$) in their molecular structure. Other chemical groups, such as methylene ($-CH_2-$) or sulfone ($-SO_2-$), may also be present. They are frequently used as elevated-temperature curing agents for epoxy resins.

aromatic compounds. A term referring to benzene (C_6H_6) and all compounds derived from it or resembling it in chemical behavior. See also aromatic hydrocarbon.

aromatic divinyl compound. A divinyl (1,3-butadiene) compound, such as divinyl benzene $[C_6H_4(CH=CH_2)_2]$, containing one or more benzene rings in its structure.

aromatic heterocyclic compound. A *heterocyclic compound* containing one or more benzene rings and one or more other groups in its structure.

aromatic hydrocarbon. Any of a group of unsaturated hydrocarbons containing one or more closed benzene rings of six carbon atoms each, or benzene-resembling groups in their molecular structures. Examples include naphthalene ($C_{10}H_8$), toluene ($C_6H_5CH_3$), and phenol (C_6H_5OH).

aromatic isocyanates. A group of compounds, such as toluene diisocyanate, diphenylmethane diisocyanate and naphthalene diisocyanate, containing one or more benzene groups ($-C_6H_4$) and one or more isocyanate groups ($-NCO$) in their structures. Other chemical groups, such as methyl ($-CH_3$), may also be present. Used in the manufacture of polyurethanes.

aromatic polyethers. A family of essentially amorphous, hydrophobic engineering thermoplastics including polyaryl ether and methyl-substituted phenylene oxide resins that are made by the oxidative coupling of phenolic monomers, e.g., dimethylphenol. They have excellent thermo-oxidative and dimensional stability, high creep resistance, good flame resistance and dielectric properties, and low water absorption. Used for electrical and electronic applications.

aromatic polyarylates. A family of completely aromatic and amorphous engineering thermoplastics derived from aromatic dicarboxylic acids and diphenols. They have outstanding weatherability, excellent low-temperature and ultraviolet-light resistance, good mechanical properties at elevated temperatures, and good impact and radiation resistance. Used for electrical connectors, switch and fuse covers, relay housings, fire hel-

mets, face shields, automotive headlight housings, light reflectors, exterior and interior trim, outdoor construction, and lighting components.

aromatic polyesters. A family of engineering plastics obtained from monomers in which all the carboxyl ($-COOH$) and hydroxyl ($-OH$) groups are directly attached to aromatic nuclei. They do not melt like conventional plastics, but can be made to flow at temperatures above 427°C (800°F). They have excellent high-temperature performance, high thermal conductivity, very high compressive strength, high tensile strength and moduli of elasticity, good wear resistance, and excellent resistance to solvents, oils and corrosive chemicals. Abbreviation: ARP; ArP. Also known as *polyoxybenzoates*.

aromatic polymer composites. A group of composite materials consisting of aromatic polymer matrices (e.g., polyether ether ketone) with fiber reinforcements.

aromatic polysulfones. A family of transparent, amorphous polyaryl sulfone polymers with good resistance to inorganic acids, alkalies, and aqueous salt solutions.

aromatic substance. Any of a large group of organic compounds containing a benzene ring, or benzene-resembling groups. Examples include benzene (C_6H_6), naphthalene ($C_{10}H_8$) and toluene (C_7H_8).

aromatic sulfones. A family of amorphous engineering thermoplastics including polysulfones, polyether sulfones and polyphenylene sulfones that have molecular chains consisting of partially or fully aromatic building units interlinked with sulfonyl groups. They have high glass-transition temperatures, excellent dimensional stability, good flame resistance, high creep resistance, good impact strength and hydrolytic stability, and high thermal and thermo-oxidative stability. Used for microwave cookware, printed circuit boards, electrical and electronic connectors, battery parts, switches, fuse housings, pumps, meter components, valves, seals, supercharger and turbocharger components, automotive heater fans and bearing cages, aircraft radomes, medical and dental instruments, and lighting fittings.

aromatic thermoplastic polyesters. A family of structural thermoplastics produced by the polycondensation of ethylene glycol and terephthalic acid. The most important member of this group is *polyethylene terephthalate* (PET).

aromatic thermoplastic polyimides. A family of fully imidized, linear polymers produced by polycondensation reactions of aromatic dianhydrides with aromatic diamines or aromatic diisocyanates in a suitable reaction medium. They have exceptional thermomechanical properties, outstanding high-temperature resistance, high toughness, good dielectric properties, high radiation resistance, and low flammability. Used for structural parts, adhesives, bearings for the aerospace and automotive industries, thermal and electrical insulators, and printed circuit boards.

Aron Alpha. Trade name of Aron Alpha, Division of Elmer's Products (USA) for cyanoacrylate adhesives.

Arophlex. Trademark of Ashland Oil, Inc. (USA) for sheet molding compounds used for automotive and industrial applications.

Aroplaz. (1) Trade name of Ashland Oil, Inc. (USA) for a series of alkyd resins including several oil-modified grades. Used for coating applications.

(2) Trademark of Reichhold Chemicals, Inc. (USA) for hard, linear baking polyesters with good gloss and color retention, used for coatings.

Aropol. (1) Trademark of Ashland Oil, Inc. (USA) for single-component and glass-reinforced unsaturated polyester resins that can be formulated for room- and high-temperature use. They have excellent electrical properties, and are used for automotive and aerospace components, and electronic applications.

(2) Trade name of Ashland Oil, Inc. (USA) for fast-drying vinyltoluene-modified short-oil alkyds used for coating applications.

Aropol WEP. Trademark of Ashland Oil, Inc. (USA) for water-extended polyester resins.

Aroset. Trademark of Ashland Oil, Inc. (USA) for a series of thermosetting acrylic resins.

Arosperse. Trademark of Engineered Carbons, Inc. (USA) for a series of oil-pelleted furnace- and thermal-grade carbon blacks (92-95% pure) with a specific gravity of 1.7-1.9. The iodine content and average particle size of furnace grades ranges from 26 to 129 mg/g and 19 to 91 nm, respectively. For thermal grades the iodine content is typically 10 mg/g, and the average particle size 290 nm. Used as fillers and reinforcers in rubbers and plastics, and as pigments.

Arostit. Trade name of Hoffmann Elektrogusstahl (Germany) for a series of corrosion- and heat-resistant austenitic, ferritic and martensitic stainless steels.

Arotech. Trademark of Ashland Oil, Inc. (USA) for sheet molding compounds for automotive and industrial applications.

Arothane. Trade name of Reichhold Chemicals, Inc. (USA) for linseed-oil-modified urethane resins used for coating applications.

Arothix. Trade name of Reichhold Chemicals, Inc. (USA) for thixotropic isophthalic alkyd resins used for coating applications.

Arotran. Trademark of Ashland Oil, Inc. (USA) for resin transfer molding materials used for automotive and industrial applications.

Arpak. Trade name of Arco Chemical Company (USA) for chemical-resistant expanded polyethylene beads used for foam-type packaging, and furniture.

Arpocalloy. Trade name of Arpocalloy Company (USA) for an alloy cast iron containing varying amounts of carbon, iron, nickel and molybdenum. Used for tool shanks.

Arpro. Trade name of Arco Chemical Company (USA) for heat-resistant expandable polypropylene beads used for packaging applications, and in the manufacture of molded foam products.

arquerite. A soft, malleable mineral from Chile composed of silver amalgam containing approximately 87% silver and 13% mercury.

Arradur. Trade name of Atofina (France) for acrylonitrile-butadiene-styrene resins used in the manufacture of electronic products, and sporting and leisure goods.

Arrestite. Trade name of Republic Steel Corporation (USA) for a nondeforming tool steel containing 0.9% carbon, 1.5% manganese, 0.2% chromium, and the balance iron.

arrojadite. A dark yellowish green mineral composed of potassium sodium calcium iron manganese aluminum fluoride phosphate hydroxide, $KNa_4CaMn_4Fe_{10}Al(PO_4)_{12}(OH,F)_2$. Crystal system, monoclinic. Density, 3.56 g/cm³; refractive index, 1.670. Occurrence: Brazil, USA (South Dakota).

Arrowblast. Trade name of Norton Company (USA) for high-pressure aluminum oxide blasting abrasives. Available in particle sizes from 16 to 80 mesh, they are used for blast cleaning of metals.

arsenate. (1) A salt or ester of (ortho)arsenic acid.

(2) A compound containing the AsO_4^{3-} radical.

arsenbrackebuschite. A light brown to yellow mineral of the brackebuschite group composed of lead iron zinc arsenate hydrate, $Pb_2(Fe,Zn)(AsO_4)_2 \cdot H_2O$. Crystal system, monoclinic. Density, 6.54 g/cm³; refractive index, 2.0. Occurrence: Namibia.

arsendescloizite. A pale yellow mineral of the descloizite group composed of lead zinc arsenate hydroxide, $PbZn(AsO_4)(OH)$. Crystal system, orthorhombic. Density, 6.56 g/cm³; refractive index, 2.030. Occurrence: Namibia.

arsenic. A metalloid element belonging to Group VA (Group 15) of the Periodic Table. It is usually a steel-gray brittle, crystalline solid with metallic luster, but a black, amorphous (β-arsenic), and a yellow cubic form are also known. It occurs combined in minerals, such as realgar, orpiment, arsenopyrite and arsenolite, and in the free state as native arsenic. Crystal system, hexagonal; crystal structure, rhombohedral. Density, 5.6-5.9 g/cm³; melting point, 814°C (1497°F); boiling point, sublimes at 613°C (1135°F); atomic number, 33; atomic weight, 74.922; divalent, trivalent and pentavalent; hardness, 57-69 Vickers; low thermal conductivity; semiconductive properties. Metallic *arsenic* is used as an alloying element in lead and copper alloys (e.g., lead shot, arsenical brasses and bronzes, battery grids, cable, sheaths, boiler tubes, etc.), and in glass manufacture. High-purity grades are used in the manufacture of gallium arsenide semiconductors, as doping agents in germanium and silicon solid-state devices (e.g., light-emitting diodes), and in solders and lasers. Symbol: As.

arsenical admiralty brass. An embrittlement-free, corrosion- and dezincification-resistant brass containing 71% copper, 28% zinc, 1% tin, and 0.03-0.04% arsenic. Used for heat exchangers, condensers, condenser tubing, and marine parts.

arsenical aluminum brass. An embrittlement-free aluminum brass containing 76-77.5% copper, 20.5-22% zinc, 2% aluminum, and 0.03-0.1% arsenic. Commercially available the form of tubing, it has good to excellent corrosion resistance, excellent cold workability, and high elongation. Used for distiller, evaporator and heat-exchanger tubing, condensers, ferrules, etc. Also known as *inhibited aluminum brass*.

arsenical aluminum bronze. A corrosion-resistant bronze containing 94.25% copper, 5.5% aluminum, and 0.25% arsenic. It has good resistance to seawater corrosion, and is used for condenser tubes, heat exchangers, and marine parts.

arsenical antimony. See allemontite.

arsenical babbitt. A lead-base babbitt containing up to about 3% arsenic. The arsenic minimizes softening at elevated temperatures (95-150°C or 203-302°F), improves the fatigue strength, and increases the hardness. Used for rolling-mill bearings, automobile bearings, and diesel engines.

arsenical bronze. (1) A free-machining brass containing about 55-65% copper, 30-40% zinc, 0.5-3.0% lead, 1-2% nickel, 1-2% iron, and 0.5-0.75% arsenic. It has excellent wear and corrosion resistance, and is used for machine elements, such as gears and pinions, and for screw-machine parts.

(2) A heavy-duty bronze containing 80% copper, 10% tin, 9.2% lead, and 0.8% arsenic. Used for bearings.

arsenical copper. A corrosion-resistant, wrought copper containing 0.2-0.6% arsenic, and 0.02% phosphorus. The arsenic has a slightly increasing effect on the hardness and strength, while raising the recrystallization temperature. Used for condensers, heat exchangers, boiler tubes, firebox stays, etc.

arsenical nickel. See nickeline.

Arsenic-Antimony. Trade name of Blackwells Metallurgical Limited (UK) for antimony containing up to 10% arsenic.

arsenic bisulfide. See arsenic disulfide.

arsenic bloom. See arsenolite.

arsenic bromide. A moisture-sensitive yellow powder or yellowish-white crystals (99+% pure) usually obtained by the interaction of bromine and arsenic. It has a density of 3.54 g/cm³, a melting point of 32.8°C (91°F), and a boiling point of 221°C (429°F). Used in ceramics, analytical chemistry, medicine, and organometallic synthesis. Formula: AsBr₃. Also known as *arsenic tribromide; arsenious bromide.*

arsenic chloride. A colorless or yellowish, oily liquid (99+% pure) usually obtained by the interaction of chlorine and arsenic. It has a density of 2.163 g/cm³, a melting point of -18°C (-0.4°F), and a boiling point of 130.2°C (266°F). Used in ceramics, as an intermediate for arsenic compounds, and in organometallic synthesis. Formula: AsCl₃. Also known as *arsenic trichloride; arsenious chloride; caustic arsenic chloride; fuming liquid arsenic.*

Arsenic-Copper. Trade name of Blackwells Metallurgical Limited (UK) for an alloy of arsenic containing 50% copper. Used as a primary metal.

arsenic disulfide. See arsenic sulfide (i).

arsenic fluoride. An oily liquid with a density of 2.67 g/cm³, a melting point of -8.5°C (17°F), a boiling point, 63°C (145°F)/752 mm. Used as a catalyst, as an ion-implantation source, as a dopant, and as a fluorinating agent. Formula: AsF₃. Also known as *arsenic trifluoride; arsenious fluoride.*

arsenic hydride. A colorless gas with a melting point of -113.5°C (-172°F) and a boiling point of -62°C (-80°F). It decomposes above 230°C (446°F). Used as a doping agent in solid-state electronics. Formula: AsH₃. Also known as *arsine.*

arsenic iodide. A red powder (98+% pure) colorless or yellowish, oily liquid obtained usually by the interaction of chlorine and arsenic. It has a density of 4.39 g/cm³, a melting point of 146°C (294°F), and a boiling point of 403°C (757°F). Used in ceramics and organometallic synthesis. Formula: AsI₃. Also known as *arsenic triiodide; arsenious iodide.*

Arsenic-Iron. Trade name of Blackwells Metallurgical Limited (UK) for an arsenic alloy containing 45% iron. Used as a primary metal.

Arsenic-Lead. Trade name of Blackwells Metallurgical Limited (UK) for a lead alloy containing 25% arsenic. Used as a primary metal.

Arsenic-Nickel. Trade name of Blackwells Metallurgical Limited (UK) for a nickel alloy containing 40% arsenic. Used as a primary metal.

arsenic oxide. See arsenic pentoxide; arsenic trioxide.

arsenic pentafluoride. A gaseous compound with a melting point of -79°C (-110°F) and a boiling point of -52.8°C (-63°F). Used as a doping agent in electroconductive polymers. Formula: AsF₅.

arsenic pentasulfide. See arsenic sulfide (ii).

arsenic pentoxide. A white, amorphous powder (99+% pure), obtained from arsenious oxide, that forms arsenic acid (H₃AsO₄·0.5H₂O) in water. It has a density of 4.32 g/cm³, and decomposes at 315°C (599°F). Used in the manufacture of adhesives and colored glass. Formula: As₂O₅. Also known as *arsenic oxide.*

arsenic phosphide. A brownish red powder used in ceramics, semiconductors, and electronics. Formula: AsP. Also known as *arsenious phosphide.*

arsenic selenide. A compound of arsenic and selenium available in the form of black, high-purity crystals. It has a density of 4.75 g/cm³ and a melting point of approximately 360°C (680°F).

Used as a semiconductor and in electronics. Formula: As₂Se₃. Also known as *arsenic triselenide.*

arsenic sesquioxide. See arsenic trioxide.

arsenic sulfide. Any of the following compounds of arsenic and sulfur: (i) *Arsenic disulfide.* A red or orange-red powder (98+% pure) obtained by roasting iron pyrites and arsenopyrite. It occurs naturally as the mineral *realgar.* Density, 3.4-3.6 g/cm³; melting point, 307°C (585°F). Used as paint pigment, in the manufacture of shot, in pyrotechnics, and in the leather and textile industries. Formula: As₂S₂; AsS. Also known as *red arsenic glass; red arsenic sulfide; ruby arsenic;* (ii) *Arsenic pentasulfide.* A yellow or orange crystalline powder (99.99% pure) used as a paint pigment, and in light filters. Formula: As₂S₅; and (iii) *Arsenic trisulfide.* A yellow or orange-red, crystalline powder (99.9+% pure) obtained from arsenious acid. It occurs in nature as the mineral *orpiment.* Density, 3.43 g/cm³; changes from yellow to orange-red form at 170°C (338°F); melting point, 300°C (572°F); boiling point, 707°C (1305°F). Used as a pigment, in glass for infrared lenses, as a semiconductor, and in pyrotechnics. Formula: As₂S₃. Also known as *arsenic tersulfide; arsenic yellow; arsenious sulfide; king's yellow; yellow arsenious sulfide.*

arsenic telluride. A compound of arsenic and tellurium available in the form of black crystals or lumps (99+% pure). It has a density of 6.50 g/cm³ and a melting point of 621°C (1150°F). Used as a semiconductor and in electronics. Formula: As₂Te₃.

arsenic tersulfide. See arsenic sulfide (iii).

Arsenic-Tin. Trade name of Blackwells Metallurgical Limited (UK) for a tin alloy containing 5% arsenic. Used as a primary metal.

arsenic tribromide. See arsenic bromide.

arsenic trichloride. See arsenic chloride.

arsenic trifluoride. See arsenic fluoride.

arsenic triiodide. See arsenic iodide.

arsenic trioxide. A white, crystalline powder (99+% pure) obtained from the flue dust resulting from the smelting of lead and copper concentrates. It occurs in nature as the minerals *arsenolite* and *claudetite.* Density, 3.738 g/cm³; melting point, 315°C (599°F); boiling point, sublimes at 465°C (869°F). Used in the manufacture of pigments, as a fining agent and decolorizer in glass, as an opacifier in glazes, and in the manufacture of ceramic enamels. Formula: As₂O₃. Also known as *arsenic sesquioxide; arsenic oxide; arsenious acid; arsenious oxide; white arsenic.*

arsenic triselenide. See arsenic selenide.

arsenic trisulfide. See arsenic sulfide (iii).

arsenic yellow. See arsenic sulfide (iii).

Arsenic-Zinc. Trade name of Blackwells Metallurgical Limited (UK) for a zinc alloy containing 40% arsenic. Used as a primary metal.

arsenides. A group of binary compounds of arsenic with a more electropositive metallic element, such as cadmium, gallium, indium, iron, nickel, thallium or zinc. Many arsenides are semiconductive and thus used as semiconductors, phosphors, and in electronics and optoelectronics.

arseniosiderite. A golden-yellow mineral composed of calcium iron arsenate hydroxide trihydrate, Ca₃Fe₄(AsO₄)₄(OH)₆·3H₂O. Crystal system, monoclinic. Density, 3.60 g/cm³; refractive index, 1.898. Occurrence: Mexico.

arsenious acid. See arsenic trioxide.

arsenious bromide. See arsenic bromide.

arsenious chloride. See arsenic chloride.

arsenious fluoride. See arsenic fluoride.

arsenious iodide. See arsenic iodide.

arsenious oxide. See arsenic trioxide.

arsenious phosphide. See arsenic phosphide.

arsenious selenide. See arsenic selenide.

arsenious sulfide. See arsenic sulfide (iii).

arsenite. (1) A salt or ester of arsenious acid (As_2O_3)

(2) A compound containing the AsO_3^{-3} radical.

arsenobismite. A yellowish brown to yellowish green mineral composed of bismuth arsenate hydroxide, $Bi_2(AsO_4)(OH)_3$. Density, 5.70 g/cm³; refractive index, above 1.86. Occurrence: USA (Utah).

arsenoclasite. A red mineral composed of manganese arsenate hydroxide, $Mn_5(AsO_4)_2(OH)_4$. Crystal system, orthorhombic. Density, 4.16 g/cm³; refractive index, 1.810. Occurrence: Sweden.

arsenocrandallite. A blue to bluish green mineral of the alunite group composed of calcium strontium barium aluminum hydrogen arsenate phosphate hydroxide, $(Ca,Sr,Ba)Al_3H[(As,P)-O_4]_2(OH)_6$. Crystal system, rhombohedral (hexagonal);. Density, 3.25 g/cm³. Occurrence: Germany.

arsenolamprite. A lead-gray mineral composed of arsenic, As, often containing small quantities of other elements, such as iron, bismuth and sulfur. It can also be made synthetically. Crystal system, orthorhombic. Density, 5.55-5.72 g/cm³.

arsenolite. A colorless, or white mineral composed of arsenic trioxide, As_2O_3. It can also be made synthetically. Crystal system, cubic. Density, 3.87 g/cm³. Also known as *arsenic bloom*.

arsenopalladinite. A white mineral with yellowish creamy tint composed of palladium antimony arsenide, $Pd_8(As,Sb)_3$. Crystal system, triclinic. Density, 10.40 g/cm³. Occurrence: Brazil.

arsenopyrite. A silver-white to steel-gray mineral with metallic luster of the marcasite group composed of iron arsenic sulfide, FeAsS. Crystal system, monoclinic. Density, 6.08 g/cm³; hardness, 5.5-6.0 Mohs. Occurrence: Canada (Ontario), Germany, USA (South Dakota). Used as the chief ore of arsenic, as a source of arsenic trioxide, in pigments, and as a leather preservative. Also known as *mispickel*.

arsenosulvanite. A yellowish gray mineral of the sphalerite group composed of copper arsenic sulfide, Cu_3AsS_4. It can also be made synthetically. Crystal system, cubic. Density, 4.11 g/cm³. Occurrence: Russia, Mongolia.

arsentsumebite. A green mineral of the brackebuschite group composed of lead copper sulfate arsenate hydroxide, $Pb_2Cu-(OH)(SO_4)(AsO_4)$. Crystal system, monoclinic. Density, 6.46 g/cm³; refractive index, 1.992. Occurrence: Southwest Africa.

arsenuranospathite. A pale yellow or almost white mineral of the autunite group composed of hydrogen uranyl aluminum arsenate hydrate, $HAl(UO_2)_4(AsO_4)_4\cdot40H_2O$. Crystal system, tetragonal. Density, 2.54 g/cm³; refractive index, 1.538. Occurrence: Germany.

arsenuranylite. A yellow mineral of the phosphuranylite group composed of calcium uranyl arsenate hydroxide hexahydrate, $Ca(UO_2)_4(AsO_4)_2(OH)\cdot6H_2O$. Crystal system, orthorhombic. Density, 4.25 g/cm³; refractive index, 1.761.

arsine. See arsenic hydride.

Arsura. Trademark of Toyobo Company (Japan) for modal (regenerated cellulose) fibers.

Artal 23. Trade name of AGS Minéraux SA (France) for cordierite-based *chamotte* calcined in a rotary-type kiln. A typical chemical analysis (in wt%) is 53.9% silica (SiO_2), 30.5% alumina (Al_2O_3), 10.5% magnesia MgO), 1.9% ferric oxide (Fe_2O_3), 1.3% titania (TiO_2), 0.8% potassium oxide (K_2O), 0.5% lime (CaO), and 0.1% sodium oxide (Na_2O). It has a bulk density of 2.1g/cm³ (0.076 lb/in.³), a water absorption of 3.8%, an apparent porosity of 8%, a pyrometric cone equivalent (Seger cone) of 14 (1410°C or 2570°F), low thermal expansion, and an upper service temperature of about 1300°C (2370°F). Used as a refractory for bricking and lining industrial furnaces, retorts and crucibles, and for steel foundry applications.

Artalloy. Trade name for a dental amalgam alloy.

A.R.T. Bond. Trade name for a multi-purpose dental adhesive.

art bronze. See architectural bronze.

Art Casters' Brass. Trade name of Belmont Metals Inc. (USA) for brass casting alloys used for jewelry and ornaments.

Art Die. Trade name of Columbia Tool Steel Company (USA) for a deep-hardening die steel containing 0.95% carbon, 0.3% chromium, 0.2% vanadium, 0.25% manganese, and the balance iron. Used for jewelry dies.

Artecoll. See Artefill.

Artefill. Trade name used in the USA for a polymethyl methacrylate (PMMA) based bipolymer for cosmetic applications including plastic surgery and body and skin reconstruction. This product is known in Canada as "Artecoll."

Artem. Trade name of Arco Chemical Company (USA) for a series of engineering resins based on styrene-maleic anhydride (S/MA).

Artfiber. Trade name of Artfiber Corporation (USA) for a glass product consisting of two sheets of clear glass with an interlayer of glass felt that have been hermetically sealed.

Artform. Trademark of Enthone-OMI, Inc. (USA) for electroformed decorative 14- and 18-karat gold coatings.

Artglass. Trade name of Heraeus Kulzer Inc. (USA) for a *ceromer*-type dental composite for esthetic crown and bridge restorations.

Arthrex. Trade name of Arthrex Company (USA) for a biodegradable polymer based on poly(L-lactide) used for tissue engineering and orthopedic applications.

arthurite. An apple green mineral composed of copper iron arsenate phosphate hydroxide tetrahydrate, $CuFe_2(AsO_4,PO_4)_2O_2\cdot4H_2O$. Crystal system, monoclinic. Density, 3.20 g/cm³; refractive index, 1.78. Occurrence: Europe (UK).

Artic bronze. A series of leaded bronzes produced by rapid cooling in chill-cast metal molds, and used for bearings and bushings.

Artifex. Trade name of Artifex-Dr. Lohmann GmbH & Co. KG (Germany) for flexible abrasives for metals, wood, plastics and glass. They are supplied as grinding wheels in various shapes and sizes, wheel dressers and truing sticks, sanding discs, buffing and polishing wheels, mounted wheels and points, lapping abrasives, and honing and sharpening stones.

artificial abrasives. Abrasives, such as aluminum oxide, silicon carbide, diamond, boron carbide and boron nitride, made synthetically, e.g., in an electric furnace. They are usually of better quality than natural abrasives (e.g., flint, garnet or emery). Also known as *manufactured abrasives; synthetic abrasives*. See also abrasives; natural abrasives.

artificial aggregates. Synthetic inorganic materials in granular form used with a suitable cementing medium to make concrete or mortar.

artificial anhydrite. A colorless *anhydrite* $(CaSO_4)$ made from calcium oxide (CaO) by treating with sulfuric acid (H_2SO_4), or from potassium sulfate (K_2SO_4) and calcium chloride $(CaCl_2)$ by precipitating and heating at 700°C (1290°F) for 16 hours.

Crystal system, orthorhombic. Density, 2.98 g/cm³; melting point, 1450°C (2640°F); hardness, 3-3.5 Mohs; refractive index, 1.575. Used as a drying agent, as a gypsum substitute for cement, as a source of sulfuric acid, and in plaster. See also calcium sulfate.

artificial barite. A white or yellowish, fine-grained, precipitated grade of barium sulfate made by adding sulfuric acid (H_2SO_4) to a barium salt solution. Density, 4.47-4.50; melting point, 1580°C (2876°F); hardness, 3 Mohs. Used as an extender pigment in paints and coatings, in porcelain enamel to improve workability, in paper coatings, and as a filler for rubber, textiles, etc. Formula: $BaSO_4$. Also known as *artificial heavy spar; blanc fixe; permanent white; terra ponderosa; tiff.* See also barite; barium sulfate.

artificial basalt. Basalt made synthetically either from natural basalt, or a mixture of raw materials in a proportion similar to that of natural basalt by fusing (melting) in an electric furnace at high temperatures and casting into suitable shapes, e.g., floor slabs, paving stones, or pipes. Depending on the starting material used, it more or less resembles the natural rock. *Artificial basalt* has very high hardness, excellent resistance to abrasion and crushing, and is used in civil engineering and building construction for flooring, paving, lining, and other applications. Also known as *cast basalt; fusion-cast basalt.* See also basalt.

artificial cotton. See high-wet-modulus modal fibers.

artificial fibers. Staple fibers and filaments made by the transformation of natural organic polymers and including: (i) cellulosic fibers, e.g., cuprammonium and viscose rayon, cellulose acetate and cellulose triacetate; (ii) alginate fibers based on alginates derived from seaweed; and (iii) protein fibers obtained from animal and plant sources, e.g., ardein, casein, peanut, soybean or zein. See also artificial man-made fibers; man-made fibers; synthetic fibers.

artificial gold. A yellow to brown powder that may be obtained by adding sulfur to stannic chloride solution. It has a density of 4.45-4.55 g/cm³, and decomposes at 600°C (1110°F). Used as a yellow pigment, and for imitation gilding. Formula: SnS_2. Also known as *mosaic gold; stannic sulfide; tin disulfide.*

artificial graphite. See synthetic graphite.

artificial heavy spar. See artificial barite.

artificial latex. Reclaimed rubber dispersed in water. See also reclaimed rubber.

artificial leather. See synthetic leather.

artificial malachite. See copper carbonate.

artificial man-made fibers. Staple filaments or fibers made by the transformation (regeneration) of natural organic polymers, e.g., casein or cellulose fibers. See also artificial fibers; man-made fibers; regenerated fibers.

artificial mullite. See synthetic mullite.

artificial oilstone. A fine-grained abrasive stone manufactured from aluminum oxide, and used for sharpening tools.

artificial pumice. A material made by treating molten blast-furnace slag with water, and used as a lightweight aggregate, and in thermal insulation. Also known as *pumice slag.* See also foamed slag; pumice.

artificial refractory. A refractory material, such as aluminum oxide, silicon dioxide, silicon carbide, zirconium carbide, etc., made in an electric furnace at high temperatures.

artificial sand. See manufactured sand.

artificial scheelite. See calcium tungstate.

artificial silk. A name formerly used for *rayon.*

artificial spider silk. See spider silk (2).

artificial stone. See cast stone.

Artilana. Trade name for rayon fibers and yarns used for textile fabrics.

artinite. A white mineral composed of magnesium carbonate hydroxide trihydrate, $Mg_2CO_3(OH)_2 \cdot 3H_2O$. Crystal system, monoclinic. Density, 2.02 g/cm³; refractive index, 1.534. Occurrence: Italy, USA (New Jersey).

Artisan. Trademark of J.F. Jelenko & Company (USA) for a ceramic dental alloy composed of 69% gold, 18.5% palladium, and 8% silver. It provides high strength and hardness, and is used for fusing porcelain to metal.

Artistolite. Trade name of Vidrieria Argentina SA (Argentina) for a decorative laminated glass.

art linen. See embroidery linen.

art paper. See supercalendered paper.

art silk. Artificial silk, a name formerly used for *rayon.*

Artuff. Trademark of Advanced Composite Materials Corporation (USA) for a series of silicon carbide whisker-reinforced composites with ceramic matrices (e.g., alumina or silicon carbide). They have excellent abrasion and wear resistance, and high compressive strength and toughness. Used for cutting and forming tools, and dies.

arylalkyl. See aralkyl.

aryl compound. A chemical compound whose molecules have either the full six-carbon ring structure of benzene, or the condensed ring structure indicative of aromatic derivatives.

aryl halide. A halogen derivative of a hydrocarbon in which one hydrogen has been replaced by a halogen bonded to an aromatic ring. Examples include chlorobenzene (C_6H_5Cl), fluorobenzene (C_6H_5F) and o-dibromobenzene ($C_6H_4Br_2$).

arylide. An organometallic compound formed from a metal and one or more aryl radicals, e.g. phenyllithium, LiC_6H_5.

arylmagnesium chloride. An organometallic compound prepared by reacting metallic magnesium with an aryl chloride in tetrahydrofuran, e.g., phenylmagnesium chloride (C_6H_5MgCl).

Arylon. Trademark of E.I. DuPont de Nemours & Company (USA) for a family of high-performance thermoplastic engineering polymers based on polyarylate resins. Commercially available in various grades including general-purpose, unreinforced, flame retardant and glass-reinforced, they have excellent impact resistance, high heat resistance, excellent electrical properties, and good weatherability and processibility. Used for lawn and garden equipment, home appliances, office equipment, consumer electronics, automotive components, and power tools.

Arzade. Trade name of Arcade Malleable Iron Company (USA) for a malleable cast iron containing 3% carbon, 1.5% silicon, and the balance iron.

Arzite. Trade name of Arcade Malleable Iron Company (USA) for a malleable cast iron used for housings, fittings, and gears.

Arzite Metal. Trade name of Arcade Malleable Iron Company (USA) for a corrosion-resistant malleable cast iron containing 3.2% carbon, 1.5% silicon, 1.5% nickel, 0.8% chromium, and the balance iron. Used for corrosion-resistant malleable-iron castings.

Arzon Metal. Trade name of Arcade Malleable Iron Company (USA) for a whiteheart malleable cast iron containing 3.2% carbon, 0.9% silicon, 0.85% manganese, 0.09% sulfur, 0.16% phosphorus, and the balance iron. Used for castings, fittings, and plumbing applications.

Asahi. Trade name of Asahi Kasei Corporation (Japan) for rayon and nylon fibers and yarns used for textile fabrics.

Asahiblue. Trade name of Asahi Glass Company Limited (Japan) for a blue-tinted heat-absorbing glass.

Asahi Bemberg. Trade name of Asahi Kasei Corporation (Japan) for cuprammonium rayon fibers and yarns used for textile fabrics.

Asahibronze. Trade name of Asahi Glass Company Limited (Japan) for a bronze-tinted heat-absorbing glass.

Asahigray. Trade name of Asahi Glass Company Limited (Japan) for a gray-tinted heat-absorbing glass.

Asahikasei Cupro. Trade name of Asahi Kasei Corporation (Japan) for cuprammonium rayon fibers and yarns used for textile fabrics.

Asahikasei Ester. Trade name of Asahi Kasei Corporation (Japan) for polyester fibers and yarns used for textile fabrics.

Asahikasei Nylon. Trade name of Asahi Kasei Corporation (Japan) for nylon 6 and nylon 6,6 fibers and yarns used for textile fabrics.

Asahi Rayon. Trade name of Asahi Kasei Corporation (Japan) for viscose rayon fibers and yarns used for textile fabrics.

Asaori. Trade name of Central Glass Company Limited (Japan) for a patterned glass with weave pattern.

Asarcolo. Trade name of American Smelting & Refining Company (USA) for a series of fusible bismuth-, indium- or lead-base alloys used for soldering and sealing applications, fire-protection devices, sprinklers, foundry core, patterns and molds, pipe bending, etc.

Asarcoloy. Trade name of American Smelting & Refining Company (USA) for a series of white bearing metals containing up to 3% nickel, and the balance cadmium. They have rather high compressive strength and hardness, and low melting points and coefficients of friction. Used for heavy-duty bearings.

Asarcon. Trade name of American Smelting & Refining Company (USA) for a series of cast leaded and unleaded bronzes, red brasses and leaded red brasses, and aluminum bronzes. Used for bearings, bushings, sleeves, liners, hardware, and casings.

asbecasite. A yellow mineral composed of calcium beryllium titanium tin arsenic silicate, $Ca_3(Ti,Sn)As_6Si_2Be_2O_{20}$. Crystal system, hexagonal. Density, 3.70 g/cm^3; refractive index, 1.86. Occurrence: Switzerland.

Asbestall. Trade name of Uniroyal Inc. (USA) for asbestos fabrics.

asbestine. A white, soft, fibrous variety of *talc* (magnesium silicate) with properties similar to asbestos, used as an extender in paper, rubber and plastics, and as an extender pigment in paint. Also known as *French chalk.*

Asbeston. Trademark of United States Rubber Company (USA) for asbestos yarns and fabrics used for fireproof clothing, cable insulation, etc.

Asbestone. Trademark of National Gypsum Company (USA) for corrugated asbestos sheets, asbestos-cement flat sheets and asbestos shingles used for roofing and siding applications.

asbestos. A group of heat-resistant, chemically inert, and noncombustible minerals and including (1) *chrysotile*, a serpentine asbestos mineral, and (2) the *amphibole* minerals: (i) grunerite asbestos (or amosite); (ii) riebeckite asbestos (or crocidolite); (iii) anthophyllite asbestos; (iv) tremolite asbestos, and (v) actinolite asbestos. They are all silicates of magnesium, calcium, sodium and/or iron, have crystalline structures, and consist of strong, thermally stable fibers. Other properties include low thermal conductivity, a density range of 2.5-3.5 g/cm^3 (0.09-0.13 $lb/in.^3$), and a melting range of 1280-1310°C (2335-2390°F). Their colors range from white to gray, green and brown. Used for general fireproofing, fireproof textiles, clutch and brake linings, gaskets and filters, asbestos cement products, paper and boards, building materials, roofing compositions, electrical and thermal insulation, coatings and heat shields, paint fillers, and as reinforcements for cement, rubber and plastics. Also known as *earth flax. Note:* Asbestos is a carcinogen that is highly toxic when dust particles are inhaled. See also asbestos substitutes.

asbestos blanket. Asbestos fibers made into a flexible blanket for high-temperature insulation and fireproofing applications.

asbestos board. A fire-resistant sheet made by mixing asbestos and Portland cement. Used for fireproofing and thermal insulating applications.

asbestos cement. A mixture of asbestos and Portland cement bonded with water, and used in the production of fire-resistant flat and corrugated sheets, tiles, shingles, piping, siding, wallboard, etc. Abbreviation A/C.

asbestos cord. A uniform cord consisting of a strand of asbestos fibers. Used in the electrical industries for wrapping small cables or as a cord for resistance wires, in the glass industry, and in the manufacture of metallic gaskets.

asbestos fabrics. Flameproof fabrics woven from asbestos fibers, or asbestos fibers mixed with cotton or metallic fibers. Used for fireproof clothing, gloves, welding curtains and shields, cable insulation, outside pipe covering, high-temperature insulation, packing, brake and clutch linings, etc. Also known as *asbestos textiles.*

asbestos felt. Felted asbestos impregnated with asphalt or synthetic rubber, and used as a vapor barrier for concrete, and in roofing.

asbestos fiber. Milled and screened asbestos in fiber form having a minimum length-to-maximum transverse dimensional ratio of 10:1, or above. Used for high-temperature insulation applications.

asbestos insulation. Asbestos in fiber form used as thermal insulation at temperatures in excess of 815°C (1500°F). It can be bonded with clay and sodium silicate.

asbestos lumber. Asbestos-cement board having a grained surface imitating wood. Used for flooring and partitions, siding, etc.

asbestos paper. A strong, flexible building paper made of asbestos fibers bonded with sodium silicate or clay. It will not burn, and has good thermal insulating characteristics. Used in the insulation of pipes and walls, as electrical insulation, for gaskets and packing, in building construction as a fire retardant between subfloor and finished flooring, as a pipe wrap for air ducts, etc.

asbestos plaster. Asbestos in fiber form bonded with bentonite clay, and used for thermal insulation and fireproofing applications.

asbestos replacements. See asbestos substitutes.

asbestos roofing. Asbestos bonded with asphalt, cement or rubber and made into sheets for roofing applications.

asbestos rope. Several strands of asbestos fibers braided or twisted into a rope for use as a caulking material.

asbestos substitutes. Materials introduced over the past five decades as viable alternatives to *asbestos* fibers which have been found to be highly toxic and carcinogenic materials posing many risks to human health (especially respiratory diseases, such as asbestosis) and whose mining, production and use have been banned in many countries around the world. These alternative materials have many properties in common with asbestos including excellent fire resistance and/or good thermal or electrical insulation properties. Alternative materials for fire protection include bleached polytetrafluoroethylene fibers (e.g., *Teflon*), heat-resistant nylon fibers (e.g., *Nomex*) and glass fabrics reinforced with ceramic fibers. High-alumina cement products with vermiculite fillers as well as ceramic, glass and mineral fibers are used for thermal insulation applications. Alternatives for electric insulation include thermosetting plastics, such as polyethersulfones and polyimides, and fabrics made of, or reinforced with glass fibers. Also known as *asbestos replacements*.

asbestos tape. A fireproof tape made from asbestos fibers and supplied in typical widths from ranging from 13-75 mm (0.5-3.0 in.) and length of from 30 to over 45 m (100 to over 150 ft.). Used for electrical and thermal insulating applications, e.g., for wrapping hot pipes and tubes, and electrical wires and cables.

asbestos textiles. See asbestos fabrics.

asbestos tubing. Heat-resistant flexible sleeving made from asbestos fibers. It is often of the expandable or shrinkable type, and supplied with inside diameters ranging from less than 10 to over 25 mm (less than 0.4 to over 1.0 in.). Used for insulating electrical cables, wires, pincers, tongs, thermocouple rods, etc.

asbestos yarn. A yarn consisting of plain asbestos fibers, or asbestos fibers mixed with cotton or fine metal wires. It has excellent thermal and fire resistance, and good electrical insulating properties. Used in the manufacture of asbestos cord, tape, tubing, tape, and fabrics.

asbolane. A black, soft, impure, earthy mixture of hydrated cobalt and manganese oxides used in the manufacture of underglaze blue colors. Also known as *asbolite; black cobalt; earthy cobalt.*

asbolite. See asbolane.

Ascarite. Trademark of Arthur H. Thomas Company (USA) for a solid substance composed of sodium hydroxide-coated silica and available in mesh sizes from about 8 to 30. Used as an absorbent for carbon dioxide.

aschamalmite. A lead-gray mineral composed of lead bismuth sulfide, $Pb_6Bi_2S_9$. Crystal system, monoclinic. Density, 7.35 g/cm^3. Occurrence: Austria.

Ascend. Trademark of Solutia Inc. (USA) for a nylon 6,6 extrusion polymer supplied in several grades and product forms. Used for a wide range of applications from fibers and filaments for wearing apparel, home furnishings and industrial fabrics to food packaging and specialty films.

Asco. Trade name of Pasminco Europe (Mazek) Limited (UK) for a water-hardening steel containing 1.2-1.7% carbon, and the balance iron. Used for general tools.

Ascoloy. Trade name of Allegheny Ludlum Steel (USA) for a group of quenched-and-tempered martensitic stainless steels with a nominal composition of 0.14% carbon, 12.0-13.0% chromium, 2.0% nickel, 0-1.8% molybdenum, small amounts of nitrogen and vanadium, and the balance iron. They have good mechanical properties, high strength up to 590°C (1094°F), and good corrosion resistance.

ash. The strong, tough, straight-grained wood of any of several trees of the genus *Fraxinus* belonging to the olive family, especially the American or white ash (*F. americana)*, the black ash (*F. nigra)*, and the European ash (*F. excelsior)*. See also American ash; black ash; European ash.

ashanite. (1) A pitch-black mineral of the columbite group composed of iron manganese niobium tantalum uranium oxide, $(Nb,Ta,U,Fe,Mn)_4O_8$. Crystal system, orthorhombic. Density, 6.61 g/cm^3; refractive index, 2.35. Occurrence; China.

(2) A pitch-black mineral of the columbite group composed of iron manganese niobium tungsten oxide, $(Nb,W,Fe,-Mn)O_2$. Crystal system, monoclinic. Density, 6.55 g/cm^3. Occurrence: Russia.

Ashberry. (1) British trade name for a series of pewters containing 79-80% tin, 14-15% antimony, 2-3% copper, 1-2% nickel, 1-2% zinc, and 0-1% aluminum. Used for tableware and utensils.

(2) British trade name for a series of tin babbitts containing 78-80% tin, 14-19% antimony, 0-3% copper, 0-3% nickel, and 0-2.8% zinc. Used for bearings and utensils.

ashcroftine. A pink mineral composed of potassium sodium calcium yttrium silicate hydroxide tetrahydrate, $KNaCaY_2Si_6O_{12}$·$(OH)_{10}$·$4H_2O$. Crystal system, tetragonal. Density, 2.61 g/cm^3; refractive index, 1.536. Occurrence: Greenland.

Ashland Furane. Trade name of General Polymers (USA) for a series of furane resins.

Ashland P'Est. Trade name of General Polymers (USA) for polyesters supplied as flexible and rigid casting resins.

ashlar. A thin square or rectangular stone or brick with a rough-hackled surface used for facing walls. Also known as *ashlar brick.*

ashlar brick. See ashlar.

Ash Casting Silver. Trade name for a dental casting alloy based on *sterling silver.*

ashleaf maple. See box-elder.

Ashlene. Trademark of Ashley Polymers Inc. (USA) for a series of thermoplastic resins based on acrylonitrile-butadiene-styrene, and supplied in the following grades: transparent, low-gloss, high- and medium-impact, high-heat, fire-retardant, UV-stabilized, plating, glass fiber-reinforced and structural foam.

Ashlene Nylon 6. Trademark of Ashley Polymers Inc. (USA) for a series of resins based on polyamide 6 (nylon 6) and supplied in the following grades: standard, casting, stampable sheet, elastomer copolymer, glass fiber- or carbon fiber-reinforced, silicone-, polytetrafluoroethylene- or molybdenum disulfide-lubricated, glass bead- or mineral-filled, high-impact, fire-retardant and UV-stabilized.

Ashlene Nylon 66. Trademark of Ashley Polymers Inc. (USA) for a series of resins based on polyamide 6,6 (nylon 6,6), and supplied in the following grades: standard, glass fiber- or carbon fiber-reinforced, silicone-, polytetrafluoroethylene- or molybdenum disulfide-lubricated, glass bead- or mineral-filled, high-impact, supertough, fire-retardant and UV-stabilized.

Ashlene SAN. Trademark of Ashley Polymers Inc. (USA) for a series of acrylonitrile-butadiene-styrene resins supplied in the following grades: standard, 30% glass fiber-reinforced, high-impact, fire-retardant, high-heat and UV-stabilized.

A Shot. Commercially pure nickel shot (97.7+%) containing 0.75% carbon, 0.9% iron, and 0.07% sulfur. Used as a virgin metal.

ashtonite. See mordenite.

A-silicone. A silicone resin that has been cured (hardened) by an addition-type chemical reaction. Also known as *addition-cured silicone.* See also silicones.

Asota. Trademark of Asota GmbH (Austria) for polypropylene and nylon 6 fibers.

asparagolite. A greenish-yellow transparent variety of the mineral *apatite* found in Spain. Used chiefly as a gemstone. Also known as *asparagus stone.*

asparagus stone. See asparagolite.

Aspen. Trade name for a palladium-based bonding alloy used in dentistry for porcelain bonding.

aspen. The soft, light wood of any of several trees of the genus *Populus* belonging to the willow family, especially the quaking or trembling aspen *(P. tremuloides)*, and the bigtooth aspen *(P. grandidentata)*. See also American aspen.

aspergillin. A black, graphite-like pigment material produced by spores of molds of the genus *Aspergillus*, especially *A. niger*. It is composed of about 53% carbon, 36% oxygen and a total of 4% hydrogen and nitrogen, and yields mellitic acid and oxalic acid on oxidation with 10% hydrogen peroxide.

asphalt. An amorphous, solid, or semisolid, brownish-black pitch or bitumen produced from the higher-boiling mineral oils by the action of oxygen. It occurs naturally, and is used for pavements, roofing, and waterproofing. Also known as *mineral pitch*. See also bitumen; pitch.

asphalt cement. A specially prepared asphalt having a penetration value at 25°C (77°F) of about 5-300 under specified conditions. Used for direct application in making bituminous pavements, and in asphalt-rubber compounds. Also known as *asphaltic cement; paving asphalt; road asphalt*.

asphalt concrete. A special concrete made with asphaltic cement using coarse or fine aggregate (e.g., gravel or sand). Used as a road paving material. Also known as *asphaltic concrete; bituminous concrete*.

asphalt emulsion. A suspension of fine globules or particles of liquid asphalt and water usually stabilized with a selected emulsifier. The mixture is of pourable consistency, and does not require heating prior to application. Used in road construction. Also known as *cold asphalt*.

asphaltene. A polynuclear hydrocarbon fraction of high molecular weight obtained from asphalt by precipitation. It is soluble in carbon tetrachloride and carbon disulfide, but insoluble in paraffinic naphthas.

asphalt felt. A sheet or felt-like substance saturated with asphalt, and used in roofing and for waterproofing applications.

asphalt filler. Viscous liquid bitumen or pitch used for paving and joint sealing.

asphaltic cement. See asphalt cement.

asphaltic concrete. See asphalt concrete.

asphaltic limestone. Limestone that is impregnated with asphalt.

asphaltic mortar. A mixture of sand and stone flour, typically with an asphalt content of 12 wt% or more, and a softening point of 50-85°C (122-185°F).

asphaltic sand. See sand asphalt.

asphalt laminate. A laminate made by bonding sheets of paper or felt with asphalt.

asphalt mastic. A blend of solid asphalt and graded mineral aggregate that can be made pourable by heating. Also known as *bituminous mastic; mastic asphalt*.

asphalt paint. A paint with an asphaltic base in a volatile solvent that may or may not have fillers, pigments, resins or drying oils added. Used for waterproofing and roofing applications.

asphalt paper. Paper coated or saturated with asphalt.

asphalt rock. A calcareous rock, such as limestone or dolomite, or a sandstone naturally impregnated with asphalt. Used for flooring and paving applications. Also known as *asphalt stone; bituminous rock*. See also rock asphalt.

asphalt roofing. Roofing materials based on asphalt, and including asphalt-saturated felts, roll roofing, and shingles. See also asphalt felt; asphalt shingles.

asphalt-rubber. A mixture of up to 85 wt% asphaltic cement and selected additives and 15 wt% or more reclaimed tire rubber in which the reaction of the rubber with the asphalt and additives has resulted in a significant swelling of the rubber particles.

asphalt stone. See asphalt rock.

asphalt tile. Flooring blocks and tiles made from mixtures of asphalt, mineral pigments and inert fillers.

asphaltum varnish. A black quick-drying varnish with good resistance to heat and acids.

Aspire. Trade name of Degussa-Ney Dental (USA) for a white, high-noble ceramic alloy containing 52% gold, 25.6% palladium, and 17% silver. It has a high yield strength, and a hardness of 255 Vickers, and is used in restorative dentistry for crowns, implant cases and long-span bridgework.

Asplit. Trademark of Atofina Chemicals, Inc. (USA) for acid-resisting self-hardening phenolic resin-type cements and mortars.

Aspun. Trademark of Dow Chemical Company (USA) for fiber-grade linear-low-density polyethylene resins with melt indices of 17-30 g/10 min., and densities ranging from 0.920 to 0.955 g/cm³ (0.033 to 0.035 lb/in.³). They can be processed into yarn, bicomponent fibers, bonding fibers, and nonwoven fabrics by several processes including spin bonding and melt blowing.

assacu. See ochoo pine.

Assam silk. A wild silk obtained from a species of Indian moth, and used for the manufacture of liht, rough-surfaced fabrics. Also known as *muga silk; munga silk*.

asselbornite. A brown to lemon-yellow mineral composed of lead barium uranyl bismuth arsenic phosphate trihydrate, $(Pb,Ba)(UO_2)_6(BiO)_4((As,P)O_4)_2(OH)_{12} \cdot 3H_2O$. Crystal system, cubic. Density, 5.60 g/cm³; refractive index, 1.9. Occurrence: Germany.

Assemblofil. Trademark of Nuovo Italtess (Italy) for glass fibers.

assembly adhesive. An adhesive used in the manufacture of a finished structure, such as a boat, or a piece of furniture. Also known as *assembly glue*.

Assie mahogany. The wood of the mahogany tree *Entandrophragma utile* which is similar to sapele mahogany, but not as heavy, and with a more open texture. It works well, and has an average weight of 660 kg/m³ (41 lb/ft³). Source: Ivory Coast, Cameroon. Used for general construction applications, furniture, and joinery. Also known as *Sipo mahogany; utile*.

Assurance. Tradename of E.I. DuPont de Nemours & Company (USA) for strong, lightweight nylon 6,6 fibers and yarns used for the manufacture of textile fabrics.

Assurex. Trade name of CIVE SA (Argentina) for a toughened glass.

A-stage resin. A soluble, fusible, linear thermosetting resin of the phenolic type in an early stage of polymerization. Also known as *one-step resin; resole*.

ASTAR-811C. A high-strength high-temperature refractory alloy containing 8% tungsten, 1% rhenium, 1% hafnium, 0.025% carbon, and the balance titanium. Developed under NASA sponsorship, it has good compatibility with liquid alkali metals, good time-dependent deformation properties, and good weldability and fabricability. Used for space nuclear-power systems.

astatine. A rare, radioactive, nonmetallic element belonging to Group VIIA (Group 17) of the Periodic Table. It is the heaviest member of the halogen group obtained by bombarding bismuth with alpha radiation. Atomic number, 85; atomic weight, 210; half-life of most stable isotope, about 8 h. Used chiefly in research. Symbol: At.

A Steel. Trade name of British Steel Corporation (UK) for a plain-carbon steel containing 0.5% carbon, 0.8-0.9% manganese, traces of sulfur and phosphorus, and the balance iron. It has a medium high hardness, low shock resistance, and is used for general engineering applications and tools.

Astel. Trade name for polysulfone.

ASTM spring bronzes. A group of spring bronzes that have been standardized by the American Society for Testing and Materials (ASTM). *Grade A* calls for 95% copper, 4.75% tin and 0.25% phosphorus, and is used for general-purpose springs. *Grade B* contains 92% copper, 7.75% tin and 0.25% phosphorus. It has better physical properties than Grade A, and is used for coil and flat springs.

ASTM steels. A large group of steel products that have been standardized by the American Society for Testing and Materials (ASTM). The basic ASTM specification consists of the letter "A" (for ferrous materials) and an arbitrary serial number, e.g. A611 refers to a cold-rolled steel sheet, and A714 to welded or seamless high-strength alloy-alloy (HSLA) steel pipe.

Astra. Trade name of George Cook & Company Limited (UK) for a tough, oil-hardened tool steel containing 0.55% carbon, 0.5% silicon, 2% tungsten, 1% chromium, and the balance iron. Used for piercers, shears, and cold punches.

astrachan. See astrakhan.

Astraglas. Trademark of Dynamit Nobel AG (Germany) for polyester and other resins supplied in the form of foils, sheets, tubes and rods.

Astrakan. Trade name of SA Glaverbal (Belgium) for patterned glass.

astrakanite. See bloedite.

astrakhan. A pile fabric, usually knit or woven from wool, having a distinctive curled or looped finish made to imitate the fleece of young lambs from Astrakhan, Russia. Used especially for coats and trimming. Also spelled astrachan. Also known as *poodle cloth*.

Astra. Trade name of Drake Extrusion Limited (UK) for a triangular polypropylene fiber (8-20 denier/filament) that can be made into a high-class, bulky yarn for the manufacture of floor coverings, especially for bathrooms and bedrooms.

Astral. Trade name of Verreries de la Gare et A. Belotte Réunies SARL (France) for a patterned glass available in four versions. Versions 1 and 2 have diamond patterns, and versions 3 and 4 a pattern combining diamonds and squares.

Astradur. Trade name of Dynamit Nobel AG (Germany) for modified polyvinyl chlorides (PVCs).

Astraglas. Trade name of Dynamit Nobel AG (Germany) for flexible polyvinyl chloride (PVC) supplied in film form.

Astralene C. Trade name for polyester fibers and yarns used for textile fabrics.

Astralit. (1) Trade name of Dynamit Nobel AG (Germany) for a polyvinyl chloride copolymer supplied in several grades including films.

(2) Trade name for a dental silicate cement.

Astralloy. Trade name of Lukens Steel (USA) for hardenable ultrahigh-strength steels containing 0.23-0.33% carbon, 0.7-1.0% manganese, 3.25-3.75% nickel, 1.25-1.75% chromium, 0.2-0.3% molybdenum, up to 0.02% sulfur, up to 0.02% phosphorus, and the balance iron. Available in the form of bars and plates, they have high toughness and ductility, high abrasion resistance, and good weldability. Used for mining equipment, and pulp and paper mill equipment.

Astraloft. Trade name for a series of nylon fibers.

Astralon. Trademark of Dynamit Nobel AG (Germany) for a transparent polyvinyl chloride (PVC) copolymer supplied in many grades including films for the manufacture of relief and topographic maps, in general cartography and surveying, and in offset and intaglio printing. *Astralon-C* polyester and nylon fibers for textile fabrics are also available.

Astraloy. Trademark of J.P. Stevens & Company, Inc. (USA) for high-temperature woven glass-fiber fabrics.

Astranit. Trade name of Krupp Stahl AG (Germany) for a series of standard and high-manganese austenitic stainless steels. The standard grades contain 0.08-0.12% carbon, 18% chromium, 9-10% nickel, and the balance iron, and the high-manganese grades contain about 0.12% carbon, 18% manganese, 12% chromium, 2% nickel, 0.5% molybdenum, and the balance iron. *Astranit* steels have very high corrosion resistance and low to moderate strength. Used for chemical plant equipment, food-processing equipment, and household appliances. High-manganese types are also used for wear-resistant parts and equipment operating in corrosive atmospheres.

Astratherm. Trademark of Dynamit Nobel AG (Germany) for polyvinyl chloride (PVC) resins and film materials.

Astrel. Trade name of 3M Company (USA) for polyaryl sulfone resins.

Astro. Trade name of Astro (USA) for corrosion-resistant titanium supplied in billet, bar, and sheet form, and used for chemical and marine applications.

astrochanite. See bloedite.

Astro Fibers. Trade name of Astro-Tex Refining Company (USA) for metal fibers.

Astro-Foam. Trademark of Astro-Valcour, Inc. (USA) for polyethylene foam supplied in sheet and roll form for protective packaging applications.

AstroGrass. Trademark of Southwest Recreational Industries, Inc. (USA) for tufted nylon 6,6 fabrics designed for sand-filled turf applications in sports arenas.

Astrolon. Trademark of Flexsteel Industries Inc. (USA) for nylon fibers used for the manufacture of durable upholstery fabrics.

Astroloft. Trade name for a series of lofty nylon fibers used for textile fabrics, e.g., clothing.

Astrolon. Trade name for a series of nylon fibers.

Astroloy. (1) Trade name of Cannon-Muskegon Corporation (USA) for a nickel-base superalloy containing 15% chromium, 15-18% cobalt, 5-5.2% molybdenum, 3.5-5.0% aluminum, 3.5-5.5% titanium, 0.06% zirconium, 0.05-0.06% carbon, and traces of iron and boron. It has high heat and corrosion resistance, and high stress-rupture and creep strength, and is used for turbine wheels and blades, nuclear reactors, petrochemical equipment, and aerospace and aircraft applications, e.g., jet engine components, nozzles, compressor disks, etc.

(2) Trade name of Ludlow-Saylor (USA) for a high-tensile stainless steel.

(3) Trademark of Reichhold Chemicals, Inc. (USA) for polyester resins.

Astrolure. Trade name for a series of nylon fibers used for textile fabrics.

Astron. Trade name for a dual-cure denture resin.

astrophyllite. A bronze- to golden-yellow mineral composed of potassium iron manganese titanium silicate hydroxide, $(K,Na)_3(Fe,Mn)_7Ti_2Si_8O_{24}(O,OH)_7$. Crystal system, triclinic. Density, 3.35 g/cm^3; refractive index, 1.703. Occurrence: Norway.

Astroquartz. Trademark of J.P. Stevens & Company, Inc. (USA) for nonhygroscopic, water-insoluble fused silica (99.9-99.95%). Commercially available in the form of continuous filaments, rovings, yarns, mats and fabrics, it has a density of 2.2 g/cm^3 (0.079 lb/in.^{3}), an average fiber diameter of 10 μm (394 μin.), excellent chemical stability, a maximum service temperature of 1050°C (1920°F), high tensile strength, and high strength-to-weight ratio. Used as reinforcements in high-performance epoxy, phenolic, polyimide and other composites, and in thermal insulation.

Astroquartz II. (1) Trademark of J.P. Stevens & Company, Inc. (USA) for high-purity (99.95%) fiber grade of Astroquartz with outstanding ablative properties, a low coefficient of thermal expansion, a low dielectric constant, and excellent mechanical properties in composites.Used as reinforcements in high-performance epoxy, phenolic, polyimide and other composites, and in thermal insulation.

(2) Trademark of J.P. Stevens & Company, Inc. (USA) for epoxy prepregs with *Astroquartz II* fused silica fiber reinforcement used for compression molding applications.

AstroTurf. (1) Trademark of Southwest Recreational Industries, Inc. (USA) for nylon 6,6 fibers used in the manufacture of synthetic turf for sports arenas.

(2) Trademark of Southwest Recreational Industries, Inc. (USA) for knitted nylon 6,6 fabric surfaces with factory-applied polyurethane foam cushioning components. Used as synthetic turf in sports arenas, e.g., hockey and soccer installations.

(3) Trade name of Solutia Inc. (USA) for nylon 6,6 fibers and products including doormats and poultry nest pads.

Astro-Wrap. Trademark of Tenneco Packaging-AVI (USA) for protective cellulose wadding used for packaging applications.

Astryn. Trade name of Himont USA Inc. for a series of impact-resistant polypropylene/polyethylene copolymers with or without mineral fillers (e.g., calcium carbonate or talc). Used for containers, automotive parts, appliances, and aerospace parts.

atacamite. A bright green mineral composed of copper oxychloride, $Cu_2Cl(OH)_3$. Crystal system, orthorhombic. Density, 3.76 g/cm^3; refractive index, 1.861. Occurrence: Chile, Peru.

atactic polymer. A polymer with a random arrangement of side groups or atoms on either side of the polymer chain or backbone. This irregularity results in a polymer with decreased crystallinity, but higher melting point and rigidity. Examples include atactic polypropylene and atactic polystyrene.

atelestite. A yellow mineral composed of bismuth oxide arsenate hydroxide, $Bi_8(AsO_4)_3O_5(OH)_5$. Crystal system, monoclinic. Density, 6.82 g/cm^3; refractive index, 2.15. Occurrence: Germany.

A-Tell. Trade name for a polyester fiber based on polyethylene oxybenzoate (PEB), and formerly made in Japan.

Atephen. Trademark of Hoechst AG (Germany) for phenol-formaldehyde resins and plastics.

Atercliffe. Trade name of Sanderson Kayser Limited (UK) for a series of plain-carbon tool steels with carbon contents of 0.6-1.5% that are available in a wide range of tempers.

Aterite. Trade name of Barber Asphalt Company (USA) for a corrosion-resistant copper nickel containing 35-65% copper, 10-44% nickel, 2-20% iron and 0-23% zinc. Used for valves, cocks, plumbing fixtures, hardware, and chemical equipment.

Atex. Trade name of Atex-Werke Wilhelm Holzhäuer GmbH & Co. KG (Germany) for chipboard.

athabascaite. A light gray to bluish gray mineral composed of copper selenide, Cu_5Se_4. Crystal system, orthorhombic. Density, 6.63 g/cm^3. Occurrence: Canada (Saskatchewan).

Atha Pneu. Trade name of Crucible Specialty Steel (USA) for a shock-resisting tool steel (AISI type S1) containing 0.50% carbon, 2.50-2.75% tungsten, 1.25-1.50% chromium, 0.25% vana-

dium, and the balance iron. It has high hardness and strength, excellent toughness, good wear and abrasion resistance, and good hot hardness. Used for hand and pneumatic chisels, forming tools, headers, piercers, rivet busters, and shear blades.

atheneite. A white mineral composed of palladium mercury arsenide, $(Pd,Hg)_3As$. Crystal system, hexagonal. Density, 10.20 g/cm^3. Occurrence: Brazil.

Athenium. Trademark of Williams Dental Company Inc. (USA) for high-palladium dental alloys used for casting dental restorative appliances, such as dentures, inlays and crowns.

Athermal. Trade name of Schott DESAG AG (Germany) for green-tinted uncoated and mirrored protective glass used for spectacles and welder's goggles.

Athermane. Trade name of SA Glaverbel (Belgium) for heat-absorbing sheet glass.

Athermic. French trade name for green, heat-absorbing glass.

Athos. (1) British trade name for a heat- and corrosion-resisting steel containing up to 0.5% carbon, 22% nickel, 8% chromium, 1.75% silicon, 1% copper, 0.7% manganese, and the balance iron. Used for stainless parts.

(2) Trademark of Drama Marble Inc. (USA) for a white natural marble. Also known as *Athos White.*

(3) Trade name of Athos Steel & Aluminum, Inc. (USA) for carbon, galvanized and stainless steel, and aluminum supplied in the form of angles, bars, coils, plates, rounds, sheets, structural shapes, expanded metals, and perforated sheets.

Athos White. See Athos (2).

Atlac. Trademark of Reichhold Chemicals Company (USA) for a series of synthetic resins available in several basic types including polyesters, vinyl esters, bisphenol fumarates, and isophthalic and terephthalic polyesters, often reinforced with glass fibers, eg., *Atlac BP* glass-filled bisphenol polyester laminates, or *Atlac VE* vinyl ester resin. They have good chemical and/or flame/smoke resistance, and are used for ductwork, chemical storage tanks, chemical piping, gasoline tanks, fume ducts, sewer pipes, wall and roofing systems, pulp washer drums, pickling and plating equipment, and structural applications.

Atlantaloy. Trade name of Atlantic Casting & Engineering Company (USA) for a series of cast alloys including beryllium coppers, yellow brasses, aluminum, manganese and silicon bronzes, and several aluminum-silicon and aluminum-zinc alloys.

Atlantic. (1) Trade name of Pilkington Brothers Limited (UK) for patterned glass.

(2) Trade name of Atlantic Steel Corporation (USA) for a standard tungsten-type (18W-4Cr-1V) high-speed tool steel used for cutting tools, milling cutters, etc.

(3) Trade name of JacksonLea (USA) for a greaseless buffing compound used for satin finishing.

Atlantic Polybead. Trademark of Atlantic Gypsum Limited (Canada) for gypsum wallboard and polystyrene foam insulation.

Atlantic white cedar. The soft, durable, aromatic wood of the swamp tree *Chamaecyparis thyoides.* The heartwood is light brown tinged with pink, and the sapwood whitish-brown. It has a straight grain, low weight, low to medium strength, high durability and decay resistance, works well and holds paint well. Average weight, 375 kg/m^3 (23 lb/ft^3). Source: United States (Atlantic coast from southern Maine to central Florida, and Gulf coast from Florida to Louisiana). Used for lumber, construction work, shipbuilding, poles, ties, lumber, posts, fencing, and woodenware. Also known as *southern white cedar.*

Atlantis Marine. Trademark of Bruin Plastics (USA) for ultraviolet-, mildew- and shrink-resistant fabrics available in 3-ply vinyl-laminated polyester and 2-ply fabric-backed grades. Used for protective covers for harsh marine environments.

Atlas. (1) Trade name AL Tech Specialty Steel Corporation (USA) for an extensive series of steels including various high-strength, wear-resistant and heat-resistant grades as well as several austenitic, ferritic and martensitic stainless steel grades, and a wide range of tool steel grades including water-hardening, shock-resisting, hot- and cold-work and high-speed types.

(2) Trade name of AL Tech Specialty Steel Corporation (USA) for tungsten-type high-speed steels including *Atlas A* (AISI type H21) with about 0.3% carbon, up to 0.4% manganese, up to 0.5% silicon, up to 3.7% chromium, up to 10.0% tungsten, and 0.6% vanadium, respectively, and the balance iron. *Atlas B* (AISI type H22) is modified version with higher carbon and tungsten contents. Used for cutting, finishing and shaping tools.

(3) Trade name of Eyre Smelting Company (UK) for a series of tin-base bearing alloys. See also Atlas Admiralty A; Atlas Admiralty B; Atlas Amacol; Atlas Infranga; Atlas Tenaxas.

(4) Trade name of Eyre Smelting Company (UK) for cast copper-tin alloys.

(5) Trade name of Degussa AG (Germany) for unfilled and aluminum-filled polymethacrylate resins used for casting applications, thermoforming molds, and laminating and rim mold coatings.

(6) Trademark of Degussa AG (Germany) for polyethylene terephthalate (PET) resins.

(7) Trade name for a dual-cure dental luting resin.

atlas. (1) A warp-knit fabric made by knitting a set of yarns for several couses shifting one wale per course and then returning to the initial position.

(2) An elegant fabric, usually of silk or synthetic fibers, in a satin weave. Used chiefly for making dresses, especially evening wear.

(3) A satin-weave fabric similar to (2) made with cotton weft (filling) yarn. Used for lining fabrics.

Atlas Admiralty A. Trade name of Eyre Smelting Company (UK) for a tin-base bearing alloy containing varying amounts of zinc. It has good compressive strength, and a melting range of 200-425°C (390-800°F). Used for bearings.

Atlas Admiralty B. Trade name of Eyre Smelting Company (UK) for a lead-free tin-base bearing alloy with good compressive strength and a melting range of 240-325°C (465-615°F). Used for steam turbines.

Atlas Amacol. Trade name of Eyre Smelting Company (UK) for a leaded tin-base bearing alloy. It has good compressive strength, and a melting range of 190-370°C (375-700°F). Used for reciprocating engines.

Atlas Brake Die. See Brake Die (2).

Atlas cedar. The yellow, durable, fragrant wood of the coniferous tree *Cedrus atlantica*. It takes a fine polish, and has an average weight of 575 kg/m³ (36 lb/ft³). Source: Northern Africa and southern Europe. Used for construction applications. Also known as *silver cedar*.

Atlas Cement-Cote. Trade name of Atlas Chemical Company (USA) for a cement water paint.

Atlas Commando. Trade name of Atlas Specialty Steels (Canada) for a die steel containing 1.2% carbon, 0.25% manganese, 0.2% silicon, and the balance iron. Used for drill rod, dies, reamers, punches, twist drills, and dowels.

Atlas Die Casting Steel. Trade name of Atlas Specialty Steels (Canada) for an oil-hardening die steel containing 0.4% carbon, 0.7% manganese, 0.6% chromium, 0.15% molybdenum, varying amounts of nickel, and the balance iron. It has good machinability, good nondeforming and nonwarping properties, and good wear and erosion resistance after heat treatment. Used for die-casting dies for lead, tin and aluminum alloys.

Atlas Hobbing Iron. Trade name of Atlas Specialty Steels (Canada) for a high-quality electric-furnace iron containing 0.05% carbon, 0.2% manganese, 0.15% silicon, and the balance iron. It may be carburized, and has a high grade finish, good hardenability, and is free from porosity. Used for plastic molding dies with hobbed cavities.

Atlas Infranga. Trade name of Eyre Smelting Company (UK) for a lead-free tin-base bearing alloy with good compressive strength and a melting range of 240-330°C (465-625°F). Used for high-speed diesel engines.

Atlas KK. See Sioux.

Atlas Mold Base. Trade name of Atlas Specialty Steels (Canada) for a medium-carbon, resulfurized mold steel with good toughness, stability and strength. Used for plastic die-casting mold bases, die sets, and fixture bases.

Atlas Mold Special. Trade name of Atlas Specialty Steels (Canada) for a mold steel (AISI type P20) containing 0.3% carbon, 0.8% manganese, 0.5% silicon, 1.65% chromium, 0.4% molybdenum, and the balance iron. It has good machinability, good wear resistance, high cleanliness, and moderate toughness. Used for plastic molds and die-casting dies.

Atlas Refined. Trade name of Atlas Specialty Steels (Canada) for a series of water-hardening tool steels (AISI type W1). *Atlas Refined-8* contains 0.8% carbon, 0.25% manganese, 0.2% silicon, and the balance iron. It has high toughness and shock resistance, good resistance to decarburization, moderate machinability, and low red hardness and wear resistance, and is used for blacksmith tools, hand and quarry chisels, pneumatic rivet sets, shear blades, drift pins, and clamping dies. *Atlas Refined-10* contains 1% carbon, 0.25% manganese, 0.2% silicon, and the balance iron. It has good resistance to decarburization, high to moderate shock resistance and toughness, moderate wear resistance and machinability, and low red hardness. Used for lathe centers, mandrels, arbors, cones, blanking, drawing and forming dies, coining, embossing and trimming dies, die plates, knife blades, knurls, and springs.

Atlastacrete. Trade name of Atlas Minerals & Chemicals, Inc. (USA) for polymer concrete containing epoxy, vinyl or furan resins.

Atlas Tenaxas. Trade name of Eyre Smelting Company (UK) for a leaded tin-base bearing alloy with good compressive strength and a melting range of 186-340°C (367-644°F). Used for reciprocating engines.

Atlastic. Trademark of Atlas Minerals & Chemicals, Inc. (USA) for protective corrosion-resistant coatings based on *cutback asphalt*. They are waterproof and resistant to organic and inorganic acids, alkalies, and salts. Used on ferrous and nonferrous metals, concrete, and wood.

Atlastik. Trademark of Degussa AG (Germany) for flexible polyvinyl chloride film materials.

Atmodie. Trade name of Columbia Tool Steel Company (USA) for a series of air-hardening high-carbon, high-chromium cold-work tool steels (AISI type D2) containing 1.5% carbon, up to 0.4% manganese, up to 0.5% silicon, 11.6-12.0% chromium, 0.8-1.0% molybdenum, 0.2-1.0% vanadium, and the balance iron. They have high toughness, high abrasion and wear resistance, high hardness, good nondeforming properties, and good corrosion resistance. Used for blanking, trimming and stamping dies, shear blades, punches, bending and seaming tools, drawing and forming tools, thread-rolling dies, solid gages, and structural components.

Atmodie Smoothcut. Trade name of Columbia Tool Steel Company (USA) for a series of air- or oil-hardening free-machining cold-work tool steels (AISI type D2) containing 1.5% carbon, 12% chromium, 0.8% molybdenum, 0.9% vanadium, and the balance iron. It has high toughness, high abrasion and wear resistance, high hardness, good nondeforming properties, and good corrosion resistance. Used for blanking, trimming and stamping dies, shear blades, punches, solid gages, and mandrels.

AT Nickel. Trade name of Inco Alloys International Inc. (USA) for a commercially pure wrought nickel (99.0-99.5%) containing 0.15% carbon, 0.4% iron, 0.25% copper and 0.01% sulfur. It is commercially available in sheet, plate and bar form in the annealed or cold-rolled condition. Used for electrical and electronic applications.

atokite. A light cream-colored mineral of the gold group composed of palladium platinum tin, $(Pd,Pt)_3Sn$. Crystal system cubic. Density, 14.19 g/cm^3. Occurrence: South Africa.

Atom. German trade name for a patterned glass.

Atom Arc. Trade name of Chemtron Corporation (USA) for a series of low-alloy, high-strength low-alloy and quenched-and-tempered structural steel arc-welding electrodes with low-hydrogen iron powder and low hydrogen coverings.

atomic clusters. Small clusters (typically 2-1000 atoms) constructed from metal, diamond or other nanoparticles, e.g., by laser ablation, molecular beam epitaxy, nanosphere lift-off lithography or other nanomanufacturing techniques.

atomic-cluster-assembled materials. A relatively recent class of materials which have unique electronic, optical, magnetic, mechanical and/or other properties, and are synthesized (assembled) from small *atomic clusters*. Present and potential future applications include the manufacture of unique electronic, photoluminescent and optoelectronic devices, semiconductor laser diodes for nonlinear optics, magnetic storage devices with ultrahigh storage densities, photocatalysts, and chemical sensor for biochemistry and DNA research. Also known as *cluster-assembled materials*.

Atomet. Trade name of Quebec Metal Powders Limited (Canada) for ferrous powders used in the manufacture of powder-metallurgy and powder-forged parts. The powders available include reduced iron powders, atomized steel powders in non-alloy and prealloyed grades, and ferromagnetic composite powders.

atomic fuel. See reactor fuel.

Atomite. Trademark of Cyprus Mines Corporation (USA) for natural water-ground calcium carbonate essentially free from particles larger than 15 μm (590 μin.). Used as a filler for rubber, paint, paper, etc.

atomized powder. A powder consisting of small particles of varying size and produced from a meltable solid material, e.g., a metal, metal alloy or ceramic, by dispersion in a rapidly moving continuous stream of fluid (gas or liquid), or by mechanical means. An example is atomized gold powder.

Atprime. Trademark of ICI Americas Inc. (USA) for two-component polyester-base resins used for bonding reinforced plastics.

Atrix. Trade name of Saarstahl AG (Germany) for a series of plain-carbon and low-alloy machinery steels including several case-hardening grades.

Atrisorb. Trade name of Atrix Laboratories (USA) for a biodegradable, bioabsorbable polymer based on poly(DL-lactide), and used for tissue engineering applications and periodontal guided-tissue-regeneration (GTR) membranes.

Atryl. Trademark of Alpha/Owens-Corning LLC (USA) for low-profile polyester thermoset resin systems that provide low curing times, optimum surface finishes, and improved productivity. They have high strength, low shrinkage, acceptable shelf life, and good paint adhesion. Used for sheet-molding compound applications in the automotive industry, e.g., body panels.

Atsil. Trade name of Atlantic Steel Corporation (USA) for a shock-resisting tool steel (AISI type S5) containing 0.5% carbon, 0.6% manganese, 0.5% tungsten, 0.3% molybdenum, 1.3% silicon, and the balance iron. It has outstanding toughness, excellent shock resistance, a high elastic limit, a great depth of hardening, and good ductility. Used for cutting tools, pneumatic tools, shear blades, punches, and rivet sets and busters.

Atsina. Trade name of Chemalloy Electronics Corporation (USA) for a steel containing 0.7% carbon, 0.5% chromium, 5% tungsten, and the balance iron. Used for electrical machinery and magnets.

ATTA. Trade name of Btech Corporation (USA) for carbon fiber-filled epoxy film adhesives with good processibility and thermal conductivity. Used in the manufacture of electronic equipment to bond integrated circuits to leadframes.

Attachment Bond. Trade name for a dual-cure dental luting composite resin used in bonding restorations.

attakolite. A light pink mineral composed of calcium manganese aluminum phosphate silicate hydroxide trihydrate, $(Ca,Mn,Sr,Fe)_3Al_6(PO_4)_5(SiO_4)_2 \cdot 3H_2O$. Crystal system, orthorhombic. Density, 3.23 g/cm^3; refractive index, 1.664. Occurrence: Sweden.

Attane. Trademark of Dow Chemical Company (USA) for a series of linear and ultra-low-density polyethylene resins available in standard, blown and cast film, and impact-modified grades. They have a density range of 0.904-0.913 g/cm^3 (0.032-0.033 lb/in.^{3}), good low-temperature flexibility and flex-crack resistance, and high tear resistance. Used for heavy-duty sacks, turf and consumer bags, food and detergent packaging, silage wrap, mulch films, heating and water pipes, and injection-molded products.

attapulgite. A light green to grayish, magnesium-rich, fibrous clay mineral composed of aluminum-magnesium silicate tetrahydrate, $(Mg,Al)_2Si_4O_{10}(OH) \cdot 4H_2O$. It is the chief ingredient of *fuller's earth*. Crystal system, monoclinic; density, 2.36-2.40 g/cm^3. Occurrence: USA (Florida and Georgia). Used as a suspension agent in ceramic slips, as an emulsifier, as a flatting agent and extender in paints, in drilling fluids, and as a filter medium.

Atvantage. Trademark of National Nonwovens (USA) for a series of high-performance fiber composites that are supplied as composite structural core (CSC) and composite insulating core (CIC) products. CSC products are isotropic, mechanically interlocked fiber arrays with chemical enhancement and provide high strength and stiffness, excellent thermal, acoustic and vibration attenuation in a multitude of applications. CIC products are self-supporting structured fiber arrays used for applications requiring outstanding high-temperature insulating properties.

Au-ABA. Trade name Wesgo Division of GTE Products Corporation (USA) for a gold-based active brazing alloy. See also ABA.

Aubel. Trade name for a dental alloy containing a minimum of

92% gold. It has a melting range of 970-995°C (1780-1825°F).

Aubert & Duval. Trade name of Aubert & Duval (France) for an extensive series of steels including (i) alloy machinery steels, such as chromium, chromium-molybdenum, chromium-molybdenum-vanadium, chromium-vanadium, nickel-chromium, nickel-chromium-molybdenum, silicon-manganese and chromium-aluminum-molybdenum grades as well as several nitriding and case-hardening types, (ii) plain-carbon, cold-work and hot-work tool steels, (iii) austenitic, ferritic and martensitic stainless steels, and (iv) several nickel-chromium and nickel-chromium-cobalt superalloys.

aubertite. An azure-blue mineral composed of copper aluminum chloride sulfate hydrate, $CuAl(SO_4)_2Cl \cdot 14H_2O$. Crystal system, triclinic. Density, 1.82 g/cm^3; refractive index, 1.482. Occurrence: Chile.

Aucrom. Trade name for a gold solder used in dentistry.

Auden Wire. Trade name of Johnson Matthey plc (UK) for a hardened alloy wire composed of 75% gold and 25% platinum. It has a melting range of 965-1000°C (1770-1830°F), and is used for dental applications.

Audio. Trade name for a series of soft magnetic iron alloys containing up to 4.75% silicon. They have high magnetic permeability and are used for laminations in electrical equipment.

Audiocote. Trademark of US Gypsum Company (USA) for acoustical plaster.

Audiolloy. Trademark of Colt Industries (UK) for a magnetic alloy of 52% iron and 48% nickel. It has high magnetic permeability, moderately high magnetic saturation, and is used for magnets, sensitive relays, shut-off valves, etc.

AudioSeal. Trademark for acoustical or sound-barrier materials.

Auer metal. A pyrophoric alloy composed of 65% mischmetal (chiefly cerium, lanthanum, ytterbium and erbium), and 35% iron. Formerly used for gas and cigarette lighters.

Aufriloy. Trade name of Wellman Dynamics Corporation (USA) for an alloy of copper, lead and tin, used for heavy-duty bearings and bushings.

augelite. A colorless mineral composed of aluminum phosphate hydroxide, $Al_2(PO_4)(OH)_3$. Crystal system, monoclinic. Density, 2.68 g/cm^3; refractive index, 1.576. Occurrence: Canada, Sweden.

Auger. (1) Trade name of Ziv Steel & Wire Company (USA) for a water-hardening steel containing 0.75% carbon, and the balance iron. Used for machine-tool parts, and mining equipment.

(2) Trade name of Bethlehem Steel Corporation (USA) for a tool steel containing 0.8% carbon, up to 0.4% manganese, up to 0.4% silicon, and the balance iron. Used for mining equipment and augers.

augite. A dark green to black, or brown mineral of the pyroxene group composed of calcium iron magnesium silicate, $Ca(Fe,Al,Mg)(Al,Si)_2O_6$. Crystal system, monoclinic. Density, 3.2-3.4 g/cm^3; refractive index, 1.70-1.71; hardness, 5-6 Mohs. Occurrence: USA.

Aumet. Trade name for a wire composed of 83.3% gold and 16.7% platinum. It has a melting range of 1000-1085°C (1830-1985°F) and is used for dental applications.

Auracem. Trade name for an universal luting cement used in restorative dentistry.

Aurafil. Trade name for a light-cure filled hybrid composite resin used in restorative dentistry.

Auralite. Trade name of Asbestos Building Supply Limited (Canada) for translucent fiberglass building sheets and panels for siding, roofing, etc.

Aurall. Trade name of LeaRonal, Inc. (USA) for high-purity gold (24 karat) used for plating.

Aurastan. Trade name of PAX Surface Chemicals Inc. (USA) for mat and bright tin and tin-lead electroplates.

Auratone. Trade name of Smith Precious Metals Company (USA) for a low-karat preplate gold base.

Aurea. (1) German trade name for a dental casting alloy containing 50% gold.

(2) Trademark of American Structural Systems, Inc. (USA) for prefabricated concrete flooring panels.

Aureocem. Trade name of Inter-Africa Dental (Zaire) for a water-resistant, resin-based dental cement with very high adhesion to gold, dentin and enamel. Used for permanent luting of crowns, bridges, inlays, onlays, posts, etc.

aureolin. See Indian yellow.

Auresin. Trade name of Flachglas AG Delog-Detag (Germany) for a special glass coated with gold using an evaporation technique.

Auribond-GP. Trade name of Aurident, Inc. (USA) for a durable, high-strength dental ceramic alloy containing 86.5% gold, 5.4% platinum, and 4.55% palladium. It has a light golden color, excellent bonding properties, excellent porcelain shade control, outstanding workability, and good polishability. Used for porcelain-to-metal restorations, especially long-span bridges and single units.

auric bromide. See gold tribromide.

Auric brown. An intimately ground, lightfast hydrated ferric oxide pigment used in papermaking.

auric chloride. See gold trichloride.

auric cyanide. Colorless, hygroscopic crystals that decompose at 50°C (122°F). Used as an electrolyte in the electroplating industry. Formula: $Au(CN)_3 \cdot 3H_2O$. Also known as *gold cyanide; cyanoauric acid*.

aurichalcite. A greenish to bluish mineral composed of zinc copper carbonate hydroxide, $(Zn,Cu)_5(CO_3)_2(OH)_6$. Crystal system orthorhombic. Density, 3.96 g/cm^3. Occurrence: Mexico. Also known as *brass ore*.

auric hydroxide. See gold hydroxide.

auric iodide. See gold triiodide.

auric oxide. See gold oxide.

auric trioxide. See gold oxide.

auricupride. A rose mineral of the gold group composed of copper palladium gold, $(Cu,Pd)_3Au_2$. Crystal system, orthorhombic. Density, 13.77 g/cm^3. Occurrence: Russian Federation.

Auridium. Trade name for a high-gold dental casting alloy with rich yellow color.

Auriga. Trade name of David Brown Foundries Company (USA) for a series of wear-resistant, high-strength or high-temperature steel castings, plain-carbon, low-carbon and medium-carbon steel castings, austenitic, cold-work manganese steel castings, standard and high-manganese austenitic stainless steel castings, and corrosion-resistant nickel-iron-chromium alloy castings.

Auritex. Trade name of Aurident, Inc. (USA) for a series of gold-palladium and palladium-silver dental ceramic alloys. Gold-palladium grades include silver-free *Auritex-66* (66Au-24Pd), *Auritex WP* (51.8Au-38Pd), *Auritex XP* (51.8-38Pd) and silver-containing *Auritex-HP* (62.2Au-23.4Pd). They are compatible with most porcelains, have excellent mechanical properties, and are used for porcelain-to-metal restorations including long-span bridges, single units, and implant cases. *Auritex-ZP* is a whitish, low-density palladium-silver grade (60Pd-29Ag) that is compatible with high-expansion porcelains, has excellent porcelain bonding properties, and is used for porcelain-to-metal restorations.

Aurocast. Trade name for a series of high-gold dental casting alloys with a rich yellow color.

Aurofluid. Trade name of Metalor Technologies SA (Switzerland) for several dental alloys including *Aurofluid CPF*, a palladium- and copper-free universal alloy containing 71.5% gold and 11.5% platinum, used for medium-fusing ceramic restorations, and *Aurofluid Plus*, a dental bonding alloy containing 72.8% gold and 9% palladium, used for low-fusing ceramic restorations.

Auromet. Trade name of Aurora Industries (USA) for a series of copper alloys including several aluminum bronzes, leaded bronzes and brasses, yellow brasses, tin bronzes, silicon bronzes, and manganese bronzes.

Auronal. Trade name of Shipley Ronal (USA) for gold finishes used for semiconductor applications.

Auropal. Trademark of Wieland Dental + Technik GmbH & Co. KG (Germany) for a series of yellow gold-silver dental alloys. *Auropal 60* is a hard-type alloy, and *Auropal 1, Auropal 2* and *Auropal 4* are extra-hard types.

aurorite. A mineral composed of silver calcium manganese oxide trihydrate, $(Mn,Ag,Ca)Mn_3O_7 \cdot 3H_2O$. Crystal system, triclinic. Density, 3.88 g/cm³. Occurrence: USA (Nevada).

aurosmiridium. A silver-white, cubic mineral consisting of a solid solution of gold and osmium in iridium. Occurrence: Russia.

aurostibite. An opaque mineral of the pyrite group with gray metallic luster composed of antimony gold, $AuSb_2$. It can also be made synthetically. Crystal system, cubic. Density, 9.98 g/cm³. Occurrence: Canada, Germany.

AuRoTech. Trademark of Lucent Technologies (USA) for gold and gold alloy electrodeposits and plating processes.

aurous bromide. A yellowish-gray mass that has a density of 7.9 g/cm³ and decomposes at approximately 165°C (330°F). Formula: AuBr. Also known as *gold bromide*.

aurous chloride. Yellowish crystalline compound (99.9% pure) that has a density of 7.57 g/cm³ and decomposes at 289°C (552°F). Formula: AuCl. Also known as *gold chloride*.

aurous cyanide. A moisture-sensitive, pale yellow, crystalline powder with a density of 7.14 g/cm³, used in chemistry and materials research. AuCN. Also known as *gold cyanide*.

aurous iodide. Greenish-yellow, hygroscopic powder (99.9% pure) that has a density of 8.25 g/cm³ and decomposes at 120°C (248°F). Formula: AuI. Also known as *gold iodide*.

aurous oxide. A gray-violet compound that loses its oxygen at approximately 205°C (400°F). Formula: Au_2O. Also known as *gold oxide*.

Aurowhite. Trade name of Auromet Corporation (USA) for imitation rhodium used in electroplating.

Auro-Yellow. Trade name of Auromet Corporation (USA) for imitation gold used in electroplating.

Aurum. Trade name of Mitsui Toatsu Chemical Company (Japan) for melt-processable thermoplastic polyimide resins. They have excellent chemical and radiation resistance, low outgassing properties and an upper service temperature of up to 240°C (465°F) in the amorphous state, and up to 288°C (550°F) in the annealed, semicrystalline state. Used for analytical and medical equipment, aircraft and automotive components, business machines (e.g., copiers and laser printers), compressors, construction machinery, conveyors, electric motors, electronic and

solid-state applications, farm equipment, industrial machinery, off-road equipment, pumps, textile equipment, and transmissions.

Auruna. Trade name of Degussa-Huls Corporation (USA) for a gold electrodeposit and electroplating process.

ausformed steels. Steels, often of the high-alloy type, that have been hot formed while in the metastable austentic range. Such steels attain high strengths.

Austalon. Trade name of Howmet Corporation (USA) for an austenitic stainless steel containing 0.08% carbon, 18% chromium, 8% nickel, 2% molybdenum, and the balance iron. Used for orthodontic and dental applications.

austempered ductile irons. A group of easily cast, high-strength irons produced from cast ductile iron with small additions of copper, molybdenum or nickel by a heat treatment process that involves initial heating to dissolve carbon followed by rapid quenching to prevent pearlite formation, and subsequent holding at the appropriate austempering temperature for a prescribed period of time. The final microstructure consists of austempered bainite and spheroidal graphite, and the fracture surface appears silver-gray. *Austempered ductile iron* has excellent toughness and wear properties, and high strength, and is used for chain links, hooks, clevis pins, engine mounts, drive pinions, sprockets, log splitters, and truck equalizers. Abbreviation: ADI irons.

austempered steels. Steels that have been quenched in heated baths from the austentization temperature, held at a temperature just above the martensitic transformation range for sufficient time to allow the austenite to completely transform into bainite, and then cooled at any rate. Such steels typically have hardness above 50 Rockwell and enhanced toughness.

Austenal Chairside. Trade name of Austenal Limited (UK) for a self-cure dental acrylic resin.

Austenit. Trade name of Austenit Gesellschaft mbH & Co. KG (Germany) for corrosion- and acid-resistant steels including various austenitic stainless grades.

Austenite. Trade name of T. Inman & Company Limited (UK) for a tungsten-type high-speed steel containing 0.65% carbon, 14% tungsten, 4% chromium, 0.5% vanadium, and the balance iron. Used for lathe and drilling tools, cutters, etc.

austenite. (1) A soft, ductile, nonmagnetic, face-centered cubic (fcc) solid solution of carbon in gamma iron (γ-iron) which is very unstable below its critical temperature. The maximum carbon content is 2.11% at 1147°C (2097°F). In high-alloy steels (e.g., stainless steels) stable austenite may be obtained at room temperature, since alloying additions of nickel and manganese lower the critical transformation temperature. See also gamma iron.

(2) Any solid solution of one or more elements in face-centered cubic (fcc) iron. For example, if the solute is nickel, the solid solution is known as "nickel austenite."

austenitic alloys. See austenitic steels.

austenitic cast irons. A group of gray and ductile cast irons that have been alloyed with considerable amounts of nickel, chromium, copper, manganese and silicon to improve corrosion and heat resistance. See also austenitic ductile irons; austenitic gray irons.

austenitic ductile irons. A group of nonmagnetic ductile cast irons that have been alloyed with up to 36% nickel and up to 5% chromium. They have a density range of 7.3-7.7 g/cm³ (0.26-0.28 lb/in.³), excellent heat and corrosion resistance, good resistance to saltwater and seawater corrosion, good dimensional stability, and good weldability. Used for chemical and paper-mill equipment, e.g.. for pumps, valves, compressors, impellers, bushings and pipes. Also known as *high-alloy ductile irons.*

austenitic gray irons. A group of gray cast irons containing about 18-20% nickel, up to 4% chromium, and up to 7% copper. They have good corrosion and heat resistance, good seawater and saltwater resistance, good scaling and growth resistance up to about 820°C (1508°F), high toughness, good thermal shock resistance, and good wear and abrasion resistance. Used for pumps, valves, flood gates, filtering equipment, pipes, stove tops, and general heat- and corrosion-resistant castings.

austenitic manganese steels. See Hadfield steels.

austenitic precipitation-hardenable stainless steels. A group of wrought stainless steels with stable, austenitic microstructures, and chromium and nickel contents ranging from 12 to 16% and 4 to 9%, respectively. Hardening is achieved by a final aging treatment which precipitates very fine second-phase particles (usually carbides) from a supersaturated solid solution. Also known as *precipitation-hardenable austenitic stainless steels.* See also precipitation-hardenable stainless steels.

austenitic stainless steels. A group of tough, ductile, nonmagnetic steels whose microstructure is austenitic at room temperature. They contain significant additions of chromium (16-30%) and nickel (6-20%) together with manganese, molybdenum, copper, titanium and/or niobium (columbium). The carbon contents range from about 0.03 to 0.15%. The so-called 18-8 stainless steels are typical austenitic stainless steels containing approximately 18% chromium and 8% nickel. *Austenitic stainless steels* have very high strength and corrosion resistance even at elevated temperatures, good low-temperature properties, good weldability, and moderate machinability. Used for structural applications, such as aircraft and transport equipment, cutlery, chemical equipment, food-processing equipment and surgical instruments. Abbreviation: ASS. See also stainless steels.

austenitic steels. A group of extremely corrosion-resistant alloy steels that essentially contain carbon, chromium (18-22%), nickel (5-10%) and iron, and have austenitic microstructures at room temperature. Also known as *austenitic alloys.* See also austenitic stainless steels; Hadfield steels.

austenitized alloys. Ferrous alloys which have been heated to or above their upper critical temperatures to form face-centered-cubic iron and thus austenite.

austinite. A colorless, or pale yellow mineral of the descloizite group composed of calcium zinc arsenate hydroxide, $CaZnAsO_4(OH)$. Crystal system, orthorhombic. Density, 4.13 g/cm^3; refractive index, 1.763; hardness, 4.5 Mohs. Occurrence: Bolivia, USA (Utah).

Austinox. Trade name of Société Nouvelle des Acieries de Pompey (France) for a series of austenitic stainless steels (AISI types 304 through 321) containing 0.006-0.1% carbon, 18% chromium, 5-9% nickel, 0.2% molybdenum, 0.2% sulfur, and the balance iron. They have good to excellent acid and corrosion resistance, and are used for chemical equipment, process equipment, tanks, agitators, pulp and paper mill equipment, etc.

Austral. Trade name of Vidrieria Argentina SA (Argentina) for a patterned glass.

Australian blackwood. The durable, reddish to dark-brown wood of the locust tree *Acacia melanoxylon.* It has a low weight, and an attractive grain. Source: Australia, Tasmania. Used chiefly for cabinetmaking. Also known as *Australian locust; Tasmanian blackwood.*

Australian gold. See coinage gold (2).

Australian locust. See Australian blackwood.

Australian merino wool. A high-quality wool obtained from Austrlian hybrid sheep, and valued for making clothing.

Australian pine gum. A white, brittle resinous substance obtained from the Australian coniferous tree *Callitris arenosa.* Used for lacquers and varnishes.

Australian red mahogany. The hard, dark-red wood of the tree *Eucalytus resinifera.* It has good durability, and a coarse, open grain. Source: Australia. Used for general construction work.

Australian teak. The hard, orange-colored wood from the tree *Flindersia australis.* It has a close grain, poor workability, and a greasy feel. *Australian teak* is not a true teak. Source: Australia (New South Wales). Used for general construction, furniture, etc. See also teak.

Australite. Trade name of Australian Window Glass Proprietary Limited for patterned glass.

Austrian alloy. An alloy containing 90-94% zinc, 4-6% copper, and 2-3.5% aluminum. Used for die castings.

Austrian journal box. A gilding metal of 92.5% copper and 7.5% zinc. Used for bearings, bushings, etc.

Austrian oak. See chestnut oak (2).

Austwin. Trade name of Australian Window Glass Proprietary Limited for double glazing units with metal spacer.

Autan. Trademark of Acla-Werke GmbH (Germany) for cellular and solid elastomers with good shock-absorbing properties available in the form of bars, blocks, rings, and strips. Used in the manufacture of buffers, fenders and shock absorbers for air and land vehicles and ships.

Auto. Trade name of Westa-Westdeutsche Edelstahlhandelsgesellschaft (Germany) for an oil- or water-hardening, shock-resisting steel containing 0.41% carbon, 1.1% chromium, 0.7% manganese, 0.25% silicon, and the balance iron. Used for bolts, pins, shafts, axles and gears.

Auto A. Trade name for a lead-base bearing alloy containing 16% antimony and 1% copper.

Auto C. Trade name for a tin-base bearing alloy containing 11% antimony, 4% copper and 4% lead.

autoclaved lime. A *dolomitic lime* that has been highly hydrated under pressure in an autoclave. Used for structural applications. Also known as *pressure-hydrated lime.*

autoclaved Portland cement. Portland cement that has been expanded by heating under pressure in an autoclave.

Autocrat. Trade name of United American Metals Corporation (USA) for a bearing alloy containing varying amounts of copper and tin.

Autocrat Bushing Bronze. Trade name of United American Metals Corporation (USA) for a bronze containing copper, tin, and phosphorus. Used for bushings and bearings.

auto-crosslinked polysaccharides. A relatively new class of engineered polysaccharides derived from the natural polymer *hyaluronan* (hyaluronic acid) by creating crosslinking bonds and stabilizing by direct esterification of some of the carboxyl groups of *glucoronic acid* along the chain backbone with hydroxyl groups obtained from similar or dissimilar hyaluronan molecules. They possess excellent biodegradability and cytocompatibility, low cell adhesiveness, and are used as biomaterials for foul-resistant coatings on medical devices, and in surgery for the prevention of surgical adhesions. Abbreviation: ACP.

Auto-grade. Trademark for an alpha-beta rolled titanium alloy for automotive applications.

Autogrip. Trade name of Pechiney/Electrométallurgie (France) for an antiskid and wear-resistant tungsten carbide-base alloy used for automotive snow tires, antiskid studs, and ice chains with studded straps.

automolite. A dark green to nearly black variety of the mineral *gahnite* ($ZnAl_2O_4$).

Automotive Jetcoat. Trade name of Bethlehem Steel Corporation (USA) for a galvanized sheet with an intermediate alloy coating. Used for automotive applications.

Autonic. Trade name of Stapleton Technologies (USA) for electroless nickel coatings.

Auto-Pak. Trademark of 3M Company (USA) for a series of abrasives for automotive applications including fine, medium and coarse aluminum oxide and silicon carbide sandpapers, adhesive-backed sanding disks and disk pads, mechanical rust strippers, grinding and finishing products, etc. Also included under this trademark are autobody putties.

Autopan. Trade name of VDM Nickel-Technologie AG (Germany) for a free-cutting aluminum alloy containing 0.6-1% manganese, 0.6-1.2% silicon, up to 0.3% chromium, and a total of 0.5-2.5% lead, tin, cadmium and bismuth. Used for screw-machine products, and fasteners.

autophoretic paint. A water-reducible paint that has been deposited on a metallic surface by the catalytic action of a (usually ferrous) metal on the paint material in the bath. Used for automotive components.

Auto Sumus. Trade name of Stahlwerke R. & H. Plate (Germany) for a high-speed tool steel containing 0.74% carbon, 4.1% chromium, 1.1% vanadium, 18.5% tungsten, and the balance iron. Used for lathe and planer tools, drills, taps, reamers, and broaches.

Autothane. Trade name of Hyperlast Limited (UK) for a microcellular polyurethane elastomer with excellent fatigue resistance, high impact, shock and vibration resistance, and exceptional compression and flex properties. Used for automotive suspension systems.

Autothermic. (1) British trade name for a heat-resistant aluminum-coated steel used for fire walls.

(2) Trademark of Bohn Aluminum & Brass Corporation (USA) for nonferrous metal castings used for engine pistons.

Autovalve Steel. Trade name of Carpenter Technology Corporation (USA) for a corrosion-resistant steel containing carbon, iron, and nickel. It has a low thermal coefficient of resistivity, and is used for valves and spark plugs.

Autowine. Trade name for polyolefin fibers and products.

Autumn. Trade name of Pilkington Brothers Limited (UK) for a patterned glass.

autunite. A canary-yellow, radioactive mineral with a dark green core and light yellow margin composed of calcium uranyl phosphate hydrate, $(Ca,Sr)(UO_2)_2(PO_4)_2 \cdot 10H_2O$. Crystal system, tetragonal. Density, 3.05-3.23 g/cm^3; refractive index, 1.582; hardness, 2-2.5 Mohs. Occurrence: France, USA (Washington). Used as a minor ore of uranium.

auxetic materials. See negative Poisson's ratio materials.

auxiliary pigment. See extender.

Avalex. Trade name of Lilly Jamestown Inc. (USA) for varnishes, lacquers and other finishes.

Avalloy. See Cavex Avalloy.

Avantige. Trademark of E.I. DuPont de Nemours & Company (USA) for synthetic copolymer fibers composed of 88% polyamide (nylon 6,6) and 12% *Lycra*, used for transparent, elastic fabrics, especially hosiery.

Avanto. Trade name of Voco (USA) for a glass-ionomer dental luting cement.

Avantra. Trade name of BASF Corporation (USA) for a series of polystyrene resins.

Avcarb. (1) Trademark of Textron Specialy Materials (USA) for high-strength carbon fibers used as reinforcements in phenolic- and carbon-matrix engineering composites.

(2) Trademark of Textron Specialy Materials (USA) for engineering composites consisting of phenolic or carbon matrices reinforced with carbon fabrics made from Avcarb (1) carbon fibers. Used for high-temperature applications including heating elements, brake disks, aerospace components, etc.

Avceram. Trademark of FMC Corporation (USA) for a series of engineering fibers including *Avceram CS* heat-resistant high-strength carbon-silica fibers, and *Avceram RS* rayon-silica fibers. Also included under this trademark are high-temperature fabrics containing these fibers.

Aventurine. Trade name of Asagi Glass Company Limited (Japan) for a patterned glass.

aventurine. (1) A glass or glaze containing colored, opaque spangles of nonglassy materials, such as copper oxide, chromic oxide or ferric oxide, which give it a glittering appearance. Copper and ferric oxides produce golden colours while chromic oxide results in greenish shades. Also known as *aventurine glass; aventurine glaze.*

(2) A transparent or translucent feldspar that has inclusions of hematite which give it a golden shimmer. Used as a gemstone. Also known as *sunstone.*

(3) A yellow, brown, or red, transparent to opaque variety of quartz, spangled with scales of mica, hematite, or some other mineral.

Avesta. Trade name of Avesta AB (Sweden) for an extensive series of corrosion- and/or heat-resistant austenitic, ferritic, superferritic and duplex stainless steels.

Avial. Trade name of Bidault-Elion SA (France) for a heat-treatable aluminum alloy containing 2.5% copper, 1% nickel, 0.5% silicon, 0.6% magnesium and 0.7% chromium. Used for light-alloy parts.

Avialite. Trade name of Anaconda Company (USA) for an aluminum bronze containing 8-10% aluminum, 1% iron, and the balance copper. Used for die-pressed parts, and heat-resistant parts for aircraft engines.

Avibest. Trademark of FMC Corporation (USA) for microcrystalline asbestos.

avicennite. A black mineral of the bixbyite group composed of thallium oxide, Tl_2O_3. Crystal system, cubic. Density, 10.35 g/cm^3. Occurrence: Germany. It can also be made synthetically.

Avicolor. Trade name for cellulose acetate fibers used for wearing apparel.

Avicron. Trade name for rayon fibers and yarns used for textile fabrics.

Avila. Trade name for rayon fibers and yarns used for textile fabrics.

Aviloc. Trade name for rayon fibers and yarns used for textile fabrics.

Avimid. Trade name of E.I. Du Pont de Nemours & Company, Inc. (USA) for a series of condensation-type polyimide resins including *Avimid K* and *Avimid N* high-temperature thermoplastic polyimide resins with glass transition temperatures of 210°C (410°F) and 340°C (645°F), respectively.

Aviol. Russian trade name for an aluminum alloy containing 0.6% magnesium and 0.7% silicon. Used for aircraft parts.

Avional. Trademark of Alusuisse (Switzerland) for a series of lightweight, high-strength wrought aluminum alloys containing up to 5% copper, 1-1.5% magnesium, up to 1% manganese, and sometimes up to 1% silicon. Traces of lead, chromium and titanium may also be present. They have a density of 2.8 g/cm^3 (0.10 lb/in.^{3}), moderate weldability, and poor corrosion resistance. Used for aircraft structures and parts, motor vehicle parts, machinery, bus and truck bodies, machine-tools, etc.

Avisco. Trademark of Avisco (USA) for an extensive range of acetate, vinyon and viscose rayon fibers used for the manufacture of textile fabrics. The rayon fibers include standard and modified and high-tenacity grades including *Avisco Super L* and *Avisco XL.* The standard (low- and medium-tenacity) grades are used for clothing, and blended with other fibers for suitings, furnishings, carpets, linings and medical fabrics. The high-tenacity grades are used for industrial fabrics, including tire cords.

Aviscose. Trademark of Avisco (USA) for cellulose acetate fibers.

Avisun. Trade name of Avisco (USA) for polyolefin fibers and products.

Avlin. Trademark of Avlin Fibers Inc. (USA) for polyester and multicellular rayon fibers used for textile fabrics.

avodire. See African satinwood.

avogadrite. A colorless natural mineral of the barite group composed of potassium fluoborate, (K,Cs)BF$_4$, that can also be made synthetically. Crystal system, orthorhombic. Density, 2.49-2.51 g/cm^3; refractive index, 1.3245. Occurrence: Italy.

Avonbond. Trade name of Avonlea Mineral Industries Limited (Canada) for a bentonite clay used in foundry moldings.

Avonlite. Trademark of Avon Aggregates Limited (Canada) for *lightweight aggregates.*

Avonmouth. (1) Trade name of Imperial Smelting Company Limited (UK) for zinc ingots (98.5% pure) used as primary metals.
(2) Trade name of British Metal Corporation Limited (UK) for an aluminum alloy used to make light-alloy parts.

Avora FR. Trademark of KoSa (USA) for an inherently flame-resistant polyester fibers, yarns and fabrics used for clothing, drapes, home furnishings, and industrial applications.

Avox. Trademark of Textron Specialy Materials (USA) for a stabilized acrylic (polyacrylonitrile) reinforcing fiber with a density of 1.35 g/cm^3 (0.049 lb/in.^{3}) and a maximum tensile strength of 302 MPa (43.8 ksi). Used for engineering composites, e.g., conveyor belts, protective clothing and fire-blocking applications.

Avricon. Trade name of Avisco (USA) for rayon fibers and yarns used for textile fabrics.

Avril. (1) Trade name of G.A. Avril Company (USA) for an extensive series of copper- and lead-base alloys.
(2) Trade name of Avisco (USA) for high-modulus rayon fibers with excellent breaking strength and abrasion and pilling resistance.

Avron. Trade name of Avisco (USA) for high-modulus rayon fibers.

Avtel. Trademark of Phillips Chemical Company (USA) for a series of composites based on polyphenylene sulfide resins that are available as unidirectional E-glass tapes, E-glass continuous swirl mats and tapes, and fabric laminates.

Awa. Trademark of Delta Metal Company (UK) for a nickel silver composed of 62% copper, 20% zinc, and 18% nickel. Usually supplied in foil and tube form, it has a density of 8.72 g/cm^3 (0.32 lb/in.^{3}), a melting point of 1060-1110°C (1940-2030°F), excellent weldability and cold workability, good ma-

chinability, and poor hot formability. Used for fasteners (rivets, screws, bolts, etc.), costume jewelry, and optical components.

Awarnite. British corrosion- and heat-resistant alloy containing 75% nickel and 25% iron.

awaruite. A native alloy composed of 57.7% nickel and 42.3% iron, and found in New Zealand and the USA (California).

Awco. Trade name of British Alcan Wire Limited (UK) for a series of aluminum wire products including commercially-pure aluminum, and aluminum-silicon, aluminum-magnesium, aluminum-magnesium-silicon, aluminum-copper, and aluminum-zinc alloys. Used for welding wire, conductors, rivets, costume jewelry, etc.

A/W Glass-Ceramic. A dense, nonporous, two-phase bioactive glass-ceramic developed at Kyoto University, Japan. It is essentially composed of polycrystalline *apatite* [Ca$_{10}$(PO$_4$)$_6$(OH, F)$_2$] and *wollastonite* (CaSiO$_3$) embdded in a silica glass matrix. The nominal composition (in wt%) is 44.9% calcium oxide (CaO), 34.2% silicon dioxide (SiO$_2$), 16.3% phosphorus pentoxide (P$_2$O$_5$), 4.6% magnesium oxide (MgO) and 0.5% calcium difluoride (CaF$_2$). *A/W Glass-Ceramic* forms strong bonds with human bone, and is used in vertebral and illiac crest surgery. Addition of transformation-toughened zirconia (about 0.5 wt%) to these glass-ceramic somewhat inhibits bone attachment, but results in a composite with high bend strengths.

awning duck. A strong, heavy cotton fabric woven with a colored stripe.

awning stripe. A heavy canvas fabric with an evenly striped design.

Axaloy. Trade name of Timken-Detroit Axle Company (USA) for a water-hardening steel containing 0.4% carbon, and the balance iron. Used for truck axles.

axinite. A clove-brown, gray, green, blue or violet mineral with vitreous luster composed of calcium aluminum manganese borosilicate, H$_2$(Ca,Fe,Mn)$_4$(BO)Al$_2$(SiO$_4$)$_5$. Crystal system, triclinic. Density, 3.22-3.31 g/cm^3; hardness, 6.5-7.0. Occurrence: France, Switzerland, Japan, USA (New Jersey, Pennsylvania). Used chiefly as a gemstone. Also known as *glass schorl.*

Axiom-Eutekt. Trade name of Braun-Loetfolien (Germany) for brazing fluxes.

Axite. Trade name of Axelson Manufacturing Company (USA) for a wear-resistant alloy containing varying amounts of cobalt, tungsten, and chromium. Used for hardfacing electrodes and valve seats.

axle steel. An oil- or water-hardening carbon steel used for axles, gears, shafts, etc.

Axloy. Trade name of Axelson Manufacturing Company (USA) for a series of corrosion- and/or wear-resistant cast irons used for engine parts, valves, etc.

Axpandcrete. Trademark of Anti-Hydro Canada Inc. for nonshrink grouts.

AXXEL. Trademark of Fiber Innovations Technology, Inc. (USA) for specialty staple fibers available in single fiber (e.g., polyethylene terephthalate or poly-1,4-cyclohexanedimethanol terephthalate), sheath/core bicomponent fiber and segmented splittable bicomponent fiber forms.

AXXIS PC. Trademark for polycarbonate sheeting.

Ayrlyn. Trade name for a series of nylon fibers.

Azdel. Trade name of PPG Industries Inc. (USA) for a series of strong, rigid, impact-resistant engineering thermoplastics based on polypropylene or polyethylene terephthalate and reinforced with continuous glass-fiber strands. Usually supplied in sheet form, they are used for automotive interior and exterior components (e.g., instrument panels, knee bolsters, bumper beams, steering-column covers and glove-box doors) and for lawnmower shrouds and tractor parts.

Azecron. Trademark of Kimex SA (Mexico) nylon 6 fibers and filament yarns.

azide. (1) Any of numerous compounds containing an azide group ($-N_3$).

(2) A salt or ester of hydrazoic acid (HN_3).

azlin. A colored cotton fabric in a plain weave, used for furnishings.

azlon fiber. A smooth man-made fiber with a soft hand in which the fiber-forming substance is composed of a regenerated natural protein, such as corn, peanut, soybean or milk. It blends well with other fibers, and is used for garments, coats, and sportswear. Also known as *azlon.*

Azloy. Trade name for amorphous thermoplastic composites.

Azmet. Trade name for crystalline thermoplastic composites.

azobe. See African ironwood.

azo compound. A strongly colored chemical compound containing the azo group ($-N=N-$), e.g., azobenzene, $C_6H_5N=NC_6H_5$, and *p*-hydroxyazobenzene, $C_6H_5N=NC_6H_4OH$.

azo dye. Any of a group of strongly colored synthetic dyes containing the azo group ($-N=N-$). Used for dyeing cotton, paper, wool, etc.

Azolite. Trade name for *barite* containing 71% barium sulfate and 29% zinc sulfide. It is available as a 325-mesh powder, and used as a filler in rubber and paints. See also artificial barite.

Azolone. Belgian trade name for phenolic plastics.

azoproite. A black mineral of the ludwigite group composed of magnesium iron titanium borate, $Mg_2(Fe,Ti,Mg)BO_3O_2$. Crys-

tal system, orthorhombic. Density, 3.63 g/cm³; refractive index, 1.822. Occurrence: Russian Federation.

azoton. A textile fiber with enhanced properties produced by reacting cotton with acetonitrile. Used especially for apparel.

Azowit. Trade name of Thyssen Edelstahlwerke AG (Germany) for a series of chromium-aluminum, chromium-molybdenum, chromium-aluminum-molybdenum, chromium-molybdenum-vanadium, and chromium-aluminum-nickel nitriding steels.

p-azoxyanisole. A crystalline compound (98+% pure) with a melting point of 118-121°C (244-250°F). It is credited to be the first synthetic liquid crystal, prepared by F.A.L. Gatterman in 1890. Formula: $C_{14}H_{14}N_2O_3$. Also known as *4,4'-azoxyanisole; p-azoxydianisole; 4,4'-azoxydianisole.*

AZS Grog. Trade name of BPI Inc. (USA) for an alumina-zirconia-silicate (AZS) grog obtained from recycled AZS refractories. It contains about 46% alumina (Al_2O_3), 39.5% zirconia (ZrO_2), 12% silica (SiO_2), and a total of 2.5% sodium, hafnium, titanium, ferric and calcium oxides and carbon. Supplied in several mesh sizes, it is used for special ceramic and refractory applications.

Azure. Trade name for a dental die stone with high surface hardness.

azure blue. See cobalt blue.

azurite. A bright blue mineral with vitreous luster composed of copper carbonate hydroxide, $Cu_3(CO_3)_2(OH)_2$. Crystal system, monoclinic. Density, 3.77-3.83 g/cm³; hardness, 3.5-4.0 Mohs; refractive index, 1.758. Occurrence: Australia, France, Russia, South Africa, USA. Used for blue pigments, as an ore of copper, and as an ornamental stone. Also known as *blue copper; blue copper carbonate; blue copper ore; blue malachite; chessy copper; chessylite.*

Azurose. Trade name of Sovirel (France) for a color-filtered glass.

B

B-Alloy. A non-heat-treatable aluminum alloy composed of 25% zinc and 3% copper, and used for light-alloy parts.

babbitts. A class of soft, tin-, lead- or tin-lead-base white metal alloys. *Tin-base babbitts* or *tin babbitts* contain 65-95% tin, 8-12% antimony, 1-8% copper, and small additions of zinc, aluminum, arsenic, bismuth and iron. They are corrosion-resistant, have fair fatigue resistance, and are used for the manufacture of bearings for heavy-duty service applications. *Lead-base babbitts*, or *lead babbitts* contain 72-94% lead, 5-15% antimony, 1-10% tin, 0-3% copper, and small additions of other metals, such as alkaline earths, and/or arsenic. They are corrosion-resistant, moderately fatigue-resistant, and are used for the manufacture of bearings for light service applications. *Intermediate babbitts* contain 20-50% tin with the balance usually being lead and antimony. They are used for bearings designed for medium-duty service applications. *Babbitts* have a microstructure consisting of discrete hard particles in a soft matrix, are usually cast on steel, bronze or brass bases, or directly in the bearing housing, and machine well. In small bearings, they are commonly electrodeposited as thin layers. They readily bond with the substrate metal, maintain oil films on their surfaces, and have nonseizing and antifriction properties. Also known as *babbitt metals; white bearing metals; white metal alloys; white metal bearing alloys.*

babefphite. A white mineral composed of barium beryllium oxide phosphate, $BaBePO_4(O,F)$. Crystal system, tetragonal. Density, 4.31 g/cm^3; refractive index, 1.629. Occurrence: Siberia.

babingtonite. A greenish to brownish black mineral of the pyroxenoid group composed of calcium iron silicate hydroxide, $Ca_2Fe_2Si_5O_{14}(OH)$. Crystal system, triclinic. Density, 3.36 g/cm^3; refractive index, 1.731. Occurrence: Italy.

Babosil. Trade name for a frit for pottery glazes that contains varying amounts of barium oxide, boron oxide and silicon dioxide.

baby cord. See pincord.

baby flannel. A lightweight, washable flannel in plain or twill weave used for blankets and robes.

Bachite. Trade name of Bachite Development Corporation (USA) for an austenitic stainless steel containing 0.2% carbon, 18% chromium, 8% nickel, and the balance iron. Used for stainless parts and chemical equipment.

back case metal. An alloy containing 62% copper, 20% zinc, and 18% nickel. Used for heat- and corrosion-resistant parts.

back coating. An adhesive coating applied to the back of a textile fabric to increase its body and stiffness, lock pile yarn tufts into carpets, or bond secondary backings to primary backings.

backed fabrics. Textile fabrics woven with additional sets of warp or weft yarns to enhance their weight and/or strength.

back filling. A filler material applied to the back of a textile fabric to increase its weight and/or enhance its feel.

backing brick. See common brick.

backing fabrics. Textile fabrics that have an extra reinforcing fabric bonded to the reverse side.

backing sand. Reconditioned foundry sand that is rammed uniformly in the flask around the pattern on top of the facing sand. It supports the latter and forms the major portion of the total sand volume in the flask. See also facing sand; foundry sand.

back-sizing. See filler (8).

Bacon EP. Trade name of Bacon Industries (USA) for a series of epoxies available in general-purpose, flexible and high-heat grades, and as aluminum-, glass-, mineral- and silica-filled casting resins.

bacor. Alumina-zirconia refractories made from 15-30% *baddeleyite* (ZrO_2), 50-65% *corundum* (Al_2O_3), and 15% *silica* (SiO_2). Used in glass manufacture.

Bactekiller. Trademark of Kanebo Limited (Japan) for a durable, antimicrobial polyester fiber for the manufacture of clothing and medical products.

bactericide paper. Paper that has been treated with a chemical agent or solution that destroys bacteria. Used chiefly in biochemistry, biotechnology, medicine, and hygienics.

bacteriorhodopsin. A purple protein pigment or photochrome that occurs naturally in the cell membranes (purple membranes) of the bacteria *Halobacterium halobium* and *H. salinarum*. It is chemically similar and structurally related to rhodopsin (a purple pigment found in the rod cells of mammalian retinas). It has a typical molecular weight of approximately 26784 daltons, contains about 248 amino acids, and has a purple membrane protein-to-lipid ratio of 3:1 by molecular weight. *Bacteriorhodopsin* functions in converting solar energy (sunlight) into chemical and electrical energy. It can also be synthesized in the laboratory (supplied as wild-type or genetically modified products). The light-induced electric signals in the cell membrane can be utilized in the manufacture of intelligent electronic, optical, photonic and bioengineered materials and devices. Typical applications in electronics, optics, electrooptics, photonics and life sciences include optical computing and holographic recording devices, associative optical memories, spatial light modulators, neural networks, and artificial retinas. See also bacteriorhodopsin film; lyophilized bacteriorhodopsin; rhodopsin.

bacteriorhodopsin film. An optical information recording and processing material based on *bacteriorhodopsin* and supplied in the form of a thin film with normal or slow thermal relaxation, high spatial resolution and outstanding thermal and photochemical stability. The bacteriorhodopsin-containing layer has a typical thickness of 30-100 μm (0.001-0.004 in.) and is sealed between two windows of high-quality optical glass. Upon photochemical excitation by yellow light the bacteriorhodopsin gradually changes from the initial state (B-state) to the final state (M-state). This photochemically reversible B-M transition is utilized for optical data recording and processing applications (holography, pattern recognition, optical filtering, etc.). The reverse M-B transition is suited for data recording and information processing applications.

BactiShield. Trademark of Wellman Inc. (USA) for a durable, antimicrobially-treated polyester fiber that inhibits bacterial odors, fungi and mildew. Used for fiberfill and dry batting applications including clothing and home furnishings.

Badell. Trade name of Badell Company Inc. (USA) for heat-treated, wear-resistant steel plate used for severe service applications, pressure vessels, and wear plates.

baddeleyite. A colorless, yellow, brown, or black mineral with a white streak and submetallic to vitreous luster composed of zirconium dioxide, ZrO_2. Crystal system, monoclinic. Density, 5.5-6.0 g/cm³; melting point, 2300-2950°C (4172-5342°F); refractive index, 2.19; high resistance to heat and corrosion. Occurrence: Brazil, India, Sri Lanka. Used as a source of zirconium, as a refractory, e.g., for furnace linings and muffles, and as an ingredient in low-expansion ceramic bodies.

Badger. Trade name of Latrobe Steel Company (USA) for an oil-hardening, nondeforming cold-work tool steel (AISI type O1) containing 0.94% carbon, 1.2% manganese, 0.3% silicon, 0.5% chromium, 0.5% tungsten, and the balance iron. It has a great depth of hardening, good machinability, and low tendency to shrinking and warping. Used for blanking, forming and trimming dies, and punches.

Badger Cast. Trademark of Badger Mining Corporation (USA) for industrial silica sand, limestone, and zeolite.

Badin metal. A foundry alloy containing about 19% silicon, 9% aluminum, 5% titanium, and the balance iron. Used as a deoxidizer for steel.

Bafa. German trademark for scrim fabrics of rayon staple yarn, jute or synthetic fibers, or glass-fiber yarn.

bafertisite. A yellow, red, or brown mineral of the seidozerite group composed of barium iron titanium silicate, $BaFe_2TiSi_2O_9$. Crystal system, orthorhombic. Density, 3.96 g/cm³. Occurrence: Mongolia.

bagasse. The crushed fibrous cellulosic residue left after the juice is expressed from sugar cane. It is used as reinforcement and filler in plaster products, e.g., acoustic tile, in the manufacture of lightweight refractories, for low-grade paper, in compressed form as an insulating board in construction, and as a fuel.

bag cloth. A heavily sized fabric woven from inferior-grade yarns, and used for bags and sacks to hold dry goods.

bagging cloth. See gunny.

baggings. Textiles made from bast fibers, chiefly jute, and used primarily in the manufacture of bags, sacks, packaging materials, and for wall coverings.

Bagheera velvet. A largely uncrushable *velvet* with a rough surface texture that has been piece-dyed, and is used for outerwear.

bag leather. A staple leather made from the hides of cattle, sheep, goats and seals, and used for suitcases and traveling bags. Also known as *case leather.*

bagtikan. The reddish gray to pale brown wood of the lauan trees *Parashorea plicata* and *P. malagnonan*. It is stronger and heavier than most other lauans, but has poor durability, and must be treated with preservatives. Source: Southeast Asia. Used as a veneer for plywood. See also lauan.

bahianite. A tan- to cream-colored mineral composed of aluminum antimony oxide hydroxide, $Sb_3Al_5O_{14}(OH)_2$. Crystal system, monoclinic. Density, 4.89 g/cm³; refractive index, 1.87. Occurrence: Brazil.

Bahn-Aluminium. A German wrought aluminum alloy containing 6% copper. Used for light-alloy parts and electric pantographs.

Bahn-Metall. A German alkali-lead alloy containing 0.68-0.76% calcium, 0.62-0.72% sodium, 0.03-0.05% lithium, 0.02-0.04% potassium, 0.2% aluminum, and the balance lead. It retains its high hardness even at high temperature, and is used for locomotive and railroad-car bearings. Also known as *B-Metall*. Also spelled "Bahnmetall."

Bahrain sand. A type of sand that occurs in the State of Bahrain and contains predominantly calcite ($CaCO_3$) and quartz (α-SiO_2) with small quantities of dolomite ($CaMg(CO_3)_2$) and feldspars, e.g., albite and microcline.

Baily's metal. A British corrosion-resistant alloy containing about 82% copper, 13% tin, and 5% zinc. Used for bearings and castings.

bainite. A transformation product of *austenite* formed in some steels and cast irons at temperatures higher than those where *martensite* starts to form on cooling. The microstructure consists of alpha *ferrite* and a fine dispersion of *cementite*. If formed in the upper part of the transformation temperature range, the bainite has a feathery appearance and is known as "upper bainite;" if formed in the lower part, the bainite has an acicular (needle-like) appearance, somewhat resembling *tempered martensite*, and is known as "lower bainite."

Bainitex. Trademark of Kaltwalzwerk Brockhaus GmbH (Germany) for texture-rolled steel strip used for spings.

baize. A coarse, thick woolen or cotton cloth in a plain weave, usually dyed green and having a short nap. Used especially as a table, wall and screen covering.

BAK. Trademark of Bayer Corporation (USA) for biodegradable polyamides used for bottles, garbage bags, flower ties, and films.

Bakadie. (1) Trade name of CCS Braeburn Alloy Steel (USA) for a low-carbon die steel containing varying amounts of nickel and molybdenum. Used for bakelite molds and dies.

(2) Trade name of CCS Braeburn Alloy Steel (USA) for an austenitic stainless steel containing 0.2% carbon, 18% chromium, 8% nickel, and the balance iron. Used for colored stainless steel parts.

bake-hardening steels. A group of high-strength sheet steels with ultralow carbon contents (typically 50 ppm or less) that have been strengthened by bake hardening, i.e., by yield strength-increasing age hardening during the paint-bake treatment. They have an optimal combination of high yield and tensile strengths and good ductility, drawability and formability. Used especially for structural applications, and in the automotive and domestic appliance industries for parts which require good appearance.

Bakelite. Trademark of Union Carbide Corporation (USA) for a series of phenolic resins made of phenol and furfural, or phenol and formaldehyde. They have good heat and electrical resistance, high rigidity, strength, hardness and stability, good to excellent ductility, creep resistance and low-temperature properties, low moisture absorption, and excellent thermal stability to over 150°C (300°F). They are easy to mold, cast and laminate, highly resistant to organic solvents, poorly resistant to strong acids and alkalies, and moderately resistant to weathering and moisture. Used for handles, pulleys, wheels, distributors for cars and trucks, electrical fixtures, fuse blocks, plugs, coil forms, electric devices, television and radio cabinets, computer components, insulators, housings, containers, buttons, toilet seats, piping, conduits, ducts, etc. *Note:* The trademark *Bakelite* now also includes other plastic materials, such as acrylics, acrylonitrile-butadiene styrenes, epoxies, ethylene copolymers, parylenes, polyethylenes, polypropylenes, polystyrenes, polysulfones and vinyl resins and compounds.

bakelites. See phenol-formaldehyde resins.

Baker. Trade name of Engelhard Corporation (USA) formerly used for various gold, palladium and platinum dental alloys, now sold under the trade name *Engelhard.*

Baker Cast Resin. Trade name of Baker Oil Tool Company (USA) for cast phenolic resins.

Baker Clasp Wire. Trade name of Engelhard Corporation (USA) for a wrought gold alloy wire used for dental applications. See also Baker.

Baker Four. Trade name of Engelhard Corporation (USA) for a medium-gold dental casting alloy. See also Baker.

bakerite. A white mineral composed of calcium borosilicate hexahydrate, $Ca_8B_{10}Si_6O_{35}·6H_2O$. It resembles unglazed porcelain. Occurrence: USA (California).

baking enamel. The top or finish coating enamel that incorporates catalysts and crosslinking agents which require heat for polymerization, and must be cured by baking at a temperature above 65°C (150°F) in an oven, under infrared lamps, or by induction heating. Often used on automobiles and household appliances (refrigerators, etc.). See also enamel; enamel paint.

baking japan. An enamel that has to be cured by baking at elevated temperatures to attain maximum hardness and toughness of film.

baking soda. See sodium bicarbonate.

Bakos Iron. Trade name of Webster Industries, Inc. (USA) for a machinable white iron that does not require an annealing treatment. Used for chains and castings.

Balacron. Trademark of BN International BV (Netherlands) for artificial leather.

balanced fabrics. Textile fabrics with an equal number of warp and weft (filling) yarns per inch. Both yarns are of the same count number.

balanced laminate. A structural laminate in which all plies at angles other than 0° and 90° occur only in ± pairs, and are symmetrical on both sides of the laminate principal axis.

balanced yarn. A yarn that will not double, kink or twist on itself when permitted to hang in an open loop.

balancing paper. A resin-impregnated *kraft paper* used as a warp-free core material in plastic laminates for structural applications.

balangeroite. A brown mineral composed of iron magnesium manganese silicate hydroxide, $(Mg,Fe,Mn)_{42}Si_{15}O_{54}(OH)_{36}$. Crystal system, orthorhombic. Density, 2.98 g/cm^3; refractive index, 1.680. Occurrence: Italy.

balas. See balas ruby.

balas ruby. A red variety of the mineral *spinel* from Sri Lanka and Burma, used as a gem and for ornamental purposes. The term is a misnomer, since this mineral is not a true ruby. Also known as *spinel ruby.*

balata. A hard, thermoplastic, nonelastic, rubberlike substance obtained by drying the milky juice of trees of the Sapotaceae family, especially the South American bully tree *(Manilkara bidentata).* It is a *trans*-isomer of isoprene and resembles *gutta-percha.* Source: Venezuela, Brazil, Guianas. Used for golf ball covers, for impregnating conveyor and power transmission belts, and as a replacement for gutta-percha. Also known as *gutta-balata.*

Balatum. Trade name for wool felt impregnated with balata, or a rubber solution, and used as floor covering material.

Balco. Trademark of Wilbur B. Driver Company (USA) for a magnetic alloy containing 29-31% iron and 69-71% nickel. It has a density of 8.46 g/cm^3 (0.306 $lb/in.^3$), high magnetic permeability, good heat resistance up to 590°C (1094°F), and good strength properties. Used for thermometer bulbs, ballast tubes, voltage resistors, and cores for magnetic amplifiers.

Balco Alloy. Trade name of BALCO–Bahrain Saudi Aluminum Marketing Company for a series of aluminum ingot alloys.

baldcypress. The soft, light wood of the large coniferous tree *Taxodium distichum* belonging to the redwood family (Taxodiaceae). The heartwood varies from yellowish-brown to brown or dark reddish-brown to chocolate, and the sapwood is rather narrow and almost white. *Baldcypress* has a fine, straight grain, exceptional resistance to decay and insects, high resistance to warping, high durability even under damp conditions, and moderate stiffness, strength and hardness. Average weight, 510 kg/m^3 (32 lb/ft^3). Source: Swamps and wetlands of southeastern United States and Mexico. Used for construction work including docks, warehouses, bridges and boats, and for shingles, interior paneling, railroad ties, posts, piling, tanks, vats, and cooperage. Also known as *cypress; gulf cypress; marsh cypress; red cypress; southern cypress; white cypress.*

Balder. Trade name of Uddeholm Corporation (USA) for a soft-magnetic iron containing 0.05% carbon, 0.1% silicon, 0.1% manganese, and the balance iron. Used for electrical equipment.

Baldolux. Trade name of Esperanza SA (Spain) for a solid glass block with a reeded pattern on one surface.

Baldwins. Trade name of British Steel Corporation (UK) for an extensive series of austenitic, ferritic and martensitic stainless steels.

Bale-lok. Trade name for polyolefin fibers used for textiles and cordage.

baler twine. A heavy strand, usually 5 mm (0.2 in.) or less in diameter, consisting of fibers or yarns of cotton, flax, jute, sisal, rayon or nylon, compacted into a twisted structure. Used for tying or binding parcels, bundles, bales, newspapers and lumber.

Balfosteel. Trade name of Ekstrand & Tholand Company (USA) for a water-hardening tool steel containing 0.7-1.2% carbon, and the balance iron. Used for saws, punches, reamers, and dies.

Balfour. Trade name of Darwins Alloy Castings (UK) for an extensive series of cold-work, hot-work and plain-carbon tool steels.

Balfour Darwins. Trade name of E.A. Balfour Steel (UK) for a series of acid- and corrosion-resistant stainless steels.

Balinet. Trade name of Balzers Tool Coating Inc. (USA) for a series of physical-vapor deposition coatings including (i) *Balinet A* wear-resistant titanium nitrides for ferrous metals and plastics, (ii) *Balinet B* titanium carbonitrides for milling, forming and punching tools, (iii) *Balinet C* low-friction tungsten carbide/carbon composites for gears, gear drives, engine parts, pumps, compressors, etc., (iv) *Balinet Cast* and *Balinet D* chromium carbide and chromium nitride coatings respectively for die-cast mold components, and for machining copper, (v) *Balinet Futura* multilayer titanium-aluminum nitride coatings for carbides, cermets and high-speed steel tooling, (vi) *Balinet Xtreme* single-layer titanium-aluminum nitride coatings for carbide end mills for machining hardened steel, and (vii) *Balinet HardLube* cutting-tool coatings based on titanium-aluminum nitride and tungsten carbide particles in carbon matrices.

baling paper. A kraft or Manila paper that has asphalt-coated cheesecloth pasted to one side. Used for packaging applications. See also kraft paper; Manila paper

balipholite. A white mineral of the carpholite group composed of lithium barium magnesium aluminum fluoride silicate hydroxide, $BaMgLi_{1.5}Al_{3.5}Si_4O_{12}(OH,F)_8$. Crystal system, orthorhombic. Density, 3.32 g/cm³; refractive index, 1.5943. Occurrence: China.

balk. A large square-sawn or hewn softwood timber used in the building and construction trades. Also spelled "baulk."

balkanite. A gray mineral composed of copper silver mercury sulfide, $Cu_9HgAs_2S_8$. It can also be made synthetically. Crystal system, orthorhombic. Density, 6.32 g/cm³. Occurrence: Bulgaria.

ballas. A hard, spherical industrial diamond consisting of numerous tiny crystals arranged radially around a central point. Used for drilling applications.

ballast. Crushed stone or coarse gravel used in railroad beds, road construction, and as an aggregate in the manufacture of concrete. See also railroad ballast.

ballast concrete. A *heavy concrete* containing crushed stone or coarse gravel as the aggregate.

Ballast. Trade name of Gilby-Fodor SA (France) for a heat-resistant, commercially pure nickel (99.8%) used for cathodes and for electronic tube filaments.

Ballast Nickel. Trade name of Wilbur B. Driver Company (USA) for a commercially pure nickel (99.6+%) used for current-limiting controls in electrical equipment.

ball-bearing stainless steel. A stainless steel containing 0.4% carbon, 0.25% silicon, 0.3% manganese, 11.5% chromium, and the balance iron. Used for corrosion-resistant ball bearings.

ball-bearing steel. A chromium steel, such as AISI-SAE 52100, usually through- or case-hardened and containing about 0.5-1% carbon and 1-4% chromium. Its microstructure after oil quenching and stress relieving is composed of a fine acicular martensite matrix with uniform undissolved carbides. It has good machinability, dimensional stability and wear resistance, and high hardness and compressive strength. Used for ball-bearing balls and races. Also known as *bearing steel.*

ball clay. A fine-grained, sedimentary lignite-bearing clay consisting chiefly of aluminum silicate, and characterized by high plasticity, high dry and wet strength, high refractoriness, high tendency to balling, strong bonding power, a long vitrification range, and a clean, white, ivory or buff color after firing. Used for ceramic bodies to provide plasticity during forming and induce vitrification during firing, as a bonding agent or plasticizer in whiteware, porcelains, stoneware, terra cotta, glass refractories, floor and wall tile, and as a suspension agent in porcelain enamels and glazes.

Ballit. German trade name for a plastic wood made into a soft, malleable paste.

ball-milled powder. An alloyed (blended), mixed and/or ground metal or nonmetal powder produced in a rotating cylindrical container filled with hard metal or nonmetal balls.

ballnut hickory. See mockernut hickory.

balloon cloth. A tightly woven fabric made from fine, superior-quality cotton yarn in a plain weave. Used for balloons, shirts, and formerly for typewriter ribbons.

Ballotini. Trade name of Potters Industries Inc. (USA) for glass beads used for stress relief and blast cleaning.

balsa. The strong, buoyant wood of the trees *Ochroma lagopus, O. pyramidale* and *O. velutina,* that is the lightest wood in general use. Its average weight varies from 40 to more than 320 kg/m³ (2.5 to more than 20 lb/ft³). It is relatively soft, white to pale gray, and has good heat-insulating and sound-absorbing properties. Source: West Indies, Central and South America. Used for general modelbuilding, model airplanes, airplane fairings, floats and rafts, life-saving equipment, sound and heat insulators, cold storage, etc. Also known as *balsawood; corkwood.*

balsa fiber. A dark-colored fiber obtained from the Central American balsa tree *Ochroma velutina.* Used for insulation and padding applications.

balsam fir. The soft, light, perishable wood of the fir tree *Abies balsamea.* It is creamy white to pale brown with straight grain and indistinguishable heartwood and sapwood. Average weight, 417 kg/m³ (26 lb/ft³). Source: Eastern and central Canada, and northern United States (mainly from New England and Lake states). Used as a pulpwood, for lumber, and as a Christmas tree. *Note:* A resin known as *Canada balsam* is obtained from the bark blisters of this fir. It is used in making varnish, cementing lenses, waterproofing cement, and mounting specimens on microscope slides. See also Canada balsam.

balsam of fir. See Canada balsam.

balsam poplar. The soft, weak wood of the poplar tree *Populus balsamifera.* Average weight, 450 kg/m³ (28 lb/ft³). Source: Canada; northeastern United States. Used for pulpwood, lumber, veneer, containers, and excelsior. Also known as *tacamahac.* See also cottonwood; excelsior; poplar.

balsa sawdust. Particles of balsawood produced in sawing, and used as lightweight fillers for plastics.

balsawood. See balsa.

Baltic. (1) Trade name of Pittsburgh Corning Corporation (USA) for a glass block with linear pattern.

(2) Trade name of Joseph Beardshaw & Son Limited (UK) for a series of cold-work, hot-work and plain-carbon tool steels, and several die and mold steels.

Baltic pine. See northern pine.

Baltic redwood. See northern pine.

Baltic whitewood. See Norway spruce.

Baltoc. Trade name of Crucible Specialty Metals (USA) for a cobalt-tungsten high-speed tool steel containing 0.7% carbon, 18% tungsten, 4% chromium, 5% cobalt, 1% vanadium, and the balance iron. Used for hogging tools.

balyakinite. A mineral composed of copper tellurium oxide, $CuTeO_3$. Crystal system, orthorhombic. Density, 5.62 g/cm³.

Bamberko. Trademark of Claude Bamberger Molding Compounds (USA) for a thermoplastic purging compound used in the injection molding and extrusion of cellulosic, polyolefin, polyvinyl chloride and styrene resins at 176-275°C (°349-527°F).

bambollaite. A black mineral composed of copper selenide telluride, $Cu(Se,Te)_2$. Crystal system, tetragonal. Density, 5.64 g/cm³. Occurrence: Mexico.

Bamboo. Trade name of Libbey-Owens-Ford Company (USA) for patterned glass.

bamboo. Tropical grasses of the order *Graminaceae* with hollow, tree-like stems. They are native to Southeast Asia (mainly Indonesia, Philippines, India, Java and Sri Lanka), but also found in the southern United States, and Central and South America. Used for furniture, fishing poles, window blinds, arrows and lances, baskets, walking sticks, etc., for making specialty paper, and as a source of cellulose.

Bamilex. Trademark of Bay Mills Corporation (Canada) for a composite fabric of polyester, nylon, glass fibers, etc., used chiefly as backing for upholstery, as a reinforcing membrane for roof coatings, and as a coating substrate for cargo parachutes and pond liners.

banak. The pinkish-brown to brownish-gray wood of various tropical American trees especially *Virola koschnyi, V. surinamensis,* and *V. sebifera.* It has a straight grain, a uniform texture, moderate strength, and poor resistance to decay and insects, but can be treated with preservatives. Source: Central and South America. Used for lumber, plywood, and veneer.

banalsite. A white mineral of the feldspar group composed of sodium barium aluminum silicate, $Na_2Ba(Al_2Si_2O_8)_2$. Crystal system, orthorhombic. Density, 3.07 g/cm^3; refractive index, 1.571. Occurrence: Sweden, UK (Wales).

banana fibers. The fibers obtained from the leafstalks any of various treelike, herbaceous tropical banana plants (genus *Musa*), especially the plantain (*M. sapientum*) and Manila hemp (*M. textilis*). Used for making cordage and paper.

Banard. Trade name for a ball clay.

banca tin. A high-grade tin from Banca and Malacca. See also Banka.

Bandalasta. Trade name of Brookes & Adams (UK) for urea-formaldehyde resins and products.

bandana. See bandanna.

bandanna. A fabric, usually of cotton, that has a bright or dark ground with a printed design depicting white or brightly colored motifs. It is originally from India, and used for napkins, etc. Also known as *bandana*.

Bandit. Trade name of Triplex Safety Glass Company Limited (UK) for a laminated glass that consists of two sheets of glass with a polyvinyl butyral interlayer.

Bandlite. Trade name of C-E Glass (USA) for glass with a heavily banded pattern.

Band-Rite. Trademark of Pulpdent Corporation (USA) for a strong, fluoride-releasing self- or light-cure dental glass-ionomer cement with excellent mixing and handling characteristics. Used for the cementation of orthodontic bands.

bandylite. A deep blue mineral composed of copper borate-chloride tetrahydrate, $CuB_2O_4 \cdot CuCl_2 \cdot 4H_2O$. Crystal system, tetragonal. Density, 2.81 g/cm^3; refractive index, 1.691. Occurrence: Chile.

Bangor limestone. A limestone from Bangor, Alabama, USA that consists of small rounded grains cemented together, and is used as a building stone.

Banka. Brand name for high-grade tin (99.935% pure) containing 0.03% lead, 0.006% copper, and 0.014% arsenic. See also banca tin.

banknote paper. A strong, durable, crease-resistant safety paper with watermark.

bank sand. A sand that is found in pits or banks, and whose total content of clay and silt is usually less than 12%. Used in casting for making cores, and for synthetic molding sands.

banner cloth. See bunting.

bannermanite. A black mineral composed of sodium vanadium oxide, NaV_6O_{15}. It can also be made synthetically. Crystal system, monoclinic. Density, 3.50-3.58; refractive index, 2.2. Occurrence: Central America.

bannisterite. A dark brown mineral composed of potassium manganese iron aluminum silicate hydroxide hydrate, $(K,Ca)_{0.5}(Mn,Fe,Zn)_4(Si,Al)_7O_{14}(OH)_8$. Crystal system, monoclinic. Density, 2.92 g/cm^3; refractive index, 1.586. Occurrence: USA.

Banrock. Trade name for mineral wool made from high-silica limestone, and used as oven wall insulation.

baotite. A black mineral of the axinite group composed of barium titanium niobium chloride silicate, $Ba_4Ti_7NbSi_4O_{28}Cl$. Crystal system, tetragonal. Density, 4.71; refractive index, 1.944. Occurrence: USA (Montana), Mongolia.

Bapolan. Trademark of Bamberger Polymers, Inc. (USA) for a series of high-impact polystyrene and acrylonitrile-butadiene-styrene resins used for appliances, housewares, toys, electronic instruments, telephones, etc.

Bapolan PS. Trademark of Bamberger Polymers, Inc. (USA) for a series of polystyrenes supplied in the following grades: standard, medium-impact, 30% glass fiber-reinforced, 2% silicone-lubricated, UV-stabilized, extrusion, injection molding, and structural foam.

Bapolan HIPS. Trademark of Bamberger Polymers, Inc. (USA) for a series of high-impact polystyrenes available in standard, fire-retardant, and UV-stabilized grades.

Bapolene EVA. Trade name of Bamberger Polymers, Inc. (USA) for ethylene-vinyl acetates supplied in standard and stabilized grades.

Bapolene PE. Trade name of Bamberger Polymers, Inc. (USA) for low-density and linear-low-density polyethylenes supplied in standard and UV-stabilized grades. Used for housewares, toys, pens, closures, appliances, medical devices, instrument panels, etc.

Bapolene PP. Trade name of Bamberger Polymers, Inc. (USA) for polypropylene homopolymers supplied in standard and UV-stabilized grades. Used for housewares, toys, pens, closures, appliances, medical devices, instrument panels, etc.

Bapolon. Trademark of Bamberger Polymers, Inc. (USA) for a series of nylon 6 and nylon 6,6 plastics available in powder and pellet form, and used for general industrial applications.

bar. A round, square, rectangular or other polygonal piece of rolled, drawn or extruded metal having a length greater than its width or thickness. Also known as *barstock*.

Baral. Trade name of Calloy Limited (UK) for an aluminum alloy containing up to 50% barium. It has good stability in air, and is used as a getter in electrical discharge devices and vacuum tubes.

bararite. A white mineral composed of ammonium fluosilicate, $(NH_4)_2SiF_6$. Crystal system, hexagonal. Occurrence: India.

barathea. A silk, silk-wool, wool, or worsted fabric with a pronounced broken filling effect, used for neckties, evening wear and women's suits and coats. See also silk; worsted.

baratovite. A white mineral composed of lithium potassium calcium titanium zirconium fluoride silicate, $Li_3KCa_7(Ti,Zr)_2(Si_6O_{18})_2F_2$. Crystal system, monoclinic. Density, 2.92 g/cm^3; refractive index, 1.672. Occurrence: Tadzhikistan.

barbed wire. A twisted wire with sharp hooks or points fixed to it at short intervals, or a single or twisted wire with double-pointed wire barbs. It is usually made of galvanized steel, and used for fences. Also known as *barbwire*.

Barberite. Trade name of Barber Asphalt Company (USA) for a corrosion-resistant alloy containing containing 5% nickel, 5% tin, 1.5% silicon, 0.5% manganese, 0.5% iron, 0.04% carbon, and the balance copper. It has a density of 8.8 g/cm^3 (0.32 lb/in.^{3}), a melting point of 1070°C (1960°F), and good resistance to sulfuric acid, seawater, sulfurous atmospheres, and mine waters. Used for marine parts, mining equipment, ornaments, and fixtures.

barbertonite. A violet to light pinkish red mineral composed of magnesium chromium hydroxide carbonate tetrahydrate, $Mg_6Cr_2(OH)_{16}CO_3 \cdot 4H_2O$. Crystal system, hexagonal. Occurrence: Southern Africa (Transvaal).

barbosalite. A black mineral of the lazulite group composed of iron phosphate hydroxide, $FeFe_2(PO_4)_2(OH)_2$. Crystal system, monoclinic. Density, 3.60 g/cm^3; refractive index, 1.5943. Occurrence: Brazil.

barbwire. See barbed wire.

Bardempal. Trade name of Saint-Gobain (France) for exterior wall panels composed of rigid glass fiberboard between two surfacing layers.

Barden. Trade name for a series of yellowish white hydrous aluminum silicates (sedimentary kaolins) with fine particle sizes. They have a density of 2.60 g/cm³ (0.094 lb/in.³), and are used for flooring and tile-caulking compounds, putties, roofing granules, boxboard, and as fillers or extenders in engineering plastics.

Bareco. Trademark of Petrolite Corporation (USA) for microcrystalline waxes.

bare glass. Glass yarns, rovings and fabrics prior to the application, or after the removal of sizings or finishes.

barentsite. A colorless mineral composed of hydrogen sodium aluminum carbonate fluoride, $Na_7AlH_2(CO_3)_4F_4$. Crystal system, triclinic. Density, 2.56 g/cm³; refractive index, 1.479. Occurrence: Russian Federation.

Barex. Trademark of BP Amoco (USA) for impact- and chemical-resistant acrylonitrile-methacrylate copolymers used for household products, toys, and packaging applications.

bariandite. A mineral composed of vanadium oxide hydrate, $V_{10}O_{24} \cdot 12H_2O$. Crystal system, monoclinic. Density, 2.70 g/cm³; refractive index, above 1.85. Occurrence: Africa (Gabon).

baricite. A colorless to pale blue mineral of the vivianite group composed of iron magnesium phosphate octahydrate, $(Mg,Fe)_3$-$(PO_4)_2 \cdot 8H_2O$. Crystal system, monoclinic. Density, 2.42 g/cm³; refractive index, 1.564. Occurrence: Canada (Yukon).

Bario. (1) British trade name for a stainless and corrosion-resisting alloy of 57.4% nickel, 21.4% chromium, 15.4% tungsten, 1% iron, 0.3% carbon, and the balance iron. Used for resistor elements.

(2) Trade name for a stainless and corrosion-resisting sheet material containing 90% nickel, 4.3% chromium, 1.2% tungsten, 0.3% silicon, and traces of cobalt, copper, and iron. Used for tools and corrosion-resistant parts.

Bario Hard. Trade name for a corrosion- and heat-resistant alloy containing 30% cobalt, 30% chromium, 25% tungsten, 10% manganese, and 5% titanium. Used for tools and high-temperature applications.

bariomicrolite. A pink, reddish, yellowish-brown, or colorless to white mineral of the pyrochlore group composed of barium tantalum oxide hydrate, $BaTa_2(O,OH)_7$. Crystal system, cubic. Density, 5.68 g/cm³. Occurrence: Brazil.

bariopyrochlore. A yellowish gray mineral of the pyrochlore group composed of barium strontium niobium oxide hydrate, $(Ba,Sr)Nb_2O_6(OH)$. Crystal system, cubic. Density, 4.00 g/cm³. Occurrence: Tanganyika.

Bario Soft. Trade name for a corrosion- and heat-resistant alloy containing 60% cobalt, 20% chromium, and 20% tungsten. Used for tools and heat- and corrosion-resistant parts.

bar iron. Wrought iron in the form of bars.

Barite. Trade name for a series of coarse, medium, fine and superfine ground barites (80, 100, 120 and 140 mesh respectively). Used as extenders for rubber, plastics, paper, paints, textiles, etc.

barite. A colorless, white, or light yellow mineral with vitreous luster composed of barium sulfate, $BaSO_4$. It can also be made synthetically. Crystal system, orthorhombic. Density, 4.3-4.5 g/cm³; melting point, 1580°C (2876°F); hardness, 2.5-3.5 Mohs; refractive index, 1.637. Occurrence: Europe, USA (Arkansas, California, Connecticut, Georgia, Idaho, Missouri, New York, Tennessee). Used as an important source of barium, as an ex-

tender in rubber, plastics, paint, paper and textiles, as a flux in glasses, in ceramic bodies, glazes and porcelain enamels, and as an aggregate in high-density radiation-shielding concrete. Also known as *baryte; heavy spar.* See also artificial barite.

Barium. Trade name of Barium Stainless Steel Corporation (USA) for a die steel containing 0.4-0.6% carbon, 1-2% nickel, 0.5-1% chromium, 0.2-0.3% molybdenum, and the balance iron. Used for forging dies.

barium. A soft, silvery-white metallic element of Group IIA (Group 2) of the Periodic Table (alkaline-earth group). It is commercially available in the form of rods, sticks, billets, plates, lumps, powder and wire. The commercial barium ores are *barite* and *witherite.* Crystal system, cubic. Crystal structure, body-centered cubic. Density, 3.6 g/cm³; melting point, 704°C (1317°F); boiling point, 1640°C (2980°F); atomic number, 56; atomic weight, 137.327; divalent, heptavalent; good extrudability and machinability. Used as a getter in vacuum tubes, in fluorescent lamps, as a deoxidizer for copper, in Frary's metal and spark-plug alloys, in pyrotechnics, in superconductors, and in biochemistry and medicine. Symbol: Ba.

barium acetate. A white crystalline powder (99+% pure) obtained by treating a solution of barium carbonate or barium sulfate with acetic acid. Density, 2.47 g/cm³; melting point, decomposes. Used in paint and varnish driers, and in organometallic and superconductor research. Formula: $Ba(C_2H_3O_2)_2$.

barium aluminate. Any of the following compounds of barium oxide and aluminum oxide: (i) *Tribarium aluminate.* A gray powder. Melting point, 2000°C (3632°F). Used as a source of barium oxide in glassmaking, and in cathode coatings for vacuum tubes. Formula: $Ba_3Al_2O_6$; (ii) *Barium monoaluminate.* An off-white powder. Density, 3.99 g/cm³; melting point, 1998°C (3628°F). Used in ceramics. Formula: $BaAl_2O_4$. Also known as *barium aluminum oxide (BAO);* and (iii) *Barium hexaaluminate.* An off-white powder. Density 3.64 g/cm³; melting point, 1860°C (3380°F). Used in ceramics. Formula: $BaAl_{12}O_{19}$.

barium aluminate silicate. A compound of barium oxide, aluminum oxide and silicon dioxide. Found in nature as the mineral *celsian.* Density, 3.2-3.3 g/cm³; melting point, 1716°C (3121°F). Used in ceramics and refractories. Formula: $BaAl_2Si_2O_8$.

barium aluminum oxide. See barium aluminate (ii).

barium arsenate. A black powder used in ceramics. Density, 5.1 g/cm³; melting point 1604°C (2919°F). Formula: $Ba_3(AsO_4)_2$.

barium binoxide. See barium dioxide.

β-barium borate. A synthetically grown crystal of barium borate with high nonlinear optical coefficient and damage threshold, and a transparency range of 190-350 nm. Used in nonlinear optics, and for laser applications especially Nd:YAG, Ti:sapphire and Q-switched types. Formula: $β-BaB_4O_7$. Abbreviation: BBO.

barium boride. A black crystalline substance. Density, 4.32 g/cm³; melting point, 2270°C (4118°F); hardness, approximately 3000 Vickers; low thermal expansion; high resistivity. Used in ceramics and materials research. Formula: BaB_6. Also known as *barium hexaboride.*

barium bromide. A white or colorless crystalline substance (99+% pure) obtained by interaction of barium sulfide and hydrobromic acid. Density, 3.58 g/cm³; melting point, loses H_2O at 75°C (167°F); boiling point, loses $2H_2O$ at 120°C (248°F). Used in photographic compounds and in phosphors. Formula: $BaBr_2 \cdot 2H_2O$. Also known as *barium bromide dihydrate.*

barium bromide dihydrate. See barium bromide.

barium calcium aluminate. Any of the following compounds of barium oxide, calcium oxide and aluminum oxide used in ceramics and materials research: (i) *Barium calcium monoaluminate*. Formula: $BaCaAl_2O_5$; and (ii) *Barium tetracalcium dialuminate*. Formula: $BaCa_4Al_4O_{11}$.

barium calcium monoaluminate. See barium calcium aluminate (i).

barium calcium silicate. A compound of barium oxide, calcium oxide, and silicon dioxide used in ceramics and materials research. Melting point, 1320°C (2408°F). Formula: $BaCa_2Si_2O_7$.

barium carbide. A gray crystalline substance. Density, 3.75 g/cm^3; melting point, above 1760°C (3200°F). Used in ceramics. Formula: BaC_2.

barium carbonate. A white powder (99+% pure) made by a reaction of sodium carbonate on carbon dioxide with barium sulfide. It is also found in nature as the mineral *witherite*. Density, 4.43 g/cm^3; melting point, 1360°C (2480°F). Used as a flux in porcelain enamels and glazes, as an ingredient in flint glass, pressed tableware, radiation-resistant television tubes, laboratory glassware and structural clay products, in steatite, forsterite and zircon porcelain, in titanate electronic components, hard-core permanent magnets, ferrites, ceramic superconductors and case-hardening baths, and in oil-well drilling. Formula: $BaCO_3$.

barium carbonate-nickel cermet. A cermet consisting of barium carbonate in a nickel matrix. It has high hardness and toughness and good thermal properties, and is used in the manufacture of electronic components, and for high-temperature applications.

barium chloride. A white powder (99.9+% pure). Density, 3.9 g/cm^3; melting point, 963°C (1765°F); boiling point, 1560°C (2840°F). Used in ceramics, superconductors, and metal-surface treatment. Formula: $BaCl_2$.

barium chloride dihydrate. A white crystalline substance (99+% pure). Density, 3.097 g/cm^3; melting point, loses $2H_2O$ at 113°C (235°F). Used as a set-up agent and scum preventive in porcelain enamels, in barium salts, in lubrication oil additives and boiler compounds, in pigments, in electronics, and in the manufacture of white leather. Formula: $BaCl_2 \cdot 2H_2O$.

barium chromate. A heavy, yellow, crystalline powder (98+% pure) obtained by the interaction of barium chloride and sodium chromate. Density, 4.498 g/cm^3; melting point, decomposes above 1000°C (1830°F). Used in the manufacture of yellow and pale green overglaze colors, as a yellow paint pigment (*lemon yellow*), in metal primers, and in fuses, ignition-control devices and pyrotechnics. Formula: $BaCrO_4$.

barium columbate. See barium niobate.

barium crown glass. An *optical crown glass* in which the calcium oxide has been partially replaced by barium oxide.

barium cyanide. A white, crystalline powder made by treating barium hydroxide with hydrocyanic acid. Used in metallurgy and electroplating. Formula: $Ba(CN)_2$.

barium cyanoplatinite. See barium-platinum cyanide.

barium dioxide. A white, corrosive powder obtained by heating barium oxide in oxygen or air. Density, 4.96 g/cm^3; melting point, 450°C (840°F); boiling point, decomposes at 800°C (1470°F). Used as a glass decolorizer, in aluminothermy, and as an oxidizer and bleaching agent. Formula: BaO_2. Also known as *barium peroxide; barium superoxide*.

barium di-o-phosphate. See barium hydrogen phosphate.

barium disilicate. See barium silicate (ii)

barium dititanate. See barium titanate (ii).

barium diuranate. A yellow substance used in coloring porcelain. Formula: BaU_2O_7. Also known as *uranium barium oxide*.

barium ferrite. (1) A ferrimagnetic ionic material composed of

barium oxide (BaO) and ferric oxide (Fe_2O_3). Used in electronics. Formula: $BaFe_2O_4$.

(2) An anisotropic, hard-magnetic ceramic material usally made by pressing micro-sized blended powder composed of ferric oxide and barium oxide in a magnetic field and sintering at prescribed temperatures to obtain the desired magnetic properties. It has a hexagonal crystal structure, a Curie temperature of approximately 450°C (840°F), a high coercive force and magnetic energy product, and a maximum service temperature of 400°C (750°F). Used for permanent magnets. Formula: $BaFe_{12}O_{19}$, or $BaO \cdot 6Fe_2O_3$. Also known as *barium hexaferrite*.

barium flint glass. An *optical flint glass* in which considerable quantities of barium oxide and lead oxide are used as fluxing ingredients. Also known as *lead-barium crown glass*.

barium fluoride. A white, corrosive, crystalline powder (98+% pure) made by treating barium sulfide with hydrofluoric acid. Vacuum-grown nonhygroscopic barium fluoride single crystals (99.9+% pure) in optical and scintillator grades are also available. Crystal system, cubic. Density, 4.890 g/cm^3; melting point, 1280°C (2336°F); boiling point, 2137°C (3879°F); hardness, 3 Mohs; refractive index, 1.4741. Used as an opacifier and flux in porcelain enamels, in glassmaking, as a primary modifier in fluorozirconate glasses, and in the form of crystals for spectroscopy, in lasers and dry-film lubricants, and in electronics and superconductivity studies. The single crystals are used especially for optical windows, objective lenses, and as mirror substrates in infrared and ultraviolet optical mirror systems. Formula: BaF_2.

barium fluosilicate. A white, crystalline powder. Density, 4.29 g/cm^3; melting point, decomposes at 300°C (572°F). Used as a flux and opacifier in porcelain enamels and glazes. Formula: $BaSiF_6$. Also known as *barium silicofluoride*.

barium germanium fluoride. See barium hexafluorogermanate.

barium glass. A *soda-lime glass* in which the calcium oxide has been partially replaced by barium oxide.

barium hexaaluminate. See barium aluminate (iii).

barium hexaboride. See barium boride.

barium hexaferrite. See barium ferrite (2).

barium hexafluorogermanate. A compound of barium fluoride and germanium fluoride available as a white crystalline powder with a density of 4.56 g/cm^3 and a melting point of approximately 665°C (1229°F). Used in electronics and materials research. Formula: $BaGeF_6$. Also known as *barium germanium fluoride*.

barium hydrate. See barium hydroxide (2).

barium hydride. Gray, moisture-sensitive crystals (99.5+% pure) with a density of 4.21 g/cm^3 and a melting point of 675°C (1247°F) (decomposes). Used in organic and inorganic research. Formula: BaH_2.

barium hydrogen phosphate. A white powder with a density of 4.16 g/cm^3, used for phosphors, in flame retardants and in ceramics. Formula: $BaHPO_4$. Also known as *dibasic barium phosphate; secondary barium phosphate*.

barium hydroxide. (1) A white, corrosive powder (95+% pure). Density, 3.743 g/cm^3; melting point, above 300°F (572°F). Used in the manufacture of barium ferrite magnets, as a sulfate-controlling agent in ceramics, in glassmaking, in the manufacture of phenol-formaldehyde resins, in vulcanization, as a steel carbonizing agent, and as a boiler scale remover. Formula: $Ba(OH)_2 \cdot H_2O$. Also known as *barium hydroxide monohydrate; barium monohydrate*.

(2) White powder or colorless crystals usually made by

precipitation from an aqueous barium sulfate solution by caustic soda, or by dissolving barium oxide in water. Density, 2.180 g/cm^3; melting point, loses 8H$_2$O at 78°C (172°F); boiling point, 103°C (217°F). Used for barium salts, in the fusion of silicate compounds, in superconductivity studies, and in ceramics as a source of high-purity barium oxide. Formula: Ba(OH)$_2$·8H$_2$O. Also known as *barium hydrate; barium hydroxide octahydrate; barium octahydrate; baryta; caustic baryta.*

barium hydroxide monohydrate. See barium hydroxide (1).

barium hydroxide octahydrate. See barium hydroxide (2).

barium hypophosphite. A white, crystalline powder with a density of 2.90 g/cm^3 used in nickel plating. Formula: BaH$_4$(PO$_2$)$_2$.

barium hyposulfite. See barium thiosulfate.

barium iodide dihydrate. White or colorless, hygroscopic, light-sensitive crystals. Density, 5.15 g/cm^3; melting point, 740°C (1364°F) (loses 2H$_2$O). Used in the preparation of iodide compounds and as a chemical reagent. Formula: BaI$_2$·2H$_2$O.

barium manganate. A gray to emerald-green crystalline powder (90+% pure) with a density of 4.85 g/cm^3. Used as a paint pigment, as an oxdizing agent, for the selective oxidation of diols, and in organometallic research. Formula: BaMnO$_4$. See also Cassel green; manganese green; Rosenstiehl's green.

barium metaphosphate. A white, crystalline powder with a melting point of 849°C (1560°F). Used as an opacifier in glasses and glazes, as an ingredient in porcelains and enamels, and as a metal precoating treatment to eliminate primary boiling in sheet steel enamels. Formula: Ba(PO$_3$)$_2$.

barium metasilicate. See barium silicate (i).

barium mica. A pink *muscovite* mica [KAl$_2$(AlSi$_3$)O$_{10}$(OH)$_2$] in which part of the potassium has been replaced by barium. Crystal system, monoclinic. Density, 2.83 g/cm^3; refractive index, 1.59. Used in ceramics and for electrical insulation applications. See also mica.

barium molybdate. A white, crystalline powder (99.9% pure). Density, 4.65 g/cm^3; melting point, approximately 1600°C (2910°F). Used as an opacifier and adherence promoter in porcelain enamels, as a paint pigment, as a pigment in protective coatings, in electronic and optical equipment, and in organometallic research. Formula: BaMoO$_4$.

barium monoaluminate. See barium aluminate (ii).

barium monohydrate. See barium hydroxide (1).

barium monosulfide. See barium sulfide.

barium monoxide. See barium oxide.

barium naphthenate. The barium salt of naphthenic acid containing about 22.6% barium. Used as a drier, binder and/or hardener in adhesives and linoleum.

barium niobate. A compound of barium oxide and niobium (columbium) oxide. Density, 5.98 g/cm^3; melting point, 1927°C (3500°F). Used in ceramics and superconductor research. Formula: Ba$_6$Nb$_2$O$_{11}$ (Ba$_6$Cb$_2$O$_{11}$). Also known as *barium columbate.*

barium nitrate. A white, crystalline powder (99+% pure) made by treating barium carbonate or sulfide with nitric acid. It is also available as single crystals, and occurs in nature as the mineral *nitrobarite.* Density, 3.230 g/cm^3, melting point, 592°C (1098°F); strong oxidizer. Used as an ingredient in optical glasses, as a homogeneity and opacity improver in porcelain enamels, in ceramic glazes, and in electronics, superconductivity studies, explosives and pyrotechnics. The single crystals are used especially in optical filters and in nonlinear frequency converters for tunable lasers. Formula: Ba(NO$_3$)$_2$.

barium octahydrate. See barium hydroxide (2).

barium oxalate. A white, crystalline powder (97+% pure) with a density of 2.658 g/cm^3 and a melting point of 400°C (750°F) (decomposes). Used as an analytical and organometallic reagent, and in pyrotechnics. Formula: BaC$_2$O$_4$.

barium oxide. A colorless, white or yellowish-white crystalline powder (97+% pure) obtained from calcined barium carbonate. Density, 5.72 g/cm^3; melting point, 1923°C (3493°F); boiling point, about 2000°C (3630°F); refractive index, 1.98. Used as a fluxing ingredient in glass, in superconductors, as a detergent for lubricating oils, and as a desiccant. Formula: BaO. Also known as *barium monoxide; barium protoxide; baryta; calcined baryta.*

barium permanganate. Brownish-violet crystals used as oxidizing agent, and in the depolarization of dry cells. Formula: Ba(MnO$_4$)$_2$.

barium peroxide. See barium dioxide.

barium pharmacosiderite. A red-brown or green mineral of the pharmacosiderite group composed of barium iron arsenate oxide hydroxide pentahydrate, BaFe$_4$(AsO$_4$)$_3$O$_2$(OH)$_5$·5H$_2$O. Crystal system, tetragonal. Density, 3.19-3.24 g/cm^3. Occurrence: Germany.

barium phosphate. White crystals with a density of 4.1 g/cm^3 and a melting point of 1727°C (3141°F). Used in ceramics. Formula: Ba$_3$(PO$_4$)$_2$.

barium phosphide. A compound of barium and trivalent phosphorus. It has a density of 3.18 g/cm^3 and is used in ceramics. Formula: Ba$_3$P.

barium plaster. A gypsum plaster to which selected barium salts have been added during milling. Used on the walls of X-ray rooms.

barium-platinum cyanide. A yellow or greenish crystalline substance. Density, 2.08 g/cm^3; melting point, loses 2H$_2$O at 100°C (212°F). Used in X-ray screens. Formula: BaPt(CN)$_4$·4H$_2$O. Also known as *barium cyanoplatinite; platinum-barium cyanide.*

barium-potassium chromate. A light yellow pigment obtained by heating barium carbonate and potassium dichromate in a kiln at 500°C (932°F). It has a density of 3.65 g/cm^3 and is used as a component of protective organic coatings for ferrous and light-metal alloys. Formula: BaK(CrO$_4$)$_2$.

barium potassium bismuth oxide. A superconductor composed of atomic planes of bismuth and oxygen, and barium and potassium ions located between these planes. It has been found that it does not exhibit any measurable specific heat change while going through up to 3 different critical magnetic fields (e.g., T$_c$, H$_c$ and J$_c$). Formula: Ba$_{0.6}$K$_{0.4}$BiO$_3$.

barium protoxide. See barium oxide.

barium selenide. A compound of barium and divalent selenium available in the form of a crystalline powder with a density of 5.0 g/cm^3. Used as a semiconductor, in photocells, and in electronics. Formula: Ba$_2$Se.

barium selenite. A black powder with a density of 4.4, used as a pigment in glass, and in ceramics. Formula: BaSeO$_3$.

barium silicate. Any of the following compounds of barium oxide and silicon dioxide used in ceramics and materials research: (i) *Barium metasilicate.* A colorless or white, crystalline powder. Density, 4.40 g/cm^3; melting point, 1604°C (2919°F); refractive index, 1.67. Formula: BaSiO$_3$; (ii) *Barium disilicate.* A crystalline powder. Density, 3.73 g/cm^3; melting point, 1420°C (2588°F). Formula: BaSi$_2$O$_5$; (iii) *Dibarium silicate.* A crystalline powder. Density, 5.20 g/cm^3; melting point, above 1755°C (3190°F). Formula: Ba$_2$SiO$_4$; and (iv) *Dibarium trisilicate.* A crystalline powder. Density, 3.93 g/cm^3; melting point, 1450°C (2642°F). Formula: Ba$_2$Si$_3$O$_8$.

barium silicide. A compound of barium and silicon available in the form of light-gray lumps. Melting point, 1180°C (2156°F). Used as a deoxidizer and desulfurizer for steels, and for ceramic applications. Formula: $BaSi_2$.

barium silicofluoride. See barium fluosilicate.

barium sodium columbate. See barium sodium niobate.

barium sodium niobate. An artificial electrooptical crystal that undergoes no optical damage during high-power laser irradiation. It is used in the production of coherent green light, and in the manufacture of electro-optical modulators and optical oscillators. Also known as *barium sodium columbate.*

barium stannate. A white, crystalline powder. Density, 7.6 g/cm^3; melting point, loses $3H_2O$ at 280°C (536°F). The anhydrous form can contain up to 50-51% barium oxide. Used in the manufacture of special ceramic insulators, as a Curie peak modifier for barium titanate capacitors, and in glass enamels to improve alkali resistance. Formula: $BaSnO_3 \cdot 3H_2O$ (trihydrate); $BaSnO_3$ (anhydrous).

barium stearate. A white crystalline solid, or waxy, nontacky white powder. Density, 1.145 g/cm^3; melting point, 160°C (320°F). Used in waterproofing agent, as a lubricant in wire drawing, as a dry lubricant for plastics and rubber, in wax compounding, in grease manufacture, as a heat and light stabilizer in plastics. Formula: $Ba(C_{18}H_{35}O_2)_2$.

barium sulfate. A white or yellowish powder (99+% pure) made by treating a solution of a barium salt with sulfuric acid or sodium sulfate. It occurs in nature as the mineral *barite*, and is commercially known as *blanc fixe.* Density, 4.500 g/cm^3; melting point, 1580°C (2876°F); hardness, 3 Mohs. Used as a workability improver in porcelain enamels, as an extender pigment in paints, in the preparation of lake pigments, as a filler for fabrics, inks, plastics, rubber and coatings, and as a standard reflecting agent in the measurement of the whiteness and reflectance of papers. Formula: $BaSO_4$. See also artificial barite.

barium sulfide. Yellow-green, gray, or black powder or lumps (85+% pure) obtained from crude *barite* (barium sulfate) and coal by roasting in a furnace. Crystal system, polymorphous. Density, 4.25-4.50 g/cm^3; melting point, above 1660°C (3020°F). Used in the manufacture of cerium and uranium melting crucibles, in the manufacture of barium salts and hydrogen sulfide, as a flame retardant, as a paint pigment, and in luminous paints. Formula: BaS. Also known as *barium monosulfide; black ash.*

barium superoxide. See barium dioxide.

barium telluride. A compound of barium and tellurium with a melting point of 1527°C (2781°F). Used in ceramics and electronics. Formula: BaTe.

barium tetracalcium dialuminate. See barium calcium aluminate.

barium tetracyanoplatinate hydrate. A crystalline compound (99.9% pure) with a density of 3.050 g/cm^3, and a melting point of 100°C (210°F) (decomposes). Used in organometallic synthesis, and in ceramics. Formula: $BaPt(CN)_4 \cdot 4H_2O$.

barium tetratitanate. See barium titanate (iv).

barium thiosulfate. A white crystalline powder (99% pure). Density, 3.5 g/cm^3; melting point, 220°C (428°F) (decomposes upon heating). Used in luminous paints, varnishes, explosives and matches. Formula: $BaS_2O_3 \cdot H_2O$. Also known as *barium hyposulfite.*

barium thorate. A compound of barium oxide and thorium oxide. Density, 7.66 g/cm^3; melting point, 2300°C (4172°F). Used in ceramics. Formula: $BaThO_3$.

barium titanate. Any of the following compounds of barium oxide and titanium dioxide: (i) *Barium titanate.* White or light-gray powder, or white sintered lumps made by die-pressing barium carbonate and titanium dioxide and sintering at high temperature. Crystal system, tetragonal, or cubic. Density, 5.85-6.08 g/cm^3; melting point, 1618-1654°C (2944-3009°F); high dielectric constant (above 1000). *Barium titanate* crystals have photorefractive properties and large electrooptical coefficients, and low absorption losses in the visible and infrared range. Used for piezoelectric and ferroelectric ceramics, electrostrictive transducers, electronic and communication equipment, vacuum deposition coatings, and in ultrasonic cleaning. The crystals are used in electronics and optoelectronics especially in optical information processing and for computer applications. Formula: $BaTiO_3$; (ii) *Barium dititanate.* A compound of barium oxide and titanium dioxide. Melting point, 1320°C (2408°F). Used in ceramics. Formula: $BaTi_2O_5$; (iii) *Barium trititanate.* A powder. Density, 4.7 g/cm^3; melting point, 1356°C (2473°F). Used in ceramics. Formula: $BaTi_3O_7$; (iv) *Barium tetratitanate.* A powder. Density, 4.6 g/cm^3; melting point, 1420°C (2588°F). Used in ceramics. Formula: $BaTi_4O_9$. See also barium titanate ceramics; cerium-doped barium titanate.

barium titanate ceramics. A group of piezoelectric and ferroelectric ceramics, such as barium strontium titanate, based on barium titanate ($BaTiO_3$). They have high dielectric constants (about 1250), and are used for capacitors in electronic equipment, such as television, radio receivers, and storage devices, and for transducers, accelerometers, communication equipment, dielectric amplifiers, digital calculators, guided missiles, measuring instruments, miniature electronics, storage devices, underwater sonar, depth sounders, hydrophones, etc.

barium titanate disilicate. See barium titanium silicate (ii).

barium titanate silicate. See barium titanium silicate (i).

barium titanate silicate. Any of the following compounds of barium oxide, titanium dioxide, and silicon dioxide: (i) *Barium titanate silicate.* Melting point, 1398°C (2548°F). Used in ceramics and materials research. Formula: $BaTiSiO_5$; and (ii) *Barium titanate disilicate.* Melting point, 1250°C (2282°F). Used in ceramics and materials research. Formula: $BaTiSi_2O_7$.

barium trititanate. See barium titanate (iii).

barium tungstate. A white powder (99.5+%) available in standard (99.9% pure) and optical grades (99.95% pure). It has a density of 5.040 g/cm^3 and is used as a white pigment, in the manufacture of phosphorescent and intensifying screens, and in X-ray photography. Formula: $BaWO_4$. Also known as *barium wolframate; barium white; tungstate white; wolfram white.*

barium white. See barium tungstate.

barium wolframite. See barium tungstate.

Barium XA. Trademark of Barium and Chemicals, Inc. (USA) for a degasifier and cleanser used in the manufacture of high-grade tool steels.

barium zirconate. A light-gray to dull-yellow powder. Density, 5.52 g/cm^3; melting point, 2510°C (4550°F). Used as an additive in barium titanate or zirconate ceramics, in electronics, and in the manufacture of certain silicone elastomers. Formula: $BaZrO_3$.

barium zirconium silicate. A compound of barium oxide, zirconium dioxide, and silicon dioxide available in the form of a white powder. It has a melting point of 1538°C (2800°F), and is used in the manufacture of electrical resistors, glaze opacifiers, and as a stabilizer for colored ground coat enamels. Formula: $BaZrSiO_5$.

bark cloth. A textile fabric that has a surface texture imitating tree bark.

bark-tanned leather. Leather that has been tanned in a water extract of wood bark (e.g., chestnut, hemlock, mangrove, oak, quebracho, sumac, wattle or willow). Sole or heavy-duty leather is often tanned with bark extracts.

Barmag. Trade name of Calloy Limited (UK) for an alloy of 65% magnesium and 35% barium used as a getter in electrical discharge devices and vacuum tubes.

barnesite. A dark red mineral composed of sodium vanadium oxide trihydrate, $Na_2V_6O_{16}\cdot3H_2O$. Crystal system, monoclinic. Density, 3.15 g/cm^3; refractive index, above 2.0. Occurrence: USA (Utah).

Barnite. A British age-hardenable aluminum alloy containing 5.5% copper, and small additions of magnesium, chromium, silicon and titanium. Used for light-alloy parts.

Baroid. Trade name of Baroid Division of N.L. Petroleum Services Inc. (USA) for bentonite clays and barium sulfates.

Barolite. Czech trade name for semitransparent patterned glass with rib-like pattern.

Baroque. Trade name of C-E Glass (USA) for wired glass having a Georgian-type mesh.

Baros. Trade name of Creusot-Loire (France) for a heat-resisting alloy of 90% nickel and 10% chromium. Used for balance weights and pen points.

barras. A coarse linen fabric resembling sackcloth, originally made in Holland.

Barrday Guard. Trademark of Barrday Inc. (Canada) for *Kevlar* fabrics used for soft body armor and flak jackets.

barrel-galvanized material. See hot-galvanized material.

barrerite. A white, or pink mineral of the zeolite group composed of sodium potassium calcium aluminum silicate heptahydrate, $(Na,K,Ca)_2(Si,Al)_9O_{18}\cdot7H_2O$. Crystal system, orthorhombic. Density, 2.13 g/cm^3; refractive index, 1.485. Occurrence: Italy.

BarriCut. Trademark of Honeywell Performance Fibers (USA) for a tough, durable, air-jet textured sheath-core bicomponent polyester fiber produced by a proprietary hard-particle fiber technology. It consists of a polyester core fiber with numerous embedded microscopic ceramic platelets, and a smooth polyester sheath. *BarriCut* provides significantly enhanced cut resistance as compared to ordinary polyester, excellent abrasion and snag resistance, excellent dyeability, coatability, launderability and bleachability, and a comfortable hand. Used for heavy duty industrial gloves and sleeves, and knitted and woven industrial fabrics.

barrier. See barrier material.

barrier coatings. A group of protective coatings applied to keep moisture, oxygen and/or corrosives away from a structure, or prevent or reduce the permeation of gases and the diffusion of certain species. They may consist of plastics, rubber, waxes, tars, etc., and can vary in thickness from a few microns to several centimeters. See also thermal barrier coatings.

barrier fabrics. A term refering to fabrics that are suitable for use as dust, dust mite and allergen barriers.

barrier material. (1) A material that restricts the passage of solids, semisolids, fluids, vapors and certain types of energy through itself, or through another material. Also known as *barrier*. See also air barrier; diffusion barrier; grease-resistant barrier; thermal barrier; vapor barrier; water-resistant barrier.

(2) A material that impedes the transmission of water or water vapor, prevents or impedes the transmission of grease or

oil, or is impervious to liquids, vapors and/or gases including air. Also known as *barrier*. See also air barrier; grease-resistant barrier; vapor barrier; water-resistant barrier.

(3) A material, such as concrete or lead, which is used in radiographic installations as a protection against gamma and X-rays. Also known as *barrier*.

barrier paper. A dark-colored *kraft paper* impregnated with asphalt, gilsonite or wax, and used in building construction and civil engineering.

barrier plastics. A class of rigid, lightweight, transparent plastics, usually based on acrylonitrile copolymers, that have good impact properties and restrict the passage of gas, flavor and aroma. Also known as *barrier resins*.

barrier resins. See barrier plastics.

barringerite. A synthetic mineral composed of iron phosphide, Fe_2P. Crystal system, hexagonal. Density, 6.91 g/cm^3.

barringtonite. A colorless mineral composed of magnesium carbonate dihydrate, $MgCO_3\cdot2H_2O$. Crystal system, triclinic. Density, 2.46 g/cm^3; refractive index, 1.473. Occurrence: Australia.

barronia. The common name of a bronze containing 83% copper, 12.5% zinc, 4% tin, and 0.5% lead. Used for superheated steam equipment, heat exchangers, evaporators, condenser tubes, and fittings.

bar solder. Soft solder in bar form commonly used for hand soldering.

Barshot. Trade name of BEI PECAL, Division of Stake Technology Limited (Canada) for specular hematite-based abrasives for abrasive blast-cleaning applications. See also specularite.

barstock. See bar.

bartelkeite. A colorless to very pale green mineral composed of iron lead germanium oxide, $PbFeGe_3O_8$. Crystal system, monoclinic. Density, 4.97 g/cm^3; refractive index, 1.910. Occurrence: Namibia.

Barto. Trade name of Carpenter Technology Corporation (USA) for a tough, shock-resistant, oil-hardening alloy steel containing 0.5% carbon, 0.5% manganese, 0.25% silicon, 1% chromium, 1.75% nickel, and the balance iron. Used for expander punches, chuck jaws, vise jaws, clutch parts, and feeder rolls.

bartonite. A blackish-brown mineral composed of potassium iron sulfide, $K_3Fe_{10}S_{14}$. Crystal system, tetragonal. Density, 3.31 g/cm^3. Occurrence: USA (California).

barwood. The reddish hardwood of the tree *Pterocarpus santalinus*. It has a coarse grain and high hardness. Source: West Africa. Used for machine bearings and tool handles. *Note:* The tree contains "santalin," a red coloring matter used to dye textiles.

Barworth. Trade name of Barworth Flockton Limited (UK) for a series of cold- and hot-work tool and die steels, tungsten and cobalt-tungsten high-speed steels, mold steels, and plain-carbon tool steels.

Baryfloor FRC. Trademark of Van Styn Sandwich Panels BV (Netherlands) for sandwich panels with fire-resistant surface coatings, used especially for flooring applications.

barylite. A white mineral composed of barium beryllium silicate, $BaBe_2Si_2O_7$. It can also be made synthetically. Crystal system, orthorhombic. Density, 4.07 g/cm^3; refractive index, 1.70. Occurrence: USA (New Jersey).

barysilite. A pink mineral composed of lead manganese silicate, $Pb_8Mn(Si_2O_7)_3$. Crystal system, rhombohedral (hexagonal). Density, 6.72 g/cm^3. Occurrence: Sweden.

baryta. See barium hydroxide octahydrate; barium oxide.

baryte. See barite.

barytocalcite. A colorless, white, grayish or yellowish mineral of the aragonite group composed of barium calcium carbonate, $BaCa(CO_3)_2$. Crystal system, monoclinic. Density, 3.68 g/cm^3; refractive index, 1.684. Occurrence: Europe (UK). Used as a source of barium.

barytolamprophyllite. A dark brown mineral of the seidozerite group composed of sodium barium iron titanium silicate hydroxide, $(Ba,K,Ca,Sr,Na)_2(Na,Mn,Fe,Mg)_3(Ti,Fe)_3(Si,Al)_4O_{16}$-$(OH,O,F,Cl)_2$. Crystal system, monoclinic. Density, 3.62 g/cm^3; refractive index, 1.754. Occurrence: Russian Federation.

Bary Vam. Trade name of H.L. Blachford Limited (Canada) for plywood with good acoustic insulating properties.

Basal. Trademark of Didier Refractories Corporation (USA) for refractory bricks.

basaluminite. A white mineral composed of aluminum sulfate hydroxide pentahydrate, $Al_4SO_4(OH)_{10}\cdot5H_2O$. Crystal system, hexagonal. Density, 2.12 g/cm^3; refractive index, 1.519. Occurrence: Europe (UK).

basalt. A hard, fine-grained dark-brown or dark-green to black, crystalline volcanic rock composed primarily of soda-lime feldspar, pyroxene, magnetite, olivine, magnesite and ilmenite. It has a density of 2.8-3.0 g/cm^3 (0.10-0.11 lb/in.³), a hardness of 5-9 Mohs, a refractive index of 1.62, a melting point of 1450 (2640°F), and a high content of iron and magnesium. Used chiefly as a paving and building stone, as railroad ballast, in the manufacture of rock wool, glass wool, thermal insulating fiber materials, high-grade textile fibers, acid-resistant industrial equipment and floor tiles, and as a concrete aggregate. Basalt laminates are also employed as protective coatings. See also artificial basalt; traprock.

basalt glass. A black, green, or brown volcanic glass that takes a fine polish, and is used for ornaments. Also known as *basalt obsidian; tachylite.*

basalt obsidian. See basalt glass.

basalt ware. A hard, black, highly vitreous stoneware having a dull gloss and resembling basalt rock in appearance.

Bascodur. Trademark of Raschig GmbH (Germany) for modified and unmodified phenol-formaldehydes.

BaseLine. Trade name for a glass-ionomer dental cement.

BaseLine VLC. Trade name for a light-cure glass-ionomer dental cement.

base paper. Paper used as feedstock for further treatment or refinement operations. Also known as *basis paper.*

basic aluminum acetate. (1) An unstabilized, water-insoluble, white, crystalline compound used chiefly in the preparation of pigments and lakes, as a flame retardant and waterproofing agent, and as a mordant in textile dyeing. Formula: $(CH_3CO_2)_2$-AlOH. Also known as *aluminum diacetate; aluminum subacetate; dihydroxyaluminum acetate.*

   (2) A water-soluble form of (1) stabilized with boric acid and containing about 19% aluminum. Formula: CH_3CO_2Al-$(OH)_2\cdot0.33H_3BO_3$. Also known as *dihydroxyaluminum acetate.*

basic beryllium acetate. See beryllium acetate oxide.

basic Bessemer pig iron. A pig iron that contains 2-3% phosphorus, and is used in making basic Bessemer steel. In Europe this pig iron is known as "Thomas pig iron."

basic anhydride. A compound, such as an ionic oxide, that reacts with water to form hydroxide (OH^-) ions.

basic beryllium acetate. See beryllium acetate oxide.

basic Bessemer steel. A steel made in a pear-shaped, basic-refractory-lined vessel, known as a "basic Bessemer converter," by blowing air through a molten bath of pig iron whereby most of the carbon and impurities are removed by oxidation. *Note:* In Europe this steel is known as *Thomas steel* or *Thomas-Gilchrist steel.* See also acid Bessemer steel; Bessemer steel.

basic bismuth carbonate. See bismuth oxycarbonate.

basic bismuth chloride. See bismuth oxychloride.

basic brick. Firebrick made from basic refractory materials, such as chrome, lime or magnesia. Also known as *basic firebrick.* See also acid brick; firebrick

basic cobalt chromate. See cobaltous chromate.

basic cobaltous carbonate. See cobaltous carbonate (2).

basic copper acetate. Green or greenish-blue powder, or shiny crystals obtained by the action of acetic acid on copper in the presence of air. The green variety [$2Cu(C_2H_3O_2)_2\cdot CuO\cdot6H_2O$] is also known as *green verdigris,* and the blue variety [$(C_2H_3O_2)_2\cdot Cu_2O\cdot6H_2O$] as *blue verdigris.* Blue and green verdigris are *true verdigris* based on basic copper acetate and must not be confused with artificial malachite (basic copper carbonate) or patina (basic copper sulfate, or basic copper chloride). Both artificial malachite and patina are also referred to as *false verdigris.* Used as a pigment, in antifouling paints, as a mildew preventive, and as a mordant. Also known as *copper subacetate.*

basic copper carbonate. See copper carbonate.

basic firebrick. See basic brick.

basic hydroxide. A metallic hydroxide, such barium hydroxide, that when reacted with an acid yields a salt and water.

basic lead carbonate. See white lead.

basic lead chloride. See Turner's yellow.

basic lead chromate. A red, crystalline powder used as a paint pigment (chrome red), and in anticorrosive coatings on steel substrates. Formula: $PbCrO_4\cdot PbO$. See also chrome red.

basic lead silica chromate. A compound of lead monoxide (litharge), chromium trioxide and silicon dioxide, used as a paint pigment. Formula: $PbO\cdot CrO_3\cdot SiO_2$.

basic lead silicate. A compound made by fusing lead monoxide (litharge) with silica sand and hydrating the resulting intermediate product by ball milling with water. The general formula is $3PbO\cdot2SiO_2\cdot H_2O$. It has good corrosion-inhibiting properties, and is used in protective paints for metal parts and structures exposed to water, in ceramics, and as a vinyl stabilizer. Also known as *white lead silicate.* See also litharge.

basic lead sulfate. See basic white lead.

basic open-hearth steel. A steel produced by melting selected pig iron and malleable scrap iron with the addition of pure iron ore in an open-hearth furnace constructed of basic refractories covered with magnesite or burnt dolomite. See also acid open-hearth steel; open-hearth steel.

basic oxide. A metallic oxide, such as calcium oxide, potassium oxide or sodium oxide, that will form a hydroxide when combined with water, and will enter into a chemical reaction with acidic materials.

basic-oxygen-furnace steel. A steel made from a charge of molten pig iron, scrap steel and fluxes in a basic-refractory-lined furnace by blowing high-purity oxygen at supersonic speed into the furnace through a lance onto the top of the metal bath accelerating the burning-off of unwanted elements and greatly reducing the impurities. Examples include *Kaldo steel* and *LD steel.* Abbreviation: BOF steel. Also known as *oxygen-furnace steel.*

basic phosphate slag. A finely ground basic slag that contains 12% or more phosphorus pentoxide (P_2O_5). It is obtained as a byproduct in the manufacture of basic Bessemer steel, and is used as a fertilizer. In Europe this slag is known as *Thomas slag*. Also known as *slag flour*.

basic pig iron. A low-silicon pig iron containing less than 1% phosphorus and used in the manufacture of basic open-hearth steel.

basic refractory materials. See basic refractories.

basic refractories. Refractory materials composed chiefly of lime (CaO) and/or magnesia (MgO) which undergo a reaction with acidic slags and fluxes at high temperatures. They may also contain calcium, chromium and iron compounds. Used for high-temperature applications, e.g., in linings for open-hearth furnaces. Also known as *basic refractory materials*. See also acid refractories; neutral refractories; silica refractories; siliceous refractories.

basic rock. An igneous rock containing 52% or less silica, and considerable amounts of calcium, magnesium or iron. Used in civil engineering and building construction. See also acid rock.

basic slag. A slag that contains variable amounts of tricalcium phosphate, calcium silicate, lime, and metallic oxides, especially those of iron, magnesium and manganese. It is obtained as a byproduct in the manufacture of basic Bessemer and basic open-hearth steel. Used chiefly as a fertilizer. Also known as *phosphatic slag*. See also acid slag.

basic steel. A steel made in a melting furnace with basic refractory lining (usually containing magnesite or lime). It may be a basic Bessemer, basic open-hearth or basic-oxygen-furnace steel. See also acid steel.

basic white lead. A compound of lead sulfate and lead monoxide available in the form of white, monoclinic crystals. It occurs in nature as the mineral *lanarkite*. Density, 6.92 g/cm^3; melting point, 977°C (1791°F). Used in ceramics, and as a paint pigment. Formula: $PbSO_4 \cdot PbO$. Abbreviation: BSWL. Also known as *basic lead sulfate; sublimed white lead; white lead sulfate*.

basic zirconium carbonate. A white, amorphous, water-insoluble powder obtained by the addition of sodium carbonate to a zirconium salt solution. Used in the peparation of zirconium dioxide. Formula: $ZrOCO_3$. Also known as *zirconium carbonate; zirconyl carbonate*. See also zirconium basic carbonate; basic zirconium carbonate hydrate.

basic zirconium carbonate hydrate. A white hydrous powder containing zirconium oxide and carbon dioxide. Formula: $3ZrO_2 \cdot CO_2 \cdot xH_2O$. See also zirconium basic carbonate; basic zirconium carbonate.

Basifrit. Trademark of Canadian Refractories Limited for a quick-setting *magnesite* refractory used for lining and resurfacing ladles, furnaces, etc.

basil. Uncolored sheepskin or lambskin tanned with wood or bark.

Basimag. Trademark of Premier Services Corporation (USA) for magnesite-based refractories supplied as dry particulate mixes.

basis brass. A term used in the United Kingdom for a yellow brass containing 58-66% copper, and 34-42% zinc. It has good cold workability, and is used for sheathing, bolts, nuts, pins, rivets, etc. See also Bobierre's metal.

basis paper. See base paper.

basis quality brass. A brass containing 76.5% copper and 23.5% zinc.

basketane. A polycyclic compound whose three-dimensional structure is characterized by six carbon rings arranged in the form of an open cubical box, or basket.

basket ash. See black ash (1).

basketweave fabrics. Loose-textured woven fabrics in which two or more warp (longitudinal) threads or yarns interlace over and under two or more weft (transverse) threads or yarns. They resemble the weave in a basket, and are used as composite reinforcements and for the manufacture of clothing.

Basofil. Trademark of Basofil Fibers, LLC (USA) for a heat- and fire-resistant melamine fiber that can be blended with other fibers, and manufactured into woven and nonwoven fabrics. Used for protective clothing for firefighters, steelworkers, factory workers, etc., in preventive fire protection, for safety curtains in commercial and public buildings, in fire-blocking protective layers for automobile and train upholstery, as insulation for automotive engine compartments, aircraft seat coverings, etc.

Basopor. Trademark of BASF Corporation (USA) for urea-formaldehyde resin precondensates.

Basotect. Trademark of BASF Corporation (USA) for a flexible, open-celled melamine-resin foam with three-dimensional structure. It can be combined with woven fabrics, etc., and has high heat resistance, very good flexibility and soundproofing performance, good flame retardancy, and low water absorption. Applications include soundproofing of automobiles and rail vehi-cles, heat- and flame-proofing of aircraft, and the construction of buildings and industrial plants.

basra locus. See angelique.

bassanite. A colorless, or whitish mineral composed of calcium sulfate, $CaSO_4$. It can also be made synthetically, but the synthetic material contains $0.5H_2O$. Crystal system, orthorhombic. Density, 2.70 g/cm^3; refractive index, 1.558. Occurrence: Italy.

bassetite. A yellow to olive-green mineral of the meta-autunite group composed of iron uranyl phosphate octahydrate, $Fe(UO_2)_2(PO_4)_2 \cdot 8H_2O$. Crystal system, monoclinic. Density, 3.68 g/cm^3; refractive index, 1.684. Occurrence: Europe (UK).

basswood. The soft, light wood of a group of hardwood trees (genus *Tilia*) including chiefly the American basswood (*T. americana*) and the white basswood (*T. heterophylla*). Its yellowish-brown heartwood merges gradually into the white sapwood. *Basswood* has a fine, even texture with straight grain, is somewhat porous, works well, has good gluing and nailing properties, and possesses low to medium strength and durability, large shrinkage and high resistance to warpage in use. Average weight, 420 kg/m^3 (26 lb/ft^3). Source: Canada (from New Brunswick to Manitoba) and northern United States. Used for furniture, cabinetwork, sash and door frames, paneling, Venetian blinds, pianoforte manufacture, drawing boards, woodenware, containers, boxes, crates, cooperage and veneer, as corestock for plywood, in carving, and as a pulpwood. See also American basswood; white basswood.

bast fibers. Strong, tough, flexible fibers obtained from the phloem or inner bark of plants, such as abaca, flax, hemp, jute, and ramie. Used for textiles (matting, etc.), ropes and paper. Also known as *basts*. See also soft vegetable fibers.

bastnaesite. A wax-yellow, reddish-brown, or brick-red mineral that is a fluorocarbonate of cerium, lanthanum or yttrium $(Ce,La,Y)CO_3F$, and may also contain other rare earths, calcium and/or thorium. Crystal system, hexagonal. Density, 3.90-4.99 g/cm^3; refractive index, 1.72. Occurrence: Russia, Kazakhstan, Sweden, USA (California, Colorado). Used as a source of rare earths especially cerium, dysprosium, gadolinium, lanthanum, neodymium and yttrium.

Bastoni. Trade name of Saint-Gobain (France) for solid glass lenses with linear design.

Bataan mahogany. See tangile.

Batalbra. Trade name of Birmingham Battery & Metal Company Limited (UK) for an inhibited brass containing 76% copper, 22% zinc, 2% aluminum, and 0.03% arsenic. It has good corrosion resistance, excellent cold workability, and is used for chemical equipment, condenser tubes, heat exchangers, and seawater pipelines.

Batavia dammar. A pale yellow to nearly colorless high-grade resin obtained from trees of the species *Hopea*. Density, 1.04-1.06 g/cm³. Source: Batavia, Sumatra. Used in the manufacture of lacquers, varnishes, inks, paints and adhesives.

batched aggregate. See dry-batched aggregate.

bath metal. (1) A red brass containing 83% copper and 17% zinc, and used for pipes and plumbing applications.

(2) A yellow brass containing 55% copper and 45% copper, and used for bathroom fixtures and buttons.

batisite. A dark to reddish brown mineral composed of sodium barium titanium silicate, $Na_2BaTi_2(Si_2O_7)_2$. Crystal system, orthorhombic. Density, 3.43 g/cm³; refractive index, 1.735. Occurrence: Russian Federation.

batiste. (1) A fine, thin, plain-woven cotton, linen, or polyester-cotton fabric used for blouses, dresses, nightwear, lingerie, and children's wear.

(2) A fine, thin, plain-woven fabric similar to (1), but made of polyester, rayon, silk, or wool.

Batnaval. Trade name of Birmingham Battery & Metal Company Limited (UK) for a naval brass containing 60-62% copper, 37-39% zinc, and 1% tin. Used for marine and power-station condensers, and heat exchangers.

Batnickon. Trade name of Birmingham Battery & Metal Company Limited (UK) for copper nickels containing 5-30% nickel, up to 1.2% iron, up to 0.5% manganese, and the balance copper. Commercially available in tube and sheet form, they are used for pipelines, condenser tubes, and fittings.

Batonnet. Trade name of Saint-Gobain (France) for hollow glass blocks with irregular decorative designs on the inner faces.

Batonnets. Trade name of Boussois Souchon Neuvesel SA (France) for a patterned glass.

Bats. Trade name of Shieldalloy Metallurgical Corporation (USA) for a series of foundry alloys containing 20-58% titanium, 11-17% aluminum, 3.5-8% sulfur, 3.5-6% zirconium, 0-8% manganese, 0.5-2% boron, and the balance iron. Used as ladle additions.

batt. See batt insulation.

batten. A piece of square-sawn softwood timber 50-100 mm (2-4 in.) thick, and 100-200 mm (4-8 in.) wide. Used to reinforce or cover joints.

Batterium. Trade name of Batterium Metal & Vislok Limited (UK) for an acid-resisting aluminum bronze containing 89% copper, 9% aluminum, 1% nickel, and 1% other metals, such as iron and tin. It has exceptional corrosion resistance, and is used for plate terminals, plug cocks, valves, chemical equipment, etc.

battery copper. An alloy of 94% copper and 6% zinc, used for batteries and gilding metal.

battery grid alloys. A group of lead alloys (UNS L50760 to L50780) hardened with 0.01-0.1% calcium, and sometimes up to 0.5% tin. Used for storage battery grids.

battery manganese. See manganese dioxide.

battery paint. A paint consisting of asphalt or gilsonite dissolved in petroleum. Used for thick, acid-, corrosion- and water-resistant coatings on batteries.

battery-plate lead. Lead containing about 7-12% antimony, 0.25%

tin, and small quantities of copper and arsenic. Used for the positive plate grids of storage batteries.

Battery Plates. Trade name for a hard alloy of 94% copper and 6% antimony, used for battery storage plates.

batting. Continuous sheets or webs of natural (e.g., cotton or wool) or synthetic fibers (e.g., polyester) used for lining quilts, stuffing mattresses, and for packing applications.

batt insulation. Thermal insulation composed of loosely felted mats of mineral or vegetable fibers, such as rock, slag, fiberglass, wood or cotton, and supplied in the form of blankets, typically in lengths of 610 and 1220 mm (24 and 48 in.), widths of 406 and 610 mm (16 and 24 in.) and thicknesses of 76-304 mm (3-12 in.). It is sometimes faced with kraft or other paper, and used especially for insulating exposed wall cavities and attics.

battleship linoleum. A resilient, heavy, washable floor covering made by rolling a mixture of ground cork or wood flour, hot-oxidized linseed oil, an oleoresinous binder (e.g., a rosin or fossil resin), a mineral filler and pigments (e.g., lithopone) on a backing of burlap. It is coated with a cellulose or alkyd base lacquer and has a hard, glossy surface.

Batt/Z. Trade name for high-purity zinc sheet used for battery boxes.

Baturnal. Trade name of Birmingham Battery & Metal Company Limited (UK) for a free-machining brass containing 63% copper, 35% zinc, and 2% lead. Used for tubing.

batu. An East Indian dammar gum with a density of 1.00-1.05, and a melting point of 180°C (356°F). Used as a flatting agent in paints and oleoresinous varnishes, and in spirits, adhesives, and oilcloth.

Baudoin alloy. An early French nickel silver composed of 16% nickel, 10.2% zinc, 1.8% cobalt, and the balance copper. Used for silverware.

Bäuerle. Trade name of J. Adolf Bäuerle Präzisionsziehwerk (Germany) for bright-finished steels.

Baudrin's metal. A British corrosion-resistant alloy containing 72% copper, 16-16.6% nickel, 2.25-7.1% zinc, 0-2.75% tin, 0-2.5% iron, 1.8-2.0% cobalt, and 0-0.5% aluminum. Used for ornamental and corrosion-resistant parts.

baulk. See balk.

baumhauerite. A lead to steel-gray, or black mineral composed of lead arsenic sulfide, $Pb_3As_4S_9$. Crystal system, triclinic. Density, 5.33 g/cm³. Occurrence: Switzerland.

baumite. A black mineral composed of iron magnesium manganese zinc aluminum silicate hydroxide, $(Mg,Mn,Fe,Zn)_3(Si,-Al)_2O_5(OH)_4$. Crystal system, orthorhombic. Density, 2.90 g/cm³; refractive index, 1.598. Occurrence: USA (New Jersey).

bauranoite. A reddish brown mineral of the becquerelite group composed of barium uranium oxide hydrate, $BaU_2O_7 \cdot xH_2O$. Density, 5.28 g/cm³. Occurrence: Russian Federation.

bauxite. A noncrystalline red, reddish-brown, brown, yellow, white, cream, or grayish rock or earthy soil that consists largely of aluminum hydroxides together with varying amounts of impurities, such as iron and titanium oxides, silica, clay and silt. A typical composition is 30-75% aluminum oxide (Al_2O_3), 9-31% water (H_2O), 3-25% ferric oxide (Fe_2O_3), 2-9% silicon dioxide (SiO_2), and 1-3% titanium dioxide (TiO_2). Density, 2.0-3.25 g/cm³ (0.07-0.12 lb/in.³); melting point, above 1800°C (3270°F); hardness, 1-3 Mohs. Occurrence: (i) *White bauxite*: Brazil, Greece, India, Surinam, USA (Arkansas); (ii) *Red bauxite*: France, Hawaii, Indonesia, Italy, Malaya, Pacific Islands; (iii) *Phosphatic bauxite*: Brazil; (iv) *Diaspore*: Europe, Russia, USA (Arkansas, Massachusetts, Missouri, Pennsylvania); (v) *Gibbsite:* Australia, Brazil, Jamaica, Guyana, Norway, USA (Massachusetts); (vi) *Laterite* (ferruginous bauxite): Europe; USA (Oregon). *Bauxite* derives its name from the town Les Baux in southern France, where it was originally found. Used as the most important aluminum ore, in the manufacture of aluminum oxide (alumina), alum, firebricks, aluminous cement, aluminum oxide abrasives, grinding wheels, refractories and electroceramics, as an extender in paints, plastics and rubber, as a catalyst, as a decolorizer, and as a filter medium.

bauxite brick. A firebrick that consists largely of aluminum hydroxide and ferric oxide. Used for chemically neutral furnace linings.

bauxite cement. A quick-setting cement made from bauxite and lime by burning in an electric furnace. See also aluminous cement.

bauxite clay. See bauxitic clay.

bauxite residue. An inorganic extender composed of iron oxide.

bauxitic clay. A natural mixture of bauxitic minerals, such as *diaspore* and *gibbsite*, with clay minerals, such as *attapulgite*, *montmorillonite* and *kaolinite*, containing between 47 and 65% alumina (Al_2O_3) on a calcined basis. Also known as *bauxite clay.*

bavenite. A white fibrous mineral composed of beryllium calcium aluminum silicate hydroxide, $Ca_4Be_2Al_2Si_9O_{26}(OH)_2$. Crystal system, orthorhombic. Density, 2.73 g/cm³. Occurrence: Italy, Poland, USA (California). Also known as *duplexite.*

Baxboard. Trade name of Canadian Gypsum Company Limited for gypsum backing boards.

Bayblend. Trademark of Bayer Corporation (USA) for a series of acrylonitrile-butadiene-styrene/polycarbonate alloys with good mechanical properties, high rigidity, superior toughness, good surface finish, low shrinkage, high melt-flow rate, and good smoke and flame resistance. Used in the manufacture of electrical and electronic equipment, business machines, computer housings, automotive parts including wheel covers.

Baycarb. Trademark of Bayer Corporation (USA) for glass-fiber-reinforced polycarbonates.

Bayclad. Trade name of Bay Mills Limited (Canada) for clad aluminum wire.

Baycoll. Trademark of Bayer Corporation (USA) for adhesives used for bonding plastic films and wood, and for food packaging applications.

Baycryl. Trademark of Bayer Corporation (USA) for a series of polymethyl methacrylate resins and plastics.

Baydur. Trademark of Bayer Corporation (USA) for a series of high-density polyurethanes processed by the reaction-injection molding (RIM) process. Supplied as hard-cast, hard, soft and reinforced microcellular elastomers and as structural and semi-rigid foams, they have good strength and toughness, good load-bearing properties, high durability, good resistance to weathering, high-quality surface finish, low flammability, and good moldability. Used for automotive components, agricultural equipment, workstations, utility carts, consumer electronics, office equipment, housings, etc.

Baydur GMV. Trademark of Bayer Corporation (USA) for a hard, lightweight, glass-mat-reinforced polyurethane integral foam with excellent mechanical properties, used for automotive interior parts.

Bayer alumina. A white, smelter-grade alumina of 99.5% Al_2O_3 used for the manufacture of metallic aluminum, and produced by the Bayer extraction process which involves leaching of pul-

verized *bauxite* ore with a hot, concentrated solution of sodium hydroxide to dissolve the aluminum by forming sodium aluminate. This clear aluminate solution is then separated from the residue (known as "red mud") by filtration, and allowed to cool. On cooling, *alumina trihydrate* is precipitated from the solution, filtered out of the sodium hydroxide solution, and calcined to anhydrous alumina.

Bayer Cement. Trade name of Bayer Corporation (USA) for a zinc polycarboxylate dental cement.

bayerite. A white mineral composed of aluminum hydroxide, $Al(OH)_3$. It is also made synthetically, e.g., in the Bayer process. Crystal system, monoclinic. Density, 2.53 g/cm³. Occurrence: Croatia, Hungary, Israel.

Bayester. Trademark of Bayer Corporation (USA) for a series of polyester resins including several grades of polybutylene terephthalate and polyethylene terephthalate.

Bayferrox. Trademark of Mobay/Bayer Corporation (USA) for iron oxide pigments.

Bayfide. Trademark of Bayer Corporation (USA) for polyphenylene sulfide resins.

Bayfill. Trademark of Bayer Corporation (USA) for a series of pour-in-place polyurethane foam materials available in regular and semi-rigid foam grades with good energy absorption properties. Used for automotive components, e.g., bumpers.

Bayfit. Trademark of Bayer Corporation (USA) for flexible molded polyurethane foams available in high- and low-density grades. Most have high sound absorption and high resilience. Used for furniture, automotive components, etc.

Bayflex. Trademark of Bayer Corporation (USA) for a series of flexible polyurethane integral skin foams and elastomers processed by the reaction-injection molding (RIM) process. They have hardnesses ranging from Shore A 55 to Shore D 61, good impact resistance, and are used for business machines, automotive components, sports equipment, packaging, etc.

Bayfol. Trademark of Bayer Corporation (USA) for polycarbonate resins used for sports equipment.

Baygal. Trademark of Bayer Corporation (USA) for low-cost polyurethane plastics with excellent electrical properties that can be formulated for room- and high-temperature applications. Used for automotive components, furniture, appliance housings, and electrical and electronic equipment.

Bayhydrol. Trademark of Bayer Corporation (USA) for aqueous, two-component polyurethane coatings for automotive finishes, and for coating wood, plastics and furniture.

Baykisol 30. Trademark of Bayer Corporation (USA) for an aqueous solution of colloidal silica.

bayldonite. A green mineral composed of copper lead arsenate hydroxide, $Cu_3Pb(AsO_4)_2(OH)_2$. Crystal system, monoclinic. Density, 5.35-5.50 g/cm³; refractive index, 1.754; hardness, 4.5 Mohs. Occurrence: Southern Africa, UK

bayleyite. A yellow, radioactive mineral composed of magnesium uranyl carbonate hydrate, $Mg_2(UO_2)(CO_3)_3 \cdot 18H_2O$. Crystal system, monoclinic. Density, 2.05 g/cm³; refractive index, 1.490. Occurrence: USA (Arizona).

baylissite. A colorless mineral composed potassium magnesium carbonate tetrahydrate, $K_2Mg(CO_3)_2 \cdot 4H_2O$. It can also be made synthetically. Crystal system, monoclinic. Density, 2.01 g/cm³; refractive index, 1.485. Occurrence: Switzerland.

Baylon. Trademark of Bayer Corporation (USA) for a series of polyethylene resins with a density range of 0.92-0.96 g/cm³, high tensile strength, a maximum service temperature of 120°C (248°F), good resistance to dilute acids and alkalies, fair resistance to mineral oils and gasoline, and low resistance to trichloroethylene and tetrachlorocarbon. Used for containers, pipes, tubing, etc. *Baylon V* is a polyethylene copolymer supplied in film form.

Bayloy. Trademark of Bay Bronze Industries Limited (Canada) for high-strength permanent-mold casting alloys of aluminum, zinc, and other metals

Baymal. Trademark of Bayer Corporation (USA) for a water- and oil-soluble aluminum hydroxide powder used in the production of adhesives, coatings, paints, adhesives, paints, and ceramics.

Baymer. Trademark of Bayer Corporation (USA) for rigid polyurethane foams used for automotive components, sports equipment, refrigeration containers, appliances, and construction applications.

Baymidur. Trademark of Bayer Corporation (USA) for isocyanate resins and foams used in the manufacture of molded polyurethane products, e.g., electrical and electronic equipment.

Baynat. Trademark of Bayer Corporation (USA) for rigid polyurethane foams used for automotive components, sports equipment, refrigeration containers, appliances, and construction applications.

Baypreg. Trademark of Bayer Corporation (USA) for polyurethane elastomers used for automotive components, sports and leisure equipment, electrical components, appliances, and for applications in the construction and transportation industries.

Baypren. Trademark of Bayer Corporation (USA) for chloroprene rubber used in the manufacture of adhesives.

Baysilex. Trademark of Bayer Corporation (USA) for silicone elastomer dental impression materials.

Baysilon. Trademark of Bayer Corporation (USA) for a series of silicone resins and elastomers.

Baystal. Trademark of Bayer Corporation (USA) for rubber latex used in building construction, as a backing for textiles, carpets and paper, and in the manufacture of molded foams.

Baytal. Trademark of Bayer Corporation (USA) for acetal resins.

Baytec. Trademark of Bayer Corporation (USA) for polyurethane elastomer systems and prepolymers used for automotive components, electrical equipment, sports and leisure equipment, machine parts, and construction and transportation applications.

Baytherm. Trademark of Bayer Corporation (USA) for rigid polyurethane foams used for automotive components, sports equipment, refrigeration containers, appliances, and construction applications.

Baytron. Trademark of Bayer Corporation (USA) for conductive polymers supplied in resin and coating grades for electronic applications.

Bazar. Trade name of Barker & Allen Limited (UK) for a nickel silver containing 10% nickel, 20-30% zinc, and the balance copper. Used for domestic utensils and ornaments.

Bazar metal. Trade name for an alloy of 91% silver and 9% nickel that is commercially available in sheet and wire form.

bazirite. A colorless mineral of the benitoite group composed of barium zirconium silicate, $BaZrSi_3O_9$. It can also be made synthetically. Crystal system, hexagonal. Density, 3.85 g/cm³; refractive index, 1.6751. Occurrence: Scotland.

bazzite. An azure-blue mineral of the beryl group composed of beryllium iron scandium silicate, $Be_3(Sc_{1.75}Fe_{0.25})Si_6O_{18}$. It can also be made synthetically. Crystal system, hexagonal. Density, 2.77 g/cm³; refractive index, 1.627. Occurrence: Kazakhstan.

BB-Blok. Trade name of BNZ Materials, Inc. (USA) for large

blocks of insulating firebrick with ceramic or stainless steel anchors. Used in the construction of kiln roofs.

BD nickel. See beryllia-dispersed nickel.

beach sand. Smooth, separate spherical or ovaloid grains or particles of rock materials that appear on beaches, and are the result of the abrasive action of waves. *Beach sand* is usually composed chiefly of silica (SiO_2), but may also contain minerals, such as garnet, monazite, zircon, etc. Used as source of chemical elements, as a foundry sand, and in the manufacture of hydrophobic sand. See also foundry sand; hydrophobic sand.

bead board. See expanded polystyrene foam board.

beaded fabrics. Textile fabrics that are decorated with clear or colored beads.

Beadex. Trade name of ASG Industries Inc. (USA) for patterned glass having a beaded or water-drop linear pattern imitating a beaded screen.

beading enamel. Any of several special types of porcelain enamel applied as a beading on ware for decoration and edge protection purposes.

beading lace. A lace, usually machine made, with a row of worked-in holes through which a decorative ribbon can be run.

beam. A structural member, usually a bar or straight girder, subjected primarily to flexure by transverse forces.

Beamette. Trademark of Wayn-Tex, Inc. (USA) for polyethylene and polypropylene monofilaments used for textile fabrics.

Bearcat. Trade name of Latrobe Steel Company (USA) for an air-hardening, shock-resisting tool steel (AISI type S7) containing 0.50% carbon, 0.70-0.75% manganese, 0.25% silicon, 3.25% chromium, 1.40% molybdenum, and the balance iron. It has excellent toughness and good nondeforming properties. Used for cold- and hot-tooling applications, chisels, punches, blades, rivet sets, swaging and gripper dies, and die-casting dies.

Bearcomo. Trade name of British Steel Corporation (UK) for normalized, creep-resisting steel plate containing 0.16% carbon, 1.5% manganese, 0.25% molybdenum, 0.5% copper, and the balance iron. It has good weldability and hardenability, and is used for large welded structures, boilerplate, etc.

bearing alloys. A group of alloys used in the manufacture of bearings and including babbitt metals, cadmium-base alloys, copper-lead alloys, bronzes, brasses, aluminum-base alloys, and cast irons. Also known as *bearing metals*.

bearing bronze. Any bronze that contains up to 11% tin and 30% lead, respectively, and is used mainly for bearings and bushings. Such bronzes have good castability and machinability, good antiscoring properties, good fatigue strength, good high-load capacity, and good antifriction properties.

bearing jewels. Hard synthetic sapphire used in the bearings of certain scientific instruments.

bearing materials. A group of materials used in the manufacture of bearings and including bearing alloys, porous metals, plastics (e.g., phenolics, nylons and polytetrafluoroethylenes), wood (e.g., lignum vitae, oak and maple), rubber, and carbon-graphite materials. *Bearing materials* should have good antifriction and antiscoring properties, good fatigue and corrosion resistance, good malleability, good embeddability, good high-load capacity, good self-lubricating properties, and low thermal expansion.

bearing metals. See bearing alloys.

Bearingoy. Trade name of Studebaker Chemical Company (USA) for an alloy of lead, copper, antimony and tin, used for machine bearings with long service life even under poor lubrication conditions.

bearing steel. See ball-bearing steel.

Bearite. Trade name of A.M. Cadman Manufacturing Company (USA) for a bearing alloy containing 80.1% lead, 16.75% antimony, 0.37% copper, 0.13% bismuth, and 2.65% hardener. It has high compressive strength, and a low coefficient of friction, and is used for medium-duty bearings.

Bearium. Trade name of Bearium Metals Corporation (USA) for high-lead bronzes containing about 70% copper, 17.5-28% lead, and the balance tin. Used for bearings, bushings, and thrust washers.

bearsite. A white mineral composed of beryllium arsenate hydroxide tetrahydrate, $Be_2AsO_4(OH)\cdot4H_2O$. Crystal system, monoclinic. Density, 1.90 g/cm^3; refractive index, 1.502. Occurrence: Kazakhstan.

Beau-Grip. Trademark of North American Corporation (USA) for viscose rayon fibers and yarns.

Beauknit. Trademark for a series of nylon, polyester and rayon fibers and yarns.

Beautywood. Trade name for a plastic laminate with grained surface imitating wood grains. Used for wall paneling.

Beaver. (1) Trade name of Atlas Specialty Steels (Canada) for a cold-work tool steel containing 0.68% carbon, 8.25% chromium, 1% silicon, 1% vanadium, 1.4% molybdenum, 1.5% nickel, and the balance iron. It possesses a high wear resistance, and is used for slitter knives, blanking dies, and cold-work tools.

(2) Trade name for water-hardening nickel-chromium steel containing 0.55% carbon, 0.60% manganese, 1.50% nickel, 0.75% chromium, and the balance iron. Used for drop-forging dies, forging-die blocks, tools, and forgings.

(3) Trade name for a series of plastic coatings.

(4) Trade name of Domtar Inc. (Canada) for bristol board.

Beaver Babbitt. Trade name of United American Metals Corporation (USA) for an alloy of antimony, lead and tin, used for machinery bearings.

beaver cloth. See melton.

beaverite. A yellow mineral of the alunite group composed of lead copper iron aluminum sulfate hydroxide, $Pb(Fe,Cu,Al)_3$-$(SO_4)_2(OH)_6$. Crystal system, rhombohedral (hexagonal). Density, 4.36 g/cm^3; refractive index, 1.85. Occurrence: USA (Utah).

Bechgaard salts. Electron-transfer salts that consist of two tetramethyltetraselenafulvalene (TMTSF) cations and one nonorganic anion, such as PF_6^- or ClO_4^-. The first of these organic superconductors were tetramethyltetraselenafulvalene phosphorus hexafluoride [$(TMTSF)_2PF_6$] and tetramethyltetraselenafulvalene perchlorate [$(TMTSF)_2ClO_4$], synthesized by K. Bechgaard et al. between 1979 and 1981. They have relatively low superconductivity transition temperatures of just above 1K at ambient pressure, and their general formula is $(TMTSF)_2X$. See also electron-transfer salts; organic conductors; organic superconductors.

Beckacite. Trademark of Reichhold Chemicals, Inc. (USA) for fumaric and maleic resins and modified phenolic resins.

Beckacrylat. Trademark of Reichhold Chemicals, Inc. (USA) for alkyd resins.

Beckamine. Trademark of Reichhold Chemicals, Inc. (USA) for urea- and melamine-formaldehyde resins and solutions used for fast-curing, chemical-resistant coatings.

Beckaminol. Trademark of Reichhold Chemicals, Inc. (USA) for urea-formaldehyde plastics and compounds.

Becket alloy. A corrosion-resistant alloy containing 1.5-3% carbon, 3% silicon, 25-30% chromium, and the balance iron. Used

for stainless castings.

Beckocoat. Trademark of Reichhold Chemicals, Inc. (USA) for polyurethane coating resins.

Beckoform. Trademark of Reichhold Chemicals, Inc. (USA) for polyurethane foams and foam products.

Beckol. Trademark of Reichhold Chemicals, Inc. (USA) for alkyd resins.

Beckophen. Trademark of Reichhold Chemicals, Inc. (USA) for heat-setting phenolic plastics.

Beckopox. Trademark of Reichhold Chemicals, Inc. (USA) for a series of epoxy resins.

Beckosol. Trademark of Reichhold Chemicals, Inc. (USA) for a series of pure or modified alkyd resins or solutions used in paints and coatings.

Beckton white. A white lithopone pigment composed of co-precipitated barium sulfate ($BaSO_4$), and zinc sulfide (ZnS). The mixing ratio is about 2:1. Used in paints, white rubber and leather goods, paper, textiles, etc.

Beckurol. Trademark of Reichhold Chemicals, Inc. (USA) for urea-formaldehyde plastics, compounds and laminates.

Beclawat. Trade name of Beclawat (Ontario) Limited (Canada) for safety glass.

becquerelite. A yellow or amber-yellow, radioactive mineral composed of calcium uranyl hydroxide octahydrate, $Ca(UO_2)_6O_4$-$(OH)_6\cdot8H_2O$. Crystal system, orthorhombic. Density, 5.12 g/cm^3; refractive index, 1.825. Occurrence: Central Africa.

Bedford cord. A heavy, durable, clear-finished worsted fabric with longitudinal ribs resembling corduroy, used for coats, suits, hats, uniforms, upholstery, etc. It is named after Bedford, a town in southern England.

Bedford limestone. See Indiana limestone.

beech. The hard, heavy, strong, fine-grained wood of any of various hardwood trees of the genus *Fagus*, the most important of which are the American beech *(F. grandfolia)* and the European beech *(F. sylvatica)*. Also known as *beechwood*. See American beech; European beech.

beechwood. See beech.

Bee-Mix. Trade name of Aardvark Clay & Supplies (USA) for a smooth, easy-to-throw, off-white clay (cone 10) that works well with glazes.

beeswax. A white, yellow or brown amorphous wax from the honeycombs of bees. Density, 0.95 g/cm^3; melting range, 62-65°C (144-149°F). Used in foundry pattern coatings, in art metal work, and in transparent paper, textile sizes and finishes, leather dressings, polishes, candles, etc.

Beethane. Trademark of Rohm & Haas Company (USA) for self-texturing, protective urethane coatings.

Beetle. Trademark of BIP Plastics (UK) for a series synthetic resins including (i) *Beetle Acetal* acetal copolymers, (ii) *Beetle Nylon 6* (polyamide 6) resins supplied in various grades: standard, glass bead-filled, glass fiber-reinforced, fire-retardant, high-impact, UV-stabilized, and molybdenum disulfide-lubricated; (iii) *Beetle Nylon 66* (polyamide 6,6) resins supplied in various grades: standard, glass bead-filled, glass fiber-reinforced, fire-retardant, high-impact, supertough, UV-stabilized, and molybdenum disulfide-lubricated; (iv) *Beetle Polyester* flexible and rigid polyester casting resins and bulk-molding compounds (BMCs). The BMCs are supplied in electrical, fire-retardant, high-heat and low-profile grades; (v) *Beetle Melamine* melamine-formaldehyde resins; (vi) *Beetle UF* urea-formaldehyde resins available as foams or filled with cellulose; and (vii) *Beetle Urea* urea-formaldehyde resins, which are similar to

Bakelite, and supplied in translucent, opaque, and colored grades. They have excellent rigidity, strength, hardness and stability, good heat and electrical resistance, excellent mold-ability, high resistance to organic solvents, alcohols, esters, hydrocarbons and oils, poor resistance to strong acids and alkalies, and moderate resistance to weathering and moisture, and are used for electric components and devices, and for containers, cabinets, buttons, toilet seats, etc.

beetled fabrics. Fabrics, usually cotton or linen, with a hard, flat, shiny surface finish obtained by pounding.

Beetleware. Trade name of American Cyanamid Corporation (USA) and BIP Plastics (UK) for colored urea-formaldehyde resins and products.

Begoform. Trade name of BEGO (Germany) for a dental laminating and veneering porcelain.

Begolloyd. Trade name of BEGO (Germany) for a series of precious and nonprecious dental alloys including several gold-, silver-, platinum- and palladium-based types.

behierite. A grayish pink mineral of the zircon group composed of tantalum borate, $TaBO_4$. It can also be made synthetically. Crystal system, tetragonal. Density, 7.86 g/cm^3; refractive index, above 2.0. Occurrence: Madagascar.

behoite. A colorless mineral composed of beryllium hydroxide, β-$Be(OH)_2$. It can also be made synthetically. Crystal system, orthorhombic. Density, 1.92 g/cm^3; refractive index, 1.544. Occurrence: USA (Texas).

beidellite. A greenish-yellow mineral of the smectite group composed of calcium aluminum silicate hydroxide hydrate, $Ca_{0.2}Al_2$-$Si_4O_{10}(OH)_2 \cdot xH_2O$. Crystal system, orthorhombic. Density, 1.51 g/cm^3. Occurrence: USA (Colorado, Idaho).

Beira-Mar. Trade name of Vidrobrás (Brazil) for patterned glass.

Bel. Trade name of Uyemura International Corporation (USA) for electroless nickel-boron.

Belais white gold. A corrosion-resistant alloy containing 75-85% gold, 8-18% nickel, 2-14% zinc, and sometimes small additions of platinum or manganese. Used for jewelry, ornaments, etc.

Belastraw. Trade name for rayon fibers and yarns used for textile fabrics.

Belectric. Trade name of Belle City Malleable Company (USA) for a series of cast irons containing 2.90-3.25% carbon, 2.0% silicon, and the balance iron. Used for pumps, machinery castings, gears, pinions, brake shoes, and brake drums.

B Electromal. Trade name of Belle City Malleable Company (USA) for a malleable cast iron containing 2.3-2.4% carbon, 1.2-1.3% silicon, 4% nickel, 1% copper, and the balance iron.

Belgian silex. A hard, tough *quartzite* imported from Belgium in rectangular blocks, and used for mill linings.

Belgrey. Trade name of SA Glaverbel (Belgium) for a gray, heat-absorbing plate glass.

Belimat. Trade name for rayon fibers, yarns and products.

Belira. Trademark of Incel (Bosnia-Hercegovina) for polyester fibers.

belite. A colorless or gray mineral composed of beta dicalcium silicate, β-Ca_2SiO_4. It can also be made synthetically by heating calcium carbonate and silicon dioxide with 0.5% boron oxide under prescribed conditions. Crystal system, monoclinic. Density, 3.28 g/cm^3; refractive index, 1.715. Occurrence: Northern Ireland. Used in the manufacture of cement. Also known as *larnite*.

belite cement. A cement that, unlike ordinary *Portland cement*, contains *belite*, a dicalcium silicate mineral (β-Ca_2SiO_4), as

the principal constituent. Used in the manufacture of concrete and cement products.

Bell Brand. Trade name of Veerman International Company (USA) for an antique, colored sheet and rondel glass.

bell brass. A brass containing 64% copper, 35% zinc, and the balance tin. Used for bells.

bell bronze. A bronze containing 80% copper and 20% tin, used for bells.

Bellcombi. Trademark of Kanebo Limited (Japan) for polyester fibers used for textile fabrics.

Belleek china. A thin, highly translucent chinaware that is composed of a body containing considerable amounts of frit, and is usually coated with a soft, pearly luster glaze. It has nil water absorption, and is named after Belleek, Ireland, where it was first produced. Also known as *Belleek porcelain*.

Belleek porcelain. See Belleek china.

Belleglass. Trade name of Bell de St. Clair/Kerr (USA) for a laboratory-processed dental restorative resin composite.

bellidoite. A creamy white mineral composed of copper selenide, Cu_2Se. Crystal system, tetragonal. Density, 7.03 g/cm^3. Occurrence: Czech Republic.

bellingerite. A light green mineral composed of copper iodate dihydrate, $Ca_3(IO_3)_6 \cdot 2H_2O$. Crystal system, triclinic. Density, 4.89 g/cm^3; refractive index, 1.900. Occurrence: Chile.

bell metal. A casting bronze containing about 75-85% copper and 15-25% tin. It has a density of 8.6-8.7, an uniform, fine-grained microstructure, and good fusibility. Used for bells and bearings.

bell-metal ore. See stannite.

Belmalloy. Trade name of Belle City Malleable Company (USA) for a pearlitic malleable iron containing 2.15-2.35% carbon, 1% silicon, 1% manganese, and the balance iron. Used for axles and tractor parts.

Belmont. (1) Trade name of Belmont Metals Inc. (USA) for a series of low-melting alloys including zinc-tin, tin-zinc, lead-antimony, bismuth-tin, tin-bismuth, and bismuth-lead-tin types. Used as solders, bearing metals, and for sprinklers, tube-bending applications, etc.

(2) Trade name of Wesgo Division of GTE Products Corporation (USA) for advanced engineering ceramics available in five rugged alumina bodies (purities ranging from 94-99+%) in the glazed or metallized condition.

belovite. A pale green mineral of the apatite group composed of sodium strontium calcium cerium phosphate hydroxide, $Sr_3(Ce,-Na,Ca)_2(PO_4)_3OH$. Crystal system, hexagonal. Density, 3.84 g/cm^3; refractive index, 1.641. Occurrence: Russian Federation.

Belseta. Trade name for a polyester fiber.

belt duck. A cotton fabric, either loose- or hard-woven, used for conveyor and transmission belts.

Beltec. Trademark of Honeywell Performance Fibers (USA) for a durable polyester reinforcement fiber used in the cap plies (overlays) of high-performance automotive tires.

belting. (1) A collective term for all sturdy fabrics (e.g., nylon, polyester, rayon, etc.), natural and synthetic rubbers, and heavy leathers (e.g., cattle hide) used for making belts for industrial and mechanical applications including those employed for conveying and lifting goods and transmitting power.

(2) The various belts made from the fabrics, rubbers and leathers mentioned in (1), e.g., drive belts, V-belts, etc.

belting leather. A chrome- or oak-tanned leather, usually made from cattle hides, that has an average weight of about 970 kg/m^3 (61 lb/ft^3), and is used for power transmission belting.

Bemal. Trade name of Yorkshire Imperial Metals Ltd. (UK) for a brass containing 70% copper, 29-30% zinc, and small amounts of lead and/or phosphorus. Used for condenser tubes.

Bembella. Trade name of American Bemberg Corporation (USA) for rayon fibers and yarns used for textile fabrics.

Bemberg. Trade name of American Bemberg Corporation (USA) for cuprammonium rayon used for the manufacture of chiffons, net, sheer, satin and similar fabrics for warp-knit underwear, dresses, linings, and hosiery.

bementite. A light-gray, gayish-brown, or brown mineral of the pyrosmalite group composed of manganese silicate hydroxide, $Mn_5Si_4O_{10}(OH)_6$. Crystal system, orthorhombic. Density, 2.98 g/cm^3; refractive index, 1.650. Occurrence: USA (New Jersey).

Bemit. German trade name for an aluminum alloy containing 2% copper, 0.45% manganese, and 0.27% tungsten. Used for welded, drawn or stamped light-alloy parts.

benavidesite. A mineral composed of iron manganese lead antimony sulfide, $Pb_4(Mn,Fe)Sb_6S_{14}$. Crystal system, monoclinic. Density, 5.60 g/cm^3. Occurrence: Peru.

Benchmark I. Trade name of J.F. Jelenko & Company (USA) for a soft, microfine palladium-free gold-platinum dental casting alloy (ADA type I) with deep yellow color, high elongation, low yield strength, good workability, and high polishability and biocompatibility. Used for low-stress inlays.

Benchmark III. Trade name of J.F. Jelenko & Company (USA) for a yellow-colored, hard, microfine palladium-free gold-platinum dental casting alloy (ADA type III) with high strength and biocompatibility. Used for hard inlays, crowns and fixed bridgework.

Benchmark IV. Trade name of J.F. Jelenko & Company (USA) for a rich gold-colored, extra-hard, microfine heat-treatable gold-platinum dental casting alloy (ADA type IV) that does not contain palladium. It has high strength, resiliency and biocompatibility, and good resistance to tarnish and discoloration. Used for hard inlays, thin crowns and long-span bridges.

Benchmark C. Trade name of J.F. Jelenko & Company (USA) for a rich gold-colored, microfine palladium- and silver-free high-gold/high-platinum porcelain-to-metal dental alloy with high tensile strength.

Bendalite. Trade name of Bend-A-Lite Plastics (USA) for a series of polystyrenes.

Bend-All. Trademark of Fabian Furniture Manufacturing Company Limited (Canada) for molded plywood.

Bendalloy. See Cerrobend.

Bend-flex. Trademark of Rogers Corporation (USA) for a formable, copper-clad epoxy laminate designed for computer and electronic applications requiring multiplanar circuitry. It has good bending characteristics and processibility, and good flame and smoke resistance.

bending board. Paperboard made up of one or more plies that usually consist of refined cellulosic fibers. It can be readily scored on the surface and subsequently bent along this scores without much fracture of the long, strong fibers used in the bending areas. The board surface may be coated or specially treated to facilitate scoring and bending. Used for automotive applications (e.g., trunk liners, glove boxes, etc.).

Bendurlen. Trademark for a series of polyethylene copolymer products.

Bendurplast. Trademark for a series of unsaturated polyester products.

Benecor. Trademark of J.H. Benecke GmbH (Germany) for flexible polyvinyl chloride film materials.

Benedict Metal. (1) Trade name of Riverside Metals Corporation (USA) for a nickel silver containing 75-80% copper, 20-25% nickel, and small amounts of iron and manganese. It has excellent corrosion resistance, and is used for feedwater heaters, condensers, and coinage.

(2) Trade name of Anaconda Company (USA) for a nickel silver containing 57-59% copper, 20-29.5% zinc, 12-12.5% nickel, 0-9% lead, and 0-2% tin. Used for hardware and plumbing fixtures.

Benedict plate. A nickel silver containing 57% copper, 28% zinc, and 15% nickel. It has excellent corrosion resistance, and is used as a white metal for flatwork.

Benedur. Trademark of J.H. Benecke GmbH (Germany) for polyvinyl chloride film materials.

bengaline. (1) A durable fabric in a plain weave with characteristic lateral cords or ribs that are formed by coarse weft (filling) yarns and fine warp (longitudinal) yarns.

(2) A corded or ribbed silk or rayon fabric that also contains wool, worsted or cotton.

benge. The yellow to medium-brown wood of the tree *Guibourtia arnoldiana*, used for furniture, veneer, and lumber.

benitoite. A blue, white or colorless mineral composed of barium titanium silicate, $BaTiSi_3O_9$. Crystal system, hexagonal. Density, 3.64 g/cm^3; refractive index, 1.756. Occurrence: USA (California). Used chiefly as a gemstone.

benjaminite. A grayish white or light blue mineral composed of copper silver bismuth lead sulfate, $(Ag,Cu)_3(Bi,Pb)_7S_{12}$. Crystal system, monoclinic. Density, 6.70 g/cm^3; refractive index, 1.825. Occurrence: Canada.

Beno-Therm. Trade name of Benolec (Canada) for cellulose insulation.

Bensilkie. Trade name for rayon fibers and yarns used for textile fabrics.

Benson. Trade name of Hewitt Metals Corporation (USA) for a babbitt containing 75% lead, 15% antimony, and 10% tin. Used for bearings and bushings.

benstonite. A white mineral of the calcite group composed of calcium barium carbonate, $Ca_7Ba_6(CO_3)_{13}$. Crystal system, rhombohedral (hexagonal). Density, 3.60 g/cm^3; refractive index, 1.691. Occurrence: USA (Arkansas).

Bentokol. Trademark of Foseco Minsep NV (Netherlands) for a series of additives for synthetic, semisynthetic and green sand which improve the casting finish and enhance molding characteristics. Used in the manufacture of castings of iron, copper, nickel and their alloys.

Bentone. Trademark of Rheo, Inc. (USA) for a quaternary amine-treated *montmorillonite* clay available as a fine, white powder for use as a gelling agent for adhesives and paints, cellulose lacquers, etc.

bentonite. A colloidal clay derived from volcanic ash and composed of 85% or more *montmorillonite* minerals and 5-10% alkalies or alkaline-earth oxides. It is characterized by an extremely fine grain size and high adsorption. There are two general types: (i) *Sodium bentonite* also known as *swelling bentonite*, which is cream-colored, has a great affinity for water and will swell enormously; and (ii) *Calcium bentonite* with negligible swelling capacity. *Bentonite* is named after Fort Benton in Montana, USA, where it was originally found. Used for refractory linings, as a suspension agent in porcelain-enamel slips and pottery clays, as a bonding agent in molding sand, as a pelletizer for iron ores, as a filler in paper and paper coatings, in abrasives, grouting fluids and oil-well drilling muds, in water softening, in the decolorization of oils, as a thickener in greases, in fireproofing agents, and as a filtering agent.

bentonitic clay. A montmorillonite-type clay having high swelling or wetting capacity.

bentorite. A bright violet mineral composed of calcium aluminum chromium sulfate hydroxide hydrate, $Ca_6(Cr,Al)_2(SO_4)_3$-$(OH)_{12}·26H_2O$. Crystal system, hexagonal. Density, 2.03 g/cm³; refractive index, 1.478. Occurrence: Israel.

Benum. Trade name of George Cook & Company Limited (UK) for a tough, air-hardening steel containing 0.3% carbon, 4.2% nickel, 1.3% chromium, 0.5% manganese, 0.3% silicon, 0.3% molybdenum, and the balance iron. Used for dies and molds.

Benvic. Trademark of Solvay Polymers, Inc. (USA) for polyvinyl chloride resins, compounds, plasticizers and fillers.

1,2-benzanthracene. See tetraphene.

2,3-benzanthracene. See naphthacene.

Benz Aviatek. Trade name for a non-heat-treatable aluminum alloy containing 12% zinc, 6% copper, and 2% iron. Used for pistons.

benzene. A clear, colorless, volatile, flammable aromatic liquid hydrocarbon that has a characteristic, pleasant odor. It is obtained chiefly from coal tar. Density, 0.879 g/cm³; melting point, 5.5°C (42°F); boiling point, 80.1°C (176°F); flash point, 12°F (-11°C); autoignition temperature, 1044°F (562°C); refractive index, 1.5011; slightly soluble in water; soluble in all proportions in alcohol, ether, acetone, chloroform, acetic acid, carbon tetrachloride and carbon disulfide. Used as a solvent, and in the production of many organic compounds including chlorobenzene, cyclohexane, dodecylbenzene, ethylbenzene, nitrobenzene, maleic anhydride, and diphenyl. Formula: C_6H_6. Also known as *benzol*.

1,2-benzenediol. See catechol.

1,3-benzenediol. See resorcinol.

1,4-benzenediol. See hydroquinone.

benzhydrylamine resin. A resin consisting of benzhydrylamine ($C_{13}H_{13}N$) and 1-2% divinylbenzene ($C_{10}H_{10}$) crosslinked polystyrene. It is supplied in mesh sizes from 80 to 200, and also available as the hydrochloride. Used for the solid-phase synthesis of peptide amines. Also known as *p-methylbenzhydrylamine resin; 4-methylbenzhydryl amine resin.*

benzil. An aromatic compound obtained from *benzoin* by oxidation with nitric acid. It is available in the form of yellow needles (98% pure) with a density of 1.52 g/cm³, a melting point of 94-97°C (201-207°F), and a boiling point of 346-348°C (655-658°F). Used as photoinitiator for polymers, and in chemical synthesis. Formula: $C_6H_5COCOC_6C_5$. Also known as *dibenzoyl; diphenylethanedione.*

benzocyclobutane polymers. See polybenzocyclobutanes.

benzoic acid. An aromatic carboxylic acid commercially made by the chlorination of toluene ($C_6H_5CH_3$) and hydrolysis of the resulting benzotrichloride ($C_6H_5CCl_3$). It is a colorless solid, slightly soluble in hot water, with a melting point of 122°C (252°F) and a boiling point of 249°C (480°F). Used in organic synthesis, in the manufacture of benzoates, alkyd resins, dentifrices, flavors, etc., and in biochemistry, biotechnology and medicine. Formula: C_6H_5COOH.

benzoin. An aromatic compound obtained by the condensation of benzaldehyde (C_6H_5CHO) in an alkaline cyanide solution. It is available in the form of white or yellowish crystals (98+% pure) with a melting point of 134-137°C (273-279°F), a boiling point of 194°C (382°F)/12 mm, and a specific optical rotation of 0° (at 25°C/77°F). Used as a photopolymerization catalyst, as a

ultraviolet catalyst initiator, as an intermediate, and in organic synthesis. Formula: $C_6H_5CH(OH)COC_6H_5$. Also known as *2-hydroxy-2-phenylacetophenone.*

benzol. See benzene.

1,12-benzoperylene. A crystalline aromatic compound usually supplied with a purity of 98%. Melting point, 277-279°C (531-534°F); boiling point, above 500°C (932°F). Used in the manufacture of organic conductors and semiconductors. Formula: $C_{22}H_{12}$. Also known as *benzo[ghi]perylene.*

1,2-benzopyrone. See coumarin.

p-benzoquinone. A dienophile available in the form of yellow crystals (98+% pure) with a density of 1.307 g/cm³ and a melting point of 113-116°C (235-241°F). Used in the preparation of hydroquinone and dyes, in Diels-Alder cycloaddition to form naphtoquinones and 1,4-phenanthrenediones, as a free radical inhibitor, as an oxidizer, as an analytical reagent, and in photography. Formula: $C_6H_4O_2$. Also known as *1,4-benzoquinone; quinone; chinone.*

o-benzosulfimide. See saccharin.

benzoylaminoacetic acid. See hippuric acid.

benzoylglycine. See hippuric acid.

benzoylglycocoll. See hippuric acid.

benzoyl peroxide. A white, crystalline solid (97% pure) with a density of 1.334 g/cm³, and a melting point of 103-106°C (217-223°F). It is also available with a purity of 70+% (remainder water), and as a blend in dibutyl phthalate, tricresyl phosphate or butyl benzyl phthalate plus water. Used as a curing and cross-linking agent in polymerization reactions, as a polymerization initiator for acrylic and polyester resins, in the vulcanization of rubber, as a drying and bleaching agent, etc. Formula: $(C_6H_5$-$CO)_2O_2$. Also known as *dibenzoyl peroxide.*

Beraco. Trade name of Paul Bergsoe & Son (Denmark) for a bearing alloy containing 76% lead, 13.5% antimony, 10% tin and 0.5% copper. It has a melting point of 243-374°C (470-705°F), and is used for transmission bearings.

Beralcast. Trademark of Nuclear Metals, Inc. (USA) for a series of lightweight beryllium-aluminum investment casting alloys with high strength and rigidity, high thermal conductivity, and good ductility. The two standard types are *Beralcast-363*, used for structural components, and *Beralcast-191*, used for electronic heat-exchanger parts.

Beraloy. Trademark of Wilbur B. Driver Company (USA) for a series of beryllium bronzes available in four grades: (i) *Beraloy A* which contains 1.8-2.0% beryllium, 0.25-0.50% cobalt, and the balance copper; (ii) *Beraloy B* with 1.9-2.2% beryllium, 0.25-0.50% nickel, and the balance copper; (iii) *Beraloy C* composed of 0.4-0.5% beryllium, 2.5-2.7% cobalt, and the balance copper; and (iv) *Beraloy D* which contains 1.6-1.8% beryllium, 0.25-0.5% cobalt, and the balance copper. *Beraloy* has high corrosion and fatigue resistance, high electrical conductivity, good hardenability, and high strength and rigidity in heat-treated condition. Used for springs, diaphragms, resistance wire, electric contacts, and bearings.

beraunite. A red or reddish-brown mineral composed of iron phosphate hydroxide tetrahydrate, $Fe_6(PO_4)_4(OH)_5·4H_2O$. Crystal system, monoclinic. Density, 3.01 g/cm³; refractive index, 1.786. Occurrence: USA (New Hampshire).

berborite. A colorless artificial mineral composed of beryllium borate hydroxide monohydrate, $Be_2(BO)_3OH·H_2O$. It can be made from beryllium hydroxide [$Be(OH)_2$] gel and boric acid at 230°C (446°F). Crystal system, hexagonal. Density, 2.05 g/cm³; refractive index, 1.580.

Berdan. Trade name of Bertex GmbH (Germany) for cotton yarn and twine.

Bergadur PB. Trade name of Bergmann GmbH (Germany) for polybutylene terephthalates.

Bergal. Trade name of Bergmann Elektrizitätswerke (Germany) for an age-hardenable aluminum alloy containing 4% copper, 0.8% magnesium, 0.7% manganese, and 0.4% silicon. Used for light-alloy parts.

Bergaform C. Trade name of Bergmann GmbH (Germany) for a polyoxymethylene (POM) resin.

Bergamid. Trade name of Bergmann GmbH (Germany) for polyamide (nylon) plastics including nylon 6,6 (*Bergamid A*), nylon 6,66 (*Bergamid AB70*) and nylon 6 (*Bergamid B*).

bergenite. A yellow mineral of the phosphouranylite group composed of barium uranyl phosphate hydroxide octahydrate, $Ba(UO_2)_4(PO_4)_2(OH)_4 \cdot 8H_2O$. Crystal system, orthorhombic. Density, 4.10 g/cm^3; refractive index, 1.690. Occurrence: Germany.

Bergerac. Trade name of Permacon (Canada) for paving stone with rustic appearance supplied in rectangular, square and circular units of varying size in red, charcoal, gray and beige colors. Used for walkways, patios, steps, etc.

bergslagite. A colorless, whitish or grayish mineral of the datolite group composed of calcium beryllium arsenate hydroxide, $CaBeAsO_4OH$. Crystal system, monoclinic. Density, 3.40 g/cm^3; refractive index, 1.681. Occurrence: Sweden.

Bergstahl. Trade name of Bergische Stahl Industrie (Germany) for a series of machinery steels, case-hardening steels, and various cast corrosion-resistant and/or heat-resistant steels.

Bergla-Tel. Trade name of Glasfaser Gesellschaft mbH (Germany) for glass wool.

berkelium. A synthetic, radioactive, metallic element of the actinide series of the Periodic Table. It was first produced at the University of California, Berkeley in 1949 as the 243 isotope (half-life, 4.5 hours) by helium-ion bombardment of americium-241 (^{241}Am) in a cyclotron. Density, 14.8 g/cm^3; melting point, 986°C (1807°F); atomic number, 97; atomic weight of most stable isotope, 249; trivalent and tetravalent; eight isotopes are known ranging from 243 to 250; chemically similar to other transuranium elements; forms compounds with oxygen, chlorine and fluorine. There are two allotropic forms: (i) *Alpha berkelium* (close-packed double hexagonal) present below 910°C (1670°F); and (ii) *Beta berkelium* (face-centered cubic) stable between 910 and 986°C (1670 and 1807°F). Symbol: Bk.

berkeyite. See lazulite.

Berlin. German trade name for a nickel silver containing containing 48-56% copper, 24-29% zinc, 16-24% nickel, and 0-4% iron. It has excellent corrosion resistance, and is used for electrical resistances, ornaments, and drawn, stamped and spun parts.

Berlin blue. See Prussian blue.

berlinite. A colorless, rose-red, or grayish mineral of the quartz group composed of aluminum phosphate, $AlPO_4$. It can also be made synthetically from aluminum metal and phosphoric acid, or by crystal growing techniques. Crystal system, hexagonal. Density, 5.64 g/cm^3; refractive index, 1.524; low water solubility; good chemical inertness; high acoustic wave propagation velocity. Occurrence: Sweden, Central Africa. Synthetic berlinite is used for bulk and surface acoustic wave devices.

Berlin porcelain. A German hard paste porcelain made from fine clay and feldspar by firing at 1000°C (1830°F), glazing and refiring at 1400°C (2550°F).

Berlin red. A red pigment composed mainly of ferric oxide (Fe_2O_3).

Berlin white. A type of *white lead* used as a paint pigment.

Berlox. Trade name for beryllium oxide powder used for heat and wear-resistant coatings applied by flame spraying techniques. It is supplied in particle sizes ranging from 80 to 325 mesh.

bermanite. A reddish brown mineral composed of manganese phosphate hydroxide tetrahydrate, $Mn_3(PO_4)_2(OH)_2 \cdot 4H_2O$. Crystal system, monoclinic. Density, 2.84 g/cm^3; refractive index, 1.725. Occurrence: USA (Arizona).

Bermax. Trade name of Federal-Mogul Corporation (USA) for an alloy containing 9-11% antimony, 0-0.5% copper, and the balance lead. It has excellent antifriction properties, and is used for bearings.

Bermudent. Trade name for a high-gold dental bonding alloy.

berndtite. A deep yellow-brown mineral of the brucite group composed of beta tin sulfide, β-SnS_2. It can also be made synthetically. Crystal system, hexagonal. Density, 4.46 g/cm^3; refractive index, 1.825. Occurrence: Central Africa.

BERNER. Trade name of Berner Ges.mbH (Austria) for a range of glues and adhesives including *BERNER KSK 310* instant construction adhesives and *BERNER Plast-O-Fix* plastic repair adhesives.

Berolith. German trade name for thermoplastic casein plastics.

berryite. A bluish gray mineral composed of lead silver copper bismuth sulfide, $Pb_2(Ag,Cu)_3Bi_5S_{11}$. Crystal system, monoclinic. Density, 6.70 g/cm^3. Occurrence: Sweden. A variety with slightly different formula $[Pb_3(Cu,Ag)_5Bi_7S_{16}]$ is found in Greenland.

Berry Metal. Trade name of Berry Metal Company (USA) for a wear-resistant alloy containing 80% copper, 10% lead, 8% nickel and 2% antimony. Used for bearings.

Bersch metal. A German bearing alloy of 93% aluminum and 7% nickel.

Bertex. Trade name of Bertex GmbH (Germany) for cotton yarn and twine.

berthierine. A green mineral of the kaolinite-serpentine group composed of iron aluminum silicate hydroxide, $(Fe,Al)_3(Si,Al)_2O_5(OH)_4$. Crystal system, hexagonal. Density, 3.03-3.04 g/cm^3. Occurrence: Scotland, UK.

berthierite. A dark steel-gray to black mineral composed of iron antimony sulfide, $FeSb_2S_4$. It can also be made from iron sulfide (FeS), antimony and sulfur at 300-530°C (572-986°F). Crystal system, orthorhombic. Density, 4.64 g/cm^3. Occurrence: France.

Berthier's alloy. (1) A corrosion-resistant copper nickel containing 68% copper and 32% nickel. Used for stills and evaporators.

(2) A free-cutting brass containing 72% copper, 25% zinc, 2% lead, and 1% tin. Used for screw-machine products.

berthonite. See bournonite.

bertrandite. A colorless, or yellowish mineral composed of beryllium silicate hydroxide, $Be_4Si_2O_7(OH)_2$. Crystal system, orthorhombic. Density, 2.60 g/cm^3; refractive index, 1.602. Occurrence: USA (Virginia).

Beruda alloy. A British corrosion-resistant, high-tensile brass used for strong, corrosion-resistant parts.

Berydur. Trade name for a wrought copper alloy containing 1% beryllium and 0.25% cobalt. Used for springs, contacts, nonsparking tools, etc.

beryl. A mineral that is usually colorless, white, pale-yellow or bluish green with vitreous luster, but may also be light-green, golden-yellow, or pink. It is composed of beryllium aluminum silicate, $Be_3Al_2(SiO_3)_6$, and may contain up to 15% beryllium oxide (BeO). Crystal system, hexagonal. Density, 2.63-2.80 g/cm³; melting point, 1420°C (2588°F); hardness, 7.5-8.0 Mohs; refractive index, 1.579. Occurrence: Argentina, Brazil, Canada (Ontario), Colombia, India, Russia, South Africa, USA (Colorado, Maine, Massachusetts, Nevada, New Hampshire, North Carolina, South Dakota, Utah), Zimbabwe. Used as an important ore of beryllium, as a dielectric, in superconductors, as a green colorant in glazes, as a gemstone (aquamarine and emerald), and for spark plugs, and X-ray windows and tubes.

Berylco. Trademark of NGK Metals Corporation (USA) for metallic beryllium, cast beryllium bronzes, beryllium nickels and various other ferrous and nonferrous beryllium-bearing alloys. Used for springs, welding electrodes, contacts, switches, hardware, marine parts, cams, bearings, gears, dies, tools, etc.

Beryldur. Trade name of NGK Metals Corporation (USA) for age-hardenable beryllium-copper alloys containing 0.8-1.2% beryllium, 0-0.2% nickel and 4.0% other additions. They have high strength, good formability, good electrical conductivity, and excellent corrosion and erosion resistance. Used for connectors, springs, clips, and switchgear.

beryllia. A white, highly toxic powder (99+% pure) made by heating beryllium nitrate or hydroxide. It occurs in nature as the mineral *bromellite*. Density, 3.008 g/cm³; melting point, 2575°C (4667°F); boiling point, 4300°C (7770°F); hardness, 9 Mohs; can be fabricated into finished shapes. Used in the preparation of beryllium compounds, as a catalyst, in the manufacture of ceramics, glass, plastics and refractories, and for high-strength fibers. Formula: BeO. Also known as *beryllium oxide*.

beryllia ceramics. A group of ceramic products composed chiefly of beryllia (BeO). They have superior dielectric characteristics, high electrical and thermal conductivity, good wetting resistance, good physical and mechanical properties, high heat-stress resistance, high electrical resistance, a maximum service temperature of 2400°C (4352°F), poor resistance to acids and alkalies, and are transparent to microwave radiation and undamaged by nuclear radiation. Used as moderators, reflector materials, matrices for fuel elements in nuclear reactors, in electronic components, electron tubes, klystron-tube windows, transistor mountings, solid-state devices, microwave parts, gyroscopes, resistor cores, uranium and thorium-melting crucibles, aircraft and missile equipment, and as whisker materials. Also known as *beryllium oxide ceramics*.

beryllia-dispersed nickel. A dispersion-hardening alloy produced by the addition of several volume percent of beryllia (BeO) as finely dispersed particles to nickel. The beryllia significantly enhances the strength and elevated temperature properties of the nickel matrix. Abbreviation: BD nickel. Also known as *beryllium oxide-dispersed nickel*.

beryllia fibers. High-strength low-density single-crystal fibers composed of beryllia (BeO), and used as reinforcements in engineering composites. Also known as *beryllium oxide fibers*.

beryllide coating. A hard, pore-free, heat-resistant coating applied to various substrates by diffusion coating techniques.

beryllides. A group of intermetallic compounds made by chemically combining beryllium with metals, such as chromium, cobalt, hafnium, iron, manganese, molybdenum, nickel, niobium, palladium, platinum, plutonium, tantalum, thorium, titanium, tungsten, uranium, vanadium, yttrium and zirconium. The general formula is Me_xBe_y, where Me can represent any of the previous metals. *Beryllides* have high strength and strength reten-

tion even at elevated temperatures, excellent thermal-shock resistance, excellent oxidation resistance up to about 1260-1540°C (2300-2805°F), high melting temperatures ranging from about 1427-2080°C (2600-3775°F), a hardness range of 500-1300 kgf/mm², and high thermal conductivities. Used for structural applications, special ceramics, spark-resistant tools, and coatings.

beryllite. A white mineral composed of beryllium silicate hydroxide monohydrate, $Be_3SiO_4(OH)_2 \cdot H_2O$. Crystal system, orthorhombic. Density, 2.20 g/cm³; refractive index, 1.553. Occurrence: Russian Federation, Greenland.

beryllium. A brittle, gray-white metallic element of Group IIA (Group 2) of the Periodic Table (alkaline-earth group). It is commercially available in the form of flakes, lumps, hot-pressed, cold-pressed and sintered blocks, and in sheets, flake, foil, rods, tubes, powder and wire. The commercial beryllium ores are *beryl* and *chrysoberyl*. Density, 1.85 g/cm³; melting point, 1280°C (2336°F); boiling point, 2770°C (5018°F); superconductivity critical temperature, 0.026K; hardness 55-60 Brinell; atomic number, 4; atomic weight, 9.012; divalent. It has high electrical conductivity (52% IACS), high thermal conductivity, a high heat absorption rate, high strength-to-weight ratio, high permeability to X-rays, good machinability and formability, low impact strength, poor brazeability and weldability, and is the lightest structural metal. There are two forms of beryllium: (i) *Alpha beryllium* with hexagonal close-packed crystal structure, present at room temperature; and (ii) *Beta beryllium* with body-centered cubic structure, present at high temperatures. *Beryllium* is used as an alloying element in copper and nickel to increase elasticity and strength characteristics, as a structural material in missiles and aircraft, in watch springs, spark-free tools, windows for X-rays and counter tubes, in rockets, in nuclear reactors as a neutron reflector and moderator, in inertial guidance systems, gyroscopes and computer components, in superconductors, and in the melting and casting of magnesium to reduce burning tendency.

beryllium acetate. White crystals with a melting point of 285-286°C (545-546°F) and a boiling point of 330-331°C (626-627°F). Used as a source of pure beryllium salts. Formula: $Be(C_2H_3O_2)_2$.

beryllium acetate oxide. A white, moisture-sensitive powder with a density of 1.36 g/cm³ and melting point of 200°C (392°F) (sublimes). Used as a source of pure beryllium salts and in organometallic research. Formula: $Be_4O(C_2H_3O_2)_6$. Also known as *basic beryllium acetate*.

beryllium alloys. A group of alloys containing varying amounts of beryllium, and including beryllium-aluminum alloys (up to 62% beryllium), beryllium-copper alloys (up to 2.5% beryllium), beryllium-gold alloys (up to 5.0% beryllium), beryllium-nickel alloys (up to 2.7% beryllium), and beryllium-silver alloys (up to 0.9% beryllium). Those containing only a few percent of beryllium are usually of the precipitation-hardening type.

beryllium aluminate. A compound of beryllium oxide and aluminum oxide available as orthorhombic crystals. It occurs in nature as the mineral *chrysoberyl*. Density, 3.5-3.8 g/cm³; melting point, 1870°C (3398°F); hardness, 8.5 Mohs. Used in ceramics and materials research. Formula: $BeAl_2O_4$. Also known as *beryllium aluminum oxide (BAO)*.

beryllium-aluminum. A master alloy containing 94% aluminum, 5% beryllium and 1% magnesium. Used in the remelting of alloys.

beryllium-aluminum composites. Low-density engineering composites consisting of ductile aluminum matrices reinforced with beryllium particles. They have high moduli of elasticity and high yield strength, and are used for structural aerospace parts, computer memory drums and disks, nuclear fuel canning, etc.

beryllium aluminum oxide. See beryllium aluminate.

beryllium aluminum silicate. A compound of beryllium oxide, aluminum oxide and silicon dioxide available in the form of hexagonal crystals. It occurs in nature as the mineral *beryl*. Density, 2.63-2.80 g/cm^3; melting point, 1420°C (2588°F); hardness, 7.5-8.0 Mohs. Used as a dielectric, in ceramics and glazes, and for spark plugs, and X-ray windows and tubes. Formula: $Be_3Al_2-(SiO_3)_6$.

beryllium boride. Any of the following hard, strong compounds formed by heating mixtures of beryllium and boron: (i) *Beryllium diboride*. A refractory solid. Melting point, at least 1,970°C (3,578°F). Used in ceramics and materials research. Formula: BeB_2; (ii) *Diberyllium boride*. Used in ceramics and materials research. Formula: Be_2B; and (iii) *Beryllium hexaboride*. Formula: BeB_6. Used in ceramics and for aerospace applications.

beryllium bronzes. A group of tough, strong, precipitation-hardenable copper alloys containing 1-3% beryllium. Small amounts of silicon, nickel and iron may be added to increase hardness, ductility and strength, refine the grain, and/or improve the overal microstructure. They are usually supplied in the cast, hot-worked or cold-worked condition, and are costly because of the beryllium additions. *Beryllium bronzes* have a density range of 8.2-8.9 g/cm^3 (0.30-0.32 $lb/in.^3$), a hardness of 100-360 Brinell, high strength, excellent electrical properties, relatively high electrical conductivity, excellent corrosion and fatigue resistance, and good wear resistance. Used for ship and boat propellers, valves, automotive parts, nonsparking tools, bearings and bushings for aircraft landing gears, locomotive bearings, shims, cams, springs, diaphragms, electrical switch parts, dental and surgical instruments, golf club heads, optical alloys, parts subjected to frictional wear, plastic molds, electronic equipment, and spot-welding electrodes. Also known as *beryllium copper; copper-beryllium*.

beryllium carbide. Yellow, hexagonal crystals obtained from beryllium and carbon, or beryllium oxide and carbon. Density, 1.9 g/cm^3; melting point, decomposes above 2950°C (5342°F); unstable in oxygen above 982°C (1800°F); hardness, 8.5-9.5 Mohs; high toughness and elasticity; good corrosion resistance at elevated temperatures. Used as a neutron moderator in nuclear applications, and for nuclear reactor cores. Formula: Be_2C.

beryllium chloride. A compound of beryllium and chlorine available in the form of white or pale yellow, water-soluble, moisture-sensitive crystals, or an off-white powder (99+% pure). Density, 1.90 g/cm^3; melting point, 440°C (824°F); boiling point, 520°C (968°F). Used in dry form as a catalyst for organic and organometallic reactions. Formula: $BeCl_2$.

beryllium-cobalt. A master alloy of 50% beryllium and 50% cobalt, used in the remelting of alloys.

beryllium-columbium. See beryllium-niobium.

beryllium copper. (1) A master alloy containing about 96% copper and 4% beryllium used to introduce beryllium into copper alloys.

(2) See beryllium bronzes.

beryllium diboride. See beryllium boride (i).

beryllium fluoride. A compound of beryllium and fluorine usually supplied in the form of small, moisture-sensitive, amorphous pieces (99+% pure). Density, 1.986 g/cm^3; melting point, sublimes at 800°C (1472°F). Used in the manufacture of beryllium metal, in nuclear reactors, in organometallic research, and

in the manufacture of optical glass. Formula: BeF_2.

beryllium fluoride glass. An optical crown glass in which beryllium fluoride replaces all or most of the silica. It has a low refractive index, low dispersion and high light transmission, and is used for optical equipment.

beryllium gold. A hardened gold alloy that contains up to 5% beryllium, and is used in dentistry for inlays and as a gold solder.

beryllium hexaboride. See beryllium boride (iii).

beryllium-iron. A master alloy containing 50% beryllium and 50% iron used to introduce beryllium into iron-copper alloys.

beryllium-magnesium-aluminum. A master alloy containing 5% beryllium, 5% magnesium and 90% aluminum. Used in the foundry industries for remelting applications.

Beryllium Malleable. Trade name for an alloy of 99.8% beryllium and 0.2% titanium with very high malleability, used for X-ray windows and camera shutters.

beryllium-manganese bronze. A corrosion-resistant bronze composed of 89-95.5% copper, 3-10% manganese, and 1-1.5% beryllium. Used for corrosion-resistant parts.

beryllium metaphosphate. White, porous, high-melting powder or granules used as a raw material for special ceramics, and as a catalyst carrier. Formula: $Be(PO_3)_2$.

beryllium monel. A *monel*-type nickel-copper alloy containing a small percentage of beryllium.

beryllium-nickel. A master alloy containing 50% beryllium and 50% nickel. Used to introduce beryllium into copper-nickel alloys.

beryllium-nickel alloys. A group of wrought and cast age-hardenable nickel alloys containing about 1.8-2.8% beryllium. Chromium, titanium, carbon, and manganese may also be added to control grain size and improve corrosion resistance, hot workability and/or machinability. *Beryllium-nickel alloys* have high strength, and good thermal conductivity, and are used for springs, valves, turbines blades, and glass molds.

beryllium-niobium. A hard, low-density refractory compound of beryllium and niobium (columbium) usually supplied in powder form. It has a high strength, and good oxidation and corrosion resistance. Used for sintered parts and refractory coatings. Also known as *beryllium-columbium*.

beryllium nitride. A compound of beryllium and nitrogen obtained in the form of white refractory crystals by heating beryllium metal powder in a controlled nitrogen atmosphere. Density, 2.71 g/cm^3; melting point, 2200±50°C (3992±90°F); oxidizes in air above 600°C (1110°F); high hardness, elasticity, corrosion resistance and toughness at elevated temperatures. Formula: Be_3N_2. Used for incandescent mantles, in nuclear applications, in the manufacture of the carbon-14 isotope, and in rocket fuels.

beryllium oxide. See beryllia.

beryllium oxide ceramics. See beryllia ceramics.

beryllium oxide-dispersed nickel. See beryllia-dispersed nickel.

beryllium oxide fibers. See beryllia fibers.

beryllium phosphide. A compound of beryllium and trivalent phosphorus available in crystalline form. Density, 2.06 g/cm^3. Used in ceramics and semiconductor research. Formula: Be_3P_2.

beryllium polonide. A compound of beryllium and polonium. Crystal system, cubic. Crystal structure, sphalerite. Density, 7.3 g/cm^3. Used as a semiconductor. Formula: $BePo$.

beryllium-potassium sulfate. A crystalline substance used in electroplating (e.g., chromium and silver). Formula: $BeSO_4 \cdot K_2SO_4$.

beryllium selenide. A compound of beryllium and selenium. Crystal system, cubic. Crystal structure, sphalerite. Density, 4.315 g/cm^3; refractive index, 3.61. Used as a semiconductor. Formula: BeSe.

beryllium silicate. A compound of beryllium oxide and silicon dioxide. It occurs in nature as the mineral *phenacite*. Density, 2.99 g/cm^3; melting point, 1560°C (2840°F). Used in ceramics and materials research. Formula: Be_2SiO_4.

beryllium silver. An alloy of silver and 0.41-0.90% beryllium with good resistance to sulfur-laden atomspheres.

beryllium-silver. A master alloy containing 50% beryllium and 50% silver. Used in the foundry industry for remelting applications.

beryllium-sodium fluoride. A white, crystalline substance composed of beryllium fluoride and sodium fluoride. It has a melting point of approximately 350°C (660°F), and is used in the production of pure beryllium metal. Formula: $BeNa_2F_4$. Also known as *sodium-beryllium fluoride*.

beryllium sulfide. A crystalline compound of beryllium and sulfur. Crystal system, cubic. Crystal structure, sphalerite. Density, 2.36-2.47 g/cm^3; refractive index, 4.17. Used in ceramics, and as a semiconductor. Formula: BeS.

beryllium-tantalum. A hard, refractory compound of beryllium and tantalum usually supplied in powder form. It has a low weight, high strength, and good oxidation and corrosion resistance. Used for sintered parts and refractory coatings.

beryllium telluride. A crystalline compound of beryllium and tellurium. Crystal system, cubic. Crystal structure, sphalerite. Density, 5.09; refractive index, 1.45. Used as a semiconductor. Formula: BeTe.

beryllium-zirconium. A hard, refractory compound of beryllium and zirconium usually supplied in powder form. It has a low weight, high strength, and good oxidation and corrosion resistance. Used for sintered parts and refractory coatings.

beryllonite. A colorless, white, or pale yellow mineral composed of sodium beryllium phosphate, $NaBePO_4$. It can also be made synthetically. Crystal system, monoclinic. Density, 2.85 g/cm^3; refractive index, 1.558; hardness, 5.5-6 Mohs. Occurrence: USA (Maine).

Berylloy. Trade name of Telcon Metals Limited (UK) for a beryllium bronze containing 2.7% beryllium, 0.5% cobalt, and the balance copper. Used for springs, nonsparking tools, contacts, etc.

Beryvac. Trademark of Vacuumschmelze GmbH (Germany) for a series of age-hardenable copper-beryllium, nickel-beryllium and copper-nickel-manganese-beryllium alloys characterized by high strength combined with good thermal conductivity. The copper-beryllium alloys typically contain about 1.6-2.05% beryllium, up to 0.2% nickel and up to 2.5% cobalt, the nickel-beryllium alloys 2% beryllium, and the copper-nickel-manganese-beryllium alloys contain typically about 63.8% copper, 18% nickel, 18% manganese and 0.2% beryllium. Used for regulating or relay springs, diaphragms, mechanically-stressed spectacle parts, plugs, connectors, and resistance-welding electrodes.

berzelianite. A dusty black to silver-white mineral composed of copper selenide, $Cu_{2-x}Se$. Crystal system, cubic. Density, 6.65 g/cm^3. Occurrence: Sweden.

berzeliite. A cream-colored mineral of the garnet group composed of magnesium manganese calcium sodium arsenate, $(Ca,Na)_3(Mg,Mn)_2(AsO_4)_3$. Crystal system, cubic. Density, 4.08 g/cm^3. Occurrence: Sweden.

Berzelius. Trade name of Berzelius Metallhütten-GmbH (Ger-

many) for a tin, zinc and their alloys, and including commercially pure and high-purity tins, mixed tin, various alloys of tin, and commercially pure and high-grade zinc, special coating-grade zinc, and several zinc alloys. They are supplied in wire, sheet, and strip form, or as semifinished products.

Besfight. Trademark of Teijin Corporation (Japan) for a carbon fiber produced from a polyacrylonitrile (PAN) precursor material. It has high tensile strength and elastic modulus, extraordinary heat resistance, and is used in the aerospace, civil engineering, sports and leisure and many other industries.

Beslon. Trademark of Toho Rayon Company Limited (Japan) for acrylic fibers used for the manufacture of textile fabrics.

Besniflec. Trademark of McGean (USA) for bright nickel deposits and plating processes.

Besplate. Trade name of McGean (USA) for copper, nickel, chrome and brass electrodeposits and plating processes.

Bessemer iron. A low-sulfur pig iron with less than 0.10% phosphorus used for making Bessemer and acid open-hearth steel. A typical composition is 3.0-4.0% carbon, 1.5-2.5% silicon, 0.5-2.0% manganese, 0.07-0.10% phosphorus, 0.01-0.05% sulfur, and the balance iron. Also known as *Bessemer pig iron*.

Bessemer pig iron. See Bessemer iron.

Bessemer steel. A steel made by blowing air through molten pig iron, contained in a pear-shaped vessel with refractory lining, whereby most of the carbon and impurities, such as silicon, phosphorus and magnanese, are removed by oxidation. *Acid Bessemer steel* is produced in a converter lined with acid refractories and *basic Bessemer steel* in a converter lined with basic refractories.

Best. Trade name of Bethlehem Steel Corporation (USA) for a water-hardening tool steel (AISI type W2) containing 0.70-1.10% carbon, 0.2% vanadium, small amounts of chromium, manganese and silicon, and the balance iron. It has good wear resistance and toughness, low hot hardness, and good machinability. Used for shear blades and knives, broaches, reamers, punches, hand tools, threading dies, drawing and heading dies, and forming dies.

best bronze. A gilding brass that contains 90% copper and 10% zinc, and used as a jewelry base and for primers and ammunition. See also gilding metal.

best cokes. Coke tinplate carrying a slightly heavier tin coating than standard coke tinplate. See also coke tinplate.

Bestem. Trademark of Walter Somers Limited (UK) for high-carbon die steels containing 3% nickel and varying amounts of molybdenum.

Bestlite. Trade name of Corning Glass Works (USA) for photochromic spectacle glass.

Best-Lok. Trademark of Central Precast Products Limited (Canada) for interlocking paving stones.

best patent wire. Steel wire with a tensile strength of 1.10-1.24 GPa (160-180 ksi).

best plow wire. Steel wire with a tensile strength of 1.38-1.52 GPa (200-220 ksi).

best selected copper. A commercial copper of lower purity (99.75+%) than electrolytic tough-pitch copper.

best type metal. A typemetal alloy of 50% lead, 25% tin and 25% antimony.

Best Yorkshire. A name used in the United Kingdom for a commercially pure iron containing 0.06% manganese and 0.16% phosphorus. Used as a primary metal.

Best Warranted. Trade name of E.A. Balfour Steels Limited (UK) for a plain-carbon tool steel containing 0.9% carbon, and the balance iron. Used for cutting tools, press tools, fixtures, shear blades, and knives.

Beta. Trade name of Owens-Corning Fiberglass Corporation (USA) for glass-fiber yarns.

Beta-III. Trademark of RMI Titanium Company (USA) for a heat-treatable beta-titanium alloy containing 11.5% molybdenum, 6% zirconium, and 4.5% tin. It has a microstructure composed of alpha particles in a transformed beta matrix. Processed by extrusion, it has high strength, excellent resistance to oxidation and corrosion, good fatigue resistance, and a beta-transus temperature of 760°C (1400°F). Used mainly for high-strength aerospace applications. See also beta-titanium alloys.

beta americium. See americium.

beta arsenic. See arsenic.

beta berkelium. See berkelium.

beta beryllium. See beryllium.

beta boron. See boron.

Betabrace. Trademark of Essex Specialty Products, Inc. (USA) for a structural reinforcing film matrix consisting of a high-strength glass fabric and an adhesive, heat-curing polymer. It is used to strengthen sheet metals or plastics, improve distortion of thin substrates and/or deaden sound and control vibration. Used for automotive applications, such as quarter-panel or deck-lid reinforcements.

beta brass. A copper-zinc alloy with a nominal composition of about 37-57% copper and 63-43% zinc. At low temperatures, it is relatively brittle and hard, and has a body-centered-cubic (bcc) crystal structure. However, ductility increases significantly at temperatures above about 460°C (860°F), thus providing good hot-working properties. At temperatures between about 250 and 460°C (482 and 860°F), the so-called "beta prime phase" (β') is present which has a nominal composition of about 45% copper and 55% zinc, is very ductile, and possesses a simple cubic structure. Abbreviation: β-brass.

Beta-C. Trademark of RMI Titanium Company (USA) for a heat-treatable beta-titanium alloy containing 8% vanadium, 6% chromium, 4% molybdenum, 4% zirconium, and 3% aluminum. It has a microstructure composed of alpha particles in a transformed beta matrix. Processed by forging or hot rolling, it has high strength, excellent resistance to oxidation, sulfide stress cracking and hot-chloride crevice corrosion, and a beta-transus temperature of 732°C (1350°F). Used in metal-matrix composites, and in the form of extruded piping for geothermal and sour-gas environments. See also beta-titanium alloys.

beta calcium. See calcium.

beta cadmium iodide. See cadmium iodide (ii).

beta-cellulose. That fraction of the total carbohydrate component of a wood substance that is empirically defined as soluble in 17.5% aqueous sodium hydroxide (NaOH) at 20°C (68°F). It is subsequently re-precipitated on neutralization of the sodium hydroxide solution. Abbreviation: β-cellulose.

beta cerium. See cerium.

Beta-CEZ. Trademark of RMI Titanium Company (USA) for a heat-treatable beta-titanium alloy containing 5% aluminum, 2% tin, 4% molybdenum, 4% zirconium, and 1% iron. It has a microstructure composed of equiaxed coarse primary alpha particles and fine secondary alpha particles in a transformed beta matrix. Processed by solution treatment, it has high strength, excellent resistance to oxidation and corrosion, good fatigue resistance, and a beta-transus temperature of 890°C (1634°F). Used for high-strength aerospace applications. See also beta-titanium alloys.

beta cobalt. See cobalt.

Betacote. (1) Trademark of Lilly Industries, Inc. (USA) for urethane coatings, and chlorinated rubber-based maintenance paints. The former are used for flameproofing wood, and the latter for the protection of metals, wood, masonry, etc., against acids, alkalies and solvents.

(2) Trademark of E.C.C. America Inc. (USA) for a specialty clay for coating paper products.

(3) Trademark of Essex Chemical Corporation (USA) for special coatings for terrazzo and cement floors.

Betacryl II. Trade name for a heat-cure denture acrylic.

beta curium. See curium.

beta dysprosium. See dysprosium.

betafite. A black, brown, greenish, or yellow mineral of the pyrochlore group composed of calcium uranium niobium oxide hydroxide, $(Ca,Na,U)_2(Nb,Ta,Ti)_2O_6(O,OH)$. Crystal system, cubic. Density, 4.15-4.35 g/cm^3; refractive index, above 1.9. Occurrence: Canada (Ontario), Madagascar. Also known as *ellsworthite; hatchettolite*.

Betaflex. Trademark of Cork Manufacturing Company (Canada) Inc. for cellulose fibers with synthetic binders.

beta gadolinium. See gadolinium.

Betaglas. Trademark for rigid polyvinyl chloride films.

beta hafnium. See hafnium.

beta iron. See iron.

beta lanthanum. See lanthanum.

beta manganese. See manganese.

Betamate. Trademark of Essex Specialty Products Inc. (USA) for a series of structural epoxy adhesives used for bonding precoated metals, thermoplastics, and fiberglass-reinforced plastics.

beta molybdenum monoboride. See molybdenum boride (ii).

beta neodymium. See neodymium.

beta neptunium. See neptunium.

beta plutonium. See plutonium.

beta polonium. See polonium.

beta praseodymium. See praseodymium.

beta promethium. See promethium.

beta quartz. Quartz formed at temperatures between 573 and 870°C (1063 and 1598°F). Abbreviation: β-SiO$_2$. See also high quartz.

Betaseal. Trademark of Essex Specialty Products, Inc. (USA) for urethane adhesive used for bonding car windshields.

beta samarium. See samarium.

beta scandium. See scandium.

beta selenium. See selenium.

beta sodium. See sodium.

beta strontium. See strontium.

beta sulfur. See sulfur.

beta terbium. See terbium.

beta thorium. See thorium.

beta thallium. See thallium.

Betathane. Trademark of Essex Specialty Products, Inc. (USA) for high-quality castable urethane elastomers with high toughness and resilience, excellent abrasion, corrosion and impact resistance, and excellent resistance to solvents and oil. Also included under this trademark are polyurethane prepolymers and curing agents.

beta tin. A pure solid, silver-white room-temperature modification of *tin* that at 13°C (55°F) undergoes a crystallographic change ("tin pest") transforming to a gray powder (alpha tin). Abbreviation: β-Sn. Also known as *white tin*.

beta titanium. An allotropic form of *titanium* with body-centered cubic crystal structure that is stable above 880°C (1616°F). Abbreviation: β-Ti.

beta-titanium alloys. A class of hardenable titanium alloys containing varying amounts of elements, such as aluminum, chromium, iron, molybdenum, tin, vanadium and/or zirconium. Their beta phase is completely retained on water quenching, and their crystal structures at room temperature are body-centered cubic. Some of the most common of these alloys are: Ti-13V-11Cr-3Al, Ti-11.5Mo-6Zr-4.5Sn and Ti-8Mo-8V-2Fe-3Al. Important properties of *beta-titanium alloys* include very high strength up to approximately 310-320°C (590-608°F), high hardenability, low toughness and fatigue strength, increased brittleness at temperatures below -73°C (-100°F). In the solution-treated condition, *beta-titanium alloys* have good ductility and toughness, excellent forgeability and good formability, but relatively low strength. Following solution treating, they are usually aged at about 450-650°C (850-1200°F) to improve strength properties. Used for aerospace applications, such as rocket motor cases, airframe springs, and aircraft fasteners, and for structural applications. Abbreviation: β-Ti alloys. See also alpha-titanium alloys.

beta uranium. One of the three allotropic forms of *uranium* (the other two being alpha and gamma uranium) which has a tetragonal crystal structure, and is stable between 667°C and 775°C (1232°F and 1427°F). Abbreviation: β-U.

beta ytterbium. See ytterbium.

beta yttrium. See yttrium.

beta zirconium. A high-temperature form of *zirconium* with body-centered cubic crystal structure. Abbreviation: β-Zr.

betekhtinite. A pinkish gray mineral composed of copper iron lead sulfide, $Cu_{10}(Pb,Fe)S_6$. Crystal system, orthorhombic. Density, 5.96 g/cm³. Occurrence: Turkey.

Bethadur. Trade name of Bethlehem Steel Corporation (USA) for an extensive series of austenitic, ferritic and martensitic stainless steels, and several heat-resisting steels.

Bethalloy. Trade name of Bethlehem Steel Corporation (USA) for a tough, oil-hardening tool steel (AISI type L6) containing 0.75% carbon, 0.75% manganese, 0.3% silicon, 0.9% chromium, 0.4% molybdenum, 1.8% nickel, and the balance iron. It has excellent resistance to decarburization, a great depth of hardening, and good wear resistance. Used for tools, reamers, taps, broaches, and slitting cutters.

Bethalon. Trade name of Bethlehem Steel Corporation (USA) for a series of *austenitic stainless steels* of the 18-8 type. They have excellent corrosion resistance, good machinability, and are used for screw-machine parts, pump parts, valve components, etc.

bethanized product. An iron or steel product with a surface layer of very pure zinc produced by electrogalvanizing.

bethanized steel. A steel wire coated with very pure zinc in a process involving electrodeposition and subsequent passing through a die. The resulting wire can be worked and deformed, and is well suited for severe weather and handling conditions.

Bethcon Jetcoat. Trade name of Bethlehem Steel Corporation (USA) for a gray, spangle-free hot-dip galvanized steel sheet in which the zinc is combined with the steel. It has good high-temperature properties, weldability, corrosion resistance, and paintability, and is used for corrosion-resistant parts, roofing, structural parts, etc.

Beth-Cu-Loy. Trade name of Bethlehem Steel Corporation (USA) for a carbon steel containing 0.2% carbon, 0.20% or more copper, and the balance iron. It has high resistance to atmospheric and seawater corrosion, and is used for roofing, siding, and sheetmetal construction.

Beth-Cu-Loy PC. Trade name of Bethlehem Steel Corporation (USA) for polymer-coated *Beth-Cu-Loy* sheet steel.

Beth-Led. Trade name of Bethlehem Steel Corporation (USA) for a resulfurized free-machining steel containing 0.15-0.35% lead. Used for screw-machine parts.

Bethlehem. Trade name of Bethlehem Steel Corporation (USA) for an extensive series of steels including various high-speed, hot-work, shock-resisting and water-hardening tool steels, numerous free-machining, magnet, spring and gear steels, several manganese steels, and a wide range of wear-resistant and stainless steels.

Bethnamel. Trade name of Bethlehem Steel Corporation (USA) for a low-carbon steel with good drawing properties, used for porcelain enameled articles.

BethStar. Trademark of Bethlehem Steel Corporation (USA) for a family of high-strength low-alloy (HSLA) plate steels with low carbon and sulfur contents, high toughness, excellent weldability and formability, and good abrasion, impact and fatigue resistance. Used for machine parts, construction machinery, mining equipment, frame parts, large crane and offshore structures, crane and excavator booms, outriggers, off-road hauler frames, and dump bodies and canopies.

betpakdalite. A lemon-yellow mineral composed of calcium iron hydrogen molybdenum oxide arsenate decahydrate, $CaFe_2H_8(MoO_4)_5(AsO_4)_2 \cdot 10H_2O$. Crystal system, monoclinic. Density, 3.00 g/cm³; refractive index, 1.821. Occurrence: Kazakhstan.

beudantite. A black, brownish-yellow, or dark green mineral of the alunite group composed of lead iron arsenate sulfate hydroxide, $PbFe_3(AsO_4)(SO_4)(OH)_6$. Crystal system, rhombohedral (hexagonal). Density, 4.05 g/cm³; refractive index, 1.957. Occurrence: Germany.

beusite. A reddish brown mineral composed of manganese iron phosphate, $(Mn,Fe)_3(PO_4)_2$. Crystal system, monoclinic. Density, 3.70 g/cm³; refractive index, 1.703. Occurrence: Argentina.

Bevelite. Trade name of C-E Glass (USA) for a patterned glass with alternate angular planes, similar to louvers in Venetian blinds.

Bexan. (1) Trade name for vinyon (vinyl chloride) fibers with excellent acid, alkali and mildew resistance. Used for textile fabrics and carpets.

(2) Trade name for saran (polyvinylidene chloride) fibers with good chemical resistance, good weatherability, and good resistance to mildew and insects. Used for carpets and rugs, draperies, upholstery, clothing, industrial fabrics, etc.

Bexfoam. Trade name of BX Plastics (UK) for polystyrene foam.

Bexloy. Trademark of E.I. DuPont de Nemours & Company (USA) for a series of reinforced thermoplastic polyethylene terephthalate (PET) products used for automotive components (e.g. fenders). Also included under this trademark are various ionomers and several impact- and chemical-resistant polyester/thermoplastic elastomer blends for automotive components, such as door frames, bumpers and fenders.

Bexoid. Trademark of BX Plastics (UK) for cellulose acetate and cellulose nitrate plastics used for molded articles, such as toys and spectacle frames, and film products.

Bexone. Trade name of BX Plastics (UK) for polyvinyl chloride plastics.

BeXor. Trademark of Kusan Inc. (USA) for a family of biaxially oriented thermoformable, thick-gauge plastic sheet products based on polypropylenes.

Bexphane. Trademark of BIP Limited (UK) for chemically inert polypropylene plastics that have excellent electrical and fatigue properties and good resistance to heat distortion, but are susceptible to ultraviolet radiation. Used for television cabinets, bottles, luggage, etc. Also included under this trade name are plastic packaging and wrapping films and sheets and molded goods made of various thermoplastic and thermosetting resins.

Bexthene. Trade name of BX Plastics (UK) for polyethylene and polypropylene film materials.

Bextrene. Trade name of Bakelite Xylonite Limited (UK) for polystyrenes.

beyerite. A white or yellow mineral of the bismuthite group composed of calcium bismuth oxide carbonate, $CaBi_2O_2(CO_3)_2$. Crystal system, tetragonal. Density, 6.51 g/cm^3; refractive index, 2.13. Occurrence: Germany, India, USA (California).

bianchite. A colorless, or white mineral of the hexahydrite group composed of zinc iron sulfate hexahydrate, $(Zn,Fe)SO_4 \cdot 8H_2O$. Crystal system, monoclinic. Density, 2.03 g/cm^3. Occurrence: Poland.

Biasill. Trademark of E.I. DuPont de Nemours & Company (USA) for a fine, less aggressive grade of *Starblast* blasting abrasives made from *staurolite* sand. Used for blasting thin metal pieces of soft substrates such as aluminum.

biaxial crystal. See optically biaxial crystal.

biaxial fabrics. Essentially warp-knitted fabrics with extra warp threads in vertical direction and weft threads in horizontal direction added over the entire width and length. See also multiaxial fabrics.

biaxial film. A polymer film with a molecular orientation along two axes (i.e., lateral and longitudinal), i.e., it has high strength in both directions.

bibiru. See greenheart (1).

bible paper. A thin, strong, opaque printing paper usually containing linen or cotton rags. Also known as *India paper*.

Bicaloy. Trade name of British Insulated Callender's Cables (UK) for a wear- and deformation-resistant copper alloy used for resistance welding electrodes.

bicarburetted hydrogen. See ethylene.

bicchulite. A white gray mineral composed of calcium aluminum silicate hydroxide, $Ca_2Al_2SiO_6(OH)_2$. It can be made synthetically by hydrothermal treatment of *gehlenite* ($Ca_2Al_2SiO_7$) at 260-650°C (500-1200°F). Crystal system, cubic. Density, 2.75 g/cm^3; refractive index, 1.628. Occurrence: Japan.

Biciflex. Trade name for biaxially-oriented polystyrene sheeting.

bicomponent fiber. A synthetic fiber made from two chemically and/or physically different polymers by spinning and joining them in a concurrent process from one spinneret. Although composed of various different polymers, a bicomponent fiber is commonly very fine, has a unique cross section, and is thermal bonding and self bulking. Also known as *bigeneric fiber; bilateral fiber; conjugate fiber; matrix fiber.*

bicomponent yarn. A yarn that consists of two different components. Examples include yarns spun from physically or chemically different polymers, yarns made from a staple fiber yarn and a filament yarn, plied yarns, and core-spun yarns.

biconstituent fiber. (1) A composite fiber containing a dispersion of fibrils of one synthetic material within, and axially to another. Also known as *bicomponent fiber; matrix fiber.*

    (2) A fiber made up of two or more materials, e.g., a polymer fiber with a metal filament.

Bicop. Trade name for a commercially pure, oxygen-free high-conductivity copper used for electrical motors and generators.

Bicor. Trademark for biaxially oriented polypropylene film.

Bideny metal. An alloy containing 88.5% zinc and a total of 11.5% copper and lead. Used for domestic utensils.

bidalotite. See anthophyllite.

bideauxite. A colorless mineral composed of silver lead chloride fluoride, $Pb_2AgCl_3(F,OH)_2$. Crystal system, cubic. Density, 6.27; refractive index, 2.192. Occurrence: USA (Arizona).

bidery. A free-cutting alloy of 48.5% copper, 33.3% zinc, 6.05% tin, and 12.15% lead. Used for buttons and ornaments.

bidirectional fabrics. Fabrics made of fibers, filaments or yarns with a bidirectional (two-directional) weave. Also known as *two-directional fabrics.*

bidirectional laminate. A plastic laminate whose fibers are oriented in two directions. Also known as *cross laminate.*

bieberite. A pink to red mineral of both the melilite and melanterite group composed of cobalt sulfate heptahydrate, $CoSO_4 \cdot 7H_2O$. It can also be made synthetically. Crystal system, monoclinic. Density, 1.96 g/cm^3; melting point, 96.8°C (206°F); refractive index, 1.482. Used as a source of cobalt, and in ceramics, pigments and glazes. Also known as *cobalt vitriol; red vitriol; rose vitriol.*

Biermann bronze. German trade name for a strong, corrosion-resistant tungsten bronze containing 95% copper, 3.4% tin and 1.6% tungsten.

Bifil. Trade name for a series of nylon filaments and fibers.

Bifix. Trade name for a dual-cure dental cement.

Bifix DC. Trade name for a dual-cure dental resin cement for luting applications.

bigeneric fiber. See biconstituent fiber.

Big J. Trade name of Joseph Jackman & Company Limited (UK) for a high-speed steel containing 0.7% carbon, 18% tungsten, 4% chromium, 1% vanadium, and the balance iron. Used for high-speed tools, such as cutters, reamers, broaches, taps, drills, tool bits, etc.

bigleaf mahogany. The reddish-brown wood of the West African mahogany tree *Khaya grandifolia.* Used for furniture, paneling, and veneer. See also African mahogany; khaya.

bigleaf maple. The reddish-brown wood of the maple tree *Acer macrophyllum.* Average weight, 545 kg/m^3 (34 lb/ft^3). Source: Western United States. Also known as *broadleaved maple; Oregon maple.* See also maple.

Big Porcelain. Trade name of Aardvark Clay & Supplies (USA) for a porcelain clay (cone 10) that has been enhanced for throwing larger forms.

Bijohai. Trade name for rayon fibers and yarns used for textile fabrics.

bijvoetite. A yellow mineral composed of lanthanum uranyl carbonate hydroxide hydrate, $(La)_2(UO_2)_4(CO_3)_4(OH)_6 \cdot 11H_2O$. Crystal system, orthorhombic. Density, 3.95 g/cm^3; refractive index, 1.650. Occurrence: Zaire.

Bikalith. Trademark of American Potash & Chemical Corporation (USA) for a series of lithium silicate minerals including *eucryptite* (α-LiAlSiO$_4$), *lepidolite* [K(Li,Al)$_3$(Si,Al)$_4$O$_{10}$(OH,F)$_2$], and *spodumene* (α-LiAlSi$_2$O$_6$). Used in glass manufacture and in ceramics.

bikitaite. A colorless, or white mineral of the zeolite group composed of lithium aluminum silicate monohydrate, $LiAlSi_2O_6 \cdot H_2O$. Crystal system, triclinic. Density, 2.29 g/cm^3; refractive index, 1.521. Occurrence: Zimbabwe.

Bilacryl PU. Trade name of Monopol AG (Switzerland) for an extensive series of polyurethane coating and painting products including enamel and one-coat paints, glossy, matte and textured finishes, and primers.

Biladur EP. Trade name of Monopol AG (Switzerland) for an extensive series of epoxy coating and painting products including several primers and hardeners, and clear and metallic gloss finishes.

Bilafloor. Trade name of Monopol AG (Switzerland) for an epoxy floor and wall paint.

Bilame. Trade name of Creusot-Loire (France) for a series of nickel-alloy bimetals used for fire detectors and thermostats.

bilateral fiber. See bicomponent fiber.

Bilgen bronze. A German corrosion-resistant, electrical bronze containing 1.9% tin, 0.5% iron, 0.2% lead, and the balance copper.

bilibinskite. A light brown to rose-brown mineral composed of copper gold lead telluride, $Au_3Cu_2PbTe_2$. Crystal system, cubic. Density, 14.27 g/cm^3. Occurrence: Russian Federation, Far East.

bilinga. The medium-heavy, yellow or orange-brown wood of the tree *Nauclea diderrichii*. It has very high durability, works fairly well, has a slight tendency to surface checking, and poor nailing quality (splits easily). Average weight, 740 kg/m^3 (46 lb/ft^3). Source: West Africa. Used for construction work, flooring, piling, etc. Also known as *kusia; opepe*.

bilinite. A white or yellowish mineral of the halotrichite group composed of iron sulfate hydrate, $Fe_3(SO_4)_4 \cdot 22H_2O$. Crystal system, monoclinic. Refractive index, 1.49. Occurrence: Czech Republic.

bilirubin. A pigment that is responsible for the characteristic brownish-yellow color of bile. It is available as an orange-red powder with a melting point of 192°C (378°F), and used in analytical chemistry, biochemistry, biotechnology, and medicine. Formula: $C_{33}H_{36}N_4O_6$.

biliverdin. A pigment produced by the conversion of hemoglobin and responsible for the greenish color of bile in certain animals. In man and many animals, it is reduced to bilirubin. It is usually commercially supplied as the dihydrochloride, and used in analytical chemistry, biochemistry, biotechnology, and medicine. Formula: $C_{33}H_{34}N_4O_6$.

bilayer membrane. A row or layer two molecules thick such as that formed in some phospholipids at the aperture between two aqueous solutions in which the two layers are lined up with their polar ends projecting into water.

bilayer photoresist. A photoresist that consists of a patternable, high-resolution thin-film top layer (typically 100-200 nm or 3.9-7.8 μin. thick), e.g., an acid-cleavable 4SiMA silicon monomer, and a thicker, etch-resistant underlayer on a silicon wafer. The top layer is imaged and the image is transferred to the underlayer by a process that selectively etches the underlayer, but neither the top layer nor the substrate. *Bilayer photoresists* are ideally suited for modern photolithography technologies which utilize excimer lasers having typical imaging radiation energies in the deep ultraviolet range (below 250 nm).

billet. (1) A piece of metal that has a cross-sectional area of 105-230 cm^2 (16 to 36 $in.^2$) and is usually longer and smaller in cross section than a *bloom*. In the preliminary stages of working steel, an *ingot* is gradually reduced to a bloom, billet and *slab* in the rolling mill. It is the wrought starting stock for making forgings or extrusions.

(2) A powder-metallurgy compact, green or sintered, suitable for subsequent working. Also known as *ingot*.

billet steel. Steel, continuously cast or reduced directly from ingots, that has been made by the open-hearth, basic-oxygen or electric-furnace process.

billiard cloth. A smooth, full, high-quality merino wool fabric, usually dyed green, in a plain or twill weave and with a fibrous surface finish. Used for billiard and card table tops.

billietite. A deep golden yellow mineral of the becquerelite group composed of barium uranyl hydroxide octahydrate, $Ba(UO_2)_6O_4(OH)_6 \cdot 8H_2O$. It can also be made synthetically. Crystal system, orthorhombic. Density, 5.28 g/cm^3; refractive index, 1.800. Occurrence: Zaire.

billingsleyite. A dark lead-gray mineral composed of silver arsenide sulfide, Ag_7AsS_6. It can also be made synthetically. Crystal system, orthorhombic. Density, 5.96 g/cm^3; refractive index, 1.825. Occurrence: USA (Utah).

Bi-Lof. Trademark for acrylic fibers used for the manufacture of textile fabrics.

Bi-Loft. Trademark of Monsanto Company (USA) for bicomponent fibers used for textile fabrics.

Biloy. Trade name of Enthone-OMI Inc. (USA) for a multilayer nickel-iron electrodeposit and plating process.

Biltmore ash. See American ash.

Bimax. Trademark of A&P Technology (USA) for braided biaxial *broad goods*.

Bi-Met. Trade name of Cleveland Black Oxide (USA) for a black oxide conversion coating.

bimetal. A laminate composed of two thin strips of dissimilar metals having different coefficients of thermal expansion (e.g., copper and zinc, zinc and iron, brass and iron, etc.) joined together by rolling, riveting, brazing, soldering, welding or plating. When heated the metal strip with the greater coefficient will expand more causing the entire laminate to bend or curl. Used in temperature indication, regulation and control, e.g., for thermocouples, thermometers, switches, contacts, thermostats, thermographs and relays. Also known as *bimetallic strip; thermometal*.

bimetallic strip. See bimetal.

Bimoco. Trademark of Bildstein, Mommer & Co. KG (Germany) for various semifinished lead products and lead goods including seals, balls, ball strip for curtains, fishing net weighting, fishing lead, lead weight for dresses, lead wire, fine lead shot, etc.

Binal. (1) Trade name of Alloy Technology International, Inc. (USA) for a sintered cast boron carbide.

(2) Trade name for an aluminum foil with 2% boron produced from the metal powders by blending, compacting, sintering, and hot rolling.

binary alloy. (1) An alloy composed of two principal components with at least one being a metal, e.g., iron and carbon, aluminum and silicon, copper and nickel, etc.

(2) See binary polymer blend.

binary compound. A compound, such as sodium chloride or zinc oxide, that consists of only two elements.

binary granite. (1) A granite containing both light *muscovite* mica and dark *biotite* mica. Also known as *two-mica granite*.

(2) A granite containing both *feldspar* and *quartz*.

binary polymer alloy. See binary polymer blend.

binary polymer blend. A miscible, immiscible, or isomorphic polymer blend resulting from the combination of two polymers, e.g., a blend of a polycarbonate and a polyester resin. Also known as *binary alloy; binary polymer alloy*.

binche lace. A durable lace with a machine-made mesh ground and randomly appliqued motifs usually depicting snowflakes. It is named after Binche, a town in Belgium where it was originally made.

binder. (1) A material added to a metal powder to facilitate bonding of the particles and enhance the strength of the resulting compact. Also known as *binding agent*.

(2) A substance, such as clay, natural or synthetic resins, linseed oil, molasses, or cereals, added to a foundry sand to hold the grains together and give the core or mold strength. Also known as *binding agent; bonding agent*.

(3) The nonvolatile portion, or film former, of a paint that serves to bind the pigment particles. Also known as *binding agent*.

(4) That constituent of an adhesive that is the major cause for the cohesive forces between two bodies. Also known as *binding agent*.

(5) The resin or cementing component of a plastic or polymer that binds the other constituent together. Also known as *binding agent*.

(6) The agent applied to preforms or glass mats to bond the fibers prior to molding or laminating. Also known as *binding agent*.

(7) The continuous phase of a reinforced plastic that binds the reinforcement together.

(8) The organic or inorganic bonding agent (usually a urea-formaldehyde or phenolic resin) added to the flakes, chips, sawdust, or other wood particles to bind them together into a strong particleboard. Also known as *binding agent*.

(9) A mixture of organic compounds, such as tallow, fatty acids, fish oil, waxes, etc., that is used as a carrier for the abrasive in buffing compounds. Also known as *binding agent*.

(10) A cementing medium added to a powder or granular material to give items formed from it workability and green or dry strength adequate for handling and machining in all stages prior to firing. It is usually expelled during sintering or firing. Also known as *binding agent*.

(11) A cementing material, such as hydrated cement or a product of cement or lime and reactive siliceous materials, used in building construction. Also known as *binding agent*.

binder twine. A strong, coarse string of sisal or jute used for binding and tying applications.

bindheimite. A mineral of the pyrochlore group composed of lead antimony oxide, $Pb_2Sb_2O_6(O,OH)$. It can also be made synthetically. Crystal system, cubic. Density, 7.32 g/cm^3; refractive index, 1.84. Occurrence: USA (Nevada).

binding agent. See binder.

binding brass. A free-cutting brass containing 63.25-63.5% copper, 35% zinc, and 1.5-1.75% lead. Used for automatic and screw machine products.

binding metal. An alloy containing about 3.7% antimony, 2.8% tin, and the balance zinc. Used for wire-rope slings.

B-Inlay. Trademark of Degussa-Ney Dental (USA) for a medium-hard, microfine dental alloy (ADA type II) containing 77% gold, 12% silver, and 1% palladium. It has a yellow color and a hardness of 135 Vickers. Used in restorative dentistry for inlays and single crowns.

Binormat. Trade name for a dental alloy that contains 21.7% platinum and 78.3% gold, and has a melting range of 925-945°C (1695-1735°F).

Bio 72. Trade name of World Alloys & Refining, Inc. (USA) for a biocompatible, palladium-free dental porcelain alloy (ADA type IV) containing 72.1% gold, 10.0% silver, and 9.2% platinum.

It has a pale yellow color, a density of 16.6 g/cm^3 (0.60 $lb/in.^3$), a melting range of 905-990°C (1650-1795°F), and low elongation. Used for porcelain-bonding, and crown and bridge restorations.

Bio 74. Trade name of World Alloys & Refining, Inc. (USA) for a biocompatible palladium- and copper-free dental porcelain alloy (ADA type III) containing 74.0% gold, 10.0% platinum, and 12.0% silver. It has a yellow color, a density of 17.0 g/cm^3 (0.61 $lb/in.^3$), a melting range of 950-1060°C (1740-1940°F), and a hardness of 155 Vickers. Used for porcelain-bonding, and crown and bridge restorations.

bioabsorbable polymers. Polymeric materials, such as polyhydroxybutyrate, polyethylene oxide, polycaprolactone, polylactic/glycolic acid copolymers, etc., that can be made to dissolve in the human body, and are gradually absorbed over time into the tissue. This makes them attractive for applications, such as sutures, vascular grafts, artificial ligament components, drug delivery systems, etc.

bioadhesives. A group of biocompatible polymeric adhesive materials including polymethyl methacrylates, cyanoacrylates, resorcinols, fibrin and molluscan glues, gelatin foams, oxidized regenerated celluloses and succinylated amyloses. Used to fix or secure bone structures or join tissue, especially in dentistry, orthopedics, ophthalmology and wound management.

bioactive ceramics. Specialized compositions of glasses, ceramics, glass-ceramics, ceramic composites, etc., that, due to the formation of a biologically active hydroxylcarbonate apatite surface layer, will bond to bone and soft tissue. See also bioceramics.

bioactive glasses. A group of dense, nonporous, surface-reactive soda-lime silica glasses that bond to bone and soft tissue. Typically, they contain 30-60% silica (SiO_2), 19-25% sodium oxide (Na_2O), 14-25% calcium oxide (CaO) and 6% phosphorus pentoxide (P_2O_5), and may also contain significant additions of calcium difluoride (CaF_2) or boron oxide (B_2O_3). The molar ratio of calcium to phosphorus is usually kept between 4-5:1 to facilitate chemical bonding to bone.

bioactive materials. Biomaterials that invoke specific biological responses at the material-tissue or material-bone interfaces resulting in bonds between the two materials. Many of the glasses, glass-ceramics, etc., used in body-part reconstruction or repair are bioactive and enhance the bone and soft-tissue adhesion.

Bio-Alcamid. Trademark of Polymekon Research Inc. (USA) for a polyalkylamide gel used in prosthetic and reconstructive plastic surgery.

Biobon. Trademark of Biomet Merck (Germany) for a resorbable microcrystalline calcium phosphate cement for the filling or reconstruction of bone defects. It sets endothermically at body temperature and mimics natural bone in chemical composition and crystal structure.

Bio C&B. Trade name for a medium-gold dental casting alloy used for crown and bridge restorations.

Bio-Cast. Trade name for a corrosion-resistant, biocompatible cobalt-chromium dental bonding alloy.

biocatalyst. See enzyme.

bioceramic coatings. Coatings of pyrolytic carbon, hydroxyapatite, bioactive glass, tricalcium phosphate, alumina, etc., applied to prostheses, bone replacements, artificial heart valves made of stainless steel (e.g., AISI type 316L), and cobalt-chromium or titanium-aluminum-vanadium alloys to improve biocompatibility, fatigue life, bond strength, and/or other properties.

bioceramics. A group of ceramic materials including alumina (e.g., sapphire and polycrystalline alumina), hydroxyapatite, glass, glass-ceramics, etc., that have been specifically engineered to reconstruct or replace damaged or diseased human body parts. See also bioactive ceramics; medical ceramics.

biocide. An antimicrobial agent that is sometimes added to a paint or other coating to slow down the rate of attack by microorganisms.

BiOcclus 4. Trademark of Degussa-Ney Dental (USA) for a microfine-grained, extra-hard dental alloy containing 85.8% gold and 11% platinum. It has a rich yellow color, high yield strength, and a hardness of 210 Vickers, polishes to a high luster, and is suitable for use with most conventional dental porcelains. Used in restorative dentistry, especially for bridges.

biocompatible materials. Biomaterials that when introduced into the human body produce a minimum degree of rejection. This biocompatibility is a prerequisite for implants, prostheses, sutures, vascular grafts, etc.

biocomposites. A group of composites that are suitable for application in the repair and reconstruction of injured or diseased body parts. Depending on the matrix and/or reinforcing phase, they may essentially be: (i) nearly inert, e.g., carbon fiber-reinforced polymers (polyethylene, polysulfone, etc.), or alumina- or stainless steel-reinforced epoxies; (ii) bioactive, e.g., titanium or stainless steel fiber reinforced glasses, hydroxyapatite-reinforced polyethylene or collagen, or transformation-toughened glass-ceramics (e.g., zirconia); or (iii) resorbable, e.g., hydroxyapatite-reinforced polyhydroxybutyrate or polylactic/glycolic acid copolymers. See also bioactive ceramics; bioactive materials; resorbable biomaterials.

Biocryl C. Trade name of Scheu-Dental (Germany) for break-resistant, monomer-free round blanks of pure polymethyl methacrylate (PMMA) that readily bond to acrylics, and are supplied in clear, rose-opaque, rose-clear, rose-transparent, red, blue and yellow colors with 125 mm (5 in.) in diameter, and 1.5, 2.0 or 3.0 mm (0.06, 0.08 or 0.1 in.) in thickness. Used for denture bases, immediate dentures, orthodontic retainers and splints.

Biocryl M. Trade name of Scheu-Dental (Germany) for hard elastic, monomer-free round plates of laminated acrylic copolymers supplied with a diameter of 125 mm (5 in.) and a thickness of 2.0 mm (0.08 in.). They readily bond to acrylics, and are available in the following multicolor grades: camouflage, polka dots, rainbow, tiger, and zebra. Used for orthodontic plates and retainers.

Biocryl-Resin. Trade name of Scheu-Dental (Germany) for a special acrylic resin used in orthodontic plates and retainers for attaching clasp arms and expansion screws.

biodegradable polymers. Natural or synthetic polymers, such as polycaprolactones, polyglycolates, polyhydroxybutyrates, polyorthoesters, or polylactic/glycolic acid copolymers, that are susceptible to being broken down by microorganisms and enzymes. Their physical integrity and useful properties are lost during the degradation process. These polymers are used as environmentally friendly alternatives to regular polymers. Also, the property of "degradability" makes them useful in the biomedical field, e.g., for soft-tissue reconstruction, vascular graft coatings, sutures, bone plates and other implants, sustained release devices, medical packaging, etc.

biodendrimers. A group of biocompatible or biodegradable polymers that are tree-like, globular-shaped substances made by reacting acids with alcohols to form esters. They are consid-

ered for use in biomedical implants for orthopedic, pharmaceutical and surgical applications. See also dendrimers.

BioDur. Trademark of Carpenter Technology Corporation (USA) for nonmagnetic, biocompatible vacuum-arc-remelted austenitic stainless steels (AISI type 316LS) containing up to 0.08% carbon, 17.0-23.5% chromium, 0-15.5% nickel, 0.5-3.5% molybdenum, 0-24.0% manganese, 0-1.0% silicon, 0-0.5% copper, 0-1.1% nitrogen, 0-0.8 niobium, 0-0.3% vanadium, 0-0.03% sulfur, 0-0.04% phosphorus, and the balance iron. Supplied in as-annealed and as-cold-worked form, they have excellent long-term corrosion and wear resistance, high strength and fatigue resistance, superior formability, and are not affected by X-rays and magnetic resonance imaging. Used for hip and knee replacement prostheses and other medical implants. A low-nickel grade, *BioDur 108*, with up to 0.08% carbon, 19-23% chromium, 21-24% manganese, up to 0.05% nickel, and 0.85-1.1% nitrogen, is also supplied for medical applications, such as implanted orthopedic devices (e.g., bone screws, bone plates, etc.) to prevent allergic reactions to nickel.

BioDur CCM Plus. Trademark of Carpenter Technology Corporation (USA) for nonmagnetic, biocompatible, wrought cobalt-base powder-metallurgy superalloys containing 26.0-30.0% chromium, 5.0-7.0% molybdenum, 0.2-0.3% carbon, and 0.15-0.20% nitrogen. Supplied in as-annealed and as-hot-worked form, they have excellent long-term corrosion and wear resistance, high strength and fatigue resistance, superior formability, and are not affected by X-rays and magnetic resonance imaging. Used for medical implants.

bioengineered materials. A group of materials that have been made by applying the principles of bioengineering–a discipline that combines biology, genetics and/or medicine with engineering. Examples of such materials include modified biological substances, e.g., crosslinked collagen for tissue engineering, genetically engineered materials, and fully synthetic biological substances, e.g., synthetic hydroxyapatite for body part repair or replacement. The term *bioengineered materials* also includes plants and crops that have been genetically or otherwise modified to produce bigger and better yields.

bioengineered polymers. A group of polymers that includes polysaccharides and proteins which have been modified by applying the principles of biology and genetic engineering to polymer science. They are used in tissue engineering, body part repair and replacement and drug delivery.

bioengineered spider silk. See spider silk (2).

Bio Ethic. Trade name of Cendres & Métaux (Switzerland) for a biocompatible, yellow-colored gold-platinum dental alloy used for metal-ceramic restorations. Also known as *CM Bio Ethic*.

Biofix. Trademark of Bionx Corporation (USA) for biodegradable polymers that are based on poly(L-lactide) or poly(glycolide), and are used for medical fixation devices, such as surgical pins and rods.

Bio-Fresh. Trade name of Sterling Fibers, Inc. (USA) for an acrylic fiber that contains "Triclosan," a chlorinated phenoxy antimicrobial compound which inhibits bacteria, fungi and yeast growth. Available in a range of deniers and colors, it is soft, comfortable and colorfast, wicks away moisture, and has good shape retention and excellent resistance to wrinkles and stains. Used for wearing apparel, e.g., shirts, pants, outerwear, underwear, sweaters, scarves, socks and gloves, sleeping bags, household textiles, and a wide range of industrial, medical and technical applications.

Bioglass. Trademark of the University of Florida at Gainesville

(USA) for a series of bioactive glasses composed of about 30-55% silicon dioxide (SiO_2), 19-25% sodium oxide (Na_2O), 14-25% calcium oxide (CaO), and 6% phosphorus pentoxide (P_2O_5). Some may also contain calcium difluoride (CaF_2) or boron oxide (B_2O_3), and/or can be reinforced with stainless steel fibers. Used in human body part reconstruction and repair (e.g., surgical and dental implants, bone replacements, bone screws, rods and pins, orthopedic coatings, etc.). Also included under this trademark are bioactive glass-coated or laminated ceramic and metals.

Biogran. Trade name for bioactive glass granules used in dentistry.

Bio-Herador. Trade name of Heraeus Kulzer Inc. (USA) for biocompatible dental casting alloys.

Bio Heragold B. Trade name of Heraeus Kulzer Inc. (USA) for a biocompatible dental casting alloy containing 86.2% gold, 11.5% platinum and 1.5% zinc. Used for crowns and bridges.

biohybrid graft. A hybrid implant that contains living and non-living materials.

Bio-Krome. Trade name BL Dental Company (USA) for a chromium-cobalt alloy used for dental applications.

Biokryl. Trademark for acrylic fibers used for the manufacture of textile fabrics.

Bio-Lastic. Trade name Tnemec Company, Inc. (USA) for waterborne industrial and architectural acrylate coatings containing mildewcides.

Bioline. Trade name of Sandvik Steel (USA) for a molybdenum-alloyed stainless steel supplied in AISI 316LVM and High-Nitrogen grades. The latter are ultraclean steels with high yield strength and corrosion resistance. *Bioline* stainless steels are used for surgical tools, coronary stents, bone fracture management systems, forms for making permanent orthopedic implants, etc.

biological materials. Materials, such as bone and dentin, that occur in living organisms.

Bio-Maingold. Trade name of Heraeus-Kulzer, Inc. (USA) for a biocompatible gold-base dental casting alloy.

biomaterials. (1) Natural substances, such as carbohydrates, collagen, proteins, bone, dental enamel, casein, starch, cellulose, vegetable gums, etc., that are produced in biological systems, i.e., plants and animals.

(2) Engineered organic or inorganic materials, such as aluminum and titanium and their alloys, stainless steel, cobalt-base superalloys, carbon and certain ceramic and polymeric materials (e.g., aluminum oxide, polyamide, polyurethane, polycarbonate, etc.), that are noncorrosive, biocompatible and usually nondegradable, and can thus be used for biological applications, and in the manufacture of surgical implants for the human body, e.g., hip and knee replacements, artificial limbs, and mastectomies.

Biomax. Trademark of E.I. DuPont de Nemours & Company (USA) for a family of soft, pliable hydro- and biodegradable polyester resins based on polyethylene terephthalate (PET). They can be made into resins, films and fibers, and processed by standard processes including injection molding and thermoforming. Used for disposable cutlery, bottles, cups and trays, blister packs, yard waste bags, disposable diaper sheets, coated paper products, lidding and aluminized food-packaging films, agricultural films, plant pots and bags, hot-melt adhesives, etc.

biomedical materials. A group of synthetic or natural materials including various compositions of metals, polymers, ceramics, composites, modified animal or human tissue, etc., that have been specially designed to be suitable for application in medical devices (e.g., heart valves and parts, catheters, contact lenses, pacemakers, etc.), dental and orthopedic implants, prostheses, sutures, orthopedic wire, etc.

BioMEMS. A hybrid implant material or drug-delivery device that is essentially a microelectromechanical system (MEMS) incorporating micro- and nano-scale features. Examples include (i) cell encapsulation devices with nanometer-sized pores, fabricated by micromachining, that allow small molecules like hormones to pass but prevent large molecules like antibodies to do so; and (ii) micro-textured tissue scaffolds for growing cells in culture. See also MEMS materials.

Biomer. (1) Trademark of Ethicon, Division of Johnson & Johnson (USA) for a biocompatible segmented polyurethane similar in properties to *spandex*, and used in artificial heart implants, and self-cure dental resin cements. This material is no longer available.

(2) Trade name of Dentsply Caulk (USA) for a dental resin cement used for luting applications.

biomimetic ceramics. A group of engineered ceramic materials obtained by imitating natural or biological fabrication principles (e.g., the formation of sea shells, bone or dental enamel) by low-temperature, aqueous synthesis of carbonates, oxides, phosphates, sulfides, etc.

biomimetic gels. A relatively new class of polymer gels that due to the movement of individual gel loops in an electric field with varying polarity can shrink or expand and imitate biological activities. See also polymer gel.

biomimetic materials. Engineered ceramics, polymers and gels that either mimic the natural environment (human or animal body) in which cells grow in the regeneration or repair of organs and tissues, or imitate the natural or biological fabrication principles, e.g., the formation of bone or dental enamel.

biomimetic polymers. A group of novel biodegradable polymers, such as polyesters (e.g., polylactic and polyglycolic acids and their copolymers) that mimic the natural environment in which cell growth occurs.

biomineral. A mineral that either occurs naturally in or on humans and animals (e.g., bone, teeth, coral, etc.), or is made synthetically to repair or replace damaged human or animal body parts (e.g., hydroxyapatite).

biomolecular material. An engineered biomaterial based on organic macromolecules found in a living organism, e.g., a material engineered from deoxyribonucleic acid (DNA) and protein molecules.

Bionate. Trademark of Polymer Technology Group, Inc. (USA) for a series of thermoplastic elastomers that are products of a reaction between an aromatic diisocyanate, a hydroxyl-terminated polycarbonate, and a glycol chain extender. They have good oxidative stability, good biocompatibility and biostability, excellent mechanical strength, and good abrasion resistance, and are usually supplied in liquid, solid, pellet or crumb form. Used as biomaterials for stents, ventricular assist devices, pacemaker leads, catheters, etc.

Bioney. Trademark of Degussa-Ney Dental (USA) for a copper and palladium-free dental alloy containing 90% gold, 7.25% platinum, and 1% silver. It has a brilliant gold color, a light oxide layer, good strength, and a hardness of 180 Vickers. Used for porcelain restorations.

Bionic Bubble. Trademark of Sphere Services Inc. (USA) for strong alumina-silica spherical fillers and extenders with good flowability supplied in particle sizes of 75-125 μm (3-5 mils).

Used particularly for spray paint and coating applications. See also Cenospheres.

Bio-Oss. Trademark of Osteohealth Company (USA) for a bioengineered graft material that has a structure similar to human bone, and is used for dental implants. Once implanted it supports new bone growth and preserves bone and overlying soft tissue.

Biopaque. Trade name for a biocompatible dental porcelain for metal-to-ceramic bonding.

Bioplast. Trade name of Scheu-Dental (Germany) for a soft, pliable rubber supplied in round and square shapes with a diameter of 125 mm (5 in.), or a size of 125 mm × 125 mm (5 in. × 5 in.), respectively, in thicknesses of 1.0, 1.5, 2.0, 3.0 and 4.0 mm (0.04, 0.06, 0.08, 0.12 and 0.16 in.). Used in dentistry for night guards, mouth protectors, model duplications, etc.

Bioplast Color. Trade name of Scheu-Dental (Germany) for *Bioplast* supplied in multicolored rounds with a diameter of 125 mm (5 in.) and a thickness of 3.0 (0.12 in.). Used in dentistry for night guards, mouth protectors, model duplications, etc.

Biopol. Trade name of Monsanto Company (USA) for biodegradable copolymers that are based on polyhydroxybutyrate (PHB) or polyhydroxyvalerate (PHV), and are used for medical devices.

biopolymers. (1) Polymeric substances, such as carbohydrates and proteins, that are produced in biological systems (i.e., plants and animals). See also natural polymers.

(2) A group of synthetic polymers, usually derived by the action of bacteria on carbohydrates, that can be highly fluid or viscous. Fluid biopolymers are used in oil-well drilling muds and as thickeners, and the more viscous grades in orthopedic and surgical devices, slow-release systems for drugs, packaging, etc.

(3) A group of biodegradable or biostable synthetic polymers including poly(lactic acid) and poly(glycolic acid), poly(dioxanone), poly(ε-caprolactone) homopolymers and copolymers, poly(trimethylene carbonate) copolymers, polyorthesters, polyanhydrides and polyphosphazenes. They are used in tissue engineering for sutures and wound dressings, and in various medical devices.

BioPorta G. Trademark of Wieland Dental + Technik GmbH & Co. KG (Germany) for a yellow-colored, extra-hard gold-platinum dental ceramic alloy.

BioPortadur. Trademark of Wieland Dental + Technik GmbH & Co. KG (Germany) for a biocompatible, yellow, extra-hard gold-based dental casting alloy.

Bioram-M. Trade name for a dental ceramic for CAD/CAM (computer-aided design/computer-aided manufacturing) inlays.

bioresorbable polymers. A class of biocompatible, synthetic polymers, such as glycolic acid, L-lactic acid or their copolymers, that degrade biologically in the human body (e.g., by hydrolysis) over time, and are gradually replaced by tissue.

Bioscrew. Trade name of Linvatex (USA) for a biodegrable polymer based on poly(L-lactide), and used for medical fixation devices, such as surgical screws.

Bio-Sil. Trade name of Sil-Med Corporation (USA) for high-quality silicone rubber products used in medicine and diagnostics.

Biosil. (1) Trademark of Degussa-Ney Dental (USA) for hard and extra-hard cobalt-chromium-molybdenum alloys for partial dentures.

(2) Trademark of BioPlexus Corporation (USA) for elastomeric silicone sheeting for subcutaneous implantation.

BioSpan. Trademark of Polymer Technology Group, Inc. (USA) for a series of translucent, highly elastomeric segmented polyurethanes (SPUs) based on aromatic polyetherurethaneurea with a hard segment of diphenylmethane diisocyanate and mixed diamines, and a soft segment of polytetramethylene oxide. They have an average molecular weight of 180000 daltons, and exhibit excellent mechanical properties including high strength and flexibility, a Shore hardness of about 70 A, excellent wear and fatigue resistance, excellent biocompatibility, biostability and thromboresistance, and low water absorption. Used for left ventricular assist devices (LVADs), vascular grafts and prostheses, and total artificial hearts.

BioSpan C. Trademark of Polymer Technology Group, Inc. (USA) for a biocompatible segmented polyurethane (SPU) that has an aromatic urethane hard segment and an aliphatic polycarbonate soft segment. It has excellent mechanical strength and elasticity, high flex life, and very good oxidative stability, and is supplied as a solution in dimethyl acetamide (DMAc). Used for long-term implants, such as stents, pacemaker leads, or catheters which require maximum oxidative stability.

BioSpan CS. Trademark of Polymer Technology Group, Inc. (USA) for a biocompatible, biostable segmented polyurethane (SPU) that has an aromatic urethane hard segment, an aliphatic polycarbonate soft segment, and silicone surface-modifying end groups. It has excellent mechanical strength and elasticity, high flex life, and very good oxidative stability, and is supplied as a solution in dimethyl acetamide (DMAc). Used for stents, pacemaker leads, etc.

BioSpan S. Trademark of Polymer Technology Group, Inc. (USA) for a biocompatible, biostable segmented polyurethane (SPU) that has an aromatic urethane hard segment, a polyether soft segment, and silicone surface-modifying end groups. It has excellent mechanical strength, flex life and elasticity and a hydrophobic surface, and is supplied as a solution in dimethyl acetamide (DMAc). Used for stents, pacemaker leads, etc.

Biosyn. Trademark of US Surgical Corporation (USA) for a biodegradable terpolymer of *p*-dioxanone, glycolide and trimethylene carbonate that is used for tissue engineering applications, e.g., surgical sutures.

Biotech 2000. Trade name of The Argen Corporation (USA) for a 24-karat dental gold.

Bio-Therm. Trade name of BL Dental Company (USA) for acrylics used for dentures.

biotite. A common, dark green, dark brown, dark red, or black mineral of the *mica* group composed of potassium magnesium iron aluminum silicate with the general composition, $K(Mg,Fe)_3AlSi_3O_{10}(OH)_2$. It is a frequent impurity in *feldspar* and *nepheline syenite*. Crystal system, monoclinic. Density, 2.80-3.25 g/cm^3; hardness, 2.5-3.0; refractive index, 1.627. Occurrence: Canada, Central and Southern Africa, South America, USA. Used as a minor source of mica. Also known as *black mica; iron mica.*

Biotrey. Trade name of Dentsply Corporation (USA) for a dental silicate cement.

BioUniversal PDF. Trademark of Ivoclar Vivadent AG (Liechtenstein) for a palladium-free universal dental alloy containing 71.1% gold, 11.7% silver, 9.19% platinum, 4.45% copper, 1.3% indium, and less than 1.0% iridium, iron, tantalum, and zinc, respectively. It has a yellow color, a density of 16.2 g/cm^3 (0.59 lb/in.^{3}), a melting range of 910-970°C (1670-1780°F), an as-cast hardness of 450 Vickers, low elongation, and excellent biocompatibility and good compatibility with special dental ceramics, and composite veneers. Used for crowns, inlays, onlays, posts, and bridges.

Bipalit. Trade name of Bauglasindustrie AG (Germany) for decorated U-shaped glass channels.

biphenyl. See diphenyl.

biphenyltetramine. See diaminobenzidine.

biphosphammite. A colorless mineral that is composed of ammonium hydrogen phosphate, $(NH_4)H_2PO_4$, and may also contain potassium. It can also be made synthetically. Crystal system, tetragonal. Density, 1.80 g/cm^3; refractive index, 1.529. Occurrence: Australia.

Biplate. Trade name of E.L. Yencken & Company Proprietary Limited (Australia) for sealed double-glazing units.

Bi-Ply. Trademark of Owens-Corning Fiberglass (USA) for a strong fabric reinforcement consisting of one ply of fast-wetting, bidirectionally-oriented woven fiberglass roving stitch-bonded to an isotropically oriented chopped-fiberglass strand mat using the proprietary "Kyntex" knitting process. Used for tanks, tooling, marine applications, and for hand layup composites.

Biradur. Trademark of Bisterfeld + Stolting for phenol-formaldehyde plastics.

Birakrit. Trademark of Bisterfeld + Stolting for phenol- and melamine-formaldehyde and epoxy resins.

Biralit. Trademark of Bisterfeld + Stolting for melamine-formaldehyde plastics.

Biramin. Trademark of Bisterfeld + Stolting for melamine-formaldehyde plastics.

Biratex. Trademark of Bisterfeld + Stolting for phenol-formaldehyde plastics and compounds.

Birax. Trademark of Bisterfeld + Stolting for phenol- and melamine formaldehyde and epoxy resins.

birch. The hard, strong, close-grained wood from any of various trees of the genus *Betula*. There are over a dozen commercially important varieties of birch in North America including yellow birch *(B. alleghaniensis)*, sweet birch *(B. lenta)*, paper birch *(B. papyrifera)*, river birch *(B. nigra)*, silver birch *(B. lutea)*, gray birch *(B. populifolia)* and Virginia birch *(B. uber)*. Two European species of commercial interest are white birch *(B. pubescens)* and silver birch *(B. pendula)*.

Birdsboro. Trade name of Birdsboro Corporation (USA) for a series of plain-carbon and alloy steels. The alloy steels include various molybdenum, nickel, chromium-molybdenum, nickel-chromium-molybdenum and nickel-molybdenum types. Used for machinery parts and structural applications.

bird's-eye. (1) A fabric of cotton, linen or synthetic fibers knit or woven with a novelty pattern consisting of small diamonds, each having a center dot. Used especially for towels and diapers.

(2) A clear-finished worsted fabric knit or woven with small indentations imitating the eye of a bird. Used especially for making suits.

birefringent material. (1) A material that has more than one index of refraction.

(2) A transparent anisotropic crystalline material, such as calcite, in which an unpolarized ray of light is seperated into two rays with different directions and relative velocities. Also known as *doubly refracting material*.

biringuccite. A mineral composed of sodium borate hydroxide monohydrate, $Na_2B_5O_8(OH)\cdot H_2O$. It can also be made synthetically by hydrothermal treatment at 250°C (480°F). Crystal system, monoclinic. Density, 2.30 g/cm^3; refractive index, 1.539. Occurrence: Italy.

Birmabright. Trade name of Birmetals Limited (UK) for aluminum alloys containing 1-7% magnesium and up to 0.75% manganese. Commercially available as-cast, and in sheet and strip form, they have good corrosion resistance and fair strength. Used for fasteners, builder's hardware, tubes, architectural work, and marine and lightweight construction.

Birmag. Trade name for a free-machining magnesium alloy containing varying amounts of aluminum, zinc, and manganese. Used for machine parts, automotive components, etc.

Birmal. Trade name of Birmingham Aluminum Casting Company (UK) for a series of aluminum-silicon and aluminum-tin alloys.

Birmalite. Trade name of Birmingham Aluminum Casting Company (UK) for heat-treatable aluminum casting alloys containing 10% copper, 0.3% magnesium, and 0-1% iron. Used for gear boxes, brackets, and low-stressed components.

Birmasil. Trade name of Birmingham Aluminum Casting Company (UK) for a die-casting alloy containing 86-89% iron, and 11-14% silicon.

Birmasil Special. Trade name of Birmingham Aluminum Casting Company (UK) for a corrosion-resistant aluminum casting alloy containing 10-13% silicon, 2.5-3.5% nickel, 0.2% titanium, and 0.6% magnesium. Used for automotive engine parts, such as cylinder blocks and heads, and for automotive trim and fittings.

Birmastic. Trade name for an aluminum casting alloy containing 12% silicon and 3% nickel.

Birmetal. Trade name of Birmetals Limited (UK) for a series of wrought aluminum products including commercially pure aluminum (99.0+%), aluminum-manganese, aluminum-silicon and aluminum-magnesium-silicon alloys, aluminum-copper and aluminum-zinc alloys, and aluminum-clad aluminum-copper alloys.

Birmid. Trade name of Birmingham Aluminum Casting Company (UK) for a series of aluminum-copper, aluminum-magnesium, and aluminum-silicon cast alloys. Used for gear cases, housings, and strong castings.

Birmidal. Trade name of Stirling International Technology Limited (UK) for age-hardenable aluminum casting alloys containing 3-6% silicon, and 0.3-0.8% magnesium. Used for pump casings, gear boxes, instrument housings, etc.

Birmid Qualcast. Trade name of Birmingham Aluminum Casting Company (UK) for lightmetal alloy castings (e.g., aluminum and its alloys), chill and precision castings, ferrous and nonferrous automobile castings, gray cast irons, etc.

Birmingham. (1) A British trade name of a very corrosion-resistant nickel silver containing 50-62% copper, 20-32% zinc, and 12-30% nickel. Used for cutlery, flatware, household utensils, etc.

(2) Trade name of Birmingham Aluminum Casting Company (UK) for a series of aluminum-copper, aluminum-magnesium, and aluminum-silicon cast alloys. Used for gear cases, housings and strong castings.

Birmingham platina. A British corrosion-resistant zinc alloy containing 20.2-46.6% copper and 0.2-0.4% iron. Used for hardware, ornaments, and buttons. Also known as *platina*.

Birmingham platinum. A British alloy composed of 53-79% zinc, 20-47% copper and 0.3% iron. Used for ornamental parts and castings. Also known as *platinum lead*.

birnessite. A black mineral composed of sodium manganese oxide nonahydrate, $Na_4Mn_{14}O_{27}\cdot 9H_2O$. It can also be made synthetically. Crystal system, orthorhombic. Density, 3.00 g/cm^3; refractive index, 1.73. Occurrence: Scotland.

Birox. Trademark of E.I. DuPont de Nemours & Company (USA) for thick-film compositions supplied in various grades.

birunite. A white mineral composed of calcium carbonate silicate sulfate hydrate, $Ca_{15}(CO_3)_{5.5}(SiO_3)_{8.5}SO_4 \cdot 15H_2O$. Crystal system, orthorhombic. Density, 2.36 g/cm^3; refractive index, 1.527. Occurrence: Uzbekistan.

Bisbo Arma. Trade name of General Motors Corporation (USA) for a pearlitic malleable cast iron containing 0.7% combined carbon, 2.5% total carbon, 1.4% silicon, 0.4% manganese, and 0.025% bismuth. Used for automotive components, such as gears, rocker arms, crankshafts, universal-joint yokes, etc.

bischofite. A colorless, or white mineral composed of magnesium chloride hexahydrate, $MgCl_2 \cdot 6H_2O$. It can also be made synthetically. Crystal system, monoclinic. Density, 1.59 g/cm^3; refractive index, 1.505.

Bis-Core. Trademark of BISCO, Inc. (USA) for a highly filled two-component dental composite supplied in natural and opaque shades. The high-viscosity base paste composite is light-curing, and the low-viscosity catalyst is self curing. Used for core-build-ups in fixed restorations.

biscuit. (1) Unglazed ceramic ware, such as tile, fired in a biscuit or bisque oven.

(2) Ceramic ware, such as pottery or china, fired once, but not yet glazed.

(3) A small cake of virgin metal, such as uranium.

(4) See preform (3).

bis(cyclopentadienyl)chromium. See chromocene.

bis(cyclopentadienyl)cobalt. See cobaltocene.

bis(cyclopentadienyl)iron. See ferrocene.

bis(cyclopentadienyl)manganese. See manganocene.

bis(cyclopentadienyl)nickel. See nickelocene.

bis(cyclopentadienyl)osmium. See osmocene.

bis(cyclopentadienyl)ruthenium. See ruthenocene.

bis(cyclopentadienyl)vanadium. See vanadocene.

Bisecurit. Trade name of Boussois Souchon Neuvesel SA (France) for toughened glass with two fracture zones.

Bi-Seal. Trade name of Thermo Insulating Glass Company (USA) for insulating glass.

bis(ethylenedithio)tetrathiafulvalene. A light- and air-sensitive, orange powder (98+% pure) that decomposes at 238-244°C (460-471°F). Used in the manufacture of organic superconductors. Formula: $C_{10}H_8S_8$. Abbreviation: BEDT-TTF.

Bisfil I. Trademark of BISCO, Inc. (USA) for a light-cured dental hybrid composite.

Bisfil II. Trademark of BISCO, Inc. (USA) for a self-cured, highly filled hybrid composite with high strength and low water absorption. Supplied in universal and cervical shades, it is used in dentistry for posterior cavity fillings.

Bisfil 2B. Trademark of BISCO, Inc. (USA) for a fast-setting, radiopaque hybrid composite used as base increment in dental restorations. It cures internally and has low viscosity and optimal flowability.

Bisfil-Core. Trademark of BISCO, Inc. (USA) for a blue, light-cured hybrid composite with optimal viscosity, used in dentistry for core build-ups and metal-based, fixed restorations.

Bisfil M. Trademark of BISCO, Inc. (USA) for a light-cure microfine dental composite.

Bishilite. Trade name of Mitsubishi Metals Corporation (Japan) for a series heat- and wear-resistant cobalt-chromium-type hardfacing alloys supplied in the form of bare and covered electrodes.

bismaleimide 4,4'-diphenyl methane. See polyimide.

bismaleimides. Addition-type thermosetting polyimides that are the imidized products from the reaction of two moles of maleic anhydride with one mole of the diamine. They can be converted to homopolymers, copolymers and terpolymers, and are supplied under various trademarks and trade names, such as Kerimid, Compimide, Matrimid, etc. Used as matrix resins for composites for high-temperature applications, e.g., in the aircraft and aerospace fields, and as crosslinking agents. Abbreviation: BMIs.

bismaleimide-triazine resins. The reaction products of bismaleimides (e.g., bismaleimide-4,4'-methylene dianiline) and 0,0'-dicyanobisphenol A. The mixture ratio is usually from 10:90 to 60:90, and the dicyanobisphenol A component is trimerized. They have a density ranging from 1.20 to 1.35 g/cm^3 (0.043-0.049 lb/in.^{3}), and a glass-transition temperature of 230-320°C (446-608°F). Used as matrix resins for engineering composites, and as intermediates.

bis(4-maleimidodiphenyl)methanes. Addition type thermosetting polyimide resins that are the products from reactions of methylene dianiline (MDA) with maleic anhydride. Abbreviation: MDA BMI.

Bismanol. Trade name for a magnetic alloy or compound of about 20-21% manganese and 79-80% bismuth, usually made by powder metallurgy methods. It has a very high coercive force, high magnetic energy product, and low oxidation resistance. Used for permanent magnets.

bismite. A light yellow mineral composed of bismuth trioxide, Bi_2O_3, and containing theoretically about 80.6% bismuth. It can also be made synthetically. Crystal system, monoclinic. Density, 8.64-9.37 g/cm^3; refractive index, above 2.42. Used as an ore of bismuth. Also known as *bismuth ocher.*

bismoclite. A creamy white, grayish, or yellowish brown mineral of the matlockite group composed of bismuth oxide chloride, BiOCl. It can also be made synthetically. Crystal system, tetragonal. Density, 7.72 g/cm^3. Occurrence: South Africa.

bismuth. A silvery-white, diamagnetic, brittle metallic element of Group VA (Group 15) of the Periodic Table. It is commercially available in the form of ingots, foil, rods, wire, lumps, powder, shot, needles, granules and single crystals, and occurs native and in the mineral *bismuthinite*, but is usually obtained as a byproduct in the smelting of silver, gold, etc. Crystal system, hexagonal (rhombohedral). Density, 9.80 g/cm^3; melting point, 271.3°C (520°F); boiling point, 1560°C (2840°F); hardness, 16-19 Vickers; atomic number, 83; atomic weight, 208.98; trivalent, pentavalent; low electrical conductivity (1.5% IACS); very low thermal conductivity. Used in the manufacture of low-melting alloys, as a carbide stabilizer in malleable iron, as an additive in low-carbon steel and aluminum to improve machinability, as a catalyst in acrylonitrile production, in glassmaking and ceramics, for coating selenium, as a heat detector in fire protection sprinkler systems, for safety plugs in boilers, in electric fuses, and in thermoelectric materials, permanent magnets, semiconductors and superconductors. Symbol: Bi.

bismuth alloys. A group of alloys based on bismuth and containing significant amounts of lead, tin, cadmium, indium, antimony, zinc, silver and/or indium. They are fusible alloys, i.e., alloys with relatively low melting points, usually between 50 and 250°C (120 and 480°F). Used for automatic safety devices, such as sprinklers, boiler plugs, furnace controls, and in special solders, and type metal.

bismuth amalgam. A shiny, highly liquid alloy of varying amounts of bismuth and mercury, used to increase the plasticity of white-metal alloys, and for silvering mirrors.

bismuth antimonide. A compound of bismuth and antimony supplied as single crystals for semiconductor applications. Formula: BiSb.

bismuth blende. See eulytite.

bismuth brass. (1) A corrosion-resistant brass containing 47% copper, 20-21% zinc, 30-31% nickel, 1% tin, and 0.1% bismuth. Used for ornaments and hardware.

(2) A corrosion-resistant brass containing 52% copper, 30% nickel, 12% zinc, 5% lead, and 1% bismuth. Used for ornaments and hardware.

bismuth bromide. A yellow, moisture-sensitive powder (98+% pure) with a density of 5.72 g/cm^3, a melting point of 218°C (424°F), and a boiling point of 453°C (847°F). Used for bismuth salts, and as a catalyst. Formula: $BiBr_3$. Also known as *bismuth tribromide.*

bismuth bronze. A corrosision-resistant bronze containing 45-53% copper, 10-33% nickel, 20-22% zinc, 15-16% tin, 1% bismuth, and up to 0.1% aluminum. Used for ornaments and hardware.

bismuth chloride. White, corrosive, crystals (98+% pure) obtained by treating bismuth with hydrochloric acid. Density, 4.75 g/cm^3; melting point, 230-232°C (446-450°F); boiling point, 447°C (837°F). Used in bismuth salts, and as a catalyst. Formula: $BiCl_3$. Also known as *bismuth trichloride.*

bismuth chloride oxide. See bismuth oxychloride.

bismuth chromate. An orange-red, amorphous powder obtained from the interaction of potassium chromate and bismuth nitrate. Used as a pigment in porcelain enamels and glazes. Formula: $Bi_2O_3 \cdot CrO_3$.

bismuth fluoride. Moisture-sensitive, off-white powder (98-99% pure), or white-gray crystals (99.999% pure). Density, 5.32 g/cm^3; melting point, 727°C (1341°F); refractive index, 1.74. Used in superconductivity studies, and as a source of fluoride for doping applications. Formula: BiF_3. Also known as *bismuth trifluoride.*

bismuth flux. A mixture of about 50% sulfur and 50% potassium oxide, or 50% sulfur, 25% potassium bisulfate and 25% potassium iodide. Used as a flux in brazing and soldering bismuth.

bismuth germanate. A nonhygroscopic compound of bismuth oxide (Bi_2O_3) and germanium oxide (GeO_2) that is commercially available in the form of high-purity (99.99+%) single crystals. Crystal system, cubic. Density, 7.13 g/cm^3; hardness, 5 Mohs; melting point, 1050°C (1922°F); high photo-peak efficiency and stopping power, and low afterglow. Used as an intrinsic scintillator for positron CAT scanners, electromagnetic shower calorimeters, oil-well logging applications, and as a gamma-ray detector in high-energy physics. Formula: $Bi_4Ge_3O_{12}$. *Bismuth germanate* is also supplied as single crystals ($BiGeO_{20}$) with outstanding piezoelectric properties for use in electronics and optoelectronics. Also known as *bismuth germanium oxide (BGO).*

bismuth germanium oxide. See bismuth germanate.

bismuth glance. See bismuthinite.

bismuth hydrate. See bismuth hydroxide.

bismuth hydroxide. A white, amorphous powder obtained by treating a solution of bismuth nitrate with sodium hydroxide. It has a density of 4.36 g/cm^3 and is used in the separation of plutonium, and as an absorbent. Formula: $Bi(OH)_3$. Also known as *bismuth hydrate; bismuth oxyhydrate; bismuth trihydrate; bismuth trihydroxide; hydrated bismuth oxide.*

bismuthinite. A dark gray mineral of the stibnite group with metallic luster and yellowish tarnish. It is composed of bismuth trisulfide, Bi_2S_3, and contains theoretically 81.2% bismuth. It can also be made synthetically. Crystal system, orthorhombic. Density, 6.78-6.81 g/cm^3; hardness, 2 Mohs. Occurrence: Australia, Bolivia, Central Europe (Germany), Mexico, Peru, USA (Utah). Used as a source of bismuth. Also known as *bismuth glance.*

bismuth iodide. Moisture-sensitive, grayish-black crystals or gray powder (99+% pure). Density, 5.778 g/cm^3; melting point, 408°C (766°F); boiling point, approximately 500°C (930°F). Used in analytical chemistry and organometallic research, and in the manufacture of bismuth oxyiodide. Formula: BiI_3. Also known as *bismuth triiodide.*

bismuth-lead alloys. A group of fusible alloys of bismuth and lead used for casting master patterns for making molds. The eutectic alloy 55.5Bi-44.5Pb has a melting temperature of 124°C (255°F), and exhibits net expansion during solidification.

bismuth lead strontium calcium copper oxide. A fine superconducting powder or thin film used for high-temperature superconductor applications. Formula: $Bi_{1.6}Pb_{0.4}Sr_{1.6}Ca_{2.0}Cu_{2.8}O_{9.2+x}$ (x = 0.45). Abbreviation: BPSCCO.

bismuth-manganese alloys. Magnetic alloys of bismuth and manganese that have high energy products and coercivities, and are used for permanent magnets. See also Bismanol.

bismuth monoselenide. See bismuth selenide (ii).

bismuth monosulfide. See bismuth sulfide (ii).

bismuth monotelluride. See bismuth telluride (ii).

bismuth nitrate. Colorless to white crystals, or white crystalline powder (98+% pure) obtained by treating bismuth with nitric acid. Density, 2.830 g/cm^3; melting point, 30°C (86°F); boiling point, loses $5H_2O$ at 75-80°C (167-176°F). Used in the production of bismuth luster on tin, in the preparation of bismuth salts, in luminous paints, in enamels and glazes, and in the preparation of bismuth-containing superconductors. Formula: $Bi(NO_3)_3 \cdot 5H_2O$. Also known as *bismuth ternitrate; bismuth trinitrate; bismuth nitrate pentahydrate.*

bismuth nitrate oxide. See bismuth oxynitrate.

bismuth nitrate pentahydrate. See bismuth nitrate.

bismuth nitride. A gray, refractory crystalline compound of bismuth and nitrogen. Crystal system, cubic. Density, 2.71 g/cm^3; melting point, 2200°C (3992°F). Used in ceramics and as a semiconductor. Formula: Bi_2N_3.

bismuth ocher. See bismite.

bismuth orthophosphate. White crystals with a density of 6.32 g/cm^3, that do not melt on heating. Used as an ingredient in optical glass, and in the recovery of plutonium. Formula: $BiPO_4$. Also known as *bismuth phosphate.*

bismuth oxide. A heavy, yellow crystalline powder, or yellow sintered pieces (99.9+%) obtained by heating bismuth nitrate in air. It also occurs in nature as the mineral *bismite*. Density, 8.90 g/cm^3; melting point, 817-825°C (1503-1517°F); boiling point, 1890°C (3434°F). Used as a yellow pigment in porcelain enamels, glazes and other ceramics, in ceramic colors, as a fluxing component in optical glasses, as a flux in cast-iron porcelain enamels, as a flux and bonding agent for metallic components in ceramic glazes, as an ingredient in fluxes for fired-on conductive silver paints, in the production of bismuth salts, and as a starting material in superconductivity studies. Formula: Bi_2O_3. Also known as *bismuth trioxide; bismuth yellow.*

bismuth oxycarbonate. A white, light-sensitive powder with a density of 6.86 g/cm^3, obtained by adding ammonium carbonate to a bismuth salt solution. Used as a flux and opacifier in glass and porcelain enamels, in ceramic glazes, and in the manufacture of other bismuth compounds. Formula: $(BiO)_2CO_3$ (anhydrous); $Bi_2O_3 \cdot CO_2 \cdot 0.5H_2O$ (hydrated). Also known as *basic bismuth carbonate; bismuth subcarbonate.*

bismuth oxychloride. White, crystalline powder (99+% pure) with a density of 7.72 g/cm^3, obtained by treating bismuth chloride with water. Used as a paint pigment (*pearl white*), and in dry cell cathodes. Formula: $BiOCl$. Also known as *basic bismuth chloride; bismuth chloride oxide; bismuth subchloride.*

bismuth oxyhydrate. See bismuth hydroxide.

bismuth oxynitrate. A white, heavy, hygroscopic powder obtained by hydrolysis of bismuth nitrate. Density, 4.930 g/cm^3; melting point, decomposes at 260°C (500°F). Used as constituent in high-refractive glass, in low-temperature porcelain enamels and colorants, in the production of pearly luster on glasses and glazes, in the production of bismuth luster on metals, in burning gold on ceramic ware, in the manufacture of bismuth salts, and in microscopy. Formula: $4BiNO_3(OH)_2 \cdot BiO(OH)$. Also known as *basic bismuth nitrate; bismuth nitrate oxide; bismuth subnitrate; bismuth white; Spanish white.*

bismuth phosphate. See bismuth orthophosphate.

bismuth salicylate. See bismuth subsalicylate.

bismuth selenide. Any of the compounds of bismuth and selenium: (i) *Bismuth triselenide.* Black crystals. Crystal system, hexagonal (rhombohedral). It occurs in nature as the mineral *guanajuatite.* Density, 6.82-7.51 g/cm^3; melting point, 706°C (1303°F); hardness, 167 Knoop. Used for thermoelectric applications, and in semiconductor technology. Formula: Bi_2Se_3; and (ii) *Bismuth monoselenide.* Crystal system, cubic. Crystal structure, halite. Density, 7.98 g/cm^3; melting point, 607°C (1125°F). Used as a semiconductor. Formula: $BiSe$.

bismuth silicate. Any of the following compounds containing bismuth, silicon and oxygen: (i) A compound of bismuth oxide (Bi_2O_3) and silicon dioxide (SiO_2), available as high-purity (99.99+%) single crystals. It has good electro-optical, photoconductive and piezoelectric properties, and is widely used in image amplifiers and modulators, and for beam combination and matrix inversion applications. Formula: $Bi_{12}SiO_{20}$; and (ii) A compound of bismuth dioxide (BiO_2) and silicon dioxide (SiO_2). Used in ceramics and materials research. Formula: $BiSiO_4$. Also known as *bismuth silicon oxide (BSO).*

bismuth silicon oxide. See bismuth silicate (ii).

bismuth solders. See fusible solders.

bismuth spar. See bismutite.

bismuth stannate. A light-colored crystalline powder that loses $5H_2O$ above 140°C (284°F), and has a melting point above 1300°C (2372°F). Used as an additive in the production of relatively temperature-insensitive barium titanate capacitors with medium dielectric constants. Formula: $Bi_2(SnO_3)_3 \cdot 5H_2O$.

bismuth strontium calcium copper oxides. Black superconducting powders with particle sizes of less than 125 μm (0.005 in.), and purities of over 99.6% including: (i) $Bi_2Sr_2CaCu_2O_x$ which contains about 47% bismuth, 19.5% strontium, 4.5% calcium, and 14.0% copper, and is usually dry processed from selected carbonates and oxides, and also available as layered structures. It has a density of 6.4 g/mL, and a superconductivity critical temperature of 80-85K. Abbreviation: Bi-2212; (ii) BiSrCa-Cu_2O_x. A powder, usually dry processed from selected carbonates and oxides. Abbreviation: Bi-1112; and (iii) $Bi_2Sr_2CaCu_2O_x$. A powder, usually dry processed from selected carbonates and oxides. Abbreviation: Bi-2223. All powders and layered struc-

tures are used as high-temperature superconductors. Abbreviation: BSCCO.

bismuth subcarbonate. See bismuth oxycarbonate.

bismuth subchloride. See bismuth oxychloride.

bismuth subnitrate. See bismuth oxynitrate.

bismuth subsalicylate. A white, crystalline, light-sensitive powder obtained by treating bismuth hydroxide with salicylic acid. Used in coating of plastics, and in copying paper. Formula: $Bi(C_7H_5O_3)_3Bi_2O_3$. Also known as *basic bismuth salicylate.*

bismuth sulfide. Any of the following compounds of bismuth and sulfur: (i) *Bismuth trisulfide.* A black powder (98+% pure). It occurs in nature as the mineral *bismuthinite.* Density, 7.70 g/cm^3; melting point, 685°C (1265°F). Used in the manufacture of bismuth compounds. Formula: Bi_2S_3; and (ii) *Bismuth monosulfide.* A gray powder. Density, 7.7 g/cm^3; melting point, 685°C (1265°F). Used in ceramics and semiconductors. Formula: BiS.

bismuth telluride. Any of the following compounds of bismuth and tellurium: (i) *Bismuth tritelluride.* Gray platelets, lumps, or single crystals available in high-purity grades (99.999%). It occurs in nature as the mineral *tellurobismuthite.* Crystal system, hexagonal. Density, 7.3-7.7 g/cm^3; melting point, 585°C (1085°F); hardness, 1.5-2.0 Mohs, or 155 Knoop; service temperature range, -45 to +204°C (-49 to +400°F). Used as a semiconductor, and as a thermoelectric material in cooling devices and power-generation equipment. Formula: Bi_2Te_3; and (ii) *Bismuth monotelluride.* Crystal system, cubic. Crystal structure, halite. It is available as polycrystalline and single crystalline material in n- and p-types, and has good thermal and electrical properties which make it suitable for use in thermoelectric coolers, and for various semiconductor applications. Formula: $BiTe$. See also tellurobismuthite.

bismuth ternitrate. See bismuth nitrate.

bismuth tetroxide. A yellowish-brown powder with a density of 5.6 g/cm^3 and a melting point of 305°C (581°F). Used as a lubricant for metal extrusion dies. Formula: Bi_2O_4.

bismuth-tin alloys. A group of fusible alloys containing varying amounts of bismuth and tin.

bismuth titanate. A compound of bismuth trioxide (Bi_2O_3) and titanium dioxide (TiO_2) available as a white powder (98+% pure). Used in ceramics, electronics, and in inorganic and organometallic research. Formula: $Bi_2Ti_2O_7$.

bismuth tribromide. See bismuth bromide.

bismuth trichloride. See bismuth chloride.

bismuth trifluoride. See bismuth fluoride.

bismuth trihydrate. See bismuth hydroxide.

bismuth trihydroxide. See bismuth hydroxide.

bismuth triiodide. See bismuth iodide.

bismuth trinitrate. See bismuth nitrate.

bismuth trioxide. See bismuth oxide.

bismuth triselenide. See bismuth selenide (i).

bismuth trisulfide. See bismuth sulfide (i).

bismuth tritelluride. See bismuth telluride (i).

bismuth white. See bismuth oxynitrate.

bismuth yellow. See bismuth oxide.

Bismuth-Zinc. Trade name of Blackwells Metallurgical Limited (UK) for a zinc alloy containing 40% bismuth. Used as a primary metal.

bismutite. A white, straw-yellow, yellowish-green, or gray amorphous mineral composed of basic bismuth carbonate, $(BiO)_2CO_3$, and containing theoretically 78.3% bismuth. Density, 7.40-8.26 g/cm^3; refractive index, 2.12-2.30. Occurrence: East Africa. Used as an ore of bismuth. Also known as *bismuth spar.*

bismutoferrite. A dull olive-green mineral composed of bismuth iron silicate hydroxide, $Fe_2Bi(Si_2O_4)_2(OH)$. Density, 4.47 g/cm³; refractive index, 1.97. Occurrence: Germany.

bismutomicrolite. A dark gray to black mineral of the pyrochlore group composed of bismuth calcium tantalum niobium oxide hydrate, $(Bi,Ca)(Ta,Nb)_2O_6(OH)$. Crystal system, cubic. Density, 6.83 g/cm³. Occurrence: Brazil.

bismutotantalite. A light brown to pitch black mineral of the columbite group composed of bismuth tantalum oxide, $Bi(Ta,Nb)O_4$. It can also be made synthetically. Crystal system, orthorhombic. Density, 8.84 g/cm³; refractive index, 2.403. Occurrence: Uganda.

Bison. Trade name of Time Steel Service Inc. (USA) for an oil-hardening shock-resisting tool steel (AISI type S1) containing 0.5% carbon, 2.5% tungsten, 0.75% silicon, 1.15% chromium, 0.2% vanadium, and the balance iron. It has high hardness and strength, and high toughness and hot hardness. Used for hot-work tools, forming tools, and hand and pneumatic chisel.

Bisonal. Trade name of Flake Board Company Limited (Canada) for overlaid particleboard.

bisphenol A. The product formed by a condensation reaction of two molecules of phenol with acetone catalyzed by hydrochloric acid at 65°C (149°F). It is available in the form of white flakes (97+% pure) with a density of 1.195 g/cm³, a melting point of 153-156°C, a boiling point of 220°C (428°F)/4.0 mm, and a flash point of 175°F (79°C). Used as an intermediate in the manufacture of epoxy, phenolic, phenoxy, polycarbonate, polysulfone and polyester resins. Formula: $(CH_3)_2C(C_6H_4OH)_2$. Abbreviation: BPA.

bisphenol A fumarates. Unsaturated polyesters made by treating propoxylated or ethoxylated bisphenol A (BPA) with fumaric acid. They have high degree of hardness and rigidity, and good chemical, electrical and thermal stability. Used for electrical components, industrial equipment, automotive parts, and marine equipment.

bisphenol A vinyl ester resin. An unsaturated ester of epoxy resin that is the reaction product of methacrylic acid and bisphenol A (BPA) epoxy resin dissolved in styrene monomer. It has excellent corrosion resistance, and good mechanical and handling properties. Used for corrosion-resistant reinforced plastics used for pipes and tanks, electrical equipment, and equipment for the chemical-process, pulp and paper, wastewater and mining industries.

bisphenol E. The reaction product of bisphenol A and ethylene. It has a melting point of 123-127°C (253-261°F), and is used in the manufacture of various synthetic resins. Formula: $CH_2(C_6H_4OH)_2$. Abbreviation: BPE.

bisphenol F. The condensation product of bisphenol A and formaldehyde. It has a melting point of 162-164°C (324-327°F), and is used in the manufacture of various synthetic resins. Formula: $CH_2(C_6H_4OH)_2$. Abbreviation: BPF.

bisphenol M. A bisphenol compound with a melting point of 135-139°C (275-282°F) used in the manufacture of certain synthetic resins. Formula: $C_6H_4[C(CH_3)_2C_6H_4OH]_2$. Abbreviation: BPM.

bisphenol P. A bisphenol compound with a melting point of 193-195°C (379-383°F) used in the manufacture of certain synthetic resins. Formula: $C_6H_4[C(CH_3)_2C_6H_4OH]_2$. Abbreviation: BPP.

bisphenol S. A bisphenol compound with a melting point of 243-248°C (469-478°F) used in the manufacture of certain synthetic resins. Formula: $O_2S(C_6H_4OH)_2$. Abbreviation: BPS.

bisphenol Z. A bisphenol compound with a melting point of 190-192°C (374-378°F) used in the manufacture of certain synthetic resins. Formula: $C_6H_{10}(C_6H_4OH)_2$. Abbreviation: BPZ.

bisque. (1) Unglazed ceramic ware that has been fired once.
(2) A dried, but unfired coating of wet-process porcelain enamel.

Bistite II SC. Trademark of J. Morita Company (Japan) for a self-etching, self-cured all-purpose dental resin cement.

Bitestone. Trade name of Whip Mix Corporation (USA) for a dental stone with a very short setting time (about 1-3 minutes) supplied as a white powder. It is especially suitable for mounting casts, as model stone for splint and temporary appliances, and for occlusal registration.

Bi-Triplex. Trade name of Société Industrielle Triplex SA (France) for laminated glass made up of three sheets of glass separated by two plastic interlayers.

bitternut hickory. The tough, hard, heavy wood of the medium-sized to tall pecan tree *Carya cordiformis*. The tree is named after the bitter, astringent kernels of its nuts. The heartwood is reddish, and the sapwood white. *Bitternut hickory* has high strength, high shock resistance, and large shrinkage. Source: Eastern and Central United States. Used for tool handles, flooring, agricultural implements, and furniture. See also hickory.

bitter spar. A relatively pure crystalline form of *dolomite* composed of about equal proportions of calcium carbonate ($CaCO_3$) and magnesium carbonate ($MgCO_3$).

bitudobe. A sun-baked brick (adobe) with a bituminous binder. See also adobe.

Bitumastic. Trademark of Koppers Company, Inc. (USA) for a specially refined coal-tar pitch and filler. It provides useful protection up to about 430°C (806°F), and is used for hot and cold-applied corrosion- and water-resistant protective pitch primers and coatings on metals, and for industrial coatings and for sealing spray coatings.

bitumen. A generic term for any of several amorphous, black or dark-colored solid or liquid hydrocarbon polymers that are obtained naturally (e.g., asphalt and tar), or by distillation from coal, petroleum, wood, etc. They are mixtures of fats, waxes, resins, lignin, proteins and carbohydrates, and are used for hot-melt adhesives, road-surface materials, roofing compositions, and paints and sealants.

bituminous cement. A black solid, semiliquid, or liquid hydrocarbon material with good cementing properties that occurs in nature, and is also obtained as a residue in the distillation of petroleum, coal and wood. Used as a binder.

bituminous coal. A soft, dark brown to black coal which burns with a smoky luminous flame, and tends to break up into small pieces. It contains 10-50% volatile matter and is high in carbonaceous matter. Used as an industrial fuel, and in the preparation of coke, coal tar, liquid fuels and petrochemicals. Also known as *soft coal*.

bituminous coating. A protective coating based on coal tar or asphalt, and used as a road-surfacing material, and in the building trades for water-repellent barrier coatings.

bituminous concrete. A special concrete made without cement. It consists of aggregates, such as crushed stone, gravel, sand and/or slag, combined with a bituminous binder. The stone content may be as high as 65%. It is usually available in slab form. See also asphaltic concrete.

bituminous enamel coating. A flexible, tough, highly adhesive coating system that combines bitumen with thermoplastic elastomers for use on pipeline surfaces.

bituminous macadam. Asphalt made by bonding grit or crushed stone with bitumen.

bituminous material. A material containing a considerable proportion of bitumen, or resembling bitumen in consistency, odor, etc.

bituminous mastic. See asphalt mastic.

bituminous paint. A dark-colored, weatherable paint usually composed of asphalt or coal tar dissolved in mineral spirits. Used for waterproofing of concrete, and for protective coatings on outdoor tanks and piping.

bituminous rock. See asphalt rock.

bituminous sand. A sand that has been naturally impregnated with asphalt or tar.

Bitumuls. Trademark of Chevron Asphalt Limited (Canada) for asphalt road emulsions.

Bituplastic. Trademark for a quick-drying refined coal-tar pitch used for coating pipes and waterproofing buildings, and in flooring, roofing, and insulation.

Bituvia. Trade name for a road tar.

bityite. A yellowish or white mineral of the mica group composed of lithium calcium aluminum beryllium silicate hydroxide, $CaAl_2Li(AlBeSi_2)O_{10}(OH)_2$. Crystal system, monoclinic. Density, 3.07 g/cm^3; refractive index, 1.659. Occurrence: Madagascar.

bivoltine silk. Silk produced by silkworms that have two broods per season.

bixbyite. An opaque black powdery mineral composed of manganese iron oxide, $(Mn,Fe)_2O_3$. It can also be made synthetically. Crystal system, cubic. Density, 4.95 g/cm^3. Occurrence: USA. An iron-free variety with orthorhombic crystal structure is found in South Africa. Also known as *partridgeite*.

bjarebyite. A mineral composed of barium iron manganese aluminum phosphate hydroxide, $Ba(Mn,Fe)_2Al_2(PO_4)_3(OH)_3$. Crystal system, monoclinic. Density, 3.90 g/cm^3; refractive index, 1.727. Occurrence: Central Africa, USA (New Hampshire).

Bla-Caloy. Trade name of Black-Clawson Company (USA) for a ductile cast iron containing 3.3% carbon, 0.7% manganese, 2% silicon, 0.05% magnesium, and the balance iron. Used for gears, cams, shafts, and housings.

black alder. The soft, light reddish-brown wood of the tree *Alnus glutinosa*. It closely resembles poplar, has a smooth, fine grain, and low durability. Source: Native to Europe including British Isles, but introduced and established in the northern United States and eastern Canada. Average weight, 530 kg/m^3 (33 lb/ft^3). Used for plywood, turnery, cabinetwork, and toys. Also known as *common alder; European alder*.

Black Armor. Trade name of Atotech USA Inc. (USA) for black chromate conversion coatings for automotive applications.

black arsenic. An allotrope of *arsenic* (β-arsenic) derived by controlled condensation of arsenic vapor. It is a black, amorphous solid substance.

black ash. (1) The hard, stiff, moderately strong wood of the tree *Fraxinus nigra*. Average weight, 545 kg/m^3 (34 lb/ft^3). Source: Eastern Canada, northeastern United States (New England and Lake states). Used for furniture, veneer, barrel hoops, and basketweaving. Also known as *basket ash; brown ash; hoop ash*.

    (2) See soda ball.

Black Beauty. (1) Trade name of Empire Sheet & Tin Plate Company (USA) for a black-oxide steel sheet used for construction applications.

    (2) Trademark of Harsco Corporation/Reed Minerals Division (USA) for black, chemically inert coal slag abrasives

composed of hard, sharp, angular particles of uniform density providing fast cutting action and low moisture contents. They have a density of 2.7 g/cm^3 (0.10 $lb/in.^3$), a hardness of 6-7 Mohs, and low free-silica contents (less than 1%). A typical chemical composition is 47.2% silicon dioxide (SiO_2), 21.4% aluminum oxide (Al_2O_3), 19.2% ferric oxide (Fe_2O_3), 6.8% calcium oxide (CaO), 1.6% potassium oxide (K_2O), 1.5% magnesium oxide (MgO), 1% titanium dioxide (TiO_2), and 0.6% sodium oxide (Na_2O). Supplied in four grades: coarse, medium, fine, and extra fine. Used for blasting and finishing applications.

black birch. The hard, heavy, strong wood of the tall birch tree *Betula lenta*. The heartwood is dark-brown and the sapwood light brown or yellowish. *Black birch* has a fine texture, and high strength and shock resistance. Source: USA (Northeastern and Lake states, and Appalachian Mountains). Used for lumber and veneer, furniture, boxes, crates, woodenware, and doors. Also known as *cherry birch; mahogany birch; sweet birch*.

Blackbird. Trademark of Unitika Limited (Japan) for viscose rayon fibers and yarns used for textile fabrics.

Black Blast. Trade name of Virginia Materials & Supplies, Inc. (USA) for coal slag blasting abrasives.

blackboard enamel. A matte porcelain enamel of varying color that is used in the manufacture of blackboards (chalkboards) to provide a slightly roughened writing surface. Also known as *chalkboard enamel*.

black boy gum. See yellow acaroid.

black boy resin. See yellow acaroid.

black cherry. See American cherry.

Black Chromium. Trade name of Atotech USA Inc. (USA) for black chromium electroplates and processes for automotive applications.

black-coated steel. An aluminum-deoxidized low-carbon steel coated with a spongy layer of hydrogen-reduced nickel that has subsequently been impregnated with a carbonaceous slurry to give a black surface.

black cobalt. See asbolite.

black copper. See tenorite.

black copper oxide. A black, or brownish-black crystalline powder (99+% pure) obtained from copper carbonate or copper nitrate. It occurs in nature as the mineral *tenorite*, and contains on average 78-80% copper. Crystal system, monoclinic. Density, 6.31-6.49 g/cm^3; melting point, 1362°C (2484°F). Used as a source of copper, as a flux in welding and metallurgy, for batteries and electrodes, in the production of blue or green colors on glass, faience, porcelain, stoneware and other ceramics (when fired in an oxidizing atmosphere), and red colors (when fired in a reducing atmosphere), in electroplating, as a solvent for chromic iron ores, in antifouling paints, as a phosphor and catalyst, and in copper oxide superconductors. Formula: CuO. Also known as *copper monoxide; copper oxide; cupric oxide*.

black cottonwood. The soft, light-colored wood of the large poplar tree *Populus trichocarpa*. The heartwood is light-brown to gray and the sapwood whitish. *Black cottonwood* as an uniform texture, a fine, straight grain, moderate strength, and works moderately well. Average weight, 384 kg/m^3 (24 lb/ft^3). Source: Alaskan Pacific Coast; western Canada; northwestern United States. Used as a pulpwood, and for boxes, crates, and excelsior. See also cottonwood.

Black Diamond. (1) Trade name of Crucible Materials Corporation (USA) for a water-hardening tool steel (AISI type W1) containing 1% carbon, and the balance iron. It has excellent machinability, good wear resistance and toughness, and low hot hardness. Used for hand tools such as hammers, axes and chisels, and for cutting tools, drills, drawing and heading dies, and forming dies.

(2) Trade name of TemTex Temperature Systems (USA) for a corrosion-resistant diamond-like coating.

black diamond. An imperfect, compact, dark gray to black, cryptocrystalline variety of *diamond* used as an industrial diamond for cutting and grinding tools and machines, and for borehole drilling. Source: South America especially Brazil; South Africa. Also known as *carbonado; carbon diamond.*

black East Indian gum. A black variety of *East Indian gum,* that is a fossil or semirecent resin with a density of 1.04 g/cm³ (0.038 lb/in.³). Used in dark-colored oleoresinous varnishes, and in gloss paints. See also East Indian gum.

black ebony. See African ebony.

black factice. Brown factice, i.e., vulcanized oil, that contains some mineral rubber (e.g., gilsonite or grahamite). Used for rubber goods. See also brown factice; factice.

black ferric oxide. Black crystals or amorphous powder occurring in nature as the mineral *magnetite,* or made synthetically. Crystal system, cubic. Density, 5.18 g/cm³; melting point, 1538°C (2800°F); exhibits ferrimagnetic properties. Used as a paint pigment, in buffing and polishing compounds, in the decarbonization of steel, and for magnetic inks, ferrites, and magnetic tape coatings. Formula: Fe_3O_4. Also known as *black magnetic rouge; black rouge; ferroferric oxide; ferrosoferric oxide; magnetic iron oxide.*

Black Flux. Trade name for high-temperature fluxes for silver brazing of ferrous metals, stainless steel, nickel and nickel alloys, copper, brass and bronze.

Blackfriars. Trade name of James Clark & Eaton Limited (UK) for toughened glass and mirrors.

Black Giant. Trade name for a silicon steel containing about 1.4% carbon, 2.2% silicon, 0.9% manganese, 0.5% tungsten, 0.45% chromium, and the balance iron. It has high wear and abrasion resistance, and a low coefficient of friction. Used for bending, drawing and stamping dies.

black glass. Common glass containing additions of manganese or ferric oxides.

Black Gold. Trade name of Deveco Corporation (USA) for chromate coatings and processes used on cadmium or zinc.

black granite. See diorite.

black gum. See sourgum.

blackheart malleable iron. A malleable cast iron with a matrix of ferrite with interspersed nodules of temper carbon (nominal composition about 2-3%) made by prolonged annealing of white cast iron during which graphitization is the predominant reaction. The resulting product has a velvety black fracture surface with a mouse-gray rim and is called "blackheart malleable" iron. A typical composition is 1.5-2.5% carbon, 0.1-0.35% manganese, 0.5-1.2% silicon, 0.2% phosphorus, 0.1% sulfur, and the balance iron. *Blackheart malleable iron* has an average density of 7.2-7.4 g/cm³ (0.26-0.27 lb/in.³), high ductility and shock resistance, and good machinability, brazeability, solderability, hardenability and coatability. Used for railroad, aircraft and automotive castings, boiler, tank and engine castings, building hardware, conveyor systems and other handling equipment, electrical and industrial power equipment, pneumatic and portable tools, hand tools, machine-tool parts, industrial machinery, road and off-road machinery castings, marine and high-

way construction equipment, mining and oilfield castings, pipe fittings and plumbing supplies. Also known as *blackheart malleable cast iron.* Formerly known in North America as *American malleable (cast) iron.* See also cupola malleable iron.

black hickory. The tough, hard, heavy wood of the tall pignut hickory *Carya texana.* The heartwood is reddish and the sapwood white. It has high strength, high shock resistance, and large shrinkage. Source: USA (Louisiana, Texas, Oklahoma, Arkansas, Missouri and Illinois). Used for tool handles, rungs, agricultural implements, furniture, and pallets. See also pignut hickory; hickory.

blacking. A carbonaceous material, such as graphite or carbon powder, often suspended in water, alcohol, oil, etc., for application to the working surface of a core or mold as a parting material to prevent the casting from sticking, and improve its surface finish.

black iron oxide. A black, antiferromagnetic powder that can be obtained by heating iron oxalate. Density, 5.7 g/cm³; melting point, 1420°C (2588°F). Used as a colorant in glass and glazes, as a paint pigment, in steelmaking, in electronics, and as a catalyst. Formula: FeO. Also known as *ferrous oxide; iron monoxide; iron oxide.*

black jack. See sphalerite.

blackjack staple fiber. A glossy, even-surfaced, dark-colored fiber obtained from the leafstalks of a species of Sri Lankan palms (genus *Caryota*) and used as a replacement for horsehair. See also kittul fiber.

Black Knight. Trademark of U.S.E. Hickson Products Limited (Canada) for a series of bituminous roof and foundation coatings supplied in cans and spray cartridges, and several durable, all-weather coal-tar base sealers for airport runways, driveways and sidewalks, etc.

Black Label. Trade name of Peninsular Steel Company (USA) for a water-hardening tool steel containing 1.1% carbon, 0.5% chromium, 0.2% vanadium, and the balance iron. Used for tools and dies.

black lead. See graphite.

black lead oxide. See lead suboxide.

black locust. The strong, heavy, durable wood from the tall hardwood tree *Robinia pseudoacacia.* The heartwood is greenish-yellow to dark-brown and darkens on exposure, and the narrow sapwood is creamy-white. *Black locust* has a coarse grain, a satiny, lustrous surface, high hardness, high strength, stiffness, and shock resistance, high decay and termite resistance, and moderate shrinkage. Average weight, 760 kg/m³ (47 lb/ft³). Source: USA (Pennsylvania along Appalachian Mountains to northern Georgia, and northwestern Arkansas). Used for timbers, wheel spokes, fence posts, stakes, poles, railroad ties, furniture, general construction, agricultural implements, and as a fuelwood. Also known as *acacia; false acacia; green locust; post locust; red locust; white locust; yellow locust.* See also locust.

Black Magic. (1) Trade name of Hubbard-Hall Inc. (USA) for room-temperature and hot-applied black oxide conversion coatings for ferrous and nonferrous metals.

(2) Trademark of Abrasives Inc. (USA) for low free-silica coal slag abrasives composed of hard, angular particles for fast cleaning action.

black magnetic rouge. See black ferric oxide.

Black Magnum. Trademark of Fairmount Minerals (USA) for coal slag abrasives.

black maple. The heavy, strong wood of the large tree *Acer nigrum* belonging to the hard maple group. The heartwood is light reddish-brown, and the sapwood white with reddish-brown tinge. *Black maple* has a fine, uniform texture, high hardness and stiffness, high shock and abrasion resistance, and large shrinkage. Average weight, 641 kg/m³ (40 lb/ft³). Source: USA (Central United States, Lake states and Middle Atlantic states); Canada (southern Ontario). Used for lumber, flooring, furniture, veneer, boxes, pallets, crates, crossties, woodenware, and as a pulpwood. See also hard maple.

black mercury oxide. A black powder. Density, 9.8 g/cm³; melting point, decomposes at 100°C (212°F). Formula: Hg_2O. Also known as *mercurous oxide.*

black mercuric sulfide. A black powder. It occurs in nature as the mineral *metacinnabarite*. Crystal system, cubic. Crystal structure, sphalerite. Density, 7.60-7.73 g/cm³; hardness, 3 Mohs; melting point, 583.5°C (1082°F). Used as a semiconductor and pigment. Formula: HgS. Also known as *black mercury sulfide.*

black mica. See biotite.

Black Mountain. Trade name of Aardvark Clay & Supplies (USA) for a deep brown-black clay (cone 10) with medium texture.

black nickel oxide. A gray-black powder made by heating nickel oxide (NiO) above 400°C (752°F). Crystal system, cubic. Density, 6.67 g/cm³; melting point, reduced to nickel oxide (NiO) at 600°C (1112°F). Used in storage batteries, as a source of nickel oxide, and as a colorant in porcelain enamels, glass and glazes. Formula: Ni_2O_3. Also known as *nickelic oxide; nickel peroxide; nickel sesquioxide.*

black nickel plate. A black nickel electrodeposit (about 1-1.5 μm, or 0.04-0.06 mil thick) produced on a workpiece to obtain a decorative effect, or provide a nonreflecting surface. Black nickel plating baths are usually sulfate or chloride baths containing zinc, ammonium and nickel ions. The deposits are brittle and chip or flake easily on bending or impact.

black oak. The hard, moderately strong wood of the tree *Quercus velutina* belonging to the red oak group. The heartwood is brown with a reddish tinge, and the sapwood white. *Black oak* is similar to *northern red oak*, and has a coarse texture, moderate durability, and is difficult to saw and machine. Average weight, 705 kg/m³ (44 lb/ft³). Source: Eastern and central United States from Maine to Georgia, and from Texas to Wisconsin, southern Ontario. Used for lumber, flooring, veneer, furniture, boxes, pallets, crates, caskets, agricultural implements, and handles. See also northern red oak.

Black Onyx. Trade name of Cleveland Black Oxide (USA) for black oxide coating and process for heat-treated parts.

Blackor. Trade name of Blackor Company (USA) for a ditungsten carbide (W_2C) hardfacing compound containing 87% tungsten, 9% carbon, and 4% of a binder. Used for oilwell core bits, cutters, tools, etc.

Black Pearl. Trade name of Enthone-OMI Inc. (USA) for a decorative black nickel electroplate and process.

black phosphorus. An allotropic form of *phosphorus* in the form of a black, electrically conducting solid made by heating white phosphorus under pressure. It resembles graphite in appearance. Crystal system, monoclinic, Density, 2.25-2.69 g/cm³. Also known as *violet phosphorus.*

black pigment. Lampblack produced by incomplete combustion of coal tar in a closed system with insufficient supply of air. Used in paints, inks, rubber, carbon paper, lead pencils, crayons, ceramics, linoleum, coatings, polishes, soaps, etc. See also lampblack.

black pine. See matai.

black plate. A thin sheet steel, 305 to 915 mm (12 to 36 in.) wide, produced in a tin mill by cold reduction, prior to any cleaning operation. It often has a special lacquer or baked-enamel finish applied, or is chemically treated to resist rust and corrosion. The term *black plate* is a misnomer, since it is not black in color. Used for cans and containers. Also known as *black sheet iron.*

black poplar. The comparatively light and soft, yellowish white wood of the large tree *Populus nigra* that derives its name from the blackish bark. It has a fine, straight grain, low durability, high toughness, withstands rough usage without splintering, and is easy to work. Average weight, 450 kg/m³ (28 lb/ft³). Source: Europe including British Isles. Used for paneling, packaging, inlays, carpentry, cart and wagon bottoms, brake blocks, and paper pulp. Also known as *English poplar.*

black rouge. See black ferric oxide.

black shellac varnish. A black solution made by first dissolving shellac in grain alcohol or, sometimes, wood alcohol, and then adding high-quality lampblack. Used for coating foundry patterns.

black sheet iron. See black plate.

black silicon carbide. A black to bluish-black, tough *silicon carbide* (98.5+% pure) made by heating carbon (e.g., coke) and silicon dioxide (e.g., silica sand) in an electric furnace at about 2000°C (3630°F). It forms hard, brittle, iridescent crystals, and is used in the form of abrasive powder and grains for cutting and grinding metals, and in the manufacture of grinding wheels, hones, etc.

black silver. See stephanite.

Blackskin Admiralty. Trade name of Phelps Dodge Industries (USA) for an admiralty brass containing 70% copper, 29% zinc, and 1% tin. Used for pump equipment, and fixtures on marine equipment.

black solder. A brazing alloy of copper, zinc, and a small amount of tin, used for brazing blackheart malleable iron.

black spruce. The soft, moderately heavy, light-colored wood of the tree *Picea mariana* that is one of the most common trees of the northern forests in Canada and the United States. It is usually marketed together with red and white spruce. The heartwood and sapwood are almost indistinguishable. *Black spruce* has high stability and moderate strength. Source: Northern United States (New England and Lake states); Eastern Canada (from Manitoba to Nova Scotia and Newfoundland). Used mainly as pulpwood and framing lumber, and for millwork, boxes, crates, and piano sounding boards. Also known as *blue spruce; bog spruce.*

black stainless steel. A stainless steel with a lustrous black surface produced by dipping into a fused bath of potassium dichromate ($K_2Cr_2O_7$) and sodium dichromate ($Na_2Cr_2O_7$). Used for electronic components.

black thermit. Thermit consisting of a mixture of finely divided aluminum and black ferric oxide (Fe_3O_4). Used in welding iron and steel. See also thermit.

black tellurium. See nagyagite.

Black Topaz. Trade name of Sur-Fin Chemical Corporation (USA) for a cold jet-black finish for ferrous metals.

black tupelo. See sourgum.

black varnish. (1) A solution of coal-tar pitch used in the paint and coating industries.

(2) Shellac with lampblack used in foundries for pattern-making.

Black Velvet. Trade name of Henkel Surface Technologies (USA) for black-oxide coatings for steel surfaces.

black walnut. See American walnut.

black wash. A coating that consists of 1 wt% graphite in water for application to the surfaces of foundry molds and cores to serve as a barrier between the molten metal and the mold and core surfaces.

black willow. (1) See American willow.

(2) See western black willow.

blakeite. A yellow-brown to reddish-brown mineral composed of iron tellurium oxide, Fe-Te-O. Density, 3.10 g/cm³; refractive index, 2.16. Occurrence: USA (Nevada).

Blakodize. Trademark of Luster-On Products Inc. (USA) for an abrasion- and corrosion-resistant black finish for steel and stainless steel applied by a chemical treatment.

Blanc de Blancs. Trade name for a series of white-colored nylon fibers.

blanc de chine. A bright, white and glazed porcelain from China.

Blanc Fixe. Trade name of Sachtleben Corporation (USA) for precipitated barium sulfate (*artificial barite*) supplied in several grades including *Blanc Fixe N, Blanc Fixe Super F* and *Blanc Fixe Micro*. Used as extenders and pigments in paints and coatings.

blanc fixe. See artificial barite.

Blanco. Trade name of Robert-Leyer-Pritzkow & Co. (Germany) for an extensive series of austenitic, ferritic and martensitic stainless steels.

blanket insulation. Thermal insulation consisting of loosely felted mats of mineral or vegetable fibers, such as rock, slag, fiberglass, wood or cotton, usually enclosed in kraft or other paper, and somtimes surfaced with aluminum or other reflective foil. It is supplied in rolls or strips of varying lengths and widths in thicknesses of 19-304 mm (0.75-12 in.), and used in building construction, especially for insulating exposed wall cavities and attics.

Blanko-Blech. Trade name of Robert-Leyer-Pritzkow & Co. (Germany) for a corrosion-resistant alloy of 80% copper and 20% nickel. Used for condenser tubes and turbine blades.

BlasMaster. Trade name of F.J. Brodmann & Company LLC (USA) for shot-peening and abrasive blasting powders.

blast-furnace coke. A strong, metallurgical coke with a low sulfur, phosphorus and ash content. It is produced by the destructive distillation, in the absence of air, of special grades of bituminous coal. Used in blast furnaces to furnish the heat necessary to attain the desirable chemical equilibria and adequate rates of reaction, and provide the carbon monoxide that is largely responsible for the reduction of the iron ore.

blast-furnace iron. The final product of blast-furnace reduction of iron oxide in the presence of limestone. It consists of about 4-4.6% carbon, 0.8-2% silicon, up to 2.4% manganese, 0.01-0.03% sulfur, 0.25-0.5% phosphorus, and the balance iron. Used as a basic raw material for steel and cast-iron manufacture.

blast-furnace lead. The final product of blast-furnace reduction of lead sinter containing lead oxide, coke and fluxes. It contains many impurities, such as copper, antimony, arsenic, zinc, tin, bismuth, sulfur, and some noble metals, and must be refined by subsequent metallurgical processes.

blast-furnace refractories. The refractory brick and linings used in the construction of blast furnaces. The type of brick used depends upon the particular zone of the furnace, e.g., refractory bricks used at the top are hard-fired, superduty bricks with optimum abrasion resistance, while those in the inwall zone, bosh and hearth are resistant to high temperatures, erosion and slags.

blast-furnace slag. The nonmetallic byproduct that is obtained in the molten condition simultaneously with iron in a blast furnace. It consists primarily of silicates and aluminosilicates of calcium and other alkaline materials. Once solidified, it has a porous structure, and is either crushed and used in the manufacture of concrete aggregate, iron Portland cement, ballast, slag wool and slag bricks, or pulverized and used as a total or partial replacement for cement in the manufacture of concrete products.

Blastikote. Trade name of Metal Coatings International Inc. (USA) for mechanical impact zinc-iron alloy plating.

blasting abrasives. A group of abrasives, such as metallic grit and shot, sand, glass, alumina, silica, etc., used in dry and wet blasting cleaning of ferrous and nonferrous castings, forgings, and weldments, and nonmetallic parts, e.g., rubber, plastics, wood, glass, leather, etc.

blasting shot. Small spherical or elongated particles of chilled white cast iron, cast steel, aluminum, or cut steel wire used in dry abrasive blast cleaning to remove surface contaminants (scale, sand, rust, dry soils, paint, etc.) by impact. See also shot.

blasting sand. A sharp-grained, abrasive sand suitable for blasting cleaning operations. See also abrasive sand; blasting abrasive.

Blastite. Trademark of Washington Mills Electro Minerals Corporation (USA) for aluminum oxide blasting media.

Blastkote. Trade name of Metal Coatings International Inc. (USA) for mechanical-impact zinc-iron alloy plating.

Blast-O-Lite. Trade name of Flex-O-Lite Division (USA) for glass beads used for surface-finishing operations.

Blastox. Trademark of The TDJ Group, Inc. (USA) for a granular, cutting, chemical additive for blending with abrasives used in the removal of lead-based paint.

Blaupunkt. Trade name of SWB Stahlformguss Gesellschaft mbH (Germany) for a high-speed tool steel containing 0.82% carbon, 4.1% chromium, 0.85% molybdenum, 1.6% vanadium, 8.7% tungsten, and the balance iron. Used for tools and dies.

Blaw Knox. Trade name of Blaw-Knox Corporation (USA) for a series of machinery steels, cast steels and nickel-chromium-iron castings, and austenitic cast chromium-nickel and nickel-chromium steels. Used for general machinery, gears, casings, fittings, etc. The austenitic grades are used for corrosion-resistant parts.

blazer cloth. An all-wool or wool blend fabric, either plain-colored or striped, and usually with a short nap on the face side. Used for blazers.

bleaching clay. See activated clay.

bleaching earth. See activated clay.

bleeder cloth. A woven or nonwoven fabric used in vacuum-bag molding of polymers and polymer-matrix composites, usually under the breather cloth and a release film, to allow excess gases and resin to escape. See also breather cloth.

blend. (1) A general term denoting a homogeneous combination of two or more materials, such as metal powders, paints, plastics, solvents, textiles, etc.

(2) In powder metallurgy, an intimate mixture of metal powders.

(3) In the construction industry, the combination of various cementitious materials, such as cement, lime, or mortar.

(4) In the textile industry, a textile material, such as a yarn

or fabric, made up of more than one type of fiber whereby each of the fibers may require a different type of dye.

(5) In the plastics industry, an alloy of two or more immiscible polymers or copolymers with other polymers or elastomers obtained by melting and mechanical mixing. The resulting plastic resin usually has improved performance properties. Alloying permits the combination of resin polymers that cannot be polymerized. Also known as *alloy; polymer alloy; polymer blend.*

Blend-a-gum. Trade name for a silicone elastomer used for dental impressions.

Blendalloy. Trade name of Spang Specialty Metals (USA) for a series of powder-metallurgy materials including numerous cobalt, nickel, iron-nickel, nickel-iron-molybdenum, nickel-copper, and cobalt-nickel alloys.

Blend Base. Trade name of Akzo Coatings America, Inc. (USA) for a series of high-solids acrylic urethane coatings.

blende. See sphalerite.

blended cement. A ceramic product made by mixing Portland cement with granulated blast-furnace slag, pozzolan, or slaked lime, during or after the final grinding of the cement.

blended fabrics. Textile fabrics made wholly or in part from blended fibers or yarns.

blended fibers. A mixture of fibers of a different nature (e.g synthetic and natural fibers), appearance (e.g., coarse and fine fibers), color (e.g., light- and dark-colored fibers), or physical properties (e.g., hard and soft fibers).

blended hydraulic cement. A hydraulic cement composed of Portland cement or Portland cement clinker mixed with two or more inorganic ingredients by blending or intergrinding at the mill to enhance its strength.

blended sand. A mixture of sands having different clay contents and grain sizes. Used chiefly in sand casting and construction.

blended yarn. A yarn composed of different fibers or different component yarns.

Blendex. Trademark of Borg-Warner Corporation (USA) for an acrylonitrile-butadiene-styrene (ABS) powder used in the manufacture of ABS alloys.

Blendur. Trademark of Bayer Corporation (USA) for high-temperature polyurethane casting resins and molding compounds based on polyetherpolyol/isocyanate blends. They are supplied in the form of granules, powders, pastes, emulsions and liquids, and used for electrical and appliance parts, molds, sealing and potting compounds, elastic coating and lining materials, etc.

Blin-Art. Trade name of Santa Lucia Cristal SACIF (Argentina) for toughened float, plate and sheet glass. It is decorated by sand blasting, or special incrustations of copper, steel, wood, etc.

Blindex. Trade name of Santa Lucia Cristal SACIF (Argentina) for toughened float, plate and sheet glass.

Blindovis. Trade name of Fabbrica Pisana SpA (Italy) for bulletproof, multi-laminated plate glass.

Blinduch. Trade name of Santa Lucia Cristal SACIF (Argentina) for toughened float, sheet, rolled, and colored glass used for shower enclosures.

Blin-Lux. Trade name of Santa Lucia Cristal SACIF (Argentina) for toughened and enameled float and plate glass.

Blin-Mur. Trade name of Santa Lucia Cristal SACIF (Argentina) for toughened and enameled float and plate glass.

Blin-Plac. Trade name of Santa Lucia Cristal SACIF (Argentina) for a prefabricated cladding panel having an outer layer of enam-

eled toughened glass.

Blin-Tro. Trade name of Santa Lucia Cristal SACIF (Argentina) for laminated sheet glass with a polyvinyl butyral (PVB) interlayer decorated with a special paint.

Blisan. Trade name of Santa Lucia Cristal SACIF (Argentina) for laminated sheet glass.

blister bar. See blister steel.

blister copper. An intermediate copper (96-99% pure) obtained by blowing copper matte (essentially cuprous sulfide, Cu_2S, and iron sulfide, FeS) in a converter. It contains many impurities including precious metals as well as nickel, bismuth, selenium, tellurium and sulfur, and is cast into pigs for further refining. The name refers to the porous, blistered surface due to the evolution of gases (mainly sulfur dioxide) during solidification.

blister fabric. See cloque.

blister knit. A fabric knit from a special yarn on selected cylinder needles to produce blister-like raised areas on the surface at irregular intervals.

blister steel. A steel made by completely surrounding wrought iron bars with a carbonaceous material, such as charcoal, in a steel container hermetically sealed with clay, and heating to 850-1000°C (1560-1830°F) for a prescribed period of time. The name refers to the porous, blistered surface due to the evolution of gases from the steel during cooling. Also known as *blister bar; cementation steel; cement steel; cemented bar; converted bar; converted steel.*

blitz. A light- or medium-weight fabric, usually made of acetate/rayon or polyester/rayon blends, that has a delicate lateral rib, and is woven with spun weft (filling) yarns and filament warp (longitudinal) yarns.

Blitz Fee. Trade name of Fakirstahl Hoffmanns GmbH & Co. KG (Germany) for a high-speed tool steel containing 0.86% carbon, 4.3% chromium, 0.85% molybdenum, 2.1% vanadium, 12% tungsten, and the balance iron. Used for lathe and planer tools, drills, taps, and hobs.

blixite. A pale yellow mineral composed of lead oxide chloride, $Pb_2Cl(O,OH)_{2-x}$. Crystal system, orthorhombic. Density, 7.35 g/cm³. Occurrence: Sweden.

bloating clay. See expanded clay.

block. (1) See concrete block.
(2) See glass block.

block brick. A brick, larger in size than a standard or jumbo, used to bond adjacent or intersecting walls. See also jumbo brick; standard brick.

block brass. A brass containing about 66.5% copper, 32% zinc, and 1.5% lead. Used for free-machining brass parts.

block copolymer. An essentially linear copolymer in which identical mer units are clustered in blocks along the molecular chain, e.g., –A–A–B–B–A–A–. It is usually referred to as poly(A-*block*-B), e.g., a copolymer of ethylene and propylene glycol may be referred to as poly(ethylene glycol)-*block*-poly(propylene glycol). Also known as *block polymer.* See also copolymer.

blocked polyurethane. The reaction product of phenol and diisocyanate.

block insulation. Rigid thermal insulating materials in the form of rectangular units.

block mica. Mica in the form of sheets or blocks with at least 0.18 mm (0.007 in.) in thickness and an overall usable area of at least 645 mm² (1 in.²).

block polymer. See block copolymer.

Block's alloy. A corrosion- and heat-resistant alloy containing

54% cobalt, 45% nickel, and 1% silicon. Used for tools.

block steatite. A compact, massive variety of *steatite* composed chiefly of *talc*. Used as a ceramic insulator in electronic devices.

block tin. Refined commercial tin, cast into ingots or bars.

Blo-Cote. Trade name of W.S. Rockwell Company (USA) for powder coating products.

bloedite. A colorless mineral composed of sodium magnesium sulfate tetrahydrate, $Na_2Mg(SO_4)_2 \cdot 4H_2O$. Crystal system, monoclinic. Density, 2.23 g/cm^3; refractive index, 1.488. Occurrence: USA. Also known as *astrakanite; astrochanite; blödite.*

Blombit. Trade name of Allgemeine Deutsche Elektrizitätsgesellschaft (Germany) for an alloy containing 93-98% copper and 2-7% silver. It has high electrical conductivity (approximately 83-90% IACS), and is used for electrodes and tips for spot welding.

bloodstone. (1) A dark-green variety of *chalcedony* quartz with small spotty inclusions of red *jasper*. It occurs in India, and is used chiefly as a gemstone. Also known as *heliotrope.*

(2) See hematite.

bloom. A semifinished metal product that has a rectangular cross-sectional area of 232 cm^2 (36 $in.^2$) and is made from an ingot by hot rolling on a blooming mill.

blotting paper. A thick, soft, unsized *absorbent paper* with a spongy surface, used to dry writing by soaking up excess ink.

blowing agent. A chemical substance incorporated into plastics and rubber to produce inert gases by thermal, chemical and/or mechanical action. Common blowing agents include air, ammonium carbonate, sodium bicarbonate, halocarbons, methylene chloride, pentane and hydrazine. Used in the manufacture of foamed plastics and rubber. Also known as *foaming agent.*

blow-molded plastics. Hollow, rigid plastic products of simple or complex shape and varying size formed by first extruding hollow tubes of molten or semi-molten plastics known as "parisons," placing the latter in the cavities of two-piece molds of desired shape and size, and then blowing pressurized air into the parisons, whose temperature and viscosity are controlled, forcing them to expand and assume the contours of the molds. Blow-molded plastics are usually of the thermoplastic type, such as acrylonitrile-butadiene styrenes, nylons, high-density polyethylenes, polycarbonates and polyphenylene ethers.

blown asphalt. A black, friable, solid compounding material obtained by blowing air or steam at high temperature through petroleum asphalt with subsequent cooling. It is a variety of mineral rubber used as a tackifier, softener or extender in rubber products, and in paints and coatings. Also known as *oxidized asphalt.*

blown film. A flat film of polyethylene or other plastic made by extruding a tube through the ring-shaped opening of a die and expanding it into a bubble, followed by cooling below the softening point and collapsing of the bubble.

blown foam. A plastic that has been made porous and spongy by the incorporation of blowing or foaming agents, i.e., chemicals, such as ammonium carbonate or sodium bicarbonate, that facilitate the formation of air or gas bubbles.

blown glass. Glass made by the use of pneumatic air, or mouth blowing.

blown tubing. A thermoplastic film formed from an extruded tube by first expanding the molten tube by the application of a slight internal pressure, followed by cooling to room temperature and flattening by means of suitable mechanisms, and finally reeling the flattened tube up onto a roll.

BLOX. Trademark of Dow Chemical Company (USA) for epoxy thermoplastics supplied in two types: (i) adhesive resins; and (ii) high-adhesion barrier resins. The ideal combination of the flexibility and ease of processing of thermoplastics with the adhesion and durability of epoxy resins makes them suitable for a myriad of applications.

Blue Anchor Drill Rod. Trade name of Teledyne Vasco (USA) for a water-hardening carbon tool steel (AISI type W1) containing 1.2-1.35% carbon, and the balance iron. It has a uniformly spheroidized microstructure, high wear resistance, high hardenability, good machinability, and good dimensional stability. Used for threading taps, twist drills, reamers, shear blades, punches, jeweler's and engraver's tools, dental instruments, and dies.

Blue & White Label. British trade name for a tool steel containing 0.97% carbon, 0.5% vanadium, and the balance iron. It has high hardness and good fatigue properties, and is used for cutting and press tools.

blue asbestos. See crocidolite.

blue ash. The strong, stiff and durable wood of the tree *Fraxinus quadrangulata* which derives its name from the blue dye obtained from the inner bark. It has a straight grain, high shock resistance, and works fairly well. Source: Central United States (chiefly from Missouri, Illinois, Indiana, Ohio, Tennessee and Kentucky). Used for sporting goods, such as tennis rackets, baseball bats, etc., for millwork, furniture, cabinetwork, agricultural implements, tool handles, and as a fuelwood. See also ash.

blue basic lead sulfate. See blue lead.

Bluebell. Trademark of Belfast Ropework plc (USA) for polypropylene fibers and yarns used for cordage.

blue-black porcelain enamel. An alkali borosilicate ground-coat enamel for sheet iron and steel that fires to a dark blue color. The enamel frit is composed of about 33.7% silicon dioxide (SiO_2), 20.2% boric oxide (B_2O_3), 16.7% sodium oxide (Na_2O), 9.2% barium oxide (BaO), 8.5% calcium oxide (CaO), and small adherence-promoting additions of aluminum oxide (Al_2O_3), fluorine (F), manganese dioxide (MnO_2), nickel oxide (NiO), phosphoric oxide (P_2O_5), potassium oxide (K_2O), and cobaltocobaltic oxide (Co_3O_4). Also known as *blue ground coat.*

Blue C. Trade name for a range of nylon 6,6, polyester and spandex (segmented polyurethane) fibers.

Blue Chip. Trade name of Teledyne Firth-Sterling (USA) for a high-speed steel containing 0.7% carbon, 0.25% manganese, 4% chromium, 18% tungsten, 1.15% vanadium, and the balance iron. It is commercially available in the form of rods, bars, flats, and tool shapes, and used for tools, cutters, reamers, punches, drills, shears, taps and dies.

Blue Chip Superior. Trade name of Teledyne Firth-Sterling (USA) for a high-speed steel containing 0.67% carbon, 0.2% manganese, 4.1% chromium, 15.9% tungsten, 0.7% vanadium, and the balance iron. Used for high-speed cutting tools.

blue copper. See azurite.

blue copperas. See copper sulfate pentahydrate.

blue copper carbonate. See azurite.

blue copper ore. See azurite.

Blue Core Build-up. Trade name for a composite paste used in restorative dentistry for core build-ups.

blue cotton. A derivative of trisulfonated copper phthalocyanine cellulose consisting of cotton bearing trisulfocopper phthalocyanine residues. Used in medicine, biochemistry, hygienics and biotechnology to absorb mutagenic, condensed aromatic compounds. See also copper phthalocyanines.

Blue Cross. Trade name of American Biltrite, Inc. (USA) for pressure-sensitive tape.

Blue-Cut. Trade name of Kokour Company Inc. (USA) for a stainless steel buffing compound.

blued sheet steel. A cold-rolled sheet steel whose surface has been oxidized by a special annealing treatment to produce a bluish, corrosion-resistant, very thin surface layer of magnetite (Fe_3O_4), usually less than 1 μm (40 μin.) thick. Supplied in commercial and drawing qualities, it has good yield and tensile strengths, high elongation, and good formability. Used for stove pipes, and food-processing equipment, such as cake and pastry molds and baking trays.

Bluedac. Trade name of Champion Rivet Company (USA) for an all-position arc-welding electrode for low-carbon steel. It is composed of 0.08% carbon, 0.6% manganese, 0.2% silicon, and the balance iron.

Blue Devil. Trade name of Champion Rivet Company (USA) for a series of carbon and low-alloy steel welding electrodes.

blue dogwood. The heavy, close-grained wood of the hardwood tree *Cornus alternifolia*. Source: Eastern United States. Used for shuttles for weaving looms, pulleys, mallet heads, and bobbins. See also dogwood.

Blue Flag. Trade name for fiberglass reinforcing mat used for underground pipeline coatings.

Blue Flash. Trade name of Dresser Industries Inc. (USA) for abrasives and grinding wheels.

blue glass. Ordinary soda-lime glass to which a small amount of cobalt oxide or cupric oxide has been added to give a blue or greenish-blue color, respectively.

blue gold. A corrosion-resistant alloy containing 66.7-75% gold and 25-33.3% iron. Used for jewelry and ornaments.

Blue Grit. Trade name of 3M Company (USA) for coated abrasives.

blue ground coat. See blue-black porcelain enamel.

blue iron earth. See vivianite.

blue john. A compact violet to purple variety of the mineral *fluorite* found in the UK, and used chiefly as an ornamental stone.

Blue Label. Trade name of Peninsular Steel Company (USA) for a series of water-hardening tool steels (AISI types W1 and W2) containing 1-1.4% carbon, up to 0.25% silicon, up to 0.3% manganese, 0.1-0.3% vanadium, and the balance iron. They have excellent machinability, good wear resistance and toughness, and low hot hardness. Used for hand tools, cutting tools, such as drills, hand taps and cutters, cold-swaging dies, forming rolls, cold punches, knives, pipe cutters, drill bushings, and collets.

Blue Label Extra. Trade name of Wallace Murray Corporation (USA) for a water-hardening tool steel (AISI types W1) containing about 0.6-1.4% carbon, 0.25% silicon, 0.25% manganese, and the balance iron. Used for punches, drills, taps and cutters.

blue lead. (1) A bluish-gray pigment composed of 45+% lead sulfate ($PbSO_4$), 30+% lead oxide (PbO), 12+% lead sulfide (PbS), up to 5% lead sulfite ($PbSO_3$) and zinc oxide (ZnO), respectively, and up to 3% carbon and other materials. It is obtained as a byproduct in the smelting of lead ores in special furnaces, and is used as a corrosion inhibitor in paints, plastics, rubber and lubricants, and in priming-coat paints for iron and steel. Also known as *blue basic lead sulfate; sublimed blue lead.*

(2) In the lead industry, a general term for metallic lead as distinguished from certain lead compounds with other than the characteristic bluish-gray color, e.g., orange lead, red lead, white lead, etc.

(3) A general term referring to any lead product that has not been changed chemically during manufacture.

(4) See galena.

blue mahoe. See mahoe.

blue malachite. See azurite.

Blue Mountain. Trademark of IMC Industry Corporation (USA) for *nepheline syenite* used in ceramic products, such as porcelain, pottery, and tile.

Bluenose. Trade name of Nor-Var Paints Limited (Canada) for spar varnish for exterior wood surfaces.

blue ocher. See vivianite.

Blue-Pane. Trade name of Nippon Sheet Glass Company Limited (Japan) for a pale-blue, heat-absorbing glass.

Bluephase P. Trade name for a silicone elastomer for dental die applications.

blue porcelain enamel. A weather-resistant, blue porcelain enamel that is used in cover coats on sheet iron and steel. It is produced from a frit containing mainly silicon dioxide (SiO_2), lead monoxide (PbO), sodium oxide (Na_2O), boric oxide (B_2O_3), titanium dioxide (TiO_2), fluorine (F), lithium oxide (Li_2O), calcium oxide (CaO), and cobaltocobaltic oxide (Co_3O_4).

Blueral. Trade name of Central Glass Company Limited (Japan) for light greenish-blue, heat-absorbing glass.

Blue Ridge. Trade name of ASG Industries Inc. (USA) for patterned glass.

blue salt. See nickel sulfate hexahydrate.

Blue Seal. Trade name of Sanderson Kayser Limited (UK) for a water-hardening tool steel containing 0.9-1% carbon, and the balance iron. Used for springs, gages, and tools.

blue spar. See lazulite.

blue spruce. See black spruce.

blue steel plate. Rolled steel plate with a bright, grayish blue surface scale.

bluestone. (1) A greenish or bluish-gray feldspathic sandstone that splits readily into thin, smooth slabs. It is hard, compact, and fine-grained, and is used as flagstone. Occurrence: Eastern United States.

(2) A sandstone of even texture and bedding that contains a substantial amount of clay or clay minerals.

(3) See chalcanthite.

Blue Streak. Trade name of Diehl Steel Company (USA) for a series of cobalt-tungsten, tungsten and molybdenum-type high-speed tool steels.

Blue Tip. Trade name of Mueller Brass Company (USA) for a highly corrosion-resistant naval brass containing 60% copper, 0.75% tin, and the balance zinc. Used for marine hardware, valve stems, bolts, and nuts.

blue verdigris. Bright blue powder or shiny crystals obtained by the action of acetic acid on copper in the presence of air. Both blue verdigris and green verdigris (the green variety of basic copper acetate) are known as *true verdigris*. Used as a pigment, in antifouling paints, and as a mildew preventive and mordant. See also basic copper acetate; copper subacetate; green verdigris; false verdigris; true verdigris.

blue vitriol. See chalcanthite.

bluing salt. A solution of 0.9 kg/L (7.5 lbs/gal) sodium hydroxide and 0.3 kg/L (2.5 lbs/gal) sodium nitrate heated to 150°C (300°F), and used in the formation of oxidized blue surfaces on steel.

Blu-Mousse. Trade name for a silicone elastomer used for dental impressions.

Blu-Sil. Trade name of Perma Flex Rubber Company (USA) for silicone rubber.

blythite. A mineral of the garnet group composed of manganese silicate, $Mn_3^{2+}Mn_2^{3+}Si_3O_{12}$. Crystal system, cubic. Used in ceramics.

B metal. Trade name for steel castings containing 0.25% carbon, 0.3% silicon, 0.7% manganese, and the balance iron. Used for general-purpose engineering applications.

B-Metall. See Bahn-Metall.

BMG formers. See bulk metallic glass formers.

B-Monel. Trademark of Inco Alloys International Inc. (USA) for a series of cast alloys containing 61-68% nickel, 27-33% copper, 2.5% iron, 2.7-3.7% silicon, 1.5% manganese, and 0.30% carbon. They have excellent corrosion resistance as well as high strength and wear resistance. Used for rotating parts, wear rings, and stainless parts. See also monel.

board. (1) In softwood terminology, a square-sawn timber 50 mm (2 in.) or less thick, and 100 mm (4 in.) or more wide.

(2) In hardwood terminology, a square-sawn or unedged timber 50 mm (2 in.) or less in thickness, and of varying width.

(3) In ceramics terminology, a ceramic material supplied in the form of a flat, thin square or rectangular slab.

(4) In the construction trades, a flat, usually rectangular piece of acoustical or thermal insulating material.

board insulation. Lightweight insulation manufactured from glass-fiber or plastic-foam materials (e.g., expanded or extruded polystyrene, phenolics or polyurethane) in the form of rigid or semi-rigid boards in a wide range of sizes. It has a high thermal insulating value per unit thickness (also known as R-value or heat-flow resistance value) and, depending on the particular material, can be used for the acoustical and thermal insulation of interior and exterior building walls, attics and floors, and for air and/or vapor barrier applications. See also rigid board insulation; semi-rigid board insulation.

bobbinet. A hexagonal-mesh net fabric, usually of cotton, nylon, rayon or silk, made by machines, and used especially for garments.

bobbin lace. A lace made by hand using a fabric pillow to hold pins around which a thread, held and fed by bobbins (i.e., reels or spools), is wound. Also known as *pillow lace.*

Bobierre's metal. A yellow brass containing 63% copper and 37% zinc. It has good cold workability, and is used for sheathing, bolts, nuts, pins, rivets, etc. See also basis brass.

bobierrite. A colorless or white transparent mineral of the vivianite group composed of magnesium phosphate octahydrate, $Mg_3(PO_4)_2 \cdot 8H_2O$. It can also be made synthetically. Crystal system, monoclinic. Density, 2.19 g/cm^3; refractive index, 1.513.

Bobina. Trade name for rayon fibers and yarns used for textile fabrics.

Bobina-Perlon. Trade name for a series of nylon fibers and yarns used for textile fabrics including clothing.

Bocato. Trademark of J.H. Benecke GmbH (Germany) for flexible polyvinyl chloride film materials.

Bochum. Trade name of SWB Stahlformguss Gesellschaft mbH (Germany) for an extensive series of machinery, case-hardening, high-manganese and plain-carbon tool steels, and a wide range of austenitic, ferritic and martensitic stainless steels.

Bochumer. Trade name of Krupp Stahl AG (Germany) for an extensive series of cold-work, hot-work, plain-carbon and high-speed tool steels, machinery steels, case-hardening steels, rail steels and spring steels.

Bodana. Trade name for rayon fibers and yarns used for textile fabrics.

Bodanita. Trade name for rayon fibers and yarns.

Bodannella. Trade name for rayon fibers and yarns used for textile fabrics.

Bodanyl. Trade name for nylon fibers.

Bodvar. Trade name of Uddeholm Corporation (USA) for a series of permanent magnet materials containing 0.9-1% carbon, 3.5-5.5% chromium, 0.5% tungsten, 2-3.5% cobalt, and the balance iron.

body filler. See body putty.

bodying agent. Any substance used to enhance the viscosity of a paint.

Body Lite. Trade name of Marson Corporation (USA) for polymer-based autobody fillers.

body putty. A pasty mixture of epoxy or polyester resins and talc used in rebuilding holes and gaps in metallic surfaces, especially automobile and truck bodies. Also known as *body filler.*

Bodyrite. Trade name of American Smelting & Refining Company (USA) for a lead-tin solder for automobile and truck bodies.

body solder. A solder, usually of the 80Pb-20Sn type, used for filling dents or seams in automobile and truck bodies.

Böhler. Trade name of Vereinigte Edelstahlwerke/Böhler Gesellschaft mbH (Austria) for an extensive series of carbon and alloy structural and machinery steels (including various free-machining, case-hardening and nitriding grades), bearing, spring and valve steels, austenitic, ferritic and martensitic stainless steels, heat-resistant steels, tool steels (including numerous plain-carbon, cold-work, hot-work and high-speed grades), several nickel and cobalt-base superalloys, and several wrought special-purpose aluminum alloys.

Böhlerit. Trade name of Böhler Gesellschaft mbH (Austria) for a series of sintered carbides used for cutting tools and dies. They are available as straight grades composed of tungsten carbide in cobalt binders, or complex grades consisting of titanium carbide plus tungsten carbide in cobalt binders.

boehmite. A grayish, brownish, or reddish mineral composed of aluminum oxide hydroxide, $AlO(OH)$. It can also be made synthetically. Crystal system, orthorhombic. Density, 3.01-3.07 g/cm^3; melting point, decomposes at 360°C (680°F); refractive index, 1.645. It is a major constituent in certain bauxites and bauxite clays.

Boeing. Trade name of The Boeing Company (USA) for fiber-reinforced plastics used for aircraft applications.

Bofors. Trade name of Bofors AB (Sweden) for an extensive series of plain-carbon, cold-work and hot-work, high-speed and nondeforming tool steels, austenitic, ferritic and martensitic stainless steels, heat-resisting steels, and numerous plain-carbon and alloy constructional and machinery steels including several case-hardening and nitriding grades.

bogdanovite. A rose-brown to bronze-brown mineral composed of copper iron lead gold tellurium, $Au_5(Cu,Fe)_3(Te,Pb)_2$. Crystal system, cubic. Occurrence: Kazakhstan and eastern regions of Russian Federation.

boggildite. A salmon-colored mineral composed of sodium strontium aluminum oxide fluoride phosphate, $Na_2SrAl_2F_9(PO_4)$. Crystal system, monoclinic. Density, 3.66 g/cm^3; refractive index, 1.466. Occurrence: Greenland.

bog iron ore. A low-grade ore of iron containing considerable amounts of clay and other impurities. It is a soft, spongy, porous type of *limonite* composed of impure hydrous iron oxides, and having the approximate formula, $Fe_2O_3 \cdot xH_2O$. It occurs in bogs, swamps, marshes, peat mosses and shallow lakes. Also known as *lake ore; limnite; marsh ore; meadow ore; morass ore; swamp ore*.

bog spruce. See black spruce.

bohdanowiczite. A creamy yellow mineral composed of silver bismuth selenide, $AgBiSe_2$. Crystal system, hexagonal. Density, 7.87 g/cm^3; refractive index, 1.645. Occurrence: Poland.

Bohemian garnet. See pyrope.

Bohemian glass. A hard glass, usually of the lime-potash type, with high silica content. Used for making brilliant glassware for chemical and domestic ware.

Bohemian topaz. See citrine.

Bohler Welding. Trade name of Bohler Thyssen Welding USA, Inc. for an extensive series welding alloys including electrodes, wires, rods, and powders.

Bohn. (1) Trade name of Bohn Aluminum & Brass Corporation (USA) for an extensive series of aluminum alloys including numerous aluminum-silicon-copper types.

(2) Trade name of Bohn Aluminum & Brass Corporation (USA) for a series of low-melting lead-tin-antimony solders.

Bohnalite. Trade name of Bohn Aluminum & Brass Corporation (USA) for an extensive series of commercially pure aluminums and various aluminum-copper, aluminum-magnesium, aluminum-silicon, aluminum-tin, aluminum-zinc, magnesium-aluminum and magnesium-manganese alloys.

Bohn Alloy. Trade name of Bohn Aluminum & Brass Corporation (USA) for a series of brasses and bronzes including numerous manganese bronzes and leaded and unleaded brasses and bronzes.

Bohnalloy. Trade name of Bohn Bearing Division of BOHN/ADEC Group (USA) for an extensive series of copper-lead, copper-tin-lead, lead-antimony-tin, copper-zinc and copper-zinc-lead alloys. Used for bearings, bushings, and liners.

boiled wool. A knit or woven fabric of wool or a wool blend that has been heavily felted or immersed in a hot chemical solution to produce a coarse, crinkled surface texture.

boiler plate. A flat-rolled plate of carbon or alloy steel, typically 6.4-12.7 mm (0.25-0.50 in.) in thickness, used for making pressure vessels, steam boilers, tanks, chemical reaction equipment, and as cladding on ships and military equipment. Plain-carbon steel containing about 0.15-0.30% carbon and 0.3-0.6% manganese is used for low- and medium-duty applications, and alloy steels, usually manganese, molybdenum or chromium-molybdenum types, for high-duty boiler plate requiring higher tensile and creep strength, and improved resistance. Also known as *boiler steel*.

boiler steel. See boiler plate.

Bokebit. Trade name of H. Boker & Company (USA) for a high-speed tool steel containing 0.8% carbon, 18.8% tungsten, 8% molybdenum, 9% cobalt, 4.2% chromium, 2% vanadium, and the balance iron. Used for tools and cutters.

bokite. A black mineral composed of potassium aluminum iron vanadium oxide hydrate, $Fe_6KAl_3V_6(V_{20}O_{76}) \cdot 30H_2O$. Density, 3.04 g/cm^3; refractive index, 2.03; hardness, 3 Mohs. Occurrence: Kazakhstan.

Bolatron. Trade name of GenCorp Inc. (USA) for fiber-reinforced polypropylenes.

bole. See bolus.

boleite. A deep blue mineral composed of lead silver copper chloride hydroxide trihydrate, $Pb_{26}Ag_{10}Cu_{24}Cl_{62}(OH)_{48} \cdot 3H_2O$. Crys-

tal system, cubic. Density, 5.05 g/cm^3; refractive index, 2.04-2.09. Occurrence: Mexico.

bolivia. A soft, firm, tightly woven wool fabric, often blended with mohair or alpaca fibers, having a cut pile in narrow ribs running in the warp (lengthwise) direction. Used for high-priced coats and cloaks.

Boloney. Trade name of Oman Non-Friction Metal Company (USA) for a bearing alloy of copper, tin, and lead. Used for heavy-duty bearings.

bolster silver. A nickel silver containing 65.5% copper, 16% zinc, 18% nickel, and 0.5% phosphorus. Used for ornaments and hardware.

Bolta. Trade name of GenCorp Inc. (USA) for saran (polyvinylidene chloride) fibers with good chemical resistance, good weatherability, and good resistance to mildew and insects. Used for carpets and rugs, draperies, upholstery, clothing, and industrial fabrics.

Boltaflex. Trade name of GenCorp Inc. (USA) for a vinyl-coated rayon cloth with a grain imitating leather. Used in the manufacture of upholstery.

Boltaflex Saran. Trade name of GenCorp Inc. (USA) for saran (polyvinylidene chloride) fibers used for carpets and rugs, draperies, upholstery, clothing, industrial fabrics, etc.

Boltaron. Trade name of GenCorp Inc. (USA) for polyvinyl chloride sheet and rigid film supplied in various colors and with smooth and leather-imitating surface finishes. Used for credit cards, luggage, maps, wall panels and ceiling tiles, furniture, decorative coverings, etc.

Boltathene. Trade name of GenCorp Inc. (USA) for polyolefin fibers used for woven fabrics.

bolting cloth. An open-mesh fabric woven with uniformly spaced wire, silk, hair or other warp and weft threads to specified mesh sizes. Used for sifting flour or meal, in screen printing, and for needlework.

Boltomet. Trademark of Thomas Bolton Limited (UK) for a series of coppers and copper alloys including various commercially pure coppers (oxygen-free high-conductivity, electrolytically-refined, fire-refined, arsenical, tough-pitch and other grades), chromium coppers, aluminum bronzes and brasses, silicon and tin bronzes, and alpha and beta brasses.

Boltomet L. Trademark of Thomas Bolton Limited (UK) for a brass containing 63% copper and 37% zinc. Supplied in foil, rod, wire and tube form, it has a density of 8.45 g/cm^3 (0.305 $lb/in.^3$), a melting point of 900-920°C (1650-1690°F), excellent brazeability, formability and hot workability, and good machinability. Used for fasteners (e.g., rivets and screws), chains, architectural applications, and reflectors.

Bolton. Trade name of Thomas Bolton Limited (UK) for a series of phosphor bronzes.

boltwoodite. (1) A yellow mineral of the uranophane group composed of potassium oxonium uranyl silicate, $K(H_3O)(UO_2)(SiO_4)$. Crystal system, monoclinic. Density, 4.37 g/cm^3. Occurrence: USA (California).

(2) A pale yellow mineral of the uranophane group composed of potassium oxonium uranyl silicate monohydrate, $K(H_3O)UO_2SiO_4 \cdot H_2O$. Crystal system, monoclinic. Density, 3.60 g/cm^3; refractive index, 1.69. Occurrence: USA (Utah).

bolus. A fat, reddish-brown to yellow clay composed mainly of hydrous aluminum silicates. Also known as *bole*.

bolus alba. See kaolin.

Bombay hemp. See sunn hemp.

bombiccite. See hartite.

bonaccordite. A reddish brown mineral of the ludwigite group of nickel iron oxide borate, $Ni_2FeBO_3O_2$. Crystal system, orthorhombic. Density, 5.19 g/cm^3. Occurrence: South Africa.

Bonafil. Trademark of Bonar Textiles Limited (UK) for polypropylene fibers and yarns used for textile fabrics.

Bonaril. Trademark of Dow Chemical Company (USA) for a hydrolyzed polyacrylamide used in foundry sand.

bonattite. A blue mineral composed of copper sulfate trihydrate, $CuSO_4 \cdot 3H_2O$. Crystal system, monoclinic. Density, 2.67 g/cm^3. Occurrence: Canada (British Columbia), Italy.

Bondalcap. Trade name for a polycarboxylate dental cement.

Bondalox. Trademark of Exolon-ESK Company (USA) for brown fused abrasive alumina grains supplied in bulk, and as bonded abrasive products (e.g., paper and cloth).

BondArc. Trademark of Hobart Tafa Technologies, Inc. (USA) for a bond coat for thermal arc-spraying of stainless steels, alloy steels, cast iron, titanium, tantalum, aluminum, nickel, etc.

Bond-A-Strike. Trade name of MacDermid Inc. (USA) for an alkaline electroless nickel strike.

Bondbronze. Trademark for a copper-aluminum alloy coating which is arc-sprayed on annealed or hardened carbon steels, hardened alloy steels, stainless steels, and cast iron to provide corrosion and wear resistance. It has excellent impact, sharp-edge and bond resistance.

bond clay. A clay of high dry-strength and plasticity used to bond nonplastic materials, e.g., molding sand.

bond coating. A thin prime or intermediate coat applied to a substrate to improve the adhesion of subsequent thermal spray coats. Also known as *bonding coat*.

bonded abrasives. Abrasive grains that are sized, graded, bonded and pressed or molded into solid shapes, such as disks (e.g., grinding wheels), sticks, segments, cups, rings, cylinders, mounted points, etc., using bonding media, such as clay-feldspar, sodium silicate, synthetic resins, hard rubber or shellac. Also known as *bonded products*.

Bonded Carbide. Trade name of CCS Braeburn Alloy Steel (USA) for cobalt-tungsten high-speed steels (AISI types T5 and T6) containing 0.7% carbon, 0.2% manganese, 4-4.5% chromium, 1.5-2% vanadium, 18-20% tungsten, 0.7% molybdenum, 7.5-12% cobalt, and the balance iron. They have excellent red hardness and wear resistance, and are used for lathe and planer tools, form tools, cutoff tools, cutters, drills, taps, reamers, and dies.

bonded fabrics. Fabrics composed of a thin face fabric joined to a backing fabric of equal or different material by a suitable bonding agent, e.g., an adhesive resin or foam. See also laminated fabrics.

bonded-fiber fabrics. A group of nonwoven fabrics consisting of one or several webs of natural and/or synthetic fibers bonded together by chemical means, e.g., with adhesives, or by fusion. Used for protective clothing, cleaning pads, furnishings and disposables.

bonded friable alumina. Sized and graded grains of friable alumina, bonded and pressed, or molded into solid shapes using a suitable medium. See bonded abrasives; friable alumina.

bonded leather. An artificial leather made from ground leather by mixing with polyvinyl alcohol, rolling under pressure, and subsequently applying a texture that imitates real leather grains.

bonded magnets. Permanent magnets composed of barium or strontium ferrite, neodymium-iron-boron, samarium cobalt, alnico, etc., bonded with a polymeric or elastomeric material,

such as polyamide, polyphenylene sulfide, epoxy or nitrile rubber. Depending on the particular binder, they can be made into rigid forms by injection molding (polyamides and polyphenylene sulfides) or compression bonding (epoxies), or flexible forms by calendering (nitrile rubbers).

bonded products. See bonded abrasives.

bonded refractories. Refractories in which the components are held or bonded together by a bonding agent.

bonded wood. A composite consisting of wood particles, paper fibers, or other cellulosic particulate or fibers in an organic (e.g., polymer) or inorganic (e.g., gypsum, or Portland cement) matrix. Used in the manufacture of building products, such as wallboard, roofing shingles and shakes.

Bonderite. Trademark of Henkel Surface Technologies (USA) for a series of corrosion-resistant phosphatic paint bases and conversion coatings.

bonderized steel. A steel to whose surface a protective phosphatic coating, usually 5-10 µm (0.2-0.4 mil) thick, has been applied by treating with an aqueous solution of metallic phosphates using immersion, spray or flooding methods. The coating provides a rough, tough, chemical primer base used for decorative or protective finishes.

Bonderlube. Trade name of Parker Chemical Company (USA) for a composition used to lubricate surfaces that have been previously immersion-coated with zinc phosphate. Used as a rust inhibitor, especially in cold-drawing operations.

Bondermetic. Trade name of Libbey-Owens-Ford Company (USA) for metal-to-glass seal used in *Thermopane* double glazing units.

Bondex. (1) Trademark of Bondex International, Inc. (USA) for ready-mix swimming-pool paints for aluminum, concrete and other surfaces.

(2) Trade name of McInerney Plastics Company (USA) for phenol-formaldehyde resins and products.

Bondfast. Trademark of Lepage's Limited (Canada) for a polyvinyl resin emulsion glue supplied ready-to-use in plastic squeeze bottles. It bonds leather, paper, wood, etc., in minutes, and dries transparent.

bond fireclay. See plastic fireclay.

Bondi-Loy. Trade name for a corrosion-resistant cobalt-chromium dental bonding alloy.

bonding additive. See adhesion promoter (2).

bonding agent. (1) A coating or paint applied to hardened concrete to improve the bonding of subsequently applied concrete or mortar.

(2) Any substance added to a concrete or mortar to facilitate bonding.

(3) See binder (2).

bonding coat. See bond coating.

bonding material. Any of various organic materials used to impart or improve adhesion, strength, chemical resistance, weatherability, electrical properties, etc., to ceramic and glass fibers, sheets and molded shapes. Used in the manufacture of fabrics, laminates, electronic and electrical components, insulators, etc.

Bond-It. (1) Trademark of Cotronics Corporation (USA) for high-temperature- and room-temperature-curing epoxies for high-strength bonding of dissimilar materials, e.g., metals, ceramics, composites, plastics, glass, etc. Supplied in tube form, they have high electrical, thermal-shock and corrosion resistance.

(2) Trademark of Jeneric/Pentron Inc. (USA) for primer/adhesive bonding agents used for dental restorations.

Bondlite. Trade name for a *dentin* bonding agent.

Bondo. Trademark of Dynatron/Bondo Corporation (USA) for a series of autobody repair products (e.g., plastic body fillers, liquid hardeners and fiberglass resins) and marine products (e.g., fiberglass woven rovings, pourable polymer foams, gel-coat, epoxy and general-purpose finishing resins, and epoxy fairing compounds for boat and ship bottoms and keels).

Bond On. Trademark of Degussa Dental (USA) for a series of dental bonding alloys.

Bond-1. Trademark of Jeneric/Pentron Inc. (USA) for a light-cured *dentin* primer/adhesive used for bonding dental restorations.

Bond-On 4. Trademark of Degussa Dental (USA) for a silver-free palladium dental bonding alloy.

bond plaster. A calcined *gypsum plaster* suitable for use on rough monolithic concrete as a bond coat for additional gypsum plaster coats.

Bondrite. Trade name of Plasma Powders & Systems Inc. (USA) for composite wires consisting of solid nickel cores with aluminum sheaths, and used for applying thermal-arc spray bond coats. The wires produce enhanced bond strength, and effect homogeneous spray patterns.

Bond-set. Trademark of Fairey & Company Limited (Canada) for firebrick mortar.

Bond 3 Plus. Trade name for a silicone elastomer used for dental impressions.

Bondtie. Trademark of American Manufacturing Company (USA) for polyethylene and polypropylene fibers and monofilaments.

Bondur. Trade name of Vereinigte Aluminium-Werke AG (Germany) for an age-hardenable high-strength alloy containing 3.5-5.5% copper, 0.3-0.5% silicon, 0.25-1% manganese, 0.2-0.7% magnesium, and the balance aluminum. Used for structural components, aircraft parts, transportation equipment, fittings, and fasteners.

Bondurplate. Trade name of Vereinigte Leichtmetallwerke GmbH (Germany) for an age-hardenable clad alloy composed of a plate made from an aluminum alloy containing 2.5-5% copper, 0.2-1.8% magnesium, 0.3-1.5% manganese and 0-1.2% silicon, coated (or clad) with an aluminum alloy of different composition (0-1.5% manganese, 0-1% magnesium and 0-1.2% silicon). Used for aircraft applications, e.g., wings, fuselages, and in building construction.

Bondwich. Trade name of Texas Instruments Inc. (USA) for a shim metal with silver brazing alloy bonded (or clad) to both sides. The shim absorbs stresses during cutting and brazing, and is used for sandwich brazing of carbide-tipped cutting tools.

bone. The hard substance that forms vertebrate bone. It consists of an organic matrix of collagen, gelatin and traces of elastin and cellular materials and fats in which large amounts (about 65-70 wt%) of mineral matter are deposited. The composition of the mineral matter of dried, fat-free bone can be given by the formula $3Ca_3(PO_4)_2 \cdot CaX_2$, where X_2 is usually CO_3, but may also be F_2, SO_4, O or OH. *Note:* From an engineering standpoint, *bone* is a natural composite material composed of a strong, yet soft organic matrix and hard, brittle mineral matter. In addition to the natural substance, modern bioengineering has made possible the synthesis of bone or bone-like materials in the laboratory.

bone ash. A white, porous ash containing 67-85% tribasic calcium phosphate, $Ca_3(PO_4)_2$, and small amounts of magnesium phosphate, calcium carbonate and calcium fluoride. It is usually obtained by calcining bones in air, although tribasic calcium phosphate made by synthetic means is also referred to as *bone ash*. Used as an opacifier and fluxing ingredient in milk glass, porcelain, porcelain enamel and pottery, in cleaning and polishing compounds for metals and nonmetals, in mold coatings, and for jewelry cleaning purposes.

bone black. See animal black.

bone cement. A polymeric cement that may or may not contain antibiotic additives, and is used in the repair or bonding of bone.

bone char. See animal black.

bone charcoal. See animal black.

bone china. A soft, translucent chinaware of high quality made from a whiteware body containing a minimum of 25% bone ash as a fluxing agent. A typical composition is 50% *bone ash*, 25% *kaolin* and 25% *Cornish stone*. It has a water absorption of 0.3-2%, a relatively low-firing temperature, and is used chiefly for dinnerware.

bone clay. A high-quality *kaolin* used in the manufacture of high-strength porcelain.

bone glue. A hard gelatin obtained from animal bones (green bone glue). Also, the pale-amber adhesive obtained from this substance by boiling in water. During boiling, the bone proteins mix with water, swell and, upon cooling, solidify to a horn-like mass. *Bone glues* are applied hot, and bind on cooling. Used chiefly in carpentry.

BoneSource. Trademark of Stryker Leibinger (USA) for a self-setting bone cement composed of a calcium phosphate salt mixture that when mixed with water forms a putty-like substance which hardens to a strong microcrystalline *hydroxyapatite* in the body and adheres well to the surrounding bone. It is biocompatible, partly resorbed and replaced by natural bone, and used as a bone graft substitute for filling cranial defects, such as craniotomy cuts, neurosurgical burr holes, etc.

Bondilene. Trade name of Matador Converters Company Limited (Canada) for bonded polyester battings.

Bongrip. Trade name of Pechiney/Eurotungstène (France) for a tungsten carbide with cobalt binder, used for antiskid tire studs, stutted straps, and snow and ice chains.

Bonpoly. Trademark of Bongaigaor Refineries & Petrochemicals Limited (India) for polyester staple fibers.

bonshtedite. A colorless mineral of the bradleyite group with a rose, greenish or yellowish tint composed of sodium iron carbonate phosphate, $Na_3Fe(PO_4)(CO_3)$. Crystal system, monoclinic. Density, 2.95 g/cm³; refractive index, 1.568. Occurrence: Russian Federation.

Bontile Super Red. Trade name of Bonar Packaging Limited (Canada) for red plastic stretch film used for packaging applications.

Bontite. Trade name of Bonar Packaging Limited (Canada) for plastic shrink-wrap and stretch film.

Bontone. Trademark of Bondex International, Inc. (USA) for interior and exterior acrylic latex paints.

book cloth. A usually colored, filled textile fabric of varying composition in a plain weave, and often with a calendered or embossed surface. Used in bookbinding.

bookform splittings. Sheets of mica from the same block arranged in individual books, or bunches. See also mica; splittings.

book mica. See sheet mica.

book paper. A bulky, machine-coated, usually high-grade paper used in book and magazine printing.

Book Tex. Trade name of Atlas Powder Company (USA) for cellulose nitrate plastics.

Booster. Trademark of DuPont Canada Inc. (USA) for polyethylene concentrates.

Booth. Trade name of Alcan-Booth Industries Limited (UK) for a series of aluminum- and copper-base alloys.

boothite. A blue mineral composed of copper sulfate heptahydrate, $CuSO_4 \cdot 7H_2O$. Crystal system, monoclinic. Occurrence: USA (California).

Booth-Kote. Trade name of Klem-Seco Division of Stan Sax Corporation (USA) for a spray-booth coating.

Boothstrip. Trade name of Spraylat Corporation (USA) for strippable spray-booth coatings.

Bora. Trade name of Thyssen Edelstahlwerke AG (Germany) for a series of tool and die steels.

boracic acid. See boric acid.

boracite. A colorless, blue, yellow, green, or grayish white, pyroelectric mineral composed of magnesium chloride borate, $Mg_3B_7O_{13}Cl$. Crystal system, orthorhombic. Density, 2.90-2.95 g/cm^3; refractive index, 1.662; hardness, 7 Mohs. Occurrence: Germany, UK.

Boral. (1) Trade name of Brooks & Perkins Inc. (USA) for a composite made by adding up to 50 wt% boron carbide (B_4C) crystals or powder to molten aluminum and subsequent rolling. The resulting sheet product is then clad on both sides with commercially pure aluminum. It has high thermal neutron absorption, and is used for nuclear reactor shields, safety rods, neutron curtains, shutters for thermal curtains, and spent-fuel storage containers.

(2) Trade name of KB Alloys Inc. (USA) for lightweight aluminum-boron alloys with high electrical conductivity.

Boralectric. Trademark of Advanced Ceramics Corporation (USA) for graphite and boron nitride heating elements.

Boralloy. Trademark of Advanced Ceramics Corporation (USA) for nonporous, exceptionally pure pyrolytic boron nitride (BN) synthesized by high-temperature low-pressure chemical vapor deposition. It is a refractory ceramic with an apparent density of 1.95-2.22 g/cm^3, high strength even at high temperatures, high electrical resistivity and outstanding dielectric properties, negligible outgassing, high oxidation resistance, good directional thermal conductivity and thermal-shock resistance, and a maximum use temperature of 2500°C (4530°F). Its thermal, electrical and mechanical properties are anisotropic. Used for applications, such as electrical insulation, heating elements, crucibles and boats for growing compound semiconductor crystals, insulators, crucibles and hardware for high-temperature vacuum processes, and as a coating on graphite susceptors, heat shields, nozzles, evaporation boats, etc.

Boralyn. Trade name of Alyn Corporation (USA) for a powder-metallurgy aluminum-matrix composite reinforced with 15% boron carbide (B_4C) particles. Produced by cold-isostatic pressing, it can be fabricated by casting, rolling, extrusion and forging, and has high specific stiffness and good weldability. Used for aircraft landing gears, sporting goods, data-storage disk substrates, etc.

borane. (1) Any of a class of compounds of boron and hydrogen, such diborane, pentaborane, etc. Most boron hydrides are unstable, react readily with air or water, and are used as boron fuels, and in boron plastics. There are also several organoboranes that are suitable for use in nickel plating of metals and plastics. Many *boranes* form complexes with solvents, such as dimethylamine, dimethylsulfide, morpholine, butylamine, pyridine and tetrahydrofuran. Also known as *boron hydride*.

(2) A derivative of a boron hydride, e.g., triethylboron, boron fluoride, etc.

borate glass. A glass with little or no silica in which boric oxide (B_2O_3) is the chief glass-forming ingredient. It may contain tantalum and lanthanum oxides, has a low dispersion, a relatively high refractive index, and is used for camera lenses, and eyepieces for wide-angle binoculars.

Borawire. Trade name of Johnson Matthey plc (UK) for a hardened alloy of 61% gold and 39% platinum. It has a melting range of 1080-1180°C (1975-2155°F), and is used in wire form for dental applications.

borax. A colorless, white, gray, yellow, green, or bluish mineral composed of sodium borate decahydrate, $Na_2B_4O_7 \cdot 10H_2O$. Crystal system, monoclinic. Density, 1.71-1.73 g/cm^3; hardness, 2-2.5 Mohs; refractive index, 1.468; melting point, loses $10H_2O$ between 75 and 320°C (167 and 608°F); fuses to a glassy mass at red heat (borax glass). Occurrence: Found in salt lakes and alkali soils, e.g., in the USA (California). Used as an ore of boron, as a powerful flux and glass-forming agent in glass, glazes, porcelain enamel, etc., as a flux in smelting, welding and brazing, in metal fluxing, as a chemical reagent, in photography, and as a buffer in biology, biochemistry and medicine. Also known as *borax decahydrate; sodium borate; sodium borate decahydrate; sodium pyroborate; sodium tetraborate decahydrate; tincal*.

borax decahydrate. See borax.

borax glass. A glass in which *borax* is the principal glass-forming component in combination with silica.

Borazon. Trademark of General Electric Company (USA) for *cubic boron nitride* produced at extremely high pressure and temperature from ordinary boron nitride, or from mixtures of boron and nitrogen. It has a cubic structure, and is supplied in the form of small crystals. The color ranges from colorless or white for chemically-pure grades to reddish or black. It has a density of 3.5 g/cm^3 (0.13 lb/in.³), a hardness of 10 Mohs (as hard as diamond), and high stability up to about 1925°C (3500°F). Used as an abrasive and cutting-tool material. See also cubic boron nitride.

borcarite. A mineral composed of calcium magnesium borate carbonate hydroxide, $Ca_4Mg(B_4O_6(OH)_6)(CO_3)_2$. Crystal system, monoclinic. Density, 2.77 g/cm^3. Occurrence: Siberia.

Borcher's alloy. Any of a group of heat- and corrosion-resistant chromium-iron-nickel, nickel-cobalt-chromium and nickel-chromium-gold-silver alloys. The Cr-Ni-Fe alloys contain about 32.5-65% chromium, 35-60% iron, 0-35% nickel, 1.8-4% molybdenum and 0-0.5% silver, and are used for chemical apparatus, pyrometer tubes and crucibles. The Ni-Co-Cr alloys contain about 34-35% nickel, 34-35% cobalt, 30% chromium, 0-2% silver, 0-1% molybdenum, and are used for chemical apparatus, heat-treating parts and annealing pots. The Ni-Cr-Au-Ag alloys contain about 65-68% nickel, 30% chromium, 0.15-5% gold and 0.15-1.5% silver, and are used for electrical resistances, and for heat- and corrosion-resistant parts. Also known as *Borcher's metal*.

Borcher's metal. See Borcher's alloy.

Borcoloy. Trade name of General Aircraft Equipment Company (USA) for a series of iron-cobalt casting alloys containing varying amounts of boron. They have high heat and corrosion resistance, excellent cutting properties, and good abrasion and wear resistance. Used for general and cutting tools.

Boreal. Trade name of Vitrobrás (Brazil) for patterned glass.

Borealis. Trade name of Pilkington Brothers Limited (UK) for a patterned glass.

Borel. Trade name of Electroverre Romont SA (Switzerland) for an electromelted glass.

Borgolon. Trademark of Torcituradi Borgomanero SpA (Italy) for nylon 6 fibers and yarns.

borickite. A reddish-brown, isotropic mineral composed of calcium iron phosphate hydroxide trihydrate, $CaFe_5(PO_4)_2(OH)_{11}\cdot 3H_2O$.

boric acid. White, water-soluble powder, or colorless to white crystals or scales (99+% pure) usually made by treating a solution of borax with hydrochloric or sulfuric acid and crystallizing. Density, 1.435 g/cm³; melting point, loses water to form first metaboric acid (HBO_2), then pyroboric acid ($H_2B_4O_7$) and finally boric oxide (B_2O_3). Used as a flux in glazes, porcelain enamels, glass pastes, borosilicate and other special glasses, glass fibers and cements, as a flux in the manufacture of cast-iron and sheet-steel porcelain enamel frits, as a flux in welding and brazing, in nickel electroplating baths, and in chemistry, biochemistry, medicine and electrophoresis as a component (pH 9.0) in tris-EDTA-borate buffers. Formula: H_3BO_3. Also known as *boracic acid; orthoboric acid.*

boric acid gel. A crosslinked polymer of gel-like consistency formed by a reaction of dihydroborylanilino-substituted methacrylic acid with 1,4-butanediol dimethacrylate. It is available in particle sizes ranging from 0.1 to 0.4 mm (0.004 to 0.015 in.) for the column chromatographic separation of mixtures which contain ingredients that form more or less stable complexes with boric acid.

boric anhydride. See boric oxide.

boric oxide. Colorless or white powder or vitreous crystals (99+% pure) obtained by heating boric acid (H_3BO_3). Crystal system, polymorphous. Density, 2.46 g/cm³; melting point, 450°C (842°F); boiling point, approximately 1860°C (3380°F). Used in the manufacture of cements, heat-resistant glass and porcelain enamels, as a fluxing agent, as a flame retardant for paints, in electronics, materials research and liquid encapsulation, and as a thermal-neutron absorber. Formula: B_2O_3. Also known as *boron oxide.*

boric oxide glass. (1) A glass in which boric oxide in combination with silica is the principal glass-forming component. It is transparent to ultraviolet radiation. Also known as *boron oxide glass.*

(2) A clear, colorless glass or amorphous powder composed of boric oxide. It has a density of 1.86 g/cm³, a boiling point above 1500°C (2730°F), and high hardness and brittleness. Used as a fire-resistant ingredient in paint, and as an ingredient in heat-resistant glassware.

boride-alumina composites. A group of porous, highly electrically conducting composites, such as TiB_2-Al_2O_3 and ZrB_2-Al_2O_3, obtained by aluminothermic reaction in sintering titanium or zirconium dioxide, and boron oxide and aluminum powder under prescribed conditions of pressure and temperature. Used as electrodes for fused-salt electrolysis, and for filters and separators for corrosive metals and salts.

boride coatings. Hard, oxidation-resistant coatings produced on alloy and tool steels, cobalt- and nickel-base superalloys and refractory metals by vapor-phase deposition or diffusion coating. Used on rocket nozzles, turbine blades and vanes, aerospace and marine parts, etc.

borides. A group of nonstoichiometric intermetallic compounds of boron and an alkaline-earth, rare-earth or transition metal having compositions ranging from Me_3B to MeB_{12} (where Me represents the metal). Usually made by vapor-phase deposition, fused-salt electrolysis or high-temperature sintering of mixtures of metal powder and boron, they have high hardnesses (from 8 to 10 Mohs), tensile strengths and elastic moduli, high melting points (up to about 3260°C or 5900°F), excellent strength retention and oxidation resistance at high temperatures, high thermal conductivity, good thermal-shock resistance, high chemical stability, good resistance to acids, poor resistance to hot alkalies, and low electrical resistivity. Used chiefly for high-temperature ceramics, e.g., in rocket nozzles, turbine blades and vanes and aerospace parts, and in structural ceramics, and refractory coatings.

Borite. Trade name of Swedish American Steel Corporation (USA) for a water-hardening tool steel containing 1.0-1.2% carbon, and the balance iron. Used for tools and drills.

Borium. Trade name of Stoody Corporation (USA) for a hard, abrasion-resistant material composed of tungsten carbide (WC) and ditungsten carbide (W_2C), and used for hardfacing and welding rods and hard inserts.

Borneo camphorwood. See kapur.

Borneo cedar. The reddish wood of several species of trees of the genus *Shorea.* It has medium strength, and good workability and nailing qualities. Source: Borneo. Used as a mahogany substitute, and for superior joinery and interior construction work. Also known as *Borneo mahogany.*

Borneo mahogany. See Borneo cedar.

Borneo rosewood. See ringas.

bornemanite. A pale yellow mineral composed sodium barium titanium niobium fluoride phosphate silicate hydroxide, Na_4Ba-$Ti_2NbSi_4O_{17}(F,OH)\cdot Na_3PO_4$. Crystal system, orthorhombic. Density, 3.50 g/cm³; refractive index, 1.687. Occurrence: Russian Federation.

bornhardtite. A rose-colored mineral of the spinel group composed of cobalt selenide, Co_3Se_4. Crystal system, cubic. Density, 6.17 g/cm³. Occurrence: Germany.

bornite. A copper-red to pinchbeck-brown mineral that turns purple or blue on exposure. It has a metallic luster, a grayish-black streak, and is composed of copper iron sulfide, Cu_5FeS_4, containing theoretically 63.3% copper. Crystal structure, cubic. Density, 5.03-5.09 g/cm³; hardness, 3 Mohs. Occurrence: Canada, Chile, Europe, Peru, South Africa, USA (Arizona, California, Montana). Used as an important ore of copper. Also known as *erubescite; horseflesh ore; peacock ore; purple copper ore; variegated copper ore.*

boroaluminate. See aluminum borate (1) and (2).

Borobest. Trade name of Krupp Stahl AG/Bochumer Verein (Germany) for a series of hardenable martensitic stainless steels containing 0.15-0.4% carbon, 13% chromium, and the balance iron. Used for cutlery, turbine blades and surgical instruments.

borocarbides. A class of compounds which are composed of boron, carbon, a rare-earth metal, such as erbium, holmium, thulium or yttrium, and a transition metal, such as iron, nickel or palladium, and exhibit unusual superconductive behaviors in that they are normal materials (i.e., non-superconductors) below their critical temperatures. The latter may range from as low as 7.5K for holmium nickel borocarbide ($HoNi_2B_2C$) to over 23K for yttrium palladium borocarbide (YPd_2B_2C).

borocarbon. A compound manufactured from boron and carbon in an electric furnace, and used in particular for electrical resistors.

Borod. Trade name of Stoody Company (USA) for a sintered material composed of 62% tungsten carbide and 38% steel. It has excellent heat and abrasion resistance, and is used for hardfacing and welding rods.

Borofloat. Trademark of Schott Glas AG (Germany) for a temperature-resistant borosilicate float glass.

Boroflux. Trade name for boron carbide containing flake graphite, and used as a fluxing agent in casting.

Borolite. Trade name of Borolite Corporation (USA) for a sintered boron carbide used for cutting tools.

Borolon. Trade name for aluminum oxide made by fusing bauxite, and used as an abrasive and refractory.

boron. A highly reactive, nonmetallic element of Group IIIA (Group 13) in the Periodic Table. It exists in the form of a black, hard solid, a soft, yellow to brown amorphous powder, and a yellow crystalline substance. It is commercially available in the form of lumps, pieces, powder, microfoil, zone-refined rods, filaments, whiskers, single crystals and the isotopes ^{10}B and ^{11}B (various enrichments). The commercial boron ores are *borax, colemanite, kernite* and *ulexite*. Density, 2.34-2.45 g/cm³; melting point, 2300°C (4170°F); hardness, 9.5 Mohs (3300 Knoop); refractive index, 3.4; atomic number, 5; atomic weight, 10.811; trivalent; amphoteric; high neutron absorption capacity. It exists in three forms: (i) *Alpha boron* (rhombohedral crystal structure); (ii) *Beta boron* (rhombohedral crystal structure); and (ii) *Gamma boron* (tetragonal crystal structure). *Boron* is used in alloy steels, in the cementation of iron, as a neutron absorber in reactor controls, as a regulator in nuclear plants, in high-temperature brazing alloys, as an oxygen scavenger for copper and other metals, as a catalyst, as an abrasive, in welding and brazing fluxes (borax), for boron-coated tungsten wires, heat-resistant glassware, fibers and filaments in metal or ceramic-matrix composites, in semiconductors, and in rocket propellants. Symbol: B. See also amorphous boron.

boron-10. A stable isotope of boron with mass number 10. It constitutes about 18.5-19% of natural boron, and is available in the form of crystalline and amorphous powders. It efficiently absorbs slow neutron, emitting high-energy alpha particles in the process. Used in neutron counters, and in neutron and radiation shielding. Symbol: ^{10}B.

boron-11. A stable isotope of boron with mass number 11. It constitutes about 81-81.5% of natural boron, and is transparent to neutron. Abbreviation: ^{11}B.

boron alloys. Uniformly dispersed mixtures of boron with one or more metals. *Ferroboron* is a boron-iron master alloy containing 15-25% boron used as a hardening, degasifying, and deoxidizing agent in special steels, and to increase the high-temperature strength of alloy steels. *Manganese-boron* is a boron-manganese master alloy containing up to 25% boron, and small amounts of iron, silicon and aluminum used in the manufacture of nonferrous alloys, such as brass and bronze, and for hardening and deoxidizing applications.

boron-aluminum composites. A group of lightweight composite materials consisting of matrices of aluminum reinforced with high-modulus continuous boron fibers. They have anisotropic mechanical properties, high strength-to-weight ratios, high stiffness, high thermal conductivity, low coefficients of thermal expansion, and a maximum use temperature of 510°C (950°F). Used for structural tubular struts for the NASA Space Shuttle, and for jet-engine parts, aircraft-wing skins, landing-gear components, as a heat-dissipating or cold-plate materials for microchip carriers, as a neutron shielding materials, for bicycle frames, etc.

boron arsenide. A compound of boron and arsenic. Crystal system, cubic. Crystal structure, sphalerite. Density, 5.22 g/cm³; hardness, 1900 Knoop; melting point, 2027°C (3681°F); band gap, 1.5 eV. Used as a semiconductor. Formula: BAs.

boronatrocalcite. See ulexite.

boron carbide. A compound of boron and carbon produced by reduction of boron oxide with carbon in an electric furnace. It is commercially available in the form of hard, black crystals, gray-black powder, hot-pressed sheets or rods, fibers, whiskers and sintered lumps. Crystal system, rhombohedral. Density, 2.45-2.52 g/cm³; melting point, 2350°C (4260°F); boiling point, above 3500°C (6330°F); hardness, 9.3 Mohs (almost as hard as diamond). It has a high capture cross section for thermal neutrons, fair resistance to concentrated acids, alkalies, halogens and metals, and good resistance to dilute acids. Used as a grinding and lapping powder, as an abrasive in form of grinding wheels, belts, papers and powders, for abrasion-resisting parts, refractories, nuclear reactor nozzles and control rods, electrical-resistance heating elements for high-temperature furnaces, as reinforcement in composites, and as an deoxidizer for casting copper. Formula: B_4C. Also known as *carbon tetraboride; tetraboron carbide.*

boron carbide powder. A gray-black powder composed of boron carbide with a density of 2.52 g/cm³, used for grinding and lapping. It is available in sizes from 220 mesh (for grinding) to 800 mesh (for lapping), and is also used as an alloying addition to aluminum to produce sheets with high neutron absorption capacity (e.g., Boral).

boron carbide whiskers. Very fine single crystal fibers of boron carbide with high tensile strength and elastic moduli used as reinforcements in composites.

boron-carbon composites. Composite materials consisting of metallic or organic matrices reinforced with boron-carbon fibers, i.e., boron filaments deposited on carbon. Due to the high reactivity of boron with most metals, and the fact that boron fibers, whose outer surfaces are mainly boron oxide, decompose at temperatures above 500°C (930°F), the fibers must be specially conditioned (e.g., coated with silicon carbide, boron carbide, or boron nitride) before use in composites.

Boron-Cu. Trade name of Belmont Metals Inc. (USA) for a deoxidizing alloy used in the manufacture of boron-deoxidized copper.

boron-deoxidized copper. A high-purity copper (99.99%) containing 0.01% boron. It has high resistance to grain growth, oxide penetration and thermal stress cracking, and good resistance to corrosion. Used for electrical and electronic parts, magnetrons, synchrotrons, and vacuum switchgears.

boron-epoxy composites. Composite materials consisting of epoxy matrices reinforced with high-strength, high-modulus boron fibers. They have relatively low maximum service temperatures (only about 120°C or 248°F), but good mechanical properties. Used for structural aerospace applications, helicopter rotor blades, sporting goods, etc.

boron fibers. A group of synthetic inorganic, high-strength, high-modulus fibers produced as rather large, continuous monofilament fibers or "wires" by chemical vapor deposition of elemental boron onto tungsten or pyrolyzed carbon substrates. They have a low density (about 2.35-2.57 g/cm³), excellent thermal stability and stiffness, superior tensile, compressive and flexural strengths, good oxidation resistance up to 500°C (930°F), large fiber diameters (approximately 100-140 μm or 0.004-0.005 in.), and smooth outer surfaces. Used in the production of boron-epoxy preimpregnated tape or prepregs, and as high-performance reinforcements in advanced organic- and metal-matrix composites for structural aerospace and aircraft applications. Also known as *boron filaments*.

boron filaments. See boron fibers.

boron hydride. See borane.

Boronized coating. A relatively hard, boron-rich surface layer produced on iron and steel.

boron-loaded concrete. A special heavy concrete composed of a boron-containing aggregate or ingredient, such as a boron frit, a boron alloy or a mineral, e.g., colemanite or kernite. Used as in nuclear applications as neutron attenuator.

boron mica. A soft, flexible, synthetic *mica* with the approximate composition, $KMg_3BSi_3O_{10}F_2$, and a melting point of 1150°C (2100°F). Used for electrical insulating application. Also known as *boron phlogopite*.

boron nitride. A compound of boron and nitrogen produced by heating a mixture of boric acid and tricalcium phosphate in an ammonia atmosphere in an electric furnace. It is commercially available in the form of white, crystals or powder, hot-pressed sheets or rods, fibers, or whiskers. Crystal system, hexagonal. Density, 1.80-2.25 g/cm³; melting point, sublimes at 3000°C (5430°F); hardness, 2 Mohs (powder). It has a upper continuous-use temperature of 950-1200°C (1740-2190°F), high thermal and electrical resistance, excellent heat-shock resistance, low mechanical strength, good resistance to metals, fair resistance to acids and alkalies, and poor resistance to halogens. Used as a refractory in crucibles, as furnace insulation, as a high-temperature lubricant (for glass molds), in parts of pumps for molten metals, in chemical equipment, as an abrasive in special metal-grinding operations, as a high-temperature dielectric, for heat-resistant and high-strength fibers, as a matrix material for engineering composites, for heat shields, rocket nozzles, vacuum-tube separators, seals and gaskets, self-lubricating bushings and machine tools, as a neutron absorber in nuclear applications, in rectifying tubes, transistor and rectifier mounting wafers, and as a filler in encapsulating and potting compounds. Formula: BN. See also cubic boron nitride; hexagonal boron nitride; wurtzite-type boron nitride.

boron nitride fibers. Heat-resistant, high-strength fibers composed of boron nitride. They have good thermal stability up to 870°C (1600°F) in oxidizing atmospheres, and good chemical and electrical resistance. Used as reinforcements in engineering plastics and composites, and in filters for hot chemicals.

boron nitride fullerene. See boron nitride nanotube.

Boron Nitride Hardcoat. Trade name of ZYP Coatings, Inc. (USA) for abrasion- and wear-resistant high-temperature boron nitride coatings for ladles, crucibles, etc.

Boron Nitride Lubricoat. Trademark of ZYP Coatings Inc. (USA) for wear-resistant protective coatings with excellent resistance to molten aluminum, magnesium and zinc.

boron nitride nanotube. A relatively new material analogous to a carbon nanotube, but composed of a single sheet of boron nitride rolled into a tube with ends sealed with spherical fullerene molecules. It has semiconductive properties, better oxidation resistance than carbon nanotubes, and is used for electronic and high-temperature applications. Abbreviation: BN nanotube. Also known as *boron nitride fullerene*. See also carbon nanotubes; inorganic nanotubes.

Boron Nitride Releasecoat. Trade name of ZYP Coatings Inc. (USA) for release-type boron nitride top coatings for permanent molds.

Boron Nitride TPC. Trade name of Orpac, Inc. (USA) for a boron nitride transition plate coating used on billet casters.

boron oxide. See boric oxide.

boron oxide glass. See boric oxide glass.

boron phlogopite. See boron mica.

boron phosphate. A white crystalline substance (99.99+% pure). Density, 2.520 g/cm³; melting point, vaporizes at 1400°C (2550°F). Used as a constituent in special ceramic bodies and special glasses, and as an acid cleaner. Formula: BPO_4.

boron phosphide. A compound of boron and trivalent phosphorus available in low and high-temperature forms. *Low-temperature boron phosphide* is a refractory, noncorrosive, maroon powder. Crystal system, cubic. Crystal structure, sphalerite. Density, 2.97 g/cm³; melting point, approximately 2527°C (4580°F); hardness, 9.5 Mohs. Used as a semiconductor and electroluminescent materials. *High-temperature boron phosphide* is available in the form of hexagonal crystals with a wurtzite (zincite) structure, and used as a semiconductor. Formula: BP.

boron polymer. A polymer that contains boron-arsenic, boron-phosphorus or boron-nitrogen compounds.

boron powder. Boron in the form of an amorphous or crystalline powder. Amorphous powders (typically 99.995% pure) are available in particle sizes from submicron to over 5 μm (less than 40 to over 200 μin.). Crystalline powders are supplied in various purities (from 90% to over 99.9%) and particle sizes from 35 μm to over 4 mm (0.0014 to over 0.16 in.).

Boron Pyralloy. Trade name for a pyrolytic graphite containing some boron, and used for nuclear shielding applications.

boron silicide. Any of the following compounds of boron and silicon: (i) *Tetraboron silicide*. A black, free-flowing crystalline powder. Density, 2.46 g/cm³; melting point, decomposes at 1093°C (2000°F). Formula: B_4Si; (ii) *Hexaboron silicide*. A black crystalline powder. Density, 2.43 g/cm³; melting point, 1946°C (3535°F). Formula: B_6Si; and (iii) *Triboron silicide*. A crystalline powder. Density, 2.64 g/cm³; melting point 1927°C (3500°F). Formula: B_3Si. All are used in the manufacture of high-temperature ceramics and protective coatings.

boron steels A group of alloy steels containing small amounts of boron (approximately 0.0005-0.003%) to increase hardenability. They are identified in the AISI-SAE designation system by the letter "B" after the first two digits, e.g., 86B45 or 94B30. Such small amounts of boron may also be added to (i) structural steels to improve deep-hardening properties, (ii) case-hardening steels to increase core hardness, (iii) chrome-nickel steels to increase elastic limits, and (iv) molybdenum steel (with 0.50% or more molybdenum) to increase yield and tensile strengths. Since boron absorbs high amounts of neutrons, it may be added in amounts up to 2% to steels for nuclear applications.

Boron-T. Trade name of British Steel Corporation (UK) for an oil-hardening, impact-resistant steel containing 0.4% carbon, 1.6% manganese, 0.003% boron, 0.8% other elements (e.g., nickel, chromium, and molybdenum), and the balance iron. Used for crankshafts, axles, shafts, gears, pinions, and fasteners.

boron-tungsten composites. Composite materials consisting of metallic or organic matrices reinforced with boron-tungsten fibers, i.e., boron filaments deposited on tungsten. Due to the high reactivity of boron with most metals, and the fact that boron fibers, whose outer surface is mainly boron oxide, decompose at temperatures above 500°C (930°F), the fibers must be specially conditioned (e.g. coated with silicon carbide or, boron carbide or boron nitride) before use in composites.

Borosil. Trade name of SiMETCO (USA) for an alloy of 38-42%

silicon, 3-4% boron, and the balance iron. Used to introduce boron into steel.

borosilicate crown glass. An optical-grade crown glass containing substantial amounts of silicon dioxide (SiO_2) and boron oxide (B_2O_3). They are very corrosion-resistant, and exhibit high light transmissivity. Abbreviation: BSC glass.

borosilicate glass. A silicate glass containing 5% or more boron oxide (B_2O_3). It commercially available in six grades: low-expansion, low-electrical-loss, optical, ultraviolet light-transmitting, sealing (glass-to-metal sealing), and laboratory-apparatus. The composition varies greatly from grade to grade with *low-expansion glasses,* such as *Pyrex,* containing about 80.5% silicon dioxide (SiO_2), 12.9% boron oxide (B_2O_3), 3.8% sodium oxide (Na_2O), 2.2% aluminum oxide (Al_2O_3), and 0.4 potassium oxide (K_2O), and *low-electrical-loss glasses* having 70% silicon dioxide (SiO_2), 28% boron oxide (B_2O_3), 1.2% lead oxide (PbO), 1.1% aluminum oxide (Al_2O_3), and 0.5% potassium oxide (K_2O). *Borosilicate glass* has a low coefficient of expansion, high softening point of about 593°C (1100°F), high heat and thermal-shock resistance, a continuous use temperature of about 480°C (900°F), and excellent chemical durability. Used for heat-resistant glassware, in the chemical industry, for insulators, glass-to-metal seals, and for glass fibers for composite reinforcement. Abbreviation: BS glass. See also low-expansion glass.

Boro-Silicon. Trademark of Reactor Experiments, Inc. (USA) for a solid, self-extinguishing, heat-resistant, field-castable elastomer.

Borotex. Trade name of Surmet Corporation (USA) for a duplex cubic boron nitride-based coating deposited in hard vacuum to molds and dies to impart hard, chemically stable, lubricating release surfaces that are not wet by molten metals.

borovskite. A dark gray mineral composed of palladium antimony telluride, Pd_3SbTe_4. Crystal structure, cubic. Density, 8.12 g/cm³. Occurrence: Russian Federation.

borsic-aluminum composites. Composite materials consisting of aluminum or aluminum-alloy matrices reinforced with continuous high-strength borsic fibers. They have high specific strength and moduli, and low density. Used for structural applications.

borsic fibers. Continuous, high-strength fibers prepared by the vapor deposition of a layer of boron on thin tungsten wires (average diameter, 10 μm or 394 μin.). If used in metal-matrix composites, they must subsequently be coated with a thin layer of silicon carbide to retard undesirable reactions that occur between boron and many metals.

Borsten. Trade name for vinyon (vinyl chloride) fibers with excellent acid, alkali and mildew resistance. Used for textile fabrics and carpets.

bort. Impure, imperfect, off-colored diamond or diamond fragments unsuitable for cutting into gemstones. It is an industrial diamond of which black diamond, or carbonado is a variety. Used as an abrasive for grinding or polishing hard materials, as a bonded tip on cutting tools, and in borehole drilling. See also black diamond; carbonado; synthetic diamond.

Bortal. Trademark of Galt Alloys, Inc. (USA) for ferroboron used as a degasifier and deoxidizer in steelmaking.

Bortam. Trade name of NL Industries (USA) for an alloy of 16-18% titanium, 13-15% aluminum, 22-24% manganese, 20-25% silicon, 0-1% carbon, 1.5-2% boron, and the balance iron. Used for adding deep-hardening properties to steel.

Bosch. Trade name of Robert Bosch Metallwerk GmbH (Germany) for a series of aluminum- and copper-base alloys.

Bo-Stan. Trade name of Alloy Technology International, Inc. (USA) for a sintered austenitic stainless steel (AISI type 304) containing 1-2% boron. Available in the form of foil and sheet, it is used in neutron shielding for pressurized-water reactors.

bostwickite. A dark brownish red mineral composed of calcium manganese silicate hydrate, $CaMn_6Si_3O_{16}·7H_2O$. Refractive index, 1.798. Occurrence: USA (New Jersey).

BOTACEM. Trade name of BOTAMENT Systembaustoffe GmbH & Co. KG (Austria) for drywall and spackling compounds and concrete and mortar repair products.

BOTACT. Trade name of BOTAMENT Systembaustoffe GmbH & Co. KG (Austria) for concrete and mortar joint and crack sealing cements and compounds, and various building boards.

botallackite. A pale green mineral composed of copper chloride hydroxide, $Cu_2Cl(OH)_3$. Crystal structure, monoclinic. Density, 3.60 g/cm³; refractive index, 1.800. Occurrence: UK (Cornwall).

Botany wool. A fine merino wool used for the manufacture of high-quality yarns and fabrics. It is named after Botany Bay, Australia, near which it was originally grown.

BOTAZIT. Trade name of BOTAMENT Systembaustoffe GmbH & Co. KG (Austria) for bituminous roof sealing courses and various building repair cements and compounds.

BOTON. Trade name of BOTAMENT Systembaustoffe GmbH & Co. KG (Austria) for a range of concrete and mortar additives.

botryogen. An orange-red mineral composed of magnesium iron sulfate hydroxide heptahydrate, $MgFe(OH)(SO_4)_2·7H_2O$. Crystal structure, monoclinic. Density, 2.14 g/cm³; refractive index, 1.530. Occurrence: Chile.

botting clay. A plastic fireclay of high refractoriness used to plug tapholes of cupolas and melting furnaces.

Bottle. Trade name of SA Glaverbel (Belgium) for patterned glass depicting linked, long-necked bottles in relief.

bottle brick. A hollow building product of clay, 305 mm (12 in.) in length, 76 mm (3 in.) in outer diameter and 13 mm (0.5 in.) in wall thickness, that resembles a bottomless bottle in appearance. Used in the construction of arches, beams or flat slabs.

bottle glass. A *soda-lime glass* usually composed of about 72-76% silicon dioxide (SiO_2), 14-16% sodium oxide (Na_2O), 8-10% calcium oxide (CaO) and up to 1% aluminum oxide (Al_2O_3). The SiO_2 acts as a network former, and the other ingredients facilitate the formation of glass. Green bottles are due to intentionally or unintentionally introduced iron impurities. Sparkling bottle glass contains some barium, and ultraviolet absorbing bottle glass some cerium.

Bottomkote. Trademark of International Paints (Canada) Limited for a copper-base antifouling paint for wood and steel boat hulls.

bouclé. (1) Any crimped, looped yarn. See also bouclé yarn.

(2) A woven or knitted fabric with a spongy feel, made with bouclé yarn (1). It has small loops and curls on its surface.

bouclé yarn. A plied yarn, usually acrylic, nylon, polyester, viscose, wool, worsted, or a blend from these, having a rough, buckled appearance produced either by varying the twisting tension between two yarns, or by unevenly winding a gimping yarn around a core yarn to leave the latter sporadically exposed. See also bouclé; gimped yarn.

boulangerite. A bluish-gray mineral of the lillianite group composed of lead antimony sulfide, $Pb_5Sb_4S_{11}$. Crystal structure, monoclinic. Density, 6.23 g/cm³. Occurrence: Canada, France.

boule. A pure single crystal of silicon, sapphire, etc., often bottle-

or pear-shaped. It is grown in a special furnace by rotating a small seed crystal while slowly pulling it out of the melt. Used in the manufacture of microchips, bearings and thread guides, and formerly phonograph needles.

Bounce-Back. Trade name of Solutia Inc, (USA) for a water-reversible crimp bicomponent acrylic fiber used for the manufacture of colorfast, high-quality knitting and crochet yarns with excellent shape retention after laundering.

Boundary. Trademark of Pope & Talbot Limited (Canada) for softwood lumber.

Bound Brook. Trade name of GKN Powder Met Inc. (USA) for a sintered alloy of copper, tin, phosphorus and graphite. Used for bearings and bushings.

Bourbounes. British trade name for an alloy of 51% tin and 49% aluminum, with traces of iron and copper.

Bourdon lace. A lace on a net ground usually having a scroll-like design with a heavy cord outline.

bourette. (1) A fancy, plied novelty yarn that has differently colored nubs and knots.

   (2) A hairy fabric having irregularly spaced nubs and knots of fiber, and produced with a yarn spun from short, carded waste silk.

bourrelet. A weft-knitted fabric with a horizontally ribbed and rippled texture made by double knitting.

Bourne Fuller. Trade name of Republic Steel Corporation (USA) for an air-hardening steel containing 0.5% carbon, 3.5% tungsten, and the balance iron. Used for bulldozing and gripper dies, piercers, and extrusion punches.

bournonite. A steel-gray to black mineral composed of copper lead antimony sulfide, $CuPbSbS_3$. Crystal structure, orthorhombic. Density, 5.83 g/cm^3. Occurrence: Czech Republic. Also known as *berthonite; cogwheel ore.*

boussingaultite. A colorless, or yellowish-pink mineral of the picromerite group composed of ammonium magnesium sulfate hexahydrate, $(NH_4)_2Mg(SO_4)_2 \cdot 6H_2O$. It can also be made synthetically by room-temperature evaporation of an aqueous solution of $(NH_4)_2SO_4$ and $MgSO_4$. Crystal structure, monoclinic. Density, 1.72 g/cm^3; refractive index, 1.4705.

Bower. Trade name of Republic Steel Corporation (USA) for a case-hardened chromium-nickel steel used for jet-engine bearings.

bowieite. A mineral composed of rhodium sulfide, Rh_2S_3. Crystal structure, orthorhombic. Density, 6.40 g/cm^3.

bowstring hemp. Hemp-like fibers obtained from the leaves of a perennial herb (genus *Sansevieria*) of the lily family, native to Sri Lanka, but also found in Africa and East India. Used as a substitute for true hemp. Also known as *sansevieria.*

bow wire. A corrosion-resistant wire containing 93% copper, 5% tin and 2% zinc.

boxboard. Paperboard used to manufacture boxes and cartons including cereal, shoe and tissue boxes, and detergent and milk cartons, but excluding corrugated cardboard. See also corrugated cardboard; paperboard.

boxcalf. A fine-grained, chrome-tanned, aniline-dyed calfskin, glossy or ground, usually finished with the grain side boarded or stamped with irregular rectangular crosslines, and used for shoe uppers.

box cloth. A firm wool fabric of any weave having a fibrous surface concealing the threads. Used for billiard table covers and gaiters.

box-elder. The soft, white wood of a medium-sized maple tree *Acer negundo*. The heartwood is pale reddish-brown, and the sapwood white with a light reddish-brown tinge. Source: Eastern and central United States, southern Canada. Used for lumber, veneer, crossties, boxes, pallets and crates, and as a pulpwood. Also known as *ashleaf maple; Manitoba maple.*

box leather. A high-quality leather with parallel surface creases or wrinkles produced by "boarding," a process that usually involves heating and pressing.

box metal. A metal or alloy, such as antifriction alloy, brass or bronze, used for the journal boxes of axles and shafts.

boxwood. The hard, fine-grained, durable light-yellow wood of various box trees, especially the European or Turkish box tree (*Buxus sempervirens*), used for making tool handles, musical instruments, rulers, mathematical instruments, engraving blocks, inlays, etc. Other commercial boxwoods are West Indian or Maracaibo boxwood (*Gossypiospermum praecox*), and kamassi or knysna boxwood (*Gnioma kamassi*) and African or Cape boxwood (*B. macowani*), both from Africa. See also Maracaibo boxwood; kamassi boxwood; African boxwood.

Boyd. Trade name of Time Steel Service Inc. (USA) for a tool steel containing 0.75% carbon, 0.75% manganese, 0.9% chromium, 1.75% nickel, 0.35% molybdenum, and the balance iron. Used for blanking and forming dies, shears, shear blades, and woodworking tools.

boyleite. A white mineral of the starkeyite group composed of magnesium zinc sulfate tetrahydrate, $(Zn,Mg)SO_4 \cdot 4H_2O$. Crystal structure, monoclinic. Density, 2.41 g/cm^3; refractive index, 1.531. Occurrence: Germany.

BP Polystyrene (HIPS). Trade name of BP Chemicals (UK) for high-impact polystyrenes supplied in standard, fire-retardant, and UV-stabilized grades.

BP Polystyrene (PS). Trade name of BP Chemicals (UK) for high-impact polystyrenes supplied in standard, medium-impact, and UV-stabilized grades.

brabantite. A white or green mineral of the monazite group composed of calcium thorium phosphate, $CaTh(PO_4)_2$. It can also be made synthetically. Crystal structure, monoclinic. Density, 5.28 g/cm^3; refractive index, 1.780. Occurrence: Namibia.

bracewellite. A dark red-brown to black mineral of the diaspore group composed of chromium oxide hydroxide, $CrO(OH)$. Crystal structure, orthorhombic. Density, 4.47 g/cm^3. Occurrence: Guyana.

brackebuschite. A dark brown to black mineral composed of lead manganese vanadium oxide monohydrate, $Pb_2(Mn,Fe)(VO_4)_2 \cdot H_2O$. Crystal structure, monoclinic. Density, 6.05 g/cm^3. Occurrence: Argentina.

bradleyite. A light gray mineral composed of sodium magnesium carbonate phosphate, $Na_3Mg(PO_4)(CO_3)$. Crystal structure, monoclinic. Density, 2.72 g/cm^3; refractive index, 1.546. Occurrence: USA (Wyoming).

Braeburn. Trade name of CCS Braeburn Alloy Steel (USA) for a series of molybdenum- or tungsten-type high-speed tool steels. Used for cutting tools, high-quality hand tools, shears, punches, and dies.

Braecut. Trade name of CCS Braeburn Alloy Steel (USA) for a molybdenum-type high-speed tool steel (AISI type M44) containing 1.15% carbon, 4.25% chromium, 2.2% vanadium, 5.2% tungsten, 12% cobalt, 6.25% molybdenum, and the balance iron. It has high hot hardness, excellent wear resistance, and good resistance to decarburization. Used for cutting tools for high temperature alloys, e.g., lathe tools, form and milling cutters, end mills, drills, broaches, etc.

Braefour. Trade name of CCS Braeburn Alloy Steel (USA) for a molybdenum-type high-speed steel (AISI type M4) containing 1.25% carbon, 4.5% chromium, 4% vanadium, 5.5% tungsten, 4.5% molybdenum, and the balance iron. It has excellent abrasion and wear resistance and high hot hardness. Used for lathe and planer tools, milling cutters, end mills, reamers, broaches, hobs, chasers, gages, and cutting, forming and finishing tools.

Braemax. Trade name of CCS Braeburn Alloy Steel (USA) for a molybdenum-type high-speed tool steel (AISI type M42) containing 1.1% carbon, 9.5% molybdenum, 8% cobalt, 3.7% chromium, 1.5% tungsten, 1.1% vanadium, and the balance iron. It has high abrasion and wear resistance and high hot hardness. Used for lathe tools, milling cutters, end mills, reamers, broaches, hobs, gear shapers, drills, taps, and form tools.

Braemow. Trade name of CCS Braeburn Alloy Steel (USA) for a series of high-speed steels containing 0.6-0.8% carbon, 5% molybdenum, 5-6.5% tungsten, 4-4.2% chromium, 1-2% vanadium, 0-0.2% manganese, and the balance iron. It has excellent wear resistance, good toughness and high hot hardness. Used for planing and shaping tools, reamers, broaches, taps, drills, chasers, milling cutters, and cutting tools.

Braemow Special. Trade name of CCS Braeburn Alloy Steel (USA) for a series of high-speed steels containing 0.66% carbon, 5% molybdenum, 6.5% tungsten, 4% chromium, 1.7% vanadium, and the balance iron. Used for planing and shaping tools.

Braetuf. Trade name of CCS Braeburn Alloy Steel (USA) for a high-speed steel (AISI type M44) containing 1.15% carbon, 7.25% cobalt, 5-6.75% tungsten, 5-6.7% molybdenum, 4% chromium, 2% vanadium, and the balance iron. It has high hot hardness, excellent wear resistance, and good resistance to decarburization. Used for cutting tools for tough superalloys, milling cutters, end mills, broaches, reamers, drills, taps, and cold-extrusion punches.

Braetwist. Trade name of CCS Braeburn Alloy Steel (USA) for a cobalt-molybdenum high-speed steel (AISI type M33) containing 0.9% carbon, 0.3% manganese, 0.3% silicon, 9.5% molybdenum, 1.6% tungsten, 3.75% chromium, 1.15% vanadium, 8% cobalt, and the balance iron. Used for broaches, end mills, milling cutters, reamers, taps, and drills.

Braevan. Trade name of CCS Braeburn Alloy Steel (USA) for a series of molybdenum-type high-speed steels (AISI type M3) containing 1-1.2% carbon, 5-5.7% molybdenum, 5.6-6.25% tungsten, 4% chromium, 2.5-3.3% vanadium, and the balance iron. It has excellent wear resistance, high hot hardness, and good toughness. Used for lathe and planer tools, reamers, counterbores, broaches, hobs, drills, taps, end mills, milling cutters, and form tools.

Brage. Trade name of Uddeholm Corporation (USA) for a series of water-hardening tool steels containing 1.05-1.10% carbon, 0.25-0.50% chromium, 0-1% tungsten, 0.1% vanadium, and the balance iron. Used for threading tools and twist drills.

braggite. A steel-gray mineral composed of platinum palladium sulfide, $(Pt,Pd)S$. Crystal structure, tetragonal. Density, 8.90 g/cm^3. Occurrence: South Africa.

braid. A ribbon or cord formed by weaving together three or more strands of natural or synthetic fibers.

braided composites. Braided structures with two, three or more braiding-yarn systems with or without laid-in yarns, with the yarns consisting of different materials. Typical examples of such composites include graphite-epoxy, fiberglass-epoxy and glass-polyester composites. They have good damage tolerance, and the ability to limit the damage area. Used in the manufacture of sporting goods (e.g., golf clubs, squash and tennis rackets, ski equipment, fishing rods, etc.), radomes, automotive and marine parts, and aerospace and aircraft parts. See also braid.

braided fabrics. Flat or tubular fabrics consisting of interwoven or interlaced strands.

braided rope. A rope made by weaving together three or more strands of natural or synthetic fibers. It is slightly superior in strength to *twisted rope*, has less tendency to elongate under tension and unlay under load, and is used largely for small-diameter lines, e.g., sash cords, clothesline, etc.

braitschite. A white to reddish pink or colorless mineral composed of calcium rare-earth cerium borate heptahydrate, $(Ca,Na_2)_7(Ce,Ln)_2B_{22}O_{43} \cdot 7H_2O$. Crystal structure, hexagonal. Density, 2.90 g/cm^3; refractive index, 1.646. Occurrence: USA (Utah).

Brake Die. (1) Trade name of Peninsular Steel Company (USA) for hardened and tempered die steels containing 0.51% carbon, 0.87% manganese, 0.95% chromium, 0.2% molybdenum, and the balance iron. Used for press-brake dies, fixtures, tools, and dies.

(2) Trade name of Atlas Specialty Steels (Canada) for a machinery steel containing 0.5% carbon, 1.1% manganese, 0.65% chromium, 0.15% molybdenum, 0.03% phosphorus, 0.08% sulfur, and the balance iron. Usually supplied in the heat-treated condition, it has high tensile strength, good machinability, and good toughness and wear resistance. Used for press-brake, spring-pad and gooseneck dies, hemming and curling dies and fixtures, differential forgings, and machinery parts, such as shafts, gears, drive gears, speed gears, worm gears, gear shafts, driveshafts, hoisting engine parts, motor parts, etc. Also known as *Atlas Brake Die*.

brammallite. A white mineral of the mica group composed of sodium aluminum silicate hydroxide, $NaAl_2(Si,Al)_4O_{10}(OH)_2$. Crystal structure, monoclinic. Density, 2.88 g/cm^3; refractive index, 1.531. Occurrence: Kazakhstan, UK.

branched polyethylene. A low- or medium-density polyethylene having repeating units both on the principal chain and in the branches.

branched polymer. A thermoplastic or network-type polymer having a molecular structure of side or secondary chains that extend from the main or primary chains. See also hyperbranched polymer; star-branched polymer.

Brandalen. Trade name of Höhn & Höhn GmbH (Germany) for a series of polyethylenes.

Brandalit. Trademark of Höhn & Höhn GmbH (Germany) for acrylonitrile-butadiene-styrene resins.

Brandobutyrad. Trademark of Höhn & Höhn GmbH (Germany) for cellulose acetate butyrate plastics.

brandtite. A colorless or white mineral of the roselite group composed of calcium manganese arsenate dihydrate, $Ca_2Mn(AsO_4)_2 \cdot 2H_2O$. Crystal structure, monoclinic. Density, 3.67 g/cm^3; refractive index, 1.711. Occurrence: Sweden.

Brandur. Trade name of Höhn & Höhn GmbH (Germany) for a series of polyvinyl chlorides.

brannerite. A black, opaque mineral composed of titanium uranium oxide, $(U,Ca,Fe,Y,Th)Ti_2O_6$. It can also be made synthetically. Crystal structure, monoclinic. Density, 6.35 g/cm^3; refractive index, 2.30. Occurence: USA (Idaho).

brannockite. A colorless mineral of the osumilite group composed of lithium potassium tin silicate, $Li_3KSn_2Si_{12}O_{30}$. Crystal structure, hexagonal. Density, 2.98 g/cm^3; refractive index, 1.567. Occurrence: USA (North Carolina).

BRASS. See Micarta BRASS.

brass. An alloy of copper and zinc that may also contain varying amounts of other metals, such as lead, iron, manganese, aluminum, nickel, silicon, mercury and/or tin. The proportion of copper varies according to use from 55-95%, and the color ranges from bright yellow for high-copper brasses to reddish for high-zinc brasses. Some common brasses are yellow brass (about 35% zinc), red brass (about 15-25% zinc), naval brass (about 37-39% zinc), cartridge brass (about 30% zinc), muntz metal (about 40% zinc) and gilding metal (5% zinc). *Brass* has good castability and machinability, good corrosion resistance and work-hardening properties, good hot workability, good to medium cold workability and good to medium strength. Used for condenser-tube plates, water-heater and evaporator tubes, automotive radiators, housings, fittings, fasteners, locks, piping, hose nozzles and couplings, valves, drain cocks, air cocks, marine equipment, oil gages, flow indicators, electrical components, plating anodes, worm wheels, bearings, bushings, deep-drawn parts, sheets, foil, wire, powder, springs, musical instruments, jewelry, fine arts, architectural applications, clocks, stamping dies, ammunition components, cartridge casings, coins, and solders. See also alpha brass; beta brass.

brass-clad steel. A low-carbon sheet steel coated (or clad) on one side with corrosion-resistant commercial brass (90% copper and 10% zinc), and used for shell cases and bullet jackets.

brass foil. Brass rolled out to a thickness of about 0.005-2.0 mm (0.0002-0.08 in.), and supplied in various tempers (e.g., as-rolled, half-hard or hard) and sizes.

brass ingot metal. Brass used in the manufacture of ingots for subsequent casting or further working. Eight grades are specified by the ASTM containing between 63.5 and 88% copper, 6.5-34% zinc, the balance being tin and/or lead.

brassite. A white synthetic mineral composed of magnesium hydrogen arsenate tetrahydrate, $MgHAsO_4 \cdot 4H_2O$. Crystal structure, orthorhombic. Density, 2.28 g/cm^3; refractive index, 1.546.

Brasslux. Trade name of MacDermid Inc. (USA) for a bright brass electroplate and plating process.

Brassoid. Trade name of American Nickeloid Company (USA) for brass-coated zinc sheet that is easily formed, drawn and stamped into finished parts.

brass ore. See aurichalcite.

brass steel. Brass-coated steel that is easily formed, drawn and stamped into finished parts.

brass tin. Brass-coated tin that is easily formed, drawn and stamped into finished parts.

Brastil. Trade name of NL Industries (USA) for a corrosion-resistant, cast alloy containing 14-15% zinc, 4-5% silicon, and the balance copper. Used for general castings and bearings.

brattice cloth. A coarse, heavy, fire-resistant cotton ot jute fabric in a plain weave that is often coated or impregnated with chemicals to enhance its gas and vapor absorption properties. Used in the mining and other industries, especially as a screen or to confine and direct air.

braunite. A reddish brown mineral composed of manganese silicate, Mn_7SiO_6, or calcium manganese silicate, $CaMn_{14}SiO_{24}$. Crystal structure, tetragonal. The disordered variety (Mn_7SiO_6) has a density of 4.80-4.84 g/cm^3, and can also be made synthetically. The ordered variety ($CaMn_{14}SiO_{24}$) has a density of 4.73 g/cm^3. Occurrence: Europe (Sweden), India, South Africa, South America, USA.

bravoite. A yellowish mineral of the pyrite group composed of iron nickel sulfide, $(Fe,Ni)S_2$. Crystal structure, cubic. Density, 4.66 g/cm^3. Occurrence: Germany, Peru.

Braze. Trade name of Handy & Harman (USA) for an extensive series of brazing filler metals containing varying amounts of copper, silver and/or zinc, and one or more of the following elements: manganese, nickel and tin.

Braze 650. See Handy Silver Solder.

Braze 700. See Handy Silver Solder Medium.

Braze 750. See Handy Silver Solder Hard.

BrazeCoat. Trade name of Degussa AG (Germany) for flexible sheets composed of metallurgically bonded tungsten and/or chromium carbides in metal binders. They have excellent wear resistance, and are used for brazing to parts for application in aggressive environments, such as mining, oil drilling, gas exploration, excavation, etc.

braziers' copper. See coppersmiths' copper.

Brazilian. Trade name of John L. Armitage & Company (USA) for textured powder coatings.

Brazilian cork. A soft, medium-grade cork obtained from the bark of several Brazilian trees (genera *Angico* and *Piptadenia*).

brazilianite. A pale yellow mineral composed of sodium aluminum phosphate hydroxide, $NaAl_3(PO_4)_2(OH)_4$. Crystal structure, monoclinic. Density, 2.98 g/cm^3; refractive index, 1.609. Occurrence: Brazil, USA (New Hampshire). Used chiefly as a gemstone.

Brazilian jute. The strong, durable woody fiber from several Brazilian plants (genus *Hibiscus)* used for making burlap, rope, and caulking.

Brazilian pine. See araucarian pine.

Brazilian rosewood. The hard, heavy wood of the tree *Dalbergia nigra* which is now in very short supply. It is available in colors ranging from cream, brown and red to violet shades with black streaks, and takes a smooth and high polish. Source: Eastern Brazil. Used for veneer for decorative plywood, inlays in cabinetwork, pianos, cutlery handles, and turned items.

Brazilian walnut. The tough, strong, straight-grained wood of the tree *Cordia goeldiana* used for cabinetwork, furniture and cooperage. Also known as *frejo*.

brazilite. A variety of the zirconium ore *baddeleyite* mined exclusively in Brazil.

Brazil wax. See carnauba wax.

brazilwood. The bright-red wood of any of several tropical trees, especially *Caesalpinia brasiliensis*, *C. echinata* and *C. crista* of Tropical America, and *C. sappan* of India, Sri Lanka, Malaysia, Borneo, and the nearby islands. It takes a fine, brilliant polish, and is used for cabinetwork, fine furniture, violins, and as a source of red or purple dye (brazilwood extract). See also Pernambuco wood.

Brazinal. Trade name of A.E. Ullman & Associates (USA) for a brazing alloy of aluminum and silicon used for brazing aluminum and aluminum alloys.

brazing alloys. A group of solders composed of metal alloys that melt at temperatures above 370°C (700°F), and have soldering temperatures above 450°C (840°F). They are used to produce high-strength soldered or brazed joints, or solder or braze metals, such as precious metals, heavy metals, steels, etc., that cannot be joined by soft (tin-lead) solders. *Brazing alloys* include copper, copper-phosphorus, copper-zinc, copper-zinc-nickel and copper-tin alloys, and magnesium, nickel, aluminum-silicon and silver alloys. They are available in the form of strips, wires, rods, powder, sheets, pastes, strips, and transfer tape. Also known as *brazing filler metals; hard solders*.

brazing brass. A brass containing approximately equal amounts of copper and zinc, and small amounts of other elements, such as tin, iron, manganese, etc. Used widely in manual torch brazing and braze welding of low-alloy and low-carbon steels, for brazing copper-base alloys to themselves and to cast iron and steel, and for brazing nickel-base alloys to themselves and to steel, cast iron and copper alloys. Also known as *spelter solder*.

brazing filler metals. See brazing alloys.

brazing fluxes. Materials that are used during brazing to remove existing oxides from base and filler metals, and prevent the formation of new oxides. They commonly contain chemical compounds, such as alkalies, borates, boric acid, chlorides, fluorides, fluoborates or fused borax, and wetting agents.

brazing metal. A tough, ductile, low-melting brass composed of about 85% copper and 15% zinc. Used for brazed joints on steel and copper-alloy parts.

brazing rod. Brazing filler metal in rod form.

brazing sheet. A flat-rolled metal sheet clad on one or both sides with brazing alloy, or a brazing alloy in sheet form.

brazing solder. A solder of about equal parts of copper and zinc, sometimes with small percentages of lead and iron. Supplied in sheet, strip, rod, wire and powder form, it has a solidus temperature of 868°C (1595°F), a liquidus temperature of 877°C (1610°F), and a brazing range of 877-941°C (1610-1725°F). Used for brazing ferrous and nonferrous metals.

brazing wire. Brazing filler metal in wire form.

brea. Sand or soil saturated with petroleum tar used for dressing roads.

Brearley. Trade name of Dunfield Hadfields Limited (UK) for a series of ferritic and martensitic stainless steels containing up to 0.4% carbon, 11-18% chromium, and the balance iron. Used for boiler and furnace fittings, cooking utensils, bearings, kitchen sinks, engine valves, machinery, and chemical equipment.

breather cloth. A loosely woven fabric used in vacuum-bag molding of polymers and polymer-matrix composites, usually as the top layer over bleeder cloths and release film, to establish a continuous vacuum path. See also bleeder cloth.

breathable coating. A synthetic resin coating applied to textile fabrics to make them resistant or proof to wetting by water, but allow water vapor (i.e., perspiration) to pass through. Used especially on fabrics for outdoor clothing and garments for winter sports.

Breather Paper. Trademark of J.H. McNairn Limited (Canada) for building paper.

Breda. (1) Trade name of Breda Company (Italy) for a series of austenitic, ferritic and martensitic stainless steels.

(2) Trade name for rayon fibers and yarns used for textile fabrics.

Bredanese. Trade name for rayon fibers and yarns used for textile fabrics.

bredigite. A mineral composed of calcium magnesium silicate, $Ca_{1.7}Mg_{0.3}SiO_4$. It can also be made by heating calcium silicate (Ca_2SiO_4) and magnesium silicate $(MgSiO_4)$. Crystal system, orthorhombic. Density, 3.29 g/cm³. Occurrence: Northern Ireland, Scotland. Found also in slags and Portland cement.

breeze. (1) A rather imprecise term including any type of furnace residue from disintegrated clinkers to fine, incompletely sintered ashes with a large proportion of combustible matter.

(2) The finely divided residue from coke and charcoal manufacture used as a concrete filler and for making bricks. Also known as *breeze coal*.

breeze coal. See breeze (2).

breithauptite. A copper-red mineral of the nickeline group composed of antimony nickel, NiSb. Crystal structure, hexagonal. Density, 8.23 g/cm³. Occurrence: Germany.

Bremen blue. See copper carbonate.

Bremen green. See mineral green.

Brenka. Trade name of Enka BV (Netherlands) for rayon fibers and yarns used for textile fabrics.

brenkite. A colorless mineral composed of calcium fluoride carbonate, $Ca_2F_2CO_3$. Crystal structure, orthorhombic. Density, 3.13 g/cm³; refractive index, 1.590. Occurrence: Germany.

Brenkona. Trade name for rayon fibers and yarns used for textile fabrics.

Breon. Trademark of BP Chemicals (UK) for a series of polycarbonates and polyvinylidene chloride copolymers.

breton lace. A hand- or machine-made lace with an open net ground having designs embroidered with heavy, lustrous, often colored yarn. It is named after Brittany, a region in northwestern France where it was originally made. Also known as *bretonne lace*.

breunnerite. A grayish or yellowish brown mineral composed of magnesium iron manganese carbonate, $(Mg,Fe,Mn)CO_3$. Crystal system, rhombohedral. Density, 3.2-3.5 g/cm³; hardness, 4-5 Mohs. Occurrence: Central Europe (Austria, Germany), Canada, India.

Brevolite. Trade name for a transparent cellulose lacquer used for protective coatings on metals and other substrates.

brewsterite. A white mineral of the zeolite group composed of strontium barium aluminum silicate pentahydrate, $(Sr,Ba,Ca)Al_2Si_6O_{16}\cdot5H_2O$. Crystal structure, monoclinic. Density, 2.45 g/cm³; refractive index, 1.512. Occurrence: Scotland.

brezinaite. A brownish gray mineral of the the wilkmanite group composed of chromium sulfide, Cr_3S_4. Crystal structure, monoclinic. Density, 4.11 g/cm³. Occurrence: USA (Arizona).

brianite. A colorless mineral composed of sodium calcium magnesium phosphate, $Na_2CaMg(PO_4)_2$. Crystal structure, monoclinic. Density, 3.00 g/cm³; refractive index, 1.605. Occurrence: USA (Ohio).

briartite. A gray-black mineral of the chalcopyrite group composed of copper iron germanium sulfide, Cu_2FeGeS_4. It can be synthesized from copper, iron, germanium and sulfur in vacuo at 700°C (1290°F). Crystal structure, tetragonal. Density, 4.16-4.27 g/cm³.

Bricbond. Trade name of The Ceilcote Company (USA) for acidproof cement.

Brichrome. Trade name of British Piston Ring Company (UK) for a wear-resistant cast iron containing 1.75% total carbon, 30% chromium, 1.5% silicon, 0.8% manganese, and the balance iron. Used for piston rings and cylinder liners.

brick. (1) A ceramic product in the form of a molded block of building material made from clay, shale or marl, and hardened by firing or drying in a kiln, or sun baking. The American standard size building brick is 216 × 114 × 57 mm (8.50 × 4.50 × 2.25 in.), and the standard size firebrick 229 × 114 × 64 mm (9.00 × 4.50 × 2.50 in.). Used as a masonry unit in building and other construction.

(2) A bonded abrasive in the form of a block.

brick cement. A waterproof masonry cement used for bricklaying.

brick clay. A clay or earth suitable for the manufacture of brick for construction purposes. It contains loam or pure clay, flint or quartz sand, a high amount of fluxes, and up to 10% iron oxide. It is resistant to warping and cracking during firing, and usually fires to a red color and an adequate hardness at low temperature. Also known as *brick earth; building clay*.

Brick-Cote. Trademark of Phillips Paint Products Limited (Canada) for masonry paint.

brick dust. Brick that has been finely divided by mechanical means, e.g., crushing, grinding, etc.

brick earth. See brick clay.

brick hardcore. Aggregate consisting of crushed bricks. Also known as *brick rubble*.

Brickmaster Periclase. Trade name for *periclase* composed of 99+% magnesium oxide (MgO), and used as a refractory material for lining steelmaking furnaces (especially open-hearth or basic oxygen furnaces), and in the manufacture of glass.

Brick Red. Trade name of Maruhachi Ceramics of America, Inc. (USA) for glazed and unglazed clay roofing tile.

brick rubble. See brick hardcore.

bridge bronze. (1) A group of phosphor bronzes containing 80-88% copper, 9-20% tin, 0-2% zinc, and 0.1-0.3% phosphorus. Used for tubes, springs, electrical parts, gears, worm wheels, and bearings.

(2) A heavy-duty phosphor bronze containing 80% copper, 9-10% tin, 9-10% lead, and 0.7-1% phosphorus. Used for bearings.

Bridgeport. Trade name of Olin Brass, Indianapolis (USA) for an extensive series of coppers and copper-base alloys including cadmium, silver-bearing, high-conductivity, tough pitch and deoxidized coppers, and leaded and unleaded brasses and bronzes, free-cutting bronzes, and aluminum, manganese and silicon bronzes.

Bridgit. Trade name of J.W. Harris Company, Inc. (USA) for a lead-free solder containing 0-2% silver, 3% copper, 0-1% nickel, 5% antimony, and the balance tin. It has a solidus temperature of 238°C (460°F), and liquidus temperature of 332°C (630°F). Used for soldering drinking water systems.

bright alloy. An alloy containing varying amounts of copper, zinc, and tin. It derives its name from the bright surface appearance.

bright annealed products. (1) Metal or metal alloy products with bright surfaces that have been heated to a red heat or above in an inert or reducing atmosphere to inhibit or prevent oxidation and discoloration.

(2) Enameling iron or steel that has been heated to a red heat or above in a reducing atmosphere to produce a clean, bright surface preparatory to subsequent porcelain-enamel coating.

bright electrodeposited copper. A bright coating of copper produced by electrolytic deposition, and used as an undercoating for subsequent coatings, or as an overcoating protected by a thin layer of transparent lacquer.

brightener. (1) An agent or compound that when added to an electroplating bath will yield a smooth, bright, reflective deposit. Also known as *brightening agent*.

(2) A material, such as gypsum, kaolin or talc, added to paper pulp to improve its brightness or degree of whiteness. Also known as *brightening agent*.

(3) A colorless compound, such as a coumarin or diaminostilbenesulfonic acid derivative, that converts ultraviolet light into visible fluorescent light and, thus, is added to paper, plastics, textiles and other substances to improve their brightness or whiteness. Also known as *brightening agent; fluorescent brightener; optical brightener; optical whitener; fluorescent whitening agent (FWA)*.

brightening agent. See brightener.

bright extruded bronze. An alloy of 58.5% copper, 38.7% zinc and 2.8% lead, used for hinges, ornaments and architectural applications.

bright gold. A mixture of gold resinate with other metal resinates and fluxing agents used as a decoration on glass, glazes, and vitreous enamels.

bright glaze. A white, colored, or clear ceramic glaze having a shiny or lustrous appearance.

bright nickel plate. A bright to brilliant, reflective nickel electrodeposit produced on metallic surfaces in a Watts bath modified with organic or organic/inorganic brightening agents. It is used to provide decorative finishes on metals which otherwise tarnish or corrode.

bright plate. A brilliant highly reflective electrodeposit used to provide decorative finishes on certain metals and alloys.

Brightray. Trademark of Inco Alloys International Inc. (USA) for a series of electrical resistance alloys of the nickel-chromium and iron-nickel-chromium type. Used for heating elements, resistors, electric furnace elements, etc.

Brightside. Trademark of International Paint (USA) Inc. for marine paint finishes.

Brightside Polyurethane. Trademark of International Paint (USA) Inc. for a polyurethane paint system for topcoats with good gloss and color retention used for boat hulls and decks previously finished with alkyd enamels.

bright steel products. Steel products, such as tubes, wires, sheets and structural shapes, having a bright surface appearance, usually due to either the forming process itself (e.g., bright drawing or rolling), or subsequent heat treatments (e.g., bright annealing) or finishing processes (e.g., polishing).

Bright Trim. Trade name of Textron Specialty Materials (USA) for a series of bright corrosion-resistant coatings for automobile parts with an as-applied appearance of chrome finish.

Briglo. Trade name for lustrous rayon fibers and yarns used for textile fabrics.

Brillex. Trade name of Brilliant Safety Glass Company Limited (UK) for toughened safety glass.

Brilliant. (1) Trade name of Libbey-Owens-Ford Company (USA) for patterned glass.

(2) Trade name of Swedish American Steel Corporation (USA) for a series of tungsten and cobalt-tungsten high-speed tool steels.

(3) Trade name for a light-cure dental hybrid composite.

Brilliant Dentin. Trade name for a dental composite used for inlays.

Brillum. Trade name of Alcoa of Great Britain (UK) for a wrought aluminum alloy containing 2% nickel and 1.5% copper. Used for pistons.

Brilybdenum. Trade name of British Piston Ring Company (UK) for a heat-resistant cast iron containing 3.2% carbon, 2% silicon, 0.4% nickel, 0.2% chromium, and the balance iron. Used for piston rings.

Brimco. Trade name of Bridgeport Rolling Mills Company (USA) for a corrosion-resistant bronze containing 87-88% copper, 2.35-3.25% tin, 0-0.06% phosphorus, 0-0.23% manganese, and the balance zinc. It has high strength and good forming properties, and is used for contacts, fuse clips, electrical components, fasteners, lock washers, and springs.

Brimol. Trade name of British Piston Ring Company (UK) for an austenitic cast iron containing 2.8% carbon, 2% silicon, 14% nickel, 7% copper, 3% chromium, and the balance iron. Used for cylinder liners and pistons.

brin. An individual filament of degummed silk.

Brinalloy. Trade name of Lunkenheimer Company (USA) for a series of corrosion-, wear- and galling-resistant cast nickel alloys containing 15-16% chromium and 8-10% silicon. Used for valve seats and disks.

brindled brick. A building brick of high crushing strength made of ferriferrous sedimentary clays that are reduced, in part, during the firing process.

brindleyite. A dark yellowish-green mineral of the kaolinite-serpentine group composed of nickel aluminum silicate hydroxide, $(Ni,Al)_3(Si,Al)_2O_5(OH)_4$. Crystal structure, monoclinic. Density, 3.17 g/cm^3; refractive index, 1.63. Occurrence: Greece.

Bri-Nylon. Trade name of Courtaulds Limited (UK) and SANS Fibres Proprietary Limited (South Africa) for non-absorbent nylon 6,6 fibers and filament yarns used for the manufacture of carpets, and clothing, e.g., nightwear and knitwear.

Brioude. Brand name for *regulus* metal containing 98.3+% antimony, 1.3% arsenic, and 0.25% lead.

briquet. See briquette (1), (2).

briquette. (1) In general, a block of solid material made by compressing a raw material, such as coal dust or charcoal, with or without a binder. Also known as *briquet*.

(2) In the construction industries, a block or slab of artificial stone. Also known as *briquet*.

(3) In the lumber industries, a round log or rectangular block made by compressing sawmill shavings and dry sawmill waste. Used as a clean-burning, ash-free fuel.

(4) See compact.

Bristahl. Trade name of Nihon Jyokiki Seikoshi Goshi (Japan) for an austenitic stainless steel containing 0.2% carbon, 18% chromium, 8% nickel, and the balance iron. Used for stainless steel parts.

Bristol. (1) British trade name for a nitriding steel containing 0.3% carbon, 0.9% chromium, 0.65% nickel, 1% molybdenum, 0.6% aluminum, and the balance iron. Used for nitrided parts, gears, cams, etc.

(2) Trade name of Pittsburgh Corning Corporation (USA) for light-diffusing glass building blocks.

(3) See Bristol brass.

Bristol board. A fine, smooth cardboard, usually available in various colors in sheets 559 × 711 mm (22 × 28 in.) in size and 0.15 mm (0.006 in.) or more in thickness. It is used for posters, and suitable for painting, writing and printing.

Bristol brass. A high brass containing 61-76% copper and 24-39% zinc. Used for hardware, clocks, fixtures, condenser tubes, pipes, and ornamental applications. Also known as *Bristol; Bristol metal*.

Bristol brick. A rectangular block of abrasive (e.g., sand or alumina) used for polishing and scouring.

Bristol glaze. An unfritted glaze that contains zinc oxide, and is used on stoneware, terra cotta, etc.

Bristol metal. See Bristol brass.

Bristrand. Trade name for vinyon (vinyl chloride) fibers with excellent acid, alkali and mildew resistance. Used for textile fabrics and carpets.

Britanica. Trade name for a medium-gold dental casting alloy.

Britannia metal. A silvery white, corrosion-resistant alloy, quite similar to *pewter*, containing 80-94% tin, 5-11% antimony, 1-3.5% copper and, sometimes, lead, zinc and/or bismuth. In the wrought form, it can be easily fabricated by stamping, rolling, or spinning. It is also available in cast form Used for tableware, domestic utensils, bearings, and pewter.

Britbem. Trade name of British Bemberg Company (UK) for rayon fibers and yarns used for textile fabrics.

Britecarb. Trademark of Wilbur B. Driver Company (USA) for carbonized nickel.

Brite-Lite. Trade name of Laminated Glass Corporation (USA) for laminated safety glass mirrors.

Britenka. Trade name of British Enka (UK) for rayon fibers and yarns used for textile fabrics.

Britest. Trade name of British Piston Ring Company (UK) for a wear-resistant cast iron containing 3.1% total carbon, 2% silicon, 0.8% manganese, 1% chromium, and the balance iron. Used for piston rings and cylinder liners.

Brite-X. Trade name of Pacer Corporation (USA) for mica supplied in several powder grades of varying particle size. It has a density of 2.75 g/cm^3 (0.099 $lb/in.^3$), a hardness of 2.5 Mohs, a refractive index of 1.65, excellent resistance to hydrofluoric and concentrated sulfuric acid, water and air, and good resistance to high and low temperatures. Used as a filler and extender in paints, coatings, plastics, rubber, etc., and as a raw material in the ceramic industry.

britholite. (1) A red-brown mineral of the apatite group composed of calcium lanthanum silicate hydroxide, $Ca_4La_6(SiO_4)_6$-$(OH)_2$. It can also be made synthetically. Crystal structure, hexagonal. Density, 4.69 g/cm^3; refractive index, 1.78-1.79. Occurrence: Russia.

(2) A black mineral of the apatite group composed of calcium yttrium silicate hydroxide, $Ca_2Y_3Si_3O_{12}(OH)$. Crystal structure, hexagonal. Density, 4.25 g/cm^3; refractive index, 1.732. Occurrence: Russia.

(3) A clove-brown mineral of the apatite group composed of calcium cerium phosphate silicate hydroxide, $(Ca,Ce,Na,-Th)_5[(P,Si)O_4]_3(OH,F)$. Crystal structure, hexagonal;. Density, 3.86-3.95 g/cm^3; refractive index, 1.72. Occurrence: Canada (Quebec), Russia.

British Canadian. Trade name of Asbestos Corporation Limited (Canada) for crude asbestos and asbestos fibers.

British gold. See coinage gold (3).

British oak. See English oak.

brittle material. A material, such as a ceramic, that breaks without appreciable plastic deformation under stress. Also known as *nonductile material*.

brittle mica. A group of minerals including *chloritoid, margarite* and *ottrelite*, that resemble true mica in appearance, but cleave to brittle flakes.

brittle silver ore. See stephanite.

Brix. (1) Trade name for an alloy of 60-75% nickel, 15-20% chromium, 5% copper, 4% silicon, 3% titanium, 2% aluminum, 1-4% tungsten, and 1% boron. Used for heat- and corrosion-resistant parts and heating elements.

(2) Trade name of Foseco Minsep NV (Netherlands) for fluxing briquettes used in the iron and steel industry. They are added to the cupola to protect the molten metal against oxidation, improve overall fluxing conditions, and effect a more efficient refining of the slag.

Brixil. Trade name of Foseco Minsep NV (Netherlands) for fluxing briquettes similar to those sold under the trade name *Brix*, but containing some silicon. They are added to the cupola to protect the molten metal against oxidation, improve overall fluxing conditions and machinability, and raise the silicon content of the iron.

Broaching. Trade name of Bethlehem Steel Corporation (USA) for a water-hardening tool steel (AISI type W1) containing 0.8%

carbon, 0.4% manganese, 0.3% silicon, and the balance iron. Used for cutting tools for hard stone, granite, quartz, etc.

Broadcast. Trademark of Vernon Plastics (USA) for a range of fabrics with enhanced printability and ink adhesion used for billboards and banners.

broadcloth. (1) A smooth wool fabric, generally in a plain or twill weave, having a napped and polished surface. Used especially for suits and coats.

(2) A smooth, closely woven fabric in a plain weave and having a fine, crosswise rib, and made of cotton, rayon, silk, synthetics, or fiber blends thereof. Used for dresses, shirts, pyjamas, etc.

broad goods. Fabrics, 457 mm (18 in.) or more in width, woven from glass or synthetic fibers.

Broadleaf. Trade name for a crosslinked copolymer of acrylamide and acrylic acid that can absorb large amounts of water (about 400 times its weight). Supplied in granule form, it is used for laboratory applications, and in horticulture.

broadleaved maple. See bigleaf maple.

broadleaved tree. See decidious tree.

Broadline. Trade name of Australian Window Glass Proprietary Limited for patterned glass.

Broadlite. Trade name of C-E Glass (USA) for sheet glass with large concave bands on one side and various other patterns on the reverse.

Broad Reeded. Trade name of Pilkington Brothers Limited (UK) for a patterned glass.

Broad Reedlyte. Trade name of Pilkington Brothers Limited (UK) for a patterned glass.

broachanthite. See broachantite.

brocade. A moderately heavy silk or velvet fabric in jacquard weave with a multicolor, raised design, used chiefly for dresses, ties, draperies, and upholstery. See also silk; velvet.

brocatelle. A heavy satin- or twill-weave fabric with crosswise ribs, similar to brocade, but with jacquard design with a raised appearance. Used especially for draperies and upholstery.

brochantite. An emerald to dark green mineral composed of copper sulfate hydroxide, $Cu_4SO_4(OH)_6$. Crystal structure, monoclinic. Density, 3.97 g/cm³. Used as minor ore of copper. Also known as *brochanthite; warringtonite.*

Brockhouse. Trade name of Brockhouse Casting Company (UK) for a heat- and corrosion-resistant steel containing 0.2% carbon, 18% chromium, 8% nickel, and the balance iron. Used for steam boiler parts and case-hardening boxes.

brockite. A red-brown or yellow mineral of the rhabdophane group composed of calcium thorium phosphate monohydrate, $(Ca,Th,-Ln)(PO_4)\cdot H_2O$. Crystal structure, hexagonal. Density, 3.90-4.28 g/cm³; refractive index, 1.680. Occurrence: USA (Colorado).

broken slag. See crushed slag.

broken stone. See crushed stone.

Brolunick. Trade name used in France and the UK for a tough, corrosion-resistant aluminum bronze containing 81.5% copper, 7% aluminum, 5.5% nickel, 4% iron, and 2% manganese. Used for marine hardware, housings, gears, cams, and propellers.

bromargyrite. A yellowish mineral of the halite group composed of silver bromide, AgBr. Crystal structure, cubic. Density, 6.47 g/cm³; melting point, 432°C (810°F); refractive index, 2.253.

bromellite. A colorless, or white mineral of the wurtzite group composed of beryllium oxide, BeO. It can also be made synthetically. Crystal structure, hexagonal. Density, 3.02 g/cm³; refractive index, 1.719. Occurrence: Sweden.

brominated butyl rubber. See brominated isobutylene-isoprene rubber.

brominated epoxy resin. An epoxy resin made by reacting bisphenol A (BPA) with the diglycidyl ether of tetrabromobisphenol A (BrDGEBPA). Used for electrical laminates, e.g., printed-circuit boards.

brominated isobutylene-isoprene rubber. Isobutylene-isoprene rubber (butyl rubber) modified with a small amount (about 1.2%) of bromine. It has better ozone and environmental resistance than unmodified grades, good stability at high temperatures, and high compatibility with other rubber types in blends. Abbreviation: BIIR. Also known as *bromobutyl rubber; brominated butyl rubber.* See also butyl rubber.

bromine. A heavy, dark- or reddish-brown, nonmetallic element of Group VIIA (Group 17) of the Periodic Table (halogen group). It is derived from seawater, underground brines, salt beds, etc. Density, 3.11 g/cm³; melting point, 7.2°C (19°F); boiling point, 58.8°C (138°F); refractive index, 1.647; atomic number, 35; atomic weight, 79.904; monovalent, trivalent, pentavalent and heptavalent; attacks most metals. Used as a fire retardant, in dyes and in photographic films, in the manufacture of antiknock gasoline (ethylene dibromide), in water purification, in the manufacture of plastics and dyes, in organic and organometallic synthesis, and in biochemistry, biotechnology, photography and medicine. Symbol: Br.

bromine-82. A radioactive isotope of bromine with a mass number of 82 and a half-life of 36 hours. It emits beta and gamma rays, and is used in chemical, biological and medical research, e.g., as a tracer. Symbol: ⁸²Br. Also known as *radiobromine.*

bromlite. See alstonite.

bromoauric acid. See hydrogen tetrabromoaurate.

bromobutyl rubber. See brominated isobutylene-isoprene rubber.

bromohydrin. Any of a class of compounds, such as ethylene bromohydrin, obtained from glycols or polyhydroxy alcohols by partial substitution of the (−OH) groups by bromine atoms.

bromoplatinic acid. Moisture-sensitive, red-brown crystals (99.9% pure) with a platinum content of about 24-26%. They are soluble in water and alcohol, and have a melting point below 100°C (212°F). Used in electroplating, in the manufacture of catalysts, production of color effects on porcelain and other ceramics, and in microscopy. Formula: $H_2PtClBr_6\cdot H_2O$. Also known as *dihydrogen hexabromoplatinate hydrate; hydrogen hexabromoplatinate hydrate.*

bromotrifluoroethylene polymers. A group of polymers obtained from high-purity bromotrifluoroethylene gas ($BrFC=CF_2$). They are flammable, clear, oily liquids at room temperature, and used as flotation fluids in accelerometers and gyroscopes of inertial guidance systems. Abbreviation: BFE.

bromyrite. A yellowish to greenish, or gray mineral composed of silver bromide, AgBr, and containing about 57% silver. Crystal system, cubic. Density, 3.8-3.9 g/cm³; hardness, 2-3 Mohs. Occurrence: USA. Used as a minor ore of silver.

Bronco. Trade name of Texas Instruments Inc. (USA) for a copper strip clad (or coated) on both sides with about 25 wt% phosphor bronze. It has a high current-carrying capacity and good resiliency, and is used for springs.

Bronwite. Trade name of American Smelting & Refining Company (USA) for a corrosion-resistant bronze containing 59% copper, 20% zinc, 20% manganese, and 1% aluminum. Available as die and sand casting, it is used for general die castings, hardware, fixtures, and die-casting copper alloys.

Bronzac. Trade name of Zinex Corporation (USA) for a bronze finish and noncyanide plating process for nickel, copper, steel, tin, lead, zinc castings, and zincated aluminum.

Bronzalum. Trade name of American Abrasive Metals Company (USA) for a wear-resistant, antislip aluminum bronze containing 85% copper and 15% aluminum. It has abrasive grains cast in the metal, and is used for car steps, stair treads, floor plates, and doors saddles.

Bronzark. Trade name of Westinghouse Electric Corporation (USA) for copper-tin electrodes used for welding cast iron, copper and copper alloys.

bronze. An alloy of about 60-98% copper and several other elements including tin, aluminum, lead, zinc, silicon and/or nickel. Although the term usually refers to an alloy of copper and up to 20% tin, known as *tin bronze*, there are also tin-free bronzes including aluminum bronze (with about 5-10% aluminum), leaded bronze (with about 0.5-30% lead), phosphor bronze (with traces of phosphorus), silicon bronze (with up to about 3% silicon), nickel bronze (with about 5-8% nickel) and beryllium bronze (with about 1-3% beryllium). Ordinary (tin) bronze has a face-centered cubic crystal structure, high corrosion resistance, high abrasion and wear resistance, good tensile properties, high toughness, good antifriction properties, and good castability. Used for fittings, valves, drain cocks, valve seats, water gages, flow indicators, pump and turbine housings, bells, wheels, gears, helical and worm gears, bearings, bushings, bellows, springs, diaphragms, fasteners, window and door seals, welding rods, spark-resistant tools, wear plates, fourdrinier wire, electrical components, vacuum dryers, blenders, chemical equipment, and fine arts.

Bronze Blast. Trade name of The Mitchell Company (USA) for a metallic abrasive blasting material.

bronze blue. A blue pigment composed of ferric ammonium or sodium ferrocyanide.

bronze casting alloys. A group of bronzes containing about 70-85% copper, 5-15% tin, and frequently additions of lead (1.5-25%), zinc, nickel, antimony, etc. They have good antifriction properties, good wear resistance, good corrosion and seawater resistance, and good strength properties, and are used for bushings, worm gears, fittings, wheels, pump and turbine housings, bells, acid-resisting high-strength castings, bevel gears, control levers, bearings, machine-tool parts, and wear plates.

bronze-clad copper. Copper clad on one or both sides with bronze.

bronze-clad steel. Sheet steel clad on one or both sides with strong, corrosion-resistant bronze. Used for chemical equipment and storage vessels.

Bronze Devil. Trade name of Champion Rivet Company (USA) for arc-welding rods made of phosphor bronze containing 91.25% copper, 8.5% tin, and 0.25% phosphorus.

bronze filter powder. A prealloyed bronze powder composed of 90% copper and 10% tin. Commercially available in 10 to 325 mesh grades, it is used to make sintered bronze filters.

bronze green. An inorganic pigment composed of chrome green (a mixture of lead chromate and ferric ferrocyanide) modified with other pigments, such as iron oxides.

bronze leather. A leather made from kid or calfskin and finished with a cochineal (insect-derived) dye to provide a bronze-colored metallic finish.

Bronzeless Gold. Trade name of Rohm & Haas Company (USA) for a metallic-effect coating.

Bronzelite. Trade name of Australian Window Glass Proprietary Limited for bronze sheet glass.

bronze mica. See amber mica.

Bronze-on-steel. Trade name for strip steel with a sintered layer of leaded bronze. Used for bearings, bushings, etc.

Bronzepane. Trade name of Nippon Sheet Glass Company Limited (Japan) for heat-absorbing glass with a reddish-bronze color.

bronze paper. See glazed paper.

bronze plate. A gold-colored electrodeposit of bronze, usually with 10-15% tin, produced on the surface of a metal, such as cast iron or steel, to provide a more attractive appearance, and improve corrosion and wear resistance. It is sometimes used as an undercoat for bright nickel and chromium plate.

bronze P/M parts. See bronze powder-metallurgy parts

bronze powder. A powdered alloy of copper and tin, usually in flaky form, used as a paint pigment, in japanning, powder metallurgy, and as a dusting powder in printing.

bronze powder-metallurgy parts. Structural parts made from prealloyed powders, or elemental premixes of tin and copper by sintering at prescribed temperatures. The most common powder used contains 90% copper and 10% tin. Bronze P/M parts are rather difficult to press, and have moderate strength. Used for bearings, filters, and electrical friction products. Abbreviation: bronze P/M parts.

Bronzeral. Trade name of Central Glass Company Limited (Japan) for light-brown heat-absorbing glass.

bronze steel. An alloy composed of copper, tin and iron, and used in the manufacture of gunmetal.

Bronze Wabbler. Trade name for a tough, hard copper-tin alloy used for rolls.

Bronzex. Trade name of Enthone-OMI Inc. (USA) for a bronze electroplate and plating process.

bronzite. A yellowish-gray variety of the mineral *enstatite* belonging to the pyroxene group. It has a bronze-like luster, and is composed of iron magnesium silicate, $(Mg,Fe)SiO_3$. Crystal system, orthorhombic. Density, 3.37-3.40 g/cm^3; refractive index, 1.67; hardness, 5-6 Mohs. Occurrence: Norway, Czech Republic.

Bronzochrom. Trademark of Eutectic Corporation (USA) for a series of powder-metallurgy alloys, and bronze and brass welding and hardfacing electrode materials for copper, iron, steel, etc., supplied in the form of rods and wires.

Bronzstox. Trade name of A.W. Cadman Manufacturing Company (USA) for an alloy of copper, tin and lead. Used for heavy-duty bearings and bushings.

brookite. A black, brown, or reddish mineral composed of titanium dioxide, TiO_2. It is one of the three polymorphic forms of titanium dioxide, the other two being *rutile* and *anatase*. Crystal system, orthorhombic. Density, 3.87-4.14 g/cm^3; hardness, 5.5-6 Mohs; refractive index, 2.584; dielectric constant, 14-110. Occurrence: Mozambique; USA (Arkansas, Georgia, Massachusetts, Virginia). Used as a source of titanium. Also known as *pyromelane.*

Brookville. Trade name of Brookville Manufacturing Company (Canada) for crushed stone and dolomite.

Brookville Dolomitic. Trade name of Brookville Manufacturing Company (Canada) for pulverized dolomite.

broomcorn fibers. The fibers obtained from the tall broomcorn plant *Sorghum vulgare* of the sorghum family cultivated in East India, Mexico and the southern United States. They are usually processed in lengths ranging from 0.3 to over 0.6 m (1 to over 1.8 ft.) for use in the manufacture of brooms and brushes. Also known as *broom fibers.*

broom fibers. See broomcorn fibers.

Broternal. Trade name of IMI Kynoch Limited (UK) for a corrosion-resistant aluminum bronze containing 7% aluminum, 1% manganese, and the balance copper. Used for periscope tubes, and paper-mill and dyeing equipment.

brown ash. See black ash (1).

brown coal. A very soft, brownish to black low-quality coal of relatively recent origin. It contains less carbon and more moisture (30-50%) than bituminous coal, often has a fibrous, or woody texture, and ranks between peat and bituminous coal with a calorific value of only about 14-20 MJ/kg (6020-8600 Btu/lb). Ocurrence: Australia, Europe (Germany, Netherlands, Eastern Europe), USA (including Alaska). Used as a fuel, and for the preparation of lignite wax. Also known as *lignite*.

brown clay. See red clay.

brown fused alumina. See fused brown alumina.

Brownie Extra. Trade name of Raven Steel & Tool Company (USA) for a water-hardening tool steel (AISI type W1) containing 0.8% carbon, and the balance iron. Used for cutters, drills, taps, and reamers.

brown hematite. See limonite.

brown iron ore. See limonite.

brown iron oxide. A reddish-brown powder that is not a true oxide, but obtained from a solution of ferrous sulfate and sodium carbonate. It contains varying amounts of ferric carbonate, ferric hydroxide and ferrous hydroxide, and is used as a pigment in paint, glass and rubber. Also known as *iron subcarbonate; precipitated iron carbonate*.

brown ironstone clay. Limonite (brown hematite) containing considerable amount of clay or clayey minerals. See also limonite.

Brown Label. (1) Trade name of Wallace Murray Corporation (USA) for a hot-work tool steel containing 0.85-1% carbon, 3-4% chromium, and the balance iron. Used for hot-work tools and dies, hot shear blades, and bull dies.

(2) Trade name of Peninsular Steel Company (USA) for a tough, oil-hardening, shock-resistant tool steel (AISI type S1) containing 0.5% carbon, 0.2% manganese, 1.15% chromium, 2.5% tungsten, 0.2% vanadium, 0.75% silicon, and the balance iron. Used for coining and swaging dies, and chisels.

brown lead oxide. See lead dioxide.

brown mahogany. See Tiama mahogany.

brown metal. The common name of a copper alloy containing 15% zinc. It has a density of 8.75 g/cm³ (0.316 lb/in.³), a melting point of 1000-1025°C (1830-1875°F), good brazeability and machinability, and is used for fasteners, heat-exchanger tubing, architectural trim, electrical conduits, etc.

brown mica. See amber mica.

brownmillerite. See tetracalcium aluminoferrite.

brown ocher. A naturally occurring brown earth consisting chiefly of hydrous ferric oxide with varying amounts of clay and sand. It is a variety of *limonite* and used as a paint pigment.

brown ore. See limonite.

brown oxide. A naturally occurring, metallic brown pigment containing about 28-95% ferric oxide.

brownskin. A strong, waterproof brownish paper made by saturating building paper with bituminous material.

brownstone. A dark brown to reddish brown, dense, medium-grained iron-bearing *sandstone* occurring in the northeastern United States and used as a building stone.

brucite. A colorless, white, pale gray, greenish mineral with waxy or pearly luster composed of magnesium hydroxide, $Mg(OH)_2$. It can also be made synthetically. Crystal system, hexagonal. Density, 2.37-2.40 g/cm³; hardness, 2.5 Mohs; refractive index, 1.581. Occurrence: Canada (Ontario), USA (Nevada, New York). Used for basic refractories, and as a source of dead-burnt magnesite for furnace linings and welding-rod coatings.

brueggenite. A colorless to yellow mineral composed of calcium iodate monohydrate, $Ca(IO_3)_2 \cdot H_2O$. Crystal system, monoclinic. Density, 4.24 g/cm³; refractive index, 1.7985. Occurrence: Chile.

Brüggamid. Trade name of L. Brüggemann KG (Germany) for a series of polyamides (nylons).

brugnatellite. A light pink or yellow mineral composed of magnesium iron carbonate hydroxide tetrahydrate, $Mg_6FeCO_3(OH)_{13} \cdot 4H_2O$. Crystal system, hexagonal. Density, 2.14 g/cm³; refractive index, 1.5365. Occurrence: Italy, USA (Colorado).

Brulon. Trade name of ICI Limited (UK) for a series of polyamides (nylons) supplied in several grades including glass-fiber-reinforced.

Brunium. Trade name of Sovirel (France) for a brown-tinted spectacle glass.

brunogeierite. A gray mineral of the spinel group composed of iron germanium oxide, Fe_2GeO_4. It can also be made synthetically. Crystal system, cubic. Density, 5.55 g/cm³. Occurrence: Southern Africa.

Brunner's yellow. An antimony yellow made by calcining and washing a mixture of about 14.3% antimony potassium tartrate [$K(SbO)C_4O_6 \cdot 0.5H_2O$], 28.6% lead nitrate [$Pb(NO_3)_2$], and 57.1% sodium chloride (NaCl). Used in ceramics. See also antimony yellow.

Brunswick. Trade name of Hall & Pickles Limited (UK) for a series of tungsten-type hot-work tool steels, high-carbon and high-carbon high-chromium tool steels, low-carbon hot-work tool steels, medium-carbon die steels, etc. Used for cutting and press tools, dies, shear blades, ana various other tools.

Brunswick black. A black solution of *gilsonite* and *rosin* used for roof coatings.

Brunswick blue. A blue inorganic pigment composed of *Prussian blue* and an extender.

Brunswick green. See chrome green.

Brun-Tuff. Trade name of Bruin Plastics (USA) for tough, flame-retardant, heat-sealable vinyl-reinforced nylon- and vinyl-reinforced polyester-based industrial fabrics with excellent dimensional stability and durability.

Brush. Trade name of Brush Wellman Corporation (USA) for a series of beryllium coppers, beryllium aluminums, beryllium magnesiums, beryllium nickels, copper nickels, chromium coppers, etc.

Brush Bronze. Trademark of Brush Wellman Corporation (USA) for a series of aluminum-nickel bronzes with high strength and ductility, and good abrasion resistance. Used in the oil and gas industries, and the aerospace industry for bushings, bearings, wear strips, rub pads, etc.

brushed acrylics. Light- to medium-weight woven or knitted acrylic fabrics, usually printed, having a fuzzy or furry surface on one or both sides consisting of fibers that have been raised by brushing.

brushed cotton. An extremely flammable, plain or printed fabric of cotton having a fuzzy or furry surface on either side consisting of fibers that have been slightly raised by brushing. Used blouses, shirts and cildren's daywear.

brushed nylon. A strong, warm, crease-resistant nylon *jersey* with a napped face consisting of fibers that have been raised by brushing. It tends to built up static electricity, and is used chiefly for sheets and nightwear.

brushed plywood. Plywood whose surface has been treated to accent the grain pattern and create an interesting and attractive texture. Used for building interiors and exteriors. See also plywood.

brushed wool. A woven or knitted woolen fabric with a napped or teaseled finish used especially for coats.

brushite. A colorless, or ivory-yellow mineral of the gypsum group composed of dicalcium phosphate, $CaHPO_4 \cdot 2H_2O$. Crystal system, monoclinic. Density, 2.30 g/cm³; refractive index, 1.545. Occurrence: USA (Virginia). Used in bioceramics.

Brushlon. Trademark of 3M Company (USA) for man-made fibers with a synthetic-resin backing available in the form of strips, sheets, disks and continuous rolls. Used as coated abrasives.

Brush'n Seal. Trade name of King Packaged Products Company (Canada) for a brush-applied waterproof masonry coating.

brush plaster. Plaster of a consistency making it suitable for brush application.

brush wire. (1) Drawn steel wire used for rotary-powder brushes in metal cleaning and finishing. Hard brushes are made of hard-drawn wire with a diameter of 0.13 mm (0.005 in.) and soft brushes are made of soft-drawn wire with a diameter of 0.06 mm (0.002 in.).

(2) An alloy composed of 64.25% copper, 35% zinc and 0.75% tin, used for making wire products.

Brussels lace. A general term for any type of heavy bobbin or needlepoint lace with a very elaborate design. It is named after the city of Brussels in Belgium where it was originally made. See also bobbin lace; needlepoint lace.

Bryan. Trade name of Time Steel Service Inc. (USA) for a water-hardening tool steel (AISI type W4) consisting of 1.1% carbon, 0.3% manganese, 0.5% silicon, 0.25% chromium, and the balance iron. It has good hardness and high wear resistance. Used for punches, reamers, forming tools and dies, and wear plates.

Bry-Cad. Trade name of Enthone-OMI Inc. (USA) for a bright cadmium electroplate and plating process.

Bryiron. Trade name of Fillmore Foundry Inc. (USA) for a nickel cast iron containing 2.96% total carbon, 1.2% silicon, 0.73% manganese, 1.5% nickel, and the balance iron. It has high strength and good hardness, and is used for pistons and liners.

BSC glass. See borosilicate crown glass.

BS glass. See borosilicate glass.

α-BSM. Trademark of ETEX Corporation (USA) for a crystalline apatic calcium phosphate bioceramic with a composition similar to that of natural bone. Used in orthopedics as a bone graft substitute.

B-stage resin. A thermosetting resin of the phenolic type in an intermediate stage of polymerization. It softens when heated, but is not fully fusible even at 150-180°C (300-355°F), and only partially soluble in ordinary solvents. Also known as *resitol*.

Bubble Alumina. Trade name of Zircar Products Inc. (USA) for a rigid, low-density, high-alumina insulating refractory consisting of hollow, thin-wall ceramic spheres and refractory cement. Used as hot-face refractories in furnaces, for kiln furniture and radiant burners, and for high-temperature load-bearing insulation applications.

bubble brick. A ceramic product made by pouring a molten mix through a blast of air, and pressing and baking. The resulting product is a lightweight refractory brick with numerous small air bubbles, and high thermal-shock resistance. Used for non-load-bearing applications in basic furnaces.

bubble-coated paper. A paper whose surface coating contains numerous pores or cavities caused by minute air bubbles that have been intentionally introduced into the coating material.

bubble glass. A decorative glassware containing numerous bubbles of prescribed size and arrangement.

Bubblekup. (1) Trademark of Malvern Minerals Company (USA) for ceramic and/or polymeric compositions supplied in standard, bulk and sheet-molding grades for the manufacture of building products and automotive components.

(2) Trade name for surface-modified glass micro-spheres.

Bubble Wrap. Trademark of Sealed Air Corporation (USA) for cushioning materials consisting of cellular plastics with air bubbles. Used for protective packaging applications.

Bubblfil. Trade name for a tough, resilient cellular cellulose acetate used as thermal insulation, and for floats.

Bucholith. German trade name for casein plastics.

buchwaldite. A white mineral composed of sodium calcium phosphate, $NaCa(PO_4)$. Crystal system, orthorhombic. Density, 3.26; refractive index, 1.610. Occurrence: USA.

Buckeye. Trade name of Time Steel Service Inc. (USA) for a series of molybdenum- and tungsten-type high-speed tool steels.

buckeye. The light, soft, tough wood of any of various deciduous trees of the genus *Aesculus* belonging to the horse-chestnut family. The native Ohio buckeye (*A. glabra*) and the sweet or yellow buckeye (*A. octandra*) grow from Pennsylvania to Iowa, south to Oklahoma and Texas, and east to North Carolina and Virginia. An imported relative, the horse-chestnut (*A. hippocastanum*), is widely grown in Europe as an ornamental tree. The heartwood of *buckeye* is yellowish-white, and the sapwood white. It has an uniform texture, a straight grain, low shock resistance, and is difficult to machine. Used for lumber, boxes, planing-mill products, and for pulpwood.

buckeye bronze. A babbitt containing 41.6% tin, 37.8% zinc, 15.6% lead, 4.3% aluminum, and 0.6% copper. Used for bearings. See also babbitts.

Buckingham. Trademark of Canadian Forest Products for panelwood.

bucklandite. See allanite.

buckle brass. A free-cutting brass containing 65-90% copper, 9-34% zinc, and 1% lead. Used for cartridges, shells, bullets, and tubes.

buckminsterfullerene. A form of solid carbon with a hollow, soccer-ball-shaped molecular structure, each molecule usually being composed of sixty to seventy carbon atoms. It is commercially available in the form of a black powder. Crystal structure, face-centered cubic. Density, 1.7 g/cm³; refractive index, 2.2 (at λ = 630 nm); superconductive properties; good electrical insulator. The 60-carbon-atom molecule (C_{60}) is nicknamed "bucky ball," the 70-carbon-atom molecule (C_{70}) is known as "rugby ball" and the osmylated (OsO_4 treated) 60-carbon-atom is referred to as "bunny ball." *Buckminsterfullerene* molecules with more than 500 carbon atoms have also been synthesized. *Note:* Recently, noncarbon buckministerfullerenes have been discovered. They are composed of nearly round cages of either 70 or 74 atoms of indium supported on the inside by two successively smaller cages of sodium and indium atoms respectively, and one single, central atom of nickel, palladium or platinum. See also carbon nanotubes; fullerene-like carbon nitrides; fullerene polymers; fullerenes; fullerene superconductors; fullerids; fullerites.

buckram. A strong, coarse, plain-woven linen or cotton fabric made stiff with glue, and used in bookbinding, for stiffening of hats and other clothing, and for waistbands and insoles.

buckskin. (1) A soft, strong, yellowish or grayish leather made from the skin of elk or male deer. It usually has a suede finish and is tougher and coarser than deerskin. Used for clothing, gloves, shoes, belts, etc.

(2) A pliable, suede-finished leather made from sheepskin or goatskin, and resembling both *chamois leather*, and the leather described under (1).

(3) A heavy, smooth-faced fabric, usually made of a fine wool, such as merino, in a satin weave. Also known as *buckskin fabric*.

buckskin fabric. See buckskin (3).

bucky ball. See buckminsterfullerene.

bucky tubes. See carbon nanotubes; inorganic nanotubes.

Budd. Trade name of Budd Company (USA) for molding powders based on *urea resins*.

Budd Cast 6. Trade name of Budd Company (USA) for a series of resins based on polyamide 6 (nylon 6) supplied in various grades: standard, casting, elastomer, copolymer, glass fiber- or carbon fiber-reinforced, glass bead-reinforced, silicone-, polytetrafluoroethylene- or molybdenum disulfide-lubricated, UV-stabilized, stampable sheet, glass bead- or mineral-filled, high-impact, and fire-retardant.

buddingtonite. A colorless mineral of the feldspar group composed of ammonium aluminum silicate hydrate, $NH_4AlSi_3O_8 \cdot 0.5H_2O$. Crystal system, monoclinic. Density, 2.32; refractive index, 1.531. Occurrence: USA (California).

Budd SMC. Trade name of Budd Company (USA) for polyester sheet-molding compounds supplied in fire-retardant and low profile grades.

Budene. Trademark of Goodyear Tire & Rubber Company (USA) for a stereospecific *cis*-1,4-polybutadiene rubber (butadiene rubber) made by solution polymerization using an organometallic catalyst.

Buderus. Trade name of Buderus AG (Germany) for a series of plain-carbon, cold-work, hot-work and high-speed tool steels.

buergerite. A brown mineral of the tourmaline group composed of sodium iron aluminum fluoride borate silicate, $NaFe_3Al_6(BO_3)_3Si_6O_{18}(O,F)_4$. Crystal system, rhombohedral (hexagonal). Density, 3.31 g/cm³; refractive index, 1.735. Occurrence: Mexico.

buetschliite. A bluish gray mineral of the calcite group composed of potassium calcium carbonate hexahydrate. $K_6Ca_2(CO_3)_5 \cdot 6H_2O$. It can also be made synthetically. Crystal system, rhombohedral (hexagonal). Density, 2.61 g/cm³; refractive index, 1.605.

Bufcom. Trade name of Stan Sax Corporation (USA) for buffing compounds.

buffalo cloth. A heavy twill-weave fabric with a thick nap formerly used for winterwear.

buffalo leather. A heavy, coarse leather made from the hides of domestic water buffalo of Southern Asia. Used for boots and protective clothing, buffing wheels, gears and mallets.

Buff Aloy. Trade name of Buffalo Wire Works Company Inc. (USA) for an abrasion-resistant high-carbon steel used for wire cloth.

buffing compositions. See buffing compounds.

buffing compounds. Abrasives, such as tripoli, silica, fused or calcined alumina, red rouge and green chrome oxide, immersed in a suitable binder carrier, such as wax. They are applied to buffing wheels used to produce smooth, reflective surfaces on workpieces. Also known as *buffing compositions*.

Buff'n Lime. Trade name of Kokour Company Inc. (USA) for lime-based buffing compounds.

Buflokast. Trade name of Duraloy/Blaw Knox Corporation (USA) for a cast iron containing 3.2% carbon, 2.4% silicon, 0.7% manganese, and the balance iron. Used for chemical equipment, kettles, drums, and dryers.

Bugnato. Trade name of Vetereria di Vernante SpA (Italy) for a patterned glass.

builder fabric. A square-woven cotton duck produced from heavy ply yarns, formerly used for building up the foundation structure of rubber tires.

building block. See concrete block; glass block.

building board. A fibrous-felted structural insulating board with a natural surface finish.

building brick. A brick formed from ground and tempered clay, and fired to a stable unit at temperatures of 900-1250°C (1652-2282°F), but not primarily produced for color, texture, or other decorative effects. Used in building construction. See also brick.

building clay. See brick clay.

building lime. A slaked lime or quicklime that has chemical and physical properties which make it useful for building construction. Also known as *construction lime*.

building materials. See construction materials.

building paper. A heavy, waterproof plain or rosin-sized paper made essentially from waste paper. Used as sheathing for walls and subfloors, and as underlayment. Also known as *construction paper*.

building sand. A silica sand that consists of fine spherical grains, and is used for making concrete and mortar, and for laying bricks and building blocks.

building stone. Natural rock, such as granite or limestone, suitable for use in building construction.

Build-It. Trade name for a dental composite resin for core-build-ups.

built-up mica. Small pieces of mica combined with a bonding agent and compressed into sheets or blocks.

bukovite. A brownish gray to grayish blue mineral of the chalcopyrite composed of copper thallium iron selenide, $Cu_3Tl_2FeSe_4$. It can also be made synthetically. Crystal system, tetragonal. Density, 7.36 g/cm³. Occurrence: Czech Republic.

bukovskyite. A green mineral composed of iron arsenate sulfate hydroxide heptahydrate, $Fe_2(AsO_4)(SO_4)(OH) \cdot 7H_2O$. Crystal system, monoclinic. Density, 2.33 g/cm³; refractive index, 1.582. Occurrence: Czech Republic.

Bulana. Trademark of Yambolen (Bulgaria) for acrylic fibers.

bulk cement. A cement that is delivered to the job site in bulk rather than in bags.

bulk concrete. See mass concrete.

bulked yarn. See textured yarn.

bulking agent. A general term for a chemically inert material that is used to increase the bulk of a substance. It may or may not improve its properties. See also filler.

bulking paper. A thick, yet relatively lightweight paper formed by a process that allows the fibers to arrange loosely.

bulk metallic glass formers. A family of multicomponent metallic glassy alloys in bulk form, such as lanthanum-aluminum-nickel, magnesium-copper-yttrium, zirconium-nickel-aluminum-copper or zirconium-titanium-copper-nickel-beryllium. They are dense, deeply supercooled liquids produced by slow cooling from the melt, or mechanical alloying followed by powder consolidation in the supercooled state (solid-state processing). They exhibit strong glass-forming properties, good room-temperature mechanical properties, high thermal stability to crystallization, high melt viscosity, and a large tendency to chemical short-range ordering. Abbreviation: BMG formers. See also glass former.

bulk molding compound. A mixture of viscous, putty-like consistency composed of a thermosetting resin, fiber reinforcements, inert fillers, catalysts, plasticizers, thickeners and additives. It is usually made into a rope, sheet or other shape for compression, injection or transfer molding applications. Abbreviation: BMC. Also known as *fiber-reinforced thermoset molding compound.*

bulk polyvinyl chloride. See mass polyvinyl chloride.

bulk rope molding compound. A *bulk molding compound* composed of a thickened polyester resin and chopped fiberglass (typically 12.7 mm or 0.5 in. long) made into a rope with excellent flow and surface appearance.

bulky yarn. (1) A yarn with increased loft, usually produced from bulked or textured fibers.

(2) A yarn produced from bulky fibers, e.g., hollow or partially hollow synthetic fibers.

bull alloy. An antifriction alloy containing 81% lead, 18.3% antimony and 0.7% iron used for bearings.

Bulldog. Trade name of Pennsylvania Steel Corporation (USA) for a shock-resisting tool steel (AISI type S1) containing 0.5% carbon, 1.2% chromium, 0.25% vanadium, 2.5% tungsten, and the balance iron. It has a great depth of hardening, high hardness and strength, excellent toughness, good wear and abrasion resistance, and good hot hardness. Used for punches, chisels, crimpers, upsetters, and dies.

Bulldog Grip. Trademark of Canadian Adhesives, Division of Rexnord Canada Limited for a series of construction adhesives for use on wall panels, tiles, carpets, bathtubs, etc., and several adhesive sealants for exterior and interior applications. They are supplied in cans and/or cartridges.

bullet brass. A brass containing 90% copper, 9-10% zinc and 0-1% lead used for bullets, shells and cartridges.

bulletproof glass. A special safety glass made by cementing two sheets of plate glass together with an interlayer of transparent plastic sheet (e.g., polyvinyl butyral). The thickness ranges from less than 25 mm (1 in.) to up to 76 mm (3 in.). Used in banks, for military equipment, armored cars, airplanes, automobiles, pressure cookers, etc. Also known as *bullet-resisting glass.* See also laminated glass; safety glass.

bullet-resisting glass. See bulletproof glass.

Bull Head Rail. British trade name for a plain-carbon steel containing 0.55% carbon, 1% manganese, 0.06% sulfur, 0.06% phosphorus, and the balance iron. It has medium to high hardness and fair shock resistance, and is used for bull head rails.

bullion. Refined gold or silver in the form of an ingot or bar.

Bull's Eye. Trademark of Zinsser Corporation (USA) for a quick-drying shellac wood finish and sealer supplied in white and orange colors.

Bulls Head. Trade name of Marsh Brothers & Company Limited (UK) for a series of plain-carbon steels containing 0.6% or more carbon. Used for rails, and general engineering applications.

Bull's Metal. Trade name of Bull's Metal & Marine Company (UK) for a corrosion-resistant alloy containing 57-60% copper, 0.5-1.5% aluminum, 0.5-1% manganese, 0.5-1% tin, 0.8-1.2% iron, and the balance zinc. Used for marine hardware, propellers, etc.

Bull's White Metal. Trade name of Bull's Metal & Marine Com-

pany (UK) for a wear-resistant white metal containing 80% tin, 10% antimony, 6% lead, and 4% copper. Used for bearings and marine linings.

bultfonteinite. A pink mineral composed of calcium fluoride silicate hydroxide, $Ca_2SiO_2(OH,F)_2$. Crystal system, triclinic. Density, 2.73 g/cm^3; refractive index, 1.590. Occurrence: South Africa.

Bulwark. Trade name of Redfern's Rubber Works (UK) for hard rubber.

Bumper Nickel. Trade name of Atotech USA Inc. (USA) for a bright nickel used for replating bumpers.

Buna. (1) Trademark of Bayer Corporation (USA) for synthetic rubbers originally made from butadiene using finely divided sodium catalysts.

(2) Trademark of Bayer Corporation (USA) for an extensive series of synthetic rubbers of varying composition including several grades of butadiene, acrylonitrile-butadiene, ethylene-propylene, styrene-butadiene and vinyl-butadiene-styrene-butadiene rubber.

Buna A. Trademark of Bayer Corporation (USA) for acrylonitrile-butadiene copolymers with excellent resistance to vegetable and animal oils and petroleum, poor low-temperature properties, moderate electrical properties, and a service temperature range of -50 to +150°C (-58 to +300°F). Used for gasoline, chemical and oil hoses, seals, gaskets, O-rings, heels, soles, etc.

Buna BL. Trademark of Bayer Corporation (USA) for light-colored high-purity styrene-butadiene diblock copolymers made by solution polymerization. Supplied in several grades, they are used in the manufacture of acrylonitrile-butadiene-styrene copolymers and high-impact polystyrenes.

Buna CB. (1) Trademark of Bayer Corporation (USA) for light-colored high-purity butadiene rubber made by solution polymerization using a cobalt or lithium catalyst. Supplied in several grades, it is used as a modifier in the manufacture of acrylonitrile-butadiene-styrene copolymers and high-impact polystyrenes.

(2) Trademark of Bayer Corporation (USA) for abrasion-resistant, resilient butadiene rubber made with a cobalt, lithium or neodymium catalyst. It has good aging and flex-cracking resistance, and good low-temperature flexibility. Available in standard and masterbatch grades, it is used for tires, conveyor belting, footwear soles, V-belts, seals, profiles, injection-molded goods, etc.

Buna EP. Trademark of Bayer Corporation (USA) for ethylene-propylene rubber copolymers and terpolymers. The copolymers are supplied in several grades with ethylene contents of 52-68%, and the terpolymers are available in four unsaturation grades: (i) low unsaturation with ethylene contents of 61-72% and ethylidene norbornene contents of 1.5-3%, (ii) medium-unsaturation with ethylene contents of 48-71% and ethylidene norbornene contents of 4-4.5%, (iii) high unsaturation with an ethylene content of 53% and ethylidene norbornene contents of 6-6.5%, and (iv) very high unsaturation with ethylene contents of 48-66% and ethylidene norbornene contents of 8-11%. *Buna EP* possesses good to excellent weathering and ozone resistance, good aging resistance, low temperature flexibility and electrical conductivity, and fair resistance to polar chemicals. Used for hoses, sheet, sponge rubber, moldings, extrusions, automotive and building profiles, roll covers, and low-voltage cable insulation.

Buna N. Trademark of Bayer Corporation (USA) for an acrylonitrile-butadiene rubber with excellent oil and gasoline resistance, good chemical resistance, fair mechanical properties, low elasticity, and poor low-temperature properties. Used for carburetor and gasoline tanks, pump parts, gaskets, printing rolls, etc.

Buna NB. Trademark of Bayer Corporation (USA) for a modified acrylonitrile-butadiene rubber.

Buna S. Trademark of Bayer Corporation (USA) for styrene-butadiene copolymers made by solution polymerization. They have good physical properties (especially reinforced), good to moderate electrical properties, excellent abrasion and heat resistance, good water resistance, and poor oil, ozone and weather resistance. Its elasticity is somewhat lower than that of natural rubber, and its service temperature range is -60 to +120°C (-76 to +248°F). Used for pneumatic tires and tubes, heels and soles, gaskets, seals, conveyor belts, etc.

Buna SL. Trademark of Bayer Corporation (USA) for abrasion-resistant, resilient butadiene rubber made by solution polymerization. Supplied in several grades with styrene contents of 18-25%, it exhibits good aging and reversion resistance, and good low-temperature flexibility. Used in the manufacture of rubber goods, and blended with nitrile rubber or emulsion styrene-butadiene rubber for automotive tires.

Bunatex. Trademark of Bayer Corporation (USA) for a styrene-butadiene rubber latex used in building construction, as a backing for textiles, carpets and paper, and in the manufacture of molded foams.

Buna VSL. Trademark of Bayer Corporation (USA) for reversion-resistant solution vinyl-butadiene-styrene-butadiene rubber supplied in several grades with vinyl contents of 20-55% and styrene contents of 25-34%. The abrasion resistance, low-temperature flexibility and resilience vary with the vinyl and styrene contents. Used in the manufacture of automotive tires, and in blends with nitrile rubber, butadiene rubber, emulsion styrene-butadiene rubber and solution styrene butadiene rubber.

Bundy. Trademark of Bundy Corporation (USA) for tubular steel products.

Bundyweld. Trade name of Bundy Corporation (USA) for hydrogen-welded, copper-coated rolled carbon steel tubing used for gasoline and oil lines, and refrigerator coils.

Bu-Nite. British trade name for an alloy of copper, nickel and aluminum with good strength retention up to 370°C (698°F). Used for pistons.

bunny ball. See buckminsterfullerene.

bunsenite. A light yellowish green mineral of the halite group composed of nickel monoxide, NiO. It can also be made synthetically. Crystal system, cubic. Density, 6.90 g/cm³; refractive index, 2.73; hardness, 5.5 Mohs. Occurrence; USA.

Bunting. Trade name of NL Industries (USA) for a series of powder-metallurgy materials including several sintered ferrous materials (e.g., irons, steels, copper irons, copper steels, nickel steels, etc.), and sintered copper-base alloys (e.g., brasses and bronzes).

bunting. A thin, durable plain-woven cotton or woolen cloth, usually colored, and resembling cheesecloth in texture. Used for flags and decorations. Also known as *banner cloth.*

Bur-A-Loy. Trade name of Burton Rubber Processing, Inc. (USA) for elastomeric blends of nitrile-butadiene rubber and polyvinyl chloride.

burangaite. A bluish to bluish-green mineral of the dufrenite group composed of sodium calcium iron magnesium aluminum phosphate hydroxide tetrahydrate, $(Na,Ca)_2(Fe,Mg)_2Al_{10}(PO_4)_8$-

$(OH,O)_{12}\cdot 4H_2O$. Crystal system, monoclinic. Density, 3.05 g/cm³; refractive index, 1.635. Occurrence: Rwanda.

burbankite. A pale to grayish yellow mineral composed of sodium calcium strontium carbonate, $Na_2Ca_2Sr_2(CO_3)_5$. It can also be made by mixing the carbonates of calcium, sodium and strontium, adding a small amount of water and subjecting the resulting mixture to a hydrothermal treatment. Crystal system, hexagonal. Density, 3.50 g/cm³; refractive index, 1.615. Occurrence: USA (Montana).

Burberry. Trade name of Burberry Corporation (UK) for durable textile fabrics made from *worsted yarn.*

burckhardtite. A carmine to violet-red, paramagnetic mineral composed of lead aluminum iron manganese tellurium silicate oxide hydroxide monohydrate, $Pb_2(Fe,Mn)Te(AlSi_3)O_{12}(OH)_2\cdot H_2O$. Crystal system, monoclinic. Density, 3.20 g/cm³; refractive index, 1.85. Occurrence: Mexico.

Burgate. Trade name of Teijin Fibers Limited (Japan) for polyester staple fibers used for the manufacture of textile fabrics.

Burgess. A British aluminum solder containing 76% tin, 21% zinc and 3% aluminum.

burkeite. A white, grayish, or buff mineral composed of sodium carbonate sulfate, $Na_6CO_3(SO_4)_2$. It can also be made by mixing boiling aqueous solutions of sodium carbonate and sodium sulfate in a molar ratio of 1:2. Crystal system, orthorhombic. Density, 2.57; refractive index, 1.490. Occurrence: USA (California).

Burlap. Trade name of C-E Glass (USA) for a patterned glass with fabric texture.

burlap. A coarse, heavy, plain-woven fabric made of jute, hemp, flax or cotton, and sometimes impregnated with hot-melt adhesives. Used in the manufacture of sacks and bags, wrappings, wall coverings, linings or backing in upholstery, carpets, linoleum, etc., and also for draperies, clothing, inexpensive laminated composites, and as a water-retaining covering in curing concrete. See also hessian.

burlap-lined paper. A waterproof laminate composed of *kraft paper* with *burlap* bonded to it using an asphaltic binder. Used as a wrapping material.

burley clay. A *fireclay* containing aluminous and/or ferruginous nodules. Also known as *burley flint.*

burley flint. See burley clay.

Burlington. Trade name of Slater Steels Corporation (Canada) for grinding balls.

Burloy. Trade name of Canada Electric Steel Castings Limited for a series of steels containing up to 0.5% carbon, 12-30% chromium, 0-15% nickel, and the balance iron. Used for heat- and/or corrosion-resistant parts.

burnable neutron absorber. See burnable poison.

burnable poison. A neutron absorber or nuclear poison, such as boron, intentionally included in the fuel or fuel cladding of a reactor to assist in the compensation of long-term reactivity changes during progressive burnup. Also known as *burnable neutron absorber.* See also nuclear poison.

Burndy. Trade name of Burndy Corporation (USA) for an extensive series of wrought and cast coppers and copper alloys including various aluminum and silicon bronzes, gunmetals, phosphor and gilding bronzes, high brasses, lithium coppers, etc.

burnishing sand. A fine-grained silica sand of round particle shape and uniform size ranging from 65 to 100 mesh. Used in metal polishing operations.

Burns. Trade name of Matchless Metal Polish Company (USA) for rouge buffing compounds.

burnt bauxite. Bauxite from which the water has been removed by heating (burning). It contains on average about 75-85% alumina, the balance being silica, titania and ferric oxide. Used as an abrasive, and in gunning and ramming mixes, castables and refractory bricks and shapes. Also known as *calcined bauxite*. See also bauxite.

burnt borax. See anhydrous borax.

burnt clay. See calcined clay.

burnt lime. See calcium oxide.

burnt magnesia. See dead-burnt magnesite.

burnt ocher. A paint pigment that has been made by calcining raw ocher. See also ocher.

burnt-out fabrics. See etched-out fabrics.

burnt sienna. A reddish-brown inorganic pigment made by calcining raw sienna. It has a density of 3.95 g/cm^3 (0.14 lb/in.^{3}), and is used in marine, grease and oil paints, metal primers, and rubber, and in polishing compounds. See also sienna.

burnt umber. A reddish-brown pigment containing ferric oxide together with silica, alumina, manganese oxides and lime, and made by calcining raw umber at low heat. Used as a paint pigment, and in lithographic inks, wallpaper (pigment), and artists' color. See also umber.

Burr. British trade name for a brass containing 62-90% copper and 10-38% zinc. Used for hardware, tubes, pipes, and radiator and window-screen wire.

bur oak. The strong, durable wood of the tall tree *Quercus macrocarpa* belonging to the American white oak group. It is similar to overcup and white oak. Source: Canada (southern Quebec and Ontario); northern and central United States (from Maine to North Dakota and south to Texas). Also known as *mossycup oak*. For properties and use, see American white oak and overcup oak.

burrstone. A yellowish building stone composed of *chalcedony*, a variety of quartz.

bursaite. A gray mineral of the lillianite group composed of lead bismuth sulfide, Pb$_5$Bi$_4$S$_{11}$. Crystal system, orthorhombic. Density, 7.62 g/cm^3. Occurrence: Turkey.

Burundum. Trademark of US Stoneware Company (USA) for ultrahigh-fired, white-colored alumina-based ceramic materials available in purities of 87, 96 and 97%. They have high density, hardness (9+ Mohs) and toughness, excellent mechanical and thermal shock resistance, and good resistance to chipping, and most commercial acids and alkalies. Supplied in the form of cylinders with outside diameters of 13-32 mm (0.5-1.25 in.) and lengths of 13-32 mm (0.5-1.25 in.), and as rods, beads and spheres. Used as media for fine grinding and wet and dry milling, and for the rapid dispersion of liquids, and general laboratory applications.

Bush Hammer. Trade name of Colt Industries (UK) for a tool steel containing 1% carbon, 0.1% vanadium, and the balance iron. Used for tools and cutter.

bush metal. A bearing metal composed of 72% copper, 14% tin, and 14% yellow-brass ingot metal. Used for journals and bearings for rolling stock.

Buster Alloy. Trade name of Columbia Tool Steel Company (USA) for a series of oil-hardening, shock-resistant tool steels (AISI type S1) containing 0.5-0.6% carbon, 0.8-1% silicon, 2.2-2.25% tungsten, 1.25-1.35% chromium, 0-0.25% manganese, 0.25% vanadium, and the balance iron. They have high hardness and strength, excellent toughness, good abrasion and wear resistance, and good hot hardness. Used for rivet sets, rivet busters, pneumatic chisels, punches, heading dies, shear blades, forging dies, dies, and hot work tools.

bustamite. A pink mineral of the pyroxenoid group composed of calcium manganese silicate, (Ca,Mn)$_3$Si$_3$O$_9$. It can also be made synthetically. Crystal system, triclinic. Density, 3.42 g/cm^3; refractive index, 1.697. Occurrence: UK (Wales).

Butacite. Trademark of E.I. DuPont de Nemours & Company (USA) for transparent polyvinyl butyral supplied in the form of sheeting, and used in the manufacture of laminated safety glass for automotive windshields and commercial and residential atriums, doors, curtain walls, partitions, roofs, and skylights.

Butaclor. Trade name of EniChem SpA (Italy) for polychloroprene elastomers and latexes.

butadiene-acrylonitrile copolymers. See acrylonitrile-butadiene copolymers.

butadiene rubber. An oil-extendable synthetic thermoplastic polymer made from 1,3-butadiene (C$_4$H$_6$) using catalysts, such as peroxides, organometallics (e.g., butyl lithium), titanium tetrachloride, aluminum iodide, or cobalt, lithium, sodium or nickel salts. The stereospecific *cis*-isomer (*cis*-1,4-polybutadiene) is of great commercial importance, and the gutta-percha-like *trans*-isomer is of limited commercial use. *Butadiene rubber* has excellent abrasion and crack resistance, high resilience, good low-temperature flexibility, good oxidation resistance, a low glass-transition temperature of -73°C (-100°F), a service temperature range in air -70 to +80°C (-94 to +175°F), low heat-buildup, low hysteresis, poor chemical and weathering resistance, and poor UV and gas-permeability resistance. Used chiefly in blends with natural rubber and styrene-butadiene rubber for wear-resistant tires and other rubber products, and for shoe heels and soles, gaskets and belting. Abbreviation: BR. Also known as *polybutadiene elastomer; polybutadiene rubber*. See also gutta-percha; 1,4-polybutadiene.

butadiene-styrene plastics. A family of synthetic resins that are copolymerization products of 1,3-butadiene gas (C$_4$H$_6$) and liquid styrene (C$_8$H$_8$).

butadiene-styrene rubber. A synthetic rubber that is the copolymerization product of butadiene (C$_4$H$_6$) and styrene (C$_8$H$_8$).

Butakon. Trademark of ICI Limited (UK) for a series of styrene-butadiene and acrylonitrile-butadiene rubbers.

Butaprene. Trademark of Firestone Tire and Rubber Company (USA) for synthetic rubber based on copolymers of butadiene with various other monomers. Used as reinforcing agents, and in paints and coatings.

Butcher P'Ester Lam's. Trade name of Alan Butcher Associates (USA) for polyester laminates supplied in chopped-glass filled and woven-glass roving grades.

butcher's linen. (1) Formerly, a strong, heavy linen fabric in a plain weave having yarns of varying thickness in both warp and weft. Used especially for aprons and tablecloths.

(2) Now, any stiff, heavy, plain-woven fabric resembling butcher's linen (1), but made with synthetic yarns (e.g., rayon). Used for overalls and protective coats.

cis-butenedioic acid. See maleic acid.

Butler Finish. Trade name of Chas. F. L'Hommedieu & Sons, Co. (USA) for a greaseless, satin-finishing composition.

butlerite. A deep orange mineral composed of iron sulfate hydroxide dihydrate, Fe(OH)SO$_4$·2H$_2$O. It can also be made synthetically. Crystal system, monoclinic. Density, 2.55 g/cm^3; refractive index, 1.674. Occurrence: Argentina.

Buton. Trade name of Esso-Petrol Limited (UK) for a nonelastic copolymer of styrene and butadiene.

butt brass. A high brass composed of 64% copper, 35% zinc, and 1% lead. Used for hardware, hinges, etc.

buttercup yellow. A yellow pigment whose composition approximates $4ZnO \cdot K_2O \cdot 4Cr_2O_3 \cdot 3H_2O$. It is available in the form of a crystalline powder with a density of 3.4 g/cm³, low tinting strength, good rust-inhibitive properties, high light stability, and good resistance to staining and discoloration. Used for rust-inhibiting paints and varnishes. Also known as *citron yellow; zinc chrome; zinc potassium chromate; zinc yellow.*

butternut. The light, soft, weak wood of the tree *Juglans cinerea* of the walnut family. The heartwood is light brown, and the sapwood almost white. *Butternut* has a coarse texture, a close grain, large open pores, and darkens upon exposure to air. It has moderate toughness, and works and polishes well with hand tools and machines. Average weight, 432 kg/m³ (27 lb/ft³). Source: Canada (New Brunswick); USA (Maine to Minnesota, south to Arkansas, and east to North Carolina). Used for lumber, cabinetwork, furniture, paneling, interior trim, and instrument cases. Also known as *oilnut; white walnut.*

butter rock. See halotrichite.

buttgenbachite. An azure-blue mineral composed of copper chloride nitrate hydroxide dihydrate, $Cu_{19}Cl_4(NO_3)_2(OH)_{32} \cdot 2H_2O$. Crystal system, hexagonal. Density, 3.42 g/cm³; refractive index, 1.738. Occurrence: Central Africa (Zaire).

Button. Trade name for a red brass containing 80% copper and 20% zinc. Used for die castings and ornaments.

button alloy. A corrosion-resistant brass containing 43-60% copper, 30-57% zinc, and 0-10% tin. Used for buttons and ornaments. Also known as *white button metal.*

buttonball. See American sycamore.

button brass. A brass containing 89-90% copper, 9-10% zinc, and 0.5% tin. Used for buttons and ornaments.

button metal. A brass composed of 80% copper and 20% zinc, and used for buttons. See also button alloy.

buttonwood. See American sycamore.

Butvar. Trademark of Solutia Inc. (USA) for transparent polyvinyl butyral resins that are reaction products of polyvinyl alcohol and butyraldehyde, and provide enhanced flexibility, toughness and adhesion in solvent-borne adhesives, bonding agents, coatings and inks. Used for insulation applications, as interlayers in laminated safety glass, for film, for structural adhesives and sealing compounds, for coating metals, ceramics, textiles, wood, etc., as wash primers, in gravure or flexographic inks, and for composites and molded articles.

butylated hydroxyanisole. An organic compound that is commercially available as a mixture of 90+% 3-*tert*-4-methoxyphenol and 9% 2-*tert*-4-methoxyphenol. It is a white or pale yellow solid with a melting range of 48-63°C (118-145°F), and used as an antioxidant. Formula: $(CH_3)_3CC_6H_3OH(OCH_3)$. Abbreviation: BHA.

butylated hydroxytoluene. A white, crystalline solid (99+% pure) with a density of 1.048 g/cm³, a melting point of 69-70°C (156-158°F), a boiling point of 265°C (509°F), a flash point of 275°F (135°C), and a refractive index of 1.486. Used as an antioxidant for plastics, rubber, food packaging, petroleum products, aviation gasoline and other products. Formula: $[C(CH_3)_3]_2CH_3C_6H_2OH$. Abbreviation: BHT.

butylene plastics. Copolymers of butylene and one or more unsaturated compounds or polymers of butylenes. An ethylene-butylene copolymer is an example.

Butylin. Trademark of Chemtron Manufacturing Limited (Canada) for a butyl rubber sealant.

butyl rubber. A copolymer of 97+% isobutylene (C_4H_8) and about 1-3% isoprene (C_5H_8) usually made using an aluminum chloride catalyst. Additions of about 1.2% bromine or chlorine results in bromobutyl or chlorobutyl rubbers with modified cure characteristics, and enhanced adhesion. *Butyl rubber* has a density of 0.92-1.15 g/cm³ (0.03-0.04 lb/in.³), very high impermeability to air and gases, excellent shock resistance, high vibration damping at ambient temperatures, good abrasion, tearing and flexing resistance, excellent resistance to aging and sunlight, good ozone, weathering and heat resistance, high dielectric constant, good resistance to many chemicals, fair resistance to oils and greases, good low-temperature flexibility, good heat resistance up to approximately 150°C (300°F), and a service temperature range of -55 to +204°C (-65 to +400°F). Used for inner tubes, curing bladders, steam hose, tubing, diaphragms, mechanical rubber goods, electric wire insulation, encapsulation compounds, weatherstripping, coated fabrics, curtain wall gaskets, machinery mounts, seals for food jars and medicine bottles, and as pond and reservoir sealant, and as a general sealant in building and construction. Also known as *isobutylene-isoprene rubber (IIR); isobutylene-isoprene copolymer.* See also brominated isobutylene-isoprene rubber; chlorinated isobutylene-isoprene rubber.

butyl rubber sealants. Durable, synthetic sealants based on butyl rubber. They are supplied in a variety of colors, paintable after curing, and bond well to most materials, especially metals and masonry. Used in construction to seal joints between building components.

Butynol. Trademark for polyisobutylene-based film materials.

Bynel. Trademark of E.I. DuPont de Nemours & Company (USA) for adhesive resins supplied in a wide range of standard and custom grades. They can be co-extruded or extrusion-coated substrates (including barrier, heat-seal and structural materials) to produce strong interlayer bonds between dissimilar films, tubes, packaging material, etc. Also included under this trademark are co-extrudable reactive polymer films used for packaging and industrial applications.

Byrd cloth. A closely woven fabric in a plain or twill weave made from combed ply mercerized cotton yarns. It is named after the explorer Richard Byrd, and was formerly used for raincoats, suits, aviation garments, and windbreakers.

bystromite. A blue-gray mineral of the rutile group composed of magnesium antimonate, $MgSb_2O_6$. It can also be made synthetically. Crystal system, tetragonal. Density, 5.70 g/cm³; refractive index, 1.885. Occurrence: Mexico.

bytownite. A plagioclase (soda-lime) feldspar composed of 70-90% *anorthite* ($CaAl_2Si_2O_8$) and 10-30% *albite* ($NaAlSi_3O_8$).

C

cabal glass. A glass composed of calcium oxide (CaO), boric oxide (B_2O_3) and aluminum oxide (Al_2O_3).

cable. A strong, thick rope of large size, usually made by twisting together wires or strands. See also wire rope.

cabled cord. A term used in the automotive industries to refer to *cabled yarn* employed for the reinforcement of automotive tires.

cabled yarn. A yarn made by twisting together two or more plied yarns. See also plied yarn.

cable laid rope. A *laid rope* made by twisting together three or more ropes to form a helix around a common central axis. The twist of the finished cable is opposite to that of the rope forming the secondary strands.

cable lead. See sheathing lead.

cable paper. A heavy kraft or Manila paper impregnated with insulating varnish. It has high electrical breakdown strength, great reliability, a long service life, good elevated temperature resistance, and poor moisture resistance. Used for paper-insulated cables. See also kraft paper; manila paper.

Cab-O-Sil. Trademark of Cabot Corporation (USA) for *fumed silica* (SiO_2) available in the form of a finely divided powder. Used as a filler, as a pigment, and as a carbon-black substitute in light colored rubber products.

Cab-O-Sperse. Trademark of Cabot Corporation (USA) for *colloidal silica* dispersions.

Cabot. Trade name of Cabot Corporation (USA) for a series of thermoplastic resins.

Cabot Alloy. Trade name of Haynes International, Inc. (USA) for a series of wrought nickel-chromium, nickel-iron-chromium and iron-nickel-chromium superalloys used for corrosion-resistant and high-temperature applications in the chemical and petroleum industries.

Cabra. Trade name for cast copper-beryllium alloys containing small amounts of cobalt. Used for springs, contacts and nonsparking tools.

cabretta. Tanned skin obtained from sheep raised in Brazil and India, and used for glove and garment leather, and shoe uppers.

cacoxenite. A yellow or brownish mineral composed of iron phosphate hydroxide dodecahydrate, $Fe_4(PO_4)_3(OH)_3 \cdot 12H_2O$. Crystal system, hexagonal. Density, 2.26 g/cm^3; refractive index, 1.575. Occurrence: USA (Tennessee).

Cadco. Trademark of Cadillac Plastic, Division of Dayco (Canada) Limited for acrylonitrile-butadiene-styrene plastics with good physical, mechanical, chemical and electrical properties and a maximum service temperature of 121°C (250°F). Used for appliance housings, piping, cases, automotive trim, helmets, impellers, knobs, handles and levers.

Cad Glo. Trade name of Alchem Corporation (USA) for a cadmium electroplate and plating process.

Cadmet. Trade name of Castings Development Company (USA) for a sintered tungsten carbide with cobalt binder. Used for cutting tools.

cadmia. See cadmium sulfide.

Cad-Mir. Trade name of Plating Resources Inc. (USA) for a cadmium electrodeposit and cyanide plating process.

cadmium. A soft, bluish-white, malleable metallic element of Group IIB (Group 12) of the Periodic Table. It is commercially available in the form of bars, rods, sheets, foil, wire, sticks, lumps, shot, powder, granules, balls and single crystals. The single crystals are usually grown by the Bridgeman technique. A commercial cadmium ore is *greenockite*, and it also occurs in small amounts in zinc-bearing copper and lead ores. Crystal system, hexagonal. Crystal structure, hexagonal close-packed. Density, 8.64 g/cm^3; melting point, 320.9°C (609.6°F); boiling point, 767°C (1412°F); refractive index, 1.13; hardness, 2.0 Mohs; embrittlement temperature, 80°C (176°F); atomic number, 48; atomic weight, 112.411; divalent. *Cadmium* takes a high polish, emits a crackling sound when bent, and is a high absorber of slow neutrons. Used in protective electroplating of iron and steel, dipped coatings and mechanical plating, as an alloying addition to copper to improve hardness, in bearing metals, low-melting alloys and brazing alloys, in the manufacture of pigments, phosphors, enamels and plastic stabilizers, in semiconductor compounds, for contacts and terminals in electronic circuits, in nickel-cadmium (nicad) batteries, as a regulator in nuclear plants, for selenium rectifiers, electrodes for cadmium-vapor lamps, and in photocells, fire protection systems, and power transmission systems. Symbol: Cd.

cadmium acetate. Colorless or white crystalline compound (99.9+% pure) usually made by treating cadmium oxide with acetic acid. Density, 2.01 g/cm^3; melting point, loses $3H_2O$ at 130°C (266°F). Used in the production of iridescent glazes, as an assistant in dyeing and printing textiles, and in electroplating baths. Formula: $Cd(C_2H_3O_2)_2 \cdot 3H_2O$. Also known as *cadmium acetate trihydrate*.

cadmium acetate dihydrate. A white powder (98+% pure). Density, 2.341 g/cm^3; melting point, 256°C (493°F); dehydrates above 135°C (275°F). Used in the production of iridescent effects in porcelain. Formula: $Cd(C_2H_3O_2)_2 \cdot 2H_2O$.

cadmium acetate trihydrate. See cadmium acetate.

cadmium alloys. A group of alloys of cadmium and one or more other metallic elements, and including cadmium-bismuth for low-melting alloys and solders, cadmium-lead for low-melting alloys, cadmium-silver for high-temperature solders, cadmium-zinc for soldering aluminum, cadmium-nickel for bearings, and cadmium copper for wires and contacts.

cadmium amalgam. A silvery-white compound of cadmium and mercury that becomes soft like wax when moderately heated. Used in modeling, for fusible plugs, and for filling holes in metals.

cadmium antimonide. A hard, brittle compound of cadmium and antimony available in the form of orthorhombic crystals with a density of 6.92 g/cm^3 and a melting point of 456°C (853°F). Used as a semiconductor, and in thermoelectric devices. Formula: CdSb.

cadmium-base babbitts. A group of cadmium-base bearing alloys containing 1-15% nickel, or 0.4-0.75% copper plus 0.5-2.0% silver. They have excellent antiscoring properties, good malleability, a maximum service temperature of 260°C (500°F), and poor corrosion resistance. Used for sliding bearings.

cadmium blende. See greenockite.

cadmium bromide. A white to yellowish crystalline powder (99.9+% pure) usually made by heating cadmium in bromine vapor. Density, 5.192 g/cm³; melting point, 567°C (1052°F); boiling point, 863°C (1585°F). Used in lithography, photography and process engraving, and as a chemical reagent. Formula: $CdBr_2$.

cadmium bronze. See cadmium copper.

cadmium carbonate. A white, crystalline powder (98+% pure) usually made by adding an alkali carbonate to a solution of a cadmium salt. Density, 4.258 g/cm³; melting point, decomposes below 500°C (930°F). Used to improve the stability of cadmium reds, and in the manufacture of other cadmium salts. Formula: $CdCO_3$.

cadmium chloride. A white, crystalline powder (99.9+% pure) made by treating cadmium with hydrochloric acid. Density, 4.047 g/cm³; melting point, 568°C (1054°F); boiling point, 960°C (1760°F). Used as an addition agent for tinning solutions, in the manufacture of special mirrors, as an additive in electroplating baths, in the preparation of cadmium sulfide, in photography, and for the pigment *cadmium yellow*. Formula: $CdCl_2$.

cadmium chloride hydrate. A white crystals (98+% pure). Density, 4.047 g/cm³; melting point, 568°C (1054°F); boiling point, 960°C (1760°F). Used as an addition agent for tinning solutions, in the manufacture of special mirrors, as an additive in electroplating baths, in photography, in the preparation of cadmium sulfide, and for the pigment *cadmium yellow*. Formula: $CdCl_2 \cdot xH_2O$.

cadmium columbate. See cadmium niobate.

cadmium copper. A group of copper alloys containing up to 1.25% cadmium. Commercially available in the form of flat products, rods and wires, they have good to excellent corrosion resistance, excellent cold workability, good hot formability, high strength in the cold-drawn condition, and good conductivity. Used for telephone, telegraph and trolley wires, heating pads, spring contacts, transmission lines, connectors, cable wrap, switch gear components, and waveguide cavities. Also known as *cadmium bronze*.

cadmium cyanide. A white precipitate obtained by treating concentrated solution of a cadmium salt with potassium or sodium cyanide. It decomposes above 200°C (390°F) in air, and is used in cadmium and copper plating. Formula: $Cd(CN)_2$.

cadmium ferrite. A ceramic product with a cubic spinel crystal structure (*ferrospinel*). It is ferrimagnetic, has excellent magnetic properties at high frequencies, and very high resistivity. Used as a soft magnetic material for various electrical, electronic and magnetic applications. Formula: $CdFe_2O_4$.

cadmium fluoride. White powder or high-purity crystals (98+% pure). Density, 6.64 g/cm³; melting point, 1100°C (2012°F); boiling point, 1758°C (3196°F); refractive index, 1.56. Used for electronic and optical applications, in dry-film lubricants for high-temperature applications, as a starting material for laser crystals, and in phosphors. Formula: CdF_2.

cadmium foil. Cadmium (99.7-99.99+% pure) in the form of foils of varying length ranging from 0.01 to 3.0 mm (0.0004 to 0.120 in.) in thickness, and from 50 to 100 mm (2 to 4 in.) in width. Used in electronics and neutron shielding.

cadmium gallium sulfide. A compound of cadmium, gallium and sulfur. Crystal system, cubic. Density, 5.37 g/cm³ (theoretical); hardness, 670-750 Vickers; melting point, 1990°C±30°C

(3614°F±54°F). Used as a semiconductor. Formula: $CaGd_2S_4$.

cadmium gallium telluride. A compound of cadmium, gallium and tellurium. Crystal system, tetragonal. Crystal structure, "defect" chalcopyrite. Density, 5.9 g/cm³; melting point, 787°C (1449°F). Used as a semiconductor. Formula: $CdGa_2Te_4$.

cadmium germanium arsenide. A compound of cadmium, germanium and arsenic. Crystal system, tetragonal. Crystal structure, chalcopyrite. Density, 5.6 g/cm³; melting point, 665°C (1229°F); hardness, 4700 Knoop. Used as a semiconductor. Formula: $CdGeAs_2$.

cadmium germanium phosphide. A compound of cadmium, germanium and phosphorus. Crystal system, tetragonal. Crystal structure, chalcopyrite. Density, 4.48 g/cm³; melting point, 776°C (1429°F); hardness, 5650 Knoop. Used as a semiconductor. Formula: $CdGeP_2$.

cadmium gallium telluride. A compound of cadmium, gallium and tellurium. Crystal system, tetragonal. Crystal structure, "defect" chalcopyrite. Density, 5.9 g/cm³; melting point, 787°C (1449°F). Used as a semiconductor. Formula: $CdGa_2Te_4$.

cadmium hydrate. See cadmium hydroxide.

cadmium hydroxide. White powder made by treating a cadmium salt solution with sodium hydroxide. Density, 4.79 g/cm³; melting point, loses H_2O at 300°C (570°F). Used in the electrodeposition of cadmium, and in nickel-cadmium storage battery electrodes. Formula: $Cd(OH)_2$. Also known as *cadmium hydrate*.

cadmium indium selenide. A compound of cadmium, indium and selenium. Crystal system, tetragonal. Crystal structure, "defect" chalcopyrite. Used as a semiconductor. Formula: $CdIn_2Se_4$.

cadmium iodide. White powder or crystals (99+% pure) made by treating cadmium oxide with hydriodic acid. It occurs in two allotropic forms: (i) *Alpha cadmium iodide*. Density, 5.670 g/cm³; melting point, 387°C (729°F); boiling point, 796°C (1465°F). Used in electrodeposition, in phosphors and lubricants, and in biology and biochemistry as a negative stain; and (ii) *Beta cadmium iodide*. Density, 5.30 g/cm³; melting point, 404°C (759°F). Used in electrodeposition, and in phosphors and lubricants. Formula: CdI_2.

cadmium lithopone. A family of alkali-resistant, light-permanent pigments based on cadmium sulfide (for yellow shades), or cadmium selenide (for red shades), extended with barium sulfate. Used in paints and high-gloss baking enamels.

cadmium maroon. Red, non-fading, inorganic pigment composed of co-precipitated cadmium sulfide (CdS) and cadmium selenide (CdSe). Used in paints and enamels.

cadmium metasilicate. See cadmium silicate.

cadmium molybdate. White powder, or yellow crystals. Density, 5.35 g/cm³; melting point, 1250°C (2282°F). Used in electronics and for optical applications. Formula: $CdMoO_4$.

cadmium-nickel alloys. A group of white cadmium-base bearing alloys containing up to 15% nickel. They have high compressive strength and hardness, and low coefficients of friction. Used for heavy-duty bearings.

cadmium niobate. An antiferroelectric material with low-loss properties at high frequencies, and a Curie temperature of -103°C (-153°F). Used for electroceramics. Formula: $Cd_2Nb_2O_7$ ($Cd_2Cb_2O_7$). Also known as *cadmium columbate*.

cadmium nitrate. White, hygroscopic crystals (98+% pure) made by treating cadmium metal, oxide or carbonate with nitric acid. Density, 2.45 g/cm³; melting point, 59.4°C (138.9°F); boiling point, 132°C (270°F). Used in the manufacture of cadmium yellow and fluorescent pigments, as a reddish-yellow colorant in porcelain enamels and glass, as a catalyst, and in cadmium salts. Formula: $Cd(NO_3)_2 \cdot 4H_2O$. Also known as *cadmium nitrate tetrahydrate.*

cadmium ocher. See greenockite.

cadmium orange. A brilliant reddish-yellow pigment made by calcining selenium with cadmium sulfide (CdS), and used in paints and baking enamels, and as a ceramic colorant.

cadmium orthophosphate. A colorless, amorphous compound with a melting point of 1500°C (2730°F), used in ceramics and materials research. Formula: $Cd_3(PO_4)_2$.

cadmium oxide. A brown or red powder (99+% pure). Crystal system, cubic. Crystal structure, halite. Density, 8.15 g/cm³; melting point, 1430°C (2606°F). High-purity grades (99.99-99.9999%) are also available. Used as an additive in cadmium electroplating baths, in electrodes for nickel-cadmium (nicad) storage cells, in the manufacture of ceramic pigments and glazes, in semiconductors and phosphors, as a catalyst, and for cadmium salts. Formula: CdO.

cadmium pigments. A group of yellow or red inorganic pigments based on cadmium sulfide (yellow) and/or cadmium selenide (red). They have high color retention, good lightfastness and alkali resistance, and are used in paints and high-gloss baking enamels. See also cadmium lithopone; cadmium maroon; cadmium orange; cadmium red; cadmium yellow.

cadmium plate. A corrosion-resistant, silvery-white electrodeposit of cadmium, usually less than 25 μm (1 mil) thick, produced on iron or steel in a cyanide bath (e.g., a solution of cadmium oxide and sodium cyanide) or a noncyanide bath (e.g., cadmium fluoborate). It provides anodic protection against atmospheric corrosion, and offers good lubricity, excellent electrical conductivity and low contact resistance.

cadmium polonide. A compound of cadmium and polonium. Crystal system, cubic. Crystal structure, sphalerite. Used as a semiconductor. Formula: CdPo.

cadmium propionate. The cadmium salt of propionic acid available as a white powder for use in scintillation counters. Formula: $Cd(O_2C_3H_5)_2$.

cadmium red. A family of alkali-resistant, light-permanent, red inorganic pigments composed of co-precipitated cadmium sulfide (CdS) and cadmium selenide (CdSe), often extended with barium sulfate ($BaSO_4$). Used in paints and high-gloss baking enamels. See also cadmium lithopone; cadmium maroon; cadmium orange.

cadmium selenide. A compound of cadmium and selenium with either cubic (sphalerite) or hexagonal (wurtzite or zincite) structure. It is available in high-purity electronic grades (99.99-99.999%) and in the form of seeded vapor-phase crystals. The cubic (sphalerite) form has a white color, a density of 5.57 g/cm³, a melting point of 1239°C (2262°F), and a hardness of 1380 Knoop. The hexagonal (wurtzite or zincite) form is a moisture-sensitive red to reddish-brown crystalline powder (99.9+% pure) with a density of 5.66 g/cm³, a melting point of 1239°C (2262°F), and a band gap of 1.74 eV. Used for red pigments (cadmium red and cadmium maroon), in the production of red ceramic colors, as a semiconductor and phosphor, and in photoelectric cells. The crystals are also used in polarizers, infrared windows, and as substrates for A_{III}-B_{VI} epitaxy. Formula: CdSe.

cadmium selenide lithopone. An alkali-resistant, light-permanent, red pigment based on cadmium selenide (CdSe), and extended with barium sulfate ($BaSO_4$). It provides high color retention, and is used in paints and high-gloss baking enamels.

cadmium silicate. A colorless crystalline compound with a density of 4.93 g/cm³ and a melting point of 1242°C (2268°F). Used in ceramics and materials research. Formula: $CdSiO_3$. Also known as *cadmium metasilicate.*

cadmium silicon arsenide. A compound of cadmium, silicon and arsenic. Crystal system, tetragonal. Crystal structure, chalcopyrite. Hardness, 6850 Knoop. Used as semiconductor. Formula: $CdSiAs_2$.

cadmium silicon phosphide. A compound of cadmium, silicon and phosphorus. Crystal system, tetragonal. Crystal structure, chalcopyrite. Density, 4.0 g/cm³; melting point, approximately 1197°C (2187°F); hardness, 10500 Knoop. Used as semiconductor. Formula: $CdSiP_2$.

cadmium-silver. An alloy of cadmium and silver used to make corrosion-resistant parts.

cadmium-silver solder. A group of cadmium-base solders with about 5-10% silver used primarily in applications where high service temperatures are required. Joints made with cadmium-silver solders have high strength retention up to 218°C (425°F).

cadmium sulfate. A white powder (98+% pure) with a density of 4.691 g/cm³ and a melting point of 1000°C (1830°F). Used in ceramics and materials research. Formula: $CdSO_4$.

cadmium sulfate hydrate. A white powder (98+% pure). Density, 3.09 g/cm³; refractive index, 1.565. Used in electrodeposition, as an electrolyte in Weston standard cells, in fluorescent screens, in pigments, and in the biosciences. Formula: $3CdSO_4 \cdot 8H_2O$.

cadmium sulfate tetrahydrate. White crystalline powder with a density of 3.05 g/cm³. Used in electrodeposition and ceramics. Formula: $CdSO_4 \cdot 4H_2O$.

cadmium sulfide. A yellow-orange crystalline powder (99+% pure) available in cubic (sphalerite structure) and hexagonal (wurtzite or zincite structure) form. The hexagonal form occurs in nature as the mineral *greenockite*, and is also available in high-purity grades (99.99-99.999%) and in the form of seeded vapor-phase crystals. Important properties of cadmium sulfide include: Density, 4.820 g/cm³; melting point, 1476°C (2689°F); hardness, 3-3.5 Mohs; refractive index, 2.3; n-type semiconductor. Used in the production of orange (cadmium orange) and yellow (cadmium yellow) porcelain enamels, ceramic glazes and inks, in the manufacture of ruby glass, in semiconductors, rectifiers, transistors, scintillation counters, photoconductors, photovoltaic cells, solar batteries, phosphors and fluorescent screens, in xerography, in fireworks, and as a source of cadmium. The high-purity crystals are used in polarizers, infrared windows and as substrates for A_{III}-B_{VI} epitaxy. Formula: CdS. Also known as *cadmia; cadmium yellow; orange cadmium.*

cadmium sulfide selenide. A mixed crystal of cadmium, sulfur and selenium. Crystal system, hexagonal. Used for electronic and semiconductor applications. Formula: $CdS_{(x)}Se_{(1-x)}$.

cadmium sulfoselenide. A brilliant, inorganic pigment, such as cadmium orange or cadmium red, made from cadmium sulfide and either selenium or cadmium selenide.

cadmium telluride. A compound of cadmium and tellurium available in the form of black lumps, powder (99.9+% pure) or high-purity crystals (99.99-99.999%) in cubic (sphalerite structure) and hexagonal (wurtzite or zincite structure) forms. Density, 5.86-6.20 g/cm³; melting point, 1092°C (1998°F); hardness, 600 Knoop; refractive index, 2.5; photorefractive properties. Used as a semiconductor, in phosphors, rectifiers, solar cells, special windows, infrared detectors, optical and optoelectronic systems, and for growing epitaxial layers of mercury cadmium telluride. Formula: CdTe.

cadmium telluride selenide. A mixed crystal of cadmium, tellurium and selenium. Crystal system, cubic. Used for electronic and semiconductor applications. Formula: $CdTe_{(x)}Se_{(1-x)}$.

cadmium tetrafluoroborate. A colorless liquid (99.99% pure) available as a 45-50 wt% solution in water with a density of 1.485 g/cm³ and a flash point above 230°F (110°C). Used in chemistry and materials research. Formula: $Cd(BF_4)_2$.

cadmium tin arsenide. A compound of cadmium, tin and arsenic. Crystal system, tetragonal. Crystal structure, chalcopyrite. Density, 5.72 g/cm³; melting point, 607°C (1125°F); hardness, 3450 Knoop. Used as semiconductor. Formula: $CdSnAs_2$.

cadmium tin phosphide. A compound of cadmium, tin and phosphorus. Crystal system, tetragonal. Crystal structure, chalcopyrite. Hardness, 5000 Knoop. Used as a semiconductor. Formula: $CdSnP_2$.

cadmium titanate. A compound of cadmium oxide and titanium dioxide with an ilmenite crystal structure at room temperature. It exhibits ferroelectricity and has a Curie temperature of approximately 220°C (430°F). Used in ferroelectric ceramics. Formula: $CdTiO_3$.

cadmium tungstate. White or yellowish-green crystals or powder made by the interaction of ammonium tungstate and cadmium nitrate. Refractive index, 2.25. Used as a scintillator, e.g., for radiation counting, spectrometry, radiometry and computer tomography, and in fluorescent paint pigments, X-ray screens, phosphors, optoelectronic systems, and ceramics. Formula: $CdWO_4$. Abbreviation: CWO.

cadmium yellow. See cadmium sulfide.

cadmium-zinc solder. A group of solders with varying amounts of cadmium (10-85%) and zinc (15-90%) used for soldering aluminum. Cadmium-zinc soldered joints possess intermediate strength and corrosion resistance.

cadmium zinc telluride. A cubic crystal of cadmium, zinc and tellurium used in solar cells, optical windows and infrared detectors, and for growing epitaxial layers of mercury cadmium telluride. Formula: $Cd_{(1-x)}Zn_{(x)}Te$.

cadmium zirconate. A compound made from cadmium oxide (CdO) and zirconium dioxide (ZrO_2), and used as an additive to barium-titanate capacitors to depress the dielectric constant at Curie temperature. Formula: $CdZrO_3$.

Cadmolith. Trademark of Glidden-Durkee Division of SCM Corporation (USA) for nonbleeding, chemical-resistant brilliant cadmium lithopones in red and yellow shades. Used as pigments in plastics and rubber. See also cadmium lithopone.

cadmopone. A red inorganic pigment composed of co-precipitated cadmium sulfide (CdS) and barium sulfate ($BaSO_4$).

cadmoselite. A black, opaque mineral of the wurtzite group composed of cadmium selenide, CdSe. It can also be made synthetically. Crystal system, hexagonal. Density, 5.66 g/cm³.

Cadon. Trademark of Bayer Corporation (USA) for a series of engineering thermoplastics based on styrene-maleic anhydrides or acrylonitrile-butadiene-styrene/styrene-maleic anhydride alloys. Available in general-purpose, injection-molding, extrusion and plating grades, they have good high-temperature stability, and good impact resistance. Used for automotive components, appliance housings, and electrical components. Also included under this trademark are nylon fibers and plastics.

Cadvert. Trade name of MacDermid Inc. (USA) for a cadmium electroplate and plating process.

cafarsite. A brown mineral composed of calcium iron manganese titanium arsenate hydrate, $Ca_6Mn_2Fe_3Ti_3O(AsO_3)_{12}\cdot4.5H_2O$. Crystal system, cubic. Density, 3.90 g/cm³. Occurrence: Swit-

zerland.

Cafco. Trade name of US Mineral Products Company (USA) for an extensive series of sprayable and/or trowelable asbestos- and other mineral-based building products including acoustical plasters (*Cafco Sound Shield*), fireproofing products (*Cafco Blaze Shield*) and thermal insulation products (*Cafco Heat Shield*).

Cafe Cinco. Trade name of Aardvark Clay & Supplies (USA) for a medium coarse, warm buff to medium brown clay (cone 5).

cafetite. A pale yellow mineral composed of calcium iron titanium oxide tetrahydrate, $(Ca,Mg)(Fe,Al)_2Ti_4O_{12}\cdot4H_2O$. Crystal system, orthorhombic. Density, 3.28 g/cm³; refractive index, 2.08. Occurrence: Russian Federation.

cage zeolite. Any of several natural or synthetic zeolites which consist of cage-like arranged sodium aluminosilicate tetrahedral frameworks with the sodium atoms located at the intersections and the oxygen atoms at the midpoints. They are extremely effective catalysts. Also known as *sodalite*.

cahnite. A colorless mineral composed of calcium boron arsenate hydroxide, $Ca_2BAsO_4(OH)_4$. Crystal system, tetragonal. Density, 3.16 g/cm³; refractive index, 1.662. Occurrence: USA (New Jersey).

Caicara. Trademark of Manap (Brazil) for nylon 6 monofilaments.

cake alum. See aluminum sulfate octadecahydrate.

cake copper. Refined copper cast in a round, cake-shaped mass. Also known as *tough cake*.

cake of gold. See sponge gold.

Cal-Al. Trademark of Timminco Limited (Canada) for a calcium-aluminum alloy used as an addition in the manufacture of calcium alloys.

calamine. (1) A white, bluish, greenish, yellowish or brown mineral composed of zinc silicate hydroxide monohydrate, $Zn_4Si_2O_7(OH)_2\cdot H_2O$, and containing 67.5% zinc oxide, but only about 3% metallic zinc. Crystal system, orthorhombic. Density, 3.4-3.5 g/cm³; hardness, 4.5-5 Mohs; refractive index, 1.619; pyroelectric properties. Occurrence: Europe, USA (Arkansas, Missouri, New Jersey, Pennsylvania). Used as an ore of zinc, and in electronics. Also known as *hemimorphite*.
  (2) An alloy of zinc, lead and tin.

calamitic liquid crystal. A *liquid crystal* which consists of thin rodlike molecules in contrast to a *discotic liquid crystal* in which they are flat and disk-shaped. Both nematic and smectic liquid crystals can be of the calamitic type. See also nematic liquid crystal; smectic liquid crystal.

Calan. Trade name for an electrical insulating material.

calaverite. A brass-yellow to silver-white mineral with metallic luster composed of gold silver telluride, $(Au,Ag)Te_2$, and containing on average 40-44% gold and 1-3% silver. Crystal system, triclinic. Density, 9.0-9.29 g/cm³; hardness, 2.5 Mohs. Occurrence: Australia, Canada, USA (Colorado, California). Used as an important source of gold.

Calcarb. Trademark of Calcarb Limited (USA) for a rigid, strong, lightweight, high-temperature insulation material made from discontinuous lengths of carbon-bonded carbon fibers, vacuum-formed and bonded together with a carbonized resin. It has high purity, high toughness and rigidity, good resistance to shock and vibration, excellent chemical resistance, low bulk density, low water absorption, low thermal conductivity, good machinability, and an operating temperature up to 3000°C (5430°F). Used as thermal insulation for industrial furnaces, and for heat-treating furnace fixtures.

calcareous clay. A clay, such as *marl*, containing considerable quantities of calcium-bearing minerals, especially calcium sulfate and/or carbonate. Used in cement, bricks and stoneware.

calcareous coating. A coating composed of calcium carbonate and magnesium hydroxide applied to a surface that, due to the increase in pH near the surface, is cathodically protected. Also known as *calcareous deposit.*

calcareous glaze. A glaze in which the principal flux is lime, limestone or a similar calcium compound.

calcareous marl. Marl that contains 75-90% calcium carbonate ($CaCO_3$), and is used in the production of cement and building brick, and as an anticrazing ingredient in stoneware Also known as *lime marl.* See also marl.

calcareous sandstone. Sandstone containing considerable quantities of calcium carbonate ($CaCO_3$) as a bonding material for the detrital grains. Used in building construction.

Calcene. Trademark of PPG Industries, Inc. (USA) for a special precipitated calcium carbonate ($CaCO_3$) used in compounding paints, plastics and rubber.

calcia. See calcium oxide.

calcia-stabilized zirconia. See lime-stabilized zirconia.

calciborite. A mineral composed of calcium borate, CaB_2O_4. Crystal system, orthorhombic. Density, 2.88 g/cm^3; refractive index, 1.654. Occurrence: Russia.

calciclase. See anorthite.

calcimine. An inexpensive, temporary, white or colored paint composed chiefly of a mixture of whiting or chalk, water and glue. Formerly used on plastered ceilings and walls. Also known as *kalsomine.*

Calcimol. Trade name for a calcium hydroxide dental cement.

Calcine. Trademark of Hoechst Celanese Corporation (USA) for acetal copolymers with excellent flexibility and high impact resistance. Used for automotive components, and parts of consumer appliances and electronic devices.

calcine. A ceramic material or mixture of materials, usually refractory in nature, such as fireclay, that has been heated to a high temperature without fusion to remove volatile matter and produce a material with desired physical properties for use in ceramic compositions.

calcined alumina. Alumina (Al_2O_3) that has been subjected to one or more thermal treatments above 1093°C (2000°F). The resulting product is at least 99.1% pure, and contains 0.5% residual water, and a total of about 0.4% iron, silicon, sodium and titanium oxides. It is commercially available in several grades depending on the degree of heat treatment, and has a density of 3.4-4.0 g/cm^3 (0.12-0.14 $lb/in.^3$), a melting point of approximately 2040°C (3705°F), a refractive index of 1.765, a hardness of 9 Mohs, high friability, good thermal conductivity, good thermal and mechanical shock resistance, and high electrical resistivity at high temperatures. Used in abrasive products for grinding and polishing, in the manufacture of glass, porcelains, refractories, spark plugs, electrical insulators and similar products, and for metallurgical and melting applications.

calcined aluminum silicate. A compound of alumina (Al_2O_3) and silica (SiO_2) consisting essentially of 95% *mullite.* It has a density of 3.15 g/cm^3 (0.11 $lb/in.^3$), a melting point of 1810°C (3290°F), and a softening temperature of 1650°C (3000°F). Used in the manufacture of porcelains, refractories, vitreous ware, and laboratory ware. Formula: $Al_6Si_2O_{13}$.

calcined baryta. See barium oxide.

calcined bauxite. See burnt bauxite.

calcined borax. See anhydrous borax.

calcined clay. A nonplastic clay made by heating ball or china clay to remove the volatile materials. Used as a natural abrasive, and as an aggregate. Also known as *burnt clay.* See also ball clay; china clay.

calcined dolomite. Dolomite rock calcined at a temperature of about 1700°C (3090°F). It consists of a mixture of calcium oxide (CaO) and magnesium oxide (MgO), and has an upper service temperature of 1650°C (3000°F). Used as a refractory. Also known as *dolime; doloma; single-burnt dolomite.* See also dolomite.

calcined gypsum. See plaster of Paris.

calcined kaolin. A white, grayish, or reddish heat-treated *kaolin* composed essentially of mullite ($3Al_2O_3 \cdot 2SiO_2$) and amorphous siliceous materials. It has a density of 2.67 g/cm^3 (0.096 $lb/in.^3$), a melting point of 1770°C (3220°F), a deformation temperature of 1750-1770°C (3180-3220°F), high refractoriness, mechanical strength and thermal-shock resistance, good load-bearing properties, and good resistance to corrosion by molten glasses, fritted glazes, porcelain-enamel frits and slags. Used in refractories, castables, porcelains, kiln furniture, low-expansion and insulating bodies, investment molds, and high-temperature ceramics.

calcined limestone. Limestone that has been heat-treated in a kiln to remove carbon dioxide. See also limestone.

calcined magnesia. See calcined magnesite.

calcined magnesite. A ceramic product made by calcining (burning) *magnesite* at a temperature between 700 and 1450°C (1290 and 2640°F). It is composed principally of magnesium oxide (MgO) with 2-10% carbon dioxide (CO_2), and has good adsorptive properties. Used in the manufacture of refractory cements (e.g., magnesium oxychloride and oxysulfate types), welding rod coatings, refractories, glass, abrasives, rubber, etc. Also known as *calcined magnesia; caustic-calcined magnesia; caustic calcined magnesite.*

calcined refractory. A refractory that has been heated to a high temperature without fusion to eliminate volatile matter and materials that effect changes in volume.

calcined refractory dolomite. Refractory dolomite that has been heated to a temperature below its melting point for a specified period of time to decompose carbonates, and remove volatile constituents. See also calcined dolomite; refractory dolomite.

calcined soda. See sodium carbonate.

calcinite. A silicon carbide abrasive made by a calcination process.

calciocopiapite. A gray to brownish yellow mineral of the copiapite group composed of calcium iron sulfate hydroxide hydrate, $CaFe_4(SO_4)_6(OH)_2 \cdot 19H_2O$. Crystal system, triclinic. Density, 2.22 g/cm^3. Occurrence: Azerbaidjan.

calcioferrite. A yellow or green mineral composed of calcium iron phosphate heptahydrate, $Ca_2Fe_2(PO_4)OH \cdot 7H_2O$. Crystal system, monoclinic. Density, 2.53 g/cm^3; hardness, 2.5 Mohs.

calciogadolinite. (1) A light brown synthetic mineral composed of beryllium calcium lanthanum iron silicate, $CaBe_2LaFeSi_2O_{10}$. Crystal system, monoclinic. Density, 4.16 g/cm^3; refractive index, 1.75-1.80.

(2) A colorless synthetic mineral composed of beryllium calcium gallium yttrium silicate, $CaBe_2YGaSi_2O_{10}$. Crystal system, monoclinic. Density, 4.07 g/cm^3; refractive index, 1.75-1.80.

calciotantite. A light brown synthetic mineral composed of calcium tantalate, $CaTa_4O_{11}$. Crystal system, hexagonal. Density, 7.54 g/cm^3; refractive index, above 2.0.

calciouranoite. A brown to orange-brown, amorphous mineral of the becquerelite group composed of calcium uranium oxide hydrate, $CaU_2O_7 \cdot 11H_2O$. Density, 4.62 g/cm^3; refractive index, 1.726.

calciovolborthite. A greenish gray mineral of the descloizite group composed of calcium copper vanadate hydroxide, $CaCu(VO_4)$-(OH). Crystal system, orthorhombic. Density, 3.75 g/cm^3; refractive index, 2.05. Occurrence: Turkestan, USA (Colorado).

calcite. A colorless, white, or colored mineral with vitreous to earthy luster that is composed essentially of calcium carbonate, $CaCO_3$, but may contain small quantities of iron, magnesium, manganese and zinc. It is the principal constituent of chalk, limestone and marble. Dogtooth spar, Iceland spar, nailhead spar and satin spar are varieties of calcite. Crystal system, hexagonal (rhombohedral). Density, 2.71-2.72 g/cm^3; hardness, 2-3 Mohs; refractive indices, $\varepsilon = 1.487$, $\omega = 1.659$ (doubly refractive). Used as a flux in the manufacture of pig iron and steel, as a major component in Portland cement, soda-lime glass and pottery bodies, in insulating coatings for capacitors and printed circuits, as a pigment and extender, and as a phosphor. The Iceland spar variety is used in optical instruments e.g., nicol prisms. Also known as *calcspar.*

calcite dolomite. A rock consisting of 10-50% calcite ($CaCO_3$), and the balance dolomite [$CaMg(CO_3)_2$]. Also known as *calcitic dolomite.*

calcite limestone. A limestone containing 5% or less magnesium carbonate ($MgCO_3$). Also known as *calcitic limestone.* See also limestone.

calcite marble. A crystalline limestone containing 5% or less magnesium carbonate ($MgCO_3$). Used as a concrete aggregate. Also known as *calcitic marble; crystalline limestone.* See also marble.

calcitic dolomite. See calcite dolomite.

calcitic limestone. See calcite limestone.

calcitic marble. See calcite marble.

Calcitite. Trademark of Sulzer Calcitek Inc. (USA) for a dense *hydroxyapatite* used as an bioactive ceramic in bone replacement and repair, implantable dental prostheses and, in the form of cones, cylinders and shapes, for tooth-root replacement.

calcium. A silvery-white, soft metallic element of Group IIA (Group 2) of the Periodic Table (alkaline earth group). It is commercially available in the form of granules, shot, crowns, nodules, lumps, ingots, extruded form, sheets, foils, microleaf, powder, turnings and crystals (99.9+% pure). It is an essential constituent of bones, teeth, shells, limestone, chalk, milk, etc. Density, 1.57 g/cm^3; melting point, 845°C (1553°F); boiling point, 1480°C (2696°F); hardness, 17 Vickers; atomic number, 20; atomic weight, 40.078; divalent. Two forms are known: (i) *Alpha calcium* that is present at room temperature and has a face-centered cubic crystal structure; and (ii) *Beta calcium* that exists at temperatures between 464 and 845°C (867 and 1553°F) and has a body-centered cubic crystal structure. *Calcium* is used as a deoxidizer, decarburizer and/or desulfurizer for various ferrous and nonferrous alloys, as an alloying or modifying agent for aluminum, beryllium, copper, lead, tin and magnesium alloys, as a reducing agent in the preparation of chromium, thorium, uranium, vanadium, zirconium and rare earths, in alkali bearing alloys, as a getter for residual gases in high vacuums and vacuum-tube applications, in reagents for the purification and scavenging of inert gases, for cable insulation and batteries, in compounds for making concrete and plaster, in coatings for phototubes, and in biochemistry, biotechnology and medi-

cine. Symbol: Ca.

calcium-45. A radioactive isotope of calcium with mass number 45 obtained by neutron bombardment of scandium, or reactor irradiation of calcium carbonate. It has a half-life of 165 days, and emits beta rays. Used in the study of calcium exchange in clays, in ion exchange, in the diffusion of calcium in glass, and in water purification. Abbreviation: ^{45}Ca. Also known as *radiocalcium.*

calcium abietate. See calcium resinate.

calcium acetate. A brown, gray, or white amorphous or crystalline powder (99+% pure) used as an additive to calcium-soap lubricants, in metallic soaps, as a stabilizer in plastics, and as a corrosion inhibitor. Formula: $Ca(O_2C_2H_3)_2 \cdot H_2O$. Also known as *calcium diacetate; gray acetate; lime acetate.*

calcium acrylate. The calcium salt of acrylic acid available as a free-flowing, white powder. Used as a binder for clay products and foundry molds, as a clay soil stabilizer, in ion exchange, and in oil-well sealing. Formula: $Ca(O_2C_3H_3)_2$.

calcium aluminate. Any of the following compounds of calcium oxide (CaO) and aluminum oxide (Al_2O_3): (i) *Tricalcium aluminate.* White crystals or powder. Density, 3.04 g/cm^3; melting point, decomposes at 1538°C (2800°F). Used in refractories, as an ingredient of cements, especially of aluminous cements, and for fused calcium aluminate (glass). Formula: $Ca_3Al_2O_6$; (ii) *Calcium monoaluminate.* A powder. Density, 3.67 g/cm^3; melting point, 1605°C (2920°F). Used as a constituent in high-alumina cements, and in ceramics. Formula: $CaAl_2O_4$; (iii) *Calcium dialuminate.* A powder. Density, 2.90 g/cm^3; melting point, 1760°C (3200°F) (incongruent). Used in ceramics. Formula: $CaAl_4O_7$; (iv) *Tricalcium pentaaluminate.* Melting point, 2230°C (4045°F). Used in ceramics. Formula: $Ca_3Al_{10}O_{18}$; (v) *Calcium hexaaluminate.* A powder that forms corundum and a liquid phase while melting incongruently at 1850°C (3360°F). Formula: $CaAl_{12}O_{19}$.

calcium-aluminate cement. See aluminate cement.

calcium aluminoborosilicate. A compound of silicon dioxide, calcium oxide, aluminum oxide and boric oxide. It has high electrical resistivity, and is used as the principal ingredient in electric glass (*E-glass*).

calcium-aluminum disilicate. See calcium-aluminum silicate (ii).

calcium-aluminum hydride. See aluminum-calcium hydride.

calcium-aluminum monosilicate. See calcium-aluminum silicate (i).

calcium-aluminum silicate. Any of the following compounds consisting chiefly of calcium oxide, aluminum oxide and silicon dioxide: (i) *Calcium aluminum monosilicate.* A slag-like material used in the manufacture of amber, green, and other glasses. Formula: $CaAl_2SiO_6$; (ii) *Calcium aluminum disilicate.* A powder. Density, 2.77 g/cm^3; melting point, 1549°C (2820°F). It is found in nature as the mineral *anorthite.* Used in ceramics. Formula: $CaAl_2Si_2O_8$; and (iii) *Dicalcium aluminum silicate.* A powder. Density, 3.04 g/cm^3; melting point, 1596°C (2905°F). Used in ceramics. Formula: $Ca_2Al_2SiO_7$.

calcium-aluminum-silicon. A foundry alloy composed of 50-52% silicon, 10-14% calcium and 8-12% aluminum, and the balance iron. Used as a degasifier and deoxidizer in steelmaking.

calcium antimonate. A compound of calcium oxide and antimony pentoxide used as an opacifier in certain porcelain enamels and glazes. Formula: $CaSb_2O_6$.

calcium-base grease. See lime grease.

calcium borate. A white powder used as a metallurgical flux, in the manufacture of porcelain, in fire-retardant compositions, and in antifreezes. Formula: CaB_4O_7.

calcium boride. A compound of calcium and boron available in the form of black crystals or powder. Density, 2.3 g/cm^3; melting point, 2235°C (4055°F); hardness, 2740 Vickers. Used as a deoxidizer and degasifier for nonferrous metals and alloys (e.g., copper), and in ceramics. Formula: CaB_6. Also known as *calcium hexaboride*.

calcium bromide. A white, moisture-sensitive powder (99+% pure). Density, 3.353 g/cm^3; melting point, decomposes at 730°C (1345°F); boiling point, 806-812°C (1483-1495°F). Used in freezing mixtures, sizing compounds, presevatives for wood treatment, in photography, biochemistry and medicine, and as a fire retardant. Formula: $CaBr_2$.

calcium bromide hexahydrate. White crystals (98+% pure). Density, 2.295 g/cm^3; melting point, 38.2°C (101°F); boiling point, decomposes at 149°C (300°F). Used in freezing mixtures, sizing compounds, presevatives for wood treatment, in photography, biochemistry and medicine, and as a fire retardant. Formula: $CaBr_2 \cdot 6H_2O$.

calcium carbide. A hard, grayish-black crystalline compound that reacts with water to form acetylene gas. It is made by heating powdered limestone or quicklime with carbon (usually in the form of anthracite or crushed coke) in an electric furnace. Density, 2.22 g/cm^3; melting point, approximately 2300°C (4170°F). Used in the generation of acetylene for welding purposes, in the manufacture of calcium cyanamide, as a vinyl acetate monomer, and as a reducing agent. Formula: CaC_2.

calcium carbimide. See calcium cyanamide.

calcium carbonate. White powder or colorless crystals (99+% pure). It can be made synthetically, and occurs widely in nature as aragonite, calcite, chalk, limestone, lithographic stone, marble, marl, oyster shells, and travertine. Density, 2.7-2.95 g/cm^3; melting point, decomposes at 825°C (1517°F). Used as a metallurgical flux, as a component in Portland cement, soda-lime glassware and pottery bodies, in insulating coatings for printed circuits and capacitors, in superconductor research for the preparation of bismuth, calcium, copper, strontium oxides with very high transition temperatures, in whiting, as a filler and extender in plastics and rubber, as an extender pigment, in the manufacture of rubber tires, a source of quicklime and calcium metal, as an opacifier in paper, as a separator in glass firing, and in biochemistry, bioengineering and medicine. Formula: $CaCO_3$.

calcium caseinite. A colloidal aggregate that occurs as a heterogeneous complex of calcium, phosphorus, and numerous proteins in milk, and can be removed from the milk by any of several fractionation techniques. See also casein.

calcium chloride. A white, porous solid or powder (97+% pure). It occurs in nature as the mineral *hydrophilite*. Density, 2.15 g/cm^3; melting point, 782°C (1440°F); boiling point, above 1600°C (2910°F). Used as a desiccant and dehumidifier, in ceramics, and as a dustproofing agent. Formula: $CaCl_2$.

calcium chloride dihydrate. Colorless crystals (98+% pure). Density, 0.835 g/cm^3 (25°C/77°F); melting point, decomposes at 176°C (349°F). Used in ceramics. Formula: $CaCl_2 \cdot 2H_2O$.

calcium chloride monohydrate. Solid, colorless flakes, or aqueous solution. Melting point, 260°C (500°F). Used in dust control, for melting snow and ice on roads, in freezeproofing and freezing mixtures, in refrigeration brine, and in concrete as an accelerator or curing aid. Formula: $CaCl_2 \cdot H_2O$.

calcium chloride hexahydrate. Colorless crystals (98+% pure).

Density, 1.71 g/cm^3 (at 25°C or 77°F); boiling point, loses $4H_2O$ at 30°C (86°F) and $6H_2O$ at 200°C (392°F). Used as a desiccant and dehumidifier, as a source of calcium metal produced by electrolysis, as a mill addition in porcelain-enamel slips, a flocculant in glazes, an accelerator in Portland cement, a waterproofing agent in concrete, in the pulp and paper industry, as a deicing and dustproofing agent, for freezeproofing and thawing coke, coal, sand, stone and ore, in refrigerants, and in tire weighting. Formula: $CaCl_2 \cdot 6H_2O$.

calcium chromate. Bright yellow powder. Melting point, dihydrate loses $2H_2O$ at 200°C (392°F). Used as a yellow colorant, in the preparation of pigments, in coatings for light-metal alloys, as a corrosion inhibitor, as an oxidizer, and as a battery depolarizer. Formula: $CaCrO_4$ (anhydrous); $CaCrO_4 \cdot 2H_2O$ (dihydrate).

calcium chromate dihydrate. See calcium chromate.

calcium chromite. A compound made by reacting calcium oxide with chromic oxide. Density, 4.8 g/cm^3; melting point, 2161°C (3922°F). Used in ceramics. Formula: $CaCr_2O_4$.

calcium columbate. See calcium niobate.

calcium cyanamide. The calcium salt of *cyanamide* available as colorless crystals or powder, and made by heating powdered calcium carbide in an electric furnace in an nitrogen atmosphere. It contains about 45% calcium, and has a density of 1.083 g/cm^3 and a melting point of 1200°C (2190°F). Used in the hardening of iron and steel, in nitrogen products, and as a fertilizer. Formula: $CaCN_2$. Also known as *calcium carbimide; cyanamide; lime nitrogen*.

calcium-deoxidized steel. An alloy or carbon steel whose oxygen content has been reduced by the addition of calcium, usually in the form of calcium-silicon (calcium silicide, $CaSi_2$) or calcium-manganese-silicon to the ladle. The calcium also degasifies the steel and increases its scaling resistance. *Calcium-deoxidized steels* usually possess good machinability, and are often used for carburized or through-hardened machine elements, such as worms, gears, pinions, etc. Also known as *calcium-killed steel*.

calcium diacetate. See calcium acetate.

calcium dialuminate. See calcium aluminate (iii).

calcium dichromate. Reddish-brown crystalline compound available in two hydrated forms: (i) $CaCr_2O_7 \cdot H_2O$; and (ii) $CaCr_2O_7 \cdot 4.5H_2O$. Used as a corrosion inhibitor, in the manufacture of chromium compounds, and as a catalyst.

calcium dihydrogen phosphate. See calcium monophosphate.

calcium disilicide. See tricalcium disilicide.

calcium-doped lanthanum manganites. See lanthanum calcium manganites.

calcium drier. A drier, such as calcium naphthenate or calcium resinate, that is used in combination with other metal driers in the conversion of paint to hard films.

calcium 2-ethylhexanoate. A compound available in the form of a white powder (98+% pure) containing about 12.0-12.5% calcium, and in superconductor grades as a viscous liquid containing about 3-4% calcium. Used in superconductivity research, and in the preparation of high transition temperature superconducting phases in bismuth, calcium, copper, lead, strontium oxide thin films. Formula: $Ca(O_2C_8H_{15})_2$. Also known as *calcium octoate*.

calcium feldspar. See anorthite.

calcium ferrite. Any of the following three compounds: (i) *Monocalcium ferrite*. Density, 5.08 g/cm^3; melting point, 1215°C (2220°F). Used in high-alumina cement and ceramics. Formula:

$CaFe_2O_4$; (ii) *Dicalcium ferrite*. Density, 3.98 g/cm³; melting point, 1438°C (2620°F). Used in ceramics. Formula: $Ca_2Fe_2O_5$; and (iii) *Tetracalcium ferrite*. A powder used in ceramics. Formula: $Ca_4Fe_2O_7$.

calcium fluophosphate. See fluoroapatite.

calcium fluorophosphate. See fluoroapatite.

calcium fluoride. White powder or crystals (99+% pure) made synthetically by the action of sodium fluoride and a soluble calcium salt, and occurring in nature as the mineral *fluorite* (fluorspar). It is also available in optical grades (99.99% pure), phosphor grades (99.995% pure), and in the form of vacuum-grown crystals in three grades: infrared, ultraviolet and visible. Crystal system, cubic. Density, 3.18 g/cm³ (density range 2.97-3.25 g/cm³); melting point, 1360°C (2480°F); boiling point, about 2500°C (4530°F); hardness, 4 Mohs; refractive index, 1.434. Used as an important constituent of opal glass, as an opacifier and flux in porcelain enamels, glass and glazes, as a fluxing agent in whiteware bodies, as a glass etchant, as an ingredient in certain cements, as a component in uranium melting crucibles, as a source of fluorine and its compounds, as a fluxing agent in open-hearth steel furnaces, as a flux in metal smelting, in electric-arc welding equipment, in the manufacture of carbon electrodes and emery wheels, in optical equipment, in phosphors (99.95% pure) and synthetic cryolite, in bioactive bone cements, as a paint pigment, as a catalyst in wood preservatives, and in biochemistry, biotechnology and medicine. The single crystals (99.93+% pure) are employed in electronics, laser technology, prisms and lenses, spectroscopy, thermoluminescent dosimeters, and dry-film lubricants for high-temperature applications. Formula: CaF_2. See also europium-doped calcium fluoride.

calcium fluosilicate. A white crystalline powder with a density of 2.66 g/cm³, used in ceramics. Formula: $CaSiF_6$. Also known as *calcium silicofluoride*.

calcium fluosilicate dihydrate. Colorless, fine ground solid. Density, 2.25 g/cm³. Used in rubber compounding, as a flotation agent, and in ceramic glazes. Formula: $CaSiF_6 \cdot 2H_2O$. Also known as *calcium silicofluoride dihydrate*.

calcium formate. A white crystalline powder. Density, 2.015 g/cm³; melting point, above 300°C (570°F). Used in lubricants, as a briquette binder, in drilling fluids, and in leather tanning. Formula: $Ca(CHO_2)_2$.

calcium glass. An optical crown glass containing a substantial proportion of calcium oxide (lime) together with silicon dioxide (silica) and sodium oxide (soda). See also lime crown glass.

calcium grease. See lime grease.

calcium hafnate. A compound of calcium oxide and hafnium dioxide. Density, 5.73 g/cm³; melting point, approximately 2470°C (4480°F). Used in ceramics. Formula: $CaHfO_3$.

calcium hexaaluminate. See calcium aluminate (v).

calcium hexaboride. See calcium boride.

calcium hydrate. See calcium hydroxide.

calcium hydride. Moisture-sensitive, coarse or fine ground, white-gray powder, or white-gray granules or lumps (90+% pure). Density, 1.9 g/cm³; melting point, 816°C (1500°F). Used in the manufacture of chromium, titanium and zirconium by the Hydromet process, and in ceramics. Formula: CaH_2.

calcium hydrogen phosphate. See calcium phosphate (ii).

calcium hydroxide. Soft, white, corrosive, crystalline powder obtained by the interaction of calcium oxide and water. Density, 2.08-2.34 g/cm³; melting point, loses H_2O at 580°C (1075°F). Used in metallurgy, in ceramics for mortar, plaster and cement, in paint, whitewash and dental cements, in water softening, and as an accelerator in certain rubber compounds. Formula: $Ca(OH)_2$. Also known as *calcium hydrate; caustic lime; hydrated lime; lime hydrate; slaked lime.*

calcium iodide. Off-white, hygroscopic powder or beads (97+% pure). Density, 4.0 g/cm³; melting point, 783°C (1440°F); boiling point, above 1100°C (2010°F). Used in ceramics, photography, medicine, and in organic synthesis. Formula: CaI_2.

calcium-killed steel. See calcium-deoxidized steel.

calcium lanthanum sulfide. A compound of calcium, lanthanum and sulfur. Crystal system, cubic. Density, 4.53 g/cm³; melting point, 1810°C±30°C (3290°F±54°F); hardness, 548-648 Vickers; infrared transmission, 74%. Used in ceramics, electronics, and optoelectronics. Formula: $CuLa_2S_4$.

calcium-lead alloys. (1) Bearing alloys composed of lead hardened with small amounts of calcium (0.4-0.75%), sodium (0.15-0.70%) and lithium (0.04%), and made by electrolysis of the fused alkali salts using a cathode of molten lead. See also lead alkali metals.

(2) Lead hardened with about 0.04-0.1% calcium, and used for cable sheathing, battery grids, and as substitute for antimonial lead.

calcium magnesium disilicate. See calcium magnesium silicate (ii).

calcium magnesium monosilicate. See calcium magnesium silicate (i).

calcium-magnesium pyrophosphate. A green powder used in porcelains and enamels. Formula: $Ca_2Mg_2(P_2O_7)_2$.

calcium magnesium silicate. Any of the following four compounds of calcium oxide, magnesium oxide and silicon dioxide: (i) *Calcium magnesium monosilicate*. Density, 3.2 g/cm³; melting point, 1500°C (2730°F), but incongruently. It is found in nature as the mineral *monticellite*. Used in ceramics. Formula: $CaMgSiO_4$; (ii) *Calcium magnesium disilicate*. Density, 3.28 g/cm³; melting point, 1390°C (2535°F). Used in ceramics. Formula: $CaMgSi_2O_6$; (iii) *Dicalcium magnesium disilicate*. Density, 2.94 g/cm³; melting point, 1460°C (2660°F). Used in ceramics. Formula: $Ca_2MgSi_2O_7$; and (iv) *Tricalcium magnesium disilicate*. Density, 3.15 g/cm³; melting point, 1574°C (2865°F). Used in ceramics. Formula: $Ca_3MgSi_2O_8$.

calcium manganese oxide. See calcium manganite.

calcium-manganese-silicon. A master alloy containing 17-22% calcium, 8-12% manganese, 10-11% iron, and 55-65% silicon. Used as a deoxidizer and desulfurizer for steels and cast iron.

calcium manganite. A compound that can be made from calcium oxide and manganese dioxide, and is supplied as crystals or powder. Crystal system, body-centered cubic. Crystal structure, perovskite. It is an antiferromagnetic insulator, exhibits magnetoresistance, and is used for electrical, electronic and magnetic-storage applications. Formula: $CaMnO_3$. Also known as *calcium manganese oxide (CMO)*. See also manganites.

calcium metaborate. Colorless crystals with a melting point of 1100°C (2010°F). Used in ceramics. Formula: $Ca(BO_2)_2$.

calcium metasilicate. See calcium silicate (i).

calcium metatitanate. See calcium titanate.

calcium metazirconate. See calcium zirconate.

calcium mica. See margarite.

calcium molybdate. White, crystalline powder (99+% pure) made by interaction of molybdenum trioxide and calcium oxide. It contains about 39.3% calcium oxide, with the balance being molybdenum trioxide. It occurs in nature as the mineral *powellite*. Density, 4.38-4.53 g/cm³; melting point, 965°C (1770°F).

Used as an adherence-promoting agent in some antimony-bearing porcelain-enamel ground coats, for introducing molybdenum into iron and steel made by the open-hearth, air-furnace or electric-furnace process, as crystals for electronic and optical applications, and in phosphors. Formula: $CaMoO_4$.

calcium monoaluminate. See calcium aluminate (ii).

calcium monocolumbate. See calcium niobate (iii).

calcium mononiobate. See calcium niobate (iii).

calcium monophosphate. See calcium phosphate (i).

calcium neodecanoate. The calcium salt of neodecanoic acid. It is a waxy solid compound available in superconductor grades with 9-11% calcium. Used in electronics, superconductor research, and organometallic synthesis. Formula: $Ca(O_2C_{10}H_{19})_2$.

calcium neodymium aluminate. A compound of calcium, neodymium, aluminum and oxygen available in the form of high-purity single crystals. Crystal structure, cubic. Crystal structure, perovskite. It has a low dielectric constant, and is used for superconductor, microwave and high-frequency applications. Formula: $CaNdAlO_3$.

calcium neodymium sulfide. A compound of calcium sulfide and neodymium sulfide. Crystal system, cubic. Density, 4.88 g/cm^3; melting point, 1800-1860°C (3270-3380°F); hardness, 592-716 Vickers. Used in ceramics and materials research. Formula: $CaNd_2S_4$.

calcium niobate. Any of the following three compounds of calcium oxide and niobium pentoxide (columbium pentoxide): (i) *Tricalcium niobate.* Density, 4.23 g/cm^3; melting point, 1560°C (2840°F) (incongruently). Used for ceramics. Formula: Ca_3-Nb_2O_8 ($Ca_3Cb_2O_8$). Also known as *tricalcium columbate;* (ii) *Dicalcium niobate.* Density, 4.39 g/cm^3; melting point, 1565°C (2849°F). Used in ceramics. Formula: $Ca_2Nb_2O_7$ ($Ca_2Cb_2O_7$). Also known as *dicalcium columbate;* and (iii) *Calcium mononiobate.* Density, 4.72 g/cm^3; melting point, 1560°C (2860°F). Formula: $CaNb_2O_6$ ($CaCb_2O_6$). Also known as *calcium monocolumbate.*

calcium nitrate. Colorless or white, hygroscopic crystals. Density, 2.36 g/cm^3; melting point, 561°C (1042°F). Used in incandescent gas mantles, fireworks and explosives. Formula: $Ca(NO_3)_2$.

calcium nitrate tetrahydrate. White crystals (99+% pure). Density, 1.82-1.86 g/cm^3; melting point, 42°C (108°F); boiling point, decomposes at 132°C (270°F). Used as an oxidizing agent in zirconia and titania opacified porcelain enamels and in fireworks. Formula: $Ca(NO_3)_2 \cdot 4H_2O$. Also known as *lime nitrate; lime saltpeter; nitrocalcite; Norwegian saltpeter.*

calcium nitride. A compound of calcium and nitrogen available in the form of a brown, crystalline powder. Density, 2.63 g/cm^3; melting point, 900°C (1690°F). Used in ceramics. Formula: Ca_3N_2.

calcium nitrite. Colorless, or yellowish hygroscopic crystals. Density, 2.23; melting point, loses H_2O at 100°C (212°F). Used as a corrosion inhibitor in steel-reinforced concrete and lubricating oils and greases, and in ceramics. Formula: $Ca(NO_2)_2 \cdot H_2O$.

calcium octoate. See calcium 2-ethylhexanoate.

calcium orthophosphate. See calcium phosphate (iii).

calcium orthoplumbate. See calcium plumbate.

calcium orthosilicate. See calcium silicate (ii).

calcium orthotungstate. See calcium tungstate.

calcium oxalate. The calcium salt of oxalic acid available as a white, crystalline powder with a density of 2.2 g/cm^3. Used in glazes, rare-earth metal separations, and organic oxalates. Formula: CaC_2O_4.

calcium oxide. An oxide of calcium usually obtained by roasting limestone (calcium carbonate) in kilns to drive off the carbon dioxide. It is a white or grayish-white, alkaline substance that crumbles on exposure to moist air, and reacts with water to form calcium hydroxide (hydrated lime) with evolution of heat. It is available as a white powder in several grades: standard (98+% pure), phosphor (99.95+ pure) and high-purity (99.99-99.995%). Density, 3.40 g/cm^3; melting point, 2570°C (4660°F); boiling point, 2850°C (5160°F). Used as a source of calcium metal, as a flux in steelmaking, in refractories, mortars and cements, as a setting accelerator for Portland cement, in glassmaking, as an absorbent, in the pulp and paper industry, in the manufacture of calcium carbide, calcium hydroxide and sodium carbonate. Formula: CaO. Also known as *anhydrous lime; burnt lime; calcia; calx; common lime; fluxing lime; lime; pebble lime; quicklime; unslaked lime.*

calcium permanganate. A reddish-blue crystalline compound with a density of 2.4 g/cm^3. Used as an additive in certain liquid rocket propellants, and as binders for welding-electrode coatings. Formula: $Ca(MnO_4)_2 \cdot 4H_2O$.

calcium phosphate. Any of the following compounds of calcium containing oxygen and phosphorus, and sometimes hydrogen: (i) *Monocalcium phosphate.* Colorless, scaly crystals or powder. Density, 2.2 g/cm^3; melting point loses H_2O at 100°C (212°F). Used as a stabilizer for plastics, and in the manufacture of glass. Formula: $CaH_4(PO_4)_2 \cdot H_2O$. Also known as *calcium dihydrogen phosphate; calcium monophosphate; monobasic calcium phosphate; primary calcium phosphate;* and (ii) *Dicalcium phosphate.* A white, crystalline powder (99+% pure) made by treating a suspension of hydrated lime (slaked quicklime) in water with fluorine-free phosphoric acid. It is also available in phosphor grades (99.95% pure) and in anhydrous form and as dihydrate. The dihydrate occurs in nature as the mineral *brushite.* Density, 2.306 g/cm^3; melting point, loses $2H_2O$ at 109°C (228°F). Used in glassmaking, in phosphors, as a stabilizer in plastics, and in bioceramics. Formula: $CaHPO_4$ (anhydrous); $CaHPO_4 \cdot 2H_2O$ (dihydrate). Also known as *calcium hydrogen phosphate; dibasic calcium phosphate; dicalcium orthophosphate; secondary calcium phosphate;* and (iii) *Tricalcium phosphate.* A white powder that can be made by treating hydrated lime with phosphoric acid, and also occurs in nature as the minerals *apatite* and *phosphorite.* Density, 3.18 g/cm^3; melting point, 1670°C (3038°F); refractive index, 1.63. Used in ceramics for porcelain enamels, potteries and milk glass, as a polishing powder, as a plastics stabilizer, and in bioceramics. Formula: $Ca_3(PO_4)_2$. Also known as *calcium orthophosphate; tertiary calcium phosphate; tribasic calcium phosphate; tricalcium orthophosphate.*

calcium phosphate ceramics. A group of microporous and macroporous bioceramics based in particular on dicalcium phosphate, tricalcium phosphate or *hydroxyapatite.* They are made from powder precursors by initial compaction into the desired shape, and sintering at temperatures between 1000 and 1500°C (1830 and 2730°F). Used in particular for percutaneous access devices, dentures, periodontal treatment, maxillofacial surgery, spinal surgery, orthopedics and otolaryngology, and in bioactive and resorbable biocomposites.

calcium phosphide. Red-brown crystals or gray granular lumps or pieces (97+% pure). Density, 2.51 g/cm^3; melting point, approximately 1600°C (2915°F). Used in electronics and semiconductor research, as phosphors, and in pyrotechnics, torpedoes and signal fires. Formula: Ca_3P_2. Also known as *photophor.*

calcium plumbate. Reddish-brown crystalline powder with a density of 5.71 g/cm³. Used as a fluxing agent in glass manufacture, as an oxidizer, and in organic coatings, storage batteries, and fireworks. Formula: Ca_2PbO_4. Also known as *calcium orthoplumbate.*

calcium potassium silicate. A compound of calcium oxide, potassium oxide and silicon dioxide. It has a melting point of 1631°C (2968°F), and is used in ceramics. Formula: CaK_2SiO_4.

calcium pyrophosphate. A fine, white powder (99.9% pure) with a density of 3.09 g/cm³, a melting point of 1230°C (2246°F). A phosphor grade (99.95% pure) is also available. Used as a mild abrasive for metal polishing, and as a phosphor. Formula: $Ca_2P_2O_7$.

calcium resinate. A metallic soap in the form of a yellowish-white, amorphous powder or lumps, produced by boiling slaked lime with rosin, and subsequent filtering. Used as a paint drier, as a binder in ceramic inks and pastes, in waterproofing, in the manufacture of porcelains and enamels, as a coating for fabrics, wood and paper, and in tanning. Formula: $Ca(C_{44}H_{62}O_4)_2$. Also known as *calcium abietate; calcium rosinate.* See also limed rosin.

calcium rosinate. See calcium resinate.

calcium scandate. A compound of calcium oxide and scandium oxide with a density of 3.89 g/cm³, used in ceramics. Formula: $CaSc_2O_4$.

calcium silicate. Any of the following compounds of calcium oxide and silicon dioxide: (i) *Calcium metasilicate.* A white powder. It occurs in nature as the mineral *wollastonite.* Density, 2.8-2.9 g/cm³; melting point, 1544°C (2810°F); hardness, 4.5-5 Mohs. Used for pottery bodies, wall tile, wallboard, mineral wool, special low-loss electroceramics, in the manufacture of glass and Portland cement, as an absorbent, as an extender for paper coatings, as a reinforcing agent in rubber, as a filler in plastics, ceramics, paints and coatings, as a viscosity controller in liquids, in welding-rod coatings, a gloss reducer for certain coatings, and in electrical insulators. Formula: $CaSiO_3$. Abbreviation: CS; (ii) *Dicalcium silicate.* A colorless crystalline compound obtained as a byproduct in electric-furnace operation. Density, 3.28 g/cm³; melting point, 2130°C (3866°F). Used as a constituent of Portland cement, and certain dolomite refractories. Formula: Ca_2SiO_4. Abbreviation: C_2S. Also known as *calcium orthosilicate; dicalcium orthosilicate*; (iii) *Tricalcium disilicate.* Melting point, decomposes at 1900°C (3450°F). Used in ceramics. Formula: $Ca_3Si_2O_7$. Abbreviation: C_3S_2; and (iv) *Tricalcium silicate.* Melting point, decomposes at 1465°C (2670°F). Used as the principal cementing constituent of Portland cement, and for stabilized dolomite refractories. Formula: Ca_3SiO_5. Abbreviation: C_3S.

calcium silicate brick. A brick made from a mixture of silica sand and lime, and cured in an autoclave. Used for fancy walls.

calcium silicide. A compound of calcium and silicon available in moisture-sensitive pieces of varying size, and as a fine powder with a density of 2.5 g/cm³. Used in metallurgy (e.g., in iron and steelmaking), and in semiconductor research. Formula: $CaSi_2$.

calcium silicofluoride. See calcium fluosilicate.

calcium silicofluoride dihydrate. See calcium fluosilicate dihydrate.

calcium-silicon. A master alloy of calcium, silicon and iron supplied in three grades: (i) *High-iron* with 18-22% calcium, 57-60% silicon and 15-20% iron; (ii) *Low-iron* with 22-28% calcium, 65-70 silicon and up to 5% iron; and (iii) *High-calcium*

containing 28-35% calcium, 60-65% silicon and up to 6% iron. Used as a deoxidizer, desulfurizer and degasifier for steel and cast iron.

calcium soap. A metallic soap based on calcium resinate, $Ca(C_{44}H_{62}O_4)_2$, and used as a binder in ceramic inks and pastes.

calcium stannate. White, crystalline powder that has a melting point above 1200°C (2190°F), and loses $3H_2O$ at about 350°C (660°F). Used as an additive to ceramic capacitors, in the preparation of ceramic colors (pink to maroon colors), in barium titanate bodies to lower Curie temperature, and as a base for phosphors. Formula: $CaSnO_3$ (anhydrous); $CaSnO_3 \cdot 3H_2O$ (trihydrate).

calcium stearate. White, soft, crystalline powder with a melting point of 150°C (302°F) used as a flatting agent in paints, in waterproofing concrete, cements, stucco, textiles, plastics and wood, as a stabilizer for vinyl resins, as a softener in lead pencils, as a mold-release agent for sheet-molding compounds, as a lubricant for rubber and plastic molds, and in drawing compounds for steel wires. Formula: $Ca(C_{18}H_{35}O_2)_2$.

calcium strontium sulfide. A grayish powder consisting of calcium sulfide and strontium sulfide, and used as a phosphor and phosphorescent pigment. Formula: $CaSrS_2$.

calcium sulfate. Calcium sulfate dihydrate (*gypsum*) from which all the water of crystallization has been removed. It is available as white powder or crystals (99.9+% pure), and also occurs in nature as the mineral *anhydrite.* Density, 2.93-2.96 g/cm³; melting point 1450°C (2642°F); hardness, 3-3.5 Mohs; refractive index, 1.575. Used as a drying agent, as a gypsum substitute for cement, as a source of sulfuric acid, in the manufacture of plaster, as a paper filler, and doped with dysprosium or manganese for use in thermoluminescent dosimeters. Abbreviation: $CaSO_4$. Also known as *anhydrous calcium sulfate; anhydrous gypsum plaster; dead-burnt gypsum; dead-burnt plaster.*

calcium sulfate dihydrate. A hydrous sulfate of calcium in the form of a white powder (98+% pure). It occurs in nature as the mineral *gypsum.* Density, 2.32 g/cm³; melting point, loses $2H_2O$ at 163°C (325°F); hardness, 1.5-2.0 Mohs. Used in ceramics and building construction. Formula: $CaSO_4 \cdot 2H_2O$.

calcium sulfate hemihydrate. A hydrated form of calcium sulfate formed from calcium sulfate dihydrate above a temperature of 128°C (262°F). It forms anhydrous calcium sulfate above 163°C (325°F), and is available as a moisture-sensitive crystalline powder (98+% pure). Also known as *hemihydrate.* Formula: $CaSO_4 \cdot 0.5H_2O$. See also calcined gypsum.

calcium sulfide. An off-white powder (99.9+% pure). It occurs in nature as the mineral *oldhamite.* It has a density of 2.5-2.6 g/cm³, and is used in phosphors and luminous paint, in electronics and ceramics, as an additive in lubricants, and as an ore dressing and flotation agent. Formula: CaS. See also sulfurated lime.

calcium titanate. A ceramic powder (99+% pure). It can be made synthetically, and occurs in nature as the mineral *perovskite.* Density, 4.10 g/cm³; melting point, 1975°C (3587°F); high dielectric constant. Used in high-potassium bodies, as an addition to barium, in titanate and titanate-lead piezoelectric compositions, in temperature-compensating capacitors and electronics. Formula: $CaTiO_3$. Also known as *calcium metatitanate.*

calcium titanium silicate. A compound of calcium oxide, titanium dioxide and silicon dioxide. Density, 3.4-3.6 g/cm³; melting point, 1382°C (2520°F). Used in ceramics. Formula: $CaTiSiO_5$.

calcium tungstate. A white powder (99+% pure) made by treating calcium oxide with tungstic acid. It is also available in the form of high-purity crystals, and occurs in nature as the mineral *scheelite*. Density, 6.062 g/cm³; melting point, 1620°C (2948°F); good mechanical strength and chemical stability. Used as a phosphor in luminous paints and fluorescent lamps. The crystals are employed in scintillation counters and lasers. Formula: CaWO₄. Also known as *artificial scheelite; calcium orthotungstate; calcium wolframate*.

calcium uranate. A compound of calcium oxide and uranium trioxide. Density, 7.45 g/cm³; melting point, 1800 (3270°F). Used in ceramics. Formula: CaUO₄.

calcium wolframate. See calcium tungstate.

calcium zinc silicate. A compound of calcium oxide, zinc oxide, and silicon dioxide. Melting point, 1427°C (2600°F). Used in ceramics. Formula: Ca₂ZnSi₂O₇.

calcium zirconate. A white refractory powder. Density, 4.78 g/cm³; melting point, 2550°C (4620°F); good stability under highly reducing conditions up to 1750°C (3180°F); low firing shrinkage. Used in titanate dielectrics, and in refractories. Formula: CaZrO₃. Also known as *calcium metazirconate*.

calcium-zirconium silicate. A white solid compound of calcium oxide, zirconium oxide and silicon dioxide. Melting point, 1587°C (2889°F). Used in electrical resistor ceramics, electroceramics, and as an opacifier in glazes. Formula: CaZrSiO₅.

calcjarlite. A white mineral composed of sodium calcium strontium aluminum fluoride hydroxide, Na(Ca,Sr)₃Al₃(F,OH)₁₆. Crystal system, monoclinic. Density, 3.51 g/cm³; refractive index, 1.428. Occurrence: Russian Federation.

calclacite. A white natural mineral composed of calcium acetate chloride pentahydrate, C₂H₅CaClO₂·5H₂O. It can also be made synthetically. Crystal system, monoclinic. Density, 1.55; refractive index, 1.484.

Calcoat. Trademark of Calcarb Limited (USA) for special protective coatings for *Calcarb* furnace insulation materials to prevent erosion during gas quenching processes.

calcrete. A mixture of gravel and sand bonded by calcium carbonate.

calcspar. See calcite.

calcurmolite. A honey-yellow mineral composed of calcium molybdenum uranyl oxide hydroxide octahydrate, Ca(UO₂)₃(MoO₄)₃(OH)₂·8H₂O. Refractive index, 1.8215. Occurrence: Russian Federation.

calderite. A dark yellowish mineral of the garnet group that is composed of manganese iron silicate, Mn₃Fe₂Si₃O₁₂, and may also contain calcium and aluminum. Crystal system, cubic. Density, 4.08 g/cm³; refractive index, 1.872. Occurrence: Southwest Africa; India.

Caldria. Trade name of Union Carbide Corporation (USA) for raw asbestos fibers and pellets.

Caldur. Trade name of Uddeholm Corporation (USA) for a hot-work tool steel containing 0.25% carbon, 10-12% chromium, 7% tungsten, 4% cobalt, and the balance iron. It has high hardness and toughness, and good resistance to heat checking. Used for die-casting dies for aluminum and brass, and for forging and extrusion dies.

Caledonian brown. A brown pigment made from a British variety of umber.

caledonite. A bluish green mineral composed of copper lead carbonate sulfate hydroxide, Cu₂Pb₅(SO₄)₃CO₃(OH)₆. Crystal system, orthorhombic. Density, 5.76 g/cm³; refractive index, 1.866. Occurrence: UK.

Caleidoscopio. Trade name of Vidrobrás (Brazil) for a figured glass.

calender bowl paper. A paper containing cotton, wool and/or asbestos. Used in the manufacture of calender rollers with elastic surfaces.

calender-bonded nonwovens. Nonwoven fabrics that have been bonded by softening and melting a batt or web of heat-sensitive fibers by the application of heat and pressure by means of plain or embossed calender rolls.

calendered fabrics. Textile fabrics that have been given a flat, smooth, shiny surface finish by passing through a machine with two or more rollers known as a "calender".

calendered paper. A paper that has been made smooth and glossy and/or reduced in thickness by running through a machine with two or more rollers known as a "calender".

calendered plastics. Plastic sheet or foil materials with good surface appearance and uniform thickness produced by passing the starting material through a series of pressure rollers. For the processing of thermoplastic sheets and foils these rollers are usually heated.

Cal-Fab. Trade name of Flexfab, LLC (USA) for self-fusing silicon rubber tape.

Cal Flux. Trade name of BPI Inc. (USA) for a series of calcium aluminates of varying compositions used as a synthetic slags for metallurgical and melting applications.

calf leather. A light, fine-grained, pliable leather made from the skins of calves. Used for lining garments, for patent leather, and for gloves, shoes, suede, bookbinding, etc. Also known as *calfskin*.

Calfoil. Trademark of Calcarb Limited (USA) for a foil material used for bonding *Calcarb* furnace insulation materials to provide a shiny, graphite finish and improved thermal efficiency.

calfskin. See calf leather.

Calibra. Trade name Dentsply Caulk (USA) for a dental resin cement.

Calibre. Trademark of Dow Chemical Company (USA) for a series of polycarbonate resins with high impact and heat resistance available as general-purpose, sterilization, branched, 30% carbon fiber-reinforced, 20 or 30% glass fiber-reinforced, fire-retardant, high flow, impact-modified, UV-stabilized, 15% polytetrafluoroethylene-lubricated and structural foam grades. Used for appliances, business machines, electronics, medical housewares, recreational equipment, and automotive components.

Calicel. Trade name for a porous aggregate composed of calcium and aluminum silicates, and used in the manufacture of lightweight concrete.

calico. A lightweight cotton or cotton blend fabric in a plain weave that usually has colored patterns printed on one side. Used for aprons, dresses, curtains, and quilts.

Calido. Trade name of Abex Corporation (USA) for a heat-resistant alloy of 60-64% nickel, 8-16% chromium, up to 3% manganese, and 24-25% iron. It has excellent oxidation resistance up to 1000°C (1830°F). Used for electrical resistance wire, resistors, and dipping baskets.

California. Trade name of Cristales California SACIF (Argentina) for toughened glass.

California laurel. See laurel (2).

California red fir. See red fir.

California redwood. See redwood.

californite. A compact green variety of the mineral *idocrase* [Ca₁₀Al₄(Mg,Fe)₂Si₉O₃₄(OH)₄], found in Switzerland and California and used as a gem.

californium. A synthetic radioactive element of the actinide series of the periodic table. Californium-251 (^{251}Cf) was first produced by bombarding curium-242 (^{242}Cm) with high-energy helium isotopes in a cyclotron. Density, 9.31 g/cm³; atomic number, 98; atomic weight of most stable isotope, 251. The most important isotopes are californium-249 (^{249}Cf), californium-251 (^{251}Cf) and californium-252 (^{252}Cf). It forms compounds with oxygen (californium trioxide and sesquioxide), chlorine (californium trichloride), fluorine (californium trifluoride) and several other elements. Two allotropic forms are known: (i) *Alpha californium* (close-packed double hexagonal); and (ii) *Beta californium* (face-centered cubic) below 940°C (1724°F). Californium-252 (^{252}Cf) is used in neutron-activation analysis, as a neutron source, in mineral prospecting, and in oil-well logging. Symbol: Cf.

Caligen. Trade name of Calligen Foam Limited (UK) for polyurethane foam for automotive, electronic and health-care applications.

Calipso. Trade name of Circeo Filati S.r.l. (Italy) for a line of 100% pure polyester yarn products, supplied in various counts, and including low-pilling grades (*Calipso Low Pill*), microfiber grades (*Calipso Micro*), and various differently colored grades (*Calipso Melange*). Used for textile fabrics.

Calite. (1) Trade name of Calorizing Company (USA) for a heat-resistant alloy containing 40% nickel, 5.5% chromium, 4.5% aluminum, and the balance iron.

(2) Trade name of Calorizing Company (USA) for an extensive series of cast and wrought stainless alloys including various iron-chromium-nickel, nickel-chromium-iron, and nickel-iron-chromium stainless steels with good high-temperature strength, corrosion resistance and weldability. Used for heat-treating equipment, furnace parts, fan blades, chemical-plant equipment, pulp and paper mill equipment, fasteners, etc.

calite. (1) A heat-resisting alloy composed of aluminum, iron, and nickel. It has outstanding corrosion resistance under normal conditions, and excellent oxidation resistance up to about 1200°C (2190°F).

(2) Iron or steel that has been calorized. See also alumetized steel.

Calix. Trade name of Uddeholm Corporation (USA) for an oil-hardening hot-work steel containing 0.4% carbon, 1.3% chromium, 4% tungsten, 0.25% vanadium, and the balance iron. Used for punches, upsetters, mandrels, and extrusion dies.

calkinsite. A pale yellow mineral composed of cerium lanthanum carbonate tetrahydrate, $(Ce,La)_2(CO_3)_3 \cdot 4H_2O$. Crystal system, orthorhombic. Density, 3.28 g/cm³; refractive index, 1.657. Occurrence: USA (Montana).

callaghanite. An azure-blue mineral composed of copper magnesium carbonate hydroxide dihydrate, $Cu_2Mg_2(CO_3)(OH)_6 \cdot 2H_2O$. Crystal system, monoclinic. Density, 2.71 g/cm³; refractive index, 1.653. Occurrence: USA (Nevada).

Calliflex. Trade name of GTE Sylvania (USA) for a tungsten-base thermostat alloy.

Callinite. Trade name of GTE Sylvania (USA) for a series of tungsten, tungsten-copper and silver-tungsten alloys with high electrical and thermal conductivity. Used for facing electrical contacts, and welding electrodes.

Calloy. Trade name of Calloy Limited (UK) for a series of cadmium- and lead-base bearing alloys containing considerable quantities of alkaline-earth metals, such as barium, calcium and strontium. Also included under this trade name are aluminum-cadmium alloys used as deoxidizers in steelmaking, and alumi-

num-strontium alloys used as getters in vacuum tubes.

Calmalloy. Trade name of General Electric Company (USA) for magnetically soft alloys containing 68-88% nickel, 10-30% copper and 2% iron. They have high permeability, and are used for electrical and magnetic equipment.

Cal-Max. Trade name of Ash Grove Cement Company, Inc. (USA) for pulverized and chemical quicklime.

Calmax. Trade name of Uddeholm Corporation (USA) for an air- or oil-hardening hot-work tool steel containing 0.28% carbon, 11-12% chromium, 7-7.5% tungsten, 9-9.5% cobalt, 0.4-0.5% vanadium, and the balance iron. It has high hot hardness, and is used for extrusion and forging dies, die-casting die cores, and mandrels.

Calmet. Trade name of Caloriz Corporation of Great Britain (UK) for an austenitic stainless steel containing 25% chromium and 12% nickel. It has excellent heat resistance up to 1050°C (1920°F). Used for thermocouples and resistances wire.

Calmolloy. Trade name of GTE Sylvania (USA) for a tungsten alloy used for grid wire.

Calobar. Trade name of American Optical Corporation (USA) for a green-tinted spectacle glass supplied in various grades.

calomel. A secondary mineral composed of mercurous chloride, Hg_2Cl_2, and occurring in colorless, white, grayish, yellowish, green, blue, reddish or brown masses usually associated with *cinnabar*. Crystal system, tetragonal. It can also be made synthetically. Density, 6.48-7.16 g/cm³; hardness, 1-2 Mohs; refractive index, 1.973. Occurrence: Germany, Italy, Spain, USA. Used as a source of mercuric chloride, and for electrodes. Also known as *calomelite; calomelano; horn quicksilver; mercurial horn ore*.

calomelano. See calomel.

calomelite. See calomel.

Calo-MER. Trademark of Polymer Technology Group, Inc. (USA) for a shape-memory thermoplastic alloy consisting of a block copolymer and an soft-segment antiplasticizer. It has a glass-transition temperature of 20-60°C (68-140°F), and is usually supplied in the form of pellets or configured shapes.

Calomic. Trade name of Telcon Metals Limited (UK) for a resistance alloy containing 60-65% nickel, 15-16% chromium and 20-24% iron. It has an operating temperatures up to 1000°C (1830°F), and is used for electrical resistances and heating elements.

Calor. (1) Trade name of Haeckerstahl GmbH (Germany) for a series of cold-work and hot-work die and tool steels.

(2) Trade name of BNZ Materials, Inc. (USA) for silica-alumina-based firebrick for kiln and furnace insulating applications.

Calorex. (1) Trade name of Pilkington Brothers Limited (UK) for textured translucent blue-green tinted glass that absorbs a high percentage of infrared radiation while simultaneously transmitting a relatively high proportion of visible light.

(2) Trademark of Schott Glas AG (Germany) for a coated, sun-reflecting sheet glass.

Calorite. (1) Trade name for a resistance alloy containing 65% nickel, 15-23% iron, 12% chromium, and 0-8% manganese. It has high heat resistance, and is used in the form of resistance wire for electrical heating devices.

(2) Trademark of Delachaux SA (France) for a heat-producing mixture of oxides of aluminum, magnesium, silicon and other metals for aluminothermy and thermit-welding applications.

calorized steel. See alumetized steel.

calotropis fiber. A durable fiber obtained from the silky down of the milkweed seeds (*Calotropis gigantea*) native to India but also cultivated in other tropical or subtropical countries. Used for the manufacture of carpets, ropes, fishing nets and sewing thread.

Caloxo. Trade name of Warman Steel Casting Company (USA) for a series of cast corrosion- and heat-resistant chromium and chromium-nickel stainless steels.

Cal-Seal. Trade name of California Insulating Glass Company (USA) for insulating glass products.

Calsifer. Trade name of Cyprus Foote Mineral Company (USA) for a *ferrosilicon* containing small additions of both aluminum and calcium. Used as an inoculant for ductile and gray cast iron, and as a silicon addition to iron and steel.

Cal-Sil. Trademark of Betcon Division, Graybec Inc. (Canada) for building bricks.

Calsilite. Trade name of Ruberoid Company (USA) for calcium silicate-based thermal insulation cements, blocks and pipe coverings.

Calspar. Trade name for a dental impression plaster.

Calsun. Trade name of Anaconda Company (USA) for a strong, corrosion-resistant aluminum bronze containing 95.5% copper, 2% tin and 2.5% aluminum. It is available in soft and hard-drawn form for use in electrical conductors and wires.

Cal-Tab. Trade name of ALCOA, Inc. (USA) for castable refractories.

calumetite. A blue mineral composed of copper chloride hydroxide dihydrate, $Cu(OH,Cl)_2 \cdot 2H_2O$. Crystal system, orthorhombic. Refractive index, 1.690. Occurrence: USA (Michigan).

calumite. A slag-like material composed chiefly of calcium oxide, aluminum oxide, magnesium oxide and silicon dioxide. Used in the manufacture of amber, green and other glasses.

calx. See calcium oxide.

Calxyl. Trade name for a calcium hydroxide paste for dental applications.

Calypso. Trade name of G.O. Carlson Inc. (USA) for a series of die- or permanent-mold-cast aluminum-copper, aluminum-silicon and aluminum-zinc alloys. Used for tools, machine parts, electrical and electronic equipment, and aircraft and automotive components.

Calypto. Trade name of Chance Brothers Limited (UK) for decorative table glassware with beautiful leaf designs.

Calyx. Trade name for rayon fibers and yarns used for textile fabrics.

calziritite. A dark brown to black mineral composed of calcium titanium zirconium oxide, $CaTiZr_3O_9$. Crystal system, tetragonal. Density, 5.01 g/cm^3; refractive index, 2.23. Occurrence: Siberia.

Camacari. Trademark of Suramericana de Fibras SA (Peru) for acrylic fibers.

Camadil. Trademark of Global Stone James River (USA) for dolomitic limestone supplied in fine and ultrafine particle sizes. Used as a filler and extender in adhesives, coatings, plastics and rubber, and as a chemical fluxing stone.

Camalon. Trademark of Camalon for nylon 6 fibers and yarns.

Cambrelle. Trademark of E.I. DuPont de Nemours & Company (USA) for nylon 6 fibers, yarns and fabrics.

cambric. (1) A fine, thin, lustrous, canvas-like cotton fabric in a plain weave used for aprons, tablecloths, handkerchiefs, shirts dresses, blouses, nightwear, and underwear.

(2) A strong, fine-woven cotton cloth, sometimes calendered on one side, and used as *varnished cambric* in cable insulation.

cambric paper. A water- and abrasion-resistant paper with leather-like surface. It may be white or colored, and is often coated and embossed. Colored cambric paper is made from colored base paper.

Camceram. Trademark of CAM Implants BV (Netherlands) for biocompatible, synthetic *hydroxyapatite* ceramic coatings with good bonding properties. Used on dental and orthopedic implants.

camel cloth. (1) A soft fabric, usually tan or brown in color, made from camel hair or a blend of this hair and wool or other fibers. Used for dresses, suits, coatings, etc. Also known as *camel hair*.

(2) A soft tan-colored fabric with a slight pile imitating camel hair, but made of wool or wool/synthetic yarn blends. Used for rugs, scarves, coats, etc.

Camelia metal. A free-cutting brass containing 70% copper, 15% lead, 10% zinc, 4.2% tin, and 0.5% iron. Used for bearings and hardware.

camel hair. (1) Fine, soft natural protein fibers obtained from the bactrian camel (*Camelus bactrianus*) of the highlands of central Asia (especially China, Mongolia and western Russia). For commercial use, they are usually scoured and dehaired to remove coarser hairs. Used chiefly for weaving into fabrics. Also spelled camel's hair.

(2) See camel cloth.

CAM Elyaf. Trademark of CAM Elyaf (Turkey) for glass fibers.

Cameo. Trade name of J.F. Jelenko & Company (USA) for a white, microfine ceramic dental alloy containing 52.5% gold, 27% palladium, and 16% silver. It provides good strength, thermal stability and elasticity. Used for fusing dental porcelain to metal.

Cameolite. Trade name for J.F. Jelenko & Company (USA) for a low-gold dental casting alloy.

Camite. Trade name of Cleveland Automatic Machinery Company (USA) for an alloy of carbon, tungsten and iron. Used for cutting-tool tips.

camlet. A glossy fabric of silk, hair or wool in a plain weave, used for suits and furnishings.

Camloy. Trade name of Cameron & Son Limited (UK) for a heat- and corrosion-resistant stainless alloy of 25-35% nickel, 10-20% chromium and 45-65% iron. Used for marine parts, valves and tubing for superheated steam equipment, turbine blading, fittings, and tanks.

Camper-Tex. Trademark of Bruin Plastics (USA) for a tear-, water- and mildew-resistant polymer-coated textile fabric for pop-up campers.

Campholoid. Japanese trade name for cellulose nitrate plastics.

camphorwood. (1) The wood of the East African tree *Octea usambarensis*. It darkens on exposure to a deep brown color, and has an average weight of 590 kg/m^3 (37 lb/ft^3). Used for fine cabinetwork, interior decoration work, and fittings. Also known as *East African camphorwood*.

(2) The strong, light reddish-brown wood of the tree *Dryobalanops camphora*. It resembles *apitong* and *keruing*, has an interlocking grain, a fine texture, and the scent of camphor. Average weight, 770 kg/m^3 (48 lb/ft^3). Source: Borneo, Malaysia, Sumatra. Used for exterior joinery, fenceposts, piers, piles, marine installations, and cabinetwork. Also known as *Borneo camphorwood; kapur*.

campigliaite. A light blue mineral composed of copper manganese sulfate hydroxide tetrahydrate, $Cu_4Mn(SO_4)_2(OH)_6 \cdot 4H_2O$. Crystal system, monoclinic. Density, 3.00; refractive index,

1.645. Occurrence: Italy.

camwood. The hard, heavy wood of the tree *Baphia nitata*. It has a coarse, dense grain, and contains a red dye (santalin). Average weight, 1040 kg/m³ (65 lb/ft³). Source: West Africa. Used for tool handles, and machine bearings.

Canada balsam. A tacky, transparent, yellowish resin with a pine-like odor obtained from the North American balsam fir *Abies balsamea*. It has a density of 0.98 g/cm³ (0.035 lb/in.³), a refractive index of 1.530-1.545, and a softening point below 100°C (212°F). Used as a mounting medium in microscopy, as a cement for optical elements, prisms and lenses, and in fine lacquers and varnishes. Also known as *Canada turpentine*.

Canadac. Trade name for ductile and gray cast irons, and wear-resistant low-alloy and corrosion- and heat-resistant steel castings.

Canada-Lafarge. Trademark of Canada Cement Lafarge Limited for standard, high-early-strength, low-alkali, oil-well and sulfate-resisting types of Portland cements, and various masonry cements.

Canada Metal. Trademark of The Canada Metal Company Limited for a series of babbitts, solder alloys, lead castings, etc.

Canada turpentine. See Canada balsam.

Canadian asbestos. A term referring to *chrysotile*, a dark green, fine, fibrous variety of *asbestos* composed of hydrous magnesium silicate, $Mg_3Si_2O_5(OH)_4$, occurring throughout the Province of Quebec, Canada.

Canadian ash. See American ash.

Canadian spruce. See red spruce.

Canadize. Trademark of General Magnaplate Corporation (USA) for a synergistic coating for titanium made by the *Canadize* process, a proprietary electrochemical coating process that involves the production of a light anodized, or heavy hard-coated overall surface which is subsequently treated with a coating of polytetrafluoroethylene (PTFE), molybdenum disulfide, graphite, adhesives, or colored pigments. The film is locked into and over the surface resulting in extremely high hardness. The process also imparts a high-fatigue, low-friction, corrosion-resistant surface with fracture toughness superior to steel.

Canalé. Trade name of Glasindustrie Pietermann NV (Netherlands) for a figured glass.

Canalloy. Trademark of Canalloy (Canada) for alloy-tool steel products.

Canary whitewood. See American whitewood.

canasite. A greenish yellow mineral composed of sodium calcium silicate hydroxide, $(Na,K)_6Ca_5Si_{12}O_{30}(OH,F)_4$. Crystal system, monoclinic. Density, 2.71 g/cm³; refractive index, 1.538. Occurrence: Russian Federation.

canavesite. A white mineral composed of magnesium carbonate borate pentahydrate, $Mg_2(CO_3)(HBO_3)\cdot 5H_2O$. Crystal system, monoclinic. Density, 1.80 g/cm³; refractive index, 1.494. Occurrence: Italy.

cancrinite. A white, gray, yellow, green, or pinkish red mineral composed of sodium aluminum silicate carbonate, $Na_3CaAl_3Si_3O_{12}(CO_3)(OH)_2$. Crystal system, hexagonal. Density, 2.44 g/cm³; refractive index, 1.525. Occurrence: Canada (Ontario).

candelilla wax. A yellowish-brown, opaque to translucent amorphous solid obtained from the stems of the shrubs *Pedilanthus pavonis* and *Euphorbia antisyphilitica*. Density, 0.983 g/cm³; melting point, approximately 68°C (154°F). Used in varnishes, electric insulating composition, waterproofing and insect-proofing compositions, in leather dressing and polishes, as a sealing wax, soft-wax stiffener and paper size, for candles, and as a substitute for beeswax and carnauba wax.

candlewick. (1) A soft, loosely twisted cotton yarn resembling that used for candlewicks.

(2) A cotton fabric with short, cut loops and a fluffy surface appearance similar to that of *chenille*. Used especially for bedspreads, draperies and robes.

Candok. Trade name of Canbensan Industry & Trade Company (Turkey) for a foundry-grade, brown- to cream-colored natural *sodium bentonite* containing at least 75% of the mineral *montmorillonite*. It has excellent swelling characteristics, a moisture content up to 9.5%, moderately high dry strength, and a sintering temperature of 1050-1150°C (1920-2100°F). Used in iron foundries in greensand molding to impart exceptional green strength.

cane ash. See American ash.

cane clay. A fireclay of reduced refractoriness that may be sandy.

Canelado. Trade name of Vitrobrás (Brazil) for patterned glass.

Canexel. Trade name of Abitibi Building Products (Canada) for residential siding profiles made from 100% wood fiber and having five separate baked-on finishing coats. Supplied in a variety of colors, they are extremely tough, crack-, dent- and warp-resistant and can withstand harsh climates.

Canfelzo. Trade name of Pigment & Chemical Inc. (USA) for French-process zinc oxide.

canfieldite. A black mineral composed of silver tin sulfide, Ag_8SnS_6. It can also be made synthetically. Crystal system, cubic. Density, 5.95-6.28 g/cm³. Occurrence: Bolivia, Germany.

Canfor. Trademark of Canadian Forest Products Limited for fir, pine and spruce lumber, plywood, hardboard, panelwood and other wood products.

Canins. Trade name of Canbensan Industry & Trade Company (Turkey) for a *sodium bentonite* with excellent thixotropic and filter-cake forming properties. It is used in the civil engineering and construction industries for dam sealing, piling, tunneling, diaphragm wall construction, waste containment, and in the manufacture of bricks and cement to improve water resistance and green and dry strengths, and to increase the yield point.

Cankim. Trade name of Canbensan Industry & Trade Company (Turkey) for a *bentonite clay* used in the manufacture of ceramics, rubber, textiles, paper, fertilizers, cosmetics, etc.

can metal. A lead-alkali bearing alloy containing 1.75% calcium, 1.25% copper, 1% strontium and 1% barium.

cannizzarite. A light to silvery gray mineral composed of lead bismuth sulfide, $Pb_4Bi_6S_{13}$. It can also be made synthetically. Crystal system, monoclinic. Density, 6.70 g/cm³. Occurrence: Italy.

canoe birch. See American white birch.

Cannon. Trade name of Darwin & Milner Inc. (USA) for a high-speed steel containing 0.7% carbon, 16% tungsten, 3.5% chromium, 1% vanadium, and the balance iron. Used for twist drills, punches, and planing and lathe tools.

Cannonite. Trade name of Textron Inc. (USA) for a cast iron containing 2.7% carbon, 1.5% silicon, and the balance iron. Used for cylinder brake drums, pistons, crankshafts, diesel engine components, and refrigerator parts.

Cannon-Muskegon. Trade name of Cannon-Muskegon Corporation (USA) for a series of abrasion- and heat-resistant cobalt-chromium-tungsten and nickel-chromium-iron-boron hardfacing alloys used on industrial equipment and parts, e.g., gear teeth, bushings, tools, valves, scraper blades and trimming dies.

Cannon Special. Trade name of Darwin & Milner, Inc. (USA) for a tool steel containing 0.7-0.75% carbon, 18-20% tungsten, 4% chromium, 2-2.25% vanadium, 0.5% molybdenum, and the balance iron. Used for tools and cutters.

Cannon Vanadium. Trade name of Darwin & Milner, Inc. (USA) for a high-speed steel containing 1% carbon, 18-20% tungsten, 4% chromium, 3.5% vanadium, 0.5-0.8% molybdenum, and the balance iron. Used for tools and cutters.

Canoe. Trademark of Federated Co-operatives Limited (Canada) for plywood and lumber products.

Canon. Trade name of Creusot-Loire (France) for a series of oil-hardened steels used for machine-tool parts.

can solder. A lead solder that contains 2% tin, and has a solidus temperature of 270°C (518°F) and a liquidus temperature of 312°C (594°F). Used for soldering side seams of cans.

Canson. Trade name of Canbensan Industry & Trade Company (Turkey) for a *sodium bentonite* for borehole-drilling applications (drilling mud).

cantala. See cantala fibers.

cantala fibers. The hard vegetable fibers obtained from the agave plant *Agave cantala*, and used especially for binder twine and cordage. Also known as *cantala; Cebu maguey; Manila maguey.* See also agave fibers; maguey fibers.

canted antiferromagnetic material. A material exhibiting weak ferromagnetism and a net moment that are the result of canting of the antiferromagnetically oriented spins. Also known as *canted antiferromagnet.* See also antiferromagnetics.

Cantona. Trade name for silky rayon fibers and yarns used for textile fabrics.

canton blue. A bluish-purple ceramic color obtained by the addition of barium carbonate to *cobalt blue* (a mixture of cobalt oxide and aluminum oxide).

Canton crepe. A soft fabric with a wrinkled surface, originally of silk, but now also made of rayon or polyester. Named for Canton, a city in southern China where it was originally made. Used for blouses and dresses. See also crepe.

Canton flannel. A strong, heavy cotton or cotton-blend fabric with a twill face and a soft, fleecy, heavily combed back used for gloves, nightwear, underwear and linings, and for light-duty buffing wheels. Named for Canton, a city in southern China where it was originally made. See also flannel.

Central. Trade name of Société des AFC (France) for a heat-treatable aluminum alloy containing 12% silicon, 2.5% copper, 1.25% magnesium, 1.25% manganese, 2.5% nickel, and 0.25% titanium. Used for light alloy parts and castings.

Cantrece. Trade name of E.I. Du Pont de Nemours & Company (USA) for strong, lightweight nylon 6,6 fibers with high resistance to abrasion, wrinkling, mildew and moths. Used for clothing including hosiery.

canvas. A strong, durable, closely woven fabric of cotton, linen, hemp, flax or jute usually in a plain weave. It may be water-proofed for outdoor use, and is used for awnings, tarpaulins, bags, sails, tents, welding curtains, upholstery, footwear, and protective clothing, e.g., aprons and gloves. See also duck.

Canzler. Trade name of Perry Equipment Corporation (USA) for brass and bronze welding rods used for welding copper and copper alloys.

caoba. See American mahogany.

caoutchouc. (1) The gummy, coagulated *latex* obtained from various tropical rubber trees and shrubs of the genera *Hevea* and *Ficus,* especially *H. brasiliensis.* It consists of an aqueous dispersion of *cis*-1,4-polyisoprene $(C_5H_8)_n$, a high-molecular-weight, unsaturated hydrocarbon. Also known as *India rubber.*

See also polyisoprene; natural rubber; rubber.

(2) A term sometimes used to refer to the solid, dark, elastic substance obtained from the coagulated *latex* described in (1) by preparing as sheets and subsequent drying. Used as a starting material for natural rubber production.

capacitor paper. A strong, tough, high-grade paper, usually less than 10 μm (0.4 mil) thick. Used as a dielectric in the manufacture of capacitors. *Note:* Paper capacitors are manufactured by rolling alternate layers of paper and metal foil in ribbon form, or by depositing a thin metal coating by any of several processes.

capacitor tissue paper. An electrical insulating paper used as a dielectric in capacitors.

Capaloy. Trade name of Oscap Manufacturing Company (USA) for a platinum alloy used for laboratory ware.

Caparol. Trademark of Amphibolin for a series of acrylic resins and compounds.

cap copper. A high-copper alloy containing 3-3.5% zinc. Usually supplied in the form of strips, it has excellent cold workability, good hot workability and good corrosion resistance, and is used for cartridge and fuse caps.

cape asbestos. See crocidolite.

cape blue. See crocidolite.

cape boxwood. See African boxwood.

cape leather. A glove leather finished with a glossy surface and originally obtained from South African sheep (Cape of Good Hope area), but now also imported from other countries.

cape wool. A high-quality wool obtained from sheep raised in southern Africa.

cap gilding. A commercial bronze containing 90% copper and 10% zinc. It has good corrosion resistance, and is used for ornamental window screens, grillwork, and screen cloth.

Capilene. Trade name for polyester fibers and yarns used for textile fabrics.

Capima. Trade name for nylon fibers and yarns.

capirona. The strong, straight-grained yellowish wood of several species of trees, the best known of which is the degami (*Calycophyllum candidissimum*). It has a fine dense grain, high elasticity, and poor resistance to stains, decay and insects. Source: Cuba; West Indies. Used for textile industry equipment, and archery equipment. See also degami.

Capital. Trade name of Eagle & Globe Steel Limited (Australia) for a series of high-speed steels containing 0.8-1.2% carbon, 4.0-4.5% chromium, 1.7-6.4% tungsten, 5.0-8.7% molybdenum, 1.1-4.0% vanadium, 0-5.0% cobalt, and the balance iron. They have good wear resistance and red hardness, and high toughness. Used for cutting tools, e.g., tool bits, reamers, twist drills and milling cutters.

Capitor. Trade name of Cendres & Métaux SA (Switzerland) for a pale yellow gold dental alloy used for special ceramic restorations. Also known as *CM Capitor.*

Caplana. Trademark of AlliedSignal, Inc. (USA) for nylon 6 fibers, yarns and monofilaments.

Caposite. Trade name for braided asbestos rope, 13-51 mm (0.5 to 2 in.) in diameter and up to 30.5 m (100 ft.) in length, consisting of long white asbestos fibers enclosed in a braided asbestos yarn cover. Used as thermal insulation for steam or hot water pipes, valves, and furnace doors.

Cappagh brown. A brown pigment made from a British variety of *umber.*

capped steel. A type of steel intermediate in most characteristics between rimmed and semikilled steel. As with rimmed steel, the surface (or rim) is relatively soft, easily machinable, and virtually free of voids. However, it contains numerous blowholes close to the surface and from the top to the bottom. These holes are due to the application of a heavy metal cap over the top of the ingot causing the top metal to solidify. See also rimmed steel; semikilled steel.

cappelenite. A white, yellowish or greenish-brown mineral composed of barium yttrium borosilicate, $Ba(Y,La)_6(Si_3B_6O_{24})$. Crystal system, hexagonal. Density, 4.20 g/cm^3; refractive index, 1.760. Occurrence: Kazakhstan.

Capran. Trademark of Allied Chemical Corporation (USA) for a flame-retardant antistatic nylon film that reduces losses from static electricity. Used as a packaging material.

Capricorn. Trademark of Ivoclar Vivadent AG (Liechtenstein) for an extra-hard ceramic dental alloy containing 78.1% palladium, 6.5% indium, 6.0% gold, 6.0% gallium, 3.0% silver, and less than 1.0% ruthenium, rhenium and lithium, respectively. It has a white color, a density of 11.0 g/cm^3 (0.40 lb/in.^{3}), a melting range of 1170-1335°C (2140-2435°F), a hardness of 260 Vickers, moderate elongation, and excellent biocompatibility. Used for crowns, onlays, posts and bridges.

caproester-urethane. A linear thermoplastic urethane polymer.

caprolactam. A cyclic amide with six carbon atoms, usually supplied in the form of a water-soluble fused solid or white flakes (99+% pure). It has a melting point of 70-72°C (158-162°F), a boiling point of 180°C (356°C)/50 mm, and a refractive index of 1.496 (30°C/86°F). A 70% solution with a density of 1.050 g/cm^3 is also available. Used in the manufacture of polycaprolactam (nylon 6) coatings, fibers and resins, and for synthetic leather, plasticizers, paint vehicles, etc., as a crosslinker for polyurethanes, and in the synthesis of lysine. Formula: $C_6H_{11}NO$. Also known as ε-*caprolactam*.

caprolactones. Crystalline compounds (98+% pure) available in two forms: (i) γ-caprolactone with a density of 1.027 g/cm^3, a boiling point of 219°C (426°F), a flash point of 209°F (98°C) and a refractive index of 1.439; and (ii) ε-caprolactone which is the product of a reaction of peracetic acid with cyclohexanone. It has a density of 1.030 g/cm^3, a boiling point of 96-97.5°C (205-208°F), a flash point of 229°F (109°C) and a refractive index of 1.463. *Caprolactones* are used as intermediates in the manufacture of caprolactam, as intermediates in adhesives and urethane coatings and elastomers, in the manufacture of polymers, plasticizers, films and synthetic fibers, as diluents for epoxy resins, and in organic synthesis. Formula: $C_6H_{10}O_2$.

Caprolan. Trademark of Honeywell International (USA) for polyamide fibers, yarns and monofilaments made from high-molecular-weight polymerized caprolactam (nylon 6). It has outstanding dimensional stability after heat setting, exceptional mechanical properties, and takes dyes well. Used for tire cord.

Capron. Trademark of AlliedSignal Inc. (USA) for polyamide (nylon 6 and nylon 6,6) resins and blends available as homopolymers with glass-fiber or carbon-fiber reinforcement for high impact, filled with glass beads or minerals, or lubricated with silicone, polytetrafluoroethylene or molybdenum disulfide. They are also supplied in UV-stabilized, fire-retardant, stampable sheet, elastomer copolymer and casting grades. *Capron* polyamides have high mechanical strength and creep resistance, high impact strength, good elevated temperature properties, a heat deflection temperature of 210°C (410°F), and are used for appliances, fuel tanks, automotive air ducts and resonators, marine parts, lawn and garden equipment, packaging, fasteners, switches, connectors, and multilayer films. *Capron UltraTough*

refers to series of polyamide (nylon 6) resins with good low-temperature impact strength and ductility to -40°C (-40°F). Used for automotive components, power-tool and small-engine components, casings, and sports gear.

capsule metal. An alloy of 92% lead and 8% tin used for bearings.

Captek. Trade name of Unikorn Limited for high-gold dental alloys for crown and bridge restorations.

Captiva. Trade name for nylon fibers and yarns.

caracolite. A colorless mineral composed of sodium lead chloride sulfate, $Na_3Pb_2(SO_4)_3Cl$. Crystal system, monoclinic. Density, 5.10 g/cm^3; refractive index, 1.754. Occurrence: Chile.

Caraloft. Trade name for acrylic fibers used for the manufacture of clothing.

carapa. See andiroba.

Carat. Trade name for a dental porcelain for metal-to-ceramic bonding.

Carawisp. Trade name for acrylic fibers used for the manufacture of textile fabrics.

Carbaloy. Trade name of General Electric Company (USA) for various grades of tungsten carbide.

carbamide. See urea (2).

carbamide peroxide. See urea peroxide.

carbazole. A white, crystalline tricyclic hydrocarbon containing nitrogen. It is available in technical grades (95+% pure) with a melting point of 243-246°C (469-475°F) and a boiling point 352-355°C (665-670°F). Used in the manufacture of dyes, lubricants, rubber antioxidants, as a UV-sensitizer in photography, and in biochemistry and biotechnology. Formula: $(C_6H_4)_2NH$.

Carbdi. Trade name of Eagle & Globe Steel Limited (Australia) for a steel containing 0.2% carbon, 1% manganese, 1.2% chromium, 0.2% molybdenum, 0.3% silicon, and the balance iron. It has high hardenability, and is used for case-hardened molds, and for compression, transfer or injection molds for abrasive plastics.

Carbex. Trade name for a refractory brick made from silicon carbide.

carbide. (1) A binary compound of carbon with a more electropositive metallic or nonmetallic element, e.g., boron carbide (B_4C) or molybdenum monocarbide (MoC).
  (2) See cemented carbides.

carbide-base cermets. A group of cermets made by powder-metallurgy techniques from a ceramic carbide, such as chromium, titanium or tungsten carbide, with a metal or alloy matrix binder, such as cobalt, molybdenum, nickel, or a nickel alloy. They have high rigidity, strength, hardness and wear resistance, and good strength retention at elevated temperatures. Used for gages, turbine parts, valves, bearings, pumps, cutting tools, and nuclear equipment. See also cermets; chromium carbide cermets; titanium carbide cermets; tungsten carbide cermets.

carbide ceramic coatings. Ceramic coatings prepared from carbides, such as boron, hafnium, niobium (columbium), silicon, tantalum, titanium or tungsten carbide, and applied by plasma-arc or detonation-gun techniques. They have high hardness and excellent wear resistance, and are used for wear and seal applications, e.g., jet-engine seals, in knives for rubber-skiving machines and paper machines, for plug gages, etc.

carbide fibers. Continuous and discontinuous ceramic fibers made of silicon carbide, or carbide mixtures, such as silicon carbide with titanium or zirconium carbide. They have excellent mechanical properties and good oxidation resistance, and are used as reinforcing fibers in ceramic, metallic or plastic composites.

carbide fuel. A strong, oxidation-resistant composition made by mixing a nuclear fuel with a metal and a carbon compound. Also known as *carbide nuclear fuel.*

carbide nuclear fuel. See carbide fuel.

carbide tool materials. See cemented carbides.

Carbidie. Trade name of Carbidie Division of Aiken Industries Inc. (USA) for a series of hard, wear-resistant sintered carbides composed of tungsten carbide or carbide mixtures and a cobalt or nickel binder. Used for blanking, drawing and forming dies, compacting and stamping dies, header and coining dies, lamination and extrusion tooling, slitter knives, mandrels, punches, boring bars, drills, liners, pump bearings and plungers, gauges, wear parts, nozzles, seal-ring faces, valve balls and seats, mill rolls, crusher rolls, and pulverizing hammers.

Carbium. Trade name of A. Gayer Company (France) for a heat-treatable aluminum alloy containing 4-5% copper used for light-alloy parts.

Carbo. British trade name for a cemented tungsten carbide with cobalt matrix binder, used for cutting tools.

CarboAccucast. Trademark of CARBO Ceramics Inc. (USA) for ceramic media supplied in various grades for casting and reclamation applications. They possess high flowability, strength and durability, and low coefficients of thermal expansion.

Carbo-Alkor. Trade name of Atlas Minerals & Chemicals, Inc. (USA) for acid- and alkali-resistant cements, coatings and linings for floors, tanks, etc.

Carboalumina. Trade name for fused alumina.

Carbo-Atlasiseal. Trade name of Atlas Minerals & Chemicals, Inc. (USA) for corrosion-resistant linings.

carboborite. A colorless mineral composed of calcium magnesium carbonate borate tetrahydrate, $Ca_2Mg(CO_3)_2B_2(OH)_8 \cdot 4H_2O$. Crystal system, monoclinic. Density, 2.12 g/cm^3; refractive index, 1.5459. Occurrence: China.

Carbobrant. Trade name for silicon carbide abrasives.

Carbobronze. A British trade name for an acid- and alkali-resistant phosphor bronze containing about 92% copper, 8% tin, and 0.3% phosphorus. Used for tubing.

carbocernaite. A mineral of the aragonite group composed of calcium strontium cerium fluoride carbonate, $(Ca,Sr,Ce)_4(CO_3)_4F$. Crystal system, orthorhombic. Density, 3.53 g/cm^3; refractive index, 1.679. Occurrence: Canada (Ontario).

Carb-O-Cite. Trade name of Shamokin Graphite (USA) for granular *anthracite* supplied in several particle sizes.

Carboco. Trade name for a zinc polycarboxylate dental cement.

Carbo-Cor. Trade name of USFilter/Permutit (USA) for activated carbon.

Carbo-Cut. Trade name of Carborundum Abrasive Company (USA) for nonwoven finishing abrasives.

carbodiimide. See cyanamide.

Carbodis. Trade name for a series of carbon blacks.

Carbo-Dur. Trade name of USFilter/Permutit (USA) for activated carbon products.

Carb-O-Fil. Trade name of Shamokin Graphite (USA) for powdered *anthracite* supplied in various particle sizes. Used as a carbonaceous filler, and as a substitute for carbon black in certain phenolic resins.

Carboflex. Trademark of Ashland Oil, Inc. (USA) for multi-purpose carbon fibers supplied in the form of chopped fibers and mats.

Carbofol. Trademark of Niederberg-Chemie GmbH (Germany) for roofing, roofing membranes and sealing courses.

Carboform. Trade name of Cytec Inc. (USA) for epoxy laminates available as *Kevlar-* and carbon-fiber prepregs.

Carbofrax. Trademark of The Carborundum Corporation (USA) for silicon-bonded silicon carbide for the manufacture of highly refractory products. The silicon carbide content of the products is 85+% and the porosity about 13%. Used for refractory cements and mortars, bonded refractory bricks and shapes for furnace linings, muffle walls, checkers, domes, radiant tubes, and hearths.

Carbo-Glas. Trade name of Fred A. Wilson Company (USA) for roof coatings and patches, and roof-patch reinforcing glass membranes.

carbohydrates. A group of organic compounds of carbon, hydrogen and oxygen including monosaccharides (e.g., fructose and glucose), disaccharides (e.g., sucrose and maltose), and polysaccharides (e.g., starches and celluloses). Used in medicine, biochemistry, biotechnology, and in the synthesis of novel biopolymers.

carboirite. A pale to dark green mineral composed of iron aluminum germanium oxide hydroxide, $FeAl_2GeO_5(OH)_2$. Crystal system, triclinic. Density, 4.10 g/cm^3; refractive index, 1.735. Occurrence: France.

Carbo-Korez. Trademark of Atlas Minerals & Chemicals, Inc. (USA) for an acid-resisting, carbon-filled phenol-formaldehyde resin used as a mortar cement.

Carbolac. Trademark of Cabot Corporation (USA) for a series of carbon blacks, used as pigments in lacquers, varnishes and paints.

Carbolap. Trademark of Exolon-ESK Company (USA) for silicon carbide grains for lapping and tumbling applications.

Carbo-Lastomeric. Trade name of Fred. A. Wilson Company (USA) for elastomeric roof coatings.

CarboLex. Trade name for single-walled *carbon nanotubes* available in several grades with diameters of 120-150 nm (5-6 μin.).

Carboline. Trade name of Corro Therm, Inc. (USA) for a series of polymer-based protective coatings and linings including high-gloss polyurethanes and high-temperature modified silicones.

Carbolite. (1) Trade name for silicon carbide.

(2) Trademark of Mitsubishi Chemical Industries Limited (Japan) for carbon fibers used in insulating materials to enhance oxidation resistance and thermal diffusivity.

Carbolon. (1) Trademark of The Exolon-ESK Company (USA) for silicon carbide (SiC) supplied in black (97.8% pure) and green (99% pure) colors. Used as an abrasive for grinding, lapping, polishing and blasting applications, in wire-sawing of hard materials, and as a metallurgical additive.

(2) Trademark of Nippon Carbon Company Limited (Japan) for raw carbon fibers used as reinforcement in composites, and for packing and insulating applications. Also included under this trade name are carbon fiber threads and yarns.

(3) Trademark of Nippon Carbon Company Limited (Japan) for carbonized and graphitized man-made fiber cloth, felt, tape and sheeting.

Carbolox. Trade name for silicon carbide.

Carboloy. Trademark of General Electric Company (USA) for a composite material made by sintering a powdered tungsten or chromium carbide with cobalt or nickel. It has extremely high hardness and excellent wear and abrasion resistance. Used in the form of cemented carbides for cutting tools, drawing dies, gages, and wear-resistant parts.

Carbolux. Trade name of Rütgerswerke AG (Germany) for acetylene cokes including *Carbolux S* composed of 98.5% carbon,

0.5% ash, 0.2% hydrogen, 0.15% nitrogen and 0.1% sulfur, and *Carbolux SK* composed of 99% carbon, 0.2% hydrogen, 0.15% ash, 0.06% nitrogen, and 0.06% sulfur. Supplied in the form of pellets, they are used as carburizers in the manufacture of mass-produced steel and cast iron, and as carbon sources in the production of high-carbon high-grade steels.

Carbomang. Trade name of Detroit Alloy Steel Company (USA) for an oil-hardenable, cast-to-shape tool steel containing 0.9-1% carbon, 1-1.25% manganese, 0.45-0.60% chromium, 0.4-0.6% tungsten, and the balance iron. Used for tools and dies, machine parts, and castings.

Carbomet. Trade name for a series of carbon blacks.

Carbomix. Trademark of DSM Copolymers (USA) for a black styrene-butadiene rubber master batch used in the manufacture of solid rubber goods.

Carbon. (1) Trade name of Latrobe Steel Company (USA) for a plain-carbon steel containing 1.05% carbon, 0.2% manganese, 0.2% silicon, 0.08% chromium, and the balance iron. It has high hardness, and is used for cutting tools, press tools, header and forming dies, etc.

(2) Trade name of US Steel Corporation (USA) for a low-carbon steel containing 0.1-0.3% carbon, 0.3-0.6% manganese, 0.1-0.2% silicon, 0.4-0.7% molybdenum, and the balance iron. It has good high-temperature properties, and excellent creep strength. Used for steam and oil piping, and for still tubes for oil-cracking equipment.

carbon. A nonmetallic element of Group IVA (Group 14) of the Periodic Table. There are three basic allotropic forms one amorphous (e.g., charcoal, coal, coke and carbon black), and two crystalline (graphite and diamond). It occurs in combination with other elements in all plants and animals, some minerals, oil, coal and natural gas deposits, and can also be made synthetically in varying degrees of purity. Two more recently discovered special forms are *buckminsterfullerene* and *carbon nanotubes*. Carbon is commercially available in the form of rods, sheets, foils, microleaves, tubes, electrodes, granules, powder, soot, single crystals, fibers, yarns and fabrics. Density, 1.88-2.1 g/cm³ (amorphous), 2.25 g/cm³ (graphite), 3.5 g/cm³ (diamond); melting point, sublimes above 3500°C (6330°F); boiling point, 4827°C (8721°F); hardness, 0.5-1.0 Mohs (graphite), 10 Mohs (diamond); atomic number, 6; atomic weight, 12.011; tetravalent. *Carbon* has a low coefficient of thermal expansion, high thermal-shock resistance, high thermal conductivity, fair electrical conductivity, high strength with increasing temperature, excellent abrasion, erosion and corrosion resistance, and a low neutron-capture cross section (about 0.0034 barns). Used in the form of graphite, diamond and carbon black, as a reducing agent, in pencils, polymers, pigments, electrodes, electrical equipment and molds, in nuclear reactors as a moderator and neutron deflector, in steelmaking, and in biochemistry, bioengineering, biology and medicine. See also *amorphous carbon; crystalline carbon.*

carbon-12. A natural isotope that constitutes 98.89% of the element carbon. Used as the standard of atomic mass and defined as exactly 12u (relative nuclidic mass unit). 12 grams of carbon-12 contain exactly 1 mole of carbon atoms (6.02 × 10²³). Symbol: ^{12}C.

carbon-13. A natural isotope of carbon with a mass number of 13, which constitutes 1.11% of the element. Used in nuclear magnetic resonance spectroscopy, and for synthetic diamonds. Symbol: ^{13}C.

carbon-13 diamonds. Synthetic diamonds composed of 99+%

carbon-13 (^{13}C), and produced by a two-step process involving the low-pressure chemical vapor deposition (LPCVD) of an aggregate of small diamonds using high-purity methane gas as the carbon source, and a subsequent high-pressure technology to dissolve and recrystallize the diamond aggregates into gem-quality diamonds which are highly abrasive, harder than natural diamonds (10+ Mohs), and useful for grinding, machining and polishing applications.

carbon-14. A heavy, radioactive isotope of carbon with a mass number of 14 that is produced in nature by collisions between nitrogen atoms and neutrons from cosmic radiation, and from atmospheric nitrogen. It can also be produced synthetically by irradiation of calcium nitrate in a nuclear reactor. It emits beta rays and has a half-life of 5780 years. Used as a radiation source in thickness gauges and other instruments, in metallurgical research, and in radiocarbon dating in geology and archaeometry. Symbol: ^{14}C. Also known as *radioactive carbon; radiocarbon.*

carbonado. See black diamond.

carbon aerogels. A class of highly porous, low-density, electrically conductive aerogels derived by sol-gel polymerization of organic monomers, such as resorcinol-formaldehyde, in solution. They consist of three-dimensional networks of nanometer-sized, covalently bonded carbon particles. Available in the form of powders, monoliths and thin films, they are used in double-layer capacitors, electrochemistry, composite structures, high-temperature insulation, printer pigments, etc. See also aerogel.

carbon-aramid composites. Hybrid fabrics woven from aramid and carbon fibers in a plain or twill weave pattern. They provide laminate reinforcements which are uniform in thickness and free of holes, facilitate lay-up, and have high resin compatibility. These fabrics are made up of alternate warp and weft picks (threads) of aramid (*Kevlar*) and carbon fibers. Also known as *carbon-polyaramid composites.* See also plain-weave fabrics; twill-weave fabrics.

carbonate. (1) In chemistry, an ester or salt of carbonic acid (H_2CO_3). The reaction of an organic compound with H_2CO_3 produces an ester (e.g., diethyl carbonate) and that of a metal yields a salt (e.g., calcium carbonate).

(2) In ceramic terminology, any salt composed of a metallic element in combination with a CO_3 radical. Examples include calcium carbonate ($CaCO_3$) and strontium carbonate ($SrCO_3$). Used as a source of metal oxides in ceramic bodies.

carbonate-cyanotrichite. A sky blue to azure blue mineral composed of copper aluminum carbonate sulfate hydroxide dihydrate, $Cu_4Al_2(CO_3,SO_4)(OH)_{12} \cdot 2H_2O$. Density, 2.65 g/cm³. Occurrence: Russian Federation.

carbonate-fluoroapatite. A colorless mineral of the apatite group composed of calcium fluoride carbonate phosphate hydroxide, $Ca_{10}(PO_4)_5CO_3F_{1.5}(OH)_{0.5}$. Crystal system, hexagonal. Density, 3.12 g/cm³; refractive index, 1.627. Occurrence: Germany, South Africa.

carbonate-hydroxylapatite. A mineral of the apatite group composed of calcium carbonate phosphate hydroxide, $Ca_{10}(PO_4)_3(CO_3)_3(OH)_2$. It can also be made synthetically by adding calcium acetate in the form of drops to solutions of phosphate and carbonate at 95-100°C (203-212°F). The natural form may contain some fluorine. Crystal system, hexagonal. Density, 2.87-3.05 g/cm³. Occurrence: Germany.

carbon base paper. A *base paper* that is wood-free and sometimes contains cotton or linen rags. It is ready to be coated with pressure-transferable pigments, and used in the manufacture of carbon and self-copying paper. Base papers used for making one-time carbon paper may contain some wood.

carbon black. A general term for any of various colloidal black substances of extremely fine particle size consisting essentially of amorphous elemental carbon prepared by partial combustion or thermal decomposition of hydrocarbons, such as oil, gas, etc. They have high oil absorption and low specific gravity, and are used as colorants in concrete, as pigments for paint and printing inks, as fillers, as ultraviolet light absorbers and reinforcing agents in rubber products and plastics, as electrical conductivity improvers (e.g., in plastics and composites), in molding compounds intended for outside weathering applications, as opacifiers, for stove polishes, for typewriter and printer ribbons, as expanders in battery plates, and as solar energy absorbers. See also channel black; furnace black; gas black; jet black.

Carbon Bonded Carbon Fiber. Trade name of Calcarb Limited (USA) for a low-density rigid thermal insulation consisting of carbon fibers bonded with carbon. It is produced by vacuum molding, firing to 2000°C (3630°F), and subsequent outgassing under vacuum. Supplied in standard and custom shapes, it is used in the heat treating, sintering and graphitizing industries for vacuum and controlled-atmosphere applications.

carbon brick. A brick usually manufactured from a carbonaceous material, such as carbon, graphite, calcined anthracite coal or metallurgical or petroleum coke, using a bituminous binder, such as pitch or tar.

carbon bronze. A *plastic bronze* containing 75% copper, 15% lead and 10% tin, used for bearings.

carbon-carbon composites. See carbon-matrix composites.

carbon-ceramic refractories. See ceramic-carbon refractories.

Carbon Chisel. Trade name of Bissett Steel Company (USA) for a water-hardening steel containing 0.7-1% carbon, and the balance iron. Used for tools, chisels, punches, etc.

Carbon Cold Header. Trade name of Bethlehem Steel Corporation (USA) for a water-hardening tool steel containing 0.9% carbon, up ro 0.1% chromium, 0.18% vanadium, and the balance iron. Used for hammers, and cold-heading, gripper, forming and swaging dies.

Carbondale. (1) Trade name of Robinson Brick Company (USA) for a red clay.

(2) Trade name of Robinson Brick Company (USA) for building bricks.

Carbondale silver. A nickel silver containing 66% copper, 18% nickel, and 16% zinc. Used for spinning and drawing applications, and for flatware, spoons, forks and knives to be plated.

carbon diamond. See black diamond.

carbon-epoxy composites. Composite materials consisting of epoxy-resin matrices reinforced with carbon fibers. Usually supplied in sheet, rod or tube form, they have high strength-to-weight ratios, and good thermal and electrical properties, and are used for external structural panels for aircraft and space structures, and in filament winding of components, such as tanks, pipes, pressure vessels, and shapes of revolution.

carbon fabrics. Fabrics woven from carbon fibers obtained from high-tenacity rayon, pitch or polyacrylonitrile precursors. They usually have have high tensile, compressive and flexural strengths. Woven carbon fabrics provide a laminate reinforcement that is uniform in thickness, free of holes, and has high resin compatibility. See also carbon fibers; graphite fabrics.

carbon-fiber ceramic composites. Composite materials consisting of ceramic matrices (e.g., silicon carbide, silicon nitride,

glass, or glass-ceramics) reinforced with aligned, continuous carbon fibers. They have low specific densities, high strength and stiffness, good high-temperature characteristics in inert atmospheres, and rapid degradation in oxidizing atmospheres above 500°C (930°F). Used for structural applications.

carbon-fiber composites. See carbon-fiber-reinforced composites.

carbon fiber-reinforced carbon. See carbon-matrix composites.

carbon fiber-reinforced cement. A durable, high-strength composite that has a cement-based matrix reinforced with carbon fibers. A thin nickel coating on the fibers provides strong fiber-matrix bonding that significantly improves both mechanical strength and fracture toughness. Used for building construction and civil engineering applications. Abbreviation: CFRC.

carbon fiber-reinforced composites. Composite materials consisting of metal or nonmetal matrices reinforced with continuous or discontinuous carbon fibers. Examples include carbon-aramid, carbon-carbon and carbon-epoxy composites as well as carbon fiber-reinforced ceramics and plastics. Abbreviation: CFRC. Also known as *carbon-fiber composites.*

carbon fiber-reinforced plastics. Composite materials consisting of plastic matrices (e.g., polyester, epoxy, vinylester or nylon) reinforced with continuous or discontinuous carbon fibers. They have high strength and stiffness, improved friction and wear properties, low energy dissipation, and good weatherability. Used for mechanical energy-storage devices, bearings and electrical waveguides. Abbreviation: CFRP. Also known as *carbon fiber-reinforced polymers.*

carbon fibers. A class of fibers consisting nongraphitic carbon obtained by thermal decomposition of natural or synthetic organic fibers, or fibers drawn from organic precursors (e.g., rayon, polyacrylonitrile or pitch) and subsequently carbonized (heat-treated) at temperatures up to 2725°C (4937°F). The content of elemental carbon after heat-treatment is about 93-98%. *Carbon fibers* have an average density range of 1.75-2.15 g/cm³ (0.06-0.08 lb/in.³), and an average diameter range of 0.1-11 μm (3.9-433 μin.). Used as reinforcements for engineering plastics and composites. Abbreviation: CF. See also graphite fibers.

carbon fiber, type I. A type of carbon fiber that has been heat-treated at temperatures above 2500°C (4530°F) and exhibits high elastic modulus and medium strength.

carbon fiber, type II. A type of carbon fiber that has been heat-treated a temperatures up to about 1400°C (2550°F) and exhibits high tensile strength and medium elastic modulus.

carbon fiber, type A. A type of carbon fiber that has been heat-treated at up to about 1000°C (1830°F) and exhibits relatively low elastic modulus and tensile strength.

carbon-filled polyethylenes. A family of reinforced polymers composed of butyl polyethylene filled with carbon black. They have a density of 0.96 g/cm³ (0.03 lb/in.³), good electrical properties, good ultraviolet-light resistance, high tensile strength, good resistance to dilute acids and alcohols, and fair resistance to concentrated acids and ketones. Used for plastic parts with enhanced electrical conductivity.

Carbon-Ford. Trade name of Republic Steel Corporation (USA) for a steel containing 0.15-0.45% carbon, 0.3-0.85% manganese, 0-0.09% silicon, and the balance iron.

carbon-graphite. A graphitic composition used in the manufacture of sliding bearings, carbon brushes, packing rings, seals, etc. Bearing materials molded from carbon-graphite do not require lubrication, and may contain a metal or metal alloy to improve compressive strength. *Carbon-graphite* has high dimensional stability, excellent resistance to chemicals, a maximum operating temperature in air of 400°C (750°F), good wear resistance, a low coefficient of friction, good strength, and fair impact properties.

Carbon Gun. Trade name of Wellsville Fire Brick Company (USA) for carbon-base gunning refractories.

Carbonite. (1) Trademark of Anchor Packing Division, Robco Inc. (Canada) for bulk asbestos fiber packings.

(2) Trademark of Dresser Industries, Inc. (USA) for silicon carbide abrasives.

(3) Trademark of Hill & Griffith Company (USA) for carbonaceous metal-casting additives.

carbonitrided steel. A steel, usually a plain- or low-carbon grade, that has been subjected to a surface hardening process involving heating at 700-870°C (1290-1600°F) for several hours in a gaseous hydrocarbon and ammonia atmosphere. Carbon and nitrogen are introduced into the surface layer (or case), increase the hardenability, and permit hardening by oil quenching. The case depth is about 0.5 mm (0.02 in.). *Carbonitrided steel* is often used for machine elements, such as gears, nuts, bolts, etc. Also known as *gas-cyanided steel; nicarbed steel.*

Carbonized Nickel. Trade name of Harrison Alloys Inc. (USA) for a commercially pure nickel (99+%) used for anode plates.

carbonized rag fiber. A vegetable fiber from which cellulosic matter has been removed by treating with an acid solution (e.g., a sulfuric acid solution) or an acid gas (e.g., hydrogen chloride), followed by heating. This treatment significantly increases the friability of the cellulosic matter.

carbonless paper. A paper coated or treated with chemicals to make smudge-free copies without requiring carbon paper interleaves. Used for cash-register rolls, business forms, etc. Also known as *no-carbon-required paper; NCR paper.*

carbon-manganese steels. A group of carbon steels containing about 0.18-0.48% carbon, 1.00-2.00% manganese, 0.15-0.50% silicon, traces of sulfur and phosphorus, and the balance iron. This term usually refers to the steels in the AISI-SAE 13xx series, which have high abrasion and wear resistance, high tensile strength and good weldability. Used for structural applications, rails, railway equipment, carburized parts, and machinery parts. Also known as *intermediate manganese steels; manganese steels; pearlitic manganese steels.*

carbon-matrix composites. Biocompatible composites consisting of carbonaceous matrices (e.g., carbon or graphite) reinforced with carbon or graphite fibers. They have high strengths and moduli, high creep resistance, high fracture toughness, high thermal conductivity, low coefficients of thermal expansion, low thermal-shock sensitivity, and an upper service temperature above 2800°C (5070°F). Used for aircraft and automotive brakes, structural components and thermal protective panels for spacecraft, gas turbine engine parts, hot-pressing molds, and prosthetic devices, e.g., knee and hip joints. Abbreviation: CMC. Also known as *carbon-carbon composites; carbon fiber-reinforced carbon.*

carbon micro-trees. A class of relatively new micron-sized, tree-like carbon structures produces by chemical vapor deposition (CD) of methane gas.

Carbon Moly Steel. Trade name of Bonney-Floyd Company (USA) for a heat-treated abrasion- and wear-resistant steel containing 0.25-0.45% carbon, 0.8-1.1% chromium, 0.15-0.25% molybdenum, and the balance iron. Used for shafts, gears, and castings.

carbon nanofiber. A carbon fiber with a diameter in the nanometer size region (typically 2-100 nm or 0.08-3.94 μin.), and a length of 5-100 μm (200-4000 μin.). It can be prepared in bulk quantities by the decomposition of a mixture of hydrocarbon, hydrogen and inert gas on metal foil, gauze or powder catalysts (e.g., cobalt, iron, nickel or selected alloys thereof) in a flow reactor system. It possesses a high degree of flexibility, a high aspect ratio and good mechanical properties, and is used as a reinforcing fiber in various engineering materials. See also graphite nanofiber.

carbon nanohorn. A relatively new, essentially defect-free, high-purity carbon material that is similar in its structure to carbon nanotubes, but can form an aggregate, or secondary particles (about 100 nm or 3.94 μin. in size) by the grouping together of individual fibers. It has very high stiffness, strength and toughness, a thermal conductivity comparable to that of diamond, and the extremely high electrical conductivity. Current and proposed uses include fuel-cell electrodes, hydrogen-storage applications, molecular electronics, electron emitters for cold cathodes and flat-panel displays, cell-phone batteries, supercapacitors, lithium-ion batteries, EMI shielding, conductive coatings and composites, and artificial organs. Also known as *single-wall carbon nanohorn (SWNH).* See also carbon nanotubes.

carbon nanotubes. A relatively new class of materials composed of a single sheets of graphite rolled into tubes, with ends sealed with spherical fullerene molecules containing millions of carbon atoms. Single-walled nanotubes (SWNTs) are about 1 nm (0.04 μin.) in diameter and 1-100 μm (40-4000 μin.) in length. Multi-walled nanotubes (MWNTs) are composed of one nanotube embedded within a larger tube. Produced by electric-arc, laser ablation or chemical vapor deposition methods using a metallic catalyst, carbon nanotubes are assembled as low-density carbon fibers with superior tensile strength, and high electrical and thermal conductivities. Used as reinforcements in composite materials, in the formation of nanoribbons, and for molecular quantum wires. Also known as *bucky tubes.* See also carbon nanohorn; fullerenes; buckminsterfullerene; inorganic nanotubes.

Carbon P. Trade name of Rütgerswerke AG (Germany) for calcined petroleum coke composed of 98% carbon, 0.2% ash, 0.2% hydrogen, 0.9% nitrogen and 0.3% sulfur. It is supplied in the form of granules for use as a carburizer in the manufacture of mass-produced steel and cast iron, and as a carbon source in the production of high-carbon high-grade steels.

carbon paper. A thin paper coated on one or both surfaces with a composition consisting of a pigment, such as carbon black or Prussian blue, mixed with a medium, such as oil or wax. It is interleaved between two sheets of paper and, upon application of pressure on the top sheet, the coating will transfer to the bottom sheet at the point of pressure contact. Used for making copies of written, typed or printed material.

carbon-phenolic prepregs. Carbon fibers that are preimpregnated with phenolic resin in the uncured state. They have excellent heat resistance, exceptional ablative properties, and good flame resistance and ablative properties. Used for aircraft interior panels, etc.

carbon-pitch fibers. Carbon or graphite fibers produced from pitch or tar precursors, e.g., coal tar or petroleum asphalt, by a process that involves thermal treatment at about 350-450°C (660-840°F) to form a mesophase (containing both anisotropic and isotropic phases), extrusion at about 380°C (715°F), and thermofixing below 300°C (570°F) to make the isotropic phase unmeltable. This is followed by a final carbonization treatment at either 1000°C (1830°F) to produce carbon fibers, or at 2000°C (3630°F) to produce graphite fibers. See also carbon fibers.

carbon-polyacrylonitrile fibers. See polyacrylonitrile carbon fibers.

carbon-polyaramid composites. See carbon-aramid composites.

carbon-polyetheretherketone composites. Engineering composites consisting of polyetheretherketone (PEEK) matrices reinforced with carbon fibers. Used especially for aerospace and automotive components.

carbon-rayon fibers. Carbon fibers produced from rayon precursors by heat treatment and pyrolysis. See also carbon fibers.

carbon refractories. Refractories manufactured substantially or entirely from carbon and/or graphite, and used for crucibles, stopper nozzles in steelmaking furnaces, etc.

carbon-reinforced aluminum. A composite with good thermal properties consisting of an aluminum or aluminum-alloy matrix reinforced with discontinuous carbon/graphite fibers.

carbon-reinforced silicon carbide. A ceramic composite consisting of a silicon carbide matrix reinforced with carbon fibers. It has a density of about 2.3-2.8 g/cm³ (0.08-0.10 lb/in.³), and is used for high-strength and high-temperature applications.

Carbon Solid Drill. Trademark of Atlas Specialty Steel (Canada) for a plain-carbon solid drill steel used for concrete breakers, bull and crow bars, scaling bars, moils, hand tools, etc. for mining, quarrying and excavating applications.

carbon soot. A black powder that can be obtained by resistive heating of graphite and typically contains about 5-20% buckminsterfullerene (C_{60}/C_{70}). Used as precursor to fullerenes. See also buckminsterfullerene; fullerenes.

carbon-steel castings. Castings made from low-, medium-, or high-carbon steels. Low-carbon steel castings contain less than 0.20% carbon, 0.50-0.80% manganese, 0.35-0.70% silicon, up to 0.05% phosphorus, up to 0.06% sulfur, and the balance iron. Medium-carbon steel castings have 0.20-0.50% carbon, 0.50-1.50% manganese, 0.35-0.80% silicon, up to 0.05% phosphorus, up to 0.06% sulfur, and the balance iron. High-carbon steel castings contain more than 0.50% carbon, 0.50-1.50% manganese, 0.35-0.70% silicon, up to 0.05% phosphorus and sulfur, respectively, and the balance iron. See also cast steels; steel castings.

carbon steels. See plain-carbon steels.

carbon tetraboride. See boron carbide.

carbon tetrachloride. A chlorinated hydrocarbon available in the form of colorless crystals (99% pure) with a density of 1.594 g/cm³, a melting point of -23°C (-9°F), a boiling point of 76-77°C (168-170°F) and a refractive index of 1.460. It is also available in electronic and chromatography grades. Used as a chlorinating agent, as a solvent for rubber, oils, fats, greases, waxes, etc., as a regrigerant, in metal degreasing, in the preparation of semiconductors, and in high-pressure liquid chromatography (HPLC). Formula: CCl_4. Also known as *tetrachloromethane*.

carbon tool steels. See plain-carbon tool steels.

Carbon Vanadium Drill Rod. Trade name of Teledyne Allvac (USA) for a water-hardening tool steel (AISI type W2) containing 0.9-1.1% carbon, 0.15-0.25% vanadium, and the balance iron. It has a fine-grained structure, high toughness and good hardenability. Used for punches, shear blades, chisels, threading taps, masonry drills, keys, pins, bushings, heading and swaging dies, and dental tools.

carbon-vinylester composites. Composites consisting of vinyl-ester resin matrices reinforced with carbon fibers. They have high strength and stiffness, good friction and wear properties, and good weatherability. Used for appliances, structural parts for aircrafts and spacecraft, automotive components, and equipment for the chemical industry.

carbon white. A very finely divided white powder of amorphous silica made from ethyl silicate by calcining. It does not contain carbon, and the name was derived in reference to "carbon black." Used as a filler, pigment and carbon-black substitute in light-colored rubber products. Also known as *fumed silica; silica white; white carbon*. See also carbon black.

carbonyl diamide. See urea (2).

carbonyl iron powder. A high-purity powder with microscopically fine, spherical particles prepared by treating finely divided iron with carbon monoxide in the presence of a catalyst, such as ammonia, and vaporizing and depositing the resulting yellow to orange liquid composed of iron pentacarbonyl (Fe-$(CO)_5$). Used for magnetic coils and cores for high-frequency equipment.

carbonyl powder. A powder made by the thermal decomposition of a metal carbonyl, such as iron pentacarbonyl, $Fe(CO)_5$, or nickel tetracarbonyl, $Ni(CO)_4$.

Carbopack. Trade name of Supelco, Inc. (USA) for graphitized carbon black used as packing for chromatography columns.

Carbo-Plastic. Trade name of Fred A. Wilson Company (USA) for plastic roof patching and high-temperature cements.

Carbo-Plastic Aluminum. Trade name of Fred A. Wilson Company (USA) for aluminized plastic roof patching and flashing cements.

Carbo-Plastomeric. Trade name of Fred A. Wilson Company (USA) for neoprene rubber-base roof patching and flashing cements.

Carboprene. Trade name of The Polymer Group (USA) for carbon-fiber-reinforced polypropylenes.

Carborax. Trademark of The Carborundum Company (USA) for silicon-bonded silicon carbide.

Carborite. Trade name for silicon carbide.

Carbortam. Trademark of NL Industries (USA) for a master alloy containing 15-20% titanium, 2.4-4.0% silicon, 6-8% carbon, 1-2.25% boron, traces of phosphorus and sulfur, and the balance iron. Used in steelmaking as a deoxidizer, and as an addition to impart deep-hardening properties.

Carborundum. Trademark of The Carborundum Corporation (USA) for an extensive series of highly refractory silicon carbide (SiC) products produced by fusing amorphous carbon (coke) and silica sand in an electric furnace. Supplied in colors from greenish to dark blue or black, they have densities ranging from 3.06 to 3.20 g/cm³ (0.111 to 0.116 g/cm³), a hardness of 9.2 Mohs, excellent resistance to acids, good resistance to oxidization up to approximately 1000°C (1830°F), outstanding resistance to corrosion, erosion, wear, impact, thermal shock and heat, and good heat dissipation. Used as abrasive grains and powders for cutting, grinding and polishing aluminum, brass, copper, cast iron and other hard or low-strength alloys, in valve-grinding compounds, grinding stones and wheels, in whetstones, rubbing bricks, coated abrasives and antislip tiles and treads, refractory grains, bonded refractories for incinerators, blast furnaces, chemical reactors, mineral processing equipment, etc., as high-temperature insulation and protection against thermal shock and corrosion, and as a semiconductor. Also included under this trademark are several fused alumina products, and specialty graphite products, such as carbon and graphite felt for high-temperature insulation and corrosion and thermal-shock protection.

carbosand. A term referring to a fine sand that has been treated with an organic solution and roasted for spray application onto oil slicks to sink or disperse them.

Carbo-Sil. Trademark for Polymer Technology Group, Inc. (USA) for a series of hydrophobic, optically clear, biocompatible thermoplastic silicone polycarbonate urethane copolymers consisting of hard segments of an aromatic diisocyanate, usually 4,4'-methylene diisocyanate (MDI), a glycol chain extender, and soft segments of silicone (e.g., polymethylsiloxane). They exhibit high tensile strength, outstanding oxidative stability, and low-energy silicone surfaces, and are supplied in the form of free-flowing pellets, and as granules. Used for medical implants including artificial heart components and ventricular assist devices (VADs).

Carbosil. (1) Trade name for a pyrogenic silica.

(2) Trademark of Carbone-Lorraine North America Corporation (USA) for silicon carbide used in the electronic and semiconductor industries for manufacturing and processing semiconductor materials and electronic components by coating, crystal growing, etching, heating, implanting, masking, sealing and sputtering.

(3) Trademark of Polymer Technology Group, Inc. (USA) for synthetic high polymers used in the manufacture of coated, dipped, extruded and molded goods for consumer, industrial and medical applications. They are supplied in the form of solid pellets and granules, and as a liquid.

Carbo-Silicone. Trade name of Fred A. Wilson Company (USA) for water-repellent silicone coatings for roof and floor repair.

CarboSpan. Trademark of Polymer Technology Group, Inc. (USA) for a segmented polycarbonate urethane similar to *CarboSil* with surface-modifying end groups.

Carbo-Spheres. Trademark of Versar, Inc. (USA) for thin-walled carbon microspheres and macrospheres. They are commercially available in various sizes and compositions including regular hollow carbon microspheres, low-impedance hollow carbon spheres, ferromagnetic hollow carbon microspheres, and solid carbon products. Used as low-density fillers, carbon particles for various applications, and as aggregates.

Carbotam. Trade name of Shieldalloy Metallurgical Corporation (USA) for a ferrotitanium master alloy containing 15-20% titanium, 2.5-4% silicon, 6.5-7.5% carbon, 0-3% aluminum, 1.5-2% boron, 0-1% calcium, and the balance iron. Used as a hardness-increasing ladle addition of boron to cast steels.

CarboTherm. Trademark of The Carborundum Corporation (USA) for a series of boron nitride powders.

Carbo Tool. Trade name of Bissett Steel Company (USA) for a water-hardening tool steel containing 1% carbon, and the balance iron. Used for tools, dies, drills, and taps.

Carbo-Vitrobond. Trade name of Atlas Minerals & Chemicals Inc. (USA) for corrosion-resistant cements.

Carbowax. Trademark of Union Carbide Corporation (USA) for polyethylene glycols (PEGs) and methoxypolyethylene glycols (MPEGs). The PEGs are supplied in a wide range of molecular weights ranging from viscous liquids with 200-800 and waxy solids with 1000-18000 to hard solids with more than 18500. Used as solvents for dyes, proteins, resins, cosmetics and pharmaceuticals, as plasticizers for caseins, gelatins, cork, glues, zein, printing inks, etc., as water-soluble lubricants, and as intermediates in the preparation of nonionic surfactants and alkyd resins.

carboxylic acid. An organic acid obtained by the oxidation of a primary alcohol or aldehyde, and containing one or more carboxyl (–COOH) groups in the molecule. Examples include acetic, butyric, formic, phthalic and terephthalic acid.

carboxylic rubber. A synthetic rubber in whose synthesis one or more carboxylic acids (e.g., phthalic or terephthalic acid) have been used. Abbreviation: COX.

Carboxylon. Trademark of 3M Company (USA) for dental adhesive cements.

carboxymethylated cotton. Stiff, absorbent, crease-resistant cotton fibers, yarns or fabrics that have been treated first with chloroacetic acid and then with a strong solution of sodium hydroxide.

carboxymethylcellulose. A completely water-soluble polymer that is an acid ether derivative of cellulose. It is usually available as the sodium salt with molecular weights ranging from as low as 20000 to over 700000 in the form of colorless granules or powder with a density of 1.590 g/cm^3, a refractive index of 1.510, a melting point above 300°C (572°F). Aqueous solutions with high, low or medium viscosity are also supplied. *Carboxymethylcellulose* has excellent colloidal and thixotropic properties, and is used as a thickener, polyelectrolyte, suspending agent, protective colloid, emulsion stabilizer, paper and paperboard coating, in emulsion paints, as a cation exchanger in chromatography, etc. Abbreviation: CMC; CM cellulose. Also known as *sodium carboxymethyl cellulose.*

Carbo Zinc. Trademark of Corro Therm, Inc. (USA) for a series of self-curing primers for the galvanic protection of steel. *Carbo Zinc 11* is a solvent-based inorganic silicate primer, *Carbo Zinc 11HS* a solvent-based inorganic zinc primer and *Carbo Zinc 11VOC* a high-solids inorganic zinc primer.

car brass. A high-lead brass containing 77% copper, 15% lead and 8% tin used for automotive bearings.

Carbrax. Trade name of W. Canning Materials Limited (USA) for a steel-buffing composition.

Carburit. Trade name for a carburizing paste used for selective carburizing of iron and steel parts.

Carburite. Trade name for an alloy containing 47-48% carbon, 28% iron, 0.3% silicon and 0.2% phosphorus. Used as a recarburizer in steelmaking.

carburized steel. A steel, usually a low-carbon or low-alloy grade, that has been subjected to a process producing a high-carbon surface by heating at 870-930°C (1600-1705°F) in contact with carbonaceous gaseous, solid or liquid substances for several hours. The high-carbon steel surface is then hardened by quenching. Depending on the type of steel and the carburizing time and temperature, the hardened surface layer (or case) may be about 0.15-4.0 mm (0.006-0.15 in.) deep. Used for machine elements, such as gears, shafts, camshafts, axles, crank pins, levers, etc.

carburizing compound. A solid compound containing about 10-20% alkali or alkaline-earth metal carbonates, such as barium carbonate ($BaCO_3$), calcium carbonate ($CaCO_3$) or sodium carbonate (Na_2CO_3), bound to charcoal (usually hardwood or animal based) or coke by oil or tar. Used as an energizer or catalyst in pack carburizing of steel.

Carbyl. Trademark of Inquitex SA (Spain) for nylon 6 fibers and yarns.

cardboard. A stiff, compact pasteboard composed of two or more webs of paper available in various qualities and thicknesses. Used for cards, signs, printed material, cartons and boxes. See also boxboard; pasteboard.

carded cotton yarn. A cotton yarn that has been carded, i.e., subjected to a preparation process that roughly aligns the fibers.

carded fiberglass yarn. A yarn that has been made from continuous glass fiber yarn by chopping into lengths of about 40 mm (1.6 in.) followed by directional alignment in mat form and conversion to a yarn of spinnable length. Used for automotive brake linings, and as a fiber reinforcement.

carded wool. Scoured wool processed through a carding machine to partially straighten its fibers.

carded yarn. Yarn made from fibers that have been partially straightened and cleaned by processing through a carding machine.

Cardinal. Trade name of Cardinal Insulated Glass Company (USA) for an insulating glass.

Cardinal Rapid. Trade name of Thyssen Edelstahlwerke AG (Germany) for a high-speed steel containing 0.7% carbon, 18% tungsten, 4% chromium, 1% vanadium, and the balance iron. Used for high-speed cutting tools.

Cardolite. Trade name of British Resin Products (UK) for protein plastics derived from cashew nut shells.

Cards of Wood. Trade name of Lenderink, Inc. (USA) for plywood made in a wide range of thicknesses by bonding together several hardwood veneers, each 0.1-4.2 mm (0.005-0.167 in.) thick, using a special adhesive. Hardwoods commonly used as veneers include aspen, basswood, birch, hard maple, oak, and poplar. Used for building and construction applications.

Cardura. Trademark of Shell Chemical Company (USA) for a series of polyesters.

Care. Trade name for a calcium hydroxide dental cement.

Careco. Trade name for a white bearing alloy of lead, tin and antimony.

Carend. Trade name of Arcos Alloys (USA) for an Armco-type iron containing 0.03% carbon. Used for welding electrodes for sheetmetal. See also Armco iron.

Caress. See Anso Caress.

Carex. Trade name of Monsanto Chemical Company (USA) for polystyrenes.

Carey. Trade name of Philip Carey Manufacturing Company (USA) for crysotile asbestos fibers and a wide range of asbestos-, magnesia- or calcium silicate-based products including high-temperature blocks and pipe coverings as well as asbestos felts, papers, ropes and wicks.

Careycell. Trade name of Philip Carey Company (USA) for asbestos insulation blocks and pipe covering materials.

Carey Millboard. Trade name of Philip Carey Manufacturing Company (USA) for asbestos millboard.

Carey Firefoil. Trademark of Philip Carey Manufacturing Company (USA) for an asbestos-laminated insulation board. It has good strength retention after prolonged soaking in water, and is used in ship and boat construction, building partitions, and insulating linings for stacks.

Careystone. Trademark of Philip Carey Manufacturing Company (USA) for asbestos-cement boards and corrugated sheeting for roofing, siding, etc.

Careytemp. Trademark of Philip Carey Manufacturing Company (USA) for asbestos blocks and pipe coverings and fibrous high-temperature adhesives and cements.

Cargan. Trade name for regenerated protein (azlon) fibers.

Caribbean pine. The heavy, strong, yellow wood of the tree *Pinus caribaea*. It has high hardness, strong resinous odor and greasy feel, and machines well. Source: Caribbean Central America

(Belize, Bahamas, Cuba, British Honduras, Nicaragua, etc.). Used for constructional work, paper pulp, cooperage, and interior fittings. Also known as *Cuban pine*.

Cariflex. Trademark of Shell Chemical Company (USA) for styrene-butadiene, styrene-butadiene-styrene and isoprene based thermoplastic elastomers.

Carilan. Japanese trade name for a vinyl acetate fiber.

Carilloy. Trade name of US Steel Corporation for an extensive series of carbon and alloy machinery and structural steels including various carbon-manganese, carbon-molybdenum, carbon-vanadium, nickel-chromium-molybdenum, high- and medium-nickel, chromium-molybdenum, manganese-molybdenum and chromium-vanadium grades.

Carilon. Trade name of Shell Chemical Company (USA) for aliphatic linear thermoplastic polyketone resins produced by single-stage catalysis from ethylene, propylene and carbon monoxide building blocks. Available in regular, fiber-reinforced, extrusion, blow molding, injection molding, rotomolding, lubricated and flame-retardant grades, they have a melting point of 220°C (430°F), a heat-deflection temperature of 210°C (410°F), high stiffness, good impact resistance, strength, friction and wear characteristics, good low- and high-temperature modulus retention, outstanding wear and friction properties, and good resistance to corrosive chemicals, industrial solvents, fuels and automotive fluids. Used for automotive components, gears, consumer appliances, and business and office equipment.

Carina. Trademark of Shell Chemicals (UK) for polyvinyl chloride plastics.

Carinex. Trademark of Shell Chemicals (UK) for modified and unmodified polystyrene resins with excellent electrical properties and optical clarity and good dimensional and thermal stability. Used for panels, housings, casings, etc.

Cariron. Trade name of Fillmore Foundry Inc. (USA) for a nickel cast iron containing 2.96% total carbon, 1.2% silicon, 0.75% manganese, 1.55% nickel, and the balance iron. Used for pistons and liners.

Carise. Trademark of DuPont Australia Limited (Australia) for nylon 6,6 yarns and monofilaments.

carletonite. A pink or blue mineral composed of potassium sodium calcium carbonate silicate hydroxide monohydrate, $KNa_4Ca_4Si_8O_{18}(CO_3)_4(OH,F)\cdot2H_2O$. Crystal system, tetragonal. Density, 2.45 g/cm^3; refractive index, 1521. Occurrence: Canada (Quebec).

Carlex. Trade name of Verreries de la Gare & A. Belotte Réunies SARL (France) for enameled sheet or wired cast glass for cladding applications.

carlfriesite. A yellow mineral composed of calcium hydrogen telluride, $CaH_4(TeO_3)_3$. Crystal system, monoclinic. Density, 6.30 g/cm^3; refractive index, 2.095. Occurrence: Mexico.

carlhintzeite. A white to colorless mineral composed of calcium aluminum fluoride monohydrate, $Ca_2AlF_7\cdot H_2O$. Crystal system, monoclinic. Density, 2.86 g/cm^3; refractive index, 1.416. Occurrence: Germany.

carlinite. A dark gray mineral composed of thallium sulfide, Tl_2S. It can also be made synthetically. Crystal system, rhombohedral (hexagonal). Density, 8.39 g/cm^3. Occurrence: USA (Nevada).

Carlite. Trade name of Ford Motor Company (USA) for toughened and laminated safety glass for automotive applications.

Carlon. Trade name of Carnegie Fabrics, Inc. (USA) for polyvinyl chloride used for treating fabrics.

Carlona. Trademark of Shell Chemical Company (USA) for tough, electrically-insulating, chemically-resistant low-density polyethylenes supplied in the form of sheets, rods, tubes, granules, powders and fibers. UV-stabilized grades are also available. Used for electrical applications, containers, linings, packaging film, etc. Also included under this trademark are several polypropylenes.

Carloy. Trade name of Benedict-Miller, Inc. (USA) for a water-hardening tool steel containing varying amounts of carbon, molybdenum, and iron.

carlsbergite. A light gray mineral of the halite group composed of chromium nitride, CrN. It can also be made synthetically. Crystal system, cubic. Density, 5.90 g/cm³; refractive index, 1.653. Occurrence: USA.

Carlson Alloys. Trade name of G.O. Carlson Inc. (USA) for an extensive series of austenitic, duplex, ferritic, martensitic and precipitation-hardening stainless steels, titanium, nickel alloys and copper-nickel alloy products supplied in standard plate form and as heads, rings, disks, cut bars and special shapes.

Carlton. Trade name of Time Steel Service, Inc. (USA) for a free-machining, air-hardening tool steel (AISI type A4) containing 0.95% carbon, 2% manganese, 2.2% chromium, 1.1% molybdenum, a trace of lead, and the balance iron. It has a great depth of hardening, good wear resistance, and good to moderate machinability and toughness. Used for forming tools, forming, bending, trimming and blanking dies, and punches.

Carmelia bronze. A heavy-duty bronze that contains varying amounts of copper, zinc and lead, and is used for machinery bearings.

Carmet. Trademark of Carmet Materials Division of Allegheny Ludlum Corporation (USA) for a series of cemented and coated cemented carbides including various straight-carbide types (e.g., tungsten carbide, titanium carbide, tantalum carbide or chromium carbide) and complex carbide mixtures. Used for blanks inserts and holders for cutting tools for ferrous and nonferrous metals and certain nonmetals, and for gauges and dies.

carminite. A carmine-red mineral composed of lead iron arsenate hydroxide, PbFe₂(AsO₄)₂(OH)₂. Crystal system, orthorhombic. Density, 5.18-5.42 g/cm³. Occurrence: Mexico.

Carmo. Trade name of Uddeholm Corporation (USA) for a cold-work tool steel.

Carnalite. Trade name for a benzyl cellulose formerly used for denture bases.

carnallite. A white, brownish, or reddish mineral with white streak and greasy luster composed of potassium magnesium chloride hexahydrate, KMgCl₃·6H₂O. Crystal system, orthorhombic. Density, 1.59-1.62 g/cm³; melting point, 265°C (509°F); hardness, 1 Mohs; refractive index, 1.475; strongly phosphorescent. Source: France, Germany, USA (New Mexico). Used as an ore of potassium, as a source of manufactured potash salts, and as a flux in magnesium refining.

carnauba wax. Hard, vitreous yellow to greenish-brown lumps obtained from the leaves of the Brazilian wax palm (*Copernica cerifera*). Density, 0.995 g/cm³; melting point, 84-86°C (183-187°F). Used in electric insulating compositions, furniture and floor polishes, leather finishes, varnishes, waterproofing and hardening compositions, shoe wax, carbon paper and candles, and as a substitute for beeswax. Also known as *Brazil wax*.

carnelian. A red, orange, brownish, or brownish-yellow variety of *chalcedony* quartz that contains iron impurities. It occurs in Brazil and India, and is used chiefly as a gemstone.

carnival glass. A multicolored glass product made by applying metallic salts to a colored glass and firing.

Carnolia. Trade name of Charles Carr Limited (UK) for a babbitt metal containing lead, tin and antimony in varying amounts. Used as an antifriction alloy for bearings.

carnotite. A bright yellow to lemon- or greenish-yellow, strongly radioactive mineral composed of potassium uranyl vanadium oxide hydrate, K₂(UO₂)₂(VO₄)₂·xH₂O, containing about 2-5% uranium oxide and up to 6% vanadium oxide. Crystal system, monoclinic. Density, 4.70-4.91 g/cm³; refractive index, 2.06. Source: Australia, USA (Arizona, Colorado, New Mexico, Utah). Used as a source of radium, uranium and vanadium.

Caro. Trade name of Carobronze Limited (UK) for copper and copper alloys (chiefly bronzes and brasses) supplied as semifinished products, rods, rounds, squares, structural shapes, tubing, etc. *Caro-A* is a corrosion-resistant phosphor bronze containing 91.2% copper, 8.5% tin and 0.3% phosphorus, and is also known as *Carobronze*. *Caro-B* is a tin- and nickel-free wrought copper-zinc alloy with less than 88% copper. *Caro-C* is a wrought specialty copper alloy for valve seats, and *Caro-E* a wrought copper-aluminum alloy supplied in the form of rods and finished parts.

caroa fibers. The hard vegetable fibers obtained from a Brazilian plant (*Neoglazovia variegata*). They are chemically and physically similar to *sisal* and *cantala fibers* but not of great commercial importance.

caroba. See jacaranda.

carobbiite. A colorless mineral of the halite group composed of potassium fluoride, KF. It can also be made synthetically. Crystal system, cubic. Density, 2.52 g/cm³; refractive index, 1.352.

Carobronze. Trade name of Carobronze Limited (UK) for a corrosion-resistant phosphor bronze containing 91.2% copper, 8.5% tin and 0.3% phosphorus. Supplied in cast and hard-drawn form, it is used for bushings with or without steel backing for heavy-duty bearings, solid cold-drawn tubes, and wear parts. Also known as *Caro A*.

carob wood. See jacaranda.

Carocel. Trade name of Celotex/Philip Carey Company (USA) for asbestos pipe covering materials.

Caroga. See Albany Caroga.

Carolan. Trade name for cellulose acetate fibers used for wearing apparel and as industrial fabrics.

Carolina ash. See water ash.

Carolina stone. A kaolinized feldspar, similar to *Cornwall stone*, found in North Carolina, USA. Used as a building stone.

Carolith. Trade name of OMYA GmbH (Austria) for a range of fillers for paper, paints, lacquers, varnishes, plastics and adhesives.

carotenes. Intensely colored hydrocarbons (lipochromes) corresponding to the formula C₄₀H₅₆. They are pigments that occur in plants and animals, exist in the alpha, beta and gamma form, and can be converted into vitamin A. Used in medicine, pharmacology, nutrition, biochemistry, and biotechnology.

carotenoids. A class of easily oxidizable yellow, orange, red or purple pigments found in plants or animals. Examples include *carotene* and *xanthophyll*.

carotenol. See xanthophyll.

Carpefin. Trademark of Carpol SpA (Italy) for nylon 6 fibers and yarns.

Carpental. Trade name of Alumetal (Italy) for a heat-treatable, high-strength aluminum alloy containing 4-6% zinc, and 2.5-3.5% magnesium. Used in the aircraft and automotive industries.

Carpenter. Trade name of Carpenter Technology Corporation

(USA) for an extensive series of engineering materials including several plain-carbon, free-machining, case-hardening, carburizing and spring steel grades, various stainless and permanent-magnet steel grades and several tool steel grades (e.g., high-speed, cold-work, hot-work, shock-resisting, water-hardening and mold types) as well as a wide variety of superalloys.

Carpenter Consumet. See Consumet.

Carpenter Glass Sealing. Trademark of Carpenter Technology Corporation (USA) for a series of iron-nickel, iron-nickel-chromium and iron-chromium alloys used for glass-to-metal sealing applications.

Carpenter Super Speed Star. See Super Speed Star.

Carpenter's Wood Filler. Trade name of Aron Alpha, Division of Elmer's Products, Inc. (USA) for a line of polymer-based wood fillers.

Carpenter's Wood Glue. Trade name of Aron Alpha, Division of Elmer's Products, Inc. (USA) for a line of polymer-based wood glues.

carpholite. A yellow mineral composed of manganese aluminum silicate hydroxide, $MnAl_2Si_2O_6(OH)_4$. Crystal system, orthorhombic. Density, 2.93-3.04 g/cm^3; refractive index, 1.629. Occurrence: Japan.

carpincho leather. A washable, chrome-tanned leather resembling pigskin in appearance. It is made from the skin of the carpincho, a medium-sized South American mammal of the order Rodentia. Used for gloves.

Carrara. (1) Trade name of PPG Industries Inc. (USA) for colored, opaque glass with hard, brilliant finish supplied in thicknesses up to 38 mm (1.5 in.). It has good resistance to crazing and checking, and is used as a structural glass for storefronts, countertops, tiling, and paneling.

(2) Trade name for a dental porcelain for jacket crowns.

Carrara marble. A hard, fine-textured, white to bluish *marble* from Carrara, Italy, used for statuaries, and in building construction.

Carrara M+K. Trade name for a high-gold dental bonding alloy.

carrboydite. A yellowish green to blue-green mineral composed of nickel aluminum sulfate hydroxide heptahydrate, $Ni_{14}Al_9(SO_4)_6(OH)_{43} \cdot 7H_2O$. Crystal system, hexagonal. Density, 2.50 g/cm^3; refractive index, 1.55. Occurrence: Western Australia.

carrier cloth. See scrim cloth.

carrier fiber. A fiber added to the principal fiber in a yarn to enhance overall processibility.

carrollite. A light gray mineral of the spinel group composed of copper cobalt sulfide, $CuCo_2S_4$. Crystal system, cubic. Density, 4.65 g/cm^3. Occurrence: Zimbabwe. A variety from China contains 38.2% platinum.

Carr's Quality. Trade name of Richard W. Carr & Company Limited (UK) for an extensive series of steel products including several carbon-manganese steels (AISI-SAE 13xx series), plain-carbon tool steels (AISI group W), cold-work tool steels (AISI groups D, O and A), hot-work tool steels (AISI group H), shock-resisting tool steels (AISI group S), carbon-tungsten-type special-purpose tool steels (AISI group F), low-alloy-type special-purpose tool steels (AISI group L), mold steels (AISI group P), various austenitic, ferritic and martensitic stainless steels (AISI 2xx, 3xx and 4xx series), and a wide range of alloy steels of the nickel-chromium, nickel-molybdenum, nickel-chromium-molybdenum or chromium-molybdenum-vanadium types including several carburizing grades.

Carsil. Trade name of Foseco Minsep NV (Netherlands) for a complete range of silicate binders used in the foundry for core-making and molding.

Carta. Trademark of Isola Werke AG (Germany) for laminated paper. It is modified with phenol-formaldehyde resin and supplied in the form of sheets, rods, tubing and machined and punched parts.

Carta Textil. Trademark of Isola Werke AG (Germany) for laminated fabrics supplied in the form of sheets, rods, tubing and machined and punched parts.

Carter white gold. A corrosion-resistant gold alloy containing 16.7% nickel, used for ornaments and jewelry.

Carthane. Trade name for a series of polyurethane and isocyanate resins.

Cartoglue. Trade name of Ato Findley (France) for *dextrin* and starch adhesives for packaging applications.

carton boards. Paper pulp boards coated with synthetic resin for use in packaging.

cartridge brass. A highly ductile, yellow copper alloy with a nominal composition of 70% copper and 30% zinc. Commercially available in the form of tubes, rods, flat products and wires, it has a density of 8.53 g/cm^3 (0.31 $lb/in.^3$), excellent cold workability, good deep-drawing qualities, and good brazeability. Used for fasteners, springs, spring clips, eyelets, rivets, screws, hinges, locks, plumbing accessories, automotive tanks, ammunition components (e.g., cartridge cases, primer cups and shot shells), automotive radiator cores, lamp fixtures, hardware, and deep-drawn and spun parts. Abbreviation: CB.

cartridge gilding. A *gilding metal* containing 90% copper and 10% zinc. It has excellent cold workability, good hot workability, and good corrosion resistance. Used for cartridge shells, ornaments, and as a base for fire enameling.

Cartun. Trade name of Delsteel Inc. (USA) for a water- or oil-hardening tool steel containing 1.35% carbon, 2.75% tungsten, and the balance iron. Used for cutters and high-speed tools.

Carulean blue. A blue, inorganic pigment composed of mixed oxides of cobalt and aluminum.

Carvacraft. Trade name of J. Dickinson & Company (UK) for phenol-formaldehyde plastics.

Car-Van. Trade name of Allied Steel & Tractor Products Inc. (USA) for a water-hardening tool steel containing 0.75-1.1% carbon, 0.2% vanadium, and the balance iron. Used for tools, dies, shear blades, and punches.

Carvan. Trade name of Bethlehem Steel Corporation (USA) for a water-hardening tool steel containing 0.8% carbon, and the balance iron. Used for cutting tools.

Car-Van Special. Trade name of Allied Steel & Tractor Products Inc. (USA) for a water-hardening tool steel containing 0.8-1.35% carbon, 0.2% vanadium, and the balance iron. Used for tools, dies, cutters, and drills.

Carvite. Trade name of Columbia Tool Steel Company (USA) for a tungsten-type high-speed tool steel (AISI type T9) containing 0.7-0.9% carbon, 4% chromium, 18% tungsten, 4% vanadium, and the balance iron. It has high hot hardness and wear resistance, and is used for cutting tools.

caryinite. A brown mineral composed of calcium sodium manganese arsenate hydroxide, $(Ca,Na,Mn)_3(Mn,Mg)_2(AsO_4)_{3-y}(OH)_x$. Crystal system, monoclinic. Density, 4.29 g/cm^3; refractive index, 1.780. Occurrence: Scandinavia.

caryopilite. A mineral of the kaolinite-serpentine group composed of manganese silicate hydroxide, $(Mn,Mg)_3Si_2O_5(OH)_4$. Crystal system, monoclinic. Density, 2.87 g/cm^3; refractive index, 1.6325. Occurrence: Japan.

Casalegno. Trade name of Casalegno Giovanni & Figli SpA (Italy) for a toughened glass.

Casalloy. Trade name of British Steel Corporation (UK) for a case-hardening steel containing 0.1% carbon, 1.0% nickel, and the balance iron. Supplied in the annealed condition, it has good core ductility and toughness, and is used for machine parts with hard, wear-resistant surfaces and tough, ductile cores, e.g., gears, shafts, pins, etc.

Casar. Trade name of Fakirstahl Hoffmanns GmbH & Co. KG (Germany) for a high-speed steel containing 1.3-1.4% carbon, 0.85% molybdenum, 4.2-4.3% chromium, 3.8-4% vanadium, 12% tungsten, and the balance iron. Used for lathe and planer tools, reamers, broaches, and taps.

Cascade. Trade name of Latrobe Steel Company (USA) for a mold steel (AISI type P21) containing 0.2% carbon, 0.25% chromium, 4.1% nickel, 0.2% vanadium, 0.3% silicon, 0.3% manganese, 1.2% aluminum, and the balance iron. Usually supplied in the solution-treated-and-aged condition, it has high core strength, a great depth of hardening, outstanding resistance to decarburization, good toughness, machinability and wear resistance, and can be surface-hardened by nitriding. Use for plastic molds, and zinc die-casting dies.

Cascamid. Trade name of Borden Inc. (USA) for a series of urea-formaldehyde resins.

cascandite. A pale pink mineral composed of calcium scandium silicate hydroxide, $CaScSi_3O_8(OH)$. Crystal system, triclinic. Density, 3.01-3.02 g/cm^3. Occurrence: Italy.

Cascaphene. Trade name of Cascelloid Limited (USA) for cellulose nitrate plastics.

Cascelloid. Trade name of Cascelloid Limited (USA) for celluloid-type cellulose nitrate plastics.

Casco. (1) Trademark of Borden Inc. (USA) for casein-base adhesives, and various formaldehyde condensation resins including urea-formaldehydes.

(2) Trade name for a self-cure acrylic for dental applications.

Cascobond. Trademark of Borden Inc. (USA) for liquid contact adhesives for industrial applications.

Cascophen. Trademark of Borden Inc. (USA) for a series of synthetic resins based on phenol-formaldehyde.

cased glass. (1) Glassware whose surface composition is different from that of the core.

(2) A glass product composed of two or more sheets of different colors.

Case Die. Trademark of Wallace Murray Corporation (USA) for a shock-resisting tool steel (AISI type S3) containing 0.45% carbon, 1% tungsten, 0.2% molybdenum, 0.9% chromium, and the balance iron. It has high strength and toughness, fair wear resistance, and is used for tools and dies.

case-hardened steel. A steel, usually of the low-carbon type, whose surface has been enriched with carbon, nitrogen or a mixture thereof, followed by appropriate thermochemical treatment. It has a hard, wear-resistant surface layer (or case), and a soft, ductile, tough core. Common case-hardening processes includes carburizing, carbonitriding, cyaniding and nitriding. Abbreviation: CHS.

case-hardening bone. Specially treated bone black used as a carburizing compound in the case hardening of steel. See also bone black.

case-hardening materials. A group of materials used to introduce carbon into the surface layer (or case) of an iron-base alloy, such as a low-carbon steel, to obtain a hardened, wear-resistant case upon subsequent quenching. Common case-hard-

ening materials include charcoal obtained from hardwood, animal bones, leather and prepared carburizing compounds composed of charcoal mixed with barium carbonate, etc.

casein. A white, or pale-yellow, tasteless, odorless, hygroscopic, amorphous phosphoprotein material that contains calcium, and is precipitated from skimmed milk by the enzyme rennet or a dilute acid. It has a density of 1.25-1.31 g/cm^3 (0.045-0.047 lb/in.^{3}), a melting point of 280°C (536°F) (decomposes), and a refractive index of 1.55-1.56. Used in the manufacture of adhesives, paints, paper and plastics, as a foundry-sand binder, in textile fibers, in cheese-making, foods, feeds and dietary supplements, in biochemistry, and in the preparation of biopolymers.

casein adhesives. Colloidal dispersions of casein (milk curd) in water, often containing certain additives. They may be prepared at room temperature or elevated temperatures. See also casein glues.

casein fibers. The soft, silky, white-colored regenerated protein fibers made by spinning chemically treated casein through an extrusion die (spinneret). Used for knitting yarns.

casein-formaldehyde resin. A modified thermosetting polymer formed by the reaction of formaldehyde with casein. See also casein plastics.

casein glues. Water-resistant adhesives made from casein (milk curd), hydrated lime and sodium salts. They are supplied in the form of dry powders, and mixed with cold water for use. *Casein glues* have good joint-filling qualities, fairly good strength, poor water resistance, and a high tendency to stain certain woods, such as oak, maple and redwood. Used for gluing oily woods, such as teak, padouk and lemonwood, for laminating wood that has a high moisture content, and for general plywoods and assembly work. See also casein adhesive.

casein-latex adhesives. Fairly strong adhesives made by mixing casein with rubber latex. Used for joining wood to metals or plastics.

casein paint. An organic coating made by adding pigments and other ingredients to a casein solution.

casein plastics. A class of soft, nonflammable, thermoplastic materials commonly made by the action of formaldehyde on rennet casein. They have a density of 1.3-1.4 g/cm^3 (0.04-0.05 lb/in.^{3}), a refractive index of 1.554, excellent moldability and machinability, fair to poor mechanical properties, high water absorption, poor resistance to alkalies, an upper service temperature of 150°C (300°F), and can be colored to imitate amber, agate, ivory, etc. Used for buttons, beads, ornamental parts, knitting needles, fountain-pen holders, and novelty items. Abbreviation: CS.

case leather. See bag leather.

casement cloth. Very light and thin fabrics, usually in an open weave, used especially as backing for heavy draperies and for curtains.

casha. A soft twill-weave fabric resembling *flannel* in appearance, originally made from vicuna wool, but now woven from fine sheep's wool or a blend of sheep's and *cashmere* goat wool. Used especially for overcoats and other clothing.

cashgora. Hybrid protein fibers obtained from the downy undercoat of the cashgora, a crossbreed of cashmere and angora (mohair) goats, and produced chiefly in Turkey, Mongolia, New Zealand, and western and southern Russia. For commercial use, it is usually scoured and dehaired to remove coarser hairs. Used for the manufacture of textile fabrics.

cashmere. (1) A soft, downy fiber obtained from the undercoat hair of the Cashmere goat *(Capra hircus)* of the Himalayas.

Used for the manufacture of textiles.

(2) A soft, downy fiber, quite similar to cashmere (1), obtained from the hair of Feral goats of Scotland, Australia and New Zealand.

(3) A soft, silky, lightweight fabric made from cashmere fiber (1), and used for fine garments, sweaters, worsted, etc.

(4) A generic term for any fine, soft, silky woolen fabric.

Cashmilon. Trademark of Asahi Chemical Industry Company (Japan) for acrylic fibers used for the manufacture of textile fabrics.

Casidiam. Trade name of Anatech Limited (USA) for a diamond-like carbon coating (1-4 μm or 0.04-0.16 mil thick) composed of carbon, hydrogen and selected doping elements deposited at 100°C (212°F). It has a density of 1.8-2.1g/cm³ (0.07-0.08 lb/in.³), high hardness, a low coefficient of friction, high thermal conductivity, and high electrical resistivity and dielectric strength. It is generally inert to water, salts, acids, alkalies and solvents, and forms an impermeable barrier to hydrogen and other gases. Used on automotive engine components, pump parts, spray nozzles, ceramic faucet washers, injection molds, sewing and knitting needles, outboard-engine components, and marine parts. Owing to its good biocompatibility, it is also used on hip replacements.

Casino. Trade name for a high-speed steel containing 0.7% carbon, 18% tungsten, 4% chromium, 1% vanadium, and the balance iron. It has good hot hardness and wear resistance, a high cutting ability, and is used for cutters, reamers, shapers, etc.

Caslen. Trade name for casein fibers obtained from milk protein.

Casolith. Dutch trade name for casein plastics.

Casona. Trade name of Osborn Steel Limited (UK) for a case-hardened steel containing up to 0.1% carbon, 0.25% silicon and 0.25% manganese, respectively, and the balance iron. It is suitable for carburizing, and used for deep-hobbing dies.

Casowyte. Trade name for a medium-gold dental casting alloy.

Cassair. Trademark of Brinco Mining Limited (Canada) for asbestos fibers.

Cassel brown. See Vandyke brown.

Cassel earth. See Vandyke brown.

Cassel green. An emerald-green crystalline powder composed of barium manganate (BaMnO₄), and used as a paint pigment. Also known as *manganese green; Rosenstiehl's green.*

Cassel yellow. See Turner's yellow; yellow ocher.

cassidyite. A green mineral of the the fairfieldite group composed of calcium nickel phosphate dihydrate, Ca₂(Ni,Mg)(PO₄)₂· 2H₂O. Crystal system, triclinic. Density, 3.10 g/cm³. Occurrence: Western Australia.

cassimere. A fabric usually woven in a herringbone weave from a blend of wool and worsted yarns, and used especially for making suits.

cassiterite. A brown, black, yellow, or white mineral of the rutile group with a white streak and an adamantine or dull submetallic luster. It is composed of tin dioxide, SnO₂, and contains 78.6% tin when pure. It can also be made synthetically. Crystal system, tetragonal. Density, 6.8-7.2 g/cm³; hardness, 6-7 Mohs; refractive index, 2.00. Occurrence: Australia, Bolivia, East Indies, Malaya, Uganda, Zaire, UK, USA (California, Nevada, South Dakota). Used as the most important ore of tin. The transparent grades are used as gemstones. Also known as *tin stone.*

Cassius Basaltic. Trade name of Aardvark Clay & Supplies (USA) for a clay (cone 5) with a very black finish resembling ebony.

Cast. Trade name of Samsung/OmegaDent (South Korea) for dental casting alloys for crowns, bridges and inlays.

castable refractories. See castables.

castables. Mixtures of specially selected refractory aggregates and suitable binders, such as aluminous hydraulic cement, that when combined with water can be sprayed, poured or rammed into place to form shapes or structures which set and develop considerable structural strength. Used in the construction and repair of cupolas, ladles and furnaces. Also known as *castable refractories; hydraulic-setting refractories.*

Casta-Crete. Trade name of Brouk Company (USA) for expanded perlite and vermiculite roof slabs.

Castadur. Trade name for a low-gold dental casting alloy.

castaingite. A yellow mineral composed of copper molybdenum sulfide, CuMo₂S₅₋ₓ. Occurrence: Germany.

Castaloy. (1) Trade name of Fisher Scientific Company (USA) for a zinc die-casting alloy containing 4.1% aluminum and 0.04% magnesium. Used for chemical equipment, clamps, and holders.

(2) Trade name of Detroit Alloy Steel Company (USA) for a die steel with excellent machinability containing 1.5-1.6% carbon, 12-14% chromium, 0.7-0.8% molybdenum, 0.45-0.55% manganese, and the balance iron.

cast basalt. See artificial basalt.

cast brass. See casting brass.

cast bronze. See casting bronze.

Cast Composite. Trade name of A. Milne & Company (USA) for a water-hardening tool steel containing 0.6-1.05% carbon, and the balance iron. Used for tools, drills, taps, and blanking dies.

cast-coated paper. High-gloss *coated paper* whose coating has been dried by pressing against a polished cylinder.

cast copper. A soft, low-strength, corrosion- and oxidation-resistant copper (99.7-99.95% pure) containing small additions of silver and other elements. Its electrical and thermal conductivity are about 10-15% lower than those for wrought copper of the same composition.

cast copper alloys. See copper casting alloys.

Castdie. Trade name of Columbia Tool Steel Company (USA) for an oil-hardened steel containing 0.35% carbon, 5.25% chromium, 1.35% molybdenum, 0.95% silicon, 0.5% vanadium, and the balance iron. Used for die-casting dies.

Castell. Trade name for a low-gold dental casting alloy.

cast film. A polymer film of varying thickness produced by pouring a layer of liquid polymer dispersion or solution onto a surface and either allowing it to solidify, stabilizing it by solvent evaporation, or fusing it to the surface. See also cast plastics.

CastForm PS. Trade name of Accelerated Technologies Inc. (USA) for a polystyrene-based material for the selective laser-sintering production of defect-free investment casting patterns. It has a very low ash content (less than 0.02%), and can be used with many castable alloys including aluminum, magnesium, nickel, steel, stainless steel, titanium and zinc.

cast glass. (1) Glass made by pouring the melt into a mold and allowing it to cool and solidify.

(2) Glass made by centrifugal casting in which the melt is forced against the sides of a rapidly revolving mold.

casting alloys. Ferrous or nonferrous alloys that can only be cast into a final shape, but not forged or rolled.

casting brasses. A group of brasses that can be divided into the following three general categories: (i) *Copper-zinc-tin alloys* including various leaded and unleaded red, semi-red and yellow brasses; (ii) *Manganese* and *leaded manganese bronze alloys;* and (iii) *Copper-zinc-silicon alloys* including various silicon brasses and bronzes. The properties of *cast brasses* depend greatly on the alloying additions and impurities, the microstructure resulting from the particular melting and casting process, e.g. sand, die or centrifugal casting, and from the wall thickness of the casting. Typically, they cast and machine well, and have higher strength and toughness than ordinary cast iron. Also, they are frequently made from brass ingot metal. Also known as *cast brass*. See also red casting brass; semi-red casting brass; yellow casting brass; silicon brass.

casting bronze. A group of bronzes that can be divided into the following general categories: (i) *Copper-tin alloys (or tin bronzes)*; (ii) *Leaded copper-tin alloys (or leaded tin bronzes)*; (iii) *Copper-tin-nickel alloys (or nickel-tin bronzes)*; and (iv) *Copper-aluminum alloys (or aluminum bronzes)*. See also aluminum bronze; leaded tin bronze; tin bronze; nickel-tin bronze. Also known as *cast bronze*.

casting copper. A fire-refined tough pitch copper cast into ingots for the manufacture of foundry castings.

casting paper. See anti-adhesive paper.

casting-pit refractories. Specially shaped refractories into which molten ferrous metals are cast.

casting plaster. A white, high-grade gypsum plaster used in casting and carving.

castings. Object produced by pouring liquids or suspensions, such as molten metals, a liquid monomer-polymer solutions, ceramic body slips or hot glasses, into suitably shaped containers or molds and allowing them to solidify. See also cast metals; cast plastics.

Castingweld. Trade name of Westinghouse Electric Corporation (USA) for a carbon steel used for electrodes to make non-machinable welds on cast iron.

cast-in-place concrete. Concrete that is poured in the same place where it is required to set as part of a structure. It is the opposite of precast concrete. Abbreviation: CIP concrete. Also known as *in-situ concrete; pour-in-place concrete*. See also precast concrete.

cast iron. See cast irons.

cast-iron enamel. A relatively low-melting *porcelain enamel* that is designed specially for cast iron, applied over an adherence-promoting ground coat, and fired to maturity over a relatively long-firing time.

cast irons. A group of iron-carbon alloys that contain more carbon than can be dissolved in austenite. Most commercial cast irons contain between 1.8 and 4.5% carbon, up to 3.5% silicon, and up to 2% phosphorus. The carbon may exist either as free carbon in the form of graphite, or as combined carbon in the form of cementite (iron carbide). *Cast iron* is obtained by remelting pig iron and casting to desired shapes, and is characterized by good castability and wear resistance, relatively high brittleness, and the onset of the melting range at about 1427°C (2600°F). Cast irons can be divided into the following general categories: (i) *White cast iron;* (ii) *Gray cast iron;* (iii) *Malleable cast iron* (including blackheart and whiteheart malleable irons); (iv) *Ductile cast iron;* (v) *Mottled cast iron*; and (vi) *Special cast iron* (including alloy iron, chilled iron, austempered ductile iron and compacted-graphite iron).

cast-iron thermit. Thermit to which ferrosilicon and steel punchings have been added. Used for welding iron castings. See also thermit.

cast metals. Objects produced by pouring molten metals or alloys, such as aluminum, copper or iron, into suitably shaped containers or molds and allowing them to solidify. Metal casting molds may consist of clay- or resin-bonded sand (sand casting), or metal (permanent-mold and die casting). See also castings.

Castomatic. Trademark of Fry's Metals Inc. (USA) for antifriction and white metals, babbitts, solder alloys, type metals and zinc-base alloys supplied in the form of bars, ingots, rods and sticks.

Castor. Trade name of Uddeholm Corporation (USA) for a series of tungsten and cobalt-tungsten high-speed tool steels used for lathe and planer tools, reamers, broaches, cutters, etc.

Castordag. Trade name of C-I-L, Inc. (USA) for a suspension of graphite in castor oil, used as a lubricant.

Castoro. Trade name for a high-gold dental casting alloy.

castor oil. A pale-yellowish, greenish or colorless nondrying oil which contains ricinoleic and dihydroxystearic acids. It has a density of 0.945-0.967 g/cm³ (0.034-0.035 lb/in.³), a solidification point of about -10 to -12°C (10 to 14°F), and a boiling point of about 313°C (595°F). Used in polyurethane adhesives, coatings and elastomers, as a plasticizer in lacquers, as a leather preservative, in lubricants, waxes, soaps and surface-active agents, in electrical insulating compounds, in the synthesis of thixotropic materials, and in medicine, pharmaceuticals and cosmetics.

Castorwax. Trade name of Parchem (USA) for a solvent-resistant wax composed of hydrogenated castor oil, and supplied in the form of pearl-white flakes. It has a melting point of 85-88°C (185-190°F). Used as electrical insulation, for candles, in carbon paper, and as a substitute for carnauba wax. See also Opalwax.

cast plastics. Plastic products of varying shape and size usually produced by pouring liquid monomer-polymer solutions into open molds and allowing them to solidify. Cast plastics may be of the thermosetting type, such as epoxies, phenolics, silicones or unsaturated polyesters, or the thermoplastic type, such as acrylics, nylons or certain cellulosics. See also cast film; castings; cast sheet; centrifugal castings (2); rotational castings.

Cast-rite. Trademark of American Colloid Company (USA) for *bentonite clay* products.

Cast Seal. Trade name for a sodium silicate based sealant for filling porous metals.

cast sheet. A polymer sheet, such as an acrylic, usually produced by heating a catalyzed monomer between two pieces of polished glass and allowing it to cure (cell casting process), or by pouring a liquid polymer between moving stainless steel belts (continuous casting process).

cast stainless steels. See corrosion-resistant cast alloys.

cast steels. See steel castings.

cast stone. A concrete or mortar block or slab shaped or cast to resemble natural building stone. Also known as *artificial stone; concrete ashlar*.

CasTuf. Trademark of Advanced Cast Products, Inc. (USA) for a tough austempered ductile iron (ADI) produced by the lost foam (or evaporative pattern) casting process.

Cast Well. Trade name for a silver-free palladium dental bonding alloy.

caswellsilverite. A synthetic mineral composed of sodium chromium sulfide, $CrNaS_2$. Crystal system, rhombohedral. Density, 3.21-3.30 g/cm³.

Catabond. Trade name of Catalin Corporation (USA) for phenol-formaldehyde plastics.

Catalex. Trade name of Catalin Corporation (USA) for phenol-formaldehyde plastics and laminates.

Catalin. Trademark of Catalin Corporation (USA) for thermosetting phenol- and melamine-formaldehyde resins with a density of about 1.3 g/cm³ (0.05 lb/in.³), excellent heat and electrical resistance, good strength and hardness, good colorability and color stability, low moisture absorption, and low flammability. They are supplied in the form of blocks, sheets and rods for the manufacture of moldings, lighting fixtures, switchgear, switch panels, and kitchen and dinnerware, and as varnishes and lacquers.

Catalox. Trademark of Condea Vista Company (USA) for a series of aluminas supplied in several grades.

Cataloy. Trade name for a series of cast copper-lead and lead-copper alloys for heavy-duty bearing applications.

Catapal. Trademark of Condea Vista Company (USA) for a series of aluminas supplied in several grades.

catapleiite. A yellow to brown, or bluish mineral composed of sodium zirconium silicate dihydrate, $(Na_2,Ca)ZrSi_3O_9 \cdot 2H_2O$. Crystal system, hexagonal. Density, 2.7-2.8 g/cm³; hardness, 6 Mohs. Occurrence: Norway.

catalyst. (1) In general, a reusable substance that brings about a change in the rate of chemical reaction without being changed or consumed itself. Examples include aluminum chloride, aluminum oxide, ammonia, cobalt, ferric chloride, manganese dioxide, platinum metals, silver, triethylaluminum, zeolites, etc.

(2) In plastics terminology, a material that is utilized to activate resins and initiate curing. In polyester systems, it involves chiefly organic peroxides, and in epoxy systems amines and anhydrides.

catalyst carrier. A solid, porous material, such as activated alumina or kieselguhr, often supplied in pellet form, and used to support a catalyst.

catalyzed asphalt. A weather-resistant asphalt treated with phosphorus pentoxide (P_2O_5), and used in roadbuilding.

Catavar. Trademark of Catalin Corporation (USA) for phenol-formaldehyde-bonded paper laminates.

Catawba. Trade name of Jessop Steel Company (USA) for a shock-resisting tool steel containing carbon, silicon, manganese, and iron. Used for tools, punches and upsetters.

catechol. A phenolic compound available as a white to faint tan crystalline powder (99+% pure) that may discolor on exposure to air and light. It has a density of 1.371 g/cm³, a melting point of 104-106°C (219-223°F), and a boiling point of 245°C (473°F), and a flash point of 279°F (137°C). Used in organic synthesis, as photographic additive, as a dyestuff, in antioxidants and light stabilizers, in electroplating, and in medicine (antiseptic). Formula: $C_6H_4(OH)_2$. Also known as *1,2-benzenediol; pyrocatechol.*

Cathaloy. Trade name of Superior Tube Company (USA) for a series of cathode materials including several high-purity nickels and nickel-tungsten alloys. Used for cathodes in electronic tubes.

cathedral glass. A translucent sheet glass that is usually shaped by rolling, not polished and may or may not have a texture on one surface. Used for cathedral and church windows.

Cathoclear. Trade name of Enthone-OMI, Inc. (USA) for a cathodically applied polymer coating.

Cathode Alloy. Trade name of Wilbur B. Driver Company (USA) for a nickel alloy that contains up to 4% tungsten, and is used for electron-tube cathodes.

cathode copper. Copper (99.9% pure) that has been refined electrolytically, i.e., deposited on the cathode of an electrolytic bath of an acidified copper sulfate solution from a black copper anode. It is sold as-is, or as-remelted. Abbreviation: CC.

cathode nickel. An alloy of 96-99.5% nickel and 0.5-4% selected additives, used for cathodes in electron tubes.

cathodic coatings. Coatings that have been deposited on the cathode in an electrolytic bath.

cathodic material. Any material that acts as the cathode in an electrolytic cell.

cathodic metal. A metal that, in reference to other metals, does not have the tendency to corrode, dissolve or oxidize preferentially.

Cathogard. Trade name of BASF Corporation (USA) for intermediate-film electrocoating products.

Cathospheres. Trade name of Plastic Methods Company (USA) for precision copper-plated spheres used in barrel plating.

cation exchanger. An *ion exchanger*, usually an organic compound, whose negatively charged functional groups are attached to the support by covalent bonds. It has the capacity to reversibly bind cations, i.e., to enter into a reversible process involving the exchange of positively charged atoms. See also anion exchanger.

cationic polymer. A polymer produced by a chain-reaction polymerization initiated by an acid, such as sulfuric acid, aluminum trichloride or boron trifluoride. See also anionic polymer; ionic polymer.

cativo. The wood of the tree *Prioria copaifera* having a texture resembling that of *mahogany.* It seasons rapidly, but has a tendency to bleed resinous material. Source: Tropical America. Used for furniture, cabinetwork, interior trim, veneer for patterns, plywood, etc.

catlinite. See pipestone.

Catolux. Trade name of Esperanza SA (Spain) for solid glass blocks of elongated form that have parallel grooves which are specially adaptable to curved walls.

catoptrite. See katoprite.

cat's eye. A honey-yellow to yellowish-green variety of the mineral *chrysoberyl* $(BeAl_2O_4)$ that when properly cut shows a beam of light across the curved surface. Obtained from Sri Lanka, it is used chiefly as a gemstone. Also known as *cymophane.*

cattierite. A pinkish metallic mineral of the pyrite group composed of cobalt sulfide, CoS_2. Crystal system, cubic. Density, 4.80 g/cm³. Occurrence: Zaire.

cattle hides. The hides of domesticated bovine animals including cows, bulls, steers and oxen used for making leather for protective clothing, linings, suitcases, traveling bags, upholstery, and shoe soles, uppers and insoles. See also cowhides; steerhides.

caulk. See caulking compound (1).

caulking. See caulking compound (1).

caulking compound. (1) A material, usually of solid or viscous consistency, that has good adhesive and cohesive properties. Used in building construction to make seams or joints air-, steam- or waterproof, seal between window or door frames, or fill crevices, holes, etc. Also known as *caulk; caulking.*

(2) A pasty mass composed of a drying oil (e.g., linseed) and whiting, or a drying oil, asbestos fibers, inert fillers and pigments. It is now often made with synthetic resins or elastomers (e.g., polysulfide). Used for setting window and door frames. Also known as *caulking putty; glazier's putty; glazing compound.* See also putty (1).

caulking putty. See caulking compound (2).

causal metal. An austenitic gray cast iron with excellent corrosion resistance to atmospheric conditions, many acids and all alkalies. Used for marine applications, e.g., pump bodies.

Caurite. Trade name of Atofina SA (France) for urea- and melamine-formaldehyde adhesives and binders

Causplit. Trade name of Atofina Chemicals Inc. (USA) for an acid- and alkali-resisting resin type cement.

caustic arsenic chloride. See arsenic chloride.

caustic baryta. See barium hydroxide (2).

caustic-calcined magnesia. See calcined magnesite.

caustic-calcined magnesite. See calcined magnesite.

caustic lime. See calcium hydroxide.

caustic potash. See potassium hydroxide.

caustic silver. See silver nitrate.

caustic soda. See sodium hydroxide.

Cavalon. (1) Trademark of E.I. DuPont de Nemours & Company (USA) for polyacrylic resins used for hard, abrasion-resistant coatings and enamels.

(2) Trademark of Cavalier Carpets (USA) for bulk continuous polyolefin yarn used for pile carpets.

cavalry twill. A durable fabric with a double diagonal stripe, usually made of cotton, wool or worsted in a steep twill weave. Used for coatings, suitings, and army uniforms. Also known as *elastiqué*.

cavansite. A blue mineral composed of calcium vanadyl silicate tetrahydrate, $Ca(VO)Si_4O_{10} \cdot 4H_2O$. Crystal system, orthorhombic. Density, 2.21 g/cm^3; refractive index, 1.544. Occurrence: USA (Oregon).

Cavex. Trade name of Glaceries de Saint Roch SA (Belgium) for a patterned glass.

Cavex Avalloy. Trade name of Cavex (Netherlands) for a durable dental amalgam alloy composed of lathe-cut particles. It contains 45% silver, 30.5% tin, 24% copper, 0.5% zinc, and does not contain the corrosive tin-mercury (or gamma-2) phase. Also known as *Avalloy*.

Cavex Clearfil. Trade name of Cavex (Netherlands) for a series of dental hybrid composites (e.g., *Cavex Clearfil Lustre* and *Cavex Clearfil CR Inlay)*, and dental adhesives and bonding agents (e.g., *Cavex Clearfil Activator* and *Cavex Clearfil Newbond)*.

Cavex Non Gamma 2. Trade name of Cavex (Netherlands) for a durable dispersed dental amalgam alloy of lathe-cut and spherical particles. It contains 69.2% silver, 18.6% tin, 11.9% copper, 0.3% zinc, and does not contain the corrosive tin-mercury (or gamma-2) phase. Also known as *Non Gamma 2*.

Cavex Octight. Trade name of Cavex (Netherlands) for a gamma-2-phase-free dental amalgam alloy supplied in the form of capsules, and containing 45% silver, 30.5%, tin 24%, and the balance mercury. Also known as *Octight*.

Caviar. Trade name of SA Glaverbel (Belgium) for patterned glass.

Cavinol. Trade name for a zinc oxide/eugenol dental cement.

Cavitec. Trademark of Sybron Corporation (USA) for a unmodified zinc oxide/eugenol dental cement.

Cavitile. Trademark of Cindercrete Products Limited (Canada) for concrete tiles.

cavity brick. See hollow brick.

caysichite. A colorless to white mineral composed of yttrium calcium carbonate silicate tetrahydrate, $(Y,Ca)_4(CO_3)Si_4O_{10} \cdot 4H_2O$. Crystal system, orthorhombic. Density, 3.03 g/cm^3; refractive index, 1.614. Occurrence: Canada (Ontario).

cazin. An alloy containing 82.6% copper and 17.4% zinc. It has a melting point of 263°C (505°F), and is used for brazing steel, e.g., cables.

C&B-Metabond. Trade name of Parkell, Inc. (USA) for a self-curing polymer-based dental adhesive cement that forms a strong bond between the tooth structure (i.e., dentin and enamel)

and metal, porcelain or composite-resin restorations. It contains a base [4-methacryloxyethyl trimellitate anhydride (4-META) and methyl methacrylate (MMA)], a dentin activator solution (citric acid/ferric chloride), an enamel etchant (phosphoric acid), a catalyst (tributyl borane), and tooth colored and clear powders [polymethyl methacrylate (PMMA)]. Used for the cementation of crowns, bridges, inlays, onlays and posts, for bonding splints, and as a protective lining of vital teeth.

C-Cast. Trademark of Cominco Limited (Canada) for large semi-continuously cast zinc ingots.

CCM. Trademark of Carpenter Technology Corporation (US) for a nonmagnetic, wrought alloy composed of 26% chromium, 6% molybdenum, 1% iron, 1% silicon, 1% manganese, 1% nickel, 0.5% copper, 0.5% tungsten, 0.18% nitrogen, 0.05% carbon, 0.015% sulfur, 0.015% phosphorus, and the balance cobalt. It has high strength, excellent corrosion resistance, good biocompatibility and wear resistance, and a highly polished surface finish. Used in medical implants including hip-joint femoral stems and caps, and artificial-knee-joint stabilizing posts and locking screws.

CD Tow. Trade name of Pepin Associates Inc. (US) for commingled fiber tow composed of a mixture of discontinuous structural man-made filaments, such as glass fibers, and continuous, meltable thermoplastic filaments, such as recycled polyethylene terepthalate (PET). It can be made into braids, fabrics, etc., and molded under heat and pressure into structural composites in which the molten thermoplastic filaments form the matrix, and the discontinuous (glass) fibers the reinforcement.

Ceamarc. Trade name of Air Products & Chemicals Inc. (USA) for a series of corrosion- and wear-resistant inert-gas arc-spray coatings.

ceba. See kapok.

cebollite. A colorless, or greenish to white mineral composed of calcium aluminum silicate hydroxide, $Ca_5Al_2(OH)_4Si_3O_{12}$. Crystal system, orthorhombic. Density, 2.96 g/cm^3; refractive index, 1.60; hardness, 5 Mohs. Occurrence: USA (Colorado).

Cebu maguey. See cantala fibers.

Cecaperl. Trade name of Atofina Ceca (France) for expanded *perlite* for cryogenic insulators.

Cecarbon. Trade name of Atofina Chemicals Inc. (USA) for granular activated carbon.

Cecatherm. Trade name Atofina Ceca (France) for *perlite* used for foundry applications.

cechite. A black mineral of the descloizite group composed of iron lead manganese vanadium oxide hydroxide, $Pb(Fe,Mn)(VO_4)(OH)$. Crystal system, orthorhombic. Density, 5.88 g/cm^3. Occurrence: Czech Republic.

Ceco. A British trade name for an alloy containing 62.5% copper, 32% lead, 4.6% tin, and 0.9% nickel. Used for heavy-duty bearings and bushings.

Cecolloy. Trade name of Chambersburg Engineering Company (USA) for a series of shock-resistant cast irons containing 2.8-3% total carbon, 0-0.9% manganese, 0-1.3% silicon, 0-1.5% nickel, 0.5% molybdenum, 0-0.35% chromium, and the balance iron. Used for castings, valves, frames, steam cylinder liners, forming dies, and anvils.

cedar. (1) The yellow, fragrant wood of any of several coniferous trees of the genus *Cedrus* belonging to the pine family. *Cedars* grow throughout Southern Europe, North Africa and Asia (Himalayas). Three species are of commercial interest: (i) Lebanon cedar *(C. libani)*; (ii) Himalayan cedar *(C. deodora)*; and

(iii) Atlas cedar *(C. atlantica)*. Cedarwood has high durability, and takes a fine polish. Average weight, 575 kg/m³ (36 lb/ft³). Used for general construction. For more information, see individual species.

(2) The wood of any of a large group of North American coniferous trees of the genera *Thuja, Juniperus,* and *Cupressus,* especially the eastern red cedar *(J. virginiana),* western red cedar *(T. plicata),* northern white-cedar *(T. occidentalis)* and Alaska cedar *(Cupressus sitkaensis)*. North American cedars are not true cedars. For property and application data, see individual species.

(3) The soft, durable, reddish wood of various coniferous trees of the genus *Cedrela* found throughout Africa, Asia and Tropical America, especially the *Spanish cedar* of Central and South America that is not a true cedars, but somewhat resembles the lighter grades of true mahogany.

cedar pine. See spruce pine.

cedarwood. See cedar (1), (2) and (3).

cedro. See Spanish cedar.

cedro macho. See andiroba.

Ced R-Tex. Trademark of Canexel Hardboard Inc. (Canada) for textured hardboard used as siding for residential construction.

Cefor. Trade name of Shell Chemical Company (USA) for polypropylene resins supplied in the form of pellets, sheets and film, and used in the manufacture of plastic goods.

Cegeite. Trade name of La Compagnie Générale d'Electricité (France) for phenolic resins and plastics.

Cehadent. Trade name for a medium-gold dental bonding alloy.

Cehadentor. Trade name for a medium-gold dental casting alloy.

ceiba. See kapok.

Ceilcoat. Trademark of The Ceilcote Company (USA) for a wide range of polymer coatings and linings, and epoxy grouts.

Ceilcrete. Trademark of The Ceilcote Company (USA) for a range of glass-reinforced polyester lining materials.

CeilGuard. Trademark of Sandoz Limited (Switzerland) for a series of corrosion-resistant paints and coatings.

Cekas. Trade name of C. Kuhbier & Sohn (Germany) for corrosion-resistant, hardenable steels containing 0.1-0.2% carbon, 13% chromium, 1.0% molybdenum, and the balance iron. Used for ball bearings, gears, springs, tableware, surgical instruments, and chemical and oil-refinery equipment.

Cekol. Trademark of Noviant Inc. (USA) for carboxymethylcellulose (CMC).

Celacloud. Trade name of Cotton Felts, Limited (Canada) for cellulose acetate fibers used for textile fabrics.

Celacrimp. Trade name for cellulose acetate fibers and crimped yarns used for wearing apparel.

celadon. A light bluish green, semi-opaque glaze fired in a reducing atmosphere using iron as the colorant. Also known as *celadon glaze.*

celadon glaze. See celadon.

celadonite. A blue-green mineral of the mica group composed of potassium magnesium iron aluminum silicate hydroxide, $K(Mg,Fe,Al)_2(Si,Al)_4O_{10}(OH)_2$. Crystal system, monoclinic. Density, 2.95-3.00 g/cm³. Occurrence: USA (Washington).

Celafibre. Trade name for a cellulose acetate fiber usually supplied in cut staple form. It blends well with other fibers, and is used for the manufacture of blankets and other textile fabrics.

Celafil. Trade name for cellulose acetate fibers used for textile fabrics.

Celairese. Trade name of Hoechst Celanese Corporation (USA) for soft, light cellulose acetate staple fibers used especially for

underlinings.

Celaloft. Trade name for cellulose acetate fibers used for clothing.

Celanar. Trademark of Hoechst Celanese Corporation (USA) for thermoplastic polyester made from polyethylene terephthalate (PET). Available in the form of a tough, transparent, biaxially-oriented all-plastic film, it has a density of 1.4 g/cm³ (0.05 lb/in.³), excellent fatigue and tear strength, outstanding dimensional stability, high dielectric strength, a service temperature range of -60 to +150°C (-75 to +300°F), and good resistance to humidity, acids, greases, oils and solvents. Used for magnetic recording tapes, pressure-sensitive tapes, automotive tire cords, drafting materials, and as a dielectric, clothing and packaging material.

Celanese. (1) Trademark of Hoechst Celanese Corporation (USA) for a series of cellulose acetate fibers and yarns used for the manufacture of textile fabrics with good moth and mildew resistance, e.g., garments with silky luster and good drape.

(2) Trademark of Hoechst Celanese Corporation (USA) for engineering resins based on polyamide 6,6 (nylon 6,6). They are available in standard and various specialty grades including glass-reinforced, graphite-reinforced, glass bead- or mineral-filled, conductive, molybdenum disulfide-lubricated, UV-resistant, impact-modified, supertough, fire-retardant and food grade. They have good abrasion, chemical and temperature resistance, excellent strength, toughness and lubricity, and good flame and smoke resistance. Used for automotive components, appliances, and electronic and electrical components.

Celanese Nylon 6. Trademark of Hoechst Celanese Corporation (USA) for engineering resins based on polyamide 6 (nylon 6). They are available in standard and glass fiber-reinforced grades for automotive and electrical applications.

Celanese Nylon 66. Trademark of Hoechst Celanese Corporation (USA) for engineering resins based on polyamide 6,6 (nylon 6,6). They are available in standard and glass fiber-reinforced grades for automotive and electrical applications.

Celanex. Trademark of Hoechst Celanese Corporation (USA) for a family of thermoplastic polyester resins based on polybutylene terephthalate (PET) or polybutylene terephthalate (PBT). They are supplied in various grades including neat, unfilled, glass bead- and/or mineral-filled, glass fiber- or carbon fiber-reinforced, high-impact, low-warp, improved surface finish, fire-retardant, UV-stabilized, silicone- or polytetrafluoroethylene-lubricated, and structural foam. A PBT/PET alloy grade is also available. *Celanex* resins are processed by resin-injection molding techniques, and have excellent wear resistance, excellent thermal and chemical resistance, high toughness and rigidity, high impact strength, outstanding physical properties, exceptional dimensional stability, good dielectric properties, good surface finish, good moldability, and high flame resistance. Used for high-performance applications, electrical and electronic components, lighting equipment, cassette and tape cases and boxes, household appliances, sports and recreational equipment, and automotive applications.

Celaperm. Trade name for cellulose acetate fibers used for textile fabrics.

Celara. Trade name for cellulose acetate fibers used for wearing apparel.

Celarandom. Trade name for cellulose acetate fibers and products.

Celaspun. Trade name for spun cellulose acetate fibers and yarns used for wearing apparel.

Celastar. Trademark of Hoechst Celanese Corporation (USA) for a series of polyester resins.

Celastics. Trade name of Celastic Corporation (USA) for cellulose nitrate plastics.

Celastoid. Trademark of Hoechst Celanese Corporation (USA) for cellulose acetate plastics supplied in film, sheet, strip, rod and tube form.

Celastraw. Trade name for cellulose acetate fibers used for textile fabrics.

Celatom. Trademark of Eagle-Picher Industries, Inc. (USA) for a series of high-quality amorphous lightweight siliceous materials based on *diatomite*. Used as filter aids, catalyst supports, paper and paint fillers, and extenders in concrete and asphalt.

Celatow. Trade name for cellulose acetate fibers and tow.

Celatress. Trade name for cellulose acetate fibers and yarns used for wearing apparel.

Celatron. Trade name of Hoechst Celanese Corporation (USA) for modified and unmodified polystyrene resins with excellent electrical properties and optical clarity, and good dimensional and thermal stability. Used for panels, housings, casings, etc.

Celaweb. Trade name for cellulose acetate fibers and fabrics.

Celawrap. Trade name of Hoechst Celanese Corporation (USA) for cellulose acetate film materials for wrapping and packaging applications.

Celazole. Trademark of Hoechst Celanese Corporation (USA) for ultrahigh-performance thermoplastics based on polybenzimidazole (PBI) polymers. They are commercially available in powder form, stock shapes, as finished parts, and in several melt-processible grades including unreinforced, 30% glass-reinforced, 30% carbon-reinforced and self-lubricating. The powder grades are processed by sintering, and the melt-processible grades can be used for making high-performance parts by injection molding or extrusion. *Celazole* thermoplastics have outstanding wear resistance, high compressive strength, excellent tensile and flexural strength, good dimensional stability, high volume resistivity, very high thermal stability, low coefficients of thermal expansion, low coefficients of friction, low hysteresis losses, and excellent resistance to many chemicals. Typical properties of molded *Celazole* parts include excellent wear and friction properties, a maximum heat-deflection temperature of 326°C (619°F), good compressive and flexural strengths of 220 MPa (32 ksi) and 315 MPa (46 ksi), respectively. Used for automotive, electronics and/or office equipment industries for bushings, bearings, carburetor links, electrical and electronic connectors, valve components, high-temperature seals, packings, electrical insulators, thrust washers, piston rings, hard disk carriers, etc.

Celbar Spray. Trade name of National Cellulose Company (USA) for cellulose-based spray-on acoustical and thermal coatings.

Celbond. Tradename of KoSa (USA) for sheath-core bicomponent staple fibers consisting of strong, resilient polyester core and a low-melting polymer sheath (e.g., polyethylene terephthalate/polyester) which upon heating can thermally bond with other fibers. Used for the manufacture of nonwovens and high-loft textiles including air-laid, wet-formed and dry-formed products, interliners, filters, automotive components, packaging, and medical and protective fabrics.

Celcon. Trademark of Hoechst Celanese Corporation (USA) for a series of thermoplastic linear acetal copolymers based on polyoxymethylene, and used as molding materials. They are commercially available in standard and various special grades including carbon fiber-reinforced, supertough, food, electro-platable, antistatic, weather-resistant and UV-resistant. They have a density of 1.41 g/cm³ (0.051 lb/in.³), high crystallinity, high strength, stiffness and endurance, high toughness and stability, high flexural strength, low coefficients of friction, high dielectric strength, good resistance to corrosion and electrolysis, good heat resistance, inherent lubricity, good resistance to solvents, fair to poor resistance to radiation, weathering and strong acids and bases, poor flame resistance, and a maximum service temperature of 85-104°C (185-220°F). Used for automotive trim, housings, windshield washer pump housings, piping, impellers, bearings, gears, cams, levers, fans, and medical equipment.

Celcos. Trade name of Celanese Corporation (USA) for a partially saponified acetate staple fiber composed of a core of cellulose acetate and a skin or wrapping of rayon. Used for textile fabrics.

Celebrate. Trade name for cellulose acetate fibers used for textile fabrics.

Celecrome. Trade name for cellulose acetate fibers used for fabrics.

Celero. Trade name of Disston Inc. (USA) for an oil-hardening tool steel containing 1.35% carbon, 0.25% chromium, 2.75% tungsten, and the balance iron. Used for plug gages and finishing tools.

Celesta. Trademark of Hoechst Celanese Corporation (USA) for polyester yarn used for the manufacture of carpets.

celeste blue. (1) Any of several iron-blue pigments that usually contain considerable amounts of extenders, such as barites (e.g., barium sulfate, $BaSO_4$).

(2) Any cobalt-blue pigment softened by additions of zinc oxide (ZnO). Also known as *celestial blue*.

celestial blue. See celeste blue (2).

celestine. A colorless, white, yellow, or sky-blue natural mineral of the barite group with a vitreous to pearly luster. It is composed of strontium sulfate, $SrSO_4$, and can also be made synthetically. Crystal system, orthorhombic. Density, 3.94-3.97 g/cm³; melting point, decomposes at 1580°C (2875°F); hardness, 3-3.5 Mohs; refractive index, 1.624. Occurrence: USA (California, Ohio, Texas, West Virginia). Used as a source of strontium and its compounds, in ceramics to impart iridescence to glasses and pottery glazes, and as a fining agent in crystal glass. Also known as *celestite*.

celestite. See celestine.

Celestra. Trademark of Celanese Acetate AG for solution-dyed cellulose acetate yarns used for woven and knitted fabrics.

CelFlow. Trademark of Noviant Inc. (USA) for a series of cellulose ethers including ethylcellulose, methylcellulose and carboxymethylcellulose.

Celfor. (1) Trade name of Sanderson Kayser Limited (UK) for a case-hardening steel containing 0.2% carbon, and the balance iron. Used for shafts, pinions, and gears.

(2) Trade name of Sanderson Kayser Limited (UK) for a plain-carbon tool steel containing 1.0% carbon. Used for forging, heading and trimming dies, shear blades, etc.

Celfort. Trademark of Celfort Construction Materials Inc. (USA) for extruded rigid polystyrene foam used as thermal building insulation.

Celgard. (1) Trademark of Hoechst Celanese Corporation (USA) for a series of chemical- and impact-resistant polypropylene materials used for microporous membranes, filters, food-processing equipment, and biological filtration applications.

(2) Trademark of Hoechst Celanese Corporation (USA)

for microporous and hollow synthetic films and fibers used in the manufacture of control-release and medical devices, batteries, sterile packaging, etc.

Celion. Trademark of Celanese/ToHo (USA) for a series of high-strength polyacrylonitrile (PAN) based carbon fibers used as reinforcements in high-performance composites for aircraft, aerospace and military applications.

Celipal. Trademark of CWH (Germany) for unsaturated polyesters.

Celite. (1) Trademark of Celite Corporation (USA) for *diatomaceous earth* and related products composed of about 93% silica (SiO_2), 4% alumina (Al_2O_3), 1.5% ferric oxide (Fe_2O_3), 1.0% lime (CaO) and magnesia (MgO), and 0.5% potash (K_2O) and soda (Na_2O). Used as ingredients in cements, as abrasives in glass and metal polishing, as filter aids, and as flatting agents in paints and paper finishes.

(2) Trade name for an extremely fine glass fiber used mainly in paper manufacture.

celite. A solid-solution constituent in Portland-cement clinker composed of tetracalcium aluminoferrate ($4CaO \cdot Al_2O_3 \cdot Fe_2O_3$) and hexacalcium dialuminoferrate ($6CaO \cdot 2Al_2O_3 \cdot Fe_2O_3$).

Cell-Aire. Trademark of Sealed Air Corporation (USA) for polyurethane foam used for protective packaging applications.

Cellanite. Trade name of Continental-Diamond Fiber Company (USA) for phenol-formaldehyde resins and plastics.

Cellasta. Trade name for cellular polyurethane elastomers used for casting applications.

Cellastine. Trademark of Celanese Corporation (USA) for cellulose acetate plastics.

Cellasto. Trademark for thermoplastic polyester and polyether urethane elastomers..

Cell-Cast. Trade name of Three Sixty Corporation (USA) for molded plastics.

Cellene. Trade name of Ato Findley (France) for neoprene-based adhesives for floor coverings and building applications.

Cellestron. Trade name for cellulose acetate fibers used for wearing apparel and as industrial fabrics.

Cellidor. Trademark of Albis Corporation (USA) for transparent, flexible, thermoplastics made from cellulose nitrate and camphor, and supplied as molding compositions. They have good resistance to water, gasoline and dilute acids and alkalies, moderate flammability, can be processed by injection molding, extrusion and machining, and are supplied in several grades including glass-fiber-reinforced. Used in the manufacture of spectacle frames, photographic film, etc. *Cellidor B* refers to a series of tough, transparent, amorphous thermoplastics based on cellulose acetate butyrate plastics supplied in the form of sheet and tubes. They have a density of 1.2 g/cm³ (0.04 lb/in.³), good dielectric properties, excellent weathering properties, good dimensional stability, good resistance to oils and greases, and relatively low moisture absorption (0.9-2.2%). Used for lenses, goggles, consumer products, packaging, signs, plastic film and sheeting, piping and tubing, etc. *Cellidor CP* refers to cellulose propionate plastics.

Cellini. Trade name of Bethlehem Steel Company (USA) for a tool steel containing 0.8% carbon, 0.9% manganese, 0.5% chromium, and the balance iron. Used for tools and dies.

Cellit. Trademark of Bayer Corporation (USA) for a cellulose acetate and cellulose acetate butyrate plastics.

Cello. Trade name of Hoechst Celanese Corporation (USA) for cellulose acetate film materials.

Cello-Bond. Trade name of Cello-Foil Products, Inc. (USA) for

cellophane laminations for flexible packaging applications.

Cellobond. Trade name of BP Chemicals (USA) for urea-formaldehyde resins and products. *Cellobond K* refers to phenolic-based foams.

Cellocel. Trade name for cellulose compounds.

Cel-Lok. Trademark of Owens Corning Fiberglass (USA) for moisture-resistant extruded polystyrene insulation supplied in pink-colored, pregrooved, 0.6 × 2.4 m (2 × 8 ft.) sheets with a thickness of 38 or 51 mm (1.5 and 2 in.). It has a heat flow resistance value (R-value) of R-5 per inch (25 mm) of thickness, and is used for basement insulation.

Cellokyd. Trade name of Reichhold Chemicals, Inc. (USA) for a series of short-oil alkyds used in paints and coatings.

Cellomold. Trade name for a cellulose-acetate molding powder.

Cellon. (1) Trade name of Albis Corporation (USA) for a transparent, flexible, thermoplastic made from cellulose nitrate and camphor. It has good resistance to water, gasoline and dilute acids and alkalies, moderate flammability, and can be processed by injection molding, extrusion and machining. Used for spectacle frames and photographic film.

(2) Trademark of Dynamit Nobel AG (Germany) for cellulose acetate plastics.

(3) Trademark of Koppers Company, Inc. (USA) for pressure-treated wood.

Cellonex. Trademark of Dynamit Nobel AG (Germany) for cellulose acetate plastics.

Cellophane. Trade name of E.I DuPont de Nemours & Company (USA) for regenerated cellulose film and sheeting materials. See also cellophane.

cellophane. A regenerated cellulose, i.e., a transparent, clear substance obtained by treating cellulose with sodium hydroxide and carbon disulfide. It is supplied in the form of thin, strong, flexible film or sheeting, typically 0.02-0.04 mm (0.0008-0.0016 in.) thick, and frequently modified with softeners, flame retardants, coatings, etc., and sometimes colored or embossed. It is impermeable to nonaqueous substances, can be coated with cellulose lacquer and laminated with synthetic resin for moisture-proofing, and has moderate mechanical properties, and excellent resistance to greases, oils and air. Untreated film softens on exposure to heat at about 150°C (300°F), decomposes at 170-205°C (338-400°F), is flammable, not self-extinguishing, and chars at about 190°C (375°F). Used for wrapping foods, candy, tobacco, etc., and for general packaging applications.

Cellothane. Trade name of Reichhold Chemicals, Inc. (USA) for oil-modified urethane resins used for abrasion-resistant coatings.

Cellothene. Trademark for a very strong, heat-sealable film composed of *cellophane* laminated with a thin layer of polyethylene. It has excellent resistance to oils, greases, and many chemicals, and is used for wrapping foods, etc.

Cello Vanadium. Trade name of McInnes Steel Company (USA) for a nondeforming, oil-hardening tool steel containing 0.9% carbon, 1.2% manganese, 0.5% chromium, 0.5% tungsten, 0.15% vanadium, and the balance iron. Used for dies, taps, reamers, hobs, and broaches.

Cellovar. Trade name of Reichhold Chemicals, Inc. (USA) for phenolic-modified varnish resins for coating applications.

Celltate. Trade name for cellulose acetate fibers used for textile fabrics.

Cellufix. Trademark of Noviant Inc. (USA) for carboxymethylcellulose (CMC).

Cellulac. (1) Trademark of C-I-L Paints (Canada) for cellulose lacquers.

    (2) Trade name of Continental Diamond Fiber Company (USA) for shellac plastics.

    (3) Trade name of British Plastoids Company (UK) for cellulose nitrate plastics.

cellular adhesives. Adhesives of low apparent density containing numerous small, closed or interconnected bubbles or cells of air or gas evenly dispersed throughout their masses. Also known as *foamed adhesives.*

cellular cellulose acetate. See cellulose acetate foam.

cellular ceramics. See ceramic foams.

cellular concrete. See aerated concrete.

cellular elastomer. A flexible, semirigid or rigid foam composed of a natural or synthetic elastomer containing numerous closed cells, and used chiefly as an insulating material.

cellular glass. A rigid expanded glass with an essentially closed-cell structure. It is noncombustible, moistureproof and buoyant, and has a high thermal insulating value. *Cellular glass is made by adding powdered carbon or other gas-forming materials to crushed glass and heating in a manner that entraps the evolving gas bubbles. Available in the form of blocks or sheets, it is used as thermal insulation for walls, floors, roofing and domestic and industrial appliances, and for piping, low-temperature equipment, etc. Also known as *cellulated glass; expanded glass; foamed glass; glass foam.*

cellular fabrics. Textile fabrics that have been woven, knit or otherwise combined so as to have closely and evenly spaced holes. For example, weaves, such as honeycomb and leno, can produce such fabrics.

cellular material. A material that contains numerous small, closed or interconnected cavities or cells. See also cellular solids.

cellular nylon. See nylon foam.

cellular plastics. Thermoplastic or thermosetting resins processed into flexible or rigid foams with numerous small, closed or interconnected cells by the thermal, chemical or mechanical action of a blowing or foaming agent, such as air, ammonium carbonate, sodium bicarbonate, halocarbons, methylene chloride, pentane, hydrazine, etc. Rigid foams have fair compressive strength and good machinability, and are used as core materials for sandwich constructions, and for light construction (e.g., boats, airplanes, etc.) and insulation applications. Flexible foams have high resiliency and softness, and are used for furniture, mattresses, and automobile interiors. Also known as *expanded plastics; foamed plastics; plastic foams.*

cellular polystyrene. See polystyrene foam.

cellular polyurethane. Polyurethane that has been processed with a blowing agent (e.g., carbon dioxide, trifluoromethane, etc.) to form a rigid foam containing numerous closed cavities or cells. It has a density range of about 0.03-0.80 g/cm³ (0.001-0.029 lb/in.³), excellent thermal and acoustic insulating properties, high impermeability to air and water, and fair to poor flame resistance. Used for acoustic and thermal insulation, as a filling material, for boat hulls, surfboards, skis, etc., in packaging, for furniture, buoyancy and flotation devices, and for automobile bumpers. See also polyurethane foam; rigid polyurethane foam; flexible polyurethane foam.

cellular rubber. See sponge rubber.

cellular solids. A very broad class of solid materials with sponge-like structures containing numerous, minute, closed or interconnected cavities, pores or cells, and including natural materials, such as bone, cork, sponge and wood, and synthetic (engineered) materials, such as ceramic, metallic, polymeric and vitreous

foams. See also cellular glass; cellular plastics; ceramic foams; metal foams; honeycomb.

Cellulate. Trade name of National Plastic Products Company (UK) for cellulose acetate plastics.

cellulated ceramics. See ceramic foams.

cellulated glass. See cellular glass.

Cellulith. Trade name for vulcanized fiber.

Celluloid. Trademark of Hoechst Celanese Corporation (USA) for a strong, flexible, thermoplastic made from cellulose nitrate and a plasticizer, such as camphor, and supplied in the form of films, sheets, rods and tubes. It is available in transparent, translucent, opaque, colored and colorless forms, and has a density of 1.3-1.8 g/cm³ (0.05-0.07 lb/in.³), a refractive index of 1.50, good resistance to water, gasoline, oils and dilute acids and alkalies, good molding properties, and can be processed by injection molding, extrusion and machining. The untreated material is highly flammable, but frequently contains flame-retardant, such as ammonium phosphate to reduce flammability. Used for spectacle frames and photographic film.

celluloid. A generic name for colorless, transparent plastics obtained from cellulose nitrate by mixing with fillers and pigments in a solution of camphor in alcohol. They have high durability and dimensional stability, good moisture resistance, good polishability, poor flame resistance, and can be colored, rolled and molded. Used for denture bases, spectacle frames, photographic film, combs, brushes, buttons, etc.

cellulose. A natural long-chain, linear carbohydrate polymer (polysaccharide) composed of β-D-1,4 linked-anhydroglucose units. It is a major chemical constituent of higher plant cell walls (e.g., wood and cotton), has a high degree of polymerization ranging from 1000 for wood pulp to 3500 for cotton fiber, and thus a high molecular weight (typically 160000-560000). It is available in the form of fibers, microgranules and microcrystalline powders of varying purity with a density of approximately 1.5 g/cm³ (0.054 lb/in.³). Used in the manufacture of paper, plastics, textiles, explosives, cotton fabrics, and insulation and soundproofing materials, and as a fuel. High-purity fibers are also used in chromatography, biochemistry and medicine. Formula: $(C_6H_{10}O_5)_n$.

cellulose acetate. (1) An ester of cellulose and acetic acid obtained by reacting wood pulp or cotton linters with acetic acid (ethanoic acid) or acetic anhydride (ethanoic anhydride) in the presence of a perchloric or sulfuric acid catalyst. It is available in the form of white flakes or powders with a typical acetyl (ethanoyl) content of 39-40%, a density of 1.3 g/cm³ (0.05 lb/in.³), and a refractive index of 1.475. Used in thermoplastic molding compositions for the manufacture of plastics and fibers, and in biochemistry and biotechnology. Abbreviation: CA. Also known as *acetylated cellulose; ethanoylated cellulose.* See also primary cellulose acetate; secondary cellulose acetate.

    (2) A tough, flexible, combustible thermoplastic material made by compounding a cellulose acetate ester with plasticizers and other ingredients. Supplied in sheet, film, fiber and molded form, it has a density of 1.3 g/cm³ (0.05 lb/in.³), a heat-deflection temperature of 48-105°C (118-220°F), a softening temperature of approximately 60-97°C (140-207°F), a continuous-use temperature of -20 to +95°C (-4 to +203°F), a melting point of approximately 260°C (500°F), high dielectric strength, good resistance to ultraviolet light, a high moisture absorption of 1.9-7.0%, high impact strength, high tensile strength and modulus, rather low dimensional stability to heat and cold flow, good moldability, good resistance to aromatic hydrocarbons,

greases, oils and halogens, fair resistance to alcohols and dilute acids, poor resistance to alkalies, ketones and concentrated acids. Used for magnetic tapes, photographic film, transparent sheeting, acetate fibers, protective coating solutions and lacquers, binders, and molded articles. Also known as *cellulose acetate plastic*.

cellulose acetate butyrate. (1) A mixed ester of acetic acid and butyric acid with cellulose. It is available in the form of white pellets or granules with typical acetyl and butyryl contents of 2-30 wt% and 17-52 wt%, respectively. Used in thermoplastic molding compositions for the manufacture of plastics, and in biochemistry and biotechnology. Abbreviation: CAB. Also known as *cellulose acetobutyrate*.

(2) A very tough, clear, combustible thermoplastic material made by compounding a cellulose acetate butyrate ester with plasticizers and other ingredients. Supplied in film, sheet, tube and pipe form, it has a density of 1.2 g/cm³ (0.04 lb/in.³), a refractive index of 1.475, a continuous-use temperature from below -40 to over +100°C (-40 to +210°F), low heat conductivity, high dielectric strength, high impact resistance, excellent weathering properties, good resistance to ultraviolet light, low water absorption (0.9-2.2%), good dimensional stability, good resistance to oils and greases, poor resistance to alcohols, ketones, organic acetates and lactates, methylene, ethylene, propylene chlorides and high-boiling solvents. Used for automotive taillight lenses, dial covers, outdoor signs, television screen shields, consumer products, tool and brush handles, typewriter keys, packaging, hydrometers, plastic film and sheeting, piping and tubing, toys, photographic film, protective coating solutions and lacquers, and covering for aluminum fibers. Also known as *cellulose acetate butyrate plastic*.

cellulose acetate fibers. See acetate fibers.

cellulose acetate film. See acetate film.

cellulose acetate foam. Cellulose acetate expanded into a material with cellular structure and reduced density, usually by combined chemical and mechanical action. Also known as *cellular cellulose acetate*.

cellulose acetate propionate. (1) A mixed ester of acetic acid and propionic acid with cellulose. It is available in the form of white pellets or powder with typical acetyl and propionyl contents of 0.6-2.5 wt% and 42.5-46 wt%, respectively. Used in thermoplastic molding compositions for the manufacture of plastics, and as an enteric coating material. Abbreviation: CAP. Also known as *cellulose propionate*.

(2) A thermoplastic material made by compounding a cellulose acetate propionate ester with plasticizers and other ingredients. It has slightly higher strength and modulus of elasticity than *cellulose acetate butyrate*, low heat conductivity, high dielectric strength, high impact resistance, moderate weathering properties, good resistance to ultraviolet light, good dimensional stability, good moldability, good resistance to oils and greases, and poor resistance to alcohols, ketones and aromatic hydrocarbons. Used for automotive components, such as steering wheels and fuel filter bowls, and for appliance housings and consumer products. Also known as *cellulose acetate propionate plastic*.

cellulose acetate rayon. See acetate fibers.

cellulose acetobutyrate. See cellulose acetate butyrate.

cellulose diacetate. (1) An ester of cellulose and acetic acid in which the latter has esterified two hydroxyl groups of the cellulose molecules. Its theoretical acetic acid content is 48.8%.

(2) A term sometimes used as a synonym for *acetone-soluble cellulose acetate*.

cellulose ether. See ethylcellulose; methylcellulose.

cellulose ethyl ether. See ethylcellulose.

cellulose-fiber insulation. Loose-fill thermal insulation of small particle size, often made from shredded newsprint and treated with fire-, fungus- and corrosion-resistant chemicals. It can be blown or poured, and is used in building construction for filling irregular and inaccessible spaces in walls, floors, roofs and attics.

cellulose fibers. Fibers based on cellulose or its derivatives (e.g., esters or ethers). Cotton fibers are almost pure cellulose. Cellulose acetate fibers are made from esters of cellulose and acetic acid, and paper pulp contains processed cellulose fibers obtained from wood. Also known as *cellulosic fibers*.

cellulose lacquer. A liquid coating formulation comprised of an ester or ether of cellulose (e.g., cellulose acetate or nitrate), plasticizers and pigments dissolved in an organic solvent.

cellulose methyl ether. See methylcellulose.

cellulose mixed ester. An ester of cellulose and mixed acids or anhydrides, e.g., *cellulose acetate butyrate* is a mixed ester of cellulose with acetic acid and butyric acid, and *cellulose acetate propionate* a mixed ester of cellulose with acetic acid and propionic acid.

cellulose nitrate. See nitrocellulose.

cellulose nitrate acetate. See cellulose nitroacetate.

cellulose nitroacetate. A nonflammable nitrated cellulose acetate used for textile finishing compounds, protective coatings, and plastic molding composition. Also known as *cellulose nitrate acetate*.

cellulose plastics. A group of plastics based on cellulose, or cellulose derivatives (e.g., esters or ethers). Cellulose acetate plastics are based on esters, while ethyl cellulose plastics are based on ethers. Also known as *cellulosic plastics*.

cellulose polymers. (1) Natural biopolymers based on *cellulose*, or cellulose derivatives.

(2) A class of synthetic (engineered) biopolymers obtained from *cellulose* by introducing modifying functional groups and/or substituting existing native functional groups. They have relatively high thermal stability and good biocompatibility, and are used for medical products, such as dialyzer membranes, orthopedic prostheses, and as adhesive coatings for biomedical applications.

cellulose propionate. See cellulose acetate propionate.

cellulose triacetate. A type of cellulose acetate in which the cellulose is almost completely esterified by acetic acid. It is available in the form of white flakes or pellets with a density of 1.2 g/cm³ (0.04 lb/in.³) and a melting point of approximately 300°C (570°F). Used in the manufacture of protective coatings, packaging, textile fibers, and magnetic tapes. Abbreviation: CTA.

cellulose xanthate. A viscous, soluble compound obtained by first treating cellulose with strong sodium hydroxide solution to form soda cellulose, and then reacting the latter with carbon disulfide. Used in the manufacture of regenerated cellulose (viscose rayon).

cellulosic fiberboard. A building material manufactured from lignocellulosic fibers (e.g., wood, cane, bagasse or straw) with or without additives and modifiers by interfelting, but without consolidation (i.e., pressing or rolling). The density ranges from about 0.15 to 0.50 g/cm³ (0.005-0.018 lb/in.³).

cellulosic fibers. See cellulose fibers.

cellulosic plastics. See cellulose plastics.

cellulosic resin. A resin based on cellulose or its derivatives, and used in the manufacture of cellulosic plastics.

cellulosics A term referring to cellulose derivatives, such as cellulose acetete, cellulose nitrate or cellulose propionate.

Celluvarno. Trade name of Sillcocks-Miller Company (USA) for cellulose nitrate plastics.

Cellux. Trademark of Cellux AG (Switzerland) for *regenerated cellulose* and derivatives supplied in the form of sheets and films, and used for pressure-sensitive adhesive tapes, etc.

Celmar. Trademark of Courtaulds Advanced Polymers (UK) for chemical-resistant, high-strength polypropylene laminates used for structural applications, duct channels and beams.

Celon. Trade name for nylon fibers and yarns.

Celoron. Trademark of Diamond State Fibre Company (USA) for laminated and molded phenol-formaldehyde plastics made by impregnating paper, or canvas or other fabrics. They have good resistance to oil, water and chemicals, high impact strength, and an upper continuous-use temperature of 120°C (250°F). Used for machine parts, e.g., bearings and automobile timing gears, and electrical insulation.

Celotex. (1) Trademark of Celotex Corporation (USA) for building products made from *bagasse* or wood fibers. They have excellent sound insulation properties, and are used in the manufacture of structural building and insulation board, acoustic tiles and panels, wallboard, hardboard, plasters, etc.

(2) Trade name of Celotex Corporation (USA) for asbestos millboard, rollboard, paper and roof coatings.

Celramic. Trademark of Pittsburgh Corning Corporation (USA) for cellular, all-glass materials and cellular glass-bonded crystalline materials used for insulation applications. Supplied in the form of nodules or beads.

celsian. A colorless mineral of the feldspar group composed of barium aluminum silicate, $BaAl_2Si_2O_8$, sometimes with a small quantity of potassium. Crystal system, monoclinic. It can also be made synthetically. Density, 3.23-3.39 g/cm^3; melting point, 1780°C; (3236°F); refractive index, 1.58-1.59. Occurrence: Australia, Sweden, USA (California). Used in ceramics, and in special refractories for kilns and electric furnaces.

Celso. Trademark of Montecatini Edison SpA (Italy) for carboxymethylcellulose polymers.

Celstar. Trademark of Celanese Acetate AG for dull and bright cellulose acetate yarns used for the manufacture of textile fabrics.

Celstran. (1) Trademark of Hoechst Celanese Corporation (USA) for a long-strand fiber-reinforced thermoplastic polypropylene. It has greater impact resistance and fatigue strength than conventional fiber-reinforced polypropylene, excellent impact resistance and mechanical strength, and good fracture toughness and stiffness. Used for automotive applications, e.g., structural parts, such as battery trays, floor pans and seat shells.

(2) Trademark of Hoechst Celanese Corporation (USA) for an extensive series of polymers including polyamides (e.g., nylon 6,6), polyurethanes, polyethylenes, polybutylene terephthalates, polypropylenes, polyphenyl sulfides and polyethylene terephthalates.

Celtik. Trade name of Permacon (Canada) for architectural stone with an attractive surface resembling natural stone. It is supplied in various shapes, sizes and colors for retaining walls, planters, curbs, etc.

Celta. Trade name for rayon fibers and yarns used for textile fabrics.

Celto. Trade name of Keasby & Matteson Company (USA) for a carbon steel containing 0.2% carbon, and the balance iron. Used

for electrodes for shielded-arc welding processes.

Celtral. Trademark of Fairway Filamentos SA (Brazil) for nylon 6,6 fibers and yarns.

Celufibre. Trademark of Kohler International Limited (Canada) for cellulose blowing wool for thermal insulation applications.

Cem-Fil. Trademark of Saint-Gobain Vetrotex (France) for alkali-resistant glass fibers for cement and mortar reinforcement.

Cemkote. Trademark of International Chemical Technologies, Inc. (USA) for a nickel-cobalt-boron metal coating with very high wear resistance. It can be applied by electroless coating and electroplating.

cement. (1) A fine, gray powder made from a mixture of calcined clay and limestone that, when mixed with water, sand, gravel or crushed stone, forms concrete or mortar, and hardens into a stone-like mass. See also hydraulic cement; Portland cement.

(2) A ceramic or nonceramic adhesive used to bind solids together.

(3) A hard material, quite similar to bone in composition, that forms a relatively thick layer covering the roots of teeth. It contains about 50% mineral matter in an organic matrix consisting mainly of *collagen*. Also known as *cementum*.

(4) An adhesive or nonadhesive dispersion or solution of vulcanized rubber or a plastic in a volatile vehicle.

cement aggregate. Coarse particles (e.g., sand, gravel, or crushed stone or rock), lightweight materials (e.g., blast-furnace slag, shale, clay or carbon), or synthetic inorganic materials used with a cementing medium to make concrete or mortar. See also concrete aggregate.

cementation coating. A hard, corrosion and/or oxidation-resistant coating produced on a metallic surface by heating in intimate contact with another metal in powder, liquid or gaseous form. It may also consist of a boride, carbide or silicide of the base metal. Used on refractory metals or cobalt-, nickel- and vanadium-based alloys, and applied by pack cementation, fluidized-bed cementation or vapor streaming. See also diffusion coating.

cementation steel. See blister steel.

cement-base paint. See cement paint.

cement block. A half-solid building unit that is made from a mixture of 1 part cement and 4 parts aggregate, and has a cross-sectional compressive strength of about 6.9 MPa (1000 psi). A standard cement block is 203 × 203 × 406 mm (8 × 8 × 16 in.) in size.

cement board. An engineering composite consisting of a mixture of 50% cement and 50% wood or similar lignocellulosic fibers. Used in the manufacture of building products, e.g., roofing shingles and shakes.

cement brick. A molded brick made from a mixture of Portland cement and sand, and usually steam cured at 93°C (200°F) after pressing. Used as a backing brick.

cement clinker. See clinker (2).

cemented bar. See blister steel.

cemented carbides. Sintered mixtures of one or more powdered carbides of refractory metals bonded together in metallic matrices. Tungsten carbide is a common powdered carbide used for this purpose, and common matrix binders include cobalt, nickel and iron. The mixtures may also contain small amounts of titanium, niobium (columbium) and/or tantalum carbide. They have high hardness (8-9 Mohs), high melting points, high toughness, good shock resistance, low electrical and thermal conductivities, excellent thermal resistance, high abrasion and wear resistance, high compressive strength, and high moduli of elasticity. *Cemented carbides* are often sold under trademarks or trade names, such as *Carboloy, Kennametal. Talide* or *Valenite*, and used for abrasive products including grinding wheels, belts and papers, and for electrical-resistance heating elements for kiln and furnaces, machining and cutting tools, drill bits, sawteeth, sandblast nozzles, machine parts, tire studs, wire-drawing dies, balls for the tips of ball-point pens, and hardfacing and welding rods. Also known as *carbide tool materials; sintered carbides*.

cemented oxides. See oxide ceramics.

cement grout. A plaster or mortar of troweling or pouring consistency composed of cementitious materials, such as Portland cement and lime, together with a suitable aggregate and water, and sometimes sand. Used for painting-up and finishing mortar joints, filling crevices, and coating building walls. Also known as *cement-water grout*.

Cement-It. Trademark of Jeneric/Pentron Inc. (USA) for a self-curing, radiopaque, fluoride-releasing composite cement with 68 wt% barium borosilicate glass filler. Used in dentistry for the cementation of precious-metal restorations and posts.

cementite. The very hard and brittle compound *iron carbide*, Fe_3C, containing 6.67 wt% carbon and 93.33 wt% iron. It has an orthorhombic crystal structure and occurs as a phase in steels and cast irons.

cementitious material. A material, such as cement, that when mixed with a liquid, such as water, forms a plastic paste with adhesive and cohesive properties and, upon placement, hardens into a solid mass. An aggregate may or may not be added to the paste.

cementitious mixture. A mixture, such as concrete, grout or mortar, containing a cementitious material, such as hydraulic cement.

cement-lime mortar. See lime-cement mortar.

cement mortar. A plastic mixture of sand and Portland cement (mixing ratio, 4:1) with water. Lime may be added to facilitate spreading.

cement paint. A dry powder composed of Portland cement, hydrated lime, pigments, fillers, accelerators and water repellents. When mixed with water it can be applied as a waterproof coating to brickwork, concrete, masonry, etc. Also known as *cement-base paint*.

cement paste. A more or less plastic mixture of Portland cement, water and entrained air.

cement plaster. A plaster consisting of Portland cement and sand, and used as a finish coat for plastering interior surfaces.

cement rock. A natural high-calcium limestone that contains at least 18% clay, and is used in cement manufacture. Also known as *cement stone*.

cement-sand grout. See sanded grout.

cement steel. See blister steel.

cement stone. See cement rock.

cementum. See cement (3).

cement-water grout. See cement grout.

Cemofoam. Trade name of IFP Enterprises Inc. (USA) for a rigid, fire-resistant polyimide foam available in densities from 48 to 496 kg/m³ (3 to 31 lb/ft³). It has high strength and heat resistance, and is used for casings for electronic equipment, sandwich composites, and thermal or cryogenic insulation.

Cemper. Trade name for a self-cure dental resin cement.

Cemtex. (1) Trademark of Macnaughton-Brooks Limited (Canada) for a latex bonding agent.

(2) Trademark of Detroit Graphite Company (USA) for protective and fire-retarding paints.

Cenco. Trade name of Central Brass & Aluminum Foundry Company (USA) for an alloy of copper, antimony and tin, used for bearings and bushings.

Cendré. Trade name of Société Industrielle Triplex SA (France) for a laminated glass with tinted interlayer.

Cenospheres. Trade name of Sphere Services Inc. (USA) for alumina-silica spherical fillers and extenders supplied in particle sizes of 75-300 µm (3-12 mils). They have a low density (0.5-0.8 g/cm³), high compressive strength, high inertness to acids, alkalies, solvents, organic chemicals and water, high thermal stability above 980°C (1800°F). Used as fillers and extenders in cements, composites and synthetic resins, as thermal-spray coating powders, and as filler and extender replacement for glass spheres and calcium carbonate, clay, talc, and silica particles.

Centanin. Trade name of Isabellenhütte Heusler GmbH (Germany) for a resistance alloy containing 65-67% copper, 27-29% manganese, 4-6% nickel and 0-1% aluminum. It has an upper service temperature of 300°C (570°F), and is used for resistances, and electrical equipment and instruments.

Centari. Trademark of E.I. DuPont de Nemours & Company (USA) for a series of coatings and enamels.

Centaur. Trade name of Jessop-Saville Limited (UK) for a water-hardened tool steel used for tools, drills, and taps.

Central American cedar. The soft, reddish, fragrant wood of any of various large trees of the genus *Cedrela*, principally *C. odorata*. The heartwood is light to dark reddish brown. *Central American cedar* has medium strength, good resistance to decay and insects, works and seasons very well, takes a beautiful polish, has good gluing qualities, and somewhat resembles the lighter grades of true mahogany. Average weight, 480 kg/m³ (30 lb/ft³). Source: Central America, West Indies. Used in construction, for furniture, cabinetwork, joinery, interior trim, boatbuilding, cigar boxes, foundry patterns, and as a mahogany replacement.

Centralloy. Trade name of Central Iron & Steel Company (USA) for a high-strength steel containing 0.15% carbon, 0.8% nickel, 0.2% chromium, and the balance iron. Used for bus and railroad bodies.

centrally mixed concrete. Concrete that is blended in a stationary mixer and then delivered in agitators to the building site. Also known as *central-mixed concrete*.

Centrard. Trade name of Sheepbridge Engineering Limited (UK) for a nitrogen-hardened alloy cast iron containing 2.7% carbon, 1.5-1.75% chromium, 1.5-1.75% aluminum, and the balance iron. It has excellent abrasion and wear-resistance, and is used for cylinder liners, and general parts to resist severe abrasion.

Centra Steel. Trade name of General Motors Corporation (USA) for a steel containing 1.7% carbon, 2.25% silicon, 0.4% magnesium, 0.1% sulfur, 0.05% phosphorus, 0.01% boron, and the balance iron. It has good wear resistance, and is used for crankshafts, gears, agricultural equipment, and castings.

Centrex. Trademark of Bayer Corporation (USA) for a series of acrylic-styrene-acrylonitriles, acrylic ethylene acrylonitriles, and various blends of these resins. They are supplied in general-purpose, extrusion, weather-resistant and automotive grades.

Centrex Inducto-Chrome. Trade name of Ludlow Steel Company (USA) for induction-hardened hard chrome plates.

Centricast. Trade name of Sheepbridge Engineering Limited

(USA) for an extensive series of cast irons shaped by centrifugal casting, and used for piston rings, cylinder liners, shafts, rollers, and automotive engine components.

centrifugal castings. (1) Accurate, good-quality cylinderical ferrous or nonferrous castings made by a process in which molten metal is poured into rapidly revolving or rotating molds.

(2) Cylindrical thermoplastics, such as pipes, made by introducing granular resins into heated, rotating or revolving containers.

(3) Cylindrical composites, such as pipes, made by introducing synthetic resins into hollow, heatable, rotating or revolving mandrels containing chopped strand mats.

centrifugally cast concrete. See spun concrete.

centrifugally spun fibers. Synthetic fibers produced by centrifugally throwing a molten or dissolved fiber-forming polymer off the edge of a high-speed rotating disk.

centrifuged rubber latex. Rubber latex that has an increased concentration due to the complete or partial removal of the dispersion medium by rotating or revolving in a suitable container.

Century. (1) Trade name of CCS Braeburn Alloy Steel (USA) for a tool steel containing 1.0% carbon, 4.0% chromium, 8.0% molybdenum, 0.8% tungsten, 2.0% vanadium, and the balance iron. Used for cutting tools, dies, shear blades, etc.

(2) Trade name of LTV Steel Corporation (USA) for a high-strength steel supplied in the form of cold-drawn bars, e.g., flats, hexagons, rounds and squares.

(3) Trade name of G.J. Nikolas & Company Inc. (USA) for an air-drying lacquer for brass, bronze and silver parts.

Ceot. Trade name of Saarstahl AG (Germany) for a hot-work tool steel containing 0.45% carbon, 1.4% chromium, 0.7% molybdenum, 0.3% vanadium, and the balance iron. Used for extrusion-press tools.

Cerablanket. Trademark of Morgan Crucible Company plc (UK) for a refractory ceramic fiber insulation used in heat-treating furnaces.

CeraBond. Trademark of Manville Corporation (USA) for a spray-applied refractory-fiber insulation coating for furnace lining applications to 1350°C (2460°F).

Cerac. Trade name of Cerac Incorporated (USA) for an extensive range of high-quality products supplied in commercial and high purities. Included are various metals and alloys, rare earths, refractories, intermetallics, salts, nitrides, oxides, phosphides, selenides, sulfides and tellurides. Also included are sputtering targets of varying size and shape, and evaporation materials supplied as coarse or fine powders, granules, tablets, pellets and various other forms.

Cerachrome. Trade name of Manville Corporation (USA) for ceramic fibers composed of 42.5% alumina (Al_2O_3), 55% silica (SiO_2), and 2.5% chromia (Cr_2O_3), and supplied in discontinuous, bulk and mat forms.

Cer-A-Cote. Trade name of Amortek Industries, Inc. (USA) for ceramic coatings used on printing equipment.

Ceradelta. Trade name of Metalor Technologies SA (Switzerland) for a series of precious metal dental alloys.

Cerafiber. Trade name of Manville Corporation (USA) for ceramic fibers composed of 47% alumina (Al_2O_3) and 53% silica (SiO_2), supplied in discontinuous, bulk, and mat form.

Ceraflex. Trade name of Ceradyne Inc. (USA) for a series of *yttria-stabilized zirconia* ceramics available in sheet form for use in sensors, solid-oxide fuel cells (SOFCs), as substrates, and for applications requiring thin ceramics.

Ceraglas. Trade name of London Sandblast Decorative Glass

Works Limited (UK) for a cladding glass with a fired ceramic enamel fused onto its reverse side.

Ceralbond-BN. Trade name of Orpac Inc. (USA) for an alumina-bonded boron nitride with high electrical resistance and good thermal conductivity. It machines like graphite, is inert and nonwetting to molten aluminum and magnesium, and can be modified by adding reinforcing fibers, etc. Used in handling equipment for molten metals, glasses and salts.

Ceralloy. (1) Trademark of Ceradyne, Inc. (USA) for a series of tough, wear-resistant, hot-pressed ceramics based on aluminum and beryllium oxides, boron and silicon carbides, silicon nitride, thorium oxide, titanium diboride, etc. They have good elevated-temperature properties, and high tensile moduli and compressive strength. Used for turbine components, friction plates, bearing seals, hydraulic valves, wire guides, sandblast nozzles, rocket nozzles, microwave absorbers, tooling and hot-pressing dies, metallurgical refractories, nuclear ceramics, and other technical ceramics.

(2) Trade name of Ronson Metals Corporation (USA) for a series of materials composed of cerium-rich, mischmetal-type rare-earth metal mixtures alloyed with iron, aluminum or magnesium. Used as additives, desulfurizers, nodulizers, etc., in the manufacture of steel and ductile cast iron, in the manufacture of magnesium and other nonferrous alloys, and as getters in vacuum tubes.

Ceralox. Trademark of Ceralox Corporation (USA) for an extensive range of aluminas and alumina products including alpha alumina, gamma alumina, magnesium aluminate spinel and yttrium aluminate, supplied in various forms including corase and fine powders, and granules.

Ceralumin. Trade name of Stone Manganese–J. Stone & Company Limited (UK) for aluminum casting alloys containing 1-4.5% copper, 0.7-2.5% silicon, 0.25-2% nickel, 0.1-2.5% magnesium, 0-1.5% iron, 0-0.3% niobium, 0.2% titanium, and 0-0.3% cerium. They have good castability, and improved mechanical properties, and are used for cylinder heads, impellers, gear casings, and low-stressed parts.

Cerama-bond. Trademark of Aremco Products Inc. (USA) for alumina, silica and magnesia ceramic coatings.

Cerama-cast. Trademark of Aremco Products Inc. (USA) for ceramic oxide and carbide products including alumina, magnesia and zirconia.

Cerama-Dip. Trademark of Aremco Products, Inc. (USA) for ceramic silica coatings.

Cerama-Fab. Trademark of Aramco Products, Inc. (USA) for ceramic alumina and silica coatings.

Ceramag. Trademark of Stackpole Corporation (USA) for a ferrimagnetic *barium ferrite* ceramic with excellent soft magnetic properties, available in the form of coated and uncoated toroids, E-cores, pot cores, and special shapes. It has high permeability and low loss, and is used for EMI/RFI filter applications, wideband transformers, pulse transformers, antenna cores, etc.

ceramagnet. A hard ceramic magnet material composed of ferrimagnetic barium ferrite ($BaO \cdot 6Fe_2O_3$) with a Curie temperature of approximately 450°C (840°F), high coercive force and energy product, and a maximum service temperature of 400°C (750°F). Used for permanent magnets.

ceramals. See cermets.

Ceramalite. Trade name of C-E Glass (USA) for a ceramic-coated, toughened glass for spandrels.

Ceramalloy. Trade name for a corrosion-resistant cobalt-chromium dental bonding alloy.

Ceramapearl. Trade name for a castable glass-ceramic for dental applications.

Ceramapot. Trademark for an extensive series of ceramic potting materials used in the high-temperature encapsulation of electronic and electrical components. It includes dispensable aluminum oxide ceramic compounds, such as *Ceramapot 575* with outstanding thermal conductivity, magnesium oxide materials, such as *Ceramapot 583,* that maintain good dielectric properties at elevated temperatures, and fast-setting magnesium oxide compounds, such as *Ceramapot 584.*

Ceramarc. Trademark of Air Products and Chemicals, Inc. (USA) for abrasion- and wear-resistant coatings deposited onto metal and nonmetal substrates by arc-thermal spraying. *Ceramac 2000* is an amorphous iron-chromium material for arc-spray coatings produced by a process that utilizes a controlled inert atmosphere to atomize molten droplets. The coatings have reduced oxide levels, higher density and lower porosity than those produced with conventional compressed-air atmospheres and greater film thicknesses than high-velocity oxyfuel (HVOF) and plasma-sprayed coatings. Other properties include outstanding wear resistance, a very high hardness (1200-1400 Vickers). Used for boiler tubes, hammer-mill screens, impact-mill liners, extruder barrels, and wear plates.

Ceramcast 646. Trademark of Aremco Products, Inc. (USA) for a series of castable ceramics with excellent high-temperature properties, and good dielectric properties. Used for tooling applications, e.g., brazing fixtures, and in the encapsulation of thermocouples and electrical feedthroughs.

Ceramcem. Trade name for a glass-ionomer dental cement.

Ceramco. (1) Trade name of Degussa-Ney Dental (USA) for a dental porcelain for jacket crowns.

    (2) Trade name of Degussa-Ney Dental (USA) for a high-gold dental bonding alloy.

Ceramco II. Trade name of Degussa-Ney Dental (USA) for a dental porcelain for metal-to-ceramic bonding.

Ceramco White. Trade name of Degussa-Ney Dental (USA) for a microfine-grained dental bonding alloy containing 50.3% gold, 30.2% palladium and 14.4% silver. It has good yield strength, high corrosion and tarnish resistance, a hardness of 220 Vickers, and is suitable for use with most dental porcelains. Used in restorative dentistry for single crowns, and short-span bridges.

Ceram-Core. Trade name for a glass-ionomer dental cement.

Ceramelec. Trademark of Erie Technological Products, Inc. (USA) for barium titanate and other piezoelectric ceramics.

ceramers. A group of inorganic-organic ceramic/polymer hybrid materials used for porous membranes, porous insulators for noise reduction, in the synthesis of biomaterials, in dentistry, as biosensors, and for the control of electrical, optical, mechanical and/or thermal properties of devices.

ceramets. See cermets.

CERAMETiL. Trademark of Lucas-Milhaupt Inc. (USA) for a 100 mm (4 in.) wide strip product consisting of a titanium-bearing coating bonded to high-purity brazing filler metal of vacuum-tube grade. Used for sealing or bonding ceramic and other nonmetallic materials, such as aluminum oxide, silicon nitride, titanium carbide, zirconium dioxide, diamond, sapphire or graphite, to themselves and other materials.

Ceramfil. Trade name for a series of glass-ionomer dental cements including *Ceramfil ART, Ceramfil B* and *Ceramfil Seasons.*

Ceram-Guard. Trademark of A.O. Smith Corporation (USA) for

oxidation-resistant ceramic frits and coatings for metals, and steel tools and dies providing controlled scaling, increased yield, die-lubricant and penetration control, and controlled decarburization.

ceramic adhesives. A group of adhesives based on ceramic materials, such as porcelain enamel, alumina, or zirconia, but excluding cementitious materials. They often consist of refractory fibers and inorganic binders, and have good corrosion, erosion and oxidation resistance, good high-temperature bond strength, and good electrical resistance. Used for high-temperature-resistant metal joints, for bonding ceramics to ceramics, metals, glass, plastics, etc., for bonding electrical components, appliances, heaters and lamps, and in potting compounds.

ceramic aggregate. Concrete containing either porous clay or lumps of ceramic materials.

Ceramicast. Trademark of Lebanon Steel Foundry (USA) for steel and alloy castings with high dimensional accuracy and good surface finish shaped by pouring into a ceramic mold.

ceramic balls. Hard, wear-resistant spheres of alumina, zirconia, porcelain, etc., used chiefly as media in grinding mills, and as fillers in composites. Also known as *ceramic spheres.*

ceramic board. A ceramic material, such as alumina, zirconia or plaster, supplied in the form of a flat, thin square or rectangular slab. Used chiefly in building construction.

ceramic-carbon refractories. Refractories manufactured from a mixture of carbon or graphite and a refractory ceramic material, such as fireclay, silicon carbide, etc. Also known as *carbon-ceramic refractories.*

ceramic-ceramic composites. Composite materials consisting of ceramic matrices reinforced with ceramic fibers, such as aluminum oxide or silicon carbide. Used for high-temperature engineering applications. See also ceramic-matrix composites.

ceramic coatings. Inorganic, nonmetallic coatings based on borides, carbides, nitrides, oxides, silicides, cermets superporcelains or special glazes bonded to ceramic or metallic substrates to protect them against high temperatures, oxidation and corrosion. They may also provide chemical, electrical and wear resistance, and/or high reflectivity and protection against hydrogen diffusion.

ceramic colorant. See color oxide.

ceramic color glaze. An opaque glaze produced on a clay body by first spraying with a compound consisting of metallic oxides, clay and selected chemicals, and then fusing by firing at high temperatures. The resulting glaze has a satin or glossy finish, and strongly adheres to the body.

ceramic composites. (1) Composite materials consisting of ceramic matrices reinforced with metal fibers or wires, carbon or graphite fibers, or ceramic fibers or particulate. See also ceramic-matrix composites.

    (2) Composites, such as concrete, composed of aggregate, cement, water, and sometimes admixtures and/or additives.

ceramic enamel. See ceramic glass enamel.

ceramic felt. A material made by interfelting ceramic fibers, and usually available in the form of mats or sheets. Used for thermal insulation and high-temperature applications.

ceramic-fiber ceramic-matrix composites. Engineering composites with greatly enhanced toughness and good strength properties at elevated temperatures consisting of ceramic matrices (e.g., aluminum oxide, magnesium oxide, silicon carbide, silicon nitride, glass or glass-ceramics) reinforced with ceramic fibers or whiskers (e.g., silicon carbide, silicon nitride or zirconium oxide). Used for cutting tools, and for applications requiring resistance to severe stress and temperature conditions.

ceramic fibers. Fibers that are formed from ceramic materials and can be subdivided into the following two broad categories: (i) Continuous and discontinuous oxide fibers, such as alumina, alumina-silica, alumina-boria-silica, zirconia-silica, fused-silica and leached-glass fibers; and (ii) Continuous and discontinuous carbide or nitride fibers, such as silicon or zirconium carbide fibers, silicon nitride fibers, and fibers containing carbide mixtures, such as silicon carbide with titanium carbide or zirconium carbide, etc. *Ceramic fibers* generally have excellent mechanical properties and thermal stability, and good electrical, acoustic and thermal insulation properties, and are used for acoustic and thermal insulation applications, as reinforcements for metal, plastic and ceramic materials, in filtration, packing, etc. Abbreviation: CF.

ceramic foams. Ceramic materials, such as alumina, mullite or silica, having spongelike, cellular structures due to gaseous cells purposely introduced and distributed throughout. They are often available in various degrees of porosity, and may be readily machinable. *Ceramic foams* have excellent high-temperature resistance, outstanding thermal-shock resistance and good thermal insulating properties. Used for high-temperature filters, catalyst supports, gas diffusion, thermal insulation, fixtures, plates, boats, etc. Also known as *cellular ceramics; cellulated ceramics.*

ceramic fuel element. A fuel rod made of uranium or plutonium oxide, boron compounds, rare-earth oxides, etc., and used in nuclear reactors.

ceramic glass enamel. A finely ground mixture of low-melting fluxes, calcined ceramic pigments and a suitable vehicle that may be applied to glassware and fired at a temperature between 500 and 760°C (930 and 1400°F) to a smooth, hard coating. Also known as *ceramic enamel; glass enamel.*

ceramic glaze. A ceramic coating applied to a ceramic article and matured to a vitreous state. Depending on the coating composition and application, the finish may be glossy or mat.

ceramic hard materials. A broad class of ceramic materials and coatings with exceptionally high hardness and usually a good combination of corrosion, friction and wear resistance. Examples include natural and synthetic diamond, diamond-like carbon, tungsten carbide/cobalt cemented carbides, cubic boron nitrides, aluminum oxide materials, silicon carbide- and silicon nitride-based materials and various nanostructured materials. Used for cutting tools and wear parts, as coatings and thin films on electronic equipment, etc.

ceramic insulator. An electrically insulating material composed of one, or a combination of several ceramic materials, such as alumina, beryllia, cordierite, electrical porcelain, silica, steatite, zircon, etc.

ceramic magnets. Engineering ceramics with low-conductivity or superconducting characteristics. Low-conductivity magnetic ceramics include ferrites and garnets. *Ferrites* are fixed mixtures of Fe_2O_3 and appropriate compounds of divalent metals, such as barium, cobalt, copper, lead, magnesium, manganese, nickel, strontium or zinc, and have crystal structures that can be represented by the general formula $MO \cdot Fe_2O_3$, where M represents a metal. They exhibit ferromagnetic, ferrimagnetic and antiferromagnetic, magneto-optical and magnetostrictive properties, and are used in antennas, computer-memory cores, computer disks, recording tapes, telecommunications systems, etc. *Garnets* are ferrimagnetic ceramic materials having very complicated crystal structures that can be represented by the gen-

eral formulas $M_3Fe_5O_{12}$ and $M_3Al_5O_{12}$, respectively where M usually represents a rare-earth element, such as dysprosium, erbium, europium, gadolinium, holmium, samarium, terbium, thulium, ytterbium or yttrium. They are used in solid-state electronics, lasers, microwave devices, etc. *Superconducting magnetic ceramics,* such as yttrium barium copper oxide (YBCO), bismuth strontium calcium copper oxide (BSCCO), lanthanum copper oxide (LCO), etc., are used for thin-film devices, and in computers and magnetic detectors. Also known as *ceramic magnet materials; magnetic ceramics.* See also ceramic permanent magnets.

ceramic mass-finishing media. Abrasive media, such as preformed porcelain, fused and sintered alumina, etc., used in mass-finishing processes, such as barrel, vibratory, centrifugal-disk and spindle finishing.

ceramic materials. See ceramics.

ceramic-matrix composites. A group of high-temperature composites consisting of ceramic matrices (e.g., aluminum oxide, silicon carbide, silicon nitride, glass or glass-ceramics) reinforced with metal fibers or wires (e.g., molybdenum, tungsten, tantalum or niobium), carbon or graphite fibers, or ceramic fibers or particulate (e.g., alumina or zirconia.). Abbreviation: CMC. See also ceramic composites.

ceramic-metal coatings. See cermet coatings.

ceramic mosaic tile. An unglazed tile composed of clay or porcelain, shaped by plastic forming or dust pressing, and usually mounted on paper sheets. The facial area is less than 232 cm² (6 in.²), and the thickness ranges from 6.4 to 9.5 mm (0.250 to 0.375 in.).

ceramic nuclear fuel. See cermet nuclear fuel

ceramic oxides. Compounds that are composed of one or more metal oxides and correspond to any of the various ceramic crystal structures. Examples include oxides with (i) rock salt (halite) structures, such as CaO, MgO, FeO, NiO and SrO; (ii) fluorite structures, such as CeO_2, SiO_2, TeO_2, ThO_2 and UO_2; (iii) corundum structures, such Al_2O_3, Cr_2O_3 and α-Fe_2O_3; (iv) perovskite structures, such as $CaTiO_3$, $BaTiO_3$, $SrZrO_3$ and $SrSnO_3$; or (v) spinel structures, such as $MgAl_2O_4$, $FeAl_2O_4$, $NiAl_2O_4$, $ZnAl_2O_4$ and $ZnFe_2O_4$.

ceramic paper. A paper made by forming ceramic fibers (e.g., quartz, alumina or sapphire) into mats or sheets with or without the addition of a bonding agent. It can easily be cut and formed into complex shapes, and is supplied in various widths (typically up to 1270 mm or 50 in.) and thicknesses (typically from 0.5 to 9.5 mm (0.020 to 0.375 in.). It has a low weight, good resistance to thermal fatigue, high flexibility, high tensile strength, and continuous-use temperatures up to 1650°C (3000°F). Used for electrical insulation, high-temperature insulation and gaskets, in combustion chambers, heat-treating and metallurgical furnaces, and for automotive applications including air bags, hoods, heat shields for exhaust pipe insulation and gas tanks.

ceramic pebbles. Hard, dense, tough rounded stones of a ceramic material, such as alumina or porcelain, used in mills as media for grinding cement, ores, minerals, etc.

ceramic permanent magnets. Ceramic magnet materials, such as barium, lead or strontium ferrite, made by sintering blended powders. They usually have hexagonal crystal structures, and high remanence, coercivity and saturation flux density as well as low initial permeabilities, and high hysteresis energy losses. Used for small dc motors, loudspeakers, and magnetic door latches.

ceramic putty. A substance of putty-like consistency made of ceramic materials, such as alumina, and selected ceramic binders that on drying, produces a strong ceramic body.

Ceramicrete. Trade name of Bindan Corporation (USA) for a ceramic material composed of a mixture of metal oxides and phosphate salts that cures exothermically in 1-2 hours without firing. Being harder and denser than Portland cement, it is used for repairing roads, protecting high-rise steel beams from fires, and making terrazzo tiles.

ceramics. A class of products composed of metallic and nonmetallic elements with mixed ionic/covalent bonding. Traditionally, this includes ware formed from earthy raw materials, such as clay or other silicate-bearing materials, and hardened by firing in a kiln. However, the term now also includes a variety of materials that do not contain silicates. *Ceramics* may classified into the following general categories: (i) *Structural clay products* including brick and tile; (ii) *Refractory materials* including fireclay, magnesite and dolomite; (iii) *Cementitious materials* including cement, lime, plaster and gypsum; (iv) *Pottery;* (v) *Glass* and related products; (vi) *Porcelain enamels and glazes;* (vii) *Abrasives* including corundum, flint, emery and silicon carbides; (viii) *Whitewares* including dinnerware, and chemical and electrical porcelains; and (ix) *Advanced ceramics* including aluminum oxide, aluminum nitride, boron carbide, boron nitride, silicon carbide, titanium diboride and zirconium oxide. In general, ceramics are relatively hard and brittle, and more resistant to high temperatures and severe environments than either metals or polymers. Also known as *ceramic materials.*

ceramic sealing alloys. See sealing alloys.

ceramic spheres. See ceramic balls.

ceramic stain. A ceramic color composed of a transition metal and one or more other elements and applied to a body, enamel or glaze as an addition to the body, enamel or glaze composition.

ceramic superconductors. See high-temperature superconductors.

ceramic veneer. A type of architectural *terracotta* held in place by the adhesion of a mortar, or by anchors connected to the backing wall.

ceramic whiteware. See whiteware.

Ceramiseal. Trade name of Wilbur B. Driver Company (USA) for an iron-base superalloy containing 48% iron, 27% nickel, and 25% cobalt. It has a melting point of 1420°C (2590°F), and is used for high-temperature applications, and ceramic-to-metal seals.

Ceram-Kote 54. Trademark of Freecom, Inc. (USA) for a barrier coating composed of submicron ceramic particles in a synthetic resin system with very high adhesion to a wide range of substrates, outstanding chemical and sliding abrasion resistance, and high surface lubricity.

Ceramlin. Trade name for a glass-ionomer dental cement.

Ceramlite. Trade name for a resin-modified glass-ionomer dental cement.

Ceramol. Trademark of Foseco Minsep NV (Netherlands) for core and mold dressings in paste or solution form.

ceramoplastic. A high-temperature ceramic insulator made by bonding synthetic mica (e.g., muscovite or phlogopite) with glass. Also known as *ceramoplastic insulator.*

CeramSave. Trade name for a glass-ionomer dental cement.

Ceram-T. Trade name of Accumet Materials Company (USA) for alumina-titania composites that can be slip-cast into standard and complex shapes. They have excellent thermal-shock resistance up to 1800°C (3270°F), and good mechanical properties. Used for thermal insulation applications.

Ceram-Tuff. Trademark of A.O. Smith Corporation (USA) for a thin, inorganic protective coating with a matte finish for metals providing corrosion and abrasion resistance up to 650°C (1200°F) for long periods, and up to 816°C (1500°F) for short periods. It can be applied to a wide range of substrates, such as aluminum, copper, iron, steel and titanium. Used on appliances, fume control ducts, stoves, pipes, exhaust manifolds, mufflers, ferrous castings, etc.

Ceramvar. Trade name of Wilbur B. Driver Company (USA) for an alloy of 48% iron, 27% nickel, and 25% cobalt. It has expansion properties similar to those of alumina, and is used for ceramic-to-metal seals.

Ceran. Trademark of Schott Glas (Germany) for glass-ceramics and specialty heat-resistant glass used for range and stove tops, heating plates, etc.

Cerap. Trade name of Metalor Technologies SA (Switzerland) for a series of precious-metal dental alloys.

Cerapall. Trade name of Metalor Technologies SA (Switzerland) for a series of palladium-based precious metal dental alloys.

cerargyrite. A colorless, white or pale gray mineral of the halite group with adamantine luster that turns violet-brown in light. It is composed of silver chloride, AgCl, and contains about 75% silver. Crystal system, cubic. Density, 5.55-5.57 g/cm^3; hardness, 2-3 Mohs; refractive index, 2.071. Occurrence: Chile, Mexico, Peru, USA (Colorado, Idaho, Nevada). Used as an ore of silver. Also known as *chlorargyrite; horn silver.*

Cerasil. (1) Trademark of GBC Materials Corporation (USA) for aluminum oxide ceramics.

(2) Trademark of Unimin Specialty Minerals, Inc. (USA) for ground, microcrystalline silica (SiO_2) used in the manufacture of whitewares, and in ceramic body and glaze formulations.

Cerasouple. Trade name of Ato Findley (France) for paste adhesives.

Cerastar. Trade name of The Carborundum Company (USA) for reaction-bonded silicon carbide ceramics.

Ceratherm. Trade name of Coors Ceramics Company (USA) for a family of high-performance ceramic products based on alumina, mullite or silicon carbide, and supplied as molten-metal immersion tubes, pump and chemical-equipment linings, recuperators, radiant tubes, oxygen sensors, and high-temperature electrical insulators.

CeraTrex CVD. Trade name of Ceradyne Inc. (USA) for fully-dense silicon carbide (SiC) produced by chemical vapor deposition at much higher deposition rates than conventional products of this type.

Ceravital. Trademark of Wild Leitz GmbH (Germany) for a series of bioactive glass-ceramics composed of about 38-47% silicon dioxide (SiO_2), 20-33% calcium oxide (CaO), 13-26% calcium metaphosphate [$Ca(PO_3)_2$)], 4-5% sodium oxide (Na_2O) and sometimes additions of aluminum, magnesium or potassium oxide. Used in human body-part reconstruction and repair, e.g., bone replacements, dental implants, dentures, coatings for dental or orthopedic prostheses, etc.

Cerawool. Trade name of Manville Corporation (USA) for ceramic fibers composed of 50% silicon dioxide (SiO_2), 40% aluminum oxide (Al_2O_3), 5% calcium oxide (CaO), 3.5% magnesium oxide (MgO) and 1.5% titanium dioxide (TiO_2), and supplied in discontinuous, bulk and mat form. Used chiefly acoustic and thermal insulation applications.

CERBEC. Trademark of St. Gobain Industrial Ceramics (France) for hard, tough, corrosion-resistant silicon nitride ceramic balls used in high-precision bearings.

Cercast. Trademark of Cercast-Howmet Corporation (USA) for aluminum-alloy precision investment castings.

Cercodim. Trademark of Institut Straumann AG (Switzerland) for bioceramic coatings based on alumina and zirconia, and used for knee prostheses and dental implants.

Cercor. Trademark of Corning Glass Works (USA) for cellular ceramic structures based on glass-ceramics.

cereal. See cereal binder.

cereal binder. An organic binder, usually corn flour, used in foundry-core mixtures. Also known as *cereal*.

cereal flour. A finely milled corn flour used as an organic binder in foundry-core mixtures.

cerebrose. See galactose.

Cerec Dicor MGC. Trade name of Dentsply Corporation (USA) for a machinable glass-ceramic used for dental restorations.

Cerec Duo. Trade name of Dentsply Corporation (USA) for a dual-cure dental resin cement used for luting restorations.

ceresin. A white to yellow waxy cake composed of high-molecular-weight hydrocarbons and obtained from purified *ozocerite* treated with sulfuric acid. It has a density of 0.92-0.94 g/cm³ (0.033-0.034 lb/in.³), a melting point of 68-72°C (154-162°F), and is used in lubricants, polishes, antifouling paints, waxed paper, candles, sizing, waterproofing and impregnating agents, electrical insulation, and floor waxes. Also known as *ceresine; ceresin wax; purified ozocerite*.

Cerex. (1) Trademark of Cerex Inc. (USA) for polystyrene resins with good electrical, heat and stain resistance, high hardness, good stability and moldability, poor resistance to ultraviolet radiation, fairly high tendency to load cracking, good tensile strength, impact strength and moduli of elasticity, and a maximum service temperature of 50-100°C (120-210°F). Used for automotive components, appliance parts, piping, battery boxes, dials, knobs, high-frequency insulation, and dinnerware.

(2) Trademark of Cerex Advanced Fabrics, LP for (USA) spun-bond nonwoven nylon fabrics.

Cergogold. Trademark of Degussa-Ney Dental (USA) for a pressable dental ceramic with high abrasion resistance, good polishability, a smooth surface, and a hardness of 450 Vickers. Used for hydrothermal all-ceramic restorations, especially crowns and inlays.

ceria. See cerium oxide.

ceria catalysts. See cerium oxide catalysts.

ceria coatings. See cerium oxide coatings.

cerianite. A light gray, greenish-yellow, or yellowish-brown mineral of the fluorite group composed of cerium oxide, CeO_2. It can also be made synthetically Crystal system, cubic. Density, 7.22 g/cm³. Occurrence: Canada (Ontario).

ceric hydroxide. A hydrated oxide that contains about 85-90% ceric oxide (CeO_2), and is usually supplied in the form of a dry white or yellowish-white powder. Used as an opacifier in glasses and enamels, as a yellow colorant for glasses and ceramics, and in the manufacture of shielding glass. Formula: $Ce(OH)_4$. Also known as *cerium hydrate; cerium hydroxide*.

ceric oxide. See cerium oxide.

ceric rare earths. See light rare earths.

cerics. See light rare earths.

ceric sulfide. See cerium sulfide.

ceric titanate. A compound of ceric oxide and titanium dioxide used as colorant and opacifier for ceramics and glasses (e.g., to impart a golden-yellow color). Formula: $CeTi_2O_6$.

ceric zirconate. A compound of ceric oxide and zirconium oxide available as an off-white powder for use in ceramics, and as colorant and opacifier for ceramics and glasses. Formula: $CeZrO_4$.

Cerinate. Trademark of Den-Mat Corporation (USA) for a translucent feldspathic porcelain-composite with high strength and low thermal expansion that does not exhibit cracking or crazing due to polymer shrinkage. Used in dentistry for porcelain laminates, inlays, onlays, crowns, bridges, and splints.

cerite. A brown mineral composed of calcium lanthanum silicate hydroxide hydrate, $(Ca,Mg)_2(La,Ce)_8(SiO_4)_7(OH,H_2O)_3$. Crystal system, rhombohedral (hexagonal). Density, 4.78-4.86 g/cm³; refractive index, 1.806; hardness, 5.5 Mohs. Occurrence: USA (California), Sweden. Used as an ore of cerium and lanthanum.

Ceri-Temp. Trade name for a dental resin for temporary crown and bridge restorations.

cerium. A steel-gray, ductile metal belonging to the lanthanide series (rare-earth group) of the Periodic Table. It is commercially available in the form of chips, turnings, granules, powder, lumps, ingot, wire, sheets, foils and rods. The commercial cerium ores are *monazite, cerite* and *bastnaesite*. Density, 6.75 g/cm³; melting point, 799°C (1470°F); boiling point, 3426°C (6199°F); hardness, 25-30 Vickers; atomic number, 58; atomic weight, 140.116; trivalent, tetravalent; readily oxidizes at room temperature in air. Four phases are known: (i) *Alpha cerium* with face-centered cubic crystal structure, present at room temperature; (ii) *Beta cerium* with hexagonal close-packed crystal structure, present at temperatures below 250K (-23.15°C); (iii) *Gamma cerium* with face-centered cubic crystal structure, present below 110K (-163.15); and (iv) *Delta cerium* with body-centered cubic crystal structure, present at temperatures between 730 and 799°C (1346 and 1470°F). *Cerium* is used as a deoxidizer and desulfurizer in ferrous alloys, as an alloying addition to heat-resisting steel to enhance nonscaling properties and to high-alloy steel to improve hot working properties, as a nodulizer in cast irons, as a reducing agent (scavenger), in superalloys to improve resistance to high-temperature oxidation, as a constituent in ceralumin (an aluminum-base alloy), with iron in pyrophoric alloys and pyrophoric ignition and ordnance devices, in lighter flints, glass-polishing compounds and glass-decolorizing agents, in the production of cerium oxides for incandescent gas mantles, in high-current-density arc carbons, jet-engine alloys, catalytic converters, petroleum cracking catalysts, as an illuminant in photography, in fluorescent tubes, solid-state devices and ceramic capacitors, as a getter in vacuum tubes, in cerium-cobalt ($CeCo_5$) permanent magnets, superconductors and rocket propellants, and as a diluent in plutonium nuclear fuels. Symbol: Ce.

cerium-140. A natural or stable isotope of cerium with a mass number of 140 that constitutes 88.4% of the element cerium. Used in the manufacture of cerium-141. Symbol: ^{140}Ce.

cerium-141. A radioactive isotope of cerium with a mass number of 141 obtained from cerium-140 by neutron capture and gamma photon emission. It has a half-life of about 21.5 days, and emits beta and gamma rays. Used in research. Symbol: ^{141}Ce.

cerium-142. A natural or stable isotope of cerium with a mass number of 142 that constitutes 11.1% of the element cerium. Symbol: ^{142}Ce.

cerium aluminate. A compound of ceric oxide and aluminum oxide with a density of 6.17 g/cm^3 and a melting point of 2704°C (4900°F). Used in ceramics. Formula: CeAl$_2$O$_5$.

cerium bismuthide. Any of the following compounds of cerium and bismuth used in ceramics and materials research: (i) CeBi. Melting point, 1527°C (2781°F); and (ii) Ce$_4$Bi$_3$. Melting point, 1604°C (2781°F).

cerium boride. Any of the following compounds of cerium and boron used in ceramics and materials research: (i) *Cerium tetraboride*. Density, 5.75 g/cm^3. Formula: CeB$_4$; and (ii) *Cerium hexaboride*. Density, 4.82 g/cm^3; melting point, 2190°C (3974°F). Formula: CeB$_6$.

cerium bromide. See cerous bromide.

cerium carbide. Any of the following compounds of cerium and carbon used in ceramics and materials research: (i) *Cerium dicarbide*. Red crystalline powder. Density, 5.2-5.5 g/cm^3; melting point, 2540°C (4604°F). Formula: CeC$_2$; and (ii) *Cerium sesquicarbide*. Density, 6.95-6.99. Formula: Ce$_2$C$_3$.

cerium carbonate. A white crystalline powder that is usually supplied in hydrated form (xH$_2$O), e.g., as the pentahydrate. Used in inorganic and organic synthesis. Formula: Ce$_2$(CO$_3$)$_3$ (anhydrous); Ce$_2$(CO$_3$)$_3$·xH$_2$O (hydrated). Also known as *cerous carbonate*.

cerium chloride. See cerous chloride.

cerium chromite. A compound of ceric oxide and chromic oxide with a melting point of 2438°C (4420°F) used in ceramics. Formula: CeCr$_2$O$_5$.

cerium-cobalt. A permanent-magnet material consisting of a compound of cobalt and cerium in a ratio of 5:1 (CeCo$_5$). Commercially available in cast and sintered form, it has high permanency, high induction values, high coercive force, and a very high magnetic energy product. Used in the manufacture of cerium-cobalt magnets for computers and other electronic devices.

cerium copper. An alloy composed of mischmetal (a mixture of rare-earth metals) and copper.

cerium dicarbide. See cerium carbide (i).

cerium dioxide. See cerium oxide.

cerium-doped barium titanate. A high-purity crystalline material made by doping barium titanate (barium titanium oxide) with cerium (Ce^{3+}) ions. It has excellent photorefractive properties owing to the formation of a self-pumped phase conjugate (SPPC) wave. Used in SPPC mirrors. Formula: Ce:BaTiO$_3$. Abbreviation: BTO:Ce. See also barium titanate.

cerium-doped gadolinium silicate. Gadolinium silicate (gadolinium silicon oxide) doped with cerium (Ce^{3+}) ions and supplied as high-purity single crystals. Crystal system, monoclinic. Density, 6.71 g/cm^3; hardness, 5-7 Mohs. Used as a scintillator crystal in spectroscopy, computer tomography, and gamma radiometry. Formula: Ce:Gd$_2$SiO$_5$. Abbreviation: GSO:Ce. See also gadolinium silicate.

cerium-doped yttrium-aluminum garnet. Yttrium-aluminum garnet (Y$_3$Al$_5$O$_{10}$) doped or activated with cerium (Ce^{3+}) that has improved chemical and mechanical properties over the unactivated material, and is used for ultrathin scintillation screens, imaging screens, etc. It may also contain other trivalent dopants, such as chromium, holmium, erbium or neodymium. Crystal system, cubic. Density, 4.57 g/cm^3; melting point, 1970°C (3578°F); hardness, 8.5 Mohs. Abbreviation: YAG:Ce. See also yttrium-aluminum garnet.

cerium-doped yttrium-aluminum perovskite. Yttrium-aluminum perovskite (YAlO$_3$) doped or activated with cerium (Ce^{+3}) ions. It is nonhygroscopic, and has improved chemical and mechanical properties over the unactivated material. Crystal sys-

tem, rhombohedral. Density, 5.37 g/cm^3; melting point, 1875°C (3407°F); hardness, 8.5+ Mohs. Used for entrance windows, scintillator crystals in gamma and X-ray counters, electron and X-ray imaging screens, and tomography units. It may also contain other trivalent dopants, such as cerium, erbium, neodymium or thulium. Abbreviation: YAP:Ce. See also yttrium-aluminum perovskite.

cerium fluoride. See cerous fluoride.

cerium-free mischmetal. A mixture of rare-earth metals, such as lanthanum, neodymium, praseodymium, samarium, dysprosium, holmium, etc., but, unlike standard *mischmetal*, without cerium. Abbreviation: CFM.

cerium hexaboride. See cerium boride (ii).

cerium hydrate. See ceric hydroxide; cerous hydroxide.

cerium hydroxide. See ceric hydroxide; cerous hydroxide.

cerium iodide. See cerous iodide.

cerium metal. (1) A relatively pure cerium used for alloying purposes.

(2) See light rare earths.

cerium monoselenide. See cerium selenide (ii).

cerium monotelluride. See cerium telluride (ii).

cerium naphthenate. The cerium salt of naphthenic acid commercially available as a viscous mixture of rare-earth soaps. Used as a paint drier and flatting agent, in lubricants, etc.

cerium nitrate. See cerous nitrate.

cerium nitrate hexahydrate. See cerous nitrate.

cerium nitride. A compound of cerium and nitrogen made by the interaction of ammonia (NH$_3$) and cerium at 500°C (930°F), or nitrogen (N$_2$) and cerium at 800°C (1470°F). It has a density of 8.09 g/cm^3, and is used in ceramics. Formula: CeN.

cerium oxalate. A hygroscopic, yellowish-white, crystalline powder obtained from *monazite* sand by extraction and crystallization. Commercial cerium oxalate is a mixture of cerium, lanthanum and didymium oxalates. It decomposes upon heating, and is used in the isolation of cerium metals and the manufacture of cerium oxide. Formula: Ce$_2$(C$_2$O$_4$)$_3$·9H$_2$O. Also known as *cerium oxalate nonahydrate; cerous oxalate*.

cerium oxide. A white or yellow powder (99.9+% pure) obtained by heating cerium oxalate. Crystal system, cubic. Crystal structure, face-centered cubic. Density, 7.132 g/cm^3; melting point, approximately 2600°C (4710°F); hardness, 6 Mohs. Used as an opacifier in porcelain enamels and photochromic glass, as a glass decolorizer and brightener, in ceramic coatings, as an abrasive for polishing glass, marble and optical surfaces, in refractory oxides, phosphors, semiconductors and electronic components, as a catalyst for environmental redox reactions, and as a diluent in nuclear fuels. Formula: CeO$_2$. Also known as *ceria; ceric oxide; cerium dioxide*.

cerium oxide catalysts. A group of cerium oxide (CeO$_2$) based materials used as active catalysts in environmental redox reactions, e.g., in the production of hydrogen in fuel cells, the conversion of hydrogen sulfide and sulfur dioxide, and the combination with other catalysts in the conversion of hydrocarbons, carbon monoxide and nitric oxide. Also known as *ceria catalysts*.

cerium oxide coatings. Oxidation-resistant high-temperature coatings based on cerium oxide (CeO$_2$) and used on austenitic stainless steels (especially AISI types 304, 316, 321 and 304) for reheater tubes, superheaters, turbine blades, vanes, etc. Also known as *ceria coatings*.

cerium phosphide. A compound of cerium and trivalent phosphorus with a density of 5.56 g/cm^3 used in ceramics and electronics. Formula: CeP.

cerium ruthenium. An intermetallic compound of cerium and ruthenium with a melting point of 1538-1571°C (2800-2860°F), used in ceramics. Formula: $CeRu_2$.

cerium selenide. Any of the following two compounds used in ceramics and electronics: (i) *Cerium sesquiselenide.* Melting point, 1593-2051°C (2900-3724°F). Formula: Ce_2Se_3; and (ii) *Cerium monoselenide.* Melting point, 1816°C (3300°F). Formula: CeSe.

cerium sesquicarbide. See cerium carbide (ii).

cerium sesquiselenide. See cerium selenide (i).

cerium sesquitelluride. See cerium telluride (ii).

cerium silicide. A compound of cerium and silicon with a density of 5.41 g/cm³. Used in ceramics and materials research. Formula: $CeSi_2$.

cerium standard alloy. A *mischmetal* containing 50-55% cerium, 22-25% lanthanum, 15-17% neodymium, 0.5-1.0% iron, with the balance being a mixture of the rare-earths praseodymium, promethium, samarium, terbium and yttrium. Used in the manufacture of magnetic and pyrophoric alloys, ferrous and nonferrous alloys, cigarette lighters, and as a getter in electronic tubes.

cerium stearate. See cerous stearate.

cerium sulfide. A compound of cerium and sulfur available as red crystals, or brown to dark purple powder. Melting point, 2450°C (4440°F); good stability and thermoelectric properties at temperatures to 1100°C (2010°F); good thermal-shock resistance; high chemical and thermal resistance. Used for high-temperature thermoelectric devices, high-stage units in conversion devices, and metallurgical melting crucibles. Formula: CeS. Also known as *ceric sulfide.*

cerium telluride. Any of the following compounds of cerium and tellurium used in ceramics and electronics: (i) *Cerium sesquitelluride.* Melting point, 1665-1800°C (3029-3272°F). Formula: Ce_2Te_3. and (ii) *Cerium monotelluride.* Melting point, 1587-1888°C (2889-3430°F). Formula: CeTe.

cerium tetraboride. See cerium boride (i).

cerium trifluoride. See cerous fluoride.

cerium vanadate. A compound of ceric oxide and vanadium tetroxide. It has a melting point of 1832°C (3330°F) and is used in ceramics and materials research. Formula: $Ce_2V_2O_8$.

cermet ceramic coatings. See cermet coatings.

cermet coatings. Electrodeposited or flame-sprayed coatings consisting of mixtures of metal and ceramic borides, carbides or oxides, such as metal-bonded carbides (e.g., tungsten carbide with 8-15% cobalt or chromium carbide and with nichrome), or aluminum oxide or refractory carbides and an oxidation-resistant metallic binder. Used for the protection of metallic substrates against corrosion, erosion and oxidation. Also known as *ceramic-metal coatings; cermet ceramic coatings.*

CermeTi. Trade name of Dynamet Technology, Inc. (USA) for a series of particulate-reinforced titanium-matrix composites available in three grades: (i) *CermeTi A* composites consisting of titanium-alloy matrices reinforced with titanium aluminide (TiAl); (ii) *CermeTi B* composites with titanium-alloy matrices with titanium diboride (TiB_2) reinforcement; and (iii) *CermeTi C* composites having titanium-alloy matrices reinforced with titanium carbide (TiC). *CermeTi* composites possess high strength, hardness and stiffness as well as excellent high-temperature performance. All are used as high-performance engineering composites for aerospace applications.

cermet nuclear fuel. A special nuclear fuel made by mixing a fissionable material with a ceramic and a high-temperature metal or alloy to increase refractoriness and damage tolerance. Used in nuclear reactors. Also known as *ceramic nuclear fuel.*

cermets. A class of composite materials consisting of ceramic materials, such as metal borides, carbides, nitrides, oxides, silicates or silicides, and metallic binders, such as aluminum, cobalt, chromium, iron, nickel, tantalum, titanium or zirconium. They are made at high temperatures under controlled atmospheres using powder-metallurgy techniques involving pressing and sintering. They have outstanding high-temperature strength and wear resistance, high resistance to oxidation and intergranular corrosion at elevated temperatures, good corrosion resistance, excellent stress-to-rupture properties, and high toughness. Used for high-temperature applications, super high-speed cutting tools, turbojet engines, gas turbines, rocket motors, nuclear reactors, nuclear fuel elements, chemical equipment, pumps for severe service, rocket-fuel handling systems, brake linings, high-temperature-resistant coatings, electrical components, seals, bearings, etc. Also known as *ceramals; ceramets; metal-ceramics; metallic ceramics.*

cernyite. A steel-gray mineral of the chalcopyrite group composed of copper cadmium tin sulfide, Cu_2CdSnS_4. Crystal system, tetragonal. Density, 4.95 g/cm³. Occurrence: USA (South Dakota).

Cero-Crystide. Trade name T. & W. Ide Limited (UK) for rough-cast glass containing additions of cerium oxide. It is uniquely white, and has high ultraviolet light absorption. Used in galleries and museums as protection against fading.

ceromers. Composite materials composed of ceramic matrices reinforced with polymers, e.g., in the form of fibers or particulates. Frequently used in dentistry for the preparation of anterior and posterior restorations including crowns.

cerotungstite. An orange-yellow mineral composed of cerium neodymium tungsten oxide hydroxide, $(Ce,Nd)(WO_3)_2(OH)_2$. Crystal system, monoclinic. Density, 6.25 g/cm³; refractive index, 1.95. Occurrence: Uganda.

cerous bromide. A hygroscopic, orange, crystalline powder (99+%) with a melting point of 730°C (1346°F). It is also available in hydrated form as white crystals. Used as a polymerization catalyst, and in the preparation of cerium metal. Formula: $CeBr_3$ (anhydrous); $CeBr_3 \cdot xH_2O$ (hydrate). Also known as *cerium bromide.*

cerous carbonate. See cerium carbonate.

cerous chloride. A hygroscopic, white crystalline powder made by treating cerium carbonate or hydroxide with hydrochloric acid. It is also available as the heptahydrate in the form of white crystals. Density, 3.88 g/cm³ (anhydrous), 3.92 g/cm³ (heptahydrate); melting point, 848°C (1558°F) (anhydrous); boiling point, 1727°C (3141°F) (anhydrous). Used in incandescent gas mantles, as a polymerization catalyst, and in the preparation of cerium metal. Formula: $CeCl_3$ (anhydrous); $CeCl_3 \cdot 5H_2O$ (heptahydrate). Also known as *cerium chloride.*

cerous fluoride. A white crystalline powder (99+% pure) made by treating cerous oxalate with hydrofluoric acid. It occurs in nature as the mineral *fluocerite.* Density, 6.16 g/cm³; melting point, 1460°C (2660°F); boiling point, approximately 2300°C (4170°F). Used in arc carbons to increase brilliance, in the preparation of cerium metal, in the fluorination of hydrocarbons, and for laser crystals. Formula: CeF_3. Also known as *cerium fluoride.*

cerous hydroxide. A white, gelatinous, precipitated powder obtained from *monazite* sand. It may be yellow, brown, or pink when impurities are present. Used in pure form to color glass yellow, to prepare cerium salts, and as an opacifier in porcelain enamels and glazes. The crude form is used in flaming arc lamps. Formula: $Ce(OH)_3$. Also known as *cerium hydrate; cerium hydroxide.*

cerous iodide. A yellow, crystalline powder (99.9+%) with a melting point of 750-760°C (1380-1400°F). It also available as the nonahydrate in the form of reddish-white crystals. Used as a polymerization catalyst, and in the preparation of cerium metal. Formula: CeI_3 (anhydrous); $CeI_3 \cdot 9H_2O$ (nonahydrate). Also known as *cerium iodide; cerous triiodide.*

cerous nitrate. A colorless or white crystalline compound (99+% pure) made by treating cerous carbonate with nitric acid. Melting point, loses $3H_2O$ at 150°C (300°F); boiling point, decomposes at 200°C (390°F). Used in gas mantles, in the separation of cerium from other rare earths, and as an oxidizer. Formula: $Ce(NO_3)_3 \cdot 6H_2O$. Also known as *cerium nitrate; cerium nitrate hexahydrate; cerous nitrate hexahydrate.*

cerous oxalate. See cerium oxalate.

cerous oxide. A gray to greenish powder. Density, 6.9-7.0 g/cm^3; melting point 2040°C (3704°F); refractive index 2.19. Used as an opacifier in porcelain enamels and other ceramics, and in glass to enhance absorption of ultraviolet radiation. Formula: Ce_2O_3.

cerous stearate. The cerium salt of stearic acid commercially available as a white powder (95+%) with a melting point of 120-124°C (248-255°F). Used as a paint drier and flatting agent, in lubricants, etc. Formula: $Ce(O_2C_{18}H_{35})_3$. Also known as *cerium stearate.*

cerous triiodide. See cerous iodide.

Cerox. (1) Trademark of Babcock & Wilcox Company (USA) for a series of hard, white aluminum silicate refractories containing 50-90% aluminum oxide, and varying amounts of silicon dioxide and other oxides. Commercially available in molded shapes, they have excellent corrosion, erosion, abrasion and wear resistance, good to excellent resistance to thermal shock and reducing atmospheres, and a minmum melting temperature of 1980°C (3596°F). Used in the manufacture of furnace and electronic components, abrasives, powder metallurgy, and high-temperature gas reactors.

(2) Trade name of Babcock & Wilcox Company (USA) for a fine, fast-cutting cerium oxide polishing powder used to produce smooth surfaces on optical lenses and automobile windshields.

Cerpass-XTL. Trademark of St. Gobain Industrial Ceramics (USA) for nonfused macro- and submicron-sized aluminum oxide grains.

Cerpress. Trademark of Leach & Dillon Company (USA) for a dental porcelain.

Cerrobase. Trade name of Cerro Metal Products Company (USA) for a series of fusible alloys containing 5-58% bismuth, 1-51.5% lead, 0-60% tin, and 0-10% cadmium. The melting point of the 55.5Bi-44.5Pb alloy is 124°C (255°F), it has no shrinkage or expansion on cooling, and can be readily cast. Used for molds and tube fillers, as a pattern metal and for proof-casting forging dies.

Cerrobend. Trade name of Cerro Metal Products Company (USA) for a fusible alloy containing 50% bismuth, 26.7% lead, 13.3% tin, and 10% cadmium. It has a melting point of 70°C (158°F) and expands on solidification. Used as a filler in tube bending, as a low-melting solder, and in glass-to-metal seals. Formerly known as *Bendalloy.*

Cerrocast. Trade name of Cerro Metal Products Company (USA)

for a noneutectic fusible alloy of 60% tin and 40% bismuth. It has a yield temperature of 150°C (302°F), a melting-temperature range of 38-170°C (280-338°F), and negligible shrinkage. Used for wax pattern molds.

Cerrodent. Trade name of Cerro Metal Products Company (USA) for a dental alloy containing 38.14% bismuth, 26.42% lead, 31.67% tin, 2.64% cadmium, 0.06% copper, and 1.07% antimony. It has a melting range of 75-118°C (167-244°F), and is used for dental models.

CerroGroove. Trade name of Cerro Copper Tube Company (USA) for internally grooved copper tubing used for air-conditioning equipment.

Cerrolow. Trade name of Cerro Metal Products Company (USA) for a series of fusible alloys based on bismuth and containing varying amounts of lead, tin, cadmium, and indium. Used for bismuth and indium solders, fuses, thermal safety devices, fusible elements, etc.

Cerromatrix. Trade name of Cerro Metal Products Company (USA) for a noneutectic fusible alloy containing 48% bismuth, 28.5% lead, 14.5% tin, and 9% antimony. It expands on solidification, has a yield temperature of 116°C (240°F) and a melting-temperature range of 102-227 (216-440°F). Used as a dental alloy, as a fusible alloy for die mounting, and for short run dies.

Cerrosafe. Trade name of Cerro Metal Products Company (USA) for a noneutectic fusible alloy of 42.5% bismuth, 37.7% lead, 11.3% tin, and 8.5% cadmium. It has a yield temperature of 72.5°C (163°F), a melting range of 70-90°C (158-194°F), and will not burn wood or cause fires. Used for toy casting sets and patterns. Formerly known as *Saffalloy.*

Cerroseal. Trade name of Cerro Metal Products Company (USA) for an alloy of 50% indium and 50% tin. It has a melting range of 117-127°C (243-260°F), and will adhere to ceramics and glass. Used for sealing applications in vacuum and low-vapor pressure environments, soldering glass to metal, and for fusible links.

Cerrotric. Trade name of Mining & Chemical Products Limited (UK) for a fusible alloy composed of 58% bismuth and 42% tin.

Cerrotru. Trade name of Cerro Metal Products Company (USA) for an eutectic fusible alloy of 58% bismuth and 42% tin. It has a melting temperature of 137.5°C (280°F), and does not exhibit any volume change on cooling. Used for castings, molds, and patterns.

Cer-Seal. Trademark of ZYP Coatings Inc. (USA) for an oxidation- and wear-resistant water-based ceramic coating for graphite that is applied by conventional painting techniques, and does not require curing. Used for applications below 1400°C (2550°F).

Certain. Trade name for a light-cure dental hybrid composite.

Certainteed. Trade name of Certainteed Products Corporation (USA) for an extensive series of asbestos and non-asbestos building products including asbestos cement pipe, flexible trim-line, joint treatment compounds, sound-absorbing plasters, plastic cements and various glass and fiberglass products.

certified malleable irons. Malleable cast irons produced according to specifications of the American Society for Testing and Materials (ASTM), such as ASTM A47, A220, A338 or A602.

certified oxygen-free nickel. An oxygen-free high-conductivity nickel (99.99% pure) produced according to generally accepted specifications. It has good corrosion resistance, excellent hot and cold workability, and is used for electrical and electronic applications.

certified reference material. A reference material whose composition and physical and chemical properties have been certified by a recognized testing laboratory or standardizing authority. Abbreviation: CRM.

Certus. Trade name of Ato Findley (France) for casein adhesives for bonding champaign corks.

cerulean blue. A light- to sky-blue pigment made from stannous oxide, copper sulfate and chalk. It is essentially cobaltous stannate, $CoO·x(SnO_2)$.

ceruleite. A deep blue mineral composed of copper aluminum arsenate hydroxide hydrate, $Cu_2Al_7(OH)_{13}(AsO_4)_4·11.5H_2O$. Crystal system, triclinic. Density, 2.70 g/cm^3; refractive index, 1.60. Occurrence: Bolivia, Chile.

cerusa. A *white lead* (basic lead carbonate) used as a paint pigment.

cerussite. A colorless, white, yellow, or grayish mineral of the aragonite group. It has an adamantine luster and is composed of lead carbonate, $PbCO_3$. It can also be made synthetically. Crystal system, orthorhombic. Density, 6.55 g/cm^3; hardness, 3-3.5 Mohs; refractive index, 2.07. Occurrence: Australia, Europe, USA (Arizona, Colorado, Idaho, New Mexico). Used as an ore of lead. Also known as *white lead ore.*

cervantite. A white, or yellow naqtural mineral of the columbite group composed of antimony tetroxide, Sb_2O_4. It can also be made synthetically. Crystal system, orthorhombic. Density, 6.64 g/cm^3.

Cer-Vit. Trademark of Owens-Corning Fiberglass (USA) for glassceramics composed of minute ceramic crystals in a glass matrix. They have a linear expansion near zero, and negligible deflection and distortion of light rays. Used as optical glass for telescope mirrors, and molded electronic parts.

Cerwool. Trade name of Combustion Engineering (USA) for ceramic fibers composed of 52-55% alumina (Al_2O_3) and 41-44% silica (SiO_2), and supplied in discontinuous, bulk and mat form.

cesanite. A colorless mineral of the apatite group composed of sodium calcium sulfate hydroxide, $Ca_2Na_3(SO_4)_3(OH)$. Crystal system, hexagonal. Density, 2.79 g/cm^3; refractive index, 1.570. Occurrence: Italy.

cesarolite. A steel-gray mineral composed of lead manganese oxide monohydrate, $PbMn_3O_7·H_2O$. Density, 5.29 g/cm^3. Occurrence: Tunisia.

cesbronite. A green mineral composed of copper tellurate hydroxide dihydrate, $Cu_5(TeO_3)_2(OH)_6·2H_2O$. Crystal system, orthorhombic. Density, 4.45 g/cm^3; refractive index, 1.928. Occurrence: Mexico.

cesium. A silvery-white metallic element of Group IA (Group 1) of the Periodic Table (alkali group). It is commercially available as lumps or ingots sealed in glass ampules or steel container under vacuum, inert gas (argon) or oil. The chief cesium ore is *pollucite*. Crystal system, cubic. Crystal structure, bodycentered cubic. Density, 1.90 g/cm^3; melting point, 28.5°C (83.3°F); boiling point, 705°C (1301°F); hardness, 0.2 Mohs; atomic number, 55; atomic weight, 132.905; monovalent. *Cesium* has high ductility below its melting point, high affinity for oxygen, halogens, sulfur and phosphorus, ignites immediately on contact with air, reacts explosively with water, and is the most reactive element in the electrochemical series. Used for photocells, phototubes, photocathodes, rectifiers, infrared and vapor lamps and thermionic diodes, as a semiconductor, as a getter in vacuum tubes, in atomic clocks and ion engines, in plasma for thermoelectric conversion, as a gamma radiation

source, rocket propellant and heat transfer fluid in power generators, and in thermochemistry. Symbol: Cs.

cesium-134. A radioactive isotope of cesium with a mass number of 134 that emits beta particles and has a half-life of about 2.2 years. Used in photocells and ion engines. Symbol: [134]Cs.

cesium-137. A radioactive isotope of cesium with a mass number 137. It is recovered from the waste of nuclear reactors, and emits beta particles. The beta decay of cesium-137 produces radioactive barium-137 that emits the characteristic gamma radiation of 0.662 MeV. It has a half-life of 33 years, and is used as a gamma-radiation source in radiography. Symbol: [137]Cs. Also known as *radiocesium.*

cesium acid phthalate. The cesium salt of phthalic acid available in the form of single crystals with high plasticity and fissionability for use as an analyzing crystal in X-ray spectral analysis. Formula: $CsC_8H_5O_4$. Abbreviation: CsAP. Also known *cesium biphthalate; cesium hydrogen phthalate.*

cesium alum. See cesium-aluminum sulfate.

cesium-aluminum sulfate. A colorless, water-soluble crystalline substance obtained by the addition of a solution of cesium sulfate to a solution of potassium aluminum followed by concentration and crystallization. Density, 1.202 g/cm^3; melting point, decomposes at 110°C (230°F). Used in water purification, in the manufacture of cesium salts, in the purification of cesium by fractional crystallization, in mineral waters, and in materials research. Formula: $AlCs(SO_4)_2·12H_2O$. Also known as *aluminum-cesium sulfate; aluminum-cesium sulfate dodecahydrate; cesium alum; cesium-aluminum sulfate dodecahydrate.*

cesium antimonide. A compound of cesium and antimony used as a high-purity semiconductor, e.g., for photocathodes. Formula: Formula: Cs_3Sb.

cesium arsenide. A compound of cesium and arsenic used as a high-purity semiconductor. Formula: Cs_3As.

cesium biphthalate. See cesium acid phthalate.

cesium bismuth telluride. A semiconductor material with a layered anisotropic needle-like crystal structure composed of anionic bismuth telluride (Bi_4Te_6) slabs with alternating layers of positive cesium ions. It exhibits thermoelectric properties, high electrical conductivity, low thermal conductivity, and melts at 545°C (1013°F). It can be used in electronic devices, such as laser diodes or infrared detectors to cool components operating at 100°C (212°F). Formula: $CsBi_4Te_6$.

cesium bromide. Colorless or white, hygroscopic, crystalline powder (99+% pure). Density, 4.440 g/cm^3; melting point, 636°C (1177°F); boiling point, 1300°C (2372°F); refractive index, 1.6984. Used in scintillation counters and fluorescent screens. The crystals are also employed in infrared spectroscopy. Formula: CsBr.

cesium carbonate. White, hygroscopic, crystalline powder (99+% pure) made by introducing carbon dioxide into a cesium hydroxide solution. Melting point, decomposes at 610°C (1130°F). Used in specialty glasses, and as a polymerization catalyst for ethylene oxide. Formula: Cs_2CO_3.

cesium chloride. White or colorless, hygroscopic crystals (99+% pure) made by treating cesium oxide with hydrochloric acid. Crystal system, cubic. Density, 3.988 g/cm^3; melting point, 646°C (1195°F); boiling point, sublimes at 1290°C (2354°F); refractive index, 1.6418. Used in photocells, infrared signaling lamps, in the evacuation of radio tubes, in fluorescent screens, in biochemistry and molecular biology, and for optical applications. Formula: CsCl.

cesium fluoride. White, hygroscopic, crystalline powder (99+%

pure). Density, 4.115 g/cm³; melting point, 682°C (1260°F); boiling point, 1251°C (2284°F). Used in specialty glasses, in optics, and as a catalyst. Formula: CsF.

cesium fluoroaluminate. A powder (99.9+% pure) with a density of 3.70 g/cm³ and a melting point of 423-436°C (793-817°F), used in materials research, ceramics and electronics. Formula: $(CsF)_x(AlF_3)_{3y}$.

cesium formate. The cesium salt of formic acid available as hygroscopic crystals (98% pure) with a melting point of 265°C (509°F). Formula: $CsCHO_2$.

cesium formate hydrate. A hydrated form of cesium formate that decomposes at 41°C (105°F). Used to form density gradient solutions. Formula: $CsCHO_2 \cdot H_2O$.

cesium germanium fluoride. See cesium hexafluorogermanate.

cesium hexafluorogermanate. A compound of cesium fluoride and germanium fluoride available as a white crystalline powder with a density of 4.1 g/cm³ and a melting point of 675°C (1247°F) that is readily soluble in hot water. Used in electronics and materials research. Formula: Cs_2GeF_6. Also known as *cesium germanium fluoride*.

cesium hexatitanate. See cesium titanate.

cesium hydrate. See cesium hydroxide.

cesium hydrogen phthalate. See cesium acid phthalate.

cesium hydroxide. The technical grade is a colorless or yellowish, fused, crystalline substance made by treating a cesium sulfide solution in water with barium hydroxide. It has a density of 3.675 g/cm³ and a melting point of 272.3°C (522°F). Used as an electrolyte in alkaline storage batteries at subzero temperatures and as a polymerization catalyst for siloxanes. Formula: CsOH. Also known as *cesium hydrate*.

cesium iodide. White or colorless, hygroscopic powder or crystals (99+%). Density, 4.510 g/cm³; melting point, 621-626°C (1150-1159°F); boiling point, 1280°C (2336°F); refractive index, 1.7876; very low hardness. Used in scintillation counters, as crystals and precipitated powders for infrared spectroscopy, spectrophotometry, and fluorescent screens. The high-purity crystals also find application in satellite-borne radiation detectors, and as beamsplitters or interferometer plates in lasers for far-infrared applications. Formula: CsI.

cesium iodide thallide. A slightly hygroscopic cesium iodide single crystal doped or activated with thallium (Tl^{3+}) ions. Crystal system, cubic. Density, 4.51 g/cm³; hardness, 2 Mohs; melting point, 621°C (1150°F). Used in the discriminator circuits of scintillators for gamma-ray spectroscopy, in silicon photodiodes and lasers, and for infrared applications. Formula: CsI:Tl.

cesium kupletskite. A gold-brown mineral of the astrophyllite group composed of cesium manganese iron titanium niobium silicate, $Cs_3(Mn,Fe)_7(Ti,Nb)_2Si_8O_{24}(O,OH,F)_7$. Crystal system, triclinic. Density, 3.62-3.68 g/cm³; refractive index, 1.713. Occurrence: Russian Federation.

cesium lithium borate. A compound of cesium, lithium, boron and oxygen that melts congruently and can be grown directly from the melt in the form of high-purity single crystals. It has excellent nonlinear optical properties, high transparency down to 190 nm, and generates fourth and fifth harmonics of fundamental wavelength in neodymium-doped yttrium aluminum garnet lasers. Formula: $CsLiB_6O_{10}$. Abbreviation: CLB; CLBO.

cesium metavanadate. The cesium salt of metavanadic acid supplied in the form of a fine, high-purity powder (99.9+%). Used in optics, ceramics and electronics. Formula: $CsVO_3$.

cesium molybdate. The cesium salt of molybdic acid supplied in the form of a fine, high-purity powder for use in optics, ceram-

ics and electronics. Formula: Cs_2MoO_4.

cesium nitrate. A crystalline powder (99+% pure) with a density of 3.687 g/cm³ and a melting point of 414°C (1150-1159°F) (decomposes). Used in the manufacture of cesium salts, and as an oxidizing agent. Formula: $CsNO_3$.

cesium oxide. Moisture-sensitive, orange-red crystals, or yellow-brown crystalline powder (99+% pure) usually supplied as a mixture of oxides. Crystal system, hexagonal. Density, 4.36 g/cm³; melting point, 490°C (914°F). Used in the manufacture of cesium salts, and in ceramics and materials research. Formula: Cs_2O.

cesium perchlorate. A white crystalline substance (97+% pure) . Density, 3.327 g/cm³; melting point, 250°C (482°F); strong oxidizer. Used as an oxidizer, in specialty glasses, and in optics, catalysis and power generation. Formula: $CsClO_4$.

cesium phosphide. A compound of cesium and trivalent phosphorus used as a high-purity semiconductor. Formula: Cs_3P.

cesium silicate. Yellow, crystalline substance of cesium oxide and silicon dioxide used in ceramics. Formula: Cs_2SiO_3.

cesium sulfate. Colorless, hygroscopic crystals (99+% pure) made by treating cesium carbonate with sulfuric acid. Crystal system, cubic. Density, 4.243 g/cm³; melting point, 1010°C (1850°F). Used as a density gradient in ultracentrifuge separation, for optical applications, and in the biosciences. Formula: Cs_2SO_4.

cesium superoxide. A yellow crystalline compound. Crystal system, tetragonal. Density, 3.77 g/cm³; melting point of 432°C (810°F). Formula: CsO_2.

cesium titanate. A compound consisting of cesium oxide and titanium dioxide with a melting point above 300°C (570°F). Used in ceramics and materials research. Formula: $Cs_2Ti_6O_{13}$. Also known as *cesium hexatitanate*.

cesium trichlorogermanate. A compound (95% pure) that contains cesium chloride. Density, 3.45 g/cm³. Used as an auxiliary catalyst with platinum for the hydroformylation of olefins, and in materials research. Formula: $CsGeCl_3$.

cesium triiodide. White or colorless, hygroscopic powder or crystals (99.9+%). Density, 4.510 g/cm³; melting point, 207.5°C (406°F). Used in materials research. Formula: CsI_3.

cesium tungstate. The cesium salt to tungstic acid supplied in the form of a fine, white high-purity powder with a melting point above 350°C (570°F). Used in ceramics and materials research. Formula: Cs_2WO_4.

Cessato. Trademark of Fillatice SpA (Italy) for spandex fibers and yarns.

cesstibtantite. A colorless to gray mineral of the pyrochlore group composed of cesium sodium antimony tantalum oxide, (Cs,Na)-SbTa$_4$O$_{12}$. Crystal system, cubic. Density, 6.40 g/cm³; refractive index, above 1.8. Occurrence: Russian Federation.

Cestidur. Trademark of DSM Engineering Plastics (USA) for a series of polyethylene resins.

Cestilene. Trademark of DSM Engineering Plastics (USA) for a series of polyethylene resins.

Cetal. Trade name of Metallgesellschaft Reuterweg (Germany) for a non-heat-treatable aluminum alloy containing 80.5% aluminum, 10% zinc, 6.5% silicon, and 3% copper. Used for light-alloy parts and castings.

Ceto. Trade name of Philips NV (Netherlands) for an alloy containing 80% thorium, 15% aluminum and 5% *mischmetal*. It has high gas absorption properties, and is used as a getter.

Ceva-Crete. Trade name of E-Poxy Industries, Inc. (USA) for elastomeric concrete.

Cevian. (1) Trade name of Hoechst Celanese Corporation (USA) for acrylonitrile-butadiene-styrene resins.

(2) Trademark of Daicel Chemical Industries, Limited (Japan) for synthetic resins available in liquid, paste and powder form. Used for making automotive components, electrical equipment, cosmetic containers, etc.

ceylon. A colored fabric woven with cotton warp yarn and cotton-wool filling yarn. Used for blouses and shirts.

ceylonite. A dark-green, brown, or black variety of *spinel* ($MgAl_2O_4$) containing some iron. It is found in Sri Lanka (formerly Ceylon), and used as a gemstone. Also known as *pleonaste.*

CFF Fibrillated Fiber. Trade name of Sterling Fibers, Inc. (USA) for fibrillated acrylic staple fibers with excellent environmental resistance, high mechanical strength, low moisture sensitivity and good web formation properties. Used as high-efficiency binders in nonwovens, technical and special papers, composites, and pulp-molded products, for high-performance absorptive and particulate filtration media, etc.

C-Fill MH. Trade name for a light-cure dental hybrid composite.

C-Flex. Trademark of Concept, Inc. (USA) for opaque to white silicone-modified thermoplastic elastomer polyblends formulated from styrene-ethylene-butadiene-styrene (SEBS) block copolymers. Available in molding, forming and extrusion grades, they have good strength and processibility, good moisture and temperature stability, good flexibility, optimum product clarity, smooth surface finish, biocompatibility, and an useful temperature range of -100 to +100°C (-148 to +212°F). Used for medical products, laboratory tubing, and peristaltic pumps.

CG alloy. A ferroalloy containing 48-52% silicon, 8.5-10.5% titanium, 4-5% magnesium, 4-5.5% calcium, 0.2-0.35% cerium, 1-1.5% aluminum, and the balance iron. Used in the manufacture of compacted graphite cast iron.

CG iron. See compacted graphite cast iron.

C-glass. A soda-lime-borosilicate glass with excellent chemical stability used for highly corrosive environments, and as glass fiber reinforcement in composites that either contain, or are in contact with acidic substances.

chabazite. A colorless, white, or salmon-pink mineral of the zeolite group composed of calcium aluminum silicate hexahydrate, $CaAl_2Si_4O_{12} \cdot 6H_2O$. Crystal system, rhombohedral (hexagonal). Density, 2.05-2.16 g/cm³; hardness, 4-5 Mohs. Occurrence: Canada (Nova Scotia). Used in water treatment and ion exchange processes.

chabourneite. A black mineral composed of thallium antimony arsenic sulfide, $(Tl,Pb)_5(Sb,As)_{21}S_{34}$. Crystal system, triclinic. Density, 5.10 g/cm³. Occurrence: France, Japan.

Chace. (1) Trade name of W.M. Chace Company (USA) for a series of iron-nickel bimetals used for thermostats.

(2) Trade name of W.M. Chace Company (USA) for a series of copper-manganese-nickel and manganese-copper-nickel alloys used for springs and electrical resistances.

Chadolene. Trade name for polyolefin fibers and products.

Chadolon. Trade name for nylon fibers and yarns.

chagrin paper. A black *cambric paper.*

chain bronze. A corrosion-resistant, high-strength phosphor bronze containing 95% copper, 4.9% tin and 0.1% phosphorus. Used for diaphragms, springs, and chains.

chainette yarn. See loop-wale yarn.

chain iron. A wrought iron containing varying amounts of iron, carbon, silicon, and manganese. Used for crane chains, slings, hoists, steam shovels, marine uses, etc.

Chainlon. Trademark of Chain Yarns Company (Taiwan) for nylon fibers and yarns.

chain-reaction polymer. See addition polymer.

chalcanthite. A sky-blue, or greenish mineral composed of copper sulfate pentahydrate, $CuSO_4 \cdot 5H_2O$. It can also be made synthetically. Crystal system, triclinic. Density, 2.12-2.30 g/cm³; refractive index, 1.5368; hardness, 2.5. Occurrence: Chile, Spain, UK, USA. Used as a source of copper. Also known as *bluestone; blue vitriol; copper vitriol.*

chalcedony. A transparent, or translucent white, gray, blue, brown, or red, cryptocrystalline variety of *quartz* with waxy luster. There are several forms including carnelian, heliotrope and moss agate. Used as a gemstone, and for ornamental purposes.

chalcoalumite. A turquoise-green to pale blue mineral composed of copper aluminum sulfate hydroxide trihydrate, $CuAl_4SO_4 (OH)_{12} \cdot 3H_2O$. Crystal system, monoclinic. Density, 2.29 g/cm³; refractive index, 1.525. Occurrence: USA (Arizona).

chalcocite. A black or lead-gray mineral with lead-gray steak, and metallic luster that tarnishes dull black on exposure to air. It is composed of copper sulfide, Cu_2S, and contains about 80% copper. It can also be made synthetically. Crystal system, orthorhombic. Density, 5.5-5.8 g/cm³; melting point, 1130°C (2065°F); hardness, 2.5-3 Mohs. Occurrence: Bolivia, Chile, Europe, Mexico, Peru, USA (Alaska, Arizona, Connecticut, Montana, Nevada, Utah). Used as an important ore of copper. Also known as *chalcosine; copper glance; glance; redruthite; vitreous copper.*

chalcogenide glass. A special glass containing considerable amounts of one of the chemical elements of Group VI (Group 16) of the Periodic Table in particular sulfur, selenium or tellurium. Used for glass switches.

chalcogenides. Binary compounds composed of a chalcogen, i.e., an element of Group VI (Group 16) of the Periodic Table, and a more electropositive element or radical.

chalcogens. A collective term for the chemical elements of Group VI (Group 16) of the Periodic Table, i.e., oxygen, sulfur, selenium, tellurium and polonium.

chalcocyanite. A colorless, or white mineral composed of copper sulfate, $CuSO_4$. Crystal system, orthorhombic. It can also be made synthetically. Density, 3.65 g/cm³; refractive index, 1.733. Occurrence: USA (Colorado).

chalcolite. See torbernite.

chalcomenite. A bright blue mineral composed of copper selenite dihydrate, $CuSeO_3 \cdot 2H_2O$. Crystal system, orthorhombic. Density, 3.31 g/cm³; refractive index, 1.732. Occurrence: Canada (Saskatchewan), USA.

chalconatronite. A greenish blue mineral composed of sodium copper carbonate trihydrate, $Na_2Cu(CO_3)_2 \cdot 3H_2O$. Crystal system, monoclinic. Density, 2.27 g/cm³; refractive index, 1.530. Occurrence: Egypt.

chalcophanite. A purplish black mineral composed of zinc manganese oxide hydrate, $(Zn,Fe)Mn_3O_7 \cdot xH_2O$. Crystal system, triclinic. Density, 3.98 g/cm³; refractive index, above 2.72. Occurrence: USA (New Jersey).

chalcophyllite. An emerald-green mineral composed of copper aluminum arsenate sulfate hydroxide hydrate, $Cu_{18}Al_2(AsO_4)_3 (SO_4)_3(OH)_{27} \cdot 36H_2O$. Crystal system, rhombohedral (hexagonal). Density, 2.67 g/cm³; refractive index, 1.618. Occurrence: Chile. Also known as *copper mica.*

chalcopyrite. A brass-yellow mineral with greenish-black streak, and metallic luster. It is composed of copper iron sulfide, $CuFeS_2$, and contains 34.5% copper and sometimes small quantities of gold or silver. It can also be made synthetically. Crystal system, tetragonal. Density, 4.1-4.3 g/cm³; hardness, 3.5-4 Mohs. Occurrence: Africa, Canada, Chile, Europe (Germany, Sweden, Spain), UK, USA (Arizona, Montana, Tennessee, Utah, Wisconsin). Used as an important ore of copper, and in semiconductor research. Also known as *copper pyrite; yellow copper; yellow pyrite.*

chalcopyrites. A group of ternary compounds with chalcopyrite structure and corresponding to the general formula $M'M''X_2$, where M' is a transition metal, e.g., cadmium, copper, zinc or silver, M'' an element of Group III or IV (Group 13 or 14), e.g., gallium, germanium, silicon or tin, and X_2 an element of Group V or VI (Group 15 or 16), e.g., arsenic, phosphorus, selenium or sulfur. All *chalcopyrites* are semiconductors and used essentially in electronics, optoelectronics, nonlinear optics, etc. Examples include cadmium germanium arsenide ($CdGeAs_2$), silver gallium sulfide ($AgGaS_2$) and zinc germanium phosphide ($ZnGeP_2$).

chalcosiderite. A dark green mineral of the turquoise group composed of copper iron phosphate hydroxide tetrahydrate, $CuFe_6(PO_4)_4(OH)_8 \cdot 4H_2O$. Crystal system, triclinic. Density, 3.22 g/cm³; refractive index, 1.840. Occurrence: UK.

chalcosine. See chalcocite.

chalcostibite. A gray-black mineral composed of copper antimony sulfide, $CuSbS_2$. It can also be made synthetically from the elements by hydrothermal treatment at 200°C (392°F). Crystal system, orthorhombic. Density, 4.95 g/cm³; hardness, 283-309 Vickers; refractive index, 1.618. Occurrence: Morocco.

chalcothallite. A lead-gray mineral composed of copper thallium antimony sulfide, $Cu_6Tl_2SbS_4$. Crystal system, tetragonal. Density, 6.60 g/cm³. Occurrence: Greenland.

Chalet. Trademark of Canadian Forest Products Limited for plywood used as siding.

chalk. (1) A soft, fine-grained, white, gray or dull yellow limestone made up chiefly of fossil sea shells. It is a variety of the mineral *calcite* ($CaCO_3$) possesses high porosity and friability, and is used in the preparation of lime.

(2) A white, very fine preparation obtained by levigating and washing naturally occurring chalk (1). One of the finest grades is Paris white, while coarser grades include commercial whiting and gilder's whiting. Used as a pigment for paints and inks, in putties (with linseed oil), as an extender in rubber, plastics and paper coatings, in whitewashes, as a sealant, as a polishing and scouring material, in earthenware and vitreous sanitaryware bodies, glazes, glasses and porcelain enamels, and as a constituent in refractories. Also known as *drop chalk; prepared chalk; prepared calcium carbonate; precipitated chalk; whiting.* See also precipitated calcium carbonate.

chalkboard enamel. See blackboard enamel.

Chalkelle. Trade name for rayon fibers and yarns used for textile fabrics.

Challenge. Trade name for a bonding resin used in orthodontics.

challis. A soft, lightweight, plain-woven fabric usually made of wool, worsted, cotton or synthetics in solid color or with distinctive, printed designs. Used for blouses, dresses, skirts, scarves, and children's wear.

chalmersite. See cubanite.

chalybite. See siderite (1).

chambersite. A colorless to deep purple mineral of the boracite group composed of manganese chloride borate, $Mn_3B_7O_{13}Cl$. Crystal system, orthorhombic. Density, 3.49 g/cm³; refractive index, 1.737. Occurrence: USA (Texas).

chambray. A fine, lightweight cotton fabric in a plain weave combining colored warp (longitudinal) threads with white weft (filling) threads in a variety of designs. Used for blouses, dresses, shirts, sportswear, pyjamas and linings.

chameanite. A dark gray mineral of the tetrahedrite group composed of copper iron arsenic selenium sulfide, $(Cu,Fe)_4As(Se,S)_4$. Crystal system, cubic. Density, 6.17 g/cm³. Occurrence: France.

Chamel. Trademark of Winter & Co. GmbH (Germany) for synthetic *chamois* leather substitutes.

Chameleon. Trade name for a dental laminate veneer porcelain.

chameleon. A varicolored fabric made from a colored warp yarn and two differently colored filling yarns.

Chamet. Trade name of Chase Brass & Copper Company (USA) for a series of unleaded and leaded bronzes. The unleaded bronzes contain about 60-62.25% copper, 37-39.25% zinc and 0.5-0.75% tin, and are used for fasteners and welding rods. The high-strength, free-cutting leaded bronzes contain 60% copper, 38.5% zinc, 0.75% lead and 0.75% tin, and are used for pipes, fittings and hardware.

Chamex. Trademark of Meyer & Myer Limited (Canada) for synthetic *chamois* leather substitutes.

chamois. Originally, a very soft pliant suede leather of yellowish-brown color made from the hides of chamois (*Rupicapra rupicapra*) which belong to the same family (Bovidae) as goats and antelopes and are native to the mountains of Europe and Southwest Asia. This term now refers to a soft, oil-tanned suede-finished leather made from the hides of deer, lamb, sheep or goats. Also known as *chamois leather.*

chamois paper. A yellowish brown copying paper used in photography. It does not contain *chamois*, the name being in reference to its general appearance.

chamosite. A green to greenish-gray mineral of the chlorite group composed of iron magnesium aluminum silicate hydroxide, $(Fe,Al,Mg)_6(Si,Al)_4O_{10}(OH)_8$. Crystal system, monoclinic. Density, 3.12 g/cm³. Occurrence: Czech Republic.

chamotte. A coarsely graded refractory, such as dead-burnt fireclay grog bonded with plastic fireclay, used in the manufacture of refractory bricks for industrial furnaces, lining of retorts and crucibles, and for steel-foundry applications.

Champaloy. Trade name of Crucible Materials Corporation (USA) for an oil-hardening, special-purpose tool steel (AISI type L6) containing 0.75% carbon, 1.50% nickel, 0.75% chromium, 0.30% molybdenum, and the balance iron. It has high toughness, good wear resistance, and is used for tools, rolls, shear blades, jigs, and collets.

Champion. Trade name of Champion Rivet Company (USA) for a series of stainless, low-alloy, high-carbon, manganese and tool steels.

Champlain. Trade name of North American Steel Corporation (USA) for a tough, nondeforming, oil-hardening tool steel containing 0.9% carbon, 0.5% tungsten, 0.5% chromium, 1.2% manganese, 0.2% vanadium, and the balance iron. Used for dies, forming rolls, knives, guides, shear blades, and vise jaws.

Chance. Trade name of Chance Brothers Limited (UK) for a dark-blue glass that selectively transmits ultraviolet light.

Chance-Crookes. Trade name of Pilkington Brothers Limited (UK) for a series of tinted ophthalmic glasses including light pink *Crookes Alpha*, light blue *Crookes A2*, and dark neutral *Crookes B* and *Crookes B2*. Used for sunglasses.

changbaiite. A yellow mineral composed of lead niobate, $PbNb_2O_6$. It can be made synthetically from a stoichiometric mixture of lead monoxide (PbO) and niobium dioxide (NbO_2). Crystal system, rhombohedral (hexagonal). Density, 5.50 g/cm^3.

channel black. A variety of *carbon black* formerly made by the impingement of a smoky natural or methane gas flame on iron channels. This preparation method is now virtually obsolete. Formerly used as a pigment for ink and rubber goods, and in stove polishes. See also gas black.

channel bronze. A copper alloy containing 39% zinc and 1% tin. It has improved corrosion resistance, and is used for marine applications, and for fasteners and sections.

channel compounds. A subclass of *inclusion compounds* in which the host lattices form nonintersecting channels which the guest species can occupy. An example is an urea compound that includes *n*-alkanes.

channel groove plywood. Plywood panels that have longitudinal channel grooves machined into their surfaces. Used particularly as house siding material.

Channel Side. Trademark of Canexel Hardboard Inc. (Canada) for hardboard with longitudinal surface channels, used as house siding material.

chantalite. A colorless mineral composed of calcium aluminum silicate hydroxide, $CaAl_2SiO_4(OH)_4$. Crystal system, tetragonal. Density, 2.80 g/cm^3; refractive index, 1.653. Occurrence: Turkey.

chantilly lace. A type of *bobbin lace* having a fine, netlike ground with a pattern of scrolls or florals, outlined by flat, heavy yarn. Named after Chantilly, a town in northern France, where it was originally made. Also known as *chantilly*.

chapmanite. An olive-green mineral composed of iron antimony silicate hydroxide, $Fe_2SbSi_2O_8(OH)$. Density, 3.58 g/cm^3. Occurrence: Canada (Ontario).

charcoal. A black, brittle, porous material containing about 85-90% carbon. It is the solid residue of the destructive distillation of animal or vegetable matter and can be obtained by charring wood, cellulose, peat or bones in a retort or kiln from which the air is shut out. *Charcoal* can be divided into the following classes: (i) *Activated charcoal;* (ii) *Animal charcoal;* (iii) *Bone charcoal;* (iv) *Vegetable charcoal;* (v) *Wood charcoal;* and (vi) *Charcoal blacking*. Used as a fuel for heating and cooking, as a decolorizing and filtering medium, as a gas adsorbent, in the manufacture of gunpowder, in arc light electrodes, in decolorizing and purifying oils, in solvent recovery, and in biochemistry, biotechnology and medicine.

charcoal blacking. Charcoal powder, either dry or suspended in clay, applied to the working surface of a core or mold to improve the overall surface finish.

charcoal briquet. A small cake or block made by compressing charcoal with a starch binder. Used chiefly as a fuel for domestic heating.

charcoal iron. A high-quality pig iron with a close, dense structure containing significantly reduced amounts of sulfur and phosphorus. It is made in a blast furnace using charcoal as fuel. Also known as *charcoal pig iron*. See also pig iron.

charcoal pig iron. See charcoal iron.

charcoal plate. See charcoal tinplate.

charcoal tinplate. High-grade *tinplate* made from charcoal iron. It has a rather heavy coating of tin whose thickness is indicated by a letter ranging from A (or 1A) for the lightest coat to AAAAAA (or 6A) for the heaviest coat. Even the lightest coat is still heavier than that of *coke tinplate*. Also known as *charcoal plate*.

Chardonize. Trade name for rayon fibers and yarns used for tex-

tile fabrics.

chardonnet silk. A name formerly used for *rayon*.

Chardon S. Trade name of Compagnie Ateliers et Forges de la Loire (France) for a water-hardening tool steel (AISI type W1) containing 0.6-1.1% carbon, and the balance iron. Used for cutters, hammers, chisels, punches, and blacksmith tools.

Charisma. (1) Trademark of Heraeus Kulzer Inc. (USA) for a light-cure ionomer microhybrid composite resin containing ultrafine (0.7 μm or 28 μin.) *Microglass* particles. It provides high strength, wear resistance, and takes a high lustrous polish. Supplied in several shades, it is used posterior dental restorations.

(2) Trademark of Plasticisers Limited (UK) for polypropylene staple fibers.

(3) Trade name of Fifield Inc. (USA) for imitation suede.

charlesite. A colorless mineral of the ettringite group composed of calcium aluminum borate silicate sulfate hydroxide hydrate, $Ca_6(Al,Si)_2(SO_4)_2(B(OH)_4)(OH,O)_{12}·26H_2O$. Crystal system, hexagonal. Density, 1.77 g/cm^3; refractive index, 1.492. Occurrence: USA (New Jersey).

Charles Leonard. Trade name of Jessop-Saville Limited (UK) for a water-hardening tool steel (AISI type W1) containing 0.7-1% carbon, 0.3% manganese, 0.3% silicon, and the balance iron. Used for drills, cutters, and blacksmith tools.

Charlton white. See lithopone.

Charme. Trademark of Saint-Gobain (France) for an acid-etched glass.

charmeuse. A soft, lightweight cotton, rayon or silk fabric in a satin weave, usually with a smooth, semilustrous face and a dull back. Used for blouses, women's apparel, nightwear and linings.

Charmour. Trade name of Celanese Corporation (USA) for cellulose acetate plastics.

charoite. (1) A lilac to violet mineral composed of potassium sodium calcium silicate fluoride hydroxide monohydrate, $(Ca,K,Na)_3Si_4O_{10}(OH,F)·H_2O$. Crystal system, monoclinic. Density, 2.54 g/cm^3; refractive index, 1.553. Occurrence: Russian Federation.

(2) A purple mineral composed of potassium sodium barium calcium strontium silicate fluoride hydroxide hydrate, $(Ca,Na)_4(K,Sr,Ba)_2Si_9O_{22}(OH,F)_2·xH_2O$. Crystal system, monoclinic. Density, 2.54 g/cm^3; refractive index, 1.553. Occurrence: Russian Federation.

Char-Pac. Trade name of US Steel Corporation for a heat-treated carbon steel plate containing 0.17% carbon, 1.25% manganese, 0.35% silicon, 0.25% copper, 0.15% nickel, 0.12% chromium, 0.04% molybdenum, and the balance iron. Used for pressure vessels, bridges and buildings, storage tanks, and penstocks.

Charpy alloy. A babbitt metal containing 83.3% tin, 11.1% antimony and 5.6% copper, used for bearings.

Charpy phosphor bronze. A heavy-duty phosphor bronze containing 86.2-87.4% copper, 12.2-13.4% tin and 0.4% phosphorus. Used for bushings, bearings, and gears.

Chartek. Trade name of Epoxy Technology Inc. (USA) for a fire-resistant, intumescent epoxy coating.

chart paper. See graph paper.

charvet. A soft acetate or silk fabric, usually having alternating lateral stripes of dull and lustrous finish, woven in a diagonal rib pattern. Used for neckties.

Chassis Blue. Trade name of Pavco Inc. (USA) for a blue bright chromate coating.

Chateau. Trademark of Iko Industries Limited (Canada) for asphalt shingles.

Chateauguay iron. A *pig iron* with low phosphorus content made from low-iron *magnetite* ore found in the vicinity of Chateauguay in northern New York state. It contains about 4.0% carbon, 0.7-4.1% silicon, 0.1-0.2% phosphorus, up to 0.035% sulfur, up to 0.035% phosphorus, and the balance iron. Used in the manufacture of iron and steel, and for casting rolls, cylinders, gears, and machine parts. Also known as *Chateauguay pig iron; Chateauguay low-phosphorus pig iron.*

Chatillon. Trade name of Chiers-Chatillon (France) for a series of cold-work, hot-work and high-speed tool steels, several high-temperature structural steels, and various austenitic, ferritic and martensitic stainless steels.

chatkalite. A pale rose colored mineral of the chalcopyrite group composed of copper iron tin sulfide, $Cu_6FeSn_2S_8$. Crystal system, tetragonal. Density, 4.97 g/cm³; refractive index, 1.553. Occurrence: Uzbekhistan.

chat sand. Sand tailings from lead and zinc ores used in making concrete pavements, and for sawing building stone.

Chauvel. Trade name of Vereinigte Glaswerke (Germany) for a plate glass with parallel wire strands.

checker bricks. See checkers.

checkers. Refractory bricks of special design stacked openly to permit the passage of hot air or gases through a regenerative furnace. Used in metallurgical furnaces (open-hearth design). Also known as *checker bricks.*

Checo. Trade name for a dental alloy containing 89% silver, 10% tin and 1% platinum.

cheesecloth. A thin, loosely woven cotton fabric in a plain weave formerly used chiefly for wrapping cheese, but now also used for lining, filtering and polishing applications.

chelkarite. A colorless mineral composed of calcium magnesium chloride borate heptahydrate, $CaMgB_2O_4Cl_2 \cdot 7H_2O$. Crystal system, orthorhombic. Density, 2.44 g/cm³. Occurrence: Russia.

chelate. See chelate compound.

chelate compound. A coordination compound formed by the combination of a chelating agent, such as ethylenediamine (en) or ethylenediaminetetraacetic acid (EDTA), and a metal atom or ion, such as Co^{2+}, Ni^{2+}, Cu^{2+}, Zn^{2+} or a rare-earth. Structurally, it consists of two or more nonmetal atoms (ligands) in the same molecule attached to the central metal atom by coordinate linkages forming a heterocyclic ring. Used in rare-earth chelate lasers, as catalysts, in ion exchange applications, in medicine to distroy viruses, bacteria, etc., and in biochemistry and biotechnology.

chelating resin. An *ion-exchange resin* that has a very high selectivity for specific metal cations.

Chelon-Fil. Trade name for a glass-ionomer dental filling cement.

Chelon-Silver. Trade name for a silver-filled glass-ionomer dental cement.

Chemaco. Trade name of Chemaco (USA) for a series of vinyls including polyvinyl acetate, polyvinyl alcohol and polyvinyl chloride supplied as flexible to rigid polymers. They are available in various colors and grades for optimum abrasion, chemical, electrical, flame, oil and/or weather resistance, and have good processibility, high stiffness at low temperatures, and a maximum service temperature of 138°C (280°F). Used for floor and wall coverings, upholstery, rainwear, tubing, insulation, toys, phonographic products, and safety glass layers.

Chemalloy. Trade name of Abex Corporation (USA) for a series of iron-nickel-chromium casting alloys with excellent corro-

sion resistance used for chemical and refinery equipment, pulp and paper mills, retorts, turbines, valves, pumps, condensers, etc.

Chemax. Trade name of Pax Surface Chemicals, Inc. (USA) for conversion coatings for zinc, cadmium and brass.

Chembond. Trademark of Chembond Limited (Canada) for a construction adhesive.

Chemcalk. Trademark of Chemtron Manufacturing Limited (Canada) for caulking compounds.

Chemcast. (1) Trademark of Clayburn Refractories Limited (Canada) for a chemically-bonded gunning castable.

(2) Trade name of Chemtech International Corporation (USA) for cell-cast acrylic sheeting.

Chemclad. Trademark of Enecon Corporation (USA) for chemically resistant polyvinyl chloride and other polymeric coating compounds for metal, concrete and masonry surfaces.

Chemcor. Trade name of Corning Glass Works (USA) for chemically strengthened glass.

Chem Cote. Trade name of Brent America Inc. (USA) for iron and zinc phosphate conversion coatings.

Chemelec. Trade name for a series of polytetrafluoroethylene resins.

Chemester. Trade name of Atlas Minerals & Chemicals Inc. (USA) for a chemically resistant vinyl ester resin for floor topping applications.

ChemFil. Trade name for glass-ionomer composite resins for dental restorative applications.

ChemFil Superior. Trade name for a glass-ionomer dental filling cement.

Chemfilm. Trademark of Chemfab Corporation (USA) for a series of polytetrafluoroethylene films with one side modified for adhesive or heat bonding. Used for wire and cable wrap, low-noise coaxial cables, printed circuits, and in the manufacture of aerospace and automotive composites.

Chemfilm MR. Trademark of Chemfab Corporation (USA) for conformable, high-temperature parting, release and vacuum-bagging films based on advanced fluoropolymers. They have excellent drapability, excellent chemical inertness and release characteristics, and a maximum use temperature to 400°C (752°F). Used for thermoset and thermoplastic composites.

Chemfit. Trade name for nylon fibers and yarns used for textile fabrics.

Chemflon. Trademark of C&K Trading Company (South Korea) for fluoropolymer-based nonstick coatings for cookware.

Chemfloor. Trademark of Norton Chemplast (USA) for impact- and wear-resistant polytetrafluoroethylene in fibrous-porous form for flooring applications.

Chemfluor. Trademark of Norton Performance Plastics (USA) for a series of fluoropolymers including nonporous, chemically inert polytetrafluoroethylene (PTFE) polymers with a maximum service temperature of 93°C (200°F), and perfluoroalkoxy (PFA) polymers with superior chemical resistance and a maximum service temperature of 120°C (248°F). Supplied in the form of sheets, pipes, tubes and rods, they are used for chemical equipment, fittings, and diaphragm valves.

Chemfos. Trade name of Pax Surface Chemicals Inc. (USA) for a phosphate coating and process.

Chemgard Epoxy. Trademark of Shamrock Construction Chemicals (Canada) for acid-, alkali-, and salt-resistant epoxy coatings for metals.

Chemgard Platecoat. Trademark of Shamrock Construction Chemicals (Canada) for electroplating coatings.

Chemgard Stripcoat. Trademark of Shamrock Construction Chemicals (Canada) for a strippable coatings for radiation-control applications.

Chemgard Topcoat. Trademark of Shamrock Construction Chemicals (Canada) for acrylic coatings for external industrial applications.

Chemglas. (1) Trade name of Chemfab Corporation (USA) for oil- and water-resistant polytetrafluoroethylene (*Teflon*)- or silicone-coated glass fibers used in the manufacture of fabrics, and various industrial applications, e.g., packaging and insulation. Also included under this trade name are *Chemglas-UVR* ultraviolet-resistant *Teflon*- or silicone-coated composite fabrics.

(2) Trademark of Reichhold Chemicals Inc. (USA) for glass rovings.

Chemglaze. (1) Trademark of Lord Corporation (USA) for a series of high-performance polyurethane coatings for metal, plastics, leather, glass, wood, concrete, plastics, textiles, and elastomers including neoprene, natural and styrene-butadiene rubbers. They have good resistance to abrasion, wear, and chemicals. Also included under this trademark are various oil-free, moisture-curing polyurethane coatings, wood fillers and metal primers.

(2) Trademark of Lord Corporation (USA) for a series of decorative or protective paint-like coating compositions as well as fillers, sealers, primers, overlayments, etc.

(3) Trademark of Chemtron Manufacturing Limited (Canada) for glazing compounds and putties.

Chemgrate. Trademark of Grating Pacific, LLC (USA) for molded gratings made by combining specially formulated resin mixtures with continuous fiberglass stands or roving.

Chemiace II. Trade name of Sun Medical Company Limited (Japan) for a dual-cure, water-resistant, radiopaque dental resin cement containing the adhesive monomer 4-methacryloxyethyl trimellitate anhydride (4-META). It provides high bond strength, and is used in restorative dentistry for crowns, bridges, inlays and onlays.

chemical brick. Chemical stoneware in brick form used for lining chemical process equipment, such as reaction chambers, tanks, storage vessels, etc.

chemical cellulose. See alpha cellulose.

chemical conversion coatings. Protective or decorative adherent surface layers of low-solubility oxide, phosphate or chromate compounds produced *in situ* by the chemical reaction of suitable reagents with a metallic surface. They modify several surface properties including appearance, electrochemical potential, electrical resistivity, hardness and absorption, and are often used as preparatory coatings form improving the adhesion of supplementary organic finishes or films.

chemical conversion pulp. Wood pulp consisting of cellulose from which most of the non-cellulosic matter has been removed by any of the following chemical processes: (i) the sulfate process in which sodium sulfate is added to the caustic soda digestion liquor; (ii) the sulfite process in which wood chips are digested with a solution of magnesium, ammonium or calcium disulfite containing free sulfur dioxide; (iii) the soda process in which wood chips are digested by caustic soda; or (iv) the holopulping process in which wood fibers are delignified by alkaline oxidation of extremely thin wood chips followed by solubilization of the lignin fraction. Chemical pulp is suitable for further chemical processing. Also known as *chemical pulp; chemical wood pulp.* See also wood pulp.

chemical cotton. Alpha cellulose prepared from cotton or cotton linters. See also alpha cellulose.

chemical glass. A special glass, usually of the soda-lime-borosilicate type, with excellent chemical durability and resistance to corrosive (acidic) environments. Used for laboratory apparatus, chemical glassware, etc. See also C-glass.

chemical grout. A *grout* that is a true solution and contains no suspended particles.

chemical lead. A grade of *pig lead* containing 99.90+% lead, 0.040-0.080% copper, 0.002-0.020% silver, up to 0.005% bismuth, up to 0.002% iron, up to 0.001% zinc, and up to a total of 0.002% arsenic, antimony and tin. Available in cast and rolled form, it is used for storage battery plates, chemical piping, tank linings, sheet, chemical valves and piping, fuses, and cable coverings.

chemical lime. A high-purity hydrated lime containing less than 3% magnesium and less than 2% insoluble matter, and having chemical and physical properties which make it suitable for use in the chemical industries.

chemically active pigment. (1) Any pigment that reacts with the oil of the vehicle to form a soap in order to influence film toughness, increase durability, etc.

(2) A pigment, such as red lead, that reacts with acids formed at metallic surfaces to prevent corrosion.

chemically bonded brick. A brick made by a process in which mechanical strength is not developed by firing, but by the action of chemical bonding agents added to the raw batch.

chemically bonded ceramics. Ceramic products obtained from raw batches, or mixtures by the addition of one or more selected chemical bonding agents. Firing is not required, since the agent causes hardening and setting at room or moderately elevated temperatures resulting in the desired mechanical strength. Abbreviation: CBC.

chemically bonded refractory cement. A finely divided air-setting (or air-hardening) refractory cement containing chemical agents that cause setting or hardening at room temperature or above, but below the vitrification temperature.

chemically deposited metal coating. A coating produced by the deposition of a metal from a solution of its salts, or by the precipita-tion of a metal from a gaseous compound.

chemically foamed plastics. Thermoplastic or thermosetting resins processed into flexible or rigid foams with numerous small, closed or interconnected cells produced by gases generated by the chemical reaction of the components (e.g., by thermal decomposition). Also known as *chemically foamed polymers.*

chemically precipitated metal powder. A fine powder produced by the reduction of a metal from a solution of its salts by a less noble metal, or other reducing agent.

chemically reactive coatings. Two-component coatings that cure on combination by chemical reaction. Examples are polyamide- or amino-cured epoxies, amine-cured coal-tar epoxies, urethanes, inorganic zinc, etc. They provide an exceptional combination of durability, and water, solvent and chemical resistance.

chemically strengthened glass. A special glass whose surface has been treated by an ion-exchange process to alter its composition or structure and produce a layer of high compressive stresses (often exceeding 690 MPa, or 100 ksi).

chemical porcelain. A high-quality, hard-glazed, vitreous porcelain produced from a blend of kaolin, clay, feldspar and quartz. It is of white color, usually transparent, and highly resistant to chemicals and acids, except hydrofluoric acid. Used for chemical and laboratory equipment and ware, e.g., tanks, storage vessels, reaction equipment, pumps, valves, fittings, and pipes.

chemical pulp. See chemical conversion pulp.

chemical reaction paper. A paper that produces a visible image by chemical reaction of certain colorless additives upon application of pressure.

chemical reduction coating. A coating produced by depositing a metal from a chemical solution onto metal or plastic substrates to form a mirror-like film whose thickness is between those formed by vapor deposition and electroplating.

chemical-resistant cement. A Portland cement (especially ASTM types II, IV and V) having high tetracalcium aluminoferrate (C_4AF) and low tricalcium aluminate (C_3A) contents, and containing special additives, such as water glass, that make it more resistant to chemicals than standard grades. It has relatively low heat of hydration, usually good sulfate resistance, and is used in the manufacture of chemical-resistant concrete.

chemical-resistant concrete. A concrete containing a Portland cement of high tetracalcium aluminoferrate (C_4AF) and low tricalcium aluminate (C_3A) contents together with special additives, such as calcium soaps, water glass, etc., that make it resistant to chemicals.

chemical-resistant material. A material that possesses the ability to resist degradation by chemical reactivity or solvent action.

chemical-resistant polymer concrete. A chemically resistant composite consisting of an aggregate and a polymeric binder. Used as special concrete for applications requiring high chemical resistance. See also polymer concrete.

chemical-resistant resin grout. A mixture of a liquid resin, a hardener and a filler that sets by chemical action.

chemical-resistant resin mortar. A trowelable mixture of a liquid resin, a hardener and a filler that sets by chemical action.

chemical stoneware. Ceramic products made from blends of selected iron- and lime-free clays having low sand, kaolin, feldspar and quartz contents. They have low firing shrinkage, low water absorption (up to 0.4%), and excellent resistance to acids (except hydrofluoric acid) and alkalies (except strong hot caustics). *Chemical stoneware* has a density of 2.2 g/cm³ (0.08 lb/in.³), a Scleroscope hardness of 100, low tensile strength, good compressive strength, and low thermal-shock resistance. Used for utensils, pipe, stop cocks, pumps, valves, tubes, fittings, laboratory ware, and chemical apparatus.

chemical-vapor-deposited carbon. Carbon deposited on a material by thermal decomposition of a hydrocarbon compound, such as methane. Abbreviation: CVD carbon.

chemical vapor deposition coatings. Coatings deposited from the gaseous phase by the chemical vapor deposition (CVD) process. In this process a gaseous mixture is passed over the parts to be coated at high temperature. A chemical reaction takes place between the gases whereby a dense, even layer of uniform thickness and excellent adhesion is built up on the surface of the part. The composition of the coating can be controlled by modifying the gas mixture. Abbreviation: CVD coatings.

chemical wood pulp. See chemical conversion pulp.

Chemigum. Trademark of Goodyear Tire & Rubber Company (USA) for a series of thermoplastic elastomers and latexes including acrylonitrile-butadiene and styrene-butadiene copolymers and polyester rubbers.

Chemiko. Trade name of CHEMIKO (Germany) for one- and two-component epoxy, polyester, cyanoacrylate, anaerobic and dispersion adhesives used in the metal, marble, furniture and electrical industries.

chemiluminescent material. A luminous substance that emits electromagnetic radiation (e.g., light) as the result of an internal chemical reaction.

Chemline. Trade name for nylon fibers and yarns.

Chemlok. Trademark of Hughson Chemical Company (USA) for a series of one-coat, two-coat and water-borne cyanoacrylate adhesives used for bonding unvulcanized and vulcanized rubber to metal and other substrates.

Chemlon. Trademark of Chemlon AS (Slovak Republic) for nylon 6 fibers and yarns.

Chemlux. Trade name for nylon fibers and yarns.

Chemocure. Trademark of Chemor Inc. (Canada) for concrete curing membranes.

Chemodent. Trade name of Chemodent Dental Laboratory (USA) for a precious-metal dental ceramic alloy.

Chemoplast. Trademark for cresol-formaldehyde and phenol-formaldehyde plastics and compounds.

Chemopoxy. Trade name of Chemor Inc. (Canada) for epoxy ester coatings.

Chemorane. Trademark of Chemor Inc. (Canada) for clear polyurethane finishes for wood flooring and other interior wood surfaces.

Chemorciment. Trademark of Chemor Inc. (Canada) for fast-setting mortars.

Chemorclad. Trademark of Chemor Inc. (Canada) for protective epoxy coatings.

Chemorcrete. Trademark of Chemor Inc. (Canada) for emulsion-type concrete adhesives.

Chemorpatch. Trademark of Chemor Inc. (Canada) for epoxy mortar used for concrete repairs.

Chemorplex A. Trademark of Chemor Inc. (Canada) for expansive chemical agents for grouts and mortars.

Chemorquartz. Trademark of Chemor Inc. (Canada) for epoxy floor toppings.

Chemorset E. Trademark of Chemor Inc. (Canada) for epoxy adhesives for metal, wood, ceramic tile, stone and concrete.

Chemortar. Trademark of Chemor Inc. (Canada) for cementitious mortars.

Chemortop. Trademark of Chemor Inc. (Canada) for a series of cementitious adhesive mortars, coal-tar epoxy coatings and waterproof floor surfacings.

Chemortex. Trademark of Chemor Inc. (Canada) for coatings and finishes including *Chemortex C* cementitious coatings and *Chemortex S* textured finishes.

Chem-Peel. Trademark of Chemco Manufacturing Company Inc. (USA) for strippable spray booth coatings.

Chempol. Trademark of Cook Composites & Polymers Company (USA) for a series of unmodified and modified polyurethanes used in protective coatings and varnishes for plastics and paper. Supplied as hard-cast, hard, soft and reinforced microcellular elastomers, and semi-rigid and structural foams.

Chempruf. Trade name of Atlas Minerals & Chemicals, Inc. (USA) for fiberglass flake-reinforced high-build polymer coatings, usually 0.5-1.5 mm (20-60 mils) thick, applied by brushing, rolling or spraying to floors, structural steel and interior and exterior tank surfaces. They are supplied in a wide range of polymer resins including bisphenol epoxy, novolac bisphenol F epoxy, bisphenol A fumarate, vinyl ester, novolac vinyl ester, and chlorendic anhydride polyester.

chempure tin. A grade of commercial tin, 99.90-99.95% pure, obtained by electrolytic refining.

ChemRex. Trademark of ChemRex Inc. (USA) for a series of polyurethane adhesives, sealants, and coatings.

Chem-Rock. Trade name of Rock-Tred Corporation (USA) for epoxy coatings with 100% solids.

Chemset. Trademark of Sandoz AG (Switzerland) for chemical-resistant epoxy grouting and setting compounds for floor brick, tiles, and construction joints.

Chemsil. (1) Trademark of Chemtron Manufacturing Inc. (Canada) for a silicone sealant used for porous surfaces.

(2) Trademark of Chemfab Corporation (USA) for glass fabrics coated with silicone available in sheet and roll form.

Chemstik. Trade name of Chemfab Corporation (USA) for a series of polytetrafluoroethylene (*Teflon*) coated fiberglass fabrics with adhesive backing on one side.

Chemstone. (1) Trademark of Chemstone Veneering Company Limited (Canada) for precast concrete.

(2) Trademark of Manville Corporation (USA) for chemical- and fire-resistant, impregnated asbestos-cement slabs and sheets for siding, wallboard, etc.

Chemstrand. Trade name of Chemstrand Corporation (USA) for acrylic, nylon and polyester fibers used for the manufacture of textile fabrics.

Chem-Thane. Trade name of Rock-Tred Corporation (USA) for water-based urethane polymer coatings.

Chemtile. Trademark of Chemstone Veneering Company Limited (Canada) for precast concrete.

Chemtool. Trade name of Catarpillar Inc. (USA) for industrial-grade contact adhesives.

Chemtree. Trademark of Chemtree Corporation (USA) for a series of metallic mortars supplied as dry powders composed of high percentages of lead and an organic binder and, sometimes, neutron-absorbing isotopes. They are mixed with water for application as neutron attenuators in nuclear shielding.

Chemtrol. Trade name of Precision Finishing Inc. (USA) for mass-finishing compounds.

Chem-X. Trade name of Chemtron Manufacturing, Limited (Canada) for sealants based on a styrene-butadiene rubber/block polymer.

chenevixite. A dark green, olive green to greenish yellow mineral composed of copper iron arsenate hydroxide monohydrate, $Cu_2Fe_2(AsO_4)_2(OH)_4 \cdot H_2O$. Crystal system, monoclinic. Density, 4.38 g/cm^3; refractive index, 1.96. Occurrence: Mexico.

Chenite. British trade name for an air-hardening alloy of titanium, vanadium, chromium and hafnium. Used for cutting tools and dies.

chenille. (1) A yarn, usually of cotton, wool, silk, rayon or synthetic fibers, that has a soft, fuzzy pile protruding from its surface.

(2) A textile fabric woven from chenille yarn (1) and used for bedspreads, curtains, coverings, rugs, robes, dresses, coats, etc.

cheralite. A pale to dark green mineral of the monazite group composed of rare-earth thorium calcium uranium phosphate silicate, $(Ln,Th,Ca,U)(PO_4,SiO_4)$. Crystal system, monoclinic. Density, 5.30 g/cm^3; refractive index, 1.780. Occurrence: India.

Cheratolo. Trademark of Montecatini Edison SpA (Italy) for phenol-formadehyde compounds and plastics.

chernovite. A colorless to pale yellow mineral of the zircon group composed of yttrium arsenate, $YAsO_4$. It can also be made synthetically. Crystal system, tetragonal. Density, 4.85 g/cm^3; re-

fractive index, 1.783. Occurrence: Russia.

chernykhite. An olive-green mineral of the mica group composed of sodium barium aluminum vanadium silicate hydroxide, $(Ba,Na)_{0.55}(V,Al,Mg)_{2.28}(Si,Al,V)_4O_{10}(OH)_2$. Crystal system, monoclinic. Density, 3.14 g/cm^3; refractive index, 1.691. Occurrence: Kazakhstan.

Cherokee. Trade name of AL Tech Specialty Steel Corporation (USA) for a high-carbon, high-chromium cold-work tool steel (AISI type D5) containing 1.5% carbon, 12% chromium, 0.5% nickel, 0.7% molybdenum, 3% cobalt, and the balance iron. It has a great depth of hardening, good abrasion and wear resistance, and good nondeforming properties. Used for cold-forming dies, blanking and coining dies, draw and trimming dies, punches, etc.

cherry. The wood of any of various trees of the genus *Prunus* belonging to the rose family, especially the strong, stiff, moderately heavy wood of the black cherry (*P. serotina*) growing throughout the southeastern Canada and eastern United States, and the English cherry (*P. cerasus*), native to Europe. See also black cherry; English cherry.

cherry birch. See black birch.

cherry mahogany. See African cherry.

chert. A light gray, fine-grained silica rock resembling *flint*, and containing cryptocrystalline or microcrystalline *quartz* and hydrated silica (*chalcedony* and/or *opal*). It has a splintery fracture, and is sometimes used as an abrasive, and in building and paving. Also known as *hornstone*.

chervetite. A colorless mineral composed of lead vanadium oxide, $Pb_2V_2O_7$. Crystal system, monoclinic. Density, 6.49 g/cm^3; refrac-tive index, 2.2-2.6. Occurrence: Africa (Gabon).

chessexite. A white mineral composed of sodium calcium magnesium zinc aluminum silicate sulfate hydroxide hydrate, $Na_4Ca_2(Mg,Zn)_3Al_8(SiO_4)_2(SO_4)_{10}(OH)_{10} \cdot 40H_2O$. Crystal system, orthorhombic. Density, 2.21 g/cm^3; refractive index, 1.460. Occurrence: France.

chessy copper. See azurite.

chessylite. See azurite.

chesterite. A colorless to very light pinkish brown mineral of the amphibole group composed of iron magnesium silicate hydroxide, $(Mg,Fe)_{17}Si_{20}O_{54}(OH)_6$. Crystal system, orthorhombic. Density, 3.23 g/cm^3; refractive index, 1.6320. Occurrence: USA (Vermont).

Chesterton. Trade name of A.W. Chesterton Company (USA) for asbestos gasketing and packing materials.

chestnut. The wood of any of various trees of the genera *Castanea* and *Aesculus* belonging to the beech family, especially the American chestnut (*C. dentata*), the European or sweet chestnut (*C. sativa*), and both the American horse chestnut or buckeye (*A. glabra*) and the European horse chestnut (*A. hippocastanum*). See American chestnut; sweet chestnut; horse chestnut.

chestnut oak. (1) The wood of the medium-sized upland oak tree *Quercus prinus* found in southern Ontario, Canada, and throughout most of the eastern United States, especially along the Appalachian Mountains from Southwestern Maine to Northern Georgia and Alabama. Used for hardwood lumber and construction work, it is similar to *Chinkapin oak* and *swamp oak*.

(2) The strong, light-brown wood of the oak tree *Quercus petraea*. It has a coarse, open grain, a firm texture, and good strength and durability. Average weight, 720 kg/m^3 (45 lb/ft^3). Source: Europe including British Isles. Used for constructional applications. Also known as *Austrian oak; mountain oak.* See also English oak.

Chetra. Trademark of CHETRA GmbH (Germany) for cyanoacrylate, anaerobic and several specialty adhesives.

Cheviot. Trade name for rayon fibers and yarns used for textile fabrics.

cheviot. (1) A rough, coarse, heavily napped cotton or cotton blend fabric used for shirts and coats.

(2) A rough woolen fabric closely woven in a twill weave and having a nap. Used for overcoats, suits and casual wear. It was originally made from wool obtained from a breed of sheep native to the Cheviot Hills in Scotland.

Chevis. Trade name of George Cook & Company Limited (UK) for a water-hardening tool steel used for punches, and dies for sheetmetal blanking and forming.

Chevisol. Trade name for rayon fibers and fabrics.

chevkinite. (1) A colorless or pink natural mineral composed of calcium cerium iron titanium silicate, $(Ca,Ce)_4(Fe,Mg)_2Ti-Fe_3Si_4O_{22}$. Crystal system, monoclinic. Density, 4.53 g/cm³. Occurrence: USA (Kansas).

(2) A colorless or pink synthetic mineral composed of cerium iron titanium silicate, $Ce_4Fe(Fe,Ti)Ti_2Si_4O_{22}$. Crystal system, monoclinic. Density, 5.13 g/cm³.

Chevre. Trade name of Chiers-Chatillon (France) for a series of water-hardening tool steels containing 0.9-1.1% carbon, and the balance iron. Used for tools, taps, drills, dies and springs.

chevreau leather. A fine, chrome-tanned, aniline-dyed goatskin used for shoe uppers.

Chevron. Trade name for high-density polyethylene plastics.

Cheyenne. Trade name of AL Tech Specialty Steel Corporation (USA) for a general-purpose high-speed steel (AISI type M2) containing 0.8% carbon, 4% chromium, 6% tungsten, 5% molybdenum, 2% vanadium, and the balance iron. It has a great depth of hardening, excellent wear resistance, good toughness, and high hot hardness. Used for blanking and forming dies, punches, and cutting tools.

Chiaro. Trade name of Pittsburgh Corning Corporation (USA) for hollow glass blocks having raised surface patterns with fired-on black ceramic material.

chiastolite. A white variety of the mineral *andalusite* (Al_2SiO_5) containing carbonaceous impurities forming a black cross. Used as a gemstone.

chiavennite. An orange or reddish-orange mineral composed of calcium manganese beryllium silicate hydroxide dihydrate, $CaMnBe_2Si_5O_{13}(OH)_2\cdot 2H_2O$. Crystal system, orthorhombic. Density, 2.56-2.64 g/cm³; refractive index, 1.596. Occurrence: Italy, Norway.

Chicago. Trade name of Engelhard Corporation (USA) for high-platinum gold alloys used in dentistry, and supplied in the annealed and age-hardened condition. Depending on the particular composition, they have melting temperatures ranging from 890 to 940°C (1635 to 1725°F).

Chickasaw. Trade name of AL Tech Specialty Steel Corporation (USA) for an oil-hardening, shock-resisting tool steel (AISI type S5) containing 0.6% carbon, 0.85% manganese, 2% silicon, 0.45% molybdenum, 0.25% vanadium, and the balance iron. It has a high elastic limit and good ductility, excellent toughness and shock resistance, and moderate wear resistance. Used for dies, punches, stamps, shear blades, and machinery parts.

Chiemlon. Trademark of Chiem Patana Synthetic Fibers Company (Thailand) for polyester fibers and yarns.

chiffon. A delicate fabric made of silk, rayon, polyester, nylon, cotton, worsted, etc., usually in a plain weave. It is very thin and light, usually soft, and used for blouses, dresses, scarves, lingerie, veils, etc.

Chilac. (1) Trademark of Winter & Co. GmbH (Germany) for artificial leather and other leather substitutes.

(2) Trademark of James River Corporation of Virginia (USA) for vinyl-coated materials for bookbinding applications.

Chilcote. Trade name of Foseco Minsep NV (Netherlands) for a series of coatings for application to metal chills that are inserted in sand molds. It provides improved surface finish of the casting, and increases the service life of the chills.

childrenite. A yellowish-brown, brown, or pink mineral composed of iron manganese aluminum phosphate hydroxide monohydrate, $(Fe,Mn)AlPO_4(OH)_2\cdot H_2O$. Crystal system, orthorhombic. Density, 3.11-3.24 g/cm³; refractive index, 1.66; hardness, 4.5-5 Mohs. Occurrence: Germany.

Chili bar. A free-machining grade of commercially pure copper containing 1% sulfur, and traces of other impurities. Used for automatic screw-machine products.

chill coating. (1) A dressing for application to a chill that is inserted in a mold cavity to prevent the molten metal from adhering to it.

(2) A dressing for application to the surface of a sand mold to effect undercooling of the molten metal.

chilled cast iron. A *white cast iron* that has been poured into molds faced with iron or steel to effect rapid solidification and produce castings with hard, abrasion- and wear-resistant surface layers of white iron, and tough interiors of gray iron. It can only be machined by grinding or with carbide tools. Thin-walled castings are through-hardening. Fully hardened solid castings are obtained by lowering the silicon content, or by adding manganese until it outweighs the effect of silicon. A typical composition is 2.8-4.0% carbon, 0.2-1% silicon, 0.6-1.5% manganese, 0.2-0.5% phosphorus, 0.08% sulfur, and the balance iron. Used for hydraulic pistons, crusher parts, railroad wheels, rolls for rolling mills and for printing, paper, rubber and plastics machinery, for grinders, slurry pumps, brake shoes, and various other industrial components. Also known as *chilled iron*.

chilled shot. (1) Shot made from chilled cast iron. It has high hardness (58-65 Rockwell C), and is frequently used for shot peening of carbon and low-alloy steels, and for abrasive blast cleaning of ferrous and nonferrous parts. See also shot.

(2) Lead shot hardened with 3-6% antimony. See also lead shot.

Chimitex. Trade name of Chimitex Cellchemie GmbH (Germany) for carboxymethylcellulose.

china. A fine, vitreous translucent ceramic whiteware, glazed or unglazed, made of high-grade *kaolin* and baked at high temperatures. It may have colored designs baked in. Examples include dinnerware, sanitary ware, artware, and other nontechnical products.

China brass. A Chinese brass containing 56.6% copper, 27-37% zinc, 0.2-1% tin and 0-0.8% lead. It has good workability, and is used for hardware and sheet products

china clay. See kaolin.

China grass. See ramie.

china sanitary ware. Glazed, vitrified whiteware used for plumbing fixtures and other sanitary applications.

China silk. See habutai.

china stone. A weathered *granite* whose chief constituents are quartz, feldspar and fluorine minerals. *Cornish stone* found in the United Kingdom is a variety. Used as a flux in the production of ceramic whitewares.

chinchilla. (1) The soft, fine silvery-gray, highly valued fur of the chincilla (*Chinchilla laniger*) native to the South American mountains.

(2) A thick, durable, twill-weave wool fabric that has a long nap or small closely set tufts or nubs imitating the fur of the chinchilla. Used for overcoats and sportswear. Also known as *chinchilla cloth.*

Chinese art metal. A wrought copper alloy containing 17.5% lead, 10% zinc and 1% tin. Used for ornamental and architectural applications.

Chinese blue. (1) A bluish black mineral aggregate composed of hydrated oxides of cobalt and manganese. Used as an underglaze for porcelain. Also known as *Chinese cobalt.*

(2) A term used for any of the following sky- to grayish-blue pigments: (i) a mixture of *ultramarine* (mixture of kaolin, soda ash, sulfur, and carbon, or charcoal) and *flake lead* (basic lead carbonate); and (ii) a mixture of *white lead* (basic lead carbonate) and *cobalt blue* (a mixture of cobalt oxide and aluminum oxide).

Chinese bronze. (1) A heavy-duty leaded bronze containing 53-88% copper, 10-20% lead, 1-14% zinc, and 1-13% tin. Used for bearings.

(2) A corrosion-resistant nickel bronze containing at least 10% nickel. It has a characteristic white color, and is used for decorative parts and hardware.

Chinese cinnabar. A fine, light, brilliant red pigment made from *cinnabar* by grinding, and used in paints and inks

Chinese cobalt. See Chinese blue (1).

Chinese German silver. A Chinese *nickel silver* containing 40.4-41% copper, 25.4-26.5% zinc, 30.8-31.6% nickel, and 2.6-2.7% iron. It has excellent corrosion resistance, and is used for electrical resistances and ornaments.

Chinese insect wax. See Chinese wax.

Chinese lacquer. A glossy black finishing material made from the gum exudation of the sumac plant (*Rhus vernicifera*) by mixing with oils. It dries by evaporation of the thinner or solvent, and is used as a varnish to give a shiny appearance to metals, wood, paper, etc.

Chinese red. Any of various red and orange pigments made by mixing lead chromate ($PbCrO_4$), and lead oxide (PbO) in various proportions. It blends well to the chrome red pigments.

Chinese scarlet. (1) A red to orange pigment composed chiefly of basic lead chromate ($PbCrO_4 \cdot PbO$). It is one of the *chrome red* pigments. Also known as *American vermilion.*

(2) A light to brilliant red pigment usually composed of lead molybdate, lead chromate and lead sulfate. Also known as *scarlet chrome.*

(3) A light to brilliant red pigment composed of an organic dye, lead oxide and barium sulfate.

Chinese silver. An alloy containing 58% copper, 17.5% zinc, 11.5% nickel, 11% cobalt, and 2% silver. It has good corrosion resistance, and is used for imitating silver, for ornaments, and for electrical parts.

Chinese speculum. A reddish alloy of 81% copper, 2-11% tin and 8-8.5% antimony. A special leaded grade containing 81% copper, 9% lead, 8% antimony and 2% tin is also produced. Used for mirrors.

Chinese vermilion. See red mercuric sulfide.

Chinese wax. A colorless, crystalline wax obtained from the secretion of certain Chinese plant lice. It is composed principally of ceryl cerotate ($C_{26}H_{53} \cdot OOCC_{25}H_{51}$), and used for candles, and

leather and furniture polishes. Also known as *Chinese insect wax.*

Chinese white. See zinc white.

Chinese yellow. A raw (unburnt) yellow *ocher* used as a pigment in paints, paper, etc.

chinkapin oak. The wood of the medium-sized upland tree *Quercus muehlenbergii* belonging to the white oak group. It is quite similar to the chestnut and live oak. Also spelled "chinquapin." See also American white oak; white oak.

Chinlon. Trade name for nylon fibers and yarns.

chino. A strong cotton or cotton-blend twill fabric of medium weight, often with a smooth, lustrous finish and light to medium brown or yellowish brown color. Used especially for sportswear, workwear and military uniforms.

Chinon. Trademark of Toyobo Company (Japan) for azlon fibers and yarns produced by grafting milk protein with an acrylic polymer.

chinone. See p-benzoquinone.

Chinsang. Trade name for rayon fibers and yarns used for textile fabrics.

chintz. A firm, plain-woven cotton, cotton-blend or linen fabric, usually printed with colorful designs, and having a bright, glazed or wax-coated surface. Used for dresses, draperies, slipcovers, etc.

chiolite. A colorless, or snow-white mineral composed of sodium aluminum fluoride, $Na_5Al_3F_{14}$. It can also be made by treating sodium hydrogen carbonate ($NaHCO_3$) and aluminum powder with hydrofluoric acid (HF) solution. Crystal system, monoclinic. Density, 3.00 g/cm^3; refractive index, 1.3486. Occurrence: Russia.

chipboard. (1) A building board made in large rigid panels by compressing wood chips and fibers, and bonding with synthetic resin.

(2) A lightweight building board of low density, strength and quality made from waste paper, woodpulp, chips, etc., usually on a cylinder-type papermaking machine. Used for book covers, boxes, paneling, and gaskets.

Chippaway. Trade name of Jessop Steel Company (USA) for a water-hardening tool steel containing 0.6-1.4% carbon, and the balance iron. Used for pneumatic hammers, chisels, punches, vise jaws, and rock drills.

chipped glass. A glass product whose surface has been purposely chipped to produce special surface effects.

Chipper. Trade name of Uddeholm Corporation (USA) for a cold-work tool steel containing 0.5% carbon, 1% silicon, 0.5% manganese, 8% chromium, 1.5% molybdenum, 0.5% vanadium, and the balance iron. Used for knives, blades, shears, and dies.

Chipper Knife. (1) Trade name of Disston Inc. (USA) for a tool steel of 0.81% carbon, 11.5% chromium, 0.4% molybdenum, and the balance iron. It has good to fair corrosion resistance, and high hardness. Used for chipper knives and dies.

(2) Trade name of Disston Inc. (USA) for a tool steel of 0.7% carbon, 1.15% chromium, 0.7% molybdenum, 0.25% vanadium, and the balance iron. Used for chipper knives, shears, dies, etc.

Chippewa. Trade name of Atlas Specialty Steels (Canada) for a hollow shank steel containing 0.33% carbon, 0.6% manganese, 0.25% silicon, 0.4% chromium, 3% nickel, 0.25% molybdenum, 0.02% phosphorus, 0.02% sulfur, and the balance iron. It has a good combination of high toughness and strength, and is used for detachable mining bit shanks, and hollow shanks for sectional or long-hole drilling.

chiral liquid crystal. See chiral nematic liquid crystal; chiral smectic liquid crystal.

chiral nematic liquid crystal. A *liquid crystal* whose molecules are chiral, i.e., not symmetric when reflected, and preferably lie side by side in a slightly skewed orientation. The induced director (i.e., the preferred molecular orientation direction) configuration gives this material a helical internal structure. This structure can be destroyed by the application of a voltage eliminating its light transmissability. Chiral nematics are used for twisted nematic liquid crystal displays with the zero-voltage light-transmitting state being white and the non-zero non-transmitting state black. See also chiral smectic liquid crystal.

chiral smectic liquid crystal. A *liquid crystal* that exhibits a chiral smectic C (SmC*) phase in which the director (i.e., the direction of preferred molecular orientation) is tilted and gradually rotates from layer to layer forming a helical structure. See also chiral nematic liquid crystal.

chirimen. A coarse crepe fabric, usually of silk or polyester, having a dull finish.

Chirulen. Trade name for a tough semicrystalline ultrahigh-molecular-weight polyethylene with a density of 0.94 g/cm^3 (0.034 lb/in.^{3}), excellent wear and abrasion resistance, high impact resistance, high melt viscosity, low coefficient of friction, and good resistance to most chemicals except halogens and aromatic hydrocarbons. Used for industrial equipment and machine components.

chisel steel. (1) A steel containing 0.3-0.4% carbon, 0-0.6% chromium, 0-3% nickel, and the balance iron. It has a high surface hardness and abrasion resistance, and is used for hand tools, chisels, hammers, etc., and pneumatic tools.

(2) A steel that contains 0.9-1% carbon, has good forgeability, and is used for cold chisels, punches, mining drills, etc.

(3) A low-alloy medium-carbon steel that contains varying amounts of tungsten and chromium, and is used for chisels, boring and chipping tools, etc.

Chisso Polypro. Trademark of Chisso Polypro Fiber Company (Japan) for polypropylene fibers and yarns.

chitin. A glucosamine polysaccharide that is a constituent of the shells or integuments of crustaceans, arachnids and insects, such as crabs, shrimp, lobsters and beetles. It is available in practical and purified grades as a white, amorphous, semitransparent mass or powder, and is used as a source of *chitosan*, and in biology, biochemistry and biotechnology. Formula: $(C_8H_{13}NO_5)_n$.

Chitonal. Trade name of Lavorazione Leghe Leggere SpA (Italy) for a series of wrought heat-treatable aluminum-copper alloys with small additions of silicon, magnesium, and/or manganese. Used for strong lightweight and structural applications, e.g., aircraft components.

chitosan. A polymer derivative of *chitin* obtained by deacetylation, and available with high, medium and or low molecular weights. Used as a biomaterial in wound dressings, in photographic emulsions, as a dyeing assistant, as a heavy metal absorber in industrial wastewater treatment, and as coatings in manufacture of certain paper and sheet products.

chitosan-coated fibers. Fibers of ceramics, glass, metals, wood, or polymers (e.g., polypropylene) that have been coated with *chitosan* by first suspending them into an agitated acidic solution of the latter, and then adding an alkaline substance (e.g., NaOH). Many of these coated fibers can be made into useful products, e.g., sheet-like materials, that have properties far superior to those made from uncoated fibers.

Chi-Vit. Trademark of Chi-Vit Corporation (USA) for porcelain enamel frits for coating metals.

Chiz-Alair. Trade name of Kloster Steel Corporation (USA) for an air-hardening, shock-resistant tool steel (AISI type S7) containing 0.5% carbon, 0.7% manganese, 0.25% silicon, 3.25% chromium, 1.4% molybdenum, and the balance iron. It has a great depth of hardening, and excellent toughness. Used for dies, chisels, punches, chucks, etc.

Chiz-Alloy. Trade name of Kloster Steel Corporation (USA) for an oil-hardening, shock resisting tool steel (AISI type S1) containing 0.5% carbon, 1.15% chromium, 2.5% tungsten, 0.2% vanadium, 0.75% silicon, and the balance iron. It has a great depth of hardening, high hardness and strength, excellent toughness, good wear and abrasion resistance, and good hot hardness. Used for punches, and hand and pneumatic chisels.

chkalovite. A white mineral composed of sodium beryllium silicate, $Na_2BeSi_2O_6$. It can also be made synthetically. Crystal system, orthorhombic. Density, 2.66 g/cm^3; refractive index, 1.549. Occurrence: Russia.

chloraluminite. A colorless mineral composed of aluminum chloride hexahydrate, $AlCl_3 \cdot 6H_2O$. It can also be made synthetically. Crystal system, rhombohedral (hexagonal). Density, 1.67 g/cm^3; refractive index, 1.560.

chlorapatite. A dark red to dark green mineral of the apatite group composed of calcium chloride phosphate, $Ca_5(PO_4)_3Cl$. It can be made synthetically by heating stoichiometric amounts of calcium chloride and calcium phosphate under prescribed conditions. Crystal system, monoclinic. Density, 3.17 g/cm^3; refractive index, 1.668. Occurrence: Canada (Ontario).

chlorargyrite. See cerargyrite.

chlorendics. A generic term for polyester resins made from blends of chlorendic anhydride ($C_9H_2Cl_6O_4$) or chlorendic acid ($C_9H_4Cl_6O_4$) and maleic anhydride ($C_4H_2O_3$) or fumaric acid ($C_4H_4O_4$). They exhibit excellent chemical resistance, and good fire retardancy.

Chlorene. Trade name for saran (polyvinylidene chloride) fibers with good chemical resistance, good weatherability, and good resistance to mildew and insects. Used for carpets and rugs, draperies, upholstery, clothing, industrial fabrics, etc.

Chlorimet. Trademark of Duriron Company, Inc. (USA) for a series of corrosion-resistant cast nickel-base alloys. *Chlorimet 2* contains 60-66% nickel, 30-33% molybdenum, 2-3% iron, 1% silicon, 1% manganese, 1% chromium, and 0-0.07% carbon. *Chlorimet 3* contains 58-60% nickel, 17-20% molybdenum, 17-20% chromium, 3% iron, 1% silicon, 1% manganese, and 0-0.07% carbon. *Chlorimet* alloys have excellent corrosion resistance, good resistance to hydrochloric acid, hot sulfuric and phosphoric acid, brines, bleaches and chlorides, and are superior to stainless steels in many severe environments. Used for pumps, valves, fluid handling equipment, e.g., the for chemical, pulp and paper, and food-handling industries.

Chlorin. (1) Trade name for vinyon (vinyl chloride) fibers used for textile fabrics and carpets.

(2) Trade name for saran (polyvinylidene chloride) fibers used for carpets and rugs, draperies, upholstery, clothing, industrial fabrics, etc.

chlorinated butyl rubber. See chlorinated isobutylene-isoprene rubber.

chlorinated fluorocarbon. See chlorofluorocarbon.

chlorinated hydrocarbon. An organic compound in which one or more of the hydrogen atoms have been replaced by chlorine. Examples include carbon tetrachloride, chlorobenzene, chloroform, hexachloroethane, methyl chloride, methyl chloroform, methylene chloride and trichloroethylene. Also known as *chlorohydrocarbon (CHC)*.

chlorinated isobutylene-isoprene rubber. Isobutylene-isoprene rubber modified with a small amount (typically about 1-2%) of chlorine. It has better ozone and environmental resistance than unmodified butyl rubber, good stability at high temperatures, and high compatibility with other rubber types in blends. Abbreviation: CIIR. Also known as *chlorobutyl rubber; chlorinated butyl rubber.* See also butyl rubber.

chlorinated low-density polyethylenes. The products of reactions of *low-density polyethylenes* with chlorine. Abbreviation: CLDPE.

chlorinated paraffin. An oily or waxy chlorine derivative of paraffin or a paraffin compound, used chiefly as a flame retardant in plastics and textiles, as a plasticizer for polyvinyl chloride, as a high-pressure lubricant, and in polyethylene sealants.

chlorinated polyethers. A family of highly crystalline polyether thermoplastics containing high concentrations of chlorine. They are commercially available as molding powders for injection-molding and extrusion, and in the form of blocks, sheets, rods, tubes and pipes, and have high molecular weights, outstanding corrosion resistance, excellent resistance to most chemicals even at elevated temperatures, exceptional resistance to thermal degradation, good mechanical properties, good chemical resistance, good dielectric properties over a wide range of frequencies and temperatures, high dielectric strength, good electrical resistance, and good processibility and fabricability with conventional equipment and techniques. Used for laboratory and chemical-processing equipment, piping, valves, lining tanks and other equipment, and protective coatings produced by fluid-bed processes.

chlorinated polyethylene elastomer. A thermosetting elastomer that is the reaction product of high-density polyethylene and varying amounts of chlorine, or chlorine plus chlorosulfonyl, cured with a peroxide catalyst. It has good resistance to hydrocarbon fluids, oils and fuels and elevated temperatures, but relatively poor mechanical strength. Used for hose linings and tires. Abbreviation: CPE. Also known as *chlorinated polyethylene rubber.*

chlorinated polyethylene rubber. See chlorinated polyethylene elastomer.

chlorinated polyolefins. A class of polyolefin elastomers to which high concentrations chlorine have been added to produce solid, film-forming resins. They are available as white, amorphous powders, and used chiefly in paints for marine environments, swimming pools, masonry, etc. See also polyolefin copolymers; polyolefin resins.

chlorinated polypropylenes. A family of film-forming polymers available in the form of white incombustible powders. They have high abrasion and chemical resistance, and good light stability. Used for coatings, paper sizing, adhesives, and inks.

chlorinated polyvinyl chlorides. A family of plastics based on resins, such as polyvinyl dichloride (PVDC). They are similar to polyvinyl chloride (PVC) with good chemical resistance and high strength, but increased softening temperatures. Used for pipes and fittings for hot corrosives. Abbreviation: CPVC; PVC+.

chlorinated rubber. The reaction product of chlorine and natural rubber or a polyolefin usually available a white, amorphous powder containing about 65-67% chlorine, and in various viscosities. *Chlorinated rubber* is a nonrubbery, incombustible,

film-forming resin with a density of 1.64 g/cm³ (0.059 lb/in.³), good corrosion resistance, good resistance to acids, alkalies, lacquer solvents and alcohols, good solubility in aromatic and aliphatic hydrocarbons and turpentine, and good compatibility with most natural and synthetic resins. It decomposes at 125°C (257°F), and is often sold under trademarks and trade names, such as *Alloprene, Dartex, Hypalon, Parlon, Pergut, Roxprene, Tegofan* or *Tornesit.* Used for marine, swimming-pool and masonry paints, varnishes, adhesives, inks and plastics.

chlorinated rubber paint. A paint containing 60% or more chlorinated rubber. It is cured by air drying (i.e., solvent evaporation), and has excellent water, alkali and acid resistance. Used for maintenance coatings, ship bottoms, swimming pools, and chemical processing equipment.

chlorine. A nonmetallic element of Group VIIA (Group 17) of the Periodic Table (halogen group). At normal temperatures it is a dense, diatomic, strongly electronegative, greenish-yellow gas with density of 3.21 g/L (at 0°C or 32°F), and a critical temperature of 144°C (291°F). Liquid *chlorine* is clear amber, and has a density of 1.56 g/cm³ (-35°C or -31°F), a freezing point of -101°C (-150°F) and low electrical conductivity. *Chlorine* occurs in nature only in combined form, e.g., in the minerals *halite* (rock salt), *sylvanite* and *carnallite*, and in seawater. Atomic number, 17; atomic weight, 35.453; monovalent; trivalent; tetravalent; pentavalent; heptavalent. Used in the manufacture of chlorinated hydrocarbons, polychloroprene, polyvinyl chloride and metallic chlorides, in dyes and explosives, as a powerful bleaching and oxidizing agent, as a disinfectant in water purification, as a stain remover, in flame-retardant compounds, in special batteries, and in medicine, biochemnistry and biotechnology.

chlorine-36. Radioactive chlorine with a mass number of 36. It emits beta radiation, and has a half-life of about 300000 years. Used as a tracer in salt water corrosion studies of metals (steels, cast irons, etc.). Symbol: ^{36}Cl.

chloritoid. A dark green mineral composed of iron aluminum silicate hydroxide, $FeAl_2SiO_5(OH)_2$. Crystal system, monoclinic. Density, 3.58-3.61 g/cm³; refractive index, 1.728-1729. Used as a brittle mica. Occurrence: USA (Rhode Islands); Canada (Quebec).

chlormanasseite. A colorless to greenish brown mineral of the sjogrenite group composed of iron magnesium aluminum chloride carbonate hydroxide trihydrate, $(Mg,Fe)_2Al_3(OH)_{16}(Cl,OH,-(CO_3)_{0.5})_3\cdot3H_2O$. Crystal system, hexagonal. Density, 1.98 g/cm³; refractive index, 1.540. Occurrence: Siberia.

chlormanganokalite. See chloromanganokalite.

chloroauric acid. Yellow to red, hygroscopic crystals (99.9% pure) with a density of 3.90 g/cm³. Used in gold plating, photography, as a metal stain in microscopy, in medicine, in ceramics and glass, in special inks, and in the preparation of finely divided gold and purple of Cassius. Formula: $HAuCl_4\cdot xH_2O$ (x = 3-4). Also known as *acid gold trichloride; hydrogen tetrachloroaurate hydrate; hydrochloroauric acid.*

chlorobutyl rubber. See chlorinated isobutylene-isoprene rubber.

chlorocalcite. A colorless, or white mineral composed of potassium calcium chloride, $KCaCl_3$. It can also be made from a mixture of potassium chloride and anhydrous calcium chloride. Crystal system, orthorhombic. Density, 2.16 g/cm³; refractive index, 1.568. Occurrence: Italy.

chlorocarbon. A compound, such as carbon tetrachloride, chloroform or tetrachloroethylene, that consists of carbon and chlorine, or carbon, hydrogen and chlorine. The latter is also referred to as a *chlorohydrocarbon*. Abbreviation: CC.

chloro fibers. (1) A generic term for man-made fibers whose fiber-forming substance is any long-chain synthetic polymer made up of more than 50% vinyl chloride (–CH$_2$–CHCl–) or vinylidene chloride (–CH$_2$–CCl–) units.

(2) A generic term for man-made fibers composed of either vinyl chloride (CH$_2$=CHCl) or vinylidene chloride (CH$_2$=CCl) and a second comonomer, e.g., vinyon or saran fibers.

chlorofluorocarbon. A compound, such as trichlorofluoromethane or trichlorodifluoromethane, that consists of carbon, hydrogen, fluorine and chlorine. Abbreviation: CFC. Also known as *chlorinated fluorocarbon*.

chlorofluorocarbon plastics. Plastics based on resins, such as chlorotrifluoroethylene, made by the polymerization of monomers consisting solely of chlorine, fluorine and carbon atoms.

chlorofluorohydrocarbon plastics. Plastics based on resins, such as ethylene chlorotrifluoroethylene, made by the polymerization of monomers consisting solely of chlorine, fluorine, hydrogen and carbon atoms. Also known as *hydrochlorofluorocarbon plastics*.

chloroform. An organic halide obtained commercially by the reduction of carbon tetrachloride (CCl$_4$) using iron and steam as the reducing agents. It is a sweet-smelling volatile liquid (99+% pure) that is commercially available as a solution in an ethyl alcohol stabilizer. It has a density of 1.485 g/cm^3, a melting point of -63 to -64°C (-81 to -83°F), a boiling point of 61°C (142°F), and a refractive index of 1.446. Used in the manufacture of fluorocarbon plastics, as a solvent, in analytical chemistry, in pesticide residue analysis, in capillary gas chromatography and high-pressure liquid chromatography, in spectrophotometry, and in biochemistry and biotechnology. It was formerly widely used in medicine and dentistry as an anesthetic. Formula: CHCl$_3$.

chlorohydrin. A heavy, hygroscopic, colorless liquid that is commercially available as a mixture (98% pure) of the a and b isomers with a density of 1.322 g/cm^3, a melting point of -40°C (-40°F), a boiling point of 213°C (415°F) (decomposes), a flash point above 230°F (110°C), and a refractive index of 1.4800. Used chiefly as an intermediate in organic synthesis, as a solvent for cellulose acetate and glyceryl phthalate resins, and in biochemistry and biotechnology. Formula: C$_3$H$_7$ClO$_2$.

chlorohydrin rubber. See epichlorohydrin elastomer.

chlorohydrocarbon. See chlorinated hydrocarbon.

chlorohydroquinone. White to light tan, water-soluble crystals available in technical grades (85% pure) and photographic grades (90+% pure) with a melting point of 101-102°C (214-216°F), and a boiling point of 263°C (505°F). Used as a polymerization inhibitor, as an organic intermediate, as a photographic developer, in dyestuffs, and in the biosciences. Formula: ClC$_6$H$_3$(OH)$_2$.

5-chloro-2-hydroxybenzophenone. A yellow, crystalline compound (99% pure) with a melting point of 93-98°C (199-208°F), and an optimum light absorption wavelength of 320-380 nm. Used as a light absorption medium. Formula: ClC$_6$H$_3$(OH)CO-C$_6$H$_5$.

chloromagnesite. Colorless crystals. Crystal system, rhombohedral (hexagonal). Density, 2.325 g/cm^3; melting point, 714°C (1317°F); boiling point, 1412°C (2574°F). Used in ceramics and materials research. Formula: MgCl$_2$. Also known as *chloromagnesite*.

chloromanganokalite. A yellow mineral composed of potassium manganese chloride, K$_4$MnCl$_6$. It can also be made synthetically by adding an excess of manganese chloride to fused potassium chloride. Crystal system, rhombohedral (hexagonal). Density, 2.31 g/cm^3; refractive index, 1.59. Occurrence: Italy. Also known as *chlormanganokalite*.

chloroosmic acid. Black, deliquescent, alcohol and water-soluble crystals obtained by dissolving osmium in aqua regia followed by evaporation and crystallization. Used in electroplating, in the manufacture of catalysts, and in microscopy. Formula: H$_2$OsCl$_6$·6H$_2$O. Also known as *dihydrogen hexachloroosmate hexahydrate; hydrogen hexachloroosmate hexahydrate*.

chlorophoenicite. A green to grayish-green mineral composed of manganese zinc arsenate hydroxide, (Mn,Zn)$_5$(AsO$_4$)(OH)$_7$. Crystal system, monoclinic. Density, 3.46 g/cm^3; refractive index, 1.690. Occurrence: USA (New Jersey).

chlorophyll. A porphyrin derivative that contains a central magnesium atom, is found in most plants and some marine organisms, and acts as a catalyst in photosynthesis. It is a photoreceptor for wavelengths up to about 700 nm, and can transfer radiant energy. There are three basic forms: (i) *chlorophyll a* (C$_{55}$H$_{72}$MgN$_4$O$_5$)–a bluish green microcrystalline wax; (ii) *chlorophyll b* (C$_{55}$H$_{70}$MgN$_4$O$_6$)–a light-sensitive, yellowish-green microcrystalline wax, and (iii) *chlorophyll c*–a substance obtained from marine organisms. *Chlorophyll* can also be made synthetically, and is used in dentistry, as a colorant in oils, soaps, fats, waxes, etc., as a photographic sensitizer, as an energy converter in synthetic photovoltaic cells, and in optoelectronics, biochemistry and biotechnology.

chloroplast. A particle, 3-6 μm (118-236 μin.) in diameter, that is found in the cells of plants, contains the green pigment *chlorophyll*, and plays a major role in photosynthesis.

chloroplatinic acid. Heat-, light- and moisture-sensitive, brownish-yellow crystals (99.9% pure) containing about 38-40% platinum. Density, 1.421 g/cm^3; melting point, 60°C (140°F). Used for platinum mirrors, in electroplating, in etching of zinc for printing, in the manufacture of catalysts, in the production of color effects on porcelain and other ceramics, in indelible ink, and in microscopy as a metal stain. Formula: H$_2$PtCl$_6$·6H$_2$O. Also known as *dihydrogen hexachloroplatinate hexahydrate; hydrogen hexachloroplatinate hexahydrate*.

chloroprene elastomer. See chloroprene rubber.

chloroprene resin. A resin made by the polymerization of chloroprene (C$_4$H$_5$Cl), and used in the manufacture synthetic rubber goods.

chloroprene rubber. A thermosetting rubber made from chloroprene (C$_4$H$_5$Cl), and vulcanized with metallic oxides. Commercially available in solid form, as latex, and as flexible foam, it has a density of 1.23-1.35 g/cm^3 (0.044-0.049 lb/in.^{3}), high cohesive strength, excellent weathering, heat, flame, oil, solvent, oxygen, ozone and corona-discharge resistance, good resistance to alkalies, dilute acids, petroleum oils and greases, fair resistance to water and antifreezes, poor resistance to oxidizing agents, high flame resistance (only the isocyanate-modified grades), electrical properties inferior to natural rubber, and a service temperature range of -50 to +105°C (-58 to +220°F). It is often sold under trademarks or trade names, such as *Neoprene* or *Perbunan*. The solid grades are used for mechanical rubber products, conveyors belts, V-belts, tires, hose covers, brake diaphragms, motor mounts, rolls, seals and gaskets, adhesive cements, linings for chemical reaction equipment, tanks, oil-loading hoses, etc., wire and cable, coatings for electric wiring, roofing membranes and flashing, footwear, and as binders for rocket fuels. The flexible foam grade is used for adhesive tape, automotive components, seat cushions and carpet backing, as a metal fastener substitute, and in sealants. The latex grade is used for specialty items. Also known as *chloroprene elastomer; polychloroprene rubber.*

chlorosulfonated polyethylene elastomer. The reaction product of polyethylene with chlorine and sulfur dioxide. It is white in color and has excellent resistance to oxygen, ozone and most chemicals, fair oil and low-temperature resistance, good ultraviolet and weather resistance, low gas permeability, good rigidity and abrasion resistance, and good to fair mechanical properties. Used for protective coatings, wire and cable insulation, flexible hose, tubing, connectors, linings, gaskets, seals, diaphragms, automotive goods, building products, shoe soles and heels, flooring, high-temperature conveyor belts, automotive components, spark-plug boots, coatings, and polymer blends. Abbreviation: CSM. Also known as *chlorosulfonated polyethylene rubber.*

chlorosulfonated polyethylene rubber. See chlorosulfonated polyethylene elastomer.

chlorothionite. A bright blue mineral composed of potassium copper chloride sulfate, $K_2CuSO_4Cl_2$. Crystal system, orthorhombic. Density, 2.69 g/cm^3; refractive index, 1.760. Occurrence: Italy.

chlorotrifluoroethylenes. A family of colorless, melt-processible thermoplastic fluoropolymers characterized by the repeating structure CF_2–CCl, and obtained by the polymerization of chlorotrifluoroethylene (ClFC=CH$_2$). They are available in the form of transparent films and thin sheets, and have a high density (2.10-2.15 g/cm^3 or 0.076-0.078 $lb/in.^3$), refractive index of 1.43, good corrosion resistance, good resistance to most chemicals, good resistance to most organic solvents, poor resistance to chlorinated solvents, good lubricity, excellent flame retardancy, good low and high-temperature properties, high heat resistance, a melting point of 210-215°C (410-420°F), a maximum service temperature of 175°C (347°F), good electrical properties, zero moisture absorption, high impact strength, high resilience, and good moldability and extrudability. Used for chemical piping, fittings, connectors, valves, gaskets, tank linings, diaphragms, wire and cable insulation, and electronic components. Abbreviations: CTFE. Also known as *chlorotrifluoroethylene polymers; fluorothenes; polychlorotrifluoroethylenes; polyfluorochloroethylene resins; polytrifluorochloroethylene resins.*

Chlorovene. Trade name of F.A. Hughes & Company (UK) for polyvinyl chloride plastics.

Chlorowax. Trademark for a chlorinated paraffin wax used in adhesives and in fireproofing and waterproofing compositions.

chloroxiphite. A dull olive green mineral composed of copper lead oxide chloride hydroxide, $Pb_3CuCl_2(OH)_2O_2$. Crystal system, monoclinic. Density, 6.93 g/cm^3. Occurrence: UK.

Chockfast. Trademark of Illinois Tool Works, Inc. (USA) for pourable resinous compositions based on epoxy or other polymers. Used for shimming and chocking industrial and marine machinery and equipment.

Choice. Trademark of BISCO, Inc. (USA) for light- and dual-cured composite luting cements supplied in translucent and opaque shades. Used in dentistry for porcelain veneers, inlays, and onlays.

cholesteric liquid crystal. A *liquid crystal,* such as *cholestryl benzoate,* that is in a birefringent fluid state, and composed of optically active, elongated organic molecules that are arranged parallel within a layer plane, but have a preferred direction of alignment (known as "cholesteric director") rotated slightly from layer to layer forming a continuous helix or coil. See also nematic liquid crystal; smectic liquid crystal.

cholesterol. A sterol that is the precursor of many other steroids in animal and human tissue including bile acids, sex hormones, adrenal corticosteroids, vitamin D, saponin, and cardiac glycosides. It is a white or pale yellow, light-sensitive crystalline monohydric secondary alcohol of the cyclopentenophenanthrene system with a density of 1.067 g/cm^3, a melting point of 148-150°C (298-302°F), a boiling point of 360°C (680°F), and a specific optical rotation of -40° (levorotatory). Used as an emulsifier in cosmetics and pharmaceuticals, in medicine, as a source of estradiol, in biotechnology, and in liquid-crystal technology. Formula: $C_{27}H_{45}OH$. Also known as *cholesterin.*

cholesteryl acetate. A liquid crystal compound with a melting point of 112-114°C (233-237°F) and a specific rotation of -44° (24°C/75°F). Formula: $C_{29}H_{48}O_2$. Abbreviation: CA.

cholesteryl benzoate. A crystalline substance that has two distinct melting points. When heated to the first melting point at 145°C (293°F), this originally solid material reversibly changes into a turbid liquid, and upon further heating reaches its second melting point at 179°C (354°F) and reversibly changes to a clear, transparent isotropic liquid. It is thus considered to be a thermotropic liquid crystal. This phase change in *cholestryl benzoate* was first observed in 1888 by the Austrian botanist Friedrich Reinitzer. *Cholesteryl benzoate* has a specific optical rotation of -15.0°, and is used as a liquid crystal in chemistry, biochemistry, polymer science, medicine, electronics and optoelectronics, e.g., for electronic displays, thermometers and other temperature-indicating devices and optoelectronic devices including waveguides, windows and filters. Formula: $C_{34}H_{50}O_2$. Abbreviation: CB. See also liquid crystal; thermotropic liquid crystal.

cholesteryl oleate. A monotropic liquid crystal compound with a melting point of 44-47°C (111-117°F) and a specific optical rotation of -24°. Used in electronics and optoelectronics. Formula: $C_{45}H_{78}O_2$.

cholesteryl palmitate. A monotropic liquid crystal compound with a melting point of 75-77°C (167-171°F), and a specific rotation of -24° (27°C/81°F). Used in electronics and optoelectronics. Formula: $C_{29}H_{48}O_2$.

cholesteryl stearate. A liquid crystal compound with a melting point of 79-83°C (174-181°F), and a specific optical rotation of -21° (25°C/77°F). Used in electronics and optoelectronics. Formula: $C_{45}H_{80}O_2$.

chololaite. A forest-green mineral composed of copper lead tellurium oxide monohydrate, $CuPb(TeO_3)_2 \cdot H_2O$. Crystal system, cubic. Density, 6.40 g/cm^3; refractive index, 2.04. Occurrence: USA (Arizona).

chondrodite. A yellow, brown or red mineral of the humite group with resinous luster composed of magnesium fluoride silicate, $Mg_5F_2(SiO_4)_2$. It may also contain iron and/or titanium, and can be made synthetically. Crystal system, monoclinic. Density, 3.16-3.23 g/cm^3; refractive index, 1.594. Occurrence: USA (New York, Arizona).

Chop-Pak. Trademark of Johns Manville Corporation (USA) for glass-fiber products.

chopped fibers. See discontinuous fibers.

chopped glass. A generic term for glass fiber products cut to short lengths including chopped fibers, mats and fabrics. Also known as *chopped-strand glass fiber products.*

chopped glass fibers. Glass fiber strands cut into short lengths, typically 3.2-12.7 mm (0.125-0.500 in.), and used as reinforcement in injection-molding and bulk-molding compounds, and as fillers for molded plastics. Also known as *discrete glass fibers.*

chopped mat. A fiberglass mat made by randomly placing fibers, cut to short lengths, onto a belt or chain and joining them together with a thermoplastic resin binder. Also known as *chopped-strand mat.*

chopped-strand glass fiber products. See chopped glass.

chopped-strand mat. See chopped mat.

Chopvantage. Trademark of PPG Industries, Inc. (USA) for chopped glass fibers.

christite. A crimson to deep red mineral composed of thallium mercury arsenic sulfide, TlHgAsS$_3$. It can also be made synthetically. Crystal system, monoclinic. Density, 6.37 g/cm^3. Occurrence: USA (Nevada).

Christofle metal. A silver-plated nickel silver with excellent corrosion resistance, used for cutlery and hardware.

chromadized aluminum. Aluminum or aluminum alloy whose paint adhesion properties have been improved by treating with a solution of chromic acid. It is used chiefly for aircraft skins. Also known as *chromatized aluminum; chromodized aluminum.*

Chromador. Trade name of British Steel plc (UK) for a corrosion-resistant high-tensile steel containing up to 0.3% carbon, 0.7-1% manganese, 0.7-1% chromium, 0.25-0.5% copper, and the balance iron. Usually supplied in the rolled condition, it is used for structural work, tanks, rivets, etc.

ChromaFusion. Trademark of Cesar Color, Inc. (USA) for colored laminated glass panels.

Chromal. (1) Swedish trade name for a non-heat-treatable wrought alloy containing 2-4% chromium, small additions of nickel and manganese, and the balance aluminum. Used for airplane parts, propellers, cooking utensils, and milk and oil separators.

(2) British trade name for a water-hardening tool steel containing 0.3% carbon, 0.8% chromium, 0.8% manganese, 0.8% molybdenum, and the balance iron. Used for gears, pinions, and shafts.

Chromalloy. Trade name of General Electric Company (USA) for a steel containing 0.2% carbon, 1% chromium, 1% molybdenum, 0.75% silicon, 0.5% manganese, 0.1% vanadium, and the balance iron. It has good high-temperature properties and good weldability, and is used for structural applications.

Chromaloid. Trade name of American Nickeloid Company (USA) for zinc sheet bonded with chromium, and used for hardware and plumbing fixtures.

Chromaloy O. Trademark of A.C. Scott & Company Limited (UK) for a resistance heating alloy of 75% iron, 20% chromium and 5% aluminum. Supplied in the form of fine annealed wire with a diameter of 0.25-0.5 mm (0.01-0.02 in.), it has a density of 7.1 g/cm^3 (0.26 lb/in.^{3}), a melting point of approximately 1510°C (2750°F), a maximum service temperature (in air) of 1250°C (2280°F), a Curie temperature of 600°C (1110°F), high strength at ordinary temperatures, excellent resistance to oxidation at elevated temperatures, and high electrical resistivity. Used for resistors in instruments and controls.

Chromaloy V. Trademark of A.C. Scott & Company Limited (UK) for a resistance alloy of 80% nickel and 20% chromium. It has good high-temperature resistance to 620°C (1150°F), and high strength and resistivity. Used for heating and resistance elements.

Chroman. Trade name of Vacuumschmelze GmbH (Germany) for corrosion-resistant nickel-chromium-manganese, nickel-iron-chromium-manganese, and nickel-chromium-manganese alloys used for heating elements.

Chromang. Trade name of Arcos Alloys (USA) for a stainless steel containing 0.1% carbon, 4% manganese, 19% chromium, 9% nickel, and the balance iron. Used for welding electrodes for armor and high-alloy steels.

Chromar. Trademark of Kurz-Hastings, Inc. (USA) for a metallized plastic film.

ChromaScreen. Trademark of Cesar Color, Inc. (USA) for unidirectional laminated glass panels.

chromate coatings. Thin, usually amorphous inorganic protective films of complex chromium compounds obtained on metal surfaces, e.g., steel, magnesium, aluminum, copper, zinc or silver, by chemical reactions resulting from the application of aqueous solutions of chromic acid, or sodium or potassium chromate or dichromate, with or without electrolytic assistance. The coatings are extremely abrasion resistant, provide excellent bases for organic coatings, but only limited protection against corrosion in water. Used on parts and equipment operating in marine atmospheres or high-humidity environments. Also known as *chromate conversion coatings; chromium chromate coatings.*

chromate conversion coatings. See chromate coatings.

chromated protein films. See chromate-protein coatings.

chromate-protein coatings. Thin, hard, yellowish corrosion-resistant coatings produced by first immersing ferrous objects into a solution of a *protein*, such as albumin or casein, and then immersing the dried product into a dilute solution of chromic acid. Used as undercoatings for iron and steel. Also known as *chromated protein films.*

chromate red. See chrome red.

chromatite. A yellow mineral of the zircon group composed of calcium chromate, CaCrO$_4$. It can also be made synthetically. Crystal system, tetragonal. Density, 3.14 g/cm^3; refractive index, 1.84-1.88. Occurrence: Israel.

chromatized aluminum. See chromadized aluminum.

Chromax. (1) Trade name of Harrison Alloys, Inc. (USA) for a resistance heating alloy of 45% iron, 35% nickel, 20% chromium, and 0-1% silicon. Used in wire, rod or ribbon form for heating elements up to 925°C (1697°F).

(2) Trade name of Empire Steel Castings Company (USA) for an abrasion-resistant steel containing 0.3-1% carbon, 1.25-1.5% chromium, 0.35-0.45% molybdenum, and the balance iron.

(3) Trade name of Driver Harris Company (USA) for an abrasion-, impact-, strain-, heat-, and corrosion-resistant cast alloy of 45-65% iron, 20-35% nickel, and 15-20% chromium. Used for furnace parts, carburizing boxes, and enameling fixtures.

(4) Trade name for a bronze containing 15.2% nickel, 12% zinc, 3% aluminum, 3% chromium, and the balance copper.

(5) Trade name of Chemtech Finishing Systems Inc. (USA) for conversion coatings for zinc plate.

(6) Trade name of MacDermid Inc. (USA) for trivalent iridescent chromate coatings for zinc plate.

Chrombal. Trade name of Foseco Minsep NV (Netherlands) for a series of chromium-bearing covering fluxes used during copper alloy melting to decrease chromium oxide formation and gas pick-up.

Chromdie. Trade name of Faitout Iron & Steel Company (USA) for a tough, nondeforming, air-hardening tool steel (AISI type D2) containing 1.5% carbon, 12% chromium, 1% molybdenum, 0.4% vanadium, and the balance iron. It has high abrasion and wear resistance, a great depth of hardening, high hardness, good resistance to decarburization, and good nondeforming properties. Used for blanking, drawing and thread-rolling dies, wire-drawing and stamping dies, punches, broaches, hobs, and gages.

chromdravite. A dark green, or nearly black mineral of the tourmaline group composed of sodium magnesium iron chromium borate silicate hydroxide, $NaMg_3(Cr,Fe)_6(BO_3)_3Si_6O_{18}(OH)_4$. Crystal system, rhombohedral (hexagonal). Density, 3.40 g/cm^3; refractive index, 1.778. Occurrence: Russian Federation.

chrome. (1) A collective term for chrome-bearing pigments.
(2) A term sometimes used to refer to *chromium*.

chrome alum. See chromium potassium sulfate.

chrome-alumina pink. A class of pink ceramic colors consisting of mixtures of chromic oxide (Cr_2O_3), aluminum oxide (Al_2O_3) and zinc oxide (ZnO).

chrome ball-bearing steel. See chrome bearing steel.

chrome bearing steel. A water-hardening steel containing 1% carbon, 1.2% chromium, and the balance iron. Used for ball-bearing races and balls. Also known as *chrome ball-bearing steel*. See also ball-bearing steel.

Chromeblue. Trade name of Shatterprufe Safety Glass Company Proprietary Limited (South Africa) for a mirror of laminated glass made by depositing a chrome alloy on the inner surface of one of the two sheets of glass.

Chrome Brass. Trade name of American Nickeloid Company (USA) for chromium-coated brass which is easily formed, drawn and stamped. Used for fabricated parts.

chrome brick. A refractory brick produced from chrome ore, and having excellent resistance to chemical reaction with basic and acidic oxides at temperatures above 2040°C (3705°F). Used as a replacement for magnesia brick in furnace and kiln linings, as a spacer between acidic and basic walls of open-hearth and similar metallurgical furnaces. Also known as *chrome refractory*.

Chrome Cast. Trade name of Atlantic Steel Casting Company (USA) for a heat-, wear- and corrosion-resistant cast steel containing up to 0.1% carbon, 16-23% chromium, and the balance iron. Used for heat-, wear-, and corrosion-resistant parts.

Chrome Castable. Trade name of National Refractories & Minerals Corporation (USA) for ram-cast refractory gunning mixes.

Chrome Coat. Trade name of Marson Corporation (USA) for bright aluminum paint for automotive applications.

chrome coatings. See chromium coatings.

chrome copper. See chromium copper.

Chrome Core. Trademark of Carpenter Technology Corporation (USA) for a series of iron-chromium ferritic alloys supplied in standard and free-machining grades. *Chrome Core 8* and *Chrome Core 12* have 8% and 12% chromium, respectively. Both grades have good corrosion resistance, good electrical properties, high saturation induction, and are used for shunts, magnetic cores, solenoids, relays, automotive fuel injectors, and motor laminations. *Chrome Core 13-FM* is a soft magnetic alloy with high magnetic permeability and saturation induction, and low coercive field strength as well as good electrical and corrosion resistance. It is used for fuel injection parts, antilock braking systems, solenoid valves and other electromechanical and automotive applications. *Chrome Core 29* is a corrosion-resistant solenoid quality soft magnetic ferritic stainless steel.

Chrome Die. Trade name for a hot-work steel containing 0.5% carbon, 5% chromium, and the balance iron. Used for hot dies, punches, and shears.

Chromeflex. Trade name for metallic fibers.

chrome green. Any of a class of brilliant green pigments made by mixing *chrome yellow* and *Prussian blue* in various proportions. They have high brightness, excellent light-fastness, good opacity, and excellent strength. Used in paints, rubber, plastics and vitreous enamels, and as green stains or oxides for ceramic products. Also known as *Brunswick green*. *Note:* Do not confuse with *chrome oxide green*.

Chrome Hot Die. Trade name of Wallace Murray Corporation (USA) for an air-hardening die steel containing 0.8-1% carbon, 3-4% chromium, and the balance iron. Used for bolt and rivet dies, forging mandrels and piercing dies.

Chrome Hot Work. Trade name of A. Milne & Company (USA) for a hot-work steel containing 0.9% carbon, 3.75% chromium, and the balance iron. Used for tools, dies and mandrels.

chrome iron ore. See chromite.

Chromel. Trade name of Hoskins Manufacturing Company (USA) for a series of strong, heat- and corrosion-resistant nickel-chromium alloys of varying composition. *Chromel A* contains 20% chromium, 1% silicon, 0.5% iron, and the balance nickel. It has a melting point of 1420°C (2590°F), and is used in the form of heating and resistance wire for appliances. *Chromel AA* contains 20% chromium, 8.3% iron, 2% silicon, and the balance nickel. It has a continuous service temperature of 1230°C (2250°F), and is used as a heating and resistance wire. *Chromel B* contains 85% nickel and 15% chromium, and is used for heating elements. *Chromel C* contains 60% nickel, 16% chromium, and 24% iron. It has a continuous service temperature of 1010°C (1850°F), and is used as heating wire for household appliances and industrial equipment. *Chromel D* contains 35-36% nickel, 18.5% chromium, 0-1.5% silicon, and the balance iron. It has a continuous service temperature of 649°C (1200°F), and is used for heating and resistance wire. *Chromel P* contains 90% nickel and 10% chromium. It has a maximum operating temperature of 1315°C (2400°F), and is used as a thermocouple wire. *Chromel R* contains 80% nickel and 20% chromium. It has a continuous-use temperature of 1150-1200°C (2100-2190°F), and is used for precision-wound resistors, potentiometers, and heating elements. *Chromel 70/30* contains 70% nickel and 30% chromium. It has a continuous-use temperature of 1175°C (2150°F), and is used for heating elements.

chrome leather. See chrome-tanned leather.

chrome-magnesite brick. A refractory brick, either fired or chemically bonded, made from a mixture of refractory-grade *chrome ore* and *dead-burnt magnesia*, in which the chrome ore, by weight, is predominant. Also known as *chrome-magnesite refractory*.

chrome-magnesite refractory. See chrome-magnesite brick.

chrome-magnesite cement. A quick-setting refractory cement made from a mixture of refractory-grade *chrome ore* and *magnesite*. Used for patching metallurgical furnace linings.

chrome-manganese steels. See chromium-manganese steels.

Chrome-Moly. Trade name of Detroit Alloy Steel Company (USA) for a tough, close-grained cast iron containing 2.75-3% carbon, 2-2.6% silicon, 0.8-1% manganese, 0.4% chromium, 0.4-0.6% molybdenum, and the balance iron. Used for heavy dies.

chrome-molybdenum steels. See chromium-molybdenum steels.

chrome-molybdenum-vanadium steels. See chromium-molybdenum-vanadium steels.

Chrome Moly Roll. Trade name of Crucible Specialty Metals (USA) for a tough, deep-hardening steel containing 1% carbon, 1.2% chromium, 0.3% molybdenum, and the balance iron. Used for straightening and reducing rolls.

Chrome Nickel. Trade name of VDM Nickel-Technologie AG (Germany) for a resistance alloy of 78% nickel, 20% chromium and 2% manganese. It has high heat and corrosion resistance, and a maximum service temperature of 1150°C (2100°F). Used for heating elements, resistances, rheostats, and furnace parts. See also Chromnickel I.

Chrome Nickel Iron. Trade name of VDM Nickel-Technologie AG (Germany) for a resistance alloy of 70% nickel, 20% chromium, 8% iron and 2% manganese. It has good heat and acid resistance, and a maximum service temperature of 1150°C (2100°F). Used for heating elements, rheostats, and measuring instruments. See also Chromnickel II.

chrome-nickel-molybdenum steels. See nickel-chromium-molybdenum steels.

Chrome Nickel Silver. Trade name of American Nickeloid Company (USA) for a corrosion-resistant nickel silver sheet bonded with chromium. Used in automatic fabrication, and for sheet and strip.

chrome-nickel steels. See chromium-nickel steels.

chrome ocher. A brilliant greenish clay material containing 2-10.5% chromic oxide (Cr_2O_3).

chrome orange. An orange pigment composed principally of basic lead chromate ($PbCrO_4 \cdot PbO$). Used in paints, rubber and plastic products. See also chrome red.

chrome ore. See chromite.

chrome oxide green. A green inorganic pigment consisting essentially of chromic oxide (Cr_2O_3). It is supplied in the form of a dry powder, or ground in oil, and the pure grade consists of 99% chromic oxide. It has extremely high color permanence and stability, good resistance to strong alkalies and acids, and good heat resistance. Used in paints and lacquers, as a rubber colorant, and in paints and finishes for cement and concrete surfaces. Also known as *chromium oxide green. Note:* Do not confuse with *chrome green.*

chrome pigments. A group of light-fast inorganic chromium-containing pigments that have high stability to sunlight, weathering and chemical action. The most important types are: (i) chrome oxide green; (ii) chrome green; (iii) chrome red; (iv) chrome yellow, and (v) molybdate orange and (vi) zinc yellow.

chrome plates. See chromium coatings.

chrome refractory. See chrome brick.

chrome red. A group of light orange to red pigments obtained by mixing lead chromate ($PbCrO_4$) and lead oxide (PbO) in varying proportions. Used in paints, rubber and plastic products. Also known as *chromate red.* See also American vermilion; Chinese red; Chinese scarlet; chrome orange.

chrome refractory. See chrome brick.

chrome retan. Leather tanned first with chrome and then with plant-derived extracts.

Chrome Roll. Trade name of Midvale-Heppenstall Company (USA) for a water-hardening steel containing 0.9% carbon, 1% chromium, and the balance iron. Used for tools, cutters, and dies.

Chromesco. Trade name of Vallourec SA (France) for a series of steels containing up to 0.15% carbon, 0.5-2.6% chromium, 0.5-1.2% molybdenum, and the balance iron. Used for boilers, steam pipes, superheaters, and oil refinery equipment.

Chromeset. Trademark of Dresser Industries, Inc. (USA) for refractory cement containing a considerable amount of *chrome ore.*

Chrome Silicon Spring. Trade name of US Steel Corporation (USA) for a spring steel containing 0.5-0.6% carbon, 0.5-0.8% manganese, 1.2-1.6% silicon, 0.5-0.8% chromium, and the balance iron. Used for springs.

chrome spinel. See picotite.

Chrome Stainless. Trade name for a low-carbon steel containing 17% chromium, and the balance iron. Used for stainless parts.

Chrome Steel. Trade name of American Nickeloid Company (USA) for a chromium-bonded to steel with good heat resistance up to 750°C (1380°F), used for floor plates, reflectors and trim and molding.

chrome steels. See chromium steels.

chrome-tanned leather. Leather that has been tanned with a salt (e.g., an acetate or sulfate) of trivalent chromium or a chromium compound, e.g., chromium-ammonium sulfate or chromium-potassium sulfate. Shoe uppers and linings, and clothing are often made from chrome-tanned leather. Also known as *chrome leather.*

Chromet. German trade name for an alloy of 90% aluminum and 10% silicon used for bearings for hard shafts.

Chrome Tin. Trade name of American Nickeloid Company (USA) for tin products coated with chromium.

chrome-tin pink. A colorant for ceramic glazes which produces a pink color, and is composed chiefly of chromic oxide and tin oxide, with some lime (CaO). Also known as *English pink.* See also chrome-zircon pink.

Chrometough. Trade name of British Steel plc (UK) for oil-hardening general-purpose tool steel containing 0.55% carbon, 0.5% manganese, 0.5% chromium, 0.2% vanadium, and the balance iron. Used for chisels, boring bars, and rivet sets.

Chrome Tungsten. Trade name of Latrobe Steel Company (USA) for a hot-work tool steel containing 0.55% carbon, 7.5% chromium, 7.4% tungsten, 0.16% nickel, and the balance iron. It has good toughness and hot hardness, and is used for press tools, jigs and fixtures, hand tools, dies, punches, and shear blades.

chrome-vanadium steels. See chromium-vanadium steels.

Chromevert. Trade name of American Chemical & Equipment (USA) for chromate conversion coatings.

Chromewear. Trade name of Teledyne Vasco (USA) for an air-hardening medium-alloy cold-work tool steel (AISI type A7) containing 2.3% carbon, 0.4% silicon, 0.7% manganese, 5.25% chromium, 4.75% vanadium, 1.1% molybdenum, 1.1% tungsten, and the balance iron. It has good deep-hardening properties, excellent abrasion and wear resistance, and high hardness in the heat-treated condition. Used for forming and drawing dies for wires and other shapes, liners for sand slingers, shot-blasting equipment, refractory brick molds, tube-making rolls, blanking rings and punches for cans, and extrusion dies for ceramic components.

Chromex. (1) Russian trade name for a series of heat-resistant cast iron containing 1.2-2.8% carbon, 32-35% chromium, 1.2-2.5% silicon, and the balance iron. Used for furnace equipment.

(2) Russian trade name for steel welding electrodes containing 0.07-0.7% carbon, 13-30% chromium, and the balance iron. Used for welding low and medium-carbon high-chromium steels.

chrome yellow. A class of inorganic pigments composed chiefly of lead chromate ($PbCrO_4$), and made by mixing solutions of lead acetate and potassium dichromate. They are available in light greenish-yellow to lemon-yellow shades. Medium-yellow shades contain essentially lead chromate, while light shades contain lead chromate and varying amounts of lead sulfate. Used in paints, enamels, glasses, glazes, rubber, plastics, etc. Also known as *Leipzig yellow; Paris yellow.*

chrome-zircon pink. A colorant for ceramic glazes which produces a pink color, and is composed of chromic oxide, zircon, and some tin oxide in the presence of lime. See also chrome-tin pink.

Chromfast. Trade name for color and lightfast cellulose acetate fibers and yarns used for textile fabrics.

chromia. See chromic oxide.

chromia-alumina. A ceramic material composed of chromia (Cr_2O_3) and alumina (Al_2O_3), and supplied in the form of a fine powder. It has a density of 4.26 g/cm^3, high hardness, good resistance to molten metals, and is used for the manufacture of advanced ceramics.

Chromic. Trade name of Riverside Metals Corporation (USA) for a series of resistance alloys containing 60-80% nickel, 18-20% chromium and 0-22% iron. They have high electrical resistance, and are used for heating elements and resistors.

chromic acetate. Grayish-green or bluish-green pasty mass obtained by reacting chromium hydroxide with acetic acid. Used as a polymerization and oxidation catalyst, and as an emulsion hardener. Formula: $Cr(C_2H_3O_2)_3 \cdot H_2O$. Also known as *chromium acetate.*

chromic acid. A name commonly used for chromium trioxide (CrO_3). The true chromic acid (H_2CrO_4) exists only in solution. *Chromium trioxide* is a moisture-sensitive, oxidizing, dark purplish-red crystalline compound (98+% pure) with a density of 2.70 g/cm^3 and a melting point of 195°C (383°F). Used in hard chromium plating baths, in the purification of plating wastes, in acid pickling of magnesium alloys, in etching baths for steel, in electropolishing of aluminum, brass, copper and ferrous metals, in acid cleaning baths for cast iron and steel, in phosphate coating and decorative chromium plating, in anodizing, in ceramic glazes and colored glass, in paints, and as an etchant for plastics. Formula: CrO_3. Also known as *chromic anhydride; chromium trioxide.* See also polymer-supported chromic acid.

chromic anhydride. See chromic acid.

chromic anodized aluminum. Aluminum or aluminum alloy on whose surface a corrosion-resistant, protective coating of aluminum oxide (Al_2O_3) has been produced by converting the aluminum surface to aluminum oxide by making the aluminum or alloy the anode in an electrolytic cell, and using a chromic acid electrolyte. Abbreviation: CAA. See also anodized aluminum.

chromic boride. See chromium boride (i).

chromic bromide. Black crystals with a density of 4.25 g/cm^3 used as catalyst in olefin polymerization. A hexahydrate in the form of green crystals (99% pure) with a density of 5.4 g/cm^3 is also available. Formula: $CrBr_3$ (anhydrous); $CrBr_3 \cdot 6H_2O$ (hexahydrate). Also known as *chromium tribromide.*

chromic carbide. See chromium carbide.

chromic chloride. Purple, hygroscopic crystals or flakes (99+% pure) made by passing chlorine over a mixture of chromic oxide and carbon. Density, 1.76 g/cm^3; melting point, 1150°C (2100°F); boiling point, sublimes at approximately 1300°C (2370°F); antiferromagnetic properties. Used in chemistry, electronics and materials research. Formula: $CrCl_3$. Also known as *chromium chloride; chromium trichloride; chromium sesquichloride.*

chromic chloride hexahydrate. Green crystals made by treating chromium hydroxide with hydrochloric acid. Density, 1.76 g/cm^3; melting point, 83°C (181°F). Used in chromium plating, vapor plating and flame metallizing, in the preparation of sponge chromium, in alloying steel powders, as a polymerization catalyst, and in waterproofing. Formula: $CrCl_3 \cdot 6H_2O$. Also known as *chromium chloride hexahydrate.*

chromic fluoride. Green powder or crystals (98+% pure) obtained by treating chromium hydroxide with hydrogen fluoride. It is also available as the tetrahydrate and nonahydrate. Density, 3.78-3.80 g/cm^3; melting point, above 1000°C (1830°F); boiling point, sublimes at approximately 1100-1200°C (2010-2190°F); antiferromagnetic properties. It is also available as the tetrahydrate and nonahydrate. Used in ceramics and in the preparation of mixed-metal fluoride catalysts. Formula: CrF_3 (anhydrous); $CrF_3 \cdot 4H_2O$ (tetrahydrate); $CrF_3 \cdot 9H_2O$ (nonahydrate). Also known as *chromium fluoride; chromium trifluoride.*

chromic iron. See chromite.

chromic iron ore. See chromite.

chromic nitrate. Purple, or green-black, hygroscopic crystals (99+% pure) made by treating chromium hydroxide with nitric acid. Melting point, 60°C (140°F); boiling point, decomposes at 100°C (212°F). Used as a corrosion inhibitor, and as a catalyst. Formula: $Cr(NO_3)_3 \cdot 9H_2O$. Also known as *chromium nitrate; chromium nitrate nonahydrate.*

Chromicoat. Trademark of Chemetall, Oakite Products Inc. (USA) for a conversion coating for aluminum and zinc.

chromic octadecasulfate. Violet cubes with a density of 1.70 g/cm^3, used in chrome plating, in chromium alloys, in green paints, in varnishes and inks, and in the production of green effects on glazes and ceramics. Formula: $Cr_2(SO_4)_3 \cdot 18H_2O$. Also known as *chromium octadecasulfate.*

chromic oxide. A bright-green, hard, crystalline powder, or black, fused substance (98+% pure). Crystal system, hexagonal. Density, 5.21 g/cm^3; melting point, 2435°C (4415°F); boiling point, 4000°C (7230°F); refractive index, 2.551. The green powder is used as a green pigment (chrome green), in metallurgy, as an abrasive, as a constituent of refractory brick, and as green granules in asphalt roofing. The black, fused grade is used for high-temperature spraying applications. Synthetically grown crystals are used for magnetic applications. Formula: Cr_2O_3. Also known as *chromia; chromium oxide; chromium sesqui-oxide.*

chromic nitride. A compound of chromium and nitrogen supplied in the form of an amorphous or crystalline powder. Density, 5.9-6.1 g/cm^3; melting point, decomposes at 1500°C (2730°F); hardness, about 7.5 Mohs. Used in ceramics and materials research. Formula: CrN. Also known as *chromium mononitride; chromium nitride.*

chromic octadecasulfate. See chromic sulfate.

chromic pentadecasulfate. See chromic sulfate.

chromic phosphate. See chromic phosphate hexahydrate; chromic phosphate tetrahydrate.

chromic phosphate hexahydrate. Violet crystals with a density of 2.12 g/cm^3 (at 14°C or 57°F). Used as a paint pigment (Plessy's green), and as a catalyst. Formula: $CrPO_4 \cdot 6H_2O$. Also known as *chromium orthophosphate hexahydrate; chromium phosphate hexahydrate.*

chromic phosphate tetrahydrate. Green crystals. Formula: $CrPO_4 \cdot 4H_2O$. Also known as *chromium orthophosphate tetrahydrate; chromium phosphate tetrahydrate.*

chromic phosphide. See chromium phospide (ii).

chromic sulfate. A violet or red powder with a density of 3.012. Hydrated forms are also available including chromic pentadecasulfate (green-black powder or dark-green amorphous scales with a density of 1.867 g/cm^3 and a melting point of 100°C or 212°F), and chromic octadecahydrate (violet cubic crystals with a density of 1.70 g/cm^3). Used in chrome plating, ceramics (glazes), green paints, varnishes and inks, in chromium alloys, and as catalysts. Formula: $Cr_2(SO_4)_3$ (anhydrous); $Cr_2(SO_4)_3 \cdot 15H_2O$ (pentadecasulfate); $Cr_2(SO_4)_3 \cdot 18H_2O$ (octadecasulfate). Also known as *chromium sulfate*.

chromic sulfide. See chromium sesquisulfide (ii).

Chromidium. Trade name of Midland Motor Cylinder Company (USA) for a series of wear-resistant cast irons containing 3.1-3.5% total carbon, 2.0-2.2% silicon, 0.20-0.45% chromium, 0.75.1.0% manganese, and the balance iron. Used for automotive castings, e.g., cylinders, engine blocks, brake drums, etc.

Chromindur. Trademark for a series of alloys containing about 28% chromium, 10.5-15.% cobalt, small amounts of other elements, and the balance iron. They have hard magnetic properties, good formability, and are used for ductile permanent magnets, e.g. for telephone receivers.

Chro-Mir. Trade name of Plating Resources Inc. (USA) for chromate conversion coatings.

chromite. An iron-black, brownish-black, brown, or dark reddish brown mineral of the spinel group with a brown streak and a metallic to submetallic luster. It is composed of iron chromium oxide, $FeCr_2O_4$, sometimes with some magnesium and/or aluminum. When pure, it contains about 68% chromic oxide (Cr_2O_3). Crystal system, cubic. Density, 4.65-5.09 g/cm^3; melting point, 1850°C (3360°F); hardness, 5.5 Mohs. Occurrence: Cuba, Greece, India, Morocco, New Caledonia, Philippines, Russia, South Africa, Turkey, USA (California, Maryland, Montana, Oregon, Pennsylvania), Zimbabwe. *Chromite* (or *chrome ore*) is commercially available in three grades: (i) *Metallurgical* with at least 48% Cr_2O_3 and a Cr-Fe ratio of 3:1; (ii) *Refractory* that is high in Cr_2O_3 and Al_2O_3 and low in iron; and (iii) *Chemical* that is low in SiO_2 and Al_2O_3, and high in Cr_2O_3. Used as an important ore of chromium, as a source of chromium and its compounds, in the manufacture of refractories, chromium alloys, chromium steels, stainless steels, resistance wire and high-speed cutting tools, in pigments, and in chromium plating. Also known as *chrome iron ore; chrome ore; chromic iron; chromic iron ore; iron chromite.* See also aluminian chromite.

chromite brick. Refractory brick made essentially of refractory-grade *chrome ore*. It will withstand temperatures up to 2040°C (3705°F), and is used for lining industrial furnaces.

chromite flour. A finely divided *chromite* (average particle size, 45μm or 0.002 in.) used chiefly for foundry applications, and in pigments.

chromite sand. A sand based on *chromite* and typically containing about 46-47% chromic oxide (Cr_2O_3), 24.5% ferrous oxide (Fe_2O_3), 15% alumina (Al_2O_3), 9.5% magnesia (MgO), and up to 1% silica (SiO_2). Supplied in mesh sizes from -16 to +200 for use as a foundry sand and in refractory mortars.

chromium. A steel-gray metallic element of Group VIB (Group 6) of the Periodic Table. It is commercially available in the form of granules, pellets, flakes, lumps, chips, chunks, pieces, foil, microfoil, disks (arc-melted), rod (hot-pressed), tubules, composites, single crystals, crystal bars, high-purity crystals, crystallites, degassed and regular powders and high-purity powders. The crystals are usually grown by the Bridgeman or the float-zone technique. The only important chromium ore is *chro-*

mite. Crystal system, cubic. Crystal structure, body-centered cubic. Density, 7.138 g/cm^3; melting point, 1875°C (3407°F); boiling point, 2665°C (4829°F); hardness, 130-220 Vickers; atomic number, 24; atomic weight, 51.996; divalent, trivalent, hexavalent, less commonly tetravalent. Commercially pure chromium is hard and brittle, while high-purity chromium is soft. Used in stainless steels to increase corrosion resistance, high-temperature stability, scaling resistance and strength, in tool steels to improve high-temperature stability, wear resistance and cutting edge life, in high-strength steels to improve high-temperature stability, nitridability, hardness, tensile and yield strength and scaling and corrosion resistance, in magnetic steels to improve coercivity and remanence, and as an alloying element in nickel-base and cobalt-base superalloys, aluminum-base alloys and electrical resistance alloys. Other applications include hardfacing, electroplating of metals and plastics (to produce corrosion- and wear-resistant as well as decorative coatings), chromizing, protective coatings for automotive and industrial equipment, nuclear and high-temperature research, ferrite compositions, stereo and video tapes, lasers, inorganic pigments, and camouflage paints.

chromium-51. A radioactive isotope of chromium with a mass number of 51 that emits gamma rays, has a half-life of 26.5 days, and is used in the life sciences and medicine, e.g., as a tracer for blood volume studies. Symbol: ^{51}Cr. Also known as *radiochromium*.

chromium acetate. See chromic acetate.

chromium alloys. A group of alloys of chromium and one or more other elements. The most common chromium alloys are: chromium-aluminum, chromium-antimony, chromium-cobalt, chromium-iron, chromium-molybdenum, chromium-nickel, chromium-tin, chromium-tungsten, and chromium-vanadium alloys.

chromium-alloy steels. A group of alloy steels containing chromium as the predominant, or one of the predominant alloying elements, and including plain chromium, chromium-molybdenum, chromium vanadium, nickel-chromium and nickel-chromium-molybdenum steels.

chromium-alumina cermet. A composite material consisting of aluminum oxide (Al_2O_3) particles bonded with chromium, and made by powder-metallurgy techniques. It has good high-temperature properties, and is used for nuclear applications.

chromium aluminide. A compound of chromium and aluminum with a melting point of 2160°C (3920°F) and excellent oxidation resistance. Used in ceramics and materials research. Formula: CrAl.

chromium beryllide. A compound of chromium and beryllium. Density, 4.34 g/cm^3; melting point, 1840°C (3344°F); hardness, 7-8 Mohs. Used in ceramics, and for structural applications. Formula: CrBe.

chromium boride. Any of the following hard, corrosion-resistant compounds of chromium and boron: (i) *Chromium monoboride*. A silvery crystalline powder. Density, 6.1-6.2 g/cm^3; melting point, 2760°C (5000°F); hardness, 8.5 Mohs. Used in ceramics, metallurgy, cermets and refractories. Formula: CrB. Also known as *chromic boride*; (ii) *Chromium diboride*. Fine, crystalline powder. Density, 4.36 g/cm^3; melting point, 1850°C (3360°F); hardness, above 8 Mohs; high tensile strength; good resistance to oxidation and thermal shock up to 1100°C (2010°F). Used in ceramics, metallurgy, cermets and refractories. Formula: CrB_2; (iii) *Dichromium boride*. Density, 6.5 g/cm^3; melting point, 1830°C (3325°F). Used as an alloying addition to austenitic steels and as an absorber material or shielding for nuclear reactors. Formula: Cr_2B; (iv) *Chromium pentaboride*. Melting point, 2000°C (3630°F). Used in ceramics and materials research. Formula: Cr_2B_5; (v) *Trichromium diboride*. Crystalline powder. Density, 6.1 g/cm³; hardness, above 9 Mohs. Used in ceramics, metallurgy, cermets and refractories. Formula: Cr_3B_2; (vi) *Chromium tetraboride*. Density, 5.76 g/cm³; melting point, 1927°C (3500°F). Used in ceramics and materials research. Formula: Cr_3B_4; (vii) *Tetrachromium boride*. Density, 6.2 g/cm³; melting point 1650°C (3000°F). Used in ceramics and materials research. Formula: Cr_4B; and (viii) *Chromium triboride*. Density, 6.12 g/cm³; melting point 1900°C (3450°F). Used for jet and rocket engines, as a metallurgical additive, in high-temperature electrical conductors, in cermets and refractories, and in coatings resistant to attack by molten metals. Formula: Cr_5B_3.

chromium-boron. A master alloy of 80% chromium and 20% boron used in the manufacture of nonferrous alloys.

chromium bromide. See chromous bromide.

chromium bronze. A high-strength, cast bearing bronze containing 2-10% tin, 1% iron, 1% chromium, and the balance copper.

chromium buffing compounds. Buffing compounds that contain fine unfused alumina (Al_2O_3) abrasives, and are used for color buffing of chromium-plated parts, and in secondary coloring compounds for stainless steel. See also buffing compounds.

chromium carbide. Gray crystals or powder (99.5+% pure). Crystal system, rhombohedral. Density, 6.65-6.68 g/cm³; melting point, 1890°C (3434°F); boiling point, approximately 3800°C (6870°F); hardness, above 9 Mohs; very high oxidation resistance at high temperatures; good resistance to acids and alkalies. Used for bearings, seals, pump components, valve seats, orifices, chemical equipment, gage blocks, and hot-extrusion dies. It is also used in powder form as a spray-coating material. Formula: Cr_3C_2. Also known as *chromic carbide*. Other chromium carbides include: (i) Cr_4C. Density, 6.99 g/cm³; melting point, 1520°C (2768°F); (ii) Cr_7C_3. Melting point, 1780°C (3236°F); (iii) Cr_7C_6. Melting point, 1665°C (3029°F); and (iv) $Cr_{23}C_6$. Melting point, 1250°C (2282°F).

chromium carbide cermet. A cermet containing 80-90% chromium carbide (Cr_3C_2) and 10-20% nickel or nickel alloy as matrix binder. It has a density of 7.0 g/cm³ (0.253 lb/in.³), high tensile strength, excellent corrosion and oxidation resistance, and high hardness, stiffness and abrasion resistance. Used for gages, bearings, pump rotors, valve components, seal rings, spray nozzles, and cutting tools.

chromium carbonyl. See chromium hexacarbonyl.

chromium cast irons. A group of alloy cast irons containing high amounts of chromium and including (i) *Abrasion-resistant white irons* with about 2.3-3.0% total carbon and about 23-28% chromium, martensitic microstructures, and good toughness and corrosion resistance; (ii) *Corrosion-resistant cast irons* containing 1.2-4.0% total carbon and 12-35% chromium, and having austenitic or martensitic microstructures and good resistance to dilute oxidizing acids, salt solutions and marine atmospheres; and (iii) *Heat-resistant gray cast irons* with 1.8-3.0% carbon and 15-35% chromium, having ferritic or coarse pearlitic microstructures, and good resistance to scaling. Also known as *chromium irons*.

chromium chloride. See chromic chloride; chromous chloride.

chromium chromate coatings. See chromate coatings.

chromium coatings. Hard, bright layers of chromium produced on metallic substrates by electrodeposition. The metallic sub-

strates are often given very thin initial copper and/or nickel underlayers. Thin chromium coatings provide significantly enhanced corrosion resistance, while somewhat thicker coatings improve abrasion and wear resistance. Very thin coatings may also be used for decorative purposes, however they do not hide surface defects, and the surface must be fully finished prior to plating. Used on automotive components and trim, aircraft components, domestic equipment, chemical equipment, bearing surfaces, and bathroom fixtures. Also known as *chrome coating; chrome plate; chromium plate*. See also chromium-copper-nickel plate.

chromium columbium. See chromium niobium.

chromium copper. (1) A wrought high-copper alloy containing about 0.7-0.9% chromium. Commercially available in the form of flat products, rods, structural shapes, tubes and wires, it has good to excellent corrosion resistance, excellent cold workability, good hot workability, high impact strength, and high electric and thermal conductivity. Used for welding electrodes, switchgear, cable connectors, circuit breaker components, and electrical and thermal conductors. Also known as *chrome copper*.

(2) A master alloy containing 89-92% copper, and 8-11% chromium. Used in the manufacture of hard steels, as an addition of chromium to chrome copper, copper-chromium alloys and other nonferrous alloys. Also known as *chrome copper*.

(3) A commercially pure copper surface-coated first with nickel and then with chromium, and is used for stampings, display cases, refrigerators, and kitchen equipment. Also known as *chrome copper*.

chromium-copper plates. Chromium coatings applied over undercoatings of copper by electrodeposition. The initial bright copper plate is usually 20 μm (0.8 mil) and the final chromium plate only 0.3 μm (0.012 mil) thick. Used for decorative applications. Also known as *copper-chromium plates*.

chromium-copper-nickel plates. Chromium coatings applied over undercoatings of copper and nickel by electrodeposition. The initial bright copper plate is usually 20 μm (0.8 mil) thick, and is followed by a bright nickel plate that is usually 30 μm (1.2 mils) thick. The final chromium plate is only 0.3 μm (0.012 mil) thick. The copper and nickel undercoatings impart bright, semibright or satin appearance to the chromium. Used for decorative applications, and for corrosion protection. Also known as *copper-nickel-chromium plates*.

chromium diboride. See chromium boride (ii).

chromium dichloride. See chromous chloride.

chromium dioxide. A compound of chromium and oxygen available in the form of a brownish-black powder, or black, acicular crystals and obtained by heating chromic acid. It has a density of 4.9 g/cm³, strong ferromagnetic properties, and high coercivity. Used for magnetic recording tapes, as a semiconductor, and as a catalyst. Formula: CrO_2.

chromium diphosphide. See chromium phosphide (i).

chromium disilicide. See chromium silicide (ii).

chromium fluoride. See chromic fluoride.

chromium hafnium. A compound of chromium and hafnium with a melting point of 1485°C (2705°F). Used in ceramics. Formula: Cr_2Hf.

chromium hexacarbonyl. White crystals with a density of 1.770 g/cm³, and a melting point of 150-155°C (270-311°F). It explodes above 204°C (399°F), and is used as a catalyst. Formula: $Cr(CO)_6$. Also known as *chromium carbonyl*.

chromium hot-work steels. A group of hot-work tool steels (AISI subgroup H1 through H19) containing about 0.30-0.45% carbon, 0.20-0.70% manganese, 0.20-1.20% silicon, 3.00-5.50% chromium, up to 5.25% tungsten, 3.00% molybdenum, 2.20% vanadium and 4.50% cobalt, respectively, and the balance iron. They have good deep-hardening properties, high hardenability, excellent toughness, high ductility, good hot hardness, and moderate wear resistance. Often sold under trade names or trademarks, such as *Alcodie, Chro-Mow, Crodi, Dart, Dievac, Firedie, Lumdie* or *Potomac*. Used for hot extrusion, trimming and die-casting dies, hot gripper and header dies, hot punches, hot shear blades, and die inserts. Also known as *chromium-type hot-work tool steels*.

chromium irons. See chromium cast irons.

chromium isopropoxide. A green metal alkoxide supplied in the form of a 10-12% solution in isopropanol/tetrahydrofuran (THF) with a density of 0.80. Used in the stabilization of sol-gel derived titania ceramics. Formula: $Cr(OC_3H_7)_3$.

chromium magnet steels. A group of magnet steels that contain about 1% carbon, 3-5% chromium and small additions of manganese, and are used for permanent magnets. A chromium magnet steel with 3.5% chromium and 1% carbon has a Curie temperature of approximately 745°C (1370°F), a density of 7.70-7.80 g/cm³ (0.278-0.282 lb/in.³), good hard magnetic properties including high coercivity and remanent magnetization. See also magnet steel.

chromium manganese antimonide. A brittle, gray, solid semiconductor compound that exhibits magnetism above its Curie temperature of about 250°C (480°F).

chromium-manganese steels. (1) A group of austenitic stainless steels (AISI types 201, 202 and 205) containing about 0.12-0.25% carbon, 0.5-1.0% silicon, 5.5-15.5% manganese, 16-19% chromium, 1.6% nickel, traces of nitrogen, phosphorus and sulfur, and the balance iron. Used for automobile parts, flatware, kitchen utensils, corrosion-resistant parts, etc. Also known as *chrome-manganese steels*.

(2) A group of case-hardening steels containing 0.1-0.2% carbon, 1-1.5% chromium and 1-4% manganese. Used for gears, shafts, piston pins, camshafts, gages, etc. Also known as *chrome-manganese steels*.

chromium-molybdenum steels. (1) A group of alloy steels containing chromium and molybdenum as the chief alloying elements, usually in the ranges of 0.50-0.95% chromium and 0.08-0.30% molybdenum, respectively. This term usually refers to the steels in the AISI-SAE 41xx series that have high strength and toughness, high ductility, good forgeability and machinability, and are used for forgings, carburized parts, axles, shafts, gears, propeller shafts, etc. Also known as *chrome-molybdenum steels*.

(2) Alloy steels containing 0.5-1% molybdenum, and 1-10% chromium. They have good to excellent corrosion resistance, good high-temperature properties, high toughness, good yield and creep strengths even at high temperatures, good weldability, forgeability and machinability, and high hardenability. Used for camshafts, piston pins, gauges, gears, and shafts. Also known as *chrome-molybdenum steels*.

chromium-molybdenum-vanadium steels. Alloy steels similar to *chromium-molybdenum steels*, but containing 0.5-1.3% molybdenum, 1-5% chromium and 0.2-1.0% vanadium. They have good to excellent corrosion resistance, excellent high-temperature properties, good fatigue and yield strength even at high temperatures, and good weldability. Used for camshafts, piston pins, gauges, gears, and shafts. Also known as *chrome-mo-*

lybdenum-vanadium steels.

chromium-molybdenum white irons. A class of *martensitic white cast irons* containing typically about 2.0-3.6% total carbon, 1.0% silicon, 0.5-1.5% manganese, 11-23% chromium, 0.5-3.5% molybdenum, 0.10% phosphorus, 0.06% sulfur, 1.5% nickel, 1.2% copper, and the balance iron. They have excellent abrasion and wear resistance, high hardness, good strength and toughness, and good oxidation and corrosion resistance at ordinary and elevated temperatures. Also known as *chromium-molybdenum white cast irons*. See also chromium white cast irons; high-carbon chromium-molybdenum white irons; low-carbon chromium-molybdenum white irons; martensitic white irons; white irons.

chromium monoboride. See chromium boride (i).

chromium mononitride. See chromic nitride.

chromium monophosphide. See chromium phosphide (ii).

chromium monosilicide. See chromium silicide (i).

chromium monosulfide. See chromium sulfide (i).

chromium-nickel-molybdenum steels. See nickel-chromium-molybdenum steels.

chromium-nickel steels. A group of highly corrosion-resistant stainless steels containing considerable amounts of chromium and nickel. They can be divided into two general categories: (i) *Ferritic and pearlitic case-hardening and heat-treatable steels* with about 0.1-0.4% carbon, 0.4-5.0% chromium, and 0.5-4.0% nickel; and (ii) *Austenitic and ferritic stainless steels* with about 0.05-0.2% carbon, 12-18% chromium and 1-10% nickel, including the so-called "eighteen-eight steels." Also known as *chrome-nickel steels*.

chromium niobium. A compound of chromium and niobium (columbium) with a melting range of 1480-1710°C (2695-3110°F) used in ceramics and materials research. Formula: Cr_2Nb (Cr_2Cb). Also known as *chromium columbium*.

chromium nitrate. See chromic nitrate.

chromium nitride. See chromic nitride.

chromium octadecasulfate. See chromic sulfate.

chromium orthophosphate. See chromic phosphate hexahydrate; chromic phosphate tetrahydrate.

chromium oxide. See chromic oxide.

chromium oxide green. See chrome oxide green.

chromium pentaboride. See chromium boride (v).

chromium pentadecasulfate. See chromic sulfate.

Chromium Permalloy. Trade name of Bell Telephone Laboratories (USA) for a soft-magnetic alloy containing 78.5% nickel, 17.5% iron and 4% chromium. It has high magnetic permeability, and is used for high-frequency equipment.

chromium phosphate. See chromic phosphate hexahydrate; chromic phosphate tetrahydrate.

chromium phosphide. Any of the following compounds of chromium and phosphorus used in ceramics and materials research: (i) *Chromium diphospide*. Density, 4.5 g/cm³. Formula: CrP_2; (ii) *Chromium monophosphide*. Gray-black crystals. Density, 5.49 g/cm³; melting point, 1360°C (2480°F); hardness, above 632 Vickers. Formula: CrP. Also known as *chromic phosphide*; (iii) *Dichromium phosphide*. Formula: Cr_2P; and (iv) *Trichromium phosphide*. Density, 6.51 g/cm³. Formula: Cr_3P.

chromium plates. See chromium coatings.

chromium platinum. A compound of chromium and platinum with a melting point of 1500°C (2730°F), used in ceramics and materials research. Formula: Cr_3Pt.

Chromium Plus. Trademark for a chromium electroplating deposit containing polytetrafluoroethylene (PTFE). Used on tooling and plastic molding dies to enhance lubricity, and corrosion and wear resistance.

chromium-potassium sulfate. Deep violet-red crystalline compound with a density of 1.81 g/cm³, and a melting point of 89°C (192°F). It loses 12H₂O at 400°C (750°F). Used as a red or green ceramic colorant, as a mordant, in tanning, and in photographic fixing baths. Formula: $CrK(SO_4)_2 \cdot 12H_2O$. Also known as *chrome alum; chromium-potassium sulfate dodecahydrate; potash chrome alum; potassium-chromium sulfate.*

chromium powder. A grayish powder which contains 99.0+% chromium and is available in regular and degassed grades with particle sizes ranging from 2 to 250 µm (0.00008 to 0.0098 in.). Used in powder metallurgy, hardfacing, and pigments.

chromium ruthenium. A compound of chromium and ruthenium with a melting point of 1540-1570°C (2804-2858°F), used in ceramics and materials research. Formula: $CrRu_2$.

chromium sesquichloride. See chromic chloride.

chromium sesquioxide. See chromic oxide.

chromium sesquisulfide. See chromium sulfide (ii).

chromium silicide. Any of several compounds of chromium and silicon possessing excellent oxidation resistance, good resistance to thermal shock, moderate strength, poor resistance to impact loading. Many of these compounds are used in wear-resistant components for high-temperature applications, and for hardfacing and high-temperature coatings. The following compounds are of commercial importance: (i) *Chromium monosilicide.* Density, 5.43 g/cm³; melting point 1543°C (2809°F); hardness, 1000 Vickers. Formula: $CrSi$; (ii) *Chromium disilicide.* Gray powder (99+% pure). Crystal system, hexagonal. Density 4.7-5.5 g/cm³; melting point 1490°C (2714°F); hardness, 100-1600 Vickers. Used as reinforcement for structural ceramic composites and as a matrix material for structural silicide compounds. Formula: $CrSi_2$; (iii) *Trichromium silicide.* Density, 6.45 g/cm³; melting point, 1710°C (3110°F); hardness, 1000 Vickers. Formula: Cr_3Si; and (iv) *Trichromium disilicide.* Density, 5.6 g/cm³; melting point, 1560°C (2840°F); hardness, 1280 Vickers. Formula: Cr_3Si_2.

chromium stearate. A metallic soap available in the form of a dark-green powder with a melting point of 95-100°C (203-212°F). Used in ceramics, plastics, waxes and greases, and as a catalyst. Formula: $Cr(C_{18}H_{35}O_2)_3$.

chromium steels. (1) A generic term for steels containing chromium as their principal alloying element. Also known as *chrome steels.*

    (2) A group of extremely hard, strong, wear-resistant steels containing 1.00% or more carbon and 0.27-1.50% chromium. This term usually refers to the steels in the AISI-SAE 50xx, 51xx, 50xxx, 51xxx and 52xxx series. Also known as *chrome steels.*

    (3) A large group of case-hardening, heat-treatable and tool steels containing 0.3-30% chromium. They have high tensile strength and hardness, and good high-temperature properties, and corrosion resistance. Also known as *chrome steels.*

chromium sulfate. See chromic sulfate.

chromium sulfide. Any of the following compounds of chromium and sulfur used in ceramics, metallurgy and materials research: (i) *Chromium monosulfide.* A black powder. Density, 4.1 g/cm³. Formula: CrS. Also known as *chromous sulfide*; (ii) *Chromium sesquisulfide.* A moisture-sensitive, brown-black powder (99+% pure). Density, 3.77 g/cm³. Formula: Cr_2S_3. Also known as *chromic sulfide*; and (iii) *Chromium tetrasulfide.* A gray-black powder. Density, 4.16 g/cm³. Formula: Cr_3S_4.

chromium tantalum. A compound of chromium and tantalum with a melting point of approximately 1980°C (3595°F). Used in ceramics and materials research. Formula: Cr_2Ta.

chromium tetraboride. See chromium boride (vi).

chromium tetrasulfide. See chromium sulfide (iii).

chromium triboride. See chromium boride (viii).

chromium tribromide. See chromic bromide.

chromium trichloride. See chromic chloride.

chromium trifluoride. See chromic fluoride.

chromium trioxide. See chromic acid.

chromium-type hot-work tool steels. See chromium hot-work steels.

chromium-vanadium steels. A group of alloy steels containing 0.45-0.55% carbon, 0.50-1.10% chromium and 0.10-0.25% vanadium. This term usually refers to the steels in the AISI-SAE 61xx series that have high strength, ductility and hardness, and exceptional resistance to wear and fatigue. Used for high-duty forgings, locomotive frames, gears and shafts, crankshafts, propeller shafts, and coil and leaf springs. Also known as *chrome-vanadium steels.*

chromium white irons. A class of *martensitic white cast irons* with a hardness of 550 Brinell typically contain 2.4-2.86% carbon, 11-14% chromium, up to 0.5% nickel, 0.5-1.5% manganese, up to 1.2% copper, up to 0.5-1.0% molybdenum, up to 1.0% silicon, up to 0.1% phosphorus, up to 0.06% sulfur, and the balance iron. Also known as *chromium white cast irons.* See also high-chromium white irons; chromium-molybdenum white irons.

chromium zirconium. A compound of chromium and zirconium with a density of 6.8 g/cm³, and a melting point of 1677°C (3050°F). Used in ceramics and materials research. Formula: Cr_2Zr.

Chromix. Trademark of Quigley Company, Inc. (USA) for plastic refractory plastic and ramming mixes containing a considerable amount of *chromite* (chrome ore).

chromized iron and steel. Iron and steel with hard chromium-rich surface layers produced by heating in intimate contact with a chromium salt (e.g., chromium dichloride) in powder, liquid, or gaseous form at elevated temperatures typically 950-1100°C (1740-2010°F). The highly corrosion- and high-temperature-resistant chromium layers are formed by the reaction of the salt with the substrate while diffusing inward. Used for fittings, containers, screws, bolts, valve parts, bearings, pins, pipes, etc.

Chromnickel I. Trade name of VDM Nickel-Technologie AG (Germany) for a resistance alloy of 78% nickel, 20% chromium and 2% manganese. It has high heat and corrosion resistance, and a maximum service temperature of 1150°C (2100°F). Used for heating elements, resistances, rheostats, and furnace parts. See also Chrome Nickel.

Chromnickel II. Trade name of VDM Nickel-Technologie AG (Germany) for a resistance alloy of 70% nickel, 20% chromium, 8% iron and 2% manganese. It has good heat and acid resistance, and a maximum service temperature of 1150°C (2100°F). Used for heating elements, rheostats, and measuring instruments. See also Chrome Nickel Iron.

chromocene. An organometallic coordination compound (molecular sandwich) composed of a chromium atom (Cr^{2+}) and two five-membered cyclopenediene (C_5H_5) rings, one located above and the other below the chromium atom plane. *Chromocene* is supplied in the form of scarlet crystals with a melting point of 172-173°C (342-343°F), and used as a catalyst, and in organometallic synthesis. Formula: $(C_5H_5)_2Cr^{2+}$. Also known as *dicyclopentadienylchromium; bis(cyclopentadienyl)chromium.*

chromodized aluminum. See chromadized aluminum.

Chromodur. Trade name of Krupp Stahl AG (Germany) for a series of stainless steels containing 0.2% carbon, 12% chromium, 0.3% nickel, 1% molybdenum, 0.5% tungsten, 0.3% vanadium, and the balance iron. Used for tableware, household ornaments, chemical equipment, and oil-refinery equipment.

Chromodur M. Trade name for a corrosion-resistant cobalt-chromium dental casting alloy.

chromogenic materials. A group of materials, such as certain ceramics (e.g., tungsten oxide) and liquid crystals usually produced as thin films, that exhibit a change in optical properties in response to electrical, photo or thermal stimulation. Used for chromogenic mirrors and windows. See also electrochromic materials.

Chromo-Loy. Trade name of Grand Northern Products Limited (USA) for a tough, wear-resistant cast iron containing 3% carbon, 2% silicon, 0.43% manganese, 21.6% chromium, 4.2% molybdenum, 3.4% nickel, and the balance iron. Used for coated hardfacing electrodes for producing overlays or buildups on roll crushers, shovel teeth, bucket lips, and oil-field tool joints.

Chromoloy. Trade name for a low-carbon alloy steel containing 0.2% carbon, 1% chromium, 1% molybdenum, 0.1% vanadium, and the balance iron. It has good elevated temperature properties, and good creep and fatigue strength. Used for automotive and aircraft components, and structural applications.

chromo paper. A smooth, white or colored printing paper coated on one side with chalk, and having a dull or glossy surface. It can be painted, bronzed and laminated. Moisture- and alkali-resistant grades are also available. Used primarily for offset printing and lithography.

chromophores. A class of organic pigments and dyes which contain a chromophore functional group, e.g., –NO (ni-troso dyes), –NO$_2$ (nitro dyes) or –N=N– (azo dyes), that causes absorption of ultraviolet or visible light. Some chromophore dyes are suitable as laser dyes and for imparting semicon-ductivity to certain polymers. Also known as *chromophore pigments.*

Chromosorb. Trademark of Manville Corporation (USA) for an extensive series of diatomite-based compounds of varying color, density and mesh size used as gas chromatography supports. See also diatomite.

chromous bromide. White crystals with a density of 4.356 g/cm^3 and a melting point of 842°C (1547°F). Used as a catalyst and reducing agent. Formula: $CrBr_2$. Also known as *chromium bromide.*

chromous chloride. Hygroscopic, air-sensitive, white needles or off-white powder (95+% pure) with a density of 2.90 g/cm^3, and a melting point of 824°C (1515°F). Used in the chromizing steel, as a reducing agent, and as a catalyst. Formula: $CrCl_2$. Also known as *chromium chloride; chromium dichloride.*

chromous formate. Reddish, acicular crystals obtained by treating chromous chloride with sodium formate. Used in chromium electroplating baths and as a catalyst in organic synthesis. Formula: $Cr(CHO_2)_2$.

chromous sulfide. See chromium sulfide (i).

Chromovan. Trade name of Teledyne Firth Sterling (USA) for a nondeforming, wear-resistant die steel containing 1.6% carbon, 12.5% chromium, 0.80% molybdenum, 1% vanadium, and the balance iron. Used for coining and thread-rolling dies.

Chro-Mow. Trade name of Crucible Materials Corporation (USA)

for a chromium-type hot-work tool steel (AISI type H12) containing 0.35% carbon, 1.05% silicon, 5.00% chromium, 1.25% tungsten, 1.35% molybdenum, 0.35% vanadium, and the balance iron. It has high toughness, good machinability, high hot hardness, and good abrasion resistance. Used for hot punches, hot shear blades, piercing tools, mandrels, die holders, and forging and extrusion tools.

Chrom/Slik. Trademark of Chromium Industries, Inc. (USA) for a composite coating of chromium and *Teflon* (polytetrafluoro-ethylene) polymer applied to machine components, such as cylinders, drums, and rolls for reduced friction applications.

Chromsol. Trade name of Union Carbide Corporation (USA) for a master alloy containing 62% chromium, 5% manganese, 1.5% silicon, 5.25% carbon, and the balance iron. Used to introduce chromium into steel melts.

Chromspun. Trade name of Eastman Chemical Company (USA) for cellulose acetate fibers with good moth and mildew resistance used for the manufacture of garments with silky luster and good drape.

Chromva-W. Trade name of Firth Brown Limited (UK) for an alloy steel containing 0.18% carbon, 2.7% chromium, 0.5% molybdenum, 0.5% tungsten, 0.7% vanadium, and the balance iron. It has a maximum service temperature of 580°C (1075°F), and is used for machine parts, tools, and structural applications.

Chronit. Trade name of J.C. Soding & Halbach (Germany) for an extensive series of cast and wrought corrosion- and heat-resistant austenitic, ferritic and martensitic stainless steels including several free-cutting and welding grades.

Chronite. Trade name of J.C. Soding & Halbach (Germany) for a heat- and corrosion-resistant nickel alloy containing 10% iron, 13.5% chromium, 1% aluminum, 1% manganese, and 0.4% silicon. Used for burners and valves.

Chronix. Trade name of ThyssenKrupp VDM GmbH (Germany) for resistance alloys supplied in several grades including *Chronix 70* with about 68% nickel, 30% chromium, 1.5% silicon and small amounts of other elements, and *Chronix 80* with about 78% nickel, 20% chromium, 1.5% silicon, and small amounts of other elements. Used for resistors and heating elements.

ChronoFlex. Trademark of CT Biomaterials-Division of Cardio-Tech International (USA) for a family of ether-free, biodurable, segmented polyurethane elastomers that are resistant to environmental stress cracking. They are supplied in three grades: *ChronoFlex AL* and *ChronoFlex C* which are available in the form of hard and soft pellets (hardness range from 80 Shore A to 75 Shore D) for use in the manufacture of long-term implanted medical devices, such as pacemaker leads, catheters, orthopedic devices and cardiovascular products, and *ChronoFlex AR* which is a highly flexible aromatic thermoplastic elastomer supplied in solution form for the manufacture of artificial-heart diaphragms, vascular grafts and stent coverings.

Chronothane. Trademark of CT Biomaterials-Division of Cardio-Tech International (USA) for aromatic, ether-based, thermoplastic polyurethane elastomers available in the form of hard and soft pellets (hardness range from 80 Shore A to 75 Shore D). They have high dimensional stability, low coefficients of friction, excellent chemical inertness and biostability, and good processibility by injection molding and extrusion. Used for medical devices, and as replacements for other medical-grade polyurethanes.

chrysene. A tetracyclic hydrocarbon obtained in the distillation of coal tar. It is commercially available in the form of standard

(95% pure), zone-refined (98+% pure) and high-purity zone-refined (99.9+%) crystals with a density of 1.274 g/cm³, a melting point of 252-254°C (486-489°F), and a boiling point of 448°C (838°F). Used in organic synthesis. The high-purity, zone-refined crystals (99.9+%) are used as aromatic reference standards. Formula: $C_{18}H_{12}$.

Chrysiode. Trade name for an alloy containing 92% silver and 8% aluminum.

Chrysite. British trade name for a copper-based dental alloy containing 37% zinc and 0.24% lead.

chrysoberyl. A green or greenish-yellow mineral with a vitreous streak belonging to the olivine group. It is composed of beryllium aluminate, $BeAl_2O_4$. Crystal system, orthorhombic. Density, 3.5-3.8 g/cm³; melting point, 1870°C (3398°F); hardness, 8.5 Mohs; refractive index, 1.75. Occurrence: Brazil, Russia, Sri Lanka, USA (Colorado). Used as a source of beryllium, beryllia and alumina, and as a gemstone. Also known as *chrysopal; golden beryl; gold beryl.*

chrysocolla. An emerald-green, greenish-blue, or sky-blue mineral with a white streak and vitreous luster. It is composed of copper silicate dihydrate, $CuSiO_3 \cdot 2H_2O$, and may contain cobalt and manganese. Crystal system, orthorhombic. Density, 2.10-2.60 g/cm³; refractive index, 1.6; hardness, 2-4 Mohs. Occurrence: Canada, Europe, USA (Arizona, New Mexico, Nevada, Utah). Used as minor ore of copper, and as a gemstone.

Chrysokalk. (1) British leaded brass containing 86-95% copper, 4.5-8% zinc, and 0.5-6% lead. It has poor resistance to tarnishing, and is used for cheap jewelry.

(2) British leaded brass containing 59% copper, 40% zinc and 1% lead, used for tubes.

chrysolite. An olive-green to yellowish brown mineral of the olivine group composed of magnesium iron silicate, $(Mg,Fe)_2SiO_4$, and occurring in igneous and metamorphic rocks, meteorites, and blast furnace slags. Crystal system, orthorhombic. Density, 3.26-3.40 g/cm³; hardness, 6.5-7.0 Mohs. Occurrence: Brazil, Egypt, USA (Arizona, New Mexico). The crude grades are crushed and used in refractories, cement, foundry sand, and floors. The purer grades are used in the manufacture of electronic components and ceramic-metal seals. The clear yellow-green variety, known as *peridot*, is used as a gemstone.

chrysopal. See chrysoberyl.

chrysoprase. An apple-green variety of *chalcedony* quartz containing nickel oxide, and found in Germany and the USA (California, Oregon). Used as a semiprecious gemstone.

Chrysorin. British trade name for a high-strength brass containing 63-72% copper and 28-37% zinc, used for plumbing, pipes, and tubes.

chrysotile. A dark green, fine fibrous asbestos mineral of the kaolinite-serpentine group. It is composed of hydrous magnesium silicate, $Mg_3Si_2O_5(OH)_4$, and may contain some aluminum, iron, manganese oxide, calcium oxide, potassium oxide and/or sodium oxide. Microstructurally, it consists of SiO_4 tetrahedra arranged in sheets. Crystal system, monoclinic, or orthorhombic. Density, 2.53-2.62 g/cm³; hardness 3-4 Mohs; strong, flexible fibers suitable for spinning and weaving; loses strength at elevated temperatures. Occurrence: Canada (Quebec), South Africa, USA (Arizona, California, Vermont). Used for textiles, paperboard and fireproof fabrics. See also asbestos; serpentine.

chrysotile felt. A product made by interfelting of *chrysotile* asbestos fibers, and used for the absorption and dissipation of heat from aerospace parts.

chudobaite. A pink mineral composed of magnesium zinc hydrogen arsenate decahydrate, $(Mg,Zn)_5H_2(AsO_4)_4 \cdot 10H_2O$. Crystal system, triclinic. Density, 2.94 g/cm³; refractive index, 1.608. Occurrence: Southwest Africa.

Chugal. Russian trade name for a cast iron containing 2.5-3.2% carbon, 1.6-2.3% silicon, 5.5-20% aluminum, and the balance iron. Used for furnace equipment.

chukhrovite. A colorless or white mineral composed of calcium aluminum yttrium fluoride sulfate decahydrate, $Ca_3Al_2(Y,Ce)(SO_4)F_{13} \cdot 10H_2O$. Crystal system, cubic. Density, 2.35 g/cm³; refractive index, 1.440. Occurrence: Kazakhstan.

chunk glass. Optical glass in the form of rough, irregular pieces obtained in breaking open a melting pot of transfer glass.

churchite. A smoke-gray mineral of the gypsum group composed of yttrium lanthanum phosphate dihydrate, $(Y_{0.9}La_{0.1})PO_4 \cdot 2H_2O$. Crystal system, monoclinic. Density, 3.27 g/cm³; refractive index, 1.631. Occurrence: UK.

Cibanoid. Trademark of Ciba Products Corporation (USA) for urea- and melamine-formaldehyde plastics.

Cico. Trademark of Denison Avery (USA) for one- and two-part industrial epoxy adhesives, cyanoacrylate adhesives and anaerobic adhesives.

Cicron. Trade name of Forjas Alavesas SA (Spain) for a series of hot-work tool steels containing 0.60-0.65% carbon, 0-1.2% silicon, 0.8% chromium, 0.3% molybdenum, and the balance iron. Used for hot punches and chisels, hot-work tools, etc.

C-I-L Nylon. Trade name of C-I-L Inc. (Canada) for nylon fibers and yarns.

Cilux. Trademark of C-I-L Inc. (Canada) for alkyd enamel paints.

Cimacell. Trade name of Cimar for cellulose acetate plastics.

Cimalen. Trade name of Cimar for polyethylene plastics.

Cimasit. Trade name of Cimar for cresol-formaldehyde and phenol-formaldehyde plastics.

Cimavil. Trade name of Cimar for polyvinyl chloride plastics.

Ciment De La Farge. Trade name for a rapid-hardening *aluminate cement* made by fusing in an electric furnace or kiln.

Ciment Fondu. Trade name for a slow-setting, rapid-hardening *aluminate cement* containing 40% alumina (Al_2O_3), 40% calcia (CaO), 10% silica (SiO_2), and 10% impurities.

Cimet. Trade name of Driver Harris Company (USA) for a corrosion- and heat-resistant iron alloy containing 22-29% chromium and 10-14% nickel. It has good resistance to acid mine waters and most chemicals, and is used for mine-water pumps, mining machinery, valves, screens, and furnace parts.

cimolite. A white, grayish, or reddish, rather soft and clayey mineral composed of aluminum silicate hexahydrate, $Al_4Si_9O_{33} \cdot 6H_2O$.

Cinch. Trademark of Parkell, Inc. (USA) for a no-slump addition-reaction polyvinylsiloxane-based dental impression material supplied in several viscosities.

Cindal. Trade name of D. & J. Tullus Limited (UK) for a heat-treatable aluminum alloy containing 0.8% zinc, 0.2% magnesium, and 0.3% chromium. Used for light-alloy parts.

cinder. See slag (1).

cinder block. A hollow building block made from concrete in which cinder (slag) is used as the aggregate.

cinder concrete. A composite made from a mixture of cement, water and cinder (slag) aggregate, and used for making cinder blocks.

cinder wool. Slag wool made from molten blast-furnace slag by blowing a jet of steam or compressed air through it. Used for packing, acoustic and thermal insulation of walls, in fireproofing, as a filter medium, and for compressing into mats, blankets and boards. See also slag wool.

Cinidur. Trade name for a heat-resistant steel containing 0.25% carbon, 24% nickel, 19% chromium, 2% molybdenum, 1% tungsten, 2.25% titanium, 1% aluminum, and the balance iron. Used for furnace parts, and oil-refinery equipment.

cinnabar. A reddish or brownish mineral with a scarlet streak and an adamantine to dull earthy luster. It is composed of mercury sulfide, HgS, and contains 86.2% mercury when pure. It can also be made synthetically. Crystal system, hexagonal. Density, 8.10 g/cm^3; hardness, 2-2.5 Mohs; refractive index, 2.905. Occurrence: China, Italy, Japan, Mexico, Spain, USA (Arkansas, California, Nevada, Oregon, Texas). Used as the most important ore of mercury, and as a red pigment (vermilion) in making paints, dyes, etc. Also known as *cinnabarite; liver ore; natural vermilion.* See also vermilion.

cinnabarite. See cinnabar.

cinnamon stone. See hessonite.

Cinemoid. Trade name of British Celanese (UK) for cellulose acetate films.

Cinseal. Trade name of ITT Components Group Europe (UK) for an alloy of 53% iron, 29% nickel, 17.5% cobalt, and 0.5% manganese. Used for metal-to-glass seals, hermetic seals, and telecommunications equipment.

CIPed products. See cold-isostatically-pressed products.

Circeo. Trade name of Circeo Filati S.r.l. (Italy) for a yarn blend composed of 65% polyester and 35% cotton, and supplied in several counts. Used for textile fabrics.

Circle. Trade name of Teledyne Vasco (USA) for a series of cobalt-tungsten high-speed steels containing 0.75-1.3% carbon, 4-4.5% chromium, 1.0-5.1% molybdenum, 6-18.5% tungsten, 1.5-2.0% vanadium, 5.0-9.0% cobalt, and the balance iron. Used for tools, cutters, drills, reamers, broaches, taps, and slotting and forming tools.

circuit materials. A group of engineering and engineered materials used in the manufacture of integrated and printed circuits and related components. Included are materials, such as copper, gold, low-expansion alloys, p- and n-type semiconductors, calcite, sapphire, ceramics, glass-ceramics and various polymers, such as epoxies, polyethylene terephthalate, polyetheretherketone, polyethersulfone and polyimide.

circular knit fabrics. Tubular textile fabrics that have been knitted on a machine with circularly arranged knitting needles.

Circuposit. Trademark of Shipley Company LLC (USA) for a series of stable electroless copper products for circuit applications.

ciré. A wax-treated fabric having an extremely smooth and lustrous surface.

Cirlex. Trademark of E.I. DuPont de Nemours & Company (USA) for polyimide laminates.

CIROM. Trademark of CERAC, Inc. (USA) for a family of specially prepared thin-film deposition materials. CIROM stands for "CERAC Infra-Red Optical Material." *CIROM-2* is an ultradense, polycrystalline zinc sulfide (ZnS) produced from a high-purity (99.99%) powder by a proprietary high-pressure vacuum pressing technique. It is deposited by electron-beam or thermal evaporation, and used in electrooptics and optics to produce high-index films on beamsplitters and bandpass filters, and as an antireflective coating for semiconductors. *CIROM-IRX* is a non-radioactive modified cerous fluoride (CeF_3) supplied in the form of lumps, granules and grains with a purity of 99.9%. It is deposited as a relatively thin film by electron-beam evaporation on substrates, such as germanium,

zinc sulfide or zinc selenide. The deposit has good mechanical durability and moisture resistance and good antireflection properties. It can be used as a replacement for thorium fluoride (ThF_4), e.g., for low-loss laser mirrors, filters, and antireflective coatings on electronic components.

Cisalfa. Trade name for rayon fibers and yarns used for textile fabrics.

Cita. Trade name of The Wilkinson Company, Inc. (USA) for gold-based dental and medical alloys.

Citadur. Trademark for aluminate cement.

citric acid. A carboxylic acid that occurs in fruits, such as lemons and limes, and is also found in all animal and plant cells. It is available in the form of colorless or white, translucent, water-soluble, sour-tasting powders or crystals with a density of 1.542 g/cm^3, and a melting point of 152-154°C (306-309°F). It is also supplied in the form of the monohydrate and various salts including diammonium, sesquicalcium tetrahydrate, sesquimagnesium hydrate, lead trihydrate, monosodium, disodium, trisodium, trilithium, tripotassium and trisilver salts. Used in cleaning and polishing metals and alloys, such as steel, in the synthesis of citrates, as an acidifier, as a sequestering agent, as a water-conditioning agent, as a detergent builder, as a flavor or antioxidant in foods and soft drinks, in alkyd resins, in the manufacture of dyes, in waste-gas sulfur removal, and in biochemistry, molecular biology, biotechnology, medicine, electrophoresis, materials testing and microscopy. Formula: $C_6H_8O_7$.

Citricon. Trade name for a silicone elastomer used for dental impressions.

citrine. A light-yellow to brown transparent variety of *quartz* resembling true *topaz*, and found in India, Spain, Madagascar, and eastern Europe. Used chiefly as a gemstone. Also known as *Bohemian topaz; false topaz; quartz topaz.*

Citrix. Trade name for a self-cure dental hybrid composite.

Citroen. Trade name of Citroen SA (France) for a cast aluminum alloy containing 12% copper, 0.1% magnesium, 0.5% manganese, 0.5% iron, 0.5% silicon, and the balance iron. Used for pistons.

citron yellow. See buttercup yellow.

Civona. Trade name for rayon fibers and yarns used for textile fabrics.

Cladboard. Trademark of Domtar Construction Materials (Canada) for melamine panels used in building construction.

clad brazing sheet. A sheet of metal clad on one or both sides with brazing alloy.

Clad Foam. Trademark of Cor Tec (USA) for fiberglass-reinforced plastic foam products.

Cladmate. Trademark of Dow Chemical Company (USA) for blue-colored, rigid *Styrofoam* polystyrene home insulation with excellent air-barrier and thermal insulation properties. It is supplied in 0.6 × 2.4 m (2 × 8 ft) sheets of varying thickness, and is used as exterior wall sheathing.

clad metal. A composite metal, usually in sheet, plate, tube or wire form, made by adding one or more layers of an acid-, corrosion-, and/or heat-resistant or more electrically conductive or resistive metal to a base metal by (i) heating and rolling together, (ii) welding, (iii) application of a powder coating and heating to effect diffusion, or (iv) heavy chemical deposition or electrodeposition. Typical examples include aluminum-, copper-, bronze- and nickel-clad steel, stainless steel-clad copper, and copper-clad aluminum. Also known as *laminated metal.*

clad steel. A plain-carbon or low-alloy steel, usually in sheet or plate form, having a layer of another metal or alloy, such as stainless steel, nickel, aluminum, copper, Monel or Inconel, firmly bonded to one or both sides.

Clair de Lune. Trade name of Celanese Corporation (USA) for cellulose acetate plastics.

Clamer's alloy. A heat- and corrosion-resistant alloy of 50-90% iron, 5-25% nickel and 5-25% cobalt. Used for electrical machinery.

Claradex. Trade name of Shin-A (USA) for acrylonitrile-butadiene-styrene resins.

claraite. A bluish green mineral composed of copper zinc carbonate hydroxide tetrahydrate, $(Cu,Zn)_3(CO_3)(OH)_4 \cdot 4H_2O$. Crystal system, rhombohedral (hexagonal). Density, 3.35 g/cm^3; refractive index, 1.751. Occurrence: Germany.

Clarene. Trademark of Solvay Polymers, Inc. (USA) for ethylene vinyl alcohol copolymers.

Clarex DR-III. Trade name of Astra Products Inc. (USA) for light-diffusion polymethyl methacrylates with additives available in clear, white and various colors. Depending on the particular grade, they are supplied in thicknesses from 0.2 to 2.0 mm (0.008 to 0.2 in.) and light transmissions from 45 to 93%. Used in automotive and electronic applications, e.g., avionic systems, computers and video cameras.

Clarglas. Trade name of Cristaleria Espanola SA (Spain) for a patterned glass.

Claricast. Trade name of Vereinigte Glaswerke (Germany) for a colored, rough cast glass.

Clarifoil. Trademark of Acordis Acetate Chemicals Limited (UK) for a clear, glossy cellulose diacetate thermoplastic supplied in film and foil form. It has a density of 1.3 g/cm^3 (0.05 lb/in.^{3}), a service temperature range of -20 to +95°C (-4 to +203°F), rather high moisture absorption (1.9-7.0%), moderate ultraviolet stability, and fair to poor chemical resistance. Used for print laminations, photo sleeves, graphic film, window cartons, labels, invisible tape, and packaging applications.

Clarilux. Trade name of Fabbrica Pisana SpA (Italy) for glass blocks.

claringbullite. A blue mineral composed of copper chloride hydroxide hemihydrate, $Cu_4Cl(OH)_7 \cdot 0.5H_2O$. Crystal system, hexagonal. Density, 3.90 g/cm^3; refractive index, 1.782. Occurrence: Zambia.

Clarite. Trade name of Columbia Tool Steel Company (USA) for a series of tungsten-type hot-work and high-speed tool steels.

Clarity. Trademark of H.B. Fuller Company (USA) for advanced bottle labeling adhesives.

clarkeite. A dark brown, reddish-brown, or brownish-yellow, radioactive mineral of the becquerelite group composed sodium calcium lead uranium oxide hydroxide, $(Na,Ca,Pb)_2U_2(O,OH)_7$. Density, 6.29 g/cm^3; refractive index, 2.098. Occurrence: USA (North Carolina).

Clark's alloy. A corrosion-resistant alloy containing 75% copper, 14% nickel, 7.2% zinc, 1.9% cobalt and 1.9% tin. Used for chemical equipment and corrosion-resistant parts.

Clarus metal. A strong aluminum alloy containing 4% silicon and 1.5% copper. It has good oxidation resistance, and is used for light-alloy parts.

Classic. Trade name of Owens-Illinois Inc. (USA) for hollow glass blocks.

Classic-II. Trade name of Ivoclar Vivadent AG (Liechtenstein) for a bonding resin used in orthodontics.

Classic III. Trade name of Ivoclar Vivadent AG (Liechtenstein) for a high-gold dental bonding alloy.

Classic IV. Trade name of Ivoclar Vivadent (Liechtenstein) for a

high-noble dental ceramic alloy.

Classic Cote. Trade name of Perry & Derrick Company (USA) for one-coat house paints.

Classic Visions Pisces. Trademark of Ivoclar Vivadent AG (Liechtenstein) for a ceramic dental alloy containing 73.7% nickel, 12.6% chromium, 8.0% molybdenum, 3.3% aluminum, 1.7% beryllium, and less than 1.0% silicon and iron, respectively. It has a white color, a density of 7.6 g/cm^3 (0.27 lb/in.^{3}), a melting range of 1165-1270°C (2130-2320°F), a hardness after ceramic firing of 415 Vickers, moderate elongation, high strength, good processibility, and excellent biocompatibility and resistance to oral conditions. Used for single crowns, posts and bridges.

clathrate compounds. (1) Inclusion compounds in which the molecules of one compound are wholly enclosed within the lattice of another. Also known as *clathrates*. See also inclusion compounds.

(2) Nearly transparent, crystalline solids that occur naturally in seabed rocks and sediments under suitable conditions of temperature and pressure, and are stable to about 35°C (95°F). They are essentially composed of gas molecules (principally methane, CH_4) wholly enclosed within a hydrogen-bond-stabilized cage of water molecules and correspond to the formula $CH_4 \cdot 6H_2O$. Clathrate compounds are potential sources of energy. Also known as *clathrates; gas hydrates*.

clausthalite. A metallic blue-gray mineral of the halite group composed of lead selenide, PbSe. It can also be made synthetically. Crystal system, cubic. Density, 7.80-8.28 g/cm^3; refractive index, 1.92. Occurrence: Germany.

Clavo. German trade name for glass with nailhead pattern.

clay. A fine-grained sedimentary deposit obtained from weathering rocks, and composed principally of hydrous aluminum silicate minerals containing other impurities (e.g., ferric and ferrous oxides, magnesia, lime, etc.). It becomes plastic and moldable when mixed with water, and sets to a rigid, hard, brittle mass when dried. Its color may vary from dark reddish-brown to pale dull yellow. The most important clay minerals are *attapulgite, bentonite, halloysite, illite, kaolinite* and *montmorillonite*. Used in the manufacture of ceramic products, such as whiteware, pottery, brick, tile, stoneware, refractories, foundry molds and cores, mortars and cements, as a lightweight aggregate, as a filler in plastics and elastomers, in paper coatings, in colloidal suspensions, as a filtering media, in ion exchange, as catalyst support, and in drilling muds.

clay aggregate. Expanded or sintered clay or shale used as an aggregate in *lightweight concrete*. See also lightweight aggregate.

clay brick. A building brick of varying shape and size made from clay. Also known as *clay building brick*. See also brick clay.

Clayburn. Trademark of Clayburn Industries Limited (Canada) for firebrick, and fired clay and refractory products.

Claycast. Trademark of Clayburn Industries Limited (Canada) for clay-base castable refractories.

Claycrete. Trademark of Clayburn Industries Limited (Canada) for clay-base gunning mixes and castables.

clay ironstone. See argillaceous hematite.

Clay-Loy. Trade name of Colorado Fuel & Iron Company (USA) for a high-strength low-alloy (HSLA) constructional steel containing 0.22% carbon, 1.25% manganese, 0.35% manganese, 0.35% silicon, 0.5% copper, 0.2% vanadium, and the balance iron. Used for railroad and bus bodies.

clay-mortar mix. Clay in finely divided form added to a masonry mortar to improve plasticity.

clay refractories. Refractories made from clays, such as *china clay, ball clay* or *fireclay*, that are able to withstand temperatures up to 1700°C (3090°F). Used for linings in furnaces, ladles, and other molten-metal containment components.

clay rock. A rock containing or composed of fine, worn clayey particles (clay, silt, etc.). Also known as *argillaceous rock; clay stone.*

clay roofing tile. A roofing product in the form of overlapping or interlocking structural units composed of molded, hard-burnt shale, or mixtures of shale and clay. It is hard, fairly dense, durable, and usually unglazed, although glazed tile is sometimes used. A variety of shapes, colors, and textures are available.

clay slip. A thin suspension of finely divided clay in water. Also known as *clay slurry.*

clay slurry. See clay slip.

Clayspar. Trade name for a ceramic raw material based on silicon dioxide (SiO_2) and containing 2.5% aluminum oxide (Al_2O_3), 1.5% potassium oxide (K_2O), and 0.5% sodium oxide (Na_2O), and 0.5% other oxides.

clay sponge. A fired ceramic product that has a porous, spongy structure, and is composed of calcined clay foamed by the introduction of gas.

clay stone. See clay rock.

clayware. Ceramic products made of *clay*, and traditionally including pottery and allied whitewares as well as structural clay products, such as brick, tile and pipe.

clay wash. A clay slip or slurry used as a coating on molds, flasks, etc.

Clean-Cut. Trademark of Lukens Steel Company (USA) for a series of alloy plate steels with sulfur contents of about 0.02-0.05%. They are subjected to a calcium resulfurization treatment that provides a finished product with uniformly distributed sulfide inclusions, improved machinability, and ultimately in superior machined surfaces and reduced wear. Used for structural applications.

Cleangard. Trade name of C.E. Thurston & Sons, Inc. (USA) for asbestos fabrics, ropes and tapes.

Cleanlube. Trade name for high-temperature dry-film lubricants used in the forming of metals and polymers.

Clean'n Glaze. Trademark of 3M Company (USA) for a glazing compound.

Clean'n Strip. Trademark of 3M Company (USA) for abrasives supplied in disk and wheel form.

clear ceramic glaze. An inseparable, translucent or tinted glaze that has a glossy finish and is bonded to a ceramic product (e.g., a facing tile) by firing.

clear chromate conversion coating. A thin, clear conversion coating produced on the surface of cadmium-plated parts by treating in a bath containing chromic acid, sulfuric acid and water. It is a protective coating providing good passivation and a pleasing appearance.

Clearclad. Trade name of Clearclad Coatings Inc. (USA) for an electrophoretically-applied clear coating.

clear clay. A *clay* of the kaolin-type used as a mill addition in the manufacture of certain enamels. It is free from deleterious impurities, such as organic matter, and produces high gloss and deep, clear colors.

Clear-Cut. Trade name of Composition Materials Company Inc. (USA) for plastic media used for blast cleaning soft metals.

Clearfil. (1) Trademark of J. Morita Company (Japan) for an ex-

tensive range of restorative dental hybrid composites and composite resins including *Clearfil AP-X, Clearfil CR Inlay, Clearfil Photo Core, Clearfil Posterior* and *Clearfil Ray-Posterior,* and several dental adhesives and bonding agents and cements including *Clearfil Bond, Clearfil Liner Bond, Clearfil New Bond, Clearfil Photo Bond, Clearfil Porcelain Bond* and *Clearfil SE Bond.*

(2) See Cavex Clearfil.

Clearform. Trade name of Corning Glass Works (USA) for sintered glasses used for glass-to-metal seals.

clear frit. A frit that produces a clear, transparent porcelain enamel. *Clear frits* for sheet steels and cast irons usually contain 8-11% titanium dioxide (TiO_2) and produce relatively strong colors.

clear glaze. A transparent, colored or colorless ceramic glaze.

Clearlite. Trade name of Fourco Glass Company (USA) for sheet glass.

Clearlyte. Trade name of Enthone-OMI, Inc. (USA) for a cathodically applied polymer coating.

Clearsite. Trade name of Celluplastic Corporation (USA for cellulose acetate plastics.

Clearshield. Trade name of Corning Glass Works (USA) for radiation-shielding windows in which a proprietary connective agent replaces the oil between the panes.

Cleartemp. Trade name of Fourco Glass Company (USA) for toughened sheet glass.

Cleartran. Trademark of Rohm & Haas Company (USA) for a water-clear, high-purity zinc sulfide (ZnS) prepared by chemical vapor deposition and modified by subsequent hot-isostatic pressing. It has high chemical inertness and theoretical density, good machinability, and is used for infrared applications.

Cleaver Cast. Trade name of National Refractories & Minerals Corporation (USA) for castable refractories.

cleavelandite. A snow-white lamellar high-purity variety of *albite* (soda feldspar).

Clebrium. British trade name for a series of heat- and corrosion-resistant cast irons containing 2.0-2.6% carbon, 2-4.6% nickel, 13.1-18.3% chromium, 0.75-2.8% manganese, 0-3.6% molybdenum, 1.5% silicon, 0-2% copper, and the balance iron.

Cleerspan. Trademark of Globe Manufacturing Company (USA) for a *spandex* (segmented polyurethane) fiber used for the manufacture of textile fabrics with excellent stretch and recovery properties.

Clep-A-Laq. Trade name of MacDermid Inc. (USA) for water-based lacquers.

Clepox. Trade name of Frederick Gumm Chemical Company Inc. (USA) for black oxide coatings.

Clepoxy. Trade name of DFI Pultruded Composites, Inc. (USA) for pultruded carbon-epoxy composite rods with diameters down to 0.25 mm (0.01 in.). Used for molded carbon prepreg reinforcements, twisted structural cables, etc.

Cleremont. Trade name of Crucible Materials Corporation (USA) for a tool steel containing 1.05% carbon, and the balance iron. Used for drill rod, pivots and tools.

Cletaloy. Trade name of Chase Brass & Copper Company (USA) for a series of sintered copper-tungsten and silver-tungsten alloys used for electrical contacts and welding electrodes.

Clevite. (1) Trade name of Gould Inc. (USA) for a series of moderately hard copper-lead-tin bearing alloys that are usually cast or sintered on steel backs. Also included under this trade name are several aluminum-base bearing alloys, and soft lead-tin and lead-tin-copper bearing overlays.

(2) Trade name for lead zirconate titanate piezoelectric ceramics.

Clevyl. Trademark of Rhovyl SA (France) for a series of flame-resistant vinyon fibers with excellent resistance to water, sunlight, bacteria and moths. Used for clothing, drapes, furnishings, etc.

Clichier metal. An antifriction metal with any of the following compositions: (i) 80% tin, 15% bismuth and 5% lead; (ii) 48% tin, 32% lead, 11% antimony and 9% bismuth; and (iii) 50% lead, 36% tin and 14% cadmium. Used for fuses and bearings.

Clicker. Trade name for an enamel/dentin bonding agent.

Clicking Die. (1) Trade name of Disston Inc. (USA) for a water hardening die steel containing 0.55% carbon, 0.2% molybdenum, and the balance iron.

(2) Trade name of Disston Inc. (USA) for a water hardening die steel containing 0.4% carbon, 0.7% chromium, and the balance iron.

cliffordite. A yellow mineral composed of uranium tellurite, UTe_3O_9. Crystal system, cubic. Density, 6.57 g/cm^3; refractive index, above 2.11. Occurrence: Mexico.

cliftonite. A black mineral composed of carbon, C. Crystal system, hexagonal. Density, 2.26 g/cm^3. Occurrence: USA (Arizona).

Climax. Trade name of Climax Performance Materials Corporation (USA) for a series of cast steels and alloy cast irons.

Climelt. Trade name of Climax Performance Materials Corporation (USA) for high-purity molybdenum and molybdenum alloys for high-temperature applications. They have good strength and corrosion resistance, and are used for chemical equipment, machine parts, grinding equipment, furnace parts, heat exchangers, engines, welding tips, electrical and electronic parts, etc.

clinker. (1) A very hard, dense brick with somewhat distorted shape that has been burnt at very high temperatures, and is almost completely vitrified. Also known as *clinker brick.*

(2) A ceramic product made by fine grinding and intimate mixing of clay, lime, silica, alumina, magnesia and ferric oxide in the proper proportions, and heating (calcining) to about 1400°C (2550°F) in a rotary or shaft kiln. It is the raw product for the manufacture of Portland cement. Also known as *cement clinker; Portland cement clinker.*

clinkered dolomite. See double-burnt dolomite.

clinkstone. See phonolite.

clinobisvanite. An orange-yellow mineral composed of bismuth vanadium oxide, $BiVO_4$. It can be made synthetically by heating stoichiometric mixtures of bismuth oxide and vanadium oxide. Crystal system, monoclinic. Density, 6.95 g/cm^3.

clinochlore. A white, pink, light green, olive-green, yellowish, purple or dark brown mineral of the chlorite group composed of magnesium aluminum silicate hydroxide with or without iron and/or chromium, $(Mg,Cr,Al,Fe)_6(Si,Al)_4O_{10}(OH)_8$. Crystal system, monoclinic, or triclinic. Density, 2.60-3.10 g/cm^3; hardness, 2.0-2.5 Mohs. Used for porcelain enamels, glazes and welding-rod coatings. Also known as *clinochlorite.*

clinochlorite. See clinochlore.

clinoclase. A dark to bluish green mineral composed copper arsenate hydroxide, $Cu_3AsO_4(OH)_3$. Crystal system, monoclinic. Density, 4.33 g/cm^3; refractive index, 1.87. Occurrence: Japan. Also known as *clinoclasite.*

clinoclasite. See clinoclase.

clinoenstatite. A colorless mineral of the pyroxene group composed of magnesium silicate, $MgSiO_3$. It can be made synthetically from silicon dioxide and basic magnesium carbonate by

a solid-state reaction process. Crystal system, monoclinic. Density, 3.21 g/cm^3; refractive index, 1.654. Occurrence: In meteoritic stones.

clinoferrosilite. A greenish mineral of the pyroxene group composed of iron silicate, $FeSiO_3$. Crystal system, monoclinic. Density, 4.07 g/cm^3; refractive index, 1.767. Occurrence: Africa, Iceland, USA (California, Wyoming).

clinohedrite. A colorless, or white mineral composed of zinc calcium hydroxide silicate, $ZnCa(OH)_2SiO_3$. Crystal system, monoclinic. Density, 3.33 g/cm^3; refractive index, 1.667; hardness, 5.5 Mohs. Occurrence: USA (New Jersey).

clinoholmquistite. A mineral of the amphibole group composed of lithium sodium iron magnesium aluminum fluoride silicate, $(Li,Na)_{2.5}(Al,Mg,Fe)_5Si_8O_{22}(O,F,OH)_2$. Crystal system, monoclinic. Density, 3.00 g/cm^3; refractive index, 1.627. Occurrence: Siberia.

clinohumite. A yellow or brown mineral of the humite group composed of magnesium fluoride silicate, $Mg_9(SiO_4)_4(F,OH)_2$, with or without titanium. Crystal system, monoclinic. Density, 3.21 g/cm^3; refractive index, 1.618. Occurrence: Finland, USA.

clinojimthompsonite. A colorless to light pinkish brown mineral of the amphibole group composed of iron magnesium silicate hydroxide, $(Mg,Fe)_5Si_6O_{10}(OH)_2$. Crystal system, monoclinic. Density, 3.02 g/cm^3. Occurrence: USA (Vermont).

clinophosinaite A pale lilac mineral composed of sodium calcium phosphorus silicate, $Na_3CaPSiO_7$. Crystal system, monoclinic. Density, 2.88 g/cm^3; refractive index, 1.561. Occurrence: Russia.

clinoptilotite. A colorless or white mineral of the zeolite group composed of potassium sodium calcium aluminum silicate hydrate, $(Na,K,Ca)_6(Si,Al)_{36}O_{72} \cdot 20H_2O$. Crystal system, monoclinic. Density, 2.10 g/cm^3; refractive index, 1.479. Occurrence: USA (California).

clinosafflorite. A gray mineral of the marcasite group composed of cobalt arsenide, $CoAs_2$. It can be made synthetically. Crystal system, monoclinic. Density, 7.48 g/cm^3.

clinotyrolite. An emerald green mineral composed of calcium copper oxide arsenate sulfate hydroxide decahydrate, $Cu_9Ca_2[(As,S)O_4]_4(O,OH)_{10} \cdot 10H_2O$. Crystal system, monoclinic. Density, 3.22 g/cm^3; refractive index, 1.6862. Occurrence: China.

clinozoisite. A colorless, light yellow, gray, green, or pink mineral of the epidote group composed of calcium aluminum iron silicate hydroxide, $Ca_2(Al,Fe)Al_2(SiO_4)_3(OH)$. Crystal system, monoclinic. Density, 3.21 g/cm^3; refractive index, 1.711. Occurrence: Czech Republic.

Clinpro. Trade name of 3M ESPE (USA) for a fluoride-containing dental pit and fissure sealant supplied as a pink-colored resin that turns white after light activation.

clintonite. A reddish-brown, or yellowish-brown mineral of the mica group composed of calcium magnesium aluminum silicate hydroxide, $Ca(Mg,Al,Fe)_3(Al,Si)_4O_{10}(OH)_2$. Crystal system, monoclinic. Density, 3.10 g/cm^3; refractive index, 1.658. Occurrence: Russia.

Clipper. Trade name for a high-speed steel (AISI type T1) containing 0.73% carbon, 18% tungsten, 4% chromium, 1.1% vanadium, and the balance iron. Used for cutting tools.

Clipsonic. Trademark of Alsapan (France) for coatings with waterproof thermosetting aminoplastic resin wearing layers used for laminating wood flooring, paneling and veneers.

clock brass. A free-cutting brass containing 62-65% copper, 33-37% zinc and 2% lead. It is commercially available in bar and sheet form, and is used for clock frames and plates, clock and watch backs, gears, wheels, and meter parts.

Clomo. Trade name of Uddeholm Corporation (USA) for an oil-hardening tool steel containing 0.97% carbon, 1.15% chromium, 0.32% molybdenum, and the balance iron. Used for hollow mine drills.

cloque. A term referring to a textile fabric having a blister or wrinkle on the surface intentionally produced by printing with a chemical, such as caustic soda, or by using yarns with different degrees of shrinkage or different tension in the weave. Also known as *blister fabric; relief fabric.*

closed-cell foam. A thermoplastic or thermosetting plastic, or an elastomer in the form of a flexible or rigid foam with numerous small, predominately closed (non-interconnected), spherical, polyhedral or honeycomb-shaped cells. Also known as *closed-cell cellular polymer; closed-cell foamed polymer.* See also open-cell foam.

closed-cell foamed polymer. See closed-cell foam.

closed-cell material. An organic or inorganic material containing numerous small cells, most of which are closed, i.e., not interconnected. See also open-cell material.

closed-face fabrics. Face fabrics that when joined to another material do not reveal the substrate. It is the opposite of open-face fabrics.

close-grained wood. Wood, such as birch, maple or pine, that has very small, and closely spaced pores, i.e., whose fibers are fine, and held closely together. Usually, such woods do not require fillers before finishing. Also known as *fine-textured wood.* See also open-grained wood.

Closeloy. Trade name of Armstrong-Whitworth Limited (UK) for a medium-carbon alloy steel containing varying amounts of carbon, nickel, chromium, molybdenum and iron. Used for machine parts.

closely graded sand. See uniform sand.

cloth. A generic term for *fabrics* of cotton, silk, linen, wool, animal hair, synthetic fibers, etc., made by weaving, knitting, rolling, pressing, or other processes.

clothing wool. Uncombed, short-fiber wool used especially for making woolens.

Cloverleaf. Trade name of E.A. Williams & Sons (USA) for an alloy of copper, tin and lead used for bushings and bearings.

Clunise. British trade name for an alloy containing 40% copper, 32% nickel, 25% zinc and 3% iron. Used for decorative parts.

Cluny lace. A *bobbin lace* made of heavy linen or cotton yarns, and usually having geometric designs. Used especially for curtains, tableware and apparel trim. Named after Cluny, a town in eastern France.

Clupak. Trade name for a strong, tough, strechable kraft paper.

cluster-assembled materials. See atomic cluster assembled materials.

Clydall. Trade name of Osborn Steels Limited (UK) for a series of cobalt-tungsten-type high-speed steels containing about 0.8% carbon, 4% chromium, 5-12% cobalt, 18.5-22% tungsten, 1.35% vanadium, and the balance iron. Used for lathe and planer tools, broaches, drills, cutters, tool bits, and roll-turning tools.

Clyde Alloy. Trade name of Steel Company Limited (UK) for a water hardening steel containing 0.3-0.6% carbon, and the balance iron. Used for machinery parts.

Cly-Die. Trade name of Osborn Steels Limited (UK) for a corrosion-resistant die steel containing 0.1-0.3% carbon, 13% chromium, and the balance iron. Used for plastic mold dies.

Clydmo. Trade name of Osborn Steels Limited (UK) for a high-speed steel containing 0.83% carbon, 4% chromium, 5% molyb-

denum, 1.8% vanadium, 6.2% tungsten, and the balance iron. Used for lathe and planer tools, reamers, hobs, and drills.

Clysar. Trademark of E.I. DuPont de Nemours & Company (USA) for thin, tough, elastic, highly transparent polyvinyl chloride shrink-on film for food and nonfood packaging applications including meat trays, video tapes, etc. It is supplied in various grades for specific requirements.

CM Bio Ethic. See Bio Ethic.

CM Capitor. See Capitor.

CM Esteticor. See Esteticor.

CM Neocast. See Neocast.

CM Opticast. See Opticast.

CM Protor 3. See Protor 3.

C Metal. Trade name of Edgar Allen & Company Limited (UK) for steel castings composed of 0.32% carbon, 0.3% silicon, 0.7% manganese, and the balance iron. Used for general engineering applications.

C-Monel. Trademark of Inco Alloys International Inc. (USA) for a series of cast nickel-copper alloys containing 60-70% nickel, 27-31% copper, 0-2.5% iron, 3.3-4.3% silicon, 1.5% manganese, and 0.20% carbon. They have good corrosion resistance, and are used for water-meter components, and machined parts. See also Monel.

coal. A general term for a firm, brittle, black or brownish-black combustible substance containing varying amounts of carbon. It is a sedimentary rock formed during the course of many millions of years by partial to complete decomposition of vegetable matter away from air and under varying degrees of pressure. Depending on the particular type (e.g., brown coal, bituminous coal or anthracite coal), it has a heating value ranging from less than about 19.5 MJ/kg (8300 Btu/lb.) to over 32.5 MJ/kg (14,000 Btu/lb.). Used as a fuel, and in the manufacture of coal tar, hydrocarbon gases, pigments and foundry facings. See also anthracite; bituminous coal; brown coal.

coal blacking. Pulverized coal applied to the working surface of a foundry core or mold to improve the surface finish of castings.

coal briquettes. Powdered coal or charcoal compressed into rectangular, cubic, pincushion-shaped, cylindrical, or ovoid pieces used primarily as fuel. A binder of pitch, asphalt or starch is commonly used. Also known as *fuel briquettes.*

coalesced copper. Oxygen-free copper in massive form made from brittle *cathode copper* by compacting, sintering in a reducing atmosphere at high pressure and temperature, and subsequent hot working.

coalingite. A red brown mineral of the sjogrenite group composed of magnesium iron carbonate hydroxide dihydrate, $Mg_{10}Fe_2(OH)_{24}(CO_3)\cdot 2H_2O$. Crystal system, rhombohedral (hexagonal). Density, 2.26 g/cm³; refractive index, 1.594. Occurrence: USA (California).

Coalni. Italian trade name for a steel containing 25.5% nickel, 11.5% aluminum, 4% cobalt, 4% copper, and the balance iron. Used for permanent magnets.

Coalnimax. Italian trade name for a steel containing 24% cobalt, 14% nickel, 8% aluminum, 3% copper, and the balance iron. Used for permanent magnets.

coal slag. An amorphous mixture of iron, aluminum and calcium silicates used in the manufacture of blasting and finishing abrasives.

coal tar. A dark brown or black, heavy, viscous mixture consisting chiefly of aromatic compounds. It is obtained as a residue in the destructive distillation of bituminous coal, and has a calorific value of about 38-40 MJ/kg (16300-17200 Btu/lb.). Used in the manufacture of dyes, plastics, solvents, explosives, etc., in waterproofing compositions and organic coatings, for road surfacing, and roofing applications, as electrical insulation, and as a fuel.

coal-tar coating. A black nonmetallic, water-resistant, two-component coating consisting of an epoxy resin to which coal tar pitch has been added.

coal-tar creosote. See creosote.

coal-tar enamel. A corrosion- and water-resistant organic coating based on coal tar, and used as cathodic protection for oil and gas pipelines.

coal-tar felt. Felt impregnated with purified coal tar.

coal-tar epoxy coating. A black nonmetallic, water-resistant, two-component coating consisting of an epoxy resin to which coal tar pitch has been added.

coal-tar pitch. A dark brown to black amorphous combustible residue obtained by the partial evaporation or distillation of coal tar. Depending on the degree of distillation, it is either a cementitious solid, or a sticky mass. It is available in several grades including roofing grades that soften at 65°C (149°F) and electrode grades that soften at 110-115°C (230-239°F). Density, 1.07 g/cm³ (0.039 lb/in.³); melting point, 60-120°C (140-248°F). Used in waterproofing, as a base for waterproof and acidproof paints and coatings, in roofing compositions, in paving compounds, as insulation, for covering of underground pipes, in bonded refractory brick, fillers, coal briquettes and foundry-core compounds, as a binder for carbon electrodes, as a fuel, in sealants and thermoplastics, and as a plasticizer for elastomers and polymers.

coal-tar resins. Thermosetting synthetic resins, such as phenolic or coumarone-indene resins, produced from the lighter fractions of coal tar. Available in the form of liquids and chips, they have good resistance to acids, alkalies and water.

Coanailium. Trade name of R.W. Coan Limited (UK) for a corrosion-resistant aluminum alloy used for marine parts.

coarse aggregate. Aggregate graded to a diameter exceeding 4.76 mm (0.187 in.) and including crushed stone and gravel. Used with a suitable cementing material and water to form concrete.

coarse aggregate bituminous concrete. A special concrete made without cement, but with *coarse aggregate* (e.g., crushed stone, gravel or slag) combined with a bituminous binder.

coarse concrete. A concrete made with coarse aggregate.

coarse-grained metal. A polycrystalline metal that contains primarily large grains (crystallites) that are more than 0.5 mm (0.02 in.) in size. Owing to its smaller total grain boundary area, a coarse-grained metal is softer and weaker than a fine-grained material. Occasionally, coarse-grained microstructures may be preferred and purposely produced, e.g., steel may be subjected to a special annealing treatment (at high temperature for long periods of time) to produce a coarse-grained material with lower toughness and improved machinability, or improved magnetic properties.

coarse gravel. Gravel that has an average grain diameter of 25-75 mm (1-3 in.). Used especially as a concrete aggregate, and in road construction.

coarse paper. Paper that ranges intermediate in appearance and composition between newsprint and the lighter weight paperboards and includes high-grade sulfite bond writing paper as well as groundwood, kraft, jute and parchment paper, and glassine. See also newprint; paperboard.

coarse-textured wood. See open-grain wood.

Coast. Trade name of Coast Metals Inc. (USA) for a series of tool-steel, stainless-steel, cast-iron, cobalt-base and nickel-base welding and hardfacing electrodes, rods and wires.

coast redwood. See redwood.

coated abrasives. Abrasive products, such as sandpaper or emery cloth, made by coating backings of paper or cloth on one or both sides with abrasive grains using a glue or synthetic resin. They are usually available in the form of sheets, strips, belts or disks. Common *coated abrasives* include aluminum oxide, flint, quartz, garnet, silicon carbide and tungsten carbide. *Coated abrasives* that are densely covered with abrasive grains are known as "close-coated." They provide fast-cutting action, but tend to clog quickly. "Open-coated" abrasives have spaces between the grains (coverage about 50-70% of backing) resulting in slower cutting action and reduced risk of clogging. Also known as *coated abrasive products; coated products.*

coated cemented carbides. Cemented carbides on whose surface a thin layer (usually 5 μm or 0.2 mil) of a material, such as titanium carbide, titanium nitride or aluminum oxide, has been produced by chemical or physical vapor deposition. They have a tough carbide core with an exceptionally hard and wear-resistant surface, and good to excellent shock resistance, good to excellent high-temperature machining characteristics, good resistance to oxidation and diffusion, and allow for faster cutting speeds than uncoated carbides. Used for indexable inserts and tool bits.

coated cotton. A soft, plain or printed cotton fabric with a thin waterproof polyurethane coating. Used for raincoats, jackets, coveralls, and other outdoor clothing.

coated fabrics. Textile fabrics that have oils, lacquers, varnishes or synthetic resins (e.g., acrylics, polyurethanes or polyvinyl chlorides) applied to their surfaces by calendering, brushing, dipping, spraying or other processes to impart or enhance properties, such as water resistance or repellency, appearance, and/or feel. See also protective fabrics.

coated felt. (1) A felt composed of glass fibers, and coated or impregnated on both sides with asphalt or coal tar.

(2) A sheet- or felt-like substance coated on both sides with asphalt. See also asphalt felt.

coated paper. A highly enameled paper made by coating with clay, sizes or other materials. It comes in dull and glossy finishes, and is especially useful for high-grade halftone work. Also known as *enamel paper.* See also dull-coated paper.

coated products. See coated abrasives.

coating clay. A finely divided clay made into a thin, watery suspension (or slip) suitable for application to ceramic bodies for masking the color and texture, or imparting color and/or opacity.

coating resins. Synthetic resins, such as acrylics, alkyds, phenolics or polyurethanes, used as film formers or binders in organic coatings (paints). See also synthetic resins; film formers; organic coatings.

coatings. Films or thin layers of solid, liquid or semiliquid materials applied to substrates by electrolysis, vapor deposition, immersion, brushing, spraying, calendering, troweling or other coating processes. After drying/curing, they serve to prevent corrosion, wear and/or oxidation, or improve adhesion, lubrication, surface finish and/or chemical, mechanical, thermal, optical and electrical properties.

Cobalt. (1) Trade name of Latrobe Steel Company (USA) for tungsten-type high-speed tool steels (AISI type T4) containing 0.7% carbon, 18% tungsten, 5% cobalt, 4% chromium, 1% vanadium, and the balance iron. They have high hot hardness, excellent wear resistance, and good machinability. Used for lathe and planer tools, hobs, and shaper tools.

(2) Trade name of Thyssen Edelstahlwerke AG (Germany) for a series of iron-cobalt permanent magnet materials.

(3) Trade name of Arcos Alloys (USA) for a series of bare and coated cobalt-chromium hardfacing electrodes.

cobalt. A nickel-white to silvery gray, brittle, lustrous metallic element of Group VIII (Group 9) of the Periodic Table. It is commercially available in the form of rondels, sheet, foil, microfoil, platelets, pieces, cakes, rod, wire, shot, sponge, granules, powder, anodes, ductile strips (95% cobalt, 5% iron), high-purity strips, and pure, single crystals. The commercial cobalt ores are *cobaltite, linnaeite* and *smaltite*. Density, 8.85 g/cm^3; melting point, 1495°C (2723°F); boiling point, 2870°C (5198°F); hardness, 170-320 Vickers; atomic number, 27; atomic weight, 58.933; divalent, trivalent. It has ferromagnetic properties, high toughness and ductility, relatively good electrical conductivity (about 27% IACS), and takes a high polish. There are two forms of cobalt: (i) *Alpha cobalt* with hexagonal close-packed crystal structure, present at ordinary temperatures; and (ii) B*eta cobalt* with face-centered cubic crystal structure, stable between about 450 and 1495°C (842 and 2723°F). *Cobalt* is used in the manufacture of stainless steels, cobalt steels for permanent and soft magnets, cobalt-chromium and cobalt-tungsten high-speed tool steels, as an alloying element in permanent and soft magnetic materials and magnetostrictive, hardfacing and wear-resistant alloys, cobalt-base tool materials, superalloys, sintered carbide cutting tools and cermets, in high-temperature spring and bearing alloys, electrical resistance alloys and special expansion and constant-modulus alloys, in electroplating, for lamp filaments, as an oxidizer, in ceramics, in catalytic converters, as a trace element in glass, as a drier in printing inks, paints and varnish-es, and as a gamma radiation source. The salts are used to produce brilliant blue colors in porcelain, glass and pottery. Symbol: Co.

cobalt-60. A radioactive isotope of cobalt with a mass number of 60, obtained by neutron irradiation of cobalt metal or cobaltic oxide. It has a half-life of 5.3 years and emits beta and gamma rays. Used in industrial radiography (e.g., for testing welds, castings and materials for flaws), as a gamma radiation source, in the determination of the height of molten metal in cupolas, as a radiation source in liquid-level gages, as a source of ions in gas-discharge devices, in the study of the oil consumption in internal-combustion engines and the permeability of pourous media to flow of oil, etc. Symbol: ^{60}Co. Also known as *radioactive cobalt; radiocobalt.*

cobalt acetate. See cobaltous acetate.

cobalt acetylacetonate. See cobaltic acetylacetonate.

cobalt alloys. Alloys containing cobalt as the predominant element and including soft magnetic cobalt-iron alloys, permanent magnet cobalt alloys, cobalt steels, cobalt-base superalloys, and wrought cobalt-base wear and hardfacing alloys.

cobalt aluminate. A blue mixture of aluminum oxide and cobalt oxide with a density of 4.37 g/cm^3, a melting point of 1960°C (3560°F), and good resistance to weathering and chemicals. It is used as a blue ceramic colorant, and as a very durable bright-blue to bluish-green paint pigment. The greenish shades may contain some zinc oxide. The blue pigments are known as *cobalt blue*. Formula: $CoAl_2O_4$. Also known as *cobaltous aluminate.*

cobalt aluminide. A compound of cobalt and aluminum with a

density of 6.04 g/cm^3, a melting point of 1630°C (2966°F), a hardness of about 440 Vickers, and good strength and oxidation resistance. Used for protective coatings. Formula: CoAl.

cobalt aluminide coating. An oxidation-resistant diffusion coating composed of cobalt aluminide, and produced on cobalt-base superalloys by reacting with aluminum.

cobalt aluminum bronze. A wear- and corrosion-resistant aluminum bronze containing 10.5% aluminum, 3.5% iron, 2.5% cobalt, and the balance copper. Used for pump bodies, impellers, shafts, heavy-duty bearings, screwdown nuts, and steel-mill slippers.

cobalt-ammonium sulfate. See cobaltous ammonium sulfate.

cobalt antimonide. A compound of cobalt and antimony. Crystal system, cubic. Crystal system, cubic. Crystal structure, skutterudite. Melting point, 850°C (1562°F). Used as a semiconductor. Formula: $CoSb_3$.

cobalt arsenate. See cobaltous arsenate.

cobalt arsenide. A compound of cobalt and arsenic. Crystal system, cubic. Crystal structure, skutterudite. Density, 6.73 g/cm^3; melting point, 957°C (1755°F); hardness, 3450 Knoop. Used as a semiconductor. Formula: $CoAs_3$.

Cobalt Ascoloy. Trade name of Allegheny Ludlum Steel (USA) for a heat- and corrosion-resistant steel containing 0.2% carbon, 12.25% chromium, 5% cobalt, 3% tungsten, 0.25% vanadium, and the balance iron. Used for high-temperature applications.

cobalt-base superalloys. A group of wrought and cast heat-resisting alloys containing about 35-65% cobalt, 19-30% chromium, up to 30% nickel, 15% tungsten, 11% molybdenum, 9% tantalum, 4.5% aluminum, 4% niobium (columbium) and 4% titanium, respectively, and small amounts of iron and carbon. They have excellent high-temperature properties, and high tensile fatigue strengths. Usually sold under trademarks and trade names, such as *Eliloy, J-Alloy, Stellite* or *Vitallium*, they are used for gas turbines, jet-engine parts, aircraft skins, spacecraft structures, nuclear reactors, pressure vessels, chemical-process equipment, and orthopedic and dental implants and prostheses. Also known as *cobalt-chromium alloys; high-temperature cobalt alloys.*

cobalt beryllide. A compound of cobalt and beryllium with a melting point of 1505°C (2741°F), good oxidation resistance, and high strength and strength retention at elevated temperatures. Used for structural applications. Formula: CoBe.

cobalt black. See cobaltic oxide.

cobalt bloom. See erythrite

cobalt blue. A very durable bright-blue to bluish-green pigment made from a mixture of aluminum oxide and cobalt oxide, and approximating *cobalt aluminate* ($CoAl_2O_4$) in composition. Greenish shades may contain some zinc oxide. It has good resistance to weathering and chemicals, and is used as a ceramic colorant and paint pigment. Also known as *azure blue; cobalt ultramarine; king's blue; Leyden blue; Thenard's blue.*

cobalt boride. Any of the following compounds of cobalt and boron used in ceramics and materials research: (i) *Cobalt monoboride*. Crystalline prisms. Density, 7.29 g/cm^3; melting point, above 1400°C (2550°F). Formula: CoB. Also known as *cobaltic boride;* (ii) *Cobalt diboride*. Formula: CoB_2; and (iii) *Dicobalt boride*. Melting point, 1260°C (2300°F). Formula: Co_2B; and (iv) *Tricobalt boride*. Melting point, 1093°C (2000°F). Used in ceramics. Formula: Co_3B.

cobalt-boron-silicon alloy. A permanent magnet amorphous alloy (metallic glass) composed of 69% cobalt, 12% boron, 12%

silicon, 4% iron, 2% molybdenum and 1% nickel. It has a density of 7.80 g/cm³ (0.282 lb/in.³), a hardness of 900 kgf/mm², excellent magnetic properties, a Curie temperature of 365°C (690°F), a maximum service temperature of 80°C (175°F) in air, a crystallization temperature of 520°C (970°F), and high strength and modulus. Used for magnetic and electrical equipment.

cobalt bromide. See cobaltous bromide.

cobalt carbonate. See cobaltous carbonate.

cobalt carbonyl. See cobalt tetracarbonyl.

cobalt chloride. See cobaltous chloride; cobaltous chloride hexahydrate.

cobalt chromate. See cobaltous chromate.

Cobalt Chrome. Trade name of Latrobe Steel Company (USA) for a series of high-carbon, high-chromium tool steels containing 1.3-1.5% carbon, 12-12.5% chromium, 3.0% cobalt, 0.8-0.9% molybdenum, up to 0.5% silicon, up to 0.5% manganese, and the balance iron. They have good abrasion and wear resistance and good nondeforming properties. Used for broaches, punches, burnishing tools, and valves.

cobalt-chromium alloys. (1) Alloys of 70% cobalt and 30% chromium that have a melting point of 1470°C (2680°F), and are used for high-temperature applications.

(2) See cobalt-base superalloys.

cobalt-chromium-aluminum-yttrium coatings. Cobalt-base overlay coatings containing rather large amounts of chromium, intermediate amounts of aluminum, and small amounts of yttrium. They are used to provide cobalt-base superalloys with improved resistance to oxidation and hot corrosion, and are applied by physical-vapor deposition, or plasma spraying. Abbreviation: CoCrAlY coatings. See also MCrAlY coatings.

cobalt-chromium steel. A high-temperature steel containing 1.5% carbon, 0.4% silicon, 0.4% manganese, 13.3% chromium, 3.7% cobalt, 0.7 % molybdenum, and the balance iron. It has good high-temperature properties, good resistance to pitting and oxidation. Used for internal-combustion engine valves.

cobalt columbium. See cobalt niobium.

Cobaltcrom. Trade name of Darwin & Milner Inc. (USA) for a series of alloy and high-speed steels containing about 1-1.5% carbon, 12.5-18% chromium, 3.0-3.7% cobalt, up to 0.75% molybdenum, 0.25% vanadium and 0.3% manganese, respectively, and the balance iron. They have good abrasion, corrosion and scaling resistance, and good nondeforming properties. Used for tools, such as blanking and trimming dies, shear blades, milling cutters, broaches, press tools, valves, welding rods, and special cutlery.

cobalt diboride. See cobalt boride (ii).

cobalt dichromate. A compound consisting of cobaltous oxide and chromium trioxide. It is available in the form of black crystals for use in ceramics, inorganic synthesis and materials research. Formula: $CoCr_2O_7$.

cobalt dichromium tetroxide. A compound consisting of cobaltous oxide and chromic oxide. It is available in the form of a black powder for use in ceramics, inorganic synthesis and materials research. Formula: $CoCr_2O_4$.

cobalt difluoride. See cobaltous fluoride.

cobalt disilicide. See cobalt silicide (ii).

cobalt disulfide. See cobalt sulfide (ii).

cobalt drier. A very efficient, powerful paint and varnish drier that contains cobalt linoleate, cobalt naphthenate, cobalt octoate, or cobalt resinate and that is soluble in virtually all vegetable and animal drying or semidrying oils.

cobalt 2-ethylhexanoate. See cobaltous 2-ethylhexanoate.

cobalt ferrate. A compound of cobaltous oxide and ferric oxide with a density of 5.30 g/cm³, a melting point of 1570°C (2860°F). Used in ceramics. Formula: $CoFe_2O_4$.

cobalt ferrite. A ferrimagnetic ceramic product with inverse cubic spinel crystal structure (*ferrospinel*) produced from mixtures of cobaltous oxide and ferric oxide powders. It has outstanding soft magnetic properties at high frequencies, high magnetic permeability, low coercive force, low magnetic saturation, very high resistivity, and high corrosion resistance. Used as a soft magnetic material for various applications. Formula: $CoFe_2O_4$. Also known as *cobaltous ferrite*.

cobalt fluoride. See cobaltic fluoride; cobaltous fluoride.

cobalt fluosilicate. Pink or orange-red crystalline powder with a density of 2.09-2.11 g/cm³ used in ceramics and materials research. Formula: $CoSiF_6 \cdot 6H_2O$. Also known as *cobaltous fluosilicate; cobalt silicofluoride; cobaltous silicofluoride*.

cobalt glance. See cobaltite.

cobalt-gold alloy. An alloy containing 40-75% cobalt and 25-60% gold, and used for vapor-deposited magnetic films.

cobalt green. A pale to dark yellowish green, permanent pigment made by mixing oxides of cobalt and zinc. Also known as *Rinman's green; Swedish green*.

cobalt hafnium. A compound of cobalt and hafnium with a melting point of 1570°C (2858°F) used in ceramics and materials research. Formula: Co_2Hf.

Cobalt High Speed. Trade name of McInnes Steel Company (USA) for a high-speed steel containing 0.7% carbon, 4% chromium, 18% tungsten, 5% cobalt, and the balance iron. Used for high-speed cutting tools and dies.

cobalt high-speed steels. See cobalt-tungsten high-speed steels.

cobalt hydrate. See cobaltic hydroxide; cobaltous hydroxide.

cobalt hydroxide. See cobaltic hydroxide; cobaltous hydroxide.

cobaltic acetylacetonate. A metal alkoxide available in the form of dark green or black, moisture-sensitive crystals made by treating cobaltous carbonate with acetylacetone and peroxide. Density, 1.43 g/cm³; melting point, decomposes at 211-216°C (412-421°F); boiling point, 340°C (644°F); soluble in water and toluene. Used in the vapor plating of cobalt, as a co-catalyst for the polymerization of dienes, as a catalyst for the polymerization of propylene oxide, as a photoinitiator, and as a component in preparation of light-sensitive photographic materials. Formula: $Co(O_2C_5H_7)_3$. Also known as *cobaltic 2,4-pentanedionate; cobalt acetylacetonate*.

cobaltic boride. See cobalt boride (i).

cobaltic fluoride. A moisture-sensitive, tan powder (99+% pure) with a density of 3.88 g/cm³. Used as a fluorinating agent especially for hydrocarbons. Formula: CoF_3. Also known as *cobalt fluoride; cobalt trifluoride*.

cobaltic hydroxide. A black or dark-brown powder that has a density of 4.46 g/cm³ and loses $3H_2O$ at 100°C (212°F). Used as a catalyst and in cobalt salts. Formula: $Co(OH)_3$. Also known as *cobalt hydrate; cobalt hydroxide*.

cobaltic oxide. Steel-gray to black crystalline powder (99+% pure) made by low-temperature heating of cobalt compounds with an excess of air. Density, 4.8-5.6 g/cm³. melting point, decomposes at 895°C (1643°F). Used as a pigment, as a colorant for enamels, and for glazing earthenware. Formula: Co_2O_3. Also known as *cobalt black; cobalt oxide*.

cobaltic 2,4-pentanedionate. See cobaltic acetylacetonate.

cobaltic potassium nitrite. See cobalt potassium nitrite.

cobaltic sulfide. See cobalt sulfide (iii).

cobalt iodide. See cobaltous iodide.

cobalt-iron alloys. A group of anisotropic and isotropic hard- and soft-magnetic alloys including those sold under the trademarks *P-6 Magnet Alloys* (45-50% iron, 44-45% cobalt, 6% nickel and 4-5% vanadium) and *Vicalloy* (52% cobalt, 38% iron and 10% vanadium) as well as the magnetic semihard alloys sold under the trade name *Remendur* (46-50% iron, 48-49% cobalt and 2-5% vanadium). Hard magnetic *cobalt-iron alloys* are used for hysteresis motors, magnetic equipment, magnetic memories, recording tapes, magnetic clutches and other magnetic devices, and telephone and reed switches. Soft-magnetic *cobalt-iron alloys* contain about 30-40% cobalt, have high magnetic saturation, and are used for both ac and dc magnetic and electrical applications. See also iron-cobalt alloys.

cobaltite. A silver-white, reddish or grayish mineral with a metallic luster. It belongs to the pyrite group and is composed of cobalt iron arsenic sulfide, (Co,Fe)AsS, containing about 35.5% cobalt. Crystal system, cubic. Density, 6.0-6.34 g/cm^3; hardness, 5.5 Mohs. Occurrence: Canada (Ontario), Germany, Sweden, Zaire. Used as an important ore of cobalt, and in ceramics. Also known as *cobalt glance; gray cobalt; white cobalt.*

cobaltkoritnigrite. A deep purple mineral composed of cobalt zinc arsenite hydroxide monohydrate, (Co,Zn)(AsO$_3$)(OH)·H$_2$O. Crystal system, triclinic. Refractive index, 1.668.

cobalt linoleate. See cobaltous linoleate.

Cobaltloy. Trade name of W.A. Zelnicker (USA) for an abrasion-resistant steel containing 1.5% carbon, 12% chromium, 0.9% molybdenum, 0.5% vanadium, 3.25% cobalt, and the balance iron. Used for hot-forming dies.

Cobalt Major. Trade name of Jessop Steel Company (USA) for a cobalt-tungsten high-speed steel containing 0.7% carbon, 4% chromium, 19% tungsten, 8% cobalt, 2% vanadium, and the balance iron. Used for hobs, broaches, reamers, and lathe and planer tools.

cobalt magnet steels. A group of *magnet steels* containing about 0.5-0.8% carbon, 15-40% cobalt; 3-9% tungsten, 1.5-6% chromium, and the balance iron. They have a density range of 8.0-8.4 g/cm^3 (0.29-0.30 lb/in.^{3}), a hardness of 60-65 Rockwell C, and hard magnetic properties, such as high remanent magnetization and coercivity. Used for permanent magnets.

cobalt molybdenum. An intermetallic compound of cobalt and molybdenum with a melting point of 1494°C (2721°F). Used in ceramics and materials research. Formula: Co$_7$Mo.

cobalt monoboride. See cobalt boride (i).

cobalt monophosphide. See cobalt phosphide (i).

cobalt monosilicide. See cobalt silicide (i).

cobalt monoxide. See cobaltous oxide.

cobalt naphthalocyanine. A naphthalocyanine derivative with a dye content of about 85%, and a maximum absorption wavelength of 731 nm. Used as a dye and pigment. Formula: Co(C$_{48}$H$_{24}$N$_8$). See also naphthalocyanine.

cobalt naphthenate. See cobaltous naphthenate.

cobalt neodecanoate. A dark blue, pasty compound containing about 12% cobalt. Used as an additive in the drying of printing inks. Formula: Co(O$_2$C$_{10}$H$_{19}$)$_2$.

cobalt niobium. An intermetallic compound of cobalt and molybdenum with a melting point of 1570°C (2858°F). Used in ceramics and materials research. Formula: Co$_2$Nb. Also known as *cobalt columbium.*

cobalt nitrate. See cobaltous nitrate.

cobaltocene. An organometallic coordination compound (molecular sandwich) consisting of a cobalt atom (Co^{2+}) and two five-membered cyclopenediene (C$_5$H$_5$) rings, one located above and the other below the cobalt atom plane. *Cobaltocene* is supplied in the form of air-, light- and heat-sensitive purple-black crystals with a melting point of 173-180°C (343-356°F). It is also available as a 7.5% solution in diethyl benzene. Used as a paint drier, as an oxygen stripping agent, as a polymerization inhibitor, and in Diels-Alder reactions. Formula: (C$_5$H$_5$)$_2$Co^{2+}. Also known as *dicyclopentadienylcobalt; bis(cyclopentadienyl)cobalt.*

cobalt ocher. See erythrite.

cobaltocobaltic oxide. Black or gray-black, hygroscopic, crystalline powder (99+% pure). Crystal system, cubic. Density, 6.07-6.45 g/cm^3; melting point, decomposes at 895°C (1643°F). Used for ceramics, in the manufacture of metallic cobalt, in pigments, as a catalyst, and in semiconductors. Formula: Co$_3$O$_4$. Also known as *tricobalt tetroxide.*

cobalt octoate. See cobaltous 2-ethylhexanoate.

cobalt oleate. See cobaltous oleate.

cobaltomenite. A red, synthetic mineral composed of cobalt selenite dihydrate, CoSeO$_3$·2H$_2$O. Crystal system, monoclinic. Density, 3.39 g/cm^3; refractive index, 1.728.

cobaltonickelous sulfate. See nickel-cobalt sulfate.

cobalt orthoarsenate. See cobaltous arsenate.

cobalt orthophosphate octahydrate. See cobaltous phosphate.

cobaltous acetate. Pink hygroscopic crystals made by treating cobalt carbonate with acetic acid. Density, 1.705 g/cm^3; melting point, loses 4H$_2$O at 140°C (284°F); refractive index, 1.542. Used in driers for paints and varnishes, in sympathetic inks, in anodizing, as a foam stabilizer, and as a catalyst for esterification and oxidation reactions. Formula: Co(O$_2$C$_2$H$_3$)$_2$ (anhydrous); Co(O$_2$C$_2$H$_3$)$_2$·4H$_2$O (tetrahydrate). Also known as *cobalt acetate.*

cobaltous aluminate. See cobalt aluminate.

cobaltous ammonium phosphate. Violet to deep red crystals used as paint pigments (cobalt violet) and as colorants in glasses, glazes and enamels. Formula: CoH$_4$NO$_4$P. Also known as *ammonium cobaltous phosphate.*

cobaltous ammonium sulfate. Deep, glowing red crystals obtained by treating cobaltous sulfate crystals with ammonium sulfate. Density, 1.902 g/cm^3. Used in ceramics, cobalt plating and catalysis. Formula: CoSO$_4$(NH$_4$)$_2$·SO$_4$·6H$_2$O. Also known as *cobalt-ammonium sulfate.*

cobaltous arsenate. A violet-red crystalline powder with a density of 2.95 g/cm^3. Used as a colorant to produce light blue colors on glass and porcelain, and in blue ceramic inks. Formula: Co$_3$(AsO$_4$)$_2$·8H$_2$O. Also known as *cobalt arsenate; cobalt orthoarsenate; cobaltous orthoarsenate.*

cobaltous carbonate. (1) A red or pink crystalline powder with a density of 4.13 g/cm^3, used in ceramics, as a temperature indicator, in pigments, and as a catalyst. Formula: CoCo$_3$. Also known as *cobalt carbonate.*

   (2) The commercial cobalt carbonate available as violet-red crystals made by treating a solution of cobaltous acetate with sodium carbonate. It decomposes on heating and is used in the manufacture of cobalt monoxide, cobalt salts and cobalt pigments, and in the production of blue and black ceramic colorants. Formula: 2CoCO$_3$·3Co(OH)$_2$·H$_2$O. Also known as *basic cobaltous carbonate.*

cobaltous bromide. Red-violet, moisture-sensitive crystalline powder (99+% pure) with a density of 2.46 g/cm^3 and a melting point of 47-48°C (116-188°F). It loses 6H$_2$O at 130°C (266°F) turning bright green. Used as a catalyst, and in barometers and hygrometers. Formula: $CoBr_2$ (anhydrous); $CoBr_2 \cdot 6H_2O$ (hexahydrate). Also known as *cobalt bromide*.

cobaltous chloride. Blue, moisture-sensitive crystalline powder (97+% pure) with a density of 3.356 g/cm^3, a melting point of 724°C (1335°F), and a boiling point of 1049°C (1920°F). Used in electroplating, barometers and hygrometers, as a flux for magnesium refining, as a catalyst, in sympathetic inks, and as a solid lubricant. Formula: $CoCl_2$. Also known as *cobalt chloride*.

cobaltous chloride hexahydrate. Deep red crystals (98+% pure) that have a density of 1.924 g/cm^3, a melting point of 87°C (188°F), and lose $6H_2O$ at 100°C (212°F). Used as a decolorizer in iron-tinted glass, in electroplating, in barometers and hygrometers, as a flux for magnesium refining, as a catalyst, e.g., for diaminobenzidine polymers, as a metal stain, in sympathetic inks, and in solid lubricants. Formula: $CoCl_2 \cdot 6H_2O$. Also known as *cobalt chloride; cobalt chloride hexahydrate*.

cobaltous chromate. Gray-black high-purity crystals, or brownish to yellowish-brown powder of varying composition. Used chiefly as a ceramic colarant, and in porcelain enamels (usually with zinc oxide and alumina) to produce light-colored stains. Formula: $CoCrO_4$. Also known as *cobalt chromate; basic cobalt chromate*.

cobaltous 2-ethylhexanoate. The cobaltous salt of 2-ethylhexanoic acid available as a bluish 65 wt% solution in mineral spirits containing up to 12% divalent cobalt and having a density of 1.002 g/cm^3 and a flash point of 104°F (40°C). It is also available as a solution in 6% naphtha. Used as a polymerization accelerator and initiator, as a catalyst, as a whitener, and as a paint and varnish drier. Formula: $Co(O_2C_8H_{15})_2$. Also known as *cobalt 2-ethylhexanoate; cobalt octoate; cobaltous octoate*.

cobaltous ferrite. See cobalt ferrite.

cobaltous fluoride. Pink crystals or powder (98+% pure) with a density of 4.46 g/cm^3, a melting point of approximately 1200°C (2190°F) and a boiling point of 1400°C (2550°F). Used as a catalyst. Formula: CoF_2. Also known as *cobalt fluoride; cobalt difluoride*.

cobaltous fluosilicate. See cobalt fluosilicate.

cobaltous hydroxide. Pink, hygroscopic, crystalline powder (95+% pure). Density, 3.597 g/cm^3; melting point, decomposes. Used in driers for paints and varnishes, in cobalt salts, as a catalyst, and in electrodes for storage batteries. Formula: $Co(OH)_2$. Also known as *cobalt hydrate; cobalt hydroxide*.

cobaltous iodide. Moisture-sensitive black crystals or powder (95+% pure) with a density of 5.68 g/cm^3, a melting point of 515°C (960°F), and a boiling point of 570°C (1058°F). Used as a catalyst and in barometers and hygrometers. Formula: CoI_2. Also known as *cobalt iodide*.

cobaltous iodide dihydrate. Green, deliquescent crystals with a melting point of about 100°C (210°F) (decomposes). Formula: $CoI_2 \cdot 2H_2O$.

cobaltous iodide hexahydrate. Brownish-red crystals that have a density of 2.90 g/cm^3, and lose $6H_2O$ at 27°C (80°F). Formula: $CoI_2 \cdot 6H_2O$.

cobaltous linoleate. A brown, amorphous powder obtained by boiling a cobalt salt with sodium linoleate. Used as a drier for paints and varnishes, and in enamels and white paints. Formula: $Co(O_2C_{18}H_{31})_2$. Also known as *cobalt linoleate*.

cobaltous naphthenate. The cobalt salt of naphthenic acid. It is available as a brown, amorphous powder, a bluish-red solid, or a dark-colored solution in mineral spirits. Used as a catalyst accelerator, as a drier for paints and varnishes, and in bonding steel and other metals to rubber. Also known as *cobalt naphthenate*.

cobaltous nitrate. Red or purple crystals (98+% pure). Density, 1.87 g/cm^3; melting point, 56°C (133°F); boiling point, 75°C (167°F). Used in metal treatment to promote adherence of porcelain enamels to iron and steel, and in porcelain decoration, in cobalt pigments, as a catalysts and in sympathetic inks. Formula: $Co(NO_3)_2 \cdot 6H_2O$. Also known as *cobalt nitrate; cobalt nitrate hexahydrate*.

cobaltous octoate. See cobaltous 2-ethylhexanoate.

cobaltous oleate. A brown, amorphous powder. Melting point, 235°C (455°F). Used as drier for paints and varnishes. Formula: $Co(O_2C_{18}H_{33})_2$. Also known as *cobalt oleate*.

cobaltous orthoarsenate. See cobaltous arsenate.

cobaltous orthophosphate octahydrate. See cobaltous phosphate.

cobaltous oxide. A cobalt compound that is usually supplied as a grayish, hygroscopic powder, but also as greenish brown crystals. Crystal system, cubic. Density, 6.45 g/cm^3; melting point, 1935°C (3515°F). Used as a pigment in ceramics and paints, in porcelain enamels, as a colorant in glass, glazes and enamels, in ground-coat enamels, as a decolorizer in glass and enamels, as a source for cobalt metal powder, as a catalyst, and in the preparation of cobalt salts. The crystals are also used for magnetic applications. Formula: CoO. Also known as *cobalt monoxide; cobalt oxide*.

cobaltous phosphate. Reddish or purple powder. Density, 2.769 g/cm^3; melting point, loses $8H_2O$ at 200°C (392°F). Used in the preparation of cobalt pigments, as a colorant for glasses, and in the production of light blue colors on porcelain. Formula: $Co_3(PO_4)_2 \cdot 8H_2O$. Also known as *cobalt orthophosphate octahydrate; cobaltous orthophosphate octahydrate; cobalt phosphate; cobalt phosphate octahydrate*.

cobaltous resinate. A brownish red powder used as a drier in paints and varnishes. Formula: $Co(O_4C_{44}H_{62})_2$. Also known as *cobalt resinate*.

cobaltous selenide. See cobalt selenide.

cobaltous silicate. See cobalt silicate.

cobaltous silicofluoride. See cobalt fluosilicate.

cobaltous sulfate. A red crystalline powder obtained by treating cobaltous oxide with sulfuric acid. Density, 3.47 g/cm^3; melting point, 735°C (1355°F). Used in ceramics, pigments and glazes. Formula: $CoSO_4$. Also known as *cobalt sulfate*.

cobaltous sulfate heptahydrate. A red-pink crystalline powder. Density, 1.92 g/cm^3; melting point, 97°C (206°F); boiling point, loses $7H_2O$ at 420°C (788°F). Used in ceramics, pigments and glazes, in whiteware bodies to impart blue and blue-white colors, as a drier for paints and inks, in plating solutions for cobalt, as a catalyst, and in storage batteries. Formula: $CoSO_4 \cdot 7H_2O$. Also known as *cobalt sulfate heptahydrate*.

cobaltous sulfide. See cobalt sulfide (i).

cobaltous tungstate. Red-brown or purple powder with a density of 8.42 g/cm^3, used as a drier for paints, varnishes and inks, in enamels, pigments and antiknock agents, and in electronics. Formula: $CoWO_4$. Also known as *cobalt tungstate; cobaltous wolframate; cobalt wolframate*.

cobaltous wolframate. See cobaltous tungstate.

cobalt oxide. See cobaltic oxide; cobaltous oxide; cobaltocobaltic oxide.

cobalt 2,4-pentanedionate. See cobaltic acetylacetonate.

cobalt pentlandite. A mineral of the pentlandite group composed of cobalt iron nickel sulfide, $(Co,Fe,Ni)_9S_8$. Crystal system, cubic. Density, 5.22 g/cm^3. Occurrence: Russian Federation,

Finland.

cobalt phosphate. See cobaltous phosphate.

cobalt phosphide. Any of the following compounds of cobalt and phosphorus used in electronics and materials research: (i) *Cobalt monophosphide*. Density, 6.24 g/cm^3. Formula: CoP; (ii) *Dicobalt phosphide*. Density, 7.55 g/cm^3; melting point, 1385°C (2525°F). Formula: Co_2P; and (iii) *Cobalt triphosphide*. Crystal system, cubic; skutterudite structure. Density, 4.26 g/cm^3; lower melting point, 997°C (1827°F). Used as a semiconductor material. Formula: CoP_3.

cobalt phthalocyanine. A purple, crystalline phytalocyanine dye that contains a divalent central cobalt ion (Co^{2+}). It has a dye content of about 97%, and a maximum absorption wavelength of 677 nm. Used as a dye and pigment. Formula: $Co(C_{32}H_{16}N_8)$. Abbreviation: CoPc. See also phthalocyanines.

cobalt-platinum. A compound containing 76.8 wt% platinum and 23.2 wt% cobalt. It has a density of 15.5 g/cm^3 (0.560 $lb/in.^3$), high coercive force and residual induction, a Curie temperature of approximately 480°C (900°F), and a maximum service temperature of 350°C (660°F). Used as a permanent magnet material for small magnets, electric instruments, etc. Formula: CoPt.

cobalt-potassium cyanide. Yellow high-purity crystals with a density of 1.87 g/cm^3, used in electronics. Formula: $K_3Co(CN)_6$.

cobalt-potassium nitrite. Yellow, crystalline powder made by adding potassium nitrite to a solution of a cobalt salt. It has a density of 5.18 g/cm^3, and is used as a yellow pigment (cobalt yellow), as a colorant for rubber, glass and ceramics, and for painting on glass and porcelain. Formula: $K_3Co(NO_2)_6$. Also known as *cobaltic potassium nitrite; Fischer's salt; potassium cobaltinitrite*.

cobalt powder. Cobalt in powder form available in purities from 99.6 to 99.995%, and particles sizes from less than 2 to over 700 μm (0.00008 to over 0.0276 in.). Used in powder metallurgy, as a catalyst, and for cobalt salts and chemicals.

cobalt pyrites. See linnaeite.

cobalt resinate. See cobaltous resinate.

cobalt selenide. Yellow crystals or pieces. Crystal system, hexagonal. Density, 7.65 g/cm^3; melting point, 1055°C (1931°F). Used in ceramics and semiconductor research. Formula: CoSe. Also known as *cobaltous selenide*.

cobalt silicate. See cobaltous silicate.

cobalt silicide. Any of the following compounds of cobalt and silicon used as semiconductors and high-temperature structural ceramics: (i) *Cobalt monosilicide*. Semiconductive powder or crystals. Density, 5.5-6 g/cm^3. Formula: CoSi; (ii) *Cobalt disilicide:* Gray crystals. Crystal system, cubic. Density, 4.95 g/cm^3; melting point, 1326°C (2419°F). Suitable as reinforcement for structural ceramic composites and as matrix materials for structural silicide compounds. Formula: $CoSi_2$; and (iii) *Dicobalt silicide*. Formula: Co_2Si.

cobalt silicofluoride. See cobalt fluosilicate.

cobalt-silicon-boron alloys. A group of amorphous alloys (metallic glasses) composed of 66-69% cobalt, 12-16% silicon, 12-14% boron, 4% iron, 0-2% molybdenum, and 0-1% nickel. They have a density of 7.6-7.8 g/cm^3 (0.27-0.28 $lb/in.^3$), a hardness of 900 kgf/mm^2, a maximum service temperature of 80°C (176°F), crystallization temperature range of 500-550°C (930-1020°F), excellent magnetic properties, high strength and ductility, and high chemical and electrical resistance. Used for magnetic applications.

cobalt-silicon-boron-manganese alloys. A group of amorphous alloys (metallic glasses) composed of 70% cobalt, 23% combined silicon and boron, 5% manganese, and 2% combined iron and molybdenum. They have a density of approximately 7.6 g/cm^3 (0.27 $lb/in.^3$), a hardness of 900 kgf/mm^2, a maximum service temperature of 80-120°C (176-248°F), excellent magnetic properties, high strength and ductility, and high chemical and electrical resistance. Used for magnetic applications.

cobalt soap. A water-insoluble metallic soap based on cobaltous linoleate, naphthenate, oleate, resinate or tallate. Used as drier for paints and varnishes and in flatting agents.

cobalt-sodium nitrate. See sodium hexanitritocobaltate.

Cobalt Special. Trade name of Thyssen Edelstahlwerke AG (Germany) for a heavy-duty high-speed steel containing 0.65% carbon, 4.2% chromium, 0.65% molybdenum, 1.7% vanadium, 18% tungsten, 15.5% cobalt, and the balance iron. Used for lathe and planer tools, form cutters, etc.

cobalt stearate. The cobalt salt of stearic acid ($C_{18}H_{36}O_2$) available in the form of purple pellets for use as a paint and varnish drier.

Cobalt Steel. Trade name of CCS Braeburn Alloy Steel (USA) for a high-speed steel containing 0.73% carbon, 0.15-0.3% manganese, 3.75-4.25% chromium, 0.85-1% vanadium, 17.5-18.5% tungsten, 4-4.5% cobalt, 0.75-1% molybdenum, and the balance iron. Used for high-speed cutting tools and general tools.

cobalt steels. See cobalt magnet steels; cobalt-tungsten high-speed steels.

cobalt sulfate. See cobaltous sulfate.

cobalt sulfate heptahydrate. See cobaltous sulfate heptahydrate.

cobalt sulfide. Any of the following compounds of cobalt and sulfur used in ceramics and materials research: (i) *Cobaltous sulfide*. Red powder, or brownish crystals. It occurs in nature as the mineral *syeporite*. Density, 5.45 g/cm^3; melting point, above 1100°C (2010°F). Formula: CoS; (ii) *Cobalt disulfide*. Black, cubic crystals. Density, 4.27 g/cm^3. Formula: CoS_2; and (iii) *Cobaltic sulfide*. Black crystals. Density, 4.8 g/cm^3. Formula: Co_2S_3.

cobalt tallate. A cobalt soap made with refined tall oil and used as a drier for paints and varnishes.

cobalt tantalum. A compound of cobalt and tantalum with a melting point of 1604°C (2919°F). Used in ceramics and materials research. Formula: Co_2Ta.

cobalt tetracarbonyl. Air- and heat-sensitive dark orange or dark brown crystals obtained by combining finely divided cobalt with carbon monoxide under pressure. Density, 1.78 g/cm^3; melting point, decomposes above 51-52°C (124-126°F). Used in the manufacture of high-purity cobalt salts, and in the preparation of cobalt powder for use in powder metallurgy. Formula: $Co_2(CO)_8$. Also known as *cobalt carbonyl; dicobalt octacarbonyl*.

cobalt tetrathiocyanatomercurate. See mercury tetrathiocyanatocobaltate.

cobalt trifluoride. See cobaltic fluoride.

cobalt triphosphide. See cobalt phosphide (iii).

cobalt tungstate. See cobaltous tungstate.

cobalt tungsten. A compound of cobalt and tungsten with a melting point of about 1750°C (3180°F), used in ceramics and materials research. Formula: Co_2W.

cobalt-tungsten alloys. Alloys of cobalt with up to 50% tungsten used in the production of hard, heat-resistant electrodeposits.

cobalt-tungsten high-speed steels. A group of high-speed tool steels (AISI types T4-T15) containing 0.7-1.5% carbon, 0.1-0.4% manganese, 0.2-0.4% silicon, 5-13% cobalt, 10-21% tungsten, 4-5% chromium, 1-5% vanadium, up to 1.3% molybdenum, up to 0.3% nickel, and the balance iron. They have excellent wear resistance, high hot hard-ness, good nonwarping properties, poor toughness, and allow for high cutting speeds and feeds. Often sold under trademarks or trade names, such as *Braeburn Cobalt, Gold Star, King Cobalt, Maxite, Nipigon, Red Cut Cobalt* or *Super Cobalt*, they are used for tools employed in the cutting and working of hard metallic materials, single-point tools and inserts, twist drills, lathe-tool bits, milling cutters, and forming and blanking dies. Also known as *cobalt high-speed steels*.

cobalt ultramarine. See cobalt blue.

cobalt violet. A permanent light red to deep blue violet, semi-opaque paint pigment composed of *cobaltous ammonium phosphate* (CoH_4NO_4P).

cobalt vitriol. See bieberite.

cobalt wolframate. See cobaltous tungstate.

cobalt yellow. A yellow paint pigment composed of *cobalt potassium nitrite* [$K_3Co(NO_2)_6$].

cobalt zincate. A reaction product of zinc dioxide and cobalt available as a green powder and used as a pigment.

cobalt-zippeite. An orange to tan mineral of the zippeite group composed of cobalt uranyl sulfate hydroxide hydrate, $Co_2(UO_2)_6(SO_4)_3(OH)_{10} \cdot 16H_2O$. It can be synthesized from cobalt sulfate and uranyl sulfate. Crystal system, orthorhombic. Density, 3.20 g/cm^3; refractive index, 1.779.

cobalt zirconium. A compound of cobalt and zirconium with a density of 8.46 g/cm^3, and a melting point of 1560°C (2840°F). Used in ceramics and materials research. Formula: Co_2Zr.

Cobanic. Trade name of Wilbur B. Driver Company (USA) for a series of heat- and corrosion-resistant alloys containing about 54.5-55% nickel, 44.5-45% cobalt, and up to a total of 1% iron and carbon. Used for electron tube cathodes and filaments, and vacuum-tube filament wire.

Cobar. Trade name of Arcos Alloys (USA) for a cobalt-base weld metal containing 1% carbon, 30% chromium, 4.5% tungsten, 2% nickel, and 2% iron. Used for cored-wire hardfacing electrodes.

cobble. (1) A fully or partly rounded stone with a nominal size of 64-305 mm (2.5 to 12 in.) used in paving streets, roads, sidewalks, etc. Also known as *cobblestone*.

(2) A coarse aggregate with a nominal diameter of 76-152 mm (3-6 in.). Used in concrete.

cobble mix. A concrete containing coarse, cobble-based aggregate.

cobblestone. See cobble (1).

Cobend. Trade name of Arcos Alloys (USA) for a series of cobalt-chromium alloys used for hardfacing electrodes.

Cobenium. Trademark of Wilbur B. Driver Company (USA) for a heat-treatable, corrosion-resistant superalloy containing 40% cobalt, 20% chromium, 15% nickel, 7% molybdenum, 2% manganese, 0.04% beryllium, 0.15% carbon, and the balance iron. It has excellent high-temperature properties, high strength and hardness, and is suitable for sub-zero temperatures and temperatures up to about 540°C (1000°F). Used for watch and instrument springs, and high-temperature applications. See also Durapower and Elgiloy.

Cobex. Trade name of BIP Plastics (UK) for polyethylene and polyvinyl chloride plastics.

Cobite. Trade name of Columbia Tool Steel Company (USA) for a series of cobalt-tungsten high-speed steels (AISI types T5 and T6) containing about 0.8% carbon, 18-20% tungsten, 8.75-

12% cobalt, 4.25-4.50% chromium, 1.50-2.05% vanadium, and the balance iron. They have very high red hardness, great depths of hardening, and excellent wear resistance. Used for lathe and planer tools, milling cutters, cutoff tools, boring tools, form tools, hobs, etc.

Coblac. Trademark for a series of black dispersions composed of about 10 parts cellulose nitrate, 3 parts dibutyl phthalate, and 2 parts carbon black. Used in the production of black laquers for automotive and industrial applications, black leather finishes, etc.

Cobra. (1) Trade name of Ludlow Steel Corporation (USA) for a high-speed tool steel (AISI type T1) containing 0.7% carbon, 18% tungsten, 4% chromium, 1% vanadium, and the balance iron. It has high red hardness, and high hardness in the heat-treated condition. Used for cutters, lathe and planer tools, hobs, form cutters, reamers, and broaches.

(2) Trade name Cobra Bandstahl GmbH (Germany) for refined strip steel.

Cobrasox. Trademark of A&P Technology (USA) for hybrid sleevings braided with carbon fibers having a tensile modulus of 3.86 GPa (560 ksi), and either aramid or E-glass fibers. Supplied in light-, medium- and heavy-fabric grades, they are used as reinforcements in composites.

Cobre. Trade name of Zinex Corporation (USA) for a solderable, high-adhesion, fine-grained copper electrodeposit and alkaline, noncyanide plating process.

cocal. See sande.

Cochrome. (1) Trade name of Cochrane Corporation (USA) for a high-strength steel containing 0.3% carbon, 1.2% chromium, 0.3% molybdenum, and the balance iron. Used for valves, valve fittings, and high-pressure castings.

(2) British trade name for a heat-resistant alloy of 60% cobalt, 24% iron, 12% chromium, and 2% manganese. It has high heat resistance, and is used for filaments and heating elements.

cochromite. A dark bluish-green mineral composed of cobalt chromium oxide, $CoCr_2O_4$. It can be made by heating cobaltous oxide (CoO) and chromic oxide (Cr_2O_3). Crystal system, cubic. Density, 5.22 g/cm^3.

cock bronze. A corrosion-resistant bronze of 84-90% copper, 8-10% tin and 2-6% zinc. Used for cocks, fittings and hardware.

cock metal. A soft copper alloy containing 65-70% copper and 30-35% lead. Used for taps and cocks.

Co-Co. Trade name of Teledyne Vasco (USA) for a cobalt-tungsten high-speed steel containing 0.7% carbon, 4% chromium, 1% vanadium, 18% tungsten, 5% cobalt, and the balance iron. Used for high-speed cutters, and lathe, planer and single-point cutting tools.

cocoawood. The tough, hard wood of several trees of the genus *Brya*. The heartwood is brown with yellow streaks, and the sapwood pale yellow. *Cocoawood* has a fine, dense, even grain. Source: Central and South America. Used for inlays and turnery. Also known as *cocos wood; West Indian ebony*.

cocobola. The hard, oily wood of the hardwood tree *Dalbergia retusa*. It is a fine wood with intriguing colors, usually having orange and red stripes alternating with darker streaks. It takes a beautiful polish, but has low gluing quality. Average weight, 1280 kg/m^3 (80 lb/ft^3). Source: Central America. Used for turnery, inlays, ornamental work, knife handles, canes, and instrument cases. Also known as *Honduras rosewood*.

coconinoite. A yellow mineral composed of aluminum iron uranyl phosphate sulfate hydroxide hydrate, $(Fe,Al)(UO_2)_4(PO_4)_2$-

$(SO_4)_2(OH)·22H_2O$. Crystal system, orthorhombic. Density, 2.90 g/cm³; refractive index, 1.58. Occurrence: China, USA (Utah).

coconut charcoal. An amorphous, porous, highly carbonaceous substance obtained by the destructive distillation of coconut shells in a closed retort, and subsequent heating with steam or carbon dioxide to develop high adsorptive properties. Used in gas masks.

coconut fiber. See coir.

coconut shell flour. A powder obtained by the repeated grinding of coconut shells and supplied in coarse, fine and very fine grades with mesh sizes ranging from 50 to 600. Used as a mild blasting abrasive for cleaning historic buildings and delicate objects, as a filler and extender in phenolic molding powders, as a degradable additive for plastics, as a filler in resin glues, in hand cleaner pastes, in oil well drilling, and as an adsorbent.

coconut shells. The crushed shells of coconuts used for the manufacture of activated charcoal, in abrasive blasting, and as an extender in plastics.

cocos wood. See cocoawood.

CoCrAlY coatings. See cobalt-chromium-aluminum-yttrium coatings; MCrAlY coatings.

Cocuman. Trade name of GTE Products Corporation (USA) for a brazing alloy of 58.5% copper, 31.5% manganese and 10% cobalt. It is commercially available in the form of foil, flexibraze, powder, wire, extrudable paste and preform, and has a liquidus temperature of 999°C (1830°F) and solidus temperature of 896°C (1645°F).

CodeBord. Trademark of Owens Corning Fiberglass (USA) for a strong, lightweight extruded polystyrene insulation for above-grade applications.

co-deposit. Two or more materials simultaneously or sequentially applied to a substrate or base material by any of several processes, such as chemical or physical vapor deposition, electrodeposition, electroless deposition, etc.

Coe Bronze. Trade name of Anaconda Company (USA) for a tough tin bronze containing 89.5% copper and 10.5% tin. Available in the form of hard and soft sheets, it is used for gears.

Coedal. Trade name of GC America Inc. (USA) for a phenol-formaldehyde resin formerly used for denture bases.

Coe-Flex. Trade name of GC America Inc. (USA) for a flexible polysulfide-based dental impression material.

Co-Elinvar. Trade name of Wallace Murray Corporation (USA) for an alloy containing 57-63% cobalt, 22-35% iron and 6-15% chromium. It has low thermal expansion, and a constant modulus of elasticity over a wide range of temperatures. Used for instruments, chronometers and hair springs. See also Elinvar.

Coe-Rect. Trade name of GC America Inc. (USA) for an acrylic resin for relining dentures.

coeruleolactite. A white to sky-blue mineral of the turquoise group composed of calcium aluminum phosphate hydroxide octahydrate, $CaAl_6(PO_4)_4(OH)_8·8H_2O$. Crystal system, triclinic. Density, 2.57-2.99 g/cm³; refractive index, 1.58. Occurrence: Germany, USA (Pennsylvania).

coesite. A colorless mineral composed of silicon dioxide, SiO_2. It can be made synthetically by a method involving high-pressures and elevated temperatures. Crystal system, monoclinic. Density, 2.90-3.01 g/cm³; refractive index, 1.6. Occurrence: Meteor impact craters.

Coe-Soft. Trade name of GC America Inc. (USA) for a soft, resilient self-cure acrylic resin-based denture relining material containing zinc undecylenate $(ZnO_4C_{22}H_{38})$.

Co-EVA. Trade name of Agriplast Srl (Italy) for a direct-light thermal film composed of a synthetic resin with high ethylene-vinyl acetate (EVA) copolymer content, and selected stabilizers. It is manufactured by co-extrusion, and provides high visible light transmission (above 90%), excellent mechanical properties, and high resistance to photo-oxidation. Supplied in three grades, it is used in greenhouses to provide vegetables and flowers with high luminosity, and protection against thermal inversion.

co-extrusions. Plastic products formed by the simultaneous extrusion of two or more polymers with different properties and/or colors, often through the same extruder. The starting polymers used in the manufacture of such products may either have similar compositions or dissimilar, but compatible compositions (e.g., different polymers, but with similar melting points and/or molecular structures). Examples of co-extrusions include (i) corrugated tubing composed of polymers with different properties, e.g., a smooth, chemically resistant inner pipe extruded along with a strong, tough outer pipe; (ii) window sashes made by co-extruding a soft, rubbery vinyl with a more rigid vinyl; and (iii) multi-ply plastic sheeting products manufactured by co-extruding several layers of similar or dissimilar polymers with different properties and/or colors.

coffinite. A black, radioactive mineral with an adamantine luster. It belongs to the zircon group and is composed of uranium silicate, $USiO_4$. A bluish green variety can be made synthetically. Crystal system, tetragonal. Density, 5.1 g/cm³; refractive index, 1.83-1.85; hardness, 230-300 Vickers. Occurrence: UK, USA (Arizona, Colorado, New Mexico, Utah, Wyoming). Used as an ore of uranium.

cog bronze. A corrosion-resistant bronze containing 85% copper, 11% tin and 4% zinc. Used for cogs, worms, worm wheels, etc.

Cogemica. Trade name of Cogebi Inc. (USA) for foil and paper products based on *muscovite* and *phlogopite* mica and used for electrical and metallurgical applications.

Cogemicafoil 504. Trade name of Cogebi Inc. (USA) for mica-based foil products used in the metals melting industry for the separation of refractory linings from the grout of coreless induction furnaces.

Cogemica High-Temp. Trade name of Cogebi Inc. (USA) for an asbestos-free gasket material consisting of mica paper, with high *phlogopite* content, impregnated with a silicone resin binder. It has excellent high-temperature properties up to 1000°C (1830°F), and good resistance to most acids, bases, solvents, and mineral oils. Used for automotive exhaust manifolds, heat exchangers, gas turbines, gas and oil burners, bolted flanged connections, etc.

Cogemicanite. Trade name of Cogebi Inc. (USA) for a series of mica products including *Cogemicanite 400* foils consisting of epoxy resin-impregnated *Cogemica* muscovite mica paper with close thickness tolerances and good flexural strength for use in the insulation of commutators of electric motors. *Cogemicanite 505* refers to rigid mica laminates with excellent high-temperature electrical insulating properties for use in appliance heating elements, and *Cogemica 132* to a flexible, formable asbestos-free, electrically insulating *Cogemica* muscovite or phlogopite mica paper impregnated with a special heat-resistant polymeric binder for use in heating elements and components.

Cogemicatape. Trade name of Cogebi Inc. (USA) for mica tapes composed of reinforced *Cogemica* mica paper. They have high electrical strength and thermal conductivity, low loss factors, and good moisture resistance. They are supplied in low-resin porous, and high-resin forms for the electrical insulation of high voltage coils, bars and slots of rotating machinery. The low-resin form is particular suitable for vacuum pressure insulation.

Cogetherm. Trademark of Cogebi Inc. (USA) for asbestos-free, temperature-resistant laminate insulation available in several grades, and consisting of reconstituted mica paper impregnated with synthetic resin and pressed into 2-76 mm (0.08-3.00 in.) thick plates. *Cogetherm P* withstands temperatures up 1000°C (1830°F) and *Cogetherm M* resists pressure to 400 MPa (58 ksi). Used for electrical and thermal barriers, heating elements, industrial heaters, etc.

Cogne. Trade name of Cogne Company (USA) for a series of corrosion-resistant austenitic, ferritic and martensitic stainless steels including several stabilized and welding grades.

Cogwheel. Trade name of Phosphor Bronze Company Limited (UK) for a series of chill-cast phosphor bronzes containing 10-14% tin, 0-0.2% phosphorus, 0-10% lead, and the balance copper. Used for gears, cog and worm wheels, bearings, bushings, connecting rods, pumps, piston rings, wear plates, gib keys, etc.

cogwheel ore. See bournonite.

Cohardite. Trade name of Connecticut Hard Rubber Company (USA) for hard rubber.

cohenite. (1) A tin-white meteoritic mineral composed of iron carbide, $(Fe,Ni,Co)_3C$. Crystal system, cubic.

(2) A dark grayish-brown synthetic mineral that is composed of iron carbide, Fe_3C, and can be extracted from steel as *cementite*. Crystal system, orthorhombic. Density, 7.20-7.67 g/cm³.

COHRlastic. Trademark of Saint Gobain Performance Plastics (USA) for a series of flame-retardant, low-density silicone foam sheet products, and silicone-coated asbestos and polymer fabrics. The foam products have low density, moisture absorption, compression-set and smoke evolution, excellent thermal and acoustical properties, good resistance to outgassing and softening by many chemicals, good resistance to various thermal conditions, and are used for aircraft, aerospace, automotive and electronics applications, e.g., gaskets, seals, diaphragms, tapes, ducting, etc. The coated fabrics are used for applications requiring resistance to acids, alkalies and moderate temperatures.

coil-coated steel. A steel sheet or strip to which a liquid organic coating (organosol, lacquer, varnish, etc.) has been applied on a continuous coil coating line to enhance its resistance to marine, rural, industrial and/or chemical atmospheres. The sheet or strip is usually galvanized or electrogalvanized prior to coating.

Coilex. Trade name of Murex Limited (UK) for a plain-carbon steels with 0.08% carbon, 0.4-0.8% manganese, up to 0.2% silicon, and the balance iron. Used for steel welding electrodes.

coinage bronze. A corrosion-resistant wrought bronze containing 94-96% copper, 2-4% tin and 1-2% zinc, and used for coins. Also known as *coinage copper.*

coinage copper. See coinage bronze.

coinage gold. (1) An alloy of 90% gold and 10% copper used for coins in the United States. Also known as *American gold; coin gold; standard gold.*

(2) An alloy of 91.67% gold and 8.33% silver used for coins in Australia. Also known as *Australian gold; coin gold.*

(3) An alloy of 91.67% gold and 8.33% copper used for coins in the United Kingdom. Also known as *British gold; coin gold.*

coinage metal. A durable metal, such as gold, silver, niobium (columbium) or nickel, that is used for coinage. It may or may not be a noble or precious metal.

coin bronze. A British high-copper alloy containing 2.5% zinc and 0.5% tin. It is available in the form of strips, and used in the United Kingdom, Australia and New Zealand for stamping coins.

coin gold. See coinage gold.

co-injection-molded plastics. Plastic products made from two chemically or structurally similar or dissimilar plastic materials (thermoplastics or thermosets) by injecting the two materials through a common manifold into the cavities of a mold and allowing them to solidify. The plastic material forming the surface layer usually has properties that are different from those of the material forming the core, e.g., the surface material may be a plastic having high chemical, ultraviolet and/or impact resistance, or high surface gloss, and the core material a low-cost commodity plastic that lacks these properties. See also injection-molded plastics.

coin silver. See sterling silver.

coin steel. A stainless steel used in some countries for coins.

coir. The tough, coarse, durable, reddish-brown- to buff-colored cellulose fiber obtained from the fruit of the tropical coconut palm (*Cocos nucifera*). It has low to moderate tensile strength, but fairly high elasticity and decay resistance, and is used for ropes, fishing nets, brooms, brushes, and door and hall mats. Also known as *coconut fiber.*

coke. The porous, gray, solid residue remaining after the destructive distillation of bituminous coal, petroleum, coal-tar pitch, or other carbonaceous material in ovens or retorts in the absence of air. It has a high fixed-carbon content (about 88%), a low volatile content (about 2-7%), and burns with much heat and little smoke. High-temperature coke is difficult and low-temperature coke relatively easy to ignite. The calorific value of *coke* is about 23-30 MJ/kg (9900-12900 Btu/lb.). It is used chiefly as a fuel for residential and industrial heating, and in metallurgy to reduce metallic oxides to metal, e.g., in blast furnaces and cupolas. It is also used as a source of synthesis gas, town gas, pitch and petroleum-derived coke for refractory furnace linings, in electrofining of aluminum, in the production of silicon carbide and calcium carbide, and for electrodes in the electrolytic reduction of alumina to aluminum. See also blast-furnace coke; metallurgical coke.

coke breeze. The fine residue from screening after all graded coke has been removed. It consists of pieces and particles below about 12.7-19.0 mm (0.5-0.75 in.) in size. Used in blacking mixes for foundry applications, and for making briquettes for cupolas.

coke iron. An ordinary pig iron made in a blast furnaces using coke as a fuel. Also known as *coke pig iron.*

coke oven tar. A brownish or blackish cementitious substance obtained as a byproduct in the destructive distillation of bituminous coal in coke ovens.

coke pig iron. See coke iron.

coke plate. See coke tinplate.

coke tinplate. A bright *tinplate* made from coke iron. It has a very light coating of tin, usually 1.5 to 4 µm (0.06 to 0.16 mil). The tin of the coating tends to form various compounds (e.g., $FeSn$, $FeSn_2$ or Fe_2Sn) with the iron of the plate. *Best coke* is coke tinplate carrying a slightly heavier tin coating than standard coke tinplate. Used for food containers and oil canning. Also known as *coke plate.* See also charcoal tinplate.

Colacril. Trade name for acrylic fibers used for the manufacture of textile fabrics.

Colalloy. Trade name of Colonial Alloys Inc. (USA) for a corrosion-resistant aluminum alloy containing 2% magnesium, 1% manganese, and 1% silicon. It has medium strength and hardness, and is used for chemical and food-handling equipment.

cola mahogany. See niangon.

Colas. Trademark for cold asphalt containing 60% asphalt and 40% water. Used in road paving.

Colastex. Trademark for a road construction material made by first blending 90% asphalt with 10% rubber latex, and then adding an *asphalt emulsion* (cold asphalt) containing 60% asphalt and 40% water.

colcather. See colcothar.

Colcessa. Trade name for rayon fibers and yarns used for textile fabrics.

Colclad. Trade name of British Steel plc (UK) for a series of low carbon steels (mild steels) in sheet or plate form, having a layer of stainless steel firmly bonded to one or both sides.

Colclad Inconel. Trade name of British Steel plc (UK) for a series of clad steels composed of sheets or plates of mild steel having a layer of Inconel firmly bonded to one, or both sides.

Colclad Nickel. Trade name of British Steel plc (UK) for a series of clad steels composed of sheet or plates of mild steel having a layer of nickel firmly bonded to one, or both sides.

Colcord. Trade name for rayon fibers, yarns and fabrics used for industrial applications including tire cords.

colcothar. A reddish-brown oxide of iron (Fe_2O_3) obtained as a residue by heating ferrous sulfate at high temperatures. It was formerly used as a pigment and polishing agent. Also spelled colcather. Also known as *English red*. See also crocus; rouge.

Colcrete. Trademark for colloidal concrete used for cement-bound surfacing.

cold asphalt. See asphalt emulsion.

cold-blast pig iron. A blast-furnace pig iron made without the use of a heated blast of air. Also known as *cold-blast iron.*

Cold Cure. Trademark of Industrial Formulator of Canada Limited for cold-setting bonding agents, and grouting materials used in building construction.

cold-cured rubber. Rubber vulcanized at ordinary temperatures by treating with a solution of a sulfur compound (e.g., sulfur chloride), or by exposure to its vapors.

cold-drawn sheet. A thermoplastic sheet that has been formed by pulling or drawing through a die of suitable cross section at room temperature.

cold-drawn steel. A round, rectangular, square, hexagonal or other shape of steel bar, wire or tube whose diameter has been gradually reduced by pulling or drawing through a die, or a series of dies at room temperature. The resulting product has a smooth, bright surface finish, is very accurate, and has a yield point about twice that of hot-rolled steel. See also hot-drawn steel.

cold-drawn product. A metal rod, tube, or wire that has been pulled or drawn through a die, or series of dies at room temperature. See also hot-drawn product.

cold-drawn wire. Wire that has been pulled or drawn through a die, or a series of dies at room temperature. See also hot-drawn wire.

cold-drawn-wire reinforcement. Steel wire, usually between 2 and 16 mm (0.080 and 0.625 in.) in diameter, made from rods obtained from hot-rolling mills by drawing through a die at room temperature. Used as a concrete reinforcement.

cold-finished steel. A steel bar, rod, plate, sheet or structural shape that has been shaped by cold rolling, cold drawing, smooth turning, or centerless grinding. It has a fine finish, and a uniform temper.

Cold Header Die. Trade name of Jessop Steel Company (USA) for a water-hardening tool steel (AISI type W1) containing 0.9-1% carbon, 0.3-0.4% manganese, and 0.2-0.4% silicon, and the balance iron. It has excellent machinability, good to moderate wear resistance and toughness, and low hot hardness. Used for coining and forming dies, cold-heading dies, and knurls.

Cold Header Vanadium. Trade name of Jessop Steel Company (USA) for a water-hardening tool steel (AISI type W2) containing 1% carbon, 0.25% manganese, 0.2% silicon, 0.2% vanadium, and the balance iron. Used for punches, pneumatic tools, taps, dies, etc.

Coldie. Trade name of CCS Braeburn Alloy Steel (USA) for a water-hardening tool steel (AISI type W2) containing 0.9% carbon, 0.25% silicon, 0.25% manganese, 0.2% vanadium, and the balance iron. It has excellent machinability, good to moderate wear resistance and toughness, and low hot hardness. Used for nail sets, blanking, drawing and forming dies, rivet sets, etc.

cold-isostatically-pressed products. Ceramic, metal or metal-alloy products of high and uniform density formed by first sealing the powdered starting material into a flexible metal or nonmetal container (or mold) and then applying equal fluid pressure from all directions (isostatic pressure) at or below room temperature. Cold-isostatically-pressed products usually require subsequent firing or sintering. Abbreviation: CIPed products. See also hot-isostatically-pressed products.

cold-laid plant mix. A composition made from cutback asphalt, bituminous emulsion and mineral aggregate in a central or stationary mixing plant. It is delivered to, and spread and compacted at the construction site when the composition is at ambient temperature. Also known as *cold mix.* See also hot-laid plant mix.

Coldlite. Trade name of Australian Window Glass Proprietary Limited for a heat-absorbing figured glass with bluish-green tint.

cold mix. See cold-laid plant mix.

cold-molded plastics. A group of plastic materials consisting of asbestos, either finely ground or in fiber form, bonded with inorganic materials, such as Portland cement or silica-lime cement, or organic materials, such as asphalt, pitch, phenolics or melamines. See also inorganic cold-molded plastics; organic cold-molded plastics.

Coldo. Trade name of A/S Drammens Glasverk (Norway) for an enameled, toughened glass used for cladding applications.

cold-painted glass. Fine glassware that has been ornamented by the application of oil paints or lacquer colors without subsequent firing. Used especially for architectural applications and in the arts.

cold-pressed materials. Ceramic, metal or metal-alloy products formed by compacting the powdered starting material at or below room temperature by the application of equal uniaxial or equiaxial (isostatic) pressure. Cold-pressed materials therefore require subsequent sintering. See also hot-pressed materials.

cold press moldings. Reinforced thermosetting plastic moldings with good surface appearance formed by a process in which glass-fiber fabrics, mats or prepregs are first placed on one side of two-piece molds or tools and then premixed, liquid thermosetting resin is poured onto them and the molds are closed by the application of pressure. Excess resin is squeezed out of the mold, and the products are allowed to cure.

cold-process cement. See slag cement.

cold-resistant steels. See cryogenic steels.

cold-rolled sheet. A rolling-mill product, usually of carbon or high-strength low-alloy steel, that has good surface finish, improved mechanical properties, and relatively uniform thickness. It is manufactured by cold (i.e., room-temperature) reduction of hot-rolled pickled and oiled coils. See also hot-rolled sheet.

cold-rolled stainless steel sheet. A rolling-mill product produced from cold-rolled stainless strip in numerous finishes by light cold rolling on a so-called skin-pass or planishing mill between highly polished rolls.

cold-rolled stainless steel strip. A cold-reduced rolling-mill product obtained from hot-rolled strip by pickling and subsequent rolling to the required thickness and/or strength on a either a two-high, or four-high reversing mill.

cold-rolled steel. (1) A steel that has been shaped by rolling while cold (usually at room temperature). Abbreviation: CRS. See also hot-rolled steel; rolled steel.

(2) A low-carbon steel that has been cold-reduced, and subsequently annealed.

cold rubber. A tough, strong, abrasion-resistant synthetic rubber, usually of the styrene-butadiene type, formed by polymerization at a relatively low temperature (4°C or 40°F).

cold-setting adhesive. An adhesive, such as hemlock-bark extract, resorcinol-formaldehyde, epoxy polyaminoamide or a methacrylate, that hardens at temperatures below 20°C (68°F).

cold solder. A mixture of a metal powder with a catalyst-cured synthetic resin (e.g., epoxy), or a pyroxylin cement that may or may not contain additives and fillers. Used for filling cracks in metals at ambient temperatures.

cold spray coatings. Coatings deposited on metallic or dielectric substrates by exposing them to high-velocity jets (300-1200 m/s or 90-3900 ft/s) of metal, alloy or composite powder particles (typically about 1-50 μm or 40-1970 μin. in size) accelerated by superconic jets of compressed gas at temperatures below the particle melting temperatures.

cold tar. A road tar dissolved in a light oil (distilled below about 210°C or 410°F), and used for road paving.

cold-water paint. A paint whose vehicle is composed of casein glue or similar material dissolved in water.

cold-worked metals. Metals that have been deformed plastically by subjecting to processes, such as rolling, upsetting, stamping, twisting, forging, extrusion or drawing, at a temperature below the recrystallization temperature.

cold-worked steel. Steel that has been deformed plastically at a temperature below the recrystallization temperature. For carbon-steels this temperature is between 450 and 600°C (840 and 1110°F), and for medium- and high-alloy steels between 600 and 800°C (1110 and 1470°F).

cold-worked steel reinforcement. Steel wire that has been shaped by cold working, usually involving rolling or drawing at ordinary temperatures. Used as concrete reinforcement.

cold-work tool steels. A group of tool steels that are extensively used for tools whose regular service does not involve elevated temperatures. They can be divided into the following groups: (i) *High-carbon, high-chromium types* (AISI group D); (ii) *Medium-alloy air-hardening types* (AISI group A); and (iii) *Oil-hardening types* (AISI group O). *Cold-work tool steels* have good nondeforming properties, good machinability, high toughness, and good wear resistance. Used for cold-working press tools (e.g., dies, punches, shears, etc.), plug and ring gages, blanking and forming punches, trimming dies, gages, press tools, etc. Also known as *cold-work steels.*

colemanite. A colorless to white, or yellowish mineral with vitreous luster composed of calcium borate pentahydrate, $Ca_2B_6O_{11} \cdot 5H_2O$. Crystal system, monoclinic. Density, 2.43 g/cm^3; hardness, 4.0-4.5 Mohs; refractive index, 1.59; exhibits ferroelectric properties below 266K. Occurrence: Turkey; USA (California, Nevada). Used as an ore of calcium borate (the principal natural source of borax), as a source of calcium oxide and boric oxide in pink and maroon raw-lead glazes, as a substitute for boric acid in the manufacture of glass fibers, and for ferroelectric crystals.

Coleman Porcelain. Trade name of Aardvark Clay & Supplies (USA) for porcelain clay (cone 10), translucent when thin, with a true *grolleg* body. It works well with glazes.

Colesco. Trade name for a lead babbitt composed of 77% lead, 14% antimony, 8% tin and 1% copper, used for bearings.

Colflex. Trade name of Ato Findley (France) for polyurethane adhesives for packaging, pasting and laminating applications.

Col-Graph. Trade name of Columbia Tool Steel Company (USA) for an oil-hardening cold-work tool steel (AISI type O6) containing 1.45% carbon, 0.8% manganese, 1.2% silicon, 0.2% chromium, 0.25% molybdenum, and the balance iron. It has a great depth of hardening, good machinability, good wear and abrasion resistance, and good toughness. Used for wear plates, blanking, trimming and forming dies, gages, taps, and machine parts.

Colgrout. Trademark for a colloidal grout used in the building trades.

Colgunite. Trademark for colloidal grouts and concrete products and the related wet-mix application process. Used in the building and road construction trades.

Collacral. Trademark of BASF AG (Germany) for a copolymer of polyvinylpyrrolidone and polyacrylonitrile/polymethyl methacrylate.

Collagraft. Trademark of NeuColl, Inc. (USA) for a synthetic material composed of *collagen* and a mineral composite of *tricalcium phosphate* and *hydroxyapatite* that is chemically very similar to natural bone. It has excellent biocompatibility, osteoconductivity and osteoinductivity, and is used in the restoration and repair of human bone.

collagen. A protein that forms most of the white fibrous portion of connective tissues in the tendons, muscles and skin, and helps to hold cells and tissues together. It is composed of a high proportion of proline ($C_5H_9NO_2$) and hydroxyproline ($C_5H_9NO_3$) amino acid residues, and has a structure intermediate between an alpha helix and a pleated-sheet structure. Obtained in various physical forms including solutions, powders, films and sponges, *collagen* from animal skins and hides is used in the manufacture of fibers, glues, and in wound dressings and biomedical coatings. Crosslinked collagen fibers are used in the reconstruction of damaged knee ligaments. See also cross-linked collagen.

collagen fibers. The dimensionally stable protein fibers obtained from split animal hides. Used for brushes and biomedical applications.

collaurin. See colloidal gold.

Collegiate. Trade name of J.F. Jelenko & Company (USA) for a microfine, gold-palladium (50% gold) porcelain-to-metal dental alloy with characteristics similar to *Olympia*, but more compatible with high-coefficient porcelains.

collet brass. A free-cutting brass containing 61% copper, 37% zinc and 2.5% lead, used for collets, and hot-forged parts.

Collet Steel. Trade name of Crucible Materials Corporation (USA) for a medium-carbon steel containing 0.6% carbon, 0.8% chromium, 0.4% nickel, 0.2% molybdenum, and the balance iron. Used for machine parts.

collimated roving. Roving that has been specially processed to improve the parallelism of the strands. It is usually of the parallel-wound type.

collinsite. A white mineral of the fairfieldite group composed of calcium magnesium phosphate dihydrate, $Ca_2Mg(PO_4)_2 \cdot 2H_2O$, and sometimes contains also iron or zinc. Crystal system, triclinic. Density, 2.93 g/cm³; refractive index, 1.642. Occurrence: Western Australia.

Collocarb. Trade name for colloidal carbon blacks.

collodion. A flammable, pale yellow viscous solution of cellulose nitrate (*pyroxylin*) in a mixture of ether and alcohol. It is deposited as a tough, waterproof, transparent film on evaporation of the ether, and has a flash point of approximately -18°C (0°F). Used for fibers, film, cements, lithography, photography, and as a replica-forming material in electron microscopy.

collodion cotton. See pyroxylin.

colloid. A solid, liquid or gaseous substance, e.g., albumin, glue, starch, gelatin, or swelling bentonite, made up of particles ranging in size from 1 to 1000 nm (0.04 to 40 μin.), e.g., single large molecules, groups of smaller molecules, nanocrystals or amorphous powders, that will remain suspended without dissolving in a different medium. A *colloid* may be suspended in a solid, liquid or gas, exhibits Brownian movements and, due to its electrical charge, is subject to cataphoresis. See also nanocolloid.

colloidal antimony oxide. Finely divided particles of antimony oxide (Sb_2O_3) dispersed in a medium, such as a water solution or polypropylene. It contains about 20% solids, and is used as a calibration aid in microscopy.

colloidal clay. A very fine clay, such as *bentonite*, that usually swells when mixed with water forming a jelly-like liquid. Used as a binder for nonplastic materials.

colloidal concrete. A concrete composed of an aggregate bonded with colloidal grout.

colloidal crystals. Closely packed crystals assembled from 5 nm (0.2 μin.) to 5 μm (0.2 μin.) organic particles (e.g., emulsion droplets, or polystyrene latex microspheres), inorganic particles (e.g., silica, titania, or metals) or biomolecules (e.g., DNA, proteins, or viruses). They exhibit a wide range of interesting properties including high surface-to-volume ratios, high structural stability, light diffraction properties, photonic band gaps, and high catalytic throughput, and are suitable for various applications in electronics, optoelectronics, photonics, catalysis, medicine and biotechnology.

colloidal gold. Finely divided particles of gold uniformly dispersed in a water solution. The color of the solution changes with particle size from blue to red. Used in electron microscopy and immunochemistry. Also known as *collaurin*.

colloidal graphite. Extremely fine flakes of pure synthetic graphite suspended in water, petroleum oil, castor oil, glycerin or other liquid. Used as a lubricant, and in lubricating films, mold-release agents, foundry facings, black coatings on glass, and conductive shields on the inside or outside surfaces of electron tubes.

colloidal grout. A *grout* that has the ability to hold the dispersed solid particles in permanent suspension.

colloidal metal. Finely divided metal particles dispersed in an aqueous suspension, e.g., colloidal gold or colloidal silver.

colloidal microgel. A *polymer gel*, such as poly(*N*-isopropylacrylamide), which consists of minute cells, typically 1-1000 nm (0.04-40 μin.) in size, and is maintained in a suitable solvent (e.g., water). It can reversible shrink on heating and may also be sensitive to changes in the pH and to other ion or molecule species. Used as a viscosity modifier, in oil recovery, and for drug delivery applications. See also microgel.

colloidal photonic crystals. Photonic crystals with three-dimensional (3D) lattice structure produced by several techniques including (i) the accurate arrangement of micromachined silicon wafers in successive layers, (ii) the self-arrangement of submicron-sized silica spheres in colloidal suspension and filling of the air voids with a high-refractive index titania-dioxide solution, or from a suspension of polymer spheres in a hydrogel film; and (iii) the formation of an face-centered cubic (fcc) structure by the sedimentation of monodisperse colloidal silica on a template, followed by imbibition of molten selenium into the air voids and etching away of the silica. The resulting replica has a high dielectric constant. See also photonic crystals; 3D photonic crystals.

colloidal silica. A colloidal suspension of silicon dioxide (SiO_2) used in the semiconductor industry as a fine abrasive for polishing silicon, in sol-gel production, in ceramics, textile treatment, etc. Monodisperse colloidal silica is used in the preparation of photonic crystals.

colloidal silver. Finely divided particles of silver dispersed in a quick-drying aqueous or organic liquid. Used to increase the surface area on terminals of electronic components, and to provide low-resistance contact points for mounting specimens for scanning-electron microscopy.

collophane. A white, or almost white cryptocrystalline variety of the mineral *apatite* containing calcium. Density, 2.7 g/cm³; hardness, 2-2.5 Mohs. Also known as *collophanite*.

collophanite. See collophane.

Colmirit. Trade name of A. Michelsens Glashaerderei (Denmark) for toughened glass products with fired-on colors, decorations and letters.

Colmonoy. Trademark of Wall Colmonoy Corporation (USA) for an extensive series of nickel- and nickel-chromium-type hardfacing alloys in powder and rod form, tungsten-nickel alloys for thermal spraying, nickel-alloy welding powders for cast irons and filling blowholes, hard, wear-resistant nickel-tungsten coatings, and copper-phosphorus powders for braze welding of copper and its alloys.

Colnova. Trade name for rayon fibers and yarns used for textile fabrics.

Colocrete. British trademark for a colored cement.

Cologne brown. A dark-brown, naturally occurring pigment similar to *Cassel brown* that consists of mixtures of iron oxide and decomposed vegetable matter. It is found in peat and lignite beds, and was originally obtained from lignitic ocher occurring near the German city of Cologne. Also known as *Cologne earth; Cologne umber.* See also Vandyke brown.

Cologne earth. See Cologne brown.

Cologne umber. See Cologne brown.

Colomat. Trade name for rayon fibers, yarns and fabrics.

Colomo. Trade name of Uddeholm Corporation (USA) for a tough, oil-hardening tool steel containing 0.95% carbon, 1.1% chromium, 0.33% molybdenum, and the balance iron. Used for mining drills, tools, and cutters.

Colonial. (1) Trade name of Teledyne Vasco (USA) for a series of tool steels including *Colonial No. 6*, an oil-hardening cold-work tool steel (AISI type O1), and several water-hardening tool steels, such as *Colonial No. 14* (AISI type W1) and *Colonial No. 7* (AISI type W2).

(2) Trade name of Colonial Metals Company (USA) for an extensive series of leaded and unleaded cast and wrought brasses, bronzes, nickel silvers, and other copper-base alloys. Used for bearings, fittings, hardware, machine parts, valves, pumps, and castings.

Colonial Pine. Trademark of Canadian Forest Products Limited for panelboards.

Colonial High Speed. Trade name of Teledyne Vasco (USA) for a high-speed tool steel containing 0.52% carbon, 0.3%, manganese, 3.8% chromium, 18.24% tungsten, 0.25% vanadium and the balance iron. Used for cutting tools.

Colonnade. Trade name of Libbey-Owens Ford Company (USA) for a patterned glass.

colophony. A light yellow to black amorphous residue that is left after the volatiles have been removed by distillation or heating of crude turpentine obtained from the wood of various species of pine. Density, 1.08 g/cm³ (0.039 lb/in.³); melting point, 70-85°C (158-185°F). Used as a drier for paints, varnishes and lacquers, and in putties, floor covering, synthetic resins, cements and printing inks, for gluing writing paper, in steel hardening, in solders, and in cold bending of pipes.

Colora. Trade name of Saint-Gobain (France) for a hollow glass block with smooth external surfaces and internal surfaces permanently colored by a vitrified enamel applied by spraying.

Colorado. Trade name for a corrosion-resistant nickel silver containing 57% copper, 25% nickel and 18% zinc. Used for corrosion-resistant parts and electrical resistances.

coloradoite. A black mineral of the sphalerite group composed of mercury telluride, HgTe. It can also be made synthetically. Crystal system, cubic. Density, 8.1-8.6 g/cm³; hardness, 3 Mohs. Occurrence: USA (Colorado).

Colorado Rose. Trademark of Robinson Brick Company (USA) for building bricks.

Colorado Water Hardening. Trade name of Sanderson Kayser Limited (UK) for a water-hardening tool steel containing 0.6-1.5% carbon. Its carbon content is specified by temper number. Used for tools, drills and taps.

colorant. A substance, such as a dye, pigment or other agent, that imparts a specific color to another material or a mixture of materials. It can be a dry pigment that is admixed with the substance by mechanical means, or a dispersion of pigments in a liquid. Used in paints, coatings, ceramics, plastics, rubber, fibers, etc. Also known as *coloring agent.*

Coloray. Trade name for colored rayon fibers and yarns used for textile fabrics.

Colorbel. Trade name of SA Glaverbel (Belgium) for a thick, toughened, smooth or figured glass that has a coating of colored enamel fired onto one side at high temperatures.

Colorbestos. Trade name for a flameproof hybrid fabric in a plain weave composed of cotton and asbestos. Used for draperies.

color black. A finely ground carbon black used as a pigment.

Colorblock. Austrian trade name for a cladding unit composed of a sheet of *Colorbrunit* (an enameled toughened glass), a layer of cellular glass or pressed rock wool, and a plasterboard backing.

Colorbond. Trade name of Ivoclar North America/Williams Alloys (USA) for a ceramic gold coating used on precious and nonprecious porcelain-bonding dental alloys to enhance the porcelain-to-metal bond.

Colorbrunit. Austrian trade name for an enameled, toughened glass for wall-cladding applications.

Colorclad. Trade name of Pilkington Brothers Limited (UK) for a ceramic-enameled and toughened glass for cladding applications.

Colorcoat. Trade name of British Steel plc (UK) for a prepainted low-carbon strip steel.

Colorcomp. Trade name of LNP Engineering Plastics (USA) for acrylic thermoplastic resins supplied in ivory and black colors.

Color-Cote. Trade name of U-C Coatings Corporation (USA) for wax paints.

colored fancy paper. Generic term for a colored paper whose surface has been treated by coating, printing, painting, marbling, patterning, bronzing, embossing, or other processes

colored frit. A frit to which a pigment or colorant has been added to produce a colored porcelain enamel, glaze or other ceramic coating.

colored glass. A glass that has been made by adding metal oxides, such as iron oxide or pigments to the melt. It transmits only part of the incident light.

Colorfine. Tradename of Martin Color-Fi, Inc. (USA) for colored polyester fibers used for the manufacture of textile fabrics.

Colorglas. Trade name of Glasbauelemente GmbH (Germany) for a cladding glass with colored enamel fused onto one side.

Colorglaze. Tradename of Guardian Industries Corporation (USA) for a colored, toughened spandrel-type glass for inside and outside installations.

coloring agent. See colorant.

Colorit. Trade name of Glaceries Réunies SA (Belgium) for a toughened glass with ceramic enamel coating.

Colorite. Trademark of Vicwest (Canada) for prepainted protective finishes on galvanized sheet steel profiles.

color lake. See lake.

Colorlux. Trade name of John Healey Limited (UK) for a square glass slab supplied in a various colors. It is usually 203 × 203 mm (8 × 8 in.) in size and has a thickness of 22-25 mm (0.875 to 1.000 in.).

color oxide. A material, usually a metallic oxide, or combination of oxides, used to impart color to ceramic ware, porcelain enamels, glazes, glasses, etc. Examples of such materials include oxides of chromium, cobalt, iron, titanium and zinc. Also known as *ceramic colorant.*

Colorpave. Trademark of Mulco Inc. (Canada) for acrylic pavement coatings.

color pigments. Pigments of blue, red or other shades that absorb a portion of the incident light and reflect to the eye certain groups of light bands, which make the recognition and discernment of various colors possible.

Colorsealed. Trade name for colored rayon fibers and yarns used for textile fabrics.

Colorspan. Trade name of Flachglas AG (Germany) for a laminated glass with opaque interlayer. Used for architectural applications, e.g., spandrels.

Colorspun. Trade name for spun rayon fibers and yarns.

Colortemper. Trade name Asahi Glass Company Limited (Japan) for a toughened cladding glass produced by fusing ceramic pigments onto a rough, or twin-polished plate glass surface.

Colorthane. Trade name of Spatz Paints (USA) for a high-gloss acrylic-urethane coating.

Colortrans. Trade name of Flachglas AG (Germany) for a laminated glass with transparent colored interlayer.

Colortread. Trademark of Tremco Inc. (USA) for a gasoline-, oil- and detergent-resistant concrete floor paint with durable high-gloss finish for use on exterior and interior concrete and wood surfaces. It is available in gray, silver gray, sand and terra cotta colors.

Colortrend. Trademark of Degussa AG (Germany) for a several aqueous, non-aqueous and VOC-free dispersions of pigments (colorants).

colossal magnetoresistance materials. A class of materials with perovskite crystal structure, such as lanthanum strontium manganite ($La_{1-x}Sr_xMnO_3$) and lanthanum calcium manganite ($La_{1-x}Ca_xMnO_3$), that exhibit increases or decreases in electrical resistivity by several orders of magnitude with the application of a relatively weak magnetic field. They are useful for applications in magnetic data processing and storage, e.g., for computer hard drives. Abbreviation: CMR materials.

Colourbond. Trademark of Stelco Inc. (Canada) for galvanized steel sheets with wiped coatings.

Colourcast. Trade name of Glass Tougheners Proprietary Limited (Australia) for a ceramic fired and toughened cast glass for cladding applications.

Colourclad. Trade name of Glass Tougheners Proprietary Limited (Australia) for a ceramic-fired and toughened plate glass for cladding applications.

Colourseal. Trade name of Hollow Seal Glass Company Limited (UK) for a double glazing unit, one sheet of which has an inner surface with a fired-on, usually translucent, ceramic color. Used chiefly for wall cladding applications.

Colphene. Trade name for roof sealing courses and dampproofing membranes.

colquiriite. A white mineral composed of lithium calcium aluminum fluoride, $LiCaAlF_6$. Crystal system, hexagonal. Density, 2.94 g/cm^3; refractive index, 1.388. Occurrence: Bolivia.

ColSpray. Trade name of Wall Colmonoy Corporation (USA) for a series of two-step thermal spray powders based on metallic and ceramic materials (e.g., steel, bronze, nickel, etc.). Used to provide metal parts with improved wear resistance.

Coltex. Trade name for a silicone elastomer used for dental impressions.

Colt Hot. Trade name of A. Finkl & Sons Company (USA) for a hot-work steel containing carbon, chromium, nickel, molybdenum, and iron. Used for dies and tools.

Coltoflax. Trade name of Coltene-Whaledent (USA) for a condensation-type silicone dental impression material with excellent physical properties, optimal elastic recovery, and adjustable setting and working times.

Coltosol. Trade name of Coltene-Whaledent (USA) for an eugenol-free, radiopaque resin-based dental filling material for temporary restorations.

Coltra. Trade name of Coltra Inc. (USA) for decorative glass panels made by leading a solid sheet of glass with ceramic colors and fire fusing for permanence.

coltskin. A split leather made from colt or horsehide.

Coltuf. Trade name of British Steel Corporation (UK) for plain-carbon steel plate containing up to 0.22% carbon, 0.2-1.25% silicon, 1.2-1.6% manganese, up to 0.5% nickel, a total of 0.05% sulfur and phosphorus, and the balance iron. It has high ductility, good resistance to embrittlement, and exhibits low work hardening. Used in ship construction.

Columax. Trade name of Permanent Magnet Association (UK) for a series of heat-treatable permanent-magnet steels containing 24-25% cobalt, 13.5-14% nickel, 8% aluminum, 3.0% copper, 0-2% niobium, and the balance iron. It is a columnar crystal form of *Alcomax*.

Columbia. Trade name of Columbia Tool Steel Company (USA) for an extensive series of cold-work, hot-work, shock-resisting, water-hardening and special-purpose tool steels.

Columbia Carbon. Trademark of Columbia International Chemicals Company (USA) for a series of carbon blacks.

Columbia Electrex. Trade name of Columbia Tool Steel Company (USA) for a water-hardening tool steel containing 0.9-1.6% carbon, 0-0.25% silicon, 0.35% manganese, and the balance iron. Used for drills, hobs, reamers, taps, punches, shafts, arbors, brake dies, etc. Also known as *Electrex*.

Columbia Extra. Trade name of Columbia Tool Steel Company (USA) for a water-hardening tool steel (AISI type W1-2) containing 1.06% carbon, 0.25% manganese, 0.25% silicon, and the balance iron. It has excellent machinability, good to moderate depth of hardening, good wear resistance and toughness, and low hot hardness. Used for threading dies, drawing and heading dies, shear blades and knives, hand stamps, and punches.

Columbia Extra Headerdie. Trade name of Columbia Tool Steel Company (USA) for a water-hardening tool steel (AISI type W1-2H) containing 0.95% carbon, 0.35% manganese, 0.25% silicon, and the balance iron. It has good wear resistance and toughness. Used for stamping and heading dies, hand stamps and punches.

Columbian pine. See Douglas fir.

Columbia Special. Trade name of Columbia Tool Steel Company (USA) for a water-hardening tool steel (AISI type W1-1) containing 1.06% carbon, 0.3% manganese, 0.25% silicon, 0.2% chromium, 0.05% vanadium, and the balance iron. It has excellent machinability, good to moderate depth of hardening, good wear resistance and toughness, and low hot hardness. Used for hand taps, reamers, shear and paper knives, and gages.

Columbia Standard. Trade name of Columbia Tool Steel Company (USA) for a water-hardening tool steel (AISI type W1-3) containing 1.06% carbon, 0.3% manganese, 0.25% silicon, and the balance iron. It has excellent machinability, good wear resistance and toughness, and low hot hardness. Used for hand tools, blacksmith tools, and dowel and drift pins.

columbite. See niobite.

columbium. Entries related to columbium and columbium alloys and compounds can be found under "niobium."

Columbus. Trade name of Schmidt & Clemens Edelstahlwerke (Germany) for a series of tungsten and cobalt-tungsten high-speed steels used for lathe and planer tools, milling cutters, broaches, reamers, drills, taps, hobs, etc.

columnar liquid crystal. A *liquid crystal* composed of disk-like organic molecules stacked up in columns that are organized in two dimensions. Also known as *columnar mesophase liquid crystal*.

Columns. Trade name SA Glaverbel (Belgium) for patterned glass.

colusite. A bronze-colored mineral of the tetrahedrite group composed of copper arsenic tin vanadium sulfide, $Cu_3(As,Sn,V,Fe,Te)S_4$. Crystal system, cubic. Density, 4.44-4.50. Occurrence: USA (Montana).

Colva. Trade name for rayon fibers and yarns used for textile fabrics.

Colvadur. Trade name for rayon fibers and yarns.

Colvalon. Trade name for rayon fibers and yarns used for textile fabrics.

Co Major. Trade name of Jessop Steel Company (USA) for a high-speed steel containing 0.7% carbon, 18% tungsten, 4% chromium, 1% vanadium, 5% cobalt, and the balance iron. Used for lathe and planer tools, cutters, and other tools.

Comalco. Trade name of Comalco Limited (Australia) for a series of cast aluminum-silicon, aluminum-silicon-magnesium and aluminum-silicon-copper alloys. They have good castability, and are used for intricate castings.

Comalloy. British trade name for a precipitation-hardenable alloy of 71% iron, 17% molybdenum and 12% cobalt. It has high magnetic permeability, and is used for permanent magnets, and electrical and magnetic equipment.

ComAlloy. Trade name of ComAlloy International Corporation (USA) for an extensive series of synthetic resins. *ComAlloy 6/12* refers to polyamide 6,12 (nylon 6,12) resins supplied in standard, glass or carbon fiber-reinforced, fire-retardant and silicone- or polytetrafluoroethylene-lubricated grades. *ComAlloy ABS* embraces a series acrylonitrile-butadiene-styrene resins supplied in glass-fiber-reinforced, fire-retardant, high-heat, UV-stabilized, high- and low-impact, low-gloss, transparent, plating and structural-foam grades. *ComAlloy Nylon 66* refers to polyamide 6,6 (nylon 6,6) resins supplied in standard, glass or carbon fiber-reinforced, glass-bead and/or mineral-filled, fire-retardant, high-impact, supertough, UV-stabilized and molybdenum disulfide- or polytetrafluoroethylene-lubricated grades. *ComAlloy PP* are polypropylene resins filled with 20 or 40% calcium carbonate or talc, reinforced with 20% glass fibers, or coupled with 30% glass fibers. *ComAlloy PPS* refers to polyphenylene sulfide resins reinforced with 40% glass fibers, 30% carbon fibers or glass fibers or beads, or lubricated with 20% polytetrafluoroethylene. *ComAlloy PS* are polystyrene resins supplied in standard, glass-fiber-reinforced, medium-impact, UV-stabilized, structural-foam, fire-retardant and silicone-lubricated grades. *ComAlloy PSUL* is a series of polysulfone resins supplied in standard, glass fiber- or carbon fiber- reinforced, or polytetrafluoroethylene-lubricated grades, and *ComAlloy SAN* refers to styrene-acrylonitriles supplied in standard, glass fiber-reinforced, fire-retardant, high-impact, high-heat and UV-stabilized grades.

Comanche. Trade name of Pyramid Steel Company (USA) for an oil-hardening tool steel (AISI type O1) containing 0.9% carbon, 1.2% manganese, 0.5% chromium, 0.2% vanadium, 0.5% tungsten, and the balance iron. Used for dies, knives, and punches.

comancheite. An orange-red to yellow mineral composed of mercury bromide chloride oxide, $Hg_{13}(Cl,Br)_8O_9$. Crystal system, orthorhombic. Density, 7.70-8.00 g/cm^3; refractive index, 1.78-1.79. Occurrence: USA (Texas).

Comat. Trade name for a bonded fiberglass mat/weft knitted fabric combination that provides a quick laminate thickness build-up, and has a low weight (0.64-1.07 kg/m^2 or 2.11-3.5 oz/ft^2) and good strength properties. Used in composites for marine applications.

Combarloy. Trade name of Thomas Bolton Limited (UK) for a commercially pure copper containing 0.1% silver. It has high electrical conductivity, and is used for electrical applications, e.g., commutator bars.

Combat. Trademark of The Carborundum Company (USA) for machinable, engineered ceramics based on boron nitride. Supplied in machinable solids, and powders, coatings and aerosols, they have good electrical resistance, low dielectric con-

stants, high dielectric strengths, very low coefficients of thermal expansion, high thermal conductivity, high lubricity, good non-wetting and free-machining properties, and a maximum service temperature of 2775°C (5027°F). The solid products are used in various shapes and sizes for dielectric components, electrical insulators, thermocouple protection tubes, low-loss insulators, nozzles for high-temperature applications, friction modifiers, heat sinks, fixtures, microwave windows, dies and crucibles for molten metals, refractory linings, etc. The powders are used as additives and fillers in thermally conductive elastomers and plastics, and the coatings for mold-release and antioxidant applications in the metal, glass and plastics industries.

combed cotton fibers. Carded cotton fibers that have been combed, i.e., subjected to a preparation process that removes shorter fibers and further aligns the remaining longer fibers. See also carded yarn.

combed cotton yarns. A yarn with a relatively smooth appearance, made from combed, carded cotton fibers. See also carded fibers.

combed finish tile. A tile whose back has been scored or scratched to improve its bond with mortar or plaster.

combed ware. Ware with a combed (e.g., scored or scratched) surface finish.

combed yarn. Yarn made from carded and combed fibers, i.e., fiber that have been cut to length and directionally aligned.

combeite. A colorless mineral composed of sodium calcium silicate hydroxide, $Na_2(Ca,Al,Fe)_3Si_6O_{16}(OH,F)_2$. Crystal system, rhombohedral. Density, 2.84 g/cm^3; refractive index, 1.598. Occurrence: Zaire.

comber leather. A specially treated leather made from steerhide, and used in the textile industry for combing machines.

comber noils. See noils.

combination coating. A mechanical coating that is obtained by adding individual powders of metals, such as cadmium, tin or zinc to a coating barrel. The revolving action of the barrel generates sufficient dispersion of the powders to obtain a uniform mixture in the coating. Also known as *combination mechanical coating*.

combination fabrics. Textile fabrics made from two or more different single yarns which in turn are composed of only one type of fiber. For example, a fabric made from cotton and polyester yarns.

combination yarn. A yarn made by twisting together two or more yarns of different materials. For example, a yarn made by combining fibers of cotton and rayon, or acetate and rayon.

combined rope. A rope consisting of central strands of steel whose outer portions are made from natural or synthetic fibers.

Combinervo. Trademark of Rippenstreckmetall-Gesellschaft mbH (Germany) for galvanized or black-coated ribbed expanded metal used in building interiors.

Combirip. Trademark of Rippenstreckmetall-Gesellschaft mbH (Germany) for black lacquered or galvanized ribbed expanded metal.

comblainite. A turquoise-blue mineral of the sjogrenite group composed of cobalt nickel carbonate hydroxide hydrate, $(Ni_{6.10}Co_{2.90})(OH)_{18.27}(CO_3)_{1.315} \cdot 6.7H_2O$. Crystal system, rhombohedral (hexagonal). Density, 3.05 g/cm^3. Occurrence: Zaire.

Combomat. Trademark of PPG Industries (USA) for a series of electrical-grade glasses (E-glasses) used in marine composites.

combustible material. Any solid, liquid or gaseous substance that will ignite and burn.

combustible textiles. Textile materials that will ignite and burn when subjected to an ignition source. Most textiles support combustion, unless properly treated.

Comet. (1) Trade name of Carpenter Technology Corporation (USA) for a water-hardening tool steel (AISI type W1) containing 1.05% carbon, 0.25% manganese, and the balance iron. It has good machinability, good to moderate wear resistance and toughness, and low hot hardness. Used for blanking, drawing, forming, heading, threading and trimming dies, taps, and reamers.

(2) Trade name of Driver Harris Company (USA) for a resistance alloy containing 65% iron, 30% nickel, and 5% chromium. It has high strength, high electrical resistivity, high heat resistance, and a maximum service temperature of 700°C (1290°F). Used for resistance wire, rheostats, and dipping baskets.

(3) Trade name of SA des Verreries de Fauquez (Belgium) for a patterned glass.

Comforel. Trade name of E.I. DuPont de Nemours & Company (USA) for polyester fibers used for the manufacture of textile fabrics including pillows, comforters and pads.

Comforlast. Trade name of E.I. DuPont de Nemours & Company (USA) for strong, durable nylon 6,6 fibers used for the manufacture of textile fabrics.

Comfor-Lite. Trade name of Amerada Glass Company (USA) for a double glazing unit with a louvred screen hermetically sealed between the two glass panes.

ComforMax. Trademark of E.I. DuPont de Nemours & Company (USA) for a series of barrier fabrics.

Comfort Fiber. Trade name for polyester fibers and yarns used for textile fabrics, especially clothing.

Comfortfil. Trademark of KoSa (USA) for durable high-loft continuous filament polyester fiberfill with excellent compression recovery and insulation performance, used for furniture and bedding.

Comfort Foam. Trademark of Foam Enterprises, Inc. (USA) for polyurethane foam for spray or froth application into window sashes and main frames to prevent heat flow.

Comfort-Pane. Trade name of Comfort Products Company (USA) for an insulating glass.

ComFortrel. Trademark of Wellman Inc. (USA) for a soft polyester fiber with a natural hand, excellent wrinkle-, pill-, shrink- and stretch resistance, and outstanding shape retention. Used for comfortable, breathable, wickable clothing.

ComFortrel Plus. Trademark of Wellman Inc. (USA) for a breathable, wickable, colorfast polyester fiber with excellent wrinkle and shrink resistanc, and outstanding shape retention. It has a softer feel, a different luster, a higher dye absorption, better moisture management properties, better antistatic properties, and higher strength than the original *ComFortrel* fiber. It blends well with other fibers including spandex, and is used for action-wear, sportswear, swimwear, workwear, lingerie, and other apparel.

Comfort Ti-R. Trademark of AFG Industries (US) for a low-emissivity (low-E) window glass coated with a thin, invisible titanium layer that blocks heat flow, reduces sun damage and fading of furnishings, carpets, paint and artwork, and significantly reduces condensation and drafting.

Cominco. Trade name of Cominco Limited (Canada) for high-purity and commercially pure bismuth, cadmium, gold, indium, lead, silver and zinc, several wrought zinc alloys, and numerous lead-antimony and lead-arsenic alloys.

Comiso. Trade name for rayon fibers and yarns used for textile fabrics.

Commando. (1) Trade name of Wallace Murray Corporation (USA) for a shock-resistant hot-work tool steel (AISI type S1) containing 0.45% carbon, 1.4% chromium, 0.4% silicon, 2.1% tungsten, up to 0.2% vanadium, and the balance iron. It has high hardness and strength, excellent toughness, good wear and abrasion resistance, and good hot hardness. Used for chisels, headers, piercers, shear blades, hot punches, and tools.

(2) Trade name of Atlas Specialty Steels (Canada) for a die steel containing 1.2% carbon, 0.25% manganese, 0.2% silicon, and the balance iron. Used for drill rod, dies, reamers, punches, twist drills, and dowels.

Command Ultrafine. Trade name for a light-cure dental hybrid composite.

Commell Special. British trade name for a high-speed steel containing 1.2% carbon, 2.9% tungsten, 0.2% manganese, and the balance iron. Used for tools, dies, and cutters.

Commend. Trademark of World Alloys & Refining, Inc. (USA) for a series of corrosion-resistant copper-free nickel-chromium and nickel-chromium-beryllium dental ceramic alloys. *Commend* and *Commend NB* are nickel-chromium-beryllium grades with a density of 8.4 g/cm³ (0.30 lb/in.³), high yield strength and low elongation. The former has a melting range of 1230-1290°C (2250-2350°F) and a hardness of 220 Vickers and the latter a melting range of 1290-1345°C (2250-2350°F) and a hardness of 190 Vickers. *Commend V* is a nickel-chromium grade with a density of 7.8 g/cm³ (0.28 lb/in.³), a melting range of 1230-1290°C (2250-2350°F), a hardness of 235 Vickers, high yield strength and low elongation. *Commend* dental ceramic alloys are used for metal-to-ceramic restorations.

commercial ammonium molybdate. See ammonium heptamolybdate.

commercial brass. A wrought or cast *brass* containing 60-65% copper, 30-35% zinc and 0-3.75% lead. Used for fixtures, radiators, and ornaments.

commercial bronze. A corrosion-resistant *bronze* containing 88.5-90% copper and 10-12.5% zinc. It is available in the form of tubes, rods, flat products and wires, and has a density of 8.4-8.5 g/cm³ (0.31 lb/in.³), high electrical conductivity (about half that of pure copper), excellent cold workability, good hot workability, and excellent to good corrosion resistance. Used for weatherstripping, grillwork, screen cloth, marine hardware, screws, rivets, window screen wire, automobile radiators, ornaments, slide fasteners, costume jewelry, lipstick cases, bases for gold plate, for etching bronze, and for forgings.

commercial fine gold. Cast or wrought *gold* with a standard purity of 99.95+%, a density of 19.32 g/cm³ (0.698 lb/in.³), and a melting point of 1064°C (1947°F). Used for coinage, jewelry, electronics, etc.

commercially pure aluminum. A non-heat-treatable *aluminum* with a purity of 99.00+%. Typical impurities include silicon, iron, copper, manganese, magnesium, zirconium and titanium. The term usually refers to aluminum in the 1000 series of the AA (Aluminum Assiciation) designation system. It is commercially available in the form of foils, impacts, sheets, plates, extruded rods, bars and wire, extruded shapes, extruded tubes, cold-finished rods, bars and wires and drawn tubes, and has a density of 2.70-2.71 g/cm³ (0.097-0.098 lb/in.³), a melting range of 643-657°C (1190-1215°F), excellent resistance to general and stress-corrosion cracking, excellent brazeability, solderability and weldability, good cold workability, good machinability, high thermal and electrical conductivity, and moderate strength. Used for railroad tank cars, chemical equipment, electrical conductors, electrolytic capacitor foil, packaging foil, fin stock, sheetmetal work, and spun hollowware.

commercially pure copper. Copper having a purity of 99.88-99.99%, and containing small amounts of impurities, such as phosphorus, silver, oxygen and cadmium. The term usually refers to the coppers with numbers C10100 through C13000 in the Unified Numbering System (UNS) and includes the following types: oxygen-free electronic; oxygen-free; oxygen-free, extra-low phosphorus; oxygen-free, silver-bearing; oxygen-free, low-phosphorus; electrolytic tough-pitch; silver-bearing tough-pitch; high residual phosphorus-deoxidized; and fire-refined tough-pitch. Available in the form of flat products, rods, wires, tubes, pipes and shapes, it has good strength, good to excellent corrosion resistance, high electrical conductivity, excellent cold and hot workability, and good forgeability. Used for electrical conductors, busbars, waveguides, switches, terminals, lead-in wires, transistor components, coaxial tubes, rectifiers, electrical power transmission equipment, radio parts, gaskets, radiators, chemical-processing equipment, condenser and heat exchanger tubing, gasoline and oil lines, and hydraulic lines. See also copper.

commercially pure nickel. Nickel with a purity of 99.5+%, typically containing impurities, such as manganese, copper, iron, sulfur, silicon and carbon. Available in the form of wires, rods and strips, it has outstanding corrosion resistance, good resistance to caustics and halogens, good high-temperature properties, high thermal and electrical conductivities, and good magnetic and magnetostrictive properties. Used for anodes and passive cathodes, fluorescent-lamp parts, vacuum-tube plates, lead wires, base pins, cathode shields, electronic and electrical parts, transducers, spun and cold-formed parts, and food-processing equipment. See also nickel

commercially pure palladium. Palladium with a purity of at least 99.99%, typically supplied in foil, wire or powder form. See also palladium.

commercially pure platinum. Platinum with a purity of at least 99.0-99.9%, typically supplied in foil, gauze, powder, shot, sponge, wire, rod or tube form. See also platinum.

commercially pure tin. Tin with a purity of 99.00-99.98%. ASTM B339 lists seven grades: (i) *Grade AAA* is extra-high purity electrolytic tin (99.98% pure); (ii) *Grade AA* is high-purity electrolytic tin (99.95% pure); (iii) *Grade A* is high-purity commercial tin (99.80% pure); (iv) *Grade B* is general-purpose tin (99.80%); (v) *Grade C* is intermediate-grade tin (99.65% pure); (vi) *Grade D* is lower-intermediate-grade tin (99.50% pure); and (vii) *Common tin* (99.00% pure). Typically supplied in foil, wire, rod, bar, powder, granules, shot and wire form, it is used for lining equipment and vessels in water distillation plants, in the manufacture of plate glass, general-purpose alloys, tinfoil and tinplate. Also known as *unalloyed tin*. See also tin.

commercially pure titanium. Titanium with a purity of 98.9-99.5%. Typical impurities include palladium, oxygen, iron, carbon, nitrogen and hydrogen. Available in the form of sheets, bars, billets, tubes, rods, wires and sponge, it has excellent corrosion resistance, good strength (increases with increasing iron and oxygen contents), good to medium weldability, and good ductility and formability. Also known as *unalloyed titanium*. See also titanium.

commercial prepreg. A *prepreg* that has been made to meet specific customer requirements.

commercial zinc. A crude *zinc* (99.6% or less pure) obtained in the smelting zinc ores, and used for galvanizing applications. See also spelter.

commingled plastics. Mixed waste plastics obtained from discarded household products, such as high-density polyethylene milk jugs and polyethylene terephthalate soda bottles. They can be reprocessed and either made into new plastic products, or used in fibrous or particulate form in the manufacture of plastic or composite materials. See also recycled plastics.

comminuted powder. A powder made by mechanical reduction (e.g., crushing or grinding) of a solid.

commodity polymers. A group of general-purpose polymers, such as acrylonitrile-butadiene-styrene, polyethylene, polystyrene, polyvinyl chloride, etc., that, unlike engineering polymers, usually have relatively poor thermal properties.

common alder. See black alder.

common boards. In lumber grading, boards 25.4 mm (1 in.) thick and up to 305 mm (12 in.) wide. Also known as *commons*.

common brass. See high brass.

common brick. A hard, strong, reddish-brown brick with low susceptibility to warping and cracking, made from common brick clay. Used for rough work, filling, or backing purposes. Also known as *backing brick*.

common brick clay. A plastic clay that has a high content of fluxing ingredients, and fires to a reddish brown color at relatively low temperatures.

common cottonwood. See eastern cottonwood.

common feldspar. See orthoclase.

common formula. A British nickel silver containing 55% copper, 25% zinc and 20% nickel. Used for electrical resistances.

common iron. Iron manufactured from re-rolled scrap iron, or a mixture of steel scrap and iron.

common juniper. See juniper.

common lead. (1) A constant lead mixture of 1.4% ^{204}Pb, 24.1% ^{206}Pb, 22.1% ^{207}Pb and 52.4% ^{208}Pb whose overall atomic weight is 207.24. See also lead.

   (2) A grade of *pig lead* containing 99.94+% lead, up to 0.050% bismuth, up to 0.005% silver, up to 0.002% iron, up to 0.0015% copper, up to 0.001% zinc, and a total of up to 0.002% arsenic, antimony and tin. Used for plumbing and general applications. See also common lead A; common lead B; lead.

common lead A. A grade of lead (99.85% pure) from which silver has been removed by either the Parkes or the Pattinson process. See also common lead (2); common lead B.

common lead B. A grade of lead with a purity of 99.73%. See also common lead (2); common lead A.

common lime. See calcium oxide.

common lumber. (1) A term for nonstress graded softwood lumber that is an appearance grade lower in quality than select grades. There are five grades of common lumber: (i) *Grade 1* refers to a sound material that will contain tight knots and a limited number of blemishes; (ii) *Grade 2* that should not have splits or distortions like warp, but can have defects, such as loose knots, end checks and some blemishes and discoloration; (iii) *Grade 3* that can contain various types of defects, some of which should be removed before usage, and is usually used as a medium-quality construction lumber; (iv) *Grade 4* that is a low-quality construction lumber which may contain many defects, even open knot holes; and (v) *Grade 5* that is the lowest quality material and can be used as a filler only.

   (2) A term referring to hardwood lumber of any of the following grades: (i) *Grade 1* that is often used for interior and

less demanding cabinetwork, and must have one side at least 65% clear; and (ii) *Grade 2* that is frequently used for paneling and flooring, and has one side about 50% clear.

common mica. See muscovite.

common pear. See pear.

commons. See common boards.

common salt. See halite.

common solder. See wiping solder.

common tin. A commercially pure *tin* (ASTM B339 Grade E) with a purity of 99.00% used in lead-base alloys, bearing metals and cast bronzes.

common tombac. A ductile, corrosion-resistant wrought alloy of 72% copper and 28% zinc used for brazing applications, condenser and evaporator tubing, hardware, and cartridge cases. See also tombac.

common typemetal. A cast alloy of 55-60% lead, 10-40% tin and 4.5-30% antimony. It has a low melting point and good castability, and is used for typewriters. See also typemetal.

Commonwealth. Trade name of General Steel Industries (USA) for a steel containing 0.3% carbon, 2% nickel, and the balance iron. Used for truck frames.

Como. (1) Trade name of CCS Braeburn Alloy Steel (USA) for a series of molybdenum-type high-speed steels (AISI type M30) containing 0.7-0.8% carbon, 8.5-9% molybdenum, 1.5% tungsten, 4% chromium, 1-1.2% vanadium, up to 5% cobalt, and the balance iron. They possess high hot hardness, excellent wear resistance, and are used for lathe and planer tools, drills, hobs, milling cutters, taps, and reamers.

(2) Japanese trade name for a permanent-magnet steel containing 17% chromium, 12% cobalt, and the balance iron.

Comokut. Trade name of Bethlehem Steel Corporation (USA) for a tungsten-type high-speed steel (AISI type T4) containing 0.75% carbon, 4% chromium, 18% tungsten, 1% vanadium, 5% cobalt, and the balance iron. It has excellent hot hardness, good wear resistance and high cutting ability. Used for general-purpose cutting tools.

Comol. Trade name of Colt Industries (UK) for a series of permanent-magnet materials containing 12% cobalt, 17-20% molybdenum, and the balance iron. They have high permeability, high residual flux density, and are used for electrical and magnetic equipment.

comonomer. Any of two or more monomers that simultaneously polymerize to form a copolymer, e.g., styrene and butadiene are the two comonomers that form a styrene-butadiene copolymer.

COMP. Trade name for a crystalline, hard magnetic neodymium-iron-boron nanocomposite developed by the Idaho National Engineering and Environmental Laboratory (INEEL). It is made from an amorphous spherical powder precursor by heat treating at 800°C (1470°F), and has a microstructure consisting of nanosized $Nd_2Fe_{14}B$ grains (30-80 nm or 1.2-3.2 μin.) and titanium, zirconium and carbon (TiZrC) precipitates with a grain size of 5-10 nm (0.2-0.4 μin.). *COMP nanocomposite* has a Curie temperature of 435°C (815°F) and can be processed by injection molding into near-net-shaped magnets.

CompaAdhesiv. Trade name for a dual-cure dental resin cement.

compact. In powder metallurgy, the compressed shape of metal powder prior to sintering. Also known as *briquet; briquette; powder compact*.

compacted graphite cast irons. A group of cast irons made by treating molten iron with carefully controlled amounts of cerium or magnesium, usually in the form of ferroalloys, using titanium as an inhibitor to form fully spheroidal graphite. The chemical composition is similar to that of ductile iron, and the microstructure, consisting essentially of short, coarse, rounded interconnected graphite flakes, is intermediate between that of ductile iron (with spheroidal graphite) and gray iron (with flaky graphite). It combines the high strength and good impact resistance of ductile irons with the good thermal conductivity, damping capacity and machinability of gray iron. Also known as *CG irons; compacted graphite irons; vermicular irons*. See also CG alloy.

compacted powder. A metal powder that has been compressed to form a compact or briquet.

CompaFill MH. Trade name for a light-cure microfine dental filling cement.

Compalox. Trademark of Martinswerk GmbH (Germany) for activated alumina (Al_2O_3) used in the cleaning of organic liquids, and in wastewater treatment.

CompaMolar. Trade name for a light-cure dental hybrid composite.

Compar. Trade name of European Vinyls Corporation for a series of vinyls (polyvinyl acetate, polyvinyl alcohol, polyvinyl chloride, etc.) supplied in flexible and rigid grades, and in various colors. They have good abrasion, chemical, electrical, flame, oil and weather resistance, good processibility, high stiffness at low temperatures, and a maximum service temperature of 138°C (280°F). Used for floor and wall coverings, upholstery, rainwear, tubing, insulation, toys, phonographic products, and safety glass layers.

compatibilizer. (1) A chemical agent that is added or otherwise applied to normally incompatible materials to enhance their miscibility.

(2) A chemical agent used to modify the surface of a material, such as a clay, so that this material becomes attracted to and will disperse in a synthetic resin.

(3) In polymer engineering, an additive that promotes the miscibility of two immiscible polymers.

Compax. Trade name of Uddeholm Corporation (USA) for a chromium-molybdenum type shock-resisting tool steel containing 0.5% carbon, 3.2% chromium, 1.4% molybdenum, 0.7% manganese, 0.3% silicon, and the balance iron. Used for shear blades, cropping tools, and compression, injection and transfer molds for plastic molding applications.

compensated semiconductor. An *extrinsic semiconductor* containing both acceptor (p-type) and donor (n-type) impurities with one partially cancelling the electrical effects on the other.

compensator alloys. A group of temperature-sensitive soft-magnetic alloys of 70% iron and 30% nickel whose flux density depends strongly on the temperature, and which therefore experience a proportional loss in magnetic permeability with increasing temperature in the range between approximately -30 and +55°C (-22 and +130°F). They have low initial permeability, low coercivity, high saturation induction and a Curie temperature range of 30-120°C (85-250°F). Often sold under trademarks and trade names, such as Mutemp, Thermoflux, Thermoperm or Varioperm, they are used for compensating shunts for electrical equipment, temperature compensation of permanent magnets, and temperature-dependent switches. Also known as *temperature-compensator alloys*.

Compimide. Trademark of Rhône-Poulenc SA (France) for a series of heat-curable amorphous maleimide resins that are Michael-addition-reaction polymerization products of 4,4'-bis-maleimidodiphenylmethane (4,4'-MDA-BMI) or eutectic mixtures of bismaleimides. They have a melting range of 100-140°C (210-220°F), and are used for advanced organic-matrix composite structures.

Complac. Trade name of Poinsetta Inc. (USA) for shellac plastics.

complex compound. See coordination compound.

complex English metal. A British bearing alloy containing 87% tin, 6% antimony, 2% nickel, 2% copper, 1.5% tungsten, 1% zinc and 0.5% bismuth.

complex-phase steels. A group of precipitation-hardened high-strength steels with good formability and weldability, and microstructures composed of ferrite, bainite and martensite. Abbreviation: CP steels.

complex steel. An alloy steel composed of iron, carbon and more than two alloying elements. Examples include high-speed steels, hot-work steels and stainless steels.

Com-Ply. Trademark of APA–The Engineered Wood Association (USA) for engineered composite panels essentially consisting of a core of compressed wood-fiber strands bonded to conventional face and back wood veneers. It is available in 3-ply panels with wood-fiber core and a facing and backing veneer, and 5-ply panels with a central wood veneer crossband and a facing and backing veneer. Used for structural building applications and for industrial applications, such as bins, hoppers and pallets.

Compo. Trade name of GKN Powder Met Inc. (USA) for a series of powder-metallurgy materials used to produce porous metal bearings that usually possess good self-lubricating properties and high load-carrying capacities. Examples include various bronzes with 87-90% copper, 9-10% tin and 0-4% graphite, and graphited iron-copper alloys.

compo board. See composition board.

Compobond. (1) Trade name of The Ceilcote Company (USA) for an acidproof cement.

    (2) Trade name of Inter-Africa Dental (Zaire) for a chemically curing dental adhesive for bonding self-cure composites to the tooth structure.

compocastings. See composite castings.

Compoglass. Trademark of Ivoclar Vivadent AG (Liechtenstein) for a *compomer* consisting of a matrix of amorphous silica (SiO_2) that is 80% filled with barium glass particles. It has very high wear resistance, high tensile strength, and is used in restorative dentistry. *Compoglass Flow* is a flowable compomer and *Compoglass SCA* a dental bonding agent for bonding tooth structure.

Compolite. Trademark of Carlisle Dental (USA) for a self-cure dental hybrid composite. *Compolite II* is a light-cure dental hybrid composite.

Compolute. Trade name for a dual-cure dental resin cement for luting restorations.

compomer. A composite that has a polymeric matrix reinforced with inorganic spherical particles (e.g., barium or strontium) and is frequently used in restorative dentistry.

Composan LCM. Trade name of Inter-Africa Dental (Zaire) for a light-cure, translucent filled microhybrid dental composite. It has high radiopacity, high compressive and flexural strengths, low volume shrinkage, and good polishability to a high luster. Used for anterior and posterior restorations, and for inlays.

Composeal. Trade name of Inter-Africa Dental (Zaire) for a light-cure, fluoride-releasing ionomer dental sealing cement with very high compressive strength and acid resistance. *Composeal 1* is a light-cure, fluoride-releasing one-part dental adhesive

for dentin/enamel bonding.

Compo-Site. Trade name of Compo-Site, Inc. (USA) for shellac plastics.

Composite. Trade name of H. Boker & Company (USA) for a water-hardening tool and die steel.

composite board. A board or panel whose thin face and back veneers are bonded to a reconstituted, or solid wood core. Also known as *composite panel*.

composite castings. Composite materials formed by first injecting a thixotropic mixture composed of a metal alloy and suitable fillers into dies at high pressures, and then allowing it to solidify. Also known as *compocastings*.

composite coating. A coating on a metallic or nonmetallic substrate consisting of two or more phases of different compositions. For example, a composite nickel coating may consist of two layers of nickel, one bright and the other semibright, and a nickel-ceramic coating may consist of a nickel or nickel-phosphorus matrix with dispersed particles of silicon carbide, hexagonal boron nitride or silicon nitride. The different phases can also be arranged in alternating layer structures. Also known as *duplex coating*.

composite compact. A powder-metallurgy compact composed of two or more different materials (e.g., metals or metal alloys) with each material retaining its own identity.

Composite Die. Trade name of Jessop-Saville Limited (UK) for a water hardening tool and die steel.

composite dielectrics. Dielectric materials consisting of combinations of two or more distinct materials, often arranged in layered structures. Examples include laminates of paper and synthetic resins, or polystyrene and barium titanate. *Composite dielectrics* have low losses and high permittivity, and are used for electronic and communication components.

composite fiberglass. Glass fibers impregnated with a synthetic resin and combining the impact resistance of resins with the inertness and strength of glass. Used as reinforcement.

composite fibers. Fibers made from two or more chemically different polymers, or from two chemically similar, but not identical polymers. For example, fibers composed of a nylon core and an outer covering of polyethylene, or a polyester homopolymer and a polyester copolymer.

composite materials. See composites.

composite panel. (1) A panel consisting of a core of cellular plastic or reconstituted or solid wood to which plywood veneers or metal sheets have been adhesively bonded.

    (2) See composite board.

composite plate. An electroplate made up of two or more layers of different metals or alloys that have been deposited separately.

composite plywood. Plywood whose middle ply is a core made of oriented wood fibers bonded together by gluing. See also plywood.

composite powder. A powder whose particles are made up of two or more different materials that are not alloyed.

composite refractory coating. A coating on a metallic or nonmetallic substrate consisting of two or more layers of different heat-resistant ceramics whose production may or may not require a heat treatment. Also known as *refractory composite coating*.

composites. A group of structural or functional materials made by bonding together two or more distinct materials without changing their individual properties, but which exhibit property sets that cannot be achieved for any of the individual phases alone. They usually do not dissolve into one another, can be readily identified and are joined by processes, such as soldering, brazing, casting, cladding, sintering, spraying, vapor deposition, laminating, or electrodeposition. Examples include bimetallic strips made by soldering one metal foil onto another, single-point tools made by brazing a piece of expensive cutting material onto a shank of soft steel, bearing alloys cast on steel, brass or bronze backs, fiber- or particulate-reinforced ceramic, metallic or plastic materials, reinforced concrete consisting of steel rods in a concrete matrix, clad metals (e.g., aluminum-copper alloys clad with pure aluminum), powder-metallurgy materials, steels with sprayed-on highly corrosion-resistant coatings, cemented carbides with vapor-deposited coatings, metal foils bonded to paper, plastic materials bonded to wood or woven fabrics, and sandwich and honeycomb structures. Also known as *composite materials*.

composite steel. Steel, usually in bar or rod form, composed two or more different steel types, e.g., tool steel joined to low-carbon steel, or stainless steel joined to tool steel.

composite superconductor. A conductor comprising a superconductor material, e.g., a multifilamentary wire composed of superconducting niobium-titanium alloy filaments in a copper matrix, a core of tin encircled by superconducting niobium filaments in a copper matrix, or an individual filament of superconducting niobium-titanium alloy clad with copper. *Composite superconductors* are also available in the form of flat ribbons or strips See also filamentary superconducting alloy; multifilamentary superconducting alloy.

composite tape. See tape (2).

composite tool steel. A machined bar steel having an insert made by welding tool steel to a backing of low-carbon steel. Used for shear blades and die parts.

Composite Welder FS. Trade name of Devcon (USA) for structural adhesives for bonding plastics, concrete, brick and honeycomb materials.

composite yarn. A yarn, such as *core-spun yarn*, that contains both staple and continuous constituents.

composition board. A building product made by rolling and pressing a mechanical or chemical pulp containing fibers of wood, bagasse or similar cellulosic materials into a rigid sheet or panel of varying size. Also known as *compo board*.

composition brass. A cast leaded brass containing 84-85% copper, 4-6% zinc, 4-6% tin and 4-6% lead. It has good castability and machinability, and takes a high polish. Used for bearings, hardware, screws, nuts, hydraulic castings, such as pump parts, carburetors, valves, and pipe fittings. Also known as *composition metal*. See also ounce metal.

composition metal. See composition brass.

composition wood. Building materials made by pressing wood fibers into thin sheets.

compound. (1) A substance formed by the combination of the atoms or ions of two or more chemical elements in definite ratios. It has characteristic properties that are different from those of its constituent elements. Examples of chemical compounds include aluminum oxide, gallium arsenide, silicon dioxide and sodium chloride.

(2) An intimate mixture of a polymeric material with one or more additives, such as catalysts, colorants, curing agents, fillers, flame retardants, lubricants, modifiers, plasticizers, reinforcements or stabilizers.

compound compact. A powder-metallurgy compact composed of a mixture of two or more different metal powders joined by pressing and/or sintering, with each powder particle retaining its own composition.

compound fabrics. A generic name for fabrics comprising two or more individual layers (or plies) of textile material which have been produced at the same time and are woven together in one process.

compounding material. A material other than the polymer or rubber itself used in forming a polymer or rubber mix. Also known as *compounding ingredient*.

compound semiconductors. See semiconductor compounds.

compound sugar. A sugar composed of two or more monosaccharide units linked together by the elimination of water. Disaccharides, trisaccharides and polysaccharides are compound sugar.

Compzitex. Trademark for plastic products containing an inorganic-fiber reinforcement consisting of nonwoven materials composed of vitreous fibers. Laminates incorporating *Compzitex* exhibit excellent tensile, compressive, shear and flexural strength. The reinforcement is compatible with several reinforced-plastics manufacturing processes including pultrusion and contact, compression, vacuum-bag, pressure-bag and resin-transfer molding. The products can also be used as structural component substrates, and as structural-core materials.

compreg. See compregnated wood.

compregnated wood. A plywood material made by first impregnating two or more regular hardwood veneers with phenolic resin (e.g., phenol-formaldehyde) and then bonding, curing, and compressing them. It has increased density, strength and hardness, good dimensional stability, an attractive glossy finish, and low swelling and shrinking properties. Used for furniture and structural applications. Also known as *compreg; compressed wood; densified wood; impregnated wood (2)*.

compreignacite. A yellow mineral of the becquerelite group composed of potassium uranium oxide octahydrate, $K_2(UO_2)_6O_4(OH)_6 \cdot 8H_2O$. It can also be synthesized by room-temperature reaction of a solution of potassium hydroxide and uranyl acetate. Crystal system, orthorhombic. Density, 5.03 g/cm^3; refractive index, 1.798. Occurrence: France.

compressed wood. See compregnated wood.

compression-molded plastics. Thermoplastic or thermosetting plastics formed by first placing the proper amounts of a usually preheated, thoroughly mixed plastic resin or molding compound in a heated mold cavity, and then closing the mold and applying heat and pressure to force the plastic material to flow and conform to the mold shape.

Compresto. Trade name of Olin Chemicals (USA) for a conductor made up of a round core wire with several layers of pure aluminum wire concentrically wound around it.

Comshield. (1) Trade name of ComAlloy International Corporation (USA) for impact-resistant, high-strength polypropylene used for electrical housings.

(2) Trademark of ComAlloy International Corporation (USA) for electrically conductive acrylonitrile-butadiene-styrene resins filled with carbon black, or stainless-steel fibers. Used for EMI/RFI antistatic applications. See also EMI/RFI materials.

Comsol. Trade name of Johnson Matthey plc (UK) for a cast lead-base alloy containing silver and tin. It retains its strength at elevated temperatures, has a melting point of 296°C (565°F), and an electrical conductivity of 8% IACS. Used as a soft solder.

Comspan. Trade name of Dentsply Caulk (USA) for a self-cure dental resin cement.

Comtuf. Trade name of ComAlloy International Corporation (USA) for a series of impact-resistant thermoplastics including mineral- or glass-reinforced nylon 6, nylon 6,6 or nylon 6,12 resins as well as polypropylenes, polybutylene terephthalates, polyethylene terephthalates and polyamides. Uses range from sporting goods and appliances to automotive and electrical components.

Comymex. Trade name for rayon fibers and yarns used for textile fabrics.

Con. Trade name of McKechnie Brothers Limited (UK) for a wrought copper-zinc alloy (alpha brass) containing 34% zinc, 2.75% manganese, 1.75% nickel, 1.5% aluminum, 1% iron, 0.5% tin and 0.5% lead. It is available in the as-drawn condition, and used for extrusions.

Conap. Trade name of Cytec Industries Inc. (USA) for a series of modified polyurethane elastomers with good impact resistance used for prototype parts, metal stamping pads, foundry patterns, etc. Also included under this trade name are epoxies and acrylic conformal coatings.

Conapoxy. Trade name of Conap Corporation (USA) for a series of epoxy casting resins for adhesives, coatings, sealants and encapsulation and potting compounds.

Conastic. Trade name of Ctyec Industries Inc. (USA) for several polyurethane plastics.

Conathane. (1) Trade name of Conap Corporation (USA) for a family of conformal coatings used in the manufacture of printed-circuit boards, and including single-component, water-base polyurethanes, two-component polyurethanes and single-component acrylic materials. Also included under this trade name are several polyurethane casting systems.

(2) Trade name of Cytec Industries Inc. (USA) for several impact- and wear-resistant polyurethane, polybutadiene urethane and polyether elastomers. Used for industrial belts, pulleys, wheels, washers, gaskets, diaphragms, flexible couplings, etc.

Concept. Trade name for dental composites for inlays.

Concise. Trademark of 3M ESPE Dental (USA) for a self-cure dental hybrid composite containing a quartz filler. Used for anterior restorations.

Concise Crown Build-up. Trademark of 3M Dental (USA) for a white, light- or self-cure composite paste used in restorative dentistry for crown build-ups, and as a pit and fissure sealant.

Concorde. French trade name for film-faced glass wool ceiling boards.

concrete. A homogeneous mixture of a mortar or a hydraulic cement (e.g., Portland cement), fine aggregate (e.g., sand), coarse aggregate (e.g., gravel, crushed stone or blast-furnace slag) and water that hardens to a stone-like mass as it dries. *Concrete* can be divided into various classes according to density: (i) *Lightweight concrete* with less than 1.7 g/cm³ (0.06 lb/ft³), (ii) *Regular or normal concrete* with 1.7-2.7 g/cm³ (0.06-0.10 lb/ft³) and (iii) *Heavy concrete* with 2.7 g/cm³ (0.10 lb/ft³) or more. It can also be classed according to the method of delivery into: (i) centrally-mixed concrete, (ii) ready-mixed concrete, (iii) shrink-mixed concrete and (iv) transit-mixed concrete. *Concrete* has high compressive strength (about 14-70 MPa or 2-10 ksi depending on its manufacture) and relatively low tensile strength (about 1.4-1.8 MPa or 0.2-0.4 ksi). Used for a wide range of applications in bridge, building and road construction, in radiation shielding and in the manufacture of reinforced concrete.

concrete aggregate. Aggregate used in making regular and heavy, but not lightweight concrete, and including materials, such as sand, gravel, crushed stone and slag. See also cement aggregate.

concrete ashlar. See cast stone.

concrete block. A building unit of concrete, either solid or a with hollow core. The standard hollow-core block is made from Portland cement using aggregates, such as sand, fine gravel or crushed stone, and weighs about 18-23 kg (40-50 lbs.). Lightweight units weighing only 11-16 kg (25-35 lb.) are also available. Blocks are usually supplied in 101, 152, 203, 254 and 305 mm (4, 6, 8, 10 and 12 in.) widths and 101 and 203 mm (4 and 8 in.) heights. Used for foundation walls and other masonry construction.

concrete bonding plaster. A gypsum plaster with good adhesion properties used on smooth concrete surfaces usually as a prime or intermediate coat.

concrete brick. A conventional building brick made from a mixture of cement, aggregate and water, and supplied in various shapes and sizes. It has high compressive strength and good resistance to weathering. Used for building construction, and other structural applications.

concrete facing slab. A plate of concrete, often having a smooth, fine, medium or coarse surface texture, designed for use on the exterior or facing of a structure or wall.

concrete hardcore. Aggregate consisting of crushed concrete.

concrete hardener. A substance, such as calcium chloride, sodium chloride or sodium hydroxide, added to a concrete mix to hasten the set. Also known as *concrete-hardening agent.*

concrete masonry unit. A building unit, usually precast, made of concrete.

concrete pipe. A porous ceramic product in the form of a conduit or pipe made of reinforced or prestressed concrete, and often used for water mains, gravity lines and drainage purposes.

concrete products. Precast concrete produced at a central plant and including bricks, blocks, pipes, sills, tiles, etc.

concrete reinforcing bar. See reinforcing bar.

concrete retarder. An additive introduced in small quantities into a concrete mix to prolong the setting time and lower the rate of strength development. After setting, the retarder must have no adverse effects on the concrete.

concrete roofing tile. A roofing product of concrete supplied in the form of an overlapping or interlocking structural unit. It is available in various shapes and textures.

concrete slab. A flat horizontal concrete plate of uniform thickness and varying shape used for floors, roofing sections, bridge decks, well and pit covers, stepping stones, sidewalks, etc.

(2) A pavement made of concrete.

concrete steel. A carbon or alloy steel suitable for the manufacture of rods, wires, bars or mesh and used in reinforced and prestressed concrete.

concrete subgrade paper. In road construction, a strong paper used in long strips to reduce the friction between the subgrade of crushed stone and gravel and the base course of cement concrete. Also known as *concreting paper; underlay paper.*

concreting paper. See concrete subgrade paper.

Condal. Trademark of Continental Gummi-Werke (Germany) for vinyl rubber foam products.

condensation-cured silicone. See C-silicone.

condensation polyimide. A thermoplastic polyimide derived from polyamic acids by a chemical or thermal treatment in which the these acids are produced by a series of step-growth (condensation) reactions at room temperature from a dianhydride or dianhydride derivative and a diamine. It has excellent processibility, good high-temperature properties, and good physical and mechanical properties. Used as laminating resin for resin-matrix applications, aerospace prepregs, radomes on aircraft, and sound suppression panels.

condensation polymer. A high-molecular-weight polymer formed by an intermolecular reaction involving two or more monomer species, usually with the production of a byproduct of low molecular weight, such as water. Also known as *step-reaction polymer; step-growth polymer.*

condensation resin. A synthetic resin formed by individual chemical reactions between two or more reactive monomers with the separation of a simple substance, such as water or alcohol. Examples include alkyd, melamine-formaldehyde, phenol-formaldehyde and urea-formaldehyde resins.

Condenser Foil. Trade name for a tin alloy containing 15% lead and 2% antimony, used as a dielectric material in electronics.

condenser-spun yarn. A yarn spun from relatively thick fiber strands known as "slubbings."

conditioned sinter. Sinter that contains up to 10% lime, and is used as an iron-blast furnace charge. See also sinter.

Condor Special. Trade name of Farrelloy Company (USA) for a water-hardening tool steel (AISI type W1) containing 0.7-1.4% carbon, 0.25% silicon, 0.25% manganese, and the balance iron. Used for pneumatic tools, shear knives, chisels, drills, drill rod, and dies.

Conductal. Trade name for a series of French aluminum alloys used for busbars and insulated conductors.

Conductex. Trade name for conductive carbon blacks.

conducting material. See conductor.

conducting polymers. See conductive plastics.

conductive carbon paint. A high-quality paint or paste based on carbon (micrographite) in a suitable solvent, such as isopropanol. It dries rapidly at room temperature, and is used in sample preparation for scanning electron microscopy.

conductive ceramic tile. A ceramic tile made conductive by using special additives in its manufacture, or by special manufacturing methods that are designed to increase its electrical conductivity.

conductive coating. (1) A very thin coating of gold, gold-palladium or carbon (usually 5-100 nm or 0.2-4 μin.) applied to a surface to prevent the buildup of static electric charges.

(2) A metallic coating, porcelain enamel, or glaze capable of conducting electricity.

conductive concrete. A concrete made conductive by introducing a special aggregate to the mix. Used in the grounding of electronic equipment.

conductive elastomer. An elastomer, such as silicone or styrene-butadiene rubber, thermoplastic rubber, neoprene or ethylene-propylene-diene monomer, that has been made electrically conductive by incorporating carbon black or metal particles (e.g., aluminum or silver flakes or spheres). Used for conveyor belts, tabletops, contact sensor, electronic components, etc. Also known as *conductive rubber; electrically conductive elastomer; electrically conductive rubber.*

conductive glass. (1) Plate glass made electrically conductive by applying an ultrathin, clear coating of tin oxide at high temperatures.

(2) A glass that has been made electrically conductive by coating with a clear lacquer containing finely divided metal powder.

conductive glass coatings. Very thin, transparent, durable coatings of metallic oxides applied to the surfaces of borosilicate or lime glasses, usually by spraying. Such coatings make the otherwise electrically insulating glass conductive. Used in the manufacture of heaters and for aerospace applications.

conductive plastics. Thermoplastic or thermosetting materials that are either inherently conductive, such as the polyacetylenes, polyparaphenylenes, polyparaphenylene sulfides, polyanilines and polypyrroles, or have been made conductive by incorporating conductive additives, such as antistatic agents (e.g., quaternary ammonium compounds, alkyl amines or ethoxylated glycol ester derivatives), fillers (e.g., carbon black, aluminum flakes or fibers, or metal powders) or reinforcements (e.g., carbon fibers, metallized glass fibers, or stainless steel fibers). Used in the elimination of static buildup, for electromagnetic shielding applications, for mounting materials in metallography, in electrochemistry and solid-state physics, and for electronic, optoelectronic and photonic applications. Also known as *conducting polymers; conductive plastic materials; conductive polymers; electrically conductive plastics; electrically conductive polymers.* See also conjugated polymers; semiconducting polymers.

conductive polymers. See conductive plastics.

conductive rubber. See conductive elastomer.

conductive silver paint. A high-quality, high-purity, quick-drying paint or paste containing pure silver pigment. Used in scanning electron microscopy for affixing samples to specimen mounts.

conductive silver paste. (1) A paste composed of silver powder suspended in a suitable vehicle. It is bonded to ceramics and other insulating materials (e.g., titanates, mica, porcelain, steatite, wood, paper, etc.) to provide a conductive surface or produce ceramic-to-metal seals, or as a base for electroplating.

(2) See conductive silver paint.

conductor. (1) A material, such as a metal or strong electrolyte that offers a small resistance (or opposition) to the passage of an electric current. In a metallic conductor the energy levels of the conduction and valance bands overlap. Also known as *conducting material; electrical conductor.*

(2) Any solid, liquid or gaseous medium that easily transmits heat, i.e., kinetic energy. Metals, such as silver, copper and aluminum, are excellent conductors and most polymers and ceramics are relatively poor conductors. Also known as *conducting material.*

Conductron. Trademark of W.C. Richards Company (USA) for a conductive paint used in the electronic and automotive industries.

Conductrol. Trademark of Sterling Fibers, Inc. (USA) for a stable acrylic-based, carbon-containing conductive fiber available in spun yarn and short staple form. It is used for the control of electrostatic discharges in adhesives, rubber, epoxy flooring, carpets, woven or nonwoven apparel, upholstery, filtration products, etc.

Conduloy. Trade name of Brush Wellman (USA) for an age-hardenable high-copper alloy containing 0.23-0.32% beryllium and 1.4-1.5% nickel. Used for instrument parts, springs, and diaphragms.

Conduron. Trademark of The Duffy Company (USA) for carburizing, nitriding and oxidation preventives used in selective steel hardening.

Condursal. Trademark of The Duffy Company (USA) for paintable carburizing, nitriding and oxidation preventives (stop-offs) used in selective steel hardening.

Cone Five Porcelain. Trade name of Aardvark Clay & Supplies (USA) for a low-firing porcelain clay (cone 5) made from a modified cone 10 clay.

Conel. Trade name of Waltham Precision Instruments (USA) for an iron alloy containing 38.5% nickel, 4.6% chromium, 3.3% manganese and 3.3% silicon. Used for springs.

Co-Netic. Trade name of Magnetic Shield Corporation (USA) for magnetic alloys with high initial permeability and attenuation. Supplied in sheet and foil form for magnetic shielding applications.

Conex. Trade name for polyaramid fibers and fabrics.

confined concrete. A type of reinforced concrete that contains numerous closely spaced transverse reinforcements to restrain or restrict the concrete in a direction horizontally to the applied stress.

Conflex. Trade name of Metal & Controls and General Plate (USA) for a series of composite metals including various plain-carbon and spring steels, carbon-vanadium alloy steels, precipitation-hardening stainless steels, etc., clad with copper on both sides. They have high elasticity, good spring properties and good electrical conductivity (10-40% IACS). Used for current-carrying springs and diaphragms.

conformal coating. A coating that fully covers and exactly fits the contour of an article.

Conformat. Trademark of Nicofibers Company (USA) for glass fibers and yarns.

Confort. Belgian trade name for hollow glass blocks having one plain and one stippled side.

Congo. Trade name of CCS Braeburn Alloy Steel (USA) for an oil-hardening molybdenum-type high-speed steel (AISI type M6) containing 0.8% carbon, 4% chromium, 1.4-1.5% vanadium, 4% tungsten, 5% molybdenum, 12% cobalt, and the balance iron. It has excellent wear resistance, high hot hardness, and good resistance to decarburization. Used for cutting tools for hard materials, such as boring, form, lathe and planer tools, milling cutters, cutoff knives, reamers, and shear blades.

Congo copal. A very hard fossil resin available in the form of a yellowish, amorphous solid. It has a density of 1.06-1.07 g/cm³ (0.038-0.039 lb/in.³). Used as a replacement for amber, in the manufacture of high-gloss lacquers and varnishes, in paints for metallic substrates, and in textile finishes. Also known as *Congo gum; Congo resin.*

Congo Hot Work. Trade name of CCS Braeburn Alloy Steel (USA) for an age-hardenable, air-quenched hot-work steel containing 0.1% carbon, 3.5-4.0% chromium, 4.0% tungsten, 5.0% molybdenum, 23-25% cobalt, 0.5-0.75% vanadium, and the balance iron. Used for brass casting dies, and master hobs.

Congo jute. See urena fiber.

congolite. A light red mineral of the boracite group composed of iron chloride borate, $Fe_3B_7O_{13}Cl$. Crystal system, rhombohedral (hexagonal). Density, 3.58 g/cm³; refractive index, 1.755. Occurrence: Zaire.

Congo resin. See Congo copal.

conichalcite. A green mineral of the descloizite group composed of calcium copper arsenate hydroxide, $CaCu(AsO_4)(OH)$. Crystal system, orthorhombic. Density, 4.33 g/cm³; refractive index, 1.831. Occurrence: USA (Arizona).

Conico. Trade name for a ferromagnetic material containing 50% copper, 29% cobalt and 21% nickel. Used for soft magnets and electrical equipment.

Conicro. Trade name of ThyssenKrupp VDM GmbH (Germany)

for cobalt-base superalloys containing 20-22% chromium, 14-15% tungsten, 10-12% nickel, 2% iron and 1.2-1.5% manganese. Supplied in the form of semifinished products, they are used for heat-resistant parts and equipment.

conifer. A tree or shrub belonging to the order Pinales, most species having small, needle-shaped or scale-like, evergreen leaves and bearing their seeds in cones. Conifers produce wood commonly known as "softwood" (although not necessarily an indication of their hardness). They can be subdivided into the following genera: (i) *Abies* (firs), (ii) *Cedrus* (cedars), (iii) *Larix* (larches), (iv) *Picea* (spruces), (v) *Pinus* (pines) and (vi) *Tsuga* (hemlocks).

coniferous wood. See softwood.

conjugated polymers. A relatively new class of semiconductive, mainly planar extended organic polymers with frameworks of alternating single and double carbon-carbon or carbon-nitrogen bonds that, especially when chemically doped, have optical and electrical properties including occupied valance and unoccupied conduction bands separated by energy gaps, which make them suitable for optoelectronic applications, such as polymer light-emitting devices (P-LEDs), emissive and electroluminescent displays, electrochromic devices, capacitors and rechargeable batteries, and as polymer actuators, corrosion-protection agents and biomaterials. Unsubstituted conjugated polymers are usually insoluble, do not melt and cannot be evaporated. Examples include polyacetylene (PA), polyaniline (PAn) poly(*p*-phenylene vinylene) (PPV), poly(*N*-vinylcarbazole) (PVCZ) and polypyrrole (PPy). See also light-emitting polymers; conductive polymers; photovoltaic polymers; semiconductive polymers.

conjugated protein. A protein, such as nucleoprotein, phosphoprotein, glycoprotein or chromoprotein, that contains a prosthetic (or non-protein) group.

conjugate fiber. See bicomponent fiber.

Conlo. Trade name of British Steel Corporation (UK) for aluminum- or silicon-deoxidized plain-carbon steel plate containing 0.16% carbon, 1-1.5% manganese, 0.2% silicon, a total of 0.05% sulfur and phosphorus, and the balance iron. Usually supplied in the normalized condition, it has good brazeability and weldability, and is used for ship and boiler plate and structural applications.

Conloc. (1) Trademark of EGO Dichtstoffe GmbH & Co. Betriebs KG (Germany) for cyanoacrylate, epoxy and silicone adhesives including various quick-setting and anaerobic types. Used for joining various materials, such as metals, glass, rubber, plastics, etc.

(2) Trade name of PRO Technologies Limited (Canada) for a family of gray, fibrillated polypropylene fibers supplied in lengths up to 50 mm (2 in.) for dispersion into concrete. A nonwetting, chemically inert, lightly fibrillated grade (*Conloc*ᵗᵐ) for dispersion into shotcrete and ready-mix concrete is also available.

Conloy. Trade name of Constrictor Limited (UK) for an aluminum alloy containing 4-5% copper. Used for tubes and bicycles.

connellite. A deep-blue mineral of the buttgenbachite group composed of copper chloride sulfate hydroxide trihydrate, $Cu_{19}Cl_4(SO_4)(OH)_{32}\cdot3H_2O$. Crystal system, hexagonal. Density, 3.41 g/cm³; refractive index, 1.738. Occurrence: USA (Arizona, Utah), South Africa. Also known as *footeite.*

Conoco. Trade name of Vista Chemical (USA) for plasticized polyvinyl chlorides available in three elongation ranges: 0-100%, 100-300%, and more than 300%. *Conoco UPVC* refers to unplasticized polyvinyl chlorides available in standard, cross-linked, high-impact, UV-stabilized and structural-foam grades.

Conomet. Trade name for metallic fibers.

Con-Pac. Trade name of US Steel Corporation (USA) for a series of heat-treated high-strength structural carbon steels and quenched-and-tempered high-strength low-alloy structural steels used for construction and mining equipment, bridges, buildings, penstock, trucks, industrial machinery, ship hulls, etc.

Conpernik. Trade name of Westinghouse Electric Corporation (USA) for an alloy containing 50% iron, 50% nickel and a trace of manganese. It has high magnetic permeability and negligible permeability variation. Used for choke coils and transformer cores.

Conpol. Trademark of E.I. DuPont de Nemours & Company (USA) for additive resins made from ethylene-methacrylic acid carrier resins. Supplied in several grades, they contain additives, such as antiblock and/or slip agents, anti-fog agents or odor-absorbing agents. Used in the manufacture of food and non-food packaging materials to modify the surface properties of coatings or films of Dupont's *Nucrel* and *Surlyn* resins.

ConQuest. Trademark of DSM BV (Netherlands) for a series of conductive polymer-supported polypyrole polymers that may either be supplied as water-dispersible undoped polypyrrole shells on polyurethane cores, water-dispersible, organic acid doped conductive polypyrrole shells on polyurethane core binder resins, or organic acid doped conductive polypyrrole shells on waterborne polyurethane core resin binders.

Conquest. Trade name of Jeneric/Pentron (USA) for dental filling composites including *Conquest DFC,* an all-purpose grade.

Conrex. Trademark of Continental Gummi-Werke (Germany) for flexible polyvinyl chlorides.

Consafis. German trade name for double glazing units.

Conseal F. Trade name of Southern Dental Industries Limited (Australia) for a light-cured, fluoride-releasing, opaque dental resin composite 7% filled with a submicron special filler. It provides low viscosity and optimal flow characteristics, high surface-wear resistance and low water solubility. Used in restorative dentistry as a pit and fissure sealant.

Conservaloy. Trade name of Lehigh Steel Corporation (USA) for a corrosion and heat-resistant steel containing 0.6% carbon, 22% chromium, 8.5% manganese, 0.35% nickel, and the balance iron. Used for valve parts.

Consil. Trade name of Handy & Harman (USA) for a series of silver-cadmium, silver-cadmium oxide and silver-magnesium electrical contact alloys.

Constahl. Trade name of Nihon Jyokiko Seikosho Goshi (Japan) for an austenitic stainless steel containing 0.1% carbon, 19% chromium, 9% nickel, and the balance iron. Used for corrosion-resistant parts.

Constant. Trade name of Midvale-Heppenstall Company (USA) for an oil-hardening, nondeforming tool steel containing 0.9-0.95% carbon, 1.2-1.25% manganese, 0.5% chromium, 0.5% tungsten, and the balance iron. Used for tools, dies, cutters, hobs, reamers, taps, chasers, etc.

constantans. A generic name for a group of resistance alloys containing 40-55% nickel, 45-60% copper, and traces of iron and manganese. Commercially available in foil, rod, wire and powder form, they have a density of 8.9 g/cm³ (0.32 lb/in.³), a melting range of 1225-1300°C (2235-2370°F), a high electrical resistivity that remains relatively constant over a wide range of temperatures, low temperature coefficients of resistance, low

coefficients of expansion, high heat resistance, a maximum service temperature in air of 500°C (930°F), and a hardness of 100-300 Brinell. Used for electrical resistances, precision wire-wound resistors, thermocouples, specialized heat-measuring devices, and rheostats.

Constantin. Trade name of Driver Harris Company (USA) for a heat-resistant alloy containing 44-46% nickel, 54% copper, 0-0.4% iron, and up to 1.3% manganese. Used for electrical resistances, rheostats, and thermocouples.

constituent material. (1) Any of the components of which an alloy is made, e.g., copper and tin are the constituent materials of bronze.

(2) Any of the individual materials of which a composite is made including reinforcing elements, fillers and matrix binders, e.g., glass fibers and epoxy resin are the constituent materials of a glass-epoxy composite.

constructional steels. See structural steels.

construction glass. Any glass suitable for use in building construction including in particular insulating and safety glass, window glass, glass blocks and bricks, glass roofing tiles and glazed reinforced concrete.

construction-grade lumber. See construction lumber.

construction lime. See building lime.

construction lumber. (1) High-quality lumber with good strength, serviceability and surface appearance, and without any deleterious defects. It is suitable for a wide range of structural applications. Also known as *construction-grade lumber.*

(2) One of the two categories of softwood lumber (the other being remanufactured lumber). It is a best-quality structural material and includes: (i) *Stress-graded lumber,* (ii) *Non-stress-graded lumber* (or *yard lumber*) and (iii) *Appearance lumber.* In general, construction lumber includes timbers, beams, posts, stringers, decking, boards and planks. Also known as *construction-grade lumber.*

construction materials. A large group of natural or synthetic materials that have been processed to make them suitable for use in building and road construction, civil engineering, carpentry, masonry, architecture, etc. Included are in particular lumber, plywood and other wood panel materials, metals (e.g., steel, aluminum, lead and copper), plastics, ceramics, glass, composites, concrete, adhesives and sealants, and road construction materials (e.g., macadam, tar, bitumen, gravel, or crushed stone). Also known as *building materials.*

construction paper. See building paper.

Constructit. Italian trade name for a building element of shaped glass reinforced with steel wires embedded in the curved part of the product.

Consumet. Trade name of Carpenter Technology Corporation (USA) for a molybdenum-type high-speed steel (AISI type M50) containing 0.5-0.8% carbon, 0.25% manganese, 0.25% silicon, 4.0% chromium, 4.5% molybdenum, 1.0% vanadium, 0.1% nickel, and the balance iron. Used for aircraft and gas-turbine-engine bearings operating at temperatures up to 430°C (805°F). Also known as *Carpenter Consumet.*

Con-Tact. Trade name of Decora Industries, Decora Manufacturing Division (USA) for self-adhesive vinyl film used for decorative applications.

contact adhesives. A group of adhesives that are essentially dry to the touch and applied to each of two properly aligned surfaces. When pressed together, bonding takes place immediately upon contact. Used in the application of floor covering and automotive interiors, in bonding plastic laminates to wood, etc.

Also known as *contact bond adhesives; dry bond adhesives.*

Contact Bronze. Trade name of Criterion Metals Inc. (USA) for a series of corrosion-resistant bronzes containing 88-89% copper, 9-10% zinc, 2% tin, 0-0.05% iron, 0-0.05% lead and 0-0.15% phosphorus. Used for switches, terminals, connectors, springs, fuse clips, pen clips, weatherstripping and electrical contacts.

contact cement. A ready-to-use neoprene rubber-base adhesive that is applied to each of the surfaces to be bonded and allowed to dry for a specified period of time. The surfaces are then properly aligned and pressed firmly together. Bonding takes place immediately upon contact. Used for bonding plastic laminates to wood (e.g., countertops and plywood edging), joining parts that cannot be easily clamped together, and for joining wood, cloth, leather, rubber, and plastics.

contact materials. A group of materials used for electrical contacts particularly make-and-break and sliding contacts. They may be classified as: (i) *Contact metals,* and (ii) *Composite contact materials* including refractory, carbide-base, silver-base, and copper-base powder metallurgy products. *Contact materials* have high electrical and thermal conductivity, high arc-transfer resistance, high melting points, minimum sticking or welding tendencies, low contact closing resistance, high corrosion resistance, good wear resistance, and a good balance of hardness and ductility. Also known as *electrical contact materials.*

contact metals. A group of metals and metal alloys that are used for electrical contacts and include: (i) *Copper and copper alloys,* such as beryllium, cadmium and electrolytic-tough-pitch coppers, red, yellow and low brasses, and phosphor bronzes; (ii) *Silver and silver metals* including pure silver, and silver-cadmium, silver-copper, silver-gold, silver-graphite, silver-molybdenum, silver-palladium, silver-tungsten and silver-zinc alloys; (iii) *Gold and gold metals,* such as pure gold, and gold-copper and gold-silver alloys; (iv) *Other precious metals* including in particular platinum-group metals and alloys, such as platinum and palladium metal, and platinum-palladium, palladium-copper, palladium-ruthenium, palladium-silver, platinum-iridium, platinum-osmium, platinum-ruthenium and osmium-rhodium alloys; (v) *Pure aluminum;* and (vi) *Tungsten and molybdenum.* For properties of contact metals, see contact materials. Also known as *electrical contact metals.*

contact moldings. Reinforced plastic moldings formed by placing the reinforcement over a mold, brushing or spraying with liquid resins and then curing, either without heat and pressure using a catalyst-promoter system, or by baking in an oven.

contact resin. A liquid thermosetting resin that requires a catalyst and heat for curing, but relatively little pressure when used for bonding laminates. Also known as *contact pressure resin; impression resin; low-pressure resin.*

containerboard. Paperboard used in the manufacture of solid and corrugated fiberboard shipping containers. See also paperboard.

containment material. (1) A material, such as steel-reinforced concrete, used in the form of a shell or other enclosure around a nuclear reactor to prevent fission products from escaping into the atmosphere.

(2) A material that either encloses or is in contact with the heat-transfer or heat-storage medium of a solar-energy system.

Contan. Trademark of Continental Gummi-Werke (Germany) for flexible polyvinyl chloride film materials.

Contex. Trademark of Continental Gummi-Werke (Germany) for flexible polyurethane foams and foam products.

Conticell. Trademark of Continental Gummi-Werke (Germany) for modified polyvinyl chlorides.

Contico. Trademark of CP Rubber Inc. (Canada) for molded and extruded rubber goods.

Conrex. Trademark of Continental Gummi-Werke (Germany) for flexible polyvinyl chlorides.

Conthan. Trademark of Continental Gummi-Werke (Germany) for a polyurethane elastomer.

Contiduct. Trademark of Continental Gummi-Werke (Germany) for silicone-based products.

Contilan. Trademark of Continental Gummi-Werke (Germany) for vinyl rubber products.

Contimatic. Trademark of Continental Gummi-Werke (Germany) for flexible polyvinyl chlorides.

Continental. Trade name of CP Rubber Inc. (Canada) for molded and extruded rubber goods.

Continental Lime. Trade name of Continental Lime, Inc. (USA) for quicklime, limestone and hydrated lime products.

Continex. Trademark of Witco Chemical Company, Inc. (USA) for a furnace black used in rubber compounding.

Continol. Trademark of Continental Gummi-Werke (Germany) for polyvinyl chloride film materials.

continuous alumina fibers. Long, thin, continuous strands that are composed of up to 99.5% aluminum oxide (Al_2O_3) and produced by dry or slurry spinning with subsequent heat treatment. They have high moduli of elasticity, high melting points, and exceptional corrosion resistance. Used as reinforcing fibers in organic-, ceramic- and metal-matrix composites. Also known as *continuous aluminum oxide fibers.*

continuous boron fibers. Single, long, thin, continuous filaments of boron produced by chemical vapor deposition on tungsten or carbon substrate wires, thermal decomposition of diborane, or drawing from molten boron. They have high strength, high stiffness, high moduli of elasticity, and low density. Used for boron-epoxy preimpregnated tape or prepregs, and as reinforcing fibers for organic and metal-matrix composites.

continuous carbon fibers. Single, long, thin, continuous filaments that are composed of 93-95% carbon, have an average diameter of about 4-11 μm (157-433 μin.), and are produced from organic precursors, such as rayon, polyacrylonitrile or certain pitches. The conversion of these precursors into carbon fibers involves carbonization in the range of 1315°C (2400°F). Continuous carbon fibers have outstanding strength and moduli of elasticity even at high temperatures. Used as reinforcing fibers for ceramic-, carbon- and metal-matrix composites, fiber reinforced plastics, and prepregs.

continuous castings. Products made by a casting process that involves pouring molten metal (e.g., aluminum, copper, copper-tin and copper-zinc alloys, cast iron, steel, etc.) into one end of a mold, open at both ends, followed by rapid cooling, and extraction in a continuous length from the other end. Also known as *strand castings.*

continuous cast steel. A semifinished mill-product, such as a bloom, billet, ingot or slab, made by a casting process that involves pouring the molten, usually deoxidized (or killed) steel into one end of a mold, open at both ends, followed by rapid cooling, and removal of the continuous strand from the other end. Continuous cast steel is of high quality owing to its uniform microstructure and mechanical properties. Also known as *continuously cast steel.*

continuous-fiber ceramic composites. A class of strong, stiff composite materials consisting ceramic matrices (e.g., alumina, or silicon carbide) that have been toughened by the incorporation of continuous fibers of materials, such as carbon/graphite or silicon carbide. They have excellent resistance to high temperatures, typically 650-1200°C (1200-2190°F), and are used for applications, such as structural components, gas-turbine fan blades, high-speed aircraft fins, diesel-engine piston ring parts, components of carburizing furnaces, etc. Abbreviation: CFCC.

continuous fibers. Long, thin, continuous filaments of natural or synthetic materials, usually 50 mm (2 in.) or more in length. Examples of natural continuous fibers include hemp, jute, flax, cotton and silk. Synthetic continuous fibers include alumina, aramid, boron and carbon. Abbreviation: CF.

continuous fiber-reinforced composites. Composite materials consisting of metallic, polymeric or ceramic matrices reinforced with continuous fibers. Abbreviation: CFRC.

continuous filament nonwoven fabrics. Textile fabrics made by continuously spinning and mechanically or thermally bonding endless natural or synthetic fibers. Used for making clothing, packaging, and as wraps in building construction. Also known as *continuous filament nonwovens.*

continuous filament yarn. A yarn made by twisting together two or more mono- or multifilaments that extend substantially throughout the entire length. Also known as *filament yarn.*

continuous glass fibers. Long, thin, continuous filaments of glass obtained by melting and subsequent drawing and controlled cooling. Commercial continuous glass fibers are available in several grades including high-alkali (A-glass), electrical (E-glass), high-strength (S-glass), and modified, chemically resistant E-glass (ECR-glass). While most of their properties depend largely on the particular type of glass, common characteristics include good handleability and processibility, and high strength. Used as reinforcing fibers in plastics, and for filament-wound components. Abbreviation: CGF.

continuous graphite fibers. Long, thin, continuous filaments that are composed of 99+% carbon, have diameters normally ranging from 4 to 11 μm (157 to 433 μin.), and are produced from organic precursors, such as rayon, polyacrylonitrile and certain pitches. The conversion of these precursors to graphite fibers involves graphitization at 1900-2480°C (3450-4495°F). They have outstanding strength and moduli of elasticity even at high temperatures. Used as reinforcing fibers for ceramic-, graphite/carbon-, and metal-matrix composites, fiber-reinforced plastics, and prepregs.

continuously cast steel. See continuous cast steel.

continuous oxide fibers. A continuous man-made fiber consisting of one or several ceramic oxides, e.g., alumina, alumina and silica, or zirconia and silica.

continuous silicon carbide fibers. Long, thin, continuous filaments of (i) *Monolithic* (single structure) type, such as standard silicon carbide fibers consisting of silicon, carbon and oxygen, and modified silicon carbide fibers containing silicon, titanium and carbon, or silicon, zirconium and carbon; or (ii) *Bicomponent* type made by chemical vapor deposition of silicon carbide on a carbon core. *Continuous silicon carbide fibers* have high strength and high moduli of elasticity even at high temperatures, and are used as reinforcing fibers for metal-matrix composites.

continuous-strand mat. A fibrous material, such as fiberglass, consisting of swirled filaments loosely held together with a synthetic binder, such as a thermoplastic resin. Used in the manufacture of reinforced plastics by the closed-mold, resin-transfer molding or pultrusion process.

continuous tungsten fibers. Long, thin, continuous filaments of tungsten, tungsten-thorium dioxide ($W-ThO_2$), tungsten-hafnium-carbon (W-Hf-C) or tungsten rhenium (W-Rh). They have high strength and stiffness, excellent high temperature properties, and are used as a reinforcing fibers for metal-matrix composites and fiber-reinforced superalloys.

Contiplast. Trademark of Continental Gummi-Werke (Germany) for flexible polyvinyl chloride film materials.

Contiplastex. Trademark of Continental Gummi-Werke (Germany) for flexible polyvinyl chlorides.

Contipren. Trademark of Continental Gummi-Werke (Germany) for polyurethane foams and foam products.

Contitec. Trademark of Continental Gummi-Werke (Germany) for chloroprene rubber film materials.

Conti-PUR. Trademark of Continental Gummi-Werke (Germany) for polyurethanes supplied in a wide range of grades.

Contivyl. Trademark of Continental Gummi-Werke (Germany) for flexible polyvinyl chlorides (PVCs) and PVC copolymers.

Contour. (1) Trade name of Fairhope Fabrics, Inc. (USA) for glass tapes.

(2) Trade name of Georgia Bonded Fibers, Inc. (USA) for a coated paper.

(3) Trade name of GOEX Corporation (USA) for rigid plastic sheeting.

(4) Trademark of Saint-Gobain (France) for a curved annealed glass.

(5) Trade name of Kerr Dental (USA) for regular- and fast-set dental amalgam alloy powders.

Contour Securit. Trade name of Saint-Gobain (France) for a curved toughened glass.

Contour-Wall. Trademark (Canada) for a glass-based building cladding system.

Contracalor. German trade name for a heat-absorbing polished plate glass with light greenish-blue color.

Contracid. Trade name of Vacuumschmelze GmbH (Germany) for a series of alloys containing 58-61% nickel, 12-20% iron, 15-18% chromium, 0-10% tungsten, 0-10% molybdenum, 0-2% manganese and 0-3% cobalt. They have high heat, fatigue, acid and corrosion resistance. Used for springs, chemical apparatus, surgical instruments, jet engine parts, pumps, and valves.

Contraflam. Trademark of Saint-Gobain (France) for a fire-resistant insulating glass.

Contralloy. Trade name of Michiana Products Corporation (USA) for a heat- and corrosion-resistant alloy containing varying amounts of carbon, 28% chromium, 15% chromium, and the balance iron. Used for parts and equipment that must be resistant to sulfurous gases at high temperatures and molten salts.

Contralux. British trade name for a glass consisting of two panes with a glass-fiber interlayer.

Contrasol. German trade name for a phototropic laminated glass having a plastic interlayer with active chemicals.

Contrast. Trade name for a silicone elastomer used for dental impressions.

Contro. Trade name for rubber fibers used in the manufacture of elastic yarn for clothing, elastic bands and tapes, etc.

controlled-expansion alloys. A group of alloys including *Invar* (64Fe-36Ni) and constantan (55Cu-45Ni) that, owing to their compositions and crystal structures, either exhibit predictable high or low thermal expansion, or match the thermal expansion behavior of ceramics or glasses. See also high-expansion alloys; low-expansion alloys; matching-expansion alloys.

Controlled Zone. South African trade name for a glass that has been selectively toughened by a proprietary process. Used for windshields.

Controltac. Trademark of 3M Company (USA) for polyvinyl film with an adhesive applied to one side.

Conturan. Trademark of Schott DESAG AG (Germany) for an antireflective glass used in the manufacture of computer monitors, TV receivers, operating panels, display panels, etc.

conventional hard-magnetic materials. Hard-magnetic materials, such as tungsten steel, alnico or cunife alloys and hexagonal ferrites, e.g., barium hexaferrite ($BaO \cdot 6Fe_2O_3$) or lead hexaferrite ($PbO \cdot 6Fe_2O_3$), with magnetic energy products of about 2-80 kJ/m^3 (0.25-10 MG·Oe). See also hard-magnetic materials.

conversion coatings. Inorganic coatings or films, usually less than 25 µm (1 mil) thick, formed on a metal or metal-alloy substrate by chemical or electrochemical reaction with a suitable chemical solution. They are produced from the original surfaces and thus adhere tightly to the substrates. Examples include phosphate coatings on iron, steel and zinc, chromate coatings on steel, zinc, cadmium, aluminum, etc., black oxide coatings on steel, and anodic coatings on aluminum, zinc, magnesium, etc.

converted bar. See blister steel.

converted fabrics. Textile fabrics that have received one or more finishing treatments, such as dyeing, printing, bleaching or waterproofing.

converted steel. See blister steel.

convertible coating. An organic coating that is crosslinked or cured on the substrate after deposition.

convertible film. An organic (polymeric) film that can form crosslinks.

converting paper. A paper suitable for subsequent treatment and conversion into paper products, e.g., containers and napkins.

Convey-O-Kote. Trade name of Turco Products Inc. (USA) for a masking material used on conveyors.

conveyor duck. A loose- or hard-woven cotton fabric used for conveyor and transmission belts. See also duck.

Conweld. Trademark of Thermit Alloys Limited for special soldering and brazing alloys, special brazing and welding fluxes, electrodes, rods and filler wires, and flame-spraying powders.

Conyma. Trade name for rayon fibers and yarns used for textile fabrics.

cookeite. A white mineral of the chlorite group composed of lithium aluminum silicate hydroxide, $LiAl_4Si_3AlO_{10}(OH)_8$. Crystal system, monoclinic. Density, 2.65 g/cm^3; refractive index, 1.58. Occurrence: Australia.

Cook's alloy. An alloy of 56-69% antimony and 31-44% zinc.

Cook Silver Label. Trade name of George Cook & Company Limited (UK) for a water-hardening tool steel containing 1.4% carbon, 0.4% manganese, 0.6% chromium, and the balance iron. Used for lathe and planer tools, dies, punches, reamers and taps.

Coolaray. Belgian trade name for a heat-absorbing glass.

Coolblue. South African trade name for a laminated glass with light bluish vinyl interlayer.

Cool-Lite. Trademark of Saint-Gobain (France) for a solar control glass having a characteristic blue-green color. Two special grades, *Cool-Lite K* and *Cool-Lite SK*, with enhanced thermal insulation properties are also available.

CoolMax. Trademark of E.I. DuPont de Nemours & Company (USA) for a high-performance fabrics made from proprietary *Dacron* polyester fibers. They have good wicking and moisture evaporation properties, and are used for sportswear.

CoolMelt. Trademark of Dexter Corporation (USA) for a high-viscosity ethylene-vinyl-acetate *Hysol*-type hot-melt adhesive for bonding paper, styrofoam, plastics and softwoods.

CoolPoly. Trade name of Cool Polymers Inc. (USA) for thermally conductive polymer composites supplied in injection-molding grades. Their thermal conductivity, ranging from about 10 to 100 W/m·K (0.69 to 6.9 Btu·in./h·ft^2·°F), is due to the addition of 10-70 vol% nonmetallic, thermally conductive materials uniformly distributed throughout the matrix. Electrically conductive and nonconductive grades are also supplied. They have low thermal expansion and mold shrinkage, and are designed for low service temperatures. Used for automotive components, business machines, computers, portable and power electronics, heat sinks; medical, optical, mechanical and micromechanical parts.

Cool-View. Trade name of Shatterproof Glass Corporation (USA) for a glass product comprising several sheets of glass combined with a reflective coating. Used for oven windows.

Cooper. (1) British trade name for a corrosion-resistant dental alloy containing 40-50% gold and 50-60% palladium.

(2) Trade name of Cooper Alloys Corporation (USA) for an extensive series of alloys including several corrosion-resistant cast steels of the iron-chromium-nickel, iron-nickel-chromium and precipitation-hardening type, and numerous austenitic, ferritic and martensitic stainless steels, some of which are also heat- and/or abrasion-resistant.

Cooperite. British trade name for an extremely hard alloy of 80% nickel, 14% tungsten and 6% zirconium. Used for cutting tools.

cooperite. A steel-gray mineral composed of platinum sulfide, PtS. It can also be made synthetically. Crystal system, tetragonal. Density, 9.5 g/cm^3. Occurrence: South Africa.

coordination complex. See coordination compound.

coordination compound. A compound that comprises a central metal atom or ion to which two or more groups of nonmetallic ions or molecules, called ligands or complexing agents, are attached by coordinate covalent bonds. The most common metal ions include transition metals, such as cobalt, copper, iron, platinum, nickel and zinc, and the most common ligands are ammonia, water, chlorine, and hydroxyl groups. Also known as *coordination complex; complex compound; transition-metal complex.*

Copaco. Trademark of Twinpak Inc. (USA) for splicing film flexible packaging materials, such as polyethylene-coated and laminated paper.

Copa-Griltex. Trade name of EMS Grilon (USA) for hot-melt adhesives based on polyamide copolymers. See also Griltex.

Copaloy. Trade name of Michigan Smelting & Refining Company (USA) for an antifriction alloy of tin, antimony and lead. Used for bearings.

copals. A group of yellow to red, hard, lustrous natural resins (recent or fossil) exuded from various trees native to Africa, South America or the East Indies. Examples of copals include amber, congo, kauri, Manila, pontianak and Zanzibar. Used chiefly in the manufacture of varnishes and lacquers. Also known as *copal resins.* See also fossil resin; recent resin.

Copan. Trade name for a babbitt metal containing 80-90% tin, 10-15% antimony, 2-5% copper and 0.2% lead. Used for bearings.

Copastar. Trade name of Mitsubishi Corporation (Japan) for an aluminum bronze containing 7.5% aluminum, 2.5% nickel, 3.5% iron, 0.5% manganese, and the balance copper. It has excellent resistance to discoloration, and is used for hardware, materials for the building and home-construction industries, tableware, fixtures, and musical instruments.

Copel. Trade name of Hoskins Manufacturing Company (USA) for a resistance alloy containing 55% copper and 45% nickel. It has uniform resistivity, a very low temperature coefficient of resistance, and a service temperature up to 427°C (800°F). Used for rheostats, resistors, electrical resistances, thermocouples, and electrical instruments.

Copelmet. Trademark of Metro Cutanit Limited (UK) for a series of sintered materials containing varying amounts of copper and tungsten, or copper and tungsten carbides. A typical composition of the former is 72% tungsten and 28% copper. Supplied in bar, rod and sheet form, they have excellent thermal conductivity, low thermal expansion, and excellent resistance to mechanical wear and electrical erosion. Used for welding electrodes, electrical-contact materials, spark-erosion electrodes and contacts, and riveting dies.

Copernick. Trade name of Western Electric Company (USA) for a resistance alloy containing 50% iron and 50% nickel. Used for transformer cores.

Copes-Griltex. Trade name of EMS Grilon (USA) for hot-melt adhesives based on polyether sulfone copolymers. See also Griltex.

copiapite. A yellow mineral composed of iron sulfate hydroxide hydrate, $Fe_{14}O_3(SO_4)_{18} \cdot 20H_2O$. Crystal system, triclinic. Refractive index, 1.531. Occurrence: Canada (British Columbia), USA. Also known as *yellow copperas.*

Coply. Trademark of C-I-L Paints (Canada) for coextruded polyolefin films.

Copo. Trademark of DSM Copolymers (USA) for a styrene-butadiene elastomer.

copolymer. A long-chain polymer that consists of two or more mer units in combination along its molecular chain. Usually, it is a linear or nonlinear macromolecule formed by a chemical reaction in which the molecules of two or more monomers are linked together. The molecular weight of a copolymer is higher, and its physical properties are different from those of the original substances (monomers). Depending on the arrangement of the monomers, a copolymer can be of alternate, block, graft or random type. See also alternating copolymer; block copolymer; graft copolymer; random copolymer. Abbreviation: CP.

Coppen Light. Trade name of H.C. Spinks Clay Company (USA) for a ball clay.

copper. A reddish, malleable, ductile metallic element of Group IB (Group 11) of the Periodic Table. It is commercially available in the form of ingots, bars, rods, sheet, foil, microfoil, microleaves, wire, wire mesh, tubing, shot, powder, granules, turnings, and high-purity single crystals and whiskers. The single crystals are usually grown by the Czochralski or the Bridgeman technique. The commercial copper ores are *azurite, chalcocite, chalcopyrite (copper pyrite), covellite, cuprite* and *malachite.* Crystal system, cubic. Crystal structure, face-centered cubic. Density, 8.96 g/cm³; melting point, 1083°C (1981°F); boiling point, 2595°C (4703°F); hardness, 49-87 Vickers; atomic number, 29; atomic weight, 63.546; monovalent, divalent. It has a bright metallic luster, develops a greenish surface tinge (patina) when exposed to the atmosphere for a period of time, has high electrical conductivity (up to 103% IACS) and high thermal conductivity, good corrosion and fatigue resistance, good workability, machinability and strength, and takes a good finish. Used for electrical conductors, wires and cables, wave-guides, switches, terminals, commutator seg-

ments, electrical apparatus, electronic components, glass-to-metal seals, thermal conductors, corrosion-resistant piping, plumbing pipe and tube, condenser and heat-exchanger tubing, fuel lines, printing rolls, refrigerators, radiators, air conditioners, hardware, fasteners, in the manufacture of alloys (e.g., brass, bronze, monel, and various clad metals), in electroplating for protective coatings and undercoats for nickel, chromium and other metals, for coinage, statuaries, art metalwork, cooking utensils, building construction, roofing, heating equipment, chemical-process and pharmaceutical equipment, in antifouling paints, and as a catalyst. The powder and shot forms are used in superconductor research, and the whiskers in the manufacture of thermal and electrical composites. Symbol: Cu.

copper-64. Radioactive copper with a mass number of 64 obtained by irradiating metallic copper in a nuclear reactor. It decays by the following processes: (i) orbital electron capture (42%); (ii) negative beta-particle emission (39%); and (iii) positive beta-particle emission (19%). *Copper-64* has a half-life of 12.8 hours, and is used as an aid in studying corrosion, diffusion and friction wear in metal alloys. Symbol: ^{64}Cu.

Copper-ABA. Trade name of Wesgo Metals (USA) for copper-based active brazing alloys. See also ABA.

copper abietate. A green crystalline compound used as a preservative in metal paint, and as a fungicide. Formula: $Cu(C_{20}H_{29}O_2)_2$. Also known as *cupric abietate.*

copper acetate. Fine, greenish-blue powder, or blue-green crystals (98% pure) made by treating copper oxide with acetic acid. Density, 1.882 g/cm³; melting point, 115°C (239°F); boiling point, decomposes at 240°C (464°F). Used in pigments, as a raw material in the manufacture of Paris green, as a catalyst, and as a precursor to superconducting materials using solution methods. Formula: $Cu(C_2H_3O_2)_2 \cdot H_2O$. Also known as *copper acetate monohydrate; cupric acetate; crystals of Venus; crystallized verdigris; neutral verdigris.*

copper acetoarsenite. An emerald-green powder made by reacting sodium arsenite with copper sulfate and acetic acid. Used as a green pigment, and in antifouling paints and wood preservative. Formula: $(CuO)_3 \cdot As_2O_3 \cdot Cu(C_2H_3O_2)_2$. Also known as *cupric acetoarsenite; imperial green; king's green; Paris green; royal green; Schweinfurt green.*

copper alloys. A group of wrought and cast alloys of copper and one or more other metals. They can be classified into several general categories according to composition: (i) *High-copper alloys;* (ii) *Copper-zinc alloys* (or brasses); (iii) *Copper-tin alloys* (or bronzes); (iv) *Copper-nickel alloys* (or copper nickels); (v) *Copper-nickel-zinc alloys* (or nickel silvers); (vi) *Leaded coppers;* (vii) *Copper-lead bearing alloys;* (viii) *Copper-manganese resistance alloys;* (ix) *Copper-nickel radio alloys;* and (x) *Special copper alloys.* In general, *copper alloys* have excellent electrical and thermal conductivities, outstanding corrosion resistance, good fabricability, and a broad range of obtainable strengths.

Copper Alnico. Trade name of Harsco Corporation (USA) for an alloy of 9-11% aluminum, 16-18% nickel, 12-13% cobalt, 5.5-6.5% copper, and the balance iron. Used for magnets.

copper-aluminum. A master alloy that is composed of 50% copper and 50% aluminum, melts at 577°C (1071°F), and is used in the manufacture of aluminum alloys.

copper aluminum selenide. A compound of copper, aluminum, and selenium. Crystal system, tetragonal. Crystal structure, chalcopyrite. Density, 4.7 g/cm³; melting point, 1987°C (3609°F). Used as a semiconductor. Formula: $CuAlSe_2$.

copper aluminum sulfide. A compound of copper, aluminum, and sulfur. Crystal system, tetragonal. Crystal structure, chalcopyrite. Density, 3.47 g/cm^3; melting point, 2227°C (4041°F). Used as a semiconductor. Formula: $CuAlS_2$.

copper aluminum telluride. A compound of copper, aluminum, and tellurium. Crystal system, tetragonal. Crystal structure, chalcopyrite. Density, 5.5 g/cm^3; melting point, 2277°C (4131°F). Used as a semiconductor. Formula: $CuAlTe_2$.

copper amalgam. An alloy of about 3 parts mercury and 1 part copper available as hard, brown leaflets for use in dental cements. Also known as *copper cement.*

copper aminoacetate. See copper glycinate.

copper-antimony alloys. A group of copper-base alloys containing small amounts (usually less than 1.0%) of antimony to prevent dezincification. Examples include certain inhibited admiralty brasses containing mainly copper, zinc and tin, with traces of antimony, and some copper-base alloys containing mainly copper, lead and tin with small amounts of antimony (about 0.5%).

copper antimony selenide. A compound of copper, antimony, and selenium. Crystal system, tetrahedral. Density, 6.0 g/cm^3. Used as a semiconductor. Formula: Cu_3SbSe_4.

copper antimony sulfide. A purplish to dark gray compound of copper, antimony, and sulfur. It occurs in nature as the mineral *famatinite*. Crystal system, tetrahedral. Density, 4.9 g/cm^3. Used as a semiconductor. Formula: Cu_3SbS_4.

Copper Arc. Trade name for a carbon steel containing 0.2% carbon, 0.5% copper, and the balance iron. Used for welding electrodes for cast iron.

copper arsenic selenide. A compound of copper, arsenic, and selenium. Crystal system, tetrahedral. Density, 5.61 g/cm^3. Used as a semiconductor. Formula: Cu_3AsSe_4.

copper arsenic sulfide. A dark gray compound of copper, arsenic, and sulfur. It occurs in nature as the mineral *enargite*. Crystal system, orthorhombic. Density, 4.47 g/cm^3. Used as a semiconductor. Formula: Cu_3AsS_4.

copper arsenite. A fine, light green or yellowish-green powder that decomposes on melting, and is used as a pigment and as a wood preservative. Formula: $CuHAsO_3$. Also known as *copper orthoarsenite; cupric arsenite.* See also Scheele's green.

copperas. See copper sulfate; ferrous sulfate; zinc sulfate.

copper-base powder-metallurgy materials. A group of powder-metallurgy materials based on copper and including (i) bronzes (usually with 90% copper and 10% tin) for self-lubricating bearings; (ii) sintered copper-base materials for friction components, such as automotive clutches and aircraft brakes, and (iii) sintered copper-base materials including brasses, nickel silvers and bronzes for structural applications. Also included in this group are sintered copper-tungsten alloys containing about 70-75% copper and 25-30% tungsten. The latter are used as electrical contact materials.

copper-bearing lead. A grade of pig lead (99.90% pure) containing 0.04-0.08% copper, and traces of bismuth, iron, zinc, silver, arsenic, antimony and tin.

Copper Bearing Low-Metalloid. Trade name of US Steel Corporation (USA) for a carbon steel containing 0.2% or more copper. Used for hot rolled, annealed and galvanized sheets for culverts, building construction, and general fabrication.

copper-bearing steel. A low-carbon steel that contains 0.05-0.25% copper to improve its resistance to atmospheric corrosion. It is commercially available in the form of bars, sheets, plates, shapes and wires, and has a corrosion resistance that although moderate is better than that of ordinary carbon steels. Used for building and general fabrication, culvert pipes, ducts, piping, boilers, wire fencing, and barbed wire. Also known as *copper steel.*

copper-beryllium. See beryllium bronze.

copper blue. A term used for the bright blue mineral *azurite* [Cu_3-$(CO_3)_2(OH)_2$] when in ground form and used as a paint pigment. Also known as *mountain blue.*

copper borate. See copper metaborate.

copper bromide. See cupric bromide; cuprous bromide.

copper tert-butoxide. A fine sol-gel powder (0.01 μm or 0.4 μin.) used in the preparation of certain superconducting ceramics. Formula: $Cu(OC_4H_9)_2$.

copper-cadmium alloys. See cadmium copper.

copper cadmium tin sulfide. A compound of copper, cadmium, tin and sulfur with a density of 10.83 g/cm^3, used as a semiconductor. Formula: Cu_2CdSnS_4.

copper carbonate. A blue-green powder made by adding sodium carbonate and a copper sulfate solution. It occurs in nature as the mineral *malachite*. Density, 4.0 g/cm^3; melting point, decomposes at 200°C (392°F). Used as a green copper paint pigment, as a blue and green colorant in glazes, as a colorant to produce black coating on brass, in pyrotechnics, and for copper salts. Formula: $CuCO_3 \cdot Cu(OH)_2$. *Note:* The green pigment is also referred to as *Bremen green* and the blue pigment as *Bremen blue*. Also known as *artificial malachite; basic copper carbonate; cupric carbonate.*

copper carbonate hydroxide. See azurite.

copper casting alloys. A group of copper alloys produced by sand, continuous, centrifugal, die, investment, permanent-mold or plaster-mold casting, and including several coppers, high-copper alloys, red, semi-red and yellow brasses, silicon brasses and bronzes, tin and leaded tin bronzes, aluminum bronzes, copper nickels, and leaded coppers. Also known as *cast copper alloys.*

copper cement. See copper amalgam.

copper chloride. See cupric chloride; cuprous chloride.

copper chloride dihydrate. See cupric chloride dihydrate.

copper chromite. A black powder containing 82% cupric oxide and 17% chromic oxide. Formula: $CuCr_2O_4$. It is also available with a cupric oxide to chromic oxide ratio of 2:1 corresponding to the formula $2CuO \cdot Cr_2O_3$, and as a barium-promoted grade containing 62-64% copper chromite ($CuCr_2O_4$), 22-24% cupric oxide, 6-10% barium oxide, 1% chromic oxide, 1% chromium dioxide and 0-4% graphite. Used as a catalyst.

copper-chromium alloys. (1) A commercially pure copper coated first with nickel and then chromium. Used for stampings, display cases, refrigerators, and kitchen equipment.

(2) See chromium copper.

copper-chromium plates. See chromium-copper plates.

Copperclad. (1) Trade name of GTE Products Corporation (USA) for a 42% nickel steel core with pure copper sleeve. Used for sealing vacuum tubes.

(2) Trade name for electrosheet copper bonded to asbestos felt, and used for roofing applications.

(3) Trademark of Ferro Corporation (USA) for protective antifouling coatings applied to the bottom of ships and other marine vessels.

copper-clad aluminum. A rod, sheet, strip or tube of aluminum with a metallurgically bonded copper coating (e.g., by rolling) that forms a minimum of 10% of the total material thickness on one side, 5% on each side or, in the case of rods or tubes, a minimum of 10% of the cross-sectional area. Often sold under trademarks or trade names, such as *Alcuplate* or *Cupal*, it has good working properties, high electrical and thermal conductivities, and is used for formed and stamped parts, panels for railroad and subway cars, and for electrical conductors.

copper-clad steel. A low-carbon sheet steel that has copper bonded (or clad) to both sides. The copper cladding is equal to 10% of the total sheet thickness. Used for stamped, formed and drawn parts. See also copper-clad steel wire.

copper-clad steel wire. High-tensile steel wire having an uniform and continuous copper cladding bonded to its surface. It has good electrical conductivity, and is used for line and guy wires, and for screens. Abbreviation: CCSW.

copper-cobalt-beryllium alloy. An alloy (UNS C17500) containing 2.5% cobalt, 0.6% beryllium, and the balance copper. Supplied in the form of rods and flat products, it has a density of 8.75 g/cm³ (0.316 lb/in.³), a solidus temperature of 1030°C (1886°F), a liquidus temperature of 1070°C (1958°F), excellent cold workability, good to excellent corrosion resistance, and good formability. Used for electrical conductors, fasteners, fuse clips, springs, switch and relay components, and welding equipment.

copper cyanide. See cupric cyanide; cuprous cyanide.

coppered steel wire. Steel wire with a smooth, lustrous finish that has been dipped in a solution of copper sulfate or copper-tin sulfate, and subsequently finished by wet drawing. The copper or copper-tin plate provides a good bearing surface for the drawing die.

copper enamel. A porcelain enamel, usually with high thermal expansion, specifically intended for use as a decorative and protective coating on copper.

copper ferrite. A ferrimagnetic ceramic product with a cubic-spinel crystal structure (*ferrospinel*) produced from copper powder. It has outstanding magnetic properties at high frequencies, very high resistivity, and high corrosion resistance. Used as a soft-magnetic material for various applications. Formula: $CuFe_2O_4$.

copper ferrocyanide. A reddish-brown powder used in pigments for paints and enamels. Formula: $Cu_2Fe(CN)_6 \cdot 7H_2O$. Also known as *cupric ferrocyanide.*

Copper-Flo. Trade name of Johnson Matthey plc (UK) for a series of phosphorus-bearing silver brazing alloys containing about 90-94% copper. They have a melting range of 700-900°C (1290-1650°F).

copper fluoride. See cupric fluoride.

copper fluoride dihydrate. See cupric fluoride dihydrate.

copper foil. Copper beaten, rolled, hammered or otherwise made into a very thin sheet. Regular copper foil is available in purities from 99.0 to 99.999%, thicknesses from 0.0005 mm to 3.25 mm (0.00002 to 0.128 in.), and various tempers (hard, half hard, annealed, as rolled, etc.). It is often supported on aluminum foil that may be removed if required. Disk-shaped microfoil with a purity of 99.99+% in thicknesses ranging from 0.001 to 1.0 μm (0.04 to 39.4 μin.) is also available. Microfoil is often supplied on permanent polymer support.

copper formate tetrahydrate. A blue-green powder (97+% pure). Density, 1.831 g/cm³; melting point, decomposes at 130°C (266°F); dehydrated at 100°C (212°F) at reduced pressure over calcium chloride ($CaCl_2$); antiferroelectric below 235.5K; antiferromagnetic below 17K. Used in the production of superconducting yttrium barium copper oxide (YBaCuO) powders by spray drying. Formula: $Cu(CHO_2)_2 \cdot 4H_2O$.

copper gallium selenide. A compound of copper, gallium and

selenium. Crystal system, tetragonal. Crystal structure, chalcopyrite. Density, 5.56 g/cm³; hardness, 4200 Knoop; melting point, 1697°C (3087°F). Used as a direct semiconductor, and in solar cells. Formula: $CuGaSe_2$.

copper gallium sulfide. A compound of copper, gallium and sulfur. Crystal system, tetragonal. Crystal structure, chalcopyrite. Density, 4.35 g/cm³; melting point, 2027°C (3681°F). Used as a semiconductor. Formula: $CuGaS_2$.

copper gallium telluride. A compound of copper, gallium and tellurium. Crystal system, tetragonal. Crystal structure, chalcopyrite. Density, 5.99 g/cm³; hardness, 3500 Knoop; melting point, 2127°C (3861°F). Used as a semiconductor. Formula: $CuGaTe_2$.

copper germanium phosphide. A compound of copper, germanium and phosphorus. Density, 4.318 g/cm³; hardness, 8500 Knoop; melting point, 840°C (1544°F). Used as a semiconductor. Formula: $CuGe_2P_3$.

copper germanium selenide. A compound of copper, germanium and selenium. Density, 5.57 g/cm³; hardness, 3840 Knoop; melting point, 757°C (1395°F). Used as a semiconductor. Formula: Cu_2GeSe_3.

copper germanium sulfide. A compound of copper, germanium and sulfur available in high and low-temperature form. Density, 4.45-4.46 g/cm³; hardness, (high-temperature form) 4550 Knoop; melting point, (high-temperature form) 937°C (1719°F). Used as a semiconductor. Formula: Cu_2GeS_3.

copper germanium telluride. A compound of copper, germanium and tellurium. Crystal system, tetrahedral. Density, 5.92 g/cm³; hardness, 2890 Knoop; melting point, 757°C (1395°F). Used as a semiconductor. Formula: Cu_2GeSe_3.

copper glance. See chalcocite.

copper glycinate. Blue, triboluminescent crystals with a melting point of 130°C (266°F). Used in electroplating baths, as a catalyst in biochemistry, and in photometry. Formula: $Cu(NH_4C_2O_2)_2$. Also known as *copper aminoacetate; cupric glycinate.*

Copper Glo. Trade name of Alchem Corporation (USA) for a copper electroplate and plating process.

copper-graphite. A copper-base material that contains varying amounts of graphite, and is made by powder-metallurgy methods. Used as an electrical-contact material for starter brushes and auxiliary motor brushes in the automotive and related industries, and for friction components, e.g., clutches and brakes.

Copper Hardened. Trade name of Belmont Metals Inc. (USA) for a babbitt metal.

copper-hardened rolled zinc. A group of hot- or cold-rolled zinc alloys (UNS Z44330) hardened with 1% copper. They are supplied in sheet and plate form, and have good corrosion resistance, weldability and solderability. Their mechanical properties vary with the rolling direction.

copper hemioxide. See red copper oxide.

copper hexafluoroacetylacetonate. Moisture-sensitive, blue-green crystals that have a melting point of 85°C (185°F), and decompose at 220°C (428°F). A high-purity electronic grade (99.999%) is also available. Used in the chemical vapor deposition of superconductors, and in the deposition of copper thin films. Formula: $Cu(O_2F_6C_5H)_2$. Also known as *copper hexafluoropentanedionate.*

copper hexafluoroacetylacetonate hydrate. Green crystals with a melting point of 130-134°C (266-273°F) containing approximately 13% copper and 2-4% water. Used in the chemical vapor deposition of superconductor thin films. Formula: $Cu(O_2F_6C_5H)_2 \cdot xH_2O$. Also known as *copper hexafluoropentanedionate* *hydrate.*

copper hydrate. See copper hydroxide.

copper hydroxide. A blue, corrosive, hygroscopic powder that has a density of 3.368 g/cm³ and decomposes on melting. Used as a pigment, in staining paper, as a catalyst, in cuprammonium rayon, and in copper salts. Formula: $Cu(OH)_2$. Also known as *copper hydrate; cupric hydroxide; hydrated copper oxide.*

copper indium selenide. A compound of copper, indium and selenium. Crystal system, tetragonal. Crystal structure, chalcopyrite. Density, 5.77 g/cm³; hardness, 2050 Knoop; melting point, 1327°C (2421°F). Used as a direct semiconductor, and in solar cells. Formula: $CuInSe_3$.

copper indium sulfide. A compound of copper, indium and sulfur. Crystal system, tetragonal. Crystal structure, chalcopyrite. Density, 4.75 g/cm³; hardness, 2550 Knoop; melting point, 1127°C (2061°F). Used as a direct semiconductor, and in solar cells. Formula: $CuInS_3$.

copper indium telluride. A compound of copper, indium and tellurium. Crystal system, tetragonal. Crystal structure, chalcopyrite. Density, 6.1 g/cm³; hardness, 400 Knoop; melting point, 1387°C (2529°F). Used as a semiconductor. Formula: $CuInTe_3$.

copper-infiltrated steel. A high-density powder-metallurgy iron-carbon alloy containing about 0.3-1.0% carbon and 8-25% copper prepared by infiltrating a copper alloy into the porous steel matrix and sintering for a specified period of time at about 1120°C (2050°F) in an endothermic gas atmosphere. The final density is about 7.3 g/cm³ (0.26 lb/in.³). It is nearly fully dense, and has high strength, enhanced mechanical properties, and good machinability.

copper iodide. See cuprous iodide.

Copperior. Trade name of Becker Stahlwerk AG (Germany) for a corrosion-resistant steel containing 0.1% carbon, 0-0.2% copper, and the balance iron. Used for culvert pipes and roofing.

copper iron. An alloy (UNS No. C19200) containing 98.7% or more copper, 0.8-1.2% iron, 0-0.5% zinc, 0.01-0.04% phosphorus and up to 0.025% lead. Commercially available in the form of flat products and tubes, it has excellent cold and hot workability, good to excellent corrosion resistance especially stress corrosion, and good softening resistance. Used for electrical terminals and fuse clips, air-conditioner and heat-exchanger tubing, flexible hose, hydraulic brake lines, and thin gage sheet.

copper-iron alloys. See copper-bearing steel; copper-infiltrated steel; copper iron; sintered copper iron.

copper iron selenide. A compound of copper, iron and selenium. Crystal system, tetrahedral. Crystal structure, chalcopyrite. Melting point, 577°C (1071°F). Used as a semiconductor. Formula: $CuFeSe_2$.

copper iron sulfide. A metallic-yellow compound of copper, iron and sulfur. It occurs in nature as the mineral *chalcopyrite.* Crystal system, tetragonal. Density, 4.09 g/cm³; hardness, 3.5 Mohs. Used as a semiconductor. Formula: $CuFeS_2$.

copperized soft lead. An alloy (UNS L51125) containing 99.9+% lead, and the balance copper. It has a density of 11.34 g/cm³ (0.410 lb/in.³), excellent ductility and corrosion resistance, and is used for chemical process equipment, electroplating equipment and corrosion-resistant parts.

copper lactate. Dark blue or bluish-green crystals or granular powder used as source of copper in copper plating. Formula: $Cu(C_3H_5O_3)_2 \cdot 2H_2O$. Also known as *cupric lactate.*

copper lanthanum sulfide. A compound of copper, lanthanum and sulfur. Crystal system, tetragonal. Crystal structure, chalcopyrite. Used as a semiconductor. Formula: $CuLaS_2$.

copper lead. (1) Lead with a purity of 99.85% made by introducing copper into fully refined lead.

(2) Any of a group of copper-base bearing metals that contain 5-50% lead, and are available in the sintered, cast and continuously cast condition. They have good fatigue strength, high load capacities, good high-temperature performance, a maximum service temperature of 177°C (350°F), and are used for heavy-duty main and connecting-rod bearings, turbine and electric-motor bearings, and aircraft and automotive components. Also known as *copper-lead alloys.*

copper-lead alloys. See copper lead; high-leaded bronzes; high-leaded brasses.

Copperlok. Trademark of Copperlok, Inc. (USA) for a resin bond coat containing hollow microballoons that selectively fracture to provide undercuts and cavities which increase the adhesion of thermal-spray coatings. Used for printed circuits, EMI/RFI shielding, and architectural facades.

Copperlume. Trade name of Atotech USA Inc. for a copper electroplate and plating process.

copper-manganese. A series of master alloys that contain 25-30% manganese, and the balance copper. They are commercially available in the form of notched slabs, or shot. Grades made of *ferromanganese* contain up to 5% iron, and those made of manganese metal are free of iron and/or carbon. A 70Cu-30Mn has a melting point of about 870°C (1600°F). Used for deoxidizing and hardening brasses, bronzes, nickel silvers, copper nickels, and nickel and its alloys. Also known as *manganese-copper.*

copper-manganese alloys. (1) A group of electrical resistance alloys containing about 30-90% copper, 10-40% manganese, and 0-35% nickel. They have high electrical resistivity, and are used for resistances, resistance wire, and control equipment and measuring instruments.

(2) A group of heat- and corrosion-resistant alloys containing 29-90% copper, 8.5-52% manganese, 0-10% iron, 0-6.25% aluminum, 0-3.3% carbon, 0-1.1% silicon, 0-2.1% zinc, 0-0.4% tin, and 0-0.5% lead. Used for resistances, and corrosion- and heat-resistant parts.

(3) A group of copper-based sound-damping alloys having high percentages of manganese. Used for power tools, machinery, jackhammers, etc.

copper-manganese-lithium. A series of master alloys containing 60-70% copper, 27-30% manganese, 0.5-5.0% lithium, and 0-7% calcium. Used in the manufacture of steel, cast iron and nickel.

copper-mercury iodide. See mercuric cuprous iodide.

copper metaborate. A blue-green crystalline powder obtained by the action of copper sulfate and sodium borate. It has a density of 3.86 g/cm³, and is used as a pigment in inks for painting on porcelain and other ceramics, as a paint pigment, as a wood preservative, and as a fire retardant. Formula: $Cu(BO_2)_2$. Also known as *copper borate; cupric borate.*

copper mica. See chalcophyllite.

copper molybdate. High-purity crystals with a density of 3.4 g/cm³, and a melting point of approximately 500°C (930°F). Used as a paint pigment, in protective coatings, as a corrosion inhibitor, and in electronic and optical equipment. Formula: $CuMoO_4$.

copper-molybdenum alloys. A group of alloys containing about 30-50% molybdenum, and the balance copper. They have excellent thermal conductivity, low thermal expansion, and good machinability and platability. Used for electronic components, such as heat sinks, circuit board cores, etc.

copper monoxide. See black copper oxide.

copper nickel. See nickeline.

copper-nickel alloys. (1) A group of electrical resistance alloys including (i) *Radio alloys* containing about 75-98% copper and 2-25% nickel; and (ii) *Constantan* alloys containing 45-60% copper and 40-55% nickel.

(2) A group of master alloys containing either 70% copper and 25% nickel or 50% copper and 50% nickel, sometimes with small additions of other elements (zinc, iron, etc.). They are commercially available in the form of ingots, slabs and shot. A 70Cu-30Ni alloy melts at approximately 1195°C (2185°F), and a 50Cu-50Ni copper alloy at approximately 1270°C (2320°F). Used to introduce nickel into brasses, and in the manufacture of ferrous alloys.

(3) See copper nickels; nickel silvers.

copper-nickel-chromium plates. See chromium-copper-nickel plates.

copper-nickel-phosphorus alloy. A high-strength alloy (UNS C19000) containing 1.1% nickel, 0.25% phosphorus, and the balance copper. Fabricated by coining, drawing, upsetting, hot forging, pressing, spinning, swaging or stamping, it is supplied as rods, wires and flats, and possesses excellent cold and hot workability, good to excellent corrosion resistance, good forgeability, high electrical and thermal conductivity, high creep and fatigue resistance, and moderate machinability. Used for bolts, screws, cotter pins, nails, springs, clips, electrical components, and power- and electron-tube parts.

copper-nickel-phosphorus-tellurium alloy. A high-strength alloy (UNS C19100) containing 1.1% nickel, 0.50% tellurium, 0.25% phosphorus, and the balance copper. Supplied as flat products, it has a density of 8.78 g/cm³ (0.317 lb/in.³), good cold and hot workability, good to excellent corrosion resistance, high hardenability, good conductivity, and good machinability. Used for screw-machine parts, forgings, bolts, gears, pinions, bushing, nuts, tie rods, welding-torch tips, marine hardware, and electrical connectors.

copper-nickel powder. A powder composed of 70-90% copper and 10-30% nickel and available in various particles sizes. Used prealloyed in powder metallurgy, for the manufacture of coatings, etc.

copper nickels. A group of corrosion-resistant alloys of 69.5-88.7% copper, 5-30% nickel, 0-1.4% iron, and sometimes small amounts of beryllium, zirconium, and/or aluminum. They are commercially available in the form of flat products, rods, tubes, wire and powder. *Copper nickels* have a pinkish-white color, a density range of 8.6-8.95 g/cm³ (0.310-0.323 lb/in.³), a melting range of about 1100-1200°C (2010-2190°F), a hardness of 2-3 Mohs, high ductility and malleability, excellent corrosion resistance also to seawater, good hot and cold workability, and a good combination of strength and toughness. Used for condensers, condenser plates, condenser and heat-exchanger tubes, distiller and evaporator tubing, valve and pump components, ferrules, electrical springs, resistance wire, resistors, corrosion-resistant equipment, communication relays, coinage, and bullet jackets. Also known as *cupronickels*.

copper-nickel-titanium alloy. A heat-treatable copper alloy containing 5% nickel and 2.5% titanium. It has good high-temperature properties, and an electrical conductivity at room temperature about half that of copper. Used for electronic applications, current-carrying springs, heat sinks, switch parts, and radar components.

copper nitrate. See copper nitrate hexahydrate; copper nitrate

hydrate; copper nitrate trihydrate.

copper nitrate hexahydrate. Blue, oxidizing, deliquescent crystals. Density, 2.074 g/cm³; melting point, loses $3H_2O$ at 26.4°C (79.5°F). Used in light-sensitive papers, in electroplating, in coloring copper black, in paints, varnishes and enamels, and in the production of burnished effect on iron. Formula: $Cu(NO_3)_2 \cdot 6H_2O$. Also known as *cupric nitrate hexahydrate.*

copper nitrate hydrate. Blue, oxidizing crystals, or a blue 1M aqueous solution. It is also supplied in high purity (99.999%). The degree of hydration may vary from 2-5 H_2O. Used as a starting material for coprecipitation methods of generating precursors to superconducting powders. Formula: $Cu(NO_3)_2 \cdot xH_2O$. Also known as *cupric nitrate hydrate.*

copper nitrate trihydrate. Blue, oxidizing, deliquescent crystals (99+% pure). Density, 2.32 g/cm³; melting point, 114.5°C (238°F); boiling point, decomposes at 170°C (338°F). Used in light-sensitive papers, in electroplating, in coloring copper black, in paints, varnishes and enamels, and in the production of burnished effect on iron. Formula: $Cu(NO_3)_2 \cdot 3H_2O$. Also known as *cupric nitrate trihydrate.*

Copperoid. (1) Trade name of Youngstown Steel (USA) for a carbon steel containing at least 0.2% copper. It has good resistance to rusting and atmospheric corrosion, and is used as a sheet material for building and general fabrication.

(2) Trade name of American Nickeloid Company (USA) for a copper-coated zinc used for stamped, drawn and formed parts.

copper oleate. A brown powder or greenish-blue lumps made by the action of copper sulfate and sodium oleate. Used in ore flotation, in lube oils, and as a fuel-oil ignition improver. Formula: $Cu(C_{18}H_{33}O_2)_2$. Also known as *cupric oleate.*

copper orthoarsenite. See copper arsenite.

copper oxalate. A bluish-green powder that decomposes at approximately 300°C (570°F), and is used as a catalyst in organic and organometallic synthesis. Formula: CuC_2O_4. Also known as *cupric oxalate.*

copper oxide. See black copper oxide; red copper oxide.

copper phenolsulfonate. Blue-green crystals used in electroplating. Formula: $Cu[C_6H_4(OH)SO_3]_2 \cdot 6H_2O$. Also known as *copper sulfocarbolate.*

copper phosphide. See cupric phosphide; cuprous phosphide.

copper phosphorus sulfide. A compound of copper, phosphorus and sulfur. Hardness, 3 Mohs. Used as a semiconductor. Formula: Cu_3PS_4.

copper phthalocyanine blue. A purple phtalocyanine powder (Color Index C.I. 74160) that contains a divalent central copper ion (Cu^{2+}). It is made by heating phthalonitrile with cuprous chloride, and has a typically dye content of about 90-97% and a maximum absorption wavelength of 694 nm. Used as a bright-blue pigment in paints, lacquers, inks, paper, rubber, synthetic resins, colored chalks and pencils, in tinplate printing, and in the manufacture of blue cotton. Formula: $CuC_{32}H_{16}N_8$. Abbreviation: CuPc. Also known as *copper phthalocyanine; Pigment Blue 15.*

copper phthalocyanine green. A bright green pigment (Color Index C.I. 74260) made by heating phthalocyanine in sulfur dichloride under pressure. Used as a pigment in paints, lacquers, inks, leather and book cloth, in surfacing paper, in colored chalks and pencils, and in tinplate printing. Also known as *Pigment Green 7.*

copper phthalocyanine pigments. See copper phthalocyanine blue; copper phthalocyanine green.

copper plate. An electrodeposit of copper frequently used as an undercoating in multiple-plate systems (e.g., copper-nickel-chromium plating), in electroforming, as resists for heat transfer, and in plating printed-circuit boards.

copperplate printing paper. A rag-containing or wood-free paper, usually with high softness and absorptive capacity, having a special surface texture that makes it suitable for copperplate printing.

Copperply. Trademark of National Standard Company (USA) for an annealed or hard-drawn high-tensile steel wire with an electrodeposit of 5-10 wt% copper. Used for electrical applications.

copper-potassium cyanide. See potassium-copper cyanide.

copper powder. Copper in the form of a powder supplied in various purities, particle shapes and sizes. The high-purity grade contains 99.99-99.999% and the commercially pure grade 99+% copper. The particle shapes range from spheres and dendrites to flakes, and the particle sizes from less than 1 μm (40 μin.) to over 800 μm (0.03 in.). Flaked copper powder is made by electrolysis, and used as paint and ink pigment, and as an insulation for liquid fuels. Noncrystalline copper powder is a bronzing powder made by chemical reduction, and frequently used as a liquid vehicle for copper coating, and for powder-metallurgy parts (e.g., bearings, filters, and electrical and friction products).

copper protoxide. See red copper oxide.

copper pyrite. See chalcopyrite.

copper resinate. A green powder that is obtained from copper sulfate and rosin oil, and is used in antifouling paints for metals. Formula: $Cu(C_{20}H_{29}O_2)_2$. Also known as *cupric resinate.*

copper-rich brass. A commercial bronze containing 90% copper and 10% zinc. It has good to excellent corrosion resistance, high electrical conductivity, excellent cold workability, and good hot workability. Used for window-screen wire, screen cloth, grillwork, and ornamental applications. Also known as *rich gold metal.*

copper ruby glass. A ruby glass made by adding copper or copper oxide to the melt. When melted in a reducing atmosphere, the addition effects the formation of colloidal copper resulting in a characteristic red color with a slight greenish tinge. Used for decorative purposes, taillights, etc.

copper selenate. Bluish crystals that have a density of 2.56 g/cm³, lose $4H_2O$ at 50-100°C (122-212°F), and are used in coloring copper black. Formula: $CuSeO_4 \cdot 5H_2O$. Also known as *cupric selenate.*

copper selenide. See cupric selenide; cuprous selenide.

Copper's Gold. British trade name for a corrosion-resistant alloy of 67-81% copper, 19-30% platinum and 0-4% zinc. Used for ornaments and jewelry.

copper shot. Copper in the form of small spherical particles of varying diameter. Ultrahigh purity grades (99.9995%) are supplied in particles sizes from 2 to 4 mm (0.08 to 0.16 in.), high-purity grades (99.99+%) in particle sizes from 2 to 10 mm (0.08 to 0.4 in.) and commercially pure grades (99+%) in particle sizes from -3 to +14 mesh. Used as alloying addition in gold and silver.

copper silicide. See silicon-copper (2).

copper-silicon alloys. A group of copper-base alloys containing varying amounts of silicon and including the following: (i) *Copper silicide* (or silicon-copper) used in the manufacture of silicon bronze; (ii) *Silicon bronze,* a wrought or cast copper-base alloy that contains up to 4% silicon, and is used for electrical applications, welding rods for welding copper to steel, fasteners, hardware, castings, bearings and pumps; (iii) *Silicon brass,* a wrought or cast copper-base alloy that contains 14-34% zinc and 1-4% silicon, and is used for castings, bearings and valves; and (iv) *Silicon copper* that contains 98.75% copper, 0.75% tin, 0.3% silicon and 0.2% manganese, and is used for welding rod and wire for copper and its alloys.

copper-silicon-lithium. A master alloy containing 80-85% copper, 10-11% silicon, 2.5-10% lithium and 0-2.5% calcium. Used in the treatment of steel, cast iron and nickel.

copper silicon phosphide. A compound of copper, silicon and phosphorus. Crystal system, tetrahedral. Used as a semiconductor. Formula: $CuSi_2P_3$.

copper silicon sulfide. A compound of copper, silicon and sulfur. Crystal system, tetrahedral. Melting point, 927°C (1701°F); density, 3.6-3.8 g/cm³; hardness, 3.5 Mohs. Used as a semiconductor. Formula: $CuSiSe_2$.

copper silicon telluride. A compound of copper, silicon and tellurium. Crystal system, tetrahedral. Density, 5.47. Used as a semiconductor. Formula: Cu_2SiTe_3.

copper-silver alloys. Copper-base alloys containing varying amounts of silver, and including in particular: (i) *Silver copper* containing up to 0.11% manganese, 0.06% phosphorus and 0.034% silver, respectively, used for electronic components, electrical contacts, fasteners, fittings, and resistance welding electrodes; (ii) *Copper-silver contact alloys* containing varying amounts of silver and copper, and having good electrical properties and relatively high hardness; and (iii) *Copper-silver conductor alloys* that contain several percent of silver, and are used for high-strength good-conductivity copper wire.

copper silver iodide. A yellow compound that is commercially available as a powder, and also occurs in nature as the mineral *miersite.* Crystal system, cubic. Density, 5.45-5.64 g/cm³; refractive index, 2.2. Used as a semiconductor. Formula: (Ag,Cu)I.

copper-silver wire. High-strength wire containing about 92-95% copper and 5-8% silver, and having an electrical conductivity of about 70-85% IACS. Used for electrical conductor cables.

Copper's Mirror. British trade name for a corrosion-resistant alloy of 58% copper, 28% tin, 9.5% platinum, 3.5% zinc and 1.5% arsenic. Used for mirrors and reflectors.

coppersmiths' copper. Heavy, hot-rolled copper sheets in soft temper used by coppersmiths. Also known as *braziers' copper.*

copper soap. A water-insoluble soap of copper and oleic, linoleic or linolenic acid used in oil paints to catalyze the absorption of oxygen and facilitate paint or coating drying, and in antifouling paints to prevent algae, fungus or mold growth.

copper-sodium cyanide. See sodium-copper cyanide.

Copper's Pen Metal. British trade name for corrosion-resistant alloys for pen points that contain: (i) 50% copper, 25% silver and 25% gold, or (ii) 50% platinum, 38% silver and 12% copper.

copper sponge. A copper-bearing material with a porous, sponge-like structure produced by molding and sintering copper powder and a volatile organic compound, and then saturating the resulting intermediate with lead to provide self-lubricating characteristics. See also sintered copper, powder-metallurgy copper.

copper steel. See copper-bearing steel; copper-clad steel; copper-infiltrated steel.

copper stearate. Bluish powder made from copper sulfate and sodium stearate. It has a melting point of 125°C (257°F), and is used in antifouling paints, as a preservative for cellulose derivatives, and as a catalyst. Formula: $Cu(C_{18}H_{35}O_2)_2$. Also known as *cupric stearate*.

copper subacetate. See basic copper acetate.

copper suboxide. See red copper oxide.

copper sulfate. A green to greenish-white powder (98+% pure). Density, 3.603 g/cm³; melting point, decomposes at 200°C (392°F). Used in textile dyeing, copper plating baths, in electroforming, and as a fungicide. Formula: $CuSO_4$. Also known as *cupric sulfate*.

copper sulfate pentahydrate. Sky-blue crystals, or blue, crystalline granules or powder (98% pure) made by treating copper or copper oxide with dilute sulfuric acid. It occurs in nature as the mineral *chalcanthite*. Density, 2.284 g/cm³; melting point, loses $4H_2O$ at 110°C (230°F); boiling point, loses $5H_2O$ at 150°C (300°F). Used as a source of copper, in the manufacture of copper salts, as a colorant in production of ruby-red glass, in textile dyeing, leather tanning and pigments, in electroplating, as a wood preservative, in steel manufacture, in the treatment of natural asphalts, in synthetic rubber, and in electric batteries. Formula: $CuSO_4 \cdot 5H_2O$. Also known as *blue copperas; cupric sulfate pentahydrate*.

copper sulfide. See cupric sulfide; cuprous sulfide.

copper sulfocarbolate. See copper phenolsulfonate.

copper sulfocyanide. See cuprous thiocyanate.

copper telluride. See cuprous telluride.

copper tetraiodomercurate. See mercuric cuprous iodide.

copper thallium selenide. A compound of copper, thallium and selenium. Crystal system, tetragonal. Crystal structure, chalcopyrite. Density, 7.11 g/cm³; melting point, 627°C (1161°F). Used as a semiconductor. Formula: $CuTlSe_2$.

copper thallium sulfide. A compound of copper, thallium and sulfur. Crystal system, tetragonal. Crystal structure, chalcopyrite. Density, 6.32 g/cm³. Used as a semiconductor. Formula: $CuTlS_2$.

copper thiocyanate. See cuprous thiocyanate.

Copper Tin. Trade name of American Nickeloid Company (USA) for copper-coated tin used for stamped, formed and drawn parts.

copper-tin alloys. A group of alloys in which copper and tin are the predominant elements including (i) *Tin brasses* containing 60-95% copper, 2-38% zinc and 0.25-2.5% tin; (ii) *Tin bronzes* containing 80-93% copper, 7-20% tin and 0-4% zinc; and (iii) *Phosphor bronzes* containing 89-99% copper, 1-10% tin, and trace amounts of phosphorus. *Copper-tin alloys* have high tensile and fatigue strengths, good corrosion resistance also to seawater, good wear resistance, and good antifriction and antiscoring properties. Used for gears, bearings, bushings, springs, bellows, valves and valve seats, water gages, flow indicators, drain cocks, welding rods, spark-resistant tools, fourdrinier wire, paint, electrical hardware, vacuum dryers, blenders, and in the fine arts. See also tin brass; tin bronze; phosphor bronze.

copper-tin plate. A coating of tin applied over an undercoating of copper and produced by electrodeposition. The final tin plate must be heat-flowed in hot oil. Used on magnesium alloys for corrosion protection, and in electronics to improve the solderability of magnesium alloys.

copper tin selenide. A compound of copper, tin and selenium. Crystal system, tetrahedral. Density, 5.94 g/cm³; melting point, 687°C (1269°F); hardness, 2510 Knoop. Used as a semiconductor. Formula: Cu_2SnSe_3.

copper tin sulfide. A compound of copper, tin and sulfur. Crystal system, tetrahedral. Density, 5.02 g/cm³; hardness, 2770 Knoop; melting point, 837°C (1539°F). Used as a semiconductor. Formula: Cu_2SnS_3.

copper tin telluride. A compound of copper, tin and tellurium. Crystal system, tetrahedral. Density, 6.51 g/cm³, hardness, 1970 Knoop; melting point, 407°C (765°F). Used as a semiconductor. Formula: Cu_2SnTe_3.

copper silicon sulfide. A compound of copper, tin, and tellurium. Crystal system, tetrahedral. Density, 6.51 g/cm³; melting point, 407°C (765°F); hardness, 1970 Knoop. Used as a semiconductor. Formula: Cu_2SnTe_2.

copper titanate. A gray powder used as a semiconductor, and as a small addition to barium titanate ($BaTiO_3$) to promote the fired density. Formula: $CuTiO_3$.

copper-titanium alloys. A group of alloys in which copper and titanium are the predominant elements. Examples of include (i) *Electrical contact alloys* with 95.7% copper and 4.3% titanium for nonsparking tools, diaphragms, electrical contacts and springs, and (ii) *Titanium bronzes* with 25-90% copper and 10-75% titanium.

copper tungstate. The copper salt of tungstic acid available in anhydrous form as a brown powder, and in form of the dihydrate as a light-green powder. Used in semiconductors, nuclear reactors, and as a catalyst for polyesters. Formula: $CuWO_4$ (anhydrous); $CuWO_4 \cdot 2H_2O$ (dihydrate). Also known as *cupric tungstate*.

copper uranite. See torbernite.

copper vitriol. See chalcanthite; copper sulfate pentahydrate.

Copperweld. Trade name of British Insulated Callender's Cables (UK) for a wire consisting of a steel core with a copper covering permanently welded to it. It is commercially available in three grades: (i) extra-high strength with an electrical conductivity of 40% IACS; (ii) high-strength with an electrical conductivity of 30% IACS; and (iii) high-strength with an electrical conductivity of 40% IACS. *Copperweld* wire has high tensile strength, good resistance to electrogalvanic action, good thermal-shock resistance, good conductivity, good formability by bending, twisting, hot rolling, cold drawing, and forging. Used as electrical conductor wire for high-voltage transmission spans, and as telephone, telegraph and signal wire.

copper wire. Copper supplied in wire form with different purities, diameters, lengths and hardnesses. Commercially pure wire contains 99.9+% and ultrahigh-purity wire 99.999% copper. Wire diameters range from 0.01 to 2.0 mm (0.4 to 79 mils), and lengths, depending on purity and diameter, from 0.1 to 2000 m (4 to 78740 in.). It is made in three hardness grades: hard-drawn, medium-hard-drawn and soft (or annealed). Insulated copper wire in purities from 99.9 to 99.99+% is also available and consists of a copper conductor, 0.01-0.1 mm (0.4 to 4 mils) in diameter, with a thin special coating of polyurethane or trimel (typically 0.005-0.015 mm or 0.2-0.6 mil thick). Insulated wire is sold in lengths from 1 to 250 m (39 to 9843 in.). Used for electric conductors, transmission lines, trolley wires, telephone wires, and various other wire products.

copper wool. Fine copper threads or shavings made into a mat or pad for use in cleaning operations.

copper-zinc alloys. A group of alloys in which copper and zinc are the predominant elements, although small amounts of other metals, such as lead, iron, manganese, aluminum, nickel, silicon, mercury and tin are often present. The concentration of copper varies according to use from 55-95% and the color ranges from bright yellow for high-copper brasses to reddish for high-zinc brasses. Some common brasses are yellow brass (about 35% zinc), red brass (about 15-25% zinc), naval brass (about 37-39% zinc), cartridge brass (about 30% zinc), muntz metal (about 40% zinc) and gilding metal (5% zinc). *Copper-zinc alloys* have good castability and machinability, good corrosion resistance and work-hardening properties, good hot workability, good to medium cold workability, and good to medium strength. Used for condenser-tube plates, water-heater and evaporator tubes, automotive radiators, housings, fittings, fasteners, locks, piping, hose nozzles and couplings, valves, drain cocks, air cocks, marine equipment, oil gauges, flow indicators, electrical components, plating anodes, worm wheels, bearings, bushings, deep-drawn parts, sheets, foil, wire, powder, springs, musical instruments, jewelry, fine arts, architectural applications, clocks, stamping dies, ammunition components, cartridge casings, coins, and solders. See also brass; alpha brass; beta brass.

copper-zirconium alloys. (1) Amorphous alloy powders of copper and zirconium made by rapid solidification techniques. Some contain considerable amounts of iron, phosphorus and other elements. A powder of 70% copper and 30% zirconium has a melting point of about 1140°C (2085°F). Copper-zirconium alloy powders are used for powder-metallurgy parts with improved mechanical properties and corrosion resistance.

(2) A master alloy of 65-87.5% copper and 12.5-35% zirconium used to introduce zirconium into bronzes and brasses.

(3) See zirconium copper.

Coppralyte. Trademark of E.I DuPont de Nemours & Company (USA) for copper electroplating products.

Coppro. Trade name of Copperweld Steel Company (USA) for a series of nitrided steels containing 0.2-0.5% carbon, 0.9-1.8% chromium, 0.15-0.45% molybdenum, 0.85-1.2% aluminum, and the balance iron. Used for crankshafts, crank pins, fittings, cylinders, etc.

Coprex. Trade name of Wakefield Corporation (USA) for a sintered copper alloy used for bearings.

Cop-R-Loy. Trade name of Wheeling-Pittsburgh Steel Corporation (USA) for corrosion-resistant ingot iron and low-carbon steel with small percentages of copper. Used for roofing, wire fencing, piping, boiler tubes, and structural parts.

Copr-Trode. Trade name of Ampco Metal Inc. (USA) for a weld metal containing 98.0+% copper and up to 1% tin, 0.5% manganese, 0.5% silicon, 0.15% phosphorus and 0.50% other elements, respectively. Used for gas metal-arc and gas tungsten-arc welding of deoxidized copper castings.

Cop-Sil-Loy. (1) Trade name of Cop-Sil-Loy Inc. (USA) for an alloy of copper, silver and lead. It has a low coefficient of friction, and is used for wear reduction applications on friction surfaces.

(2) Trade name of Cop-Sil-Loy Inc. (USA) for a sintered alloy containing 47-53% lead and 47-53% copper. Used for special parts.

copy paper. A white or colored typewriter paper, usually wood-free, hard-sized and with a non-calendered surface, having a basis weight of about 30-40 g/m² (0.9-1.2 oz/yd²).

copying carbon paper. See self-copying paper.

Copyplast. Trade name of Scheu-Dental (Germany) for a tough elastic material supplied in round and square blanks. Used in dentistry for molds for temporary crowns and bridges, bleaching and bracket transfer trays, and model duplications.

Coquilles. French trade name for rigid sections of fine glass fibers, impregnated with silicon resin. Used as thermal insulation for piping up to 300°C (570°F).

coquimbite. A transparent, colorless or white mineral composed of iron sulfate nonahydrate, $Fe_2(SO_4)_3 \cdot 9H_2O$. Crystal system, hexagonal. Density, 2.11 g/cm³; refractive index, 1.54-1.55. Occurrence: USA (California), Hungary.

coquina. A coarse-textured, porous, friable limestone composed essentially of sea shells, including the small coquina shell of the genus *Donax*, and shell and/or coral fragments cemented together by calcite. Used as a calcareous raw material.

Cora. (1) Trade name of Rheinische Schmirgelwerke GmbH (Germany) for alumina-based abrasive paper and cloth, and grinding wheels.

(2) Trade name of Rhône-Poulenc SA (France) for nylon 6 fibers and yarns used for apparel.

Coradent. Trade name for a dental composite paste for core-build-ups.

Coral. Trademark of Firestone Synthetic Rubber and Latex Company (USA) for a stereospecific *cis*-1,4-polyisoprene rubber used as a substitute for natural rubber.

coral. The hard, porous, skeletal material of certain marine polyps (class Anthozoa) especially those of the genus *Corallium* that form coral reefs and islands, and exist in white, red, pink and black varieties. *Coral* is composed essentially of calcium carbonate. Occurrence: Eastern Australian coast, southwestern Pacific Ocean, Indian Ocean, and coastal areas of Northeastern Africa. Used for building stone, cement, road and airfield surfacing, and in biotechnology. The pink and red varieties are used for jewelry, beads, buckles, etc.

Coralite. Trade name for a vulcanized rubber formerly used for denture bases.

Coralli. Italian trade name for patterned glass.

coral limestone. Limestone obtained from the skeletons of any of several marine polyps (genus *Corallium*) that form coral reefs and islands. Used in building and construction.

Coralox. Trade name for fused aluminum oxide.

coral red. A low-temperature color produced in glazes and porcelain enamel by lead chromate.

Corax. Trade name for a series of carbon blacks.

Coray. Trademark of Wonjin Rayon Company (Japan) for viscose rayon fibers and yarns.

Corbin. British trade name for a non-hardenable aluminum alloy containing 12.5% copper. Used for pistons.

Corblanit. Trademark for a series of polystyrene foams and foam products.

Corcel. Trademark of Interpace Corporation (USA) for cellular glass microspheres used as fillers in plastics, concrete, etc.

Cor-Cote. Trade name of General Polymers Corporation (USA) for corrosion-resistant coatings.

cord. (1) A long, heavy, flexible, relatively thick string made by braiding, twisting, knitting, plaiting or weaving together several strands of natural or synthetic fibers.

(2) A term used in the automotive industries for a yarn made by twisting together two or more plied (or folded) yarns, and used for the reinforcement of automotive tires. See also plied yarn.

cordage. A general term used for fibrous materials made into ropes, cords, cables, twine, etc., by braiding, twisting, plaiting or weaving together two or more strands. *Cordage* is used for towing, power transmission, baling, tying and packing applications.

Cordamey. Trade name for rayon fibers and yarns used for textile fabrics.

corded fabrics. Fabrics with one or more surface ribs or ridges, usually produced by using heavier and finer yarns.

corded yarn. A yarn produced by twisting together two or more fine yarns.

Cordelan. Trademark of Kohjin Company (Japan) for polychlal bicomponent fibers. See also polychlal fibers.

Cordenka. Trademark of Acordis Group for a high-tenacity rayon filament yarn with high, permanent dimensional stability and excellent chemical resistance and thermal stability. Used as a reinforcing material for tires, hoses, twines, and strapping.

corderoite. A faint greenish white mineral composed of mercury chloride sulfide, α-$Hg_3S_2Cl_2$. It can also be made synthetically. Crystal system, cubic. Density, 6.89 g/cm^3. Occurrence: USA (Nevada).

Cordfil. Trade name of Cordfil (Canada) for cotton cord chopped into short lengths, and used as a filler in molding compounds for reinforced plastics.

cordierite. A grayish blue, lilac-blue or dark-blue, transparent to translucent mineral with a vitreous luster. It is composed of magnesium aluminosilicate, $Mg_2Al_4Si_5O_{18}$, and sometimes contains small amounts of iron. Crystal system, orthorhombic. Density, 2.10-2.67 g/cm^3; hardness, 7-7.5 Mohs; refractive index, 1.549. It is commercially available as a high-purity powder with good dielectric and thermal properties, and can also be made synthetically from stoichiometric mixtures of the oxides. Occurrence: Finland, Germany, Greenland, Madagascar, Norway, Sri Lanka, USA (Connecticut, New Hampshire). Used as an insulating material, in electronic ceramics, stoneware and porcelain and vitreous-china bodies to enhance the thermal-shock resistance, for refractory electronic parts, as a composite material, as a substrate in the packaging industries, and as a gemstone.

cordierite ceramics. A group of ceramic whitewares in which the mineral *cordierite* is the essential crystalline phase. They possess low coefficients of thermal expansion and superior thermal-shock resistance, and are used for thermocouple insulators, burner tips, and heating elements. Also known as *cordierite whiteware*.

cordierite porcelain. A hard-fired, vitrified *cordierite* whiteware used in the manufacture of spark plugs and electrical insulators.

cordierite whiteware. See cordierite ceramics.

Cordino. Trade name for a fine wale combed cotton fabric resembling *Bedford cloth* and having excellent wash-and-wear properties.

Cordley. Trade name of Teijin Fibers Limited (Japan) for artificial leather based on a nonwoven fabric made from Tetoron polyester stable fibers and polyurethane. Used for apparel, sportswear and interior furnishings.

Cordoglas. Trade name for a series of glass-fiber materials coated with thermoplastic resins or elastomeric materials.

Cordonsiv. Italian trade name for glass-fiber cords used for insulating tubing and ducts of heating installations.

cordon yarn. A yarn made by twisting together a cotton yarn and a wool or worsted yarn.

Cordopreg. Trade name for glass-fiber and other reinforcements that have been preimpregnated with stabilized thermosetting resins.

cordovan. A soft, smooth, durable leather made from split horsehide. It is finely porous, vegetable-tanned and has a distinctive waxy finish. It was originally made of goatskin from Cordoba, Spain. Used for shoes, boots, coats, jackets, and fancy leather articles.

Cordtex. Trademark of Paris Brick Company Limited (Canada) for calcite bricks.

Cordulla. Trade name for rayon fibers and yarns used for textile fabrics.

Cordura. Trademark of E.I. DuPont de Nemours & Company (USA) for abrasion-resistant, durable, lightweight, air-textured high-tenacity nylon 6,6 fibers available in 160, 330, 500 and 1000 denier grades. Used in the manufacture of luggage, business and computer cases, backpacks, sport bags, boots, sports apparel, workwear, uniforms, webbing, etc. It was formerly also used for tire cord. Also included under this trademark are a series of rayon fibers and yarns.

Corduron. Trade name of The Duffy Company (USA) for a paste used in selective case carburizing to cover those regions of a workpiece that should not be carburized.

corduroy. A strong, durable cotton, cotton-blend, wool or other fabric in a plain or twill weave and with a thick velvety cut pile in variably spaced ridges. Used chiefly for clothing and upholstery.

cordylite. A pale yellow mineral of the bastnaesite group composed of barium cerium fluoride carbonate, $BaCe_2(CO_3)_3F_2$. Crystal system, hexagonal. Density, 4.07 g/cm^3; refractive index, 1.773. Canada (Quebec).

core binder. A substance used to hold the grains of a foundry *core sand* together, e.g., flour, pitch and rosin compositions, dextrin, raw linseed oil, commercial core oils, and aqueous solutions of resins, such as urea-formaldehyde.

Corebond. Trade name of E.I. DuPont de Nemours & Company (USA) for polyester fibers used for textile fabrics.

Core-Cell Foam. Trademark of ATC Chemicals Inc. (USA) for stiff, linear styrene-acrylonitrile-based foam materials with excellent thermal properties. Used as core materials for sandwich panels, e.g., in boatbuilding.

cored brick. A brick that is at least three-quarters solid in any plane parallel to the load-bearing surface.

cored solder. A solder wire or bar with a core of flux.

core filler. A substance, such as coke, cinder or sawdust, used as a substitute for sand in large foundry cores. It normally facilitates core collapsibility.

Core-Flo. Trademark of BISCO Inc. (USA) for a fast-setting, self-cure resin with a creamy consistency and high radiopacity. Supplied in natural and white shades, it is used in restorative dentistry for core-build-ups.

Coreform. Trade name for a dental composite paste for core-build-ups.

Corega. Trademark of Block Drug Company, Inc. (USA) for a denture adhesive.

core iron. Soft iron with high magnetic permeability, usually supplied in sheet form, and used for laminated cores of electromagnets, chokes, transformers and relays. It typically contains about 0.06% carbon, up to 1% vanadium, and the balance iron.

Coreline. Trademark of Rubico Inc. (Canada) for spandex and other elastic yarns.

Corelite. Trade name for a dental composite paste for core-build-ups.

Corelox. Trademark of Dofasco Inc. (Canada) for cold-rolled steel coils, sheets and strip.

Coremat. Trademark Lantor NL (Netherlands) for a series of honeycomb structural materials that contain microspheres arranged in a hexagonal pattern. They are made into flexible mats with excellent drapability, and used for boatbuilding and in composite reinforcement.

Corepal. Trademark of Bayer AG (Germany) for a series of alkyd resins.

core paper. A paper, usually a waste paper grade, used in the manufacture of core tubes for paper, tissue, textiles, etc. Also known as *core tube paper.*

Core Paste. Trade name of Den-Mat Corporation (USA) for a creamy, radiopaque, chemical- or dual-cure composite paste with high strength and flow strength used in restorative dentistry for core and crown build-ups, post cementations, and split-roots repair.

Corephen. Trademark of Bayer AG (Germany) for phenol-form-aldehyde plastics.

core sand. Silica sand, usually with little or no clay, to which a suitable binding material has been added. It must have high refractoriness and permeability and make the core friable and collapsible after cooling. Ready-to-use commercial molding-sand mixtures are also available. Also known as *foundry core sand.*

CoreShade. Trade name of Shofu Dental Corporation (Japan) for a gray-colored, fluoride-releasing glass-ionomer dental cement belonging to the *GlasIonomer* product range. It chemically bonds to the tooth structure and provides high radiopacity, biocompatibility and compressive strength. Used in restorative dentistry for metal-free core-build-ups.

core-spun yarn. A compound yarn made by wrapping fibers (e.g., cotton) around a usually readily separable monofilament or multifilament core yarn. For example, spandex is a core-spun yarn. See also core yarn.

Corethane. Trademark of Corvita Corporation (USA) for a polycarbonate-urethane elastomer used as a biomaterial for ventricular assist devices, etc. *Note:* This material is now sold by The Polymer Technology Group, Inc. (USA) under the trademark *Bionate.*

core tube paper. See core paper.

core-type particleboard. See particleboard corestock.

core wash. A mixture applied to foundry cores to improve the surface finish of the cored area of the casting.

Corex. (1) Trademark of Corning Glass Works (USA) for a glass that is transparent to ultraviolet light.

     (2) Trade name of Hobart Brothers Company (USA) for welding filler metals supplied in the form of tubular wires.

     (3) Trade name for a silicone elastomer used for dental impressions.

Corex II. Trademark of Corning Glass Works (USA) for a chemically-strengthened borosilicate glass for laboratory glassware.

core yarn. The usually readily separable inner monofilament or multifilament yarn of a *compound yarn* around which other fibers are wrapped.

Corfam. Trademark of E.I. DuPont de Nemours & Company (USA) for a poromeric nonwoven hybrid fabric, usually impregnated with synthetic resin, composed of urethane and polyester fibers. Used as a leather substitute and for gaskets, packings, seals, vibration dampers, bridge pads, buffing and polishing pads, etc.

Corfix. Trade name of Foseco Minsep NV (Netherlands) for a series of self- and oven-drying foundry core jointing compounds.

Corfloat. Trade name of Coral International (USA) for a water-wash spray-booth compound.

Corgard. Trademark of Corning Glass Works (USA) for borosilicate glass reinforced with a plastic laminate made by impregnating continuous glass fibers with a modified polyester resin and winding on a mandrel.

Corguard. Trademark of Norton-Pakco Industrial Ceramics (USA) for a high-performance ceramic composed of about 50 wt%

alumina (Al_2O_3), 35 wt% zirconia (ZrO_2) and 15 wt% silica (SiO_2). It has a density of 3.49 g/cm^3 (0.126 $lb/in.^3$), a hardness of 2000 Vickers, a maximum service temperature of 1593°C (2900°F), low porosity (only 1.15%), high strength and modulus, good thermal-shock properties, and very low water absorption and gas permeability. Used for structural components and nonstructural parts.

Corialgrund. Trademark of BASF AG (Germany) for polyacrylic compounds.

Corian. Trademark of E.I. DuPont de Nemours & Company (USA) for a wear-resistant cast polymer made to resemble fine natural stone in richness and translucence. It is used in the manufacture of counters, sinks, and vanity tops.

Coriandoli. Italian trade name for a patterned glass.

Coriglas. Trademark of Pittsburgh Corning (Switzerland) AG for a cellular glass used for thermal insulating boards.

Corihyde. Trade name of Duralac Inc. (USA) for a textured surface finishes.

Corindite. Trade name for a synthetic *corundum* used as an abrasive.

Corinth. Trade name of Allegheny Ludlum Steel (USA) for a water-hardening tool steel (AISI type W1) containing 0.9% carbon, 0.3% manganese, and the balance iron. Used for cutters and tools.

Coritex. Trade name of Duralac Inc. (USA) for a coating with a surface texture resembling leather.

cork. The light, thick outer bark of the cork oak (*Quercus suber*) growing throughout southern Europe (especially Spain, Portugal and Italy) and northern Africa (especially Morocco and Tunesia), but now also cultivated in the southwestern United States. It is a form of cellulose with a cellular structure containing in excess of 50 vol% of air cells. With a density of 0.1-0.3 g/cm^3 (0.004-0.011 $lb/in.^3$) it is extremely light, relatively impervious to water, and has high porosity and resilience, and low thermal conductivity. Used for bottle stoppers, sound-deadening products, vibration pads, thermal insulation, corkboard, wallboard, floor coverings, floats for fishing lines, polishing wheels, inner soles of shoes, life preservers, gaskets, and oil retainers. See also cork oak.

corkboard. A thick sheet made from a mixture of ground cork and paper pulp by compressing and subjecting to heat. A binder may or may not be used. *Corkboard* has a low weight, a low thermal conductivity, high porosity, and is relatively impervious to water. Used for the thermal insulation of buildings, as acoustic insulation, for placemats, office memo boards, bulletin boards, etc.

cork elm. (1) The strong, tough wood of the elm tree *Ulmus alata* that derives its name from the corky "wings" on some branchlets. The heartwood is light brown, and the sapwood almost white. *Cork elm* has a fine texture and high hardness. Source: Southeastern United States from Virginia to Florida and westward to Texas. Used chiefly as a shade tree, but some wood is used for vehicle parts and construction work. Also known as *winged elm; wahoo.*

     (2) See rock elm.

corkite. A yellow to dark green mineral of the alunite group composed of lead iron phosphate sulfate hydroxide, $PbFe_3(PO_4)$-$(SO_4)(OH)_6$. Crystal system, rhombohedral (hexagonal). Density, 4.30 g/cm^3; refractive index, 1.93. Occurrence: Ireland.

cork oak. An evergreen oak (*Quercus suber*) growing throughout southern Europe (especially Spain, Portugal and Italy) and northern Africa (especially Morocco and Tunesia), but now also cultivated in India and the southwestern United States. Its light, thick bark yields *cork*.

Corkoustic. Trade name for a corkboard used as a sound-deadening material for walls and ceilings.

cork paint. A special paint that contains fine granules of cork, and is used on steel parts to prevent sweating.

cork powder. Cork ground to various degrees of fineness for use in the manufacture of corkboard, floor and wall surfacing (supplied in sheet or tile form), gaskets, etc.

corkscrew yarn. A fancy yarn in which the individual plies are twisted around each other in a smooth spiral or helix.

cork tile. A tile made of compressed cork bonded with a synthetic resin (e.g., phenolic), and used as floor surfacing material, and for the decoration of plywood and hardboard.

corkwood. See balsa.

Corlar. Trademark of E.I. DuPont de Nemours & Company (USA) for epoxy enamels and primers.

Corlite. Trade name of Aerospace Composite Products (USA) for a stiff, lightweight, black satin-finished composite plate consisting of an epoxy matrix with hollow ceramic microspheres, sandwiched between unidirectional carbon-fiber laminates. It is available in four- and five-ply grades, and used for aerospace applications.

Corlok. Trade name of ATOFINA Chemicals, Inc. (USA) for potassium silicate with high bond strength and good resistance to strong acids. Used as an acid-tank cement/mortar.

Cormet. Trade name of Corning Glass Company (USA) for a porous nickel made by powder-metallurgy techniques. It has a maximum service temperature of 300°C (570°F), and is used for non-contacting conveyors.

Cormin bronze. A corrosion-resistant leaded bronze containing 44% copper, 37% nickel, 11% tin and 8% lead. Used for hardware.

corncob media. Corncobs crushed and ground for use as soft-grit abrasives in blast cleaning, deflashing and finishing of plastics and delicate items. They have multi-faceted grains with angular shapes, a density of about 1.2 g/cm³ (0.04 lb/in.³), and a hardness of 4.5 Mohs.

cornetite. A peacock-blue mineral composed of copper phosphate hydroxide, $Cu_3PO_4(OH)_3$. Crystal system, orthorhombic. Density, 4.10 g/cm³; refractive index, 1.81. Occurrence: Zimbabwe.

Corning. Trademark of Corning Inc. (USA) for an extensive series of glass products and glassware of varying compositions and physical properties. Also included under this trademark are optical and photo-optical materials including optical fibers, photonic products as well as polymer products, ceramic substrates and advanced materials.

Corningware. Trade name of Corning Inc. (USA) for glass-ceramics with excellent thermal-shock resistance and high strength used as ovenware and tableware.

Cornish bronze. A heavy-duty bearing bronze containing 73.82% copper, 16.5% lead, 9.6% zinc, and 0.08% phosphor.

Cornish clay. A variety of *china stone* found in the United Kingdom. It is a weathered granite whose chief constituents are quartz, feldspar, and fluorine minerals. Available in various grades including hard white, soft white, hard purple and mild purple, it is used as a flux in the production of ceramic whitewares, and for high-grade porcelain glazes. Also known as *Cornish stone*.

Cornish Refined. British trade name for refined tin containing 99.82% tin, 0.34% copper, 0.065% lead and 0.03% arsenic.

Used in the manufacture of alloys and tin products.

Cornish stone. See Cornish clay.

Cornix. Trade name of Röchling Burbach GmbH (Germany) for a series of chromium and chromium-nickel stainless steels with excellent corrosion resistance. Used for mixers, agitators, digesters, tanks, acid vessels, chemical-process plant, oil-refining equipment, surgical instruments, and structural applications.

cornstarch. A starch obtained from any of various types of corn (*Zea mays*). It is a carbohydrate polymer of about 75% amylopectin and 25% amylose, and is available as a white powder with high swelling capacity in water. Used as a source of glucose, in biodegradable plastics, in adhesives and coatings, as a thickener, and in food products.

corn sugar. See dextrose.

cornubite. A green mineral composed of copper arsenate hydroxide, $Cu_5(AsO_4)_2(OH)_4$. Crystal system, triclinic. Density, 4.64 g/cm³. Occurrence: UK.

cornwallite. A bluish to blackish green mineral composed of copper arsenate hydroxide monohydrate, $Cu_5(AsO_4)_2(OH)_4 \cdot H_2O$. Crystal system, monoclinic. Density, 4.52 g/cm³. Occurrence: USA (Utah).

Cornwall kaolin. A high-grade *china clay* with negligible amount of iron oxide found in the United Kingdom, and used in ceramics, white paper coatings, and as a paper extender.

Cornwall stone. A weathered, kaolinized feldspar containing about 2% lime. Used for ceramic applications.

Coro. Trademark of Nylstar SpA (Italy) for nylon 6 and nylon 6,6 fibers and filaments.

Corodent. Trade name of Johnson Matthey plc (UK) for a hardened gold alloy containing 29% platinum. It has a melting range of 935-995°C (1715-1823°F), and is used for dental purposes.

Coro-Dyed. Trade name for a glass fiber yarn that has special dyes applied prior to weaving.

Coro-Gard. Trademark of Coroplast, Inc. (USA) for antistatic, conductive corrugated plastic sheeting.

Coroline. Trade name of The Ceilcote Company (USA) for reinforced epoxy lining and flooring materials.

Co-Ro-Lite. Trade name of Columbian Rope Company (USA) for phenol-formaldehyde plastics.

Coromant. Trade name of Sandvik Steel Company (USA) for a series of cemented carbides including tungsten carbides in cobalt binders, or mixtures of tungsten carbide, titanium carbide and/or tantalum carbide in cobalt binders. Used for cutting tools for metallic and nonmetallic materials.

Coromat. (1) Dutch trade name for glass-fiber roofing materials.
(2) Trade name for glass-fiber reinforcements used to protect underground pipelines against corrosion.

Corona. Trade name of Stahlwerke Kabel (Germany) for tungsten-type high-speed steels used for lathe and planer tools, cutters, drills, hobs, reamers, broaches, form tools, etc.

Corona 5. Trade name for high-strength alpha-beta titanium alloys containing 4.5% aluminum, 5% molybdenum, and 1.5% chromium. They are supplied in the beta annealed, beta worked, and alpha-beta worked condition.

Corona Tex. Trade name of Polygenex International, Inc. (USA) for conductive microfibers for specialty gloves.

coronadite. A dark gray to black mineral of the rutile group composed of lead manganese oxide, $MnPbMn_6O_{14}$. Crystal system, tetragonal. Density, 5.44 g/cm³. Occurrence: USA (Arizona), Morocco.

Coronado White. Trade name of Aardvark Clay & Supplies (USA) for a clay (cone 10) having a white body with medium-grained sand.

Coronation. Trade name of George Morrell Corporation (USA) for casein plastics.

Coronel. British trade name for an alloy containing 65-70% nickel and 30-35% copper. It is quite similar to *Monel*, and used for pumps, valves, and turbine blades.

Coronet. Trade name of AB Tilafabriken (Sweden) for melamine-formaldehyde plastics.

coronized alloy. A steel or copper-base alloy to whose surface a zinc-nickel electrodeposit has been applied. It is subsequently heat-treated at about 370°C (698°F), and extremely resistant to corrosion by sulfur dioxide and sulfur trioxide.

Coronze. Trade name of Olin Corporation (USA) for a corrosion- and oxidation-resistant high-strength aluminum bronze containing 95% copper, 2.8% aluminum, 1.8% silicon and 0.4% cobalt. Used for heat exchangers, glass-to-metal seals, springs, and electrical contacts.

Coroplast. Trademark of Coroplast Inc. (USA) for flat, extruded corrugated plastic sheeting.

Cororesist. Trade name of Eutectic Corporation (USA) for a series of low-carbon stainless alloy powders applied by spraying.

Corosoloy. Trade name of US Steel Corporation for a corrosion-resistant steel for hardfacing electrodes.

Corotex. Trademark of Degussa AG (Germany) for silicone-based film materials.

Coro Tint. Trademark of Dominion Corrugated Paper Inc. (Canada) for paper laminates.

Corovac. Trademark of Vacuumschmelze GmbH (Germany) for a series of soft-magnetic composite materials based on powders of iron alloys that receive an insulating coating before being die-pressed into cores and other shaped parts. They have a density range of 4.5-7.5 g/cm^3 (0.16-0.27 lb/in.^{3}), high electric resistance, good magnetic permeability, high coercivity and saturation flux density, good moldability and machinability, good high-temperature stability, high strength and stability, and a maximum service temperature of 150°C (300°F). Used for inductive components, shaped parts in electric motors, and structural parts in magnetic equipment.

Coro-Wrap. Trade name for fibrous glass wraps used for piping and construction applications.

Corrasil. Trade name of Atotech USA Inc. for topcoats for chromates.

Corrax. Trade name of Uddeholm Corporation (USA) for a precipitation-hardening stainless mold steel.

Correct-Fit. Trade name for a silver-free dental bonding alloy.

corrensite. A brown mineral composed of magnesium iron aluminum silicate hydroxide hydrate, $(Mg,Fe)_9(Si,Al)_8O_{20}(OH)_{10}\cdot xH_2O$. Crystal system, orthorhombic. Occurrence: New Zealand.

Corresist. Trade name of Pose-Marre Edelstahlwerk GmbH (Germany) for a series of wrought and cast austenitic, ferritic and martensitic stainless steels used for cutlery, cookware, household appliances, surgical instruments, chemical-processing equipment, pulp and paper machinery, dairy, food-processing and oil-refining equipment, fittings, tanks, machine parts, pump and valve parts, aircraft parts, and turbine blades.

Corrix-Metall. German trade name for wrought and cast aluminum bronzes containing 8.7% aluminum, 3.2% iron, and the balance copper. They have good resistance to acids, alkalies and corrosion, and good physical properties. Commercially available in the form of ingots, bars (rounds, squares, hexagons, flats), sheets, strips, wires, forgings and castings, they are used for marine and mining applications, fittings, worm

wheels, gears, etc.

CorroBan. Trademark of Pure Coatings Inc. (USA) for a smooth, continuous, level alloy electrodeposit of 82-89% zinc, and 11-18% nickel, produced in a slightly acidic electrolyte bath. It is free from hydrogen embrittlement and provides high plating efficiency. Used as a replacement for cadmium coatings in protecting steel aerospace components, such as aircraft structures, bearings, fasteners, mechanisms, etc., against corrosion.

Corrochrom. Trade name of J.C. Soding & Halbach (Germany) for a series of ferritic and martensitic low-carbon high-chromium stainless steels including several hardenable and welding grades. Used for valves, pump and turbine parts, cutlery, scissors, knives, shears, furnace equipment and parts, chemical and petrochemical equipment, oil-refining equipment, agitators, tanks, valve, architectural and automotive trim, fittings, surgical instruments, and hardware.

Corrochroni. Trade name of J.C. Soding & Halbach (Germany) for a series of corrosion-resistant austenitic chromium-nickel stainless steels including several free-machining and welding grades. Used for valve and pump parts, chemical and petrochemical equipment, oil-refining equipment, pharmaceutical equipment, pulp and paper mill equipment, agitators, tanks, fittings, exhaust manifolds, aircraft engine parts, fasteners, hardware, and machine parts.

corroding lead. A commercial grade of pig lead (99.94+% pure) containing up to about 0.050% bismuth, up to 0.002% iron, up to 0.001% zinc, up to 0.0025% silver plus copper, and a total of 0.002% arsenic, antimony and tin. It is suitable for the manufacture of white lead.

Corrodur. Trade name of Bergische Stahl Industrie (Germany) for an extensive series of corrosion- and/or heat-resistant cast and wrought austenitic, ferritic and martensitic stainless steels, and several corrosion-resistant nickel-molybdenum and nickel-molybdenum-chromium alloys.

Corrofest. Trade name of Gusstahl-Handels GmbH (Germany) for a series of corrosion- and/or heat-resistant wrought and cast austenitic, ferritic and martensitic stainless steels.

Corrofestal. Trade name of Deutsche Messingwerke (Germany) for a corrosion-resistant, heat-treatable aluminum alloy containing 0.5-2% magnesium, 0.3-1.5% silicon, and 0-1.5% manganese. It has good formability, medium strength, and is used for decorative and structural parts.

Corro-Foam. Trademark of Sealed Air Corporation (USA) for a ribbed polyethylene foam used for protective packaging applications.

Corrofond. Trade name of Montecatini Settore Alluminio (Italy) for a series of corrosion-resistant cast and wrought aluminum-magnesium and aluminum-silicon alloys. Used for automotive, marine and aircraft parts.

Corroglaze. British trade name for corrugated glass.

Corrolite. Trade name of Reichhold Chemical Company (USA) for vinyl ester resins.

Corronel. Trade name of Inco Alloys International Limited (UK) for corrosion-resistant nickel alloys containing 0-37% chromium, 0-28% molybdenum, 0-6% iron, 0-2.25% vanadium, 0-1% copper, 0-1% manganese, and sometimes small amounts of silicon, aluminum, titanium and/or carbon. Used for acid- and corrosion-resistant vessels, chemical plant and equipment, petroleum plant equipment, pickling tanks, etc.

Corronil. Trade name of Inco Alloys International Limited (UK) for a heat- and corrosion-resistant alloy of 70% nickel, 26% copper and 4% manganese. Used for corrosion-resistant vessels, chemical plant and equipment, and electrical resistances.

Corronium. British trade name for cast or wrought copper alloys containing 15% zinc, and 5% tin.

corronized steel. A steel on whose surface a coating has been produced by first electroplating a composite coating of nickel, followed by tin or zinc, and a final heat treatment to effect the diffusion of the tin or zinc into the nickel.

Corronon. Trade name of Stahlwerk Stahlschmidt GmbH & Company (Germany) for a series of heat- and/or corrosion-resistant low-carbon chromium, chromium-aluminum, chromium-nickel and chromium-molybdenum steels including various stainless, welding and free-machining grades. Used for furnace equipment, parts and fixtures, heat-treating equipment, turbine blades, cutlery, knives, surgical, chemical and petroleum-refining equipment, tanks, vessels, mixers, machine parts, and bearings.

Corrosalloy. Trade name of Sheepbridge Alloy Castings Limited (UK) for a series of cast corrosion- and/or heat-resistant austenitic, ferritic, martensitic and precipitation-hardening stainless steels including various welding grades. Used for chemical equipment, petroleum refining equipment, welded structures, tanks, and pressure vessels.

CorroShield. Trade name of CHN Anodizing Inc. (USA) for an abrasion- and impact-resistant, nonconductive finish produced on zinc-alloy or zinc-coated parts by an anodizing process. The semifused compositional buildup of zinc chromates, oxides and phosphates provides outstanding resistance to corrosion. Used for electrical components, and marine and pump parts.

Corrosil. Trade name of Centrifugal Products Inc. (USA) for a corrosion- and wear-resistant cast iron containing 0.7% carbon, 14% silicon, and the balance iron. Used for corrosion control anodes.

corrosion inhibitor. (1) An organic or inorganic substance that dissolves in a corroding medium, but is capable of forming a protective layer of some kind at the anodic or cathodic areas.

(2) A chemical substance that, when added in relatively low concentrations, retards the rate of metallic corrosion in the environment. For example, phosphates or chromates are added to closed water circulation systems (e.g., automotive engine cooling systems, or industrial heating systems).

corrosion-resistant cast alloys. A group of stainless steel made by pouring into a mold of the desired configuration and then allowed to solidify. They usually contain more than 11% chromium, 1-30% nickel and, depending on the desired degree of corrosion resistance and strength, about 0.03-0.2% carbon. All of these high-alloy cast steels resist attack by aqueous solutions at or near room temperature and high-boiling liquids and hot gases up to 650°C (1200°F). They have good resistance to many acids, chemicals and general corrosion, good weldability, are almost always heat treated, and used in the food-processing, pulp and paper and chemical industries. *Cast stainless steels* can be divided into three categories: (i) *Iron-chromium alloys* containing about 10-30% chromium, 0-8% nickel, and small amounts of molybdenum and copper; (ii) *Iron-chromium-nickel alloys* with about 14-32% chromium, 8-22% nickel, and up to 4% molybdenum and/or copper; (iii) *Precipitation-hardening alloys* with significantly higher strength; and (iv) *Iron-nickel-chromium alloys* with about 8-25% nickel and 18-22% chromium. Also known as *cast stainless steels.*

corrosion-resistant high-strength alloys. A group of vacuum- or electroslag-remelted nickel- and cobalt-base alloys that are resistant to many acids, alkalies, chemicals, hot gases and liquids, and general corrosion. They are used for oil and gas drilling and exploration equipment, aerospace and marine component, aircraft fasteners, chemical and food-processing equipment, and in medical implants.

corrosion-resistant materials. Materials that are capable of withstanding contact with an ambient corroding medium without deterioration or change in properties. Corrosion-resistant metals include stainless steels, and nickel and titanium and their alloys. Corrosion-resistant organic materials include most engineering plastics.

Corrosiron. Trade name of Pacific Foundry Company (USA) for a corrosion-resistant cast iron containing 0.8-1% carbon, 13.5-14.5% silicon, and the balance iron. It is made by casting and has poor forgeability. Used for pumps, valves, drains, cocks, and acid pans.

Corrostite. Trade name of STD Services Limited (UK) for a plain carbon steel containing 0.11% carbon, 0-0.3% silicon, 0.4% manganese, 0.3% copper, a total of 0.05% sulfur and phosphorus, and the balance iron. Commercially available in tube form, it has improved resistance to atmospheric corrosion due to the copper addition. Used for boiler feed tubes, tubing in feedwater systems, piping, etc.

corrosive flux. A soldering flux with a residue that acts chemically with the base metal, and therefore must be removed completely after soldering. It may be composed of inorganic salts and acids, such as the chlorides of ammonium, sodium and zinc, hydrochloric acid, or activated rosins or resins. *Corrosive fluxes* are effective chiefly on low-carbon steel, cadmium, copper, copper alloys (e.g., brass and bronze), lead, nickel, monel, silver, and zinc.

corrosive sublimate. See mercuric chloride.

Corrostop. Trademark of Sico Inc. (Canada) for anti-rust paint.

Corr-Paint. Trademark of Aremco Products Inc. (USA) for a series corrosion-resistant ceramic oxide coatings.

corrugated asbestos. Boards or sheets of asbestos cement, usually 9.5 mm (0.375 in.) thick, shaped into curved or wavelike ridges. Used as roofing and siding for industrial installations, warehouses, hangars, etc. Also known as *corrugated asbestos board; corrugated asbestos-cement sheet.*

corrugated bar. A steel bar with numerous crosswise ridges used in reinforced concrete.

corrugated cardboard. Cardboard that is formed into a row of straight, parallel, alternate, wavelike ridges and valleys. Used for wrapping packages, etc. See also cardboard.

corrugated-core sandwich. A sandwich that contains a honeycomb core consisting of a multiplicity of corrugated ribbon sheets joined together at the nodes and stacked into blocks.

corrugated fiberboard. Fiberboard composed of one, two or three corrugated boards (C) and one, two, three or four flat facings (F), glued together in any of the following combinations: (i) C-F; (ii) F-C-F; (iii) F-F-C; (iv) F-C-F-C-F; or (v) F-C-F-C-F-C-F. See also fiberboard.

corrugated glass. A sheet of glass rolled into a wavelike, furrowed, or corrugated form.

corrugated iron. A sheet of iron or steel, usually zinc-coated, that is formed to produce a wavy or corrugated contour.

corrugated paper. Paper that is formed into a row of straight, parallel, alternate, wavelike ridges and valleys. Used for wrapping packages, etc.

corrugated sheet. (1) Sheetmetal shaped into a row of straight, parallel ridges and valleys by means of specially designed rolling-mill rollers or by the use of a press brake with suitable die-and-punch sets.

(2) A polycarbonate, polyvinyl chloride or other plastic sheeting material shaped into a row of straight, parallel ridges and valleys similar to corrugated iron or steel sheet. Used especially for roofing and other building construction applications.

Corsair. Trade name of Latrobe Steel Company (USA) for a molybdenum-type high-speed steel (AISI type M3) containing 1.00-1.05% carbon, 6.1% tungsten, 4% chromium, 2.4% vanadium, 5-6% molybdenum, and the balance iron. It has excellent wear resistance, high hot hardness and good toughness, and is used for form tools, broaches, boring tools, etc.

Cor-Seal. Trade name of General Polymers Corporation (USA) for chemical-resistant polymer sealants.

Corseal. (1) Trade name for Foseco Minsep NV (Netherlands) for a special compound available in ready-to-use air-drying form, or as a pasty mass diluted with water. Used to seal and repair foundry cores.

(2) Trade name of Tronex Chemical Corporation (USA) for rust and corrosion inhibitors used in metal finishing operations.

Corsican pine. The hard, yellowish wood of the softwood tree *Pinus nigra* (subspecies *laricio*). It is similar to Scotch pine (*P. sylvestris*), but with a wider sapwood zone, and slightly coarser texture. Average weight, 510 kg/m³ (32 lb/ft³). Source: Europe. Used as timber for telegraph, telephone and transmission poles, scaffolding, and home building.

Corteccia. Italian trade name for patterned glass.

Cor-Ten. Trademark of Algoma Steel Corporation Limited (Canada) for a series of high-strength low-alloy (HSLA) steels typically containing 0.1-0.2% carbon, 0.2-1.25% manganese, 0.15-0.75% silicon, 0-1.25% chromium, 0-0.6% copper, 0-0.7% nickel, 0-0.1% vanadium, 0.04-0.15% phosphorus, 0-0.05% sulfur, and the balance iron. They have good weldability and good resistance to atmospheric corrosion. Usually supplied in plate form, they are used for bus and truck bodies, mine cars, derricks, booms, cranes, bridges, towers, buildings, and various other structural applications.

Corterra. Trademark of Royal Dutch/Shell Group (USA) for multi-purpose fibers and yarns spun from polytrimethylene terephthalate (PTT) polymers (*Corterra Polymers*). They have the stain resistance of polyester, the softness of nylon and the bulk of acrylic fibers, and blend well with natural and synthetic fibers including acetate, cotton and wool. Used for the manufacture of durable, hardwearing, easy-care apparel with excellent drape, stretch and feel, e.g., blouses, dresses, skirts, suits, casual wear, sportswear, nightwear and underwear.

Corterra Polymers. Trade name of Royal Dutch/Shell Group (USA) for a polytrimethylene terephthalate (PTT) polymer belonging to the polyester family. Produced from a 1,3-propanediol ($C_3H_8O_2$) starting material, it exhibits high strength, rigidity and heat resistance, good processibility, low moisture absorption, and can be blended with thermoplastics, such as polycarbonate, polyethylene terephthalate and polybutylene terephthalate. Used for molded and extruded products and for spinning into fibers and yarns for the manufacture of clothing.

Cortico. Trade name of Teijin Fibers Limited (Japan) for polyester filaments used for the manufacture of textile fabrics.

Corton. Trade name of PolyPacific (USA) for a series of mineral-filled synthetic resins.

Cortran. Trade name of Corning Glass Works (USA) for a calcium-aluminosilicate glass that transmits infrared radiation.

Corval. Trade name for rayon fibers and yarns.

Corvala. Trade name for rayon fibers and yarns used for textile fabrics.

Corvinair. Trade name for rayon fibers and yarns.

Corum. Trade name of Ivoclar Vivadent AG (Liechtenstein) for dental porcelains.

corundum. An extremely hard mineral occurring in various shades of blue, yellow to golden, and sometimes violet, green, pink or deep red. It is composed of alpha aluminum oxide, α-Al_2O_3, and sometimes contains iron, magnesia or silica. *Corundum* can also be prepared synthetically with high purity. There are three varieties, namely, *ruby*, *sapphire* and *emery*. Ruby and sapphire are transparent gem varieties, and emery is an impure variety mixed with magnetite and hematite. Crystal system, rhombohedral (hexagonal). Density, 3.98-4.05 g/cm³; melting point, 2050°C (3720°F); boiling point, 2700°C (4890°F); hardness, 9 Mohs; refractive index, 1.768. It is commonly coarsely crystalline, but sometimes microcrystalline, and has high strength, good nonmagnetic and dielectric properties, good infrared and ultraviolet transmission properties, good resistance to acids, alkalies, halogens, and metals, and a upper continuous-use temperature of 1800-1950°C (3270-3540°F). Occurrence: Brazil, Burma, Greece, India, Sri Lanka, South Africa, Turkey, USA (Georgia, Montana, New Jersey, New York, North Carolina, Pennsylvania, South Carolina). Used in the manufacture of abrasives for polishing and grinding tough, high-strength materials, such as tool steels, cast steels, malleable iron, bronze, etc., and for grinding wheels, abrasive tools and refractory bricks. Ruby and sapphire are also used as gemstones and in lasers.

corundum brick. A refractory brick composed essentially of alpha aluminum oxide (α-Al_2O_3), and used for high-temperature applications, such as furnace and kiln linings. See also alumina; corundum.

Corvel. Trademark of Morton Powder Coatings (USA) for powder coatings based on plastic materials, such as acrylics, epoxies, polyester, nylon, etc.

Corvex. Trade name of Grating Pacific, LLC (USA) for a corrosion- and flame-resistant, polyester grating used for light industrial and water/wastewater applications.

Corvic. (1) Trade name of Chase Brass & Copper Company (USA) for a corrosion-resistant bronze containing 98.2-98.5% copper, 1.5% tin, and up to 0.3% phosphorus. It has good electrical conductivity, and is used for chemical plant and equipment.

(2) Trade name of ICI Limited (UK) for stiff, strong unplasticized polyvinyl chlorides supplied in rod, sheet, tube, powder and film form. They are amorphous inherently flame-retardant thermoplastic resins with a density of 1.4 g/cm³ (0.05 lb/in.³), low water absorption (0.03-0.4%), good barrier properties and ultraviolet stability, an upper continuous-use temperature of 50°C (122°F), and moderate chemical resistance. Used for building products, containers, and bottles.

corvusite. A bluish-black to brownish red mineral composed of vanadium oxide hydrate, $V_7O_{17} \cdot xH_2O$. Density, 2.82 g/cm³. Occurrence: Kazakhstan.

Cosal. German trade name for a synthetic resin made from vinyl isobutyl ether.

cosalite. A lead to steel gray mineral of the lillianite group composed of lead bismuth sulfide, $Pb_2Bi_2S_5$. Crystal system, orthorhombic. Density, 6.86-7.13 g/cm³. Occurrence: Japan, Sweden.

Cos-Ar-Cor. Trade name of COSIPA–Companhia Siderurgica Paulista (Brazil) for an alloy steel containing up to 0.16% carbon, up to 1.2% manganese, up to 0.03% phosphorus, 0.015%

sulfur, up to 0.5% silicon, 0.2-0.5% copper, 0.4-0.7% chromium, a total of 0.15% niobium, vanadium and titanium, and the balance iron. It has good resistance to atmospheric corrosion, and good weldability, and is used for structural members, buildings, bridges, and steel structures.

coslettized product. An iron or steel product with a thin, hard, rust-resisting coating of gray basic iron phosphate that has been formed by immersing for 3-4 hours in a hot bath composed of a dilute solution of iron phosphate and phosphoric acid.

Cosmesil SNM4. Trade name for a soft, silicone-based polymer used in dentistry for maxillofacial applications

Cosmic. Trade name of Cosmic Plastics (USA) for a series of reinforced or filled synthetic resins and molding compounds including *Cosmic DAIP* glass fiber-reinforced or mineral-filled diallyl isophthalate resins, *Cosmic DAP* glass or synthetic fiber-reinforced or mineral-filled diallyl phthalate resins, and *Cosmic EP* glass fiber-reinforced or mineral-filled epoxy molding compounds.

Cosmos. (1) Trade name of Lumen Bearing Company (USA) for a lead babbitt containing 15% antimony, 9% tin, and the balance lead. Used for bearings.

(2) Belgian trade name for patterned glass with a crescent moon and star design.

Cosmotech. Trade name of GC Dental (Japan) for a dental porcelain used for metal-to-ceramic restorations.

Cosmotech Porcelain. Trade name of GC Dental (Japan) for a dental laminating and veneering porcelain.

co-spun yarn. A yarn made by the simultaneous extrusion of two polymers (e.g., two chemically different polymers, such as polyester and nylon, or a homopolymer and a copolymer of the same polymeric substance) through a spinneret whereby the exiting filaments form a single strand.

costibite. A gray mineral of the marcasite group composed of antimony cobalt sulfide, CoSbS. Crystal system, orthorhombic. Density, 6.90 g/cm³. Occurrence: Australia (New South Wales).

Cosyfelt. Trade name of D. Anderson & Son Limited (UK) for glass-wool roof insulation.

Cosywrap. Trade name of Fiberglass Limited (UK) for a 25-mm (1-in.) thick glass-fiber insulation product used for domestic attic insulation applications.

Cote-A-Cor. Trade name of Westvaco Corporation (USA) for coated corrugated paperboard.

Cote d'Or. Trade name of Technic Inc. (USA) for a heavy gold electrodeposit.

Cothias alloy. An aluminum alloy containing 6.5% copper and 0.5% zinc. Used for lightweight castings.

Cotin. Trade name of Zinex Corporation (USA) for a bright nickel-palladium-rhodium electroplate used as an alternative undercoat for gold and other precious metals, and as an end finish on jewelry.

CotMaster. Trade name of F.J. Brodman & Company LLC (USA) for powder coatings based on plastics, such as polyamide, polyethylene, polyimide, polyvinyl chloride, polyethylene terephthalate, polyphenylene sulfide, various fluoropolymers, ethylene acrylic acid or ethylene vinyl acetate.

Cotron. Trade name for cellulose acetate fibers and yarns.

Cotronics. Trade name of Cotronics Corporation (USA) for a series of molded ceramic fibers (e.g., alumina silicate) used as high-temperature insulation, e.g., for pipes, burners, turbines, brazing equipment, and thermocouples.

Cotswold. Trade name of Pilkington Brothers Limited (UK) for patterned glass.

Cottaer Sandstein. Trademark of Sächsische Sandsteinwerke GmbH (Germany) for a high-grade sandstone from Saxony. Used in building and construction.

cotton. The soft, downy white or yellowish fibers obtained from the seed pods of various species of *Gossypium* belonging to the mallow family. The cotton surrounding the seeds in a fluffy mass contains about 88-96% cellulose. *Cotton* has a density of 1.52-1.54 g/cm³ (0.054-0.056 lb/in.³), high tensile strength, low permanent set, poor resistance to acids, decomposes at approximately 150°C (300°F), and swells in caustics, but is not damaged. Source: Brazil, Central Asia, Egypt, India, USA. Used as an important textile fiber for apparel, household and industrial fabrics, upholstery, cordage, padding, thread, etc., for absorbent cotton and cottonseed oil, and a source of cellulose. See also American cotton; Egyptian cotton; Indian cotton; Sea Island cotton.

cotton fabrics. Textile fabrics made from cotton fibers and available in many weave types and weights. Examples include batiste, canvas, corduroy, and duck. Used for cable insulation, tarpaulins, bags, sails, tents, awnings, welding curtains, apparel, protective clothing (e.g., aprons and gloves), conveyor belts, filter cloths, machine covers, etc.

cotton fillers. Cotton fibers or fabrics chopped into short lengths, and used as fillers in molding compounds for reinforced plastics.

cotton flock. Waste cotton used as a filler in plastics, and as a stuffing material for mattresses and cushions.

cottongrass. The silky fibers obtained from the fruits of several species of a genus (*Eriophorum*) of sedges, especially the narrow-leaved cottongrass (*E. angustifolium*), used as a stuffing for pillows, cushions and upholstered furniture.

cotton gum. See water tupelo.

cotton linters. See linters.

cotton-phenolic composite. A structurally anisotropic composite manufactured in tube form by winding phenolic resin-impregnated cotton cloth on a mandrel, and subsequent curing at elevated temperatures. It can absorb significant amounts of lubricating oils, and is used for ball bearing cages.

cotton pulp. Chemically purified cotton *linters* supplied in the form of sheets or pressed bales as starting materials for the manufacture of fibers, lacquers and other products.

cotton-spun yarn. A yarn other than cotton that has been made with cotton-yarn making machinery.

cottonwood. The soft, yellowish-white wood of any of various North American poplars including especially the eastern cottonwood (*Populus deltoides*), the black cottonwood (*P. trichocarpa*), the swamp cottonwood (*P. heterophylla*), and the balsam poplar (*P. balsamifera*). The name "cottonwood" refers to the cottony tufts of hairs on the seeds of the tree. *Cottonwood* has grayish-white to light-brown heartwood merging gradually with the whitish sapwood. It has a fine, open grain, a uniform texture, low to moderate strength, excellent gluing and nailing qualities, high warping tendency, and works fairly well. Average weight, 480 kg/m³ (30 lb/ft³). Source: Eastern United States and Canada. Used for lumber and veneer for crates, pallets, boxes, etc., for paneling, and as a pulpwood and fuelwood. See also balsam poplar; black cottonwood; eastern cottonwood; swamp cottonwood.

cotton wool. (1) A term used for raw (unprocessed) cotton.

(2) A term referring to webbing or batting composed of cotton and/or viscose rayon, and used especially for medical or

cosmetic applications.

cotunnite. A colorless, or pale white to yellowish mineral composed of lead chloride, $PbCl_2$. It can also be made synthetically. Crystal system, orthorhombic. Density, 5.91; refractive index, 2.217.

Cougar. Trade name of Pyramid Steel Company (USA) for a cold-work tool steel (AISI type D2) containing 1.55% carbon, 0.04% manganese, 11.5-12% chromium, 0.9% vanadium, 0.8% molybdenum, and the balance iron. Used for dies, shear blades, forming rolls, punches, broaches, etc.

coulsonite. A bluish gray mineral of the spinel group composed of iron vanadium oxide, FeV_2O_4. Crystal system, cubic. Density, 5.17 g/cm^3. Occurrence: India, USA (Nevada).

coumarin. A lactone available in the form of colorless crystals, flakes or powders with a melting point of 69-73°C (156-163°F), and a boiling point of 298°C (568°F). Used in biochemistry, biotechnology, medicine, pharmacology, electroplating, and in the preparation of coumarin dyes. Formula: $C_9H_6O_2$. Also known as *cumarin; 1,2-benzopyrone.*

coumarin dyes. A large group of coumarin derivatives available in the form of crystals and/or powders with various compositions and melting points ranging from about 70°C to 240°C (158 to 464°F), and including *coumarin 1* ($C_{14}H_{17}NO_2$), *coumarin 4* ($C_{10}H_8O_3$), *coumarin 6* ($C_{20}H_{18}N_2O_2S$), *coumarin 7* ($C_{20}H_{19}N_3O_2$), *coumarin 120* ($C_{10}H_9NO_2$), *coumarin 151* ($C_{10}H_6F_3NO_2$), *coumarin 152* ($C_{12}H_{10}F_3NO_2$), *coumarin 311* ($C_{12}H_{13}NO_2$), *coumarin 314* ($C_{18}H_{19}NO_3$), *coumarin 334* ($C_{17}H_{17}NO_3$), *coumarin 337* ($C_{16}H_{14}N_2O_2$) and *coumarin 343* ($C_{16}H_{15}NO_2$). Used in biochemistry and medicine (e.g., as fluorescent labeling reagents), and in biotechnology, pharmacology, electroplating and electronics. Many coumarin dyes are suitable as laser dyes.

coumarone-indene resins. A class of thermosetting resins made by treating a mixture of coumarone (C_8H_6O) and indene (C_9H_8), both of which are compounds derived from coal-tar naphtha, with sulfuric or phosphoric acid to induce polymerization. Often sold under trademarks or trade names, such as Cumar, Nevidene, Paradene, etc., they are soft and tacky at room temperature, and have high dielectric strength. Used for adhesives, paints, varnishes, lacquers, enamels, and printing inks. Also known as *polycoumarone resins; polyindene resins.*

coumarone resins. A class of thermosetting resins made by treating coumarone (C_8H_6O) with sulfuric or phosphoric acid to induce polymerization. They have densities of 1.05-1.15 g/cm^3 (0.038-0.042 $lb/in.^3$), high dielectric strength, and are used for molded parts, paints, lacquers, varnishes, adhesives, and waterproofing compounds.

courbaril. The strong, durable wood of any of over 30 species of trees of the genus *Hymenaea*, with *H. courbaril* being the most important one. The heartwood is brown to violet, and the sapwood grayish-white. *Courbaril* has good turning, finishing and gluing qualities, and good shock resistance. Source: Brazil (Amazon Basin and Valley). Used for sporting goods, handles, furniture, and veneer.

Courlene. Trade name for cellulose acetate fibers used for wearing apparel and as industrial fabrics.

Courlene-Duracol. Trade name for polyolefin fibers used for textile fabrics.

Courlene-PY. Trade name for polyolefin fibers and yarns used for textile fabrics.

Coursier. Trade name of Compagnie Ateliers et Forges de la Loire (France) for a water-hardening steel containing 0.6-0.7% carbon, and the balance iron. Used for knives, hammers, chisels,

cutting tools, and drills.

Courtaulds Lyocell. Trademark of Courtaulds Limited (UK) for lyocell textile fibers used for textile fabrics including clothing.

Courtek M. Trademark of Courtaulds Limited (UK) for acrylic staple fibers.

Courtelle. Trademark of Courtaulds Limited (UK) for strong, washable acrylic fibers that are resistant to moths, mildew and creasing, and can be blended with various other fibers. Used for clothing, furnishings, blankets, and carpets.

CourtGard. Trademark of Solutia Inc. (USA) for a clear polyester film impregnated with UV-blocking chemicals. It is highly resistant to flaking, peeling, rubbing and scratching, and is used for the protection of light-sensitive materials and cultural treasures including legal documents.

Cousoff. Trade name of Teijin Fibers Limited (Japan) for polyester filaments used for the manufacture of textile fabrics.

Coussinal. French trade name for a series of aluminum alloys containing 3-4% tin, 0-4% copper, 0-3% magnesium, 0-3% antimony, 0-2% lead, and 0-1% manganese. Used for antifriction bearings.

Coustic-Aire. Trade name for multi-ply acoustical units made from fiberglass.

Cova. Trade name for rayon fibers and yarns.

covalent ceramics. A group of ceramic compounds including silicon carbide, copper indium diselenide, magnesium silicon dinitride, etc., in which the dominant interatomic bonding is of the covalent type. See also covalent compounds.

covalent compounds. A group of compounds, such as gallium arsenide, indium antimonide, silicon carbide, etc., having covalent interatomic bonding in which valence electrons from the atoms involved in the bonding are mutually shared by adjacent atoms.

Covan. Trade name of CCS Braeburn Alloy Steel (USA) for a molybdenum-type high-speed steel (AISI type M34) containing 0.9% carbon, 4% chromium, 8% molybdenum, 8% cobalt, 2% vanadium, and the balance iron. It has good hot hardness, and is used for cutting tools.

Covan 55. Trade name of Atlantic Steel Company (USA) for a high-strength low-alloy cobalt-vanadium structural steels for buildings and bridges.

Covandur. Trade name of Carpenter Technology Corporation (USA) for an oil-hardening steel used for machine-tool parts.

Covar. Trade name for a cobalt-vanadium-rhodium alloy for glass feedthroughs.

covellite. An indigo-blue mineral composed of copper sulfide, CuS. Crystal system, hexagonal. It can also be made synthetically. Density, 4.67 g/cm^3; refractive index, 1.45; hardness, 1.5-2 Mohs. Used as a minor source of copper. Also known as *indigo copper.*

Coveral. Trade name of Foseco Minsep NV (Netherlands) for covering and protecting fluxes for aluminum alloys.

cover-coat enamels. A class of clear, opaque or semiopaque porcelain enamels that are either applied over a ground coat or directly onto the properly prepared metal substrate. They may or may not be pigmented to take on various colors, and improve the appearance and properties of the substrate. *Cover-coat enamels* are often used on sheet steel, cast iron, aluminum, and aluminum alloys.

covered yarn. A multiple yarn made from a wrapping yarn contained on a revolving spindle and a tensioned core yarn fed through the axis of this spindle.

Coverlac. Trade name of Spraylat Corporation (USA) for conductive coatings for plastics.

Coverlight. Trade name of Reeves, Industrial Coated Fabrics (USA) for a textile fabric coated with an polymeric material, such as neoprene, nylon, chlorosulfonated polyethylene or vinyl chloride.

cover paper. A strong, durable paper that can be easily folded and printed. It is generally heavy and used for covers of catalogs, periodicals, etc.

covert. A medium- to light-weight fabric of acate, cotton, silk, rayon, or wool in a twill weave and with a two-color warp and a smooth finish. Used for general clothing and raincoats and overcoats.

Covertex. Trademark of J.H. Beneckw GmbH (Germany) for polyvinyl chloride film materials.

Cover-Up II. Trade name for a dental adhesive resin cement.

Covington. Trade name for rayon fibers and yarns used for textile fabrics.

Covinite. Portuguese trade name for a toughened glass.

Cowaplast. Trademark of Coswig for flexible polyvinyl chloride film materials.

cow hair. The long, relatively stiff protein fibers obtained from the protective skin hair of domestic cows. They are sometimes spun as yarn but more often used for the production of felt products.

cowhide. (1) Leather made from the hides of cows.

(2) A general term describing any tanned leather obtained from hides of bovine animals including cows, steers, bulls and oxen.

Cowles. (1) British trade name for a series of tough, corrosion-resistant aluminum bronzes containing 1.25-11% aluminum, 0-1.5% silicon, 0-0.5% iron, and the balance copper. They have high strength, and good wear resistance. Used for hardware, worm wheels, gears, and trolley wires.

(2) Trade name for a corrosion-resistant high-manganese brass containing 67-80% copper, 15-18% manganese, 5-13% zinc, and 0.1% aluminum and silicon. It has excellent cold formability, high strength, and is used for strong, corrosion-resistant parts.

cowlesite. A gray to white mineral of the zeolite group composed of calcium aluminum silicate hydrate, $CaAl_2Si_3O_{10} \cdot xH_2O$. Crystal system, orthorhombic. Density, 2.14 g/cm^3; refractive index, 1.515. Occurrence: USA (Oregon).

coyoteite. A black mineral composed of sodium iron sulfide dihydrate, $NaFe_3S_5 \cdot 2H_2O$. Crystal system, triclinic. Density, 2.88 g/cm^3. Occurrence: USA (California).

CP Evalux. Trade name of Agriplast Srl (Italy) for thermal films composed of a synthetic resin with 14% ethylene-vinyl acetate (EVA). They are extremely thin (only 25-30 μm or 1.0-1.2 mils), provide high transparency, excellent tear and puncture resistance, and are antidrop-treated to minimize fungal diseases and control condensation. Supplied in three grades, they are used in greenhouses to protect vegetables and flowers.

"CPM" Steel. Trademark of Crucible Specialty Metals (USA) for powder-metallurgy high-speed and cold-work tool steels with microstructures containing fine, uniform distributions of carbides, produced by the proprietary CPM (Crucible Particle Metallurgy) process, a powder-metallurgy technique based on gas atomization and hot-isostatic pressing. Supplied in the form of powder and barstock, the high-speed steels (e.g., *CPM Rex 76*, *CPM Rex 121* and *CPM Rex T15*) have a superior combination of toughness and wear resistance, high red hardness and good machinability, and the cold-work steels (e.g., *CPM 3V*

and *CPM 10V*) contain high additions of vanadium (typically 3-15%) resulting in remarkably high wear resistance for tooling applications.

CP steels. See complex-phase steels.

Crabiyon. Trademark of Omikenshi Company (Japan) for chitosan fibers and filament yarns.

crabwood. See andiroba.

Crackled. British trade name for an antique glass.

crackled glass. A glass that has been intentionally cracked by immersion in water while hot, and subsequent reheating and shaping.

crackleware. Ceramic ware made with a glaze exhibiting extensive crazing.

Craig gold. A copper-base alloy containing 10% zinc and 10% nickel. Used for jewelry and ornaments.

Cralclad. Japanese trade name for clad *duralumin*.

Cralfer. British trade name for a cast iron containing 3% carbon, varying amounts of manganese, silicon and aluminum, and the balance iron. Used for machinery castings.

Cramco. Trademark of Cramco Solder Alloys (Canada) for a series of soldering alloys.

Cramp Alloy. Trade name of Baldwin-Lima-Hamilton Corporation (USA) for a series of gray and ductile cast irons, chromium and nickel cast irons, and several cast and wrought coppers and copper alloys including high-conductivity coppers, composition and ounce metals, high-strength copper-zinc alloys, aluminum, manganese, nickel and silicon bronzes, tin-base alloys, and copper-lead-tin alloys. *Cramp Alloy* is mainly used for bearings, bushings and other machine parts.

Cramp's Super-Strength Bronze. Trade name of Baldwin-Lima-Hamilton Corporation (USA) for a series of manganese bronzes containing 57% copper, 28% zinc, and a total of 15% hardeners of manganese, iron and aluminum. Used for bushings, bearings, shafts, gears, worms, valve stems, pump impellers, die castings, tubes, and sheets.

crandallite. A yellow to light gray mineral of the alunite group composed of calcium aluminum phosphate hydroxide monohydrate, $CaAl_3(PO_4)_2(OH)_5 \cdot H_2O$. Crystal system, rhombohedral (hexagonal). Density, 2.85 g/cm^3; refractive index, 1.622. Occurrence: Guatemala, USA (Utah).

Crane. Trade name of Crane Company (USA) for a water-hardening steel containing 0.7-1.2% carbon, and the balance iron. Used for tools, cold chisels, and dies.

craquelé. Crepe fabrics with a conspicuously cracked and grained surface. See also crepe fabrics.

crash. A coarse linen fabric in a plain or twill weave, used for dresses, tablecloths, curtains, upholstery, towels, and backings.

Crastin. Trade name of E.I. Du Pont de Nemours & Company (USA) for stiff, tough, heat-resistant thermoplastic polyesters based on polybutylene terephthalate. They have excellent fatigue and tear strength, excellent electrical insulation characteristics, outstanding resistance to acids, oils, solvents, greases and humidity, and good flow properties and processibility. Available in regular, glass-reinforced, mineral-filled, flame-retardant and arc-resistant grades, they can be welded, glued, painted, hot-stamped and laser-marked. Used for plastic film, beverage containers, injection-molded products, automotive, electronic and electrical applications, sporting goods, etc. *Crastin S* is a polybutadiene terephthalate with excellent fatigue and tear strength and outstanding resistance to acids, oils, solvents, greases and humidity, *Crastin SK* a polybutadiene terephthalate available in regular, glass-fiber-reinforced, glass-bead filled and UV-stabilized grades, and *Crastin SO* a fire-retardant polybutadiene terephthalate reinforced with 30% glass fibers.

CRATEC. Trademark of Owens-Corning Fiberglass (USA) for glass fibers supplied in chopped strand form in several grades for use as reinforcements for unsaturated polyester bulk molding compounds, epoxy or phenolic molding compounds, polycarbonate and polyethylene composites, and thermoplastic polymers, such as acetals, acrylonitrile-butadiene-styrenes, styrene-acrylonitriles, styrenic maleic anhydrides and polybutylene terephthalates. Also available are *CRATEC Plus* chopped glass-fiber strands suitable as reinforcements in polymer-matrix composites for automotive and electrical applications.

Cratère. Belgian trade name for figured glass.

Cravenette. Trademark for a water-repellent fabric used for raincoats, topcoats, etc.

crazeproof enamel. An enamel specially formulated to prevent the development of minute surface cracks when subjected to thermal shocks.

Crea-Lite. Trademark of Saint-Gobain (France) for heat-formed glass.

cream of tartar. See potassium bitartrate.

crease-resistant fabrics. Textile fabrics, such as cotton, flax and linen, whose wash-and-wear resistance to and recovery from creasing, wrinkling and folding has been enhanced by impregnating with suitable chemicals, e.g., synthetics. Also known as *wrinkle-resistant fabrics*.

creaseyite. A green mineral composed of copper lead iron aluminum silicate hexahydrate, $Cu_2Pb_2(Fe,Al)_2Si_5O_{17}\cdot6H_2O$. Crystal system, orthorhombic. Density, 4.10 g/cm^3; refractive index, 1.747. Occurrence: USA (Arizona).

Creation. Trade name of Jensen Dental (USA) for a dental porcelain for metal-to-ceramic restorations.

Creation Bond. Trade name for a bonding agent used in restorative dentistry for bonding dentin.

crednerite. A iron-black mineral composed of copper manganese oxide, $CuMnO_2$. Crystal system, monoclinic. Density, 5.34. Occurrence: UK.

creedite. A colorless to white mineral composed of calcium aluminum fluoride sulfate dihydrate, $Ca_3Al_2(SO_4)(F,OH)_{10}\cdot2H_2O$. Crystal system, monoclinic. Density, 2.71 g/cm^3; refractive index, 1.478; hardness, 2 Mohs. Occurrence: USA (Colorado, Nevada).

Creel-Pak. Trademark of Owens-Corning Corporation (USA) for glass fibers.

creep-resistant steels. A group of steels suitable for parts, such as gas and steam turbine parts, boiler parts, automotive engine exhaust valves, valve springs, etc., exposed to elevated temperatures (350-650°C or 660-1200°F) and repeated mechanical stresses. Chromium and chromium-nickel austenitic stainless steels, and several low-alloy steels of the chromium, chromium-molybdenum-vanadium, molybdenum, tungsten and tungsten-chromium type are examples.

creep-resistant martensitic stainless steels. A group of *martensitic stainless steels* with excellent high-temperature oxidation resistance and creep strength. They have a typical composition (in wt%) of about 0.07% carbon, 9.5% chromium, 3% cobalt, 1% nickel, 0.6% molybdenum, 0.3% titanium, and the balance manganese, silicon and iron. Contrary to conventional corrosion-resistant martensitic steels, which contain chromium-rich precipitates (e.g., $Cr_{23}C_6$ and Cr_7C_3), creep-resistant grades have uniform dispersions of very fine coarsening-resistant titanium carbide particles, and the addition of austenite stabilizers (e.g.,

carbon, nickel and cobalt) has been balanced with additions of ferrite stabilizers (e.g., chromium, molybdenum and titanium). Used especially for steam-turbine components, tubes and pipes for power stations and oil refineries, oilfield downhole casings, automotive engine valves, etc.

Crelan. Trademark of Bayer Corporation (USA) for powder-coating resins.

Cremnitz white. A white inorganic pigment composed of lead carbonate and hydrated lead oxide.

Cremona. Trade name for a vinal (polyvinyl alcohol) fiber used for textile fabrics.

Creoserve. Trademark of Domtar Inc. (Canada) for coal tar-based protective coatings.

creosote. A pungent colorless, yellowish or greenish oily liquid consisting of a mixture of phenols distilled from coal tar. It has a density of 1.03-1.10 g/cm^3 (0.037-0.040 $lb/in.^3$), a distilling range of 200-400°C (390-750°F), a flash point of 74°C (165°F), and is used chiefly as a wood preservative and for protective coatings. Also known as *coal-tar creosote; pitch oil; tar oil.*

creosote stain. A *wood stain* made by mixing creosote obtained from wood and coal tars with linseed oil and a drier, and subsequent thinning with benzene or kerosene.

crepe. See crepe fabrics; crepe paper; crepe rubber.

crepe-back satin. A fabric with one crepe and one satin side. Also known as *satin-back crepe*. See also crepe fabrics; satin.

creped duplex paper. A paper laminate consisting of two sheets of *crepe paper* bonded together with asphalt, rubber latex, etc.

crepe de chine. A soft, thin, washable, light- to medium-weight crepe fabric made with silk or synthetic yarns in a plain weave. Used especially for blouses and dresses.

creped paper. See crepe paper.

crepe fabrics. Fabrics with a crinkled, knotty surface usually obtained by using special cotton, wool, polyester, rayon, silk, or nylon yarns, by twisting alternately right and left in the filling, or by treating with certain chemicals. Used especially for making dresses, coats and suits. Examples include crepe georgette, crepe de chine and crepe marocain. Also known as *crepe*.

crepe georgette. See georgette.

crepe marocain. See marocain.

crepe paper. A thin, lightweight, strong *kraft paper* with a crinkled surface texture resembling crepe fabrics. It has high stretch and conformability, and is available in a range of colors. Used for decorations, displays, wrapping, floats, and bag and barrel liners. Also known as *crepe; creped paper.*

crepe rubber. Originally, rubber latex coagulated with acid and formed into light-colored sheets with a crinkled surface. *Crepe rubber* is now made from synthetic rubber. Used for shoe soles and heels. Also known as *crepe*.

Crepeset. Trade name for nylon fibers and yarns used for manufacture of crepe and other fabrics.

Crepesly. Trade name for rayon fibers and yarns used for crepe and other fabrics.

Crepesoft. Trade name for polyester fibers and yarns used for clothing, especial dresses, blouses and suits.

crepe tape. Adhesive tape made of crepe and having an adhesive backing that sticks to metal, glass, fabrics, etc. Used for wrapping and autoclaving applications.

crepe yarn. A high-twist yarn sometimes used in the manufacture of crepe fabrics.

Crepi. Belgian trade name for a patterned glass.

crepon. A fabric made from crimped high-twist yarn and having a warp effect resembling pleat-like crepe. Used for blouses, dresses, shirts and nightwear.

Crescent. (1) Trade name of Spencer Clark Metal Industries Limited (UK) for water-hardening carbon steels supplied in six carbon ranges and tempers: (i) 1.20-1.40%, (ii) 1.05-1.15%, (iii) 0.90-1.00%, (iv) 0.80-0.90%, (v) 0.70-0.80%, and (vi) 0.60-0.70%. They have excellent machinability, and good to fair toughness and shock resistance. Used for cutting tools, hand tools, and dies.

(2) Trade name of Colt Industries (UK) for a hot-work tool steel containing 0.95% carbon, 3.5-3.9% chromium, and the balance iron. Sometimes small amounts of vanadium, molybdenum, or nickel are added. It has good hardness and toughness, and good to fair resistance to heat checking. Used for header dies, gripper dies, bloom shears, and tools.

(3) Trade name for a tool steel containing 2.1% carbon, 2.7% manganese, 0.1% silicon, 6.7% tungsten, and the balance iron. Used for tools and dies.

(4) Trade name for a contoured glass block having a sloping concave shape on its surface which provides a constantly changing pattern as light falls upon it.

Creslan. Trademark of Sterling Fibers Inc. (USA) for an acrylic fiber supplied in staple and tow form. It has good flame resistance, and good blendability with natural and synthetic fibers including cotton and wool. Used for wrinkle- and shrinkage-resistant clothing, sweaters, socks, blankets, draperies, home furnishings, and industrial and technical fabrics.

Cres-Lite. Trade name of Crescent Bronze Powder Company (USA) for bronzing powders and liquids, and aluminum paints.

Cresloft. Trade name of Sterling Fibers, Inc. (USA) for acrylic fibers used for the manufacture of textile fabrics.

Crest. (1) Trade name of St. Lawrence Steel Company (USA) for an oil-hardening tool steel (AISI type O1) containing 0.9% carbon, 1.15% manganese, 0.65% chromium, 0.25% vanadium, 0.5% tungsten, and the balance iron. Used for punches, knurling tools, gages, and blanking and bending dies.

(2) Trade name of St. Lawrence Steel Company (USA) for an air-hardening tool steel (AISI type A2) containing 1% carbon, 0.7% manganese, 5.5% chromium, 1.15% molybdenum, 0.3% vanadium, and the balance iron. Used for broaches, rolling dies, shear blades, and blanking and trimming dies.

Crestalkyd. Trademark of Scott Bader & Company, Limited (USA) for a series of oil-modified alkyd resins used for lacquers and varnishes, enamels, paints, and plasticizer.

Crestaloy. Trade name of Crescent Tool Company (USA) for an oil-hardening tool steel containing 0.5% carbon, 1.5% chromium, 0.2% vanadium, and the balance iron. Used for hand tools.

Creston. Trade name of Applied Steel & Tractor Products Inc. (USA) for a series of water-hardening tool steels containing 0.75-1.4% carbon, and the balance iron. Used for tools, dies, cutters, and battering tools.

cretonne. A strong, moderately heavy cotton fabric in a plain or twill weave, usually having designs printed on one or both sides. Used for curtains, draperies, furniture covers, and bedspreads.

Creusabro. Trade name of Creusot-Loire (France) for a series of strong, abrasion- and wear-resistant steels with good weldability used for structural applications and machinery parts.

Creuselso. Trade name of Creusot-Loire (France) for a series of weldable high-strength structural carbon steels and heat-treated structural low-alloy steels used for mining equipment and structures, gas containers, pressure vessels, low-temperature structures, marine platforms, derricks, bridges, penstock, etc.

Creusem. Trade name of Creusot-Loire (France) for a series of

soft-magnetic materials containing 0-0.08% carbon, 0-0.4% manganese, 0.25% silicon, and the balance iron. Used for relays.

Creusot. Trade name of Creusot-Loire (France) for an extensive series of structural and machinery carbon and alloy steels including several case-hardening grades and various plain-carbon, cold-work, hot-work, high-speed and shock-resisting tool steels.

crichtonite. A black mineral composed of iron titanium oxide, $(Fe,Ti)_{1.71}O_3$. Crystal system, pseudohexagonal. Density, 4.46 g/cm^3. Occurrence: France.

Crilicon. Trade name for a series of acrylic resin emulsions used in paints, adhesives, and textile finishes.

Crilley Metals. Trade name of Metal Sales Corporation (USA) for a series of self-lubricating bearing alloys composed of mercury, copper and tin.

Crilobal. Trademark of Sumar SA (Chile) for nylon 6 fibers and yarns.

crimped yarn. A filament yarn with permanent crimps, coils, curls, folds or loops in direction of the filament axis.

Crimplene. Trade name for polyester fibers and yarns used for textile fabrics, especially clothing.

Crinex. Trade name for nylon fibers and yarns.

crinkled fabrics. Fabrics with an uneven, wrinkled, rippled or puckered surface produced by chemical or mechanical means, or by the use of high-twist yarns.

Crinofil. Trade name for rayon fibers and yarns.

Crinol. Trade name for rayon fibers and yarns used for textile fabrics.

crinoline. A stiff fabric, usually heavily sized and woven in an open plain weave. Used as a lining, interlining and stiffening material.

Crinovyl. Trade name for vinyon (vinyl chloride) fibers with excellent acid, alkali and mildew resistance. Used for textile fabrics and carpets.

Cripmore. Trade name of Bethlehem Steel Corporation (USA) for a hot-work tool steel containing 0.35% carbon, 1% silicon, 0.35% manganese, 5% chromium, 1.75% molybdenum, 1.35% tungsten, and the balance iron. Used for hot-work dies.

Crisil. Trade name of Johnston Steel & Wire Company Inc. (USA) for a hard-drawn wire of steel containing 0.55% carbon, 0.5-0.8% chromium, 1.2-1.6% silicon, 0.5-0.8% manganese, and the balance iron. Used for highly stressed mechanical springs.

Crispella. Trade name for rayon fibers and yarns used for textile fabrics.

Crispline. Trade name of Libbey-Owens-Ford Company (USA) for a patterned glass.

Cristal B. Trade name for a corrosion-resistant nickel-chromium dental bonding agent.

Cristallino. Trade name of Vernante SpA (Italy) for patterned glass.

Cristamid. Trade name of Atofina SA (France) for polyamide 11 and 12 (nylon 11 and 12) resins usually supplied in the form of granules.

Cristanola. Trade name of Cristaleria Espanola SA (Spain) for a polished plate glass.

Cristobal. Trademark of Degussa-Ney Dental (USA) for a dental composite resin supplied in two-component form (powder plus liquid) and used for metal, composite or ceramic restorations including crowns, bridges, inlays, onlays and implants.

Cristoll. Trade name of Vidrierias de Lllodio SA (Spain) for a thick sheet glass.

Cristite. Trade name of Commercial Alloys Company (USA) for a series of abrasion- and wear-resistant alloys composed of 17% tungsten, 10-12% chromium, 3-3.5% carbon, 0-6% nickel, 2-2.5% molybdenum, 0-0.8% titanium, 0-0.3% hafnium, and the balance iron. They have good high-temperature resistance, and are used for hardfacing electrodes for facing dredgers, shovel teeth and bucket lips, agricultural implements, knives, hand and cutting tools, etc.

cristobalite. A colorless mineral that is a crystalline allotropic form of silica (SiO_2) formed by the inversion of quartz at 1470°C (2678°F). Crystal system, tetragonal. Density, 2.30-2.34 g/cm³; melting point, 1713°C (3115°F); hardness, 2.5-3.0; refractive index, 1.487. Used as a major component of silica refractories, in investment casting of metals, and sometimes in siliceous ceramic bodies.

Criterion. Trade name of Taskem Inc. (USA) for a nickel electrodeposit and plating process.

Critnic. Trade name for a nickel alloy containing 2.8% titanium and 0.2% carbon. Used for corrosion- and heat-resisting parts.

Crobalt. Trade name of Crobalt Inc. (USA) for a series of wear-resistant cast alloys containing 40-50% cobalt, 25-33% chromium, 14-20% tungsten, 0-9% iron, and sometimes carbon. Used for cutters, tools, tool bits, and punches.

Crobrit. Trade name for a dental resin for temporary crown and bridge restorations.

Crocar. Trade name of Teledyne Vasco (USA) for a high-carbon, high-chromium cold-work tool steel (AISI type D4) containing 2.2% carbon, 0.6% manganese, 12% chromium, 1% molybdenum, 0.9% vanadium, and the balance iron. It has good wear resistance and good resistance to decarburization. Used for blanking, extrusion, trimming and stamping dies, punches, gages, and forming tools.

Crocem. Trade name of Forjas Alavesas SA (Spain) for a chromium-manganese carburizing steel containing 0.16% carbon, 1.2% manganese, 1% chromium, and the balance iron. Used for camshafts, transmission shafts, gears, piston pins, and gages.

crochet lace. A fine or heavy *lace* that is hand-made by interlocking loops of a single yarn with a hooked needle. Used for sweaters, shawls, doilies, tablecloths, etc.

crocidolite. An indigo-blue, lavender-blue or leek-green variety of *riebeckite* asbestos belonging to the amphibole mineral group. It is a silicate of iron and sodium with the approximate chemical formula $Na_2Fe_5Si_8O_{22}(OH)_2$, and occurs in fibrous, massive, and earthy forms. The fibrous form is preferred for commercial purposes. Asbestos fibers of this type have diameters up to 2.5 mm (0.1 in.) and lengths up to 76 mm (3 in.), high tensile strength, high heat resistance up to about 650°C (1200°F), good insulating properties, and good resistance to most chemicals. Occurrence: South Africa. Used for spinning and weaving fireproof fabrics, as a reinforcing agent in plastics and rubber, as a molding powder, and for electrical and heat insulation applications. Also known as *blue asbestos; cape asbestos; cape blue.*

crocoisite. See crocoite.

crocoite. A hyacinth-red mineral of the monazite group composed of lead chromium oxide, $PbCrO_4$. Crystal system, monoclinic. Density, 5.99 g/cm³; refractive index, 2.36. Also known as *crocoisite; red lead ore.*

crocus. A fine natural or synthetic ferric oxide (Fe_2O_3) powder with a bright red color. It is very soft, and used for cleaning and polishing, when a minimum of stock is to be removed, e.g., in buffing cutlery and some nonferrous metals (e.g., tin).

crocus martis. Impure purple, yellowish, or brownish red ferric oxide (Fe_2O_3) used as a pigment in decalcomanias and glazes, and in polishing powders.

Crodi. Trade name of Atlas Specialty Steels (Canada) for a chromium-type hot-work tool steel (AISI type H12) containing 0.35% carbon, 0.5% manganese, 5% chromium, 1.4% molybdenum, 1.2% tungsten, 0.3% vanadium, and the balance iron. It has excellent resistance to decarburization, high toughness, good machinability, high hot hardness, and good abrasion resistance. Used for hot punches, hot shear blades, header, trimmer and extrusion dies, etc.

Crodon. Trade name of Chromium Corporation of America (USA) for an electroplating alloy of chromium and iron. Used for anodes in electrolytic copper cells.

Crodur. Trade name of SWB Stahlformguss Gesellschaft mbH (Germany) for a series of nondeforming high-carbon high-chromium type cold-work tool steels used for punches, and blanking and forming dies.

Crofer. Trade name of TyssenKrupp VDM GmbH (Germany) for a series of corrosion- and/or heat-resistant cast and wrought austenitic, ferritic and martensitic stainless steels including several welding grades. Used for petroleum-refining, chemical-process, dairy and heat-treating equipment, tanks, mixers, furnace parts and fixtures, turbine blades, jet engine, valve and pump parts, cutlery, and marine hardware.

Crofil. Trademark of Sinteticos Slowak SA (Uruguay) for nylon 6 fibers and monofilaments.

Croflex. Trade name for carbon black.

Croform. British trade name for a corrosion-resistant dental casting alloy containing 60% cobalt, 30% chromium, 5% iron and 5% molybdenum.

Croform Excel. British trade name for a corrosion-resistant cobalt-chromium dental casting alloy.

Croform Exten. British trade name for a heat-cure acrylic resin for dentures.

Croform Repair. British trade name for a self-cure acrylic resin for denture repair.

Croform Springhard. British trade name for a corrosion-resistant cobalt-chromium dental casting alloy.

croisé. A fabric made of cotton or worsted yarn in a twill weave. Used especially for making clothing.

Crolac. Trade name for carbon black used as a pigment in lacquers, varnishes and paints.

Crolan. Trademark of Celanese Mexicana SA (Mexico) for polyester fibers and filaments.

Croloy. Trade name of Babcock & Wilcox Company (USA) for an extensive series of austenitic, ferritic, martensitic and precipitation-hardening stainless steels, heat- and oxidation-resistant steels, and several low-carbon chromium-molybdenum steels for welding rods and high-strength applications.

Croma. Trade name of Lehigh Steel Corporation (USA) for a carbon steel containing 0.35% carbon, 0.8% manganese, 1% chromium, and the balance iron. Used for machinery parts.

Cromadur. Trade name of Krupp Stahl AG (Germany) for a heat-resistant stainless steel containing up to 0.15% carbon, 18% manganese, 12.5% chromium, 1% vanadium, 0.2% nickel, and the balance iron. Used for gas-turbine blades and jet-engine parts.

Cromal. Russian trade name for a corrosion-resistant cast iron containing 1.8% carbon, 20-22% chromium, 2.2% silicon, 3-4% aluminum, and the balance iron. Used for furnace equipment.

Cromansil. Trade name of Union Carbide Corporation (USA) for a medium-carbon steel containing 0.1-0.3% carbon, 1.1-1.4% manganese, 0.7-0.8% silicon, 0.4-0.6% chromium, and the balance iron. Used for boilers, pressure vessels, tanks, ship plate, buildings, bridges, and seamless tubing. Heat-treated grades are also used for gears, shafts, etc.

Cromargan. Trade name for a corrosion-resistant austenitic stainless chromium-nickel steel.

Cromax. Trade name of Creusot-Loire (France) for a series of wear-resistant steel castings containing 0.5-0.75% carbon, 0.2-0.35% silicon, 0.5-0.8% manganese, 0.7-2.0% chromium, and the balance iron. Used for machinery parts, gears, shafts, housings, and wear plates.

Cromaz. Trade name of Creusot-Loire (France) for a series of heat-resisting steels with high thermal resistance up to 955°C (1750°F). Used for furnace parts and heat-treating boxes.

Crome. Trade name of Bethlehem Steel Corporation (USA) for a shock-resistant tool steel containing 0.5% carbon, 0.5% manganese, 1% chromium, 0.35% molybdenum, and the balance iron. Used for battering tools.

Crometex. Trade name of Grayborn Steel Company (USA) for a shock-resistant steel containing 0.38-0.43% carbon, 0.7% manganese, 0.3% silicon, 0.8% chromium, 0.25% molybdenum, 1.8% nickel, and the balance iron. Used for arbors, axles, gears, and piston rods.

Cromex. Russian trade name for a heat-resistant steel containing 0.5-1% carbon, 25-32% chromium, varying amounts of silicon, and the balance iron. Used for furnace parts and equipment.

cromfordite. See phosgenite.

Cromimphy. Trade name of Creusot-Loire (France) for an extensive series of ferritic and martensitic stainless steels containing about up to 0.25% carbon, 0-1% manganese, 0-1% silicon, 12-30% chromium, 0-1% molybdenum, 0-2.5% nickel, 0-0.3% vanadium, 0.04% phosphorus, 0.03% sulfur, and the balance iron. Used for machine parts, valve and pump components, fasteners, cutlery, hardware, surgical instruments, aircraft-engine parts, mining equipment, heat-treating equipment, furnace parts, household appliances, and turbine blades.

Cromin. Trade name of Gilby-Fodor SA (France) for a resistance alloy containing 9-11% nickel, 19-21% chromium, and the balance iron. It has a maximum service temperature of 343°C (650°F), and is used for heating elements.

Cromino. Trade name of Ambo-Stahl-Gesellschaft (Germany) for an extensive series of corrosion- and heat-resistant ferritic and martensitic stainless steels including several free-machining and welding grades. The nominal composition is up to 0.9% carbon, 0-1% manganese, 0-1% silicon, 0-1% molybdenum, 0-2.5% nickel, 0-0.3% vanadium, 0.04% phosphorus, 0.03% sulfur, and the balance iron. Used for machine parts, chemical and oil-refining equipment, valve and pump components, bearings, fasteners, cutlery, hardware, surgical instruments, aircraft-engine parts, mining and heat-treating equipment, furnace parts, household appliances, sinks, and food-processing equipment.

Cromo. Trade name of Bethlehem Steel Corporation (USA) for a series of steels including several hot- and cold-work tool steels.

Cromoco. Trade name of Teledyne Firth-Sterling (USA) for an air-hardening tool and die steel containing 1.6% carbon, 12% chromium, 1% cobalt, 1% molybdenum, and the balance iron. Used for coining, blanking and forming dies, swages, and metal shears.

Cromodi. Trade name of Pyramid Steel Company (USA) for a nondeforming, preheat-treated steel used for brake dies.

Cromodie. (1) Trade name for a series of high-carbon die steels containing 1-2.4% carbon, 5.25% chromium, 1.1-1.2% molybdenum, 0.25-4.1% vanadium, and the balance iron. They have good hardenability, and are used for hot-work dies.

(2) Trade name of Firth Brown Limited (UK) for a tough, oil-hardening tool and die steel containing 0.31-0.35% carbon, 0.9-1% silicon, 0.3% manganese, 5% chromium, 0-1.75% molybdenum, 0-1.1% tungsten, 0-1.5% magnesium, 0.25-1% vanadium, and the balance iron. Used for punches, crimpers and dies.

Cromol. Trade name of Continental Foundry & Machine Company (USA) for alloy steel castings containing 0.4% carbon, 1.1% chromium, 0.3% molybdenum, and the balance iron. Used for rams, sow blocks, and dies.

Cro-Mo-Loy. Trade name of Atlas Specialty Steels (Canada) for an air-hardening cold-work tool steel (AISI type A2) containing 1.0% carbon, 1.0% manganese, 5.0% chromium, 0.25% vanadium, 1.0% molybdenum, 1.0% cobalt, and the balance iron. It has good nondeforming properties, and good wear resistance and toughness. Used for blanking, forming and trimming dies, tools, shear blades, punches, etc.

Cromonite. Trade name of Continental Foundry & Machine Company (USA) for a wear-resistant steel used for chills rolls.

CromoPal. Trade name for a dental gold solder.

Cromopal. Trade name of Saint Gobain (France) for a pot-color glass mosaic whose surface has been obscured by sand blasting.

Cromotung. Trade name of Ziv Steel & Wire Company (USA) for a tough, oil-hardening tool steel containing carbon, tungsten and iron. Used for taps, drills, reamers, punches, and dies.

Cromovan. Trade name of Teledyne Firth Sterling (USA) for a heat- and wear-resistant tool and die steel containing 1.4-1.7% carbon, 12-14% chromium, 0.5-1% molybdenum, 1-1.5% vanadium, and the balance iron. It has good nondeforming properties and high abrasion resistance. A free-machining grade (*Cromovan F.M.*) with 0.12% sulfur is also available. Used for blanking, extrusion, pressing, swaging, thread-rolling and trimming dies, punches, shear blades, brick-mold liners, reamers, broaches, and taps.

Cromtube. Trade name of Cromweld Steels Limited (UK) for low-carbon high-chromium stainless steel tubing containing about 0.03% carbon, 1.5% manganese, 1% silicon, 10-12% chromium, 1.5% nickel, 0.6% titanium, 0.04% phosphorus, 0.03% sulfur, 0.03% nitrogen, and the balance iron. Used for food and chemical-plant equipment, building exterior components, vehicle chassis and frames, and power transmission structures.

Cromva. (1) Trade name of Bisset Steel Company (UK) for a series of water- or oil-hardening tool steels containing 0.35% carbon, 1% chromium, 0.2% vanadium, and the balance iron. Used for punches, dies, upsetters, and crimpers.

(2) Trade name of Bissett Steel Company (UK) for an oil-hardening steel containing 0.5% carbon, 0.6% manganese, 1% chromium, 0.2% vanadium, and the balance iron. Used for tools and die-casting dies.

(3) Trade name of Bethlehem Steel Corporation (USA) for a hot-work steel containing 0.35-0.5% carbon, 0.6% manganese, 1% chromium, 0.2% vanadium, and the balance iron. Used for hot-work dies.

Cromylite. Trade name of Enthone-OMI Inc. (USA) for a decorative chromium electroplate and plating process.

Cronelka. Trade name for rayon fibers and yarns used for textile fabrics.

Croni. Trade name of Forjas Alavesas SA (Spain) for a series of nickel-chromium carburizing steels containing 0.13-0.19% carbon, 0.5-0.9% manganese, 0.25% silicon, 1-2.9% nickel, 0.7-1% chromium, and the balance iron. Used for shafts, pins, bolts, gears, axles, etc.

Cronidur. Trade name of Krupp Stahl AG (Germany) for an extensive series of strong, corrosion- and/or heat-resistant austenitic, ferritic and martensitic stainless steels.

Cronifer. Trade name of ThyssenKrupp VDM GmbH (Germany) for an extensive series of corrosion- and heat-resistant wrought austenitic chromium-nickel stainless steels used for furnace and chemical-process equipment, acid digesters and tanks, pressure vessels, storage vessels for acids and gases, chemical and nuclear reactor equipment, agitators and mixers, high-temperature applications, transportation and bulk-handling equipment, and seawater piping and structures.

Cronimo. Trade name of Colt Industries (UK) for an oil-hardening die steel containing 0.5% carbon, 1.1% chromium, 3.25% nickel, 0.25% molybdenum, and the balance iron.

Croniro. Trade name of GTE Products Corporation (USA) for a brazing alloy containing 72% gold, 22% nickel and 6% chromium. It is commercially available in powder, wire, foil, preform, flexibraze and extrudable paste form. It has a liquidus temperature of 1000°C (1830°F), and a solidus temperature of 975°C (1787°F). Used for high-temperature brazing applications.

Cronit. British trade name for a resistance alloy of 60% nickel and 40% chromium used for electrical resistances.

Cronite. Trade name of Cronite Foundry Company Limited (UK) for age-hardenable, heat-resistant nickel casting alloys and cast stainless steels. The nickel casting alloys contain about 14.5-30% iron, 13-25% chromium, 0-1% tungsten, 0-0.7% carbon, 0-0.3% molybdenum and 0-0.2% cobalt, and the stainless steel castings about 0.5% carbon, up to 25% chromium, up to 20% nickel, varying amounts of molybdenum and/or tungsten, and the balance iron. *Cronite* alloys have high strength at elevated temperatures, and are used for heat-treating equipment, furnace parts, fire doors, grids, etc.

Cronix. Trade name of ThyssenKrupp VDM GmbH (Germany) for a nickel-chromium alloy supplied in the form of semifinished products.

cronstedtite. A black to brownish black mineral of the kaolinite-serpentine group composed of iron silicate hydroxide, $Fe_3(Si,Fe)_2O_5(OH)_4$. Crystal system, monoclinic. Density, 3.34 g/cm³. Occurrence: UK.

Crookes glass. A glass that contains rare earths, such as cerium, and has a low transmission for infrared and ultraviolet radiation. Used for sunglasses.

crookesite. A lead-gray mineral composed of copper thallium silver selenide, $(Cu,Tl,Ag)_2Se$. Crystal system, tetragonal. Density, 6.90 g/cm³. Occurrence: Sweden.

Cro-Sil. Trade name of Colt Industries (UK) for a series of corrosion-resistant steels containing 0.08-0.2% carbon, 3% silicon, 10% chromium, and the balance iron. Used for corrosion-resistant parts.

Crospane. Japanese trade name for a figured glass.

cross-blended fibers. A single yarn spun from a blend of natural and synthetic fibers (e.g., wool and rayon), or animal and vegetable fibers (e.g., wool and cotton).

cross-creped paper. A thin, lightweight, strong *kraft paper* with a crinkled surface texture produced by creping in different directions.

cross-dyed fabrics. Textile fabrics with contrasting color effects produced by differently dyeing the cotton, rayon, silk or wool yarns used in their manufacture.

crossite. (1) A blue mineral of the amphibole group composed of sodium iron aluminum silicate hydroxide, $(Na,Ca)_2(Fe,Mg,Al)_5(Si,Al)_8O_{22}(OH)_2$. Crystal system, monoclinic. Density, 3.21 g/cm³; refractive index, 1.670. Occurrence: USA (California), France.

(2) A lavender-blue or gray mineral of the amphibole group composed of sodium magnesium iron aluminum silicate hydroxide, $Na_2(Fe,Mg)_5(Si,Al)_8O_{22}(OH)_2$. Crystal system, monoclinic. Density, 3.21 g/cm³; refractive index, 1.656. Occurrence: UK (Wales).

cross laminate. See bidirectional laminate.

crosslinked cellulose. A cellulose that has been modified by the addition of a chemical substance (crosslinker) that initiates the formation of chemical links between the cellulose molecules. Crosslinking enhances several properties including the crease resistance and drying properties.

crosslinked collagen. An insoluble polymer produced from *collagen* obtained from the extracellular matrix by crosslinking with agents, such as formaldehyde, glutaraldehyde and several isocyanates. Used as a biomaterial in tissue engineering.

crosslinked high-density polyethylenes. A class of high-density grades of crosslinked polyethylene with off-white natural color, excellent stress-crack and impact resistance, outstanding weatherability, very good chemical resistance, good abrasion resistance, high rigidity, good resistance to acids, alkalies and chemical compounds, poor resistance to strong oxidizing agents, aromatic and halogenated aliphatic hydrocarbons, liquefied petroleum gas and solvents, a maximum service temperature of 65°C (150°F), and a brittleness temperature of -118°C (-180°F). Used for piping and molded fittings, tanks, chemical equipment, laboratory ware, storage tanks and equipment for sulfuric, hydrochloric and hydrofluoric acids, sodium hydroxide and other corrosives, for boiler treatment equipment and storage vessels, and for water and sewage treatment equipment. See also crosslinked polyethylenes.

crosslinked polyethylenes. A class of previously thermoplastic *polyethylenes* within which chemical links have been set up between the molecular chains, e.g., by high-energy radiation, or chemically with an organic peroxide as crosslinking agent, to produce three-dimensional, translucent structures that are thermoset in nature. They are available in high-density, low-density and low-resistivity grades, and have high creep, impact and tensile strength, high thermal and electrical resistance, good stress-crack resistance, a service temperature range of -118 to +65°C (-180 to +150°F), excellent resistance to weak and dilute acids, good resistance to strong and weak bases and many solvents, poor resistance to strong and concentrated acids, good resistance to ozone and gamma radiation, low permeability to carbon dioxide, oxygen and nitrogen, and poor ultraviolet resistance. Used for wire and cable coatings and insulation, pipe and molded fittings, tanks, tubing, laboratory ware, and in semiconductor applications. Abbreviation: XLPE.

crosslinked polyimides. A class of *polyimides* derived from an addition reaction between unsaturated groups of imide monomers or oligomers. Examples are bismaleimides, reverse Diels-Alder polyimides and acetylene end-capped polyimides. *Crosslinked polyimides* have a continuous service temperature range of 177-232°C (351-450°F), and are used as matrix resins for laminates and composites, e.g., for the aircraft and aerospace industries. Abbreviation: XLPI.

crosslinked polymers. See network polymers.

crosslinked polystyrenes. A class of previously thermoplastic *polystyrenes* within which chemical links between the molecular chains have been set up chemically with a cross-linking agent, such as divinylbenzene (DVB), and subsequent exposure to heat in order to produce three-dimensional structures that are thermoset in nature. They have a density of 1.05 g/cm³ (0.038 lb/in.³), good tensile strength, high thermal and electrical resistance, good stress-crack resistance, good resistance to alpha, beta, gamma and X-ray radiation, good resistance to ultraviolet radiation, a maximum service temperature of 93°C (200°F), good resistance to dilute acids and alcohols, fair resistance to concentrated acids, alkalies, greases and oils, and poor resistance to aromatic hydrocarbons, halogens and ketones. Used for packaging, containers and molded household items, machine housings, electrical equipment, as dielectrics in X-ray equipment, radiation detectors, etc. It is also available in bead form for chromatographic applications. Abbreviation: XLPS.

crosslinked rubber. See vulcanized rubber.

crosslinker. See crosslinking agent.

crosslinking agent. A chemical substance that is added to a polymer to promote the setting up of chemical links between the molecular chains, or to a rubber to promote a chemical reaction in which the rubber's physical properties are changed by causing it to react with the crosslinking agent. *Crosslinking agents* for polymers include organic perioxides, such as benzoyl peroxide for polyethylene and divinylbenzene for polystyrene. The most common crosslinking agents for rubber are sulfur, organic peroxides, metallic oxides, chlorinated quinines, and nitrobenzenes. Also known as *crosslinker.*

Crosslite. Trade name of C-E Glass (USA) for patterned sheet glass with large raised bands or ribbons crossing each other on opposite sides.

Crossnet. Trade name of Libbey-Owens-Ford Company (USA) for patterned glass.

Cross Pipes. Trade name of Osborn Steels Limited (UK) for a cold-work tool steel containing 1.5% carbon, 11% chromium, 2% cobalt, and the balance iron. Used for press tools and forging dies.

cross-plied laminate. A nonparallel laminate in which adjacent plies or layers are oriented at right angles to each other. Also known as *cross-ply laminate.*

Crossral. Trade name of Central Glass Company Limited (Japan) for a patterned glass.

Cross Reeded. Trade name of Pilkington Brothers Limited (UK) for a glass with a cross-reeded pattern on one surface.

Crossweld. Trade name of Libbey-Owens-Ford Company (USA) for a wired glass with square mesh.

Crostar. Trade name of Société Nouvelle des Acieries de Pompey (France) for a cold-work tool steel containing 2.0% carbon, 12.5% chromium, 0.1% vanadium, and the balance iron. Used for dies and press tools.

crotch wood. Those sections of wood cut from the joint between the limbs, branches and trunks of a tree. They have beautiful grain patterns, and are used for fine furniture, construction, and veneer.

Crotorite. Trade name of Delta Metal (BW) Limited (UK) for alloys containing up to 10% aluminum, 0-6.3% nickel, 0-3% iron, 0-0.6% manganese, and the balance copper. They have

good strength and corrosion resistance, and are used for machine parts.

Crouse Fusible Alloy. British trade name for a series of white metal alloys containing 45% bismuth, 25% tin, 25% lead, and 5% cadmium. They have a melting point of 88°C (190°F), and are used for fire extinguishers, fire sprinklers, etc.

Crovac. Trade name of Vacuumschmelze GmbH (Germany) for a series of malleable, hard-magnetic alloys containing varying amounts of iron, chromium, cobalt and molybdenum. Commercially available as strips and wires, they can be deep drawn and punched, and have a density of 7.6 g/cm³ (0.27 lb/in.³), a hardness of 330 Vickers (rolled), a Curie temperature of 640°C (1185°F), a maximum service temperature of 480°C (900°F), good magnetic properties comparable to those of high-grade aluminum-nickel-cobalt alloys, and high magnetic energy products. Used for moving permanent magnetic systems, rotational frequency pick-ups, tachometers, polarized relays, and switching magnets for reed relays.

Crovan. Trade name of Atlas Specialty Steels (Canada) for a hot-work tool steel (AISI type H13) containing 0.35% carbon, 5% chromium, 1% silicon, 1.4% molybdenum, 0.9% vanadium, and the balance iron. It has excellent thermal-shock resistance and good nondeforming properties. Used for die-casting dies, plastic molds, and extrusion tools.

Crovani. Trade name of Haeckerstahl GmbH (Germany) for an oil-hardening, shock-resistant tool steel containing 0.58% carbon, 1.05% chromium, 0.1% vanadium, and the balance iron. Used for springs, bolts, studs, and gears.

Crow. Trade name of AL Tech Specialty Steel Corporation (USA) for a water-hardening tool steel (AISI type W5) containing 1.2% carbon, 0.5% chromium, varying amounts of vanadium, manganese and silicon, and the balance iron. It has excellent machinability, good cutting and wear properties, and low toughness. Used for drills, reamers, broaches, tools, cutlery dies, and drawing and bending dies.

Crowelon. Trademark of Crowe Rope Industries Company (USA) for polypropylene monofilaments.

Crown. (1) Trade name of Fibreglass Limited (UK) for a series of glass-fiber insulation products.

(2) Trade name of Crown Crystal Glass (Australia) for pressed and blown glassware.

(3) Trade name of Latrobe Steel Company (USA) for a special-purpose tool steel (AISI type L2) containing 0.5-1.1% carbon, 0.1-0.9% manganese, 1% chromium, 0.2% vanadium, and the balance iron. It has excellent toughness, good to moderate machinability, and moderate wear resistance. Used for forming tools, and forming and trimming dies.

(4) Trade name of Jessop-Saville Limited (UK) for an oil-hardening, shock-resisting tool steel containing 0.4% carbon, 1.3% chromium, 2.3% tungsten, 0.15% vanadium, and the balance iron. Used for hot dies, pneumatic tools, chipping chisels, and punches.

Crown Brand. Trade name of Robinson Ransbottom Pottery Company (USA) for stoneware and pottery.

Crown Cast. Trademark of Pentron Laboratory Technologies (USA) for a corrosion-resistant nonprecious dental casting alloy composed of 63% cobalt, 30% chromium, 3% molybdenum, and the balance manganese, silicon and boron. Used for crown and bridge restorations.

Crown Clay. Trade name of Southeastern Clay Company (USA) for hard rubber clay.

crown ethers. A group of cyclic polyethers that contain four or more oxygen atoms and can hold a metal cation in a hole in the middle of the ether molecule through ion-dipole bonds. They are named by the total number of atoms in the puckered heterocyclic ring and the corresponding oxygen atoms, e.g., 18-crown-6 ($C_{12}H_{24}O_6$) contains 18 atoms in the ring of which 12 are carbon and 6 oxygen. Used as catalysts to transfer ionic compounds into organic phases.

crown flint glass. See lead crown glass.

crown glass. A hard, white, easily polished, lead-free optical glass typically containing 72% silica (SiO_2), 15% soda (Na_2O), and 13% lime (CaO), and sometimes small amounts barium oxide (BaO), alumina (Al_2O_3), boria (B_2O_3) and/or fluorine, phosphorus, arsenic and zinc. According to their characteristic composition, they are called barium, borosilicate, fluor, phosphate, or zinc crown. *Crown glass* has high refraction, an average refractive index of 1.5, a low-dispersion value, a nu-value (v value) above 55.0, high transparency, and relatively low melting temperature. Used for electric lamp bulbs, windows, containers, high-grade bottles, and optical lenses. Abbreviation: CG. Also known as *optical crown glass*.

Crowngold. Trade name of Engelhard Electro Metallics Division (USA) for a gold strike for plating onto stainless steels.

Crown Hill Stone. Trademark of Crown Hill Stone, Inc. (USA) for manufactured stone used as building veneer.

Crownlite. Trade name of Crown Crystal Glass (Australia) for painted and fired glass used for cladding applications.

Crown Max. Trade name of Firth-Vickers Stainless Steels Limited (UK) for a wrought austenitic stainless steel containing 0.2% carbon, 23% chromium, 11.5% nickel, 2.8% tungsten, and the balance iron. Used for stainless parts, valves, pumps, etc.

Crown Superb. Trade name of Latrobe Steel Company (USA) for a high-carbon tool steel (AISI type L2) containing 0.75% carbon, 0.95% chromium, 0.18% vanadium, and the balance iron.

Crownwrap. Trade name of Fibreglass Limited (UK) for fiberglass insulation supplied in batts or blankets with a thickness of 63 mm (2.5 in.). Used for the thermal insulation of domestic attics.

Crozat. Trade name of American Tooth Industries (USA) for a wrought cobalt-chromium-nickel-iron wire for dental applications.

CR-PAA. Trademark of Hexcel Corporation (USA) for phosphoric acid-anodized aluminum honeycomb materials.

CRS. Trade name of Tata Iron and Steel Company Limited (India) for corrosion-resistant reinforcing bars of high-strength steel that contains 0.12% phosphorus and 0.50% copper, and may also contain 0.8% chromium. They form passive protective surface oxides, and are suitable for a wide range of applications in reinforced-concrete construction.

CRS Wrap. Trade name of Lydall, Inc. (USA) for nonwoven thermal fiber insulation supplied in the form of rolls.

Crucast. Trade name of Colt Industries (UK) for a series of cast, heat- and corrosion-resistant chromium and chromium-nickel stainless steels.

Crucia. Trade name of Republic Steel Corporation (USA) for a steel containing 0.8-0.95% carbon, 0.3-0.5% manganese, 0.2-0.4% chromium, and the balance iron. Used for cutters, springs, tools, and dies.

Crucia Steel. Trade name of Republic Steel Corporation (USA) for a steel containing 0.75-1.15% carbon, 0.3-0.5% manganese, 0.1-0.5% silicon, 0.15-3.25% chromium, and the balance iron.

Used for tools, dies, taps, reamers, and cutters.

Crucible. Trade name of Crucible Materials Corporation (USA) for an extensive series of austenitic, ferritic and martensitic stainless steel, high-carbon high-chromium steels, heat-resisting steels, hot-work, cold-work and shock-resisting tool steels, wear-resisting austenitic high-manganese steels, high-strength structural steels, iron-nickel superalloys, commercially pure titanium, titanium alloys, etc.

crucible clay. A *ball clay* that is refractory enough to be used in the manufacture of high-temperature crucibles and pots.

Crucible Kapstar. Trade name of Crucible Materials Corporation (USA) for a plastic mold and die-casting steel.

Crucible Marlok. Trade name of Crucible Materials Corporation (USA) for a maraging ultrahigh-strength die steel containing up to 0.01% carbon, 18% nickel, 11% cobalt, 5% molybdenum, up to 0.3% titanium, and the balance iron. It has an excellent heat check resistance, and is used for long-run, aluminum die-casting dies.

crucible steel. A high-quality steel with low phosphorus and sulfur contents made in an electric induction furnace from a charge of ferroalloys, pig iron, steel scrap, or ingot iron.

Crucin. Trade name of Vereinigte Deutsche Nickelwerke AG (Germany) for a corrosion- and wear-resistant alloy of 56% copper and 44% nickel, used for kettles and kitchen utensils.

Crucore. Trademark of Crucible Materials Corporation (USA) for *samarium cobalt* permanent magnet materials.

crude-dressed mica. Crude mica from which contaminants have been removed.

crude mica. Mica in the as-mined condition with contaminants, such as dirt, rock, etc., still present.

crude rubber. A soft, tacky, crosslinkable (vulcanizable) natural or synthetic rubber in bale or package form. It is thermoplastic and has low resiliency and strength. Used as a starting material for the manufacture of rubber goods. Also known as *raw rubber*.

crude tar. A viscous or liquid bituminous material derived by the destructive distillation of organic substances.

Crumax. Trademark of Crucible Materials Corporation (USA) for neodymium-iron-boron permanent-magnet materials.

Crumb Rubber. Trade name of Rubberecycle (USA) for coarse, brittle chips of recycled rubber about 20 mm (0.75 in.) in size, obtained by cryogenic freezing of used rubber goods, e.g., automotive tires.

crumb rubber. Coarse, porous particles of rubber used in adhesives and plastics.

Crumeron. Trademark of Zoltek Magyar Viscosa RT (Hungary) for acrylic fibers.

Crusader. Trade name of Latrobe Steel Company (USA) for a high-speed steel (AISI type M3) containing 1.2% carbon, 4% chromium, 6.1% tungsten, 2.4-3.2% vanadium, 5% molybdenum, and the balance iron. It has good machinability, red-hardness, wear resistance and edge toughness, and is used for roll-turning cutters, planer and lathe tools, and form cutters.

Crusca. Trade name of Colt Industries (UK) for a series of plain carbon tool steels containing 0.8% carbon, 0.15% silicon, 0.3% manganese, and the balance iron. Used for hollow, solid and twist drills, cutting and blanking dies, tools, and molds and dies for plastics.

Crusco Steel. Trade name of Colt Industries (UK) for a tungsten-type high-speed steel containing 0.7% carbon, 18% tungsten, 4% chromium, 1% vanadium, 5% cobalt, and the balance iron. It has high abrasion resistance at elevated temperatures, and is used for forming rolls, dies, piercing points, and rolling-mill plugs.

crushed gravel. Gravel whose particle size has been reduced by a primary crushing operation. Its particle size ranges usually from about 7 to 30 mm (0.25 to 1.20 in.). Used as an aggregate in concrete.

crushed gypsum. The gypsum product resulting from a primary crushing operation.

crushed slag. A blast-furnace slag with an average weight of 1.10-1.25 g/cm³ (0.040-0.045 lb/in³) that has been crushed for use in cement and lightweight concrete. Also known as *broken slag.*

crushed steel. An abrasive made from *crucible steel* by heating, quenching in cold water, crushing the resulting fragments, and subsequent sizing and grading. Used for grinding, polishing and cleaning metals, stone, brick, glass, etc.

crushed stone. The product obtained by crushing or grinding quarried rocks, boulders, cobblestones, etc., into smaller irregular fragments usually about 30-70 mm (1.20-2.75 in.) in size. Also known as *broken stone.*

Crusher. Trade name of Jenney Cylinder Company (USA) for a lead-base bearing bronze containing varying amounts of tin and copper. Used for heavy-duty bearings.

crusher-run aggregate. Aggregate as it comes from a mechanical crusher and prior to screening.

crush-resistant fabrics. Pile fabrics, such as corduroy or velvet, that have received a special finishing treatment to enhance the ability of the pile to recover after being crushed or flattened by pressure.

CrushRegister. See Anso CrushRegister.

Crutanium. Trade name for a corrosion-resistant cobalt-chromium dental casting alloy.

Cru-Wear. Trade name of Crucible Specialty Steel (USA) for air-hardening wear-resistant tool steels.

Cruxite. Trade name of Aearo Company (USA) for an anti-glare spectacle glass with a pinkish to brownish tint.

Crylon. Trademark for acrylic fibers used for the manufacture of textile fabrics.

Crylor. Trademark of Rhodiaceta (France) for acrylic (polyacrylonitrile) polymers supplied in fiber and other forms.

Cryoflux BE. Trade name of BPI Inc. (USA) for a *cryolite* byproduct containing alumina and various fluorides. A typical chemical composition is 50% alumina, 22% sodium fluoaluminate, 22% aluminum fluoride, 3% carbon, 2% aluminum metal, the balance being lithium, magnesium and calcium fluoride. Used as a metallurgical flux in steelmaking.

cryogenic conductors. See superconductors.

cryogenic steels. A class of steels including ferritic nickel steels with about 5-9% nickel, and austenitic stainless steels containing about 15-26% chromium, 5-22% nickel and 2-10% manganese. They have high toughness and medium to high strength at temperature as low as -269°C (-452°F), and excellent resistance to low-temperature embrittlement. Used for tanks, vessels and containers for storing and handling cryogenic liquids, for pipes, tubing, valves and pumps used in the transfer of cryogenic liquids, and for Dewar flasks, bubble chambers, liquefaction equipment, cylindrical magnet tubes for superconducting magnets, cryogenic components for superconducting equipment, magnets and transmission systems, and structural components for aircraft, space vehicles and missiles. Also known as *cold-resistant steels; low-temperature steels.*

Cryogenic Tenalon. Trade name of US Steel Corporation for an austenitic stainless steel containing 0.1% carbon, 15.1% man-ganese, 0.7% silicon, 17.5% chromium, 5.5% nickel, 0.38% nitrogen, and the balance iron. It has high strength at low temperatures, and is used for cryogenic structures.

Cryolite. Trademark of Cyro Industries (USA) for impact-resistant acrylic molding and extrusion compounds used for medical devices, medical packaging, and food containers.

cryolite. A colorless to white, occasionally reddish brown, or black mineral with a white streak and a greasy to vitreous luster. It is composed of sodium fluoroaluminate, Na_3AlF_6, containing about 54.3% fluorine and 12.9% aluminum. It can also be made synthetically from fluorite, sodium carbonate, sulfuric acid and hydrated aluminum oxide. Crystal system, monoclinic. Density, 2.95-3.00 g/cm³; hardness, 2.5 Mohs; refractive index, 1.3377; melting point, 1020°C (1868°F). Occurrence: Greenland. Used as a flux and opacifier in opal glass and porcelain enamels, as a flux in whiteware bodies, as a constituent in dental cements, light bulbs and welding-rod fluxes, as a flux in producing aluminum from bauxite (electrolysis process), as a filler in grinding wheels, as a binder for abrasives, in the manufacture of soda, and as an electrical insulator. Also known as *Greenland spar; icestone.*

cryolite glass. A translucent or nearly opaque white or colored glass made by adding cryolite (Na_3AlF_6) and zinc oxide (ZnO), or fluorspar (CaF_2) and alumina (Al_2O_3) to an ordinary soda-lime glass batch. Used for windows and doors, and optical applications. Also known as *fusible porcelain; milk glass; milky glass.*

cryolithionite. A colorless mineral of the garnet group composed of lithium sodium fluoroaluminate, $Li_3Na_3Al_2F_{12}$. Crystal system, cubic. Density, 2.77 g/cm³; refractive index, 1.3395. Occurrence: Greenland, Russia.

Cryoperm. Trade name of Vacuumschmelze GmbH (Germany) for a soft-magnetic alloy containing 77% nickel and 23% iron. It has high initial permeability and low saturation induction, and is used for cryogenic engineering applications.

Cryovac. Trademark of W.R. Grace & Company (USA) for low-temperature-resistant polyolefins (e.g., polyethylene) with good tensile and seal strengths. It is available in the form of shrink-film with good heat-sealing properties for packaging food items.

cryptocyanine. A light-sensitive solid compound (96% pure) with a melting point of 250.5-253°C (483-487°F) and a maximum absorption wavelength of 703 nm. Used as an organic photosensitizing dye, and as a laser dye. Formula: $C_{25}H_{25}N_2I$. Also known as *1,1'-diethyl-4,4'-carbocyanine iodide; kryptocyanine.*

cryptohalite. A colorless mineral composed of ammonium silicon fluoride, $(NH_4)_2SiF_6$. It can also be made synthetically. Crystal system, cubic. Density, 2.01 g/cm³; refractive index, 1.3696.

cryptomelane. A steel-gray, bluish-gray, or black mineral of the rutile group. It is composed of potassium manganese oxide, KMn_8O_{16}, and found in high-grade manganese ores. Crystal system, monoclinic, or tetragonal. Density, 4.35-4.58 g/cm³. Occurrence: Gabon, Ghana, India, USA (Montana). Used as a source of manganese and battery manganese.

Cryptone. Trade name for a series of white pigments containing high quantities of zinc sulfide, and varying amounts of barium sulfate, calcium sulfide, titanium dioxide, and zinc oxide.

Crysel. Trademark of Celulosay Derivados SA (Mexico) for acrylic fibers.

Crys-Glas. Trade name of Dearborn Glass Company (USA) for solid-glass wall tiles with fired-on colors.

Crystacel. Trade name for die stones used in dentistry.

Crystal. Trade name of Walter Wurdack, Inc. (USA) for silicone-base waterproofing compounds for masonry surfaces.

Crystal 2100. Trade name of Reade Advanced Materials (USA) for an ion-exchange material for filtration and water-purification applications. It can remove the following cationic metals simultaneously: silver, cadmium, cobalt, copper, iron, mercury, manganese, nickel, lead and zinc.

crystal. (1) A transparent, colorless natural or synthetic crystal of quartz (SiO_2) with high purity and piezoelectric properties. Used for lenses and prism components in optical instruments, and for jewelry and ornaments. Also known as *quartz crystal; rock crystal.*

     (2) A hand-cut, or engraved blown glassware that takes a high polish.

     (3) See crystalline materials.

crystal bar. A pure, solid metal, such as hafnium, titanium or zirconium, produced by the reaction of the impure starting metal with iodine to form a volatile tetraiodide that is subsequently decomposed on a hot wire at high temperatures to form the high-purity metal and iodine.

Crystal Bay. Trade name of 3M Company (USA) for pressure-sensitive adhesive tapes and emery cloth-coated abrasives.

crystal carbonate. See sodium carbonate monohydrate.

Crystal Clear. (1) Trade name of ATW Manufacturing Company, Inc. (USA) for shrink wrapping film.

     (2) Trade name of Loctite Corporation (USA) for a glass adhesive.

Crystal-Core. Trade name of Crystal Research (USA) for crystal-cored glass fibers.

Crystalcrete. Trade name of James Clark & Eaton Limited (UK) for panels made up of thick textured glass, assembled to any design and set in reinforced concrete.

Crystalex. Trade name of Rohm & Haas Company (USA) for for acrylic resins and plastics.

crystal glass. A water-clear, highly-transparent high-quality glass without bubbles or schlieren, often with lead oxide (PbO) to further increase its refractive index. It is made from high-purity raw materials, such as crystallized soda or rock crystal, and takes a fine polish. Used for art- and tableware. Also known as *lead crystal.*

Crystal Glaze. Trade name of Porcelanite, Inc. (USA) for glazed wall tiles.

crystal glaze. See crystalline glaze.

Crystalgrit. Trade name of Virginia Materials & Supplies, Inc. (USA) for *rouge* and dustfree *specular hematite* blasting abrasives.

Crystalite. (1) Trade name of Crystalite Corporation (USA) for diamond and cubic boron nitride products.

     (2) Trademark of Georgia-Pacific Corporation (USA) for weather-resistant stucco cement for exterior finishing applications.

     (3) Trade name of Rohm & Haas Company (USA) for a synthetic acrylate resin.

     (4) Trade name of Rohm & Haas Company (USA) for an acrylic molding powder.

Crystallin. Trade name for cast phenolic plastics with good thermal and electrical resistance.

crystalline alumina. An abrasive grade of alumina that is composed predominantly of the mineral corundum (α-Al_2O_3) and whose grains may range from microcrystalline to coarsely crystalline in size.

Crystallined. Trade name of Pilkington Brothers Limited (UK) for a glass with irregular, frost-like pattern obtained by coating one obscured surface with glue followed by drying under controlled conditions of temperature and humidity.

crystalline calcium. An air- and moisture-sensitive free-flowing crystalline powder (94-97% pure) supplied under argon. It has a density of 1.54 g/cm³ (0.056 lb/in.³), and a melting point of 851°C (1564°F). Used in the reduction of light-metal ores.

crystalline glaze. A devitrified glaze in which crystallization has taken place during the cooling period following firing. Also known as *crystal glaze.*

crystalline graphite. A soft, crystalline allotropic form of carbon that occurs naturally, but can also be produced synthetically. It is available in various grades as crystals, flakes, powder, rods, plates and fibers. See also graphite.

crystalline limestone. See calcite marble.

crystalline materials. Solid materials that have a regular arrangement of the atoms, ions or molecules in space, i.e., that are characterized by long-range order. Also known as *crystals.* See also amorphous materials; liquid crystals; microcrystalline materials; nanocrystalline materials.

crystalline plastics. See semicrystalline plastics.

crystalline polymers. See semicrystalline plastics.

crystalline silicon. The form of silicon in which the atoms have the cubic-diamond crystal structure arrangement. It is available in the form of dark-colored crystals including large single crystals for semiconductor applications. Commercially pure crystalline silicon (96-98%) is obtained by heating silica with carbon in an electric furnace, followed by zone refining. The ultrahigh-purity, semiconductor-grade (99.97%) is made by reduction of purified tetrachlorosilane or trichlorosilane with purified hydrogen. *Crystalline silicon* has a density of 2.34 g/cm³ (0.085 lb/in.³), a melting point of 1410°C (2570°F), a boiling point of 2355°C (4270°F), a hardness of 7.0 Mohs, and high electrical resistivity. See also amorphous silicon; silicon.

crystallized verdigris. See copper acetate.

crystallizing lacquer. A novelty lacquer that when applied to a suitable substrate crystallizes forming unusual crystal and floral patterns during drying.

Crystalloy. Trade name of Transformer Steels Limited (UK) for an alloy containing 97% iron and 3% silicon. It has high magnetic permeability, and is used for transformers, and laminations for motors and generators.

Crystalon. (1) Trademark of Phillips Paint Products Limited (Canada) for polyurethane varnishes.

     (2) Trade name of Nano Film, Limited (USA) for non-stick, molecular film coatings for glass and plastics.

     (3) Trade name for ayon fibers and yarns used for textile fabrics.

Crystalor. Trademark of Phillips Chemical Company (USA) for engineering resins based on polymethylpentene, and supplied in unfilled and glass-reinforced grades. They have high arc resistance, a low dielectric constant, a low dissipation factor, excellent resistance to acids and alkalies, and good resistance to most automotive fuels and fluids. Used for automotive component including engine components, power trains, distributor caps, sensors, and connectors.

crystals of Venus. See copper acetate.

crystal whisker. A single crystal that has such a high degree of growth anisotropy that it forms fine, short fibers or filaments. See also whisker.

Crystar. (1) Trademark of E.I. DuPont de Nemours & Company (USA) for specialty resins based on polyethylene terephthalate. Supplied in clear and white grades, they are used in the manufacture of polyester films, engineering resins, and fibers.

(2) Trademark of Norton Company (USA) for a recrystallized silicon carbide used for hot-surface igniters in gas-fired appliances.

Crystic. Trademark of Scott Bader & Company Limited (USA) for a series of unsaturated polyesters used for low-pressure plastic laminates, potting compounds, coatings, etc.

Crystic Impel. Trademark of Scott Bader & Company Limited (USA) for a general-purpose polyester molding compound supplied in unfilled and mineral-filled grades.

Crystle. Trade name of Marblette Corporation (USA) for phenolic resins and plastics and molded products.

Crystolex. Trademark of Kerr Dental (USA) for acrylic resins used in dentistry.

Crystolon. Trademark of Norton Company (USA) for a series of silicon carbide (SiC) materials including (i) regular SiC used as an abrasive for grinding and polishing, and in refractories; (ii) stabilized SiC used for refractories and thermal-shock-resistant parts; and (iii) self-bonding SiC used for coating graphite components. Also known as *Norton Crystolon*.

Cryston. Trade name of Norton-Pakco Industrial Ceramics (USA) for high-performance silicon nitride-bonded silicon carbide based ceramics supplied in various of shapes. They have a bulk density of 2.61 g/cm^3 (0.094 $lb/in.^3$), a hardness of 2700 Knoop, an apparent porosity of 18%, and a maximum service temperature of 1590°C (2900°F). Used for abrasion- and wear-resistant applications.

C-Silicone. Trade name for a condensation-cured silicone for dental impressions.

C-silicone. A silicone resin that has been cured (hardened) by a condensation-type chemical reaction. Also known as *condensation-cured silicone*.

CSIOX. Trade name of Catalytic Solutions Inc. (USA) for catalytic coating materials based on metal oxides, such as perovskites, that owing to their crystal structures can reduce and convert carbon monoxide, nitrogen oxides and hydrocarbons. Used in automotive emission control systems.

C Special. Trademark of Degussa-Ney Dental (USA) for an extra-hard dental alloy (ADA type IV) containing 51.8% gold, 24.2% silver, and 5% palladium. It has high yield strength, excellent physical properties, and a hardness of 280 Vickers. Used in restorative dentistry for crown and bridgework.

C-stage resin. A thermosetting resin of the phenolic type that is in the final stage of polymerization. It is infusible and insoluble in virtually all ordinary solvents. Also known as *resite*.

C-Temp. Trade name of G.O. Carlson Inc. (USA) for corrosion-resistant nickel-alloy and stainless steel plate products with excellent high-temperature strength and stability, and very good thermal fatigue and shock resistance, and good formability and weldability. Supplied in standard plate form in thicknesses from 4.76 mm (0.1875 in.), widths of 3.2 m (10.5 ft.) and lengths of 12.2 m (40 ft.), and as cut bars, rings, disks, welded cylinders, tubesheets, heads, and special shapes. Used for thermal-processing applications.

CTI Nylon. Trade name of Compounding Technology Inc. (USA) for a series of polyamide (nylon) resins including CTI Nylon 6, CTI Nylon 6/10, CTI Nylon 6/12 and CTI Nylon 11. They are supplied in various grades including standard, glass or carbon fiber-reinforced, silicone-, molybdenum disulfide- or polytetrafluoroethylene-lubricated and fire-retardant. Some CTI nylons are also supplied in glass bead- and/or mineral-filled, high-impact, UV-stabilized, stampable-sheet, elastomer-copolymer, semiflexible, flexible, and/or casting grades.

CTI PBT. Trade name of Compounding Technology Inc. (USA) for a series of polybutylene terephthalate resins supplied in various grades including standard, glass or carbon fiber-reinforced, silicone-, molybdenum disulfide- or polytetrafluoroethylene-lubricated, glass bead- and/or mineral filled, structural foam, UV-stabilized and fire-retardant.

CTI PC. Trade name of Compounding Technology Inc. (USA) for a series of polycarbonate resins supplied in various grades including standard, glass or carbon fiber-reinforced, polytetrafluoroethylene-lubricated, structural foam, high-flow, UV-stabilized and fire-retardant.

CTI PEEK. Trade name of Compounding Technology Inc. (USA) for a series of polyetheretherketone resins supplied in unreinforced, 10, 20 or 30% glass fiber-reinforced and 30% carbon fiber-reinforced grades.

CTI PES. Trade name of Compounding Technology Inc. (USA) for a series of polyether sulfone resins supplied in various grades including unreinforced, 20 and 30% glass fiber-reinforced and 30% carbon fiber-reinforced grades

CTI PI. Trade name of Compounding Technology Inc. (USA) for a series of polyimide resins supplied in various grades including standard, 40% glass fiber-reinforced and graphite-, molybdenum disulfide- or polytetrafluoroethylene-lubricated.

CTI PPS. Trade name of Compounding Technology Inc. (USA) for a series of polyphenylene sulfide resins supplied in various grades including 40% glass fiber-reinforced, 30% carbon fiber-reinforced, glass bead-reinforced and 20% polytetrafluoroethylene-lubricated.

CTI PSUL. Trade name of Compounding Technology Inc. (USA) for a series of polysulfone resins supplied in various grades including standard, 10 and 30% glass fiber-reinforced, 30% carbon fiber-reinforced and 15% polytetrafluoroethylene-lubricated.

CTI SAN. Trade name of Compounding Technology Inc. (USA) for a series of styrene-acrylonitriles supplied in various grades including standard, 30% glass fiber-reinforced, fire-retardant, high-heat, high-impact and UV-stabilized.

cubane. A molecular compound that is composed of 8 carbon atoms located at the corners of a cube and 1 symmetrical hydrogen atom linked to each carbon atom. It can be prepared in solid molecular crystal form. Applications include fuels, drugs, and chemical explosives. Formula: C_8H_8.

cubanite. A brass to bronze-yellow mineral composed of copper iron sulfide, $CuFe_2S_3$. Crystal system, orthorhombic. Density, 4.10 g/cm^3. Occurrence: Canada (Ontario). Also known as *chalmersite*.

Cuban mahogany. See Spanish mahogany.

Cuban pine. See Caribbean pine.

Cuba wood. See fustic.

Cu-Be. Trademark of Telcon Metals Limited (UK) for a series of age-hardenable, corrosion-resistant alloys containing 0.25-3.0% beryllium, 0.25-2.5% cobalt, and the balance copper. Used for connectors, strips, springs, wire, etc.

Cube-Alloy. Trade name of Handy & Harman (USA) for a beryllium oxide-strengthened copper with high creep resistance used for motor windings, magnet wire, cables, welding electrode tips, and springs.

Cube Injection Mold. Trade name of Telcon Metals Limited (UK) for an age-hardenable, corrosion-resistant alloy of 2.7% beryllium, 0.5% cobalt, and the balance copper. Used for plastic injection molds and zinc die-casting dies.

Cubelloy. Trade name of Imperial Smelting Corporation Limited

(UK) for a master alloy containing 96% copper and 4% beryllium. Used in the manufacture of beryllium-copper alloys.

cube ore. See pharmacosiderite.

cube spar. See anhydrite.

Cubex. Trademark of Westinghouse Electric Corporation (USA) for an oriented silicon-iron electrical steel (about 3% silicon) in rolled sheet form having cubic grains with faces parallel to the surface, and two directions of easy magnetization (preferred directions) parallel to the surface, one parallel and the other perpendicular or transverse to the rolling direction. *Cubex* steels have high magnetic permeability and low energy losses, and are used for electrical and magnetic equipment, motors, transformers, and inductive devices.

cubic aggregate. An angular aggregate consisting of particles which are mostly cubic or near-cubic in shape.

cubic boron nitride. Boron nitride produced at extremely high pressure and temperature from ordinary *boron nitride*, or from mixtures of boron and nitrogen. It has a cubic crystal structure and is supplied in the form of small crystals. The color ranges from colorless or white for chemically-pure grades to reddish or black. It has a density of 3.5 g/cm^3 (0.12 $lb/in.^3$), a hardness of 10 Mohs (as hard as diamond), and high stability up to about 1925°C (3500°F). Used as an general abrasive, and as an abrasive for grinding wheels used for grinding hardened tool steels and high-speed steels. It is also used as a cutting tool material, e.g., for indexable inserts and cutting tool blanks that consist of polycrystalline cubic boron nitride coatings firmly bonded to cemented carbide substrates and are employed for high-speed machining of very hard and tough materials, such as hardened plain carbon and alloy steels, hardened tool steels, hard cast irons, and superalloys. Abbreviation: CBN; cBN. Also known as *polycrystalline boron nitride*.

Cubicut. Trademark of 3M Company (USA) for coated abrasives.

cubic ferrites. A group of magnetically hard ceramic materials having inverse spinel crystal structures with cubic symmetry, and a general chemical composition of $MO \cdot Fe_2O_3$, or MFe_2O_4, where M is a divalent metal cation, such as cadmium, cobalt, copper, iron, magnesium, nickel or zinc. Magnesium ferrite ($MgO \cdot Fe_2O_3$, or $MgFe_2O_4$) and nickel ferrite ($NiO \cdot Fe_2O_3$ or $NiFe_2O_4$) are examples. See also inverse spinels.

cubic spinels. See synthetic spinels.

cubic zirconia. See yttrium-stabilized zirconia.

Cubitron. Trademark of 3M Company (USA) for coated abrasives.

CuBond. Trademark of Glidden-Durkee Division of SCM Corporation (USA) for a series of copper brazing pastes composed of copper powder on high-purity cuprous oxide pigments in petroleum or organic vehicles. Used for brazing ferrous components in a furnace with reducing atmosphere.

CuBraz. Trademark of Wall Colmonoy Corporation (USA) for a powdered copper brazing filler metal (99.0+% copper) suspended in a gel-type binder and supplied in disposable cartridges for use in air-powered applicators. It has a brazing temperature of 1093-1149°C (2000-2100°F), and is used for furnace brazing carbon and alloy steels to irons.

Cu-Brite. (1) Trade name of Heatbath Corporation (USA) for a deoxidizing agent and chemical polish for copper and its alloys.

(2) Trade name of Univertical Corporation (USA) for electrolytic copper supplied in the form of roll-forged anodes for cyanide copper plating.

Cu Clad. Trademark of 3M Company (USA) for copper-based

integrated-circuit materials.

Cudo. Trade name of Flachglas AG (Germany) for double glazing units with profiles fastened to the ground glass edges and soldered at the corners.

Cudo-Auresin. German trade name of Flachglas AG (Germany) for a double glazing unit reflecting about 60% of the incident solar energy. It is composed of one sheet of plate glass, and one sheet of glass covered with gold by an evaporation process.

Cudo-Bronze. German trade name of Flachglas AG (Germany) for a double glazing unit that appears yellowish brown when viewed from the outside, and warm gray from the inside. It is composed of one pane of plate glass, and one pane of metal-coated glass.

Cudo-Gold. German trade name of Flachglas AG (Germany) for a double glazing unit reflecting about 74% of the incident solar energy. It is composed of one sheet of plate glass and one sheet of gold-coated glass.

Cudo-Grau. German trade name of Flachglas AG (Germany) for a double glazing unit that appears neutral in color whether viewed from the outside or inside. It is composed of one pane of plate glass and one pane of metal-coated glass.

Cudo Infrastop. German trade name of Flachglas AG (Germany) for a double glazing unit that has a heat- and sunlight-reflecting metallic coating on the inner surface of one sheet.

Cufenium. Trade name for a corrosion-resistant nickel silver containing 60-72% copper, 20.5-22% nickel and 6-19.5% iron. Used as a base metal for tableware.

Cufenloy. Trade name of Phelps Dodge Industries (USA) for a high-strength, corrosion-resistant alloy of 29-41% nickel, 0.5-2% iron, 0.3-1% manganese, and the balance copper. Used for heat-exchanger tubes, high-pressure feedwater heaters, high-temperature alloys, and castings.

Cuferco. Trade name of Westinghouse Electric Corporation (USA) for a hardenable alloy containing 2% iron, 2% cobalt, and the balance copper. It has an electrical conductivity about 0.7 times that of copper, and is used for spot-welding tips.

Cu-Flex. Trademark of 3M Company (USA) for copper-coated plastic sheeting used for circuit-board laminates.

Cu-High Ten. Trade name of Nippon Steel Corporation (Japan) for a hot-rolled sheet steel with extra-low carbon content (0.0013%) and 1-1.6% copper. When heat treated at about 475-675°C (890-1250°F) following cold forming, it develops high mechanical strength due to copper precipitation hardening. Used especially for automotive applications, such as cross members and side frames.

Cuivral. French trade name for copper-clad aluminum.

Cuivre Poli. A French cartridge brass containing 70% copper and 30% zinc. It has high ductility, excellent cold workability, and is used for cartridges, shell cases, and condenser tubes.

Culimeta. Trade name of Culimeta Textilglas-Technologie GmbH & Co. KG (Germany) for textile glass-fiber yarn and glass-fiber fabrics, ribbons, tubing, etc.

Culminal. Trademark of Henkel KGaA (Germany) for methylcellulose plastics.

CuLox. Trade name of Sylvania Chemicals and Metals (USA) for low-oxidizing copper powders available in average particle sizes from 3-10 μm (120-395 μin.) at oxygen levels of about 5000-3000 ppm. Used in thick-film pastes for hybrid circuits.

Cultured Brick. Trade name of Cultured Stone Corporation (USA) for manufactured stone veneer.

cultured granite. A building material composed of a cast polymer (e.g., unsaturated polyester) usually containing considerable quantities of fillers and pigments and made to resemble natural *granite* in appearance. Supplied in both polished and matte surface finishes, it is used in the manufacture of kitchen countertops, vanity tops, sinks, shower enclosures, whirlpools, soaking tubs, etc.

cultured marble. A building material composed of a cast polymer (e.g., unsaturated polyester) usually containing considerable quantities of fillers and pigments and made to resemble natural *marble* in appearance. Supplied in several surface finishes, it is used in the manufacture of kitchen countertops, vanity tops, sinks, shower enclosures, whirlpools, soaking tubs, etc.

cultured onyx. A building material composed of a cast polymer (e.g., unsaturated polyester) usually containing considerable quantities of fillers and pigments and made with high surface brightness to resemble natural *onyx* in appearance. Supplied in both polished and matte surface finishes, it is used in the manufacture of kitchen countertops, vanity tops, sinks, shower enclosures, whirlpools, soaking tubs, etc.

Cumanite. Trade name of Abex Corporation (USA) for a wear-resistant steel containing carbon, manganese and iron. Used for machine-tool parts.

Cumar. Trademark of Neville Chemical Company (USA) for a series of stable, neutral, synthetic *coumarone-indene resins* made from selected distillates of tar. Used as softeners and tackifiers in varnishes, in rubber compounding to enhance tear resistance, in adhesives and waterproofing compounds, and in floor tile and printing ink.

cumar gum. A synthetic resin formed by copolymerization of coumarone (C_8H_6O) and indene (C_9H_8). Supplied in several grades ranging from hard solids to soft, tacky gums, it has good alkali resistance and a melting range of 5-140°C (41-284°F). Used especially in adhesives, varnishes, and rubber compounds. Also known as *paraindene; paracoumarone.*

cumarin. See coumarin.

Cumberland. Trade name of Brown-Wales Company (USA) for a case-hardening steel containing 0.18-0.23% carbon, and the balance iron. Used for machine parts, such as gears, shafts, etc.

cumengite. An indigo-blue mineral composed of copper lead chloride hydroxide hydride, $Pb_4Cu_4Cl_8(OH)_8 \cdot H_2O$. It can also be made synthetically. Crystal system, tetragonal. Density, 4.70; refractive index, 2.041. Occurrence: Mexico.

cummingtonite. A beige mineral of the amphibole group composed an iron magnesium silicate hydroxide, $(Fe,Mg)_7Si_8O_{22}(OH)_2$. Crystal system, monoclinic. Density, 3.37 g/cm³; refractive index, 1.671. Occurrence: Canada (Labrador).

Cumloy. Trade name of West Steel Casting Company (USA) for a tough steel containing 0.3-0.4% carbon, 1.25-1.5% manganese, 0.3-0.4% silicon, 0.2-0.3% molybdenum, 0.12% vanadium, and the balance iron. Used for shafts, gears, and machinery parts.

Cumuloft. Trade name for a strong, lightweight nylon fiber with high resistance to abrasion, wrinkling, mildew and moths. Used for lofty clothing.

Cunic. Trade name of George W. Prentiss & Company (USA) for a resistance alloy containing 55% copper and 45% nickel. It has good low-temperature resistance, and is used for thermocouples, shunts, and rheostats.

Cunico. Trade name of General Electric Company (USA) for a hard-magnetic material containing 50% copper, 21% nickel, and 29% cobalt. It has a density of 8.3 g/cm³ (0.30 lb/in.³), a Curie temperature of 860°C (1580°F), a hardness of 95 HRB, high coercive force, high magnetic energy product, and a maximum service temperature of 500°C (930°F). Used for permanent magnets for magnetic and electrical equipment.

Cunife. Trademark for Hoskins Manufacturing Company (USA) for a ductile hard-magnetic material containing 60% copper, 20% nickel and 20% iron. Commercially available in the form of cast and/or rolled wires, strips and finished magnets, it has a density of 8.6 g/cm³ (0.31 lb/in.³), a Curie temperature of 410°C (770°F), a hardness of 95 Rockwell B, a magnetic orientation developed by rolling or other mechanical working, high coercive force, high magnetic energy product, and a maximum service temperature of 350°C (660°F). Used for permanent magnets for speedometers, instruments, electrical equipment, control systems, and magnetic recording wire.

Cunifer. Trademark of ThyssenKrupp VDM GmbH (Germany) for a series of copper-nickel-iron alloys used in the chemical, petrochemical and marine industries. Also includes several welding alloys for submerged-arc and electroslag overlay welding.

Cunilate. Trademark of Scientific Chemicals Inc. (USA) for an antifouling, fungicidal paint based on copper quinolate, and used on ship bottoms and marine structures.

Cuniloy. Trade name for a corrosion-resistant alloy containing 25% copper, 3.8% manganese, 1% lead, and the balance nickel. Used for valve parts, pump rods, etc.

Cunimat. Trade name of ThyssenKrupp VDM GmbH (Germany) for a copper-nickel alloy supplied in the form of semifinished products.

Cunip. Trade name of Handy & Harman (USA) for a heat-treatable copper alloy containing 1.1% nickel and 0.2-0.3% phosphorus. Commercially available in wire and strip form, it has high tensile strength and electrical conductivity, and is used for electron-tube components, cathode supports, tuning fingers, and spring clips.

Cunisil. Trade name of Anaconda Company (USA) for a corrosion-resistant, precipitation hardening copper alloy containing 97.5% copper, 1.9% nickel and 0.6% silicon. It has high strength, good electrical conductivity (about 35% IACS), and good to fair machinability. Used for electrical apparatus and equipment, electrical hardware, and mechanical fasteners.

Cupa. Trade name of Vereinigte Silberhammerwerke Hetzel (Germany) for copper-clad aluminum.

Cupal. Trade name of Ambolt Machine Tool Company (USA) for corrosion-resistant copper-clad aluminum sheet and strip products. The coating forms 10% of the total material thickness on each side. *Cupal* has good cold workability, good solderability and coatability, and high electrical and thermal conductivities. Used as paneling for railroad and subway cars, in electrical engineering, and in machine and instrument construction. See also copper-clad aluminum.

Cupalloy. Russian trade name for a high-conductivity copper containing 0.5% chromium, and used for electrical equipment and motors.

Cupaloy. Trade name of Westinghouse Electric Corporation (USA) for an age-hardenable, high-conductivity copper containing 0.5% chromium and 0.1% silver. Commercially available in wrought and cast form, it has high strength, high electrical conductivity (about 85% IACS), and is used for commutators, slip rings, terminal studs, welding electrodes, and electrical components.

Cupersil. Trade name for silky cuprammonium rayon fibers and yarns used for textile fabrics.

cup grease. See lime grease.

Cu-Phos. Trade name of Univertical Corporation (USA) for copper supplied in the form of roll-forged anodes for copper acid plating.

Cupioni. Trade name for cuprammonium rayon used for the manufacture of chiffons, net, sheer, satin and similar fabrics for warp-knit underwear, dresses, linings, and hosiery.

Cuplat. Trade name of Western Gold & Platinum Company (USA) for a corrosion-resistant alloy of 60% copper and 40% platinum. It has a melting range of 1185-1216°C (2165-2220°F) and is used for brazing cathode structures.

cupola iron. An iron produced by melting a charge composed of pig iron, foundry cast iron and steel scrap and a flux in contact with metallurgical coke in a cylindrical vertical refractory-lined furnace known as "cupola." Used in the manufacture of castings.

cupola malleable iron. Blackheart malleable iron produced from white cast iron by melting in a cylindrical vertical refractory-lined furnace known as a "cupola," or in a cupola in conjunction with an air furnace. The chemical composition of white iron suitable for this purpose is 2.80-3.30% carbon, 0.60-1.10% silicon, 0.40-0.65% manganese, less than 0.25% sulfur and 0.20% phosphorus, respectively, and the balance iron. *Cupola malleable iron* is highly fluid and may be used to make sound castings and castings with very light sections. It has very good machinability, relatively low strength, and its average density (7.15 g/cm³ or 0.258 lb/in.³) and shrinkage are lower than those of standard malleable grades. Used for pipe fittings, valves, building hardware, and small tools. See also blackheart malleable iron.

cupola refractories. Refractories that are highly resistant to abrasion, erosion, chemicals and wear, and are used in the manufacture of cupola furnaces, i.e., cylindrical vertical refractory-lined furnaces for melting cast iron and remelting pig iron and scrap. The cupola lining is usually high or medium heat-duty fireclay brick. Patching mixes for lining repairs are composed of broken refractory brick and plastic fireclay or silica sand and high-grade plastic fireclay.

Cupolloy. Trademark of Foseco Minsep NV (Netherlands) for briquetted *ferroalloys* containing a refractory binder to reduce oxidation losses of the alloying additions. They are usually made in unit quantities, and used as a cupola addition to cast iron.

Cupolux. Italian trade name for glass blocks.

Cuposit. Trade name of Shipley Company Inc. (USA) for electroless copper.

cupra. See cuprammonium rayon.

Cupracolor. Trade name for colored cuprammonium rayon fibers and yarns used for textile fabrics.

Cuprakote. Trade name of Heatbath Corporation (USA) for a copper coating produced by immersion.

Cupralan. Trade name for cuprammonium rayon fibers and yarns used for textile fabrics.

Cupralinox. Trade name of Le Bronze Industriel (France) for a series of cast and wrought aluminum, aluminum-nickel and aluminum-manganese bronzes.

Cupralith. Trade name for a series of copper-lithium alloys containing 1-10% lithium.

Cupralium. German trade name for a non-heat-treatable aluminum alloy containing 7-12.5% copper. Used for light-alloy parts.

Cupralon. Trademark of Bayer AG (Germany) for a yarn made from spun rayon (*Cuprama*) and nylon (*Perlon*).

Cupraloy. Trade name for a dental amalgam alloy.

Cupralum. Trade name of Knapp Mills Inc. (USA) for an acid-resistant material consisting of lead-clad copper. Used for chemical equipment.

Cuprama. Trademark of Bayer AG (Germany) for spun rayon produced by dissolving cotton linters in an ammoniacal copper solution and reconverting it to cellulose by treating with acid. Used for the manufacture of chiffons, net, sheer, satin and similar fabrics for warp-knit underwear, dresses, linings, and hosiery.

cuprammonium rayon. Fibers or filaments of rayon produced by dissolving cotton linters in an ammoniacal copper solution and reconverting to cellulose by treating with acid. *Cuprammonium rayon* is attacked by mildew and disintegrated by acids, swell in alkalies, and burns quickly. Used for dresses, shirts, shear fine-denier fabrics, knitted fabrics, draperies, upholstery, etc. Also known as *cuprammonium cellulose; cupra; cupro; cupro silk*. See nitrocellulose rayon; rayon fibers; viscose rayon.

Cupranium. British trade name for a nickel silver used for corrosion-resistant parts.

cuprates. A class of layered perovskite-type compounds that contain atomic planes of copper and oxygen with rare-earth and other metal ions (e.g., yttrium, lanthanum, barium, calcium, strontium, thallium, lead, etc.) located between these planes. *Cuprates* exhibit superconductivity at relatively high transition temperatures (Tc > ~ 35K). Examples of high-temperature superconductor cuprates include yttrium barium copper oxide (Y-$Ba_2Cu_3O_7$) and thallium barium calcium copper oxide (Tl_2Ba_2-$Ca_2Cu_3O_{10}$). See also high-temperature superconductors.

Cuprel. Trade name for cuprammonium rayon fibers and yarns used for textile fabrics.

Cupresa. Trade name for cuprammonium rayon used for the manufacture of chiffons, net, sheer, satin and similar fabrics for warp-knit underwear, dresses, linings, and hosiery.

Cuprex. (1) Trade name of Foseco Minsep NV (Netherlands) for a series of degassing and covering fluxes for copper and nickel alloys used to provide full or partial oxidizing conditions required to limit hydrogen pick-up during melting.

(2) Trade name of Murex Limited (UK) for a plain carbon steel containing 0.07% carbon, 0.35% manganese, 0.5% copper, and the balance iron. Used for welding electrodes.

(3) Trade name of Atofina Chemicals, Inc. (USA) for a semi-bright acid copper plating system for circuit boards.

cupric abietate. See copper abietate.

cupric acetate. See copper acetate.

cupric acetoarsenite. See copper acetoarsenite.

cupric arsenite. See copper arsenite.

cupric borate. See copper metaborate.

cupric bromide. Gray, hygroscopic powder, or black crystals (99+% pure) available in cubic and hexagonal form. Density, 4.77 g/cm³; melting point, 498°C (928°F). Used as a battery electrolyte, as a semiconductor, as an intensifier in photography, and as a wood preservative. Formula: $CuBr_2$. Also known as *copper bromide.*

cupric carbonate. See copper carbonate.

cupric chloride. Hygroscopic, brownish-yellow powder, or yellow crystals (97+% pure). Density, 3.386 g/cm³; melting point, 620°C (1148°F); boiling point, decomposes to cuprous chloride at 993°C (1819°F); refractive index, 1.93. Used for refining copper, gold and silver, in the recovery of mercury from its ores, in electroplating copper on aluminum, and for semiconductors and superconductors. Formula: $CuCl_2$. Also known as *copper chloride*.

cupric chloride dihydrate. Green, deliquescent crystals (99+% pure). It occurs in nature as the mineral *eriochalcite*. Density, 2.51 g/cm^3; melting point, loses $2H_2O$ at 100°C (212°F). Used in refining copper, gold and silver, in electroplating baths, as a pigment for glass and ceramics, as a wood preservative, in sympathetic ink, in the manufacture of acrylonitrile, and in fireworks, photography and water treatment. Formula: $CuCl_2 \cdot 2H_2O$. Also known as *copper chloride dihydrate*.

cupric cyanide. A green or yellow-green powder made by treating solution of copper sulfate with potassium cyanide. Used in electroplating copper on iron and steel, and in organic synthesis to introduce the cyanide group in aromatic compounds. Formula: $Cu(CN)_2$. Also known as *copper cyanide*.

cupric ferrocyanide. See copper ferrocyanide.

cupric fluoride. White to light-gray, hygroscopic powder (98+% pure) turning blue in moist air. Crystal system, cubic. Crystal structure, sphalerite. Density, 4.23 g/cm^3; melting point, 908°C (1666°F); boiling point, decomposes at 950°C (1742°F). Used in the investigation of halogen reactions with superconductors, and as a semiconductor. Formula: CuF_2. Also known as *copper fluoride*.

cupric fluoride dihydrate. Blue-green crystals (97+% pure) with a density of 2.93 g/cm^3. Used as a metallurgical flux, as a flux and colorant in porcelain enamels and glazes, in ceramics, in high-energy batteries, and as a fluorinating agent. Formula: $CuF_2 \cdot 2H_2O$. Also known as *copper fluoride dihydrate*.

cupric glycinate. See copper glycinate.

cupric hydroxide. See copper hydroxide.

cupric lactate. See copper lactate.

cupric nitrate. See copper nitrate hexahydrate; copper nitrate hydrate; copper nitrate trihydrate.

cupric nitrate hexahydrate. See copper nitrate hexahydrate.

cupric nitrate hydrate. See copper nitrate hydrate.

cupric nitrate trihydrate. See copper nitrate trihydrate.

cupric oleate. See copper oleate.

cupric oxalate. See copper oxalate.

cupric oxide. See black copper oxide.

cupric phosphide. Gray-black, metallic powder made by heating copper and phosphorus. Density, 6.67 g/cm^3. Used in the manufacture of phosphor bronze. Formula: Cu_3P_2. Also known as *copper phosphide*.

cupric resinate. See copper resinate.

cupric selenate. See copper selenate.

cupric selenide. Fine, black powder, or black, high-purity crystals. Density, 5.99 g/cm^3; melting point, decomposes at 550°C (1020°F). Used in ceramics, electronics, and as a semiconductor. Formula: $CuSe$. Also known as *copper selenide*.

cupric stearate. See copper stearate.

cupric sulfate. See copper sulfate.

cupric sulfate pentahydrate. See copper sulfate pentahydrate.

cupric sulfide. Black, crystalline powder, or lumps. It occurs in nature as the mineral *covellite*. Crystal system, hexagonal. Density, 3.9-4.6 g/cm^3; melting point, decomposes at 220°C (428°F); refractive index, 1.45. Used in antifouling paints for ships and marine structures, and in dyeing. Formula: CuS.

cupric tungstate. See copper tungstate.

Cuprifil. Trade name for cuprammonium rayon fibers and yarns used for textile fabrics.

Cuprit. Trade name of Foseco Minsep NV (Netherlands) for a series of covering and protecting fluxes added to copper and cop-

per-zinc alloys to significantly decrease zinc losses due to fuming during melting.

cuprite. A deep red or orange-red secondary mineral with a brownish-red streak and a dull or adamantine luster. It is composed of cuprous oxide, Cu_2O. Crystal system, cubic. Density, 5.85-6.15 g/cm^3; hardness, 3.5-4 Mohs. Occurrence: Australia, Bolivia, Chile, China, France, Germany, Peru, Russia, UK, USA (Arizona). Used as a source of copper. Also known as *red copper ore; red oxide of copper; ruby copper ore*.

cupro. See cuprammonium rayon.

cupro-aluminum. An aluminum bronze containing 10% aluminum, 0-2% iron, and the balance copper. Used for gears, worm wheels, pinions, etc.

cuprobismutite. A dark bluish gray mineral composed of copper bismuth sulfide, $Cu_4Bi_5S_{12}$. It can also be made synthetically. Crystal system, monoclinic. Density, 6.47-6.81 g/cm^3. Occurrence: USA (Colorado).

Cuprochrome. Trade name of Wilbur B. Driver Company (USA) for a high-conductivity copper containing 1-2% chromium, and the balance copper. It has good hot and cold workability, and is used for electron tubes.

cuprocopiapite. A sulfur-yellow mineral of the copiapite group composed of copper iron sulfate hydroxide hydrate, $CuFe_4(SO_4)_6(OH)_2 \cdot 20H_2O$. Crystal system, triclinic. Density, 2.13 g/cm^3; refractive index, 1.575. Occurrence: Greenland.

Cuprodie. Trade name of A. Finkl & Sons Inc. (USA) for a tool steel containing 0.5-0.8% carbon, 0.65-0.9% manganese, 1.4-1.75% nickel, 0.8-1.1% chromium, 0.25-0.35% molybdenum, 0.6-0.9% copper, and the balance iron. It has high resistance to heat checking, and mechanical properties that depend largely on its hardness. Used for drop-hammer dies, forging press dies, and hammer dies.

Cuprodine. Trade name of Henkel Surface Technologies (USA) for a copper immersion coating for steel.

cupromagnesium. An alloy containing 90% copper and 10% magnesium used as a cast iron inoculant and graphite spheroidizer.

cupromanganese. (1) A heat-resistant alloy containing 90% copper and 10% manganese used for staybolts and heat-resistant parts.

(2) A corrosion-resistant alloy of 96% copper and 4% manganese used for tubes.

Cupron. Trade name of Gilby-Fodor SA (France) for a resistance alloy, similar to constantan, containing 55% copper and 45% nickel. It has a low temperature coefficient of resistance, a maximum service temperature of 500°C (930°F), and an electrical conductivity of 4% IACS. Used for strain gauges, rheostats, voltmeters, shunts, resistances, thermocouples, pyrometers, etc.

cupronickel-manganese. A hard, nonmagnetic alloy containing 60% copper, 20% nickel and 20% manganese. It is difficult to machine, and is used for gears and escapement wheels for watches.

cupronickels. See copper nickels.

Cuprophan. Trademark of Enka AG (Germany) for a transparent, clear substance obtained by treating cellulose with ammoniacal copper solution. Available as a thin strong film or sheeting, it is used for packaging applications, and flat, tubular and hollow-fiber membranes for dialysis and artificial kidneys.

Cupror. Trade name for a corrosion-resistant alloy of 94.2% copper and 5.8% aluminum used for valve stems, pump rods, etc.

cuprorivaite. A blue mineral composed of calcium copper silicate, $CaCuSi_4O_{10}$. Crystal system, tetragonal. Density, 3.06 g/cm^3; refractive index, 1.636. Occurrence: Italy.

Cuprosal. German trade name for copper plating salts for zinc, cast iron, steel, etc.

cuprosilicon. An alloy of 55% silicon and 45% copper used as a hardener for copper alloys.

cupro silk. See cuprammonium rayon.

cuprosklodowskite. A yellowish green, grass-green or emerald green mineral of the uranophane group composed of copper uranyl silicate hydroxide hexahydrate, $Cu(UO_2)_2(SiO_3OH)_2 \cdot 6H_2O$. Crystal system, triclinic. Density, 3.50-3.85 g/cm³; refractive index, 1.664. Occurrence: Czech Republic; Central Africa.

cuprospinel. A black mineral of the spinel group composed of copper iron oxide, $CuFe_2O_4$. Crystal system, cubic. Density, 5.42 g/cm³. Occurrence: Canada (Newfoundland).

cuprostibite. A gray mineral composed of antimony copper thallium, $Cu_2(Sb,Tl)$. Crystal system, tetragonal. Density, 8.42 g/cm³. Occurrence: Greenland.

Cuprotec. Trade name of Eutectic Corporation (USA) for a series of copper-base alloy powders for brazing thin-walled copper alloy sections.

Cuprothal. Trademark of Kanthal Corporation (USA) for a series of copper alloys containing 2-45% nickel. Supplied in wire and strip form, they have service temperatures of up to 540°C (1000°F), and are used for heating elements, and resistors.

Cuprotherm. Trade name of Wieland-Werke AG (Germany) for a chromium copper containing 0.3-1.2% chromium, 0.04-0.1% zirconium, and the balance copper. It has good corrosion resistance, high electrical conductivity, good formability, high strength, and good tempering properties. Used in electrical engineering, and for welding electrodes, wheels and tips.

cuprotungstite. A green mineral composed of copper tungsten oxide hydroxide, $Cu_2WO_4(OH)_2$. Crystal system, tetragonal. Density, 6.98 g/cm³; refractive index, 2.18. Occurrence: USA (Arizona), Germany.

cuprous bromide. A hygroscopic, light-sensitive, white to off-white, crystalline powder (98+% pure). Density, 4.71 g/cm³; melting point, 500°C (932°F); boiling point, 1345°C (2453°F); refractive index, 2.116. Used as a catalyst in organic and organometallic synthesis. Formula: $CuBr$; Cu_2Br_2. Also known as *copper bromide*.

cuprous chloride. White, hygroscopic, light- and air-sensitive cubic crystals, or light brown to light gray powder (97+% pure). It occurs in nature as the mineral *nantokite*. Density, 4.14 g/cm³; melting point, 430°C (806°F); boiling point, 1490°C (2714°F); refractive index, 1.93. Used as a catalyst, as an absorbent for carbon monoxide, as a fungicide and preservative, and in decolorizing and desulfurizing of petroleum. Formula: $CuCl$; Cu_2Cl_2.

cuprous cyanide. A cream-colored powder (99+% pure). Density, 2.92 g/cm³; melting point, 473°C (883°F); boiling point, decomposes. Used in electroplating, in antifouling paints, as a catalyst, and in the biosciences. Formula: $CuCN$; $Cu_2(CN)_2$. Also known as *copper cyanide*.

cuprous iodide. White crystals or white to brownish-yellow powder (98+% pure) available in (i) cubic and (ii) hexagonal form. The cubic form occurs in nature as the mineral *marshite*. Crystal structure: (i) cubic form: sphalerite; (ii) hexagonal form: wurtzite (or zincite). The cubic form has a density of 5.63 g/cm³, a melting point of 605°C (1121°F), a boiling point of 1290°C (2354°F), and a refractive index of 2.346. Used as a semiconductor, as a catalyst, in cloud seeding, as a feed additive and in table salt. Formula: CuI. Also known as *copper iodide*.

cuprous meruric iodide. See mercuric cuprous iodide.

cuprous oxide. See red copper oxide.

cuprous phosphide. Gray-black powder. Density, 6.4-6.8 g/cm³. Used in chemistry and materials research. Formula: Cu_3P.

cuprous potassium cyanide. See potassium-copper cyanide.

cuprous selenide. Dark bluish black crystals. Density, 6.84 g/cm³; melting point, 1100°C (2012°F). Used in semiconductor research. Formula: Cu_2Se.

cuprous sulfide. Black powder, or crystals. It occurs in nature as the mineral *chalcocite*. Crystal system, orthorhombic. Density, 5.6 g/cm³; melting point, 1100°C (2012°F). Used for pigments, antifouling and luminous paints, in solar cells and electrodes, as a solid lubricant, and as a catalyst. Formula: Cu_2S.

cuprous telluride. Fine, high-purity powder, or small pieces. Density, 4.6-7.27 g/cm³; refractive index, 1.45-1.86. Used in ceramics, as a semiconductor, and in electronics and materials research. Formula: Cu_2Te. Also known as *copper telluride*.

cuprous thiocyanate. A yellow-white, air- and light-stable powder with a density of 2.843 g/cm³ and a melting point of 1084°C (1983°F). Used in antifouling paints, as a source of monovalent copper, and in textile printing. Formula: $CuSCN$. Also known as *copper thiocyanate; copper sulfocyanide*.

Cuprovac. Trade name of Crucible Materials Corporation (USA) for a high-conductivity copper used for electrical equipment.

Cuprussah. Trade name for silky cuprammonium rayon fibers and yarns used for textile fabrics.

Cupten. Trade name of NKK Corporation (Japan) for a series of weather-resistant steels containing up to 0.12% carbon, 0-0.75% silicon, 0-0.6% manganese, 0.06-0.15% phosphorus, 0-0.04% sulfur, 0.2-0.6% copper, 0-0.45% nickel, 0-0.1% vanadium, and the balance iron. Used for structural parts.

Cur-adex. Trademark of Currie Products Limited (Canada) for fire-retardant adhesives.

curative. See curing agent (1) and (2).

CuRay-Dual Match. Trade name for a dual-cure dental cement.

Curb. Trade name of Novamex Technologies Inc. (USA) for a water-displacing corrosion preventive.

Curel. Trade name for a spandex (segmented polyurethane) fiber used for the manufacture of textile fabrics with excellent stretch and recovery properties.

Cure-on-touch. Trademark of Sci-Pharm, Inc. (USA) for a light-cure composite resin for dental bonding applications.

curetonite. A yellow-green mineral composed of barium aluminum titanium oxide phosphate hydroxide, $Ba_4Al_3Ti(PO_4)_4(O,OH)_6$. Crystal system, monoclinic. Density, 4.42 g/cm³; refractive index, 1.680. Occurrence: USA (Nevada).

curienite. A yellow, radioactive mineral of the carnotite group composed of lead uranyl vanadium oxide pentahydrate, $Pb(UO_2)_2V_2O_8 \cdot 5H_2O$. Crystal system, orthorhombic. Density, 4.88 g/cm³; refractive index, above 2. Occurrence: Gabon.

curing agent. (1) An agent that when added to a linear polymer converts it to a crosslinked polymer. Also known as *curative; hardener*.

(2) A chemical substance or mixture of substances that when introduced into an adhesive facilitates or controls the curing or hardening reaction. Also known as *curative; hardener*.

(3) An additive that when added to cement or asbestos-cement products increases the chemical activity between the cementitious ingredients resulting in an increase or decrease in the curing rate.

curing compound. A liquid composition sometimes sprayed on the surface of fresh concrete as a sealant to prevent loss of moisture.

curite. A deep orange-red, radioactive mineral of the becquerel group composed of lead uranium oxide tetrahydrate, $Pb_2U_5O_{17}\cdot 4H_2O$. Crystal system, orthorhombic. Density, 7.40 g/cm^3; refractive index, 2.08. Occurrence: Australia.

curium. A silvery-white, synthetic radioactive metallic element of the actinide series of the Periodic Table named after Pierre and Marie Curie. Curium-242 was originally produced by the bombardment of plutonium-239 with helium ions. Curium metal can be prepared by reduction of curium fluoride (CmF$_3$) with barium metal vapor at 1275°C (2327°F). Density, 13.5 g/cm^3; melting point, 1330°C (2426°F); atomic number, 96; atomic weight, 247; trivalent, tetravalent. Two allotropic forms are known: (i) *Alpha curium* (close-packed double hexagonal) below 1176°C (2149°F) and (ii) *Beta curium* (face-centered cubic) between 1176 and 1330°C (2149 and 2426°F). *Curium* is an alpha emitter, forms compounds with oxygen (CmO$_2$, Cm$_2$O$_3$, Cm(OH)$_3$) and halides (CmBr$_3$, CmCl$_3$, CmF$_3$, CmF$_4$, CmI$_3$, Cm$_2$(C$_2$O$_4$)$_3$), and is used a thermoelectric power source and as a dopant for thoria. Symbol: Cm.

curium-242. A radioactive isotope of curium with mass number of 242 that emits alpha particles and has a half-life of about 165.2 days. It is was originially obtained from plutonium-239 by helium-ion bombardment, the daughter being americium-242. It may also be obtained by neutron irradiation of americium-241, the daughter being californium-246. Symbol: ^{242}Cm.

curium-244. A radioactive isotope of curium with mass number of 244 that emits alpha particles and has a half-life of about 16.6 years. Used as a thermoelectric power source. Symbol: ^{244}Cm.

curled yarn. See loop yarn.

Curlon. Trademark of RK Carbon Fibers Limited (UK) for carbon fibers.

Curoloy. Trade name of Birdsboro Corporation (USA) for an alloy cast iron containing 3% carbon, 1.2% silicon, 0.9% manganese, and the balance iron. Used for rolls.

Curon. Trademark of Reeves Brothers Inc. (USA) for polyurethane foam.

Curseal. Trademark of Chemor Inc. (Canada) for a concrete curing membrane and sealer.

Curtisol. Trade name of Curtis-Wright Corporation (USA) for a silver alloy with a melting range of 704-760°C (1300-1400°F) used as a solder for titanium and titanium alloys.

Curtiss. British trade name for a heat-treatable aluminum alloy containing 2.5% copper and 1.5% magnesium. Used for pistons.

curupay. The very hard, reddish wood of the tree *Piptadenia cebil*. It has an attractive, wavy grain, and an average weight of 1190 kg/m^3 (74 lb/ft^3). Source: Argentina, Brazil, Paraguay. Used for construction work and ornaments.

Curveil. Trademark of Schmelzer Industries Inc. (USA) for flexible fiberglass surfacing veils.

cushioning material. A material, such as cork, rubber, felt or fiberboard, used to isolate or lessen the effect of mechanical shocks and/or vibrations. See also vibration isolators.

Cusil. Trade name of GTE Products Corporation (USA) for an eutectic brazing alloy containing 72% silver and 28% copper. Commercially available in powder, wire, extrudable paste, foil, preform and flexibraze form, it has a melting point of 780°C (1435°F), a brazing range of 780-900°C (1435-1650°F), excellent flow characteristics, high vapor pressure, and good brazing properties. Used for joining ferrous and nonferrous metals.

Cusil-ABA. Trade name of GTE Products Corporation (USA) for an active brazing alloy containing 63% silver, 35.25% copper,

and 1.75% titanium. It has a liquidus temperature of 816°C (1500°F), a solidus temperature of 779°C (1435°F), and is used for brazing protective devices, e.g., surge arrestors. See also ABA.

Cusiloy. Trade name for a corrosion-resistant series of copper alloy containing 1-3% silicon, 0.5-1.5% tin and 0.7-1% iron. Used for wires.

Cusiltin. Trade name of GTE Products Corporation (USA) for a series of brazing alloys containing 60-70% silver, 25-30% copper and 5-10% tin. They are commercially available in the form of powder, wire, extrudable paste, foil, flexibraze and preform. The 60Ag-30Cu-10Sn alloy has a solidus of 602°C (1115°F), a liquidus of 718°C (1325°F), and a brazing range of 718-843°C (1325-1550°F). Used for joining most ferrous and nonferrous metals.

Cusio. Trade name for rayon fibers and yarns used for textile fabrics.

cuspidine. A rose-red to colorless mineral of the wohlerite group composed of calcium fluoride silicate, Ca$_4$F$_2$Si$_2$O$_7$. Crystal system, monoclinic. Density, 2.95; refractive index, 1.596. Occurrence: Italy.

CusterMica. Trade name of Pacer Corporation (USA) for a translucent gray, dry-ground *muscovite* mica supplied in several powder grades of varying particle size. Used as a filler and extender in paints, coatings, plastics, rubber, etc.

Custom. Trademark of Carpenter Technology Corporation (USA) for a series of precipitation-hardening stainless steel. They have good corrosion resistance, high strength and hardness, and good fabricability. Used for chemical-processing equipment, papermill equipment, oilfield equipment, oilfield valve components, aircraft and missile fittings, pump impellers and shafts, pump gears, fasteners, valve bodies, valve stems, seals, retaining rings, fittings, and rods.

Custom Age. Trademark of Carpenter Technology Corporation (USA) for a series of nickel-base alloys with high strength and excellent resistance to stress-corrosion cracking, sulfide stress cracking, pitting and crevice corrosion.

Custom Cast. Trade name for a series of dental casting alloys including *Custom Cast I* and *Custom Cast II* medium-gold alloys and *Custom Cast III* silver-palladium alloys.

Custom Ceram. Trade name for a palladium-based dental bonding alloy.

Custom Flo. Trade name of Carpenter Technology Corporation (USA) for a series of high-quality nonmagnetic chrome-nickel stainless steels with excellent resistance to corrosion and rusting, low work-hardening rates, and good cold headability. Used for tools, machine components, fasteners, etc.

Custom Resins. Trade name of Custom Resins (USA) for a series of wear-resistant nylon (polyamide) extrusion resins used for thick rod stock, heavy cross sections, film, sheet, pipe, and profiles.

Custom Rolled. Trade name of United Aluminum Corporation (USA) for coiled aluminum sheet supplied in the annealed condition in various gauges, tempers, and widths.

cut-and-color compound. See double-duty compound.

Cutanit. Trade name of Jessop-Saville Limited (UK) for a series of cemented carbides containing tungsten carbide, titanium carbide, and a cobalt or molybdenum binder. Used for tools, cutters, taps, and dies.

cutback asphalt. Asphalt that has been rendered soft or liquid by the addition of a suitable petroleum-based diluent, such as spirit, kerosine or creosote. Used for protective coatings, in waterproofing building walls, in road surfacing, and as a cementing medium for floor and wall coverings.

cut glass. Fine glassware that has been decorated or ornamented by first grinding figures, designs or patterns into the surface by means of special abrasives, and then smoothing and polishing the cut surface.

Cutin. Trade name of GTE Products Corporation (USA) for a brazing alloy containing 85% copper and 15% tin. Commercially available in the form of powder, wire, extrudable paste, foil, flexibraze and preform, it has a liquidus temperature of 960°C (1760°F), and a solidus temperature of 798°C (1468°F).

Cutlery. Trade name of Dunford Hadfields Limited (UK) for a corrosion-resistant steel containing 0.15-0.4% carbon, 12-14% chromium, and the balance iron. Used for high-strength engineering applications and cutting tools.

Cutrite. (1) Trademark of 3M Company (USA) for coated abrasives.

(2) Trademark of Scott Paper Company (USA) for waxed paper.

Cutter Alloy. Trade name of Columbia Tool Steel Company (USA) for a water-hardening tool steel containing 1.05% carbon, 0.25% manganese, 0.25% chromium, 0.25% tungsten, and the balance iron. Used for cutting tools for hard materials.

cutting alloys. A class of intrinsically hard-cast cobalt-chromium-tungsten, nickel-tungsten and iron-cobalt base cutting tool materials usually containing varying amounts of one or more of the following elements: carbon, niobium, molybdenum, silicon, boron, manganese and zirconium. They are often sold under trademarks and trade names, such as *Borcoloy, Cooperite, Crobalt, Rexalloy, Speedaloy* or *Stellite*. Commercially available in the form of tool bits and shear blades, they have excellent wear resistance and good thermal and corrosion resistance. They are now largely replaced by tungsten and cobalt-tungsten base superhigh-speed steels, cermets, cemented carbides and coated cemented carbides, but still used for hardfacing applications.

cutting powder. A powder, e.g., iron or aluminum, used in metal-powder cutting, an oxygen cutting process in which the introduction of the powder in the orifice of an oxyfuel gas torch raises the total heat of the flame and thus improves the cutting performance.

cutting sand. A granular material that is composed of sharp un-graded quartz grains, and used as an abrasive for sawing stone and rock.

cutting steels. A group of tool steels that are suitable for the manufacture of cutting tools, such as lathe, planer and boring tools, milling cutters, drills, reamers, taps, threading dies, form cutters, etc. Examples include high-speed (AISI types M and T), cold-work (AISI types D, A and O) and water-hardening (AISI type W) tool steels. Also known as *cutting-tool steels.*

cutting-tool steels. See cutting steels.

cuttlefish bone. See sepia.

Cutwire. Trade name of Premier Shot Company (USA) for cut wire shot produced from drawn steel and supplied in various sizes and hardnesses for the abrasive blast cleaning of steel and cast iron.

(2) Trade name of Sumi-Pac Corporation (Japan) for curl- and twist-free copper-clad low-carbon steel wire for electrical applications including wire electro-discharge machining (WEDM). It is also supplied in tin or tin-alloy plated grades (TCP wire), and standard and special brass grades.

Cuyo. Trade name of Hoover Ball & Bearing Company (USA)

for a heat-resisting stainless steel containing 0.2% carbon, 18% chromium, 8% nickel, and the balance iron. Used for furnace parts, and stainless parts.

Cuzinal. Trade name of Olin Brass, Indianapolis (USA) for an embrittlement-free arsenical aluminum brass of 77% copper, 20.87% zinc, 2.1% aluminum, and 0.03% arsenic. It has good corrosion resistance and excellent cold workability. Used for condensers, condenser and heat-exchanger tubes, heat exchangers, ferrules, etc.

cuzticite. A yellow to brown mineral composed of iron tellurium oxide trihydrate, $Fe_2TeO_6 \cdot 3H_2O$. Crystal system, hexagonal. Density, 3.90; refractive index, 2.06. Occurrence: Mexico.

CVD carbon. See chemical vapor deposited carbon.

CVD coatings. See chemical vapor deposition coatings.

CVD Silicon Carbide. Trademark of Rohm & Haas Company/ Advanced Materials (USA) for solid, porosity-free high-purity silicon carbide (SiC) produced by chemical vapor deposition. It has good high-temperature performance (above 1500°C or 2730°F), and excellent thermal conductivity, stiffness and chemical resistance. Supplied in plates and near-net shapes, it is used for optical, wear and semiconductor applications.

CVD Zinc Selenide. Trademark of Rohm & Haas Company/Advanced Materials (USA) for chemically inert, high-purity zinc selenide (ZnSe) produced by chemical vapor deposition. It has high thermal-shock resistance, and is supplied in the form of blanks, rectangles, prisms, sheets, and near-net domes. Used for optical components in high-powered carbon-dioxide lasers and high-resolution forward-looking infrared thermal imaging equipment, in infrared windows and lenses for medical and industrial applications, and as an evaporative source material.

CVD Zinc Sulfide. Trademark of Rohm & Haas Company/Advanced Materials (USA) for chemically inert, high-purity zinc sulfide (ZnS) produced by chemical vapor deposition. It has high hardness and fracture strength, and is supplied in the form of blanks, rectangles, prisms, sheets, and near-net domes. Used for infrared windows, missile domes, optical elements, and as an evaporative source material.

CV-than. Trademark of Continental Gummi-Werke (Germany) for flexible polyvinyl chlorides.

CX-Plus. Trade name of Shofu Dental Corporation (Japan) for a glass-ionomer dental cement belonging to the *GlasIonomer* product range. The glass particles are coated with silicon oxide using the "Silicon-Oxide Layer Coating" (SLC) technique that significantly increases the allowable working time. Used in dentistry for luting restorations.

cyanamide. See also calcium cyanamide; lead cyanamide.

Cyanaprene. Trademark of Air Products and Chemicals Inc. (USA) for toluene-2,4-diisocyanate (TDI) prepolymers, and thermoplastic urethane elastomers.

cyanate. An ester or salt of cyanic acid (HCNO).

cyanate esters. See cyanate resins.

cyanate resins. A family of thermosetting resins obtained from polyphenols or bisphenols. Commercially available as monomers, oligomers, blends and solutions, they have high dimensional stability, excellent corrosion resistance, low dielectric loss, and low moisture absorption. Used as matrix materials for printed wiring boards, structural composites, etc. Also known as *cyanate esters; cyanates; cyanic esters; triazine resins.*

cyanates. See cyanate resins.

cyanic esters. See cyanate resins.

cyanide. An ester or salt of hydrocyanic acid (HCN), e.g., potassium cyanide (KCN).

cyanide copper. Copper electrodeposited from an electrolytic solution containing a complex ion composed of monovalent copper and the cyanide radical, e.g., a solution of potassium or sodium cyanide.

cyanided steel. A low-carbon steel that has been subjected to a surface-hardening process producing a hard, wear-resistant carbon- and nitride-containing surface (case depth about 0.3 mm, or 0.01 in.) by heating at 870°C (1600°F) in a molten 30% sodium cyanide bath for about one hour. Quenching in oil or water from this bath hardens the surface of the steel. Used for screws, nuts, bolts, and small gears.

cyanine dyes. A class of dyes including dicyanines, cryptocyanines, isocyanines, merocyanines, neocyanines, pseudocyanines and xenacyanines. *Cyanine dyes* consist of two heterocyclic groups, which are essentially quinoline nuclei, linked by a chain of conjugated double bonds having an odd number of carbon atoms. *Cyanine dyes* are photosensitizing, and are used for photographic emulsion, in electronics, and in organic synthesis.

cyanite. See kyanite.

cyanoacrylate adhesives. A group of rapid-curing one-part adhesives based on thermoplastic monomers, such as the alkyl 2-cyanoacrylates (e.g., 2-cyanoethyl acrylate, 2-cyanoethyl methacrylate, etc.), and having very good polymerizing and bonding characteristics. They cure in the presence of surface moisture, or by surface activation, and have excellent adhesion to most substrates, fair heat and chemical resistance, and poor peel strength. Used for rubber printing plates, tools and rubber swimming masks, as tissue adhesives in wound closure and microsurgery, and as structural adhesives for engineering applications.

cyanoauric acid. See auric cyanide.

cyanocarbons. A class of organic compounds including tetracyanoethylene $(CN)_2C=C(CN)_2$ in which the cyanide radical (-CN) replaces hydrogen. They are used with aromatic hydrocarbons to prepare colored complexes.

Cyanocel. Trademark of American Cyanamid Company (USA) for cyanoethylated cellulose used as a dielectric for capacitors.

cyanochroite. A greenish blue mineral of the picromerite group composed of potassium copper sulfate hexahydrate, $K_2Cu(SO_4)_2 \cdot 6H_2O$. It can also be made synthetically from an aqueous solution of potassium copper sulfate hexahydrate. Crystal system, monoclinic. Density, 2.23 g/cm³; refractive index, 1.4838. Occurrence: Italy.

cyanoethylated cellulose. Cotton fibers that have been passed through a caustic bath, treated with acrylonitrile, and subsequently neutralized with acetic acid. The treatment forms cyanoethyl ether groups in the fiber, increases the strength retention after exposure to heat and abrasion and the stretch resistance, and improves the resistance to bacteria, mildew and rot, and the receptiveness to dyes. It has a glass-transition temperature of 180°C (356°F). Also known as *cyanoethylated cotton.*

cyanoethylated cotton. See cyanoethylated cellulose.

cyanoethylated paper. A very thin, electrical insulating paper treated with acrylonitrile, and used as a dielectric in capacitors.

cyanoethylated starch. Starch that has been treated with acrylonitrile.

cyanoguanidine. See dicyanodiamide.

cyanophillite. A mineral composed of copper aluminum antimony oxide hexahydrate, $Cu_{10}Al_4Sb_6O_{25} \cdot 6H_2O$. Crystal system, orthorhombic. Density, 3.10 g/cm³; refractive index, 1.664. Oc-

currence: Germany.

cyanotrichite. A sky-blue, or azure-blue mineral composed of copper aluminum sulfate hydroxide dihydrate, $Cu_4Al_2SO_4(OH)_{12} \cdot 2H_2O$. Crystal system, orthorhombic. Density, 2.85 g/cm³; refractive index, 1.617. Occurrence: USA (Arizona).

cyanurtriamide. See melamine.

cyclic anhydride. An organic anhydride that contains one or more closed rings in its structure, e.g., benzoic, maleic or phthalic anhydride.

cyclic compound. An organic compound that contains one or more closed rings in its structure. It can be an alicyclic, aromatic or heterocyclic compound.

cyclic olefin copolymers. A family of amorphous engineering thermoplastics obtained by further reacting the bicyclic olefin 2-norbornene (a reaction product of ethylene and a cyclopentadiene catalyst) with ethylene. They have a density of 1.02 g/cm³ (0.037 lb/in.³), high stiffness and dimensional stability, high tensile strength and modulus, high surface hardness, high heat-deflection temperatures, high transparency and optical clarity, excellent electrical insulation properties, excellent water-vapor barrier properties, good resistance to aqueous acids and bases and polar solvents, fair to poor resistance to aliphatic and aromatic hydrocarbons, low moisture absorption, and good processibility by most conventional methods including injection and blow molding and cast and blown film extrusion and co-extrusion. Uses include capacitor films, optical lenses, lightguides, display panels, medical and laboratory ware including syringes, test tubes, petri dishes and pipettes, containers, packaging applications, electronic components, etc. Abbreviation: COC.

cyclized rubber. A tough, thermoplastic elastomer that has been cyclized (or isomerized) by treating natural rubber with a Lewis acid, such as chlorostannic or sulfonic acid, or with sulfonyl chloride. Its structure consists of the characteristic polyisoprene molecules with cyclic (ring-shaped) linkages. Used for paints, lacquers, adhesives, waterproofing compositions, etc. Also known as *isomerized rubber.*

Cyclo. Trade name of Cytemp Specialty Steel Division (USA) for an oil-hardening tool steel containing 0.7% carbon, 1.2% manganese, and the balance iron. Used for punches, dies, and tools.

cycloaliphatic epoxy resin. See cycloalkenyl epoxy resin.

cycloalkane. Any of a class of hydrocarbons that like alkanes contain only single carbon-carbon bonds, but have the carbon atoms arranged in a ring. The general formula is C_nH_{2n}. Examples include cyclobutane, cyclohexane and cyclopropane.

cycloalkene. An unsaturated hydrocarbon containing one closed ring in its structure. The general formula is C_nH_{2n-2}. Examples include cyclobutane, cyclohexane and cyclopentane. Also known as *cycloolefin.*

cycloalkenyl epoxy resin. A high-temperature-resistant polymer made by epoxidation of certain polycyclic aliphatic compounds (multicycloalkenyls) with organic peracids (e.g., peracetic acid). Used for adhesives with good heat resistance, outdoor electrical equipment, etc. Also known as *cycloaliphatic epoxy resin.*

cyclohexane. A cycloalkane available as a clear, colorless, mobile liquid (99+% pure) with density of 0.779 g/cm³, a melting point of 6.5°C (44°F), a boiling point of 80.7°C (177°F), a flash point of -1°F (-18°C), a refractive index of 1.4260, and an autoignition temperature of 473°F (245°C). It is also available in special grades: spectrophotometry, high-performance liquid chromatography, deuterated (99.5 at% deuterium). Used in nylon manufacture, as a paint and varnish remover, as a solvent for bitumens, cellulose ethers, fats, oils, synthetic resins, crude rubber and waxes, in glass substitutes, in solid fuels, and in spectrophotometry and chromatography. Formula: C_6H_{12}. Also known as *hexamethylene; hexanaphthene; hexahydrobenzene.*

Cycloid. Trade name of Carpenter Technology Corporation (USA) for a silico-manganese tempering steel containing 0.4% carbon, 1.5% silicon, 0.9% manganese, and the balance iron. Used for gears, pinions, and shafts.

cyclol acrylate. See 2-hydroxymethyl-5-norbornene acrylate.

Cyclone. Trade name of British Steel Corporation (UK) for a series of tool steels containing about 0.8-1.25% carbon, 3.7-4.3% chromium, 5-8.7% molybdenum, 1.5-7% tungsten, 2-3% vanadium, 0-8% cobalt, and the balance iron. Used for cutting tools of all types, drills, and blanking and thread-rolling dies.

Cyclone Extra. Trade name of British Steel Corporation (UK) for a series of cobalt-tungsten high speed steels containing about 0.8% carbon, 4.4% chromium, 1% molybdenum, 18.5% tungsten, 1.6% vanadium, 10% cobalt, and the balance iron. Used for cutting tools.

Cyclone Supercut. Trade name of British Steel Corporation (UK) for a series of tool steels containing about 1.5% carbon, 4.7% chromium, 12.5% tungsten, 5% vanadium, 5% cobalt, and the balance iron. Used for cutting tools.

cycloolefin. See also cycloalkene.

cyclopean. A mass concrete that contain aggregate consisting of large stones usually 152 mm (6 in.) or more in size. It is used in dams or other thick structures. Also known as *cyclopean concrete.*

Cyclope. Trade name of TradeARBED Inc. (USA) for a hot-work tool steel (AISI type H21) containing 0.35% carbon, 3.5% chromium, 9% tungsten, and the balance iron. Used for hot-blanking and trimming dies, extrusion and die casting dies, and gripper dies.

Cyclops. Trade name of Cytemp Specialty Steel Division (USA) for a series of tool steels including the following types: (i) low-alloy special-purpose tool steels, such as *Cyclops L2* (AISI type L2) and *Cyclops L6* (AISI type L6); (ii) molybdenum-type high-speed steels, such as *Cyclops M4* (AISI type M4) and *Cyclops M42* (AISI type M42); (iii) tungsten-type high-speed steels, such as *Cyclops T15* (AISI type T15); (iv) water-hardening tool steels, such as *Cyclops W1* (AISI type W1) and *Cyclops W2* (AISI type W2); and (v) shock-resisting tool steels, such as *Cyclops S2* (AISI type S2) and *Cyclops S5* (AISI type S5).

Cyclo-Rubber. Trade name of Elementis Performance Polymers (USA) for cyclized rubber used in the manufacture of adhesives.

Cycloset. Trade name for cellulose acetate fibers used for wearing apparel and as industrial fabrics.

Cyclosit. Trademark of Bayer AG (Germany) for cyclized rubber.

cyclosilicates. Silicate minerals, such as beryl ($Be_3Al_2Si_6O_{18}$), in which three SiO_4 tetrahedra are linked to form rings, such as $Si_3O_9^{6-}$ or $Si_6O_{18}^{12-}$, with a Si:O ratio of 1:3. Also known as *ring silicates.*

Cyclotene. Trademark of Dow Chemical Company (USA) for a series of advanced high-purity resins derived from B-stage bisbenzocyclobutene (BCB) monomers. Supplied in dry-etch and photosensitive grades, they are formulated as high-solids low-viscosity solutions. They have high optical clarity, low moisture absorption, cure temperature, low dielectric constants, excellent chemical and corrosion resistance, and good thermal stability and compatibility with a wide range of metallization systems. Used for microelectronic applications.

cyclyd. A cyclic alkyd coating. See also alkyd coatings.

Cycogel. Trade name of Nova Chemicals Inc. (USA) for acrylonitrile-butadiene-styrene resins.

Cycolac. Trademark of GE Plastics (USA) for a series of flame- and smoke-resistant acrylonitrile-butadiene-styrene (ABS) plastics with outstanding strength and toughness, excellent impact resistance, good dimensional stability, high heat-distortion resistance and high-temperature stability, and good electrical properties. They are supplied in pellet form in more than 100 different grades including unfilled, filled, blow-molding, injection-molding, blending, extrusion, plating, glass fiber-reinforced, fire-retardant, high-heat, high- and medium-impact, low-gloss, transparent, UV-stabilized and structural foam. *Cycolac* plastics are used for business machines, computers, appliance housings, highway safety devices, lawn and garden equipment, piping, household appliances, refrigerator linings, toys, cases, automotive trim, helmets, impeller, knobs, handles, levers, etc. Also included under this trademark are several ABS polymer blends, such as acrylonitrile-butadiene-styrene/polybutylene terephthalate and polycarbonate/acrylonitrile-butadiene-styrene.

Cycolin. Trade name of GE Plastics (USA) for acrylonitrile-butadiene-styrene and polybutylene terephthalate (ABS-PBT) alloys.

Cycoloy. Trade name of GE Plastics (USA) for a series of impact-resistant acrylonitrile-butadiene-styrene/polycarbonate (ABS-PC) and ABS/nylon alloys, some of which are lubricated with polytetrafluoroethylene (*Teflon*). Supplied in unfilled injection molding and plating grades, they are used for automotive trim, wheel covers, etc.

Cycom. Trademark of American Cyanamid Company (USA) for a series of modified bismaleimide laminating materials, polyester structural resins, high-strength modified resins, heat-resistant phenolic resins, and modified epoxy laminating materials.

Cycom MCG. Trademark of American Cyanamid Company (USA) for metal-coated graphite fibers supplied as continuous and precision chopped fibers, woven fabrics, and nonwoven mats. The metal coatings are based on nickel, copper or silver. Used in composites for electromagnetic shielding applications.

Cycom NCG. Trademark of American Cyanamid Company (USA) for nickel-coated graphite fibers supplied as continuous and precision chopped fibers, woven fabrics, and nonwoven mats. Supplied in filament counts of 3000, 6000 and 12000, they are used for lightning-strike protection in structures, and in composites for electromagnetic shielding applications.

Cycopol. Trademark of Koppers Company, Inc. (USA) for a series of copolymers including acrylic, vinyl, toluene, styrenated and modified styrenated alkyd resins, etc., used for coating metals.

Cycor. Trade name of Cytec Company (USA) for a series of urea-formaldehyde resins.

Cycovin. Trademark of GE Plastics (USA) for acrylonitrile-butadiene-styrene/polyvinyl chloride (ABS-PVC) alloys supplied in high- and low-gloss grades.

Cycowood. Trade name for a modified acrylonitrile-butadiene-styrene (ABS) resin that is specially formulated for easy machining and dimensional stability. The abrasion-resistant extruded engineering plastic board does not change dimensionally when exposed to moisture, temperature fluctuations and rough handling. It remains unaffected by most chemicals, and may be joined into laminations like wood.

Cyglas. Trademark of American Cyanamid Company (USA) for a series of chemically resistant polyester composite bulk molding compounds with or without mineral or glass-fiber reinforcement, that can be processed by compression, injection or transfer molding. Supplied in standard, electrical, fire-retardant and high-heat grades, they are used for circuit breakers, transformers, microwave cookware, automotive components, and heavy industrial parts.

Cy-Less. Trade name of Technic Inc. (USA) for a cyanide-free bright silver bath, gold bath and copper strike.

Cylight. Trademark of Sterling Fibers (USA) for a tough, colorfast, weather-resistant acrylic fiber with excellent UV, chemical, mold and mildew resistance. Used for outdoor fabrics.

cylinder iron. Cast iron containing about 3-3.25% carbon, 2-2.25% silicon, 0.4-0.75% manganese, and the balance iron. It has a dense microstructure, excellent castability, high strength, good wear resistance, and low coefficient of friction. Used for cylinders of compressors and engines, and for pistons, piston rings, and other machinery parts.

cylinder kraft. Special *containerboard* made from sulfate process (kraft) pulp on a cylinder-type papermaking machine containing several rotary cylindrical filters.

cylindrite. A lead-gray mineral composed of iron lead tin antimony sulfide, $FePb_3Sn_4Sb_2S_{14}$. Crystal system, triclinic. Density, 5.46 g/cm^3. Occurrence: Bolivia.

Cylpeb. Trademark of Hop Mac Inc. (Canada) for cylindrical grinding media used in ball and vibratory mills for ore grinding applications.

Cymbal metal. A red brass containing 78% copper and 22% zinc. Used for ornamental and architectural parts.

Cymel. Trade name of Cytec Company (USA) for cellulose-filled melamine-formaldehyde plastics. They have excellent heat and electrical resistance, good mechanical properties including tensile strength, relatively high hardness, good colorability and color stability, low moisture absorption and good mechanical properties. Used for moldings, lighting fixtures, switchgear, switch panels, kitchen and dinnerware, etc.

cymophane. See cat's eye.

cymrite. A colorless to dark green, or brown mineral composed of barium aluminum silicate monohydrate, $BaAl_2Si_2O_8 \cdot H_2O$. Crystal system, hexagonal. Density, 3.41 g/cm^3; refractive index, 1.614. Occurrence: USA (Alaska).

cypress. (1) The wood of any of several related species of evergreen trees (genus *Cupressus)* in the order Pinales found in North America, Europe and Asia especially the aromatic, light brown wood of the Italian cypress (*C. sempervirens)* native to southern Europe and the Mediterranean, but now also planted in the southern United States from Florida to Texas, and in California. It has high durability, relatively low weight, and is used for construction work, furniture, chests, and doors. See also Italian cypress.

(2) The lightweight, moderately heavy, strong and stiff wood of medium-large evergreen trees of the cypress family especially *Chamaecyparis nootkaensis* and *Cupressus sitkaensis*. See also Alaska cedar.

(3) The wood of any of several species of coniferous trees of the family Cupressaceae including true cypresses, junipers, and arborvitae.

(4) See baldcypress.

Cyprus bronze. A soft bronze containing 65% copper, 30% lead and 5% tin. It has excellent machinability, good castability, and is used for hardware, high-speed bearings, bushings, fittings, and castings.

Cyprus pine. A coniferous tree of the genus *Callitris* found throughout the Mediterranean region. It exudes a white, brittle solid resin, known as "sandarac," that is used in the manufacture of varnishes, lacquers and paints.

Cyprus umber. A medium- to dark-brown earth containing iron and manganese oxides. It is found chiefly in Cyprus, and has high inertness and stability, good covering power, and excellent resistance to heat, light, moisture and alkalies. Used as a pigment. See also umber.

Cyrene. Trade name of Neville Company (USA) for polystyrene resins and products.

Cyrex. Trade name of Cyro Industries (USA) for high-impact acrylic/polycarbonate alloys with good toughness and chemical resistance.

cyrilovite. A bright yellow mineral of the wardite group composed of sodium iron phosphate hydroxide dihydrate, $NaFe_3(PO_4)_2(OH)_4 \cdot 2H_2O$. Crystal system, tetragonal. Density, 3.08 g/cm^3; refractive index, 1803. Occurrence: Brazil.

Cyrolite. Trademark of Cyro Industries (USA) for a series of acrylic-based multipolymer compounds for injection molding applications. *Cyrolite G* refers to methacrylate-butadiene-styrene (MBS) terpolymers and *Cyrolite HP* to polycarbonate sheeting.

Cyrolon. Trade name of Cyro Industries (USA) for polycarbonate sheeting.

Cyroloy. Trade name of Cyro Industries (USA) for acrylic resins.

Cyrovu. Trademark of Cyro Industries (USA) for high-performance multipolymer compounds.

Cystar. Trademark of Cytec Industries, Inc. (USA) for acrylic fibers.

Cystrip. Trade name of AC Molding Compounds (USA) for plastic media used for blast cleaning applications.

cytochromes. A class of oxidation-reduction pigments consisting of iron-porphyrin protein complexes, and occurring in the cells of most animals and plants. Used in medicine, biochemistry, biotechnology, electronics and optoelectronics. See also porphyrins.

D

0D nanostructures. See nanostructured materials.

1D nanostructures. See nanostructured materials.

1D photonic crystals. A class of photonic crystals, usually produced by thin-film deposition, that reflect and refract light in one direction, i.e., exhibit one-dimensional photonic band gaps. In their simplest form they consists of multilayer stacks of alternating high-k and low-k materials, i.e., materials with high and low dielectric constants. Other 1D photonic crystals are composed of silicon structures with lithographically etched holes. *1D photonic crystals* are useful for applications, such as optical waveguides, vertical-cavity surface-emitting lasers (VCSELs), distributed-feedback devices, optical fiber gratings and filters, and dielectric mirrors. Also known as *one-dimensional photonic crystals*. See also high-k materials; photonic crystals.

2D nanostructures. See nanostructured materials.

2D photonic crystals. A class of photonic crystals that reflect and refract light in two directions, i.e., exhibit two-dimensional photonic band gaps. There are many examples of such materials including nanometer-sized lithographically produced silicon pillars, cubic arrays of dielectric cylinders purposely designed with defects (e.g., a missing cylinder rows and columns), microstructured optical fibers drawn from structured fiber preforms, and structures composed of III-V (13-15) semiconductors, such as gallium arsenide (GaAs). In spite of the lack of optical control over the third dimension, they are useful for optical filters and networks, lasers and planar optical-waveguide applications. Also known as *two-dimensional photonic crystals*. See also microstructured optical fibers; photonic crystals; semiconductors.

2D polymers. Molecular structures produced by chemically linking one-dimensional monomers at two molecular locations, incorporating a hydroxy acid in the vicinity of the molecule center and terminating one end with a *p*-pentoxy biphenyl. This causes the chains to self-organize into layers and bond, forming a molecular polymer sheet. *2D polymers* have stable nonlinear optical properties and better long-term thermal stability than ordinary (one-dimensional) polymers. Uses include layered filter membranes, reinforcement of ordinary polymers, and the manufacture of nonlinear optical devices. Also known as *two-dimensional polymers*.

3D nanostructures. See nanostructured materials.

3D photonic crystals. A class of photonic crystals that reflect and refract light in three directions, i.e., exhibit three-dimensional photonic band gaps, and have ideal crystal structures based on three-dimensional lattices. Examples of such materi-

als include the so-called "Yablonovite", a photonic crystal produced by drilling small holes into a block of dielectric material, structures with micrometer-scale features prepared by two-photon polymerization of photoresists, and structures based on III-V (13-15) semiconductors, such as gallium arsenide (GaAs), produced by high-resolution semiconductor processing techniques. There are also colloidal photonic structures produced by accurate arrangement of micromachined silicon wafers in successive layers, self-arrangement of submicron-sized silica spheres in colloidal suspension and subsequent filling of the air voids with a high-refractive index titania-dioxide solution, or from a suspension of polymer spheres in a hydrogel film. Used for lasers, optical waveguides and filters and other telecommunication applications. Also known as *three-dimensional photonic crystals*. See also colloidal photonic crystals; photonic crystals; Yablonovite.

DAB-Am. Trade name of DSM NV (Netherlands) for a series of polypropylenimine (PPI) based dendrimer polymers, such as PPI tetramine (*DAB-Am-4*), PPI octamine (*DAB-Am-8*), PPI hexadecamine (*DAB-Am-16*), PPI dotricontaamine (*DAB-Am-32*) and PPI tetrahexacontaamine (*DAB-Am-64*). *Note:* The numbers appended to the trade name "DAB-Am" indicate the number of surface primary amino groups.

Dacar. Trade name of Jessop Steel Company (USA) for a water-hardening steel containing 0.7-1.2% carbon, and the balance iron. Used for machinery parts.

dachiardite. (1) A colorless to snow-white mineral of the zeolite group composed of potassium calcium aluminum silicate hexahydrate, $(Ca,K_2)Al_2Si_{12}O_{28} \cdot 6H_2O$. Crystal system, orthorhombic. Density, 2.17 g/cm^3; refractive index, 1.482. Occurrence: Bulgaria.

(2) A colorless mineral of the zeolite group composed of potassium sodium calcium aluminum silicate hydrate, $(Ca,Na,-K,Mg)_4(Si,Al)_{24}O_{48} \cdot 13H_2O$. Crystal system, monoclinic. Density, 2.16 g/cm^3; refractive index, 1.496. Occurrence: Italy.

Dacromet. Trade name of Metal Coatings International Inc. (USA) for water-based chromium and zinc dispersion coatings.

Dacron. (1) Trademark of E.I. DuPont de Nemours & Company (USA) for thermosetting polyester based on polyethylene terephthalate (PET). Available in the form of fibers and films, it has outstanding dielectric properties, low moisture absorption, good heat resistance up to 204°C (400°F), good flame and weather resistance, good resistance to humidity, acids, greases, oils and solvents, good color stability, and high transparency to high-frequency radio waves. Plastic films based on *Dacron* are very tough, and have excellent fatigue and tear strength. Used for textile fabrics (e.g., sports- and leisurewear, curtains and furnishings), reinforced structural shapes, boat and car bodies, fire hoses, aircraft glaze, television, video and radio parts, radar domes, magnetic recording tapes, and automotive tire cords.

(2) Trademark of E.I. DuPont de Nemours & Company (USA) for soft synthetic polyester fibers based on polyethylene terephthalate (PET). Supplied as filament, yarn, staple, tow and fiber fill, they have a density of 1.4 g/cm^3 (0.05 $lb/in.^3$), a melting point of 250°C (482°F), high tensile strength, high elastic recovery, excellent mildew resistance, good chemical, flame and insect resistance, and good self-extinguishing properties. Used for textile fabrics, cordage, fire hose, fabric covers for paint rollers, etc.

dadsonite. A lead gray mineral composed of lead antimony chloride sulfide, $Pb_{21}Sb_{23}S_{55}Cl_{28}$. Crystal system, monoclinic. Density, 5.51 g/cm^3. Occurrence: Germany.

Daehalon. Trademark of Daeha Synthetic Fibers Inc. (South Korea) for polyester fibers and yarns.

Dag. Trademark of Acheson Colloids Company (USA) for a series of dispersions used as coatings, lubricants and mold-release agents, and usually applied by spraying or immersion. Most dispersions consist of molybdenum disulfide, colloidal graphite or graphite powder in water, oils, resins (e.g., alkyd, epoxy, phenolic or silicone) or solvents. Graphite dispersions are frequently used to produce black coatings, as electrically conductive coatings, and as lubricant for dies, tools, and molds. Molybdenum disulfide dispersions are used chiefly as lubricants and mold-release-agents.

dagger fibers. The coarse fibers obtained from the stem and foliage of the dagger plant (genus *Yucca*) belonging to the lily family and found in Mexico, the West Indies and the southern United States. Used especially in the manufacture of cordage.

dahoma. See greenheart (2).

Daido. Trade name of Daido Steel Company Limited (Japan) for a series of tough, constructional steels containing 0.13-0.2% carbon, 1.2-1.85% manganese, 0.3-0.65% silicon, 0-0.35% copper, and the balance iron. Used for bridges, buildings, structural members, agricultural equipment, gears, etc.

Daido Super Gear. Trade name of Daido Steel Company Limited (Japan) for a series of carburizing gear steels with high strength, toughness and fatigue resistance. DSG1 has 0.2% carbon, 0.7% manganese, 0.15% or less silicon, 1% chromium, 0.4% molybdenum, 0.015% sulfur, 0.015% or less phosphorus, and the balance iron. DSG2 contains 0.2% carbon, 0.6% manganese, 0.15% or less silicon, 1.0% nickel, 0.8% chromium, 0.3% molybdenum, 0.015% sulfur, 0.015% or less phosphorus, and the balance iron. DSG3 has 0.2% carbon, 0.3% manganese, 0.15% or less silicon, 2.0% nickel, 0.3% chromium, 0.75% molybdenum, 0.015% sulfur, 0.015% or less phosphorus, and the balance iron. Used for automotive transmission gears. Abbreviation: DSG.

Daiflon. Trade name of Daikin Kogyo (Japan) for a series of fluoropolymers.

Daifuki. Japanese trade name for rayon fibers and yarns used for textile fabrics.

Daimler. British trade name for a corrosion-resistant bearing metal containing 76% copper, 20% zinc, 3% tin, and 1% lead. Used for bushings and bearings.

Dai-Nippon Film. Japanese trade name for cellulose nitrate film materials.

dairy bronze. A cast nickel silver containing 63-68% copper, 20-25% nickel, 1-5% tin, 2-5% lead, and 2-8% zinc. It has good strength, machinability and corrosion resistance, and is used for dairy equipment.

Dairy Metal. Trade name of Amesbury Brass & Foundry Company (USA) for a corrosion-resistant alloy containing 68% copper, 28% nickel, 1.5% tin, 2% aluminum, and 0.5% lead. Used for stainless filters.

Dairywhite. Trade name for a nickel silver with excellent corrosion resistance, used for dairy equipment.

Daisies. Trade name of Chance Brothers Limited (UK) for decorative glass tableware with floral pattern.

Daiwabo. Trade name of Asahi Kasei Corporation (Japan) for rayon fibers and yarns used for textile fabrics.

Daiwabo Polypro. Trade name of Daiwabo Company (Japan) for polypropylene fibers.

Daiya. Trade name of Asahi Glass Company Limited (Japan) for patterned glass.

Daka-Ware. Trade name of Harvey Davies Molding Company (USA) for urea-formaldehyde resins.

Dakota Gold. Trademark of Abrasives Inc. (USA) for silica abrasives composed of hard, angular particles for fast cleaning action.

Dalcobel. Trade name of SA Glaverbel (Belgium) for colored patterned glass. One side is slightly dimpled, and the other has a wavy surface with striated pattern.

Daltoflex. Trade name of ICI Polyurethane (UK) for reinforced microcellular polyurethane elastomers.

Dalton alloy. A fusible alloy containing 60% bismuth, 25% lead and 15% tin. It has a melting point of 92°C (198°F), and is used for fire extinguishers and sprinklers.

Dalvin. Trademark of Novacor Chemicals Limited (Canada) for polyvinyl chloride resins.

dalyite. A colorless mineral composed of potassium zirconium silicate, $K_2ZrSi_6O_{15}$. It can also be made synthetically. Crystal system, triclinic. Density, 2.84 g/cm^3; refractive index, 1.590. Occurrence: Ascension Island.

Dama. Trade name of Fabbrica Pisana SpA (Italy) for patterned glass with diamond design.

Damar. British trade name for a heavy-duty bearing alloy containing 76% copper, 13% lead and 11% tin.

Damascite. Trade name of Chrysler Corporation (USA) for an iron with a strength comparable to that of carbon steel. Used for machine pads, and molded iron parts.

Damascus. (1) Trade name of Damascus Steel Casting Company (USA) for a bronze containing 77% copper, 10% tin and 13% lead. Used for bearings and ornaments.

(2) Trade name of Latrobe Steel Company (USA) for an oil-hardening tool steel containing 0.55% carbon, 2% silicon, 0.9% manganese, 0.2-0.3% vanadium, 0.25% chromium, and the balance iron. Used for chisels, punches, shears, cold cutters, and stamps.

Damascus steel. A hard, hypereutectoid or hypoeutectoid steel made at Damascus, Syria, during the Middle Ages, and used for the manufacture of sword blades. Hypereutectoid Damascus steel has a microstructure composed of a matrix of *tempered martensite* with a dispersion of particles of spheroidized hypereutectoid *cementite*. The microstructure of hypoeutectoid Damascus steel usually consists of tempered martensite and small particles of cementite. It is believed that some Damascus steel was *crucible steel* (probably *wootz steel*) or a steel made by a combined cementation and crucible process. See also cementation steel; hypereutectoid steel; hypoeutectoid steel.

Damask. Trade name of SA Glaverbel (Belgium) for a patterned glass.

damask. (1) A linen fabric, usually in one color, with woven decorative designs (e.g., flowers or animals) and usually combining plain and satin weave. Used especially for towels, upholstery, and tablecloths. It is named after Damascus, Syria, where it was originally made.

(2) A reversible cotton, linen or silk fabric with woven decorative design, used especially for towels, upholstery, and tablecloths.

dammar. A colorless or pale yellowish-white solid resinous exudation of various species of trees of the genera *Shorea, Balanocarpus, Hopea,* and *Agathis* growing in the East Indies. It is a recent (non-fossil) resin with a density of 1.04-1.12 g/cm^3 (0.038-0.040 lb/in.^{3}), and a melting point of 120°C (248°F). Used in the preparation of varnishes and lacquers, alkyd baking enamels, paper and textile coatings, and printing ink. Also known as *gum dammar.*

damourite. See sericite.

Damping. Trade name for an alloy containing 87% manganese and 13% copper. It has a low electrical temperature resistance coefficient, and is used for measuring instruments.

dampproofer. A substance, such as sodium silicate, or a fluosilicate of aluminum or zinc, that is introduced into a batch of concrete or mortar, or applied as a coating to the surface of hardened concrete to decrease the passage or absorption of water or water vapor.

dampproofing. An impervious material employed to prevent moisture from seeping through exterior building walls. This may include one or more of the following: Portland cement plaster, hot bituminous coatings, polyethylene sheeting, tar paper, or plastic foam boards.

Dana. Trade name of Paul Bergsoe & Son (Denmark) for a series of babbitts including *Dana Auto Special* with 90.5% tin, 6.5% antimony and 3% copper used for engine bearings, *Dana Common* with 76% lead, 13.5% antimony, 10% tin and 0.5% copper for transmission bearings, *Dana Diesel* with 83.5% tin, 8% antimony, 6.5% copper and 2% lead for diesel engine bearings, and *Dana Steam* with 80% tin, 11.5% antimony, 5.5% copper and 3% lead for steam turbine and generator bearings.

danaite. A variety of the mineral *arsenopyrite* (FeAsS) in which cobalt replaces part of the iron.

Danaklon. Trademark of Danaklon Americas, Inc. (USA) for polypropylene fibers.

danalite. A reddish brown to gray mineral composed of iron manganese beryllium silicate sulfide, $(Fe,Mn,Zn)_4Be_3Si_3O_{12}S$. Crystal system, cubic. Density, 3.31 g/cm^3; refractive index, 1.753. Occurrence: Canada (British Columbia).

Danalloy. Trade name of Inland Electronics Products Corporation (USA) for a series of nonmagnetic, heat-treatable alloys containing varying amounts of silver, gold, nickel and magnesium. They have high electrical and thermal conductivities and are used for relay contacts, circuit-board retainers, and microcircuit backup plates.

Danamid. (1) Trade name of Zoltek Corporation (USA) for a series of nylon-6 based thermoplastics supplied in unreinforced, glass fiber-reinforced and mineral-filled grades for injection molding of automotive and electric components, and furniture.

(2) Trade name of Zoltek Corporation (USA) for nylon-6 fibers and yarns.

Danar. Trademark of Dixon Industries Corporation (USA) for a series of amorphous high-performance thermoplastic polyetherimide film and sheet materials available in varying thicknesses down to 6 μm (236 μin.). They can be fabricated by melt extrusion, vacuum forming, thermoforming and fusion bonding, and have good high-temperature resistance up to 204°C (400°F), good chemical and radiation resistance, high dielectric strength, stable dielectric constants, low dissipation factors, good melt adhesion to most products, good thermoformability and stability, low shrinkage, very low flammability, and low smoke evolution. Used in electronics, packaging, etc., and as bonding media for metallic, fibrous or cellulosic substrates.

danburite. A white, brown-yellow, or orange-yellow translucent to transparent mineral of the feldspar group composed of calcium borosilicate, $CaB_2Si_2O_8$. Crystal system, orthorhombic. Density, 2.98 g/cm^3; refractive index, 1.633; hardness, 7-7.5. Occurrence: Czech Republic. Used chiefly as an ornamental stone.

Dandelion metal. A high-strength babbitt containing 72% lead,

18% antimony and 10% tin. Used for heavy-duty machine bearings and locomotive crosshead linings.

Danish flint pebbles. See Danish pebbles.

Danish Mint. British trade name for a corrosion-resistant copper alloy containing 92% copper, 6% aluminum and 2% nickel. Used for coinage.

Danish pebbles. Hard, tough, uniformly rounded stones of flint with smooth surfaces ranging in size from about 25-203 mm (1-8 in.). Occurrence: Canada (Newfoundland), Denmark, Greenland. Used as grinding media in ball, pebble and tube mills for grinding cement, minerals and ores. Also known as *Danish flint pebbles.*

Dannemora iron. See Swedish iron.

dannemorite. A dark-green, yellowish-brown or greenish-gray mineral of the amphibolite group composed of iron magnesium manganese silicate hydroxide, $(Fe,Mg,Mn)_7Si_8O_{22}(OH)_2$. Crystal system, monoclinic. Density, 3.3-3.4 g/cm^3; refractive index, 1.672. Occurrence: Sweden.

d'ansite. A mineral composed of sodium magnesium chloride sulfate, $Na_{21}Mg(SO_4)_{10}Cl_3$. Crystal system, cubic. Density, 2.63 g/cm^3; refractive index, 1.488. Occurrence: China.

danta. The reddish-brown wood of the tree *Nesogordonia papaverifera.* It has a fine, even texture, an interlocked grain giving a striped effect, high durability, works fairly easily, and turns well. Average weight, 740 kg/m^3 (46 lb/ft^3). Source: West Africa. Used for shipbuilding, carriages, wagon work, flooring, and crossarms.

Danufil. Trademark of Acordis Kelheim GmbH (Germany) for viscose rayon fibers used for clothing, and blended with other fibers for suitings, furnishings, bedding, tablecloths, curtains, carpets, linings, sanitary and medical fabrics, e.g., sanitary and cosmetic pads, OP garments and medical dressings, and filter fleeces.

Danuflor. Trademark of Acordis Kelheim GmbH (Germany) for viscose rayon fibers.

Danulon. Trademark of Acordis Kelheim GmbH (Germany) for nylon fibers and yarns used for textile fabrics including clothing.

Danzig pine. See northern pine.

daomanite. A steel-gray mineral composed of copper platinum arsenic sulfide, $(Cu,Pt)_2AsS_2$. Crystal system, orthorhombic. Density, 7.06 g/cm^3. Occurrence: China.

Dapex. Trademark of CPC International Inc. (USA) for a diallyl phthalate (DAP) molding compound used for electrical applications.

Daplan. Trade name for polyolefin fibers and yarns.

Daplen. Trade name of ÖSL (Austria) for polyolefin plastics including several polyethylenes and polypropylenes.

Dapon. Trademark of FMC Corporation (USA) for a thermoplastic diallyl phthalate (DAP) polymer supplied in the form of a fine, white powder.

dapsone. See tetramethylene sulfone.

DAPtex. Trade name of DAP, Inc. (USA) for a water-repellent latex foam sealant for exterior and interior applications.

Daraco. German trade name for graphited white-metal bearing alloys.

Daraloy. Trade name for aluminum bronzes.

Daran. (1) Trademark of W.R. Grace & Company (USA) for saran (polyvinylidene chloride) latex coatings used as barrier coatings for paper, paperboard, plastics, etc.

(2) Trademark of W.R. Grace & Company (USA) for saran (polyvinylidene chloride) fibers with good chemical resis-

tance, good weatherability, and good resistance to mildew and insects. Used for carpets and rugs, draperies, upholstery, clothing, and industrial fabrics.

darapiosite. A colorless to white mineral of the osumilite group composed of lithium potassium sodium manganese zinc zirconium silicate, $LiKNa_2MnZnZrSi_{12}O_{30}$. Crystal system, hexagonal. Density, 2.92 g/cm^3; refractive index, 1.580. Occurrence: Tadzhikistan.

darapskite. A colorless mineral composed of sodium nitrate sulfate monohydrate, $Na_3(NO_3)(SO_4) \cdot H_2O$. Crystal system, monoclinic. Density, 2.20 g/cm^3; refractive index, 1.480.

Daratak. Trademark of W.R. Grace & Company (USA) for a series of emulsions based on polyvinyl acetate homopolymers and used in special coatings and adhesives.

Darc Cure. Trademark of Dymax Corporation (USA) for an adhesive, conformal coating which cures to a tough, abrasion-resistant solid in air, or when subjected to ultraviolet radiation.

D'Arcet alloy. A fusible alloy containing 50% bismuth, 25% lead and 25% tin. It has a melting point of 93°C (200°F), and is used for fire extinguishers, fire sprinklers, and boiler safety plugs.

Darco. Trademark of American Norit Company Inc. (USA) for a series of chemically activated wood-based carbons including acid-washed *lignite* and high-purity grades. Available in several mesh sizes, they are used for minimizing contamination in products, such as electroplating baths, vegetable and animal oils, waxes, greases, and fats.

Darelle. Trade name for viscose rayon used for clothing, and blended with other fibers for suitings, furnishings, carpets, linings, and medical fabrics.

Darex. Trademark of W.R. Grace & Company for styrene-butadiene rubber latexes used for adhesives, textile coatings, carpet backings, paint and coating resins.

Darex AEA. Trademark of W.R. Grace & Company (USA) for an air-entraining agent for concrete that is essentially a triethanolamine salt of sulfonated hydrocarbon.

Dargraph. Trade name of Darwin & Milner Inc. (USA) for an oil-hardening, wear-resistant graphite steel containing 1.45% carbon, 1-1.15% silicon, 0.25% molybdenum, and the balance iron. Used for punches, dies, cams, wear plates, cutters, and gages.

dark plaster. Plaster made from gypsum that has been calcined, but not ground.

Dark Red Gold. British trade name for a corrosion-resistant alloy of 50% gold and 50% copper, used for ornaments.

dark-red silver ore. See pyrargyrite.

dark-ruby silver. See pyraragyrite.

Darleen. Trade name for rubber fibers used in the manufacture of elastic yarn for clothing, and elastic bands and tapes.

Darlspan. Trade name for a *spandex* (segmented polyurethane) fiber used for the manufacture of textile fabrics with excellent stretch and recovery properties.

Darmstadt bell metal. A German corrosion-resistant leaded tin bronze containing 72.5-74% copper, 21.1-21.7% tin, 2.1% lead, 0.05-0.2% iron and 2-2.6% nickel. Used for bells.

Dart. (1) Trade name of Latrobe Steel Company (USA) for a chromium-type hot-work tool steel (AISI type H10) containing 0.41% carbon, 1% silicon, 0.55% manganese, 3.3% chromium, 2.4% molybdenum, 0.35% vanadium, and the balance iron. It has good hardness and wear resistance, good machinability, high toughness, and excellent ductility. Used for forging, header and press dies, hot forming and extrusion dies, and die holders

and inserts.

(2) Trademark of E.I. DuPont de Nemours & Company (USA) for a series of impact-resistant, medium- to high-strength polystyrenes used for toys, housewares, containers, and packaging applications.

Dartek. Trademark of E.I. DuPont de Nemours & Company (USA) for nylon films that provide good stretchability, low-temperature formability, good oxygen-barrier properties, and good oil, grease and moisture resistance. They can be processed by thermoforming, and are used for packaging applications.

Dartex. (1) Trademark of DuPont Canada Inc. for chlorinated rubber.

(2) Trademark of Courtaulds Textiles (Holdings) Limited (UK) for coated fabrics.

Darvan. Trademark of Celanese Chemical Company (USA) for a nytril fiber based on a vinylidene dinitrile polymer. It has high tenacity, outstanding resistance to dry heat degradation and outdoor weathering in direct sunlight, good resistance to mildew, good to excellent resistance to hot and cold acids, fair resistance to cold dilute alkalies, and poor resistance to strong alkalies. Used for apparel, yarns, pile fabrics, and industrial fabrics.

Darvic. Trademark of ICI Limited (UK) for low-cost general-purpose rigid polyvinyl chloride plastics with relatively moderate heat distortion properties. They can be made flexible with plasticizers, and are used for floor covering, electrical insulation, piping, etc.

Darwin. (1) Trade name of Darwin Alloy Castings (UK) for a series of age-hardenable acid-resistant nickel-chromium-iron, nickel-molybdenum-iron, and nickel-molybdenum-chromium alloy castings.

(2) Trade name of Darwin & Milner Inc. (USA) for a series of steels including several cobalt-tungsten high-speed tool steels used for cutting tools, hobs, borers, milling cutters, etc., and various stainless steels for surgical instruments, rollers, and dies.

Darwins. Trade name of Darwins Alloy Castings (UK) for a series of corrosion- and/or heat-resistant austenitic, ferritic and martensitic stainless steels, tungsten and tungsten-cobalt high-speed steels, cold- and hot-work steels, and water-hardening plain-carbon tool steels. Also included under this trade name are several iron-nickel controlled-expansion alloys, and corrosion-resistant nickel-molybdenum-iron and nickel-molybdenum-chromium-iron alloys.

Dasag. Trade name of DASAG-Deutsche Naturasphalt GmbH (Germany) for slabs made of natural asphalt.

dash-bond coat. A thick slip made by mixing Portland cement and sand with adequate amounts of water and applied to a surface by machine spraying, or dashing with a broom or paddle to provide a base for subsequent plaster or stucco coats.

datolite. A colorless, white, yellowish, or reddish mineral with greenish tinge and vitreous luster composed of calcium borosilicate, $CaBSiO_4(OH)$. Crystal system, monoclinic. Density 2.8-3.0 g/cm^3; hardness, 5-5.5 Mohs. Occurrence: USA (Connecticut, Massachusetts, New Jersey). Used as a flux in glazes, and as a gemstone.

daubreelite. A black mineral of the spinel group composed of iron chromium sulfide, $FeCr_2S_4$. It can also be made synthetically. Crystal system, cubic. Density, 3.81 g/cm^3. Occurrence: Found in meteorites.

Dauphinox. Trade name of Forges et Acieries de Bonpertuis (France) for an extensive series of austenitic, ferritic and martensitic stainless steels used for chemical equipment, corrosion-resistant welded structures, household appliances, equipment for the pulp and paper industries, food-processing equipment, dairy and brewing equipment, aircraft fittings, bearings, valve and pump parts, cutlery, surgical instruments, machine parts, fasteners, fittings, turbine blades, and wheels.

davanite. A colorless mineral composed of potassium titanium silicate, $K_2TiSi_6O_{15}$. Crystal system, triclinic. Density, 2.76 g/cm^3. Occurrence: Russian Federation.

Davisil. Trademark of W.R. Grace & Company (USA) for a series of silica gels available in a variety of mesh sizes for chromatography and other adsorption applications. They are usually supplied as 5% slurries with a pH of 6.5-7.0.

davidite. A gray-black, or black mineral of the crichtonite group composed of iron cerium uranium titanium oxide, $(Fe,Ce,U)_2(Ti,Fe)_5O_{12}$. Crystal system, rhombohedral (hexagonal). Density, 4.46 g/cm^3; refractive index, 2.3; hardness, 803-907 Vickers. Occurrence: Mozambique, Australia, USA (Arizona).

Davignon. British trade name for an alloy containing 58% gold, 37% copper, and 5% aluminum. It has excellent corrosion resistance, and is used for ornaments and jewelry.

Davis metal. Trade name for a series of hard, corrosion- and heat-resistant alloys of 67% copper, 25-29% nickel, 2-6% iron, 0.3-1.5% manganese, 0.2-0.5% carbon, 0-0.8% lead, and a trace of silicon. Used for valves, fittings, and turbine blades.

davreuxite. A colorless mineral composed of manganese aluminum silicate hydroxide, $Mn_2Al_{12}Si_7O_{31}(OH)_6$. Crystal system, monoclinic. Density, 3.00 g/cm^3; refractive index, 1.680. Occurrence: Belgium.

davyne. A colorless mineral of the cancrinite group composed of sodium calcium potassium aluminum silicate chloride sulfate, $(Na,Ca,K)_8(Si,Al)_{12}O_{24}(Cl,SO_4)_3$. Crystal system, hexagonal. Density, 2.42 g/cm^3; refractive index, 1.519. Occurrence: Italy.

Dawbarn. Trade name of Dawbarn Brothers, Inc. (USA) for saran (polyvinylidene chloride) fibers.

Dawnilux. Czech trade name for a glass with random, diaphanous, rippled and finely granulated pattern.

Dawson. Trade name of Time Steel Service Inc. (USA) for a water-hardening shock resisting tool steel (AISI type S4) containing 0.65% carbon, 0.9% manganese, 2% silicon, and the balance iron. Used for hand and pneumatic tools.

dawsonite. A colorless, or white mineral composed of sodium aluminum carbonate hydroxide, $NaAlCO_3(OH)_2$. Crystal system, orthorhombic. Density, 2.40-2.44 g/cm^3; refractive index, 1.542. Occurrence: Canada.

Dawson's bronze. A bronze containing 83.9% copper, 15.9% tin, 0.1% lead and 0.05% arsenic. Used for journal bearings and castings.

DAW Steel. Trade name of Mitsui Engineering & Shipbuilding Company Limited (Japan) for an iron-nickel-manganese alloy with a very high damping ratio. The material absorbs vibrational energy by expanding or contracting laminated defects in its structure. Used for products requiring noise and vibration attenuation.

Dayan. Trademark of La Seda de Barcelona SA (Spain) for nylon 6 fibers and yarns used for textile fabrics including clothing.

Day-Glo. Trademark of Day-Glo Color Corporation (USA) for fluorescent pigments, dyes and inks. The pigments are used for coatings and plastics, and in the graphic arts, the dyes for plastics, and the inks in the graphic arts. Also included under this trademark are various tracer and textile dyes, UV brighteners, and luminescent/phosphorescent colorants for "glow-in-the-

dark" applications.

(2) Trade name of A.R. Monteith Limited (Canada) for a series of fluorescent inks and paints.

daylight glass. A special glass that absorbs red light and produces a light that resembles daylight. Used for incandescent lamps.

Daylite. Trademark of Day & Campbell Limited (Canada) for concrete building blocks.

Dazzle PS. Trade name of Whip Mix Corporation (USA) for a fine-grain synthetic silica powder used as a pumice substitute for polishing dental acrylics.

DB-Float. Trade name of Dry Branch Kaolin Company (USA) for a clay with fine particle size made from soft Georgia kaolin.

dc-fix. Trademark of Konrad Hornschuh AG (Germany) for flexible polyvinyl chloride film materials.

DC Resins. Trade name of Dow Corning Corporation (USA) for silicone resins with excellent electrical properties, chemical inertness (but susceptible to attack by steam), and outstanding heat resistance. Used for laminates, terminal strips, and high-temperature insulation.

DC Xtra. Trade name of A. Finkl & Sons Company (USA) for a double-vacuum-refined tool steel.

deacetylated acetate. Regenerated cellulose fibers made by removing virtually all acetyl groups from cellulose acetate. Also known as *deacetylated acetate fibers*.

dead-burnt dolomite. A granular refractory dolomite, $CaMg(CO_3)_2$, with or without additives, that has been calcined to remove carbonates and other volatile constituents, and form calcium oxide (CaO) and periclase (MgO). It is resistant to atmospheric hydration and recombination with carbon dioxide. Also known as *dead-burnt refractory dolomite*.

dead-burnt gypsum. See calcium sulfate.

dead-burnt magnesia. See dead-burnt magnesite.

dead-burnt magnesite. Granular magnesia (MgO) or other magnesium-bearing material convertible to magnesia that has been produced by calcining magnesite ($MgCO_3$) at temperatures above 1450°C (2640°F) to form a stable material suitable for use as a refractory, or refractory component in metal-melting furnaces and kilns. It can also be made from magnesium hydroxide or chloride. Abbreviation: DBM. Also known as *burnt magnesia; dead-burnt magnesia; refractory magnesia; sinter magnesia*.

dead-burnt plaster. See calcium sulfate.

dead-burnt refractory dolomite. See dead-burnt dolomite.

deadlime. A *chalk* that has or has been decomposed.

dead mild steel. A low-carbon steel containing 0.07-0.15% carbon. See also mild steel.

dead soft steel. (1) Steel whose carbon content is lower than that of mild or soft steel (below 0.05%). See also soft steel.

(2) Steel, usually a medium- or high-carbon type, that has been fully annealed, i.e., austenitized by heating to a temperature just above the critical temperature until equilibrium is achieved, and then cooled to room temperature in the furnace. It has a coarse pearlitic structure, and is very soft.

DEAE cellulose. See diethylaminoethyl cellulose.

de-aired brick. A brick that has been densified by removing air during forming by the application of a vacuum.

deal. (1) A piece of square sawn softwood timber, 50-100 mm (2-4 in.) thick, less than 230 mm (9 in.) wide, and up to 1.8 m (6 ft.) long.

(2) European name for the wood of the northern pine (*Pinus*

silvestris), also known as red deal, and the Norway spruce (*Picea abies)*, also known as white deal. See also northern pine; Norway spruce.

decaborane. Flammable, colorless or white crystals (95+% pure) with a density of 0.94 g/cm^3, a melting point of 98-100°C (208-212°F), and a boiling point of 213°C (415°F). It decomposes into boron and hydrogen at 300°C (572°F), and is used as a catalyst, reducing agent, fluxing agent, stripping agent, stabilizer, fuel additive, propellant, corrosion inhibitor, oxygen scavenger, and as a boron source for ion-implantation processes. Formula: B$_{10}$H$_{14}$. Also known as *decaboron tetradecahydride*.

decaboron tetradecahydrate. See decaborane.

decacarbonyl dimanganese. See dimanganese decacarbonyl.

decalcomania simplex paper. A unsized, wood-free paper with a water-soluble coating on which a printed design or picture can be produced, and which, after moistening, allows transfer of this design or picture.

Decaplast. Trade name of Rhodiatoce (France) for a series of polyamides (nylons).

Decapol. Trade name of Monopol AG (Switzerland) for paint strippers.

decarburized steel. A steel from whose surface layer carbon has been removed by heating in a medium that acts chemically with carbon.

decarburized enameling steel. A sheet steel with very low carbon content used in porcelain enameling especially for single-coat coverage, i.e., for the application of a white or colored finish coat of porcelain enamel directly to the steel. Also known as *direct enameling steel; direct-on cover coat steel; direct on steel; zero-carbon steel*.

Deccan fibers. See kenaf fibers.

Decelith. Trademark of Orbitaplast for a series of thermosetting plastics including flexible, modified and unmodified polyvinyl chlorides.

deciduous tree. Any tree or shrub that usually sheds its soft, broad leaves annually. Deciduous trees produce wood commonly known as *hardwood* (although not necessarily an indication of hardness). Ashes, basswoods, beeches, birches, cherries, elms, hickories, mahoganies, maples, oaks and poplars are deciduous trees. Also known as *broadleaved tree*. See also hardwood.

Decobondglas. Trade name of G.F. Fleischer (Germany) for strips of figured glass, 3.2 mm (0.125 in.) thick and approximately 51 mm (2 in.) wide, bonded together to form a rigid sheet. Used for "false" ceilings.

Decobra. British trade name for a corrosion-resistant copper alloy containing 74.4% copper, 5.4% zinc, 19% nickel, 0.8% iron and 0.4% manganese. Used for hardware and ornamental parts.

Deco-Coat. Trademark of Epoxies, Etc. (USA) for epoxy resins used to form crystal clear, glossy decorative lens effects on photographs, decals, labels, medallions, magnets, etc.

Decofalt. Trade name of Westerwald AG (Germany) for a flexible glass made from narrow strips of cast glass joined in longitudinal direction with synthetic resin and tissue film.

Decoflexglas. Trade name of G.F. Fleischer (Germany) for a flexible product made from strips of figured glass, 3.2 mm (0.125 in.) thick and approximately 51 mm (2 in.) wide, strung together.

decolorizer. A material added to a glass melt to improve the appearance of the resulting glass by removing or hiding color in finished products. Examples of such materials include antimony oxide, arsenic trioxide, barium dioxide, cerium dioxide, cobalt chloride, cobalt monoxide, manganese dioxide, neodymium oxide, neodymium sulfate and nickel monoxide. Also known as *glass decolorizer.*

Decoltal. Trade name of Alusuisse (Switzerland) for wrought aluminum-copper alloys available in regular and free-machining grades (containing lead). They have high strength, and are used for architectural and structural applications. The free-machining grades are also used for automatic lathe products.

Decolux. Trademark of DECOLUX GmbH (Germany) for a range of decorative plastic and metallic film and foil products diffraction, mirror, 3D-, holographic, optic and sun-protection films and foils.

Decora. (1) Trade name of Pittsburgh Corning Corporation (USA) for a nearly transparent, non-light-directing glass block with smooth outside surfaces, and a random design pressed into the inside surface.

(2) Trade name for rayon fibers and yarns used for textile fabrics.

Decoral. Trade name of Aluminium Laufen AG (Switzerland) for a wrought alloy containing 5.5% zinc, 1% magnesium, and the balance aluminum. Used in the aircraft industries.

decorated glass. Glass that has been ornamented by cutting, engraving, etching, gilding, painting, sandblasting, or other methods. Used especially for architectural applications and in the arts.

Decorative Art. Trade name of SA de Verreries de Fauquez (Belgium) for patterned glass.

decorative brass plate. A thin, bright electrodeposit of yellow brass or gold-colored brass applied over a bright nickel plate for decorative purposes. Used on wire goods, picture frames, builder's hardware, etc.

decorative chromium plate. A very thin electrodeposit of hexavalent or trivalent chromium applied over undercoatings of copper, nickel or copper plus nickel for decorative purposes. Depending on the undercoating, the plate can be bright, semibright or satin, and corrosion-resistant. Used on automotive components and ornaments, hardware, etc.

decorative gold plate. A very thin electrodeposit of gold or gold alloy applied to bronze or brass parts to obtain a particular color. Used chiefly on jewelry and ornaments.

decorative laminates. High-pressure laminates comprising a core of phenolic paper and a top sheet of melamine paper with a decorative pattern.

decorative nickel plate. A very thin electrodeposit of nickel applied to provide decorative finishes on ferrous and nonferrous metals. It may be used over undercoatings of cadmium, copper, nickel or zinc, or as a undercoating for chromium or precious metals. The most common deposits are: (i) thin, black, nonreflecting, nonprotective deposits derived from baths containing zinc chloride or zinc sulfate, and (ii) bright, corrosion-resistant deposits containing varying amounts of sulfur, and produced in modified Watts baths.

decorative plate. An electrodeposit of brass, chromium, nickel or precious metals applied chiefly for decorative purposes, i.e., to obtain a surface finish with a certain color, appearance or other optical properties.

decorative rhodium plate. A thin, white electrodeposit of rhodium applied to provide decorative finishes on platinum products, and impart non-tarnishing properties to sterling silver.

decorative thermosetting laminates. A group of plastic laminates produced by bonding together, under heat and pressure, sheets of paper or fabrics impregnated with thermosetting resins, such as phenolics, ureas or melamines. Used mainly for decorative applications (usually in the form of panel materials).

decorative vacuum coating. A very thin vacuum-deposited coating, usually of aluminum, applied to a metallic or plastic substrate solely for decorative purposes. It does not provide protection against corrosion or wear, and is widely used on automotive components, home appliances, hardware, jewelry, and novelty items.

Decorex. Trade name of Enthone-OMI Inc. (USA) for a decorative palladium electrodeposit and plating process.

Decor-Glass. Trade name of Amerada Glass Company (USA) for a patio door glass with insignias, designs or permanent laminations between two sheets of glass.

Decorglass. Trademark of Saint-Gobain (France) for a patterned glass with clear or tinted body.

Decorglass Wired. Trade name of Saint-Gobain (France) for a patterned wired glass.

Decra-Brite. Trade name of Manville Corporation (USA) for a white fiberglass decorative blanket.

Decra-Glo. Trade name of Manville Corporation (USA) for fiberglass used for sculptured, decorative ornaments.

Decra-Loc. Trademark of D. Barnett & Company Limited (Canada) for interlocking paving stones.

Decraloy. Trademark of Cominco Limited (Canada) for zinc-color coatings for steels.

decrystallised cotton. Cotton with a low degree of crystallization obtained by reacting with concentrated sodium hydroxide solution, zinc chloride, or certain amine-based compounds (e.g., anhydrous liquid ethylamine). It has enhanced absorbency and dyeability.

Dederon. Trade name for nylon fibers and yarns used for textile fabrics including clothing.

Dedikanol. Trademark of Rheinpreussen AG (Germany) for a series of unsaturated polyesters.

Deefive. Trade name of Handy & Harman (USA) for a hard gold casting alloy containing a total of 15-40% copper, platinum, zinc and silver. Used for dentures and dental inlays.

Deeformat. British trade name for a mechanically bonded mat of chopped glass-fiber strands.

Deefour. Trade name of Handy & Harman (USA) for a hard gold casting alloy containing a total of 15-40% copper, platinum, zinc and silver. Used for dentures.

Deefourteen. Trade name of Handy & Harman (USA) for a hard gold casting alloy containing a total of 15-40% copper, platinum, zinc and silver. Used for dentures and dental inlays.

Deeglas. Trade name of Deeglas Fibres Limited (UK) for glass-fiber reinforcements.

Deelite. Trade name of Handy & Harman (USA) for a hard gold casting alloy containing a total of 15-40% copper, platinum, zinc and silver. Used for dentures and dental inlays.

Deeone. Trade name of Handy & Harman (USA) for a soft dental casting alloy containing 60-85% gold, and a total of 15-40% copper, platinum, zinc and silver. Used for dental inlays.

deep-drawing brass. A commercial wrought brass containing about 67-70% copper and 30-33% zinc. It has high elongation and excellent cold workability. Used for seamless tube, condenser and evaporator tubing, and deep-drawn mechanical parts.

Deepep. Trade name of Handy & Harman (USA) for a wrought gold alloy containing a total of 15-40% copper, platinum, zinc and silver. Used for dentures and wire.

Deepex. Trade name of Murex Limited (UK) for a plain-carbon

steel containing 0.15% carbon, 0.35% manganese, and the balance iron. Used for welding electrodes.

Deep Hardening Berkshire. Trade name of Carpenter Technology Corporation (USA) for a cold-work tool steel containing 1.2% carbon, 1% manganese, 0.5% manganese, 0.2% vanadium, and the balance iron. Used for thread-chasing dies.

Deepreg. Trade name of Deeglas Fibres Limited (UK) used for polyester preimpregnated glass strands.

deep-textured fabrics. Textile fabrics with 3-dimensional effects produced either by using two or more differently colored yarns, or curled or subbed yarns of different thickness, or by combining different weave patterns.

deerite. A black mineral composed of iron aluminum silicate hydroxide, $Fe_6(Fe,Al)_3Si_6O_{20}(OH)_5$. Crystal system, monoclinic. Density, 3.84 g/cm^3. Occurrence: USA (California).

Deeseven. Trade name of Handy & Harman (USA) for a hard gold casting alloy containing a total of 15-40% copper, platinum, zinc and silver. Used for dental inlays.

Deesix. Trade name of Handy & Harman (USA) for a hard gold casting alloy containing a total of 15-40% copper, platinum, zinc and silver. Used for dentures and dental inlays.

Deethree. Trade name of Handy & Harman (USA) for a hard gold casting alloy containing a total of 15-40% copper, platinum, zinc and silver. Used for dental inlays and dentures.

Deetwo. Trade name of Handy & Harman (USA) for a gold casting alloy containing a total of 15-40% copper, platinum, zinc and silver. Used for dental inlays.

defect semiconductor. A nonstoichiometric transition metal oxide that gains its conductivity from ions that have more than one valence, e.g., Fe^{2+} and Fe^{3+}, or Ni^{2+} and Ni^{3+}.

Defender. (1) Trade name of Keystone Steel & Wire Company (USA) for barb wire.

(2) Trade name of Magnolia Metal Corporation (USA) for babbitts.

(3) Trade name of Lantor Inc. (USA) for coated nonwoven fabrics for filtration applications.

Defender Polyester. Trade name of Lantor Inc. (USA) for a series of polyester-based air filtration media with improved particulate collection efficiencies and air flows. They have very high air-throughputs, outstanding filtration efficiencies, and are used for pollution-control applications, and in the chemical, pharmaceutical, food-processing and mining industries.

defernite. A colorless mineral composed of calcium chloride carbonate hydroxide monohydrate, $Ca_3(CO_3)(OH,Cl)_4 \cdot H_2O$. Crystal system, orthorhombic. Density, 2.50 g/cm^3; refractive index, 1.572. Occurrence: Turkey.

Defiheat. Trade name of Armco Steel Corporation (USA) for a corrosion- and heat-resistant stainless steel containing 0.3% carbon, 0.5% manganese, 0.75% silicon, 25-30% chromium, and the balance iron. Used for stainless articles, and furnace parts.

Definite. Trade name of Degussa Dental (USA) for a light-curing hybrid ormocer composite used in restorative dentistry. See also ormocers.

Defistain. Trade name of Armco Steel Corporation (USA) for an austenitic stainless steel containing about up to 0.2% carbon, 0.5% manganese, 0.5% silicon, 18% chromium, 8% nickel, and the balance iron. It has excellent corrosion resistance, and is used for chemical equipment, valves and trim. Free-machining (with higher sulfur contents) and welding grades are also available.

Deflex. Trade name for a low-temperature-resistant thermoplastic olefin elastomer used for automotive applications.

deflocculant. A basic material, such as a carbonate, phosphate or silicate of sodium, that when added to a clay slurry or slip will disperse the agglomerates or flocs of fine clay or soil particles, and form a more fluid colloidal or near-colloidal suspension. Also known as *deflocculating agent*.

deflocculating agent. See deflocculant.

defoamer. See antifoamer.

defoaming agent. See antifoamer.

deformed bar. A reinforcing steel rod or bar having several ribs or indentations to improve its bond with concrete.

Deft. Trademark of DEFT, Inc. (Canada) for wood stains used for finishing surfaces.

Defthane. Trademark of DEFT, Inc. (Canada) for a series of polyurethane finishes supplied in gloss, satin and semi-gloss grades for sealing and finishing exterior and interior wood surfaces.

Degalan. Trademark of Degussa AG (Germany) for acrylics based on polymethyl methacrylate polymers, and available in various grades and forms. They have a density of 1.2 g/cm³ (0.04 lb/in.³), good to moderate mechanical properties, a maximum service temperature of 60-93°C (140-200°F), good optical properties, excellent resistance to gasoline and mineral oils, and moderate resistance to dilute acids and alkalies and trichloroethylene. Used for glazing, signs, enclosures, light fixtures, molding resins, etc.

degami. The strong, straight-grained yellowish wood of the tree *Calycophyllum candidissimum* belonging to the capironas. It has a fine, dense grain, high elasticity, and poor resistance to stains, decay and insects. Source: West Indies. Used for textile and archery equipment. Also known as *degame; degami lancewood; degami wood*. See also capirona.

Degamid. Trademark of Degussa AG (Germany) for polyamide (nylon) film materials.

degami wood. See degami.

Degaplast. Trademark of Degussa AG (Germany) for a series of polyacrylic resins and compounds.

Degaser. Trade name of Foseco Minsep NV (Netherlands) for a series of degassing and/or grain-refining agents supplied in the form of tablets, and used to remove hydrogen from aluminum and aluminum-alloy melts.

degasifier. A substance added to a molten metal or alloy to draw off soluble gases before they are trapped in the material upon solidification.

degasser. See scavenger (2).

degenerate semiconductor. A semiconductor in which the number of conduction electrons approaches that of a metal.

Deglas. Trademark of Degussa AG (Germany) for plexiglass-type polymethyl methacrylate plastics.

degradable plastics. A class of plastics that are susceptible to decomposition by (i) microorganisms, known as *biodegradable plastics*; (ii) ultraviolet radiation, known as *photodegradable plastics*; or (iii) water, known as *water-soluble plastics*.

Degubond. Trademark of Degussa-Ney Dental (USA) for a series of noble palladium-gold (e.g., Degubond E), palladium-silver (e.g., Degubond 60) and palladium-silver-gold (e.g., Degubond Ultra) dental ceramic alloys that bond well to most commercial dental porcelains. Used in restorative dentistry for single crowns, and bridges.

Degubond Lite. Trademark of Degussa-Ney Dental (USA) for a highly biocompatible, fine-grained, silver-free dental ceramic alloy containing 79.5% palladium and 5.5% gold. It has high corrosion and tarnish resistance, high mechanical strength, a

hardness of 250 Vickers, bonds well to most dental porcelains, and is used in restorative dentistry for long-span bridges and single crowns.

Degubond Ultra. Trademark of Degussa-Ney Dental (USA) for a highly biocompatible, microfine-grained dental ceramic alloy containing 74.4% palladium, 12.5% silver, and 2% gold. It has high corrosion and tarnish resistance, high mechanical strength, a hardness of 230 Vickers, and is suitable for use with high-expansion dental porcelains. Used in restorative dentistry for single crowns, short- and long-span bridges, and milled restorations.

Degucast U. Trade name of Degussa-Ney Dental (USA) for a medium-gold dental bonding alloy.

Degudent. Trademark of Degussa-Ney Dental (USA) for a series of dental bonding alloys including gold-platinum (e.g., Degudent G), gold-platinum-palladium (e.g., Degudent GS and Degudent H), gold-platinum-palladium-silver (e.g., Degudent U), and gold-palladium-platinum (e.g., Degudent LS and Degudent SF). Used in restorative dentistry for crowns, inlays, dentures, etc.

Degufill. Trademark of Degussa Dental (USA) for synthetic substances and filling materials for dental applications. Degufill H, Degufill-M and Degufill SC Micro Hybrid are light-cure hybrid composites, self-cure hybrid composites and light-cure microfine composites, respectively. Used in restorative dentistry.

Deguflex. Trademark of Degussa Dental (USA) for dental impression and duplicating materials.

Deguform. Trademark of Degussa Dental (USA) for elastic duplicating and impression materials for dental applications.

Degulfit. Trade name of Degussa-Ney Dental (USA) for a soldering alloy.

Degulcor. Trade name of Degussa-Ney Dental (USA) for a precious metal alloy.

Degulor. Trademark of Degussa-Ney Dental (USA) for a series of high-gold dental casting alloys and solders.

degummed silk. Silk yarns, fibers or waste that have all or most of the silk gum (*sericin*) removed by treating with a hot mildly alkaline solution.

Degunorm. Trademark of Degussa-Ney Dental (USA) for an extra-hard microfine-grained, low-copper dental bonding alloy (ADA type IV) containing 73.8% gold, 9.2% silver, and 9% platinum. It has a deep-golden color, excellent corrosion and tarnish resistance, high yield strength, a hardness of 230 Vickers, and can be veneered with most commercial resins and *Duceragold* porcelain. Used for long-span bridges, single crowns, inlays, partial dentures, milled restorations, and implant superstructures.

Degupal. Trademark of Degussa-Ney Dental (USA) for palladium-based dental alloys.

Deguplus. Trademark of Degussa-Ney Dental (USA) for a series of precious-metal dental alloys used especially for bridges, milled restorations, and cast partial dentures.

Degupress. Trademark of Degussa Dental (USA) for self-cure acrylic dental resins used in denture manufacture and repair.

Degusint. Trademark of Degussa-Ney Dental (USA) for precious-metal dental alloys and materials.

Degussa. Trade name of Degussa AG (Germany) for an extensive series of precious metals (e.g., gold, platinum, palladium, rhodium, iridium, osmium, ruthenium, rhenium, etc.) supplied raw and processed in various purities. Also included under this trade name are bimetal strips for contacts, various nonprecious metals and nonmetals (e.g., mercury, thallium, bismuth and selenium).

Degussa alloy. A yellow brass containing 66% copper, 34% zinc, and traces of iron. Used for lamp fixtures, hardware and ornamental parts.

Degussa Black. Trade name of Degussa AG (Germany) for a series of carbon blacks.

Degussit. Trademark of Degussa AG (Germany) for sintered aluminum oxide (Al_2O_3) materials used as electrical insulation materials, thermal insulating materials (available in brick, cement and paste form), for sintered bearings, for ceramic applications, and as abrasives in the manufacture of grinding tools, e.g., grinding wheels, mounted points, files and whetstones.

Degustar. Trademark of Degussa-Ney Dental (USA) for fine-grained and microfine-grained palladium-silver, palladium-silver-gold and silver-palladium-gold ceramic alloys used in restorative dentistry for crowns, bridges, implant superstructures, etc.

Dehesive. Trademark of Wacker Silicones Group (USA) for silicone-base paper release coatings.

dehrnite. A colorless, greenish-white, or gray mineral composed of calcium sodium potassium phosphate hydroxide, $(Ca,Na,K)_5$-$(PO_4)_3(OH)$. Crystal system, hexagonal. Occurrence: Germany, USA (Utah).

dehydrated borax. See anhydrous borax.

dehydrated tar. See refined tar.

Deimitech. Trademark of Hogen Industries (USA) for a tungsten-based material containing a total of 7.5% nickel, iron and molybdenum, and having excellent erosion and wear resistance, good thermal conductivity, thermal-shock resistance, hardness, and strength at high working temperatures. Used for high-temperature tooling, die-casting and extrusion dies, and electro-brazing applications.

deKhotinsky cement. A thermoplastic cement made by mixing *shellac* with *pine tar*. It has excellent resistance to hydrochloric, nitric and sulfuric acid, gasoline and many solvents, good to fair resistance to chloroform, ether and alkalies, and poor resistance to ethanol. Used as an adhesive for joining metals, ceramics and glasses.

Dekorit. Trademark of Raschig GmbH (Germany) for phenol-formaldehyde plastics and laminates.

Dekorsil. Trade name of Vereinigte Metallwerke Ranshofen-Berndorf (Austria) for an aluminum casting alloy containing 5% silicon. It has good castability, weldability, polishability, and machinability.

delafossite. A black mineral composed of copper iron oxide, $CuFeO_2$. Crystal system, rhombohedral (hexagonal). Density, 5.41 g/cm³. Occurrence: Russian Federation.

delafossites. A class of materials that are oxides of copper and another metallic element, usually iron or a rare earth, such as lanthanum or yttrium. The general formula is $CuMO_{2-x}$ (x < 1). They have unique electronic and magnetic properties, and are used for layered superconductors.

delaine. A light, usually printed fabric made from fine worsted wool yarn in a plain weave, and used for luxurious dresses.

Delair. Trade name of Delsteel Inc. (USA) for a tough, wear-resistant, air-hardening tool steel (AISI type A2) containing 1% carbon, 5% chromium, 1% molybdenum, 0.4% vanadium, and the balance iron. Used for blanking, trimming and thread-rolling dies, cutters, engraving tools, shear blades, and broaches.

Delaron. Trade name of De La Rue Plastics (UK) for phenol-formaldehyde plastics.

Delcar. (1) Trade name of Delsteel Inc. (USA) for a water-hardening tool steel (AISI type W1) containing 0.6-1.4% carbon, and the balance iron. Used for punches, reamers, drills, taps, springs and hobs.

(2) Trade name of Deloro Stellite Limited (UK) for tube carbide rods and electrodes supplied in three grades for hardfacing applications. They are neither machinable nor grindable, and are used on ore-crusher rolls, grab bucket teeth and coal plow picks.

Delceram. Trade name of Delceram-Risius GmbH (Germany) for heat-resistant textiles, yarns, cords, sheet packing, tapes, cloth, wool and tubing based on ceramic materials reinforced with glass or *Inconel*. Used as asbestos substitutes in high-temperature insulation.

Delcondex. Trade name of Alcan-Booth Industries, Limited (UK) for a series of corrosion-resistant alloys containing 10-31% nickel, 0.7-2% manganese, 0.5-2% iron, and the balance copper. They have excellent resistance to corrosion in seawater and brackish water, and are used for condenser tubes, heat exchangers and fittings, boiler feedwater heaters, and marine condenser tubes.

Delconion. Trade name of Alcan-Booth Industries, Limited (UK) for an alloy containing 5.5% nickel, 1.2% iron, 0.6% manganese, and the balance copper. It has good formability, and is used for seawater pipes and condensers.

Delcro. Trade name for a series of cobalt-chromium-iron alloys used for heat-resistant parts and furnace equipment.

Delcrome. Trade name of Haynes International (USA) for abrasion-resistant white cast iron containing 2.7% carbon, 27% chromium, 0.9% manganese, 0.8% silicon, 0.8% vanadium, and the balance iron. It has good to fair corrosion resistance, good oxidation resistance up to 980°C (1796°F), a density of 7.59 g/cm³ (0.27 lb/in.³), a hardness of 670 Brinell, a melting point of 1255-1308°C (2292-2387°F). Used for hardfacing and welding electrodes, wear plates, dry bearings, valve slides, pump barrel liners, and other wear-resistant parts.

Delcron. Trademark of Nylon de Mexico SA (Mexico) for polyester fibers and yarns.

Deletot's alloy. An exceptionally ductile alloy containing 80% copper, 18% zinc and 2% manganese. Used as a brass solder, and for cartridge cases.

Delfer. Trade name of Deloro Stellite Limited (UK) for wear-resistant alloy castings containing 2.5% carbon, 16.5% cobalt, 13.5% chromium, 9% molybdenum, 5.5% tungsten, and the balance iron. Used for hardfacing electrodes.

Delfion. Trade name for nylon fibers and yarns used for textile fabrics including clothing.

delftware. (1) A kind of calcareous earthenware that has an opaque white glaze and is usually decorated in a characteristic blue. It is named for Delft, a city in the Netherlands where it was originally made.

(2) Any earthenware or pottery having a similar glaze or color as that defined under (1).

delhayelite. A colorless mineral composed of potassium sodium calcium aluminum silicate chloride hydrate, $(K,Na)_{10}Ca_5Al_6Si_{32}$-$O_{20}Cl_6 \cdot 18H_2O$. Crystal system, orthorhombic. Density, 2.60 g/cm³; refractive index, 1.532. Occurrence: Central Africa (Zaire).

Delhi Iron. Trade name of Associated Steel Corporation (USA) for a steel containing 0.1-0.11% carbon, 0.75-1% silicon, 16.5-18% chromium, and the balance iron. Used as a substitute for galvanized iron for roofing, autobody sheets, etc.

Delhi Hard. Trade name of Associated Steel Corporation (USA) for a steel containing 1-1.1% carbon, 16.5-18% chromium, up to 0.1% nickel, up to 0.5% manganese, 0.75-1.1% silicon, and the balance iron. It has excellent resistance to mine and seawater, cold ammonium hydroxide and moist sulfurous atmospheres.

Delhi Special. Trade name of Associated Steel Corporation (USA) for a heat- and corrosion-resistant steel containing 0.14% carbon, 0.5% manganese, 0.6% silicon, 17% chromium, and the balance iron. Used for furnace parts, furnace linings, and conveyors.

Delhi Tough Iron. Trade name of Associated Steel Corporation (USA) for a corrosion-resistant steel containing 0.07% carbon, 0.3% manganese, 1.25% silicon, 17% chromium, and the balance iron. Used for corrosion-resistant parts.

Delignit. Trademark of Blomberger Holzindustrie B. Hausmann GmbH & Co. KG (Germany) for standard- and aircraft-quality plywood, laminated wood, and multi- and fine-ply paneling.

Delignit-Panzerholz. Trademark of Blomberger Holzindustrie B. Hausmann GmbH & Co. KG (Germany) for high-strength wood laminates.

Deliplan. Trade name for flexible polyvinyl chloride film materials.

Deliplast. Trademark for flexible polyvinyl chloride film materials.

dellaite. A synthetic mineral composed of calcium silicate hydroxide, $Ca_6(SiO_4)(Si_2O_7)(OH)_2$, and made by a 3-week hydrothermal treatment of β-Ca_2SiO_4 at 350-370°C (660-700°F). Crystal system, triclinic. Density, 2.97 g/cm³; refractive index, 1.661.

Delloy. Trade name of Delloy Metals (USA) for a series of cobalt-chromium-tungsten alloys, tungsten-type high-speed steels, and cobalt-bonded cemented (tungsten) carbides. Used for cutting tools and hardfacing electrodes.

Delodur. Trade name of Flachglas AG (Germany) for a toughened safety glass.

Delogcolor. Trade name of Flachglas AG (Germany) for a colored, ceramic-coated toughened glass.

Delog-Hartglas. Trade name of Flachglas AG (Germany) for a toughened glass.

Deloro. Trade name of Deloro Stellite Limited (UK) for an extensive series of heat- and wear-resistant cobalt- and nickel-base alloys used for hardfacing electrodes and furnace equipment. See also Stellite.

Deloro Alloy. Trade name of Deloro Stellite Limited (UK) for a series of abrasion-, corrosion- and/or oxidation-resistant hardfacing materials including various nickel-chromium-silicon and nickel-chromium-molybdenum-silicon alloys containing considerable amounts of boron (typically 1-3%), and numerous nickel-molybdenum-iron and nickel-molybdenum-chromium types containing about 0.1-0.5% carbon.

Deloro Stellite. See Stellite.

Delovisca. Trade name for viscose rayon fibers and yarns used for textile fabrics.

Delphic. Trade name for a self-cure dental hybrid composite.

Delram AF. Trade name of National Refractories & Minerals Corporation (USA) for ram-cast refractory mixes.

Delray. Trade name for rayon fibers and yarns.

Delrin. Trademark of E.I. DuPont de Nemours & Company (USA) for thermoplastic homopolymer acetal resins based on polyoxymethylene (POM) and supplied in white and various colors in glass-fiber-reinforced, UV-stabilized, unfilled and polytetrafluoroethylene-filled grades. An acetal/elastomer alloy is also available. *Delrin* acetals have a density of 1.42 g/cm³ (0.05 lb/ in.³), a high tensile and impact strength, high toughness, stiffness and endurance, high creep resistance, high flexural fatigue strength, good durability and dimensional stability, excellent wear resistance, smooth surface finishes, low coefficients of friction, good moldability, low moisture absorption, excellent water resistance, good solvent and heat resistance, good resistance to oils, greases and gasoline, fair to poor resistance to burning, radiation, weathering and strong acids and bases, a maximum service temperature of 85-104°C (185-220°F), and high dielectric strength. Used for automotive components and trim, housings, tubing and pipes, underground pipes, impellers, door handles, bushings, bearings, gears, cams, couplings, levels, fans, industrial equipment, sports and recreational equipment, injection-molded and extruded parts, electrical parts, and fasteners. *Delrin II* refers to acetal resins available in various grades including UV-resistant, tough and supertough and glass-reinforced with high melt stability and impact resistance. *Delrin ST* refers to thermoplastic copolymer and elastomer acetal resins supplied in carbon and glass fiber-reinforced, UV-stabilized, supertough and silicon-lubricated grades.

delrioite. A yellow-green mineral composed of calcium strontium vanadium oxide hydroxide trihydrate, $CaSrV_2O_6(OH)_2 \cdot 3H_2O$. Crystal system, monoclinic. Density, 3.10 g/cm³; refractive index, 1.834. Occurrence: USA (Colorado).

Delsteel Alloy. Trade name of Delsteel Inc. (USA) for a series of shock-resisting tool steels (AISI type S5) containing various amounts of carbon, manganese, vanadium, molybdenum and iron. Used for chisels, mandrels, shear blades, punches, and stamps.

Delta. (1) Trade name Saint-Gobain (France) for triangular glass blocks made of two half blocks bonded at high temperature and enclosing an insulating air cushion.

(2) Trade name of Sainte des Acieries de Micheville (France) for antifriction metals composed of varying amounts of copper, tin and antimony. Used for bearings, linings, castings, bolts, and nuts.

(3) Trade name of Delta Extruded Metals Company Limited (UK) for an extensive series of cast and wrought coppers and copper-base alloys including numerous electrolytic tough-pitch and oxygen-free high-conductivity coppers, tellurium coppers, copper-nickel-phosphorus alloys, aluminum and silicon-aluminum bronzes, high-tensile and cold-forming brasses, high-speed machining brasses, manganese and naval brasses, nickel silvers, etc.

(4) Trade name of GPC (Taiwan) for acrylonitrile-butadiene-styrene resins.

Deltabestos. Trade name of General Electric Company (USA) for asbestos-insulated magnet wire.

Deltabra. Trade name of Alcan-Booth Industries Limited (UK) for an arsenical aluminum brass containing 77% copper, 2% aluminum, 21% zinc and 0-0.03% arsenic. It has excellent cold workability, and good resistance to seawater corrosion. Used for condenser and heat-exchanger tubes, and tanker heater tubes.

Delta Cast. (1) Trademark of Acheson Industries Inc. (USA) for graphite dispersions used for die-casting applications.

(2) Trade name of Wakefield Engineering Inc. (USA) for a pourable, thermally-conductive synthetic resin for electronic applications.

delta cerium. See cerium.

Deltachop. Trademark of PPG Industries, Inc. (USA) for chopped glass fibers.

Delta Coll GZ. Trade name of Bodycote Metallurgical Coatings Limited (UK) for a black or opaque, solvent-based organic coating applied by dip spinning to specially prepared substrates, and cured at 80-100°C (175-212°F). It has excellent corrosion resistance, especially to salt-spray corrosion, and is used on automotive components including fasteners.

delta dysprosium. See dysprosium.

Delta Forge. Trademark of Acheson Industries Inc. (USA) for dispersions of graphite, molybdenum disulfide, etc., used for die-forging applications.

Delta Immadium. Trade name of Delta Metal (BW) Limited (UK) for a series of wrought or cast, corrosion-resistant high-strength alloys of copper, zinc, aluminum and manganese. Used for pump rods, valve spindles, staybolts, and propeller shafts.

delta iron. See iron.

Deltal. Trade name of Deutsche Delta Metallgesellschaft (Germany) for an aluminum alloy containing 0.6-1.4% magnesium, 0.6-1.6% silicon, 0.6-1% manganese and 0-0.3% chromium. Used for structural members.

Deltalloy. Trademark of Alcoa–Aluminum Company of America for wrought aluminum-silicon alloys (AA 4000 Series) with outstanding wear resistance, high strength, good machinability and drilling characteristics, good weldability, good stress-corrosion cracking resistance, good surface finish, and low thermal expansion. Used for automotive components, such as transmission valves, brake cylinders and bushings for rack and pinion steering systems, and for various bearings and hydraulic components, copier parts, sound recording devices, and forged pistons.

Delta Maid. Trade name of Rockwool Manufacturing Company (USA) for high-temperature refractory cements.

delta manganese. See manganese.

Deltamax. Trade name of Allegheny Ludlum Steel (USA) for a grain-oriented soft-magnetic alloy containing 50% iron and 50% nickel. It has high saturation induction, good permeability, and a square hysteresis loop, and is used for choke cores and magnets.

delta metal. A high-strength brass containing about 55-61% copper, 36-42% zinc, up to 2% iron, and small percentages of manganese and lead.

Delta-Mu. Trade name of Magnetic Metals Company (USA) for an alloy of 50% iron, 45% nickel, and 5% chromium. It has high electrical resistivity, and is used for magnetic cores.

delta plutonium. See plutonium.

Delta Seal. Trade name of Bodycote Metallurgical Coatings Limited (UK) for organic resin coatings with about 2% polyetrafluoroethylene (PTFE) applied to a thickness of 8-10 μm (315-394 μin.) by spraying or dip spinning. They have excellent corrosion resistance including salt-spray corrosion, and are used on automotive components, especially fasteners. *Delta Seal GZ* coatings are similar to *Delta Seal*, but have higher PTFE contents and thus lower coefficients of friction, and are used on self-tapping and self-drilling screws and other fasteners.

delta terbium. See terbium.

Deltatovis. Trade name for rayon fibers and yarns used for textile fabrics.

Deltavit. Trade name of Deutsche Delta Metallgesellschaft (Germany) for heat-treatable high-strength aluminum alloy containing 2.5-5.0% copper, 0.2-1.8% magnesium, and 0.3-1.5% manganese. Used for aircraft and automobile parts.

Deltokal. Trade name of Deutsche Delta Metallgesellschaft (Germany) for a corrosion-resistant aluminum alloy containing 2.0-4.0% magnesium, 0-0.4% manganese, and 0-0.3% chromium. Used in the aircraft, shipbuilding and food-processing industries, and for automotive parts, trim and covers.

Delton. Trade name for self-cure dental pit and fissure sealants.

Deltoxal. Trade name of Deutsche Delta Metallgesellschaft (Germany) for corrosion-resistant aluminum alloy containing 2-4% magnesium, 0-0.4% manganese, and 0-0.3% chromium. It has excellent resistance to seawater corrosion, and is used for aircraft, missile and marine parts.

Deltumin. Trade name of Deutsche Delta Metallgesellschaft (Germany) for heat-treatable aluminum alloy containing 3.5-5.5% copper, 0.2-1.5% silicon, 0.1-1.5% manganese, and 0.2-2% magnesium. It has high tensile and fatigue strength, and is used for hardware, fittings, and aircraft-engine parts.

delustered fibers. Man-made fibers whose luster or sheen has been reduced by the addition of a delustrant to the spinning solution, or the finish coating. Their degree surface luster ranges from clear and bright to dull and mat.

Delustra. Trade name for delustered rayon fibers and yarns used for textile fabrics.

delustrant. A substance, such as titanium dioxide (anatase), barium sulfate, clay, chalk, etc., added to a man-made fiber prior to extrusion to reduce the luster or sheen. It may be added to the spinning solution, or the finish coating.

De Luxe. Trade name of Federal Foundries & Steel Company Limited (UK) for tools steels containing 0.9-1.2% carbon, 0.25% manganese, 0.2% silicon, and the balance iron. Used for dies and forming tools, header dies, reamers, and cutters for nonferrous metals.

Delwood. Trade name for molded products made by bonding wood chips and discontinuous glass fibers with polyester resin. Used for furniture, picture frames, cabinetwork, etc.

demantoid. A green variety of *andradite* garnet with brilliant luster from Russia. Used chiefly as a gemstone.

Demark. Trade name of Pipe Machinery Corporation (USA) for a sintered product composed of titanium carbide with cobalt binder. Used for thread-gauging dies.

Demerara greenheart. See greenheart (1).

demesmaekerite. A mineral composed of lead copper uranyl selenite hydroxide dihydrate, $Pb_2Cu_5(UO_2)_2(SeO_3)_6(OH)_6 \cdot 2H_2O$. Crystal system, orthorhombic. Density, 2.17 g/cm³; refractive index, 1.482. Occurrence: Central Africa (Zaire).

Demilan. Trademark for melamine-formaldehyde plastics.

Demilon. Trademark of De Millus SA (Brazil) for nylon 6 fibers and yarns.

Demo bronze. A high-leaded bronze containing 61% copper, 33% lead, 4-6% tin and 1-2% nickel. Used for bearings and utensils.

Demon. Trademark of Plibrico Company (USA) for mortars.

Denavis. Trade name of Denman & Davis Company (USA) for a water-hardening tool steel containing 0.7-1.2% carbon, and the balance iron. Used for tools and cutters.

Denavis High Speed. Trade name of Denman & Davis Company (USA) for a water-hardening tool steel containing 0.7% carbon, 4% chromium, 18% tungsten, and the balance iron. Used for tools, cutters, and reamers.

dendrimers. A class of branched, three-dimensional synthetic macromolecules with dendritic substituents and uniform molecular weights. They are build up step-by-step from simple branched monomers, and have chemical, physical and mechanical properties that are superior to those of conventional linear polymers. Dendrimeric macromolecules typically have amorphous structures and spherical shape, high solubility and compatibility, and low viscosity and compressibility. They can have *phthalocyanines* incorporated, can be produced as glassy films for optical applications, and can also be produced in other forms for use as nano-sized building blocks for the manufacture of molecular devices and machines, and drug delivery systems. They are also suitable for the preparation of adhesives, coatings, additives, etc., and in the manufacture of advanced microelectronic and magnetic storage devices. Also known as *dendritic macromolecules*.

dendritic macromolecules. See dendrimers.

dendritic powder. See arborescent powder.

Denertia. Trademark of Cabot Corporation (USA) for dental casting alloys.

Denflex. Trade name of Dennis Chemical Company (USA) for plastisols, plastisol primers, and strippable coatings.

denim. A heavy, coarse, durable *twill* fabric, usually of cotton or a blend of cotton and synthetic fibers and woven with blue- or black-colored warp (longitudinal) and white weft (filling) threads. Used for work and casual clothes, upholstery, etc.

Denine. Trade name of Atlas Specialty Steels (Canada) for a water-hardening steel containing 1.2% carbon, 1.5% tungsten, and the balance iron. Used for taps, and drawing dies.

Denka. Trade name of Showa Company/Denka Polymers (Japan) for acrylonitrile-butadiene-styrene resins.

Den-Mat Core Paste. Trade name of Den-Mat Corporation (USA) for a creamy, radiopaque, chemical- or dual-cured composite paste with high strength and flow strength used in restorative dentistry for core and crown build-ups, cementation of posts, and for split-roots repair.

Den-Mat Crown Reline. Trade name of Den-Mat Corporation (USA) for self-cure dental resin cement for relining crowns.

Den-Mat Paste Laminate. Trade name of Den-Mat Corporation (USA) for a light-cure dental opaquing composite.

Den-Mat Rembrandt. Trade name of Den-Mat Corporation (USA) for a light-cure dental coloring composite.

Den-Mat Resto-Seal. Trade name of Den-Mat Corporation (USA) for a light-cure resin-based dental fissure sealant.

Den-Mat TetraPaque. Trade name of Den-Mat Corporation (USA) for a dual-cure dental opaquing composite.

denningite. A pale-green or colorless mineral composed of a manganese calcium zinc tellurate, $(Mn,Ca,Zn)Te_2O_5$. Crystal system, tetragonal. Density, 5.05 g/cm^3; refractive index, 1.89. Occurrence: Mexico.

Densalloy. Trade name of Welded Carbide Company Inc. (USA) for a sintered heavy-metal alloy containing 90% tungsten, 6% nickel, and 4% copper. Used for balance weights and centrifugal clutches.

DensArmor. Trademark of Georgia-Pacific (USA) for an interior building board consisting of a noncombustible gypsum core board with a moisture-, mold- and mildew-resistant coated glass-mat backing and a smooth heavy-duty paper facing. It is supplied with a width of 1.2 m (4 ft.) and a length of 2.4-3.6 m (8-12 ft.) in thicknesses of 12.5 mm (0.500 in.) and 9.5 mm (0.375 in.). Special grades, such as fire-resistant *DensArmor Fireguard*, and *DensArmor Plus* with glass fiber-reinforced gypsum core for higher moisture resistance, are also available.

Denscast. Trade name of Driver Harris Company (USA) for an alloy of 80% nickel and 20% chromium. Used for thin-walled castings.

Dens-Cote. Trademark of Georgia-Pacific Corporation (USA) for glass fiber- and powdered plaster-reinforced gypsum composi-

tion boards for building applications.

Dense Alloy. Trade name of GTE Sylvania (USA) for a series of sintered tungsten-cobalt-silver, tungsten-cobalt-nickel, and tungsten-copper-nickel alloys. Used for balancing weights.

dense concrete. A general term for concrete in which the number of void spaces is kept to a minimum. It usually refers to a concrete of a density greater than or equal to 1.9 g/cm^3 (0.07 lb/in^3).

dense-graded aggregate. Graded mineral aggregate that when compacted produces relatively small voids or spaces between the aggregate particles. Abbreviation: DGA. See also open-graded aggregate; aggregate.

DensGlass. Trade name of Georgia-Pacific (USA) for glass-reinforced gypsum wallboard.

DensGuard. Trademark of Georgia-Pacific (USA) for moisture-, mold- and mildew-resistant exterior building panels.

Densified Cru-Die. Trade name of Slater Steels Corporation (USA) for a tough alloy steel supplied in various hardness grades. It contains 0.55% carbon, 0.9% manganese, 0.3% silicon, 0.8% nickel, 1.05% chromium, 0.4% molybdenum, and the balance iron. Used for die blocks, dies and tools.

densified hardboard. See high-density hardboard.

densified wood. See compregnated wood.

densifier. (1) An alloy added to a metal or alloy to impart a fine, close-grained homogeneous structure.

(2) A chemical compound used to impart water repellency to concrete and masonry.

Densilay. Trademark of Degussa-Ney Dental (USA) for a hard, microfine dental alloy (ADA type III) containing 60% gold, 27.2% silver, and 3.6% palladium. It has high yield strength, a hardness of 240 Vickers, and is used in restorative dentistry for crowns, short-span bridges, inlays, and onlays.

Densimag. Trade name for a product composed of a thin stainless steel tape with a magnetic-powder coating applied to one side. It has a total product thickness of 26.7 μm (1 mil), an upper service temperature of 315°C (600°F), and is used for electronic tapes.

Densite. (1) Trademark of Atlas Turner Inc. (Canada) for asbestos-cement panels and boards.

(2) Trade name of Foseco Minsep NV (Netherlands) for a series of master alloys composed of nickel and nickel-tin. Used to introduce nickel into gunmetal to enhance its strength properties, and into leaded bronzes to refine the lead distribution.

(3) Trade name of Jamison Steel Corporation (USA) for air-hardening tool and die steels containing 1.4-1.5% carbon, 11-13% chromium, 0.8-1% molybdenum, 0-0.2% vanadium, and the balance iron.

Denstar. Trade name of Denmat Corporation (USA) for a high-copper dental amalgam alloy.

Denstone. (1) Trade name of Denmat Corporation (USA) for die stone used in dentistry.

(2) Trademark of St. Gobain Industrial Ceramics (France) for catalyst-bed support media.

Dentabond. Trade name for a medium-gold dental bonding alloy.

Dentagiene. Trade name of C-I-L Inc. (Canada) for cellulose nitrate plastics used in dentistry.

dental alloys. A group of alloys suitable for use in dentistry and including gold- and silver-base alloys with/or without palladium and/or platinum, nickel-titani-um shape-memory alloys, nickel-chromium alloys and cobalt-base superalloys. Included in this group are dental amalgams for fillings and dental casting alloys for bridges, braces, prostheses, inlays, crowns, bars, clasps and implants.

dental amalgam. A dental material in the form of a plastic mass made by mixing finely divided spherical or lathe-cut particles of a silvery-white alloy, usually consisting of about 20-70% silver, 12-28% tin, 2-15% copper and 0-1% zinc, with liquid mercury. It has good wear and abrasion properties, high strength and good corrosion resistance, and a short setting time. Gallium-based amalgams are now employed as replacements for mercury. Used for dental fillings.

dental casting alloys. A group of dental alloys processed by casting and including gold casting alloys containing 50-80% gold, and silver casting alloys containing 25-70% silver and varying amounts of palladium. *Dental casting alloys* are classified by the American Dental Association (ADA) according to their hardness into (i) *Type I* – soft alloys with a hardness of 50-90 Vickers; (ii) *Type II* – medium-hard alloys with a hardness of 90-120 Vickers; (iii) *Type III* – hard alloys with a hardness of 120-150 Vickers; and (iv) *Type IV* – extra-hard alloys with a hardness greater than 150 Vickers. Used for inlays, crowns, bridges, bars, claps, dentures, and porcelain-to-metal restorations. See also dental alloys.

dental cements. Materials used in restorative dentistry for cementing, luting or lining crowns, inlays, onlays, bridges, prostheses and posts. Examples of such materials include zinc oxide with or without eugenol, zinc phosphate, zinc oxide with ethoxybenzoic acid or zinc polycarboxylate, and modified or unmodified glass-ionomer and resin-ionomer composites.

Dental D. Trademark of Biodent New Zealand/Australia for biocompatible technolopymer based on semi-crystalline polyoxymethylene (POM). It has excellent physical and mechanical properties, and is used for removable prosthodontic applications, and as a replacement for metals and acrylic resins in many prosthetic applications.

dental enamel. The hard, white calcareous substance that covers the exposed surface of teeth. It contains about 96% mineral matter consisting chiefly of tricalcium phosphate hydroxide $[3Ca_3(PO_4)_2 \cdot Ca(OH)_2]$ and an organic matrix composed essentially of *keratin*. Also known as *enamel*.

dental floss. A thin, strong, smooth, specially treated silk or synthetic yarn used to remove plaque and food particles from between the teeth.

dental foil. Soft, high-purity platinum or palladium foil used in orthodontics and for dentures, or a plastic foil (e.g., acrylic) used for dentures, or as an articulating foil for bite registration.

dental gold. A group of white or yellowish wrought or cast dental alloys containing 50-90% gold, 5-12% silver, 0-10% copper, and usually platinum and/or palladium. See also dental casting alloys.

Dentalloy. Trade name of Cerro Metal Products Company (USA) for a dental casting alloy containing 38.2% bismuth, 26.4% lead, 31.7% tin, 2.6% cadmium, 1% antimony, and the balance copper. Used for dental castings and models.

dental materials. A large group of materials including metals, alloys, plastics, ceramics and combinations thereof used to make durable and esthetically pleasing dental restorations including bars, braces, bridges, clasps, crowns, fillings, inlays, etc. Abbreviation: DM.

Dentalon Plus. Trade name for a dental resin for temporary bridge and crown restorations.

Dentalor. Trade name of Degussa Dental (USA) for a high-gold dental casting alloy.

dental porcelain. A dense, bubble-free, usually feldspathic or aluminous porcelain with superior strength that is shaped, tinted

and fired for oral prosthetic use.

dental resin cement. See resin cement.

dental spar. Potash feldspar used for making dental porcelain.

DenTastic. Trademark of Pulpdent Corporation (USA) for multi-purpose dental bonding agents. *DenTastic Uno* is a single-component, light-cure dental adhesive that forms strong bonds with tooth and restorative surfaces including composites, resins and resin cements. *DenTastic Uno-Duo* is a hydrophilic dental adhesive system consisting of the dental adhesive *DenTastic Uno* and the catalyst *DUO*, and is used for indirect restorations including crowns and inlays, and for bonded amalgams.

Dentecon. Trade name of Mallory Metallurgical Products Limited (UK) for a hardened dental alloy containing 63% gold and 37% platinum. It has a melting range of 885-945°C (1625-1733°F).

Denthesive. Trade name for a *dentin* bonding agent used in dentistry.

Denthesive II. Trade name for a multi-purpose dental bonding agent.

dentin. A calcareous material that comprises the principal mass of a tooth. It contains about 70% mineral matter and 30% organic matter (mainly collagen) and water, and is similar to bone, but harder and denser. It can also be made synthetically. Also known as *dentine*.

Dentitan. Trade name for a corrosion-resistant cobalt-chromium dental bonding alloy.

Dentobond. Trade name for a corrosion-resistant cobalt-chromium dental bonding alloy.

Dentolith. Trade name for a cellulose acetate formerly used for denture bases.

Dentormat. Trade name of Mallory Metallurgical Products Limited (UK) for a hardened dental alloy containing 76% gold and 24% platinum. It has a melting range of 880-925°C (1616-1697°F).

Denture Clasp. Trade name of J.M. Ney Company (USA) for a gold-colored dental alloy containing 56% copper and 5% palladium. It has a fusing temperature of 900°C (1650°F), and is used for clasps and orthodontic wire.

Denturlac. Trade name for benzyl cellulose formerly used for denture bases.

deoxidant. See deoxidizer.

Deoxidising Tubes. Trade name of Foseco Minsep NV (Netherlands) for a variety of deoxidizers supplied in exact and controlled amounts required for a given melt. Used to deoxidize copper and nickel alloys and/or increase their fluidity.

deoxidized copper. A pure copper from which cuprous oxide (Cu_2O) has been removed by introducing a deoxidizer, such as aluminum, beryllium, boron, carbon, lithium, magnesium, silicon, sodium or phosphorus, to the molten bath. *Deoxidized copper* is much less porous than oxygen-bearing copper (or tough-pitch copper). See also deoxidized high-residual phosphorus copper; deoxidized low-residual-phosphorus copper.

deoxidized high-residual-phosphorus copper. A commercially pure copper (99.90%) containing about 0.02-0.4% phosphorus. Available in the form of flat products, rods, tube, plate and sheet, it has excellent hot and cold workability, good forgeability, and good to excellent corrosion resistance. Used for tubing for oil burners, condensers, evaporators, heat exchangers, distillers, air, gasoline, hydraulic lines, plumbing pipe, etc. Abbreviation: DHP Cu. Also known as *DHP copper; high-phosphorus copper; high-residual-phosphorus copper.*

deoxidized leaded copper. A free-machining grade of deoxidized copper containing up to about 0.8-1.2% lead. It has high electrical and thermal conductivity, good corrosion resistance and machinability, good cold workability, and is used for screw-machine parts, contact pins and inserts, connectors, and motor and switch parts.

deoxidized low-residual-phosphorus copper. A commercially pure copper (99.90-99.99%) containing 0.004-0.12% phosphorus. Commercially available in the form of flat products, rods, tubes, pipes and structural shapes, it has high electrical conductivity, excellent cold and hot workability, and good corrosion resistance. Used for busbars, electrical apparatus, electrical conductors, tubular bus, waveguides, welded and brazed components, etc. Abbreviation: DLP Cu. Also known as *DLP copper; low-phosphorus copper; low-residual-phosphorus copper.*

deoxidized steel. A steel treated with a deoxidizing agent, such as silicon, aluminum, manganese, vanadium or titanium, in order to reduce the oxygen content. Fully deoxidized steel is also referred to as *killed steel.*

deoxidizer. (1) In general, a material that has a strong affinity for oxygen, i.e., that will reduce the oxygen content of another material. Also known as *deoxidant.*

(2) For metals and compounds, a substance that, when added to a compound or molten metal or alloy, removes free or combined oxygen. Common deoxidizers include silicon, aluminum and manganese for steel, and phosphorus for copper and copper-tin alloys. Also known as *deoxidant.*

deoxidizing alloy. An alloy that when added to a molten metal removes free or combined oxygen. Ferroalloys are common deoxidizing alloys for steel.

Deoxo. Trade name for general-purpose fluxes for medium-temperature silver brazing of ferrous metals, stainless steel, nickel and nickel alloys, copper, brass, bronze, and precious metals.

Deoxoloy. Trade name of Belmont Metals Inc. (USA) for a deoxidizer for copper alloys.

deoxy-D-erythropentose. See D-deoxyribose.

deoxyribonucleic acid. A nucleic acid occurring as a major component of the genes located on the chromosomes in the nucleus of all living cells. It genetically codes amino acids and their peptide chain pattern, and determines the type of life form into which a cell will develop. It is thus the 'genetic blueprint' essential for protein synthesis, cellular reproduction and heredity transfer. The DNA molecule is comprised of two polynucleotide chains arranged in a right-handed double helix containing D-deoxyribose, phosphoric acid and the nitrogenous bases adenine and guanine, which contain the purine-ring system, and cytosine and thymine, which contain the pyrimidine-ring system. The two complementary chains are antiparallel and held together at intervals by linear hydrogen bonds between adenine and thymine and between guanine and cytosine. Abbreviation: DNA. Also known as *deoxyribose nucleic acid.*

D-deoxyribose. A pentose sugar that is an essential component of DNA and has a melting point of 89-90°C (192-194°F) and a specific optical rotation of -59° (at 22°C/72°F). Used in chemistry, biochemistry, biotechnology and carbohydrate research. Formula: $C_5H_{10}O_4$. Also known as *deoxy-D-ribose; deoxy-D-erythropentose.*

Depal. Trade name of Duval et Poulain (France) for an age-hardenable casting alloy containing 2% copper, 2% nickel, 2% manganese, and the balance aluminum. Used for cylinder heads and pistons.

Deparex. Trade name for plastic sheeting consisting of a rigid core of expandable polystyrene foam coated on both sides. Used in exhibition contracting for partition walls.

depitched wool. Wool that has tar and other deleterious substances removed by solvent extraction or other processes.

depleted uranium. A grade of metallic uranium containing less than 0.720 wt% of the fissile isotope uranium-235 (^{235}U) normally found in natural uranium. It is primarily used for its high density (18.9 g/cm^3, or 0.683 lb/in^3). Abbreviation: DU.

deposit. The material applied to a substrate or base material by any of several processes, such as chemical or physical vapor deposition, electrodeposition, electroless deposition, chemical deposition, thermal spraying, welding, hardfacing or screening.

Dera. Trade name of Detlef Rave KG (Germany) for glass wool.

Derakane. (1) Trade name of Dow Chemical Company (USA) for a series of vinyl ester resins of the bisphenol-A epoxy and epoxy-novolac types. They have good resistance to acids, alkalies, bleaches and organic solvents, and are available in several grades providing varying degrees of chemical resistance, strength, elongation and high-temperature properties. A brominated grade for fire-retardant applications is also supplied. Used for corrosion-resistant composite products and structures, and as filament-winding resins.

(2) Trade name of Dow Chemical Company (USA) for low-cost polyester plastics with excellent electrical properties that can be formulated for room- and high-temperature applications. Used for automotive components, furniture and appliance housings.

Derakane Momentum. Trade name of Dow Chemical Company (USA) for epoxy vinyl ester resins with high reactivity, high strength, good workability and corrosion resistance and, depending on the particular grade, varying degrees of chemical resistance, strength, elongation and high-temperature properties. A brominated grade for fire-retardant applications is also supplied. Used in the manufacture of fiber-reinforced plastics.

Deranox. Trademark for a series of alumina ceramics.

Derapex. Trademark of Crown Diamond Paints Limited (Canada) for nonslip paint.

derbylite. A black, or brown mineral composed of iron titanium antimony oxide hydroxide, $Fe_4Ti_3SbO_{13}(OH)$. Crystal system, monoclinic. Density, 4.62 g/cm^3. Occurrence: Italy.

derived protein. A protein that is produced from other proteins by partial hydrolysis, denaturation, or heat. Examples include proteoses, peptones, metaproteins and coagulated proteins.

Dermagraft. Trade name for an artificial skin grafting material based on a synthetic matrix made of a polyglycolic acid mesh suitable for seeding with human dermal fibroblasts.

Dermagrain. Trade name of MarkeTech International Inc. (USA) for high-purity, microcrystalline alumina (corundum) for microdermabrasion.

Dermaloy. Trade name of Atotech USA Inc. for a white bronze electrodeposit and plating process.

Dermatoid. German trademark for a durable, water-washable synthetic leather used primarily for bookbinding.

D.E.R. Resins. Trademark of Dow Chemical Company (USA) for an extensive series of epoxies including solid and liquid resins as well as solutions. Examples include *D.E.R. 332* bisphenol A diglycidyl ethers, *D.E.R. 732* polyethylene glycol diglycidyl ethers and *D.E.R. 736* epichlorohydrin-polyglycol epoxy resins. The solids are useful as engineering resins and the liquid resins are used in electron microscopy as embedding media.

derriksite. A green mineral composed of copper uranyl selenite hydroxide, $Cu_4(UO_2)(SeO_3)_2(OH)_6$. Crystal system, orthorhombic. Density, 4.62 g/cm^3; refractive index, 1.85. Occurrence: Zaire.

Derustal. Trade name of Heatbath Corporation (USA) for alkaline derusting compounds.

deruster. An alkaline compound, usually sodium- or sulfonate-based, used in the removal of scale and rust from workpieces by electrolytic or immersion methods. Also known as *descaler; derusting compound.*

derusting compound. See deruster.

Derustit. Trademark of Deutsche Derustit GmbH (Germany) for pickling baths and pastes for steel products.

dervillite. A white mineral composed of silver arsenic sulfide, Ag_2AsS_2. Crystal system, monoclinic. Density, 5.61 g/cm^3. Occurrence: France.

Desagnat 'Mural'. Trade name of Desagnat (France) for a glass that is usually supplied on cloth support for use in wall-covering applications.

Desamin. Trademark for melamine-formaldehyde plastics.

desautelsite. A bright orange mineral of the sjogrenite group composed of magnesium manganese carbonate hydroxide tetrahydrate, $Mg_6Mn_2(CO_3)(OH)_{16}\cdot4H_2O$. Crystal system, rhombohedral (hexagonal). Density, 2.13 g/cm^3; refractive index, 1.569. Occurrence: USA (Pennsylvania, California).

descaler. See deruster.

descloizite. A brownish-red mineral composed of zinc copper lead vanadium oxide hydroxide, $(Zn,Cu)PbVO_4(OH)$. Crystal system, orthorhombic. Density, 6.25 g/cm^3; refractive index, 2.265. Occurrence: Argentina. A greenish black variety with about 8.8% copper oxide and a density of 5.90 g/cm^3 is found in Angola. The latter is known as "cuprian descloizite."

Descoglaze. Trademark of Macnaughton-Brooks Limited (Canada) for epoxy wall coatings.

desiccant. An absorbent or adsorbent substance, such as activated alumina, bauxite, calcium chloride, zinc chloride, silica gel or glycerol, used to remove water or water vapor. Used for drying applications. Also known as *drying agent; siccative.*

Desilube. Trade name of Desilube Technology Inc. (USA) for graphite lubricants that have a service temperature range up to 870°C (1598°F), and can be used in inert and reducing atmospheres, and in vacuum.

desilverized lead. Lead (99.85+% pure) from which silver has been removed either by the Parkes process that utilizes a small percentage of zinc to form a zinc-lead compound that can be removed, or the Pattinson process that employs slow cooling of the melt to solidify and remove lead crystals poor in silver.

Desimpel. Trade name of Hanson Desimpel (Belgium) for building bricks.

desized textiles. Textile yarns and fabrics that have sizes and sizing compounds, such as starch, gums, oils, waxes, etc., removed.

Deslustra. Trade name for viscose rayon used for clothing, and blended with other fibers for suitings, furnishings, carpets, linings, and medical fabrics.

Desmalkyd. Trademark of Bayer AG (Germany) for alkyd resins and polyurethanes.

Desmocoll. Trademark of Bayer AG (Germany) for hot-melt and solvent-borne polyol (polyester) adhesives.

Desmodur. (1) Trademark of Bayer Corporation (USA) for flexible and rigid polyurethane foams, integral skin foams and reaction-injection-molding elastomers. Used for automotive compo-

nents, furniture, appliances, mattresses, sports equipment, and electrical and electronic equipment.

(2) Trademark of Bayer AG (Germany) for two-component polyurethane coatings used as automotive finishes, and for coating wood, furniture and plastics.

Desmodut. Trademark of Bayer AG (Germany) for a series of isocyanate resins.

Desmoflex. Trademark of Bayer AG (Germany) for hard, microcellular polyurethane elastomers used for machine parts, automotive components, electrical equipment, sports equipment, shoes, etc.

Desmomelt. Trademark of Bayer AG (Germany) for hot-melt and solvent-borne adhesives.

Desmopan. Trademark of Bayer AG (Germany) for polyol-based thermoplastic polyurethane elastomers with excellent chemical stability, hydrolysis resistance, low-temperature flexibility, and mechanical properties. Used for automotive components, machine parts, sports equipment, hoses, and films.

Desmophen. (1) Trademark of Bayer AG (Germany) for flexible and rigid polyurethane foams used for automotive components, furniture, appliances, mattresses, sports equipment, and insulation.

(2) Trademark of Bayer AG (Germany) for two-component polyurethane coatings used as automotive finishes, and for coating wood, furniture and plastics.

Desoto. Trade name of PRC-Desoto International Inc. (USA) for epoxy-based coating primers and base coats.

despujolsite. A lemon-yellow mineral of the fleischerite group composed of calcium manganese sulfate hydroxide trihydrate, $Ca_3Mn(SO_4)_2(OH)_6\cdot3H_2O$. Crystal system, hexagonal. Density, 2.46 g/cm^3; refractive index, 1.656. Occurrence: Morocco.

Desulco. Trademark of Superior Graphite Company (USA) for a series of natural and synthetic high-purity carbon and graphite products used in the ferrous and nonferrous metal industries as additives to ductile and gray iron, steel and copper-base alloys. They are also used in the manufacture of advanced composites, electrical conductors, brakes, clutches, and gaskets.

desulfurizer. A material, such as calcium or magnesium, that is sometimes added to a molten metal or alloy, such as steel or cast iron, to reduce the sulfur content.

desulfurizing alloy. A metal alloy, such as sodium-copper or sodium-tin, used to remove sulfur from copper-tin and copper-zinc alloys.

detackifier. A chemical agent added to ground rubber to prevent the particles from sticking together.

Detaclad. Trade name of DuPont Company (USA) for a series of explosion-bonded clad metals available in a variety of metal combinations.

Detag-Auresin. Trade name of Flachglas AG (Germany) for a glass coated with gold by means of an evaporation technique. Used for incorporation into *Cudo-Auresin* or *Sigla* laminated glasses.

Detag-Gold. Trade name of Flachglas AG (Germany) for a gold-coated glass for inclusion in *Cudo-Gold* or *Sigla* laminated glasses.

Detalux. Trade name of Flachglas AG (Germany) for polymethyl methacrylate film and sheeting.

Detaseal E. Trade name for a silicon elastomer for dental impressions.

Dethermal. Czech trade name for a heat-absorbing glass.

De Trey Gutta Percha. Trade name of Dentsply International (USA) for gutta-percha used in dentistry for temporary fillings.

De Trey Orthoresin. Trade name of Dentsply International (USA) for a self-cure acrylic for dental applications.

De Trey RR. Trade name of Dentsply International (USA) for a self-cure acrylic for dental applications.

De Trey's Zinc. Trade name of Dentsply International (USA) for a zinc phosphate dental cement.

De Trey Triad System. Trade name of Dentsply International (USA) for a light-cure denture resin.

Deurance metal. A British babbitt metal containing 44.5% antimony, 33.3% tin and 22.2% copper. Used for locomotive bearings.

deuterated compound. A compound in which one or more of the hydrogens is a deuterium isotope, e.g., deuterated ethylene-d_2 (CH_2CD_2) and deuterobenzene (C_6D_6).

deuterium. An isotope of hydrogen with a mass number of 2 that unlike ordinary hydrogen has one proton and one neutron in its nucleus. It is obtained from high-purity *heavy water* by electrolysis or from liquid hydrogen by fractional distillation, and occurs in natural hydrogen in the proportion of 1:6500. It has a melting point of -254.4°C (-426°F), a boiling point of -249.5°C (-417°F), an autoignition temperature of 585°C (1085°F), and forms compounds with halides and oxygen. Used in the bombardment of atomic nuclei, in thermonuclear reactions, in chemistry, biochemistry, biotechnology and materials research, and in nuclear magnetic resonance spectroscopy. Symbol: D; d; H²; ²H. Also known as *double-weight hydrogen; heavy hydrogen; hydrogen 2.*

deuterium oxide. Water composed of one atom of oxygen and two atoms of deuterium (heavy hydrogen). It can be produced by the electrolysis of ordinary water, and is available in purities ranging from 90 to 100 at%. The 100% grade has a density of 1.107 g/cm³, a melting point of 3.8°C (39°F), a boiling point of 101.4°C (215°F) and a refractive index of 1.328. Used in nuclear power plants to moderate and control nuclear reactions, and in chemistry, biochemistry and biotechnology. Formula: D_2O. Abbreviation: DOD. Also known as *heavy water.*

deuteroporphyrin IX dihydrochloride. A synthetic porphyrin derivative available in the form of purple crystals (96+% pure) for use as a pigment and dye. It is also obtained from bovine blood. Formula: $C_{30}H_{30}N_4O_4 \cdot 2HCl$.

deuteroporphyrin IX dimethyl ester. A synthetic porphyrin derivative available in the form of purple, polymorphic crystals (90+% pure) with a melting point of 217-220°C (422-428°F), and a maximum absorption wavelength of 399 nm. It is also obtained from bovine blood. Used as a pigment and dye. Formula: $C_{32}H_{34}N_4O_4$.

Deutro. Trade name of Deutro GmbH (Germany) for a series of austenitic stainless steels.

Deutsche Nickel. Trade name of Vereinigte Deutsche Nickel-Werke AG (Germany) for semifinished nickel products (bars, sheets, wires, tubes, etc.), nickel and nickel-copper (Monel) tubing, nickel-type thermometals, nickel sand and chill castings, and beryllium-copper, copper-nickel and nickel-silver strip and wire for electronics, and drawing and stamping applications.

Deva. Trademark of Degussa-Ney Dental (USA) for a wide range of gold-palladium dental alloys, including *Deva M* bonding alloy for bridgework, and *Deva Plus* tarnish and corrosion-resistant ceramic alloy for crowns, bridges and restorations.

Devarda's alloy. An alloy of 50% copper, 45% aluminum, and 5% zinc supplied in the form of a fine, gray, flammable pow-

der with a melting point of 550°C (1022°F). Used for analysis applications. Also known as *Devarda's metal.*

Devcon Metal Welder. Trade name of Norris & Company (USA) for a methacrylate structural adhesive with excellent impact and peel strength, and good resistance to thermal shocks, elevated temperatures, humidity and weathering. Used for bonding aluminum, brass, copper, steel, stainless steel and other metals to dissimilar substrates including various plastics.

devilline. A dark blue-green mineral composed of calcium copper sulfate hydroxide trihydrate, $CaCu_4(SO_4)_2(OH)_6 \cdot 3H_2O$. Crystal system, monoclinic. Density, 3.13 g/cm³; refractive index, 1.651. Occurrence: Slovakia. Also known as *devillite.*

devillite. See devilline.

Devils Iron. Trade name of Allegheny Ludlum Steel (USA) for a soft magnetic, grain-oriented iron-silicon alloy used for core transformers and telephone components.

devitrified ceramics. See glass-ceramics.

devitrified glasses. See glass-ceramics.

devitrite. A crystalline product, $Na_2O \cdot 3CaO \cdot 6SiO_2$, obtained by crystallizing (devitrifying) a commercial glass. It decomposes to *wollastonite* and a liquid at 1045°C (1913°F).

Dev-Kote. Trade name of Deveco Corporation (USA) for a zinc phosphate coating.

Devlac. Trade name of Deveco Corporation (USA) for air-drying and thermosetting water-soluble lacquers.

devoré fabrics. See etched-out fabrics.

Devstar. Trade name of Deveco Corporation (USA) for electroless nickel, and chromium and brass electrodeposits and plating processes.

Deward. Trade name of AL Tech Specialty Steel Corporation (USA) for an oil-hardening cold-work tool steel (AISI type O2) containing 0.9% carbon, 1.55% manganese, 0.3% molybdenum, and the balance iron. It has good nondeforming properties, good machinability, and good to moderate wear resistance. Used for punches, dies, gages, taps, rolls, cutters, bushings, tools, reamers, and saws.

dewindtite. A yellow, radioactive mineral of the phosphuranylite group composed of lead uranyl phosphate hydroxide trihydrate, $Pb(UO_2)_3(PO_4)_2(OH)_2 \cdot 3H_2O$. Crystal system, orthorhombic. Density, 4.35 g/cm³; refractive index, 1.73-1.74. Occurrence: Central Africa, Australia.

Dewlite. Trade name of C-E Glass (USA) for a patterned glass featuring small, slightly elevated lenses resembling dewdrops.

Dewoglas. Trademark of Degussa AG (Germany) for plexiglass-type acrylics based on polymethyl methacrylate.

Dewopor. Trademark of Degussa AG (Germany) for flexible polyvinyl chloride film materials.

Dexel. Trademark of Courtaulds Chemicals (USA) for clear, glossy thermoplastic cellulose acetate resins with a density of 1.3 g/cm³ (0.05 lb/in.³), a service temperature range of -20 to +95°C (-4 to +203°F), high moisture absorption (1.9-7.0%), moderate ultraviolet stability, and fair to poor chemical resistance. Used for spectacle frames, tool and brush handles, and packaging and graphics applications.

Dexflex. Trademark of Solvay Engineered Polymers (USA) for an extensive series of thermoplastic olefins (e.g., polypropylenes) with good rigidity, dimensional stability, flowability, and impact resistance. They are available in standard, extrusion and blow-molding grades, and used for automotive bumper parts.

Dexil. Trademark of Foseco Minsep NV (Netherlands) for dextrin-based organic breakdown additives for core sand mixes, especially for sand bonded by the CO_2-process. It produces excellent "shake-out" properties, and has very low tendency to contaminate the re-used sand.

Dexlon. Trade name of Dexter Corporation (USA) for a series of high-impact unfilled or glass-filled nylon 6 and nylon 6,6 plastics used for injection-molded products.

Dexocor. Trade name for a dextrin binder used in foundry core oils.

Dexon. Trademark of Davis & Geck, Inc. USA) for a biodegradable, bioabsorbable *polyglycolide* used for surgical sutures.

Dexonite. Trade name of Dexine Limited (UK) for hard rubber.

Dexpro. Trade name of Dexter Corporation (USA) for a series of unfilled, glass fiber-reinforced or calcium carbonate- or mineral-filled polypropylene plastics used for injection-molded products. The glass-reinforced grades are also used for high-impact applications.

dextrin. A class of colloidal products that are carbohydrate mixtures of varying composition made by the treatment of corn or potato starch with enzymes or dilute acids, or by heating dry starch. They are polymers of D-glucose available in the form of yellow or white granules or powders, and used in the manufacture of adhesives, printing inks and felt, as binders in glazes, as carriers or binders for ceramic inks and decorating colors, as sizes for paper and textiles and glass-silvering compositions, in metal casting, and in biology, biochemistry, biotechnology and medicine. Formula: $(C_6H_{10}O_5)_n$. Also known as *starch gum.*

dextrose. A dextrorotatory monosaccharide that is found in most plants and in animal and human blood. It is synthetically obtained by the hydrolysis of cornstarch with acids or enzymes, and available as a white, crystalline, water-soluble, sweet-tasting powder with a melting point of 153-156°C (307-313°F) and a specific optical rotation of +52.5-53.0° (25°C/77°F). Used as a blood-volume extender, in medicine as a diuretic, in nutrition, in food products, in biochemistry, biotechnology, bioengineering and molecular biology. Formula: $C_6H_{12}O_6$. Also known as *D-glucose; D-glucopyranose; corn sugar; grape sugar.* See also glucose.

Dex-Tung. Trade name of North American Steel Corporation (USA) for an oil-hardening tool steel containing 0.32% carbon, 0.93% chromium, 0.48% molybdenum, 0.47% tungsten, and the balance iron. Used for blacksmith tools, punches, collets, clutch parts, shafting, and gears.

D-Film. Trademark of Cryovac Division W.R Grace & Company (USA) for plastic shrink film used for packaging food and non-food products.

D-Glass. Trademark of Nuova Italtess (Italy) for D-glass fibers.

D-glass. A glass that contains a high percentage of boron, and is often used in fiber form in composites that require low or controlled dielectric constants.

DHP copper. See deoxidized high-residual-phosphorus copper.

Diabestos. Trade name of General Refractories Company (USA) for diatomite-based filter aids.

Diabolique Satan. Trade name of Creusot-Loire (France) for a series of fast-finishing tool steels containing 1.15% carbon, 1.9% tungsten, 0.45% chromium, and the balance iron. Used for cutters and tools.

diaboleite. A light to sky blue mineral composed of lead copper chloride hydroxide, $Pb_2CuCl_2(OH)_4$. It can also be made synthetically. Crystal system, tetragonal. Occurrence: UK, USA (Arizona).

Diabon. Trademark of Great Lakes Carbon Corporation (USA) for an acid-, heat- and corrosion-resistant material made from porous graphite.

DiaBond. Trademark of Powdermet, Inc. (USA) for diamond grit with enhanced retention for use in the manufacture of metal-bonded abrasives.

diacetate. (1) An ester or salt that contains two acetate groups in its structure.

(2) A fiber manufactured from acetone-soluble cellulose acetate. Also known as *diacetate fiber.* See also acetone-soluble cellulose acetate.

diacetate fiber. See diacetate (2).

diacetone acrylamide. A hygroscopic, white, crystalline solid (99+% pure) that is soluble in water and many organic solvents, polymerizes readily, and has a melting point of 54-57°C (129-135°F), a boiling point of 120°C (248°F)/8.0 mm and a flash point above 230°F (110°C). Used as a vinyl monomer in production of hydrophilic polymers, in latex and water-based coating compositions, as a crosslinker in polyester resins, as an adhesion promoter for glass, concrete and cellulosics, and in color photography. Formula: $C_9H_{15}NO_2$. Abbreviation: DAA.

diacetoxytin. See stannous acetate.

Dia-Clust. Trade name of Bales Mold Service Inc. (USA) for a composite electrodeposit composed of a chromium matrix with a dispersion of nanometer-sized, spherical diamond particles. It has a very high toughness, a hardness exceeding 80 Rockwell, high corrosion resistance, and relatively low coefficient of friction.

Diacro. Trade name of Weatherly Casting & Machine Company (USA) for an abrasion- and corrosion-resistant cast alloy containing 1.4-1.8% carbon, 0-2% silicon, 1.75-2.5% nickel, 27-29% chromium, 0-1.5% manganese, 1-2.25% molybdenum, and the balance iron.

diadic nylon. A nylon made by the condensation of a diamine and a dicarboxylic acid. Also known as *diadic polyamide.*

diadic polyamide. See diadic nylon.

diadochite. A yellow, yellowish brown, or brown mineral composed of iron phosphate sulfate hydroxide pentahydrate, $Fe_2(PO_4)(SO_4)(OH)\cdot5H_2O$. Crystal system, triclinic. Density, 2.00 g/cm³; refractive index, 1.64. Occurrence: USA (Nevada), Czech Republic.

Diafil. Trade name for rayon fibers and yarns used for textile fabrics.

Diafoil. Trade name for polethylene terephthalate resins supplied in film form.

Diagonal. Trade name of Vereinigte Glaswerke (Germany) for a rhomboid glass block having a diagonal stripe creating a zig-zag effect when glazed.

Diakon. Trademark of ICI Chemicals & Polymers Limited (UK) for an amorphous thermoplastic acrylate resin available in colorless and colored transparent sheets, tubes, rods, films, lacquers and coatings. It has a density of 1.2 g/cm³ (0.04 lb/in.³), good machinability, moldability and extruding properties, good resistance to acids and alkalies, excellent optical clarity and good ultraviolet stability. Used for displays, signs, sinks and glazing. *Diakon TD* is a high-impact grade thermoplastic acrylate resin.

Dialead. Trademark of Mitsubishi Kasei Corporation (Japan) for chopped, pitch-derived carbon fibers used in road barriers, curtain walls, etc.

dialkyllithium. An organometallic compound prepared by adding a cuprous halide (CuX) to alkyllithium (RLi). The general formula is R_2CuLi. Examples of such compounds include lithium diethylcopper [$(C_2H_5)_2CuLi$] and lithium dimethylcopper [$(CH_3)_2CuLi$].

Diall. Trademark of Allied Chemical Company (USA) for a series of diallyl phthalate (DAP) thermosetting molding compounds supplied in various grades including unmodified, and glass or synthetic fiber-reinforced and mineral-filled. They have high strength, good physical properties, good dimensional stability, excellent electrical resistance, good flame resistance, exceptional colorfastness in sunlight and heat, good resistance to almost all solvents and many chemicals, and good resistance to fungi.

Diallist. British trade name for a magnet material of nickel, aluminum, cobalt and iron. It has a high coercive force, and is used for permanent magnets.

Dialloy. Trade name of Gorham Tool Industries Inc. (USA) for a chromium-nickel-molybdenum tool steel with high impact and fatigue resistance, used for battering tools, and punches.

diallyl isophthalates. A family of thermosetting resins that are meta-resins of diallyl phthalate [$C_6H_4(CO_2CH_2CH=CH_2)_2$]. They have excellent electrical-insulating properties, high compressive strength, high tensile moduli, low impact resistance, excellent heat resistance, and a continuous use temperature of approximately 190°C (375°F). Used for molding compounds for electrical and electronic applications, e.g., connectors, potting cups, switches, bobbins, etc. Abbreviation: DAIP.

diallyl phthalates. A family of thermosetting resins that are the products of the reaction of allyl alcohol and phthalic anhydride. They are available in two types, namely, the ortho-resin of diallyl phthalate (diallyl-*o*-phthalate, C_6H_4-*o*-$CO_2CH_2CH=CH_2)_2$) with superior electrical properties, and the heat-resistant meta-resin (diallyl-*m*-phthalate, C_6H_4-*m*-$CO_2CH_2CH=CH_2)_2$). *Diallyl phthalates* have high dielectric strength, excellent insulating properties, high compressive strength, high tensile moduli, low impact resistance, and a continuous-use temperature of approximately 180°C (355°F). Used as mounting materials in metallography, as molding compounds for electrical and electronic applications, e.g., connectors, potting cups, switches, bobbins, etc. Abbreviation: DAP.

Dialox. Trade name of Plansee Metallwerk-Gesellschaft (Austria) for a ceramic aluminum oxide (Al_2O_3) used for wear parts and cutting tools.

Dialoy. Trade name of J.M. Ney Company (USA) for a bismuth-lead dental alloy containing some tin and cadmium. It expands on solidification, and is used for dental tooth impressions, dental bridgework, and models.

diamagnetic materials. A class of materials, such as bismuth, copper, zinc, gold, silver, nitrogen, barium sulfate and water, that have relative permeabilities less than that of a vacuum, and therefore experience repulsion of magnetic flux. They will position themselves at right angles to the magnetic flux lines. Also known as *diamagnetics; diamagnetic substances.*

Diamalloy. Trade name of Sulzer Metco (USA) for nickel-, copper-, cobalt-, iron-, chromium carbide-, and tungsten carbide-base powders used for thermal-spray coatings applied by the high-velocity oxyfuel (HVOF) process, and providing excellent resistance to abrasive and sliding wear, and corrosion by aggressive media (e.g., acids and alkalies) at temperatures up to 875°C (1605°F).

Diamant. (1) Trade name of Vereinigte Glaswerke (Germany) for decorative glass blocks.

(2) Trade name of Styria-Stahl Steirische Gussstahlwerke AG (Austria) for a series of tool steels.

(3) Trade name of Eisenwerk Würth GmbH & Co. (Germany) for abrasive blast-cleaning media based on chilled white cast iron.

(4) Trade name of Saint-Gobain (France) for an extra-clear glass with diamond pattern.

Diamantato. Trade name of Fabbrica Pisana SpA (Italy) for a patterned glass.

Diamantbronze. Trade name of Ostermann GmbH & Co. (Germany) for a corrosion-resistant bronze containing 58% copper, 38% zinc, 1% nickel, 1% aluminum, 1% iron, and 1% manganese. Used for valve gage fittings.

Diamante. Trade name of Vitrobrás (Brazil) for a patterned glass.

Diamax. Trade name of Eutectic Corporation (USA) for a series of alloy powders for the metal spraying of hard final coatings.

diamine polyimide. A thermoplastic polyimide that is the condensation product of a reaction of anhydrides or anhydride derivatives with diamines.

diaminobenzidine. A light-sensitive solid compound (97+% pure) with a melting point of 175-180°C (347-356°F), used in analytical chemistry and biochemistry as a nucleic-acid stain and peroxidase substrate, as a reagent for the spectrophotometric determination of selenium, and copolymerized with diphenyl isophthalate to produce polybenzimidazoles. Formula: $(H_2N)_2$-$C_6H_3C_6H_3(NH_2)_2$.

Diamite. (1) Trade name of Weatherly Casting & Machine Company (USA) for an abrasion-resistant white cast iron containing 3-3.5% carbon, 1.1-1.5% silicon, 0.6-0.9% manganese, 1-3% chromium, 3-5% nickel, and the balance iron. It has good strength and hardness, and is used for chute liners, pulverizer hammers, sand pump parts, and hardfacing rods.

(2) Trade name of Rennie Tool Company Limited (UK) for a sintered tungsten carbide used for cutting tools.

Diamite Mo. Trade name of Weatherly Casting & Machine Company (USA) for an abrasion-resistant white cast iron containing 3-3.5% carbon, 15-18% chromium, up to 3% molybdenum, and the balance iron.

Diamolith. Trade name of Innovative Coatings Company (France) for a flexible, wear-resistant carbon coating composed of a mixture of diamond- and graphite-like structures and applied by physicochemical plasma processes. It has a thickness of 3-5 μm (0.1-0.2 mil), a hardness of 4000 Vickers, a very low coefficient of friction, good thermal stability, good resistance to acids, alkalies, solvents and salts, and good biocompatibility.

DiaMon. Trade name of Diamond Tool Coating (USA) for a chemical-vapor-deposited (CVD) polycrystalline diamond coating for carbide cutting tools with excellent wear resistance and a hardness of about 10000 Vickers.

Diamond. (1) Trade name used in various countries for patterned glass products.

(2) Trade name of Pilkington Brothers Limited (UK) for a wired plate and cast glass with diamond mesh.

(3) Trade name of Jessop Saville Limited (UK) for a series of tool steels containing 0.65-1.3% carbon, 0.25% vanadium, and the balance iron. They have good machinability and high toughness. Used for general-purpose tools.

(4) Trade name of J.F. Jelenko & Company (USA) for a silver- and palladium-free high-gold/high-platinum porcelain-to-metal dental alloy that has a rich gold color and provides high strength and durability, and outstanding thermal stability and sag resistance.

diamond. A crystalline allotrope of carbon that is the hardest known mineral. It occurs naturally in various colors ranging from colorless, slightly yellowish and yellow to red, green, blue or black. It has a three-dimensional structure consisting of covalent bonds in which each carbon atom is at the center of a tetrahedron with its vertices at its nearest four neighbors. *Diamond* can also be made synthetically, e.g., by heating a carbonaceous material (e.g., carbon or graphite) with a metallic catalyst (e.g., chromium, cobalt or nickel) in an electric furnace at about 1200-2400°C (2190-4350°F) under high pressure. The synthetic form is usually colorless. Crystal system, cubic. Crystal structure, diamond-cubic. Density, 3.50-3.53 g/cm³; melting point, 3700°C (6690°F); boiling point, 4200°C (7590°F); hardness, 10 Mohs; refractive index, 2.419. It is an excellent electrical insulator, transparent to infrared radiation, and has high stability, a low coefficient of friction, and very high thermal conductivity. Occurrence: Brazil, Borneo, India, South Africa, USA (Arkansas), Venezuela. Used as a gemstone when clear and without flaws, otherwise for polishing powders, abrasive wheels, glass cutters, drill bits, coatings for cutting tools and surgical knives, windows of space probes and high-capacity transmitters. See also industrial diamond; synthetic diamond.

diamond abrasives. A group of very hard (Mohs 10) abrasives composed of diamond dust or powder in a viscous material, such as a paste, slurry or aerosol. Used as a polishing medium for metallographic specimens.

Diamond Alloy. Trade name of Brown Alloy Works (USA) for an abrasion- and wear-resistant alloy of 45% cobalt, 40% molybdenum and 15% chromium. Used for tools and cutters.

diamond ballas. A hard, spherical *industrial diamond* consisting of numerous tiny crystals arranged radially around a central point. Used for drilling applications.

Diamond Black. Trade name of Carolina Coating Technologies Inc. (USA) for a microthin, hard boron-carbide coating with excellent wear resistance, a very smooth surface, a very high hardness (above 9 Mohs), and a deposition temperature of 120°C (248°F).

Diamond Bond. Trade name of Biodent New Zealand/Australia for an universal dental adhesive system with synthetic collagen fixers.

Diamond Bronze. Trade name for a corrosion-resistant, high-strength bronze containing 88% copper, 10% aluminum and 2% silicon. Used for marine parts, hardware, pump parts, and valves.

Diamond Carbide. Trademark of Protective Metal Alloys, Inc. (USA) for a family of composite alloys composed of a nickel-chromium-silicon-boron (*VERSAlloy*) matrix with dispersed tungsten, tungsten-cobalt or vanadium-tungsten carbide particles of varying shape and mesh size. They are supplied in several grades with hardnesses of (i) 55-60 Rockwell C for corrosion-resistant applications at extreme abrasion and relatively low impact conditions, such as drill and mining bits, conveyor screws, digging tools, etc.; (ii) 50+ Rockwell C for corrosion-resistant applications at severe abrasion and moderate impact conditions, such as augers, pipe elbows, food-processing screws, etc.; and (iii) 40+ Rockwell C for corrosion-resistant applications at low to medium abrasion and high impact conditions, such as mill hammers, debarkers, etc.

diamond compound. A micronized diamond powder incorporated into a specially formulated paste or lubricant, and used for lapping, polishing and superfinishing applications.

Diamond-Cool. Trade name of Asahi Glass Company Limited (Japan) for heat-absorbing glass.

Diamond Crown. Trade name of Biodent New Zealand/Australia for a direct and indirect, microcrystalline poly-ceramic dental restorative for crowns.

diamond dust. See diamond powder.

Diamond Grit. Trademark of Carborundum Company (USA) for coated abrasives.

DiamondHard. Trademark for an extremely hard amorphous diamond coating.

diamond-like carbon. A hard, low-friction, amorphous form of carbon with a structure lacking crystalline diamond order. Depending on its manufacture, it can consist of carbon only, or of hydrogenated amorphous carbon (a-C:H). It is often supplied in the form of a coating or film.

diamond-like nanocomposites. Thin-film coatings with amorphous microstructures consisting of two random interpenetrating molecular networks–one based on carbon and the other on silicon–stabilized with hydrogen and oxygen. They can also be doped with a third ceramic or metal network to produce the required mechanical, optical or electrical (superconductive, conductive, or dielectric) properties. These coatings is are usually produced by plasma- or ion-beam vacuum deposition, and contain carbon in a primarily diamond bonding configuration. They have exceptional corrosion and erosion resistance, good wear resistance, good thermal stability, low coefficients of friction, and adhere well to many substrates. Used as conductive or insulating coatings, optical coatings, and on chemical and electrochemical components, automotive, marine and microelectronic components, surgical instruments, etc.

DiamondLink. Trade name of Biodent New Zealand/Australia for dental luting composite.

DiamondLite. Trade name of Biodent New Zealand/Australia for a direct microcrystalline microhybrid crosslinked polyethlene (PEX) dental composite.

Diamondlite. Trade name of Australian Window Glass Proprietary Limited for wired glass with diamond mesh.

Diamondlube C11. Trade name of Richter Precision Inc. (USA) for a hard, corrosion-resistant, low-friction diamond-like carbon coating for plastic molds.

diamond mesh. A steel sheet that has been slit and expanded to form diamond-shaped openings. Used as a base for plaster or tile on building walls and ceilings.

Diamond Ninety. Trade name for a glass-polyphosphonate dental cement. Also spelled *Diamond 90*.

diamond paste. Diamond dust dispersed in a viscous material, such as a paste or slurry, and used as an abrasive for grinding or polishing. See also diamond compound; diamond slurry.

diamond powder. Finely divided powder obtained either by crushing industrial diamonds, or by cutting gem diamonds. It is available in particle sizes ranging from submicron to 85 μm (less than 40 μin. to over 0.003 in.), and used as an abrasive for grinding, polishing and lapping of steel, stone, etc., and in the manufacture of diamond grinding wheels. Abbreviation: DP. Also known as *diamond dust*.

Diamond Raindrops. Trade name of SA des Verreries de Fauquez (Belgium) for patterned glass.

Diamondshield. Trademark of Diamonex (USA) for a water-clear diamond-like carbon coating with superhigh abrasion resistance for application to metallic and plastic substrates.

diamond slurry. A micronized diamond powder dispersed in a liquid, such as water, or a suspension, and used for polishing, lapping, sectioning or texturing applications.

Diamondstone. Trade name of Permacon (Canada) for triangular paving stones supplied in charcoal, dark brown and gray colors, and used in conjunction with square or rectangular pavers to create various patterns.

Diamond Weld. Trade name of Libbey-Owens-Ford Company (USA) for polished wired glass with diamond mesh.

diamond willow. Willow wood with a conspicuous diamond-like grain pattern resulting from an abnormal growth of the stems that can occur in any species of willow. Used in the manufacture of lamps, walking sticks, ornaments, etc.

Diamonite. Trade name of John A. Crowley Inc. (USA) for a superhard alloy of tungsten carbide and ditungsten carbide containing 95.65% tungsten, 3.91% carbon and 0.44% iron. Used for cutting tools.

dianite. See niobite.

diaphanous. A thin transparent or semitransparent film.

diaphorite. A steel-gray mineral composed of silver lead antimony sulfide, $Ag_3Pb_2Sb_3S_8$. Crystal system, orthorhombic. Density, 6.04 g/cm^3. Occurrence: Czech Republic. Also known as *ultrabasite*.

diaphragm brass. A corrosion-resistant brass of 95% copper, 3% tin and 2% zinc used for diaphragms.

Diaplex. Trade name for a mica substitute made from bentonite, and supplied in thin, hard, transparent sheets.

Dia-Plus. Trademark of Walter Messner GmbH (Germany) for diamond pastes and powders used as abrasives for grinding, polishing and lapping of steel, ceramics, stone, and other materials.

diaspore. A colorless, white, gray, yellowish, or greenish mineral with vitreous to pearly luster composed aluminum oxide hydrate, β-AlO(OH), and occurring with *corundum* and *dolomite*, and in *bauxite*. Crystal system, orthorhombic. Density, 3.35-3.45 g/cm^3; hardness, 6.5-7 Mohs. Occurrence: Czech Republic, Russia, Slovakia, Switzerland, USA (Arkansas, Massachusetts, Missouri, Pennsylvania). Used as an abrasive, for refractories, and as a source of aluminum.

diaspore brick. A high-alumina refractory brick made substantially from *diaspore clay*.

diaspore clay. A rock composed essentially of *diaspore* bonded with *fireclay*. Calcined diaspore clay contains about 70-80% alumina (Al_2O_3), and is used in the manufacture of refractories.

diastereoisomer. One of two pairs of stereoisomers that are not mirror images of each other, have similar chemical properties, but differ in physical properties. Also known as *diastereomer*. See also stereoisomer.

Diatherm. Trademark of Clayburn Refractories Limited (Canada) for refractory insulating brick.

Diato. Trademark of Fairey & Company Limited (Canada) for insulating fireclay brick.

diatomaceous earth. A white, yellowish, or light-gray, friable, highly siliceous material derived from the shells of diatoms, i.e., microscopic, unicellular, aquatic algae. It contains various impurities, such as silica sand, argillaceous minerals, iron, alkalies and alkaline earths. The typical composition range is 86-92% silica (SiO_2), 3-9% alumina (Al_2O_3), 1-2% ferric oxide (Fe_2O_3), 0.5-1% calcia (CaO), and the balance other oxides. It is available as light-colored blocks, bricks, powder, aggregate, cement and mild abrasive, has a density of 1.9-2.35 g/cm^3 (0.07-0.08 lb/in.^{3}), a bulk density of 80-240 kg/m^3 (5-15 lb/ft^3); a melting point of 1715°C (3120°F), a hardness of 4-6.5 Mohs (unprocessed), high water absorption capacity (up to four times its weight), high oil absorption capacity, and is a poor conductor of electricity, heat and sound. Used as a filter medium, as an absorbent, as a filler in paints, rubber and plastics, in refractories, ceramics and paper coatings, as a mild abrasive for polishing operations, for acidproof liners, in thermal and acoustical insulation, in asphalt compositions, and as a catalyst. Also known as *diatomite; infusorial earth; kieselguhr; tripolite*.

diatomaceous-earth-base blocks. An insulating material made from uncalcined or calcined silica, asbestos fibers and clay by molding under heat and pressure. It has good high-temperature resistance up to 950°C (1740°F) and is used chiefly for boiler walls. Also known as *diatomite blocks*.

diatomaceous silica. An insulating material composed of diatomaceous earth with reinforcing fibers, and with or without binders.

diatomite. See diatomaceous earth.

diatomite blocks. See diatomaceous-earth-base blocks.

Diaweld. Trade name of Krupp Stahl AG (Germany) for a corrosion-resistant cast iron containing 4.2-4.4% carbon, 1.8% silicon, 5-6% manganese, 30% chromium, and the balance iron. Used for hardfacing electrodes.

diazonium salt. A salt produced when a primary amine reacts with nitrous acid (HNO_2). For example, benzenediazonium chloride [$C_6H_5N(N)Cl$] is a salt produced by the reaction of anisole with nitrous acid.

dibarium silicate. See barium silicate (iii).

dibarium trisilicate. See barium silicate (iv).

dibasic barium phosphate. See barium hydrogen phosphate.

dibasic calcium phosphate. See calcium phosphate (ii).

dibasic lead phthalate. See lead phthalate.

dibasic sodium orthophosphate. See disodium phosphate.

dibasic sodium phosphate. See disodium phosphate.

dibenz[*a,c*]anthracene. An aromatic compound that is available in the form of yellow crystals (99+% pure) with melting point of 205-207°C (401-405°F) and a boiling point of 518°C (964°F). Used in electronics, biochemistry and biotechnology. Formula: $C_{22}H_{14}$.

dibenz[*a,h*]anthracene. A yellow, crystalline aromatic compound with semiconducting properties that is available with a purity of 99+%. It has a melting point of 266-267°C (511-513°F), a boiling point of 524°C (975°F), and is used as an organic semiconductor for electronic and microelectronic applications. Formula: $C_{22}H_{14}$. See also pentacene.

dibenz[*de,kl*]anthracene. See perylene.

2,5-dibiphenylyloxazole. A crystalline solid compound (98% pure) with a melting point of 237-240°C (459-464°F) used as a scintillation counter or wavelength shifter in liquid scintillation spectrometry. Formula: $C_{27}H_{21}NO$. Abbreviation: BBO.

diberyllium boride. See beryllium boride (ii).

diblock copolymer. A block copolymer in which there are two identical mer units clustered in blocks along the molecular chain. See also block copolymer.

diborane. A colorless, flammable, volatile organic compound derived by reacting lithium hydride with boron trifluoride in the presence of an ether catalyst at 25°C (77°F). It is available in technical grades (95% pure), high-purity grades (99+%), electronic grades (10% in hydrogen), and deuterated electronic grades (10% in deuterium or helium). It has a density of 0.180 g/mL (17°C/63°F), a melting point of -165°C (-265°F) and a boiling point of -92.5°C (-134.5°F). Used in the synthesis of organic boron compounds, as a p-type dopant for semiconductor, as a reducing agent, as a polymerization catalyst for ethylene, and as a fuel for rocket engines. Formula: B_2H_6.

Dibronze. Trade name of Ampco Pittsburgh Corporation (USA) for an aluminum bronze containing copper, aluminum and iron.

It has good wear and galling resistance, and is used for bending and forming dies, and deep-drawing dies.

Dica. Trade name of Jessop Steel Company (USA) for a series of hot-work tool steels.

Dical. Trade name of Die Casting Limited (UK) for a die-casting alloy of 88% copper and 12% silicon.

dicalcium aluminum silicate. See calcium aluminum silicate (iii).

dicalcium ferrite. See calcium ferrite (ii).

dicalcium columbate. See calcium niobate (ii).

dicalcium magnesium disilicate. See calcium magnesium silicate (iii).

dicalcium niobate. See calcium niobate (ii).

dicalcium orthophosphate. See calcium phosphate (ii).

dicalcium orthosilicate. See calcium silicate (ii).

dicalcium phosphate. See calcium phosphate (ii).

dicalcium silicate. See calcium silicate (ii).

Dicalite. Trademark of Great Lakes Carbon Corporation (USA) for a series of products made from *diatomaceous earth* or *perlite*. They are usually supplied in the form of lumps or fine powders of varying particle size, and used as filters and filter aids, absorbents, fillers, thermal insulation for walls, and in heat-insulating cement, concrete admixtures, and polishes.

dicarbadodecaborane. See carboranes.

dicarboxylic acid. A carboxylic acid containing two substituent carboxyl (–COOH) groups, e.g., adipic, fumaric, glutaric, isophthalic, maleic and oxalic acid.

Dicel. Trade name of Novaceta SpA (Italy) for bright to dull cellulose acetate fibers and filament yarns having good dyeability, wrinkle resistance, moth and mildew resistance, good draping qualities, and very good blendability with other fibers, e.g., viscose rayon. They may have flameproofing agents (e.g., *Lo-Flame Elastomerics*) and stretch fibers (e.g., *Lycra*) added to further improve the drape. Used especially for clothing, linings, furnishings, and knit lingerie.

Dicelesta. Trade name of Novaceta S.p.a. (Italy) for cellulose acetate fibers and yarns used for the manufacture of wearing apparel.

dicetyl ether. A crystalline compound with a density of 0.811 g/cm^3, a melting point of 54°C (129°F) and a boiling point of 300°C (572°F) (decomposes). Used as a chemical intermediate, in electrical insulators, antistatic substances, water repellents and as lubricants in plastics. Formula: $(C_{16}H_{33})_2O$. Also known as *dihexadecyl ether*.

dichlorodimethyltin. See dimethyltin dichloride.

Dichrofil. Trade name of Sovirel (France) for a Crookes-type tinted spectacle glass. See also Crookes glass.

dichroic glass. A glass that will display certain colors when viewed from one angle and different colors when viewed from a different angle, or that will transmit some colors and reflect others.

dichromate coating. A chromate conversion coating for magnesium alloys obtained by immersing in a hot solution of sodium dichromate ($Na_2Cr_2O_7$). It is often used as a decorative finish or a base for subsequent coatings on die-cast cameras, chain saws, and satellites and missiles.

dichromium boride. See chromium boride (iii).

dichromium phosphide. See chromium phosphide (iii).

Dichroyal. Trade name of Sovirel (France) for a Crookes-type tinted spectacle glass.

dickinsonite. A green or yellowish green mineral composed of sodium manganese iron calcium magnesium phosphate monohydrate, $H_2Na_6(Mn,Fe,Ca,Mg)_{14}(PO_4)_{12} \cdot H_2O$. Crystal system, monoclinic. Density, 3.334 g/cm^3; refractive index, 1.67. Occurrence: USA. See also arrojadite.

dickite. A colorless mineral of the kaolinite-serpentine group composed of aluminum silicate hydroxide, $Al_2Si_2O_5(OH)_4$. Crystal system, monoclinic. Density, 2.60; refractive index, 1.566. Occurrence: USA (Pennsylvania).

Dick's bronze. A bronze containing 80% copper, 10% lead, 9.2% tin and 0.8% phosphorus used for bearings and bushings. Also known as *Dick's bearing bronze*.

Dico. Trade name of Platt Metals Limited (UK) for zinc castings with a melting point of 390°C (734°F) used for press tools.

dicobalt boride. See cobalt boride (iii).

dicobalt octacarbonyl. See cobalt tetracarbonyl.

dicobalt phosphide. See cobalt phosphide (ii).

dicobalt silicide. See cobalt silicide (iii).

dicolumbium aluminide. See niobium aluminide (ii).

dicolumbium carbide. See niobium carbide (ii).

dicolumbium germanide. See niobium germanide (ii).

dicolumbium nitride. See niobium nitride (ii).

Dicor. Trademark of LD Caulk (USA) for castable ceramics used as dental porcelains, and glass-ceramics used for dentures, artificial teeth, inlays, bridges and crowns. Examples include *Dicor MGC* machinable glass-ceramic for inlay restorations, and *Dicor Plus* substrate ceramics.

Dicrome. Trade name of Firth Brown Limited (UK) for a tool steel containing 0.6% carbon, 0.25% silicon, 0.6% manganese, 0.6% chromium, and the balance iron. Used for punches, heading dies, and axes.

Dicut. Trade name of Kokour Company (USA) for buffing compounds for zinc-base die castings.

dicyanodiamide. A white, crystalline compound (99% pure) with a density of 1.400 g/cm^3 (25°C/77°F) and a melting point of 208-211°C (406-412°F). Used chiefly as a catalyst for epoxies, as an accelerator, as a starch modifier and detergent stabilizer, in case-hardening, cleaning, fireproofing and soldering compositions, in dyestuffs, for guanidine salts, in organic synthesis, as a nitrocellulose stabilizer, and in biochemistry and biotechnology. Formula: $H_2NC(NH)(NHCN)$. Also known as *cyanguanidine; dicyanodiamide*.

dicyclopentadiene. A flammable liquid that is the dimer of cyclopentadiene, and usually supplied stabilized with BHT (2,6-D-*tert*-butyl-4-methylphenol). It has a density of 0.979 g/cm^3, a melting point of 33°C (91°F), a boiling point of 170°C (338°F) and a flash point of 114°F (45°C). Used as an intermediate in the manufacture of metallocenes, EPDM elastomers, paints and varnishes, and as a flame retardant for plastics. Formula: $C_{10}H_{12}$.

dicyclopentadienylchromium. See chromocene.

dicyclopentadienylcobalt. See cobaltocene.

dicyclopentadienyliron. See ferrocene.

dicyclopentadienylmanganese. See manganocene.

dicyclopentadienylnickel. See nickelocene.

dicyclopentadienylosmium. See osmocene.

dicyclopentadienylruthenium. See ruthenocene.

dicyclopentadienylvanadium. See vanadocene.

Didby's alloy. An alloy containing 64-72% iron, 14-18% chromium and 14-18% copper. Also known as *Didby's cupritic alloy*.

didymium. (1) A natural mixture of the rare-earth elements neodymium and praseodymium. Symbol: Di (quasichemical).

(2) A mixture of rare-earth elements obtained from the mineral *monazite* by extraction and subsequent removal of cerium and thorium. The average composition is 46% lanthanum

oxide (La_2O_3), 32% neodymium oxide (Nd_2O_3), 10% praseodymium oxide (Pr_2O_3), 5% samarium oxide (Sm_2O_3), 3% gadolinium oxide (Gd_2O_3), 1% cerium oxide (Ce_2O_3), 0.4% yttrium oxide (Y_2O_3) and 2.6% other rare-earth oxides. Used in the production of rare-earth metals, for coloring and decolorizing glass (protective eyewear), and as alloying additions in metallurgy.

didymium metal. An alloy of neodymium and praseodymium used as an alloying addition.

didymium oxide. A brown powder obtained from didymium and used as glass colorant and decolorizer, in carbon arc cores, as an additive to stainless steel, in metallurgical research, and as a component in temperature-compensating capacitors. Symbol: Di_2O_3.

didymium salts. Rare-earth salt mixtures used as colorants and decolorizers in glass, in temperature-compensating capacitors, and in metallurgical research.

Die Blanking. Trade name of Midvale-Heppenstall Company (USA) for a tool steel containing 2.2% carbon, 10% chromium, 0.2% vanadium, and the balance iron. Used for tools and dies.

Die Casting. Trade name of CCS Braeburn Alloy Steel (USA) for a hot-work steel containing 0.45% carbon, 2.5% chromium, 0.6% manganese, 0.3% silicon, 0.25% vanadium, and the balance iron. Used for die-casting dies.

die-casting alloys. Alloys, such as those of aluminum, copper, lead, magnesium, tin and zinc, that are suitable for the manufacture of die castings. Also known as *die-casting metals*.

die-casting brasses. A group of brasses suitable for die casting and including (i) leaded yellow brass (UNS No. C85800) containing 58% copper, 40% zinc, 1% tin and 1% lead; (ii) silicon brass (UNS No. C87900) containing 65% copper, 34% zinc and 1% silicon, and (iii) high silicon brasses (UNS Nos. C87500 and C87800) containing 82% copper, 14% zinc and 4% silicon, and UNS No. C87400 with 83% copper, 14% zinc and 3% silicon. Used for die-cast marine, plumbing and electrical parts, bearings, brackets, gears, impellers, valve parts, clamps, and fasteners.

die-casting metals. See die-casting alloys.

die castings. Ferrous or nonferrous castings made by forcing molten metal under high pressure into steel molds or dies, usually by means of hydraulic rams. They usually have medium strength and hardness, excellent dimensional accuracy and uniformity, and can be made into intricate forms. See also die-casting alloys.

Diecrat. Trade name of Marshall Steel Company (USA) for an air-hardening cold-work tool steel (AISI type A6) containing 0.75% carbon, 2% manganese, 0.3% silicon, 1% chromium, 1.35% molybdenum, and the balance iron. Used for blanking, forming, coining and trimming dies.

Die Epoxy. Trade name for an epoxy resin for dental model and die work.

Die Flex. Trade name of Wallace Murray Corporation (USA) for a tough, hot-work die steel containing 0.4% carbon, 1.5% chromium, 0.8% molybdenum, 4.25% nickel, and the balance iron.

Diehard. Trade name of Firth Brown Limited (UK) for a series of nondeforming die steels containing 0.85-1.5-2.1% carbon, 0.3% silicon, 0.3% manganese, 12-13% chromium, 0-0.5% molybdenum, 0-1.15% vanadium, 0-1.3% cobalt, and the balance iron.

Diehard Standard. Trade name of Firth Brown Limited (UK) for a die steels containing 1.5% carbon, 0.3% silicon, 0.3% manganese, 12% chromium, 0.9% molybdenum, 0.9% vanadium, and the balance iron.

dielectric coating. (1) An electrically insulating coating composed of one or more dielectric materials.

(2) A multilayered coating with high reflectivity applied over a substrate in alternate layers of identical or different thickness consisting of quarter-wave film having either a lower or higher refractive index than the substrate. Such coatings may or may not be wavelength-specific.

dielectric crystal. A crystal in which the positive and negative electrically charge entities are separated, i.e., which has an electric dipole structure, and is electrically nonconductive. See also dielectric materials.

dielectric film. A film or foil used as a dielectric between the plates of a capacitor. Examples include oil-filled paper foil, silicon monoxide film, and films of polyester, polypropylene, polystyrene, polysulfone, polycarbonate and polytetrafluoroethylene.

dielectric materials. Insulators in which an electric field can be established with little or no leakage current. They may be solid (e.g., mica, quartz, glass, rubber, ceramics, plastics, wood, etc.), liquid (e.g., hydrocarbon oils, askarel, silicone oils, etc.) or gaseous (e.g., air, sulfur hexafluoride, etc.). Also known as *dielectrics*. See also high-k materials.

dielectrics. See dielectric materials.

Dielectrite. Trade name of Industrial Dielectrics Inc. (USA) for an extensive series of polyester bulk molding compounds with glass fiber reinforcements. They have good electrical properties and are used for automotive and electrical applications. Conductive grades for EMI/RFI shielding applications are also available.

die lubricant. A lubricating oil or compound, such as graphite or molybdenum disulfide, applied to the working surfaces of a die or added to the material or product being shaped to facilitate movement, reduce friction and die wear, and simplify removal of the finished product from the die. Such lubricants are commonly used in powder metallurgy, forging, forming, pressing and deep drawing.

DI-ELYTE. Trademark of Glastic Corporation (USA) for a high-strength E-glass/S-2 glass-reinforced epoxy resin supplied in wire and rod for composite and fiberoptic cable reinforcement.

Diemac. (1) Trade name of Macauley Foundry Company (USA) for a close-grained electric furnace cast iron containing 3.1-3.3% carbon, 1.7-1.9% silicon, 0.6-0.8% manganese, 0.8-1% molybdenum, 1-1.2% nickel, 0-0.05% phosphorus, and the balance iron. Used for permanent molds and casting dies.

(2) Trade name of Macauley Foundry Company (USA) for a cast iron containing 3.2% carbon, 2.6% silicon, 1% nickel, 0.6% molybdenum, 0.1% sulfur, 0.1% phosphorus, and the balance iron. Used for casting dies and permanent molds for aluminum, zinc, lead, etc., and plastic materials.

Die-Mate. Trade name of US Metalsource for a standard carbon steel containing 0.2% carbon, 0.2% silicon, 0.8% manganese, 0.015% phosphorus, and the balance iron. Used for cold-finished flats and squares for noncritical components of tooling and die sets.

Diemitech. Trademark of Mi-Tech Metals Inc. (USA) for a powder-metallurgy heavy-metal material composed of tungsten, nickel, iron and molybdenum, and manufactured by compaction of powders at pressures of 180-415 MPa (26-60 ksi) followed by liquid-phase sintering at 1300-1500°C (2400-2700°F). The final material is fully dense and homogeneous, and can be used for high-temperature die-casting, extrusion and upsetting tooling.

Dien. Trade name for nylon fibers and filaments used for textile fabrics.

Diene. Trademark of Firestone Synthetic Rubber & Latex Company (USA) for stereospecific solution-polymerized polybutadiene rubber available in high-cis (96%) and medium-cis (40%) grades, and vulcanized and unvulcanized form. Used as an addition to other elastomers to improve the low-temperature characteristics and mechanical hysteresis, as an addition to high-impact polystyrene and acrylonitrile-butadiene styrene to improve impact resistance, and for mechanical rubber goods, golf balls and tires.

diene. Any of a group of unsaturated aliphatic hydrocarbons containing two carbon-carbon double bonds. Examples include butadiene (C_4H_6) and hexadiene (C_6H_{10}). The general formula is C_nH_{2n-2}. Also known as *alkadiene; diolefin.*

diene polymer. A polymer derived from a diene (diolefin), or a diene (diolefin) and one or more other monomer species.

Dienett German silver. A free-cutting nickel silver (German silver) containing 51% copper, 32% zinc, 9.5% lead, 6.4% nickel and 1.6% tin. Used for hardware, novelties, and ornaments.

dierbium trisilicate. See erbium silicate (ii).

Diesel Bearings. British trade name for a babbitt of 80% lead, 15% tin and 5% copper, used for diesel-engine bearings.

Diesel Oil Engine Babbitt. Trade name of NL Industries (USA) for a babbitt containing 83-87% tin, 9-11% antimony and 4-6% copper. It has high impact loading resistance, and is used for diesel engine bearings.

die steels. A group of tool steels suitable for making dies for forging, hot and cold forming, extrusion, blanking, piercing, trimming, punching, stamping, swaging, coining and heading operations, and for die inserts, mandrels, dummy blocks, shear blades, punches, chisels, blacksmith tools, etc. Steels suitable for these applications include several cold-work, hot-work, shock-resisting and water-hardening tool grades.

Dietemper. (1) Trade name of Gulf Steel Corporation (USA) for an oil-hardening hot-work tool steel containing 0.4% carbon, 1.5% chromium, 1% molybdenum, 0.65% manganese, 0.25% vanadium, and the balance iron. Used for forging and die-casting dies.

(2) Trade name of Gulf Steel Corporation (USA) for an oil-hardening hot-work tool steel containing 0.3% carbon, 9.5% tungsten, 2.75% chromium, 1.6% nickel, 0.3% molybdenum, and the balance iron. Used for forging and die-casting dies.

diethylaminoethyl cellulose. A cellulose ether in which the diethylaminoethyl $[(C_2H_5)_2NCH_2CH_2-]$ group is attached to the cellulose in an ether linkage. It can be obtained by the addition of β-chloroethyldiethylamine hydrochloride to a sodium hydroxide solution of cellulose, and is commercially available in fibrous and microgranular form for use as an anionic ion-exchange material in chromatography, and as a catalyst for epoxide reactions. Abbreviation: DEAE cellulose.

1,1'-diethyl-4,4'-carbocyanine iodide. See cryptocyanine.

1,1'-diethyl-2,4'-cyanine iodide. A light-sensitive, crystalline compound with a dye content of about 97%, a melting point of 158-160°C (316-320°F), and a maximum absorption wavelength of 558 nm. Used as a dye and pigment, and for electronic and optoelectronic applications. Formula: $C_{23}H_{23}N_2I$. Also known as *isocyanine iodide.*

1,1'-diethyl-4,4'-dicarbocyanine iodide. A light-sensitive, crystalline compound with a dye content of about 90%, a melting point of 158°C (316°F), and a maximum absorption wavelength of 814 nm. Used as a dye and pigment, and for electronic and optoelectronic applications. Formula: $C_{25}H_{25}N_2I$.

diethylene glycol. A colorless, hygroscopic liquid (99+% pure) obtained as a byproduct in the production of ethylene glycol. It has a density of 1.118 g/cm^3, a melting point of -80°C (-112°F), a boiling point of 245°C (473°F), a flash point of 290°F (143°C), an autoignition temperature of 444°F (229°F) and a refractive index of 1.446. Used as a solvent, e.g., for nitrocellulose, dyes and oils, in the manufacture of unsaturated polyesters, polyurethane and triethylene glycol, in petroleum solvent extraction, in the dehydration of plasticizers, surfactants and natural gas, in adhesives and synthetic sponges, and in antifreezes. Formula: $C_4H_{10}O_3$. Abbreviation: DEG. Also known as *oxydiethanol.*

diethylsilane. An organosilicon compound available as a flammable, colorless liquid (99% pure) with a density of 0.681 g/cm^3, a boiling point of 56°C (132°F), a flash point of -20°F (-28°C) and a refractive index of 1.391. Used as a reagent in the chemical vapor deposition (CVD) of silicon dioxide for microelectronic applications. Formula: $(C_2H_5)_2SiH_2$.

1,3-diethylthiourea. A buff solid compound (98% pure) with a melting point of 76-78°C (169-172°F) used as an accelerator, as an activator in elastomers, and as a corrosion inhibitor in metal pickling baths. Formula: $C_2H_5NHCSNHC_2H_5$.

diethylzinc. An air- and moisture-sensitive, colorless, pyrophoric liquid with a density of 1.205 g/cm^3, a melting point of -28°C (-18°F), a boiling point of 118°C (244°F), a flash point of -1°F (-18°C) and a refractive index of 1.498. It is also available in electronic grades with a purity of 99.999%, as a 1.0M solution in heptane or hexane, and as a 1.1M solution in toluene. Used as an organometallic reagent, in organic synthesis, e.g., as a catalyst for olefin polymerization and in the preparation of ethyl mercuric chloride, in electronics, and as a high-energy fuel for aircraft and missile applications. Formula: $(C_2H_5)_2Zn$. Also known as *ethylzinc; zinc ethyl; zinc diethyl.*

dietrichite. A white mineral of the halotrichite group composed of zinc aluminum sulfate hydrate, $ZnAl_2(SO_4)_4 \cdot 22H_2O$, and may also contain some iron and/or manganese. Crystal system, monoclinic. Density, 1.86 g/cm^3; refractive index, 1.480. Occurrence: Hungary.

dietzeite. A golden-yellow mineral composed of calcium chromium oxide iodate, $Ca_2(CrO_4)(IO_3)_2$. Crystal system, monoclinic. Density, 3.70 g/cm^3; refractive index, 1.842. Occurrence: Chile.

Dievac. Trademark of Atlas Specialty Steels (Canada) for a chromium-type hot-work tool steel (AISI type H13) containing about 0.3-0.4% carbon, 0.2-0.5% manganese, 1% silicon, 5% chromium, 1.5% molybdenum, 1% vanadium, and the balance iron. It has high hot hardness, good abrasion resistance, good resistance to heat checking, good machinability and toughness, and high hot hardness. Used for die-casting, extrusion, forging, header, plastic-molding and trimmer dies.

Dievar. Trade name of Uddeholm Corporation (USA) for a hot-work die steel with high ductility, toughness and fracture toughness, and good thermal fatigue resistance. Used for die-casting dies.

Diewear. Trade name of Gorham Tool Industries Inc. (USA) for an oil-hardening, cold-work tool steel (AISI type O1) containing 0.9% carbon, 1% manganese, 0.5% chromium, 0.5% tungsten, and the balance iron. It has good nondeforming properties, and is used for blanking, forming, bending and drawing dies, and punches.

Dieweld. Trade name of Edgcomb Metals Corporation (USA) for a high-speed tool steel containing 0.7% carbon, 18% tungsten, 4% chromium, 1% vanadium, and the balance iron. Used for welding electrodes.

differential coating. A unequally coated product, i.e., a product that has a heavy coating on one side and a much lighter coating on the other. Examples include electrolytic tinplate and hot dip galvanized products.

differential tinplate. Unequally coated electrolytic tinplate, i.e., tinplate made by electrodeposition of unequal amounts of tin on the two surfaces of a continuous strip of rolled steel. The heavier coating is less than 2.0 μm (0.08 mil), while the lighter coating is in the range of 0.4-1.5 μm (0.02-0.06 mil).

Diffuglas. Trade name SA Glaverbel (Belgium) for a glass product consisting of two sheets of glass with an interlayer of glass fibers and sealed all around with a supple joint.

Diffulite. Czech trade name for a semitransparent glass with a somewhat irregular pattern.

diffused stainless steel. A stainless sheet steel that has a soft, chromium-free, low-carbon core and a corrosion-resistant chromium-rich surface produced by heating in intimate contact with a chromium compound at temperatures above 1000°C (1830°F). The chromium-rich layer is formed by diffusion of chromium from the chromium compound. See also chromized steel.

diffusion aid. A solid filler metal applied to surfaces to be welded to assist in diffusion welding.

diffusion-alloyed steel. A steel made by first adding the alloying elements, e.g., nickel, chromium, molybdenum, etc., in the form of finely divided elemental or oxide powders to the iron powder, and then pressing and sintering. Co-reduction with the iron powder produces a steel in which the elemental additions are firmly attached and partially diffused into the iron.

diffusion barrier. A material usually applied as a protective coating to prevent or reduce the diffusion of certain species. See also barrier.

diffusion coatings. A class of corrosion and/or oxidation-resistant alloy coatings produced on base-metal or alloy substrates by heating in intimate contact with another metal or alloy in powder, liquid or gaseous form at elevated temperatures. The coatings are formed by the reaction of the coating material with the base-metal or alloy by diffusing inward. Examples include aluminum, chromium and zinc coatings on steel.

Diffusopane. Trade name of Dearborn Glass Company (USA) for light-diffusing laminated glass with a light transmission of 63%.

Difulit. Trade name of former Gerresheimer Glas AG (Germany) for patterned glass.

digadolinium disilicate. See gadolinium silicate (ii).

digenite. A blue to black mineral composed of copper sulfide, Cu_9S_5. Crystal system, rhombohedral or cubic. It transforms from the rhombohedral to the cubic crystal form at 81°C (178°F). Density, 5.60 g/cm³. Occurrence: Scotland. It can also be made synthetically ranging in composition from $Cu_{1.8}S$ to Cu_9S_5 and densities from 5.60 to 5.71 g/cm³.

digermane. A flammable gas supplied in 10% hydrogen. It has a melting point of -109°C (-164°F), a boiling point of 31.5°C (89°F), and is used in electronics, and as a precursor to germanium and germanium-silicon alloy thin films. Formula: Ge_2H_6.

Diglaze. Trade name for a diamond polishing paste used in dentistry.

diglycidyl ether of bisphenol A. A nonflammable epoxide that is the reaction product of *epichlorohydrin* (C_3H_5ClO) and *bisphenol A* ($C_{15}H_{16}O_2$) in the presence of an alkali (e.g., sodium hydroxide). It has excellent mechanical properties, good fatigue

resistance, good moisture resistance and environmental properties, good electrical properties, a density of about 1.16 g/cm³, and a service temperature of 80-88°C (176-190°F). Used for epoxy resins, in dental resins and composites, as a matrix composite for pipes, automotive parts, sporting goods and printed-wiring boards, and as an embedding medium in microscopy. Abbreviation: DGEBPA. See also bisphenol A; epoxide.

diglycidyl ether of bisphenol F. A nonflammable epoxide that is the reaction product of *epichlorohydrin* (C_3H_5ClO) and *bisphenol F* ($C_{13}H_{12}O_2$). It has a viscosity of about 2000-3000 centipoise, a density of 1.190 g/cm³, a melting point of -15°C (-5°C), a flash point above 230°F (110°C) and a refractive index of 1.579. Used for epoxy resins and composites. Abbreviation: DGEBPF. See also bisphenol F; epoxide.

Di Hard. Trade name of Certified Alloy Products Inc. (USA) for a wear-resistant cast iron containing 3.5% carbon, 4.2% nickel, 1.1% boron, and the balance iron. Used for fusible coating for lined cylinders.

Di-Hard. Trade name of Xaloy Inc. (USA) for a hard, wear-resistant white cast iron containing 3.5% carbon, and varying amounts of boron, nickel, silicon, manganese, and the balance iron. Used for extruder and injection molding barrels, cylinder liners, pump parts, etc. See also Xaloy.

diheptylazoxybenzene. A nematic liquid crystal with a flash point above 230°F (110°C), used in electronics and optoelectronics. Formula: $CH_3(CH_2)_6C_6H_4N=N(O)C_6H_4(CH_2)_6CH_3$.

dihexadecyl ether. See dicetyl ether.

dihexylazoxybenzene. A nematic liquid crystal with a flash point above 230°F (110°C). Used in electronics and optoelectronics. Formula: $CH_3(CH_2)_5C_6H_4N=N(O)C_6H_4(CH_2)_5CH_3$.

dihexyloxacarbocyanine. A light-sensitive solid compound (98% pure) with a melting point of 219-221°C (426-430°F) and a maximum absorption wavelength of 485 nm. Used in biology, biochemistry and biotechnology as a fluorescent probe for determining cell membrane potentials. Formula: $C_{29}H_{37}N_2O_2I$.

dihydrated lime. Dolomitic lime that contains less than 8% unsaturated oxides, and is made by high-pressure hydration. Also known as *double-hydrated lime*. See also monohydrated lime.

dihydrogen hexachloroosmate hexahydrate. See chloroosmic acid.

dihydrogen hexachloroplatinate hexahydrate. See chloroplatinic acid.

m-dihydroxybenzene. See resorcinol.

p-dihydroxybenzene. See hydroquinone.

Di-Iron. Trade name of Aluminium Union Limited (UK) for a cast iron containing 3.2% carbon, 2.2% silicon, 1.5% nickel, 0.8% chromium, 0.15% molybdenum, and the balance iron. Used for gears, shafts; housings, and dies.

diiron nickel tetraoxide. See nickel ferrite.

diiron titanium. See iron titanium (ii).

diisocyanate. An organic compound obtained by treating a diamine, such as hexamethylenediamine or toluene-2,4-diamine, with phosgene ($COCl_2$). It has two isocyanate groups (–NCO) and is used in the production of polyurethane elastomers and foams, in rubber-to-nylon and rubber-to-rayon bonding, and as an alkali and water resistance improver in phenolic resins.

diisocyanate foam. See polyurethane foam.

dilanthanum silicide. See lanthanum silicide (ii).

dilanthanum trisilicate. See lanthanum silicate (ii).

Dilar. Trademark of Norman Wade Company Limited (Canada) for sensitized film.

dilatational materials. See negative Poisson's ratio materials.

Dilatherm. Trade name of Vereinigte Edelstahlwerke (Austria) for a series of alloys containing 0-36% nickel, 0-25% chromium, 0.1% carbon, 0-0.1% sulfur, and the balance iron. They have low coefficients of expansion, and are used for glass-to-metal seals, hot conductors, etc.

Dilaton. Trade name of Vereinigte Deutsche Nickel-Werke AG (Germany) for a series of alloys including iron-nickel, iron-nickel-cobalt, nickel-iron and nickel-iron-cobalt alloys supplied as strip, bar, wire, or forged and turned parts for use in electronics and glass-sealing applications.

Dilecto. Trade name of Continental-Diamond Fiber Company (USA) for phenolic resins and laminates.

Diline. Trademark of Norman Wade Company Limited (Canada) for a sensitized paper.

dilithium phthalocyanine. See lithium phthalocyanine.

dilithium silicate. See lithium silicate (ii).

Dilon. Trade name of Dilon Limited (Pakistan) for nylon 6 fibers and yarns.

Dilphy. Trade name of Creusot-Loire (France) for a nickel-iron alloy with controlled coefficient of expansion, used for thermostats.

diluted magnetic semiconductors. A group of electronic materials composed essentially of nonmagnetic host materials, such as arsenides, selenides, sulfides or tellurides of cadmium, mercury or zinc, into which magnetic ions of metallic materials, such as cobalt, iron or manganese, have been introduced to replace a carefully controlled fraction of cations. Abbreviation: DMS. Also known as *magnetic semiconductors; semimagnetic semiconductors.*

Dilver. Trademark of Creusot-Loire (France) for a series of iron-chromium, iron-nickel and iron-nickel-cobalt alloys. Usually supplied in rod, sheet, tube, wire, foil and powder form, they have low coefficients of thermal expansion, and good high-temperature properties. Used for electronic equipment, hard-glass sealing applications, heating elements, etc.

dimagnesium columbate. See magnesium niobate (ii).

dimagnesium dialuminum pentasilicate. See magnesium aluminum silicate (i).

dimagnesium germanate. See magnesium germanate (ii).

dimagnesium niobate. See magnesium niobate (ii).

dimagnesium silicate. See magnesium silicate (iii).

dimagnesium stannate. See magnesium stannate (ii).

dimagnesium sulfide. See magnesium sulfide (ii).

dimagnesium titanate. See magnesium titanate (ii).

dimanganese antimonide. See manganese antimonide (ii).

dimanganese arsenide. See manganese arsenide (ii).

dimanganese boride. See manganese boride (iii).

dimanganese decacarbonyl. See manganese carbonyl.

dimanganese dialuminum pentasilicate. See manganese aluminum silicate (i).

dimanganese phosphide. See manganese phosphide (ii).

dimanganese silicate. See manganese silicate (ii).

dimanganese titanate. See manganese titanate (ii).

Dimension. Trade name of AlliedSignal Inc. (USA) for a series of modified polyamide (nylon 6) alloys for automotive applications. They are supplied in various grades including extrusion, blow-molding, glass-reinforced and low-moisture absorption. A polphenylene ether/nylon alloy grade is also available.

dimension lumber. Lumber that is from 50 mm (2 in.) to, but not including, 125 mm (5 in.) thick and 50 mm (2 in.) or more wide, and graded for strength rather than appearance. Used in almost any type of building construction.

dimension stock. Timber sawn to specific dimension for a special application.

dimension stone. A large, relatively undamaged natural stone cut to shape or size. It may or may not have mechanically trimmed surfaces, and is used especially as a building and paving stone.

dimer. A polymer made from two basic molecules, or a polymer consisting of two repeating units.

Di-Metal. Trade name of American Smelting & Refining Company (USA) for a series of die-casting alloys containing 3.5-4.5% aluminum, 0.75-3.5% copper, 0.02-0.1% magnesium, up to 1% iron, and the balance zinc. Used for general-purpose die castings.

Dimetcote. Trademark of Ameron International Corporation (US) for a series of inorganic zinc and zinc silicate primers for the protection of steel structures, tanks, equipment, piping, ships, barges, offshore structures, and other surfaces.

dimethyldichlorotin. See dimethyltin dichloride.

(3S)-cis-3,6-dimethyl-1,4-dioxane-2,5-dione. See L-lactide.

3,6-dimethyl-1,4-dioxane-2,5-dione. See lactide.

dimethylenemethane. See allene (2).

dimethylhydantion-formaldehyde polymer. A brittle, water-soluble, light-colored polymer resin with a density of 1.30 g/cm^3 and a softening point of 59-80°C (138-176°F). It is prepared from crystalline 5,5-dimethylhydantoin ($C_5H_8N_2O_2$) and contains up to 0.3 wt% formaldehyde. Used in the manufacture of adhesives and sizes, and as a blending agent.

dimethyl silicones. A class of polysiloxanes obtained by the hydrolysis of dimethyldichlorosilane, and corresponding to the general formula $[(CH_3)_2SiO]_n$, where n is typically 3-9. They are colorless oils with boiling points between 134 and 188°C (273 and 370°F)/20 mm. Used chiefly in brake fluids and as transformer oils.

dimethyltin dichloride. Moisture-sensitive, white crystals (97% pure) with a melting point of 103-105°C (217-221°F) and a boiling point of 185-190°C (365-374°F) used as an intermediate for tin oxide deposition on glass, as a polyvinyl chloride stabilizer, as a catalyst, and in the preparation of electroluminescent materials. Formula: $(CH_3)_2SnCl_2$. Also known as *dichlorodimethyltin; dimethyldichlorotin.*

dimethylzinc. An air- and moisture-sensitive, colorless, pyrophoric liquid available in standard and high-purity (99.999%) grades. It has a density of 1.386 g/cm^3, a melting point of -42°C (-43°F) and a boiling point of 46°C (114°F). It is also available as a flammable 1.0M solution in heptane with a density of 0.724 g/cm^3, a boiling point of 44-46°C (111-114°F), and a flash point of 30°F (-1°C), and as a pyrophoric, moisture-sensitive 2.0M solution in toluene with a density of 0.931 g/cm^3 and a flash point 1°F (-17°C). Used as organometallic reagent, in organic synthesis, as a catalyst for polymerization reactions, and in electronics. Formula: $(CH_3)_2Zn$. Also known as *methylzinc; zinc methyl; zinc dimethyl.*

dimity. A thin, white or colored cotton fabric plain-woven with heavy threads at intervals usually in a striped arrangement. Used for dresses, blouses, nightwear and curtains.

Dimlon. Trademark of Tekstiplik Sanayii AS (Turkey) for nylon 6 fibers and yarns.

Di-Mol. Trade name of Disston Inc. (USA) for a high-speed steel containing 0.8% carbon, 4% chromium, 9% molybdenum, 1.5% tungsten, 1% vanadium, and the balance iron. Used for cutting tools.

dimolybdenum boride. See molybdenum boride (iv).

dimolybdenum carbide. See molybdenum carbide (ii).

dimolybdenum nitride. See molybdenum nitride (ii).

dimolybdenum trioxide. See molybdenum sesquioxide.

Dimondite. Trade name of Teledyne Firth-Sterling (USA) for a sintered carbide of 87-95% tungsten carbide and 5-13% cobalt. It has an extremely high hardness (nearly that of diamond), and is used for tips of high-speed tools and cutters, drawing dies, wear parts, and gages.

dimorphite. An orange to yellow mineral composed of arsenic sulfide, As_4S_3. Crystal system, orthorhombic. Density, 3.58 g/cm^3; refractive index, 1.66. Occurrence: Italy.

Dimpalloy. (1) Trade name of Sheffield Smelting Company (UK) for a copper-zinc alloy used for brazing brasses and bronzes.

    (2) Trade name of Sheffield Smelting Company (UK) for a silver solder sheet with flux available in various compositions but usually with 22-23% cadmium, 15-20% copper, 15-20% zinc, and the balance silver.

dimyristyl ether. A liquid compound (95+% pure) with a density of 0.813 g/cm^3, a melting point of 38-40°C (100-104°F), and a boiling point of 238-248°C (460-478°F)/4.0 mm. Used as a chemical intermediate, in electrical insulators, antistatic substances, water repellents, and as lubricants in plastics. Formula: $(C_{14}H_{29})_2O$. Also known as *ditetradecyl ether.*

Dinabase. Trade name for a thermoplastic acrylic resin used in dentistry for base relining applications.

Dinas brick. An acid refractory brick that contains more than 95% silica (SiO_2) and is used for furnace linings. See also silica brick.

Dinas rock. A natural rock or sand consisting almost entirely of silica (SiO_2). It has a melting point of 1680°C (3056°F), and is used as acid refractory brick for furnace lining applications.

Dinertia. Trade name of Deloro Stellite, Limited (UK) for a corrosion-resistant, reactive refractory alloy of cobalt, chromium, molybdenum and vanadium that is available in cast and rolled form, and used for dental and surgical tools.

dinickel boride. See nickel boride (ii).

dinickel phosphide. See nickel phosphide (ii).

diniobium aluminide. See niobium aluminide (ii).

diniobium carbide. See niobium carbide (ii).

diniobium germanide. See niobium germanide (ii).

diniobium nitride. See niobium nitride (ii).

dinnerware. Glassware and ceramic whiteware used in dinner service.

dioctyl ether. A liquid compound (99+% pure) with a density of 0.806 g/cm^3, a melting point of -7°C (19°F), a boiling point of 292°C (557°F), a flash point above 230°F (110°C), and a refractive index of 1.432. Used as a chemical intermediate, in electrical insulators, antistatic substances, water repellents, and as lubricants in plastics. Formula: $(C_8H_{17})_2O$. Also known as *octyl ether.*

Diofan D. Trademark of BASF AG (Germany) for polvinylidene chlorides.

diolefin. See diene.

Diolen. Trademark of Acordis Industrial Fibers Inc. (USA) for high-tenacity polyester fibers and filament yarns used for industrial fabrics and protective clothing.

Dion. Trade name of Reichhold Chemicals, Inc. (USA) for an extensive series of synthetic resins including bisphenol fumarates, bisphenol-epoxy vinyl esters, vinyl esters, and isophthalic and orthophthalic polyesters. They have good chemical and/or flame/smoke resistance, and are used in ductwork, chemical-storage tanks, chemical piping, gasoline tanks, fume ducts, sewer pipes, wall and roofing systems, pulp-washer drums, pick-ling and plating equipment and structural applications. *Dion ISO* is a line of rigid, high-molecular-weight isophthalic resins with excellent corrosion and heat resistance and outstanding mechanical properties. *Dion VER* are strong, heat-resistant ultrahigh-density crosslinked vinyl ester resins containing 30% styrene monomer and a divinyl benzene comonomer. They have superior resistance to various solvents, hot caustics, and hot concentrated acids. *Dion VPE* refers to a line of strong, tough vinyl polyester resins with low styrene content and good corrosion and heat resistance.

Dioplast. Trademark of Chemtron Manufacturing Limited (Canada) for a polymercaptan sealant.

diopside. A white to light-green mineral of the pyroxene group with vitreous luster that is composed of calcium magnesium silicate, $CaMg(SiO_3)_2$, and may contain manganese and iron. The iron-rich varieties have a deep-green color. It can also be made synthetically. Crystal system, monoclinic. Density, 3.2-3.3 g/cm^3; melting point, 1392°C (2538°F); hardness, 5-6 Mohs; refractive index, 1.671. Occurrence: Austria, Italy, Russia, Switzerland, USA (Connecticut, New York). Used for refractories, in welding-rod coatings, as a component in whiteware bodies, glazes and glass, and in ceramics and materials research. The transparent varieties are used as gemstones. Also known as *alalite; malacolite.* See also synthetic diopside.

dioptase. An emerald-green mineral composed of copper silicate monohydrate, $CuSiO_3 \cdot H_2O$. Crystal system, rhombohedral (hexagonal). Density, 3.28 g/cm^3; refractive index, 1.667. Occurrence: Namibia.

Diorid. Trade name for a modacrylic fiber used chiefly for clothing and industrial fabrics.

diorite. Any of a group of crystalline, coarse-grained, usually grayish igneous rocks that are composed chiefly of *plagioclase feldspar* and *hornblende*, but may also contain *quartz, biotite* and/or ferromagnesian silicates. Used as a building and ornamental stone, and as a concrete aggregate. Also known as *black granite.*

2,5-p-dioxanedione. See glycolide.

Dioxsil. Trade name for an acid-resistant material made by the fusion of silica sand or crushed natural quartz. See also fused quartz.

dipalmitoyl phosphatidylcholine. A chiral lipid-based compound that is a lyotropic liquid crystal [$C_{15}H_{31}$-COO-(di)phosphatidylcholine] available in R- and S-configurations. Used for medical and biomedical applications. Abbreviation: DPPC.

dip coating. See immersion coating.

dipentaerythritol. An off-white, hygroscopic powder obtained from pentaerythritol and usually supplied in technical grades with a density of 1.33 g/cm^3 and a melting point of 215-218°C (419-424°F) for use in coatings and paints. Formula: $C_{10}H_{22}O_7$.

dipentylazoxybenzene. A nematic liquid crystal with a flash point above 230°F (110°C), used in electronics and optoelectronics. Formula: $C_{22}H_{30}N_2O$.

diphenyl. White, water-insoluble crystalline compound (99.5% pure) consisting of two phenyl rings and having a density of 0.992 g/cm^3, a melting point of 69-72°C (156-161°F), a boiling point of 255°C (491°F), and a flash point of 235°F (113°F). Used in organic synthesis, as a heat-transfer agent, and in the manufacture of benzidine. High-purity, zone-refined grades (99.9+%) are used as aromatic reference standards. Formula: $C_{12}H_{10}$. Also known as *biphenyl.*

diphenylacetylene. Monoclinic white crystals or powders (98% pure) with a density of 0.990 g/cm^3, a melting point of 59-

61°C (138-141°F), and a boiling point of 300°C (572°F). Used in organic and organometallic research. The purified grades are used as primary fluor or wavelength shifters in soluble scintillators. Formula: $C_{14}H_{10}$. Also known as *tolan*.

diphenylanthracene. A crystalline compound (98% pure) with a melting point of 245-253°C (473-488°F), used as primary fluor or wavelength shifters in soluble scintillators. Formula: $C_{26}H_{18}$.

α-diphenylene. See fluorene.

trans-1,2-diphenylethylene. See *trans*-stilbene.

1,6-diphenylhexatriene. A crystalline compound (98% pure) with a melting point of 199-203°C (390-397°F) used as monomer, as a wavelength shifter in liquid scintillation spectrometry, and in biochemistry as a fluorescent membrane probe and assay for phospolipid vesicles. Formula: $C_{18}H_{16}$. Abbreviation: DPH.

2,5-diphenyloxazole. A white, crystalline compound supplied in scintillator grades (99% pure). It has a melting point of 72-74°C (162-165°F), a boiling point of 360°C (680°F), and a peak lasing wavelength of 357 nm. Used as a primary fluor or wavelength shifter in liquid scintillation spectrometry, and as a laser dye. Formula: $C_{15}H_{11}NO$. Abbreviation: PPO; DPO.

diphenyl oxide resins. Thermosetting resins that are based on diphenyl oxide, $(C_6H_5)_2O$, and have excellent heat resistance and handling properties.

4,4'-diphenylstilbene. A crystalline compound with a melting point of 308-310°C (586-590°F). The purified grade is used in liquid scintillation spectroscopy. Formula: $C_{26}H_{20}$. Abbreviation: DPS. Also known as *trans-p,p'-diphenylstilbene*.

diphenylzinc. An organozinc compound available in the form of white, air- and moisture-sensitive crystals (95+% pure) with a melting point of 102-107°C (215-224°F) and a boiling point of 280-285°C (536-545°F)/760 mm. Used as an organometallic reagent, in organic synthesis, and as a catalyst for polymerization reactions. Formula: $(C_6H_5)_2Zn$. Also known as *phenylzinc; zinc phenyl; zinc diphenyl*.

dip metal. A free-cutting alloy of 70% copper, 28.5% zinc, 0.5% nickel, 0.5% lead, and 0.5% antimony. Used for hardware.

dipoxy resins. Epoxy resins manufactured by the epoxidation of olefins with a peracid, such as peracetic acid. See also epoxies.

Dip-Pak. Trade name of Fidelity Chemical Products Corporation (USA) for hot-melt strippable plastic coating.

dipping black. A solution of *gilsonite*.

dipping brass. A highly ductile brass containing 67% copper and 33% zinc. It has excellent cold workability, moderate strength, fair machinability, and is used for spun, stamped or deep-drawn parts.

Dippit. Trade name of United American Metals Corporation (USA) for a babbitt of 94.5% lead, 5% tin and 0.5% copper. Used for bearings and bushings.

direct-band-gap semiconductor. A semiconductor in which an electron can be promoted from the valence band to the conduction band without changing the magnetic moment of the electron. When the excited electron falls back into the valance band, it recombines with a hole to produce light. This is known as radiative recombination. Examples are gallium arsenide and solid solutions of gallium arsenide and aluminum arsenide. Used for the manufacture of light-emitting diodes (LED's) of different colors. See also indirect-band-gap semiconductor.

direct-bonded basic brick. A fired basic brick in which the refractory grains are bonded by a solid-state diffusion mechanism.

direct enameling steel. See decarburized enameling steel.

directionally solidified alloy. A cast alloy that has been solidi-

fied such that molten metal was always available for that portion about to solidify. It possesses a structure in which all grain boundaries are parallel to the direction of heat flow. Abbreviation: DS alloy.

direct metal. The molten pig iron from a blast furnace that is collected in car ladles and transported directly to the mill for steelmaking.

direct-on cover coat enamel. See direct-on enamel.

direct-on cover coat steel. See zero-carbon steel.

direct-on enamel. A white or colored porcelain-enamel cover or finish coat applied directly to a steel base without a ground coat. Also known as *direct-on cover coat enamel*.

direct-on steel. See zero-carbon steel.

direct-reduced iron. Metallic iron, usually in the form of pellets, produced from iron ores by removing most of the oxygen with suitable reducing agents at temperatures above 650°C (1200°F), but below the melting temperature of about 1200°C (2190°F). Below 1000°C (1830°F), the iron will be porous, when the reducing temperature is between 1000 and 1200°C (1830 and 2190°F), the material begins to sinter. Direct reduction is carried out by fluidized-bed, fixed-bed retort, moving-bed shaft, or rotary-kiln processes. Used as a charge material for blast furnaces and other melting furnaces, and in the manufacture of clean high-grade steel in electric furnaces. Abbreviation: DRI; DR iron.

diresorcinolphthalein. See fluorescein.

dirhenium decacarbonyl. See rhenium carbonyl.

dirhenium phosphide. See rhenium phosphide (iii).

Dirigold. Trade name of Dirigold Corporation (USA) for a strong copper casting alloy containing 9-10% aluminum, 1.1-2% nickel, and 0-1.7% tin. Used for ornamental castings and corrosion resisting parts.

Dirilyte Metal. Trade name of American Art Metals, Inc. (USA) for a gold-colored aluminum bronze containing 95% copper and 5% aluminum. It has good corrosion resistance and cold workability, and fair hot workability. Used for tableware.

Diron. Trade name of British Steel plc (UK) for a low-carbon steel with single-coat porcelain enamel.

Dirostahl. Trade name of Karl Diederichs Stahl-, Walz- und Hammerwerk (Germany) for an extensive series of steel products including various cold-work, plain-carbon and tungsten and cobalt-tungsten high-speed tool steels, various standard carbon steels, and a wide range of high-grade alloy steels, e.g., chromium, chromium-manganese, chromium-molybdenum, chromium-molybdenum-vanadium, chromium-nickel, molybdenum, chromium-nickel-molybdenum and chromium-vanadium grades.

Dirt Fighter. Trademark of Dutch Boy Inc. (USA) for an exterior acrylic latex enamel paint.

disaccharide. A compound sugar produced when two monosaccharides are linked together with the elimination of a water molecule. Examples include cellobiose, sucrose, maltose and lactose.

disamarium trisilicate. See samarium silicate (iii).

Dis-a-paste. Trade name of Aptek Laboratories (USA) for urethane adhesives.

Discaloy. Trademark of Westinghouse Electric Corporation (USA) for a series of wrought iron-base superalloys and austenitic stainless steels containing 24-28% nickel, 12-15% chromium, 2.5-3.5% molybdenum, 1.35-1.85% titanium, 0.6-1% silicon, 0.6-1% manganese, 0-0.35% aluminum, 0-0.08% carbon, and the balance iron. They have excellent heat and oxidation resistance, good high-temperature strength, good corrosion resistance, and good rupture strength. Used for turbine blades, turbine disks, gas turbine parts, jet engine components, extrusion dies, nonmagnetic gears and pistons, bolts, furnace parts, retorts, combustion chambers, salt pots, and heat treating equipment.

Discolith. Trade name of Fabbrica Pisana SpA (Italy) for glass blocks.

discontinuous carbide fibers. Thin, short ceramic fibers consisting of ceramic carbides, e.g., silicon carbide or mixtures of carbides, such as silicon carbide, titanium carbide and zirconium carbide. Used as reinforcing fibers for engineering composites.

discontinuous fiber composites. A group of composites that contain discontinuous (short) reinforcing fibers, and have properties varying with the fiber length. They are supplied in various product forms including bulk- and injection-molding compounds and dry preforms. Common examples include polymers reinforced with particulate fillers or chopped-glass, aramid and carbon fibers. Also known as *discrete-fiber composites; short-fiber composites.*

discontinuous-fiber-reinforced thermoplastics. A group of polymer composites consisting of thermoplastic resin matrices (e.g., polyamide, polycarbonate, polystyrene, polyphenylene sulfide or polyetheretherketone) reinforced with discontinuous (short) fibers of glass, carbon, graphite, etc. Also known as *short-fiber-reinforced thermoplastics.*

discontinuous fibers. Thin filaments of natural or synthetic materials, such as cotton, flax, alumina, boron, carbon, aramid, polyamide, glass, etc., cut to short lengths, usually 50 mm (2 in.) or less, and used as reinforcing fibers for engineering composites. Abbreviation: DF. Also known as *chopped fibers; discrete fibers; short fibers.* See also continuous fibers.

discontinuous oxide fibers. Thin, short fibers composed of ceramic oxides. Examples are discontinuous alumina, alumina-silica or zirconia-silica fibers.

discotic liquid crystal. A *thermotropic liquid crystal* in which the flat disk-shaped molecules self-assemble by stacking on top of each other forming columns. The latter are then packed on a two-dimensional lattice. Possible starting materials for the preparation of this type of liquid crystal are various hexakis(*n*-alkoxy)triphenylenes (HATn). Abbreviation: DLC. See also calamitic liquid crystal; liquid crystal.

Discovery. Trademark of World Alloys & Refining, Inc. (USA) for a corrosion-resistant chromium-cobalt dental ceramic alloy with a density of 8.2 g/cm³ (0.30 lb/in.³), a melting range of 1330-1357°C (2425-2475°F), a hardness of 300 Vickers, high yield strength, and low elongation. Used for metal-to-ceramic restorations.

discrete fibers. See discontinuous fibers.

discrete-fiber composites. See discontinuous fiber composites.

discrete glass fibers. See chopped glass fibers.

Discus. (1) Trade name of British Steel plc (UK) for a mild steel used for high-strength galvanized corrugated sheet.

(2) Trade name of Pilkington Brothers Limited (UK) for a patterned glass.

Disil. Trade name for a series of high-temperature oxidation-resistant silicide and modified silicide coatings for molybdenum and niobium. The oxidation-resistant coatings contain only silicon, and the modified coatings contain silicon as well as vanadium, chromium and titanium. *Disil* coatings are deposited by a fluidized-bed or pack cementation processes.

disilane. A pyrophoric compound of silicon and hydrogen available in standard and electronic grades (up to 99.998% pure). It has a melting point of -132.6°C (-207°F), a boiling point of -14.5°C (6°F), and is used as a precursor for the fast low-temperature deposition of epitaxial silicon and silicon-based dielectrics. Formula: Si_2H_6.

disilicates. Silica materials in which three of the four oxygen atoms of the SiO_4 tetrahedra are shared with adjacent tetrahedra, i.e., they have only one nonbridging oxygen atom per tetrahedron. The general formula is XSi_2O_5, where X represents one or more metals, such as aluminum, barium or calcium.

disodium dihydrogen pyrophosphate. See sodium acid pyrophosphate.

disodium diphosphate. See sodium acid pyrophosphate.

disodium edetate. See disodium EDTA.

disodium EDTA. The disodium salt of ethylenediaminetetraacetic acid ($C_{10}H_{16}N_2O_8$), usually supplied as the dihydrate. It is available as a white, water-soluble, crystalline powder (99+% pure) and as a 0.1M solution in water. The pH typically ranges between 4.0 and 6.0. Used as a chelating and sequestering agent. Formula: $C_{10}H_{14}N_2Na_2O_8 \cdot 2H_2O$. Abbreviation: EDTA·Na₂. Also known as *disodium edetate.*

disodium hydrogen phosphate. See disodium phosphate.

disodium orthophosphate. See disodium phosphate.

disodium phosphate. A term used for any of the following compounds which are usually available as colorless, translucent crystals or white powders:: (i) *Anhydrous dibasic sodium phosphate.* A compound that is converted to sodium pyrophosphate at 240°C (464°F). Formula: Na_2HPO_4; (ii) *Dibasic sodium phosphate dihydrate.* Density, 2.066 g/cm³; melting point, loses $2H_2O$ at 92.5°C (198°F). Formula: $Na_2HPO_4 \cdot 2H_2O$; (iii) *Dibasic sodium phosphate heptahydrate.* Density, 1.679 g/cm³; melting point, loses $5H_2O$ at 48°C (118°F). Formula: $Na_2HPO_4 \cdot 7H_2O$; and (iv) *Dibasic sodium phosphate dodecahydrate.* Density, 1.524 g/cm³; melting point, 35°C (95°F); loses $5H_2O$ on exposure to air at ordinary temperature; loses $12H_2O$ at 100°C (212°F). Formula: $Na_2HPO_4 \cdot 12H_2O$. The anhydrous compound and the di- and heptahydrate are used in textiles, in fireproofing wood and paper, in soldering enamels, in ceramic glazes, in tanning, in boiler water treatment, and as components of buffer solutions in chemistry, biology, biochemistry and medicine. The dodecahydrate is employed in the production of opalescent glass, in the purification of clays, as a water conditioner, as a deflocculant in porcelain enamels and glazes, and as a cooling agent. Also known as *dibasic sodium orthophosphate; dibasic sodium phosphate; disodium hydrogen phosphate; disodium orthophosphate; secondary sodium orthophosphate.*

disodium pyrophosphate. See sodium acid pyrophosphate.

Disovern. Trade name of Monopol AG (Switzerland) for a series of epoxy primers, epoxy textured finishes and epoxy knifing fillers.

Disperal. Trademark of Condea Visa Company (USA) for alumina supplied in sol and powder forms.

Dispercoll. Trademark of Bayer Corporation (USA) for aqueous adhesives used for bonding automotive components, furniture, shoes, and in the building trades.

dispergator. See emulsifier (1).

Dispersalloy. Trademark of Dentsply/Caulk (USA) for a corrosion-resistant dental amalgam mixture composed of silver-copper eutectic spherical and lathe-cut particles. Supplied in capsule, powder and tablet form for restorative work.

dispersed-phase alloys. A group of plastically deformable two- or multi-phase alloys that have been hardened and strengthened by the dispersion of a second phase of particles (e.g., metallic or nonmetallic oxides) whereby the matrix phase completely surrounds the latter. See also dispersion-strengthened metals; oxide-dispersion-strengthened alloys.

dispersed zirconia ceramics. See transformation-toughened zirconia.

disperser. See emulsifier (1).

dispersing agent. See emulsifier (1).

dispersion coatings. Water-borne coatings consisting of very small, finely divided particles of resin, less than 0.1 μm (2.5 mils) in diameter, suspended or dispersed in an aqueous vehicle. Upon evaporation of the aqueous vehicle the coatings form continuous adherent layers on the substrate. They have high gloss and toughness, and good to moderate weathering properties.

dispersion-hardened aluminum alloys. See sintered aluminum powder alloys.

Dispersionsfarbe. Trade name of Monopol AG (Switzerland) for a series of emulsion paints for interior and/or exterior applications.

dispersion-spun fibers. Synthetic fibers made from rather insoluble and infusible polymers, such as *Teflon*, by first dispersing as a polymer powder into a solution of sodium alginate or sodium xanthate, extruding into fibers, and then heating to induce polymer coalescence.

dispersion-strengthened composites. A group of metal-matrix composites strengthened and hardened by the uniform dispersion of several volume percent of very fine metallic or nonmetallic particles of a very hard and inert material. Examples include thoria-dispersed nickel (containing 3 vol% of thoria, ThO_2), and sintered aluminum powder alloys.

dispersion-strengthened copper. Copper that contains a fine dispersion of a refractory oxide, usually alumina (Al_2O_3). A typical composition is 98.0-99.8 wt% Cu and 0.2-2.0 wt% Al_2O_3. It is usually made by internal oxidation of a copper-aluminum alloy powder, and has excellent elevated-temperature stability, and good electrical conductivity. Used as leads for incandescent lamps, and for relay blades, welding electrode caps, and fuel pumps. Abbreviations: DSC; DS copper.

dispersion-strengthened lead. A fine-grained, corrosion-resistant lead containing a fine dispersion of lead monoxide (PbO). A typical compositions 98.5-99.9 wt% Pb and 0.1-1.5 wt% PbO. It has good strength and stiffness, and is used for chemical construction, chemical piping and fittings, storage batteries, counterweights, roofing, and gutters. Abbreviation: DS lead.

dispersion-strengthened materials. Powder-metallurgy materials composed of metallic matrices and one or more finely dispersed metallic and/or nonmetallic phases, such as alumina, magnesia, silica, thoria, yttria, zirconia, etc., that are virtually insoluble in the matrix. Dispersion strengthening increases the high-temperature strength of metallic matrices. Abbreviation: DS material.

dispersion-strengthened metals. Aggregate or particulate composites consisting of metal matrices and small concentrations (less than 15 vol%) of one or more finely dispersed small-diameter (typically 0.01-0.1 μm, or 0.0004-0.004 mil) metallic and/or nonmetallic oxides. The particles serve to strengthen the metal by blocking dislocation motion. Examples include dispersions of alumina in aluminum, copper or iron, thoria in nickel, lead monoxide in lead, etc. See also dispersion-strengthened copper; dispersion-strengthened lead; dispersion-strengthened nickel; dispersion-strengthened silver; dispersion-strength-

ened tungsten.

dispersion-strengthened nickel. Nickel containing a fine dispersion of thoria (ThO_2.) A typical composition is 98 wt% Ni and 2 wt% ThO_2. It is commercially available in the form of a powder for sintering, and in sheets or bars. It has an as-sintered density of 8.9 g/cm^3 (0.32 lb/in.^{3}), and good high-temperature properties. Used for high-temperature structural applications. Abbreviations: DSN, DS nickel.

dispersion-strengthened silver. Silver containing a uniform dispersion of nanometer-sized aluminum oxide (Al_2O_3) particles. Developed by OMG Americas Metal Powders Division, Research Triangle Park, North Carolina, USA, it is produced by a special powder metallurgy technique involving an internal oxidation process, and supplied in powder, strip, bar and wire form in two grades–OMAg5 with 0.5 wt% Al_2O_3 and OMAg10 with 1.0% Al_2O_3. It can be processed by various techniques including hot extrusion and press-sinter repressing, and can also be combined with various metal oxides (e.g., cadmium, tin, zinc, and cobaltocobaltic oxides) to produce high-strength and high-conductivity products. Used for electrical contacts, etc. Abbreviation: DS silver.

dispersion-strengthened tungsten. Tungsten containing fine dispersions of thoria (ThO_2) or hafnium carbide (HfC) and, sometimes, additions of rhenium. It has exceptionally high strength at temperatures above 2000°C (3630°F), and can be produced by powder-metallurgy methods. Used for filaments, high-temperature structural applications, aerospace applications and as reinforcement for nickel-base superalloys. Abbreviation: DS tungsten.

dispersion-toughened alumina. See transformation-toughened alumina.

dispersion-toughened zirconia. See transformation-toughened zirconia.

Dispersite. Trademark of Uniroyal Chemical Division of Uniroyal Inc. (USA) for water dispersions of crude or reclaimed natural and synthetic rubbers and resins. Applied by spraying, saturation, spreading or impregnation, they have high softness and tackiness, and are used in adhesives for paper, leather and textiles, in coatings for metal, paper, textiles, felt, jute, carpets, book covers, tape, and for dipping tire cords.

Disque. Trade name of Société Nouvelle des Acieries de Pompey (France) for a series of oil-hardening, shock-resistant tool steels containing 0.45-0.55% carbon, 0.7% manganese, 1.75% silicon, and the balance iron. Used for agricultural equipment and plows.

dissolving pulp. A term referring to a wood tissue-derived cellulose of high purity.

Dissolvo. Trademark of CMS Gilbreth Packaging Systems, Inc. (USA) for water-soluble paper and tape made from cellulose polymer, and available in sheet and roll form. Used for purge dams employed in the gas-tungsten-arc welding of stainless steel tubes and pipes for the aerospace, oil and gas, offshore-drilling, nuclear and paper-mill industries, and for packaging, printing and labeling application.

Distaloy. Trade name of Hoeganaes Corporation (USA) for a series of diffusion-alloyed powder-metallurgy low-alloy steels containing additions of molybdenum, nickel and copper, and having good hardenability characteristics. The starting powders are also referred to as *Distaloy*.

distemper. A water paint that is an emulsion of pigments in casein, egg yolk, egg white, or glue.

disthene. See kyanite.

Di-strate. Trade name of Aptek Laboratories (USA) for composite dielectrics.

Distrene. Trade name of British Resin Products (UK) for polystyrene plastics.

distrontium silicate. See distrontium silicate (ii).

Ditac. Trademark of E.I. DuPont de Nemours & Company (USA) for a series of engineering adhesives for integrated circuit applications.

ditantalum boride. See tantalum boride (iii).

ditantalum carbide. See tantalum carbide (ii).

ditantalum nitride. See tantalum nitride (ii).

ditantalum silicide. See tantalum silicide (ii).

diterbium carbide. See terbium carbide (ii).

ditetradecyl ether. See dimyristyl ether.

dititanium nickelide. See titanium nickelide (ii).

dittmarite. A colorless mineral composed of ammonium magnesium phosphate monohydrate, $NH_4MgPO_4 \cdot H_2O$ Crystal system, orthorhombic. Density, 2.15 g/cm^3; refractive index, 1.569. Occurrence: Australia.

ditungsten boride. See tungsten boride (ii).

ditungsten carbide. See tungsten carbide (ii).

ditungsten nitride. See tungsten nitride (iii).

Ditzler. Trademark of PPG Industries Inc. (USA) for synthetic resin coatings for automotive refinishing jobs.

divanadium carbide. See vanadium carbide (ii).

divanadium nitride. See vanadium nitride (ii).

Divco. Trade name of Division Lead Company (USA) for a series of lead-tin solders for general applications.

Diverfos. Trade name of Novamax Technologies (US) Inc. for a zinc phosphate coating and coating process.

Divinylcell. Trade name of Diab Group (USA) for lightweight structural polyvinyl chloride foam core materials.

dixenite. (1) A deep red brown mineral composed of copper iron manganese arsenite arsenate silicate hydroxide, $CuMn_{14}Fe(AsO_3)_5(SiO_4)_2(AsO_4)(OH)_6$. Crystal system, rhombohedral (hexagonal). Density, 4.37 g/cm^3; refractive index, 1.651. Occurrence: Sweden.

(2) A black mineral composed of manganese arsenate silicate hydroxide, $Mn_{15}(AsO_3)_6(SiO_4)_2(OH)_8$. Crystal system, hexagonal. Density, 4.20 g/cm^3; refractive index, 1.96. Occurrence: Sweden.

Dixie. Trademark of R.T. Vanderbilt Company (USA) for a fine high-quality *kaolin* clay used as a rubber extender and reinforcement.

Dixiolbronze. Trade name of Seitzinger's Inc. (USA) for a corrosion-resistant bronze containing 88% copper, 10% tin and 2% zinc. Used for bearings, bushings, liners, and gears.

Dixon. (1) Trademark of Albion Industrial Products (Canada) for graphite lubricants.

(2) Trade name for a series of water-hardening plain-carbon tool steels containing about 0.7-1.4% carbon, and the balance iron.

diyttrium aluminate. See yttrium aluminate (ii).

diyttrium pentaaluminate. See yttrium aluminate (iii).

dizinc titanate. See zinc titanate (ii).

dizirconium germanide. See zirconium germanide (ii).

dizirconium silicide. See zirconium silicide (iii).

djerfisherite. (1) An olive-green mineral composed of potassium copper iron sulfide, $K_3(Cu,Na)(Fe,Ni)_{12}S_{14}$. Crystal system, cubic. Density, 3.90 g/cm^3. Occurrence: India.

(2) An olive-brown mineral composed of potassium copper iron nickel chloride sulfide, $K_6(Cu,Fe,Ni)_{25}S_{26}Cl$. Crystal system, cubic. Density, 3.90 g/cm^3. Occurrence: Russian Federation.

DJ Metal. Trade name for a corrosion-resistant nickel-chromium dental bonding alloy.

djurleite. A dark lead-gray to bluish-black mineral composed of copper sulfide, $Cu_{1.93}S$. Crystal system, monoclinic. Density, 5.63 g/cm^3. Occurrence: Japan, USA (Arizona). An orthorhombic form, $Cu_{31}S_{16}$, can be made synthetically from high-purity copper and sulfur.

DLP copper. See deoxidized low-residual-phosphorus copper.

D Metal. Trade name of Edgar Allen & Company Limited (UK) for steel castings composed of 0.45% carbon, 0.3% silicon, 0.85% manganese, and the balance iron. Used for general engineering applications.

D-Monel. Trade name of Inco Alloys International Inc. (USA) for a series of cast nickel-copper alloys containing 60-66% nickel, 27-31% copper, 2.5% iron, 3.5-4.5% silicon, 1.5% manganese, and 0.25% carbon. It has good corrosion resistance, exceptional galling resistance, good strength, and is not weldable. See also Monel.

D-Nickel. Trade name of Inco Alloys International Inc. (USA) for a medium-strength nickel alloy containing about 4-5% manganese and 0-0.1% carbon. It has medium hardness and excellent resistance to corrosion, heat and spark-erosion, and is used for spark-plug electrodes, grid wires, electronic tubes, and electrical and electronic applications.

Do Alloy. Trade name of DoAll Company (USA) for a cast alloy of cobalt, chromium and tungsten with high red hardness and shock resistance, used for general-purpose machining applications.

dobby. Fabrics with small, repetitive woven designs of checks, dots, lozenges or squares. Used especially for shirts and dresses. Also known as *dobby-weave fabrics*.

dobby-weave fabrics. See dobby.

Dobeckan. Trademark of BASF Farben + Fasern AG, Beck Elektroisolier-Systeme (Germany) for unsaturated polyester resins.

Dobeckot. Trademark of BASF Farben + Fasern AG, Beck Elektroisolier-Systeme (Germany) for epoxy resins.

Doblin alloy. A German corrosion-resistant cobalt-silicon alloy used for chemical equipment.

DockGARD. Trademark of Vintex Corporation (Canada) for high-strength, tear-resistant composite fabrics consisting of textile substrates with vinyl extrusion coating. They have good resistance to low temperatures, peeling, chipping and delamination, and are used to protect docks.

Doco. Trade name of Saarstahl AG (Germany) for a cold-work tool steel containing 1.75% carbon, 2.5% cobalt, 12.5% chromium, 0.9% molybdenum, 0.25%, and the balance iron. Used for heavy-duty cutting and stamping dynamo and transformer sheets.

Doctor metal. A brass containing 88% copper, 9.5% zinc and 2.5% tin. Used for corrosion-resisting parts.

Document. Trade name for an ultraviolet-absorbing glass used for the protection of documents and valuable papers against discoloration and fading.

document glass. A glass with high ultraviolet-absorbing properties used as a cover to protect documents and valuable papers against deterioration from strong light.

dodecacarbonyliron. See triiron dodecacarbonyl.

dodecenylsuccinic anhydride. A branched chain compound available in technical grades (90% pure) as a mixture of isomers in the form of a pale yellow, moisture-sensitive, viscous liquid with a density of 1.005 g/cm³, a boiling point of 150°C (302°F)/ 3.0 mm, a flash point above 230°F (110°C), and a refractive index of 1.479. A high-purity (95%) grade with a melting point of 41-43°C (106-109°F), a boiling point of 180-182°C (356-360°F)/5.0 mm and a flash point of 352°F (177°C) is also available. Used as an epoxy hardener, in alkyds and other synthetic resins, as a plasticizer, as a wetting agent for bituminous compounds, and as an anticorrosive. Formula: $C_{16}H_{26}O_3$. Abbreviation: DDSA.

Dodge. Trade name of Dodge Foundry & Machine Company (USA) for a series of plain-carbon, cold- and hot-work tool steel castings, heat-resistant steel castings, and various alloy steel castings (e.g., molybdenum, chromium-molybdenum and nickel-chromium-molybdenum types).

Do-Di. Trade name of NL Industries (USA) for a free-cutting brass containing 60% copper, 0.65% tin, 0.82% lead, 0.35% iron, 0.53% aluminum, and the balance zinc. Used for hardware and marine parts.

Doe Run Copperized Lead. Trade name for a lead containing 0.05% copper, 0.0006% silver and 0.0003% cadmium. It has a density of 11.3 g/cm³ (0.41 lb/in.³), a melting point of 326°C (620°F), and is used for chemical equipment, and cable sheathing.

Doe Run Lead. Trade name for a lead containing 0.0005% silver and 0.0003% cadmium. It has a density of 11.3 g/cm³ (0.41 lb/in.³), a melting point of 327°C (620°F), and is used as a battery oxide, and in solders, chemical pigments and antiknock compounds.

doeskin. (1) A very soft leather made from the skin of female deer.

(2) A soft, white leather obtained by tanning the skin of sheep or lambs with aluminum sulfate and/or formaldehyde. Used especially for gloves.

(3) A smooth, woolen fabric with a soft, short nap and a kidglove handle, used for suits, dress uniforms, sportswear, coats, etc. Also known as *doeskin fabric*.

Dofasco. (1) Trade name of Dofasco Inc. (Canada) for a steel containing 0.3% carbon, 2-3% nickel, 0.9-1.1% chromium, 0.3-0.4% molybdenum, and the balance iron. It has good high temperature properties up to 593°C (1100°F), and is used for valves, fittings, and oil crusher roll shells.

(2) General trademark identifying products supplied by Dofasco Inc. (Canada) and including carbon and alloy steel plate, cold-rolled flat steel, flat-rolled steel coils and sheets either coated with chromium, zinc-aluminum alloy or tin or galvanized, galvannealed, painted or surface-textured, hot- and cold-rolled steel tubing, hot-rolled flat steel coils and strips, vitreous enameling steel, cast pig iron, and various metallurgical products, blast-furnace additives and byproducts, (e.g., iron oxides, cokes, slags and dross).

Dofascoloy. Trade name of Dofasco Inc. (Canada) for a series of high-strength low-alloy structural steels with excellent resistance to atmospheric corrosion and good formability and weldability. Used for structural members of bridges, buildings, cranes, derricks, mining equipment, railway and agricultural equipment, automobile components, etc.

Dofasco Pre-Coat. Trade name of Dofasco Inc. (Canada) for prepainted cold-rolled steel sheet and plate.

dogskin. A strong leather made from sheepskin.

dogwood. The hard, heavy wood of any of a genus (*Cornus*) of trees and shrubs found chiefly in the United States. The most important species are: roughleaf dogwood (*C. drummondii*), flowering dogwood (*C. florida*), stiff dogwood (*C. stricta*), alternate-leaf dogwood (*C. alternifolia*) and Pacific dogwood (*C. nuttalli*). *Dogwood* grows also in Southern Europe and China. It has a close grain and an uniform texture, and is used for weaving shuttles, bobbins, pulleys, jewelers' blocks, skate rollers, tool handles, mallets, and golf-club heads.

Doha sand. A type of sand that occurs in the vicinity of Doha, Qatar and contains predominantly gypsum ($CaSO_4 \cdot 2H_2O$) with considerable quantities of dolomite ($CaMg(CO_3)_2$) and small amounts of calcite ($CaCO_3$) and quartz (α-SiO_2).

Dohlen. Trade name of Dohlen-Stahl Gusstahl-Handels GmbH (Germany) for a series of machinery, tool and die steels.

Do-It. Trade name of Ziv Steel & Wire Company (USA) for an oil-hardening tool steel containing 0.7% carbon, 0.8% manganese, 2% silicon, 0.2% molybdenum, and the balance iron. Used for shear blades, punches, and chisels.

Dolan. Trademark of Hoechst AG and Acordis Kelheim GmbH (Germany) for an acrylic (polyacrylonitrile) fiber used for chemical protection clothing, outdoor fabrics, e.g., awnings and boat covers, and car soft tops.

Dolanit. Trademark of Acordis Kelheim GmbH (Germany) for a high-tenacity acrylic fiber used in fiber cement, for filtration applications and for frictional materials.

Doler. Trade name of NL Industries (USA) for a series of aluminum-, copper-, magnesium- and zinc-base casting and die-casting alloys.

dolerophanite. A brown mineral composed of copper oxide sulfate, $Cu_2O(SO_4)$. Crystal system, monoclinic. Density, 4.17 g/cm³; refractive index, 1.820. Occurrence: Italy.

Dolfil. Trademark of Canada Talc Industries Limited for dolomite products used as fillers.

dolime. See calcined dolomite.

Dollar Blue Chip. British trade name for a high-speed tool steel containing 0.62% carbon, 4.3% chromium, 17.8% tungsten, 1.7% vanadium, 3.9% cobalt, and the balance iron. Used for cutters, reamers, and tools.

doloma. See calcined dolomite.

dolomite. A colorless or white mineral of the calcite group composed of calcium magnesium carbonate, $CaMg(CO_3)_2$, and containing 54.3 wt% calcium carbonate ($CaCO_3$) and 45.7 wt% magnesium carbonate ($MgCO_3$). Small amounts of iron may replace some of the magnesium. Crystal system, rhombohedral (hexagonal). Density, 2.85-2.99 g/cm³; hardness, 3.5-4.0 Mohs; refractive index, 1.680. Occurrence: Austria, Canada (Ontario), Germany, Italy, UK, USA (Midwestern States). Used for building and ornamental stone, in the manufacture of certain cements, refractories, glass, tile and pottery bodies, as a fluxing ingredient in glazes, in the manufacture of magnesium compounds and magnesium metals, in papermaking, and for mineral wool. See also calcined dolomite; dead-burnt dolomite; double-burnt dolomite; pearl spar; raw refractory dolomite.

dolomite brick. A refractory brick that is manufactured essentially from *dead-burnt dolomite* and used for furnace linings.

dolomite limestone. See dolomitic limestone.

dolomite-magnesite brick. A refractory brick composed of a mixture of *dead-burnt dolomite* and *dead-burnt magnesite*, but in which the dolomite content exceeds that of the magnesite.

dolomite marble. See dolomitic marble.

dolomitic lime. A lime that contains 50-70% calcium oxide (CaO) and 30-50% magnesium oxide (MgO).

dolomitic lime putty. A product of varying consistency made by slaking *dolomitic lime* with water. Used on masonry walls, etc.

dolomitic limestone. A limestone composed of calcium carbonate ($CaCO_3$) and magnesium carbonate ($MgCO_3$), with the $MgCO_3$ content ranging between 35 and 46%. Used as a refractory or component in refractories, as an ingredient in cement, as a source of calcium in certain glazes, as a filler and extender in rubber, latex, plastics, adhesives, paints, fiberglass, asphalt roofing, etc., and for chemical fluxing applications. Also known as *dolomite limestone; dolostone.*

dolomitic marble. (1) A crystalline variety of limestone containing 60% or less calcium carbonate ($CaCO_3$), and 40% or more magnesium carbonate ($MgCO_3$).

(2) A limestone containing between 60 and 95% calcium carbonate ($CaCO_3$), and 5 and 40% magnesium carbonate ($MgCO_3$). Used as a concrete aggregate. Also known as *magnesian marble.*

doloresite. A dark brown to nearly black mineral composed of hydrous vanadium oxide, $H_8V_6O_{16}$. Crystal system, monoclinic. Density, 3.44 g/cm^3; refractive index, 1.90. Occurrence: USA (Colorado, Arizona).

dolostone. See dolomitic limestone.

Dolphin. Trade name of Vereinigte Glaswerke (Germany) for patterned glass with fish design.

Dom. Trade name of Vallourec SA (France) for steel containing 0.37% carbon, 0.95% chromium, 0.2% molybdenum, and the balance iron. It has high elongation, and is used for mechanical and tool joints.

Domal. (1) Trade name of Timminco Limited (Canada) for an extensive series of wrought and cast magnesium products including various magnesium-aluminum, magnesium-aluminum-zinc and magnesium-manganese alloys. Used for high-strength castings, housings, tanks, cylinder heads, pistons, valve and pump bodies, compressor parts, aircraft parts including engines, casings and structures.

(2) Trade name of Timminco Limited (Canada) for calcium metal (98.5+% pure) containing about 0-0.4% aluminum and 0-1% magnesium. Used as an alloying addition and grain refiner for aluminum and magnesium alloys.

dome brick. A brick of tapered or curved shape suitable for use in the construction of a dome.

Domestic. Trade name of Edgar T. Ward's Sons Company (USA) for a water-hardening tool steel containing 0.7-1.2% carbon, and the balance iron. Used for punches, taps, and tools.

domeykite. A tin-white to steel-gray mineral composed of copper arsenide, Cu_3As. Crystal system, cubic. Density, 7.20-7.92 g/cm^3. Occurrence: USA (Michigan).

Dominator. Trade name of Boehler GmbH (Austria) for a series of high-carbon, high-chromium tool steels containing 1.5-1.9% carbon, 11.5-12% chromium, 0-0.8% molybdenum, up to 2% tungsten, up to 1% vanadium, and the balance iron. Used for cutting tools, reamers, and dies.

Dominial. Trade name of Kind & Co. Edelstahlwerk (Germany) for an extensive series of steels including various alloy tool steels, wear-resistant manganese steels, heat-resistant steels, and stainless steels as well as several nickel- and cobalt-base superalloys.

Domite. Trade name of Canron, Inc. (Canada) for a series of nickel cast irons containing 3.0-3.5% carbon, 0-1.5% nickel, 1.5-2.5% silicon, and the balance iron. Used for pulleys, gears, stamping dies, pumps, die shoes, evaporators, flywheels, impellers, crankshafts, and forming dies.

Domite Ni-Hard. Trade name of Canron, Inc. (Canada) for a cast iron containing 3.0-3.6% carbon, 0.5% silicon, 4.2-4.7% nickel, 1.4-2.5% chromium, 0.4-0.6% manganese, and the balance iron. Used for abrasion- and wear-resistant castings.

Domite Noduloy. Trade name of Canron, Inc. (Canada) for nodular cast irons containing 3.3% total carbon, 1.9% silicon, and the balance iron. Used for high-strength castings.

Donal. Trade name of Constrictor Limited (UK) for a corrosion-resistant, non-heat-treatable aluminum alloy containing 1-2% manganese and 0.3-0.5% silicon. Used for general structures, containers, heat exchangers, and trim.

donathite. A blackish brown mineral composed of iron magnesium chromium oxide, $(Fe,Mg)(Cr,Fe)_2O_4$. Crystal system, tetragonal. Density, 5.00 g/cm^3. Occurrence: Norway.

Donegal. Trade name of Donegal Steel Foundry Company (USA) for a series of steels including several cast, corrosion- and heat-resistant steels of the iron-chromium-nickel type, several wrought stainless steels, and various case-hardening and shock-resistant steels.

donegal. A white-colored woolen fabric resembling tweed and made with colored filling yarns and brightly colored nubs usually in a herringbone or plain weave. It is now also made from cotton, acrylics, viscose rayon or blends of these fibers. Used for coats, dresses, and suits.

donnayite. A pale yellow to yellow mineral of the mckelveyite group composed of sodium calcium strontium yttrium carbonate trihydrate, $NaCaSr_3Y(CO_3)_6 \cdot 3H_2O$ Crystal system, triclinic. Density, 3.30 g/cm^3; refractive index, 1.646. Occurrence: Canada (Quebec).

dopant. (1) In powder metallurgy, a material introduced in small quantity into a metallic powder to stop or control grain growth or recrystallization.

(2) In powder metallurgy, a substance added in minute amounts to a powder or to the sintering atmosphere to improve the densification rate. Also known as *activator.*

(3) In polymer engineering, a substance introduced into a polymer to change one or more physical properties.

(4) In semiconductor technology, a small, controlled amount of donor or acceptor impurities (foreign atoms) or alloying elements, such as antimony, arsenic, nitrogen, phosphorus, etc., intentionally added to a semiconductor crystal in order to change its electrical and/or physical properties. The resulting semiconductor is called an "extrinsic" or "n-type" semiconductor. Also known as *dope; doping agent.*

dope. (1) A solution of cellulose acetate in acetone used to produce an adhesive or coating.

(2) A term used in the textile industries as a synonym for "spinning solution," i.e., a solution of a fiber-forming polymer suitable for extrusion through a spinneret.

(3) A lubricant, such as graphite, used on glass molds to decrease friction and eliminate sticking during the forming of glass products.

(4) See dopant (4).

(5) See adhesion promoter (2).

doped binder. In concrete manufacture, a cementing material containing an adhesion-promoting agent.

doped cutback. Cutback asphalt containing an adhesion promoter. See also cutback asphalt.

doped semiconductor. A semiconductor to which a small, controlled amount of donor or acceptor impurities (foreign atoms) or alloying elements have been added. See also dopant (4); extrinsic semiconductor.

doped solder. A solder into which small quantities of an element have been purposely introduced to provide it with certain properties or characteristics of the base metal.

dope-dyed fibers. Synthetic fibers that have a dye or pigment added prior to filament formation.

doping agent. See dopant (4).

Dopploy. Trade name of Sowers Manufacturing Company (USA) for a series of corrosion-resistant cast irons containing 3.0% total carbon, 2% silicon, 0.8% manganese, 18.5% nickel, 2.5% chromium, and the balance iron. Used for agitators, and for casting jacketed kettles.

Doquat. Russian trade name for a hard, sintered alloy of iron and tungsten, used for cutting tools.

Dorcasine. Trade name of O. Murray & Company (UK) for casein plastics.

Dorcasite. Trade name of Charles Horner Limited (UK) for cellulose acetate plastics.

Dordent. Trade name of Johnson Matthey plc (UK) for a hardened dental alloy containing 72.5% gold and 27.5% platinum. It has a melting range of 870-920°C (1600-1690°F).

Dorex. Trade name for rayon fibers and yarns used for textile fabrics.

dorfmanite. A white mineral composed of sodium hydrogen phosphate dihydrate, $Na_2HPO_4 \cdot 2H_2O$. Crystal system, orthorhombic. Density, 2.07; refractive index, 1.461.

Doric. (1) Trade name of Libbey-Owens-Ford Company (USA) for figured glass.

(2) Trade name for a dental porcelain for metal-to-ceramic bonding applications.

Doric Acrylic. Trade name for a heat-cure denture acrylic.

Doric Bonding Gold. Trade name for a high-gold dental bonding alloy.

Doric Dy-Rok. Trade name for a hard die stone used in dentistry.

Doric GII. Trade name for a silver-free palladium dental bonding alloy.

Doric HTZ. Trade name for a zirconia ceramic for dental die work.

Doric MDS. Trade name for a silicone elastomer for dental laboratory applications.

Doric M-K. Trade name for a dental porcelain for metal-to-ceramic bonding applications.

Doric Non-Precious. Trade name for a corrosion-resistant beryllium-free nickel-chromium dental bonding alloy.

Doric Porcelain. Trade name for a dental porcelain for jacket crowns.

Doric Silicone. Trade name for a silicone elastomer used for dental impressions.

Doric SP. Trade name for a low-gold dental casting alloy.

Dorium. Trade name for a series of copper-zinc alloys (brasses) used for condenser tubes.

Dorix. Trade name of Bayer Antwerpen BV (Belgium) for nylon 6 fibers.

Dorlan. Trade name for acrylic fibers and yarns used for textile fabrics.

Dorlastan. Trademark of Bayer Corporation (USA) for spandex (segmented polyurethane) fibers and yarns used for elastic outerwear, hosiery, lingerie and swimwear.

Dorlon. Trade name for nylon fibers and yarns used for textile fabrics.

Dörrenberg. Trade name of Dörrenberg Edelstahl GmbH (Germany) for an extensive series of austenitic, ferritic and martensitic stainless steels, heat-resistant steels, alloy steels (including case-hardening and heat-treatable grades), and several plain-

carbon, hot-work and cold-work tool steels.

Dorrlmate. Trade name of Magni Group Inc. (USA) for a black zinc-rich coating.

Dorrltech. Trade name of Magni Group Inc. (USA) for an aluminum-based organic coating system.

dotted swiss. A fine, sheer, crisp fabric, usually of cotton or a cotton blend, having a pattern of woven, printed or flocked dots. Used especially for dresses, children's wear and curtains. See also swiss.

doublé. A French *duplex metal* composed of copper or brass with a thin cladding of a noble metal, such as gold, silver or platinum. Used for costume jewelry.

Double-Bond. Trade name of Calgon Corporation (USA) for iron phosphate coatings.

double-braided rope. A fiber rope that consists of two components–a core formed of several braided fiber strands and an outer cover or sheath also made of several braided fiber strands. The core lies coaxially within the outer cover. *Double-braided rope* has higher strength, and lower tendency to elongate when under tension, or rotate and unlay when under load than ordinary twisted rope. Used for sash cord, clotheslines, and medium-duty towing and lifting applications. See also braided rope; rope materials.

double brick. A brick that is twice as high as a standard brick, i.e., 135 × 102 × 203 mm (5.33 × 4.00 × 8.00 in.) as compared to only 68 × 102 × 203 mm (2.66 × 4.00 × 8.00 in.).

double bronze. A wire containing copper and aluminum. Its skin is made from pure copper and its core from a copper-aluminum alloy. Used for electrical equipment.

double-burnt dolomite. A dolomite that contains considerable amounts of iron oxide and is burnt at very high temperatures. Also known as *clinkered dolomite*.

Double Chrome Vanadium. Trade name of Allegheny Ludlum Steel (USA) for a tool and die steel containing 0.5% carbon, 1.5% chromium, 0.2% vanadium, and the balance iron. Used for tools, drills and dies.

double cloth. A compound woven fabric in which the two component fabrics are secured in position either with stitches or by a weaving technique that allows them to more or less interchange with each other.

Double Conqueror. Trade name of Joseph Beardshaw & Son Limited (UK) for a water-hardening, wear-resistant tool steel containing 1.1-1.4% carbon, and the balance iron. Used for roll-turning tools, rock drills, and punches.

Double Diamond. Trade name of Crucible Materials Corporation (USA) for a tool steel containing 0.9% carbon, 0.95% chromium, 0.3% molybdenum, and the balance iron. Used for drills.

double-duty compound. A grease-bonded mixture of *tripolite* and amorphous silica powder used for combined fast cutdown and color buffing applications. Also known as *cut-and-color compound*.

doubled yarn. See plied yarn.

Double Echo Cobalt. Trade name of British Steel Corporation (UK) for a cobalt-tungsten high-speed tool steel containing 0.83% carbon, 4.4% chromium, 0.5% molybdenum, 18% tungsten, 1.6% vanadium, 10% cobalt, and the balance iron. Used for cutting tools.

Double Extra. Trade name of Midvale-Heppenstall Company (USA) for an oil-hardening tool steel containing 0.8% carbon, 0.65% chromium, and the balance iron. Used for dies, tools, and mandrels.

double-faced fabrics. Textile fabrics whose two sides are of equally good surface appearance, texture, finish, etc., and thus either side can be used as the face side.

double-face ware. Porcelain-enameled ware having a finish coat on both surfaces of the metal base.

double-fagoted iron. A wrought-iron bar made by repeated heating of a *fagot* to welding heat, followed by rolling down to a solid bar.

Double-Flex. Trademark of Hexcel Corporation (USA) for aluminum honeycomb core that is supplied as corrosion-resistant alloy foil and allows the formation of sharp curves without compromising mechanical properties or structural integrity.

Double Flygo. Trade name of Turton Brothers & Mathews Limited (UK) for a tungsten-type high speed tool steel containing 0.73% carbon, 4% chromium, 18% tungsten, 1% vanadium, and the balance iron. Used for cutting tools.

double-frit glaze. A glaze made from frits of two different compositions, usually to complement and enhance the chemical and physical properties and extend the firing range.

Double Geant. Trade name of Société des Acieries de Longwy (France) for a high-speed steel containing 0.7% carbon, 18% tungsten, 9% cobalt, 4% chromium, 2% vanadium, 1% molybdenum, and the balance iron. Used for tools, dies and cutters.

double glazing. (1) Glazing that incorporates two or three layers of glass separated by spacers and argon gas or dehydrated air. Used in energy-efficient windows to reduce heat losses and drafts. See also low-E glass.

(2) A glaze composed of two coats applied one over the other.

Double Griffin. Trade name of Darwins Alloy Castings (UK) for a shock-resisting steel containing 0.35% carbon, 0.3% manganese, 1.7% chromium, 3.5% nickel, 0.2% silicon, and the balance iron. Used for chisels, punches, shear blades, wedges, and snaps.

double-hydrated lime. See dihydrated lime.

double jersey. A firm, heavy two-layer jersey fabric, usually acrylic or wool, and often textured, plain-colored and/or reversibly knitted. It has two identical sides and is usually made by using a double set of needles. Used especially for jackets, dresses, pants and skirts. Also known as *double knit.*

double knit. See double jersey.

double metal alkoxides. Complexes of metal alkoxides having two metal atoms within each molecular species, e.g., magnesium aluminum isopropoxide, $MgAl_2(OC_3H_7)_8$ and sodium tin ethoxide, $NaSn_2(OC_2H_5)_9$. The general formula is $M'M''(OR)_x$, where M' and M'' represent two different metals. Abbreviation: DMA.

Double Musket. Trade name of Osborn Steels Limited (UK) for a high speed steel containing 0.6-0.8% carbon, 18% tungsten, 4% chromium, 1% vanadium, and the balance iron. Used for tools and cutters.

Double Niagara. Trade name of Perma Paving Stone Company (Canada) for paving stones, 198 × 198 × 60 mm (7.9 × 7.9 × 2.4 in.) in size, supplied in several colors including natural, red, brown and charcoal, and various color blends. Used for driveways, walkways, patios, etc.

double nickel salt. See ammonium-nickel sulfate.

Double-O-Two. Trademark of Lukens Steel Company (USA) for a plate steel that contains no more than 0.002% sulfur, and is used for structural applications.

Double Rapid. Trade name of George Cook & Company, Limited (UK) for a high-speed steel containing 0.8% carbon, 4.5% chromium, 18.5% tungsten, 5.5% cobalt, 0.3% molybdenum,

1.2% vanadium, and the balance iron. Used for hobs, drills, reamers, and lathe and planer tools.

double-refracting spar. A pure, transparent colorless variety of the mineral *calcite* found in Iceland. It is readily cleaved along its crystal faces in rhombohedrons, each face being a parallelogram with angles of 78.5° and 101.5°. It exhibits strong birefringence (double refraction), and is used for light polarizers (e.g., Nicol prisms) and other optical instruments. Also known as *double-refraction calcspar.* See also Iceland spar.

double-refraction calcspar. See double-refracting spar.

Double Satan. Trade name of Creusot-Loire (France) for oil-hardening hot-work steel containing 0.55% carbon, 3.5% chromium, 5% tungsten, and the balance iron. Used for hot work tools and dies.

double-screened ground refractory material. A refractory material that has been graded only once, but has been screened twice to remove particles that are either coarser or finer than a specified size range.

Double Seven. Trade name of Edgar Allen Balfour Limited (UK) for a series of nondeforming tool steels containing 1.2-1.3% carbon, 0.3% manganese, 12.5-13.5% chromium, 2.5-4.5% cobalt, 0-1.5% molybdenum, 0-1% nickel, and the balance iron. Used for tools, broaches, shear blades, punches, cold forging dies, blanking dies, and lamination dies.

double-side-prepared self-copying paper. A self-copying paper that exhibits a chemical or physical color reaction. It is composed of two sheets having different layers on the front and reverse surface. An image appears when pressure is applied by means of writing utensils, typewriters, printers, etc.

Double Six. (1) Trade name of Edgar Allen Balfour Limited (UK) for a nonshrinking, air- or oil-hardening tool steel containing 1.9% carbon, 12.5% chromium, 0.8% molybdenum, 0.25% vanadium, and the balance iron. Used for dies, punches, press tools, and gauges.

(2) Trade name of A. Milne & Company (USA) for a nondeforming tool steel containing 2.25% carbon, 14% chromium, and the balance iron. Used for gauges, and blanking and drawing dies.

Double Six Super. Trade name of Edgar Allen Balfour Limited (UK) for a nonshrinking, air-hardening tool steel containing 1.9% carbon, 12.5% chromium, 0.8% molybdenum, 0.25% vanadium, and the balance iron. Used for dies, press tools, gauges, press tools, reamers, broaches, etc.

Double Special. Trade name of Sanderson Kayser Limited (UK) for a water-hardening tool steel containing 1.4% carbon, 0.8% chromium, 5% tungsten, and the balance iron. Used for turning tools, and cutters.

double-standard brick. A brick, usually of the refractory type, whose width is twice that of a standard square brick. Used in building construction.

double-strength glass. A clear sheet glass that has a strength (i.e., nominal thickness) of 3-6 mm (0.125-0.250 in.) and smooth, truly parallel surfaces produced e.g., by floating a ribbon of glass on molten tin. Used for window panes. Abbreviation: DS glass.

Double Super Express. Trade name of Leadbeater & Scott Limited (UK) for a high speed steel containing 0.7% carbon, 5% chromium, 22% tungsten, 1% vanadium, and the balance iron. Used for cutters for rubber and asbestos.

Double Super Hydra. Trade name of Osborn Steels Limited (UK) for a high-speed steel containing 0.8% carbon, 4.5% chromium, 18% tungsten, 1.75% vanadium, 5% cobalt, and the balance iron. Used for lathe tools, reamers, broaches, and drills.

Double Twelve. Trade name of Edgar Allen Balfour Limited (UK) for an oil-hardening tool steel containing 0.32-0.35% carbon, 12% chromium, 12% tungsten, 1% vanadium, and the balance iron. Used for brass die-casting molds, extrusion dies, mold inserts for the plastic industry, and master hobs for beryllium copper.

Double Two. Trade name of Multicore (USA) for a lead solder containing 2% tin and 2% antimony. It has a melting point of 305°C (580°F), and a fusion point of 315°C (600°F). Used for high-temperature joining applications.

double-weave fabrics. Woven fabrics made with two sets of warp and weft (filling) yarns, e.g. double-weave satin.

double-weight hydrogen. See deuterium.

Doublex. Trade name of ASG Industries Inc. (USA) for a glass with a 279 mm (11 in.) chequered pattern with long, rounded grooves on one surface running at right angles to wavelike ridges on the other.

Double You Chrome. Trade name of Wardlows Limited (UK) for a nonshrinking tool steel containing 2.25% carbon, 13% chromium, and the balance iron. Used for dies and tools.

Double You Die. Trade name of Wardlows Limited (UK) for an air-hardening tool steel containing 0.9% carbon, 3.75% chromium, and the balance iron. Used for tools and dies.

Double You Hot Stuff. Trade name of Wardlows Limited (UK) for a hot-work tool steel containing 0.3-0.5% carbon, 2.5% chromium, 9.5% tungsten, 0.1% vanadium, and the balance iron. Used for hot dies and tools.

Double Zebra. Trade name of Sanderson Kayser Limited (UK) for a cold-work tool steel containing 0.4% carbon, 0.3% chromium, 3.5% nickel, and the balance iron. Used for chisels, drifts, etc.

doubly refracting material. See birefringent material (2).

Douglas fir. The hard, strong, pale to medium red-brown wood of the large softwood tree *Pseudotsuga menziesii* belonging to the pine family. It has a straight, close grain, and is moderately resinous with heavy contrast between springwood and summerwood. It dries quickly, splinters easily, machines and sands poorly with hand tools, and works easily with power tools. Average weight, 530 kg/m³ (33 lb/ft³). Source: Western North America from Rocky Mountains to Pacific Coast, and from Mexico to central British Columbia. Used for lumber, wall and roof framing, general house building, heavy construction work, interior fittings, joinery, piling and plywood, railroad ties, cooperage, poles, mine timbers, and fencing. It is seldom used for finish, but the tree is popular as a Christmas tree. Also known as *Columbian pine; Douglas spruce; Oregon pine; yellow fir.* See also red fir (2).

Douglas spruce. See Douglas fir.

doupion. See dupion.

doupioni. See dupion fabrics.

Doux. Trade name of Vallourec SA (France) for a low-carbon steel containing 0.2% carbon, 0.25% silicon, 0.8% manganese, and the balance iron. Used for oil refinery tubes.

Dover. Trade name of Time Steel Service Inc. (USA) for a water-hardening tool steel (AISI type W1) containing 1% carbon, 0.25% manganese, 0.25% silicon, and the balance iron. Used for tools, dies, mandrels, etc.

Doverite. Trade name of Dover Limited (UK) for cellulose acetate plastics.

Dow ABS. Trade name of Dow Plastics (USA) for a series of acrylonitrile-butadiene-styrene resins supplied in various grades

including transparent, low-gloss, high- and medium-impact, glass fiber-reinforced, fire-retardant, high-heat, UV-stabilized, plating and structural foam.

Dow Corning. Trademark of Dow Corning Corporation (USA) for an extensive series of emulsions, greases, heat-resistant coatings, insulating varnishes, laminating polymers, lubricants, and mold-release agents based on silicone or polysiloxane polymers.

Dow CPE. Trade name of Dow Plastics (USA) for chlorinated polyethylene elastomers with good elevated temperature properties and good resistance to hydrocarbon fluids, oils and fuels.

Dow EP. Trade name of Dow Plastics (USA) for a series of epoxy resins supplied in various grades including general-purpose, flexible and high-heat. They are also available as aluminum-, glass-, mineral-, or silica-filled casting resins.

Dowex. Trademark of Dow Chemical Company (USA) for an extensive series of anionic and cationic ion-exchange resins made from styrene-divinylbenzene copolymers. Used in water softening, and as catalysts.

Dow LDPE. Trade name of Dow Plastics (USA) for low-density polyethylenes.

Dowlex. Trademark of Dow Plastics (USA) for polymers and copolymers based on olefinic substances in particular hybrids made from linear-low-, low- and high-density polyethylenes. Used for molded and extruded parts, and for plastic films.

Dowlex LLDPE. Trademark of Dow Plastics (USA) for linear-low density polyethylenes with excellent toughness and puncture resistance, and good processibility and tear resistance. They are supplied in various grades including injection-molding, blown-film, cast-film, extrusion, coating/laminating, and roto-molding. Used for wrapping pallet loads, for food and detergent packaging, agricultural silage wrap and mulch films.

Dow LLDPE. Trade name of Dow Plastics (USA) for linear low-density polyethylenes.

Dow Magnum. Trademark of Dow Plastics (USA) for a series of acrylonitrile-butadiene-styrene resins.

Dow MDPE. Trade name of Dow Plastics (USA) for medium-density polyethylenes.

Dowmetal. Trade name of Dow Chemical Company (USA) for a series of wrought and cast magnesium alloys containing varying amounts of aluminum, manganese, copper, cadmium, nickel, zinc and silicon. Used for castings, light-alloy parts, valve and pump parts, superchargers, housings, motors, aircraft parts, etc.

downeyite. A colorless mineral composed of selenium oxide, SeO_2. Crystal system, tetragonal. Density, 4.16 g/cm³. Occurrence: USA (Pennsylvania).

Downspun. Trademark of PFE Limited (UK) for polypropylene fibers used for fabrics.

Dow PE. Trade name of Dow Plastics (USA) for low-density polyethylene supplied in standard and UV-stabilized grades.

Dow Silicone. Trade name of Dow Plastics (USA) for silicone resins supplied in standard and glass-fiber and mineral-filled grades.

DP steel. See dual-phase steel.

DQ steel. See drawing-quality steel.

drafting paper. A heavy, high-grade paper, usually white or cream-colored, used for making drawings. It is supplied in the form of rolls and sheets in standard lengths including slightly rough grades for pencilwork and smooth, hard-sized and coated grades for inkwork. Also known as *drawing paper.*

dragline silk. The silk spun by spiders, such as the golden orb-weaving spider (*Nephila clavipes*) and the golden garden spider (*Argiope aurantia*), to produce draglines and make the scaffolding of their webs. It is the strongest of the various types of natural spider silk–stronger than steel, lighter and stronger than *Kevlar* aramid, and tougher, stretchier and more waterproof than true silk produced by the silkworm. See also spider silk.

Dragon. Trade name of Allegheny Ludlum Steel (USA) for a water-hardening steel containing 0.35% carbon, 0.55% manganese, 0.7% chromium, 0.25% molybdenum, and the balance iron. Used for bucket teeth, bolts, studs, and keys.

Dragonite. (1) Trade name of British Steel plc (UK) for low-carbon steel plate and sheet electrocoated with zinc. See also *Zintec*.

(2) Trade name of Jaygo Inc. (USA) for solid glass spheres for surface finishing applications.

dragon's blood. A deep-red, water-insoluble resin obtained from tropical trees and shrubs of the genus *Daemonorops*. It has a melting point of 120°C (248°F), and is used as a pigment for coloring lacquers, varnishes, paints, polishes, paper, plaster, earthenware, marble, etc., in photoengraving, and to protect zinc plates from acid.

Dragonzin. Trade name of British Steel plc. (UK) for hot-dip galvanized mild steel. See also *Galvatite*.

Drahtokulit. See Okulit.

DrainBoard. Trademark of Roxul Inc. (Canada) for moisture-resistant mineral wool insulation supplied in boards 76 × 102 mm (3 × 4 in.) and 102 × 152 mm (4 × 6 in.) in size with an R-value (heat flow resistance) of 4.3 per inch (25 mm) of thickness. Used for the thermal insulation of below-grade basement foundations. See also Roxul.

drain tile. A circular, unglazed tile of burnt-clay, concrete, etc., used to collect and remove ground water.

Draion. Trade name for acrylic fibers and yarns used for textile fabrics.

Draka. Trade name for saran (polyvinylidene chloride) fibers with good chemical resistance, good weatherability, and good resistance to mildew and insects. Used for carpets and rugs, draperies, upholstery, clothing, and industrial fabrics.

Dralon. Trademark of Bayer Corporation (USA) for acrylic fibers and continuous filaments based on polyacrylonitrile. They have good resistance to acids, alkalies, sunlight and moths, very low shrinkage and water absorption, and can blended with other fibers. Used in the manufacture of woven and knitted fabrics for upholstery, furnishings, carpets, bedspreads, blankets, curtains, tablecloth, napkins and flags, and as reinforcing fibers.

Dramex. (1) Trademark of Expanded Metal Corporation (Canada) for expanded metal based on aluminum, steel or stainless steel.

(2) Trademark of Bondex International Inc. (USA) for ready-mixed textured paint.

Drapespun. Trade name for spun rayon fibers and yarns used for textile fabrics.

Dravel. Trademark of Saint-Gobain (France) for noninsulated, fire-resistant, wired glass used in building construction and civil engineering.

dravite. A light-brown mineral of the tourmaline group composed of sodium magnesium aluminum borate silicate hydroxide, $NaMg_3Al_6(BO_3)_3Si_6O_{18}(OH)_4$. Crystal system, rhombohedral (hexagonal). Density, 3.03 g/cm³; refractive index, 1.634. Occurrence: Austria. A variety containing chromium is found in Finland. Used chiefly as a gemstone.

Drawalloy. Trade name of Welding Equipment & Supply Company (USA) for a series of nonmagnetic austenitic steel con-

taining 0.2% carbon, 18% chromium, 8% nickel, and the balance iron. Used for hardfacing welding rods.

Drawcote. Trade name of Atofina Chemicals, Inc. (USA) for dry-film lubricants for drawing and ironing applications.

Draw-ex. Trademark of Humble Oil & Refining Company (USA) for a series of oil- and water-soluble drawing compounds used in the cold and hot working of metals.

Drawinella. Trade name for cellulose acetate fibers used for wearing apparel and as industrial fabrics.

drawing brass. A *deep-drawing brass* containing 67-70% copper, and the balance zinc. Used for seamless tubes, condensers, and evaporators.

drawing compound. A composition of graphite, talc, molybdenum disulfide, grease, oil, etc. applied to the working surfaces of drawing dies or workpieces to serve as a lubricant for the reduction of friction and prevention of galling, scoring, draw marks and tearing. Also known as *drawing lubricant*.

drawing lubricant. See drawing compound.

drawing paper. See drafting paper.

drawing-quality steel. A carbon or low-alloy steel suitable for the manufacture of parts requiring severe deformation, especially deep-drawn parts. It possesses more uniform properties and greater ductility than commercial-quality steels, and is supplied in the form of sheets, plates and strips. Special drawing grades, e.g., rimmed (DQR), aluminum-killed (DQAK) and special-killed (DQSK) are also available. Abbreviation: DQ steel.

drawn fibers. Directionally oriented synthetic fibers with nearly circular cross sections formed by drawing through a die.

drawn glass. Glass drawn continuously from a melting tank and processed by rolling or shaping. Used for the manufacture of glass rods, tubing, sheets, fibers, etc.

drawn metals. Metal rods, tubes or wires that have been pulled, or drawn through a die, or series of dies either at elevated temperature (hot-drawn metals) or room temperature (cold-drawn metals).

drawn tube. A metal tube made by drawing a tube bloom through a die or over a mandrel.

drawn yarn. An extruded man-made yarn that has been oriented by stretching or drawing in the direction of the filament axis.

draw-spun filaments. Single continuous man-made fibers that have been partially or highly oriented by a drawing process prior to entering the spinning machinery.

draw-wound yarn. A filament yarn that has been oriented by drawing and then winding onto a package.

Dreadnaught. Trade name of Hawkridge Brothers Company (USA) for a high-speed steel containing 0.7% carbon, 4% chromium, 1.15% vanadium, 18% tungsten, and the balance iron. Used for tools, cutters and hacksaw blades.

Dreadnought. Trade name of Colt Industries (UK) for an oil- or air-hardening high-speed steel containing 0.8% carbon, 0.3% manganese, 4% chromium, 1% vanadium, 18% tungsten, and the balance iron. Used for tools, dies, punches, and general-purpose tools.

Dream Slub. Trade name for rayon yarns and slubbings.

dressed crude mica. Mica from which dirt, rock and other contaminants have been removed.

dressed flax. Flax consisting of fibers with parallel strands and square ends and without naps or entanglements, and produced by a special combing process.

dressed lace. Lace that has been stretched to size, stiffened with suitable chemicals and then dried.

dressed lumber. See surfaced lumber.

dresserite. A white mineral of the dundasite group composed of barium aluminum carbonate hydroxide trihydrate, $Ba_2Al_4(CO_3)_4(OH)_8 \cdot 3H_2O$. Crystal system, orthorhombic. Density, 1.96 g/cm³. Occurrence: Canada (Quebec).

dressing. A coating applied to casting chills, cores, dies or molds for the prevention of metal penetration, extension of tool life, elimination of surface defects on, or improvement of overall surface finishes of castings, or other purposes. It is usually a refractory powder for dry application, a refractory or graphite paste for dilution with alcohol or water, or a suspension of graphite in oil. See also facing; mold coating.

Dreve Gum Quick. Trade name for a silicone resin used in dental laboratory work.

Drew Foam. Trade name for polyurethane foam materials for flotation, insulating and packaging applications.

Drewosta. Trade name of Stahlwerk Stahlschmidt GmbH & Co. (Germany) for a wear-resistant, water-hardening tool steel containing 1.4% carbon, varying amounts of tungsten and vanadium, and the balance iron. Used for forming and blanking dies, and engravers tools.

Drexflex. Trade name of D&S Plastics (USA) for thermoplastic elastomers.

dreyerite. An orange-yellow mineral composed of bismuth vanadate, $BiVO_4$. Crystal system, tetragonal. Density, 6.25 g/cm³. It can be made synthetically from solutions of sodium vanadate and bismuth nitrate, and changes to the stable monoclinic form on heating to 400-500°C (750-930°F).

Dri-Dek. Trade name of Dri-Dek Corporation (USA) for interlocking, acid-resistant vinyl floor tiles, and interlockable self-draining vinyl liners for interior and exterior applications, e.g., decks and balconies.

drier. A compound of certain metals that promotes oxidation or drying of oils when added to paints, varnishes, printing inks, etc. Most driers are solutions of metallic soaps (e.g., linoleates, naphthenates or resinates of metals, such as cobalt, magnesium, cerium, lead, chromium, iron, nickel, uranium or zinc) in oils and volatile solvents. Several types are available including oil driers, japan driers, and liquid driers. See also paint drier; varnish drier.

Drierite. Trademark of W.A. Hammond Drierite Company (USA) for an anhydrous calcium sulfate ($CaSO_4$) available as granules and powders of varying mesh size. It has highly porous granular structures, and high affinity for water. Used as a drying agent for solids, liquids and gases.

Dri Fast. Trade name of Foamex (USA) for drainable polyurethane foam used as a cushioning material for outdoor furniture.

Dri-Film. Trademark of Allied Chemical Corporation (USA) for an extensive series of silicone resins used to impart water and weather resistance to textiles, and metallic and ceramic surfaces.

Dri-Ink. Trade name of Paxar Systems Division (USA) for hot stamping foil.

Drilac. Trademark of THM Biomedical (USA) for a biodegradable polymer based on poly(DL-lactide), and used for dental applications.

Drill. Trade name of US Steel Corporation for a water-hardening tool steel containing 0.85% carbon, 0.5% manganese, 0.25% silicon, and the balance iron. Used for hammers, axes, shear blades, chisels, and drills.

drill. A strong, twilled cotton fabric similar to *denim* and lighter than *duck*. Used for overalls, uniforms, linings, etc.

Drillalloy. Trade name of Delsteel Inc. (USA) for a water-hardening tool steel containing 0.65% carbon, 0.7% manganese, 0.4% molybdenum, and the balance iron. Used for tools, drills and mining drills.

Drillex. Trade name of Rankin Manufacturing Company (USA) for a sintered tungsten carbide screened from -60 to +100 mesh size powders, and subsequently alloyed with small amounts of nickel, manganese and silicon to develop a strong impact-resistant matrix for welding.

Drill Rod. Trade name of Peninsular Steel Company (USA) for a water-hardening tool steel containing 1.2% carbon, and the balance iron. Used for drills and tools.

drill rod. A polished, round rod of oil- or water-hardening high-speed or high-carbon tool steel manufactured to close tolerances in sizes equivalent to twist drill diameters. *Drill rod* can be heat-treated, and is supplied in diameters as large as 82.550 mm (3.250 in.) and as small as 0.343 mm (0.0135 in.), and in lengths up to 0.9 m (3 ft.). Used in the manufacture of drills, taps, punches, dowel pins and shafts, in inspection-hole alignment and location, and in checking the size of drilled holes (e.g., plug gages).

drill-rod brass. A brass that contains 62% copper, 35.5% zinc and 2.5% lead, drills and turns well, and is used for automatic screw-machine products.

drill steel. A carbon steel containing about 0.8-0.9% carbon, 0.15-0.30% manganese, 0.01-0.03% sulfur and phosphorus, respectively, up to 0.3% silicon, and the balance iron. It is usually made by the electric-furnace process, and supplied in the form of squares, octagons, rounds, and solid or hollow bars. Used for mine and quarry drills. Abbreviation: DS.

Dril-Tec. Trade name of Eutectic Corporation (USA) for an abrasion- and wear-resistant material consisting of hard carbides in a nonferrous matrix. Used for hardfacing of earthmoving equipment, shovels, etc.

Dri-Nylon. Trade name of E.I. DuPont de Nemours & Company (USA) for strong, tough polyamide (nylon) fibers similar to *Perlon*.

Drioil. Trade name of Pax Surface Chemicals, Inc. (USA) for a protective dip coating.

drip-dry fabrics. Textile fabrics that do not require squeezing or wringing after washing but are dried by being let drip and then require little or no ironing. Fabrics for curtains are often of the drip-dry type.

Driscopipe. Trademark of Phillips Petroleum Company (USA) for a polyethylene resin with high corrosion resistance, durability, and flexibility, used for pipes and fittings.

Dri-So-Mar. Trade name of Marcus Paint Company (USA) for powder coatings for industrial applications.

Dri-Surf. Trade name of Contact Paint & Chemical Corporation (USA) for waterproofing paints.

Dri Touch. Trade name of Birchwood Casey (USA) for a water-displacing rust preventive used on iron and steel parts.

Drittelsilber. See drittel silver.

drittel silver. A German alloy of two parts (66.7%) aluminum or nickel silver, and one part (33.3%) silver used for ornamental applications. Also known as *Drittelsilber; tiers-argent. Note:* The words "Drittel" and "tiers" are German and French, respectively, for "one third."

Driver. Trade name of Wilbur B. Driver Company (USA) for an extensive series of resistance alloys containing about 78-98% copper and 2-22% nickel. They have a density of about 8.9 g/

cm^3 (0.32 lb/in.^{3}), moderate to good strength, and a maximum service temperature in air of 316-538°C (600-1000°F). Used for resistances, rheostats, precision resistors, instruments, etc.

Drop. Trade name of SA Glaverbel (Belgium) for patterned glass with pear-shaped spots.

drop black. A black pigment composed of washed and ground *bone black*, and used for paints and enamels.

drop chalk. See chalk (2).

drop machine brick. A brick formed by automatically dropping a prepared mix from a considerable height into a mold. The force of the drop forces the mix in all corners to fully fill the mold. Also known as *drop mold brick.*

drop mold brick. See drop machine brick.

Dropnyl. Trademark of Fairway Filamentos SA (Brazil) for nylon 6,6 fibers used for textile fabrics.

Dropsin. Trade name for a modified zinc phosphate dental cement.

drop tin. Granules of tin made by pouring molten tin into water.

drop zinc. Zinc in the form of globules made by pouring molten zinc into water.

Dr. Smooth Stone. Trade name of Mikel & Company, LLC (USA) for a non-clumping mixture of TriCo Polymers, wetting agents, smoothing agents and water, used as a water substitute in the preparation of smooth, accurate dental stones.

DRW copper. An electrolytic tough-pitch copper containing 99.90% copper and 0.04% oxygen. It has excellent hot and cold workability, and good forgeability and corrosion resistance. Used for electrical applications including power-transmission components, radio parts, busbars, and in rod and wire form for various general applications.

drugmanite. A colorless mineral composed of lead aluminum iron phosphate hydroxide monohydrate, $Pb_2(Fe,Al)(PO_4)_2(OH) \cdot H_2O$. Crystal system, monoclinic. Density, 4.10 g/cm^3; refractive index, 1.88. Occurrence: Belgium.

dry-batched aggregate. A batch of concrete that contains aggregate (e.g., sand or gravel), cement and admixtures in the proper proportion, but does not contain water. Also known as *batched aggregate.*

dry blend. See powder blend.

dry body. (1) A ceramic body from which essentially all moisture has been removed.

(2) An unglazed ceramic body, e.g., stoneware.

Dry Bond. Trademark of Laticrete International Inc. (USA) for grouts, mastics and sealers for glass, marble, masonry, slate, tile, etc.

dry bond adhesives. See contact adhesives.

Dry Coat. Trade name for a light-cure dental sealing resin.

Drycrete. Trademark for a concrete waterproofing compound.

Dryden. Trademark of Great Lakes Forest Products Limited (Canada) for sulfate-bleached softwood.

dry developer. In liquid penetrant inspection, a fine dry absorbent powder developer applied to the test piece by dusting, spraying or immersion. Also known as *dry powder developer.*

Dry Film. Trade name of Atofina Chemicals, Inc. (USA) for dry-film lubricants for drawing and ironing.

dry-film lubricant. See solid-film lubricant.

dry-film rust-preventive compound. A compound, usually composed of a blend of various basis materials, additives and inhibitors, that produces a thin, hard, strippable coating on iron or steel surfaces protecting them against corrosive environments during fabrication, storage and use.

drying agent. See desiccant.

dry-laid fabrics. Textile fabrics made by first carding and/or air laying fibers into a batt or web and then bonding by any suitable process. See also air-laid fabrics.

dry-laid nonwoven fabrics. Nonwoven fabrics produced by loosely combining natural or synthetic fibers by mechanical means, e.g., carding and/or air laying. Also known as *dry-laid nonwovens.* See also nonwoven fabrics; wet-laid nonwoven fabrics.

Drylark. Trade name for a structurally foamed copolymer with good moldability, dimensional stability and heat resistance, and high stiffness. Used for automotive components.

Drylene. Trademark of Plasticisers Limited (UK) for polyethylene fibers and monofilaments used for textile fabrics.

Drylok. Trademark for a range of concrete and masonry repair and treatment products including concrete cleaners, degreasers and etchants, latex-base concrete floor paints, latex- and oil-base masonry waterproofers and masonry joint and crack sealers.

dry lubricant. See solid-film lubricant.

dry-milled fireclay. Fireclay ground in a pan-type rotating grinding mill and subsequently passed over a screen.

dry mix. A ceramic mix containing batch ingredients blended in the dry state. Liquid is added as required during subsequent processing.

dry-mix shotcrete. Pneumatically applied shotcrete containing little or no water. The necessary mixing water is added at the gun nozzle. Also known as *air-placed concrete; gun-applied concrete; jetcrete; pneumatically placed concrete.* See also shotcrete; gunite.

dry pack. (1) A mixture of cement and sand containing only a small amount of water. Used as a filler to repair imperfections in concrete.

(2) A damp mixture of Portland cement and aggregate placed by forcibly ramming into a confined area. Also known as *dry-packed concrete.*

dry-packed concrete. See dry pack (2).

dry powder. (1) Finely pulverized particles of ferromagnetic material used in magnetic particle inspection.

(2) A finely pulverized substance or mix containing little or no moisture.

dry powder developer. See dry developer.

dry-pressed brick. A brick formed in metal molds under high pressure from a clay powder with a moisture content of only 5-7%.

dry-pressed ceramics. Ceramic products, usually of simple shape, formed by filling free-flowing ceramic powders with very low moisture contents into the cavities of dies, and compacting them with a punch under high pressure, followed by firing or sintering at high temperatures.

dry-process enameled steel. A steel that has been heated to a temperature above the maturing temperature of the *porcelain enamel.* Dry, powdered enameling materials are then applied to the hot steel, and fired-on. Dry powder cover coats may be applied over dry powder ground coats and matured in a single firing.

dry-process hardboard. Hardboard produced from wood fibers and old newspapers (ratio 50:50) by adding 7% thermosetting resin, and forming into boards by mechanical means. Used for thermal and acoustical insulation applications.

drysdallite. A grayish black mineral of the molybdenite group composed of molybdenum diselenide, $MoSe_2$. Crystal system, hexagonal. Density, 6.90 g/cm^3. Occurrence: Zambia.

dry-spun fibers. Man-made fibers produced by dissolving the fiber-forming chemical in a solvent that evaporates in warm air once the filaments exit the extrusion die or spinneret.

dry-spun yarn. (1) A coarse flax yarn that has been spun from air-dry roving.

　　(2) A worsted yarn made from a dry-combed top.

　　(3) A synthetic yarn made with dry-spinning machinery.

Dry Step. Trade name of Honeywell International (USA) for nylon-6 fibers.

drywall. See gypsum wallboard.

dry-zone mahogany. See oganwo.

DSC Aluminum. Trade name of Chesapeake Composites Corporation (USA) for strong, readily machinable aluminum alloys strengthened by the incorporation of 30-40 vol% micro-sized alumina particles. They can withstand temperatures of 600°C (1110°F) and have improved ductility, and controlled thermal expansion. Supplied in the form of billets, they can be forged or squeeze-cast into finished products, such as automobile pistons, connecting rods, drive shafts, valves, brake parts, etc.

DS copper. See dispersion-strengthened copper.

DSD Autofine. Trade name for a low-copper dental amalgam.

DSD Fine Grain. Trade name for a fine-grained low-copper dental amalgam.

DSG. See Daido Super Gear.

DS glass. See double-strength glass.

DS lead. See dispersion-strengthened lead.

DS nickel. See dispersion-strengthened nickel.

DSP. Trademark of Allied Signal/Honeywell Performance Fibers (USA) for dimensionally stable high-performance yarns based on high-modulus, low-shrinkage polyester fibers. They have excellent tenacity, toughness and abrasion resistance, excellent thermal and chemical stability, and nearly zero moisture absorption. Used for reinforcements including hoses, mechanical belts, au-tomotive tire cords and transmission belts, and conveyor belts.

D-steel. A British steel containing 0.33% carbon, 1.1-1.4% manganese, 0.12% silicon, and the balance iron. Used for shipbuilding.

DS tungsten. See dispersion-strengthened tungsten.

Dual Cement. Trade name Ivoclar/Vivadent North America (USA) dual-cure, microfill composite resin cement used in restorative dentistry for luting applications.

Dualcoat. Trade name for continuous galvanized steel sheets and coils.

Dual Cure. Trade name for a dual-cure dental resin cement.

Dual Glaze. Trade name of Thermal Industries Inc. (USA) for insulating glass.

Dualite. Trademark of Pierce & Stevens Corporation (USA) for expanded polymeric microspheres embedded with calcium carbide particles, and used to enhance the tensile and flexible strength, thermal shock resistance and workability and reduce the weight of unreinforced and fiber-reinforced polyesters.

Duallan. Trade name of Duall Division Met Pro Corporation (USA) for laminates of polyvinyl chloride or other plastics and fiber-reinforced plastics.

Duallor. Trademark of Degussa-Ney Dental (USA) for medium-gold dental casting alloys. *Duallor R* is a hard, fine-grained dental casting alloy (ADA type III) containing 40.5% silver, 40% gold and 6% palladium. It has high yield strength, a hardness of 190 Vickers, and is used in restorative dentistry for full crowns and short-span bridges.

Dualoy. Trade name of Duke Steel Company Inc. (USA) for a

water-hardening tool steel used for dies and tools.

dual-phase steels. A class of high-strength steels that have been heat-treated to produce a microstructure consisting of more than one phase. They usually consist of a continuous phase (the ferrite matrix) and isolated particles of a second phase that appear as small islands (approximately 20 vol%) of dispersed martensite. Unlike most conventional high-strength low-alloy (HSLA) steels, they possess good formability. Used for structural and nonstructural applications. Abbreviation: DP steel.

Dual Seal. Trademark of RPM, Inc. (USA) for protective coatings.

Dual-Ten. Trade name of US Steel Corporation for a dual-phase steel with high strength and strain rate sensitivity, greatly enhanced formability, and good fatigue properties. Its strength is significantly enhanced through rapid work-hardening during the stamping and bake hardening processes.

Dual Wave. Trade name of Acoustical Surfaces, Inc. (USA) for an urethane foam for acoustical insulation applications.

Dubai sand. A type of sand that occurs in the vicinity of Dubai, United Arab Emirates and contains predominantly calcite ($CaCO_3$) and quartz (α-SiO_2) with small quantities of other minerals, such as dolomite [$CaMg(CO_3)_2$].

Dublfilm. Trade name of Coating Sciences Inc. (USA) for double-faced high-strength acrylic-adhesive tape. It has high peel strength, good non-yellowing properties, good resistance to plasticizers, and good long-term aging properties.

Dublisil. Trade name for a silicone elastomer for dental laboratory work.

Ducera LFC. Trademark of Degussa-Ney Dental (USA) for a low-fusing dental porcelain with high strength and stability. It fuses at only 668-676°C (1235-1250°F), and is compatible with most dental porcelains. Used for metal-to-ceramic bonding.

Duceragold. Trademark of Degussa-Ney Dental (USA) for a hydrothermal dental porcelain matching *Degunorm* extra-hard dental porcelain alloy (ADA type IV). It has a self-healing protective layer, high abrasion resistance, excellent resistance to acids and oral conditions, and is used for porcelain-to-metal restorations.

Ducera-Lay. Trademark Degussa-Ney Dental (USA) for a dental laminate veneer porcelain.

Duceralloy U. Trademark Degussa-Ney Dental (USA) for a corrosion-resistant cobalt-chromium dental bonding alloy.

Duceram Plus. Trademark Degussa-Ney Dental (USA) for a dental porcelain with a natural white fluorescence and good handling properties. Used for metal-to-ceramic bonding of restorations with brilliant finish.

Duceratin Porcelain. Trademark Degussa-Ney Dental (USA) for a dental porcelain used for metal-to-ceramic bonding.

duchesse lace. A machine-made lace that features floral and leaf motifs with very little ground, and in which a raised texture is obtained by intertwining heavier threads. Used especially for gowns and bridal veils.

Duchrome. Trade name of MacDermid Inc. (USA) for chromate conversion coatings.

Ducilo. Trade name for nylon fibers and yarns used for textile fabrics.

duck. A strong, heavy, durable fabric of cotton, linen or synthetic fibers with a lighter and finer weave than canvas. Used for sailcloth, conveyor belts, filter fabrics, sailor's clothes, awnings, tents, machine covers, bags, etc. See also awning duck; conveyor duck; canvas.

Duco. Trade name for a cellulose nitrate lacquer.

Ducol. Trade name of British Steel Corporation (UK) for a series of plain-carbon and low-carbon alloy steels containing up to 0.3% carbon, 1.6% manganese, 0-0.4% silicon, a total of 0.05% sulfur and phosphorus, and the balance iron. The low-carbon alloy grades contain small amounts of nickel, chromium and vanadium. *Ducol* steels have good weldability, and are used for pressure vessels, bridges, structures, and rail cars.

Ductal. French trade name for an impermeable, high-performance concrete made of reactive powders and organic or inorganic fibers. It has good ductility in bending and tension, high resistance to acids, chlorides and sulfates, high abrasion resistance, and very low shrinkage and creep. Used for building structures, floors, platforms, civil engineering applications, architectural components, and structural components for nuclear applications.

Ductalloy. Trade name of Abex Corporation (USA) for a ductile cast iron containing 3-3.5% total carbon, 2-2.5% silicon, 0.15% magnesium, 0.8% manganese, and the balance iron. Used for machinery castings, gears, and shafts.

Ductaluminum. Trade name of Abex Corporation (USA) for a series of age-hardenable aluminum alloys containing 0.2-0.4% magnesium, 6.5-7.5% silicon, 0-0.2% copper, 0-0.5% manganese, 0-0.2% titanium, and 0-0.5% iron. Used for aircraft castings.

Ductilend. Trade name of Arcos Alloys (USA) for a series of alloy-steel welding electrodes including various molybdenum, nickel, nickel-molybdenum, nickel-molybdenum-chromium types.

ductile cast irons. A class of cast irons that have their free graphite in the form of balls, nodules or spheroids rather than in lamellae or flakes as gray cast irons. This form of graphite is obtained by adding small amounts of magnesium, cerium, silicon or other elements to the molten iron just prior to pouring. The carbon content of *ductile cast iron* ranges from about 2.5-3.6%, and the microstructure may be predominantly ferritic, pearlitic, ferritic-pearlitic or martensitic. Their ductility and flexural and tensile strengths are significantly higher than those of conventional gray cast irons. Furthermore, they have high wear resistance and impact strength, good elevated-temperature strength, pressure tightness under stress, corrosion resistance, chemical resistance, machinability, brazeability, weldability, and can be surface-hardened by flame and induction methods. *Ductile cast iron* has a density of about 7.1-7.3 g/cm³ (0.257-0.264 lb/in.³), and a melting point of about 1400°C (2550°F). The American Society for Testing and Materials (ASTM) uses a three-number designation to specify five grades (60-40-18, 65-45-12, 80-55-06, 100-70-03 and 120-90-02) in which the numbers refers to the minimum tensile strength in ksi, the 0.2% yield strength in ksi, and the minimum percent elongation in 2 in. (50 mm), respectively. *Ductile cast irons* are used for crankshafts, connecting rods, bearings, pistons, cylinder heads, gearboxes, gears, pinions, rollers, slides, disc-brake calipers, chains, axles, couplings and clutches, rocker arms, steering knuckles, levers, cylinders, guides, forging hammer anvils, pumps, turbines, valves and pump bodies, fittings, piping for industrial furnaces and chemical plant, farm machine and tractor parts, mining machinery, clamp frames, wrenches, dies, and faceplates. Also known as *ductile irons; nodular cast irons;, nodular irons; SG irons; spheroidal graphite cast irons; spheroidal graphite irons; spheroidal irons; spherulitic cast irons; spherulitic irons; spherulitic graphite cast irons; spherulitic graphite irons.*

Ductile Cecolloy. Trade name of Chambersburg Engineering Company (USA) for a ductile cast iron containing 3.3-3.6 total carbon, 2.2-2.6% silicon, 0.5% magnesium, and the balance iron. It has good machinability, and is used for castings, and forging hammer parts.

ductile iron pipe. A relatively strong and ductile pipe made from ductile cast iron. It has fairly good corrosion resistance when coated, and is used for the industrial and residential transmission and distribution of water, gas, etc. Abbreviation: DIP.

ductile irons. See ductile cast irons.

Ductiliron. Trade name for a series of ductile cast irons containing 3.3% carbon, 0.7% manganese, 2.2% silicon, 0.05% magnesium, and the balance iron. Used for gears, shafts, and machine-tool housings.

Ductilite. Trade name of Holt Equipment Company (USA) for a ductile cast iron containing 3.3-3.5% carbon, 2.2-2.5% silicon, 0.7% manganese, 0.05% magnesium, and the balance iron. Used for earthmoving equipment.

Ductillite. Trade name of Wheeling-Pittsburgh Steel Corporation (USA) for tinplate.

Ductiloy. Trade name of National Intergroup Inc. (USA) for a structural steel containing up to 0.18% carbon, 0.5-0.9% manganese, 0.6-0.9% silicon, 0.4-0.8% chromium, 0.03-0.12% zirco-nium, and the balance iron. It has good weldability and corro-sion resistance.

Ductimet. Trade name of Alloy Metal Products Inc. (USA) for a series of inoculants used in the production of ductile cast iron. The following compositions are supplied: (i) 80-95% nickel and 5-20% magnesium; and (ii) 50% nickel, 20% silicon, 15% magnesium and 15% iron.

Dudley's bearing metals. A series of British bearing alloys containing about 98% tin, 1.6% copper, and 0.4% lead.

Dudley's bronze. A series of British bronzes containing 77% copper, 8-10.5% tin, and 12.5-15% lead. Used for heavy-duty bearings and bushings.

Dudley's phosphor bronze. A British phosphor bronze containing 80% copper, 10% tin, 9.6% lead and 0.8% phosphorus. Used for heavy-duty bearings.

Duelplast. Trade name of Duelplast GmbH (Germany) for a series of plastic products including tubes, rods and sheets made of acetal, polyamide (nylon), polyethylene terephthalate or poly-vinylidene fluoride plastics, and sheets and rods made of polyethylene, polypropylene, polystyrene or polyvinyl chloride.

Duett. Trade name for high-copper dental amalgam alloys.

Duex. Trade name of Compagnie Ateliers et Forges de la Loire (France) for an oil-hardening high-speed steel containing 0.88% carbon, 17.5% tungsten, 8% cobalt, 1.5% vanadium, 1.2% molybdenum, and the balance iron. It has high hardness and red-hardness, and is used for heavy-duty lathe and planer tools, boring tools, hobs, and broaches.

Dufay-Chromex. Trade name of Dufay-Chromex Limited (UK) for cellulose acetate plastics.

duffel. A coarse, heavy, inexpensive woolen fabric having a thick nap on both sides. Used for coats (duffel coats) and socks. It is named for Duffel, a town near Antwerp, Belgium, where it was originally made.

Dufour white gold. A corrosion-resistant alloy containing 75% gold, 21.5% lead and 4.5% platinum, used for ornaments.

dufrenite. (1) A greenish-black mineral composed of calcium iron phosphate hydroxide tetrahydrate, $CaFe_{12}(PO_4)_8(OH)_{12} \cdot 4H_2O$. Crystal system, monoclinic. Density, 3.41 g/cm³. Occurrence: UK.

(2) A dark green to olive-green mineral composed of iron phosphate hydroxide dihydrate, $Fe_5(PO_4)_3(OH)_5 \cdot 2H_2O$. Crystal system, monoclinic. Density, 3.33 g/cm³; refractive index, 1.837. Occurrence: Germany.

dufrenoysite. A lead-gray mineral composed of lead arsenic sulfide, $Pb_2As_2S_5$. Crystal system, monoclinic. Density, 5.53 g/cm³. Occurrence: Switzerland.

duftite. An olive-green to gray-green mineral composed of lead copper arsenate hydroxide, $PbCuAsO_4(OH)_6$. Crystal system, orthorhombic. Density, 6.40 g/cm³; refractive index, 2.08. Occurrence: Southwest Africa.

dugganite. A colorless to green mineral composed of lead zinc tellurate arsenate hydroxide, $Pb_3Zn_3(TeO_6)(AsO_4)(OH)_3$. Crystal system, hexagonal. Density, 6.33 g/cm³; refractive index, 1.977. Occurrence: USA (Arizona).

Dugulor. Trade name for a series of high-gold dental casting alloys including *Dugulor M, Dugulor MO* and *Dugulor S*.

duhamelite. A green mineral composed of copper lead bismuth vanadium oxide hydroxide octahydrate, $Cu_4Pb_2Bi(VO_4)_4(OH)_3 \cdot 8H_2O$. Crystal system, monoclinic. Density, 3.13 g/cm³; refractive index, 1.651. Occurrence: USA (Arizona).

Duka Glas. Dutch trade name for a glass that has been made water-repellent by a special surface treatment.

Dukane. Trade name of Denman & Davis Company (USA) for a water-hardening tool steel containing 0.4% carbon, and the balance iron. Used for shafts and gears.

Duke-Kidd. Trade name of Duke Steel Company Inc. (USA) for a water-hardening tool steel containing 0.9% carbon, and the balance iron. Used for tools, drills, and taps.

Duke's Metal. (1) British trade name for an alloy of 40% nickel, 30% copper and 30% iron. Used for heat- and corrosion-resistant parts.

(2) British trade name for a nondeforming tool steel containing 1.5% carbon, 12% chromium, 4% cobalt, 0.6% silicon, 0.4% tungsten, 0.2% manganese, and the balance iron. Used for corrosion- and heat-resistant parts, dies, and tools.

Dukex. Trade name of Duke Steel Company Inc. (USA) for a water-hardening tool steel that contains varying amounts of carbon, tungsten and iron, and is used for cutting tools.

Dulamel. Trademark of Benjamin Moore & Company Limited (Canada) for semi-gloss enamel paint.

Dularit. Trademark of Henkel KGaA (Germany) for epoxy resins and adhesives.

Dulesco. Trade name for dull rayon fibers and yarns.

Dul-fast. Trade name for rayon fibers, yarns and fabrics.

Dulin. Trade name of Empire Rubber Company (UK) for a series of synthetic rubbers.

Dulkona. Trade name for rayon fibers and yarns.

dull-coated paper. Paper with a dull (or mat) surface finish produced by the application of a low-gloss surface coating. See also coated paper.

Dullray. Trade name of Inco Alloys International Limited (UK) for a resistance alloy containing 33-35% nickel, 60-64% iron and 3-5% chromium. It is suitable for temperatures below dull red, and used for motor-starters and other electrical resistances. Formerly known as *Ferrozoid*.

Dul-tone. Trade name for dull rayon fibers and yarns used for textile fabrics.

Dulux. Trademark of DuPont Canada Inc. for an extensive series of enamels, finishes, paints, thinners and varnishes.

Dumet. Trade name of Carpenter Technology Corporation (USA) for a composite metal made by enclosing a core of 54-58%

iron and 42-46% nickel in a sheath of copper and drawing into a thin wire. It has a very low thermal expansion similar to that of soft glass. Used for glass-to-metal seals, lead-in wires, seal-in wire in incandescent lamps, and vacuum tubes.

Dumold. Trade name of Cascelloid Limited (USA) for cellulose nitrate plastics used for making molded products.

dumontite. A pale yellow to golden-yellow, radioactive mineral of the phosphuranylite group composed of lead uranyl phosphate hydroxide trihydrate, $Pb_2(UO_2)_3(PO_4)_2(OH)_4 \cdot 3H_2O$. Crystal system, monoclinic. Density, 5.65; refractive index, 1.87. Occurrence: Zaire (Katanga).

Dumont's blue. See smalt.

Dumore. Trade name of Ziv Steel & Wire Company (USA) for an air-hardening tool steel containing 0.95-1.05% carbon, 5-5.5% chromium, 0.95-1.25% molybdenum, and the balance iron. Used for blanking, coining, drawing, forming and trimming dies.

dumortierite. A bright smalt-blue, greenish-blue, violet or pink mineral with vitreous luster composed of aluminum iron oxide borate silicate, $(Al,Fe)_7BO_3(SiO_4)_3O_3$. Crystal system, orthorhombic. Density, 3.26-3.41 g/cm³; hardness, 7 Mohs. Occurrence: Brazil, France, Mexico, Madagascar, Norway, USA (Arizona, California, Nevada, New York). Used for spark-plug porcelain, in the manufacture of high-grade porcelains to improve thermal-shock resistance and physical damage, and in the manufacture of special refractories. The blue or violet grades are valued as gemstones.

Dun-Chrome. Trademark of Dunmore Corporation (USA) for metallic silver films.

dundasite. A white mineral composed of lead aluminum carbonate hydroxide hydrate, $PbAl_2(CO_3)_2(OH)_4 \cdot xH_2O$. Crystal system, orthorhombic. Density, 3.41 g/cm³. Occurrence: Italy, Tasmania, UK.

Dundie. Trade name of Allegheny Ludlum Steel (USA) for a steel used for tools, gears, shafts, etc.

Dunelt. Trade name of Dunford Hadfields Limited (UK) for an extensive series of low-, medium- and high-carbon steels containing 0.1-1.1% carbon, and including various carburizing, case-hardening and free-machining grades. Used for machine parts, such as gears, shafts, bolts, axles, connecting rods, valves, fasteners, etc. Also included under this trade name are various corrosion-resistant chromium and chromium-nickel steels used for chemical equipment, roller bearings, etc.

dungaree. A coarse cotton fabric for workwear, sails, etc., that is principally blue *denim*.

dunite. An olive-green, ultrabasic rock composed of up to 90% magnesium-rich *olivine*, and a total of about 10% *picotite* and *chromite*. 85-95% of the olivine is *forsterite* (Mg_2SiO_4) and 5-15% *fayalite* (Fe_2SiO_4). Occurrence: Australia, New Zealand, USA (North Carolina, Washington). Used in the manufacture of forsterite refractories, and as a source of chromium.

Dunn. Trade name of Old Hickory Clay Company (USA) for clay and clay products.

Duo-Andralyt. Trade name of Rasselstein AG (Germany) for double-reduced electrolytic tinplate supplied in sheet, ring and strip form.

Duo-Bethcolite. Trade name of Bethlehem Steel Corporation (USA) for double-reduced tin-mill products.

Duo-Bond. Trade name for a dual-cure dental resin.

Duocarb. Trademark of Wilbur B. Driver Company (USA) for carbonized nickel with both surfaces polished.

Duocast. Trade name for a high-gold dental casting alloy.

Duocel. Trademark of ERG Materials & Aerospace Corporation (USA) for a series of lightweight open-celled foams composed of metals and ceramics. Typical metals and ceramics used include nickel, aluminum, carbon, aluminum oxide, silicon carbide and silicon nitride. Available in monolithic or composite form, they have large surface areas, low flow resistance, outstanding rigidity, and good high-temperature capabilities. Used for aerospace structures, composite panels, heat exchangers, energy absorbers, flame arrestors, filter, diffusers, optical mirrors, and porous electrodes.

Duo-Cement. Trade name for a dual-cure dental resin cement.

Duocor. Trade name of Amorphous Metals Technologies Inc. (USA) for hardfacing materials containing a dual carbide dispersion in a tough crystalline matrix. They provide abrasion and wear resistance for a multiplicity of substrates, and are supplied as metal powder-cored wires for application by arc-spraying and arc-welding processes.

Duofoam. (1) Trade name for a series of urethane foam products.

(2) Trademark of Scapa Tapes North America, Inc. (USA) for a pressure-sensitive foam tape with adhesive applied to both sides.

Duo-Glass. Trade name of 3M Company (USA) for coated abrasives.

Duoline. Trademark of Rice Engineering & Operating Limited (Canada) for steel tubing that is internally lined with polyvinyl chloride, and used for oilfield applications.

Duo-Link. Trademark of BISCO, Inc. (USA) for a smooth, highly filled dental composite luting cement with ultrafine particles. It provides high strength, high wear and stain resistance, and low solubility. *Duo-Link* light-cures for immediate finishing, and can be dual-cured for complete polymerization. Used for bonded porcelain or composite inlays, onlays and crowns.

Duolite. (1) Trademark of PPG Industries, Inc. (USA) for flat laminated safety glass.

(2) Trademark of Rohm & Haas Company (USA) for a series of ion-exchange resins including weakly basic anion exchangers with amine functionalities, and strongly acidic cation exchangers with sulfonic acid functionalities.

Duolith. Trade name for a white lithopone pigment containing about 15% titanium dioxide (TiO_2), 25% zinc sulfide (ZnS), and the balance barium sulfate ($BaSO_4$).

Duon. Trade name for polyolefin fibers used for textile fabrics.

Duo Pal 6. Trademark of Wieland Dental + Technik GmbH & Co. KG (Germany) for a white-colored, extra-hard palladium-base dental ceramic alloy.

Du-O-Pane. Trade name of National Du-O-Pane Corporation (USA) for insulating glass.

Duoplast. Trade name of Monopol AG (Switzerland) for an extensive series of coating products including polyurethane top coats, plastic finishing coats and plastic textured finishes, finishing varnishes and plastic fillers.

Duo Plus B. Trademark of Wieland Dental + Technik GmbH & Co. KG (Germany) for a yellow-colored, extra-hard gold-platinum dental ceramic alloy.

Duopol. Trade name of Monopol AG (Switzerland) for an extensive series of coating products including epoxy and epoxy/zinc dust primers, epoxy paints with micaceous iron oxide, epoxy floor sealing paints, glossy polyurethane finishes and polyurethane paints with micaceous iron oxide.

Duo-Prime. Trade name of Kwal-Howells Paint (USA) for an exterior alkyd primer.

Duoterm. Trade name of A/S Stormbull (Norway) for double glazing units.

Duotherm. Trade name of Ragnar Bergstedt AB (Sweden) for double glazing units.

Duo-Twist. Trade name of Lake Shore Cryotronics, Inc. (USA) for a single, twisted pair of phosphor bronze wires.

Duozinc. Trade name of E.I. DuPont de Nemours & Company (USA) for a cast zinc with 0.25% mercury. Used for zinc plating anodes, and electric resistance applications.

Dupan. Trademark of Produit Forestier MacLaren Inc. (Canada) for particleboard.

Duphos. Trade name of MacDermid Inc. (USA) for iron and zinc phosphate coatings.

dupion. A thick, rough, uneven silk fiber reeled from two cocoons. Also known as *doupion*.

dupion fabrics. (1) Fabrics with a rough, uneven surface, woven from *dupion* silk fiber yarns. Also known as *doupioni*.

(2) Fabrics made to imitate *dupion* fabrics (1), but woven from synthetic fiber yarns, e.g., acetate or viscose rayon. Also known as *doupioni*.

Duplacryl. Trade name of Coralite Dental Products (USA) for acrylic resins for dental applications.

Duplalux. Trade name of Atotech USA Inc. (USA) for bright and semibright nickel electrodeposits and plating processes.

Duplate. Trademark of Duplate Division of PPG Canada Inc. (USA) for nonshattering laminated plate glass.

Duplex. Trade name of Faitout Iron & Steel Company (USA) for a water-hardening tool steel (AISI type W1) containing 0.9-1.1% carbon, 0.2% silicon, 0.2% manganese, and the balance iron. Used for taps, drills, reamers, punches, stamps, knurls, and mandrels.

Duplex Central. Trade name of Central Glass Company Limited (Japan) for polished plate glass.

duplex coating. See composite coating.

duplexed malleable iron. Malleable cast iron exposed to a two-furnace melting and refining procedure. It is first melted in a cupola, arc or induction furnace at temperatures usually between 1425 and 1500°C (2600 and 2730°F), and subsequently refined in an air, arc or induction furnace at temperatures between 1475 and 1600°C (2690 and 2910°F).

Duplex Gear. Trade name of Colt Industries (UK) for a series of oil-hardening steels containing 0.35-0.5% carbon, 1.5-3.5%, 0.75-1.25% chromium, and the balance iron. Used for gears, axles, shafts, crankshafts, etc.

duplex iron. Cast iron exposed to a two-furnace melting and refining process. It is first melted in a cupola furnace at relatively low temperatures, and subsequently refined in an arc, induction or air furnaces at superheating temperatures.

duplexite. See bavenite.

duplex metal. A composite metal made by applying one or more layers of different metals to a base metal by rolling, welding or electrodeposition. The layers usually serve to improve the corrosion resistance and/or appearance. See also clad metal.

duplex nickel plate. An electrodeposit consisting of two layers of nickel, one bright and the other semibright. Used as an undercoating for chromium plates.

duplex paper. A wall paper composed of two separate sheets of paper pasted together and used to create highly embossed effects.

duplex stainless steels. A class of superplastic stainless steels that contain about 24-30% chromium and 4-9% nickel, and are produced by thermomechanical processing. They have very fine-grained microduplex structures in which ferrite and austenite are distributed uniformly. *Duplex stainless steels* have high strength and toughness, good hot workability and good corrosion resistance. Abbreviation: DSS.

duplex steel. A steel that is first produced by one process (e.g., the Bessemer process) and subsequently refined by another (e.g., the electric-furnace process). Abbreviation: DS.

Dupliver. Trade name of Soliver SA (Belgium) for a safety glass.

Duplover. Trade name of Vetreria di Vernante SpA (Italy) for a laminated glass.

Dupont Metal. Trade name of Cutler Hammer Inc. (USA) for a corrosion-resistant alloy containing about 96% copper, 3% silicon, 1% manganese and 0.1% iron. It has good strength and weldability, and is used for bolts, screws, nuts, turnbuckles and welding wire.

DuPont Nylon. Trademark of E.I. DuPont de Nemours & Company (USA) for nylon 6,6 fibers and yarns.

DuPont XTC. Trademark of E.I. DuPont de Nemours & Company (USA) for a thermoplastic polyethylene terephthalate (PET) sheet material reinforced with long glass fibers. Available in ready-to-use rolls, it has a lower weight than conventional sheet-molding composites, high melt recyclability, and can be assembled and coated with the same equipment as steel panels. Used for horizontal autobody panels, such as trunk lids, hoods and roofs.

Duprene. Trademark for an oil-resistant synthetic rubber.

Dura-20. Trade name of Metallic Building Company/Division of NCI Building Systems, Inc. (USA) for a silicone polyester paint coating.

Dura-Bar. Trademark of Dura-Bar, Division of Wells Manufacturing Company (USA) for continuous-cast iron bar stock supplied as-cast or cold-finished in ductile, gray, *Ni-Resist* and special alloyed iron grades in the form of rectangles, rounds, squares, tubes and custom shapes. It has an uniform, fine-grain microstructure free from gas holes, shrinkage and tool-wearing inclusions, such as sand. Other important properties include high strength and wear resistance, excellent surface finish, good machinability, and good noise- and vibration-dampening properties. Used for pumps, compressors, automotive components, machine tools, and agricultural and construction equipment.

Durabil. Trade name of Duke Steel Company Inc. (USA) for a water hardening steel for tool and dies.

Durabilit. German trade name for protective surface coatings based on crosslinked natural and synthetic rubber.

durable-press fabrics. See permanent-press fabrics.

durable-press resins. See permanent-press resins.

Durablu. Trade name of Performance Ceramic Inc. (USA) for a blue, hard, wear-resistant glaze with high-gloss surface that is impervious to water, and has good electrical properties.

Dura Bond. Trade name of National Refractories & Minerals Corporation (USA) for refractory mortars.

Durabond. (1) Trademark of US Gypsum Company (USA) for drywall patching compounds for gypsum wallboard, plaster, etc.

(2) Trademark of Cotronics Corporation (USA) for metallic putties that dry in air to form machinable composites with continuous-use temperatures to 650°C (1200°F). They have good ductility and impact resistance, and are used for bonding, casting, potting or repairing aluminum, copper, cast iron, steel, high-expansion metals, and ceramics.

Dura Bright. Trademark of Cuprinol Korzite Limited (Canada) for polyurethane transportation coatings.

Duracast. Trade name of West Steel Casting Company (USA) for a series of tough steels containing 0.3-0.4% carbon, 0.6-0.8% manganese, 0.3-0.4% silicon, 0.15-0.4% molybdenum, and the balance iron. Used for tools and dies.

Duracel. Trade name of Celanese Mexicana SA (Mexico) for nylon 6 fibers and yarns used for textile fabrics.

Duracid. Trade name of Maschinenbau AG (Germany) for a corrosion-resistant cast iron containing 3% carbon, 16% silicon, and the balance iron. Used for chemical equipment.

Duracoat. (1) Trade name of Heatbath Corporation (USA) for chromate conversion coatings for copper, brass, cadmium and zinc.

(2) Trade name of Leybold Technologies, Inc. (USA) for hard titanium nitride coatings.

(3) Trademark of Norman Wade Company Limited (Canada) for sensitized paper.

Duracoat Star Black. Trade name of Heatbath Corporation (USA) for two-component chromate conversion coatings for producing durable, corrosion-resistant black finishes on cadmium electrodeposits or zinc-plated parts.

Duracoat Star Blue. Trade name of Heatbath Corporation (USA) for liquid coating products used to produce clear to light-blue chromate conversion finishes on zinc-plated parts.

Duracol. Trade name for rayon fibers and yarns used for textile fabrics.

Duracon. Trade name for acrylic resins based on polyoxymethylene (POM).

Duracor. Trade name of The Ceilcote Company (USA) for a synthetic plastic compound for ducts, hoods, tanks, etc.

Duracorr. Trade name of Lukens Inc. (USA) for a strong, corrosion-resistant 10.5-12.5% chromium dual-phase stainless steel with good formability and weldability. Supplied in standard plate and sheet gages with hot-rolled and pickled finishes, it is used for various applications in the agricultural, mining, transportation, and cement industries.

Duracron. Trademark of PPG Industries, Inc. (USA) for durable thermoset acrylic coil and extrusion coatings for aluminum and galvanized steel. They have excellent chemical, stain and dirt resistance, outstanding flow characteristics and weatherability, and are used for architectural and automotive applications, such as wall panels, recreational-vehicle stock, truck trailer sheets, awnings, canopies, and door and window frames.

Dura-Decor. Trademark of Duracote Corporation (USA for textile fabrics laminated or reinforced with vinyl resin. Used for curtains, drapery, and partitions.

Duradek. Trademark of Strongwell Company (USA) for lightweight, flame-retardant, yellow or gray composite panels composed of premium-grade vinyl ester resin with glass-roving or glass-mat reinforcements. Used for various construction applications.

Duradene. Trademark of Firestone Synthetic Rubber & Latex (USA) for solution styrene-butadiene rubbers with excellent abrasion and flex-cracking resistance, excellent processibility, good resilience and wet-skin resistance, and inherent hysteresis. Used for molded rubber goods, shoe soles, flooring, cove bases, automotive tire compounds, conveyor belting, and as replacements for emulsion styrene-butadiene rubber.

Dura-Die. Trade name for an epoxy resin for dental model and die work.

Durafab. (1) Trademark of Dominion Fiber Textile Inc. (Canada) for industrial-grade textile fabrics.

(2) Trademark of Advanced Flexible Composites, Inc.

(USA) for industrial fabrics coated with fluoroelastomer or silicone elastomer, or impregnated with polymeric resins. Used for conveyor belts, insulation, gaskets, seals, filters, release fabrics and liners, architectural materials, protective covers, clothing, curtains, etc.

Duraface Foamglas. Trade name of Pittsburgh Corning Corporation (USA) for a cellular glass insulation material with hard ceramic surface.

Dura-Fibers. Trade name of Durawool, Inc. (USA) for chopped fibers of metals and alloys, such as copper, brass, steel, stainless steel, etc.

Durafibre. Trademark of Cargill Limited (UK) for wheat chaff obtained by refinement from the *bast fibers* of oilseed flax straw (an agricultural waste product). It has a high *holocellulose* content, a low *lignin* content, low bulk density, excellent tensile strength and modulus, good durability, and is supplied in several grades in purities of 55%, 80% or 97% bast fiber, with the balance being shive (woody core material). It has an average fiber length of 100-150 mm (4-6 in.), but some grades can be chopped to shorter lengths. Used as a natural reinforcement in thermoplastics and thermosets. See also Durafill (2).

Durafil. Trade name for a strong viscose rayon staple fiber used for clothing, and blended with other fibers for suitings, furnishings, carpets, linings, and medical fabrics.

Durafill. (1) Trademark of Heraeus-Kulzer, Inc. (USA) for anterior, photo-curing, microfine composite resins for dental filling and restorative applications.

(2) Trademark of Cargill Limited (UK) for wheat chaff obtained by refinement from the woody core (shive) of oilseed flax straw (an agricultural waste product). It has a high *holocellulose* content, a low *lignin* content, low bulk density, and is supplied in four particle sizes: 10, 18, 35 and 60 mesh. Used as a natural filler in thermoplastics and thermosets. See also Durafibre.

Dura-Flex. Trademark of Building Products of Canada Limited for phenolic roof insulation products.

Duraflex. (1) Trade name of Anaconda Company (USA) for a phosphor bronze containing 1-10% tin, 0.01-0.3% phosphorus, and the balance copper. Commercially available in the form of strip and wire, it has a high endurance limit, and is used for springs, clips, and diaphragms.

(2) Trade name of Shell Chemical Company (USA) for polybutylene resins.

(3) Trade name for a corrosion-resistant cobalt-chromium dental casting alloy.

Duraflox. Trade name for rayon fibers.

Dura/Flute. Trade name of Brown Products, Inc. (USA) for molded corrugated paper for protective packaging applications.

Dura-Foam. Trade name of Foam Plastics of New England (USA) for durable plastic foams for building insulation and packaging applications.

DuraFoil. Trade name of Allegheny Ludlum Corporation (USA) and Texas Instruments (USA) for an aluminum-clad stainless steel foil for automotive catalytic converters. It possesses excellent resistance to high-temperature oxidation, and good mechanical strength, ductility, formability, and thermal shock resistance.

Dura-Form. Trademark of Carpenter Technology Corporation (USA) for a high-carbon tool steel composed of 0.65% carbon, 0.5% manganese, 1.4% silicon, 4% chromium, 2.5% molybdenum, 1.5% vanadium, and the balance iron. It has good toughness, high hardness and resistance to chipping and cracking,

high hot hardness, and good abrasion and wear resistance. Used for cold- and hot-working applications, such as hot shear blades and gripper dies, cold forming dies, and forging dies.

Durafrax 2000. Trademark of Norton-Pakco Industrial Ceramics (USA) for a high-performance ceramic composed of about 90% aluminum oxide. It has a density of 3.52 g/cm^3 (0.127 $lb/in.^3$), a hardness of 918 Brinell, a maximum service temperature of 1250°C (2280°F), high elastic and shear modulus, good wear and thermal-shock properties, and nil water absorption and gas permeability. Used for wear-resistant parts.

Dura-Gard. Trade name of DuBois Chemicals, Division of Chemed Corporation (USA) for iron phosphating compounds used for ferrous metal products.

Dura Glass. Trademark of Johns Manville Corporation (USA) for glass fibers and glass-fiber products.

Dura-Glaze. Trademark of Porter Paints (USA) for a line of water-based epoxy paints and coatings for industrial and architectural applications.

Duragrate. Trademark of Strongwell Company (USA) for strong, skid-resistant, molded mesh grating panels composed of thermosetting matrix resins, such as vinyl ester, isophthalic or orthophthalic polyester, reinforced with fiberglass rovings. Supplied in panel widths of 0.9, 1.2 and 1.5 m (3, 4 and 5 ft.), and lengths of 2.4, 3.0 and 3.6 m (8, 10 and 12 ft.), they are used for various construction applications, e.g., stairtreads, panel connectors, panel holddown clips, etc.

Dura-Grid. Trademark of Alcore, Inc. (USA) for aluminum honeycomb core materials for structural applications.

Duragrid. Trademark of Strongwell Company (USA) for lightweight, heavy-duty grating supplied in bar shape with or without antiskid surface grit. It is a pultruded product composed of fiberglass with a choice of polymer-matrix resins. Phenolic grades have excellent fire resistance. Used as metal-grate replacements, e.g., for stair treads, landings, panel holddowns, etc.

Dura Guard. Trade name of Flexfab LLC (USA) for self-fusing silicone rubber tape.

Durak. (1) Trade name for a series of high-temperature oxidation-resistant modified silicide coatings (containing silicon and selected additives) for molybdenum and niobium, deposited by pack-cementation diffusion.

(2) Trademark of Rohm & Haas Company (USA) for synthetic plastic film and sheet materials with or without adhesive coatings.

Durakapp Babbitt. Trademark of Kapp Alloy & Wire Inc. (USA) for a high-strength tin-based babbitt used as bearing lining for high-stress and shock applications, such as mill and production bearings, and for turbine components. It has a density of 7.6 g/cm^3 (0.27 $lb/in.^3$), a liquidus temperature of 260°C (500°F), a solidus temperature of 176°C (350°F), and high ductility.

Dura-Kote. Trade name of Universal Metal Finishing Company (USA) for a series of hard anodic oxide coatings (hardcoats) for aluminum and aluminum alloys supplied in standard and nonfriction grades. Also included under this trade name are anodized coatings for aluminum and other alloys.

Durakote. Trademark of Dural Products Limited (Canada) for a line of paint products.

Dural. (1) Trademark of Williams & Wilson Limited, Industrial Ceramics Division (Canada) for wear-resistant ceramic compounds.

(2) Trademark of Alcan Aluminium Limited (UK) for an age-hardenable high-strength aluminum alloy containing 4%

copper and 1% manganese. Usually supplied in rod, tube and powder form, it has a density of 2.8 g/cm³ (0.10 lb/in.³), a hardness of 115-135 Brinell, and good resistance to corrosion by acids and seawater. Used for aircraft structures, boats, machinery, etc.

(3) Trade name of Alpha Chemical (USA) for a series vinyl resins including polyvinyl chloride used for medical applications.

(4) Trade name of Dexter Corporation (USA) for unmodified and acrylic-modified polyvinyl chlorides used for electrical and electronic components, appliance and medical parts, hardware, packaging and various industrial and consumer applications.

(5) Trademark of Hoechst Celanese Corporation (USA) for a high-temperature polyarylate resin.

Dural 65. Trade name of Wellsville Fire Brick Company (USA) for a refractory bonding mortar.

dural. See duralumin.

Duralak. Trade name of Heatbath Corporation (USA) for a series of clear, lacquer-based protective coatings.

Dura-Lar. Trade name of Grafix Plastics/Division of Graphic Art Systems, Inc. (USA) for oriented polyester films (*Mylar, Melinex*, etc.) for the graphic arts.

Duralbrite. Trade name of Alcan-Booth Industries Limited (UK) for a series of hard-drawn aluminum products including commercially pure aluminums (99+% pure) containing varying amounts of magnesium and manganese as well as several wrought aluminum-magnesium and aluminum-magnesium-silicon alloys. Used for extrusions.

Duralcan. (1) Trademark of Alcan Wire & Cable (Canada) for metallizing wire based on an aluminum composite. It has excellent wear, abrasion and corrosion resistance, and good antiskid properties.

(2) Trademark of Dural Aluminum Composites Corporation (USA) for castable metal-matrix composites produced by mixing fine ceramic particles into molten aluminum while stirring constantly. The particles are uniformly distributed throughout the matrix of the cast products (e.g., foundry ingots, rolling slabs and extrusion billets). Casting can be carried out by almost any commercial method including sand, investment, permanent-mold, and high-pressure die casting. Used in the aerospace, automotive, industrial-equipment, and sporting-goods industries.

Duralco. Trademark of Cotronics Corporation (USA) for a series of high-temperature two-component epoxies for temperatures of 260-315°C (500-600°F). Available as adhesive, sealant, coating, thermally-conductive and rubber-like high-flexibility grades, they have excellent chemical, electrical and radiation resistance, high bond strength, and low moisture absorption and shrinkage. Used for bonding, potting, sealing and repairing metals, plastics, ceramics and glass, and as protective coatings for electronic components, strain gages, instruments, motors, and appliances.

Duralex. Trade name of Saint-Gobain (France) for toughened glassware.

Duralfa. Trade name of Alfa Romeo (Italy) for an aluminum alloy used for light-alloy parts.

Duralimin. Russian trade name for an age-hardenable alloy of 96% aluminum and 4% copper. Used for light-alloy parts.

Duralin. (1) Trade name of Buesing & Fasch KG (Germany) for epoxy resins.

(2) Trade name for high-density polyethylenes.

Dura-lin. Trademark of Dural Products Limited (Canada) for linoleum tile cements.

Duraline. Trade name of Atlas Specialty Steels (Canada) for a steel containing 0.35% carbon, 1.3% silicon, 3.5% chromium, 0.85% vanadium, 4.25% molybdenum, and the balance iron. It has high hot strength and hardness, and is used for extrusion tools and dies.

Dura-Liner II. Trade name for a self-cure acrylic resin used as a liner in restorative dentistry.

Duralinox. Trade name of Société du Duralumin (France) for a series of corrosion-resistant aluminum alloys containing 4.5-12% magnesium and 0-0.5% manganese. Used for light-alloy parts, partitions, floors, etc.

Duralit. (1) French trade name for an age-hardenable, wrought aluminum alloy containing 3% copper, 0.5% nickel, 0.5% magnesium, 0.7% silicon, and 1% titanium. Used for connecting rods.

(2) Trademark of Degussa Dental (USA) for dental stones used for modeling applications.

Dura-Lite. Trade name of North Texas Marble (USA) for cultured marble for kitchen countertops.

Duralite. (1) Trade name of Alluminio SA (Italy) for a series of age-hardenable, high-strength aluminum alloys containing 3-3.8% copper, 0.5-0.7% silicon, 0-0.8% nickel, 0.5-0.75% magnesium, 1.3-1.6% iron, and 0.05-0.2% titanium. Used for automotive and aircraft parts, pistons, general engineering components, structures, fittings, fasteners, and bolts.

(2) Trademark of Apex Concrete Products Limited (Canada) for concrete blocks.

(3) Trade name of Washington Mills Ceramics Corporation (USA) for lightweight ceramic mass-finishing media.

Duralium. Trade name for a corrosion-resistant cobalt-chromium dental casting alloy.

Durall. Trademark of Phillips Paint Products Limited (Canada) for exterior and interior enamel paints.

Duralloy. (1) Trade name of West Steel Casting Company (USA) for a cast steel containing 1-2% chromium, 1-3.5% nickel, and the balance carbon and iron.

(2) Trade name of Degussa-Ney Dental (USA) for a silver alloy used for dental amalgams.

(3) Trade name of Hogen Industries (USA) for pure elemental molybdenum and tungsten, and tungsten-copper powder metallurgy composites.

Duralo. Trade name of Detroit Paper Products Corporation (USA) for phenolic resins and laminates.

Duralon. Trade name of The Thermoclad Company (USA) for plastic coating powders.

Duraloy. (1) Trade name of Duraloy Blaw-Knox Corporation (USA) for abrasion- and wear-resistant manganese steel containing 1.3% carbon, 0-0.5% silicon, 14% manganese, 1.6% chromium, and the balance iron. Used for wear plates and tread plates.

(2) Trade name of Celanese Corporation (USA) for impact-modified polyesters.

(3) Trade name of Detroit Paper Products Corporation (USA) for phenol-formaldehyde plastics.

Duralplat. Trade name of Dürener Metallwerke (Germany) for a highly corrosion-resistant composite metal consisting of a copper-base core containing *duralumin* coated with copper-free duralumin.

Duralum. (1) Trademark of Washingtom Mils Electro Minerals Corporation (USA) for fused alumina powders.

(2) German trade name for a non-hardenable alloy containing 77-79% aluminum, 10-11% magnesium, 10% copper, and 0.5% phosphorus. Used for light-alloy parts.

duralumin. A group of age-hardenable, high-strength and low-density aluminum alloys composed of 3.5-6% copper, 0.5-0.8% magnesium, 0.25-1.0% manganese, and up to 0.8% iron, 0.1% chromium and 0.7% silicon, respectively. They have good resistance to corrosion by acids and seawater, and can be cast, forged, hot-rolled and cold-rolled. Used for aircraft parts, railroad cars, boats and machinery. Also known as *dural*.

Duralux. (1) Trade name of Fabbrica Pisana SpA (Italy) for toughened glass blocks.

(2) Trademark of Loctite Corporation (USA) for a series of adhesives, encapsulants and sealants.

(3) Trademark of Process Paint Manufacturing Company (USA) for a series of paints.

Duramat. Trademark of Power Marketing Group, Inc. (USA) for bitumen-coated reinforced glass-fiber mats used in the construction trades as lining materials for ponds and drainage ditches, supporting bases for drain tiles, and pipeline wrap.

Dura Max. Trademark of Kaiser Tool Company, Inc. (USA) for submicron-grain carbides used for tools.

Duramesh. Trade name of Dura Undercushions Limited (Canada) for bitumen-coated fiber glass mesh and fabrics.

Duramic. Trade name of Wesgo Division of GTE Products Corporation (USA) for precision-engineered high-temperature ceramics based on aluminum oxide, aluminum silicate, silicon carbide, graphite composites, boron nitride, or machinable glass-ceramics. They are made to very close tolerances and with smooth surface finishes, and used for electrical and electronic components, etc.

Duramium. Trade name of Duke Steel Company Inc. (USA) for a high-speed steel containing 0.7% carbon, 18% tungsten, 4% chromium, 1% vanadium, and the balance iron. Used for taps, drills, reamers, and hobs.

Duramix. (1) Trademark for epoxy-resin molding compounds.

(2) Trademark of Richard J. Gallagher Company (USA) for wear-resistant elastomeric bars, sheets, strips and other products used in the manufacture of conveyors, chutes, guides, rail covers, pusher dogs, locators, etc.

(3) Trademark of Sylvax Chemical Corporation (USA) for a concrete mix.

(4) Trademark of Polymer Engineering Corporation (USA) for autobody repair compositions based on urethane polymers.

Duramold. Trade name of Bethlehem Steel Corporation (USA) for a series of low-carbon mold steels (AISI types P2, P3, P4 and P6) containing 0.07-0.1% carbon, 0.1-0.7% manganese, 0.1-0.4% silicon, 0-4.5% chromium, 0-3.25% nickel, 0-1% molybdenum, and the balance iron. They possess high strength at elevated temperatures, good wear properties and toughness, and low hardness. Used for plastic molding dies.

Duran. (1) Trade name of Schott Glas AG (Germany) for borosilicate sheet glass and laboratory glassware with exceptional chemical resistance and low thermal expansion.

(2) Trademark of W.R. Grace & Company (USA) for polyvinylidene chloride (PVDC) used for packaging applications.

(3) Trade name of Scheu-Dental (Germany) for highly transparent, hard-elastic round blanks supplied with a diameter of 125 mm (5 in.) in thicknesses of 0.5, 0.75, 1.0, 1.5, 2.0 and 3.0 mm (0.02, 0.03, 0.04, 0.06, 0.08 and 0.1 in.). Used in dentistry for temporary crowns, bridges and splints, periodontal and retainer splints, invisible retainers, and bite guards.

Durana. German trade name for a series of corrosion-resistant high-strength brasses containing 59-78% copper, 29.5-40% zinc, 0-2.2% tin, 0-1.7% aluminum, 0.3-1.5% iron, and 0.4-2% lead. They possess excellent hot workability and forgeability, and good machinability. Used for chemical-plant equipment, forgings and pressings.

Duranal. German trade name for an alloy containing 5-10% magnesium, 0.6% manganese, 0.2% silicon, and the balance aluminum. Used for light-alloy parts.

Duranalium. Trade name of Dürener Metallwerke (Germany) for a series of corrosion-resistant aluminum-magnesium and aluminum-magnesium-manganese alloys used for aircraft, automotive and marine applications, and for fasteners, fittings and hardware.

Duranar. Trade name of PPG Industries, Inc. (USA) for pigmented fluoropolymer coil and extrusion coatings with excellent weather resistance used for architectural applications, e.g., on building and roof panels, curtainwalls, louvers, windows, etc. Special grades such as *Duranar-Plus* for enhanced resistance to industrial and seawater atmospheres are also supplied.

Durance. German trade name for a babbitt containing 44.5% antimony, 33.3% tin and 22.2% copper, used for bearings.

Durand's alloy. A tough, strong, non-hardenable alloy of 67% aluminum and 33% zinc, used for lightweight parts.

durangite. An orange-red mineral composed of sodium aluminum fluoarsenate, $NaAlF(AsO_4)$. Crystal system, monoclinic.

Durango Red. Trade name of Aardvark Clay & Supplies (USA) for a red-colored earthenware clay.

Duranic. French trade name for a non-heat-treatable aluminum alloy composed of 2% manganese and 2% nickel, and used for light-alloy parts.

Duranickel. Trademark of Inco Alloys International, Inc. (USA) for heat-treatable alloys containing a total of 93.7-96.5% nickel and cobalt, and 0.2% carbon, 0.2-0.3% manganese, 0.15-0.35% iron, 0.5% silicon, 0-0.1% copper, 0.5-0.6% titanium, 4.4-4.5% aluminum, and 0.005% sulfur. Commercially available in the form of wire, rod and strip, they possess high strength and hardness, a maximum service temperature of 350°C (660°F), high fatigue resistance, excellent spring properties, low-sparking properties, good corrosion resistance, and an electrical conductivity of 12% IACS. They are slightly magnetic after heat treatment, and are used for springs, diaphragms, clips, bellows and other flexing parts, valve disks, glass molds, extrusion and injection dies for plastics, instrument parts, thermostat contacts, electronic parts, etc. Formerly known as *Z-Nickel*.

Duranit. (1) Trade name of Thyssen Edelstahlwerke AG (Germany) for a series of coated welding electrodes based on iron alloys (e.g., steels) or nickel alloys. Used for overlaying applications on machinery, cutting tools, forging tools, dies, etc.

(2) Trademark for styrene-butadiene rubber.

Duranmium. Trade name of Duke Steel Company Inc. (USA) for a tungsten-type high-speed steel used for cutting tools.

Duranoid. Trade name of Specialty Insulation Manufacturing Company (USA) for phenolic and shellac plastics.

duranusite. A pale gray mineral composed of arsenic sulfide, As_4S. Crystal system, orthorhombic. Density, 4.53 g/cm³. Occurrence: France.

Durapatite. Trade name for a dense *hydroxyapatite* composed of calcium phosphate hydroxide, $3Ca_3(PO_4)_2 \cdot Ca(OH)_2$, and used as a bioactive ceramic in bone replacement and repair.

Duraperm. Trade name of Hamilton Technology Inc. (USA) for a hard, brittle, soft-magnetic alloy containing 84.5% iron, 9.5% silicon, and 6% aluminum. It has high magnetic flux, high permeability at low flux density, and high resistivity. Used for pole tips in magnetic recording heads operating in the audio or video range.

Durapid. Trade name for a calcium aluminate cement.

DuraPlast. Trade name of EnviroSafe Products, Inc. (USA) for a series of high- and low-density recycled plastics.

Duraplast. Trade name of Duraplast GmbH (Germany) for polyethylene shrink films.

Duraplen. Trademark of Filpersa (El Salvador) for polyethylene and polypropylene monofilaments.

Duraplex. Trademark of Rohm & Haas Company (USA) for fast-drying, oil-modified alkyd resins used to produce durable, tough, glossy coatings on metal parts, automotive parts, furniture, etc.

Durapol. Trademark of Isola Werke AG (Germany) for fiberglass mats and reinforced laminates available in the form of plates, sheets, rods and tubes. Used for dry transformers, switch levers, and insulating walls.

Durapolymer. Trademark of Creative Extruded Products, Inc. (USA) for bulk plastic materials available in the form of sheets and rolls.

Duraposit. Trade name of Shipley Ronal (USA) for electroless nickel.

Durapower. Trade name of Durapower (USA) for a nonmagnetic superalloy containing 40% cobalt, 20% chromium, 15% nickel, 7% molybdenum, 2% manganese, 15% iron, 0.15% carbon, 0.05% beryllium, and the balance iron. Commercially available in the form of wire, strip, rods and tubes, it has high strength and good corrosion resistance, good sub-zero temperature properties, good elevated temperature properties up to about 540°C (1005°F), and can be precipitation-hardened. Used for seal components, instrumentation, and watch and instrument springs. See also *Cobenium* and *Elgiloy*.

Durapox. (1) Trade name for glass-fiber-reinforced epoxy-resin molding compounds.

(2) Trademark of Palmer Products Inc. (USA) for resinous coating compositions for patching non-skid coated surfaces.

Durapreg. Trademark for epoxy-resin molding compounds.

Durapress. (1) Trade name for glass-fiber-reinforced phenolic and melamine resin molding compounds.

(2) Trademark of UCAR Carbon Technology Corporation (USA) for synthetic carbon used in the hot pressing of metallic powders.

(3) Trademark of FiberMark, Inc. (USA) for pressboard used for document covers and folders.

Dura-Pro. Trademark of Duracote Corporation (USA) for heat-sealable textile fabrics laminated with vinyl resin, and used for commercial and outdoor applications.

Durapso. Trade name of Société des Acieries de Pompey (France) for a series of low-carbon chromium steels containing about 0.2% carbon, 0.4% chromium, 0.4-0.5% copper, and the balance iron. Used for structural applications requiring improved corrosion resistance and paint adherence.

Dura-Seal. Trade name of Century Industries Inc. (USA) for insulating glass.

Dura-Shade. Trademark of Duracote Corporation (USA) for fiberglass fabrics laminated with vinyl resin, and used for roller shades.

Durashield. Trademark of Strongwell Company (USA) for a series of strong, lightweight, flame-retardant panels comprising a fiberglass skin of vinyl ester or isophthalic polyester pultruded over a rigid, closed-cell urethane foam core. They are weather- and corrosion-resistant, nonconductive and transparent to electromagnetic radiation. The tongue-and-groove panels are supplied in thicknesses of 25 and 76 mm (1 and 3 in.) and widths of 0.3 and 0.6 m (1 and 2 ft.) in typical lengths of 3.6-9.7 m (12-32 ft.). Used for structural applications, e.g., roofs, walls, covers, and electrical and chemical enclosures.

Durasint. Trade name of Sintered Products Limited (UK) for a series of powder-metallurgy materials including various sintered iron, iron ceramic, bronze, bronze ceramic, and bronze graphite alloys. Used as friction materials for aircraft and automotive clutches, brakes, transmissions, etc.

Dura-Skrim. Trade name of Raven Industries, Inc. (USA) for polyethylene sheeting reinforced with scrim fabrics and used for pond liners and construction enclosures.

Dura-Slide. Trade name of Precision Coatings, Inc. (USA) for wear-resistant nonstick *Teflon* coatings containing durable ceramics or metals. Used on metallic and nonmetallic substrates.

Durasoft. (1) Trade name of Scheu-Dental (Germany) for abrasion- and break-resistant plastic sandwich foil materials with one hard and one soft side supplied in thicknesses of 1.8 and 2.5 mm (0.07 and 0.1 in.). The 1.8-mm (0.07-in.) foil consists of a 0.8 mm (0.03 in.) hard side and a 1.0 mm (0.04 in.) soft side, and the 2.5-mm (0.1-in.) foil has a 1.0 mm (0.04 in.) hard side, and a 1.5 mm (0.06 in.) soft side. The hard side bonds well to acrylics. *Durasoft* is used in dentistry for periodontal splints, bite splints, and night guards.

(2) Trade name of Solutia Inc. (USA) for strong, supple nylon 6,6 fibers and yarns used for the manufacture of textile fabrics.

Dura-Sonic. Trademark of Duracote Corporation (USA) for sound deadening materials based on lead-free, mass-loaded vinyl with or without reinforcement and/or decoupling foams.

Duraspan. Trade name for a spandex (segmented polyurethane) fiber used for the manufacture of textile fabrics with excellent stretch and recovery properties.

Duraspun. (1) Trade name of Duraloy Blaw-Knox (USA) for an austenitic stainless steel containing 0.2% carbon, 18% chromium, 8% nickel, and the balance iron. Used for corrosion-resistant centrifugally cast tubes, sleeves, liners, etc.

(2) Trade name of Solutia Inc. (USA) for a soft, strong, durable, colorfast spun acrylic fiber used for friction- and abrasion-absorbent dress socks that draws perspiration away from the foot.

DuraStar. Trade name of Eastman Chemical Company (USA) for high-clarity injection-molding copolymers with excellent chemical resistance, and good elongation and toughness properties. Used for appliance parts, toys, packaging of cosmetic and personal-care items, etc.

Durastic. Trade name of Eyre Smelting Company (UK) for a series of bearing alloys containing varying amounts of tin, lead and copper. Used for bearings for reciprocating engines.

Dura-Stran. Trade name for metallic fibers.

Dura-Strand. Trade name of International Wire Products (USA) for a high-strength, nonmagnetic core wire plated first with copper and then with silver. Used for coaxial transmission applications.

Durastrength. Trade of Atofina Ceca (USA) for acrylic copolymers used as impact modifiers in methyl methacrylates.

Dura-Surf. Trademark of Crown Plastics Company (USA) for pressure-sensitive tape made of ultrahigh-molecular-weight polyethylene.

Dura Tab. Trade name of Wellsville Fire Brick Company (USA) for special high-alumina refractories.

Duratape. (1) Trademark of Manville Corporation (USA) for bitumen-coated fiber-glass mats for underground pipeline joints and fittings.

(2) Trademark of Ideal Stencil Machine & Tape Company (USA) for pressure-sensitive tape used for carton-sealing applications.

Duratex. Trademark of Drytex, Division of JWI Limited (Canada) for all-synthetic and semisynthetic open-weave felts for paper-machine dryer sections.

Dura-Thane. Trade name of Temp Paint & Varnish Company (USA) for polyurethane-base finishes for aircraft applications.

Dura-Therm. Trademark of Duracote Corporation (USA) for plastic bubbles laminated with aluminum foil, and used for high-reflection and thermal insulation applications.

Duratherm. (1) Trademark of Vacuumschmelze GmbH (Germany) for a series of age-hardenable spring alloys with 41% cobalt, 26% nickel and 12% chromium that are hardenable with additions of beryllium and titanium, or aluminum and titanium. Commercially available in the form of bands, strips, wires, foils, rods, slabs and finished parts, they have a density of 8.4-8.5 g/cm³ (0.30-0.31 lb/in.³), a melting range of 1350-1390°C (2460-2535°F), a maximum service temperature of 565°C (1050°F), high strength and moduli of elasticity, high ductility and good formability in the solution-treated condition, and good high-temperature and corrosion resistance. Used for thermally-stressed springs, corrosion-resistant membranes for manometers, nonmagnetic, wear-resistant spacer foils, temperature- and corrosion-resistant cup springs, etc., and for heat-resistant springs in regulators and switches.

(2) Trade name of Atofina Ceca (USA) for an acrylic copolymer used as a heat modifier in polyvinyl chlorides.

Duratite. Trademark of DAP, Inc. (USA) for wood-dough surfacing putties.

Duratlas. Trade name of Firth Brown Limited (UK) for nickel-chromium cast iron.

Duratop. Trade name of The Thermoclad Company (USA) for high-solids, low volatile-organic-compounds, low-bake coatings.

Duratrek. Trademark of Solutia Inc. (USA) for a durable, wear-resistant nylon 6,6 fiber used for activewear, e.g., hiking, trekking, climbing apparel.

Dura-Trim. Trademark of Duracote Corporation (USA) for flexible polyvinyl chloride with a covering of woven glass fibers used for wall coverings.

Duratron. Trademark of DSM Engineered Plastics (USA) for a series of durable polyimide resins available in unfilled, graphite-filled and compression-molded grades, and used for bearings.

Dura-Tube. Trade name of Dura-Bar, Division of Wells Manufacturing Company (USA) for continuous-cast iron tubing.

Duraver. Trademark of Isola Werke AG (Germany) for fiberglass-reinforced fabrics.

Duravin. Trade name of The Thermoclad Company (USA) for a series of polyvinyl chlorides used for electrostatic-spray coating of metal parts.

Durawax. Trade name of Heatbath Corporation (USA) for a non-flammable, water-soluble dip- or spray-applied synthetic wax emulsion used to provide corrosion protection on parts and equipment for indoor applications, and as a dry-film lubricant in metalworking.

Durax. (1) Trademark for a series of glass fiber-reinforced polyester-resin molding compounds.

(2) Trade name for phenolic resins.

(3) Trade name of Schott Glas AG (Germany) for gage glass having a very low coefficient of expansion. See also *Suprax*.

(4) Trade name of Swedish American Steel Corporation (USA) for a shock-resisting tool steel (AISI type S1) containing 0.5% carbon, 2.5% tungsten, 1.5% chromium, and the balance iron. It has high hardness and strength, excellent toughness, good wear and abrasion resistance and good hot hardness. Used for chisels, tools, dies, and shear blades.

Durazit. (1) Trade name of Plansee Metallwerk Gesellschaft (Austria) for a series of tungsten carbide and cobalt-chromium-tungsten hardfacing alloys.

(2) Trademark of Quarzit-Werke (Germany) for *quartzite* used as a concrete hardener.

Durbar. (1) Trade name of Buffalo Bronze Die Casting Company (USA) for a high-leaded tin bronze containing 70-72% copper, 20-24% lead, and 4-10% tin. Used for bearings and ornaments.

(2) Trade name of British Steel plc (UK) for a slip-resistant steel floor plate with raised pattern.

Durbar-Hard. Trade name of Buffalo Bronze Die Casting Company (USA) for a tin bronze containing 70% copper, 20% lead, and 10% tin. Used for heavy-duty bearings, e.g., for cranes, excavating machines, and steel-mill machinery.

Durbar-Standard. Trade name of Buffalo Bronze Die Casting Company (USA) for a tin bronze containing 72% copper, 24% lead and 4% tin. Used for bearings, e.g., for machine tools, electric motors, pumps, cranks and gas engines.

Durcilium. Trade name of Trefileries & Laminoirs du Havre (France) for an extensive series of wrought aluminum products including commercially pure aluminums and various aluminum-magnesium, aluminum-silicon and aluminum-copper alloys. Used for aircraft parts, drawn wire, tubing, and structural members.

Durco. Trade name of Duriron Company, Inc. (USA) for heat- and corrosion-resistant nickel-base alloys and austenitic stainless steels.

Durcon. Trademark of Duriron Company, Inc. (USA) for a series of epoxy resins with high toughness and impact resistance, and low moisture absorption.

Durcoton. Trade name for laminated phenolics.

Durcupan. Trademark of Fluka Chemie AG (Germany) for water-soluble epoxies used as embedding media in biochemical electron microscopy.

Dureco. Trademark of Sico Inc. (Canada) for paints and wood stains.

Duredge. (1) Trade name of Swedish American Steel Corporation (USA) for a hard, tough tool steel containing 0.55% carbon, 1.6% silicon, 0.4% molybdenum, 0.6% vanadium, and the balance iron. It maintains a keen cutting edge, and is used for cutting tools, chisels, rivet sets, shear blades, etc.

(2) Trade name of Boyd Wagner Company (USA) for water- and oil-hardening tool steels containing 0.55% carbon, 0.75% manganese, 0.35% molybdenum, 2% silicon, and the balance iron. Used for punches, shears, and dies.

Durehete. Trade name of British Steel Corporation (UK) for a series of low- and medium-carbon alloy steels containing 0.1-0.45% carbon, 0-0.5% silicon, 0.9-1.5% chromium, 0.5-1.1% molybdenum, 0-0.7% nickel, 0-0.8% vanadium, 0-0.2% titanium, and the balance iron. They have good high-temperature properties, and are used for fasteners, bolts, studs, and steam-plant parts.

Durel. (1) Trademark of Hoechst Celanese Corporation (USA) for a series of thermoplastic polyarylates available in transparent and opaque grades. They have high optical clarity, high heat-distor-tion temperatures, excellent flame resistance and ultraviolet stability, and good impact strength.

(2) Trade name for polyolefin fibers.

Durelon. Trademark of 3M Espe Dental (USA) for zinc polycarboxylate dental cements used for the cementation of bridges, crowns and inlays, and for lining and filling tooth cavities.

Dureta. Trade name for rayon fibers and yarns used for textile fabrics.

Durethan. (1) Trademark of Bayer Corporation (USA) for linear polyurethanes available in soft and rigid grades. They have softening temperatures ranging from 75 to 175°C (167 to 347°F), and properties that are largely dependent on the particular grade. Used for machine elements, e.g., screws, bolts, bushings, bearings, gears, tubing, and hoses.

(2) Trademark of Bayer Corporation (USA) for a series of polyamide (nylon 6 and nylon 6,6) resins supplied in unreinforced and mineral-reinforced grades for injection molding. The reinforced grades possess high strength, high toughness, good surface characteristics, and low warpage, and are used for automotive components, such as wheel covers, door handles and fuel-filler doors, household appliances, and electrical and electronic equipment. *Durethan A* refers to a thermoplastic polyamide (nylon 6,6) resins available in standard, glass fiber-reinforced and UV-stabilized grades. *Durethan B* is a thermoplastic polyamide (nylon 6) available in transparent and opaque shades and in many grades including standard, glass fiber-reinforced, fire-retardant, glass bead-filled, UV-stabilized, and elastomer copolymer. They have high strength and rigidity, high-impact strength, excellent resistance to gasoline, mineral oils, tetrachlorocarbon and dilute alkalies, moderate resistance to trichloroethylene and dilute acids, and a maximum service temperature of 100-150°C (210-300°F). Used for machine elements, e.g., bolts, screws, bearings, bushings, gears, cams, etc., and for hoses and tubes. *Durethan KU & RM KU* refers to polyamide 6 and 6,6 (nylon 6 and 6,6) and *Durethan T* to transparent polyamide 6 and polyamide 6,6/6T (nylon 6 and nylon 6,6/6T) resins.

Durex. (1) Trade name of Darwins Alloy Castings (UK) for a high-speed steel containing 0.7% carbon, 4.5% chromium, 1% vanadium, 18.5% tungsten, and the balance iron. Used for shaping and planing tools, broaches, hobs, and dies.

(2) Trade name of Moraine Manufacturing, Inc. (USA) for iron and steel made by blending iron powder with a small amount of carbon, pressing, and sintering for a prescribed period of time in a special furnace under controlled atmospheres. It has a density range of 5.5-7.6 g/cm³ (0.20-0.27 lb/in.³), a porosity of 5-35 vol%, and is used for mechanical parts, and porous and impregnated bearings.

(3) Trade name of Moraine Manufacturing, Inc. (USA) for a sintered bearing bronze composed of 95% copper and 5% tin.

(4) Trade name of Moraine Manufacturing, Inc. (USA) for a porous material containing 83.2% copper, 10% tin, 0.6% impurities, and 4.4-4.7% total carbon (in the form of graphite). It can absorb 25 vol% oil, and is used for bearings.

(5) Trade name of Manifattura Specchi e Vetri Felice Quentin (Italy) for toughened sheet glass.

(6) Trade name for reinforced phenolic resins.

Durexon. Trade name of Durexon GmbH& Co. (Germany) for a range of polymer coating products.

Durez. Trademark of Occidental Chemical Corporation (USA) for a series of synthetic resins including *Durez Alkyd* alkyd resins supplied in long and short glass fiber-reinforced, fire-retardant, mineral-filled,and synthetic fiber-filled grades, *Durez DAIP* diallyl isophthalate resins supplied in long and short glass fiber-reinforced and mineral-filled grades, *Durez DAP* diallyl phthalate resins supplied in long and short glass-fiber-reinforced, fire-retardant, mineral-filled and synthetic fiber-filled grades, and *Durez GMC* general-purpose polyester molding compounds supplied in standard, high-impact and mineral-filled grades. Also included under this trademark *Durez Urea* urea-formaldehyde resins, and *Durez PF* phenolic resins and molding compounds based on phenol-furfural, phenol-formaldehyde, diallyl phthalate, furfural alcohol and other resins, and available in various grades including compression-molding, injection-molding, mineral- and cellulose-filled, impact-resistant, glass-filled, wood flour-filled, and as varnishes and powders. They have good heat and electrical resistance, excellent thermal stability to over 150°C (302°F), high rigidity, strength, hardness and stability, low moisture absorption, good moldability and castability, and may be compounded with a large number of resins, fillers, etc., and made fairly resistant to chemicals, except strong acids and alkalies. *Durez* resins are used for laminates, shell forms, handles, pulleys, wheels, television and radio cabinets, plugs, fuse blocks, coil forms, automotive applications, such as drive pulleys, pump housings, disk brake pistons, distributors, electrical fixtures, etc., electrical and appliance applications, electronic components, and motor housings.

Durface. Trade name of Eutectic Corporation (USA) for a hard-facing alloy of carbon, nickel, molybdenum and iron.

Durflex. French trade name for a thick, toughened sheet glass.

Durichlor. Trademark of Duriron Company, Inc. (USA) for an alloy steel composed of 0.9-1.1% carbon, 14.5% silicon, 4-5% chromium, up to 0.65% manganese, and the balance iron. It has a relatively high hardness, moderate strength, and excellent resistance to chlorine, ferric chloride, sulfurous and sulfuric acid, nitric acid, chlorinated solutions and hydrochloric acid. Used for acid-handling pumps and impressed current anodes.

Durimet. Trademark of Duriron Company, Inc. (USA) for a series of alloys containing 15-35% nickel, 2.5-22% chromium, 1-5% silicon, 0.50-1% manganese, 1-3.5% molybdenum, 0.25-4% copper, up to 0.6% carbon, and the balance iron. Available in various grades including austenitic stainless steel. *Durimet* alloys have good strength, medium hardness, good machinability and weldability, good resistance to many acids and caustics, and good heat and corrosion resistance. Used for chemical equipment, pickling tanks, pumps, valves, corrosion-resistant parts, and fume ducts.

Durinval. Trade name of Creusot-Loire (France) for a corrosion-resistant, precipitation-hardening alloy of 54% iron, 42% nickel, 2% aluminum, and 2% titanium. Used for chronometer springs.

durionized material. A material on whose wearing surfaces has been produced an electrodeposit of hard chromium as a protection against frictional wear.

Duriron. Trademark of Duriron Company, Inc. (USA) for a high-silicon cast iron composed of 0.4-1.1% total carbon, 0.3-0.7% manganese, 14.2-14.7% silicon, and the balance iron. It has a ferritic microstructure, a density of 7.0 g/cm³ (0.25 lb/in.³), good resistance to abrasion, corrosion and erosion, fair to poor castability, low mechanical and thermal shock resistance at room temperatures, moderate mechanical and thermal shock resistance at elevated temperatures, moderate strength, high hardness, and poor machinability, but fairly good grindability. Used for chemical-processing equipment for handling corrosive media, such as sulfuric and nitric acids, and for pickling tanks, vessels, pumps, and valves.

Durisite. Trademark of US Stoneware Company for a furan resin with good acid and alkali resistance used in cement mortars.

Durisol. Trade name of Henkel Surface Technologies (USA) for an iron phosphate coating for steel products.

Duritas. Trade name of Jessop-Saville Limited (UK) for a cemented tungsten carbide used for nibs and dies.

Durite. (1) Trademark of Borden Chemical (USA) for a series of synthetic resins including phenol-formaldehydes and phenol furfurals with excellent thermal stability to above 150°C (300°F) used in the manufacture of brake linings, clutch facings, and grinding wheels.

   (2) Trade name of Columbia Tool Steel Company (USA) for a tungsten-type hot-work steel used for tools and dies.

   (3) Trademark of Federated Genco Limited (Canada) for white-metal bearing alloys.

   (4) Trade name for rough and semifinished iron-alloy castings.

   (5) Trade name for a wear-resistant nickel cast iron.

   (6) Trademark of Norton Company (USA) for abrasive cloth and paper.

   (7) Trademark of Durite USA, Inc. (USA) for premixed stucco.

Dur-Lab-Sil. Trade name of WP Dental/William & Pein GmbH (Germany) for a very hard, non-stick, heat-resistant, dimensionally stable condensation crosslinked silicone used for dental laboratory applications.

Durlux. Trade name of Saint-Gobain (France) for a toughened cast glass.

Durmax. Trademark of M.A. Hanna Engineered Materials (USA) for polyacetals and thermoplastic polyesters.

Durmes. Trade name of Dürener Metallwerke (Germany) for a free-cutting aluminum alloy containing 2.5-5% copper, 0.2-1.8% magnesium, 0.3-1.5% manganese, 0.5-2.5% lead, and small amounts of tin, cadmium and bismuth. Used for screw-machine products.

Dur-Ni. Trade name of Enthone-OMI Inc. (USA) for a bright nickel electroplate and process.

Durni-Coat. Trademark of AHC-Oberflächentechnik (Germany) for a nickel electroplate and plating process.

Duro. (1) Trade name of Electro Chemical Engineering & Manufacturing Company (USA) for corrosion resistant cements, coatings and linings.

   (2) Trade name of Thyssen Edelstahlwerke AG (Germany) for a water-hardening plain-carbon tool steels.

Duro-Antifriction. Trade name of Federal-Mogul Corporation (USA) for a tin babbitt containing 3-5% copper, 7-8% antimony, and the balance tin. Used for bushings, liners, and bearings.

Durobax. Trademark of Schott Rohrglas AG (Germany) for a special glass tubing with exceptional chemical resistance used for pharmaceutical and medical applications, e.g., test tubes, syringes and pipettes.

Duro-Bond. Trade name of Electro Chemical Engineering (USA) for natural rubber sheet linings supplied in hard (e.g., *Duro-Bond H-160*) and soft (e.g., *Duro-Bond S-103*) grades.

Durobor. Trade name of Durobor SA (Belgium) for high-strength glassware.

Durobronze. Trade name for a bronze containing 97% copper, 2% tin and 1% silicon. It has high electrical conductivity and high toughness.

Duro-Chip. Trade name of Allied Steel & Steel Tractor Products Inc. (USA) for an oil-hardening, shock-resistant tool steel containing 0.6% carbon, 0.7% manganese, 1.85% silicon, 0.5% molybdenum, 0.25% vanadium, and the balance iron. Used for hand and pneumatic chisels, and caulking tools.

Durochrome. Trade name of TRW Inc. (USA) for a wear-resistant cast iron used for valve parts.

Durock. Trade name of Canadian Gypsum Company for extremely durable and water-resistant cement board panels made of glass fiber-wrapped Portland cement and used inside shower stalls or around bathtubs.

Durocyl. Trade name of Midland Motor Cylinder Company (USA) for a centrifugally-cast cast iron containing 3% carbon, 2% silicon, 0.8% manganese, 0.3% chromium, and the balance iron. Used for wear plates and cylinder liners.

Durode. Trade name of Thomas Bolton Limited (UK) for a copper alloy with high electrical conductivity, used for spot-welding electrodes.

Durodet. Trademark of Flachglas AG (Germany) for thermosetting, glass fiber-reinforced unsaturated polyester prepregs and premixes, and molded articles.

Durodi. Trade name of A. Finkl & Sons Company (USA) for a tough, air-hardening hot-work tool steel containing 0.5-0.6% carbon, 0.7-0.9% molybdenum, 0.5-0.6% manganese, 0.85-1.15% chromium, 1.4-1.75% nickel, and the balance iron. It has good nondistorting properties, and is used for punches, shear knives, and upsetting and gripping dies.

Durodie. Trade name of Jessop-Saville Limited (UK) for a hot-work steel containing 0.4% carbon, 3.3% chromium, 2.7% tungsten, 0.3% vanadium, and the balance iron. Used for bolt dies, hot extrusion dies, and swaging dies.

Duro-Flex. Trademark of RPC, Inc. (USA) for rubber-based roof coatings.

Duroflint. Trademark of Sterling Limited (Canada) for concrete floor enamel paints.

Duro-Form. Trade name of Gulf States Steel Inc. (USA) for a series of high-strength low-alloy (HSLA) steels containing 0.1% carbon, 0.8% manganese, 0.5% silicon, 0.5% copper, 0.9% chromium, 0-0.02% niobium (columbium), and the balance iron. They have excellent atmospheric corrosion resistance, and are used for structural members, bridges, buildings, etc.

Duro-Glas. Trade name of former Spiegelglaswerke Germania AG (Germany) for a toughened polished or rough-cast plate glass. Also spelled "Duroglas."

Duro High Speed. Trade name of McInnes Steel Company (USA) for a high-speed steel containing 0.8% carbon, 18.5% tungsten, 4% molybdenum, 4% chromium, 7% cobalt, and the balance iron. Used for heavy-duty cutters.

Duroid. (1) Trade name of Duroid Covering Company (UK) for cellulose acetate plastics.

   (2) Trade name for polytetrafluoroethylenes supplied in unfilled and glass-, graphite- and bronze-filled grades.

Durolam. Trademark of Durolam Inc. (Canada) for high-pressure laminates.

Durolith. Trade name of Durex GmbH (Germany) for a zinc die-casting alloy.

Duromax. Trade name of Ambo-Stahl-Gesellschaft (Germany) for a series of high-strength low-alloy (HSLA) steels containing 0.1-0.2% carbon, 0.8-1% manganese, 0.15-0.35% silicon, 0.4-0.65% chromium, 0.4-0.6% molybdenum, 0.5-1% nickel, up to 0.05% vanadium, a trace of boron, and the balance iron. Used for structural members, bridges, buildings, etc.

Duron. Trade name of Drake Extrusion Limited (UK) for hard-wearing, rot- and stain-resistant polypropylene fibers used for carpet tiles, and nonwovens for automotiove applications.

Duronze. Trade name of Olin Brass, Indianapolis (USA) for a series of high-strength silicon bronzes containing 0-7% aluminum, 1-3% silicon, up to 0.95% manganese, 0.1% iron and 1.5% tin, respectively, and the balance copper. They have good corrosion resistance, and good cryogenic and elevated temperature properties. Used for forgings, condenser tubes, pump rods, boat shafting, gears, pinions, valve parts, liquid oxygen valves, marine hardware, pole-line hardware, water storage tanks, chemical vessels, range boilers, fasteners, and cable.

Duropal. Trademark of Duropal-Werk Eberhard Wrede GmbH & Co. KG (Germany) for high-pressure plastic laminates used in for furniture, and for finishing interior walls and ceilings.

Durophen. Trademark of Hoechst AG (Germany) for modified phenol-formaldehyde plastics.

Duroplast. Trademark of Duochem Inc. (USA) for pressed and extruded parts made of thermosetting plastics.

Duroplat. German trade name for an aluminum-coated *duralumin* alloy for bimetal parts.

Duro-Plate. Trade name of Gulf States Steel Inc. (USA) for a series of high-strength low-alloy (HSLA) steels containing 0.17%, 0.5-1.4% manganese, 0.25-0.5% silicon, 0.3-0.5% copper, 0.4-0.7% chromium, 0.005% niobium (columbium), up to 0.07% vanadium, 0.1% molybdenum and 0.4% nickel, respectively, and the balance iron. They have good resistance to atmospheric corrosion, and are used for structural members, bridges, buildings, etc.

Duro-Ply. Trade name of Electro Chemical Engineering & Manufacturing Company (USA) for neoprene-rubber sheet linings.

Duropool. Trademark of Sterling Limited (Canada) for synthetic rubber-based swimming-pool paints.

Duroprene. Trade name for chlorinated rubber.

Durostone. Trademark of Röchling Haren AG (Germany) for a series of fiber-reinforced plastics.

Duro-Tak. Trade name of Specialty Adhesives, National Starch and Chemical Corporation (USA) for solvent-borne, pressure-sensitive acrylic adhesives with high cohesive strength used for mounting tapes, decals, and nameplates.

Dur-O-Wal. Trademark of Dur-O-Wal Limited (Canada) for masonry wall reinforcements.

Durox. (1) Trade name for a lightweight, cellular concrete supplied in the form of blocks, panels and precast units.

(2) Trademark of Sonoco Products Company (USA) for strong paper and paperboard used for cores, tubes and similar shapes.

(3) Trade name of Ferromet Praha (Czech Republic) for ferrite materials.

Duroxyn. Trademark of Hoechst AG (Germany) for epoxy resins.

Dursico. Trademark of Sico Inc. (Canada) for an interior and exterior acrylic latex texture coating for masonry surfaces.

Dursil. Trade name of Dürener Metallwerke (Germany) for a series of aluminum alloys containing up to 10% silicon. Used in the automotive and aircraft industries, and for light-alloy parts.

Dursilium. French trade name for an age-hardenable aluminum

alloy that contains 4-5% copper, and is used for light-alloy parts.

Durvit. Trade name of Flachglas AG (Germany) for a toughened sheet glass.

Dusklite. Trade name of Duplate Canada Limited for a laminated sheet glass with tinted interlayer of vinyl plastic.

dussertite. A green mineral of the alunite group composed of a barium iron arsenate hydroxide monohydrate, $BaFe_3(AsO_4)_2$-$(OH)_5 \cdot H_2O$. Crystal system, rhombohedral (hexagonal). Density, 4.09 g/cm³; refractive index, 1.87. Occurrence: Algeria.

dust. (1) Particles of solid matter usually 1-150 μm (40-5900 μin.) in size.

(2) In powder metallurgy, a powder whose particles are mostly of submicron size, e.g., gold dust.

(3) In ceramics, very fine particles of minerals, e.g., mica, or ceramic compounds.

dust coat. A thin, sprayed coating of porcelain enamel or glaze with a dry, dusty appearance.

dust gold. A term referring to very fine gold, usually those pieces lighter than 3-5 g (0.11-0.18 oz.).

dusting agent. Graphite, talc, mica, slate, flour or clay powder used as an abherent and mold-release agent in the plastics and rubber industries.

Dustlon. Trade name of Nippon Mineral Fiber Manufacturing Company Limited (Japan) for a filter material consisting of fiberglass wool with a fiber diameter of 0.5 mm (0.04 in.), finished in felt shape.

Dutch Boy. Trade name of NL Industries (USA) for a series of lead-tin and tin-lead solders, and several lead-tin-antimony-copper and tin-antimony-copper bearing metals.

Dutch gold. A bright low brass composed of 76-80% copper and 20-24% zinc. It is usually supplied in the form of leaves or foils, and used as imitation gold leaf and for cheap jewelry. Also known as *Dutch leaf; Dutch metal; Dutch metal leaf.*

Dutch leaf. See Dutch gold.

Dutch metal. See Dutch gold.

Dutch silver. A corrosion-resistant alloy containing 9.6% copper, 8.8% antimony, and the balance tin. Used for tableware, novelties, and art metal objects.

Dutch white. A white paint pigment containing 3 parts of lead sulfate ($PbSO_4$) and 1 part of lead carbonate ($PbCO_3$).

Dutch white metal. A babbitt containing 9.5-9.6% copper, 8.5-8.8% antimony, and the balance tin. Used for bearings and antifriction metals.

Dutral. Trademark of EniChem SpA (Italy) for a series of olefinic thermoplastic elastomers.

duttonite. A pale brown mineral composed of vanadium oxide hydroxide, $VO(OH)_2$. Crystal system, monoclinic. Density, 3.00 g/cm³; refractive index, 1.900. Occurrence: USA (Colorado).

Duval. Trade name of Aubert & Duval (France) for a series of plain-carbon, cold-work, hot-work and high-speed tool steels.

duvetine. An imitation velvet made of wool, cotton or synthetic fibers.

duvetyn. (1) A soft, tightly woven fabric of cotton, wool, a wool blend or other fibers having a velvety finish and a short nap covering the weave. Used for suits, coats, etc.

(2) A close-faced, napped spun-silk fabric resembling duvetyn (1).

Dux. Trade name of British Steel Corporation (UK) for a series of tool steels containing about 0.5-1.2% carbon, 0.75-1.1% chromium, 1.4-2.25% tungsten, and the balance iron. Used for shear blades, punches, chisels, drills, taps, and burrs.

DWL-Lot. German trade name for a dental gold solder.

dwornikite. A snow-white mineral of the kieserite group composed of iron nickel sulfate monohydrate, $(Ni,Fe)SO_4 \cdot H_2O$. Crystal system, monoclinic. Density, 3.34 g/cm^3; refractive index, 1.63. Occurrence: Peru.

Dyal. Trade name for alkyd resin finishes.

Dycal. Trademark of Dentsply Corporation (USA) for calcium hydroxide dental cements.

Dycast. (1) Trade name of Latrobe Steel Company (USA) for a series of chromium-type hot-work tool steels (AISI type H11) containing 0.35-0.4% carbon, 0.95-1% silicon, 0-0.3% manganese, 5-5.4% chromium, 0.8-1.4% molybdenum, 0-1.25% tungsten, 0-0.5% vanadium, and the balance iron. They have high hot hardness, good resistance to heat checking, good toughness and machinability, and good wear resistance. Used for die-casting dies and hot punches.

(2) Trade name for a polyurethane resin used in dentistry for model and die work.

Dycastal. Trade name of Foseco Minsep NV (Netherlands) for compounds that are sometimes added in controlled amounts to aluminum-alloy melts for die castings to eliminates cracks and reduce localized shrinkage porosity.

Dycote. Trade name of Foseco Minsep NV (Netherlands) for a series of dressings for die-casting molds used for nonferrous alloys. See also dressing.

Dycro. Trade name of Pennsylvania Steel Corporation (USA) for an air-hardening, nondeforming tool steel containing 1.5% carbon, 12% chromium, 0.8% molybdenum, 0.25% vanadium, and the balance iron. Used for broaches, and forming and blanking dies.

dye. (1) An organic coloring matter that is used to impart color to other materials. Natural dyes, such as cochineal, indigo, logwood and madder, are derived from animal or vegetable sources, while synthetic dyes, such as acetate, anthraquinone, azo, alizarin and aniline dyes, are derived from petroleum and coal-tar intermediates. Used for dyeing or staining textiles, glazes, porcelain enamels, in coatings, in biochemistry and biotechnology, and in electronics and optoelectronics. Also known as *dyestuff*.

(2) The liquid or bath containing a dyestuff (1).

(3) A coloring agent in brush-on solution used to assist in the detection of surface imperfections, such as cracks or pinholes, in metals and ceramics.

dyestuff. See dye (1).

Dyflor. Trade name of Degussa-Huels/Creanova Inc. (USA) for polyvinylidene fluoride (PVDF) resins supplied in unreinforced and carbon fiber-reinforced grades. They have a density of about 1.78 g/cm^3 (0.064 $lb/in.^3$), nil water absorption, and an upper continuous-use temperature of 70°C (158°F). Processible by injection and compression molding and extrusion, they are used for chemical equipment, electrical cables, automotive components, and industrial machinery.

Dy-Krome. Trade name of Allied Steel & Tractor Products Inc. (USA) for an air-hardening tool steel containing 1.5% carbon, 12% chromium, 0.8% molybdenum, 0.2% vanadium, and the balance iron. Used for blanking and cold-extrusion dies, broaches, and punches.

Dy-Krome Special. Trade name of US Steel Corporation for an oil-hardening, nondeforming tool steel containing 2.15% carbon, 12.5% chromium, 1% vanadium, and the balance iron. Used for forming and drawing dies, punches, etc.

Dylan. Trademark of Koppers Company, Inc. (USA) for high-density linear polyethylene with a density of 0.95 g/cm^3 (0.034

$lb/in.^3$), high elongation, and a softening point of 123°C (253°F). Used for fibers, paper and cloth laminates, piping, film packaging, and wire insulation.

Dylark. Trademark of Nova Chemicals Inc. (USA) for a series of engineering resins based on styrene-maleic anhydride (SMA) copolymers and available in many grades including transparent, high-impact and glass-reinforced. An SMA/polystyrene alloy is also supplied. *Dylark* SMAs can be injection molded to produce automotive and other components.

Dylene. Trademark of Arco Chemical Company (USA) for polystyrene resins supplied in standard and high-impact grades.

Dylex. Trademark of Arco Chemical Company (USA) for acrylic styrene-butadiene latexes for paint applications.

Dylite. (1) Trade name of Arco Chemical Company (USA) for polystyrene beads and pellets that expand on application of heat into closed-cell foams for thermal insulation, packaging, flotation equipment, etc.

(2) Trade name of Arco Chemical Company (USA) for polystyrene plastics with plywood or gypsum facings.

Dy-lok. Trade name for rayon fibers and yarns used for textile fabrics.

Dylyn. Trademark of Advanced Refractory Technologies, Inc. (USA) for thin-film coatings of the diamond-like nanocomposite type. The amorphous microstructure consists of two random interpenetrating molecular networks—one based on carbon and the other on silicon—stabilized, respectively with hydrogen and oxygen. The coatings are produced by a special plasma/ion beam vacuum deposition technique and contain carbon in a primarily diamond bonding configuration. They have outstanding corrosion and erosion resistance, good wear resistance, low coefficients of friction, good thermal stability, good physical flexibility, low stress, and adhere well to many metal, ceramic and plastic substrates. They can also be doped with a third ceramic or metal network to produce the required mechanical, optical or electrical (superconductive, conductive or dielectric) properties, and deposited in multiple layers. Used for conductive or insulating coatings and optical coatings, and on chemical and electrochemical components (e.g., valves, probes, and sensors), automotive powertrain components, microelectronic components, marine components (e.g., valves), and surgical instruments.

Dymax. Trademark for a series of UV-cure adhesives for assembling rotors or solid fields made of ferrites, flexible magnets, alnico magnets, or rare-earth magnets.

Dymerex. Trade name of Hercules, Inc. (USA) for polymer-modified *rosin* used in the manufacture of coatings.

Dymix. Trade name of Saint-Gobain (France) for a neodymium-lead optical glass.

Dymo. Trade name for an extremely fine diamond powder with uniform particles supplied in sizes down to 0.5 μm (20 μin.). Used as a polishing powder.

Dymonhard. Trade name of Dymonhard Corporation of America (USA) for a series of hardfacing electrodes based on chromium-nickel-molybdenum, chromium-nickel-tungsten, chromium-tungsten-molybdenum or iron-chromium-nickel alloys.

Dynablast. Trademark of Norton Company (USA) for a high-purity aluminum oxide abrasive supplied in grit sizes from 8 to 240 mesh. Used for pressure blasting of metals, for cutting rock, and for polishing metals and glass.

Dynabond. Trademark of C-I-L Paints (Canada) for autobody fillers.

Dynacast. (1) Trade name of Dynacast, Inc. (USA) for zinc-base precision die castings.

(2) Trademark of Dynamit Nobel AG (Germany) for electrically fused metallic oxides and oxide mixtures used in the manufacture of fusion-cast refractory bricks and pipes.

Dynacor. Trade name for rayon fibers.

Dyna Core. Trade name of Conneaut Industries, Inc. (USA) for twisted aramid used in high-performance automotive ignition wires.

Dynacut. Trade name of Latrobe Steel Company (USA) for a wear-resistant high-speed tool steel (AISI type M-43) containing 1.2% carbon, 3.75% chromium, 2.7% tungsten, 1.6% vanadium, 8% molybdenum, 8.2% cobalt, and the balance iron. It has high red hardness, and is used for broaches, chasers, drills, and cutters for heavy-duty machining.

Dynafilm. Trademark of National Distillers & Chemical Corporation (USA) for a series of very thin polypropylene films and sheets coated or laminated on one or both sides with other polymers, such as vinyl acetate or polyethylene. They have good heat-sealing properties, and are used for packaging applications.

Dynaflex. (1) Trademark of Woodbridge Foam Corporation (Canada) for flexible polyurethane foam used for seat cushions.

(2) See Dynaflex Vac-Arc.

Dyna-Flex. Trade name for high-temperature resistant ceramic sheets made by interfelting aluminum silicate fibers.

Dynaflex Vac-Arc. Trade name of Latrobe Steel Company (USA) for a hot-work tool steel (AISI type H11) containing 0.4% carbon, 0.3% manganese, 0.9% silicon, 1.3% molybdenum, 5% chromium, 0.45% vanadium, and the balance iron. Used for upsetters, punches, extrusion and forging dies, and mandrels. Formerly known as *Dynaflex*.

Dyna-Gel. Trade name of Dyna-Dri (USA) for silica gel supplied in clear and humidity-indicating beads.

Dynakote. Trademark of C-I-L Paints (Canada) for acrylic enamel paints used in automotive refinishing.

Dyna-Kut. Trade name of Stanadyne (USA) for a series of resulfurized and rephosphorized free-machining grades of standard carbon steel. Depending on the particular grade, the composition range is: up to 0.15% carbon, 0.6-1.15% manganese, 0.07-0.12% phosphorus, 0.1-0.35% sulfur, and the balance iron. Some grades contain 0.15-0.35% lead. Supplied in bar form, they are used for screw-machine products.

Dynalite. (1) Trademark of C-I-L Paints (Canada) for acrylic lacquer for automotive refinishing.

(2) Trademark of Arco Chemical Company (USA) for expandable polystyrene used for various applications in the chemical industry.

Dyna-Loy. Trade name of Stanadyne (USA) for a series of carbon and alloy steels containing considerable amounts of lead (typically 0.15-0.35%) to improve machinability.

DynaMax. Trade name of Johns Manville Company (USA) for modified bitumen roofing.

Dynamax. (1) Trade name for an alloy of 65% nickel, 33% iron and 2% molybdenum. It has high permeability, and is used for toroidal cores.

(2) Trade name of Timken Latrobe Steel Company (USA) for a high-speed tool steel (AISI type M42) containing 1.08% carbon, 3.75% chromium, 9.5% molybdenum, 1.5% tungsten, 1.1% vanadium, 8% cobalt, and the balance iron. It has very high wear resistance and good red hardness, and is used for cutting tools for long production runs or heavy-duty machining.

Dynamet. Trade name of Dynamet Inc. (USA) for a wrought cobalt alloy containing 20% chromium, 17% iron, 13% nickel and 2.5% tungsten. It has good high-temperature properties, and is used for springs.

Dynamo. (1) Trade name of Patriarche & Bell (USA) for a water-hardening tool steel containing 0.7-0.9% carbon, and the balance iron. Used for dies, jigs, and tools.

(2) Trade name of Johnson Matthey plc (UK) for a low-carbon steel containing 0.12% carbon, 0.2% silicon, 0.1% manganese, and the balance iron. It has high magnetic permeability, and is used for electrical and magnetic equipment.

dynamo sheet steel. A silicon steel containing up to 0.1% carbon, 3-5% silicon, up to 0.3% manganese, up to 0.03% phosphorus and sulfur, respectively, and the balance iron. Its microstructure consists of very coarse, homogeneous ferrite grains, and it has high electrical resistivity, high magnetic permeability, and low hysteresis losses. Used for electrical equipment and laminated sheets for dynamos. Also known as *dynamo steel sheet*.

dynamo steel. A low-alloy steel containing about 0.05-0.15% carbon, 0.20% manganese, up to 2.0% silicon, and the balance iron. It has high magnetic permeability, and is used for dynamos.

dynamo steel sheet. See dynamo sheet steel.

Dynamullit. Trademark of Dynamit Nobel AG (Germany) for high-temperature-resistant *mullite* based ceramics for furnace insulation applications.

Dynaplan. Trademark of Dynamit Nobel AG (Germany) for polyvinyl chloride film materials.

Dyna-Plus. Trade name of Dynatron/Bondo Corporation (USA) for a stain-free, nonbleeding autobody filler that dries tackfree in less than 20 minutes.

Dynapro. Trade name of Dynapro-Products (USA) for waterproofing coatings for roofing repairs, and roof-seam sealing applications.

Dyna-Quartz. Trade name for silica fiber felt (99% pure) used for lightweight, semirigid tiles and blocks.

Dynarohr. Trademark of Dynamit Nobel AG (Germany) for polyvinyl chloride used for pipes.

Dyna Seal. Trade name of Williams Products, Inc. (USA) for polyurethane sealants.

Dynaspinell. Trademark of Dynamit Nobel AG (Germany) for spinel (magnesium aluminate) based ceramics used for high-temperature furnace insulation applications.

Dynasta. Trade name of Fiberglass Limited (UK) for fiberglass-reinforced materials.

Dynasty. Trade name for a silver-free palladium dental bonding alloy.

Dynavan. Trade name of Latrobe Steel Company (USA) for a high-speed steel (AISI type T-15) containing 1.6% carbon, 12.5% tungsten, 4% chromium, 5% vanadium, 5% cobalt, and the balance iron. It has good heat and abrasion resistance, and high red hardness. Used for cutting tools, blanking dies, broaches, drills, and milling cutters.

Dynavar. Trade name of Hamilton Technology Inc. (USA) for a wrought, age-hardenable nonmagnetic alloy of 40% cobalt, 20% chromium, 15% nickel, 7% molybdenum, 2% manganese, 0.15% carbon, 0.04% beryllium, and the balance iron. It is suitable for sub-zero temperatures and elevated temperatures up to about 400°C (750°F), and has good corrosion resistance. Used for electronic components, watch and instrument springs, flapper valves, and valve parts.

Dynazirkon. Trademark of Dynamit Nobel AG (Germany) for a series of high-purity zirconia (ZrO_2) powders stabilized with yttria (Y_2O_3) or magnesia (MgO), and available in fine particle sizes down to 0.6 μm (24 μin.) and low silica contents. Used in the production of ceramics with tetragonal zirconia polycrystal (TZP) structures, in electronics, and as coatings for gas turbine engines. Also included under this trademark are various electrically molten metallic zirconium-based or zirconium-bearing oxides and oxide mixtures.

Dyneema. Trademark of DSM High Performance Fibers (Netherlands) for an ultrahigh-molecular-weight polyethylene fiber produced by gel spinning. It has a density of 0.97 g/cm^3 (0.035 $lb/in.^3$), and high ultimate tensile strength (2.70 GPa) and elastic modulus (87 GPa). Used as a reinforcing fiber for ballistic, medical and other applications.

Dynel. Trademark of Dynel Division, Union Carbide Corporation (USA) for a modified acrylic fiber (modacrylic fiber) spun from a copolymer of 40% acrylonitrile and 60% vinyl chloride. It has excellent water resistance, good resistance to most acids, alkalies and solvents, good mildew resistance, high tenacity and excellent drape, and does not support combustion. Used for pile and napped fabrics, work clothing, anode bags, fabric covers for paint rollers and other industrial goods, and blended with cotton, rayon, wool or other fibers for dresses, suits and underwear.

Dynelec. Trade name of Empire Sheet & Tin Plate Company (USA) for an alloy containing 97.5% iron and 2.5% silicon. It has good magnetic permeability, and is used for armatures, electric motors, and generators.

Dyno. Trademark of Dyno Overlays Inc. (USA) for a series of unprocessed aminoplast and phenoplast resins used for saturated overlays and glue films.

Dynobond. Trade name of Dynobond (USA) for an acrylic polymer cement used in building construction as an exterior/interior coating and for concrete repair and restoration.

Dynopas. Trademark of Dynamit Nobel AG (Germany) for vulcanized fiber products.

Dynoplast. Trade name of Dynoplast AB (Sweden) for a series of plastics, foams and rubbers.

Dynos. Trademark of Dynamit Nobel AG (Germany) for vulcanized fiber products.

Dynyl. Trademark of Rhodiaceta (France) for polyamide (nylon) plastics and products.

Dyphene. Trademark of Sherwin-Williams Chemicals (USA) for a series of phenolic resins with excellent chemical resistance, high hardness, and good drying properties.

dypingite. (1) A white mineral composed of magnesium carbonate hydroxide pentahydrate, $Mg_5(CO_3)_4(OH)_2 \cdot 5H_2O$. Crystal system, monoclinic. Density, 2.15 g/cm^3; refractive index, 1.510. Occurrence: Norway.

(2) A snow-white or gray mineral composed of magnesium carbonate hydroxide octahydrate, $Mg_5(CO_3)_4(OH)_2 \cdot 8H_2O$. Refractive index, 1.521. Occurrence: Japan.

Dypro Polyallomer. Trade name of Arco Chemical Company (USA) for an ethylene propylene copolymer.

Dypro PP. Trade name of Arco Chemical Company (USA) for a series of polypropylene resins supplied in homopolymer, UV-stabilized and talc-filled grades.

Dyract. Trademark of Dentsply/Caulk (USA) for a fluoride-releasing *compomer* consisting of a matrix of amorphous silica filled with barium glass particles. It has high wear resistance, good luster and polishability, high tensile strength, and is used in dentistry for restorations and cervical erosions. *Dyract Cem* is a polyacid-modified dental composite cement, and *Dyract PSA* a pressure-sensitive dental primer/adhesive for tooth structure (i.e., dentin and enamel) bonding.

dyscrasite. A silver-white mineral composed of antimony silver, Ag_3Sb. Crystal system, orthorhombic. Density, 9.71-9.74 g/cm^3. Occurrence: Germany, Australia.

Dysoid. British trade name for a free-cutting alloy containing 63% copper, 18% lead, 10% tin and 10% zinc used for hardware.

dysprosia. See dysprosium oxide.

dysprosium. A lustrous, silvery metallic element of the lanthanide series (rare-earth group) of the Periodic Table. It is commercially available in the form of ingots, rods, lumps, turnings (chips), sponge, wire, sheet, foil, powder and single crystals. Density, 8.536 g/cm^3; melting point, 1412-1505°C (2574-2741°F); boiling point, 2562°C (4644°F); hardness, 55 Vickers; atomic number, 66; atomic weight, 162.50; trivalent. It reacts slowly with water and halogen gases, and has a high cross section for thermal neutrons, outstanding corrosion resistance, and good magnetic properties. Four forms of dysprosium are known: (i) *Alpha dysprosium* with hexagonal close-packed crystal structure, present between room temperature and 950°C (1742°F); (ii) *Beta dysprosium* with body-centered cubic crystal structure, present at high temperatures (above 950°C or 1742°F); (iii) *Gamma dysprosium* with orthorhombic crystal structure, present below 86K (-187.15°C); and (iv) *Delta dysprosium* with rhombohedral crystal structure, present at room temperature and 75 kbar pressure. *Dysprosium* is used for control rods in nuclear reactors, reactor fuels, in permanent magnets and optical memories, as a fluorescence activator in phosphors, as a dopant for laser crystals, in garnets, ceramics, superconductors and semiconductors, in catalysts and garnet microwave devices, and in neutron flux measurement. Symbol: Dy.

dysprosium aluminate. A crystalline compound of dysprosium oxide and aluminum oxide. Density, 6.05 g/cm^3; melting point, 1816°C (3300°F). Used in ceramics. Formula: $Dy_2Al_4O_9$.

dysprosium antimonide. A compound of dysprosium and antimony supplied in high purity and used as a semiconductor. Formula: DySb.

dysprosium arsenide. A compound of dysprosium and arsenic supplied in high purity and used as a semiconductor. Formula: DyAs.

dysprosium boride. Any of the following compounds of dysprosium and boron used in ceramics and materials research: (i) *Dysprosium tetraboride*. Tetragonal crystals. Density, 6.74-6.98 g/cm^3; melting point, 2500°C (4530°F). Formula: DyB_4; and (ii) *Dysprosium hexaboride*. Density, 5.49 g/cm^3. Formula: DyB_6.

dysprosium bromide. White, hygroscopic crystals with a melting point of 879°C (1614°F). Used in ceramics, electronics, superconductor research, and organic synthesis. Formula: $DyBr_3$.

dysprosium carbide. Any of the following compounds of dysprosium and carbon used in ceramics and materials research: (i) *Tridysprosium carbide*. Density, 9.21 g/cm^3. Formula: Dy_3C; (ii) *Dysprosium sesquicarbide*. Formula: Dy_2C_3; and (iii) *Dysprosium dicarbide*. Density, 7.45 g/cm^3. Formula: DyC_2.

dysprosium chloride. An off-white powder (99.9+% pure) with a density of 3.67 g/cm^3 and a melting point of 718°C (1324°F), used in ceramics, electronics, superconductor research, and organic synthesis. Formula: $DyCl_3$.

dysprosium chloride hexahydrate. Light yellow, hygroscopic crystals (99+% pure) used in ceramics, electronics, superconductor research, and organic synthesis. Formula: $DyCl_3 \cdot 6H_2O$.

dysprosium columbate. See dysprosium niobate.

dysprosium dicarbide. See dysprosium carbide (iii).

dysprosium dichloride. Black crystals with a decomposition temperature of 721°C (1330°F), used in chemistry and materials research. Formula: $DyCl_2$.

dysprosium diiodide. Purple crystals with a melting point of 659°C (1218°F), used in chemistry and materials research. Formula: DyI_2.

dysprosium disilicate. See dysprosium silicate (ii).

dysprosium disilicide. See dysprosium silicide (i).

dysprosium disulfide. See dysprosium sulfide (iii).

dysprosium fluoride. Green crystals or white, hygroscopic powder (99.9+% pure) with a melting point above 1154°C (2109°F) and a boiling point above 2200°C (3990°F). Used in ceramics, electronics, superconductor research, and organic synthesis. Formula: DyF_3.

dysprosium heptasulfide. See dysprosium sulfide (i).

dysprosium hexaboride. See dysprosium boride (ii).

dysprosium hydride. Hexagonal crystals with a density of 7.1 g/cm^3, used in chemistry and materials research. Formula: DyH_3.

dysprosium iodide. Green crystals, or yellow, hygroscopic powder (99.9+% pure) with a melting point of 955°C (1750°F) and a boiling point of 1320°C (2410°F). Used in ceramics, electronics, superconductor research, and organic synthesis. Formula: DyI_3.

dysprosium monosilicate. See dysprosium silicate (i).

dysprosium niobate. A compound of dysprosium oxide and niobium pentoxide with a density of 5.49 g/cm^3, a melting point of approximately 1950°C (3540°F). Used in ceramics. Formula: $Dy_2Nb_2O_8$ ($Dy_2Cb_2O_8$). Also known as *dysprosium columbate.*

dysprosium nitrate. Yellow, hygroscopic crystals (99.9% pure) with a melting point of 88.6°C (191°F), used in ceramics, electronics, material research, chemistry, and as a strong oxidizing agent. Formula: $Dy_2O_3 \cdot 5H_2O$. Also known as *dysprosium nitrate pentahydrate.*

dysprosium nitride. A compound of dysprosium and nitrogen. Crystal system, cubic. Density, 9.93 g/cm^3. Used in ceramics, electronics, and materials research. Formula: DyN.

dysprosium oxide. A white, hygroscopic crystalline powder (99.9+% pure). Crystal system, cubic. Density, 7.81 g/cm^3; melting point, 2330-2408°C (4225-4365°F); more magnetic than ferric oxide. Used in cermets, component in control rods for nuclear reactors, as a phosphor activator and neutron-density

indicator, and in dielectric compositions. Formula: Dy_2O_3. Also known as *dysprosia.*

dysprosium pentasilicide. See dysprosium silicide (ii).

dysprosium phosphide. A compound of dysprosium and phosphorus usually supplied in high purity and used as a semiconductor. Formula: DyP.

dysprosium sesquicarbide. See dysprosium carbide (ii).

dysprosium sesquisulfide. See dysprosium sulfide (ii).

dysprosium silicate. Any of the following compounds of dysprosium oxide and silicon dioxide used in ceramics and materials research: (i) *Dysprosium monosilicate.* Melting point, 1930°C (3505°F); hardness, 5-7 Mohs. Formula: Dy_2SiO_5; (ii) *Dysprosium disilicate.* Melting point, 1721°C (3130°F); hardness, 5-7 Mohs. Formula: $Dy_2Si_2O_7$; and (iii) *Dysprosium trisilicate.* Melting point, 1920°C (3488°F); hardness, 5-7 Mohs. Formula: $Dy_2Si_3O_9$.

dysprosium silicide. Any of the following compounds of dysprosium and silicon used in ceramics and materials research: (i) *Dysprosium disilicide.* Orthorhombic crystals. Density, 5.2-6.8 g/cm^3; melting point, 1248°C (2278°F). Formula: $DySi_2$; and (ii) *Dysprosium pentasilicide.* Crystalline compound. Formula: Dy_3Si_5.

dysprosium sulfate. Bright yellow, hygroscopic crystals (99.9+% pure) that lose $8H_2O$ at 360°C (680°F) turning light yellow. Used in atomic-weight determinations, and in chemical synthesis. Formula: $Dy_2(SO_4)_2 \cdot 8H_2O$. Also known as *dysprosium sulfate octahydrate.*

dysprosium sulfate octahydrate. See dysprosium sulfate.

dysprosium sulfide. Any of the following compounds of dysprosium and sulfur: (i) *Dysprosium heptasulfide.* Density, 6.35 g/cm^3; melting point, 1540°C (2804°F). Used in ceramics and materials research. Formula: Dy_5S_7; (ii) *Dysprosium sesquisulfide.* Red-brown crystals. Crystal structure, orthorhombic. Density, 6.1-6.5 g/cm^3; melting point, 1479°C (2694°F); band gap, 1.77 eV. Used in ceramics and electronics. Formula: Dy_2S_3; and (iii) *Dysprosium disulfide.* Density, 6.48 g/cm^3. Formula: DyS_2.

dysprosium tetraboride. See dysprosium boride (i).

dysprosium trisilicate. See dysprosium silicate (iii).

Dytron XL. Trade name of Advanced Elastomer Systems (USA) for a flame-retardant thermoplastic elastomer used for electronic components.

dzhalindite. A yellow to brown mineral of the sohngeite group composed of indium hydroxide, $In(OH)_3$. It can also be made synthetically. Crystal system, cubic. Density, 4.45 g/cm^3; refractive index, 1.716. Occurrence: Russian Federation.

E

Eagle & Globe. Trade name of Eagle & Globe Steel Limited (Australia) for an extensive series of steels including several austenitic, ferritic, martensitic and precipitation-hardening stainless steels, hot-die steels, and chromium roller-bearing steels.

eagle wood. The aromatic reddish wood of large tree *Aquilaria agallocha*. Source: Southern Asia (especially Burma, India and Java.). Used for ornamental applications.

eakerite. A colorless mineral composed of calcium aluminum tin silicate hydroxide dihydrate, $Ca_2SnAl_2Si_6O_{18}(OH)_2 \cdot 2H_2O$. Crystal system, monoclinic. Density, 2.93 g/cm^3; refractive index, 1.586. Occurrence: USA (North Carolina).

E Alloy. Trade name for a non-heat-treatable aluminum alloy containing 20% zinc, 2.5-3% copper, 0.2-1% silicon, 0.5% magnesium, 0.5% manganese, and traces of iron. Developed by the National Physical Laboratory in the United Kingdom, it is used for lightweight structures.

E-Alloy. Trade name of NL Industries (USA) for a non-heat-treatable aluminum casting alloy containing 2% copper and 2% nickel.

earlandite. A white to pale yellow mineral composed of calcium citrate tetrahydrate, $Ca_3(C_6H_5O_7)_2 \cdot 4H_2O$. It can also be made synthetically. Crystal system, monoclinic. Density, 1.95 g/cm^3; refractive index, 1.56. Occurrence: Antarctica (Weddell Sea).

Earlumin. Spanish trade name for an age-hardenable alloy containing 95-96% aluminum and 4-5% copper. Used for light-alloy parts.

earth. (1) A term referring to any argillaceous or siliceous mixture or compound, such as diatomaceous earth, fuller's earth, etc.

(2) A natural metallic oxide. See also earth color.

(3) Any of a series of chemically related metallic elements, such as the alkaline earths (e.g., barium or strontium), or the rare earths (e.g., dysprosium or samarium).

earth color. Any pigment of mineral origin, such as umber, ocher, red iron oxide, etc.

earthenware. An opaque, nonvitreous, glazed or unglazed type of pottery consisting of clay, silica, kaolin and feldspar, fired at a relatively low temperature, and having a water absorption greater than 3%. It must be glazed to become nonporous.

earthenware clay. A fine-textured plastic clay with very low lime and gypsum contents that is quick-slaking, has low air and fire shrinkage, and is suitable for earthenware.

earth flax. See asbestos.

earth materials. See geological materials.

earth metals. Metals, such as barium, calcium or strontium, whose oxides are classified as earths.

earth oxides. Oxides of alkaline earths, such as barium, calcium and strontium. See also alkaline earths; barium oxide; calcium oxide; strontium oxide.

earth pigments. Natural or mineral pigments, such as yellow ocher, red and yellow iron oxides, raw and burnt siennas and umbers, that occur as deposits in the earth.

earth wax. See ozocerite.

earthy cobalt. See asbolane.

Easilite. Trade name of Libbey-Owens-Ford Company (USA) for heat-absorbing glass.

Eastacoat. Trade name of Eastman Chemical Company (USA) for extrusion coating resins based on polyester and polyethylene resins.

East African camphorwood. See camphorwood.

Eastalloy DA. Trade name of Eastman Chemical Company (USA) for polycarbonate/polyethylene terephthalate (PC/PET) alloys.

Eastapak. Trade name of Eastman Chemical Company (USA) for polyethylene terephthalate resins for injection stretch-blow molding. Used in food and nonfood packaging, and beverage containers.

Eastar. Trade name of Eastman Chemical Company (USA) for a series of copolyester resins in particular glycol-modified polycarbonate terephthalates (*Eastar DN*) and glycol-modified polyethylene terephthalates (*Eastar GN* and *Eastar PETG*). Available in standard and specialty grades, they can be injection-molded, extruded into film or thin sheeting, and blow-molded into containers.

Eastar Bio. Trade name of Eastman Chemical Company (USA) for biodegradable polyester available in the form of flexible film for food packaging, and lawn and garden bags.

eastern ash. See American ash.

eastern cottonwood. The wood of the tall tree *Populus deltoides* whose grayish-white to light-brown heartwood merges gradually with the whitish sapwood. It has a uniform texture, a straight grain, moderate strength, and is somewhat difficult to work with tools, because of its fuzzy surface. Average weight, 450 kg/m^3 (28 lb/ft^3). Source: Canada (southern Ontario), USA (eastern states). Used for lumber and veneer for crates, boxes, baskets, pallets, woodenware and paneling, and as a pulp- and fuelwood. Also known as *common cottonwood*.

eastern hemlock. The wood of the tall conifer *Tsuga canadensis*. The heartwood is pale brown with a reddish hue, and the sapwood is not distinctly separated, but may be lighter. *Eastern hemlock* has a coarse, uneven texture, and a bark rich in tannin. Average weight, 450 kg/m^3 (28 lb/ft^3). Source: United States (Northeastern and Lake states, and from Appalachian Mountains down to northern Alabama) and Canada (New Brunswick, Nova Scotia, southern Quebec and southern Ontario). Used as construction lumber, railroad ties, boxes and crates, and as pulp- and fuelwood.

eastern larch. See American larch.

eastern red cedar. The moderately strong, light, durable wood of the conifer *Juniperus virginiana*. The aromatic heartwood is bright to dull red or rose-brown, and the thin sapwood nearly white. *Eastern red cedar* has a straight, close grain, contains numerous knots, possesses good decay resistance, high durability, high stability in use, and works well. Average weight, 545 kg/m^3 (34 lb/ft^3). Source: Canada (Southern Ontario) and United States (eastern and central states). Used as lumber for cedar chests, wardrobes, cabinets, closet linings, novelty items, lead pencils and fenceposts, and as fuelwood. Also known as *pencil cedar; red juniper.*

eastern spruce. The light-colored wood of either the black spruce *(Picea mariana),* red spruce *(Picea rubens)* or white spruce *(Picea glauca),* available under the same name. There is only a slight difference between the heartwood and the sapwood of these trees. *Eastern spruce* has good stability and moderate strength. Average weight, 450 kg/m³ (28 lb/ft³). Source: Eastern United States and Canada. Used as pulpwood, framing lumber, and for millwork, boxes, crates, and piano sounding boards. See also black spruce; red spruce; white spruce.

eastern sycamore. See American sycamore.

eastern white cedar. The light, soft, durable wood of the arborvitae *Thuja occidentalis.* The heartwood is light colored and decay-resistant. *Eastern white cedar* has low strength, good painting qualities, and works well. Average weight, 352 kg/m³ (22 lb/ft³). Source: Canada (from Nova Scotia to Manitoba) and United States (northeastern and Lake states). Used for poles, posts, fencing, ties, lumber, tanks, boatbuilding, woodenware and shingles. Also known as *northern white cedar; swamp cedar.* See also arborvitae.

eastern white pine. The soft, light wood of the conifer *Pinus strobus.* The heartwood is light brown and darkens on exposure. *Eastern white pine* has a fine, uniform texture, a straight grain, very low porosity, is generally not as resinous as other pines, easily kiln-dried, and has a small shrinkage, high stability, works and glues well, and takes a beautiful finish. Average weight, 380 kg/m³ (24 lb/ft³). Source: Eastern North America from Maine to Georgia, also from Great Lake region and St. Lawrence forests. Used as lumber for foundry patterns, window frames, doors, millwork, furniture, trim, paneling, caskets, toys, containers, and packaging. Also known as *northern white pine; Weymouth pine.*

East India gum. Any of several fossil or semirecent dammar resins, especially the following varieties: (i) *East India Macassar gum,* a fossil dammar with a density of 1.03; (ii) *Singapore gum,* a brown or reddish fossil dammar with a density of 1.04; and (iii) *Batu,* a dammar gum with a density of 1.00-1.05 and a melting point of 180°C (356°F). Used in spirit and oleoresinous varnishes, as flatting agents in paints, in printing inks, in plastics, and in adhesives and oilcloth. See also dammar.

East India Macassar gum. See East India gum (i).

East Indian teak. The beautiful fragrant wood of the tall tree *Tectona grandis* belonging to the verbena family. It is golden-yellow to deep-brown in color, somewhat resembling oak in appearance, and has a coarse, open grain, high strength, hardness, stability and durability, low shrinkage, good resistance to insect attack, a moist, greasy feel, and moderate workability (i.e., the silica content tends to dull tools). Average weight, 640 kg/m³ (40 lb/ft³). Source: Burma, East Indies, India, Java, Thailand. Used for shipbuilding, bridges, fine furniture, cabinetwork, boxes, chests, flooring and paneling, decorative objects, and plywood. See also teak.

East Indian walnut. The hard, dense wood of the medium-sized tree *Albizzia lebbek.* It has a fine texture, close grain, and is dark brown with grayish stripes. Average weight, 800 kg/m³ (50 lb/ft³). Source: Southern and Southeast Asia, Central and Southern Africa. Used for furniture, cabinetwork, paneling, furnishings, etc.

Eastlene. Trademark of Far Eastern Textile Limited (Taiwan) for polyester filament yarns.

Eastlon. Trademark of Far Eastern Textile Limited (Taiwan) for polyester staple fibers.

Eastman. Trade name of Eastman Chemical Company (USA) for polyester-based thermoplastic resins and fibers including polypropylenes, polyethylene terephthalates, polycarbonate terephthalates, glycol-modified polycarbonate terephthalates, and polycarbonate/polyester alloys. Also included under this trade name are several grades of cellulose acetate fibers including *Eastman 50* and *Eastman 75.*

Eastman Hifor. Trade name of Eastman Chemical Company (USA) for high-strength linear polyethylene used for packaging applications.

Eastobond. Trademark of Eastman Adhesives (USA) for hot-melt adhesives for bonding board, film, foil and paper. They are now sold by H.B. Fuller Company (USA) under the trade name *Opt-E-Bond.*

eastonite. A mineral composed of potassium magnesium aluminum silicate hydroxide, $K_2Mg_5AlSi_5Al_3O_{20}(OH_4)$.

Eastpac. Trade name of Eastman Chemical Company (USA) for a series of polyester resins including in particular polyethylene terephthalates for packaging applications.

Easy. Trade name of Engelhard Corporation (USA) for an alloy containing 65% silver, 20% copper, and 15% zinc. It has a melting range of 693-718°C (1280-1325°F), and is used as a silversmithing solder for sterling silver.

EasyBond. Trade name of Parkell, Inc. (USA) for a light-cure dental bonding system consisting of the adhesive monomer 4-methacryloxyethyl trimellitate anhydride (4-META) and a surface conditioner composed of citric acid and ferric chloride. It forms a strong bond between the composite resin, metal or ceramic restoration and the tooth substrate (i.e., dentin and enamel).

easy-care fabrics. Textile fabrics, usually of the machine washable and tumble drying type, that regain their original appearance after washing and require little or no ironing.

Easy Dye. See Accepta.

Easy-Flo. Trademark of Handy & Harman (USA) for a series of silver-based brazing filler metals containing varying amounts of copper, zinc and/or cadmium. They have a brazing range of 600-750°C (1112-1382°F).

Easy Flow. Trade name for a dental gold solder.

Easy Seal. Trade name of Combustion Engineering Company (USA) for an insulating cement.

Easy Silicone. Trade name of DAP Inc. (USA) for silicone-base sealants, adhesives, cements, grouts and caulking, and spackling compounds.

Eatonite. Trade name of Eaton Corporation (USA) for an alloy containing 2-2.75% carbon, 37-41% nickel, 14% tungsten, 27-31% chromium, 9-11% cobalt, 0-8% iron, and 0-1% silicon. It retains its hardness at high temperature, and has good corrosion resistance to hot gases and antiknock fuels. Used for valve facings and valve seats.

EAZALL. Trademark of Eastern Alloys (USA) for a series of zinc alloys including *Zamak* and *ACuZinc* die-casting alloys, *ZA* sand, pressure-die casting and permanent-mold alloys, *Galfan* and *Galvalume* alloys for coating steels, *Kirksite* alloys, and several slush-casting alloys.

EBA cement. See ethoxybenzoic acid cement.

Ebalta. Trade name of Ebalta Kunststoff GmbH (Germany) for a series of specialty and custom-made casting resins for the manufacture of foundry patterns, tools, molds, gages and prototypes.

EB copolymers. See ethylene-butylene copolymers.

Eberle. Trade name of J.N. Eberle & Cie. GmbH (Germany) for cold-rolled specialty strip steel.

Ebonit. German trade name for a hard material made from rubber by vulcanizing with large quantities of sulfur.

ebonite. A hard, rigid, black substance made by compounding crude rubber with 30-35 wt% sulfur and curing at prescribed temperatures. It has high toughness, is several times stiffer than normal rubber, and takes a high polish. Used for tank linings, battery boxes, acid and alkali-resistant equipment, combs, buttons, ornaments, and electric insulation. *Ebonite* dust is used as a filler for other rubber goods. Also known as *hard rubber; vulcanite.*

ebonized asbestos. Asbestos building panels bonded with asphalt.

ebonized nickel. A nickel or nickel-base alloy with a black, glossy oxide finish.

Ebontex. Trade name for emulsified asbestos.

ebony. The heavy, brittle, scarce wood of any of various tropical persimmon trees of the genus *Diospyrus,* especially *D. denta* of West Africa, and *D. melanoxylos* of India. Its color varies from brown, streaked with gray and black, to jet-black, and it has a close texture, high hardness, good wear resistance and high durability. Average weight, 1000 kg/m³ (62 lb/ft³). Source: India, Sri Lanka, East Indies, Tropical Africa. Used especially for decorative woodwork, carvings, ornamental inlays, turnery, tool and knife handles, etc., and for veneer in cabinetwork. It was originally used for courtly furniture in ancient Egypt, Persia and India. See also black ebony; kaki.

Ebony-Blend. Trade name for a pigmented, heat-cure acrylic resin for dentures.

E-Brite. Trademark for AL Tech Specialty Steel Corporation (USA) for vacuum-melted, low-interstitial, high-chromium ferritic stainless steel containing 0.01% carbon, 25-27.5% chromium, 0.75-1.5% molybdenum, 0.5% nickel, 0.4% manganese, 0.4% silicon, and small amounts of copper, niobium (columbium), phosphorus, sulfur and nitrogen. It develops minimum hardness and maximum toughness, ductility and corrosion resistance in the annealed and quenched condition, and has excellent resistance to pitting and stress-corrosion cracking, high thermal conductivity, and low thermal expansion. Used for automotive parts, fasteners, and heat-treating and chemical equipment.

EC aluminum. See high-conductivity aluminum.

Ecarit. German trade name for cellulose acetate plastics.

Eccentric Ring. British trade name for an alloy of 84% copper, 14% tin and 2% zinc used for bells and bearings.

eccentric yarn. A specialty yarn with a marked tendency to form into a spiral, made either by twisting together a single and a doubled yarn, or two single yarns having different counts, or degrees and/or directions of twist or tension. Also known as *spiral yarn.*

Eccobond. Trademark of Emerson & Cuming (USA) for clear, fire-retardant, electrically conductive high-strength adhesives supplied in silver, low-cost silver, and non-silver grades. Used for wire bonding, as a cold solder, in EMI/RFI shielding, and general industrial applications. See also EMI/RFI shielding materials.

Eccocoat. Trademark of Emerson & Cuming (USA) for electrically conductive coatings for EMI/RFI shielding, circuit boards, resistor and capacitor dip coatings, and for general-purpose applications. See also EMI/RFI shielding materials.

Eccofoam. Trademark of Emerson & Cuming (USA) for polyurethane foam-in-place resins used for microwave and other applications.

Eccolite. Trademark of Emerson & Cuming (USA) for lightweight, high-strength syntactic foam adhesives, sealants and coatings with good thermal properties for aerospace applications.

Eccoshield. Trademark of Emerson & Cuming (USA) for coatings and sealants for EMI/RFI and electrostatic shielding applications. See also EMI/RFI shielding materials.

Eccosil. Trademark of Emerson & Cuming (USA) for clear, low-density silicone casting resins. Electrically conductive and other special grades are also available. Used for moldmaking and encapsulation applications.

Eccosphere Microballoon. Trademark of Emerson & Cuming (USA) for lightweight hollow glass microspheres used as reinforcements for plastics.

Eccospheres. Trademark of Emerson & Cuming (USA) for lightweight glass microspheres and macrospheres used as reinforcements in plastics.

Eccothane. Trademark of Emerson & Cuming (USA) for a family of filled and unfilled urethane casting resins specifically designed for electronic/electrical casting and cable splicing. They have good toughness and abrasion resistance, high tear strength, variable hardness and flexibility, good stability in high moisture environments, good electrical insulation properties, and excellent thermal shock resistance. Used for potting of electrical modules, devices and connectors, and for motor housings, telecommunication modules and devices, optical lenses, power supplies, and sensors.

Ecdel. Trade name of Eastman Chemical Company (USA) for thermoplastic copolyester-ether elastomers available in several grades including blow- and injection-molding, film-extrusion, sheet and tubing. Used for flexible packaging.

ecdemite. A greenish yellow to yellow mineral composed of lead arsenate chlorate, $Pb_6As_2O_7Cl_4$. Crystal system, tetragonal. Density, 7.14 g/cm³. Occurrence: Sweden.

Ecepolen. Trade name of Orbitaplast for polyethylene film materials.

ECG iron. See enhanced compacted-graphite iron.

Echo. Trade name of Hall & Pickles Limited (UK) for a tool steel containing 0.65% carbon, 4% chromium, 14.5% tungsten, 0.6% vanadium, and the balance iron.

Echo-Alufilm. German trademark for aluminum-metallized plastic sheeting.

Echo Molybdenum. Trade name of Hall & Pickles Limited (UK) for an alloy steel containing 0.82% carbon, 4.1% chromium, 5% molybdenum, 6.4% tungsten, 1.9% vanadium, and the balance iron.

Ecka. Trade name of Eckart-Werke Carl Eckart GmbH & Co. (Germany) for nonferrous metal powders and pastes.

eckermannite. A dark bluish-green mineral of the amphibole group composed of sodium magnesium aluminum silicate hydroxide, $Na_3Mg_4AlSi_8O_{22}(OH)_2$, and sometimes small amounts of lithium, iron and/or fluorine. It can also be made synthetically from a gel consisting of sodium oxide, magnesium oxide, aluminum oxide, silicon dioxide and water. Crystal system, monoclinic. Density, 3.00 g/cm³; refractive index, 1.6385. Occurrence: Sweden.

Eclair. Trade name of Creusot-Loire (France) for a series of tungsten hot-work, and tungsten or cobalt-tungsten high-speed tool steels.

eclarite. A whitish gray mineral of the lillianite group composed of copper iron lead bismuth sulfide, $(Cu,Fe)Pb_9Bi_{12}S_{28}$. Crystal system, orthorhombic. Density, 6.85 g/cm³. Occurrence: Austria.

Eclipsalloy. Trade name of Eclipse-Pioneer Foundries (USA) for aluminum-copper alloys.

Eclipse. (1) Trademark of Degussa-Ney Dental (USA) for a white, hard, silver-free dental bonding alloy (ADA type III) of 52% gold, 37.5% palladium and 3.6% zinc. It has high yield strength, excellent resistance to sagging and porcelain greening, and a hardness of 254 Vickers. Used in restorative dentistry for implants, long-span bridges, and high-stress restorations.

(2) Trade name of Honeywell International (USA) for nylon-6 fibers.

Eclipse Bronze. Trade name of Sargent & Company (USA) for a white nickel bronze containing varying amounts of copper, tin and nickel. Used for hardware.

Eco. (1) Trade name of Johnson Matthey plc (UK) for yellow dental casting alloys containing 40-41% silver, 20% gold, 20% palladium, 15-18% indium, and the balance zinc.

(2) Trade name of Wieland Dental + Technik GmbH & Co. KG (Germany) for hard and extra-hard silver-palladium dental casting alloys.

(3) Trade name for a series high-temperature alumina-based furnace insulation products available in various densities. They have good high-temperature stability up to 1700°C (3090°F), very low shrinkage, and good resistance to many chemicals and atmospheres.

Eco-Borne. Trademark of G.J. Nikolas & Company, Inc. (USA) for a line of EPA-compliant coatings with low volatile organic compound (VOC) content.

Ecobrom. Trademark of Agriplast Srl (Italy) for a barrier-film consisting of a top layer of low-density polyethylene, an intermediate layer of *Orgalloy* polyamide alloy, and a bottom layer of low-density polyethylene. Supplied in four grades, it is used for soil disinfecting with methyl bromide.

Ecocer. Trade name of Johnson Matthey plc (UK) for a white dental bonding alloy of 79% palladium, 10% copper, 9% gallium, and 2% gold.

Ecoceramics. Trade name of NASA Glenn Research Center (USA) for strong, tough net-shape silicon carbide ceramics and composites made by infiltrating pyrolized wood or wood sawdust with molten silicon or silicon alloys.

EcoClear. Trademark of Wellman Inc. (USA) for a solid-state polymerized copolymer resin that is a blend of new polyethylene terephthalate (PET) resin and reprocessed postconsumer PET (from recycled bottles). Used for molded and blow molded products, especially for food packaging applications.

Ecocryl. Trade name of Atofina SA (France) for acrylic emulsions and dispersions.

Ecodex. Trade name for powdered ion exchange resins.

Ecoflex. Trade name of BASF AG (Germany) for polymer blends of polybutylene terephthalate and adipic acid resins.

Eco-Foam. Trademark of National Starch & Chemical Company (USA) for a loosefill material made from extruded hybrid cornstarch.

Ecolan. Trade name of Porvair (USA) for starch-modified polyurethanes.

EcoLon. Trademark of Wellman Inc. (USA) for a series of heat-stabilized, mineral/glass-fiber-reinforced polyamides based on recycled nylon 6,6 (25% postconsumer nylon from carpet fibers). Used for automotive components. See also Wellamid.

Ecomass. Trademark of Ideas to Market, LP (USA) for nontoxic high-strength thermoplastic composites (e.g., polyamides, polyurethanes or polyphenylene sulfides filled with tungsten powder). They have high yield strength and good processibility, provide high levels of radiation shielding, and can be formu-

lated with various fillers and binders with densities ranging from about 6-11 g/cm^3 (0.22-0.40 lb/in^3). Other properties, such as impact and tensile strengths, heat deflection and flexural modulus, can also be tailored by proper formulation. *Ecomass* composites can be processed by compression and injection molding, and are used as replacements for lead (density, 11.35 g/cm^3 or 0.410 lb/in.^{3}) in military projectiles, and for radiation-shielding, soundproofing, vibration-damping applications. Other possible applications include X-ray shielding of medical and laboratory equipment, counterweights and inertia brakes, and as a replacement for lead wool and paste.

ecomaterials. A large group of different materials including metals, alloys, ceramics, polymers and composites that have been designed or modified to reduce harmful effects on the environment. An ideal *ecomaterial* is one that does not only minimize adverse environment effects, but also provides an affordable alternative with performance characteristics comparable to those of traditional materials. Typically, such a material is nonhazardous ("environmentally friendly") and contains natural, renewable or recyclable constituents, e.g., lead-free solders, ultralight steels, biodegradable plastics with nonhazardous plasticizers, recyclable plastics and elastomers, ceramics incorporating wood or soil, fly-ash concrete, low-temperature cements, and environmental sorbents and catalysts. By extension the term *ecomaterials* also includes materials used for the treatment and purification of water and wastewater, and the removal and/or destruction of volatile organic compounds (VOCs), and air pollutants from exhaust system and flue stack emissions. Also known as *environmentally conscious materials*. See also green engineered materials.

EcoMer. Trademark of of EcoSynthetix (USA) for a family of sugar-based macromers. The glucose used in the synthesis of these hybrid polymers is obtained from corn starch and other renewable sugar resources. *EcoMer* polymers can be copolymerized with acrylic monomers, and are available in a wide variety of viscosities ranging from solid to liquid. Used as bio-based building blocks for the manufacture of waterborne sugar-acrylic adhesives, resins, inks and toners.

Econo. (1) Trade name of CCS Braeburn Alloy Steel (USA) for a hot-work tool steel containing 0.4% carbon, 0.2% manganese, 3.75% chromium, 5.7% molybdenum, 1.1% tungsten, 0.7% vanadium, and the balance iron. Used for hot-work dies, cutters and tools.

(2) Trade name of Permacon (Canada) for paving slabs with embossed surfaces resembling natural stone. Supplied in square units, 295 × 295 × 40 mm (11.750 × 11.750 × 1.625 in.) in gray, red/black and beige/black colors. Used for walkways, back yards, and patios.

Econo-Chrome. Trademark of McGean (USA) for chromium electroplates and plating processes.

Economedia. Trade name of Almco Inc. (USA) for ceramic mass-finishing media.

economic mineral. A mineral of interest or value to commerce, e.g., corundum, dolomite, hematite, magnetite, pyrite, quartz, talc, etc.

economy brick. A brick employed as a closure unit and having nominal dimensions of 102 × 102 × 203 mm (4 × 4 × 8 in.).

Economy Bronze. Trade name of Anaconda Company (USA) for an alloy containing copper, tin and zinc, used for welding rods for steel and cast iron.

economy-grade lumber. See economy lumber.

Economy Hardface. Trade name of Abex Corporation (USA) for a series of abrasion- and wear-resistant steels containing 0.5-1% carbon, 3-5% chromium, 1.7% molybdenum, and the balance iron. Used for hardfacing electrodes.

economy lumber. Low-quality lumber with many defects. It is the lowest commercially available grade of lumber and is not suitable for structural framing members. Also known as *economy-grade lumber.*

ECOPET. Trade name of Teijin Fibers Limited (Japan) for a polyester fiber produced from recycled polyethylene terephthalate (PET) bottles. Used for apparel, and household and industrial fabrics.

EcoPLA. Trade name of Cargill Dow Polymers (USA) for biodegradable polylactic resins composed of lactic acid chains derived from agricultural products, such as corn and sugar beets. Readily processed by most melt-fabrication techniques, they possess relatively high strength, toughness, dimensional and temperature stability, and good resistance to grease and oil.

EcoSphere. Trademark of EcoSynthetix (USA) for a family of starch adhesives based on biopolymer nanospheres synthesized from molecularly redesigned starch molecules. They form high solids dispersions in water, have high tack and long shelf lifes, activate at low temperatures and require no additional chemicals for gelling.

EcoSpun. Trademark of Wellman Inc. (USA) for a soft, colorfast, spun polyester fiber belonging to the *Fortrel* range of fibers and made from 100% recycled plastic bottles. Used for designer clothes, home furnishings and outdoor fabrics.

ECOSS. Trade name of Figla Company Limited (Japan) for a sandwich material composed of glass sheets, aluminum honeycomb or louvers and acrylic bars.

EcoStix. Trademark of EcoSynthetix (USA) for family of low VOC, repulpable pressure-sensitive adhesive (PSA) products synthesized by a proprietary sugar macromer technology which copolymerizes *EcoMer* building blocks with acrylic monomers in an aqueous emulsion polymerization process. Used for paper label, stamp, film and specialty applications, their adhesive properties are designed to turn off in response to specific external conditions, and their copolymer structure can be modified resulting in a variety of sugar-acrylic PSA copolymers.

Eco-Vapor Cote. Trade name of Mon-Eco Industries, Inc. (USA) for solvent-based paints.

ecrasé leather. A colored, vegetable-tanned, coarse-grained leather made from goatskin.

E-CR Glas. Trademark of Fibre Glasty Developments Corporation (USA) for corrosion-resistant electrical glass (E-CR glass).

E-CR glass. Electrical grade glass (E-glass) that has been modified for improved resistance to corrosion by many acids. It contains chiefly silica, calcium oxide and alumina, and is used in the form of reinforcing fibers for engineering composites.

Ecsaine. Trademark of Toray Industries, Inc. (Japan) for a synthetic imitation suede used for clothing and shoes.

Ecusit. Trademark of DMG Hamburg (Germany) for light-cure universal dentin/enamel hybrid composite used in restorative dentistry for bonding ormocers, compomers, and composites.

Edal. German trade name for a heat-treatable alloy containing 95% aluminum and 5% magnesium used for light-alloy parts and welding rods.

Edco. Trade name of Eccles & Davis Machinery Company (USA) for a phosphor bronze containing 91.5% copper, 8.25% tin, and 0.25% phosphorus. It has a melting point of 1010°C (1850°F), and is used for welding rods.

Edelbronze. German trade name for a corrosion-resistant tin bronze used for chemical equipment, textile machinery, marine hardware, turbine wheels, machine parts, and fasteners.

Edelit. Trade name of Flachglas AG (Germany) for figured glass. See also Sigla Edelit.

Edelmessing. German trade name for high-strength brass that typically contains about 60% copper and 30-40% zinc, and varying amounts of iron, manganese, and/or aluminum. It has good hardness, and is used for marine forgings and castings, hydraulic cylinders, gears, cams, bushings and bearings, and valve parts.

Edelstahl. German trade name for a group of electric-furnace steels with low phosphorus and sulfur contents, high cleanliness, and uniform quality. This group comprises a wide range of different steel grades including case-hardening, heat-treatable, stainless and tool steels containing varying amounts of alloying elements, such as chromium, nickel, tungsten, molybdenum, and vanadium.

Edelweiss. Trade name of Vereinigte Edelstahlwerke (Austria) for a series of corrosion- and/or heat-resistant stainless steels.

edenite. A green to black mineral of the amphibole group composed of sodium calcium magnesium aluminum silicate hydroxide, $NaCa_2Mg_5AlSi_7O_{22}(OH)_2$. Crystal system, monoclinic. Density, 3.06 g/cm^3; refractive index, 1.6656. Occurrence: USA (New Jersey), Canada (Ontario).

Ederol. Trademark J.C. Binzer Papierfabrik GmbH & Co. KG (Germany) for nonwovens used in the automotive and building industries, in the hotel and catering trades, for medical applications, and for liquid- and air-filtration applications.

edestan. A protein obtained from *edestin* by treatment with dilute hydrochloric acid and subsequent neutralization of the acid.

edestin. A globulin with a molecular weight of about 310000 found in hempseed.

Edge. Trade name for a dental shoulder porcelain.

edge-grained lumber. Lumber cut from a log that was first sawed into quarter sections. Each quarter is then sawn into boards at right angles to the annual rings and parallel to the medullary rays. This method produces beautiful wood grain effects. Also known as *quarter-sawn lumber; rift-sawn lumber; vertical-grained lumber.*

Edgetek. Trade name of M.A. Hanna Engineered Materials (USA) for polyethersulfones reinforced with carbon or glass fibers, polyetheretherketones filled with carbon fibers, and filled or unfilled polytetrafluoroethylenes. Used for high-performance applications.

Edimet. Trademark of Enichem (Italy) for acrylics based on polymethyl methacrylate.

edinam. See Tiama mahogany.

edingtonite. A pink, white or grayish mineral of the zeolite group composed of barium aluminum silicate tetrahydrate, $BaAl_2Si_3O_{10}\cdot4H_2O$. Crystal system, orthorhombic. Density, 2.78 g/cm^3; refractive index, 1.5528. Occurrence: Sweden, Scotland.

Edistir. Trademark of Enichem SpA (Italy) for polystyrene resins supplied in standard, medium-impact, UV-stabilized, silicone-lubricated, glass fiber-reinforced, and structural foam grades.

Edistir HIPS. Trade name of Enichem SpA (Italy) for high-impact polystyrene resins supplied in standard, fire-retardant and UV-stabilized grades.

Edlon. Trademark of Edlon Products, Inc. (USA) for a series of high-performance adhesives for bonding fluoropolymer products.

Edwards speculum. A corrosion-resistant alloy containing 63-70% copper, 25-32% zinc, 16-2.4% arsenic, and 0-2.6% zinc.

Used for mirrors and reflectors.

EEA copolymers. See ethylene-ethyl acrylate copolymers.

Efbecol. Trade name of Friedrich Branding GmbH & Co. (Germany) for an extensive series of synthetic and natural adhesives and glues including hot-melt, latex, solvent and water-based, dextrin and other types. Depending on the particular composition, they may be used for bonding tiles, cardboard, wood, ceramics, metals, plastics, building products, etc., and for coating, packaging, labeling, bookbinding, envelope-sealing, and antislip applications.

EF-Extra. Trade name for a high-purity zirconia (ZrO_2) powder used primarily in the piezoelectric ceramics industry.

Effektsteine. Trade name of former Glas- und Spiegel-Manufactur (Germany) for decorative glass blocks.

Eftrelon. Trade name of Schwarz GmbH (Germany) for nylon fibers and yarns used for textile fabrics.

Efylon. Trade name for nylon fibers and yarns used for textile fabrics.

Egalit. Trade name Société Nouvelle des Acieries de Pompey (France) for an alloy steel containing 0.85% carbon, 2.1% manganese, 0.07% chromium, 0.16% vanadium, and the balance iron.

Egalite. Trade name of Egal Metal Products Company (USA) for an aluminum alloy that is subjected to a special process while in the molten state. Used for light-alloy parts.

EG Bond. Trade name of Sun Medical Company Limited (Japan) for a light-cure dental bonding system consisting of the adhesive monomer 4-methacryloxyethyl trimellitate anhydride (4-META) and a surface conditioner composed of citric acid and ferric chloride. It forms a strong bond between the composite resin, metal or ceramic restoration and the tooth substrate (i.e., dentin and enamel).

egg albumin. A water-soluble globular protein found as a colorless viscous fluid in egg white. Also known as *ovalbumin.*

eggshell. (1) A very thin, highly translucent porcelain. Also known as *eggshell porcelain.*

(2) Fired glaze or porcelain enamel with a semimat, egg-shell-like texture.

eggshell glazed tile. A tile that has a glazed coating with a semimat, eggshell-like texture.

eggshell porcelain. See eggshell (1).

Eglas. Trademark of Saint-Gobain (France) for heatable glass.

E-Glass. Trademark of Nuova Italtess (Italy) for E-glass fibers.

E-glass. An electrical-grade glass fiber composed chiefly of silicon dioxide (SiO_2), calcium oxide (CaO), aluminum oxide (Al_2O_3) and boron oxide (B_2O_3), and an alkali content of less than 2.0%. It has a density of about 2.6 g/cm³ (0.09 lb/in.³), excellent electrical properties, high electrical resistivity, an ultimate tensile strength and elastic modulus of about 3.45 GPa (500 ksi) and 72 GPa (10.4×10^3 ksi), respectively, good durability and dimensional stability, and good moisture resistance. Used as reinforcing fiber for textiles and plastics, e.g., in printed-circuit boards and electrical laminates.

E-glass/epoxy composites. A class of high-performance sheet composites that have epoxy resin matrices reinforced with *E-glass* fibers. They have high tensile strength, good elevated-temperature strength, high tensile moduli, and good thermal and electrical properties. Used for electrical laminates for circuit-board manufacture, and structural aerospace and construction components.

eglestonite. A yellow, orange yellow or brownish yellow mineral composed of mercury oxide chloride hydroxide, $Hg_6Cl_3O_2H$.

Crystal system, cubic. Density, 8.33; refractive index, 2.49. Occurrence: USA (Texas).

eguëite. A brownish-yellow mineral composed of calcium iron phosphate hydroxide hydrate, $CaFe_{14}(PO_4)_{10}(OH)_{14} \cdot 21H_2O$. Occurrence: Sudan.

Egyptian blue. Blue frit, or powdered synthetic pigment that is a double silicate of calcium and copper ($CaO \cdot CuO \cdot 4SiO_2$) contained in a glassy matrix. It has good resistance to chemicals, produces permanent colors, and was already used by the ancient Egyptians.

Egptian cotton. (1) A fine, lustrous, naturally colored high-quality cotton yarn or fiber, originally obtained from Egypt. It is of the extra-long staple type (fiber length, approximately 38-44 mm, or 1.5-1.7 in.), and can be bleached. Used for the manufacture of textile fabrics.

(2) A strong, durable cotton fabric with a soft feel, woven from Egyptian cotton fibers (1) in a plain weave. It dyes well, can be printed, and is used for high-priced dresses, blouses, nightwear, and infant's clothes.

Egyptianized clay. A clay that has been made more plastic by adding tannin.

Egyptian paper. See papyrus.

EH copolymers. See ethylene-hexylene copolymers.

Ehret. Trade name of Baldwin-Ehret-Hill Company (USA) for 85% magnesia high-temperature blocks and pipe covering products.

Ehrhardt's bearing metal. A bearing alloy containing 84.4% zinc, 10-11% copper, 2.5% aluminum, 1-1.2% lead and 0.2% antimony. It has poor resistance to heat or live steam, and is used for bearings.

Ehrhardt's metal. A babbitt containing 89% zinc, 4% copper, 4% tin and 3% lead used for bushings and bearings.

Ehrhardt's type metal. An alloy containing 89% zinc, 6% tin, 3% copper and 2% lead, used for type metal.

eifelite. A colorless to very light yellow or green mineral of the osumilite group composed of potassium sodium magnesium silicate, $KNa_3Mg_4Si_{12}O_{30}$. Crystal system, hexagonal. Density, 2.67 g/cm³; refractive index, 1.5445. Occurrence: Germany.

eighteen-eight steels. See chromium-nickel steels.

Eighteen Per Cent. British trade name for a nickel silver containing 65% copper, 18% nickel and 17% zinc. Used as a base for silver-plated flatware.

eight-strand rope. A rope usually consisting of a braided core around which 4 pairs of strands, each composed of 2 yarns, have been twisted.

Einheitsmetall. A German babbitt containing 79% lead, 14% antimony, 5.5% tin, and 1.5% copper. Used for bushings and bearings.

einsteinium. A rare, radioactive, synthetic element of the actinide series of the Periodic Table. It is produced either in a cyclotron by bombardment of uranium-238 (^{238}U) with accelerated nitrogen ions, or in a nuclear reactor by neutron irradiation of californium or plutonium, and is also obtained as a byproduct of nuclear fission. It is named for Albert Einstein. Density, 8.84 g/cm³; atomic number, 99; half-life, 276 days; divalent; chemically similar to holmium. Isotopes with mass numbers ranging from 246 to 254 have also been prepared. Two allotropic forms are known: (i) *Alpha einsteinium* (close-packed double hexagonal) and (ii) *Beta einsteinium* (face-centered cubic) below 820°C (1508°F). Symbol: Es.

Eirenglass. Trade name of Mallow Industries Limited (Ireland) for woven glass cloth, woven rovings, and composite cloth-mat products.

Eisenbronze. German iron bronze containing 57.5-82.5% copper, 4.4-39.5% zinc, 1-8.6% tin, 0-0.3% lead, 0-0.3% aluminum, and 1.3-4% iron. Used for marine parts, and hardware.

Eisenheiss. Trademark of Mameco International (USA) for metallic surface coatings.

Eisennickel. German trade name for a soft magnetic iron-nickel alloy containing 78-80% nickel and 20-22% iron. It has high initial permeability, low saturation induction, low coercive force, a square hysteresis loop, and is used for magnetic and electrical equipment including motors.

Eisler's bronze. A high-conductivity bronze containing 94% copper and 6% zinc. Used for springs and electrical contacts.

eitelite. A colorless mineral of the calcite group composed of sodium magnesium carbonate, $Na_2Mg(CO_3)_2$. Crystal system, rhombohedral (hexagonal). Density, 2.74 g/cm^3; refractive index, 1.605.

Eka-Color-Reliefglas. German trade name for a flat glass that is first laminated with colorants inside and then fired.

Ekadur. Trade name for polyvinyl chloride film materials.

Ekadure 2001. Trade name of Ashland Specialty Chemical Company (USA) for a UV-stable pigmentation resin.

Ekafluvin. Trade name for a series of fluoropolymers.

Ekalit. Trade name for flexible polyvinyl chloride film materials.

Ekalon. Trade name for polyvinyl chloride resins and products.

Ekanate. Trade name for isocyanate foams and foam products.

ekanite. A straw-yellow to dark-red mineral composed of calcium thorium silicate, $ThCa_2Si_8O_{20}$. Crystal system, tetragonal. Density, 3.08; refractive index, 1.580. Occurrence: Canada (Yukon, Quebec), Sri Lanka.

EKasic. Trademark of Elektroschmelzwerk Kempten GmbH (Germany) for sintered silicon carbide (SSiC) products.

EKasic D. Trademark of Elektroschmelzwerk Kempten GmbH (Germany) for 98.5+% pure sintered silicon carbide (SSiC) containing 1.0% free carbon, 0.3% aluminum, and traces of oxygen and nitrogen. It has a density of 3.1 g/cm^3 (0.11 $lb/in.^3$), a porosity of 3.5%, excellent resistance to aqueous media, acids, alkalies and organic solvents, excellent wear resistance, and high hardness and strength. Used for pump sealing rings, shaft sleeves, sliding bearings, precision balls, pump nozzle casings, and discharge sleeves.

EKasin. Trademark of Elektroschmelzwerk Kempten GmbH (Germany) for sintered silicon nitride (Si_3N_4) products.

ekaterinite. A white mineral composed of calcium chloride borate hydroxide dihydrate, $Ca_2B_4O_7(Cl,OH)_2 \cdot 2H_2O$ Crystal system, hexagonal. Density, 2.44 g/cm^3. Occurrence: Russian Federation.

Ekatit. Trade name of Ekatit Stahl GmbH (Germany) for an extensive series of austenitic, ferritic and martensitic stainless steels.

Ekavin. Trademark for polyvinyl chlorides.

Ekazell. Trademark for flexible polyvinyl chloride foams and foam products.

ekhimi. See greenheart (2).

ekki. See African ironwood.

Eko. Trademark of Saint-Gobain (France) for a glass with enhanced thermal insulation properties.

Ekonol. Trademark of Harbinson-Carborundum Corporation (USA) for a series of engineering plastics based on aromatic polyester (polyoxybenzoate). They have excellent high-temperature performance, an upper continuous-use temperature above 315°C (600°F), high thermal conductivity, very high compressive strength, high tensile strength and modulus of elas-

ticity, self-lubricating surface, good wear resistance, and excellent resistance to solvents, oils and corrosive chemicals. Used for pumps and chemical-process equipment, protective coatings on metal parts, disk brakes, etc.

Eko Plus. Trademark of Saint-Gobain (France) for a glass with enhanced thermal insulation properties.

Ekosil. Trade name for a condensation-curing silicone paste for dental applications.

Ektafil. Trade name for polyester fibers and products.

Ektar. Trademark of Eastman Chemical Company (USA) for a series of thermoplastic resins including polyethylene terephthalates, polybutylene terephthalates, and unmodified or glycol-modified polyethylene terephthalates and polycyclohexane terephthalates. Several glass fiber-reinforced grades (10-30% glass fibers) and melt blends are also available. *Ektar CG* refers to a series of glass fiber-reinforced polycyclohexane terephthalates (PCTs) for injection-molded automotive and electrical components. *Ektar DN* are glycol-modified polycyclohexane terephthalates (PCTGs). *Ektar FB* refers to a series of glass- or mineral-reinforced polyethylene terephthalates (PETs), polypropylenes (PPs), polycyclohexane terephthalates (PCTs) and glycol-modified polycyclohexane terephthalates (PCTGs) with good high-temperature properties and excellent molding characteristics, used for automotive, electrical and electronic applications. *Ektar GN* are transparent, amorphous, glycol-modified polyethylene terephthalates. *Ektar MB* are melt-blend polymers including polyethylene terephthalate/polycarbonate, glycol-modified polycyclohexane terephalate/polycarbonate and other polyester alloys with good impact and chemical resistance, and high clarity. *Ektar TPO* refers to thermoplastic polyolefin elastomers used for automotive components, golf carts, and sporting and recreational equipment.

Elana. Trademark of Chemitex-Elana (USA) for polyester spun fibers and yarns used for textile fabrics.

ElastaGard. Trade name of Neogard, Division of Jones-Blair Company (USA) for a durable, flexible elastomeric roof coating system. Supplied in several colors, it is used for the direct application over new and existing bitumen or metal roofs.

Elastalloy. Trade name of GLS Corporation (USA) for thermoplastic elastomers.

Elastamax. Trade name of M.A. Hanna (USA) for an extensive series of polymers including various grades of styrene-butadiene-styrene thermoplastics, thermoplastic elastomers based on polyvinyl chloride/nitrile, styrene, ethylene-propylene-diene-monomer, olefin and polyurethane polymers as well as several polyurethane alloys and polypropylene compounds.

Elastane. Trademark of Polymer Technology Group, Inc. (USA) for biocompatible polyether urethanes.

elastane. See elastane fiber.

elastane fiber. A term used by the ISO (International Organization for Standardization) for a synthetic man-made fiber composed of linear macromolecules with 85wt% or more segmented polyurethane groups in the chain. It can be stretched to 3 times its original length and, upon release, recovers instantly and substantially returns to its initial length. Used especially for garments. This fiber is known in North America as "spandex." Also known as *elastane*. See also spandex fiber.

Elast-Eon. Trademark of Elastomedic Inc. (USA) for a family of durable, tear-resistant rigid and flexible polyurethanes and polyurethane-silicone biopolymer alloys that can be easily processed into various shapes. *Elast-Eon 1* is a modified polyurethane with enhanced environmental resistance. Siloxane has been incorporated into the soft and hard segments of *Elast-Eon 2* to improve biostability and flexibility. *Elast-Eon 3* is a modified version of *Elast-Eon 2* that has siloxane incorporated into the hard segments to further increase flexibility. *Elast-Eon 4* is a biostable, rigid polyurethane with significantly enhanced use characteristics. *Elast-Eon* biopolymers are used for heart valves, arterial grafts and wound dressings.

Elastex. Trade name for an emulsified asphalt.

elastic bitumen. See elaterite.

elastic fabrics. Woven or nonwoven textile fabrics, wholly or partially composed of elastomeric materials, and including hybrid fabrics woven from elastomeric and textile yarns, e.g., fabrics of cotton and rubber (spandex) fibers. They can be stretched to several times their original lengths and, upon release, recover instantly and substantially return to their initial lengths and shapes. The elasticity or stretchability usually increases with the elastomeric content. Used especially for garments, and industrial belting.

elastic mineral pitch. See elaterite.

Elasthane. Trademark of Polymer Technology Group, Inc. (USA) for a series of high-strength, aromatic thermoplastic polyetherurethane elastomers that are products of a reaction of a polyol, usually polytetramethylene oxide (PTMO), an aromatic isocyanate, usually 4,4'-methylene diisocyanate (MDI), and a glycol-chain extender, such as 1,4-butanediol. They have a microstructure consisting of soft, rubbery polyether segments and hard urethane segments. *Elasthane* polyetherurethanes have good biocompatibility, biostability and hydrolytic stability, excellent mechanical properties and stability, good low-temperature resistance, and smooth fungus-resistant surfaces. They are supplied in the form of hygroscopic pellets and granules, and used as biomaterials for implantable devices, such as artificial hearts, heart valves, pacemaker leads, ventricular assist devices (VADs) and intraaortic balloons.

Elasticon. Trademark of Kerr Dental (USA) for silicone dental impression materials.

Elastic-Kote. Trademark of Ferox Coatings Inc. (Canada) for elastomeric swimming-pool paints.

Elasti-Glass. Trade name of S. Buchsbaum Company (USA) for polyvinyl chloride and other vinyl products.

Elastileum. Trademark of Canadian Elastileum Limited (USA) for protective coatings.

elastin. A collagen-like albuminoid protein found in elastic fibers of ligaments and tendons. The reconstituted form is used in biomedical coatings.

Elastinol. Trade name for a nickel-titanium wire for dental applications.

elastiqué. See cavalry twill.

Elastique Spring. Trade name of Republic Steel Corporation (USA) for a spring steel containing 0.4-0.6% carbon, 0.6-1.3% manganese, 0.1-2% silicon, and the balance iron.

elastodiene. See elastodiene fiber.

elastodiene fiber. An elastomeric fiber, filament or yarn composed of natural or synthetic polyisoprene, or one or more dienes, sometimes polymerized with vinyl monomers. It can be stretched to 3 times its original length and, upon release, recovers instantly and substantially returns its initial length. Used especially for garments. Also known as *elastodiene*. See also elastomeric yarn; elastic fabrics.

Elastoglas. Trademark of Elastogran GmbH (Germany) for a series of plastic powders, granules, liquids and pastes used in the manufacture of industrial goods, sports equipment, and roller skates.

Elastogran. Trade name of Elastogran GmbH (Germany) for thermoplastic polyurethane elastomers.

Elast-O-Lene. Trademark of Kirkhill Rubber Company (USA) for thermoplastic rubber compounds used in the manufacture of extruded and molded goods.

Elastollan. Trade name of Elastogran GmbH (Germany) for transparent polyester-type thermoplastic polyurethane elastomers with excellent abrasion resistance and toughness. They are also supplied in wire and cable grades.

Elast-O-Life. Trademark of Elp Products Limited (Canada) for a polyurethane elastomer.

Elastomeric. Trademark of Rexnord Canada Limited for a series of elastomeric adhesives.

elastomeric adhesives. A class of adhesives based on natural or synthetic rubbers of high molecular weight, e.g., polyisoprene, polychloroprene, polyurethane, silicone or polysulfide. They are usually supplied as rubber solutions dispersed in water or hydrocarbon solvents, as latexes, or as hot-melt adhesives. Tackifiers and antioxidants are usually incorporated in these solutions. *Elastomeric adhesives* can be crosslinked or vulcanized to improve both shear strength and creep resistance. Important properties include high peel strength, relatively low shear strength, poor resistance to solvents and fuels, and fair to poor resistance to weathering. Used for bonding paper, textiles, engineering composites, etc. See also rubber-base adhesive, rubber adhesive.

elastomeric yarn. A nontextured yarn, such as *elastane* or *elastodiene*, made from an elastomer (natural or synthetic rubber) that can be repeatedly stretched to at least twice its original length. Used for elastic fabrics. See also elastane fiber; spandex fiber; elastic fabrics; elastodiene fiber.

elastomers. A class of rubberlike polymeric materials that may experience large and reversible elastic deformations before fracture. At room temperature, they can be stretched to at least twice their original lengths, usually do not conform to Hooke's law and, upon removal of the deforming stress, return to approximately their original lengths and shapes. Examples include butyl and chloroprene, ethylene-propylene, isoprene, natural, nitrile, polyvinyl, silicone and styrene-butadiene rubber as well as chlorosulfonated polyethylene, ethylene-vinylacetate copolymers, and fluorocarbon, polysulfide and polyurethane elastomers.

Elaston. Trademark of Chemitex (Poland) for spandex fibers and yarns used for elastic fabrics.

elasto-optic materials. A group of materials whose optical properties (e.g., light propagation) are changed by mechanical deformation.

Elastophene. Trademark of Soprema (France) for bituminous roof-sealing courses, and dampproof roofing membranes.

elastoplastics. See thermoplastic elastomers.

elastoresistance materials. A group of materials whose electrical resistance is changed due to the application of a mechanical stress within its limits of elasticity.

Elastoshield. Trademark of Heveatex Corporation (USA) for elastomeric protective roof coatings.

Elastsil. Trademark of Wacker Silicones Corporation (USA) for a series of silicone elastomers including liquid silicone rubber (LSR), heat-curable (silicone) rubber (HCR), and room-temperature vulcanizing (RTV) silicone rubber. Used in the manufacture of mechanical rubber goods

Elastuf. Trade name of Horace T. Potts Company (USA) for a series of low- and medium-carbon steels and low-alloy steels used mainly for machine parts, such as gears, pinions, worms, axles, shafts, spindles, arbors, crankshafts, bolts, and studs.

elaterite. A massive, dark brown, amorphous material composed of petroleum asphalt and having moderate elasticity and softness. Occurrence: USA (Colorado and Utah). Used for compounding protective coatings and varnishes, rubber and paints, and for insulating, paving and waterproofing applications. Also known as *elastic bitumen; elastic mineral pitch; mineral caoutchouc.*

Elba. Trade name of Circeo Filati S.r.l. (Italy) for a yarn blend composed of 50% polyester and 50% viscose rayon, and supplied in several counts. Used for textile fabrics.

elbaite. A pale green, or red mineral of the tourmaline group composed of lithium sodium aluminum borate silicate hydroxide fluoride, $NaLi_3Al_6(BO_3)_3Si_6O_{18}(OH,F)_4$. Crystal system, rhombohedral (hexagonal). Density, 3.02 g/cm³; refractive index, 1.637. Occurrence: Italy, USA (Maine).

Elbama. Trade name for rayon fibers and yarns used for textile fabrics.

El-Chem. Trade name of Electro Chemical Engineering & Manufacturing Company (USA) for an extensive series of corrosion-resistant organic coatings and linings, and corrosion-resistant, fire-retardant and/or insulating cements and mastics. Included are several grades of epoxy and coal tar/epoxy, neoprene, hypalon and vinyl coatings, hydraulic, silicate, sulfur and carbon sulfur cements, epoxy, furan and phenolic resin cements, epoxy and polyester floor toppings, antigalling compounds, concrete adhesives, epoxy machinery grouts, and unreinforced and carbon-fabric reinforced *Kynar* (polyvinylidene fluoride) laminates.

Elcomet. Trade name of La Bour Pump Company (USA) for a series of alloys containing 0.15% carbon, 10-22% nickel, 20-23% chromium, 0-5% silicon, 0-4% copper, 0-3% molybdenum, 0-0.3% manganese, and the balance iron. Used for valves, pumps, and spinner heads.

Eleclite. Trade name of Nippon Sheet Glass Company Limited (Japan) for a glass with electroconductive coating.

Electalloy. Trade name of McQuay-Norris Manufacturing Company (USA) for a series of wear-resistant cast irons containing 2.7-3.85% total carbon, 2-2.9% silicon, 0-1.2% nickel, 0.6-1.3% molybdenum, 0.3-1.2% chromium, and the balance iron. Used for piston rings.

Electem. Trademark of Walter Somers Limited (UK) for a high-carbon tool steel used for making die blocks.

Electorspec. Trademark of Cominco Limited (Canada) for a series of zinc and zinc alloys.

Electra. Trade name of Cytemp Specialty Steel Division (USA) for a series of water- or oil-hardening tool steels containing 1.2% carbon, 0.6-0.85% manganese, 0.5-0.6% chromium, 0-0.6% molybdenum, and the balance iron. Used for taps and edge tools.

Electrafil. Trademark of DSM Engineering Plastics (USA) for an extensive series of conductive acetal, ethylene vinyl acetate, fluoropolymer, nylon 6, nylon 6,6 and nylon 6,12, polycarbonate, polyphenylene ether, polyphenylene sulfide, polyether sulfone, polyetheretherketone, polyethylene terephthalate, polybutylene terephthalate, polyimide, polypropylene, polysulfone, polystyrene, and acrylonitrile-butadiene-styrene resins incorporating low loadings (typically less than about 7 wt%) of long stainless-steel fibers or carbon fibers or fillers (typically less than about 5 wt%). Supplied in pellet form for injection molding, they have good mechanical properties, and are used for electromagnetic and radio-frequency interference (EMI/RFI) shielding applications, tote bins, electronic housings, and packaging.

Electralloy. Trademark of Alcan Wire and Cable (Canada) for aluminum-alloy-based electrical conductors.

Electrapane. Trade name of Libbey-Owens-Ford Company (USA) for an electrically conductive coated glass.

electret. Any of a group of solid substances, such as the titanate ceramics and certain low-dielectric-constant (low-k) polymers, in which a permanent state of electrical polarization exists without the constant supply of electrical charges. They are dielectrics that possess either permanent or semipermanent polarity in a manner analogous to a permanent magnet. See also polymer electrets.

Electrex. See Columbia Electrex.

electrical conductor. See conductor (1).

electrical conductor aluminum. See high-conductivity aluminum.

electrical contact materials. See contact materials.

electrical contact metals. See contact metals.

electrical glass. Any glass with good electrical insulating properties, e.g., lead glass.

electrical gold alloy. An electrical contact alloy composed of 70% gold, 25% silver and 5% nickel.

electrical-grade glass fibers. A class of general-purpose glass fibers that are based on electric glass (E-glass), and have excellent electrical properties. They are used in reinforced plastics and textiles, and in electrical laminates, e.g., printed circuit boards. See also electric glass; E-glass.

electrical-grade plastics. A class of plastics with excellent electrical properties including good insulating properties, high dielectric strength and arc resistance, and low dissipation factors. Examples are diallyl isophthalate (DAIP), diallyl phthalate (DAP), polybutylene terephthalate (PBT), polyethylene terephthalate (PET), polyphenylene sulfide (PPS), and polyamide (nylon).

electrical heating alloys. See heating alloys.

electrical insulating paper. See insulating paper.

electrical insulating varnish. See insulating varnish.

electrical insulator. See insulator (1).

electrical laminate. A flat plastic laminate usually made from dry lay-up prepregs of epoxy resin by co-laminating with copper foil facings. Used for circuit boards.

electrically conductive adhesive. An adhesive, usually based on an epoxy, polyimide or silicone resin, that has been made electrically conductive by incorporating metallic fillers, such as aluminum, copper, nickel, silver, gold, etc., or carbon powder. Abbreviation: ECA.

electrically conductive elastomer. See conductive elastomer.

electrically conductive plastics. See conductive plastics.

electrically conductive polymers. See conductive polymers.

electrically conductive rubber. See conductive elastomer.

electrical porcelain. See insulation porcelain.

electrical resistance alloys. See resistance alloys.

electrical resistance materials. See resistance materials.

electrical sheet. See electrical steels.

electrical sheet steels. See electrical steels.

electrical steels. A group of hot-rolled alloy or low-carbon steels in strip or sheet form that have 0.3-4.5% silicon and low manganese, sulfur and phosphorus contents, and exhibit good soft magnetic properties. Examples are magnetic lamination steels, nonoriented electrical steels and grain-oriented electrical steels. Used in the manufacture of armatures, dynamos, motors, and transformers. Also known as *electrical sheet; electrical sheet steels; silicon-iron electrical steel.*

electrical tape. See insulating tape (1).

electrical transformer steels. See transformer steels.

electrical vacuum coatings. Vacuum-deposited metallic or ceramic films ranging in thickness from 0.01 to 100 μm (0.0004 to 4 mils) used for electronic circuit applications, e.g., for conductors, resistors and capacitors. Typical vacuum-deposited electrical coating materials include aluminum, bismuth, chromium, germanium, gold, indium, platinum, silver, tin, and their alloys as well as aluminum oxide, cadmium sulfide, cerium oxide, titanium oxide, and zinc oxide.

electrical varnish. See insulating varnish.

electric-arc thermal-spray coatings. Coatings produced by a process using a controlled electric arc between two consumable electrodes or wires to heat a metallic or nonmetallic coating material to a molten state. The molten material is then atomized and propelled onto the substrated by a stream of compressed air or gas.

electric cement. A cement made by bonding red ocher and beeswax with resin. Used for joining brass to glass.

electric furnace steels. See electric steels.

electric glass. See E-glass.

electrician's solder. A rosin-fluxed low-melting solder containing about 60-63% tin and 37-40% lead. Used for electronic and electrical applications. See also rosin.

Electric Railway Babbitt. Trade name of Hoyt Metal Company of London Limited (UK) for a lead babbitt containing tin and zinc. Used for bushings and bearings especially for rail vehicles.

Electric Spindle. Trade name of Colt Industries (UK) for a water- or oil-hardening tool steel containing 0.7% carbon, 1.15% manganese, 0.5% chromium, and the balance iron. Used for machine-tool spindles.

Electric Star. Trade name of Carpenter Technology Corporation (USA) for a high-speed tool steel containing 1.3% carbon, 0.25% silicon, 0.25% manganese, 4.5% molybdenum, 4.5% chromium, 5.5% tungsten, 4% vanadium, and the balance iron. Used for cutting tools.

electric steels. A group of steels with high and uniform quality made in an electric furnace, e.g., the direct- or indirect-arc furnace, or the induction furnace. Electric furnaces permit close temperature control and the direct addition of alloying elements; also the high melt temperatures allow the addition of refractory metals, such as tungsten, tantalum and molybdenum. All high-alloy steels including tool and high-strength steels are electric steels. Abbreviation: ES. Also known as *electric furnace steels.*

Electriplex. Trade name of Triplex Safety Glass Company Limited (UK) for electrically heated laminated glass. It has fine resistance wires incorporated between one of the glass sheets and the transparent plastic interlayer.

Electrite. Trade name of Latrobe Steel Company (USA) for an extensive series of tungsten, tungsten-cobalt and molybdenum-type high-speed tool steels. They have excellent hot hardness, good deep-hardening properties, good abrasion and wear resistance, and good to fair machinability. Used for cutting tools, such as bits, drills, taps, chasers, reamers, broaches, milling cutters, lathe and planer tools, and for form tools and blanking, shearing, thread-rolling and trimming dies.

Electrite Kelvan. See Kelvan.

Electrite Lacomo. See Lacomo.

Electrite Stark. See Stark.

Electrite Super. Trade name of Latrobe Steel Company (USA) for a tungsten-type high-speed tool steel (AISI type T5) containing 0.85% carbon, 4.1% chromium, 0.8% molybdenum, 18.7% tungsten, 9% cobalt, 1.9% vanadium, and the balance iron. It has excellent red hardness and wear resistance, and is used for cutting tools.

Electrite Super Cobalt. See Super Cobalt.

Electrite Tatmo. See Tatmo.

Electrite Ultra Cobalt. See Ultra Cobalt.

Electrite Ultravan. See Ultravan.

Electro. (1) Trade name of Electro-Steel Company (USA) for a high-speed steel containing 0.7% carbon, 6% molybdenum, 6% tungsten, and the balance iron. Used for cutting tools.

    (2) Trademark of Electro Abrasives (USA) for high-purity, high-density boron carbide and silicon carbide powders with excellent flow rates. Used in the manufacture of ceramic and metal composites.

electroacoustic materials. A class of materials that are used for both their electrical and acoustic properties. For example, they can change electric energy and waves into acoustic energy and waves, and vice versa.

electroactive materials. A class of inorganic and organic materials that exhibit unique electrical, electrochemical and opto-electronic properties that make them suitable for use in computers, flat-panel displays, sensors, and energy-storage devices.

electroactive organic materials. See electroactive polymers.

electroactive polymers. A class of polymers, such as polysilicon, polyaniline, poly(*p*-phenylene vinylene) and polypyridine, that have reversible redox properties. They are suitable for use as active components in the manufacture of solar cells, electronic circuits and displays, computer memory elements, actuators, biological and chemical sensors, lasers, fuel cells, and solar cells and batteries. Abbreviation: EAP. Also known as *electroactive organic materials.*

Electroblack. Trade name of Enequist Chemical Company Inc. (USA) for black electroplates and plating processes used on antiques.

Electro-Brite. Trade name of Electrochemicals Company (USA) for conversion coatings and plating processes.

Electrocarb. Trademark of Electro Abrasives Company (USA) for a medium-high-density black powder composed of 98.5% silicon carbide, 0.5% silicon dioxide, 0.3% silicon, 0.3% carbon, and small additions of aluminum and iron. Supplied in various bulk densities and particle sizes and shapes, it is used as an abrasive for various blasting, grinding, polishing, lapping, tumbling, and wire-sawing applications.

electrocast brick. See fusion-cast brick.

electrocast refractories. See fusion-cast refractories.

electrocatalytic coatings. Metal coatings produced by an electrocatalytic process that accelerates the half-cell reactions at the cathode surface.

electrocement. A special cement made in an electric furnace by adding lime to molten slag.

electroceramic composites. Composite materials composed of polymer matrices (e.g., polyurethanes, silicone elastomers or epoxies) filled with a ceramic powder, such as barium or lead titanate, zirconium oxide, or titanium dioxide. Depending on the filler used they may have insulating, semiconductive or conductive properties. Used in electronics and electrical engineering.

electroceramics. A group of ceramic materials, such as electrical and zircon porcelains, titanate ceramics, steatite and cordierite, that are used in the manufacture of spark plugs, insulators for power lines, and other electrical components.

electrochemical coatings. A group of coatings formed on metal surfaces by a chemical reaction in which there is a transfer of electrons from one chemical species to another. This reaction involves oxidation and reduction, and is carried out in an electrochemical cell containing an anode, a cathode and an aqueous organic or molten electrolyte.

electrochemical composites. Corrosion-resistant composites suitable for use in electrochemical applications (e.g., for fuel-cell electrodes), and typically consisting of glass, ceramic or carbon matrices with organic or inorganic material additions.

electrochromic materials. Solid, organic or inorganic insulating materials that change color when injected with positive and negative charges, which makes them suitable for use in electrochromic and other passive solid-state displays.

electrocoatings. See electrophoretic coatings.

electroconductive elastomers. See conductive elastomers.

electroconductive polymers. See conductive plastics.

Electrodag. Trade name of Acheson Colloids Company (USA) for nickel-based conductive coatings used in metal finishing.

electrode. See arc-welding electrode.

electrode carbon. Carbon or graphite made into various electrode shapes by compressing and molding. Used in arc furnaces employed for the manufacture of metallic and nonmetallic products.

electrodeposit. A metal, ceramic, semiconductor or composite material produced by an electroplating process from an aqueous, organic or molten-salt bath electrolyte which contains the material to be plated in ionic form. Electroplating is carried out by applying a voltage between a cathode, on which the deposit forms, and a consumable or dimensionally stable anode such that the ions migrate from the electrolyte to the cathode where they are reduced and incorporated into the growing electrodeposit. Electrodeposits are used on many engineering components to provide hardness, corrosion resistance, wear resistance, fatigue properties, appearance and functionality of their surfaces. Also known as *electrodeposited coating; electrolytic deposit.*

electro-exploded nanopowders. See exploded nanopowders.

Electrofil. (1) Trade name for a composite consisting of 5 wt% stainless-steel fibers in a polycarbonate matrix. It has high strength, impact resistance and durability, and good moldability.

(2) Trade name for a carbon-powder-filled polyolefin for blow and injection molding and thermoforming. It provides good electromagnetic interference shielding, and good protection against electrostatic discharge. Used for electronic components. See also EMI/RFI shielding materials.

electroformed nickel. Nickel that has been nonadherently electrodeposited as a relatively thick coating onto a mandrel, mold, or matrix contained in a plating bath, and can subsequently be removed as a formed, integral part or product. See also electroformed products.

electroformed products. Products or parts formed by electrodepositing a metal, such a copper, nickel, silver or iron, onto a mandrel, mold, or matrix, which may be soluble for removal from the product or part, or be coated with a parting agent. Electroformed parts have good dimensional accuracy and are often produced for aerospace, automotive, electronic, telecom-

munication and various other applications. See also electroformed nickel.

electrofused ceramics. Ceramic materials, such as alumina, magnesia or silica, made by heating the starting materials to high temperatures in an electric furnace.

electrogalvanized steel. A steel, usually a plain-carbon steel in sheet or wire form, on whose surface a usually thin, uniform corrosion-resistant electrochemical deposit of pure zinc has been produced by immersion in an electrolytic bath. In application, the zinc preferentially corrodes and protects the steel. Also known as *electrozinc-plated steel; zinc-plated steel.*

Electroglass. Trade name for a specially formulated protective porcelain coating based on a dry powder frit containing bonding additives for good adhesion. It provides improved resistance to discoloration, flaking and thermal shock. Used for metal barbecue grills, gas-fired range grates, etc.

Electro High Speed. British trade name for a high speed steel containing 0.76% carbon, 0.45% manganese, 2.95% chromium, 13.2% tungsten, 1.5% vanadium, and the balance iron. Used for drills, cutters, and tools.

electroless coating. A protective metal or metal-alloy coating deposited on a substrate by chemical reduction in a suitable bath, without the application of an electric current. Electroless copper, cobalt, nickel and precious metal coatings are common examples. Also known as *electroless deposit; electroless plate.*

electroless deposit. See electroless coating.

electroless nickel. An engineering coating deposited by autocatalytic reduction of nickel ions by aminoborane, borohydride and hypophosphite compounds without the application of an electric current. Electroless nickel coatings offer many favorable properties including excellent corrosion and wear resistance and outstanding uniformity. Electroless nickel-boron, nickel-phosphorus, nickel-cobalt-phosphorus and nickel composite coatings are also produced. Abbreviation: EN.

electroless nickel-boron. An engineering coating deposited onto ferrous or nonferrous substrates by autocatalytic reduction of nickel ions by aminoborane or borohydride compounds without the application of an electric current. In addition to its corrosion- and wear resistance, a borohydride-reduced electroless nickel containing about 5% boron has a density of 8.25 g/cm^3 (0.298 lb/in.^{3}), a melting point of 1080°C (1980°F), and an as-deposited hardness of about 700 Vickers.

electroless nickel-phosphorus. An engineering coating deposited onto metallic or nonmetallic substrates by autocatalytic reduction of nickel ions by hypophosphite compounds without the application of an electric current. In addition to its corrosion- and wear resistance, a hypophosphite-reduced electroless nickel containing about 10% phosphorus has a density of 7.75 g/cm^3 (0.280 lb/in.^{3}), a melting point of 890°C (1630°F), and an as-deposited hardness of about 500 Vickers.

electroless plate. See electroless coating.

electroluminescent materials. A class of materials including various p- and n-type semiconductors, such as germanium, silicon and gallium arsenide, and several organic materials, such as anthracene and poly(p-phenylene vinylene), that emit light when excited by an applied electric field or alternating current. Used in electroluminescent devices, e.g., panels, displays, lights, etc. See also electroluminescent phosphor; organic electroluminescent materials.

electroluminescent phosphor. A phosphor, usually a p or n-type semiconductor, that gives off light when excited by an electric field or alternating current. Also known as *electroluminor.*

electroluminor.   See electroluminescent phosphor.

electrolytic aluminum.   (1) A term that usually refers to ultrapure aluminum (about 99.99%) which has been electrolytically refined utilizing an anode of aluminum-copper alloy in a fused fluoride bath. The refining process involves three liquid layers and is referred to as the "Hoope's process."

(2) Aluminum as produced by the standard Hall-Heroult process involving the electrolytic reduction of a fused bath of alumina dissolved in *cryolite* (Na_3AlF_6).

electrolytic bismuth.   Bismuth (99.5+% pure) obtained as a byproduct in the debismuthizing of lead by an electrolytic process known as the "Betts process." The bismuth is obtained from the scrap anodes by special treatment.

electrolytic cadmium.   Cadmium (99.5% pure) obtained in zinc-ore roasting by collecting cadmium dust in an electrostatic precipitator, followed by leaching and fractional precipitation and distillation, or by recovery in the electrolytic zinc process.

electrolytic cobalt.   Pure cobalt (99.9+%) obtained from ore concentrations processed by roasting followed by electrolytic reduction of metal solutions.

electrolytic copper.   Metallic copper refined by the electrolytic deposition process. This process produces a metal of high purity (99.9+%), and enables precious metals, such as gold and silver, to be recovered. See also electrolytic tough-pitch copper.

electrolytic copper powder.   Copper powder made by electrolysis using anodes of electrolytic copper, cathodes of lead-alloy sheet, and an electrolyte of sulfuric acid, with or without addition agents. It is subsequently washed and usually subjected to a reduction treatment in a furnace at relatively low temperature to further control the size and shape of the particles. With a purity of 99.5+%, and an apparent density range of 1-4 g/cm^3 (0.04-0.15 lb/in^3), it is used for electrical parts, and friction applications.

electrolytic deposit.   See electrodeposit.

electrolytic iron.   A high-purity iron (99.9+%) produced by electrolytic deposition from solutions of a ferrous salt. It contains traces of carbon (0.006%), sulfur (0.004%), silicon (0.005%) and copper (0.015%), is brittle as deposited, and has excellent magnetic properties. Used for thin seamless tubing, magnetic cores, and electrical instruments.

electrolytic iron powder.   A fine, high-purity ferrous powder metallurgy product (99+%) made from electrolytic powder. It has a density of 2.5 g/cm^3 (0.09 lb/in^3), high compressibility, high green strength, and irregular particle shape. Used for soft magnetic sintered parts with final densities of 7.2-7.7 g/cm^3 (0.26-28 lb/in^3).

electrolytic lead.   High-purity lead (99.995-99.998%) refined by an electrolytic process (the Betts process) in which the electrolyte contains lead fluosilicate and fluosilicic acid.

electrolytic magnesium.   High-purity magnesium (99.8+%) produced by electrolysis of fused magnesium chloride.

electrolytic manganese.   Manganese (99.9% pure) made electrolytically from sulfate or chloride solutions of low-grade ores. Used for pyrotechnics, and metallurgical applications. Also known as *electromanganese.*

electrolytic nickel.   A virgin nickel (99.5+%) produced by refining sulfide-type nickel ores by flotation and roasting to sintered nickel oxide, followed by electrolytic refining.

electrolytic paper.   A paper that changes its color due to an electrochemical reaction taking place in the current-conducting areas.

electrolytic powder.   A metal powder produced directly by electrodeposition, or indirectly by pulverization of an electrodeposited coating.

electrolytic tantalum.   Tantalum (99.5% pure) obtained from tantalum potassium fluoride by fused salt electrolysis.

electrolytic tinplate.   Continuous strip or sheet of rolled steel (usually 0.15-0.60 mm, or 0.006-0.024 in. thick) on one or both surfaces of which a thin, protective coating of tin has been produced by electrolytic deposition. The average coating thickness per surface is 0.375-2.286 μm (15 to 90 μin.). Also known as *electro-tinplate.*

electrolytic titanium.   Commercially pure titanium (99.9+%) obtained by electrolysis of titanium tetrachloride in a bath of fused salts (alkali or alkaline-earth chlorides).

electrolytic tough-pitch copper.   A commercially pure copper (99.90%) that has been refined electrolytically. It contains small additions of oxygen (0.02-0.04%) and less than a total of 50 ppm metallic impurities. It has high electrical conductivity (above 100% IACS), good corrosion resistance, excellent hot and cold workability, and good forgeability. Used for electrical conductors, anodes, busbars, radio parts, automobile radiators, pipes, tubes, roofing, gutters, downspouts, flashing, fasteners, rivets, gaskets, architectural shapes, etc. Abbreviation: ETPC, ETP Cu. Also known as *high-conductivity copper.* See also fire-refined tough-pitch copper.

electrolytic vanadium.   A pure vanadium (99.99%) obtained by electrolytic refining using a molten salt electrolyte containing vanadium chloride.

electrolytic zinc.   A high-grade zinc (99.9+%) made by a hydrometallurgical or electrolytic process in which zinc oxide is leached from roasted or calcined zinc ores with sulfuric acid to form a zinc sulfate solution which is then electrolyzed in cells to deposit the zinc on cathodes.

Electro-Magma.   Trade name for a corrosion-resistant beryllium-free nickel-chromium dental bonding alloy.

electromagnetic composites.   Composites that due to the combination of their individual components (matrices and inclusions) have unique properties making them useful for electromagnetic applications, i.e., for parts that absorb, reflect, dissipate or control electromagnetic radiation (e.g., cores, toroids, shields, radar-deception equipment, etc.). Depending on the desired composite properties, the properties of the individual components may either support or oppose each other, e.g., dielectric matrices and dielectric phases, or magnetic matrices and nonmagnetic inclusions.

electromagnetic interference/radio-frequency interference shielding materials.   See EMI/RFI shielding materials.

electromanganese.   See electrolytic manganese.

Electromatic.   Trade name for a steel containing 0.7% carbon, and the balance iron. Used for springs and oil-tempered wire.

Electromet.   Trade name of Union Carbide Corporation (USA) for a series of ferroalloys (e.g., ferroboron, ferrochromium, ferromanganese, ferroniobium, ferrosilicon, ferrovanadium and ferrotungsten) used in the manufacture of steel and cast iron. Also included under this trade name are a wide range of calcium, chromium, copper, manganese, nickel, niobium, silicon, tantalum, titanium and vanadium-base materials used as deoxidizers, degasifiers, cleansers, slag reducers and/or alloying additions in the manufacture of ferrous and nonferrous alloys.

Electrometal.   Trade name of Magnesium Elektron Limited (UK) for an alloy containing 94% magnesium and 6% aluminum.

Used for light-alloy parts.

electron beam-cure adhesives. A class of adhesives based on synthetic resins, such as acrylics, epoxies or polyesters, that are applied to the substrates (e.g., glass, metal, plastic, etc.) in the liquid state, and subsequently cured to dry films by subjecting to an electron beam. They possess good dimensional stability and toughness, good abrasion, chemical and heat resistance, and improved substrate adhesion.

electron compounds. See intermetallic compounds.

electronic ceramics. Ceramic products, usually of high purity, displaying dielectric, semiconductive, superconductive, piezoelectric, ferroelectric, electro-optic, magnetic or similar properties that make them useful as substrates and for the manufacture of electronic and solid-state devices, such as insulators, transistors, capacitors, varistors, thermistors, piezoelectric components, electron tubes, transducers, sensors, magnetic amplifiers, integrated circuit packages, etc. Examples of dielectric ceramics include titanate ceramics, porcelain, mica, amber, steatite, alumina, zircon, soda-lime glass and E-glass. Semiconductor ceramics include silicon carbide and zinc sulfide. Piezoelectrics and ferroelectrics include titanate ceramics and quartz, and magnetic ceramics include magnetite, ferrites and garnets, manganous oxide and nickel oxide. Superconductors include yttrium-barium and various other cuprates.

electronic gold. Special compositions of gold used to produce conductive, corrosion-resistant coatings on ceramics for use as semiconductors, capacitors, and integrated circuits.

electronic materials. A class of engineering materials used primarily for their electrical and electronic properties (e.g., conductivity, resistivity, polarization, dielectric constant and strength, volume resistivity, etc.). Included are conductors (aluminum, copper, gold, certain polymers, etc.), superconductors (mercury, lead, niobium stannide, yttrium barium copper oxide, etc.), semiconductors (germanium, silicon, gallium arsenide, etc.), and insulators (aluminum oxide, glass, ceramics, most plastics, etc.).

Electronics Grade Precision Alumina. Trade name of St. Gobain Industrial Ceramics (France) for a single-crystal calcined alumina (Al_2O_3) made by a proprietary process. Used for optical applications, and for precision lapping of silicon wafers. Abbreviation: EGPA.

electron-transfer salts. A class of single-crystal organic salts that consist of large organic donor molecules (cations), such as tetramethyltetraselenafulvalene (TMTSF) or bisethylenedithiotetrathiafulvalene (BEDT-TTF), which can stack in chains or planes, and smaller inorganic acceptor molecules (anions), such as ClO_4^-, PF_6^-, ReO_4^-, etc., which can locate themselves between these stacks. In all of these compounds there occurs a transfer of one electron from two donor molecules (cations) to one acceptor molecule (anion). Depending on the particular cation-anion combination, the type of stacking, and the existing pressure, temperature and/or applied magnetic field, they can exhibit insulating, magnetic, superconducting and metallic states, and a bulk quantum Hall effect. The general formula is $(ET)_2X$, in which ET represents the cation and X the anion. Abbreviation: ET salts. Also known as *organic charge-transfer salts*. See also Bechgaard salts; organic conductors (2); organic superconductors.

electro-optical materials. A group of materials, such as a dielectrics and liquid crystals, in which optical properties (e.g., the index of refraction) are changed by the application of an electric field. See also optoelectronic materials.

Electro-OX. Trade name of Electro Abrasives Company (USA) for aluminas including white fused alumina and white calcine alumina supplied in several grit sizes. See also fused alumina; calcine alumina.

Electro-OX BZ. Trade name of Electro Abrasives Company (USA) for a gray, durable, fused zirconia-alumina powder composed of about 72% alumina (Al_2O_3), 25% zirconia (ZrO_2), 1.2% magnesia (MgO), 0.5% titania (TiO_2), and small additions of silica (SiO_2), calcia (CaO), and ferric oxide (Fe_2O_3). Supplied in several grit sizes, it is used for bonded abrasives.

electropainted coatings. See electrophoretic coatings.

electrophoretic coatings. A class of thin coatings, usually less than 40 μm (1.5 mils) thick, produced by a process in which a metal substrate with a positive potential is dipped into a negatively charged tank. Colloidal particles of paint, ceramics or metals, suspended in the water in the tank, are attracted to the substrate and adhere firmly to its surface. These coatings are commonly used for automobile bodies and household appliances. Also known as *electrocoatings; electropainted coatings*.

electrophoretic paint. A special water-reducible paint used to produce electrophoretic paint coatings.

electrophotographic material. A photoconductive material, such as cadmium selenide, that is suitable for use in photographic processes, e.g. photocopying.

electrophotographic paper. A paper having a thin, photoreceptive surface layer on which an electrostatic image is created by exposure to electromagnetic radiation (light).

electrophotoluminescent material. An electroluminescent material that emits light upon electric excitation, e.g., by an applied electric field. See also electroluminescent materials.

Electroplate. British trade name for an alloy containing 50-70% copper, 10-20% nickel, and 5-30% zinc. Used as a base for silver plate and tableware.

electroplate. A thin layer or coating of electroplated metal, e.g., a nickel or copper electroplate.

electropolymer. A protective, polymeric coating produced on the surface of a metal by electrolytic or electrophoretic means. Abbreviation: EP.

electrorheological fluid. A fluid that exhibits the so-called "electrorheological effect," i.e., a significant increase in apparent viscosity (flow resistance) and elastic modulus (stiffness) in the presence of a strong electric field. For example, the application of such a field to an originally fluid, slightly water-activated, nonaqueous silica suspension results in rapid solidification yielding a more or less rigid gel. Also known as *ER fluid*.

Electrose. Trade name of Specialty Insulation Manufacturing Company (USA) for shellac plastics used for electrical applications.

electrosensitive paper. A paper that has a conductive surface coating, which evaporates when an electric current is sent through it resulting in darkening of the paper. The coating consists either of metallic materials, such as aluminum or zinc, or a dielectric. However, the latter requires an additional conductive interlayer.

electro-sheet copper foil. A commercially pure copper (99.5%) available in the form of foils on rolls in weights ranging from 0.15 to 2.14 kg/m² (0.5 to 7 oz/ft²). It has high electrical conductivity (94.3-99.5% IACS), and is used for electrical-coil and transformer windings, flexible cable, electrostatic shielding, die-stamped circuits, roofing, and dampproofing.

electroslag-remelted steel. Refined steel produced by immersing a consumable electrode composed of the steel to be refined into a resistively heated molten slag, and passing an electric current through the slag to melt the electrode. Molten steel drips from the end of the electrode and falls through the slag, which enters into a chemical reaction with the impurities and dissolved gases in the steel, forming a high-purity ingot with uniform microstructure in a water-cooled crucible. ESR steels are of high quality, have very low gas contents, and good mechanical properties, and are used in the manufacture of turbine, engine, aerospace and aircraft parts, and other highly stressed parts. Abbreviation: ESR steel.

Electro Solder. Trade name of Temrex Corporation for precious metal dental alloys and solders.

electrostatic coatings. Nonmetallic coatings produced by spraying electrically charged atomized particles of pigments, such as phthalocyanine blue, onto a metal substrate and subsequent baking. Abbreviation: ESC.

electro-tinplate. See electrolytic tinplate.

electrotype metal. A type metal alloy containing 93-94% lead, 3-4% antimony and 3-4% tin. It has excellent antifriction properties, a low melting point, and is used for electrotypes, i.e., metal plates for printing applications.

Electroweld. Trademark of Irving Industries Limited (Canada) for welded-wire fabrics for structural applications.

electro-zinc plated steels. See electrogalvanized steels.

Electrum. (1) British trade name for a nickel silver containing 52% copper, 26% nickel, and 22% zinc. Used for corrosion-resisting parts.

(2) British trade name for a corrosion-resistant alloy containing 55-58% gold and 42-45% silver. Used for jewelry and ornaments.

electrum. (1) A natural alloy of gold with up to 26% silver. Also known as *gold argentide; electrum metal.*

(2) A corrosion-resistant alloy containing 55-58% gold and 42-45% silver. Used for jewelry and ornaments.

electrum metal. See electrum (1).

Elefant. (1) Trade name of Dörrenberg Edelstahl GmbH (Germany) for a series of water-hardening tool steels (AISI type W1) containing 0.7-1.3% carbon, 0-0.25% silicon, 0-0.25% manganese, and the balance iron. They are supplied in various grades including extra-tough, hard, medium-hard and tough. Used for drills, taps, reamers, hobs, broaches, milling cutters, punches, engravers tools, springs, hammers, axes, and rails.

(2) Trade name of Wanfrieder Schmirgelwerk Gottlieb Gries KG (Germany) for emery cloth and paper, and waterproof abrasive paper.

Elektra. Trademark of Ivoclar Vivadent AG (Liechtenstein) for an extra-hard dental casting alloy (ADA type IV) containing 58.3% silver, 25.0% palladium, 14.67% copper, 2.0% indium, and less than 1.0% ruthenium, rhenium and lithium, respectively. It has a white color, a density of 10.4 g/cm³ (0.376 lb/in.³), a melting range of 865-990°C (1589-1814°F), an as-cast hardness of 185 Vickers, low elongation, and excellent biocompatibility. Used for crowns, onlays, posts, and bridges.

Elektro. Trade name of Friedrich Lohmann GmbH (Germany) for a water-hardening tool steel (AISI type W1) containing 0.85%, 0.3% silicon, 0.6% manganese, and the balance iron. Used for reamers, taps, drills, springs and tools.

Elektrobronze. Trade name of Ostermann GmbH & Co. (Germany) for a corrosion-resistant bronze containing 81% copper, 10% aluminum, and a total of 9% iron and manganese. Used for valve gages, and fittings.

Elek-Tro-Cut. Trademark for coated abrasives.

Elektroglas. Trade name of NV Syncoglas SA (Belgium) for glass fabrics used in electrical laminates.

Elektron. Trademark of Magnesium Elektron, Limited (UK) for a series of lightweight, wrought or cast magnesium alloys containing small additions of other elements, such as aluminum, manganese, thorium, zinc, zirconium, and/or rare earths. Supplied in the form of sand, chill and die castings, forgings, extrusions, tubing, structural shapes, rods, sheets and strips, they have a low density (1.8 g/cm³, or 0.07 lb/in.³), good mechanical properties, and good machinability. Used for aircraft and automotive parts, lightweight structural parts, portable machinery, machine parts, fixtures and jigs, and dies.

element 104. See rutherfordium.

elemental semiconductor. A semiconductor, such as germanium or silicon, in which the bonding is essentially covalent, i.e., in which the valence electrons are shared by adjacent atoms and have very limited mobility. This bonding is highly directional giving rise to a structure of high strength and low ductility.

Elementis. Trademark of V.O. Baker Inc. (USA) for a series of potting and encapsulating compounds.

elemi. A soft, white, yellow, or green resin, rich in essential oils, obtained from various tropical trees of the family Burseraceae in the order Sapindales, especially the trees *Canarium communis* and *C. luzonicum*. It has a density of 1.02-1.08 g/cm³ (0.037-0.039 lb/in.³), a melting point of 120°C (248°F), and is used as an addition to certain varnishes, lacquers, lithographic inks, wax compositions and cements, in waterproofing and engraving, and in textile and paper coatings. Also known as *elemi gum.*

Elemid. Trade name of GE Plastics (USA) for a polyamide/acrylonitrile-butadiene-styrene alloy.

elemi gum. See elemi.

Elephant Brand. Trade name of Seymour Products Company (USA) for a series of phosphor bronzes, phosphor-copper master alloys, and phosphorized copper-base bearing alloys.

Elfur. Trade name of Baldwin-Lima-Hamilton Corporation (USA) for a series of nickel cast irons containing 3.2-3.45% carbon, 1.3-1.5% silicon, 0-0.8% manganese, 0-0.3% chromium, 1.5-3.0% nickel, traces of phosphorus, and the balance iron. Used for pistons, cylinders, cylinder heads, engine liners, turbine casings, valve bushings, brake drums, and camshafts.

Elftex. Trade name for a series of carbon blacks.

Elga. Trade name of Saint-Gobain (France) for a thin, decorative hollow glass block with smooth external surfaces and an irregular chequerboard-type design on its internal surfaces.

Elgiloy. Trademark of Elgiloy Limited Partnership (USA) for a corrosion-resistant, nonmagnetic cobalt-base superalloy containing 20% chromium, 15% nickel, 15% iron, 7% molybdenum, 2% manganese, 0.15% carbon, and 0.05% beryllium. Commercially available in wire, strip, rod and tube form, it can be precipitation-hardened at 480°C (900°F) to produce high hardness. It has high strength and elastic modulus, excellent fatigue life and dimensional stability, good high-temperature corrosion resistance, excellent resistance to stress-corrosion cracking, excellent resistance to hydrogen sulfide and chlorides, good performance in marine and other corrosive environments, and excellent performance at sub-zero and elevated temperatures ranging from about -180 to +540°C (-300 to +1000°F). Used for seal components, coil and flat springs, watch springs, instrumentation, marine equipment, oil rigs, chemical and petrochemical plant and equipment, and aerospace, automotive and general industrial applications. *Elgiloy* wire is also used for dental applications. See also Cobenium and Durapower.

Elginite. Trade name of Elgin Watch Company (USA) for an alloy of chromium, nickel, iron, molybdenum and cobalt. It has constant flexibility from -37 to +50°C (-35 to +122°F), and is used for temperature-compensating hairsprings.

Elhanco. British trade name for a water-hardening tool steel containing 0.7-1.2% carbon, 0.2% vanadium, and the balance iron.

Elianite. Italian trade name for a brittle, acid-resistant alloy containing 81.4-84.4% iron, 15.0% silicon, 0-2.2% nickel, 0-0.82% carbon, and 0.5-0.6% manganese. Used for pumps, valves, drains, and chemical equipment.

Eliminal. Trade name of Foseco Minsep NV (Netherlands) for a series of fluxing agents introduced into copper and nickel alloys to make them more fluid and thus improve their overall castability by eliminating small quantities of aluminum or silicon.

Elinvar. Trade name of Telcon Metals Limited (UK) for a slightly magnetic alloy containing 33-36% nickel, 4-12% chromium, 0.5-2% silicon, 0.5-2% carbon, 0-2% manganese, 0-3% tungsten, and the balance iron. It may also contain small additions of titanium and molybdenum, and has a low coefficient of thermal expansion, a constant modulus of elasticity over a considerable range of temperatures, and good strength properties. Used for resistances, instrument springs, hairsprings, chronometer balances, and watches.

Elinvar Extra. Trade name of International Nickel Inc. (USA) for a precipitation-hardening nonmagnetic alloy containing 48% iron, 42-43% nickel, 5-6% chromium, 2.5-2.75% titanium, 0-0.5% manganese, 0.04-0.06% carbon, 0.3-0.5% aluminum, 0.5% silicon, and 0-0.35% cobalt. It has a low thermal expansion and good corrosion resistance. Used for springs, flexures, orthodontic wires, diaphragms, etc.

Elisha EMC. Trade name of Elisha Technologies Co., LLC (USA) for a chromate-free passivation coating used as a tiecoat over zinc.

Elite. (1) Trade name of Stahlwerk Stahlschmidt GmbH & Co. (Germany) for a wear-resistant, oil-hardening tool steel containing 1.35% carbon, varying amounts of tungsten and cobalt, and the balance iron. Used for punches, mandrels, and blanking and forming dies.

(2) Trademark of Dow Chemical Company (USA) for enhanced polyethylene resins produced by a proprietary processing technology (known as INSITE). Supplied in blown-film, cast-film and extrusion-coating/laminating grades, they possess high impact and puncture resistance, high stiffness and stretchability, and good sealability and processibility. Used for packaging dry, liquid and frozen foods, and for heavy-duty shipping sacks, diaper backsheets and agricultural silage wrap, and for rotomolded products, such as toys and playground equipment.

(3) Trade name of Aurident, Inc. (USA) for a dental ceramic alloy containing 88.5% gold and 8.3% platinum. It has a deep yellow color, outstanding shade control, excellent bonding properties, and is used for porcelain-to-metal restorations.

(4) Trademark of Libbey-Owens-Ford Company (USA) for tempered safety glass.

(5) Trade name of Nylstar SpA (Italy) for polybutylene terephthalate fibers and yarns.

Elite II. Trade name of Aurident, Inc. (USA) for a biocompatible dental ceramic alloy containing 88.0% gold and 9.5% platinum. It has a deep yellow color, excellent physical and mechanical properties, and takes a lustrous polish. Used for porcelain-to-metal restorations.

Elite Cement. Trade name of Aurident, Inc. (USA) for a zinc phosphate dental cement.

Elite Fast. Trade name of Aurident, Inc. (USA) for a silicone elastomer used for dental impressions.

Elkaloy. Trade name of CMW Inc. (USA) for coppers and copper alloys including cadmium copper for resistance welding electrodes, and several aluminum bronzes for bushings, bearings, forming dies, fixtures, and welding electrodes.

Elkem. Trade name of Elkem Metals Company (USA) for a range of high-quality ferrosilicon, silicon metal, silica fume, calcium carbide and carbon products used in the aluminum, steel, foundry or chemical industries, and/or electronics.

elk leather. (1) Leather made from elkskin, preferably known as *buckskin*.

(2) Durable, flexible, chrome-tanned cattlehide used for pocketbooks, and shoe and boot uppers, e.g., for heavy-duty work boots and shoes, hunting boots, and sport shoes.

Elkonite. Trademark of CMW Inc. (USA) for an extensive series of sintered materials based on tungsten, molybdenum or their carbides, and containing copper or silver in varying proportions. Used for electrical contacts, resistance brazing and welding electrodes and dies, projection welding tips, and hot-riveting electrodes.

Elkonium. Trademark of CMW Inc. (USA) for an extensive series of electrical contact materials including numerous copper-, silver-, palladium- and platinum-based metals and alloys.

Elkro. Trade name of Disston Inc. (USA) for an oil-hardening tool steel containing 0.5% carbon, 0.95% chromium, 0.2% vanadium, and the balance iron. Used for dies.

ellisite. A dark gray mineral composed of thallium arsenic sulfide, Tl_3AsS_3. Crystal system, hexagonal. Density, 7.10 g/cm³. Occurrence: USA (Nevada).

ellsworthite. See betafite.

Elm. Trade name of Elm Engineering Limited (UK) for wrought aluminum-magnesium and aluminum-manganese alloys supplied in strip and tube form.

elm. The hard, durable wood of any of a genus *(Ulmus)* of tall hardwood trees native mainly to North America and Europe. There are numerous North American species including American elm *(U. americana)*, slippery or red elm *(U. rubra)*, rock or hickory elm *(U. thomasii)*, winged elm or wahoo *(U. alata)*, cedar elm *(U. crassifolia)* and September or southern red elm *(U. serotina)*. Common European elms include wych elm *(U. glabra)*, and English elm *(U. procera)*.

Elma. Trade name for rayon fibers and yarns used for textile fabrics.

Elmarid. British trade name for a cemented carbide containing both tungsten carbide (WC) and ditungsten carbide (W_2C). Used for dies, and tips for high-speed tools.

Elmax. Trade name of Uddeholm Corporation (USA) for a specialty stainless mold steel made by powder metallurgy techniques. It contains 1.7% carbon, 0.3% manganese, 0.4% silicon, 17.0% chromium, 1.0% molybdenum, 3.0% vanadium, and the balance iron. *Elmax* has high wear resistance, good toughness, good to moderate corrosion resistance, and high hardness. Used for molding electronic and electrical components.

Elmedur. Trade name of Plansee Metallwerk-Gesellschaft (Austria) for a sintered tungsten alloy used for resistance welding electrodes.

Elmer's. Trademark of Borden Company (USA) for carpenter's glues, contact cements, and wood fillers.

Elmet Rotung. Trade name of Plansee Metallwerk-Gesellschaft (Austria) for a copper-tungsten electrical contact alloy.

Elmet Silno. Trade name of Metro Cutanit Limited (UK) for a series of sintered materials containing silver and nickel with silver and cadmium oxides. Used for electrical contacts.

Elmet Siltung. Trade name of Plansee Metallwerk-Gesellschaft (Austria) for a silver-tungsten electrical contact alloy.

Elner's German silver. A nickel silver (German silver) of 57.4% copper, 26.6% zinc, 13% nickel and 3% iron, used for ornaments.

Elnic. Trade name of MacDermid Inc. (USA) for an electroless nickel plate and process.

Elo. Trade name of Birkby Limited (UK) for phenolic resins.

Eloma. Trade name for rayon fibers used for textile fabrics.

Elomag. German trade name for hard, corrosion and wear-resistant protective coatings of magnesium oxide produced on magnesium and aluminum alloys by anodic oxidation. See also anodized aluminum; anodized magnesium.

Elongatable Dow Fiber. Trade name of Dow Chemical Company (USA) for a flexible, curly carbon fiber with good flame resistance, insulation properties, and moisture resistance. Used in the manufacture of insulation and fire-block products for aircraft, ships, recreational equipment (e.g., tents), and clothing. Abbreviation: EDF.

elongated single-domain magnet. A magnet manufactured from fine aligned particles (e.g., iron-cobalt) of a size below the domain wall size, and having an elongated shape in the preferred direction of magnetization. Also known as *ESD magnet.*

Eloxal. German trade name for hard protective coatings of aluminum oxide produced on aluminum and aluminum alloys by anodic oxidation, i.e., by conversion of the aluminum surface to aluminum oxide by making the metal or alloy the anode in an electrolytic cell, and using an oxalic, sulfuric and chromic acid electrolyte. These coatings are usually from 3 to 60 μm (0.1 to 2.4 mils) thick, as compared to the natural oxide film with less than 0.1 μm (0.004 mil). They have excellent chemical resistance, high hardness, good electrical insulating properties, good abrasion and wear resistance, and good corrosion and oxidation resistance.

elpasolite. A colorless mineral composed of potassium sodium aluminum fluoride, K_2NaAlF_6. It can also be made synthetically. Crystal system, cubic. Density, 2.99 g/cm³; refractive index, 1.376. Occurrence: USA (Colorado).

Elphal. Trade name of British Steel plc (UK) for an aluminum-coated mild steel.

elpidite. A white to brick red mineral of the pyroxenoid group composed of sodium zirconium silicate hydroxide, $Na_2ZrSi_6O_{15} \cdot 3H_2O$. It can also be synthesized from zirconium dioxide, silicon dioxide and hydrous sodium carbonate by a hydrothermal process. Crystal system, orthorhombic. Density, 2.58 g/cm³; refractive index, 1.561. Occurrence: Greenland, Russia.

Elprene. Trademark of Atlas Minerals & Chemicals, Inc. (USA) for self-curing neoprene rubber used for general maintenance coatings.

Elran. Trade name of Vereinigte Metallwerke Ranshofen-Berndorf (Austria) for a heat-treatable, corrosion-resistant aluminum alloy containing 0.3-0.7% silicon and 0.4-0.8% magnesium. Used for structural and decorative parts.

Elstrama. Trade name for rayon fibers and yarns used for textile fabrics.

Eltex. Trademark of Solvay Polymers, Inc. (USA) for thermoplastic polyolefins and copolymers based on high-density poly-

ethylene. Usually supplied in plates, sheets and rods.

Eltex P. Trademark of Solvay Polymer Inc. (USA) for polypropylene homopolymers and copolymers supplied in unstabilized and UV-stabilized grades.

Elura. (1) Trade name for a modacrylic fiber used chiefly for clothing and industrial fabrics.

(2) Trade name for a spandex (segmented polyurethane) fiber used for the manufacture of textile fabrics with excellent stretch and recovery properties.

Elustra. Trademark of Hercules, Inc. (USA) for polypropylene fibers and yarns.

Elvace. Trademark of Reichhold Chemicals, Inc. (USA) for vinyl acetate and ethylene vinyl acetate emulsion copolymers, terpolymers and comonomers used in the manufacture of building products, nonwovens, carpets, paper, adhesives, sealants, caulking compounds, paints, and coatings.

Elvace CPS. Trademark of Reichhold Chemicals, Inc. (USA) for vinyl acetate ethylene copolymers supplied in various grades and forms including powders and flexible and tough films, and used as binders in caulking compounds, plaster, concrete, tile grouts and mortars, and as hydraulic cement additives.

Elvacet. Trademark of E.I. DuPont de Nemours & Company (USA) for a series of vinyl acetate molding resins and polymerized vinyl esters.

Elvacite. Trademark of E.I. DuPont de Nemours & Company (USA) for acrylic resins and adhesives based on polymethyl methacrylate.

Elvamide. Trademark of E.I. DuPont de Nemours & Company (USA) for a family of tough, high-elongation, alcohol-soluble polyamide (nylon) multipolymer resins and copolymers for molding, extrusion, and bonding applications. They adhere well to nylon yarns, are strong and abrasion-resistant, and can be easily processed.

Elvandi. Trade name of Teledyne Vasco (USA) for water-hardening tool steels (AISI type W2) containing 0.6-1.4% carbon, 0.1-0.4% silicon, 0.1-0.4% manganese, 0.18-0.22% vanadium, and the balance iron. They have fine-grained microstructures, excellent machinability, shallow depths of hardening, and high toughness. Used for punches, caulking tools, blanking and forging dies, cold heading dies, cold chisels and cutters, battering and beading tools, stamps, and shear blades.

Elvanol. Trademark of E.I. DuPont de Nemours & Company (USA) for a series of strong, flexible or rigid, water-soluble polyvinyl alcohol resins supplied in three grades: fully hydrolyzed, partially hydrolyzed, and copolymer. Available in various colors, they possess high stiffness at low temperatures, excellent adhesive, film-forming and emulsifying properties, good flame and electrical resistance, good chemical, oil and weather resistance, good abrasion resistance, good processibility, and a maximum service temperature of 138°C (280°F). Used for floor and wall coverings, upholstery, tubing, insulation, safety glass interlayers, phonographic products, toys, and rainwear, in adhesives for water-absorbent substrates, such as leather, paper, textiles and wood, as binders for paper coatings, ceramics and nonwovens, and in textile sizing and finishing.

Elvax. Trademark of E.I. DuPont de Nemours & Company (USA) for ethylene-vinyl-acetate copolymer resins containing 12, 25, or 33% vinyl acetate. Supplied in several packaging and film grades, they provide excellent clarity, impact and puncture resistance, high toughness, and good low-temperature sealability.

Elverite. Trade name of Babcock & Wilcox Company (USA) for series of abrasion- and wear-resistant cast irons available in four grades: (i) *Elverite A* which is a wear-resistant chilled cast iron that contains 3.4-3.7% carbon, 0.5-1% silicon, and the balance iron, and is used for jaw crusher sprockets, wheels and tube mill linings; (ii) *Elverite B* which is a wear-resistant chilled cast iron with 3.3% carbon, 2.4% silicon, 1.5% nickel, 0.7% chromium, and the balance iron, and is used for pulverizer, crusher, stamp mill and conveyor parts; (iii) *Elverite C* which is an abrasion-resistant cast iron that contains 3.5% carbon, 0.6% silicon, 1.5% chromium, 4.5% nickel, 0.6% molybdenum, and the balance iron, and is used for crushers, mixers, pulverizers in the coal and cement industries; and (iv) *Elverite D* which is a wear-resistant cast iron with 3.3% carbon, 2.2% silicon, 0.4% molybdenum, 0.7% manganese, and the balance iron, used for castings.

elyite. A violet mineral composed of copper lead sulfate hydroxide, $CuPb_4(SO_4)(OH)_8$. Crystal system, monoclinic. Density, 6.32 g/cm^3; refractive index, 1.993. Occurrence: USA (Nevada).

elysian. A thick fabric, usually of wool, having a deep surface nap arranged in a diagonal or rippled pattern. Used especially for making coats.

Emac. Trade name of Chevron Chemical Company LLC (USA) for ethylene-vinyl-acetate copolymers in pellet form.

EMA copolymers. See ethylene-maleic anhydride copolymers.

Emalit. Trade name of Saint-Gobain (France) for ceramic-coated, toughened glass for cladding applications.

Emarex. (1) Trade name of MRC Polymers (USA) for impact-modified polyphenylene ether/nylon 6,6 alloys available in unreinforced and mineral- or glass-reinforced grades. Used for injection-molded products.

(2) Trademark of Ziegler Chemical & Mineral Corporation (USA) for uintaite-based natural asphalt. See also uintaite.

Ematal. German trade name for hard, dense, nontransparent, gray, opaque, extremely abrasion-resistant protective coatings of aluminum oxide produced on aluminum and aluminum alloys by anodic oxidation, i.e., by converting the aluminum surface to aluminum oxide while the alloy is the anode in an electrolytic cell containing an electrolyte composed of oxalic acid, citric acid, boric acid, a titanium salt, and water.

Émauglas. Trade name of Boussois Souchon Neufesel SA (France) for ceramic-coated, toughened plate glass.

Emaye Cam. Trade name Türkiye Sise ve Cam Fabrikalari AS (Turkey) for enameled glass used in building construction for decorative applications.

Embacoid. Trade name of May & Baker (UK) Limited for cellulose acetate plastics.

Embafilm. Trade name of May & Baker (UK) Limited for cellulose acetate plastics supplied in film form.

Embeebush. Trade name of Delta Extruded Metals Company Limited (UK) for a corrosion-resistant cast copper alloy.

Embeesh. Trade name of Delta (Manganese Bronze) Limited (UK) for strong, corrosion-resistant copper alloys containing 40% zinc, and varying amounts for aluminum, iron, manganese and tin. Used for machine parts, bushings and bearings, marine parts, valve stems, and fasteners.

Emblazia. Trademark of KoSa (USA) for solution-dyed polyester yarns supplied in a wide range of brilliant colors. The colors are uniformly distributed throughout the yarns, and the latter blend well with other fibers. Used for textile fabrics including wearing apparel and home furnishings.

embolite. A yellow-green mineral composed of native silver bromide and silver chloride, Ag(Cl,Br). Crystal system, cubic. Occurrence: Chile. Used as an ore of silver.

Embonte. Trade name for a zinc-oxide/eugenol dental cement for temporary restorations.

embossed fabrics. (1) Textile fabrics that have been decorated with raised designs or patterns by passing through a calender.

(2) Textile fabrics that have raised designs or patterns produced by pressing an engraved metal cylinder onto them, usually with heat and pressure.

embossed hardboard. Decorative hardboard panel with a pattern resembling basketweave, leather, woodgrain, etc., pressed into the surface during manufacture.

embossed leather. Leather decorated with a raised pattern or design by pressing or stamping in a special machine.

embossed plywood. Plywood panel that has received a special surface treatment to effect a texture or pattern. This treatment usually consists of pressing the heated panel against a master pattern.

embossed satin. A usually heavy satin-weave fabric with a permanent, impressed pattern. Used especially for wedding gowns and ceremonial clothing. See also satin-weave fabrics.

Embrace. Trade name of Eastman Chemical Company (USA) for copolyester supplied in the form of shrink film used for shrink labels for containers, and high-contour bottle labels.

Embrace WetBond. Trade name of Pulpdent Products, Inc. (USA) for light-activated, lightly-filled, fluoride-releasing hydrophobic dental resin used as a pit and fissure sealant and lesion restorative.

embreyite. An orange mineral composed of lead chromium oxide phosphate monohydrate, $Pb_5(CrO_4)_2(PO_4)_2 \cdot H_2O$. Crystal system, monoclinic. Density, 6.45 g/cm^3; refractive index, 2.36. Occurrence: Siberia.

embroidered fabrics. Textile fabrics ornamented with raised designs or patterns made with needle and thread or yarn either by hand or by machine.

embroidery crash. See embroidery linen.

embroidery linen. A well-balanced textile fabric usually made from smooth, round cotton or cotton/linen yarns in a plain weave. Used for the manufacture of table linen, draperies, slipcovers, pillowcases and apparel to be embroidered. Also known as *art linen; embroidery crash.*

Emburite. Trademark of Embury Company (Canada) for fire-labeled anodized aluminum building panels.

emeleusite. A colorless mineral of the osumilite group composed of a lithium sodium iron silicate, $Li_2Na_4Fe_2Si_{12}O_{30}$. Crystal system, orthorhombic. Density, 2.76 g/cm^3; refractive index, 1.597. Occurrence: Greenland.

Emera. Trade name for rayon fibers and yarns used for textile fabrics.

emerald. A brilliant-green to grass-green variety of the mineral *beryl* ($Al_2Be_3Si_6O_{18}$), which contains sufficient chromium to impart the green color. Crystal system, hexagonal. Density, 2.7-2.9 g/cm^3; hardness, 8 Mohs. Occurrence: Colombia, Egypt, Russia, Siberia, USA (North Carolina). Used chiefly as a high-purity laser crystal, and as a gemstone.

emerald glass. A bright-green glass obtained by fusing *beryl* ($Al_2Be_3Si_6O_{18}$). It has a density of 2.5 g/cm^3 (0.09 $lb/in.^3$), and a refractive index of 1.52.

emerald green. A name for two bright-green pigments, namely Paris green (a copper acetoarsenite) and Guignet's green (a chrome green type pigment). See also copper acetoarsenite; chrome green.

Emerald Ram. Trade name of Premier Refractories and Chemicals Inc. (USA) for alumina-chrome plastic refractory mixes.

Emerez. Trademark of National Distillers & Chemical Corporation (USA) for polyamide and polyester resins used in the manufacture of thixotropic paints and hot-melt adhesives.

emerized fabrics. Fabrics, such as *serge* and *worsted*, with a chamois- or suede-type surface texture produced by a roller covered with emery.

emery. A dark gray to black, hard, impure variety of the mineral *corundum* (Al_2O_3) mixed with magnetite, hematite, quartz and/or spinel. Used in pulverized form as a slow-cutting abrasive for grinding, smoothing and polishing.

emery cake. Powdered emery made into cake form using a suitable binder.

emery cloth. A strong fabric of cotton or linen to which emery powder has been bonded. Used as a bonded abrasive for grinding and polishing metals, wood and plastics. See also emery paper.

emery flour. A very fine dust obtained by crushing and grinding of emery powder.

emery paper. Emery powder tightly bonded to a strong paper or cloth backing. Emery paper and cloth are usually available in grit sizes ranging from very coarse (24 mesh) to very fine (280 mesh) on sheets typically 230 × 280 mm (9 × 11 in.) in size for use as abrasive paper for cleaning, grinding and polishing metals, wood and plastics. See also emery cloth.

emery paste. A tallow or grease stick impregnated with fine emery powder, and used for polishing and buffing applications.

emery powder. Impure *corundum* in the form of a powder containing 50-60% alumina (Al_2O_3), 30-40% magnetite (Fe_3O_4), with the balance being chiefly ferric oxide (Fe_2O_3), silica (SiO_2) and chromium. Used in metal finishing for cutting down and preliminary operations. See also abrasive powder.

emery stone. A mixture of emery powder and a shellac or clay binder used for molding grinding wheels.

EMI/RFI shielding materials. A broad class of engineering and engineered materials that have or have been given high electrical conductivity, low electrical resistivity or surface resistivity, or low dielectric strength and, thus, suppress changing magnetic fields or electromagnetic radiation at electronic or communication systems (EMI shielding), and/or prevent radio-frequency electromagnetic radiation from leaving or entering the systems (RFI shielding). Materials used for these applications include various metals and alloys, carbon, and conductive plastics and composites, such as nylon (e.g., nylon 6,6) reinforced with nickel-coated carbon fibers, polycarbonates reinforced with polyacrylonitrile, and urethane hybrid composites reinforced with conductive carbon mats or metal fabrics. Also known as *electromagnetic interference/radio-frequency interference shielding materials*.

Emisaloy. Trademark of Wilbur B. Driver Company (USA) for a heat- and corrosion-resistant alloy containing 80% nickel and 20% cobalt. It has a density of 8.84 g/cm³ (0.32 lb/in.³), a melting point of 1500°C (2730°F), and good strength. Used for electron tubes.

Emlie. Trademark of Toho Rayon Company (Japan) for regenerated cellulose (modal) fibers.

Emmembrucke. Trade name for nylon fibers and yarns used for textile fabrics.

emmonsite. A yellow-green mineral composed of iron tellurate dihydrate, $Fe_2(TeO_3)_3 \cdot 2H_2O$. Crystal system, triclinic. Density, 4.53 g/cm³; refractive index, 2.09. Occurrence: USA (Nevada).

E-Monel. Trade name of Inco Alloys International Inc. (USA) for a series of cast nickel alloys containing 26-33% copper, 3.5%

iron, 1-2% silicon, 1.5% manganese, 0.3% carbon, and a total of 1-3% niobium (columbium) and tantalum. They have excellent corrosion resistance, good strength, and are used for pumps, valves, and fittings. See also Monel.

Empedur. Trade name of Plansee Metallwerk-Gesellschaft (Austria) for a sintered alloy of tungsten carbide and iron, used for hardfacing electrodes.

Empee. Trademark of Monmouth Plastics Inc. (USA) for an extensive series of flame-retardant polyolefin resins. *Empee PE* refers to high, medium and low-density polyethylenes with a density range of 0.91-1.4 g/cm³ (0.33-0.51 lb/in.³), used for electric wire and cable, injection and blow-molded parts, extruded sheet, and shapes. *Empee PP* are fire-retardant high- and low-density polypropylenes with a density range of 0.90-0.99 g/cm³ (0.32-0.36 lb/in.³), used for injection- and flow-molded parts, thermoformed parts, extruded sheet, shapes, and for various electrical applications. *Empee PS* refers to transparent polystyrenes with good physical properties used for office machines, dust covers, consumer electronics, light diffusers, decorative products, and clear film for packaging.

Emperor. (1) Trade name of Sutcliff, Speakman & Company (UK) for a heat- and corrosion-resistant steel containing 0.2% carbon, 20% chromium, 10% nickel, and the balance iron. Used for furnaces, crucibles, carburizing pans, hardening boxes, and glass molds.

(2) German trade name for an antifriction alloy containing lead, tin and antimony, used for bearings.

emperor brass. A corrosion-resistant brass of 60% copper, 20% aluminum and 20% zinc.

Emperor Chrome Nickel Alloy. Trade name of Sutcliff, Speakman & Company (UK) for an austenitic stainless steel containing 0.2% carbon, 19% chromium, 9% nickel, and the balance iron. It has good heat and corrosion resistance, and is used for furnaces and furnace components.

Empire. Trade name of Empire Steel Castings Company (USA) for an extensive series of cast corrosion- and heat-resistant steels for various chemical, petrochemical and engineering applications.

emplectite. A grayish white mineral composed of copper bismuth sulfide, $CuBiS_2$. Crystal system, orthorhombic. Density, 6.38 g/cm³. Occurrence: Germany.

Empress. Trade name of Ivoclar Vivadent AG (Liechtenstein) for a dental porcelain based on a glass-ceramic.

empress. See kiri; paulownia.

empressite. A pale-bronze mineral composed of silver telluride, AgTe. Crystal system, orthorhombic. Density, 7.61 g/cm³. Occurrence: USA (Colorado).

Emralon. Trademark of Acheson Colloids Company (USA) for a series of dry-film lubricants composed of colloidal tetrafluoroethylene resins dispersed in epoxy, phenolic, thermoplastic or other resin solutions.

Emro. Trade name of Pyramid Steel Company (USA) for a shock- and fatigue-resistant steel used for gears, shafts, hooks, chains, and chisels.

Emsodur. Trademark of EMS Grilon (USA) for cubical-shaped polyamide shot blasting media for efficient deflashing and deburring of thermoset moldings and rubber articles. *Emsodur S* is designed for cryogenic deflashing of rubber components at temperatures as low as -200°C (-328°F). It has a natural color, and is supplied in 1.0 mm (0.04 in.) cubical size. *Emsodur GV* is glass-reinforced for added weight to provide efficient deflashing at lower speeds. It is suitable for difficult-to-deflash parts as well as for metal parts and roughing up metal surfaces to promote better adhesion of coatings. It is has a natural color, and is supplied in 1.0 mm (0.04 in.) cubical size. *Emsodur Micro* is a micron-size medium for deflashing electrical and electronic parts. The natural color product is supplied in granules with an average size of 200-400 µm (0.008-0.016 in.) that further reduce the risk of slot-and-hole-blocking in small parts.

Emu. Trademark of Boral Kinnaers Proprietary Limited (Australia) for polypropylene fibers and yarns.

emulsified asphalt. An asphalt emulsion in water used for waterproofing concrete, painting steam and hot-water pipes, and surfacing floors.

emulsifier. (1) A surface-active agent that intimately mixes with and disperses dissimilar, generally immiscible materials, such as oil and water, or water and fine particles, to produce a stable emulsion. Also known as *dispergator; disperser; dispersing agent; emulsifying agent.*

(2) In liquid penetrant inspection, an agent that is added to a liquid penetrant to make it soluble and thus water-washable.

emulsifying agent. See emulsifier (1).

emulsion. (1) A stable dispersion of two immiscible liquids. In an emulsion minute droplets of one of the liquids (e.g., oil) are evenly distributed throughout the other (e.g., water).

(2) The photosensitive coating on a camera film, or photographic plate.

emulsion butadiene rubber. A white, viscous aqueous suspension of butadiene rubber. Abbreviation: E-BR.

emulsion-coated paper. A paper coated with polymers or synthetic resins that are in the form of an emulsion any of various processes.

emulsion coating. An aqueous suspension of latex, oil or resin. See also emulsion paint.

emulsion-derived foams. A group of engineered, low-density polymeric foams (e.g., polystyrene, polymethacrylonitrile and resorcinol-formaldehyde) with open-celled structure produced by high internal phase emulsion. Used as absorption, separation and filter media, in supercapacitors and rechargeable batteries, and as supports for biochemical synthesis.

emulsion paint. A paint made by emulsifying the film former or binder, usually a latex, oil or resin, in a volatile liquid, usually water. See also latex paint.

emulsion polymer. A polymer, often a synthetic rubber, such as styrene butadiene, that is the end product of a polymerization reaction carried out at standard pressure (101.325 kPa or 14.696 psi) and temperatures between -20 and +60°C (-4 and +140°F) with the reacting species in emulsified form. Abbreviation: EP.

emulsion rust-preventive compound. A compound of emulsifying or soluble oils, usually oil-in-water emulsions or wax emulsions, used as temporary protective coatings on metal surfaces.

emulsion-spun fibers. Synthetic fibers that have been spun from a polymer emulsion.

Emultex. Trade name polyvinyl acetates.

Emu-Pulver. Trademark of BASF AG (Germany) for polystyrene copolymer powders.

enamel. (1) See enamel paint.
(2) See porcelain enamel.
(3) See dental enamel.
(4) See enamel leather.

enamel brick. A brick with a hard, smooth, glazed or enamellike surface produced by a fired-on, usually colored, wash-type coating. Also known as *enameled brick.*

enamel brick clay. A clay similar in composition and consistency to that used in the production of buff-colored face brick. It can be used as in coatings, and fired to vitrification.

enamel clay. A fat clay, such as *ball clay*, suitable for use as a suspension promoter for porcelain enamels and glazes in aqueous slips. Also known as *enameling clay.*

enamel colors. Inorganic compositions used to impart color to porcelain enamels, ceramic bodies, glass and glazes.

enameled brick. See enamel brick.

enameled sheet steel. See porcelain enameled sheet steel.

enamel film insulator. A magnetic wire consisting of a core of a metallic conductor coated with a thin, flexible enamel film based on epoxy, polyester or polyvinyl acetate resins.

enamel frit. See porcelain enamel frit.

enameling clay. See enamel clay.

enameling iron. See porcelain enameled iron.

Enamelize. Trade name for alumina polishing paste for dental applications.

enamel leather. Leather having a shiny finish on the grain side. Also known as *enamel.*

Enameloid. Trade name of Gemloid Corporation (USA) for acrylic and cellulose acetate plastics and products.

enamel oxides. See porcelain enamel oxides.

enamel paint. A paint made by grinding or mixing pigments with varnishes, lacquers, synthetic resins or combinations thereof. It dries by oxidation or baking, and is used to form smooth, hard, glossy or semiglossy surfaces on metals or ceramics. Also known as *enamel.* See also baking enamel.

enamel paper. See coated paper.

enamelware. Articles of steel, cast iron, aluminum or copper coated with porcelain enamel and including various products of the kitchenware industry, such as pots, pans, etc.

Enant. Trade name for nylon fibers.

enantiomer. One of a pair of stereoisomers, whose molecular configurations have left-handed (levo) and right-handed (dextro) forms and differ only in being mirror images of each other. For example, the two glyceraldehydes L-glyceraldehyde and D-glyceraldehyde are enantiomers. See also stereoisomer.

enantiotropic liquid crystal. A type of thermotropic liquid crystal that exhibits the liquid crystalline state when the temperature of a solid is raised, or the temperature of a liquid is lowered. See also liquid crystal; monotropic liquid crystal; thermotropic liquid crystal.

enargite. A dark to metallic gray mineral of the wurtzite group composed of copper arsenide sulfide, Cu_3AsS_4. It can also be made synthetically. Crystal system, orthorhombic. Density, 4.43-4.45 g/cm³; hardness, 3 Mohs. Occurrence: Europe (Balkan peninsula), USA (Colorado, Montana, Utah), Peru. Used as an ore of copper and arsenic.

Enathene. Trademark of Quantum Chemical Corporation (USA) for ethylene butyl acrylate copolymers used for adhesives and polymer films.

Enbiron. Trade name for vinyon (vinyl chloride) fibers with excellent acid, alkali and mildew resistance. Used for textile fabrics and carpets.

Enbo. Trade name of Surface Technology Inc. (USA) for electroless nickel-boron coatings.

encapsulant. An insulating material in the form of a conformal or thixotropic polymer coating, or a molten polymer or elastomer film or foam applied to an object, such as a small electronic component or device (e.g., a capacitor or circuit board module) by brushing, dipping or spraying. Polymers frequently used for this application include epoxies and silicones.

encapsulated adhesive. A special adhesive in which the particles of one of the reactants are enclosed in a protective film to suppress the curing reaction until the film is intentionally destroyed by mechanical or chemical means.

Encaptrax. Trademark of Glotrax Polymers Inc. (USA) for a line of epoxy-based potting compounds and encapsulants.

Encasa. Trade name for regenerated protein (azlon) fibers.

encaustic. A paint based on hot, pigmented wax.

encaustic tile. A ceramic tile that has a pattern or design set in its surface with clays of a color different than that of the body, followed by firing.

Encel. Trade name for rayon fibers and yarns used for textile fabrics.

ENCo-CC. Trademark of Engineered Coatings (USA) for abrasive and abradable thermal-spray composite coatings supplied in metallic, ceramic, organic and inorganic grades. They have excellent corrosion, erosion and oxidation resistance, and are applied to rotating parts, such as compressors, expanders and turbochargers, mainly for clearance control.

Encon. Trade name of Delta Enfield Metals Limited (UK) for continuously cast copper-tin alloys.

Encore. Trademark of Degussa-Ney Dental (USA) for a hard, white copper and silver-free dental bonding alloy containing 48% gold and 40.5% palladium. It has a light oxide layer, high yield strength, good resistance to margin warping, a mid-range thermal expansion coefficient compatible with most porcelain, and a hardness of 250 Vickers. Used for porcelain-to-gold restorations subject to stress.

Encore Plus. Trade name of Degussa-Ney Dental (USA) for a silver-free palladium dental bonding alloy.

Encron. Trade name for polyester fibers and yarns used for textile fabrics.

end-and-end. A plain-weave fabric made with alternating warp (longitudinal) yarns of white and a color, and alternating weft (filling) yarns of two different colors producing a small check design. Used especially for shirts. Also known as *end-to-end*.

end construction tile. A hollow concrete or fired ceramic tile designed to receive compressive stresses parallel to the axes of the cells.

end-cut brick. An extruded brick having both ends cut by a wire.

end-grained wood. A tree or log cut at right angle to the direction of the fibers. The resulting piece has a cross section revealing pitch, cambium, annual rings, etc.

Endobon. Trademark of Biomet Merck (Germany) for a highly biocompatible *hydroxyapatite* ceramic with an interconnecting pore system (pore size, 100-1500 μm, or 0.004-0.06 in.). Supplied in blocks, granules and cylinders, it is used as a bone graft material in total joint replacement and in the repair of bone defects.

Endofix. Trade name Acufex Microsurgical Inc. (USA) for biodegradable polymers based on poly(L-lactide) or poly(glycolide-*co*-trimethylene carbonate) and used for medical and surgical fixation devices, e.g., screws and tacks.

EnDOtec. Trademark of Eutectic Corporation (USA) for continuous gas-shielded welding alloys for maintenance, repair and wear resistance. They have good abrasion resistance under high pressure and moderate impact.

Endox. Trade name of Enthone-OMI, Inc. (USA) for alkaline compounds for removing rust and scale from iron and steel. They are supplied in powder form for addition to water.

end-to-end. See end-and-end.

Endur. (1) Trademark of Rogers Corporation (USA) for a series of thermoplastic elastomers with good tear strength, good resistance to ultraviolet light, ozone and oils, and a long wear life. Used for business equipment, industrial equipment, machine elements, etc.

(2) Trademark of Ormco Corporation (USA) for dental adhesives.

(3) Trade name of Corhart Refractories Corporation (USA) for fused cast magnesite-chrome refractories.

Endura. Trademark of Endura Coatings Inc. (USA) for special non-stick, dry-film polyurethane coatings with excellent wear resistance, good anti-galling properties, good chemical and corrosion resistance, and low friction coefficients.

Enduralac. Trade name of Egyptian Lacquer Manufacturing Company Inc. (USA) for a clear lacquer.

Enduran. Trademark of GE Plastics (USA) for a series of polybutylene terephthalate (PBT) resins, and polybutylene terephthalate/polyethylene terephthalate (PBT/PET) alloys available in unfilled extrusion, and glass- and/or mineral-filled injection grades.

Endurance. (1) Trade name of A. Cohn Limited (UK) for a series of tin bronzes supplied in various grades: *Endurance AA* is a wear-resistant bronze containing 87% copper, 12% tin, and 1% phosphorus. It has a low coefficient of friction, and is used for automotive and aircraft bearings. *Endurance BB* is a wear-resistant bronze containing 89% copper, 10.5% tin and 0.5% phosphorus, and is used for bushings. *Endurance CC* is a corrosion-resistant bronze containing 65-89.5% copper, 10-24.65% tin and 0.35% phosphorus used for bearings, collars and pumps. *Endurance DD* is a heavy-duty bronze containing 80% copper, 10% tin, 9.75% lead and 0.25% phosphorus used for connecting-rod and railroad bearings. *Endurance EE* is a heavy-duty bronze containing 84% copper, 9.5% tin, 6.3% lead and 0.2% phosphorus used for connecting-rod and railroad bearings.

(2) Trade name of American Hard Rubber Company (USA) for hard rubber.

Endur-C. Trademark of Rogers Corporation (USA) for a microcellular urethane with good vibration-damping properties, wear resistance and chemical resistance, used to control vibration and shock in computer peripherals, office equipment, telecommunications equipment, appliances and electronic components.

Endur-Ever. Trade name of Kelly Moore/Preservative Paints (USA) for a white baking enamel.

Enduria. Trade name of Bethlehem Steel Corporation (USA) for a spring steel containing 0.7% carbon, and the balance iron.

Enduro. Trade name of Republic Steel Corporation (USA) for an extensive series of heat-resistant and stainless steels including various low-nickel austenitic, standard chromium-nickel, and ferritic chromium grades.

Enduron. Trade name of Sheepbridge Engineering Limited (UK) for a heat- and abrasion-resistant cast iron containing 1.9% carbon, 16% chromium, 1.9% silicon, 0.06% phosphorus, and the balance iron. Used for cylinder liners and piston rings.

Endweldur. Trade name of American Chain & Cable (USA) for a series of steels containing 0.2-0.4% carbon, 0-3.5% nickel, and the balance iron. Used for chains.

Enebra. Trade name of Eisenwerke Neubrandenburg GmbH (Germany) for an extensive series of cast bronzes and brasses including several highly corrosion resistant tin bronzes for chemical and textile equipment, various high-strength bearing bronzes for pumps and automotive gasoline and diesel engines, miscellaneous copper-zinc alloys for general construction applications, and diverse special bronzes for electrical applications.

EnerBond. Trademark for an expanding foam adhesive for subfloor applications.

EnerGraph DBX. Trade name of BP Amoco (USA) for high-purity, mesophase pitch-based graphite materials used as active anode materials in fuel cells, lithium batteries, and super-capacitors.

Energy Foam. Trademark of Bradford Industries, Inc. (USA) for a lead-free loaded polyvinyl chloride foam with high energy absorption used to dampen shocks and vibrations, and absorb airborne noise and energy.

EnerFoam. Trade name for expanding foam sealants.

Energy-Shield. Trademark of Bruin Plastics (USA) for highly reflective textile fabrics with thermal barrier properties used for coolers, panel ceilings, aircraft windshield covers, and spa cover liners.

EnerWrap. Trademark of Roxul Inc. (Canada) for a series of flexible, lightweight mineral wool insulation products. They have a typical density of about 96 kg/m^3 (6.0 lb/ft^3), excellent moisture and thermal resistance, good fire resistance, and an upper service temperature of 650°C (1200°F). Used in the form of blankets for the thermal insulation of industrial boilers, furnaces, vessels, pipes, structural members, and other mechanical equipment. See also Roxul.

Enfinity. Trade name of Stapleton Technologies (USA) for electroless nickel.

Enforce. Trade name of Dentsply Corporation (USA) for a dual-cure dental resin cement.

Engage. Trade name of Dow Chemical Company (USA) for thermoplastic polyolefin elastomers.

Engaloy. Trade name of Engelhard Corporation (USA) for a series of brazing alloys based on copper (e.g., Cu-Au, Cu-Au-Ni and Cu-Pd-Ni-Mn), gold (e.g., Au-Cu, Au-Cu-Ag, Au-Cu-Ni, Au-Ni and Au-Ni-Cr), palladium (e,g., Pd-Ni and Pd-Ni-Mn), or silver (e.g., Ag-Cu-Pd, Ag-Pd and Ag-Pd-Mn). Used for brazing electrical and electronic components.

Engelbond Elite. Trade name of Engelhard Industries (USA) for a high-gold dental bonding alloy.

Engelbond Hi. Trade name of Engelhard Industries (USA) for a dental gold solder.

Engelbond PGX. Trade name of Engelhard Industries (USA) for a dental bonding alloy containing 45% gold.

Engelhard. Trade name of Engelhard Minerals & Chemicals Corporation (USA) for an extensive series of alloys containing varying amounts of precious metals, such as gold, silver, platinum, iridium, etc.

Engild. Trade name of Engelhard Electro Metallics Division (USA) for decorative gold electrodeposits and plating processes.

engineered brick. See engineering brick.

engineered mineral fibers. A class of mineral fibers made by a controlled production process from a melt of high-grade raw materials of specific size and chemical composition from which all non-fibrous material (or shot) has been removed using high-velocity spinning wheels. They are surface treated for the application of coatings reactive within the matrix of a specific composite material. Used as reinforcements for engineering composites.

engineered materials. Materials, such as metals, alloys, polymers, ceramics, composites, etc., that have been purposely designed and manufactured to exhibit well defined, predetermined properties and characteristics. See also engineering materials.

engineered wood. A class of structural and non-structural wood products made by bonding together wood fibers, strands, veneers or lumber with a suitable adhesive. The four main groups of structural engineered wood products are: (i) *Structural wood panels;* (ii) *Structural composite lumber;* (iii) *Glued laminated timber;* and (iv) *I-joists* or *I-beams.* Nonstructural engineered wood products include *particleboard* and *medium-density fiberboard.* Used as building materials.

Engineering. Trade name of Dunford Hadfields Limited (UK) for a martensitic stainless steel (AISI type 420) containing 0.15-0.25% carbon, 12-14% chromium, and the balance iron. Used for valves, valve seats, and hydraulic and steam equipment.

engineering adhesives. Adhesives with high bond strength used to join metals, ceramics, glass, plastics, rubber and wood which often have structural applications. The term does not include glues and mucilages used for joining paper. See also adhesives.

engineering alloys. Engineering materials used in the manufacture of products for structural applications, and including ferrous materials, such as cast irons and steels, and nonferrous materials, such as aluminum, magnesium, titanium, nickel, zinc, copper and their alloys. *Engineering alloys* are characterized by strength, hardness, toughness, formability, machinability, ductility, metallic luster, and corrosion and oxidation resistance. See also alloys (1).

engineering brick. A brick whose nominal dimensions are 81 × 102 × 203 mm (3.2 × 4.0 × 8.0 in.). It is denser and harder than ordinary brick, and has high strength and frost resistance. Used in bridge and building construction. Also known as *engineered brick.*

engineering ceramics. A group of silicate and nonsilicate ceramics used in the manufacture of technical products, and including advanced ceramics, glasses, glass-ceramics, glazes, porcelain enamels, refractories, clays, cement, lime, plaster, and gypsum. Abbreviation: EC. See also advanced ceramics.

engineering coatings. Coatings produced for engineering applications, e.g., to improve corrosion protection, lubricity or adhesion, and increase service life, but not for decorative purposes. Abbreviation: EC. See also coatings.

engineering materials. A large class of materials used in the manufacture of technical products including machinery, plant, equipment, installations and systems, structural members, machine parts, building products, device and instruments, tools, and auxiliary products. Important types of engineering materials include ferrous and nonferrous metals, ceramics, concrete, glass, plastics and elastomers, composite materials, semiconductors, superconductors, and natural substances, such as minerals, wood, stone, animal skin or hide and their derivatives. *Engineering materials* are characterized by strength, hardness, toughness, dimensional stability, formability and machinability, chemical, electrical and thermal properties, heat and flame resistance, and resistance to corrosion, acids, and solvents.

engineering plastics. A term including all reinforced and unreinforced thermoplastic and thermosetting polymers that have performance characteristics (e.g., mechanical, chemical, electrical and/or thermal properties) that make them suitable for applications which were traditionally reserved for metallic or ceramic materials, e.g., the manufacture of products, such as machinery, equipment, construction materials, machine parts, etc., that require definite and predictable structural performance over a wide range of temperatures. Abbreviation: EP. Also known as *superpolymers.* See also engineering thermoplastics; engineering thermosets.

engineering polymers. A class of high-performance polymers that can replace metals and alloys in many applications. Abbreviation: EP. Also known as *high-performance polymers*. See also engineering plastics.

engineering resins. A class of high-performance high-melting synthetic resins. See also engineering plastics.

engineering thermoplastics. A term used to refer to a group of thermoplastics typically having a tensile strength greater than or equal to 35 MPa (5 ksi) and a tensile modulus of more than 2.07 GPa (0.3 × 10^6 psi). Examples include acetals, acrylic plastics, acrylonitrile-butadiene-styrenes, high-density and ultrahigh molecular weight polyethylenes, high-impact polystyrene, polyamides, polyamideimides, polyarylates, polyaryletherketones, polyaryl sulfones, polybenzimidazoles, polybutylene terephthalates, polycarbonates, polyetherimides, polyether sulfones, polyethylene terephthalates, polyphenylene sulfides, polysulfones, polyvinyl chlorides, styrene-acrylonitriles, styrene-maleic anhydrides, thermoplastic fluoropolymers, and thermoplastic polyimdes. Abbreviation: ET. Also known as *thermoplastic engineering plastics*. See also engineering plastics.

engineering thermosets. A group of thermosets, usually reinforced with particulate (wood flour, cellulosic and mineral fillers, mica, etc.) or fibers (cotton, asbestos, glass, etc.), having a unique combination of mechanical, thermal, chemical and electrical properties that allow them to be used for the same engineering applications as ceramics or metals. Such plastics include in particular the aminos, epoxies, phenolics, unsaturated polyesters, polyurethanes, silicones and vinyl esters. Also known as *thermoset engineering plastics*. See also engineering plastics.

English. (1) British trade name for a nickel silver containing 61.8% copper, 19.1% zinc and 19.1% nickel. Used for silver plating, knives, and forks.

(2) British trade name for a free-cutting brass containing 70.3% copper, 29.3% zinc, 0.17% tin and 0.26% lead. Used for hardware.

English alloy. A British babbitt containing 53% tin, 33% lead, 11% antimony, 2.4% copper and 1% zinc. Used for bushings and bearings.

English bearing metal. A British bearing alloy of 80% copper, 14.5% tin and 5.5% copper. It has poor resistance to heat or live steam, and is used for bearings.

English brass. A British free-cutting brass containing 70% copper, 29% zinc, 0.3% lead, and 0.2% tin. Used for hardware and novelties.

English elm. The fairly strong, tough wood of the tree *Ulmus procera*. It is cross-grained, must be treated to reduce its tendency to warp, and has good durability under water. Average weight, 550 kg/m^3 (34 lb/ft^3). Source: British Isles. Used for piling, dock and wharf construction, barge and boatbuilding, sea groynes, weatherboarding for agricultural buildings, ends of packing cases and boxes, carts, wheelbarrows and wagons, furniture, chairs, interior fittings, and coffin boards.

English German silver. A British corrosion-resistant nickel silver of 59.4% copper, 25% zinc, 13% nickel and 3% iron, used for ornamental parts.

englishite. A colorless, or white mineral composed of potassium sodium calcium aluminum phosphate hydroxide hydrate, $K_2Na_2Ca_{10}Al_{15}(PO_4)_{21}(OH)_7 \cdot 26H_2O$. Crystal system, monoclinic. Density, 2.65 g/cm^3. Occurrence: USA (Utah).

English linotype. A British alloy containing 83% lead, 12% antimony, and 5% tin. Used as type metal for linotype machines.

English penny bronze. A British corrosion-resistant alloy containing 95.5% copper, 3% tin and 1.5% zinc. Used for springs.

English pewter. A British corrosion-resistant alloy containing 81.2% tin, 11.5% lead, 5.7% antimony and 1.6% copper. Used for domestic utensils, ornaments, and bearings.

English phosphor bronze. A British heavy-duty phosphor bronze containing 79.4% copper, 10% tin, 9.6% lead and 1% phosphorus. Used for bearings, gears, gear wheels, and pinions.

English oak. The strong, light-brown wood of the oak tree *Quercus robur*. It has a coarse, open grain, a firm texture, and a good combination of strength and durability. Average weight, 720 kg/m^3 (45 lb/ft^3). Source: Europe including British Isles. Used especially for constructional applications, and furniture. Also known as *British oak*.

English pink. See chrome-tin pink.

English poplar. See black poplar.

English red. See colcothar.

English speculum. A British alloy containing 67% copper and 33% zinc, used for deep-drawn, spun and stamped parts.

English stereotype. A British alloy containing 82.5% lead, 4.5% tin and 13% antimony. Used as type metal for stereotypes (i.e., one-piece printing plates).

English type metal. A series of British type alloys containing 60.5-77.5% lead, 4.5-14.5% tin, 16-26% antimony, and 0.8-2% copper. Used as type metal for printing plates.

English vermilion. A bright red pigment consisting of precipitated mercury sulfide (HgS) prepared directly by heating mercury and sulfur. It becomes darker on exposure to light, and is used in paints. Also known as *vermilion red*. See also vermilion.

English walnut. See European walnut.

English white. A fine, whitish powder derived by grinding English chalk (calcium carbonate) and agitating with water. Used in the manufacture of polishing powders, cements, and certain ceramics.

English yew. The reddish-brown, heavy and strong wood from the coniferous tree *Taxus baccata*. It has a fine grain, exceptional durability, very high hardness, and is very difficult to work. Source: British Isles. Average weight, 670 kg/m^3 (42 lb/ft^3). Used for furniture, cabinetmaking, archery bows, gateposts, fencing, and turned items.

Englo. Trade name for rayon fibers and yarns used for textile fabrics.

Engold. Trade name of Engelhard Electro Metallics Division (USA) for acid gold electrodeposits and high-speed plating processes.

Engraver Plate. Trade name of Colt Industries (UK) for a steel containing 0.35% carbon, and the balance iron. Used for engraver plates.

engraver's brass. A free-cutting leaded brass containing 62.5-66% copper, 33-36% zinc and 1-2.5% lead. Used for gears, watch and clock backs, engravings, engraved items, and dials.

enhanced compacted-graphite iron. A modified compacted graphite cast iron with a nodularity of up to 50%, which combines the high stiffness, toughness and strength of ductile iron with the high damping capability and easy machinability of gray iron. Used for automotive bedplates. Abbreviation: ECG iron. See also compacted graphite cast irons.

Enhancer. Trade name for a dental porcelain for metal-to-ceramic bonding.

E Nickel. Trade name of Inco Alloys International Inc. (USA) for an alloy containing 2% manganese, 0.1% carbon, 0.05% iron, 0.05% silicon, 0.03% copper, 0.005% sulfur, and the balance nickel. Used for electron tube supports. Now known as *Nickel*

212.

enigmatite. See aenigmatite.

Enjay. Trade name for saran (polyvinylidene chloride) fibers.

Enka. Trade name of Acordis Group for a series of rayon and nylon fibers.

Enkadrain. Trademark of Colbond Geosynthetics/Acordis Group for a strong, lightweight nylon geocomposite comprising a three-dimensional core bonded to one or two nonwoven filter layers. Used for civil engineering applications, e.g., as gas collection, leakage control or drainage layers in landfill sites and for the drainage of road edges, cellars, tunnels, etc.

Enkaflex. Trademark of Enka Corporation for textile fabrics used for carpets and rugs.

Enkafort. Trademark of NV Silenka AKU-Pittsburgh (Netherlands) for glass-fiber reinforcements.

Enkagrid. Trademark of Colbond Geosynthetics/Acordis Group for a family of geogrids used for civil engineering applications including road stabilization, base reinforcement of embankments, reinforced slopes and reinforced block walls. *Enkagrid MAX* is a biaxial grid made from polypropylene strips, *Enkagrid PRO* is an uniaxial grid made from polyethylene terephthalate strips, and *Enkagrid TRC* consists of a composite reinforcement geogrid and a nonwoven separation layer.

Enkaire. Trademark of Acordis Group for rayon fibers.

Enkalene. Trade name of Acordis Group for nylon fibers used for textile fabrics.

Enkaloft. Trade name of Acordis Goup for nylon fibers used for textile fabrics.

Enkalon. Trademark of Acordis Industrial Fibers for high-tenacity polyamide 6 (nylon 6) fibers and filament yarns.

Enkalure. Trade name of Acordis Group for nylon fibers used for textile fabrics.

Enkamat. Trademark of Colbond Geosynthetics/Acordis Group for a three-dimensional, non-biodegradable nylon polymer matting used in civil engineering and landscaping for erosion control applications on embankments, slopes, riverbanks, etc.

Enka Nylon. Trademark of Acordis Industrial Fibers for polyamide 6,6 (nylon 6,6) fibers and filament yarns.

Enkasa. Trade name for regenerated protein (azlon) fibers derived from casein and used for textile fabrics.

Enkasheer. Trade name of Acordis Group for nylon fibers and yarns used for textile fabrics including clothing.

Enkasonic. Trade name of Acordis Group for nylon fibers and plastics.

Enkatron. Trade name of Acordis Group for nylon fibers and products.

Enkaturf. Trade name of Acordis Group for nylon fibers used in the manufacture of synthetic turf for sports arenas.

Enka Viscose. Trademark of Acordis Group for viscose rayon fibers and yarns.

Enkona. Trade name of Acordis Group for rayon fibers and yarns used for textile fabrics.

Enkrome. Trade name of Acordis Group for rayon fibers.

Enkron. Trade name of Acordis Group for rayon fibers and yarns used for textile fabrics.

Enloy. Trade name of Enthone-OMI, Inc. (USA) for a cobalt alloy electroplate and process.

Enlyte. Trade name of Enthone-OMI, Inc. (USA) for bright electroless nickel.

Enocol. Trade name of Ato Findley (France) for resorcinol-formaldehyde and resorcinol-phenol-formaldehyde adhesives for building and automotive applications.

Enpax. Trade name of Pax Surface Chemicals Inc. (USA) for electroless nickel-phosphorus.

Enplate. Trademark of Enthone-OMI, Inc. (USA) for an extensive series of electroless copper and nickel coatings and plating solutions providing excellent corrosion and wear resistance on a wide range of substrates including carbon and low-alloy steels, stainless steels, aluminum and copper alloys, powdered metals, plastics, ceramics and glasses. Depending on the particular coating material and composition, other property enhancements may be obtained, such as resistance to erosion, galling and high temperatures, controlled or improved electrical and/or optical properties, and/or improved solderability or brazeability.

Enseal. Trade name of Enthone-OMI, Inc. (USA) for corrosion-resistant sealants for zinc and zinc alloys.

enriched fuel. Nuclear reactor fuel composed of uranium having a concentration of uranium-235 (^{235}U) greater than the concentration in natural uranium (i.e., 0.720%). It is necessary to enrich uranium in order to produce a fuel that will sustain a chain reaction in a water-cooled core. Abbreviation: EF.

enriched material. A material or mixture of fissile and nonfissile isotopes in which the fissile fraction is increased by any of various standard methods including gaseous diffusion, gas centrifuging and electromagnetic separation.

enriched uranium. Uranium whose content of the fissile isotope uranium-235 (^{235}U) has been raised from the 0.720% normally found in nature to a higher level. *Low-enriched uranium* (LEU) contains 2-4% ^{235}U and is is used as fuel in many types of nuclear reactors. *High-enriched uranium* (HEU) often contains more than 90% ^{235}U and is used in some special nuclear reactors and in the manufacture of nuclear weapons. Standard methods for enriching uranium include gaseous diffusion, gas centrifuging, and electromagnetic separation. Abbreviation: EU.

Enshield. Trademark of Enthone-OMI, Inc. (USA) for a series of electroless nickel coatings for electromagnetic interference/radio-frequency interference (EMI/RFI) shielding applications. See also EMI/RFI shielding materials.

Ensifide. Trade name of Ensinger, Inc. (USA) for a polyphenylene sulfide (PPS) resin with excellent chemical resistance and good mechanical and thermal properties, low ionic impurity, low water absorption (0.02%), and an upper continuous-use temperature (in air) of 104°C (220°F). Supplied as extruded product, it is used for automotive components, machine parts, industrial equipment, appliances, electrical and electronic components, and in semiconductor manufacture.

Ensifone. Trade name of Ensinger, Inc. (USA) for an engineering thermoplastic based on polysulfone (PSU). Supplied in the form of transparent, extruded products, it has a density of 1.24 g/cm³ (0.04 lb/in.³), low water absorption (0.3%), excellent chemical and hydrolytic resistance, excellent high-temperature mechanical properties, very good rigidity, and an upper continuous-use temperature of 171°C (285°F). Used for electronic, food service, medical and pharmaceutical equipment, for semiconductor processing equipment, and for equipment and devices operating in autoclave environments.

Ensikem. Trade name of Ensinger, Inc. (USA) for a polyvinylidene fluoride (PVDF) resin supplied in the form of extruded products. It has a density of 1.78 g/cm³ (0.06 lb/in.³), a melting point of 178°C (350°F), low water absorption (0.04%), excellent resistance to aging and oxidizing environments, exceptional resistance to many aliphatic and aromatic hydrocarbons, mineral and organic acids, halogenated solvents and alcohols, and good processibility. Used for chemical, petrochemical, pharmaceutical and pulp and paper equipment, hydrometallurgical and nuclear equipment, and for semiconductor processing applications.

Ensilon. Trade name of Ensinger, Inc. (USA) for extruded polyamide 6,6 supplied in unreinforced and 30% glass-fiber-reinforced grades. Both grades have outstanding resistance to abrasion, wear and organic chemicals. The unreinforced grade, *Ensilon 6/6*, has a density of 1.14 g/cm³ (0.04 lb/in.³), a water absorption of about 1.2%, and a maximum continuous-use temperature of 99°C (210°F). The 30% glass fiber-reinforced grade, *Ensilon 6/6 GF30*, possesses very high tensile and flexural strength, and excellent toughness, rigidity and heat-deflection properties. It has a density of 1.35 g/cm³ (0.05 lb/in.³), low water absorption (0.7%), a maximum continuous-use temperature of 110°C (230°F), and is used for automotive components, railway tie insulators, valves, etc.

Ensipro. Trade name of Ensinger, Inc. (USA) for an engineering thermoplastic based on polypropylene homopolymer, and supplied in the form of extruded products. It has a density of 0.9 g/cm³ (0.03 lb/in.³), low water absorption (0.02%), a melting point of 160°C (320°F), excellent chemical and moisture resistance, good weldability, and high purity. Used for laboratory equipment, food-processing equipment, tanks, and pump and valve parts.

Ensital. Trade name of Ensinger, Inc. (USA) for semicrystalline acetals based on polyoxymethylene (POM) and supplied as copolymers and homopolymers in the form of extruded products. The copolymer acetals have a density of 1.41 g/cm³ (0.05 lb/in.³), low water absorption (0.2%), a maximum continuous-use temperature of 141°C (285°F), high strength, toughness and dimensional stability, good wear resistance and bearing properties, good resistance to hydrocarbons, solvents and hot water. It is used for gears, bushings, rollers, fittings, wear parts, pump components, material and fluid handling equipment, etc. The homopolymer acetals are internally lubricated with polytetrafluoroethylene, and have a density of 1.54 g/cm³ (0.06 lb/in.³), a melting point of 175°C (347°F), an upper continuous-use temperature (in air) of 85°C (365°F), outstanding chemical resistance, low friction, excellent wear resistance, toughness, flexural fatigue properties and machinability. Used for food-processing equipment, automotive and aerospace components, business equipment, valves, and precision and measuring devices.

Ensitep. Trade name of Ensinger, Inc. (USA) for thermoplastic polyesters based on polybutylene terephthalate (*Ensitep PBT*) and polyethylene terephthalate (*Ensitep PET*), respectively, and supplied in the form of extruded products. They have a density of about 1.38 g/cm³ (0.05 lb/in.³), a melting point of 254°C (490°F), low water absorption (about 0.1%), outstanding wear resistance, excellent impact resistance, high flexural modulus, good dimensional stability, low friction, and good resistance to chemicals, solvents, and food products. Used for mechanical and electromechanical parts, food-processing equipment, valves, textile machinery, printing equipment, and water purification systems.

Enslip. Trade name of Plating Process Systems Inc. (USA) for an electroless nickel composite coating with polytetrafluoroethylene (*Teflon*).

Ensolite. Trademark of Ensolite Inc. (USA) for foamed vinyl plastisols and cellular sheet plastics.

enstatite. A grayish, or white mineral of the pyroxene group composed of magnesium silicate, MgSiO₃. Iron-rich varieties, such as *bronzite*, may be yellowish, greenish, brownish, or bronze. Crystal system, orthorhombic. Density, 3.10-3.43 g/cm³; hardness, 5-6 Mohs; refractive index, 1.652. Occurrence: Norway, Czech Republic, Slovakia, USA (New York, North Carolina, Pennsylvania, Texas). Used in electronic ceramics as a replacement for talc to minimize shrinkage.

Entecrod. Trade name of Eutectic Corporation (USA) for a series of cadmium-silver alloys with a bonding temperature range of 343-371°C (650-700°F), used as hard solders.

Enterprise. Trade name of The Enterprise Glass Company Limited (UK) for laminated safety glass.

Enthox. Trademark of Enthone-OMI, Inc. (USA) for chromate conversion coatings.

Enuran. Trade name of GE Plastics (USA) for a polybutylene terephthalate alloy.

EnVe. Trade name of NSI Dental Pty. Limited (Australia) for a radiopaque bisphenol glycidyl methacrylate (BisGMA) based universal composite used in dentistry for anterior and posterior and veneering application.

Envex. Trademark of Rogers Corporation (USA) for a series of polyimide resins supplied in extruded rods, profiles or tubes. Available in unfilled and graphite-, molybdenum disulfide- or polytetrafluoroethylene-filled grades, they have good cryogenic and elevated-temperature performance, good strength and dimensional stability, low wear rates, and good chemical properties. Used for insulators, valve seats, seals, gaskets, bushings, bearings, clamps, transducer heads, machine components, and automotive and aircraft components.

Envilon. Trademark of Toyo Chemical Company (Japan) for vinyon (vinyl chloride) fibers with excellent acid, alkali and mildew resistance. Used for textile fabrics and carpets.

Envirez 5000. Trade name of Ashland Chemical Company (USA) for a thermosetting polyester sheet molding compound containing a total of about 25 wt% soybeans and corn. Used in the manufacture of automotive components, such as hoods and body panels.

Enviroacryl. Trade name of PPG Industries Inc. (USA) for an acrylic powder coating material.

Enviroboard. Trade name of EnviroSafe Products, Inc. (USA) for a composite board material made from recycled plastics and wood dust.

EnviroBrass. Trade name of Copper Development Association (USA) for a series of brass casting alloys containing only trace amounts of lead, and small amounts of a selenium-bismuth mixture added to improve castability and free-machining properties. Supplied in the form of castings and ingots, *EnviroBrass I* (UNS C89510) is a cast brass used for potable water applications, such as valves, water meters and plumbing fixtures, and *EnviroBrass II* (UNS C89520) is a leadfree yellow brass used for permanent-mold casting applications. Formerly known as *SeBiLOY*.

Envirochrome. Trade name of MacDermid Inc. (USA) for a trivalent chromium electrodeposit and plating process.

Enviro-Crete. Trade name of Tnemec Company (USA) for elastomeric masonry coatings.

Enviroline. Trademark of Industrial Environmental Coatings Corporation (USA) for green-colored, fast-curing 100%-solids epoxy coatings and linings with excellent chemical resistance. Used for single coat application on industrial equipment, such as tanks, pipes, and on concrete surfaces.

Envirolloy. Trade name of Aarco Inc. (USA) for a selenium-bismuth master melt used in the manufacture of *EnviroBrass*.

Envirolloy ACS. Trade name of A Brite Company (USA) for alloy chrome substitutes available in the form of a decorative, bright, corrosion-resistant chrome-like electrodeposits plated over nickel and steel. The plating process is also referred to as "Envirolloy ACS."

Environ. Trade name of PPG Industries, Inc. (USA) for durable, waterborne, pigmented acrylic latex resin coil coatings for aluminum building products.

environmentally conscious materials. See ecomaterials.

environmentally friendly materials. A group of materials specifically designed to minimize the adverse effects on the environment. Examples include biodegradable plastics, polymers that contain natural products (e.g., corn or soybean), lead-free solders, solvents with low volatile organic compound (VOC) contents, chromium-free coatings, and water-base adhesives and paints. Also known as *environmentally preferable materials*.

environmentally preferable materials. See environmentally friendly materials.

environmentally sensitive biopolymers. A class of polymeric materials, such as polyethylene oxide, polyvinyl methyl ether, hydroxypropyl cellulose, polyvinyl alcohol and certain hydrogels, e.g., *N*-isopropyl acrylamide (NIPAAm), that exhibit marked changes in behavior (e.g., phase separations, changes in shape, hardness, permeability, optical properties, etc.) in response to rather small changes in conditions, such as temperature, pH, chemical agents, mechanical stress, etc. Such polymers are suitable for use in biomedical applications, e.g., drug delivery, phase separations, cell binding, and cell and protein attachment or detachment.

Enviro-Prep. Trademark of Hoffer's Coatings, Inc. (USA) for water-base coatings used for removing lead-based paints.

Envirostrip. Trademark of Canning-Gumm, Inc. (USA) for an environmentally safe, solvent-free blasting abrasive based on a modified wheat starch polymer. It is supplied in the form of granules in several size ranges, and is used for stripping coatings, paints and primers from metal or composite surfaces.

EnviroTex Lite. Trademark of EnviroTex (USA) for a high-gloss plastic finish for wood surfaces.

Envirothane. Trademark of Industrial Environmental Coatings Corporation (USA) for a series of urethane coatings for exterior applications. Applied by brushing, rolling or spraying, they provide enhanced resistance to ultraviolet light, chemical media, chipping and marring.

Envirozin. Trade name of MacDermid Inc. (USA) for an alkaline zinc electrodeposit and plating process.

Envision. Trademark of Dow Chemical Company (USA) for high-performance foams custom-made by laminating any of several *Ethafoam Brand Foam Plank* core materials on one side of green, blue, or natural-colored, tough, low-abrasion *Synergy Soft Touch Foam* sheets. Supplied in high- and low-density sheets from 3.2 to 13 mm (0.125 to 0.5 in.) in size, and high- and low-density planks 13 mm to 108 mm (0.5 to 4.25 in.) in size. Used for protective packaging and consumer goods.

enzyme. Any of numerous protein substances that are formed by living cells and catalyze a thermodynamically possible reaction by lowering the activation energy so that the rate of reaction is compatible with the conditions in the cell. There are six major classes of enzymes: (i) oxidoreductases; (ii) transferases; (iii) hydrolases; (iv) lyases; (v) isomerases; and (vi) ligases. They are of great importance in biology, biochemistry, biotechnology, and medicine. Also known as *biocatalyst*.

enzyme-washed fabrics. Textile fabrics, usually of cotton or denim, that have been washed with an enzyme (cellulase) that hydrolyzes and decomposes their cellulose producing a material with a used, worn appearance and a soft hand.

eolienne. A delicate, lightweight, airy fiber with a high luster, produced with a low thread count.

eosphorite. A pink to rose-red mineral composed of manganese aluminum phosphate hydroxide hydrate, $(Mn,Fe)Al(PO_4)(OH)_2 \cdot H_2O$.

EPAlloy. Trademark of Copper Development Association (USA) for a series of free-machining lead-free brass casting alloys with small additions of selenium and bismuth.

Epco-Oil Hard. Trade name of Williamson Brothers Inc. (USA) for an oil-hardening nondeforming tool steel containing 1% carbon, 0.5% chromium, 1% manganese, 0.5% tungsten, and the balance iron. Used for dies, gages, and master tools.

Epco-Water Hard. Trade name of Williamson Brothers Inc. (USA) for a water-hardening tool steel containing 1.05% carbon, 0.25% manganese, and the balance iron. Used for blanking, forming and trimming dies.

EPDM rubber. See ethylene-propylene-diene monomer.

Eperan. Trade name of Kaneka Texas Corporation (USA) for molded polyethylene and polypropylene foams used for automotive and packaging applications, sporting goods and foam pads.

EPG. Trademark of Solutia Inc. (USA) for an automotive glass product composed of two sheets of glass and an interlayer of polyvinyl butyral (PVB) laminated together under heat and pressure. It provides protection from ultraviolet rays, interior heat build-up, intrusion and outside noise, and helps prevent injury in case of accident. Used for side and rear automobile window applications.

ephesite. (1) A pink to pinkish brown mineral of the mica group composed of sodium aluminum silicate hydroxide, $Na_2Al_2(Al_2Si_2)O_{10}(OH)_2$. Crystal system, monoclinic. Density, 3.00 g/cm³; refractive index, 1.625. Occurrence: South Africa.

(2) A synthetic mineral of the mica group composed of lithium sodium aluminum silicate hydroxide, $Na(LiAl_2)Al_2Si_2O_{10}(OH)_2$. It can be made by hydrothermal treatment under prescribed conditions. Crystal system, monoclinic. Density, 2.95 g/cm³.

E-Phos. Trade name of Electrochemical Products Inc. (USA) for iron, manganese and zinc phosphate coatings.

Epiall. Trademark of Plaskon Products, Inc. (USA) for an extensive series of thermosetting molding compounds including various epoxy-based grades. Available in regular, and mineral-, glass- and long glass fiber-filled grades, they have exceptional elevated-temperature resistance up to 260°C (500°F), excellent specific resistance, good physical properties, good dimensional stability, very low flammability, excellent resistance to solvents and many chemicals, and low water absorption.

Epibond. Trademark of Ciba-Geigy Corporation (USA) for a series of epoxy adhesives, coatings and sealants used in the aircraft and aerospace industry, and for electrical and electronic applications.

Epic-AP. Trade name of Parkell, Inc. (USA) for a highly flowable, light-cure, no-slump hybrid composite consisting of the tough, hydrophilic proprietary resin "RDMA" (a non-Bis-GMA resin), a reactive organic filler (trimethylolpropane trimethacrylate, or TMPT), and an inorganic barium glass filler. It has high radiopacity, mechanical strength and wear resistance, excellent polishability, and low water absorption. Used for anterior and posterior dental restorations.

Epic-Cast. Trade name of Epic Resins (USA) for a series of low-viscosity epoxy and polyurethane casting resins for tooling applications.

Epic-Lam. Trade name of Epic Resins (USA) for a series of filled or unfilled epoxies available in high-temperature and impact-resistant grades for automotive and aircraft applications.

Epic-TMPT. Trademark of Parkell Inc. (USA) for a tough, resilient, nonsticky, light-cure composite resins with a chemically reactive trimethylolpropane trimethacrylate (TMPT) organic filler. It provides high wear resistance, very good polishability, and is used in restorative dentistry.

epichlorohydrin. An epoxide that is highly reactive with polyhydric phenols, such as bisphenol A. It is available as a flammable, highly volatile, unstable liquid (99% pure) with a density of 1.183 g/cm^3, a melting point of -57°C (-71°F), a boiling point of 115-117°C (239-243°F), a flash point of 93°F (33°C), a viscosity of 1.12 centipoise and a refractive index of 1.438. Used as the basic epoxidizing resin intermediate in the production of epoxy resins, as a raw material for phenoxy resins, as a monomer for the manufacture of epichlorohydrin rubber, in the curing of propylene-based rubbers, as a solvent for cellulose derivatives, in the preparation of polyglycidyl ethers, as a crosslinking agent for polymers, and in affinity chromatography. Formula: CH_2OCHCH_2Cl.

epichlorohydrin elastomer. An elastomer that results from the homopolymerization of epichlorohydrin, or the copolymerization of epichlorohydrin with ethylene oxide. It has high resistance to aging, oxidation, ozone and hot oil, good resistance to hydrocarbon solvents, low gas permeability (homopolymer), good low-temperature flexibility (copolymer), and a service temperature range of -40 to +130°C (-40 to +266°F). Used for molded products for automotive and industrial applications, tires, and vibration damping. Also known as *chlorohydrin rubber.*

Epiclon. Trade name of Dinippon (Japan) for epoxy prepreg resins of the diglycidyl ether of bisphenol F type used in aerospace applications.

epididymite. (1) A colorless mineral composed of sodium beryllium hydrogen silicate, $NaBeHSi_3O_8$. Crystal system, orthorhombic. Density, 2.56 g/cm^3. Occurrence: Greenland, Norway.

(2) A mineral composed of sodium aluminum silicate hydroxide, $NaBeSi_3O_7OH$. Crystal system, monoclinic. Density, 2.55 g/cm^3. Occurrence: Greenland.

epidote. A dark green mineral with a vitreous luster composed of calcium aluminum iron silicate hydroxide, $Ca_2(Al,Fe)_3Si_3O_{12}$-(OH). Crystal system, monoclinic. Density, 3.35-3.50 g/cm^3; refractive index, 1.768; hardness, 6-7 Mohs. Occurrence: Austria, Switzerland, USA (Alaska, California, Connecticut). Used as a gemstone.

epigenite. A steel-gray mineral composed of copper iron arsenic sulfide, $(Cu,Fe)_2AsS_6$. Crystal system, orthorhombic. Hardness, 4.5 Mohs.

EpiHBT. Trademark of EpiWorks, Inc. (USA) for epitaxial gallium arsenide (epi-GaAs) and epitaxial indium phosphide (epi-InP) wafers for heterojunction bipolar transistors. The epi-GaAs wafers consist of indium gallium phosphide (InGaP) deposited on gallium arsenide (GaAs), and the epi-InP wafers of indium gallium arsenide (InGaAs) deposited on indium phosphide (InP) by metal-organic chemical vapor deposition. See also epitaxial coating.

EpiFET. Trademark of EpiWorks, Inc. (USA) for epitaxial gal-

lium arsenide (epi-GaAs) and epitaxial indium phosphide (epi-InP) wafers for field-effect transistors. The epi-GaAs wafers consist of aluminum gallium arsenide (AlGaAs) indium gallium phosphide (InGaP) deposited on gallium arsenide (GaAs), and the epi-InP wafers of indium aluminum arsenide (InAlAs) or indium gallium arsenide (InGaAs) deposited on indium phosphide (InP) by metal-organic chemical vapor deposition. See also epitaxial coating.

Epikote. Trademark of Shell Chemical Corporation (USA) for epoxy resins with good adhesion properties used for protective coatings.

Epilox. Trade name for a series of epoxy resins and products.

epimer. One of a pair of diastereomers that differ only in the arrangement of the hydrogen and hydroxyl attached to the last asymmetric carbon atom.

epinglé. (1) A heavy fabric of wool, silk, cotton, etc. with alternating long or short ribs in the weft (filling) threads.

(2) An upholstery fabric with uncut loops.

Epi-Rez. Trademark of Interez Inc. (USA) for a series of epoxy resins with good adhesion and strength properties, good chemical and corrosion resistance, good dimensional stability, and good electrical properties. Used for electrical moldings and components, sinks, adhesives, protective coatings, and fiberglass laminates.

epistilbite. A colorless, or white mineral of the zeolite group composed of calcium aluminum silicate hydrate, $(Ca,Na)(Si,Al)_6O_{12}$·$4H_2O$. Crystal system, monoclinic. Density, 2.25 g/cm^3; refractive index, 1.510.

epistolite. A white or yellow mineral composed of sodium niobium titanium silicate hydrate, $Na_2(Nb,Ti)_2Si_2O_9·xH_2O$. Crystal system, monoclinic. Density, 2.89 g/cm^3; refractive index, 1.650. Occurrence: Greenland.

epitaxial coating. A thin-film layer composed of a metal, semiconductor or insulator deposited onto a substrate by any of various vapor deposition techniques (e.g., physical or chemical vapor deposition, molecular beam epitaxy, etc.) or electrodeposition, while maintaining a particular crystallographic relationship or orientation between the deposited layer and the substrate. Epitaxial films are arranged in specific orientations to achieve certain physical, chemical or mechanical properties of the substrate-film system, and are often used in semiconductor devices, lasers, high-speed networks, etc. Examples of epitaxial films include epitaxial silicon (epi-Si), epitaxial gallium arsenide (epi-GaAs) and epitaxial indium phosphide (epi-InP). Abbreviation: EC.

epitropic fibers. Synthetic fibers (e.g., nylon and polyester) that have fine, electrically conductive carbon particles embedded in their surfaces. Used for the manufacture of antistatic and low-static fabrics, e.g., carpets and industrial clothing.

EPK Kaolin. Trade name of The Feldspar Corporation (USA) for high-quality water-washed *kaolin* with a mineral content of about 97% *kaolinite*. It fires to a white color, has a high green strength, and very good forming properties. Used mainly for ceramic applications including refractories, tiles, frits, glazes and tableware.

E Plus. Trademark of Repla Limited (Canada) for insulating glass.

Epocase. Trade name of Furane Products (USA) for a series of epoxy resins.

Epocast. Trademark of Ciba-Geigy Corporation (USA) for an extensive series epoxy products used for aircraft and aerospace applications. It includes various fairing compounds, syntactic foam compounds and corefills, potting, molding and tooling compounds, and laminating and bonding systems.

Epocolor. Trade name of Ato Findley (UK) for epoxy joint fillers.

Epok. Trade name of Firth Brown Limited (UK) for austenitic 13%-manganese steel castings.

(2) Trade name of British Resin Products (UK) for phenolic resins and plastics.

Epo-Kote. Trademark of Sternson Limited (Canada) for decorative, fire-retardant, water-repellent, weather-resistant and protective epoxy coatings used for painting interior walls. They can be used over bricks, blocks and stucco.

Epolene. Trademark of Tennessee Eastman Company (USA) for a series of high- and low-density polyethylene waxes.

Epolite. Trade name of Hexcel Corporation (USA) for a series of epoxies supplied as aluminum-, glass-, mineral-, silica-filled casting resins, and in high-heat, and general-purpose grades.

Epomik. Trademark of Mitsui Chemicals, Inc. (Japan) for special epoxy resins used in the manufacture of electrical, semiconductor and printed circuits.

Epon. Trademark of Shell Chemical Company (USA) for a series of epoxy resins obtained by condensation polymerization of *epichlorohydrin* and *bisphenol A*. They have a density of 1.18 g/cm^3 (0.04 lb/in.^{3}), an excellent combination of mechanical properties and corrosion resistance, good adhesion, strength, chemical resistance, good dimensional stability, and good electrical properties. Used for electrical moldings, sinks, adhesives, fiberglass laminates, protective coatings, as embedding media for electron microscopy, and as a thermoset matrix material for reinforcing with woven quartz.

Epon HPT. Trademark of Shell Chemical Company (USA) for a series of epoxy resins and curing agents with excellent hot/wet performance (hot/wet flexural modulus) at elevated temperatures up to 177°C (350°F), high glass transition temperatures of 227-246°C (440-475°F), and low water absorption. Used for prepreg and structural adhesive applications in the aircraft, aerospace and military industries, and for tooling.

Eponal. Trade name of Ato Findley (France) for an extensive series of epoxy resins, adhesives, cements, mortars, coatings and sealants. The resins are formulated for injection-molding, sealing and repair applications. The adhesives, cements, coatings and sealants are used for building construction and flooring applications.

éponge. A rather lightweight cotton or wool fabric with a spongy feel, used especially for desses, shirts and robes.

Eponol. Trademark of Shell Chemical Company (USA) for a series of linear copolymers of *bisphenol A* and *epichlorohydrin*. They have high molecular weights and are used for surface coatings.

Epophen. Trademark for epoxy resins.

Epon Resin. Trademark of Shell Chemical Company (USA) for a series of epoxy resins.

Epo-TeK. (1) Trademark of Epoxy Technology Inc. (USA) for high-temperature polyimides with excellent adhesion resistance, good resistance to fluxes, solders and solvents, long room-temperature shelf life, short curing times, degradation temperatures above 550°C (1022°F), low coefficients of thermal expansion, and excellent screen printability and definition. Used for passivation coatings on thick- and thin-film resistors, wafer coatings for microcircuit packages, alpha particle barriers on memory chips, and general-purpose microchip protection.

(2) Trademark of Epoxy Technology Inc. (USA) for an extensive series of unfilled or metal- or ceramic-filled epoxy materials for optical and microelectric applications.

Epotuf. Trademark of Reichhold Chemicals, Inc. (USA) for multi-functional liquid epoxy resins with high heat-distortion temperatures and good chemical resistance, good adhesion properties, and relatively low viscosity. Used for maintenance coatings, tank linings, chemically resistant flooring, electrical encapsulation compounds, and composites.

Epox. Trademark of Mitsui Chemicals, Inc. (Japan) for high-performance epoxy adhesives with low outgassing, and excellent electrical properties, and high heat resistance. Used in the manufacture of relays and printed circuits.

Epoxal. Trademark of Niagara Protective Coatings (Canada) for epoxy-base protective coatings.

Epoxall. Trademark of Allied Compositions Company, Inc. (USA) for epoxy compounds for concrete surfacing applications.

Epoxalloy. Trademark of Ashland Oil, Inc. (USA) for sheet and resin-transfer molding compounds.

Epox-EEZ. Trademark of Cotronics Corporation (USA) for machinable room-temperature-curing aluminum putties that form adhesion-, chemical- and corrosion-resistant composites. Used in the repair of pumps, machinery and equipment, in rebuilding worn housings, linings and shafts, in patching leaky fittings, valves, and pipes, and in filling holes prior to powder coating.

Epoxibond. (1) Trademark of Canada Wire & Cable Limited for epoxy-resin bonded magnet wire.

(2) Trademark of Atlas Minerals & Chemicals, Inc. (USA) for epoxy-based putties, glues, and wood-restoration products.

Epoxical. Trademark of Epoxical, Inc. (USA) for a series of urethane casting resins for vacuum-forming molds and models, foam fixtures, hand layup of high-temperature molds for polyester and prepregs, and injection molds for the manufacture of prototype parts.

Epoxicote. Trademark of Harris Steel Group Inc. (Canada) for epoxy-coated concrete reinforcing steel bars.

Epoxicrete. Trademark of Coast Pro-Seal & Manufacturing Company (USA) for epoxy-based composites containing sand, silicon carbide and/or other abrasion-resistant aggregates. Used in building construction.

epoxide. An organic compound that is formed by the attachment of a reactive group in which an oxygen atom is singly bonded to each of two carbon atoms that are already united in some other way. Examples of such compounds include epichlorohydrin, ethylene oxide and propylene oxide.

epoxidized phenol novolacs. See epoxy novolacs.

epoxidized polybutadiene. See polybutadiene oxide.

epoxidized polyolefin. A polymer that is prepared by the epoxidation (i.e., attachment of the epoxide group) of polyolefins with organic peracids, such as peracetic acid, and has more epoxide groups than epoxies made from bisphenol and epichlorohydrin. Abbreviation: EPO.

epoxidized rubber. The reaction product of natural rubber and peracetic acid. Abbreviation: ER.

epoxies. A class of thermosetting resins formed either by the condensation reaction of epichlorohydrin and a dihydric phenol, such as bisphenol A, or by the epoxidation of olefins with peracetic acid. Common curing agents or co-reactants include acids, anhydrides, alcohols, amides, amines, and phenols. *Epoxies* can be viscous liquids or brittle solids. The latter have good strength, toughness, dimensional stability and adhesion properties, excellent resistance to corrosion and chemicals, including acids and alkalies, and excellent dielectric properties. *Epoxies* are used for protective surface coatings, in structural adhesives for composites, metals, plastics, glass and ceramics, as binders in mortars and concretes, in low-temperature mortars, as matrix resins for reinforced composites, as bonding agents for fiberglass laminates, in filament winding, electrical moldings, printed circuit boards, encapsulation of electrical and electronic components, as neutron-shielding materials, in casting and molding of tools and dies, as floor- and road-surfacing materials, and as rigid foams. Abbreviation: EP. Also known as *epoxy resins.*

Epoxi-Mica. Trademark of Motor Coils Manufacturing Company (USA) for electrical insulation materials that consist of epoxy resins and filled with mica, and are used for the insulation of field coils in electric motors and other rotating electrical machinery.

Epoxin. Trademark of BASF AG (Germany) for an epoxy resin with good chemical and electrical resistance, low shrinkage, good resistance to most organic solvents, excellent combination of mechanical properties and corrosion resistance, good electrical properties, and good dimensional stability. Used for molded products, electrical casings and components and tools.

Epoxize. Trade name of Egyptian Lacquer Manufacturing Company Inc. (USA) for several clear and transparent colored epoxy finishes used on polished metals.

Epoxolon. Trade name for an epoxy resin formerly used for denture bases.

epoxy. See epoxies.

epoxy adhesives. Adhesives based on epoxy resins and usually supplied in room temperature- or heat-curing grades. The former are usually two-part liquids and pastes with a maximum service temperature of about 80°C (180°F), while the latter are one-part films curing at about 120°C (250°F). *Epoxy adhesives* are often compounded with toughening agents to enhance peel strength. In general, they have relatively good peel/tensile strength, excellent cohesive strength, good resistance to heat and many chemicals, low shrinkage, and are used for gluing wood to nonporous materials (metals, glass, plastic), for the assembly and repair of composite structures, and as structural adhesives for composites, metals, plastics, glass, and ceramics. Also known as *epoxy cements; epoxy resin cements.*

epoxy alkyd. A polymer alloy containing an epoxy and an alkyd resin.

Epoxybond. Trademark of Atlas Minerals & Chemicals, Inc. (USA) for all-purpose epoxy adhesive and filling putties supplied in the form of sticks.

epoxy cements. See epoxy adhesives.

epoxy coating. An organic coating whose nonvolatile vehicle is an epoxy resin. It has good water, chemical, abrasion and solvent resistance, and is used on floors, walls, ceilings, concrete surfaces, metals, plastics, wood, etc.

epoxy composites. Composite materials consisting of epoxy resin matrices reinforced with glass, carbon, graphite or aramid fibers and/or fabrics, and made chiefly by wet or dry lay-up methods. They have high strength-to-weight ratios, high tensile strength, good elevated-temperature strength, high tensile moduli, good flexural and compressive properties, and good thermal and electrical properties. Used for commercial pressure vessels, tanks and pipe, electrical laminates for circuit-board manufacture, structural components for spacecraft and aircraft including fuselages, wing and control surface panels and rocket nozzle structural shells. Also known as *epoxy-matrix composites.*

Epoxydent. Trade name for a dental model and die epoxy resin.

epoxy ester. The reaction product of an epoxy resin and a fatty acid.

epoxy film adhesives. A group of adhesives made from epoxy resins, and available as thin films on paper support. They are used with high-temperature aromatic amine or catalytic curing agents, and are supplied with a variety of flexibilizing and toughening agents, usually on polyester, polyamide (nylon) or glass fiber supports. The maximum service temperature range is about 118-178°C (244-352°F).

epoxy foam. A rigid foam produced when an epoxy resin powder is mixed with a suitable co-reactant, and expands and cures with the application of heat.

Epoxylite. Trade name of Epoxylite Corporation (USA) for a series of high-temperature epoxy resins supplied in the form of adhesives, coatings, encapsulating and potting compounds, and sealers for electrical, electronic and other industrial applications.

epoxy-matrix composites. See epoxy composites.

epoxy microspheres. Spheres of microscopic size based on epoxy resins, and used as fillers in ceramics, plastics and composites.

Epoxyn. Trade name of The North American Group (USA) for catalyst-curing epoxy resins filled with aluminum powder, and made into soft putty-like compounds for filling surface defects in metals.

epoxy novolacs. A class of epoxy resins made by the reaction of epichlorohydrin with phenolic novolac resins (e.g., phenol-formaldehyde). They have better elevated-temperature performance than epichlorohydrin-bisphenol A-type epoxy resins, and are capable of maintaining their chemical and mechanical integrity even at temperatures over 200°C (400°F) for extended periods. Used for high-performance aerospace applications, and for adhesives. Also known as *epoxidized phenol novolacs; novolac epoxies.*

epoxy-nylon adhesives. Heat-curing structural adhesives made from epoxy and polyamide (nylon) resins. They have excellent lap-shear strengths at low and elevated temperatures up to 80°C (180°F), and high room-temperature peel strengths.

epoxy-phenolic adhesives. Heat-curing structural adhesives made from epoxy and phenolic resins. They have low peel strengths, and good-lap shear strengths at temperatures between -73°C and +260°C (-100°F and +500°F).

epoxy plastics. Plastics based on epoxy resins. See also epoxies.

Epoxy Plus 25. Trademark of Devcon Corporation (USA) for epoxy structural adhesives used for bonding ferrous and nonferrous metals, fiberglass, many plastics, wood, brick, concrete, ceramics, and honeycomb materials.

epoxy polymer concrete. A *polymer concrete* in which an epoxy resin, hardened with a polyamine-, polyamide- or polysulfide-based curing agent, is used as the binder. It has good creep and fatigue resistance, high strength, low setting shrinkage, good adhesion to most construction materials, and high chemical resistance (in particular with polyamine-based curing agents). Reinforced with boron, carbon or glass fibers, it can be used for boat hulls, autobodies and translucent panels. Unreinforced epoxy polymer concrete is used as a mortar for skid-resistant industrial flooring and highway overlays.

1,2-epoxypropane. See propylene oxide.

epoxy resin cements. See epoxy adhesives.

epoxy resins. See epoxies.

EPR copolymers. See ethylene-propylene rubber.

EPR rubber. See ethylene-propylene rubber.

epsilon carbide. A hexagonal close-packed (hcp) type of *cementite* (iron carbide, Fe_3C) that occurs during the initial primary *martensite* tempering stage. Abbreviation: ε-carbide.

epsilon plutonium. See plutonium.

epsomite. A colorless mineral with vitreous to milky luster composed of magnesium sulfate heptahydrate, $MgSO_4 \cdot 7H_2O$, and mined in a high degree of purity. Crystal system, orthorhombic. Density, 1.678 g/cm^3; melting point, loses $6H_2O$ at 150°C (302°F); loses water at 200°C (392°F); hardness, 2-2.5 Mohs; Used in ceramics as a set-up agent to adjust the viscosity and improve the application and flow properties of slips, in fireproofing compositions, as a catalyst carrier, in papermaking, etc. Also known as *epsom salt; Seidlitz salt; magnesium sulfate heptahydrate.*

epsom salt. See epsomite.

equiaxed metals. Polycrystalline, single-phase metals and alloys that have microstructures in which the grains have approximately equal dimensions in all three directions.

Era. (1) Trade name of Dunford Hadfields Limited (UK) for a series of steels including various carburizing, free-machining, plain-carbon, low-carbon alloy, nitriding, low- and medium-carbon high-chromium, and wrought and cast austenitic chromium-nickel grades. Also included under this trade name are nickel-iron-chromium superalloys for heat-resistant parts.

(2) Trade name of Gustav Ernstmeier GmbH & Co. KG (Germany) for textile fabrics coated on one or both sides with a polymer, such as acrylate, polyvinyl chloride, polyurethane, or silicone.

Eraclene. Trade name of Enichem SpA (Italy) for high-density polyethylene supplied in standard and UV-stabilized grades.

Eradene. Trademark of LIBA SpA (Italy) for a series of polyethylenes.

Eraydo. Trade name of Illinois Zinc Company (USA) for a nonmagnetic, corrosion-resistant zinc alloy containing 1-3% copper and 0.03-0.1% silver. Used for panels, drawn case, indoor trim, and electrical appliances.

erbia. See erbium oxide.

erbium. A soft, silvery, metallic element of the lanthanide series (rare-earth group) of the Periodic Table. It is commercially available in the form of ingots, rods, lumps, turnings (chips), sponge, wire, sheet, foil, powder and single crystals. The commercial erbium ores are *euxenite, gadolinite, fergusonite* and *xenotime.* Crystal system, hexagonal. Crystal structure, hexagonal close-packed. Density, 9.051 g/cm^3; melting point, 1529°C (2784°F); boiling point, 2863°C (5185°F); hardness, 70 Vickers; atomic number, 68; atomic weight, 167.26; trivalent; high electrical resistivity (about 86 $\mu\Omega cm$). Used in nuclear controls, special alloys, vanadium steels, glass coloring compounds, phosphors, garnet microwave devices, ferrite bubble devices and catalysts, room-temperature lasers, superconductors and ceramics, and as a dopant for laser crystals and garnets. Symbol: Er.

erbium acetylacetonate. Off-white to light pink crystals (97+% pure) with a melting point of 125-132°C (257-270°F). Used as a dopant for sol-gel fiber optics. Formula: $Er(O_2C_5H_7)_3$. Also known as *erbium pentanedionate.*

erbium boride. Any of the following compounds of erbium and boron used in ceramics and materials research: (i) *Erbium tetraboride.* Crystal system, tetragonal. Density, 6.96 g/cm^3; melting point, 2450°C (4442°F). Formula: ErB_4; and (ii) *Erbium hexaboride.* Density, 5.58 g/cm^3. Formula: ErB_6.

erbium bromide. Violet, hygroscopic crystals with a melting point

of 923°C (1693°F). Used in ceramics, materials and superconductor research, and electronics. Formula: $ErBr_3$.

erbium carbide. Any of the following compounds of erbium and carbon used in ceramics and materials research: (i) *Trierbium carbide.* Density, 4.71 g/cm^3. Formula: Er_3C; (ii) *Erbium sesquicar-bide.* Formula: Er_2C_3; (iii) *Erbium dicarbide.* Density, 7.95 g/cm^3. Formula: ErC_2.

erbium chloride. A pink, hygroscopic powder (99+% pure) or violet, hygroscopic crystals. Crystal system, monoclinic. Density, 4.10 g/cm^3; melting point, 774°C (1425°F). Used in ceramics, materials and superconductor research, and electronics. Formula: $ErCl_3$.

erbium dicarbide. See erbium carbide (iii).

erbium disilicate. See erbium silicate (iii).

erbium disilicide. See erbium silicide (i).

erbium fluoride. A pink, hygroscopic, crystalline powder (99+% pure). Crystal system, orthorhombic. Density, 7.81 g/cm^3; melting point, 1350°C (2462°F); boiling point, 2200°F (3992°F). Used in ceramics, materials and superconductor research, and electronics. Formula: ErF_3.

erbium halide. A compound of erbium and bromine, chlorine, fluorine or iodine, e.g., erbium chloride ($ErCl_3$) or erbium fluoride (ErF_3).

erbium heptasulfide. See erbium sulfide (ii).

erbium hexaboride. See erbium boride (ii).

erbium iodide. A red, hygroscopic powder (99+% pure), or violet, hygroscopic crystals. Crystal system, hexagonal. Density, 5.50 g/cm^3; melting point, above 1014°C (1857°F); boiling point, 1280°C (2336°F). Used in ceramics, materials and superconductor research, and electronics. Formula: ErI_3.

erbium monosilicate. See erbium silicate (i).

erbium monosulfide. See erbium sulfide (i).

erbium nitrate. Pink, hygroscopic crystals (99.9% pure). Melting point, loses $4H_2O$ at 130°C (266° F). Used in chemistry and materials research. Formula: $Er(NO_3)_3 \cdot 5H_2O$. Also known as *erbium nitrate pentahydrate.*

erbium nitrate pentahydrate. See erbium nitrate.

erbium nitride. A crystalline compound of erbium and nitrogen used in ceramics and materials research. Crystal system, cubic. Density, 10.35-10.6 g/cm^3. Formula: ErN.

erbium oxalate. A pink, microcrystalline powder (99.9% pure). Density, 2.64 g/cm^3; melting point, decomposes at 575°C (1067°F). Used in the separation of rare-earth metals from common metals. Formula: $Er_2(C_2O_4)_3 \cdot 10H_2O$. Also known as *erbium oxalate decahydrate.*

erbium oxalate decahydrate. See erbium oxalate.

erbium oxide. A rare-earth oxide usually supplied as a rose-red, hygroscopic powder (99.9+% pure). Density, 8.640 g/cm^3; melting point, 2345°C (4253°F). Used as an activator for phosphors, in microwave ferrites, in superconductors, in laser research, as a nuclear poison, and as an ingredient in infrared absorbing glass. Formula: Er_2O_3. Also known as *erbia.*

erbium pentanedionate. See erbium acetylacetonate.

erbium pentasilicide. See erbium silicide (ii).

erbium selenide. A compound of erbium and selenium used in ceramics, electronics and materials research. Density, 6.96 g/cm^3; melting point, approximately 1520°C (2770°F). Formula: Er_2S_3.

erbium sesquicarbide. See erbium carbide (ii).

erbium sesquisulfide. See erbium sulfide (iii).

erbium silicate. Any of the following compounds of erbium oxide and silicon dioxide used in ceramics and materials research:

(i) *Erbium monosilicate.* Density, 6.80 g/cm³; melting point, 1979°C (2770°F); hardness, 5-7 Mohs. Formula: Er_2SiO_5;(ii) *Dierbium trisilicate.* Density, 6.22 g/cm³; melting point, 1900°C (3450°F); hardness, 5-7 Mohs. Formula: $Er_4Si_3O_{12}$; and (iii) *Erbium disilicate.* Melting point, 1800°C (3270°F); hardness 5-7 Mohs. Formula: $Er_2Si_2O_7$.

erbium silicide. Any of the following compounds of erbium and silicon used in ceramics and materials research: (i) *Erbium disilicide.* Formula: $ErSi_2$; and (ii) *Erbium pentasilicide.* Formula: Er_3Si_5.

erbium sulfate. Rose-red, hygroscopic crystals (99.9% pure). Density, 3.217 g/cm³; melting point, loses $8H_2O$ at 400°C (752°F). Used in the determination of atomic weight of the rare-earth elements. Formula: $Er_2(SO_4)_3 \cdot 8H_2O$. Also known as *erbium sulfate octahydrate.*

erbium sulfate octahydrate. See erbium sulfate.

erbium sulfide. Any of the following compounds of erbium and sulfur used in ceramics, electronics and materials research: (i) *Erbium monosulfide.* Formula: ErS; (ii) *Erbium heptasulfide.* Density, 6.71 g/cm³; melting point, 1620°C (2948°F). Formula: Er_2S_7; (iii) *Erbium sesquisulfide.* Light-tan, semiconductive crystals. Crystal system, monoclinic. Density, 6.07-6.21 g/cm³; melting point, 1730°C (3146°F); band gap, 2.58 eV. Formula: Er_2S_3.

erbium telluride. A crystalline compound of erbium and tellurium. Crystal system, orthorhombic. Density, 7.11 g/cm³; melting point, 1213°C (2215°F). Used in ceramics, electronics and materials research. Formula: Er_2Te_3.

erbium tetraboride. See erbium boride (i).

Erco. Trade name of Erie Steel Company (USA) for a water-hardening steel containing 0.8% carbon, and the balance iron. Used for thermal-shock resistant dies.

Ercusol. Trademark of Bayer AG (Germany) for acrylic resins.

erdite. A copper-red mineral composed of sodium iron sulfide dihydrate, $NaFeS_2 \cdot 2H_2O$. Crystal system, monoclinic. Density, 2.30 g/cm³. Occurrence: USA (California).

Eref. Trade name of Solvay Polymers, Inc. (USA) for a polyamide/polypropylene alloy.

ER fluid. See electrorheological fluid.

Erftal. Trademark of Vereinigte Aluminium-Werke AG (Germany) for high-gloss mill aluminum (99.9% pure) containing 0.04% iron. Supplied as pigs, slabs, cast and continuously cast ingots. Used for reflectors.

Erftwerk. German trade name for light-gray oxide conversion coatings produced on copper-containing alloys by a proprietary process that makes use of a bath composed of sodium carbonate, sodium chromate and sodium silicate. Abbreviation: EW.

Erges. Trade name of Schmole GmbH (Germany) for a corrosion-resistant aluminum alloy containing 0.6-1.4% magnesium, 0-1.6% silicon, 0.6-1% manganese, and 0-0.3% chromium. Used for chemical equipment.

Ergolith. Trade name of McLeod Limited (UK) for tough, hard, flexible, odorless casein plastics with good resistance to greases, oils and dilute acids used for buttons, beads, knitting needles, novelties, and decorative items.

Ergste. Trade name of Stahlwerk Ergste GmbH & Co. KG (Germany) for a series of austenitic, ferritic and martensitic stainless steels, case-hardening steels, spring steels, and several alloy tool steels.

Erhard bronze. A German aluminum bronze used for ornamental fittings.

ericaite. A red mineral of the boracite group composed of iron chloride borate, $Fe_3B_7O_{13}Cl$. It can also be made synthetically Crystal system, orthorhombic. Density, 3.22 g/cm³; refractive index, 1.75. Occurrence: Germany.

ericssonite. A deep reddish black mineral of the seidozerite group composed of barium manganese iron silicate hydroxide, $BaMn_2Fe(Si_2O_8)(OH)$. Crystal system, monoclinic. Density, 4.21 g/cm³. Occurrence: Sweden.

Erilan. Trade name for regenerated protein (azlon) fibers.

Erinite. Trade name of Erinoid Limited (UK) for phenol-formaldehyde plastics.

Erinofort. Trade name of Erinoid Limited (UK) for cellulose acetate plastics.

Erinoid. (1) Trademark of Erinoid Limited (UK) for a synthetic resin made from casein and formaldehyde. Used as a replacement for amber, ivory, ebony, tortoise shell, etc.

(2) Trademark of BP Limited (UK) for chlorosulfonated polyethylene plastics.

Erinoplast. Trade name of Erinoid Limited (UK) for phenol-formaldehyde plastics.

eriochalcite. A brilliant bluish green mineral composed of copper chloride dihydrate, $CuCl_2 \cdot 2H_2O$. Crystal system, orthorhombic. Density, 2.39 g/cm³; refractive index, 1.685. Occurrence: Chile.

erionite. A white mineral of the zeolite group composed of potassium sodium aluminum silicate hydrate, $(Na,K)_8(Si,Al)_{36}O_{72} \cdot 23H_2O$. Crystal system, hexagonal. Density, 2.02 g/cm³; refractive index, 1.458. Occurrence: USA (California).

Erium. Trade name of Eriez Magnetics (USA) for rare-earth magnet materials including samarium-cobalt. Used in the manufacture of rolls, grates and tapes.

Erisys. Trademark of CVC Specialty Chemicals, Inc. (USA) for a thermoplastic polyurethane-modified bisphenol A epoxy coating resin with enhanced peel and thermal shock resistance. It bonds to polyvinyl chloride, and is used on flooring and surfaces such as those found in sports arenas.

Erkenzweig. Trade name of Erkenzweig & Schwemann (Germany) for a series of steel products including various stainless steels, and plain-carbon and alloy tool steels (e.g., high-speed, cold-work and hot-work types).

Erkopon. Trade name for resins that contain the ethoxide ($-OC_2H_5$) reactive group, and are made with ethylene diamine catalysts. Used for protective coatings for industrial and chemical equipment.

Erlanger blue. A variety of iron blue (Prussian blue) used as a pigment.

erlichmanite. A gray mineral of the pyrite group composed of osmium sulfide, OsS_2. It can also be made synthetically. Crystal system, cubic. Density, 9.52 g/cm³.

Erm. Trade name of Delta Enfield Metals Limited (UK) for a series of coppers and copper alloys including chromium and tellurium coppers, sulfur-bearing coppers, and copper-cobalt, copper-silver and copper-nickel-phosphorus alloys. Used for flash and resistance welding electrodes, welding equipment, and switchgear.

Ermal. (1) Trade name of Erie EMI Company (USA) for a pearlitic cast iron containing 2.2-2.4% carbon, 0.9-1% silicon, 0.7-0.8% manganese, and the balance iron. Used for fittings, pipes, and valves.

(2) Trade name of Enfield Rolling Mills Limited (UK) for several wrought aluminum products supplied in sheet and strip form including commercially pure aluminum (99.0+%) and aluminum-magnesium alloys.

Ermalite. Trade name of Erie EMI Company (USA) for a wear- and shock-resistant, high-strength malleable cast iron containing 2.4-2.6% total carbon (1.6-2.0% graphitic carbon and 0.6-0.8% combined carbon), 1.9-2.0% silicon, 0.4-0.45% manganese, 0.35-0.45% nickel or molybdenum, and the balance iron. Used for brake drums.

ernstite. A yellow brown mineral composed of iron manganese aluminum oxide phosphate hydroxide, $(Mn,Fe)AlPO_4(OH,O)_2$. Crystal system, monoclinic. Density, 3.07 g/cm^3; refractive index, 1.706. Occurrence: Southwest Africa.

Erodur. Trade name of Bergische Stahl Industrie (Germany) for a series of abrasion- and wear-resistant cast steels including various high-carbon chromium, chromium-molybdenum, chromium-molybdenum-vanadium, and intermediate and high-manganese steels.

Ersta. Trade name of Starcke KG Schleifmittelwerke (Germany) for abrasive cloth and paper of alumina and silicon carbide with resin or glue binders, supplied in sheet, roll, wheels and bands for wood, metal, glass and leather sanding and grinding.

Ertacetal. Trademark of DSM Engineering Plastics (USA) for a series of homopolymer and copolymer acetals available in the form of sheets, plates, blocks, bands, rods, and tubing.

Ertalon. Trademark of DSM Engineering Plastics (USA) for dry and conditioned nylon 6 and nylon 6,6 available in a wide variety of formulations including natural, blue, cast, molybdenum disulfide-filled, internally lubricated, and heat-resistant grades.

Ertalyte. Trademark of DSM Engineering Plastics (USA) for a series of polyester thermoplastics. *Ertalyte PET-P* refers to unreinforced, semicrystalline polyethylene terephthalate-based thermoplastic polyesters with high strength, excellent dimensional stability, good wear resistance, low coefficients of friction, good chemical and abrasion resistance, low moisture absorption, and a continuous use temperature of 100°C (210°F). Used for food-processing equipment, valves, gears, and precision mechanical parts. *Ertalyte TX* is an internally lubricated thermoplastic polyester with enhanced wear resistance.

erubescite. See bornite.

Ervadiene. Trademark of Plastimer (France) for an alkyd resin.

Ervamine. Trademark of Plastimer (France) for a series of modified melamine-formaldehyde plastics and compounds.

Ervamix. Trademark of Plastimer (France) for a series of glass-fiber-reinforced unsaturated polyesters.

Ervaphene. Trademark of Plastimer (France) for phenol-formaldehyde plastics and compounds.

Ervapon. Trademark of Plastimer (France) for a series of unsaturated polyesters.

erythrine. See erythrite.

erythrite. A deep red, reddish, yellowish-pink, pink or pearl gray mineral of the vivianite group composed of cobalt arsenate octahydrate, $Co_3(AsO_4)_2 \cdot 8H_2O$, and containing about 37.5% cobalt oxide. It can also be made synthetically. Crystal system, monoclinic. Density, 2.91-3.14 g/cm^3; hardness, 1.5-2.5 Mohs; refractive index, 1.661. Occurrence: Canada (Ontario), Germany, Morocco, USA (California, Colorado, Idaho, Nevada). Used as a source of cobalt, and as a colorant for glass and ceramics. Also known as *cobalt bloom; cobalt ocher; erythrine; peachblossom ore; red cobalt.*

erythrosiderite. A deep reddish mineral composed of potassium iron chloride monohydrate, $K_2FeCl_5 \cdot H_2O$. It can also be made synthetically from an acid solution of potassium chloride in ferric chloride by slow room-temperature evaporation. Crystal system, orthorhombic. Density, 2.37 g/cm^3; refractive index, 1.75.

Esa. Trade name of Latrobe Steel Company (USA) for a tool steel containing 1.45% carbon, 0.5% chromium, 4% tungsten, 0.25% vanadium, and the balance iron.

Esbrilith. German trade name for thermoplastic casein plastics.

Escalloy. (1) Trade name of Eastern Stainless Steel Company (USA) for a series of austenitic nickel-chromium steel castings. They have excellent corrosion resistance even to most hot and/or con-centrated acids, and are used for chemical-process equipment, plating equipment, etc.

(2) Trademark of ComAlloy International Corporation (USA) for chemical- and impact-resistant glass- or hybrid-reinforced polypropylenes for high-strength applications.

Escarit. German trade name for cellulosics.

Esclad. Trademark of Shamrock Construction Chemicals (Canada) for vinyl siding, fiberboard sheathing, and siding backerboard.

Esco. (1) Trade name of ESCO Corporation (USA) for an extensive series of alloys including various cast and wrought abrasion-, corrosion-, and/or heat-resistant steels (e.g., austenitic, ferritic and martensitic stainless and high-temperature grades) as well as corrosion-resistant cast irons, and nickel and nickel-chromium alloys.

(2) Trade name of Eyre Smelting Company (UK) for several cast phosphor bronzes containing 7-10% tin, 0-3.5% lead, 0-2% zinc, 0-1% nickel, 0.3-0.5% phosphorus, and the balance copper.

Escon. Trademark of Exxon Corporation (USA) for polypropylene resins with good rigidity, good resistance to heat, chemicals, moisture and electricity, good processibility, good strength and moduli, good impact resistance, and a maximum service temperature of 120°C (250°F). Used for piping, tubing, packaging, television cabinets, automotive trim, and hinges and other hardware.

Escor. Trade name of Exxon Chemical Company (USA) for ionomers and ethylene-maleic anhydride terpolymers that have good heat-sealing properties, and excellent adhesive properties in contact with aluminum, polyamide, paper and polyethylene. Used for flexible packaging, and in polymer modification.

Escorene. Trademark of Exxon Chemical Company (USA) for an extensive series of linear-low-density, linear high-density and linear-medium-density polyethylenes, high-density polyethylenes and ethylene-vinyl acetate (EVA) resins and copolymers. The polyethylenes are available in injection, rotomolding, extrusion coating and laminating grades for automotive and electrical insulation, packaging and compounding and various specialty applications. *Escorene LDPE* and *Escorene LLDPE* are low-density polyethylenes and linear-low-density polyethylenes, respectively, supplied in standard and UV-stabilized grades. The EVAs are supplied in film and compounding resin, stretch film, shrink film and adhesive/sealant grades for food packaging, sealing, bonding, electrical and various specialty applications. Examples include *Escorene EVA* and *Escorene Ultra EVA* resins and copolymers with vinyl acetate contents of 12, 25 or 33%.

Escoy. Trade name of E.S. Coy Limited (UK) for high-purity tin (99.90% pure) containing 0.03% lead, and 0.031% arsenic.

Escusa. Trade name of Teijin Fibers Limited (Japan) for polyester staple fibers used for the manufacture of textile fabrics.

ESD magnet. See elongated single-domain magnet.

ES Fiber. Trade name of Chisso Corporation (Japan) for polyethylene and polypropylene fibers.

Esgard. Trademark of Building Products of Canada Limited for asphalt shingles, industrial roofing, and fiberboard roof insulation.

Eshalit. German trade name for phenolic molding compounds and cellulosics.

eskebornite. A creamy yellow to yellowish brown mineral of the chalcopyrite group composed of copper iron selenide, $CuFeSe_2$. Crystal system, cubic. Density, 5.45. Occurrence: Germany, Czech Republic, Slovakia.

eskimo cloth. A satin- or twill-weave fabric with a thick nap, produced from natural or synthetic fibers, and usually dyed and having a striped pattern. Used especially for overcoats.

eskimoite. A grayish white mineral of the lillianite group composed of silver lead bismuth sulfide, $Ag_7Pb_{10}Bi_{15}S_{36}$. Crystal system, monoclinic. Density, 7.10 g/cm^3. Occurrence: Greenland.

eskolaite. A black mineral of the corundum group composed of chromium oxide, Cr_2O_3. It can also be made synthetically. Crystal system, rhombohedral (hexagonal). Density, 5.24 g/cm^3. Occurrence: Guyana.

Eslon. Trademark of Cheil Synthetics Company (South Korea) for polyester fibers and yarns.

Esmo. Trade name of W. Ossenberg & Cie. Edelstahlwerke (Germany) for high-carbon cold-work tool steel containing 1.65% carbon, 0.3% manganese, 0.3% silicon, 12% chromium, 0.6% molybdenum, 0.5% tungsten, 0.1% vanadium, and the balance iron. Used for broaches, coining and punching dies for thin sheet, cold extrusion dies, punches, metal saws, and bending tools.

E-Solder. Trademark of IMI Acrylics (UK) for electrically conductive epoxy adhesives.

ESP. Trademark of KoSa (USA) for a textured polyester filament stretch yarn with recovery properties. It blends well with most other fibers including acetate, cotton, rayon, wool, and other polyesters, and can produce special effects, such as crepe or crinkle. Used for athletic and casual warp-knit and circular-knit fabrics to produce two-way stretch, and athletic and casual woven fabrics as a warp (filling) yarn to produce primarily one-way stretch. ESP stands for "Extra Stretch Performance."

Espa. Trademark of Toyoba Company (Japan) for spandex fibers and yarns used for textile fabrics.

Espandy. Trademark of Kanebo Limited (Japan) for polyester fibers.

esparto. A tough, hard cellulose fiber obtained from the stems of speargrasses (especially *Stipa tenacissima* and *Lygeum spartum*) native to Spain and northern Africa, and used for weaving hats and matting, and in the manufacture of baskets, ropes and paper. Also known as *esparto fiber; alfa grass.*

esperite. A colorless or white mineral composed of calcium lead zinc silicate, $(Ca,Pb)ZnSiO_4$. It can also be made synthetically. Crystal system, monoclinic. Density, 4.28-4.42 g/cm^3; refractive index, 1.769.

Espet. Trade name for polyethylene terephthalate resins.

Espresso. Trade name of Teijin Fibers Limited (Japan) for polyester filaments used for the manufacture of textile fabrics.

Esscalloy. Trade name of Eastern Stainless Steel Company (USA) for cobalt-chromium-tungsten alloys with good high temperature properties.

Essera. Trade name of American Fibers & Yarns Company (USA) for polyolefin fibers and yarns used for textile fabrics.

Essex. (1) Trade name of Pittsburgh Corning Corporation (USA) for light-diffusing glass blocks.

(2) Trade name for a series of carbon blacks.

Esshete. Trade name of United Engineering Steels Limited (UK) for a series of steels including several molybdenum, molybdenum-chromium and molybdenum-chromium-vanadium grades, low-carbon chromium-molybdenum grades, and austenitic stainless grades.

Estaloc. Trademark of BF Goodrich (USA) for stable, engineering thermoplastics based on fiber-reinforced thermoplastic polyurethane (TPU). They have good impact and chemical resistance, good processibility, low moisture absorption, and coefficients of thermal expansion similar to those of metals.

Estalon. Trademark of Erwin Strinz Estalon-Lote (Germany) for copper-phosphorus and silver brazing filler metals, and silver-based soft solders.

Estamid. Trade name of Dow Chemical Company (USA) for polyamide-based chlorinated thermoplastic elastomers.

Estane. Trade name of BF Goodrich (USA) for thermoplastic polyester and polyether urethane elastomers. They have outstanding toughness and abrasion resistance, extraordinary low temperature flexibility, high tensile strength, high ultimate elongation, low air permeability, good physical and chemical properties, and good resistance to solvents including gasoline. Used for wire and cable coatings, automotive fuel hoses, tanks, belting, shoe heels, coated fabrics, medical devices, and adhesives.

Estar. Trademark of Mitsui Chemicals, Inc. (Japan) for a series of unsaturated polyesters.

Estax. Trademark of Schott Rohrglas AG (Germany) for special glass tubing with low background radiation used for pharmaceutical and medical applications, e.g., test containers, actinometers, and scintillation vials.

E-Steel. Trade name of Jones & Laughlin Steel Corporation (USA) for a free-cutting steel containing up to 0.6% carbon, 0.3% manganese, 0.25% silicon, and the balance iron. Used for screws, nuts and bolts, and screw-machine products.

ester. The product of a reaction of an alcohol or other organic compound rich in hydroxyl (OH) groups with an organic or inorganic acid. The functional group is RCOOR', where R is an alkyl group (e.g., CH_3 or C_2H_5) or hydrogen atom, and R' is an alkyl group that may or may not be the same as R. For example, the reaction product of methanol with acetic acid is an ester called *methyl acetate* (CH_3COOCH_3) and that of ethanol with benzoic acid is an ester known as *ethyl benzoate* (C_6H_5-$COOC_2H_5$).

Estera. Trade name for cellulose acetate fibers used for textile fabrics.

Estercore. Trade name of Diab Group (USA) for polyester-based semistructural foam core.

ester gum. See rosin ester.

Esterwald. Trade name for polyester fibers and yarns used for textile fabrics.

Esthet X. Trade name for a light-cure universal dental composite for esthetic restorations.

Esteticor Trade name of Cendres & Métaux (Switzerland) for an extensive series of dental bonding alloys including several white or yellow universal high-, medium- or low-gold alloys (e.g., *Esteticor Royal, Esteticor Swiss, Esteticor Concorde* and *Esteticor Ecologic*) for special ceramic restorations, and rich to pale yellow gold-platinum alloys (e.g., *Esteticor Avenir, Esteticor Cosmor H, Esteticor Ideal H, Esteticor Prema H, Esteticor Topas* and *Esteticor Vision*), white gold-palladium alloys (e.g., *Esteticor Economic, Esteticor Opal* and *Esteticor Plus*) and white palladium-base alloys (e.g., *Esteticor Actual* and *Esteticor Prisma*) for metal-ceramic restorations. Also known as *CM Esteticor.*

Estilux. Trade name of Heraeus Kulzer Inc. (USA) for a light-cure microfine methacrylate-based dental composite.

Estiseal LC. Trade name of Heraeus Kulzer Inc. (USA) for a light-cure methacrylate-based dental fissure sealant.

Estochrome. Trade name for polyester fibers and yarns used for textile fabrics.

Estolit. Estonian trade name for casein plastics (artificial horn).

Estra. Trade name of Eastman Chemical Company (USA) for cellulose acetate fibers used for wearing apparel and as industrial fabrics.

Estralon PET. Trade name of The Polymer Corporation (USA) for a chemical- and wear-resistant polyethylene terephthalate resin with good electrical properties used for electrical parts, pumps, valves, seals and chemical- and food-processing equipment.

Estrell. Trademark of Aquafil SpA (Italy) for polyester fibers and yarns.

Estron. Trademark of Eastman Chemical Company (USA) for a series of cellulose ester yarns and cellulose acetate staple fibers and filament yarns used for textile fabrics.

Etad. Trade name of Société Nouvelle des Acieries de Pompey (France) for a series of low- and medium-carbon alloy steels containing 0.3-0.4% carbon, 1.3-2.5% chromium, 3.8-4.6% nickel, 0-0.5% molybdenum, 0-0.5% vanadium, and the balance iron. Used for structural applications and machinery parts.

etamine. A lightweight, twill-weave worsted fabric with a short nap, formerly also made from cotton, linen or wool.

Etan. Trade name of Aerospace Composite Products (USA) for engineering composites consisting of carbon fibers embedded in carbon matrices. They have high thermal-shock resistance, can withstand temperatures above 2000°C (3600°F), and are used for aircraft brakes, engine pistons, metallurgical and vacuum furnaces, and crystal-growing equipment.

Etano. Trade name of Société Nouvelle de Acieries de Pompey (France) for a series of hardened and tempered low-carbon alloy steels containing 0.35% carbon, 1.7% chromium, 3.75% nickel, and the balance iron. Used for structural applications and machinery parts.

etched glass. Fine glassware with rough to mat finish that has been ornamented by treating with acid to produce an intaglio design. Used especially for architectural applications and in the arts.

etched-out fabrics. Plain-colored, patterned fabrics woven from mixed fibers (e.g., cotton and polyester) and treated with acid to selectively etch or burn out fibers in certain areas of the patterns or designs. After the acid treatment these etched out pattern appear transparent while the remainder of the fabrics appears opaque. Used for draperies, dresses and blouses. Also known as *burnt-out fabrics; devoré fabrics.*

etched wood. Plywood panels that have been surface-treated by wire brushing to accent the grain pattern, and create an interesting and attractive texture.

etch primer. See wash primer.

Eternit. Trademark of Eternit AG (Germany) for an asbestos cement made from a mixture of asbestos, Portland cement and water. It has good flame resistance and excellent weatherability, and is supplied in the form of flat and corrugated sheets, shingles, tiles, drain and ventilation pipe, siding, roofing products, and wallboard.

Eterplast. Trademark of Eternit AG (Germany) for glass-fiber-reinforced unsaturated polyesters.

Eternos. Trademark of Remystahl (Germany) for a series of corrosion- and/or heat-resistant steels containing high amounts of chromium, or chromium and nickel.

Ethafoam. Trademark of Dow Chemical Company (USA) for a series of strong, resilient, closed-cell low- and medium-density polyethylene foams supplied in several grades and colors including low-abrasion (*Ethafoam LA*), high-strength (*Ethafoam HS*) and fire-retardant (*Ethafoam CI*). They possess excellent shock-absorbing, vibration-dampening, and insulating properties. Their density ranges from as low as 20 kg/m³ (1.25 lb/ft³) for *Ethafoam Nova* to as high as 154 kg/m³ (9.61 lb/ft³) for *Ethafoam M5* and *Ethafoam HS 900*. Used for cargo cushioning and protective packaging applications, automotive applications, and in museums and art conservation. The flame-retardant grades are also used as interior compartment components for commercial aircraft.

Ethafoam Nova. Trademark of Dow Chemical Company (USA) for a strong, resilient, green-colored low-density polyethylene foam with closed-cell structure. It has a density of 20 kg/m³ (1.25 lb/ft³), and possesses excellent shock-absorbing, vibration-dampening, and insulating properties. Used for cargo cushioning and protective packaging applications.

Ethafoam Select. Trademark of Dow Chemical Company (USA) for a strong, resilient, blue-colored low-density polyethylene foam with closed-cell structure. It has a density of 28 kg/m³ (1.75 lb/ft³), and possesses excellent shock-absorbing, vibration-dampening and insulating properties. Used for cargo cushioning and protective packaging applications.

ethanedial. See glyoxal.

ethanoylated cellulose. See cellulose acetate (1).

ethanoylated cotton. See acetylated cotton.

ethene. See ethylene.

ether. An organic compound containing an oxygen atom connected to two identical or different alkyl groups (e.g., CH_3, C_2H_5, C_6H_5, etc.). The generic formula is ROR'. Examples include *diethyl ether*, $(C_2H_5)_2O$, and *methylphenyl ether* (or *anisole*), $C_6H_5OCH_3$.

Ethocel. Trademark of Dow Chemical Company (USA) for thermoplastic ethylcellulose resins supplied in translucent, transparent, and opaque colors. They have good strength, toughness and shock resistance, high transparency and surface gloss, good moldability, a maximum service temperature of -40 to +93°C (-40 to +199°F), good resistance to water, and poor resistance to most organic solvents. Used for automotive components, tools for the aircraft industry, appliance housings and trim, household articles, piping, packaging, eye shades, glazing, handles, and knobs.

ethoxide. A compound that is the product of a reaction in which the hydroxyl hydrogen of ethyl alcohol (C_2H_5OH) has been substituted by a metal. Examples include *aluminum ethoxide*, $Al(OC_2H_5)_3$ and *sodium ethoxide*, $NaOC_2H_5$. Also known as *ethylate.*

ethoxybenzoic acid. An aromatic compound (98% pure) that at room temperature is usually a liquid with some crystals present. It has a density of 1.105 g/cm³, a melting point of 19-19.5°C (66-67°F), a boiling point of 174-176°C (345-349°F)/15 mm, a flash point above 230°F (110°C), and a refractive index of 1.540. Used chiefly in biochemistry, biotechnology, medicine, and in the preparation of dental cements. Formula: $C_9H_{10}O_3$. Abbreviation: EBA.

ethoxybenzoic acid cement. A dental cement composed of ethoxybenzoic acid ($C_9H_{10}O_3$) alone, or in combination with zinc oxide, and used for temporary lining and luting, and for sealing lateral perforations. Also known as *EBA cement.*

Ethron. Trade name for tough, electrically insulating, chemically resistant polyethylenes used for packaging film, and molded articles.

Ethyl. Trade name of Georgia-Pacific Company (USA) for plasticized polyvinyl chlorides available in three elongation ranges: 0-100%, 100-300%, and above 300%.

ethylate. See ethoxide.

ethylcellulose. An ethyl ether of cellulose obtained either from cellulose and ethanol in the presence of dehydrating agents, or from alkali cellulose and ethyl chloride or sulfate. It is a thermoplastic solid supplied in the form of white granules or powder typically containing 45-49% ethoxyl. While its properties depend largely on the ethoxyl content, the standard grades have a density range of 1.07-1.17 g/cm³ (0.039-0.042 lb/in.³), a softening range of 155-162°C (311-324°F), a refractive index of 1.47, high dielectric strength, good resistance to alkalies and dilute acids, and good film-forming properties. Used for hot-melt adhesives and coatings for cables, paper, textiles, etc., molding powders, extrusion of wire insulation, lacquers, protective coatings, as a binder for technical ceramics and pigments, as a parting agent for thin-sheet ceramics made by the doctor-blade process, as a hardening and toughening agent for plastics, as a thin wrapping material, in printing inks, and as a casing for rocket propellants. Formula: $(C_6H_8O_3(OC_2H_5)_2)_n$. Abbreviation: EC. Also known as *cellulose ethyl ether.*

ethylcellulose plastics. Lightweight thermoplastics based on ethylcellulose that have high toughness, moderate strength, good moldability, good dielectric properties, low water absorption, good low- and elevated-temperature flexibility from -57 to +66°C (-71 to +151°F). They are frequently sold under trade names, such as *Ethocel, Hercocel* or *Lumarith.*

ethylene. A flammable, colorless gas (96+% pure) with a melting point of -169°C (-272°F), a boiling point of -104°C (-155°C), a flash point of -213°C (-135°C), a critical temperature of 9.5°C (49°F), and an autoignition temperature of 1009°F (543°C). In its liquid state, it has a density of 0.610 g/cm³ (0°C/32°F). Used in the manufacture of many polymers and elastomers including polyesters, polyethylenes, polypropylenes, polystyrenes, polyvinyl chlorides and styrene-butadiene rubbers, in the manufacture of various chemicals, such as acetaldehyde, ethylene oxide, vinyl acetate and vinyl chloride, in the synthesis of aluminum alkyls, as a refrigerant, and in metal cutting and welding. Formula: $H_2C=CH_2$. Also known as *ethene; bicarburetted hydrogen.*

ethylenebisiminodiacetic acid. See ethylenediaminetetraacetic acid.

ethylene-butene copolymers. See ethylene-butylene copolymers.

ethylene-butylene copolymers. A family of high-density polyethylene products obtained by the copolymerization of ethylene and butylene (butene-1). They have good strength and rigidity, a useful temperature range of 65-90°C (149-194°F), relatively high susceptibility of environmental stress cracking, and good to fair resistance to chemicals and solvents. Used for molded articles, coatings, sheeting, etc. Abbreviation: EB copolymers. Also known as *ethylene-butene copolymers.*

ethylene chlorotrifluoroethylenes. A family of fluoropolymer resins produced by copolymerizing ethylene and chlorotrifluoroethylene. They have a density of 1.68 g/cm³ (0.06 lb/in.³), high strength, high wear and creep resistance, good impact strength, a upper service temperature of 150-170°C (300-340°F), a lower service temperature of -75°C (-103°F), low water absorption (less than 0.1%), low coefficients of friction, good resistance to acids, alcohols, alkalies, aromatic hydrocarbons, greases, oils and ketones, fair resistance to halogens, good electrical properties (dielectric), good ultraviolet resistance, and good processibility. Used for filter cartridges in cleanrooms, cables, electronic circuit manufacture, pipes, valves, tubing, pumps, clamps, adapters, fittings, etc. Abbreviation: ECTFE, E-CTFE. Also known as *ethylene chlorotrifluoroethylene copolymers.*

ethylene copolymers. Reaction products of the copolymerization of ethylene and either butene, hexene, ethyl acrylate, maleic anhydride, propylene or vinyl acetate. See also ethylene-butylene copolymers; ethylene-hexylene copolymers; ethylene-ethyl acrylate copolymers; ethylene-maleic anhydride copolymers; ethylene-propylene copolymers; ethylene-vinyl acrylates.

ethylene diamine tartrate. Crystalline organic compound used in the manufacture of piezoelectric crystals for electric frequency control in televisions, and as wavelength-dispersive spectrometer crystals. Abbreviation: EDT.

ethylenediaminetetraacetic acid. An organic chelating compound available in the form of colorless crystals (99+% pure) that decompose at approximately 245°C (473°F). Used as a metal chelating agent, as an eluting agent in ion exchange, in the decontamination of radioactive surfaces, in metal cleaning and plating, as an antioxidant, as a detergent, and in medicine, biochemistry, analytical chemistry and spectrophotometric titration. It is also used in the form of its salts (e.g., disodium EDTA, trisodium EDTA, tripotassium EDTA, etc.). Formula: $C_{10}H_{16}N_2O_8$. Abbreviation: EDTA. Also known as *(ethylenedinitrilo)tetraacetic acid; ethylenebisiminodiacetic acid.*

(ethylenedinitrilo)tetraacetic acid. See ethylenediaminetetraacetic acid

ethylene-ethyl acrylate copolymers. A family of soft, tough, highly flexible copolymers of ethylene and ethyl acrylate which can be processed by blow, compression, injection, rotational or transfer molding, and extrusion. They have good to moderate electrical properties, low thermal resistance, poor resistance to aliphatic and aromatic hydrocarbons, chlorinated hydrocarbons and most solvents, excellent resistance to environmental stress cracking and good resistance to ultraviolet radiation. Used for molded articles including electrical products, dishwasher trays, garbage containers, flexible hoses and tubing, water pipes, and packaging films. Abbreviation: EEA copolymers.

ethylene-hexene copolymers. See ethylene-hexylene copolymers.

ethylene-hexylene copolymers. A family of high-density polyethylene products obtained by the polymerization of ethylene and hexylene (hexene-1). They have good strength and rigidity, a service temperature range of 65-90°C (149-194°F), relatively high susceptibility to environmental stress cracking, and good to fair resistance to chemicals and solvents. Used for blow-molded articles, coatings for cables, wires and piping, sheeting, monofilament, etc. Abbreviation: EH copolymers. Also known as *ethylene-hexene copolymers.*

ethylene-maleic anhydride copolymers. Water-soluble resins made by the copolymerization of ethylene and maleic anhydride, and usually supplied in the form of fine powders as both straight-chain and crosslinking polymers in a wide range of molecular weights. Used for paints, textile sizes, printing inks, suspension agents, emulsifiers, and ceramic binders. Abbreviation: EMA copolymers.

ethylene plastics. See polyethylene plastics.

ethylene-propylene copolymer. See ethylene-propylene rubber.

ethylene-propylene-diene monomer. An elastomer that is based on stereospecific linear ethylene and propylene terpolymers with a small addition of an aliphatic or cyclic diene, such as hexadiene, dicyclopentadiene or ethylidene norbornene. It can be crosslinked (vulcanized) with sulfur, and has excellent ozone resistance, excellent resistance to weathering, oxidation and sunlight, good mechanical properties when reinforced, good resilience and flexing characteristics, good mechanical hysteresis and compression-set resistance, good chemical resistance, also to acids and alkalies, fair to poor oil resistance, good electrical resistance, an upper continuous-use temperature of approximately +175°C (+350°F), and a lower continuous-use temperature of approximately -50°C (-60°F). Used for electrical insulation, wire and cable coating, automotive parts, automotive hoses and belts, tire tubes, safety bumpers, gaskets, weatherstripping, mechanical rubber goods, footwear, coated fabrics, and as a thermoplastic resin modifier. Abbreviation: EPDM. Also known as *EPDM rubber; ethylene-propylene-diene monomer rubber; ethylene-propylene terpolymers (EPT)*.

ethylene-propylene elastomer. See ethylene-propylene rubber.

ethylene-propylene rubber. An elastomer that is the reaction product of the stereospecific copolymerization of ethylene and propylene, or terpolymerization of ethylene, propylene and an aliphatic or cyclic diene. It cannot be vulcanized, but can be cured (crosslinked) with peroxides. It has good resiliency and flexibility, an upper service temperature of 177°C (350°F), good resistance to alkalies and diluted acids, better resistance to atmospheric aging, oxidation and ozone than styrene-butadiene rubber, good to fair resistance to hydraulic fluids, ketones and hot and cold water, and poor resistance to most oils, kerosene, gasoline, aromatic aliphatic hydrocarbons, halogenated solvents, and concentrated acids. Used for automotive components, mechanical rubber goods, footwear, coatings, and coverings. Abbreviation: EPR rubber. Also known as *EPR rubber; EPR copolymers; ethylene-propylene copolymer; ethylene-propylene elastomer*.

ethylene-propylene terpolymer. See ethylene-propylene-diene monomer.

ethylene resins. See polyethylenes.

ethylene-tetrafluoroethylenes. A family of melt-processible fluoropolymer resins produced by copolymerizing ethylene and tetrafluoroethylene. They are frequently made conductive by adding up to 30% carbon fibers, and have a low density (1.7 g/cm³ or 0.06 lb/in.³), low water absorption (0-0.03%), good resistance to abrasion, cut-through, impact and low-temperature impact, good weather and radiation resistance, excellent chemical resistance, good stress-crack resistance, uniform electrical properties, low dielectric constants, and an upper service temperature of 150°C (300°F). Used for wire harnesses for automobiles and trucks, electric wires and cables, medical parts, hoses and tubing for hydraulic, oil, fuel, pneumatic and oxygen systems, printed circuit board laminates, sealing applications, and for fluidized-bed, rotomolding and electrostatic-coating applications. Abbreviation: ETFE.

ethylene-vinyl acetates. See ethylene-vinyl acetate copolymers.

ethylene-vinyl acetate copolymers. A family of soft, tough, highly flexible copolymers of ethylene and vinyl acetate that can be processed by blow, compression, injection, rotational or transfer molding, and extrusion. They have good to moderate electrical properties, low thermal resistance, a softening temperature of 64°C (147°F), poor resistance to aliphatic and aromatic hydrocarbons, chlorinated hydrocarbons and most solvents, and good resistance to environmental stress cracking and ultraviolet radiation. Used as an adhesion improver for hot-melt and pressure-sensitive adhesives, in conversion coatings and thermoplastics, for molded articles including electrical products, appliance bumpers, bushings, seals, gaskets, hoses for pneumatic tools and paint-spray equipment, for extruded tubing for vending machines, and for drug delivery systems. Abbreviation: EVA copolymers. Also known as *ethylene-vinyl acetates*.

ethylhydroxyethylcellulose. A water-soluble cellulose ether available in the form of a white, granular solid in extra-low viscosity types (10-20 centipose), low viscosity (20-35 centipose) and high-viscosity types (125-250 centipoise). Used in organic and aqueous-organic solvent systems, in protective coatings, as a film former, and as a binding, stabilizing and thickening agent. Abbreviation: EHEC.

ethyllithium. A highly reactive organometallic compound prepared by reacting ethyl chloride with metallic lithium. It is available as transparent crystals and as a 2.0M solution in benzene. Used as a Grignard reagent in organometallic synthesis. Formula: C_2H_5Li. Also known as *lithium ethyl*.

ethylorthosilicate. See tetraethyl orthosilicate.

Ethyl Rubber. Trade name of Hercules Company (USA) for a series of synthetic rubbers.

ethyl silicate. See tetraethyl orthosilicate.

Ethyl UPVC. Trade name of Georgia-Pacific Company (USA) for unplasticized polyvinyl chlorides available in standard, high-impact, crosslinked, UV-stabilized, and structural foam grades.

Ethylux. Trade name of Westlake Plastics Company (USA) for a series of low-density polyethylene plastics with good toughness over its entire service temperature range of -57 to +93°C (-70 to +200°F), good chemical, moisture and electrical resistance, low coefficient of friction, good processibility, and good tensile and impact strength. Used for piping, duct, containers, housings, insulation, houseware, toys, packaging, coatings, and films.

ethyl zinc. See diethylzinc.

ethyne. See acetylene.

etioporphyrin III. A *porphyrin* derivative available in the form of purple crystals, and used as dye and pigment. Formula: $C_{32}H_{38}N_4$.

etioporphyrin I dihydrobromide. A synthetic *porphyrin* derivative available in the form of purple crystals and used as a dye and pigment. Formula: $C_{32}H_{38}N_4 \cdot 2HBr$.

etioporphyrin I nickel. See nickel etioporphyrin.

Etiral. Trade name of Creusot-Loire (France) for a tough, hard steel with 10% chromium, used for wire drawing dies.

Etisol. Trade name of Salgotarjan (Hungary) for insulating glass.

Etocel. Trademark of Dow Plastics (USA) for ethyl cellulose plastics.

Etofil. Trade name of Fiberfil (USA) for glass-fiber-reinforced polyethylene.

ETP copper. See electrolytic tough pitch copper.

ET salts. See electron transfer salts.

ET-semicon. Trade name for conductive polyethylene resins with butyl side chains that can be processed by blow or injection molding, vacuum forming, and extrusion. They have good flame resistance, and improved electrical conductivity, and are used for packaging applications (e.g., boxed pallets), and in resistance heating of molded products.

ettringite. (1) A colorless mineral composed of calcium aluminum carbonate silicate sulfate hydroxide, $Ca_6Al_2(SO_4,SiO_4,CO_3)_3(OH)_{12} \cdot 26H_2O$. Crystal system, hexagonal. Density, 1.77-1.85 g/cm^3; refractive index, 1.49-1.85. Occurrence: USA (New Jersey, California).

(2) A colorless high-sulfate calcium sulfoaluminate mineral, $3CaO \cdot Al_2O_3 \cdot 3CaSO_4 \cdot 30\text{-}32H_2O$. It occurs naturally, e.g., in California or New Jersey, and is also formed on mortar and concrete when attacked by acids.

Etylon. Trade name for polyolefin fibers and products.

eucairite. A silver-white to lead-gray mineral composed of copper silver selenide, CuAgSe. Crystal system, tetragonal. Density, 7.50-7.70 g/cm^3; hardness, 2.5 Mohs. Occurrence: Sweden, Argentina, Chile.

eucalyptus. Any of a genus *(Eucalyptus)* of evergreen trees of the myrtle family native to Australia and New Guinea. Although there are over a hundred species, the two most important Australian eucalyptus trees are jarrah *(E. marginata)* and karri *(E. diversicolor)*. They are valued for their wood, gum, resin and oil that can be extracted from the leaves, and are now also grown in the Mediterranean, and the southern and western United States. Eucalyptus wood is very hard, heavy, strong and decay-resistant and used for heavy construction, shipbuilding, flooring, furniture, posts, tool handles, etc. American eucalyptus species are known as *gumwood* and include the blue gum, red gum and salmon gum. See also blue gum; gumwood; ironwood; jarrah; karri; red gum; salmon gum.

euchroite. A green mineral composed of copper arsenate hydroxide trihydrate, $Cu_2AsO_4(OH) \cdot 3H_2O$. Crystal system, orthorhombic. Density, 3.45 g/cm^3; refractive index, 1.698. Occurrence: Hungary.

euclase. A colorless, bluish, or green mineral with vitreous luster composed of beryllium aluminum silicate hydroxide, $BeAlSiO_4(OH)$. Crystal system, monoclinic. Density, 3.10 g/cm^3; refractive index, 1.655; hardness, 7.5 Mohs. Occurrence: Austria, Brazil, Peru, Russia, Tasmania. Used as a gemstone.

eucryptite. A colorless or white mineral of the phenakite group composed of lithium aluminum silicate, α-LiAlSiO$_4$, and containing up to 4.8% lithium oxide. Crystal system, rhombohedral (hexagonal). Density 2.66-2.67 g/cm^3; refractive index, 1.572. Occurrence: Africa. Used in glass manufacture, and as a source of lithium in ceramic bodies of low-thermal expansion.

eudialyte. A pink, brown, or red mineral composed of sodium calcium zirconium silicate hydroxide, $Na_4(Ca,Fe)_2ZrSi_6O_{17}(OH,Cl)_2$. Crystal system, rhombohedral (hexagonal). Density, 2.93 g/cm^3; refractive index, 1.613. Occurrence: Greenland, Canada (Quebec).

eudidymite. A colorless, or glassy white mineral composed of sodium beryllium silicate hydroxide, $NaBeSi_3O_7(OH)$. Crystal system, monoclinic. Density, 2.55 g/cm^3; refractive index, 1.546; hardness, 6 Mohs. Occurrence: Norway.

Eugene Vaders. British trade name for a corrosion-resistant copper alloy containing 37.7% zinc, 0.3% iron, 2.5% manganese, 1.5% aluminum, and 0.5% silicon. Used for propellers, gears, and high-strength castings.

eugenol. A colorless to light yellow liquid extracted from clove oil. It has a density of 1.066 g/cm^3, a melting point of -11°C (12°F), a boiling point of 254°C (489°F), a flash point above 230°F (110°C) and a refractive index of 1.541. Used in dental cements and resins, in the production of perfumery, essential oils, and flavors, as a substrate for peroxidase, and in medicine. Formula: $C_{10}H_{12}O_2$.

eugsterite. A colorless mineral composed of sodium calcium sul-

fate hydrate, $Na_4Ca(SO_4)_3 \cdot 2H_2O$. Crystal system, monoclinic. Occurrence: Turkey, Africa (Kenya).

eulytine. See eulytite.

eulytite. A pale yellow to dark brown mineral composed of bismuth silicate, $Bi_4(SiO_4)_3$. It can also be prepared from bismuth oxide (Bi_2O_3) and silicon dioxide (SiO_2) by firing under prescribed conditions. Crystal system, cubic. Density, 6.1-6.6 g/cm^3; refractive index, 2.05. Occurrence: Germany. Also known as *agricolite; bismuth blende; eulytine.*

eumelanins. A group of *melanins* (black pigments) produced by radical-type polymerization of a nitrogenous *melanogen*. They contain a graphitic core, and yield nitrogenous polycarboxylic acids (e.g., 2,3,4,5-pyrroletetracarboxylic acid) on oxidative cleavage.

euphorbia rubber. The gum resin exudation from the vines of various species of African plants of the genus *Euphorbia* used in the manufacture of natural rubber.

Euredur. (1) Trademark of Ciba-Geigy Corporation (USA) for cycloaliphatic amines.

(2) Trademark of Witco GmbH (Germany) for synthetic resins used in two-component resin systems.

Eureka. (1) Trademark of Eureka Welding Alloys, Inc. (USA) for an extensive series of ferrous alloys used for coated electrodes and welding rods.

(2) Trade name of J.F. Jelenko & Company (USA) for a white, silver-free gold-palladium ceramic dental alloy for fusing porcelain to metal.

(3) German trade name for an electrical resistance material of 60% copper and 40% nickel.

(4) Trade name CAE Industries Limited (Canada) for babbitt bearing alloys for railway journal boxes.

Eurekamatic. Trade name of Eureka Welding Alloys, Inc. (USA) for an extensive series of ferrous and nonferrous hardfacing alloys usually supplied in wire form.

Eureka Plus. Trade name of J.F. Jelenko & Company (USA) for a white-colored gold-palladium-silver ceramic dental alloy with high strength, and good workability, finishability, and polishability. Used for fusing porcelain to metal.

Euro. Trademark of Wieland Dental + Technik GmbH & Co. KG (Germany) for white-colored, extra-hard gold-palladium dental ceramic alloys, such as *Euro 45* and *Euro 50*.

Euroacril. Trademark of EniChem Fibre SpA (Italy) for acrylic fibers used for the manufacture of textile fabrics.

Eurodrain. Trade name of Hegler Plastik GmbH (Germany) for flexible plastic drainage pipes.

Eurolastic. Trade name of Euroteam AG (Germany) for several jointing and caulking compounds.

Eurolistral. Trade name of Saint-Gobain (France) for cast glass that may or may not be wired or colored.

Europalloy Duett. Trade name for dental amalgam alloys.

European alder. See black alder.

European ash. The tough, elastic, nearly white wood of the tall hardwood tree *Fraxinus excelsior*. It has a straight grain, exceptional toughness and shock resistance, and good bending properties. Average weight, 680 kg/m^3 (42 lb/ft^3). Source: Europe including British Isles. Used for furniture, handles for tools, such as picks, shovels, etc., agricultural implements, sporting goods, such as hockey sticks, tennis rackets, skis, etc. It was formerly widely used for motor vehicle framing, underframing, in aircraft construction, and for bent work.

European beech. The strong, reddish-brown wood of the hardwood tree *Fagus sylvatica*. It has a straight grain, close even texture, moderate hardness, good to moderate durability, and good bending properties. Average weight, 680-720 kg/m³ (42-45 lb/ft³). Source: Europe including British Isles. Used for furniture and cabinetmaking, floorboards, staircases, wood-block paving, tools, tool handles, general turnery, bent work, machine parts, pianoforte manufacture, and woodenware. It was formerly also used for autobodies. Also known as *red beech*.

European birch. The tough, white to light brown wood of the hardwood trees *Betula pubescens* and *B. pendula*. It has a fairly straight grain, and fine texture. Average weight, 600 kg/m³ (37 lb/ft³). Source: Europe including British Isles. Used for turnery, furniture, house construction, wagons, and plywood. See also silver birch.

European boxwood. See Turkish boxwood.

European chestnut. See sweet chestnut.

European Cobble. Trade name of Perma Paving Stone Company (Canada) for strong, durable paving stones that resemble original European cobblestone, and are supplied in several shapes and sizes in terra cotta, fieldstone, granite and antique buff colors.

European hornbeam. See white beech (2).

European horse chestnut. The creamy-white wood of the tall hardwood tree *Aesculus hippocastanum*. It has a fine, uniform texture, and low strength and durability. Average weight, 510 kg/m³ (32 lb/ft³). Source: Europe including British Isles. Used for turnery, dairy and kitchen utensils, etc. See also American horse chestnut; buckeye.

European larch. The strong, durable, brownish-red wood of the coniferous tree *Larix decidua*. Average weight 570 kg/m³ (36 lb/ft³). Source: Europe including British Isles. Used for lumber, gates and fencing, mine timber, and outdoor work.

European lime. The soft, white or light-yellow wood of the tall hardwood tree *Tilia vulgaris*. It has a fine texture, cuts cleanly, and has moderate durability. Average weight, 550 kg/m³ (34 lb/ft³). Source: Europe including British Isles. Used for carving, turnery, parts of musical instruments including pianos, and for charcoal. Also known as *European linden*.

European linden. See European lime.

European oak. See chestnut oak (2); English oak.

European porcelain. See hard paste porcelain.

European silver fir. See silver fir (1).

European spruce. See Norway spruce.

European walnut. The hard, finely figured wood of the hardwood tree *Juglans regia*. It is grayish-brown, available in a wide variety of colors, figures and textures, and machines and finishes well. Average weight, 640 kg/m³ (40 lb/ft³). Source: Europe including British Isles. Used for furniture, cabinetwork, paneling, interior decoration, decorative objects, veneer, turnery, and gun and rifle stocks. Also known as *English walnut*.

europia. See europium oxide (i).

europium. A soft, steel-gray metallic element of the lanthanide series (rare-earth group) of the Periodic Table. It is commercially available in the form of ingots, lumps, wire, sheet, foil, powder and single crystals. The commercial europium ores are *monazite, gadolinite, samarskite* and *xenotime*. Crystal system, cubic. Crystal structure, body-entered cubic. Density, 5.243 g/cm³; melting point, 822°C (1512°F); boiling point, 1597°C (2907°F); hardness, 20 Vickers; atomic number, 63; atomic weight, 151.964; divalent, trivalent. It oxidizes rapidly in air at room temperature, reduces metallic oxides, and is an efficient neutron absorber. Used for control rods in nuclear reactors, as a neutron scavenger, for phosphors in color television tubes, in

the electronic recognition of first-class mail, for X-ray screens and mercury lamps, in superconductors, semiconductors and ceramics, and as a dopant for laser crystals. Symbol: Eu.

europium acetylacetonate. A hygroscopic powder with a melting point above 187-189°C (368-372°F). It induces fluorescent emission upon irradiation, and is used in organic and organometallic synthesis. Formula: $Eu(O_2C_5H_7)_3$. Also known as *europium 2,4-pentanedionate*.

europium boride. A crystalline compound of europium and boron Crystal system, cubic. Density, 4.91-4.94 g/cm³; melting point, 2600°C (4712°F). Used in ceramics, electronics and materials research. Formula: EuB_6. Also known as *europium hexaboride*.

europium carbide. A compound of europium and carbon that is unstable at 21°C (70°F), and is used in ceramics and materials research. Formula: EuC_2.

europium chalcogenides. Compounds of europium and any of the elements of Group VIA (Group 16) of the Periodic Table (e.g., oxygen, sulfur, etc.), used for ferromagnetic ceramics, electronic components, computer memories, etc. See also europium oxide; europium sulfide.

europium chloride. A hygroscopic, white-yellow powder or green-yellow needles (99.9+% pure) with a density of 4.89 g/cm³ and a melting point of 623°C (1153°F). Used in ceramics, chemistry and materials research. Formula: $EuCl_3$.

europium chloride hexahydrate. Hygroscopic, white-yellow crystals (99.9+% pure) used in ceramics, chemistry and materials research. Formula: $EuCl_3 \cdot 6H_2O$.

europium dichloride. White crystals. Crystal system, orthorhombic. Density, 4.9 g/cm³; melting point, 731°C (1348°F). Used in ceramics, chemistry and materials research. Formula: $EuCl_2$.

europium difluoride. Greenish-yellow crystals. Crystal system, cubic. Density, 6.5 g/cm³; melting point, 1380°C (2516°F). Used in ceramics, chemistry and materials research. Formula: EuF_2.

europium diiodide. Green crystals or off-white powder with a melting point of 580°C (1076°F). Used in ceramics, chemistry and materials research. Formula: EuI_2.

europium disilicide. See europium silicide.

europium-doped calcium fluoride. A nonhygroscopic, chemically resistant compound of *calcium fluoride* doped (or activated) with europium (Eu^{3+}) ions and supplied as high-purity single crystals. Crystal system, cubic. Density, 3.18 g/cm³; melting point, 1360°C (2480°F); hardness, 4 Mohs; refractive index, 1.43. Used as a light scintillator for beta-ray and soft gamma-ray detection, and in spectroscopy. Formula: CaF_2:Eu.

europium fluoride. Hygroscopic, white crystals or powder (99.9+% pure) with a melting point of 1390°C (2534°F) and a boiling point of 2280°C (4136°F). Used in the preparation of mixed-metal fluorides and in fluoride glasses. Formula: EuF_3.

europium halides. Compounds of europium and a halogen, such as bromine, chlorine, fluorine or iodine. Examples include europium chloride ($EuCl_3$) and europium fluoride (EuF_3).

europium hexaboride. See europium boride.

europium monosulfide. See europium sulfide (i).

europium monoxide. See europium oxide (ii).

europium nitrate. Colorless, white, or pale pink crystals (99.9+% pure) with a melting point of 85°C (185°F). Used in chemistry and materials research. Formula: $Eu(NO_3)_3 \cdot 6H_2O$. Also known as *europium nitrate hexahydrate*.

europium nitrate hexahydrate. See europium nitrate.

europium nitride. A crystalline compound of europium and nitrogen. Crystal system, cubic. Density, 8.77 g/cm³. Used in ceramics, electronics and materials research. Formula: EuN.

europium oxide. Any of the following compound of europium and oxygen: (i) *Europium sesquioxide.* A pale rose, hygroscopic rare-earth oxide powder (99.9+% pure). Density, 7.42 g/cm³; melting point, above 1300°C (2370°F); hardness, 435 Knoop. Used for nuclear-control rods, fluorescent glass, red and infrared-sensitive phosphors, e.g., in color television tubes, in rare-earth magnets and magnetooptic magnets for computer memories, and in superconductor research. Formula: Eu_2O_3. Also known as *europia;* (ii) *Europium monoxide.* Density, 8.16 g/cm³. Formula: EuO; and (iii) *Europium tetroxide.* Density, 8.07 g/cm³. Formula: Eu_3O_4. Another europium oxide ($Eu_{16}O_{21}$) with a density of 6.74 g/cm³ is used for phosphors sensitive to red and infrared radiation.

europium 2,4-pentanedionate. See europium acetylacetonate.

europium selenide. A brown crystalline compound of europium and selenium. Crystal system, cubic. Crystal structure, halite. Density, 6.45 g/cm³; melting point, 2027°C (3681°F). Used as a semiconductor. Formula: EuSe.

europium sesquioxide. See europium oxide (i).

europium silicide. A crystalline compound of europium and silicon. Crystal system, tetragonal. Density, 5.46 g/cm³; melting point, approximately 1500°C (2730°F). Used in ceramics, electronics and materials research. Formula: $EuSi_2$. Also known as *europium disilicide.*

europium sulfate. Colorless, white, or pale pink crystals. Density, 4.95 g/cm³ (anhydrous); melting point, loses 8H₂O at 375°C (707°F). Formula: $Eu_2(SO_4)_3$ (anhydrous); $Eu_2(SO_4)_3 \cdot 8H_2O$ (octahydrate).

europium sulfide. Any of the following compounds of europium and sulfur used in ceramics, electronics and materials research: (i) *Europium monosulfide.* Crystal stystem, cubic. Density, 5.75 g/cm³; melting point, 1667°C (3033°F); semiconductive properties. Formula: EuS; and (ii) *Europium tetrasulfide.* Density, 6.27 g/cm³. Formula: Eu_3S_4.

europium telluride. A black crystalline compound with a density 6.48 g/cm³ and melting point of 1526°C (2779°F). Used in ceramics, electronics and materials research. Formula: EuTe.

europium tetrasulfide. See europium sulfide (ii).

europium tetroxide. See europium oxide (iii).

Europox. Trademark of Schering AG (Germany) for a series of epoxy resins.

Europrene. Trademark of EniChem SpA (Italy) for an extensive series of general-purpose and thermoplastic elastomers and latexes supplied in several grades for various applications.

Eurosil. Trade name for silicone elastomer used for dental impressions.

Eurotek. (1) Trade name of Euroteam AG (Germany) for polysulfide, silicone and polyurethane-based joint sealing strips.

(2) Trademark of Eurotek Finishes, Inc. (USA) for automotive paints, primers, and thinner.

Eurromatic. Trade name of Techne Inc. (USA) for hollow plastic spheres.

EutecDur. Trade name of Eutectic Corporation (USA) for ultrahard wear coatings composed of coarse tungsten-carbide particles. Used for severe high-stress and abrasion applications.

Eutecsil. Trademark of Eutectic Corporation (USA) for a series of silver brazing alloys.

Eutectal. Trade name of Alais Forges et Camargue (France) for a heat-treatable aluminum alloy containing 1.6% magnesium, 1.5% copper, 0.8% manganese, 0.35% titanium, and 0.25% silicon. Used for light-alloy parts.

Eutectic. Trade name for a series of eutectic fusible alloys including the following compositions: (i) 50Bi-27Pb-13Sn-10Cd with a melting point of 70°C (158°F); (ii) 52Bi-40Pb-8Cd with a melting point of 91°C (196°F); (iii) 53Bi-32Pb-15Sn with a melting point of 95°C (203°F), (vi) 54Bi-26Sn-20Cd with a melting point of 102.5°C (217°F); and (v) 50Sn-18Cd-32Pb with a melting point of 142°C (288°F). *Eutectic* alloys are used for fuses, fire sprinklers, furnace controls, etc.

Eutectic Alloy. Trade name of Johnson Matthey plc (UK) for a brazing alloy containing 71% silver and 29% copper. It has a melting point of 778°C (1432°F), and a brazing range of 778-900°C (1432-1650°F).

eutectic alloy. An alloy containing two or more pure metals in such proportions that the melting point of the alloy is lower than the melting point of any of the pure metals in the alloy at which it goes through a so-called "invariant reaction" in which the solid constituents with different compositions transform to a single-phase liquid with eutectic composition. It melts at one definite temperature, the so-called "eutectic temperature." For example, an alloy containing 11.1% antimony and 88.9% lead at 252°C (486°F) is eutectic. See also hypereutectic alloy; hypoeutectic alloy.

eutectic cast iron. Cast iron that contains 4.3% carbon, and has a microstructure consisting of *ledeburite* whose gamma crystals exist in the form of dendrites or globules.

eutectic solder. A solder containing two or more pure metals in such proportion that the melting point of the solder is lower than the melting point of any of the pure metals in the solder. It melts at one definite temperature, the so-called "eutectic temperature," while all other solders containing the same metals as the eutectic solder, but at other composition ratios, melt over a range of temperatures (pasty range) from solidus to liquidus. Examples of eutectic solders include 63Sn-37Pb, 96.5Sn-3.5Ag, 94.5Pb-5.5Ag, 52In-48Sn, and 80Au-20Sn.

eutectoid steel. A steel that contains 0.77% carbon and has a microstructure consisting of alternating layers (lamellae) of *alpha ferrite* and *cementite* (iron carbide). This structure is referred to as "pearlite" since it resembles mother-of-pearl in appearance. Hypoeutectoid steels have less than 0.77% carbon, while hypereutectoid steels have more. Also known as *pearlitic steel.* See also hypereutectoid steel; hypoeutectoid steel.

Eutector. Trade name of Eutectic Corporation (USA) for brazing and welding fluxes.

Eutekt. Trade name of Braun Lötfolien (Germany) for brazing foils and fluxes.

euxenite. A mineral of the columbite group that is usually black, but sometimes has a brownish or greenish tinge. It is composed of yttrium cerium calcium uranium niobium tantalum titanium oxide, $(Y,Ce, Ca,U)-(Nb,Ta,Ti)_2O_6$, and can also be made synthetically. Crystal system, orthorhombic. Density, 4.7-5.1 g/cm³; refractive index, 2.24; hardness, 6.5 Mohs. Occurrence: Japan, Madagascar, Uganda, Norway, Canada (Ontario), USA (Pennsylvania).

Eva. Trade name of Everseal International Sales Company, Inc. (USA) for heat-resisting aluminum coatings.

Evacote. Trademark of International Group, Inc. (Canada) for hot-melt coatings based on petroleum wax compounds with plastic additives.

EVA copolymers. See ethylene-vinyl acetate copolymers.

Eval. Trade name of EVAL Company of America Company (USA) for ethylene-vinyl alcohol copolymer resins and blown films.

The blown copolymer films are biaxially oriented and have a melting point of 181°C (358°F) and excellent gas barrier properties. Used especially for packaging applications.

Evalux. Trade name of Agriplast Srl (Italy) for thermal films composed of a synthetic resin with 14% ethylene-vinyl acetate. They provide high transparency, good infrared-barrier properties, and high visible-light transmission (above 90%). Supplied in three grades, they are used in greenhouses to protect vegetables and flowers against frost and light variations.

E-Van. Trade name of Midvale-Heppenstall Company (USA) for a water-hardening steel containing 0.95% carbon, 0.25% manganese, 0.25% vanadium, 0.3% silicon, and the balance iron. Used for cold-work dies.

Evanite. Trademark of Evanite Fibers Company (USA) for glass fibers.

Evanohm. Trademark of Wilbur B. Driver Company (USA) for a series of nonmagnetic resistance alloys containing 72-75% nickel, 20% chromium, 2.5-3% aluminum, and 2-2.75% copper. Commercially available in the form of foils, wire, and insulated wire, they have a density of 8.1 g/cm³ (0.29 lb/in.³), a melting point of 1340-1390°C (2445-2535°F), a low change in resistance with temperature, good tensile strength, good corrosion resistance, and a maximum service temperature (in air) of 300°C (570°F). Used for precision-wound resistors.

evansite. A colorless to milky white mineral composed of aluminum phosphate hydroxide hexahydrate, $Al_3(PO_4)(OH)_6 \cdot 6H_2O$. It may also contain some iron and uranium.

Evans Peerless. Trade name of Evans Steel Company (USA) for a water-hardening tool steel containing 0.9-1.2% carbon, and the balance iron. Used for dies, jigs, and tools.

Evansteel. Trade name of Chicago Steel Foundry (USA) for a series of abrasion- and wear-resistant steels containing 0.3% carbon, 1.5-2% nickel, 0.75-1% chromium, and the balance iron. They are available in the following grades: (i) *Grade I* for gears, sprockets, and high-pressure valves; (ii) *Grade II* for sheaves, tractor shoes, bucket lips, and dipper teeth; and (iii) *Grade III* with good heat resistance up to about 540°C (1004°F) for pulverizer hammers, guard rails, liner plates, and dipper teeth.

Evapcast. Trade name for austempered ductile iron (ADI) produced by the lost foam casting process.

evaporated coating. See evaporation vacuum coating.

evaporated film. See evaporation vacuum coating.

evaporated rubber latex. Latex with increased rubber concentration due to the partial evaporation of water.

evaporation vacuum coating. A film or coating, usually 0.1-100 μm (0.004-4 mils) thick, produced on a metallic or nonmetallic substrate by evaporating a material under high vacuum and condensing it on an exposed surface. Such films or coatings are used to improve the surface properties of various components. Also known as *evaporated coating; evaporated film; vacuum-evaporated coating; vacuum-metallized coating; vacuum-metallized film.*

Eva-Pox. Trade name of E-Poxy Industries, Inc. (USA) for epoxy adhesives.

Evatane. Trade name of Atofina SA (France) for ethylene-vinyl-acetates with vinyl acetate contents of 12, 25, or 33%. Used for packaging, as adhesives for bonding, and in the manufacture of automotive components, and sporting and leisure goods.

Evazote. (1) Trade name of BXL Limited (UK) for polyethylene resins.

(2) Trademark of Zotefoams plc (UK) for block-type polyolefin foams and foam products.

eveite. A green mineral of the adamite group composed of manganese arsenate hydroxide, $Mn_2AsO_4(OH)$. Crystal system, orthorhombic. Density, 3.67 g/cm³; refractive index, 1.715. Occurrence: Sweden.

Evenglo. Trademark of Koppers Company, Inc. (USA) for a durable lightweight polystyrene used for light fixtures.

evenkite. A colorless, or slightly yellow mineral composed of tetracosane, $C_{24}H_{50}$. Crystal system, orthorhombic. Density, 0.87 g/cm³; refractive index, 1.50. Occurrence: Siberia.

Everard. Trade name of Everard Tap & Die Company (USA) for a water-hardening tool steel containing 1-1.2% carbon, and the balance iron. Used for dies and tools.

Everbrite. Trade name of American Manganese Bronze Company (USA) for a series of white alloys composed of 56.7-65% copper, 29.5-33.5% nickel, 1-8% iron, 0-3% silicon, 3.75% chromium, 0-2.5% manganese, 0-0.25% carbon, and 0-0.35% cobalt. They have good corrosion resistance, good resistance to moisture and steam, good strength, and moderate hardness. Used for valve disks and seat rings, chemical equipment, steam turbine parts, and castings.

Everclad. Trade name of British Steel plc (UK) for a mild steel-based, galvanized, plastic-coated steel product used for roofing applications.

Evercoat. Trade name of International Technical Ceramics Inc. (USA) for refractory coatings used for quick, inexpensive repairs of refractory materials, such as brick, fibers, etc.

Everdur. Trademark of Anaconda Company (USA) for a series of copper-silicon alloys containing about 90.75-98.25% copper, 1.50-4.00% silicon, 0-1.10% manganese, 0-0.40% lead, and 0-7.25% aluminum. Available in cast and wrought form, they have good acid and corrosion resistance, excellent machinability, good fabricating and fusing qualities, and high strength and toughness. The lead-bearing grades are free-cutting. Used for plates for unfired pressure vessels, and for rods, studs, bolts, springs, valve parts, marine fittings, and tanks, seamless tubes for electrical metallic tubing and rigid conduit, switchgear, electrical hardware, screw-machine products, castings, hot forgings, and welding rods.

Everflex. Trademark of W.R. Grace & Company (USA) for polyvinyl acetate copolymer emulsions used for paints, coating resins, paper coatings, and acoustic tile coverings.

Everglaze. (1) Trade name for fabrics, usually cotton-based, made crease-resistant by a special treatment, and having a prominent pattern.

(2) Trademark of W.L. Jackson Manufacturing Company, Inc. (USA) for tank linings.

Everit. Trade name of Thyssen Edelstahlwerke AG (Germany) for a series of coated hardfacing alloys based on high-chromium cast irons and steels.

Everlast. (1) Trademark for room-temperature curing two-component epoxy-resin compounds used to produce protective coatings on metallic or nonmetallic substrates.

(2) Trademark of CAE Fiberglas Limited (Canada) for fiberglass products, such as large-diameter reinforced fiberglass pipe, and process tanks and vessels.

Ever-Lock. Trade name of Reichhold Chemicals, Inc. (USA) for a series of adhesives for automotive applications, e.g., body and door panels, instrument panels, headliners, interior trim, etc. It includes various reactive and general-purpose hot melts, two-part structural adhesives, and thermoformable adhesives.

Everlube. Trade name of E/M Engineered Coating Solutions

(USA) for a bonded solid-film lubricant.

Everon. Trade name of Shipley Ronal (USA) for regenerative electroless nickel.

Evershield. Trade name of E/M Engineered Coating Solutions (USA) for copper, nickel and graphite conductive paints.

Everstone. Trademark of SAFAS Corporation (USA) for a durable, fire-retardant, scratch-resistant solid surface material bonded to a wood core, and usually available in granite- or solid-color sheets with a size of 1.5 × 2.4 m (5 × 8 ft.) and a thickness of 19 mm (0.75 in.). Used furniture and architectural applications.

Evershyne. Trade name of Osborn Steels Limited (UK) for a stainless steel containing 0.2% carbon, 18% chromium, 8% nickel, and the balance iron.

Eversoft. Trade name for a resilient, plasticized acrylic for dental applications.

Evicom. Trade name of EVC for polyvinyl chlorides.

Evilon. Trade name of Dietzel GmbH (Germany) for polyvinyl chlorides.

Evipol. Trade name for stiff, strong unplasticized polyvinyl chloride. It is an amorphous inherently flame-retardant thermoplastic resin supplied in rod, sheet, tube, powder and film form. It has a density of 1.4 g/cm³ (0.05 lb/in.³), low water absorption (0.03-0.4%), good barrier properties and ultraviolet stability, an upper continuous-use temperature of 50°C (122°F), and moderate chemical resistance. Used for building products, containers, and bottles.

Evlan. Trademark of Courtaulds Limited (UK) for a relatively tough viscose rayon fiber that dyes well, readily accepts finishes, and can be blended with various other fibers (e.g., cotton, nylon, silk, etc.). Used especially for the manufacture of carpets.

Evolutia. Trademark of Solutia, Inc. (USA) for an acrylic fiber used for high-pile fabrics, especially wearing apparel.

Evolution. Trademark of Ivoclar Vivadent AG (Liechtenstein) for an extra-hard ceramic dental alloy containing 40.0% gold, 37.9% palladium, 12.5% indium, 6.3% silver, 3.1% copper, and less than 1.0% ruthenium, rhenium and lithium, respectively. It has a white color, a density of 12.6 g/cm³ (0.46 lb/in.³), a melting range of 1155-1235°C (2075-2255°F), a hardness after ceramic firing of 300 Vickers, low elongation, and excellent biocompatibility and resistance to oral conditions. Used for crowns, onlays, posts, and bridges.

Exabite. Trademark of GC America Inc. (USA) for a thixotropic platinum-catalyzed addition-reaction vinyl polysiloxane (VPS) impression material of creme-like consistency. It has high strength, hardness and carvability, and a short setting time. Used in dentistry for occlusal bite registration.

Exact. Trademark of Exxon Chemical Company (USA) for a series of polyethylene-based hexene plastomers supplied as monolayer and coextrusion coatings and cast films. They have a density of 0.9 g/cm³ (0.03 lb/in.³), and a melting point of 89°C (192°F).

Exact-O-Board. Trade name of Manville Corporation (USA) for fiberglass panels used for the insulation of appliances.

Exaduro. Trade name for a hard plaster used in the ceramic industries for case molds.

Exafast. Trademark of GC America Inc. (USA) for a fast-setting, hydrophilic no-slump vinyl polysiloxane (VPS) dental impression material with exceptional dimensional stability, and high resistance to tearing and distortion.

Exaflex. Trademark of GC America Inc. (USA) for a hydrophilic, thixotropic, no-slump, platinum-catalyzed addition-reaction vinyl polysiloxane (VPS) dental impression material with exceptional dimensional stability, high resistance to tearing and distortion, outstanding physical properties, and a very short setting time.

Examix. Trademark of GC America Inc. (USA) for a hydrophilic, thixotropic, no-slump, platinum-catalyzed addition-reaction vinyl polysiloxane (VPS) dental impression material with exceptional dimensional stability, high resistance to tearing and distortion, outstanding physical properties, and a very short setting time.

ExB metal. A high-leaded bronze containing 77% copper, 15% lead, 8% tin, and a trace of phosphorus. Used for bearings and bushings.

Excalibar. (1) Trade name of Copperweld Steel Company (USA) for fine-grained, calcium-treated alloy steel bars developed for machining with carbide tooling. It provides for good chip breakage.

(2) Trademark of Alcoa–Aluminum Company of America for high-strength, wrought aluminum alloys (AA 6000 Series) containing small amounts of magnesium, copper, manganese, silicon, iron, zinc, titanium, and chromium. They have excellent weldability and brazeability, excellent compressive properties, good coatability and machinability, good corrosion resistance, a density of 2.71 g/cm³ (0.098 lb/in.³), and a hardness of 130 Brinell. Used for machine parts, valves, roller-blade components, ABS braking systems, hydraulic components, and munitions.

Excel. (1) Trade name of Degussa-Ney Dental (USA) for a copper-, palladium- and silver-free dental alloy containing 82% gold and 15% platinum. It has a yellow color, a light oxide layer, high yield and bond strength, and a hardness of 220 Vickers. Used for metal-to-porcelain restorations

(2) Trademark of Excel Technologies Inc. (USA) for an extensive series of products for metallographic cutting, grinding, mounting and polishing including coated abrasives, abrasive cut-off wheels, abrasive and polishing powders, polishing compounds, diamond disks, wheels and wafering blades as well as cold and compression mounting systems.

(3) Trade name for a bonding resin used in orthodontics.

(4) Trademark of E.I. DuPont de Nemours & Company (USA) for nylon 6,6 filaments and yarns.

(5) Trade name of The Schindler Company of Canada Limited (Canada) for nylon monofilaments.

Excelclad. Trademark of Graham Products Limited (Canada) for ultraviolet-resistant film finishes for glass fiber-reinforced plastic panels.

Excelco. Trade name for feldspathic dental porcelains.

Excelite. (1) Trade name of Excelite Company (USA) for a cast alloy of 45% cobalt, 32% chromium, 18% tungsten, 2.8% iron, 2% carbon, and 0.2% boron. Used for cutting tools.

(2) Trademark of Graham Products Limited (Canada) for reinforced plastic sheets and panels used in the manufacture of glazing and skylights.

Excello. Trade name of H. Boker & Son (USA) for an oxidation-resistant wrought heating alloy containing 85% nickel, 14% chromium, 0.5% iron, and 0.5% manganese. It has relatively high strength, a maximum service temperature of 1093°C (2000°F), and is used for heating elements and resistance wire.

Excelo. Trade name of Carpenter Technology Corporation (USA) for a shock-resisting tool steel (AISI type S1) containing 0.5% carbon, 0.3% manganese, 0.3% silicon, 1.5% chromium, 2.6%

tungsten, 0.25% vanadium, and the balance iron. It has high hardness and strength, excellent toughness, good wear and abrasion resistance, and good hot hardness. Used for hand and pneumatic chisels, punches, piercers, heading dies, impact tooling, and hot shears.

Excelsior. (1) Trade name of Riverside Metals Corporation (USA) for a resistance alloy containing 55% copper and 45% nickel. It has good heat resistance, and a maximum service temperature of 427°C (800°F). Used for rheostats, resistors, and heating elements.

(2) Trademark of Degussa-Ney Dental (USA) for a fluorescent, opaque universal dental porcelain supplied in 16 shades. Used for base- and precious-metal restorations.

(3) Trade name for a series of carbon blacks.

excelsior. See wood wool.

Excite. Trade name of Ivoclar Vivadent AG (Liechtenstein) for a light-cure, finely filled, acrylate-based, multi-purpose dental adhesive.

Exelac. Trademark of Graham Products Limited (Canada) for ultraviolet-resistant exterior clear coatings.

Exelloy. Trade name of Crane Company (USA) for a series of corrosion-resistant cast and wrought alloys containing up to 0.15% carbon, 0-1% manganese, 0-1% silicon, 0-1% nickel, 11.5-14% chromium, 0-1% molybdenum, and the balance iron. Used for valves and fittings.

exfoliated lamellar nanocomposites. A subclass of lamellar nanocomposites in which the polymer-to-inorganic layer ratio is fixed and the number of intralamellar polymer chains is continuously variable. They have excellent mechanical properties and are suitable for the manufacture of lightweight, mechanically reinforced structures. See also intercalated lamellar nanocomposites; lamellar nanocomposites.

exfoliated vermiculite. Vermiculite expanded by heating at a temperature above 950°C (1742°F). It expands from 6 to 20 times its original size, and extends its crystals into curving, worm-like aggregates of sheets or plates. Used in building construction, as a filler and extender, etc. Also known as *expanded vermiculite*. See also vermiculite.

ExK metal. A high-leaded bronze containing 77% copper, 12.5% lead, 10.5% tin and a of trace phosphorus. Used for bearings and bushings.

Exl-Die. Trade name of Columbia Tool Steel Company (USA) for a nondeforming, oil-hardening cold-work tool steel (AISI type O1) containing 0.95% carbon, 1.3% manganese, 0.5% chromium, 0.5% tungsten, 0.1% vanadium, and the balance iron. It has good machinability, and a low tendency to shrinking and warping. Used for form tools, blanking, engraving, extrusion, forming and trimming dies, broaches, reamers, and taps.

Exlan. Trademark of Japan Exlan Company (Japan) for polyacrylonitrile (PAN) fibers used for padding and stuffing applications.

Exlo. Trade name of Foote Mineral Company (USA) for a silicon-chromium-iron master alloy containing 0-0.025% carbon. Used to introduce silicon and chromium into stainless steel melts.

Exnor. Trademark of Norton Performance Plastics (USA) for polyvinyl chloride foam profiles.

Exoblast. Trademark of Exolon-ESK Company (USA) for tough, hard fused alumina abrasive grains for blast cleaning, deburring, etching and finishing applications.

Exobond. Trademark of Exolon-ESK Company (USA) for brown fused abrasive alumina grains supplied in bulk, and as bonded products, e.g., paper and cloth.

Exocut. Trade name of AL Tech Specialty Steel Corporation (USA)

for a heavy-duty molybdenum-type high-speed steel (AISI type M42) containing 1.09% carbon, 3.75% chromium, 8% cobalt, 9.5% molybdenum, 1.6% tungsten, 1.15% vanadium, 0.2% molybdenum, 0.3% silicon, and the balance iron. It has high hot hardness, and excellent wear resistance. Used for lathe tools, milling cutters, end mills, reamers, broaches, etc., for machining high-temperature alloys and titanium and prehardened alloy steels.

Exofine. Trademark of US Granules Corporation (USA) for fine aluminum granules produced from recovered aluminum foil products. Used for exothermic applications including thermit and exposives manufacture.

Exohard. Trade name of AL Tech Specialty Steel Corporation (USA) for a molybdenum-type high-speed steel (AISI type M47) containing 1.1% carbon, 0.2% manganese, 0.3% silicon, 3.75% chromium, 5% cobalt, 9.5% molybdenum, 1.6% tungsten, 1.25% vanadium, 0.015% phosphorus, 0.02% sulfur, and the balance iron. Used for cutting tools for hard and superhard metals, lathe and planer tools, and broaches.

Exolon. Trade name of Exolon-ESK Company (USA) for a series of alumina and silicon carbide abrasives.

Exolon WP. Trade name of Exolon-ESK Company (USA) for white alumina grit (99.5% pure) supplied in grit sizes from 12-1200 for blasting electronic components and investment casting molds, and for metallurgical lapping applications.

Exon. Trademark of Firestone Plastics Company (USA) for a series of resins, compounds and latexes made by the polymerization of vinyl chloride, or the copolymerization of vinyl chloride with one or more monomers.

Exonite. Trade name of Dover Limited (UK) for cellulose nitrate plastics.

Exo-Sphere. Trade name of Exolon-ESP Company (USA) for spherical abrasive blast cleaning media.

Expamet. Trademark of Expanded Metal Company (USA) for expanded ferrous and nonferrous metal products supplied in the form of beads, sheets and lath.

expandable plastics. Plastics that can be made into flexible or rigid foams with open- or closed-cell structures by mechanical (agitation), chemical (blowing agents), or thermal (heating) means.

expandable polystyrenes. See expanded polystyrenes.

Expandal. Trade name of Latrobe Steel Company (USA) for an austenitic alloy containing 0.6% carbon, 5.7% manganese, 0.2% silicon, 10% nickel, and the balance iron. It has a high coefficient of thermal expansion, and good strength.

expanded aggregate. A lightweight cellular product made by heating a material, such as clay or shale, to a temperature sufficient to effect swelling and bloating. Used in the manufacture of lightweight concrete and cement, and acoustic and thermal insulation.

expanded blast-furnace slag. A lightweight cellular material obtained by treating molten blast-furnace slag with water, high-pressure steam, compressed air, or a combination thereof. It has a honeycomb-like structure, and is supplied in the form of crushed pieces 9.5-12.7 mm (0.375-0.500 in.) in size. It has a density range of 480-800 kg/m³ (30-50 lb/ft³) and is used in the manufacture of lightweight concrete and concrete products, and lightweight, heat-insulating blocks. Also known as *expanded slag; foamed slag; honeycomb slag*. See also blast-furnace slag.

expanded clay. A ceramic product made from ordinary brick clay by grinding, screening, and subsequent heating at about 1480°C (2700°F) to effect a change of ferric oxide (Fe_2O_3) to ferrous oxide (FeO) that liberates carbon dioxide and oxygen, and produces small bubbles which in turn cause the clay to swell or bloat. Also known as *bloating clay*.

expanded ebonite. A closed-cell material based on *ebonite* (hard rubber) and used chiefly for thermal insulation applications. It has very low thermal conductivity, good moisture resistance, good ultraviolet light resistance, and good machinability.

expanded glass. See cellular glass.

expanded metal. A low-carbon sheet steel mesh made by first piercing, stamping, cutting or slitting a steel sheet to a pattern, and then gripping the edges of the sheet, and applying pressure to pull or expand it to up to ten times its original width. The open mesh formed is either of diamond or rectangular shape, and enhances rigidity and strength by setting the strands at a sharp angle to the plane of the sheet. *Expanded metal* may also be made from iron, aluminum, lead, and other metals. Used in reinforced concrete and plaster walls, for vents and grilles, and for screening, guarding and partitioning applications. Abbreviation: XPM.

expanded metal lath. See lath (iii).

expanded plastics. See cellular plastics.

expanded polystyrene foam boards. Insulating boards made by bonding coarse polystyrene beads into rigid plastic foams. They are supplied in high- and low-density grades with the former having higher thermal insulating values and better resistance to moisture. In application, expanded polystyrene boards must be protected from prolonged exposure to sunlight, solvents and some sealants and covered with a fire-resistant material. They can be used for acoustical and thermal building insulation applications. The moisture-resistant high-density grades can also be used for subgrade insulation of foundation wall exteriors. Also known as *bead board*. See also expanded polystyrenes; rigid board insulation.

expanded polystyrenes. A group of flexible and rigid cellular thermoplastic polymers produced by foaming (expanding), injection of inert gases (i.e., blowing agents), such as nitrogen, carbon dioxide or methyl chloride into the melt, or addition of special monomers or volatile solvents that initiate foaming upon heating. Flexible polystyrene foam is available in the form of extruded sheets, and as beads made by treating a polystyrene suspension with pentane. The beads expand from 30-50 times on heating, and are used in automobile radiator sealants. Rigid foams are widely used for building and boat construction, filtration, as fillers in shipping containers, and for thermal insulation applications. Abbreviation: EPS. Also known as *expandable polystyrenes*.

expanded rubber. A cellular rubber composed predominantly of closed (non-interconnecting) cells, and produced directly from a solid starting material. Its stiffness is much higher than that of ordinary foam rubber. Abbreviation: ER.

expanded slag. See expanded blast-furnace slag.

expanded vermiculite. See exfoliated vermiculite.

expanding alloy. An alloy containing 67% lead, 25% tin and 8% bismuth, used for mounting applications and type metals.

expanding cement. See expansive cement.

expanding metal. See expansive metal.

Expandofoam. Trademark of Armstrong Cork Company (USA) for rigid, lightweight urethane foams used as thermal insulation in freezers, refrigerators, etc.

expansive cement. A cement or mixture of cements, usually high in calcium sulfate and alumina, which when mixed with water forms a paste that during setting and initial hardening tends to

significantly increase in volume. Used in concrete to compensate for shrinkage during drying, or induce stress in post-tensioning. Also known as *expanding cement; high-expansion cement*.

expansive-cement concrete. A concrete, mortar or grout with expansive cement as the matrix material.

expansive clay. A clay, such as sodium bentonite, that is composed chiefly of the mineral *montmorillonite*, and has a high capacity to expand or swell in water.

expansive metal. An alloy, usually containing antimony, bismuth, etc., in combination with lead, tin, etc., that expands when cooled from the molten state. Examples include 66.7Sb-33.3Bi and 75Pb-16.7Sb-8.3Bi alloys. Used for plugging holes or cracks in metals and castings, and in the manufacture of accurate castings (e.g., type metals). Also known as *expanding metal*.

exploded nanopowders. Highly reactive nano-sized particles of aluminum, copper, iron, nickel, tin, tungsten, zirconium and other metals produced by the electro-explosion technique. Used in rocket fuels, lubricants and protective coatings, and for other applications. Also known as *electro-exploded nanopowders*. See also nanopowders.

Exporit. Trademark of Gruenzweig + Hartmann und Glasfaser AG (Germany) for polystyrene foams and foam products.

Export. Trade name of Paul Bergsoe & Son (Denmark) for a shock-resistant alloy containing 92% tin, 3.9% antimony, 3.9% copper, and 0.2% nickel. It has a melting range of 227-316°C (440-600°F), and is used for castings, engine bearings, and bushings.

exposed finish tile. A tile or hollow building block that has a combed, roughened, or smooth surface which may be painted or left exposed after installation.

Express. (1) Trademark of E.B. Eddy Forest Products Limited (Canada) for manila paper.

(2) Trademark of 3M Espe Dental (USA) for a non-slumping addition-polymerizing vinyl polysiloxane (VPS) dental impression material supplied in light and regular body grades.

EX steels. A group of experimental alloy steels developed to either minimize the nickel content, or improve a particular attribute of a standard grade. They are indicated in the AISI-SAE designation scheme by the prefix "EX-" followed by a number, e.g. EX-40 or EX-53.

Ex-Ten. Trade name of US Steel Corporation for a series of high-strength low-alloy (HSLA) structural steels supplied in bar, sheet, strip, plate and structural shape forms.

Extend-A-Bond. Trade name for a bonding resin used in orthodontics.

extended pigment. An organic pigment diluted with an extender, such as barite, celestite, clay, diatomaceous earth, dolomite, gypsum, quartz, etc.

extender. (1) A pigment, usually an inert mineral substance, used to extend or increase the bulk of paint and reduce its cost per unit volume. It provides very little hiding power, but may be useful in stabilizing suspensions, increasing moisture resistance, improving flow or brushing qualities, lowering gloss, and providing other desirable qualities. Extender pigments include asbestos, blanc fixe, diatomaceous earth, gypsum, mica, talc, tripoli, and vermiculite. Also known as *auxiliary pigment; extender pigment*.

(2) An inexpensive, inactive or inert material added to a polymer or resin formulation to increase its volume.

(3) An additive whose only function is to reduce cost.

(4) A somewhat adhesive material introduced into an ad-

hesive to minimize the need for primary binder per unit area.

extender pigment. See extender (1).

Extendo-Die. Trade name of Carpenter Technology Corporation (USA) for an air-hardening hot-work tool steel (AISI type H13) containing about 0.45% carbon, 1% silicon, 0.45% manganese, 6% chromium, 2% molybdenum, 0.8% vanadium, and the balance iron. It has a high hot hardness and toughness, and long service life. Used for extrusion dies, and mandrels.

Extendosphere. Trade name of PQ Corporation (USA) for ceramic spheres used as additives in plastic compounds.

exterior-type particleboard. Particleboard treated with special moisture- and heat-resistant binders, generally phenolic resins. It is used for exterior applications, but is not waterproof. It is supplied in two density grades: (i) *High-density* with at least 800 kg/m³ (50 lb/ft³); and (ii) *Medium-density* with between 593 and 800 kg/m³ (37 and 50 lb/ft³). See also interior-type particleboard; particleboard.

exterior-type plywood. A waterproof plywood manufactured from selected veneers by bonding with a glue that is resistant to weathering, low and high temperatures, microorganisms, boiling water, steam and dry heat. It is suitable for use outdoors, in boatbuilding, for marine structures, and in damp places. There are six grades of exterior-type plywood: (i) *Grade A-A:* Sanded plywood panels with A-grade veneer on face and back, and A-grade veneer on inner plies. Used where good appearance of both sides is called for, e.g., built-ins, cabinets, shipping containers, tanks, ducts, signs, boats, and fences; (ii) *Grade A-B:* Sanded plywood panels with A-grade veneer on face and B-grade veneer on back. Used where good appearance of one side is important; (iii) *Grade A-C:* Plywood panels with A-grade veneer on face, C-grade veneer on back, and usually C-grade veneer on inner plies. Used where good appearance of only one side is called for, e.g., fences, soffits, structural uses, boxcar and truck linings, farm buildings, tanks, trays, commercial refrigerators, building exteriors; (iv) *Grade B-B:* Plywood panels with B-grade veneer on both surfaces used as paintable utility panels; (v) *Grade B-C:* Plywood panels with B-grade veneer on face, and C-grade veneer on back. Frequently used for farm service, work buildings, boxcar and truck linings, containers, tanks, agricultural equipment, and as a base for exterior coatings; and (vi) *Grade C-C (plugged):* Touch-sanded plywood panels with improved (plugged) C-grade veneer on face, and regular C-grade veneer on inner plies and back. Frequently used for severe moisture conditions, e.g., as underlayment over structural subfloor, in refrigerated or controlled atmosphere storage rooms, and for pallet fruit bins, tanks, boxcar and truck linings, and open soffits. See also interior-type plywood; plywood; veneer.

Externit. Trademark of Niagara Protective Coatings (Canada) for protective coatings for masonry surfaces.

Extir. Trademark of Enichem SpA (Italy) for an extensive series of expandable polystyrenes supplied as standard, fast-molding, flame-retardant and low-foaming-agent grades.

Extoral VLC Reline. Trade name for a dual-cure denture resin.

Extra. (1) Trade name of CCS Braeburn Alloy Steel (USA) for a water-hardening tool steel (AISI type W1) containing 0.8-1.2% carbon, 0.25% manganese, 0.25% silicon, and the balance iron.

(2) Trade name of Carpenter Technology Corporation (USA) for a water-hardening tool steel containing 0.6-1.2% carbon, and the balance iron. Used for drills, taps, reamers, and dies.

Extrad. Trade name of Midvale-Heppenstall Company (USA)

for an air-hardening steel containing 0.35% carbon, 1% silicon, 5% chromium, 1.5% tungsten, 1.65% molybdenum, 0.2% vanadium, and the balance iron. Used for hot-forming dies.

extra-duty glazed tile. A ceramic floor or wall tile, usually white or colored, with a glaze rendering it sufficiently durable provided abrasion, impact and wear are not excessive.

extra-high-leaded brass. A wrought brass containing 63.0% copper, 34.5% zinc and 2.5% lead. It has excellent machinability, good corrosion resistance, and good to moderate cold workability. Used for gears, wheels, channel plate, clock and watch backs, and clock hardware.

Extra-L. Trade name of Teledyne Vasco (USA) for water-hardening tool steels (AISI type W2) containing 0.6-1.4% carbon, 0.1-0.4% silicon, 0.1-0.4% manganese, 0.18-0.22% vanadium, and the balance iron. They have excellent machinability, shallow depths of hardening, and fine-grained microstructures. Used for punches, chuck jaws, keys, caulking tools, blanking, cold-heading and forging dies, cold chisels and cutters, battering and beading tools, stamps, hot rivet sets, shear blades, mining drills, wire cutters, lathe centers, arbors, mandrels, expander rolls, thread cutters, broaches, and countersinks.

Extra Nap Superior. Trade name of Vulcan Steel & Tool Company, Limited (UK) for a cobalt-tungsten tool steel containing 0.8% carbon, 4.5% chromium, 0.5% molybdenum, 20% tungsten, 1.25% vanadium, 10% cobalt, and the balance iron. It has high hot hardness and excellent wear resistance, and is used for lathe and planer tools, boring tools, single-point cutting tools, and tool bits.

Extrard. Trade name of Midvale-Heppenstall Company (USA) for an air- or oil-hardening tool steel (AISI type H12) containing 0.35% carbon, 5% chromium, 1.4% molybdenum, 1.25% tungsten, 0.2% vanadium, 1% silicon, and the balance iron. Used for hot-work tools, and heat-resistant parts.

Extra Superior Spur. Trade name of T. Turton & Sons Limited (UK) for a cobalt-tungsten high-speed steel containing 0.72% carbon, 4.5% chromium, 18.5% tungsten, 3.7% cobalt, 1.5% vanadium, and the balance iron. It has excellent red hardness, and excellent wear resistance. Used for cutting tools.

Extra Tough and Hard. Trade name of Vereinigte Edelstahlwerke (Austria) for a series of water-hardening, general-purpose tool and die steels containing 0.9-1.1% carbon, 0-0.15% vanadium, and the balance iron.

Extra Triple Griffin. Trade name of Darwins Alloy Castings (UK) for a tool steel containing 1.5-1.7% carbon, 0.35% manganese, 0.2-0.4% chromium, 6% tungsten, and the balance iron. It has high surface hardness and high core ductility and toughness. Used for milling cutters and tube drawing dies.

Extra Vanadium. Trade name of Vulcan Steel & Tool Company Limited (UK) for a tungsten-type high-speed steel containing 0.7% carbon, 4% chromium, 18% tungsten, 1.25% vanadium, and the balance iron. Used for high-speed cutting tools.

Extrema. Trade name for rayon fibers and yarns used for textile fabrics.

Extren. Trademark of Strongwell Company (USA) for strong, lightweight, corrosion-resistant, electrically and thermally nonconductive composites consisting of thermosetting polyester or vinyl ester matrix resins with fiberglass reinforcements. They are supplied as plates and structural shapes in three series: (i) *Series 500* based on UV-stabilized isophthalic polyester resin and available in olive green colors; (ii) *Series 525* based on a fire-retardant, UV-stabilized isophthalic polyester, and supplied in standard slate-gray colors; and (iii) *Series 625* based on UV-stabilized, fire-retardant and highly corrosion-resistant vinyl ester resin matrices and supplied in beige colors.

extrinsic ionic conductor. An electrical conductor in which conduction results from the influence of impurity atoms. See also intrinsic conductor.

extrinsic semiconductor. A semiconductor crystal to which small, controlled amounts of donor or acceptor impurities (foreign atoms) or alloying elements have been intentionally added to modify its electrical conductivity by introducing excess electrons or holes. Also known as *doped semiconductor; impurity semiconductor.* See also dopant (4); intrinsic semiconductor.

extrinsic sol. A colloidal dispersion whose stability is ascribed to charges on the surfaces of the colloidal particles.

Extrudal. Trade name of Aluminum Industrie AG (Switzerland) for a series of wrought aluminum-silicon-magnesium and aluminum-silicon-magnesium-copper alloys.

Extrudawood. Trade name of CPI Plastics Group Limited (USA) for an extruded thermoplastic composite produced in various shapes, and wood colors and textures. It combines the look and feel of natural wood (e.g., color, wood-grain effect, etc.) with the durability and flexibility of plastics. Used for interior and exterior building applications.

Extrude. Trade name of Coltene-Whaledent (USA) for a range of addition-polymerizing silicone elastomers used for dental impressions.

extruded ceramics. Ceramic products of varying shape and cross-sectional areas, such as blocks, bricks, pipes, and tiles, produced from stiff plastic ceramic masses by forcing them through the orifice of a die of the desired cross section, followed by firing or sintering at high temperatures. Also known as *extrusions.*

extruded metals. Metal products of varying shape and cross-sectional areas, such as bars, rods, tubing and structural shapes, produced from billets or ingots by forcing them to flow plastically through and take on the shape of the orifice of a die. Steel as well as copper, lead, magnesium, tin, titanium, zinc, and their alloys are commonly supplied in this form. Also known as *extrusions.*

extruded fibers. Synthetic fibers that consist of continuous filaments and have been produced by forcing molten or dissolved fiber-forming substances (e.g., polymers) through minute holes of a die or spinneret.

extruded metal powder. (1) A metal powder that has been given a desired shape by forcing through the orifice of a die of the appropiate shape.

(2) A metal powder that has been formed by first enclosing into a soft metal sheathing and then forcing it through a die orifice.

extruded particleboard. Particleboard made by combining wood particles with a bonding agent into a mass, and forcing the latter through a heated die with the applied pressure parallel to the faces and in the extrusion direction. See also particlebeard.

extruded plastics. Plastics that have been made by compacting and melting solid particles, e.g., pellets or powder, and forcing the melt through the orifice of a die to form products of varying size and shape including boards, sheets, pipes, and profiles. Many plastics can be processed by extrusion including polyether sulfones, polyvinyl chlorides, polystyrenes and styrene-maleic anhydrides. Also known as *extrusions.*

extruded polystyrene. Polystyrene foam made by an extrusion process, i.e., by forcing molten polystyrene through the orifice of a die. It is available in many shapes and sizes including pipes, blocks, sheets and boards.

extruded polystyrene foam boards. Rigid extruded polystyrene foam products in board form having numerous fine, closed cells containing a mixture of air and gases (e.g., fluorocarbons). They are supplied in many sizes in high- and low-density grades with the former having a higher thermal insulating value. Extruded polystyrene foam boards are strong and lightweight, but must be protected from prolonged exposure to sunlight and solvents, and are usually not fire-resistant. Used for above- and below-grade acoustical and thermal insulation of buildings and structures. See also extruded polystyrene; extruded plastics.

Extrudex. Trade name of BIP Plastics (UK) for extruded polyvinyl chloride and other vinyl products.

Extrudicote. Trade name of Dextrex Corporation (USA) for a phosphate coating for cold forming and cold extrusion of steel.

extrusion brass. A wrought brass containing about 60% copper, 40% zinc, and small amounts of lead. It has good formability and machinability, and high elongation. Used for architectural extrusions, building products, profiles, etc.

extrusion coating. (1) The plastic coating resulting when a thin film of molten synthetic resin is extruded and pressed into or onto a substrates to which it firmly adheres.

(2) The product resulting from an extrusion process in which a molten material adheres firmly to the surface of another solid material forming a continuous coating upon cooling. Used for flexible barrier coatings.

extrusion metals. Wrought metals or alloys, such as steel and aluminum, copper, lead, magnesium, tin, titanium and zinc and their alloys, suitable for processing by hot or cold extrusion.

extrusions. See extruded ceramics; extruded metals; extruded plastics.

Exuro. Trade name of Schott DESAG AG (Germany) for infra-red-absorbing sheet glass.

Exxtral. Trade name of Exxon Chemical Company (USA) for a lightweight, thermoplastic polyolefin supplied in varying degrees of stiffness. It eliminates fogging, and can be used as a replacement for polyvinyl chloride, e.g., in the automotive industry for instrument and door panels, body cladding, bumper fascia and rocker panels.

Eybrel. Trade name for rayon fibers and yarns.

eyelet brass. A high-ductility brass containing 65-68% copper and 32-35% zinc. It has excellent cold workability, and is used for eyelets, wire products, and drawn shells.

eylettersite. A cream-white mineral of the alunite group composed of thorium aluminum phosphate hydroxide, $(Th,Pb)Al_3(PO_4,SiO_4)_2(OH)_6$. Crystal system, rhombohedral (hexagonal). Density, 3.38 g/cm^3; refractive index, 1.61. Occurrence: Zaire.

Eymyd. Trademark of Ethyl Corporation (USA) for a wide range of synthetic resins supplied in solid, solution, emulsion and dispersion form. Also included under this trademark are pre-pregs with thermoplastic polyimide matrices and reinforcements based on various fibers, e.g., carbon, quartz, etc.

Eyrite. Trade name for a bentonite clay.

EZ Cast. (1) Trade name of National Refractories & Minerals Corporation (USA) for self-flowable castable refractories.

(2) Trade name of Aremco Products, Inc. (USA) for plastic molding compounds.

ezcurrite. (1) A colorless mineral composed of sodium borate tetrahydrate, $Na_4(B_5O_7(OH)_3)_2 \cdot 4H_2O$. Crystal system, triclinic. Density, 2.05 g/cm^3. Occurrence: Argentina.

(2) A colorless mineral composed of sodium borate heptahydrate, $Na_4B_{10}O_{17} \cdot 7H_2O$. Crystal system, triclinic. Density, 2.05

g/cm³; refractive index, 1.507. Occurrence: Argentina.

Ezda. Trademark of Pasminco Metals Proprietary Limited (UK) for copper-free zinc-alloy die castings.

E-Z-Die. Trade name of Columbia Tool Steel Company (USA) for an air-hardening medium-alloy cold-work tool steel (AISI type A2) containing 1% carbon, 5.25% chromium, 1.1% molybdenum, 0.2% vanadium, and the balance iron. It has a great depth of hardening, good nondeforming properties, and good wear resistance and toughness. Used for blanking, coining, forming and trimming dies.

E-Z-Drill. Trade name of Sandmeyer Steel Company (USA) for a free-machining austenitic stainless steel plate supplied in AISI types 304, 304H, 304L, 316 and 316L. The improvement of machinability is obtained by enhanced melting practices and ladle metallurgy allowing for higher speeds when drilling or reaming holes with high-speed tools.

Ezeform. (1) Trademark of Atlas Stainless Steel (Canada) for austenitic stainless steels of the 18% chromium and 8% nickel type.

(2) Trade name for a high-strength stretch-resistant thermoplastic used in the manufacture of protective, supportive and corrective splints.

E-Z Eye. Trade name of Libbey-Owens-Ford Company (USA) for bluish-green plate and sheet glass.

E-Z Fill. Trade name of International Technical Ceramics Inc. (USA) for refractory patches used for application to refractory materials, such as brick, furnace linings, fibers, etc., to eliminate rebricking.

E-Z Gold. Trade name Ivoclar North America, Williams Alloys Division (USA) for a modified version of *Goldent* pure gold used in restorative dentistry for repairing crowns, pit and fissure cavities, endodontic access openings, etc.

EZ-Lam Epoxy. Trade name of Aerospace Composite Products (USA) for finishing, laminating and tooling resins and vacuum bagging systems.

Ezolith. Trademark of Emil Zorn AG (Germany) for polystyrene foams and foam products.

eztlite. A blood-red mineral composed of iron lead tellurium oxide hydroxide octahydrate, $Fe_6Pb_2(TeO_3)_3TeO_6(OH)_{10}\cdot4H_2O$. Crystal system, monoclinic. Density, 4.5-4.6 g/cm³. Occurrence: Mexico.

F

Fabceram 45. Trade name of Hottec (USA) for a gray-colored, machinable ceramic with a maximum service temperature of about 600°C (1110°F) used for kiln and oven fixtures, glass, plastic and rubber molds, insulators, standoffs, guides, and wear parts.

Fabelmat. Trademark of Fabelta Fibres NV (Belgium) for rayon fibers and fabrics.

Fabelnyl. Trademark of Fabelta Fibres NV (Belgium) for nylon 6,6 fibers and filament yarns.

Fabelta. Trademark of Fabelta Fibres NV (Belgium) for rayon and nylon fibers and yarns used for textile fabrics including clothing.

fabianite. A colorless mineral composed of calcium borate hydroxide, $CaB_3O_5(OH)$. Crystal system, monoclinic. Density, 2.77 g/cm^3. Occurrence: Germany.

Fabis. Trade name of Sanderson Kayser Limited (UK) for water-hardening plain-carbon tool steels containing 0.6-1.5% carbon. Used for punches, chisels, shear blades, and dies.

Fablon. Trade name of Commercial Plastics (UK) for polyvinyl chlorides and other vinyl plastics.

Fabmat. Trademark of Fiber Glass Industries Inc. (USA) for fiberglass fabrics.

Fabmet. Trademark of James Fibre-Manufacturing Company Limited (Canada) for a combined chopped strand glass mat and cloth reinforcements.

Fabrene. Trademark of Fabrene Inc. (Canada) for woven polyolefin fabrics.

Fabrialloy. Trade name of Industrial Overlay Metals Corporation (USA) for steel electrodes containing 0.5-0.6% carbon, and the balance iron. Used for shielded metal-arc welding applications.

fabric-base laminate. A composite made by bonding a woven cloth of cotton, fiberglass or synthetic fibers with a synthetic resin under heat and pressure. Used chiefly for electrical insulating applications. See also laminated fabrics.

fabric-reinforced prepreg. A form of fiber-reinforced resin in which the reinforcement is a planar structure produced by interlacing (e.g., weaving, or chemical bonding) of fibers of textile, glass, carbon, graphite, and/or aramid.

fabrics. (1) Planar structures produced from yarns, fibers or filaments by weaving, braiding, knitting, pressing, or other processes. Sometimes nonwoven fabrics are included in this term. Also known as *textile fabrics*.

(2) A knitted, woven, or pressed cloth, e.g., canvas, felt, flannel, linen or velvet. Also known as *textile fabrics*.

Fabricut. Trademark of 3M Company (USA) for coated alumina and silicon carbide abrasives.

Fabrifil. Trade name for cotton fibers, or fabrics chopped into short lengths, and used as fillers in molding compounds for reinforced plastics.

Fabriflex. Trademark of Tredegar Corporation (USA) for non-woven fabrics composed of synthetics laminated to plastic or other films.

Fabrikoid. Trademark of E.I. DuPont de Nemours & Company (USA) for waterproof, washable textile fabrics (often made of cotton) with cellulose nitrate coating. It is made in various grades, colors, widths and grains, and can have the appearance and feel of any desired natural leather. Used as a replacement for rubber and leather in automobile seating, upholstery, book covers, etc.

Fabri Kraft. Trade name of Bradford Company (USA) for fabric-coated fiberboard used for packaging applications.

Fabrolite. Trade name of British Thomson-Houston Company (UK) for phenol resins and plastics.

Fabtuf. Trademark of Hobart Welding Products (USA) for abrasion- and wear-resistant steel hardfacing alloys.

Facade. Trademark of Atlas Turner Inc. (USA) for molded asbestos-cement panels.

face brick. See facing brick.

face brick clay. A good-grade refractory or semirefractory clay that fires at moderate temperatures to a uniform red, white, or buff color, has low absorption, and is free from cracks. Used in the manufacture of facing bricks.

face-finished fabrics. Napped or otherwise treated fabrics, such as chincilla, fleece or broadcloth, with distinctive surface textures.

Facellglass. Trade name of Vitromedia Limited (UK) for cellular glass.

face veneer. A wood veneer, usually of good quality (i.e., A- or B-grade), used for exposed surfaces of exterior- and interior-type plywood, and for panels overlaid with wood veneer. Used where good appearance is the main consideration.

Faceweld. Trade name of Lincoln Electric Company (USA) for a series of irons containing 4.0-4.5% carbon, 1-6% manganese, 0.5-1% silicon, 0.25-6% molybdenum, 18-23% chromium, 0.1-0.7% vanadium, and the balance iron. They have good resistance to severe abrasion, and are used for arc-welding and hardfacing electrodes.

facing. A material in fine pulverized form (e.g., a sand mixture), or an emulsion or suspension of several materials applied to a core or mold face to improve smoothness, facilitate the removal of the casting, and improve the surface of the casting. Also known as *foundry facing*.

facing brick. A building brick, often with a smooth, fine, medium or coarse surface texture, designed for use on the exterior or facing of a structure or wall, and made from selected clays, and in various sizes and colors. Also known as *face brick*.

facing clinker. A special clinker brick that due to its good appearance and surface texture is suitable for use on the exterior of a building.

facing leather. (1) A smooth, lightweight calfskin, lambskin, or skiver used for the interiors of wallets, billfolds, and other folding cases.

(2) Any lightweight leather used for binding the edges of shoes, or for facing seams of leather goods.

facing silk. A fine lustrous silk fabric used exclusively for facing, e.g., the lapels of coats and jackets.

facing stone. A cut stone, such as ashlar or dimension stone, usually square or rectangular in shape, used for facing building walls.

facing tile. Tile whose faces are to be left exposed in interior and exterior walls, partitions, and the like.

faconne. A term used to describe textile fabrics, usually of the jacquard type, having a design of small scattered motifs. Also known as *faconné*.

factice. See vulcanized oil.

factory-finished boards. Boards with surface finishes, such as liquid or powder coatings or overlays, applied at the factory or mill, and ready for use at the job site.

factory lumber. (1) Hardwood lumber intended to be cut up for use in further manufacture. The National Hardwood Lumber Association (NHLA) specifies the following grades: (i) *Firsts:* Top of the line lumber for high-grade cabinetwork which should be over 90% clear on both sides; (ii) *Seconds:* Good quality lumber for most cabinetwork which should be at least 80% clear on both sides; (iii) *Firsts and Seconds* (FAS): A combination of "Firsts" and "Seconds" which should contain a minimum of 20% "Firsts;" (iv) *Common No. 1:* Lumber which is frequently used for interior and less demanding cabinetwork and, one side of which should be at least 65% clear; and (v) *Common No.2:* Lumber which is used for paneling and flooring, and one side of which should be about 50% clear.

(2) Softwood lumber used by millwork plants in the fabrication of windows, doors, moldings, and other trim items.

factory-primed boards. Boards that have a primer applied at the factory or mill, but require subsequent finishing at the job site.

factory sole leather. A tanned and finished sole leather that has high compressibility and flexibility and can be readily processed with shoemaking machinery.

Factrolite. Trade name of C-E Glass (USA) for patterned glass with 900 prisms per square inch (about 140 prisms per square centimeter).

Fadeban. Trade name of Shatterprufe Safety Glass Company Limited (South Africa) for laminated glass with an interlayer of vinyl plastic, and coated with a barrier material for blocking out ultraviolet radiation.

Fade-Shield. Trade name of Amerada Glass Company (USA) for laminated glass that screens out over 90% of ultraviolet rays.

Fagersta. Trade name of Fagersta Stainless AB (Sweden) for a series of austenitic, ferritic and martensitic stainless steels, various heat-resisting, mining, machinery, case-hardening steels, a wide range of plain-carbon, cold-work, hot-work, shock-resisting and tungsten- and molybdenum-type high-speed tool steels, and several low-alloy mold steels.

faggot. A bundle of rods or pieces of iron or steel to be heated, forged, and welded into a single bar. Also known as *fagot*.

faggoted iron. An iron or wrought-iron bar made from a faggot. Also known as *fagoted iron*.

fagot. See faggot.

fagoted iron. See faggoted iron.

faheyite. A white, bluish or brownish mineral composed of magnesium manganese beryllium iron phosphate hexahydrate, $(Mn,Mg)Be_2Fe_2(PO_4)_4 \cdot 6H_2O$. Crystal system, hexagonal. Density, 2.66 g/cm³; refractive index, 1.631. Occurrence: Brazil.

fahlore. See tetrahedrite (1).

Fahlun brilliants. A German alloy that contains 60% tin and 40% lead, and is used for ornaments and theater jewelry. Also known as *Fahluner Diamanten*.

Fahluner Diamanten. See Fahlun brilliants.

Fahluner Metall. See Fahlun metal.

Fahlun metal. A white alloy of 40% tin and 60% lead used in making cheap jewelry. Also known as *Fahluner Metall*.

Fahr. Trade name of Fahralloy Company (USA) for a heat-resistant alloy containing iron, nickel and chromium used for heat-resistant parts, e.g., furnace components.

Fahralloy. Trade name of Fahralloy Company (USA) for a series of casting alloys including several cast heat-resisting steels of the iron-chromium, iron-chromium-nickel and nickel-iron-chromium type, and numerous cast corrosion-resistant steels of the iron-chromium-nickel and iron-nickel-chromium type. Depending on the particular composition and properties, they are used for centrifugally cast tubes, static and shell molded components for steel mills, furnace and heat-treating equipment and parts, and petrochemical equipment and parts, also for valves, and pump bodies.

Fahrenwald gold alloy. A British white gold that contains 60-90% gold and 10-40% palladium, has high acid resistance, and is used for chemical equipment.

Fahrig metal. A British antifriction metal that contains 90% tin and 10% copper, and is used for bearings.

Fahrite. Trade name of Teledyne Ohiocast (USA) for a series of corrosion- and heat-resistant nickel-chromium cast steels.

Fahry's alloy. A bearing alloy composed of 90% tin and 10% copper.

faience. A soft, porous fine-quality earthenware or porcelain that may or may not be coated with a transparent, or opaque glaze. Used for figurines, pottery, tile, mosaics, etc. Also known as *faience ware*.

faience tile. A glazed or unglazed earthenware tile having a hand-crafted, decorative appearance.

faience ware. See faience.

faille. A soft, narrowly cross-ribbed fabric made of acetate, rayon, silk or wool yarn, usually in a plain weave. Used especially for dresses and suits.

fairbankite. A colorless mineral composed of lead tellurium oxide, $PbTeO_3$. Crystal system, triclinic. Density, 7.45 g/cm³; refractive index, 2.31. Occurrence: USA (Arizona).

fairchieldite. A colorless mineral of the calcite group composed of potassium calcium carbonate, $K_2Ca(CO_3)_2$. It can also be made synthetically from a mixture of potassium carbonate and calcium carbonate under prescribed conditions. Crystal system, hexagonal. Density, 2.46 g/cm³; refractive index, 1.532.

fairfieldite. A white mineral composed of calcium manganese phosphate dihydrate, $Ca_2(Mn,Fe)(PO_4)_2 \cdot 2H_2O$. Crystal system, monoclinic. Density, 3.08 g/cm³; refractive index, 1.641. Occurrence: USA (Connecticut, Maine).

Fairleys. Trade name of Jas. Fairley & Sons Limited (UK) for a high-speed steel containing 0.7% carbon, 18% tungsten, 4% chromium, 1% vanadium, and the balance iron. Used for high-speed tools and cutters.

Fairtex. Trade name for metallic fibers used for textile fabrics.

Fakir. Trade name of Fakirstahl Hoffmanns GmbH & Co. KG (Germany) for a series of plain-carbon, cold-work, hot-work and high-speed tool steels.

Falcon. Trade name of Atlas Specialty Steels (Canada) for a series of shock-resisting tool steels (AISI type S1) including the following: (i) *Falcon-4* containing 0.45% carbon, 0.25% manganese, 0.3% silicon, 2% tungsten, 1.5% chromium, 0.25% vanadium, and the balance iron. It has excellent toughness, good wear and abrasion resistance and good hot hardness, and is used for pneumatic tools, chisels, shear blades, punches, trimmers, swage rollers, and beading tools; and (ii) *Falcon-6* containing 0.55% carbon, 0.25% manganese, 0.3% silicon, 2% tungsten, 1.5% chromium, 0.25% vanadium, and the balance iron. It has excellent toughness, good wear and abrasion resistance and good hot hardness, and is used for chisels, shear blades, rotary shears, flaring tools, punches, trimmers, swage rollers, beading tools, tube expanding mandrels, rivet busters, and caulking and scarfing tools.

falcondoite. A whitish green mineral of the sepiolite group composed of magnesium nickel silicate hydroxide hexahydrate, $(Ni,Mg)_4Si_6O_{15}(OH)_2 \cdot 6H_2O$. Crystal system, orthorhombic. Density, 1.90; refractive index, approximately 1.55. Occurrence: Dominican Republic.

falkmanite. A gray mineral of the lillianite group composed of lead antimony sulfide, $Pb_{5.4}Sb_{3.6}S_{11}$. Crystal system, monoclinic. Density, 6.25 g/cm³. Occurrence: Germany.

F Alloy. (1) Trade name of Otis Elevator Company (USA) for an aluminum die-casting alloy containing 20% zinc, 2.5% copper, 0.5% magnesium, 0.5% manganese, and 0.75% silicon. Used for aircraft parts and structural parts.

(2) Trade name for a steel containing 0.06% carbon, 0.4% manganese, 28% nickel, 17.5% cobalt, and the balance iron. Used for magnetic applications.

false acacia. See black locust.

false galena. See sphalerite.

false lapis. See lazurite.

false topaz. See citrine.

false-twist yarn. A yarn, usally of the synthetic type, that has been given certain stretch properties either by highly twisting, heat setting and untwisting, or by continuous folding two yarns together, heat setting and unfolding. Used in the manufacture of fabrics with desired degrees of elasticity.

false verdigris. A term that is used for both *artificial malachite* (basic copper carbonate) and *patina* (basic copper sulfate, or basic copper chloride). Not to be confused with either blue or green verdigris which are based on basic copper acetates and known as "true verdigris." See also basic copper acetate; blue verdigris; copper subacetate; green verdigris; true verdigris.

Fama. Swedish trade name for a series of high-permeability alloys composed of 50-60% iron, 20-24% nickel, 10-13% aluminum, 4-6% copper, and 0-12% cobalt. Used for permanent magnets, and electrical and magnetic equipment.

famatinite. A purplish dark gray mineral of the chalcopyrite group composed of copper antimony sulfide, Cu_3SbS_4. It can also be made synthetically from high-purity copper, antimony and sulfur at elevated temperatures. Crystal system, tetragonal. Density, 4.58 g/cm³; refractive index, 1.641.

Famco. Trade name of American Air Filter Company Inc. (USA) for decorative and surfacing glass fiber mats.

Fan Blades. British trade name for a free-cutting leaded brass containing 61% copper, 37% zinc and 2% lead, and used for fan blades.

fancy. A term used for any woven or textured novelty fabric.

fancy leather. Leather having an appealing grain pattern or distinctive finish, either natural (e.g., alligator leather) or resulting from processing by graining, embossing, ornamenting or decorating with metal foil or powder, and the like.

fancy yarn. A single or plied novelty yarn that either has curls, knots, loops, slubs, etc., worked in at irregular intervals, or has threads of a different nature (e.g., metal) or construction incorporated. Also known as *novelty yarn.*

Fanite. Trade name of Fansteel Metals (USA) for an alloy containing 55% zinc and 45% copper used as a brazing metal.

Fansteel. Trade name of Fansteel Metals (USA) for a series of nonferrous alloys including miscellaneous high-temperature molybdenum, niobium (columbium) and tantalum-base alloys, several copper-base electrical contact materials, sintered tungsten-base materials, etc.

Fantasia. Trademark of Flexsteel Industries Inc. (USA) for nylon fibers used for the manufacture of durable upholstery fabrics.

Fanweld. Trade name of Fansteel Metals (USA) for a wear-resistant alloy composed of tungsten, chromium, cobalt, tantalum carbide and niobium carbide (columbium carbide). It has high red-hardness, and is used for hardfacing and welding rods, and hot-work dies.

Faradex. Trade name of DSM Engineering Plastics (USA) for an extensive series of engineering plastics with antistatic properties making them suitable for use in the electronic, semiconductor and various other industries for electromagnetic interference/electrostatic discharge (EMI/ESD) applications.

Farbond. Trade name of Farboil Company (USA) for powder coatings.

Farbruss. German trade name for a series of carbon blacks used as pigments in coatings, paints, lacquers and varnishes.

Farm-Alloy. Trade name of Resisto-Loy Company, Inc. (USA) for a wear-resistant alloy containing 2% carbon, 12% chromium, 4% molybdenum, 1% vanadium, 3% nickel, 2% manganese, 0.5% boron, and the balance iron. Used for hardfacing electrodes.

Farmface. Trade name of Abex Corporation (USA) for an austenitic, wear-resistant alloy containing 4.5% carbon, 6% manganese, 2% silicon, 30% chromium, and the balance iron. Used for hardfacing rods.

Farrell-Cheek. Trade name of Farrell-Cheek Steel Company (USA) for a series of carbon and low-alloy steels used for machinery, e.g., gears, pinions, shafts, rollers, housings, fasteners, and machine-tool castings.

farringtonite. A colorless, wax-white, or yellow mineral composed of magnesium phosphate, $Mg_3(PO_4)_2$. It can also be made synthetically. Crystal system, monoclinic. Density, 2.74 g/cm³; refractive index, 1.544. Occurrence: Canada (Saskatchewan).

Fasadex. Trade name of Ragnar Bergstedt AB (Sweden) for a cladding glass.

Fasal. Trade name of Forjas Alavesas SA (Spain) for a series of nitriding steels containing 0.4% carbon, 0.6% manganese, 0.25% silicon, 1.5% chromium, 0.25% molybdenum, 1.1% aluminum, and the balance iron. Used for shafts, spindles, gages, etc.

Fasaloy. Trade name of Fansteel Metals (USA) for an extensive series of electrical contact materials including various gold-, silver-, palladium- and platinum-base alloys.

fasciated yarn. A staple fiber yarn made by wrapping fibers over a core usually consisting of parallel fibers.

fassaite. A white mineral of the pyroxene group composed of calcium aluminum silicate, $CaAl_2SiO_6$. Crystal system, monoclinic. Density, 3.43 g/cm³; refractive index, 1.714. Occurrence: USA (Maine). A grass-green variety containing iron and magnesium and having a density of 2.96 g/cm³ and a refractive index of 1.701 is found in the West Indies; and a deep green titanium-bearing variety occurs in Mexico.

Fassonal. German trade name for free-cutting aluminum alloys.

FAST. Trademark of Sinterama Tespiana Associates (Italy) for polybutylene terephthalate fibers and filament yarns.

FasTape. Trade name of Avery Dennison, Specialy Tape Division (USA) for high-temperature, solvent-resistant adhesive tape used for automotive applications.

Fastblast. Trademark of Exolon-ESK Company (USA) for tough, hard fused alumina abrasive grains for blast cleaning, deburring, etching, and finishing applications.

Fastbond. Trademark of 3M Company (USA) for water-based contact adhesives available in various color grades. They provide high strength and can be applied by brushing, rolling, or spraying.

Fastbor Hollow Drill. Trade name of Hoytland Steel Company (USA) for a tough, water hardening steel containing 0.8% carbon, and the balance iron. Used for hollow drills.

Fast Cast. Trade name of Goldenwest Manufacturing Inc. (USA) for liquid urethane casting resins.

Fast-Cut. Trade name of Lukens Steel Company (USA) for free-machining carbon steel plates containing 0.24-0.33% sulfur. Used for rubber and tire molds.

Fastell. Trade name of Fansteel Metals (USA) for an extensive series of sintered electrical contact materials including various silver-tungsten, copper-tungsten, tungsten-silver, tungsten-copper, molybdenum-silver, silver-cadmium oxide, silver-palladium and gold-silver alloys.

Fastemp. Trade name of C-E Glass (USA) for a toughened glass.

Fastex. Trade name of Murex Limited (UK) for a series of low-carbon steels containing 0.06-0.08% carbon, 0.4-0.6% manganese, 0-0.4% silicon, 0-0.05% molybdenum and the balance iron. Used for welding electrodes.

Fast Finishing. Trade name of Carpenter Technology Corporation (USA) for a tool steel containing 1.3% carbon, 3.5% tungsten, and the balance iron. Used for tools, fast-finishing cutters, reamers, and arbors.

Fastwork. Trade name of Darwins Alloy Castings (UK) for a tool steel containing 1.5% carbon, 12.5% chromium, 0.75% molybdenum, 0.25% vanadium, and the balance iron.

fat clay. A clay, such as *ball clay*, having relatively high plasticity and green strength.

fat concrete. Concrete that contains a substantial percentage of mortar.

Fatigue-Proof. Trademark of Stelco Inc. (Canada) for high-strength, cold-drawn steel bars containing 0.4-0.48% carbon, 1.35-1.65% manganese, 0-0.04% phosphorus, 0.24-0.33% sulfur, 0.15-3% silicon, and the balance iron. Used for structural parts, studs, shafts, axles, bolts, gears, and machined parts.

fat lime. Lime that contains 95% or more calcium oxide (CaO), and hydrates very quickly with the evolution of a great quantity of heat. Also known as *high-calcium lime; rich lime.*

fat mortar. Mortar composed of a high quantity of cement. It is sticky and plastic, and adheres to a trowel.

Fat Red. Trade name of Aardvark Clay & Supplies (USA) for a red clay (cone 10) with a rich smooth throwing body.

fatty acid. Any of a group of solid, semisolid or liquid long-chain carboxylic acids contained in, or obtained from animal or vegetable fats and oils. They are classed among the lipids and include substances, such as stearic, palmitic, linoleic, linolenic, eicosapentaenoic and decahexanoic acid. Used in biotechnology and medicine, and in special soaps including heavy-metal soaps, lubricants, detergents and emulsifiers, paints and lacquers (drying oils), candles, etc. Formula: $C_nH_{2n}O_2$.

Faturan. Trademark of NYH (Germany) for modified and unmodified phenol- and cresol-formaldehyde plastics.

faujasite. Colorless or white mineral of the zeolite group composed of sodium calcium aluminum silicate hexahydrate, $(Na_2,-Ca)Al_2Si_4O_{12} \cdot 6H_2O$. Crystal system, cubic. Density, 1.92-2.09 g/cm³; refractive index, 1.48. Occurrence: Central Europe. Used as a zeolite, and for molecular sieves.

faustite. An apple-green mineral of the turquoise group composed of zinc aluminum phosphate hydroxide tetrahydrate, $(Zn,Cu)-Al_6(PO_4)_4(OH)_8 \cdot 5H_2O$. Crystal system, triclinic. Density, 2.92 g/cm³; refractive index, 1.613. Occurrence: USA (Nevada).

faux fur. A knit or woven pile fabric, usually of acrylic or mod-acrylic fibers, made to imitate animal fur.

faux leather. A textile fabric, usually a polyurethane laminate, that has been made to look like natural leather.

faux linen. An imitation linen made with slubbed yarns consisting of one or more other fibers.

faux shearling. An imitation *shearling* made from yarns consisting of one or more other fibers.

faux silk. A soft, drapey, lustrous imitation silk fabric made from synthetic fibers, such as polyester.

Favicur. Trade name of Favicur SA (Argentina) for a toughened glass.

Favorit. Trade name of Bohler Gesellschaft mbH (Austria) for a high-carbon tool steel containing 0.9% carbon, 1.9% manganese, 0.2% chromium, 0.1% vanadium, and the balance iron.

fayalite. A grayish olive or yellowish-black mineral of the olivine group that is composed of iron silicate, Fe_2SiO_4, and may also contain magnesium or manganese. It can also be made synthetically. Crystal system, orthorhombic. Density, 3.91-4.40 g/cm³; hardness, 6-6.5; refractive index, 1.8642. Occurrence: Greenland, Japan. Used for especially for refractories, cement, and in ceramics. Also known as *iron-olivine.*

F Clay. Trade name of AGS Minéraux SA (France) for dried, ground clays supplied in various particle sizes. They are obtained from kaolinitic raw clays mined in southwestern France. A typical chemical analysis (in wt%) of the dried product is 48% silica (SiO_2), 34.8% alumina (Al_2O_3), 1.5% ferric oxide (Fe_2O_3), 1.5% titania (TiO_2), 0.6% potassium oxide (K_2O), 0.3% lime (CaO), 0.2% magnesia (MgO), 0.1% sodium oxide (Na_2O), and up to 0.07% carbon. Their loss on ignition is about 13%, and their pyrometric cone equivalent (Seger cone) 34 (1750°C, or 3182°F). Used in the manufacture of refractory products.

Feal. Trademark of Feal SpA (Italy) for a neutron-transparent alloy containing 75% iron and 25% aluminum. It has excellent strength at high temperatures, outstanding resistance to oxidation, and is used for nuclear reactor applications.

Featalak. Trade name of Featly Products Limited (UK) for phenolic resins.

Featalite. Trade name of Featly Products Limited (UK) for phenol-formaldehyde plastics.

Featheray. Trade name for rayon fibers and yarns used for textile fabrics.

feather brick. A molded brick cut diagonally from one end or side to the opposite end or side to form a shape of triangular cross section.

feathered tin. A pure tin in granulated form made by quenching molten tin in cold water.

Feather Glass. Trade name of Paramount Glass Manufacturing Company Limited (Japan) for glass fiber thermal and acoustical insulation.

Featherglass. Trade name of Eastern Gypsum Limited (Canada) for glass fiber insulating materials.

Featherlite. (1) Trademark for lightweight aggregate used in the manufacture of concrete.

(2) Trademark of PUPI Enterprises, LLC (USA) for fiberglass products used for electrical applications.

(3) Trademark of Aquashed Technologies, Inc. (USA) for lightweight polyurethane building block.

feather ore. See jamesonite.

Febolit. Trademark of Febolit for flexible polyvinyl chloride film materials.

Fechral. Russian trade name for heat-resistant alloy containing 0.06-0.15% carbon, 17-25% chromium, 4-7% aluminum, and the balance iron. Used for resistance wire.

Fecraloy. (1) Trade name of Wilbur B. Driver Company (USA) for a stainless steel containing 0.2% carbon, 13-15% chromium, 4-5% aluminum, and the balance iron. It has high heat resistance, and is used for rheostats and heating elements.

(2) Trademark of Wilbur B. Driver Company (USA) for a series of resistance alloys containing 22% chromium, 4-5% aluminum, 0-0.3% silicon, 0.1-0.3% yttrium, 0-0.1% zirconium, and the balance iron. Commercially available in the form of foil, rod, sheet and wire, they have a density of 7.22 g/cm^3 (0.26 lb/in.^{3}), a melting range of 800-1490°C (2515-2715°F), high electrical resistivity, a maximum service temperature (in air) of 1100-1300°C (2010-2370°F), excellent resistance to oxidation at elevated temperatures, and good tensile strength. Used for resistances and heating elements.

FeCrAlY coatings. See MCrAlY coatings.

Federalloy. Trade name of The Federal Metal Company (USA) for a lead-free brass casting alloy for plumbing applications. It contains small additions of bismuth (less than 1%), and can be used as a substitute for leaded red brasses (UNS C83600 with 85% copper, 5% tin, 5% lead and 5% zinc) and leaded semired brasses (UNS C84400 with 81% copper, 3% tin, 7% lead and 9% zinc).

Federal High Speed. Trade name of Swedish Iron & Steel Corporation (USA) for a high speed steel containing 0.7% carbon, 4% chromium, 18% tungsten, 1% vanadium, and the balance iron. Used for high-speed cutting tools.

Federal-Mogul. Trade name of Federal-Mogul Corporation (USA) for an extensive series of copper, cadmium, lead and tin-base bearing alloys.

Federaloy. Trade name of Federal-Mogul Corporation (USA) for a series of aluminum-, copper-, tin-, and lead-base bearing alloys.

Federal White. Trademark of Federal White Cement Limited (USA) for Portland cement.

Federated. Trade name of American Smelting & Refining Company (USA) for a series of strong aluminum-copper, aluminum-silicon and aluminum-zinc casting alloys. Used for pistons, gear cases, brake shoes, cylinder blocks, cylinder heads, oil pans, propellers, housings, reflectors, hardware, fittings, etc. Also included under this trade name are various lead- and tin-base antifriction alloys for bearing applications.

Fedol. Trade name of Swedish Iron & Steel Corporation (USA) for a water-hardening tool steel containing 1.2% carbon, 0.2% vanadium, and the balance iron. Used for tools and dies.

fedorite. A colorless mineral composed of potassium sodium calcium silicate hydroxide fluoride monohydrate, $(K,Na)_{2.5}(Ca,Na)_7Si_{16}O_{38}(OH,F)_2 \cdot H_2O$. Crystal system, triclinic. Density, 2.58 g/cm^3; refractive index, 1.530. Occurrence: Russian Federation.

fedorovskite. A brownish mineral composed of calcium magnesium manganese borate hydroxide, $Ca_2(Mg,Mn)_2B_4O_7(OH)_6$. Crystal system, orthorhombic. Density, 2.65 g/cm^3; refractive index, 1.632. Occurrence: Russian Federation.

feebly magnetic material. A nonmagnetic substance having a maximum normal magnetic permeability of 1-4.

Feedex. Trademark of Foseco Minsep NV (Netherlands) for moldable exothermic antipiping compounds used in making iron, steel, aluminum, copper, magnesium and nickel castings.

feed materials. Refined fuel material (e.g., uranium or thorium) introduced into a nuclear reactor to replace used fuel that is removed.

Feedol. Trademark of Foseco Minsep NV (Netherlands) for exothermic hot-topping compounds used in making aluminum, copper, nickel or magnesium castings for application to the surface of the molten metal in risers.

Fefassea. Trade name for nylon fibers and yarns.

feitknechtite. An iron-black to brownish-black mineral composed of manganese oxide hydroxide, β-MnO(OH). It can also be made synthetically. Crystal system, tetragonal. Density, 3.80 g/cm^3.

Felax. Spanish trade name for a series of silicon-manganese spring steels containing 0.46-0.6% carbon, 0.7-0.8% manganese, 1.75% silicon, and the balance iron.

feldspar. A group of crystalline, rock-forming minerals consisting of aluminum silicate and potassium, sodium or calcium, and sometimes barium. The general formula is (K,Na,Ca,Ba)-$(Al,Si)_4O_8$. Members of this group include albite, anorthite, celsian, danburite, labradorite, microcline, orthoclase, paracelsian and sanidine. Density range, 2.5-3.4 g/cm^3; melting range, 1100-1532°C (2010-2790°F); hardness range, 6-6.5 Mohs. Occurrence: One of the most common and widespread minerals. Used in porcelain, tile, dinnerware, other whiteware bodies, glass, glazes, porcelain enamels, and similar ceramic products generally as a flux. Also spelled "felspar." See also calcium feldspar; potassium feldspar; sodium feldspar.

feldspathoids. A group of rock-forming minerals, such as leucite, lazurite, nepheline and sodalite, containing a lower percentage of silica (SiO_2) than the feldspars.

Felex. Trade name of The Feldspar Corporation (USA) for a series of inert, white mineral fillers consisting of 68.4% silicon dioxide (SiO_2), 18.6% aluminum oxide (Al_2O_3), 7.2% sodium oxide (Na_2O), 3.8% potassium oxide (K_2O), 1.9% calcium oxide (CaO), and small additions of ferric oxide (Fe_2O_3) and magnesium oxide (MgO). Their loss on ignition is about 3.1%. *Felex* fillers have a density of 2.62 g/cm^3, a specific surface area of 1.41-3.60 m^2/g, a hardness of 6.0-6.5 Mohs, a refractive index of 1.53, and excellent abrasion, weathering and chemical resistance. Supplied in several grades with average particle sizes of 4.5-12 μm (177-472 μin.), they are used in paint coatings, adhesives, caulks, sealants, plastics, and elastomers.

felsic tuff. A light-colored *tuff* composed of feldspar and feldspathoid or silica minerals.

felsöbányaite. A yellow-white mineral composed of aluminum sulfate hydroxide pentahydrate, $Al_4SO_4(OH)_{10} \cdot 5H_2O$. Crystal system, hexagonal. Density, 2.35 g/cm^3; refractive index, 1.518. Occurrence: Rumania.

felt. (1) A porous, densely matted, nonwoven fabric made by rolling and pressing together fibers of wool, hair, fur, cotton, rayon, asbestos, glass, etc. It possesses high moisture absorption, softness and resilience. Used for hats, sportswear, table coverings, thermal insulation, as a caulk, as a sound and vibration damper, in roofing, in padding and lining of instrument cases, and in the drying section of fourdrinier papermaking machines. See also wool felt; synthetic fiber felt.

(2) A fulled woven fabric that has been compressed and shrunk by a process involving heat and pressure to obscure the

weave and entangle the fibers.

felted metal. A porous sheet made by interfelting ferrous or nonferrous metal fibers and subsequent compressing and sintering. Used for filtration applications.

Feltex. Trade name for an asphalt-saturated felt used in roofing.

felt fabrics. Woven fabrics made of surface-milled cotton, wool or other fibers and having felt-like surfaces that obscure the weaves.

Feltmetal. (1) Trade name of Brunswick Technetics, ECS (USA) for an acoustic product made by diffusion bonding randomly oriented metal fibers. It is supplied in a number of fiber types and tensile strengths.

(2) Trade name of Technetics Corporation (USA) for an aluminum-fiber sheet material consisting of diffusion-bonded aluminum alloy 6061 (a wrought, heat-treatable aluminum-magnesium-silicon alloy) reinforced with alloy 3003 (a wrought, non-heat-treatable aluminum-manganese-copper alloy) or alloy 6061 perforated sheet. Used for lightweight, acoustical and noise-abatement parts, and aluminum honeycomb structures.

FeltMetal. Trademark of Brunswick Corporation (USA) for fiber-metal materials made by sintering fibers of metals and alloys, such as iron, stainless steels, aluminum, copper, cobalt, nickel, titanium, etc., to produce metallic bonds at the fiber contact points. Available in the form of bars, blocks, discs, plates, rings, rods, sheets and tubing, they have excellent resistance to high temperature and corrosion, good oxidation resistance, and a low weight. Used for mold liners, gas burners, pipe insulation, sound absorbers, turbine engine seals, and electrode substrates.

felt paper. A heavy kraft paper of organic or mineral fibers impregnated with asphalt or tar, or treated with chemical compounds. Used as sheathing tapes on roofs, and on sidewalls for protection against moisture, heat and cold.

felt reusable surface insulation. A thermal insulation material composed of *Nylon* felt blanket, 4 or 8 mm (0.16 or 0.32 in.) thick, with an outer coating of silicone rubber. It has a maximum service temperature of approximately 400°C (752°F), and is used for the upper sidewalls, upper wing surfaces and cargo bay doors of the NASA Space Shuttle. Abbreviation: FRSI.

Feltrosiv. Trade name of Società Italiana Vetro SpA (Italy) for long glass-fiber mats for thermal and acoustical insulating applications.

Felzodox. Trademark of Pigment & Chemical Company Limited (Canada) for French-process zinc oxide.

fenaksite. A light rose mineral composed of potassium sodium iron silicate hydroxide, $(K,Na)_4Fe_2(Si_4O_{10})_2(OH,F)$. Crystal system, monoclinic. Density, 2.74 g/cm^3; refractive index, 1.560. Occurrence: Russian Federation.

Feneshield. Trade name of PPG Industries Inc. (USA) for glass fiber yarn used for the manufacture of drapery fabrics.

Fenicoloy. Trade name of Molecu-Wire Corporation (USA) for an alloy containing 53.8% iron, 29% nickel, 17% cobalt, and 0.2% manganese. It has a maximum service temperature of 450°C (842°F), and is used for resistance wire.

Fenilon. Trade name for polyaramid fibers and fabrics.

Fennolith. Trade name of A. Ahlström Osakeyhtiö (Finland) for glass blocks.

Fenochem. Trade name of Chemiplastica for phenol-formaldehyde resins and plastics.

Fenton's alloy. An English bearing metal containing 80% zinc, 11.5-14% tin, and 6-8.5% copper.

FEP resins. See fluorinated ethylene-propylene resins.

Fer. Trade name of A. Milne & Company (USA) for a water-hardening tool steel containing 0.7-1.2% carbon, and the balance iron. Used for punches, tools, and dies.

Feral. German trade name for a low-carbon steel containing 0.06% carbon, 0.6% manganese, and the balance iron, clad with aluminum alloy. Used for aircraft firewalls.

Feralite. Trade name of Chapman Valve Manufacturing Company (USA) for a malleable cast iron containing 3% carbon, varying percentages of silicon and manganese, and the balance iron. Used for gears, shafts, fittings, and housings.

Feralsi. Trade name of Unexcelled Manufacturing Company (USA) for an alloy of iron, aluminum and silicon. It has high magnetic permeability, and is thus used for magnetic applications.

Feralun. Trade name of American Abrasive Metals Company (USA) for a wear-resistant special iron base with embedded abrasive alumina grains. Used for stairs, platforms, and castings.

Feran. (1) Trade name of Massillon Steel Casting Company (USA) for sheet iron or strip steel coated with aluminum and rolled together to combine the good corrosion resistance of aluminum with the strength of iron. Used in engineering construction.

(2) Trade name of Almetals Company (USA) for a strip material made by roll cladding aluminum foils onto one or both sides of an aluminum-free, low-nitrogen unalloyed deep-drawing steel followed by a recrystallization heat treatment. The final product possesses good corrosion and oxidation resistance, high surface quality, and easy recyclability, and can be processed by various cold-forming operations including stamping, folding, pressing, stretch forming, deep drawing, etc. Used for automotive components, such as cylinder-head gaskets, and heat shields.

(3) Trade name of Hösch Stahl AG (Germany) for a bimetal manufactured by passing strips of aluminum and iron through rolls at a temperature of 350°C (660°F) followed by cold rolling to sheet. Used in the canning industries.

ferberite. A black mineral with submetallic luster belonging to the wolframite group. It is composed of iron tungstate, $FeWO_4$, and contains about 73.6-77.8% tungsten trioxide (WO_3). Crystal system, monoclinic. Density, 7.47-7.61 g/cm^3; hardness, 4.0-4.5 Mohs; refractive index, 2.305. Occurrence: Argentina, Germany, USA (Colorado). Used as a source of tungsten.

Ferbond. Trade name of General Refractories Company (USA) for refractory bonding and patching materials.

Ferchromit. Trade name of Skoda Works National Corporation (Czech Republic) for a stainless steel containing 0.1-1.5% carbon, 22-30% chromium, 0.5-2.5% silicon, and the balance iron. It has excellent oxidation resistance at high temperatures.

ferdisilicite. A dark gray mineral composed of iron disilicide, $FeSi_2$. It can also be made synthetically. Crystal system, tetragonal. Density, 4.99 g/cm^3. Occurrence: Russian Federation.

Ferex. (1) Trade name for a plain-carbon steel containing 0.2% carbon, and the balance iron. Used for coated welding rods.

(2) Trademark for fire-retardant polyurethane foams.

ferghanite. A sulfur-yellow, radioactive mineral composed of uranium vanadate hexahydrate, $U_3(VO_4)_2 \cdot 6H_2O$. Crystal system, orthorhombic. Used as a source of vanadium.

fergusonite. A black, gray, brown, or red to reddish-brown mineral composed of yttrium erbium cerium niobium tantalum oxide, $(Y,Er,Ce)(Nb,Ta)O_4$, and may also contain other rare earths, such as neodymium as well as titanium, uranium, thorium, zirconium, calcium, and/or iron. Crystal system, monoclinic. Density, 5.3-5.8 g/cm³; refractive index, 2.07, hardness, 5.5-6.5 Mohs. Occurrence: Africa, China, Norway, Sweden, USA (North Carolina, South Carolina, Virginia).

Ferholzer. Trade name of Creusot-Loire (France) for a very soft, magnetic material containing 0.005% carbon, the balance being high-purity iron. Used for solenoids, electromagnets, and relays.

Ferimphy. Trade name of Creusot-Loire (France) for an iron-base magnetic alloy used for soft magnets for strong or moderate fields, e.g., in armatures and transformers.

Ferinox. Trade name of Sandvik Steel Company (USA) for an austenitic stainless steel composed of 0.09% carbon, 1.2% silicon, 1.3% manganese, 17% chromium, 8% nickel, 0.7% molybdenum, and the balance iron. It has high fatigue strength, and is used for watch mainsprings.

Fermet. British trade name for a stainless steel containing 0.35% carbon, 18% nickel, 4% chromium, 2.2% manganese, 1% tungsten, 0.3% copper, and the balance iron.

Fermit. (1) Trade name of Nissen & Volk (Germany) for all-purpose jointing and packing compounds, refractory furnace and boiler putties, oil-resistant jointing compounds, and plastic jointing and sealing compounds

(2) Trademark of Ivoclar North America, Inc. (USA) for a light-cure dental filling material.

fermium. A radioactive, metallic element belonging to the actinide series of the Periodic Table. It can be produced in a cyclotron by bombardment of uranium with accelerated oxygen ions, in a nuclear reactor by neutron irradiation of californium, plutonium or einsteinum, or by other nuclear reactions. It is named for Enrico Fermi. Atomic number, 100; atomic weight, 254; half-life, approximately 80 days; isotopes with mass numbers ranging from 253 to 257 are known; trivalent; chemically similar to erbium. Used in nuclear engineering and physics. Symbol: Fm.

fermorite. A pinkish white mineral of the apatite group composed of calcium strontium potassium phosphate arsenate fluoride hydroxide, $(Ca,Sr)_5[(As P)O_4]_3(OH,F)$. Crystal system, hexagonal. Density, 3.52 g/cm³; refractive index, 1.660. Occurrence: India.

Fermo Special. Trade name of British Steel Corporation (UK) for a carbon steel containing 0.2% carbon, and the balance iron, used for case-hardened parts.

fernandinite. A dull green mineral composed of calcium vanadyl vanadate hydrate, $CaO \cdot V_2O_4 \cdot 5V_2O_5 \cdot 14H_2O$. Occurrence: Peru. Used as a vanadium ore.

fernico. An alloy composed of iron, nickel and cobalt, and used for glass-to-metal seals, and for sealing vacuum tubes.

Fernite. Trade name of Allegheny Ludlum Steel (USA) for a series of heat- and/or corrosion-resistant chromium-nickel and nickel-chromium steels.

Ferno. Trade name of Lehigh Steel Corporation (USA) for a tool steel containing 0.7% carbon, 3.75% chromium, 0.55% vanadium, 0.7% molybdenum, and the balance iron. It has a good red hardness, and is used for shear blades, punches, and hot-work dies.

Ferno Extra. Trade name of Lehigh Steel Corporation (USA) for a tough, abrasion-resistant tool steel containing 0.45% carbon, 3.75% chromium, 0.4% manganese, 0.7% vanadium, 12% tungsten, and the balance iron. Used for shear blades, punches, and hot-work dies.

Feroba. Trade name of Edgar Allen Balfour Limited (UK) for a series of permanent magnetic alloys composed of sintered barium ferrite ($BaO \cdot 6Fe_2O_3$). They have a Curie temperature of approximately 450°C (840°F), high coercivity and magnetic energy products, high remanent magnetization, and a maximum service temperature of 400°C (752°F). Used for permanent magnets, and electric and magnetic equipment.

Ferobestos. Trademark of Franklin Fiber–Lamitex Corporation (USA) for asbestos fabrics or felts impregnated with phenolic or melamine resins and sometimes graphite. Supplied in rod, tube and/or sheet forms, they are used machine parts, such as bearings, bushings, gears, rotor blades, pump and compressor parts and electrical components.

Fero-Brass. Trademark of Ferox Coatings Inc. (Canada) for brass and copper coatings.

Fero-Clair. Trademark of Ferox Coatings Inc. (Canada) for polyurethane finishes.

Ferovan. Trade name of Shieldalloy Metallurgical Corporation (USA) for a master alloy containing 42-48% vanadium, 0-8% nickel, 0-7.5% chromium, 0-5.5% manganese, 0-0.1% aluminum, 0-7% silicon, 0-1% carbon, 0-0.4% sulfur, 0-0.4% phosphorus, and the balance iron. Used for the ladle addition of vanadium to steels.

Ferox. Trademark of Ferox Coatings Inc. (Canada) for paints, enamels, and zinc-rich and epoxy coatings.

Fer-Prim. Trademark of Ferox Coatings Inc. (Canada) for a series of zinc-rich coatings.

Ferral. Russian trade name for steel containing up to 0.25% carbon, 12-15% chromium, 3.5-5.5% aluminum, and the balance iron. Used for electrical and heat-resistant elements.

Ferralium. Trade name of Haynes International, Inc. (USA) for cast and wrought ferritic-austenitic duplex stainless steels containing 0.08% carbon, 24-27% chromium, 4.5-6.5% nickel, 2.0-4.0% molybdenum, 1.3-4.0% copper, 1.5% manganese, 0.1-0.2% nitrogen, and small additions of silicon, phosphorus and sulfur, and the balance iron. They are available in the annealed or solution heat-treated condition, either rapidly cooled or hardened. They have good strength, improved corrosion resistance, and good sulfide stress-cracking resistance. Used for applications involving seawater immersion, oil-field applications, etc.

Ferrari cement. Sulfate-resistant cement composed chiefly of dicalcium silicate ($2CaO \cdot SiO_2$), tricalcium silicate ($3CaO \cdot SiO_2$), and tetracalcium aluminoferrate ($4CaO \cdot Al_2O_3 \cdot Fe_2O_3$). Used especially for marine and underwater structures.

ferrarisite. A colorless mineral composed of calcium hydrogen arsenate nonahydrate, $Ca_5H_2(AsO_4)_4 \cdot 9H_2O$. Crystal system, triclinic. Density, 2.63 g/cm³; refractive index, 1.572. Occurrence: France.

ferredoxin. A pigment containing iron to which excited electrons flow from light-stimulated chlorophyll during photosynthesis. It transfers some the electrons to flavoproteins (conjugated protein dehydrogenases that contain the coenzyme flavin), and is used in biochemistry, biotechnology, and has potential applications in optics and optoelectronics.

Ferrene. Trademark of Ferro Corporation (USA) for filled and/or reinforced polyolefins, such as polyethylene, supplied in the form of particulate. They are available in injection moldable and thermoformable grades for the manufacture of electrical appliance housings and other plastic goods.

Ferrex. Trademark of Ferro Corporation (USA) for an extensive range of filled and/or reinforced polypropylenes including mineral-, talc- and calcium carbonate-filled grades, fiberglass-reinforced grades, mica-reinforced grades, and mineral- plus fiberglass-reinforced grades. Used for appliances.

ferric acetylacetonate. A metal alkoxide available in the form of a red-brown crystalline powder or orange-red crystals (97+% pure), and containing about 15.5-15.8% trivalent iron. Density, 5.24 g/cm^3; melting point, 176-184°C (349-363°F); refractive index, 1.33; soluble in water, ethanol and toluene. Used as bonding agent and curing accelerator, in combination with polyamic acids to form magnetic films, as a moderating and combustion catalyst, as a solid fuel catalyst, and as an intermediate for ferrites by sol-gel processing. Formula: $Fe(O_2C_5H_7)_3$. Also known as *iron acetylacetone; iron pentanedionate; ferric pentanedionate.*

ferric bichromate. See ferric dichromate.

ferric bromide. Red-black, hygroscopic, hexagonal crystals (98+% pure) with a density of 4.5 g/cm^3. Used as a bromination catalyst. Formula: $FeBr_3$. Also known as *ferric tribromide; iron bromide.*

ferric chloride. A black-brown, hygroscopic, crystalline powder (97+% pure). Crystal system, hexagonal. Density, 2.898 g/cm^3; melting point, 306°C (583°F) (decomposes); boiling point, 319°C (606°F). It occurs in nature as the mineral *molysite.* Used as a pigment (flores martis), in the development of gold lusters in glass, glazes and porcelain enamels, as an oxidizing, chlorinating and condensing agent, in water and sewage treatment, and in the detection of phenols and phenolic acids. Formula: $FeCl_3$. Also known as *ferric perchloride; ferric trichloride; iron chloride; iron perchloride; iron trichloride.*

ferric chloride hexahydrate. Yellow, hygroscopic lumps (97+% pure) with a density of 1.82 g/cm^3, a melting point of 37°C (98°F), and a boiling point of 280-285°C (536-545°F). Used as a pigment, in ceramics, in chemistry, and in materials research. Formula: $FeCl_3 \cdot 6H_2O$. Also known as *ferric trichloride hexahydrate; iron chloride hexahydrate; iron trichloride hexahydrate.*

ferric chromate. A yellow powder with a density of 5.08 g/cm^3, and a melting point above 1770°C (3218°F). Used in metallurgy, as a paint pigment, in ceramics, and as a brown and black pigment in engobes, glazes, glasses and porcelain enamels. Formula: $Fe_2(CrO_4)_3$. Also known as *iron chromate.*

ferric dichromate. Reddish-brown granules used in the preparation of pigments. Formula: $Fe_2(Cr_2O_7)_3$. Also known as *iron dichromate.*

ferric ethoxide. A metal alkoxide that contains trivalent iron and is soluble in benzene and hot ethanol. It has a melting point of 120°C (248°F), a boiling point of 155°C (311°F), and is used as an intermediate for sol-gel derived ferrites with magnetic properties. Formula: $Fe(OC_2H_5)_3$. Also known as *iron ethoxide.*

ferric ferrocyanide. Dark blue crystals that decompose on heating. Used as pigment (*Prussian blue*), and in blue inks. Formula: $Fe_4[Fe(CN)_6]_3$. Also known as *iron ferrocyanide.*

ferric fluoride. Hygroscopic, green crystals or green-brown powder (99+% pure). Crystal system, hexagonal. Density, 3.52 g/cm^3 [also reported as 3.87 g/cm^3]; melting point, sublimes above 1000°C (1830°F). Used in porcelain and pottery chiefly as a flux and slight opacifier, and as a catalyst. Formula: FeF_3. Also known as *iron fluoride.*

ferric hydrate. See ferric hydroxide.

ferric hydroxide. A reddish-brown to brown flocculent precipitate obtained by treating a solution of ammonia with a solution of ferrous sulfate. It dries to ferric oxide. Density, 3.4-3.9 g/cm^3; melting point, loses $1.5H_2O$ below 500°C (932°F). Used in the preparation of pigments for paints and rubber, in water purification, and as a catalyst. Formula: $Fe(OH)_3$. Also known as *ferric hydrate; hydrated ferric oxide; hydrated iron oxide; iron hydroxide; iron hydrate.*

ferric material. A material that contains trivalent iron (Fe^{3+}). Abbreviation: FM.

ferric metavanadate. See ferric vanadate.

ferric nitrate. Light purple to yellow-green, hygroscopic crystals (98+% pure). Density, 1.684 g/cm^3; melting point, 47.2°C (117°C); boiling point, decomposes at 125°C (257°F). Used as a mordant for buffs and blacks, in tanning, and as an oxidizer. Formula: $Fe(NO_3)_3 \cdot 9H_2O$. Also known as *ferric nitrate nonahydrate; iron nitrate; iron nitrate nonahydrate.*

ferric nitrate nonahydrate. See ferric nitrate.

ferric oxalate. Yellowish-white amorphous powder or scales decomposing at about 100°C (210°F). It is available in anhydrous form and as the hexahydrate. Used in the manufacture of iron powder, as a catalyst, and for certain photographic papers. Formula: $Fe_2(C_2O_4)_3$ (anhydrous); $Fe_2(C_2O_4)_3 \cdot 6H_2O$ (hexahydrate). Also known as *iron oxalate.*

ferric oxide. See red iron oxide.

ferric oxide pigments. See iron oxide pigments.

ferric perchloride. See ferric chloride.

ferric phosphide. See iron phosphide (iv).

ferric pyrophosphate. A yellowish-white powder used in pigments, and as a catalyst and flame retardant. Formula: $Fe_4(P_2O_7)_3 \cdot xH_2O$. Also known as *iron pyrophosphate.*

ferric resinate. A reddish-brown powder used as a paint and varnish drier. Also known as *iron resinate.*

ferric stearate. The trivalent iron salt of stearic acid available as a light brown powder for use as a varnish drier and in photography. Formula: $Fe(C_{18}H_{35}O_2)_3$. Also known as *iron stearate.*

ferric sulfate. Yellow crystals or grayish-white powder with a density of 3.097 g/cm^3. Used in chemistry and materials research. Formula: $Fe_2(SO_4)_3$. Also known as *ferric trisulfate; iron persulfate; iron sulfate; iron tersulfate.*

ferric sulfate nonahydrate. Yellow crystals or grayish-white powder. Density, 2.0-2.1 g/cm^3; melting point, decomposes at 480°C (896°F). Occurs in nature as the mineral *coquimbite.* Used in pigments, in metal pickling, as an etchant for steel and aluminum, in chelated iron products, in dyeing and calico printing, and as a polymerization catalyst. Formula: $Fe_2(SO_4)_3 \cdot 9H_2O$. Also known as *ferric trisulfate nonahydrate; iron persulfate nonahydrate; iron sulfate nonahydrate; iron tersulfate nonahydrate.*

ferric tribromide. See ferric bromide.

ferric trichloride. See ferric chloride.

ferric trichloride hexahydrate. See ferric chloride hexahydrate.

ferric trioxide. See red iron oxide.

ferric trisulfate. See ferric sulfate.

ferric trisulfate nonahydrate. See ferric sulfate nonahydrate.

ferric vanadate. A grayish-brown powder used in metallurgy. Formula: $Fe(VO_3)_3$. Also known as *ferric metavanadate; iron vanadate; iron metavanadate.*

ferricopiapite. A yellow mineral of the copiapite group composed of iron sulfate hydroxide hydrate, $Fe_{4.67}(SO_4)_6(OH)_2 \cdot 20H_2O$. Crystal system, triclinic. Occurrence: Italy.

ferric oxide. See red iron oxide.

ferric trioxide. See red iron oxide.

ferridravite. A black mineral of the tourmaline group composed of sodium magnesium iron borate silicate oxide hydroxide, $NaMg_3Fe_6(BO_3)_3Si_6O_{18}(O,OH)_4$. Crystal system, rhombohedral (hexagonal). Density, 3.26 g/cm^3; refractive index, 1.800. Occurrence: Bolivia.

ferrierite. (1) A colorless to white mineral of the zeolite group composed of sodium magnesium aluminum silicate hydroxide nonahydrate, $(Na,K)_2MgAl_3Si_{15}O_{36}(OH)\cdot9H_2O$. Crystal system, orthorhombic. Density, 2.15 g/cm^3; refractive index, 1.479. Occurrence: Canada (British Columbia).

(2) A white mineral of the zeolite group composed of potassium sodium magnesium aluminum silicate nonahydrate, $(Na,K,Mg)_2(Si,Al)_{18}O_{36}\cdot9H_2O$. Crystal system, orthorhombic. Density, 2.14 g/cm^3; refractive index, 1.478. Occurrence: USA (California).

ferrihydrite. A yellow to dark brown mineral composed of iron oxide tetrahydrate, $Fe_5O_7(OH)\cdot4H_2O$. It can also be made synthetically. Crystal system, hexagonal. Density, 3.80 g/cm^3.

Ferrimag. Trademark of Crucible Materials Corporation (USA) for a series of permanent magnet materials composed of sintered barium ferrite $(BaO\cdot6Fe_2O_3)$. They have high coercivity, high permeability, high remanent magnetization, and high magnetic energy products. Used for ceramic magnets, and electrical and magnetic equipment.

ferrimagnetic garnets. A group of synthetic ferrimagnetic ceramic oxides that have garnet crystal structures and the general formula $M_3Fe_5O_{12}$, in which M represents a trivalent rare-earth ion, such as lanthanum, praseodymium, samarium, yttrium or any of the rare-earths belonging to the yttrium subgroup. Used for waveguide components in microwave communication.

ferrimagnetic materials. Materials, such as black ferric oxide (Fe_3O_4), holmium-cobalt and ceramic magnets (ferrites), that possess opposing, but unbalanced, alignments of the magnetic moments and thus have net magnetization in one direction. Also known as *ferrimagnets*.

ferrimagnets. See ferrimagnetic materials.

ferrimolybdite. A yellowish mineral with silky to earthy luster composed of ferric molybdate octahydrate, $Fe_2(MoO_4)_3\cdot8H_2O$. Density, 4.5 g/cm^3; hardness, 1.5 Mohs. Occurrence: Australia, USA (Arizona, California, Colorado, Nevada, New Mexico, Pennsylvania). Used as an ore of molybdenum. Also known as *iron molybdite*. See also molybdite.

ferrinatrite. A greenish, or whitish mineral composed of sodium iron sulfate trihydrate, $Na_3Fe(SO_4)_3\cdot3H_2O$. Density, 2.55 g/cm^3; refractive index, 1.556. Occurrence: Chile.

ferriprotoporphyrin IX chloride. See hemin.

ferrisicklerite. A dark brown mineral of the olivine group composed of lithium iron manganese phosphate, $(Li,Fe,Mn)PO_4$. Crystal system, orthorhombic. Density, 3.41 g/cm^3; refractive index, 1.805. Occurrence: Sweden, Morocco.

ferrite. (1) A *solid solution* of up to 0.022% carbon in iron that is present in all steels, cast irons and wrought irons. It is magnetic and soft, stable below 912°C (1674°F), and has a body-centered cubic (bcc) crystal structure. See also iron.

(2) An iron-rich, body-centered cubic (bcc) *solid solution* containing one or more other elements, e.g., chromium, cobalt or nickel.

(3) See ferrites.

ferrite magnet. A ceramic magnet composed of a mixture of ferric oxide and a strong basic oxide, such as barium or sodium oxide. Used in the form of a powder on recording and memory tapes, and in the manufacture of sintered rectifiers and permanent magnets.

ferrites. A class of ceramic magnetic materials that have spinel crystal structures represented by the general formula $MO\cdot Fe_2O_3$ (where M represents a metal). They are composed of fixed mixtures of ferric oxide and appropriate compounds of divalent metals, such as barium, cobalt, copper, lead, magnesium, manganese, nickel, strontium or zinc, and exhibit ferromagnetic, ferromagnetic, antiferromagnetic, magnetooptical, and/or magnetostrictive effects. They are usually magnetically soft, although some magnetically hard types (e.g., *magnetoplumbites*) have also been developed. Soft magnetic ferrites are used in antennas, computer memory cores, ferrite cores for coils, computer disks, recording tapes, telecommunications systems, etc.

ferritic cast irons. A group of cast irons with predominantly ferritic matrices and including high-silicon irons, medium-silicon, high-chromium and high-aluminum gray irons, medium-silicon ductile irons, and malleable irons. Abbreviation: FCI.

ferritic malleable cast irons. A group of malleable cast irons with microstructures consisting of nodules or spheres of temper carbon embedded in predominantly ferritic matrices. They are cast as white irons, but converted by a prolonged annealing process involving decarburization and/or graphitization reactions. Depending on the major reaction and the resulting color of the fracture surface, the product is called "blackheart malleable" (predominant graphitization resulting in dark fracture surface), or "white-heart malleable" (predominant decarburization resulting in light fracture surface). They have high toughness and ductility, good shock resistance, excellent machinability, good brazeability, solderability, hardenability and coatability. Used for railroad, agricultural and automotive castings, and for housings, valves, door locks, keys, stuffing boxes, clutches, fittings, fasteners, etc. Abbreviation: FMI. Also known as *ferritic malleable irons*. See also malleable cast irons; blackheart malleable cast irons; whiteheart malleable cast irons; pearlitic malleable cast irons.

ferritic malleable irons. See ferritic malleable cast irons.

ferritic-pearlitic cast irons. A group of low-cost, high-production gray cast irons containing 2-4% carbon and 4-5% silicon. They have microstructures consisting of pearlitic matrices with graphite flakes dispersed throughout, produced by rapid cooling. Used for general engineering applications. Abbreviation: FPCI. See also ferritic cast irons; gray cast irons; pearlitic cast irons.

ferritic stainless steels. A group of non-heat-treatable, magnetic stainless steels containing about 0.08-0.35% carbon, 10-28% chromium, and additions of aluminum. They have microstructures of alpha ferrite plus carbide, and must be hardened and strengthened by cold working. Important properties include good to moderate mechanical properties, good oxidation and corrosion resistance at elevated temperatures, high susceptibility to embrittlement at temperatures between 400 and 540°C (750 and 1000°F). Used for turbine parts, heat-resisting parts, automotive parts and trim, and transportation equipment. Abbreviation: FSS. Also known as *stainless irons*.

ferritic steels. A group of steels with very low carbon contents and fully ferritic matrices including electrolytic iron, certain sheet steels (for transformers and magnetic cores), iron-silicon and iron-nickel electrical alloys, and ferritic stainless steels. Also known as *low-carbon ferritic steels*. Abbreviation: FS.

ferritungstite. A bright yellow mineral composed of iron tungsten oxide hydroxide, $Fe_2(WO_4)(OH)_4\cdot4H_2O$. Crystal system, cubic. Density, 5.20 g/cm^3; refractive index, 2.09-2.15. Occurrence: USA (Nevada).

Ferrium. Trade name of QuesTek Innovations (USA) for a series of hardened steels. *Ferrium CS61* is a case-carburizing stainless steel with high wear life, and a ductile core used for aerospace bearings. *Ferrium CS62* is a case-hardened stainless steel with a core hardness of 48 Rockwell C and a case hardness of 62 Rockwell C used for structural applications. *Ferrium C69* is a case-hardened steel strengthened by divalent metal carbides (M_2C) and containing about 0.1% carbon (in core), 28% cobalt, 3% nickel, 5.1% chromium, 2.5% molybdenum, 0.02% vanadium, and the balance iron. It has a core hardness of 50 Rockwell C and a case hardness of 69 Rockwell C providing it with remarkable toughness and, at the same time, high ductility.

Ferro. Trade name of Ferro Corporation (USA) for an extensive series of specialty coatings including organic and inorganic powder coatings as well as porcelain enamel and glaze frit compositions.

ferro-actinolite A dark green to black mineral of the amphibole group composed of calcium iron silicate hydroxide, $Ca_2Fe_5(Si_8O_{22})(OH)_2$. Crystal system, monoclinic. Density, 3.51 g/cm³. Occurrence: South Africa, USA (Idaho).

ferroalloys. A group of alloys of iron and one or more other elements that are used as master alloys to introduce these elements into molten steel or cast iron, and neutralize harmful impurities by combining with them and separating from the steel or cast iron as flux or slag prior to solidification. Examples include ferroaluminum, ferroboron, ferromanganese and ferromolybdenum. Abbreviation: FA.

ferroalluaudite. A greenish black mineral composed of sodium calcium iron phosphate, $(Na,Ca)Fe(Fe,Mn)_2(PO_4)_3$. Crystal system, monoclinic. Density, 3.79 g/cm³. Occurrence: USA (New Jersey).

ferroaluminum. A master alloy usually composed of equal proportions of iron and aluminum, but sometimes containing up to a total of 0.5% silicon, copper and carbon. Some special grades contain up to 80% aluminum, with the balance being iron. Used to introduce aluminum into molten steel, as an alloying addition, or as a deoxidizer.

ferroaxinite. A yellow mineral of the axinite group composed of calcium iron aluminum boron silicate hydroxide, $Ca_2(Fe,Mn)Al_2BSi_4O_{15}(OH)$. Crystal system, triclinic. Density, 3.19 g/cm³; refractive index, 1.688. Occurrence: France, Japan.

ferroboron. A master alloy composed of 75-85% iron and 15-25% boron used as a hardening, degasifying and deoxidizing agent in special steels, and in alloy steels to increase the high-temperature strength.

ferrobronze. A corrosion-resistant alloy containing 8% iron, 0.3% chromium, and the balance copper. Used for hardware.

ferrobustamite. A white mineral of the dreierkette group composed of calcium iron manganese silicate, $(Ca,Fe,Mn)_3Si_3O_9$. Crystal system, triclinic. Density, 3.09 g/cm³. Occurrence: Japan.

Ferrocal. Trade name of Aluminiumwerke Maulbronn (Germany) for an aluminum casting alloy used for permanent-mold castings, general machinery parts, and automobile and motorcycle parts.

Ferrocarbo. Trademark of Harbison-Carborundum Corporation (USA) for silicon carbide briquettes and granules used as cupola additions in the manufacture of gray cast iron, and as ladle additions to steel. They decompose in the cupola or ladle adding silicon and carbon to the iron or steel, while acting as a deoxidizer and graphitizer, and improving the machinability and strength.

ferrocarbon titanium. A master alloy composed of 3-8% carbon, 15-20% titanium, and the balance iron. Used as an addition to molten low-carbon steel.

ferrocarpholite. A green to gray-green mineral composed of iron magnesium aluminum silicate hydroxide, $(Fe,Mg)Al_2(Si_2O_6)(OH)_4$. Crystal system, orthorhombic. Density, 3.04 g/cm³; refractive index, 1.630. Occurrence: France.

Ferrocart. Trade name of General Refractories Company (USA) for powdered iron used for high-frequency cores.

ferrocene. An organometallic coordination compound (molecular sandwich) composed of an iron atom (Fe^{2+}) and two five-membered cyclopenediene (C_5H_5) rings, one located above and the other below the iron atom plane. *Ferrocene* is supplied in the form of orange crystals (98% pure) with a melting point of 173-174°C (343-345°F), a boiling point of 249°C (480°F), and a maximum absorption wavelength of 358 nm. Used as a catalyst, as an antiknock agent in gasoline, as a fuel oil additive, as an intermediate for high-temperature polymers, as a coating for aerospace vehicles, missiles and satellites, as a ultraviolet absorber, in organic and organometallic synthesis, and in biochemistry and biotechnology. Formula: $(C_5H_5)_2Fe^{2+}$. Also known as *dicyclopentadienyliron; bis(cyclopentadienyl)iron*.

ferrocenylborane polymer. A ferrocene derivative that contains up to 30% borane, has good hydrolytic and oxidation resistance and long-term thermal stability up to 315°C (600°F), and is used as an ablative for aerospace applications, and in special fibers, coatings and polymer resins.

ferrocerium. A pyrophoric alloy composed of iron and mischmetal (a mixture of cerium and other rare earths) used in the manufacture of cigarette lighter flints, and ferrous and nonferrous alloys.

ferrochrome. See ferrochromium.

ferrochromium. (1) In general, any alloy composed predominantly of iron and chromium. Also known as *ferrochrome.*

(2) An alloy containing chiefly iron and chromium and used as a means of introducing chromium into low-, medium- and high-carbon steels, and cast iron. The average melting range is about 1260-1650°C (2300-3000°F). There are four types of ferrochromium: (i) *Foundry-type* composed of 5-7% carbon, 62-66% chromium, 7-10% silicon, and the balance iron. It dissolves easily and is used for metallurgical applications, and as a ladle addition of chromium to cast iron; (ii) *High-carbon type* composed of 4.5-7% carbon, 1-3% silicon, 65-70% chromium, and the balance iron. It is graded according to the carbon content, and used in the production of chromium steels and cast irons; (iii) *Low-carbon type* composed of 0.02-2% carbon, 67-73% chromium, 0.2-1% silicon, and the balance iron. It is graded according to the carbon content, and used for stainless steels; and (iv) *Nitrogen-type* composed of up to 0.1% carbon, 67-71% chromium, 0.3-1% silicon, approximately 0.75% nitrogen, and the balance iron. Also known as *ferrochrome.*

Ferrochronin. Trade name of Vereinigte Deutsche Nickel-Werke AG (Germany) for a series of alloys containing 60-75% nickel, 14-20% chromium and 6-50% iron. It has high heat and electrical resistance, and is used for electrical equipment, heating elements, resistors, regenerators, combustion liners, manifolds, and heaters.

Ferrocite. Trade name of Charles Hardy & Company (USA) for a powder metal containing 0.5% carbon, and the balance iron. Used for valves and guides.

ferrocoke. Coke made by blending selected coal with a finely crushed iron ore, such as ilmenite, limonite or magnetite, and used as iron-blast furnace burden.

ferrocolumbite. A black mineral of the columbite group composed of iron manganese niobium (columbium) tantalum oxide, $(Fe,Mn)(Nb,Ta)_2O_6$. Crystal system, orthorhombic. Density, 6.43 g/cm^3; refractive index, 2.40. Occurrence: USA (Connecticut, South Dakota).

ferrocolumbium. See ferroniobium.

Ferrocon. Trademark of Ferro Corporation (USA) for a series of modified conductive polyolefins.

ferroconcrete. A concrete strengthened or reinforced with iron or steel, usually in the form of a framework. Abbreviation: FC. See also reinforced concrete.

Ferrocor. Trade name of British Steel Corporation (UK) for a series of electrical steels used for frictional horsepower motors.

Ferrocrete. Trademark of Associated Portland Cement Manufacturers, Limited (UK) for a rapid-hardening Portland cement.

Ferrodur. Trade name of Janney Cylinder Company (USA) for a cast iron containing 3% carbon, 0.5% molybdenum, 1.2% nickel, 0.9% manganese, 2% silicon, 0.2% chromium, and the balance iron. Used for pump liners, shaft sleeves, and centrifugal castings.

ferroelectric crystal. A crystal of ferroelectric composition exhibiting ferroelectricity and piezoelectricity. Also known as *polar crystal*. See also ferroelectric materials.

ferroelectric materials. Crystalline substances, such as potassium dihydrogen phosphate, potassium-sodium tartrate (Rochelle salt), the titanates and zirconates of barium, calcium, magnesium, etc., and certain fluoropolymers, that exhibit spontaneous reversible electric-dipole alignment, electric hysteresis, and piezoelectricity. Used for capacitors, transducers, dielectric amplifiers, computers, optoelectronic devices, and similar applications. Abbreviation: FEM. Also known as *ferroelectrics; seignette electrics; seignette-electric materials.*

ferroelectric polymers. Electrically poled polymers (e.g., polyvinylidene fluoride films) that exhibit piezo- and pyroelectric effects and have potential applications in electronics and optoelectronics for sensors and actuators.

ferroelectrics. See ferroelectric materials.

ferroferric oxide. See black ferric oxide.

Ferroflo. Trade name of Ferro Corporation (USA) for lubricated thermoplastic resins including silicone-filled polystyrene for thermoformed and sheet-extruded products.

ferrofluid. A colloidal suspension of magnetic particles (e.g., superparamagnetic iron oxide nanoparticles) and selected additives in a carrier fluid. In an applied magnetic field this suspension becomes magnetized, but unlike a *magnetorheological fluid*, does not exhibit a shear yield stress. Used in seals, sensors, stepper motors, and for damping applications.

ferrogedrite. A pale green, brownish green, greenish blue mineral of the amphibole group composed of iron aluminum silicate hydroxide, $Fe_5Al_4Si_6O_{22}(OH)_2$. Crystal system, orthorhombic. Density, 3.57 g/cm^3; refractive index, 1.710. Occurrence: Japan.

Ferrogen. Trade name of Foseco Minsep NV (Netherlands) for a flux that acts as scavenger, degasser, deoxidizer and slag reducer when added to the ladle in iron and steelmaking.

Ferroglass. Trade name of Semi-Elements Inc. (USA) for optically transparent glass that is ferromagnetic at room temperature.

ferroglaucophane. A blue mineral of the amphibole group composed of sodium iron magnesium aluminum silicate hydroxide, $Na_2(Fe,Al,Mg)_5Si_8O_{22}(OH)_2$. Crystal system, monoclinic. Density, 3.23 g/cm^3; refractive index, 1.659. Occurrence: New Caledonia.

Ferro-Glo. Trade name of Electro-Glo Company (USA) for an electropolishing concentrate for carbon and alloy steels.

Ferroguard. Trade name of Alchem Corporation (USA) for a phosphate coating and coating process for ferrous metals.

ferrohexahydrite. A colorless mineral of the hexahydrite group composed of iron sulfate hexahydrate, $FeSO_4 \cdot 6H_2O$. Crystal system, monoclinic. Density, 1.93 g/cm^3. Occurrence: Russian Federation.

Ferrohone. Trade name of Aldoa Company (USA) for polishing compounds used on steel products.

ferrohornblende. A brown mineral of the amphibole group composed of sodium potassium calcium iron magnesium aluminum silicate hydroxide, $(Na,K)Ca_2(Fe,Mg)_5(Al,Si)_8O_{22}(OH)_2$. Crystal system, monoclinic. Density, 3.41 g/cm^3. Occurrence: Norway.

ferroics. A group of ferroelectric or antiferroelectric materials that have high dielectric constants. See also antiferroelectric materials; ferroelectric materials; high-k materials.

Ferrolene. Trade name of Ferro Corporation (USA) for polypropylene plastics filled with 20 and 40% talc, or calcium carbonate.

Ferrolum. Trade name of Knapp Mills Inc. (USA) for an acid-resistant, lead-clad sheet steel with good resistance to sulfuric acid. Used for chemical equipment, and tanks.

ferromagnesite. A variety of the mineral *magnesite* ($MgCO_3$) that contains some iron, and is used as a heat-bonding refractory.

ferromagnetic ceramics. A group of permanent-magnet materials made by blending magnetic (e.g., iron) and ceramic powders followed by pressing and sintering. They possess high resistivity and are extremely resistant to demagnetization. Examples include the ferrites of barium, lead and strontium. Used for magnets in high-frequency circuits, dc motors, door latches, and radio loudspeakers. See also ferrites; barium ferrite; lead ferrite; strontium ferrite.

ferromagnetic crystal. A crystal composed of ferromagnetic material.

ferromagnetic film. See magnetic thin film.

ferromagnetic materials. A large group of magnetic materials exhibiting spontaneous parallel alignment of neighboring magnetic moments in one direction. They can acquire high degrees of magnetization in relatively weak magnetic fields, have definite magnetic saturation points, appreciable residual magnetism and hysteresis, and magnetic permeabilities considerably greater than unity. Examples include several metals, such as cobalt, iron, nickel, various alloys, intermetallic compounds of rare-earth and transition elements, and numerous ceramics. Abbreviation: FMM. Also known as *ferromagnets; ferromagnetics; ferromagnetic substances.*

ferromagnetic substances. See ferromagnetic materials.

ferromagnetics. See ferromagnetic materials.

ferromagnets. See ferromagnetic materials.

ferromanganese. A master alloy that contains chiefly iron and manganese. There are four types: (i) *Low carbon type* composed of 0.07-0.5% carbon, 85-90% manganese, and the balance iron. It is graded according to the carbon content, and used as high-manganese addition to low carbon steels and for metallurgical applications; (ii) *Low-iron type* composed of 85-90% manganese, 0-3% silicon, approximately 7.0% carbon, and the balance iron. Used for metallurgical applications in nonferrous alloys, where high manganese and low iron contents are needed; (iii) *Medium-carbon type* composed of 80-85% manganese, 1.5% carbon, 0-1.5% silicon, and the balance iron. Used as high-manganese additions to medium carbon steels, in metallurgical applications, and in the manufacture of carbon-manganese steels; and (iv) *Standard type* composed of 74-76% manganese, 0-1% silicon, 0-0.3% phosphorus, approximately 7% carbon, and the balance iron. Used for various metallurgical applications, and as an alloy deoxidizer.

ferromanganese-silicon. A low-carbon low-melting ferroalloy containing about 50% silicon, 25-30% iron, 20-25% manganese and 0-0.5% carbon. Used to add manganese and silicon to iron and steel melts.

Ferromet. Trade name of Plansee Metallwerk-Gesellschaft (Austria) for a sintered iron alloy used for machine components, bearings, and magnets.

ferromolybdenum. An alloy composed predominantly of iron and molybdenum and supplied in various grades with carbon contents of 0.60%, 1.10% and 2.50%, respectively, and a molybdenum content ranging between 50 and 85%. Used to introduce molybdenum into iron or steel (e.g., cast iron, or constructional, stainless or tool steel), and as a welding-rod coating.

Ferron. (1) Trade name of Fansteel Metals (USA) for an alloy containing 50% iron, 35% nickel and 15% chromium, used for heat- and corrosion-resistant parts, and heating elements.

(2) Trade name for co-precipitated iron oxide and calcium sulfate used as a building material.

ferronickel. (1) A crude master alloy composed of iron and nickel.

(2) A nickel-steel alloy used for electrical resistance wire.

ferronickelplatinum. A silvery white mineral of the gold group composed of iron nickel platinum, Pt_2FeNi. Crystal system, tetragonal. Density, 15.39 g/cm^3. Occurrence: Russian Federation.

ferroniobium. A master alloy composed 50-60% niobium (columbium), 7% silicon, and the balance iron. Used to introduce niobium (columbium) into steels (e.g., stainless steels), and in certain welding-rod alloys. Also known as *ferrocolumbium.*

Ferro-Pak. Trade name of Cromwell-Phoenix, Inc. (USA) for corrosion-inhibiting paper supplied in rolls and sheets for storing and shipping metals.

Ferropak. Trade name of Ferro Corporation (USA) for filled and reinforced thermoplastic resins and blends (e.g., polypropylene/polyethylene) supplied in particulate form. They are available in injection-moldable, extrudable and/or thermoformable grades for use in the manufacture of microwave cookware and condiment packing applications.

ferropargasite. A mineral of the amphibole group composed of sodium calcium iron aluminum silicate hydroxide, $NaCa_2Fe_4Al-Si_6Al_2O_{22}(OH)_2$. Crystal system, monoclinic. Density, 3.44 g/cm^3; refractive index, 1.713.

Ferroperm. Trade name of NKK Corporation (Japan) for a soft-magnetic alloy containing 99% iron and 1% aluminum, and supplied in plates 1.0-4.5 m (40-175 in.) wide, and 6-100 mm (0.25 to 4 in.) thick. It has low coercivity, large saturation magnetization, high permeability, and outstanding magnetic-shielding performance (better than pure iron). Used for electrical and magnetic equipment.

ferrophosphorus. A high-phosphorus iron alloy available principally in two grades: one with about 18% phosphorus, obtained as a byproduct in making pig iron from iron, phosphorus rock, silica and coke, and the other with 23-25% phosphorus, made in the electric furnace. *Ferrophosphorus* is used as an addition to certain steels to adjust the phosphorus content, for rephosphorizing certain carbon and alloy steels, in the manufacture

of steel for tinplate, as a fluidity improver for steel used for casting, to prevent steel sheets arranged in bundles from sticking together during rolling or annealing, and to introduce phosphorus into bronzes.

ferroplatinum. See iron platinum.

Ferropol. Trade name for high-luster buffing compounds for steel, stainless steel, chromium-plated articles, and aluminum and other light-metal alloys.

Ferropox. Trademark of Kloeckner Pentaplast GmbH (Germany) for glass-fiber-reinforced epoxies.

Ferropreg EP. Trade name of Ferro Corporation (USA) for epoxy laminates available as glass fabric prepregs, and as glass, carbon graphite and *Kevlar* fiber prepregs. Used for aircraft and aerospace parts and sporting goods.

Ferropreg PF. Trade name of Ferro Corporation (USA) for a series of phenolic laminates and prepregs. The laminates are available as cotton and glass fabrics, and paper laminates, and the prepregs are supplied in the following grades: wood- or natural fiber-filled for general purposes, cellulose-filled for shock resistance, glass fiber-filled for high impact, chopped fabric-filled for medium impact, cotton-filled for medium shock resistance, mica-filled for electrical applications, mineral-filled for high heat resistance, and foam.

Ferropreg Polyester. Trade name of Ferro Corporation (USA) for polyester laminates supplied as woven-glass roving prepregs. Used for aircraft and electrical applications.

Ferropreg Polyimide. Trade name of Ferro Corporation (USA) for copper-bonded polyimide laminates supplied as woven glass fabric prepregs. Used for electrical applications including printed circuit boards.

Ferropyr. British trade name for an alloy containing 86% iron, 7% chromium, 7% aluminum, and traces of manganese and silicon. Used for electrical resistances for high temperature environments.

ferrorichterite A green mineral composed of sodium calcium iron silicate hydroxide, $Na_2CaFe_5Si_8O_{22}(OH)_2$. It can be made synthetically. Crystal system, monoclinic. Density, 3.46 g/cm^3.

Ferrosad. Trade name of Metalltechnik Schmidt GmbH & Co. (Germany) for a carbon steel containing 0.1% carbon, 0.15% silicon, 1-1.3% manganese, 0-0.03% sulfur, 0-0.03% phosphorus, and the balance iron. It has a bainitic microstructure, and is used in the form of shot for abrasive blast cleaning.

ferroselenium. A ferroalloy composed of almost equal amounts of selenium and iron, and small percentages of carbon, and used to introduce selenium into stainless and other alloy steels to improve machinability.

ferroselite. A brass yellow to white mineral of the marcasite group composed of iron selenide, $FeSe_2$. It can also be made synthetically by heating ferrous selenide and selenium under prescribed conditions. Crystal system, orthorhombic. Density, 7.21. Occurrence: Russia, USA (Utah, Wyoming).

Ferrosil. Trade name of British Steel Corporation (UK) for a series of nonoriented low-carbon iron strips used for magnetic applications.

ferrosilicon. An alloy composed entirely or predominantly of iron and silicon, and used to add silicon to steel and iron. There are five grades: (i) the *15% grade* composed of 14-20% silicon, 0-1% carbon, 74-84% iron, 0-0.05% phosphorus and 0-0.04% sulfur, used to introduce silicon into steel, and for other metallurgical applications; (ii) the *50% grade* composed of 47-51% silicon, and the balance iron, used for various metallurgical applications, and as a graphitizer and deoxidizer; (iii) the *75%*

grade composed of 73-78% silicon, and the balance iron used for metallurgical applications, particularly as a silicon addition to ferrous alloys; (iv) the *85% grade* composed of 83-88% silicon, and the balance iron, used in the manufacture of high-silicon alloys, and as an inoculant for cast iron; and (v) the *90% grade* composed of 92-95% silicon, and the balance iron, used in the manufacture of high-silicon alloys, and for other metallurgical applications.

ferrosilicon-aluminum. A master alloy containing about 44-46% silicon, 33-40% iron and 12-15% aluminum, used to deoxidize steels, and to introduce silicon into aluminum-silicon castings.

Ferrosilid. Russian trade name for an acid-resistant cast alloy containing 0.5-0.7% carbon, 12-18% silicon, and the balance iron. Used for cast conduits and pumps.

ferrosoferric oxide. See black ferric oxide.

ferrospinel. (1) Powder products with cubic spinel structures corresponding to the general formula MFe_2O_4, where M represents barium, cadmium, calcium, cobalt, copper, iron, magnesium, manganese, nickel, strontium or zinc. Used for soft magnets, and as refractories with high resistance to attack by molten glass and slags.

(2) See hercynite.

Ferrosteel. Trade name of Crane Company (USA) for a cast iron containing 3% carbon, 1.8% silicon, and the balance iron.

Ferrotherm. Trade name of Krupp Stahl AG (Germany) for a series of heat- and corrosion-resistant chromium, chromium-nickel and nickel-chromium steels including several stainless types. Used for furnace parts, recuperators, autoclaves, etc.

Ferro-TiC. Trade name of Alloy Technology International Inc. (USA) for a machinable metal-matrix composite (MMC) composed of very hard titanium-carbide particles dispersed in an alloy-steel matrix. It has high toughness, high abrasion and wear resistance, high hardness, good corrosion resistance, a low coefficient of friction, good machinability, and good thermal-shock resistance. Used for die plates, pelletizer knives, drawing rings for gas cylinders, etc. See also Superwear (2).

Ferro-Titanit. Trademark of Edelstahl Witten-Krefeld GmbH (Germany) for sintered hard alloys composed tool steel with up to 45 vol% titanium carbide (TiC). They are produced by powder-metallurgical techniques, and are available in several grades categorized according to the particular binder phase into: (i) carbon-containing alloys with as-hardened martensitic microstructures; (ii) very-low-carbon alloys having as-hardened nickel martensite structures; and (iii) corrosion-resistant alloys with austenitic microstructures. *Ferro-Titanit* alloys possess outstanding abrasion and/or corrosive wear resistance, can be distortion-free vacuum-hardened up to 70 Rockwell C, and machined by most conventional methods. Used to make tools and wear components, guide rollers, bending rollers, etc.

ferrotitanium. A master alloy composed predominantly of iron and titanium and used as a deoxidizer and scavenger and/or titanium addition to steel. It is available in three types: (i) *Low-carbon* composed of 0-0.1% carbon, 20-25% titanium, 0-7.5% aluminum, and the balance iron. Used for metallurgical applications in stainless steel; (ii) *Medium-carbon* composed of 2.5% carbon, 18% titanium, and the balance iron. Used for metallurgical applications including steelmaking; and (iii) *High-carbon* composed of 6-8% carbon, 15-18% titanium, and the balance iron. Used in steel metallurgy.

Ferrotite. Trade name of Pacific Foundry Company (USA) for a steel containing 0.2% carbon, and the balance iron. Used for welding rods.

Ferrotron. Trademark of Polypenco (USA) for fluoroploymers based on polytetrafluoroethylene.

Ferrotube. Trade name of Foseco Minsep NV (Netherlands) for a flux contained in tubes. Used as a deoxidizer in small iron and steel melts.

ferrotungsten. An alloy composed predominantly of iron and tungsten, and used in steelmaking to introduce tungsten into the ladle. It has a melting range of 1650-2750°C (3000-4980°F), and dissolves readily in the molten steel. There are three types: (i) *Standard* composed of 0.6-0.7% carbon, 76-80% tungsten, 0-0.75% silicon, and the balance; (ii) *Low melting* composed of 2.5% carbon, 65-75% tungsten, 0-1.5% silicon, and the balance iron; and (iii) *ASTM* composed of 0.6-0.7% carbon, 70-80% tungsten, 0-1% silicon, and the balance iron.

ferrotychite. A colorless or light yellowish mineral of the tychite group composed of a sodium iron carbonate sulfate, $Na_6Fe_2(SO_4)(CO_3)_4$. Crystal system, cubic. Density, 2.79 g/cm³. Occurrence: Russian Federation.

ferrouranium. An alloy of uranium (especially ^{238}U) and iron sometimes used as an alloying addition to promote the tensile strength and modulus of steels, and as a denitrogenizer, deoxidizer and carbide former in alloy steels including tool steels.

ferrous acetate. An air-and moisture-sensitive, off-white to light brown powder (95+% pure) that decomposes at 190-200°C (374-390°F), and is used in chemistry, medicine and materials research, in textile and leather dyeing, and as a wood preservative. Formula: $Fe(C_2H_3O_2)_2$. Also known as *iron acetate.*

ferrous alloys. See iron alloys.

ferrous alloying elements. A term referring to the various chemical elements introduced into steel during its manufacture to improve final properties. Such elements include aluminum, beryllium, boron, calcium, cerium, chromium, cobalt, copper, manganese, molybdenum, nickel, niobium (columbium), phosphorus, tantalum, titanium, vanadium, and zirconium. See also alloying elements.

ferrous aluminate. A spinel-type compound composed of ferrous oxide and aluminum oxide. Density, 4.35 g/cm³; melting point, 1438°C (2620°F). Used in ceramics and composites. Formula: $FeAl_2O_4$. Also known as *iron aluminate.*

ferrous ammonium sulfate. Light green crystals that have a density of 1.865 g/cm³, and decompose at 100-110°C (210-230°F). Used in chemistry and metallurgy. Formula: $Fe(SO_4) \cdot (NH_4)_2SO_4 \cdot 6H_2O$. Also known as *iron-ammonium sulfate; Mohr's salt.*

ferrous arsenide. See iron arsenide (i).

ferrous boride. See iron boride (ii).

ferrous bromide. An orange, moisture-sensitive powder (98+% pure). Crystal system, hexagonal. Density, 4.636 g/cm³; melting point, 691°C (1276°F); antiferromagnetic properties. Used as an organic catalyst, and in materials research. Formula: $FeBr_2$. Also known as *iron bromide.*

ferrous bromide hexahydrate. Deliquescent, orange crystals, or green, crystalline powder. Density, 4.636 g/cm³; melting point, 27°C (80°F). Used as a polymerization catalyst. Formula: $FeBr_2 \cdot 6H_2O$. Also known as *iron bromide hexahydrate.*

ferrous chloride. White, hygroscopic crystals or tan powder. It occurs in nature as the mineral *lawrencite.* Crystal system, rhombohedral (hexagonal). Density, 3.16 g/cm³; melting point, 674°C (1245°F); boiling point, 1023°C (1873°F); metamagnetic properties. Used in chemistry and materials research. Formula: $FeCl_2$. Also known as *iron chloride.*

ferrous chloride dihydrate. Yellowish or greenish crystals with a density of about 2.4-2.6 g/cm³, used in the manufacture of iron powder. Also known as *iron chloride dihydrate*.

ferrous chloride tetrahydrate. Light green crystals with a density of 1.93 g/cm³ and antiferromagnetic properties. Used in metallurgy, in the manufacture of ferric chloride, in the manufacture of magnets, as a mordant in dyeing, and in sewage treatment. Formula: $FeCl_2 \cdot 4H_2O$. Also known as *iron chloride tetrahydrate*.

ferrous disulfide. A gold-colored, cubic, crystalline transition-metal dichalcogenide that has an orthorhombic, metastable, semiconducting low-temperature polymorph (*marcasite*). It occurs in nature as the mineral *pyrite*, and can be manufactured by hydrothermal, flux, and chemical vapor transport methods. At 743°C (1370°F), it undergoes a peritectic transition into the more stable FeS and liquid sulfur. It exhibits both n- and p-type conduction, and is used for thin-film photovoltaic applications, e.g., solar cells. Formula: FeS_2. Also known as *iron disulfide*.

ferrous fluoride. White, hygroscopic crystals or off-white, hygroscopic powder (98+% pure). Crystal system, tetragonal. Density, 4.09 g/cm³; melting point, 1100°C (2012°F). Used in ceramics and materials research, and as a catalyst. Formula: FeF_2. Also known as *iron fluoride*.

ferrous fluoride octahydrate. Green crystals. Melting point, loses $8H_2O$ at 100°C (212°F). Used in ceramics. Formula: $FeF_2 \cdot 8H_2O$. Also known as *iron fluoride octahydrate*.

ferrous iodide. Red-violet, hygroscopic crystals or gray, hygroscopic powder (99.5+%). Crystal system, hexagonal. Density, 5.315 g/cm³; melting point, 587°C (1089°F). Used in chemistry and materials research. Formula: FeI_2. Also known as *iron iodide*.

ferrous iodide tetrahydrate. Dark violet to black hygroscopic leaflets. Density, 2.873 g/cm³; melting point, decomposes at 90-98°C (194-208°F). Used in the manufacture of alkali-metal iodides, and as a catalyst. Formula: $FeI_2 \cdot 4H_2O$. Also known as *iron iodide tetrahydrate*.

ferrous material. A material that contains divalent iron (Fe^{2+}). Abbreviation: FM.

ferrous metals. One of the two classes of metallic materials (the other being *nonferrous metals*) that includes all metals that have iron as their base material, e.g., steels, cast irons and wrought iron.

ferrous nitride. See iron nitride (ii).

ferrous oxalate. A yellow powder used in photography. Formula: FeC_2O_4. Also known as *iron oxalate*.

ferrous oxalate dihydrate. Pale yellow, crystalline powder. Density, 2.28 g/cm³; melting point, decomposes at 190°C (374°F). Used as a photographic developer, and as a pigment in glass, plastics, paints, etc. Formula: $FeC_2O_4 \cdot 2H_2O$. Also known as *iron oxalate dihydrate*.

ferrous oxide. See black iron oxide.

ferrous phosphate. See iron phosphate.

ferrous phosphate octahydrate. See iron phosphate.

ferrous phosphide. See iron phosphide (iii).

ferrous phthalocyanine. See iron phthalocyanine.

ferrous powder. See iron powder.

ferrous selenide. Black crystals or gray-black, moisture-sensitive lumps with high purity (99.99+%). Crystal system, hexagonal. Density, 6.7 g/cm³. Used in ceramics, electronics and materials research. Formula: FeSe. Also known as *iron selenide*.

ferrous stearate. The divalent iron salt of stearic acid available in the form of dark brown pills containing about 9% iron. Formula: $Fe(C_{18}H_{35}O_2)_2$. Also known as *iron stearate*.

ferrous sulfate. Green or yellow-brown crystals. Density, 1.89 g/cm³; melting point, 64°C (147°F); boiling point, loses $7H_2O$ at 300°C (572°F). It occurs in nature as the mineral *melanterite*. Used in iron oxide pigments, in dyeing, in the manufacture of ferrites, inks and iron salts, in water and sewage treatment, as a wood preservative, in photography, and in process engraving. Formula: $FeSO_4 \cdot 7H_2O$. Also known as *ferrous sulfate heptahydrate; green copperas; green vitriol; iron sulfate; iron sulfate heptahydrate; iron vitriol; sal chalybis*.

ferrous sulfate heptahydrate. See ferrous sulfate.

ferrous sulfide. Colorless, hygroscopic crystals, or dark brown or black, hygroscopic metallic sticks, pieces, granules or powder (90+% pure). Crystal system, hexagonal, or tetragonal. Density, 4.7-4.84 g/cm³; melting point, 1196°C (2185°F); boiling point, decomposes; hardness, 4 Mohs. The hexagonal form occurs in nature as the mineral *troilite*, and the tetragonal form as the mineral *mackinawite*. Used as a semiconductor, and in metallurgy. Formula: FeS. Also known as *iron sulfide; iron monosulfide; iron protosulfide; iron sulfuret*.

ferrous superalloys. See iron-base superalloys.

ferrous telluride. See iron telluride (i).

Ferrovac. (1) Trade name of Crucible Materials Corporation (USA) for a series of vacuum-melted steels of the plain-carbon type (AISI-SAE 1020), nickel-chromium-molybdenum type (AISI-SAE 4340), or chromium type (AISI-SAE 52100). Used for bolts, gears, shafts, crankshafts, plug and ring gages, and aircraft bearings.

(2) Trade name of Crucible Materials Corporation (USA) for a gas-free high-purity iron containing only 0.007% carbon. Used for magnetic applications, e.g., pole pieces, solenoids, armatures, and relays.

ferrovanadium. An alloy composed predominantly of iron and vanadium, and used to introduce vanadium into steel. It is supplied in various grades containing 30-80% vanadium, 0-15% silicon, 3-6% carbon, with the balance being iron. *Ferrovanadium* has a melting range of approximately 1480-1520°C (2700-2770°F).

Ferroweld. (1) Trade name of Pacific Foundry Company (USA) for a low-carbon steel used for coated welding rods for cast irons.

(2) Trade name of Lincoln Electric Company (USA) for a low-carbon steel containing 0.15% carbon, 0.03% silicon, 0.3% manganese, 0.04% phosphorus, 0.04% sulfur, and the balance iron. Used for arc-welding electrodes.

Ferroxcube. Trademark of Philips Electronics North America Corporation (USA) for a series of soft magnetic materials that are cubic ferrites of the composition MFe_2O_4, where M is a divalent metal, such as manganese, zinc or nickel. They have low eddy-current losses at high frequencies and high electrical resistivity. *Ferroxcube A* is composed of 48 wt% manganese ferrite ($MnFe_2O_4$) and 52 wt% zinc ferrite ($ZnFe_2O_4$), and *Ferroxcube B* contains 36 wt% nickel ferrite ($NiFe_2O_4$) and 64 wt% zinc ferrite ($ZnFe_2O_4$).

Ferroxdure. Trademark of Philips Electronics North America Corporation (USA) for a series of hard magnetic materials that are sintered ceramics based principally on barium ferrite with hexagonal crystal structure. The principal grades are *Ferroxdure I* and *Ferroxdure II* composed of barium hexaferrite ($BaFe_{12}O_{19}$, or $BaO \cdot 6Fe_2O_3$). They have a density range of 4.5-4.9 g/cm³ (0.16-0.18 lb/in.³), high resistance to demagnetization, and high coercivity and electrical resistivity. Used for permanent magnets for motors and oil filters.

Ferrozell. Trade name of Ferrozell-Gesellschaft Sachs & Co. mbH (Germany) for paper/phenolic and fabric/phenolic laminates and sheet-molding compounds.

ferrozirconium. An alloy composed of iron, zirconium and silicon and used to introduce zirconium into steel. There are two basic types: (i) *Low zirconium* with 12-15% zirconium, 39-43% silicon and 40-45% iron, used to introduce zirconium into steel of high silicon content; and (ii) *High zirconium* composed of 35-40% zirconium, 47-52% silicon and 8-12% iron, and used to introduce zirconium into steel of low silicon content.

Ferrozoid. See Dullray.

ferruccite. A colorless mineral of the anhydrite group composed of sodium fluoborate, $NaBF_4$. It can also be made synthetically. Crystal system, orthorhombic. Density, 2.50 g/cm^3; refractive index, 1.3012. Occurrence: Italy.

ferruginous sandstone. A reddish-brown to brown sandstone containing iron oxide minerals, such as *hematite* and *limonite*, as cementing materials. Also known as *iron sandstone*. See also brownstone.

Ferrul. British trade name for a free-cutting alloy containing 54.6% copper, 40% zinc, 5% lead, and 0.4% aluminum. Used for intricate castings, and for bearings.

Ferrux. Trademark of Foseco Minsep NV (Netherlands) for exothermic antipiping compounds used on the surface of feeding heads in castings of iron and steel.

Ferry. (1) Trademark of Inco Alloys International Limited (UK) for an annealed nonmagnetic alloy containing 55-56% copper and 44-45% nickel. Commercially available in the form of foil, rod and wire, it has a density of 8.9 g/cm^3 (0.32 $lb/in.^3$), high electrical resistivity, good corrosion resistance, a low temperature coefficient of resistance, a low coefficient of expansion, and a maximum service temperature of 440°C (825°F). Used for thermocouples and electrical resistances.

(2) See Ferry metal.

Ferry Alloy. Trade name of Inco Alloys International Inc. (USA) for a resistance alloy containing 55% copper, 42.4-55% nickel, 0-1% iron, 0-0.1% carbon, 0-1% manganese, and 0-0.5% silicon. It has good electrical properties, a medium-range electrical resistivity, a low temperature coefficient of resistance, and a maximum service temperature of 400°C (750°F). Used for electrical resistances and wire-wound precision resistors.

Ferrydur. Trade name of Ferry-Capitain, Usines de Bussy (France) for a cast iron containing 1.8-2% carbon, 12-13% chromium, and the balance iron. It has excellent wear resistance and good hardness, and is used for wear-resistant castings.

Ferry metal. British trade name for a lead alloy that contains 2% barium, 1% calcium and 0.35% mercury, and is used for solders and bearings. Also known as *Ferry*.

Ferrynox. Trade name of Ferry-Capitain, Usines de Bussy (France) for a series of corrosion- and heat-resistant alloys including several austenitic chromium-nickel stainless steels, and nickel-iron-chromium and cobalt-chromium-iron alloys. Used for furnace parts, heat-treating equipment, chemical and petrochemical equipment, and food-processing equipment.

fersilicite. A tin-white mineral of the halite group composed of iron silicide, FeSi. It can also be made synthetically. Crystal system, cubic. Density, 5.95 g/cm^3. Occurrence: Ukraine.

fersmanite. A dark brown mineral composed of sodium calcium titanium niobium silicate fluoride hydroxide, $(Ca,Na)_4(Ti,Nb)_2$-$Si_2O_{11}(F,OH)_2$. Crystal system, triclinic. Density, 3.44 g/cm^3; refractive index, 1.930. Occurrence: Russian Federation.

fersmite. A black mineral of the columbite group composed of calcium cerium niobium titanium oxide fluoride, $(Ca,Ce)(Nb,Ti)_2(O,F)_6$. It can also be made synthetically. Crystal system, orthorhombic. Density, 4.72 g/cm^3; refractive index, 2.10.

Fertene. Trademark of Montecatini Edison SpA (Italy) for a series of polyethylenes.

fertile material. A material that can be converted into a nuclear fuel by a nuclear reaction, e.g., neutron absorption. Uranium-238 (^{238}U) and thorium-232 (^{232}Th) are important fertile materials that can be converted to fissionable plutonium-239 (^{239}Pu) and uranium-233 (^{233}U), respectively.

fervanite. A golden brown mineral composed of iron vanadate pentahydrate, $Fe_4(VO_4)_4 \cdot 5H_2O$. Crystal system, monoclinic. Refractive index, 2.224. Occurrence: Gabon. Used as a vanadium ore.

Fervetro. Trade name of Saint-Gobain (France) for thin toughened sheet glass.

Festellan. German trade name for phenol-formaldehyde plastics and products.

Festel Metal. Trade name of Cabot Corporation (USA) for weld metal composed of 55% iron, 23% cobalt, 21% chromium, and 0.7% carbon.

Festel Stellite. Trade name of Cabot Corporation (USA) for an alloy of cobalt, chromium and tungsten, used for welding rods and in the manufacture of cutlery.

Festival. Trade name of Chance Brothers Limited (UK) for a patterned glass.

Feuerfestbinder. Trade name of Chemetall GmbH (Germany) for refractory building materials.

Feuiltex. French trade name for a laminated glass.

Feurex. Trade name of ASG Industries Inc. (USA) for flat borosilicate glass that is resistant to heat and chemicals.

Feutre-Sol. Trade name of Saint-Gobain (France) for acoustical glass-wool floor insulation materials.

Fiammingo. Trade name of Fabbrica Pisana SpA (Italy) for a patterned glass.

Fibacrete. Trademark of Bay Mills Limited (Canada) for a glass-reinforcing fabric used for outdoor wall systems.

FibaTape. Trademark of Bay Mills Limited (Canada) for fiberglass metal corner tape with twin reinforcing strips, and gypsum wallboard joint tape.

fiber. (1) A general term for a slender, elongated filament. Natural fibers are of vegetable (e.g., cotton, flax, jute, kapok, and sisal), animal (e.g., wool, silk and hair), or mineral (e.g., asbestos and rock wool) origin. Other fibers include semisynthetic fibers (e.g., rayon), synthetic fibers (e.g., nylon and polyacrylonitrile), ceramic fibers (e.g., glass and aramid), and metal fibers (e.g., aluminum, copper and tungsten).

(2) A term used to refer to a synthetic fiber made by drawing a ceramic, metal or polymer into a long, thin filament. It is characterized by an exceptionally high ratio of length to diameter (usually at least 100:1), and considerably high tenacity. Such a fiber can be polycrystalline or amorphous, and continuous or discontinuous.

Fiber B. Trade name for polyolefin fibers and yarns used for textile fabrics.

Fiberbel. Trade name of SA Glaverbel (Belgium) for a glass-fiber mat used as a bituminous roofing mat and for reinforcing coatings and paints.

fiberboard. A broad term for building materials of widely varying densities, manufactured from long, refined or partly refined vegetable fibers (e.g., wood, bagasse, straw, old newspapers or groundwood paper) by pressing or rolling into flat, firm, strong semirigid sheets. Bonding agents are optional, although they are often added to enhance strength and resistance to decay, fire and/or moisture. Used for the construction of interior walls (e.g., as acoustical and thermal insulation), inside of exterior walls, as sheathing, and for containers, partitions, counters, etc. Also known as *fibrous-felted board*.

Fiberbrite 2000. Trade name of Honeywell International (USA) for polyester fibers and yarns used for textile fabrics.

fiber bundle. A package of numerous long, thin, flexible transparent fibers, parallel to each other and usually composed of glass, used for the transmission of images from one end of the bundle to the other in fiber-optic applications.

Fiber C. Trade name for cellulose acetate fibers used for textile fabrics.

Fibercast. Trademark of Fibercast (USA) for glass-fiber-reinforced epoxies.

fiber cement composites. A group of composite materials consisting of metallic or nonmetallic fibers (e.g., steel, glass or carbon) embedded in cement matrices. The fiber-matrix bond is usually enhanced by the addition of chemical admixtures, silica fume, and/or polymers. Also known as *fiber-reinforced cement composites*. See also reinforced concrete.

Fiberclad. Trademark of Ultralux Plastic Industries Limited (Canada) for glass fiber-reinforced plastics for pipes, tanks, etc.

fiber composites. A group of composite materials consisting of metallic or nonmetallic matrices reinforced with continuous or discontinuous metallic or nonmetallic fibers. In these composites the matrix and the reinforcement exist as two discrete physical phases. Abbreviation: FC. Also known as *fiber composite materials; fiber-reinforced composites; fibrous composites*. See also metal-matrix composites, ceramic-matrix composites, fiber-reinforced plastics.

Fiber D. Trademark for acrylic fibers used for the manufacture of textile fabrics.

Fiberdri. Trademark of Camelot Superabsorbents Limited (Canada) for superabsorbent fibers used for textile fabrics.

Fiberdux. Trade name of Ciba Geigy Corporation (USA) for epoxy laminates supplied as *Kevlar* or carbon fiber prepregs.

Fiber E. Trade name for rayon fibers and yarns used for textile fabrics.

fiber-epoxy composites. Composite materials consisting of epoxy resin matrices reinforced with continuous or discontinuous fibers (e.g., carbon, graphite, aramid, glass, etc.). Used for aircraft, spacecraft and helicopter components, ship hulls, offshore structures, printed wiring boards, etc.

fiber-epoxy prepreg tape. Continuous or discontinuous fibers of aramid, carbon, fiberglass, etc., preimpregnated with epoxy resin and made into tape form.

Fiberez. Trademark of Fiberez Corrosion Molding Products Limited (USA) for reinforced plastics used for plumbing fixture applications.

Fiber Face. Trademark of Georgia-Pacific Corporation (USA) for hardboard panels for in building construction.

Fiberfil. Trademark of DSM Engineering Plastics (USA) for reinforced and filled thermoplastics. *Fiberfil G* is a series of chemical-resistant, long glass fiber-reinforced polycarbonates and styrene-acrylonitriles for automotive, electronic and other industrial applications. *Fiberfil J* refers to a series of chopped or long glass fiber- and/or mineral-filled nylon 6, nylon 6,6, nylon 6,10, nylon 6,12, polycarbonate, polyether sulfone, polyetheretherketone, polyphenylene sulfide, polyethylene terephthalate and polypropylene resins for automotive, electrical, elec-

tronic and various other applications. *Fiberfil NY* refers to a series of impact-resistant mineral-filled or toughened nylon 6,6 resins used for automotive, electrical, electronic and other industrial applications. *Fiberfil PP* are mineral-reinforced polypropylenes with excellent impact resistance used for automotive and electronic components. *Fiberfil TN* refers to composites consisting of nylon 6,6 matrices reinforced with glass beads. They have excellent toughness, stiffness and wear resistance, high dimensional stability and durability, excellent resistance to corrosion from fuels, and good heat resistance. Used for automotive components. *Fiberfil Foam* is a chemical- and flame-resistant glass fiber-reinforced polycarbonate foam material.

Fiberfill. Trade name for polyester fibers.

fiberfill. Fibers of jute, sisal, kapok or synthetic materials that are of suitable density, length, crimp and other properties to be useful as filling materials in pillows, mattresses, furniture battings, textiles, sleeping bags, etc.

Fiberfilm. Trade name of American Machine & Foundry Company (USA) for a flexible, nonwoven sheet of matted submicron glass fibers, randomly oriented and impregnated with tetrafluoroethylene resins. They are available in porous and nonporous grades in various thicknesses with a maximum service temperature of 250°C (480°F). The porous grade is used as a filter material. The nonporous grade has excellent dielectric properties, and is used in capacitors and transformers.

fiber-forming material. A term referring to any synthetic manmade material that is capable to form and/or be formed into fibers or filaments.

Fiberfrax. Trademark of The Carborundum Company (USA) for a series of high-temperature ceramic fiber products based on alumina (Al_2O_3) and silica (SiO_2). They are commercially available in bulk as blown, chopped and washed fibers, discontinuous and continuous engineered fibers, mats, blankets, textiles, paper, ropes, rovings and blocks. They have outstanding insulating properties, excellent high temperature stability up to about 1260°C (2300°F), low thermal expansion, excellent chemical and fire resistance, good resistance to most acids and hydrogen atmospheres, good mechanical properties, and high strength and resilience. Used as insulating products for commercial buildings and nuclear and chemical plants, for the high-temperature insulation of kilns, furnaces and burner blocks, for the insulation of electrical wire and motors, in jet-motor insulation, for packing expansion joints and heating elements, for transportation equipment, in rolls for roller-hearth furnaces and piping, as sound-deadening materials, for fine filtration, for the reinforcement of metal-matrix composites, and for plastic-resin and ceramic systems.

Fiberfrax Sprayable. Trademark of The Carborundum Company (USA) for sprayable ceramic-fiber insulating materials designed for refit, repair and lining-over-refractory applications. They have high thermal-shock resistance, a high insulation value, and a maximum service temperature of about 1150°C (2100°F). Used for industrial furnaces (including heat-treating furnaces) and heating plant, chemical-process equipment, and for equipment used in the ceramics industry.

Fiber Gain. Trade name of Corning Inc. (USA) for optical fibers.

Fiberglas. Trademark of Owens Corning Fiberglass (USA) for an extensive series of products manufactured with or from glass flakes or fibers and including various mats, rovings, insulating wools, coarse fibers, yarns, cordage, acoustical products, electrical insulation and reinforced plastics. Used as reinforcements for organic and inorganic materials. Coarse fibers are employed in the form of packs for filtering fluids and corrosion protection of underground pipelines.

Fiberglas Pink. Trademark of Owens Corning Fiberglass (USA) for pink-colored lightweight fiberglass wall and attic insulation. The attic insulation is supplied with a heat flow resistance (R-value) of R-40, and the wall insulation with R-values of R-12 and R-20.

fiberglass. (1) Glass drawn or spun into fine threads, fibers or filaments.

(2) A thick insulating material composed of matted fiberglass.

(3) A generic term for light, relatively strong, continuous or discontinuous fibers composed of glass. The approximate composition of glass that is most commonly drawn or spun into fibers is 55% silica (SiO_2), 16% calcia (CaO), 15% alumina (Al_2O_3), 10% boria (B_2O_3), and 4% magnesia (MgO). Fiberglass is available in various grades with specific chemical, physical or other properties including: (i) *AR-glass* (alkali-resistant glass); (ii) *C-glass* (glass with excellent resistance to corrosion by most acids); (iii) *D-glass* (glass with low dielectric constant); (iv) *E-glass* (glass with excellent electrical properties); (v) *ECR-glass* (modified E-glass with improved resistance to corrosion by most acids); (vi) *R-glass* (corrosion-resistant glass with high tensile strength and modulus); and (vii) *S-glass* (glass with high tensile strength, elastic modulus and service temperature). *Fiberglass* is used in autobodies, boat hulls, plastic pipes, storage containers, industrial flooring, and as a composite reinforcement. Abbreviation: FG.

fiberglass-epoxy composites. See glass fiber-epoxy composites.

fiberglass fabrics. Fabrics made from fiberglass yarns using any of several weave patterns.

Fiberglass Fusion. Trade name of ITW Plexus (USA) for a methacrylate-based structural adhesive for bonding fiberglass products.

fiberglass insulating boards. Insulating boards manufactured from glass-fiber materials and usually supplied as high-density, semi-rigid products in above- and below-grade types with high thermal insulating value. The above-grade types are covered with water-repellent breather-type building paper for use as exterior sheathing, and the below-grade types often have moisture-resistant coverings for use on foundation wall interiors. Also known as *glass fiber insulating boards*. See also fiberglass; fiberglass insulation; glass fibers; rigid board insulation.

fiberglass insulation. Fiberglass-based insulation products supplied in several forms including batts, blankets, semi-rigid boards and loose fill. Used for the thermal insulation of attics, floors, and interior and exterior walls. Also known as *glass fiber insulation*. See also batt insulation; blanket insulation; board insulation; loose-fill insulation; fiberglass insulating boards.

fiberglass mat. A glass product manufactured from continuous-strand fibers or randomly oriented chopped fibers loosely held together with a binder. Supplied in blankets of various weights, widths and lengths, it is used as a reinforcement for engineering plastics.

fiberglass paper. See glass paper (2).

fiberglass-polyester composites. Composite materials consisting of unsaturated polyester resin matrices reinforced with glass fibers. They have high strength, good flexural properties, and good thermal and electrical properties. Used for housings, body panels of trucks and automobiles, rear lifts on wagons and vans, electrical components, etc.

fiberglass-reinforced BPADCy prepolymer. A prepolymer of bisphenol A dicyanate (BPADCy) with glass-fiber reinforcement that has good hardness and dimensional stability, a long shelf life, and a low dielectric constant and dissipation factor. Used as laminating and molding resin for printed wiring boards and electronic components.

fiberglass-reinforced plastics. Composites consisting of reinforcing fibers or filaments of fiberglass embedded in polymer matrices (e.g., epoxy, unsaturated polyester, etc.). They have high tensile strength, good flexural properties, good thermal and electrical properties and, depending on the matrix resin, other properties, such as high strength-to-weight ratios, or good chemical resistance. Used for structural and nonstructural components for aircraft, spacecraft and helicopters, body panels of trucks and automobiles, rear lifts on vans, and sports vehicles, ship and boat hulls, offshore structures, pressure vessels, tanks and pipe, pump components for chemical and food handling, and electrical laminates for circuit-board manufacture. Abbreviation: FRP. Also known as *fibrous-glass-reinforced plastics*. See also fiber-reinforced plastics.

fiberglass reinforcement. Continuous or discontinuous glass fibers embedded in a metallic or nonmetallic matrix to provide additional strengthening (e.g., improved tensile and/or impact strength or toughness). See also reinforcement fiber.

fiberglass roving. A bundle of continuous, usually untwisted, glass fibers wound on a roll. The wound product is called a "roving package." Used in the manufacture of pipes, tanks, leaf springs, bathtubs, etc.

fiberglass yarn. See glass-fiber yarn.

Fiberglas Vetrotex. Trademark of Gevetex-Textilglas GmbH (Germany) for glass fibers.

Fibergrate. Trade name of Grating Pacific, LLC (USA) for a line of molded grating products available in 5 different resin systems.

Fiber HM. Trade name for high-modulus rayon fibers.

Fiberite. Trademark of Fiberite Corporation (USA) for an extensive series of phenolic, epoxy and specialty resin laminates and molding compounds including various glass, mineral and carbon fiber grades. *Fiberite EP* are epoxy laminates and molding compounds. The laminates are supplied as glass fabrics and prepregs, and carbon and Kevlar fiber prepregs. The molding compounds are supplied in the following grades: glass fiber-reinforced, mineral-filled, mineral- and glass fiber-reinforced, and high-heat. *Fiberite PF* refers to phenolic laminates and prepregs. The laminates are available as cotton and glass fabrics and paper laminates, while the prepregs are supplied in the following grades: wood- or natural fiber-filled for general purposes, cellulose-filled for shock resistance, glass fiber-filled for high impact, chopped fabric-filled for medium impact, cotton-filled for medium shock resistance, mica-filled for electrical applications, and mineral-filled for high heat resistance. A foam grade is also available. *Fiberite PI* are polyimide resins supplied in unmodified, glass fiber-reinforced, and graphite-, molybdenum disulfide- or polytetrafluoroethylene-filled grades. *Fiberite Silicone* refers to silicone resins and plastics supplied in standard and glass fiber- or mineral-filled grades.

Fiber K. Trade name for a white, highly flexible spandex (segmented polyurethane) fiber used for the manufacture of textile fabrics with excellent stretch and recovery properties.

Fiberlac. Trade name of Monsanto Chemical Company (USA) for cellulose nitrate plastics.

fiber leather. See leatherboard.

Fiberlets. Trade name of Lancaster Fiber Technology Group (UK) for a short-strand E-glass fiber made from glass-fiber byproducts, and used as reinforcement in thermosetting and thermoplastic composites.

Fiberline. Trademark of Kunststoffe Arthur Krüger (Germany) for glass-fiber-reinforced plastic products usually supplied in the form of profiles.

Fiber-Lite. Trade name of PK Insulation Manufacturing Company, Inc. (USA) for wood-fiber cellulose used for thermal insulation applications.

Fiberlite Epoxy. Trade name of Vigilant Plastic Company (USA) for epoxy molding compounds supplied in various grades: glass fiber-reinforced for general purposes, glass fiber-reinforced for high heat resistance, and mineral- and/or glass fiber-filled for fire retardancy.

Fiberloc. Trademark of Geon Company (USA) for chemically coupled glass fiber-reinforced engineering thermoplastics based on vinyl resins, such as polyvinyl chloride. Suitable for processing by extrusion and injection molding techniques, they have excellent flow characteristics, high strength, stiffness and dimensional stability, good creep and fatigue resistance, excellent chemical, corrosion and weather resistance, low coefficients of thermal expansion, low moisture absorption, low flammability, high ultraviolet stability, and good processibility.

Fiberloid. Trade name of The Fiberloid Corporation (USA) for a thermoplastic cellulose nitrate supplied in transparent, opaque, colored and colorless grades in the form of rods, sheets, and lacquers. It has good machining qualities, and good resistance to hydrocarbons and oils.

Fiber-Lok. Trade name of Senior Aerospace Composites (USA) for honeycomb panels for aerospace applications.

Fiberlon. (1) Trade name of The Fiberloid Corporation (USA) for a thermosetting phenol-formaldehyde resin available in transparent, translucent, opaque, colored and colorless grades. It has good machining qualities and excellent resistance to ketones, esters and hydrocarbons oils, and is used for molded products.

(2) Trade name of The Fiberloid Corporation (USA) for cellulose nitrate plastics.

(3) Trade name of Georgia Bonded Fibers, Inc. (USA) for coated paper.

Fibermax. (1) Trademark of Sohio Engineered Materials (USA) for a discontinuous ceramic fiber containing 72% alumina (Al_2O_3) and 28% silica (SiO_2). Commercially available as discontinuous fiber and in bulk form, it has a density of 3.0 g/cm³ (0.11 lb/in.³), good mechanical properties, a high service temperature of 1650°C (3000°F), and an average fiber diameter of 2-3.5 µm (79-138 µin.). Used as reinforcing fiber for metals and ceramics.

(2) Trademark of Michelman, Inc. (USA) for a series of specialty coatings for strengthening paper and corrugated containers.

(3) Trademark of Aventis CropScience SA (France) for cotton and other textile fibers, yarns and threads.

(4) Trademark of The Carborundum Company (USA) for ceramic fibers, fiber batts, sheets, rolls and coating compositions. Used as thermal insulation.

Fiberock. (1) Trademark of United States Gypsum Company (USA) for fiber-reinforced gypsum wall, ceiling and floor underlayment panels.

(2) Trademark for roofing board made by impregnating asbestos fibers with asphalt.

Fiberod. Trademark of Polymer Composites, Inc. (USA) for a family of advanced thermoplastic composite materials composed of high-strength aramid, carbon, glass or hybrid fibers in thermoplastic matrix resins, such as polyphenylene sulfide, polyethylene sulfide, polyethylene terephthalate, polybutylene terephthalate, polypropylene, polycarbonate, nylon 6,6 or acetal resin. They have high stiffness and toughness, good mechanical properties and unlimited shelf life. Used for structural applications.

fiber-optics. See optical fibers.

fiber plaster. Gypsum plaster reinforced or strengthened with hair or natural fibers, such as cotton, sisal, or wood. Also known as *fibrous plaster.*

Fiber-Ply. Trademark of Georgia-Pacific Corporation (USA) for plywood panels used for building and construction applications.

fiber preform. See preform (2).

fiber-reinforced cement. See glass fiber-reinforced cement.

fiber-reinforced cement composites. See fiber-reinforced cementitious composites.

fiber-reinforced cementitious composites. A group of composites that consist of cementitious matrices (e.g., cement, mortar or concrete) reinforced with asbestos, metal, synthetic (e.g., nylon or polyethylene) or glass fibers. Also known as *fiber-reinforced cement composites.* See also glass fiber-reinforced cement; fiber-reinforced concrete.

fiber-reinforced ceramics. A group of engineering composites that consist of ceramic matrices (e.g., aluminum oxide, silicon carbide, silicon nitride, glass, or glass-ceramic) reinforced with metal fibers or wires (e.g., molybdenum, tungsten, tantalum or niobium), carbon or graphite fibers, or ceramic fibers (e.g., silicon carbide, aluminum oxide or zirconium). See also ceramic-matrix composites.

fiber-reinforced composites. See fiber composites.

fiber-reinforced concrete. A composite material composed of a concrete-based matrix reinforced with steel fibers or high-modulus fibers of asbestos, glass, nylon, polyethylene, etc. Abbreviation: FRC. See also reinforced concrete; steel-fiber-reinforced concrete.

fiber-reinforced glass. Glass or glass-ceramic composites that have been reinforced with continuous fibers of carbon or silicon carbide. They are produced by coating the fibers with a glass-powder-based slurry, and adding an organometallic sol-gel. The fibers are then wound onto reels and formed into prepregs, followed by hot pressing. Upon drying and tempering the slurry forms the glass matrix of the composite, while the sol-gel converts into an inorganic substance. For glass-ceramic composite fabrication suitable nucleating agents must be added. The finished products have high tensile strength and impact resistance, but do not have the inherent brittleness of ordinary glass. Fiber-reinforced glass and glass-ceramic composites are available in various shapes and sizes for use in structural and nonstructural applications.

fiber-reinforced metals. A group of composites consisting of metal matrices, e.g., aluminum or copper, reinforced with continuous or discontinuous fibers or whiskers of aluminum oxide, boron, graphite, silicon carbide, etc. Abbreviation: FRM. See also metal-matrix composites.

fiber-reinforced plastics. A group of composites consisting of polymeric matrices reinforced with fibers. The matrix materials are either thermosetting polymers (e.g., epoxy, polyester, phenolic or silicone) or thermoplastic polymers (e.g., nylon, polycarbonate, polystyrene, polyphenylene sulfone or polyether-etherketone), and the fiber reinforcement is usually aramid, carbon, glass, graphite, etc., in the form of continuous or chopped strands, mats, woven fabrics, and the like. Used for aircraft, spacecraft and helicopter panels and parts, boat and ship hulls, automobile panels and components, offshore structures, printed circuit boards and other electronic components, housings of equipment and machinery, bearings and other machine parts, etc. Abbreviation: FRP. Also known as *fiber-reinforced polymers; fiber-reinforced polymer composites; polymeric composites; polymer-matrix composites.*

fiber-reinforced polymer composites. See fiber-reinforced plastics.

fiber-reinforced polymers. See fiber-reinforced plastics.

fiber-reinforced superalloys. A group of engineering composites consisting of oxidation-resistant matrix alloys, usually nickel-, cobalt- or iron-based superalloys, reinforced with strong, stiff, creep-resistant fibers or wires of tungsten, tungsten-alloy, molybdenum-alloy, silicon carbide or aluminum oxide. They have good elevated-temperature properties, good thermal fatigue properties, high impact strength, and good hot corrosion resistance. Used for aircraft engines and gas turbines, turbine blades, turbopumps, flywheels, and pressure vessels. Abbreviation: FRS.

fiber-reinforced thermoplastic composites. Composites that consist of thermoplastic matrix resins, such as polybutylene terephthalate, polyethylene terephthalate, polyamide (nylon) or polysulfone, reinforced with synthetic fibers, such as glass, carbon, etc. Also known as *fiber-reinforced thermoplastics.* See also fiber-reinforced plastics.

fiber-reinforced thermoplastics. See fiber-reinforced thermoplastic composites.

fiber-reinforced thermoset molding compound. See bulk molding compound.

fiber reinforced thermosets. See fiber-reinforced thermosetting composites.

fiber-reinforced thermosetting composites. Composites that consist of thermosetting matrix resins, such as amino, phenolic, polyester or epoxy, reinforced with synthetic fibers, such as glass, carbon or aramid. Also known as *fiber-reinforced thermosets; fiber-reinforced thermosetting plastics.*

fiber-reinforced thermosetting plastics. See fiber-reinforced thermosetting composites.

fiber reinforcement. A strong fiber phase embedded in a relatively weak material in order to strengthen or reinforce it. Typical fiber materials for this application include fiberglass, graphite, aramid, and boron. Thermosets, thermoplastics, ceramics and metals can contain fiber reinforcements.

fiber-resin composites. Composites comprising thermoplastic or thermoset resin matrices reinforced with fibrous materials, and available as "ready-to-use" fiber-resin combinations in various forms, e.g., chopped-fiber molding compounds, filament-wound fiber bundles or tows, lay-up fabrics, and unidirectional tape lay-down woven fabrics.

fiber rope. A compact, flexible, continuous line or cord of natural (cotton, manila, sisal, etc.) or synthetic (nylon, polyester, polypropylene, etc.) fibers made from strands which are braided, plaited or twisted together and have a diameter of at least 5 mm (0.2 in.). Natural fiber ropes usually have short overlapped fibers, while synthetic fiber ropes have fibers that run the entire length of the rope. Used for the transmission of a force, e.g. by hauling, rigging, lifting of loads. Also known as *fibrous*

rope. See also braided rope; twisted rope; rope materials.

fiber silk. A term formerly used for rayon.

Fibersorb. Trademark of Camelot Superabsorbents Limited (Canada) for superabsorbent fibers used for woven and nonwoven textile fabrics.

Fiberstran. Trademark of DSM Engineering Plastics (USA) for engineering thermoplastics based on long glass fiber-reinforced polycarbonates, polypropylenes, polysulfones and nylon 6, nylon 6,6, nylon 6,10 and nylon 6,12 resins with high strength and excellent impact resistance for use in automotive components such as battery trays, brackets and lighting supports, and various other applications.

Fiber T. Trade name for a modacrylic fiber used chiefly for clothing and industrial fabrics.

Fiberthin. Trademark of United States Rubber Company for a thin, strong, waterproof, woven nylon fabric coated with a plastic, and used for protective coverings, tarpaulins, etc.

Fibertuff. Trade name for polystyrene reinforced with 40 wt% glass fiber and supplied in pellet form for injection-molding. It is used to make light, moderately strong molded parts with high impact resistance, and a heat distortion temperature of 104°C (219°F).

Fibestos. Trade name of Monsanto Chemical Company (USA) for a thermoplastic cellulose nitrate available in transparent, translucent, opaque, colored, and colorless grades. It has good moldability and machining qualities, good resistance to hydrocarbons and oils, and poor flame resistance, but is slow burning. Used for molded products.

Fibral. Trademark of Sanderson Industries Limited (Canada) for fiberglass products.

Fibralloy. Trade name of Thermal Ceramics Inc. (USA) for ceramic preforms consisting of discontinuous alumina-silica or high-alumina ceramic fibers or powders, or fiber/powder blends. The ceramic content can be tailored from 7 vol% for increased wear resistance to 15 vol% for enhanced hot strength, and up to 35% for improved stiffness. The coefficient of thermal expansion can be modified to customer specifications. Used for automotive components, such as pistons.

Fibramine. A trade name for viscose rayon that contains casein and is used for making textile fabrics.

Fibrana. Trade name for rayon fibers and yarns used for textile fabrics.

Fibranne. Trade name for rayon fibers and yarns.

Fibrasiv. Trade name of Società Italiana Vetro SpA (Italy) for glass fiber wool.

Fibravid. Trade name of Vidrobrás (Brazil) for insulating glass fibers.

Fibravyl. Trademark of Rhovyl SA (France) for vinyon (vinyl chloride) fibers with excellent acid, alkali and mildew resistance. Used for textile fabrics.

FibreDecor. Trademark for a wall compound used for the roller application of decorative wall effects, and for covering defects.

Fibredux. Trade name of Ciba Geigy Composites (USA) for bismaleimide (BMI) composite prepreg materials. Used especially for aircraft structures.

Fibreen. Trade name of BSK, Division of DS Smith (UK) Limited for a strong, tough laminated product made by combining two sheets of strong *kraft paper* strengthened with two crossed layers of asphalt-bonded sisal fibers under heat and pressure. Used as a case liner, lidding or wrapping paper for commercial and industrial packaging applications.

Fibreglass. Trademark of Fibreglass plc (UK) for an extensive range of glass fibers and glass-fiber products.

Fibrelam. Trade name of Ciba-Geigy Corporation (USA) for honeycomb materials consisting of epoxy or fiberglass/epoxy skins and *Nomex*/phenolic or carbon fiber-reinforced *Nomex*/phenolic cores. Used for boat hulls, and aircraft flooring and interior parts.

Fibreloft. Trade name of E.I. DuPont de Nemours & Company (USA) for discontinuous synthetic filaments.

Fibrelta. Trade name for rayon fibers and yarns used for textile fabrics.

Fibremat. Trademark of 3M Company (USA) for fiberglass mats used in the manufacture of electrical insulating tape.

Fibrenka. Trade name of Enka BV (Netherlands) for viscose rayon fibers used for clothing, and blended with other fibers for suitings, furnishings, carpets, linings, and medical fabrics.

Fibre Plus. Trademark of Domtar Inc. (Canada) for fiberglass roof insulation.

Fibrewall. Trade name for textured wall coatings.

Fibrex. (1) Trademark for a micronic nickel fiber with a nominal diameter of 20 μm (790 μin.) that is available in various aspect ratios, and used for nonwoven webs for battery electrode structures, in conductive EMI/RFI shielding, and for electrostatic discharge (ESD) thermal conductivity and filtration applications. See EMI/RFI shielding materials.

(2) Trademark of Fibrex Isolation Limitée (Canada) for cellulosic insulation products.

(3) Trade name of Oliner Fibre Company, Inc. (USA) for industrial fiberboard and paperboard.

(4) Trade name of National-Standard, Woven Filter Fiber Division (USA) for nonwoven metallic fiber fabrics.

Fibri-cord. Trade name for polyolefin fibers and yarns used for textile fabrics and cordage.

fibrids. A generic name for fibrillar or filamentary man-made polymer fibers used as as bonding elements in manufacture of wet-laid synthetic paper, and in the production of certain nonwovens known as "textryl."

Fibri-knit. Trade name for polyolefin fibers and yarns used for textile fabrics including knitted and woven clothing.

Fibrilawn. Trade name for polyolefin fibers and products.

fibrillated fibers. (1) Textile fibers that have been given a finishing treatment to cause the *fibrils* to protrude from the surface. Used for making soft fabrics with a frosted, washed appearance. See

(2) Synthetic fibers produced by mechanically cracking stretched polymer films that have been oriented in extrusion direction.

(3) Staple fiber made from fibrillated yarn or fibrillated film tow by cutting, chopping or breaking processes. Also known as *fibrillated film fibers*.

fibrillated film fibers. See fibrillated fibers (3).

fibrillated-film tow. A product consisting of several fibrillated textile films.

fibrillated-film yarn. A yarn made from a polymer film that has been given a fibrillated structure in the longitudinal direction.

fibrillated yarn. A textile yarn consisting of a network of interconnected fibers, and made by splitting an oriented film or tape in a longitudinal direction.

fibrillating fibers. Textile fibers whose ends, when wet and mechanically agitated, peel back, splinter and protrude as minute fibrils from the surface. Fiber fibrillation can either be inherent, or intentionally produced by a special finishing treatment. See also fibrils; fibrillated fibers; nonfibrillating fibers.

fibrillating film. A polymer film that has been longitudinally oriented by stretching. Used in the manufacture of fibrillated yarns.

Fibrilon. Trade name for polyolefin fibers and yarns used for textile fabrics.

Fibrils. Trademark of Hyperion Catalysis International (USA) for graphite microfibers with an average diameter of 0.01 μm (0.4 μin.) and excellent static dissipation properties, used in the manufacture of conductive elastomeric, thermoplastic and thermosetting composites.

fibrils. The minute slender elements of natural or synthetic fibers.

fibrin. A white, insoluble, fibrous protein in blood produced in the hydrolysis of fibrinogen by the action of the enzyme *thrombin*, that causes blood to coagulate by forming blood clots. Used in biochemistry, biotechnology, and in the manufacture of fibrin glue.

fibrin glue. A biomedical adhesive based on *fibrin* and used as a tissue adhesive and vascular graft coating.

Fibro. (1) Trade name of British Steel Corporation (UK) for a steel that contains 0.15-0.25% carbon, and the balance iron, and is used for case-hardened parts, gears, and shafts.

(2) Trademark of Courtaulds plc (UK) and Acordis Cellulosic Fibers Inc. (USA) for viscose rayon staple fibers used for clothing, and blended with other fibers for suitings, furnishings, carpets, linings, and medical fabrics.

Fibrocam. Trade name of Türkiye Sise ve Cam Fabrikalari AS (Turkey) for glass tissue used as pipe insulation.

Fibroceta. Trade name for cellulose acetate fibers used for wearing apparel and as industrial fabrics.

Fibrofelt. Trade name for an acoustic insulation material manufactured from flax or rye fibers.

fibroferrite. A pale yellow to white mineral composed of iron sulfate hydroxide pentahydrate, $Fe(OH)SO_4 \cdot 5H_2O$. Crystal system, rhombohedral (hexagonal). Density, 1.92 g/cm^3; refractive index, 1.532. Occurrence: Chile.

Fibroglass. Trade name of Türkiye Sise ve Cam Fabrikalari AS (Turkey) for glass tissue used for the insulation of pipes and buildings against water and humidity.

fibroin. The water-insoluble, fibrous albuminoid protein secreted by silkworms and spiders, and containing glycine and alanine. It has a pleated sheet structure in which the polypeptide chains are arranged alongside, but antiparallel to each other, and held together by hydrogen bonds. Used in biochemistry, biotechnology and bioengineering.

Fibrolane. Trade name for regenerated protein (azlon) fibers.

fibrolite. See sillimanite.

Fibrolux. Trade name of Fibrolux GmbH (Germany) for an extensive series of plastic products including (i) epoxy, polyvinyl chloride, polyethylene, polypropylene or polycarbonate pipes and profiles, (ii) aramid, carbon or glass fiber-reinforced plastic pipes and profiles, (iii) polycarbonate, polyethylene, polypropylene, polystyrene, polyvinyl chloride, or glass fiber-reinforced polyester sheeting, (iv) epoxy, phenolic and other laminates with paper or fabrics, (v) corrugated acrylic, polyester, polycarbonate and polyvinyl chloride sheeting, (vi) plasticized and unplasticized polyvinyl chloride compounds, and (vii) biaxially oriented polypropylene films and copolymer shrink films.

Fibron. Trademark of FIBRON Wolfgang Mellert GmbH (Germany) for unsaturated polyester products.

fibronectin. Any of the cell adhesion and migration promoting glycoproteins found on animal and human cell surfaces, and in connective tissue and blood plasma. Used in biochemistry and biotechnology.

fibronectin-like engineered protein polymer. A recombinant biopolymer, developed by Protein Polymers Technologies, Inc. (USA), that incorporates repeated structural peptide units with interspersed multiple copies of the human fibronectin RGD attachment ligand. It has a molecular weight of approximately 110000 (by electrophoresis), and is used in biochemistry (e.g., tissue culture) and biotechnology.

Fibropane. Trade name of Plyglass Limited (UK) for plate glass that contains a colorless or pigmented layer of plastic-bonded glass gauze.

fibrous composites. See fiber composites.

fibrous-felted board. See fiberboard.

fibrous-glass-reinforced plastics. See fiberglass-reinforced plastics.

fibrous gypsum. A snowy white, translucent, fibrous variety of gypsum with a silky sheen. Also known as *satin spar; satin stone.*

fibrous material. (1) Any tough material with threadlike structure.

(2) A metal, polymer or ceramic in fibrous form, e.g., aluminum, nylon, aramid, glass, graphite, boron, aluminum oxide, or asbestos.

fibrous plaster. See fiber plaster.

fibrous proteins. A class of water-insoluble proteins that serve as the chief structural materials in animal tissues. Their molecules are long, thread-like, and tend to form fibers. Examples include collagen in tendons, fibroin in silk, keratin in hair, nails or skin, and myosin in muscle.

fibrous refractory composite insulation. A high-strength, low-density thermal insulation material that is a composite of 78% silicon dioxide fibers and 22% aluminum borosilicate fibers. It has a density of 0.19 g/cm³ (0.007 lb/in.³), very low thermal expansion, excellent thermal-shock resistance, a maximum-use temperature of 1480°C (2700°F), and a small modulus of elasticity. *Fibrous refractory composite insulation* in the form of ceramic tiles, available under the trade name of *FRCI-12*, are used around the access panels and door areas of the NASA Space Shuttle. Abbreviation: FRCI.

fibrous rope. See fiber rope.

fibrous silica. Silica (SiO$_2$) made into batted or felted form and then either spun into fine continuous filaments, or woven into fabrics. Used primarily as high-temperature insulating fibers for aircraft and motor components, etc. See also silica; silica fiber.

fibrous wool. A fibrous material that consists of fine filaments made by spinning, drawing, etc., and resembles loose fibers of wool in texture, e.g., glass or rock wool, or wool composed of short fibers of silicon dioxide or aluminum oxide.

Fibrox. (1) Trade name for a ceramic fiber based on silicon oxycarbide (SiCO) and supplied in light fluffy fibers, batts, blocks and sheets for thermal insulation applications.

(2) Trademark of Rank Organisation Limited (UK) for lightguides composed of glass or polymer fibers in a glass or plastic sheath.

Fibrtough. Trade name of Atlas Steels Limited (Canada) for a high-ductility steel with 0.15% carbon, 0.7% manganese, 0.25% silicon, and the balance iron. Used for boiler staybolts.

Fidelity. Trade name for a series of gold-based dental casting alloys including *Fidelity 370 & 375* containing 70% and 75% gold, respectively, and *Fidelity 340, 350 & 360* with 40%, 50% and 60% gold, respectively.

Fidelity 58.5. Trade name for a 58.5% gold dental solder.

Fidelity White. Trade name for a white-gold dental casting alloy. See also white gold.

Fidion. Trademark of EniChem Fibre SpA (Italy) for polyester fibers.

fiedlerite. A colorless mineral composed of lead chloride hydroxide, Pb$_3$Cl$_4$(OH)$_2$. Crystal system, monoclinic. Density, 5.58 g/cm³; refractive index, 2.04. Occurrence: Greece.

Field. British trade name for a stainless alloy containing 85-85.5% nickel, 10% silicon, 3% copper, and 1.5-2% aluminum. Used for applications requiring corrosion resistance.

field concrete. Concrete supplied to, or mixed, placed and cured at the construction site. See also job-mix concrete.

field pine. See loblolly pine.

Field Rolled. British trade name for a corrosion-resistant alloy containing 60% nickel, 20-29% iron and 11-20% molybdenum, used for stainless parts.

Field Stripe. Trademark of Canada Talc Industries Limited for dolomite and dolomite products.

Fifteen Per Cent. British corrosion-resistant alloy containing 57% copper, 28% zinc and 15% nickel, used for ornamental applications.

Fifty-Five. Trade name of Lake Erie Engineering Corporation (USA) for a cast iron containing 2.9-3.1% carbon, 1.2-1.5% nickel, 0.4-0.6% molybdenum, 1.4-1.6% silicon, and the balance iron. Used for diesel engine and hydraulic press parts.

figue. A fiber obtained from the leaves of a perennial plant (*Furcraea macrophylla*). See also Mauritius hemp.

figured fabrics. Fabrics, usually of the dobby- or jacquard-type, that have patterns or designs made by combining various characteristic weaves.

figured glass. Flat glass decorated on one or both surfaces with an etched or ground design or pattern.

figured rolled glass. A translucent rolled glass having a pattern etched or ground on one surface that partially or entirely obscures vision.

figured velvet. Velvet that has been given a design that stands out from the surface by cutting or pressing the pile.

figured veneer. A veneer containing attractive grain formations that make it suitable for decorative applications, such as cabinet and furniture panel faces, plywood facings, etc.

Fiji silk. A crisp silk that is somewhat heavier than China silk, usually of natural or cream color, and used for luxurious blouses and dresses.

Filac. Trade name of Alfred Harris & Company (UK) for cellulose nitrate plastics.

Filabond. Trade name of unsaturated polyesters.

Filament. Trademark of Soenmez Filament Sentetic Iplik AS (Turkey) for polyester staple fibers and filament yarns.

filament. An individual, flexible, continuous fiber of small cross section, usually extruded, spun or drawn from a material, such as glass, nylon, rayon, acetate, a metal, or a metal carbide or oxide. Also known as *filament fiber.*

filamentary superconducting alloy. A composite wire composed of one or more superconducting filaments embedded in a metal matrix. See also composite superconductor; multifilamentary superconducting alloy.

filament blend yarns. Yarns made of two different types of more or less randomly blended filaments. They uniquely combine the properties of two chemically and/or physically different yarn materials. Examples include filaments of rayon and nylon or acetate and nylon twisted together. Used for the manufacture of fabrics such as crepes, twills and jerseys.

filament fiber. See filament.

filament-winding resins. Thermosetting resins, such as epoxy, unsaturated polyester or vinyl ester, suitable for coating continuous reinforcements, e.g., fiber, tape, wire or yarn, prior to or during the filament winding operation.

filament-wound composites. Large composite structures, such as pressure vessels, chemical storage tanks, piping, etc., made by winding continuous fibers of glass, boron, silicon carbide, etc., impregnated with a thermosetting matrix resin, such as epoxy, polyester or vinyl ester, before or after winding under tension onto a mandrel.

filament yarn. See continuous-filament yarn.

Filamite. Trademark of NVF Company (USA) for filament-wound glass epoxy tubing.

Fi-lana. Trade name for acrylic fibers used for the manufacture of textile fabrics.

Filanda. Trademark of Filanda Srl (Italy) for filament yarns.

Filatex. (1) Trade name of Linatex Inc. (USA) for cold-cure putty fillers based on natural rubber.

(2) Trade name of Linatex Inc. (USA) for rubber fibers used in the manufacture of elastic yarn for clothing, elastic bands and tapes, etc.

filature silk. A raw silk unwound from cocoons into skeins by means of machinery.

Filawound. Trademark of Spaulding Fibre Company Inc. (USA) for glass-filament wound tubing.

Filbrite. British trade name for an iron-carbon alloy containing 35% nickel and 17% chromium. Used for heat- and corrosion-resistant components.

Filcryl. Trade name of Portland Plastics (UK) for acrylic resins.

File. Trade name of Toledo Steel Works (USA) for a series of plain-carbon steels containing 1.2% carbon, 0.18% silicon, 0.3% manganese, 0-3% chromium, 0.05% sulfur, 0.05% phosphorus, and the balance iron. Used for hand tools, shear blades, knives, etc.

file bronze. A free-cutting bronze containing 50.5-75% copper, 18-31% tin, 7-8.5% lead and 0-10% zinc, used for hardware and plumbing.

file metal. See Genfer file metal; Vogel file metal.

filet lace. See filet net (1).

filet net. (1) Mesh or net lace with a square-mesh ground and square patterns or motifs. Also known as *filet lace*.

(2) A woven mesh or net in which the yarns are locked at the intersections.

Filgra. Trade name of Eisenwerk Würth GmbH & Co. (Germany) for steel cut-wire abrasives.

Filigrana. Trade name of Vetreria di Vernante SpA (Italy) for patterned glass. See also filigree glass.

filigree glass. An art glass decorated by incorporating white threads of *spun glass* into the melt that upon cooling form lattices and patterns.

filigree hardboard. A special hardboard panel with a geometric or other design and a lacy, delicate or fanciful touch, used for cornices and room dividers.

filigree paper. A high-quality paper having a reticular- or linear-type watermark.

Filipoff Lead-Cadmium. Trade name for an antifriction alloy containing 1.9% calcium, 1.4% copper, 1.1% barium, 1% strontium, 0.1% sodium, and the balance lead. Used for bearings.

Filkar. Trademark of SOFICAR (France) for carbon fibers.

filled composites. Composites consisting of relatively finely divided minerals or organic substances (e.g., talc, wood, silica or slate flour, etc.), or short fibers of glass, asbestos, textiles, cellulose, etc., embedded in thermosetting, thermoplastic or elastomeric matrices. Also known as *filled plastics*. See also filler (2).

filled hardboard. Hardboard having a special manufacturing treatment that reduces surface porosity and greatly improves paintability. Also known as *sealed hardboard*.

filled particleboard. Particleboard whose surface and edges have been filled or sealed and sanded for subsequent paint coating. Also known as *sealed particleboard*.

filled plastics. See filled composites.

filler. (1) In metal joining, the metal or alloy added in making welded, brazed or soldered joints. Also known as *filler metal*.

(2) In polymer engineering, a relatively inert, finely divided substance, such as blanc fixe, calcium carbonate (whiting), carbon/graphite, glass spheres or bubbles, kaolin, mica, silica, slate or wood flour, incorporated into a polymer in relatively large proportions to increase bulk and thus decrease cost (extender), or modify the physical, mechanical, thermal, electrical or other properties. This term should not be confused with the terms *reinforcing agent* and *pigment*, since fillers usually have neither reinforcing nor coloring properties.

(3) An adhesive or nonadhesive material incorporated into an adhesive to enhance properties, such as permanence, strength and workability. It may also be added to reduce cost or improve bonding.

(4) In papermaking, a usually inorganic, water-insoluble substance in finely divided form that is added to paper pulp in order to impart certain properties to the finished paper or cardboard.

(5) In woodworking, any inert material used to fill the holes and irregularities in planed or sanded surfaces and decrease the porosity of the surface before applying finish coatings. Open-grain woods like ash, hickory, mahogany, oak, poplar and walnut require fillers.

(6) In metal casting, a wet, puttylike blend of inorganic materials used to cover pores and smoothen the surface of iron castings for subsequent enameling.

(7) In building construction, an inert, pulverized material, such as limestone, silica, etc., that may be introduced into paints, asbestos cement, Portland cement, etc., to increase bulk, decrease shrinkage, and/or improve workability.

(8) In the textile industries, a substance incorporated into a textile to increase its volume, or change the appearance or handle. Substances used for these applications include insoluble clays, gypsum, starches, gums, etc.

(9) In the coating industries, a white insoluble substance, such as barium sulfate, calcined gypsum, clay, kieselguhr, whiting, etc., used as an extender for paint pigments.

filler clay. A clay, such as kaolin in crushed or finely ground form, suitable for use as a relatively inert filler in plastics to increase volume, in paper to impart color and enhance surface finish, and in textiles to increase volume or change the appearance.

filler fabrics. Fabrics, usually cross-woven and rubber-coated, that are used as reinforcements in the bead section of automotive tires.

filler metal. See filler (1).

filler rod. See welding rod.

filler stone. See keystone.

filler wire. See welding rod.

filling. See weft yarn.

filling yarn. See weft yarn.

Fillite. Trademark of Fillite (Runcorn) Limited (USA) for glass-hard, lightweight, high-strength hollow and solid ceramic microspheres based on silicates, supplied in various grades, and used as fillers in molded plastic articles.

fillowite. A yellow, yellowish, reddish brown, or colorless mineral composed of sodium calcium manganese iron phosphate, $Na_2Ca(Mn,Fe)_7(PO_4)_6$. Crystal system, rhombohedral (hexagonal). Density, 3.43 g/cm^3; refractive index, 1.672. Occurrence: USA (Connecticut).

Fillwell. Trade name of Wellman Inc. (USA) for a family of polyester fibers supplied in several types including *Fillwell, Fillwell II and Fillwell Plus,* and used for textile fabrics.

film. (1) A polymeric material, such as cellophane, polyester, polyethylene, polypropylene, polystyrene, polyvinyl chloride or polyvinylidene, in the form of a thin sheet, usually 25-250 µm (1-10 mils) thick, produced by blowing, extrusion or coextrusion. It has a low density, a high degree of flexibility, high tensile and tear strength, and good resistance to chemicals and moisture. Used for bags for wrapping and packaging food products, textiles and other products, in heat sealing, as photographic film, for lining wood, fabric, paper, masonry, etc., on slip surfaces, and in waterproofing garments.

(2) A very thin leaf or sheet of metal. See also leaf; foil.

(3) A thin coating or layer of a substance over a substrate. See also thin film.

(4) See magnetic film.

film adhesives. A group of heat- and pressure-curing adhesives supplied in the form of relatively thin, dry films of synthetic resin with or without glass, fabric or paper carriers.

FilMaster. Trade name of F.J. Brodmann & Company LLC (USA) for functional metal and plastic filler powders.

film former. A material, such as a drying oil, varnish or synthetic resin, that imparts continuity to an organic coating (paint), and is dispersed or dissolved in water, or a solvent. Also known as *filmogen.*

film-forming polymers. A group of polymer-base sizing or finishing agents, such as polyvinyl acetate or polyvinyl alcohol, used in the protection of fiber reinforcements during processing.

Filmglas. Trade name of Owens-Corning-Ford Company (USA) for thin, flat glass flakes produced by blowing and subsequent shattering of the glass bubble.

Filmite. Trademark of Dubois Chemicals, Inc. (USA) for a protective spray-booth coating.

film mica. Trimmed mica splittings, usually 0.03-0.1 mm (0.001-0.004 in.) thick, obtained from high-grade *block mica.* See also mica.

filmogen. See film former.

Filnic. Trade name of Driver Harris Company (USA) for an extensive series of nickel-base alloys used mainly for filaments and resistances. Included are pure nickel (99.9%) as well as nickel-aluminum, nickel-cobalt, nickel-silicon and nickel-cobalt-titanium-iron alloys of varying compositions.

Filo Diamante. Trade name of Owens-Corning Fiberglass (USA) for a yarn used for decorative glass fabrics.

Filomat. Trade name of Fiberglass Industries Inc. (USA) for glass fiber reinforcements.

Filon. (1) Trademark of Kemlite Company (USA) for fiberglass and/or polyamide-reinforced unsaturated polyester resins used for flat and corrugated decorative and building panels.

(2) Trademark of BP Chemicals International (USA) for chemical- and flame-resistant glass fiber-reinforced polyester panels and sheets for trucks, RVs and boats.

(3) Trademark of Sohio Chemical Company (USA) for reinforced fiberglass panels.

Filospun. Trade name of Vetreria Italiana Balzaretti Modigliani SpA (Italy) for a series of texturized glass yarns.

Filpro. Trademark of US Silica Company (USA) for unground silica-based well and filtration sand and gravel.

Fil-Solder. Trade name of Swiss Laboratory Inc. (USA) for a self-fluxing tin-lead solder for aluminum and aluminum alloys.

Filsyn. Trademark of Flipinas Synthetic Fiber Corporation (Philippines) for staple fibers and filament yarns.

Filtec. Trademark of Rhône-Poulenc–Filtec SA (France) for nylon 6 and nylon 6,6 monofilaments.

Filtek. Trademark of 3M Dental (USA) for light-cure, radiopaque universal and posterior restorative dental composites based on polymeric resins of bisphenol A diglycidyl ether dimethacrylate, bisphenol A polyethylene glycol diether dimethacrylate or urethane dimethacrylate with silica/zirconia fillers. Common grades include *Filtek A110,* an anterior restorative composite, *Filtek P60,* a posterior restorative composite, *Filtek Flow,* a flowable restorative composite, and *Filtek Z100 & Z250* universal restorative composites. A recent addition to the *Filtek* line of products is *Filtek Supreme,* a dental resin composite containing up to 79 wt% nanomer and nanocluster filler particles (e.g., non-agglomerated nano-sized silica particles and aggregated zirconia/silica nanoclusters). It is supplied in a wide range of shades and opacities, and used for anterior and posterior restorations, inlays, onlays, veneers, and core build-ups.

filter alum. See aluminum sulfate octadecahydrate.

filter bauxite. See activated bauxite.

filter carbon. See activated carbon.

filter cloth. A fabric woven from polyethylene, polypropylene, nylon, glass or metal fibers, and used as filter medium for liquids and gases. Also known as *filter fabric.*

filter fabric. See filter cloth.

Filter Foam. Trade name for a reticulated, flexible polyester urethane foam used for filtration applications.

filter glass. An optical glass, usually tinted green, brown or gold, used in welding goggles, helmets and face shields to reduce glare and protect the eyes against harmful ultraviolet, infrared and visible radiation.

filter gravel. Specially prepared gravel used in the beds of gravity filters of water purification systems.

filter material. A porous material or graded granular material, such as sand, kieselguhr, cotton duck, nylon or glass fabrics, woven wire cloth, or filter paper, that can be used to filter liquid and gases. Also known as *filter medium.*

filter medium. See filter material.

filter paper. An unsized, wood-free paper sometimes containing cotton or linen rags and having high porosity or permeability. It is used for separating particulate from liquids or gases.

filter sand. A chemically-pure sand, such as beach sand, composed of small grains of uniform size suitable for filtration applications.

Filtrabronze. Trade name of Glaceries de Saint-Roch SA (Belgium) for bronze-colored cast glass.

Filtragrey. Trade name of Glaceries de Saint-Roch SA (Belgium) for gray, heat-absorbing plate and cast glass.

Filtralux. Trade name of Boussois Souchon Neuvesel SA (France) for gray plate glass.

Filtrasol. Trade name of Boussois Souchon Neuvesel SA (France) for green-tinted, heat-absorbing plate or cast glass.

Filtrasorb. Trade name of Calgon Corporation (USA) for activated carbon used in water treatment.

Filtrete. Trademark of 3M Company (USA) for a series of filter media.

Filtrol. Trademark of Filtrol Corporation (USA) for bentonite and chemically activated clays.

Filwell. Trademark of Wellman Inc. (USA) for polyester staple fibers.

Fimalon. Trademark of Fimalon SpA (Italy) for nylon 6,6 filament yarns.

Fina. Trade name of Atofina SA (France) for polyolefin resins.

Finaclear. Trademark of Atofina SA (France) for thermoplastic resins based on polystyrene, polyethylene, polypropylene, acrylonitrile-butadiene-styrene, styrene-butadiene-styrene or styrene-acrylonitrile. They are available in various forms including blocks, pellets, powders, plates, sheets, strips and spheres. Also included under this trademark are synthetic rubbers, such as styrene-butadiene-styrene and isoprene-styrene-butadiene and their copolymers, supplied in block, plate, strip, fiber and sphere form for various applications in the packaging industries, and in the health-care sector.

Final. Trade name of Supplies Limited (USA) for a metallographic polishing compound consisting of gamma alumina (γ-Al_2O_3) in acidic suspension.

Final Plus. Trade name of Supplies Limited (USA) for a polishing compound consisting of gamma alumina (γ-Al_2O_3) in acidic suspension. It imparts both etching action and mechanical abrasion to produce clean, smooth metallographic sample surfaces.

Finaprene. Trade name of Atofina SA (France) for unprocessed, vulcanized and nonvulcanized thermoplastic styrene-butadiene-styrene elastomers available in the form of powders, pellets, blocks, spheres, sheets, spheres and fiber. Used mainly in the automotive, building, and health-care sectors, and for adhesives.

Finaprene SX. Trade name of Atofina SA (France) for unprocessed and semiprocessed synthetic resins and rubbers. The unprocessed grades are supplied in the form of liquids, dispersions, emulsions, pastes, powders and granules, while the semiprocessed and rubber grades are available as blocks, strips, sheets, plates, spheres and fibers. Used in the manufacture of plastic products, and in road and building construction. The fiber grades can be used in bitumen and concrete.

Finathene. Trade name of Atofina SA (France) for high-density polyethylene used mainly for automotive components, and packaging applications.

fine aggregate. Aggregate graded to a diameter not exceeding 4.76 mm (0.1875 in.) and including materials, such as sand, pumice and foamed slag. Used with a suitable cementing material and water to form lightweight concrete or mortar.

fine aggregate bituminous concrete. A special concrete made without cement, but consisting instead of fine aggregate (sand, pumice, etc.) combined with a bituminous binder. See also coarse aggregate.

Fine-Clad. Trade name of Reichhold Chemicals, Inc. (USA) for a series of powder coating resins and crosslinkers.

fine gold. Gold having a very high degree of purity, e.g., chemically pure gold has a purity (fineness) of up to 1000 parts per thousand (24 karat). See also gold.

Fine Grain. Trade name for fine-grained dental amalgam alloys.

fine-grained metal. A polycrystalline metal that contains primarily small grains (crystallites), less than 0.5 mm (0.02 in.) in size. Owing to its greater total grain boundary area, it is harder and stronger and, in many cases, tougher than a coarse-grained material.

fine-grained steel. A steel that contains predominantly small grains, and is made by adding a deoxidizer and grain refiner, preferably aluminum, to the melt. This eliminates the dissolved gases and reduces the ferrous oxide (FeO) content leaving minute oxide particles in the metal, which act as nuclei and, if numerous, promote the formation of fine grains. Other grain refiners used include strong carbide formers, such as vanadium and molybdenum. *Fine-grained steels* have increased impact toughness, improved machinability, and reduced quenching cracks, distortion in quenching and surface decarburization.

fine gravel. Gravel with an average grain size of 2.00-9.52 mm (0.079-0.375 in.) used as concrete aggregate, and in filtration and road construction.

Finel. Trademark of Mitsubishi Rayon Company (Japan) for acrylic staple fibers.

Fineline. Trademark of Lukens Steel Company (USA) for a series of low-sulfur plate steels (up to 0.010% sulfur) that contain calcium compounds to modify the shape of inclusions. They have 75% or more Z-direction yield and tensile strength (of longitudinal value), 25% Z-direction reduction of area, and good Charpy V-notch impact values. Used for offshore platforms, construction equipment and bridges, chemical and petrochemical plant and equipment, and industrial equipment.

Fineline Double-O-Five. Trademark of Lukens Steel Company (USA) for a series of very-low-sulfur plate steels (up to 0.005% sulfur) with 90% or more Z-direction yield and tensile strength (of longitudinal value), 40% Z-direction reduction of area, improved Charpy V-notch impact values, excellent resistance to lamellar shearing, excellent shape control and outstanding through-thickness properties. Used for highly stressed joints, e.g., in offshore platforms, process equipment, construction equipment and bridges, materials handling equipment, pipeline fittings, etc.

Fineline Double-O-Two. Trademark of Lukens Steel Company (USA) for a series of very-low-sulfur plate steels (up to 0.002% sulfur) developed for structural applications requiring highest resistance to hydrogen-induced cracking.

Finemet. Trademark of Hitachi Metals, Limited (Japan) for a nanoscale crystalline soft magnetic material composed of 73.5% iron, 13.5% silicon, 9% boron, 3% niobium and 1% copper, and produced by the primary crystallization process from a melt-spun amorphous ribbon. It has high saturation magnetic flux density, a high remanence ratio and Curie temperature, and low coercive force, core losses and magnetostriction. Used for magnetic applications, such as induction accelerators, pulse transformer cores and pulsed power switching equipment, and for EMC noise reduction components.

fine paper. A term used for a class of high-quality, wood-free papers usually containing at least 10 wt% cotton and/or linen rags. Book and printing paper are fine papers.

fine sand. Sand grains with a diameter of 0.25-0.125 mm (0.01 and 0.005 in.). See also sand.

fine silver. A pure metal containing at least 99.9% silver, equivalent to a purity (fineness) of 999, and having high electrical conductivity (higher than pure copper), and a melting point of 961°C (1762°F). Used for alloys, plating, chemicals, jewelry, electrical contacts, and powder-metallurgy parts. See also silver.

Finesse. (1) Trade name of Tropical Dental Laboratories (USA) for translucent, fluorescent, opalescent low-fusing dental porcelain with fine grain size for porcelain-to-metal restorations.

(2) Trade name of Tropical Dental Laboratories (USA) for a self-cure dental hybrid composite.

Finesse Gold Ceramic Alloy. Trade name of Tropical Dental Laboratories (USA) for a copper- and palladium-free, biocompatible, extra-hard dental ceramic alloy used with *Finesse* dental porcelain. It contains 86.65% gold and 10.5% platinum, and has a yellow color, a density of 19.0 g/cm^3 (0.69 $lb/in.^3$), a casting temperature of 1230°C (2245°F), a hardness of 190 Vickers, and high tensile strength.

Finetex. Trade name of ASG Industries Inc. (USA) for patterned glass.

fine-textured wood. See close-grained wood.

fine zinc. A high-grade of zinc with a purity of at least 99.9% used for rolled sheet, strip and foil, drawn wire and rod, in the manufacture of brass and nickel silver for deep drawing, and in the manufacture of high-grade zinc alloys, in corrosion-resistant coatings for iron and steels, and for plating anodes.

Finfix. Trademark of Noviant Inc. (USA) for carboxymethylcellulose (CMC).

Fingal. Trade name of Uddeholm Corporation (USA) for an oil- and water-hardening tool steel containing 0.5% carbon, 0.6% chromium, 0.5% tungsten, 0.2% molybdenum, and the balance iron. Used for pneumatic tools.

finger-jointed lumber. An engineered wood product made by joining together clear, defect-free lumber stock, typically 51 × 102 mm (2 × 4 in.) or 51 × 152 mm (2 × 6 in.) in size, with glue and tight-fitting interlocking joints. It is used especially for ceiling, base, door, window and chair-rail moldings, and for structural applications.

fining agent. An agent, such as arsenic trioxide (As_2O_3), used in glass manufacture to promote the expulsion of bubbles and undissolved gases.

Finis. Trade name of Apex Steel Company (USA) for a high-speed steel containing 0.7% carbon, 18% tungsten, 4% chromium, 1% vanadium, and the balance iron. Used for tools and cutters.

finish. (1) A coating of paint, lacquer, varnish, wax, etc., applied to a wood surface to enhance its appearance and/or protect it against the elements and thus increase its durability.

(2) A material (dressing) applied to the grain surface of leather to improve its overall appearance, give uniformity of color, enhance smoothness, and/or mask defects.

(3) A material or mixture of materials applied to the surface of fibers used in reinforced plastics to improve their physical properties.

finished market products. A general term used for hardwood lumber products, such as flooring, moldings, paneling, risers, siding, stair treads, stringers, ties and trim, that have been cut and finished and are ready for use.

finished steel. A thermally and/or mechanically processed steel that is ready for commerce. Thermal processing may include annealing or tempering, while mechanical processing may include cutting, rolling, drawing, grinding, polishing and/or machining.

finished thermocouple material. A metal-sheathed thermocouple material in a form ready for application.

finishing cement. A cement composed of a mixture of dry powdered and/or fibrous materials which when blended with water become highly plastic and can be applied to a surface and, af-

ter drying, form a hard, protective finish.

finishing grout. Grout whose principal cementing ingredient is Portland cement.

finishing lime. A white, highly plastic hydrated lime that is suitable for use in plastering and other finish-coat applications.

finishing plaster. (1) Plaster made of cement and sand, and applied as a final layer over a basecoat in waterproofing masonry walls.

(2) Plaster made by mixing a special sand with gypsum or lime sand cement, or by mixing lime and gypsum or cement to a putty-like consistency. It is applied as a thin layer, typically 0.06 in. (1.5 mm) thick, directly onto the masonry surface, or over a suitable base (e.g., fiberboard or expanded metal lath). Used in finishing interior walls.

finishing quicklime. A highly plastic quicklime that has been slaked to a lime putty, and is suitable for use in plastering and other finish-coat applications.

Finishing Special. Trade name of Midvale-Heppenstall Company (USA) for a tool steel containing varying amounts of carbon, chromium, tungsten and iron.

finish lumber. Lumber that can accept a high-quality surface finish, e.g., lumber used for millwork, or lumber products employed in the finishing of building interiors, such as window and door trim, paneling, flooring, stair treads, risers and stringers, railings, moldings, etc.

finish tile. A tile with one glazed surface used in the construction of exterior or interior walls.

finnemanite. A gray to black, or olive-green mineral composed of lead chloride arsenite, $Pb_5Cl(AsO_3)_3$. Crystal system, hexagonal. Density, 7.26 g/cm^3; refractive index, 2.295. Occurrence: Sweden.

Fio Banylsa Nylon. Trademark of Banylsa Tecelagem do Brasil SA (Brazil) for nylon 6 filament yarns.

Fio Banylsa Poliester. Trademark of Denyls Industrias Quimicas de Textil SA (Brazil) for nylon 6 filament yarns.

Fiocco. Trade name for rayon fibers and yarns used for textile fabrics.

Fiolax. Trademark of Schott Rohrglas AG (Germany) for neutral glass tubing with exceptional chemical resistance supplied in two grades: *Fiolax-clear* and *Fiolax-amber*. The latter has low light transmission. Used for pharmaceutical primary packaging applications, e.g., syringes, ampoules, vials, etc.

fiorite. See geyserite.

fir. The wood of any of several softwood trees of the genera *Abies* and *Pseudotsuga* belonging to the order Pinales. Fir species found in the eastern United States include balsam fir *(A. balsamea)* and Fraser fir *(A. fraseri)*. Fir species growing in the western United States include subalpine fir *(A. lasiocarpa)*, California red fir *(A. magnifica)*, grand fir *(A. grandis)*, noble fir *(A. procera)*, Pacific silver fir *(A. amabilis)* and white fir *(A. concolor)*. A fir found throughout western North America and eastern Asia is the Douglas fir *(P. menziesii)*, and silver fir *(A. alba)* is a common European species.

Firearmor. Trade name of Michiana Products Corporation (USA) for a series of heat- and abrasion-resisting cast nickel-iron-chromium alloys containing 16.7-21.6% iron, 15% chromium, 1.7-1.8% manganese, 0.5% carbon, 1-1.2% silicon, and the balance nickel. Used for carburizing boxes, grates and furnace parts.

Fireboard. Trade name of Westroc Industries Limited (Canada) for fire-retardant gypsum wallboard.

Firebond. Trade name of Multiglass Limited (UK) for glass with a coating of ceramic color fused permanently at high temperature onto one of its surfaces. Used for infill panels in curtain wall structures.

firebox plate. Medium-carbon steel plate used in the manufacture of fire-tube boilers, etc. See also boiler plate.

firebrick. A refractory brick that is capable of withstanding very high temperatures (1500-1600°C, or 2730-2910°F) without fusion. It consists essentially of aluminosilicates and silica comprising less than 78% silica (SiO_2) and less than 38% alumina (Al_2O_3). The most common size of firebrick is $222 \times 114 \times 63$ mm ($8.75 \times 4.50 \times 2.50$ in.). Used for lining furnaces, ladles, fireplaces, chimneys, etc. Also known as *fireclay brick*.

fireclay. A soft sedimentary clay that is high in hydrated aluminum silicates or silica, low in alkalies, iron and other fluxing ingredients, and can withstand high temperatures without fusion. Used in the manufacture of refractory brick, furnace and kiln linings, glassmaking pots and tanks, crucibles, saggers, heat-resistant cement, and other refractory products. See also refractory clay.

fireclay brick. See firebrick.

fireclay cement. A heat-resistant cement composed of a mixture of dry fireclay and water glass, and used in the repair of refractories, saggers, kiln cracks, etc.

fireclay goods. Ceramic products, such as brick, cement, mortar, refractory shapes, etc., predominantly composed of fireclay, often with the addition of opening materials, such as *grog*.

fireclay grog refractory. A refractory brick consisting of fireclay bonded with *grog*.

fireclay mortar. Mortar consisting of finely divided fireclay and water. Used in the construction of heat- and fire-resistant structures.

fireclay plastic refractory. A refractory material composed of fireclay made plastic by tempering with water, and used as a ramming mix in the placing of monolithic furnace linings.

fireclay refractories. Refractories composed of high-purity fireclays, alumina and silica mixtures with an average alumina (Al_2O_3) content of about 20-40 wt%. Used for furnace construction, confinement of hot atmospheres, and in the thermal insulation of structural members against excessive temperatures.

Firecrete. Trademark of Morgan Crucible Company plc (UK) for lightweight refractories and refractory cements composed of calcined high-alumina clay. They have excellent resistance to temperatures of 1320°C (2410°F). Used for floors and doors of industrial furnaces.

Firedie. Trade name of Columbia Tool Steel Company (USA) for a series of chromium-type hot-work tool steels (AISI types H11 to H19) containing 0.38-0.4% carbon, 0-1% silicon, 0.3% manganese, 3.6-5.2% chromium, 1.4-3% molybdenum, 0-1.05% vanadium, 0-2% cobalt, and the balance iron. They have good deep-hardening properties, high hot hardness, good wear resistance and toughness, and are used for die-casting dies, hot-extrusion tools and dies, forging dies and tools, etc.

Firedie Smoothcut. Trade name of Columbia Tool Steel Company (USA) for a free-machining chromium-type hot-work tool steel (AISI type H13) containing 0.38% carbon, 0-1% silicon, 5.2% chromium, 1.4% molybdenum, 1.05% vanadium, and the balance iron. It has good deep-hardening properties, high hot hardness, good wear resistance and toughness, and is used for die-casting dies, hot-extrusion tools and dies, hot-sizing dies, and forging dies and tools.

Fire-flex. Trademark of Textron Specialty Materials (USA) for a fire-resistant, intumescent epoxy tape material.

FireFlex Contour Design. Trade name of TECHNIFOAM, Inc. (USA) for fiber-free acoustical foam panels (Class 1) used for noise control applications.

Firefrax. Trademark of The Carborundum Company (USA) for a series of refractory aluminum silicate cements made from *fireclay* or *kaolin*. Supplied in air- and heat-setting grades, they have good resistance to high temperatures up to 1650°C (3000°F). Used as bonds for crushed firebrick or ganister for patching furnace linings, in patching materials for byproduct coke ovens, for laying and repairing fireclay and silica brick work, in the manufacture of monolithic or rammed-up linings, and as washes for small pouring ladles in nonferrous foundries. Also included under this trademark is refractory aluminum silicate brick.

Fire-Glass. Trade name of Manville Corporation (USA) for fiberglass and asbestos roofing.

fire opal. A translucent or transparent, orange-yellow to bright red variety of *opal* ($SiO_2 \cdot xH_2O$) that has a brilliant play of colors, and exhibits fire-like reflections in bright light. Used chiefly as a gemstone.

fire-opal glass. A highly translucent glass that exhibiting fire-like reflections imitating those of the mineral *fire opal*.

fireproof fabrics. See flameproof fabrics.

fireproofing. A material used to protect a substance against fire or combustion.

fireproofing compound. See fire retardant.

fireproofing tile. A tile used to protect exterior or interior walls, partitions, etc., against fire.

fire-protection paint. A paint that contains a substance rendering an otherwise flammable material incapable of supporting fire. See also flame-retardant paint.

fire-refined copper. See fire-refined tough-pitch copper.

fire-refined tough-pitch copper. A commercially pure copper (99.88%) obtained by deoxidizing anode copper until the oxygen content has been reduced to about 0.02-0.04%. It contains a small amount of residual sulfur, typically 10-30 ppm, and cuprous oxide, typically 500-3000 ppm, and has high ductility, good hot and cold workability, good corrosion resistance, and excellent electrical conductivity. Used for electrical conductors, busbar, electronic components, waveguides, klystrons, etc. Abbreviation: FR Cu. Also known as *fire-refined copper*.

fire retardant. A chemical substance that when applied to the surface of, or otherwise added to a material, such as a plastic, rubber, textile, etc., will reduce its rate of flame spread, resist ignition and/or significantly reduce the tendency to ignite, burn or melt. Also known as *fireproofing compound; fire-retarding agent; flame retardant; flame-retarding agent*.

fire-retardant coatings. Surface coatings that when applied to an otherwise combustible substance will reduce the rate of flame spread, resist ignition and/or significantly reduce the tendency to ignite, burn or melt. Also known as *flame-retardant coatings*.

fire-retardant fabrics. Textile fabrics which are either made from inherently nonflammable fibers (polyamides and aramids) or have been treated with fire-retardant chemicals to reduce their flammability.

fire-retardant paint. A paint containing a substance, such as casein, polyvinyl chloride or silicone, that either slows down the rate of combustion of a flammable material, or renders it incapable of supporting fire. Also known as *flame-retardant paint*.

fire-retardant wood. Wood that has been coated or impregnated with a fire-retardant chemical to reduce its rate of flame spread, resist ignition and/or significantly reduce the tendency to ignite or burn. Used in building construction, mining, and ship- and boatbuilding.

fire-retarding agent. See fire retardant.

fire sand. See furnace sand.

Firestone. Trade name of Firestone Polymers (USA) for a range of nylon, polyolefin and saran fibers.

firestone. See flint.

Firestop. Trademark of Westroc Industries Limited (Canada) for fire-resistant gypsum wallboard.

Firet. Trademark of Firet-Gesellschaft mbH (Germany) for nonwoven fabrics used in the building, machinery, clothing, shoe, artificial-leather and mattress industries, and for liquid and air filtration, surface treatment applications in the plastics industry, and in the fields of electrical engineering and electronics.

fire tile. A fire-resistant tile used in a kiln or furnace.

Firex. (1) Trade name of Darwin & Milner Inc. (USA) for an air- or oil-hardening tool steel containing 0.4% carbon, 4.25-4.5% nickel, 1.1-1.2% chromium, 0.3% silicon, 0.6% manganese, and the balance iron. Used for chisels, hot shears, shear blades, cutting tools, and shock-resisting tools.

(2) Trade name of Pfizer Pigments Inc. (USA) for polymers based on epoxy or urethane resins, and used for thermal-barrier coatings.

Firex Special. Trade name of Darwin & Milner Inc. (USA) for a tough, shock-resistant tool steel containing 0.4-0.55% carbon, 0.4-1% chromium, 2.75-4% nickel, 0.4-0.9% manganese, 0.15-0.2% vanadium, 0.4-0.7% molybdenum, and the balance iron. Used for dies, tools and punches.

Firit. Trade name of Foseco Minsep NV (Netherlands) for refractory materials that can be made plastic by mixing with water, and are used in the iron and steel industries for dressings and protective coatings on furnace tools and ladles, crucibles, etc.

Firmilay. Trade name of J.F. Jelenko & Company (USA) for a yellow-colored, hard dental casting alloy (ADA type III) with 74.5% gold, 11% silver, and 3.5% palladium. It provides high strength and good finishability, and is used for hard inlays, crowns, and fixed bridgework.

Firminy. Trade name of Creusot-Loire (France) for a series of machinery, case-hardening and heat-treating steels, austenitic, ferritic and martensitic stainless steels, austenitic manganese steels, cold-work, hot-work and shock-resisting tool steels, and several mold steels.

Firmoid. Trade name of Bluemel Brothers (UK) for cellulose acetate plastics.

Firox. Trademark of Cogebi Inc. (USA) for mica cable tapes supplied in various grades. *Firox P* and *Firox TM* consist of *Cogemica* phlogopite mica paper bonded to an electrical glass (E-glass) fabric and impregnated with a special high-temperature-resistant silicone elastomer, and *Firox PE* is a heat-resistant, high-temperature silicone elastomer-impregnated *Cogemica* phlogopite mica paper bonded to a polyethylene support film. *Firox* cable tapes have high flexibility, outstanding electrical insulation properties and excellent flame resistance. See also Cogemica; mica.

FiRP glulam. Glue-laminated wood reinforced with polymer fiber-reinforced plastics, and used in the manufacture of building products. See also glue-laminated wood.

First-fill. Trademark of Jeneric/Pentron Inc. (USA) for a translucent, light-cured, semiflexible setting polymer-based dental filling material of putty-like consistency. It is easy to handle, shape, place and remove, and is used for temporary restorations in particular for filling inlays, onlays, and endodontic access openings.

First Quality. Trade name of Houghton & Richards Inc. (USA) for a water-hardening tool steel containing 0.8-1.1% carbon, and the balance iron. Used for punches, dies, and tools.

first-quality ware. Manufactured ceramic articles or products that are free of imperfections or defects and meet specified standards.

first-split leather. Leather of moderate durability and flexibility obtained from the layer just below the top or outside layer of cattlehide. It is intermediate in quality between top-grain and second-split leather. See also second-split leather; top-grain leather.

Firthag. Trade name of Firth Brown Limited (UK) for a medium-carbon tool steel containing 0.45% carbon, 0.9% manganese, 0.95% chromium, 0.25% silicon, and the balance iron. Usually supplied in the hardened and tempered condition, it is used for bolts, gears, punches, and forgings.

Firthaloy. Trade name of Ferranti Limited (UK) for a cemented tungsten carbide used for wire drawing and extrusion dies, cutting tools, wear parts, and gages.

Firth Brown. Trade name of Firth Brown Limited (UK) for an extensive series of plain-carbon tool and spring steels, hot-work tool steels, alloy steels (e.g., chromium, chromium-manganese, chromium-vanadium, chromium-vanadium-tungsten, nickel-chromium, nickel-chromium-molybdenum and nickel-molybdenum types including various case-hardening grades), nitriding steels, maraging steels, austenitic, ferritic, martensitic and precipitation-hardening stainless steels, heat-resisting steels, and high-strength structural steels.

Firthite. Trade name of Teledyne Firth-Sterling (USA) for a bonded tungsten carbide used for drawing dies, cutting tools, gages, and wear parts.

Firthob. Trade name of Firth Brown Limited (UK) for a case-hardened, shock-resistant steel containing 0.09% carbon, 0.5% manganese, 1.25% nickel, 0.5% chromium, and the balance iron. Used for plastic molds, cams, gears, shafts, and machinery parts.

Firth-Vickers. Trade name of Firth-Vickers Stainless Steels Limited (UK) for an extensive series of corrosion- and/or heat-resistant austenitic, ferritic and martensitic stainless steels including various free-machining and welding grades.

Fischer's salt. See cobalt-potassium nitrite.

fischesserite. A pink mineral composed of gold silver selenide, $AuAg_3Se_2$. Crystal system, orthorhombic. Density, 9.10 g/cm³. Occurrence: Czech Republic.

fish-eye cloth. A fabric similar to bird's-eye but woven with a larger novelty pattern. It is usually made of wool, polyester or blended fibers. See also bird's-eye (1).

fish glue. A glue derived from solutions of the skins, or the jelly separated from the oil of commercial fish (chiefly cod). Used in gummed tape, cartons, textile finishing, blueprint paper, letterpress printing plates, photoengraving and photomounting, and paints and sizes.

fishnet. A weft-knitted, fine-mesh net fabric with an openwork effect produced by floating a thick yarn across a thin yarn.

fish paper. A strong, thin insulating paper made from chemically treated vulcanized fiber. It has high mechanical and electrical strength, and good flexibility, and is used for wrapping conductors and transformer windings.

Fisisa. Trademark of Fibras Sinteticas SA (Mexico) for acrylic, nylon 6 and polyester fibers and filaments used for textile fab-

Fisseal. Trade name of Inter-Africa Dental (Zaire) for a low-viscosity dental pit and fissure sealant with excellent flow characteristics and good enamel adhesion.

fissile materials. See fissionable materials.

fissionable materials Materials, such as the heavier isotopes of uranium (e.g., ^{235}U), plutonium (^{239}Pu) and thorium (^{232}Th), whose nuclei are capable of undergoing fission with the emission of large amounts of energy. Also known as *fissile materials.*

fission fuel. See reactor fuel.

Fissium. Trade name for an alloy used in varying amounts in spent nuclear fuel elements. It contains 40% ruthenium, 26% molybdenum, 23% palladium, and a total of 11% rhodium and zirconium.

Fiveohm. Trade name for a wrought copper-nickel alloy containing some manganese. Used for electrical resistances.

Five Point Deep Hard. Trade name of Foote Brothers Gear & Machine Corporation (USA) for an oil-hardening tool steel containing 0.4% carbon, 1.5% nickel, 0.8% chromium, 0.2% molybdenum, and the balance iron. Used for valves, axles, gears, wheels, and wrist pins.

Fivestar. Trademark of Saint-Gobain (France) for a non-insulated, fire-resistant glass.

Five Star. Trade name of Midvale-Heppenstall Company (USA) for a high-speed steel containing 0.85% carbon, 4% chromium, 10.5% cobalt, 18% tungsten, 0.55% molybdenum, and the balance iron. Used for tools and cutters.

Fixamper. Trade name of Creusot-Loire (France) for a resistance alloy composed of 72% nickel and 28% iron. It has a maximum service temperature of 600°C (1110°F), and is used for heating elements and electrical resistances.

Fixe Moquette. Trade name of Ato Findley (France) for adhesives used for carpets and similar coverings.

Fixinvar. Trade name of Creusot-Loire (France) for an alloy of 64% iron and 36% nickel, used for clocks.

Fixotin. German trade name for tinning paste used for cast iron, cast steel, ingot steel, bronze, etc.

Fixset. Trademark of Plasticisers Limited (UK) for polypropylene staple fibers.

Fix-Zement. Trademark of Sakowsky GmbH (Germany) for a quick-setting natural cement.

fizelyite. A lead gray mineral of the lillianite group composed of silver lead antimony sulfide, $(Pb,Ag)_8Sb_{11}S_{24}$. Crystal system, orthorhombic. Density, 5.56 g/cm³. Occurrence: Hungary, Rumania.

Flachrip. Trademark of Rippenstreckmetall-Gesellschaft mbH (Germany) for black lacquered or galvanized flat-ribbed expanded metal.

flagstaffite. A colorless or white mineral composed of terpin hydrate, $CH_3(OH)C_6H_9C(CH_3)_2OH \cdot H_2O$. It can also be made synthetically. Crystal system, orthorhombic. Density, 1.10 g/cm³; refractive index, 1.512. Occurrence: USA.

flagstone. A flat slab or piece of stone, usually of sandstone, shale or other rock, either cut or naturally split for use in paving sidewalks, walkways, porches, patios, etc., and for covering exterior walls.

Flaikona. Trade name for rayon fibers.

flake. See flake powder.

flakeboard. Particleboard made of wood flakes that are bonded together under pressure. Used as a building board, and for furniture components, underlayment and sheathing. See also particleboard.

flake enamel. Porcelain enamel frit supplied in very thin, flat fragments or scales.

flake-galvanized material. See hot-galvanized material.

Flakeglas. Trade name of Owens-Corning Fiberglass Corporation (USA) for thin, flat glass flakes produced by blowing, and subsequent shattering of the glass bubble. Used for spray coatings and polarizing lighting panels.

flake graphite. A grade of crystalline graphite in the form of small scales, or thin flat plates. See also crystalline graphite; graphite.

Flakeline. Trademark of The Ceilcote Inc. (USA) for corrosion-resistant epoxy coatings.

flake powder. Flat or scalelike metal powder particles of relatively small thickness. Also known as *flake; flaky powder.*

FlakeRez. Trademark of PolySpec (US) for chemical-resistant, filled coatings supplied in a wide range of grades for various applications. Some grades are also resistant to fluorides, petroleum, and corrosive acids.

Flake Slub. Trade name for rayon fibers and yarns.

flake white. A pigment composed of flaked white lead (basic lead carbonate) in a suitable drying oil.

flake yarn. A novelty yarn in which roving flakes or tufts, secured by ply twist, are arranged at regular or irregular intervals.

flaky powder. See flake powder.

Flamaloy. Trade name of Detroit Alloy Steel Company (USA) for a wear-resistant cast iron used for general casting applications.

Flamco. Trade name of Foseco Minsep NV (Netherlands) for a series of foundry mold and core dressings in powdered form, ready to use on dilution with a suitable alcohol, such as isopropanol. Used as a coating on carbon dioxide-bonded molds and cores.

Flame. Trademark of Clayburn Industries Limited Refractories Limited (Canada) for fireclay bricks.

Flame Check. Trade name of Sico Inc. (Canada) for intumescent and fire-retardant paints.

Flamecoat. Trade name of Plastic Flamecoat Systems Inc. (USA) for a thermoplastic coating that can be applied to various substrates by flame spraying, and fluidized-bed or electrostatic processes. It provides outstanding resistance to abrasion, corrosion, impact and moisture.

flame coatings. Relatively thin refractory coatings applied to objects by a plating process in which heated powder particles are shot in bursts from a gun barrel by rapidly successive oxyacetylene explosions and impacted into the object surfaces. Excellent bonding and density are obtained without excessive heating of the objects.

Flame Curb. Trademark of PABCO Gypsum, Division of Pacific Coast Products, Inc. (USA) for flame-resistant gypsum wallboard.

Flame Hard. Trade name of Induction Steel Castings Company Limited (USA) for an oil- or flame-hardening tool steel containing 0.55% carbon, 1.2% manganese, 1.25% chromium, 0.4% molybdenum, 0.1% vanadium, and the balance iron. Used for trimming dies.

Flamenol. Trademark of General Electric Company (USA) for vinyl-resin insulated electrical conductors.

flameproof fabrics. Textile fabrics, usually treated with chemicals, that char, but do not support combustion and extinguish themselves after the source of ignition is removed. Also known as *flame-resistant fabrics; fireproof fabrics; nonflammable fabrics*.

flameproof paper. Paper that has been chemically treated with ammonium sulfamate, ammonium sulfate, dibasic or monobasic ammonium phosphate, dibasic sodium phosphate, etc., to reduce its tendency to ignite, char or burn.

flameproof ware. Articles of fired clay capable of withstanding extreme thermal shock.

flame-resistant fabrics. See flameproof fabrics.

flame-resistant material. A material that is resistant to flames or combustion, either inherently or by treatment with suitable chemicals.

flame retardant. See fire retardant.

flame-retardant coatings. See fire-retardant coatings.

flame-retardant paint. See fire-retardant paint.

flame-retardant resin. A synthetic resin to which certain chemicals, such as bromine, chlorine or phosphorus-containing compounds, have been added to reduce the rate of flame spread, and/or significantly decrease the tendency to ignite, burn or melt.

flame-retarding agent. See fire retardant.

Flame-Safe. Trademark of Manville Corporation (USA) for preformed fiberglass pipe insulation.

flame-sprayed coatings. Thin protective coatings produced on metallic or refractory substrates by a thermal spraying process in which an oxyacetylene or oxyhydrogen gas flame is the source of heat for melting the coating material (usually a metal, ceramic or plastic in rod or powder form). Compressed gas may be used to propel the coating material onto the substrate. Used to produce oxidation resistant coatings on steel and other metals and alloys.

flammable material. A solid, liquid or gaseous substance that will ignite, and readily support combustion.

flammable textiles. Textile fabrics that are combustible, and will readily support combustion.

Flamruss. Trade name for a carbon black.

Flamtard. Trade name of Alcan Chemical UK Limited for flame-retardant additives supplied in several compositions. *Flamtard H* is based on zinc hydroxystannate, *Flamtard S* on zinc stannate and *Flamtard Z10 & Z15* on zinc borate. They have low specific gravity, low toxicity and reduced smoke and carbon monoxide emission. *Flamtard H* decomposes at 180°C (356°F), and *Flamtard S* is thermally stable in all polymers. Used in PVC cabling, flooring, profiles, etc.

flange metal. See French flange metal; German flange metal.

flannel. A soft, loosely woven, slightly napped fabric made of wool, worsted, cellulose or cotton usually in a plain or twill weave. Used especially for color-buffing wheels, clothing, and sleeping bags.

flannelette. A soft, warm, lightweight fabric with a fuzzy nap resembling flannel. It is usually made from cotton, viscose rayon, modal or blended fibers. Used for pyjamas, nightgowns, and blankets.

FLARE. Trademark for thin-film polymers based on polyarylene ether. They have low dielectric constants (k < 3.0), good thermal stability, good moisture and solvent resistance, high glass-transition temperatures (T_g above 450°C or 840°F), and a long shelf life. Used as interlayer dielectrics for integrated circuit devices.

flash. A very thin metal film, usually less than 2.5 µm (0.1 mil), produced by electrodeposition. Also known as *flash plate*.

Flash Alloy. Trade name of S.G. Taylor Chain Company (USA)

for a steel containing 0.17-0.22% carbon, 1.65-2% nickel, 0.2-0.3% molybdenum, and the balance iron. Used for sling chains.

Flashbreaker. Trade name of Airtech Europe SA (France) for flash tapes based on polyester film. They have a single sized pressure-sensitive adhesive, and a maximum service temperature of 180°C (355°F). Used in vacuum-bag molding. See also flash tape.

flashed brick. Brick fired in a kiln under reducing conditions to develop a desired color.

flashed glass. Transparent glass that has a thin layer of opaque or colored glass applied to its surface.

flashing. Strips of sheet metal or other material used in roof and wall construction (especially around chimneys and vents) to protect the joints and angles, and prevent rain or other water from entering.

flashing cement. See plastic cement (3); roofing cement.

flash plate. See flash.

flash-spun fabrics. Nonwoven fabrics made from highly fibrillated polymer film by a combined bonding and extrusion process.

flash-spun fibers. Synthetic fibers made by the extrusion of a polymer solution at a temperature exceeding the boiling point of the solvent resulting in rapid evaporation of the latter upon exiting the spinneret and the formation of highly fibrillar filaments.

flash tape. A polymer tape (e.g., polyester) with single-sided adhesive used in vacuum-bag molding to mask off areas to be degreased or painted, prevent surface damage, or locate and secure vacuum-consumable materials during the bagging operation.

Flask-Stone. Trade name for a model stone used in dentistry.

Flat Bottom Rails. British trade name for a steel containing 0.55% carbon, 1% manganese, 0.06% sulfur, 0.05% phosphorus, and the balance iron. Used for flat-bottom rails.

Flatesa. Trade name for rayon fibers.

flat fabrics. Woven or knitted textile fabrics of two-dimensional construction that do not have pile loops on their surfaces.

flat glass. A collective term for any form of glass that is of a flat nature, such as sheet, plate, float and rolled glass. Abbreviation: FG.

flat-grained lumber. Softwood lumber that has been sawn lengthwise and tangential to the growth rings. The grain pattern on the surface of the sawn material is the common "U" or "V" shape, and the annual rings are very apparent. Hardwood lumber so sawn is referred to as *plain-sawn lumber*.

flat metal yarn. A yarn, such as that used in banknotes, which consists of or incorporates one or more continuous metal strips.

flat slab. A plate or slab of reinforced concrete, such as that used in flooring, designed to span in two directions.

flatting agent. See gloss reducer.

Flattsea. Trade name for rayon fibers and yarns used for textile fabrics.

flat varnish. A varnish that dries with reduced surface gloss or sheen, and is obtained by incorporating flatting agents, such as finely divided heavy-metal soaps, silica, diatomaceous earth or wax.

flat wall paint. An interior wall paint that has certain flatting agents added to produce a flat, low-gloss or lusterless finish. See flatting agents; wall paint.

flat wire. Wire with rectangular or square cross section produced by drawing or rolling.

flatwork. A collective term for concrete products, such as sidewalks, and floor and flat slabs.

flat yarn. An essentially twistless, untextured, fully drawn yarn consisting of continuous filaments.

Flax. Trade name of ASG Industries Inc. (USA) for glass with a random cobble-like pattern on one surface that resembles a loosely woven fabric.

flax. A soft natural fiber obtained from the inner bark of the linseed plant (*Linum usitatissimum*). It has a density of 1.52 g/cm³ (0.05 lb/in.³), and the fibers are usually about 50 cm (20 in.) long, and stronger and more durable than *cotton*. Source: Egypt, France, Ireland, Italy, Russia, United States. It has minor applications in fiber-reinforced composites, and is used chiefly for linen yarns and fabrics, cordage, rope, twine, and cigarette and duplicating papers.

flax-spun yarn. Any staple yarn produced and spun on machinery made for spinning flax yarn.

flax tow. Shorter fiber sections separated from flax fibers during scrutching (a beating and shaking process) or hackling (a combing process).

Fle. Trade name of Vetreria Italiana Balzaretti Modigliani SpA (Italy) for an acoustic insulating felt made from long glass fibers.

Fleck's. Trade name of Mizzy Dental (USA) for a series of dental cements including *Fleck's Zinc* zinc phosphate luting cement and *Fleck's PCA* zinc polycarboxylate dement for luting and lining applications.

fleck yarn. A fancy yarn that has some fibers of different color or luster incorporated to give it a spotted look.

Flecto Iron. Trade name of Ohio Brass Company (USA) for a specially annealed malleable cast iron containing 2.4-2.6% carbon, 0.45% manganese, 1.1% silicon, and the balance iron. Used for castings to be hot galvanized. Also known as *Flecto Metal*.

Flecto Metal. See Flecto Iron.

Flectron. Trademark of Monsanto Chemical Company (USA) for nickel-coated graphite fabrics that readily adhere to polymeric matrices, and are used for lightning-strike protection applications, and in the aerospace industry.

fleece. (1) Raw wool obtained from sheep by shearing, and prior to scouring.

(2) A fabric with a napped surface and a twill weave. The fibers may come from fine wool, camel's hair, or fiber blends. Used especially for garments and linings.

Fleeceline. Trademark of Bishop Fibretek Inc. (Canada) for fibrous insulating materials.

fleece wool. A term used for wool cut from a living sheep.

fleischerite. A pale rose mineral composed of lead germanium sulfate hydroxide trihydrate, $Pb_3Ge(SO_4)_2(OH)_6 \cdot 3H_2O$. Crystal system, hexagonal. Density, 4.20 g/cm³; refractive index, 1.776. Occurrence: Southwest Africa.

Fleka. Trademark of Nippon Sheet Glass Company Limited (Japan) for granules made by bonding electrical-grade borosilicate glass flakes with a special binder and coupling agent. They have an average particle size of 0.6-1.0 mm (0.02-0.04 in.), and are used in the manufacture of electronic components, such as connectors and sockets, and precision parts for laser printers, fax machines, photocopiers, etc.

Flemish. Trade name of Pilkington Brothers Limited (UK) for a patterned glass.

Flemish brick. Hard, yellow paving brick.

Flenderstahl. Trade name of Carl Flender Kaltwalzwerk und Bandhärterei (Germany) for high-grade carbon and alloy strip steel, strip steel for springs and packaging applications, steel wire, etc., including various stainless grades.

Fletcher & Emperer Bearing. British non-heat-treatable bearing alloys containing 92% aluminum, 7.5% copper, and 0.5% tin.

fletcherite. A steel gray mineral of the spinel group composed of copper cobalt nickel sulfide, $Cu(Ni,Co)_2S_4$. Crystal system, cubic. Density, 4.76 g/cm³. Occurrence: USA (Missouri).

Fletcher's alloy. British non-heat-treatable aluminum alloy containing 3% copper, 1% tin, 0.5% antimony, and a trace of phosphorus. Used for aircraft components and lightweight parts.

Fletcher's bearing metal. British bearing metal containing 90% aluminum, 7% copper, and 3% zinc.

fletton. A pale reddish brick from the districts of Bedford and Peterborough in the southeastern United Kingdom.

Flex. Trade name for flexible polyvinyl chloride film materials.

Flexacryl. Trade name of Lang Dental Manufacturing Company (USA) for acrylic dental rebasing plastics.

Flexal. Trade name of Aluminium Norf GmbH (Germany) for aluminum and aluminum-iron alloys.

Flexaloy. Trade name of Bergstrom Alloys Corporation (USA) for a wear-resistant alloy of carbon, chromium, manganese, nickel and iron used for hardfacing electrodes.

Flexane. Trademark of Devcon Corporation (USA) for room-temperature-curing urethanes.

Flexarc. Trade name of Westinghouse Electric Corporation (USA) for a series of welding electrodes for low-, medium- and high-carbon steels, high-tensile steels, and alloy and stainless steels. Most of them have either lime or titania-lime coatings.

Flexbac. Trade name of Carborundum Abrasive Company (USA) for coated abrasives.

Flexbond. Trademark of Harbison-Carborundum Corporation (USA) for a series of nonionic polyvinyl acetate copolymer emulsions supplied in various grades with an average particle size of 0.1-1.5 μm (0.004-0.06 mil) and a solids content of 45-60%. Used as vehicles for house paints or industrial finishes, as pigment binders in paper coatings, and as adhesive bases for plastic, textile, paper, leather, etc.

Flexcalk. Trademark of Chemtron Manufacturing Limited (Canada) for oil-base caulking compounds.

Flex-Core. Trademark of Hexcel Corporation (USA) for honeycomb cellular-core materials used for structural, decorative, construction and aerospace applications.

Flex-E-Bond. Trademark of Kleen-Flo Tumbler Industries Limited (USA) for autobody fillers.

Flexglass. Trade name of United States Plywood Corporation (USA) for strips of colored or silvered glass bonded to cloth.

Flexibatt. Trademark of Roxul Inc. (Canada) for water-resistant mineral wool home insulation supplied in R-values (heat flow resistances) of R-13.5 for 2 × 4 in. (51 × 102 mm) studs, and R-21.5 for 2 × 6 in. (51 × 152 mm) studs. It is used for the thermal insulation of walls and attics.

flexible graphite. A flexible sheet with high resilience in perpendicular direction produced by compressing flakes of exfoliated graphite, and used as a gasket material for fluids.

flexible insulation. A thermal insulation material supplied in the form of flexible blankets, sheets, batts or felts to facilitate installation by nailing, wrapping or adhesive bonding.

flexible foam. A thermoplastic or thermosetting resin made into a cellular plastic having high elasticity, resiliency and softness. Used for furniture, mattresses, and automobile interiors.

flexible magnet. A plastic sheet, strip or tape impregnated or coated with magnetic powder and used in electronics.

flexible material. A material that folds, bends, flexes, twists or bows without breaking.

flexible polyurethane foam. A cellular polyurethane elastomer with high resilience and softness usually based on polyoxypropylenediols and triols and containing predominantly open cells. Used for cushions, padding, mattresses, furniture, upholstery, automotive accessories, carpet underlayment, and packaging. Also known as *flexible urethane foam*. See also polyurethane foam.

flexible urethane foam. See flexible polyurethane foam.

Flexi-Draw. Trade name for a water-soluble metal drawing compound that has high film strength and minimizes metal-to-metal contact under high pressure and temperature.

Flexi-Flow. Trade name for a titanium-reinforced composite used in dentistry.

Flexilon. Trademark of Technical Coatings Company Limited (Canada) for silicone and polyester coatings.

Fleximat. Trade name of Kitson's Insulation Products Limited (UK) for flexible pipe insulating strip made from glass fiber.

Flexi-Mica. Trade name for sheet mica bonded with silicone resin. It has high dielectric strength and good tensile strength, and is used for electrical insulation applications.

Flexiphen. Trade name for modified phenolic resins.

Flexiplast. (1) Trade name for an injection molding acrylic for dental applications.

(2) Trade name of former Foster–Grant Company (USA0 for polyvinyl acetate resins.

Flexirene. Trade name of Enichem SpA/Polimeri Europa (Italy) for linear-low-density polyethylenes.

Flexisil. Trade name for a dental silicone elastomer used in duplicating work.

Flexit. Trade name of Manifattura Specchi e Vetri Felice Quentin (Italy) for thin toughened plate glass.

Flexite. (1) Trade name of St. Lawrence Steel Company (USA) for an oil-hardening low-alloy tool steel containing 0.45% carbon, 1.1% manganese, 1.1% chromium, 1.75% nickel, 0.2% vanadium, 0.35% molybdenum, and the balance iron. Used for pins, shafts, pinions, and hand tools.

(2) Trade name of ASG Industries Inc. (USA) for laminated glass.

Flexitime. Trademark of Heraeus-Kulzer Inc. (USA) for a controlled-set dental impression materials supplied in several grades with consistencies ranging from putty-like to thixotropic.

Flexmat. Trade name of McNichols Company (USA) for acid-resistant, thermoplastic floor tile.

Flexiteek. Trademark of Viva Vinyl BV (Netherlands) for non-slip vinyl-based decking materials that clean easily and closely resemble natural teak in appearance.

Flexo. Trade name of Carpenter Technology Corporation (USA) for a spring steel containing 0.6% carbon, 0.75% manganese, 2% silicon, and the balance iron. It has high toughness and tensile strength, and good hardness. Used for automobile leaf springs, recoil springs, and shuttle springs.

Flexo-Ceram. Trade name for dental ceramics used for inlay and veneer applications.

Flex-O-Crylic. Trademark of Flex-O-Glass, Inc. (USA) for crystal-clear acrylic sheeting.

Flex-O-Glass. Trademark of Warp Brothers/Flex-O-Glass, Inc. (USA) for a durable, flexible polymer film with special additives that provides high UV-protection, optimum light transmission and reduced condensation. Used in horticulture for pro-

tecting greenhouse crops and flowers.

Flexograin. Trade name of Riverside Metals Corporation (USA) for a phosphor bronze with high fatigue life used for clips, contacts, springs, etc.

Flexoid. Trademark of Cork Manufacturing Company Inc. (Canada) for cellulose-fiber-bonded binders used in cork products.

Flex-O-Lite. Trademark of Flex-O-Lite Division–Lukens General Industries Inc. (USA) for reflective glass beads for traffic and industrial applications.

Flex-O-Loy. Trade name of Bridgeport Rolling Mills Company (USA) for a low-leaded brass containing 66.5-71.5% copper, 30% zinc, 0.03% lead and 0.05% iron. It has good formability, and is used for wiring components, and other electrical and electronic components.

Flexomer. Trade name of Union Carbide Corporation (USA) for ultralow-density polyethylene, linear-low-density polyethylene and ethylene-vinyl acetate resins.

Flexor. Trade name of Pennsylvania Steel Corporation (USA) for a steel containing 0.34% carbon, 0.85% chromium, 0.5% molybdenum, 0.45% tungsten, and the balance iron. Used for chuck jaws, arbors, motor shafts, gears, racks, and pretreated machinery steel.

Flex Plate. Trade name of Aerospace Composite Products (USA) for stiff, high-strength composites made up of isotropic layups of unidirectional carbon plies. Used for landing-gear parts, wing-spar joiners, etc.

Flexply. Trade name of Kafus Biocomposites (USA) for non-woven mats and panels composed of natural fibers, such as hemp or kenaf, blended with polymer fibers. With heat and pressure they can be made into automotive components, such as door panels and panels, deck trays, seat backs, etc.

Flexseal. (1) Trademark of PPG Industries Inc. (USA) for a laminated glass whose vinyl resin interlayer extends beyond the glass edges for sealing into the window frame. Used for aircraft windows.

(2) Trade name of Flexrock Company (USA) for cellulose nitrate plastics.

Flex-Sheet. Trade name of National Products Inc. (USA) for strips of silvered glass attached to textile fabrics.

Flexsico. Trade name of Sico Inc. (Canada) for exterior and interior epoxy ester enamel paints for floors and porches.

Flex-Span. Trademark of Jeneric/Pentron Inc. (USA) for a neutral-colored, fast-setting eugenol-free zinc oxide/organic acids-based dental cement with a creamy, low-viscosity consistency used for cementing temporary bridges and crowns.

Flextran. Trade name of Manville Corporation (USA) for pipe made from fiberglass-reinforced polyester resin with graded silica aggregate.

Flextube. Trade name of Shape Memory Applications, Inc. (USA) for flexible tubing based on *Nitinol* nickel-titanium shape-memory alloy, and supplied in a wide range of diameters and lengths.

Flexwood. Trademark of Flexible Materials, Inc. (USA) for a wood-textile composite consisting of a sheet of cotton to which very thin sheets of veneer have been bonded under heat and pressure. Used as decorative wall covering.

Flexwrap. Trademark of Fiberite Corporation (USA) for a heat-resistant cotton fabric-reinforced, elastomer-modified phenolic tape material for ablative and aerospace applications.

Flexy. Trademark of Maple Leaf Plastics Corporation (Canada) for polyethylene pipe.

Flimba. Trade name for rayon fibers and yarns used for textile fabrics.

flimsy paper. A thin, wood-free, calendered copy paper with a basis weight of about 30-39 g/m² (0.10-0.13 oz/ft²).

flinkite. A greenish-brown mineral composed of manganese arsenate hydroxide, $Mn_3(AsO_4)(OH)_4$. Crystal system, orthorhombic. Density, 3.78 g/cm³; refractive index, 1.801. Occurrence: Sweden.

Flinso. Trade name of Grant & West Limited (UK) for a lead-tin solder for aluminum and aluminum alloys.

flint. A white, yellow, gray, brown or black impure cryptocrystalline variety of quartz composed of silicon dioxide (SiO_2), and having a greasy to waxy luster. Crystal structure, hexagonal. Density, 2.60-2.65 g/cm³; hardness, 6.5-7.0 Mohs. It has high toughness, and produces sparks when struck against steel. Used as an abrasive, as a paint extender, as a filler for rubber, plastics and road asphalt, as a component in glass manufacture, for building and paving applications, in the form of balls and liners for ball mills, and in the manufacture of earthenware, pottery and porcelain. Also known as *firestone*.

Flint alloy. British corrosion-resistant alloy containing 12.5% chromium, 0.3% carbon, and 0.5% silicon, and the balance iron. Used for cutlery and stainless parts.

Flintcast. Trade name of Pacific Foundry Company (USA) for an abrasion-resistant cast iron containing 3.3% total carbon, 2.4% silicon, 1% chromium, and the balance iron.

flint clay. Hard, smooth, nonplastic, kaolin-type refractory clay that has a conchoidal fracture surface resembling *flint*. Also known as *flint fireclay*.

Flint Edge. Trade name for a dental amalgam alloy.

Flintex. Trademark of Foxboro Company (USA) for porcelain tubing.

flint fireclay. See flint clay.

Flintflex. Trademark of E.I. DuPont de Nemours & Company (USA) for epoxy linings used on freight car, truck and trailer interiors.

flint glass. (1) A soft, fusible, colorless, lustrous and brilliant optical glass containing about 54% silica (SiO_2), 37% lead oxide (PbO), 8% potassium oxide (K_2O), and 1% sodium oxide (Na_2O). It has a high density, and a high refractive index (approximately 1.65-1.9 g/cm³), dispersion and surface brilliance. Used for lenses, fine table glassware, vacuum tubes, and electrical equipment. Also known as *optical flint glass*.

(2) A clear, colorless bottle glass.

(3) A term sometimes used for any glass of high quality.

Flintkote. Trademark of Flintkote Company, Inc. (USA) for an extensive series of different products for use in the construction and marine industries including adhesives and cements, asphalt coatings and sealers, asphalt roof coatings and roll roofing, shingles, fiberglass roofing mats, building papers, lath and wallboard, plaster, limestone and lime mixtures, gypsum products, Portland cements, mortars, gravel and stone aggregate, lumber and lumber products, and paints and other protective coatings.

Flintmetal. Trade name of Morris Machine Works (USA) for a hard, corrosion-resistant white cast iron containing 3-3.6% carbon, 4-5% nickel, 1.4-3.5% chromium, 0.4-0.7% silicon, and the balance iron. Used for cams, dies, rollers, and bearings.

flint paper. A strong, grayish-white paper that has crushed flint bonded to one side. Crushed flint for this purpose is available in various grit sizes ranging from very fine (240 mesh) to very coarse (20 mesh). *Flint paper* is fairly inexpensive, but dulls and wears rapidly, and is used as sanding paper for removing paint and old finishes from wood.

flint pebbles. Hard, tough, rounded stones of flint having smooth surfaces ranging in size from about 25 to 200 mm (1 to 8 in.). Occurrence: Canada (Newfoundland), Denmark, Greenland. Used as grinding media in ball, pebble and tube mills for grinding cement, minerals, ores, etc. *Flint pebbles* are also available calcined and ground for use as a source of silica in earthenware and porcelain. See also Danish pebbles.

Flintseal. Trademark of Stewart Paint Manufacturing Company (USA) for hot-poured rubber asphalt compounds for sealing contraction and expansion joints in concrete road and airport pavements, and finish coats for wood and concrete.

Flintseal JFR. Trademark of Stewart Paint Manufacturing Company (USA) for jet-fuel-resistant joint sealing compounds used for contraction and joints of airport pavements.

flint shot. Hard, coarse, sharp-edged, relatively clean sand used in sandblasting.

Flintuff. Trade name of Teledyne Ohiocast (USA) for a hard nickel cast iron containing 3% carbon, varying amounts of nickel and silicon, and the balance iron. Used for rolls.

Flintype. Trade name of Grand Northern Products Limited (USA) for an abrasion-resistant steel containing 0.9% carbon, 1% silicon, 0.3% manganese, 9.5% chromium, 0.6% molybdenum, 3.2% nickel, and the balance iron. Used for hardfacing electrodes for overlay or buildup applications on shovels, plow points, earthmoving machinery, etc.

Flisca. Trade name for rayon fibers and yarns used for textile fabrics.

flitch. A term applied to a log that has been sawn on two or four sides. It is ready for additional cutting, but can also be used as rough timber.

float glass. Sheet glass produced by the floating process, i.e., by flowing melted glass onto the flat surface of a molten metal, such as tin, contained in a tank. As it flows over the metal, a ribbon of glass with smooth, parallel surfaces is formed. Upon cooling, the glass becomes a rigid sheet that is subsequently carried through an annealing oven on smooth rollers. Finally, the continuous sheet of glass is inspected and cut to usable sizes. *Float glass* is available in two grades: (i) *Clear* with a nominal thickness of 2.5-6 mm (0.100 to 0.250 in.); and (ii) *Heavy-duty clear* with a nominal thickness of 8-22 mm (0.315-0.875 in.). Single-strength and double-strength grades are also supplied. Used for standard windowpanes.

flocculating agent. See flocculant.

flocculant. A reagent or electrolyte, such as alum or lime, added to a colloidal suspension (e.g., clay) to cause the particles to aggregate or coalesce and settle. Also known as *flocculating agent*.

flock. (1) Finely powdered or very short textile fibers (e.g., wool, cotton, rayon, silk, etc.) usually made by cutting or grinding. Used as an extender or filler in plastics, rubber and flooring compositions, and on wallpaper or fabrics to form velvety raised patterns.

(2) Waste wool or cotton used for stuffing, padding, cushioning or upholstery.

flocked fabrics. Textile fabrics that have very short fibers (flock) of similar or dissimilar color and/or material adhesively bonded to the face side, usually as tufts in the form of dots or other patterns. See also flock.

Flock-Lok. Trademark of Lord Corporation (USA) for a resinous plastic, paint-like adhesive used for attaching flock to rigid and flexible surfaces of metals, plastics, elastomers, ceramics, glass, concrete, wood and textiles.

Flo-Kote. Trade name of Stulz Sickles Steel Company (USA) for a wear-resistant steel containing 0.2% carbon, 11-13% manganese, varying amounts of nickel, and the balance iron. Used for welding electrodes.

FloMaster. Trade name of F.J. Brodman & Company (USA) for a series of powders based on ceramic borides, carbides and nitrides, pure metals, metal alloys, polymers, cermets, porcelain frits, composites, lanthanum chromite, molybdenum disulfide and hydroxyapatite. Used for laser cladding and thermal spraying applications.

Flomat. (1) Trademark of Fothergill & Harvey Limited (UK) for chopped-strand glass mats impregnated with unsaturated polyester resin.

(2) Trade name of DSM Resins (USA) for polyester sheet-molding compounds supplied in regular, fire-retardant, high-impact and low-profile grades.

Flonac. Trademark of Eckart America LP (USA) for pearlescent pigments.

flooding control agent. An agent added to organic coatings to retard the formation of surface irregularities.

floor brick. A brick that has exceptionally high resistance to chemicals, wear, mechanical damage and thermal conditions found in industrial and commercial installations.

floor covering materials. Planar materials used in building construction and renovation as the final surface of a floor to enhance its beauty, comfort and utility. A wide range of materials are available for these purposes including hardwood and softwood (usually in the form of strips, planks, blocks or tiles), resilient plastic tiles (e.g., vinyl or urethane), flexible vinyl sheeting, plastic laminates, ceramic tiles, flagstone, slate, and carpeting.

flooring cement. See Keene's cement.

flooring plaster. See Keene's cement.

floor sand. Reclaimed sand mixed with some new sand and coal dust, and used in foundries.

FloorThane. Trademark of Sherwin Williams Company (USA) for urethane-based industrial floor coatings.

floor tile. (1) An abrasion-resistant glazed or unglazed ceramic tile used in floor construction.

(2) A tough, durable, flexible plastic tile, usually of polyvinyl chloride or polyurethane, used as a floor covering material. It comes in various colors and sizes, may have a special shiny wear surface, and is either of the self-stick type, or must be bonded to the subfloor with a special adhesive.

Flora. (1) Japanese trade name for patterned glass.

(2) Trade name of SA Glaverbel (Belgium) for a glass with floral pattern.

Floreal. Trade name of Vidrieria Argentina SA (Argentina) for a glass with floral pattern.

Florence. Trade name of Permacon (Canada) for paving stone supplied in rectangular, square and half-round units of varying sizes and colors, and used for walkways, patios, steps, etc.

florencite. A pale yellow mineral of the alunite group composed of aluminum cerium phosphate hydroxide, $CeAl_3(PO_4)_2(OH)_6$. Crystal system, rhombohedral (hexagonal). Density, 3.67 g/cm³; refractive index, 1.695. Occurrence: Brazil.

Flo-Restore. Trademark of Den-Mat Corporation (USA) for a light-cure, radiopaque low-viscosity hybrid composite containing submicron barium glass and fumed silica. It polishes to a very high, natural luster, and has high strength and durability. Used in dentistry for small restorations with difficult access, for filling or repairing cracks in teeth, as a pit and fissure seal-

ant, and in incisal repair.

Florex. Trade name of ASG Industries Inc. (USA) for a glass with floral pattern.

Florida kaolin. A ball-type, uniform, sedimentary high-purity *kaolin* with fine particle size, and white burning properties used in various ceramic products to promote refractoriness, plasticity, bonding strength, workability, and suspending power.

Floridin. Trade name of Floridin Company (USA) for off-white fuller's earth from the state of Florida, USA, supplied in various sizes from 16 to 200 mesh.

Florigel. Trade name for fuller's earth.

Florisil. Trademark of US Silica Company (USA) for activated magnesium silicate available in various grades, and mesh sizes from 30 to 200. It has a pore diameter of 2700 nm, a specific pore volume of 0.54, and a surface area of 200 m²/g. Used as a highly selective synthetic adsorbant in gas and thin-layer chromatography.

Florite. Trademark of Floridin Company (USA) for activated bauxite, $Al_2O_3 \cdot 3H_2O$ or $Al(OH)_3$, that has been crushed, screened and calcined, and is available in 20- to 60-mesh grades. It is a porous, highly adsorptive material made by heating selected bauxites under controlled conditions, and used as adsorbent in filtration.

Flosbrene. Trademark of American Rubber Company (USA) for polystyrene polymers and copolymers.

floss. (1) A soft, glossy, untwisted thread of cotton or silk used for embroidery.

(2) A short, loose fiber of waste silk.

(3) See dental floss.

Flotectic Silver. Trade name of Eutectic Corporation (USA) for silver solders for torch, furnace and induction brazing of ferrous and nonferrous alloys.

Floterope. Trademark of American Manufacturing Company (USA) for polyethylene and polypropylene monofilaments and yarns used for textile fabrics and cordage.

Flotret Process Alloy. Trade name of Globe Metallurgical, Inc. (USA) for a special magnesium ferrosilicon alloy containing 43-47% silicon, 3-3.75% magnesium, 0.7-1.1% calcium, 9-1.2% aluminum, 1.25-1.75% total rare-earth elements, and the balance iron. Used in the Flotret process.

flour. The dust obtained by fine grinding of a solid material, such as wood, metal or stone.

flour gold. See gold flour.

Flovic. Trademark of ICI Limited (UK) for polyvinyl chloride (PVC) resins and PVC-copolymer film materials.

flow blue. A deep *cobalt blue* pigment used for underglaze printing on pottery giving a flowed or blurred effect.

Flower Brand. Trade name of Magnolia Antifriction Metal Company (USA) for a graphite-impregnated lead-base bearing metal used for bearings operating under constant load and speed conditions.

flowering dogwood. The hard, heavy wood from the small to medium-sized tree *Cornus florida*. It has a close-grained, uniform texture. Source: Throughout the eastern United States from New Hampshire to central Florida and westward to Texas. Used for weaving shuttles, bobbins, pulleys, jewelers' blocks, skate rollers, tool handles, mallets, and golf-club heads. Also known as *white dogwood*.

flowers of tin. See stannic oxide.

flowers of zinc. See zinc oxide.

Flowgrip. Trademark of Creative Pultrusions, Inc. (USA) for pultruded high-strength, fiber-reinforced polymer flooring panels used for walkways, firewalls, bridge decks, truck floors, machinery guards, structural roofing, etc.

flowing varnish. A special varnish designed to produce a smooth lustrous surface without polishing or rubbing.

Flow-It. Trademark of Jeneric/Pentron Inc. (USA) for a light-cure, radiopaque, fluoride-releasing bisphenol glycidylmethacrylate (Bis-GMA) based hybrid composite resin that is 70% filled with barium glass particles with an average size of 1.0 μm (40 μm). It has a flowable consistency, high flexibility, good compressive and tensile strengths, and good wear resistance. Used in dentistry for restorations, enamel and porcelain repair, small core-build-ups and veneering, as a pit and fissure sealant, and as a liner.

FlowLine. Trademark of Heraeus Kulzer, Inc. (USA) for a light-cure, 62 wt% filled, radiopaque, fluoride-releasing, medium-viscosity hybrid composite resin supplied in several shades, and used in dentistry for pit and fissure sealing and for restorations.

FlowMaster. Trademark of F.J. Brodmann & Co. LLC (USA) for a series of thermal-spray powders including various carbide, ceramic, cermet, glass, pure-metal, intermetallic, metal-alloy, metal-composites, microcomposite, mineral, nanophase and polymer powders used for the production of high-velocity-oxyfuel (HVOF), hardfacing, cold spraying and biomedical coatings.

FlowRestore. Trade name for a light-cure flowable dental composite restorative.

Flow-Satin. Trademark of Phillips Paint Products Limited (Canada) for a latex paint with satin finish.

FlowStone. Trade name of Whip Mix Corporation (USA) for a dental stone supplied as a powder. When cast from a suspension in water it sets with low expansion developing a dense, bubble-free abrasion-resistant surface. Supplied in buff and white colors, it is especially suitable for accurate, dimensionally stable bases and master models.

Flow-Stone. Trade name of King Packaged Products Company (Canada) for a special cement mix for anchoring posts, bolts and railings in concrete.

Flox. Trade name for rayon fibers and yarns used for textile fabrics.

Floxan. Trade name for spun rayon fibers and yarns.

fluckite. A pale to deep rose mineral composed of calcium manganese hydrogen arsenate dihydrate, $CaMnH_2(AsO_4)_2 \cdot 2H_2O$. Crystal system, triclinic. Density, 3.05 g/cm³; refractive index, 1.624. Occurrence: France.

fluellite. A colorless or white mineral composed of aluminum fluoride phosphate hydroxide heptahydrate, $Al_2PO_4F_2(OH) \cdot 7H_2O$. Crystal system, orthorhombic. Density, 2.18 g/cm³; refractive index, 1.490. Occurrence: UK.

Fluginox. Trade name of Pechiney/Ugine Aciers (France) for a series of low-carbon high-chromium stainless steels containing 0.05-0.25% carbon, 12-13% chromium, 0.5% molybdenum, 0-0.2% aluminum, and the balance iron. They have excellent corrosion resistance, and are used for chemical processing equipment, oil refining equipment, hardware, and tableware.

fluid-applied elastomer. An elastomer that at ambient temperature is in the liquid state, but cures or dries upon application forming a continuous membrane. Used as waterproofing in building construction.

fluid-bed coating. A smooth, tightly adhering coating of constant thickness, usually 0.15-1.50 mm (6-60 mils), that is produced by dipping a heated metal object in an agitated swirling pool of fine powder suspended in a stream of air. The powder particles are composed of a mixture of resin, pigment, catalyst and stabilizer, and melt upon the hot object surface forming a uniform film.

fluid-compressed steel. Steel subjected to a compression procedure while in the liquid state to remove gases and increase homogeneity.

Fluks. Trade name of Paul Bergsoe & Son (Denmark) for an alloy containing 55% tin, 32.5% lead, 10% antimony and 2.5% copper. It has good castability, a melting point of 179-327°C (354-621°F), and is used for electric motor and refrigerator bearings.

fluobarite. A mixture of the minerals *fluorite* (CaF_2) and *barite* ($BaSO_4$) used as a flux in glassmaking.

fluoborite. A colorless or white mineral composed of magnesium fluoborate, $Mg_3(BO_3)(F,OH)_3$. Crystal system, hexagonal. Density, 2.98 g/cm³; refractive index, 1.502. Occurrence: Sweden, USA (New Jersey). Also known as *nocerite*.

fluocerite. A yellow mineral that is composed of cerium fluoride, CeF_3, and may also contain lanthanum and neodymium. It can also be made synthetically. Crystal system, hexagonal. Density, 6.14 g/cm³; refractive index, 1.615. Occurrence: USA (Colorado).

Fluo-Kem. Trademark of Bel-Art Products (USA) for an antistick, dry-film fluorocarbon polymer lubricant with outstanding chemical inertness and thermal properties, used on adjacent surfaces to prevent abrasion, freezing or galling.

Fluometal. Trade name of EGC Corporation (USA) for a rigid, durable laminate composed of an expanded metal (e.g., stainless steel, *Hastelloy* or *Inconel*) sandwiched between two layers of polytetrafluoroethylene combining the high strength and workability of metal with the self-lubricating and nonstick characteristics and low friction of fluoropolymers. It is available in the form of sheets of varying thicknesses and lengths, and used for bushings and sliding bearings.

Fluon. Trademark of ICI Americas Inc. (USA) for a series of essentially chemically inert polytetrafluoroethylene polymers available in regular and specialty grades filled with 15% graphite or glass fibers, 25% glass fibers, or 60% bronze. They have a density of approximately 2.2 g/cm³ (0.08 lb/in.³), excellent flame retardancy, low coefficients of friction, and good high-temperature stability. Used for chemical pipe and valves, seals, O-rings, etc.

Fluonate. Trade name of Reichhold Chemicals, Inc. (USA) for fluorinated polyolefin copolymers used for coating applications.

fluophor. See fluor (1).

fluor. (1) A solid or liquid substance that can be stimulated to emit light by absorbing incident radiation (e.g., X-rays, cathode rays, ultraviolet radiation or alpha particles). The emission ceases almost immediately after excitation. Also known as *fluophor.* See also luminophor; phosphor.

(2) See fluorite.

fluoranthene. A tetracyclic hydrocarbon compound obtained from coal tar and available in standard grades (98% pure) in the form of colored needles with a melting point of 109-111°C (228-232°F) and a boiling point of 384°C (723°F). It is also available in deuterated and zone-refined high-purity grades (99.9+%). Used in chemistry and biochemistry. The zone-refined, high-purity grade is used as an aromatic reference standard. Formula: $C_{16}H_{10}$. Also known as *idryl.*

fluorapatite. See fluoroapatite.

fluorapophyllite. (1) A colorless, white, pink or green mineral of the apophyllite group composed of potassium sodium calcium fluoride silicate octahydrate, $(K,Na)Ca_4Si_8O_{20}F\cdot8H_2O$. Crystal system, tetragonal. Density, 2.37 g/cm^3. Occurrence: USA (Michigan).

(2) A colorless, white, pink, or green mineral of the apophyllite group composed of potassium calcium fluoride silicate octahydrate, $KCa_4Si_8O_{20}F\cdot8H_2O$. Crystal system, tetragonal. Density, 2.37 g/cm^3; refractive index, 1.534. Occurrence: Germany.

fluor crown glass. An *optical crown glass* containing considerable amounts of fluorine as a flux, and having a low dispersion and refractive index.

Fluorel. Trademark of 3M Company (USA) for a fully saturated fluorinated polymer that contains over 60 wt% fluorine, and is used for insulation applications, wire and fabric coatings, gaskets, sealants, O-rings, fuel cells, containers, hoses, and diaphragms.

Fluorel II. Trademark of 3M Company (USA) for a fluoroelastomer with excellent chemical resistance to engine oils and coolants, lubricants and transmission fluids, used for automotive components.

fluorellestadite. A pale rose mineral of the apatite group composed of calcium silicate phosphate sulfate fluoride, $Ca_5[(Si,P,S)O_4]_3(F,OH,Cl)$. Crystal system, hexagonal. Density, 3.07 g/cm^3; refractive index, 1.655. Occurrence: USA (California).

Fluoresbrite. Trademark of Polysciences, Inc. (USA) for monodisperse latex microparticles that contain fluorescent dye and may or may not contain pendant carboxyl groups. They are supplied in several fluorescences including bright blue, yellow-green, new yellow-orange and polychromatic red, and are used in electron and fluorescence microscopy.

fluorescein. A crystalline yellow crystalline powder (Color Index C.I. 45350:1) with a dye content of about 95%, a melting point of 320°C (608°F), and a maximum absorption wavelength of 496 nm. Used in biochemistry, biotechnology, medicine, electronics, and for various industrial applications. Formula: $C_{20}H_{12}O_5$. Also known as *diresorcinolphthaliein; resorcinolphthalein.*

fluorescent brightener. See brightener (3).

fluorescent dye. A dye that gives off or reflects light when exposed to shortwave radiation, and is used in nondestructive inspection, in textiles to intensify color and add brilliance, and in biochemistry and biotechnology.

fluorescent fabrics. Fabrics that are impregnated or coated with fluorescent chemicals and can be stimulated to emit light by incident ultraviolet radiation. Used for clothing and signal flags.

fluorescent glass. Glass that contains compounds that induce fluorescence, e.g., rare-earth elements or compounds or calcium fluoride. Used for fluorescent tubes, mercury-vapor discharge tubes, etc.

fluorescent magnetic powder. A ferromagnetic magnetic powder composed of finely divided fluorescent metallic particles in a medium, such as air or a liquid, that when viewed under ultraviolet light (wavelength, 320-400 nm) provide high contrast and visibility. Used in magnetic particle inspection.

fluorescent metal. A metal, usually a rare earth such as cerium, praseodymium, neodymium, samarium, europium or gadolinium, that can be stimulated to emit characteristic electromagnetic radiation by absorbing electromagnetic or corpuscular radiation. The emission ceases when the stimulus is removed.

fluorescent paint. A luminous paint containing *luminescent pigments* that give off light only so long as they are stimulated or

activated by ultraviolet or black light.

fluorescent penetrant. A substance utilizing penetrants which are usually green in color and fluoresce or glow brilliantly under ultraviolet or black light. It can either be water washable, postemulsifiable lipophilic, postemulsifiable hydrophilic or solvent removable, and is used in liquid penetrant inspection for the detection of discontinuities.

fluorescent pigment. A pigment, such as cadmium sulfide, calcium sulfide or zinc sulfide, that gives off light during exposure to radiant energy, such as ultraviolet or black light.

fluorescent plastics. Polymers or plastics, such as acrylics or vinyls, to which certain fluorescent dyes or pigments have been added that will glow or emit light after exposure to light. Used for symbol, warning and safety signs, for ornamental applications, and for labeling applications. Also known *fluorescent polymers.*

fluorescent whitening agent. See brightener (3).

Fluorex. (1) Trade name for a phosphate glass containing about 75% phosphorus pentoxide (P_2O_5) and 0.5% or less silicon dioxide (SiO_2). Supplied in rods and tubes, it is resistant to hydrogen fluoride and decomposed by alkalies.

(2) Trademark of Rexam Industries Corporation (USA) for a series of weatherable fluoropolymer films and laminates for surfacing of parts used outdoors.

fluorinated amorphous carbon. A thin film material consisting of an amorphous carbon-carbon crosslinked structure with sp^3 and sp^2 bonded carbon and strong carbon-fluorine bonds. It is typically produced by plasma-enhanced chemical vapor deposition (PECVD) from fluorocarbon source materials, such as carbon tetrafluoride (CF_4), hexafluoroethylene (C_2F_6), octafluorocyclobutane (C_4F_8) or their hydrogen mixtures. *Fluorinated amorphous carbon* materials have good thermal stability and low dielectric constants (k ~ 2.1-2.7), and are suitable as interlayer dielectrics for ultra-large-scale-integrated circuits.

fluorinated ethylene-propylenes. A group of fluorocarbon resins which are copolymers of hexafluoropropylene ($CF_3CF=CF_2$) and tetrafluoroethylene ($F_2C=CF_2$). Available in the form of pellets and powder for extrusion and molding, film, monofilament fiber, aqueous dispersions for spray or dip coating and nonstick finishes, they have excellent chemical stability, outstanding electric insulating properties, low dielectric constants and loss factors, low coefficients of friction, low melt viscosity, and a maximum continuous use temperature of 205°C (400°F). Used for hoses and tubing for fuel, oil, hydraulic, pneumatic and oxygen systems, for fittings, linings for pumps, pipes and chemical-processing equipment, for sealing applications in turbine engines, alternators and rotary actuators, for wire and cable insulation, printed circuit board laminates, medical and surgical supplies, and as fibers for filtration screening and mist separators. Abbreviation: FEP. Also known as *FEP resins; fluorinated ethylene-propylene resins; fluorinated perfluoroethylene-propylenes.*

fluorinated graphite. A compound of graphite and fluorine corresponding to the general formula $(CF_x)_n$ where n is about 0.8-1.2. It is used as a high-temperature and high-vacuum lubricant, as a cathodic depolarizer, and as a stationary phase in gas chromatography.

fluorinated perfluoroethylene-propylenes. See fluorinated ethylene-propylenes.

fluorinated polymers. See fluoropolymer.

fluorinated silicone. See fluorosilicone.

fluorine. A nonmetallic element of Group VIIA (Group 17) of the

Periodic Table (halogen group) that is a light, diatomic, corrosive, pale greenish-yellow gas or liquid. Density of gas, 1.696 g/L (at 0°C or 32°F); density of liquid, 1.108 g/L (at -188.2°C or -307°F); boiling point, -188°C or -307°F; freezing point, -219.6°C (-363°F); atomic number 9; atomic weight, 18.998; monovalent. It is extremely reactive, the most electronegative (nonmetallic) element known, and forms fluorides with many elements. Used as a powerful oxidizer, in the manufacture of metallic and other fluorides, in the production of fluoropolymers, as an active constituent of fluoridating compounds used in drinking water, toothpastes, etc., in rocket fuels, as a refrigerator coolant, in uranium enrichment, as a metallurgical and ceramic flux, and in chemistry, biochemistry, biotechnology and medicine. Symbol: F.

fluorine mica. Collective term for a group of natural or synthetic *phlogopite* and *muscovite* micas containing substantial quantities of fluorite (fluorspar). Also known as *fluormica*.

fluorite. A transparent to translucent, crystalline mineral occurring in a myriad of colors including yellow, green, pink, red, blue, violet, white and brown. It is a natural calcium fluoride, CaF_2, and may also contain lanthanides, such as dysprosium, erbium and ytterbium. It can also be made synthetically and is also in metallurgical, ceramic and acid grades containing from 85 to 98% CaF_2. Crystal system, cubic. Density, 3.18-3.20 g/cm³; melting point, 1350°C (2462°F); hardness, 4 Mohs; refractive index, 1.433. It has low dispersion, is weakly radioactive, and is transparent to radiation over a very wide range including infrared radiation of wavelengths up to about 10^{-3} cm. Occurrence: Canada, Germany, Italy, Mexico, Spain, UK, USA. Used as the principal ore of fluorine, as a flux in open hearth steelmaking, in gold, silver, copper and lead smelting, as a flux and opacifier in ceramic glazes, porcelain enamels and glass, as a flux in emery-wheel binders, in the manufacture of opalescent glass, as a component in certain cements, as a major component in crucibles for melting uranium for nuclear applications, in the manufacture of synthetic cryolite and hydrofluoric acid, in the manufacture of optical equipment and prisms, in carbon electrodes, phosphors, paint pigments, electric arc welding equipment, as a catalyst in certain wood preservatives, and in biochemistry, biotechnology and medicine. Also known as *fluor, fluorspar*.

fluorites. A group of ceramic materials corresponding to the fluorite (CaF_2) crystal structure MX_2, where M denotes a metal, such as calcium, cesium, potassium, tellurium, thorium, uranium, etc., and X a nonmetal, such as chlorine, fluorine or oxygen. Examples include tellurium dichloride ($TeCl_2$) and thorium dioxide (ThO_2).

Fluorline. Trademark of Norton Performance Plastics (USA) for polytetrafluoroethylene plastics used as container lining materials.

fluormica. See fluorine mica.

fluoroapatite. A natural or synthetic material with hexagonal crystal structure. *Natural fluoroapatite* is a mineral of the apatite group composed of calcium iron phosphate chloride fluoride, $(Ca,Fe)_5(PO_4)_3(F,Cl)$. Density, 3.1-3.2; refractive index, 1.651. Occurrence: Australia. *Synthetic fluoroapatite* is made from calcium phosphate, $Ca_3(PO_4)_2$. Density, 3.1-3.2 g/cm³; hardness, 5 Mohs; refractive index, 1.633. Both the natural and the synthetic minerals are used in ceramics, biochemistry, biotechnology, medicine and dentistry. The synthetic material is also used for laser crystals. Formula: $Ca_5F(PO_4)_3$. Abbreviation: FAP. Also known as *calcium fluophosphate; calcium fluorophosphate;*

fluorapatite. See also apatite.

Fluorobase T. Trade name of Ausimont USA for room-temperature-curing, waterborne fluoroelastomer paint that provides excellent aging and UV resistance, and outstanding resistance to acids, steam, fuels and hydrocarbon oils, and can be applied to steel, rubber and concrete products.

Fluorobase Z. Trade name of Ausimont USA for perfluoropolyethylene-based high-solids coating resins providing excellent mechanical properties, chemical resistance, durability and weatherability, and good stain and grafitti resistance. Used for high-performance topcoats on buildings, and as coil coating finishes.

fluorocarbon elastomers. A group of elastomers based on copolymers of vinylidene fluoride ($H_2C=CF_2$) and hexafluoropropylene ($CF_3CF=CF_2$). Available as standard, low-temperature and reinforced grades, they have outstanding oil, solvent and chemical resistance even at high temperatures, and good heat and oxidation resistance. Unreinforced grades have moderate mechanical properties and only fair resilience. The mechanical properties of *fluorocarbon elastomers* can be improved by reinforcement, their service temperatures range between -40 and +316°C (-40 and +600°F), and their low brittleness temperature is -23°C (-9°F). They do not support combustion and are suitable for prolonged exposure to abnormal temperatures and compounded oils. Used for aircraft and automotive components, vacuum equipment, seals, O-rings, gaskets, diaphragms, hoses, and many types of industrial equipment. Also known as *fluorocarbon rubbers*.

fluorocarbon fibers. Fibers made of a fluorocarbon resins, such as polytetrafluoroethylene, polyvinylidene fluoride or fluorinated ethylene propylene.

fluorocarbon plastics. Plastics based on resins, such as polytetrafluoroethylene and fluorinated ethylene propylene, made by polymerization of monomers consisting solely of fluorine and carbon atoms. See also fluorocarbons (2).

fluorocarbon polymers. See fluorocarbon elastomers; fluorocarbon plastics; fluorocarbons (2).

fluorocarbon powder. A powder made from a fluorocarbon resin, such as polytetrafluoroethylene, and usually available in particle sizes down to about 4 μm (0.0002 in.).

fluorocarbon resins. See fluorocarbons.

fluorocarbon rubbers. See fluorocarbon elastomer.

fluorocarbons. (1) A large group of synthetic organic compounds derived from hydrocarbons in which the hydrogen atoms have been replaced entirely or in part by fluorine atoms. They are usually nontoxic, chemically inert, nonflammable and stable to heat up to about 260-315°C (500-600°F). Some *fluorocarbons* are compressed gases, while others are liquids. *Fluorocarbons* are denser and more volatile than the corresponding hydrocarbons and have low solubilities, low surface tensions, low refractive indices, low dielectric constants, and viscosities comparable to those of hydrocarbons. Used for refrigerants, in the manufacture of plastics, for electrical insulation, fire extinguishers and certain wax coatings, in flotation and damping fluids, as dielectrics, lubricants and hydraulic fluids, and as solvents, blowing agents, and air-conditioning media.

(2) A category of engineering fluoroplastics derived principally from hydrocarbons in which the hydrogen atoms have been replaced entirely or in part by fluorine atoms. Other halogen atoms (e.g., chlorine) may also be present. The two basic types of fluorocarbons are polytetrafluoroethylene (PTFE or TFE) and fluorinated ethylene propylene (FEP). They are char-

acterized by relatively high crystallinity and molecular weights, good inertness and nonadhesiveness, good to excellent thermal and chemical resistance, and low dissipation factors and dielectric constants. Other interesting properties include low friction and low mechanical strength (but can be reinforced, e.g., with glass fibers). They are supplied in the form of molding and extrusion materials, dispersions, films, tapes, sheets, rods, tubes, fibers, monofilaments, insulated wires, powders, etc., often sold under trademarks or trade names, such as Viton, Fluorel, Fluorthene, Polyfluoron, Teflon, etc. Used for high-temperature wire and cable insulation, electrical and chemical equipment, coating of cooking utensils, piping, gaskets, seals, O-rings, etc., and continuous sheet, in bonding industrial diamonds to metal (grinding wheels), in biochemistry and biotechnology, and in biomedicine for vascular grafts. Also known as *fluorocarbon resins*. See also fluorine plastics; fluoroplastics.

Fluorocomp ETFE. Trademark of LNP Engineering Plastics (USA) for a family of ethylene-tetrafluoroethylene (ETFE) composites filled with 10% glass fibers or 30% carbon, graphite or glass fibers. They have good wear resistance and high compressive strength, and are used for nonlubricated bearings.

Fluorocore. (1) Trademark for a self-rising rigid polyimide foam produced by the action of a fluorocarbon-based blowing or foaming agent. Available in a wide range of shapes and several densities, they have high flame retardancy, good heat and humidity resistance, and low smoke emissivity. Used for aerospace, marine and general industrial applications.

(2) Trademark of Densply International Inc. (USA) for fluorocarbon-based dental core buildup materials.

fluoroelastomers. A group of high-molecular-weight, thermosetting elastomeric homopolymers and copolymers containing fluorine. They have excellent resistance to strong oxidizing acids, aromatic fuels and solvents, even at elevated temperatures, excellent oxygen and ozone resistance, and poor mechanical properties. They are often sold under trademarks and trade names, such as Fluorel, Kel-F, Viton, Kalrez, etc. Used for insulation, wire and fabric coatings, gaskets, sealant, O-rings, fuel cells, containers, hoses, and diaphragms. Abbreviation: FE. Also known as *fluororubbers*.

fluorofibers. A generic term referring to synthetic fibers composed of linear macromolecules based on fluorocarbon aliphatic monomers, such as vinyl fluoride or vinylidene fluoride.

Fluorofil. Trademark of Fiberfil (USA) for glass-fiber-reinforced polytetrafluoroethylene plastics.

Fluoroflex. Trade name of Resistoflex GmbH (Germany) for a series of fluoropolymers including polytetrafluoroethylene used for flexible tubing.

fluorogen. A chemical substance which produces fluorescence when mixed with another nonfluorescent substance.

Fluoroglas. Trade name for a fluorocarbon resin reinforced with glass particles and modified with a mineral pigment. Used for self-lubricating bearings.

FluoroGrip. Trademark of Integument Technologies, Inc. (USA) for fluoropolymer coatings and linings with excellent resistance to harsh chemicals at ordinary and elevated temperatures.

fluorohydrocarbon plastics. Plastics based on polymers made by the polymerization of monomers consisting solely of carbon, hydrogen and fluorine atoms. Examples include polyvinyl fluoride and polyvinylidene fluoride.

fluorohydrocarbons. A group of synthetic organic compounds derived from hydrocarbons in which the hydrogen atoms have been replaced in part by fluorine atoms. Examples include

fluoroform (CHF_3) and fluoromethane (CH_3F). Abbreviation: FHC. Also known as *hydrofluorocarbons*.

fluoroindate glass. A heavy-metal fluoride glass available in various compositions but usually containing about 30-40 mol% indium fluoride (InF_3), 20 mol% zirconium fluoride (ZrF_4), 10-20 mol% strontium fluoride (SrF_2), 15-20% barium fluoride (BaF_2) and varying amounts of one or more of the fluorides of lead, calcium, gadolinium, sodium, lanthanum, yttrium and cadmium. They have excellent formability and optical properties, and are used for optical fibers.

fluorol. See sodium fluoride.

Fluorolast. Trademark of Lauren Manufacturing Company (USA) for a series of two-component, catalyzed, room-temperature curing fluorocarbon rubber coating materials applied by brushing, dipping or spraying. They have excellent corrosion resistance to acids, alkalies, corrosive fuels, solvents and gases, and are used on concrete.

Fluorolin. (1) Trademark of Joclin Manufacturing Company (USA) for a thin, nonshrinking aluminum foil/glass laminate combined with a high-temperature silicone pressure-sensitive adhesive. It has high flexibility, conformability, heat resistance and peel adhesion, and is used for masking tape used for thermal-spray applications providing very good resistance to grit blast, and offering high heat reflectivity for good shielding. Also included under this trademark are similar aluminum foils with adhesive on one side and polyfluorocarbon film on the other. They are used as friction-resistant, protective facings on processing equipment and bag chutes.

(2) Trademark of Quantum Inc. (USA) for plastic materials used for thermal and electrical shielding and insulation applications.

Fluoroloy. Trade name of Fluorocarbon Company (USA) for wear-resistant fluorocarbon resins and blends of fluorocarbon materials supplied as sheets, tubes and rods. Used for seals, piston rings, valve seats, bearings, wear rings, and packings.

Fluoromelt. Trademark of ICI Americas Inc. (USA) for an extensive series of filled and reinforced melt-processible ethylene-chlorotrifluoroethylene (ECTFE) composites available in various grades including high-impact, carbon-reinforced and glass-reinforced. Used in the manufacture of molded industrial products, and for coatings. *Fluoromelt FPC* is an unreinforced ECTFE resin, and *Fluoromelt FPCF* a glass-fiber-filled ECTFE resin.

fluorophenylboronic acid. A fluorinated organic compound with a melting point of 214-218°C (417-424°F) used in the manufacture of liquid crystalline fluorobiphenylcyclohexenes and difluoroterphenyl by cross couplings in the presence of a palladium catalyst. Formula: $FC_6H_4B(OH)_2$.

fluoroplastics. Plastics based on fluoropolymers. See also fluoropolymers.

fluoropolymers. A large family of polymers derived from hydrocarbons in which the hydrogen atoms have been replaced entirely or in part by fluorine and/or chlorine atoms. Included in this group are polymers, such as polytetrafluoroethylene (PTFE), fluorinated ethylene-propylene (FEP), perfluoroalkoxy alkane (PFA), ethylene-tetrafluoroethylene (ETFE), polyvinylidene fluoride (PVDF), polychlorotrifluoroethylene (PCTFE), ethylene chlorotrifluoroethylene (ECTFE), polyvinyl fluoride (PVF), poly(ethylene-*co*-tetrafluoroethylene) (PE-TFE) and poly(ethylene-*co*-chlorotrifluoroethylene) (PE-CTFE). Also known as *fluorinated polymers*.

Fluoropore. Trademark of Millipore Corporation (USA) for polytetrafluoroethylene polymers with good resistance to acids, solvents and bases also at elevated temperatures. Used for laboratory ware, and membranes for laboratory filtration assemblies.

Fluoroprene. Trademark of Freudenberg-NOK (USA) for thermoplastic fluoroelastomers used in the manufacture of sealants.

fluororubbers. See fluoroelastomers.

fluorosilicates. A group of chemical compounds that are salts of fluosilicic acid (H_2SiF_6), and used in concrete hardening, ceramics, glass, paints, weather-resistant stone, electroplating, etc. Examples include magnesium fluorosilicate ($MgSiF_6$) and zinc fluorosilicate ($ZnSiF_6$). Also known as *fluosilicates; silicofluorides.*

fluorosilicone elastomer. A silicone polymer produced by partially substituting the methyl groups attached to the siloxane polymer of dimethyl silicone with fluoroalkyd groups. This increases chemical and ozone resistance and provides higher temperature resistance. Used for gaskets, seals, O-rings, wire and cable insulation, encapsulation of electronic components, adhesives, and sealants. Abbreviation: FMQ. Also known as *fluorosilicone rubber; methyl fluorosilicone rubber.*

fluorosilicone rubber. See fluorosilicone elastomer.

fluorosilicones. Silicone elastomers that have fluorine and sometimes methyl or vinyl substituent groups on the polymer chain. They have good chemical and thermal properties, and are used for rubber goods, such as seals, gaskets, etc. Also known as *fluorinated silicones.*

Fluorosint. Trademark of DSM Engineered Plastics (USA) for a series of polytetrafluoroethylene resins available in regular and synthetic mica-filled grades. They have high dimensional stability, low coefficients of friction, and a upper service temperature of 260°C (500°F). Used for labyrinth seals and shrouds, appliance bearings, transmission and power steering seal rings, valve seats, etc.

fluorothenes. See chlorotrifluoroethylene polymers.

Fluorothin. Trade name for a zinc silicophosphate dental cement.

fluorozirconate glass. See ZBLAN.

Fluorplast. Trade name of Norton Plastics (USA) for a series of thermoplastic fluoropolymers including several polytetrafluoroethylenes, perfluoroalkoxy alkanes and tetrafluoroalkoxy alkanes.

Fluor-Quasiceram. Trade name for glass-ceramics with fluorine bonds in their composition. See also Quasiceram.

Fluor-Simrit. Trademark of Carl Freudenberg (Germany) for polytetrafluoroethylene polymers for seals, gaskets and vibration damping devices.

fluorspar. See fluorite.

Fluorthene. Trademark for fluorocarbon resins with outstanding inertness, excellent temperature and electrical resistance, low coefficient of friction, low mechanical strength (but can be reinforced), moderate impact strength, and a maximum service temperature of 204-260°C (400-500°F). Used for bearings, piping, electrical insulation, enamels, high-temperature coatings, etc.

fluosilicates. See fluorosilicates.

Fluosite. Trademark of Montecatini Edison SpA (Italy) for phenol-formaldehyde plastics, molding compounds and laminates.

Fluran. Trademark of Norton Plastics (USA) for fluorocarbon polymer products with service temperatures up to 260°C (500°F) used for solvent-recovery systems, peristaltic pump tubing, dry-cleaning fluid lines, process-monitoring equipment, etc.

Fluropon. Trademark of DeSoto Coatings Limited (Canada) for architectural coatings used in the metal construction industry.

flush plate. A free-cutting alloy of 65.75% copper, 32.75% zinc

and 1.5% lead, used for hardware.

Flutes. Trade name of SA Glaverbel (Belgium) for patterned glass.

Flutex. Trade name of ASG Industries Inc. (USA) for glass with 25 mm (1 in.) convex flutes.

Flutilite. Czech trade name for a patterned glass with broad design of long, rounded grooves.

flux. (1) A chemical mixture applied to a metal prior to brazing, soldering or welding. It cleans the metal, prevents oxidation and corrosion and promotes better fusion by increasing fusibility and reducing the melting temperature. Also known *fluxing agent.* See also brazing flux; soldering flux; welding flux.

(2) A substance that promotes the fusing of minerals, prevents the formation of oxides and removes undesirable substances. For example, in iron smelting lime or limestone is added to the blast-furnace charge to assist in its reduction by heat and absorb mineral impurities, and in copper refining sand is used to remove iron oxide . Also known *fluxing agent.* See also purifier; mold flux.

(3) A liquid bituminous material used in the processing of other bituminous materials. Also known *fluxing agent.*

(4) An additive incorporated into a plastic to enhance its flow properties. Also known *fluxing agent.*

(5) Any substance or mixture, e.g., borax, fluorspar, lead, lime, nepheline syenite, silica, etc., that improves the fusion and flow of a ceramic, glaze or glass mixture when subjected to heat. Also known as *fluxing agent.*

(6) A clear porcelain enamel composed of silicates and other materials, but containing no coloring oxides. Used on artware. Also known as *fondant.*

Fluxcor. Trade name of Airco Vacuum Metals (USA) for a series of tubular flux-cored carbon and low-alloy steel arc-welding electrodes for use with carbon-dioxide shielding gas.

fluxing agent. See flux (1), (2), (3), (4) and (5).

fluxing lime. See calcium oxide.

fluxing stone. Pure limestone, dolomite or other rock, usually in the form of lumps 100-150 mm (4-6 in.) in size, used in the melting of iron in blast furnaces and foundries. Also known as *fluxstone.*

flux paste. A soldering or brazing flux in the form of a thin paste applied either directly or by dipping the heated metal filler rod into the paste. A popular paste for brazing is composed of a mixture of 75% powdered borax and 25% boric acid.

Fluxrite. Trade name of NL Industries (USA) for a soft lead-tin solder.

fluxstone. See fluxing stone.

fly ash. The small noncombustible particles, usually containing alumina, silica, unburned carbon and several metallic oxides, that result from the combustion of solid fuels, such as ground or powdered coke, coal, etc., and are carried off and recovered from flue gases. Used as pozzolan or additive and filler in cement, concrete and plastics, in the form of cenospheres as low-density filler in polymer composites, as absorbent for oil spills, and in mechanical powder blends for continuous casting applications. Abbreviation: PFA. Also known as *pulverized fluel ash (PFA).*

fly ash cenospheres. Spherical fillers and extenders based on fly ash, a glassy aluminosilicate-rich by product of solid-fuel fired power stations, and supplied in a typical particle size range of 75-300 μm (3-12 mils). They have low density (0.5-0.8 g/cm³, or 0.01-0.03 lb/in.³), high compressive strength, high inertness to acids, alkalies, solvents, organic chemicals and water, and high thermal stability above 980°C (1800°F). Used in synthetic resins, composites, cements, etc.

fly ash concrete. A concrete that contains fly ash, a glassy alumino-silicate-rich material, as a cement substitute that reacts with free calcium and thus contributes to the formation of a cementi-tious network.

F Nickel. Trade name of Inco Alloys International Inc. (USA) for an alloy containing 5.5% silicon, 1.5% iron, 0.5% carbon, 0.2% copper, 0.06% sulfur, and the balance nickel. It is commer-cially available in ingot and shot form, and used as a primary metal.

foam. A dispersion of a gas in a solid or liquid. Solid foams, such as plastic foams, sponge rubber, glass and refractory foams and metal foams, have sponge-like, cellular structures. Liquid foams include soap bubbles, foamed oils, etc. Also known as *foam material.*

Foamalum. Trademark of Foamalum Corporation (USA) for alu-minum foam.

foam-back fabrics. See foam-laminated fabrics.

foam-bonded nonwovens. Nonwoven fabrics made by a bonding process that involves the treatment of a web of fibers with ad-hesive liquid foam. Also known as *foam-bonded nonwoven fab-rics.*

Foamcoat. Trademark of Pierce & Stevens Chemical Corpora-tion (USA) for heat-expandable coatings for paper, textiles, films, etc.

foam core. A material composed of a foamed plastic, such as cel-lulose acetate, epoxy, phenolic, polystyrene, silicone or ure-thane, or a foamed ceramic, e.g., glass or concrete, and used as a foam-in-place, low-density core in sandwich materials for thermal-insulating and architectural applications.

Foamcote. Trademark of Ludlow Composites Corporation (USA) for a composite material composed of medium open-cell sty-rene-butadiene rubber latex foam with sheet or fabric facing.

foamed adhesives. See cellular adhesives.

foamed aluminum. See aluminum foam.

foamed clay. A lightweight, cellular clay formed either by rapid heating of selected clays, or by mechanical or chemical gen-eration of gas bubbles in a slurry or slip that are retained after it has dried. Used for acoustic and thermal insulation.

foamed concrete. See aerated concrete.

foamed glass. See cellular glass.

foamed-in-place insulation. See spray-foam insulation.

foamed metal. See metal foam.

foamed nylon. See nylon foam.

foamed plastics. See cellular plastics.

foamed polystyrene. See polystyrene foam.

foamed slag. See expanded blast-furnace slag.

foamed zinc. See zinc foam.

Foamex. (1) Trademark of Foamex LP (USA) for foam rubber products made from synthetic rubber latex.

(2) Trademark of Foamed LP (USA) for polyurethane foams available as blocks, sheets, rolls and tubes for use in the manufacture of products, such as acoustical insulation for auto-body panels, seat cushions and arm rests for airplanes, cars, trucks and boats, mattresses, pillows, carpet cushions, floor mats, goggles and helmets, air filters, gaskets, wipers, paint brushes, rollers, and orthopedic and athletic cushions and pads.

Foamglas. Trade name of Pittsburgh Corning Corporation (USA) for cellular or expanded glass in block form used as thermal insulation for intermediate temperature applications (up to about 600°C or 1110°F).

foaming agent. See blowing agent.

foam injection moldings. Thermoplastic moldings of varying size and shape usually formed by injecting gas into polymer melts and forcing the latter into the cavities of molds in which the entrapped gases expand to form foams. Foam injection mold-ings can also be made by first adding heat-activated blowing agents to plastic granules, and then heating and injecting the melt into molds cavities. Typical applications of foam-injec-tion-molded plastics include automotive bumpers, business ma-chine housings and transport pallets.

foam-laminated fabrics. Textile fabrics with a thin layer of a foamed plastic or elastomer (e.g., a polyester, polyether, poly-urethane, or polyvinyl chloride). The layer can be bonded to the fabrics by means of an adhesive, or by a melting process. Used especially for carpet backings, footwear and outerwear. Also known as *foam-back fabrics.*

foam material. See foam (1).

foam metal. See metal foam.

Foampuff. Trademark of Ludlow Composites Corporation (USA) for a composite material consisting of soft open-cell styrene-butadiene rubber latex foam with sheet or fabric facing.

foam rubber. See sponge rubber.

Foam Seal. Trade name of Foam Seal Inc. (USA) for a pressure-sensitive, adhesive polyvinyl chloride foam.

Foamsil. Trademark of Pittsburgh Corning Corporation (USA) for fused silica (99% pure) available in the form of foamed blocks, disks and slabs with closed-cell structures. They have a density range of 150-240 kg/m³ (9.4-15 lb/ft³), good high-tem-perature resistance up to about 1200°C (2190°F), and high com-pressive strength. Used for insulating and refractory materials, tank linings, refractories, and structural elements.

Foam S.T.O.P. Trade name of Acoustical Surfaces, Inc. (USA) for acoustical melamine and urethane foams for noise-control applications.

Foam X. Trademark of Alusuisse Composites Inc. (USA) for poly-styrene foam-centered boards for displays, signs, etc.

Foballoy. Trade name of Plessey Inc. (USA) for an alloy contain-ing 95% tin, 3.5% silver and 1.5% antimony, used for hermetic sealing of integrated ceramic packages.

Fobes metal. A low-strength casting alloy containing 54% zinc and 46% copper, used for ornamental castings.

fob metal. A British red brass containing 87.5% copper, 12% zinc, and 0.5% tin. Used for ornaments and fobs.

Focarbo. Trade name of Forjas Alavesas SA (Spain) for a series of low- and medium carbon steels (AISI-SAE 10xx series) in-cluding various case-hardening and heat-treatable grades. Used for machinery parts, such as bolts, joints, studs, pins, levers, shafts and axles.

Fodel. Trademark of E.I. DuPont de Nemours & Company (USA) for thick-film compositions supplied in various grades, and in-corporating photosensitive polymers.

foggite. A colorless to white mineral of the zeolite group com-posed of calcium aluminum phosphate hydroxide monohydrate, $CaAl(PO_4)(OH)_2 \cdot H_2O$. Crystal system, orthorhombic. Density, 2.78 g/cm³; refractive index, 1.610. Occurrence: USA (New Hampshire).

foil. A metal, such as aluminum, copper, gold or tin, beaten, ham-mered, rolled or deposited into a very thin sheet usually less than 0.15 mm (0.006 in.) in thickness, and used in the manu-facture of electronic components, such as capacitors, in insula-tion and other electrical applications, and in wrapping, pack-aging and laminating, e.g., aluminum foil bonded to paper. See also microfoil; nanofoil.

Foilskin. Trademark of Bakor Company (Canada) for waterproof, reflective, UV-resistant aluminum repair tape for use on gutters, flashing, garages, vehicles, etc.

Folan. Trade name of Forchheim GmbH (Germany) for polyolefin/polyvinyl chloride film materials.

Folastal. Trade name of Salmax AG (Germany) for plastic-coated metal sheets and trapezoidal bars used for roofs and walls.

folded yarn. See plied yarn.

foliated graphite. Natural graphite that occurs massive and readily separates into thin leaves or plates. Used for lubricants, crucibles, etc.

Folio. Trade name of Vereinigte Glaswerke (Germany) for a glass with leaf pattern.

Folioplast. Trademark of Schwarz GmbH (Germany) for a flexible polyvinyl chloride film material.

Follansbee. Trade name of Follansbee Steel Company (USA) for a series of electrical steels supplied with silicon contents ranging from 0.5 to 4.0%. They have high magnetic permeability, and are used for transformers, dynamos, motors, generators, pole pieces, armatures, chokes, and loudspeaker components.

Fomo. Trade name of Forjas Alavesas SA (Spain) for a series of chromium-molybdenum alloy steels including several case-hardening grades. Used for structural applications, shafts, pins, gears, etc.

fondant. See flux (6).

Fonix. Trade name of Forjas Alavesas SA (Spain) for a series of austenitic, ferritic and martensitic stainless steels.

Fonox. Trade name of Ferromet Praha (Czech Republic) for ferrite materials.

Fontain Moreau. A British reverse bronze containing 0-8% copper, 0-1% iron, 0-1% lead, and the balance zinc. Used for die castings, fittings, and ornamental work.

Fool-Proof. Trade name of P.F. McDonald & Company (USA) for an oil-hardening tool steel containing 0.7% carbon, 4% tungsten, 2% chromium, and the balance iron. Used for hand and pneumatic chisels.

fool's gold. See pyrite.

footeite. See connellite.

Foraflon. Trademark of Atofina SA (France) for a series of melt-processible engineering thermoplastic resins based on polyvinylidene fluoride, and supplied in powder, pellet, film, metallized film, rod, sheet and tube form in unreinforced and carbon fiber-reinforced grades. They have a density of 1.76 g/cm^3 (0.064 lb/in.^{3}), a melting point of 155-170°C (310-340°F), high impact strength, excellent rigidity and abrasion resistance, good electrical and self-extinguishing properties, excellent resistance to radiation, excellent stress-crack resistance, good thermal stability, a service temperature range of -40 to +150°C (-40 to +300°F), very low water absorption, good weatherability, excellent resistance to ozone, good resistance to acids, alcohols, alkalies, greases, oils and halogens, fair resistance to aromatic hydrocarbons, and poor resistance to ketones. Used for wire and cable insulation, tank and drum linings, chemical tanks, and tubing and piping, electrical components and devices, electronic applications, and in paints and coatings.

Foral. Trade name of Forjas Alavesas SA (Spain) for a free-machining, low-carbon steel containing 0.1% carbon, 1.1% manganese, 0-0.6% silicon, 0-0.05% phosphorus, 0.3% sulfur, 0.2% lead, 0-0.1% selenium, and the balance iron. Used for pins, bolts, screws, etc.

Forall. Trademark of Forest Fiber Products Company (USA) for light-colored, grain-free, nonsplintering hardboard made by compressing and interfelting *lignocellulose* wood fibers (e.g., from Douglas fir) under heat. It is available in various sizes and thicknesses for construction applications.

Foraperle. Trade name of Atofina SA (France) for fluorinated acrylic resins used mainly in the leather, paper, textile and packaging industries

Forar. Trade name of Amoco Performance Products (USA) for high-density polyethylenes.

forbesite. A dull grayish-white mineral composed of nickel cobalt arsenate hydrate, $H(Ni,Co)AsO_4 \cdot 3.5H_2O$. Occurrence: Chile.

Forbon. Trademark of NVF Company (USA) for a general-purpose vulcanized fiber supplied in the form of sheets, rods and tubes. Also included under this trademark are fiber overlays for lumber, plywood, etc.

Forca. Trademark of Tonen Kabushiki Kaisha (Japan) for carbon fibers used for reinforcing plastics, and in making sheets, tubes, boards and rods. Also included under this trademark is tow sheet made by placing liquid crystalline pitch-based carbon fiber tow on a thin scrim cloth.

Forcel. Trademark of Rayon Industrial SA (Peru) for polyester staple fibers and filament yarns.

Foremost. (1) Trade name Swift Levick & Sons Limited (UK) for a series of non-aging nickel-chromium and nickel-chromium-iron casting alloys. Usually supplied in the annealed condition, they have good corrosion resistance and good high-temperature properties. Used for heat- and/or corrosion-resistant parts.

(2) Trade name of Swift-Levick & Sons Limited (UK) for a series low- and medium-carbon alloy steels for hot dies as well as numerous high-chromium and chromium-nickel stainless steel for applications requiring good corrosion resistance.

Foremost Pneumatic. Trade name of Swift Levick & Sons Limited (UK) for a medium-carbon tool steel containing 0.45% carbon, 1.7% chromium, 2% tungsten, 0.25% vanadium, and the balance iron. Used for punches, shears, etc.

Forest. Trade name for hardboard with a smooth, lustrous surface made by bonding wood chips with synthetic resin and wax, and carrying through a hydraulic pressing procedure. Used for furniture, cabinets, wall paneling, etc.

Forestacryl. Trade name of Forestadent Limited (UK) for a self-cure dental acrylic.

Forest Green. Trade name of AFG Industries, Inc. (USA) for flat glass.

Forest Pride. Trademark of Trojan Board Limited (USA) for solid wood paneling.

Forest Ridge. Trademark of Georgia-Pacific Corporation (USA) for vinyl panels used as siding.

Forest-Wool. Trade name of CDD Inc. (USA) for cellulose insulation.

Forex. Trade name of M.A. Vink GmbH (Germany) for rigid polyvinyl chloride foam products.

Forez. Trade name of Acieries de Forez (France) for a series of molybdenum- and tungsten-type hot-work, high-carbon high-chromium cold-work, molybdenum- and tungsten-type high-speed and water-hardening plain-carbon tool steels, and various nickel-chromium machinery steels.

Forge-Die. Trade name of Teledyne Vasco (USA) for a tungsten-type hot-work tool steel (AISI type H25) containing 0.23-0.28% carbon, 0.2-0.3% silicon, 0.2-0.3% manganese, 13.5-14.5% tungsten, 3.25-3.75% chromium, 0.4-0.6% vanadium, and the balance iron. It has high hot hardness and toughness, and good resistance to heat checking. Used for extrusion dies, piercing punches, upsetter punches, nut punches and piercers, extrusion and gripper dies, forging-die inserts, and hot-forming dies and rolls.

Forgemaster Die Block Steel. Trade name of Eagle & Globe Steel Limited (Australia) for a heat-treated tool steel containing 0.55% carbon, 0.3% silicon, 0.65% manganese, 0.65% chromium, 0.3% molybdenum, 1.5% nickel, and the balance iron. Used for drop forging die blocks, shear blades, die casting dies for zinc, extrusion tooling and auxiliary tools for lead, copper and aluminum and their alloys.

Forge-Well. Trade name of St. Lawrence Steel Company (USA) for an oil hardening low-alloy mold steel containing 0.27% carbon, 1% manganese, 1.2% chromium, 0.25% vanadium, 0.3% molybdenum, and the balance iron.

forging brass. A brass with a nominal composition of 58-60% copper, 38-40% zinc, and up to 2.0% lead. Commercially available in the form of rods, structural shapes and extrusions, it has excellent mechanical properties, excellent hot and cold workability (forgeability), and good corrosion resistance and machinability. Used for forgings and pressings of all kinds, tire-valve stems, hardware, bolts, fasteners, fittings, locks, gears, wire, plumbing supplies, etc. Abbreviation: FB.

Forging Russian. Russian trade name for an alloy containing 53.5% copper, 42% zinc and 4.5% manganese used for corrosion-resistant forgings.

forgings. (1) Metal or metal alloy products of varying size and shape formed by individual or intermittent applications of pressure, usually with the simultaneous application of heat. In forging, the starting material is either compressed in cross section and thus extended in length, longitudinally squeezed and thus increased in cross section, or squeezed within and made to conform to the shape of a die cavity. Forgings have excellent grain structures and a good combination of mechanical properties. Carbon and alloy steels including structural, tool and stainless grades as well as many nonferrous alloys including those of aluminum, copper, magnesium, molybdenum, nickel, niobium and titanium are available as forgings.

(2) See powder-metallurgy forgings.

Forlion. Trademark of Fortex SIDAC SpA (Italy) for nylon fibers and yarns used for textile fabrics including clothing.

Formable. Trade name of Armco International (USA) for a series of high-strength low-alloy (HSLA) steels used for structural members, bridges, and buildings.

Form-Acoustic. Trade name of Manville Corporation (USA) for molded glass-fiber acoustic panels.

Formaldafil. Trade name of Fiberfil (USA) for glass-fiber-reinforced acetal resins based on polyoxymethylene (POM).

formaldehyde. An easily polymerizable, colorless gas with a density of 1.075-1.081 g/cm³, a melting point of -92°C (-134°F), a boiling point of -21°C (-6°F), an autoignition temperature of 806°F (430°C). It is usually supplied as 37-50 wt% solution in water, and used as a hardening agent, in the manufacture of phenolic, polyacetal, urea and melamine resins, in the production of plywood, particleboard and foam insulation, as a chemical intermediate, as a disinfectant and preservative, as a reducing agent in the recovery of gold and silver, as a corrosion inhibitor. Formula: HCHO. Also known as *formic aldehyde; oxymethylene; oxomethane; methanal.*

Formalin. Trade name of National Plastic Products Company (USA) for phenolic resins and plastics.

Formaloy. (1) Trade name of American Smelting & Refining Com-

pany (USA) for a high-strength die-casting alloy containing 3% copper, 0.4% magnesium, and the balance zinc. Used in dies for forming drop hammer punches and aluminum sheet dies.

(2) Trade name of Boyd-Wagner Company (USA) for an oil-hardening tool steel used for dies and shearing tools.

formanite. A black mineral of the fergusonite group composed of rare-earth tantalum oxide, LnTaO₄, and may also contain uranium, zirconium, thorium, calcium, niobium, magnesium and/or titanium. Crystal system, monoclinic. Density, 6.17 g/cm³; refractive index, 2.14. Occurrence: Siberia.

Formanyl. Trademark of Glasurit (Germany) for phenol-formaldehyde resins and compounds.

Formasil. Trade name for silicone elastomers for dental laboratory applications.

Format. Trade name of Fiber Glass Industries Inc. (USA) for glass-fiber reinforcements.

Formatray. Trade name for a self-cure acrylic resin for dental tray applications.

Formbrite. Trade name of Anaconda Company (USA) for a zinc-copper alloy with good cold formability, used for deep-drawn parts.

Formdie. Trade name of Columbia Tool Steel Company (USA) for an air-hardening medium-alloy cold-work tool steel (AISI type A9) containing 0.51% carbon, 0.45% manganese, 1% silicon, 5.2% chromium, 1.5% nickel, 1.05% nickel, 1.4% molybdenum, and the balance iron. It has high toughness and hot hardness, good wear resistance, and is used for shear blades, extrusion dies, inserts, and heavy punches.

Formdie Smoothcut. Trade name of Columbia Tool Steel Company (USA) for a free-cutting medium-alloy cold-work tool steel (AISI type A9) containing 0.48-0.54% carbon, 0.9-1.1% silicon, 1.2-1.5% molybdenum, 5-5.4% chromium, 0.9-1% vanadium, 1.3-1.6% nickel, and the balance iron. Used for shear blades, hot upsetting and gripping dies, drop forging and hot heading dies, and punches.

formed fabrics. Textile fabrics whose fibers are held together by adhesive bonding, mechanical interlocking, fusion (application of heat), pressing or other techniques. See also woven fabrics; nonwoven fabrics.

formed yarn. See plied yarn.

Formetal. Trademark of Formet Technology Corporation (USA) for a series of thermoformed superplastic zinc alloys.

Formex. (1) Trade name for vinyl formal resins used on insulating wire.

(2) Trademark of Illinois Tool Works Inc. (USA) for flame-retardant polypropylene sheeting used as electrical insulation for office products, appliances, and consumer electronics.

(3) Trademark of Formex AB (Sweden) for optical fibers used in information-transfer systems.

Formflex. (1) Trade name for odor-free flexible silicone sealing compounds for windows and doors.

(2) New Zealand trademark for exterior and interior paints, primers and undercoatings.

(3) Trade name of Form Flex (USA) for standard and printable polyolefin sheet and film.

Formglas. Trademark of Formglas Inc. (USA) for glass-reinforced gypsum ceiling panels.

Formica. Trademark of Formica Corporation (USA) for laminated panels and sheets made from urea, melamine and phenolic plastics by a high-pressure process. They have hard, smooth, glossy, heat-resistant surfaces, and are used for counters and tabletops and other decorative surfacing applications, boat and recreational vehicle furniture, and in adhesives for bonding laminated plastic to other surfaces.

formic aldehyde. See formaldehyde.

forming board. A board, either single ply or laminated, composed of refined cellulosic fibers and, sometimes, synthetic fibers, usually with thermosetting or thermoplastic resin bonding. It can be readily formed into relatively rigid and stable three-dimensional shapes by application of heat and pressure while confined in a die.

Formion. Trademark of A. Schulman Inc. (USA) for a series of compounded or alloyed ionomers supplied as powders, pellets, beads, or granules for molding and extruding plastic goods.

Formite. Trade name of Columbia Tool Steel Company (USA) for a series of tungsten-type hot-work tool steels (AISI types H21 to H24) containing 0.33-0.51% carbon, 3-3.5% chromium, 9.2-15% tungsten, 0.5% vanadium, and the balance iron. They have great depth of hardening, excellent resistance to decarburization, good abrasion resistance, and high toughness. Used for extrusion and die-casting dies for brass, hot-extrusion dies, hot trimming and blanking dies, die and dummy blocks, hot punches, piercers, tool and die inserts, and hot nut tools.

form lacquer. Thin lacquer or varnish applied to forms to prevent concrete from adhering to them. It hardens before the concrete is placed.

Formoboard. Trademark of Lydall, Inc. (USA) for a series of high-performance thermoformable thermoplastic or thermosetting composites containing cellulosic and/or inorganic fibers. They are free of formaldehyde, have high strength-to-weight ratios, good dimensional stability, very low water absorption, good flame retardancy, and good deep-draw capabilities. Used for package shelves, door panels, trunk liners, spare-tire covers, kick panels, quarter trim panels, and head liners.

Formold. Trade name of Crucible Materials Corporation (USA) for an oil-hardening tool steel containing up to 0.7% carbon, 0.55% nickel, 1.35% chromium, 0.2% molybdenum, and the balance iron. It has high core strength, and is used for cold-hubbed plastic mold dies.

Formolit. Trademark of NYH for a series of epoxy resins and adhesives.

Formstones. Trade name of Grav-I-Flo Corporation (USA) for randomly shaped resin-bonded tumbling media.

Formula 1. Trade name of Atotech USA Inc. (USA) for a bright nickel electrodeposit and plating process mainly used for automotive applications.

FormulaFX. Trademark of Superior Graphite Company (USA) for graphite/carbon brake materials for superior friction performance.

FormulaOne. Trademark of Solutia Inc. (USA) for glare- and heat gain-reducing high-performance automotive window films supplied in neutral color in a wide range of light transmissions.

Formvar. Trademark of Monsanto Chemical Company (USA) for a series of tough thermoplastics based on polyvinyl formal resins. They have a density of 1.3 g/cm³ (0.05 lb/in.³), good dielectric properties, good resistance to alkalies, poor resistance to acids and moderate strength. *Formvar* is also available as 0.25 and 0.5 wt% solutions in ethylene dichloride. Used in metallography, biochemistry and electron microscopy for replicas or specimen supporting membranes, and for electrical insulation, wire enamels and coatings, e.g., for magnet wire, and for adhesives, films, molded materials, and impregnated compounds.

fornacite. An olive-green mineral composed of copper lead chromium oxide arsenate hydroxide, $Pb_2CuCrO_4AsO_4(OH)$. Crystal system, monoclinic. Density, 6.27 g/cm³. Occurrence: Congo; Iran.

Fornanc Special. Trade name of British Steel Corporation (UK) for an oil-hardening tool steel containing 0.15-0.25% carbon, 1.4-1.8% manganese, 0.4-0.7% nickel, 0.15-0.35% molybdenum, and the balance iron. Used for gears, shafts, and axles.

Forpan. Trademark of Forpan Inc. (Canada) for particleboard.

Forrovid. Trade name of Vidrobrás (Brazil) for glass-fiber acoustic tiles.

forsterite. A colorless, white, greenish or yellowish mineral of the olivine group composed of magnesium silicate, Mg_2SiO_4. It is also produced synthetically from magnesium carbonate and silicon dioxide for use as a ceramic raw material. Crystal system, orthorhombic. Density, 2.8-3.3 g/cm³; melting point, 1900°C (3450°F); hardness, 7.5 Mohs; refractive index, 1.660; high-thermal expansion; low-loss dielectric properties. Occurrence: Greenland, Italy, USA (Massachusetts). Used for electronic ceramics, ceramic-to-metal seals, refractories and cements. Also known as *white olivine*.

forsterite ceramics. Ceramics in which forsterite is the principal crystalline phase.

forsterite porcelain. A vitreous ceramic whiteware composed principally of forsterite. It has good dielectric properties, and is used for technical applications. Also known as *forsterite whiteware*.

forsterite refractories. Semibasic refractories composed of essentially of olivine rock and magnesia, with a general composition corresponding to the formula (Mg_2SiO_4). Used in the form of firebricks for metallurgical furnaces.

forsterite whiteware. See forsterite porcelain.

Fortafil. Trademark of Fortafil Fibers Inc. (USA) for high-strength carbon fibers made from low-cost acrylic filaments and used in the manufacture of automotive and aerospace components, and sporting goods.

Fortafil Uniweb. Trademark of Fortafil Fibers Inc. (USA) for unidirectional fabrics based on zero-crimp dry, unidirectional *Fortafil* carbon fibers oriented with microthin glass. Used as reinforcements in the manufacture of composites.

Fortal. Trade name of Compagnie Français des Métaux (France) for a heat-treatable aluminum alloy containing 4% copper, 0.5% magnesium, 0.5% manganese, and 0.7% silicon. Used for light-alloy parts.

Fortan. Trademark of Fortan Inc. (Canada) for chrome-tanned leather and splits.

Forte. Trade name of Forjas Alavesas SA (Spain) for a series of plain-carbon tool steels used for hand tools, chisels, axes, hammers, hatchets, pliers, saws, and knives.

For Ten. Trade name of Ford Motor Company (USA) for high-strength low alloy (HSLA) structural steels.

Forticast. Trade name of J.F. Jelenko & Company (USA) for a quenchable, hardenable, microfine extra-hard dental casting alloy (ADA type IV) containing 42% gold, 26% silver, and 9% palladium. Used for three-quarter and full crowns, long-span bridgework, and partial dentures.

Forticel. Trademark of Hoechst Celanese Corporation (USA) for thermoplastic cellulosic materials (e.g., cellulose propionate resins) available in pellets and crystals in translucent, metallic, and opaque colors for processing by blow and injection molding, extrusion, and rotational casting. They have a density of 1.20 g/cm³ (0.04 lb.in.³), a maximum service temperature of

49-93°C (120-200°F), good strength and toughness, high surface gloss, good chemical resistance, poor resistance to organic solvents, good resistance to mineral oils, good moldability, and good impact strength. Used for piping, sheeting, appliance housings and trim, pen and pencil barrels, spectacle frames, glazing, packaging, eyeshades, telephone bases, tool handles, knobs, steering wheels, etc. Also included under this trademark are liquid epoxy-based resins used for glass-bottle coatings.

Fortiflex. Trademark of Solvay Engineered Polymers (USA) for an extensive series of engineering thermoplastics based on high-, low-, or medium-density polyethylenes, and used for blow, injection, or rotationally molded products including chemical tanks, materials handling equipment, food containers, sporting goods, truck beds, packaging film, sheeting, chemical piping, fibers, and gasoline and oil containers.

Fortisan. Trade name for rayon fibers and yarns used for textile fabrics.

Fortify. Trademark of BISCO Inc. (USA) for an unfilled, light-cure, thin, penetrating resin composite with low viscosity and high wear resistance. Used in dentistry as a surface sealant for anterior and posterior composite restorations.

Fortify Plus. Trademark of BISCO Inc. (USA) for a light-cure, microfilled resin composite with low viscosity and high wear resistance. Used in dentistry as a surface sealant for anterior and posterior composite restorations.

Fortilene. Trademark of Solvay Polymers, Inc. (USA) for homopolymer and copolymer polyethylene and polypropylene resins including gas-phase polymerized, anti-staticized, UV-stabilized and impact-modified grades.

Fortinox. Trade name of British Steel plc (UK) for semi-austenitic stainless steel used for watch springs.

Fortisan. Trademark of Hoechst Celanese Corporation (USA) for a cellulosic fiber made from stretched cellulose acetate by partial saponification. It can be colored like cotton, and has a density of 1.5 g/cm³ (0.05 lb/in.³), extreme fineness (e.g., the monofilament is 1 denier, i.e., 1 gram per 9000 meters of fiber), very high tenacity, high dry strength, good wet strength, and rather low elongation under stress. Used for parachutes and fine fabrics.

Fortissimus. Trade name of Rudolf Schmidt Stahlwerke (Austria) for high-speed steel containing 0.7% carbon, 18% tungsten, 4% chromium, 1% vanadium, and the balance iron. Used for tools and cutters.

Fortiweld. Trade name of British Steel plc (UK) for a structural steel containing 0.1-0.16% carbon, 0.4-0.6% molybdenum, 0-0.6% manganese, 0-0.4% silicon, 0-0.005% boron, and the balance iron. Usually supplied in the normalized condition, it has good formability and weldability, and is used railroad cars, bridge cranes, materials handling equipment, and pressure vessels.

Fortoss Cema. Trademark of Biocomposites Limited (UK) for a biocompatible, porous calcium phosphate ceramic matrix material used in surgery to stabilize underlying graft and prevent tissue ingress. It is applied as a paste, sets in a few minutes, and resorbs in six weeks.

Fortoss Perma. Trademark of Biocomposites Limited (UK) for a deproteinized non-resorbable, porous natural hydroxyapatite ceramic matrix material used to support the remodeling of new bone, particularly to fill periodontal pockets and enhance soft tissue esthetics.

Fortoss Resorb. Trademark of Biocomposites Limited (UK) for a resorbable, porous stoichiometric tricalcium phosphate matrix material used in dentistry to restore lost bone and fill bony defect and extraction sockets.

Fortrel. Trademark of Wellman Inc. (USA) for strong, durable polyester fibers, filaments and yarns available in a wide range of grades including *Fortel*, *Fortrel EcoSpun*, *Fortel MicroSpun*, *Fortel Plus*, *Fortel Spunnaire*, *Fortel Spunnesse* and *Fortrel BactiShield*. Used for elegant clothing, nonwovens, bedding, household and industrial fabrics.

Fortress Blue Core. Trade name for a blue-colored resin-based dental core composite for metal restorations.

Fortress Core. Trade name for a resin-based dental core composite for porcelain.

Fortrex. Trade name of Murex Limited (UK) for a series of plain-carbon steels and low-carbon nickel steels for welding electrodes.

Fortron. Trademark of Ticon LLC/Hoechst Celanese Corporation (USA) for polyphenylene sulfides usually supplied in glass fiber-, carbon fiber- and mineral-filled or polytetrafluoroethylene-lubricated grades. They have a density of about 1.6 g/cm³ (0.06 lb/in.³), excellent thermal and electrical properties, excellent high-temperature resistance, good toughness and elasticity, good flame retardancy, excellent chemical resistance to harsh substances, hydrogen sulfide, salt water, hot water, hot oils and solvents, and good mechanical properties including impact strength and toughness. Used for electronic components, such as chi-carrier sockets and back-plane connectors, and for lamps and lighting equipment, industrial equipment, and automotive components.

Fortuna. Trade name of Stahlwerke Südwestfalen (Germany) for an extensive series of steels including plain-carbon, cold-work, hot-work and high-speed tool grades, austenitic, ferritic and martensitic stainless grades, chromium, nickel-chromium, nickel-chromium-molybdenum, chromium-molybdenum, chromium-vanadium and chromium-aluminum alloy grades as well as high-manganese grades.

Fortune. Trade name of Abrasive Finishing Inc. (USA) for preformed tumbling materials.

Forty-Two N. Trade name of US Steel Corporation for a structural steel containing 0.16% carbon, 1.2% manganese, 0.35% silicon, 0.05% niobium (columbium), and the balance iron. It has good formability and weldability, and is used for stressed structures at low temperatures, especially arctic and marine structures.

Fosbond. Trademark of Atofina Chemicals Inc. (USA) for a series of iron and zinc phosphate coatings used on ferrous metals for corrosion protection.

Foscoat. Trademark of Atofina Chemicals Inc. (USA) for a series of zinc phosphate coatings and iron and zinc conversion coatings used prior to cold working operations, e.g., drawing.

Foscote. Trade name of Pax Surface Chemicals, Inc. (USA) for iron phosphate coatings.

Foseco. Trade name of Foseco NV (Netherlands) for an extensive series of products used in casting and smelting, and in the manufacture of ferrous and nonferrous metals.

Fos-Flo. Trademark of Handy & Harman (USA) for a series of copper brazing alloys containing some phosphorus.

foshagite. A colorless or white mineral composed of calcium silicate hydroxide dihydrate, $Ca_5Si_3O_{10}(OH)_2 \cdot 2H_2O$. It can also be made synthetically from calcium hydroxide and silicon dioxide by hydrothermal treatment. Crystal system, monoclinic. Density, 2.36 g/cm³; refractive index, 1.594. Occurrence: USA (California).

Foslube. Trade name of Pennwalt Corporation (USA) for dry-film lubricants for drawing, ironing and cold extrusion applications.

fossil resin. A natural resin, such as amber, kauri, copal, retinite, or flagstaffite, exuded from ancient trees. It derived its characteristics through aging in the ground. Abbreviation: FR.

fossil wax. See ozocerite.

Fostacryl. Trade name of Hoechst AG (Germany) for a transparent styrene-acrylonitrile copolymer used for molded articles, such as dinnerware and food containers.

Fostalite. Trade name of Hoechst AG (Germany) for light-stable polystyrenes.

Fostalon. Trademark of Foster Corporation (USA) for nylon compounds used for medical applications.

Fostarene. Trade name of Hoechst AG (Germany) for crystalline general-purpose polystyrenes.

Fosta Tuf-Flex. Trade name of FostaGrent/Hoechst AG (Germany) for impact-modified styrene-butadiene elastomers.

Fosterite. Trademark of Westinghouse Electric Corporation (USA) for a series of photoelastic synthetic resins used as electric insulation varnishes for bonding asbestos sheets.

Fotoceram. Trademark of Corning Inc. (USA) for crystalline glass-ceramics produced from photosensitive glass and heated to alter their molecular structure. They are photochemically machined to very small dimensional tolerances, usually ±25 μm (±0.001 in.). Used for high-temperature electronic components including interconnects and integrated circuit packaging, and in the industrial arts.

Fotoform. Trademark of Corning Inc. (USA) for photosensitive glass photochemically machined to very small dimensional tolerances, usually ±25 μm (±0.001 in.), and used for electronic components and in the industrial arts.

foulard. A soft, thin, light- to medium-weight cotton, acetate, triacetate, rayon or silk fabric in a twill weave, generally with a printed pattern. Used for neckties, scarves, dresses, handkerchiefs, etc.

Foundation. Trade name for a high-gold dental casting alloy.

foundation board. A strong, rigid *kraft paper* laminate or *hardboard* that is suitable as structural foundation or support in trim panel assemblies. It is frequently used in combination with a subfoundation board, and may or may not be painted, embossed, perforated, patterned or otherwise surface-treated.

Foundrez. Trademark of Reichhold Chemicals, Inc. (USA) for a series of water-soluble urea-formaldehyde and phenol-formaldehyde resins used for foundry applications.

foundry alloys. See master alloys.

foundry clay. A refractory clay that can be made plastic by mixing with water, and is suitable as a bonding agent for molding sand. Used in the manufacture of foundry molds and cores.

foundry coke. A low-sulfur coke used in cupolas and blast furnaces. See also metallurgical coke.

foundry core sand. See core sand.

foundry facing. See facing.

foundry parting sand. See parting sand.

foundry pig iron. Pig iron obtained from a blast furnace or electric furnace, and used in the foundry as raw material in the production of cast irons.

foundry sand. A sand used to make sand molds for metal castings, and characterized by refractoriness, cohesiveness and durability. It contains about 85-92% silica, 0-15% alumina and 2% iron oxide. According to its particular application or composition, foundry sands may be classified as backing sand, core sand, facing sand, loam sand, molding sand, racing sand, parting sand, etc.

foundry-type ferrochromium. A master alloy composed of 5-7% carbon, 62-66% chromium, 7-10% silicon, and the balance iron. It dissolves easily in molten metal, and is used for metallurgical applications, and as a ladle addition of chromium to cast iron. Also known as *foundry-type ferrochrome*.

Fourdrinier brass. A red brass containing 83% copper and 17% zinc, used in the form of cloth and wire mesh for Fourdrinier papermaking machines.

Fourdrinier kraft. Special containerboard formed from sulfate process (kraft) pulp on a Fourdrinier-type papermaking machine containing a traveling endless wire mesh belt.

Fourdrinier wire. A British low-leaded brass containing 80-85% copper, 15-20% zinc, 0-0.4% tin and 0.01% lead. Used in wire form for the screens of Fourdrinier papermaking machines.

fourmarierite. A reddish-orange mineral of the becquerel group composed of lead uranium oxide tetrahydrate, $PbU_4O_{13} \cdot 4H_2O$. Crystal system, orthorhombic. Density, 5.74 g/cm³; refractive index, 1.885. Occurrence: Central Africa.

Four-Max. Trade name of Formax Manufacturing Corporation (USA) for polishing and buffing compositions used in metal finishing. Also known as *4-Max*.

Four-Ply. Trade name for weft-knitted fiberglass fabrics consisting of four layers of unidirectional cross-plied rovings joined into single biaxial fabrics. The products are made from a multiplicity of lightweight rovings and possess good physical properties. Used in composites for marine applications.

IV-VI compounds. A group of usually semiconducting compounds formed by combining a metallic element of Group IV (Group 14), e.g., tetravalent tin or lead, and a nonmetallic element of Group VI (Group 16), e.g., hexavalent sulfur, selenium or tellurium. Also known as *14-16 compounds*.

Four Star. (1) Trade name of Midvale-Heppenstall Company (USA) for a water-hardening tool steel containing 0.7-1.2% carbon, and the balance iron. Used for tools, drills, and taps.

(2) Trade name of Carpenter Technology Corporation (USA) for a molybdenum-type high-speed tool steel (AISI type M4) containing about 1.3-1.4% carbon, 0.2-0.4% manganese, 0.2-0.4% silicon, 5.5-6% tungsten, 4.5% molybdenum, 4% chromium, 4% vanadium, and the balance iron. It has excellent wear resistance, high hot hardness, and good deep-hardening properties. Used for lathe and planer tools, milling cutters, reamers, broaches, chasers, and form tools.

Fourteen Per Cent. British trade name for a nickel silver of 58-60% copper, 26-28% zinc, and 14% nickel. It has excellent corrosion resistance, and is used for ornamental flatware.

14-16 compounds. See IV-VI compounds.

fowlerite. A zinc-bearing variety of the mineral *rhodonite* ($MnSiO_3$) from New Jersey, USA.

Fox. Trade name of British Steel Corporation (UK) for a series of chromium, nickel, nickel-chromium, nickel-chromium-molybdenum, and nickel-molybdenum alloy steels including several case-hardening grades. Also included under this trade name are various austenitic, ferritic and martensitic stainless steels.

Foxfire Clear-Cote. Trade name of US Paint Corporation (USA) for clear urethane topcoats.

FPC PE. Trade name of Federal Plastics Company (USA) for low-density polyethylenes supplied in standard and UV-stabilized grades.

FPC PP. Trade name of Federal Plastics Company (USA) for polypropylene homopoloymers supplied in standard and UV-stabilized grades.

FPC PS. Trade name of Federal Plastics Company (USA) for polystyrene resins supplied in standard, UV-stabilized, medium-impact, glass fiber-reinforced, silicone-lubricated and structural foam grades.

FP fiber. A continuous polycrystalline fiber composed essentially of aluminum oxide (99+%). It is available as an uncoated fiber with rough surface and polycrystalline grains with an average size of 0.5 μm (20 μin.), and as a silicon dioxide coated fiber with smooth surface and somewhat higher strength. Both grades have good high-temperature resistance up to about 1370-1650°C (2500-3000°F), and are used for fiber-reinforced composites.

Fractalball. Trade name of Mattech Corporation (USA) for highly porous, high-strength ceramic spheres supplied in a wide range of sizes. The lightweight products are stable up to about 1900°C (3455°F), and are used for structural ceramic membranes, automotive filters and exhaust systems, and particle collectors.

Fracton. Trade name of Foseco Minsep NV (Netherlands) for refractory materials that can be made plastic by mixing with water, and are used for dressings and protective coatings for furnace tools and ladles and crucibles used in iron and steel-making.

fragmented powder. A powder that has been reduced into fine particles by mechanical comminution.

fraipontite. (1) A light bluish gray mineral of the kaolinite-serpentine group composed of copper zinc aluminum silicate hydroxide, $(Zn,Al,Cu)_3(Si,Al)_2O_5(OH)_4$. Crystal system, orthorhombic. Density, 3.54 g/cm^3. Occurrence: Greece.

(2) A colorless mineral of the kaolinite-serpentine group composed of zinc aluminum silicate hydroxide, $(Zn,Al)_3(Si,Al)_2O_5(OH)_4$. Crystal system, monoclinic. Refractive index, 1.624. Occurrence: Belgium.

Framdie. Trade name of Columbia Tool Steel Company (USA) for an oil-hardening low-alloy tool steel (AISI type L7) containing 1-1.1% carbon, 1.35% chromium, 0.4% molybdenum, 0.35% manganese, and the balance iron. Used for arbor, spindles, and driveshafts.

Frame Protec. Trade name of OMS Canada Inc. (Canada) for protective siloxane coatings that are resistant to oxidation, perspiration and discoloration, and are used on spectacle frames.

framework silicates. See tectosilicates.

framing lumber. A general term used for the following categories of dimension lumber: (i) *Light framing lumber* which is 50-100 mm (2-4 in.) thick and 50-100 mm (2-4 in.) wide, and used where high strength is not a requirement, e.g., for plates, studs, cripples, sills, and blocking; (ii) *Structural light framing lumber* which is 50-100 mm (2-4 in.) thick and 50-100 mm (2-4 in.) wide, and used where higher bending strength is required, e.g., for trusses and concrete pier wall forms; and (iii) *Structural framing lumber* which is 50-100 mm (2-4 in.) thick and more than 150 mm (6 in.) wide, and used for general structural applications, e.g., planks, joists, rafters, and general framing.

francevillite. A natural or synthetic yellow mineral of the carnotite group composed of barium uranyl vanadium oxide pentahydrate, $Ba(UO_2)_2V_2O_8 \cdot 5H_2O$. The natural mineral also contains lead. Crystal system, orthorhombic. Density, 4.42 g/cm^3; refractive index, 1.945. Occurrence: Gabon.

francium. A radioactive element of Group IA (Group 1) of the Periodic Table (alkali group). The principal and longest-lived isotope is francium-223 (^{223}Fr) which can be isolated from its parent, actinium-227 (^{227}Ac). Francium-223 has a half-life of about 22 minutes, emits both alpha and beta rays, and belongs

to the actinium decay series. Melting point, 27°C (81°F); atomic weight, 223; atomic number, 87; monovalent; appears to exist only as radioactive isotopes. Other isotopes, such as ^{218}Fr, ^{219}Fr, ^{220}Fr, ^{221}Fr and ^{222}Fr, have also been made artificially. Symbol: Fr.

franckeite. A grayish black mineral of the cylindrite group composed of lead tin antimony sulfide, $Pb_5Sn_3Sb_2S_{14}$. Crystal system, triclinic. Density, 6.01 g/cm^3. Occurrence: Bolivia.

Franclay. Trade name of AGS Minéraux SA (France) for hard clay obtained from kaolinitic sedimentary *ball clay* mined in southwestern France, and supplied as a powder with a particle size ranging from submicron to over 20 μm (less than 40 to over 800 μin.). A typical chemical analysis (in wt%) is 49% silica (SiO_2), 34% alumina (Al_2O_3), 1.5% ferric oxide (Fe_2O_3) and 1.5% titania (TiO_2). The loss on ignition is about 12%, and it has a density of 2.6 g/cm^3 (0.09 $lb/in.^3$), and a specific surface area of 22 m^2/g. Used in the manufacture of ceramic products.

Franco. Trade name of E. Frank Atkinson & Sons Limited (UK) for a high-speed steel containing 0.7% carbon, 18% tungsten, 4% chromium, 1% vanadium, and the balance iron. Used for high-speed cutters and tools.

francoanellite. A yellowish-white mineral composed of potassium aluminum hydrogen phosphate hydrate, $H_6K_3Al_5(PO_4)_8 \cdot 13H_2O$. Crystal system, rhombohedral (hexagonal). Density, 2.26 g/cm^3; refractive index, 1.510. Occurrence: Italy.

Frankalloy. Trade name for titanium used in ladle-vacuum-degassed ultra-low-carbon steels.

frankdicksonite. A colorless synthetic mineral of the fluorite group composed of barium fluoride, BaF_2. Crystal system, cubic. Density, 4.89 g/cm^3.

Frankfurt black. See vine black.

Frankilon. Trademark of Texpro SpA (Italy) for nylon 6 fibers and filament yarns.

Frankite. Trade name of Frank Foundries Corporation (USA) for nickel cast irons, plain-carbon steels, and heat- and corrosion-resistant steels.

Franklin Fiber. (1) Trade name of Franklin-Lamitex Corporation (USA) for a filler/polyolefin blend. The filler is a gypsum-based whisker reinforcing fiber and the polyolefin is usually polypropylene or high-density polyethylene.

(2) Trade name of Franklin-Lamitex Corporation (USA) for vulcanized fiber supplied in the form of rods, sheets and tubes.

franklinite. An iron-black to medium-brown, slightly magnetic mineral with dull metallic luster and brown streak that has some resemblance with magnetite. It belongs to the spinel group, and is composed of zinc manganese iron oxide $(Zn,Mn,Fe)(Fe,Mn)_2O_4$. Crystal system, cubic. Density, 5.07-5.32 g/cm^3; hardness, 6.0-6.5 Mohs; refractive index, 2-2.36. Occurrence: USA (New Jersey). Used as an ore of zinc and manganese, in spiegeleisen manufacture, and in ground form as a dark pigment for certain paints.

fransoletite. A colorless mineral composed of beryllium calcium hydrogen phosphate tetrahydrate, $H_2Ca_3Be_2(PO_4)_4 \cdot 4H_2O$. Crystal system, monoclinic. Density, 2.56 g/cm^3; refractive index, 1.566. Occurrence: USA (South Dakota).

Frantex. Trade name of AGS Minéraux SA (France) for a surface-treated hard clay obtained from kaolinitic sedimentary *ball clay* mined in southwestern France, and supplied as a powder with a particle size ranging from submicron to over 10 μm (less than 40 to over 400 μin.). A typical chemical analysis (in wt%)

of the untreated product is 49% silica (SiO_2), 34% alumina (Al_2O_3), 1.5% ferric oxide (Fe_2O_3) and 1.5% titania (TiO_2). The loss on ignition is about 13%, and it has a density of 2.6 g/cm^3 (0.09 $lb/in.^3$) and a specific surface area of 22 m^2/g. Used in the manufacture of ceramic products.

franzinite. A white mineral of the cancrinite group composed of sodium calcium aluminum carbonate sulfate silicate hydroxide hydrate, $(Na,Ca)_7(Si,Al)_{12}O_{24}(SO_4),OH,CO_3)_3 \cdot H_2O$. Crystal system, hexagonal. Density, 2.49 g/cm^3; refractive index, 1.510. Occurrence: Italy.

Frapimphy. Trade name of Creusot-Loire (France) for a series of wrought austenitic, ferritic and martensitic stainless steels for cold-heading applications, e.g., screws, nuts, bolts, balls, pins, and hinges.

Frappant. Trade name of Otto Wolff Handelsgesellschaft (Germany) for an oil-hardening nondeforming tool steel containing 0.9% carbon, 1.9% manganese, 0.1% vanadium, and the balance iron. Used for punches, and blanking and forming dies.

Frary metal. A bearing metal containing 96.5-99.4% lead, 0.2-2% barium, 0.1-1% calcium, 0.24-0.3% mercury, 0.1-0.3% tin, and 0-0.04% copper. It has good corrosion resistance and good antifriction properties, and is used for low-pressure bearings operating at medium temperatures.

Fraser fir. The soft, light, straight-grained, creamy-white to pale brown wood of the fir tree *Abies fraseri*. Its heartwood and sapwood are indistinguishable. Source: Appalachian Mountains of Virginia, North Carolina and Tennessee. Used chiefly for paper pulp.

Frebol. Trade name of F. Boehm (UK) for a phenol-formaldehyde resin.

freboldite. A violet-rose mineral of the nickeline group composed of cobalt selenide, CoSe. Crystal system, hexagonal. Density, 7.56 g/cm^3. Occurrence: Germany.

Freebond. Trade name for a corrosion-resistant cobalt-chromium dental bonding alloy.

Freecut. Trade name of Peninsular Steel Company (USA) for a series of free-machining steels containing about 0.2-0.5% carbon, 1.25% manganese, 0.05% silicon, 0.25% sulfur, 0.02% phosphorus, and the balance iron. It has high strength and medium hardness, and is used for plate stock.

free-cutting brass. An alpha-beta brass containing 60-65% copper, 31.5-35.5% zinc, and 2.0-3.5% lead. Commercially available in the form of flats, rods and structural shapes, it has excellent machinability and good corrosion resistance, and is used for automatic screw-machine products, fasteners, bolts, nuts, shafts, studs, etc.; drawn, cupped or formed parts, gears, pinions, and hardware. Also known as *high-speed brass; leaded high brass*.

free-cutting bronze. A free-cutting alloy of 89% copper, 10% zinc and 1.5% lead with good machinability and corrosion resistance, used for screwstock, bolts, and automobile radiators.

free-cutting copper. A general term for several coppers with good machinability including phosphorus-deoxidized tellurium-bearing copper, sulfur-bearing copper and leaded copper. All of them also have high conductivity and good corrosion resistance, and are used for automatic screw-machine products. See also tellurium copper, sulfur-bearing copper, leaded copper.

free-cutting muntz metal. A free-cutting alloy that contains 60% copper, 36.75-39% zinc and 1-3.25% lead, and is used for screw-machine products, butt hinges, lock bodies, mechanical devices, and forging rods.

free-cutting phosphor bronze. A free-cutting, wear-resistant

bronze containing 88% copper, 4% zinc, 4% tin, and 4% lead. It has excellent machinability and good workability and corrosion resistance. Used for bushings, bearings, valve and pump parts, pinions, gears, shafts, and thrust washers.

free-cutting steels. A group of low- and medium-carbon steels with excellent machining characteristics. The typical composition range is 0.06-0.65% carbon, 0.3-1.65% manganese, 0.1-0.5% silicon, 0.04-0.15% phosphorus, and/or 0.08-0.35% sulfur, 0-0.3% selenium and 0.2-0.5% lead, with the balance being iron. The term usually refers to the carbon steels in the AISI-SAE 11xx and 12xx series. The good machinability is due to the addition of one or several alloying elements, such as sulfur, selenium and/or lead that effect short, broken chips, reduced power consumption, improved surface finish and extended tool life. Used for screw-machine products, yokes, studs, universal joints, bolts, nuts, etc. Also known as *free-machining steels*.

free-cutting tube brass. A free-cutting brass available in the form of tubes and composed of 66.5% copper, 31.9% zinc, and 1.6% lead. Used for screw-machine products, ballpoint pens, plumbing fixtures, and musical instruments.

free-cutting yellow brass. A free-cutting brass containing 61.5% copper, 35.25% zinc and 3.25% lead, used for machined parts.

Freedom. (1) Trade name of J.F. Jelenko & Company (USA) for a gold- and silver-free palladium-base dental alloy with good castability, high bond strength, and excellent thermal stability/ sag resistance used for fusing porcelain to metal in implants, and for long-span bridgework.

(2) Trade name of Southern Dental Industries Limited (Australia) for a fluoride-releasing, no-slump *compomer* consisting of a matrix of amorphous silica that is 80% filled with strontium glass particles. It has very high wear resistance, outstanding luster and polishability, high tensile strength, high hydrolytic stability, a tooth-like translucency, and good color stability. Used in dentistry for restorations and cervical erosions.

Freedom Plus. Trade name of J.F. Jelenko & Company (USA) for a multi-purpose palladium-based dental alloy with good castability and excellent thermal stability/sag resistance, used for fusing porcelain to metal in implants, and for long-span bridgework.

free-fiber-end yarn. A yarn with a hairy surface produced bry cutting or breaking protruding filament loops by air-jet texturing.

Freeflo. Trade name of DSM Resins (USA) for general-purpose polyester molding compounds.

Free Kast. Trade name for a series of low-water, cement-free alumina-based castable refractories made using a special bonding system. Their density and porosity are comparable to those of superduty dry-press brick.

free-machining steels. See free-cutting steels.

Freemelt. Trade name of Ato Findley (France) for hot-melt adhesives used for packaging applications, and for bonding nonwovens.

Freemix. Trade name of DSM Resins (USA) for a series of bulk molding compounds supplied in standard and electrical, fire-retardant, high-heat and low-profile grades.

free-radical polymer. A polymer obtained when an initiator molecule is thermally or photochemically decomposed and produces free radicals that are then introduced to the long-chain free radical species starting the chain reaction. Abbreviation: FRP.

freibergite. A gray to black mineral of the tetrahedrite group with reddish streak composed of copper silver zinc antimony sulfide, $(Cu,Ag,Zn)_{12}Sb_{4.4}S_{12.6}$. It can also be made synthetically. Crystal system, cubic. Density, 5.05 g/cm^3. Occurrence: Germany, USA (Idaho, Nevada).

freieslebenite. A gray mineral composed of silver lead antimony sulfide, $AgPbSbS_3$. Crystal system, monoclinic. Density, 6.22 g/cm^3. Occurrence: Spain.

frejo. See Brazilian walnut.

Fremax. Trade name of US Steel Corporation for a series of hardenable, free-machining steels containing 0.13-0.50% carbon, 1.1% manganese, 0.25% sulfur, and the balance iron. Used for molds, dies, seals, and gears.

fremontite. A white mineral composed of sodium lithium aluminum hydrofluophosphate, $(Na,Li)Al(PO_4)(OH,F)$. Crystal system, monoclinic or triclinic. Occurrence: USA (Colorado). Also known as *natromontebrasite*.

French alloy. A French nickel silver of 50-58% copper, 25-31% zinc and 17-20% nickel, used for electrical resistors, fittings and ornaments.

French antelope lambskin. A term used in North America for suede leather made from the pickled skin of South American and New Zealand lambs.

French Auto. Trade name for a lead babbitt containing 15% antimony and 10% tin. Used for automobile bearings.

French blue. A brilliant blue, inorganic pigment made by heating a mixture of aluminum silicate (e.g., china clay or kaolin) and sodium sulfide. It is a fine powder with the empirical formula $Na_7Al_6Si_6O_{24}S_2$, formerly also made by powdering *lapis lazuli*. French blue has good alkali and heat resistance, poor acid resistance, low hiding power and poor outdoor durability. Used in paints, printing inks and rubber goods, as a colorant for machinery and toy enamels, as a white baking enamel, in textile printing, and in bluing for whitening paper and organic materials. See also ultramarine blue.

French Bronze. Trade name for a high-tensile brass (UZ19AL6) supplied in the form of extruded and forged bars and tubes.

French calf. Firm, wax-finished, high-grade calf leather imported from France.

French chalk. (1) White talc in the form of a finely ground powder used in the manufacture of ceramics, as a filler in paints, rubber, plaster, putty, etc. Also known as *powdered talc; talcum powder.*

(2) See asbestine.

French flange metal. A tough French leaded bronze that contains 94% copper, 5.6% tin and 0.4% lead, and is used for flanges, bushings, and fittings.

French kid. Soft, supple, alum- or vegetable-tanned kidskin. See also kid leather.

French silver. See argent français.

French solder. A French solder containing 24% copper, 10% zinc, and the balance silver.

French terry. A knitted, napped or unnapped *jersey* covered with loops on one side. See also terry.

French type metal. A French typemetal containing 55% lead, 23-30% antimony, and 15-22% tin.

Frentopal. Trade name of Hurlingham SA (Argentina) for opaline glass cladding used for exterior walls.

Frenzelit. Trade name of Frenzelit Werke GmbH & Co. KG (Germany) for asbestos-free building facing tiles that consist of cement, cellulose and selected inorganic fillers, and are coated with high-quality plastic materials. Also included under this trade name are numerous asbestos and asbestos-free insulating

and packing products including asbestos-rubber goods and asbestos and asbestos-free sheet packing, tape, tubing, cord, cloth, paper, board and rope.

Frequenta. Trade name for an engineering ceramic used as a radio-frequency insulating material.

Frequentite. Trademark of Morgan Advanced Ceramics (USA) for *steatite* $[Mg_3Si_4O_{10}(OH)_2]$ ceramics with high temperature resistance and mechanical strength, high corrosion and wear resistance, excellent resistance to water, fuels, oils and ultraviolet light, and excellent high-temperature insulating properties. Used for automotive components, e.g., connectors, shields, heating systems and engine-management systems.

Fresco. Trade name of Owens-Corning-Ford Company (USA) for an acoustic glass-fiber tile with a surface resembling stippled plaster.

Frescor. Trade name of Owens-Corning Fiberglass (USA) for acoustic glass-fiber tiles and ceiling boards having softly stippled white surfaces that produce a graded shadow effect.

fresh concrete. Concrete that has not attained its initial set.

Freshloft. Trade name of KoSa (USA) for high-loft polyester products with built-in antimicrobial agent, and supplied in cut lengths of 51-76 mm (2-3 in.), and 6 denier. Used especially for bedding, home furnishings and clothing.

fresnoite. A colorless mineral composed of barium titanium silicate, $Ba_2TiSi_2O_8$. Crystal system, tetragonal. Density, 4.43 g/cm^3; refractive index, 1.775.

freudenbergite. A black mineral composed of sodium iron titanium silicate hydroxide, $(Na,K)_2(Ti,Nb)_6(Fe,Si)_2(O,OH)_{18}$. Crystal system, monoclinic. Density, 4.30 g/cm^3; refractive index, 2.37. Occurrence: Germany.

Freund steel. A German high-elastic carbon steel containing 0.1-0.15% carbon, 0.7-1.3% silicon, 0.3-0.6% manganese, some chromium, and the balance iron. Used for highly stressed structural members.

friable abrasives. Abrasives whose friability (brittleness) is very high resulting in easy breakage of grains and ready formation of new sharp edges. They are most commonly used in finish and precision grinding operations. See also friable alumina; semifriable abrasives.

friable alumina. An alumina (Al_2O_3) of medium purity that is intermediate in friability (brittleness) between regular alumina and white alumina. Used as an abrasive in grinding. See also alumina; friable abrasives; semifriable alumina; white alumina.

Frialit-Degussit. Trade name of Degussa AG (Germany) for sintered aluminum oxides supplied in sphere and tube form for ceramic applications, and as abrasives for the manufacture of grinding tools.

Fricke's Harder. Trade name for a strong, corrosion-resistant alloy composed of 69% copper, 30% zinc, and 1% nickel.

Fricke's Silvery. Trade name for a corrosion-resistant alloy composed of 50% copper, 31.2% nickel and 18.8% zinc, and used for ornamental and decorative parts.

Frick's alloy. A British nickel silver containing 50-69% copper, 18-39% zinc and 5.5-31% nickel. It has excellent corrosion resistance, and is used for ornamental parts, and as a base for plated ware.

Frick's Bluish Yellow. A British trade name for a hard, corrosion-resistant alloy composed of 55.5% copper, 39% zinc and 5.5% nickel, and used for decorative parts and hardware.

Frick's Pale Yellow. A British trade name for a ductile, corrosion-resistant alloy composed of 62.5% copper, 31.2% zinc, and 6.3% nickel, and used for decorative parts and hardware.

frictional polymers. See friction polymers.

friction alloy. A lead babbitt that contains 40% tin and 10% antimony, and is used for bearings and antifriction metals.

Friction-Fit. Trade name of Owens-Corning Fiberglass (USA) for unfaced glass-fiber insulation.

Frictionless. Trade name of American Smelting & Refining Company (USA) for a lead babbitt that contains varying amounts of tin and antimony, and is used for bearings and linings.

friction materials. Powder-metallurgy parts composed of copper, copper-alloy, iron or iron-alloy-base composites with nonmetallic additions that greatly increase their coefficients of friction.

friction polymers. High-molecular-weight polymers made by a process that involves the polymerization of organic compounds on mating surfaces which move relative to each other. Also known as *frictional polymers.*

friedelite. A rose-red to colorless mineral of the pyrosmalite group composed of manganese chloride silicate hydroxide, $Mn_8Si_6O_{15}$-$(OH,Cl)_{10}$. Crystal system, monoclinic. Refractive index, 1.654. Occurrence: France, USA (New Jersey).

friedrichite. A creamy to yellowish-white mineral composed of copper lead bismuth sulfide, $Cu_5Pb_5Bi_7S_{18}$. Crystal system, orthorhombic. Density, 7.13 g/cm^3. Occurrence: Austria.

frieze. A thick, heavy fabric with more or less sheared, but uncut loops and a hard to soft feel. Used especially for slipcovers and upholstery.

Frigidal. Trade name of Vereinigte Deutsche Nickel-Werke AG (Germany) for a resistance alloy containing 35% nickel, some chromium, and the balance iron.

Frigolit. Trademark of Frigolit (Germany) for polystyrene foams and foam products.

Frilon. Trade name for nylon fibers and yarns used for textile fabrics including clothing.

Frioplast. Trademark of Kaspar Winkler & Co. (Switzerland) for air-entraining agents used for concrete, cement and plaster.

Frip. Trademark of Frip Panels (Canada) for fiberglass-reinforced panels used in the container, trailer, truck and construction industries.

frisé. A fabric made of natural or artificial silk and used especially for towels.

Frisella. Trademark of Montefibre SpA (Italy) for polyester filament yarns.

Frismuth solder. A British alloy that contains 67-70% tin, 27-30% lead and 3% aluminum, and is used as an aluminum solder.

frit. A glassy material that is composed of a mixture of ceramic materials, and has been melted and quenched in air or water to form small friable particles which are subsequently ground for use in the manufacture fritted glazes, chinaware, and porcelain enamels.

fritted china. Thin, translucent chinaware made from a body having zero water absorption and containing substantial amounts of high-fluxing frits. It is usually coated with a soft glaze.

fritted glass. Glass with controlled porosity made from powdered glass by sintering under prescribed conditions.

fritted glaze. A glaze in which all or most of the fluxing ingredients have been melted or quenched to form small friable particles prior to the addition to the glaze slip.

fritted porcelain. See soft paste porcelain.

fritzcheite. A mineral of the carnotite group composed of manganese uranyl vanadium oxide tetrahydrate, $Mn(UO_2)_2(VO_4)_2$·$4H_2O$. Crystal system, orthorhombic. Density, 4.39 g/cm^3.

Frogalloy. Trade name of Teledyne McKay (USA) for a wear-resistant, work-hardenable steel containing 0.4% carbon, 4.1% manganese, 19.5% chromium, 10% nickel, 1.4% vanadium, and the balance iron. Used for hardfacing electrodes.

frohbergite. A pink, purplish pink mineral of the marcasite group composed of iron ditelluride, $FeTe_2$. It can also be made synthetically. Crystal system, orthorhombic. Density, 3.50 g/cm^3. Occurrence: Canada (Quebec).

frolovite. A white mineral with dull luster composed of calcium borate tetrahydrate, $CaB_2O_4·4H_2O$. Crystal system, rhombohedral (hexagonal). Density, 2.14 g/cm^3; refractive index, 1.572. Occurrence: Russia.

frondelite. A dark brown mineral composed of manganese iron phosphate hydroxide, $MnFe_4(PO_4)_3(OH)_5$. Crystal system, orthorhombic. Density, 3.45 g/cm^3; refractive index, 1.880. Occurrence: Brazil.

Frontier. (1) Trade name of Frontier Bronze Corporation (USA) for a series of wrought and cast copper- and aluminum-base alloys including several commercially pure coppers, aluminum bronzes, leaded and unleaded tin bronzes, manganese bronzes, and aluminum-zinc alloys.

(2) Trade name for a self-cure dental cement for bridge and crown restorations.

Frontier Bronze. Trade name of Sterling International Technology Limited (UK) for a medium-strength, reasonably shock-resistant gravity die-cast aluminum alloy containing 4.8-5.7% zinc, 0.5-0.75% magnesium, 0.4-0.6% chromium, 0.1% nickel, 0.1% lead, 0.05% tin, and 0.15-0.25% titanium. It is moderately hot short, and ages in 21-30 days at room temperature.

froodite. A gray mineral composed of bismuth palladium, Bi_2Pd. It can also be made synthetically. Crystal system, monoclinic. Density, 11.5-12.5 g/cm^3. Occurrence: Canada (Ontario).

frosted glass. (1) Very thin crushed glass sometimes fused over the surface of a glass article to give a frosted appearance.

(2) Glass etched or otherwise treated to give a frosted appearance, and used for decorative applications.

frosted yarn. A yarn with a spotted or speckled appearance produced from fiber blends, e.g., acetate and cotton or rayon staple, whose constituent materials have different affinities for dyes.

Frost-Lite. Trade name of Amerada Glass Company (USA) for white-toned, laminated glass that, although appearing opaque, transmits almost two-thirds of all incident light. It provides the same level of privacy as etched or frosted glass.

Frostlyte. Trade name of Pilkington Brothers Limited (UK) for patterned glass.

Frostone. Trade name of President Suspender Company (USA) for polyvinyl chlorides and other vinyl plastics.

FRP Classic. Trademark of Melton Classics, Inc. (USA) for cast stone made from fiber-reinforced plastics.

FRP plywood. A strong, lightweight engineered wood composite consisting of a 6 to over 28 mm (0.250 to over 1.125 in.) thick plywood panel of varying grade, bonded between tough 0.6-1.5 mm (25-60 mils) thick fiberglass-reinforced plastic overlays with additional thin nonporous surface finish (typically a gel coat or polyvinyl fluoride coating). It has excellent bending strength, dimensional stability and weatherability. Available in panel sizes ranging from 1.2 × 2.4 m (4 × 8 ft.) to over 3 × 14 m (10 × 45 ft.), it is used for truck, trailer and van bodies, railcar linings and side panels, shipping containers, concrete forms, sewage treatment tanks, etc.

Frurnie. Trade name of Konrad Hornschuh AG (Germany) for a polyvinyl chloride film.

FTR Foam. Trade name of Fiber Materials Inc. (USA) for a fire-resistant rigid epoxy foam used for insulation applications.

F2 Dual Cure. Trade name of NSI Dental Pty. Limited (Australia) for a light- and chemical-cure, flowable, microrod-reinforced bisphenol glycidylmethacrylate (Bis-GMA) dental composite with exceptional strength and fracture resistance.

Furnit. Trademark of Konrad Hornschuh AG (Germany) for flexible polyvinyl chloride film materials.

Furset. Trade name of Raschig GmbH (Germany) for a series of furan resins.

Fuchs. Trade name of Otto Fuchs Metallwerke (Germany) for an extensive series of aluminum-, copper- magnesium-, and titanium-base alloys including various high-purity and commercially pure aluminums and aluminum-magnesium alloys, electrolytic, silver-bearing, tellurium and leaded coppers, leaded and unleaded brasses (e.g., muntz metal, yellow brasses, cartridge brasses, low brasses, commercial and architectural bronzes, aluminum and manganese brasses), high-purity and commercially pure magnesium, magnesium-aluminum and magnesium-zinc alloys, high-purity and commercially pure titanium, and titanium-aluminum-molybdenum, titanium-aluminum-vanadium and titanium-copper alloys.

Fuchs acid-resisting gold alloy. A British acid-resisting alloy containing 75% gold, 10-15% nickel and 10-15% tungsten, used for chemical plant and equipment.

Fuchs acid-resisting silver alloy. A British acid-resisting alloy containing 13.5-15% gold, 13-15% nickel, and the balance silver. Used for corrosion-resistant parts, and chemical equipment.

Fuchsal. Trade name of Otto Fuchs Metallwerke (Germany) for an alloy containing 0.6-1.4% magnesium, 0.6-1.6% silicon, 0.6-1% manganese, 0-0.3% chromium, and the balance aluminum. It has good forming and welding properties, and is used for boats, gutters, fan blades, and window frames.

Fuchsdur. Trade name of Otto Fuchs Metallwerke (Germany) for an age-hardenable aluminum alloy containing 2.5-5% copper, 0.2-1.8% magnesium, and 0.3-1.5% manganese. Used for fasteners, and aircraft structures and fittings.

fuchsite. A bright green variety of muscovite mica containing up to 5% Cr_2O_3. See also muscovite.

Fuchsman. Trade name of Otto Fuchs Metallwerke (Germany) for an aluminum alloy containing 1-1.5% manganese and 0-0.3% chromium. It has good formability and weldability, and is used for tanks, heat exchangers, cooking utensils, and furniture.

Fuego. Trade name of Hidalgo Steel Company Inc. (USA) for a water-hardening tool steel containing 0.5% carbon, 0.9% silicon, 0.5% manganese, varying amounts of chromium and vanadium, and the balance iron. Used for shears, chisels, and punches.

fuel briquettes. See coal briquettes.

fuel element. Nuclear fuel in the form of a plate, rod, tube or other shape. See also nuclear fuel; fission fuel.

Fuga. Trade name of Fabbrica Pisana SpA (Italy) for a glass with pendant pattern.

fugitive binder. An organic substance added to a metal powder to enhance the bond between the particles during compaction and thereby increase the green strength of the compact. It decomposes during the early stages of the sintering cycle.

fugitive colors. Temporary colors, i.e., colors which are subject to fading.

Fuji. Trade name of GC America Inc. (USA) for an extensive series of glass-ionomer dental cements, resins and bonding agents. Many grades, such as *FujiCem, Fuji Ortho, Fuji Plus*, and *Fuji Triage* are fluoride-releasing and radiopaque. Used for the permanent or temporary restoration of crowns and bridges, for securing inlays, onlays, posts and brackets, as fillers, liners, lutes or bases, or as bonding agents (e.g., *Fuji Bond LC*).

fuji. A light, plain-weave fabric usually made from 100% pure silk yarn, which however is sometimes substituted, wholly or in part, with acetate, triacetate, polyester or viscose rayon yarns. Used especially for blouses and dresses.

Fujibo Spandex. Trademark of Fuji Spinning Company (Japan) for spandex fibers and yarns used for elastic fabrics.

fujiette. A medium-weight fabric, usually of rayon or an acetate-rayon blend, that has a delicate widthwise rib made with a weft of spun yarn and warp of filament yarn. Used especially for blouses and dresses.

Fujirock. Trade name of GC America Inc. (USA) for die stones used in dentistry.

fukalite. A pale brown to white mineral composed of calcium carbonate silicate hydroxide, $Ca_4Si_2O_6(CO_3)(OH)_2$. Crystal system, orthorhombic. Density, 2.77 g/cm^3; refractive index, 1.605. Occurrence: Japan.

fukuchilite. A dark brownish-gray synthetic mineral of the pyrite group composed of copper iron sulfide, $(Fe,Cu)S_2$. Crystal system, cubic. Density, 4.66 g/cm^3.

Fulbond. Trade name for natural and modified fuller's earth compositions used in foundry greensand mixtures.

Ful-Fil. Trade name for light-cure hybrid composite resins used for posterior dental restorative applications.

Fulgurit. Trademark of Fulgurit Baustoffe GmbH (Germany) for asbestos cement products including corrugated sheets, facing and roofing tiles, flat sheets, extruded profiles and various other asbestos and non-asbestos building products.

Fulgupal. Trade name of Fulgurit Baustoffe GmbH (Germany) for interior building panels composed of asbestos cement.

Fulguplast. Trade name of Fulgurit Baustoffe GmbH (Germany) for translucent asbestos cement sheets for building construction.

Full Cast. Trademark of World Alloys & Refining, Inc. (USA) for a corrosion-resistant nickel-chromium-aluminum dental ceramic alloy with a density of 7.7 g/cm^3 (0.28 $lb/in.^3$), a melting range of 1280-2500°C (2280-2370°F), a hardness of 165 Vickers, high yield strength, and low elongation. Used for crown and bridge restorations.

fulled fabrics. See milled fabrics.

Fuller Alloy. Trade name of Fuller & Basche Company (USA) for a heat-treatable aluminum alloy containing 6% zinc, 1.2% magnesium and 2% iron. Used for light-alloy castings.

fullerene-like carbon nitrides. A novel class of noncubic carbon nitride (CN_x) compounds with fullerene-like structures produced by the incorporation of nitrogen atoms into the basic structural units. They are currently available in film form, can be deposited by reactive magnetron sputtering, and have found application as highly resilient, wear-resistant coating materials.

fullerene polymers. Structures obtained by incorporating buckminsterfullerene (C_{60}) or other fullerenes into polymers. For example, these polymer structures can be synthesized by reacting suitable monomers (e.g., xylylene) with fullerene molecules resulting in entirely new copolymers, or by linking existing polymers, such as polystyrene, to fullerene molecules.

Many fullerene polymer structures consist of fullerene molecule networks linked by hydrocarbon bridges. See also fullerenes; buckminsterfullerenes.

fullerenes. A relatively recent class of carbon cluster compounds typically composed of 60- or 70-carbon-atom molecules (C_{60} or C_{70}), although compounds with higher carbon atom numbers, such as C_{76}, C_{78}, C_{84} and C_{120}, also available. They have high oxidative and thermal stability, good resistance to corrosive compounds, and interesting optical, electric and magnetic properties including nonlinear variation of transparency as a function of incident light intensity, and good electron-release properties. *Fullerenes* are typically supplied as nanometer-sized agglomerated clusters. Colored fullerenes, e.g., red fullerenes (*C_{60} magenta* and *C_{70} red*) and black fullerenes (*C_{120} black*), and noble-gas fullerenes, e.g., xenon and helium fullerenes ($^{129}XeC_{60}$) and ($^{129}HeC_{60}$) are also available. Used chiefly for electronic and optoelectronic applications, and in the manufacture of nanotubes, conducting polymers, etc. See also alkali-doped fullerenes; buckminsterfullerenes; fullerene superconductors; giant fullerenes; interstitial fullerenes; carbon nanotubes.

fullerene superconductors. A fairly recent class of fullerenes with very high superconducting transition temperatures (T_c > 40K). They can be produced either by the incorporation of interhalogen molecules (e.g., ICl) into a crystal composed of buckminsterfullerene molecules with 60 or more carbon atoms each, or by doping the purified buckminsterfullerene compound with alkali metals (e.g., potassium or rubidium). See also alkali-doped fullerenes; organic superconductors.

fullerids. Fullerene-metal complexes produced either by laser vaporization of a metal, such as cesium, lanthanum or potassium, and mixed with graphite, or by doping fullerenes with rubidium, cesium or other alkali, alkaline-earth or rare-earth metals. The doped materials have superconductive properties and are used in the manufacture of electronic and optoelectronic devices. See also fullerenes.

Fuller-O'Brien. Trade name of O'Brien Corporation (USA) for thermoset powder coatings.

fulleroids. A group of diphenyl molecules (C_{61} to C_{66}) obtained by linking fullerenes to complex diazo compounds with phenyl (Ph) rings. The general formula is Ph_xC_y (x = 2, 4, 6, 8, 10 and 12; y = 61-66). See also fullerenes.

fullerites. Mixtures of fullerenes with different numbers of carbon atoms (e.g., 90% C_{60} and 10% C_{70}) used as precursor to buckminsterfullerenes. See also buckminsterfullerenes; fullerenes.

fuller's earth. A gray, yellow, greenish white or greenish brown, porous, nonplastic colloidal clay composed of hydrated aluminum silicate. A typical fuller's earth contains up to 75% silica (SiO_2), 10-20% alumina (Al_2O_3), 2-4% magnesia (MgO), 1-4% lime (CaO), and sometimes ferric oxide (Fe_2O_3). Supplied as a powder in particle sizes from 8 to 200 mesh, it has high adsorptive capacity, and is used as a pigment extender, rubber filler, talcum powder replacement, decolorizer for oils and fats, filtering medium, catalyst carrier, and in oil-well drilling muds.

full-trimmed mica. Mica that has been trimmed on all sides, with cracks or cross grains removed.

fully acetylated cotton. See acetylated cotton.

fully active flux. A highly active rosin-base soldering flux that produces superior cleaning results. See also rosin.

fully deoxidized steel. See killed steel.

fully drawn yarn. See fully oriented yarn.

fully inoculated iron. A molten or solidified cast iron containing all required alloying additions and inoculants.

fully oriented yarn. A continuous filament yarn, usually made by melt spinning, whose high and essentially complete molecular orientation is imparted by a process that either involves spinning at high wind-up speeds or drawing at high draw ratios. Abbreviation: FOY. Also known as *fully drawn yarn*.

fully-stabilized zirconia. An engineering ceramic composed of zirconium oxide (ZrO_2) stabilized with about 8-16 wt% lime (CaO), magnesia (MgO), yttria (Y_2O_3), etc. It has a cubic crystal structure, a theoretical density of 5.56-6.1 g/cm³ (0.20-0.22 lb/in.³), a melting point of approximately 2500°C (4530°F), high transverse rupture strength, good fracture toughness, good hardness, high strength even at temperatures above 1300°C (2370°F), an upper continuous-use temperature of 2200°C (3990°F), a very high modulus, a very low coefficient of expansion, low thermal conductivity, good thermal-shock resistance, good electrical-insulating properties, and good resistance to many acids and alkalies. Used for refractories, and in advanced composites. Abbreviation: FSZ. See also lime-stabilized zirconia; magnesia-stabilized zirconia; partially-stabilized zirconia; yttria-stabilized zirconia.

Ful-O-Mite. (1) Trademark of H.B. Fuller Company (USA) for structural adhesives and insulative exterior wall coatings.

(2) Trademark of TEC Inc. (USA) for polymer-modified cementitious adhesive-base coatings for bonding and priming substrates and producing decorative exterior wall finishes.

fuloppite. A lead gray mineral composed of lead antimony sulfide, $Pb_3Sb_8S_{15}$. Crystal system, monoclinic. Density, 5.22 g/cm³. Occurrence: Hungary, Rumania.

Fultalloy. Trade name of Fulton Iron Works Company (USA) for cast steel containing 0.5% carbon, varying amounts of chromium, and the balance iron.

Ful-Thane. Trademark of E.I. DuPont de Nemours & Company (USA) for urethane enamels.

Ful-Thik. Trade name of Manville Corporation (USA) for fiberglass home insulation.

Fulton. (1) Trade name of Fulton Gold Refineries Corporation (USA) for an extensive series of copper- and silver-base brazing alloys. The copper-base alloys contain varying amounts of silver, zinc, and/or phosphorus, and the silver-base alloys contain copper, zinc, tin, and/or nickel.

(2) Trade name of LNP Engineering Plastics (USA) for a series of homopolymer and copolymer acetal resins used for machine parts, fasteners, seals, etc. They are supplied in various grades including unmodified, carbon fiber-reinforced, silicone- or polytetrafluoroethylene-lubricated and UV-stabilized. Also included under this trade name are 30% glass fiber-reinforced acrylonitrile-butadiene-styrene resins, and supertough acetal/elastomer alloys.

Ful-wrap. Trade name of Qualified Innovation Inc. (USA) for a plastic film or paper sheet with cohesive coating.

fumed alumina. Submicron aluminum oxide powder produced from aluminum chloride ($AlCl_3$) by a flame reduction process, and used for ceramic ferrites, and in coatings.

fumed silica. See carbon white.

functional ceramics. A very broad class of high-performance ceramics engineered to have specific functional and performance characteristics. They are widely used for the manufacture electronic components (e.g., capacitors, sensors and resistors), automotive components (e.g., spark plugs, gaskets, catalytic converters, valves and valve seats), aerospace and medical components, superconductive and piezoelectric components, etc.

functionally graded materials. A class of composite materials that exhibit spatially inhomogeneous (or graded) microstructures and properties. Their gradients can be designed at the microstructural level so as to produce application-specific materials with tailored functional and performance characteristics. Abbreviation: FGMs.

functional paper. See graph paper.

fungicidal pigment. A pigment, such as hydrated cupric oxide or copper quinolinolate, used in antifouling paints to inhibit the growth of fungii.

Furafil. Trademark of QO Chemicals, Inc. (USA) for a dark-brown, absorbent powder based on *lignocellulose*, and used for foundry facings, as an extender for phenolic plywood glues, as an ingredient in phenolic molding resins, and as an additive for bulk or absorbency.

Furalac. Trade name of Ato Findley (France) for furane resins used for adhesives, concrete, mortars, etc.

furan plastics. A group of black or dark-colored plastics based on furan resins. Usually reinforced, they have high adhesion and chemical resistance, good resistance to solvents and many acids and alkalies, and are used for foundry sand cores, shell molding, chemical equipment, pipes and fittings, adhesives and asphaltic pavement coatings, and as corrosion-resistant construction materials. Also known as *furan polymers*.

furan polymer concrete. A polymer concrete in which a furan polymer, usually derived from furfuryl alcohol ($C_4H_3OCH_2OH$), is used as the binder. It provides excellent resistance to most aqueous acidic and basic solutions and strong solvents, and has enhanced heat and thermal-shock resistance. Used in mortar or grout form for brick floors and linings. See also polymer concrete.

furan polymers. See furan plastics.

furan resins. Dark-colored, liquid, thermoplastic resins derived from furfuryl alcohol ($C_4H_3OCH_2OH$), furfural (C_4H_3OCHO), or reaction products of furfural and a ketone. Used for furan plastics, and in the impregnation of special plaster of Paris.

Furbaloi. A British trade name for corrosion-resistant steel with 0.13% carbon, 0.05% manganese, 0.45% copper, 0.08% nickel, 13.25% chromium, and the balance iron. Used for corrosion-resistant parts.

fur fabrics. (1) Fabrics with a distinctive color and texture, made wholly or in part from fibers or yarns obtained from beaver, rabbit, mink or other fur. Used chiefly for cover, trim, and line clothing.

(2) Fabrics made in a wide range of colors and grades from synthetic fibers, such as acrylics or nylon, and used for coats, jackets, cuffs and other items of clothing to imitate genuine animal fur.

fur felt. Felt fabrics produced by adding hairs from mammals, such as rabbits, rats or beavers, to wool, rayon or other conventional felt fibers.

fur fibers. The usually soft, silky natural protein fibers obtained from the underfur of animals, such as beavers, bears, otters, raccoons, muskrats and weasels. They are composed chiefly of the scleroprotein *keratin*, and used for spinning textile fabrics and making felt products.

furfural-phenol resins. Dark-colored, glossy, self-curing resins derived from furfural (C_4H_3OCHO) and phenol (C_6H_5OH). They have excellent adhesion to other materials, good chemical and electrical properties, and good thermal resistance. Used chiefly for chemical and electrical coatings.

furfural resins. Thermosetting resins obtained by condensation

reactions of furfural (C_4H_3OCHO) with phenols, amines, urea, etc. They have excellent resistance to acids and alkalies, and are used in the manufacture of adhesives, molding materials, etc.

furfuryl alcohol resins. Dark-colored, liquid thermosetting resins produced by the autopolymerization of furfuryl alcohol ($C_4H_3OCH_2OH$) with an acid catalyst. They have excellent chemical resistance, and are used in mortar for bonding acid-proof brick and chemical stoneware, and in protective coatings and tank linings.

Furious. Trade name of Osborn Steels Limited (UK) for an oil-hardening, abrasion-resistant tool steel containing 1.25% carbon, 4.5% tungsten, 1.2% chromium, 0.25% vanadium, and the balance iron. Used for blanking and heading dies, punches, broaches, taps, etc.

Furlon. Trade name for rayon fibers and yarns.

furnace black. A finely divided type of *carbon black* produced by the partial combustion of liquid or gaseous hydrocarbons in a closed system under controlled conditions. It is available in numerous grades including general-purpose, reinforcing, semi-reinforcing, high-modulus, low-modulus, super-abrasion, superior-processing and conducting. Used as filler in rubber goods, and as reinforcing agent for plastics. See also carbon black.

furnace lining. Any acid, basic or neutral refractory material used to construct the interior part (walls, bottom, etc.) of a furnace in contact with the molten charge. Acid or silica-rich refractories include fireclay, flint and ganister, basic refractories are lime and magnesia, and neutral refractories include chromite and graphite.

furnace sand. Highly refractory silica sand composed of coarse quartz grains, and used as a foundry sand, and for furnace linings. Also known as *fire sand*.

Furnal. Trade name for a series of furnace blacks.

Furnex. Trademark of Columbian International Chemicals Company (USA) for semi-reinforcing (SRF) furnace blacks.

Furniture Wrap. Trademark of Sealed Air Corporation (USA) for polyethylene foam used for protective packaging of furniture.

Furodit. Trade name of Röchling Burbach GmbH (Germany) for a series of corrosion- and/or heat-resistant austenitic, ferritic and martensitic stainless steels supplied in wrought and cast form. They are used for chemical equipment, tanks, mixers, furnace and heat-treating equipment, heat exchangers, oil-refinery equipment, valves, turbine and jet-engine parts, etc.

Furon. Trade name for nylon fibers and yarns used for textile fabrics including clothing.

furongite. A bright yellow to lemon yellow mineral composed of aluminum uranyl phosphate hydroxide hydrate, $Al_{13}(UO_2)_7$-$(PO_4)_{13}(OH)_{14} \cdot 58H_2O$. Crystal system, triclinic. Density, 2.86 g/cm³; refractive index, 1.5655. Occurrence: China.

furring brick. A hollow brick with one surface grooved to receive and retain plaster. Used in the construction of building walls.

furring tile. A nonload-bearing tile used for lining or furring the inside of a building wall.

furutobeite. A gray mineral with creamy yellowish tint composed of copper silver lead sulfide, $(Cu,Ag)_6PbS_4$. Crystal system, monoclinic. Density, 6.74 g/cm³. Occurrence: Japan.

Fusabond. Trademark of E.I DuPont de Nemours & Company (USA) for reactive polyolefin resins supplied in pellet and powder form for incorporation into commodity base polymers as compatibilizers and tougheners, and as interlayers in coextrusions in laminated structures. They are available in several modified polyolefin grades including polyethylene, polypropylene, ethylene propylene rubber, ethylene vinyl acetate and ethylene acrylate carbon monoxide terpolymer. *Fusabond* polyolefins are used as multilayer pipeline coating adhesives, for powder coatings, and for linings and sealants.

fused alumina. Aluminum oxide (94-99% pure) made by heating a mixture of calcined bauxite or a pure grade of alumina with iron borings to a temperature above 1980°C (3600°F) in an electric furnace. The color of fused alumina is related to its titanium dioxide (TiO_2) content, e.g., the color of chemically purified grades is white, but changes with increasing TiO_2 content to pink, reddish brown and dark brown. Used for applications requiring high abrasion resistance (e.g., spindles, bearings, bushings, etc.), refractory brick and other refractory applications, for metallurgical applications, and as an abrasive. Also known as *fused aluminum oxide*. See also brown fused alumina; white fused alumina.

fused aluminum oxide. See fused alumina.

fused borax. See anhydrous borax.

fused brown alumina. An extremely tough fused alumina (Al_2O_3) with a reddish brown to dark brown color due to the presence of up to 3.5% titania (TiO_2). It is produced directly from bauxite by heating in an electric furnace and, due to the easy breakage of its grains and ready formation of new sharp edges, it is used for heavy grinding applications. Also known as *brown fused alumina*. See also fused alumina.

fused-cast brick. See fusion-cast brick.

fused-cast refractories. See fusion-cast refractories.

fused fabrics. Textile laminates produced by bonding a fusible fabric to another fabric. See also fusible fabrics.

fused grain refractory. A refractory made from materials that have solidified from a fused or molten state.

fused magnesia. Magnesia (MgO) made by fusing in an electric furnace, and used as a refractory.

fused quartz. A transparent or translucent form of silica (SiO_2) glass made by the fusion of silica sand or crushed natural quartz. It is composed of 99.9+% SiO_2, and has similar properties to the latter, but is harder, though not quite as strong. Typically supplied in the form of sheets, rods, tubes, etc., it has a density of 2.2 g/cm³ (0.079 lb/in.³), a hardness above 7 Mohs, a refractive index of 1.46, an upper continuous-use temperature of 1100-1400°C (2010-2550°F), good strength, low thermal expansivity, good thermal-shock resistance, good ultraviolet transmission in the 180-3000 nm range, good resistance to acids and halogens, and fair resistance to alkalies and metals. Used as an insulating material, in optical glass, crucibles, furnace rods and tubes, and for chemical and heat-treating apparatus and equipment. Also known as *quartz glass*.

fused silica. A transparent, glasslike material composed of 99.5+% silica (SiO_2), and made by melting silica, quartz or sand, or by flame hydrolysis of silicon tetrachloride ($SiCl_4$). Supplied in the form of rods, tubes and other shapes, it has a density of 2.15 g/cm³ (0.078 lb/in.³), a very low coefficient of thermal expansion, good thermal spalling resistance, a melting point of about 1750°C (2100°F), a upper continuous-use temperature of 1050°C (1920°F), high thermal-shock resistance, a hardness of above 6 Mohs, good strength, good dielectric strength, good infrared transmission in the 400-4000 nm range, good ultraviolet and ultrasonic transmission properties, good resistance to acids and halogens, fair resistance to metals, and poor resistance to alkalies. Used as an ablative material in rocket en-

gines, spacecraft, etc., for refractories, fibers in reinforced plastics, special camera lenses, special thermometers, chemical equipment and ware, laboratory equipment, crystal-growing crucibles, radar delay lines, and heat-resisting glassware. Abbreviation: FS. Also known as *silica glass; silicon dioxide glass; vitreous silica*.

fused silica fibers. Nonhygroscopic, water-insoluble continuous ceramic fibers composed of 99.5+% fused silica (SiO_2). They have a density of 2.20 g/cm³ (0.079 lb/in.³), good chemical stability, high service temperatures above 1000°C (1830°F), high ultimate tensile strength (3.45 GPa), high elastic modulus (69 GPa), good electrical properties, low dielectric constants, and a refractive index of 1.45. Used as reinforcing fibers and for thermal insulation applications.

fused silica refractories. Refractories in which the predominant component is fused silica (SiO_2).

fused white alumina. A white, dense aluminum oxide (99.4+% pure) made by fusing calcined alumina in an electric-arc furnace under controlled conditions and subsequent, crushing, grinding and precision screening of the resulting ingots. Its microstructure consists predominantly of alpha alumina crystals, and it is available in various grain sizes with an average bulk density of about 3.4-3.6 g/cm³ (0.12-0.13 lb/in³) and an apparent porosity of about 7-8%. Used in the manufacture of refractory shapes and monolithics, as abrasives, and for investment casting shells. Also known as *white fused alumina*. See also fused alumina.

fused zirconia. Zirconium oxide (ZrO_2) made in an electric furnace. It has a hardness of about 7-8 Mohs, and is used as an abrasive for applications requiring high rates of removal.

Fusible. A British trade name for an alloy composed of 45% bismuth, 17% tin, 30% lead and 10% mercuric sulfate. It has a melting point of 85°C (185°F), and is used for binding plugs, and solders.

fusible alloys. A large group of white metal alloys that melt at relatively low temperatures, usually in the range of 50-260°C (120-500°F). Commercial fusible alloys contain bismuth, lead, tin, antimony and cadmium, and sometimes zinc, silver, indium, gallium and/or thallium. Eutectic fusible alloys have definite minimum melting points, while noneutectic alloys have melting ranges. Most fusible alloys are heavy, bright, silvery, nontarnishing, and can be aged. Important applications include fuses, fire sprinklers, furnace controls and boiler plugs, solders for soft metals, casting patterns, tube bending applications, ornaments and toys. Also known as *fusible metals; low-melting alloys*.

fusible fabrics. Textile fabrics that either have a thermoplastic adhesive bonded to one side, or can be fused by the application of heat and pressure, and are thus used for bonding fabrics to other fabrics and materials. Also known as *fusibles*. See also fused fabrics.

fusible glass. A special glass with brilliant color made by laying out glasses of varying colors and shapes on a large glass sheet and fusing in an oven.

fusible metals. See fusible alloys.

fusible porcelain. See cryolite glass.

fusibles. See fusible fabrics.

fusible solders. A group of low-melting solders used for soldering operations at temperatures below about 180°C (360°F). They contain approximately 40-60% bismuth, 20-45% lead, 0-60% tin and, sometimes, additions of cadmium or antimony. Bismuth solders are more fluid than other solders and are used, usually with corrosive fluxes, for soldering in temperature-sensitive areas, e.g., fire sprinkler links, heat-sensitive equipment, or heat-treated areas on parts. Also known as *bismuth solders; low-temperature solders.*

Fusil 2000. Trade name of Industrial Ceramic Products Inc. (USA) for a fused silica (SiO_2) based ceramic with excellent thermal-shock resistance and zero thermal expansion. Processed by extrusion or compression molding, it is used in the metals melting industries for crucible liners and refractory shapes.

Fusion. Trade name of Fusion Inc. (USA) for an extensive series of gold-, nickel- silver-, copper-phosphorus- and aluminum-silicon-base brazing filler metals, and several tin-antimony, tin-lead, tin-lead-bismuth, tin-silver and lead-tin solders. They are supplied in paste form for joining various ferrous and nonferrous alloys.

fusion-cast basalt. See artificial basalt.

fusion-cast brick. A dense refractory brick or other shape that is made by fusing premixed refractory oxides in an electric furnace and casting the molten material into appropriate molds. Also known as *electrocast brick; fused-cast brick; molten cast brick.*

fusion-cast refractories. Hard, nonporous, vitreous refractory products made from mullite, aluminum silicate, etc., by fusing in an electric furnace, and subsequent casting into blocks or other shapes. They have low coefficients of thermal expansion, excellent high-temperature resistance, excellent resistance to corrosion and erosion from slags, fluids, glass, etc. Also known as *electrocast refractories; fused-cast refractories; molten cast refractories.*

fusion fuel. A fuel, such as deuterium or tritium used in thermonuclear reactors.

Fusite. (1) Trademark for solvent cement adhesive used for structural applications.

(2) Trademark of Redland Ohio, Inc. (USA) for sized, iron-coated dolomitic quicklime refractories.

Fusor. Trademark of Lord Corporation (USA) for a series of one- and two-component structural adhesives based on epoxy resins, and used in the manufacture and repair of vehicles and vehicle components.

fustian. (1) Originally, a coarse, rather heavy fabric made of cotton or flax, and used for clothing.

(2) Now, a thick, hard-wearing, heavily wefted cotton fabric like corduroy, moleskin or velveteen, used for making clothing.

fustic. The yellow wood of the tree *Chlorophora tinctoria* belonging to the mulberry family. It has a fine, open grain, very high hardness, and contains a liquid extract that produces a yellow dyestuff. Average weight, 650 kg/m^3 (41 lb/ft^3). Source: Tropical America. Used for cabinetwork and as a dyewood. Also known as *Cuba wood.*

Futar D'Occlusion. Trade name for a silicone elastomer used in dentistry for bite registration applications.

Futura. Trademark of Indian Organic Chemicals Limited (India) for polyester staple fibers and filament yarns.

Futurabond. Trade name for a light-cure dental adhesive for tooth structure (i.e., dentin/enamel) bonding.

Futurethane. Trademark for urethane structural adhesives.

Futurit. Trade name of Kabelwerk AG for phenolics and colored casein and other plastics.

Futurol. Trade name for phenolic resins and plastics.

Fybex. Trademark of E.I. DuPont de Nemours & Company (USA) for crystalline potassium titanate (K_2TiO_3) fibers used as reinforcing fibers in engineering thermoplastics.

Fynal. Trade name of Dentsply/LD Caulk (USA) for a polymer- and ethoxybenzoic acid (EBA) reinforced zinc oxide–eugenol dental cement used for temporary and permanent luting and restoration applications.

Fyre-Cote. Trade name of John Boyle & Company, Inc. (USA) for flame-retardant polyester/cotton fabrics for awnings, canopies and tents.

G

gabardine. A medium-weight, closely woven fabric made of cotton, rayon or wool in a twill weave. It has small, diagonal surface ribs, and is used especially for suits, dresses, slacks, shirts, rainwear, and sportswear.

gabbro. Any of a large group of heavy, dark-colored igneous rocks with granular texture formed from *basalt* and composed of *plagioclase* feldspar (commonly *labradorite* or *anorthite*), with or without pyroxenes, hornblende and olivine. Used as a concrete aggregate.

gaboon. See okoume.

Gaboon mahogany. See okoume.

Gabraster. Trademark of Montecatini Edison SpA (Italy) for a series of unsaturated polyesters.

gabrielsonite. A black mineral of the descloizite group composed of lead iron arsenate hydroxide, $PbFe(AsO_4)(OH)$. Crystal system, orthorhombic. Density, 6.67 g/cm^3; refractive index, above 2.00. Occurrence: Sweden.

Gabrite. Trademark of Montecatini Edison SpA (Italy) for urea-formaldehyde resins.

Gacoflex. Trademark of Gaco Western Inc. (USA) for neoprene-*Hypalon*-, acrylic latex- and urethane-based protective coatings.

gadolinia. See gadolinium oxide.

gadolinite. A black, greenish-black, or brown mineral with a vitreous to greasy luster. It belongs to the datolite group, is composed of beryllium iron yttrium silicate, $Be_2FeY_2Si_2O_{10}$, and may also contain other lanthanides, such as lanthanum, cerium, praseodymium, neodymium and samarium. It can also be made synthetically. Crystal system, monoclinic. Density, 4.15-4.41 g/cm^3; hardness, 6.5-7 Mohs; refractive index, 1.76-1.81. Occurrence: Scandinavia. Used as a source of rare earths, and in ceramics.

gadolinium. A rare, silvery-white metallic element of the lanthanide series (rare-earth group) of the Periodic Table. It is commercially available in the form of ingots, rods, lumps, sponge, wire, sheet, foil, turnings (chips), powder and single crystals. The commercial gadolinium ores are *monazite, gadolinite, samarskite* and *xenotime*. Density, 7.87 g/cm^3; melting point, 1312°C (2394°F); boiling point, 3266°C (5911°F); hardness, 55 Vickers; atomic number, 64; atomic weight, 157.25; trivalent. It has the highest thermal neutron absorption cross section of any known element, very high saturation magnetization, and exhibits superconductive properties. There are two forms of gadolinium: (i) *Alpha gadolinium* with hexagonal close-packed crystal structure, present at ordinary temperatures; and (ii) *Beta gadolinium* with body-centered cubic crystal structure, present between 1262 and 1312°C (2304 and 2394°F). *Gadolinium* is used as a burnable poison in shields and control rods of nuclear reactors, in garnets, microwave filters and X-ray tubes, as a phosphor activator and in host materi-als for rare-earth phosphors, in computer memories, as a catalyst, in the manufacture of chromium steel and permanent magnets, in ceramics and superconductors, and as a scavenger for oxygen in the manufacture of titanium. Symbol: Gd.

gadolinium aluminate. A compound of gadolinium oxide and aluminum oxide with a melting point of 1982°C (3600°F). Used in ceramics and materials research. Formula: $Gd_2Al_2O_6$. Also known as *gadolinium aluminum oxide (GAO)*.

gadolinium aluminum oxide. See gadolinium aluminate.

gadolinium boride. Any of the following compounds of gadolinium and boron used in ceramics and materials research: (i) *Gadolinium tetraboride*. Crystals or powder. Density, 6.48 g/cm^3. Formula: GdB_4; and (ii) *Gadolinium hexaboride*. Black-brown crystals. Crystal system, cubic. Density, 5.28-5.31 g/cm^3; melting point, 2099°C (3810°F) [also reported as 2510°C or 4550°F]; hardness, 2360 Vickers. Formula: GdB_6.

gadolinium bromide. White, hygroscopic crystals or powder available in purities of 99.9-99.99%. Crystal system, monoclinic. Density, 4.56 g/cm^3; melting point, 770°C (1418°F). Used as a source of gadolinium, and in ceramics, electronics and materials research. Formula: $GdBr_3$.

gadolinium carbide. Any of the following compounds of gadolinium and carbon used in ceramics and materials research: (i) *Trigadolinium carbide*. Density, 8.70 g/cm^3. Formula: Gd_3C; (ii) *Gadolinium sesquicarbide*. Density, 8.02 g/cm^3. Formula: Gd_2C_3; and (iii) *Gadolinium dicarbide*. Density, 6.90 g/cm^3; melting point, above 2204°C (4000°F). Formula: GdC_2.

gadolinium chloride. White, hygroscopic crystals or powder available in purities of 99.9-99.99%. Crystal system, monoclinic. Density, 4.52 g/cm^3; melting point, 609°C (1128°F). Used as a source of gadolinium, for gadolinium sponge metal by contact with a reducing metal vapor, and in ceramics, electronics and materials research. Formula: $GdCl_3$.

gadolinium chloride hydrate. Hygroscopic, colorless crystals, or white powder (99.9+% pure) with a density of 2.424 g/cm^3. Used as a source of gadolinium, for gadolinium sponge metal, and in ceramics, electronics and materials research. Formula: $GdCl_3 \cdot xH_2O$.

gadolinium dicarbide. See gadolinium carbide (iii).

gadolinium diiodide. Bronze-colored crystals with a melting point of 831°C (1528°F) used in ceramics, electronics and materials research. Formula: GdI_2.

gadolinium disilicate. See gadolinium silicate (iii).

gadolinium disilicide. See gadolinium silicide (i).

gadolinium disulfide. See gadolinium sulfide (iii).

gadolinium ferrite. A compound of gadolinium oxide and ferric oxide with a melting point of approximately 1650°C (3000°F). Used in electronics, materials research and ceramics. Formula: $Gd_2Fe_2O_6$.

gadolinium fluoride. White crystals or powder available in purities of 99.9-99.99%. Melting point, 1231°C (2248°F). Used as a source of gadolinium, in electronics and materials research, and in the synthesis of non-oxide glasses. Formula: GdF_3.

gadolinium-gallium garnet. A soft magnetic material made from gadolinium oxide and gallium oxide. It is available in the form of small pieces and crystals with a purity of 99.9+% and a density of 7.09 g/cm^3. Used for computer bubble memories and other electronic applications, and as substrates in superconductor research. Formula: $Gd_3Ga_5O_{12}$. Abbreviations: GdGG; GGG.

gadolinium hexaboride. See gadolinium boride (ii).

gadolinium iodide. Yellow crystals with a melting point of 926°C (1699°F) available in purities of 99.9+%. Used as a source of gadolinium, and in ceramics, electronics and materials research. Formula: GdI_3.

gadolinium-iron garnet. A powder-made *cubic ferrite* of the magnetic garnet type used for magnetic and electronic applications, waveguide components and microwave applications. Formula $Gd_3Fe_5O_{12}$. Abbreviation: GdIG.

gadolinium monosilicate. See gadolinium silicate (i).

gadolinium monosulfide. See gadolinium sulfide.

gadolinium monotelluride. See gadolinium telluride (i).

gadolinium nitrate hexahydrate. Colorless triclinic crystals. Density, 2.33 g/cm³; melting point, 91°C (196°F). Used as an oxidizer and chemical synthesis. Formula: $Gd(NO_3)_3 \cdot 6H_2O$.

gadolinium nitrate pentahydrate. White, hygroscopic crystals and chunks available in purities from 99.9 to 99.999%, and containing 34.9-36.5% gadolinium. Density, 2.406 g/cm³; melting point, 92°C (198°F). Used in superconductivity studies. Formula: $Gd(NO_3)_3 \cdot 5H_2O$.

gadolinium nitride. A crystalline compound of gadolinium and nitrogen. Crystal system, cubic. Density, 9.10 g/cm³. Used in ceramics, electronics and materials research. Formula: GdN.

gadolinium oxide. White to cream-colored, hygroscopic powder (99.9+%). Density, 7.407 g/cm³; melting point, 2310°C (4190°F) [also reported as 2420°C or 4388°F]; hardness, 486 Knoop; high neutron absorption. Used for special glasses, filament coatings, ceramic dielectrics, high-temperature ceramics, neutron shields, nuclear-reactor control rods, phosphor activators, catalysts, lasers, masers, and as a starting material for preparation of gadolinium-substituted superconductors. Formula: Gd_2O_3. Also known as *gadolinia*.

gadolinium oxysulfide. A compound made from gadolinium sulfide and gadolinium oxide, and used as a scintillator in general and neutron radiography. Formula: $Gd_4(SO)_3$.

gadolinium pentasilicide. See gadolinium silicide (ii).

gadolinium selenide. A compound of gadolinium and selenium. Crystal system, cubic. Crystal structure, halite. Density, 8.1 g/cm³; melting point, 2127°C (3861°F); band gap, 1.8 eV. Used in ceramics and materials research, as a semiconductor, and in thermoelectric generating devices. Formula: GdSe.

gadolinium sesquicarbide. See gadolinium carbide (ii).

gadolinium sesquisulfide. See gadolinium sulfide (ii).

gadolinium sesquitelluride. See gadolinium telluride (ii).

gadolinium silicate. Any of the following compounds of gadolinium oxide and silicon dioxide used in ceramics, materials research and electronics: (i) *Gadolinium monosilicate*. Density, 6.55 g/cm³; melting point, 1900°C (3452°F); hardness, 5-7 Mohs. Formula: Gd_2SiO_5; (ii) *Digadolinium disilicate*. Density, 6.29; hardness, 5-7 Mohs. Formula: $Gd_4Si_2O_{10}$; and (iii) *Gadolinium disilicate*. Density, 5.34 g/cm³; melting point, 1720°C (3128°F); hardness, 5-7 Mohs. Formula: $Gd_2Si_2O_7$. See also cerium-doped gadolinium silicate.

gadolinium silicide. Any of the following compounds of gadolinium and silicon: (i) *Gadolinium disilicide*. Crystal system, orthorhombic. Density, 5.9-6.4 g/cm³; melting point, approximately 1538°C (2800°F). Used in ceramics and materials research. Formula: $GdSi_2$; and (ii) *Gadolinium pentasilicide*. Crystals used in ceramics and materials research. Formula: Gd_3Si_5.

gadolinium sulfate. Colorless, hygroscopic crystals with a density of 4.2 g/cm³. Used in chemical research. Formula: Gd_2-$(SO_4)_3$.

gadolinium sulfate octahydrate. Colorless or white hygroscopic crystals (99.9+%). Density, 3.01 g/cm³; Used chiefly in cryogenic research. Formula: $Gd_2(SO_4)_3 \cdot 8H_2O$.

gadolinium sulfide. Any of the following compounds of gadolinium and sulfur: (i) *Gadolinium monosulfide*. Crystals. Density, 7.26 g/cm³. Used in ceramics and materials research. Formula: GdS; (ii) *Gadolinium sesquisulfide*. Red brown, or yellow crystals. Crystal system, orthorhombic. Density, 6.15 g/cm³; melting point, 1885°C (3425°F); band gap, 1.73 eV. Used in ceramics and materials research, and in electronics as a semiconductor. Formula: Gd_2S_3; and (iii) *Gadolinium disulfide*. Crystals. Density, 5.98 g/cm³. Used in ceramics and materials research. Formula: GdS_2.

gadolinium telluride. Any of the following compounds of gadolinium and tellurium used in ceramics, electronics and materials research: (i) *Gadolinium monotelluride*. Melting point, 1870°C (3398°F). Formula: GdTe; and (ii) *Gadolinium sesquitelluride*. Orthorhombic crystals. Density, 7.7 g/cm³; melting point, 1255°C (2291°F). Formula: Gd_2Te_3.

gadolinium tetraboride. See gadolinium boride (i).

GAF Carbonyl Iron Powders. Trademark of GAF Materials Corporation (USA) for spherical particles of high-purity carbonyl iron (up to 99.5% iron) available in various size groups ranging from 2 to 20 µm (79 to 790 µin). They have uniform particle size, outstanding electromagnetic interference (EMI) properties, excellent high-frequency absorption, and good compatibility with many plastic resins and with metals and alloys including tungsten, copper and bronze. Used in coatings and advanced composites for aerospace applications, in ferrous powder metallurgy, as alloying agents, in magnetic fluids, and in high-frequency cores for electrical and electronic equipment, such as radar equipment, radio short-wave transmitters, telephones and televisions.

Gafite. Trademark of GAF Materials Corporation (USA) for vinyl polymers including craze-resistant, transparent polymethyl α-chloroacrylates and other resins. They have good tensile and flexural strength, a heat distortion point of 127°C (260°F), and are supplied as beads, pellets, granules, molding powders, plates, sheets, films, rods, tubes and foams. Used in aircraft components, e.g., windows.

Gaflon. Trademark of Plastic Omnium GmbH (Germany) for several polytetrafluoroethylene plastics.

Gaftex. Trademark of GAF Materials Corporation (USA) for fiberglass rovings used as reinforcements for epoxy, phenolic, polyester and vinyl ester resins.

gagarinite. A creamy yellow, or rose mineral composed of yttrium calcium sodium fluoride, $(Y,Ca,Na)F_2$. Crystal system, hexagonal. Density, 4.21 g/cm³; refractive index, 1.472. Occurrence: Kazakhstan, Russian Federation.

gaged brick. See gauged brick.

gaged mortar. See lime-cement mortar.

gaged steel. See gauged steel.

gageite. A colorless mineral that is composed of magnesium manganese silicate hydroxide, $(Mn,Mg)_7Si_2O_7(OH)_8$, and may also contain some zinc. Crystal system, orthorhombic. Density, 3.58 g/cm³; refractive index, 1.734. Occurrence: USA (New Jersey).

gaging plaster. See gauging plaster.

gahnite. A dark blue-green, sometimes yellow, brown, gray or black mineral of the spinel group composed of zinc aluminate, $ZnAl_2O_4$. It can also be made synthetically. Crystal system, cubic. Density, 4.48-4.62 g/cm³; melting point, 1950°C (3542°F);

hardness, 7.5-8.0 Mohs; refractive index, 1.790. Used in refractories for lining melting pots, and in ceramics, materials research and electronics. Also known as *zinc spinel*.

gaidonnayite. A colorless to white mineral composed of sodium zirconium silicate dihydrate, $Na_2ZrSi_3O_9 \cdot 2H_2O$. Crystal system, orthorhombic. Density, 2.67 g/cm^3; refractive index, 1.592. Occurrence: Canada (Quebec).

gainesite. A pale bluish lavender mineral composed of sodium beryllium zirconium phosphate, $Na_6Zr_2Be(PO_4)_4$. Crystal system, tetragonal. Density, 2.94 g/cm^3; refractive index, 1.630. Occurrence: USA (Maine).

Gainex. Trade name of Armco International (USA) for a series of high-strength low-alloy (HSLA) steels.

gaitite. A white to colorless mineral of the fairfieldite group composed of hydrogen calcium zinc arsenate hydroxide, $H_2Ca_2Zn(AsO_4)_2(O)_2$. Crystal system, triclinic. Density, 3.81 g/cm^3; refractive index, 1.730. Occurrence: Namibia.

gaize cement. A special cement product consisting of a mixture of finely ground pozzolanic material and Portland cement, or pozzolanic material and hydrated lime. See also pozzolan.

Gala. (1) Trade name of Dunford Hadfields Limited (UK) for low-carbon high-chromium steels with 0.1% carbon, 13% chromium, and the balance iron. Used for corrosion-resistant parts, and cutlery.

(2) Trade name of George Morrell Corporation (USA) for casein plastics.

galactose. A monosaccharide hexose that is a constituent of *lactose* (milk sugar), and also occurs in raffinose, pectin, gums, mucilages and agar-agar. It exists in levorotatory (L-), dextrorotatory (D-) and racemic (DL-) forms. The commercial D-form has a melting point of about 168-170°C (334-338°F), and a specific rotation of +80.2° (at 20°C or 68°F). Used in biochemistry, biotechnology, organic synthesis, and as a diagnostic aid in medicine. Formula: $C_6H_{12}O_6$. Also known as *cerebrose*.

galacturonan. See galacturonic acid.

galacturonic acid. A monobasic acid that is a major constituent of plant pectins, and is obtained from these by hydrolysis. It occurs in alpha- and beta-form, exhibits mutarotation, and is used in biochemistry and biotechnology. Formula: $C_6H_{10}O_7$. Also known as *galacturonan*.

Galahad. Trade name of Dunford Hadfields Limited (UK) for a series of wrought and cast ferritic and martensitic stainless steels containing 0.1-0.3% carbon, 13% chromium, 0-1% nickel, up to 0.8% molybdenum, and the balance iron. Free-machining grades may contain up to 0.75% sulfur. Used for corrosion-resistant parts, turbine blades and wheels, propeller shafts, machine parts, structural parts, cutlery, and surgical instruments.

Galalith. (1) Trademark of International Galalith (Germany) for a tough, hard, flexible, noncombustible, tasteless and odorless thermoplastic molding material made from rennet *casein* and formaldehyde. It has good resistance to grease, oils and dilute acids, and good electrical properties. Used for electrical insulation, buttons, beads, knitting needles, novelties, and decorative items.

(2) Trademark of Phoenix AG (Germany) for chlorosulfonated polyethylene plastics.

galatea. A durable cotton fabric, usually made in white and stripes in a twill weave with pronounced warp effect. Used especially for uniforms and children's wear.

Galatom. Trade name for diatomaceous earth. This term is now sometimes used as a synonym for diatomaceous earth and related products.

Galavan. Trade name of US Steel Corporation for a low-carbon pre-galvanized steel sheet for truck and trailer bodies.

galaxite. A black or yellowish brown mineral of the spinel group composed of manganese aluminate, $MnAl_2O_4$. It can also be made synthetically from manganese tetroxide (Mn_3O_4) and alpha aluminum oxide (α-Al_2O_3). Crystal system, cubic. Density, 4.15 g/cm^3; refractive index, 1.856. Occurrence: USA (North Carolina).

Galaxy. (1) Trade name of BEI PECAL, Division of Stake Technology Limited (Canada) for a garnet abrasive based on the mineral *almandine*. It has a hardness of 8 Mohs and sharp grains, which make it suitable for blast cleaning and water-jet cutting.

(2) Trademark of Acordis Cellulosic Fibers Inc. (USA) for viscose rayon staple fibers.

(3) Trade name of Galaxy Dental Manufacturing Company (USA) for a medium-gold dental casting alloy.

Galaxy Granules. Trademark of SAFAS Corporation (USA) for granular, cured polymer fillers supplied in a wide range of sizes and colors. They are fire-retardant, corrosion-, stain- and UV-resistant, and used for the manufacture of cast, molded and pultruded polymer products.

galeite. A colorless mineral composed of sodium chloride fluoride sulfate, $Na_{15}(SO_4)_5F_4Cl$. Crystal system, hexagonal. Density, 2.61 g/cm^3; refractive index, 1.447. Occurrence: USA (California).

galena. A lead-gray or black mineral with a metallic luster belonging to the halite group. It is composed of lead sulfide, PbS, and may contain up to 86.6% lead. Crystal system, cubic. Density, 7.3-7.6 g/cm^3; melting point, 1114°C (2037°F); hardness, 2.5 Mohs. Occurrence: Africa, Canada, South America, USA (Colorado, Idaho, Kansas, Missouri, Oklahoma). Used as an important ore of lead, as a source of metallic lead, in ceramics, in glazings for pottery as a flux substitute for lead oxide. Formerly also used as a detector in crystal radio receivers. Also known as *blue lead; galenite; lead glance*.

galenite. See galena.

galenobismutite. A lead-gray, or tin-white mineral composed of lead bismuth sulfide, $PbBi_2S_4$. It can also be made synthetically. Crystal system, orthorhombic. Density, 6.88-7.04 g/cm^3; hardness, 3-4 Mohs. Occurrence: Sweden.

Galette. Trade name of SA Glaverbel (Belgium) for a patterned glass.

Galfan. (1) Trademark of International Lead Zinc Research Organization (ILZRO) for an alloy coating of 95% zinc and 5% aluminum/mischmetal that is one of the most formable hot dipped sacrificial coatings. It is an excellent substrate for paint adhesion with a corrosion protection superior to that of conventional zinc coatings. Used for coating steel sheets, strips, tubing, wire and cable. *Note:* The coated products are also known by this trademark.

(2) Trade name of Sollac/Usinor (France) for steel sheets with a corrosion-resistant, low-friction alloy coating of about 95% zinc and 5% aluminum. Supplied in thicknesses of 0.2-3.0 mm (0.008-0.12 in.) and widths of about 0.6-1.5 m (2.0-4.9 ft.), the sheets are suitable for deep-drawing processes. Used for automotive components, such as universal-joint sleeves, oil filter cartridges and motor stators, in building construction for gutters, window frames, metal doors, garage doors and signposts, and for casings and structural parts of domestic appliances, such as washers, dryers and refrigerators.

Galflex B. Trade name of Sollac-Usinor (France) for a sheet steel with crack-resistant galvanized coating. It has a smooth, uniform appearance without spangling, and is supplied in coils, slit strips and cut-to-length sheets. Depending on the product form, it is available in thicknesses of 0.2-2.0 mm (8-800 mil) and widths from 0.6 to 1.85 m (2.0 to 6.1 ft.). Used for domestic appliances, such as washers, dryers, dishwashers, ovens, cookers and water heaters, and for roofing, suspended ceilings, and wall cladding.

Galicar. Trade name of Charles Carr Limited (UK) for a babbitt composed of tin, lead and antimony, and used for bearings and antifriction alloys.

galkhaite. (1) A brown to reddish-black mineral of the tetrahydrite group composed of cesium thallium copper mercury zinc antimony arsenic sulfide, $(Cs,Tl)(Hg,Cu,Zn)_6(As,Sb)_4S_{12}$. Crystal system, cubic. Density, 5.36 g/cm^3. Occurrence: USA (Nevada).

(2) An orange-red mineral composed of mercury arsenic sulfide, $HgAsS_2$. Crystal system, cubic. Density, 5.40 g/cm^3; refractive index, 2.8. Occurrence: Russian Federation.

Gallery. Trade name of PPG Industries Inc. (USA) for a non-glare picture glass with special surface coating.

Gallia-Rubber. French trade name for vulcanite (hard rubber).

Gallimore Metal. Trade name of William Gallimore & Sons, Limited (UK) for a wrought alloy containing 45% nickel, 28% copper, 25% zinc, 2% iron, and traces of silicon and manganese. It has good corrosion resistance in seawater, and is used for airplane parts, and stampings.

gallite. A gray mineral of the chalcopyrite group composed of copper gallium sulfide, $CuGaS_2$. It can also be made synthetically. Crystal system, tetragonal. Density, 4.29 g/cm^3. Occurrence: Zaire, Southwest Africa.

gallium. A rare, gray-blue to silvery-white metallic element of Group IIIA (Group 13) of the Periodic System. It is commercially available in various grades with purities from 99.9% to 99.9999%, often as lumps or splatter, and occurs naturally in zinc ores, bauxite, and certain iron ores. Crystal system, orthorhombic. Density, 5.904 g/cm^3 (at 20°C or 68°F); melting point, 29.8°C (85.6°F); boiling point, 2205°C (4001°F); superconductivity critical temperature, 1.08K; hardness, 1.5-2.5 Mohs; atomic number, 31; atomic weight, 69.723; divalent, trivalent. It is liquid at ordinary temperatures, may be undercooled to almost 0°C (32°F) without solidifying, and reacts with most metals at high temperatures. Used in liquid metals, in high-temperature and quartz thermometers, as a heat-exchange medium for nuclear reactors, in combination with elements of Groups III, IV and V (Groups 13, 14 and 15) of the Periodic Table to form semiconductor materials, in combination with arsenic and/or phosphorus in light-emitting diodes, laser diodes, solar cells, transistors, etc., as a metallic coating for ceramics, as a backing for optical mirrors, and as a nonpoisonous substitute for mercury in certain dental alloys. Symbol: Ga.

Gallium Alloy GF. Trade name of Tokurike Honten (Japan) for a low-melting dental restorative amalgam alloy produced by first alloying gallium with indium and tin, and then mixing the resulting liquid ternary alloy with a silver-based spherical alloy powder containing 25.7% tin, 15% copper, 9% palladium and 0.3% zinc.

gallium antimonide. A compound of gallium and antimony supplied in small pieces, and as single crystals in p- and n-types, usually grown by horizontal zone melting or liquid-encapsulated Czochralski methods. The electronic grade is 99.9+% pure. Crystal system, cubic. Crystal structure, sphalerite. Density, 5.619 g/cm^3; melting point, 707°C (1305°F); hardness, 4480 Knoop; refractive index, 3.8. Used as a semiconductor, and for epitaxial deposition. Formula: GaSb.

gallium antimonide semiconductor. An intrinsic compound semiconductor (Group III-V or Group 13-15) with the following electrical properties at room temperature: band gap, 0.68 eV; electron mobility, 0.200-0.400 m^2/V-s; electron hole mobility, 0.100-0.140 m^2/V-s. Maximum operating temperature, below 250°C (480°F). Used in solid-state electronics.

gallium arsenide. A compound of gallium and arsenic supplied as dark gray crystals or pieces. It is available in various grades including ingots, polycrystalline, high-purity electronic (99.999%) and single crystal, and is frequently alloyed with gallium phosphide, indium arsenide, or other compounds or elements. Crystal system, cubic. Crystal structure, sphalerite. Density, 5.316 g/cm^3; melting point, 1238°C (2260°F); hardness, 7500 Knoop; electroluminescent in infrared light; low absorption; photorefractive properties. Used for microwave diodes, high-temperature rectifiers and transistors, as a high-grade semiconductor in light-emitting diodes, injection lasers, solar cells, magnetoresistive devices, thermistors, integrated circuits, optoelectronic devices, and in the fabrication of photonic crystals. The alloyed or doped grades are used in laser, microwave, and photoelectric devices. Single crystal substrates are used for III-V (13-15) compound depositions for light-emitting diodes and blue lasers. Formula: GaAs.

gallium arsenide semiconductor. An intrinsic compound semiconductor (Group III-V or Group 13-15) having the following electrical properties at room temperature: band gap, 1.47 eV; high electron mobility, 0.720-0.850 m^2/V-s; electron hole mobility, 0.020-0.040 m^2/Vs. Maximum operating temperature, 400°C (750°F) in transistors and 315°C (600°F) in rectifiers. It can be doped with aluminum, beryllium, carbon, chromium, erbium, silicon, tellurium, or zinc. Used in field-effect transistors for high-speed logic circuits, microwave devices, rectifiers, and other solid-state electronics.

gallium arsenide superconductor. A superconductive gallium arsenide semiconductor produced in thin layers by growing under prescribed conditions. It has zero electrical resistance at a temperature of about 10K, and is used in computer, laser and CD-player components.

gallium bromide. White, moisture-sensitive high-purity crystals. Density, 3.69 g/cm^3; melting point, 121.5°C (251°F); boiling point, 278.8°C (534°F). Used in electronics, and materials research. Formula: $GaBr_3$.

gallium chloride. See gallium dichloride; gallium trichloride.

gallium dichloride. White, moisture-sensitive high-purity crystals. Melting point, 164°C (327°F); boiling point, 535°C (995°F). Used in electronics and materials research. Formula: Ga_2Cl_4.

gallium-doped yttrium iron garnet. Yttrium-iron garnet ($Y_3Fe_5O_{12}$) doped or activated with gallium (Ga^{3+}) and having improved electrical and other properties over the unactivated material. Used for various applications in electronics, optoelectronics, photonics and scintillation including lasers, microwave devices, acoustic and electronic transmitters, and oscillators. Abbreviation: YIG:Ga. See also yttrium-iron garnet.

gallium ferric oxide. High-purity crystals that are piezoelectric between room temperature and -195°C (-319°F), and magnetic below -13°C (9°F). Used in electronics and materials research. Formula: $Gd_2Fe_2O_6$. Abbreviation: GFO. Also known as *gallium ferrite*.

gallium ferrite. See gallium ferric oxide.

gallium fluoride. A white, hygroscopic powder (99.9+%) with a density of 4.47 g/cm³, used in electronics and in the preparation of optical fluorogallate glasses. Formula: GaF₃.

Gallium GFII. Trade name of Tokurike Honten (Japan) for a low-melting gallium-based dental restorative amalgam alloy similar to *Gallium Alloy GF*, but with only 2% palladium.

gallium halide. A compound of gallium and any of the halogens (e.g., chlorine, fluorine, or iodine) used chiefly in electronics and materials research.

gallium monosulfide. See gallium sulfide (i).

gallium 2,3-naphthalocyanine chloride. An organometallic dye with a dye content of about 90%, and a maximum absorption wavelength of 814 nm. Used in biochemistry, biotechnology and electronics. Formula: $C_{48}H_{24}N_8GaCl$.

gallium nitride. A compound of gallium and nitrogen supplied in standard grades as a dark gray powder, and in high-purity, moisture-sensitive grades (99.99+%). Crystal system, hexagonal. Crystal structure, wurtzite (or zincite). Density, 6.1 g/cm³; melting point, 1227°C (2241°F) [also reported as 800°C or 1472°F]. Used as a semiconductor. Formula: GaN.

gallium oxide. See gallium sesquioxide; gallium suboxide.

gallium phosphide. A compound supplied as amber, transparent high-purity crystals (99.99+%), or whiskers (up to 20 mm or 0.8 in. long), often made by low-temperature reaction of phosphorus and gallium suboxide vapors. Crystal system, cubic. Crystal structure, sphalerite. Density, 4.13 g/cm³; melting point, 1477°C (2691°F) [also reported as 1348°C or 2458°F]; hardness, 9450 Knoop; refractive index, 3.2; electroluminescent in visible light. Used as a semiconductor in solid-state devices. Formula: GaP.

gallium phosphide semiconductor. An intrinsic compound semiconductor (Group III-V or Group 13-15) having the following electrical properties at room temperature: band gap, 2.25 eV; low electron mobility, 0.011-0.030 m²/V-s; electron hole mobility, 0.007-0.015 m²/V-s; maximum operating temperature, 870°C (1600°F) in transistors. Used for field-effect transistors and other solid-state electronic devices.

gallium phthalocyanine chloride. An organometallic dye with a dye content of about 97% and a maximum absorption wavelength of 694 nm. Used in biochemistry, biotechnology, electronics, chemistry and materials research. Formula: $C_{32}H_{16}N_8$-GaCl.

gallium selenide. A compound of gallium and selenium supplied in the form of high-purity crystals or powder (99.99+%). Crystal system, cubic. Crystal structure, sphalerite. Density, 4.92; melting point, 747°C (1377°F); hardness, 3160 Knoop. Used as a semiconductor, and for other electronic applications. Formula: GaSe.

gallium sesquioxide. White high-purity crystals or powder (99.99+%) available in alpha and beta form. Density, (alpha) 6.44; (beta) 5.88; melting point, 1900°C (3452°F); changes to beta form at 600°C (1112°F). Used in spectrography, vacuum deposition coatings, ceramics, materials research, and electronics. Formula: Ga_2O_3. Also known as *gallium oxide*.

gallium sesquisulfide. See gallium sulfide (ii).

gallium suboxide. A compound of gallium and oxygen supplied as a brownish-black powder. Melting point, above 660°C (1220°F). Used in ceramics, electronics and materials research. Formula: Ga_2O. Also known as *gallium oxide*.

gallium subsulfide. See gallium sulfide (iii).

gallium sulfide. Any of the following compounds of gallium and sulfur: (i) *Gadolinium monosulfide.* Small, air- and moisture-sensitive, light-yellow high-purity pieces (99.99+%). Density, 3.86 g/cm³; melting point, approximately 965°C (1769°F). Used as a semiconductor, and in electronics, materials research and ceramics. Formula: GaS; (ii) *Gallium sesquisulfide.* Yellow, moisture-sensitive crystals or white amorphous powder. Density, 3.5-3.65 g/cm³; melting point, approximately 1248°C (2278°F). Used in ceramics and materials research. Formula: Ga_2S_3; and (iii) *Gallium subsulfide.* Green crystals, or black powder. Density, 4.22 g/cm³. Used in ceramics and materials research. Formula: Ga_2S.

gallium telluride. A compound of gallium and tellurium. Crystal system, cubic. Crystal structure, sphalerite. Density, 5.75 g/cm³; hardness, 2370 Knoop; melting point, 790°C (1454°F). Used as a semiconductor, and in electronics and materials research. Formula: Ga_2Te_3.

gallium-tin alloy. A low-melting alloy of gallium and tin, sometimes used as an electron carrier in silicon semiconductor devices.

gallium trichloride. White, moisture-sensitive powder or crystals (99.99+% pure). Density, 2.47 g/cm³; melting point, 77.9°C (172.2°F); boiling point, 201.3°C (235°F). Used in electronics and materials research. Formula: $GaCl_3$.

gallium triethyl. See triethylgallium.

gallium trimethyl. See trimethylgallium.

gallopheomelanins. A group of red-brown *pheomelanin* pigments obtained from chicken feathers. They are macromolecules occurring in the uncombined state, or as protein complexes. See also *melanin*.

G-Alloy. (1) Trade name of American Smelting & Refining Company for a lead-base babbitt used for bearings.

(2) Trade name of the National Physical Laboratory (UK) for a lightweight alloy containing 18% zinc, 2.5% copper, 0.35% magnesium, 0.35% manganese, 0.2% iron, 0.75% silicon, and the balance aluminum. Used for cast truck, airplane and boat parts.

Galloy. Trade name of Southern Dental Industries Limited (Australia) for a non-palladium gallium-based dental restorative amalgam alloy produced by mixing a liquid quaternary alloy composed of 62% gallium, 25% indium, 13% tin and a trace of bismuth with a silver-based spherical alloy powder containing 28.05% tin, 11.8% copper and 0.05% platinum. It has a melting point of about 10°C (50°F), relatively low setting expansion, good creep resistance and compressive strength, and is used as a filler for posterior and other restorations.

Gall-Tough. Trademark of Carpenter Technology Corporation (USA) for a series of austenitic stainless steels with outstanding galling and metal-to-metal wear resistance, high strength and oxidation resistance, and good corrosion resistance. Used for pump and valve fittings, fasteners, and chain-link conveyors belts. *Gall-Tough Plus* refers to austenitic stainless steels with similar properties and applications, supplied in the annealed, or annealed and cold-worked (10-30%) condition.

Galorn. Trade name of George Morrell Corporation (USA) for several casein plastics.

Galvabec. Trademark of BMF Group Inc. (Canada) for hot-dip galvanized products.

Galvallia. Trade name of Sollac-Usinor (France) for a sheet steel with a mat gray appearance which after galvanizing is heated to form a zinc-iron alloy surface layer (about 10% iron). Supplied in typical thicknesses from 0.5-3.0 mm (0.02-0.12 in.) and in widths of 0.6-1.5 m (2.0-4.9 ft.), or 0.5-1.8 mm (0.02-0.07 in.) and 0.6-1.8 m (2.0-5.9 ft.), respectively. It has excellent spot weldability and good corrosion resistance and formability. Used in the automotive industries for structural and autobody applications, and for gas-pump casings and ski bindings.

Galvalloy. Trade name of Metalloy Products Company (USA) for a lead-tin alloy that contains some zinc, and is used as a flux-eliminating solder for aluminum and aluminum alloys.

Galvalume. (1) Trade name of Bethlehem International Engineering Corporation (USA) a coating containing 55% aluminum, 43.5% zinc and 1.5% silicon. It has a microstructure consisting of silicon needles in an aluminum-zinc matrix. Used as a protective coating on iron and steel.

(2) Trademark of Dofasco Inc. (Canada) for sheet steels coated with aluminum-zinc alloy.

Galvamatt. Trade name of British Steel Corporation (UK) for galvanized sheet steel containing 0.2% carbon. It has a mat surface appearance, and is used for metal signs, and stoves.

Galvan. Trademark of Cabot Corporation (USA) for electrically conductive carbon black.

Gal-Van-Alloy. Trade name for galvanized sheet steel on which a protective zinc coating has been developed by hot dipping. It is characterized by the absence of spangle, and can be worked without crazing or cracking of the coating. See also hot-dip galvanized coating.

Galvanax. Trade name of Alvin Products Inc. (USA) for an aerosol-packaged, zinc-rich cold galvanizer.

Galvanite. Trademark of Kapp Alloy & Wire Inc. (USA) for a lead-free solder that forms a metallurgical bond with steel, and is thus used for repairing galvanized steel surfaces. Supplied in rod form, it has a density of 6.95 g/cm³ (0.251 lb/in.³), a melting range of 200-300°C (390-570°F), a hardness of 100 Brinell, good ductility, and excellent cathodic and barrier protection.

galvanized coatings. A class of coatings produced by depositing zinc on the surface of metals, such as iron or steel, using any of various processes including hot-dip galvanizing, electrogalvanizing (electrodeposited zinc coating), zinc spraying, sherardizing and mechanical galvanizing. Also known as *zinc coatings*.

galvanized iron. Sheet iron with a surface coating composed of zinc applied by dipping in molten zinc, or by electrodeposition. It has good corrosion and rust resistance, and is used for culverts, roofing and sheathing. Abbreviation: GI. Also known as *zinc-coated iron*.

galvanized sheet. Sheet steel, usually of the low-carbon type and in coil form, coated with zinc, and supplied in various grades and sizes. Also known as *zinc-coated sheet*. See also hot-dipped zinc coating.

galvanized steel. Low-carbon or copper-bearing steel coated with a thin layer of zinc by dipping in a molten zinc bath, or by electrodeposition. In application, the zinc preferentially corrodes and protects the steel. Abbreviation: GS. Also known as *zinc-coated steel*.

galvanized steel wire. Plain, twisted or barbed steel wire coated with zinc by dipping in molten zinc, or by continuous electrodeposition. It has good corrosion and rust resistance. Also known as *zinc-coated steel wire*.

Galvannealed. Trade name of US Steel Corporation for a low-carbon steel sheet with remelted zinc coating.

galvannealed sheet. Sheet steel on whose surface a zinc-iron alloy coating has been produced, usually by hot dipping. It is free from spangle, and can be painted. However, the alloy may be somewhat brittle and thus may powder during subsequent forming operations. See also hot-dipped zinc coating.

galvannealed steel. Steel on whose surface a special zinc-iron coating has been deposited. The coating is being kept molten following hot dipping in order to allow the zinc to alloy completely with the substrate. See also hot-dipped zinc coating.

Galva-One. Trade name of US Steel Corporation for a low-carbon electrogalvanized steel sheet coated with zinc on one side only.

Galvaprime. Trade name of British Steel plc (UK) for a galvanized and prepainted *mild steel* used for roofing and cladding applications.

Galvaseal. Trademark of Pennwalt Corporation (USA) for a thin chromate conversion coating for galvanized steel applied by immersion, spraying, or roller coating. It has good salt-spray resistance and eliminates white rust formation. Used as an undercoating.

Galvatite. Trade name of British Steel plc (UK) for a hot-dip galvanized low-carbon steel.

Galvilite. Trademark of ZRC Worlwide (USA) for a zinc-rich galvanizing compound supplied in spray cans for repairing galvanized metal parts and rustproofing welds.

Galvobrite. Trade name of American Smelting & Refining Company (USA) for a zinc alloy used in the galvanizing industry to improve the fluidity of metallic zinc.

Galvomag. Trademark of Dow Chemical Company (USA) for extruded magnesium anodes containing up to 0.01% aluminum, 0.5-1.3% manganese, and up to 0.02% copper, 0.001% nickel and 0.03% iron, respectively.

Galvorod. Trademark of Dow Chemical Company (USA) for extruded magnesium anodes containing 2.5-3.5% aluminum, 0.7-1.3% zinc, 0.2% or more manganese, and up to 0.05% silicon, 0.01% copper, 0.001% nickel and 0.002% iron, respectively.

Galv-Weld. Trade name of Galv-Weld Inc. (USA) for a low-melting zinc-base alloy containing varying amounts of lead, tin and bismuth. Available in the form of sticks, it is used as a solder for regalvanizing welds.

Gama. Trade name of Gama Aluminium Ltd. (UK) for a low-friction aluminum alloy containing 12% copper, 2% nickel, 0.5% iron and 0.5% silicon. Used for pistons.

gamma alumina. Aluminum oxide (Al_2O_3) in the form of a white, ultrafine, high-purity powder with cubic crystals. It has an average density of 3.6 g/cm³ (0.13 lb/in³), and a hardness of 8 Mohs or above. Used as an abrasive for lapping, grinding and metallographic polishing. It is also available in the form of white or grayish pellets or powders (97.5+% pure) for use as a drying agent, as an activated or neutral catalyst support, and in special grades for hydration or dehydration reactions. Abbreviation: γ-Al_2O_3. Also known as *gamma aluminum oxide (GAO)*. See also alumina.

gamma aluminum oxide. See gamma alumina.

gamma boron. See boron.

gamma brass. A copper-zinc alloy that contains more than 45% zinc and has very poor hot and cold workability. Abbreviation: γ-brass. See also brass.

gamma bronze. See gamma metal.

gamma cellulose. A cellulose produced by acidification of a solution of *beta cellulose* in 17.5% aqueous sodium hydroxide. Abbreviation: γ-cellulose. See also cellulose.

gamma cerium. See cerium.

Gamma Columbium. Trade name of Universal Cyclops (USA) for a heat-resistant steel containing 0.4% carbon, 1% manganese, 1% silicon, 15% chromium, 25% nickel, 2-4% molybdenum, 4% columbium (niobium), and the balance iron. Used for supercharges, wheels, and aircraft-engine components.

gamma dysprosium. See dysprosium.

gamma iron. See iron.

gamma lanthanum. See lanthanum.

gamma manganese. See manganese.

gamma metal. An alloy of copper and tin. Also known as *gamma bronze.*

gamma neptunium. See neptunium.

Gamma Nickel Steel. Trade name of Midvale-Heppenstall Company (USA) for an alloy containing 64% iron and 36% nickel. It has low thermal expansion, and is used for chronometers and other instruments.

gamma plutonium. See plutonium.

gamma selenium. See selenium.

Gammasox. Trademark of A&P Technology (USA) for sleevings braided with high-modulus carbon fibers and supplied as light, medium and heavy fabric grades. Used as reinforcements in composites.

gamma terbium. See terbium.

gamma thallium. See thallium.

gamma titanium aluminide alloys. A group of intermetallic alloys composed of gamma titanium aluminide (γ-TiAl) with high melting point, low density and good to excellent specific strength, creep strength and oxidation resistance. Used for aircraft and spacecraft engine and nozzle components, missile parts, thermal-protection system panels, and fire barriers. Abbreviation: γ-TiAl alloys. See also titanium aluminide.

gamma uranium. One of the three allotropic forms of *uranium* (the other two being alpha and beta uranium) which has a body-centered cubic (bcc) crystal structure, is stable above 775°C (1427°F) and melts at 1132°C (2070°F). Abbreviation: γ-U.

gamma ytterbium. See ytterbium.

gangaw. The hard, heavy, tough, pinkish wood of the ironwood tree *Mesua ferrea.* Source: Burma. Used as lumber for construction applications. See also ironwood.

Ganisand. Trademark of Quigley Furnace Specialties Company, Inc. (USA) for *ganister* with a melting point of about 1800°C (3270°F), used as a refractory for lining furnaces.

ganister. (1) A mixture of crushed or ground silica rock (e.g., quartz) or firebrick with clay. Used for tamped furnace linings.

(2) A highly refractory, fine-grained metamorphic and sedimentary rock composed almost entirely of silica (98+% SiO_2). A low percentage of impurities, such as iron oxide (Fe_2O_3), aluminum oxide (Al_2O_3), calcium oxide (CaO) and magnesium oxide (MgO), are also present. Its microstructure consists predominantly of quartz grains bonded by silica. Used in the manufacture of silica brick, refractories, abrasives, linings for tube mills, grinding pebbles and building road-paving stone. Also known as *quartzite.*

Gannaloy. Trade name of Midvale-Heppenstall Company (USA) for a corrosion- and heat-resistant steel containing 0.03% carbon, 1.4% manganese, 0.4% silicon, 5.5% chromium, 24.5% nickel, 2.25% titanium, 0.68% aluminum, 0.003% boron, and the balance iron.

ganomalite. A light yellowish mineral of the apatite group composed of calcium lead silicate hydroxide, $Ca_4Pb_6(Si_2O_7)_3(OH)_2$. Crystal system, hexagonal. Refractive index, 1.910. Occurrence: Sweden.

ganophyllite. A light brown mineral composed of potassium sodium manganese aluminum silicate hydroxide hydrate, $(K,Na,Ca)_2Mn_8(Si,Al)_{12}(O,OH)_{32}(OH)_4 \cdot 8H_2O$. Crystal system, monoclinic. Density, 2.84 g/cm³; refractive index, 1.611. Occurrence:

Sweden.

Gantrez. Trademark of GAF Corporation (USA) for a series of alternating copolymers of methyl vinyl ether and maleic anhydride with an average molecular weight of 20000 or above. Used in coatings for metals, in adhesives, and as protective colloids and thickeners.

Gapex. Trade name of Ferro Corporation (USA) for fiberglass-reinforced polypropylene compounds with good thermal and ultraviolet stability.

gap-filling adhesive. An adhesive that, when placed between two adherends, can bridge the existing space.

gap-graded aggregate. Aggregate that contains both large and small particles, but none or almost none of certain intermediate sizes.

gar-alloy. A bluish-gray zinc-base alloy that contains small amounts of copper and silver, has high cold workability, and can be buffed to a high polish. Used as an inexpensive *pewter* substitute.

Garan. Trade name of Manville Corporation (USA) for fiberglass rovings and yarns.

Garanized. Trade name of Manville Corporation (USA) for glass fibers and glass-fiber products.

Garanmat. Trade name Manville Corporation (USA) for a fiberglass mat having a *Garan* finish.

Garant. Trade name of Haeckerstahl GmbH (Germany) for a series of tungsten and cobalt-tungsten high-speed steels used for lathe and planer tools, drills, and various other machining and finishing tools.

garavellite. A gray mineral with a brown-olive tint composed of iron antimony bismuth sulfide, $FeSbBiS_4$. Crystal system, orthorhombic. Density, 5.65 g/cm³. Occurrence: Italy.

Garba. Trade name of Ekstrand & Tholand Company (USA) for a water-hardening tool steel that contains 1% carbon, and the balance iron. Used for taps, drills, and drill rod.

Garbel. Trade name of Pechiney SA (France) for polyvinyl chloride polymers and copolymers.

Garden Blok. Trade name of Perma Paving Stone Company (Canada) for architectural stone that has one surface with a texture resembling natural stone. It is supplied in regular units, 200 × 200 × 112.5 mm (8 × 8 × 4.5 in.) in size, and in outside and inside tapered units in several colors including natural, charcoal, brown and sandstone. Used for retaining walls.

Gardenlite. Czech trade name for a patterned glass with flower design.

garden tile. A molded tile, usually square, used for stepping-stones in patios and gardens.

Gardglas. Trade name of J. Starkie Gardner Limited (UK) for a laminated glass.

Gardlite. (1) Trade name of Gardlite Limited (UK) for synthetic resin made by the condensa-tion of a toluol sulfonamide and formaldehyde.

(2) Trademark of Reichhold Chemicals, Inc. (USA) for reinforced fiberglass panels.

Gardobond. Trademark of Chemetall, Oakite Products Inc. (USA) for a zinc phosphate conversion coating.

Gar-Dur. Trademark of Garland Manufacturing Company (USA) for a wear-resistant ultrahigh-molecular-weight (UHMW) polyethylene supplied in the form of bars, rods, slabs and tubes. It has a density of 0.94 g/cm³ (0.03 lb/in.³), a low coefficient of friction, and high abrasion, wear, corrosion and impact resistance.

Garfield. Trade name of Time Steel Service Inc. (USA) for a water-hardening tool steel (AISI type W7) containing 1% carbon, 0.3% manganese, 0.25% silicon, 0.5% chromium, 0.2% vanadium, and the balance iron. Used for hand and cutting tools.

garment leather. A term referring to leather obtained from the hides of cattle, horses, pigs, goats or sheep, and used for coats, jackets, hats and other articles of clothing, but excluding footwear.

garnet. See natural garnets; synthetic garnets.

garnet abrasives. Red or reddish abrasives based on natural garnets. It is harder and sharper than *flint*, and used as a popular all-purpose abrasive for hardwood, softwood, composition board, horn and various plastics. Available in various grit sizes ranging from very fine (240 mesh) to very coarse (24-36 mesh), they can be used by hand or with machines. They however tend to fracture so that their use is often limited to softer materials. *Garnet abrasives* are usually supplied on kraft, manila or cloth backing. See also garnet paper; natural garnets.

garnet-coated paper. See garnet paper.

garnet paper. A layer of crushed natural garnet tightly bonded to a strong kraft or manila paper or a cloth backing for durability. Used as an abrasive paper for sanding and grinding wood and plastics. Also known as *garnet-coated paper*. See also garnet abrasives; natural garnets.

garnierite. An apple-green or pale-green mineral of the kaolinite-serpentine composed of nickel magnesium silicate hydroxide, $(Ni,Mg)_3Si_2O_5(OH)_4$. Crystal system, monoclinic. Density, 3.20 g/cm^3; refractive, 1.63. Occurrence: Czech Republic. Used as an ore of nickel, and as a gemstone. Also known as *nepuite; noumeite*.

Garon. Trade name of Garon Inc. (USA) for glass fibers and glass-fiber products.

garrelsite. A colorless mineral composed of sodium barium borate silicate hydroxide, $NaBa_3Si_2B_7O_{16}(OH)_4$. It can also be made synthetically. Crystal system, monoclinic. Density, 3.68 g/cm^3; refractive index, 1.633. Occurrence: USA (Utah).

garronite. (1) A whitish mineral of the zeolite group composed of sodium calcium aluminum silicate hydrate, $Na_2Ca_5Al_{12}S_{20}O_{64}\cdot 27H_2O$. Crystal system, tetragonal. Density, 2.15; refractive index, 1.507. Occurrence: Ireland.

(2) A white mineral of the zeolite group composed of calcium aluminum silicate hydrate, $Ca_3(Si,Al)_{16}O_{32}\cdot 13H_2O$. It can also be made synthetically. Crystal system, tetragonal. Density, 2.16 g/cm^3; refractive index, 1.51. Occurrence: Iceland, Ireland.

garspar. A substitute for *feldspar* composed of a mixture of finely ground glass and quartz and obtained as a byproduct in the grinding and polishing of *plate glass*. Used as filler in hard rubber and for ceramic applications.

Garware. Trademark of Garware Nylons Limited (India) for nylon 6 fibers and filament yarns.

garyansellite. A brown mineral of the phosphoferrite group composed of iron magnesium phosphate hydroxide hydrate, $(Mg,Fe)_3(PO_4)_2(OH)_{1.5}\cdot 1.5H_2O$. Crystal system, orthorhombic. Density, 3.16 g/cm^3; refractive index, 1.757. Occurrence: Canada (Yukon).

gasarite eutectics. Porous structures that have elongated pores filled with hydrogen and are produced by directional solidification of metal-hydrogen eutectic alloys. These materials have good mechanical properties at reduced weights. Also known as *gasar materials; lotus-type porous metals*. See also eutectic alloy.

gasar materials. See gasar eutectics.

gas black. A smooth, finely divided, jet-black substance composed of amorphous elemental carbon, and made by cracking or burning methane or natural gas with an insufficient supply of air. Used as a pigment and colorant for rubber, plastics and concrete, and as a reinforcing agent in rubber products. See also carbon black; channel black; jet black.

gas carbon. Carbon obtained as a byproduct in the production of coal gas by destructive distillation of bituminous coals in closed retorts. Used in the manufacture of carbon electrodes and graphitic crucibles.

gas-carburized steel. A low-carbon steel with a hard surface layer produced by a case-hardening process involving the introduction of carbon into the steel surface from hydrocarbon gases, such as natural gas or propane, hydrocarbons plus carbon monoxide, or easily vaporized hydrocarbon liquids, such as terpenes, benzenes, alcohols or glycols.

Gas-Chrom. Trademark of Applied Science Laboratories, Inc. (USA) for diatomaceous earths supplied in flux-calcined, acid-washed and screened grades, and used as supports in gas chromatography. See also diatomaceous earth.

gas concrete. See aerated concrete.

gas-cyanided steel. See carbonitrided steel.

Gas Engine Babbitt. Trade name of Hoyt Metal Company of London Limited (UK) for a babbitt composed of antimony, tin and lead, and used for gas-engine bearings.

gas-formed concrete. See aerated concrete.

gas hydrates. See clathrate compounds (2).

Gasite. Trade name of Georgia Iron Works Company (USA) for a corrosion-resistant white cast iron containing 3-3.6% carbon, 4-4.75% nickel, 1.4-3.5% chromium, 0.4-0.7% silicon, and the balance iron. Used for roller-bearing races, heavy cams, and dies.

gasket materials. Flat, relatively deformable pieces of asbestos, leather, rubber, metal or waxed paper cut, formed or molded into desired shapes and placed between mating parts to prevent leakage.

gasket leather. Highly stuffed, chrome- and/or vegetable-tanned cattlehide leather suitable for the manufacture of gaskets for pump valves, and for piston packing applications.

gas-nitrided steel. An alloy steel subjected to a case-hardening process that involves heating in an electric furnace at about 500-520°C (930-970°F) in an atmosphere containing nitrogen (usually ammonia gas, NH_3) for about 12-96 hours during which the nitrogen diffuses into the surface layer (or case) of the steel.

gaspeite. (1) A green mineral of the calcite group composed of nickel magnesium carbonate, $(Ni_{0.49}Mg_{0.43}Fe_{0.08})CO_3$. Crystal system, hexagonal. Density, 3.71 g/cm^3; refractive index, 1.83. Occurrence: Canada (Quebec).

(2) A green mineral of the calcite group composed of nickel carbonate, $NiCO_3$. Crystal system, rhombohedral (hexagonal). It can also be made synthetically. Density, 3.71 g/cm^3; refractive index, 1.83. Occurrence: Canada (Quebec).

Gastex. Trade name for gas black used as a reinforcing agent and colorant in rubber.

Gatron. Trademark of Gatron Industries Limited (Pakistan) for polyester fibers and filament yarns.

gattar. A lustrous, drapable fabric produced in solid colors with silk warp and cotton weft threads. Used for luxurious evening wear.

gatumbaite. A pure, white mineral composed of calcium aluminum phosphate hydroxide monohydrate, $CaAl_2(PO_4)_2(OH)_2\cdot H_2O$. Crystal system, monoclinic. Density, 2.92 g/cm^3; refractive index, 1.63. Occurrence: Africa (Rwanda).

gaudefroyite. A black mineral composed of calcium manganese borate carbonate hydroxide, $Ca_4Mn_{3-x}(BO_3)_3(CO_3)(O,OH)_3$. Crystal system, hexagonal. Density, 3.35 g/cm³; refractive index, 1.81. Occurrence: Morocco.

gauged brick. A brick that has been made to accurate dimensions, e.g., by grinding or cutting. Also known as *gaged brick*.

gauged mortar. See lime-cement mortar.

gauge steel. A general trade term for hardened and ground plain-carbon steel containing 1% carbon. Also known as *gaged steel*.

gauging plaster. A finish-coat plaster composed of calcined gypsum plaster and lime. Also known as *gaging plaster*.

Gaussit. A German trade name for wrought cobalt-iron-vanadium alloys used for permanent magnets, motors, and computers.

gauze. (1) A very thin, light, loosely woven fabric of cotton, silk, or other fibers used for surgical dressings.

(2) A net-like structure formed by a series of crossed metal wires or plastic fibers. See also mesh.

Gaydon. Trade name of John Dickinson & Company (UK) for phenol- and melamine-formaldehyde resins and plastics.

gaylussite. A white to yellowish-white, transparent mineral with a vitreous luster. It is composed of sodium calcium carbonate pentahydrate, $Na_2Ca(CO_3)_2 \cdot 5H_2O$. Crystal system, monoclinic. Density, 1.94 g/cm³; refractive index, 1.516. Occurrence: Central Africa.

G Babbitt. Trade name of American Smelting & Refining Company (USA) for a heavy-duty babbitt of 83.75% lead, 12.5% antimony, 3% arsenic and 0.75% tin. It has a melting range of 252-287°C (486-549°F), and is used for bearings, and in the manufacture of other babbitts.

GBE. Trademark of Integran Technologies Inc. (USA) for several *grain-boundary-engineered materials*.

GBEST. Trademark of Integran Technologies Inc. (USA) for metals and alloys whose surfaces have been treated to produce grain-boundary-engineered structures. See also grain-boundary-engineered materials.

G bronze. See admiralty gunmetal.

G-Cast. Trademark of Degussa-Ney Dental (USA) for a hard, microfine dental alloy (ADA type III) composed of 50% gold, 32% silver and 5% palladium. It has low yield strength, a hardness of 235 Vickers, and is used in restorative dentistry for inlays, single crowns, and short-span bridgework.

GC Dentin Cement. Trade name of GC America Inc. (USA) for a glass-ionomer dental cement used for dentin bonding applications.

GC Reline. Trade name of GC America Inc. (USA) for flowable acrylic resins used for relining hard dentures. They do not contain methyl methacrylate (MMA) and self-cure intra-orally.

GC Reline Soft/Extra Soft. Trade name of GC America Inc. (USA) for dimensionally stable vinyl polysiloxane-based dental relining materials.

Geadur. Trademark of AEG Aktiengesellschaft (Germany) for several thermosetting resins based on epoxies or unsaturated polyesters.

Geaplast. Trademark of AEG Aktiengesellschaft (Germany) for several epoxy and unsaturated polyester plastics.

Geapox. Trademark of AEG Aktiengesellschaft (Germany) for several epoxy resins and adhesives.

gear bronze. (1) A term referring to any casting bronze suitable for the manufacture of gears, pinions and worm wheels.

(2) A tough, leaded tin casting bronze containing about 78-91.5% copper, 8.5-13% tin, 0-3% zinc, 0.2-2% lead, and 0.1-0.3% phosphorus. It has high strength, good machinability, and a low coefficient of friction. Used for gears and worm wheels.

gear-crimped yarn. A texturized yarn produced by passing the heated yarn through a crimping device, such as a gear-wheel mechanism.

Gearite. Trade name of Swift Levick & Sons Limited (UK) for an alloy steel containing 0.31% carbon, 1.2% chromium, 4.2% nickel, and the balance iron. Used for gears, pinions, and machine parts.

gearksutite. A white mineral composed of calcium aluminum fluoride hydroxide monohydrate, $CaAl(F,OH)_5 \cdot H_2O$. Density, 2.77 g/cm³; refractive index, 1.454. Occurrence: Greenland.

Gearmac. Trade name of Swift Levick & Sons Limited (UK) for an alloy steel containing 0.31% carbon, 0.7% chromium, 2.5% nickel, 0.6% molybdenum, and the balance iron. Used for gears, pinions, and machine parts.

GearMet. Trademark of QuesTek Innovations LLC (USA) for a series of gear and bearing steels used for aerospace and automotive applications. *GearMet C-61* is a martensitic steel containing 0.16% carbon (core), 18% cobalt, 9.5% nickel, 3.5% chromium, 1.1% molybdenum, 0.08% vanadium, and the balance iron. It has high core strength and toughness, a core hardness of 53 Rockwell C, a surface hardness exceeding 60 Rockwell C, and a plane-strain fracture toughness (K_{IC}) exceeding 77 MPa-m$^{1/2}$ (70 ksi-in.$^{1/2}$). It is used especially for automotive gears (e.g., racing cars). *GearMet C-69* with slightly different composition has high core hardness (50 Rockwell C) and very high surface hardness (70 Rockwell C) for wear and fatigue resistance, but a somewhat lower plane-strain fracture toughness (K_{IC}) of 55 MPa-m$^{1/2}$ (50 ksi-in.$^{1/2}$). It is used especially for aircraft and helicopter transmissions. *GearMet* alloys are precipitation-hardened by metal carbides, such as chromium carbide (Cr_2C), molybdenum carbide (Mo_2C) and/or vanadium carbide (V_2C).

Gearol. Trade name of Swift Levick & Sons Limited (UK) for an alloy steel containing 0.4% carbon, 1.2% chromium, 1.5% nickel, 0.3% molybdenum, and the balance iron. Used for gears, pinions, and machine parts.

Gear Steel. British trade name for an oil-hardening steel containing 0.45-0.5% carbon, 3.5% nickel, 1.5% chromium, and the balance iron. Used for gears and pinions.

gear steels. A group of steels suitable for use in the manufacture of gears, transmissions and gear components and including in particular austenitic stainless steels (AISI 300 Series), ferritic and martensitic stainless steels (AISI 400 Series), plain-carbon steels (e.g., AISI 1020, 1045, 1116 and 1140), nickel alloy steels (e.g.,, AISI 2512), chromium-molybdenum alloy steels (e.g., AISI 4130), nickel-chromium-molybdenum alloy steels (e.g., AISI 4340, 8615, 8620 and 9310) and nickel-molybdenum alloy steels (e.g., AISI 4615 and 4620).

gear wheel bronze. A tin casting bronze containing 86.2-88.9% copper, 11-12% tin, up to 1.5% zinc, and 0.1-0.3% phosphorus. Used for high-duty worm wheels working with hardened steel worms.

Geax. Trademark of AEG Aktiengesellschaft (Germany) for a series of epoxy-, phenol-formaldehyde- and melamine-formaldehyde-based resins.

gebhardite. A brown mineral composed of lead chloride oxide arsenite, $Pb_8OCl_6(As_2O_5)_2$. Crystal system, monoclinic. Density, 6.07 g/cm³. Occurrence: Namibia.

Geberit. Trade name of Geberit GmbH (Germany) for polyethylene plastics.

Gecalloy. Trade name of General Electric Company (USA) for a magnetic powder containing principally cobalt and iron.

Gecesa. Trade name for rayon fibers and yarns.

Gecet. Trademark of GE Plastics (USA) for a high-memory thermoreactive polyurethane foam used especially for medical and biotechnology applications, and for bicycle helmets.

Gecor. Trade name for a hard-magnetic, cobalt-base rare-earth compound that is magnetically anisotropic, has an extremely high magnetic energy product and very high coercive force. Used for permanent magnets.

Gedelite. Trademark of Atofina SA (France) for phenol-formaldehyde resins and plastics.

Gedex. Trademark of Atofina SA (France) for a series of polystyrenes.

Gedexal. Trademark of Atofina SA (France) for polystyrene foams and foam products.

Gedge's metal. A British casting brass composed of 60% copper, 38.2% zinc and 1.8% iron. It is malleable at red heat, and used for cylinders of hydraulic presses, and ship sheathing.

Gedopal. Trademark for phenol-formaldehyde resins and plastics.

gedrite. A dark green to wine-yellow mineral of the amphibole group composed of iron magnesium aluminum silicate hydroxide, $(Fe,Mg,Al)_7Al_2Si_6O_{22}(OH)_2$. Crystal system, orthorhombic. Density, 3.37 g/cm^3; refractive index, 1.681. Occurrence: USA (Maine).

geerite. A black mineral of the sphalerite group composed of copper sulfide, $Cu_{1.60}S$. Crystal system, cubic. Density, 5.61 g/cm^3. Occurrence: USA (New York).

gedu nohor. See Tiama mahogany.

geffroyite. A brown mineral with a creamy tint belonging to the pentlandite group. It is composed of copper iron silver selenide sulfide, $(Cu,Fe,Ag)_9(Se,S)_8$. Crystal system, cubic. Density, 5.39 g/cm^3. Occurrence: France.

Gehaplast. Trade name of Gebrüder Halbert (Germany) for acrylonitrile-butadiene styrenes.

gehlenite. A colorless or green mineral of the melilite group composed of calcium aluminum silicate, $Ca_2Al_2SiO_7$. It can also be made synthetically from calcium carbonate, aluminum oxide and silicon dioxide. Crystal system, tetragonal. Density, 3.04 g/cm^3. Occurrence: USA (California).

geikielite. A colorless, or bluish to brownish-black mineral of the corundum group composed of magnesium titanate, $MgTiO_3$. Crystal system, rhombohedral (hexagonal). Density, 4.05 g/cm^3; refractive index, 2.31. Occurrence: Sri Lanka.

Geilon. Trade name for nylon fibers and yarns used for textile fabrics including clothing.

Geko. Trademark of Süd-Chemie AG (Germany) for activated *bentonite*.

gel. A substance formed from a colloidal solution by coagulation (e.g., evaporation, cooling or precipitation). It can be jelly-like and elastic (e.g., gelatin), or semirigid to rigid (e.g., silica gel).

Gelamide 250. Trade name of Polysciences Inc. (USA) for a linear polyacrylamide powder usually diluted to a 1% solution in water for use in biochemistry (e.g., electrofocusing).

gelatin. A hard, transparent, crosslinkable collagen-derived mixture of proteins obtained by boiling animal connective tissues, such as skins, hoofs and horns. It can be treated to high purity, swells in water, and forms reversible gels of high strength and viscosity. Used as a coating for artificial heart bladders, for capsules of medicinals, as a blood-plasma volume expander, in biotechnology, and in photographic film, sizing, textile and paper adhesives, cements and light filters. See also collagen.

gelatin fibers. Glossy fibers spun from a *gelatin* solution and then made water-insoluble by treating with formaldehyde.

gelatinous aluminum hydroxide. See alumina gel.

GelBond. Trademark of FMC Corporation (USA) for a 0.2 mm (0.04 in.) thick plastic support film used in polyacrylamide gel electrophoresis (PAGE).

gel cement. A cement mix with gel-like consistency made by adding a small quantity of *bentonite* clay to cement in order to increase the ratio of cement to water, improve the homogeneity, and decrease water losses.

gel coat. (1) A quick-setting synthetic resin used in the manufacture of certain laminates to improve bonding and surface characteristics. It is applied to the mold and gelled prior to wet lay-up or resin-transfer molding.

(2) A specially formulated thermoplastic polyester resin with pigments and fillers that provides a smooth, pore-free surface for fiberglass parts.

gel-dyed fibers. Wet-spun synthetic fibers, usually of the acrylic or modacrylic type, colored with soluble dyes after extrusion while in the gel state.

Gel-Kote. Trade name of Glidden Company (USA) for unsaturated polyesters used for gel-coat applications. See also gel coat.

gellant. See gelling agent.

gelling agent. A substance, such as dextran, bentonite clay, or a synthetic resin, added to a liquid, solution or slurry to aid in gel formation. Also known as *gellant*. See also gel.

Gelo. Trademark of Gelo Holz- und Kunststoff GmbH (Germany) for sawn lumber, and various lumber products.

Gelon. Trademark of GE Plastics (USA) for a series of polyamide (nylon) resins including nylon 6 (*Gelon B*), nylon 6,6 (*Gelon A*) and nylon 6,66 (*Gelon C*).

Geloy. Trademark of GE Plastics (USA) for thermoformable acrylonitrile-styrene acrylate (ASA) resins as well as ASA-polycarbonate (ASA-PC) and ASA-polyvinyl chloride (ASA-PVC) blends. The ASA resins are supplied in unfilled extrusion and blow-molding grades, and the ASA-PC and ASA-PVC blends in unfilled injection-molding and extrusion grades. *Geloy* resins and blends have good impact and heat resistance, good weatherability and good resistance to corrosive environments including chlorinated solvents and salt water. Used for sports and recreational equipment, and exterior automotive components.

gel paint. See thixotropic paint.

Gelva. Trademark of Monsanto Chemical Company (USA) for a series of vinyl acetate polymers with a melting range of approximately 65-195°C (149-383°F) and poor resistance to alcohol and toluol. Used for coatings, hot-melt adhesives, binders, slush moldings, paints, textile sizes, and in papermaking.

Gemax. (1) Trade name of Esperanza SA (Spain) for glass lenses with concave surfaces and net-like patterns used for vertical glazing.

(2) Trade name of Gemax Glass AB (Sweden) for a laminated glass.

(3) Trademark of GE Plastics (USA) for a series of thermoplastic polybutylene terephthalate/polyphenylene ether (PBT/PPE) alloys for automotive applications, such as body hardware, doors and side panels.

Gem-Base. Trade name of Dental Composite Limited (UK) for glass-ionomer dental base cements. *Gem-Base LC* is a light-cure, resin-modified version.

Gem-Bond. Trade name of Dental Composite Limited (UK) for a dual-cure dental resin cement.

Gem-CC1. Trade name of Dental Composite Limited (UK) for a self-cure dental hybrid composite.

Gem-Cem. Trade name of Dental Composite Limited (UK) for a glass-ionomer dental cement.

Gemco. Trade name of GTE Products Corporation (USA) for a high-temperature brazing alloy of 87.75% copper, 12% germanium and 0.25% nickel. It is commercially available in the form of foil, powder, flexibraze, wire, extrudable paste and preform, and has a liquidus temperature of 965°C (1769°F) and solidus temperature of 820°C (1508°F).

Gem-Core. Trade name of Dental Composite Limited (UK) for a glass-ionomer/cermet cement used in restorative dentistry for core-build-ups.

Gem-Fil. Trade name of Dental Composite Limited (UK) for a glass-ionomer dental filling cement.

Gemguard. Trade name of Safetee Glass Company Inc. (USA) for laminated glass units with heavy plastic interlayer used particularly in jewelry stores.

Geminol. Trade name of Driver Harris Company (USA) for nickel-base alloys used for nuclear-grade thermocouples. *Geminol N* is a nickel alloy containing 3% silicon, and *Geminol P* contains 79-80% nickel, 19-20% chromium and up to 1% silicon.

Gemini. (1) Trade name of MacDermid Inc. (USA) for a bright nickel electroplate and plating process.

(2) Trade name of Dentsply International (USA) for a corrosion-resistant nickel-chromium dental bonding alloy.

Gemite. Trademark of Gemite Unique Products Inc. (Canada) for fiber-reinforced cement composites.

Gemlike. Trade name of Gemloid Corporation (USA) for cellulose nitrate plastics.

Gemlite. Trade name of Gemloid Corporation (USA) for acrylic, cellulose acetate and polystyrene plastics.

Gem-Lite 1. Trade name of Dental Composite Limited (UK) for a light-cure dental hybrid composite.

Gemma. Trade name of Creusot-Loire (France) for a soft-magnetic alloy composed of 65% iron and 35% nickel. It has high initial permeability and low saturation induction.

Gemon. Trademark of GE Plastics (USA) for thermosetting polyimide molding compounds used for electrical, electronic, aerospace and other engineering applications.

Gem-Ortho. Trade name of Dental Composite Limited (UK) for a glass-ionomer cement for ortho-dontic applications.

Gempco. Trade name of General Metals Powder Company (USA) for a porous, self-lubricating powder-metallurgy alloy of tin, lead, graphite and copper. Used for friction materials, clutch disks, motors, and brushes.

Gem-Seal. Trade name of Dental Composite Limited (UK) for a glass-ionomer dental sealing cement.

Gemstone. Trade name of John A. Knoedler Company (USA) for phenolic resins and molded phenol-formaldehyde plastics.

Genal. Trade name of GE Plastics (USA) for thermosetting engineering phenol-formaldehyde plastics and molding compounds.

Genafol. Trademark of Hoechst AG, Kalle Werk (Germany) for a rigid polyvinyl chloride film.

Genalloy. Trade name of Alloy Engineering & Casting Company (USA) for a series of alloys containing up to 0.5% carbon, 31-40% nickel, 12-21% chromium, up to 1% silicon, and the balance iron. Used for heat- and corrosion-resistant parts.

Genappe yarn. A worsted yarn without surface fibers, produced by passing through a flame or over heated elements. Originally from Genappe, a town in Belgium.

Gencalloy. Trade name of Federated Genco Limited (Canada) for a corrosion-resistant lead alloy used for cable sheathing.

Gencivex. Trade name for a phenol-formaldehyde resin formerly used for denture bases.

Genco. Trade name of Federated Genco Limited (Canada) for aluminum ingots and zinc dust.

Genelite. Trade name of British Thomson Houston Company Limited (UK) for a porous, self-lubricating synthetic powder-metallurgy bronze containing 70-73% copper, 12-14% tin, 9-10% lead, and 4.5-5.5% graphite. Used for aerospace engine bearings.

GenEpoxy. Trademark of GE Plastics (USA) for a series of solid and liquid epoxy resins usually made from epichlorohydrin and a bisphenol. Used for structural adhesives, casting and laminating resins, sealants and coatings, and potting compounds for electronics.

General Alloys. Trade name of Alloy Engineering & Casting Company (USA) for a series of wrought and cast alloys including several corrosion- and/or heat-resistant wrought austenitic, ferritic and martensitic stainless steels, corrosion- and/or heat-resistant steel castings, and various *Hastelloy*-type nickel-alloy castings (e.g., nickel-molybdenum-iron, nickel-molybdenum-chromium-iron, nickel-silicon-iron and nickel-chromium-iron types). Used for chemical-process equipment, agitators, dryers, pickling equipment, pump and valve parts, furnace parts, and heat-treating equipment.

General Plate. Trade name of Texas Instruments Inc. (USA) for a series of silver brazing alloys composed of (i) silver, copper and zinc, (ii) silver, copper, zinc and cadmium, (iii) silver, copper, zinc and nickel, or (iv) silver, copper, zinc, cadmium and nickel.

Genesee. Trade name of Symington-Gould Corporation (USA) for a series of steels including various corrosion-resistant, heat-resistant, case-hardening, structural and machinery grades, and several corrosion-, heat- and/or wear-resistant cast irons.

Genesis. (1) Trade name J.F. Jelenko & Company (USA) for a light-cure silicone elastomer for dental impressions. *Genesis II* is a beryllium-free, ruthenium-bearing cobalt-chromium ceramic alloy used for bonding dental porcelains to metal.

(2) Trade name of Dentsply/Caulk (USA) for a visible-light-activated polyether urethane dimethacrylate (UDMA) dental impression material.

Genex. Trade name of Murex Limited (UK) for a low-carbon steel containing 0.02% carbon, 0.1% manganese, and the balance iron. Used for welding electrodes.

Gen-Flo-Latices. Trademark of Diversitech General (USA) for a series of styrene-butadiene emulsion polymers used for adhesives, latex paints, printing inks, and sizings.

Gen-Glaze. Trademark of Genesta Manufacturing Limited (Canada) for safety glazing.

genkinite. A pale brown or tan mineral with a yellowish tinge composed of antimony palladium platinum, $(Pt,Pd)_4Sb_3$. Crystal system, tetragonal. Density, 9.26 g/cm^3. Occurrence: South Africa.

Genoa silk. A ribbed silk fabric, somewhat resembling *corduroy* in appearance, used for upholstery and furnishings. It is named after the Italian city of Genoa.

Genolon. Trademark of Hoechst AG, Kalle Werk (Germany) for rigid polyvinyl chloride (PVC) and PVC copolymer films.

Genopac. Trademark of Hoechst AG, Kalle Werk (Germany) for a rigid polyvinyl chloride packaging film.

Genotherm. Trademark of Hoechst AG, Kalle Werk (Germany) for rigid polyvinyl chloride film with excellent chemical resistance, good resistance to water, oils and fats, and good flame retardancy. Used for packaging and furniture.

Genovita. Trade name for a phenol-formaldehyde resin formerly used for denture bases.

Gensil. Trademark of GE Plastics (USA) for several silicone rubber compounds.

Gensteel. Trade name of General Malleable Corporation (USA) for pearlitic malleable irons used for castings.

Gen-Tac Latex. Trademark of Diversitech General (USA) for a latex that contains varying amounts of vinyl pyridine, butadiene and styrene. Used as an adhesive for bonding textiles to rubber, e.g., in automotive tires.

Genthane. Trademark of Diversitech General (USA) for polyurethane elastomers used for packets, grommets and mechanical products.

genthelvite. A colorless to light green mineral of the helvite group composed of beryllium zinc sulfide silicate, $Zn_4Be_3(SiO_4)_3S$. Crystal system, tetragonal. Density, 3.66 g/cm^3; refractive index, 1.743. Occurrence: Canada (Quebec).

Gentro. Trademark of Diversitech General (USA) for a series of vulcanizable polymers that contain about 3 parts butadiene and 1 part styrene, and are frequently mixed with reinforcing agents and pigments. Used in the manufacture of tires, mechanical goods, extruded goods, and shoe heels and soles.

Gentro-Jet. Trademark of Diversitech General (USA) for styrene-butadiene rubber co-precipitated carbon blacks used in the manufacture of tires, mechanical goods, extruded goods, shoe heels and soles.

genuine babbitt. Any tin babbitt that approximates the original composition of 89.3% tin, 7.1% antimony and 3.6% copper. It has a density of 7.39 g/cm^3 (0.267 lb/in.³), high softness, a solidus temperature of 241°C (466°F), a liquidus temperature of 354°C (669°F), good antiscoring properties, conformability and corrosion resistance, and fair fatigue resistance. Used for bearings and bushings. See also babbitts.

genuine hand-made paper. A paper usually made wholly or partly from cotton or linen rags, usually laid-up, and not cut to size. It has a genuine deckle edge on all four sides.

Genuine Silverine. Trade name of Glacier Metal Company (USA) for a tin babbitt containing lead and antimony. Used for bearings.

Genuine Sovereign Babbitt. Trade name of Glacier Metal Company (USA) for a tin babbitt containing a total of 10-14% antimony and copper. It has a solidus temperature of about 306°C (583°F), and is used for internal combustion engine bearings.

Genuine Stubs. Trade name of Peter Stubs (UK) for a high-carbon steel containing 1.1% carbon, 0.35% manganese, 0.2% silicon, 0.045% sulfur and phosphorus, and the balance iron. Used for dowels and pins.

Genuine Wrought Iron. Trade name of A.M. Byers Company (USA) for a ductile, easily welded iron containing up to 0.05% carbon, up to 0.05% manganese, 0.1-0.15% silicon, 0.3% sulfur, 0.1-0.15% phosphorus, and the balance iron. Used for staybolts, rivets, roofing sheets, water and steam pipes, and boiler tubes.

Geocel. Trademark of Geocel Corporation (USA) for caulking and sealing materials based on acrylic resins, and used for sealing and waterproofing cracks and joints, as weatherproofing, and in vehicle construction.

Geocoustic. Trade name of Pittsburgh Corning Corporation (USA) for open cellular glass panels that are 50 mm (2 in.) thick, and 340 × 340 mm (13.5 × 13.5 in.) in size.

geocronite. A lead-gray mineral composed of lead antimony arsenic sulfide, $Pb_{14}(Sb,As)_6S_{23}$. Crystal system, monoclinic. Density, 6.46 g/cm^3. Occurrence: UK, USA (Utah, Idaho), Germany, Sweden, Ireland.

Geofilm. Trademark of Hughes-Owens Limited (Canada) for a drafting film.

Geolast. Trademark of Advanced Elastomer Systems (USA) for a series of black thermoplastic polypropylene/nitrile elastomers with excellent fluid resistance, good resistance to hot air and oils, good weldability, a hardness of 70 Shore A, and a processing temperature of 216°C (421°F). They can be processed into various shapes by blow and injection molding, or extrusion. Used for automotive, oilfield and industrial hose applications, as lining materials for tanks and containers, and as replacements for nitrile rubber and epichlorohydrin thermoset rubbers.

Geoline. Trade name of Agriplast Srl (Italy) for a weldable plastic film with excellent aging, tear and perforation resistance, and relatively large thickness (300-500 µm or 0.012-0.020 in.) used in building construction for waterproofing groundwork and roofs, in civil engineering, in water storage, and for vegetable crop protection.

geological materials. A broad class of materials that occur on or within the Earth. They were the first raw materials used by man to fashion tools, utensils and fuels and include stone, rock, flint, naturally occurring minerals, such as hematite, quartz, halite or fluorite, silicate glass, and mixtures, such as soil, kerogen, lignite and coal. Also known as *earth materials*.

Geomant. Trade name of Thyssen Edelstahlwerke AG (Germany) for a coated electrode of carbon steel used for overlaying coal plows.

Geomet. Trade name of Metal Coatings International Inc. (USA) for a thin, chromium-free, water-based coating consisting of overlapping zinc and aluminum flake bonded with an inorganic binder. It has good corrosion, solvent and thermal resistance, and is electrically conductive. Used as a replacement for chromium coatings, especially in automotive applications.

Geon. Trademark of The Geon Company (USA) for a series of plasticized (flexible) polyvinyl chloride (PVC) polymers available in three elongation ranges: 0-100%, 100-300% and above 300%. Used for cable insulation. Several chlorinated and unplasticized PVCs as well as a number of polyvinylidene chloride (PVDC) plastics are also included under this trademark. The chlorinated PVCs are known as *Geon CPVC* and chlorinated polyvinyl chlorides as *Geon PVDC*. *Geon UPVC* refers to the unplasticized polyvinyl chlorides which are supplied in a wide range of grades included compounded, rigid, crosslinked, high-temperature, low-gloss, opaque, transparent, UV-stabilized, glass fiber-reinforced, high-impact, structural-foam, lightweight, custom-injection molding, extrusion, latex and polyblend. *Geon UPVC* is used for automotive, building and construction applications, in the recreational and apparel industries, and in the wire and cable industries.

georgette. A light, fine, sheer crepe fabric, usually in a plain weave, having a dull pebbled or crinkled texture. Used especially for blouses and dresses. Also known as *crepe georgette*. See also crepe fabrics.

georgiadesite. A white or brownish-yellow mineral composed of lead oxide chloride arsenate hydroxide, $Pb_8(AsO_4)_2Cl_7O(OH)$.

Crystal system, monoclinic. Density, 6.30 g/cm³. Occurrence: Greece.

Georgian. Trade name of Pilkington Brothers Limited (UK) for wired plate and cast glass with square mesh.

Georgian glass. A cast or polished, fire-resistant building glass in which a wire mesh of square pattern is incorporated as reinforcement. Also known as *Georgian-wired glass.*

Georgia-Pacific. Trade name of Georgia-Pacific Corporation (USA) for crushed gypsum.

Georgia pine. See longleaf pine.

Georo. Trade name of GTE Products Corporation (USA) for an eutectic brazing alloy containing 88% gold and 12% germanium. Commercially available in extrudable paste, preform, flexibraze, foil, wire and powder form, it has a melting point of 356°C (673°F).

GeoSpan. Trademark of Plasti-Fab, Division of PFB Corporation (USA) for an insect- and vermin-resistant, compressible fill material composed of expanded polystyrene (EPS). It has a brown color with earthy tone, and is supplied precut with a standard length of 2.4 m (96 in.), in custom widths up to 1.2 m (48 in.), and custom thicknesses up to 0.6 m (24 in.). Used as a compressible medium under concrete grade beams and certain structural floor slabs, and against foundation walls.

Geotex. Trademark of Synthetic Industries, Inc. (USA) for woven and nonwoven geotextile fabrics based on synthetic materials, and used in road construction, erosion control and soil filtration.

geotextiles. A class of woven or permeable nonwoven textile fabrics manufactured from man-made synthetic fibers or yarns, and used especially in civil engineering and road construction as soil/aggregate membranes and reinforcements.

Geovel. Trade name of Hughes-Owens Limited (Canada) for drafting vellum.

GeoVoid. Trademark of Plasti-Fab, Division of PFB Corporation (USA) for an insect- and vermin-resistant compressible fill material composed of expanded polystyrene (EPS) similar to *GeoSpan*, but supplied in precut sections with a length of 2.4 m (96 in.) in custom thicknesses and depths. Used as a compressible medium under structural concrete floor slabs.

Gepolit. Trademark of CWH (Germany) for polyvinyl chloride copolymers.

Gepoxy. Trademark of GE Plastics (USA) for several epoxy-based adhesive compounds.

Gerfil. Trade name for polyolefin fibers and yarns used for textile fabrics.

Gerfus. Trade name for flexible polyvinyl chloride film materials.

Gerflex. Trade name of Gerflex Limited (UK) for flexible polyvinyl chloride film and sheeting for floor covering products.

gerhardtite. A dark to emerald green mineral with vitreous to brilliant luster composed of copper nitrate hydroxide, $Cu_2(OH)_3$-NO_3. Crystal system, orthorhombic. Density, 3.40 g/cm³; refractive index, 1.713; hardness, 2 Mohs. Occurrence: Central Africa, USA (Arizona).

Geril. Trade name for polyolefin fibers and yarns used for textile fabrics.

Geristore. Trade name of Den-Mat Corporation (USA) for a self-adhering, translucent, radiopaque hybrid ionomer-composite resin based on hydrophilic bisphenol A glycidyl methacrylate (BisGMA). It is dual-curing, polishable and biocompatible, bonds well to enamel, dentin, cementum, dental alloys and old amalgam, and has low thermal expansion and polymerization shrinkage. Used in restorative dentistry for core build-ups, cavity lining, luting, and pit and fissure sealing.

germane. A colorless combustible gas composed of germanium and hydrogen and available in electronic grades with high purity (99.997+%). It has a density of 3.43 g/L, a melting point of -165°C (-265°F), a boiling point of -88.4°C (-127°F), a critical temperature of 308K, and can ignite spontaneously in air. Used in the epitaxial chemical vapor deposition of germanium thin films for solar devices and light-emitting diodes, and as a precursor for germanium-containing thin films made by chemical vapor deposition or molecular-beam epitaxy. Formula: GeH_4. Also known as *germanium tetrahydride.*

German flange metal. A corrosion-resistant German brass containing 92% copper, 5% zinc and 3% tin, and used for flanges, fittings and pipes.

Germania bronze. A British bearing alloy composed of 79% zinc, 10% tin, 5% copper, 5% lead, and 1% iron.

germania. See germanium dioxide.

germanide. A compound of an alkaline metal (e.g., lithium, sodium or potassium) or alkaline earth metal (e.g., magnesium, calcium or strontium) with germanium.

germanium. A rare gray-white metallic element of Group IVA (Group 14) of the Periodic Table. It is commercially available in the form of semicircular, trapezoidal or circular rods, and as ingots, lumps, pieces, foil, microfoil, sheet, pure or doped powder, chips, pieces, or single crystals or slices. The single crystals are usually grown by the Czochralski method. It occurs only in a few minerals, such as *argyrodite*, and in *coal*. Crystal system, cubic. Crystal structure, diamond-cubic. Density, 5.323 g/cm³; melting point, 937.4°C (1719°F); boiling point, 2830°C (5126°F); hardness, 6.25 Mohs; atomic number, 32; atomic weight, 72.61; divalent, tetravalent. It has high brittleness, oxidizes readily at 600-700°C (1112-1292°F) and is a p-type semiconductor. Used for solid-state electronic devices, such as transistors, diodes, triodes, rectifiers and microwave detectors, for wavelength-dispersive spectrometer crystals, for high-purity semiconducting applications, in brazing alloys, phosphors and gold and beryllium alloys, in infrared-transmitting glass and infrared prisms, and for reflectors in projectors and wide-angle lenses. Symbol: Ge.

germanium bromide. See germanium dibromide; germanium tetrabromide.

germanium chloride. See germanium dichloride; germanium tetrachloride.

germanium columbide. See germanium niobide.

germanium crystals. High-purity single crystals of germanium usually produced (grown) by pulling from a pure melt (Czochralski technique), or by vaporization of germanium diiodide (GeI_2). Used for n-type and p-type semiconductors.

germanium dibromide. Yellow crystals. Crystal system, monoclinic. Melting point, 122°C (252°F); decomposes at 150°C (300°F). Used in chemistry, and materials research. Formula: $GeBr_2$. Also known as *germanium bromide.*

germanium dichloride. A white-yellow, hygroscopic powder used in chemistry, materials research and electronics. Formula: $GeCl_2$. Also known as *germanium chloride.*

germanium difluoride. White crystals. Crystal system, orthorhombic. Density, 3.64 g/cm³; melting point, 110°C (230°F); decomposes at 130°C (266°F). Used in chemistry, materials research and electronics. Formula: GeF_2. Also known as *germanium fluoride.*

germanium diiodide. Orange-yellow crystals, or an air- and moisture-sensitive yellow powder (99.9+% pure). Density, 5.4-5.73 g/cm³; melting point, decomposes at 545°C (1013°F); soluble in chloroform and cold water; oxidizes in air at 210°C (410°F) to form germanium tetraiodide and germanium dioxide. Used in the manufacture of germanium crystals, for reactions with alkyl halides to prepare alkylgermanium diiodohalides, and as a reagent to form stable ionic adducts with carbenes. Formula: GeI_2. Also known as *germanium iodide*.

germanium dinitride. See germanium nitride (i).

germanium dioxide. Colorless or white crystalline or amorphous powder (99.9+% pure). It is available in technical, semiconductor and electronic grades with purities up to 99.9999%. The crystalline form exists in two modifications: (i) high germanium dioxide with hexagonal crystal structure, and (ii) low germanium dioxide with tetragonal crystal structure. Density, 4.25-4.70 g/cm³ (crystalline), 3.64 g/cm³ (amorphous); melting point, 1081-1115°C (1978-2039°F); refractive index, 1.650. Used as a substitute for silica, as a glass former in glazes and bodies, and in glasses of high refractive index and infrared-transmitting glass, as a semiconductor in transistors and diodes, in phosphors, and as a component in optical fibers. Formula: GeO_2. Also known as *germania*.

germanium diselenide. See germanium selenide (ii).

germanium disulfide. See germanium sulfide (ii).

germanium fluoride. See germanium difluoride; germanium tetrafluoride.

germanium halide. A binary compound of a halogen, e.g., chlorine or fluorine, and germanium.

germanium iodide. See germanium diiodide; germanium tetraiodide.

germanium monoselenide. See germanium selenide (i).

germanium monosulfide. See germanium sulfide (i).

germanium monotelluride. See germanium telluride.

germanium monoxide. A black, high-purity solid (99.999%) with a melting point of 700°C (1291°F) (decomposes). Formula: GeO.

germanium mullite. A compound of aluminum oxide and germanium dioxide. Hardness, 1120 Knoop; refractive index, 1.664; infrared transmission, 70%; melting point, 1550°C (2822°F); sintering temperature, approximately 1500°C (2732°F). Used for ceramics and refactories. Formula: $Al_6Ge_2O_{13}$.

germanium niobide. A compound of germanium and niobium (columbium) containing 19.9-20.9% germanium. Crystal structure, cubic. Melting point, 1980°C (3596°F); superconductivity transition temperature, 23.6K. Used as a superconductor. Formula: $GeNb_3$ ($GeCb_3$). Also known as *germanium columbide*.

germanium nitride. Any of the following compounds of germanium and nitrogen: (i) *Germanium dinitride*. Black crystals used in ceramics and electronics. Formula: Ge_3N_2; and (ii) *Germanium tetranitride*. A compound of germanium and nitrogen supplied as orthorhombic crystals, or as a moisture-sensitive, white to light-brown powder (99.99+% pure). Density, 5.25 g/cm³; melting point, decomposes at 500°C (932°F); high electrical resistivity. Used as a dielectric in gallium arsenide MISFETs (metal-insulator-semiconductor field-effect transistors), and for electroceramics. Formula: Ge_3N_4.

germanium oxide. See germanium dioxide; germanium monoxide.

germanium selenide. Any of the following compounds of germanium and selenium: (i) *Germanium monoselenide*. Gray, high-purity crystals (99.99+%), or brown powder. Crystal system, orthorhombic. Density, 5.60 g/cm³; melting point, 667°C

(1233°F). Used in electronics. Formula: GeSe; and (ii) *Germanium diselenide*. Yellow-orange, high-purity crystals (99.99+%). Crystal system, orthorhombic. Density, 4.56 g/cm³; melting point, 707°C (1305°F) (decomposes). Used as a semiconductor for solid-state electronic devices. Formula: $GeSe_2$.

germanium sulfide. Any of the following compounds of germanium and sulfur: (i) *Germanium monosulfide*. Gray, high-purity crystals (99.99+%), or yellowish-red amorphous powder. Crystal system, orthorhombic. Density, 4.01 g/cm³ (crystalline); 3.31 g/cm³ (amorphous); melting point, 530°C (986°F). Used as a semiconductor and in electronics. Formula: GeS; and (ii) *Germanium disulfide*. White, moisture-sensitive, high-purity powder (99.999+%), or black crystals. Density, 2.94-3.01 g/cm³; melting point, approximately 530°C (985°F). Used as a semiconductor and in electronics. Formula: GeS_2.

germanium telluride. A compound of germanium and tellurium supplied as a ferroelectric powder in high-purity grades (99.999+%). Density, 6.2 g/cm³; melting point, 725°C (1337°F); Curie temperature, 400°C (752°F). Used as a very efficient semiconductor, and as a phase-change reversible alloy for erasable optical storage. Formula: GeTe. Also known as *germanium monotelluride*.

germanium tetrabromide. White, moisture-senstive high-purity crystals (99.99+%). Density, 3.132 g/cm³; melting point, 26.1°C (79°F); boiling point, 186.4°C (368°F); refractive index, 1.6269. Used as a doping source in vapor-phase epitaxy of III-V (13-15) semiconductors, such as GaAs, GaP, GaSb, InAs and InP. Formula: $GeBr_4$. Also known as *germanium bromide*.

germanium tetrachloride. Colorless, moisture-sensitive high-purity liquid (99.9+%). Density, 1.874 g/cm³; melting point, -49 to -52°C (-56 to -62°F); boiling point, 83.1°C (182°F); refractive index, 1.4644; superconductivity critical temperature, 548K. Used in the production of optical waveguides, and as a coupling agent for the production of rubbers from living polymers. Formula: $GeCl_4$. Also known as *germanium chloride*.

germanium tetrafluoride. A compound of germanium and fluorine supplied as a colorless gas in electronic grades (99.9+% pure) with a melting point of -36.5°C (-34°F). It is also available as a solid. Used in the formation of photoconductive films in combination with germane (GeH_4), in fluoride glasses, in the fabrication of SiGe crystallites on glass substrates, and in studies of the room-temperature oxidation of noble metals. Formula: GeF_4. Also known as *germanium fluoride*.

germanium tetrahydride. See germane (1).

germanium tetraiodide. Moisture-sensitive, red to orange crystals (99.9+% pure). Crystal system, cubic. Density, 4.25-4.40 g/cm³; melting point, 146°C (294°F); boiling point, 350°C (662°F). Used in organic and organometallic synthesis. Formula: GeI_4. Also known as *germanium iodide*.

germanium tetranitride. See germanium nitride (ii).

German silver foundry alloy. A nickel silver of 45% copper, 35% nickel and 20% zinc, used in the manufacture of white metal castings.

German silvers. See nickel silvers.

German silver solder. (1) A group of solders including copper-zinc base types containing 30-40% copper, 30% zinc, 10-20% silver and 20% cadmium.

(2) A German silver solder containing 10% silver, 3% copper, 2% zinc, and the balance tin.

German type metals. A series of German alloys containing 60-75% lead, 5-25% antimony and 2-35% tin.

Gerona. Trade name for a bolivia wool pile fabric.

Gerowal. Trade name of G. Robert Wilms AG (Germany) for a heat-treatable aluminum alloy containing 2-7% copper, 2-7% silicon and up to 0.6% manganese, 2.5% zinc, 1.1% iron, respectively. Used for light-alloy parts.

Gerowi Silberit. Trade name of G. Robert Wilms AG (Germany) for a series of aluminum alloys containing 2-4% magnesium, and up to 1.3% silicon, 0.6% manganese, 0.3% chromium and 0.2% titanium, respectively. They have excellent resistance to seawater corrosion, and are used for aircraft tanks and fittings, fuel lines, and marine parts.

Gerrip. Trade name of Saint-Gobain (France) for rigid insulating board products composed of glass fibers bonded with *Bakelite,* and having a perforated metal facing.

gersdorffite. A white or gray mineral of the pyrite group composed of nickel arsenic sulfide, NiAsS. It may also contain antimony or cobalt. Crystal system, cubic. Density, 5.90-6.49 g/cm³. Occurrence: Germany, Austria. Also known as *nickel glance.*

gerstleyite. A reddish mineral of the zeolite group composed of sodium antimony arsenic sulfide dihydrate, $Na_2(Sb,As)_8S_{13} \cdot 2H_2O$. Crystal system, monoclinic. Density, 3.62 g/cm³; refractive index, above 2.01. Occurrence: USA (California).

gerstmannite. A white to pale pink mineral composed of magnesium manganese zinc silicate hydroxide, $(Mg,Mn)_2Zn(SiO_4)(OH)_2$. Crystal system, orthorhombic. Density, 3.68 g/cm³; refractive index, 1.675. Occurrence: USA (New Jersey).

GE Silicone. Trade name of GE Plastics (USA) for several unfilled and glass fiber- or mineral-filled silicone plastics.

getchellite. A dark red mineral composed of arsenic antimony sulfide, $AsSbS_3$. Crystal system, monoclinic. Density, 3.92 g/cm³; refractive index, above 2.72. Occurrence: Iran, USA.

GETEK. Trademark of GE Electromaterials (USA) for naturally ultraviolet-blocking laminates and prepregs composed of epoxy and polyphenylene oxide resins and supplied in three grades: *Thin Laminates* (ML200), *Rigid Laminates* (RG200) and *Prepregs* (T-series). They provide high maximum operating and glass-transition temperatures, excellent thermal-cycling crack resistance, excellent craze resistance, low and stable dielectric constants and dissipation factors, low moisture absorption, and good dimensional stability. Used in computers, telecommunications equipment, instrumentation devices, load boards, high-speed digital boards, probe cards, wireless modems, transceiver backplanes, antennas, voltage-control oscillators and remote meter readers, and for electronic packaging applications, e.g., burn-in boards and ball-grid-array (BGA) substrates.

getter. See scavenger (2).

geversite. A light gray mineral of the pyrite group composed of antimony platinum, $PtSb_2$. It can also be made synthetically. Crystal system, cubic. Density, 10.91. Occurrence: South Africa.

Gevetex. Trade name of Gevetex-Textilglas GmbH (Germany) for glass-fiber yarns for various technical applications.

Gewa. Trade name of Krupp Stahl AG (Germany) for a series of tool steels including various cold-work, hot-work, shock-resisting and high-speed types.

geyserite. A soft, porous white or grayish type of *opal* deposited from hot springs and geysers. Also known as *fiorite; pearl sinter; siliceous sinter.*

G-Glass. Trademark of Nuova Italtess (Italy) for glass fibers.

gianellaite. A straw yellow mineral composed of mercury nitride sulfate, $Hg_4N_2SO_4$. Crystal system, cubic. Density, 7.19 g/cm³; refractive index, 2.085. Occurrence: USA (Texas).

Giant. Trade name of Clayburn Refractories Limited (Canada) for hollow, structural clay brick.

giant arborvitae. See western red cedar.

giant fullerenes. A subclass of fullerene carbon cluster compounds composed of molecules with 100 or more carbon atoms, e.g., C_{120}, C_{540}, etc. See also fullerenes.

giant magnetoresistance materials. A relatively recent class of materials including various manganites and other compounds with perovskite crystal structures, and magnetic multilayered nanostructures (e.g., Co/Cu/Fe/Cu) that exhibit increases or decreases in electrical resistivity of 20% or more in a relatively weak applied magnetic field. They are useful for applications in magnetic data processing and storage, e.g., computer hard drives and nonvolatile magnetic memory devices, and for solid-state compasses, automotive sensors, and landmine detectors. Abbreviation: GMR materials.

giant magnetostriction materials. A relatively recent class of materials that exhibit large recoverable strains by the application of magnetic fields and include in particular ferromagnetic shape-memory alloys, such as Fe-Pd, Fe_3Pt and Ni_2MnGa. They are suitable as actuator and sensor materials. Abbreviation: GMS materials.

Giant Special. Trade name of Champion Steel Company (USA) for an oil- or water-hardening steel containing 0.35% carbon, 0.9% chromium, 0.9% molybdenum, and the balance iron. Used for plastic molds.

Giasiblend. Trade name of C.G.T. (Italy) for polymer blends of polycarbonate and acrylonitrile-butadiene-styrene resins.

Giasilac. Trade name of C.G.T. (Italy) for acrylonitrile-butadiene-styrene resins.

Giasilon. Trade name of C.G.T. (Italy) for a polyamide 6,6 (nylon 6,6) resin supplied in several grades.

Giasimid. Trade name of C.G.T. (Italy) for a series of polyamide (nylon) resins including nylon 6 and nylon 6,6, supplied in a range of grades.

Giasiplen. Trade name of C.G.T. (Italy) for standard and modified polypropylenes supplied in several grades.

Giasixan. Trade name of C.G.T. (Italy) for a polyoxymethylene (acetal) resin supplied in a range of grades.

gibbsite. A colorless, white or tinted mineral composed of aluminum hydroxide, $Al(OH)_3$. It can also be made synthetically. Crystal system, monoclinic. Density 2.40-2.44 g/cm³; hardness, 2.5-3.5 Mohs; refractive index, 1.577. Occurrence: Guianas, Norway, USA (Alabama, Arkansas, Georgia, Massachusetts). It is a major component of bauxite and is used as a refractory binder for china clays, and as a bat wash. Also known as *hydrargillite.*

Gibraltar. Trade name of H. Boker & Company (USA) for a water-hardening tool steel (AISI type W1) containing 0.95-1.05% carbon, 0.3% silicon, 0.25% manganese, and the balance iron. Used for taps, drills, cutters and punches.

Gibraltar stone. See onyx marble.

Gibsiloy. Trade name of Gibson Electric Company (USA) for a series of electrical contact alloys made by powder metallurgy techniques and including various composites of silver-nickel, silver-cadmium-nickel, silver-graphite, silver-carbon, silver-molybdenum, copper-tungsten and copper-tungsten carbide.

Gibson. Trade name of Gibson Electric Company (USA) for a series of silver-cadmium, silver-copper, silver-platinum, silver-palladium and silver-zinc electrical contact alloys.

giessenite. A creamy to dull gray mineral that is composed of lead bismuth sulfide, $Pb_8Bi_6S_{17}$, and may also contain copper and antimony. Crystal system, orthorhombic. Occurrence: Switzerland, Norway.

Gigant. Trade name of Saarstahl AG (Germany) for a series of tungsten, cobalt-tungsten and molybdenum high-speed tool steels. They have high hot hardness, excellent wear resistance and good toughness, and are used for boring, forming, finishing and lathe, planer and shaper tools, and for various types of dies.

Giglio. Trade name of Circeo Filati S.r.l. (Italy) for a yarn blend composed of 65% polyester and 35% viscose rayon, and supplied in several counts. Used for textile fabrics.

Gila. Trademark of Solutia Inc. (USA) for an extensive range of automotive and home solar window films. The automotive window films are supplied in a wide range of syles, sizes and colors for car, limousine and RV tinting applications. The home window films are used on doors and windows to reduce glare, interior heat buld-up and fading of wallpaper, furniture, carpets, etc. Specialty grades, such as heat-control films, mirror tinting films, privacy films and static cling vinyl films, are also available.

gilalite. A green to pale blue green mineral composed of copper silicate heptahydrate, $Cu_5Si_6O_{17} \cdot 7H_2O$. Crystal system, monoclinic. Density, 2.82 g/cm^3; refractive index, 1.635. Occurrence: USA (Arizona).

gilded glass. Fine glassware that has been ornamented by the application of gold leaf, dust or paint, and either left unfired or fired at low temperatures to achieve permanency. Used especially for architectural applications, and in the arts.

Gildent EPR. Trade name of Guilini SpA (Italy) for a beige dental gypsum.

gilder's whiting. Coarse and medium-coarse (extra gilder's whiting) grades of naturally occurring calcium carbonate (*whiting*).

gilding brass. See gilding metal.

gilding foil. A corrosion-resistant alloy containing 2.2% copper, 0.1% iron, and the balance tin. Used in foil form for manufacture of cheap jewelry.

gilding metal. A golden-red alloy that contains 85-97% copper and 3-15% zinc, and is commercially available in the form of flat products and wires. It has a density of 8.75 g/cm^3 (0.316 $lb/in.^3$), excellent cold workability, good hot workability and corrosion resistance, good electrical conductivity (about half that of pure copper), a low temperature coefficient, and good strength. Used for coins, medals, ornamental trim, bullet jackets, fuse caps, primers, as a jewelry base for gold plate, and for angles, channels and red brass. Abbreviation: GM. Also known as *gilding brass*.

gilding metal-clad steel. Steel that has *gilding metal* bonded to both sides. The decorative cladding is about 5% of the total thickness on one side and 15% on the other. Used for coinage and ammunition.

Gilgrid. Trade name of Gilby-Fodor SA (France) for a series of corrosion- and heat-resistant magnetic alloys containing 45-67% nickel, 10-30% molybdenum, and up to 45% iron, 1% chromium, 1% silicon and 1% manganese, respectively. Used for grid wire, tubes and other electronic applications.

Gillener. Trade name of M.C. Gill Corporation (USA) for an impact-resistant polyester laminate reinforced with woven glass fibers. Used for aircraft applications.

gillespite. A rose red mineral composed of barium iron silicate, $BaFeSi_4O_{10}$. Crystal system, tetragonal. Density, 3.40 g/cm^3; refractive index, 1.621. Occurrence: USA (Alaska, California).

Gillfab. Trade name of M.C. Gill Corporation (USA) for a series

of thermosetting composites including E-glass-reinforced polyester and Kevlar-reinforced vinyl ester laminates, glass cloth-reinforced phenolic composites, carbon-reinforced phenolic composites with aramid honeycomb cores, glass fiber-reinforced phenolic composites with Nomex honeycomb cores, and S-2 glass-reinforced epoxies with aramid honeycomb cores. Used for aircraft applications and military equipment.

Gilfloor. Trade name of M.C. Gill Corporation (USA) for impact-resistant fiberglass-reinforced polyester sandwich panels with balsawood cores. Used used for aircraft and boat flooring.

Gilliner. Trade name of M.C. Gill Corporation (USA) for chemical- and impact-resistant glass fabric-reinforced polyester laminates used for cargo liners.

Gilmore Nickel Alloy. Trade name of F.F. Gilmore & Company (USA) for a nickel alloy that contains some copper, and is used for diamond setting tools and holders.

Gilmore Tool. Trade name of Bissett Steel Company (USA) for a tool steel containing 0.85% carbon, 3.5% chromium, 1% tungsten, and the balance iron. Used for hot- and cold-work tools and dies.

gilsonite. A black, pure variety of natural *asphalt* with a bright luster and a lacing of brown streaks. It is a a variety of *mineral rubber* that occurs widely in Utah, USA, has a density 1.05-1.07 g/cm^3 (at 25°C or 77°F) and a hardness of 2 Mohs, and softens and flows on heating. Used in the manufacture of black asphaltum varnish, lacquers, japans, baking enamels, acid-, alkali- and waterproof coatings, wire-insulation compounds, mineral waxes, waterproofing compounds, roofing, paving and floor tiles, storage-battery cases, and as a diluent in low-grade rubber compounds. Fine dust is used on foundry cores, as a carbonaceous addition to foundry sands, as an additive for heavy fuel oils, and as a source of gasoline, fuel oil, and *coke*. Also known as *uintaite*.

Gilsulate. Trademark of American Thermal Products, Inc. (USA) for ground, hard *bitumen* used for insulation purposes.

Giludur SK. Trade name of Guilini SpA (Italy) for a dental gypsum.

Gimap. Trade name of Pechiney/Société Français d'Electrométallurgie (France) for a series of powdered manganese alloys used for coating welding rods and wire.

Gimel. Trade name of Pechiney/Société Français d'Electrométallurgie (France) for an electrothermic manganese metal (96-98% pure) that contains about 2-4% iron, and is used as an alloying addition for steel, copper alloys and several light alloys.

gimped yarn. A *plied yarn* consisting of a core yarn with a wrapping of another filament, strip or yarn in which the core yarn does not twist with the surrounding yarn so that the latter can be separately unwrapped.

Ginga. Trade name of Asahi Glass Company Limited (Japan) for a patterned glass featuring large and small stars.

ginger pine. See Lawson cypress.

gingham. A light- to medium-weight, plain-woven fabric, usually of cotton or a cotton blend, having woven two-color checks, stripes or plaids. Used for clothing, bedspreads and curtains.

Gingoshi. Trade name of Asahi Glass Company Limited (Japan) for a patterned glass.

giniite. A blackish green to blackish brown mineral composed of iron phosphate hydroxide dihydrate, $FeFe_4(PO_4)_4(OH)_2 \cdot 2H_2O$. Crystal system, orthorhombic. Density, 3.41 g/cm^3; refractive index, 1.803. Occurrence: Namibia.

ginkgo wood. The light, white to yellowish wood of the tall maidenhair tree *Ginkgo biloba*, the only living species of the order

Ginkgoales that once included several trees with fernlike leaves. It has a fine texture and works well. Source: Central Asia, especially China. Used for chessboards and pieces.

Ginmoru. Trade name of Asahi Glass Company Limited (Japan) for a reeded glass.

Ginny. Trademark of Solutia, Inc. (USA) for a colorfast, high-wicking acrylic microfiber that looks and feels like cotton. Supplied in a wide range of colors, it is fast drying, and has excellent processibility. Used especially for wearing apparel including socks.

ginorite. A white mineral composed of calcium borate octahydrate, $Ca_2B_{14}O_{23} \cdot 8H_2O$. Crystal system, monoclinic. Density, 2.09 g/cm^3; refractive index, 1.524. Occurrence: Italy.

giorgiosite. A synthetic mineral composed of magnesium carbonate hydroxide pentahydrate, $Mg_5(CO_3)_4(OH)_2 \cdot 5H_2O$. Density, 2.15 g/cm^3.

giraudite. A light gray mineral of the tetrahedrite group composed of copper silver zinc antimony arsenic selenide sulfide, $(Cu,Zn,Ag)_{12}(As,Sb)_4(Se,S)_{13}$. Crystal system, cubic. Density, 5.75 g/cm^3. Occurrence: France.

girder. A horizontal structural member of wood, steel or reinforced concrete used to support a concentrated load.

girdite. A white mineral composed of hydrogen lead tellurium oxide, $H_2Pb_3(TeO_3)(TeO_6)$. Crystal system, monoclinic. Density, 5.50 g/cm^3; refractive index, 2.47. Occurrence: USA (Arizona).

G-Iron. Trade name of Tonawanda Electric Steel Casting (USA) for a graphitized *pig iron* containing 4.2% carbon, 2.4% silicon, 0.6% manganese, and the balance iron. Used for the manufacture of gray iron castings.

gismondine. (1) A colorless, white, or grayish mineral of the zeolite group composed of sodium calcium aluminum silicate tetrahydrate, $(Ca,Na_2)Al_2Si_2O_8 \cdot 4H_2O$. Crystal system, monoclinic. Density, 2.19 g/cm^3. Occurrence: Hawaii.

(2) A white mineral of the zeolite group composed of calcium aluminum silicate hydrate, $CaAl_2Si_2O_8 \cdot 4H_2O$. Crystal system, monoclinic. Density, 2.20 g/cm^3. Occurrence: Germany.

Gisolan. Trade name for regenerated protein (azlon) fibers.

Gizolan. Trade name for rayon fibers and yarns used for textile fabrics.

Gittermetall. A German graphitic-lead bearing alloy composed of 10% tin, 15% antimony, 1.75% copper, 0.20% graphite, and the balance lead.

gittinsite. A grayish white mineral of the thortveitite group composed of calcium zirconium silicate, $CaZrSi_2O_7$. Crystal system, monoclinic. Density, 3.63 g/cm^3; refractive index, 1.736. Occurrence: Canada (Quebec).

giuseppettite. A pale violet-blue mineral of the cancrinite group composed of potassium sodium calcium aluminum chloride silicate sulfate, $(Na,K,Ca)_8Al_6Si_6O_{24}(SO_4)_2Cl_{0.25}$. Crystal system, hexagonal. Density, 2.35 g/cm^3; refractive index, 1.491. Occurrence: Italy.

glacé leather. A very soft, delicate leather with a glossy surface made from the skin of goats or lambs.

Glacelite. Czech trade name for a patterned glass with irregular texture.

Glacetex. Trade name of Glaceries Réunies SA (Belgium) for a laminated plate glass.

Glaciale. Trade name of Vetreria di Vernante SpA (Italy) for a glass with floral pattern.

Glacier. (1) Trade name of Southern Dental Industries Limited (Australia) for a radiopaque, light-cure, no-slump hybrid den-

tal composite of micro-sized strontium glass particles and amorphous silica. Supplied in natural shades, it has high viscosity and high early strength, excellent handling and physical characteristics, and good polishability. Used for posterior-anterior restorations.

(2) Trade name of Libbey-Owens-Ford Company (USA) for a patterned glass.

gladite. A lead-gray mineral of the stibnite group composed of copper lead bismuth sulfide, $CuPbBi_5S_9$. Crystal system, orthorhombic. Density, 6.88 g/cm^3. Occurrence: Sweden.

Glakresit. Trade name of Glakresit glass-fiber-reinforced phenol-formaldehyde plastics, compounds and laminates.

Glam. Trademark of Pantasone Company (USA) for a series of polyvinyl chlorides and other vinyls.

glance. See chalcocite.

glance pitch. A pure, black, naturally occurring *asphalt* resembling *gilsonite*, but having a black instead of a brown streak, a higher density, a higher amount of fixed carbon, and a bright conchoidal fracture. Occurrence: Argentina, Mexico, Trinidad, USA (Oklahoma). Used in electrical insulation products, paints, and molding materials.

Glanzstoff. Trade name of Glanzstoff Austria AG (Austria) for nylon fibers and yarns used for the manufacture of clothing, e.g., dresses and blouses.

Glapoband. Trade name of Zehdenick for vinyl rubber products.

Glare-Check. Trade name of Polacoat Inc. (USA) for a laminated architectural glass with polarizing interlayer.

Glare-Gard. Trade name of Amerada Glass Company (USA) for a no-glare glass product composed of a polarizing coating sealed between two sheets of glass.

Glareprufe. Trade name of Shatterprufe Safety Glass Company (Proprietary) Limited (South Africa) for a laminated glass with tinted interlayer.

Glare-X. Trade name of Tyre Brothers Glass Company (USA) for a double glazing unit enclosing a metal mesh with slanting horizontal members resembling miniature Venetian blinds.

Glascofoam. Trade name of Trent Insulation (UK) for polystyrene foam boards and sheets.

Glaseal. Trade name of Hill Brothers Glass Company Limited (UK) for double glazing units.

Glasef. Trade name of Safeglas Corporation Limited (India) for a toughened glass.

Glaseta. Trade name of Vereinigte Seidenwebereien AG (Germany) for glass-fiber curtain materials.

Glasfab. Trademark of Bay Mills Limited (Canada) for woven glass fabrics used as reinforcements in plastics.

Glasflex. Trade name of Glasflex (USA) for acrylics supplied in general-purpose, casting, cast sheet and high-impact grades.

Glasfloss. Trade name of Glasfloss Industries, Inc. (USA) for glass fibers and glass fiber products for commercial, industrial and residental air filtration applications, including

Glasfoil. Trade name of Manville Corporation (USA) for fireproof, flexible glass fiber facings.

GlasIonomer. Trademark of Shofu Dental Corporation (Japan) for a product range of glass-ionomer dental cements. *GlasIonomer Base Cement* is a biocompatible, fluoride-releasing cement that chemically bonds to the tooth structure (dentin and enamel), provides high radiopacity and compressive strength, and is used as a protective and supportive base under amalgam, composite and porcelain restorations. *GlasIonomer Cement CX-Plus* is a cement that contains glass particles coated with silicon oxide using the "Silicon-Oxide Layer Coating" (SLC) technique to significantly increase the allowable working time. *GlasIonomer Cement Type II* is a biocompatible, fluoride-releasing cement that chemically bonds to the tooth structure (i.e., dentin and enamel), provides high compressive and tensile strength, a thermal expansion coefficient similar to tooth structure, and is used for various restorations including tooth repair, cervical erosions and pedodontic restorations.

Glasiosite. Trade name of Dentamed (Czech Republic) for a polyacid-modified dental composite.

Glas-Isol. Trade name of Fibres de Verre SA (Switzerland) for a glass silk.

Glaskyd. (1) Trademark of American Cyanamid Company (USA) for resinous (including polyester) molding compounds that can be processed by compression, injection and transfer molding. Used for numerous applications ranging from automotive components to heavy industrial parts.

(2) Trademark of Cytec Company (USA) for alkyd resins supplied in various grades including long glass fiber-reinforced for high impact, short glass fiber-reinforced for fire retardancy, mineral-filled and mineral- or glass fiber-reinforced.

Glas-Lok. Trade name of Innotek World Resins, LLC (USA) for thermoplastic powder coatings.

glasphalt. A special particulate composite consisting of recycled-glass aggregate in an asphalt matrix.

Glasrod. Trade name of Glastic Corporation (USA) for fiberglass-reinforced polyester bars and rods used for electrical applications.

Glasron. Trade name of Asahi Fiber Glass Company Limited (Japan) for glass wool and textile glass fiber products.

glass. An essentially hard, brittle, inorganic and predominantly oxidic solid that may be colorless or colored, and either clear or opaque due the presence of impurities. It is made by super-cooling of a melt without crystallizing, has a short-range (amorphous) atomic order, and a structure consisting of an irregular, three-dimensional network composed of certain atomic units (e.g., SiO_4 tetrahedra) having large cations (e.g., calcium or sodium) interspersed. Ordinary soda-lime glass is made from silica sand, soda ash and lime in a ratio of about 75:20:5, but may contain other compounds, such as dolomite, feldspar, potash, borax, saltpeter, metallic oxides (e.g., barium, calcium, cerium, lead, lithium, titanium and/or zinc oxide). *Glass* is available in almost any form including pastes, sheets, rods, tubing, pipes, powders, beads, flakes, solid and hollow spheres, fabrics, fibers, yarns and filaments. Used for windows, containers, building blocks, chemical equipment, laboratory ware, dinnerware, incandescent light bulbs, vacuum tubes, reinforcing fibers and microspheres for engineering composites, lasers, optical equipment and instruments, and variety of other applications. See also metallic glasses; heavy metal fluoride glasses.

GlassBase. Trademark of Pulpdent Corporation (USA) for a radiopaque, dentin-shaded dental glass-ionomer cement with high compressive strength, low thermal expansion, and controlled setting. Used in dentistry as a firm, hard base under amalgam and composite restorations and for the restoration of teeth.

glass belt. Glass-fiber reinforced synthetic resins made into belts for use in conveyors.

glass block. A translucent building block of glass that is made of two formed pieces of glass fused together to leave an insulating air space between. It has good thermal insulating properties, and may be used in outside and inside walls. Also known as *glass building block.*

glass-bonded mica. A ceramic material obtained by bonding finely divided natural or synthetic *mica* with powdered glass at high pressure and temperature. It has good dielectric properties, good transverse strength, low thermal expansion, and a maximum upper-use temperature of approximately 500°C (930°F). Used as an electrical insulating material.

glass brick. A hollow block of translucent glass that has a plain or patterned surface, and is used in the construction of walls, partitions and windows.

glass building block. See glass block.

glass cement. An adhesive or glue that is particularly suitable for bonding glass.

glass-ceramics. A group of a hard, strong, nonporous, fine-grained ceramic materials that are partly crystalline and partly glassy (amorphous), and made by adding a nucleating agent, such as titania (TiO_2) or zirconia (ZrO_2), to an ordinary glass melt followed by rolling and cooling. The resulting product is then devitrified (crystallized) by heating to a temperature at which crystal nucleation occurs. They are usually opaque white or transparent, and have a theoretical density of approximately 2.4-5.9 g/cm³ (0.09-0.21 lb/in.³), low thermal expansion (5-17 × 10⁻⁶/K), high thermal-shock resistance up to 900°C (1650°F), high resistance to breakage, high flexural strength, high elastic moduli (typically 83-138 GPa or 12-20 ksi), high rupture moduli, good impact strength, an upper continuous-use temperature of about 700-1000°C (1290-1830°F), good chemical resistance, excellent corrosion and oxidation resistance, and low electrical conductivity. They are often sold under trademarks or trade names, such as *Cervit, Macor, Nucerite* or *Pyroceram*, and used in the manufacture of high-capacitance and magnetic glasses, architectural panels, molded mechanical and electrical parts, electronic components, laboratory bench tops, heat-exchanger tubing, missile cones, cookware and dinnerware, restaurant heating and warming equipment, range and stove tops, radomes, high-temperature bearings, telescope mirrors, coatings, cellular foams, and adhesives. Abbreviation: GC. Also known as *devitrified ceramics; devitrified glasses; nucleated glasses; vitro ceramics.* See also machinable glass ceramics.

glass cloths. See glass fabrics.

glass-coated steel. Steel coated with a special type of vitreous enamel. It has high resistance to chemicals at high temperatures and pressures and good structural properties, and is used for chemical reactors and hot-water tanks, containers, vessels, tanks, piping and other equipment used in the chemical, plating, pulp and paper and metal-refining industries. Also known as *glass-lined steel; glassed steel.*

Glasscor. Trademark of Domtar Construction Materials (Canada) for fiberglass shingles.

GlassCore. Trademark of Pulpdent Corporation (USA) for a biocompatible, radiopaque, fluoride-releasing dental glass-ionomer cement that has high compressive strength, low thermal expansion and controlled setting, and is used in restorative dentistry for core-build-ups and tooth restorations.

Glass Core. Trade name of Manville Corporation (USA) for fiberglass-asbestos felt used for wrapping underground pipelines.

Glass-cote. Trade name of PolySciences Inc. (USA) for a solution of 1% dimethyldichlorosilane, $(CH_3)_2SiCl_2$, in benzene used to make electrophoresis gel tubes nonwettable.

glass decolorizer. See decolorizer.

Glasseal. Trade name of Techalloy Company Inc. (USA) for a series of iron- and nickel-base alloys used for sealing metal to glass. The typical composition range is 29.0-50.5% nickel, up to 17.0% cobalt, 0.02-0.06% carbon, 0.50-0.80% manganese, 0.20-0.30% silicon, 0.50% magnesium, and the balance iron. Used for sealing electronic tubes, industrial lamps, and sealed-beam headlights.

GlasSeal Thermopane. Trade name of Libbey-Owens Fiberglas Corporation (USA) for all-glass double glazing units.

glassed steel. See glass-coated steel.

glass enamel. See ceramic glass enamel.

glass-epoxy composites. High-performance composites having epoxy resin matrices reinforced with glass fibers or fabrics. Also known as *glass-epoxy matrix composites*. See also epoxy composites.

glasses. See glass.

Glass-Fabric. Trademark of J.H. McNairn Limited (Canada) for a glass-cloth reinforcement used for membrane waterproofing materials.

glass fabric-reinforced composites. Engineering composites consisting of *glass cloths* (glass fabrics) impregnated with synthetic resins, such as epoxies or phenolics.

glass fabrics. Fabrics woven from continuous or staple glass fibers, usually in a plain weave. They have high strength and thermal resistance, good dielectric properties, and are used as reinforcements in polymer-matrix composites, and in the manufacture of laminated plastics, especially for insulating applications. Also known as *glass cloths; glass fiber cloths; glass fiber fabrics*.

glass felt. A sheet-like material consisting of glass fibers bonded or impregnated with synthetic resins, and used in building construction as waterproofing, and for shingles and roofing membranes.

glass fiber cloths. See glass fabrics.

glass fiber-epoxy composites. Composite materials consisting of epoxy resin matrices reinforced with glass fibers. They have high-strength-to-weight ratios, high strength at elevated temperatures, high elastic moduli, good flexural and compressive properties, good thermal and electrical properties, and excellent chemical resistance. Used as structural components for spacecraft and aircraft including fuselages, wings and control surface panels, rocket nozzle structural shells, commercial pressure vessels, tanks and pipes, pump components for chemical and food handling, and electrical laminates for circuit board manufacture. Also known as *fiberglass-epoxy composites*.

glass fiber fabrics. See glass fabrics.

glass fiber insulating boards. See fiberglass insulating boards.

glass fiber insulation. See fiberglass insulation.

glass fiber-reinforced cement. A tough, high-strength composite with a cement-based matrix reinforced with glass fibers that have been coated with carbon to prevent alkali corrosion. Used in the construction trades. Abbreviation: GFRC.

glass fiber-reinforced plastics. A class of composite materials composed of continuous or discontinuous reinforcing glass fibers or filaments embedded in resin matrices (e.g., epoxy or unsaturated polyester). They have high strength, good flexural properties, and good thermal and electrical properties. Depending on the matrix resin, they may also have high strength-to-weight ratios and excellent chemical resistance. Used for structural and nonstructural components of aircraft, spacecraft and helicopters, e.g., fuselages, wing and control surface panels, rocket nozzle structural shells and exterior and interior panels, for body panels of trucks and automobiles, rear lifts on station wagons, vans and sports vehicles, ship and boat hulls, offshore structures, pressure vessels, tanks and pipes, pump components

for chemical and food handling, and electrical laminates for circuit board manufacture. Abbreviation: GFRP. Also known as *glass fiber-reinforced polymers*.

glass fibers. A group of noncombustible continuous synthetic staple fibers or filaments, usually less than 25 μm (0.001 in.) thick, made by extruding or spinning molten glass at high speed through extremely small orifices . Available in bulk and woven form, they can be colored by resin-bonded pigments, or by dyeing. *Glass fibers* have a density of 2.54 g/cm^3 (0.092 $lb/in.^3$), outstanding resistance to heat and chemicals, good strength, low elongation, no moisture regain, an upper-use temperature of 315°C (600°F), a softening temperature of approximately 815°C (1500°F), good resistance to most commercial chemicals and solvents, and fair to poor resistance to hydrogen fluoride and alkalies. Used as acoustic, electrical and thermal insulating materials, as media and fabrics in air and other filters, for decorative and utility fabrics, such as drapes, curtains, table linen, carpet backing and tenting, in nonceramic products, e.g., as reinforcements in rubber, plastic and similar products, in tire cord as belts between treads and carcasses, as reinforcements in cement products for construction use, in light transmission for communication signals, for surgical sutures, and numerous other domestic and commercial applications. Abbreviation: GF.

glass fiber yarn. Fine, continuous or staple glass fibers or filaments of varying composition twisted together into strands. Also known as *fiberglass yarn*.

glass filament. An individual, flexible, continuous glass fiber drawn to a small diameter, typically 0.8-25.4 μm (0.00004-0.001 in.).

GlassFill. Trademark of Pulpdent Corporation (USA) for a tooth-shaded dental glass-ionomer dental cement with high compressive strength, low thermal expansion, a tooth structure-matching refractive index and a short setting time. Used as a filler for the restoration of cervical erosions, root caries, cavities and deciduous teeth, and as a fissure sealant.

Glass-Fill. Trademark Mulco Inc. (USA) for fiberglass-reinforced autobody fillers.

glass-filled high-impact polystyrenes. High-impact polystyrenes with up to 40 wt% E-glass fibers added to improve tensile properties, stiffness and creep properties, impact resistance and toughness, and lower the coefficient of thermal expansion. Used for packaging materials, e.g., containers and disposables, and for toys. See also high-impact polystyrenes.

glass flake. Thin, flat scales or flakes of glass produced by blowing and subsequent shattering the glass bubbles, or by supercooling very thin sheets of glass to induce shattering. Used as reinforcement and filler in plastics and paints.

Glass-Flex. Trade name of PPG Industries Inc. (USA) for flexible pre-insulated glass-fiber duct products used for heating and air-conditioning applications.

Glassfloss. Trade name of Durez Plastics & Chemicals Limited (USA) for phenolic resins and plastics.

glass foam. See cellular glass.

glass formers. Various oxides, such as silicon dioxide (SiO_2), boric oxide (B_2O_3), germanium oxide (GeO_2), phosphoric pentoxide (P_2O_5) and arsenic pentoxide (As_2O_5), and multicomponent alloys, such as lanthanum-aluminum-nickel or magnesium-copper-yttrium alloys, which aid in the formation of glass by producing oxide polyhedra. Also known as *network formers*. See also bulk metallic glass formers.

glass-grinding sand. A screened, but ungraded abrasive sand with tough grains of uniform size, used for general grinding of glasses.

Glassgrit. Trade name of Virginia Materials & Supplies, Inc. (USA) for glass grit-based blasting abrasives.

Glas-Shot. Trade name of Cataphote, Inc. (USA) for coated and uncoated glass spheres used for abrasive blast cleaning, deburring and finishing.

glassine. A thin, highly glazed, greaseproof paper obtained by excessive beating of sulfite pulp and subsequent supercalendering. It may have urea-formaldehyde added to enhance strength. *Glassine* has high toughness and flexibility and high transparency or translucency, and is used for packaging, sanitary wrapping, envelope windows, dust covers for books and documents, insulating paper between layers of iron-core transformer windings, and for general household purposes.

glass-ionomer cements. See ionomer-glass cements.

glass-ionomers. See ionomer glasses.

Glass-Kraft. Trademark for a building product consisting of a layer of loosely felted glass fibers sandwiched between two facings of strong, tough *kraft paper*. Available in various lengths and width, it used for the thermal insulation of buildings.

GlassLine. Trademark of Pulpdent Corporation (USA) for a fluoride-releasing, radiopaque dental glass-ionomer cement with very high compressive strength, very high bonding strength and adhesion to dentin, low thermal expansion, and controlled setting. Used in restorative dentistry for the cementation of crowns, posts and orthodontic bands, and as a thin liner under composite and amalgam restorations.

glass-lined steel. See glass-coated steel.

Glasslon. Trademark of Asahi Fiber Glass Corporation (Japan) for glass fibers.

GlassLute. Trademark of Pulpdent Corporation (USA) for a fluoride-releasing, radiolucent dental glass-ionomer luting cement with very high compressive and tensile strengths, very high bonding strength and good adhesion to tooth structure and restorations, low thermal expansion, controlled setting and an ultrathin film thickness of only 15 μm (0.6 mil). Used for the cementation of crowns, bridges and posts.

glassmaker's lepidolite. Lepidolite (lithium mica) used as a flux and source of alumina (Al_2O_3) in opal and flint glasses. See also flint glass; lepidolite; opal glass.

glassmakers' soap. A glass decolorizer, such as manganese dioxide, certain selenium compounds or white arsenic, used to remove the green color produced by iron salts. See also decolorizer.

Glass Mat. Trade name of Saint-Gobain Vetrotex (France) for nonwoven glass fabrics used for the reinforcement of bituminous coverings for roofs and walls.

glass mat. A thin mat manufactured from continuous-strand glass fibers, randomly oriented chopped glass fibers or directionally oriented glass fibers, interfelted or intertwined and with or without a binder. Supplied in various weights, widths and lengths, it us used as a reinforcement for engineering plastics.

glass-mat thermoplastics. Laminated composites consisting of *glass mat* reinforcements contained in thermoplastic resin matrices. The most commonly used resins are polypropylene, polybutylene terephthalate, polyethylene terephthalate, polycarbonate, and alloys of polycarbonate and polybutylene terephthalate. Used for automotive, aerospace and industrial parts and equipment. Abbreviation: GMT.

glass-matrix composites. Composites having glass (e.g., borosilicate) or glass-ceramic matrices reinforced with materials, such

as alumina, carbon or silicon carbide fibers.

glass-metal seal alloys. See sealing alloys.

glass microspheres. Small solid or hollow spherical particles of glass, usually of the soda-lime type and used in plastics or rubber as extenders and reinforcing agents. Solid microspheres have a density of about 2.5 g/cm³ (0.09 lb/in.³), and are available in diameters from 1 to 5000 μm (0.00004 to 0.2 in.). Hollow microspheres range from 0.08 to 0.8 g/cm³ (0.002 to 0.029 lb/in.³) in density and from 10 to 200 μm (0.0004 to 0.008 in.) in diameters.

glass mineral wool. See glass wool.

Glass Mold Alloy. A British alloy composed of 55-65% copper, 12-18% nickel, 11-17% zinc, 8-12% iron and 0.5-1% silicon, and used for glass molds.

glassoid. A thin, highly glazed, transparent paper that is similar to *glassine*, but superior in surface finish.

glass paper. (1) A paper made of glass fibers and having high heat and chemical and weathering resistance. Used for permanent documents.

(2) A relatively thin fiberglass mat incorporating chopped strands usually 25-50 mm (1-2 in.) in length. Used as reinforcement for fiberglass roofing shingles. Also known as *fiberglass paper*.

(3) A paper that contains a bonded coating of powdered glass, and is used as an abrasive.

(4) Borosilicate glass flake bonded with a synthetic resin, such as an alkyd, phenolic or silicone.

glass-phenolic composites. A group of composites comprising phenolic resin matrices (e.g., single-stage *resoles* or two-stage *novolacs*) reinforced with glass fibers often based on *E-glass* or *S-glass*.

glass polyalkenoate cement. A dental cement made by mixing an ion-leachable fluoroaluminosilicate glass powder with a solution of a polymeric acid, such as polyacrylic acid, and used in restorative dentistry for luting and tooth-filling applications.

glass-polyester composites. A group of composites comprising polyester resin matrices (e.g., orthophthalic or isophthalic polyester resins, bisphenol A fumarates, chlorendics, or vinyl esters) reinforced with glass fibers often based on *E-glass* or *S-glass*.

glass-reinforced acrylamate composites. A group of high-performance engineering composite materials consisting of acrylamate polymer based matrices randomly or directionally reinforced with glass fibers. They have high strength and elastic moduli, good high-temperature capabilities, good chemical resistance, and are used for automotive and aerospace components, recreational products, agricultural equipment, etc. See also acrylamate polymers.

glass-reinforced epoxies. A group of high-performance engineering composites consisting of thermosetting epoxy resin matrices reinforced with glass fibers or flakes. They have high tensile strength and elastic moduli, high thermal stability, good dielectric properties and good chemical resistance. Used chiefly for electrical and electronic components.

glass-reinforced phenolics. A group of composite materials consisting of thermosetting phenolic resin matrices reinforced with glass fibers or flakes. They have low to medium tensile strength and moduli, good electrical resistivity, good chemical resistance, very low water absorption and good overall properties. Used for electrical and electronic components, such as circuit boards, electrical housings, and gears.

glass-reinforced plastics. A group of high-performance engineering materials consisting of thermoplastic or thermosetting resin matrices reinforced with glass fibers or flakes. Abbreviation: GRP.

Glassroot. Trade name of PPG Industries Inc. (USA) for a glass-fiber product used in soil-erosion control.

glass rope. A compact, flexible, continuous line or cord woven from continuous glass filaments with a typical diameter of 6.4-19 mm (0.25-0.75 in.) and used for chemical and electrical insulation applications.

glass sand. A high-grade quartz sand composed of medium-sized grains usually between 20 and 100 mesh in size. Typically, it contains 99.0+% quartz (SiO_2), along with minor amounts of alumina (Al_2O_3), ferric oxide (Fe_2O_3), calcium oxide (CaO), and magnesium oxide (MgO). It has a density of 2.2-2.6 g/cm^3 (0.08-0.09 $lb/in.^3$) , a melting point of 1710°C (3110°F), and is used in the manufacture of glass.

glass schorl. See axinite.

Glass Sealing. See Carpenter Glass Sealing.

glass sealing alloys. See sealing alloys.

glass silk. See glass wool.

glass spar. Ground *feldspar* used in the manufacture of *flat glass* (e.g., for milk bottles, mason jars, etc.), and for enamels.

glass spheres. Glass made into small solid or hollow spheres ranging in size from about 1 μm (0.00004 in.) to over 5 mm (0.2 in.) for use as fillers and reinforcements in plastics and composites. See also glass microspheres.

glass stain. A color applied to glassware by immersing in a solution of a color-forming metal salt and heating to a temperature at which the color is developed and absorbed by the glass surface, or by subjecting the glassware to the vapors of a color-forming salt at elevated temperatures in a closed furnace.

glass textiles. Glass fibers woven or spun into textile fabrics, and used in plastic laminates, fireproof draperies and curtains, high-temperature-resistant insulation, filter fabrics, and similar products. See also textile glass.

glass-to-metal seal alloys. See sealing alloys.

glassware. Articles made of glass, and used for domestic, industrial and laboratory applications.

glass wool. A randomly oriented, fleecy mass of fine spun glass fibers that resembles the loose fibers of wool in texture. Used for acoustical and thermal insulation, air filters, packing, and other applications. Abbreviation: GW. Also known as *glass mineral wool; glass silk.*

glassy alloys. See amorphous alloys.

glassy carbon. See amorphous carbon.

glassy liquid crystal. See liquid crystal glass.

glassy material. See amorphous material; noncrystalline material.

glassy plastics. See amorphous plastics.

glassy polymers. See amorphous polymers.

Glasteel. Trademark of Pfaudler Company (USA) for a corrosion-resistant engineering composite consisting of a substrate of low-carbon steel to which glass has been chemically bonded by first applying a slurry of powdered glass and subsequently firing at high temperatures. Used for piping, valves, pump equipment, process and reaction equipment, and heat exchangers.

Glastherm. Trademark of Glastic Corporation (USA) for a thermal insulating sheet for molds composed of a heat-resistant mineral-filled thermosetting resin with glass fiber reinforcement.

Glastic. Trademark of Glastic Corporation (USA) for an extensive series of polyesters including various glass-reinforced bulk

molding compounds supplied in injection and compression molded, glass- or mineral-filled, and glass mat-reinforced laminate grades. Used for standoff insulators, topsticks, and high-temperature applications. *Glastic BMC* are polyester bulk molding compounds supplied in standard, high-heat, fire-retardant and electrical grades, and *Glastic Polyester* are polyester laminates filled with chopped glass.

Glastone. Trade name of Libbey-Owens-Ford Company (USA) for colored structural glass (*Vitrolite*) bonded to a lightweight concrete (*Haydite*).

Glastrusion. Trademark of Glastrusions, Inc. (USA) for angles, bars, channels, rods and tubes composed of glass fiber-reinforced plastics, and made by a special process involving extrusion through a die. Used as reinforcements and for structural support.

Glasuld. Trade name of Aktieselskabet Dansk Glasuldfabrik (Denmark) for a glass wool.

Glasutex. Trade name of Aktieselskabet Dansk Glasuldfabrik (Denmark) for industrial fabrics based on glass-fiber rovings and yarns.

Glas-Wich. Trade name of Dearborn Glass Company (USA) for a laminated glass used for architectural applications.

Glas-Wrap. Trade name of Royston Laboratories (USA) for impregnated fiberglass mesh used for building and construction applications.

Glattantik. Trade name of former Vereinigte Glaswerke (Germany) for a colored cast glass.

glauberite. A colorless or grayish-yellow mineral composed of sodium calcium sulfate, $Na_2Ca(SO_4)_2$. It can also be crystallized from an aqueous solutions of sodium sulfate and calcium chloride at 80°C (176°F). Crystal system, tetragonal. Density, 2.78 g/cm^3; refractive index, 1.530.

Glauber's salt. See sodium sulfate decahydrate.

glaucocerinite. A sky-blue mineral composed zinc aluminum copper sulfate hydrate, $Zn_{13}Al_8Cu_7(SO_4)_2O_{30}\cdot34H_2O$. Occurrence: Greece.

glaucochroite. A bluish green mineral of the olivine group composed of calcium manganese silicate, $(Ca,Mn)_2SiO_4$. Crystal system, orthorhombic. Density, 3.48 g/cm^3; refractive index, 1.723. Occurrence: USA (New Jersey).

glaucodot. A grayish tin-white to reddish silver-white mineral of the marcasite group composed of cobalt iron arsenic sulfide, (Co,Fe)AsS. Crystal system, orthorhombic. Density, 6.04 g/cm^3. Occurrence: Sweden.

glauconite. A green or bluish green clay mineral with earthy luster belonging to the mica group. It is composed of potassium iron aluminum and magnesium silicate hydroxide, $K_2(Mg,Fe)_2Al_6(Si_4O_{10})_3(OH)_{12}$, and contains up to 25% iron. Crystal system, monoclinic. Density, 2.3-2.9 g/cm^3. Occurrence: Venezuela. Used as an ion exchanger, e.g., in in water softeners, and for foundry molds. Also known as *greensand*.

glaucophane. (1) A blue or black mineral of the amphibole group composed of sodium magnesium iron aluminum silicate hydroxide, $Na_2(Mg,Fe,Al)_5Si_8O_{22}(OH)_2$. Crystal system, monoclinic. Density, 3.17 g/cm^3. Occurrence: Ireland.

(2) A blue mineral of the amphibole group composed of sodium magnesium aluminum silicate hydroxide, $Na_2Mg_3Al_2Si_8O_{22}(OH)_2$. It can also be made synthetically. Crystal system, monoclinic. Density, 3.14 g/cm^3; refractive index, 1.644. Occurrence: USA (California).

glaukosphaerite. A green mineral of the rosasite group composed of copper nickel carbonate hydroxide, $(Cu,Ni)_2CO_3(OH)_2$. Crystal system, monoclinic. Density, 3.78; refractive index, 1.83. Occurrence: Australia.

Glava. Trade name of Glasvatt A/S (Norway) for glass wool.

Glavex. Trade name of SA Glaverbel (Belgium) for a glass that has been obscured by acid etching or sandblasting.

glaze. (1) Any smooth, glossy, vitreous coating.

(2) A glassy or vitreous coating applied to and fired onto glass or ceramic ware to provide a smooth, protective surface.

(3) The mixture from which the coating in (2) is made.

glaze clay. Fine-grained, high-purity clay, usually a type of *ball clay*, that contains considerable amounts of colloidal organic matter, and is used as a suspension agent in glaze slips.

glazed brick. A brick having a hard surface made glossy by applying and firing on a glaze.

glazed ceramic mosaic tile. A *ceramic mosaic tile* that has received a glaze either on the face, or on other exposed surfaces.

glazed fabrics. (1) Textile fabrics, such as cotton, that have been given smooth, glossy surface finishes by first treating with glue, paraffin, starch or a synthetic resin, and then running through a calender.

(2) Textile fabrics with glossy, polished surfaces produced by the intentional or unintentional application of heat, pressure or friction.

glazed kid. A black or tan leather with glazed finish made from chrome-tanned goatskin or kidskin. The glazed finish is produced by ironing with a glass cylinder. Used especially for shoes.

glazed paper. Paper whose surface has been improved by application of a metallic powder, or a nonmetallic powder with metallic color. Also known as *bronze paper*.

glazed tile. A tile consisting of a vitreous, semivitreous or nonvitreous body coated with an impervious colored or uncolored ceramic glaze, and used for building exteriors, interiors and floors. See also facing tile; floor tile.

glazed whiteware product. A clay-based ceramic product, such as china, earthenware, porcelain or tile, to which a glassy or vitreous coating has been applied and fired on to provide a smooth nonporous, protective surface. See also whiteware.

glaze stains. Calcined ceramic pigments added to the mixture used in the manufacture of a glaze.

glazier's putty. See caulking compound (2).

glazing compound. See caulking compound (2).

gleitmo. Trademark of Gleitmolybdän Schmierstoffe GmbH (Germany) for an extensive series of high-performance products based on molybdenum disulfide and other inorganic or organic solid lubricants. They are supplied in the form of pastes, liquids (suspensions), greases, lacquers and powders, and used for the high-temperature and extreme-pressure lubrication of machine elements, and in deep drawing and hot and cold forming.

Glendion. Trademark of Montecatini Edison SpA (Italy) for polyether foam and foam products.

Glennite. Trade name for engineering ceramics based on barium titanate ($BaTiO_3$).

Gliadel. Trade name of Guilford Pharmaceuticals, Inc. (USA) for biopolymers based on *polyanhydrides* exhibiting time-controllable degradation (days or years) by surface erosion and having excellent *in vivo* compatibility. Used for drug-delivery applications, e.g., for the delivery of the chemotherapeutic cancer drug BCNU (carmustine).

GlidCop. Trademark of SCM Metal Products Inc. (USA) for a dispersion-strengthened high-performance copper composed of ultrafine aluminum oxide (Al_2O_3) particles uniformly dispersed

as a strengthening phase within the copper matrix. It has high strength, stiffness and thermal softening resistance, excellent electrical and thermal conductivity, and good high-temperature and cryogenic temperature performance. Used for complex, thin and highly stressed components.

Glidden. Trade name of Glidden Company, Division of SCM Corporation (USA) for an extensive series of paint products.

Glidshield. Trade name of Glidden Company, Division of SCM Corporation (USA) for protective polyurethane enamel paints.

Glievor Bearing. A British trade name for antifriction metals containing (i) 9% tin, 7% antimony, 5% lead, 4% copper, 1.5% cadmium, and the balance zinc; or (ii) 14% antimony, 8% tin, 1.5% iron, and the balance lead. Used for bearings.

Glillon. Trade name for nylon fibers and yarns used for textile fabrics.

Glitter. Trade name for lustrous rayon fibers and yarns used for textile fabrics.

Glixey. Trade name of Die Casting Appliance Corporation, Limited (UK) for a series of antifriction metals of lead, antimony and tin, used for bearings.

Globar. Trademark of Harbison-Carborundum Corporation (USA) for silicon carbide (SiC) with high resistivity and a service temperature up to about 1500-1600°C (2730-2910°F). Used for heating elements and electrical resistances.

Globle. Trade name of Teledyne Firth-Sterling (USA) for a water-hardening tool steel containing 1.25% carbon, 0.1% chromium, 0.3% manganese, and the balance iron used for drills, cutters and reamers.

globular powder. A powder composed of particles having approximately spherical shape.

globular proteins. A class of proteins whose molecules are folded into compact units that frequently approach spheroidal shapes. They are readily soluble in aqueous solvents and include all enzymes and many antibodies and hormones including insulin, ovalbumin, hemoglobin and fibrinogen. See also protein.

globulin. A class of simple proteins that are insoluble in water, but soluble in dilute salt solutions. They are coagulated by heat, and include serum globulin in blood, myoglobulin in muscle, lactoglobulin in milk, and edestin in hemp seed.

Glomag. Trade name of Globe Metallurgical, Inc. (USA) for a series of alloys containing 43-47% silicon, 3-10% magnesium, 0.7-1.1% calcium, 0-1.2% aluminum, 0-2% rare earth metals, and the balance iron. Used in the manufacture of ductile cast iron.

Glomar. Trademark of Dry Branch Kaolin Company (USA) for a series of calcined, fractionated kaolins supplied in a wide range of grain sizes for use as extender pigments in paints, rubber compounds, plastics, ceramics, and other materials. See also kaolin.

Glospan. Trademark of Globe Manufacturing Company (USA) for *spandex* (segmented polyurethane) fibers and yarns used for the manufacture of textile fabrics with excellent stretch and recovery properties.

gloss control agent. A *flatting agent* or *gloss enhancer* used in certain organic coatings.

gloss enhancer. A substance that promotes gloss, e.g., a clear surface film. It is the opposite of a gloss reducer. Also known as *gloss-enhancing agent*. See also gloss reducer.

gloss reducer. A substance, usually a finely ground heavy-metal soap, such as aluminum, calcium or zinc stearate, or a wax, diatomaceous earth or silica, used in paints, lacquers, varnishes and other coating materials to reduce the surface gloss or sheen of the dried film or produce a "rubbed" effect. It is the opposite of a gloss enhancer. Also known as *flatting agent; gloss-reducing agent*. See also gloss enhancer.

gloss white. A suspension of co-precipitated aluminum hydroxide (about 25%) and barium sulfate (about 75%) in water, used as a white pigment for paper and printing inks, and as an extender.

glove grain. Light, soft-finished *side leather* used for shoes.

Glowray. Trade name of Inco Alloys International Limited (UK) for a heat-resistant alloy composed of 65% nickel, 20% iron and 15% chromium. Used for furnace parts and heating elements.

glucine. A mineral composed of beryllium calcium phosphate hydroxide hydrate, $CaBe_4(PO_4)_2(OH)_4 \cdot 0.5H_2O$. Density, 2.32 g/$cm^3$. Occurrence: Russia.

gluconic acid. A monocarboxylic acid obtained by the oxidation of D-glucose, and supplied as a pure crystalline product or as a 45-50% solution in water. The solution is light brown in color, has a density of 1.24 g/cm^3, and a refractive index of 1.416. Used in the manufacture of pharmaceuticals and food products, in metal cleaning, in biochemistry and biotechnology, and for various other commercial and industrial applications. Formula: $C_6H_{12}O_7$. Also known as *glyconic acid; glycogenic acid.*

D-glucopyranose. See dextrose.

glucose. A monosaccharide formed in plants by photosynthesis, and also found in the blood and tissue fluids of humans and animals. It is utilized by the cells as a source of energy, and can also be made synthetically by the hydrolysis of cornstarch and cellulosic wastes. It is available in D-, L- and DL-forms. The D-form is also known as *dextrose. Glucose* is a colorless crystalline powder with a density of 1.544 g/cm^3, a melting point of about 146-153°C (295-307°F), and is used in foods and pharmaceuticals, in medicine, biochemistry and biotechnology, and as a source of certain amino acids. Formula: $C_6H_{12}O_6$.

D-glucose. See dextrose.

glucuronic acid. An acid that is formed by the oxidation of *glucose*, and is a constituent of many polysaccharides, especially glycosaminoglycans and certain vegetable gums, such as gum arabic. It is a crystalline compound with a melting point of 159-161°C (318-322°F) and a specific optical rotation of +36° (20°C or 68°F). Used in biochemistry, biotechnology and medicine. Formula: $C_6H_{10}O_7$. Also known as *glycuronic acid.*

glucuronide. A derivative of glucuronic acid combined with a phenol, alcohol or carboxylic acid. Also known as *glycuronide.*

glucuronoxylan. A hardwood-derived hemicellulose that contains *xylose.*

glue. See glues.

Gluebond. Trademark of 3M Company (USA) for coated abrasives.

glues. (1) A group of adhesive materials made from the hoofs, organs, skins and bones of animals by boiling in water, from casein (milk), blood or rubber, or from vegetable sources, such as soybean protein. They typically have lower bond strengths than synthetic and engineering adhesives.

(2) A term now frequently used synonymously with the term *adhesives.*

glue-laminated timber. See glue-laminated wood.

glue-laminated wood. An engineered, stress-rated wood panel constructed by gluing together two or more layers (veneers or plies) of wood. It is similar to *plywood*, but with the grain direction of all layers essentially parallel instead of turned at right angles in each successive layer. Used especially for beams,

structural members, headers, purlins, ridge beams, floor girders and bridges. Also known as *glue-laminated timber; glulam.*

glulam. See glue-laminated wood.

Gluma. Trade name of Heraeus Kulzer Inc . (USA) for a range dental bonding agents. *Gluma 2000* is used for bonding dentin and *Gluma Comfort Bond, Gluma CPS* and *Gluma One Bond* for bonding dentin and enamel.

Glu-Rific. Trademark of Itoya of America, Limited (USA) for a clear-drying *white glue* supplied in a flexible applicator. It is suitable for bonding porous and semiporous materials, and used chiefly for household and stationery applications.

glushinskite. A creamy white, white or colorless mineral composed of magnesium oxalate dihydrate, β-$MgC_2O_4 \cdot 2H_2O$. It can also be made synthetically. Crystal system, monoclinic. Density, 1.85 g/cm^3; refractive index, 1.530. Occurrence: Scotland; Arctic Russia.

gluten. A water-insoluble protein mixture that contains especially gliadin, glutenin, globulin and albumin, and is found in cereal grains, especially wheat. Used in the production of adhesives and certain amino acids.

Glutofix. Trademark of Hoechst AG, Kalle Werk (Germany) for methylcellulose-based adhesives.

Glutolin. Trademark of Hoechst AG, Kalle Werk (Germany) for methylcellulose-based adhesives.

Glycene Thermolite. Trade name of Oralite Company (UK) for a *Glyptal* resin formerly used for denture bases.

glyceraldehyde. A monosaccharide, derived from *glycerol*, that is isomeric with dihydroxyacetone and exists in D-, L- and DL-form. Used in biochemical research, nutrition, and in the preparation of polyesters, adhesives and certain biopolymers. Formula: $C_3H_6O_3$. Also known as *glyceric aldehyde.*

glycerides. A large group of neutral lipids including most of the common animal fats. Glycerides found in vegetables usually exist as oils rather than fats. In simple glycerides, such as tristearin, tripalmitin and triolein, all the fatty acids in the fat molecule are the same. In mixed glycerides, often found in naturally occurring fats, there are different fatty acids found in the same molecule, and they may contain both saturated and unsaturated fatty acids. Used in the biosciences, especially biochemistry, biotechnology, nutrition and medicine.

glycerin. See glycerol.

glycerol. A trihydric alcohol obtained as byproduct of the manufacture of soap and from a synthetic process that uses propylene. It is a clear, colorless or light yellow, hygroscopic syrupy liquid (99+% pure) that is soluble in all proportions in water and alcohol. *Glycerol* has a density of 1.249-1.265 g/cm^3, a melting point of 18-20°C (64-68°F), a boiling point of 182°C (360°F)/20 mm, a flash point of 351°F (177°C), an autoignition temperature of 739°F (392°C), and a refractive index of 1.474. It is also available on 1% divinylbenzene-crosslinked polystyrene support. Used in the manufacture of alkyd resins, as a binder for abrasives, cements and mixes, as an additive to clay to increase and maintain the pliability and workability of the mass, in dynamite and ester gums, as a plasticizer for regenerated cellulose, as an emulsifier, solvent, lubricant, penetrant and softener, in polyurethane polyols, in printing and copying inks, in special soaps, hydraulic fluids and antifreeze mixtures, and in DNA/RNA analysis, biochemistry and spectrophotometry. Formula: $C_3H_8O_3$. Also known as *glycyl alcohol.*

glycidyl novolac resins. A group of epoxy resins that are glycidyl ethers of *novolac resins*, i.e., they are made by the reaction of epichlorohydrin with any of various novolacs. Examples include glycidyl ethers of bisphenol A novolacs, glycidyl ethers of cresol novolacs, glycidyl ethers of phenolic novolacs and tetraglycidyl ethers of tetraphenol ethane. They have excellent elevated temperature performance. See also epoxy novolacs.

Glyco. Trade name of Joseph T. Ryerson & Son Inc. (USA) for a series of lead-base alloys.

Glycodur. Trade name of Glyco-Metallwerke Daelen & Hofmann KG (Germany) for a series of porous tin bronzes (composed of about 90% copper and 10% tin) sintered onto steel. Their pores are filled with a polymer, such as an acetal or polytetrafluoroethylene (PTFE), and the surface layers or running-in layers are composed of the same material. Used for building machinery, hydraulic excavators and cylinders, mechanical handling equipment, transmissions, shock absorbers, pumps, and agricultural equipment.

gycogenic acid. See gluconic acid.

glycolic acid. A colorless crystalline compound derived from chloroacetic acid ($C_2H_3O_2Cl$) or oxalic acid ($C_2H_2O_4$), and commercially available with a purity of 99%, or as a 70% solution in water. The pure grade has a melting point of 75-80°C (167-176°C), and the solution is a light straw-colored liquid with a density of 1.25 g/cm³, a melting point of 10°C (50°F), a boiling point of 112°C (234°F), and a refractive index of 1.412. Used for soldering and cleaning compounds, leather and textile dyeing, polishing compositions, electroplating and chemical milling, as a chelator, and in the synthesis of biodegradable polymers. Formula: $C_2H_4O_3$. Also known as *hydroxyacetic acid*.

glycolide. A compound that is the cyclic dimer of *glycolic acid* and has a melting point of 79-82.8°C (174-181°F). Used in biochemistry and in the synthesis of biopolymers known as "poly(glycolides)." Also known as *2,5-p-dioxanedione*. See also poly(glycolide); poly(glycolic acid).

Glyco Metal. Trade name of former Glyco Metal Company (UK) for a babbitt containing 85.5% zinc, 5% tin, 4.7% lead, 0.4% copper and 2% aluminum. Used for bushings and bearings.

glyconic acid. See gluconic acid.

glycopeptide. See glycoprotein.

glycoproteins. A group of conjugated *proteins* composed of a protein and a carbohydrate, and occurring as mucin in saliva and as mucoids in tendon and cartilage. Used in biochemistry, biotechnology and medicine. Also known as *glycopeptide.*

glycosaminoglycan. Any of several amino hexose-derived polysaccharides, such as *chitin, hyaluronic acid* and *keratin,* that are essential in restoring damaged cartilage and reduce joint inflammation in the body. Used as biomaterials in the modification of natural polymers. Abbreviation: GAG. Also known as *mucopolysaccharide.* See also polysaccharide.

Glyco Turbo. Trade name of former Glyco Metal Company (UK) for a lead alloy containing 22% antimony and 8% tin, used for turbine bearings.

glycuronic acid. See glucuronic acid.

glycyl alcohol. See glycerol.

Glydag. Trademark of Acheson Colloids Company (USA) for a lubricant composed of a solution graphite in glycerin or butylene glycol.

glyoxal. A yellow crystalline solid or pale yellow liquid that easily polymerizes in the presence of water or on standing. The solid has a density of 1.140 g/cm³, a melting point of 15°C (59°F), a boiling point of 51°C (124°F) and a refractive index of 1.383. The liquid is usually supplied in the form of a 40% solution. Used as an insolubilizing agent for proteins, gelatin, animal glue and compounds with polyhydroxyl groups, e.g.,

cellulose and starch, as a reducing agent, in permanent-press fabrics, in the dimensional stabilization of fibers such as rayon, in leather tanning, and in light and electron microscopy. Formula: $C_2H_2O_2$. Also known as *ethanedial; oxalaldehyde.*

Glyptal. Trademark of GE Plastics (USA) for a series of transparent synthetic resins based on esters of glycerol ($C_3H_8O_3$) and phthalic anhydride ($C_8H_4O_3$). They are waterproof, have high adhesiveness, and used for polymers, plasticizers in paints, varnishes, lacquers and cements, as shellac substitutes, as protective coatings for wiring and insulation, and formerly for denture bases.

gmelinite. A white, yellowish, greenish white, reddish white or colorless mineral of the zeolite group composed of sodium calcium aluminum silicate hydrate, $(Na_2,Ca)Al_2Si_4O_{12}\cdot6H_2O$. Crystal system, hexagonal. Density, 2.03 g/cm³; refractive index, 1.476. Occurrence: Ireland.

G metal. See Guillaume metal.

G-Monel. Trade name of Inco Alloys International Inc. (USA) for a series of alloys containing 63-70% nickel, 25-37% copper, 0-2.5% iron, 0-2% manganese and 0.3% carbon. They have good corrosion resistance and strength, and are used for watermeter components and machined parts. See also Monel.

GMR materials. See giant magnetoresistance materials.

GMS materials. See giant magnetostriction materials.

gneiss. A metamorphic rock similar to *granite*, but with a layered structure, and usually composed of *quartz, feldspar* and *mica* or *hornblende*. Used as a building material.

goat hair. The protein fibers obtained from the protective coats of goats (genus *Capra*). They are relatively strong, and can be blended with other fibers and spun into textile fabrics, especially clothing and upholstery.

goatskin. Vegetable-, alum-, or chrome-tanned leather obtained from the hides of mature goats and used for gloves, garments, linings, suede leather, etc. The leather from young goats is usually referred to as *kidskin.*

gobbinsite. A white mineral of the zeolite group composed of sodium calcium aluminum silicate dodecahydrate, $Na_4CaAl_6Si_{10}O_{32}\cdot12H_2O$. Crystal system, tetragonal. Density, 2.19 g/cm³; refractive index, 1.494. Occurrence: Ireland.

godlevskite. A pale yellow mineral composed of nickel sulfide, Ni_7S_6. Crystal system, orthorhombic. Density, 4.10 g/cm³. Occurrence: Russian Federation.

goedkenite. A colorless to pale yellow mineral of the brackebuschite group composed of calcium strontium aluminum phosphate hydroxide, $(Sr,Ca)_2Al(PO_4)_2(OH)$. Crystal system, monoclinic. Density, 3.83 g/cm³; refractive index, 1.673. Occurrence: USA (New Jersey).

Goetheglas. Trademark of Schott DESAG AG (Germany) for an antique glass used in the restoration of historic windows, and for external protective glazing applications.

goethite. A yellow, red or dark-brown mineral of the diaspore group composed of iron oxide hydroxide, α-FeO(OH). Crystal system, orthorhombic. Density, 4.00 g/cm³; refractive index, 2.393. Occurrence: Europe. Used as a minor ore of iron. See also aluminian goethite.

Gohi iron. A rust-resistant iron containing up to 0.02% carbon, 0.25% copper, 0.025% manganese, and 0.003% silicon. Used for sheetmetal construction work, roofing, culverts, pipes, ventilators, and skylights.

Gold. Trade name of Plettenberger Gusstahlfabrik (Germany) for high-speed steels containing 0.8-1.3% carbon, 0-2.8% cobalt, 4.3% chromium, 0.85% molybdenum, 2.1-3.8% vanadium, 12%

tungsten, and the balance iron. Used for lathe and planer tools, drills, milling cutters, and engravers tools.

gold. A heavy, ductile, yellow metallic element of Group IB (Group 11) of the Periodic Table. It is commercially available in the form of sponge, powder, shot, wire, insulated wire, foil, leaf, sheet, microfoil, microleaf, rod, ingots, tubing, aqueous colloidal suspensions, and single crystals. The crystals are usually grown by the Bridgeman technique. *Gold* occurs naturally as native gold and in tellurides. Crystal system, cubic; crystal structure, face-centered cubic. Density, 19.32 g/cm^3; melting point, 1063°C (1945°F); boiling point, 3080°C (5576°F); hardness, 20-60 Vickers; atomic number, 79; atomic weight, 196.967; monovalent and trivalent. It is a precious metal and has high malleability, high electrical conductivity (about 75% IACS), good thermal conductivity, good resistance to most acids and chemicals except aqua regia, potassium cyanide solutions and hot sulfuric acid, good corrosion resistance in air, excellent reflection of infrared radiation, and extremely high light reflectivity. Used for lining chemical equipment, in the manufacture of high-melting solders, electrical contact alloys, dental alloys, amalgams, gilding, jewelry and coinage, in electronics for infrared reflectors and other devices, and for bonding transistors and diodes to wires, metallizing ceramic and mica capacitors, and in polarographic electrodes and printed circuits, in optoelectronic devices, for space vehicle equipment, for spinnerets, in ceramics for glass and ceramic decorations, as a nucleating agent in photosensitive glass, and in gold plating anodes and laboratory ware. Symbol: Au.

gold-198. A radioactive isotope of gold with a mass number of 198 made by neutron irradiation of gold and supplied as gold metal or *colloidal gold*. It has a half-life of 64.8 hours, and is used in medicine, e.g., for radiation theraphy, in metallurgy, e.g., for solidification studies, in leak detection, and as a tracer for metallic silver. Symbol: ^{198}Au. Also known as *radiogold*.

Gold-ABA. Trademark of Wesgo Metals, Division of Morgan Advanced Ceramics Inc. (USA) for several gold-base active brazing alloys. See also ABA.

Goldal. British trade name for a corrosion-resistant gold alloy used as a solder and dental alloy.

gold alloys. A group of alloys containing gold as a minor or major constituent and including (i) alloys of gold, silver and copper used as solders, in the manufacture of jewelry and in dentistry; (ii) alloys of gold and silver with or without palladium used as electrical contact materials and in dentistry; (iii) alloys of gold and copper used as solders; (iv) alloys of gold and nickel used in electroplating; (v) alloys of gold and silicon used in solid-state electronics; and (vi) alloys of gold and indium used in electronics.

gold amalgam. A native mercury alloy containing about 30-43% gold.

Gold Anchor Drill Rod. Trade name of Teledyne Allvac (USA) for drill rod manufactured from *Red Cut Superior* tungsten-type high-speed tool steel (AISI type T1) containing 0.5-0.8% carbon, 0.25-0.4% silicon, 0.1-0.3% manganese, 17.5-18.5% tungsten, 3.75-4.25% chromium, 0.95-1.10% vanadium, and the balance iron. It has excellent resistance to decarburization, excellent machinability and wear resistance, high hot hardness and good toughness and cutting ability. It is supplied in standard shapes including rounds, flats, squares and hexagons, and used for twist drills, taps, milling cutters, shear blades, punches and a large variety of other cutting, finishing or shaping tools.

gold-antimony alloy. A gold alloy with up to 0.7% antimony

used in electronics.

gold argentide. See electrum.

Gold Art. Trade name of A.P. Green Industries, Inc. (USA) for a fireclay used as a refractory and in the industrial arts.

Goldbach-Edelstahl. Trade name of A. Goldbach KG (Germany) for a series of high-grade steels including several high-alloy steels and various low-alloy constructional steels.

gold beryl. See chrysoberyl.

Gold Bond. Trade name of National Gypsum Company (USA) for an extensive series of building products including asbestos-cement boards, asbestos-cement siding shingles, flat and corrugated asbestos-cement sheeting and paneling, corrugated asbestos siding and roofing materials, asbestos ceiling tiles, canal bulkheading, asbestos and non-asbestos thermal insulation panels, flexible plastic and/or mineral-based sheeting and boards, all-weather caulking compounds, ready-mixed joint cements and compounds, sprayable and/or trowelable textured wall finishes, mortar mixes, sprayable and/or trowelable acoustical and patching plasters, exterior stucco compounds, plastic- or mineral-based topping cements and compounds, spackling pastes, and several laminating adhesives.

Gold Bond R. Trade name of Unimin Specialty Minerals Inc. (USA) for ground silica (SiO_2) used as an extender in adhesives and paints.

gold bromide. See aurous bromide; gold tribromide.

gold bronze. (1) An aluminum bronze containing 10% aluminum, and the balance copper. It can be produced in powder form by reduction from gold leaf and subsequent mechanical polishing and coating with stearic acid. Used as a pigment in paints and inks, and as a gold substitute. Also known as *aluminum bronze powder; gold bronze powder.*

(2) An aluminum bronze containing about 3-5% aluminum, and the balance copper. It has good corrosion resistance and cold workability, and is used for condenser, evaporator and distiller tubes.

(3) A corrosion-resistant bronze composed of 90.5% copper, 6.5% tin and 3% zinc, and used for ornaments and jewelry.

(4) A bronze composed of 89.5% copper, 2% tin, 5.5% zinc and 3% lead, used for architectural castings.

gold bronze powder. See gold bronze (1).

Gold Buttons. A British trade name for a free-cutting brass containing 59% copper, 33% zinc, 5% tin, and 3% lead. Used for buttons and jewelry.

gold chloride. See aurous chloride; gold trichloride.

Gold-Chrom. German trade name for a resistance alloy containing up to 10% chromium, and the balance gold. It has high electrical resistivity (about 3-115 μΩcm at room temperature), a low coefficient of expansion, and is used for rheostats.

Goldchrom. German trade name for a resistance alloy containing 97.95% gold and 2.05% chromium. It has high electrical resistivity (about 33 μΩcm at room temperature), a low coefficient of expansion, and is used for precision resistors.

gold coatings. Corrosion-resistant layers of gold of varying thickness applied to metal or nonmetal substrates by electroplating, high-vacuum evaporation, or sputtering. Used as relatively thick solderable coatings on circuits and other electronic components, as conductive coatings on specimens for scanning-electron microscopy, and as thin, decorative coatings on jewelry and ornamental items. Also known as *gold plates.*

gold-colored brass plate. A decorative coating of yellow brass (85-90% copper and 10-15% zinc) resembling gold in color, and applied by electroplating to hardware, picture frames, ornamental items and cosmetic articles.

gold-copper solders. High-temperature brazing alloys typically containing 34.5-79.5% gold and 20.5-65.5% copper and having brazing temperatures above 890°C (1634°F). Used for brazing cobalt, nickel and iron-base alloys requiring oxidation and corrosion resistance.

Goldcres. Trade name for polyolefin fibers used for textile fabrics.

gold cyanide. See auric cyanide; aurous cyanide.

gold dust. Gold powder consisting of fine particles or flakes and used for coatings, jewelry and ornaments, in the manufacture of gold ruby glass, and in solid-state electronics. See also gold flake; gold flour; gold powder.

golden antimony sulfide. See antimony sulfide (ii).

golden beryl. See chrysoberyl.

Golden Brass. Trade name of LeaRonal, Inc. (USA) for a brass electroplating formulation. See also gold-colored brass plate.

Golden Caprolan. Trade name of Allied Chemical Corporation (USA) for nylon fibers and yarns used for textile fabrics especially clothing.

Golden Ceramic. Trademark of Ivoclar Vivadent AG (Liechtenstein) for a dental ceramic alloy containing 86.9% gold, 8.0% platinum, 2.5% palladium and less than 1.0% silver, tin, indium, ruthenium, rhenium, tantalum, iron and lithium, respectively. It has a rich yellow color, a density of 18.5 g/cm^3 (0.67 $lb/in.^3$), a melting range of 1060-1140°C (1940-2084°F), a hardness after ceramic firing of 165 Vickers, moderate elongation and excellent biocompatibility. Used for crowns, onlays and bridges.

Golden Decade. Trade name of Del Paint Manufacturing Corporation (USA) for interior and exterior enamel paints.

Goldenedge. Trademark of General Magnaplate Corporation (USA) for a microthin coating with dense, fine-grain microstructure. It provides ultrahigh surface hardness and low surface tension, and is applied to the cutting edges of cutting tools (e.g., blades or knives) to maintain sharpness.

Golden Extra. Trade name of Marson Corporation (USA) for autobody fillers.

golden fir. See red fir.

Golden Glow. Trade name for polyester fibers and yarns used for textile fabrics, especial clothing.

Golden Multibond. Trade name for a high-gold dental bonding alloy.

golden ocher. A mixture of *ocher, chrome yellow* and *whiting* used as a pigment.

Goldent. Trademark of Williams Gold Refining Company (USA) for a pure gold used for dental filling applications.

Golden Touch. Trade name for polyester fibers and yarns used for textile fabrics.

goldfieldite. An iron-black mineral of the tetrahedrite group composed of copper antimony arsenic tellurium sulfide, $Cu_{12}(Te,As,Sb)_4S_{13}$. Crystal system, cubic. Density, 4.95 g/cm^3. Occurrence: USA (Nevada).

gold-filled metal. A relatively inexpensive base metal, such as *brass* or *cupronickel*, clad or plated on one or more surfaces with a thin layer of gold alloy (10 karat or above) with the amount of gold alloy being at least 5% of the total weight. Used for jewelry and ornaments.

gold filler metals. A group of filler metals used for brazing stainless steel and copper alloys. They contain varying amounts of gold, copper, palladium and nickel, and are considered high-cost compositions restricted to highly specialized applications, such as the fabrication of aerospace equipment, and for joining vacuum components.

gold-film glass. Glass that has a thin, electrically heatable film of gold incorporated. It can be demisted and deiced, and is used for windshields and windows.

gold-flash palladium nickel. Palladium nickel coatings with a very thin electrodeposited film of gold used in the electric connector industry and as a replacement for gold in the manufacture of electronic packaging. Abbreviation: GFNiPd.

gold flake. Gold in the form of small, flat, scalelike particles or platelets of varying size, but usually less than 1 μm (40 μin.) in thickness, used as radiation-control coatings on spacecraft.

gold flour. A very fine grade of *gold dust* that will float when sprinkled on water. Also known as *flour gold.*

gold foil. A sheet of gold, usually 0.00025-0.5 mm (0.00001-0.02 in.) thick, made by beating or hammering. Gold microfoil in thicknesses from 0.001 to 1 μm (0.04 to 40 μin.) on permanent *Mylar* support is also available.

gold-gallium alloys. Gold-based alloys that contain small percentages of gallium, and are used in electronics.

gold-germanium solders. A group of gold-base special-purpose solders containing varying amounts of germanium (typically 8-12%). The eutectic solder contains 88% gold and 12% germanium and melts at 356°C (673°F). Supplied in foil and lump form, gold-germanium solders have high corrosion resistance, good wettability and strength, and are compatible with semiconductor materials (e.g., silicon). Used for soldering semiconductor devices and solid-state electronics.

gold-graphite. A powder-metallurgy material made by blending gold powder with graphite, compacting at ambient or elevated temperatures, and subsequent sintering to the desired shape. The typical composition range is 85-99% gold and 1-15% graphite. Used for electrical contacts.

gold hydrate. See gold hydroxide.

gold hydroxide. A brown, light-sensitive powder used in gilding liquids, in the decoration of porcelain and other ceramics, and in gold plating. Formula: $Au(OH)_3$. Also known as *auric hydroxide; gold hydrate.*

Gold Imitation. A British trade name for a corrosion-resistant alloy resembling gold in appearance, but composed of 97.8% copper, 2% aluminum and only 0.2% gold.

gold-indium plate. A hard, bright, wear-resistant electrodeposited coating produced on a substrate by adding indium to the gold plating bath. Used on transistors and other electronic components.

gold-indium solders. A group of gold-base solders containing varying amounts of indium (typically 18-22%). They have high corrosion resistance, good wettability and strength, and are compatible with semiconductor materials (e.g., silicon). The brazing temperature range is 485-500°C (905-932°F). Used for soldering semiconductor devices and solid-state electronics.

gold iodide. See aurous iodide; gold triiodide.

Gold Label. Trade name of Hidalgo Steel Company Inc. (USA) for a series of cobalt-tungsten high-speed steels used for cutting tools.

gold latten. Gold in very thin sheets used for decorative applications.

gold leaf. Gold made into extremely thin foils or leaves by hammering or rolling. The typical thickness range is 0.001-1.0 μm (0.04-40 μin.), and it is usually available in books with interleaves, or as square or round leaves on plastic support. The 0.001 μm (0.04 μin.) leaf or foil has a weight of 1.9 $μg/cm^2$ and a purity of 99.99+%. Used for gilding books and art objects, embossing leather and fabrics, and for architectural applications.

GoldLink. Trade name for dental adhesives used for bonding synthetic resins to gold and other dental alloys.

goldmanite. A dark to brownish green mineral of the garnet group composed of calcium vanadium iron aluminum silicate, $Ca_3(V,Fe,Al)_2(SiO_4)_3$. Crystal system, cubic. Density, 3.74 g/cm^3; refractive index, 1.821. Occurrence: USA (New Mexico). Used in ceramics.

Goldmaster. Trade name of Herbert Cutanit Limited (UK) for a series of sintered carbide tool materials used for coated throwaway tips employed for light or heavy cutting of steel and nonferrous metals.

Gold Metal Leaf. A British trade name for a corrosion-resistant alloy composed of 66-84% copper, 16-34% zinc and 0-4% lead, used as a gold solder.

Gold Nugget. Trademark of Westroc Industries Limited (Canada) for expanded vermiculite.

Gold N Wear. Trade name of Advanced Chemical Company (USA) for a gold electrodeposit and plating process used on jewelry.

gold oxide. See aurous oxide; gold trioxide.

gold-palladium coatings. Thin, conductive coatings, usually only a few nanometers in thickness, frequently applied to nonconductive scanning-electron microscopy samples by sputtering or evaporation techniques to prevent surface charging.

gold plates. See gold coatings.

gold-potassium chloride. Yellow crystals used in the decoration of glass and ceramics, and in photography. Formula: $AuCl_3 \cdot KCl \cdot 2H_2O$. Also known as *gold-potassium dichloride; potassium aurichloride; potassium chloroaurate; potassium chloroaurate dihydrate; potassium-gold chloride.*

gold-potassium cyanide. A white, crystalline powder (98+% pure) with a density of 3.45 g/cm^3, used in electrogilding. Formula: $KAu(CN)_2$. Also known as *potassium cyanoaurite; potassium dicyanoaurate; potassium dicyanoaurate; potassium-gold cyanide.*

gold-potassium dichloride. See gold-potassium chloride.

gold powder. Gold in the form of a fine powder (99.99+% pure) made by atomization and chemical reduction. The average particle size range is 1-595 μm (0.00004-0.0234 in.). Atomized gold powder has spherical crystalline particles, and chemical gold powder is amorphous. Used for coatings, and for soldering applications in solid-state electronics.

Goldpunkt. Trade name of SWB Stahlformguss Gesellschaft mbH (Germany) for a high-speed steel containing 0.86% carbon, 4.1% chromium, 0.85% molybdenum, 2.5% vanadium, 12% tungsten, and the balance iron. Used for lathe and planer tools, and various other cutting and finishing tools.

Goldric. Trade name of Uyemura International Corporation (USA) for an electroless gold coating.

gold ruby glass. A deep-red *ruby glass* whose color is due to the addition of *gold dust* to the melt, and thermal treatment involving cooling and subsequent heating to high temperatures. The gold dissolves in the glass forming *colloidal gold* that in turn produces the characteristic red color.

gold salt. See gold-sodium chloride.

goldschmidtine. See stephanite.

goldschmidtite. See sylvanite.

gold-silicon alloys. Alloys of gold and silicon usually made in the form of extremely thin films (typically 10 μm or 0.4 mil thick) by rapid cooling of the molten alloy on a rotating wheel.

Used for solid-state electronics. Also known as *silicon-gold alloys.*

gold-silicon solders. A group of gold-base solders containing varying amounts of silicon (up to 5%). They have high corrosion resistance, good wettability and strength, and are compatible with semiconductor materials (e.g., silicon). The brazing temperature is above 370°C (700°F), and they are used for semiconductor devices and other solid-state electronics.

Goldsmith. Trade name of NL Industries (USA) for an extensive series of copper- and silver-based brazing alloys.

gold-sodium chloride. Yellow crystals that decompose at 100°C (212°F) and are used for staining fine glass, in the decoration of glass, porcelain and other ceramics, and in photography. Formula: $NaAuCl_4 \cdot 2H_2O$. Also known as *gold salt; sodium aurichloride; sodium chloroaurate; sodium-gold chloride.*

gold-sodium cyanide. A yellow powder that contains 46% or more gold, and is used in gold electroplating of electronic components. Formula: $NaAu(CN)_2$. Also known as *sodium aurocyanide; sodium cyanoaurite; sodium-gold cyanide.*

gold solders. A group of corrosion-resistant solders that are composed of 45-85% gold, 3-35% silver, 7-22% copper, 2-3% tin and 2-4% zinc. Their solidus and liquidus temperatures tend to increase with the gold content, and the average fusion range is about 680-870°C (1255-1600°F). Used mainly for dental and electronic applications, and in the manufacture of jewelry.

Gold Spezial. Trade name of Plettenberger Gusstahlfabrik (Germany) for a high-speed steel containing 1.35% carbon, 2% cobalt, 4.3% chromium, varying amounts of vanadium and tungsten, and the balance iron. Used for milling cutters, form cutters, and engravers tool.

Gold Spray. Trade name of Chance Brothers Limited (UK) for decorative table glassware.

Gold Star. Trade name of Carpenter Technology Corporation (USA) for a tungsten-type high-speed steel (AISI type T8) containing 0.77% carbon, 13.7-14% tungsten, 5% cobalt, 3.75-4% chromium, 2% vanadium, and the balance iron. It has high hot hardness and excellent wear resistance, and is used for tools for heavy-duty machining, and for high-speed cutting tools.

Goldstar. Trade name of J.F. Jelenko & Company (USA) for a microfine dental casting alloy composed of 60% palladium, 26% silver and 2% gold. It has a density of 10.7 g/cm^3 (0.39 lb/in.^{3}), good strength and castability, and medium hardness. Used for fusing porcelain to metals.

goldstone glaze. An *aventurine* glaze composed chiefly of basic lead carbonate, $2PbCO_3 \cdot Pb(OH)_2$, feldspar, $K_2O \cdot Al_2O_3 \cdot 6SiO_2$ and flint (SiO_2) with small additions of ferric oxide (Fe_2O_3) and whiting ($CaCO_3$).

gold-tin precipitate. See gold-tin purple.

gold-tin purple. A brown powder obtained by reacting a neutral solution of *gold trichloride* with a mixture of stannous and stannic chlorides, and available as a mixture of colloidal gold and tin oxide in varying proportions. Used in the manufacture of ruby glass, various glazes, in coloring enamels, and in painting and staining glass and porcelain. Also known as *gold-tin precipitate; purple of Cassius.*

gold-tin solders. A group of gold-base solders containing varying amounts of tin (typically up to 20%). They have high corrosion resistance, good wettability and strength, and are compatible with semiconductor materials (e.g., silicon). The brazing temperature is about 280°C (535°F), and they are used for soldering semiconductor devices and other solid-state electronics.

Gold Tip. Trade name of LaSalle Steel Company (USA) for a free-machining steel containing 0.08-0.13% carbon, 0.6-0.9% manganese, 0.24-0.33% sulfur, and the balance iron. Used for automatic screw-machine parts.

gold tissue. A soft, transparent, elegant fabric made with gold-colored metal warp threads and silk or synthetic weft (filling) threads. Used especially for fine evening wear.

gold-titanium. Any of the following intermetallic compounds of gold and titanium used in ceramics and materials research: (i) AuTi. Melting point, 1488°C (2710°F); and (ii) Au_2Ti. Melting point, 1465°C (2670°F).

gold tribromide. An orange-brownish, hygroscopic powder (99.9% pure) that decomposes at 160°C (320°F). Used in chemical analysis, biochemistry and medicine. Formula: $AuBr_3$. Also known as *auric bromide; gold bromide*.

gold trichloride. Maroon, hygroscopic crystals (99+% pure). Density, 3.9 g/cm³; melting point, decomposes at 254°C (489°F); boiling point, sublimes at 265°C (509°F). Used with a mixture of stannous and stannic chlorides to produce *gold-tin purple* (purple of Cassius), in coloring or decoration of glass and ceramic ware, in the production of ruby reds in glasses, glazes and porcelain enamels, in the manufacture of finely divided gold, in gold electroplating, and in the biosciences. Formula: $AuCl_3$. Also known as *auric chloride; gold chloride*.

gold triiodide. A green powder used in chemistry and electroplating. Formula: AuI_3. Also known as *auric iodide; gold iodide*.

gold trioxide. An orange-brown to brownish-black, light-sensitive powder typically containing 85-86% gold. It is decomposed by heat and loses oxygen at 250°C (482°F). Used in gold electroplating, gilding, and in the decoration of porcelain and other ceramics. Formula: Au_2O_3. Also known as *auric oxide; auric trioxide; gold oxide*.

gold-uranium. An intermetallic compound of gold and uranium that has a melting point above 1450°C (2640°F), and is used in ceramics. Formula: Au_3U.

gold-zirconium. An intermetallic compound of gold and zirconium that has a melting point of 1560°C (2840°F), and is used in ceramics. Formula: Au_3Zr.

Golfalloy. Trade name of Certified Alloy Products, Inc. (USA) for a corrosion-resistant, investment-cast steel containing 0.08-0.2% carbon, 15-16.5% chromium, 1.5-2.2% nickel, up to 1% manganese, 1% silicon, 0.5% copper and 0.5% molybdenum, respectively, and the balance iron. Used for golf-club heads.

Gollet Steel. Trade name of Bethlehem Steel Corporation (USA) for a water-hardening tool steel containing 0.95% carbon, 0.7% manganese, and the balance iron.

Gölzathen. Trade name of Gölzaplast (Germany) for a polyethylene film material.

Gomglas. Trade name of Splintex Belge SA (Belgium) for a laminated glass made by the synthetic resin process.

Gomak. French trade name for a series of zinc die-casting alloys containing 3.5-5% aluminum, 0-3.5% copper and 0.02-0.1% magnesium. Used for hardware, motor frames, gear housings and fuel pumps.

goncalo alves. The strong, dense wood of several trees of the species *Astronium*, especially *A. graveolens* and *A. fraxinifolium*. Its heartwood has good moisture and fungus resistance, turns very well and finishes smoothly. Source: Southern Mexico, Central America and northern South America, e.g., in the Amazon Basin. Used for turnery, carving, and specialty items, such as archery bows and billiard cue butts.

gong metal. A bronze composed of 78% copper and 22% tin, used for gongs.

gonnardite. A white mineral of the zeolite group composed of calcium sodium silicate aluminum oxide heptahydrate, $CaNa_4Si_6Al_4O_{20}\cdot7H_2O$. Crystal system, orthorhombic. Density, 2.26 g/cm³. Occurrence: France.

gonyerite. A deep brown mineral of the chlorite group composed of magnesium manganese iron silicate hydroxide, $(Mn,Mg,Fe)_6Si_4O_{10}(OH)_8$. Crystal system, orthorhombic. Density, 3.01 g/cm³. Occurrence: Sweden.

gooseberry stone. See grossularite.

goosecreekite. A colorless mineral of the zeolite group composed of calcium aluminum silicate pentahydrate, $CaAl_2Si_6O_{16}\cdot5H_2O$. Crystal system, monoclinic. Density, 2.21 g/cm³; refractive index, 1.498. Occurrence: USA (Virginia).

gopher plum. See ogeche tupelo.

gorceixite. (1) A brown mineral of the alunite group composed of barium calcium aluminum phosphate hydroxide monohydrate, $(Ba,Ca)Al_3(PO_4)_2(OH)_5\cdot H_2O$. Crystal system, tetragonal. Density, 3.10 g/cm³; refractive index, 1.625. Occurrence: Brazil, Ireland.

(2) A colorless mineral of the alunite group composed of barium aluminum phosphate hydroxide monohydrate, $BaAl_3(PO_4)_2(OH)_5\cdot H_2O$. Crystal system, tetragonal. Density, 3.10 g/cm³; refractive index, 1.625. Occurrence: Australia.

Gordon. Trade name of Latrobe Steel Company (USA) for a tough tool steel containing 0.6% carbon, 0.6% chromium, 0.3% manganese, 0.4% molybdenum, 0.2% vanadium, 0.7% silicon, and the balance iron. Used for cutters, bending dies, and tools. *Gordon Die Steel* is a tough tool and die steel containing 0.57% carbon, 0.7% silicon, 0.6% chromium, 0.35% molybdenum, and the balance iron. Used for swaging dies, chisels, and pneumatic tools.

gordonite. A colorless mineral of the paravauxite group composed of magnesium aluminum phosphate hydroxide octahydrate, $MgAl_2(PO_4)_2(OH)_2\cdot8H_2O$. Crystal system, triclinic. Density, 2.23 g/cm³; refractive index, 1.543. Occurrence: USA (Utah).

gorgeyite. A colorless mineral of the zeolite group composed of potassium calcium sulfate monohydrate, $K_2Ca_5(SO_4)_6\cdot H_2O$. Crystal system, monoclinic. Density, 2.93 g/cm³; refractive index, 1.569. Occurrence: Austria.

gormanite. A blue green mineral composed of iron aluminum phosphate hydroxide dihydrate, $Fe_3Al_4(PO_4)_4(OH)_6\cdot2H_2O$. Crystal system, triclinic. Density, 3.13 g/cm³; refractive index, 1.653. Occurrence: Canada (Yukon).

gortdrumite. A mineral composed of copper iron mercury sulfide, $(Cu,Fe)_6Hg_2S_5$. Crystal system, tetragonal. Density, 6.80 g/cm³. Occurrence: Ireland.

goslarite. A colorless mineral of the epsomite group composed of zinc sulfate heptasulfate, $ZnSO_4\cdot7H_2O$. It can also be made synthetically. Crystal system, orthorhombic. Density, 1.98 g/cm³; refractive index, 1.480. Occurrence: Ireland.

Göttingen. Trade name of Göttingen Aluminiumwerke GmbH (Germany) for a series of aluminum products including commercially pure aluminum and several wrought aluminum-magnesium and aluminum-manganese alloys. Also known as *Gottingen*.

gotzenite. A mineral composed of sodium calcium titanium fluoride silicate, $(Na,Ca,Al)_7(Si,Ti)_3O_{15}F_{3.5}$. Crystal system, triclinic. Density, 3.14 g/cm³; refractive index, 1.662. Occurrence: Central Africa.

goudeyite. A mineral of the mixite group composed of copper aluminum arsenate hydroxide trihydrate, $Cu_6Al(AsO_4)_3(OH)_6\cdot$ $3H_2O$. Crystal system, hexagonal. Density, 3.50 g/cm³; refractive index, 1.704. Occurrence: USA (Nevada).

Goulard's powder. See lead acetate.

Goutte d'Eau. Trade name of Glacerie de Franière SA (Belgium) for a patterned glass.

Government Bronze. Trade name of Century Brass Products, Inc. (USA) for a leaded bronze containing 86-89% copper, 7.5-11% tin, 1.5-4.5% zinc, 0.75% nickel and 0.3% lead. Used for valves, gears, fittings, bearings and bushings.

Government Genuine Babbitt. Trade name of United American Metals Corporation (USA) for a babbitt containing lead, antimony and tin. It has good shock resistance, and is used for bearings.

Government Rubber-Styrene. See GR-S rubber.

gowerite. A white or colorless mineral composed of calcium borate pentahydrate, $CaB_6O_{10}\cdot 5H_2O$. Crystal system, monoclinic. Density, 2.00 g/cm³; refractive index, 1.501. Occurrence: USA (California).

goyazite. A colorless or yellowish-white mineral of the alunite group composed of strontium aluminum phosphate hydroxide monohydrate, $SrAl_3(PO_4)_2(OH)_5\cdot H_2O$. Crystal system, rhombohedral (hexagonal). Density, 3.22 g/cm³; refractive index, 1.6397. Occurrence: Brazil, Canada (Yukon).

GPC Delta. Trade name of Hsimex International Inc. (USA) for a series of impact-resistant acrylonitrile-butadiene-styrene terpolymers and styrene-acrylonitrile copolymers available in ultra-high-impact, high-temperature, high-modulus, medical and food and drug grades. Used for structural applications, beverage containers, cosmetioc and pharmaceutical packaging, and medical implants.

GPX BioCoatings. Trade name of General Plastics (USA) for a series of biocompatible thermal-spray coatings supplied in various grades including cobalt-chromium and commercially pure titanium. They can be applied by various thermal spraying processes including high-velocity oxyfuel (HVOF), and are used to improve the joining of bone to prosthetic devices.

Grac. Trade name of Graphitized Alloy Corporation (USA) for an alloy of lead, tin, antimony, arsenic, cadmium, nickel and graphite. It has a melting range of 234-302°C (453-576°F), and is used for bearings.

Gracite. Trade name of Grayborn Steel Company (USA) for an oil-hardening, nondeforming tool steel containing 0.9% carbon, 1.25% manganese, 0.5% chromium, 0.5% tungsten, and the balance iron. Used for forming and beading dies, upsetters, and punches.

gradated coating. See graded coating.

graded abrasive. An abrasive powder that is available in specified particle size ranges only, e.g., in 220-360 mesh or 120-180 mesh grades, but not in intermediate grades.

graded aggregate. Aggregate classified according to size, and usually available in continuous particle sizes ranging from coarse to fine.

graded coating. A deposit composed of several layers of different materials that progressively change in composition from 100% undercoating material to 100% top layer material. For example, a deposit composed of a metallic undercoating, an intermediate cermet coating and a ceramic top layer on a metallic substrate. Graded coatings are often applied by thermal spray processes. Also known as *gradated coating; gradient coating; graduated coating.*

graded sand. Sand whose particles are of fine, medium or coarse size, but have no continuous grading.

graded standard sand. Standard *Ottawa sand* that will pass the No. 30 (600 μm) and be retained on the No. 100 (150 μm) US Standard Sieve. Used for testing cements.

gradient coating. See graded coating.

graduated coating. See graded coating.

graemite. A blue-green mineral composed of copper tellurate monohydrate, $CuTeO_3\cdot H_2O$. Crystal system, orthorhombic. Density, 4.13 g/cm³; refractive index, 1.960. Occurrence: USA (Arizona).

Grafene. Trade name for a *colloidal graphite* in oil or grease, used as a lubricant.

Grafidin. Trade name of Pechiney Electrométallurgie (France) for a cast iron *inoculant* containing 63% silicon, 19.5% iron, 9% manganese, 4.5% barium, 2.5% calcium, and 1.5% aluminum.

Grafil. (1) Trademark of Grafil Inc. (USA) for standard- and high-modulus carbon fibers used as reinforcements in engineering polymers.

(2) Trademark of Grafil Inc. (USA) for a series of carbon fiber-reinforced engineering plastics including *Grafil DG* carbon fiber-reinforced polycarbonates, *Grafil GG* carbon fiber-reinforced polysulfones, *Grafil JG* carbon fiber-reinforced polyether sulfones, *Grafil OG* carbon-fiber-reinforced polyphenylene sulfides, *Grafil PVDF* carbon-fiber-reinforced polyvinylidene fluorides and *Grafil RG* carbon fiber-reinforced polyamide 6,6 (nylon 6,6) resins.

Graflok. Trade name of Premier Refractories and Chemicals Inc. (USA) for refractory graphitic plastic mixes.

Graflon. Trade name for a modacrylic fiber with excellent moth and mildew resistance, relatively high tenacity and abrasion resistance, good resistance to acids and alkalies, and fair to poor resistance to ketones. Used chiefly for clothing and industrial fabrics.

Grafoam. Trademark of Poco Graphite (USA) for a nearly 100% graphite foam with open-celled structure, high conductivity and low density. Used for heat-transfer applications in cooling systems for satellites and computers, and as heat sinks for power electronics.

Grafoil. Trademark of UCAR Carbon Technology Corporation (USA) for graphite supplied in various forms including flexible webs, fabrics, plates, sheets, strips, tapes, ribbons, strands and filaments. The graphite plates are made from flexible graphite and used in batteries, fuel cells and electrical devices for electrodischarge (EDM) machining applications. The directional, flexible, pure graphite tapes are used for thermal insulating applications. Also included under this trademark are graphite foams and graphite foam products as well as graphite pastes for sealing and lubricating screw and pipe threads.

Grafseal. Trademark of Anchor Packing, Division Robco Inc. (Canada) for graphite filament-based braided packing.

graft copolymer. A copolymer in which the sequence of repeating units of one monomer is present as side chains or branches on the main polymer chain. It is usually referred to as poly(A)-*graft*-poly(B). For example, a copolymer of main-chain propylene and and side-chain maleic anhydride may be referred to as poly(propylene-*graft*-maleic anhydride). Also known as *graft polymer.* See also copolymer.

Graham Fiber Glass. Trademark of Graham Fiber Glass Limited (Canada) for acoustical and thermal fiberglass insulation supplied in the form of batts and blowing wool.

grahamite. A solid, brittle, pure, jet-black *bitumen* found in the USA (Colorado and Oklahoma), Argentina, Mexico and Trinidad. It has a density of 1.1-1.5 g/cm³ (0.04-0.05 lb/in.³), and is used for molding materials, electrical insulation products, and in paints.

Grainal. Trade name of Shieldalloy Metallurgical Corporation (USA) for a series of master alloys of varying composition. A typical composition is 20% titanium, 13% aluminum, 8% manganese, upto 5% silicon, 3.5-4% zirconium, 0.5-1% boron, and the balance iron. Special grades composed of 13-25% vanadium, 15-20% titanium, 10-12% aluminum and 0.2% boron with the balance being iron, are also available. *Grainal* master alloys are used as ladle additions in steelmaking·

grain board. A wall covering panel consisting of a fireproof gypsum core and an imitation wood-grain surface.

grain-boundary-engineered materials. (1) Polycrystalline materials in which the structure and frequency of special grain boundaries and triple junctions are controlled to improve mechanical, physical and chemical properties.

(2) Nanocrystalline materials in which the overall density of grain boundaries and triple junctions has been drastically increased, and which consequently exhibit greatly improved and, in many cases, unique properties as compared with their conventional crystalline or amorphous counterparts.

grained leather. Leather whose natural grain has been synthetically softened, raised or otherwise altered.

graining oxide. A finely milled mixture of ceramic pigments that may contain small amounts of fluxing ingredients and is incorporated into graining pastes.

graining paste. A well-blended oil suspension of graining oxides used in the decoration of porcelain enamels and glaze surfaces by the rubber-roll process.

grain leather. Leather made from the dehaired side of skins or hides.

Grainloy. Trade name of Birdsboro Corporation (USA) for a cast iron containing 3% carbon, 1.3% silicon, varying amounts of manganese, and the balance iron. Used for rolls.

grain magnesite. Dead-burnt magnesia in granular form that has been produced by calcining magnesite ($MgCO_3$) at temperatures above 1450°C (2640°F). See also dead-burnt magnesia.

grain-oriented electrical steel. See oriented electrical steel.

grain-oriented sheet. See oriented electrical steel.

grain-oriented silicon steel. See oriented electrical steel.

grain refiner. An alloying element added to molten metals or alloys to restrict grain growth and promote fine-grained microstructures. For example, aluminum, molybdenum, titanium and vanadium are used as grain refiners for steels.

grain-refining inoculant. A master alloy, such as a titanium or titanium-boron master alloy, added to an aluminum alloy melt to modify the solidification structure of cast ingots by promoting the growth of fine, uniform grains and reducing the formation of center cracks.

Gralur. Trade name of Duke Steel Company, Inc. (USA) for a tool steel used for drills.

Gramix. Trade name of US Graphite Inc. (USA) for a series of ferrous and nonferrous materials made by powder-metallurgy techniques, i.e., by blending, compacting and subsequent sintering. The ferrous materials include various compositions, such as Fe, Fe-C, Fe-Cu-C, Fe-Cu-Pb-C and Fe-Ni-C, with medium to high density. The nonferrous materials include compositions, such as Cu, Cu-Sn, Cu-Sn-Ni, Cu-Sn-Pb-Ni, Cu-Sn-C, Cu-Zn, Cu-Zn-Pb and Cu-Zn-Ni with low to medium density. *Gramix*

powder-metallurgy materials are used for bearings, bushings, rotors, pole pieces, valve plates, and structural parts.

Granada. Trade name of Crucible Materials Corporation (USA) for a general-purpose, water-hardening tool steel containing 1% carbon, 0.3% manganese, 0.25% silicon, and the balance iron. *Granada Vanadium* is a water-hardening tool steel containing 1% carbon, 0.3% manganese, 0.2% vanadium, and the balance iron. It is used for punches, shears, dies and forging tools.

granada. A fabric with a grainy texture made with a cotton warp and a fine alpaca or mohair weft (filling) in a broken twill weave.

Granada Tool. Trade name of Colt Industries (UK) for a water-hardening tool and die steel containing 1% carbon, and the balance iron.

Granal. Trade name of Eisenwerk Würth GmbH & Co. (Germany) for light-metal blasting abrasives. *Granal-Filgra* are abrasive blast-cleaning media based on corundum.

Granalec. Trade name of National Intergroup Inc. (USA) for an iron-silicon alloy with high magnetic permeability used for transformer laminations.

Granator. Trade name of National Intergroup Inc. (USA) for a high-permeability iron-silicon soft magnetic alloy used for transformer laminations.

Granature. Trade name of National Intergroup Inc. (USA) for an iron-silicon alloy with high magnetic permeability used for armatures and transformer laminations.

Grancon. Trade name of Grace GmbH (Germany) for several polyvinyl chlorides.

grandidierite. A greenish blue mineral composed of magnesium iron aluminum boron silicate, $(Mg,Fe)Al_3BSiO_9$. Crystal system, orthorhombic. Density, 2.98 g/cm³; refractive index, 1.618. Occurrence: Madagascar.

grandrelle yarn. A *plied yarn* made by combining two yarns of different color or luster.

Grane. Trade name of Uddeholm Corporation (USA) for a series of tough medium-carbon tool steels containing 0.55% carbon, 1-1.5% chromium, 3% nickel, up to 0.3% molybdenum, and the balance iron. Used for drop forging dies.

Granellino. Trade name of Vetreria di Vernante SpA (Italy) for a patterned glass.

Granex. Trademark of Graham Products Limited (Canada) for exposed-aggregate exterior facade panels.

Granfin Cast Iron. A British trade name for a cast iron containing 2.2-2.6% carbon, 0.7-1% manganese, 2.3-2.5% silicon, 1.7-2.1% graphite, and the balance iron. Used for frames, housings, and general castings.

Granfo. Trademark of US Granules Corporation (USA) for aluminum granules produced from recovered aluminum foil products.

Granformer. Trade name of National Intergroup Inc. (USA) for a series of soft magnetic iron materials that contain 2.5-5% silicon, have high magnetic permeability, and are used for transformer laminations.

GrA-Ni. Trademark of Inco Limited (Canada) for a lightweight, wear-resistant metal-matrix composite consisting of cast aluminum reinforced with silicon carbide (SiC) and nickel-coated graphite particles added to the aluminum melt. Used for automotive applications, e.g., brakes, cylinder liners, valve guides, pistons and sleeves.

Granicoat. Trademark of SAFAS Corporation (USA) for a durable, seamless, lightweight solid surface material available in a wide range of granite and other colors and designs. It is UV-stabilized, fire-retardant, bacteria-, fungi-, corrosion-, stain- and thermal shock-resistant, and can be spray-applied to various architectural and building products including counter and table tops, vanities and sinks, shower enclosures and wall panels.

Granimo. Trade name of National Intergroup Inc. (USA) for a high-permeability soft magnetic iron-silicon alloy used for transformer laminations and dynamos.

Granisil. Trade name of National Intergroup Inc. (USA) for a soft magnetic iron-silicon alloy with high magnetic permeability used for transformer laminations.

Granit. (1) Trade name of Saint-Gobain (France) for a patterned glass with an indeterminate cathedral-type pattern.

(2) Trade name of Bergische Stahl-Industrie (Germany) for a series of steels including various cold- and hot-work die and tool steels, several stainless steels and machinery steels and numerous case-hardening steels.

Granitan. Trade name for a mixture of granite aggregates and epoxy resin with vibration damping characteristics superior to those of cast iron. It can be cast to desired dimensions, and has good resistance to water, humidity and lubricants, and low shrinkage during curing (1% or less). Used for machine bases and noise-suppression enclosures.

Granite. Trade name of Tech Stone Floor & Wall Coatings Inc. (USA) for decorative, flexible acrylic emulsion coatings containing colored quartz compounds. They are breathable, water-repellent, durable, resistant to abrasion, bacteria, cracking, impact and salts, and add slip resistance. Used on exterior and interior concrete, drywall and masonry surfaces.

granite. A pink to light or dark gray crystalline igneous rock composed chiefly of *albite* or *orthoclase* feldspar, quartz, mica and/or hornblende, and rarely pyroxene. It has a uniform coarse- to fine-grained structure, a density of 2.63-2.75 g/cm³ (0.095-0.099 lb/in.³), a melting range of 1100-1240°C (2010-2265°F), a crushing strength of about 104-138 MPa (15-20 ksi), and high hardness (above 6 Mohs). Used for building and paving stone, monuments, surface plates for machine shop layout work, and for pulverized granite that is similar to china clay in properties and use.

Granitec. Trade name of Novocol, Inc. (USA) for a root-canal dental cement. *Granitec M5 Ortho* is a dual-cure cement used for bonding orthodontic bands.

Granite Elite. Trade name of Huber Engineered Materials (USA) for an alumina trihydrate (ATH) filler containing pigmented polymer granules. It is available in a wide range of colors and shades for the manufacture of cast polymers including gel-coated *cultured marble*.

granite granules. Small pellets of *granite* in various sizes and colors used chiefly in the manufacture of building products, as a substitute for clay in specialized applications, and in making hard terrazzo floors.

granite pebbles. Small pebbles made from *granite* and used in mills for grinding ores and minerals. See also grinding pebbles.

graniteware. An article, such as an item of kitchenware, covered with a single coat of porcelain enamel, and having a mottled appearance due to controlled corrosion of the metal base prior to application.

Granito. (1) Trade name of Fabbrica Pisana SpA (Italy) for a glass with indeterminate pattern.

(2) Trade name of Vitrobrás (Brazil) for a patterned glass.

(3) Trade name of Permacon (Canada) for durable paving slabs with unpolished granite texture supplied in square units of 610 × 610 mm (24 × 24 in.) in gray and pink colors.

Granlac. Trademark of Ensio Industries Inc. (USA) for roofing granules.

Granlar. Trademark of Granmont Inc. (USA) for an injection moldable, glass fiber-reinforced liquid crystal polymer for high-temperature and general-purpose applications.

Granmarfil. Trademark of Toho Rayon Company Limited (Japan) for high-quality combed cotton fibers used for the manufacture of wearing apparel.

Granodine. Trade name of Henkel Surface Technologies (USA) for a zinc phosphate coating.

Granodized steel. Steel with a very thin zinc phosphate coating. "Granodizing" is a proprietary zinc phosphate coating process.

Granoleum. Trade name of Parker & Amchem (USA) for a rust preventive for iron and steel products.

granolithic concrete. Concrete containing selected aggregates and having appropriate hardness, surface texture and particle shape for use as a finish on floor wearing surfaces.

Grant. (1) Trade name of Time Steel Service Inc. (USA) for a low-alloy special-purpose tool steel (AISI type L2) containing 0.45% carbon, 0.55% manganese, 0.2% silicon, 0.95% chromium, 0.2% vanadium, and the balance iron. Used for hatchets, axes, hammers and mallets.

(2) Trademark of Grant Waferboard (Canada) for wafer-board.

grantsite. A dark olive-green mineral composed of sodium calcium vanadium oxide dihydrate, $NaCa_xV_3O_8 \cdot 2H_2O$ (x < 1). Crystal system, monoclinic. Density, 2.94 g/cm³; refractive index, above 2.0. Occurrence: USA (Colorado, New Mexico).

Granubi. Trade name of Haas GmbH (Germany) for a biodegradable blend of natural raw substances, such as sugar cane, soybeans, cornstarch, jute or cotton, with minerals and selected synthetic resins. It can be processed by injection molding and most other techniques, and is used as a substitute for polypropylene and acrylonitrile-butadiene-styrene in automotive applications.

granular activated carbon. Activated carbon composed of black granules having an average particle size of 80 mesh (180 μm) or above. See also activated carbon.

granular powder. A metal powder composed of particles with nearly equidimensional and nodular, but nonspherical shapes.

Granulart. Trade name of Permacon (Canada) for paving slabs with an attractive texture that highlights the luster of the aggregates used in their manufacture. They are supplied in square units of 610 × 610 mm (24 × 24 in.) in gray, beige and glacier-blue colors.

granulated blast-furnace slag. The glassy, granular, nonmetallic product obtained by rapidly quenching molten blast-furnace slag in water, steam or air. Used as an aggregate in mortar and concrete, and in the manufacture of slag bricks. See also blast-furnace slag; slag sand.

granulated cork. Cork made into small crumbs or particles, typically less than 12.7 mm (0.5 in.) in size, and used for the thermal insulation of refrigeration equipment, and in the manufacture of corkboards and similar pressed products. See also cork.

granulated metals. Metals in the form of small pellets of spherical or near-spherical shape obtained by pouring drops of the molten metal onto a revolving disk, or through a sieve, screen or similar device, and then chilling in water. See also shot.

granulated rubber. See rubber powder.

granulated slag. The granular product obtained by directing a jet of high-pressure water at a stream of molten slag, by pouring molten slag into water, or by dropping the molten slag on a revolving disk, and then chilling in water.

granules. (1) A generic term referring to small grains, particles or pellets of material.

(2) Small pieces of gravel that will pass the No. 5 (4.0 mm) and be retained on the No. 10 (2.0 mm) US Standard Sieve.

(3) Particles of crushed slag, rock, slate, tile, porcelain, etc., approximately 8 mesh (2.4 mm) in size, used as surfacing materials on asphalt roofing and shingles. The granules protect the asphalt coating from solar radiation, add color and provide fire resistance. Also known as *roofing granules*.

(4) Small pellets or grains of ceramics or naturally colored minerals used to impart color to the surface of asbestos-cement products.

Granulite. Czech trade name for a patterned glass with a rough-grained, nearly crystalline surface.

Granuplast. (1) Trademark of SAFAS Corporation (USA) for a compounding material incorporating *Galaxy Granules* (proprietary granular, cured polymer fillers) and used in acrylonitrile-butadiene-styrene, polycarbonate, polypropylene and polystyrene and other thermoplastic resin mixes for processing into a wide range of cast, extruded or injection-molded products resembling natural granite or stone in appearance, e.g., appliances, automotive components, consumer electronics, computer housings, telecommunication equipment, furniture and housewares.

(2) Trade name for particulate composites of relatively low density consisting of wood particles bonded together under low heat and pressure with melamine, phenolic or urea resins, and made into molded parts.

grape sugar. See dextrose.

Graph-Air. Trade name of Latrobe Steel Company (USA) for an air-hardening medium-alloy cold-work tool steel (AISI type A10) containing 1.35% carbon, 1.2% silicon, 1.8-1.85% manganese, 1.5% molybdenum, 1.8% nickel, and the balance iron. It contains free graphite, and has good abrasion and wear resistance, good machinability, and good nondeforming properties. Used for forming rolls, shear blades, gages, dies, punches, and other tools.

Graph-Al. Trade name of Timken Company (USA) for a high-carbon die steel composed of 1.5% carbon, 0.2% silicon, 0.3% manganese, 0.15% aluminum, and the balance iron. It contains free graphite, and is used for dies and mandrels.

Graphallast. Trademark of Graphite Metallizing Corporation (USA) for abrasion- and corrosion-resistant, oilless materials based on graphite and hydrocarbons, and used for seals and low-friction bearings and bushings to be operated dry or subjected to harsh environments.

Graphalloy. Trade name of Graphite Metallizing Corporation (USA) for a material composed of graphite impregnated at high pressure with bronze, cadmium, copper, gold, iron, lead, lead babbitt, nickel or silver. It has a low coefficient of friction, a long service life, good self-lubricating properties and good abrasion and corrosion resistance. Used for contact shoes, electric brushes, circuit breakers, switches, relays, slip-ring brushes, contacts for controllers, seals, seal rings, and oilless, self-lubricating bearings and bushings.

graphene. A single layer of graphite that is only one atom thick. Also known as *graphene layer; graphene sheet*.

Graphex. Trade name of Wakefield Corporation (USA) for a self-lubricating bronze composed of 4.0% graphite, 9-11% tin, and the balance copper. The graphite is evenly dispersed through-out the bronze matrix. Used for bearings and bushings.

Graphi-coat. Trademark of Aremco Products Inc. (USA) for ceramic coatings based on titanium dioxide (TiO_2).

graphic tellurium. See sylvanite.

Graphidox. Trade name of Cyprus Foote Mineral Company (USA) for a ferroalloy containing 50-55% silicon, 9-11% titanium, 5-7% calcium, 1-1.3% aluminum, 0.15-0.25% carbon, and the balance iron. Used as an inoculant for cast iron and for the deoxidation of steel.

Graphilm. Trademark of Graphite Metallizing Corporation (USA) for liquid lubricants composed of graphite suspended in suitable media for application by brushing, dipping or spraying. They dry to thin lubricating films, and are used for vacuum, high-temperature and cryogenic equipment.

Graphitar. Trade mark of US Graphite Inc. (USA) for molded products made from mixtures of amorphous and other allotropic forms of carbon. They have high hardness and strength, high crushing strength and good acid and corrosion resistance, and are used for piston and sealing rings, thrust washers, machinery bearings, bushings, and pump vanes.

graphite. An allotropic crystalline form of carbon that is soft, steel-gray to black in color and has a metallic luster and a greasy feel. It is found in nature, and can also be produced synthetically, e.g., from petroleum coke in an electric resistance furnace. An amorphous form is also found in nature. Properties of crystalline graphite: Crystal structure, hexagonal (rhombohedral). Density, 1.9-2.28 g/cm³; hardness, 0.5-2 Mohs; melting point, 3650°C (6600°F) (sublimes); boiling point, 4200°C (7590°F). It has high thermal conductivity, good electrical conductivity (about 10^5 $(\Omega\text{-m})^{-1}$), a low coefficient of friction, good resistance to oxidation and thermal shock, and fair resistance to chemicals. Occurrence: Austria, India, Italy, Malagasy, Mexico, Sri Lanka, USA (Alabama). Used for pencil leads, electrodes for welding and arc furnaces, cathodes in electrolytic cells, crucibles, retorts, bricks, polymers, coatings (e.g., for foundry patterns), crucibles and molds, mold and die lubricants, pigments, paints, powder glazings, chemical equipment, brushes for electric motors and generators, rocket nozzles, boiler compounds, sealing rings, lubricants for machinery and self-lubricating bearings, as a moderator in nuclear reactors, in monochromators for X-ray diffraction analysis, and for engineering fibers. Also known as *black lead; plumbago*. See also amorphous graphite.

graphite-aluminum composites. A group of lightweight composite materials consisting of reinforcing graphite fibers in aluminum-alloy (e.g., AA 2009, 6061 or 6092) matrices. They are often produced by casting, i.e., by adding molten metal to a casting mold containing the continuous reinforcing fibers, or by diffusion bonding or powder metallurgy techniques. They have high strength, elastic moduli and thermal conductivity and good machinability, and are used for high-performance applications, e.g., structural parts for aerospace. Abbreviation: Gr/Al composites.

graphite-base carbon refractories. Refractory materials manufactured substantially or entirely from graphite and used for crucibles and stopper nozzles in steelmaking furnaces. See also carbon refractories.

graphite brick. A brick manufactured from a carbonaceous material, such as petroleum coke, using a bituminous binder (e.g., pitch or tar). The brick is heat-treated to produce a graphitic structure and used as a refractory and for lining chemical-process equipment.

graphite carbon raiser. A graphite powder that is introduced into certain steel melts to increase the carbon content.

graphite composites. A group of composite materials consisting of polymeric matrices (e.g., epoxy, polyester, polyetheretherketone or polyphenylene sulfide) reinforced with graphite fibers or fabrics. See also graphite-fabric composites; graphite fiber-reinforced composite.

graphite-copper composites. A group of composite materials consisting of reinforcing graphite fibers in copper-alloy matrices. They are often produced by casting, i.e., by adding molten metal to a casting mold containing the continuous reinforcing fibers, or by diffusion bonding or powder metallurgy techniques, and have high strength and elastic moduli, very high thermal conductivity and good machinability. Used for high-performance applications, e.g., for structural parts for aerospace. Abbreviation: Gr/Cu composites.

graphited bronze. An oil-impregnated porous bronze made from a mixture of copper and tin powders and small amounts of lead and graphite by compacting and sintering. Used for self-lubricating bearings.

graphited metal. A self-lubricating porous metal, such as sintered bronze, brass, or nickel silver, containing considerable amounts of graphite dispersed evenly throughout its matrix.

graphite-epoxy composites. See graphite fiber-epoxy composites.

graphite-fabric composites. See graphite laminates.

graphite fabrics. Fabrics woven from graphite fibers. Laminates based on graphite fibers usually have high tensile, compressive and flexural strengths. Woven graphite fabrics provide laminate reinforcements that are uniform in thickness, free of holes and have high resin compatibility. See also carbon fabrics.

graphite fiber-epoxy composites. A group of composite materials consisting of reinforcing graphite fibers embedded in epoxy resin matrices. They have high strength-to-weight ratios, high stiffness, low weights, good thermal and electrical properties, and an upper service temperature of about 190°C (375°F). Used for unidirectional prepreg tapes, aircraft external structural panels, space structures, such as truss elements, waveguides, antennas and parabolic reflectors, for golf-club shafts and tennis rackets, and in filament winding of components, such as tanks, pipes, pressure vessels and other shapes of revolution. Also known as *graphite-epoxy composites.*

graphite fiber-reinforced composites. A group of composite materials consisting of metal or nonmetal matrices reinforced with continuous or discontinuous carbon fibers. Examples include graphite-aramid and graphite-epoxy composites as well as graphite-reinforced aluminum-matrix composites and graphite fiber-reinforced ceramics and plastics.

graphite fiber-reinforced plastics. A group of composite materials consisting of plastic matrices (e.g., epoxy, nylon, polyester, or vinylester) reinforced with graphite fibers. They have high strength and stiffness, improved friction and wear properties, improved weatherability and reduced energy dissipation. Used chiefly for mechanical energy-storage devices, bearings, electrical waveguides and structural components. Also known as *graphite fiber-reinforced polymers.*

graphite fibers. High-strength, high-modulus fibers made from organic precursors (e.g., rayon, polyacrylonitrile or pitch) by graphitization (heat treatment) at temperatures up to 1900-2480°C (3450-4500°F). The content of elemental carbon after heat-treatment is 99+%. Depending on the precursor, their crystal structures may be amorphous (rayon) or polycrystalline (polyacrylonitrile and pitch). *Graphite fibers* have an average density of 1.75-2.15 g/cm³ (0.06-0.08 lb/in.³) and an average diameter of 7-8 μm (275-310 μin.). Used chiefly as reinforcements for engineering plastics and composites, and in protective clothing and flameproof textiles. Abbreviation: GF; GrF. See also carbon fibers.

graphite laminates. Woven graphite fabric configurations layered with matrix polymers to form laminates. Used as reinforcements. Also known as *graphite-fabric composites.*

graphite lubricants. Lubricating oils or greases containing small additions of finely divided graphite (e.g., colloidal graphite or flake graphite). Available in the form of pastes and sprays, they have excellent lubricating power and high temperature stability, and are used for the solid-film lubrication of high-load bearings, machine beds, gears, shafts, chains, screw threads and other machine parts, for locks, typewriters, burner assemblies, pipe joints and gaskets, and as friction reducers in cold- and hot-forming operations. See also solid-film lubricant.

graphite-magnesium composites. A group of lightweight composite materials consisting of reinforcing graphite fibers in magnesium or magnesium-alloy matrices. They are often produced by casting, i.e., by adding molten metal to a casting mold containing the continuous reinforcing fibers, or by diffusion bonding techniques, and have high strength, elastic moduli and thermal conductivity, and good machinability. Used for high-performance applications, e.g., structural parts in aerospace. Abbreviation: Gr/Mg composites.

graphite-matrix composites. A group of biocompatible composite materials composed of carbonaceous matrices (carbon or graphite) reinforced with carbon or graphite fibers. They have high strength and elastic moduli, high creep resistance and fracture toughness, high thermal conductivities, low coefficients of thermal expansion, low thermal-shock sensitivity, and upper service temperatures exceeding 2800°C (5070°F). Used for aircraft and automotive brakes, structural components and thermal protective panels for spacecraft, gas-turbine engine parts, hot pressing molds, and prosthetic devices, e.g., knee and hip joints. See also carbon-matrix composites.

Graphite Metal. A British trade name for an alloy containing 68-80% lead, 17-20% antimony and up to 15% graphite. Ued for lead pencils, lubricants, crucibles, foundry facings, and electric brush carbons.

graphite nanofibers. A new class of ordered crystalline carbon fibers that are thypically 5-100 nm (0.2-4 μin.) in diameter and 5-100 μm (0.0002-0.004 in.) in length, and synthesized by the decomposition of selected hydrocarbons (e.g., mixtures of ethylene, carbon monoxide and hydrogen) at temperatures of about 400-800°C (750-1470°F) using a metal powder catalyst (e.g., an alkali metal). They are made up of graphene sheets (one-atom-thick graphite layers) arranged in various orientations with respect to the fiber axis resulting in various distinct structural types including: (i) herringbone (angle of graphene sheets to nanofiber axis, approximately 45°), (ii) platelet (angle of graphene sheets to nanofiber axis, 0°), and (iii) tubular (angle of graphene sheet to nanofiber axis, 90°). *Graphite nanofibers* have high active surface areas (typically 300-700 m²/g), can reversibly adsorb and store large amounts of molecular hydrogen, and are currently investigated for use in hydrogen-storage systems. Abbreviation: GNFs. See also nanofibers.

Graphite Nitralloy. Trade name of Bethlehem Steel Corporation (USA) for a steel containing 1.25-1.5% carbon, 1.25-1.5% silicon, 0.4-0.6% manganese, 1-1.5% aluminum, 0.2-0.4% chromium, 0.3% molybdenum, and the balance iron. It has excellent wear resistance after nitriding, and is used for wear-resistant parts.

graphite paint. A quick-drying mixture of graphite, hot linseed oil and a small amount of a drier (e.g., a linoleate, naphthenate or resinate of cobalt, lead or manganese), used as a coating on iron and steel products.

graphite-polyimide composites. A group of composite materials consisting of polyimide matrices reinforced with graphite fibers. They have high strength and stiffness, improved friction and wear properties, reduced energy dissipation and improved weatherability. Abbreviation: Gr/PI composites.

graphite-polyphenylene sulfide composites. A group of composite materials consisting of polyphenylene sulfide matrices reinforced with graphite fibers. They have high strength and stiffness, good friction and wear properties and good weatherability. Abbreviation: Gr/PPS composites.

graphite refractories. Refractory materials manufactured substantially or entirely from graphite and/or carbon, and used for crucibles and stopper nozzles of steelmaking furnaces.

graphite-reinforced composites. See graphite-fiber-reinforced composites.

graphite-silver-copper composites. Composites consisting of silver-copper matrices reinforced with graphite fibers. They have high electrical conductivity and are used for electrical and electronic applications.

graphite steel. See graphitic steel.

graphite yarns. Twisted filaments, fibers or strands of graphite with high tensile strength and elastic moduli and low density, used for structural components of high-performance aircraft, solid-propellant-rocket motor casings, reentry vehicles, deep submergence vessels, and high-strength composites.

graphitic carbon. An allotropic form of carbon with hexagonal (rhombohedral) crystal structure. See also graphite.

graphitic cast iron. Cast iron that contains graphite in the form of flakes (gray cast iron), nodules (ductile cast iron) or temper-carbon (malleable iron).

graphitic silicon carbide. Silicon carbide (SiC) containing graphitic carbon and usually available in the form of electrically conductive thin-film coatings for application to electronic, optoelectronic and microelectronic components. Metastable graphitic SiC can also be synthesized in tubular form by extreme hole injection techniques.

Graphitic Steel. Trade name for a pearlitic malleable cast iron.

graphitic steel. An alloy steel in which some of the carbon is distributed throughout the matrix in the form of graphite flakes. Also known as *graphite steel*.

graphitides. A relatively new class of graphite-base intercalation compounds containing metallic donor elements. They can be made by laser vaporization of a metal (e.g., cesium or rubidium) and subsequent mixing with graphite, or by doping the graphite with metallic ions. The doped materials have superconductive properties and are used in the manufacture of electronic and optoelectronic devices.

graphitized carbon. Carbon of submicron particle size graphitized (heat-treated) at 2700°C (4890°F). It is used in microscopy for fine-structure resolution calibration.

graphitized carbon black. A nonporous, highly inert carbon-based compound used as an adsorbent and solid support in gas chromatography.

graphitizer. A material, such as aluminum, silicon or titanium, that when added to certain cast irons promotes the formation

of graphite.

Graphlite. Trade name of Neptco (USA) for carbon-fiber rods consisting of standard, high or ultrahigh-modulus carbon fibers with diameters ranging from 0.75 to 5 mm (0.03 to 0.2 in.) locked in fully cured epoxy-resin matrices. Used as high strength-to-weight reinforcements for composites, e.g., in helicopter components, such as spar caps and longerons.

Graphlon. Trademark of Graphite Metallizing Corporation (USA) for a series of graphite materials including molded products (e.g., bearings, bushings and seals) composed of graphite in synthetic resin matrices as well as packings composed of asbestos fiber cores covered with a fluoropolymer (e.g., polytetrafluoroethylene) and an outer sheath of graphite fabrics.

Graph-Mo. Trademark of Timken Company (USA) for a series of oil-hardening cold-work tool steel (AISI type O6) containing 1.45-1.5% carbon, 0.4-1% manganese, 0.8-1.3% silicon, 0.25% molybdenum, and the balance iron. Available in the form of bars, billets, blooms, ingots, tubing, strips, plates, sheets, structural shapes and forgings, they contain free graphite, and have good wear and abrasion resistance, toughness and nondeforming properties. Used for punches, rolls, spindles, taps, and various dies. *Graph-Mo Hollow-Bar* are for turned and bored bars made from *Graph-Mo* steel.

Grapho Babbitt Metal. Trade name of Lehigh Babbitt Company (USA) for a series of babbitts containing varying amounts of tin, lead and antimony and 0.1-0.4% graphite. Used for a variety of bearings for automobiles, rolling stock, blowers, pumps and compressors, machine tools and other industrial machinery and equipment.

Graphokote. Trade name of Robco Canada Inc. for colloidal dispersions of high-purity graphite in oil, used as additives to other lubricants, such as crankcase and penetrating oils, forging compounds, glass-mold release oils and oils for high-speed precision equipment. Also includes several graphite dry-film lubricants for close-tolerance high-precision parts, such as cams and sliding mechanisms.

graph paper. A coated or uncoated paper in roll, strip or sheet form having a coordinate system printed on one side. Used for the registration of measurements, and for drawing graphs, mathematical functions, etc. Also known as *chart paper; functional paper*.

Graph-Sil. Trade name of Timken Company (USA) for an abrasion-resistant graphitic steel containing 1.5% carbon, 0.75-0.95% silicon, 0.35-0.4% manganese, and the balance iron.

Graph-Tung. Trade name of Timken Company (USA) for an abrasion-resistant, nonseizing, graphitic high-carbon tool steel containing 1.5% carbon, 0.5% manganese, 0.65% silicon, 0.5% molybdenum, 2.8% tungsten, and the balance iron. It contains free graphite, has good wear resistance, and is used for drawing, coining and blanking dies, and shear blades.

grappier cement. A finely ground cement made from over- or underburned slaked lime.

Grasscloth. Trademark of Canadian Forest Products Limited for panelwood products.

grass cloth. A loose, glossy fabric made from flax, hemp, ramie, or other vegetable fibers, usually in a plain weave. Used especially for sportswear, blouses and tablecloth.

Grassweave. Trade name of Libbey-Owens Fiberglas Corporation (USA) for a patterned glass.

Graticcio. Trade name of Fabbrica Pisana SpA (Italy) for a patterned glass.

gratonite. A lead-gray mineral composed of lead arsenic sulfide,

$Pb_9As_4S_{15}$. Crystal system, rhombohedral (hexagonal). Density, 6.22 g/cm^3. Occurrence: Peru.

gravel. Loose, rounded, or semirounded fragments of rock, usually about 2-70 mm (0.08-2.75 in.) in diameter, worn chiefly by the action of water. It is coarser than sand and includes pebbles and cobbles. *Gravel* has an average weight 1440-1920 kg/m^3 (90-120 lb/ft^3), and is used as coarse aggregate in concrete, and for backfilling, road construction and filtration.

gravel aggregate. Coarse or finely graded gravel used as aggregate in the manufacture of concrete.

gravel cement. Gravel formed into a compact mass by mixing with calcite, clay, silica or other suitable material.

gray acetate. See calcium acetate.

gray antimony. See stibnite.

gray birch. The light-colored wood of the small birch tree *Betula populifolia*, which has a chalky white bark that does not peel. Source: USA (New England) and Canada (Maritime Provinces). Used chiefly for charcoal and fuel, but also for small wooden articles.

gray cast irons. A group of cast irons that have been cooled very slowly from their critical temperature allowing some of the carbon to separate and form free graphite flakes. In addition to slow cooling, the amount and size of these flakes depends largely on the percentage of silicon present. It is this flake graphite that causes the gray crystalline appearance of the fracture. The typical composition range is 1.4-4.5% carbon, 1.8-3% silicon, 0.4-1.0% manganese, 0.2-0.9% phosphorus, 0.08-0.12% sulfur, and the balance iron. The American National Standards Institute (ANSI) groups *gray cast irons* according to tensile strength into sever-al classes ranging from 20 to 65 ksi (138 to 448 MPa) in increments of 5 ksi (34.5 MPa). *Gray cast irons* have a density range of 6.95-7.35 g/cm^3 (0.251-0.266 $lb/in.^3$), a melting range of 1150-1250°C (2100-2280°F), a casting temperature of approximately 1350°C (2460°F), high fluidity, good castability, a shrinkage of approximately 1%, very low elongation, good anti-friction properties, good machinability and vibration-dampening capacity (both of which decrease with increasing strength), good corrosion and wear resistance, moderate strength, and high brittleness. Used for bearings, gears, rolls, machine tools, machine and press frames and bases, housings, fittings, pipes, compressors, high-pressure cylinders, automotive cylinder blocks, gear boxes, crankcases, diesel-engine castings, flywheels, brake drums, and agricultural implements. Abbreviation: GCI. Also known as *gray irons*. See also ferritic cast irons; pearlitic cast irons; ferritic-pearlitic cast irons; cast irons.

gray cobalt. See cobaltite.

gray copper ore. See tetrahedrite (1).

Gray Cut Cobalt. Trade name of Teledyne Vasco (USA) for a tungsten-type high-speed steel (AISI type T6) containing 0.75-0.85% carbon, 0.25-0.35% silicon, 0.2-0.3% manganese, 4-4.5% chromium, 12-12.5% cobalt, 0.5-0.7% molybdenum, 20-21% tungsten, 1.5-1.7% vanadium, and the balance iron. It has high cutting ability, very high red hardness and high abrasion resistance, and is used for heavy-duty lathe and planer tools, shapers, milling cutters, and cutting tools for hard materials.

Graydac. Trade name of Champion Rivet Company (USA) for a carbon steel containing 0.08% carbon, 0.31% manganese, 0.29% silicon, and the balance iron. Used for low-spatter welding electrodes.

Gray Devil. Trade name of Champion Rivet Company (USA) for a carbon steel containing 0.09% carbon, and the balance iron.

Used for welding rods for steel.

Gray Diamond. Trade name of Latrobe Steel Company (USA) for an oil-hardening cold-work tool steel (AISI type O6) containing 1.45% carbon, 1.1% silicon, 0.25% molybdenum, and the balance iron. It contains free graphite in nodular form, has good machinability, and is used for structural parts, cutting tools, and dies.

gray fir. See Alaskan pine.

Gray Gold. A British trade name for corrosion-resistant gold alloy that contains 5.7-17% iron and up to 8.6% silver, and is used for jewelry.

gray goods. Textile fabrics that have either been woven in a loom or knitted in a knitting machine, but have not yet been bleached, dyed or finished. See also greige goods.

gray hematite. See specularite.

gray irons. See gray cast irons.

Gray Label. Trade name of Peninsular Steel Company (USA) for a water-hardening plain-carbon die steel containing about 0.9-1% carbon, and the balance iron.

Graylite. Trade name of PPG Industries Inc. (USA) for a neutral gray glare-control sheet glass.

gray manganese ore. See manganite.

Graypane. Trade name of Nippon Sheet Glass Company Limited (Japan) for a gray, heat-absorbing glass.

gray poplar. The tough, light yellow wood from the hardwood tree *Populus canescens*. It has a straight grain, a close texture, a comparatively low weight and only moderate durability. Source: Europe. Used for carpentry and flooring applications.

Grayral. Trade name of Central Glass Company Limited (Japan) for a heat-absorbing glass with a calming, pacific effect.

Gray Seal. Trade name of Progress Paint Manufacturing Company (USA) for paints.

gray selenium. A crystalline modification of selenium available in the form of gray crystals. Crystal structure, hexagonal. Density, 4.8 g/cm^3; melting point, 217°C (423°F); boiling point, 685°C (1265°F); boiling point, 685°C (1265°F); hardness, 2.0 Mohs. It is a p-type semiconductor, has photoconductive and photovoltaic properties, and its electrical conductivity varies with light irradiation. Used as a decolorizer in glass, in the production of rose and ruby colors in glass glazes and porcelain enamels, in the manufacture of steel, copper and *Invar* (e.g., as a degasifier and machinability enhancer), as an additive to lead-antimony battery grid metal, as a rubber accelerator, as a catalyst, in photocells, solar cells, solar batteries, metallic rectifiers, electronic components, magnetic computer cores, photocopier plates, xerographic drums and television cameras, and as a pigment in paints, plastics and ceramics.

gray tin. See alpha tin.

graywacke. Gray sandstone composed of a fine argillaceous matrix with numerous embedded mineral or rock fragments consisting chiefly of felspathic and ferromagnesian minerals. Used in civil engineering and construction, e.g., as a concrete aggregate. See also sandstone.

grayware. An article, such as an item of enamelware, covered with a single coat of porcelain enamel having mottled appearance.

grease. A solid or semisolid lubricant that is composed of emulsified mineral oils and lime or soda soap. See also lubricating greases.

greaseless compound. A buffing compound that contains no grease binders. It is composed of an abrasive or a blend of abrasives, mixed with gelatin-glue and water.

grease paint. A mixture composed of finely ground, white talc powder, a pigment, and a carrier, such as glycerin or lard.

grease-resistant barrier. A material used in packaging and containers to prevent or inhibit the transmission of grease or oils.

Great Western. Trade name of Great Western Steel Company (USA) for an oil-hardening, nondeforming tool steel containing 0.85% carbon, 1% manganese, 0.35% silicon, 0.5% chromium, 0.5% tungsten, and the balance iron. Used for milling cutters, drills, dies, punches, and mandrels.

Great Woods. Trademark of Georgia-Pacific Corporation (USA) for prefinished wood wall paneling materials.

Grecoform. Trade name of Grethe Kunststoff GmbH (Germany) for polystyrene foams and foam products.

Gredag. Trademark of Acheson Colloids Company (USA) for special lubricants composed of aqueous dispersions of aluminum, graphite and/or molybdenum disulfide. Used as release agents and die lubricants, and as lubricants for machine parts.

Greek Ascoloy. Trade name of Teledyne Firth-Sterling (USA) for a heat-resistant steel containing up to 0.17% carbon, 12-14% chromium, 1.8-2.2% nickel, 2.5-3.5% tungsten, up to 0.5% molybdenum, up to 0.5% copper, and the balance iron. Usually supplied in the hardened and tempered condition, it has good high-temperature resistance to about 540°C (1000°F). Used for compressor blades and vanes, fasteners and jet-engine components.

Greenal. Czech trade name for a green to yellowish-brown anti-sun glass.

green ash. The hard, strong wood of the tree *Fraxinus pennsylvanica* that was formerly categorized as a hairless variety of *red ash. Green ash* is quite similar to *black ash*. Average weight, 625 kg/m^3 (39 lb/ft^3). Source: Eastern and central United States, southern Canada (from Nova Scotia to Alberta). Used for furniture, veneer, containers, and cooperage.

green ceramic tapes. Flexible ceramic tapes based on aluminum oxide (Al_2O_3) or other ceramics and containing selected binders and plasticizers. They are usually stacked and laminated, and used in the manufacture of electronic products, such as capacitors, actuators, packages or multilayer circuits, for gas sensors, and in the manufacture of layered ceramic composites. Low-temperature, high-strength co-fired glass/ceramic tapes for use as insulating layers in hybrid circuits, multichip modules, single-chip packages and ceramic printed wiring boards are also available.

green chrome rouge. See green rouge.

green cinnabar. A green pigment made by mixing and calcining cobalt oxide (CoO) and zinc oxide (ZnO).

green compact. A metal powder that has been compressed, but not yet sintered. See also compact.

green copperas. See ferrous sulfate.

green copper carbonate. See malachite.

green copper ore. See malachite.

green earth. A native clay colored greenish by small quantities of iron in magnesium, and used as a light- and colorfast paint pigment.

green engineered materials. Materials that are designed, commercialized and used in processes and products which are feasible and economical, while minimizing the generation of pollutants and risks to human health and the environment at all stages of their life cycles. See also ecomaterials.

green flax. See natural flax.

Greenflex. Trade name of Enichem SpA/Polimeri Europa (Italy) for ethylene-vinyl acetates (EVAs) available with vinyl acetate

contents of 12, 25 and 33%. Also included under this trade name are polyethylene/EVA blends.

green glass. A glass tinted green by the introduction of a chromium compound (e.g., chromium oxide) into an ordinary glass batch. The introduction of copper monoxide results in blue-green color.

Green Gold. A British trade name for a corrosion-resistant jewelry alloy composed of 75% gold, 11-25% silver and 0-13% cadmium.

green gold. A greenish, corrosion-resistant jewelry alloy composed of gold and silver with little or no copper. It is available in several grades: (i) *14-karat* containing 58.33% gold, 34.38% silver and 7.29% copper; (ii) *15-karat* with 62.50% gold, 33.33% silver and 4.17% copper; (iii) *16-karat* having 66.67% gold and 33.33% silver; and (iv) *18-karat* containing 75.00% gold and 25.00% silver.

greenheart. (1) The dense, durable wood from the large tree *Octotea rodioei* of the laurel family. The heartwood is pale olive to nearly black and the sapwood pale yellow to greenish. *Greenheart* has a straight grain, outstanding strength and durability, very high hardness, very high resistance to decay, fungi, termites and marine borers, and good resistance to wear and seawater. Average weight, 1030 kg/m^3 (64 lb/ft^3). Source: Guyana, Surinam. Used for shipbuilding, docks, marine planking, marine jetties, piles, wharf construction, lock gates, bridges and trestles. Also known as *bibiru; Demerara greenheart; Surinam greenheart.*

(2) The yellowish-brown wood from the large tree *Piptadeniastrum africanum.* It is slightly coarse-textured, moderately heavy, has good nailing properties, works quite well, distorts in use, and is not as resistant to marine borers as common greenheart (1). Average weight, 690 kg/m^3 (43 lb/ft^3). Source: West Africa. Used for general construction purposes, shipbuilding, wharf and dock construction, and as an alternative to structural grades of oak. Also known as *African greenheart; dahoma; ekhimi.*

Green Label. Trade name Carpenter Technology Corporation (USA) for a water-hardening tool steel (AISI type W1) containing 0.8-1.4% carbon, 0.2% manganese, 0.2% silicon, and the balance iron. Used for tools and drill rod.

greenlandite. See niobite.

Greenland spar. See cryolite.

green lead ore. See pyromorphite.

Greenleaf. Trade name of Greenleaf Corporation (USA) for a series of sintered materials including various cemented carbides (e.g., tungsten carbide with cobalt binder), pressed and sintered aluminas, and alumina-boron tetracarbide, alumina-titanium carbide and alumina-silicon nitride composites.

Green Lightning. Trademark of Unimin Corporation (USA) for low free-silica blasting abrasives composed of hard, dense angular grains.

green locust. See black locust.

green lumber. Freshly cut lumber that has not yet been dried or seasoned.

green marble. A commercial term sometimes used for *serpentine asbestos.*

Green-Mousse. Trade name of Parkell, Inc. (USA) for a green-colored silicone elastomer used for dental impressions.

green nickel oxide. A green powder (99+% pure) obtained by heating nickel above 400°C (750°F) in the presence of molecular oxygen. At this temperature the nickel starts absorbing the oxygen forming nickelic oxide (Ni_2O_3) and is then reduced back to nickel oxide at 600°C (1110°F). It also occurs in nature as the mineral *bunsenite*. Crystal system, cubic. Density, 6.67 g/cm³; melting point, 1985°C (3614°F); antiferromagnetic properties. Used as a nickel salt, as a decolorizer in glass, as a blue, brown, green and yellow colorant for porcelain enamels and glazes, as an adherence promoter in porcelain-enamel ground coats, in fuel-cell electrodes, and as a principal ingredient in certain ferrites (e.g., nickel and nickel-zinc ferrites). Formula: NiO. Also known as *nickel monoxide; nickelous oxide; nickel oxide; nickel protoxide.*

greenockite. A yellow, green or orange mineral of the wurtzite group composed of cadmium sulfide, CdS, and having a cadmium content of about 77.7%. It can also be made synthetically. Crystal system, hexagonal. Density, 4.82 g/cm³; melting point, 780°C (1435°F); refractive index, 2.506. Occurrence: Associated with zinc ores, such as *sphalerite*. Used as the only commercial ore of cadmium. Also known as *cadmium blende; cadmium ocher; xanthochroite.*

GREEN PU-LINE. Trademark of Hanno-Werk GmbH & Co KG (Austria) for a line of green polyurethane foams and foam products for acoustical insulation applications.

green rouge. A very fine, greenish abrasive powder composed of chromic oxide (Cr_2O_3), and used for buffing and polishing chromium and stainless steels, and as a burnishing medium for soft metals. Also known as *green chrome rouge; levigated chromic oxide; levigated chrome rouge.*

greensand. See glauconite.

green silicon carbide. A green-colored *silicon carbide* of high purity (99.5%) made from silica sand and coke. It has high a hardness (about 9.4 Mohs or 2600 Knoop), a melting point of 2600°C (4710°F) and a density of 3.2 g/cm³ (0.12 lb/in³). Used for superrefractory applications, and as an abrasive for blasting, honing, lapping and polishing.

green spinel. A light green mixed metal oxide of calcined cobalt and titanium, used as a lightfast, permanent pigment.

greenstone. See nephrite.

Greenstripe. Trade name for fireclay.

GreenTape. Trademark of E.I. DuPont de Nemours & Company (USA) for a low-temperature, high-strength co-fired glass/ceramic tape used as an insulating layer in hybrid circuits, multichip modules, single-chip packages and ceramic printed wiring boards.

green verdigris. Pale green powder or shiny crystals obtained by the action of acetic acid on copper in the presence of air. It is the green variety of basic copper acetate, $2Cu(C_2H_3O_2)_2 \cdot CuO \cdot 6H_2O$. Both green verdigris and blue verdigris (the blue variety of basic copper acetate) are *true verdigris*, and used as pigments, e.g., in antifouling paints, as mildew preventives, and as mordants. See also basic copper acetate; copper subacetate.

green vitriol. See ferrous sulfate.

Grefco. (1) Trademark of General Refractories Company (USA) for a high-temperature insulating cement based on chrome iron ore, and used in the refractory industry.

(2) Trademark of General Refractories Company (USA) for a range of refractory products, such as brick, tile, mortars, castable and plastic refractories, ramming and hot-patching mixes, and coatings.

gregoryite. A synthetic mineral composed of sodium carbonate, α-Na_9CO_3. Crystal system, hexagonal. Density, 2.77 g/cm³. This alpha phase of sodium carbonate is stable above 490°C (914°F).

greige goods. A term used in North America for textile fabrics that have been woven in a loom, but have not yet been bleached, dyed or finished. See also gray goods.

greigite. A sooty-black mineral of the linnaeite group composed of iron sulfide, Fe_3S_4. Crystal system, cubic. Density, 4.05 g/cm³. Occurrence: USA (California), Mexico.

Gremopal. Trade name of Gremolith (Germany) for unsaturated polyester resins.

grenadine. A thin, openwork fabric usually made with high-twist yarns in a leno weave. It may have woven stripes, checks or other patterns, and is used especially for blouses, dresses and curtains.

Grenfell cloth. A water-repellent, wind-resistant, closely woven reversible twill cotton, polyester or fiber-blend fabric. Named after Sir Wilfred Grenfell, and formerly used for raincoats and other outdoor clothing.

Grey Dawn. British trade name for a decorative glass tableware.

Gridaloy. Trade name of Molecu-Wire Corporation (USA) for a heat-resistant nickel alloy containing up to 6% manganese and traces of other metals. Used for electrical resistors.

grid metal. A lead alloy that contains up to 12% antimony and, sometimes, small quantities of tin. Used for grids and plates of storage batteries. See also antimonial lead.

Gridnic. Trade name of Driver Harris Company (USA) for an extensive series of nickel alloys including the following compositions: nickel-chromium, nickel-iron, nickel-chromium-iron, nickel-manganese and nickel-cobalt-manganese-titanium-silicon-aluminum. Used for electrical and electronic applications.

Gridur. Trade name of Griesogen Griesheimer Autogen GmbH (Germany) for hardfacing electrodes supplied in various alloy grades including cobalt-base superalloy, high-carbon high-chromium steel, high-speed and hot-work tool steel, high-chromium cast iron, and tungsten carbide.

Griesheim. Trade name of Griesogen Griesheimer Autogen GmbH (Germany) for a series of steel hardfacing and welding electrode products available in various grades including machinery and austenitic manganese steels.

Griffin. Trade name of Adams & Osgood Steel Company (USA) for a nondeforming tool steel used for thread-rolling tools, punches and dies.

Griffine. Trademark of Griffine for polyurethane film materials.

Griffith's white. See lithopone.

Griffolyn. Trademark of Schleyer GmbH (Germany) for polyethylene film materials.

grignard. An organomagnesium halide used in acetylenic and other organic synthesis.

Grignard reagent. An organometallic halide formed by the addition of metallic magnesium to an aryl or alkyl halide in the presence of an ether or tetrahydrofuran solvent. It is capable of reacting with water, carbon dioxide, alcohols, aldehydes, ketones or amines to produce various organic compounds. The general formula is RMgX, in which R is an alkyl, aryl or other organic group, and X is a halogen, such as bromine, chlorine or iodine. Examples include ethyl magnesium chloride (C_2H_5MgCl), ethyl magnesium iodide (C_2H_5MgI), and phenyl magnesium bromide (C_6H_5MgBr). Used in the synthesis of organic compounds. See also organometallic compounds.

Grilamid. Trademark of EMS Grilon (USA) for an extensive series of polyamides (nylons) including various highly flexible thermoplastic nylon-12 elastomers, various grades of strong nylon-12 for blow and injection molding or extrusion applications, amorphous nylons, and several (glass) reinforced nylons. Used for electrical applications (e.g., cables and insulation tapes), insulators for electronic components, insulation of optical glass fibers, insulation jackets for pipes, and packaging materials.

Grilbond. Trademark of EMS Grilon (USA) for a series of bonding agents designed to improve the adhesion of polyester and aramid yarns and fabrics and used as reinforcements in rubber goods.

Grilene. Trademark of EMS Grilon (USA) for polyester staple fibers.

Grilesta. Trade name of EMS Grilon (USA) for polyester supplied in powder form and used for coating applications.

Grill. Trade name of SA Glaverbel (Belgium) for a patterned glass.

Grillo. Trade name of Zinkberatungsstelle GmbH (Germany) for zinc casting alloys containing up to 10% aluminum, 1% copper and 0.02% gallium.

Grillodur. Trade name of EMS Grilon (USA) for a series of glass fiber-reinforced polyurethane plastics supplied as sections, pressboard and resin-impregnated glass mats.

Grilon. Trademark of EMS Grilon (USA) for an extensive series of polyamides (nylons) including low-viscosity nylon 6 for injection-molding operations, impact-modified nylon 6, glass fiber-reinforced nylon 6 for injection molding, impact-modified and glass fiber-reinforced nylon 6, various nylon 6,12 resins, and various thermoplastic nylon blends. They are supplied in unprocessed form as powders, fibers, granulates, liquids, dispersions and emulsions for blow and injection molding or extrusion applications. Many of these grades are also available in processed form as blocks, sheets and rods. *Grilon A, B, EB* and *R* are nylon 6 resins supplied in a wide range of grades. *Grilon BM* and *C* refers to nylon 6,12 resins supplied in unmodified, glass or carbon fiber-reinforced, fire-retardant and silicone- or polytetrafluoroethylene-lubricated grades. *Grilon ELX* is an elastomer-copolymer grade based on nylon 6, and *Grilon AX, T* and *TV* refers to nylon 6,6 resins supplied in a wide range of grades including unmodified, glass fiber- or glass bead-reinforced, mineral-filled, fire-retardant, high-impact, supertough, UV-stabilized and molybdenum disulfide-lubricated. *Grilon TS* are modified nylon 6,6 resins supplied in several grades.

Griltex. Trade name of EMS Grilon (USA) for a series of thermoplastic copolyamide adhesives, wrapping films, pellets, extruded bars, blocks, rods, sheeting and tubing used in the manufacture of household, sports and industrial products.

grimaldite. A deep-red mineral of the delafossite group composed of chromium oxide hydroxide, $CrO(OH)$. Crystal system, rhombohedral (hexagonal). Density, 4.11 g/cm^3. Occurrence: Guyana.

Grimm. Trade name of Gustav Grimm Edelstahlwerk GmbH (Germany) for a series of high-grade steels including various plain-carbon and stainless steels, and several alloy tool steels including tungsten and cobalt-tungsten high-speed, hot-work, and water-hardening medium- and high-carbon grades.

Grimm's solder. An English alloy containing 50-69.1% tin, 25-28.8% lead, 1.44-25% zinc and 0-0.72% silver. Used as a solder for aluminum and aluminum alloys.

grimselite. A yellow mineral composed of potassium sodium uranyl carbonate monohydrate, $K_3NaUO_2(CO_3)_5 \cdot H_2O$. It can also be made from K_2CO_3, Na_2CO_3 and $C_4H_6O_6U$ in aqueous solution. Crystal system, hexagonal. Density, 3.30 g/cm^3.

Grinatal. Trade name of Anglo-Swiss Aluminium Company (UK) for a series of wrought aluminum alloys containing 4-5% silicon and 0.3% magnesium.

grinding aid. A liquid material with surface-active or lubricating properties added to a ball mill or ball-mill charge to accelerate the grinding process.

grinding balls. Hard, dense, abrasion-resistant spheres composed of alumina, dense porcelain, flint, steel, or tungsten carbide, and used as crushing media in ball or tube mills.

grinding flour. A finely divided powder used for grinding and polishing glass.

grinding media. Hard spheres, rods, rolls and the like made of porcelain, flint, alumina, bronze-alumina, alloy steel, or tungsten carbide, and used in grinding mills (e.g., ball or tube mills) to reduce the particle size of solid materials. The typical size range for spherical media is 6.4-51 mm (0.25-2.00 in.).

grinding pebbles. Hard, dense, tough, rounded stones, usually of flint, chert, quartz, feldspar, porcelain or alumina, used in ball, rod or tube mills for grinding cement, ores or minerals.

grinding-type resin. A synthetic resin, such as a vinyl, that has to be reduced to a powder prior to dispersion into an *organisol* or *plastisol*.

Grinolit. Trade name of EMS Grilon (USA) for epoxy resins and curing agents.

Grip. Trade name for polyester fibers and products.

griphite. A dark brown to brownish black mineral of the garnet group composed of sodium aluminum manganese phosphate hydroxide, $(Mn,Na,Ca,Fe)_6(Al,Mn)_4(PO_4)_5(OH)_4$. Crystal system, cubic. Density, 3.40 g/cm^3. Occurrence: USA (South Dakota).

Gripmaster. Trade name of JacksonLea (USA) for a polishing cement.

Gripmore. Trade name of Bissett Steel Company (USA) for hot-work tool steels used for punches, dies, shears and rams.

Griptex. Trade name of Combustion Engineering Company (USA) for a mineral wool block used for thermal insulation applications.

Grisuten. Trademark of Maerkische Faser AG (Germany) for polyester staple fibers and filament yarns.

grit. (1) Sharp, angular metallic particles composed of chilled white iron shot or crushed cast steel shot, and used in abrasive blast cleaning to remove heavy scale or rust.

(2) Sharp, angular, coarse-grained granules of alumina, corundum, garnet, sand or other suitable materials used primarily as abrasives in the manufacture of grinding wheels.

(3) Very small pieces or particles of gravel or sand.

Grital-K. Trade name of Vulkan Strahlverfahrenstechnik GmbH & Co. KG (Germany) for cast-steel grit used for abrasive-blast cleaning applications.

Gritite. Trade name of M.P. Iding Company Inc. (USA) for a polishing-wheel cement.

Grit-Lok. Trademark of 3M Company (USA) for coated abrasives.

gritstone. Coarse-grained sandstone composed of angular grains of different size.

Grivory. Trade name of EMS Grilon (USA) for an extensive series of polyamide (nylon) resins including modified nylon 6,6 (*Grivory XE* and *Grivory GC*), impact-resistant transparent nylon 12 (*Grivory G*) and transparent nylon 6,6/6T (*Grivory HTM-V*).

grog. Crushed refractory materials, such as firebrick, clinkers, pottery, quartz, quartzite, burnt ware, saggers or crucibles, added as raw materials to acidproof ware, high-temperature porcelain, refractories, sewer pipe, stoneware, terra cotta, vitreous china sanitaryware and similar products to improve the working and service properties.

grog-fireclay mortar. A finely ground *refractory mortar* composed of a mixture of raw *fireclay* with calcined fireclay or broken fireclay brick. See also grog.

Grolleg. Trade name for a high-quality kaolin clay from England that fires to a white color, and is therefore highly valued for the manufacture of porcelain.

grommet brass. A high-ductility brass that contains 70% copper and 30% zinc, has excellent cold workability, and is used for grommets.

grooved hardboard. Hardboard panels that have longitudinal channel grooves machined into the surface, and are used particularly as siding materials for buildings.

grosgrain. A closely woven fabric, usually of silk or rayon, having a heavy crosswise rib and a dull finish. Used especially for neckties, trim and millinery.

gros point. (1) A strong, durable pile fabric with uncut loops, used especially for upholstery.

(2) A heavy needlepoint lace with a raised floral or other design. Also known as *rose point; Venetian rose point.* See also Venetian lace.

Grossman. Trade name for an aluminum alloy used for light-alloy parts.

grossular. See grossularite.

grossularite. A colorless, white, yellow, brown, red, or green mineral of the garnet group composed of calcium aluminum silicate, $Ca_3Al_2(SiO_4)_3$. It can also be made synthetically. Crystal system, cubic. Density, 3.60 g/cm^3; refractive index, 1.737; hardness, 7 Mohs. Occurrence: USA (California). Used as a natural spinel in ceramics. Also known as *gooseberry stone; grossular.*

ground-coat enamel. Any of a class of black, blue, brown or gray porcelain enamels applied as a first coat to metal substrates, such as sheet steel, cast iron, aluminum or aluminum alloys. They contain oxides that improve the adherence to the substrate and, in the case of cast iron, are also used to fill surface irregularities.

ground fireclay. A milled fireclay or blend of fireclays subjected only to weathering.

ground fireclay mortar. A refractory mortar of workable consistency made by mixing finely ground raw fireclay with water.

groundnut fibers. The soft, wooly protein fibers obtained from groundnut plants, and used chiefly in fiber blends.

ground vulcanized rubber. Particulate composed of *vulcanized rubber* and used as filler and/or extender.

groundwood paper. A paper, such as newsprint, wallpaper or other low-grade paper, manufactured from groundwood pulp.

groundwood pulp. A coarse woodpulp (or mechanical pulp) that has been made wholly or in part by mechanical means, i.e., by physically separating the fibers, e.g., by grinding in wood grinders.

grout. A plaster or mortar of troweling or pouring consistency composed of cementitious materials, such as Portland cement and lime, together with aggregate and water, and sometimes sand. Used for painting-up and finishing mortar joints, filling crevices, and as a coating for building walls.

grouted-aggregate concrete. Concrete made by placing coarse aggregate and injecting a grout into it.

grouted macadam. See penetration macadam.

grouting sand. Sand composed of particles that will all pass the No. 20 (850 μm) sieve, and 95% of which will be retained on the No. 200 (75 μm) US Standard Sieve. Used for the preparation of grouts.

groutite. A black mineral of the diaspore group that is composed of manganese oxide hydroxide, MnO(OH), and may also contain 5-6% antimony. Crystal system, orthorhombic. Density, 4.14 g/cm^3. Occurrence: USA (Minnesota).

GR-S rubber. A designation formerly used in the USA for styrene-butadiene copolymers. The original GR-S rubber was a Buna-S type general-purpose rubber made by emulsion polymerization of about 75% butadiene and 25% styrene using soap as the emulsifier. It had excellent abrasion and heat resistance, good physical properties, especially when reinforced, good to moderate electrical properties, good water resistance, poor oil, ozone and weather resistance, an elasticity somewhat lower than natural rubber, a useful temperature range of -60 to +120°C (-75 to +250°F), and was used for pneumatic tires and tubes, heels and soles, gaskets, seals, conveyor belts, etc. "GR-S" stands for "Government Rubber-Styrene." See also Buna-S; styrene-butadiene rubber.

gruenlingite. A pale steel-gray mineral composed of bismuth sulfide telluride, Bi_4S_3Te. Density, 7.32 g/cm^3.

grunerite. A light grayish-brown asbestos mineral of the amphibole group composed of iron magnesium silicate hydroxide, $(Fe,Mg)_7Si_8O_{22}(OH)_2$. Crystal system, monoclinic. Density, 3.3-3.6 g/cm^3; refractive index, 1.689. It has coarse, long, flexible fibers, a high iron content (above 80%), high tensile strength and good chemical resistance. Occurrence: Canada (Labrador), South Africa (Transvaal). Used for thermal insulation applications. Also known as *amosite.*

Gruvex. Trade name of ASG Industries Inc. (USA) for a patterned glass.

gruzdevite. A gray-black mineral of the nowackiite group composed of antimony copper mercury sulfide, $Cu_6Hg_3Sb_4S_{12}$. Crystal system, rhombohedral (hexagonal). Density, 5.85 g/cm^3. Occurrence: Russian Federation.

GTX. Trade name of GE Plastics (USA) for polyphenylene-ether and polyamide (nylon) resins.

guaiac. The solid, brownish gum resin from certain lignum vitae trees, especially *Guaiacum santum* and *G. officinale* growing in Mexico and the West Indies. It has a melting point of 85°C (185°F), and is used in varnishes. Also known as *guaiac gum; guaiac resin; guaiacum.* See also lignum vitae.

guanaco fibers. Natural protein fibers obtained from the thick, soft, pale-brown wool of the guanaco, a South American wild mammal (*Lama guanicoe*) closely related to the alpaca and llama. Used for making clothing.

guanajuatite. A bluish-gray mineral of the stibnite group composed of bismuth selenide sulfide, $Bi_2(Se,S)_3$. Crystal system, orthorhombic. Density, 6.61 g/cm^3. Occurrence: Mexico.

guanine. A purine base contained in both DNA and RNA, and commercially supplied as colorless, rhombic, water-insoluble crystals with a melting point of 360°C (680°F) in the form of the pure base and the hemisulfate and hydrochloride salts. It is usually obtained from nucleic acids by organic synthesis, but also occurs as a natural mineral, e.g., at North Chincha Island in Peru. Used chiefly in biochemistry and DNA research. Formula: $C_5H_5ON_5$.

Guardcote. Trademark of Advanced Materials Inc. (USA) for liquid, fast-setting, two-component epoxy resins obtained by the polymerization of epichlorohydrin (C_3H_5ClO) and bisphenol A ($C_{15}H_{16}O_2$). Used for highway resurfacing and repair, and for lining of tanks.

Guardian. Trade name of Guardian Industries Corporation (USA) for a toughened and laminated safety glass.

guarea. The pale pinkish to reddish wood of hardwood trees of the species *Guarea*, especially *G. cedrata* and *G. thompsonii*. It resembles *mahogany*, but is somewhat stronger and close-grained, and has good seasoning properties, good stability, and acceptable working properties. Average weight, 600 kg/m³ (37 lb/ft³). Source: West Africa. Used as general utility timber for furniture, high-grade joinery and boatbuilding. See also Nigerian pearwood.

guar gum. A *mucilage* obtained from the seeds of the guar plant *Cyanopsis tetragonoloba*. It is a yellowish-white, free-flowing powder that dissolves completely in cold and hot water. Source: India, Pakistan, USA (Arizona, Texas). Used for coating paper and textiles, and in printing and polishing.

guayule. The shrub *Parthenium argentatum,* that is a member of the Compositae family and grows in Mexico and the southwestern United States (California, Texas). It bears a natural rubber that is nearly identical in composition with *hevea rubber* (*cis*-polyisoprene).

gudmundite. A silver-white to steel-gray mineral of the marcasite group composed of antimony iron sulfide, FeSbS. Crystal system, monoclinic. Density, 6.72 g/cm³. Occurrence: Sweden.

guerinite. A white mineral composed of calcium hydrogen arsenate nonahydrate, $Ca_5H_2(AsO_4)_4 \cdot 9H_2O$. Crystal system, monoclinic. Density, 2.68 g/cm³; refractive index, 1.582. Occurrence: Germany, UK (Wales).

guettardite. A black mineral composed of lead antimony arsenic sulfide, $Pb(Sb,As)_2S_4$. Crystal system, monoclinic. Density, 5.49 g/cm³. Occurrence: Canada (Ontario).

Guettiere's button. An alloy composed of 56.0-61.5% copper, 29-44% zinc and 0-9.7% tin, and used for buttons.

gugiaite. A colorless mineral of the melilite group composed of beryllium calcium silicate, $Ca_2BeSi_2O_7$. Crystal system, tetragonal. Density, 3.03 g/cm³; refractive index, 1.664. Occurrence: China.

Guidon. Trademark of Harbison-Walker Refractories Company (USA) for magnesia-chrome refractories manufactured from electrically fused grains. Used for the manufacture of walls and roofs of open-hearth furnaces and the sidewalls of electric furnaces.

guildite. A yellow to brown mineral composed of copper iron sulfate hydroxide tetrahydrate, $CuFe(SO_4)_2(OH) \cdot 4H_2O$. Crystal system, monoclinic. Density, 2.69 g/cm³; refractive index, 1.628. Occurrence: USA (Arizona).

Guignet's green. A paint pigment available in the form of a green, amorphous powder composed of hydrated chromium oxide, $(CrO)_2(OH)_2$. Also known as *permanent green*.

Guillaume alloy. See Guillaume metal.

Guillaume metal. An alloy with low thermal expansion composed of 64-66% iron and 34-36% nickel, and used for chemical equipment, and insulating tape. Also known as *Guillaume alloy; G metal.*

guilleminite. A yellow mineral composed of barium uranyl selenite hydroxide trihydrate, $Ba(UO_2)_3(OH)_4(SeO_3)_2 \cdot 3H_2O$. Crystal system, orthorhombic. Density, 4.88 g/cm³; refractive index, 1.798. Occurrence: Central Africa.

Guinea gold. A red brass composed of 88% copper and 12% zinc, and used for cheap jewelry and ornamental parts.

guipure lace. A machine-made lace with a heavy buttonhole stitch, and usually with a pattern or motif on a coarse net ground. Also known as *guipure.*

Guishibuichi. A Japanese alloy containing 51-67% copper, 32-49% silver, and a total of 0-1% iron and gold. Used for ornaments and jewelry.

Gulf. (1) Trade name of Gulf Steel Corporation (USA) for a series of tool steels including a wide range of tungsten- and molybdenum-type high-speed as well as plain-carbon and cold-work grades.

(2) Trade name of Chevron Chemical Company (USA) for an extensive series of synthetic resins including *Gulf HIPS* standard, fire-retardant and UV-stabilized high-impact polystyrenes, *Gulf PE* neat and UV-stabilized low-density polyethylenes, and *Gulf PS* for standard, UV-stabilized, structural-foam, medium-impact, glass fiber-reinforced and silicone-lubricated polystyrene resins.

Gulf Air. Trade name of Gulf Steel Corporation (USA) for an air-hardening nondeforming tool steel (AISI type A2) containing 1% carbon, 5% chromium, 1% molybdenum, and the balance iron. Used for dies and tools.

gulf cypress. See baldcypress.

Gulflex. Trade name of Gulf Canada Limited (Canada) for molybdenum disulfide (MoS_2) based solid-film lubricants.

Gullfiber. Trade name of Gullfiber AB (Sweden) for glass wool and tissue.

gum. (1) A natural, mucilaginous vegetable secretion given off by, or obtained from certain trees and plants. It dissolves in water, hardens in air, and is used in the manufacture of glues, varnishes, lacquers and similar products, for the preparation of protective colloids and emulsifying agents, in textile sizing and in the electrolytic deposition of metals. See also synthetic gum; natural gum; ester gum.

(2) The fine-textured, moderately heavy, strong, hard and stiff wood from any of several species of hardwood trees of the species *Nyssa* of the sourgum family (Nyssaceae) including in particular the water tupelo or cotton gum (*N. aquatica),* the sourgum or black gum (*N. sylvatica)* and the ogeche tupelo or sour tupelo (*N. ogeche).* All but the sourgum grow mainly in the southeastern United States. The sourgum is found in the eastern United States from Maine to Texas and Missouri. Average weight, 544-560 kg/m³ (34-35 lb/ft³). Used as lumber for crates, pallets, boxes and baskets, and for furniture, veneer and plywood. See also sourgum, ogeche tupelo; water tupelo.

(3) See sweetgum.

gum accroides. See acaroid resin.

gum arabic. See acacia gum.

gumbo. A term used to refer to very tough, highly plastic fine-grained clays found in the western and central United States. Due to their high shrinkage on firing, they cannot be used for brickmaking, but find application as railroad ballast.

gum compound. A rubber mix that contains all ingredients required for vulcanization in the proper proportions, and small proportions of other ingredients that facilitate processing, improve aging resistance and/or give or improve color. It does not contain fillers.

gum dammar. See dammar.

gummed paper. A coated or uncoated paper that may be colored or uncolored, has a gummy substance applied to one surface, and is essentially dry to the touch, but can be made mucilaginous or adhesive by moistening. Used for labels, stickers, stamps and tapes.

gummed paper tape. Kraft paper of varying basis weight that has a gummy substance applied to one side, and is available in rolls of varying width.

Gum Metal. Trade name of Toyota Tsusho and Toyotsu Nonferrous Sales Corporation (Japan) for high-strength titanium alloys.

gummite. A reddish, or yellow to brownish, radioactive mineral composed of varying quantities of hydrous oxides of lead, uranium and thorium. Density, 3.9-5.1 g/cm^3; hardness, 2.5-3.0 Mohs. Also known as *uranium ocher*.

gum resin. A mixture of a true gum and a natural resin obtained from certain plants. Examples of true gums include gutta-percha, olibanum and rubber, while common natural resins are rosin, dammar, mastic and lac.

gum rosin. See pine gum.

gum rubber. See natural rubber; vulcanized rubber.

gum Senegal. See acacia gum.

gum turpentine. An oleoresinous material obtained from living pine trees, and made by distillation of the crude exudation.

gumwood. The wood of any of various trees, such as *sweet gum and eucalyptus*, that yield gum.

gum Zanzibar. See animi gum.

gun-applied concrete. See dry-mix shotcrete; gunite.

guncotton. A high-nitrogen form of nitrocellulose (13.3-14% nitrogen) used mainly in the manufacture of explosives and propellants. Also known as *pyrocellulose*. See also nitrocellulose.

Gundol. Trade name of Basic Inc. (USA) for granular magnesia-base refractories and high-temperature gunning mixes.

gun iron. A fine-grained cast iron that typically contains 3% total carbon, 1.5% silicon, 0.85% manganese, 1.9% nickel, 0.4% chromium, 0.65% molybdenum, 0.075% sulfur, 0.18% phosphorus, and the balance iron. It is made with charcoal in an air furnace, has a uniform texture, and is used for castings and cylinder liners.

Gunite. Trade name of Kelsey Hayes Company (USA) for a series of alloy, high-strength and pearlitic malleable irons used for pump bodies, camshafts, brake drums, gears, cylinder sleeves, and forming dies.

gunite. A mixture of cement and sand or broken slag that contains little or no water and is applied in place pneumatically, with the necessary mixing water being added at the gun nozzle. Also known as *gunned concrete*. See also dry-mix shotcrete.

Gun Metal. A proprietary, chrome-tanned, fine-grade calfskin leather with a dull or semi-bright finish used for coats and other apparel.

gunmetal. (1) Any metal or alloy used in making cannons or guns.

(2) See admiralty gunmetal.

Gunmac. Trademark of C-E Refractories (USA) for high-temperature gunning mixes.

gun mount bronze. A high-strength bronze containing 80% copper, 17% zinc and 3% tin, used for gun mounts.

gunned concrete. See gunite.

gunningite. A colorless mineral of the kieserite group composed of zinc sulfate monohydrate, $ZnSO_4 \cdot H_2O$. It can also be made by exposing zinc sulfate hexahydrate to dry air for several days. Crystal system, monoclinic. Density, 3.36 g/cm^3. Occurrence: Canada (Yukon).

gunning mixes. Monolith-forming refractories (e.g., magnesite, dolomite or silica) applied with air-placement guns. See also monolithic refractories.

gunny. A strong, coarse fabric, usually of jute and in a plain weave. Used for sacks, bags, bailing, etc. Also known as *bagging fabric*.

Guntapite. Trade name of Quigley Company Inc. (USA) for high-temperature gunning cements.

Gupla. Trade name of Gummi-Plast (Germany) for quick-setting rubber and synthetic adhesives.

Gurimur. Trademark of Lonza-Werke GmbH (Germany) for flexible polyvinyl chloride film materials.

gurjun. The wood of any of several trees of the species *Dipterocarpus* growing in Burma and the Andaman Islands. Similar wood from Borneo and the Malay Peninsula is known as *keruing*, and that from Thailand as *yang*. The wood of these species is interchangeable for most commercial applications. Average weight, 720 kg/m^3 (45 lb/ft^3). Used as constructional timber in railroad construction, in building construction as a substitute for oak, and for parquet flooring.

Gurley's metal. A British free-cutting copper alloy containing 5.4% zinc, 5.4% tin, 2.7% lead, and the balance copper. Used for cocks, valves, and steam fittings.

Gurney's bronze. A bronze composed of 76% copper, 15% lead and 9% tin, and used for marine parts.

Guronit. Trade name of Guronitwerke Vervoort GmbH (Germany) for a series of steels including various austenitic, ferritic and martensitic stainless grades and several wrought and cast corrosion-resistant and/or heat-resistant grades.

Gussbronze. A German cast bronze composed of about 80-90% copper and 10-20% tin. It has good corrosion and wear resistance and good resistance to seawater, and is used for hardware, gears, worm wheels, bearings and bushings, piston rings, wear plates, bells, and pump and turbine housings.

gusset leather. Any soft, pliable leather suitable for gussets in bags or shoes.

Gussolit. Trademark of Gussolit Verbindungstechnik GmbH & Co. (Germany) for synthetic adhesives in building construction and industry for bonding metals, glass and rubber.

Guss-Korrofestal. Trade name of Aluminiumwerke Wutöschingen GmbH (Germany) for a corrosion-resistant aluminum casting alloy containing 9-13% silicon, 0.25-0.4% magnesium and 0.3-0.5% manganese. Used for light-alloy parts.

Gussmessing. A German cast brass composed of 60-90% copper and 10-40% zinc. It has good castability, strength and toughness, and is produced by sand, centrifugal, die or chill casting. Used for air conditioners, current-carrying equipment, electrical parts, housings, and machinery parts.

Guss Pantal. Trade name of Pinter Guss GmbH (Germany) for an aluminum alloy containing 2-4% magnesium, 1.2-1.5% manganese, and a total of 0.2% antimony and titanium. It has good resistance to seawater corrosion, and is used for light-alloy castings.

gustavite. A white mineral of the lillianite group composed of lead silver bismuth sulfide, $Bi_{11}Pb_5Ag_3S_{24}$. Crystal system, orthorhombic. Density, 6.80 g/cm^3. Occurrence: Greenland.

Gutan. Trademark for flexible polyvinyl chlorides.

Guthrie's alloy. A fusible alloy composed of about 20% tin, 19% lead, 13% cadmium, and the balance bismuth. Used for fire extinguisher plugs.

gutsevichite. A yellow-olive mineral composed of aluminum iron vanadium oxide phosphate hydroxide octahydrate, $(Al,Fe)_3$-$(PO_4,VO_4)_2(OH)_3 \cdot 8H_2O$. Crystal system, monoclinic. Density, 1.95 g/cm^3; refractive index, 1.5675. Occurrence: Kazakhstan.

gutta-balata. See balata.

Guttagena. Trademark of Hoechst AG, Kalle Werk (Germany) for flexible polyvinyl chloride film materials.

gutta-percha. A grayish-white, stiff, hard, inelastic thermoplastic substance composed chiefly of *trans*-polyisoprene, and obtained from the latex of tropical trees of the Sapotaceae family, especially the genera *Palaquium* and *Payena*. It has a softening point of 60°C (140°F), a melting point of 100°C (212°F), good water and alkali resistance, fair resistance to alcohol, ether and oils, and poor resistance to acids (except hydrofluoric acid), chloroform, turpentine, benzol, carbon disulfide, naphtha and atmospheric oxygen, and good electrical insulating properties. It resembles natural rubber and can be vulcanized with sulfur. Source: Borneo, Malaysia, New Guinea. Used for adhesives, waterproofing, electrical insulation (e.g., cable and wire), valve seats, washers, golf-ball covers, belt dressings, dental fillings, in surgery, and as a compounding ingredient in rubber.

Guttasyn. Trademark of Guttasyn–Meister-Plast GmbH (Germany) for flexible polyvinyl chlorides with a service temperature range of about -20 to +60°C (-4 to +140°F). Used for the manufacture of flexible tubing and hose products for industrial and laboratory applications.

guyanaite. A dark reddish-brown, red, brown or green mineral composed of chromium oxide hydroxide, $CrOOH$. It can also be made synthetically. Crystal system, orthorhombic. Density, 4.57 g/cm^3; refractive index, 1.9. Occurrence: Guyana.

Guy's alloy. An alloy composed of 82.5% nickel and 17.5% aluminum. It has high stress-rupture strength, and is used for turbine blades and jet-engine components.

Gymlene. Trademark of Drake Extrusion Limited (UK) for a circular polypropylene fiber supplied in various cut lengths and deniers, and used for spun and needle-punched floor coverings, automotive kick panels and package trays, and noise insulation products.

Gyproc. Trade name of Domtar Inc. (Canada) for gypsum wallboard, lath and joint fillers.

Gypsogum. Trade name for dental impression plaster.

Gypsona. Trade name Smith & Nephew (UK) for a plaster of Paris.

gypsum. A colorless, white, or gray mineral, sometimes tinted grayish, reddish, yellowish, bluish or brownish. It has a white streak, a pearly, silky or vitreous luster, and is composed of calcium sulfate dihydrate, $CaSO_4 \cdot 2H_2O$. Crystal system, monoclinic. Density, 2.31-2.33 g/cm^3; hardness, 1.5-2.0 Mohs; refractive index, 1.523; loses 1.5H_2O at 128°C (262°F) and 2H_2O at 163°C (325°F). Occurrence: Europe, Mexico, USA (California, Iowa, Michigan, Nevada, New Mexico, New York, Texas, Utah). Used in Portland cement, as a pigment and extender in certain paints, as filler for paper and cotton, calcined at 190-200°C (375-390°F) to produce plaster of Paris, for agricultural products, in metallurgy, in pottery and ceramics, and as a fining agent in glass. Also known as *natural gypsum; raw gypsum.*

gypsum acoustical tile. A thin, square sound-absorbing ceiling or wall tile made of gypsum. See also acoustical tile.

Gypsuman. Trade name of Westroc Industries Limited (Canada) for gypsum wallboard.

gypsum backing board. A laminated material in slab or sheet form, usually 6-16 mm (0.250-0.625 in.) thick, having a fireproof gypsum core and a gray liner paper on both sides. It is used as the base sheet on multilayer applications, as backing for gypsum wallboard, acoustical tile or other dry cladding, and is not suited for decorating and finishing applications.

gypsum block. See precast gypsum block.

gypsum board. See gypsum wallboard.

gypsum-bonded wood. A building product composed of wood particles in a gypsum matrix, and used in the manufacture of wallboard.

gypsum cement. See gypsum plaster.

gypsum concrete. A conglomerate mass made by mixing calcined gypsum with aggregates, or a blend of aggregates and water.

gypsum coreboard. A gypsum wallboard, 19-25 mm (0.75-1.00 in.) thick, composed either of a single, homogeneous board or a factory-laminated double or multiple board. Used as core or stud in gypsum board partitions.

gypsum fiber concrete. A concrete consisting of calcined gypsum and an aggregate of wood chips, flakes, fibers or shavings.

gypsum formboard. A panel or sheet having a fireproof gypsum core, a moisture-and fungus-resistant paper on the exposed side, and a special paper on the reverse side suitable for receiving pour-in-place gypsum concrete.

gypsum lath. See lath (ii).

gypsum lime plaster. A plaster prepared by mixing gypsum with lime, and used as finish coat for interior walls.

gypsum molding plaster. A plaster composed chiefly of calcined gypsum, and used primarily for plaster casts and molds, as a gauging plaster, for interior cornices and decorations.

gypsum partition block. A building block composed essentially of gypsum, and used for nonbearing walls in building interiors, and for fireproofing columns, and elevator shafts.

gypsum partition tile. A tile composed essentially of gypsum, and used for nonbearing walls in building interiors, and for fireproofing columns, and elevator shafts.

gypsum paste. A putty-like mixture of lime and gypsum used as a plastering material for final or finish coats on interior walls. Also known as *gypsum putty.*

gypsum plaster. Any of a group of cements made essentially from completely or partially dehydrated gypsum (calcium sulfate) by mixing with specially selected ingredients such as alum, borax or sodium or potassium carbonate, and enough water to produce a mixture of trowelable consistency. Depending on the selected ingredients, the cement may be known as *Keene's cement* (alum), *Mack's cement* (potassium or sodium sulfate), or *Parian cement* (borax). Also known as *gypsum cement.*

gypsum putty. See gypsum paste.

gypsum sheathing. A sheet or panel, usually 13 mm (0.5 in.) thick, composed of a fireproof and/or water-resistant gypsum core and a covering of water-repellent paper. Used as a backing for exterior surface materials. Also known as *gypsum sheathing board.*

gypsum spar. See selenite.

gypsum tile. An essentially flat building product of varying shape and size composed of gypsum.

gypsum-vermiculite plaster. An acoustic and fireproof plaster prepared by mixing gypsum with *vermiculite* aggregate. Used as a base coat for finishing interior walls.

gypsum wallboard. A laminated material composed of a gypsum core and a covering of paper or other fibrous material on one side. Standard-size gypsum wallboard comes in sheets 1.2 m (4 ft.) wide and 2.4 m (8 ft.) long, but it is also sold in 2.1, 2.7, 3.1, 3.7 and 4.3 m (7, 9, 10, 12 and 14 ft.) lengths. It is available in thicknesses of 6.4, 8.0, 9.5, 12.7 and 16.0 mm (0.250, 0.312, 0.375, 0.500 and 0.625 in.) in various grades including regular, fire-resistant, insulating, foil-backed, coreboard, predecorated and water-resistant. Used on interior walls and ceilings, and for plaster veneering. It is commonly referred to as "drywall." Also known as *gypsum board; plasterboard.*

Gyro. Trade name of CCS Braeburn Alloy Steel (USA) for a water-hardening tool steel containing 0.7% carbon, 4% chromium, 2% vanadium, 14% tungsten, and the balance iron.

gyrolite. A colorless or white mineral of the reyerite group composed of calcium silicate hydroxide trihydrate, $Ca_4(Si_6O_{15})$-$(OH)_2 \cdot 3H_2O$. It can also be made synthetically from silicon dioxide and calcium oxide in an autoclave. Crystal system, hexagonal. Density, 2.39 g/cm^3; refractive index, 1.549. Occurrence: India.

Gyrocast. Trade name of Youngstown Alloy Castings Company (USA) for a medium-carbon steel used for straightening rolls, centrifugal castings, and bar mill guides.

H

haapalaite. A red mineral of the valleriite group composed of iron magnesium nickel sulfide hydroxide, $(Fe,Ni)_2Mg_{1.6}S_2$-$(OH)_{3.2}$. Crystal system, hexagonal. Density, 3.57 g/cm³. Occurrence: Finland.

habit. A dark-colored, medium-weight wool fabric of good quality used for dresses and suits.

habutai. A fabric woven from low-grade *raw silk* that has not yet been thrown. It has a dense, firm, but irregular texture, and is used for clothing, umbrellas and parachutes. Also known as *China silk.*

hackberry. The wood of any of several trees of the genus *Celtis* belonging to the elm family. The principal members of this genus are the American or northern hackberry (*C. occidentalis*), and the sugarberry or southern hackberry (*C. laevigata*). *Hackberry* wood has high strength and toughness, is similar to elm, and is used mainly for furniture and shipping containers. Source: Canada (Southern Quebec to southern Manitoba), USA (Northeastern, central and southestern states). The sugarberry is also found in Texas and northern Mexico.

hackia. The hard, brown wood of the tree *Ixora ferrea*. It has a coarse, open grain and high toughness. Average weight, 880 kg/m³ (55 lb/ft³). Source: West Indies, Tropical South America. Used for furniture and lumber.

hackmarack. See American larch.

hackmatack. See western larch.

Hadesite. Trade name for refractory cement made by bonding a mixture of refractory clay, selected aggregates and mineral wool with a suitable binder.

Hadfield manganese steels. See Hadfield steels.

Hadfield silicon steels. Low-carbon steels containing 0.1-0.2% carbon, 4% silicon, and the balance iron. They have a narrow magnetic hysteresis and high permeability, and are used for transformer cores, armatures for dynamos and electric motors, and loading coils.

Hadfield steels. A group of nonmagnetic, abrasion- and wear-resistant austenitic steels containing about 1.0-1.4% carbon, 10-15% manganese, up to 0.1% phosphorus, up to 0.3% silicon, and the balance iron. They have high toughness, ductility, shock resistance and work-hardening capacity, low yield strength and corrosion resistance, and fair to poor machinability. Used for rails, wear plates, screens, journal box liners, screens, crushing equipment, mill liners, sheaves, and crane wheels. Also known as *austenitic manganese steels; Hadfield manganese steels; manganese steels.* See also carbon-man-ganese steels.

Hadura. Trade name of Dunford Hadfields Limited (UK) for a water-hardening tool steel used for cold-rolling rolls.

Haecker. Trade name of Haeckerstahl GmbH (Germany) for an extensive series of low-, medium- and high-carbon steels including various case-hardening, oil-hardening, air-hardening, stainless and hot-work grades.

Hafesta. Trade name of Hagener Feinstahl GmbH (Germany) for acid- and high-temperature-resistant stainless steels and various electrical resistance and heating-element materials supplied in the form of wires, bars, strips and structural shapes.

Hafnia. Trade name of Paul Bergsoe & Son (Denmark) for a cast alloy composed of 62% lead, 24.5% tin, 13% antimony and 0.5% copper. It has a melting range of 180-280°C (355-535°F), and is used for bearings in refrigerators and electrical motors.

hafnia. See hafnium oxide.

hafnia-stabilized zirconia. A ceramic material containing 98.5% zirconia (ZrO_2), stabilized with about 1.5% hafnia (Hf_2O_3). It is commercially available in powder form, and has a monoclinic crystal structure, a true density of 5.70 g/cm³ (0.206 lb/in.³), a melting point of 2700°C (4890°F), and a hardness of 6.5 Mohs. Used for fluid-bed applications, and in the manufacture of zirconia ceramic products. See also zirconia; hafnium oxide.

hafnium. A silvery metallic element of Group IVB (Group 4) of the Periodic Table. It is commercially available in the form of sponge, crystal bars, pieces, ingots, powder, wire, sheet, foil, rods, arc-melted buttons, zone-refined rods and single crystals, and is found in most zirconium ores. Density, 13.1 g/cm³; melting point, approximately 2227°C (4041°F); boiling point, 4602°C (8316°F); hardness, 150-180 Vickers; atomic number, 72; atomic weight, 178.49; superconductivity critical temperature, 0.128K; divalent, trivalent and tetravalent. It has a high thermal neutron cross section, high strength, good corrosion resistance, and is similar to zirconium in chemical properties. There are two forms of hafnium: (i) *Alpha hafnium* with hexagonal close-packed crystal structure; and (ii) *Beta hafnium* with body-centered cubic crystal structure, present at temperatures between about 1950 and 2227°C (3542 and 4041°F). *Hafnium* is used in the manufacture of tungsten filaments for light bulbs, special glasses, control rods for water-cooled nuclear reactors, as a scavenger in vacuum tubes, for electrodes, and in ceramics and superconductors. Symbol: Hf.

hafnium beryllide. Any of the following compounds of hafnium and beryllium used in ceramics and materials research: (i) $HfBe_{13}$. Density, 3.93 g/cm³; melting point, 1427°C (2600°F); (ii) Hf_2Be_{17}. Density, 4.78 g/cm³; and (iii) Hf_2Be_{21}. Density, 4.26 g/cm³; melting point, above 1927°C (3500°F).

hafnium boride. Any of the following compounds of hafnium and boron: (i) *Hafnium monoboride.* A crystalline powder. Density, 12.80 g/cm³; melting point, 2900°C (5250°F). Used in ceramics, refractories, and in materials research. Formula: HfB; and (ii) *Hafnium diboride.* A gray, crystalline powder. Density, 10.5-11.2 g/cm³; melting point, 3100-3250°C (5610-5880°F). Used as a refractory material, and in high-temperature resistant products for nuclear applications. Formula: HfB_2.

hafnium bromide. White, cubic crystals or moisture-sensitive, tan powder (99.7+%) with a density of 4.90 g/cm³ and a melting point of 424°C (795°F). Used in chemistry and materials research. Formula: $HfBr_4$.

hafnium carbide. Bluish black crystals or powder made by high-temperature reaction of hafnium oxide and carbon. Crystal structure, cubic. Crystal structure, halite. Density, 12-12.76 g/cm³; melting point, 4190°C (7575°F); hardness, approximately 1800-2900 Vickers. It has a high thermal-neutron absorption cross section, and is one of the most refractory binary compositions known. Used for nuclear-reactor control rods, in ceramic coatings and dispersion-strengthened tungsten, and for refractory components. Formula: HfC.

hafnium chloride. White, monoclinic crystals or moisture-sensitive, white powder (98+%) that has a density of 4.90 g/cm³ and sublimes at 320°C (608°F). Used in chemistry and materials research. Formula: $HfCl_4$.

hafnium diboride. See hafnium boride (ii).

hafnium dioxide. See hafnium oxide.

hafnium disilicide. See hafnium silicide (ii).

hafnium disulfide. See hafnium sulfide (ii).

hafnium ethoxide. A metal alkoxide supplied in the form of moisture-sensitive white crystals with a melting point of 178-180°C (352-356°F) and a boiling point of 180-200°C (356-392°F). Used in the preparation of refractory hafnia coatings. Formula: $Hf(OC_2H_5)_4$. Also known as *hafnium ethylate*.

hafnium ethylate. See hafnium ethoxide.

hafnium fluoride. White, monoclinic crystals with a density of 7.1 g/cm³, a melting point of 970°C (1778°F) or above. Used in chemistry and materials research. Formula: $HfCl_4$.

hafnium hydride. A compound of hafnium and hydrogen that is available in the form of a solid refractory, tetragonal crystals, or fine powder with a density 11.4 g/cm³ for use in the preparation of other hafnium compounds. Formula: HfH_2.

hafnium-iron. An intermetallic compound of hafnium and iron with a melting point of 1650°C (3000°F) used for refractories and ceramics. Formula: HfFe.

hafnium iodide. Yellow-orange, cubic crystals or moisture-sensitive, red powder (99+%) with a density of 5.60 g/cm³ and a melting point of 449°C (840°F). Used in chemistry and materials research. Formula: HfI_4.

hafnium-manganese. An intermetallic compound of hafnium and manganese with a melting point of 1642°C (2988°F) used for refractories and ceramics. Formula: $HfMn_2$.

hafnium-molybdenum. An intermetallic compound of hafnium and molybdenum with a melting point of about 2300°C (4170°F) used for refractories and ceramics. Formula: $HfMo_2$.

hafnium monoboride. See hafnium boride (i).

hafnium monosulfide. See hafnium sulfide (i).

hafnium-nickel. An intermetallic compound of hafnium and nickel with a melting point of 1788°C (3250°F) used in refractories and ceramics, and in materials research. Formula: $HfNi_2$.

hafnium nitride. Yellow to brown crystals or fine powder. Crystal system, cubic. Density, 13.8-14.0 g/cm³; melting point, 3305°C (5981°F); hardness, 8-9 Mohs; one of the most refractory of all known metal nitrides. Used for refractories and ceramics, and in materials research. Formula: HfN.

hafnium oxide. White crystals, off-white powder or sintered lumps (98+% pure). Crystal system, cubic. Density, 9.68 g/cm³; melting range, 2774-2812°C (5025-5094°F); boiling point, approximately 5400°C (9750°F). Used in zirconia refractories to lower thermal expansion, increase inversion temperature and reduce volume change during inversion, and as stabilizer in zirconia ceramics. Formula: HfO_2. Also known as *hafnia; hafnium dioxide*.

hafnium phosphide. A compound of hafnium and phosphorus supplied as hexagonal crystals or powder with a density of 9.78 g/cm³. Used in ceramics and materials research. Formula: HfP.

hafnium selenide. A compound of hafnium and selenium available as brown hexagonal crystals or powder with a density of 9.46-9.78 g/cm³. Used in ceramics and materials research. Formula: $HfSe_2$.

hafnium sesquisilicide. See hafnium silicide (i).

hafnium sesquisulfide. See hafnium sulfide (iv).

hafnium silicate. See hafnon.

hafnium silicide. Any of the following compounds of hafnium and silicon used in ceramics and materials research: (i) *Hafnium sesquisilicide*. Melting point, 2100°C (3812°F). Formula: Hf_2S_3; (ii) *Hafnium disilicide*. A gray powder (99% pure). Density, 8.02 g/cm³; melting point, 1680°C (3056°F); hardness, 865 Vickers. Formula: $HfSi_2$; and (iii) *Pentahafnium trisilicide*. Melting point, 2300°C (4172°F). Formula: Hf_5Si_3.

hafnium sulfide. Any of the following compounds of hafnium and sulfur: (i) *Hafnium monosulfide*. Melting point, approximately 2150°C (3900°F). Used in ceramics and materials research. Formula: HfS; (ii) *Hafnium disulfide*. Purple-brown crystals or fine powder. Crystal system, hexagonal. Density, 6.03 g/cm³. Used as a solid lubricant. Formula: HfS_2; (iii) *Hafnium trisulfide*. Density, 5.70 g/cm³; unstable above 871°C (1600°F). Used in ceramics and materials research. Formula: HfS_3; and (iv) *Hafnium sesquisulfide*. Density, 7.50 g/cm³. Used in ceramics and materials research. Formula: Hf_2S_3.

hafnium titanate. An off-white powder composed of varying proportions of hafnium dioxide and titanium dioxide. Density, 7.20-7.21 g/cm³; melting point, approximately 2200°C (3990°F). Used in the manufacture of special refractories with high rupture strength, good shock resistance, and low thermal expansion. Formula: $HfTiO_4$.

hafnium trisulfide. See hafnium sulfide (iii).

hafnium-vanadium. An intermetallic compound of hafnium and vanadium with a melting point of 1500°C (2732°F). Used in ceramics and materials research. Formula: HfV_2.

hafnon. A synthetic mineral of the zircon group made by firing pure hafnium dioxide (HfO_2) with silicon dioxide (SiO_2) at 1400-1600°C (2550-2910°F) for a precribed period of time. Crystal system, tetragonal. Density, 6.97 g/cm³; low thermal expansion. Formula: $HfSiO_4$. Also known as *hafnium silicate*.

hagendorfite. A greenish black mineral of the varulite group composed of sodium calcium iron manganese phosphate, $(Na,Ca)Mn(Fe,Mn)_2(PO_4)_3$. Crystal system, monoclinic. Density, 3.71 g/cm³; refractive index, 1.742. Occurrence: Germany.

Hagesta. Trade name of J.C. Soding & Halbach (Germany) for a series of highly corrosion-resistant casting alloys containing 0.1% carbon, up to 16% chromium, 17-30% molybdenum, 60-65% nickel, and the balance iron. Used for chemical and food-processing equipment.

haggite. A black mineral composed of vanadium oxide hydroxide, $V_2O_2(OH)_3$. Crystal system, monoclinic. Density, 3.53 g/cm³. Occurrence: USA (Wyoming).

Hagulen. Trademark of Hagusta GmbH (Germany) for a series of polyethylenes.

haidingerite. A colorless or white mineral composed of calcium hydrogen arsenate monohydrate, $CaHAsO_4 \cdot H_2O$. Crystal system, orthorhombic. Density, 2.96 g/cm³; refractive index, 1.602; hardness, 1.5-2.5 Mohs. Occurrence: Brazil, Germany, USA (California).

hair canvas. A broad term for a range of woven interfacing fabrics made of different fiber materials and available in various weights. Examples of fibers used include cotton and hair, goat hair and cotton or wool, and viscose rayon and hair. Used in coats and jackets, and as a filling material.

hair cloth. A stiff, bristly fabric in which the weft consists of plain or dyed horsehair or camel's hair and the warp of cotton,

linen polyester, or wool. Used as a garment stiffener, interlining, furniture cover, or sieve and press cloth.

haircord. A dyed or printed cotton fabric in a plain weave, having fine lengthwise cords or wales.

hair felt. Felt made from animal hair (e.g., cattle, horses, etc.), and used as thermal insulation for refrigeration equipment, water pipes, and for cushioning and padding applications.

hair fibers. The natural protein fibers obtained from the hairy coats of mammals, such as camels, cats, cows, goats, hares, horses and rabbits. They are composed chiefly of the scleroprotein *keratin*, and are used for spinning textile fabrics and making felt products.

Hairsprings. Trade name for a phosphor bronze containing 93.3-95.3% copper, 4.5-6.5% tin and 0.2% phosphorus, and used for hairsprings.

haiweeite. A pale yellow mineral of the weeksite group composed of calcium uranium silicate pentahydrate, $Ca(UO_2)_2Si_6O_{15}\cdot 5H_2O$. Crystal system, monoclinic. Density, 3.35 g/cm^3; refractive index, 1.572. Occurrence: Argentina, Brazil, USA (California, Colorado).

hakite. A gray-brown, cream-white or light brown mineral of the tetrahedrite group composed of copper mercury antimony selenide, $(Cu,Hg)_{12}Sb_4Se_{13}$. Crystal system, cubic. Density, 6.30 g/cm^3. Occurrence: Czech Republic.

Halar. (1) Trademark of Ausimont USA for translucent, melt-processible ethylene-chlorotrifluoroethylene (ECTFE) thermoplastic copolymers supplied in standard and glass-fiber-filled grades. The standard (unfilled) have a density of 1.7 g/cm^3 (0.06 lb/in.^{3}), excellent resistance to hydrocarbons, alkalies and weak and dilute acids, a maximum service temperature of 150°C (300°F), poor resistance to gamma radiation, high rigidity, and a water absorption of less than 0.1%. Used for laboratory ware, such as clamps, adapter, fittings, pipes, valves, tubing, pumps, fittings, etc. Also known as *Halar ECTFE.*

(2) Trade name of Albany International, Inc. (USA) for polytetrafluoroethylene (PTFE) monofilaments.

Halberland alloy. A German red brass composed of 87% copper and 13% zinc, and used for fittings and hardware.

Halco. Trade name of Crucible Specialty Metals (USA) for a tough heat-resistant steel containing 0.9% carbon, 3.6% chromium, and the balance iron. Used for hot-work dies, shear blades, and hot punches.

Halcomb. Trade name of Evans Steel Company (USA) for a series of tool steels including several hot-work, water-hardening, high-speed and shock-resisting grades as well as several stainless steels.

Halcut. Trade name of Crucible Special Metals (USA) for a tough, abrasion-resistant steel containing 0.5% carbon, 1.3% chromium, 2.7% tungsten, and the balance iron. Used for forging dies, chisels, shear blades, and hot-work tools.

Haldi. Trade name of Crucible Specialty Metals (USA) for an air-hardening, wear- and abrasion-resistant alloy containing 2.25% carbon, 11.5% chromium, and the balance iron. Used for drawing dies, shears, punches, and swages.

Halex. Trade name of Halex Limited (UK) for cellulose nitrate plastics.

half-and-half solder. A general-purpose solder containing 50% tin and 50% lead. It has a density of 8.89 g/cm^3 (0.321 lb/in.^{3}), a solidus temperature of 183°C (361°F), a liquidus temperature of 216°C (421°F) and good strength, and is used for plumbing applications, electrical connections, and general soldering of domestic items.

half-hard wire. An aluminum wire processed to produce a tensile strength intermediate between that of soft- and hard-drawn wire. See also soft-drawn wire; hard-drawn wire.

halftone printing paper. An uncoated, calendered, wood-containing printing paper having an ash content of about 15% wt%, and a basis weight of about 55-80 g/m^2 (0.18-0.26 oz/ft^2).

half-trimmed mica. Mica trimmed on two sides whereby two thirds or more are trimmed on two adjoining sides, and the remainder on two parallel sides. All graded areas must be free of cracks.

Halgraph. Trademark of Crucible Materials Corporation (USA) for an oil-hardening cold-work tool steel (AISI type O6) containing 1.5% carbon, 0.25% molybdenum, 1% silicon, 0.75% manganese, and the balance iron. It has good machinability, toughness and wear and abrasion resistance, and is used for machine-tool components.

halide glass. A glass composed of a halogen, such as fluorine, and a metal, such as barium, hafnium or zirconium. Used in electronics, optoelectronics and telecommunications.

halides. A class of binary compounds composed of a halogen and another element or organic radical. For example, sodium chloride (NaCl) and ethyl bromide (C_2H_5Br) are halides. See also halogenides; haloids.

halite. A colorless, yellowish, or bluish-red naturally occurring crystalline mineral with white streak composed of sodium chloride, NaCl. It can also be made synthetically. Crystal system, cubic. Density, 2.14-2.17 g/cm^3; melting point, 801-804°C (1474-1479°F); boiling point, 1413-1490°C (2575-2714°F); hardness, 2.5 Mohs; refractive index, 1.542; transparent to infrared radiation up to 14.5 μ and ultraviolet radiation up to 175 nm. Occurrence: (i) Ocean water; (ii) Salt lakes and brines, e.g., Great Salt Lake in Utah (USA), Caspian Sea and Dead Sea; and (iii) Salt deposits in the USA (Louisiana, Michigan, New York, Ohio, Texas), Austria, and other countries. Used in ceramic glazes, glassmaking, soda-ash and soap, in the preparation of metallic sodium, in metallurgy, for paper pulp, in ion exchange and water softening, in deicing of roads and sidewalks, in nuclear reactors, and in fire extinguishers, supercooled solutions and tanning agents. Single crystals are used in ultraviolet and infrared transmission, for spectroscopy, and as selective reflectors. Also known as *common salt; rock salt; table salt.*

halites. Ceramic materials with halite (NaCl) crystal structure corresponding to the general formula MX, where M denotes a metal, such as lithium, sodium, calcium, magnesium, iron or nickel, and X a nonmetal, such as chlorine, fluorine, oxygen or sulfur. Examples include lithium fluoride (LiF), calcium oxide (CaO) and magnesium sulfide (MgS).

Hallamite. Trade name of Hallamshire Steel Company (UK) for an oil-hardening high-speed steel containing 0.7% carbon, 14% tungsten, 4% chromium, 1% vanadium, and the balance iron. Used for saws, shear blades, and planing and turning tools.

Hallamitier. Trade name of Hallamshire Steel Company (UK) for an oil-hardening high-speed steel containing 0.7% carbon, 18% tungsten, 4% chromium, 1% vanadium, and the balance iron. Used for taps, saws, shear blades, and various metal finishing tools.

Hallamitiest. Trade name of Hallamshire Steel Company (UK) for an oil-hardening high-speed steel containing 0.7% carbon, 22% tungsten, 4% chromium, 1% vanadium, and the balance iron. Used for turning and planing tools, saws, and hacksaw blades.

Hallamsteel. Trade name of Hallamshire Steel Company (UK) for a series of water-hardening tool steels containing 0.5-1.2% carbon, and the balance iron. Used for hammers, chisels, scissors, awl blades, blacksmith tools, boilermaker tools, cutlery, drills, wood and circular saws, hacksaws, band saws, shear blades, turning and shaping tools, taps, and broaches.

Hall Coat. Trade name of Hubbard-Hall Inc. (USA) for a series of conversion coatings.

Hallcote. Trademark of C.P. Hall Company (USA) for anti-blocking coatings that contain small quantities of clay, and are used in rubber processing.

Halletts. A brand name for *regulus* containing 98.5% antimony.

Hallimax. Trade name of Hallamshire Steel Company (UK) for nickel-chromium and nickel-iron-chromium resistance alloys including *Hallimax I* with 65% nickel, 20% iron and 15% chromium, and *Hallimax II* with 80% nickel and 20% chromium. Both have high heat resistance, and are used for heating elements and resistances.

hallimondite. A yellow mineral composed of lead uranyl arsenate dihydrate, $Pb_2(UO_2)(AsO_4)_2 \cdot 2H_2O$. It can also be made synthetically in anhydrous form. Crystal system, triclinic. Density, 6.39 g/cm^3. Occurrence: Germany.

halloysite. A colorless clay mineral of the kaolinite-serpentine group composed of aluminum silicate hydroxide dihydrate, $Al_2Si_2O_5(OH)_4 \cdot 2H_2O$. It occurs in nature with *kaolinite*. Crystal system, monoclinic. Density, 2.12-2.14 g/cm^3; melting point, above 1500°C (2730°F); hardness, 1-2 Mohs; refractive index, 1.490. Used in the production of dinnerware and refractories, and as a catalyst support.

Halmo. Trade name of Colt Industries (UK) for an oil- or air-hardening tool steel containing 0.35% carbon, 5% chromium, 5% molybdenum, and the balance iron used for tools and dies.

Halo. Trade name of Superior Threads Company (USA) a sewing thread based on a polyester/acrylic blend and supplied in a wide range of colors having either gold or silver metallic highlights. Used for stiching, serging and embroidering.

halocarbon. See halogenated hydrocarbon.

halocarbon plastics. Plastics based on halocarbon resins made by the polymerization of monomers consisting solely of carbon and one or more halogens.

halocarbon resins. Synthetic resins, such as polychloroprenes or polytetrafluoroethylenes, made by the polymerization of monomers consisting of halogenated hydrocarbons.

halogenated hydrocarbon. A compound composed carbon, one or more halogens, e.g., bromine, chlorine or fluorine, and hydrogen. Monohalogen compounds, such as tetrafluoroethylene, contain one halogen atom, and polyhalogen compounds, such as trifluorochloroethylene, two or more halogens. Used in the manufacture of halocarbon plastics. Abbreviation: HC. Also known as *halocarbon*.

halogenides. (1) A group of binary compounds containing a halogen, i.e., an electronegative element of Group VIIA (Group 17) of the Periodic Table including bromine, chlorine, fluorine and iodine, and a more electropositive element or radical. See also halides.

(2) A salt consisting of a halogen combined with a metal, e.g., lithium fluoride (LiF) or silver chloride (AgCl). Also known as *haloids*.

halogens. The highly active, electronegative elements of Group VIIA (Group Group 17) of the Periodic Table, namely, iodine, bromine, chlorine, fluorine and astatine, that are nonmetallic and monovalent, form stable diatomic molecules, and combine directly with metals to form salts, and with hydrogen to form gaseous molecular compounds.

haloids. See halogenides (2).

Halon. Trademark of Ausimont USA for polytetrafluoroethylene (PTFE) polymers supplied in unfilled and bronze-, graphite- and glass fiber-filled grades. They have negligible moisture absorption, inertness to nearly all chemicals, low coefficients of friction, good high- and low-temperature resistance, high impact strength, excellent dielectric properties, and good self-extinguishing properties. Used in the chemical industries and for industrial parts and equipment.

halotrichite. A yellowish mineral composed of iron aluminum sulfate hydrate, $FeAl_2(SO_4)_4 \cdot 22H_2O$. Crystal system, monoclinic. Density, 1.89 g/cm^3. Also known as *butter rock; iron alum*.

Halowax. Trademark of Koppers Company Inc. (USA) for various chlorinated hydrocarbons including numerous low-viscosity oils and hard microcrystalline waxes used as electrical insulation, in cable coatings, and in fireproofing and waterproofing fabrics.

Halsic. Trade name of W. Haldenwanger Technische Keramik GmbH & Co. KG (Germany) for silicon carbide (SiC) materials used for furnace and kiln components resistant to abrasion, corrosion and high temperatures. *Halsic I* has exceptional thermal-shock resistance and structural integrity retention and high load-bearing capacity up to 1350°C (2460°F), and *Halsic R* is used for applications requiring extremely high temperatures (1600-2000°C, or 2910-3630°F) in different atmospheres or in vacuum.

halurgite. A white mineral composed of magnesium borate pentahydrate, $Mg_2B_8O_{14} \cdot 5H_2O$. Crystal system, monoclinic. Density, 2.19 g/cm^3; refractive index, 1.545. Occurrence: Russian Federation.

Halvan. Trade name of Crucible Materials Corporation (USA) for a tough, shock-resistant tool steel (AISI type L2) containing 0.5% carbon, 1% chromium, 0.7% manganese, 0.2% vanadium, and the balance iron. It has excellent toughness, good to moderate machinability, and moderate wear resistance. Used for punches, chisels, rivet sets, die parts and fixtures.

Halvic. Trade name of Halvic for flexible polyvinyl chlorides.

hambergite. A white, grayish-white or colorless transparent mineral composed of beryllium borate hydroxide, $Be_2BO_3(OH)$. Crystal system, orthorhombic. Density, 2.36 g/cm^3; refractive index, 1.5873; hardness, 7.5 Mohs. Occurrence: Madagascar.

Hamburg blue. A general term used for a wide variety of blue and bluish iron-bearing pigments.

Hamiloy. Trade name of Hamilton Die Cast (USA) for a corrosion-resistant aluminum-silicon alloy used for die castings.

Hamilton. Trade name of Hamilton Foundry (USA) for an extensive range cast irons including (i) *Hamilton Alloyed Iron* alloy cast iron for making strong castings, (ii) *Hamilton Ductile Iron* high-strength, high-ductility iron, (iii) *Hamilton Ductile Ni-Resist* strong, ductile, corrosion-resistant, austenitic cast iron with 18-25% nickel, (iv) *Hamilton Gray Iron* general-purpose gray cast iron, (v) *Hamilton Meehanite* high-quality gray cast iron, (vi) *Hamilton Ni-Hard* white, abrasion-and wear-resistant nickel-chromium cast iron, and (vii) *Hamilton Ni-Resist* corrosion- and erosion-resistant austenitic nickel and nickel-copper cast irons.

Hamilton Metal A. British trade name for a zinc casting alloy containing 3.5% copper, 3.1% lead, 1.5% antimony and a total of 0.5% tin and phosphorus, used for ornamental castings.

Hamilton Metal B. British trade name for high-ductility brass containing 66.7% copper and 33.3% zinc, used for stamped, drawn or spun parts.

Hamlon. Trade name for polyolefin fibers and products.

hammarite. A steel-gray mineral with red tint belonging to the stibnite group. It is composed of bismuth copper lead sulfide, $Bi_4Cu_2Pb_2S_9$. Crystal system, orthorhombic. Density, 6.73 g/cm³. Occurrence: Sweden, Russian Federation.

Hammered. Trade name of Pilkington Brothers Limited (UK) for a patterned glass.

hammered satin. A *satin* fabric having a surface texture imitating hammered metal.

Hammerflex. Trade name of Tenax Finishing Products Company (USA) for finishes with a hammered effect.

Hammerlite. Czech trade name for a patterned glass with a surface finish imitating hammered metal.

Hammond Alloy. Trade name of Hammond Brass Works (USA) for a series of cast copper alloys including various aluminum, silicon, manganese and leaded manganese bronzes, leaded and high-leaded tin bronzes, and red and leaded red brasses used for machinery parts, such as bearings, bushings, gears and worms. Some corrosion-resistant grades are suitable for use in the manufacture of hydraulic castings, valves, steam-valve bodies, fittings, cocks and pumps. The manganese and leaded manganese bronzes are used for applications requiring high strength and toughness and good corrosion resistance.

Hammonia metal. A German alloy containing 65% tin, 32% zinc and 3% copper, used for ornamental parts.

Hampden. Trade name of Carpenter Technology Corporation (USA) for an oil-hardening, high-carbon, high-chromium cold-work tool steel (AISI type D3) containing 2.1% carbon, 0.25-0.35% manganese, 0.3% silicon, 12-12.5% chromium, 0.25-0.5% nickel, and the balance iron. It has excellent abrasion and wear resistance, excellent nondeforming properties, great depth of hardening, and good compressive strength. Used for blanking, forming and various other dies, cutters, rolls, and spinning tools.

Hana. Trade name for rayon fibers and yarns used for textile fabrics.

Hanalac. Trade name of Miwon Company (Japan) for acrylonitrile-butadiene-styrene terpolymers.

hancockite. A brownish-red mineral of the epidote group composed of lead calcium aluminum iron silicate hydroxide, $(Pb,Ca,Sr)_2V(Al,Fe)_3Si_3O_{12}OH$. Crystal system, monoclinic. Density, 4.03 g/cm³; refractive index, 1.81. Occurrence: USA (New Jersey).

hand-blown glass. Glassware formed and shaped by hand manipulation at the end of a blowpipe to which air is supplied by mouth.

hand-formed brick. See hand-made brick.

Handi-Foam II. Trade name of FOMO Products, Inc. (USA) for two-part disposable polyurethane foam used for thermal insulation applications.

Handi-tak. Trademark of Pacer Technology Corporation (USA) for reusable, plasticine-type adhesives used as replacements for tacks and adhesive tape.

handkerchief linen. A fine, light fabric of linen or a linen blend in a plain weave. Used especially for blouses and dresses.

Handler. Trade name of Handler GmbH (Germany) for a series of steels including numerous plain-carbon and alloy machinery grades, nitriding grades, tool and die grades (e.g., plain-carbon, high-speed, hot-work, high-carbon high-chromium cold-work and nondeforming), high-chromium heat-resisting grades, and austenitic, ferritic and martensitic stainless grades.

hand-made brick. Brick shaped by hand or in a mold from a body of suitable consistency. It may or may not be subsequently pressed in a mechanical or hand-operated press. Also known as *hand-formed brick; hand mold brick.*

hand mold brick. See hand-made brick.

Hand Moulded Calcite. Trade name of Paris Brick Company (Canada) for hand-molded calcite bricks.

hand-repressed brick. Hand-made brick pressed after partial drying in a mechanical or hand-operated press.

Handy Coin Silver. Trade name of Handy & Harman (USA) for a corrosion-resistant alloy composed of 90% silver and 10% copper, and used for coins.

Handy Hi-Temp. Trade name of Handy & Harman (USA) for a series of high-temperature brazing alloys including various nickel-base alloys (e.g., nickel-chromium-silicon-boron and nickel-silicon-boron-iron) and several copper-manganese-nickel alloys. Used for brazing austenitic, ferritic and martensitic stainless steels and nickel- and cobalt-base superalloys.

Handy Silver Solder. Trade name of Handy & Harman (USA) for a silver solder (ASME BAg-9) containing 65% silver, 20% copper and 15% zinc. It has a solidus temperature of 671°C (1240°F), a liquidus temperature of 718°C (1325°F), and is used for brazing silverware, iron and nickel alloys. Now known as as *Braze 650. Handy Silver Solder Hard* is a silver solder (ASME BAg-11) containing 75% silver, 22% copper and 3% zinc. It has a solidus temperature of 741°C (1365°F), a liquidus temperature of 788°C (1450°F), and is used for brazing silverware, step brazing or subsequent enameling, and for brazing iron and nickel alloys. Now known as *Braze 750. Handy Silver Solder Medium* is a silver solder (ASME BAg-10) containing 70% silver, 20% copper and 10% zinc. It has a solidus temperature of 691°C (1276°F), a liquidus temperature of 738°C (1360°F), and is used for brazing silverware. Now known as *Braze 700.*

Hanilon. Trademark of Hanil Synthetic Fiber Industrial Company (South Korea) for acrylic staple fibers used for textile fabrics.

hanksite. A white, yellow, or colorless mineral composed of potassium sodium chloride carbonate sulfate, $KNa_{22}Cl(CO_3)_2(SO_4)_9$-Cl. Crystal system, hexagonal. Density, 2.56 g/cm³; refractive index, 1.481. Occurrence: USA (California).

hannayite. A yellowish or colorless mineral composed of ammonium magnesium hydrogen phosphate octahydrate, $Mg_3(NH_4)_2H_4(PO_4)_4\cdot8H_2O$. It can be made synthetically. Crystal system, triclinic. Density, 2.03 g/cm³; refractive index, 1.522.

Hannosil. Trademark of Hanno-Werk GmbH & Co KG (Austria) for a range of silicone-based adhesives and sealants for a wide range of construction and industrial applications.

Hanover white metal. An antifriction metal composed of 86.8-87% tin, 7.5-7.6% antimony and 5.5-5.6% copper, and used for bearings.

Hansa. (1) Trade name of Wehmeier & Olheide Glasbiegerei (Germany) for small panes of curved glass used for making leaded light windows.

(2) Trade name of Vereinigte Edelstahlwerke (Austria) for a series of tungsten- and cobalt-tungsten high-speed steels used mainly for lathe and planer tools, drills, taps, milling cutters, and other cutting and finishing tools.

Hapflex. Trade name of Hapco Inc. (USA) for castable hybrid elastomers available in a wide range of hardnesses including *Series 500* and *600* with durometer hardnesses from 50-70 D (hard and tough) to 45-95A (soft and flexible), *Series 700* and

800 colorless castable rubbers with durometer hardnesses of 65 A to 72D, and *Series 1000* water-clear castable rubbers with durometer hardnesses of 20-55 A.

haradaite. A bluish green mineral of the pyroxenoid group composed of strontium vanadyl silicate, $SrVOSi_2O_6$. It can also be made synthetically. Crystal system, orthorhombic. Density, 3.80 g/cm³; refractive index, 1.721. Occurrence: Japan.

Harbide. Trade name of Harbison-Walker Refractories Company (USA) for silicon carbide (SiC) brick made by an impact pressing process, and having high hardness, good abrasion and wear resistance, low permeability, dense impervious surfaces, and high oxidation resistance. Used as a refractory for furnace linings, ceramic kilns, recuperator tubes, radiant tubes, and retorts.

Harbronze. Trade name of Arthur Harris & Company (USA) for an alloy of copper and tin used for bearings.

Hardalloy. Trade name of Teledyne McKay (USA) for a series of covered hardfacing electrodes including plain-carbon, austenitic manganese and high-chromium stainless steel as well as high-chromium and high-chromium high-titanium cast iron products.

Hardas process coating. A light yellow to brown coating produced on aluminum and its alloys by a hard anodizing process in an oxalic acid bath.

Hard Babbitt. British trade name for an antifriction metal composed of 83.3% tin, 8.4% copper and 8.3% antimony, and used for bearings.

Hard Bearing. British trade name for a lead-base bearing alloy containing 1.3% magnesium.

hard bearing bronze. See hard bronze (1).

hardboard. A thin, strong, smooth, solid building board manufactured in large panels by compressing wood fibers and sawdust with an adhesive lignin binder under heat. Special materials may be added during the panel-forming processes to modify the rigidity, density or other properties of the final product. It is commercially available in various grades including standard, service, service-tempered, tempered, patterned, embossed, perforated, filigree, grooved, striated, sealed and filled. Panels with tough, baked-on finishes are also available. Standard panels are brownish, have a density of at least 500 kg/m³ (31 lb/ft³), one textured surface, and no moisture inhibitors added. *Hardboard* is available in various sizes and thicknesses. Standard panels are 1.2 m (4 ft.) wide, 2.4 m (8 ft.) long and 3.2 or 6.4 mm (0.125 or 0.250 in.) thick. Perforated hardboard (pegboard) comes in thicknesses of 6.4 mm (0.250 in.) and 9.5 mm (0.375 in.). Uses include furniture components, siding, underlayment and interior walls.

hardboard underlayment. Service-grade hardboard, usually 6.4 mm (0.250 in.) in thickness, and supplied with close thickness tolerances. Used to provide a smooth or regular surface for floor covering materials or as a leveling course.

hard brass. A brittle brass that has not been annealed after drawing and rolling to remove the effects of cold working.

hard bronze. (1) A tough, high-strength leaded bronze containing 88% copper, 7% tin, 3% zinc and 2% lead. Used for bushings and bearings. Also known as *hard bearing bronze*.

(2) A bronze containing 88% copper, 10% tin and 2% zinc. Used for bushings and gears. Also known as *hard gear bronze*.

hard-burnt brick. Brick fired at a higher than normal temperature.

Hardcast. Trade name of Scheu-Dental (Germany) for a hard-elastic, cast plastic materials supplied round or square with a diameter of 125 mm (5 in.) or a size of 125 mm × 125 mm (5 in. × 5 in.) in thicknesses of 0.4, 0.6 and 0.8 mm (0.0016, 0.024 and 0.030 in.). Used in dentistry for copings.

hard cast iron. See white cast iron.

Hardchrome. Trademark of Frederick Hardchrome Limited (Canada) for a hard chromium coating produced by electrodeposition.

hard chromium coatings. Coatings with a thickness up to 0.1 mm (4 mils) produced on steel parts by plating in a standard chromium bath, but without the usual undercoating of nickel. These coatings are much thicker, but not necessarily harder than decorative chromium coatings. They have good corrosion and wear resistance, and low coefficients of friction. Typical hardness values range from 800-1200 Vickers. Used on dies, tools, sliding mechanisms, etc. Also known as *hard chromium plates*.

hard coal. See anthracite.

hardcoat. A hard anodic aluminum oxide coating produced on aluminum or aluminum alloys by a process making use of a sulfuric or oxalic acid bath, with or without additives. The bath temperature is relatively low, typically 0-20°C (32-68°F), and the current density is usually 2-4 A/dm² (20-43 A/ft²). Both, the coating thickness (typically 25-50 μm or 1-2 mils) and the wear resistance are higher than those of conventional anodized coatings.

Hardcote. Trade name of Foseco Minsep NV (Netherlands) for dressings used to produce hard surfaces on foundry sand cores and molds.

HardCut. Trademark of Honeywell Performance Fibers (USA) for a continuous-filament bicomponent polyester fiber produced by a proprietary hard-particle fiber technology. It consists of a polyester core fiber with numerous embedded microscopic ceramic platelets, and a smooth polyester sheath. *HardCut* provides significantly enhanced cut resistance as compared to ordinary polyester, excellent dyeability, coatability, printability and launderability, and excellent laminating properties. Supplied in flat and air-textured yarn grades, it is used for heavy duty industrial gloves and sleeves, and knitted and woven industrial fabrics.

Hard Devil. Trade name of Champion Rivet Company (USA) for a carbon steel containing 0.85-1% carbon, and the balance iron. Used for hardfacing and welding electrodes.

hard-drawn spring wire. A grade of round cold-drawn spring wire made from open-hearth, basic-oxygen or electric-furnace steel composed of 0.45-0.85% carbon, 0.6-1.30% manganese, 0.10-0.30% silicon, up to 0.040% phosphorus, up to 0.050% sulfur, and the balance iron. The wire size ranges from 0.518-15.875 mm (0.0204 to 0.6250 in.) in diameter. It can be readily plated, and has tensile strengths ranging from 1014 to 2509 MPa (147 to 364 ksi) at ordinary temperatures. Used for general-purpose springs.

hard-drawn wire. (1) A general term used for any cold-drawn aluminum, copper or low-carbon steel wire with high tensile strength.

(2) A copper wire that usually has high tensile strength and relatively low elongation. It is stiffer and harder to bend and work than *medium-hard-drawn* or *soft-drawn copper wire*.

hardened steel. Steel that has been heated to a temperature above the hardening temperature (typically between 750 and 1200°C, or 1380 and 2190°F), held at that temperature long enough to ensure thorough heating, and then rapidly quenched in a suitable medium (e.g., water) with the objective to produce a martensitic microstructure.

hardener. See concrete hardener; curing agent (1) and (2).

hardening media. See quenching media.

Hardenite. Trade name of Hardenite Steel Company Limited (UK) for an oil-hardening steel containing 0.9% carbon, 1.2% manganese, and the balance iron. Used for dies and tools.

Hardex. (1) Trade name Vetrerie Riunite Bordoni Miva (Italy) for a glass insulating material with high thermal-shock resistance.

(2) Trademark of Nimet Industries Inc. (USA) for a hard autocatalytic coating combining nickel and phosphorus with silicon carbide.

Hard Facing. Trade name of Abex Corporation (USA) for a series of wear- and heat-resistant alloys containing carbon, chromium, manganese, molybdenum and iron. Used for hardfacing rods.

hardfacing alloys. See hardfacing materials.

hardfacing materials. Materials that consist of hard particles of borides or carbides of chromium, manganese, tungsten, molybdenum or other metals, or intermetallics, uniformly distributed in soft matrix alloys normally based on cobalt, iron or nickel. There are various types of hardfacing materials including ferrous and nonferrous alloys (e.g., high-speed, austenitic manganese and austenitic high-chromium steels, cobalt- and copper-base alloys or nickel-chromium-boron alloys), cemented carbides and sintered oxides. Depending on the type of material used, they may provide high hardness, toughness, impact and/or wear resistance, or one or more other properties. They are usually applied by spraying or welding techniques. A more recent process is based on flexible tungsten carbide fabrics impregnated with hardfacing alloys and metallurgically bonded to the surfaces or substrates to be hardfaced as composite coatings by heating in a furnace to temperatures above the melting point of the alloys. *Hardfacing materials* are often used to rebuild worn cutting tools, mill hammers, earth-moving equipment, crushers, mechanical shovels, ball mills, and shears. Also known as *hard surfacing materials.* See also hardfacings.

hardfacings. Alloy, ceramic or composite coatings or layers deposited or applied to substrates by thermal spraying, or gas or arc welding to provide them with abrasion, cavitation, corrosion, erosion, galling, heat, impact and/or wear resistance. Hardfacings are also applied to rebuild worn surfaces on tools or machinery parts. For materials used for these applications, see hardfacing materials.

hard ferrite magnets. A class of ceramic permanent-magnet materials that are mixtures of ferric oxide (Fe_2O_3) and suitable compounds of divalent metals, such as barium, cobalt, copper, lead, magnesium, manganese, nickel, strontium or zinc. Common examples include barium and strontium hexaferrite. They exhibit ferromagnetic, ferrimagnetic and antiferromagnetic, magnetooptical and magnetstrictive effects and, owing to their excellent corrosion and demagnetization resistance, are often used for DC motors, automotive sensors, magnetic separators, magnetic resonance imaging (MRI) devices, etc. See also ferrites.

hard fibers. A group of usually long, stiff, heavily lignified vegetable fibers that occur throughout the hard, coarse leaves of various plants. Examples include abaca, henequen and sisal fibers. Used especially for cordage, twine, and textile fabrics. Also known as *leaf fibers.*

hard-film rust-preventive compound. A rust-preventive compound that produces a hard, thin varnish-like surface coating on iron or steel, and is applied by the evaporation of a solvent, diluent or by chemical reaction. See also rust-preventive com-

pound.

hard-finished fabrics. Cotton, wool or worsted fabrics that do not have a raised or napped surface finish.

hard-finished plaster. Overburnt gypsum treated with an accelerator, such as aluminum-potassium sulfate or potassium sulfate, and then recalcined. Used in special cements, such as *Keene's cement.*

hard-fired clay. Clay products having high compressive strengths and low water absorption due to firing at temperatures near the vitrification point. Also known as *hard-fired ware.*

Hardflex. Trade name of Sandvik Steel Company (USA) for several prehardened, austempered medium-carbon strip steels with high tensile and yield strengths used for sewing machines, appliances, and business machine and office equipment components.

hard gear bronze. See hard bronze (2).

Hardglas. Trade name of NV Hardmaas (Netherlands) for toughened glass.

hard glass. A high-silica soda-lime glass with high viscosity at elevated temperatures, high softening point and high resistance to abrasion, scratching or other mechanical damage, or any combination of these.

hard glaze. High-silica glaze that fires at a relatively high temperature, usually above 1050°C (1920°F). It is harder and more chemically resistant than soft glazes. Also known as *high-temperature glaze.*

Hard Head. British trade name for an antifriction metal containing 90% tin, 8% antimony and 2% copper, used for bearings.

Hardinox. Trade name of Acieries Nouvelle de Pompey (France) for stainless steels containing 0.3-0.4% carbon, 13% chromium, and the balance iron. Used for corrosion-resistant parts.

Hardipanel. Trademark of James Hardie Siding Products (Canada) for durable, crack- and moisture-resistant fiber cement siding supplied primed, unprimed or finished in panel sizes of 1.2 × 2.4, 1.2 × 2.7 and 1.2 × 3.0 m (4 × 8, 4 × 9 and 4 × 10 ft.).

Hardiplank. Trademark of James Hardie Siding Products (Canada) for durable, crack and moisture-resistant fiber cement siding supplied in primed, unprimed or finished planks in lengths of 3.6 m (12 ft.) and widths of 160, 210, 240 and 300 mm (6.25, 8.25, 9.50 and 12.00 in.).

hard iron. See white cast iron.

Hardite. Trade name of Hardite Metals Inc. (USA) for a series of iron-nickel, iron-chromium and nickel-iron-chromium based heat-resistant alloys used for heating elements, heat-treating boxes and furnace parts.

Hard Kote. Trade name of Bonney-Floyd Company (USA) for a series of heat- and corrosion-resistant alloys containing 1.1% carbon, 16-30% tungsten, 16-23% chromium, 24-30% nickel, and the balance iron. Used for heat- and corrosion-resistant parts, tools and dies.

hard lac resin. A *shellac* from which the soft constituents have been removed by solvent extraction.

hard laid rope. A rope that consists of tightly twisted or laid strands. It is stiff, less flexible, but more abrasion-resistant than *soft-laid rope.*

hard lead. See antimonial lead.

Hardline. Trade name of Honeywell International (USA) for nylon-6 fibers.

hard magnetic alloys. See hard magnetic materials; permanent magnet materials.

hard magnetic materials. A group of ferrimagnetic or ferromagnetic materials that strongly retain appreciable magnetization when the applied field is removed. They are characterized by high coercive forces, large hysteresis loops and high maximum magnetic energy values. Examples include carbon, tungsten and cobalt steels, Alnico, Bismanol, Cunico, Cunife and Ferroxdur. Used for permanent magnets. Also known as *hard magnets; magnetically hard materials.* See also permanent magnets.

hard maple. A collective term used for the hard, strong, stiff wood of the sugar maple *(Acer saccharum)* and black maple *(A. nigrum).* The heartwood is light reddish-brown, and the sapwood white with reddish-brown tinge. It has a fine, uniform texture and grain pattern, good shock resistance, large shrinkage, good abrasion resistance, is difficult to work with hand tools, machines well, and has medium gluing and low nailing quality. Used for flooring, bowling alleys, interior decoration, cabinetwork, turnery, woodenware, veneer, paneling, shoe lasts, handles, quality furniture, boxes, pallets, crates, spools and bobbins, crossties, and as pulpwood. See also black maple; sugar maple.

hard metals. A generic term for sintered materials with high hardness, strength and wear resistance made from one or more hard materials, such as carbides, borides or nitrides of the refractory metals (e.g., tungsten, molybdenum and titanium) and a tough metallic binder, such as cobalt or nickel. Used for cutting tools. See also cemented carbides.

hard mica. Mica that does not tend to delaminate when slightly bent.

Hardnair. Trade name of Atlantic Steel Corporation (USA) for an air-hardening nondeforming tool steel (AISI type A2) containing 1% carbon, 5% chromium, 1% molybdenum, 0.4% manganese, 0.4% vanadium, and the balance iron. Used for cutters, gages, blanking, lamination and trimming dies, cams, and shears.

hard nickel plates. Electrodeposits of nickel, usually at least 250 μm (10 mils) thick, produced on metallic substrates. They have an average hardness of 350-500 Vickers, a tensile strength of 990-1100 MPa (115-160 ksi), and are used chiefly for salvage or buildup applications.

Hardox. Trade name of Scanglas A/S (Sweden) for a toughened glass.

hard paste porcelain. A high-grade ceramic whiteware made of china clay, feldspar and silica, and fired at relatively high temperatures, usually 1300-1445°C (2370-2630°F). It is characterized by negligible water absorption, good translucency, high strength and thermal-shock resistance, and a hard-glaze surface. It is also known by the French term "pâte dure," and is the opposite of *soft porcelain* ("pâte tendre") which refers to porcelain fired at relatively low temperatures, usually 1100-1300°C (2010-2370°F). Also known as *European porcelain; hard porcelain; true porcelain.*

hard phosphor bronze. A strong, corrosion-resistant phosphor bronze containing 92.8% copper, 7% tin and 0.2% phosphorus. Used for springs, bearings, gears, and electrical parts.

hard pine. See longleaf pine.

hard porcelain. See hard paste porcelain.

Hardrite. Trade name of Universal Cyclops Specialty Steels (USA) for a nonshrinking tool steel (AISI type O7) containing 1.1% carbon, 1.7% tungsten, 0.6% chromium, 0.25% vanadium, and the balance iron. Used for various dies, and for punches, knives and cams.

Hardrock. Trade name of Industries Trading Company (USA) for a series of high-speed steels containing 0.7-0.8% carbon, 4% chromium, 18% tungsten, 7% cobalt, 1% vanadium, and

the balance iron. Used for tool bits.

Hard Rock. Trade name of Whip Mix Corporation (USA) for dental gypsum (die stone) supplied as a powder. It is typically cast from a creamy, thixotropic aqueous suspension (21 mL H_2O per 100 g powder) and, upon setting, develops a smooth nonbrittle surface, high compressive strength and high setting expansion (0.28%). Supplied in aqua, green and ivory colors, it is compatible with all types of impression materials including polyvinyl and polyether.

hard rubber. See ebonite.

hard silk. Raw silk prior to the degumming process. See also raw silk.

hard solders. See brazing alloys.

hard-spun yarn. See high-twist yarn.

Hardsteel. Trade name of Black Drill Company Inc. (USA) for a hard alloy containing 40% cobalt, 32% chromium, 18% tungsten, 2% nickel, 2% iron, and traces of manganese and boron. Used for drills, tool bits, and wear parts.

hard superconductor. A superconductor material, such as niobium or vanadium, that is completely diamagnetic at low applied fields, but with increasing magnetic field experiences a gradual transition from the superconducting to the normal state between two critical values known as the "upper critical field" (H_{c1}) and the "lower critical field" (H_{c2}). Upon reaching H_{c2}, the material is normal (i.e., nonsuperconducting). Superconductors with applied magnetic fields between the two critical values exist in a mixed state with superconducting and normal regions. *Hard superconductors* are useful for the generation of high magnetic fields in nonferrous coils. Abbreviation: HSC. Also known as *high-field superconductor; type II superconductor.*

hard surfacing alloys. See hardfacing alloys.

Hardtem. (1) Trade name of Creusot-Loire (France) for oil-hardening hot-work steel containing 0.55% carbon, 0.75% manganese, 0.55% nickel, 1% chromium, 0.45% molybdenum, 0.05% vanadium, and the balance iron. Used for hot-forming dies.

(2) Trade name of Heppenstall Company (USA) for a hot-work steel containing 0.6% carbon, 0.7% manganese, 0.61% chromium, 0.17% vanadium, 0.28% molybdenum, up to 0.7% nickel, and the balance iron. Available in drop-forged and pressed form, it is used for die blocks for forging and die casting.

Hard-Trak. Trade name of Joseph T. Ryerson & Son Inc. (USA) for a steel containing 0.31% carbon, 1.4% manganese, 0.5% chromium, 0.24% silicon, 0.11% molybdenum, 0.0004% boron, and the balance iron. Used for wear bars and rails.

Hardtuf. Trademark of Tiodize Company, Inc. (USA) for a polytetrafluoroethylene (PTFE) impregnated aluminum oxide coating applied to provide hard, corrosion-, abrasion- and wear-resistant surfaces on aluminum and aluminum alloys.

hard vegetable fibers. A class of *vegetable fibers* that are found in the stems and leaves of plants and include henequen, manila and sisal.

hard vulcanized fiber. A product made from cotton rags by treating with a zinc chloride solution and subsequent lamination.

hardware bronze. A corrosion-resistant casting brass of 85-89% copper, 9-13.25% zinc and 1.75-2% lead used for hardware, screws, nuts, bolts, tie rods, automobile fittings, and architectural applications.

Hardwear. Trademark of Lukens Steel Company (USA) for a series of formable, weldable, abrasion-resistant alloy plate steels having optimum flatness tolerance and high toughness and hardness. Extra-low sulfur grades are also available. Used for construction and mining equipment, such as liners for crushers and hoppers, truck-body liners, buckets and bucket lips.

Hardwood. Trade name for hardboard with a smooth, hard surface manufactured in large sheets by compressing hardwood waste under heat.

hardwood. The wood produced by broad-leaved or deciduous trees, such as ash, birch, cherry, maple, oak, maple and walnut, belonging to the botanical class Angiospermae. The term should not always be taken literally, e.g., the wood of some hardwoods (e.g., basswood) is much softer than that of softwoods, such as yellow pine. Also known as *nonconiferous wood*. See also softwood.

hardwood lumber. Hardwood timber, logs, beams, boards and other product forms, roughly sawn and prepared for use. The following grades of hardwood lumber are specified by the National Hardwood Lumber Association (NHLA): (i) *Firsts.* High-quality lumber for fine cabinetwork that should be over 90% clear on both sides; (ii) *Seconds:* Good quality lumber, suitable for most cabinetwork, that should be over 80% clear on both sides; (iii) *Firsts and Seconds (FAS):* A combination of Firsts and Seconds that should contain a minimum of 20% Firsts; (iv) *Selects:* Quite acceptable lumber for cabinetwork, but some culling may have to be done. One side should be at least 90% clear, the other may be ungraded; (v) *Common No. 1:* Lumber frequently used for interior and less demanding cabinetwork. It should have one side at least 65% clear; and (vi) *Common No. 2:* Lumber frequently used for flooring and paneling, that should have one side about 50% clear.

hard yellow solder. An alloy containing about 55-58% copper, 1.3% tin, 0.3% lead, and the balance zinc. Used for brazing copper, copper alloys, and steel.

Hardyne. Trade name for pressed compound oxides bonded with thermoplastic resin, and used for permanent magnets.

Hardy Nickel Iron. Trade name Atlas Specialty Steels (Canada) for an electric-furnace steel containing 0.08% carbon, 2% nickel, 0.2% manganese, 0.2% silicon, and the balance iron. It has high toughness even at subzero temperatures, good resistance to vibratory stresses, and low notch sensitivity. Used for low-duty subzero applications, tinning and soldering rolls, bolts, links, rods and staybolts.

hardystonite. A colorless mineral of the melilite group composed of calcium zinc silicate, $Ca_2ZnSi_2O_7$. It can also be made synthetically from calcium carbonate, zinc oxide and silicon dioxide by a high-temperature treatment. Crystal system, tetragonal. Density, 3.39 g/cm³. Occurrence: USA (New Jersey).

hard zinc. A German zinc alloy containing 5.3% iron, 2.4% lead and 0.1% copper, used for washboards. Also known as *Hartzink.*

hare hair. The soft, wooly protein fibers obtained from the protective hairy coat of hares (family Leporidae). They are sometimes spun as yarn, either alone or with wool.

Hares. Trademark of Resopal Werk (Germany) for cresol- and phenol-formaldehyde plastics.

Harex. Trademark of Resopal Werk (Germany) for cresol- and phenol-formaldehyde plastics.

Hargus. Trademark of Ziv Steel & Wire Company (USA) for an oil-hardening, nondeforming tool steel containing 0.9% carbon, 1.2% manganese, 0.5% chromium, 0.5% tungsten, and the balance iron. Used for drills, reamers, punches, taps, gages, and blanking dies.

haricot. A red copper oxide (CuO) used as a background material in ceramic decoration. Also known as *harrico.*

harkerite. A colorless mineral composed of calcium magnesium aluminum borate carbonate silicate monohydrate, $Ca_{12}Mg_4Al$-$(SiO_4)_4(BO_3)_3(CO_3)_5 \cdot H_2O$. Crystal system, hexagonal. Density, 2.96 g/cm³; refractive index, 1.653. Occurrence: Scotland.

Harlequin. Trade name of Pittsburgh Corning Corporation (USA) for a sculptured glass module having a pressed-in diamond pattern on both sides.

Harlington bronze. A British high-strength bronze containing 56% copper, 43% zinc, 0.9% tin and 0.6% iron. Used for bolts, nuts, sheathing, and condenser tubes.

Harlon. Trade name for saran (polyvinylidene chloride) fibers with good chemical resistance, good weatherability, and good resistance to mildew and insects. Used for carpets and rugs, draperies, upholstery, clothing, industrial fabrics, etc.

Harmonia bronze. A British bronze containing 55.7-57% copper, 40-41.2% zinc, 0.4-0.46% lead, 1.29-1.8% iron, up to 0.5% tin and up to 0.86% aluminum. Used for hardware and fittings.

Harmony. Trademark of Ivoclar Vivadent AG (Liechten-stein) for a range of dental casting alloys. *Harmony Hard* is a hard casting alloy (ADA type III) containing 74.0% gold, 12.0% silver, 9% copper, 3.8% palladium, and less than 1.0% zinc, indium and iridium, respectively. It has a deep yellow color, a density of 14.9 g/cm³ (0.538 lb/in.³), a melting range of 895-970°C (1645-1780°F), as-cast hardness of 130 Vickers, high elongation, and good biocompatibility. Used for crowns, posts, inlays, onlays and bridges. *Harmony Medium* is a medium-hard casting alloy (ADA type II) containing 76.8% gold, 12.8% silver, 8.3% copper, and less than 1.0% palladium, zinc, indium and iridium, respectively. It has a rich yellow color, a density of 15.2 g/cm³ (0.549 lb/in.³), a melting range of 880-945°C (1615-1735°F), an as-cast hardness of 120 Vickers, high elongation, and good biocompatibility and resistance to oral conditions. Used for crowns, inlays, onlays and short-span bridges. *Harmony Soft* is a soft casting alloy (ADA type I) containing 83.4% gold, 11.5% silver, 5% copper, and less than 1.0% palladium and iridium, respectively. It has a rich yellow color, a density of 18.7 g/cm³ (0.675 lb/in.³), a melting range of 940-990°C (1725-1815°F), a hardness of 85 Vickers, high elongation, and good biocompatibility and resistance to oral conditions. Used for crowns, inlays and bridges.

harmotome. A colorless mineral of the zeolite group composed of sodium barium aluminum silicate hexahydrate, $(Ba,Na)_2(Si,$-$Al)_8O_{16} \cdot 6H_2O$. Crystal system, monoclinic. Density, 2.44 g/cm³; refractive index, 1.508; hardness, 4.5 Mohs. Occurrence: Finland, USA (Arizona).

harness leather. Vegetable-tanned, oil-finished leather made from cattlehide or pigskin, and used for saddle seats, collars and harnesses.

harness-weave fabrics. See satin-weave fabrics.

harrico. See haricot.

Harris. Trade name of J.W. Harris Company, Inc. (USA) for a series of copper-base brazing and braze-welding rods and filler metals used for welding and brazing steel, cast iron and nonferrous alloys. Also included under this trade name are several aluminum and stainless steel welding wire products.

Harris tweed. A high-quality, hand-woven woolen *tweed*, originally made on the Island of Harris in the Hebrides.

Harrison. Trade name of Harrison Steel Castings Company (USA) for several ferritic and pearlitic cast irons.

harstigite. A colorless mineral composed of calcium magnesium beryllium silicate hydroxide, $Ca_6(Mn,Mg)Be_4Si_6(O,OH)_{24}$. Crystal system, orthorhombic. Density, 3.05 g/cm³; refractive index, 1.68. Occurrence: Sweden.

Hartalumin. Trade name for a hard aluminum alloy used for light-alloy parts.

Hart-Coat. Trademark of AHC-Oberflächentechnik (Germany) for oxide coatings produced on aluminum and its alloys by hard anodizing.

Hartex. Trademark of Firestone Synthetic Rubber & Latex (USA) for centrifuged low-ammonia or ammoniated natural latex used for adhesives, coatings, dipped goods, extruded threads, and carpet backing.

Hartford. Trade name of Hartford Technologies (USA) for rayon fibers and yarns used for textile fabrics.

hartite. A colorless, or white mineral composed of α-dihydrophyllocladene, $C_{20}H_{34}$. Crystal system, triclinic. Density, 1.08 g/cm^3. Occurrence: Italy. Also known as *bombiccite*.

Hartmetall. German trade name for cemented carbides used for indexable inserts, dies, and cutting tools.

Hartzink. See hard zinc.

Harvard Cement. Trade name of Richter & Hoffmann for a zinc phosphate dental cement.

harvard brick. A common red brick overburnt on one side to produce a bluish-black surface color.

Harveyized steel. Steel plate with a relatively soft core and a very hard surface produced by subjecting the surface to a cementation process in order to increase the carbon in that portion of the plate. Used for armor plate.

Harvill. Trade name of Harville Machine Inc. (USA) for a series of copper-base casting alloys including manganese brasses, nickel brasses, yellow brasses, and aluminum bronzes.

Harvite. Trade name Siemon Company (USA) for shellac plastics used in electrical engineering applications.

Harz refined lead. A corrosion resistant commercial lead (99.9% pure) used for lead tubing and sheeting for the chemical industry.

Has-All. Trade name of William Hassall & Sons (UK) for a high-speed steel containing 0.7% carbon, 18% tungsten, 4% chromium, 1% vanadium, and the balance iron. Used for high-speed tools.

Hasberg. Trade name of Hasberg Schneider GmbH (Germany) for strip steel used for precision jigs.

Haschrome Alloy. Trade name of Haynes International, Inc. (USA) for an alloy composed of 1.25% carbon, 12% chromium, 3% manganese, and the balance iron. Used for covered electrodes for AC-DC welding and surfacing.

hashemite. (1) A dark brown mineral of the barite group composed of barium chromium oxide, $BaCrO_4$. Crystal system, orthorhombic. Density, 4.50 g/cm^3; refractive index, 1.96. Occurrence: Jordan.

(2) A dark brown synthetic mineral of the barite group composed of barium chromium oxide sulfate, $Ba(Cr,S)O_4$. Crystal system, orthorhombic. Density, 4.59 g/cm^3; refractive index, 1.960.

Hass. British trade name for an age-hardening aluminum alloy containing 4.5% copper, 1% silicon and 0.75% manganese. Used for aircraft structures.

Hastelloy. Trademark of Haynes International Inc. (USA) for an extensive series of nickel-base alloys available in wrought and cast form and including several nickel-molybdenum-iron, nickel-molybdenum-chromium-iron, nickel-chromium-molybdenum-iron, nickel-iron-chromium-molybdenum, nickel-chromium-molybdenum and nickel-silicon-copper alloys. They have high tensile strengths, good resistance to acids and general cor-

rosion, good high-temperature properties and creep resistance, and good to excellent resistance to chlorine, hydrochloric and sulfuric acids, etc. Used for chemical process equipment, agitators, autoclaves, heat exchangers, dryers, burners, blowers, pickling equipment, furnace parts, oilfield pumps and valves, turbo-superchargers for jet engines, and welding electrodes and wires. The following two *Hastelloy* alloys are of particular commercial importance: (i) *Hastelloy B*, a precipitation-hardenable, corrosion-resistant nickel alloy containing 28% molybdenum, 5% iron and small additions of chromium, manganese and silicon. Used especially for jet-engine turbo-superchargers, it has a density of 9.24 g/cm^3 (0.334 lb/in.^{3}), a melting point of 1340-1390°C (2445-2535°F), a maximum service temperature (in air) of 760°C (1400°F), high tensile strength, good resistance to acids and general corrosion, and good high-temperature properties and creep resistance; and (ii) *Hastelloy C*, a corrosion-resistant nickel alloy containing 17% molybdenum, 16% chromium and small additions of iron, tungsten and manganese. It has a density of 8.94 g/cm^3 (0.323 lb/in.^{3}), a melting point of 1270-1390°C (2320-2535°F), a maximum service temperature (in air) of 1090°C (1995°F), high tensile strength, good resistance to acids and pitting and crevice corrosion, good high-temperature properties and creep resistance, and is used for chemical-process equipment, agitators, pickling equipment, pulp and paper production equipment, and wastewater treatment equipment.

hastingsite. (1) A green mineral of the amphibole group composed of potassium sodium calcium iron magnesium aluminum chloride silicate, $(K,Na)Ca_2(Fe,Mg)_5(Si,Al)_8O_{22}Cl_2$. Crystal system, monoclinic. Density, 3.59 g/cm^3. Occurrence: Russian Federation.

(2) A green mineral of the amphibole group composed of calcium iron magnesium aluminum silicate hydroxide, $(Ca,Na)_2(Fe,Mg)_5(Si,Al)_8O_{22}(OH)_2$. Crystal system, monoclinic. Density, 3.21 g/cm^3. Occurrence: Canada (Ontario).

hastite. A reddish violet or light brownish red mineral of the marcasite group composed of cobalt selenide, $CoSe_2$. Crystal system, orthorhombic. Density, 7.23 g/cm^3. Occurrence: Germany.

hatchettine. A yellowish-white to greenish-yellow mineral composed of mineral wax, $C_{38}H_{78}$. Melting point, 79°C (174°F) (pure); 55-65°C (131-149°F) (impure). Also known as *adipocerite; hatchettite; mineral tallow*.

hatchettite. See hatchettine.

hatchettolite. See betafite.

hatchite. A lead-gray mineral composed of silver thallium lead arsenic sulfide, $PbTlAgAs_2S_5$. Crystal system, triclinic. Density, 5.81 g/cm^3. Occurrence: Switzerland.

Hathal. Trade name of Aluminium-Zentrale e.V. (Germany) for a series of heat-treatable aluminum alloys. *Hathal A* is an aluminum alloy that contains 3-5% copper, 0.4% silicon, 0.4-1% manganese and 0.3-1% magnesium, has good weldability and machinability, and is used for hardware, rivets and hydraulic fittings. *Hathal B* is a corrosion-resistant aluminum alloy that contains 0.2-1% silicon, 0.2-0.6% manganese and 0.2-0.8% magnesium, and is used in the manufacture of fasteners, marine hardware and structural members. *Hathal C* is a corrosion-resistant aluminum alloy that contains 0.5-1% silicon, 0.3-1.5% manganese and 0.8-1.5% magnesium, and is used for automatic screw-machine products, and marine and structural parts.

hatrurite. A mineral that is composed of calcium silicate, Ca_3SiO_5. It can also be made synthetically. Crystal system, rhombohedral (hexagonal). Density, 3.02 g/cm^3; stable at temperatures above 1050°C (1920°F) Occurrence: Russian Federation.

Hatsushimo. Trade name of Central Glass Company Limited (Japan) for a cathedral-type glass.

hauckite. A light orange to yellow mineral composed of magnesium manganese zinc iron carbonate sulfate hydroxide, $(Mg,Mn)_{24}Zn_{18}Fe_3(SO_4)_4(CO_3)_2(OH)_{81}$. Crystal system, hexagonal. Density, 3.02 g/cm^3; refractive index, 1.630. Occurrence: USA (New Jersey).

hauerite. A reddish-brown to black mineral of the pyrite group composed of manganese disulfide, MnS_2. It can also be made synthetically by thermal treatment of a solution of manganese sulfate and potassium pentasulfide. Crystal system, cubic. Density, 3.50 g/cm^3; refractive index, 2.69.

hausmannite. A brownish-black, opaque mineral of the spinel group composed of manganese tetroxide, Mn_3O_4, sometimes with small quantities of magnesium and iron. Crystal system, tetragonal or orthorhombic. Density, 4.43-4.84 g/cm^3; hardness, 536-566 Vickers. Occurrence: Germany.

hauyne. A pale blue mineral of the sodalite group composed of sodium calcium aluminum silicate sulfate, $(Na,Ca)_8(Si,Al)_{12}O_{24}(SO_4)_2$. Crystal system, cubic. Density, 2.4-2.5 g/cm^3; refractive index, 1.496; hardness, 5.5-6 Mohs. Occurrence: Germany. A white synthetic form with slightly different formula $[Na_6Ca_2Al_6Si_6O_{24}(SO_4)_2]$ is also available. Also known as *hauynite*.

hauynite. See hauyne.

Havaflex. Trade name of Ametek-Havag Division (USA) for two-component, trowelable phenolic ablative adhesives and coatings with a density of 1.25 g/cm^3 (0.045 lb/in.^{3}), and a hardness of 89 Shore A used on ship decks.

Havar. Trade name of Hamilton Technology Inc. (USA) for a nonmagnetic, age-hardenable alloy containing 42.5% cobalt, 20% chromium, 13% nickel, 2.8% tungsten, 2% molybdenum, 1.6% manganese, 0.2% carbon, up to 0.04% beryllium, and the balance iron. It has a density of 8.3 g/cm^3 (0.30 lb/in.^{3}), high tensile strength, excellent corrosion and fatigue resistance, and is used for electronic parts, valves, watch and power springs, and medical implants.

Haveg. Trademark of Haveg Industries, Inc. (USA) for a series of plastic molding compounds and plastic laminates used in the manufacture of acid- and corrosion-resistant products, such as tanks, piping, agitators and gaskets.

Havo. Trade name of HAVO-Strangguss (Germany) for continuously cast red-brass products.

Havoc. Trade name of Wallace Murray Corporation (USA) for a water-hardening shock-resistant tool steel (AISI type S2) containing 0.5% carbon, 1% silicon, 0.2% vanadium, 0.5% molybdenum, and the balance iron. It has outstanding toughness, good deep-hardening properties, and moderate machinability and wear resistance, and is used for gages, arbors, punches and reamers.

Hawaiian obsidian. A black *volcanic glass* from the Hawaiian island of Oahu formed from basaltic magma. It takes a high polish, and is used for ornamental purposes. See also basalt glass; obsidian.

hawleyite. A yellow mineral of the sphalerite group composed of cadmium sulfide, CdS. Crystal system, cubic. Density, 4.87 g/cm^3. Occurrence: Canada (Yukon).

hawser laid rope. A rope made by twisting three strands to form helices around a common central axis.

Hawthorne. British trade name for a commercially pure tin containing 0.029% arsenic, 0.028% copper and 0.028% lead.

haxonite. A white mineral composed of iron nickel carbide,

$(Fe,Ni)_{23}C_6$. Crystal system, cubic. Occurrence: USA.

haycockite. A metallic-yellow mineral of the chalcopyrite group composed of copper iron sulfide, $Cu_4(Fe,Ni)_5S_2$. Crystal system, orthorhombic. Density, 4.35 g/cm^3. Occurrence: South Africa.

Haydite. Trademark of Buildex Inc. (USA) for a lightweight aggregate composed of expanded clay, shale or slate, and used in the manufacture of lightweight concrete and concrete products.

Haynes. Trade name of Haynes International, Inc. (USA) for a series of wrought and cast cobalt-base heat-resistant alloys containing varying amounts of chromium, nickel and tungsten, and sometimes iron, molybdenum, silicon, carbon and/or boron. Examples of important *Haynes* alloys include (i) *Haynes Alloy 6B*, a wear-resistant cobalt-base alloy containing 30% chromium, 4% tungsten, up to 3% iron, 2.5% nickel, 1.5% manganese, 1.1% carbon, 1% molybdenum and 0.7% silicon; (ii) *Haynes Alloy 188*, a high-temperature cobalt-base alloy containing 22% chromium, 22% nickel, 14% tungsten, 1.25% manganese, up to 3% iron, 0.35% silicon and 0.05% lanthanum; and (iii) *Haynes Alloy 1233*, a corrosion-resistant cobalt-base alloy containing 25.5% chromium, 9% nickel, 5% molybdenum, 3% iron, 2% tungsten, and up to 0.1% nitrogen and 0.08% carbon, respectively. *Haynes* alloys are used for turbine parts, nuclear equipment, aircraft and aerospace engine parts, and in rod and powder form for hard-facing applications. Also included under this trade name are several iron- and nickel-base superalloys and niobium-base alloys.

Haysite. Trade name of Haysite Reinforced Plastics (USA) for a series of polyester laminates and molding compounds including (i) *Haysite BMC* polyester bulk-molding compounds, (ii) *Haysite Polyester* woven-glass-roving polyester laminates, (iii) *Haysite Reinforced Plastics* fiberglass-reinforced polyesters with good electrical, structural and chemical properties, supplied as sheet pultrusions, sheet-molding compounds, bulk molding compounds, and custom moldings, and (iv) *Haysite SMC* polyester sheet-molding compounds supplied in standard and low-profile grades.

Haystellite. Trade name of Haynes International, Inc. (USA) for a series of tungsten carbide powders that may be blended with other metal powders (e.g., cobalt, iron or nickel alloys). Used in the production of hard, abrasion-resistant surfaces by welding, flame spraying or manual torch techniques.

Hazel Bronze. Trade name of Thomas Bolton Limited (UK) for a tin bronze used for screws, nuts, bolts, and studs.

H-band steels. See H-steels.

H beam. A steel beam whose cross section resembles the letter "H." It is similar to an *I-beam*, but has wider flanges.

HCR Alloy. Trade name of Telcon Metals Limited (UK) for a soft magnetic alloy that is composed of 50% iron and 50% nickel, has a rectangular hysteresis loop, and is used for magnetic amplifiers.

HCx. Trade name of Powdrex Limited (UK) for a powder-metallurgy stainless steel that has high mechanical strength, good wear and corrosion resistance, low density, and is used for water-based hydraulic systems, water pumps, and equipment operating in severe seawater environments.

Header. Trade name of Uddeholm Corporation (USA) for a water-hardening tool steel containing 0.9% carbon, 0.2% vanadium, and the balance iron. Used for tools and cold-heading dies.

Headmore. Trade name of Bissett Steel Company (USA) for an oil-hardening steel containing 0.6% carbon, 1% chromium, 0.2% vanadium, and the balance iron. Used for header dies.

Headwell. Trade name of Atlas Steels Limited (Canada) for an austenitic stainless steel containing 0.07% carbon, 16% chromium, 18% nickel, and the balance iron. Used for corrosion-resistant parts and fasteners.

Healon. Trademark of Upjohn-Pharmacia (USA) for a viscous gel based on a *hyaluronic acid* derivative (sodium hyaluronate) and used in ophthalmics in the form of an injection into the anterior chamber for protecting the corneal endothelium and other eye tissues.

heap sand. Foundry sand mixed or blended on the foundry floor.

Heanium. Trademark of Heany Industries, Inc. (USA) for a magenta pink-colored high-purity alumina (Al_2O_3) ceramic with excellent wear and corrosion resistance and good electrical insulating properties that can be readily processed by dry pressing, extrusion, injection molding, isostatic pressing and machining. It is available in a wide variety of standard and custom shapes including eyelets, bushings and tubes for applications in the wire and cable, automotive and aerospace, and computer and telecommunications industries.

hearth-refined iron. (1) Iron, such as wrought iron, manufactured from pig iron in a reverberatory or puddling furnace, or a hearth furnace.

(2) Formerly, pig iron whose carbon content has been reduced by melting in a hearth on a charcoal fire in a blast of air.

heartwood. The inner layers of wood extending from the pith (or center) to the *sapwood* and which in the growing tree consist only of dead cells and contain deposits of extraneous substances. It is usually closer-grained, harder and darker in color and, in some species, more decay-resistant than sapwood, though not always clearly differentiated.

heat-absorbing glass. A type of glass that is capable of absorbing most of the radiation in the near infrared range of the electromagnetic spectrum.

heat-activated adhesive. An adhesive that is dry to the touch, but can be made sticky or liquid by application of heat, or heat and pressure.

heat-convertible resin. An insoluble and infusible plastic mass obtained by heating a thermosetting resin.

heat-curable rubber. A siloxane polymer that due to the addition of special catalysts (usually platinum or a peroxide) and additives cures in a relatively short time (about 10 minutes) at temperatures between 121 and 177°C (250 and 350°F). It can be processed by molding, extrusion and calendering, and has excellent chemical and weathering resistance, good high-temperature properties and good low-temperature flexibility. Used for molded parts, such as seals, gaskets, diaphragms, and wire and cable. Abbreviation: HCR.

Heatherbond. Trademark of Releasall-Target Inc. (Canada) for high-plasticity aluminum, bronze and steel products.

Heatherdine. Trade name for nylon fibers and yarns used for textile fabrics including clothing.

Heatherloft. Trade name for cellulose acetate fibers used for wearing apparel.

heating alloys. A group of electrical resistance alloys used in the form of heating elements in furnaces and appliances to generate heat. They have high electrical resistivities, low thermal expansions, good elevated-temperature properties, high melting points, good oxidation resistance and high emissivity. The following materials are most suitable for these applications: (i) pure metals, such as molybdenum, platinum, tantalum and tungsten; (ii) alloys, such as nickel-chromium, nickel-chromium-iron and iron-chromium-aluminum alloys; and (iii) some nonmetallic materials, such as graphite, molybdenum disilicide and silicon carbide. Also known as *electrical heating alloys*.

Heating Elements. Trade name for a wrought, non-aging nickel alloy that contains 23% iron, 14% chromium, 0.5% zirconium and 0.2% copper, and is used in heating elements for furnaces and appliances.

heat insulators. A group of materials that have low coefficients of thermal conductivity (k-values), and high resistance to heat flow (R-values), and thus can provide protection against hot or cold environments. Examples of such materials include cork, wool, glass wool, fiberglass, mineral fibers, kieselguhr, expanded perlite, cellular (or foam) glass and expanded polystyrenes and polyurethanes. They are used in various forms (e.g., sheets, blocks, blankets, batts, powders, granules or loosefill) in building construction. Also known as *thermal insulators*.

Heatlex. Trade name of Central Glass Company Limited (Japan) for conductive and heated glass products including laminated glass having fine electric heating wires in the interlayer, and glass made electrically conductive by glazing a special material on its surface.

heat-resistant alloys. See superalloys.

heat-resistant casting alloys. A group of casting alloys retaining high mechanical strength and resistance to corrosion, creep, warping, cracking and/or thermal fatigue at temperatures between about 650 and 1200°C (1200 and 2190°F). Included are iron-chromium, iron-chromium-nickel and iron-nickel-chromium cast steels as well as cast nickel- and cobalt-base superalloys. Used in the manufacture of metallurgical furnaces, power-plant equipment, turbines and turbochargers, chemical process equipment, equipment used in the cement, glass and rubber industries.

heat-resistant cast irons. A group of alloys of iron, carbon and silicon whose resistance to high-temperature oxidation and scaling is considerably enhanced by the addition of chromium, nickel, molybdenum and aluminum in excess of 3%. They have excellent high-temperature resistance and good resistance to softening or microstructural degradation. Included in this group are: (i) silicon gray and ductile irons; (ii) chromium gray and white irons; and (iii) high-nickel austenitic gray and ductile irons. See also cast irons.

heat-resistant coatings. See heat-resistant paints.

heat-resistant ceramics. See refractory ceramics.

heat-resistant concrete. Concrete that will not deteriorate when subjected to continuous heating or thermal shocks.

heat-resistant ductile irons. A group of *ductile cast irons* that possess excellent resistance to high-temperature oxidation and scaling. They are usually austenitic or ferritic as cast, and the carbon exists predominantly as spherulitic or nodular graphite. This group includes medium-silicon (2.5-6%), and high-nickel (18-36%) irons. Also known as *heat-resistant nodular irons*.

heat-resistant glass. A low-alkali glass, such as borosilicate glass, that is heat-treated or leached to remove alkalies so that it withstands high heat and sudden cooling without shattering. Also known as *heat-resisting glass*.

heat-resistant gray irons. A group of gray cast irons that possess excellent resistance to high-temperature oxidation and scaling. They are usually austenitic or ferritic as cast, and the carbon exists predominantly as flake graphite. This group includes high-chromium (15-35%), nickel-chromium (13-36% Ni and 1.8-6% Cr), nickel-chromium-silicon (13-43% Ni, 1.8-5.5% Cr and 5-6% Si), medium-silicon (4-7% Si) and high-aluminum (20-25% Al) irons. See also gray cast irons.

heat-resistant nodular irons. See heat-resistant ductile irons.

heat-resistant paints. Organic coatings in which the film-forming material is usually a silicone or alkyd-amine resin, and which may or may not contain reflective metal pigments (e.g., aluminum). They typically cure by air drying, or by baking. Unmodified silicone types require high baking temperatures. Used for coating metal surfaces exposed to temperatures of up to 290°C (550°F). Aluminum-pigmented paints can be used up to 650°C (1200°F). Also known as *heat-resistant coatings.*

heat-resistant steels. Steels that have high resistance to oxidation and scaling and good to moderate mechanical properties at temperatures above 500°C (930°F). They are usually alloy steels containing considerable amounts of chromium, nickel, tungsten, cobalt and/or molybdenum. Also known as *heat-resisting steels; high-temperature steels.*

heat-resisting alloys. See superalloys.

heat-resisting glass. See heat-resistant glass.

heat-resisting steels. See heat-resistant steels.

heat-sealing adhesive. A thermoplastic adhesive in the form of a film that is melted between the surfaces to be joined by the application of heat, or heat and pressure to either or both surfaces.

heat-sealing paper. A paper that develops heat-sealing characteristics by the application of heat, or heat and pressure.

heat-sensitive material. A material that will produce a visible image in a selected area to which heat has been applied.

heat-sensitive paint. A paint that undergoes several color transitions as its temperature is raised from about 40 to 1600°C (104 to 2910°F). Used in thermal inspection to indicate surface temperatures.

heat-sensitive paper. Paper that undergoes color transitions with changing temperature and finds application in temperature measurements. It is usually bonded directly to the surface.

heat-setting mortar. A finely divided refractory mortar that attains its strength at elevated temperatures.

heat-setting refractory. A finely divided refractory material that develops a ceramic bond at high temperatures.

heat-set textiles. Textile fibers, yarns or fabrics that have been given stability in form and/or texture by the application of heat.

heat shield. A protective coating, covering or layer of special material, such as a refractory, carbon composite, or reinforced plastic, used on the nose cones of spacecraft, missiles and rockets to protect them from the heat generated upon reentering the earth's atmosphere. See also ablative materials.

heat-shrinkable plastics. See shape-memory plastics.

heat-sink material. A material, such as beryllium, copper or graphite, that is capable of absorbing high amounts of heat, and has a high thermal conductivity, specific heat and melting point.

heat stabilizer. A chemical agent, such as a metal carboxylate, organotin or organophosphorus compound, or calcium or barium salt, added to certain plastics (e.g., polyvinyl chloride) or coatings to decrease the effect of heat.

heat-strengthened glass. Glass that has undergone a tempering treatment, i.e., has been heated to a temperature above the glass-transition region, yet below the softening point, and cooled to room temperature in a jet of air. It is similar to tempered glass with the resulting surface layer being slightly less compressively stressed. Abbreviation: HSG. Also known as *heat-tempered glass.* See also tempered glass.

heat storage materials. A group of materials that can absorb large amounts of thermal energy either due to their specific heat capacities (e.g., concrete, brick or water) or their phase transitions from the solid to the liquid state (e.g., Glauber salt). Used for scavenging heat from hot industrial exhaust gases, storing energy from solar collectors, or as night heaters to make use of lower off-peak electricity rates.

heat-transfer agent. See heat-transfer material.

heat-transfer material. A material that is capable of transmitting thermal energy within a body by conduction, or between a body and its surroundings by conduction, convection or radiation. It may be a solid, liquid or gas (e.g., air, carbon dioxide, oil, steam or water). Abbreviation: HTM. Also known as *heat-transfer agent.*

heat-treatable alloy. A term that usually designates an alloy whose mechanical strength can be enhanced by precipitation hardening or martensitic transformation, both of which involve specific heat-treating procedures. Examples of such as alloys include certain aluminum, copper, magnesium and nickel alloys, carbon and alloy steels, and cast irons.

heat-treatable steel. A quench-hardened carbon or alloy steel, usually with a carbon content of 0.2-0.6%, whose ductility and toughness (impact strength) is enhanced by heating to a temperature below the eutectoid temperature for a prescribed period of time, and quenching in air or oil. A *quenched-and-tempered steel* is an example.

heat-treated alloy. An alloy heated to a temperature above the critical temperature and subsequently quenched at a prescribed rate. Quenching may or may not be followed by a tempering treatment designed to produce desired changes in hardness, toughness and stiffness.

heat-treated glass. See heat-strengthened glass.

heavy aggregate. See heavyweight aggregate.

heavy alloys. See heavy metals (1).

heavy concrete. See heavyweight concrete.

Heavy Duty Babbitt. Trade name of Belmont Metals Inc. (USA) for a tin-base babbitt containing 8.33% antimony and 8.33% copper. It has a melting point of 422°C (792°F), and is used for high-speed bearings.

heavy-fermion superconductors. Superconductors whose inner-shell conduction electrons have effective masses larger than those of standard superconductors, and usually more than 100 times that of a free electron, resulting in low Fermi energies. In general, they are compounds containing rare-earth elements, such as cerium or ytterbium, or actinide elements, such as uranium. For such compounds, the superconductive behavior cannot be explained by the "normal" Bardeen-Cooper-Schrieffer (BCS) theory, and their superconductivity critical temperatures range from as low as 0.1-0.7K for $CeCu_2Si_2$ and 0.87K for UBe_{13} to 6.0K for $CeRu_2$. Also known as *heavy fermions.*

heavy heavy hydrogen. See hydrogen-3.

heavy hydrogen. See hydrogen-2.

heavy leather. Vegetable-tanned leather made from unsplit cattlehide and used for belting, straps, shoe soles and mechanical goods.

heavy metal fluoride glasses. A family of glasses prepared by mixing several metal fluoride powders, melting and subsequent refining at high temperatures. The melt is then either cast into shapes under normal cooling conditions, or rapidly quenched. Standard compositions are based on fluorozirconates, e.g., ZBLAN, a zirconium tetrafluoride–barium difluoride–lanthanum trifluoride–aluminum trifluoride–sodium fluoride glass (ZrF_4–BaF_2–LaF_3–AlF_3–NaF), but other metal fluorides, such as lithium fluoride (LiF), indium fluoride (InF_3), gadolium fluoride (GdF_3) or strontium fluoride (SrF_2), may replace some or all of the standard fluorides. Used for optical fibers in lasers and communication systems. Abbreviation: HMFG. See also ZBLAN.

heavy metals. (1) Metallic elements having a density higher than 4.0 g/cm³ (0.14 lb/in.³). Examples include zinc (7.13 g/cm³ or 0.257 lb/in.³), iron (7.87 g/cm³ or 0.284 lb/in.³), cobalt (8.85 g/cm³ or 0.320 lb/in.³), nickel (8.90 g/cm³ or 0.322 lb/in.³), copper (8.96 g/cm³ or 0.324 lb/in.³), lead (11.34 g/cm³ or 0.410 lb/in.³), tungsten (19.40 g/cm³ or 0.701 lb/in.³) and the noble metals gold (19.30 g/cm³ or 0.697 lb/in.³), silver (10.49 g/cm³ or 0.379 lb/in.³) and platinum (21.45 g/cm³ or 0.775 lb/in.³). See also light metals.

(2) Sintered tungsten alloys containing 6-10 wt% nickel and up to 4 wt% copper and/or additions of iron. They have a density of at least 16.5 g/cm³ (0.60 lb/in.³), high heat resistance, good high-temperature strengths up to 1100°C (2010°F), good electric conductivity, and are machinable and brazeable. Used for screens of X-ray tubes and radioactive shields, radium containers, counterweights and balances, gamma-ray absorbers, contact surfaces for circuit breakers, and various high-density applications. Also known as *heavy alloys.*

heavy oxygen. See oxygen-18.

Heavy Pressure. Trade name of NL Industries (USA) for a heavy-duty babbitt containing tin, antimony and copper. Used for bearings for crushers.

Heavy Pressure Mill Glyco. Trade name of Joseph T. Ryerson & Son Inc. (USA) for a lead babbitt containing tin and antimony. Used for bearings.

heavy rare earths. A group of rare-earth metals that includes yttrium (atomic number 39) and the chemical elements with atomic numbers 64 through 71, namely, gadolinium, terbium, dysprosium, holmium, erbium, thulium, ytterbium and lutetium. With the exception of yttrium, these elements form a subgroup within the lanthanide group of elements. Also known as *heavy rare-earth elements; yttric rare earths; yttrics.*

heavy spar. See barite.

heavy water. See deuterium oxide.

heavyweight aggregate. Aggregate with a high density, e.g., coarse sand, gravel, stone chippings, metal punchings, magnetite, barite, ilmenite or magnetite. Used in the manufacture of *heavyweight concrete.* Abbreviation: HWA. Also known as *heavy aggregate.*

heavyweight concrete. A high-density concrete (density usually above 3 g/cm³ or 0.11 lb/in.³) in which the conventional aggregates are replaced, wholly or in part, by heavyweight aggregates. Used in the manufacture of counterweights, radiation shielding and other specialized applications. Abbreviation: HWC. Also known as *heavy concrete; high-density concrete.*

heazlewoodite. A light bronze mineral composed of nickel sulfide, Ni_3S_2. Crystal system, rhombohedral (hexagonal). It can also be made synthetically. Density, 5.82 g/cm³. Occurrence: Australia.

Hebonite. British trade name for a heat-resistant steel.

hecatolite. See moonstone.

Hecht's porcelain. A type of refractory porcelain that is similar to *Marquardt porcelain*, and was originally made in Germany.

Hecla. Trade name of Dunford Hadfields Limited (UK) for an extensive series of steels including various plain-carbon, free-machining, low- and medium-carbon structural and case hardening steels, several low-, medium- and high-carbon alloy steels with chromium, molybdenum and/or nickel, high-chromium

stainless steels, and a number of hot-work die and tool steels.

Hecnum. Trademark of Arthur E. Heckford Limited (UK) for a resistance alloy containing 55% copper, and 45% nickel. Commercially available in foil, rod or wire form, it has a density of 8.9 g/cm³ (0.32 lb/in³), a melting point of 1225-1300°C (2235-2370°F), high electrical resistivity, a low temperature coefficient of resistance, a low coefficient of expansion, high heat resistance, a maximum use temperature in air of 500°C (930°F), and a hardness of 100-300 Brinell. Used for thermocouples, electrical resistances, and rheostats.

Hecolite. Trade name for a cellulose nitrate formerly used for denture bases.

hectorite. A white clay mineral of the smectite group composed of hydrous lithium magnesium silicate, $(Mg,Li)_3Si_4O_{10}(OH)_2 \cdot xH_2O$. A synthetic form with slightly different composition and formula $[Na_x(Mg,Li)_3Si_4O_{10}(OH,F)_2]$ is also available. Crystal system, monoclinic. Density, 2.50 g/cm³; good swelling properties; great affinity for water; good cation-exchange capacities. Occurrence: USA (California). Used for slips and clays, as a bonding agent, filler, and ion exchanger, Also known as *magnesium-bentonite.*

Heddal. (1) Trade name of VDM Nickel-Technologie AG (Germany) for a water-hardening steel containing 0.2% carbon, 0.3% manganese, 0.25% silicon, and the balance iron. Used for machinery parts.

(2) Trade name of VDM Aluminium GmbH (Germany) for a non-heat-treatable alloy containing 98.5% aluminum and 1.5% manganese. Used for light-alloy parts.

Heddenal. Trade name of VDM Nickel-Technologie AG (Germany) for a series of corrosion-resistant aluminum alloys containing 1.5-7.5% magnesium, and up to 0.8% manganese and 0.3% chromium, respectively. Used for aircraft parts, marine equipment, fuel lines, roofing, and architectural trim.

Heddronal. Trade name of VDM Aluminium GmbH (Germany) for a series of non-heat-treatable, corrosion-resistant aluminum alloys containing 0.5% manganese and 3.5-7% magnesium. Used for structural towers, pressure vessels, rocket motor parts, deck housings, overhead cranes, and heavy-duty structures.

Heddroxal. Trade name of VDM Nickel-Technologie AG (Germany) for an aluminum alloy containing 2-4% magnesium, 0.4% manganese and up to 0.3% chromium. It has good resistance to seawater corrosion. Used for aircraft tanks and fittings, fuel lines, and marine parts.

Heddur. (1) Trade name of Diado Steel Company Limited (USA) for a heat-treatable aluminum alloy containing 3.5-5% copper, 0.3-0.8% silicon, 0.2-1.2% manganese, and 0.4-1% magnesium. Used for structural applications, screw-machine products and fasteners.

(2) Trade name of VDM Nickel-Technologie AG (Germany) for heat-treatable aluminum alloys containing 2.5-5% copper, 0.2-1.8% magnesium and 0.3-1.5% manganese. Used for aircraft structures and fittings, and fasteners.

hedenbergite. A dark blackish, or brownish green mineral of the pyroxene group that is composed of calcium iron silicate, $CaFeSi_2O_6$, and may also contain magnesium. It can also be made synthetically. Crystal system, monoclinic. Density, 3.50 g/cm³; refractive index, 1.7265. Occurrence: Peru.

Hedervan. (1) Trade name of Latrobe Steel Company (USA) for a water-hardening tool steel (AISI type W2) containing 0.9% carbon, 0.15-0.35% vanadium, and the balance iron. It has good wear resistance, and is used for punches, draw dies, and cold-heading and punching dies.

(2) Trade name of Latrobe Steel Company (USA) for a tool steel containing 1.4% carbon, 0.15% chromium, 0.1% molybdenum, 3.5% vanadium, and the balance iron. Used for punches and dies.

Hedex. Trade name of Kiveton Park Steel Company (UK) for an extensive series of steels including low- and medium-carbon types, case-hardening types, low-, medium- and high-carbon alloy types with varying amounts of chromium, nickel and/or molybdenum, and several austenitic, ferritic and martensitic stainless types.

hedleyite. (1) A tin-white mineral of the tetradymite group composed of bismuth telluride, Bi_7Te_3. Crystal system, rhombohedral (hexagonal). Density, 8.91 g/cm^3. Occurrence: Canada (British Columbia).

(2) A tin-white synthetic mineral of the tetradymite group composed of bismuth telluride, Bi_5Te_3, and made from a molten solution of tellurium in bismuth by crystallization and subsequent cooling under prescribed conditions. Crystal system, rhombohedral (hexagonal). Density, 8.43.

Hedon. Trade name for polyvinyl acetate resins.

Hedrocel. Trade name of Ultramet Inc. (USA) for a synthetic biomaterial with an open-celled lattice reinforced with tantalum and resembling coral and cork in appearance. It has mechanical properties similar to those of human bone, mimics the latter in other properties, is biocompatible, and is used for hip and spine implants between vertebrae.

hedyphane. A yellowish-white mineral composed of calcium lead arsenate chloride, $(Ca,Pb)_5Cl(AsO_4)_3$. Crystal system, hexagonal. Density, 5.7 g/cm^3; hardness, 4 Mohs. Occurrence: Sweden, USA (New Jersey).

Heerdt. Trade name of Düsseldorf-Heerdt GmbH & Co. KG (Germany) for a series of steels used for dies, drills, cutters, punches, and finishing tools.

Heglas. Trade name of Heglas AG (Switzerland) for a multiple glazing unit with a spacer of light metal, and a stainless chromium-nickel frame.

Heglerflex. Trade name of Hegler Plastik GmbH (Germany) for flexible plastic insulating tubing used for electrical cables and wires.

Heglerplast. Trade name of Hegler Plastik GmbH (Germany) for rigid plastic insulating and armoring tubes used for electrical cables and wires.

heideite. A black mineral of the wilkmanite group composed of iron titanium sulfide, $FeTi_2S_4$. Crystal system, monoclinic. Density, 3.95 g/cm^3.

heidornite. A colorless mineral composed of sodium calcium chloride boride sulfate hydroxide, $Na_2Ca_3(SO_4)_2B_5O_8Cl(OH)_2$. Crystal system, monoclinic. Density, 2.75 g/cm^3; refractive index, 1.588. Occurrence: Germany.

HEIM. See Toyobo HEIM.

heinrichite. A lemon-yellow mineral of the autunite group composed of barium uranyl arsenate decahydrate, $Ba(UO_2AsO_4)_2 \cdot 10H_2O$. It can also be made synthetically. Crystal system, tetragonal. Density, 4.04 g/cm^3; refractive index, 1.605. Occurrence: Russian Federation.

Helanca. Trademark of Heberlein & Co. AG (Switzerland) for a highly elastic, hard-wearing, washable, crinkled nylon yarn used for making clothing, e.g., sweaters, tights, sports- and swimwear, and hosiery.

Helco. Trade name of Hansell-Elcock Company (USA) for a series of cast irons containing 3-3.2% carbon, 2-2.4% silicon, 1-2% nickel, and the balance iron. Used for castings and hydraulic press cylinders.

helimagnet. A rare-earth or transition-metal alloy or salt that at suitably low temperatures exhibits an atomic magnetic moment arrangement in ferromagnetic planes with uniform plane-to-plane variation of the direction of magnetism.

Helia. Trademark of Carl Freundenberg (Germany) for flexible polyvinyl chloride film.

Heliocolour. Trademark of Ivoclar North America Inc. (USA) for dental bonding agents and light-cure dental paste laminates.

Heliofil. Trademark of Feldmuehle AG (Germany) for flexible polyvinyl chloride film materials.

Helioflex. Trademark of Feldmuehle AG (Germany) for polyethylene film materials.

Helioglas. Trade name of PPG Industries Inc. (USA) for an ultraviolet-transmitting glass.

Heliogrey. Trade name of Glaceries de Saint-Roch SA (Belgium) for a heat-absorbing, gray plate glass.

Heliolink. Trademark of Ivoclar North America Inc. (USA) for a light-cure dental resin cement.

Heliomolar. Trademark of Ivoclar North America Inc. (USA) for light-cure, microfine dental composite resins used for posterior restorations. *Heliomolar Radiopaque* is a radiopaque all-purpose dental composite resin.

Helion. Trade name of E.I. Du Pont de Nemours & Company (USA) for nylon fibers and yarns used for textile fabrics including clothing.

heliophyllite. A yellow or greenish-yellow mineral composed of lead chloride arsenate, $Pb_6As_2O_7Cl_4$. Crystal system, orthorhombic. Density, 6.89 g/cm^3. Occurrence: Sweden.

HelioProgress. Trademark of Ivoclar North America Inc. (USA) for light-cure, microfine anterior composite resins used for dental restorative applications.

Helios. (1) Trade name of CCS Braeburn Alloy Steel (USA) for a series of alloy steels used for tools and dies.

(2) Trade name of Helios Limited (UK) for flexible, chrome-tanned leather with a base coat of linseed oil gel and a finish coat of urethane.

(3) Trademark of Schott AG (Germany) for a haze-reducing optical glass.

Helioseal. Trademark of Ivoclar North America Inc. (USA) for a light-cure dental fissure sealant.

Heliotital. Trade name of Cegedur Pechiney (France) for an aluminum alloy containing a total of 0.15% magnesium and titanium.

heliotrope. See bloodstone (1).

helium. A colorless, odorless, tasteless, nonmetallic element of the noble gas group (Group VIIIA, or Group 18) of the Periodic Table. It is a monatomic, noncombustible gas that forms about 1 part in 200000 parts by volume of the earth's atmosphere, and occurs in the atmosphere of the sun and stars, in natural gas, and in certain minerals. Bulk density, 0.1785 g/L at 0°C (32°F); boiling point, -268.9°C (-452°F) (1 atm); freezing point, -272.2°C (-458°F) (25 atm); atomic number, 2; atomic weight, 4.003; zerovalent (inert). It liquefies at 4.2K to form the phase "helium I." At 2.2K (lambda point) it forms the phase "helium II." Helium nuclei are alpha particles. Used as shielding gas in metal-inert-gas (MIG) and tungsten-inert-gas (TIG) welding, as an inert furnace atmosphere, in balloons and dirigibles, in diving bell atmospheres, in the pressurization of rocket fuels, in breathing equipment, in luminous signs, in geological dating, in lasers and leak detectors, in cryogenics, magnetohydrodynamics and chromatography, as a heat-transfer medium, in aerodynamic research, as a coolant for nuclear plants, and in superconducting electric systems. Symbol: He.

helix yarn. A highly curled or crimped textured yarn.

Hellefors. Trade name of Ovako Steel Hellefors AB (Sweden) for an extensive series of steels including various machinery and case-hardening grades, plain-carbon and cold-work tool grades, and austenitic, ferritic and martensitic stainless grades.

hellandite. A brown, or red mineral composed of calcium yttrium borate aluminum iron silicate hydroxide, $(Ca,Y)_6(Al,Fe)Si_4B_4O_{20}(OH)_4$. Crystal system, monoclinic. Density, 3.50 g/cm³. Occurrence: Canada, Norway, Italy.

hellyerite. A blue mineral composed of nickel carbonate hexahydrate, $NiCO_3 \cdot 6H_2O$. Crystal system, monoclinic. Density, 1.97 g/cm³; refractive index, 1.503. Occurrence: Australia (Tasmania).

Helmapor. Trademark of Helmitin GmbH (Germany) for polystyrene foams and foam products.

Helment bronze. A British high-ductility bronze containing 70% copper and 30% zinc. Used for water pipes, cartridges, shell cases, and deep-drawn parts.

Helmet metal. A British bronze containing 70-72% copper and 28-30% zinc. Used for cartridges, shell cases, helmets, and deep-drawn metal.

helmutwinklerite. A colorless to light blue mineral composed of lead zinc arsenate dihydrate, $PbZn_2(AsO_4)_2 \cdot 2H_2O$. Crystal system, triclinic. Density, 5.30 g/cm³. Occurrence: Namibia.

Helomit. Danish trade name for urea-formaldehyde resins.

Helta. Trademark of Plasticisers Limited (UK) for polypropylene staple fibers used for textile fabrics.

Helumin. Trade name of Transleteur & Co. (Germany) for a non-heat-treatable aluminum alloy containing 1.8% copper and 1.5% iron. Used for light-alloy parts.

Helve. Trade name of Hardenite Steel Company Limited (UK) for a water-hardening tool steel containing 0.7-1.26% carbon, and the balance iron. Used for dies and tools.

helvine. See helvite.

helvite. A yellow, green or dark brown mineral composed of beryllium manganese silicate sulfide, $Mn_4Be_3(SiO_4)_3S$, or $Mn_4(BeSiO_4)_3S$. Crystal system, cubic. Density, 3.20-3.23 g/cm³; hardness, 6.5 Mohs; refractive index, 1.736. Occurrence: Germany, USA (Montana). Used as a low-grade ore of beryllium. Also known as *helvine*.

hemafibrite. A brownish to garnet-red mineral composed manganese arsenate hydroxide, $(Mn,Mg)_4Al(AsO_4)(OH)_8$. Crystal system, rhombohedral (hexagonal). Density, 3.5-3.65 g/cm³; hardness, 3 Mohs. Occurrence: Sweden.

Hematite. Trademark of Hematite Manufacturing Company (Canada) for flexible polyvinyl chloride sheeting.

hematite. A brilliant black to blackish red, dark reddish brown or brick-red mineral with a brown to cherry red streak and a dull to metallic luster. It belongs to the corundum group and is composed of ferric oxide, Fe_2O_3, and, when pure, contains approximately 70% iron and 30% oxygen. Crystal system, rhombohedral (hexagonal). Density, 4.9-5.3 g/cm³; melting point, 1350-1360°C (2460-2480°F); hardness, 6 Mohs; refractive index, 3.22. Occurrence: Canada, Germany, Italy, Morocco, Spain, Switzerland, UK, USA (Alabama, Michigan, Minnesota, Wisconsin). Used as the most important ore of iron. Certain varieties are used as paint pigments (red ocher), for rouge, and for carved intaglios. Also known as *bloodstone; red hematite; red iron ore; red ocher; rhombohedral iron ore.*

hematolite. A brownish-red to black mineral composed of magnesium manganese aluminum arsenite arsenate hydroxide, $(Mn,Mg,Al)_{15}(AsO_3)(AsO_4)_2(OH)_{23}$. Crystal system, rhombohedral (hexagonal). Density, 3.49 g/cm³; refractive index, 1.733; hardness, 3.5 Mohs. Occurrence: Sweden.

hematophanite. A dark red to brown mineral composed of lead iron oxide chloride, $Pb_4Fe_3O_8Cl$. Crystal system, tetragonal. Density, 7.70 g/cm³. It exists also in hydrous form. Occurrence: Sweden.

hematoporphyrin. (1) Any of several isomeric forms of *porphyrin* obtained from hemoglobin by *in vitro* treatment with sulfuric acid. Used in biochemistry, bioengineering, and medicine. Formula: $C_{34}H_{38}O_6N_4$.

(2) The deep red, iron-free crystalline *porphyrin* derivative obtained by treating hematin or heme with sulfuric acid. Used as a dye and pigment in biochemistry, bioengineering and medicine. Two common synthetic hematoporphyrin derivatives are *hematoporphyrin IX* ($C_{34}H_{38}N_4O_6$) and *hematoporphyrin IX dihydrochloride* ($C_{34}H_{38}N_4O_6 \cdot 2HCl$). Both are available in the form of purple crystals.

heme. A *porphyrin* pigment molecule that is responsible for the red color of blood, and has a chelate structure in which an iron (Fe^{2+}) ion is located at the center of an octagon, surrounded by four pyrrole nitrogen atoms at the corners of a square, a globin molecule and a water molecule. It is the prosthetic (or nonprotein) group of hemoglobin and myoglobin. Used in biochemistry, biotechnology and medicine. Formula: $C_{34}H_{32}FeN_4O_4$.

hemicellulose. Any of the polysaccharides other than cellulose found in the cell walls of wood, grain hulls and other plants.

hemihedrite. An orange mineral composed of zinc lead chromium oxide fluoride silicate, $Pb_{10}Zn(CrO_4)_6(SiO_4)_2F_2$. Crystal system, triclinic. Density, 6.42 g/cm³; refractive index, 2.32. Occurrence: USA (Arizona).

hemihydrate. See calcium sulfate hemihydrate.

hemimorphite. See calamine.

hemin. The chloride of *heme* obtained by heating hemoglobin with acetic acid and sodium chloride. It is available in the form of black crystals that appear steel-blue in reflected light and brown in transmitted light. Used in biochemistry, biotechnology, medicine, and as a complexing agent. Formula: $C_{34}H_{32}N_4O_4FeCl$. Also known as *ferriprotoporphyrin IX chloride; hemin chloride; Teichmann's crystals.* See also hemoglobin.

Hemingray. Trade name of Owens-Illinois Inc. (USA) for glass insulators used for telephone and telegraph equipment.

hemlock. The relatively hard wood of any of a genus (*Tsuga*) of evergreen trees of the pine family found in North America and East Asia, especially the eastern hemlock (*T. canadensis*) and the western hemlock (*T. heterophylla*). Used for construction lumber, pulpwood, containers and plywood. See also eastern hemlock; western hemlock.

hemlock spruce. See Alaskan pine.

Hemlon. Trade name for nylon fibers and yarns used for textile fabrics.

hemochromogen. A pink pigment obtained by treating hemoglobin with alkali in the presence of a reducing agent. It is composed of *heme* and a nitrogenous substance, such as ammonia or pyridine.

hemocyanins. A class of blue respiratory pigments found in the blood of mollusks and many anthropods (e.g., keyhole limpets, whelks and horseshoe crabs). They are high-molecular-weight copper-containing protein-pigment complexes usually available as lyophilized powders or as solutions in glycerol. Used in biochemistry and biotechnology as SEM and TEM markers, and as antigens in immunology.

hemoglobin. The respiratory protein pigment of red blood cells that has a molecular weight of about 65000, and is composed of four helical globin polypeptide chains and four heme molecules. It has the ability of combining with gases, such as oxygen and carbon dioxide, reacting with oxygen to form oxyhemoglobin and carrying the latter by arterial circulation to the tissues. Used in biochemistry, biotechnology and medicine. Abbreviation: Hb.

hemoprotein. A conjugated *protein* (e.g., a cytochrome or hemoglobin) formed by combining iron with a porphyrin-base prosthetic group. See also porphyrin.

hemp. (1) The tough, durable, soft, white fibers obtained from the inner bark of the stalk of the tall annual plant *Cannabis sativa* belonging to the mulberry family, native to Asia, but also grown in southern Europe, India, Italy, Russia and the USA. Used in the manufacture of cordage, twine, cores of steel ropes, coarse cloth, sacking, packings, and as fillers for plastics. Also known as *true hemp*.

(2) Any of various strong vegetable fibers other than true hemp (1), such as Manila hemp (abaca), sunn hemp and sisal hemp.

hemusite. A gray mineral of the tetrahedrite group composed of copper molybdenum tin sulfide, Cu_6MoSnS_8. Crystal system, cubic. Density, 4.47 g/cm^3. Occurrence: Bulgaria.

Henckels. Trade name of Henkels Zwillingswerke AG (Germany) for a series of steels including various plain-carbon, cold-work, hot-work, nondeforming and shock-resistant tool grades as well as several machinery grades.

hendersonite. A greenish black to black mineral of the mica group composed of calcium vanadium oxide octahydrate, $Ca_2V_9O_{24} \cdot 8H_2O$. Crystal system, orthorhombic. Density, 2.79 g/cm^3; refractive index, above 2.01. Occurrence: USA (Colorado, New Mexico).

hendricksite. A copper-red to reddish black mineral of the mica group composed of potassium zinc manganese aluminum silicate hydroxide, $K(Zn,Mn,Fe)_3(Si,Al)_4O_{10}(OH)_2$. Crystal system, monoclinic. Density, 3.40 g/cm^3; refractive index, 1.686. Occurrence: USA (New Jersey).

henequen. The hard, strong, reddish vegetable fibers obtained from the leaves of an agave plant (*Agave fourcroydes*) native to Mexico and the West Indies. It is analogous to *sisal*, but stiffer and coarser, and ranges in fineness from 300-500 denier. Used in the manufacture of twine, cordage and cord. Also known as *Yucatan sisal*. See also sisal fibers.

Hennig Purifier. Trade name of PPG Industries, Inc. (USA) for a product based on *soda ash* and containing one or more other purifying agents. Supplied in the form of pellets or briquettes, it is used as a ladle addition to remove oxides and nonmetallic inclusions, and thus to produce cleaner steel.

Henricot. Trade name of Usine Emile Henricot, SA (Belgium) for an extensive series of wrought and cast steels including several heat- and/or corrosion-resistant austenitic, ferritic and martensitic types, austenitic manganese types, quenched-and-tempered alloy types, machinery types, molybdenum, tungsten and cobalt-tungsten high-speed types, plain-carbon tool types, etc. Also included under this trade name are numerous corrosion, heat- and/or wear-resistant cast irons, and several iron-chromium, iron-cobalt-chromium, iron-nickel-aluminum and iron-nickel-aluminum-cobalt permanent magnet alloys.

henrietta. A drapable twill-weave fabric, originally made with a silk warp and a worsted weft, but now also made from natural and synthetic fiber blends.

henritermierite. A brown mineral of the garnet group composed of calcium manganese aluminum silicate hydroxide, $Ca_3(Mn,Al)_2(SiO_4)_2(OH)_4$. Crystal system, tetragonal. Density, 3.34 g/cm^3; refractive index, 1.800. Occurrence: Morocco.

Heposil. German trade name for an aluminum alloy containing 21-22% silicon, 1.5% copper, 1.5% nickel, 1.2% cobalt, 0.7% manganese, and 0.5% magnesium. It has good corrosion resistance, low thermal expansion, and is used for pistons in engines and for liners.

Heppenstall. Trade name of Heppenstall Company (USA) for a series of machinery and tool steels including numerous case-hardening and heat-treatable grades as well as a wide range of tool materials, such as plain-carbon, shock-resisting, cold-work, hot-work, low-alloy special-purpose tool steels, and several mold steels.

4-(heptyloxy)-4-biphenylcarbonitrile. A nematic liquid crystal with a melting point of 54-57°C (129-135°F). Formula: $CH_3(CH_2)_6OC_6H_4C_6H_4CN$.

Herabond. Trade name of Heraeus Kulzer Inc. (USA) for a medium-gold dental bonding alloy.

Herador. Trade name of Heraeus Kulzer Inc. (USA) for high- and medium-gold dental bonding alloys.

Heraeus. Trade name of W.C. Heraeus GmbH (Germany) for an extensive series of precious metals and precious-metal alloys and compounds (based on gold, silver, iridium, osmium, palladium, rhodium, ruthenium, etc.), semifinished precious-metal products, precious-metal sputtering and vapor deposition targets, and various precious-metal based thin-film materials for electronic applications as well as catalysts, netting, pastes, micro-size powders, and anodes for electrolysis, plating and sputtering. Also included under this trade name are beryllium precision parts, bimetals for contact and solid-state applications, solid chromium metal, and various special nonferrous metals, such as niobium, tantalum and rhenium.

Heraeus Rotosil. Trade name of Heraeus Quarzschmelze GmbH (Germany) for an opaque, high-purity fused silica (SiO_2) that is chemically inert at various temperatures, pressures and pH levels. Used for nonporous transport tubes or crucibles in the manufacture of metallic and nonmetallic materials, and as an electrical insulator.

Hera GC. Trade name of Heraeus Kulzer Inc. (USA) for a medium-gold dental casting alloy.

Hera-Gold. Trade name of Heraeus GmbH (Germany) for an 18-karat gold alloy with excellent resistance to tarnishing.

Hera IS99. Trade name of Heraeus Kulzer Inc. (USA) for a dental gypsum used as a pore filler.

Heraklith. Trademark of Deutsche Heraklith AG (Germany) for building boards and plates used for the thermal and acoustical insulation of walls and ceilings.

Heranium P. Trade name of Heraeus Kulzer GmbH (Germany) for a non-precious dental casting alloy containing 59% niobium, 25% chromium, 15% silicon, 10% tungsten, 4% molybdenum and 0.8% manganese. Used for crowns and bridges.

HeraNordic 75. Trade name of Heraeus Kulzer GmbH (Germany) for a dental casting alloy containing 75% gold, 17.9% palladium, 4.5% indium, 1.3% silver and 1.0% tin. Used for crowns and bridges.

Hera O. Trade name of Heraeus Kulzer Inc. (USA) for a low-gold dental casting alloy.

Herbaplast. Trademark of Herbol for polyvinyl acetate resins.

Herberton. Brand name for commercially pure tin containing 0.037% arsenic, 0.03% copper, and 0.18% lead.

Herbohn bell metal. German corrosion-resistant bronze containing 60-71.4% copper, 26.4-35% tin, and 2.7-5% zinc. Used for bells.

Herboplex. Trademark of Herbol for unsaturated polyesters.

Herc-Alloy. Trade name of Columbus McKinnon Corporation (USA) for a wear- and impact-resistant, heat-treated alloy of nickel, chromium and/or molybdenum used for chains, slings and joining links.

Herclor. Trademark of Hercules, Inc. (USA) for a series of elastomers based on epichlorohydrin. *Herclor C* is a copolymer with ethylene oxide, and *Herclor H* a homopolymer. *Herclor* elastomers have excellent resistance to ozone, elevated temperatures, solvents and chemicals, and are used for wire and cable coatings, seals and gaskets, packings, belting, hose, automotive and aircraft parts, and coated fabrics.

Hercocel. Trademark of Hercules Inc. (USA) for molding powders based on cellulose acetate, ethylcellulose or methylcellulose.

Hercose. Trademark of Hercules Inc. (USA) for mixed esters composed of cellulose acetate butyrate or cellulose acetate propionate.

Hercosett. Trademark of Hercules Inc. (USA) for thermosetting resins.

Herculane. Trademark of Phillips Paint Products Limited (Canada) for polyurethane finishes.

Hercules. (1) Trade name of Cytemp Specialty Steel (USA) for a fatigue-resistant tool steel (AISI type W7) containing 1% carbon, 0.5% chromium, 0.2% vanadium, and the balance iron. Used for blanking, forming and drawing dies.

(2) Trademark of Hercules, Inc. (USA) for a series of engineering products including cellulose nitrate products, high-performance graphite fibers, epoxy prepreg tapes and woven fabrics.

Hercules bronze. (1) A high-strength bronze containing 86% copper, 10% tin, 2-3% aluminum and 1-2% zinc. Used for hardware, and worm gears.

(2) A bronze containing 54% copper, 36% zinc, 7.5% iron and 2.5% aluminum. Used for hardware.

Herculite. (1) Trade name of Midvale-Heppenstall Company (USA) for a tool steel containing 0.54% carbon, 2.6% nickel, 0.45% molybdenum, and the balance iron.

(2) Trademark of PPG Industries Inc. (USA) for tempered safety glass with excellent thermal-shock resistance to about 350°C (660°F). *Herculite K* is a heat-toughened glass made by the air-support process, and *Herculite II* a chemically strengthened sheet glass.

(3) Trademark of Consolidated Concrete Limited (Canada) for expanded shale aggregates.

(4) Trade name of SDS/Kerr (USA) for all-purpose, light-cure hybrid dental composite resins used for various restoration applications. *Herculite XRV* is a dental resin composite and *Herculite XR* a two-part resin composite (bonding agent plus primer).

(5) Trademark of Herculite Products Inc. (USA) for flexible plastic sheeting reinforced and/or laminated with natural or synthetic fibers.

(6) Trademark of Bendix Corporation (USA) for abrasive materials supplied in particulate form.

Herculite Plaster MC. Trade name of Foseco Minsep NV (Netherlands) for a casting plaster specially designed for the manufacture of aluminum matchplates by gravity or pressurized casting processes.

Herculoid. Trade name of Hercules, Inc. (USA) for *celluloid*-type cellulose nitrate plastics.

Herculon. Trademark of Hercules, Inc. (USA) for polypropylene fibers supplied in bulked continuous and continuous multifilament yarns, staple, and uncut tow. Used for industrial applications, apparel, and home furnishings.

Herculoy. Trademark of Revere Copper Products, Inc. (USA) for a series of silicon bronzes containing 3% silicon and up to 1% manganese, and the balance copper. Supplied as rod, wire, tube, and flat products, they have excellent hot and cold workability and good to excellent corrosion resistance. They can be fabricated by blanking, drawing, shearing, squeezing, swaging, forming, bending, hot forging, etc., and are used for machine and cap screws, bolts, marine hardware, nuts, rivets, propeller shafts, hydraulic pressure lines, cable clamps, electrical conduits, heat-exchanger tubing, and welding rods.

Hercuprene. Trade name of J-Von (USA) for several thermoplastic elastomers.

Hercuvit. Trade name of PPG Industries Inc. (USA) for glass-ceramic materials.

hercynite. A greenish-gray to black mineral of the spinel group that is composed of iron aluminum oxide, $FeAl_2O_4$, and may also contain some magnesium. It can also be made synthetically. Crystal system, cubic. Density, 4.25-4.26 g/cm^3; melting point, 2135°C (3875°F); hardness, 7.5-8.0 Mohs; refractive index, 1.800. Also known as *ferrospinel; iron spinel.*

herderite. A colorless, pale yellow or greenish-white mineral of the datolite group composed of beryllium calcium phosphate hydroxide, $CaBePO_4(OH,F)$. Crystal system, monoclinic. Density, 2.95 g/cm^3; refractive index, 1.611; hardness, 7.5-8 Mohs. Occurrence: USA (Maine).

Heresite. Trademark of Heresite-Saekaphen, Inc. (USA) for a series of corrosion-resistant coatings based on phenol-formaldehyde resins, and used for lining equipment and containers for chemicals, petroleum products, pharmaceuticals and foods, heat exchangers, etc.

Herex. Trademark of Alcan Airex AG (Switzerland) for a range of core materials for sandwich and composite reinforcement applications.

Hergermuhl brass. A British brass containing 62-72% copper, 28-38% zinc, 0.2-1% tin and up to 0.8% lead. It has good workability, and is used for fittings, sheathing and hardware.

Heritage. Trade name of Permacon (Canada) for interlocking paving stones with the appearance of stone slabs supplied in rectangular units in a wide range of color. Used for walkways and patios.

Heritage Square. Trade name of Perma Paving Stone Company (Canada) for interlocking paving stones supplied in square units, 230 × 140 × 60 mm (9.000 × 5.500 × 2.375 in.), and square edge units, 230 × 140 × 60 mm (9.000 × 5.500 × 2.375 in.). Supplied in a range of colors, they are used for walkways and driveways.

Herkules. Trade name of Westa-Westdeutsche Edelstahlhandelsgesellschaft (Germany) for a series of oil- or air-hardening, nondeforming tool steels containing 2.1% carbon, 0.3% silicon, 0.3% manganese, 11.5% chromium, and the balance iron. Used for punches, and forming and blanking dies.

Hermes. (1) Trade name of Vereinigte Edelstahlwerke (Austria) for a series of water-hardening steels containing 0.3-0.9% carbon, 0.25-0.5% silicon, 0.3-0.8% manganese, and the balance iron. Used for gears, shafts, machine-tool parts, bolts, axes, hammers, springs, crimpers, and cutting tools.

(2) Trade name of Norddeutsche Schleifmittel-Industrie Christiansen & Co. (Germany) for a series of abrasive products including wet-strong abrasive paper, resin-bonded abrasive bands, and vulcanized-fiber wheels.

Hermeseal. Trade name of Rentokil Laboratories Inc.–Hermeseal Insulation Division (UK) for double glazing units.

Hermitage. Trade name of Veerman International Company (USA) for flashed and pot opal glasses, and structural glass parts.

Herox. Trademark of E.I. DuPont de Nemours & Company (USA) for polyamide (nylon) resins and fibers.

Herringbone. Trade name of C-E Glass (USA) for a glass with herringbone design.

herringbone twill. A twill-weave fabric, usually of wool, having a herringbone-like fiber pattern. Used especially for top and sports coats and suitings.

herschelite. A white mineral of the zeolite group composed of sodium aluminum silicate trihydrate, $NaAlSi_2O_6 \cdot 3H_2O$. Crystal system, rhombohedral (hexagonal). Density, 2.08 g/cm^3; refractive index, 1.472. Occurrence: Italy.

herzenbergite. A black mineral composed of tin sulfide, SnS. It can be synthesized by heating tin and sulfur under prescribed conditions. Crystal system, orthorhombic. Density, 5.22 g/cm^3. Occurrence: Bolivia; Southwest Africa.

HES. Trade name of Atofina Chemicals Inc. (USA) for silicate mortars used in building and construction.

HES cement. See high-early-strength cement.

HES concrete. See high-early-strength concrete.

hessian. A coarse, plain-woven fabric, similar to *burlap*, made from jute, hemp or, sometimes, flax. It may be dyed and impregnated with hot-melt adhesives, and is used for laminated composites, bagging, linings, wall coverings, backings in upholstery, and water-retaining coverings for curing concrete surfaces.

hessite. A black or medium-gray mineral composed of silver telluride, Ag_2Te. It can also be made synthetically. Crystal system, monoclinic. Density, 8.41 g/cm^3; refractive index, 2.32. Occurrence: Rumania.

hessonite. A orange-yellow to reddish-brown variety of *grossularite* garnet used as a gemstone. Also known as *cinnamon stone*.

hetaerolite. A black mineral of the spinel group composed of zinc manganese oxide, $ZnMn_2O_4$. It can also be made synthetically by heating equimolar amounts of manganese oxide and zinc oxide under prescribed conditions. Crystal system, tetragonal. Density, 5.18 g/cm^3; refractive index, 2.35.

heterochain polymer. A polymer, such as silicone rubber, polyoxymethylene or nylon 6/6, that has more than one element in its backbone.

heteroepitaxial thin film. A semiconductor thin film vapor-deposited on a substrate with the film and substrate consisting of significantly different materials, e.g., an aluminum gallium arsenide (AlGaAs) film on a gallium arsenide (GaAs) substrate, or a gallium arsenide (GaAs) film on an indium gallium arsenide (InGaAs) substrate. See also homoepitaxial thin film.

heterofil fiber. A biconstituent fiber consisting of a relatively high-melting core of nylon 66 and a rather low-melting outer sheath of nylon 6, extruded through the same spinneret. The filament thus produced (heterofilaments) are usually assembled in the form of web and then suitably heated to induce inter-fiber bonding and produce a fabric. Also known as *meldable fiber*.

heterogenite. A black or steel gray mineral that is composed of cobalt oxide hydroxide, CoOOH. and may also contain some nickel. It can also be made synthetically. Crystal system, hexagonal. Density, 4.30 g/cm^3. Occurrence: Zaire, Germany.

heteromorphite. An iron-black mineral composed of lead antimony sulfide, $Pb_7Sb_8S_{19}$. Crystal system, monoclinic. Density, 5.73 g/cm^3. Occurrence: Germany.

heteropolymer. An additive copolymer that has been formed by the combination of two unsaturated substances, one of which, by itself, does not easily polymerize. Abbreviation: HP.

heterosite. A deep rose to reddish purple mineral of the olivine group composed of iron manganese phosphate, $(Fe,Mn)PO_4$. Crystal system, orthorhombic. Density, 3.40 g/cm^3; refractive index, 1.86-1.89. Occurrence: USA (New Hampshire).

heterostructures. Structures that consist of a single-crystal thin film grown on a substrate by a vacuum deposition process, such as molecular beam epitaxy (MBE). Examples include nonmagnetic heterostructures consisting of a metal thin film grown on a semiconductor substrate (e.g., aluminum on p-type silicon), a semiconductor material grown on a similar or dissimilar semiconductor substrate (e.g., silicon on silicon, or aluminum gallium arsenide on gallium arsenide), or a superconductor thin film grown on a semiconductor substrate (e.g., yttrium barium copper oxide on silicon).

Hetrofoam. Trade name for fire-resistant polyurethane foams with high chlorine content. They have low coefficients of thermal conductivity, and are used for refrigeration applications, and in the thermal insulation of buildings.

Hetron. Trade name for unsaturated polyesters.

heulandite. A colorless mineral of the zeolite group that is composed of calcium aluminum silicate hexahydrate, $CaAl_2Si_7O_{18} \cdot 6H_2O$, and may also contain strontium. Crystal system, monoclinic. Density, 2.22 g/cm^3; refractive index, 1.498. Occurrence: Faroe Islands, Czech Republic.

Heureka. Trade name for a phenol-formaldehyde resin formerly used for denture bases.

Heusler alloys. A group of magnetic copper alloys containing 16-30% manganese, 4-15% aluminum, and up to 4% lead. They crystallize with the ordered cubic Cu_2MnAl structure, exhibit ferromagnetic properties, and are used for electrical equipment.

hevea plus. A substance made by the polymerization of methyl methacrylate and hevea rubber latex.

hevea rubber. The white, viscous, milky fluid (or latex) obtained from any of various plants of the genus *Hevea*, especially the rubber tree *(Hevea brasiliensis)*. It consists of an aqueous dispersion of *cis*-1,4-polyisoprene, an unsaturated, high-molecular-weight hydrocarbon. For commercial applications, it is coagulated by the addition of acetic or formic acid, or sodium hexafluorosilicate, and subsequent concentration by evaporation or centrifugation. Source: Brazil, India, Indonesia, Liberia, Sri Lanka, Zaire. Used in the manufacture of natural rubber. See also crude rubber; natural rubber.

Heveatex. Trademark of Heveatex Corporation (USA) for raw and processed rubber latex in various grades obtained from trees of the genus *Hevea*.

Hevi-Duty. Trade name of Solar Basic Industries (USA) for high-tensile steel containing 0.3% carbon, and the balance iron, used for welding rods.

Hevimet. Trade name of General Electric Company (USA) for a sintered tungsten alloy containing 6-7.5% nickel and 2.5-4% copper. It has a density of 16.9 g/cm^3 (0.611 $lb/in.^3$), high tensile and compressive strengths, and is used for radiation shielding, vibration damping, counterbalances, and weights.

hewettite. A deep red mineral composed of calcium vanadate nonahydrate, $CaV_6O_{16} \cdot 9H_2O$. Crystal system, monoclinic. Density, 2.62 g/cm^3. Occurrence: Peru, USA (Arizona, Colorado, Utah).

Hewitt. Trade name of Hewitt Metals Corporation (USA) for several tin and lead babbitts including (i) *Hewitt Copper-Hard*, a tin babbitt that contains 4.5% copper and 4.5% antimony, has a melting range of 185-310°C (365-590°F), and is used for heavy-duty bearings and bushings; (ii) *Hewitt Genuine*, a tin babbitt that contains 7.5% antimony and 3.5% copper, has a melting range of 238-361°C (360-680°F), and is used for heavy-duty bearings and bushings; and (iii) *Hewitt Mill*, a lead babbitt that contains varying amounts of tin and antimony, and is used for steel-mill bearings.

Hewmet. Trade name of Hewitt Metals Corporation (USA) for a heavy-duty lead babbitt containing 15% antimony and 5% tin. Used for bearings and bushings.

Hex. Trade name of Libbey-Owens-Ford Company (USA) for a wired glass with hexagonal mesh.

hexaboron silicide. See boron silicide (ii).

hexacarbonylmolybdenum. See molybdenum carbonyl.

hexaferrites. See hexagonal ferrites.

hexafluoropropylene vinylidene fluorides. A group of crystalline thermoplastic copolymers of hexafluoropropylene (C_3F_6) and vinylidene fluoride ($C_2H_2F_2$). They are have a density of 1.8-2.0 g/cm^3 (0.06-0.07 $lb/in.^3$), excellent resistance to ultraviolet radiation, good self-extinguishing properties, a service temperature range of -50 to +300°C (-58 to +572°F), a refractive index of 1.41, low thermal expansivity, good strength, good resistance to acids, greases, oils and halogens, good to fair resistance to alcohols, alkalies and aromatic hydrocarbons, and poor resistance to ketones. Abbreviation: HFPVF. Also known as *poly(hexafluoropropylene-co-vinylidene fluoride)s*.

hexagonal boron nitride. Boron nitride with a hexagonal crystal structure commercially available in the form of a powder and as solid shapes. Abbreviation: HBN; hBN. See also boron nitride; cubic boron nitride; wurtzite-type boron nitride.

hexagonal ferrites. A group of magnetically hard ceramic materials having an inverse spinel crystal structure with hexagonal symmetry, and a general chemical composition corresponding to the formula $MO \cdot 6Fe_2O_3$ (or $MFe_{12}O_{19}$) where M is a divalent metal cation, such as barium (Ba^{2+}), lead (Pb^{2+}) or strontium (Sr^{2+}). Barium ferrite ($BaO \cdot 6Fe_2O_3$ or $BaFe_{12}O_{19}$), lead ferrite ($PbO \cdot 6Fe_2O_3$ or $PbFe_{12}O_{19}$) and strontium ferrite ($SrO \cdot 6Fe_2O_3$ or $SrFe_{12}O_{19}$) are examples. Used for high-coercive-force permanent magnets. Also known as *hexaferrites; magnetoplumbites*. See also barium ferrite (2); lead ferrite; strontium ferrite.

hexahydrate. A chemical compound that contains six molecules of water, e.g., chromium chloride hexahydrate ($CrCl_3 \cdot 6H_2O$).

hexahydrite. A colorless, white or greenish-white mineral composed of magnesium sulfate hexahydrate, $MgSO_4 \cdot 6H_2O$. It can also be made synthetically. Crystal system, monoclinic. Density, 1.76 g/cm^3; refractive index, 1.453. Occurrence: Canada (British Columbia), USA (Washington), Ukraine.

hexahydrobenzene. See cyclohexane.

hexahydroborite. A mineral composed of calcium borate hydroxide dihydrate, $Ca[B(OH)_4]_2 \cdot 2H_2O$. Crystal system, monoclinic. Density, 1.88 g/cm^3. Occurrence: Russia.

hexamethylene. See cyclohexane.

hexamethylindodicarbocyanine iodide. An organic compound supplied with a purity of 99%, a melting point of 264°C (507°F) (decomposes), and a maximum absorption wavelength of 637 nm. Used as a laser dye. Formula: $C_{27}H_{31}IN_2$.

hexamethylindotricarbocyanine iodide. An organic compound supplied with a purity of 98%, a melting point of 198°C (388°F) (decomposes), and a maximum absorption wavelength of 740 nm. Used as a laser dye. Abbreviation: HITC iodide. Formula: $C_{29}H_{33}IN_2$.

hexanaphthene. See cyclohexane.

hexarhodium hexadecacarbonyl. See rhodium carbonyl.

hexatestibiopanickelite. A yellowish white or pale yellow mineral of the nickeline group composed of nickel palladium antimony telluride, $(Ni,Pd)_2SbTe$. Crystal system, hexagonal. Density, 8.94 g/cm^3. Occurrence: USA (Arizona).

hexazirconium pentasilicide. See zirconium silicide (viii).

Hexcalite. Trade name of Australian Window Glass Proprietary Limited (Australia) for a sandwich panel comprising two sheets of clear glass with an interlayer of honeycombed, expanded aluminum bonded with a colored plastic adhesive.

Hexcel. (1) Trade name of Hexcel Corporation (USA) for lightweight honeycomb materials with high strength and flexibility used for various aircraft structures, and in the manufacture of the fuselage and wings of the Voyager aircraft. *Hexcel EP* are epoxy laminates supplied in glass fabric and glass prepreg grades, and *Hexcel Polyester* woven-glass-roving polyester laminates.

(2) Trademark of Hexcel Corporation (USA) for carbon fibers with 3000, 6000 and 12000 filaments.

Hexite. Trade name of Hexco Products, Inc. (USA) for acrylic plastics.

Hexogen. Trademark of Carlisle Chemical Works, Inc. (USA) for paint driers consisting essentially of solutions of metallic salts (e.g., calcium, cobalt, iron, lead, manganese and zinc) of 2-ethylhexanoic acid ($C_8H_{16}O_2$).

Hexoloy. Trademark of St. Gobain Advanced Ceramics (USA) for a series of silicon carbide (SiC) structural ceramics. *Hexoloy RB* is a fine-grained reaction-bonded SiC with a low silicon metal content (only 8-10%). *Hexoloy SA* is a high-temperature, pressureless sintered α-SiC material with low weight, good formability, high hardness, strength and thermal-shock resistance, excellent high-temperature performance, thermal conductivity and abrasion, wear, erosion, corrosion and oxidation resistance, and excellent resistance to most chemicals including strong acids and alkalies. *Hexoloy SA & RB* ceramics are used for machine components including washers, valve parts, nozzle insert, pump bearings, water-pump seals, mechanical face seals, components for heat exchangers, turbocharger rotors, turbines and internal combustion engines, for kiln furniture, and as disks and plates for silicon wafer processing. *Hexoloy SP* is a controlled-porosity SiC for specialized fluid handling applications. *Hexoloy SG* is an electrically conductive Hexoloy SA-based SiC thin-film coating containing graphitic carbon. It has high hardness, good abrasion and corrosion resistance, good temperature stability, high optical transparency, adheres well to many substrates, and can be applied by DC magnetron sputtering. It is used as a protective coating on flat glass, as a conductive or vapor-barrier coating for electronic, electrical and other applications, and as a coating for magnetic hard disks and optical data disks.

hexosan. A *polysaccharide* that gives only hexoses on hydrolysis.

hexose. A *monosaccharide*, such as glucose, fructose or galactose, that has six carbon atoms in its molecular chain and can be obtained from *hexosan* by hydrolysis. Formula: $C_6H_{12}O_6$.

HexWeb EM. Trade name of Hexcel Composites (USA) for an energy-absorbing, lightweight thermoplastic honeycomb material used for headrests, armrests, sun visors, knee bolsters, and similar applications.

1-(*trans*-4-hexylcyclohexyl)-4-isothiocyanatobenzene. A nematic liquid crystal with a flash point above 230°F (110°C), and a refractive index of 1.532. Formula: $CH_3(CH_2)_5C_6H_{10}C_6H_4NCS$.

4'-(hexyloxy)-4-biphenylcarbonitrile. A nematic liquid crystal with a melting point of 56-58°C (133-136°F). Formula: $CH_3(CH_2)_5OC_6H_4C_6H_4CN$.

Heydeflon. Trademark for fluoropolymers based on polytetrafluoroethylene (PTFE).

heyite. A yellow-orange mineral composed of iron lead vanadium oxide, $Fe_2Pb_5(VO_4)_2O_4$. Crystal system, triclinic. Density, 6.30 g/cm³; refractive index, 2.219. Occurrence: USA (Nevada).

heyrovskyite. A tin-white mineral composed of lead silver bismuth sulfide, $Pb_{10}AgBi_5S_{18}$. It can also be made synthetically. Crystal system, orthorhombic. Density, 7.17 g/cm³. Occurrence: Czech Republic, Slovakia.

HG Nickel. Trade name of Inco Alloys International Limited (UK) for a heat- and corrosion-resistant nickel containing 0.06% carbon. Used for glassy seals, terminal pins, and electronic valves.

Hi Al G Plastic. Trade name of Wellsville Fire Brick Company (USA) for high-alumina plastic refractories.

Hi Al Kastite. Trade name of Wellsville Fire Brick Company (USA) for heavy high-alumina castable refractories.

Hi-Alloy. Trade name of Dow Chemical Company (USA) for a magnesium alloy containing 5.6-6.7% aluminum, 2.5-3.5% zinc, 0.3% silicon, 0.18% manganese, and 0.05% copper. Used for cathodic protection anodes.

Hi-Almax. Trade name of National Refractories & Minerals Corporation (USA) for a high-alumina refractory mortar.

Hi-Al-Ram. Trade name of Wellsville Fire Brick Company (USA) for high-alumina ramming mixes.

Hibar. Trademark of Celufibre Industries Limited (Canada) for mineral fiber sprays.

Hibbo. Trade name of Central Brass & Aluminum Foundry Company (USA) for a series of copper-aluminum casting alloys used for bearings, bushings, gears, worm wheels, pinions, shafts, dies, pins and cams.

Hi Bond. Trade name of Matech, Inc. (USA) for a nonprecious ceramic dental alloy.

hibonite. A brown-black mineral of the hogbomite group composed of calcium aluminate, $CaAl_{12}O_{19}$. Crystal system, hexagonal. Density, 3.84 g/cm³; refractive index, 1.807. Occurrence: Russian Federation; Madagascar.

Hi-Brite. Trade name of Shannon Luminous Materials, Inc. (USA) for fluorescent paints.

Hi-Carbalon. Trade name of Asahi Kasei Corporation (Japan) for polyacrylonitrile-based carbon fibers used as reinforcements of high-performance composites for aircraft, aerospace and military applications.

Hi-Chrome. Trade name of Xaloy Inc. (USA) for an abrasion-resistant cast iron containing 2.6% carbon, 1% manganese, 1% silicon, 0.6% molybdenum, 27% chromium, 0.3% vanadium, and the balance iron. Used for pump and cylinder liners.

hickory. The strong, hard, tough, straight-grained wood of any of various species of North American and Asian hardwood trees of the genus *Carya* belonging to the walnut family. True hickories include the shagbark (*C. ovata*), the pignut (*C. glabra*), the shellbark (*C. laciniosa*) and the mockernut (*C. tomentosa*).

Pecan hickories include the bitternut (*C. cordiformis*), the pecan (*C. illinoensis*), the water hickory (*C. aquatica*), and the nutmeg hickory (*C. myristicaeformis*). Average weight, 720-820 kg/m³ (45-51 lb/ft³). Used mainly for spokes and rims of wheels, tool handles, machine parts, agricultural implements, wagons, gunstocks, chair backs, baskets, barrel hoops and skis, for work requiring bending, and as an excellent fuel that produces great heat and high-grade charcoal.

hickory cloth. A durable twill-weave fabric, usually of cotton, having warp stripes and produced in a wide range of colors.

hickory elm. See rock elm.

hickory poplar. See American whitewood.

Hi-Cobalt. Trade name of Indiana General (USA) for a hard magnetic material that is composed of 60% iron and 40% cobalt, has high coercive force and remanent magnetization, and is used for permanent magnets.

HiCond. Trade name of Aerotherm Corporation (USA) for a carbon/carbon composite material consisting of a pitch-based matrix reinforced with polyacrylonitrile fibers. Used for aerospace components.

Hicor. (1) Trademark of Stelco Inc. (Canada) for a zinc-plated steel.

(2) Trademark of Mobil Oil Corporation (USA) for oriented high-density polyethylene film used for food packaging applications.

Hicore. Trade name of British Steel Corporation (UK) for a series of case-hardening steels containing 0.1-0.2% carbon, 3-4.5% nickel, 0.9-1.2% chromium, up to 0.5% molybdenum, and the balance iron. Used for gears, pinions, and differential axes.

Hi-Core. Trade name of Matra Plast Industries, Inc. (Canada) for corrugated extruded polyethylene and polypropylene multiwall sheeting.

Hi Crome. Trade name for a corrosion-resistant cobalt-chromium dental casting alloy.

HiD. Trademark of Chevron Phillips Chemical Company (USA) for high- and medium-density polyethylenes used for piping. The high-density resins are supplied in standard and UV-stabilized grades.

Hi-D. Trade name of Westvaco Corporation (Canada) for corrugated paperboard used for boxes.

Hidalgo. Trade name of Dörrenberg Edelstahl GmbH (Germany) for a series of cold-work and high-speed tool steels.

hidalgoite. A white mineral of the alunite group composed of lead aluminum arsenate sulfate hydroxide, $PbAl_3AsO_4SO_4(OH)_6$. A light green variety containing iron is also found. Crystal system, rhombohedral (hexagonal). Density, 3.96 g/cm³. Occurrence: Mexico.

hiddenite. A clear, deep-green variety of the mineral *spodumene* ($LiAlSi_2O_6$) from North Carolina, USA, used as a gemstone.

hide. (1) The pelt of a large animal, e.g., a cow or a horse.

(2) The relatively soft and flexible outer layer of tissue of a small animal, such as a calf, pig, sheep, etc., used in the manufacture of leather. Also known as *skin*.

hide glue. A relatively strong, pale brown cementing material obtained by boiling the hides of animals.

Hi Dense. (1) Trade name of Shofu Dental Corporation (Japan) for a specially formulated, fluoride-releasing glass-ionomer composite containing spherical silver-tin particles. It has a packable and moldable consistency, and provides excellent adhesion and high compressive strength. Used in restorative dentistry for tooth repair, core build-ups under crowns, and as a base under amalgam.

(2) Trademark of Superior Graphite Company (USA) for a high-strength beta silicon carbide (β-SiC).

Hi-Dri. Trade name of Protex-A-Cote, Inc. (USA) for silicone-base waterproofing compounds.

Hidumatic. Trade name of High Duty Alloys Limited (UK) for a wrought aluminum alloy containing copper, bismuth and lead. Used for structural parts and screw-machine products.

Hiduminium. Trademark of High Duty Alloys Limited (UK) for an extensive series of wrought and cast aluminum products including commercially pure aluminum (99.0+%), various aluminum-manganese, aluminum-silicon, aluminum-magnesium, aluminum-magnesium-silicon, aluminum-copper, aluminum-tin and aluminum-zinc alloys, aluminum-clad aluminum-copper alloys as well as sintered materials of aluminum oxide and aluminum.

Hidurax. Trade name of Langley Alloys Limited (UK) for a series of wrought and cast aluminum bronzes and copper-nickel alloys used for machine parts, shafts, gears, liners, bushings, valve and pump components, and fittings.

Hidurel. Trade name of Langley Alloys Limited (UK) for a series of wrought aluminum bronzes and copper-zinc alloys used for strong, corrosion-resistant parts, e.g., condensers, fittings, valves, gears, and fasteners.

Hidurit. Trade name of Langley Alloys Limited (UK) for a series of high-strength casting brasses containing 40% zinc, 1-5% aluminum, 1.5-4% manganese, 1-2% nickel, 1% iron, up to 1% lead, up to 1.5% tin, and the balance copper.

Hiduron. Trade name of Langley Alloys Limited (UK) for several wrought and cast aluminum bronzes and copper nickels.

Hidux. Trade name for heat-curing type epoxy-phenolic adhesives with low peel strength and good lap shear strength at low and high temperatures from -73 to +260°C (-100 to +500°F). Used as structural adhesives for joining metals.

HiE-Coat. Trademark of Aremco Products Inc. (USA) for a series ceramic coatings.

hieratite. A grayish mineral composed of potassium silicon fluoride, K_2SiF_6. Crystal system, cubic. It can also be made synthetically. Density, 2.66 g/cm^3; refractive index, 1.340.

HiFax. Trade name of Himont USA Inc. for reaction-injection-moldable thermoplastic olefins based on polypropylene and supplied in rigid and semi-rigid grades. They have excellent weatherability and impact resistance. Some grades have good high-temperature and/or low-temperature properties. Used for automotive interior trim and electrical applications.

Hi-Fi. Trade name of Shofu Dental Corporation (Japan) for a glass-ionomer dental cement.

Hiflex. Trade name of Gulf Steel Corporation (USA) for a fatigue-resistant steel used for tools, dies and machinery parts.

Hi-Form. Trade name of Inland Steel Company (USA) for a series of high-strength low-alloy (HSLA) steels containing 0-0.15% carbon, up to 1.4% manganese and 0.15% niobium (columbium), respectively, and the balance iron. They have good formability and weldability, and are used for transportation and mobile equipment parts.

Hi-Frax. Trade name of Hercules, Inc. (USA) for polyethylene plastics supplied in several grades.

high-abrasion furnace black. A fine grade of *furnace black* composed of nanometer-sized particles, and produced by the partial combustion of oil in a closed system under controlled conditions. Used as filler in rubber goods and tires.

high-alkali glass. See A-glass.

high-alkali glass fibers. See A-glass fibers.

high-alloy ductile irons. See austenitic ductile irons.

high-alloy steels. A group of steels that contain considerable percentages (usually more than 8%) of alloying elements in addition to iron and carbon, and frequently exhibit special properties that plain-carbon and low-alloy steels do not have, such as excellent corrosion and chemical resistance, excellent scaling resistance at elevated temperatures, good capability of holding a sharp cutting edge, and specific electrical or magnetic properties. Abbreviation: HAS.

high-alumina brick. A refractory brick containing considerable amounts of alumina (Al_2O_3) which when fired to high temperatures reacts with silica (SiO_2) to form *mullite*. Used for ceramic kilns, boilers, and various applications in severe environments. See also mullite brick.

high-alumina cement. See aluminate cement.

high-alumina ceramics. A group of ceramics composed principally of alpha aluminum oxide (α-Al_2O_3) with compositions ranging from 75-100% Al_2O_3. They have melting points above 2000°C (3630°F), high hardness and strength, good abrasion and impact resistance, very high electrical resistivity, high dielectric strength, good resistance to many chemicals and industrial atmospheres, and close dimensional tolerances. Used for cutting tools, high-temperature roller bearings, plungers or liners in pumps, faces for mechanical rotary seals, chute linings, chemical equipment, dies, discharge orifices, nozzles, rock bits, electrical insulators, and aerospace components. Abbreviation: HAC.

high-alumina fibers. Dense, polycrystalline continuous fibers which contain 99+% alpha alumina (α-Al_2O_3) and are usually produced by a slurry spinning technique. They have excellent strength and elastic modulus, and good high-temperature stability. Used as reinforcements for metal-matrix composites, and ceramic fiber-ceramic matrix composites. Abbreviation: HAF.

high-alumina mortar. A *mortar* that contains at least 47.5% alumina (Al_2O_3).

high-alumina refractory. A refractory material composed of at least 45% alumina (Al_2O_3) and produced from raw materials, such as andalusite, bauxite, diaspore, fused alumina, gibbsite, kyanite and sillimanite.

high-aluminum gray irons A group of heat-resistant gray cast irons containing about 1.3-2.0% total carbon (predominantly in flake graphite form) and high amounts of aluminum (6-25%) and, sometimes, high amounts of chromium (up to 25%). They have ferritic matrices as cast and possess excellent scaling resistance, but have rather high brittleness and poor castability. Also known as *high-aluminum gray cast irons.*

high brass. A wrought brass that contains about 60-66% copper and 34-40% zinc, and is commercially available in the form of sheets, strips, rods, flat products, wire and rolls. It has high tensile strength, good ductility and good cold workability, and is used for drawn and spun parts, vessels, tanks, springs, screws, rivets, hardware and fixtures. Also known as *common brass; market brass.*

high-bulk yarn. A yarn whose degree of bulkiness (loft) has been markedly increased by any of several chemical, physical or mechanical treatments.

high-calcic material. Any material composed of 90% or more calcium oxide (CaO), the balance being usually magnesium oxide (MgO).

high-calcium lime. See fat lime.

high-calcium limestone. A limestone containing at least 95%

calcium carbonate ($CaCO_3$) and up to 5% magnesium carbonate ($MgCO_3$). Available as aggregate and pulverized grades, it is used as filler and extender in paint, rubber, plastics, adhesives, etc., and in the chemical industry for fluxing and/or acid desulfurization applications.

high-carbon cast steels. See carbon-steel castings.

high-carbon chromium. Chromium containing 8-11% carbon, and up to 0.5% silicon and iron, respectively.

high-carbon chromium-molybdenum white irons. A group of highly abrasion-resistant martensitic white cast irons containing 2.6-3.6% carbon, 14.0-23.0% chromium, 0.5-1.50% nickel, 0.5-1.5% manganese, up to 3.5% molybdenum, 1.2% copper, up to 1.0% silicon, 0.1% phosphorus, 0.06% sulfur, and the balance iron. Also known as *high-carbon chromium-molybdenum white cast irons* See also chromium-molybdenum white irons; martensitic white irons; white irons.

high-carbon ferrochrome. See high-carbon ferrochromium.

high-carbon ferrochromium. A *ferroalloy* containing 4.5-7% carbon, 1-3% silicon, 65-70% chromium, and the balance iron. Used in the production of chromium steels and cast irons. Also known as *high-carbon ferrochrome.*

high-carbon ferrotitanium. A *ferroalloy* containing 6-8% carbon, 15-18% titanium, and the balance iron. Used as a master alloy, deoxidizer and scavenger, and/or as a titanium addition to steel.

high-carbon high-chromium tool steels. A group of cold-work tool steels (AISI group D) containing 1.40-2.50% carbon, up to 0.60% manganese and 0.60% silicon, respectively, 11.00-13.50% chromium, 0.70-1.20% molybdenum, 1.00-4.40% vanadium, up to 3.50% cobalt, and the balance iron. They have excellent deep-hardening properties, excellent dimensional stability, good hot hardness, high wear resistance, and comparatively good machinability. They are often sold under trademarks or trade names, such as *Atmodie, Crocar, Croloy, Hampden, Huron, Neor, Ohio Die, Ontario, Olympic FM, Superdie, Triple Die* or *Truwear,* and used for cold-work press tools, thread-rolling dies, gages, lathe centers, and forming, bending and drawing dies.

high-carbon malleable irons. A *malleable cast iron* with approximately 3% carbon produced in an air furnace.

high-carbon nickel-chromium white irons. See nickel-chromium white irons.

high-carbon steel castings. See carbon-steel castings.

high-carbon steels. A group of steels containing between 0.44-1.50 wt% carbon. They are the hardest, strongest and least ductile of the carbon steels, usually contain chromium, vanadium, tungsten and molybdenum, and are almost always used in the hardened or tempered condition and thus are extremely wear-resistant and capable of holding a sharp cutting edge. Abbreviation: HCS.

High Chrome-Nickel. Trade name of Johnson Matthey plc (UK) for a tough steel containing 0.33-0.45% carbon, 0.4-0.75% chromium, 1-1.35% manganese, 0.9-1.2% nickel, 0.4-0.75% silicon, and the balance iron. Used for shafts and gears.

high-chromium gray irons. A group of heat-resistant gray cast irons containing 1.8-3.0% total carbon, 0.3-1.5% manganese, 0.5-2.5% silicon, 15.0-35.0% chromium, up to 5.0% nickel, up to 0.15% phosphorus, up to 0.15% sulfur, and the balance iron.

high-chromium high-carbon white irons. A group of white cast irons containing about 2.3-3.0% total carbon, 0.5-1.5% manganese, 1.0-2.2% silicon, 0.1% phosphorus, 0.06% sulfur, 7-23%

chromium, 1.5-7% nickel, 0.5-3.5% molybdenum, up to 1.2% copper, and the balance iron. They are usually martensitic or austenitic as cast, and their carbon exists in the form of carbides, predominantly as iron carbide (cementite, Fe_3C) and chromium carbide (Cr_7C_3). These cast irons have excellent abrasion resistance, good corrosion resistance, high hardness (450 Brinell) and toughness, negligible ductility, and are used for abrasion- and wear-resistant parts, and in the manufacture of malleable iron castings. Also known as *high-chromium high-carbon white cast irons.*

high-chromium irons. A group of corrosion-resistant cast irons containing about 1.2-4.0% total carbon, 0.3-1.5% manganese, 0.5-3.0% silicon, 12.0-35.0% chromium and up to 0.15% phosphorus, 0.15% sulfur, 5.0% nickel, 4.0% molybdenum and 3.0% copper, respectively, and the balance iron. They have martensitic or austenitic microstructures in the as-cast condition, and are used for marine applications, chemical-process equipment, pumps, valves, etc. Also known as *high-chromium cast irons.*

high-chromium white irons. A group of white cast irons containing about 2.8-3.6% total carbon, 0.5-1.5% manganese, 1.0-2.2% silicon, 0.1% phosphorus, 0.06% sulfur, 23-28% chromium, 1.5% nickel, 1.5% molybdenum, up to 1.2% copper, and the balance iron. They are usually martensitic as cast, and their carbon exists in the form of carbides, predominantly as iron carbide (cementite, Fe_3C) and chromium carbide (Cr_7C_3). They have excellent abrasion resistance, good corrosion and oxidation resistance, good stability at elevated temperatures, high hardness (450 Brinell) and toughness, negligible ductility, and are used for abrasion- and wear-resistant parts, and in the manufacture of malleable iron castings. Also known as *high-chromium white cast irons; high-chromium martensitic white irons; high-chromium martensitic white cast irons.* See also chromium white irons; martensitic white irons; white irons.

high-conductivity aluminum. A non-heat-treatable wrought aluminum (99.50+% pure) that contains 0.40% iron, 0.10% silicon, 0.05% copper, 0.05% zirconium, 0.01% manganese, 0.01% chromium, and 0.10% other elements (e.g., vanadium, titanium, boron and/or gallium). Supplied in the annealed or strain-hardened conditions in the form of sheets, plates, extruded bars, rods, shapes, tubes and wires and cold-finished bars, rods and wires, it has a density of 2.7 g/cm³ (0.10 lb/in.³), a melting range of 646-657°C (1195-1215°F), excellent resistance to corrosion and stress-corrosion cracking, good cold workability, good strength, excellent brazeability, solderability and weldability, moderate machinability, and good electrical conductivity (about 62% IACS). Used for electrical conductors. Formerly known as *EC aluminum; electrical conductor aluminum.*

high-conductivity copper. See electrolytic tough-pitch copper.

high-copper alloys. A group of alloys containing between 95.0 and 98.5% copper, and varying amounts of beryllium, cobalt, chromium, nickel, silver and silicon. They have higher hardness and strength than most other copper alloys, and are used for conductors, welding electrodes, dies, molds, bushings, cams, pump parts, etc.

high-count fabrics. Dense, close fabrics that have been woven with a comparatively high thread count. *Note:* The "thread count" is the number of warp and weft yarns per 1 in.² (645 mm²) of fabric.

high-density alumina ceramics. A group of ceramics made from high-purity (99.5+%), fine-grained polycrystalline alpha alumina (α-Al_2O_3) by powder-metallurgy techniques using up to 0.5% magnesia (MgO) as a sintering aid. They have a minimum density of 3.93 g/cm³ (0.142 lb/in.³), an average grain size of 3-6 μm (118-236 μin.), a hardness of 2300 Vickers, high strength, excellent corrosion and wear resistance, and good biocompatibility. Used for load-bearing prostheses, e.g., hip-joint balls, knee joints and bone screws, and dental applications, e.g., jawbone reconstruction, and blade-, post- and screw-type dental implants.

high-density hardboard. Hardboard with a density of at least 800 kg/m³ (50 lb/ft³). Also known as *densified hardboard*. See also hardboard.

high-density concrete. See heavyweight concrete.

high-density metals. Metals having densities above about 15.0 g/cm³ (0.54 lb/in³) These include iridium (22.65 g/cm³ or 0.82 lb/in³), osmium (22.61 g/cm³ or 0.82 lb/in³), platinum (21.45 g/cm³ or 0.77 lb/in³), rhenium (21.00 g/cm³ or 0.76 lb/in³), tungsten (19.40 g/cm³ or 0.70 lb/in³), gold (19.30 g/cm³ or 0.70 lb/in³), uranium (19.07 g/cm³ or 0.69 lb/in³) and tantalum (16.60 g/cm³ or 0.60 lb/in³).

high-density overlay. Exterior-type plywood with a resin-impregnated overlay. It has a hard, smooth, highly abrasion- and chemical-resistant surface, and can be used in severe environments. Abbreviation: HDO. See also exterior-type plywood.

high-density particleboard. Particleboard with a density of at least 800 kg/m³ (50 lb/ft³). It has a higher modulus of rupture, modulus of elasticity and internal bond strength than *low-density particleboard*, and a linear expansion of about 0.55%. Used for interior and exterior applications. See also particleboard.

high-density polyethylenes. A group of polyethylenes with natural white color produced by a catalyzed reaction resulting in polymer chains with few side chains or branches. They have a density of 0.94-0.96 g/cm³ (0.034-0.035 lb/in.³), a crystallinity of about 50%, high rigidity, good hardness, good tensile strength, good stress-crack resistance, good abrasion and impact resistance, smooth surface finish, fair resistance to ultraviolet radiation, high dielectric constant, low thermal expansivity, good (gas) weldability, an upper service temperature of 80-120°C (175-250°F), a brittleness temperature of -70°C (-94°F), low coefficients of friction, good resistance to acids, alcohols, alkalies, halogens and ketones, fair resistance to greases and oils, fair to poor resistance to aromatic and halogenated aliphatic hydrocarbons and strong oxidizing agents, liquefied petroleum gas, and solvents. Used for containers and equipment employed in the storage of organic and inorganic acids and caustics, and for water-treatment, metal-finishing and plating equipment. Abbreviation: HDPE. Also known as *linear high-density polyethylenes; high-molecular-weight polyethylenes (HMWPE); high-molecular-weight high-density polyethylenes (HMW HDPE)*. See also low-density polyethylenes; polyethylenes.

high-dielectric-constant materials. See high-k materials.

High Double Extra. Trade name of Midvale-Heppenstall Company (USA) for a high-speed steel containing carbon, tungsten, cobalt, vanadium and iron. Used for cutters.

High Duty. Trade name of Flockton, Tompkin & Company Limited (UK) for an abrasion-resistant, oil- or air-hardening tool steel containing 2.2% carbon, 0.6% tungsten, 12.8% chromium, 0.4% manganese, and the balance iron. Used for blanking and thread-rolling dies, and molds.

high-duty cast irons. A group of high-strength cast irons made by adding steel scrap or small amounts of chromium, molybdenum or nickel to molten pig iron during manufacture. They have an average tensile strength of 235 MPa (34 ksi). Also known as *high-strength cast irons; high-tensile cast irons; high-test cast irons*.

high-duty fireclay brick. A fireclay brick with a pyrometric cone equivalent between cones 31.5 and 33.

high-duty gray irons. A group of high-strength gray cast irons containing about 3.2-3.4% total carbon, 0.55-0.75% combined carbon, 2.1-2.3% silicon, 0.6-0.9% manganese, 0.2-0.6% chromium, 0.15% sulfur and phosphorus respectively, and the balance iron. Used for cylinder blocks, etc. Also known as *high-duty gray cast iron*.

High Dynamic. Trade name of Timken Company (USA) for a tough steel containing 0.4% carbon, 1.8% nickel, 0.8% chromium, 0.3% molybdenum, and the balance iron. Used for piston rods.

High Early Pozzolith. Trademark of Master Builders Inc. (USA) for an air-entraining agent that essentially has the same composition as *Pozzolith*, i.e., calcium lignosulfonate, sodium lauryl sulfate, calcium chloride and fly ash, but contains more calcium chloride. Used in concrete and mortar. See also air-entraining agent; air-entrained concrete.

high-early-strength cement. A type of Portland cement containing 53% tricalcium silicate (C_3S), 19% dicalcium silicate (C_2S), 11% tricalcium aluminate (C_3A), 9% tetracalcium aluminoferrate (C_4AF), and the balance being primarily simple oxides (magnesia, lime, alkali oxides) and calcium sulfate. It produces earlier strength in concrete and mortar, and has higher heat of hydration than standard Portland cement. Abbreviation: HES cement. Also known as *type III cement*. *Note:* In cement terminology C_3S, C_3A and C_4AF refer to tricalcium silicate ($3CaO \cdot SiO_2$), tricalcium aluminate ($3CaO \cdot Al_2O_3$) and tetracalcium aluminoferrate ($4CaO \cdot Al_2O_3 \cdot Fe_2O_3$), respectively.

high-early-strength concrete. Concrete in which high-early-strength cement and selected admixtures are used to develop a crushing strength greater than 11.7 MPa (1.7 ksi) when hardened in moist air for 24 hours, earlier than with standard concrete. Abbreviation: HES concrete.

high-energy hard magnetic materials. Hard magnetic materials, such as certain rare-earth cobalt (e.g., samarium cobalt) and neodymium-iron-boron materials, that have magnetic energy products of more than 80 kJ/m³ (10 MGOe). See also hard magnetic materials.

high-enriched uranium. See enriched uranium.

high-expansion alloys. A group of alloys including ferrous alloys with chromium and nickel, and nonferrous alloys, such as manganese-copper-nickel, that exhibit high thermal expansion coefficients (above $19 \times 10^{-6}/°C$ or $11 \times 10^{-6}/°F$). Used for thermostats and related equipment.

high-expansion cement. See expansive cement.

high-field superconductor. See hard superconductor.

high-gloss high-impact polystyrenes. A group of special high-impact polystyrenes (HIPS) whose mechanical properties are similar to those of conventional HIPS, while their brittleness is greatly reduced, and their gloss is comparable to that of acrylonitrile-butadiene-styrene (ABS). Usually processed by injection molding or extrusion, they are used for small appliances, and toys.

high-grade steels. A term used in Europe for specialty steels including all high-alloy steels and most low-alloy constructional steels. They are known in Germany as "Edelstähle," and in France as "aciers fins."

High Graphitic Iron. Trade name of Jenney Cylinder Company (USA) for a graphitic iron containing 3.5% total carbon, 0.4%

combined carbon, 3% silicon, 0.5% molybdenum, and the balance iron. Used for steam cylinder liners and centrifugal castings.

high-heat cement. Cement that liberates a large amount of heat during curing.

high-heat insulator. A refractory material used as thermal insulation in furnaces, kilns, or boilers.

high-impact polystyrene/polyphenylene ether alloys. See HIPS-PPE alloys.

high-impact polystyrene/polyvinyl chloride alloys. See HIPS-PVC alloys.

high-impact polystyrenes. A group of thermoplastic alloys and graft copolymers of styrene and natural rubber or styrene-butadiene rubber. They are commercially available in various grades including standard, glass-filled, high-gloss, ignition-resistant, refrigeration, ultraviolet-stabilized and very-high-impact. They have high to very high impact strength, also at low temperatures, high rigidity, good dimensional stability, good processibility, good resistance to alkalies, dilute inorganic acids, water and aqueous solutions of most salts, fair to poor resistance to most organic solvents, aromatic hydrocarbons and strong oxidizing agents. Sold under trade names or trademarks, such as *Replay* or *Styron*, they are used for appliances and consumer electronics, e.g., audio and video cassettes, business machines, TV cabinets, toys, furniture, cameras, plates, lids, bowls, portion cups, dairy containers, flatware, and packaging materials. Abbreviation: HIPS. *Note:* High-impact polystyrene is known in some English-speaking countries outside the US as "toughened polystyrene."

high-k dielectrics. See high-k materials.

high-k materials. A class of materials with high dielectric constants and including several ceramics, such as alumina, gadolinia, hafnia, titania, yttria and zirconia, and several titanates. They are used for microelectronic applications and in the fabrication of photonic crystals. Also known as *high-k dielectrics; high-dielectric-constant materials.*

high K_u materials. A class of granular alloy materials with higher unixial magnetic anisotropy (K_u) than conventional materials used for longitudinal magnetic data storage and recording applications. In these materials the number of grains per bit-cell is maintained very high to reduce media noise and increase data density.

Highland. Trademark of 3M Canada Inc. for pressure-sensitive adhesive tapes including clear and tan, strong, long-wearing, waterproof polypropylene packaging tape supplied in rolls for the permanent sealing of parcels, envelopes and cartons.

high-lead bronze. See high-leaded bronze.

high-leaded brass. A brass containing about 62-64.5% copper, 32.5-36.5% zinc and about 2% lead. Commercially available in the form of flats, rods, tubes, shapes and wires, it has excellent machinability, moderate cold workability, and enhanced antifriction and bearing properties. Used for clock and watch plates and backs, gears and wheels.

high-leaded bronze. A soft-matrix cast bronze that contains about 73-85% copper, 5-13% tin, and 5.5-20% lead, and is usually deoxidized with phosphorus. Commercially available in the form of sand, centrifugal, continuous and permanent-mold castings, it has good machinability and strength, and is used for high-speed bearings, bushings, pump parts, and impellers. Also known as *high-lead bronze; high-leaded tin bronze.*

high-leaded tube brass. A free-cutting tube brass containing 66% copper, 32.25-32.4% zinc and 1.6-1.75% lead. It has good machinability and strength and good to fair corrosion resistance. Used for screw-machine products.

Highlux. Trademark for semifinished products made of polymethyl methacrylate (PMMA) and polycarbonate (PC) polymers, and used for the thermal insulation of greenhouses, winter gardens, sports arenas and industrial buildings.

highly crystallized graphite. A graphite material in film form obtained from aromatic polyimide films (e.g., *Kapton* or *Novax*) by carbonization followed by a high-temperature graphitization treatment that produces a high degree of crystallization. Abbreviation: HCG. See also pyrolytic carbon.

highly oriented pyrolytic graphite. A highly crystallized and oriented graphite material obtained from pyrolytic carbon by heating to very high temperatures (about 3000°C or 5430°F) under conditions of mechanical stress. It consists of numerous highly oriented atomic layers of carbon, and is used in the calibration of scanning tunneling microscopes. Abbreviation: HOPG. See also pyrolytic carbon.

high-magnesia glass. A glass containing considerable additions (usually more than about 5%) of magnesium oxide (MgO) for increased formability. Abbreviation: HMG. See also magnesia glass.

high-magnesia hydraulic hydrated lime. A hydraulic hydrated lime containing at least 5% magnesium oxide (MgO).

high-manganese nickel. A heat-resistant alloy containing 94-98% nickel and 2-6% manganese. Used for spark-plug wire.

high-manganese steels. A group of nonmagnetic, wear-resistant steels containing 9% or more manganese. They have high strength and elongation, good abrasion resistance, high toughness and shock resistance, fair to poor machinability, and are used for rails, wear plates, screens, journal-box liners, bushings, dipper teeth, perforated screens, crushing equipment, mill liners, sheave and crane wheels, woven wire screens, pins, stokers, elevator buckets, earthmoving equipment, and welding rods. See also Hadfield manganese steels.

high-melting solder. A solder that melts above the melting range of a *low-melting solder*, but below that of a *brazing alloy*, i.e., above about 230°C (445°F), but below 450°C (840°F). Examples of such alloys include lead-tin, cadmium-zinc, silver-cadmium, tin-antimony, tin-silver, and gold-base and most lead-silver solders.

high-modulus aramid yarn. Aramid yarn with an initial modulus of 400 gf/denier (35 N/tex) or above. See also aramid.

high-modulus furnace black. A fine grade of *furnace black* composed of nanometer-sized particles, and produced by the partial combustion of oil in a closed system under controlled conditions. Used for tire carcasses.

high-modulus graphite fibers. Graphite fibers having a tensile modulus in axial direction that is much higher than that of ordinary grades, while the modulus in transverse direction is usually lower. The modulus for polyacrylonitrile precursor fibers is usually above 320 GPa (45×10^3 ksi), and that of mesophase pitch precursor fibers above 380 GPa (55×10^3 ksi).

high-modulus composite. An engineering composite with a high modulus of elasticity (Young's modulus).

high-molecular-weight high-density polyethylenes. See high-density polyethylenes.

high-molecular-weight polyethylenes. See high-density polyethylenes.

high-nickel cast irons. A group of nonmagnetic gray and ductile irons with austenitic microstructures containing 1.8-3.6% total carbon, 0.5-4.5% manganese, 1-6% silicon, 1.8-11% chromium, 1% molybdenum, up to 10% copper, 0.1-0.2% phosphorus, 0.1-0.2% sulfur, 5-43% nickel, and the balance iron. They have excellent corrosion and heat resistance, good toughness and thermal-shock resistance, and good abrasion resistance. Often sold under trademarks and trade names, such as *Ni-Resist* or *Pyrocast*, they are used for pumps, compressors, valves, gas turbines, turbocharger parts, exhaust manifolds, and equipment used for handling hot corrosives. See also nickel-chromium cast irons; austenitic cast irons.

high-nickel ductile irons. A group of heat-resistant austenitic ductile cast iron containing up to 3.0% total carbon, 18-36% nickel, 0.7-2.4% manganese, 1.75-5.5% silicon, 1.75-3.5% chromium, up to 1.0% molybdenum, 0.12% sulfur and 0-0.8% phosphorus, respectively, and the balance iron. They have austenitic matrix structures in the as-cast condition, and excellent corrosion resistance, good resistance to high-temperature scaling and growth, high toughness and thermal shock resistance, and low coefficients of expansion. Used for chemical-plant equipment, food-processing equipment, pumps, filter presses, valves, pipes, fittings, automotive engine pistons and sleeves, bearings, hydraulic turbines, turbocharger casings, manifolds, impellers, propellers, nozzles, oil burners, gages, glass and ingot molds, paper rolls, switchgear, furnace parts, stove tops, and cookware.

high-nitrogen ferrochromium. A *ferroalloy* composed of up to 0.1% carbon, 67-71% chromium, 0.3-1% silicon, approximately 0.75% nitrogen, and the balance iron. It is graded according to the nitrogen content, and used in the manufacture of high-chromium steels, for introducing chromium into steels and cast iron, and for introducing nitrogen into steels to refine the grain and enhance strength. Also known as *high-nitrogen ferrochrome.*

high-nitrogen stainless steels. A group of nonstandard *austenitic stainless steel* (e.g., 18Cr-9Ni or 19Cr-9Ni types) containing approximately 0.1-1.0% nitrogen as an austenite stabilizer for increased tensile and yield strength, and enhanced corrosion, creep and fatigue resistance. Used for shipbuilding, pressure vessels, power generator components, and chemical and cryogenic processing equipment.

high-performance alloys. A group of alloys including the iron-, nickel-, and cobalt-base superalloys, certain high-speed and hot-work tool steels and zirconium-based alloys used for equipment and components operating in severe environments and requiring high-temperature resistance, high hot hardness, high tensile, yield and creep strengths, and/or good wear, oxidation and corrosion resistance. They are used in particular for jet-engine parts, airfoil segments, chemical processing equipment, high-speed and hot-work tools, e.g., broaches, hobs, milling cutters, end mills, taps, punches and forming tools. See also superalloys.

high-performance alumina fibers. Engineering fibers composed of 85% alumina (Al_2O_3) and 15% silica (SiO_2). They have excellent mechanical properties and thermal stability, and an upper service temperature of 1250°C (2280°F). Used as reinforcing fiber for plastics, ceramics and light metals.

high-performance composites. Engineering composites consisting of high-strength, high-stiffness fibers embedded in metal, ceramic, plastic, rubber or carbon matrices. Their performance is equivalent or superior to that of traditional structural materials, such as steel and aluminum alloys, and they usually have high directionality of properties. Used for aircraft and spacecraft structures and systems, propulsion systems and automotive components. Abbreviation: HPC.

high-performance concrete. A concrete-based ceramic composite with greatly increased strength and compactness used for special construction applications. Abbreviation: HPC.

high-performance engineering thermoplastics. A group of *engineering thermoplastics* exhibiting outstanding mechanical properties at ordinary and elevated temperatures and usually good thermal and electrical properties, enhanced service lifes and good corrosion, moisture and solvent resistance. Examples include acrylonitrile-butadiene-styrene resins, acetals, acrylics, fluoropolymers, high-impact polystyrenes, polyamides, polyamide-imides, polybuty-lene terephthalates, polycarbonates, polyethylene terephthalates, polysulfones, polyurethanes, polyvinyl chlorides, styrene-acrylonitriles and styrene maleic-anhydrides.

high-performance polymers. See engineering polymers.

High Performance Tapes. Trademark of Norton Performance Plastics (USA) for self-adhesive fluoropolymer films and foils.

high-performance thermoplastic resins. A group of engineering resins with relatively good high-temperature capabilities, good moisture and solvent resistance and good mechanical and toughness performance, used as matrix materials for continuous fiber-reinforced composites.

high-permeability magnetic alloys. A group of soft magnetic materials based on iron-nickel alloys. The highest permeability is found in alloys containing about 79-80% nickel, 15-16% iron, 4-5% molybdenum and, sometimes, small additions of manganese. Used in magnetometer bobbin cores, low-noise audio-frequency transformers, magnetic shielding devices, tape recorder head laminations, etc.

high-phosphorus copper. See deoxidized high-residual-phosphorus copper.

high polymer. A solid polymeric material having a molecular weight greater than about 5000-6000. Natural high polymers include natural rubber and cellulose, and plastics and synthetic fibers are examples of synthetic high polymers. Abbreviation: HP.

high-pressure laminates. Laminates made by impregnating wood, paper or fabrics with synthetic resin and molding and curing with the application of heat at pressures typically in the range of 8.3-13.8 MPa (1.2-2.0 ksi). Also known as *high-pressure plastic laminates.* See also plastic laminates.

High Q Bond. Trade name of B.J.M. Laboratories Inc. (USA) for a multi-purpose dental bonding agent.

high quartz. A high-temperature form of *quartz* (SiO_2) that is stable at temperatures between 870°C and 1470°C (1600°F and 2680°F). It has a different crystal structure (monoclinic) and higher symmetry than ordinary *low quartz*. See also beta quartz; tridymite.

high-residual-phosphorus copper. See deoxidized high-residual-phosphorus copper.

high-silica fibers. Engineering fibers containing about 95-99% silica (SiO_2) and made from a fiberglass, usually S-glass or C-glass, by leaching and subsequent heat treatment. They have a low density (1.74 g/cm^3 or 0.063 $lb/in.^3$), moderate strength, superior temperature resistance and good electrical signal transparency. Used for ablatives, antenna windows, radomes and thermal barriers.

high-silica refractories. A group of *acid refractories* composed of substantial amounts of silica (SiO_2). They are classified into (i) *siliceous refractories* containing 78-92% SiO_2; and (ii) *silica refractories* containing more than 92% SiO_2.

high-silica molding sand. Pure *quartz sand* that is available in coarse, medium and fine grades, and used for foundry molds and cores.

high-silicon bronze. A bronze that contains 97.0% copper and 3.0% silicon, and is commercially available in the form of flats, rods, wires and tubes. It has good to excellent corrosion resistance and excellent hot and cold workability. Used for propeller shafts, hydraulic lines, tubing, pressure lines, marine and general hardware, pole-line hardware, fasteners and electrical components.

high-silicon cast irons. See high-silicon irons.

high-silicon ductile irons. A group of ductile iron alloys containing 2.8-4.0% silicon, up to 1.0% molybdenum or 0.9% vanadium, and the balance iron. The iron-silicon and iron-silicon-molybdenum grades are subcritically annealed for 4 hours at 790°C (1455°F) and then cooled in air while the iron-silicon-vanadium grades receive an additional heat treating cycle consisting of soaking for 3 hours at 900°C (1650°F), cooling to 700°C (1290°F), soaking for 5 hours, and furnace cooling to below 425°C (800°F). Also known as *high-silicon ductile cast irons.*

high-silicon irons. A group of *alloy cast irons* containing about 0.4-1.1% total carbon, 14-17% silicon, 1.5% manganese, 0.15% phosphorus, 0.15% sulfur, 5.0% chromium, 1% molybdenum, 0.5% copper, and the balance iron. They have ferritic microstructures, and good corrosion resistance, good resistance to abrasion and erosion, poor shock resistance, and very poor machinability. Often sold under trademarks and trade names, such as *Duriron, Durichlor* or *Superchlor,* they are used for chemical equipment, and other equipment for handling corrosive media. Also known as *high-silicon cast irons.*

high-silicon malleable irons. A group of special malleable cast irons containing approximately 2.00-2.50% carbon, 1.40-2.00% silicon, and less than 0.10% sulfur, 0.50% manganese, 0.12% phosphorus and 0.10% sulfur, respectively, and the balance iron. The extra-high silicon content reduces the annealing time required for complete malleabilization. Also known as *high-silicon malleable cast irons.* See also special malleable irons.

high-solids paint. A paint containing 50-vol% or more solids.

High Speed Brass. Trade name of Buckeye Brass & Manufacturing Company (USA) for a heavy-duty brass composed of 88% copper, 7% tin and 5% zinc, and used for high-speed bearings.

high-speed brass. See free-cutting brass.

high-speed cement. See aluminate cement.

high-speed steels. A group of alloy tools steels that retain their high abrasion resistance and hardness at high temperatures (up to about 700°C or 1290°F). They also have high tool lifes and allow for deeper cuts at higher machine speeds than regular tool steels. *High-speed steels* usually contain considerable amounts of chromium, tungsten, molybdenum and vanadium, and may or may not contain cobalt. The American Iron and Steel Institute (AISI) specifies two classes: (i) *Molybdenum high-speed steels* (AISI type M); and (ii) *Tungsten high-speed steels* (AISI type T). Used for high-speed cutting tools, tool bits, lathe and planer tools, drills, taps, chasers, reamers, broaches, hobs, milling cutters, lathe and form tools, lathe centers, dies and punches. Abbreviation: HSS. Also known as *high-speed tool steels.*

high-stiffness composite. A composite with a high ratio of modulus of elasticity to density.

High Strength. Trade name for a high-strength alloy containing 18.5-21.5% zinc, 2-3% iron, 1.5-2.25% nickel, 6-7% aluminum, and the balance copper. Used for strong, corrosion-resistant parts.

high-strength alloy. An engineering alloy whose outstanding characteristics are its high tensile and yield strength. Also known as *high-tensile alloy.*

high-strength aluminum bronze. An *aluminum bronze* containing 73.3% copper, 22.7% zinc, 3.4% aluminum and 0.6% nickel. It has high tensile strength, good corrosion resistance and excellent hot and cold formability. Used for relays, switches, springs, wiring devices, drawn parts and high-strength shells.

high-strength brass. Wrought or cast copper alloys containing about 60% copper and 30-40% zinc, and varying amounts of iron, manganese and/or aluminum. They have high strength, good hardness, and are used for marine forgings and castings, hydraulic cylinders, gears, sun mounts, bushings and bearings, valve stems, cams, screwdown nuts, etc.

high-strength cast irons. See high-duty cast irons.

high-strength commercial bronze. A *commercial bronze* containing 7.65% zinc, 1.75% lead, 1% nickel, 0.1% phosphorus, and the balance copper. It has good corrosion resistance, high strength and good electrical conductivity (about one third that of copper). Used for screw-machine products, pole-line hardware and cable clamps.

high-strength cast irons. See high-duty cast irons.

high-strength low-alloy steels. A group of alloy steels containing less than about 10 wt% alloying elements and having a typical composition of 0.08-0.3% carbon, 0.4-1.6% manganese, up to 0.9% silicon, 1.8% chromium, 5.25% nickel, 0.7% molybdenum, 0.035% phosphorus and 0.3% sulfur, respectively, traces of vanadium, boron, aluminum, titanium, zirconium and niobium (columbium), and the balance iron. They have very small grain sizes and contain finely distributed precipitates in ferrite matrices. Important properties include high tensile and yield strengths, high ductility, good weldability, good toughness and abrasion resistance, good low-temperature impact toughness, moderate cold workability, improved atmospheric corrosion resistance over most plain-carbon steel, and good to fair resistance to saltwater and fresh water corrosion. Used for structural members in bolted, riveted or welded buildings, bridges and industrial equipment, and for truck frames, rail cars, excavation equipment and crane booms. Abbreviation: HSLA steels.

High-Strength Pure Gold. Trade name of Mitsubishi Materials Corporation (Japan) for a readily weldable and machinable gold alloy (99.9% pure) that contains small amounts of toughness- and hardness-enhancing alloying elements, and is supplied in rod, sheet and wire form.

high-strength reinforcement. A steel reinforcement that has a minimum tensile strength of 525 MPa (76 ksi) and is used in the manufacture of reinforced concrete.

high-strength steels. A group of low-alloy *structural steels* with minimum yield strengths between 275 and 758 MPa (40 and 110 ksi). There are four general types: (i) as-rolled pearlitic structural steels; (ii) microalloyed high-strength low-alloy steels; (iii) high-strength structural steels; and (iv) heat-treated structural low-alloy steels. Also known as *high-tensile steels.*

high-strength structural steels. A group of quenched-and-tempered and normalized *structural steels* with enhanced mechanical properties containing between 0.08 and 0.20% carbon, and moderate amounts of other alloying elements, such as manganese, phosphorus, silicon, chromium and copper. They have high tensile and yield strengths, good resistance to atmospheric corrosion and good toughness and impact strength. Used for structural members, mainly in riveted or bolted constructions.

high tech ceramics. See advanced ceramics.

high tech materials. See advanced materials.

hhigh tech polymers. See advanced polymers.

igh-temperature alloys. A group of alloys that have useful properties at temperatures exceeding 500°C (930°F). The term usually refers to "heat-resistant alloys" and "superalloys." Abbreviation: HTAs. See also heat-resistant alloys; superalloys.

high-temperature bonding mortar. Refractory mixture to which nonrefractory materials have been added to improve the plasticity, reduce the bond temperature, and provide air-setting properties.

high-temperature bronze. A strong bronze containing 6.3% zinc, 2.7% tin, and the balance copper. It has good high-temperature properties, and is used for fittings and hardware.

high-temperature cement. Refractory cement that has a low tendency to softening, fusing and spalling at elevated temperatures.

high-temperature ceramics. See refractory ceramics.

high-temperature cobalt alloys. See cobalt-base superalloys.

high-temperature epoxy resins. Epoxy resins that have an upper continuous-use temperature between 205 and 230°C (400 to 450°F) and are used for prepregs.

high-temperature fibers. A group of engineering fibers including those made of ceramics, graphite, carbon and metals having upper use temperatures usually exceeding 400°C (750°F).

high-temperature glaze. See hard glaze.

high-temperature high-strength alloys. See superalloys.

high-temperature materials. A general term referring to materials, such as refractory metals and alloys, superalloys and engineering ceramics and composites that can be used in high-temperature environments, such as furnaces, kilns, smelters, roasters, nuclear equipment, jet engines and aerospace structures.

high-temperature metals. A class of metals and alloys that tend to maintain their high strength when operating for extended periods at elevated temperatures. Included are cobalt-, nickel- and nickel-iron-base superalloys, and refractory metals, such as niobium, tantalum and tungsten and their alloys.

high-temperature nickel alloys. See nickel-base superalloys.

high-temperature resins. A group of synthetic resins including polyaramids, polyphenylene sulfides, polyamide-imides, silicones and bismaleimides that exhibit high glass-transition and melting temperatures, and have continuous-use temperatures above 260°C (500°F). See also low-temperature resins; medium-temperature resins.

high-temperature reusable surface insulation. A low-density, low-expansion thermal insulation material composed of silica tiles with a high-emittance coating of *reaction-cured glass* (i.e., borosilicate glass with the addition of silicon tetraboride). It has a maximum use-temperature of approximately 1260°C (2300°F), and is used for the leading and trailing edges of the tail, sidewalls, lower surfaces and cockpit window area of the NASA Space Shuttle. Abbreviation: HRSI.

high-temperature structural silicides. A class of refractory or transition metal-based silicide compounds, such as molybdenum disilicide ($MoSi_2$), tunsten disilicide (WSi_2), titanium disilicide ($TiSi_2$) and chromium disilicide ($CrSi_2$), suitable for high-temperature structural applications under oxidizing conditions, typically in the range of 1200-1600°C (2190-2910°F). Examples of current and potential applications include oxidation resistant coatings for refractory metals, reinforcements for structural ceramic composites and matrix materials for struc-

tural silicide compounds. See also silicides.

high-temperature solders. Brazing filler metals, such as cobalt-base and nickel-base alloys, copper-zinc and copper-gold alloys, having operating temperatures above 1000°C (1830°F). See also brazing alloys.

high-temperature steels. See heat-resistant steels.

high-temperature superconductors. A class of ceramic materials including lanthanum strontium copper oxide [$(La,Sr)_2CuO_4$], yttrium barium copper oxide ($YBa_2Cu_3O_7$) and thallium barium calcium copper oxide ($Tl_2Ba_2Ca_2Cu_3O_{10}$) that are usually available as high-purity powders made into shapes by sintering, or as thin films. They exhibit superconductivity at relatively "high" temperatures (typically above 30K). The superconductivity critical temperatures for $(La,Sr)_2CuO_4$, $YBa_2Cu_3O_7$ and $Tl_2Ba_2Ca_2Cu_3O_{10}$ are 35K, 95K and 125K, respectively. Abbreviations: HTSC; HTSc. Also known as *ceramic superconductors.*

high-temperature textiles. A group of fireproof textiles that have high heat resistance and are often based on continuous-filament amorphous silica (SiO_2) fibers. Commercially available in the form of woven and knitted fabrics, coated fabrics, yarns, tapes, rope packings, sleeving, mats and paper, they have an effective service temperature of 1095°C (2000°F) and a melt temperature above 1650°C (3000°F). Used as substitutes for asbestos and other hazardous materials in high-heat environments.

high-temperature thermoplastics. A group of thermoplastics that have good resistance to elevated temperatures, usually above 200°C (390°F), and are used as matrix resins for engineering composites. Examples of such thermoplastics include polyamide-imides, polyarylene sulfides, polybutylene terephthalates, polyetheretherketones, polyetherimides, polyether sulfones, polyethylene terephthalates, polyimides and polyphenylene sulfides.

high-tenacity fiber. A synthetic fiber, such as rayon, that has an exceptionally high tensile (breaking) strength and is used especially for industrial fabrics, and as a reinforcement. Abbreviation: HTF. See also low-tenacity fiber.

high-tenacity rayon. Rayon fibers stretched to orient the molecular chains axially with respect to the filament axis. This increases the tensile (breaking) strength by 40-100%. The fineness of high-tenacity rayon for tire cord is 3-6 g/denier. Used for tire cords, and as fabric reinforcements in polymer-matrix composites. See also rayon. *Note:* The denier is a unit expressing the weight of a filament or yarn per 9000 m.

high-tenacity yarns. A class of industrial yarns with high tensile (breaking) strength usually exceeding 1100 denier. Used in tires and other rubber products, as reinforcement in automotive and appliance belts, as threads for shoes, strapping and webbing, and for ballistics fabrics employed in the manufacture of bulletproof clothing. See also low-tenacity yarns. *Note:* The denier is a unit expressing the weight of a filament or yarn per 9000 m.

High Tensile Alloy. Trade name of National Steel Corporation (USA) for a water-hardening carbon steel containing 0.4-0.5% carbon, and the balance iron. Use for gears and shafts.

high-tensile alloys. See high-strength alloys.

high-tensile brass. (1) A group of corrosion-resistant high-strength alloys of copper, zinc and tin, usually with small amounts of iron, manganese, nickel and/or aluminum. Used for castings and machine tools.

(2) A group of strong, corrosion-resistant aluminum brasses containing varying amounts of copper, zinc, and aluminum,

along with a trace of arsenic.

high-tensile bronze. A corrosion-resistant wrought bronze containing 68.5% copper, 6.5% aluminum, 2.5% iron and 2.2% manganese. It has high tensile strength, and is used for nuts, bolts and propellers.

high-tensile cast irons. See high-duty cast irons.

high-tensile reinforcement. A concrete reinforcing bar with a minimum yield strength above 414 MPa (60 ksi).

high-tensile steels. See high-strength steels.

high-tensile wire. High-quality, round, cold-drawn music steel spring wire supplied in diameters ranging from 0.102 to 6.350 mm (0.004 to 0.250 in.) with minimum tensile strengths between 3027 MPa (439 ksi) for the 0.102 mm (0.004 in.) wire and 1586 MPa (230 ksi) for the 6.350 mm (0.250 in.) wire. It is made from high-carbon steel with a typical composition of 0.70-1.00% carbon, 0.20-0.60% manganese, 0.12-0.30% silicon, up to 0.03% sulfur, up to 0.025% phosphorus, and the balance iron. In addition to its high tensile strength, it also possesses good fatigue properties.

high-test cast irons. See high-duty cast irons.

high-tin commercial bronze. A *commercial bronze* containing 88.5% copper, 9.5% zinc and 2% tin. Used for springs, clips, terminals, switches and weatherstrips.

high-transmission glass. See transparent glass.

high-twist yarn. A yarn produced with a comparatively high number of turns per inch (25 mm). Crepe and crepe-like fabrics are made with this yarn. Also known as *hard-spun yarn*.

high-velocity oxyfuel coatings. See HVOF coatings.

Highway. (1) Trade name of Apollo Steel Company (USA) for copper-bearing steel used in building construction.

(2) Trade name of Asahi Glass Company Limited (Japan) for a patterned glass.

highway materials. See road materials.

high-wet-modulus modal fibers. A subclass of *modal fibers* that like cotton and other cellulosic fibers have a high wet modulus (HWM). They are characterized by increased wet-to-dry breaking tenacity ratios, high wet extension resistance, and enhanced resistance to swelling in caustics. HWM modal fibers can be further classified into the following categories: (i) Standard HWM fibers; (ii) High-Elongation HWM fibers with high dry and wet elongation; and (iii) High-Strength HWM fibers with high dry and wet tenacity. Also known as *American cotton; HWM modal fibers*.

HiGlass. Trademark of Montech USA Inc. for polypropylenes and polyethylenes reinforced with 10-50% glass and/or other fibers. They have excellent impact resistance and good high-and/or low-temperature properties and are also available in flame-retardant grades. Used for automotive components, electronic connectors, household appliance parts, sports equipment, and pump parts.

Hi-Gloss. Trade name of Jessop Steel Company (USA) for a series of austenitic stainless steels containing about 0.1% carbon, 12-18% chromium, 8-12% nickel, 0.35-1.2% manganese, and the balance iron. Free-machining grades with up to 0.25% selenium are also available. *Hi-Gloss* steels have excellent corrosion resistance, and are used for piston rods, valves, pumps, radiator parts and cooking utensils. The free-machining grades are used for stainless nuts, bolts, screws, bushings and spindles.

Hi-Hard. Trade name of A. Finkl & Sons Company (USA) for a hard mold steel (AISI type P20) for plastic molding dies.

Hi-Hide. Trademark of Porter Paints (USA) for a line of interior wall and trim paints.

Hilan. Trademark of Hille & Müller KG (Germany) for high-luster, electrolytically nickel-plated cold-rolled strip steel with good deep-drawing properties and excellent corrosion resistance.

Hilanic. Trade name of Gilby-Fodor SA (France) for a heat-resistant alloy containing 79% nickel, 18% cobalt, 2% iron and 1% silicon. Used for cathodes and filaments in electronic tubes.

hilairite. A mineral composed of sodium zirconium silicate trihydrate, $Na_2ZrSi_3O_9 \cdot 3H_2O$. Crystal system, rhombohedral (hexagonal). Density, 2.72 g/cm³; refractive index, 1.609. Occurrence: Canada (Quebec).

Hi-Led-Aloy. Trade name of Nail City Bronze, Inc. (USA) for high-leaded bronze castings.

Hilex. Trade name of Hoechst Celanese Corporation (USA) for polyethylene plastics.

hilgardite. (1) A colorless mineral composed of calcium chloride borate hydroxide, $Ca_2B_5O_8(OH)_2Cl_2$. Crystal system, triclinic. Density, 2.71 g/cm³; refractive index, 1.636. Occurrence: USA (Louisiana).

(2) A colorless mineral composed of calcium strontium chloride borate hydroxide, $(Ca,Sr)_2B_5O_8Cl(OH)_2$. Crystal system, triclinic. Density, 2.99 g/cm³; refractive index, 1.639. Occurrence: Germany, UK.

HiLight. Trademark of General Electric Company (USA) for a ceramic material used in the manufacture of X-ray detectors for computed-tomography medical imaging equipment.

Hi-Light. Trade name of Sinnett-Elpaco Coatings Corporation (USA) for several baking enamels used on light fixtures.

Hi-Lite. (1) Trade name of Australian Window Glass Proprietary Limited (Australia) for a figured glass.

(2) Trade name of US Gypsum Company (USA) for acoustical plaster.

Hilite. Trademark of Shingkong Synthetic Fibers Corporation (Taiwan) for polyester staple fibers and filament yarns used for textile fabrics.

HiLiteCore. Trade name of Nippon Steel (Japan) for nonoriented electrical steel sheet products used for motor components. A light-gage grade, *HiLiteCore HTH*, with improved workability and higher core loss reduction is also available.

hillebrandite. A white mineral composed of calcium silicate hydroxide, $Ca_2SiO_3(OH)_2$. It can also be produced hydrothermically as an intermediate phase (between about 300 and 600°C or 570 and 1110°F) in low-temperature cements. Crystal system, monoclinic. Density, 2.66 g/cm³. Occurrence: Mexico.

hill pine. See longleaf pine.

Hills-McCanna. Trade name of Hills-McCanna Company (USA) for an extensive series of wrought and cast aluminum-, copper- and nickel-base alloys including several aluminum-copper and nickel-copper alloys as well as a wide range of copper-silicon alloys, aluminum, manganese and silicon bronzes, and unleaded, leaded and high-leaded tin bronzes.

Hilo. Trade name of Wilbur B. Driver Company (USA) for a magnetic, heat-resistant alloy composed of 75% nickel, 18% cobalt, 5% iron and 2% titanium, formerly used in radio and electronic-tube filaments, and cathodes.

Hilo-Al. Trade name of National Refractories & Minerals Corporation (USA) for high-alumina refractory mortar.

Hilox. Trademark Morgan Advanced Ceramics (USA) for alumina ceramics with high mechanical strength and temperature resistance, high corrosion and wear resistance, excellent resistance to water, fuels, oils and ultraviolet light, and excellent high-temperature insulating properties. Used for automotive components, e.g., connectors, shields, heating systems, engine management systems, and pump components, such as shafts, bearings, thrust rings, plungers and pistons.

Hiloy. Trademark of ComAlloy International Corporation (USA) for an extensive series of high-strength thermoplastic resins including glass fiber- or hybrid-reinforced polypropylenes, glass fiber- and/or mineral-reinforced polyethylene terephthalates, unreinforced and glass- and/or mineral reinforced nylon 6, nylon 6,6 and nylon 6,12 resins, and various grades of reinforced acrylonitrile-butadiene-styrene. Used for automotive, electrical and electronic applications, machine and appliance parts, sporting goods, housings and pump components.

Hilumin. Trademark of Hille & Müller KG (Germany) for highly corrosion-resistant electro-nickel-plated strip steel.

Hi-Mag. Trade name of National Refractories & Minerals Corporation (USA) for high-magnesia refractories.

Hi Mag Perm. Trade name of Erie Steel Company (USA) for a vacuum-degassed steel containing 0.03-0.05% carbon, 0.01-0.02% silicon, 0.03-0.07% chromium, 0.006-0.01% aluminum, 0.04-0.07% molybdenum, 0.005-0.009% phosphorus, and the balance iron. It has high magnetic permeability, and is used pole pieces, armatures, magnetic clutches and chucks, magnetic control devices and electrical applications.

Hi-Man. Trade name of Inland Steel Company (USA) for a series of high-strength, low-alloy constructional steels containing 0.2-0.3% carbon, 1.35% manganese, 0.3% silicon, 0.2% copper, and the balance iron. Used for autobodies, and agricultural, mining and railroad equipment.

HiMax. Trade name of Hitemco Corporation (USA) for thermally stable, chemically inert thick-film fluorocarbon coatings with excellent corrosion resistance and surface release properties and good permeation resistance. Used in the chemical processing industry for equipment, such as valve disks, neutralizing tanks and filter housings.

Hi-Milan. Trade name of DuPont/Mitsui Polychemicals (USA) for several ionomer resins.

Hi-Narco. Trade name for rayon fibers and yarns used for textile fabrics.

Hinge. British trade name for an engraving brass containing 62-63.5% copper, 34.5-39% zinc and 1-2% lead.

hingganite. A colorless mineral of the datolite group composed of beryl-lium yttrium silicate hydroxide, $BeYSiO_4(OH)$. It can also be made synthetically. Crystal system, monoclinic. Density, 3.88 g/cm^3; refractive index, 1.738. Occurrence: Russian Federation.

Hi-Nicalon. Trade name of Dow Corning Corporation (USA) for a ceramic fiber composed of homogeneously dispersed ultrafine α-silicon carbide (α-SiC) crystallites in an amorphous silicon-carbon mixture. Supplied in continuous and chopped fiber grades and as woven cloth, it has excellent tensile strength and modulus, very good high-temperature stability, and good chemical and oxidation resistance. Used as a reinforcement in ceramic-, polymer- and metal-matrix composites.

Hi-Nickel Alloy. Trade name of Wallace Murray Corporation (USA) for a corrosion- and heat-resistant high-nickel steel.

Hinomaru. Trade name of Nippon Sheet Glass Company Limited (Japan) for a range of glass products.

Hinristone. Trade name of Hinrich for a dental model stone supplied in white and golden brown colors.

hinsdalite. A colorless, dark -gray, or greenish mineral of the alunite group that is composed of lead strontium aluminum phosphate sulfate hydroxide, $PbAl_3(PO_4)(SO_4)(OH)_6$, and may also

contain strontium. Crystal system, rhombohedral (hexagonal). Density, 3.65 g/cm^3; refractive index, 1.671. Occurrence: USA (Colorado).

hiortdahlite. A light brown mineral composed of sodium calcium rare-earth yttrium zirconium silicate fluoride hydroxide oxide, $(Na,Ca,Ln,Y)_3Zr_{1-x}(Si_2O_7)(F,OH,O)_2$. Crystal system, triclinic. Density, 3.26 g/cm^3; refractive index, 1.643. Occurrence: Canada (Quebec).

HIPed products. See hot-isostatically-pressed products.

Hiperm. Trade name for a commercially pure iron with very low carbon content that has good soft magnetic properties and high magnetic permeability. Used for various magnetic applications.

Hiperco. Trademark of Carpenter Technology Corporation (USA) for a series of ductile, soft-magnetic alloys containing about 27-35% cobalt, 0.25% silicon, up to 1% chromium, 0.3% nickel, 0.3% manganese and 0.03% carbon, respectively, and the balance iron. They have high magnetic saturation, high permeabilities, low hysteresis losses, and are used for magnets in relays, motors, generators and rectifiers, and as core materials. *Hiperco 50* is a special composition of 49% cobalt, 49% iron and 2% vanadium that has high magnetic saturation and DC maximum permeability at very high magnetic flux densities, low DC coercive force, and low AC core losses. It is supplied in strip form, and is used in rotor and stator laminations, motors and generators for aircraft power generation, transformer laminations, and tape toroids.

Hipernik. Trade name of Westinghouse Electric Corporation (USA) for a soft, ductile magnetic alloy containing about 50% iron, 50% nickel, and traces of manganese. It has high and constant magnetic permeability, and is used for laminations, saturated reactors, magnetic amplifiers, and as a magnetic core material for transformer and transductor applications. *Hipernik V* is a grain-oriented version with a square hysteresis loop.

Hipernom. Trade name of Westinghouse Electric Corporation (USA) for a soft-magnetic alloy containing 79% nickel, 16-17% iron, 4-4.5% molybdenum and up to 0.05% carbon. It has high permeability, a Curie temperature of 460°C (860°F), and can be specially processed to exhibit square hysteresis loop properties. Used for sensitive saturable core devices, relays, amplifiers, audio transformers, and magnetic shielding applications.

Hipersil. Trade name of Westinghouse Electric Corporation (USA) for a soft-magnetic alloy containing 96-97% iron and 3-4% silicon. It has a large grain size, low impurities, high permeability, low core loss, and oriented crystals produced by hot- and cold rolling. Used for transformer cores.

Hi-Pflex. Trade name of Specialty Minerals Inc. (USA) for coated ground limestone used as a filler in paper, plastics and elastomers.

Hi-Phy Kirksite. Trade name of NL Industries (USA) for a grain-refined alloy of zinc, copper, magnesium and aluminum used for forming dies. See also Kirksite.

Hi-Plex. Trade name of Hi-Plex Corporation (USA) for several plastic laminates.

Hi-Ply Tissue. Trade name of Lincoln Pulp & Paper Company (USA) for a kraft paper made from hardwood and recycled softwood pulp.

hippuric acid. An acid obtained by adding glycine to benzoic acid. It is available as colorless crystals with a density 1.371 g/cm^3, and a melting point 188-191°C (370-376°F). Used in medicine, organic synthesis, biochemistry and biotechnology. Formula: $C_9H_9NO_3$. Also known as *benzoylglycine; benzoylglyco-*

coll; benzoylaminoacetic acid.

Hi-Proof. Trade name of British Steel Corporation (UK) for a series of austenitic stainless steels containing up to 0.08% carbon, up to 2% manganese, 16-20% chromium, 8-14% nickel, 0-3% molybdenum, 0.2% nitrogen, up to 0.8% niobium, and the balance iron. They have excellent corrosion resistance, and are used for corrosion-resistant parts, food-processing equipment, chemical equipment, hardware, brewing equipment, and welded constructions.

Hi-Protex. Trade name of Protex-A-Cote, Inc. (USA) for neoprene- and *Hypalon*-base protective coatings.

HIPS-PPE alloys. Alloys of about 50-80% high-impact polystyrene (HIPS) and polyphenylene ether (PPE). They have good melt micibility in all proportions, good processibility, low viscosity, low coefficients of thermal expansion, outstanding heat, and chemical resistance and excellent toughness. Used for automotive components, and chemical and industrial equipment.

HIPS-PVC alloys. Alloys of high-impact polystyrene (HIPS) and polyvinyl chloride (PVC) usually processed by extrusion, or injection molding, and having greatly improved flame-retardant properties.

hip tile. A roofing tile of special shape used to form the junction of two sloping sides of a roof, or a sloping side meeting a sloping end. It is the opposite of a valley tile. See also ridge tile; valley tile.

Hi-Q. (1) Trade name for *onyx* frit, a glassy, ceramic material obtained by smelting together selected minerals and metallic oxides at temperatures between 621 and 792°C (1150 and 1427°F). Used in the cast onyx industry.

(2) Trademark of Woodward & Dickerson Wood Products, Inc. (USA) for softwood lumber and finished softwood lumber products, such as beams, joists, posts, studs, plates, headers and sills.

(3) Trade name of Ferro Corporation (USA) for a line of powder coatings.

Hi-Qua-Led. Trade name of Alco Products (USA) for a series of high-quality, free-machining leaded steels including various plain-carbon (AISI-SAE Nos. 10L45 through 10L70), chromium-molybdenum (AISI-SAE Nos. 41L30 through 41L40) and nickel-chromium-molybdenum types (AISI-SAE No. 43L40) used for screw-machine products, fasteners, gears and shafts.

Hiralon. Trademark of Hirata Spinning Company (USA) for polyethylene and polypropylene monofilaments and filament yarns.

Hi-Ram. Trade name of National Refractories & Minerals Corporation (USA) for plastic refractories and ramming mixes.

Hiramic. Trade name of Hitemco Corporation (USA) for hard, abrasion-resistant hybrid coatings with excellent lubricity and nonstick, mold-release properties. They consist of a base coat of ceramic or other inorganic material and an infusent polymer, such as a fluorocarbon resin. Used particularly on food-processing equipment, such as bakeware, candy, cookie and cracker rollers, bakery pans and extruder parts as well as on pharmaceutical equipment, high-temperature platens, and skis.

Hircoe. Trade name for a high-impact acrylic resin for dentures.

H-Iron. Trade name for an iron powder made by hydrogen reduction of high-purity ferric oxide or oxalate, and used in ferrous powder metallurgy.

HiROC. Trade name of Hitemco Corporation (USA) for a series of wear-resistant metallic and ceramic coatings (alloys, metals and metallic or ceramic carbides) applied by a special high-velocity-oxyfuel (HVOF) process in a wide range of thicknesses. They bond well to many substrates, and have high densities,

very high hardness, good corrosion and erosion resistance, very low porosity, and relatively low oxide contents. Used on medical instruments, wear-resistant parts and high-pressure sealing surfaces.

Hirox. Trade name of Westinghouse Electric Corporation (USA) for a series of grain-refined iron alloys containing 6-10% aluminum, 3-9% chromium, up to 4% manganese, and small additions of zirconium and boron. They have high electrical resistivity and high oxidation resistance to 1260°C (2300°F), and are used for electrical resistances and heating elements.

Hi-R-Sol. Trade name of Tenax Finishing Products Company (USA) for a high-solids coating.

hirudin. A protein substance that is obtained from the salivary glands of medicinal leeches (*Hirudo medicinales*) and prevents human blood from clotting. Used in medicine, biochemistry, and as a natural biomaterial.

Hi-Run. Trademark of Kloster Steel Corporation (USA) for an air-hardening high-carbon, high-chromium tool steel (AISI type D2) containing 1.55% carbon, 11.5% chromium, 0.8% molybdenum, 0.9% vanadium, and the balance iron. Used for tools and dies.

Hishi. Trademark of Mitsubishi Chemical Company (Japan) for a range of polyethylene plastics.

Hishi Cross. Trade name of Asahi Glass Company Limited (Japan) for a patterned glass.

Hishikool. Trade name of Asahi Glass Company Limited (Japan) for a heat-absorbing sheet and plate glass.

Hishimoru. Trade name of Asahi Glass Company Limited (Japan) for a patterned glass.

Hishiwire. Trade name of Asahi Glass Company Limited (Japan) for a wired glass with diamond mesh.

Hi-Shock. Trade name of Precision-Kidd Steel Company (USA) for an air-hardening shock-resistant tool steel

Hi-Shrink Tape. Trade name of Dunstone Company, Inc. (USA) for heat-shrinkable polyester tape.

Hi-Si. Trade name for an aluminum casting alloy containing 20% silicon.

Hi-Sil. Trade name of PPG Industries Inc. (USA) for a series of finely divided silica gels supplied in various mesh sizes, and used as reinforcing pigments in plastics, rubber and other elastomers, and as extenders and brighteners in paints and paper.

Hisilon. Trademark of Hilados Sinteticos SA (Uruguay) for nylon 6,6 fibers and yarns used for textile fabrics.

hisingerite. A brownish to black mineral composed of iron silicate hydroxide dihydrate, $Fe_2Si_2O_5(OH)_4 \cdot 2H_2O$. Crystal system, monoclinic. Density, 2.67 g/cm^3; refractive index, 1.66. Occurrence: Rumania.

Hi-So-Mar. Trade name of Marcus Paint Company (USA) for high-solids industrial coatings.

Hi-Sorb. Trademark of W.R. Grace & Company (USA) for acoustical plasters.

HiStar. Trademark of ARBED SA (Luxembourg) for quenched and self-tempered high-strength low-alloy steel beams with high low-temperature toughness and excellent arc weldability. Used for structural applications including bolted, riveted and welded bridges and buildings. See also quenched and self-tempered steel.

Hi-Step. Trade name of Enthone-OMI Inc. (USA) for a semibright nickel electroplate and plating process.

Hi-Stress. Trade name of Armco Steel Corporation (USA) for hot-rolled, deformed medium-carbon steel sections for concrete reinforcement.

Hi-Strength. Trade name of National Refractories & Minerals Corporation (USA) for high-strength castable refractories.

Hi Strength. Trade name of Armco Steel Corporation (USA) for a series of low-carbon alloy and plain-carbon steels. *Hi Strength A* is a low-carbon alloy steel containing 0.12% carbon, 0.6% chromium, 0.7% nickel, 0.1% molybdenum, 0.4% copper, and the balance iron. It has good weldability, and is used for various engineering and constructional applications. *Hi Strength B* is a plain-carbon steel containing 0.22% carbon, 1.2% manganese, 0.2% vanadium, 0.2% copper, and the balance iron. It is used for various engineering and constructional applications requiring increased hardenability and fatigue strength. *Hi Strength C* is a plain-carbon steel containing 0.24% carbon, 1.3% manganese, 0.05% niobium, and the balance iron. Available in various grades, it has good weldability, and is used for structural applications, sections for building trusses, and for various other structural engineering applications.

Hi-T. Trade name of 3M Dental (USA) for orthodontic wire. *Hi-T Multibond* is a high-gold dental bonding alloy.

Hi-Talc. Trade name of Bakertalc Inc. (USA) for talc fillers.

Hi-Tek. Trade name of Graphic Industries Inc. (USA) for flexible plastic laminate sheets used as engraving stock.

Hi-Tem. Trade name of Bethlehem Foundry & Machine Company (USA) for a series of corrosion- and heat-resistant cast irons containing 3.2% carbon, 2% silicon, 0.5-1% manganese, and the balance iron. Used for chemical-plant equipment, processing vessels, retorts, and dye and paint equipment.

Hi-Temp. (1) Trade name of Rex Industrial Paint Works (USA) for heat-resistant paints.

(2) Trade name of Turner & Newall (USA) for high-temperature block and pipe insulation.

Hi-Tensile. Trade name of R. Lavin & Sons, Inc. (USA) for an aluminum alloy containing 7-8% zinc, 0.2-0.5% magnesium, 0.4-1% copper and 0.2% titanium. It has good room-temperature aging properties, and is used for aircraft and machine-tool castings.

Hitensiloy. Trade name of Cerro Wire & Cable Company (USA) for a series of corrosion-resistant bronzes containing about 57% copper, 1.5-1.75% nickel, 1% lead, up to 0.5% aluminum, and the balance zinc. Used for welding rod, pump rods, valve stems, bolts and nuts.

Hitenso. Trade name of Anaconda Company (USA) for a series of cadmium bronzes containing 0.5-1.2% cadmium, sometimes traces of tin and silicon, and the balance copper. They have high electrical conductivity (about 85% IACS), high strength, and good wear resistance. Used for electrical equipment, contact shoes, trolley wire, and welding tips.

Hiten-Speed. Trade name of British Rolling Mills Limited (UK) for a series of tough, oil-hardening steels containing 0.3-0.45% carbon, 1.3-2% manganese, up to 0.4% molybdenum, and the balance iron. Used for gears, bolts, crankshafts and axles.

Hi-Test. Trade name of Libbey-Owens-Ford Company (USA) for laminated plate and sheet glass products.

Hitest. (1) Trade name of Medart Engineering & Equipment Company (USA) for a cast iron containing 3% carbon, varying amounts of silicon and manganese, and the balance iron. Used for gears, shafts, housings and machinery parts.

(2) Trade name of Guardian Industries Corporation (USA) for a laminated glass.

Hi-Therm. (1) Trade name of John C. Dolph Company (USA) for insulating varnishes.

(2) Trademark of Carborundum Corporation (USA) for aluminum nitride (AlN) ceramics supplied as substrates for use in the hybrid circuit industry, and in various special shapes and components for thermal management applications.

Hi-Thoria. Trade name for thoriated tungsten, a tungsten alloy containing up to 2% thoria (ThO_2), used for stable-arc welding electrodes.

Hi-T-Lube. Trademark of General Magnaplate Corporation (USA) for a dry-film lubricant for extreme high and low temperatures and extreme compressive loads. It consists of four interlinked layers of metals and oxides, and when applied to a metal or alloy, the resulting coating becomes an integral part of the metallic surface imparting very low friction at temperatures between -220 and +540°C (-360 and +1000°F). Used as a synergistic coating for application to wear surfaces of metals in multilayer systems.

Hi-Top. Trade name of British Steel plc (UK) for electrolytic chromium- and chromium oxide-coated steel sheets.

Hi-Touch. Trade name of Apex Medical Technologies, Inc. (USA) for several synthetic elastomeric compounds.

Hi-T Steel. Trademark of 3M Unitek (USA) for a corrosion-resistant stainless steel (18% chromium, 8% nickel) used for orthodontic wire.

HiVal. Trademark of Ashland Inc. (USA) for thermoplastic resins including various grades of high-density polyethylene.

Hivalloy. Trade name of Montell North America (USA) for polyolefin resins.

Hi-Van. Trade name of Peninsular Steel Company (USA) for an oil-hardening, chromium-type hot-work steel used for punches, dies, upsetters and shear blades.

Hivol. Trademark for lightweight, high-strength composite materials consisting of alpha silicon carbide (α-SiC) particles embedded in aluminum-alloy matrices. Manufactured in North America under license by AMETEK Specialty Metal Products, they feature high thermal conductivity and controlled thermal expansion. They are supplied in two grades: *HIVOL B* and *HIVOL C*, either unplated or plated with electroless nickel, electrolytic nickel or copper, hard gold, silver or tin. Used in computers, telecommunications equipment, power supplies and electronic packages.

HI-WAX. Trademark of Mitsui Chemicals, Inc. (Japan) for a low-molecular-weight polyethylene wax used as a wax and coating modifier and as a filler.

Hi-White. Trademark of J.M. Huber Corporation (USA) for a series of hydrous aluminum silicates (sedimentary kaolin clays) with a density of 2.60 g/cm³ (0.094 lb/in.³) used for ceramic applications, and as roofing granules.

Hi-Wear. Trade name of Carpenter Technology Corporation (USA) for a wear-resistant tool steel composed of 1.5% carbon, 1% chromium, 1% molybdenum, 4% tungsten, and the balance iron.

Hi-Yaw-Ten. Trade name of Yawata Iron & Steel Company, Limited (Japan) for a steel containing up to 0.12% carbon, 0.25-0.75% silicon, 0.2-0.5% manganese, 0.06-0.12% phosphorus, 0.25-0.5% copper, up to 0.065% nickel, 0.4-1% chromium, up to 0.15% titanium, and the balance iron. It has excellent resistance to atmospheric corrosion, and is used for bridges, buildings, rolling stock, buses and mine cars.

Hi-Yield. Trade name of National Intergroup Inc. (USA) for a series of constructional steels containing 0.2-0.3% carbon, 0.9-1.4% manganese, 0.01% or more niobium (columbium), and the balance iron. They have good weldability and fabricability, and are used for trucks, mine cars, derricks, bridges and pressure vessels.

Hizek. Trade name for polyolefin fibers and products.

HI-ZEX. Trademark of Mitsui Chemicals, Inc. (Japan) for standard and UV-stabilized low-density polyethylenes.

HI-ZEX MILLION. Trademark of Mitsui Chemicals, Inc. (Japan) for a biocompatible ultrahigh-molecular-weight polyethylene synthesized by Ziegler polymerization. It has good abrasion, chemical and impact resistance and self-lubricating properties, and is used for sports and leisure goods, machine parts, textile and papermaking machinery, various components and equipment for the agricultural and building industries, medical equipment parts, artificial limbs and prostheses, and for lining chemical and industrial equipment.

Hi Zink Coatings. Trade name of Tresco Paint Company, Inc. (USA) for zinc-rich primers and coatings.

Hizutit. Trade name of Ekatit Stahl GmbH (Germany) for a series of wrought and cast heat-resistant steel products including several chromium-nickel and low-carbon high-chromium stainless grades. Used for heat-resisting parts.

H-Monel. Trade name of Inco Alloys International Limited (UK) for a series of nickel-copper alloys containing 63% nickel, 31-32% copper, 2.7-3.2% silicon, 2-3% iron, up to 0.12% carbon and 0.75-1% manganese. Supplied in the form of sand castings, they have good weldability, and a corrosion resistance, tensile strength and hardness higher than that of ordinary *Monel*. Used for machined parts, and chemical and petroleum refining and processing equipment.

H-Nylon. Trademark of Holeproof Mills Limited (New Zealand) for nylon 6 fibers and filament yarns.

Hob-A-Die. Trade name of Ziv Steel & Wire Company (USA) for a steel containing 0.06% carbon, 1% chromium, 0.3% manganese, 0.25% molybdenum, 0.2% silicon, and the balance iron. It has good resistance to heat checking, and is used for zinc-casting dies.

Hob-A-Form. Trade name for a water-hardening die steel containing 0-0.06% carbon, 0.15% manganese, 0.1% silicon, and the balance iron. Used for plastic molding dies.

Hoballoy. (1) Trade name of Colt Industries (UK) for a shock-resistant mold steel containing up to 0.1% carbon, 1.3% nickel, 0.6% chromium, and the balance iron. It has good machinability and high core strength, and is used for Bakelite molds.

(2) Trade name of Hoben Davis Limited (UK) for a range of dental casting alloys including high, medium and low gold, and cobalt-chromium alloys. *Hoballoy Economy* is a low-gold alloy and *Hoballoy White* a white gold alloy.

Hobart. Trademark of Hobart Welding Products (USA) for flux-cored welding wires and arc-welding electrodes made from various ferrous or nonferrous materials including plain-carbon, high-strength, stainless and tool steels, cast iron, and aluminum.

Hobb Die Steel. Trade name of Faitout Iron & Steel Company (USA) for a carburizing steel containing up to 0.07% carbon, 0.15% manganese, and the balance iron. Used for plastic molding dies.

Hobbing Die Steel. Trade name of Bethlehem Steel Corporation (USA) for a water- or oil-hardening steel containing 0.06% carbon, 0.15% manganese, 0.1% silicon, and the balance iron. Used for plastic molding dies.

hobbing steels. (1) A group of soft steels suitable for the manufacture of plastic molds. They have optimum cold workability, good machinability and overall workability in the annealed condition, high compressive strength, good abrasion and wear resistance in the heat treated condition, low distortion, and good di-

mensional stability and scaling resistance. This group of steels includes in particular low-carbon mold steels, such as AISI types P2 and P4, certain medium-carbon shock-resisting mold steels, such as AISI types S1, S5 and S7, and low-alloy special purpose tool steels, such as AISI type L6. Low-carbon mold grades contain about 0.1% carbon, 0.1-0.6% manganese, 0.1-0.4% silicon, 0.75-5.25% chromium, up to 0.5% nickel, 0.15-1% molybdenum, and the balance iron. They must be carburized after hobbing and before hardening. Medium-carbon mold grades contain about 0.5% carbon, 0.1-0.8% manganese, 0.15-2.25% silicon, 0.2-1.8% molybdenum, 0.15-0.5% vanadium, up to 3.5% chromium, 0.3% nickel and 3% tungsten, respectively, and the balance iron. They must be carefully spherodize-annealed to low hardness. Low-alloy special purpose tool steels are composed of about 0.7% carbon, 0.25-0.8% manganese, 0.6-1.2% chromium, 1.25-2% nickel, up to 0.3% vanadium, up to 0.5% molybdenum and silicon respectively, and the balance iron. They are often sold under trademarks or trade names, such as *Airmold*, *Crusca Cold Hubbing* or *Duramold*. Also known as *hubbing steels*.

(2) A special high-speed steel, often a tungsten-type, used for hobs, i.e., rotary gear cutting tools with helically arranged teeth.

Hobond. Trade name of Hoben Davis Limited (UK) for an extensive range of ceramic dental alloys including several high- and medium-gold, silver-palladium, silver-free palladium, cobalt-chromium, and nickel-chromium alloys for bonding porcelain and ceramics to metals.

Hobrite. Trade name of Swedish American Steel Corporation (USA) for a water-hardening die steel containing up to 0.05% carbon, 0.11% manganese, and the balance iron. Used for hobbing molds and plastic molding dies.

hocartite. A brownish gray mineral of the chalcopyrite group composed of silver iron tin sulfide, Ag_2FeSnS_4. Crystal system, tetragonal. Density, 4.77 g/cm³. Occurrence: Bolivia.

Hoco. Trade name of Hoffmann & Co. KG (Germany) for a series of tool steels including various plain-carbon, high-speed and hot-work types.

hodgkinsonite. A pink, reddish-pink or violet-pink mineral composed of zinc manganese silicate hydroxide, $Zn_2Mn(SiO_4)(OH)_2$. Crystal system, monoclinic. Density, 4.06 g/cm³; refractive index, 1.741. Occurrence: USA (New Jersey).

Hodi. Trade name of Atlas Specialty Steels (Canada) for a tungsten-type hot-work tool steel (AISI type H21) containing 0.28% carbon, 0.3% manganese, 0.3% silicon, 9.5% tungsten, 3.25% chromium, 0.4% vanadium, and the balance iron. It has high hot hardness and hot wear resistance, low cold wear resistance, good deep-hardening properties, high to moderate toughness, and moderate machinability. Used for extrusion molds, tools and dies, die-casting dies for copper alloys, hot trimming, swaging and blanking dies, forging die inserts, and hot punches.

hodrushite. A gray mineral composed of copper bismuth sulfide, $Cu_8Bi_{12}S_{22}$. Crystal system, monoclinic. Density, 6.18 g/cm³. Occurrence: Czech Republic, Slovakia.

Hodur. Trade name of Honsel-Werke AG (Germany) for a heat-treatable aluminum alloy containing 2.5-5% copper, 0.2-1.8% magnesium and 0.3-1.5% manganese. Used for fasteners, fittings and aircraft structures.

hoelite. A yellow mineral composed of 9,10-anthraquinone, $(C_6H_4)(CO)_2C_6H_4$. It can also be made synthetically. Crystal system, monoclinic. Density, 1.42 g/cm³. Occurrence: USA (Arizona).

hoernesite. A colorless mineral of the vivianite composed of magnesium arsenate octahydrate, $Mg_3(AsO_4)_2 \cdot 8H_2O$. It can also be made synthetically. Crystal system, monoclinic. Density, 2.61 g/cm^3; refractive index, 1.571; hardness, 1 Mohs. Occurrence: Czech Republic.

Hoesch. Trade name of Hoesch Stahl AG (Germany) for an extensive series of steel products including semifinished products (e.g., ingots, slabs, billets and sheet bars), galvanized and hot-rolled wide strip, hot-rolled bars (e.g., flats, rounds and squares), structural shapes (e.g., channels, piling, sheet piling, props and support frames for mining applications), beams, reinforcing bars, galvanized and plastic-coated sheet, commercial quality steel sheet, thin-gage and ultra-thin-gage sheet, corrugated sheet, hot-dip and electrolytic tinplate, floor, bulb and checker plate, heavy and medium plate, universal mill plate, steel castings, wire products and pipe and tubular products.

Hofors. Trade name of Hofors Steel Works (Sweden) for a series of high-, low- and medium-carbon steels used for machinery parts, tools, dies, broaches, cutters and drills.

hogbomite. (1) A black mineral composed of iron magnesium titanium aluminum oxide, $(Fe,Mg)_{1.8}(Al,Ti)_{3.7}(O,OH)_8$. Crystal system, rhombohedral (hexagonal). Density, 3.93 g/cm^3; refractive index, 1.848. Occurrence: Sweden, Southern Africa, USA (Virginia).

(2) A yellow mineral composed of titanium iron magnesium aluminum oxide, $(Mg,Fe)_2(Al,Tl)_5O_{10}$. Crystal system, hexagonal. Density, 3.70 g/cm^3; refractive index, 1.805. Occurrence: Central Africa.

hognut hickory. See mockernut hickory.

Hohenzollern brass. A German high-tensile brass composed of 60-70% copper and 30-40% zinc, and used for gears and worm wheels.

hohmannite. A brown to orange mineral composed of iron sulfate hydroxide heptahydrate, $Fe_2(OH)_2(SO_4)_2 \cdot 7H_2O$. Crystal system, triclinic. Density, 2.20 g/cm^3; refractive index, 1.643. Occurrence: Chile.

Hokotol. Trade name of Hoogovens Aluminium Walzprodukte (Germany) for wrought lightweight, high-strength aluminum-zinc-magnesium alloys (AA 7000 Series) supplied in sheet and plate form. They have excellent machinability, poor weldability and corrosion resistance, a density of 2.83 g/cm^3 (0.102 lb/in.^{3}) and a hardness of 180 Brinell. Used for machine parts, bolsters and force plates, and blow and injection molding molds for plastics.

Holdax. Trade name of Uddeholm Corporation (USA) for a vacuum-degassed alloy steel containing 0.4% carbon, 1.6% manganese, 1% chromium, 0.32% silicon, 0.2% molybdenum, 0.08% sulfur, and the balance iron. Used for molds, dies and plates.

holdenite. A clear pink to yellowish red, or deep red mineral composed of manganese zinc arsenate silicate hydroxide, $Mn_6Zn_3(OH)_8(AsO_4)_2(SiO_4)$. Crystal system, orthorhombic. Density, 4.11 g/cm^3; refractive index, 1.770. Occurrence: USA (New Jersey).

Holder. Trade name of Boehler GmbH (Austria) for a free-machining chromium-molybdenum alloyed die steel supplied in the prehardened condition, and used for die-casting dies.

Holder Block. Trade name of Crucible Materials Corporation (USA) for a prehardened steel containing 0.3% carbon, varying amounts of chromium, nickel and molybdenum, and the balance iron. Used for die-casting dies and plastic molds.

Holder Block Steel. Trade name of Slater Steels Corporation (USA) for an alloy steel that exhibits superior machinability

due to its sulfur content. A typical composition is 0.5% carbon, 1.15% manganese, 0.65% chromium, 0.18% molybdenum, 0.08% sulfur, and the balance iron. Used for forging die backers and holders, molds, frames for plastic molds, and brake dies.

Holdertem. Trade name of Heppenstall Company (USA) for a prehardened low-alloy steel containing 0.4% carbon, 0.95% chromium, 0.85% manganese, 0.2% molybdenum, and the balance iron. Used for arbors, shafts, structural parts, and special-purpose tools.

Holfos. Trade name of Holcraft Castings & Forgings Limited (UK) for an extensive series of wrought and cast copper alloys including various leaded and unleaded tin bronzes, and aluminum, phosphor, tin-nickel and silicon bronzes, and several brasses. *Holfos Spuncast* is a spun-cast alloy containing 11.5% tin and 0.7% phosphorus. It has good wear and abrasion resistance, and is used for machine parts, such as gears, shafts, bearings and fasteners.

Holistic. Trade name for a gold-platinum dental casting alloy.

holland cloth. (1) A plain-weave linen, or linen and cotton cloth, usually light brown and sometimes glazed, used for window shades, upholstery, etc. It was originally made in Holland. Also known as *holland; shadecloth*.

(2) A woven fabric, usually starch-filled cotton or linen, with a smooth glossy surface, used as interleaf for raw rubber and sheeted rubber compounds. Also known as *holland*.

hollandite. A silvery-gray mineral of the rutile group that is composed of barium manganate, $BaMn_8O_{16}$, and may also contain iron. The iron-bearing variety has often a grayish black or black color. Crystal system, monoclinic. Density, 4.95 g/cm^3. Occurrence: Sweden, India.

Hollandstone. Trademark of Tarmac America (USA) for interlocking paving stone used for commercial applications.

Hollex. Trademark of Owens-Corning Fiberglass (USA) for hollow high-performance S-2 glass fibers based on magnesium aluminosilicate. They have a density of 1.80 g/cm^3 (0.065 lb/in.^{3}), an ultimate tensile strength of 3.45 GPa (500 ksi), an elastic modulus of 67 GPa (9.72 × 10^3 ksi), good long-term performance, a dielectric constant of 3.6, good transparency, and good processibility with all conventional methods. Used as reinforcing fibers for polymer matrices, e.g., for aircraft radomes and military aircraft components. See also S-2 glass.

hollingworthite. An iron black mineral of the pyrite group composed of rhodium ruthenium platinum arsenic sulfide, $(Rh,Ru,Pt)AsS$. Crystal system, cubic. Density, 7.91 g/cm^3. Occurrence: South Africa.

Hollo. Trade name of Hidalgo Steel Company Inc. (USA) for a water-hardening tool steel used for rock drills.

Hollobar. (1) Trade name of Diehl Steel Company (USA) for an oil-hardening die steel containing 1.05% carbon, 0.4% manganese, 1.5% chromium, 0.3% silicon, and the balance iron. Used for cold-work dies.

(2) Trade name of Peninsular Steel Company (USA) for a water-hardening tool steel containing 1% carbon, 1.2% chromium, 0.3% molybdenum, and the balance iron.

Hollofil. Trademark of E.I. DuPont de Nemours & Company (USA) for a high-performance thermal insulation made from a blend of hollow-core or four-hole polyester fibers. Supplied in two primary grades, *Hollofil II* and *Hollofil 808*, it is used in outerwear and sleeping bags.

Holloseal. Trade name of Hollow Seal Glass Company Limited (UK) for insulating glass units.

hollow blocks. Building blocks of concrete or burnt clay in which the core (hollow) area in any plane parallel to the bearing surface is more than 25% of the total cross-sectional area. They are available in 100, 150, 200, 250 and 300 mm (4, 6, 8, 10 and 12 in.) widths and 100 and 200 mm (4 and 8 in.) heights, usually rectangular in shape and with some cross webs. Used for exterior walls, partition walls, and suspended floors and roofs. Also known as *hollow tile.*

Hollow Blue Band. Trade name of Hollup Corporation (USA) for a stainless steel containing up to 0.1% carbon, 25% chromium, 12% nickel, and the balance iron. Used for coated welding electrodes for stainless steel.

hollow brick. Kiln-fired masonry units composed of clay or shale molded with hollow spaces (cores) in them, usually reinforced by one or two cross webs. Also known as *cavity brick.*

hollow clay blocks. Kiln-fired, structural building blocks made from clay, and having internal air spaces. Used for exterior walls, partition walls, and suspended floors and roofs.

Hollow Drill. (1) Trade name of Agawam Tool Company (USA) for a water-hardening tool steel containing 1-1.3% carbon, 1-1.1% chromium, 0.35% manganese, 0.3% silicon, and the balance iron. Used for rock drills and hollow drills.

(2) Trade name of Ziv Steel & Wire Company (USA) for a water-hardening tool steel containing 1.2% carbon, and the balance iron. Used for hollow drills.

hollow fibers. Soft, light filaments or fibers of polyester, rayon, or other synthetics extruded as hollow, air-filled microtubules. The still air in these microtubules provides the fabrics made from them with added warmth. Used as fillings in outerwear and upholstery, and in thermal underwear.

hollow-filament yarn. A rayon yarn with small quantities of air or inert gases, produced from a viscose spinning solution that contains certain air- or gas bubble-producing chemical additives. It is duller and softer than ordinary *viscose rayon* and has lower heat transmission, greater covering power and higher flexibility. Used especially for underwear, neckties and hosiery. Sometimes referred to as a "macaroni yarn."

hollow masonry unit. A masonry unit, such as a building block or brick, in which the core (hollow) area in any plane parallel to the bearing surface is more than 25% of the total cross-sectional area.

hollow tile. See hollow blocks.

Hollup. Trade name of Hollup Corporation (USA) for a series of brazing rods and bare and coated welding electrodes. The brazing rods usually consist of copper-zinc or copper-phosphorus alloys, and the welding electrodes of low-carbon steels, chromium or nickel alloy steels, stainless or austenitic manganese steels, cast irons, nickel cast irons, or aluminum or nickel-copper alloys.

holly. The hard, heavy wood of any of a genus *(Ilex)* of deciduous and evergreen trees and shrubs especially the American holly *(I. opaca)*. See American holly.

holmia. See holmium oxide.

holmium. A bright silvery metallic element of the lanthanide series (rare-earth group) of the Periodic Table. It is commercially available in the form of lumps, ingots, rods, sheets, foil, wire, filings, chips, sponge, powder and single crystals. The commercial holmium ores are gadolinite and monazite. Crystal system, hexagonal. Crystal structure, hexagonal close-packed. Density, 8.803 g/cm³; melting point, 1474°C (2685°F); boiling point, 2695°C (4883°F); hardness, 60 Vickers; atomic number, 67; atomic weight, 164.930; trivalent. It has good mag-

netic and electrical properties including a room-temperature resistivity of about 94 μΩcm. Used as a getter in vacuum tubes, in phosphors and ferrite bubble devices, in superconductors and ceramics, and in electrochemistry and spectroscopy. Symbol: Ho.

holmium boride. Any of the following compounds of holmium and boron used in ceramics: (i) *Holmium tetraboride.* Density, 6.84 g/cm³. Formula: HoB_4; and (ii) *Holmium hexaboride.* Density, 5.52 g/cm³. Formula: HoB_6.

holmium bromide. Yellow, hygroscopic crystals or off-white, hygroscopic powder (99.9+% pure). Density, 4.85 g/cm³; melting point, 914-919°C (1677-1686°F); boiling point, 1470°C (2678°F). Used in chemistry and materials research. Formula: $HoBr_3$.

holmium carbide. Any of the following compounds of holmium and carbon used in ceramics: (i) *Holmium dicarbide.* Density, 7.70 g/cm³. Formula: HoC_2; (ii) *Holmium sesquicarbide.* Density, 8.89 g/cm³. Formula: Ho_2C_3; and (iii) *Triholmium carbide.* Density, 9.43 g/cm³. Formula: Ho_3C.

holmium chloride. Yellow, hygroscopic crystals, or off-white to bright yellow, hygroscopic powder (99.9+% pure). Crystal system, monoclinic. Density, 3.7 g/cm³; melting point, 718°C (1324°F); boiling point, 1500°C (2732°F). Used in chemistry and materials research. Formula: $HoCl_3$.

holmium-cobalt. A ferrimagnetic compound containing cobalt and holmium in a ratio of 5:1 ($HoCo_5$). It is available as a fine powder for the manufacture of holmium-cobalt permanent magnets.

holmium dicarbide. See holmium carbide (i).

holmium fluoride. Purple-yellow crystals or yellow powder (99.9+% pure). Crystal system, orthorhombic. Density, 7.664 g/cm³; melting point, 1143°C (2089°F); boiling point, above 2200°C (3990°F). Used in chemistry and materials research. Formula: HoF_3.

holmium hexaboride. See holmium boride (ii).

holmium iodide. Yellow crystals (99.9+% pure). Crystal system, hexagonal. Density, 5.4 g/cm³; melting point, 994°C (1821°F). Used in chemistry and materials research. Formula: HoI_3.

holmium nitride. A compound of holmium and nitrogen that is available as cubic crystals with a density of 10.2-10.6 g/cm³. Used in ceramics and materials research. Formula: HoN.

holmium oxide. Yellow to light yellow, hygroscopic crystals, or white powder (98+% pure). Density, 8.36-8.41 g/cm³; melting point, 2360°C (4280°F) [also reported as 2415°C or 4379°F]; a rare-earth oxide. Used for high-temperature refractories, in superconductors, and as a special catalyst. Formula: Ho_2O_3. Also known as *holmia.*

holmium sesquicarbide. See holmium carbide (ii).

holmium silicide. A compound of holmium and silicon that is available as hexagonal crystals with a density of 7.1 g/cm³, and used for refractories and in electronics and semiconductor technology. Formula: $HoSi_2$.

holmium sulfide. A compound of holmium and sulfur that is available as tan-yellow or yellow-orange crystals. Crystal system, monoclinic. Density, 5.92 g/cm³; band gap, 2.58 eV. Used in materials research, electronics and semiconductor technology. Formula: Ho_2S_3.

holmium tetraboride. See holmium boride (i).

holocellulose. The total water-insoluble carbohydrate fraction of extractive-free wood (about 60-80%) varying greatly in crystallinity, and containing chiefly *hexosan* and *pentosan* polymers.

Hol-O-Fil. Trade name for rayon fibers.

Holoplast. Trademark of Holoplast for phenol-formaldehyde plastics and compounds.

holtedahlite. A colorless mineral composed of magnesium phosphate hydroxide, $Mg_2PO_4(OH)$. Crystal system, hexagonal. Density, 2.94 g/cm^3; refractive index, 1.599. Occurrence: Norway.

Holtite. Trade name of Belmont Metals Inc. (USA) for a granular alloy of 50% copper and 50% zinc. It has a melting point of 877°C (1610°F), and is used as a brazing alloy for all metals except aluminum.

holtite. A buff mineral composed of aluminum tantalum silicate borate hydroxide, $(Ta,Sb)Al_6(SiO_4)_3BO_3(O,OH)_3$. Crystal system, orthorhombic. Density, 3.90 g/cm^3; refractive index, 1.7575. Occurrence: Western Australia.

Holto. Trade name of Compagnie Ateliers et Forges de la Loire (France) for a series of cold- and hot-work tool steels used for press tools, punches, bending, blanking, forming and trimming dies, drawing, thread-rolling, plastic molding and extrusion dies, gripper, heading and die-casting dies, punches, gauges, shear blades, and mandrels.

Homberg's alloy. A fusible alloy composed of 33.3% lead, 33.3% bismuth and 33.3% tin. It has a melting point of 122°C (251°F), used for sprinkler plugs.

Homberg's phosphorus. A phosphorescent material made by heating together 33.3% ammonium chloride with 66.7% quicklime.

Hombitec RM. Trademark of Sachtleben Corporation (USA) for white, ultrafine titanium dioxide (TiO_2) powders used as pigments in paints, coatings, polymers and other materials.

homespun. (1) A coarse, irregular fabric, originally of wool or worsted, woven or spun at home.

(2) A strong, loosely woven fabric, often of linen or cotton, in a plain weave and similar in appearance to homespun (1).

homilite. A black or blackish-brown mineral composed of iron calcium borosilicate, $Ca_2(Fe,Mg)B_2Si_2O_{10}$. Density, 3.38 g/cm^3; hardness, 5 Mohs. Occurrence: Norway.

homocyclic compound. See alicyclic compound (2).

homoepitaxial thin film. A semiconductor thin film vapor-deposited on a substrate with both film and substrate consisting essentially of the same material, e.g., a silicon thin film on a silicon substrate. See also heteroepitaxial thin film.

homogeneous material. An amorphous or crystalline material that has the composition and properties uniformly distributed throughout. Also known as *homogeneous substance*.

homogenized alloys. Alloys that have been heated to and held at a high temperature for a sufficient period of time to equalize their compositions by diffusion.

homopolymer. A macromolecule that has only one repeating unit in the polymer chain. Examples include polyethylene, rubber hydrocarbon, polychloroprene and polystyrene made by the polymerization of ethylene, isoprene, chloroprene and styrene respectively. Abbreviation: HP. See also copolymer.

Honalium. Trade name of Honsel-Werke AG (Germany) for a series of corrosion-resistant aluminum-magnesium and aluminum-magnesium-silicon alloys supplied in the sand or chill-cast condition. They have good resistance to seawater corrosion and good to fair strength, and are used for architectural and marine castings, decorative purposes, builder's hardware, and pump casings.

honan. A fine-quality *pongee* fabric made from *wild silk* in a plain weave. It is obtained from Henan (Honan), a province in eastern China.

Hon-An-Cut. Trade name of Ultramatic Equipment Company (USA) for plastic media used for tumbling operations.

Honda New. Trade name of Honda Motor Company (Japan) for an alloy composed of 45% iron, 27% cobalt, 18% nickel, 6.7% titanium and 3.7% aluminum, and used for permanent magnets.

Honduras rosewood. See cocobola.

hone. A fine-grained, prismatic block or stone composed of fine abrasive, such as silicon carbide. Used for fine grinding and cutting-tool sharpening. Also known as *honestone*.

honessite. A green, yellow, or orange mineral composed of iron nickel sulfate hydroxide tetrahydrate, $Ni_6Fe_2SO_4(OH)_{16}\cdot4H_2O$. Refractive index, 1.615. Occurrence: USA (Wisconsin).

honestone. See hone.

Honeycomb. (1) Trademark of Honeycomb Construction Service Limited (Canada) for honeycomb cores.

(2) Trade name of former Glas- und Spiegel-Manufactur AG (Germany) for a glass with honeycomb design. It was known in German-speaking countries as "Wabenmuster."

honeycomb. (1) A product consisting of thin metal sections (e.g., aluminum, nickel, stainless steel, or titanium) or resin-impregnated sheet materials (e.g., polyaramid or fiberglass fabrics, or kraft paper) bonded together to form a hexagonal cellular structure that closely resembles natural honeycomb. Used as a core material in the manufacture of lightweight *sandwich constructions*.

(2) A knit or woven fabric, usually of wool, cotton or cotton blend, having a distinctive pattern resembling a waffle iron or the cellular comb of the honey bee. Used for draperies, jackets and women's wear.

honeycomb sandwich structure. See honeycomb structure; sandwich construction.

honeycomb slag. See expanded blast-furnace slag.

honeycomb structure. A composite structure consisting of two thin, strong outer sheets joined to a lightweight honeycomb core with an adhesive, or fused by brazing or welding. It has extremely high strength-to-weight and rigidity-to-weight ratio, good fatigue properties, and good acoustical and thermal insulating properties. Also known as *honeycomb sandwich structure*. See also sandwich structure.

honeylocust. The heavy, hard, durable wood of the tall hardwood tree *Gleditsia triacanthos*. The heartwood is light red to reddish-brown, and the sapwood wide and yellowish. *Honeylocust* has an attractive figure, high strength and stiffness, and good shock resistance. Average green weight, 977 kg/m^3 (61 lb/ft^3). Source: Eastern United States except New England states, and Gulf Coast Plains. Used for lumber; railroad ties, fence posts, and agricultural implements.

Honeysuckle. Trade name of Chance Brothers Limited (UK) for decorative table glassware.

Hong Kong. A ribbed, plain-colored fabric usually made of silk or other fibers in a plain weave.

hongquiite. A bright white mineral of the halite group composed of titanium oxide, TiO. Crystal system, cubic. Density, 5.36 g/cm^3. Occurrence: China.

hongshiite. A bronze-colored mineral composed of copper platinum arsenide, CuPtAs. Crystal system, hexagonal. Density, 10.09 g/cm^3. Occurrence: China.

Honial. French trade name for a series of magnetic alloys containing varying amounts of aluminum, nickel and cobalt.

Honigum. Trademark of DMG Hamburg (Germany) for addition-cured silicone dental impression materials containing special additives for improved flow characteristics.

Honsel. Trade name of Honsel-Werke AG (Germany) for a series of fairly strong, corrosion-resistant, wrought aluminum-magnesium, aluminum-manganese and aluminum-magnesium-manganese alloys, usually supplied in sheet and plate form, and used for chemical equipment, storage tanks, architectural and automotive trim, builder's hardware, paneling, siding, appliances and traffic signs.

Hontal. Trade name of Honsel-Werke AG (Germany) for a strong, pressure-tight aluminum alloy containing about 4.5-5.2% copper and 0.1-0.3% titanium. Supplied in the chill- and sand-cast condition, it has a density of 2.75 g/cm³ (0.099 lb/in.³), good castability, and is used in the aircraft and automotive industries.

Hontron. Trade name of Honsel-Werke AG (Germany) for a series of magnesium casting alloys containing up to 6% aluminum. Commercially available in sand-cast, chill-cast and pressure-die-cast form, they have good tear resistance, and are used for motors, car wheels, covers and aircraft parts.

Hooker brass. A British free-cutting brass containing 61% copper, 37% zinc, and 2% lead. Used for hot forgings and brass parts.

Hooker Grade Fiber. Trade name of Ruberoid Company (USA) for asbestos fibers.

hoop ash. See black ash.

Hoope's aluminum. An ultrapure aluminum (about 99.99%) that has been electrolytically refined utilizing an anode of aluminum-copper alloy in a fused fluoride bath. The refining process involves three liquid layers, and is referred to as the "Hoope's process." See also electrolytic aluminum.

hopeite. A pale yellow, grayish white, or colorless mineral composed of zinc phosphate tetrahydrate, $Zn_3(PO_4)_2 \cdot 4H_2O$. It can also be made synthetically. Crystal system, orthorhombic. Density, 3.05 g/cm³; refractive index, 1.582. Occurrence: Zimbabwe. Used as an ore of zinc.

hop hornbeam. The heavy, tough wood of the small tree *Ostrya virginiana*. It has very high hardness, is relatively scarce and resembles *ironwood*. Source: Eastern United States and southern Canada. Used for handles, levers and machine parts.

Hopkinson alloy. A corrosion- and heat-resistant alloy containing 75% iron, 24.5% nickel and 0.5% silicon.

Hopkin's White. Trade name of Aardvark Clay & Supplies (USA) for a clay (cone 10) having a white body with fine-grained sand.

Horbach. Trade name of Horbach & Schmitz GmbH (Germany) for a series of austenitic, ferritic and martensitic stainless steels used for valves, cutlery, surgical and dental instruments, chemical-process equipment, turbine blades, tanks, mixers, furnace equipment, and jet-engine components.

horizontal-cell tile. A hollow structural masonry unit usually composed of burnt clay in which the cores or cells are incorporated into a wall in a plane horizontal to the core or cell axis.

Hornalite. Trade name of Hornalite Company (USA) for polyester resins.

hornbeam. The white wood from any of two genera of trees (*Carpinus* and *Ostrya*) of the birch family including the American hornbeam (*C. caroliniana*), the hop hornbeam (*O. virginiana*) and the European hornbeam or white beech (*C. betula*).

hornblende. A greenish-black, brown or yellow amphibole mineral with vitreous to silky luster, found in igneous and metamorphic rocks (e.g., granite). It is a silicate containing calcium, sodium, magnesium, iron, aluminum, fluorine and water, $(Ca,$ $Na)_2(Mg,Fe,Al)_5(Al,Si)_8O_{22}(F,OH)_2$. Crystal system, monoclinic. Density, 3.0-3.47 g/cm³; hardness, 5-6 Mohs.

Hornex. Trademark of Vulkanfiber-Fabrik Ernst Krüger & Co. KG (Germany) for a tough, hygroscopic plastic made by treating continuous sheets of paper with a zinc chloride solution and then compressing several of these sheets. It has good electrical insulating properties and dimensional stability, is very sensitive to moisture, and can be machined. Used for suitcases, insulating materials and lining and paneling.

Hornitex. Trade name of Hornitex Werke Gebrüder Künnemeyer GmbH & Co. KG (Germany) for plywood, chipboard, fiberboard and plastic sheeting.

horn lead. See phosgenite.

horn quicksilver. See calomel.

horn silver. See cerargyrite.

hornstone. See chert.

horse chestnut. See American horse chestnut; European horse chestnut.

horseflesh ore. See bornite.

horsehair. A stiff fabric made from horsehair fibers.

horsehair fibers. The long, relatively stiff protein fibers obtained from the manes or tails of horses. Used especially for the manufacture of textiles and carpets, and as upholstery stuffing.

Hoskins. Trade name of Hoskins Manufacturing Company (USA) for a series of alloys including heat-resistant *Chromel*-type nickel-chromium alloys (some with iron), corrosion-resistant *Monel*-type nickel-copper alloys, nickel-base superalloys and ferritic stainless steels used for resistance wire, heating elements, spark plugs, heat-resistant parts, and many other applications.

Hostacom. Trademark of Hoechst AG (Germany) for a series of polypropylenes reinforced or coupled with glass fibers, or filled with talc.

Hostadur. Trademark of Hoechst AG (Germany) for a series of polyethylene terephthalate plastics.

Hostaflex. Trademark of Hoechst AG (Germany) for a series of flexible polypropylenes and polyvinyl chlorides.

Hostaflon. Trademark of Hoechst AG (Germany) for a series of fluoropolymers based on polytetrafluoroethylene (PTFE), and often filled with glass fibers. Available in a variety of grades in film, rod and sheet form and in coating grades, they have a density of 2.10-2.20 g/cm³ (0.076-0.079 lb/in.³), good to moderate impact strength, good dimensional stability, good UV-stability, a maximum service temperature of 260°C (500°F) and excellent chemical resistance. Used for coatings, foils, gaskets, seals and piston rings. *Hostaflon ET* are melt-processible ethylene-tetrafluoroethylene (ETFE) copolymer with low density (1.7 g/cm³ or 0.06 lb/in.³), low water absorption (0-0.03%), good resistance to abrasion, cut-through and impact, good low-temperature impact resistance, good weather and radiation resistance, excellent chemical resistance and an upper service temperature of 150°C (300°F). Used for chemical-process equipment, vessel linings, and wire coatings. *Hostaflon FEP* refers to melt-processible fluorinated ethylene-propylene (FEP) polymers with a density of 2.15 g/cm³ (0.078 lb/in.³), low water absorption (0-0.03%), good resistance to abrasion, cut-through and impact, good low-temperature impact resistance, good weather and radiation resistance, excellent chemical resistance, and a service temperature range of -250 to +200°C (-418 to +392°F). Used for chemical-process equipment, vessel linings, coatings and electrical insulation. *Hostaflon PFA* are perfluoroalkoxy alkane (PFA) fluoropolymers with a density of 2.13-

2.16 g/cm³ (0.077-0.078 lb/in.³), excellent thermal resistance, negligible water absorption, excellent resistance to acids, alkalies, ozone and ultraviolet radiation, good abrasion resistance, good dielectric properties, low coefficients of friction, and a service temperature range of -195 to +260°C (-320 to +500°F). Used for piping, flexible tubing, valves, pumps and filter cartridges. *Hostaflon TF* refers to essentially chemically inert polytetrafluoroethylene (PTFE) polymers supplied in standard and glass fiber-, graphite- or bronze-filled grades. They have excellent flame retardancy, low coefficients of friction, good high-temperature stability, and are used for chemical pipe and valves, seals and O-rings.

Hostaform. Trademark of Hoechst AG (Germany) for acetal homopolymers and copolymers based on polyoxymethylene (POM) and supplied in various grades including unmodified, UV-stabilized, glass fiber-coupled and silicone-lubricated grades. They have a density of 1.41 g/cm³ (0.051 lb/in.³), high strength and toughness, excellent abrasion resistance, maximum service temperature of 100-150°C (210-300°F), excellent resistance to mineral oils, gasoline, tetrachlorocarbon, dilute alkalies, and moderate resistance to trichloroethylene and dilute acids. Used for machine parts, such as gears, bearings and pump components.

Hostalen. Trademark of Hoechst AG (Germany) for a series of high- and low-density polyethylenes with a density range of 0.92-0.96 g/cm³ (0.033-0.035 lb/in.³), low water absorption, a maximum service temperature of 120°C (248°F), good strength and toughness, excellent resistance to dilute acids and alkalies, fair resistance to mineral oils and gasoline, and poor resistance to trichloroethylene and tetrachlorocarbon. Used for containers, pipes and tubing, hoses, bottles, blow-molded parts (e.g., shipping drums, tool boxes and garbage containers), truck-bed liners, dunnage trays and large recreational products. *Hostalen GUR* are UV-stabilized chemically inert ultrahigh-molecular weight polyethylenes with a density of 0.94 g/cm³ (0.034 lb/in.³), excellent abrasion and wear resistance, high impact resistance, low coefficients of friction, high melt viscosity, and good resistance to most chemicals except halogens and aromatic hydrocarbons. Used for industrial and mining equipment, and machine parts. *Hostalen HD* are unmodified and UV-stabilized high-density polyethylenes supplied in rod, sheet, tube, granule and film form. They have a density of 0.95 g/cm³ (0.034 lb/in.³), very good electrical properties, relatively high strength, stiffness and impact resistance, good resistance to acids, alkalies, alcohols, halogens and ketones, and an upper service temperature of 55-120°C (131-248°F). Used for containers, bottles, pipes and pipe fittings. *Hostalen HMW* are high-molecular-weight high-density polyethylenes, and *Hostalen HT* thin polyethylene film materials with good thermal properties and good antistatic properties, used for packaging and for in-liners in flexible bulk-material containers. Hostalen PP refers to polypropylene resins available as straight copolymers or homopolymers in UV-stabilized, elastomer-modified and fire-retardant grades. They have a density of 0.91 g/cm³ (0.033 lb/in.³), excellent resistance to dilute acids and alkalies, fair resistance to mineral oils, gasoline and trichloroethylene, and poor resistance to tetrachlorocarbon. Used for fittings, and various parts with good elevated-temperature and chemical resistance.

Hostalit. Trademark of Hoechst AG (Germany) for a colorless, rigid, flame-retardant thermoplastic polyvinyl chloride supplied in rod, sheet, tube, powder and film form. It has a density of about 1.4 g/cm³ (0.05 lb/in.³), high tensile strength, a maximum service temperature of 70-80°C (160-175°F), low water

absorption (0.03-0.4%), good barrier properties and UV stability, excellent resistance to dilute acids and alkalies and mineral oils, good resistance to gasoline, and poor resistance to trichloroethylene and tetrachlorocarbon. Used for water pipes, containers, and profiles.

Hostaphan. Trademark of Hoechst AG, Kalle Werk (Germany) for strong, stiff, chemically resistant thermoplastic polyester film products based on polyethylene terephthalate.

Hostapox. Trademark of Reichhold Chemicals, Inc. (USA) for a series of epoxy film products.

Hostapren. Trademark of Hoechst AG (Germany) for chlorinated polyethylenes.

Hostalloy. Trademark of Hoechst Celanese Corporation (USA) for polyolefin alloys with high impact strength and excellent abrasion resistance.

Hostyren. Trademark of Hoechst AG (Germany) for a series of polystyrenes supplied in various grades including standard, impact-modified and high-gloss. They have a density of about 1.05 g/cm³ (0.038 lb/in.³), high tensile strength, good to excellent impact strength, an upper service temperature of 70-95°C (160-200°F), excellent resistance to dilute alkalies, good resistance to dilute acids, moderate resistance to mineral oils, and poor resistance to gasoline, trichloroethylene and tetrachlorocarbon. Used for packaging material, coolers, luggage, containers, housings, molded products, and household appliances.

Hot. Trade name of Sheffield Smelting Company Limited (UK) for a silver brazing alloy containing 60% silver, and a total of 40% copper and zinc. It has a melting range of 600-720°C (1110-1330°F), and is used for brazing in protective atmospheres.

hot-applied sealant. A sealing compound applied to a surface in the molten state and hardening by cooling to ambient temperatures.

Hot Die. (1) Trade name of Atlas Specialty Steels (Canada) for a hot-work steel containing 0.3% carbon, 3% chromium, 11% tungsten, 0.5% vanadium, and the balance iron. Used for hot punches, and hot-die tools.

(2) Trade name of Adams & Osgood Steel Company (USA) for a tungsten-type hot-work tool steel.

Hot Die Steel. Trade name of Teledyne Vasco (USA) for a hot-work steel containing 0.5% carbon, 1.5% chromium, 0.3% vanadium, 2.4% tungsten, and the balance iron. Used for punches and hot upsetters.

hot-die steels. See hot-work tool steels.

hot-dip coatings. Coatings produced by dipping a substrate into a molten metal, e.g., zinc or tin.

hot-dip galvanized coatings. Multilayered coatings produced on iron and steel products by dipping into a bath of molten zinc to provide protection against atmospheric, soil and/or water corrosion. Also known as *hot-dipped coatings; hot-dipped zinc coatings.*

hot-dip galvanized sheet. Steel sheet, usually in coil form, on whose surface a multilayered, protective coating has been produced by dipping into a bath of molten zinc. See also hot-dip galvanized coatings.

hot-dipped coatings. See hot-dip galvanized coatings.

hot-dipped zinc coatings. See hot-dip galvanized coatings

hot-dip tin coatings. Thin coatings produced on metallic articles (e.g., steel or cast iron) by dipping into molten tin usually to protect them against corrosion or provide decorative coatings.

hot-dip tinplate. A thin, pickled sheet of iron or steel dipped into molten tin to produce a thin coating (typically 24-30 g/m² or 1.05-1.32 oz/ft²). There are three grades; (i) charcoal plate; (ii) coke plate, and (iii) crystallized. *Hot-dip tinplate* has been largely replaced by electrolytic tinplate.

hot-drawn product. A metal rod, tube or wire that has been drawn while hot through a die or series of dies. See also cold-drawn product.

hot-drawn steel. A round, rectangular, square, hexagonal and other shape of steel bar, wire or tube whose diameter has been gradually reduced by pulling or drawing through a die or a series of dies at elevated temperature. The resulting product has a yield point about half that of *cold-drawn steel*.

hot-drawn wire. A wire that has been pulled or drawn while hot through a die or a series of dies. See cold-drawn wire.

hot-drawn synthetics. Synthetic filaments, fibers or films that have been drawn with the application of heat.

hot face insulation. See insulating refractory.

Hotform. Trade name of Teledyne Vasco (USA) for a series of chromium-type hot-work tool steels possessing high strength and hot hardness, toughness and erosion resistance. *Hotform No. 1* (AISI type H12) contains 0.32-0.38% carbon, 0.8-1% silicon, 0.2-0.4% manganese, 4.75-5.25% chromium, 1.3-1.5% molybdenum, 1.2-1.6% tungsten, 0.4-0.5% vanadium, and the balance iron. *Hotform No. 2* (AISI type H11) contains 0.32-0.38% carbon, 0.8-1% silicon, 0.2-0.4% manganese, 4.75-5.25% chromium, 1.2-1.4% molybdenum, 0.4-0.5% vanadium, and the balance iron. *Hotform No. 3* (AISI type H12) contains 0.52-0.58% carbon, 0.8-1% silicon, 0.2-0.4% manganese, 4.75-5.25% chromium, 1.1-1.4% molybdenum, 1-1.4% tungsten, and the balance iron. *Hot Form Drill Rod* refers to air-hardening hot-work tool steels (AISI types H11, H12 and H13) containing 0.35% carbon, 5% chromium, 1.45% molybdenum, 1.25% tungsten, 1% silicon, 0.5% vanadium, and the balance iron. They are used for high-temperature fasteners, cores and knock-out pins for die-casting dies, and punches and piercers. *Hot Form Ultra* is an AISI type H12 and *Hotform V* an AISI type H13 steel. *Hotform* tool steels are typically used for forging dies and forging die inserts, drop-forging dies, die-casting and extrusion dies for aluminum and brass, extrusion rams and press liners, gripper and heading dies, upsetters, punches, piercers, shear blades, and heavy-duty shears and slitters.

hot-galvanized material. A material with a zinc coating applied in a heated tumbling barrel containing zinc flakes. Also known as *barrel-galvanized material; flake-galvanized material; tumbler-galvanized material; wean-galvanized material.*

hot-isostatically-pressed products. Ceramic, metal or metal alloy (e.g., titanium, high-speed steel, or superalloys) products of uniform density formed by first sealing the powdered starting material into a flexible metal container (or can) and then simultaneously applying equal fluid pressure from all directions (isostatic pressure) and high (sintering) temperatures. Hot-isostatically-pressed products usually do not require subsequent firing or sintering. Abbreviation: HIPed products. See also cold-isostatically-pressed materials.

hot-isostatically-pressed silicon nitride. A low-porosity silicon nitride usually made from a preshaped cold-isostatically-pressed silicon nitride green body or a presintered silicon nitride body by hot isostatic pressing at high pressures and sintering temperatures up to 2000°C (3630°F). It has a density of about 3.24 g/cm³ (0.117 lb/in.³), a hardness exceeding 1600 Vickers, high mechanical strength and elastic modulus, excellent high-temperature resistance, good corrosion resistance and dimensional stability, good electrical insulating properties. The material made from preshaped green bodies is known as "hot-isostatical-

ly-pressed silicon nitride" (HIPSN) and that obtained from presintered bodies as "hot-isostatically-pressed sintered silicon nitride" (HIPSSN). Used for high-temperature applications.

hot-laid plant mix. A mixture of bituminous emulsion and heated mineral aggregate made in an asphalt plant or drum mixer. It is spread and compacted at the construction site when the composition is at a temperature above ambient. Also known as *hot mix; hot-mix asphalt.*

hot-melt adhesives. Solid thermoplastic materials that are heated and applied to substrates in the molten state bonding almost instantaneously upon cooling. They may be of the pressure-sensitive or the nonpressure-sensitive types, and are often compounded with additives for tack and wettability. Typical thermoplastics for this purpose include polyethylenes, polyvinyl acetates, polyamides, polyesters, polysulfones and polyurethanes. They have a melting temperature of about 190-230°C (375-450°F), a maximum service temperature of 50-170°C (120-340°F), good shear strength, moderate peel strength, poor creep and heat resistance, and indefinite shelf lifes, are fast setting and can be quickly applied. Used for packaging, food-carton bonding, side seaming of cans, bookbinding, composite structures and laminates, in electronics, in the automotive industry, and for furniture and footwear. Also known as *hot melts.*

hot mix. See hot-laid plant mix.

hot-mix asphalt. See hot-laid plant mix.

Hot Mount. Trade name of Mathison's (USA) for a thermally activated mounting adhesive without support, but having a moisture-resistant release liner. Used especially in the graphic arts and allied trades.

Hotpress. Trade name of Teledyne Vasco (USA) for a tungsten-type hot-work tool steel (AISI type H20) containing 0.33-0.38% carbon, 0.2-0.3% silicon, 0.2-0.3% manganese, 9-9.5% tungsten, 1.75-2.25% chromium, 0.4-0.6% vanadium, and the balance iron. It has high resistance to heat checking, and good hot hardness and toughness. Used for extrusion dies and dummy blocks, hot-work dies for brass and steel, and upsetter dies and tools.

hot-pressed abrasives. Bonded abrasives made in a mold by pressing at relatively high temperatures.

hot-pressed materials. Ceramic, metal or metal-alloy products formed by compacting the powdered starting material at an elevated (sintering) temperature by the application of equal uniaxial or equiaxial (isostatic) pressure. See also cold-pressed materials.

hot-pressed silicon carbide. See sintered silicon carbide.

hot-pressed silicon nitride. Fully dense silicon nitride produced by hot pressing, i.e., by compacting (pressing) the starting powder at temperatures exceeding 1700°C (3092°F). Usually supplied in sheet form, it has a density of 3.11, high tensile strength and modulus, high compressive strength, a hardness of 1700-2200 Vickers, and an upper continuous-use temperature of 1100-1650°C (2012-3002°F). Used for high-strength and high-temperature applications. Abbreviations: HPSN.

hot press moldings. Reinforced thermosetting moldings with good surface appearance formed by a process in which glass-fiber fabrics, mats or prepregs are first placed on one side of a two-piece heated mold or tool, premixed, liquid thermosetting resin is poured onto them, and then the mold is closed by the application of pressure. Excess resin is squeezed out of the mold, and the product is allowed to cure.

hot-rolled and pickled sheet. A hot-rolled steel sheet cleaned (pickled) in a bath of hydrochloric or sulfuric acid to remove the surface oxide (scale) formed during hot rolling.

hot-rolled sheet. A steel sheet rolled to the desired thickness at a temperature high enough to cause the formation of a blue-black surface oxide (hot mill scale) during the hot reduction operation. See also cold-rolled sheet.

hot-rolled steel. A steel ingot (billet, bloom or slab) that has been passed while red-hot through one or several sets of rolls to reduce its cross-sectional area. The temperature of the ingot is usually above the recrystallization temperature, and the resulting product is characterized by the blue-black surface oxide (hot mill scale) that forms during rolling. Abbreviation: HRS. See also cold-rolled steel; rolled steel.

hot rubber. Synthetic rubber, usually of the styrene-butadiene type, produced by polymerization at slightly elevated temperatures (typically 50°C or 120°F). The term is used to distinguish it from cold rubber that is produced at relatively low temperatures. See also cold rubber; styrene-butadiene rubber.

hot-setting adhesive. An adhesive that sets at a temperature of 100°C (212°F) or above.

Hotspur. Trade name of Dunford Hadfields Limited (UK) for a heat- and oxidation-resistant steel containing 0.2-0.4% carbon, 12-25% chromium, 7-25% nickel, 0-3% tungsten, and the balance iron. Used for furnace parts and equipment.

Hot Stamping Alloy Die. Trade name of Edgar T. Ward's Sons Company (USA) for an oil-hardening hot-work steel containing 0.5% carbon, 3% chromium, and the balance iron. Used for hot-stamping dies.

hot stamping foil. A decorative metal foil supplied on removable cellophane or polyester film support that is applied to a base metal with a heated die.

Hottrin. British trademark for a finely divided natural rubber used bituminous road construction.

Hotvar. Trade name of Uddeholm Corporation (USA) for a hot-work tool steel.

hot-worked metals. Metals and alloys that have been deformed by processes, such as drawing, extrusion, forging or rolling, at temperatures above those at which recrystallization occurs.

hot-work glue. An animal, bone or hide glue that is applied in the hot state and bonds on cooling to ambient temperatures. Used as a cementing medium for nonstructural applications.

hot-work tool steels. A group of medium-carbon tool steels (AISI type H) that have been developed for the manufacture of tools, such as punching, forming and shearing dies and tools exposed to abrasion, heat, pressure and shock. Depending on the principal alloying elements used, they are classified into: (i) *Chromium hot-work steels* (AISI types H1-H19); (ii) *Tungsten hot-work steels* (AISI types H20-H39); and (iii) *Molybdenum hot-work steels* (AISI types H40-H59). They have excellent hot hardness, good hardenability and toughness, and good to moderate wear resistance. Often sold under trade names, such as *Alcodie, Crodi, Cro-Mow, Dart, Dica-B, Dievac, Firedie, Howard, Hotform, Lumdie, Potomac, Red Indian* or *Thermold*, they are used for dies and die inserts, punches and punch inserts, headers, shear blades, mandrels for high-temperature applications, and various other tool and dies. Also known as *hot-die steels; hot-work steels.*

Houghton's Shaving Packing. British trade name for a lead alloy containing 5.97% tin and 0.01% copper. Used for piston-rod packings for steam engines.

hourdis. A large hollow clay block orginally from France, used in building construction, e.g., for vaults.

household china. Thin, highly translucent vitreous ceramic white-ware used for dinnerware.

Houselon. Trade name of Nitto Boseki Company Limited (Japan) for lightweight glass-fiber insulating blanket.

house paint. An organic coating either based on a binder composed of a synthetic resin (e.g., acrylic or alkyd) dissolved in a solvent or water and having high-grade pigments dispersed throughout, or containing an aqueous dispersion of a synthetic resin (e.g., acrylic, polyvinyl acetate or styrene-butadiene) produced by emulsion polymerization (latex paint). Exterior alkyd paint provides a hard, durable gloss finish to give long-lasting protection to house exteriors, garages, sheds and fences. It is dry to the touch in 6-8 hours, possesses excellent hiding qualities and resists blistering, peeling and cracking. Exterior latex paint can be used on wood siding, prefinished aluminum, vinyl and galvanized siding. Exterior acrylic latex paint is dry to the touch in 30-60 minutes, and provides a weather-resistant finish that combines protection, durability and ease of application. All house paints are available in a wide range of colors.

house-stone veneer. Building stone, usually limestone or sandstone, that is hard, low in water absorption, has a rough surface, and is broken into irregular lengths. Used as facing or veneer on residential buildings.

Hover. Trade name of Gebrüder Hover Edelstahlwerk (Germany) for an extensive series of steels including various austenitic, ferritic and martensitic stainless grades, plain-carbon, cold-work, hot-work and high-speed tool grades, and machinery case-hardening grades.

Howal. Trade name of Honsel-Werke AG (Germany) for an aluminum alloy containing 0.6-1.6% silicon, 0.6-1.4% magnesium, 0.6-1% manganese, and 0.3% chromium. It has good weldability and formability, and is used for boats, window frames, gutters, and fan blades.

Howard. Trade name of Simonds Worden White Company (USA) for a series of chromium-type hot-work tool steels (AISI types H11, H12 and H13) containing 0.35% carbon, 5.0% chromium, 1.5% molybdenum, 0.4-1% vanadium, 0-1.5% tungsten, and the balance iron. They have good deep-hardening properties, high toughness, good machinability, high hot hardness, and good abrasion and wear resistance. Used for hot dies, hot grippers and upsetters, and hot shear blades and punches.

Howard Alloy. Trade name of Howard Foundry Company (USA) for a series of magnesium casting alloys containing varying amounts of aluminum, zinc and manganese, or magnesium, zinc and zirconium, sometimes with thorium. Used for structural parts, airframes, aircraft engines, aircraft wheels and brakes, automotive components, crankcases, housings, oil pumps, and sand, permanent-mold and die castings.

Howe Brown. Trade name of Colt Industries (UK) for a series of water-hardening steel used for tools and dies.

Howeflex. Trademark of Konrad Hornshuh AG (Germany) for a flexible polyvinyl chloride film.

Howege. Trade name of Dohlen-Stahl Gusstahl-Handels GmbH (Germany) for tough, oil-hardening cold-work steels including (i) *Howege Extra* which contains 0.5-0.6% carbon, 0.7% chromium, 0.18% molybdenum, 1.85% tungsten, 0.1% vanadium, and the balance iron, and is used for cold-work tools, headers and upsetters; and (ii) *Howege Spezial* which contains 0.4% carbon, varying amounts of chromium and molybdenum, and the balance iron, and is used for gears, shafts, crankshafts, bolts, and studs.

Howelon. Trademark of Konrad Hornschuh AG (Germany) for a flexible polyvinyl chloride film.

Howes Shear Blade. Trade name of Crucible Materials Corporation (USA) for a water-hardening tool steel containing 0.9% carbon, 0.3% molybdenum, and the balance iron. Used for shear blades.

Howesol. Trademark of Konrad Hornshuh AG (Germany) for a flexible polyvinyl chloride film.

Howetex. Trademark of Konrad Hornshuh AG (Germany) for a flexible polyvinyl chloride film.

howieite. A dark green mineral composed of sodium manganese iron aluminum silicate hydroxide, $Na(Fe,Mn)_{10}(Fe,Al)_2Si_{12}O_{31}(OH)_{13}$. Crystal system, triclinic. Density, 3.38 g/cm^3; refractive index, 1.720. Occurrence: USA (California).

howlite. A whitish mineral composed of calcium boron silicate hydroxide, $Ca_2B_5SiO_9(OH)_5$. A typical composition is 28.2% calcium oxide, 14.8% silicon dioxide, 45.6% boric oxide and 11.4% water. Crystal system, monoclinic. Density, 2.60-2.62 g/cm^3; refractive index, 1.596. Occurrence: USA (California). Used in basic arc-furnace practice as an addition to treat slag and hearth.

Howmedica. Trade name of Howmedica Corporation (USA) for corrosion-resistant nickel-chromium ceramic dental alloys used for crown-and-bridge bonding applications. *Howmedica III* is a beryllium-free version.

Howmet. Trade name of Howmet Corporation (USA) for a series of cast cobalt-base superalloys used for furnace equipment and fixtures, heat-treating equipment, gas-turbine parts, afterburners, turbine blades, valve parts, bushings, and heat- and wear-resistant cutting tools.

Howord. Trade name of Wallace Murray Corporation (USA) for a series of air- or oil-hardening chromium-type hot-work tool steels (AISI types H11, H12 and H13). They have good deep-hardening properties, excellent ductility, high hot hardness, good toughness and machinability, and good abrasion and wear resistance. Used for hot-forming tools, forging, header and trimmer dies, hot grippers, etc. *Howord A* (AISI type H11) contains 0.35% carbon, 0.3% manganese, 1% silicon, 5% chromium, 1.5% molybdenum, 0.4% vanadium, up to 0.3% nickel, and the balance iron. *Howard B* (AISI type H12) contains 0.35% carbon, 0.3% manganese, 1% silicon, 5% chromium, 1.5% tungsten, 1.5% molybdenum, 0.4% vanadium, up to 0.3% nickel, and the balance iron. *Howard C* (AISI type H13) contains 0.35% carbon, 0.3% manganese, 1% silicon, 5% chromium, 1.5% molybdenum, 1% vanadium, up to 0.3% nickel, and the balance iron.

Hoyles metal. A babbitt containing 46% tin, 42% lead and 12% antimony used for bearings.

Hoyt. Trade name of Hoyt Darchem Limited (UK) for a series of tin and lead babbitts. The tin babbitts contain about 78% tin, 12% lead and 10% copper, have a melting point of 360°C (680°F), and are used for pump and internal-combustion engine bearings. The lead babbitts contain about 90-94% lead and 6-10% antimony, and are used general-purpose bearings.

HP Nickel. Trade name of Jeliff Corporation (USA) for a high purity nickel (99.9+%) with high electrical resistivity, used for ballast applications and resistance thermometers.

HPM Nickel. Trade name of Inco Alloys International Limited (UK) for an oxidation- and corrosion-resistant high-purity nickel (99.9+%), used for special valve components and electric resistance thermometers and controls.

H-Polyester. Trademark of Holeproof Mills Limited (New Zealand) for polyester staple fibers and filament yarns.

HPW Nickel. Trade name of Inco Alloys International Limited

(UK) for a heat-resistant nickel alloy with 3.5% tungsten, used for special cathodes and valves.

HR-Monel. Trade name of International Nickel Inc. (USA) for a series of free-cutting cast nickel-copper alloys containing 62-68% nickel, 2% silicon, up to 2.5% iron, 0.5-1.5% manganese, up to 0.35% carbon, traces of sulfur, and the balance copper. They have good corrosion resistance and machinability, and are used for corrosion-resistant parts, water-meter parts, fasteners, screw machine parts, and valve and pump parts. See also Monel.

HSC Silicon Carbide. Trademark of Superior Graphite Company (USA) for a family of silicon carbide powder products used in the manufacture of sintered and reaction-bonded high-performance metal- or ceramic-matrix composites, fine abrasive grit and refractory materials. *HSC Hi-Dense* is a special submicron, beta-phase silicon carbide powder having a boronated sintering aid incorporated which offers higher strength and density, improved mechanical and physical properties and lower sintering temperatures.

hsianghualite. A white mineral composed of lithium beryllium calcium fluoride silicate, $Li_2Ca_3Be_3(SiO_4)_3F_2$. Crystal system, cubic. Density, 2.98 g/cm^3; refractive index, 1.613. Occurrence: China.

HSLA steels. See high-strength low-alloy steels.

HSLA-i steels. A microalloyed, isotropic type high-strength low-alloy steel. See also high-strength low-alloy steels.

H-steels. A group of carbon or alloy steels made to required hardenability ranging between a specified maximum and minimum value, and often having a composition different from that of the corresponding regular carbon or alloy steels. In the AISI-SAE designation system, this is indicated by the suffix "H." For example, 1038H is a plain-carbon steel having guaranteed hardenability values. Also known as *H-band steels*.

HT Molybdenum. Trade name of Plansee Metallwerk Gesellschaft (Austria) for high-purity molybdenum (100%) that becomes ductile after heating to about 1480-1980°C (2700-3600°F), and is used for aerospace, chemical, electrical, electronic, nuclear and metallurgical applications and vacuum equipment.

Hualon. Trademark of Hualon Corporation (Taiwan) for polyester staple fibers and filament yarns.

huanghoite. A yellowish green or honey-yellow mineral of the bastnaesite group composed of barium cerium fluoride carbonate, $BaCe(CO_3)_2F$. Crystal system, hexagonal. Density, 4.59 g/cm^3; refractive index, 1.765. Occurrence: China.

huarizo. The fine fleece obtained from the huarizo, a hybrid of a female alpaca and a male llama.

Hubard Special. Trade name of Hubard Steel Company (USA) for a wear-resistant cast iron that contains varying amounts of carbon, nickel, chromium and iron, and is used for rolls, guides and castings.

Hubbing Die. Trade name of Bethlehem Steel Corporation (USA) for a case-hardening steel containing 0.06% carbon, 0.15% manganese, and the balance iron. Used for plastic molding dies.

hubbing steels. See hobbing steels.

Huber. Trade name of J.M. Huber Corporation (USA) for several grades of carbon black.

Huberbrite. Trademark of J.M. Huber Corporation (USA) for an extensive series of modified and unmodified barium sulfate materials obtained from natural *barite*. Used as extenders in rubber, plastics, composites, paint, paper and textiles, and as fluxes in glasses and ceramics.

Huberbrite Interlok. Trademark of J.M. Huber Corporation

(USA) for a white, organophilic surface-treated additive derived from natural *barite*. A typical chemical analysis of the untreated product is 98-99% barium sulfate, 0.5% silica, 0.4% crystalline silica, 0.04% ferric oxide, and the balance being other compounds. The loss on ignition at 950°C (1740°F) is about 1.8%. It consists of irregular, uniaxial particles with fine particle size (1.0 μm or 39 μin. on average) and excellent dry flow properties. It has high brightness, a density of about 4.3-4.5 g/cm^3 (0.15-0.16 lb/in.^{3}), a hardness of 2.5-3 Mohs, and a refractive index of 1.64. Used as an additive in thermoplastic and other polymers, and as a titanium dioxide extender.

HuberCAL. Trademark of J.M. Huber Corporation (USA) for a food-grade calcium carbonate. *HuberCAL Elite* is a pharmaceutical-grade (USP) calcium carbonate

Hubercarb. Trademark of J.M. Huber Corporation (USA) for ground calcium carbonate ($CaCO_3$) supplied in several grades and particle sizes. Used as a filler and extender in plastics and rubber, as an extender pigment, and various other applications.

Huber metal. The common name of a pyrophoric material containing 85% cerium and 15% magnesium.

Hubersorb. Trademark of J.M. Huber Corporation (USA) for calcium silicate ($CaSiO_3$) used as an absorbent.

huckaback. An absorptive, heavy, soft fabric, often of linen or cotton, with a birds-eye or waffle surface texture, used for towels.

Huckingerhütte. Trade name of Mannesmann-Hüttenwerke AG (Germany) for a series of case-hardening steels containing 0.15-0.20% carbon, 0.5-1.2% manganese, up to 0.5% silicon, 0.3% magnesium and 0.3% molybdenum, respectively, and the balance iron. Used for gears, bolts, shafts, cams, camshafts and fasteners.

Hudson Bay. Trade name of Hudson Bay Mining & Smelting Company, Limited (Canada) for zinc and an extensive series of zinc alloys.

Hudstat. Trade name for metallic fibers.

huebnerite. A black, brownish-red or yellow-brown mineral of the wolframite group composed of manganese tungstate, $MnWO_4$. It can also be made synthetically. Crystal system, monoclinic. Density, 7.18-7.24 g/cm^3; hardness, 4.5-5.5 Mohs; refractive index, 2.22. Occurrence: Europe, USA (Colorado, Nevada, New Mexico, South Dakota). Used as an ore of tungsten.

huegelite. A brown to orange-yellow mineral composed of lead uranyl arsenate hydroxide trihydrate, $Pb_2(UO_2)_3(AsO_4)_2(OH)_4·3H_2O$. Crystal system, monoclinic. Density, 5.80 g/cm^3. Occurrence: Germany.

Huels Nylon 12. Trade name of Huels America Inc. (USA) for a series of resins based on polyamide 12 (nylon 12) and marketed in unmodified, flexible, semiflexible, glass fiber-reinforced, glass bead-filled, fire-retardant and UV-stabilized grades.

huemulite. A mineral composed of sodium magnesium vanadium oxide hydrate, $Na_4MgV_{10}O_{28}·24H_2O$. Crystal system, triclinic. Density, 2.39 g/cm^3. Occurrence: Argentina.

Huetex. Trade name of ASG Industries Inc. (USA) for plain or patterned, toughened glass products with fused ceramic enamel coatings.

Huewhite. Trade name of ASG Industries Inc. (USA) for a translucent white glass with a pattern on one side that combines true color transmission and optimum glare reduction with good light diffusion.

Hughes Alloy. Trade name of Egal Metal Products Company (USA) for a lead alloy containing 17% tin and 13% antimony,

used for metallic packings.

hulsite. A black mineral of the ludwigite group composed of iron magnesium tin oxide borate, $(Fe,Mg)_2(Fe,Sn)BO_3O_2$. Crystal system, monoclinic. Density, 4.50 g/cm^3. Occurrence: USA (Alaska).

humberstonite. A colorless mineral composed of potassium sodium magnesium nitrate sulfate hexahydrate, $K_3Na_7Mg_2(SO_4)_6(NO_3)_2·6H_2O$. Crystal system, rhombohedral (hexagonal). Density, 2.25 g/cm^3; refractive index, 1.474. Occurrence: Chile.

humboldtine. A yellow mineral composed of iron oxalate dihydrate, $FeC_2O_4·2H_2O$. It can also be made synthetically from a mixture of ferrous hydroxide [$Fe(OH)_2$] and oxalic acid dihydrate [$(CO_2H)_2·2H_2O$]. It can also be made synthetically. Crystal system, monoclinic. Density, 2.28 g/cm^3; refractive index, 1.561. Occurrence: Italy. Also known as *humboldtite*.

humboldtite. See humboldtine.

Humboldt redwood. See redwood.

Humenegro. Trade name for a carbon black.

Hume-Rothery compounds. See intermetallic compounds.

humic acid. A brownish mixture of polymers that contains the pigment *melanin*, occurs in lignite, peat and soil. Used as a chelating agent, in drilling fluids and printing inks, and in biochemistry and biotechnology.

humite. A yellow to orange mineral composed of magnesium fluoride silicate, $Mg_7F_2(SiO_4)_3$. It can also be made synthetically. Crystal system, orthorhombic. Density, 3.24 g/cm^3; refractive index, 1.649. Occurrence: Finland.

hummerite. A yellow to orange mineral composed of potassium magnesium vanadium oxide octahydrate, $KMgV_5O_{14}·8H_2O$. Crystal system, triclinic. Density, 2.53 g/cm^3; refractive index, 1.81. Occurrence: USA (Colorado).

hundred metal. A British heat- and corrosion-resistant alloy of nickel and chromium used for high-temperature applications.

hungchaoite. A colorless mineral composed of magnesium borate hydroxide heptahydrate, $MgB_4O_5(OH)_4·7H_2O$. Crystal system, triclinic. Density, 1.71 g/cm^3; refractive index, 1.485. Occurrence: USA (California).

hunting. Chrome- or alum-tanned leather made from the second (or flesh) split of cowhide ground on one side.

Huntingdon. Trade name of United Clays, Inc. (USA) for a strong, plastic, high-alumina clay with very high kaolinite content, good chemical resistance and suspension properties, and very fine particle size. Used for refractory applications.

huntite. A white mineral of the calcite group composed of magnesium calcium carbonate, $Mg_3Ca(CO_3)_4$. Crystal system, rhombohedral (hexagonal). Density, 2.70 g/cm^3; refractive index, 1.615. Occurrence: USA (Nevada).

Huntsman. Trade name of Huntsman Chemical Company (USA) for an extensive series of polyolefins and polystyrenes. *Huntsman PP* are chemical-resistant polypropylene polymers and homopolymers available in blow- and injection-molding and extrusion grades. *Huntsman EPS* are standard and modified expanded polystyrenes used for insulation products, molded products and other applications. *Huntsman PS* refers to polystyrene resins supplied in unmodified, silicone-lubricated, UV-stabilized, glass-reinforced, medium-impact, and structural-foam grades. *Huntsman PS (HIPS)* are high-impact polystyrene resins supplied in unmodified, UV-stabilized and fire-retardant grades.

Hunvira. Trademark of Hung Chou Chemical Industry Limited (Taiwan) for polyester fibers and filament yarns.

hureaulite. An orange to reddish brown mineral composed of manganese iron hydrogen phosphate tetrahydrate, $(Mn,Fe)_5H_2$-

$(PO_4)_4 \cdot 4H_2O$. Crystal system, monoclinic. Density, 3.19 g/cm³; refractive index, 1.658. Occurrence: France, USA (California).

hurlburite. A colorless to greenish white mineral composed of beryllium calcium phosphate, $Be_2Ca(PO_4)_2$. Crystal system, triclinic. Density, 2.88 g/cm³. Occurrence: USA (New Hampshire), Zimbabwe.

Huron. (1) Trade name of AL Tech Specialty Steel Corporation (USA) for a high-carbon, high-chromium cold-work tool steel (AISI type D3) containing 2.1% carbon 0.32% manganese, 0.45% silicon, 12.5% chromium, 1% vanadium, and the balance iron. It has excellent abrasion and wear resistance, excellent nondeforming properties, good compressive strength and good deep-hardening properties. Used for blanking and cold-forming dies, punches, rolls and gages. *Huron V* (AISI type D7) is contains more carbon (about 2.15-2.50%) and vanadium (about 3.8-4.4%).

(2) Trademark of Alcan Wire and Cable (Canada) for several finished and semifinished aluminum products.

Hushlite. Trade name of Laminated Glass Corporation (USA) for a sound-controlling laminated glass.

Husky. (1) Trade name of Halls & Pickles Limited (UK) for a tungsten high-speed steel containing 0.8% carbon, 4.5% chromium, 22% tungsten, 1.5% vanadium, and the balance iron. Used for cutting tools.

(2) Trademark of Commonwealth Plywood Company Limited (Canada) for plywood products.

Husmann metal. A tin babbitt containing 11% antimony, 11% lead, 4% copper, 0.2% iron and 0.2% zinc. It has excellent antifriction properties, and is used for bearings and bushings.

hutchinsonite. A cherry-red mineral composed of thallium lead arsenic sulfide, $(Tl,Pb)_2As_5S_9$. Crystal system, orthorhombic. Density, 4.60 g/cm³. Occurrence: Switzerland.

huttonite. A colorless to very pale cream mineral of the monazite group composed of thorium silicate, β-$ThSiO_4$. It can also be made from an equimolar mixture of silicon dioxide (SiO_2) and thorium dioxide (ThO_2) under prescribed conditions. Crystal system, monoclinic. Density, 7.10 g/cm³; refractive index, 1.900. Occurrence: USA (Arizona).

HVOF coatings. Metallic, ceramic or composite coatings having very high densities, high bond strengths, and low porosity and internal stresses, produced by a high-velocity oxyfuel (HVOF) thermal spraying process. Also known as *high-velocity oxyfuel coatings.*

HWM modal fibers. See high-wet-modulus modal fibers.

Hy. Trademark of Danaklon A/S (Denmark) for polypropylene staple fibers.

hyacinth. A red to orange-colored variety of the mineral zircon ($ZrSiO_4$) used as a gemstone.

HYAFF. Trademark of FAB–Fidia Advanced Biopolymers for a class of biocompatible, cytocompatible and biodegradable polymers derived from highly purified *hyaluronan*. Unprocessed HYAFF polymers are soluble materials, but they can be processed into insoluble, transparent products, such as fibers, sponges, thin films, membranes or microspheres, by extrusion, lyophilization or spray drying. The two essential *HYAFF* biopolymers are *HYAFF 7*, an ethyl ester of *hyaluronic acid* (HA) and *HYAFF 11*, a benzyl ester of HA. Used as biomaterials for drug-delivery applications, tissue engineering, skin grafting, and foul-resistant coatings for medical devices.

Hyalgan. Trademark of Sanofi Pharmaceuticals (USA) for a viscous hyaluronan (hyaluronic acid) product used in the treatment of osteoarthritis as an intra-articular injection to improve joint mobility and relieve pain.

Hyalograft 3D. Trademark for a dermal substitute consisting of autologous fibroblasts on a three-dimensional matrix of *HYAFF* biopolymer.

hyalophane. A variety of *potash feldspar* ($KAlSi_3O_8$) in which barium replaces part of the potassium.

hyalotekite. A colorless mineral composed of barium calcium lead boron fluoride silicate, $(Ba,Pb,Ca,K)_6(B,Si,Al)_2(Si,Be)_{10}$-$O_{28}(F,Cl)$. Crystal system, orthorhombic. Density, 3.80 g/cm³; refractive index, 1.963; hardness, 5 Mohs. Occurrence: Sweden.

hyaluronan. See hyaluronic acid.

hyaluronate. An ester or salt of hyaluronic acid, e.g., sodium hyaluronate.

hyaluronic acid. A linear, anionic, high-viscosity structural *polysaccharide* found in all higher animals as an essential component of the extracellular matrix (ECM) of connective tissues, in the umbilical cord, and in the vitreous humor of the eye. The molecule of this natural polymer is made up of repeating units of D-glucuronic acid and N-acetyl-D-glucosamine joined by a β-1,3-linkage. Used in biochemistry and in the preparation of synthetic biomaterials for tissue engineering applications. Formula: $C_8H_{15}NO_6$. Abbreviation: HA. Also known as *hyaluronan.*

hyaluronans. A novel class of biocompatible, cytocompatible, degradable, resorbable materials prepared by the partial or total esterification, amidation or sulfatation of the polysaccharide functional groups (e.g., carboxyl, hydroxyl, or N-acetyl) of hyaluronic acid, or by crosslinking or coupling reactions. Examples of hyaluronic acid derivatives are ethyl and benzyl hyaluronan esters. Used as biomaterials in tissue engineering, skin grafting and drug delivery.

Hybnickel. Trade name of Hybnickel Alloys Company (USA) for a series of corrosion- and heat-resistant iron-chromium-nickel and iron-nickel-chromium alloys used for furnace equipment and parts, heat-treating equipment, and hearth plates.

Hybon. Trademark of PPG Industries, Inc. (USA) for a series of fiberglass roving materials. *Hybon 2006* is an abrasion-resistant grade formulated specifically for use in epoxy resin-based filament winding processes. *Hybon 2001* and *Hybon 2002* are designed specially for filament winding processes using vinyl ester and polyester resins. *Hybon 600HTX* and *Hybon 650HTX* are fiberglass gun rovings with improved performance in wet out, lay down and mold conformity. Both of these spray-up rovings also have negligible fuzz and fly. The 600HTX rovings are used for boats, recreational vehicles and reinforced plastic parts for general construction and consumer applications, and the 650HTX rovings are designed for tub and shower units.

Hy-Bond. (1) Trade name of Shofu Dental Corporation (Japan) for radiopaque polycarboxylate, zinc oxide/eugenol, and zinc phosphate restorative dental cements with tannin-fluoride additives. *Hy-Bond* is a zinc phosphate cement with high bond strength, used for seating permanent prostheses. *Hy-Bond Polycarboxylate* is a polycarboxylate cement that features an extremely low film thickness and forms a chemical bond between the tooth structure (i.e., dentin and enamel) and metal appliances. *Hy-Bond ZOE* is a quick-setting zinc oxide–eugenol cement that mixes easily and is used for cementing temporary crowns.

(2) Trademark of Fairey & Company Limited (Canada) for firebrick clays, ceramic clays and furnace cements.

Hybond. Trademark of Pierce & Stevens Chemical Corporation (USA) for a series of contact or pressure-sensitive adhesives, cements and glues. They can be applied by brushing, rolling, troweling and spraying, and bond well to many dissimilar substrates including aluminum, copper, steel, wood, fiberboard, leather, fabric, linoleum, rubber and paper.

Hy-Bor. Trademark of Avco Corporation (USA) for a series of composites consisting of ceramic, carbonaceous, polymeric or metallic matrices reinforced with mixtures of boron, carbon/graphite, ceramic, glass, metal and/or polymer fibers. They are supplied in the form of blocks, sheets and rods.

Hybor. Trade name of British Steel plc (UK) for a stainless steel available in special nuclear grades containing high-boron levels for increased neutron absorption.

Hy-bor. Trademark of Specialty Materials Inc. (USA) for continuous unidirectional boron-graphite prepreg tapes consisting of small-diameter graphite fibers between boron fibers with a diameter of 80 or 100 μm (0.003 or 0.004 in.) in a matrix of epoxy, polyimide or other engineering resins. The hybrid prepregs have higher flexural strength and stiffness than graphite-fiber epoxy prepregs, higher modulus of elasticity than boron-fiber epoxy prepregs, and interlaminar shear properties superior to either boron or graphite epoxy prepregs. Used in the manufacture of engineering composites.

Hybralox. Trademark of GBC Corporation (USA) for self-glazing, metallizable, hybrid alumina-loaded glass that is harder and stronger than standard sealing glasses, and also has a higher thermal conductivity. Used for glass-to-metal seals.

hybrid composites. A group of composites that have polymeric matrices reinforced with two or more different types of fibers (e.g., graphite and aramid, or glass and graphite). The fibers may be aligned and intimately mixed with one another, or they may be arranged into alternating laminas or layers, each of which consists of a different fiber type. They may be combined in the form of fabrics or tapes. *Hybrid composites* are stronger and stiffer than composites containing only a single fiber type. Also known as *hybrids*.

hybrid laminates. Composite materials composed of alternating layers of different materials, e.g., composites consisting of a layer of glass fiber-reinforced plastic (GFRP) and a layer of carbon fiber-reinforced plastic (CFRP) with excellent stiffness.

hybrids. See hybrid composites.

hybrid silk. A class of water-soluble polymeric biomolecular materials, currently available in the form of films and fibers, prepared by combining alternate building blocks of natural peptide molecules from silkworm silk with molecules from synthetic polymers, such as nylon, polyethylene, polyethylene oxide or polypropylene oxide. While mimicking comparable natural materials, they exceed them in flexibility and tensile strength. Although these materials have only been synthesized fairly recently, there are several potential applications including strong, elastic novel textiles, drug-delivery mechanisms, and wound dressings.

Hycar. Trademark of B.F. Goodrich Company (USA) for a series of synthetic elastomers including several nitrile, polyacrylic, polyisobutylene and styrene rubbers, and acrylic and nitrile latexes. Also included under this trademark are reactive liquid polymer compositions for modifying polymers, such as epoxies, vinyl esters and unsaturated polyesters.

Hyclad. Trade name of British Steel plc (UK) for a series of corrosion-resistant low-carbon high-chromium and low-carbon austenitic chromium-nickel steels used for cladding roofs and walls.

Hycoloid. Trade name of Celluplastic Corporation (USA) for cel-

luloid-type cellulose nitrate plastics.

Hycomax. Trade name of Gibson Electric Company (USA) for a series of cast or sintered alnico-type permanent magnet alloys that are heated to 1200°C (2192°F) and cooled in a unidirectional magnetic field to render them magnetically anisotropic. They have high permeabilities, high coercivities, high residual inductions and high magnetic energy products. Used for electrical and magnetic equipment. *Hycomax I* contains 21% nickel, 20% cobalt, 9.5% aluminum, 2% copper, and the balance iron. *Hycomax II* has 14-21% nickel, 20-29% cobalt, 7-9.5% aluminum, 2-3% copper, 0-4% titanium, 0-2% niobium (columbium), and the balance iron. *Hycomax III* contains 14% nickel, 34% cobalt, 7% aluminum, 5% titanium, 5% titanium, 4% copper, and the balance iron. *Hycomax IV* consists of 14% nickel, 40% cobalt, 3% copper, 7.5% aluminum, 8% titanium, and the balance iron.

HyComp. Trade name of Dexter Composites (USA) for polyimide sheet molding compounds.

Hycore. Trade name for a metal-filled glass-ionomer dental cement.

Hyco-Span. Trade name of US Steel Corporation (USA) for a stainless steel that contains 0.12% carbon, 17-19% chromium, 10-13% nickel, and the balance iron. Used for aircraft control cables.

Hyd-Cast. Trade name of A.L. Hyde Company (USA) for a cast nylon supplied in four primary grades: natural, heat-stabilized, molybdenum disulfide-filled, and oil-filled. It is available in the form of rods, tubes, slabs and stock shapes for use in the manufacture of industrial machinery parts.

Hydlar. Trademark of A.L. Hyde Company (USA) for impact- and wear-resistant high-modulus composites consisting of thermoplastic matrix resins, such as polyphenylene sulfide or nylon 6,6, reinforced with fibers, such as *Kevlar*, and supplied in the form of rods, slabs, strips and tubes.

Hydra. (1) Trade name of Osborn Steel Limited (UK) for a series of hot-work tool steels containing 0.25-0.4% carbon, 1.5-4.75% chromium, up to 4% nickel and 1.2% molybdenum, respectively, 4-14% tungsten, 0.25-1% vanadium, and the balance iron. Used for hot extrusion and forging tools and dies, die-casting dies for aluminum and brass, punches and mandrels.

(2) Trade name of Osborn Steels Limited (UK) for a tungsten high-speed steel containing 0.65% carbon, 4% chromium, 15% tungsten, 0.5% vanadium, and the balance iron. Used for cutting tools, drills, taps, reamers, etc. *Hydra Covan* is a tungsten high-speed steel that contains 1.5% carbon, 5% chromium, 12.5% tungsten, 5% cobalt, and the balance iron, and is used for cutting tools. *Hydra Husky* is a tungsten high-speed steel with 0.8% carbon, 4.5% chromium, 22% tungsten, 1.5% vanadium, and the balance iron, and is used cutting and finishing tools. *Hydra Vantage* is a tungsten high-speed steel that contains 1.25% carbon, 4.5% chromium, 13.5% tungsten, 4% vanadium, and the balance iron, and is used for cutting tools, miling cutters and various other cutting and finishing tools. *Hydra Multico* refers to a cobalt-tungsten high-speed steel containing 0.85% carbon, 4.5% chromium, 22% tungsten, 1.5% vanadium, 0.75% molybdenum, 11% cobalt and the balance iron. Used for cutting and finishing tools.

Hydraphos. Trade name of Maurer-Schumaker, Inc. (USA) for a phosphate conversion coating.

Hydraplex. Trade name of Strathmore Products, Inc. (USA) for water-reducible coatings.

hydrargillite. See gibbsite.

hydrate. (1) A substance containing bound water, e.g., barium chloride dihydrate ($BaCl·2H_2O$).

(2) A compound formed when water molecules combine with the molecules of certain other substances. For example, gypsum ($CaSO_4·2H_2O$) is a hydrate.

hydrated alumina. See alumina hydrate.

hydrated aluminum oxide. See alumina hydrate.

hydrated cellulose. See hydrocellulose.

hydrated copper oxide. See copper hydroxide.

hydrated lime. See calcium hydroxide.

hydrated iron oxide. See ferric hydroxide.

hydrated magnesium metasilicate. See magnesium silicate (i).

Hydratone. Trade name Spraylat Corporation (USA) for a water-based paint.

hydraulic bronze. A pressure-tight casting bronze containing 72.5-88% copper, 4-19.5% zinc, 1.75-11% tin and 0.1-6% lead. It has a dense structure, good resistance to liquid pressures, good castability and machinability, and a good surface finish. Used for valves, steam-valve bodies, fittings, cocks, pumps, gears and bearings. Also known as *red casting brass; steam bronze.*

hydraulic cement. Cement composed of a mixture of finely divided lime, alumina and silica. It sets to a hard product by chemical reaction with water. Some types also set under water.

hydraulic-cement refractory. A mixture of refractory materials that will set by admixture and reaction with water. Used as mortar in the lining of furnaces and kilns and for patching of refractory linings.

hydraulic hydrated lime. A dry powder obtained by calcining limestone containing considerable amounts of alumina and silica. It is chemically impure and has good hydraulic properties.

hydraulic lime. Calcined limestone that contains appreciable amounts of silica (SiO_2) and alumina (Al_2O_3), usually as clay, and some iron. It absorbs water without swelling or heating, and produces cement that will set and harden under water. Also known as *hydraulic limestone; water lime.*

hydraulic mortar. Mortar that will set to a hard product by chemical reaction with water. Some grades will also set under water.

hydraulic refractory cement. Cement composed of a mixture of ground refractory materials, some of which react chemically to effect setting at room temperature. It usually contains considerable amounts of alumina (typically more than 40%).

hydraulic-setting refractories. See castables.

hydrazine. A colorless, fuming, hygroscopic liquid (98+% pure) with a density of 1.021 g/cm^3, a melting point of 2°C (35°F), a boiling point of 113.5°C (236°F), an autoignition temperature of 518°F (270°C), a flash point of 126°F (52°C), and a refractive index of 1.4700. Used as polymeriza-tion catalyst, as a reducing agent for transition metals and arsenic, selenium, plutonium and uranium, in the electroplating of metals, glass and plastics, in the manufacure of fuel cells, in nuclear reactor cooling, in boiler feedwater cooling, in wastewater treatment, as a scavenger for gases, in high-energy fuels, and as a rocket propellant. Formula: H_4N_2. Also known as *hydrazine base; diamine.*

Hydrex. (1) Trade name of Osborn Steels Limited (UK) for a series of corrosion- and/or heat-resistant austenitic, ferritic and martensitic stainless steels including several free-machining types. Used for numerous applications requiring corrosion- and/or heat-resistant steels, and also available as welding electrodes.

(2) Trade name of Reichhold Chemicals, Inc. (USA) for a series of durable, high-performance laminating resins based on vinyl esters, vinyl ester/dicyclopentadiene blends or glass fiber-reinforced polyesters. Used in boatbuilding.

(3) Trademark of J.M. Huber Corporation (USA) for a sodium magnesium aluminosilicate.

hydride. A compound of hydrogen with another element or radical, e.g., calcium hydride (CaH_2) is a metal hydride. Many hydrides find application as hydrogen storage materials.

hydride powder. A powder produced by the elimination of hydrogen from metal hydrides.

hydridosesquioxanes. A class of inorganic, siloxane-based polymers with ladder structures in which the silicon atoms are attached to one hydrogen and three oxygen atoms. They can be crosslinked by a process involving baking and curing, and have high hardness, toughness and glass-transition temperatures and relatively low dielectric constants (k < 3). Usually prepared as thin films, they can be used as interlayer dielectrics for integrated circuit devices. Also known as *hydrogensilsequioxanes.*

Hydrin. Trademark of B.F. Goodrich Company (USA) for synthetic elastomers, such as epichlorohydrin, used for molded automotive and industrial products.

hydroaromatic compound. A compound that contains the carbon skeleton of an aromatic system, but has an excess of hydrogen. Examples include decalin, tetralin and 1,4-dihydronaphthalene.

hydroastrophyllite. A dark brown mineral of the astrophyllite group composed of potassium calcium oxonium iron manganese aluminum titanium silicate fluoride hydroxide, $(Ca,K)(H_3O)_2(Fe,Mn,Al)_7Ti_2(SiO_4)_5(OH)_{10}F$. Crystal system, triclinic. Density, 3.15 g/cm^3; refractive index, 1.720. Occurrence: China.

hyrobiotite. A mineral composed of potassium magnesium iron aluminum silicate hydroxide hydrate, $K(Mg,Fe)_9(Si,Al)_8O_{20}(OH)_4·xH_2O$. Crystal system, orthorhombic.

hydroboracite. A mineral composed of calcium magnesium borate hydroxide trihydrate, $CaMg[B_3O_4(OH)_3]_2·3H_2O$ Crystal system, monoclinic. Density, 2.15 g/cm^3; refractive index, 1.534. Occurrence: Turkey.

Hydrocal. Trademark of United States Gypsum Company (USA) for a plaster composed chiefly of *calcium sulfate hemihydrate* ($CaSO_4·0.5H_2O$). Used as filler for paints and plastics, and as a model stone in dentistry.

hydrocalumite. A colorless to green mineral composed of calcium aluminum hydroxide trihydrate, $Ca_2Al(OH)_7·3H_2O$. Crystal system, monoclinic. Density, 2.15 g/cm^3; refractive index, 1.553. Occurrence: Ireland.

hydrocarbon black. See lampblack.

hydrocarbon plastics. A family of plastics based on hydrocarbon resins.

hydrocarbon polymer. A thermoplastic polymer with a chemical structure formed by carbon and hydrogen atoms only. Examples include polybutadiene, polyethylene, polyisobutylene, polyisoprene, polymethylpentene, polypropylene, and polystyrene.

hydrocarbon resin. A resin made by the polymerization of monomers composed exclusively of hydrogen and carbon atoms (e.g., coal tar, petroleum, or rosin). Depending on the particular composition, they are gummy or brittle. Used for caulking and coating compositions, and asphalt and rubber formulations.

hydrocarbons. A large class of organic compounds that contain only the elements hydrogen and carbon. They are classified into aliphatic and cyclic hydrocarbons according to the arrangement of the carbon atoms in the molecule. Aliphatic hydrocarbons, such as paraffins, olefins, acetylenes and acyclic terpenes, are straight-chained hydrocarbons, and aromatic hydrocarbons, such as benzenes, cycloparaffins, cycloolefins, cyclo-acetylenes and naphthalenes, are closed-ring structures based on the benzene ring. *Hydrocarbons* are derived chiefly from petroleum, coal tar and plant sources.

Hydrocast. Trade name for a vinyl acrylate formerly used for dentures.

Hydrocel. Trademark of Acordis Speciality Fibres for a carboxymethyl cellulosic fiber and fabric used for the manufacture of wound dressing products, such as Aquacel and CarboFlex.

hydrocellulose. Cellulose that has been reacted with approximately 8-12% water yielding a gelatinous mass. It is derived either by pulverizing cellulose and mixing with water, or by reacting cellulose with strong salt solutions, acids or alkalies. Used in the manufacture of paper, mercerized cotton, vulcanized fiber and viscose rayon. Also known as *hydrated cellulose.* See also cellulose.

hydrocerrussite. A colorless, or white mineral composed of zinc carbonate hydroxide, $Pb_3(CO_3)_2(OH)_2$. It can also be made synthetically. Crystal system, hexagonal. Density, 6.80 g/cm^3; refractive index, 2.09.

hydrochlorborite. A colorless mineral composed of calcium chloride borate hydroxide hydrate, $Ca_4B_8O_{15}Cl_2 \cdot 21H_2O$. Crystal system, monoclinic. Density, 1.85 g/cm^3; refractive index, 1.502. Occurrence: Chile.

hydrochloroauric acid. See chloroauric acid.

hydrochlorofluorocarbon plastics. See chlorofluorohydrocarbon plastics.

Hydrock/Rapid Stone. Trade name of Kerr Dental (USA) for an accurate dental model stone composed of calcium sulfate, plaster and an inert-oxide colorant. Supplied in white, yellow and green colors, it provides maximum strength and minimal setting expansion, and is used in the preparation of full and partial dentures.

Hydrocryl. Trade name of Hydron Technologies (USA) for a denture acrylic.

Hydrodarco. Trademark of Darco Sales Corporation (USA) for finely divided activated carbon and carbonaceous materials. Used as adsorbents in water purification.

hydrodresserite. A colorless mineral composed of barium aluminum carbonate hydroxide trihydrate, $BaAl_2(CO_3)_2(OH)_4 \cdot 3H_2O$. Crystal system, triclinic. Density, 2.80 g/cm^3; refractive index, 1.594. Occurrence: Canada (Quebec).

hydroentangled nonwoven fabrics. See spunlaced nonwoven fabrics.

Hydrofil. Trademark of Honeywell International (USA) for nylon-6 fibers and yarns used for textile fabrics.

Hydrofit. Trademark of Carl Freudenberg (Germany) for polyurethane products.

Hydrofix. Trade name of Ato Findley (France) for a thin-bed mortar.

Hydroflex. Trade name of Ato Findley (France) for a flexible thin-bed mortar.

hydrofluorocarbons. See fluorohydrocarbons.

hydrogarnets. A group of cubic minerals corresponding to the general formula $A_3B_2(SiO_4)_{3-x}O(OH)_{4x}$, where A is usually calcium oxide, and B aluminum oxide or ferric oxide. *Hydrogrossularite* is a prominent member of this group.

Hydrogel. Trade name of Polytek (USA) for a crosslinked copolymer of acrylamide and acrylic acid that can absorb large amounts of water (about 400 times its weight). Supplied in granule form, it is used for contact lens and passive catheter coatings, *in vivo* controlled release of proteins, in hygiene prod-

ucts, such as disposable diapers, for laboratory applications, and in horticulture.

hydrogel. A jelly-like, viscous substance formed from an aqueous solution of a colloid by coagulation. See also colloid; gel.

hydrogen. A colorless, odorless, flammable, diatomic gaseous element of Group IA (Group 1) of the Periodic Table. It occurs combined with oxygen as water (H_2O), as a constituent of most organic compounds including petroleum and other hydrocarbons, and in acids, bases and alcohols. It is the lightest of all elements and is also present in many minerals and in living matter, and uncombined in the earth's atmosphere to the extent of less than 1 ppm. Density of gas, 0.0899 g/L; density of liquid, 0.070 g/L (at -252.7°C or -423°F); freezing point, -259.2°C (-435°F); boiling point, -252.7°C (-423°F); autoignition temperature, 1075°F (579°C); atomic number, 1; atomic weight, 1.008; monovalent. It can exist in crystalline state near absolute zero, and has both nonmetallic and metallic properties, two rare isotopes, hydrogen-2 (*deuterium*) and hydrogen-3 (*tritium*). Used in oxyhydrogen-gas and atomic-hydrogen welding, as a reducing agent for metallic ores, in the production of high-purity metals, in powder metallurgy, as a rocket and missile fuel, in cryogenic research, in the manufacture of chemicals (e.g., ammonia, aniline, methanol, and hydrochloric and hydrofluoric acid), in petroleum refining (e.g., hydrocracking, hydroforming and desulfurization), in instrument balloons, in nuclear magnetic resonance (NMR) spectroscopy, and in biochemistry and medicine. Molecular formula: H_2. Symbol: H. Also known as *hydrogen-1; light hydrogen; protium.*

hydrogen-1. See hydrogen.

hydrogen-2. See deuterium.

hydrogen-3. See tritium.

hydrogenated amorphous silicon. Amorphous silicon grown, for example, from silane gas (SiH_4) by plasma-enhanced chemical vapor deposition in the form of a thin film on a suitable substrate, such as glass, at a temperature of about 200-300°C (390-570°F). Based on the nature of the dopant gas added to the deposition chamber, it can exhibit n-type (phosphorus-containing gas) or p-type (boron-containing gas) semiconductivity. Used in electronic components, sensors, transistors, solar cells, imaging devices, and optical scanners. Abbreviation: a-Si:H

hydrogenated nitrile rubber. See also acrylonitrile-butadiene copolymer.

hydrogenated rosin. Pale-colored modified *rosin* produced by treating ordinary rosin with hydrogen to effect complete or partial saturation of the resin acids. Used in adhesives, protective coatings, and paper sizes.

hydrogen-autunite. A yellowish to greenish-yellow, radioactive mineral of the meta-autunite group composed of oxonium uranyl phosphate dodecahydrate, $(H_3O)_2(UO_2)_2(PO_4)_2 \cdot 12H_2O$. It can also be made synthetically. Crystal system, tetragonal. Density, 3.91 g/cm^3. Occurrence: USA (Arizona).

hydrogen cyanamide. See cyanamide.

hydrogen economy materials. Advanced materials used in an energy economy based on hydrogen as a fuel. These include materials for the production of hydrogen from water (e.g., fuel cells), hydrogen storage (e.g., intermetallics or complex hydrides), hydrogen conversion (e.g., fuel cells, reciprocating engines, turbines and heaters) and hydrogen distribution infrastructure (e.g., pipeline and compressor materials and hydrogen sensors). See also hydrogen storage materials.

hydrogen hexabromoplatinate. See bromoplatinic acid.

hydrogen hexachloroosmate hexahydrate. See chloroosmic acid.

hydrogen hexachloroplatinate. See chloroplatinic acid.

hydrogen phosphide. See phosphine.

hydrogen-reduced iron. Iron powder produced by reducing selected iron oxides in a hydrogen atmosphere. Used in ferrous powder metallurgy, and for magnets.

hydrogen-reduced powder. A metal powder produced by hydrogen reduction of selected metals or compounds, and used chiefly in powder metallurgy.

hydrogensilsequioxanes. See hydridosilsesquioxanes.

hydrogen-storage materials. A class of materials that can reversibly absorb and store high amounts of molecular hydrogen and are thus suitable for use in alternative energy-storage systems. Examples include rare-earth metal hydrides, magnesium-based alloys, nanocrystalline alloys, nanotubes and alkali-doped graphite nanofibers. See also hydrogen economy materials.

hydrogen tetrabromoaurate. Moisture-sensitive black crystals used in gold plating, ceramics and glass. Formula: $HAuBr_4 \cdot 4H_2O$. Also known as *acid gold bromide*. See bromoauric acid.

hydrogen tetrachloroaurate. Light- and moisture-sensitive yellow to red crystals derived by the action of aqua regia on gold, and used in gold plating, in ceramics (e.g., in enamels and for painting porcelain), in ceramic and glass gilding, and in ruby glass and purple of Cassius. Formula: $HAuCl_4 \cdot 4H_2O$. Also known as *acid gold chloride*. See also chloroauric acid.

hydroglauberite. A white mineral composed of sodium calcium sulfate hexahydrate, $Na_{10}Ca_3(SO_4)_8 \cdot 6H_2O$. Crystal system, orthorhombic. Density, 1.51 g/cm^3. Occurrence: Russian Federation.

hydrogrossularite. A colorless mineral of the garnet group composed of calcium aluminum silicate carbonate hydroxide, $Ca_3Al_2(SiO_4,CO_3,OH)_3$. Crystal system, cubic. Density, 3.13 g/cm^3; refractive index, 1.675. Occurrence: USA (California). Also known as *hydrogrossular*.

hydrohalite. A colorless mineral composed of sodium chloride dihydrate, $NaCl \cdot 2H_2O$. Crystal system, monoclinic. Density, 1.63 g/cm^3. Occurrence: Antarctica.

hydrohetaerolite. A dark brown to brownish-black mineral of the spinel group composed of zinc manganese oxide monohydrate, $Zn_2Mn_4O_8 \cdot H_2O$. Crystal system, tetragonal. Density, 4.64 g/cm^3; refractive index, 2.26. Occurrence: USA (Colorado, New Jersey).

Hydrolac. (1) Trademark of Iroquois Chemicals Corporation (Canada) for water-based coatings used on paper substrates.

(2) Trademark of Hood Products, Inc. (USA) for wood and metal finish lacquers.

(3) Trade name of Obron Atlantic Corporation (USA) for water-stabilized aluminum pastes used as colorants and pigments.

HydroLAST. Trademark of AST Products Inc. (USA) for a stable, biocompatible submicron-thick coating applied to the surface of hydrophobic materials (e.g., polymers and metals) by covalent surface grafting. Used for various applications in diagnostics and biotechnology.

Hydroloy. Trademark of Stoody Company (USA) for a stainless steel that contains 0.2% carbon, 17.0% chromium, 9.5% manganese, 9% cobalt, 2.5% silicon, 0.2% nitrogen, and the balance, and is available in the form of castings, wrought shapes, wire, powder and hardfacings. It has good resistance to cavitation erosion and impingement corrosion, good work hardenability, and moderate general corrosion resistance. It can be heat treated to enhance ductility, and is used for turbine and boiler components.

hydrolyzed starch. A water-soluble polymer obtained from potato starch by hydrolysis, and used in starch-gel electrophoresis. See also starch.

Hydrolux. Trade name of Tenax Finishing Products Company (USA) for a two-coat, textured water-base finish for metallic substrates.

hydromagnesite. A white mineral composed of magnesium carbonate hydroxide tetrahydrate, $Mg_5(CO_3)_4(OH)_2 \cdot 4H_2O$. Crystal system, monoclinic. Density, 2.25 g/cm^3; refractive index, 1.527. Occurrence: Iran.

hydromarchite. A white mineral composed of tin oxide hydroxide, $(Sn_3O_2)(OH)_2$. Crystal system, tetragonal. Density, 5.00 g/cm^3. Occurrence: Canada (Ontario).

hydromolysite. A deep orange yellow mineral composed of iron chloride hexahydrate, $FeCl_3 \cdot 6H_2O$. It can also be made synthetically. Crystal system, monoclinic. Density, 1.84 g/cm^3; melting point, 37°C (99°F). Occurrence: USA (New Jersey).

Hydronalium. Trade name of Westfälische Leichtmetallwerke GmbH (Germany) for a series of strong, corrosion-resistant wrought and cast aluminum-magnesium and aluminum-magnesium-silicon alloys, and wrought aluminum-magnesium-manganese and aluminum-zinc-magnesium alloys. They have low density, good resistance to seawater corrosion and weak alka-line solutions, good cold workability and high surface finish. Available in the form of sheet and strip, structural shapes, bars, tubing, forgings and sand, chill and die castings, they are used in the aircraft industry, in shipbuilding and marine applications, in the chemical and food industries, and in building construction.

Hydrone. (1) British trade name for an alloy containing 67% lead and 33% antimony. It has low to medium shock resistance, and is used for bearings.

(2) Trade name of Cutler Hammer, Inc. (USA) for an alloy of 69% lead and 31% sodium used as a deoxidizer for nonferrous metals, and in the production of pure hydrogen.

hydronium jarosite. A golden yellow mineral of the alunite group composed of oxonium iron sulfate hydroxide, $H_3OFe_3(SO_4)_2(OH)_6$. It can also be made synthetically. Crystal system, rhombohedral (hexagonal). Density, 2.50 g/cm^3; refractive index, 1.816. Occurrence: Poland.

hydrophilic sol. A sol that precipitates at a very low electrolyte concentration. See also sol.

hydrophobic sand. A pure *beach sand* coated with minute pure silica (SiO_2) particles which have been chemically treated by a reacting with to trimethylhydroxysilane [$(CH_3)_3SiOH$] vapors resulting in bonding of the latter to the particles with the formation of water. This renders the sand grains water-insoluble or hydrophobic. *Hydrophobic sand* is usually supplied in particle sizes of 150 µm (0.06 in.) or less, and is used chiefly in sculpting and as a potting soil. Also known as *magic sand*. See also beach sand.

hydroquinone. White crystals (99+% pure) obtained by first oxidizing aniline to quinone by manganese dioxide and then reducing to hydroquinone. It has a density of 1.330 g/cm^3, a melting point of 172-175°C (341-347°F), a boiling point of 285°C (545°F), a flash point of 329°F (165°C), and an autoignition temperature of 960°F (515°C). Used as a polymerization inhibitor, as paint, varnish and oil stabilizer, as a dye intermediate, as a fat and oil antioxidant, and as a photographic developer. Formula: $C_6H_4(OH)_2$. Also known as *1,4-benzenediol; p-dihydroxybenzene; quinol*.

Hydrosil. (1) Trademark of Dentsply/Caulk (USA) for dental impression materials based on addition-polymerizing silicone elastomers.

(2) Trademark of Polymer Technology Group, Inc. (USA) for synthetic resins supplied in the form of liquids, and solid crumbs and pellets for use in the manufacture of medical and industrial products.

(3) Trademark of Unimin Specialty Minerals, Inc. (USA) for microcrystalline silica used in plastics, elastomers, ceramics, paints, adhesives and sealants.

hydrosol. A *sol* whose dispersion medium is water. It is usually an aqueous colloidal dispersion of a solid, liquid or gas.

hydrotalcite. A white mineral of the sjogrenite group composed of magnesium aluminum carbonate hydroxide tetrahydrate, $Mg_6Al_2CO_3(OH)_{16} \cdot 4H_2O$. It has a *brucite*-like lamellar structure in which magnesium is octahedrally coordinated, and can also be made synthetically, e.g., by co-precipitation. Crystal system, rhombohedral (hexagonal). Density, 2.06 g/cm³; refractive index, 1.51. Occurrence: Norway. Used as an adsorbent and catalyst material for wastewater and flue-gas treatment.

Hydro-Tex. Trade name of Tenax Finishing Products Company (USA) for single-coat, water-based texture finish.

Hydrothane. Trademark of CT Biomaterials-Division of Cardio-Tech International (USA) for a series of hydrophilic thermoplastic polyurethanes with water contents from 5 to 25wt%. Supplied as clear resins with hardnesses ranging from 80 to 93 Shore A, these hydrogels have high elongation and tensile strength, inherent bacterial resistance and good processibility by molding and extrusion. Used for biomedical applications.

hydrotungstite. A greenish yellow mineral composed of tungstic acid monohydrate, $H_2WO_4 \cdot H_2O$. It can also be made synthetically. Crystal system, monoclinic. Density, 4.63 g/cm³; refractive index, 1.95. Occurrence: Bolivia.

hydrous aluminum oxide. See alumina gel.

hydrous aluminum silicate. See clay; kaolinite; montmorillonite.

hydrous mica. See illite.

hydroxide. A compound of a metallic or nonmetallic element or radical combined with one or more hydroxyl (OH⁻) radicals, e.g., lithium hydroxide (LiOH).

hydroxyapatite. A green, bluish-green or grayish-green mineral of the apatite group composed of calcium phosphate hydroxide, $Ca_5(PO_4)_3(OH)$, and forming the chief constituent of bones and teeth. It is a finely divided, nonstoichiometric powder with a hexagonal or pseudohexagonal crystal structure and density of 3.08-3.15 g/cm³ (0.111-0.114 lb/in.³). Synthetic hydroxyapatite is a surface-reactive bioceramic with outstanding biocompatibility available in regular and microcrystalline grades. *Hydroxyapatite* is used in adsorption and partition chromatography, as a coating material on metallic (e.g., steel or titanium-alloy based) prostheses and implants in bone reconstruction and regeneration, for dental implants, and for biocomposites. Abbreviation: HA. Also known as *hydroxylapatite*. See also Durapatite.

hydroxyacetic acid. See glycolic acid.

hydroxyapophyllite. A colorless to white mineral of the apophyllite group composed of potassium calcium silicate hydroxide octahydrate, $KCa_4Si_8O_{20}(OH) \cdot 8H_2O$. Crystal system, tetragonal. Density, 2.37 g/cm³; refractive index, 1.542. Occurrence: USA (North Carolina), South Africa.

hydroxyethylcellulose. A water-soluble, nonionic cellulose ether that is available in the form of a white, free-flowing powder in low-viscosity types (75-150 centipoise for a 5% solution) with molecular weights of 24000 to 27000, medium-viscosity types (4500-6500 centipoise for a 2% solution) with molecular weights of 90000 to 105000, and high-voscosity types (1500-2500 centipoise for a 1% solution). Used as a thickener and suspension agent, as a water evaporation retarder in cementitious products (e.g., cement, mortar and concrete), as a binder in glazes and films, as a paper and textile size, as a stabilizer in the polymerization of vinyls, and as a protective colloid.

hydroxyethyl starch. A starch usually prepared by converting an average of one hydroxyl (OH⁻) group per ten D-glucupyranose units of corn starch to hydroxyethyl (CH_2CH_2OH). See also starch.

hydroxylapatite. See hydroxyapatite.

hydroxylbastnaesite. (1) A yellow to dark brown mineral of the bastnaesite group composed of cerium lanthanum carbonate hydroxide, $(Ce,La)CO_3(OH)$. Crystal system, hexagonal. Density, 4.74 g/cm³; refractive index, 1.760. Occurrence: Russia.

(2) A yellow to dark brown mineral of the bastnaesite group composed of lanthanum carbonate hydroxide, $LaCO_3OH$. Crystal system, orthorhombic. Density, 4.74 g/cm³; refractive index, 1.760. Occurrence: Russia.

hydroxylellestadite. A purple mineral of the apatite group composed of calcium silicate sulfate hydroxide, $Ca_{10}(SiO_4)_3(SO_4)_3(OH,F,Cl)_2$. Crystal system, hexagonal. Density, 3.02 g/cm³; refractive index, 1.654. Occurrence: Japan.

hydroxyl-herderite. A colorless mineral of the datolite group composed of beryllium calcium phosphate hydroxide, $CaBe(PO_4)(OH)$. Crystal system, monoclinic. Density, 2.94 g/cm³; refractive index, 1.622. Occurrence: Brazil, France, USA (Maine).

α-hydroxypropionic acid. See lactic acid.

hydroxypropylcellulose. Cellulose ether some of whose hydroxyl (OH⁻) groups have been substituted with hydroxypropyl (C_3H_6OH) groups. It is a white, water-soluble powder with a molecular weight ranging from about 300000 to 1000000, used as an emulsifier, film-forming material, suspension agent and thickener. See also cellulose ether.

hydroxypropylmethylcellulose. Water-soluble cellulose ether that is available in the form of a white powder in low-viscosity types (50-100 centipoise for a 2% solution) and medium-voscosity types (4000 centipoise for a 2% solution). Used chiefly as an emulsifier, stabilizer and thickener. Also known as *propylene glycol ether*.

hydrozincite. A yellowish or white to gray mineral composed of zinc carbonate hydroxide, $Zn_5(CO_3)_2(OH)_6$. Crystal system, monoclinic. Density, 4.00 g/cm³; refractive index, 1.736; hardness, 2-2.5 Mohs. Occurrence: Europe, USA (California, Pennsylvania, Missouri, Nevada, Utah). Used as an ore of zinc. Also known as *zinc bloom*.

Hyflex. Trade name of 3M Company (USA) for extruded polyvinyl chloride tubing and tape used for electrical insulation applications.

Hyflon. Trademark of Ausimont USA, Inc. for a series of raw and semifinished fluoropolymer resins supplied in various grades for use in the chemical, pharmaceutical, electrical, electronic, painting, building and textile industries. *Hyflon ETFE* is based on ethylene-trifluoroethylene, and *Hyflon PFA* on perfluoroalkoxy alkane polymer. *Hyflon MFA* is a semicrystalline, fully fluorinated, melt-processible perfluoropolymer resin.

Hyflo Super Cel. Trademark of Celite Corporation (USA) for diatomaceous earth.

Hyflux Alnico. Trade name of Electronic Memories & Magnetics Corporation (USA) for a series of alnico-type permanent magnet alloys with high coercive force and magnetic energy product.

Hygenic. Trade name of Coltene-Whaledent (USA) for a denture acrylic with good workability.

Hy-Glass. Trade name for a light-cure dental hybrid composite.

Hygram. Trade name for rayon fibers and yarns used for textile fabrics.

hygroelastic resin matrix composite. A composite laminate whose resin matrix is capable of absorbing/desorbing moisture from/to the surroundings.

Hykro. (1) Trade name of British Steel Corporation (UK) for nitriding steels containing 0.28-0.4% carbon, 3-3.2% chromium, 0.5-0.9% molybdenum, up to 0.2% vanadium, and the balance iron. Used for machine parts, gears and crankshafts.

(2) Trade name of British Steel plc (UK) for a stainless armor plate steel.

Hykrom. Trade name of Osborn Steels Limited (UK) for low-carbon tool steels containing 0.35% carbon, 3-5% chromium, 0.5-1.5% molybdenum, 0.1-0.3% vanadium, and the balance iron. Used for casting, drop forging and extrusion dies.

hylans. A subclass of hyaluronan-based biomaterials for which modification with crosslinking agents, such as formaldehyde or vinyl sulfone, does not have a noticeable effect on the carboxylic and N-acetyl functional groups of the hyaluronan chain. They may be prepared by crosslinking the hyaluronan (hyaluronic acid) with vinyl sulfone to produce sulfonyl-bis-ethyl crosslinks between the hydroxyl groups, or by crosslinking with formaldehyde to create crosslinks between the hydroxyl groups of the hyaluronic acid molecule and imino or amino groups of linked proteins. Used in tissue engineering and skin grafting. See also hyaluronans; hyaluronic acid.

Hylar. (1) Trade name of Ausimont USA Inc. for a thermoplastic polyester film made from polyethylene terephthalate (PET). It has very high toughness, excellent fatigue and tear strength, good dielectric properties, good resistance to humidity and solvents, and good resistance to acids, alcohols, greases, oils, halogens and ketones. Used for magnetic recording tapes, electronic components, electrical insulation, and automotive tire cords.

(2) Trade name of Ausimont USA Inc. for polyvinylidene fluoride (PVDF) resins.

Hylastic. Trademark of Amsted Industries Inc. (USA) for a steel containing 0.3% carbon, 1.6% manganese, 0.4% silicon, 0.1% vanadium, and the balance iron. It is available as castings and forgings, and used for railroad and structural applications.

Hylene TPE. Trademark of E.I. DuPont de Nemours & Company (USA) for a p-phenylene diisocyanate-based thermoplastic elastomer with polycaprolactone backbone. It has a density of 1.19 g/cm³ (0.043 lb/in.³), a service temperature range of -29 to +135°C (-20 to +275°F), low water absorption, excellent flex-fatigue resistance, excellent dynamic properties, good cut and tear resistance, and good resistance to organics including hydraulic fluids. Used for power transmission couplings and belting, seals and gaskets.

Hylite. (1) Trade name of C-E Glass (USA) for a glass with indefinite pattern.

(2) Trade name of Jessop-Saville Limited (UK) for a series of titanium products including high-purity and commercially pure titanium, and several alpha, near-alpha and alpha-beta titanium alloys.

(3) Trade name of Hoogovens Hylite BV (Netherlands) for a lightweight sandwich laminate consisting of a 0.8 mm (0.03 in.) thick polypropylene core sheet covered on both sides with 0.2 mm (0.008 in.) thick layers of a non-heat-treatable aluminum-magnesium alloy (AA No. 5182-O or 5182-H18). It has high rigidity, and excellent formability and sound-damping capability. Used for automotive hoods, roofs, and trunk lids.

Hy-Lo. Trade name of Champion Rivet Company (USA) for welding electrodes that have low-hydrogen sodium coverings and are composed of low-alloy steel containing 0.12% carbon, 0.77% manganese, 0.72% silicon, 0.8% molybdenum, and the balance iron.

Hy-Lux. Trade name of Ellis Paint Company (USA) for industrial paint enamels and coatings.

Hymag. Trade name for short-fiber asbestos.

Hyman. British trade name for a nickel silver containing 58.3% silver, 16.7% copper and 8.3% nickel. Used for jewelry and ornaments.

Hyman alloy. A heat-treatable aluminum alloy containing 3% copper, 0.8% silicon, 0.5% nickel and 0.5% magnesium. Used for light-alloy parts.

Hymax. Trade name of Edgar Allen Balfour Limited (UK) for a series of cast cobalt magnet steels containing 0.85-1.15% carbon, 3-35% cobalt, 6-10% chromium, up to 1.5% molybdenum and 4.5% tungsten, respectively, and the balance iron. They have high remanent magnetization, coercivities and magnetic energy products. Used for permanent magnets.

Hymu. Trade name of Carpenter Technology Corporation (USA) for a series of soft magnetic high-nickel alloys. *Hymu 80* contains 80% nickel, 4% molybdenum, 0-0.1% carbon, and the balance iron. It has high initial permeability, low saturation induction, zero magnetocrystalline and magnetostrictive anisotropy, and is used for magnetic shields and transformer cores. *Hymu 800* contains 79% nickel, 17% iron and 4% molybdenum. It has high initial permeability and AC core losses at low magnetic flux densities, and is used for toroids for magnetic core parts, and laminations for motor cores.

Hynical. British trade name for a permanent magnet material containing 56.1% iron, 31.0% nickel, 12.5% aluminum and 0.4% titanium. It has high permeability, and is used electrical and magnetic equipment.

Hynico. Trade name of Johnson Matthey plc (UK) for a series of permanent magnet materials used for electric and magnetic equipment. *Hynico I* is a high-permeability alloy containing 19% nickel, 12% cobalt, 10% aluminum, 6% copper, and the balance iron. *Hynico II* contains 20% nickel, 20% cobalt, 8% aluminum, 4% copper, 4% titanium, 0.8 niobium (columbium), and the balance iron.

Hypalon. Trademark of DuPont Dow Elastomers (USA) for a series of chlorosulfonated polyethylene elastomers made by treating polyethylene with chlorine and sulfur dioxide. It has a density of 1.07-1.28 g/cm³ (0.039-0.046 lb/in.³), outstanding ozone, oxidation, ultraviolet and weather resistance under the most extreme conditions, good resistance to acids, alkalies, and many other chemicals, fair oil and low-temperature resistance, excellent color stability, good to fair mechanical properties, good abrasion resistance, a hardness of 50-92 Shore, and a service temperature range of -40 to +100°C (-40 to +210°F). Special carbon black-filled grades have elevated temperature resistance up to about 180°C (356°F). Used for wire and cable insulation, flexible chemical and petroleum hoses and tubes, connectors, gaskets, seals and diaphragms, shoe soles and heels, flooring, building products, tank linings, high-temperature conveyor belts, automotive components, spark-plug boots, coatings, and in polymer blends to improve oxidation and ozone resistance.

Hypar. Trade name of Universal Cyclops (USA) for tool and die steels.

Hypeak. Trade name of Swift Levick & Sons Limited (UK) for tungsten and cobalt-tungsten high-speed tool steels containing 0.8-1.5% carbon, 4-5% chromium, 6-18% tungsten, 1-5% vanadium, up to 5% molybdenum and 6% cobalt, respectively, and the balance iron. Used for cutting tools.

hyperbranched polymers. A class of polymers that have high degrees of branching and are usually prepared by single-step polymerization of A_xB monomers. See also branched polymer; star-branched polymer.

hypercinnabar. A red mineral composed of mercury sulfide, HgS. It can also be made synthetically. Crystal system, hexagonal. Density, 7.20-8.09 g/cm³; refractive index, 2.61.

Hypercor Alloy 15. Trade name of Carpenter Technology Corporation (USA) for a soft magnetic iron alloy containing 15% cobalt and 2.7% manganese, and supplied in strip, bar and plate form. It has high saturation magnetization and electrical resistivity, and its magnetic properties can be optimized by an annealing process involving heating at 1180°C (2156°F) and then holding at 700°C (1290°F) for a prescribed period of time. Used for solenoids, sensors, magnetic bearings and electromagnets.

Hypercut. Trade name of Teledyne Vasco (USA) for a molybdenum-type high-speed tool steel (AISI type M42) containing about 1.1% carbon, 0.3% manganese, 04% silicon, 9.5% molybdenum, 1.5% tungsten, 3.75% chromium, up to 0.3% nickel, 8% cobalt, 1.15% vanadium, and the balance iron. It has good deep-hardening properties, high hot hardness, excellent wear resistance, and is used for tool bits, fly cutters, and various other tools and dies.

hypereutectic alloy. In a binary A-B alloy system, any alloy that contains more B than the eutectic composition. See also eutectic alloy; hypoeutectic alloy.

hypereutectic cast iron. A cast iron that contains between 4.3 and 6.67% carbon. See also eutectic cast iron; hypoeutectic cast iron.

hypereutectoid steel. A steel that has a carbon content exceeding that of the eutectoid composition. For plain-carbon steel the hypereutectoid range lies between 0.77 and 2.11 wt% carbon, and its room-temperature microstructure is usually composed of pearlite and proeutectoid cementite. See also eutectoid steel; hypoeutectoid steel.

Hyperlast. Trade name of Kemira Polymers (USA) for polyurethane elastomers and foams. The elastomers are supplied in hard-cast, hard and soft microcellular, and reinforced microcellular grades. The foams are supplied in semirigid and structural form.

Hyperm. Trade name of Krupp Stahl AG (Germany) for a series of high-permeability magnet alloys including iron-cobalt and iron-cobalt-vanadium alloys, nickel-iron-copper and nickel-iron-carbon-molybdenum alloys, silicon steels and high-nickel alloy steels. Used for electric, electronic and magnetic machinery and equipment.

Hypernik. Trade name of Westinghouse Corporation (USA) for a hydrogen-annealed soft magnetic alloy composed of 50% iron, and 50% nickel. It has very high initial permeability at low flux densities, and is used for instrument transformers, and electrical and magnetic equipment.

Hypernom. Trade name of Carpenter Technology Corporation (USA) for high-permeability magnet alloys containing 79% nickel, 17% iron and 4% molybdenum, and used for magnetometers, instrument transformers, loading coils for telephones, and electronic shields.

Hypersil. Trademark of Shandon Southern Products, Limited (UK) for silica with numerous pores and a surface area of about 170 m²/g. Used as chromatography column packing.

Hypersilicie. British trade name for a low-density cast aluminum alloy containing 18% silicon, 3% copper, and 0.5% iron. Used for pistons.

Hypersilid. Trade name of Bradley Laboratories (UK) for an acid-resistant cast iron containing 0.4-1.1% total carbon, 14-16% silicon, and the balance iron. Used for acid-resistant castings.

hypersthene. A grayish, greenish, black or dark-brown mineral of the pyroxene group with bronze-like luster composed of iron magnesium silicate, $(Fe,Mg)SiO_3$. Crystal system, orthorhombic. Density, 3.63-3.65 g/cm³; refractive index, 1.67-1.68; hardness, 5-6 Mohs. Occurrence: Sri Lanka.

Hyplus. Trade name of British Steel plc (UK) for a series of high-strength structural steels containing up to 0.25% carbon, 0.5% silicon and 1.6% manganese, 0.2% vanadium and 0.1% niobium, respectively, and the balance iron. It has good weldability, and is used for bridges, cranes, ships, pipelines, and general structural applications.

hypoeutectic alloy. In a binary A-B alloy system, any alloy that contains less B than the eutectic composition. See also eutectic alloy; hypereutectic alloy.

hypoeutectic cast iron. A cast iron that has a carbon content than the eutectic composition, i.e., a carbon content between 2.06 and 4.3%. See also eutectic cast iron.

hypoeutectoid steel. A steel that has a carbon content less than the eutectoid composition. For carbon steel the hypoeutectoid range lies between 0.022 and 0.77 wt% carbon, and its room-temperature microstructure is usually composed of pearlite and proeutectoid alpha ferrite. See also eutectoid steel.

Hypoflex. Trade name of Superior Tube Company (USA) for needle-drawn stainless tubing.

Hypol. Trademark of W.R. Grace & Company (USA) for a series of polyurethane prepolymers made from toluene diisocyanate (TDI). They have good fire retardancy, high additive loading, and foam upon addition of water. Used for biomedical, medical, consumer and industrial applications. *Hypol Plus* refers to polyurethane prepolymers made from methylenediphenyl diisocyanate (MDI), and used for automotive, medical and surgical products.

Hy-Press. Trade name of British Steel Corporation (UK) for low-carbon steel sheet and strip used for formed and stamped parts.

Hyprez. Trade name of Engis Corporation (USA) for a finely divided diamond powder supplied in various mesh sizes for fine polishing applications.

Hy-Pro. Trade name of Allegheny Plastics, Inc. (USA) for a flame-retardant clad polypropylene sheeting.

HyproBlue. Trade name of Pavco Inc. (USA) for a blue, bright chromate coating applied by single dipping.

HyproCoat. Trade name of Pavco Inc. (USA) for a corrosion-resistant, blue, bright chromate coating.

Hyprocrode. Trade name of Sheepbridge Engineering Limited (UK) for an erosion-, corrosion- and heat-resistant austenitic cast iron containing about 3% carbon, 12.5-14.5% nickel, 6.5-7.5% copper, 4.75-5.75% chromium, and the balance iron. Used for burners, furnace, pump and hydraulic parts, impellers, plungers, and liners.

Hyproof. Trade name of British Steel plc (UK) for a series of austenitic stainless steels containing 17-19% chromium and 9-15% nickel, and having high yield strength.

Hy-Ra. Trademark of Carpenter Technology Corporation (USA) for several high-permeability, soft-magnetic nickel-iron and iron-nickel alloys used for electronic and electrical devices, magnetic amplifiers, etc. The standard *Hy-Ra 49* alloy contains 49% nickel, and the balance iron.

Hyrem Radiometal. Trade name of Telcon Metals Limited (UK) for a soft-magnetic alloy of 50% iron and 50% nickel. It has a square hysteresis loop, high remanence, and is used for transformers.

Hyresist. Trade name of British Steel plc (UK) for a series of austenitic stainless steels containing 18-22% chromium and 5-25% nickel. They have good resistance to crevice and pitting corrosion and to stress-corrosion cracking.

Hyrho Radiometal. Trade name of Telcon Metals Limited (UK) for a cold-rolled, soft-magnetic iron-nickel alloy with high resistivity, high saturation flux density and low eddy current losses, used for transformers and alternating-current applications.

Hysafe. Trademark of J.M. Huber Corporation (USA) for a synthetic *hydrotalcite* based on magnesium aluminum hydroxycarbonate.

Hyshield. (1) Trademark for a nontoxic, molded urethane foam with a durable skin that eliminates the need for any coverings or fabricated finishes. It has excellent flame resistance, and can be molded with metal, plastic or wooden inserts as an integral part of the final assembly. Used for medical accessories and motorcycle bumpers.

(2) Trademark of Scholle Corporation (USA) for plastic film for bags and containers.

Hysil. Trade name for a borosilicate glass with low coefficient of expansion, high heat and thermal-shock resistance and excellent chemical durability. Used for heat-resistant chemical glassware.

Hysol. Trademark of Dexter Corporation (USA) for an extensive series of structural adhesives and primers, composite surfacing films and syntactic-film core materials used for aerospace applications, and various grades of optoelectronic encapsulants. The series also includes various grades of epoxies, methacrylates, and ethylene-vinyl-acetate and polyamide hot melts.

Hy-Speed. Trade name of Illinois Zinc Company (USA) for a heavy-duty tin bronze containing 88% copper, 10% tin and 2% lead. Used for bearings and bushings.

Hyspeed. Trade name of American Manganese Bronze Company (USA) for a nickel bronze composed of varying amounts of copper and nickel. It has excellent corrosion resistance, good strength, and is used for housings, worm wheels, bearings and bushings.

Hyspray. Trade name of W. Canning Materials Limited (USA) for liquid buffing compositions.

Hystron. Trade name of Lenzing AG (Austria) for high-tenacity polyester filament yarns used for textile fabrics.

Hytac. Trade name of 3M Espe (USA) for a polyacid modified dental composite (com-pomer). *Hytac OSB* is a dental adhesive for tooth structure (dentin and enamel) bonding applications.

Hytel. Trade name of E.I. DuPont de Nemours & Company (USA) for strong, durable nylon 6,6 fibers used for the manufacture of textile fabrics.

Hytemp. Trade name of British Steel plc (UK) for a series of austenitic stainless steels containing 15-25% chromium and 10-20% nickel, and having good scaling resistance up to 1150°C (2100°F).

Hytempco. Trade name of Harrison Alloys Inc. (USA) for a wrought alloy composed of 70% nickel and 30% iron. It has a density of 8.4 g/cm³ (0.30 lb/in.³), good resistivity, a maximum service temperature of 500°C (930°F) and high tensile strength. Used for electrical resistance wire, heater pads, immersion heaters, and ballast resistances.

Hytempite. Trade name of F.J. Ballard & Company (UK) for an alloy of 60% iron, 25% nickel and 15% chromium used for carburizing and heat-treating boxes, annealing pots and pyrometer tubes.

Hy-Ten. (1) Trade name of Metalsource Corporation (USA) for a series of case-hardening steels and high-carbon low-alloy tool and die steels.

(2) Trade name of Metal Source Corporation (USA) for medium-carbon alloy steels containing 0.5% carbon, 1% manganese, 0.7% chromium, 0.2% molybdenum, 0.08% sulfur, and the balance iron. They have good machinability, and are used for machine parts.

Hytenal. British trade name for a corrosion-resistant bronze composed of 60% copper, 18% tin, 10% aluminum, 6% iron, 5% manganese and 1% zinc used for staybolts, valves, valve spindles and pump rods.

Hytensilite. Trade name of Hytensil Aluminum Company (USA) for a corrosion-resistant aluminum alloy containing a total of 2% silicon, magnesium and manganese. Used for light-alloy parts.

Hytex. Trade name of Mid-Mountain Materials Inc. (USA) for fiberglass textiles supplied as knit and woven ropes and tapes. They have excellent high temperature resistance up to 540°C (1000°F), high tensile strength, good resistance to vibration and thermal shock, good resistance to many chemicals, and good insulative properties. Used for seals and gaskets.

Hytrel. Trademark of E.I. DuPont de Nemours & Company (USA) for tough, resilient, lightweight thermoplastic polyester elastomers supplied in heat-stabilized, flame-retardant, UV-stabilized, and blow-molding grades. They can be processed by injection molding, rotational molding, melt casting, calendering and extrusion, and some grades also by blow molding. They have high moduli and strength, a continuous-use temperature of -40°C to +110°C (-40 to +230°F), high tolerance to temperature variations, excellent resistance to flex-fatigue and deformation, a durometer hardness of 30-82, and excellent resistance to abrasion, moisture and hydrocarbon fluids. Used for machine parts, automotive components, power tools, wire and cable covering, fittings, sports equipment, bearings, couplings, shoe soles, V-belts, hose and tubing, and consumer products.

Hy-Tuf. Trademark of Crucible Materials Corporation (USA) for an oil-hardening low-alloy steel containing 0.25% carbon, 1.3% manganese, 1.5% silicon, 0.3% chromium, 1.8% nickel, 0.4% molybdenum, and the balance iron. Available as bars, billets, blocks, rods, plates, slabs, and forgings, it has high strength and toughness, and good hardness and shock resistance. Used for engineering parts, gears, shafts and tools. Also included under this trademark are several powder-metallurgy steels supplied in the form of powders, compacts, bars, blooms, ingots, sheets, plates, rods, and tool bits.

Hy-Vin. Trade name of Hydro Polymers (USA) for a series of unplasticized and plasticized polyvinyl chlorides. The plasticized grades are available in three elongation ranges: up to 100%, 100-300%, and more than 300%.

Hyzod. Trademark of Sheffield Plastics, Inc. (USA) for extruded polycarbonate sheet products. *Hyzod (Signs)* is sign-grade sheeting and *Hyzod (Glazing)* is glazing-grade sheeting.

I

ianthinite. A black-violet, radioactive mineral composed of uranium oxide hydroxide, $U_6O_7(OH)_{20}$. Crystal system, orthorhombic. Density, 5.16 g/cm^3; refractive index, 1.91. Occurrence: Central Africa, Germany.

IBAD coatings. Metal, metalloid or nonmetal coatings and thin films deposited onto a wide range of metal and nonmetal substrates at low temperatures by the use of energetic ion beams. *Note:* IBAD stands for "ion-beam assisted deposition".

I-beam. (1) A rolled steel beam having an I-shaped cross section consisting of narrow top and bottom flanges joined by a vertical web.

(2) See wood I-joist.

ibera vera. The hard, heavy, durable wood from the tree *Caesalpinia melanocarpa*. It is similar in mechanical properties to *lignum vitae*. Source: Argentina; Paraguay. Used for constructional purposes and ornaments.

Ibis. Trade name of Phosphor Bronze Company Limited (UK) for a bearing metal composed of 55% lead and 45% tin and supplied in various grades for use in heavy-pressure bearings. *Ibis White Navy* contains 75% tin and 25% lead and is used for bearings in cold-rolling mills and stone crushing plants.

IBOLA. Trade name of Ibola GmbH Klebstoffe (Austria) for a series of building products including various adhesives for bonding flooring and wall coverings, and several spackling compounds, sealants and primers.

iBond. Trade name of Heraeus Kulzer Inc. (USA) for a light-curing, self-etching, single-step dental bonding system for direct and indirect composite, compomer, amalgam or porcelain restorations.

Icdal. Trademark for polyethylene terephthalates and several other phthalates supplied as resins and film materials.

ice. (1) The solid, crystalline form of water (H_2O). Crystal system, hexagonal. Density, 0.91-0.93 g/cm^3; melting point, 0°C (32°F); hardness, 1.5 Mohs (at 0°C or 32°F), refractive index, 1.309; melts under pressure.

(2) A clear, white or colored, coarse-grained glass enamel frit having high fluxing characteristics. It is applied to glassware by firing to produce frosted, textured or pebbled effects.

icecrete. A term referring to freshwater or seawater ice reinforced with randomly distributed continuous glass fibers or yarns, or other materials, such as sawdust, wood or paper in particulate or fibrous form. Reinforced ice has much higher stiffness than ordinary ice and is thus under consideration as a structural material for the erection of temporary constructions in arctic regions.

Iceberg. Trade name of Nitto Boseki Company Limited (Japan) for textile glass fibers.

IceGLO. Trademark of Osram Sylvania Inc. (USA) for nitride-coated electroluminescent phosphors available in two grades: (i) *Standard,* which is supplied in white, blue, green, orange, and blue/green colors, and (ii) *High Brite* supplied in blue, green, and blue/green colors. Used in electric lamps.

Iceland agate. See obsidian.

Iceland crystal. See Iceland spar.

Iceland spar. A pure, colorless, transparent variety of the mineral *calcite* ($CaCO_3$) that easily breaks along cleavage planes into rhombohedral crystals exhibiting strong double refraction. Originally obtained from Iceland, it is used for the polarization of light (e.g., for Nicol prisms and other optical instruments). Also known as *Iceland crystal; optical calcite.* See also calcite; double-refracting spar.

Iceland wool. Wool obtained from Island sheep that have outer coats of long, coarse hair and fine, soft undergrowths. The undergrowth wool is valued for making sweaters and shawls.

icestone. See cryolite.

Ichou. Trade name of Nippon Sheet Glass Company Limited (Japan) for a patterned glass.

icosahedral phases. See quasicrystalline materials.

IC silicon carbide. Silicon carbide impregnated with carbon and having a low bulk density (2.60 g/cm^3 or 0.094 $lb/in.^3$). See also silicon carbide.

Idaho white pine. See western white pine.

idaite. A brownish bronze mineral composed of copper iron sulfide, Cu_5FeS_6. It can also be made synthetically. Crystal system, hexagonal. Density, 4.20 g/cm^3. Occurrence: Southern Africa.

iddingsite. A reddish-brown mineral magnesium iron silicate tetrahydrate, $MgFe_2O_4(SiO_2)_3 \cdot 4H_2O$. Crystal system, orthorhombic. Occurrence: USA (California, Colorado).

Ideal. Trade name of Idealstahl Breidenbach KG (Germany) for a series of steels including case-hardening and nitriding steels, various low-alloy steels (including chromium, silicon-manganese, tungsten-chromium, chromium-vanadium and silicon-chromium types), a wide range of tool steels (e.g., cold-work, hot-work and plain-carbon types), and several low-carbon high-chromium stainless steels.

Idealith. German trade name for tough, hard, flexible, tasteless and odorless casein plastics with good resistance to grease, oil and dilute acids formerly widely used for buttons, beads, knitting needles, novelties and decorative items.

Idealoy. Trade name of Wellman Dynamics Corporation (USA) for a corrosion-resistant alloy composed of copper, tin, zinc and lead, and used for pressure castings.

ideal superconductor. See soft superconductor.

Ideor. Trade name of Darwins & Milner Inc. (USA) for a tough, shock-resisting tool steel (AISI type S1) containing 0.4% carbon, 1% silicon, 0.3% manganese, 2% tungsten, 1% chromium, 0.15% vanadium, and the balance iron. It has good toughness and high hardness and strength, and is used for shear blades, punches, rivet sets, and chisels.

Idiart. Trade name for a cast iron containing 3% carbon, 2% silicon, and the balance iron. Used for compressor pistons.

idigbo. The light-colored, yellowish wood of the tree *Terminalia ivorensis*. It resembles plain oak in appearance, has a straight grain, good stability, high decay resistance, fair nailing quality (tends to split) and works quite well. Average weight, 550 kg/m^3 (34 lb/ft^3). Source: West Africa. Used for general exterior and interior work.

Idilite. Trade name of M.P. Iding Company Inc. (USA) for alumina polishing abrasives.

Idilon. Trade name of M.P. Iding Company Inc. (USA) for coated abrasives.

idocrase. A brown, green, yellow or bluish mineral with vitreous to resinous luster that is composed of calcium aluminosilicate hydroxide, $Ca_{10}Al_4(Mg,Fe)_2Si_9O_{34}(OH)_4$, and may also contain boron, fluorine and/or beryllium. Crystal system, tetragonal. Density, 3.35-3.45 g/cm³; hardness, 6.5 Mohs. Occurrence: Switzerland, Italy, USA (California). Used chiefly as a gemstone. See also californite.

idrialite. A crystalline hydrocarbon mineral composed of dibenzanthracene, $C_{22}H_{14}$. Crystal system, orthorhombic. Density, 1.29 g/cm³. Occurrence: Slovenia.

Idronal. Italian trade name for aluminum alloys containing 5-7% magnesium and up to 1% silicon. They have excellent corrosion resistance especially to seawater corrosion, and are used for marine parts.

IF steels. See interstitial-free steels.

Igamid. Trade name of Bayer AG (Germany) for urethane textile fibers.

Igedur. Trade name of Farbenindustrie AG (Germany) for a heat-treatable, wrought aluminum alloy containing 3.5-5.5% copper, 0.2-2% magnesium, 0.2-1% silicon and 0.1-1.2% manganese. It has good resistance to seawater corrosion, high strength, and is used for light-alloy parts including aircraft components.

Igelit. Trade name of former IG Farben (Germany) for polyvinyl chloride plastics with good dielectric properties and good resistance to acids and alka-lies used for cable and wire insulation.

Igepal. Trademark of ISP Technologies Inc. (USA) for a series of polymers based on polyoxyethylene isooctylphenyl ether.

Igetalloy. Trade name of Sumitomo Electric Industries, Limited (Japan) for sintered tungsten carbides used for tool bits and metal-cutting tools.

igneous rock. A rock formed by the cooling and subsequent solidification of molten material that originated within the earth. It can either be a plutonic (coarse-grained) rock, such as granite, or a volcanic rock, such as basalt.

Ignidur. Trade name of Hoffmann Elektrogusstahlwerk (Germany) for a series of casting products based on carbon, low-alloy, chromium, or acid- and heat-resisting stainless steel.

Ignilux. Trademark of Montecatini (Italy) for acrylic plastics based on polymethyl methacrylate.

ignition-pin alloys. A group of *pyrophoric alloys* containing 61% cerium and 39% iron, or 70-73% cerium, 17-24% zinc, 1.6-6% iron and up to 2.4% aluminum. Used for ignition pins.

ihrigized steel. Steel coated with silicon at high temperatures in contact with silicon tetrachloride vapor. "Ihrigizing" is a proprietary process.

iimoriite. A buff or tan mineral composed of yttrium carbonate silicate, $Y_2(SiO_4)(CO_3)$. Crystal system, triclinic. Density, 4.47 g/cm³; refractive index, 1.824. Occurrence: USA (Alaska).

Ikilith. Trade name of S.R.F. Freed (UK) for casein plastics.

ikunolite. A lead gray-mineral of the tetradymite group composed of bismuth sulfide selenide, $Bi_4(S,Se)_3$. Crystal system, rhombohedral (hexagonal). Density, 7.80 g/cm³. Occurrence: Japan.

Ilacron. Trademark of Ahmedabad Manufacturing & Calico Printing Company (India) for polyester staple fibers used for textile fabrics.

ilesite. A clear green mineral of the starkeyite group composed of manganese zinc iron sulfate tetrahydrate, $(Mn,Zn,Fe)SO_4 \cdot 4H_2O$.

It can also be made synthetically. Crystal system, monoclinic. Density, 2.26 g/cm³; refractive index, 1.518. Occurrence: USA (Colorado).

ilimaussite. A brownish yellow mineral composed of potassium sodium barium rare-earth iron niobium silicate pentahydrate, $KNa_4Ba_2Ln(Fe,Ti)Nb_2Si_8O_{28} \cdot 5H_2O$. Crystal system, hexagonal. Density, 3.60 g/cm³; refractive index, 1.689. Occurrence: Greenland, Russian Federation.

Illax. Trademark of Schott Rohrglas AG (Germany) for a special glass tubing with exceptional chemical resistance and reduced transmission. Used for pharmaceutical applications, e.g., drinking ampoules, and vials.

illinium. See promethium.

Illinois Chrome-Nickel-Moly. Trade name of US Steel Corporation (USA) for a low-alloy steel containing 0.4% carbon, 0.8% chromium, 1.5% nickel, 0.2% molybdenum, and the balance iron. It has high resistance to torsion and stresses, and is used for shafts and axles.

Illinois Nickel Steel. Trade name of Joseph T. Ryerson & Son Inc. (USA) for a tough, oil-hardening nickel steel containing 0.3-0.4% carbon, 3.5% nickel, and the balance iron. Used for elevators, power shovels, and truck frames.

illite. A group of white, gray, light green or yellowish micaceous clay minerals ranging between *montmorillonite* and *muscovite* in composition and structure. They belong either to the dioctahedral, or the trioctahedral subgroup of the mica group. The crystal systems, compositions and properties found in these two subgroups are: (i) *Trioctahedral subgroup:* Orthorhombic; potassium iron magnesium aluminum silicate hydrate; density, 2.75 g/cm³; refractive index, 1.565; and (ii) *Dioctahedral subgroup:* Monoclinic; potassium aluminum silicate hydroxide, or potassium aluminum silicate hydroxide hydrate; density, 2.60-2.82 g/cm³; refractive index, 1.57-1.61. *Illites* occur in Canada (Alberta), Japan, UK (Wales), and the USA (Illinois), and are used in ce-ramics, e.g., as clay additions for certain ceramic bodies. Also known as *hydrous mica*.

Illium. (1) Trade name of Stainless Foundry & Engineering, Inc. (USA) for a series of cast and wrought nickel-base alloys containing about 21-30% chromium, 6-9% copper, 2-5% molybdenum, and sometimes small amounts of tungsten, iron, manganese, aluminum and silicon. They have good acid and corrosion resistance, and are used as a substitute for stainless steels in severe environments.

(2) Trade name of Stainless Foundry & Engineering, Inc. (USA) for several cast and wrought cobalt-base alloys containing 26-30% chromium, 0.1-0.8% carbon, and up to 20% iron, 5% molybdenum, 15% tungsten, 1% silicon and 1% nickel, respectively. They have good abrasion, corrosion and heat resistance, and are used chiefly for chemical and furnace equipment.

Illusion. Trade name of BISCO, Inc. (USA) for a dental resin cement.

ilmajokite. A yellow mineral composed of sodium titanium silicate hydrate, $(Na,Ce,Ba)_2Ti(Si,C)_3O_9 \cdot xH_2O$. Density, 2.20 g/cm³; refractive index, 1.576. Occurrence: Russian Federation.

ilmenite. A black to deep blue or red mineral of the corundum group that has a black to brownish-red streak, a submetallic luster, and is composed of iron titanate, $FeTiO_3$. Crystal system, rhombohedral (hexagonal). Density, 4.3-5.5 g/cm³; melting point, 470°C (878°F); hardness, 5.5-6 Mohs. It resembles magnetite, but is only feebly magnetic, and can also be made synthetically from alpha ferric oxide (α-Fe_2O_3) and titanium dioxide (TiO_2). Occurrence: Canada (Labrador), India, Norway, Russia, Sweden, USA (Arkansas, California, Florida, New York, North Carolina, Virginia, Wyoming). Used as a source of titanium metal, as a source of titania in special glasses, as an opacifier in glazes and enamels, in welding-rod coatings, in titanium alloys, in the production of titanium oxide, in titanium paints and enamels, as a black colorant in brick coatings, as a speckling agent on ceramic tile, as a component in some ceramic dielectrics, and as an aggregate in high-density concrete. Also known as *titanic iron ore*.

ilmenorutile. A black mineral of the rutile group composed of iron niobium tantalum titanium oxide, $Fe_x(Nb,Ta)_{2x}Ti_{1-x}O_2$. Crystal system, tetragonal. Density, 4.20 g/cm^3. Occurrence: Malaysia.

ilomba. The moderately strong wood of the tree *Pycnanthus angolensis*. Source: African rainforest. Used for plywood.

ilsemannite. A black or bluish-black mineral composed of molybdenum oxide hydrate, $Mo_3O_8 \cdot xH_2O$. Occurrence: USA (Utah).

ilvaite. A black mineral composed of calcium iron silicate hydroxide, $CaFe_3(SiO_4)_2OH$. Crystal system, monoclinic. Density, 3.99 g/cm^3; refractive index, 1.867. Occurrence: Switzerland. Used as a magnetic ferrite material.

ILZRO zinc alloys. A series of zinc die-casting alloys developed by the International Lead-Zinc Research Association, New York, USA. Examples of these alloys include *ILZRO-14* containing about 1-1.5% copper, 0.2-0.3% titanium and 0.01-0.03% aluminum, and *ILZRO-16* with about 1-1.5% copper, 0.15-0.25% titanium, 0.1-0.2% chromium and 0.01-0.04% aluminum. *ILZRO zinc alloys* are used in the automotive industry for creep-resistant, die-cast fuel pumps, valve parts and carburetors.

Imacryl-Foam. Trademark of ICI Americas Inc. (USA) for transparent, foamed acrylic sheet material used in the manufacture of insulating glass, and for the thermal insulation of single-glazing units.

Image. (1) Trademark of Saint-Gobain (France) for a decorative laminated glass.

(2) Trademark of Degussa-Ney Dental (USA) for a hard, high-gold dental bonding alloys for high-stress porcelain-to-gold restorations. *Image 2* for a copper-free dental casting alloy containing 84.5% gold, 6.9% platinum, 5% palladium and 1% silver. It has a rich gold color, a light oxide layer, high fluidity and density, and a thermal expansion coefficient compatible with most porcelains. Used in restorative dentistry for long-span bridges and implant cases.

Image Putty. Trade name of Cadco (USA) for a white and blue putty which on combination produces a hard, elastomeric solid.

imberline. A woven fabric having differently colored warp stripes, usually divided by fine gold cords. Used especially for upholstery, hangings and drapes.

Imbue. Trademark of KoSa (USA) for a polyester yarn with embedded silver-ceramic antimicrobial additive that resists the growth of odor- and mold-producing bacteria and fungi. Used for wearing apparel including activewear, sportswear and career apparel, and medical and industrial fabrics.

imbuia. The olive to brown wood of the tree *Nectandra villosa*. It closely resembles walnut, and takes a fine polish. Source: Brazil. Used for fine furniture, cabinetwork and flooring.

Imera. Trademark of Schott DESAG AG (Germany) for a smooth, machine-drawn colored glass with tinted body. It is supplied in three thickness ranges: (i) 2.5-3.0 mm (0.10-0.12 in.) with a maximum size of 1600 × 1500 mm (63 × 59 in.); (ii) 4.5-5.5 mm (0.18-0.22 in.) with a maximum size of 2100 × 1600 mm

(83 × 63 in.); and (iii) 7.5-8.5 mm (0.30-0.34 in.) with a maximum size of 2400 × 1600 mm (95 × 63 in.). Used for glazing applications, decorative objects, and in the manufacture of laminated safety glass and insulating glass.

imgreite. A pale rose mineral of the nickeline group composed of nickel telluride, NiTe. It can also be made synthetically. Crystal system, hexagonal. Density, 8.61 g/cm^3. Occurrence: Russian Federation.

imhofite. A red mineral composed of thallium arsenic sulfide, $Tl_5As_{15}S_{25}$. Crystal system, monoclinic. Density, 5.32 g/cm^3. Occurrence: Switzerland.

imide. A nitrogen-containing acid, such as succinimide or phthalimide, with two double bonds. It is usually obtained by reacting a cyclic anhydride with ammonia at elevated temperatures.

imidoyl halide. Any of a group of reactive organic compounds having a halo group attached to the carbon-atom of a carbon-nitrogen double bond.

imine. A highly reactive, organic compound containing a carbon-nitrogen double bond.

Imita Gold. British trade name for a corrosion-resistant alloy containing about 89% copper, 10% aluminum, 0.33% iron, 0.24% nickel, 0.23% tin and 0.2% manganese. It resembles gold in color, and is used for ornaments and jewelry.

imitation hand-made paper. A rag-containing, usually wood-free paper not made by hand, but in a Fourdrinier machine. It typically has a deckle edge on two sides only.

imitation silver. A silvery white nickel brass composed of 57% copper, 25% zinc, 15% nickel and 3% cobalt, and used for hardware and fittings.

imitation wood. Molded products that resemble wood in appearance, but are made by bonding selected wood fibers with a phenolic, urea-formaldehyde or other suitable synthetic resin.

Immadium. Trade name of Manganese Bronze Limited (UK) for a series of wrought aluminum and manganese brasses containing 28-40% zinc, and varying amounts of aluminum, manganese, iron and lead.

immersion coating. (1) A metallic coating produced on an object by immersing it into the coating material or a solution of thereof, e.g., a hot-dipped zinc coating. Also known as *dip coating*.

(2) A plastic coating applied to a ceramic or metal object by immersing it into a melted resin or plastisol, and then cooling the adhering melt. Also known as *dip coating*.

(3) A ceramic coating (e.g., enamel) applied to a ceramic body or steel by immersing it into a solution, slip or other bath, and then allowing it to drain to a desired thickness before firing. Also known as *dip coating*.

immiscible polymer blends. Binary polymer blends of two amorphous resins or an amorphous and a semicrystalline resin that exist as two separate phases, since they are almost completely insoluble in each other. Such polymer blends have two glass-transition temperatures. See also alloys (2).

Immix. Trademark of OsteoBiologics, Inc. (USA) for a line of bioabsorbable cartilage and bone repair products based on highly porous polymer foams. *Immix CB* is a biomimetic implant material consisting of a cartilage phase (known as *Immix Chondroflex*) and a porous or dense bone phase whose mechanical properties can be tailored to mimic those of specific human tissue. Both phases can deliver bioactive agents, such as cells and growth factors. Used for the repair of osteochondral defects.

Immunit. Trade name of SWB Stahlformguss Gesellschaft mbH (Germany) for austenitic, ferritic and martensitic stainless steels used for various applications requiring good corrosion and heat-resistance. Also included under this trade name are several case-hardening steels.

imogolite. A mineral composed of aluminum silicate hydroxide hydrate, $Al_4Si_2O_7(OH)_{10} \cdot xH_2O$. Occurrence: Japan.

Impaco. Trade name of Spring Tools Company (USA) for dip-spin coatings.

Impact. (1) Trademark of Timken Company (USA) for a strong, tough, impact-resistant alloy steel with excellent resistance to stress corrosion cracking. Used for equipment operating in harsh environments such as encountered in energy and oil exploration.

(2) Trade name for a vinyl acrylate copolymer formerly used for denture bases.

Impactex. Trade name of Thos. Bennett & Sons (Leeds) Limited (UK) for laminated glass and mirror products.

impact-modified acrylics. Acrylic resins modified with a suitable substance (e.g., styrene-butadiene rubber) to increase their toughness and impact strength. See also acrylic plastics.

Impacto. (1) Trade name of Atlas Specialty Steels (Canada) for a tough, case-hardening steel containing 0.2% carbon, 0.8% manganese, 0.55% nickel, 0.5% chromium, 0.2% molybdenum, and the balance iron. Used for carburized shafts, pins, camshafts and ratchets.

(2) Trade name of Atlas Specialty Steels (Canada) for a carburizing steel containing 0.16% carbon, 0.5% manganese, 0.2% silicon, 0.25% molybdenum, 1.75% nickel, 0.03% phosphorus, 0.03% sulfur, and the balance iron. Used for case-hardened parts, pinions, gears and cams.

impact paper. A type of carbonless paper composed of two or more sheets coated on the front. Application of pressure (e.g., with a pen, pencil, typewriter or printer) causes the image to appear on the front of the top sheet and any subsequent sheets. Also known as *mechanical transfer paper.*

Impalco. Trade name of Alcoa of Great Britain (UK) for an extensive series of wrought aluminum products including commercially pure aluminum (99+%) in bar, plate, sheet, strip, tube and wire form as well as various wrought aluminum-copper, aluminum-magnesium, aluminum-magnesium-silicon, aluminum-manganese, aluminum-silicon and aluminum-zinc alloys and several aluminum-clad aluminum alloys.

Impax. Trade name of Uddeholm Corporation (USA) for a pre-hardened general-type mold steel (AISI type P20) containing 0.36% carbon, 1.4% chromium, 1.4% nickel, 0.2% molybdenum, and the balance iron. It has good toughness, and is used for molds, dies, tools, and die-casting tools for zinc plastic molds. *Impax Supreme* is a vacuum-degassed mold steel (AISI type P20) containing 0.33% carbon, 0.3% silicon, 1.4% manganese, 1.8% chromium, 0.8% nickel, 0.2% molybdenum, 0.008% sulfur, and the balance iron. Supplied in the hardened or tempered condition, it is used for dies, molds, structural components, and forming tools.

Imperator. Trade name of Westfälische Stahlgesellschaft AG (Germany) for a series of tough, oil-hardening steels containing 0.3% carbon, 1-2.4% chromium, 0.2-0.6% vanadium, 3.7-4.3% tungsten, and the balance iron.

Imperial. (1) Trade name of Bethlehem Steel Corporation (USA) for a shock-resisting tool steel (AISI type S2) containing 0.5% carbon. It has outstanding toughness, excellent wear resistance, good deep-hardening properties and moderate machinability. Used for chisels, pneumatic tools, and dies.

(2) Trade name of Edgar Allen Balfour Limited (UK) for a wear-resistant, water-hardening tool steel containing 1.4% carbon, 0.75% chromium, 4.5% tungsten, and the balance iron. Used for cutting tools, reamers, and drills.

(3) Trade name of Pressed Prism Plate Glass Company (USA) for a figured glass with die-pressed design.

(4) Trademark of 3M Company (USA) for coated abrasives used in metal finishing operations.

(5) Trademark of Matchless Metal Polish Company (USA) for a series of lapping film and diamond-lapping film products for precision finishing and polishing operations.

imperial cloth. A firm, durable worsted fabric in twill weave, usually dyed dark blue and used for coats and capes.

imperial green. See copper acetoarsenite.

Imperial Major. Trade name of Edgar Allen Balfour Limited (UK) for a high-speed steel containing 0.7% carbon, 4-5% chromium, 13% cobalt, 21-22% tungsten, up to 1.5% vanadium and 0.5% molybdenum, respectively, and the balance iron. It has good deep-hardening properties, and high wear resistance and hot hardness. Used for lathe and planing tools and cutters.

Imperial Manganese. Trade name of Edgar Allen Balfour Limited (UK) for a strong, wear-resistant austenitic manganese steel containing 1.15% carbon, 12.5% manganese, and the balance iron. It has high toughness and shock resistance, good tensile strength, and poor machinability. Used for rails, excavation and crushing equipment, mill liners, sprockets and crane wheels.

imperial red. A group of red pigments composed of ferric oxide (Fe_2O_3).

Impero. Trade name of Vetreria di Vernante SpA (Italy) for a patterned glass with clear background.

Imperva. Trade name of Shofu Dental Corporation (Japan) for an adhesive resin-based dental bonding system. *Imperva Bond* is a for a dentin bonding agent.

impervious carbon. A dense, impermeable carbon or graphite body containing a bituminous binder, and made by pressing and sintering to an essentially pore-free brick. It has excellent chemical resistance, and is used for lining chemical process equipment, storage vessels, pumps and valves, and in the manufacture of acid-resistant parts.

Impervo. Trademark of Benjamin Moore Paints (USA) for a range of acrylic enamel paints. *Impervo Enamel* is a waterborne, low-luster, 100% acrylic interior enamel paint with satin finish and excellent flow and leveling properties. Used on primed or painted metal, wood, masonry, plaster and wallboard.

Impet. Trademark of Ticona LLC/Hoechst Celanese Corporation (USA) for a series of thermoplastic engineering polyesters based on polyethylene terephthalate (PET), and supplied in unreinforced and 30-45% glass-reinforced grades. The glass-reinforced grades are used for injection molding applications. *Impet* polyesters have excellent physical and electrical properties, excellent flame retardancy, good chemical resistance, and high-quality surface finishes. Used for automotive components, electronic components, machine and precision engineering parts, appliances, and sports and recreational equipment.

Imphram. Trade name of Creusot-Loire (France) for nondeforming tool steels used for chisels, hammers, razors, fine cutlery, gages, borers, planers and turning tools.

Imphy. Trade name of Creusot-Loire (France) for an extensive series of alloys including numerous machinery, case-hardening, maraging, electrical, heat-resisting and structural steels, several austenitic, ferritic and martensitic stainless steels, a wide range of high-speed, cold-work, hot-work and plain-carbon tool steels as well as nickel-, iron-nickel- and cobalt-base superalloys, high-iron, iron-chromium-aluminum and iron-nickel-chromium resistance alloys, iron-nickel low-expansion alloys and copper-nickel thermocouple alloys.

Implex. Trademark of Atoglas–Atofina Chemicals Inc. (USA) for a thermoplastic, high-impact acrylic molding powder supplied in natural and colored forms. Special impact-, stain-, and heat-resistant and high-gloss grades are also available. Used for keys for business machines and musical instruments, knobs, automotive components, housings, metallized parts, and shoe heels.

impreg. See impregnated wood.

Impreglon. Trademark of Impreglon Deutschland GmbH (Germany) for fluorocarbon coatings produced by the proprietary ChemCoat infusion process.

impregnated carbon. An arc-lamp carbon composed of carbon mixed with other specially selected substances to give the light emitted by the arc a particular color.

impregnated-carbon silicon carbide. See IC silicon carbide.

impregnated fabrics. Fabrics coated or saturated with a synthetic resin such that the openings between the yarns are completely filled. Impregnation usually serves one or several of the following purposes: (i) waterproofing; (ii) flameproofing; and/or (iii) increase of immunity to bacteria and fungi.

impregnated timber. Timber that has been vacuum- or pressure-treated with a flame retardant, fungicide and/or insecticide.

impregnated wood. (1) A modified wood product developed by the U.S. Forestry Products Laboratory and composed of thin veneers impregnated with phenolic resin (e.g., phenol-formaldehyde) and then dried, cured and laid up in a thick laminate. This treatment greatly reduces changes in the moisture content and thus improves dimensional stability. *Impregnated wood* has mechanical properties comparable to those of ordinary wood, but is stronger and more resistant to chemicals, decay and insects. Used in building construction. Also known as *impreg*. See also compregnated wood.

(2) See compregnated wood.

impregnated yarn. A yarn whose interstices have been filled or saturated with a synthetic resin or other suitable substance.

impregnation body paper. Unsized paper suitably prepared for impregnation.

Impregum. Trademark of 3M ESPE (USA) for polyether-based dental materials including *Impregum F, Impregum Penta* and *Impregum Penta Soft*. The latter is a soft-setting, intrinsically hydrophobic, medium-viscosity material, whereas the former two, while being quite similar in composition and most properties, have harder sets and are more rigid. Used for high-precision crown, bridge, inlay and onlay impressions.

Imprelon. Trade name of Scheu-Dental (Germany) for hard, non-elastic, round acrylic blanks that bond well to acrylics, and are supplied in white, clear and opaque grades with a diameter of 125 mm (6 in). The white and clear grades are available with thicknesses of 2.0 and 3.0 mm (0.08 and 0.1 in.), and the opaque grades with a thickness of 3.0 mm (0.1 in.). The clear and opaque grades are used in dentistry for impression trays and dressing carriers, and the white grades for bite registration blocks, chin caps and bite splints.

Imprenal. Trademark of Raschig GmbH (Germany) for phenol-formaldehyde plastics.

Impress. Trade name of Body-Wagner Company (USA) for a nondeforming tool steel containing 0.45% carbon, 1% silicon, 1.4% chromium, 2.2% tungsten, 0.3% molybdenum, 0.2% vanadium, and the balance iron.

Impressa. Trade name of American Fibers & Yarns Company

(USA) for polyolefin fibers and yarns used for textile fabrics.

impression resin. See contact resin.

Imprex. Trade name of Cadco (USA) for a silicone elastomer used for dental impressions.

Imprint. Trademark of 3M Dental (USA) for several silicone-base dental impression elastomers including *Imprint II* heavy- and light-body, slump-resistant, fast-curing, addition-polymerizing vinyl polysiloxane (VPS) with high tear strength and dimensional stability and excellent elastic recovery, and *Imprint SBR* styrene-butadiene rubber (SBR) which sets quickly to a rigid, dimensionally stable material, and is used in dentistry for occlusal registration purposes.

Improved Hi-T Multibond. Trade name for a medium-gold dental casting alloy.

Improved 1 Star. Trade name for a medium-gold dental casting alloy.

Improvite. (1) Trademark of Illingworth Steel Company (USA) for an alloy containing 66-68% nickel, 28-31% copper and 3-4% silicon. Supplied in the form of shot, it is used for the simultaneous introduction of nickel and copper into high-duty irons to enhance pressure tightness, machinability and general mechanical properties.

(2) Trademark of Foseco Minsep NV (Netherlands) for an easy dissolving nickel-copper shot used as an alloying addition in the manufacture of high-duty cast irons. It improves machinability, pressure tightness and general mechanical properties.

impurity semiconductor. See extrinsic semiconductor.

Imron. Trademark of DuPont High Performance Coatings (USA) for polyurethane coatings with excellent resistance to weathering and harsh chemicals and outstanding gloss and color retention. They are formulated for brush, roll or spray application to commercial and industrial plant and equipment.

Imseal. Trade name of Impco Inc. (USA) for an epoxy-type metal sealer.

Inafond. Trade name of Montecatini Settore Alluminio (Italy) for a series of wrought and cast aluminum-copper, aluminum-silicon and aluminum-zinc alloys.

Inalium. (1) Trade name of Société des Brevets Bethelmy (France) for an aluminum alloy containing 0.8% magnesium, 0.25% iron, 0.45% silicon, 0.18% zinc and 0.08% tungsten. Used for light-alloy parts.

(2) Trade name of Compagnie des Alliages (France) for an aluminum alloy containing 0.5% silicon, 1.2% magnesium and 1.7% cadmium. Used for light-alloy parts.

Inastillex. Trade name of Inastillex SA (Argentina) for a laminated glass.

Inamel. Trade name of Inland Steel Company (USA) for a decarburized low-carbon steel containing 0.008% carbon, up to 0.6% manganese, up to 0.4% sulfur, and the balance iron. Used for one-coat enameling products.

incaite. A mineral of the cylindrite group composed of lead silver tin iron antimony sulfide, $(Pb,Ag)_4FeSn_4Sb_2S_{13}$. Crystal system, triclinic. Occurrence: Bolivia.

Incanite. Trade name of Incanite Foundries Limited (UK) for a series of cast irons.

incense-cedar. The light, moderately strong wood of the softwood tree *Libocedrus decurrens* of the pine family. The heartwood is light brown, frequently tinged with red, and the sapwood white or cream-colored. *Incense cedar* has a close grain, a fine, uniform texture, a spicy resinous odor, low shock resistance and stiffness, small shrinkage, good decay resistance. It is easy to work, and takes a smooth finish. Source: USA (California, Nevada, Oregon). Used mainly for lumber, shingles, ship- and boatbuilding, and fence posts. Good grades are also used for Venetian blinds, chests, toys and pencils.

In Ceram. Trademark of Vita Zahnfabrik (Germany) for dental substrate ceramics including *In Ceram Alumina* alumina-based substrates and *In Ceram Spinell* spinel-based substrates.

inclusion compounds. Crystalline molecular mixtures in which the molecules of one compound are wholly or partly enclosed within the lattice of another. They are used to separate differently shaped molecules and as templates for controlling chemical reactions. Also known as *inclusion complexes*. See also clathrate compounds.

Inco. Trademark of Inco Alloys International, Inc. (USA) for an extensive series of stainless steels, commercially pure and high-purity nickels and various nickel alloys including nickel-chromium, nickel-chromium-iron, nickel-iron-molybdenum and iron-nickel-chromium types.

Incocal. Trademark of Inco Limited (Canada) for specialty nickel-base alloys. *Incocal Alloy 10* contains additions of about 4.5-6.5% calcium, and is used to add calcium to melts of low-alloy steel and stainless steels, nickel-base alloys and heat-resisting alloys.

Inco-Cored. Trademark of Inco Alloys International Inc. (USA) for nickel-based flux-cored wires for downhand and all-position welding applications. A typical as-deposited weld metal analysis contains 72% nickel, 19% chromium, 3% manganese, 2.5% iron, 2.5% niobium, 0.3% silicon, 0.25% titanium, 0.05% aluminum, 0.04% carbon, and less than 0.5% others elements (e.g., sulfur or phosphorus). Used for joining *Inconel, Incoloy* and *Inco* alloys to themselves or austenitic and ferritic stainless steels, high-strength low-alloy steels or carbon steels, and for overlay cladding of carbon and heat-treatable low-alloy steels.

Incofiber. Trademark of Inco Limited (Canada) for carbon fibers uniformly coated with nickel by a chemical-vapor deposition process, and used as conductive fiber reinforcements for composites.

Incofoam. Trademark of Inco Limited (Canada) for a high-purity nickel foam (99.98%) made by a proprietary carbonyl chemical-vapor deposition process. It has an open, uniform pore structure, high strength, good high-temperature and corrosion resistance, and a density of 400-800 g/m^2 (0.08-0.16 lb/ft^2). Used as a filter material and catalyst, in fuel cells, and in rechargeable batteries.

Incoloy. Trademark of Inco Alloys International, Inc. (USA) for an extensive series of corrosion- and/or heat-resistant alloys composed of varying amounts of nickel, iron and chromium. They have excellent oxidation resistance and strength at elevated temperatures, and are used for heat exchangers, furnace parts, heat-treating equipment, chemical process equipment, steam generators, turbine parts, jet-engine parts, sheathing for electric heating elements and resistance wire.

Incomag. Trademark of Inco Limited (Canada) for specialty nickel-base foundry alloys. *Incomag Alloy 1* contains additions of about 14-16% magnesium and 0.9-1.0% carbon, and *Incomag Alloy 3LC* has additions of 4.0-5.0% magnesium and 0.025% carbon. Both alloys are used in the foundry industry to introduce magnesium into ductile iron. *Incomag Alloy 3LC* is also used for sulfide inclusion control in iron- and nickel-base wrought alloys. *Incomag Alloy 4* contains 34% iron, 4.6% magnesium and 1.87% carbon, and is used as a direct melt addition to ductile iron.

IncoMap. Trademark of Inco Alloys International, Inc. (USA) for a series of high-strength, corrosion-resistant wrought aluminum alloys containing 4% magnesium, 1.1% carbon, up to 1.5% lithium and 0.5% oxygen.

Incomet. Trademark of Inco Alloys International, Inc. (USA) for a granular medium-quality nickel that contains 1.3% cobalt, 1.1% oxygen, 0.4% copper and 0.4% iron, and is used for alloying applications.

Incomparable. Trade name of Flockton, Tompkin & Company Limited (UK) for a high-speed steel containing 0.8% carbon, 4.5% chromium, 22% tungsten, 1.5% vanadium, and the balance iron. Used for cutters, and form, lathe and planer tools.

Inconel. Trademark of Inco Alloys International, Inc. (USA) for an extensive range of corrosion- and/or heat-resistant alloys containing varying amounts of nickel, chromium and iron. They have excellent oxidation resistance and strength at elevated temperatures, and are used for turbine parts, nuclear reactor components, marine and chemical equipment, dairy equipment, food-processing equipment, heat-treating equipment, springs, bolts, extrusion dies, forming tools and tubing. Examples of two important alloys of this product range are *Inconel 600* (14-17% chromium, 6-10% iron, 0.5% silicon and 0.5% copper respectively, up to 0.15% carbon, 0.1% manganese and 0.015% sulfur, balance nickel) and *Inconel 690* (30% chromium, 9.5% iron, 0.03% carbon, balance nickel).

Incor. Trade name of Electronic Memories & Magnetics Corporation (USA) for a permanent magnet alloy composed of 67% cobalt and 33% rhenium. It has high residual induction, coercivity and magnetic energy product.

Incoshield. Trademark of Inco Limited (Canada) for long-fiber nickel concentrates with excellent electroconductive properties made by a two-step process involving the application of a nickel coating onto a carbon fiber by chemical-vapor deposition followed by impregnation with a thermoplastic polymer. Used for EMI/RFI shielding of computers, communications and electronic equipment.

Incote. Trade name of Morton Powder Coatings (USA) for a powder coating used for in-mold applications.

Incra. Trade name of Brush Wellman (USA) for a special cast copper alloy containing 13.5-16.5% nickel, 9.5-10.5% aluminum, 0.4-1% iron and 1-2% cobalt. It has good thermal shock and fatigue resistance, good oxidation and growth resistance, and good strength. Used for glassmaking molds and plate glass rolls.

Incramet. Trade name of Copper Development Association, Inc. (UK) for cast copper alloys containing 13.5-16.5% nickel, 10-11.5% aluminum, 0.4-1% iron and 1-2% cobalt. They have good heat and corrosion resistance, and are used for glassmaking molds, bottle molds and plate glass rolls.

Incramute. Trade name of N.C. Ashton Limited (UK) for a series of alloys composed of copper and manganese, and having high damping characteristics.

Increspato. Trade name of Vetreria di Vernante SpA (Italy) for a patterned glass.

Incroloy. Trade name of Electronic Memories & Magnetics Corporation (USA) for a permanent magnet alloy containing 56% iron, 28% chromium, 15% cobalt and 1% silicon. It has a high magnetic energy product, and high residual induction and coercive force.

Incuro. Trade name of GTE Products Corporation (USA) for a series of brazing alloys containing 35-80% copper, 20-60% gold and 2-5% indium. Available in the form of foils, flexibraze, extrudable pastes, preforms, powder and wire, they have a liquidus temperature range of 900-1025°C (1650-1880°F), and a solidus temperature range of 860-975°C (1580-1790°F).

Incus. Trade name of Uddeholm Corporation (USA) for a tough die steel containing 0.55% carbon, 0.7% manganese, 0.7% chromium, 1.75% nickel, 0.7% molybdenum, and the balance iron. Used for drop forging dies.

Incusil. Trade name of GTE Products Corporation (USA) for a series of brazing alloys containing 60-63% silver, 24-27% copper and 10-15% indium. Available in the form of foils, flexibraze, extrudable pastes, preforms, powder and wire, they have a melting range of about 630-730°C (1165-1345°F).

Incut. Trade name of Inland Steel Company (USA) for a series of free-machining, oil-hardening steels containing 0.35-0.45% carbon, 0.75-1% manganese, 0.15-0.3% silicon, 0.15-0.35% lead, 0.8-1.1% chromium, 0.15-0.25% molybdenum, 0.035% or more tellurium, and the balance iron. Used for shafts, axles, fittings, collets, gears, wrenches, studs, forging hammers, and rams.

Inda. Trade name for a synthetic resin made from casein and formaldehyde.

Indalloy. (1) Trademark of Indium Corporation of America (USA) for an extensive series of special solders and alloys, many of which contain indium. Examples include lead-indium, lead-indium-silver and lead-tin-silver alloys as well as lead-free alloys. Available in the form of ingots, rods, sheets, foil, ribbon, shot, pellets, powders, spheres, tubing, pastes and preforms, they possess good thermal fatigue resistance and improved ductility, wettability and fusibility. Used chiefly in the microelectronics industry.

(2) Trade name of Indiana General (USA) for a sintered permanent magnet alloy containing about 17-20% molybdenum, 12% cobalt, and the balance iron. It has high coercive force and remanent magnetization.

Indar. Trade name of CMW Inc. (USA) for a series of powder-metallurgy materials including various high-purity coppers and irons, numerous copper-base alloys (e.g., brasses and nickel silvers), and numerous iron-base alloys (e.g., carbon and low-alloy steels, nickel steels and copper irons).

Indaten 355. Trade name of Sollac-Usinor (France) for a hot-rolled high-strength structural steel containing 0.1% carbon, up to 0.9% manganese, 0.2-0.5% silicon, 0.06-0.12% phosphorus, up to 0.015% sulfur, 0.5-1.0% chromium, 0.020% or more aluminum, 0.25-0.45% copper, and the balance iron. Due to the formation of a thin, protective surface oxide (patina), it has enhanced resistance to atmospheric corrosion. Other important properties include excellent machinability and paintability, good stampability, a yield strength of 355 MPa (51 ksi), an ultimate tensile strength of 480-580 MPa (70-84 ksi), and an elongation of 18% in 80 mm (3.1 in.). Used for architectural applications, factory chimneys, silos, bridges, pylons, freight cars, and containers.

Indefatigable. French trade name for an oil-hardening steel containing 0.26% carbon, 4% nickel, 0.6% chromium, 0.9% molybdenum, and the balance iron. Used for rail and switch parts and streetcar tracks.

Independence. Trade name of Sanderson Kayser Limited (UK) for a water-hardening spring steel containing 0.8% carbon.

inderborite. A colorless mineral composed of calcium magnesium borate hydrate, $CaMgB_6O_{11} \cdot 11H_2O$. Crystal system, monoclinic. Density, 2.00 g/cm³; refractive index, 1.512. Occurrence:

Kazakhstan.

Inderseal. Trademark of 3M Company (USA) for protective coatings used for automobile underbody applications.

Index. (1) Trade name of Westa-Westdeutsche Edelstahlhandelsgesellschaft (Germany) for a water-hardening tool steel (AISI type W2) containing 1% carbon, 0.2% silicon, 0.25% manganese, 0.1% vanadium, and the balance iron. Used for tools, cutters, reamers and springs.

(2) Trademark of Dow Chemical Company (USA) for a family of thermoplastic polymers made by the copolymerization of ethylene and styrene using a proprietary processing technology (known as *INSITE*). In addition to excellent melt strength, high elasticity, low-temperature flexibility, thermal stability and very good compatibility with amorphous and crystalline polymers (e.g., polyethylenes, polypropylenes, polystyrenes, polyvinyl chlorides and ethylene-vinyl acetates), grades with high-styrene content (*S-series*) provide enhanced stress relaxation, paintability and sound management, and grades with high ethylene content (*E-series*) offer enhanced abrasion resistance, toughness, creep resistance and flexibility. *Index* polymers are used for wire and cable, durable goods, toys, footwear, flooring, packaging, and automotive components.

index paper. Paperboard that consists of one or more plies and has been appropriately sized and finished for use as a writing material.

indialite. A colorless mineral of the beryl group composed of magnesium aluminum silicate, $\alpha\text{-}Mg_2Al_4Si_5O_{18}$. It can also be made synthetically. Crystal system, hexagonal. Density, 2.51 g/cm³; refractive index, 1.528. Occurrence: India.

indianaite. A white, waxy clay mineral composed mainly of impure *halloysite*, $Al_2Si_2O_5(OH)_4 \cdot 2H_2O$. Occurrence: USA (Indiana). Used for pottery and refractories, and in porcelain to increase translucency.

Indiana limestone. A gray or buff-colored massive rock from Indiana, USA, consisting of small rounded grains cemented together, and composed of 98% calcium carbonate ($CaCO_3$). It is available as large blocks usually up to 4.33 ft.² (0.40 m²) in cross section and up to 12 ft. (3.7 m) in length. Used as building stone. Also known as *Bedford limestone; spergenite.*

Indiana plastic clays. (1) Clean, plastic clays with good forming properties and early vitrification that fire to deep red to red-yellow colors, and are used in the manufacture of pottery and as fillers.

(2) Buff- to white-firing, low-carbon, high-potassium plastic clays with good casting properties, low surface area, low firing distortion and early vitrification. Used for floor tile, sanitaryware, fast-firing applications, and as fillers.

Indian cotton. A variety of *cotton* cultivated in India and consisting for rather short staple fibers with a typical length of about 15-20 mm (0.6-0.8 in.).

Indian gum. See karaya gum.

Indian hemp. See sunn hemp.

Indian jute. A strong bast fiber obtained from the stems of long pot jute *(Corchorus capsularis)*, a plant of the linden family growing in India. See also jute.

Indian mica. Pale red, high-grade muscovite mica from India used for capacitors and similar electronic components. Also known as *ruby mica*. See also mica; muscovite.

Indian red. A dark brownish-red pigment based on ferric oxide obtained by calcination of copperas ($FeSO_4 \cdot 7H_2O$). It has a fine particle size, excellent permanency, good nonbleeding properties and good resistance to acids and alkalies. It was originally composed of hematite and imported from Asia. Used as a pigment in paints, plastics and elastomers, and as a polishing agent for noble metals, such as gold and silver. Also known as *iron saffron.*

Indian rosewood. See palissander.

Indian silk. A crisp, plain, hand-woven *silk* fabric with a somewhat wrinkled surface. Originally made in India, it is used for evening wear, saris and furnishings.

Indian steel. See wootz.

Indian yellow. A durable, yellow pigment consisting of a double nitrite of cobalt and potassium. Also known as *aureolin.*

India oilstone. A fast-cutting, wear-resistant oilstone made from aluminum oxide blocks.

India rubber. See caoutchouc (1).

India paper. See bible paper.

Indic. Trade name of Kulzer/Jelenko (USA) for a pink, low-expansion dental die stone for making implant models

indicolite. An indigo-blue, transparent variety of *tourmaline* used as a gemstone. Also known as *indigolite.*

Indie. Trade name of Heppenstall Company (USA) for an oil-hardening mold steel (AISI type P20) containing 0.35% carbon, 0.8% manganese, 0.45% silicon, 1.65% chromium, 0.43% molybdenum, and the balance iron. Used for molds and dies.

indigo. (1) A dark-blue dyestuff obtained from several species of East Asian plants of the genus *Indigofera* especially *I. tinctoria,* but now usually made synthetically. Used for dyeing fabrics, and in paints and printing inks.

(2) A synthetic indigo (Color Index C.I. 73000) synthesized from aniline and choroacetic acid. It is a dark blue, crystalline powder with a bronze luster, a dye content of about 80-95%, a density of 1.35 g/cm^3 (0.049 $lb/in.^3$), a melting point above 300°C (570°F), and a maximum absorption wavelength of 602 nm. Used in printing inks, textile dyes and paints, and in biochemistry and biotechnology. Also known as *indigotin; indigo blue; Pigment Blue 66.*

indigo copper. See covellite.

indigolite. See indicolite.

indigotin. See indigo (2).

indigrite. A white mineral composed of magnesium aluminum carbonate hydroxide hydrate, $Mg_2Al_2(CO_3)_4(OH)_2 \cdot 15H_2O$. Density, 1.60 g/cm^3; refractive index, 1.91. Occurrence: Russian Federation.

Indiloy. Trade name of Shofu Dental Coorporation (Japan) for a dental amalgam alloy.

indite. A black mineral of the spinel group composed of iron indium sulfide, $FeIn_2S_4$. Crystal system, cubic. Density, 4.59 g/cm^3. Occurrence: Russian Federation.

indirect-band-gap semiconductor. A semiconductor in which there exists a difference in momenta between the maximum electron-energy state in the valence band and the minimum electron-energy state in the conduction band. As a result of this indirect band gap, electrons and holes produce heat upon recombination rather than light emission. This is known as nonradiative recombination. Examples of such semiconductors are silicon, germanium and galliumk phosphide. See also direct-band-gap semiconductor.

Indirect Porcelain. Trade name for a dental porcelain veneering and bonding system used for indirect restorations.

indium. A rare, silver-white, soft metallic element of Group IIIA (Group 13) of the Periodic Table. It is commercially available in various forms (purity range from 99.9 to 99.9999%) including rods, bars, ingots, pieces, shot, granules, powder, wire, sheets, foil and single crystals. The single crystals are usually grown by the Bridgeman technique. It occurs rarely in nature, but is found in small quantities in sphalerite and other ores. Crystal system, tetragonal. Crystal structure, face-centered tetragonal. Density, 7.31 g/cm^3; melting point, 156.6°C (313.9°F); boiling point, 2080°C (3776°F); hardness, less than 10 Vickers; atomic number, 49; atomic weight, 114.818; monovalent, divalent, trivalent. It has a superconductivity critical temperature of 3.41K, an electrical conductivity of about 20% IACS, good ductility and malleability even at subzero temperatures, good room-temperature corrosion resistance, and oxidizes quickly at elevated temperatures. Used for low-melting brazing alloys and solders, fusible alloys, dental alloys, electronic and semiconductor devices (e.g., transistors, photocells and solar cells), as a dopant in semiconductors, in glass-sealing alloys, in cryogenic seals for aerospace and other cold, high-vacuum environments, in control rods of nuclear reactors, radiation detectors, mirrors, electroplated coatings on silver-plated aircraft bearings and automobile bearings, for surface coatings on aluminum wire conductors, in low-pressure sodium lamps, and in surface lubricants. Symbol: In.

indium acetate. Moisture-sensitive white crystals (99.99% pure) that decompose at 270°C (518°F) and are used as chemical and organometallic reagents. Formula: $In(O_2C_2H_3)_2$.

indium acetylacetonate. Off-white crystals or powder (98+% pure) with a density of 1.41 g/cm^3, a melting point of 187-189°C (368-372°F), a boiling point of 260-280°C (500-536°F). It crystallizes in two forms, and is soluble in water. Used in the manufacture of clear electrically conductive films. Formula: $In(O_2C_5H_7)_3$. Also known as *indium 2,4-pentanedionate.*

indium antimonide. A compound of indium and antimony that is available in the form of black crystals. The semiconductor grade is of high purity (99.99+%). N-type polycrystals are made by the horizontal zone melting method, and doped grades are also available. Crystal system, cubic. Crystal structure, sphalerite. Density, 5.76-5.78 g/cm^3; melting point, 535°C (995°F); hardness, 2200 Knoop; refractive index, 3.96; dielectric constant, 17.7; band gap, 0.163 eV. Used for semiconductor electronics, photodetectors, infrared detectors, and magnetoresistive and Hall-effect devices. Formula: InSb.

indium arsenide. A compound of indium and arsenic supplied in the form of metallic gray high-purity crystals (99.99+%). Crystal system, cubic, Crystal structure, sphalerite. Density, 5.66 g/cm^3; melting point, 943°C (1729°F); hardness, 3300 Knoop; refractive index, 3.5; dielectric constant, 14.6; band gap, 0.36 eV; maximum operating temperature, 816°C (1500°F). Used for semiconductor electronics and devices, magnetoresistive and Hall-effect devices, injection lasers, infrared photoconductors and thermistors. Formula: InAs.

indium bromide. Moisture-sensitive, white to pale yellow crystals or powder (99+% pure). Crystal system, monoclinic. Density, 4.74 g/cm^3; melting point, 420°C (788°F). Used in electroplating and electronics. Formula: $InBr_3$. Also known as *indium tribromide.*

indium chloride. Moisture-sensitive, yellow or yellowish white crystals or powder (99.9+% pure). Crystal system, monoclinic. Density, 3.46-4.0 g/cm^3; melting point, 586°C (1086°F); boiling point, sublimes at 300°C (572°F). Used in electroplating, electronics and materials research. The high-purity (99.999%) grades are also used in biochemistry as metal stains for nucleic acids. Formula: $InCl_3$. Also known as *indium trichloride.*

indium coatings. Corrosion-resistant coatings of varying thickness produced on metal substrates, such as aluminum, copper alloys, or steels, by electrodepositing indium from an acid or alkaline solution. It provides excellent solderability, good anti-friction properties, and low electrical contact resistance.

indium cyclopentadienide. Heat- and light-sensitive off-white crystals supplied in high-purity (99.999+%) electronic grades with a boiling point of 50°C (122°F)/0.01 mm (sublimes). Used in electronics and organometallic synthesis. Formula: C_5H_5In.

indium dibromide. Pale yellow orthorhombic crystals used in chemistry and materials research. Formula: $InBr_2$.

indium dichloride. Colorless, hygroscopic crystals. Crystal system, orthorhombic. Density, 3.64 g/cm³; melting point, 235°C (455°F). Used in chemistry, electronics and materials research. Formula: $InCl_2$.

indium fluoride. White, hygroscopic crystals or powder (98+% pure). Crystal system, hexagonal. Density, 4.39; melting point, approximately 1170°C (2140°F); boiling point, above 1200°C (2190°F). A trihydrate and an octadecahydrate are also available. The trihydrate loses $3H_2O$ at 100°C (212°F). Used in chemistry, electronics and materials research, e.g., in the synthesis of nonoxide glasses. Formula: InF_3 (anhydrous); $InF_3 \cdot 3H_2O$ (trihydrate); $InF_3 \cdot 18H_2O$ (octadecahydrate).

indium iodide. Moisture-sensitive yellow to red crystals or powder (99.9+% pure). Crystal system, monoclinic. Density, 4.69; melting point, 210°C (410°F); boiling point, sublimes at 300°C (572°F). Used in chemistry, electronics and materials research. Formula: InI_3. Also known as *indium triiodide*.

indium-lead solders. A group of solders composed of 50-75% indium and 25-50% lead. The solidus and liquidus temperatures for a 50In-50Pb and a 75In-25Pb alloy are 180°C (356°F) and 209°C (408°F), and 129°C (264°F) and 264°C (508°F), respectively. They have excellent resistance to thermal fatigue and low base dissolution rates for soldering onto thin gold films. Used for electronic applications.

indium monobromide. Moisture-sensitive reddish brown crystals or powder (99+% pure). Crystal system, orthorhombic. Density, 4.96 g/cm³; melting point, 220°C (428°F); boiling point, 656°C (1213°F). Used in chemistry, electronics and materials research. Formula: InBr.

indium monochloride. Moisture-sensitive white to golden yellow crystals or powder (99.9+% pure). Crystal system, cubic. Density, 4.19 g/cm³; melting point, 211°C (412°F); boiling point, 608°C (572°F). Used in chemistry, electronics and materials research. Formula: InCl.

indium monoiodide. Brownish red crystals or powder (99.9+% pure). Crystal system, orthorhombic. Density, 5.32 g/cm³; melting point, 364.4°C (688°F); boiling point, 712°C (1314°F). Used in chemistry, electronics and materials research. Formula: InI.

indium monosulfide. See indium sulfide (i)

indium nitride. A compound of indium and nitrogen. Crystal system, hexagonal. Crystal structure, wurtzite (or zincite). Density, 6.88 g/cm³; melting point, 927°C (1701°F); band gap, 2 eV. Used as semiconductor, and in solid-state electronics. Formula: InN.

indium oxide. A compound of indium and oxygen that, depending on temperature, is amorphous or crystalline. It is supplied as yellow crystals and a white to light-yellow crystalline or amorphous powder (99.9+% pure) Crystal system, cubic. Density, 7.179 g/cm³; melting point 1913°C (3475°F); n-type semiconductor. Used as a yellow colorant in special glasses, for resistance elements in integrated circuitry, and in solid-state devices. Formula: In_2O_3. Also known as *indium sesquioxide; indi-*

um trioxide.

indium 2,4-pentanedionate. See indium acetylacetonate.

indium phosphide. A compound of indium and phosphorus that is available in the form of black crystals, and as a brittle metallic mass. The electronic grade (n- or p-type) is of high purity (99.999%), and doped grades are also available. N- and p-type single crystals are grown by the liquid-encapsulated Czochralski method. Crystal system, cubic. Crystal structure, sphalerite; density, 4.787 g/cm³; melting point, 1057°C (1935°F); hardness, 4100 Knoop; refractive index, 3.1; dielectric constant, 12.4; band gap, 1.27 eV; semiconductive and photorefractive properties. Used in semiconductor devices (e.g., rectifiers and transistors) particularly at intermediate temperatures, injection lasers, optoelectronic devices and solar cells. Iron-doped semi-insulating indium phosphide crystals are grown by the liquid-encapsulated Czochralski method, and used for high-tech microwave devices. Formula: InP.

indium selenide. A compound of indium and selenium supplied in the form of black crystals and moisture-sensitive lumps. Crystal system, hexagonal. Density, 5.67-5.8 g/cm³; melting point, 660°C (1220°F). Used in semiconductor technology. Formula: In_2Se_3.

indium sesquioxide. See indium oxide.

indium solders. A group of special solders containing varying amounts of indium (40-97%), lead (0-50%) and tin (0-40%), and small additions of cadmium and/or silver. They have outstanding ductility, wetting characteristics and oxidation resistance, relatively low strength, low vapor pressure (especially 50In-50Sn solder), good cryogenic strength (especially 50In-30Sn-19Pb-1Ag solder), low base desolation rates for soldering onto thin gold films (e.g., In-Pb solder), and a melting range of 120-315°C (250-600°F). Used for wetting glass, quartz and ceramics, for glass-to-metal seals, for seals in vacuum systems, for electronic assemblies, and as a minor indium addition to tin-lead solders to improve wetting and lower melting temperatures.

indium sulfate. A white or grayish, hygroscopic powder (99.9+%). Density, 3.438 g/cm³; melting point, decomposes at 600°C (1112°F). Used in electroplating. Formula: $In_2(SO_4)_3$.

indium sulfide. Any of the following compounds of indium and sulfur: (i) *Indium monosulfide.* Red-brown crystals. Crystal system, orthorhombic. Density, 5.2 g/cm³; melting point, 692°C (1278°F). Used in materials research and electronics. Formula: InS; and (ii) *Indium trisulfide.* Orange crystals or a yellow-orange powder (99.99+%). Crystal system, cubic. Density, 4.45 g/cm³; melting point, 1050°C (1920°F). Used in electronics and semiconductor technology. Formula: In_2S_3.

indium telluride. A compound of indium and tellurium supplied in the form of gray to black, friable, high-purity crystals (99.99+%). Crystal system, cubic. Crystal structure, sphalerite. Density, 5.8 g/cm³; melting point, 667°C (1233°F); hardness, 1660 Knoop; band gap, 1.04 eV. Used in electronics and semiconductor technology. Formula: In_2Te_3.

indium-tin alloys. A group of solders composed of 44-52% indium and 48-56% tin. The solidus and liquidus temperatures for a 52In-48Sn alloy are 118°C (244°F) and 118°C (244°F) respectively. Indium-tin solder alloys have low vapor pressure, and are capable to wet glass, quartz and ceramics. Used for glass- and ceramic-to-metal seals.

indium tin oxide. A light yellow to gray, transparent conductive polycrystalline thin film material composed of indium oxide (In_2O_3) doped with tin. A typical composition is 9-10 mol% tin oxide (SnO_2), with the balance being indium oxide. It has low sheet resistivity, high light transmissivity, a density of about 7.14 g/cm^3, a refractive index of about 1.95, and a melting point of about 1900°C (3450°F). Supplied in high purities (e.g., 99.99%), it is used as an electrode material for liquid crystal displays (LCDs), electrochromic and flat-panel displays, light-emitting diodes, solar cells, gas sensors, and heat-reflecting coatings. *Indium tin oxide* coatings are usually deposited by electron-beam evaporation or sputtering. Abbreviation: ITO. Also known as *tin-doped indium oxide.*

indium tribromide. See indium bromide.

indium trichloride. See indium chloride.

indium triiodide. See indium iodide.

indium trioxide. See indium oxide.

indium trisulfide. See indium sulfide (ii).

indole black. A black pigment based on indole (C_8H_7N) and belonging to the *eumelamins*. It yields tetracarboxylic acid on oxidation.

Indopol. Trademark of Amoco Chemical Company (USA) for a series of polybutenes with number-average molecular weights ranging from 320 to over 2300. Used in adhesives, caulks, sealants, and resin and rubber compositions.

Indorama. Trademark of P.T. Indo-Rama Synthetics (Indonesia) for polyester fibers and yarns used for textile fabrics.

Indox. Trade name of Electronic Memories & Magnetics Corporation (USA) for a series of hard, anisotropic ceramic permanent magnet materials including barium ferrites ($BaO \cdot 6Fe_2O_3$), oriented barium ferrites and strontium ferrites ($SrO \cdot 6Fe_2O_3$). They have high coercivity and residual induction, and high magnetic-energy products.

Indur. Trade name of Reilly Tar & Chemical Corporation (USA) for phenol-formaldehyde plastics.

Indura. Trade name of Horace T. Potts Company (USA) for a steel used for drawing dies.

Indurite. Trade name of Indurite Moulding Powders (UK) for phenol-formaldehyde plastics and molding compounds.

Industal. Trade name for asbestos millboard.

Industrex. Trade name of ASG Industries Inc. (USA) for a glass with small lenticular figures on one surface.

Industrial. Trade name of Industrial Steels Inc. (USA) for a series of corrosion-resistant steels including various austenitic, ferritic and martensitic stainless grades, and several chromium-tungsten grades.

industrial carbon. A general term for pure *carbon* products used solely for industrial purposes, but not as fuels. Included are activated carbon, carbon black, carbon fibers and whiskers, compressed carbon, graphite and industrial diamonds.

industrial diamond. A low-grade, impure, crystalline or cryptocrystalline *diamond* that, owing to its color, hardness, shape, size, imperfections or other physical properties, is deemed inferior for use as gems. It can be a *natural diamond* (e.g., bort or carbonado) or a *synthetic diamond* made at high temperatures and pressures in a furnace. Used for grinding and polishing hard materials, for grinding wheels, drill bits and wire-drawing dies, and for glass- and metal-cutting applications.

industrial fabrics. A generic term for fabrics used for technical or industrial applications as opposed to those used chiefly for apparel, upholstery or decorative applications. *Industrial fabrics* are manufactured from natural fibers (e.g., cotton, flax, hemp, jute, or sisal), or synthetic fibers (e.g., acrylic, nylon, polyester, polyvinyl chloride, aramid, glass, or carbon) by weaving, knitting, spinning, braiding, felting, twisting or pressing.

Typical applications include conveyor belting, V-belts, cable insulation, tarpaulins, machine covers, awnings, sails, filters, automobile seats and tires, balloons, parachutes, thermal insulation, and reinforcement of composites. Also known as *mechanical fabrics; technical fabrics.*

industrial floor brick. A brick that has exceptionally high resistance to chemicals, wear, mechanical damage and thermal conditions found in industrial and commercial installations.

industrial glass. Any shaped or molded glass product ready for the market.

Industrialite. Trademark of British Uralite Limited (USA) for special medium-density hardboard used for soundproofing applications.

industrial jewel. A gem or other hard stone, such as synthetic corundum, ruby or sapphire, cut to shape and used for bearings in instruments, timepieces, watch escapements, compasses, electrical instruments (meters) and dial indicators, and formerly also for recording needles. See also jewel; synthetic corundum; synthetic ruby; synthetic sapphire.

industrial nylon. A stiff, hard-wearing, wrinkle-resistant, colored nylon fabric in a plain weave, used for heavy-duty applications, e.g., for protective suits, overalls and pant pockets.

industrial thermosetting laminates. A group of plastic laminates produced by bonding together, under heat and pressure, paper, woven asbestos, or cotton or glass fabrics impregnated with a thermosetting resin, such as an epoxy, melamine, phenolic, polyester or silicone. Available in the form of sheets, rods, tubes and structural shapes, they have high mechanical and electrical insulating properties, and are used in the manufacture of industrial products, such as gears, mallets and vise-jaw grips, bearings, bushings, sleeves or liners for rolling mills and ship rudders, and acoustical, electrical and thermal insulation.

industrial talc. A commercial product composed predominantly of the mineral *talc*, $Mg_3Si_4O_{10}(OH)_2$, or of talc mixed with other, sometimes fibrous minerals, such as serpentine asbestos.

industrial yarns. A class of yarns with an average tenacity of 1100 denier or more used predominantly for industrial applications, such as belting, tires, hoses, cordage and webbing, and as reinforcments. See also yarn.

Industrolite. Czech trade name for a wired glass.

Inertex. Trade name of Foseco Minsep NV (Netherlands) for a dusting powder used to protect molten magnesium alloys during holding and pouring.

inert filler. An inert, finely divided substance, such as blanc fixe, whiting, kaolin or mica, incorporated into a polymer (thermoset, thermoplastic or elastomer) in relatively large proportions as an extender to increase bulk and decrease costs, or as a filler to modify the physical, mechanical, thermal, electrical or other properties.

inesite. A pale orange-pink to orange red-brown mineral composed of calcium manganese silicate hydroxide hexahydrate, $Ca_2(Mn,Fe)_7Si_{10}O_{28}(OH)_2 \cdot 6H_2O$. Crystal system, triclinic. Density, 3.03 g/cm^3; refractive index, 1.6384. Occurrence: Australia (New South Wales), USA.

infiltrated steel. See powder-metallurgy infiltrated steel.

Infiloy. Trade name of Pyron Corporation (USA) for a powder composed of 90-93% copper, 3.5-6% iron, 1-1.5% manganese, 0.4-0.6% nickel and 0.5-2.0% other elements. Used for copper infiltration in the manufacture of powder-metallurgy parts.

Infinity. (1) Trade name of Den-Mat Corporation (USA) for a dual-cure radiopaque hybrid resin-ionomer dental cement with high bonding strength, good fracture resistance and self-adhesive properties. Used for cementing crowns, bridges, prosthesis, cores and endodontic posts, for bonding amalgam, and as a fluoride-releasing liner or base.

(2) Trade name of Honeywell (USA) for a recyclable nylon 6 resin made from re-polymerized nylon waste including post-consumer carpeting. Used for automotive components.

infrared dyes. See IR dyes.

infrared glass. A glass that absorbs or reflects infrared radiation, e.g., due to a metallic surface coating.

infrared optical material. An optical material that transmits infrared radiation.

infrared photoconductor. A semiconductor material that changes (usually decreases) its resistivity when exposed to infrared radiation. See also photoconductor.

infrared-transparent material. An optical material, such as sapphire, cesium iodide, sodium chloride or high-density polyethylene, that transmits infrared radiation.

Infra-Rex. Trade name of Sovirel (France) for a glass that is permeable to yellow radiation from 0.56 μm, and exhibits marked absorption of the most harmful infrared radiation. Used for welders' goggles.

Infrastop. Trade name of Flachglas AG (Germany) for a double glazing unit having a metallic coating on the inner surface of one sheet to reflect heat and sunlight.

Infrax. Trademark of Harbison-Carborundum Corporation (USA) for a refractory insulation often used in the form of bricks for the primary linings of fuel-fired and electric furnaces and kilns. It is suitable for temperatures to about 1480°C (2696°F), and may be protected by a cement facing. It is also supplied in the form cements.

infusorial earth. See diatomaceous earth.

Ingaclad. Trade name of Ingersoll Steel Company (USA) for a series of heat- and/or corrosion-resistant steels consisting of a layer of stainless steel bonded to a carbon-steel plate substrate. Used for tanks, vessels, bins, and tabletops.

Ingal. Trade name of Ingal International Gallium GmbH (Germany) for gallium metal, and several gallium compounds and alloys.

Ingerin. German trade name for polyvinyl ether.

Ingersall. Trade name of Ingersoll Steel Company (USA) for an austenitic stainless steel containing 0.1% carbon, 18% chromium, 8% nickel, and the balance iron.

Ingersoll. Trade name of Ingersoll Steel Company (USA) for a series of high-, medium- and low-alloy tool steels.

ingot. (1) A cast block of metal, such as steel, aluminum, copper, gold, lead, magnesium, nickel, silver, vanadium or zinc, having a size and shape suitable for remelting, or for mechanical shaping by processes, such as rolling, forging or extrusion. Depending on the particular material, an ingot can weigh more than 300 tons.

(2) A green or sintered powder-metallurgy compact suitable for subsequent working. Also known as *billet*.

ingot iron. A commercially pure iron (99.85+%) that essentially is highly refined steel whose carbon, silicon and manganese contents are kept very low (e.g., as low as 0.02%) by remaining in the melting furnace for extended periods of time at high temperatures. It is supplied in the form of ingots for subsequent rolling into bars, plates or shapes. A well-known example of such an iron is *Armco iron*. *Ingot iron* has high ductility and good resistance to rusting, and is used for culverts, tanks, boilers, roofing, pipes, stoves, enameled ware and metal furniture.

ingot lead. Impure lead obtained from the blast furnace and containing copper, silver, zinc and other impurities that must be removed by processes, such as the Parkes process (desilvering), Betterton-Kroll process (debismuthizing), Betts process (electrolytic refining) or Harris process (pyrometallurgical refining).

ingot metal. Any metal suitable for making ingots. See also ingot.

ingot nickel. A standard grade of virgin nickel with a purity of 98.5%.

Ingotol. Trademark of Foseco Minsep NV (Netherlands) for several readily bonded dressings that are applied to chills, ingot molds and notch-bar molds for steel and nonferrous casting alloys (especially copper and nickel) to minimize surface defects, extend the service life of molds, facilitate the removal of castings, and simplify mold cleaning.

ingrain yarn. (1) A yarn that has been spun or woven from a mixture of fibers dyed in different colors. Also known as *ingrain*. See also marl yarn.

(2) A yarn that consists of differently colored filaments. Also known as *ingrain*. See also marl-effect yarn.

inhibited admiralty brass. See antimonial admiralty brass.

inhibited aluminum brass. See arsenical aluminum brass.

inhibiting pigment. A pigment, such as zinc chromate, that is added to an organic coating to impart color and slow down the rate of corrosion of the metal to which this coating is applied.

inhibitor. (1) A chemical substance that, when added in relatively low concentrations, stops or retards a chemical reaction and is consumed in the process.

(2) An organic or inorganic substance, such as a phosphate or chromate, that dissolves in a corroding medium, but is capable of forming a protective layer of some kind at either the anodic or cathodic areas, and thus reduces the rate of a chemical or electrochemical reaction.

(3) A chemical substance added to certain types of adhesives and thermosetting resins to slow down the curing reaction and extend the storage life. Also known as *retarder*.

(4) A chemical substance that, when added to cement, slows down the rate of reaction and increases the setting time.

(5) A chemical substance applied to a textile fabric to inhibit gas fading.

Inidex. Trademark of Courtaulds Limited (UK) for polyacrylate staple fibers used for textile fabrics.

initiator. An ionic or free radical species, such as benzoyl peroxide, or an azo compound, that starts a chain-reaction polymerization. Used as a crosslinking agent in polyethylene or elastomers, and for curing thermosets. See also photoinitiator.

Injectex. Trade name for a series of textile reinforcements for synthetic resins that incorporate specially engineered yarns in the weave to effectively reduce the resistance to resin flow in resin transfer molding for high-performance composites.

injection-molded ceramics. Ceramic products with simple or complex shapes and usually constant cross sections formed by first adding organic additives to ceramic powders to produce plastic masses, and then compressing these masses into mold cavities of the desired shape, followed by firing or sintering at high temperature.

injection-molded plastics. Thermoplastic or thermosetting products of varying shape and size formed by feeding the pelletized or powdered plastic starting material into a cylinder which is pushed into a heating chamber in which the plastic charge melts. The molten plastic is then injected into the cavity of a mold and allowed to solidify. Injection-molded thermoplastics solidify (cure) almost immediately after injection into cold molds, while injection-molded thermosets cure under pressure in heated molds. See also co-injection-molded plastics; injection molding compounds.

injection molding compounds. Thermoplastic or thermosetting materials that are specifically formulated for the injection-molding process. Acrylics, ABS resins, melamine-formaldehydes, polyaryletherketones, polyesters, polyether sulfones, polyurethanes, polyvinyl chlorides, styrene-maleic anhydrides and urea-formaldehydes are especially suited for this process. See also injection-molded plastics.

injector bronze. A free-cutting bronze composed of 89% copper, 8.5% tin and 2.5% lead, and used for injectors, fittings and hardware.

Inklurit. Trademark of BASF AG (Germany) for urea-formaldehyde plastics.

Inkomo. Trade name of Sanderson Kayser Limited (UK) for a shock-resistant steel containing 0.3% carbon, 2.7-3.2% nickel, 0.7-0.8% chromium, 0.5% molybdenum, and the balance iron. Usually supplied in the hardened and tempered condition, it is used for gears, shafts and crankshafts.

Ink-Pakt. Trademark of CC Chemicals Limited (Canada) for a non-shrinking premixed grout.

Inkromstahl. Trade name of Thyssen Edelstahlwerke AG (Germany) for a series of steels containing 0.1% carbon, 0.3-1% silicon, 0.3-3% manganese, up to 2.5% chromium, 0.45% titanium, 0.2% copper and 0.17% vanadium, respectively, and the balance iron. They are suitable for chromizing treatments, and used for molten salt-bath fixtures.

Inkus. Trade name of Uddeholm Corporation (USA) for a tough, oil-hardening steel containing 0.56% carbon, 0.85% chromium, 0.2% molybdenum, 1.8% nickel, 0.1% vanadium, and the balance iron. Used for punches and forging dies.

inlay casting wax. A blend of animal, vegetable, mineral and/or synthetic dental waxes, such as beeswax, ceresin and paraffin. It is usually supplied in stick form and used in the preparation of patterns for removable partial dentures. Available in various proprietary formulations.

IN-MA superalloys. A group of nickel-base superalloys strengthened by dispersions of inert oxide particles (usually yttria, Y_2O_3). They contain about 65-80% nickel, 15-20% chromium, up to 4.5% aluminum, 4% tungsten, 2.5% titanium and 2% molybdenum, respectively, as well as small additions of carbon, tantalum, zirconium, boron and yttria. They have good heat resistance and high strength, and are used for aircraft and aerospace applications, nuclear reactors and chemical and petrochemical plant and equipment.

Inmanite. Trade name of T. Inman & Company Limited (UK) for a cobalt-tungsten high-speed steel containing 0.8% carbon, 4.5% chromium, 5% cobalt, 20% tungsten, 1% molybdenum, 1.5% vanadium, and the balance iron. Used for cutting tools for very high speeds and hard materials.

Inmet. Polish trade name for a heat-treatable aluminum alloy containing varying amounts of copper, manganese, nickel and iron. Used for light-alloy parts.

innelite. A pale yellow to brown mineral composed of sodium barium magnesium titanium silicate hydroxide sulfate, $Na_2(Ba,K)_4(Mg,Ca,Fe)Ti_3Si_4O_{18}(OH,F)_{1.5}SO_4$. Crystal system, triclinic. Density, 3.96 g/cm³; refractive index, 1.737. Occurrence: Russian Federation.

Innerberg. Trade name of Vereinigte Edelstahlwerke (Austria) for a series of water-hardening steels containing 0.5-0.6% car-

bon, 0.1-0.4% silicon, 0.5-0.7% manganese, and the balance iron. Used for shafts, bolts and gears.

Innosein. French trade name for a heat-treatable alloy composed of 94-95% aluminum and 5-6% copper. Used for light-alloy parts.

Innosil. Trademark of Silicone Innovation, Inc. (USA) for silicone hoses and tubing.

Innotal. Trade name of Pechiney Electrométallurgie (France) for a cast iron nodularizer containing 45-50% silicon, 8-10% magnesium, 4% barium, 1% calcium, up to 1% aluminum, and the balance iron.

Innova. Trade name of American Fibers & Yarns Company (USA) for polyolefin fibers and yarns used for textile fabrics.

Innswool. Trade name of A.P. Green Refractories (USA) for ceramic fibers composed of 49% alumina (Al_2O_3) and 51% silica (SiO_2), and supplied in discontinuous, bulk and mat form.

Inobar. Trade name of Pechiney Electrométallurgie (France) for a cast iron inoculant containing 63% silicon, 25.5% iron, 9% barium, 1.5% aluminum and 1% calcium.

Inocarb. Trade name of Pechiney Electrométallurgie (France) for a cast iron inoculant containing 47% carbon, 32% silicon, 15.3% iron, 4.5% barium, 0.7% aluminum and 0.5% calcium.

inoculant. See inoculating alloy.

inoculated cast iron. Cast iron to which one or more inoculants, such as ferrosilicon, cerium or magnesium, have been added while it was in the molten state in order to control the size, shape and/or distribution of graphite particles. Also known as *inoculated iron.*

inoculating alloy. A master alloy, such as ferrosilicon, cerium or magnesium, introduced into molten cast iron to promote the nucleation of graphite and thus avoid the formation of primary carbides. Also known as *inoculant.*

Inoculin. Trademark of Foseco Minsep NV (Netherlands) for a cast iron inoculant for the efficient mold or ladle inoculation of cast iron melts from high-steel charges. It significantly increases both tensile and transverse strength, lowers section sensitivity and chilling tendency, and prevents undercooled graphite.

Inoculoy. Trade name of Cyprus Foote Mineral Company (USA) for a series of cast iron and ductile iron inoculants containing 55-65% silicon, 9-12% manganese, 1.5-3% calcium, 4-6% barium, 1-2% aluminum, and the balance iron. Used to reduce the chilling tendency of gray cast iron.

Inofo. Trade name of Acieries du Forez (France) for a series of austenitic, ferritic and martensitic stainless steels.

Inor. Trade name of Westinghouse Electric Corporation (USA) for a series of corrosion-resistant nickel alloys containing 7% chromium, 15-17% molybdenum, 0.06% carbon, up to 5% iron, 1% silicon, 0.8% manganese and 0.005% boron, a total of up to 0.5 aluminum and titanium, and the balance iron. It has good resistance to hot fluoride salts, oxidation, aging and embrittlement, and is used for high-temperature applications, and jet-engine and gas turbine components.

inorganic bucky tubes. See inorganic nanotubes.

inorganic coatings. A group of vitreous or porcelain coatings that are fused to sheet or cast surfaces, and are extremely hard, smooth, and easy to clean. See also organic coatings.

inorganic cold-molded plastics. A group of plastic materials consisting of asbestos, either finely ground or in fiber form, bonded with inorganic materials, such as Portland or silica-lime cement. They have good dielectric and thermal resistance and high arc resistance, and are used for heating coil supports, arc chutes, etc. See also cold-molded plastics; organic cold-molded plastics.

inorganic color. A chemical color obtained by combining two or more inorganic chemicals.

inorganic fiber papers. A group of papers made by forming inorganic materials, such as asbestos, ceramic or glass fibers, into sheets or mats with or without the use of a bonding agent. They have good electrical and thermal properties, and can be used for insulation and filtering applications. See also asbestos paper; ceramic paper; fiberglass paper.

inorganic fibers. A class of fibers including those made from boron, carbon, glass, metals and ceramics, that have properties superior to those of organic fibers including better rigidity, better corrosion and heat resistance, higher melting points and inherent flame resistance. Used especially for composite reinforcement applications.

inorganic fluxes. A group of highly active soldering fluxes, such as sodium chloride or zinc chloride, that do not contain carbon compounds. They clean surfaces better than most other fluxes and do not char or burn easily, but are highly corrosive.

inorganic-inorganic composites. A group of composite materials consisting of inorganic reinforcing fibers embedded in inorganic matrix materials. Examples include boron-reinforced aluminum and silicon carbide-reinforced titanium composites, and tungsten-fiber-reinforced superalloys.

inorganic materials. Materials composed of chemicals that do not contain carbon as their chief element. They may be solid (e.g., metals, alloys, semiconductors, glasses, ceramics, minerals, ores, etc.), liquid (e.g., water and many acids, such as sulfuric and hydrochloric acid) or gaseous (e.g., air and process gases.).

inorganic membrane. A microporous structure that is composed of inorganic materials, such as metals, ceramics, chemicals, or inorganic polymers, and used as permeable or semipermeable membrane in osmosis, and in liquid and gas separations. Many inorganic membranes have excellent thermal and structural stability, do not compact or swell, are resistant to chemicals and microbiological attack and possess the ability to be backflushed, steam sterilized and/or autoclaved.

inorganic nanotubes. A relatively new class of materials analogous to carbon nanotubes, but composed of a single sheet of an inorganic material, such as boron nitride, molybdenum disulfide, niobium disulfide or tungsten disulfide, made into a tube with ends sealed with fullerene molecules. Present and future applications include flat-panel displays, sensors and biosensors, solid-state devices and atomic force microscopes. They may also be useful for the production of new superconductors and conductive plastics. Also known as *inorganic bucky tubes*. See also carbon nanotubes.

inorganic-organic composites. Composite materials consisting of inorganic reinforcing fibers embedded in organic matrix materials (e.g., polymers). Examples include fiberglass-reinforced plastics and graphite-epoxy laminates.

inorganic pigments. Natural or synthetic metallic oxides (e.g., oxides of antimony, chromium, cobalt, iron, lead, manganese, titanium or zinc), metallic sulfides (e.g., sulfides of antimony, cadmium or mercury), other salts (e.g., barium sulfate, calcium carbonate or lead chromate), metal-powder suspensions (e.g., aluminum or gold powder), earth colors (e.g., ochers or umbers) or carbon black, added to a paint, polymer or ink to impart color and/or heat and light stability, weathering resistance and migration resistance.

inorganic polymers. A group of polymers that are not based on carbon, e.g., silicones or phosphazenes having repeating units of siloxane (–Si–O–Si–) and phosphonitrile (–N=P–), respectively.

inorganic zinc-rich paint. A two-component paint containing finely divided zinc dust in an inorganic vehicle, such as ethyl silicate. Used as a primer, or as a protective topcoat on ferrous substates.

inosilicates. A group of silicate minerals, such as *enstatite* or *tremolite*, in which three SiO_4 tetrahedra are linked together by the sharing of oxygen atoms to form chains of considerable lengths.

Inossidalfa. Trade name of Alfa Romeo (Italy) for a corrosion-resistant aluminum alloy.

Inostrong. Trade name of Pechiney Electrométallurgie (France) for a cast iron inoculant containing 66% silicon, 0.85% strontium, up to 0.5% aluminum and 0.1% calcium, respectively, and the balance iron.

Inotab. Trademark of Foseco NV (Netherlands) for a cast iron inoculant in tablet form used for the efficient mold or ladle inoculation of cast iron melts from high-steel charges. It significantly increases both tensile and transverse strength, lowers section sensitivity and chilling tendency, and prevents undercooled graphite.

Inova. Trade name of BASF Corporation (USA) for automotive clearcoats.

Inox. Trade name of Société Nouvelle des Acieries de Pompey (France) for a series of ferritic and martensitic stainless steels including several free-machining grades used for machine parts, pump and valve components, cutlery, hardware, fasteners, surgical instruments, and turbine blades.

Inoxalium. Trade name of T.L.M. Company (France) for a water-hardening carbon steel containing 0.2% carbon, and the balance iron. Used for machine parts.

Inoxargent. Trade name of Ugine Aciers (France) for a corrosion-resistant steel containing up to 0.12% carbon, 12-14% chromium, 12-14% nickel, a trace of copper, and the balance iron. Used for furniture, clocks and decorative trim.

Inoxesco. Trade name of Vallourec SA (France) for a series of austenitic and ferritic stainless steels.

Inoxium. Trade name of Chiers-Chatillon (France) for a series of austenitic and ferritic stainless steels.

Inoxyda. Trade name for corrosion-resistant bronzes containing 82-89% copper, 1-10% aluminum, up to 4% iron, 4.5% nickel and 1% zinc, respectively, and the balance tin. Used for steam fittings.

insecticide paper. A special paper that contains chemicals which kill insects on contact, or by means of evaporation.

insect-resistant paper. Paper made resistant to insects by treating with suitable insecticides.

insect-resistant fabrics. Textile fabrics that are capable of impeding damage by insects. This capability may either be inherent owing to the fibers used in their manufacture (e.g., synthetics), or may be obtained by treating them with suitable chemicals (e.g., insecticides).

Inselglas. Trade name of Vereinigte Glaswerke (Germany) for a special selectively toughened glass having a pattern of small, untempered disks distributed over the entire surface.

in-situ concrete. See cast-in-place concrete.

in-situ network composite. A composite material containing a continuous three-dimensional network as a filler. A homogeneous in-situ network composite has a network that is homogenous in composition throughout. A heterogenous in-situ network composite shows compositional fluctuations along the network phase. In-situ network composites do not suffer from delamination failure often observed in two-dimensional composites. Examples include tin-lead and aluminum-silicon-magnesium alloys which can be produced by the molten filler flow technique and the dissolution and deposition method, respectively.

insizwaite. A white mineral of the pyrite group composed of antimony bismuth platinum, $Pt_{1.03}Bi_{1.41}Sb_{0.56}$. Crystal system, cubic. Density, 12.88 g/cm^3. Occurrence: Southern Africa. A low-temperature, synthetic form composed of bismuth platinum, Bi_2Pt, is also known by this name.

INSPiRE. Trademark of Dow Chemical Company (USA) for a family of polypropylene-based performance polymers made by a proprietary processing technology (known as *INSITE*). *INSPiRE 112*, the first product in this family, is designed for blown-film extrusion providing high output rates and improved drawdown capabilities and bubble stability. Applications include shipping sacks, meat and poultry packaging pouches, soft-paper tissue overwrap, industrial bundling films and thin-gauge films.

Insta-Blak. Trademark of Electrochemical Products, Inc. (USA) for corrosion-resistant, room-temperature blackening finishes for iron, steel, aluminum and zinc.

Insta-Bond VL. Trade name of Lee Pharmaceuticals (USA) for a fast-setting bonding resin used in orthodontics.

Instant Patch. Trademark of Tremco Limited (Canada) for a trowelable asphalt-based compound used for roof repairs.

Insta-Pro Foam System. Trade name of Flexible Products Company (USA) for one-part polyurethane foam sealants.

Instaweld. Trade name of Bondmaster Products, National Starch & Chemical Corporation (USA) for elastomeric hot-melt adhesives used for industrial applications.

In-Steel. Trade name of Ishikawajima-Harima Heavy Industries Company (Japan) for an ultrahigh-strength steel containing up to 0.18% carbon, 0.15-0.35% silicon, 0.6-1.2% manganese, 0.4-0.8% chromium, up to 1.5% nickel, 0.6% molybdenum and 0.1% vanadium, respectively, and the balance iron. It has excellent weldability, and is used for heavy-duty welded structures and equipment, mine cars and bus bodies.

Instill. Trademark of Dow Chemical Company (USA) for a lightweight, open-celled polystyrene foam with excellent thermal insulating properties. Used as a core material in vacuum insulation panels for refrigerators, freezers, refrigerated trucks, cold-storage units, shipping containers and vending machines.

instrument bronze. A corrosion-resistant bronze containing 82% copper, 13% tin and 5% zinc used for utensils and instruments.

Insublock 19. Trade name of A.P. Green Refractories, Inc. (USA) for a refractory-based thermal insulating block.

Insulag. Trademark of Quigley Company Inc. (USA) for high-temperature refractory insulation cements based on fireclay, magnesia, or related materials.

Insul Aid. Trademark of Glidden Company (USA) for vapor-resistant latex barrier materials.

insulated wire. An electrical wire consisting of one or more flexible strands composed of a metal or alloy, such as aluminum, copper, gold, silver, tungsten, or stainless steel, enclosed in an insulating sheath, usually made of plastics or elastomers.

insulating boards. Lightweight boards manufactured from glass-fiber or plastic-foam materials (e.g., expanded or extruded polystyrenes, phenolics or polyurethanes) and supplied in the form of rigid or semi-rigid boards in a wide range of sizes. They have high insulating values per unit thickness (R-values or heat-flow resistance values) and may or may not have fire-resistant, moisture-resistant or decorative coverings. Depending on the particular type, they can be used for the acoustical and thermal insulation of interior and exterior building walls, attics and floors, building foundations and basement floors (moisture-resistant types), and as sheathing for air and/or vapor barrier applications. See also insulation board; rigid board insulation; semiflexible board insulation.

insulating brick. A refractory brick of low-thermal capacity and low-thermal conductivity used in fireboxes and furnace walls.

insulating cement. A dry cement product to which a considerable quantity of an insulating material, such as fibrous asbestos, diatomaceous earth, vermiculite or perlite, has been added, and that when mixed with water develops adequate plastic consistency to be placed, and when dried forms a covering providing low thermal conductivity and heat transmission. Also known as *thermal-insulating cement*.

insulating concrete. (1) A lightweight concrete of relatively low density, typically 1442 kg/m^3 (90 lb/ft^3), low-thermal conductivity and low-heat transmission, used as a thermal insulation and fire protection in building construction. It is often supplied in the form of blocks, sheets or corrugated slabs.

(2) A mixture of Portland cement and diatomaceous earth used as thermal insulation material.

insulating felt. A felt made by rolling and pressing together fibers of cotton, kapok, asbestos or glass. It has low acoustical and thermal conductance, and is used in the acoustical and thermal insulation of buildings.

insulating firebrick. A porous *refractory brick* made from fireclay. It has low thermal conductivity, low-heat capacity and is much lighter than regular *firebrick*. Used in the thermal insulation of industrial furnaces, kilns, flues and other high-temperature industrial equipment. Abbreviation: IFB.

insulating glass. A glass with excellent thermal insulating properties suitable for use in building and construction. Examples include glass blocks, flat glass for windows and doors and foam glass.

insulating material. See insulator.

insulating paper. Kraft or bond paper coated on both sides with insulating varnish. Supplied in thicknesses ranging from 0.050-0.500 mm (2-20 mils), it has excellent dielectric properties and is used as insulating material for electrical equipment and components. Also known as *electrical insulating paper; varnished paper; varnish paper*.

insulating refractory. A porous, lightweight refractory material often containing an exfoliated vermiculite or diatomaceous earth filler. It has low thermal conductivity, low-heat capacity and is much lighter than regular *firebrick*. Used in the insulation of industrial furnaces. Also known as *hot-face insulation*.

insulating tape. (1) A tape, usually with an adhesive backing, made from, or impregnated with an electrically insulating material. Most insulating tape is now made from synthetic fibers. Used in the electrical trades for insulating wires and cables. Also known as *electrical tape*.

(2) A strong, durable, dimensionally-stable, heat-resistant tape made from synthetic resins, or textile or ceramic fabrics. Used for thermal insulation applications, e.g., in pipe or hose wrapping, for electrical applications, and for the manufacture of gaskets.

insulating varnish. A *varnish*, usually epoxy-, polyester- or silicone-based, especially designed for the insulation of electric wires, coils, components and appliances against electrical, mechanical and environmental influences. It has good dielectric strength and good adhesion properties. Also known as *electrical insulating varnish; electrical varnish.*

insulating wallboard. A large rigid sheet or panel used as backing or finishing material on interior walls. Depending on its particular composition, it may also provide acoustic and/or thermal insulation. Examples of materials used for the manufacture of *insulating wallboard* include plywood, particleboard, hardboard, laminated plastics and asbestos cement.

insulation board. A lightweight board made from a pulp of wood, cane or other cellulosic fibers, and combining strength with acoustical and thermal insulating properties. Unlike *hardboard*, it is made by interfelting and does not require heat and pressure. It has a density ranging from 160 to 500 kg/m³ (10 to 31 lb/ft³) and is supplied in a wide range of sizes from tiles, 203 mm (8 in.) square, to sheets, 1.2 m (4 ft.) wide and 3.1 m (10 ft.) or more long. The standard thickness is 13-25 mm (0.5-1.0 in.). *Insulation board* is primarily used as a sheathing material for roofs and walls, and as such is often impregnated with asphalt to make it more resistant to moisture. Other uses include subflooring, interior surfaces of ceilings and walls, plaster bases, partitions, and insulation strips for foundation walls and slab floors. Also known as *structural insulating board*. See also insulating board.

insulation porcelain. A hard-fired, vitrified ceramic whiteware used in the manufacture of spark plugs and electrical insulators (e.g., power terminals and line insulators). Below a room-temperature breakdown voltage of about 8000 kV/m (203 V/mil) it will not conduct electricity. In contrast to porcelain for dinnerware, it contains more clay and less kaolin and feldspar. Examples of *insulation porcelains* include molded silica, molded steatite and ceramic insulators (e.g., alumina, cordierite, titania, zircon and magnesia). Also known as *electrical porcelain.*

insulator. (1) An ionically, covalently or van der Waals bonded material whose valence bands are completely filled at 0K and separated from their empty, outer conduction bands by a wide energy gap. Consequently, it offers very high electrical resistance (resistivity about 10^{10}-10^{16} Ω-m) to the passage of an electric current at room temperature. Examples include ceramics and silicates (e.g., porcelain, zircon, alumina, titanates, amber, mica, slate and steatite), glasses (e.g., soda-lime glass and E-glass), most traditional plastics (e.g., nylons, polyethylenes, polyvinyl chlorides and polystyrenes) as well as rubber, paraffin oil, waxed paper, cotton, wool and silk. Also known as *electrical insulator; insulating material; nonconductor.*

(2) A material that will prevent or retard the transfer of acoustic, thermal or other form of energy. Sound or acoustic insulating materials include tiles, plaster, perforated fiberboard, fiber-sheathing boards, mineral wool, hair felt, foamed plastics and sheathing papers. Thermal (or heat) insulating materials include cork, wool, glass wool, fiberglass, mineral fibers, kieselguhr, expanded perlite, cellular or foam glass, expanded polystyrene and expanded polyurethane. Also known as *insulating material; nonconductor.* See also heat insulator; sound insulator.

Insulbatte. Trade name of J.P. Stevens & Company Inc. (USA) for a mechanically bonded glass fiber mat.

Insulbox. Trademark of Quigley Company Inc. (USA) for precast products made from *Insulag* refractory insulation cements.

Insulbrix. Trade name for calcined fireclay supplied in the form of insulating bricks for lining furnaces.

Insul-Cote. Trademark of E/M Engineered Coating Solutions (USA) for parylene-type conformal polymer coatings applied by vacuum deposition.

Insulcrete. (1) Trademark of Specialty Refractories Inc. (USA) for a calcined fireclay supplied in the form of a lightweight concrete for insulating high-temperature processing vessels.

(2) Trade mark of Efficiency Structures Corporation (USA) for building blocks in wood cement.

(3) Trademark of Durisol Materials Limited (USA) for building blocks and panels made of cement and wood.

Insulex. (1) Trademark of Bognar & Company, Inc. (USA) for lightweight insulating compositions for sealing checker chambers of metallurgical and annealing furnaces.

(2) Trademark of Expanded Metal Corporation (USA) for expanded metal mesh sheets.

(3) Trade name of BSK, Division of DS Smith (UK) Limited for a reflective foil bubble home insulation.

(4) Trade name for a translucent insulating brick made from opalescent glass and used for industrial and commercial installations.

Insulgrease. Trademrk of General Electric Company (USA) for silicone compounds with grease-like consistency used as dielectrics in the electrical insulation of connectors and equipment, as antiseizing compounds, for corrosion-resistant and insulating coatings, and as heat-transfer media.

Insul-Ice. Trade name of Polyfoam Packers Corporation (USA) for an expanded polystyrene foam insulation used for protective packaging applications.

Insulight. Trade name of Pilkington Brothers Limited (UK) for (i) double glazing units and (ii) glass blocks composed of two molded sections attached to each other by a direct glass-to-glass fusion technique.

Insuline. Trademark of Quigley Company Inc. (USA) for high-temperature refractory insulation cements.

Insulite. (1) Trade name of Boise Cascade Corporation (USA) for a glass-fiber building insulation.

(2) Trade name of Insulite Inc. (USA) for a insulating glass.

(3) Trademark of Clayburn Refractories Limited (USA) for insulating and gunning castables.

Insul-Lock II. Trade name of RBX Corporation (USA) for a self-sealing, closed-cell sponge rubber for thermal insulation applications.

Insulmastic. Trade name of Insulmastic Building Products (Canada) for a pasty substance consisting of a solution of *gilsonite* into which asbestos fibers or mica flakes have been introduced. Used as a protective coating.

Insuloxide. Trademark of National Lead Company (US) for a compound of 95+% zirconia (ZrO_2) mixed with up to 5% silica (SiO_2). It has a melting point of 2485°C (4505°F), and is used for superrefractories, heat insulation applications, and for severe service conditions.

Insulspray. Trade name of Borden Chemicals & Plastics (USA) for urea-formaldehyde products supplied in cellulose-filled and foam grades.

Insulstruc. Trade name of Industrial Dielectrics Inc. (USA) for glass-fiber-reinforced polyester sheeting with good dimensional stability and flame resistance. Used for electrical applications including insulation products.

Insul Therm. Trade name of Bonded Insulation Company, Inc. (USA) for loosefill, stabilized and cavity spray cellulose used for the thermal insulation of buildings.

Insuluminum. Trade name for a steel whose surface has been impregnated with aluminum. Used for chemical equipment.

Insulux. Trade name of Owens-Illinois Inc. (USA) for hollow, partially evacuated glass blocks.

Insul-Wall. Trademark of Truefoam Limited (Canada) for structural wall insulation panels based on polystyrene foam.

Insure. Trade name of Cosmedent Inc. (USA) for a dental resin cement.

Insurok. Trademark of The Richardson Company (USA) for a group of laminated and molding plastics with low density, high durability, high dielectric strength and good resistance to many chemicals. Used as substitutes for cast aluminum in airplane components, and for electrical components, such as commutators, distributors and switches.

Intact. Trade name for a glass-ionomer dental cement.

Intacta. Trademark of Dow Chemical Company (USA) for strong, durable, tear-resistant polyurethane-based performance polymers made by a proprietary dispersion chemistry process. Used in natural rubber latex-free dental and examination gloves.

Intaglio. Trade name of Pittsburgh Corning Corporation (USA) for a hollow glass block having a recessed antique glass section, and a textured surrounding area formed by fusing gray ceramic frit into the glass surface.

Intal. Trade name of International Alloys Limited (UK) for a series of aluminum-copper and aluminum-zinc casting alloys.

intarsia. A plain or ribbed fabric knitted with an inlaid, multicolor pattern in which each differently colored section is made with a seperate yarn. Used especially for garments.

Intasil. Trade name of Chance Brothers Limited (UK) for a specialty glass used for joining glass objects to metals.

Integra. (1) Trade name of Saint-Gobain (France) for glass blocks having smooth exterior surfaces and light watermark designs on their interior surfaces.

(2) Trademark of Bethlehem Steel Corporation (USA) for a high-performance plate steel with very low sulfur content (0.006% or less). It has good formability and weldability, good resistance to brittle fracture at low temperatures, good through-thickness ductility and good resistance to lamellar tearing. Used for highly stressed industrial equipment.

(3) Trademark of Integra NeuroSciences (USA) for an artificial skin material based on a porous synthetic matrix that is made of collagen and chondroitin-6 sulfate and overlaid with a thin *Silastic* silicone sheet. Used as a scaffold for dermal regeneration.

Integra Bond. Trade name of Premier Dental (USA) for a dental bonding agent.

integral composite structure. An engineering composite structure, e.g., a reinforced plastic, not assembled by bonding or mechanical fastening. Typically, such a structure contains two or more structural elements that are laid up and cured to form a single integral component. Also known as *integral structure*.

integral skin foam. An expanded polyurethane or other thermoplastic material with a dense, integral outer skin and a porous, foamed core. See also rigid integral skin foam; structural foam.

integral waterproofer. A substance or mixture of substances added to concrete to reduce the capillary flow of water through it.

Integrity. (1) Trade name of Dentsply Corporation (USA) for a palladium-silver ceramic dental alloy used for porcelain-to-metal bonding applications.

(2) Trade name of Dentsply/Caulk (USA) for a chemical-curing bis-acryl type dental resin composite for temporary

bridge and crown restorations.

Intelin. Trade name of Intel Corporation (USA) for polystyrene products.

intelligent biomaterials. A relatively new class of engineered *biomaterials* consisting of conducting polymers or sol gels containing biological macromolecules (e.g., proteins, enzymes or DNA) that possess intelligent properties including self-assembly and self-diagnosis.

intelligent material. An engineered material that is considered "intelligent" because of its ability to recognize changes in the surroundings and respond accordingly. Used in devices for electronic, optoelectronic, photonic and biomedical applications. Also known as *actively smart material*. See also smart materials.

Intene. Trademark of EniChem SpA (Italy) for general-purpose butadiene rubber.

Intensiv. Trade name of Remystahl (Germany) for a series of cold- and hot-work tool steels used for cutting tools, broaches, thread-rolling dies, bending, forming and drawing dies, and various other dies and tools.

intercalated lamellar nanocomposites. A subclass of organic/inorganic lamellar nanocomposites in which the compositional ratio of the polymer chains alternating with the inorganic layers is fixed and the number of polymer layers in the intralamellar spaces is well defined. They can be synthesized from conjugated and saturated organic molecules by (i) insertion of the polymer chains into the inorganic matrix, (ii) *in-situ* intercalative polymerization of a monomer using the inorganic matrix as the oxidant, (iii) intercalation of the monomer and subsequent topotactic intralamellar solid-state polymerization, and (iv) direct precipitative polymer-chain encapsulation by colloidally dispersed matrix single layers. Examples of such intercalated materials include α-ruthenium trichloride/polymer, polyaniline/vanadium pentoxide, poly(ethylene oxide)/vanadium pentoxide, polyaniline/metal phosphate and polyaniline/molybdenum disulfide nanocomposites. *Intercalated nanocomposites* have unique electronic properties that make them suitable for applications in nonlinear optics, battery cathodes and ionics, and sensors. See also exfoliated lamellar nanocomposites; lamellar nanocomposites.

intercalate structure. A material in which a host lattice forms layers with adsorbed guest species (atoms or molecules) between the layers. For example, alkali metals can be intercalated into graphite.

intercalation compounds. A class of crystalline compounds in which "foreign" atoms are diffused or interspersed between the lattice planes of the host lattice and can act as either electron acceptors or donors (e.g., bromine can act as an acceptor in graphite). Some have structures in which a layer of "foreign" atoms is interspersed between several layers of the host material atoms, e.g., lithium atoms interspersed between dilithium nitride layers. Used for superconductors, solid lubricants, catalysts, rechargeable batteries and solid electrolytes.

Interclad. Trade name of Newalls Insulation & Chemical Company Limited (UK) for glass fiber mats.

Intercolour. Trade name of Goddard & Gibbs Studios (UK) for a laminated glass with an interlayer of translucent material.

interfacing. A woven or nonwoven fabric sewn or fused between the outside layer and the facing of a fabric or garment as a reinforcement or stiffener. Used in collars and cuffs.

Intergard. Trade name of Akzo Nobel (USA) for a series of zinc phosphate/epoxy primers and epoxy intermediates and finishes.

interior-type particleboard. Particleboard usually made with a urea-formaldehyde binder and used for interior building applications. It is supplied in three density grades: (i) *High-density* with at least 800 kg/m³ (50 lb/ft³); (ii) *Medium-density* with 593-800 kg/m³ (37-50 lb/ft³) and (iii) *Low-density* with less than 593 kg/m³ (37 lb/ft³). See also exterior-type particleboard, particleboard.

interior-type plywood. Plywood manufactured for indoor use or for constructions subjected to temporary moisture or where the moisture content of the panel will not exceed 20%. It is bonded with glue that maintains an adequate bond under interior or protected conditions, but is not waterproof. Good interior types of plywood are frequently bonded with exterior glue. There are six grades of *interior-type plywood* ranging from good-appearance grades (A-A) for furniture panels to fair-appearance grades (C-D) for use in hidden areas. See also exterior-type plywood; plywood.

Interlay. Trade name of Newall Insulation & Chemical Company Limited (USA) for glass-fiber home insulation products.

interlayer dielectrics. A group of materials with low dielectric constants used as interlayers for ultra-large-scale integrated (ULSI) circuit applications. Typically, the dielectric constants (k) range from as high as 3.5 to as low as 1.3. Examples of such materials include polytetrafluoroethylene (PTFE), parylenes and polynaphthalenes applied by vapor deposition, pentafluorostyrene applied by plasma polymerization, fluorinated silica glass and amorphous fluorinated carbon applied by high-density plasma chemical vapor deposition or plasma-enhanced chemical vapor deposition, fluorinated or nonfluorinated organic polymers (e.g., polyimides) applied by spin-on coating, inorganic siloxane-based polymers, inorganic-organic polymer hybrids and various porous materials including nanoporous silica and other aerogels and xerogels.

interleaving paper. A generic term for paper with special properties used to protect object surfaces, e.g., during manufacture and storage.

interleaving release paper. A paper that has an adhesive coating on one or both sides, and prevents adhesive or viscous substances from adhering to it.

interlining. A knit, woven or nonwoven fabric used as an extra lining between the outer layer of a fabric or garment and the lining. Used give shape, strength, warmth or body to a fabric or garment.

interlock. (1) A firm, tightly knit fabric with a straight, raised cord on each side.

(2) See interlocking stone.

interlocked-grained wood. Wood having grains in which the fibers put on for a number of years slope in one direction (e.g., a left-hand direction), then for several years the slope reverses to another direction (e.g., a right-hand direction) and later changes back to the inital direction. It is extremely hard to split in a radial direction, though it may be readily split in a tangential direction.

interlocking stone. A *paving stone* with specially prepared edges (e.g., rounded or beveled) designed to interlock (match) with an adjacent stone. Used for patios, walkways, driveways and swimming pool decks. Also known as *interlock; interlocking paver.* See also paving stone.

interlocking tile. A tile with specially prepared edges designed to interlock with an adjacent tile. Used in roofing and other construction.

Interlox. Trade name of McGean Company (USA) for a phosphate coating.

Interlux. Trade name of International Paints (Canada) Limited for antifouling and alkyd-enamel paints, spar varnishes, thinners and primers, used especially for marine applications, e.g., ship and boat hulls and decks.

intermediate babbitt. A soft, white-metal alloy containing 20-50% tin, with the balance usually being lead and antimony. It has moderate compressive strength, and is used for medium-duty bearings. See also babbitts.

intermediate compounds. Stable chemical compounds formed between the two components of binary systems. For example, mullite ($3Al_2O_3 \cdot 2SiO_2$) is an intermediate compound in the alumina-silica system, spinel ($MgO \cdot Al_2O_3$) in the magnesia-alumina system, and iron carbide (cementite, Fe_3C) in the iron-carbon system.

intermediate constituents. See intermetallic compounds.

intermediate cotton. Cotton staple intermediate in length between short and long staple, typically in the range of 12.7-33.3 mm (0.5-1.3 in.).

intermediate-duty fireclay brick. Fireclay brick with a pyrometric cone equivalent of at least 29, or more than 3.0% deformation at 1350°C (2460°F). See also fireclay brick.

intermediate fiberboard sheathing. High-density *fiberboard* used as a sheathing material under siding, shingles, stucco, brick or stone veneer in the construction of exterior walls.

intermediate fluxes. See mildly active fluxes.

Intermediate-Manganese. Trade name of US Steel Corporation (USA) for a series of manganese steels. *Intermediate-Manganese Abrasion Steel* is an abrasion-resistant grade containing 0.35-0.5% carbon, 1.5-2% manganese, 0.15-0.3% silicon, up to 0.05% phosphorus, up to 0.05% sulfur, and the balance iron. Used for screens, dredge pipes, and abrasion-resistant sheets, plates and billets. *Intermediate-Manganese Medium-Carbon* is a medium-carbon grade containing 0.28-0.38% carbon, 1.1-1.4% manganese, 0.15-0.3% silicon, up to 0.04% phosphorus, and the balance iron. Used for bars and rods. *Intermediate-Manganese Rail Steel* is a wear-resistant grade containing 0.5-0.65% carbon, 1.2-1.5% manganese, 0.1-0.23% silicon, up to 0.04% phosphorus, and the balance iron. Used for rails and railroad equipment.

intermediate manganese steels. See carbon-manganese steels.

intermediate oxides. Oxides, such as alumina (Al_2O_3), titania (TiO_2) or zirconia (ZrO_2), that are neither network formers nor network modifiers. They are used in the manufacture of silicate glass since their cations (Al^{3+}, Ti^{4+} or Zr^{4+}) can contribute to the stability of the glass network. Intermediate oxides are also used in the manufacture of porcelain enamels.

intermediate-temperature-setting adhesive. An adhesive that sets at intermediate temperatures usually between 30 and 100°C (86 and 212°F). Also known as *warm-setting adhesive.*

intermediate transfer paper. A generic term that includes (i) *copying carbon paper* used for handwriting; (ii) *carbon paper,* and (iii) *hectographic carbon paper.*

intermetallic compounds. A group of compounds consisting of one metallic and a second metallic or nonmetallic element, (e.g., aluminides, beryllides, borides, hydrides, nitrides and silicides of the *transition metals)* in which a progressive change in composition is accompanied by a progression of phases differing in crystal structure and properties. They have distinct chemical formulas, and on phase diagrams appear as intermediate phases that have simple stoichiometric ratios and exist over a very narrow range of compositions. The properties of intermetallic compounds are similar to both metals (at high temperatures) and ceramics (at low temperatures). Examples of intermetallic compounds are copper aluminide (Cu_3Al), iron aluminide (FeAl) and titanium copper (Ti_2Cu). Abbreviation: IMC. Also known as *electron compounds; intermediate constituents; intermetallics; Hume-Rothery compounds.*

intermetallic-matrix composites. A group of composite materials in which the matrix phase is an intermetallic compound, such as nickel aluminide or molybdenum disilicide, and the reinforcing phase is based on ceramic or metallic fibers, such as silicon carbide, or niobium. The unique combination of an intermetallic compound possessing the low-temperature properties of a ceramic and the high-temperature properties of a metal with a toughening or strengthening fibrous phase makes possible the manufacture of composites with a large variety of desired characteristics. Abbreviation: IMC.

intermetallics. See intermetallic compounds.

intermingled yarn. A yarn consisting of two or more monofilaments held together by entwining and/or twisting in an air jet

Inter-Mix. Trade name of Pittsburgh Corning Corporation (USA) for blocks of rough-textured glass with light-gray ceramic frit coatings. Used to match opaque portions to *Intaglio* glass blocks.

internally oxidized alloy. A metal alloy (e.g., an electrical contact material) in which certain phases or components have been preferentially oxidized by oxygen diffusion. The oxides reinforce (strengthen) the matrix of this material, producing improved wear resistance.

Interpane. Trade name of Interpane Isolierglas (Germany) for a multipane acoustic insulating glass.

interpenetrating elastomeric network. See interpenetrating polymer network.

interpenetrating homopolymer network. See interpenetrating polymer network.

interpenetrating polymer network. A mixture of polymers in which a crosslinked polymer permeates the other polymer. It results when two separate polymer networks or structures are intertwined. If the interpenetrating polymers are homopolymers, the mixture is known as "interpenetrating homopolymer network" (IHPN), and if they are elastomers, it is known as "interpenetrating elastomeric network" (IEN). Abbreviation: IPN. See also semi-interpenetrating network.

interply hybrid. A composite laminate in which alternating layers or plies are composed of different materials. For example, each alternating layer may consist of a different fiber type.

Interpol. Trade name of Cook Composites & Polymers (USA) for a series of impact-resistant polyester-polyurethane and high-strength isophthalate-polyester alloys with good chemical resistance. Used for automotive, construction and electrical applications, and cosumer products.

Inter-Poly. Trade name of International Paint Company, Inc. (USA) for several polyurethane enamel paints.

interpolymer. A polymeric material made by mixing two or more starting materials.

Interpon. Trademark of Courtaulds Coatings Inc. (USA) for paint-like powder materials used to produce decorative and protective coatings.

interstitial compounds. Metallic compounds that have high hardness and melting points. They are formed when a transition metal admits nonmetal atoms of small diameter (such as carbon, nitrogen, hydrogen, boron or silicon) into the interstices of its lattice. Metallurgically, they are intermediate phases in which interstitially dissolved elements (nonmetals) are at in-

terstitial positions because they are present in excess of the solubility limit. Examples of interstitial compounds include iron carbide (Fe_3C), tantalum carbide (TaC), titanium nitride (TiN) and tungsten carbide (WC). Used in cermets, cemented carbides and for strengthening alloys, such as steels. Also known as *interstitials.*

interstitial-free enameling steels. See interstitial-free steels.

interstitial-free steels. A group of aluminum-deoxidized, vacuum-decarburized, nonaging extra-low-carbon steels, fully stabilized by the addition of niobium and titanium which react with carbon and sulfur to form precipitates of carbides and sulfides. Thus, no carbon remains in the ferrite. A typical composition (in wt%) is 0.005% carbon, 0.2% manganese, 0.02% sulfur, 0.01% phosphorus, 0.09% niobium, 0.04% titanium, and the balance iron. They have good strength retention after strain and firing, outstanding formability, low distortion and sagging tendency, and do not exhibit stretcher strain and fishscaling. *Interstitial-free steels* are particular suited for the automotive industry and for applications requiring porcelain-enameled steels. Abbreviation: IF steel. Also known as *interstitial-free enameling steels.*

interstitial fullerenes. A subclass of fullerene compounds prepared by incorporating noble gas atoms into molecules composed of 70 carbon atoms. Examples include argon fullerenes (ArC_{70}), krypton fullerenes (KrC_{70}) and xenon fullerenes ($XeCr_{70}$). See also fullerenes.

interstitials. See interstitial compounds.

intertype metal. A type metal containing 11-14% antimony, 3-5% tin, and the balance lead. Used for making printing type.

Inter-Weld. Trade name for a composite metal made by first brazing a sheet of nickel to a base of brass and then welding a sheet of gold onto the nickel. Subsequent rolling produces a thin layer of gold. Used for decorative applications.

Interzinc. Trade name of Akzo Nobel (USA) for a series of zinc-epoxy primers.

Intex. Trademark of EniChem SpA (Italy) for styrene-butadiene rubber latexes.

Intol. Trademark of EniChem SpA (Italy) for general-purpose styrene-butadiene rubber.

Intolan. Trademark of EniChem SpA (Italy) for general-purpose polyolefin terpolymers.

Intra. Trade name of H. Boker & Company (USA) for a water-hardening tool steel containing 1.2% carbon, 1.3-1.5% tungsten, and the balance iron. Used for dies, reamers and other tools.

intraply hybrid. A composite laminate in which each layer or ply is composed of two or more different materials.

Intrex. Trade name of Sierra/Intrex, Division of Sierrcin/Sylmar Corporation (USA) for a series of transparent polyester film materials with good electrical and optical properties and with or without special coatings. *Intrex G* is an uncoated grade used for electrical and EMI/RFI shielding applications. *Intrex K* has an indium-tin oxide (ITO) coating and is used for EMI/RFI shielding, antistatic packaging and electrophotography. *Intrex K-M* has a multilayer metal coating produced by sputtering, and is used EMI/RFI shielding, electroluminescent screens, and various electronic and optoelectronic components and devices.

intrinsic ionic conductor. An electrical conductor in which conduction results from the movement of component ions. See also extrinsic ionic conductor.

intrinsic semiconductor. A semiconductor for which the electrical behavior is characteristic of the pure material. The electrons are excited from the valence band across the energy gap into the conduction band, producing positive electron holes in the valence band. The electrical conductivity depends only on the temperature and the band-gap energy. Also known as *i-type semiconductor.*

intumescent coating. A fire-retardant organic coating that bubbles and swells when heated to form a thick, highly insulating barrier. This volume expansion may be produced by the incorporation of borax, ammonia or an intumescent vinyl resin. Also known as *intumescent paint.*

intumescent material. A material that swells when heated and is used for fireproofing applications.

intumescent paint. See intumescent coating.

Invaleur. Trade name of Celanese Corporation (USA) for cellulose nitrate plastics.

Invar. Trademark of Carpenter Technology Corporation (USA) for a series of controlled-expansion alloys containing 50-64% iron, 35.5-50% nickel, up to 0.8% manganese and up to 0.2% carbon. Free-cutting grades contain up to 0.35% silicon and up to 0.25% sulfur or selenium. They are commercially available in foil, wire, rod, tube and powder form. The standard 64Fe-36Ni alloy has a density of 8.05 g/cm³ (0.291 lb/in.³); a liquidus temperature of 1495°C (2723°F), a hardness of 160 Brinell, a very low room-temperature coefficient of expansion (1.6×10^{-6}/°C or 0.9×10^{-6}/°F), high tensile strength and modulus of elasticity, high electrical resistivity (about 82 $\mu\Omega$cm), and a Curie temperature of 280°C (535°F). *Invar* alloys are used as electrical resistance materials, for automotive and aircraft engine parts, aircraft controls, radar and electronic devices, rod and tube assemblies for temperature regulators, temperature control and indicating devices, thermostats, bimetal strips (low-expansion side), clocks, geodetic instruments, radio devices, precision instruments, measuring tapes, weights, and structural members in precision optical laser and other measuring instruments.

Invariant. British trade name for a corrosion-resistant alloy composed of 53% iron and 47% nickel. It has a low coefficient of expansion, and is used for piston struts and surveyors tapes.

Invaro. Trade name of Teledyne Firth-Sterling (USA) for a series of oil-hardening tool steels containing 0.85-0.95% carbon, 1-1.5% manganese, 0.5-0.6% tungsten, 0.5% chromium, 0.2-0.25% vanadium, and the balance iron. Used for punches, shears, forming dies, milling cutters, and several other tools.

Invarod. Trade name of Carpenter Technology Corporation (USA) for a low-expansion alloy containing 36% nickel, 2.5% manganese, 0.75% titanium, and the balance iron. It facilitates cryogenic joining, and is used in the form of weld wire for crack-free welding of *Invar.*

Invarox. Trademark of E.I. DuPont de Nemours & Company (USA) for thick-film compositions supplied in various grades.

Invarstahl. Trade name of Vereinigte Deutsche Nickel-Werke AG (Germany) for an *Invar*-type alloy containing 64% iron and 36% nickel. Used for measuring devices, balances and clocks.

Inventor. Trade name of Dunford Hadfields Limited (UK) for a martensitic stainless steel (AISI type 420) containing 0.13-0.2% carbon, 1% manganese, 1% silicon, 12-14% chromium, and the balance iron. It has excellent strength and corrosion resistance, and is used for turbine blades and surgical instruments.

inverse spinels. A group of cubic spinels, such as magnesium ferrite ($MgFe_2O_4$) or nickel ferrite ($NiFe_2O_4$), with the divalent cations (e.g., Mg^{2+} or Ni^{2+}) in 6-fold (octahedral) interstitial sites and the trivalent cations (e.g., Fe^{3+}) equally divided between 6-fold (octahedral) and 4-fold (tetrahedral) sites. Many

commercially important ferrites or ferrimagnetic ceramics are inverse spinels. Also known as *inverted spinels.* See also synthetic spinels.

Invest Cast. Trade name of Invest Cast, Inc. (USA) for precision investment castings of various metals and alloys.

investment castings. Precision metal castings produced by investment casting, a process that involves making an expendable wax, plastic or frozen mercury pattern, surrounding it with a fluid refractory ceramic material (*investment compound*), melting or burning the pattern away after the investment material has dried and set to produce a mold cavity, pouring liquid metal (usually under air or centrifugal pressure) into the cavity and allowing the metal to solidify.

investment compound. A mixture composed of a refractory powder (usually silica sand with alumina and silicate additions), a binder (e.g., an organic silica compound such as ethyl silicate) and a liquid vehicle (e.g., alcohol or water). Used in the production of molds for investment casting. See also investment castings.

Invincible. Trade name of Joseph Bearshaw & Son Limited (UK) for a series of high-speed steels containing 0.7-0.8% carbon, 18-22% tungsten, 4% chromium, 1-1.5% vanadium, and the balance iron. Used for lathe tools, hobs, drills, taps and other cutting and finishing tools.

invisible glass. A highly transparent glass, such as *borax glass*, whose surface has been coated with a thin layer of sodium fluoride. It transmits nearly all (99.6%) of the incident light which makes it appear invisible.

inwall brick. Fireclay brick used in the refractory lining of blast-furnace stacks. See fireclay brick.

inyoite. A colorless mineral composed of calcium borate hydrate, $Ca_2B_6O_{11} \cdot 13H_2O$. It can also be made synthetically with slightly different composition, $CaB_3O_3(OH)_5 \cdot 4H_2O$. Crystal system, monoclinic. Density, 1.9-2.0 g/cm³; refractive index, 1.51; hardness, 2 Mohs.

Iochrome. Trade name of Alloy Technology International, Inc. (USA) for a vapor-deposited chromium of high purity (99.997%) used as an alloying element in high-chromium alloys and for high-temperature service applications.

iodargyrite. A yellowish or greenish mineral of the wurtzite group composed of silver iodide, AgI. It can be also made synthetically by precipitating from a silver nitrate solution with addition of an excess of sodium iodide. Crystal system, hexagonal. Density, 5.69 g/cm³; refractive index, 2.21.

iodide metal. A metal of high purity, such as hafnium, titanium or zirconium, obtained from the impure metal by combining with iodine (iodine process) to form a volatile tetraiodide that is subsequently decomposed at high temperatures to yield the refined metal.

iodine. A nonmetallic element of Group VIIA (Group 17) of the Periodic Table (halogen group). It is a violet-black to blackish-gray, corrosive solid with a metallic luster subliming at ordinary temperatures into an irritating blue-violet diatomic gas. It occurs in small quantities in seawater, rocks, soils and underground brines. Density, 4.93 g/cm³; melting point, 113.5°C (236°F); boiling point, 184.4°C (364°F); atomic number, 53; atomic weight, 126.904; monovalent, trivalent, pentavalent, heptavalent. It has semiconductive properties and is the least reactive halogen. Used in dyes, engraving lithography and halogen lamps, as a salt additive, as an alkylation and condensation catalyst, in photographic film, in water treatment, in biochemistry, biotechnology and medicine, and in pharmaceutical products. Symbol: I.

iodine-131. A radioactive iodine with a mass number of 131 obtained from the fission products of nuclear reactor fuels and by pile irradiation of tellurium. It has a half-life of 8 days and emits beta and gamma radiation. Used in film gages to measure film thicknesses in the order of 1 μm (39.4 μin.), in leak detection, as a radiation source in oilfield tests, and in medicine (e.g., as a radioactive tracer, as a diagnostic aid, and in radiation therapy). Symbol: ^{131}I; I-131. See also radioiodine.

iodopsin. A light-sensitive violet pigment found in the cones of the retina and formed from *retinol*.

iodyrite. A mineral of the halite group composed of silver iodide (AgI).

Ioline. Trade name for a glass-ionomer dental cement.

Iolon. Trademark of E.I. DuPont de Nemours & Company (USA) for a polyethylene copolymer film.

ion-beam-assisted-deposition coatings. See IBAD coatings.

Ion Bond. Trade name of Materials Research Corporation (USA) for a series of nitride coatings used on tool materials.

Ion Bond Coatings. Trade name of Multi-Arc Inc. (USA) for a series of physical and chemical vapor-deposited coatings (PVD and CVD coatings) based on aluminum-titanium nitride and chromium nitride. They retain their high hardness even at elevated temperatures, and are used to enhance the corrosion and wear properties of industrial cutting tools, metal-forming dies, plastic molds and wear parts.

Ione. Trade name of A.P. Green Company (USA) for a kaolin clay.

ion-exchange materials. A group of materials including silica and certain synthetic resins, usually supplied in particle form, that are capable of reversibly exchanging their ions (anions or cations) as functional groups for ions in a surrounding medium. These materials are used, for example, in water softening, filtration, water purification and chromatography, in the recovery of chromates from plating baths, and in the recovery of metals from wastes. Abbreviation: IEM. See also ion-exchange resin.

ion-exchange resin. A synthetic resin containing carboxylic, sulfonic, phenolic, substituted amino or other active groups that enable it to combine or exchange ions with an aqueous solution. For example, a synthetic resin with active sulfonic groups used to exchange sodium for calcium ions in water softening. Abbreviation: IER. See also ion-exchange materials.

ion-exclusion resin. A synthetic ion-exchange resin that in solution selectively absorbs non-ionized solutes (e.g., glycerin or sugar) but no ionized solutes. Used in ion-exclusion chromatography to separate electrolytes from nonelectrolytes.

ionic ceramics. A group of ceramic compounds including magnesium oxide (MgO) in which the interatomic bonding is mainly of the ionic type. See also ionic compounds.

ionic clays. A group of rare-earth ores of Chinese origin classified into the following two groups: (i) *Ionic type I* containing 62% yttrium, 33% other *heavy rare-earths* (or yttrics), 3% neodymium and 2% other *light rare-earths* (or cerics); and (ii) *Ionic type II* containing 29% neodymium, 3% cerium, 44% other cerics, 10% yttrium and 14% other yttrics. *Ionic clays* are considered an alternate source to traditional rare-earth ores, such as bastnaesite and monazite. See also heavy rare earths; light rare earths.

ionic compounds. A group of compounds in which component species are cations and anions held together primarily by strong electrostatic forces. They are solids at room temperature and have relatively high melting points. Examples of such compounds include calcium fluoride (CaF_2), magnesium oxide (MgO) and sodium chloride (NaCl).

ionic conductor. An electrical conductor in which conduction is a result of the movement of component ions (*intrinsic ionic conductor*) or the influence of impurity atoms (*extrinsic ionic conductor*). Abbreviation: IC.

ionic gel. A gel having ionic groups attached to its main structure.

ionic polymer. A polymer produced by a chain-reaction polymerization initiated by an acid or base. See also anionic polymer; cationic polymer.

ionic semiconductor. A solid, usually an essentially pure ionic crystal, in which electrical conduction does not result from the movement of electrons and holes, but from the movement of ions in the lattice.

ion-nitrided steel. See plasma-nitrided steel.

Ionobond. Trade name of Voco Chemie (Germany) for a glass-ionomer dental cement.

Ionoexpress. Trade name for a glass-ionomer dental cement.

Ionofil Molar AC. Trade name of Voco Chemie (Germany) for a glass-ionomer dental filling material.

Ionofil-U. Trade name for a glass-ionomer dental filling cement.

ionomer cement. See polyelectrolite cement.

ionomer-glass cements. Dental cements that are hybrids of polycarboxylate and silicate cements and used especially for general prosthedontic applications. Also known as *glass-ionomer cements*. See also dental cement; ionomer glasses; polycarboxylate cement.

ionomer glasses. Glass compositions with very high fluoride content, usually based on the fluoroaluminosilicates. They may be treated with silane, modified or filled with synthetic resins, or "doped" with metallic cations (e.g., cerium or silver) and are highly radiopaque and fluoride releasing. Standard grades have a typical chemical composition (in wt%) of about 30% silica (SiO_2), 20-30% alumina (Al_2O_3), 10-15% fluorine (F) and up to 10% phosphorus pentoxide (P_2O_5) and sodium oxide (Na_2O), respectively, and may also contain 0-20% strontia (SrO), 0-10% calcia (CaO) and 0-10% zinc oxide (ZnO). Standard grades have a density range of 2.6-3.1 g/cm³ (0.09-0.11 lb/in.³) and a refractive index of about 1.5. They are used in dental cements, compomers and composites. Also known as *glass ionomers*.

ionomers. A group of electrically conductive, thermoplastic resins based on copolymers of ethylene and vinyl monomer with an acid group, such as carboxylic or methacrylic acid. They are crosslinked polymers with covalent bonds between the elements of the chain, and ionic bonds between the chains with the anions hanging from the hydrocarbon chain, and the cations being metallic in nature, e.g., sodium, potassium, magnesium or zinc. Reinforced grades are also available. *Ionomers* can be processed by blow and injection molding, thermoforming, extrusion and other techniques, and have low densities, high molecular weights, high tensile strength, toughness and resiliency, excellent low-temperature flexural properties, a service temperature range of approximately -108 to +66°C (-160 to +150°F), high melt strength, high transparency, good resistance to abrasion, stress-cracking and corona attack, and relatively good resistance to oils, greases and solvents. Often sold under trade names or trademarks, such as *Hi-Milan* or *Surlyn*, they are used for blow and injection-molded containers, blow-molded packaging, packaging films, pipe and tubing, extruded film, sheet and tubing, break-resistant transparent bottles, mercury flasks, housewares, protective equipment, electric distribution elements, hand tools, tool handles and closures, machine parts, automotive components, such as bumper guards, decorative fascia and dunnage brackets, sporting goods, such as bowling machines, bowling pins, golf-ball covers, roller skates, skis, ski boots, and in dental cements. Ionomer foam is used as insulation material for fresh concrete. Also known as *ionomer resins.*

Ionosilver. Trade name for a silver-filled glass-ionomer dental cement.

Ionosit Fil P/L. Trade name of Zenith Dental (USA) for a light-cure glass-ionomer dental filling cement.

Ionosit-Seal. Trade name of Zenith Dental (USA) for a light-cure glass-ionomer resin cement used in dentistry as a pit and fissure sealant.

Ionosphere. Trade name of Henry Schein Inc. (USA) for a spherical dental amalgam alloy.

ion-plated coatings. Coatings deposited on substrates by subjecting their surfaces and/or the coating materials to a flux of high-energy particles (usually gas ions) sufficient to cause changes in the coating adhesion to the substrates, coating density, morphology or stress, or the surface coverage. Examples include optical coatings of chromium on plastics and titanium nitride and zirconium carbide on metals, electrical coatings of aluminum on gallium arsenide, copper on alumina, gold, platinum or silver on silicon, titanium or zirconium carbides or nitrides on carbides and corrosion-protective coatings of aluminum, cadmium, chromium or titanium on steel.

Ion S. Trade name for a glass-ionomer/cermet dental cement.

ion-sputtered coatings. Coatings deposited on a substrate by application of a high voltage to a target of the coating material in a medium of gas ions causing the target atoms (ions) to be dislodged and condense on the substrate.

ion-vapor-deposited coating. See Ivadized coating.

IOTA. Trademark of Unimin Corporation (USA) for high-purity quartz sands (99.98+%) supplied in several grades for various applications including the manufacture of crucibles, process tubing, high-temperature envelopes for halogen and mercury lamps, semiconductor glassware and wafer-handling systems.

Iotek. Trade name of Exxon Chemical Company (USA) for sodium and zinc ionomer resins supplied in regular and molding/extrusion grades.

iowaite. A bluish green mineral of the sjogrenite group composed of magnesium iron oxide chloride hydroxide hydrate, $Mg_4Fe(OH)_8OCl\cdot3H_2O$. Crystal system, rhombohedral (hexagonal). Density, 2.11 g/cm^3; refractive index, 1.543. Occurrence: USA (Iowa).

Ipaphon. Trade name of Interpane Isolierglas (Germany) for a multi-pane acoustical insulating glass.

Iper. Trade name of of Fidenza Vetraria SpA (Italy) for an extensive series of glass products including (i) *Iperbloc* solid glass lenses with concentric circle designs, (ii) *Ipercamera* hollow glass blocks with reeded or cross-reeded patterns, (iii) *Ipercolor* colored, hollow glass blocks, (iv) *Ipercontrol* hollow glass blocks with narrow horizontal reeds, (v) *Ipercristal* transparent hollow glass blocks, (vi) *Ipercrom* colored, hollow glass blocks, (vii) *Iperdecor* decorative hollow glass blocks, (viii) *Ipersimplex* solid glass lenses with reeded or cross-reeded patterns, (ix) *Ipersimplex Color* colored hollow glass blocks with cross-reeded patterns and (x) *Iperstagna* solid glass lenses with concentric circle designs.

Iplus Neutral. Trade name of Interpane Isolierglas (Germany) for a multi-pane thermal insulating glass.

Iporka. Trademark of BASF AG (Germany) for urea-formalde-hyde foam and foam products.

ipre brick. A paving brick shaped like the letter "I."

Ipro. Trademark for a small, rectangular concrete paving stone.

IPS. (1) Trade name of Ivoclar Vivadent AG (Liechtenstein) for a series of dental ceramics, glass-ceramics and porcelains. *IPS Classic* is a dental porcelain for metal-ceramic restorations and *IPS Corum* a dental ceramic for veneering inlays and onlays. *IPS d.SIGN* refers to a range of opalescent fluoroapatite-leucite glass-ceramic restoratives and matching alloys and cementation materials. The glass-ceramics provide excellent stability, enhanced contouring properties and light optical properties and wear behaviors similar to that of natural teeth. The matching alloys include various biocompatible and high- and medium-gold alloys as well as nickel-chromium-molybdenum alloys. *IPS Empress* are pressed dental glass-ceramics including lithium disilicate framework ceramics and fluoroapatite layering ceramics. They provide high translucency, high chemical resistance, excellent light optical properties, a wear behavior similar to that of natural dental enamel, and are used for inlays, onlays, crowns, veneers and all-ceramic bridges for anterior and premolar restorations.

(2) Trademark of International Polymer Corporation (USA) for an extensive series of engineering thermoplastics based on polyphenylene sulfide (PPS) or polyetheretherketone (PEEK). The PPS plastics are supplied as (i) unmodified compression molding compounds in wear-resistant, high-modulus, high-temperature and dielectric grades and are used accordingly for bearings, machine parts or electrical components; and (ii) modified compression molding compounds including carbon fiber-reinforced, PTFE-modified, and carbon fiber-reinforced plus PTFE-, molybdenum disulfide and graphite-modified grades. They are used according to grade for bearings, bushings, gaskets, packing and other machine parts and for automotive aerospace applications. The PEEK plastics are supplied as unreinforced, glass-reinforced and carbon fiber-reinforced PTFE-modified materials with good chemical resistance and high-temperature properties for automotive, chemical, electrical and electronic components.

Iralite. Trade name of Gulf & Western Manufacturing Company for a wear-resistant pearlitic cast iron containing 3.1% carbon, 1.2% chromium, 0.3% molybdenum, and the balance iron. Used for sheaves, sprocket, plungers, glass rolls, brake drums, and conveyor parts.

iranite. A yellow mineral composed of lead copper chromium fluoride oxide silicate hydrate, $Pb_{10}Cu(CrO_4)_6(SiO_4)_2(F,OH)_2$. Crystal system, triclinic. Density, 6.32 g/cm^3. Occurrence: Iran.

iraquite. A pale greenish yellow mineral of ekanite group composed of potassium calcium thorium cerium lanthanum silicate, $K(La,Ce,Th)_2(Ca,La)_5Si_{16}O_{40}$. Crystal system, tetragonal. Density, 3.27 g/cm^3; refractive index, 1.590. Occurrence: Iraq.

irarsite. A gray white mineral of the pyrite group composed of iridium ruthenium arsenic sulfide, $(Ir,Ru)AsS$. Crystal system, cubic. Density, 11.92 g/cm^3. Occurrence: South Africa.

Ircamet. Trade name of Ingersoll-Rand Company (USA) for a nonmagnetic, austenitic stainless steel containing 0.07% carbon, 29% nickel, 20% chromium, 4% copper, 2.75% molybdenum, 1% silicon, 0.8% manganese, and the balance iron. Used for chemical equipment and furnace parts.

Irco. Trademark of International Rustproof Company (USA) for several surface coatings. *Irco Aluminum Coatings* are chromate conversion coatings for aluminum applied by dipping, roller coating and spraying. They provide excellent corrosion resistance when used as undercoatings, and also enhance both adhesion and corrosion resistance of subsequent paint films. *Irco Bond* are crystalline zinc phosphates for steel applied by dipping, automatic coating and barrel coating. They are used as paint bases, as wearing surfaces on certain types of steels, and as a drawing and forming aids for certain steels. *Irco Iron Phosphates* are uniquely mixed compounds of acidic salts applied by spraying or immersion to produce iridescent iron phosphate coatings on cold-rolled steel surfaces. They are used as paint bases that enhance both adhesion and corrosion resistance.

Irdal. Trade name for a heat-treatable aluminum alloy containing 5% silicon, 0.7% magnesium and 0.3% boron. Used for architectural applications.

IR dyes. A large group of synthetic dyes of widely varying composition that are suitable for applications in the infrared range and have typical absorption wavelengths between about 675 and 1135 nm. Some members of this group (e.g., IR-140) are used primarily as infrared and chemical laser dyes, while others (e.g., IR-1040 and IR-1035) are suitable as Q-swiching laser dyes.

Ireson Foam. Trade name of Michigan Fiberglass Sales (USA) for fiberglass-based foam products.

irhtemite. A white mineral composed of calcium magnesium hydrogen arsenate tetrahydrate, $Ca_4MgH_2(AsO_4)_4 \cdot 4H_2O$. Crystal system, monoclinic. Density, 3.09 g/cm³. Occurrence: Morocco.

Iridal. Trade name of Istituto Sperimentali Metalli Leggeri (Italy) for a heat-treatable aluminum alloy containing 5% silicon, 0.7% magnesium and 0.3% boron. Used for architectural applications.

iridarsenite. A gray mineral of the marcasite group composed of iridium arsenide, $IrAs_2$. It can also be made synthetically. Crystal system, monoclinic. Density, 10.92 g/cm³. Occurrence: New Guinea.

Iriden. Trade name of Nuova Rayon SpA (Italy) for rayon fibers and fabrics.

iridescent fabrics. Textile fabrics, usually woven with differently colored warp and weft yarns, that change colors according to the angle of view of the observer and the lighting conditions.

iridescent glass. Glass that either has various colors or displays rainbow colors on its surface.

Iridex. Trade name for rayon fibers and yarns used for textile fabrics.

iridic bromide. See iridium tetrabromide.

iridic chloride. See iridium tetrachloride.

Iridite. Trade name of MacDermid Inc. (USA) for a conversion coating.

Iridium. British trade name for a series of special bearing alloys containing 15.7-21.7% tin, 1.1-1.25% copper, a trace of antimony, and the balance zinc. They have poor resistance to heat or live steam, and are used for bearings.

iridium. A hard, brittle, silver-white metallic element of Group VIII (Group 9) of the Periodic Table. It is commercially available in the form of sponge, powder, rods, slugs, sheet, foil, microfoil, wire and single crystals. The single crystals are usually grown by the float-zone or the radio-frequency technique. It usually occurs as a native alloy with platinum or gold, or with osmium (as in *iridosmine*). Crystal system, cubic. Crystal structure, face-centered cubic. Density, 22.65 g/cm³ (the heaviest element known); melting point, 2443°C (4429°F); boiling point, 4130°C (7466°F); hardness, 200-650 Vickers; superconductivity critical temperature, 0.11K; atomic number, 77; atomic

weight, 192.217; monovalent, divalent, trivalent, tetravalent, hexavalent. It does not tarnish in air, is the most corrosion-resistant element, highly resistant to chemical attack, and has good electrical conductivity (about 33% IACS), a very high modulus of elasticity (about 530 GPa) and a hardening effect on platinum. Used in platinum-iridium alloys for primary standards of weight and length, in ammonia fuel-cell catalysts, electric contacts and thermocouples, surgical instruments, spark-plug electrodes for helicopters and jet-engine igniters, commercial electrodes and resistance wires, extrusion dies for glass fibers, electrical and other scientific apparatus, high-temperature equipment, crucibles, laboratory ware, hypodermic needles, pen-nib points and jewelry. Symbol: Ir.

iridium-192. A radioactive isotope of iridium with a mass number of 192 obtained by pile irradiation of iridium. It has a half-life of 74.4 days and emits beta and gamma radiation. Used in radiography of light metal castings, and in biochemistry and medicine. Symbol: ^{192}Ir.

iridium antimonide. A crystalline compound of iridium and antimony. Crystal system, cubic. Crystal structure, skutterudite. Density, 9.35 g/cm³; melting point, 897°C (1647°F); band gap, 1.18 eV. Used as a semiconductor. Formula: $IrSb_3$.

iridium arsenide. A crystalline compound of iridium and arsenic. Crystal system, cubic. Crystal structure, skutterudite. Density, 9.12 g/cm³; melting point, 1197°C (2187°F) or above. Used as a semiconductor. Formula: $IrAs_3$.

iridium bromide. See iridium tetrabromide; iridium tribromide.

iridium carbonyl. An organometallic compound available in the form of a yellow powder (98%pure) that decomposes at 195°C (383°F). Used as an organometallic catalyst, and for iridium coatings. Formula: $Ir_4(CO)_2$. Also known as *tetrairidium dodecacarbonyl.*

iridium chloride. See iridium trichloride; iridium tetrachloride.

iridium dioxide. Brown crystals or brown to black powder (99.9% pure). Crystal system, tetragonal. Density, 11.66; melting point, decomposes at 1100°C (2010°F). Used in chemistry, biochemistry, ceramics and materials research. Formula: IrO_2. Also known as *iridium oxide.*

iridium disulfide. See iridium sulfide (ii).

Iridium Extra Special. Trade name of Becker Stahlwerk AG (Germany) for a high-speed steel containing 0.7% carbon, 18% tungsten, 5% cobalt, 4% chromium, 2% vanadium, and the balance iron. Used for tools and cutters.

iridium fluoride, See iridium hexafluoride; iridium trifluoride.

iridium hexafluoride. Yellow, hygroscopic crystals. Crystal system, cubic. Density, 4.8 g/cm³; melting point, 44°C (111°F); boiling point, 53°C (127°F). Used in chemistry and materials research. Formula: IrF_6. Also known as *iridium fluoride.*

iridium iodide. See iridium triiodide.

iridium monosulfide. See iridium sulfide (i).

iridium oxide. See iridium dioxide; iridium sesquioxide.

iridium phosphide. A crystalline compound of iridium and phosphorus. Crystal system, cubic. Crystal structure, skutterudite. Density, 7.36 g/cm³; melting point, 1197°C (1647°F) or above. Used as a semiconductor. Formula: IrP_3.

iridium-potassium chloride. Black, hygroscopic powder or dark-red crystals. Density, 3.546 g/cm³; decomposes on heating. Used as a black pigment in the decoration of porcelain. Formula: K_2IrCl_6. Also known as *potassium-iridium chloride; potassium chloroiridate; potassium hexachloroiridate.*

iridium-rhodium alloys. A group of high-temperature thermocouple alloys that are similar in properties to the platinum-rhodium types.

iridium sesquioxide. Blue-black crystals or black powder. It decomposes at about 1000°C (1830°F) and is used in ceramics as an underglaze black pigment, for the decoration of porcelain, and in materials research. Formula: Ir_2O_3. Also known as *iridium oxide*.

iridium sequisulfide. See iridium sulfide (iii)

iridium silicide. A compound of iridium and silicon used in ceramics and materials research. Formula: IrSi.

iridium-sodium chloride. A crystalline compound that is available in anhydrous form as greenish-brown, hygroscopic crystals, and as the hexahydrate ($Na_2IrCl_6·6H_2O$) in the form of a red-black powder that decomposes at 600°C (1110°F). Used as a reagent and pigment. Formula: $NaIrCl_6$ (anhydrous) Also known as *sodium-iridium chloride; sodium chloroiridate; sodium hexachloroiridate*.

iridium sulfide. Any of the following compounds of iridium and sulfur used in ceramics, electronics and materials research: (i) *Iridium monosulfide*. Blue-black, water-insoluble crystals. Formula: IrS; (ii) *Iridium disulfide*. Brown crystals. Crystal system, orthorhombic. Density, 9.3 g/cm^3. Formula: IrS_2; and (iii) *Iridium sesquisulfide*. Brown to black crystals. Crystal system, orthorhombic. Density, 10.2 g/cm^3. Formula: Ir_2S_3.

iridium tetrabromide. Black, hygroscopic crystals used in chemistry and materials research. Formula: $IrBr_4$. Also known as *iridic bromide; iridium bromide*.

iridium tetrachloride. Brownish-black, hygroscopic mass used in microscopy and in electroplating solution. Formula: $IrCl_4$. Also known as *iridic chloride; iridium chloride*.

iridium tribromide. A crystalline compound that is available in anhydrous form as water-soluble, olive-green, brown or black crystals, and as the tetrahydrate ($IrBr_3·4H_2O$) in the form of red-brown crystals. Crystal system, monoclinic. Density, 6.82 g/cm^3; melting point, loses $3H_2O$ at 100°C (212°F). Used in chemistry and materials research. Formula: $IrBr_3$ (anhydrous). Also known as *iridium bromide*.

iridium trichloride. Brown crystals or dark green to blue-black powder (99.9% pure) made by reacting iridium powder with chlorine at 600°C (1112°F). Density, 5.30 g/cm^3; decomposes at 763°C (1405°F). Used in chemistry, biochemistry and materials research. Formula: $IrCl_3$. Also known as *iridium chloride*.

iridium trifluoride. Black hexagonal crystals. Density, 8.0 g/cm^3; melting point, decomposes at 250°C (482°F). Used in chemistry and materials research. Formula: IrF_6. Also known as *iridium fluoride*.

iridium triiodide. Dark brown crystals. Crystal system, monoclinic. Density, 7.4 g/cm^3. Used in chemistry and materials research. Formula: IrI_3. Also known as *iridium iodide*.

iridium trisilicide. A compound of iridium and silicon used in ceramics and materials research. Formula: Ir_3Si.

iridium wire. Iridium (purity 99.9%) supplied in the as-drawn condition with a diameter usually ranging from 0.075 to 1.0 mm (0.03 to 0.04 in.) and a length of 0.05-2.00 m (0.16 to 6.50 ft.). Used in spark plug electrodes for helicopters and jet engine igniters.

iridizing compound. A highly adherent metal-oxide or other coating produced on a glass or other vitreous substrate to impart certain desired surface properties, e.g., electrical conductivity or resistance, or decorative or optical properties, such as reflectance or surface finish.

Irido-Platinum. Trade name for a wrought platinum-iridium alloy used for jewelry, and in the form of wire for dental applica-

tions.

iridosmine. A tin-white to light steel-gray natural alloy of iridium and osmium with a metallic luster. Its composition varies greatly from 10-77% iridium, 17-80% osmium, 0-18% ruthenium, 0-17% rhodium, 0-10% platinum, 0-3% iron, 0-1% copper, and traces of gold and palladium. Alloys containing more osmium than iridium are referred to as *osmiridium*. Crystal system, hexagonal. Density, 18.8-21.1 g/cm^3; hardness, 6-7 Mohs. It has good corrosion resistance and good resistance to acids including aqua regia. Occurrence: Russian Federation. Used as a source of iridium and osmium, for hardening platinum, as a primary standard of weight and length, and for jewelry, watch pivots, compass bearings, fountain-pen points and surgical needles.

Iridye. Trade name for rayon fibers and yarns used for textile fabrics.

iriginite. A yellow mineral composed of uranium molybdenum oxide hydroxide dihydrate, $U(MoO_4)_2(OH)_2·2H_2O$. Crystal system, monoclinic. Density, 3.84 g/cm^3; refractive index, 1.889. Occurrence: USA (South Dakota), Russian Federation.

Irilac. Trade name of MacDermid Inc. (USA) for protective coatings used as clear finishes on metals.

Irish lace. A lace with looped edges, usually of the crochet or needlepoint type, made in Ireland.

Irish lawn. A fine, sheer fabric made from linen yarns. See also lawn.

Irish linen. A thin fabric woven from high-quality Irish flax yarn.

Irilit. Danish trade name for cellulose acetate plastics.

IRM. Trade name of Dentsply Caulk (USA) for reinforced zinc oxide–eugenol dental cement.

Irmir. Trade name of Flachglas AG (Germany) for a coated glass.

iroko. See African teak.

iron. (1) A silver-white, ductile, metallic element of Group VIII (Group 8) of the Periodic Table. It is commercially available in the form of lumps, rods, chips, sponge, powder, wire, foil, microfoil, tube and single crystals. The crystals are usually grown by the float-zone or the radio-frequency technique. The most common ores are *hematite, magnetite, siderite, limonite* and *taconite*. Crystal system, cubic. Density, 7.87 g/cm^3; melting point, 1535°C (2795°F); boiling point, 2750°C (4982°F); atomic number, 26; atomic weight, 55.845; divalent, trivalent, tetravalent, hexavalent. There are four forms: (i) *Alpha iron* (α-iron), that exists below the Curie temperature of 768°C (1415°F) and is body-centered cubic and ferromagnetic; (ii) *Beta iron* (β-iron), that is stable between 768 and 910°C (1415 and 1670°F) and is body-centered cubic and paramagnetic; (iii) *Gamma iron* (γ-iron), that is stable between 910 and 1400°C (1670 and 2550°F) and is face-centered cubic; and (iv) *Delta iron* (δ-iron) that exists above 1400°C (2550°F) and is body-centered cubic. Iron is used in the manufacture of steel and cast iron, tin cans, tools, machinery, ships, automobiles, railroad equipment, buildings, bridges, hardware, magnets, and in numerous alloys especially with aluminum, chromium, cobalt, copper and nickel. Symbol: Fe.

(2) Any iron-base material not falling into the steel or ferrous superalloy classifications, e.g., pig iron, ingot iron, cast iron or wrought iron.

iron-59. A radioactive isotope of iron with a mass number of 59 that is obtained by irradiation of iron metal, has a half-life of 46.3 days and emits beta and gamma rays. Used in as a tracer in biochemistry and metallurgy, and in medicine and diagnostics, e.g., for blood cell and hemoglobin studies. Symbol: ^{59}Fe.

Also known as *radioiron*.

Ironac. British trade name for an acid-resistant cast iron containing 1.1% carbon, 13.2% silicon, 0.77% manganese, 0.8% phosphorus, and the balance iron. Used for valves, pumps, nozzles, ejectors and fans.

iron acetate. See ferric acetate; ferrous acetate.

iron acetylacetonate. See ferric acetylacetonate.

iron alloys. A group of alloys in which iron is the predominant component, but carbon as well as other alloying elements may also be present. The most important types of ferrous alloys are steels, cast irons, wrought irons and iron-base superalloys. Other important iron alloys include iron-aluminum, iron-chromium, iron-cobalt, iron-copper and iron-nickel alloys. Also known as *ferrous alloys*.

iron alum. See halotrichite.

iron aluminate. See ferrous aluminate.

iron-aluminum alloys. A group of alloys in which iron and aluminum are the principal components. Small additions of aluminum (0.5%) to nonoriented steels (iron-silicon alloys) improve the soft magnetic properties, increase the electrical resistivity and result in good permeability at low flux density. Higher additions (about 13-17%) to pure iron result in iron-aluminum alloys with excellent magnetic properties even at elevated temperatures, high saturation induction, high permeabilities and low core losses. Used for magnetic and electrical equipment, such as transformers and relays.

iron-ammonium sulfate. See ferrous ammonium sulfate.

iron arsenide. Any of the following compounds of iron and arsenic used as semiconductors and in materials research: (i) *Iron monoarsenide*. White or gray crystals. Crystal system, orthorhombic. Density, 7.85 g/cm³; melting point, 1020°C (1886°F). Formula: FeAs. Also known as *ferrous arsenide*; and (ii) *Iron diarsenide*. Silver-gray crystals. Occurs in nature as the mineral *loellingite*. Crystal system, rhombohedral. Density, 7.4 g/cm³; hardness, 2 Mohs. Formula: $FeAs_2$.

ironbark. The heavy, hard, grayish wood of an Australian hardwood tree of the genus *Eucalyptus*. It is extremely surface checked, has very high strength, and high durability and bending strength. Source: Western Australia. Used for structural work and flooring.

iron-base superalloys. A group of wrought iron-base heat-resisting alloys containing about 15-70% iron, 8-50% nickel, 5-25% chromium, 0-9% molybdenum, 0-5% niobium (columbium), 0-3% titanium, 0-2% aluminum, 0.05-0.5% carbon, and traces of manganese, silicon, boron, vanadium and copper. They have outstanding high-temperature mechanical properties, oxidation resistance and creep resistance, and a maximum operating temperature of about 900°C (1650°F). Used for jet-engine parts, aircraft parts, turbo-superchargers, turbines, nuclear reactors, and extreme high-temperature applications. Also known as *ferrous superalloys*.

iron beryllide. A compound of iron and beryllium used in ceramics. Melting point, 1483°C (2701°F). Formula: $FeBe_2$.

iron black. Finely divided black antimony obtained by adding zinc to an acid solution of an antimony salt and subsequent precipitation. Used to impart a polished-steel look on plaster of Paris and papier-mâché.

iron blue. A group of blue pigments based on *ferric ferrocyanide* ($Fe_4[Fe(CN)_6]_3$) available in various shades. They have good permanence, lightfastness, resistance to water, oils, alcohol, paraffin, organic solvents and dilute acids, and poor resistance to alkalies and reducing media. Used in paints, printing inks and plastics, in textile and paper dyeing, in baked enamel finishes for automobiles and appliances, and in industrial finishes.

iron-bonded graphite. A strong, nonbrittle material made by blending iron powder, calcium-silicon powder and graphite in defined proportions, compacting to the desired density and shape, and then sintering at a prescribed temperature. Used for oilless bearings.

iron boride. Any of the following compounds of iron and boron: (i) *Iron monoboride*. Refractory, crystalline solid. Crystal system, orthorhombic. Density, 7-7.15 g/cm³; melting point, 1538°C (2800°F) [also reported as 1650°C (3000°F). Used in ceramics and materials research. Formula: FeB; and (ii) *Ferrous boride*. A refractory, crystalline solid. Crystal system, tetragonal. Density, 7.3 g/cm³; melting point, 1389°C (2532°F). Used in ceramics, as an alloying addition to austenitic steels, and in absorber materials or shielding for nuclear reactors. Formula: Fe_2B.

iron-boron-silicon alloys. A group of alloys usually containing 77.5-81% iron, 13-16% boron, 3.5-9% silicon and up to 2% carbon. They have a density range of 7.18-7.32 g/cm³ (0.259-0.264 lb/in.³), a Vickers hardness of 860-940 kgf/mm², high tensile strength and modulus, good soft magnetic properties and a maximum service temperature (in air) of about 125-150°C (260-300°F). An amorphous structure is achieved by ultrafast cooling and the inclusion of a metalloid.

iron bromide. See ferric bromide; ferrous bromide.

iron bronze. A bronze containing 83% copper, 8.6% tin, 4.4% zinc and 4% iron. It has excellent seawater corrosion resistance and is used for marine parts, piston rods, shafts and screws.

iron carbide. See cementite.

iron-carbon alloys. Alloys of iron and carbon classified according to their carbon content into steels (up to 2.06% carbon) and cast irons (more than 2.06% and up to 6.67% carbon). The carbon may exist in two different forms: (i) as elemental carbon (*graphite*, or *temper carbon*) or chemically combined as iron carbide (*cementite*, Fe_3C). Carbon is the basis for the wide range of properties obtainable in irons and steels. It increases the tensile and yield strength, hardness, hardenability and wear resistance, and decreases the melting point, ductility, elasticity, elongation at fracture and low-temperature deformation. Most other alloying elements in irons or steels modify or enhance the benefits of carbon.

iron carbonate. See siderite (1).

iron-carbon powder. A powder consisting of a blend of iron and 0.3-1% carbon (graphite) and used to make powder-metallurgy steel. See also sintered steels.

iron carbonyl. See iron pentacarbonyl.

iron cement. A mixture composed of small iron chips, filings or punchings and ammonium chloride (NH_4Cl). Used for joining ferrous metals.

iron chloride. See ferric chloride; ferrous chloride.

iron chromate. See ferric chromate.

iron chromite. See chromite.

iron-chromium alloys. A large group of alloys composed predominantly of iron and chromium. Depending on composition they can be classified into the following subgroups: (i) *Plain iron-chromium alloys*, i.e., alloys containing only iron and chromium; (ii) *Iron-chromium-aluminum alloys* containing 70-85% iron, 12-25% chromium and 3.25-5.5% aluminum; (iii) *Iron-chromium-molybdenum alloys* containing 88-98% iron, 2-11% chromium and 0.5-1% molybdenum; and (iv) *Iron-chromium-cobalt alloys* typically containing 55-65% iron, 28% chromium,

10.5-15.5% cobalt, and small amounts of other elements.

iron-chromium-aluminum alloys. A group of alloys with compositions ranging from 70-85% iron, 12-25% chromium and 3.25-5.5% aluminum. They have a density range of 7.10-7.45 g/cm³ (0.257-0.269 lb/in.³), a maximum service temperature (in air) of 1050-1375°C (1920-2510°F), low resistivity, low temperature coefficients of resistance, low coefficients of thermal expansion and high tensile strength. They are often sold under trade names, such as *Aluchrom*, *Chromaloy* or *Fecralloy*, and used for electrical resistances and resistance heating elements, and with yttrium as FeCrAlY overlay coatings for superalloys.

iron-chromium cast steels. A group of cast stainless steels with about 10-30% chromium, up to 8% nickel, and small amounts of molybdenum and copper. Their carbon contents are usually below 0.5%, and they have ferritic microstructures, excellent oxidation resistance, low high-temperature strength, a maximum service temperature of 650-760°C (1200-1400°F), good chemical resistance, and excellent resistance to sulfur- and nitrogen-containing atmospheres. Used for chemical plant equipment, furnace parts, trays and fixtures.

iron-chromium-cobalt alloys. A group of ductile permanent magnet alloys containing about 55-65% iron, 28% chromium, 10.5-15.5% cobalt, and small amounts of other elements. Used for electrical resistances.

iron-chromium-molybdenum alloys. A group of alloys with about 88-98% iron, 2-11% chromium and 0.5-1% molybdenum. They have a density range of 7.70-7.87 g/cm³ (0.278-0.284 lb/in.³), low resistivity, low coefficient of thermal expansion, and high tensile strength. Used for electrical resistances.

iron-chromium-nickel cast steels. A group of cast stainless steels with 14-32% chromium, 8-22% nickel and up to 4% molybdenum and/or copper. Their carbon contents are usually below 0.08%, and they have austenitic microstructures, good high-temperature strength, good ductility, good resistance to corrosive media including sulfurous and acetic acids, mild chlorides and flowing seawater, good weldability and good to moderate machinability. Used for chemical plant equipment, marine parts and equipment, heat-treating furnaces, power-plant equipment, furnace parts, trays and fixtures.

Iron Clad. Trade name of Glen Falls Cement Company, Inc. (USA) for masonry and Portland cements.

iron clay. See argillaceous hematite.

iron-cobalt alloys. A group of soft magnetic alloys containing 45-65% iron, 25-50% cobalt and up to 3% vanadium. Alloys with about 35% cobalt have extremely high saturation induction values. Vanadium is usually added to improve the cold workability of the otherwise brittle alloys. Used for magnetic strips. See also cobalt-iron alloys.

iron columbium. See iron niobium.

iron-constantan. A *bimetal* consisting of iron and *constantan* (55Cu-45Ni) and used in thermocouples for temperature measurements up to 1000°C (1830°F) in oxidizing and reducing atmospheres.

iron-copper alloys. A group of powder-metallurgy alloys made by blending iron powder with 2-11% copper, pressing and sintering. Available in the as-sintered condition, they have a density range of about 6.0-6.7 g/cm³ (0.22-0.24 lb/in.³), good to moderate strength, good corrosion resistance, low elongation, and good hardness. See also sintered iron-copper.

iron-copper-carbon alloys. A group of powder-metallurgy alloys made by blending iron powder with 2-11% copper, and up to 0.8% carbon (graphite), pressing and sintering. Available in the as-sintered condition, they have a density range of about 6.0-6.7 g/cm³ (0.22-0.24 lb/in.³), improved solubility, good to moderate-strength, good corrosion resistance, low elongation, and good hardness. See also sintered copper iron; sintered copper steel; sintered iron-nickel.

iron-copper powder. A blend of iron powder with 9.5-10.5% copper and up to 0.3% carbon (or graphite), used in the manufacture of sintered iron-copper products. See also sintered iron-copper.

iron diarsenide. See iron arsenide (ii).

iron dichromate. See ferric dichromate.

iron diphosphide. See iron phosphide (ii).

iron disulfide. See ferrous disulfide.

iron disilicide. See iron silicide (ii).

iron ditelluride. See iron telluride.

iron dodecacarbonyl. See triiron dodecacarbonyl.

iron driers. A group of dark brown iron-base paint driers (e.g., ferrous octoate) having high tinting strength.

Ironex. Trade name of Murex Limited (UK) for plain-carbon steels containing 0.06-0.07% carbon, 0.5-0.7% manganese and the balance iron. Used for welding electrodes.

iron ferrite. A ferrimagnetic ceramic with a cubic spinel crystal structure (*ferrospinel*) made by powder-metallurgy techniques. It has excellent magnetic properties at high frequencies, very high resistivity, and is used as a soft magnetic material for antennas, computer memory cores, recording tapes and telecommunication applications. Formula: $FeO \cdot Fe_2O_3$.

iron ferrocyanide. See ferric ferrocyanide.

iron fluoride. See ferric fluoride; ferrous fluoride.

iron glance. See specularite.

iron hydrate. See ferric hydroxide.

iron hydroxide. See ferric hydroxide.

ironier's bronze. A corrosion-resistant special tin bronze containing some mercury.

iron iodide. See ferrous iodide.

Ironite. Trade name of Kinite Corporation (USA) for a wear-resistant nickel cast iron containing varying amounts of chromium, nickel and vanadium. Used for gears and cams.

iron metavanadate. See ferric vanadate.

iron mica. See biotite.

iron-molybdenum. An intermetallic compound of iron and molybdenum. It has a density of 8.53 g/cm³, a melting point of 1540°C (2804°F), a hardness of 76.5 Rockwell A, and is used in ceramics and materials research. Formula: FeMo.

iron molybdite. See ferrimolybdite.

iron monoarsenide. See iron arsenide (i).

iron monoboride. See iron boride (i).

iron mononitride. See iron nitride (i).

iron monophosphide. See iron phosphide (i).

iron monosilicide. See iron silicide (i).

iron monosulfide. See ferrous sulfide.

iron monotelluride. See iron telluride (i).

iron monoxide. See black iron oxide.

iron naphthenate. See ferric naphthenate; ferrous naphthenate.

iron-neodymium-boron magnets. See neodymium-iron-boron magnets.

iron-nickel alloys. A large group of alloys composed predominantly of iron and nickel that according to their properties can be classified into the following groups: (i) *Magnetically soft alloys* containing about 50-80% nickel, 16-50% iron and up to 4% molybdenum. High-nickel types have high initial permeability and low saturation induction, while low-nickel types have low initial permeability and high saturation induction. The normally low coefficients of expansion of these alloys increase with decreasing nickel content; (ii) *Low-expansion alloys* with a typical composition range of 48-64% iron, 31-52% nickel, and up to 0.4% manganese, 0.1% silicon and 5% cobalt, respectively. Their coefficients of thermal expansion increase with increasing nickel content; (iii) *Powder-metallurgy alloys* made by blending iron powder with 1-8% nickel, 0-2.5% copper and up to 0.9% carbon, followed by pressing and sintering; and (iv) *Stainless steels* which are corrosion-resistant alloy steels containing high percentages of chromium (usually 10% or more) and may also contain nickel and/or other alloying additions.

iron niobium. Any of the following compounds of iron and niobium (columbium) used in ceramics: (i) Fe_2Nb. Melting point, 1626°C (2959°F); and (ii) Fe_3Nb_2. Density, 7.89 g/cm^3; melting point 1655°C (3011°F); hardness, 76 Rockwell A. Also known as *iron columbium.*

iron nitrate. See ferric nitrate.

iron nitride. Any of the following compounds of iron and nitrogen: (i) *Iron mononitride.* Used in ceramics and materials research. Formula: FeN; and (ii) *Ferrous nitride.* Gray crystals. Density, 6.35 g/cm^3. Used in electronics and materials research. Formula: Fe_2N.

iron nonacarbonyl. Air- and heat-sensitive orange-yellow crystals that decompose at 100°C (212°F). Used in organometallic synthesis. Formula: $Fe_2(CO)_9$. Also known as *nonacarbonyliron.*

iron ochers. Impure oxides of iron, such as red and yellow ocher.

iron-olivine. See fayalite.

iron-ore cement. A light to chocolate-brown, slow-setting and hardening cement in which ferric oxide (Fe_2O_3) is used as a substitute for clay, shale or alumina. With a density of approximately 3.1 g/cm^3 (0.11 $lb/in.^3$) it is heavier than *Portland cement,* and has a higher resistance to corrosive environments (e.g., seawater).

iron ores. A group of iron-bearing minerals which also contain varying amounts of other elements, such as manganese, silicon, phosphorus, sulfur, clay, lime, titania, magnesia and/or silica. Depending on the particular ore, the average iron content may range from as low as 20% to over 70%. Examples of common iron ores include *magnetite* (Fe_3O_4) with about 60-72.5% Fe, *hematite* (Fe_2O_3) with about 30-50% Fe, *limonite* ($Fe_2O_3 \cdot 3H_2O$) with about 20-25% Fe, *siderite* ($FeCO_3$) with about 30-45% Fe and *taconite* (mixture of hematite and silica) with 20-30% Fe. Used for the manufacture iron and steel.

iron oxalate. See ferric oxalate; ferrous oxalate.

iron oxalate dihydrate. See ferrous oxalate dihydrate.

iron oxide. See black ferric oxide; black iron oxide; brown iron oxide; red iron oxide; yellow iron oxide.

iron oxide pigments. A group of pigments containing substantial quantities of ferric oxide (Fe_2O_3), and including *Indian red, Persian red, Tuscan red, Venetian red* and several commercial preparations sold under various trade names. Used in the manufacture of inexpensive paints, and as prime coats for steels. Also known as *ferric oxide pigments.*

iron pentacarbonyl. A mobile, air-, heat- and light-sensitive, flammable orange liquid obtained by treating fine iron powder with carbon monoxide in the presence of a catalyst, such as ammonia. Density, 1.490 g/cm^3; melting point, -20°C (-4°F); boiling point, 103°C (217°F); flash point, -15°C (5°F); refractive index, 1.5196. Used as a carbonyl iron powder for high-frequency coils and magnetic cores. Formula: $Fe(CO)_5$. Also known as *iron carbonyl; pentacarbonyliron.*

iron pentanedionate. See ferric acetylacetonate.

iron perchloride. See ferric chloride.

iron persulfate. See ferric sulfate.

iron persulfate nonahydrate. See ferric sulfate nonahydrate.

iron phosphate. Bluish-gray, hygroscopic powder that can be made synthetically, and also occurs in nature as the mineral *vivianite.* Density, 2.58 g/cm^3. Used in ceramics, as a catalyst, and in conversion coatings. Formula: $Fe_3(PO_4)_2 \cdot 8H_2O$. Also known as *ferrous phosphate; ferrous phosphate octahydrate; iron phosphate octahydrate.*

iron phosphate coatings. Amorphous coatings composed essentially of iron oxides and usually having iridescent blue to bluish-red colors. They are applied to iron, steel, galvanized steel or aluminum by immersion in, or spraying of an alkali-metal phosphate solution maintained at a pH of about 4-5. *Iron phosphate coatings* exhibit excellent adherence to substrates and good resistance to flaking from impact or flexing when painted. Used as bases for subsequent coats, and as base coats applied to steel to prepare its surface for the bonding to fabrics, wood or other materials. See also phosphate coatings.

iron phosphide. Any of the following compounds of iron and phosphorus: (i) *Iron monophosphide.* Density, 6.90 g/cm^3. Used in ceramics and materials research. Formula: FeP; (ii) *Iron diphosphide.* Density, 5.12 g/cm^3. Used in ceramics and materials research. Formula: FeP_2; (iii) *Ferrous phosphide.* Gray crystals or bluish-gray, ferromagnetic powder. Crystal system, hexagonal. Density, 6.56-6.8 g/cm^3; melting point, 1290°C (2354°F). Standard grades contains about 24-25% phosphorus. Used in iron and steelmaking. Formula: Fe_2P; (iv) *Ferric phosphide.* A gray powder. Density, 7.21 g/cm^3; decomposes at 1204°C (2200°F). Used in ceramics, and in the manufacture of iron-phosphorus powder-metallurgy alloys. Formula: Fe_3P.

iron-phosphorus alloys. A group of powder-metallurgy alloys made by blending atomized iron powder with up to 0.3% carbon, and iron phosphide (Fe_3P) to yield up to 1% phosphorus, followed by pressing and sintering. It has a density of 6.7 g/cm^3 (0.24 $lb/in.^3$), excellent soft magnetic properties, and good toughness and strength.

iron phthalocyanine. A dark purple, crystalline phthalocyanine derivative that contains a divalent central iron ion (Fe^{2+}). It has a typical dye content of about 90%, a maximum absorption wavelength of 657 nm, and is used as a pigment and dye. Formula: $Fe(C_{32}H_{16}N_8)$. Abbreviation: FePc. Also known as *ferrous phthalocyanine.*

iron plate. A relatively thick coating of iron or steel applied to a material to provide an electrically conductive surface, or to enhance its mechanical properties.

iron platinum. A ductile, magnetic, grayish black native platinum containing considerable amounts of iron (15-22%) and small additions of iridium, osmium and rhodium. Also known as *ferroplatinum.*

iron Portland cement. A mixture of 70% *Portland cement* and 30% granulated *blast-furnace slag* (lime-alumina silicate). Used in building and road construction.

iron powder. Iron in finely divided form (usually 0.02-800 μm, or 0.8 μin. to 0.03 in.) and in purities from 99.0-99.99%. It is obtained essentially by any of the following processes: (i) reduction of ferrous chloride dihydrate ($FeCl_2 \cdot 2H_2O$); (ii) thermal decomposition of iron pentacarbonyl [$Fe(CO)_5$] at 250°C (480°F); (iii) hydrogen reduction of high-purity ferric oxide or oxalate [Fe_2O_3 and $Fe_2(C_2O_4)_3$, respectively]; (iv) direct reduction of magnetite (Fe_3O_4) with carbon in the Hoeganaes process (sponge iron), or of finely ground selected iron oxides with hydrogen gas in the Pyron process; (v) gas or water atomization of a liquid metal; or (vi) electrolytic deposition from ferric salt solutions. Used for powder-metallurgy products, magnets, high-frequency cores, automotive components, appliances, office equipment, lawn and garden equipment, catalysts, and welding electrodes. Also known as *ferrous powder.*

iron protochloride. See ferrous chloride.

iron protoiodide. See ferrous iodide.

iron protosulfide. See ferrous sulfide.

iron putty. A mixture of ferric oxide and boiled linseed oil used as an acid-resistant putty.

iron pyrite. See pyrite.

iron pyrophosphate. See ferric pyrophosphate.

iron red. A group of pigments made from red varieties of iron oxide (Fe_2O_3). Examples include *Indian red, Persian red, Tuscan red* and *Venetian red.*

iron resinate. See ferric resinate.

iron rhenium. An intermetallic compound of iron and rhenium used in ceramics and materials research. Melting point, 1521°C (2770°F). Formula: Fe_3Re_2.

iron saffron. See Indian red.

iron sandstone. See ferruginous sandstone.

iron scurf. A mixture of stone and iron particles made by grinding and polishing gun barrels with siliceous abrasives and grindstones. Used as a blue pigment for coloring brick.

iron selenide. See ferrous selenide.

iron sesquioxide. See red iron oxide.

iron shot. Small, spherical particles of iron made by pouring drops of molten iron from a suitable height into water. Used as an abrasive in barrel finishing and shot peening, and for grinding stones. Shot made of cast iron or chilled iron has an as-cast hardness of 58-65 Rockwell, and is usually available in Society of Automotive Engineers (SAE) numbers ranging from No. 7 with a screen size of 2.82 mm (0.111 in.) to No. 120 with 0.124 mm (0.0049 in.). Peening shot is designated by SAE numbers from No. 70 to No. 930 according to size, with the shot number being the nominal diameter of a pellet per 2.54 μm (0.0001 in.).

iron silicate. A compound of ferrous oxide (FeO) and silicon dioxide (SiO_2) used in ceramics. Density, 4.24 g/cm³; melting point, 1198°C (2188°F); hardness, 5-7 Mohs. Formula: Fe_2SiO_4.

iron silicide. Any of the following compounds of iron and silicon used in ceramics and materials research: (i) *Iron monosilicide.* Gray crystals. Crystal system, cubic. Density, 6.1 g/cm³; melting point, 1410°C (2570°F); and (ii) *Iron disilicide.* Gray crystals or powder (99+% pure). Crystal system, tetragonal. Density, 4.74 g/cm³; melting point, 1220°C (2228°F). Formula: $FeSi_2$.

iron-silicon alloys. A group of soft magnetic iron-base alloys containing up to 5% silicon. See also silicon steels; electrical steels; electrical transformer steels.

iron sinter. See pharmacosiderite.

iron spinel. See hercynite.

iron sponge. See sponge iron.

iron stearate. See ferric stearate; ferrous stearate.

ironstone. Any iron ore with clay or other impurities from which iron can be obtained by smelting.

ironstone china. Generic term for a durable dinnerware made from a fine, strong vitrified earthenware body of high hardness.

ironstone clay. A collective name for brown and red ironstone clay. Brown ironstone clay is *limonite* (brown hematite) containing considerable amount of clay or clayey minerals, while red ironstone clay is a variety of the mineral *hematite* composed chiefly of ferric oxide with substantial amounts of clay or sand.

iron subcarbonate. See brown iron oxide.

iron sulfate. See ferric sulfate; ferrous sulfate.

iron sulfide. See ferrous sulfide; ferrous disulfide.

iron sulfuret. See ferrous sulfide.

iron tantalum. An intermetallic compound of iron and tantalum used in ceramics and materials research. Melting point, 1777°C (3231°F). Formula: Fe_2Ta.

iron telluride. Any of the following compounds of iron and tellurium used in ceramics, materials research and electronics: (i) *Iron monotelluride.* Grayish crystals or or gray lumps with high purity (99.99+%). Crystal system, tetragonal. Density, 6.8 g/cm³; melting point, 914°C (1677°F). Formula: FeTe. Also known as *ferrous telluride;* and (ii) *Iron ditelluride.* Pink crystals. Crystal system, orthorhombic. Density, 3.5 g/cm³. It occurs in nature as the mineral *frohbergite.* Formula: $FeTe_2$.

iron tersulfate. See ferric sulfate.

iron titanate. (1) A compound of ferrous oxide and titanium dioxide used in ceramics and materials research. Formula: $FeTiO_3$.
    (2) A compound of ferric oxide and titanium dioxide used in ceramics and materials research. Formula: Fe_2TiO_5.

iron titanium. Any of the following intermetallic compounds of iron and titanium: (i) *Iron titanium.* Density, 6.2 g/cm³; hydrogen-storage capacity, 1.8 wt%. Used in battery electrodes for hydrogen storage systems. Formula: FeTi; and (ii) *Diiron titanium.* Melting point, 1516°C (2761°F). Used in ceramics and materials research. Formula: Fe_2Ti.

iron trichloride. See ferric chloride.

iron trisulfate. See ferric sulfate.

iron tungstate. Brown to black crystals. It can be made synthetically and occurs in nature as the mineral *ferberite.* Density, 6.64-7.11 g/cm³; hardness, 4.0-4.5 Mohs. Used in metallurgy. Formula: $Fe_2(WO_4)_3$.

iron vanadate. See ferric vanadate.

iron vitriol. See ferrous sulfate.

Ironweld. Trade name of Edgcomb Metals Corporation (USA) for a cast iron containing 3.2% carbon, 2.2% silicon, and the balance iron. Used in the form of flux-coated welding rods for welding cast irons.

iron whiskers. Single, axially oriented, crystalline high-strength filaments or fibers of iron with an average diameter of 1 μm (40 μin.) used chiefly for electrical applications.

ironwood. See African ironwood; American hornbeam; hop hornbeam.

iron yarn. A stiff, starched and glazed cotton yarn, usually with a smooth surface and a black or white color, used especially for linings and hats.

iron zirconium. An intermetallic compound of iron and zirconium used in ceramics and materials research. Density, 7.70 g/cm³; melting point, 1642°C (2988°F); hardness, 436 Vickers. Used in ceramics and materials research. Formula: Fe_2Zr.

iron-zirconium-carbide cermet. A cermet made by powder metallurgy techniques and composed of zirconium carbide (ZrC) particles in an iron-base matrix. It has good strength, and wear resistance and is used for nuclear equipment.

iron-zirconium pink. See zirconium-iron pink.

Iropol. Trademark of Iroquois Chemicals Corporation (Canada) for a series of polyester and urethane resins.

Iroquois. Trade name of Atlas Specialty Steels (Canada) for a water-hardening tool steel containing 1.25% carbon, 0.25% manganese, 0.25% silicon, 1.4% tungsten, and the balance iron. It has good resistance to decarburization, high wear resistance, moderate to low toughness and shock resistance and moderate machinability. Used for threading taps, reamers, broaches, dental burrs, other fine-edged tools, and deep-drawing dies.

Irotherm. Trademark of Iroquois Chemicals Corporation (Canada) for fire-retardant coating resins used on wood products.

Irox. Trade name of Schott Glas AG (Germany) for a coated heat-reflecting glass.

irradiated wood. Wood whose hardness and strength has been significantly increased by impregnating it with a low-molecular-weight resin and subjecting to high-energy radiation (e.g., gamma rays) to effect crosslinking and bonding to the wood fibers.

Irrathene. Trademark of Insulating Materials Inc. (USA) for a thermosetting polyethylene made from standard polyethylene by irradiation with high-energy cathode rays. It can also be crosslinked chemically by heating with carbon black and a diperoxide. It has better resistance to acids, alkalies and solvents than standard polyethylene, very good electrical resistance, high tendency to oxidize quickly at elevated temperatures, and a melting temperature of approximately 250°C (480°F) when protected by a suitable inhibitor. Often supplied in film form, it is used for packaging, electrical insulation and encapsulation.

irregular powder. Metallic or nonmetallic powder composed of irregular-shaped asymmetric particles.

Irrubigo. Trade name of Schmidt & Clemens Edelstahlwerke (Germany) for a series of steels including various austenitic, ferritic and martensitic stainless grades, and several high-carbon high-chromium corrosion-, heat- and wear-resistant grades.

Isa. Trade name of Isabellenhütte Heusler GmbH KG (Germany) for a series of copper-base resistance alloys containing varying amounts of manganese and/or aluminum and nickel. They have a maximum service temperature of 300-500°C (570-930°F), and are used for electrical equipment and instruments.

Isabellin. Trade name of Isabellenhütte Heusler GmbH KG (Germany) for a series of resistance alloys containing 70-85% copper, 10-15% manganese, up to 20% nickel and up to 3% aluminum. They have a maximum service temperature of 400°C (750°F), and are used for electrical equipment and instruments.

Isa-Chrom. Trade name of Isabellenhütte Heusler GmbH KG (Germany) for a series of resistance alloys containing 60-80% nickel, 10-20% chromium and up to 20% iron. They have a maximum service temperature of 1150-1200°C (2100-2190°F), and are used for electrical equipment and instruments.

Isa-Nickel. Trade name of Isabellenhütte Heusler GmbH KG (Germany) for a resistance alloy containing 67% nickel, 31% copper, 1% iron and 1% manganese. It has a maximum service temperature of 600°C (1110°F), and is used for electrical equipment and instruments.

Isaohm. Trade name of Isabellenhütte Heusler GmbH KG (Germany) for a resistance alloy containing 74.5% nickel, 20% chromium, 3.5% aluminum, 1% silicon, 0.5% manganese and 0.5% iron. It has a maximum service temperature of 250°C (480°F), and is used for electrical equipment and instruments.

Isaryl. Trademark for aromatic polyarylate resins.

Isa-Spray. Trade name of Isabellenhütte Heusler GmbH KG (Germany) for a resistance alloy containing 88% copper, 10% manganese and 2% aluminum. It has a maximum service tempera-

ture of 400°C (750°F), and is used for electrical equipment and instruments.

Isatherm. Trademark of H.B. Fuller Austria Ges. mbH for isocyanurate polymers used for thermal insulation applications.

Isazin. Trade name of Isabellenhütte Heusler GmbH KG (Germany) for a resistance alloy containing 75.5% copper, 23% nickel and 1.5% manganese. It has a maximum service temperature of 400°C (750°F), and is used for electrical equipment and instruments.

Iscar. Trade name of Iscar Limited (Israel) for a series of cemented and coated cemented carbides used for cutting tools employed in the machining of carbon and alloy steels, cast irons, nonferrous alloys (e.g., aluminum-base alloys) and nonmetals (e.g., plastics and wood).

Iserlohn. (1) A highly ductile brass composed of 70% copper and 30% zinc used for condenser tubes, shell cases and cartridges. Also known as *Iserlohn brass*.

(2) A corrosion-resistant leaded brass composed of 64% copper, 34% zinc, 2.4% tin and 0.5% lead, used for hardware and fittings.

Ishigaki. Trade name of Central Glass Company Limited (Japan) for a figured glass having a pattern suggestive of castle walls.

Ishime. Trade name of Nippon Sheet Glass Company Limited (Japan) for a figured glass.

isinglass. Transparent sheet *mica*, usually in the form of single cleavage plates, formerly used in furnace and stove doors.

Islon. Trademark of Insa AS (Turkey) for nylon 6 fibers and yarns used for textile fabrics.

Iso. Trademark of Isolantite Manufacturing Company, Inc. (USA) for heat-resistant, noncorrosive fired mineral composites (e.g., steatite, mica, slate or amber) made by extruding, pressing or machining to shape. Used for low-loss electrical insulation.

ISO-25. Trade name of Apache Products Company (USA) for rigid, unfaced polyisocyanurate or polyurethane foam used for thermal insulation applications.

ISO Bar. Trade name of Corrosion Proof Composites, LLC (USA) for concrete reinforcing bars containing pultruded glass reinforcing fibers.

Isobloc. Trade name of Beohler GmbH (Austria) for electroslag-remelted hot-work die steels (AISI-SAE H13) with superior mechanical properties, excellent high-temperature strength and heat-checking resistance and good toughness used for die-casting and plastic-molding dies.

Isobrite. Trade name of MacDermid Inc. (USA) for copper, nickel and zinc electroplates and electroplating processes.

isobutyl-2-cyanoacrylate adhesive. A surgical tissue adhesive based on the isobutyl ester of 2-cyano-2-propenoic acid ($C_8H_{11}NO_2$).

isobutylene-isoprene copolymer. See butyl rubber.

isobutylene-isoprene rubber. See butyl rubber.

Isocalor. Trade name of Fabbrica Pisana SpA (Italy) for a toughened glass block containing an air chamber.

Isocarb. Trade name of Science Applications International Corporation (USA) for composite materials consisting of matrices based on pitch and synthetic resins reinforced with pitch- or polyacrylonitrile-based carbon fibers.

Isocast. Trade name of Empire Steel Castings Company (USA) for an extensive series of stainless steel castings, heat- and corrosion-resistant steels and corrosion- and acid-resistant nickel-iron-chromium and nickel-molybdenum-iron casting alloys.

Isochem EP. Trade name of Furane Products (USA) for epoxy resins available in general-purpose, flexible, high-heat and aluminum-, glass-, mineral- and silica-filled casting grades.

ISO-Chemie. Trade name of ISO-Chemie GmbH (Germany) for polyethylene foam packaging products and moldings

Isochem Silicone. Trade name of Furane Products (USA) for silicone resins available in standard and glass fiber- and mineral-filled grades.

Isochrome. Trade name of Isochrome (France) for a high-purity chromium (99.99%) made by the reduction of chromium iodide.

Isocor. Trademark of Shakespeare Monofilaments & Specialty Polymers (USA) for a series of polymers including several homo-, co- or multi-polyamides (nylons) supplied in general-purpose, extrusion, transparent and other grades.

Isocord. Trademark of Isovolta AG (Austria) for phenol-formaldehyde plastics.

isocyanate. A compound that contains the radical –NCO. It can be a monoisocyanate or a diisocyanate.

isocyanate plastics. Plastics based on *isocyanate resins*. See also polyurethane plastics.

isocyanate resins. A group of synthetic resins made by condensation reactions of organic isocyanates with any of various compounds containing hydroxyl groups, e.g., diols, polyols, polyethers or polyesters. Also known as *polyisocyanates*. See also polyurethane resins.

isocyanine iodide. See 1,1'-diethyl-2,4'-cyanine iodide.

isocyanurate foam boards. Rigid insulating boards consisting of closed-cell polyisocyanurate foam filled with gases (e.g., fluorocarbons). They are available in a wide range of sizes and usually come double-faced with foil, and sometimes bonded with interior or exterior finishes. They have high thermal insulating values, but in application must be protected from exposure to sunlight and water and covered with a fire-resistant material. *Isocyanurate foam boards* can be used on buildings and structures for acoustical, air and vapor barrier applications. Also known as *polyisocyanurate foam boards*. See also rigid board insulation.

isocyanurate foam insulation. A type of foamed-in place thermal insulation based on isocyanurate resins and catalysts. The foam is supplied in liquid form and sprayed directly onto building surfaces or poured into enclosed cavities with a pump-driven spray gun. It expands in place and sets to a semi-flexible, open-cell plastic material. Isocyanurate foam insulation can also be used for air barrier applications, but must be covered with a fire-resistant material. Also known as *polyisocyanurate foam insulation*.

isocyanurate resins. A group of synthetic resins obtained from isocyanurate, a compound associated with the isocyanates, but possessing three –NCO– groups. Used in the manufacture of rigid or semiflexible plastic foams for the acoustic and thermal insulation of buildings. Also known as *polyisocyanurates*.

Isodamp. Trademark of Cabot Safety Intermediate Corporation (USA) for a family of noise and vibration damping materials. *Isodamp MTL* refers to a series of low-weight, multidamping layer composites supplied in standard thicknesses from 7.6 to 14 mm (0.3 to 0.55 in.). They provide peak damping at specific temperatures and are used to control high-intensity vibration and sound levels in aircrafts, automobiles and boats. *Isodamp TAD* is a low-weight damping material composed of a pressure-sensitive, viscoelastic adhesive constrained by aluminum foils. Available as die-cut parts and in sheet form, it has a service temperature range of -40 to +120°C (-40 to +250°F), and is used for the reduction of noise, vibration and structural fati-

gue. *Isodamp TED* is a durable three-layer composite sheet product for extensional damping applications.

Isodisc. Trade name of U.N. Alloy Steel Corporation (USA) for a series of shock-resistant hot-work steels containing 0.33-0.35% carbon, 3-5% chromium, up to 1% vanadium, 1-3% molybdenum, up to 0.4% manganese, up to 1.5% tungsten, and the balance iron. Used for extrusion and forging dies, punches, rams, mandrels, upsetters, and rivet sets.

Isodur. (1) Trade name of Boehler GmbH (Austria) for cold-work tool steels possessing high yield strength, good toughness and wear resistance. Used for cutting, forming and stamping tools.

(2) Trademark of Isola-Werke AG (Germany) for polyvinyl chlorides.

Iso-Elastic. Trade name of John Chatillon & Sons, Inc. (USA) for nickel-base spring alloys containing varying amounts of nickel, iron, chromium and molybdenum. They have good fabricability, constant moduli of elasticity, and are used for food-weighing scales, instruments and dynamometers.

isoferroplatinum. A whitish steel- to dark-gray mineral of the gold group composed of iron platinum, Pt_3Fe. It can also be synthesized from the elements. Crystal system, cubic. Density, 16.50 g/cm³. Occurrence: Russia, Canada (British Columbia).

Isoflex. Trade name of O&M Kleeman (UK) Limited for cellulose acetate plastics.

Isofoam. Trademark of IPI International, Inc. (USA) for solid and foamed-in-place, expansible or flexible resins based on polyisocyanates and polyurethanes, and polyester and polyether derived polymers.

Isofolan. Trade name of Scheu-Dental (Germany) for a plastic foil supplied in rounds, 125 mm (5 in.) in diameter and 0.10 mm (0.004 in.) in thickness, and used in dentistry for insulating plaster models against cold-cure acrylics. Also available under this trade name is a specially formulated adhesive that enhances the bonding between *Isofolan* foil and dental plaster models.

Isofrax. Trade name of Unifrax Corporation (USA) for a soluble-fiber thermal insulation that has good high-temperature performance characteristics up to 1260°C (2300°F).

Iso-Genopak. Trademark of Hoechst AG, Kalle Werk (Germany) for a polyvinyl chloride film.

Isoglace. Trade name of Boussois Souchon Neuvesel SA (France) for a composite cladding panel whose exterior surface is composed of ceramic-coated, toughened *Émauglas*.

isokite. A white, pinkish or buff mineral of the tilasite group composed of calcium magnesium fluoride phosphate, $CaMgPO_4F$. Crystal system, monoclinic. Density, 3.24 g/cm³; refractive index, 1.594. Occurrence: Zimbabwe.

Isolac. Trademark of Polyresins Inc. (Canada) for styrenated alkyds used as protective coatings.

Isolantite. Trade name of Isolantite Manufacturing Company (USA) for a vitrified steatite ceramic with a density of 2.5 g/cm³, a hardness of 8-9 Mohs, high crushing strength and good dielectric properties. Used for electrical insulators.

Isolant Rose. Trademark of Fiberglas Canada Inc. for pink fiberglass insulating products.

Isolar-Glas. Trade name of A. Arnold KG (Germany) for a multiple glazing unit having a metal tube spacer that has been pushed slightly inwards from the edge of the glass to enlarge the space available for packing material.

isolators. Materials, such as felt, cork, fiberboard or rubber, that reduce or prevent vibrations and noise. They are often used between machine bases and foundations. Also known as *vibra-*

tion insulators.

Isolerglas. Trade name of A/S Drammens Glasverk (Norway) for a double or multiple glazing unit with a hollow anodized aluminum spacer and a polysulfide sealant.

Isolex. Trade name of Rohm & Haas Company (USA) for an acrylic shielding coating used for nonoxidizing copper.

Isolit. Trade name of Dentsply Ceramco (USA) for a dental gypsum used as a pore filler.

Isolite. Trademark of Schenectady Chemicals Inc. (USA) for electrical insulating varnishes.

Isolob. Trade name of Glasfaserwerk Steinach (Germany) for glass wool products.

Isolok. Trademark of Alcan Limited (Canada) for aluminum extrusion products.

Isoloss. Trademark of E-A-R Specialty Composites Corporation (USA) for a family of damping materials including *Isoloss VL,* a moldable urethane compound with high damping capacity from 0 to 650°C (32 to 1200°F).

Isoloy. Trade name of Riverside Metals Corporation (USA) for a series of nonmagnetic, austenitic stainless steels containing up to 0.15% carbon, 17-18% chromium, 8-12% nickel, and the balance iron. Used for wire rope and springs.

Isolsiv. Trade name of Società Italiana Vetro SpA (Italy) for long glass fiber felts used for the acoustical insulation of floors.

Isoltex. Trade name of Vetreria Italiana Balzaretti Modigliani SpA (Italy) for various glass fiber insulation products.

Isomatrix. Trade name of Boehler GmbH (Austria) for highly alloyed powder-metallurgy high-speed steels possessing high hardness, excellent compressive strength, abrasive and wear resistance, good strength retention, corrosion resistance and hardness at high service temperatures. A free-machining grade containing sulfur is also available. Used for broaches, milling cutters and other heavy-duty cutting tools, and for cold-work applications.

isomerized rubber. See cyclized rubber.

isomertieite. (1) A yellow mineral composed of palladium antimony arsenide, $Pd_{11}Sb_2As_2$. Crystal system, orthorhombic. Density, 5.70 g/cm³. Occurrence: China. A copper-bearing variety, $(Pd,Cu)_{11}(Sb,As)_4$, is found in Brazil.

(2) A synthetic mineral composed of palladium arsenide, Pd_3As. Crystal system, tetragonal. Density, 10.88 g/cm³.

Isomica. Trademark of Schweizerische Isola-Werke (Switzerland) for electrical insulating materials based on mica and supplied as sheets, tapes, rolls, tubing and shaped forms. Also included under this trademark is a special mica paper with epoxy binder.

Isomin. Trade name of Perstorp AB (Sweden) for melamine-formaldehyde plastics.

Isomolar. Trade name of Ivoclar Vivadent AG (Liechtenstein) for a self-cure, microfine dental composite restorative.

isomorphic polymer blends. Binary polymer blends of two resins existing as one semicrystalline and one amorphous phase that are mutually soluble in one another. They have both a single glass-transition temperature, and a single melting temperature.

Isonate. Trademark of Dow Chemical Company (USA) for compounds based on high-purity diphenylmethane diisocyanate and supplied as modified methylene diisocyanate (*Isonate 143L*) and prepolymers (*Isonate 181*). They provide easy handling, and are used for the manufacture of semiflexible and integral skin foams for automotive applications including doors, steering wheels and dunnage parts.

Isonel. Trademark of Schenectady Chemicals Inc. (USA) for wire coatings and insulating varnishes.

Isoperm. Trade name of Allgemeine Elektrizitäts-Gesellschaft AG (Germany) for a series of high-magnetic-permeability alloys including alloys of (i) 40-55% nickel and 45-60% iron, (ii) 35-50% nickel, 35-56% iron and 9-15% copper, and (iii) 40-60% nickel, 36-57% iron and 3-4% aluminum. Used for magnetic and electrical applications.

isophthalic acid. Colorless crystals (99+% pure) with a melting point of 341-343°C (645-649°F) used in the manufacture of high polymers, such as alkyds, polyesters and polyurethanes, and in plasticizers. Formula: $C_6H_4(COOH)_2$. Abbreviation: IPA. Also known as *m-phthalic acid.*

isophthalic polyesters. See isophthalic resins.

isophthalic resins. Unsaturated polyester resins made from blends of isophthalic acid and fumaric acid or maleic anhydride. They have good thermal and mechanical properties and good chemical resistance. Also known as *iso resins; isophthalic polyesters.* See also unsaturated polyesters.

Isoplast. (1) Trademark of Dow Chemical Company (USA) for impact-resistant thermoplastic polyurethane resins based on methylenediphenyl isocyanate. They combine the high performance and chemical resistance of semicrystalline resins with the dimensional stability of amorphous resins, and can be processed by injection molding and extrusion. Supplied in impact-modified, glass fiber-reinforced and clear amorphous grades, they are used for automotive components including fuel filters, fluid-level windows, latches, structural parts, engine compartment fasteners.

(2) Trade name of Boehler GmbH (Austria) for stainless mold steels possessing excellent corrosion resistance due to the high chromium content and molybdenum additions. Available as standard and electroslag-remelted grades, they are used for molds employed in the manufacture of plastics, lenses and optical components.

isopolyester. A polyester based on *isophthalic acid.*

Isopor. Trademark of Isopor GmbH (Germany) for polystyrene foams and foam products.

isoprene. A flammable, volatile, colorless liquid (99+% pure) usually inhibited with *p*-tert-butylcatechol to prevent polymerization. It has a density of 0.681 g/cm³, a melting point of -146°C (-231°F), a boiling point of 34°C (93°F), a flash point of -65°F (-53°C), an autoignition temperature of 802°F (427°C), and a refractive index of 1.422. It is the molecular unit of natural rubber, and used as a monomer in the preparation of polyisoprene and butyl rubber, and as a chemical intermediate. Formula: C_5H_8.

isoprene rubber. A synthetic rubber that is chemically and physically similar to natural rubber, but requires less mastication. It is isotactic and made from *cis*-1,4-polyisoprene (C_5H_8) using an organometallic catalyst. *Isoprene rubber* has good extensibility, high tear resistance, low mechanical hysteresis, low heat buildup, good flow characteristics, good electrical properties, low moisture absorption, good resistance to dilute acids, water and antifreezes, fair resistance to alkalies, weathering and ozone, and poor resistance to petroleum oils, greases and oxidizing agents. Used for automotive tires, power belts, hoses, gaskets, seals, rollers and electrical insulation. Abbreviation: IR. Also known as *synthetic cis-polyisoprene.*

Isoprofil. Trademark of Isoprofil GmbH Stahlprofil- und Warmwalzwerk (Germany) for structural steel shapes supplied in the bright-drawn, cold-drawn, cold-rolled and hot-rolled conditions.

isopropoxide. A compound that is the product of a reaction in which the hydroxyl hydrogen of *sec*-propyl alcohol (C_3H_7OH)

has been substituted by a metal. Examples are cerium isopropoxide [Ce(OC₃H₇)₄] and strontium isopropoxide [Sr(OC₃H₇)₂]. Also known as *isopropylate*.

isopropylate. See isopropoxide.

1,2-isopropylidene glyceryl 2-cyanoacrylate adhesive. A biodegradable surgical tissue adhesive based on 1,2-isopropylidene glyceryl 2-cyanoacrylate.

iso resins. See isophthalic resins.

Isorex. Trade name of Boussois Souchon Neuvesel SA (France) for glass insulators.

Isorod. Trade name of Grand Northern Products Limited (USA) for strong, tough hardfacing steels containing 0.2% carbon, 0.5% silicon, 0.2% manganese, 15% chromium, 8% molybdenum, 1% titanium, and the balance iron. Supplied as coated rods, they are used for overlays and buildups on sprockets, shovels, pump impellers and heavy hammers.

Isoschaum. Trademark of Schaum-Chemie Wilhelm Bauer GmbH & Co. KG (Germany) for a sprayable urea-formaldehyde insulating foam used for cold, heat and noise insulation applications.

Iso-Shield. Trade name of Apache Products Company (USA) for a rigid polyisocyanurate foam used for low-temperature insulating applications.

Isosol. Trade name of Isothermglas GmbH (Germany) for a building product composed of two sheets of colored, toughened glass with an insulating interlayer.

isostatically pressed materials. Ceramic, metal or metal-alloy products of uniform density formed by first sealing the powdered starting material into a flexible metal or nonmetal container and then applying equal fluid pressure from all directions (isostatic pressure) with or without the simultaneous application of heat. See also cold-isostatically-pressed materials; hot-isostatically-pressed materials.

Isotac. Trademark of 3M Company (USA) for pressure-sensitive adhesive transfer tapes.

isotactic polymer. A polymer in which all side groups (R) are located on the same side of the main chain or backbone. It is usually harder, siffer, stronger and denser than an *atactic polymer*, and readily forms crystals. Examples are isotactic polypropylene and isotactic polystyrene.

Isotan. Trade name of Isabellenhütte Heusler GmbH KG (Germany) for a resistance alloy containing 55% copper, 44% nickel and 1% manganese. It has a maximum service temperature of 600°C (1110°F), and is used for electrical equipment and instruments.

Iso-Temp. Trademark of 3M Dental (USA) for a light-cure resin-based dental composite restorative that provides low polymerization shrinkage, low heat generation, and good compatibility with many impression materials including alginates, silicones, plastics and waxes. Upon setting, it develops good flexural strength and low wear resistance. Used in dentistry for temporary restorations and for lining crowns.

Isotextil. Trade name of Isotextil GmbH (Germany) for textile wallpapers.

isothiocyanato-4(trans-4-octylcyclohexyl)benzene. A nematic liquid crystal (99% pure) with a boiling point of 190°C(374°F)/0.01 mm and a flash point above 230°F (110°C). Formula: CH₃(CH₂)₇C₆H₁₀C₆H₄NCS.

4-isothiocyanatophenyl 4-pentylbicyclo[2.2.2]octane-1-carboxylate. A nematic liquid crystal (99% pure) with a melting point of 73-113.5°C (163-236°F) and a flash point above 230°F (110°C). Formula: CH₃(CH₂)₇C₆H₁₀C₆H₄NCS.

4-isothiocyanato-4(trans-4-propylcyclohexyl)benzene. A nematic liquid crystal (99% pure) with a melting point of 41-44°C(106-111°F) and a flash point above 230°F (110°C). Formula: CH₃CH₂CH₂C₆H₁₀C₆H₄NCS.

Isotop. Trade name of Reinicke Klima GmbH & Co. (Germany) for glass-fiber insulating materials.

isotropic dielectric. A dielectric whose magnitude of electric polarization is independent of the direction of the applied electric field. In such a material the directions of polarization and applied field are parallel.

isotropic laminate. A laminate whose strength properties are equal in all directions. See also anisotropic laminate.

isotropic material. A material that exhibits the same properties in all directions. Also known as *isotropic substance*. See also anisotropic material.

isotropic pitch-based precursor fiber. A fiber obtained from isotropic petroleum pitch by melt spinning and subsequent carbonization, graphitization and sizing, and used as a precursor fiber in the manufacture of carbon fibers.

Isovac. Trademark of Vacuumschmelze GmbH (Germany) for a series of mineral-insulated electrical conductors with metal sheaths. The conductors are embedded in magnesia (MgO) or alumina (Al₂O₃) and the sheath consists of a corrosion-resistant high-grade stainless steel (maximum service temperature up to 800°C or 1470°F), or Inconel 600 (maximum service temperature up to 1100°C or 2010°F). The conductors are NiCr/Ni, Fe/CuNi, Cu, Ni or NiCr. Commercially available in the form of coils (50-500 m or 164-1640 ft. long), they are used for sheathed thermocouples, measurement cables and heating conductors.

Isover. (1) Trade name of Saint-Gobain (France) for glass fiber and wool products. *Isover Climaver* is a glass-fiber material for insulating air-conditioning conduits and *Isover Sonebel* a glass-fiber material for insulating acoustical ceilings.

(2) Trademark of Grünzweig + Hartmann und Glasfaser AG (Germany) for mineral-fiber materials for heat, cold, sound or fire insulation applications.

Isovyl. Trademark of Rhovyl SA (France) for vinyon (vinyl chloride) fibers with excellent acid, alkali and mildew resistance. Used for textile fabrics and carpets.

ISP Micropowder. Trademark of International Specialty Products (USA) for a spherical carbonyl iron powder with uniform particle size distribution used for making high-density, high-strength powder-metallurgy parts, soft magnetic, low-remanence high-permeability electronic components, and as an additive to metals, alloys and synthetic resins.

Isteg. Trade name of Isteg Steel Products Company (UK) for a prestressed low-carbon steel used for reinforcing applications in the construction industry.

istle. A stiff, but smooth, yellowish white fiber obtained from several species small tropical American agaves (chiefly from Texas, New Mecico and Mexico), and used for the manufacture of bags, carpets, rough cordage, saddle girths and nets. *Note:* The term "istle" is sometimes used synonymously with *tampico*.

Istra. Trademark of Heidelberger Zement AG (Germany) for an aluminate cement used in the manufacture of concrete.

Istrakin. Trademark of Paular SA (Spain) for acrylic staple fibers used for textile fabrics.

Istrona. Trademark of Istrochem (Slovak Republic) for polypropylene fibers and yarns used for textile fabrics.

Itabil. Italian trade name for an alloy composed of 95% aluminum and 5% silicon, used for light alloy-parts, and die and sand castings.

Italian asbestos. A term sometimes used in Europe for *tremolite* asbestos.

Italian chestnut. See sweet chestnut.

Italian cypress. The aromatic, light brown wood of the softwood tree *Cupressus sempervirens.* It has high durability, low weight, and an aromatic odor. Source: Southern Europe (Mediterranean); also planted in the southern United States from Florida to Texas, and in California. Used for construction work, furniture, chests and doors. See also cypress.

Italian silk. A strong, elastic, good-quality *silk* from Italy used for making dresses, tricots, underwear, knitwear and hosiery.

Italor. Trade name of Fabbrica Pisana SpA (Italy) for a plate glass with pink-amber tint.

Italsil. Trade name of Italsil (Italy) for an aluminum alloy containing 5% silicon used for light-alloy parts, and die and sand castings.

itauba. The yellowish-green wood of the tree *Silvia itauba* growing to a height of about 23 m (75 ft.) in the upland forests of the lower Amazon in Brazil. It has a compact texture, and a rough fiber, and is used for shipbuilding, furniture and flooring.

itauba preta. The yellowish-brown wood of the tall Brazilian tree *Oreodaphne bookeriana.* It closely resembles *teak* in appearance and is used for cabinetmaking, fine furniture, chests and boxes.

Itaver. Trade name of Vertex NP (Czech Republic) for glass insulation products.

Iteco. Trade name of Iteco SA (Italy) for a phenol-formaldehyde resin formerly used for denture bases.

Iten Politen. Tradename of Iten Industries (USA) for a series of glass-reinforced polyester laminates used for electrical components and devices.

Iten Resiten. Tradename of Iten Industries (USA) for a series of phenolic, melamine and epoxy laminates including (i) paper-phenolic laminates for electrical applications, (ii) cotton cloth-phenolic laminates for machine parts including gears, and pump and valve components, (iii) glass cloth-phenolic laminates for electrical and machine parts, (iv) nylon-cloth phenolic laminates for electrical and machine parts operating in humid conditions, (v) chemical-resistant glass cloth-melamine laminates for electrical applications, and (vi) glass cloth-epoxy laminates for electrical applications.

itoite. A white mineral of the barite group composed of lead germanium sulfate hydroxide, $Pb(S,Ge)(O,OH)_4$. Crystal system, orthorhombic. Density, 6.67 g/cm^3; refractive index, 1.84-1.85. Occurrence: Southwest Africa.

i-type semiconductor. See intrinsic semiconductor.

Iupiace. Trademark of Mitsubishi Corporation (Japan) for several polyphenylene oxides and polyphenylene ethers supplied in powder, pellet, granular, liquid and paste form.

Iupilon. Trademark of Mitsubishi Corporation (Japan) for a series of polycarbonate resins supplied in powder, pellet, granular, liquid and paste form.

Iupital. Trade name of Mitsubishi Corporation (Japan) for several acetal resins supplied in powder, pellet, granular, liquid and paste form.

Ivadized coating. A coating produced by an ion-vapor deposition (IVD) process. For example, Ivadized aluminum (IVD aluminum) is a corrosion-resistant coating applied by ion-vapor deposition on ferrous and nonferrous substrates. Abbreviation: IVD coating.

Iveberg. Trademark of Nitto Spinning Company (Japan) for spun glass fibers.

Ivorea. Trade name of SNIA SpA (Italy) for rayon fibers and yarns used for textile fabrics.

Ivoresco. Trade name for medium-carbon molybdenum-chromium alloy steels.

Ivoricast. Trade name of Plastics Research Company (USA) for medium-temperature-curing cast phenolic plastics filled with wood flour. They have good strength, toughness, shock resistance and dimensional stability.

Ivoride. Trade name of Daniel Spill Company (UK) for cellulose nitrate plastics.

Ivorine. Trade name of Columbia Dentoform (USA) for a dental molding compound.

Ivory. Trademark of Canadian Gypsum Company Limited for a hydrated finishing lime.

ivory. A hard, white, close-grained substance that constitutes the tusks and teeth of elephants and hippopotamuses. It has a density of 1.87 g/cm^3 (0.068 $lb/in.^3$) and takes a fine polish. Source: West and Central Africa, Southern Asia especially India. Formerly heavily used for ornamental parts, art objects, piano keys, and in the manufacture of ivory black.

ivory black. A high-grade *animal black* that is made by charring or burning *ivory* or ivory wastes. Formerly used as a deep black pigment.

ivory porcelain. A fine, white-colored ware whose surface resembles *ivory* in appearance. The ivory effect is produced by depolishing the vitreous surface glaze.

Iwamo. Trade name of Nippon Sheet Glass Company Limited (Japan) for a patterned glass.

Ixan. Trademark of Solvay Polymers, Inc. (USA) for several polyvinylidene chlorides and fluorides used as coating materials on polyethylene, polypropylene, polystyrene and polyvinyl chloride containers to inhibit permeation of gases, such as carbon dioxide, oxygen or inert gases. They are also used in form of aqueous dispersions for coating paper, HDPE-paper laminates, aluminum and plastic films, as solvent-base coating powders for regenerated cellulose and as binders for paints, lacquers and printing inks. The powder grades are also used for making film products by extrusion and co-extrusion processes.

Ixef. Trademark of Solvay Polymers, Inc. (USA) for a family of glass-reinforced polyarylamide resins supplied in various grades including antivibration, low-water-absorption and high-gloss. Usually processed by injection molding, they have low density, high modulus of elasticity and high tear strength. Used for auto-motive components, such as air filter housings, industrial equipment parts, such as gears, runners and rolls, electronic components, sporting goods, and medical equipment.

ixiolite. A black mineral of the columbite group composed of tantalum iron oxide, $(Ta,Fe,Sn,Nb,Mn)O_2$. Crystal system, orthorhombic. Density, 7.03 g/cm^3. Occurrence: Finland.

Ixolaine. Trade name for a phenol-formaldehyde resin formerly used for denture bases.

Izett. Trade name of Krupp Stahl AG (Germany) for a series of tough, nonaging, aluminum-deoxidized alloy steels containing 0.1-0.35% carbon, 0.48-0.56% manganese, 0.02-0.1% silicon, 0.018-0.025% phosphorus, 0.025-0.15% sulfur, and the balance iron. Used for boilers and similar equipment.

Izocam-Camyono. Trade name of Izocam Sanayi ve Ticaret AS (Turkey) for a glass wool.

Izosil. Macedonian trade name for short-fiber glass wool and long-fiber glass silk.

J

jacaranda. The hard, violet- to chocolate-brown wood of the large tree *Jacaranda copaia*. It is heavy and has numerous dark veins. Source: Brazil. Used for fine furniture and knife handles. Also known as *caroba; carob wood*.

Jackmanized Steel. Trade name of Joseph Jackman & Company Limited (UK) for a carbon steel containing 0.5% carbon, and the balance iron. Used for bushings, rolls and pins.

jack pine. The wood of the scrubby, small or medium-tall pine *Pinus banksiana*. The heartwood is light brown to orange and the sapwood nearly white. *Jack pine* has a coarse texture, a high resin content and high knottiness. Source: USA (New England from Maine, New Hampshire and Vermont to New York, and throughout the Lake states); Canada (from Nova Scotia to Saskatchewan). Used for general construction, and as a pulp- and fuelwood.

Jacksberg. Trade name of Babcock & Wilcox Company (USA) for a high-speed steel containing 0.7% carbon, 4% chromium, 18% tungsten, 1%% vanadium, and the balance iron. Used for high-speed cutters and tools.

Jackson. Trade name for polyolefin fibers.

Jackson's alloy. A high-strength alloy containing 60-68% copper, 30-35% zinc and 2-5% antimony. Used for architectural and ornamental applications, and buttons.

jacobsite. A black opaque, magnetic mineral of the spinel group composed of manganese iron oxide, $MnFe_2O_4$. It can also be made synthetically. Crystal system, cubic. Density, 5.21 g/cm³; refractive index, 2.3; hardness, 665-707 Vickers. Occurrence: USA (New Jersey).

Jacoby metal. A British babbitt containing 85% tin, 10% antimony and 5% copper, and used for liners, bushings, bearings and machine components.

Jacona metal. A British babbitt containing 70% lead, 20% antimony and 10% tin, and used for bearings and liners.

jacquard fabrics. Textile fabrics with intricate patterns or elaborate designs woven on a numerically controlled Jacquard loom. Used extensively for upholstery and draperies. Also known as *jacquard; jacquard-weave fabrics*.

jacquard paper. A tough, dimensionally stable paper used to make punched tape for the numerical control of jacquard-type looms and knitting machines.

jacquard-weave fabrics. See jacquard fabrics.

Jade. Trade name of Ugine Aciers (France) for a water-hardening cold-work steel (AISI type W1) containing 1% carbon, and the balance iron. Used for taps, punches and reamers.

jade. A hard, compact, opaque to translucent dark-green to almost white stone composed of the minerals *jadeite* or *neph-*

rite. Used as a gemstone, and for ornamental purposes. Also known as *jadestone*.

jade glass. A green, opaque or translucent glass that contains considerable quantities of lead monoxide (PbO).

jadeite. A white to grayish green mineral of the pyroxene group that is composed of sodium aluminum silicate, $NaAlSi_2O_6$, and usually contains some iron, calcium and magnesium. Crystal system, monoclinic. Density, 3.3-3.5 g/cm³; refractive index, 1.656; hardness, 6.5-7 Mohs. It is the more highly priced variety of *jade*. Occurrence: Burma, China, Tibet, USA (California). Used as a gemstone. See also nephrite.

Jade Stone. Trade name of Whip Mix Corporation for a dental gypsum (ADA Type V) supplied as a powder for mixing with water. The resulting cast die stone (typically made from a mixture of 22 mL water per 100 g powder) has a smooth jade-like surface, high compressive strength and low setting expansion (0.13%). Supplied in green and blue colors, *Jade Stone* is compatible with all types of impression materials, and provides an outstanding contrast to dental casting waxes.

jadestone. See jade.

Jadot. Trade name of Usines de Jadot (Belgium) for a series of corrosion- and/or heat-resistant austenitic, ferritic and martensitic stainless steels.

Jae Metal. Trade name of Inco Alloys International Limited (UK) for a wrought nickel alloy containing 30% copper, and used for magnetic and electrical equipment.

jaffer. A plain-woven cotton fabric with differently colored warp and weft. See also warp yarn; weft yarn.

Jägaduct. Trademark of Jäger KG (Germany) for melamine-formaldehyde plastics.

Jägalyt. Trademark of Jäger KG (Germany) for unsaturated polyester plastics.

Jägaphren. Trademark of Jäger KG (Germany) for phenol-formaldehyde plastics.

Jägapol. Trademark of Jäger KG (Germany) for unsaturated polyester plastics.

jagowerite. A green mineral composed of barium aluminum phosphate hydroxide, $BaAl_2(PO_4)_2(OH)_2$. Crystal system, triclinic. Density, 4.01 g/cm³; refractive index, 1.693. Occurrence: Canada (Yukon).

jahnsite. A yellow-brown mineral composed of calcium magnesium manganese iron phosphate hydroxide octahydrate, $CaMg_2MnFe_2(PO_4)_4(OH)_2 \cdot 8H_2O$. Crystal system, monoclinic. Density, 2.86 g/cm³; refractive index, 1.695. Occurrence: USA (New Jersey, South Dakota).

Jailene. Trademark of Swadeshi Polytex Limited (India) for polyester staple fibers used for textile fabrics.

Jalcase. Trade name of Jones & Laughlin Steel Corporation (USA) for a series of free-machining case-hardening steels used for screw-machine products, shafts, gears, bolts, cams and camshafts.

Jalcold. Trade name of Jones & Laughlin Steel Corporation (USA) for a carbon steel containing 0.2% carbon, and the balance iron. Used for welding rods.

J-Alloy. Trade name for a cobalt-base superalloy containing 23% chromium, 6% nickel, 6% molybdenum, 2% tantalum and 0.8% carbon. It has high heat resistance, and is used for gas-turbine components and jet-engine parts.

Jalloy. Trade name of LTV Steel (USA) for a series of low-carbon constructional steels with good weldability and formability used for industrial equipment, construction and mining machinery, trucks, railroad cars, cranes, crane booms, and tractors.

jalpaite. A metallic gray mineral composed of copper silver sulfide, Ag_3CuS_2. It can also be made synthetically. Crystal system, tetragonal. Density, 6.93 g/cm^3. Occurrence: Czech Republic.

Jalten. Trade name of LTV Steel (USA) for a series of high-strength steels containing up to 0.25% carbon, 1.6% manganese, 0.2% molybdenum, 0.065%% vanadium and 0.14% phosphorus, respectively, 0.2-0.3% copper, and the balance iron. They have good to moderate formability, good weldability, and are used for structural components, construction equipment, truck bodies, and railroad and mine cars.

Jalweld. Trade name of LTV Steel (USA) for a carbon steel containing 0.15% carbon, and the balance iron. Used for welding rods.

Jamag. Trade name of Blackstone Corporation (USA) for a malleable cast iron containing 2.2% carbon, 1.6% silicon, 0.3% manganese, 0.1% manganese, and the balance iron. It has high magnetic permeability, and is used for electrical parts and equipment with high speeds of rotation.

Jamaica. Trade name of Aardvark Clay & Supplies (USA) for a rich brown-black clay (cone 10) with smooth texture.

Jamalex. Trade name of Blackstone Corporation (USA) for a wear-resistant malleable cast iron containing 2.2% carbon, 1.6% silicon, 0.3% manganese, 0.05% magnesium, and the balance iron. It has high magnetic permeability, and is used for gears, cams, bearings, and other machine components.

Jamappes brass. A free-cutting brass containing 65% copper, 33.4% zinc, 1.4% lead and 0.2% tin. It has good machinability and workability, and is used for fittings, forgings and hardware.

jamesite. A red brown mineral composed of lead zinc iron oxide arsenate, $Pb_2Zn_2Fe_5O_4(AsO_4)_5$. Crystal system, triclinic. Density, 5.09 g/cm^3; refractive index, 1.995. Occurrence: Namibia.

jamesonite. A gray-black to lead-gray mineral with metallic luster composed of iron lead antimony sulfide, $FePb_4Sb_6S_{14}$. Crystal system, monoclinic. Density, 5.63 g/cm^3; hardness, 2-3 Mohs. Occurrence: Italy, UK. Used as an ore of antimony. Also known as *feather ore*.

Jamison. Trade name of Jamison Steel Corporation (USA) for several mold steels, oil-hardening cold-work tool steels, nondeforming tool and die steels and water-hardening plain-carbon tool steels.

Jamlon. Trademark of Cheng Chi Fiber Company (Taiwan) for nylon fibers and yarns.

janggunite. A black mineral composed of manganese iron oxide hydroxide, $Mn_{5-x}(Mn,Fe)_{1+x}O_8(OH)_6$. Crystal system, orthorhombic. Density, 3.59 g/cm^3. Occurrence: Korea.

janhaugite. A reddish brown mineral composed of sodium manganese titanium fluoride silicate hydroxide, $Na_3Mn_3Ti_2Si_4O_{15}(OH,F,O)_3$. Crystal system, monoclinic. Density, 3.60 g/cm^3; refractive index, 1.828. Occurrence: Norway.

Janney. Trade name of Janney Cylinder Company (USA) for a series of cast bronzes containing varying amounts of copper, tin, lead and, sometimes, nickel. It has good to moderate corrosion resistance, and is used for high-pressure bearings, crusher bearings, bushings, sleeves, and pump components.

Jano. Trade name of Hidalgo Steel Company Inc. (USA) for a high-speed steel containing varying amounts of carbon, chromium, vanadium, tungsten, molybdenum and iron. Used for tools and cutters.

Janus. Trade name of Dörrenberg Edelstahl GmbH (Germany) for an extensive series of steels including several water-harden-

ing tool steels, cold-work die and tool steels, tungsten high-speed steels, corrosion- and/or heat-resistant steels, and austenitic and martensitic stainless steels.

Janusit. Trade name of Dörrenberg Edelstahl GmbH (Germany) for a series of corrosion- and heat-resistant austenitic stainless steels containing 0.4-0.5% carbon, 1.3-2% silicon, 22-27% chromium, 4-30% nickel, and the balance iron. Used for furnace parts and heat-treating equipment.

japan. See japan lacquer.

Japan brass. A Japanese high-strength brass composed of 66.7% copper and 33.3% zinc, and used for hardware and fixtures.

japan color. A colored paste obtained by grinding high-quality colors in hard-drying varnish, and used for lettering purposes.

japan drier. A liquid varnish having a large proportion of metallic resinates (e.g., lead, cobalt or manganese resinates) added to speed drying.

japan enamel. A hard, glossy, black enamel coating produced by baking japan lacquer at relatively high temperatures. See also japan lacquer.

Japanese. Trade name of SA Verreries de Fauquez (Belgium) for a glass with floral pattern.

Japanese acrylics. A term used in dentistry for a group of methacrylate-based resin cements that contain special adhesive monomers, such as hydroxyethyl methacrylate (HEMA) or methacryloxyethyl trimellitate anhydride (4-META), that chemically bond to tooth structure (dentin and enamel) and metal appliances.

Japanese NiTi. Trade name for a nickel-titanium shape-memory alloy wire used for dental applications.

Japanese oak. The wood from the hardwood tree *Quercus mongolica*, especially the variety *Q. mongolica grosseserrata*. It has an open, porous texture, works extremely well, and can be obtained in large sizes. Average weight 660 kg/m^3 (41 lb/ft^3). Source: Japan. Used chiefly for interior work and furniture.

Japanese porcelain. Porcelain analogous to *Chinese porcelain*, but fired at a somewhat lower temperature to provide a softer, more appealing surface finish.

Japanese tallow. See sumac wax.

japan lacquer. A varnish composed of copal or kauri resin, linseed oil, litharge (lead monoxide) and pigments (e.g., Prussian blue), and thinned with a solvent, such as kerosene or turpentine. When baked at relatively high temperatures, it yields a hard, glossy, black coating (*japan enamel*). Used for baked enamel coatings on wood and metal. Also known as *japan*.

japanned leather. See patent leather.

Japan paper. A soft, pliant specialty paper, hand-made from the bast fibers of plants growing in Japan.

Japan tallow. See sumac wax.

Japan wax. See sumac wax.

Japco. Trade name of Jamestown Paint & Vanish Company (USA) for a series of paint and lacquer coatings used in metal finishing.

japonais. A light silk *poplin* fabric made from gray yarns and used especially for underwear and light summer dresses.

jappe. A fine, relatively square fabric made from continuous silk or other filament yarns in a plain weave.

jardiniere glazes. A group of unfritted, soft or hard lead glazes containing oxides of lead, aluminum, calcium, potassium, silicon and zinc. Used as a decorative glaze on pottery and earthenware (e.g., flower pots).

jarlite. A colorless mineral composed of sodium magnesium strontium aluminum fluoride hydroxide, $Na_2(Sr,Na)_{14}Al_{12}Mg_2F_{64}$-

$(OH,H_2O)_4$. Crystal system, monoclinic. Density, 3.87 g/cm^3; refractive index, 1.435. Occurrence: Greenland.

jarosite. A yellow or brown mineral of the alunite group composed of potassium iron sulfate hydroxide, $KFe_3(SO_4)_2(OH)_6$. It can also be made synthetically from ferric sulfate, potassium sulfate and sulfuric acid under prescribed conditions. Crystal system, hexagonal (rhombohedral). Density, 2.91 g/cm^3; refractive index, 1.820, hardness, 2.5-3.5 Mohs. Occurrence: Spain, Slovakia, USA (Nevada).

jarrah. The heavy, hard, dark-red wood of the tall hardwood tree *Eucalyptus marginata*. It has good mechanical properties, excellent durability, good resistance to termites and marine borers, and is stronger and more durable than oak. The heartwood can be treated with preservatives, but is rather difficult to work. Average weight, 800 kg/m^3 (50 lb/ft^3). Source: Western Australia. Used for heavy structural work, decking and underframing of piers, jetties and bridges, piling, flooring, shipbuilding and railroad freight cars.

jasmundite. A mineral composed of calcium oxide sulfide silicate, $Ca_{11}(SiO_4)_4O_2S$. Crystal system, tetragonal. Density, 3.03 g/cm^3; refractive index, 1.75. Occurrence: Germany.

jaspe. A textile fabric with faint stripes, either woven with light-, medium and dark-colored warp yarns or with a yarn consisting of two differently colored strands twisted together. Used especially for draperies, upholstery and suits.

jasper. (1) An opaque variety of *quartz* that is usually red due to hematite (iron oxide) impurities, but also occurs in brown, yellow or dark green colors. It cuts and polishes fairly well, and is used as an ornamental stone and as a building stone for decorative purposes. Also known as *jasperite; jaspis*.

(2) A shaded textile fabric made with a warp yarn that has a color pattern, and used especially curtains and bedspreads.

jasperite. See jasper.

jasperized wood. See silicified wood.

jaspis. See jasper.

Java cotton. See kapok.

Javelin. Trademark of Altlas Specialty Steels (Canada) for a hollow drill steel containing 0.26% carbon, 0.6% manganese, 0.3% silicon, 3.2% chromium, 0.15% nickel, 0.55% molybdenum, and the balance iron. It has high strength and good resistance to fatigue failure. Used for mining bit shanks.

Jayacrylic. Trademark of J.K. Synthetics (India) for acrylic staple fibers used for textile fabrics.

Jayanka. Trademark of J.K. Synthetics (India) for nylon 6 filament yarns used for textile fabrics.

Jaycord. Trademark of J.K. Synthetics (India) for high-strength nylon 6 filament yarns used for tire cords.

Jay-Glass. Trade name of Jayworth Industries Limited (Australia) for a masonry grille block containing several plate glass panels.

Jaykalon. Trademark of J.K. Synthetics (India) for nylon 6 filament yarns used for textile fabrics.

Jaykaylene. Trademark of J.K. Synthetics (India) for polyester staple fibers and filament yarns used for textile fabrics.

Jaysee. Trade name of Jonas & Culver (Nova) Limited (UK) for a medium-carbon tool steels containing 0.45% carbon, 0.5% chromium, 0.5% tungsten, and the balance iron. Used for hand tools, hammers, and chisels.

JD Alloy. Trade name of J.F. Jelenko & Company (USA) for a microfine, premium cobalt-chromium dental alloy with high tensile strength, toughness and durability and good tarnish resistance. Used for partial denture frameworks.

jean. A strong, twill-woven cotton fabric similar to *denim* used for work, sport and casual clothes.

jeanbandyite. A brown-orange mineral composed of manganese iron tin hydroxide, $(Fe,Mn)Sn(OH)_6$. Crystal system, tetragonal. Density, 3.81 g/cm^3; refractive index, 1.837. Occurrence: Bolivia.

jeanette. A light *jean* fabric used especially for linings.

Jectron. Trade name for a vinyl styrene resin formerly used for denture bases.

Jedmat. Trade name for rayon fibers and fabrics.

Jeffaloy. Trade name of Dresser Industries (USA) for a series of wear-resistant cast irons containing 2.85-3.5% total carbon, 0.3-0.7% manganese, up to 0.9% molybdenum, 1-2% silicon, and the balance iron. Used for castings, housings and pump parts.

jeffersonite. A dark-green or greenish-black mineral of the pyroxene group composed of calcium manganese zinc iron silicate, $Ca(Mn,Zn,Fe)Si_2O_6$. Crystal system, monoclinic. Occurrence: USA.

Jekster. Trademark of Orissa Synthetics Limited (India) for polyester staple fibers and filament yarns used for textile fabrics.

Jel. Trade name of J.F. Jelenko & Company (USA) for a series of dental casting and bonding alloys including (i) *Jel-2*, a microfine, medium-hard gold casting alloy (ADA type II) used for inlays, crowns and bridgework; (ii) *Jel-3*, a microfine, hard gold casting alloy (ADA type III) that has outstanding elongation and ductility, is color-matched to Jel-4, and is used for inlays, crowns and bridgework; (iii) *Jel-4*, a microfine, extra-hard gold casting alloy (ADA type IV) that has high strength and toughness, good castability and tarnish resistance, is color-matched to Jel-3, and is used for dental inlays, crowns and partial dentures; (iv) *Jel-5*, a silver-palladium ceramic alloy with excellent castability, used for fusing porcelain to metals; (v) *Jel-62*, a microfine, hard, 62%-gold alloy (ADA type III) with lustrous high-gold appearance, used for inlays, three-quarter crowns and bridgework; (vi) *Jel-65*, a white, microfine, silver-free, high-gold porcelain-to-metal alloy with high strength and durability, and outstanding thermal stability sag resistance; (vii) *Jel-71PDF*, a very hard, microfine gold-platinum porcelain-to-metal alloy; (viii) *Jel-96*, a strong, hard, microfine, white high-gold ceramic alloy containing platinum and palladium, used for fusing dental porcelain to metals; and (ix) *Jel-O 75*, a strong, hard, microfine, pale gold-colored high-gold ceramic alloy, used for fusing dental porcelain to metals.

Jelbon. Trade name of J.F. Jelenko & Company (USA) for a corrosion-resistant, beryllium-free nickel-chromium dental bonding alloy.

Jelbond PDF. Trade name of J.F. Jelenko & Company (USA) for a microfine, gold-platinum porcelain-to-metal alloy with high hardness used for long-span bridgework.

Jelcast. Trade name of J.F. Jelenko & Company (USA) for a microfine, low-gold dental casting alloy for inlays, crowns and bridgework.

Jelenko. Trade name of J.F. Jelenko & Company (USA) for an extensive series of gold- and palladium-base dental alloys for fusing porcelain to metal, and for inlays and crowns. *Jelenko O* is a microfine, high-gold ceramic dental alloy that provides high strength and durability, and is used for fusing dental porcelain to metals.

Jelliff. Trade name of Jelliff Corporation (USA) for a series of electrical resistance materials including high-purity nickels and various copper-manganese-nickel, iron-chromium-aluminum, iron-nickel-chromium, nickel-iron-chromium, copper-nickel and nickel-chromium-manganese alloys.

jelly paint. See thixotropic paint.

Jelspan. Trade name of J.F. Jelenko & Company (USA) for a corrosion-resistant, beryllium-free nickel-chromium dental bonding alloy.

Jelstar. Trade name of J.F. Jelenko & Company (USA) for a white, durable ceramic dental alloy containing 60% palladium and 28% silver, and used for fusing porcelain to metals.

Jeltrate. Trademark of Dentsply International Inc. (USA) for elastic, alginate-based dental impression materials.

jelutong. The light, soft, colored wood of the tree *Dyera costulata*. It has a fine, even texture, high stability, good nailing and staining qualities, seasons and works well, and takes a good finish. Average weight, 460 kg/m^3 (29 lb/ft^3). Source: Malaysia and Borneo. Used in patternmaking. *Note:* The tree also yields an inferior grade of rubber latex.

Jena blue glass. Jena glass to which a mixture of cobalt and ceric oxides has been added to produce a bluish color and fluorescence. See also Jena glass.

Jenaer Glas. Trademark of Schott Glas Jena (Germany) for heat-resistant household glassware.

Jena glass. An optical glass available in a wide range of refractive indices and dispersions. It is made chemically resistant by the addition of oxides of boron, aluminum or barium, and may or may not be resistant to mechanical and thermal shocks. Used for scientific apparatus and optical instruments. Named after the former location of the Schott Glas glass works in Jena, Germany. Abbreviation: JG.

Jenatherm. Trade name of former Jenaer Glaswerk, now Schott GmbH (Germany) for a heat-resistant glass.

jennite. A white mineral composed of calcium silicate hydroxide, $Ca_9H_2Si_6O_{18} \cdot 6H_2O$. Crystal system, triclinic. Density, 2.32 g/cm^3; refractive index, 1.564. Occurrence: USA (California).

jeremejevite. A colorless or yellowish mineral composed of aluminum fluoride borate hydroxide, $Al_6B_5O_{15}(F,OH)_3$. Crystal system, hexagonal. Density, 3.28 g/cm^3; refractive index, 1.646. Occurrence: Russian Federation.

jersey. A soft, machine-knitted fabric composed of wool, cotton, silk or synthetics (e.g., polyester or rayon). It is is somewhat elastic, has a slight rib on one side, and is used for undergarments, dresses, blouses and shirts.

Jersey pine. See Virginia pine.

jervisite. A light green or colorless mineral of the pyroxene group composed of sodium scandium silicate, $NaScSi_2O_6$. It can also be made synthetically. Crystal system, monoclinic. Density, 3.22 g/cm^3; refractive index, 1.715. Occurrence: Italy.

Jespan. Trademark of Cheil Synthetics Inc. (South Korea) for *spandex* fibers and yarns used for elastic fabrics.

Jess-Air. Trade name of Jessop Steel Company (USA) for an air-hardening, medium-alloy cold-work tool steel (AISI type A6) containing 0.7% carbon, 2% manganese, 0.3% silicon, 1.35% molybdenum, 1% chromium, and the balance iron. It has excellent resistance to decarburization, good nondeforming and deep-hardening properties, good wear resistance and good strength. Used for forming, blanking, notching and trimming dies, forming tools, punches, mandrels and shear blades.

Jessco. Trade name of Jessop Steel Company (USA) for a series of high-speed steels containing 0.7% carbon, varying amounts of tungsten, chromium, vanadium and cobalt, and the balance iron. Used for tools, cutters, punches, dies and shear blades.

Jessop. Trade name of Jessop Steel Company (USA) for an extensive series of steels including various grades of tool steels (e.g.,

molybdenum- and tungsten-type high-speed steels, cold- and hot-work steels, water-hardening steels and shock-resisting steels), stainless steels (austenitic, ferritic and martensitic types), heat-resistant steels, austenitic nonmagnetic steels, case-hardening steels, high-strength structural steels and magnet steels.

Jet. Trade name of Jet Moulding Compounds Limited (Canada) for several polyester bulk and sheet molding compounds supplied in several grades. *Jet B* series bulk molding compounds contain 15% glass-fiber reinforcement and are used depending on grade for general-purpose applications, microwav-able and ovenproof cookware and dishware, or office equipment and appliances. *Jet S* series sheet molding compounds contain 22 or 30% glass-fiber reinforcement and are used depending on grade for general-purpose applications, automotive components, electrical components, office equipment and appliances, or chemical-resistant pumps and pipe components. *Jet* molding compounds can be processed by compression, injection or transfer molding processes.

jet. An intensely black-colored variety of *lignite* (brown coal) with uniform texture used as a gem.

Jetalloy. Trade name of Canadian Quebec Metallurgical Corporation for a series of cobalt-base alloys containing about 20-25% chromium, 6-15% tungsten, 10-30% nickel, 0-4% titanium, 0-1% iron and 0.02-0.3% carbon. They have high creep resistance, and good oxidation and wear resistance. Used for high-temperature applications.

Jet Bite. Trade name of Coltene-Whaledent (USA) for a silicone elastomer used in dentistry for bite registration applications.

jetcrete. See dry-mix shotcrete.

jet-engine alloys. A group of alloys including cobalt-, nickel- and nickel-iron superalloys, certain iron-carbon alloys (e.g., stainless steels) and various iron-chromium, iron-chromium-nickel and iron-nickel-chromium casting alloys, and several aluminum-copper and aluminum-silicon alloys, magnesium-aluminum and magnesium-thorium alloys, and titanium- or zirconium-base alloys with good high-temperature, creep and oxidation resistance that make them useful for jet-engine parts. Abbreviation: JEA.

Jet Forge. Trade name of Teledyne Vasco (USA) for an air-hardening hot-work tool steel containing 0.45-0.5% carbon, 0.8-1% silicon, 0.2-0.4% manganese, 7.5-8% chromium, 1.2-1.5% molybdenum, 1.25-1.5% vanadium, and the balance iron. It has high toughness, good resistance to wear and heat checking and high hot hardness. Used for forging tools and dies for jet engine blades, and for forging and hot-heading dies, dummy blocks, cold and hot punches, and cold and hot shear blades.

Jethete. Trade name of British Steel Corporation (UK) for a series of ferritic and martensitic stainless steels containing up to 0.2% carbon, 11-13% chromium, 0-2.5% nickel, 0-2% molybdenum, 0-1.5% manganese, and the balance iron. They have excellent heat resistance and good corrosion and oxidation resistance. Used for turbine blades, gas-turbine components, annealing boxes, springs, and flatware.

Jet-melt. Trademark of 3M Company (USA) for a series of 100% solids thermosetting hot-melt adhesives available in a variety of grades and colors including clear, amber, tan, off-white, yellow, brown and gray. Upon heating, they become liquid, readily wet the surface to be bonded, and cure quickly upon cooling.

Jetset. Trademark for a quick-setting cement developed by the Portland Cement Association (USA).

Jetspun. Trademark of E.I. DuPont de Nemours & Company

(USA) for nylon fibers, yarns and fabrics used for clothing.

Jet Tooth Shade. Trade name of Lang Company (USA) for an acrylic dental resin supplied as a two-part system (powder plus liquid) and used for temporary crown and bridge restorations.

Jetweld. Trade name of Lincoln Electric Company (USA) for a series of carbon-steel arc-welding electrodes composed of 0.1% carbon, varying small amounts of manganese, silicon, nickel, chromium, molybdenum and vanadium, and the balance iron.

jewel. A term used in engineering to refer to a gem or other hard stone, such as synthetic corundum, ruby or sapphire, cut to shape and used as a bearing in instruments, timepieces and dial indicators, in watch escapements, compasses and electrical instruments (meters), and for recording needles. See also synthetic corundum; synthetic ruby; synthetic sapphire.

Jewel-Brite. Trade name of Atotech USA Inc. (USA) for a bright nickel electroplate used on jewelry alloys.

jeweler's bronze. See jewelry bronze; Nu-Gold.

jeweler's enamel. A specially formulated *porcelain enamel* that is usually lower melting than standard grades and is used in the manufacture of jewelry, art objects and insignia on gold, silver or iron. Also known as *jewelry enamel*.

Jeweler's Manganese Bronze. Trade name of Belmont Metals Inc. (USA) for a manganese casting bronze containing 55-65% copper, 35-40% zinc, 0.1-3% aluminum and 0.1-2% manganese. It has a melting point of 860-882°C (1580-1620°F) and is used for jewelry and decorative castings.

jeweler's putty. See putty (2).

jeweler's rouge. See red iron oxide; rouge.

Jewelite. Trade name of Pro-Phy-lac-tic Brush Company (USA) for acrylic plastics.

Jewell. Trade name of Apollo Steel Company (USA) for a series of ferritic and pearlitic malleable cast irons.

jewelry alloys. A group of ductile, malleable, corrosion-resistant nonferrous alloys, such as brass, bronze, nickel silver, or gold- or silver alloys, used in the manufacture of costume jewelry, ornaments, notions, medals, etc.

jewelry bronze. A corrosion-resistant wrought bronze containing 87.5% copper and 12.5% zinc. Commercially available in the form of wires and flat products, it has excellent cold workability and good hot workability, and is used for costume jewelry, lipstick cases, containers, channels, angles, chains, slide fasteners, and as a base material for gold-plated parts. Also known as *jeweler's bronze*. See also Nu-Gold.

jewelry enamel. See jeweler's enamel.

Jeyrock. Trade name of Jeydent (UK) for a die stone used in dentistry.

JFB Alloy. Trade name for a corrosion-resistant cobalt-chromium dental casting alloy.

J-Film. Trademark of W.R. Grace & Company (USA) for a polyethylene shrink film used for packaging food and nonfood products.

JFR compound. A jet-fuel-resistant joint sealing compound used in the construction and repair of airport runways, taxiways and hangar floors.

Jiffycrete. Trademark of Global Stone James River (USA) for ready-to-use concrete supplied in 60 lb. (27 kg) bags as coarse, sand or mortar mixes.

jimboite. A light purple or green mineral composed of manganese borate, $Mn_3(BO_3)_2$. It can also be made synthetically. Crystal system, orthorhombic. Density, 3.98 g/cm³; refractive index, 1.794.

jimthompsonite. A colorless to light pinkish brown mineral of the amphibole group composed of iron magnesium silicate hydroxide, $(Mg,Fe)_5Si_6O_{16}(OH)_2$. Crystal system, orthorhombic. Density, 3.03 g/cm³; refractive index, 1.626. Occurrence: USA (Vermont).

jinshajiangite. A black-red, brownish-red or golden-red mineral composed of potassium sodium barium calcium iron manganese niobium titanium oxide fluoride silicate, $(Na,K)_5(Ba,Ca)_4$-$(Fe,Mn)_{15}(Ti,Fe,Nb)_8(SiO_4)_{15}(F,O,OH)_{10}$. Crystal system, monoclinic. Density, 3.61 g/cm³; refractive index, 1.802. Occurrence: China.

Jiscon. Trade name of Jackson Iron & Steel Company (USA) for a cast iron containing 2.3-2.5% carbon, 6-6.25% silicon, 1.25% copper, 0.5% chromium, and the balance iron. It has good resistance to growth and scaling up to 927°C (1700°F), and is used for high-temperature castings.

Jisco Silvery. Trade name of Jackson Iron & Steel Company (USA) for a pig iron containing 0.8-3% carbon, 5-17% silicon, 1-2% manganese, and the balance iron. Used in the manufacture of foundry steel and iron castings.

jixianite. A red to brown red mineral of the pyrochlore group composed of lead iron tungsten oxide hydroxide, $Pb(W,Fe)_2$-$(O,OH)_7$. Crystal system, cubic. Density, 6.04 g/cm³; refractive index, 2.2885. Occurrence: China.

joaquinite. A brown to honey-yellow mineral composed of sodium barium cerium iron titanium silicate hydroxide, Ba_2NaCe_2-$FeTiSi_8O_{26}(OH)_2$. Crystal system, orthorhombic. Density, 3.98 g/cm³; refractive index, 1.767. Occurrence: USA (California).

job-mix concrete. A *field concrete* mixed, placed and cured at the construction site.

joesmithite. A black mineral of the amphibole group composed of calcium lead magnesium iron beryllium silicate hydroxide, $Ca_2Pb(Mg,Fe)_5(Si,Be)_8O_{22}(OH)_2$. Crystal system, monoclinic. Density, 3.83 g/cm³; refractive index, 1.765. Occurrence: Sweden.

johachidolite. A colorless mineral composed of calcium aluminum borate, $CaAlB_3O_7$. Crystal system, orthorhombic. Density, 3.37 g/cm³; refractive index, 1.717. Occurrence: North Korea.

johannite. A green, radioactive mineral composed of copper uranyl sulfate hydroxide hexahydrate, $Cu(UO_2)_2(SO_4)_2(OH)_2 \cdot 6H_2O$. Crystal system, triclinic. Density, 3.32 g/cm³; refractive index, 1.595. Occurrence: Canada.

johannsenite. A grayish, greenish, bluish, or clove-brown mineral of the pyroxene group composed of calcium manganese silicate, $CaMnSi_2O_6$. Crystal system, monoclinic. Density, 3.52 g/cm³. Occurrence: Italy, Mexico, USA (New Jersey, Oregon).

johillerite. A violet mineral composed of sodium copper magnesium zinc arsenate, $Na(Mg,Zn)_3Cu(AsO_4)_3$. Crystal system, monoclinic. Density, 4.15 g/cm³; refractive index, 1.743. Occurrence: Namibia.

johnbaumite. A white to colorless mineral of the apatite group composed of calcium arsenate hydroxide, $Ca_5(AsO_4)_3(OH)$. Crystal system, hexagonal. Density, 3.68 g/cm³; refractive index, 1.687. Occurrence: USA (New Jersey).

johnsomervilleite. A dark brown mineral composed of sodium calcium iron magnesium manganese phosphate, $Na_2Ca(Mg,Fe,Mn)_7(PO_4)_6$. Crystal system, rhombohedral (hexagonal). Density, 3.35 g/cm³; refractive index, 1.655. Occurrence: Scotland.

johnstrupite. A brownish-green mineral composed of calcium sodium cerium titanium zirconium fluoride silicate, $(Ca,Na)_3$-$(Ce,Ti,Zr)(SiO_4)_2F$. Crystal system, monoclinic. Occurrence: Norway

Johnson. (1) Trade name for a heat and corrosion-resistant alloy containing 24-80% iron, 9-36% nickel and 10-30% chromium.

(2) Trade name of Johnson Bronze Company (USA) for a series of bearing bronzes and tin- and lead-base babbitts.

Johnstown. Trade name of Johnstown Corporation (USA) for high-temperature and alloy machinery steels, and several abrasion-, corrosion- or heat-resistant alloy cast irons.

Joinite. Trade name for a corkboard used as vibration insulator between machine bases and foundations.

Joinpox. Trade name of Ato Findley (France) for an epoxy joint filler.

joint compound. (1) A compound used in the taping and/or finishing of gypsum wallboard.

(2) A compound used on pipe joints to lubricate the threads and prevent leakage.

joint filler. A building material used to fill and seal joints in concrete and exclude foreign materials. It usually allows some joint movement. See also joint sealant.

joint mortar. See masonry mortar.

joint-reinforcing metal. Welded or woven wire mesh or expanded metal, usually in the form of small strips, used in the reinforcement of plaster and lath.

joint-reinforcing tape. A material, such as metal, paper, fabric or glass mesh usually in tape form, used in the reinforcement of joints between gypsum wallboards.

joint sealant. A material used on joints in concrete to exclude water and debris. It may or may not be a *joint filler*.

jokokuite. A pale pink mineral of the chalcanthite group composed of manganese sulfate pentahydrate, $MnSO_4 \cdot 5H_2O$. Crystal system, triclinic. Density, 2.03 g/cm³; refractive index, 1.510. Occurrence: Japan.

joliotite. A lemon yellow mineral composed of uranyl carbonate dihydrate, $(UO_2)CO_3 \cdot 2H_2O$. Crystal system, orthorhombic. Density, 4.04 g/cm³. Occurrence: Germany.

Jonas & Culver. Trade name of Jonas & Culver (Nova) Limited (UK) for a water-hardening tool steel containing 1.36% carbon, 0.41% manganese, and the balance iron. Used for tools and cutters.

jonesite. A colorless mineral composed of potassium barium aluminum titanium silicate hexahydrate, $K_2Ba_4Ti_4Al_2Si_{10}O_{36} \cdot 6H_2O$. Crystal system, orthorhombic. Density, 3.25 g/cm³; refractive index, 1.660. Occurrence: USA (California).

Jonylon Nylon 6. Trade name of BIP Chemicals (USA) for polyamide 6 (nylon 6) resins supplied in unmodified, glass fiber-reinforced, fire-retardant and high-impact grades.

Jonylon Nylon 6/6. Trade name of BIP Chemicals (USA) for polyamide 6,6 (nylon 6,6) resins supplied in unmodified, mineral-filled, fire-retardant, supertough and high-impact grades.

jordanite. A lead-gray mineral composed of lead arsenic sulfide, $(Pb,Tl)_{13}As_7S_{23}$. It can also be made synthetically by heating lead sulfide and arsenic sulfide under prescribed conditions. Crystal system, monoclinic. Density, 6.44 g/cm³.

joseite. A white mineral of the tetradymite group that is composed of bismuth telluride sulfide, Bi_4TeS_2, and may also contain some selenium. Crystal system, rhombohedral (hexagonal). Density, 8.10 g/cm³. Occurrence: Canada (British Columbia), Brazil.

Joslyn. Trade name of Joslyn Stainless Steels Company (USA) for a series of austenitic, ferritic, martensitic and precipitation-hardening stainless steels including various free-machining and special-purpose grades.

jouravskite. A mineral of the ettringite group composed of calci-

um manganese sulfate carbonate hydroxide dodecahydrate, $Ca_3Mn(CO_3)(SO_4)(OH)_6 \cdot 12H_2O$. Crystal system, hexagonal. Density, 1.95 g/cm³; refractive index, 1.693. Occurrence: Morocco.

Journal. Trademark of Anchor Packing Division, Robco Inc. (Canada) for polytetrafluoroethylene (*Teflon*) and asbestos braiding packings.

journal bronze. A bronze suitable for journal bearings and having a typical composition of up to about 12% tin, 7% zinc, 3% lead, respectively, with the balance being copper. It is available in the centrifugally cast, continuously cast and sand-cast conditions.

Jouvencel. Trade name of Esperance Longdez (Belgium) for non-aging, low-carbon decarburized sheet steel.

J-Plast. Trade name of J-Von (USA) for thermoplastic elastomers. *J-Plast LoTemp* is supplied in pellet form, and is formulated to combine the elastomeric properties of vulcanized rubber with the ease of processing of thermoplastic polymers. It has high impact resistance at low temperatures (below -40°C or -40°F) and good weatherability. Used for outdoor equipment, sound-deadening fascias, and automotive splashguards.

juanite. A pistachio-green mineral composed of calcium iron magnesium aluminum silicate heptahydrate, $Ca_9(Mg,Fe)_3(Si,Al)_{12}(O,OH)_{36} \cdot 7H_2O$. Crystal system, orthorhombic. Density, 3.28 g/cm³. Occurrence: Russian Federation.

julgoldite. A mineral of the pumpellyite group composed of calcium iron silicate hydroxide monohydrate, $Ca_2(Fe,Al)_2(SiO_4)_2(Si_2O_7)(OH)_2 \cdot H_2O$. Crystal system, monoclinic. Density, 3.55 g/cm³. Occurrence: Sweden.

julienite. A blue mineral composed of sodium cobalt thiocyanate octahydrate, $Na_2Co(SCN)_4 \cdot 8H_2O$. Crystal system, tetragonal. Density, 1.65 g/cm³; refractive index, 1.556. Occurrence: Central Africa.

Juma. Trademark of Juma Natursteinwerke GmbH & Co. KG (Germany) for an extensive range of natural stone products used for building construction and civil engineering applications.

Jumbo. Trademark of Paris Brick Company Limited (Canada) for calcite bricks.

jumbo. A generic term for any building brick or block larger than standard size, sometimes produced according to specific dimensional specifications.

jumbo block. See jumbo brick.

jumbo brick. A hollow clay building brick or block that is usually 89 × 191 × 292 mm (3.5 × 7.5 × 11.5 in.) in size and has two large cells. Also known as *jumbo block*.

jumbo utility. A hollow clay building brick or block that is usually 89 × 89 × 292 cm (3.5 × 3.5 × 11.5 in.) and has three cells.

junco. A strong fiber obtained from the tropical shrub or small tree (*Koeberlinia spinosa*) and used especially for cordage and hard mats.

jungite. A dark green mineral composed of calcium zinc iron phosphate hydroxide hydrate, $Ca_2Zn_4Fe_8(PO_4)_9(OH)_9 \cdot 16H_2O$. Crystal system, orthorhombic. Density, 2.84 g/cm³; refractive index, 1.664. Occurrence: Germany.

juniper. The soft, fragrant wood of any of a genus (*Juniperus*) of evergreen shrubs and trees belonging to the cypress family especially the common or dwarf juniper (*J. communis*) and the eastern red cedar (*J. virginiana*) with the former not being a commercially important wood. Used for closet and chest linings, wardrobes and cabinets. See also eastern red cedar.

juniper gum. See sandarac gum.

junitoite. A mineral composed of calcium zinc silicate monohydrate, $CaZn_2Si_2O_7 \cdot H_2O$. Crystal system, orthorhombic. Density,

3.50 g/cm^3; refractive index, 1.664. Occurrence: USA (Arizona).

Junker. Trade name of Otto Junker GmbH (Germany) for an extensive series of corrosion-resistant steel castings of the iron-chromium-nickel or iron-nickel-chromium types as well as various heat-resistant steel castings of the iron-chromium, iron-chromium-nickel or iron-nickel-chromium types.

junoite. A mineral composed of copper lead bismuth sulfide selenide, $Cu_2Pb_3Bi_8(S,Se)_{16}$. Crystal system, monoclinic. Density, 6.77 g/cm^3. Occurrence: Australia.

Jupilon. Trade name of Mitsubishi Engineering Plastics (Japan) for polycarbonate resins used for engineering applications.

Jupiter steel. A cast steel made by melting wrought-steel scrap with about 2% ferrosilicon, up to about 0.5% ferromanganese and about 3% aluminum and casting in special molds. Its strength and ductility is similar to that of forged steel.

jurbanite. (1) A colorless mineral composed of aluminum sulfate hydroxide pentahydrate, $AlSO_4OH \cdot 5H_2O$. Crystal system, monoclinic. Density, 1.79 g/cm^3; refractive index, 1.473. Occurrence: USA (Arizona).

(2) A colorless mineral composed of aluminum oxide sulfate hydrate, $Al_2S_2O_9 \cdot 11H_2O$. It can also be made synthetically. Crystal system, monoclinic. Density, 1.79 g/cm^3; refractive index, 1.473. Occurrence: USA (Arizona).

jusite. A mineral composed of calcium silicate hydrate, $(Ca,K,Na,H_3O)(Si,Al)O_3 \cdot H_2O$. Crystal system, monoclinic. Density, 2.32 g/cm^3; refractive index, 1.558. Occurrence: Germany.

Just Like Wood. Trade name of 3M Company (USA) for a putty used for filling imperfections in wood.

Juta. Trade name of Vetreria di Vernante SpA (Italy) for a glass with cross-weave pattern.

jute. A strong bast fiber obtained from the stems of various plants of the linden family, especially the round pot jute *(Corchorus olitorius)* and the long pot jute *(C. capsularis)*. It is soft, lustrous and loses strength when wet. Source: India, Pakistan. Used in the manufacture of coarse sacks, bags, burlap, rope, cordage, twine, coarse paper, fiberboard, upholstery fabrics and carpet backing, as a filler for plastic molding materials, and as a reinforcement for polyester resins.

jute board. A fiberboard manufactured from jute fibers by pressing or rolling into flat sheets. Used in building construction, and for containers and boxes.

jute paper. A strong tan-colored paper made chiefly from jute fiber, and used for strong bags, e.g., for holding cement.

jute-reinforced plastics. Synthetic resins, such as polyesters or epoxies, reinforced with jute fibers that have better elongation and damage tolerance, and lower cost than synthetic fibers, such as aramid, carbon or glass. Used in building construction. Abbreviation: JRP.

jute-spun yarn. A staple yarn made from a material other than jute, but spun on a jute spinning machine.

Juvelite. German trade name for a phenol-formaldehyde resin.

K

kadaya gum. See karaya gum.

Kadel. Trademark of Amoco Performance Products, Inc. (USA) for a family of semicrystalline thermoplastic engineering plastics based on polyaryletherketone (PAEK) resins. Available in various grades including neat and glass or carbon fiber-reinforced, they are melt-processible by injection molding, thermoforming and extrusion processes, and have high tensile strength, good elongation and impact strength, good chemical resistance and hydrolytic and thermal stability, good integral lubricity, good overall electrical properties, good retention of mechanical properties at elevated temperatures, high temperature stability, high glass-transition temperatures and low flammability and smoke generation. Used for aircraft-engine components, air ducts, interior cabin materials, nonstructural components for aircraft and spacecraft, and for wire and cable, pump components, bearing surfaces, backup seals and monofilaments.

kaersutite. A black mineral composed of sodium calcium magnesium manganese iron aluminum titanium silicate hydroxide, $Ca_2(Na,K)(Mg,Mn,Fe)_6(Al,Ti)_2Si_6O_{22}(O,OH,F)_2$. Crystal system, monoclinic. Density, 3.21 g/cm³; refractive index, 1.692. Occurrence: Greenland, USA (Arizona).

Kae Tar. Trade name of Kansas Paint & Color Company (USA) for coal-tar epoxy coatings.

Kaethan. Trademark for polyurethane foams and foam products.

kafehydrocyanite. (1) A yellow mineral composed of potassium iron cyanide monohydrate, $K_4Fe(CN)_6 \cdot H_2O$. Density, 1.98 g/cm³; refractive index, 1.577. Occurrence: Russian Federation.

(2) A lemon-yellow mineral composed of potassium iron cyanide trihydrate, $K_4Fe(CN)_6 \cdot 3H_2O$. It can also be made synthetically. Crystal system, tetragonal. Density, 1.98 g/cm³; refractive index, 1.577. Occurrence: Russian Federation.

Kaffir. Trade name of United States Gypsum Company (USA) for die and model stones used in dentistry.

Kagero. Trade name of Nippon Sheet Glass Company Limited (Japan) for a patterned glass with curvilinear design.

kahikarea. The wood from the tall, white pine tree *Podocarpus dacrydioides*. The narrow heartwood is yellow and the sapwood white. It has a straight grain and works well, but has only low durability. Average weight, 465 kg/m³ (29 lb/ft³). Source: New Zealand. Used for boxes, crates and packing.

kahlerite. A lemon-yellow, radioactive mineral of the autunite group composed of iron uranyl arsenate dodecahydrate, $Fe(UO_2)_2(AsO_4)_2 \cdot 12H_2O$. It can also be made synthetically. Crystal system, tetragonal. Density, 3.22 g/cm³; refractive index, 1.632. Occurrence: Austria.

Ka Ho Loy. Trade name of Apollo Steel Company (USA) for a low-carbon steel that has 0.2-0.3% copper added to improve its resistance to atmospheric corrosion. Commercially available in the form of sheets and plates, it is used for roofing products.

kail. See oganwo.

Kailoc. Trade name of Klebchemie (Germany) for epoxy and polyurethane resins and adhesives.

kainite. A white, gray, reddish or colorless mineral with vitreous luster. It is composed of potassium magnesium chloride sulfate hydrate, $KMg(SO_4)Cl \cdot 2.75H_2O$. It can also be made synthetically. Crystal system, monoclinic. Density, 2.14 g/cm³. Occurrence: Germany. Used as a source of potassium salts, and as a fertilizer.

kainosite. A brown, yellow or colorless mineral of the axinite group composed of calcium rare-earth carbonate silicate monohydrate, $Ca_2Ln_2Si_4O_{12}CO_3 \cdot H_2O$. Crystal system, orthorhombic. Density, 3.52 g/cm³; refractive index, 1.685. Occurrence: Canada (Ontario).

Kaisaloy. Trade name of Kaiser Steel Corporation (USA) for a series of high-strength low-alloy (HSLA) steels available as cold- and hot-rolled strip and sheet, and hot-rolled plate, bars and structural shapes. They have good weldability, formability and corrosion resistance, and are used for structural applications including bridges, cranes, booms, derricks, buildings, mine and railroad cars, bodies and frames of cars, trucks and buses, and mining equipment.

Kaiser. Trade name of Kaiser Aluminum & Chemical Corporation (USA) for an extensive series of wrought or cast aluminum products including commercially pure aluminums, and various alloys of aluminum with one or more of the following metals: copper, manganese, magnesium, zinc, silicon and chromium.

kaki. The very hard wood of the persimmon tree *Diospyrus kaki* belonging to the ebony family. It is black in color with gray, brown or yellow stripes, and has a close, even grain. Source: China, Japan. Used as an *ebony* substitute, but is somewhat lighter than true African ebony.

kalabatun. (1) A fine embroidery thread obtained by winding a thin, drawn gold or silver wire around a core yarn. Used especially in the form of stripes as a decoration on cotton fabrics.

(2) Cotton fabrics decorated with kalabatun (1) threads.

Kaladex. Trademark of E.I. DuPont de Nemours & Company (USA) for tough, inherently UV-stable polyethylene naphthalene (PEN) resins supplied as granules and biaxially oriented and heat-stabilized films. They have a density of 1.36 g/cm³ (0.049 lb/in.³), good electrical properties, good chemical and temperature resistance, good dimensional stability, and an upper service temperature of 155°C (310°F). Used in the manufacture of film and sheeting for plastic bags, bubble packs, labels, wrapping and packaging applications, and for flexible printed circuitry and electrical insulation.

kalamein. A corrosion-resistant alloy composed of lead, tin, antimony, bismuth and nickel, and used as a coating for iron and steel.

Kalbord. Trademark of Foseco Minsep NV (Netherlands) for flexible insulating boards used around feederheads during the casting of iron, steel, and copper- and nickel-base alloys.

kalborsite. A colorless mineral composed of potassium boron aluminum chloride silicate hydroxide, $K_6BAl_4Si_6O_{20}(OH)_4Cl$. Crystal system, tetragonal. Density, 2.50 g/cm³; refractive index, 1.525. Occurrence: Russian Federation.

Kalchoids. Trade name for a bearing alloy composed of copper, tin and zinc.

Kalcolor. Trade name of Kaiser Aluminum & Chemical Corporation (USA) for a light yellow to brown or black coating produced by the proprietary "Kalcolor" hard anodizing process using a bath composed of 7-15 wt% sulfosalicylic acid, 0.3-4 wt% sulfuric acid, and water.

Kaldo steel. A carbon or low-alloy steel made from a charge of molten pig iron, scrap steel and fluxes in a large inclined rotating vessel (Kaldo furnace) by feeding high-purity oxygen through a lance at velocities somewhat lower than in the LD (Linz-Donawitz) process onto the molten bath. Most of the carbon monoxide and many of the unwanted elements in the steel are burned off. The process was developed under Prof. B. Kalling in Domnarvet, Sweden. See also LD steel.

kaliborite. A colorless, white or reddish brown mineral composed of potassium hydrogen magnesium borate hydroxide tetrahydrate, $KHMg_2B_{12}O_{16}(OH)_{10} \cdot 4H_2O$. Crystal system, monoclinic. Density, 1.98 g/cm^3; refractive index, 1.527. Occurrence: Germany, Italy.

kalicinite. A white, yellowish or white mineral composed of potassium hydrogen carbonate, $KHCO_3$. It can also be made synthetically. Crystal system, monoclinic. Density, 2.17 g/cm^3; refractive index, 1.482.

Kalidar. Trademark of E.I. DuPont de Nemours & Company (USA) for strong polyethylene naphthalene (PEN) polyester resins with outstanding UV absorption, excellent barrier properties to moisture, oxygen and carbon dioxide, and high glass-transition temperatures. They are supplied in several resin grades and are also available in fiber and film grades. Used for injection-molded products, such as food and beverage containers, and for compounding and blending applications.

Kalif. Trade name of Kalif Corporation (USA) for a steel-backed leaded copper containing 70% copper and 30% lead, and used for bearings, steel-backed bearings and wrist-pin bushings.

kalinite. A white mineral of the alum group composed of potassium aluminum sulfate hydrate, $KAl(SO_4)_2 \cdot 11H_2O$. Crystal system, monoclinic. Refractive index, 1.542. Occurrence: UK (Wales).

kaliophilite. A colorless mineral of the nepheline group composed of potassium aluminum silicate, $KAlSiO_4$. Crystal system, hexagonal. Density, 2.49 g/cm^3; refractive index, 1.587. Occurrence: Italy.

kalipyrochlore. A green mineral of the pyrochlore group composed of potassium strontium niobium oxide hydroxide hydrate, $(K,Sr)_{2-x}Nb_2O_6(O,OH) \cdot xH_2O$. Crystal system, cubic; refractive index, 1.985. Occurrence: Zaire.

kalistrontite. A colorless mineral composed of potassium strontium sulfate, $K_2Sr(SO_4)_2$. It can also be made synthetically. Crystal system, rhombohedral (hexagonal). Density, 3.20 g/cm^3; refractive index, 1.569.

Kalite. Trade name Great Lake Calcium Corporation (USA) for a fine, white precipitated calcium carbonate used as a pigment extender.

Kalitex. Trade name of Kalitex Sdn. Bhd. (Thailand) for textured vinyl panels used as wall coverings and for other interior design applications.

Kalkos. Trade name of Latrobe Steel Company (USA) for a tungsten-type hot-work tool steel (AISI type H23) containing 0.3% carbon, 0.35% manganese, 0.5% silicon, 12% chromium, 12% tungsten, 0.9-1% vanadium, and the balance iron. It has excellent hot hardness and good deep-hardening properties and wear resistance. Used for hot forming and working tools, and for

metal extrusion and forging dies.

Kallodent. Trade name of ICI Limited (UK) for an injection-molded acrylic used for denture bases.

Kallodentine. Trade name of ICI Limited (UK) for a dough (or bulk) molded acrylic resin used for denture bases.

Kallodoc. Trade name of ICI Limited (UK) for a molded acrylic resin used for denture bases.

K Alloy. British trade name for an aluminum alloy containing 11.3% copper, 1.1% manganese, 0.5% magnesium and 0.8% silicon, used for leakproof castings and light-alloy parts.

Kalloy. Trade name of Dunford Hadfields Limited (UK) for an alloy steel containing 0.4% carbon, 30% chromium, and the balance iron. It has high heat resistance, and is used for baffle plates, furnace parts, skids, retorts, and carburizing and annealing boxes.

Kalor. Trade name of Uddeholm Corporation (USA) for a series of tough, oil-hardening tool steels containing 0.3% carbon, 3.75-5.5% tungsten, 4.25% nickel, 1-1.25% chromium, 1% silicon, 0.15-0.2% vanadium, and the balance iron. Used for extrusion dies and tools, press dies, rams and liners, and punches.

Kalrez. Trademark of DuPont Dow Elastomers (USA) for a series of carbon black-filled thermoset perfluoroelastomer compounds. They have excellent resistance to very high temperatures (up to about 300°C or 570°F), excellent chemical resistance, exceptional resistance to hot-air aging, good resistance to aromatic fuels, solvents, alkalies and concentrated acids, good mechanical properties, low compression sets, low swell in organic and inorganic acids and aldehydes, fair to poor resistance to hot water and steam and hot aliphatic amines, ethylene oxide and propylene oxide, a hardness of 75 Shore A, and a maximum service temperature (in air) of 316°C (600°F). Used for O-rings, seals, diaphragms, and aircraft parts.

kalsilite. A colorless natural mineral with yellow tint belonging to the nepheline group. It is composed of potassium aluminum silicate, $KAlSiO_4$, and can also be made synthetically from potassium silicate and beta aluminum oxide at 800°C (1470°F). Crystal system, hexagonal. Density, 2.59 g/cm^3; refractive index, 1.541. Occurrence: Uganda.

Kalsogen. Trade name of Dentsply Australia for a zinc oxide–eugenol dental cement.

kalsomine. See calcimine.

Kaltron. Trademark of Joh. A. Benckiser GmbH (Germany) for fluorohydrocarbon products.

Kalvan. Trademark of R.T. Vanderbilt Company, Inc. (USA) for a precipitated calcium carbonate used as a compounding ingredient in paints, plastics and rubber.

Kalzinol. Trade name of Dentsply DeTrey for a zinc oxide–eugenol dental cement.

kamacite. (1) A black mineral composed of iron nickel, FeNi. Crystal system, cubic. Occurrence: USA.

(2) A body-centered-cubic (bcc) phase found in meteorites, and composed of about 4-7.5% nickel, and the balance iron. It is thought to be analogous to ferrite in iron-carbon alloys. See also plessite; taenite.

kamaishilite. A colorless mineral composed of calcium aluminum silicate hydroxide, $Ca_2Al_2SiO_6(OH)_2$. Crystal system, tetragonal. Density, 2.83 g/cm^3; refractive index, 1.629. Occurrence: Japan.

Kamax. Trademark of AtoHaas Company (USA) for high-modulus acrylic polyimide copolymers with excellent high-temperature properties, good optical properties and low thermal expansion. They can be processed by blow or injection molding or extrusion into optical lenses, automotive components and lighting fixtures.

Kana-Lite. Trademark of Mountain Mineral Company Limited (Canada) for slate products.

Kanamite. Trademark for an expanded clay used as a lightweight aggregate in concrete.

Kanebian. Trade name of Kanebo Limited (Japan) for a vinal (polyvinyl alcohol) fiber used for textile fabrics.

Kanebo. Trade name of Kanebo Limited (Japan) for an extensive series of synthetic fibers and yarns including *Kanebo Acryl* acrylic staple fibers, *Kanebo Ester* polyester staple fibers and filament yarns, *Kanebo Loobell* spandex fibers and yarns, *Kanebo Nylon* nylon 6 fibers and filament yarns and *Kanebo Polyester* polyester filament yarns. Used for textile fabrics including clothing and industrial products.

Kanecaron. Trade name of Kaneka Corporation (Japan) for modacrylic fibers.

Kanekalon. Trade name of Kaneka Corporation (Japan) for an extensive series of acrylic, modacrylic and vinyon fibers with excellent moth and mildew resistance, and good resistance to acids and alkalies. Used chiefly for clothing and industrial fabrics.

Kanelion. Trade name of Kanebo Limited (Japan) for rayon fibers and yarns used for textile fabrics.

Kanelight. Trade name of Kanebo Limited (Japan) for polyolefin fibers and yarns used for textile fabrics.

kanemite. A white mineral composed of sodium hydrogen silicate hydroxide dihydrate, $NaHSi_2O_4(OH)_2·2H_2O$. Crystal system, orthorhombic. Density, 1.93 g/cm^3; refractive index, 1.470. Occurrence: Central Africa.

kangaroo. A strong, supple, close-fibered, wear-resistant leather made from the hides of kangoroos. It takes a high polish, and is used for shoe uppers and gloves.

kangaroo calf. Calfskin finished to imitate kangaroo leather.

kangaroo hair. The protein fibers obtained from the protective hairy coat of kangaroos. They are sometimes blended with other fibers and used for weaving into bags.

kankite. A yellowish green mineral composed of iron arsenate hydrate, $FeAsO_4·3.5H_2O$. Crystal system, monoclinic. Density, 2.70 g/cm^3; refractive index, 1.666. Occurrence: Czech Republic.

kanoite. A light pinkish brown mineral of the pyroxene group composed of magnesium manganese silicate, $(Mn,Mg)_2(Si_2Oi_6)$. Crystal system, monoclinic. Density, 3.66 g/cm^3; refractive index, 1.717. Occurrence: Japan.

Kanox. Trademark of Kanox Corporation (Taiwan) for a thermally stable fibrous polyacrylonitrile with good corrosion and shrinkage resistance available in the form of fibers, threads, yarns, and woven and nonwoven fabrics. It is used as a precursor for carbon-carbon composites, and for insulation applications, reinforcements, gaskets, and protective clothing.

Kanthal. Trademark of Kanthal Corporation (USA) for a series of heating alloys composed of up to 0.05% carbon, 20-25% chromium, 5-5.5% aluminum, up to 3% cobalt, and the balance iron. Supplied in form of ribbons and wires, they have good resistance to high temperatures and oxidation up to about 1330-1425°C (2425-2600°F), high electrical resistivity, good resistance to many chemicals and good tensile strength. Used as resistance wire, and for furnace heating elements. *Kanthal Super* refers to a series of special refractory metals composed of molybdenum disilicide ($MoSi_2$) and silicon dioxide (SiO_2). They have good strength and hardness and a maximum service temperature of 1850°C (3360°F). Used for electrical heating and resistance elements in furnaces.

Kao-Bond. Trademark of Thermal Ceramics, Inc. (USA) for no-cement kaolin castables.

Kaocast. Trademark of Babcock & Wilcox Company (USA) for a refractory material composed of alumina and silica. It has a maximum service temperature of 1650°C (3000°F), and is used as a refractory.

Kaocrete. Trademark of Babcock & Wilcox Company (USA) for high-temperature kaolin refractory and insulating cements.

Kaolex. Trademark J.M. Huber Corporation (USA) for a series of Georgian and South Carolina sedimentary kaolins composed of hydrous aluminum silicates. Supplied in the form of air-floated or water-washed lumps and powders, they have a density of 2.6 g/cm^3 (0.09 $lb/in.^3$), a pyrometric cone equivalent of 33-35, and are used for ceramics and refractories, and as fillers in rubber and plastics.

kaolin. A high-quality refractory clay in the form of a white to yellowish or grayish powder that is composed chiefly of *kaolinite* (hydrous aluminum silicate) and small amounts of alkalies, iron and water, and fires to a white or almost white color. The approximate composition is $Al_2O_3·2SiO_2·2H_2O$. It has a density of 1.8-2.6 g/cm^3 (0.07-0.09 $lb/in.^3$), a high fusion point, a pyrometric cone equivalent of 34-35, high lubricity and low plasticity. Used in the manufacture of porcelain, as a filler or extender for paper and rubber, in coatings and paints, in the manufacture of refractories, ceramics, glazes, enamel frits, white cements and electrical insulators, and as a catalyst carrier. Also known as *bolus alba; china clay; white clay*. See also kaolinite; porcelain clay.

kaolin fibers. Fine, stable, high-temperature fibers composed of about 45-46% alumina (Al_2O_3), 51-53% silica (SiO_2) and 2-3% other elements, such as iron and titanium oxides. They have a density of 2.6 g/cm^3 (0.09 $lb/in.^3$), and excellent heat resistance, and are used as thermal insulation in heat-treating furnaces and ceramic kilns, for high-temperature filters, in pipe and joint protection, in sound absorption, and as reinforcing fibers.

kaolin firebrick. A refractory brick containing about 60-65% silica (SiO_2), 30-35% alumina (Al_2O_3) and up to 5% other ingredients. It has an upper service temperature of about 1250°C (2280°F) and is used for furnace and kiln linings.

kaolinite. A common, white, gray or yellowish clay mineral of the kaolinite-serpentine group that has an earthy luster and, sometimes, a reddish, brownish or bluish tint. It is composed of hydrous aluminum silicate, $Al_2Si_2O_5(OH_4)$, and has a relatively simple two-layer silicate crystal structure in which sheets of tetrahedrally coordinated silicon atoms are joined by an oxygen atom shared with octahedrally coordinated aluminum atoms. Crystal system, monoclinic. Density, 2.60-2.64 g/cm^3; hardness, 2-2.5; refractive index, 1.564. Also known as *kaolinite clay*.

kaolinite clay. See kaolinite.

Kaolith. Trademark of Babcock & Wilcox Company (USA) for monolithic kaolin refractories.

Kaomul. Trademark of Babcock & Wilcox Company (USA) for a mullite firebrick composed of 60-65% alumina (Al_2O_3) and 35-40% silica (SiO_2). It has a melting point of 1805°C (3280°F), and is used for high-temperature furnace applications.

Kao-Plas. Trademark of Thermal Ceramics, Inc. (USA) for plastic refractories based on kaolin.

Kaosil. Trademark of Harbison-Walker Refractories Company

(USA) for a semisilica firebrick made from low-alkali siliceous kaolin by firing at relatively high temperature. It is composed of about 76% silica (SiO_2) and 22% alumina (Al_2O_3), with the balance being titania (TiO_2), iron oxide (Fe_2O_3), magnesia (MgO), lime (CaO) and other alkalies. *Kaosil* firebrick has a maximum service temperature of about 1480°C (2700°F), high load-bearing capacity and excellent resistance to spalling and fluxing. Used for furnace applications.

Kao-Tab. Trademark of Thermal Ceramics, Inc. (USA) for dense, castable high-alumina refractories.

Kao-Tex. Trademark of Thermal Ceramics, Inc. (USA) for high-temperature textile products based on kaolin fibers.

Kao-Tuff. Trademark of Thermal Ceramics, Inc. (USA) for dense, abrasion-resistant, castable ceramics based on kaolin.

Kaowool. Trademark of Thermal Ceramics Inc. (USA) for a stable, lightweight, high-temperature ceramic kaolin fiber composed of 45-46% alumina (Al_2O_3), 51-53% silica (SiO_2), and 2-3% other elements. Supplied in the form of discontinuous fibers, bulk fibers, strips, boards, blocks, blankets, felts, paper, textiles and special shapes, it has a density of 2.56 g/cm^3 (0.09 lb/in.^{3}), an upper service temperature of 1650°C (3000°F), a melting point of 1760°C (3200°F), an average fiber diameter of 2.8 μm (110 μin.), and a fiber length up to 254 mm (10 in.). Used as insulation in heat-treating furnaces and ceramic kilns, for high-temperature filters, as pipe and joint protection, for sound absorption applications, and as reinforcing fibers and fabrics in composites. Kaowool TBM is a thermally bonded ceramic kaolin fiber with controlled pore size used in particular for high-temperature insulation.

Kapex. Trademark of Alcan Airex AG (Switzerland) for a range of core materials for sandwich and composite reinforcement applications.

kapok. The fine fibrous substance that surrounds the seeds of silk-cotton trees (genus *Ceiba*) found in Java. The white, fluffy fibers resemble cotton and are used as fillers for pillows, mattresses and life preservers, and as thermal insulation. Also known as *ceba; ceiba; Java cotton; silk cotton.*

Kapp. Trade name of Kapp Alloy & Wire Inc. (USA) for an extensive series of solder alloys and babbitts.

Kapprad. Trade name of Kapp Alloy & Wire Inc. (USA) for a solder composed of 40% tin, 27% zinc and 33% cadmium. It has a density of 7.64 g/cm^3 (0.276 lb/in.^{3}), a liquidus temperature of 260°C (500°F), a solidus temperature of 176°C (350°F), and is used for soldering aluminum, copper, brass, pewter, white metals, stainless steel and galvanized products, and for joining and repairing aluminum/copper radiators and heat exchangers.

KappTecZ. Trademark of Kapp Alloy & Wire Inc. (USA) for a general-purpose silver-alloy solder composed of 78% cadmium, 17% zinc and 5% silver. It has a density of 8.55 g/cm^3 (0.309 lb/in.^{3}), a liquidus temperature of 316°C (600°F), a solidus temperature of 249°C (480°F), and produces joints with good corrosion resistance, high strength, and good electrical properties. Used for joining ferrous and nonferrous metals and alloys including aluminum, copper and stainless steel.

Kapron. Trade name of JSC Kimbochko (Russia) for nylon 6 fibers and yarns used for automotive tire cords.

Kapton. Trademark of E.I. DuPont de Nemours & Company (USA) for engineering plastics based on condensation-type polyimide resins and available in rod, tube, sheet, laminate, film and fiber forms. They have a density of 1.42 g/cm^3 (0.051 lb/in.^{3}), good to medium tensile strength, high moduli of elasticity, high dimensional stability and toughness, good fatigue properties, excellent abrasion, creep and heat resistance, good resistance to alpha, beta, gamma and X-ray radiation, good resistance to ultraviolet radiation, good resistance to acids, alcohols, aromatic hydrocarbons, greases, oils and ketones, poor resistance to alkalies, good dielectric proper-ties, good high-temperature insulation properties, low thermal expansivity, a service temperature range of -270 to +320°C (-454 to +608°F), and a glass-transition temperature of 330°C (626°F). Used for automotive components, gaskets, seals, diaphragms and belts, insulated wire, circuit and instrument insulation, circuit boards, consumer electronics, components in instruments and devices, electronic components, mining and oilfield equipment, and EMI/RFI and radiation shielding applications. Film grades are used for capacitors, printed-circuit boards, and insulation.

kapur. The strong, light reddish-brown wood of the tree *Dryobalanops camphora*. It resembles *apitong* and *keruing*, has an interlocking grain, a fine texture, and a camphorous scent. Average weight, 770 kg/m^3 (48 lb/ft^3). Source: Borneo, Malaysia, Sumatra. Used for exterior joinery, fenceposts, piers, piles, marine installations and cabinetwork. Also known as *Borneo camphorwood.*

Karabuk. Trade name of Karabuk Steel (Turkey) for a constructional steel containing 0.12-0.65% carbon, 0.35-0.85% manganese, 0.15-1% silicon, and the balance iron. Used for rails, structures, gears, shafts, and machinery parts.

Karatachi. Trade name of Nippon Sheet Glass Company Limited (Japan) for a patterned glass.

Karatclad. Trade name of Enthone-OMI Inc. (USA) for a bright gold-alloy electrodeposit and plating process.

karaya gum. The gummy exudation from various species of the genus *Sterculia*, especially *S. camphanulata* and *S. urens* varying in color from white to dark brown or black. Supplied in flake and powder form, it is used in the manufacture of lacquers, varnishes and polishes, as a thickener in adhesives, and in paper coatings and textile finishes. Also known as *Indian gum; kadaya gum.*

Karbate. Trademark of Union Carbide Corporation (USA) for a series of carbon and graphite products made impervious to fluids under pressure by impregnation with chemically resistant resins (e.g., phenolics). Supplied as complete equipment items and as bricks, blocks, cylinders and tubes, they have a density of 1.7-2.0 g/cm^3 (0.06-0.07 lb/in.^{3}), good tensile and crushing strength, good thermal conductivity, good thermal-shock resistance, and good resistance to most nonoxidizing chemicals. Used for absorbers, heat exchangers, pumps, towers and valves for chemical processing equipment, pipes, tubing and fittings.

Karbon. Trade name of Fiberite Corporation (USA) for a series of reinforced high-temperature thermosets including polyacrylonitrile-, graphite- or pitch-fabric-reinforced phenolics and pitch-fabric-reinforced epoxies. Used for carbon/carbon composites for aerospace applications.

Kardel. Trademark of Union Carbide Corporation (USA) for a biaxially oriented polystyrene film used for window envelopes and packaging products.

karelianite. A black natural mineral of the corundum group composed of vanadium oxide, V_2O_3. It can also be made synthetically. Crystal system, rhombohedral (hexagonal). Density, 4.87 g/cm^3. Occurrence: Finland, USA (Arizona).

Karhueristeet. Trade name of A. Ahlström Osakeyhtiö (Finland) for insulation products composed of fine glass fibers. *Karhu* is the general group name for products of this type.

Karhula. Trade name of A. Ahlström Osakeyhtiö (Finland) for the following groups of glass fiber insulation products: (i) *Kar-hueristeet* insulation products composed of fine glass fibers; (ii) *Kontioeristeet* insulation products composed of silica fibers; and (iii) *Otsoeristeet* insulation products composed of coarse glass fibers.

karibibite. A brownish yellow mineral composed of iron arsenate, $Fe_2As_4O_9$. Crystal system, orthorhombic. Density, 4.07 g/cm³. Oc-currence: Southwest Africa.

Karma. Trade name of Driver Harris Company (USA) for a vacuum-melted resistance alloy composed of 20% chromium, 3% aluminum, 3% iron, 0.3% silicon, 0.15% manganese, 0.06% carbon, and the balance nickel. It has good strength, high electrical resistivity, and a melting point of 1400°C (2550°F). Used for shunts, resistances and potentiometers.

Karofil. Trade name of Kloeckner-Draht GmbH (Germany) for plastic-coated and sintered barbed wire products based on iron or steel.

Karolit. German trade name for a patterned glass.

Karolith. Trade name of American Plastics Corporation (USA) for a synthetic resin made from casein and formaldehyde.

Karoni. Trade name of Stahlwerke Kabel (Germany) for a series of austenitic, ferritic and martensitic stainless steels including several free-machining grades. They have good to excellent corrosion resistance, and are used for chemical and food processing equipment, tanks, mixers, valve and pump parts, turbine components, hardware, machine parts, cutlery, knives, sur-gical instruments, dyeing vats, furnace parts, and fittings.

karpatite. A red-violet synthetic mineral composed of coronene, $C_6H_2(C_4H_2)_4C_2H_2$. Crystal system, monoclinic. Density, 1.35-1.42 g/cm³; melting point, 450°C (840°F); refractive index, 1.78.

karri. The heavy, hard, dark-red wood of the tall hardwood tree *Eucalyptus diversicolor.* It is very similar to *jarrah* in appearance and mechanical properties. *Karri* wood has good mechanical properties, is very difficult to work, available in large sizes, and has moderate durability. Average weight, 880 kg/m³ (55 lb/ft³). Source: Western Australia. Used as high-class constructional timber for bridges, piers and wharves, for applications where high strength, large cross sections and long lengths are required, and for plywood. See also jarrah.

Karuta. Trade name of Central Glass Company Limited (Japan) for a figured glass with a pattern of irregularly arranged squares.

Kasema. Trade name for viscose rayon fibers and fabrics.

Kasha. Trademark for a woolen fabric in a twill weave with a decoration consisting of dark goat hair.

kasha. A soft, fine, usually twill-weave *flannel* fabric made from Tibetan goat hair. Its surface fibers have been slightly raised by brushing, and it has a lateral streak due to darker hairs. Used especially for dresses and jackets. Also known as *casha.*

Kashmilan. Trade name for acrylic fibers used for the manufacture of clothing.

Kasil. Trademark of PQ Corporation (USA) for a fine potassium silicate powder used in refractory cements and ceramic coatings.

Kasilga. Trade name for rayon fibers and yarns used for textile fabrics.

Kasle. Trade name of Kasle Steel Company (USA) for a series of tool steels including various water-hardening grades (AISI types W1, W2 and W5), tungsten-type high-speed grades (AISI types T1 through T5), molybdenum-type high-speed grades (M1 through M50), shock-resisting grades (AISI types S1 and S2), oil-hardening cold-work grades (AISI types O1 through O6), high-carbon high-chromium cold-work grades (AISI type D2

through D7), medium-carbon air-hardening cold-work grades (AISI types A2 through A8), chromium- and tungsten-type hot-work grades (AISI types H11 through H26) and carbon-tungsten-type special-purpose grades (AISI type F2).

Kasolid. Trade name of Synthetic Plastics Company (USA) for casein plastics.

kasolite. A yellow to orange, radioactive mineral of the uranophane group composed of lead uranyl silicate monohydrate, $Pb(UO_2)SiO_4 \cdot H_2O$. Crystal system, monoclinic. Density, 5.83 g/cm³; refractive index, 1.91. Occurrence: France, Central Africa.

Kasota stone. A pink or yellow, recrystallized dolomitic limestone from Minnesota, USA, having calcium and magnesium carbonate contents ranging between 48-50% and 38-40%, respectively. It is available in the form of blocks with high crushing strength, and used as a building stone for building interiors.

Kassel Wall. Trade name of Perma Paving Stone Company (Canada) for architectural stone that combines the durability of concrete with the attractive appearance of natural stone. It is supplied in several block sizes and in two colors (fieldstone and buff). Used for retaining walls, planters and steps.

kassite. A yellowish mineral composed of calcium titanium oxide hydroxide, $CaTi_2O_4(OH)_2$. Crystal system, orthorhombic. Density, 3.42 g/cm³; refractive index, 2.13. Occurrence: Russian Federation.

Kastek. Trade name of Kastek UB (Czech Republic) for a waterproof wrapping paper having a thin surface coating of polyethylene of vinyl resin.

Kasto-O-Lite. Trade name of A.P. Green Refractories, Inc. (USA) for refractory and insulating castables.

Kasumi. Trade name of Asahi Glass Company Limited (Japan) for a patterned glass.

Kasuri. Trade name of Asahi Glass Company Limited (Japan) for a glass with a pattern of incomplete rectangles.

Kasymilon. Trademark for acrylic fibers used for the manufacture of textile fabrics.

Katacolor. Trade name of Saint-Gobain (France) for a blue-green, heat-absorbing plate glass.

Kativo. Trade name of H.B. Fuller Company (USA) for plastic powders used for powder coating applications.

katoite. A white mineral of the garnet composed of calcium aluminum hydroxide, $Ca_3Al_2(OH)_{12}$. Crystal system, cubic. Density, 2.53 g/cm³; refractive index, 1.7045.

katoptrite. A black, almost opaque mineral composed of magnesium manganese iron aluminum antimony silicate, $(Mn,Mg,Fe)_{13}(Al,Fe)_4Sb_2Si_2O_{28}$. Crystal system, monoclinic. Density, 4.56 g/cm³. Occurrence: Sweden. Also spelled *catoprite.*

Kauramin. Trademark of BASF AG (Germany) for melamine-formaldehyde plastics.

Kauresin. Trademark of BASF AG (Germany) for cresol- and phenol-formaldehyde plastics.

kauri. A light-colored to brown, hard copal resin obtained from the kauri pine (*Agathis australis*) of New Zealand and found usually as a fossil in the ground, but also collected by tapping living trees. It has a density of 1.04-1.05 g/cm³ (0.037-0.038 lb/in.³), a melting point of 152-232°C (306-450°F), and is used for varnishes, lacquers, paints, organic cements, adhesives and linoleum, and as an amber substitute. Also known as *kauri copal; kauri gum; kauri resin.* See also kauri pine.

kauri pine. The yellowish-brown wood of the tall tree *Agathis australis* of the pine family. It has a straight grain, and can be mottled and figured. Source: New Zealand. Used for cabinetwork and lumber. See also kauri.

kauri resin. See kauri.

Kaurit. Trademark of BASF AG (Germany) for urea-formaldehyde resin glues.

Kavalier glass. A chemical-resistant glass with high potash content.

kawazulite. A silver- to tin-white mineral of the tetradymite group composed of bismuth telluride selenide, Bi_2Te_2Se. Crystal system, rhombohedral (hexagonal). Density, 7.50 g/cm^3. Occurrence: Japan.

Kayem. Trade name of Pasminco Europe (Mazak) Limited (UK) for rolled zinc sheet materials and several zinc die-casting alloys containing 3-5% aluminum, 2.5-3.5% copper and 0-0.1% magnesium.

Kaylo. Trade name of Owens-Corning Fiberglas (USA) for calcium silicate-based thermal insulating blocks and pipe covering materials.

Kaymat. Trade name of Kitson's Insulation Products Limited (UK) for flexible glass-fiber duct insulation.

Kaystrene. Trademark of Kaylis for modified polystyrenes.

kazakovite. A yellowish mineral composed of sodium titanium hydrogen silicate, $Na_6TiH_2Si_6O_{18}$. Crystal system, rhombohedral (hexagonal). Density, 2.84 g/cm^3; refractive index, 1.648. Occurrence: Russian Federation.

KB Star. Trade name of Pentron Laboratories (USA) for a medium-gold dental casting alloy.

keckite. A gray-brown, brown or yellow-brown mineral of jahnsite group composed of calcium magnesium manganese iron phosphate hydroxide dihydrate, $(Ca,Mg)(Mn,Zn)_2Fe_3(PO_4)_3 \cdot 2H_2O$. Crystal system, monoclinic. Density, 2.60 g/cm^3; refractive index, 1.692. Occurrence: Germany.

Keebush. Trademark of S. Kestner Evaporator & Engineering Company (USA) for a specialty polymer based on phenol-formaldehyde and having good resistance to hydrochloric and sulfuric acid. Used in the construction of pickling tanks, vats, stirrers, agitators, pumps and centrifuges.

Keene's alloy. An alloy composed of 75% copper, 16% nickel, 2.8% tin, 2.3% zinc, 2% cobalt, 1.5% iron and 0.4% aluminum, and used for corrosion-resistant parts and ornamental applications.

Keene's cement. An anhydrous, calcined or dead-burnt gypsum containing an accelerator, such as aluminum-potassium sulfate or potassium sulfate. Used as a hard, dense, white finish plaster. Also known as *flooring cement; flooring plaster; tiling plaster.*

Keen-Kut. Trade name of Associated Steel Corporation (USA) for several nontempering steels used for hollow drills.

Keenoil. Trademark for a zinc white (ZnO) used as a filler in lime soap greases.

KeepSafe. Trademark of Solutia Inc. (USA) for a tempered laminated residential window and door safety glass with good sound reduction and UV protection properties. *KeepSafe Maximum* is a variant that has been specially designed to resist forced entry, protect against flying debris and withstand hurricane-force winds up to 177 km/h (110 mph).

Keewatin. Trade name of Atlas Specialty Steels (Canada) for an oil-hardening cold-work tool steel (AISI type O1) containing 0.9% carbon, 1.2% manganese, 0.3% silicon, 0.5% chromium, 0.5% tungsten, 0.2% vanadium, and the balance iron. It has good deep-hardening and nondeforming properties and good machinability. Used for plastic molding dies, blanking, forming, hobbing, stamping and trimming dies, master tools, lathe centers, and gages.

kegelite. A white mineral composed of lead zinc aluminum silicate sulfate, $Pb_{12}Zn_2Al_4Si_{11}S_4S_{54}$. Crystal system, monoclinic. Refractive index, 1.81. Occurrence: Southwest Africa.

kehoeite. A mineral composed of calcium zinc aluminum phosphate hydride trihydrate, $(Zn,Ca)Al_2P_2H_6O_{12} \cdot 3H_2O$. Crystal system, cubic. Density, 2.34 g/cm^3. Occurrence: USA (South Dakota).

Keinmeyer's amalgam. An amalgam composed of 50% mercury, 25% zinc and 25% tin.

keithconnite. A cream to gray mineral composed of palladium telluride, $Pd_{3-x}Te$. Crystal system, rhombohedral (hexagonal). Occurrence: USA (Montana).

keiviite. (1) A natural mineral of the thortveitite group composed of yttrium silicate, $Y_2Si_2O_7$. Crystal system, monoclinic. Density, 4.05 g/cm^3. Occurrence: USA (Arizona).

(2) A synthetic mineral of the thortveitite group composed of ytterbium silicate, $Yb_2Si_2O_7$. Crystal system, monoclinic. Density, 6.15 g/cm^3.

Kelburon. Trade name of DSM Elastomers (USA) for reinforced ethylene-propylene diene monomer (EPDM) elastomers.

Kelcaloy. Trade name of M.W. Kellogg (USA) for a corrosion-resistant steel used for oil-refinery equipment.

Kel-Cast. Trade name of Kelly Foundry & Machine Company (USA) for a series of cast irons containing 3.2-3.6% carbon, 1.25-2.5% silicon, 0.7% manganese, 0.3% chromium, up to 0.4% molybdenum, and the balance iron. Used for gears, cams, bushings, valves and valve seats.

Kelcast. Trade name of Kelly Foundry & Machine Company (USA) for a series of alloy cast irons used for plungers, container molds, baffles, blanks, bottom plates, bars, and ring stock.

Kelco-gel. Trade name of Holliday Dyes & Chemicals, Limited (UK) for a purified sodium alginate of varying viscosity used in water-base and latex paints, welding-rod coatings, and in the manufacture of ceramics, films and pastes.

Keldax. Trademark of E.I. DuPont de Nemours & Company (USA) for several thermoformable, filled ethylene-based copolymer resins with outstanding acoustical sound-barrier performance. Used for automotive fascia and door panels, sandwiched flame-resistant architectural panels, agricultural panels, and in automotive carpets.

keldyshite. A white mineral composed of sodium hydrogen zirconium silicate, $(Na,H)_2ZrSi_2O_7$. Crystal system, triclinic. Density, 3.30 g/cm^3. Occurrence: Russian Federation.

Kel-F. Trademark of 3M Company (USA) for an extensive series of fluorocarbon products including polychlorotrifluoroethylene (PCTFE) and certain copolymers, and several fluoroelastomer homopolymers and copolymers. Commercially available in the form of greases, waxes, oils, dispersions, resins, extrusion and molding powders, resins, films and sheets, they have high thermal stability, good chemical and corrosion resistance, high dielectric strength, high compressive, impact and tensile strengths, and good low and high temperature properties. Used for chemical piping, fittings, connectors, valves, gaskets, seals, diaphragms, tank linings, diaphragms, wire and cable insulation, electronic components and electrical equipment.

Kelheim. Trade name of Acordis Kelheim GmbH (Germany) for rayon fibers and yarns used for textile fabrics.

Kellite. Trade name of Kellogg Switchboard & Supply Company (USA) for phenol-formaldehyde resins and plastics.

Kelly-Iron. Trade name of Kelly Foundry & Machine Company (USA) for a cast iron containing 3.2% carbon, 2% silicon, and the balance iron. Used for valves, rolls, gears, bearings and cams.

kellyite. A yellow mineral of the kaolinite-serpentine group composed of manganese aluminum silicate hydroxide, $(Mn,Al)_3$-$(Si,Al)_2O_5(OH)_4$. Crystal system, hexagonal. Density, 3.07 g/cm^3; refractive index, 1.646. Occurrence: USA (North Carolina).

Kelmar. Trademark of Macnaughton-Brooks Limited (Canada) for epoxy coatings used for garages and decks.

Kelmet. Trade name for a heavy-duty bearing alloy containing 68-70.5% copper, 22.5-25.5% lead, 6.4-6.6% tin, 0.03% sulfur, 0.03% zinc and 0.04% nickel. It has good fatigue strength, high load capacity, good elevated performance, and a maximum service temperature of 177°C (350°F). Used for main and connecting rod bearings.

Kelnod. Trade name of Kelly Foundry & Machine Company (USA) for a series of ductile cast irons that are supplied in various types and graded according to tensile strength, yield strength and elongation, respectively.

kelobra. A coarse-textured wood with very large pores and a large pronounced grain pattern, often with wavy lines. Source: Mexico, Guatemala, British Honduras. Used for furniture and cabinetry, chiefly in the form of veneers.

Kelock. Trade name of Sanderson Kayser Limited (UK) for a series of tungsten-type high-speed steels containing 0.7-1.1% carbon, 3-6% chromium, 1-18.5% tungsten, up to 9.5% molybdenum, 0.6-2% vanadium, up to 10% cobalt, and the balance iron. They have high red hardness, and good abrasion and wear resistance. Used for lathe and planer tools, reamers, broaches, form cutters, milling cutters, hobs, drills, chasers, die cores, shear blades and engraving tools.

Kelon. Trade name of Lati SpA (Italy) for mineral-filled polyamides supplied in two grades *Kelon A* and *Kelon B*.

Kelon F. Trademark of Busak & Shamban/Smiths Group (USA) for polytrifluorochloroethylene (PTFCE) plastics used for bearing applications.

Kelpol. Trade name of Reichhold Chemicals, Inc. (USA) for fast-drying alkyd copolymers used for coating applications.

Kelsol. Trade name of Reichhold Chemicals, Inc. (USA) for polyester dispersions, and polyester and polyester-acrylic resins used for coatings and baking enamels.

Keltran. Trade name of DSM Elastomers (USA) for a series of ethylene-propylene diene monomer (EPDM) elastomers with excellent ozone and weathering resistance, good resilience and flexing characteristics, good hysteresis and compression-set resistance, good resistance to chemicals including acids and alkalies, good electrical resistance and a lower continuous-use temperature of approximately -50°C (-60°F). Used for electrical insulation, automotive components, gaskets and mechanical rubber goods. *Keltran TP* refers to a series of olefin-based thermoplastic elastomers.

Kelvan. Trade name of Latrobe Steel Company (USA) for a molybdenum-type high-speed steel (AISI M33) containing 0.88% carbon, 3.75% chromium, 9.6% molybdenum, 1.7% tungsten, 1.15% vanadium, 8.4% cobalt, and the balance iron. It has good deep-hardening properties, high hot hardness and excellent wear resistance. Used for lathe tools, form tools, milling cutters, chasers, drills and taps. Formerly known as *Electrite Kelvan*.

kelyanite. A reddish brown mineral composed of mercury antimony bromine chlorine oxide, $Hg_{36}Sb_3(Cl,Br)_9O_{28}$. Crystal system, monoclinic. Density, 8.57 g/cm^3. Occurrence: Russian Federation.

Kem-Aqua. Trade name of Sherwin-Williams Company (USA) for several water-reducible baking enamels.

Kemat. Trademark of Fiber Glass Industries, Inc. (USA) for a glass-fiber mat used as reinforcement.

Kematal. Trademark of Hoechst AG (Germany) for a series of hard, stiff acetal copolymers based on polyoxymethylene and supplied as straight copolymers and in UV-stabilized, silicon-lubricated and glass-reinforced grades. They have a density of 1.41 g/cm^3 (0.051 lb/in.^{3}), excellent toughness and abrasion resistance, excellent processibility, good resistance to hot water, alcohols, aromatic hydrocarbons, greases and oils, poor resistance to acids and alkalies, and poor UV resistance (unless stabilized). Used for headphones, gears, cams, knobs and sprockets.

Kemcal. Trade name Kewaunee Scientific Corporation (USA) for a model stone used in dentistry.

Kemid. (1) Trademark of Norton Performance Plastics (USA) for a self-adhesive polyetherimide film.
(2) Trademark of Chemplast, Inc. (USA) for several polyimide compounds and products supplied as rods, sheets and stock shapes.

Kem Latex Gloss. Trademark of Sherwin-Williams Company (USA) for exterior latex house paints with glossy finish.

Kemler. British trade name for a zinc alloy containing 15% aluminum and 9% copper. Used for ornamental parts.

kemmlitzite. A brown mineral of the alunite group composed of strontium rare-earth aluminum arsenate sulfate hydroxide, $(Sr,Ln)Al_3(OH)_6(AsO_4)SO_4$. Crystal system, rhombohedral (hexagonal). Density, 3.63 g/cm^3; refractive index, 1.701. Occurrence: Germany.

kemp. A relatively short, coarse animal fiber, usually wool or hair, which has a long tapering tip and is difficult to dye or spin.

Kemper. Trade name of Gebrüder Kemper GmbH + Co. (Germany) for a series of brasses (e.g., red brasses) and bronzes (e.g., spring, lead, phosphor and tin bronzes) supplied in the form of semifinished products, sheet, strip, rods, wires, and sand and chill castings.

kempite. An emerald-green mineral composed of manganese chloride hydroxide, $Mn_2(OH)_3Cl$. Crystal system, orthorhombic. Density, 2.94 g/cm^3; refractive index, 1.695. Occurrence: USA (California).

Kemrock. Trademark of Kewaunee Scientific Corporation (USA) for an acid- and alkali-resistant sandstone with a hard surface finish made by impregnating ordinary sandstone with furfural resin and baking at prescribed temperatures. Used for counters, tabletops, chemical and laboratory equipment, and in dentistry as a die stone.

kenaf fibers. Brittle, brown-colored soft vegetable fibers obtained from the leaves and bast layer of the hemp plant *Hibiscus cannabinus* growing in India, Pakistan, Indonesia, Cuba and southern Russia. Used in making coarse fabrics and rope, as a substitute for jute, and in blends with other fibers. Also known as *ambary; Deccan hemp; kenaf.*

Kenbridge. Trade name for a high-gold dental casting alloy.

Kendcote. Trademark of Witco Chemical Corporation (USA) for several rustproofing and undercoating compounds for metal surfaces.

Kendex. Trademark of Witco Chemical Corporation (USA) for thermoplastic resins supplied in bulk, extruded, sheeted, powdered and molded form.

Kenface. Trade name of Kennametal Inc. (USA) for abrasion-resistant cemented tungsten carbides that contain cobalt-matrix binders. Used in powder form for hardfacing applications.

Kengard. Trade name of Kennametal Inc. (USA) for a composite material composed of a metallurgically bonded steel and a cemented carbide. It combines the advantages of steel, i.e., toughness and impact resistance, with those of cemented carbides, i.e., high hardness and wear resistance.

Kengrit. Trademark of Kennametal Inc. (USA) for abrasion-resistant cemented carbides with metal binders (usually cobalt), used in granule or powder form for hardfacing applications.

Kenite. Trade name of Celite Corporation (USA) for several ceramic products composed of synthetic *cordierite* and small quantities of silicon carbide and carbon.

Kenji. Trade name of Aardvark Clay & Supplies (USA) for a standard porcelain clay.

Kennametal. Trade name of Kennametal Inc. (USA) for an extensive series of hard sintered cemented carbides based on tungsten carbide usually with additions of niobium, tantalum and titanium carbides, and various binders (e.g., cobalt, nickel or platinum). Available in various degrees of hardness and shock resistance, they are used for abrasion- and wear-resistant parts, structural parts, cutting tools, and equipment for processing rock, refractories and metals.

kennedyite. A black mineral of the pseudobrookite group composed of iron magnesium titanium oxide, $Fe_2MgTi_3O_{10}$. Crystal system, orthorhombic. Density, 4.07 g/cm^3. Occurrence: Zimbabwe.

Kennertium. Trade name of Kennametal Inc. (USA) for a series of sintered, heavy tungsten alloys composed of 90-98% tungsten, up to 7.5% nickel and up to 2.5% copper. They have a density range of 17-19 g/cm^3 (0.61-0.69 $lb/in.^3$), good machinability and high strength, and are used for radiation shielding and high-inertia applications, and for counterbalances and weights.

Kent. Trade name of Time Steel Service Inc. (USA) for an air- or oil-hardening high-carbon, high-chromium tool steel (AISI type D3) containing 2.15% carbon, 0.25% manganese, 0.25% silicon, 12% chromium, 0.8% vanadium, and the balance iron. Used for dies, tools, shear blades, knives and metalworking rolls.

Kentanium. Trade name of Kennametal Inc. (USA) for a series of hard sintered carbides based on titanium carbide with cobalt or nickel-molybdenum binders. Available in various grades with up to 5% cobalt or 40% nickel-molybdenum, respectively, they are used for abrasion- and wear resistant parts, and metal forming and cutting tools and equipment.

kentrolite. A black or reddish-brown mineral that is composed of lead manganese silicate, $Pb_2Mn_2Si_2O_9$, and may contain some iron. It can also be made synthetically. Crystal system, orthorhombic. Density, 6.27 g/cm^3; refractive index, above 1.97. Occurrence: Sweden.

Kentucky. Trade name of Newport Steel Corporation (USA) for a constructional steel containing copper.

Kentucky jean. A strong *jean* fabric made with cotton warp and and wool weft threads. Used especially for pants.

Kenutuf. Trade name of J.F. Kenure (UK) for vinyl plastics.

Kenvert. Trade name of MacDermid Inc. (USA) for a chromate conversion coating and chromating solution.

kenyaite. A white mineral composed of sodium silicate hydroxide trihydrate, $NaSi_{11}O_{20.5}(OH)_4 \cdot 3H_2O$. Crystal system, tetragonal. Density, 3.18 g/cm^3; refractive index, 1.48. Occurrence: Kenya.

Keokuk Electro-Silvery. Trade name of Keokuk Electro-Metals Company (USA) for an addition agent composed of 79% iron, 6% carbon and 15% silicon, and used in the manufacture of iron and steel.

Kephos. Trade name of Henkel Surface Technologies (USA) for a phosphate coating.

Kepital. Trade name of Korea Engineering Plastics (South Korea) for polyoxymethylene (acetal) resins supplied in several grades.

Keradur. Trade name of BPB Formula (UK) for a series of pink-colored, formulated hard plasters based on calcium sulfate hemihydrate ($CaSO_4 \cdot 0.5H_2O$) produced from natural *gypsum*. They provide high compressive strength and abrasion resistance, and are used in the ceramic industries for pressing dies, and in the roof tile industry for the manufacture of molds for the Ram Press process.

Keralon. Trade name of Bruchtal GmbH (Germany) for split large-size stoneware tiles.

Keram. Trade name of BPB Formula (UK) for plasters supplied in several grades with different hardness and plaster-to-water ratios, and used in the ceramic industries for casting and jiggering molds. *Keram Rosa* refers to a pink-colored, formulated hard plaster based on calcium sulfate hemihydrate ($CaSO_4 \cdot 0.5H_2O$) produced from natural *gypsum*, and used in the ceramic industries for the manufacture of working molds for automatic clay forming processes.

Keramin. Trademark of Phoenix AG (Germany) for melamine-formaldehyde plastics.

Keramit. Trade name of Nobil-Metal (Italy) for a precious-metal dental bonding alloy.

Keramot. Trade name of Siemens Brothers & Company (UK) for hard rubber.

keratins. A class of fibrous proteins of varying hardness found in animal hair, skin, nails, wool, horn and feathers, and in human hair, skin and nails. They are rich in amino acids, especially cystine, arginine and serine, and insoluble in organic solvents, but tend to absorb and retain water. Used in biochemistry and biotechnology, in dissolvable drug coatings, and in protein hydrolyzates.

Keratol. Trade name of Atlas Powder Company (USA) for cellulose nitrate plastics.

Kerau. Trade name of Sanderson Kayser Limited (UK) for an oil-hardening tungsten high-speed steel containing 0.7% carbon, 4% chromium, 22% tungsten, 1% vanadium, and the balance iron. Used for knives, saws, drills and milling cutters. *Kerau Wunda* is a cobalt-tungsten high-speed-steel containing 0.75% carbon, 4% chromium, 0.5% molybdenum, 18% tungsten, 1.25% vanadium, 5.5% cobalt, and the balance iron. Used for tools, cutters and dies.

Kerbschnitt. German trade name for a patterned glass.

Kercast. Trademark of Lafarge Réfractaires Monolithiques (France) for castables composed of aluminum oxide, silicon carbide and carbon. Used for electric-arc furnace launders and blast-furnace runners.

Kergun. Trademark of Lafarge Réfractaires Monolithiques (France) for high-performance very-low rebond ceramic gunning mixes used for furnace applications.

Kerchsteel. Trade name for a constructional steel containing a small percentage of arsenic.

Kerimid. (1) Trademark of Rhône-Poulenc SA (France) for a series of thermosetting polyimides that consist of mixtures of several bismaleimides (BMI) with or without glass-fiber reinforcements or graphite fillers. *Kerimid 601* is a mixture of the following three BMIs: methylene dianiline bismaleimide (MDA BMI), chain-extended BMI and 2,5-diaminotoluene BMI, and *Kerimid FE 70003* is a mixture of two aromatic BMIs and a thermoplastic polymer. *Kerimid* polyimides are used as matrix resins for high-temperature composites, e.g., for aircraft and aerospace components, and as crosslinking agents.

(2) Trademark of Rhône-Poulenc SA (France) for a series of amorphous high-performance polyamide-imide (PAI) thermoplastics supplied in the form of rods, sheets and spheres. They have a density range of 1.42-1.46 g/cm^3 (0.051-0.053 lb/in.^{3}), high thermal stability, very good high-temperature characteristics, relatively high strength, good wear resistance, good radiation resistance, good resistance to acids, alcohols, ketones, aromatic hydrocarbons, greases and oils, poor resistance to alkalies, and inherent low flammability and smoke emission. Used for electrical and electronic components, machine elements, such as bearings and thrust washers, and for automotive and aerospace engine parts.

Keripol. Trademark of Phoenix AG (Germany) for unreinforced and glass-fiber-reinforced unsaturated polyester plastics.

Keripreg. Trade name of Phoenix AG (Germany) for glass-fiber reinforcements supplied in the form of prepregs.

Kerit. Trademark of Phoenix AG (Germany) for phenol-formaldehyde resins and plastics.

Kermel. Trademark of Kermel Snc (France) for polyaramid fibers and fabrics.

kermesite. A cherry-red to reddish brown mineral composed of antimony oxide sulfide, Sb_2S_2O. Crystal system, monoclinic. Density, 4.68 g/cm^3; refractive index, 2.72. Occurrence: Zimbabwe, Germany. Used as an ore of antimony. Also known as *antimony blende; pyrostibite; red antimony.*

kernite. A colorless to white mineral with a vitreous to pearly luster composed of sodium borate tetrahydrate, $Na_2B_4O_7 \cdot 4H_2O$. Crystal system, monoclinic. Density, 1.91-1.95 g/cm^3; hardness, 3 Mohs; refractive index, 1.473. Occurrence: Argentina; USA (California). Used as the chief source of borax, boron and boron compounds, and as a substitute for borax. Also known as *rasorite.*

Kern's bronze. A high-strength hydraulic bronze containing 78% copper, 12% tin and 10% zinc. Used for pressure castings, valves, fittings and vessels.

Keronyx. Trade name of Aberdeen Combworks Company (UK) for casein plastics.

Kerrofestal. German trade name for an aluminum alloy containing 1% silicon, 0.7% manganese, 0.65% manganese and 0.3% iron. It has good corrosion resistance, and is used in the chemical industry, in architecture, and for window frames.

Kerr's Fast Cure. Trade name of Kerr Dental (USA) for a self-cure acrylic resin for dental applications.

kersey. A coarse, ribbed, fabric, usually of wool or a wool blend, having a cotton warp and a fine nap. Used especially for overcoats.

keruing. The wood of various trees of the species *Dipterocarpus* growing in Borneo and on the Malay Peninsula. Similar wood from Burma and the Andaman Islands is known as *gurjun,* and that from Thailand as *yang.* The wood of these species is interchangeable for most practical purposes. Average weight, 720 kg/m^3 (45 lb/ft^3). Used as constructional timbers in railroad construction, in building construction as a substitute for oak, and for parquet flooring. Also known as *kruen.*

Kerus. Trade name of Sanderson Kayser Limited (UK) for a high-

speed steel containing 0.7% carbon, 14% tungsten, and the balance iron. Used for punches, lathe and boring tools, dies, taps and drills.

Kervit. Trademark of Industria Ceramica Veggia SpA (Italy) for tiles made by pouring a ceramic slip containing about 30% ground glass onto a bentonite- or lime-coated refractory former, firing at temperatures just below 1000°C (1830°F) and trimming the resulting products to size.

kesterite. A greenish black mineral of the chalcopyrite group composed of copper zinc tin sulfide, $Ca_2(Zn,Fe)SnS_4$. Crystal system, tetragonal. Density, 4.57 g/cm^3. Occurrence: Canada (British Columbia).

Ketac. (1) Trademark of ESPE GmbH & Co. KG (Germany) for an extensive series of dental materials including polymer-base pastes and powders for crowns, bridges, artificial teeth, prostheses and models, dental insulating materials including gypsum models, impression materials, and dental cements, fillers and lacquers. Examples include (i) *Ketac-Bond* and *Ketac-Cem* two-component (liquid plus powder) glass-ionomer dental cements, (ii) *Ketac-Fil* two-component (liquid plus powder) glass-ionomer composite filling resins, (iii) *Ketac-Molar* glass-ionomer dental cement and (iv) *Ketac-Silver* two-component (liquid plus powder), silver-reinforced glass-ionomer composite resins for dental restorative applications.

(2) Trademark of American Cyanamid Corporation (USA) for several ketone-formaldehyde resins.

(3) Trademark of Sikkens BV (Netherlands) for autobody repair and refinishing fillers and putties.

Ketjenblack EC. Trade name for a carbon black.

keto acid. A compound that has both acidic and ketonic characteristics and may either be a α-keto acid, such as α-ketobutyric acid ($C_2H_5COCOOH$), or a β-keto acid, such as β-acetoacetic acid (CH_3COCH_2COOH).

ketone. An organic compound containing one nonterminal carbonyl (C=O) group attached to two identical or different alkyl groups (R and R'). The simplest ketone is acetone, CH_3COCH_3. The general formula is RR'CO.

Ketos. Trademark of Crucible Materials Corporation (USA) for an oil-hardening cold-work tool steel (AISI type O1) containing 0.9% carbon, 1.3-1.4% manganese, 0.25-0.35% silicon, 0.5% chromium and 0.50% tungsten, and the balance iron. It has good deep-hardening and nondeforming properties and good machinability. Used for gages, reamers, hobs, dies and various other tools.

Ketron. Trademark of DSM Engineering Plastics (USA) for a series of thermoplastics based on polyetheretherketone (PEEK).

kettlecloth. A crisp, plain or colored, rather stiff cotton or polyester fabric in a plain weave and with a dull surface texture. Used especially for pants and summer jackets.

kettnerite. A brown to yellow mineral of the bismutite group composed of calcium bismuth oxide fluoride carbonate, $CaBi(CO_3)OF$. Crystal system, tetragonal. Density, 5.80 g/cm^3; refractive index, above 2.05. Occurrence: Czech Republic.

Kevlar. Trademark of E.I. Du Pont de Nemours & Company (USA) for a lightweight, aromatic polyamide (or aramid) fiber that is five times stronger than steel on an equal weight basis. *Kevlar* has a density of about 1.45 g/cm^3 (0.052 lb/in.^{3}), a high elastic modulus, high tensile strength, a low creep rate, excellent toughness and damage tolerance, high energy-absorption properties, high rigidity and stiffness, good flame resistance, high thermal stabil-ity, low thermal conductivity, a service temperature range of about -200 to +245°C (-325 to +475°F), good electrical sulating and dielectric properties, fair resistance to ultraviolet radiation, good resistance alcohols, alkalies, aromatic hydrocarbons, greases, oils, halogens and ketones, and fair resistance to dilute acids. Two common grades are *Kevlar 29* with an ultimate tensile strength of 3.62 GPa (525 ksi) and an elastic modulus of 58 GPa (8.4×10^3 ksi), and *Kevlar 49* with an ultimate tensile strength of about 3.62 GPa (525 ksi) and an elastic modulus of 120 GPa (17.4×10^3 ksi). Both fibers have high toughness and damage tolerance and are suitable for bulletproof vests and as a reinforcing fiber. *Kevlar 149* is an ultrahigh-modulus fiber for reinforcement applications. *Kevlar* fibers are used for broad-woven fabrics, coated fabric reinforcements, ropes, cordage products, cables, webbing, tapes and bulletproof vests, soft body armor, flak jackets, belting of radial tires, and as reinforcing fibers for engineering composites. See also aramid.

key alloy. A free-cutting copper alloy composed of 20.2% zinc, 12.7% nickel, 1.1% lead, 0.4% iron and 0.15% tin, and used for hardware and keys.

Keybak. Trademark of Johnson & Johnson Inc. (USA) for nonwoven fabrics.

key brick. A tapered or wedge-shaped brick used at the crown of an arch for closing and tightening applications.

keyite. A deep sky-blue mineral composed of cadmium copper zinc arsenate, $(Cu,Zn,Cd)_3(AsO_4)_2$. Crystal system, monoclinic. Density, 4.20 g/cm³. Occurrence: Southwest Africa.

Keykote. Trade name of Allied-Kelite Division, Witco Corporation (USA) for a phosphating compound used in metal finishing operations.

Key Largo. Trademark of Tarmac America (USA) for interlocking paving stones used for commercial applications.

Keysor. Trade name of Keyson-Century (USA) for several plasticized polyvinyl chlorides available in three elongation ranges: up to 100%, 100-300%, and above 300%. *Keysor UPVC* are unplasticized polyvinyl chlorides available in standard, crosslinked, high-impact, UV-stabilized and structural-foam grades.

keystock. A free-cutting copper alloy containing 22-26% zinc, 12% nickel and 1-2% lead used for hardware and keys.

Keystone. (1) Trademark of British Columbia Forest Products Limited (Canada) for cedar siding, plywood, and spruce and hemlock dimension stock.

(2) Trade name of US Steel Corporation (USA) for copper-bearing low-carbon steels with improved resistance to atmospheric corrosion used for structural applications, railroad cars, and roofing.

(3) Trade name of Kidd Drawn Steel (USA) for oil-hardening cold-work tool steel used for drill rod, punches, gauges and dies.

(4) Trade name of Permacon (Canada) for an architectural stone resembling natural stone in appearance. It is supplied in straight and curved blocks in brown, gray, beige and glacier-blue colors for use in retaining walls, garden walls, etc.

keystone. Small broken stone used in road construction to fill the voids between coarse aggregate prior to impregnation with bituminous material. Also known as *filler stone*.

Kézite. Trade name of Keramos (USA) for a series of advanced piezoelectric ceramics including several modified bismuth titanates, lead metaniobates and lead zirconate titanates (PZTs). Depending on their composition and properties, they are used for the manufacture of transducers, generators, accelerometers, hydrophones, thickness gages and flaw detectors.

K-Flex ECO. Trade name of RBX Corporation (USA) for a closed-cell, halogen-free elastomeric insulation material.

K-Grain. Trade name for magnesia (98% pure) supplied in the form of a white powder of varying mesh size, and with silica (SiO_2) contents of 0.4% or less.

khaki. A stout, lightweight, sand- to medium-brown-colored cotton fabric in a twill weave, used for uniforms and casual clothing.

khamrabaevite. A dark gray mineral composed of titanium carbide, TiC. It can also be made synthetically. Crystal system, cubic. Density, 4.91 g/cm³. Occurrence: USA (Pennsylvania).

khanneshite. A pale yellowish or nearly colorless mineral composed of sodium barium calcium rare-earth strontium carbonate, $(Na,Ca)_3(Ba,Sr,Ln,Ca)_3(CO_3)_5$. Crystal system, hexagonal. Density, 3.85 g/cm³; refractive index, 1.622. Occurrence: Afghanistan.

khaya. A class of light pinkish-brown to dark reddish-brown African hardwoods belonging to the genus *Khaya*. Most African mahogany is from red mahogany *(K. ivorensis)* and dry-zone mahogany *(K. senegalensis)* that closely resemble American mahogany, but have slightly coarser textures, more pronounced grain patterns and greater shrinkage. *Khaya* machines and finishes well, and has moderate stability and decay resistance. Source: West, East and Central Africa. Used for fine furniture, interior paneling, boat construction, store fixtures, art objects, veneer, and plywood. See also African mahogany, dry-zone mahogany.

khibinskite. A white mineral composed of potassium zirconium silicate, $K_2ZrSi_2O_7$. It can also be made synthetically. Crystal system, monoclinic. Density, 3.40 g/cm³; refractive index, 1.715. Occurrence: Russian Federation.

khinite. A dark green mineral composed of copper lead tellurate hydroxide, $Cu_3PbTeO_4(OH)_6$. Crystal system, orthorhombic. Density, 6.50 g/cm³; refractive index, 2.112. Occurrence: USA (Arizona).

kick-plate brass. A free-cutting brass containing 84% copper, 15% zinc and 1% lead. Used for hardware, water pipes and fittings.

kid leather. A thin, soft, supple leather with a fine, close grain made from the skin of immature goats. It is usually chrome- or vegetable-tanned, strong and wear-resistant, has a naturally glossy surface, and is used for gloves, shoes, pocketbooks and linings. Also known as *kid*.

kidney stone. See nephrite.

kidwellerite. A green to yellow mineral composed of sodium iron phosphate hydroxide pentahydrate, $NaFe_9(PO_4)_6(OH)_{10} \cdot 5H_2O$. Crystal system, monoclinic. Density, 3.04 g/cm³; refractive index, 1.800. Occurrence: USA (Arkansas).

kieselguhr. See diatomaceous earth.

kieserite. A colorless or white mineral composed of magnesium sulfate monohydrate, $MgSO_4 \cdot H_2O$. It can also be made synthetically. Crystal system, monoclinic. Density, 2.58 g/cm³; refractive index, 1.533. Occurrence: Austria, France, Germany, India. Used as a source of magnesium.

Kikansei. Trade name for rayon fibers and yarns used for textile fabrics.

Kiku. Trade name of Asahi Glass Company Limited (Japan) for a patterned glass with lace effect.

kilchoanite. A colorless mineral composed of calcium silicate, $Ca_3Si_2O_7$. Crystal system, orthorhombic. It can also be made synthetically from calcium silicate and quartz under prescribed conditions, but has a slightly different formula, $Ca_6(SiO_4)(Si_3O_{10})$. Density of natural mineral, 2.99 g/cm³. Occurrence:

Scotland.

Kilgard. Trademark of Clayburn Refractories Limited (Canada) for acid-resistant paving stones.

killalaite. A mineral composed of calcium hydrogen silicate hydroxide, $Ca_{3.2}(H_{0.6}Si_2O_7)(OH)$. Crystal system, monoclinic. Density, 2.94 g/cm³. Occurrence: Ireland.

killed steel. A steel whose oxygen content has been reduced by the ladle addition of a strong deoxidizer, such as aluminum, manganese, silicon or calcium. Deoxidizing is carried to the point where no reaction takes place between carbon and oxygen as the metal solidifies. Since there is no evolution of gas from the solidifying mass, the metal lies quietly in the ingot mold and is called "killed steel." *Killed steel* is high in carbon, has a homogeneous microstructure with few entrapped inclusions, and is very clean. Also known as *fully deoxidized steel.*

kiln-dried lumber. Lumber whose moisture content has been reduced to a specific percentage by placing it in a heated chamber, known as a kiln or oven, for a prescribed period of time. The kiln temperature is usually between 101 and 105°C (214 and 221°F). The temperature, relative humidity and air circulation in the kiln or oven are closely controlled during this procedure. Also known as *oven-dried lumber.*

Kilnoise. Trade name of Pfizer Industries (USA) for an *acoustical plaster.*

kiln wash. A coating or slurry usually composed of refractory materials, such as clay and silica, applied to the linings, shelves and furniture of kilns to protect them against combustion gases, volatile glazes or glaze drops from the ceramic ware being fired.

Kimax. Trademark of Owens-Illinois, Inc. (USA) for a borosilicate glass supplied in the form of tubes and rods. It has a low coefficient of thermal expansion, excellent chemical resistance, high toughness, good heat resistance and good thermal-shock resistance. Used for heat-resistant glassware, chemical and laboratory ware and ovenware.

kimkhab. A silk fabric woven or decorated with silver and/or gold threads.

kimzeyite. A brown mineral of the garnet group composed of calcium zirconium aluminum silicate, $Ca_3(Zr,Fe,Ti)_2(Al,Si,Fe)_3O_{12}$. Crystal system, cubic. Density, 4.00 g/cm³; refractive index, 1.94. Occurrence: USA (Arkansas).

Kind. Trade name of Kind & Co. Edelstahlwerk (Germany) for a series of steels including case-hardening and heat-treatable steels, acid-, corrosion- and heat-resistant steels, high-speed steels, water-hardening tool steels, cold-work and hot-work die and tool steels, and several mold steels for die-casting dies and plastic molds.

Kinel. Trademark of Rhône-Poulenc SA (France) for a series of bismaleimides and polyimides. The polyimides are supplied in unmodified, graphite-, polytetrafluoroethylene- and molybdenum disulfide-lubricated and glass-reinforced grades, have high thermal stability and radiation resistance, and inherent low flammability and smoke emission. Used for adhesives.

King Cobalt. Trade name of Jessop Steel Company (USA) for a tungsten-type high-speed steel (AISI type T6) containing 0.8% carbon, 4.5% chromium, 1.5% vanadium, 20% tungsten, 12% cobalt, and the balance iron. It has good deep-hardening properties and excellent wear resistance and hot hardness. Used for a variety of high-speed cutting and finishing tools.

Kinghorn metal. A British high-strength brass containing 58.5% copper, 39.4% zinc, 1.1% iron and 1% tin. Used for screws, nuts and hardware.

kingite. A white mineral composed of aluminum phosphate hydroxide nonahydrate, $Al_3(PO_4)_2(OH_3)\cdot9H_2O$. Crystal system, triclinic. Density, 2.30 g/cm³; refractive index, 1.514. Occurrence: South Australia.

kingnut hickory. See shellbark hickory.

king's blue. See cobalt blue.

king's green. See copper acetoarsenite.

kingsmountite. A white to very light brown mineral composed of calcium iron aluminum phosphate hydroxide dodecahydrate, $Ca_4FeAl_4(PO_4)_6(OH)_4\cdot12H_2O$. Crystal system, monoclinic. Density, 2.51 g/cm³; refractive index, 1.581. Occurrence: USA (North Carolina).

Kingston Bronze. Trade name of IMI Kynoch Limited (UK) for a high-strength bronze containing 83% copper, 12.5% zinc, 4% tin and 0.5% iron, and used for condenser tubes.

king's yellow. See arsenic sulfide (iii).

kinichilite. A dark brown mineral of the zeolite group composed of hydrogen sodium iron magnesium zinc tellurite trihydrate, $(Fe,Mg,Zn)_2(TeO_3)_3(H,Na)_2\cdot3H_2O$. Crystal system, hexagonal. Density, 3.96 g/cm³; refractive index, above 1.8. Occurrence: Japan.

Kinite. Trade name of Kinite Corporation (USA) for an abrasion- and compression-resistant steel containing 1.5% carbon, 12.5-14.5% chromium, 1.1% molybdenum, 0.7% cobalt, 0.55% silicon, 0.5% manganese, 0.4% nickel, up to 0.5% boron, and the balance iron. Used for mandrels, cutters, blanking, drawing and forming dies, shear blades, and various other tools.

Kinnalloy. Trade name of Kinney Iron Works (USA) for a cast iron containing 3.3% carbon, 2.5% silicon, 1.5% nickel, 0.8% chromium, and the balance iron. Used for castings.

Kinnite. Trade name of Kinney Iron Works (USA) for an abrasion- and wear-resistant cast iron containing 2.5% carbon, 1.5% chromium, and the balance iron. Used for castings.

kinoite. A blue mineral composed of calcium copper silicate dihydrate, $Ca_2Cu_2Si_3O_{10}\cdot2H_2O$. Crystal system, monoclinic. Density, 3.16 g/cm³; refractive index, 1.665. Occurrence: USA (Arizona).

Kinonglas. Trade name of Sekurit-Glas Union GmbH (Germany) for a laminated safety glass composed of two transparent sheets of glass with an interlayer of cellulose nitrate plastic. Used for protective shields on metalworking machinery, and protective covers on machines and equipment.

kinoshitalite. A yellow-brown mineral of the mica group composed of barium magnesium aluminum silicate hydroxide, $BaMg_3Al_2Si_2O_{10}(OH)_2$. Crystal system, monoclinic; density, 3.30 g/cm³; refractive index, 1.633. Occurrence: Japan.

Kintrel. Trademark of Kimex SA (Mexico) for polyester fibers and filament yarns.

kip leather. (1) Leather made from the skin of young or undersized animals.

(2) Light, supple, fine-grained, usually chrome- or vegetable-tanned leather made from the skin of young cattle (calves), and used for shoe uppers, and linings.

Kirara. Trade name of Nippon Sheet Glass Company Limited (Japan) for a patterned glass.

Kirklon Superbulk. Trade name for nylon fibers and yarns used for textile fabrics including clothing.

Kirksyl. Trade name for rayon fibers and yarns used for textile fabrics.

kischsteinite. A yellowish-gray mineral of the olivine group composed of calcium iron silicate, $CaFeSiO_4$. It can be made synthetically from calcium carbonate, ferric oxide and silicon dioxide. Crystal system, orthorhombic. Density, 3.43 g/cm³; refractive index, 1.734.

kiri. The strong, lightweight wood of the medium-sized oriental tree *Paulownia tomentosa*. This species is now naturalized in the southern United States under the names of *empress tree, paulownia* and *princess tree*. Kiri has a coarse grain and good resistance to warping. Average weight, 264 kg/m³ (16 lb/ft³). Source: Eastern Asia (especially Japan); southern United States. Used for lightweight construction, crates, floats and instruments.

Kinsalloy. Trade name of International Nickel Inc. (USA) for a series of heat-resistant alloys containing 21-24% molybdenum, 7-8% aluminum, up to 0.1% carbon, and the balance nickel. Used for high-temperature applications, such as turbine rotor blades and jet-engine parts.

Kin-Shibu-Ichi. Japanese trade name for a corrosion-resistant copper alloy composed of 33.3% of the alloy *Shaku-Do* (94-96% copper, 3.7-4.2% gold, 0.1-1.6% silver and traces of arsenic, iron and lead) and 66.7% of the alloy *Shibu-Ichi* (51.1-67.3% copper, 32.1-48.9% silver and traces of gold and iron). Used for jewelry and ornaments.

Kirkalloy. Trade name of NL Industries (USA) for a zinc die-casting alloy containing 8% copper, 3% aluminum, 0.04% beryllium and up to 0.1% iron, 0.007% lead, 0.003% tin and 0.003% cadmium, respectively. Used for forming dies.

Kirkcase. Trade name of Kirkstall Forge Engineering Limited (UK) for a free-machining case-hardening steel containing 0.1% carbon, 1% manganese, 0.2% sulfur, 0.06% phosphorus, and the balance iron. Used for machine parts, and fasteners.

Kirksite. Trade name of NL Industries (USA) for a zinc die-casting alloy containing 3.5% copper, 4% aluminum and 0.04% magnesium. It has a melting point of 380°C (717°F), and is used for stamping, blanking, trimming, drawing and forming dies, and press tools.

kish graphite. A nearly perfect crystalline form of excess carbon precipitated as a flake from carbon-rich iron melts (e.g., hypereutectic cast iron) on cooling. For many applications, it must be purified to remove iron impurities.

Kiski. Trade name of CCS Braeburn Alloy Steel (USA) for an oil-hardening cold-work tool steel (AISI type O1) containing 0.9-1% carbon, 1-1.2% manganese, 0.5% chromium, 0.2% vanadium, 0.45-0.75% tungsten, and the balance iron. It has good machinability and nondeforming properties, and is used for tools, blanking and forming dies, and gages.

Kissock. Trade name of Bonney-Floyd Company (USA) for a wear-resistant steel containing about 0.7% carbon, 2% nickel, 1.3-1.6% chromium, 0.4% molybdenum, and the balance iron. Used for wear plates and castings.

kitkaite. A pale yellow mineral of the brucite group composed of nickel selenide telluride, NiSeTe. Crystal system, hexagonal. Density, 7.22 g/cm³. Occurrence: Finland.

kittatinnyite. A bright yellow mineral composed of calcium manganese silicate hydroxide hydrate, $Ca_4Mn_6Si_4O_{16}(OH)_8·18H_2O$. Crystal system, hexagonal. Density, 2.61 g/cm³; refractive index, 1.727. Occurrence: USA (New Jersey).

kittul fiber. A fiber obtained from the leafstalks of a species of caryota palm (*Caryota urens*) and used for making brooms, baskets and whisks.

kivuite. A yellow, radioactive mineral of the phosphuranylite group composed of thorium uranyl hydrogen phosphate hydroxide heptahydrate, $(Th,Ca,Pb)H_2(UO_2)_4(PO_4)_2(OH)_8·7H_2O$. Crystal system, orthorhombic. Refractive index, 1.654. Occurrence: Central Africa.

K-Karb. Trademark of Kaiser Aerotech (USA) for lightweight carbon/carbon-fiber fabric-reinforced structural graphite composites. The fibers of the fabric may be based on rayon, pitch or polyacrylonitrile and its matrix on pitch or pitch plus a synthetic resin. They have very high impact strength, high spalling resistance, good moldability, excellent resistance to temperature fluctuations and low coefficients of thermal expansion. Used for aerospace and electronic applications.

kladnoite. A mineral composed of phthalimide, $C_6H_4(CO)_2NH$. It can also be made synthetically. Crystal system, monoclinic. Density, 1.47 g/cm³; refractive index, 1.519. Occurrence: Czech Republic.

KleanLime. Trademark of Global Stone James River (USA) for limestone supplied in pulverized and pelletized grades for acid soil treatment and soil conditioning applications.

Kleanrol. Trademark of Whitco Chemical Company, Inc. (USA) for a soldering flux made from ammonium chloride and zinc chloride.

klebelsbergite. A pale yellow to orange-yellow mineral composed of antimony oxide sulfate hydroxide, $Sb_4O_4(OH)_2SO_4$. Crystal system, orthorhombic. Density, 4.62 g/cm³; refractive index, 1.95. Occurrence: Rumania.

kleemanite. A colorless mineral composed of zinc aluminum phosphate hydroxide trihydrate, $ZnAl_2(PO_4)_2(OH)_2·3H_2O$. Crystal system, monoclinic. Density, 2.75 g/cm³. Occurrence: South Australia.

Kleenpeel. Trade name of Beck Chemicals, Inc. (USA) for a strippable polyvinyl chloride coating.

Klegecell. Trade name of Diab Group (USA) for lightweight structural polyvinyl chloride foam core materials.

Kleiberit. Trade name of Klebchemie M.G. Becker GmbH & Co. KG (Germany) for a series of adhesives, glues and liquid polymeric compounds.

kleinite. A yellow or orange mineral composed of mercury nitrogen chloride hydrate, $Hg_2Cl(N,SO_4)·0.3H_2O$. It can also be made synthetically. Crystal system, hexagonal. Density, 8.00 g/cm³; refractive index, 2.18. Occurrence: USA (Texas).

Klem Kote. Trade name of Stan Sax Corporation (USA) for a phosphate coating and coating process.

Klingenberg clay. A high-plasticity clay that contains about 60% silica (SiO_2) and is found near Klingenberg, Bavaria, Germany. Used as a binder in the manufacture of grinding wheels, graphite crucibles and pencil leads.

Klinger. Trademark of Richard Klinger GmbH (Germany) for asbestos supplied in the form of compressed sheets for gaskets and packings. Also included under this trademark are various sealing, jointing and packing materials and products based on polytetrafluoroethylene (Teflon).

Klingerit. Trademark of Richard Klinger GmbH (Germany) for heat-resistant asbestos-rubber sealing and jointing compounds. Also included under this trademark are a wide range of other sealing, jointing and packing materials and products based on metals (e.g., copper), graphite, polytetrafluoroethylene or metal-asbestos or metal-elastomer combinations.

klockmannite. A slate gray mineral composed of copper selenide, α-CuSe. It can also be made synthetically. Crystal system, hexagonal. Density, 5.99 g/cm³. Occurrence: Argentina, Germany, Sweden.

Klöckner. Trade name of Klöckner-Werke AG (Germany) for an extensive series of steels including carbon and low-alloy structural steels, high-grade carbon and alloy steels, chain and spring steels, high-temperature steels, tool steels, hot-rolled wide strip steels and tempered prestressing steel. Also included under this trade name are various semifinished steel products, structural shapes, bars, floor plate, hot-rolled heavy plate, hot- and cold-rolled medium plates, hot- and cold-rolled sheets in panel or coil forms, and electroplated and electrogalvanized sheet products.

K-Monel. Trade name of Huntington Alloys Inc. (USA) for a wrought, nonmagnetic, corrosion-resistant, precipitation-hardening alloy composed of 63-66.5% nickel, 27-33% copper, 2.3-3.2% aluminum, 0-2% iron, 0-1.5% manganese, 0.35-0.85% titanium, 0-0.5% silicon, 0-0.25% carbon and 0-0.01% sulfur. It can be hardened after finishing by a long-time age-hardening treatment to obtain a hardness and tensile strength superior to that of standard *Monel* and almost as high as stainless steel. *K-Monel* has low sparking properties, a service temperature range of -73 to +371°C (-100 to +700°F), and is used for valves, pumps, springs, pump rods and shafts, valve stems, impellers, marine propeller shafts, valve trim, oil-well components, fasteners, electronic components, doctor blades and scrapers, wearing sleeves, and aeronautical instruments.

K-Mute. Trade name of Kawasaki Steel Corporation (Japan) for a sound-dampening high-strength carbon steel containing 1-3% aluminum and up to 2% copper. It has good workability and weldability, and is used for various applications in building construction, civil engineering and shipbuilding.

Kneiss metal. An antifriction metal composed of 40-50% zinc, 25-42% lead, 15-25% tin and 0-3% copper, used for bearings.

knitted fabrics. Fabrics made by interlooping yarn or thread by hand or machine. Also known as *knit fabrics; knits.*

knit tubing. A seamless sheath-type fabric made on a circular knitting machine, and used for woman's dresses and skirts, and for sleeves.

knop yarn. See nub yarn.

knorringite. A deep green mineral of the garnet group composed of magnesium chromium silicate, $Mg_3Cr_2(SiO_4)_3$. It can also be made synthetically from magnesium oxide (MgO), chromic oxide (Cr_2O_3) and silicon dioxide (SiO_2) under prescribed conditions. Crystal system, cubic. Density, 3.76 g/cm³; refractive index, 1.803.

knotted yarn. See nub yarn.

knotty pine. See lodgepole pine.

K-N Plastic. Trade name of National Refractories & Minerals Corporation (USA) for a *chrome ore* used in the manufacture plastic refractories.

Knytex. Trademark of Owens-Corning Fiberglass (USA) for a range of glass-fiber fabrics including woven unidirectional rovings and woven and knitted multiaxial fabrics.

koa. The wood of the tall tree *Acacia koa* found exclusively on the Hawaiian Islands. It has a fiddleback figure, takes a fine finish, and has good resonant qualities. Used for musical instruments, furniture and paneling.

Koala. Trade name of Australian Window Glass Proprietary Limited for sheet and cast glass products.

koashvite. A pale yellow mineral of the combeite group composed of sodium calcium manganese iron titanium silicate monohydrate, $Na_6(Ca,Mn)(Ti,Fe)Si_6O_{18} \cdot H_2O$. Crystal system, orthorhombic. Density, 2.98 g/cm³; refractive index, 1.643. Occurrence: Russian Federation.

Kobalt. (1) Trade name of Friedrich Lohmann GmbH (Germany) for a series of cobalt-tungsten high-speed tool steels with high hot hardness, excellent wear resistance and good machinability. Used for lathe and planer tools, milling cutters, and a variety of other cutting and finishing tools.

(2) Trade name of Thyssen Edelstahlwerke AG (Germany) for a series of permanent magnet steels typically containing 17-36% cobalt.

kobeite. A dark brown mineral of the columbite group composed of yttrium iron titanium niobium oxide hydroxide, $(Y,Fe)(Ti,Nb)_2(O,OH)_6$. Crystal system, orthorhombic. Density, 5.00 g/cm³; refractive index, 2.205. Occurrence: Japan, New Zealand.

kobellite. A lead gray mineral of the lillianite group composed of lead antimony bismuth sulfide, $Pb_5(Bi,Sb)_8S_{17}$. Crystal system, orthorhombic. Density, 6.48 g/cm³. Occurrence: Sweden, Czech Republic.

Kobitalium. British trade name for several cast and wrought aluminum alloys containing 1-5% copper, 0.2-2% nickel, 0.25-2% manganese, 1-2% iron, 0.5-2% silicon, 0.4-2% magnesium and 0.08-0.12% titanium. Used for light-alloy parts.

Koblend. Trade name of Enichem SpA (Italy) for polycarbonate/acrylonitrile-butadiene-styrene alloys.

Koblum. Trademark of Maple Leaf Plastics Corporation (Canada) for a series of white-metal bearing alloys.

Kocetal. Trademark of Kolon Company (USA) for engineering thermoplastics based on acetal copolymers and supplied in general-purpose, high-impact, low-friction, antistatic and reinforced grades. They have excellent stiffness, strength and wear resistance, excellent resistance to organic chemicals, detergents, fats, oils and inorganic chemicals except strong acids, an upper continuous-use temperature of 100-120°C (210-250°F), and good and stable electrical properties.

Koch white gold. A corrosion-resistant alloy containing 75% or more gold, 3.3-24.75% nickel and 0.25-5% manganese. Used for ornaments and jewelry.

Kodachi. Trade name of Nippon Sheet Glass Company Limited (Japan) for a patterned glass with a twig design.

Kodachrome. Trade name of Eastman Chemical Company (USA) for cellulose acetate plastics used for photographic film.

Kodafilm. Trade name of Eastman Kodak Company (USA) for cellulose nitrate film materials.

Kodaloid. Trade name of Eastman Kodak Company (USA) for cellulose nitrate film materials.

Kodapak. (1) Trademark of Eastman Chemical Company (USA) for a lightweight, food-grade, crystalline thermoplastic polyester resin based on polyethylene terephthalate. It has high strength and good durability and shatter resistance, and is used for food-processing equipment and containers for storing and/or cooling food.

(2) Trademark of Eastman Kodak Company (USA) for thin, transparent cellulose acetate film and sheeting materials used for packaging applications.

Kodar. Trademark of Eastman Chemical Company (USA) for a food-grade thermoplastic polyester resin based on polyethylene terephthalate and supplied in various grades including amorphous, crystalline, glass fiber-reinforced, high-impact, supertough, fire-retardant, and mineral-filled and glycol-modified. It has high clarity, high toughness, excellent chemical and impact resistance and good thermal resistance. Used for safety equipment, household appliances, lighting equipment and fixtures, and blister packs.

Kodel. Trademark of Eastman Chemical Company (USA) for polyester fibers.

Kodofill. Trademark of Eastman Chemical Products (USA) for compression-resistant fiberfill with extra loft consisting of polyester fibers with hollow cross sections. Used for bedspreads, comforters, pillows, etc.

Kodosoff. Trade name of Eastman Chemical Products (USA) for polyester fiberfill used for pillows.

koechlinite. A yellow or greenish-yellow mineral composed of bismuth molybdenum oxide, Bi_2MoO_6. It can also be made synthetically. Crystal system, orthorhombic. Density, 8.26 g/cm^3; refractive index, 2.61. Occurrence: Germany.

Koeller's alloy. A tin-lead casting alloy that contains a small percentage of bismuth and is used for domestic utensils and ornaments.

koenenite. A soft, usually pale yellow or deep red mineral composed of magnesium aluminum hydroxide chloride, Mg_5Al_2-$(OH)_{12}Cl_4$. Crystal system, rhombohedral. Density, 1.98 g/cm^3; refractive index, 1.52. Occurrence: Germany.

Koerflex. Trade name of Krupp Stahl AG (Germany) for a series of permanent magnet alloys containing 30-52% cobalt, 36-55% iron, and up to 15% chromium and 10% vanadium, respectively. Usually supplied in strip and wire form, they possess high remanent magnetization, high external permeance and good tensile strength. Used for electromechanical devices, generators, motors, hysteresis motors, electric and magnetic equipment, speedometers and recorders.

Koerzit. Trade name of Krupp Stahl AG (Germany) for a series of permanent magnet alloys of the iron-cobalt-nickel, cobalt-iron-vanadium, cobalt-iron-vanadium-chromium, iron-nickel-aluminum or iron-nickel-cobalt-aluminum types. They have high magnetic permeability and are used for electromechanical devices, generators, motors, and electric, electronic and magnetic equipment.

koettigite. A light carmine or peach-blossom red mineral of the vivianite group that is composed of cobalt nickel zinc arsenate octahydrate, $Zn_3(AsO_4)_2 \cdot 8H_2O$, and may also contain some cobalt and nickel. It can also be made synthetically. Crystal system, monoclinic. Density, 3.33 g/cm^3; refractive index, 1.638. Occurrence: Germany.

kogarkoite. A white mineral composed of sodium fluoride sulfate, Na_3FSO_4. Crystal system, monoclinic. Density, 2.68 g/cm^3. Occurrence: Russian Federation, USA (Colorado).

Kohinor. Trade name of Rimtec (USA) for several vinyl resins.

Kohorn. Trade name of Oscar von Kohorn & Company (Germany) for viscose rayon fibers and yarns containing casein.

Kojin. Trade name of Kojin Corporation (Japan) for rayon fibers and yarns used for textile fabrics.

kokrodua. See afrormosia.

koktaite. A colorless to white mineral composed of ammonium calcium sulfate monohydrate, $(NH_4)_2Ca(SO_4)_2 \cdot H_2O$. Crystal system, monoclinic. Density, 2.09 g/cm^3; refractive index, 1.532. Occurrence: Czech Republic.

Kolamax. Trademark of Iko Industries Limited (Canada) for a kraft building paper.

Kolassal. Trade name of Hoffmann & Co. KG (Germany) for a series of air- or oil-hardening, nonshrinking tool steels containing 2.1% carbon, 11.5% chromium, 0.3% manganese, and the balance iron. Kolassal tool steels are also available in special grades including *Kolassal Extra* with 2.1% carbon, 11.5% chromium, 0.7% tungsten, and the balance iron, and *Kolassal Supra* with 1.65% carbon, 11.5% chromium, 0.1% vanadium, and the balance iron. They are used for punches, and blanking and forming dies.

kolbeckite. A colorless, blue, or gray mineral of the variscite group composed of scandium phosphate dihydrate, $ScPO_4 \cdot 2H_2O$. Crystal system, orthorhombic. Density, 2.36 g/cm^3; refractive index, 1.590. Occurrence: Germany, USA (Utah).

kolfanite. A red mineral composed of calcium iron arsenate dihydrate, $Ca_2Fe_3O_2(AsO_4)_3 \cdot 2H_2O$. Crystal system, monoclinic. Density, 3.76 g/cm^3; refractive index, 1.923. Occurrence: Russian Federation.

kolicite. A bright yellow-orange mineral composed of manganese zinc arsenate silicate hydroxide, $Mn_7Zn_4(AsO_4)_2(SiO_4)_2$-$(OH)_8$. Crystal system, orthorhombic. Density, 4.17 g/cm^3; refractive index, 1.786. Occurrence: USA (New Jersey).

Kolon. Trademark of Kolon Industries Inc. (South Korea) for nylon 6 and polyester fibers and yarns.

Kolorbon. Trade name for rayon fibers and yarns used for textile fabrics.

Kolor-N-Kote. Trade name of McGean & Rohco Inc. (USA) for several water-reducible paints.

kolovratite. A greenish yellow mineral composed of zinc nickel vanadium oxide hydroxide, Zn-Ni-V-O-OH. Refractive index, 1.577. Occurrence: Turkestan.

kolwezite. A beige mineral of the rosalite group composed of cobalt copper carbonate hydroxide, $(Cu,Co)_2(CO_3)(OH)_2$. Crystal system, monoclinic. Refractive index, 1.688. Occurrence: Zaire.

kolymite. A tin-white mineral composed of copper mercury, Cu_7-Hg_6. Crystal system, cubic. Density, 13.00 g/cm^3. Occurrence: Russian Federation.

Komacel. Trademark of Komcraft, Inc. (USA) for a rigid integral polyvinyl chloride foam supplied as plates, profiles, sections, rods and shapes.

Komadur. Trademark of Komcraft, Inc. (USA) for several rigid polyvinyl chloride plastics supplied as bands, plates, sheets, foil sections, rods and tapes.

Komapor. Trademark of Komcraft, Inc. (USA) for solid-foam polyvinyl chloride sheeting.

komarovite. A mineral composed of calcium manganese niobium hydrogen fluoride silicate hydroxide, $(H,Ca)_2Nb_2Si_2O_{10}(OH,F) \cdot H_2O$. Crystal system, orthorhombic. Density, 3.00 g/cm^3; refractive index, 1.766. Occurrence: Russian Federation.

Komatex. Trademark of Komcraft, Inc. (USA) for polyvinyl chloride foam sheeting.

Komo. Trade name of Thyssen Deutsche Edelstahlwerke AG (Germany) for a series of high-speed steels containing 0.8-1.25% carbon, 6-9.5% tungsten, 3.1-5% molybdenum, 4-4.2% chromium, 2-3.2% vanadium, 5-10% cobalt, and the balance iron. They have good to moderate wear resistance and good to excellent red hardness. Used for lathe and planer tools, reamers, hobs, broaches, drills, milling cutters, and chasers.

Komponit. Trademark for urea- and melamine-formaldehyde resins, compounds and laminates.

Konal. Trade name for an acid- and heat-resistant alloy composed of 17.5% cobalt, 6.5% iron, 2.5% titanium, 0.2% manganese, and the balance nickel. It has high strength at ordinary and elevated temperatures and is used for electrical and electronic components.

Konalloy. Trade name of Koerver & Nehring GmbH (Germany) for a series of steel products including austenitic stainless steel castings with good elevated temperature strength and scale resistance, heat-resistant steel castings and various cast nickel-, iron-nickel and cobalt-base superalloys.

Koncor. Trade name of Latrobe Steel Company (USA) for a tough, air-hardenable tool steel containing 1.1% carbon, 1% silicon, 0.35% manganese, 5.2% chromium, 4% vanadium, 1.1% molybdenum, and the balance iron. It has good nondeforming properties and good abrasion and heat resistance, and is used for precision-cast extrusion, forging, pointing and upsetter dies, die-castings inserts, and punches and chisels.

Kondo alloy. A dilute alloy composed of a magnetic material contained in a nonmagnetic matrix, and exhibiting an anomalous increase in electrical resistivity with decreasing temperature due to an exchange interaction between the magnetic moments of ions from the magnetic alloying addition and conduction electrons. Abbreviation: KA.

Konductomet. Trademark of Buehler Limited (USA) for a solid molding compound composed of 30-60 wt% fibrous glass, 10-30 wt% graphite, 1-5 wt% phenol and 1-5 wt% hexamethylenetetramine. It has a density of 1.88 g/cm³ (0.068 lb/in.³) and a boiling range of 83-260°C (181-500°F). Used as a conductive mounting resin in electron microscopy.

Konduit. Trade name of LNP Engineering Plastics (USA) for a thermally conductive polyphenylene sulfide used for disk-drive spindle motors.

Konel. Trade name of Westinghouse Electric Corporation (USA) for a heat- and corrosion-resistant alloy containing 73% nickel, 17.2% cobalt, 8.8% titanium, 0.5% silicon, 0.3% aluminum and 0.2% manganese. It exhibits hot shortness above 1250°C (2280°F) and is used for valves, lamp filaments, and as a replacement for platinum.

Konex. Trade name for polyaramid fibers and fabrics.

Konfetti. German trade name for a patterned glass.

koninckite. A yellow mineral composed of iron aluminum phosphate trihydrate, $(Fe,Al)PO_4 \cdot 3H_2O$. Crystal system, tetragonal. Density, 2.40 g/cm³. Occurrence: Belgium.

Konik. Trade name of Continental Steel Corporation (USA) for a case-hardening steel containing 0.15-0.25% carbon, 0.2% copper, 0.4% nickel, 0.3% chromium, and the balance iron. Used for cross chains, culverts and sheets.

Konit. Trade name of Koerver & Nehring GmbH (Germany) for a series of corrosion- and heat-resistant steel castings and cast nickel-base superalloys.

Konstantan. Trade name of ThyssenKrupp VDM GmbH (Germany) for a resistance alloy containing 54-58% copper, 41-45% nickel and 1% manganese. It has a density of 8.89 g/cm³ (0.321 lb/in.³), a maximum service temperature of about 600°C (1110°F), high electrical resistivity (about 50-52 mWcm), a negative temperature coefficient of resistance (about -0.00002 to -0.00003 K⁻¹) and good corrosion resistance. Used for electrical instruments, precision resistors, reducing rheostats, and shunts. See also constantan.

Kontioeristeet. Trade name of A. Ahlström Osakeyhtiö (Finland) for insulation products based on silica fibers. *Kontio* is the general group name for products of this type. See also Karhula.

konyaite. A white mineral composed of sodium magnesium sulfate pentahydrate, $Na_2Mg(SO_4)_2 \cdot 5H_2O$. It can also be made synthetically by evaporating an equimolar solution of sodium sulfate (Na_2SO_4) and magnesium sulfate heptahydrate $(MgSO_4 \cdot 7H_2O)$ under prescribed conditions. Crystal system, monoclinic. Density, 2.09 g/cm³; refractive index, 1.468. Occurrence: Turkey.

Kool Base. Trade mark of Mitsui Chemicals, Inc. (Japan) for a readily formable printed-circuit laminate composed of a metal, such as aluminum, molybdenum or stainless steel, laminated to a copper foil using a thermoplastic polyimide. It has a low dielectric constant, excellent insulating properties and high thermal resistance.

Kool Bond. Trade name of PPG Pretreatment & Specialty Products (USA) for phosphate-free pretreatment coatings.

Kool Draw. Trade name of PPG Pretreatment & Specialty Products (USA) for drawing compounds for ferrous and nonferrous metals and most commercial plastics.

Koolgray. Trade name of ASG Industries Inc. (USA) for a laminated glass with a neutral-tinted interlayer supplied in three densities.

Kooliner. Trade name of GC America, Inc. (USA) for a self-cure two-component (liquid plus powder) dental acrylic lining material.

Koolphen. Trade name of Kooltherm Insulation Products (USA) for phenolic-based foams used for thermal insulation applications.

Kooltherm Urethane. Trade name of Kooltherm Insulation Products (USA) for polyurethane-based structural foams.

Koolvue. Trade name of ASG Industries Inc. (USA) for a glass product composed of two layers of transparent plate or sheet glass with a green-tinted plastic interlayer.

Kopa. Trademark of Kolon Company (USA) for tough, abrasion- and impact-resistant nylon 6 resins with good mechanical properties, outstanding self-lubricating characteristics and excellent chemical resistance. They have low water absorption (1.8%), a density range of 1.13-1.15 g/cm³, a melting point of 220°C (428°F), an upper continuous-use temperature (in air) of 180°C (356°F), and are used for automotive and industrial components.

Kopandex. Trade name of Kohap Limited (South Korea) for spandex fibers and yarns used for elastic fabrics.

Kopel. Trademark of Kolon Company (USA) for a polyether-ester-type thermoplastic elastomer exhibiting rubber-like flexibility and elastic recovery. It has good processibility, mechanical properties and heat and weather resistance, an ultraviolet stability superior to that of rubber, low water absorption (0.3%) and a service temperature of -65 to +150°C (-85 to +300°F).

Koplac. Trade name of Kolon Company (USA) for a series of polyester resins.

Koplon. Trade name of Snia Viscosa SpA (Italy) for modacrylic and rayon fibers and yarns used for textile fabrics.

Koppers. Trade name of Koppers Company Inc. (USA) for a series of high-duty cast irons, abrasion-, corrosion- and/or heat-resistant alloy cast irons, alloy, ferritic and pearlitic ductile irons, centrifugally-cast ductile and gray irons, corrosion-resistant steel castings as well as several cast leaded tin bronzes. Used for engine parts, piston and seal rings, gears, housings, shafts, cams and propellers.

Kopr-Kote. Trademark of Jet-Lube of Canada Limited for high-temperature anti-seizing compounds.

Kora. Trade name of Sanderson Kayser Limited (UK) for a series of case-hardening steels containing 0.15-0.18% carbon, up to 0.15% silicon, 0.7-0.9% manganese, 0.06% sulfur, 0.06% phosphorus, and the balance iron. Used for gears, camshafts and machinery parts.

Korad. Trade name of PEP–Polymer Extruded Products (USA) for a durable acrylic lamination film produced from a 100% acrylic multipolymer. Supplied in a wide range of colors and surface finishes ranging from matte to glossy, it has outstanding weatherability and UV and fluorescent light resistance, and good printability and em-bossability. Used as a capping film for the protection of exteri-and interior plastic products against environmental fading, chalking and discoloration. Examples of such products include roof vents and tiles, windows, shutters, signage, labels, decals, boat hulls, construction equipment, farm tractors, RV components, snowmobile shrouds, bicycle helmets, and residential tubs, sinks and shower enclosures.

Korallin. Trade name for a cellulose nitrate plastic formerly used for denture bases.

Koralox. Trade name for alumina ceramics used in the manufacture of jet engines.

Koratron. Trademark of Koratron Inc. (USA) for a synthetic resin finish cured into a fabric at elevated temperatures to impart permanent crease to garments.

Korbgeflecht. Trade name of former Glas- und Spiegelmanufaktur AG (Germany) for a glass with basketweave pattern. It is known internationally under the trade name *Wickerweave.*

Koresin. Trademark of BASF AG (Germany) for modified phenol-formaldehyde plastics.

Korex. Trademark of DuPont Advanced Composites (USA) for a flame-retardant, high-modulus aromatic polyamide (aramid) honeycomb core material available in sheet, block and roll form. It has excellent fatigue resistance, high strength and dimensional stability, high shear and compressive modulus, good impact resistance, good dielectric properties, good hot/wet properties, good processibility, low weight, a low coefficient of thermal expansion, and negligible moisture growth at 100% relative humidity. Used for honeycomb sandwich structures, such as radomes, reflectors and optical components, and for aerospace and naval applications, e.g., flight control surfaces, and doors.

Koring. Trade name of Edelstahlhandel Koring GmbH & Co. KG (Germany) for a specialty strip steel.

koritnigite. A colorless mineral composed of zinc arsenate hydroxide monohydrate, $Zn(AsO_3)(OH) \cdot H_2O$. Crystal system, triclinic. Density, 3.54 g/cm^3; refractive index, 1.652. Occurrence: Namibia.

Korloy. Trademark of Cominco Limited (Canada) for a series of superplastic zinc-based forging, casting and extrusion alloys.

kornelite. A pink to violet mineral composed of iron sulfate hydrate, $Fe_4(SO_4)_6 \cdot 15H_2O$. Crystal system, monoclinic. Density, 2.30 g/cm^3; refractive index, 1.581. Occurrence: Hungary, USA (Utah).

kornerupine. A colorless, brown, or green mineral composed of magnesium aluminum borate silicate hydroxide, $Mg_3Al_6(S,Al,B)_5O_{21}(OH)$. Crystal system, orthorhombic. Density, 3.27 g/cm^3; refractive index, 1.677; hardness, 6.5 Mohs. Occurrence: Greenland, Kenya.

Korogel. Trade name of B.F. Goodrich Company (USA) for liquid polyvinyl and other vinyl resins.

Korolac. Trade name of B.F. Goodrich Company (USA) for liquid polyvinyl and other vinyl resins.

Koroseal. Trade name of B.F. Goodrich Company (USA) for flexible and rigid molded and extruded polyvinyl chloride products. They have outstanding resistance to ozone, nitric and chromic acid and other oxidizing agents, good resistance to oils, gases and aging, good dielectric properties, good flame resistance and low water absorption. Used for tank linings, fabric and paper coatings, draperies, and raincoats.

Korax. Trademark of BEGO (Germany) for an aluminum oxide powder supplied in particle sizes ranging from 50-250 mm (0.002-0.010 in.). Used as a dental blasting and polishing material.

Korrofestal. Trade name of Aluminium-Werke Wutöschingen GmbH (Germany) for an aluminum alloy containing 0.6-1.4% magnesium, 0.6-1.6% silicon, 0.6-1% manganese and up to 0.3% chromium. It has good formability and weldability and is used for gutters, window frames, fan blades, and boats.

Korronit. Trade name of Otto Wolff Handelsgesellschaft (Germany) for a series of corrosion-resistant austenitic, ferritic and martensitic stainless steels used for pump and valve parts, cutlery, kitchen equipment, sinks, surgical instruments, chemical equipment, food-processing equipment, oil-refining equipment, turbine blades, and machine parts, such as bearings, fasteners, and hardware.

Korrosil. Trade name of Georgsmarienwerke Selesiastahl GmbH (Germany) for a series of ferritic and martensitic stainless steels containing up to 0.22% carbon, 0.4% silicon, 13-17% chromium, 0-1.2% molybdenum, 0-0.7% titanium, and the balance iron. Free-machining grades contain additions of up to 0.2% sulfur. Used for cutlery, dental and surgical instruments, valves, machine parts, pump parts, hardware, mining equipment, heat-treating equipment, turbine blades, and fittings.

Korso. Trademark of Continental Gummi-Werke AG (Germany) for a polyvinyl chloride film.

Korton. Trademark of Norton Performance Plastics (USA) for self-adhesive ethylene chlorotrifluoroethylene (ECTFE) polymer film.

Korund. German trade name for a very hard material obtained from synthetic *corundum* (Al_2O_3) and used as an abrasive for grinding and polishing operations.

Korundal. Trademark of Dresser Industries, Inc. (USA) for a series mullite-bonded *corundum* products suitable for use in furnace linings and roofs subjected to temperatures up to about 1900°C (3450°F).

Korvex. Trademark of Norton Performance Plastics (USA) for heat-shrinkable fluoropolymer tubing used for insulation and packaging applications.

korzhinskite. A colorless mineral composed of calcium borate monohydrate, $CaB_2O_4 \cdot H_2O$. Refractive index, 1.647. Occurrence: Russia.

KoSa. (1) Trade name of KoSa (USA) for polyester fibers used for wearing apparel, sportswear, bedding, floor coverings, home furnishings, and automotive and other industrial fabrics.

(2) Trade name of KoSa (USA) for a high-tenacity polyester yarn used for tire cords, industrial fabrics (e.g., hoses, belts and automotive restraint systems), outdoor fabrics (e.g., for awnings, tents, deck furniture and car tops) and industrial webbing. Also included in this category are high-tenacity nylon-6 yarns for automotive tire cords.

(3) Trade name of KoSa (USA) for polyester polymers used for the manufacture of polyester resins, fibers and film products.

Kosciusko. Trade name of Australian Window Glass Proprietary Limited for a patterned glass.

Koseiseide. Trade name of Asahi Kosei Company (Japan) for regenerated protein (azlon) fibers.

Kosmink. Trade name for a carbon black used as a pigment in printing inks.

Kosmobil. Trade name for a carbon black.

kosmochlor. A green synthetic mineral of the pyroxene group composed of sodium chromium silicate, $NaCrSi_2O_6$. It can be made from sodium carbonate (Na_2CO_3), chromic oxide (Cr_2O_3) and silicon dioxide (SiO_2) under prescribed conditions. Crystal system, monoclinic. Density, 3.60 g/cm^3.

Kosmolac. Trade name for a carbon black used as a pigment in coatings, lacquers and other products.

Kosmos. (1) Trade name of Röchling Burbach GmbH (Germany) for a high-speed steel containing 0.8-0.9% carbon, 4.5% chromium, 12% tungsten, 0.7% molybdenum, 2.5% vanadium, and the balance iron. Used for tools, dies and cutters.

(2) Trade name for a furnace black used in rubber compounding.

Kosmotherm. Trade name for a thermal carbon black used in pigments, and as a reinforcing agent in rubber products.

Kosmovar. Trade name for a carbon black used as a pigment in in lacquers and varnishes.

Kossil. Trade name of Sanderson Kayser Limited (UK) for a tough spring steel containing 0.9% carbon, 1.3-1.6% silicon, 0.6-0.9% chromium, and the balance iron.

Kostil. Trade name of Enichem SpA (Italy) for several styrene-acrylonitriles supplied in unmodified, glass fiber-reinforced, fire-retardant, high-heat, high-impact and UV-stabilized grades. Used for appliances, bathroom furnishings, and toys.

kostovite. A white mineral composed of gold copper telluride, $AuCuTe_4$. Crystal system, orthorhombic. Density, 8.43 g/cm^3. Occurrence: Uzbekhistan.

Koto. Trade name of Nippon Sheet Glass Company Limited (Japan) for a patterned glass.

kotoite. A colorless mineral composed of magnesium borate, $Mg_3(BO_3)_2$. It can also be made synthetically. Crystal system, orthorhombic. Density, 3.04 g/cm^3; refractive index, 1.653. Occurrence: Korea.

kotulskite. (1) A cream- or pale-yellow mineral of the nickeline group composed of palladium telluride, PdTe. It can also be made synthetically. Crystal system, hexagonal. Density, 9.18 g/cm^3. Occurrence: China, South Africa, Russian Federation.

(2) A steel-gray mineral composed of palladium telluride bismuth, $Pd_{1-x}(Te,Bi)$. Crystal system, hexagonal. Density, 8.26 g/cm^3; hardness, 214-405 Vickers. Occurrence: Russian Federation.

koutekite. A dark brown mineral composed of copper arsenide, Cu_5As_2. It can also be made synthetically. Crystal system, hexagonal. Density, 8.48 g/cm^3. Occurrence: Czech Republic.

Koval. Trade name of Stupakoff Laboratories Inc. (USA) for a nickel-base superalloy containing 17% cobalt, 6.25% iron, 2.2% titanium, and traces of other elements. It has good high-temperature resistance, and is used for machine parts, molds and internal combustion valves.

Kovar. Trademark of Carpenter Technology Corporation (USA) for a series of low-expansion alloys containing 29% nickel, 17% cobalt, 0.2-0.4% manganese, 0.2% silicon, up to 0.02% carbon, and the balance iron. Commercially available in the form of rods, foils, wire, tube and powder, they have a density of 8.36 g/cm^3 (0.302 $lb/in.^3$), high electrical resistivity, a Curie temperature of 435°C (815°F), a hardness of 160 Brinell, a low temperature coefficient of resistivity, a low and constant coefficient of thermal expansion (same as glass), high tensile strength and modulus of elasticity and good processibility. Used for hermetic seals, hard glass- or ceramic-to-metal seals, X-ray, power, microwave and other electronic tubes, transistors, diodes and miniaturized integrated circuits.

kovdorskite. A colorless or pale rose mineral composed of magnesium phosphate hydroxide trihydrate, $Mg_2(PO_4)(OH)\cdot3H_2O$. Crystal system, monoclinic. Density, 2.60 g/cm^3; refractive index, 1.542. Occurrence: Russian Federation.

kozulite. A black or reddish black mineral of the amphibole group composed of sodium manganese silicate hydroxide, $(Na,K)_3(Mn,Mg,Fe,Al)_5Si_8O_{22}(OH,F)_2$. Crystal system, monoclinic. Density, 3.30 g/cm^3; refractive index, 1.717. Occurrence: Japan.

krabak. The Thai name for wood obtained from various species

of trees of the genus *Anisoptera*. The same wood is available as *mersawa* in Malaysia and as *palosapis* in the Philippines. It is a plain wood that is not decorative and very hard to work (due to its high silica content), but can be rotary-peeled. Source: Southeast Asia (from Philippines and Malaysia to Bangladesh). Used for plywood veneer.

Krafft's alloy. A Swedish fusible alloy that contains 66% bismuth, 26% lead and 8% tin, has a melting point of 100°C (212°F), and is used for fuses and safety devices.

kraftcord. A smooth, stiff, uniform *kraft yarn* used as a stuffing and/or backing material in woven carpets.

kraft paper. A tough, reddish brown, relatively inexpensive paper made from *wood pulp* by digestion with a mixture of caustic soda, sodium sulfide, sodium carbonate and sodium sulfate and subsequent conversion on a Fourdrinier-type papermaking machine. It has high strength and good puncture resistance, and is used as a wrapping paper, as a building paper for facing blanket insulation materials, as an insulating and sheathing paper, as a liner for laminated fiberboard, and for making shipping cartons and bags.

kraft yarn. A yarn produced by twisting a strip of kraft paper.

kraisslite. A deep coppery-brown mineral composed of magnesium manganese zinc arsenate silicate hydroxide, $(Mn,Mg)_{24}Zn_4(AsO_4)_4(SiO_4)_8(OH)_{12}$. Crystal system, hexagonal. Density, 3.88 g/cm^3; refractive index, 1.805. Occurrence: USA (New Jersey).

Kralastic. Trademark of Uniroyal Chemical Company, Inc. (USA) for a series of acrylonitrile-butadiene-styrene resins that can be processed by injection molding and extrusion, and have low density, high rigidity and toughness, good dimensional stability and good chemical and electrical resistance. Used for cams, gears, wheels, pulleys, cable floats, chemical pipe and cathode edge strips.

Kraloy. Trade name of Kraloy (Canada) for a series of polyvinyl chloride products.

Kramer. Trade name of Illingworth Steel Company (USA) for an extensive series of copper-base casting alloys including various copper nickels, nickel silvers, aluminum, manganese, silicon and silicon-aluminum bronzes, various corrosion- and/or acid-resistant leaded and unleaded bronzes and brasses, electrical contact alloys and several phosphorus-copper master alloys.

Krasil. Trade name of Ravi Rayon Limited (Pakistan) for cellulose acetate filaments and fibers.

kratochvilite. A white mineral with orange tint composed of fluorene, $C_{13}H_{10}$. Crystal system, orthorhombic. Density, 1.21 g/cm^3; refractive index, 1.663; melting point, 115°C (239°F). It can also be made synthetically. Occurrence: Czech Republic, Slovakia.

Kraton. Trade name of Kraton Polymers (USA) for an extensive series of high-performance thermoplastic elastomers that are block copolymers of polystyrene and polybutadiene, or polystyrene and polyisoprene, and possess distinctive molecular structures and compounding flexibilities. Used for sheets, tubes, sealants and coatings. *Kraton G* refers to an extensive series of high-performance thermoplastic elastomers based on styrene-ethylene-butylene-styrene (SEBS).

krausite. A yellow or greenish mineral composed of potassium iron sulfate monohydrate, $KFe(SO_4)_2\cdot H_2O$. Crystal system, monoclinic. Density, 2.85 g/cm^3; refractive index, 1.650. Occurrence: Mexico, USA (California).

krauskopfite. A colorless mineral composed of barium silicate dihydrate, $BaSi_2O_4(OH)_2\cdot2H_2O$. Crystal system, monoclinic.

Density, 3.14 g/cm^3; refractive index, 1.587. Occurrence: USA (California).

krautite. A pale rose mineral composed of manganese hydrogen arsenate monohydrate, $MnAsO_3(OH) \cdot H_2O$. It can also be made by a reaction of arsenic acid and manganese carbonate at 100°C (212°F). Crystal system, monoclinic. Density, 3.31 g/cm^3; refractive index, 1.639. Occurrence: Rumania.

Krayton. Trade name of Shell Chemical Company (USA) for several thermoplastic elastomers.

Krazy Glue. Trade name of Krazy Glue Inc. (USA) for a colorless, liquid cyanoacrylate adhesive supplied in several grades. It bonds well to a wide range of different substrates including metals, plastics, rubber, ceramics, glass and wood, and is used as a household and shop adhesive.

Krehalon. Trade name of Kureha Chemical Industry Company (Japan) for polyvinylidene (saran) fibers, monofilaments and yarns.

Krenit. Trade name of Chemfiber A/S (Denmark) for polypropylene staple fibers.

Kreidler. Trade name of Kreidler Werke GmbH (Germany) for a series of wrought copper-zinc alloys including various aluminum, manganese and nickel brasses, free-cutting brasses and forging brasses.

kremersite. A deep reddish orange mineral composed of potassium ammonium iron chloride monohydrate, $(NH_4,K)_2FeCl_5 \cdot H_2O$. It can also be made by slow evaporation of an equimolar solution of ammonium chloride (NH_4Cl), potassium chloride (KCl) and iron chloride ($FeCl_3$). Crystal system, orthorhombic. Density, 2.17 g/cm^3. Occurrence: Italy.

Kremnitz white. See Vienna white.

Krems white. See Vienna white.

Krene. Trademark of Union Carbide Corporation (USA) for flexible polyvinyl chloride film and sheeting materials used for packaging, laminations, seat covers and shower curtains.

krennerite. A pale yellowish white mineral of the calaverite group composed of gold silver telluride, $(Au,Ag)Te_2$. Crystal system, orthorhombic. Density, 8.63 g/cm^3. Occurrence: USA (Colorado), Rumania.

Kreo. Trademark of Politex SpA (Italy) for a polyester fiber.

K-Resin. Trademark of Phillips Chemical Company (USA) for thermoplastic elastomers based on styrene-butadiene copolymers and available in standard and film grades.

kribergite. A white mineral composed of aluminum phosphate sulfate hydroxide tetrahydrate, $Al_5(PO_4)_3SO_4(OH)_4 \cdot 4H_2O$. Density, 1.92 g/cm^3; refractive index, 1.484. Occurrence: Sweden.

Krikiln Castable. Trade name of National Refractories & Minerals Corporation (USA) for basic castable refractories based on magnesia and dolomite.

krinovite. An emerald-green mineral of the aenigmatite group composed of sodium magnesium chromium silicate, $NaMg_2CrSi_3O_{10}$. Crystal system, monoclinic. Density, 3.38 g/cm^3; refractive index, 1.725. Occurrence: USA.

Krispglo. Trade name for lustrous rayon fibers and yarns used for textile fabrics.

Kristallit. German trade name for silicon carbide grinding and polishing abrasives.

Kriston. Trade name of BF Goodrich (USA) for a series of thermosetting polyester resins with exceptional dielectric properties, low moisture absorption, good heat resistance up to 204°C (400°F), good chemical and flame resistance, good weatherability, good color stability, and good transparency to high-frequency radio waves. Used for reinforced structural shapes, boat hulls,

autobodies, aircraft glazing, television, video and radio parts, and radomes.

kroehnkite. An azure-blue mineral of the roselite group composed of sodium copper sulfate dihydrate, $Na_2Cu(SO_4)_2 \cdot 2H_2O$. Crystal system, monoclinic. Density, 2.92 g/cm^3; refractive index, 1.578.

KR-Monel. Trade name of Huntington Alloys Inc. (USA) for a wrought, nonmagnetic, heat treatable nickel alloy containing 29.5% copper, 2.8% aluminum, 0.5% titanium and 0.23% carbon. It has good machinability, strength, hardness and corrosion resistance. Used for valves, pumps, shafts, impellers, oil-well components, screw machine products, fasteners and electronic components.

Krokoloy. Trade name of Detroit Alloy Steel Company (USA) for an air-hardening chromium-cobalt tool steel used for blanking, coining, drawing and forming dies, and shear blades.

Kromair. Trade name of Republic Steel Corporation (USA) for an air-hardening tool steel (AISI type A2) containing 1% carbon, 5.25% chromium, 0.25% vanadium, 1.15% molybdenum, and the balance iron. Used for shear blades, forming rolls, and cold-work dies.

Kromal. (1) Trade name of Jessop Steel Company (USA) for a high-speed steel containing 0.7% carbon, 9% molybdenum, 4% chromium, 1.25% vanadium, and the balance iron.

(2) Trade name of Allied Steel & Tractor Products Inc. (USA) for a series of oil-hardening, nondeforming tool steels containing 0.5-0.9% carbon, 0.5-1.9% chromium, up to 1.5% manganese, 0.3% vanadium, 0.4% molybdenum and 1.25% nickel, respectively, and the balance iron. Used for tools, punches, and maintenance and repair work.

Krom-Aloy. Trade name of Chrome Alloys Manufacturing Company (USA) for several corrosion-resistant copper alloys containing varying amounts of zinc, chromium, tin, lead, aluminum, manganese and nickel. Used for marine hardware, propellers, and pump valves and seats.

Kromarc. Trademark of Westinghouse Electric Corporation (USA) for a series of austenitic stainless steels containing 0.03-0.04% carbon, 20-23% nickel, 15-16% chromium, 9-10% manganese, 2-2.5% molybdenum, 0-0.15% silicon, 0-0.2% nitrogen, 0-0.25% vanadium, traces of boron, phosphorus, sulfur, aluminum and zirconium, and the balance iron. They have excellent high-temperature tensile, creep and rupture strength, high heat resistance, good resistance to hot tearing, good weldability, good low-temperature properties and a service temperature range of -270 to +650°C (-450 to +1200°F). Used for welding electrodes, welded components in steam turbines, and structural applications in subzero environments.

Kromax. (1) Trade name for an electrical resistance alloy containing 80% nickel and 20% chromium. It has high electrical resistivity, good high-temperature properties and a melting point of approximately 1400°C (2550°F).

(2) Trademark of Thomas Waterproof Coatings Company (USA) for exterior paint coatings.

Kromex. Trade name for a glass that does not transmit ultraviolet rays and is used in the manufacture of dispensing pumps for gasoline, natural gas, etc.

Kromoglass. Trade name of SIV for glass-ionomer dental cements used for tooth filling and other restorative work.

Kromore. Trade name of Driver Harris Company (USA) for an alloy composed of 85% nickel and 15% chromium. It has excellent heat resistance, and is used for electrical resistances and heating elements.

Kromox. Trade name of Capitol Castings Inc. (USA) for an abrasion- and wear-resistant cast iron containing 2.6% carbon, 0.5% manganese, 1.25% molybde-num, 15.2% chromium, and the balance iron. Used for slurry pumps and mill liners.

Krona. Trade name of Time Steel Service Inc. (USA) for an oil-hardening tool steel (AISI type O1) containing 0.9% carbon, 1.2% manganese, 0.5% chromium, 0.2% vanadium, 0.5% tungsten, and the balance iron. Used for dies, knives, punches, and various other tools.

Kronos. (1) Trade name of Peter A. Frasse & Company (USA) for a tool steel containing 0.8% carbon, 0.4% manganese, and the balance iron.

(2) Trademark of Kronos, Inc. (USA) for titanium dioxide pigments based on the minerals *rutile* and *anatase,* and used in ceramics, concrete and glass, coatings, fibers, inks, paper, plastics, and rubber.

Kropunch. Trade name of Allied Steel & Tractor Products Inc. (USA) for an air-hardening tool steel containing 0.4% carbon, 2.5% tungsten, 0.3% vanadium, 2% molybdenum, 1.1% silicon, some chromium, and the balance iron. It has good hot and cold workability and is used for dies, mandrels, punches and shear blades.

Krosil. Trade name of Colt Industries (UK) for an oil-hardening cold-work tool steel containing 0.5% carbon, 0.9% manganese, 0.35% molybdenum, 1.5% silicon, and the balance iron. It has good shock and fatigue resistance, and is used for chisels, crimpers, punches, upsetters and dies.

Krotung. Trade name of Allied Steel & Tractor Products Inc. (USA) for a tough tool steel containing 0.3% carbon, 5% chromium, 1.5% molybdenum, 1.3% tungsten, 0.3% vanadium, and the balance iron. It has high hot hardness, and is used for hot-forging tools and dies.

Krovan. Trade name of Colt Industries (UK) for a tough, water-hardening tool steel containing 0.95% carbon, 1% chromium, 0.2% vanadium, and the balance iron. It has good deep-hardening properties and good wear and impact resistance, and is used for rolls and dies.

K/R-Ramcast 30 AF. Trade name of National Refractories & Minerals Corporation (USA) for a series of refractory ram-cast mixes.

kruen. See keruing.

Krumbhaar. Trademark of Lawter International Inc. (USA) for rosin-modified synthetic resins.

krupkaite. A dark steel gray mineral of the stibnite group composed of copper lead bismuth sulfide, $CuPbBi_3S_6$. It can also be made synthetically. Crystal system, orthorhombic. Density, 6.95-6.99 g/cm^3. Occurrence: Australia.

Krupp. Trade name of Krupp Stahl AG (Germany) for an extensive series of steels including numerous structural grades (e.g., *Krupp ST 50*), nitriding grades (e.g., *Krupp WL 1.8514*), heat-treatable grades (e.g., *Krupp 34 CrNiMo6*), case-hardening grades (e.g., *Krupp C40* and *Krupp CK45*), free-machining grades (e.g., *Krupp 9S20*), hot- and cold-work tool grades (e.g., *Krupp DF109CN* and *Krupp SKV81*), high-speed grades (e.g., *Krupp 3344*), stainless grades (e.g., *Krupp V2A* or *Krupp V1M*), specialty steel grades (*Krupp Sonderstahl*), and spring steel grades (*Krupp Federstahl*).

Kruppin. Trade name of Krupp Stahl AG (Germany) for a heat-resistant iron-carbon alloy containing 28% nickel. Used for electrical resistances.

krutaite. A gray mineral with bluish tint belonging to the pyrite group and composed of copper selenide, $CuSe_2$. It can also be made synthetically. Crystal system, cubic. Density, 6.62 g/cm^3. Occurrence: Czech Republic.

krutovite. A grayish white mineral belonging to the pyrite group and composed of nickel arsenide, $Ni_{1-x}As_2$. Crystal system, cubic. Density, 6.93 g/cm^3. Occurrence: Czech Republic.

Kruzite Castable. Trade name of A.P. Green Industries, Inc. (USA) for castable refractories.

Krylene. Trademark of Bayer Corporation (USA) for an abrasion-resistant styrene-butadiene rubber made by emulsion polymerization and supplied in several grades. It has good aging and heat resistance, good mechanical properties, good electrical properties and good resistance to polar solvents, dilute acids and bases. Used in the manufacture of rubber goods especially blended with nitrile rubber for automotive tires, and also for footwear soles and heels, seals, membranes, hose, and transmission belting.

Krylon. Trade name of Borden Company (USA) for a full line of interior and exterior paints and coatings including durable, scratch-resistant high-gloss spray paints, fast-drying spray paints, high-gloss enamel spray paints, fluorescent spray paints, and barbecue, stove and propane tank spray paints.

Krynac. Trademark of Bayer Corporation (USA) for an abrasion- and wear-resistant acrylonitrile-butadiene rubber supplied in several grades including standard, oil-extended, carboxylated and pre-crosslinked. It has very good resistance to liquid fuels, minerals and greases, good aging resistance and low permeability to gases. Used in the manufacture of moldings, seals, membranes, bellows, buffers, vibration dampers, hose, conveyor belting, brake and clutch linings, sponge rubber, footwear soles, gloves, cable sheathing, and rubberized fabrics.

Krynol. Trademark of Bayer Corporation (USA) for abrasion-resistant emulsion-polymerized styrene-butadiene rubber supplied in several grades. Many of its properties are similar to those of *Krylene* and it is used for similar applications.

Kryptex. Trade name for a zinc silicophosphate dental cement.

kryptocyanine. See cryptocyanine.

krypton. A colorless, odorless, rare, inert, gaseous element of the noble gas group (Group VIIIA, or Group 18) of the Periodic Table, and constituting about 0.0001 vol% of the atmosphere. Density, 2.818 g/cm^3; freezing point, -157.3°C (-251°F); boiling point, -153°C (243°F); atomic number, 36; atomic weight, 83.80; divalent. It forms compounds with halogens (e.g., bromine, fluorine and iodine) at about cryogenic temperatures that decompose at room temperature, and is a white crystalline solid at cryogenic temperatures melting at 116K. Used in incandescent bulbs, gas-filled electric lamps, fluorescent light tubes, flash bulbs and ultraviolet lasers, in high-speed photography, and as a primary standard of wavelength. The fluorides and iodides are used in ion lasers. Symbol: Kr.

krypton-85. A radioactive isotope of krypton with a mass number of 85 obtained as a fission product from irradiated nuclear fuel. It has a half-life of 10.3 years and emits beta radiation. Used as a tracer gas for leak detection, in the activation of phosphors, in luminous paints and markers, and as a radiation source. Symbol: ^{85}Kr.

krypton-86. A stable isotope of krypton with a mass number of 86 used in the measurement of the standard meter. Symbol: ^{86}Kr.

kryptonate. A material, such as a chemical element, compound, alloy, ceramic, glass, plastic or elastomer, impregnated with the radioactive isotope krypton-85 (^{85}Kr) to introduce tracer atoms into its structure. See also krypton; krypton-85.

Krytox. Trademark of E.I. DuPont de Nemours & Company (USA) for nonflammable, chemically inert fluorinated greases with good dielectric properties and high radiation resistance. Used for laboratory applications, e.g., in vacuum technology.

kryzhanovskite. A red brown mineral of the phosphoferrite group composed of manganese iron phosphate hydroxide monohydrate, $MnFe_2(PO_4)_2(OH)_2 \cdot H_2O$. Crystal system, orthorhombic. Density, 3.31 g/cm³. Occurrence: Russian Federation.

K-Shield. Trade name of Thermal Ceramics Inc. (USA) for ceramic paper and felt products based on discontinuous alumina-silica fibers. Available in several grades and thicknesses, they have a maximum service temperature of 1315°C (2400°F), and are used as thermal insulating materials, and as heat shields for automotive and aerospace applications.

K-Spec Fiber. Trade name of Slingmax, Inc. (USA) for a high-performance fiber core yarn used in the manufacture of rope slings.

K-Spun. Trade name of Koppers Company Inc. (USA) for a wear-resistant cast iron containing 2.85-3.5% total carbon, 0.95-1.45% silicon, 0.4-0.8% manganese, and the balance iron. Used for cylinder liners and piston rings.

K-Ten. Trade name of Kobe Steel Limited (Japan) for a series of high-strength low-alloy (HSLA) structural steels.

ktenasite. An emerald-green to bluish-green mineral composed of copper zinc sulfate hydroxide hexahydrate, $(Cu,Zn)_5(SO_4)_2(OH)_6 \cdot 6O$. Crystal system, monoclinic. Density, 2.94 g/cm³; refractive index, 1615. Occurrence: Norway, Greece.

K-Therm. Trademark of Arlon, Inc. (USA) for a boron nitride-filled, glass-reinforced silicone elastomer with excellent heat transfer properties, high resistance to torque loads, good thermal and electrical properties, good electrical insulation properties and a service temperature range of about -60 to +200°C (-75 to +390°F). Used for electronic components.

Kubax. Trade name of Röchling Burbach GmbH (Germany) for an oil-hardening tool steel containing 1.2-1.3% carbon, 0.1-0.2% chromium, 0.15-0.25% vanadium, and the balance iron. Used for tools, dies and drills.

Kühl cement. A rapid-hardening, Portland-type *hydraulic cement* in which 7% or more ferric oxide (Fe_2O_3) and alumina (Al_2O_3), respectively, replace a substantial portion of silica (SiO_2).

Kufil. Trade name of IMI Kynoch Limited (UK) for a free-flowing high-copper alloy powder containing 1% silver. It has a melting range of 1073-1078°C (1963-1972°F) and is used in the manufacture of welding rods for copper.

Kuhbier. Trade name of C. Kuhbier & Sohn (Germany) for a series of corrosion- and/or heat-resistant austenitic, ferritic and martensitic stainless steels used for chemical process equipment, food-processing equipment, oil-refining equipment, pump and valve parts, marine hardware, machine parts, turbine blades, furnace equipment, cutlery, and kitchen utensils.

Kuhne bronze. A heavy-duty phosphor bronze composed of 78% copper, 11% tin, 10% lead, and about 0.6-0.7% phosphorus and 0.3-0.4% nickel, and used for hard bearings.

Kuki. Trade name of Carl Urbach & Co. Stahlwerk KG (Germany) for a series of water-hardening plain-carbon tool and machinery steels.

kulanite. A blue to green mineral of the bjarebyite group composed of barium iron magnesium manganese aluminum phosphate hydroxide, $Ba(Fe,Mn,Mg)_2Al_2(PO_4)_3(OH)_3$. Crystal system, triclinic. Density, 3.91 g/cm³; refractive index, 1.705. Occurrence: Canada (Yukon).

Kulgrid. Trade name of GTE Sylvania (USA) for a hard-drawn

wire composed of a core of pure copper to which a cladding of nickel has been bonded that forms about 30 wt% of the total cross-sectional area. It has good oxidation resistance at high temperatures, good strength, and high electrical conductivity (70% IACS). Used for electronic components.

Kulite. Trade name of Kulite Tungsten Corporation (USA) for a series of heavy alloys made by powder-metallurgy techniques and containing about 90-97% tungsten, with the balance being either lead or copper, nickel and iron. They have a density range of 17-19 g/cm³ (0.61-0.69 lb/in.³), high heat resistance and good electrical conductivity. Used for aircraft counterweights, nuclear radiation shielding, and sinker weights.

kullerudite. A light steel-gray mineral of the marcasite group composed of nickel selenide, $NiSe_2$. Crystal system, orthorhombic. Density, 6.73 g/cm³. Occurrence: Finland.

Kulmerglas. Trade name of Fr. Trösch AG (Switzerland) for double glazing units.

Kulzer Adhesive Bond. Trade name of Heraeus Kulzer Inc. (USA) for an unfilled resin used for dental bonding applications.

Kulzer Adhesive Cement. Trade name of Heraeus Kulzer, Inc. (USA) for a dual-cure dental cement.

Kumanal. Trade name of IMI Kynoch Limited (UK) for an alloy containing 88% copper, 10% manganese and 2% aluminum usually supplied in the form of hard-drawn wire and strip. It has an electrical conductivity of 4.5% IACS, and is used for instrument shunts, resistance wire and heating elements.

Kumanic. Trade name of IMI Kynoch Limited (UK) for a corrosion-resistant alloy of 60% copper, 20% manganese and 20% nickel. Used for springs and contacts.

Kumium. Trade name of IMI Kynoch Limited (UK) for a hardenable high-conductivity copper containing 0.5% chromium. It has good elevated-temperature properties, and is used for electrical contacts, electrodes for spot-welding and welding tips.

Kumoi. Trade name of Taiwan Glass Corporation (Taiwan) for a figured glass with an irregular reeded pattern.

Kumy. Trademark for polyvinyl chlorides.

Kunheim metal. A German pyrophoric alloy containing 1-12% magnesium, 1-2% aluminum, and the balance various rare-earth metals including lanthanum, cerium and didymium. Formerly used for cigarette lighters.

Kunial. Trade name of IMI Kynoch Limited (UK) for a series of brasses, bronzes, copper nickels and nickel silvers.

Kunial Brass. Trade name of IMI Kynoch Limited (UK) for a strong, wrought copper alloy containing 20% zinc, 6% nickel and 1.5-2% aluminum. Supplied in bar, sheet and wire form usually in the solution-treated, cold-rolled or tempered conditions, it is used for springs, machine parts, valves, and keys.

Kunifer. Trade name of Imperial Metal Industries plc (UK) for a series of corrosion-resistant alloys containing 66-96% copper, 5-32% nickel, 0-2% iron and 0-2% manganese. They have good corrosion- and/or erosion resistance, good resistance to season cracking, and are used for condenser tubes, hydraulic and seawater pipelines, marine equipment, hardware, fasteners, and welding electrodes.

Kuniform. Trade name of IMI Kynoch Limited (UK) for a series of fine-grained wrought copper alloys containing 10-40% zinc.

Kunstharz. A general trade name used by several manufacturers in German-speaking countries for any of a wide range of chemically similar or dissimilar synthetic resins. For example, *Kunstharz AW* is used by BASF AG for aldehyde/ketone-based resins and *Kunstharz LTA* by CWH for unsaturated polyester resins.

kunzite. A lilac-hued variety of the mineral *spodumene* (LiAl-

Si_2O_6) from Madagascar, Brazil and California, used as a gemstone.

Kupal. German trade name for a copper-clad pure aluminum.

Kupfernickel. Trade name of VDM Nickel-Technologie AG (Germany) for a series of corrosion-resistant alloys containing 50-80% copper, 20-50% nickel and 0-5% manganese, used for electrical resistances, coinage and condenser tubes.

Kupfer-Silumin. Trade name of Metallgesellschaft Reuterweg (Germany) for aluminum casting alloys containing 12% silicon, 0.8% copper and 0.3% manganese. They have high fluidity, and are used for motor housings, engine blocks, wheels, and rolls.

kupletskite. A black or dark brown mineral composed of potassium sodium iron manganese titanium silicate hydroxide, $(K,Na)_2(Mn,Fe)_4TiSi_4O_{14}(OH,F)_2$. Crystal system, monoclinic. Density, 3.20 g/cm^3; refractive index, 1.699. Occurrence: Russian Federation.

Kupralume. Trade name of Alchem Corporation (USA) for a copper electroplate and plating process.

Kuprodur. Trade name of ThyssenKrupp VDM GmbH (Germany) for a heat-treatable, corrosion-resistant copper alloy containing 0.5% silicon and 0.7% nickel. It has good elevated-temperature performance, and is used for staybolts and propeller parts.

Kuralon. Trademark of Kuraray Company (Japan) for vinal (polyvinyl alcohol) staple fibers and filament yarns used for textile fabrics.

Kuralon Polyester. Trademark of Kuraray Company (Japan) for polyester fibers and yarns used for textile fabrics.

kuramite. A dark gray mineral of the chalcopyrite group composed of copper mercury tin sulfide, $(Cu,Hg)_{5.5}Sn_2S_8$. Crystal system, tetragonal. Density, 5.59 g/cm^3. Occurrence: Russian Federation, Uzbekistan.

kuranakhite. A brownish to nearly black mineral composed of lead manganese tellurate, $PbMnTeO_6$. Crystal system, orthorhombic. Density, 6.71 g/cm^3. Occurrence: Russian Federation.

Kurare. Trade name of Kurare Company Limited (Japan) for rayon fibers and yarns.

kurchatovite. A pale gray mineral composed of calcium magnesium manganese borate, $Ca(Mg,Mn)B_2O_5$. Crystal system, orthorhombic. Density, 3.02 g/cm^3; refractive index, 1.681. Occurrence: Siberia.

kurchatovium. See rutherfordium.

Kureha. Trademark of Kureha Chemical Corporation (Japan) for carbon fibers.

Kurehabo. Trade name of Kureha Chemical Corporation (Japan) for rayon fibers and yarns used for textile fabrics.

Kurehalon. Trade name of Kureha Chemical Corporation (Japan) for saran (polyvinylidene chloride) fibers with good chemical resistance, good weatherability, and good resistance to mildew and insects. Used for carpets and rugs, draperies, upholstery, clothing, industrial fabrics, etc.

Kuremona. Trade name of Kureha Chemical Corporation (Japan) for a vinal (polyvinyl alcohol) fiber used for textile fabrics.

Kurilac. Trade name of E.I duPont de Nemours & Company (USA) for shellac plastics.

kurnakovite. A colorless mineral composed of magnesium borate pentahydrate, $MgB_3O_3(OH)_5·5H_2O$. Crystal system, orthorhombic. Density, 1.86 g/cm^3; refractive index, 1.508. Occurrence: Russia.

Kuromi. A white, copper-base Japanese jewelry alloy containing varying amounts of tin and cobalt. Used for art objects and ornamental parts.

Kuromi-do. A Japanese jewelry alloy containing 99% copper and 1% arsenic. It has a liquidus (melting) temperature of about 1070°C (1958°F), and is used for art objects and ornamental parts.

kurumsakite. A greenish to bright yellow mineral composed of copper nickel zinc aluminum vanadium silicate hydrate, $(Zn,Ni,Cu)_8Al_8V_2Si_5O_{35}·27H_2O$. Crystal system, orthorhombic. Density, 4.03 g/cm^3. Occurrence: Russia.

Kuseta. Trade name for rayon fibers and fabrics.

kusia. See bilinga.

kusuite. An opaque yellow-brown mineral composed of cerium vanadium oxide, $CeVO_4$. It can also be made synthetically. Crystal system, tetragonal. Density, 4.76 g/cm^3.

Kutasa. Trade name for rayon fibers and yarns used for textile fabrics.

Kutern. Trade name of IMI Kynoch Limited (UK) for a tellurium copper containing 99.5% copper and 0.5% tellurium. It has free-machining properties, excellent cold and hot workability, good forgeability, good corrosion resistance, and an electrical conductivity of 90% IACS. Used for machined parts, and electrical apparatus and components.

kutinaite. A gray-white to gray-blue mineral composed of copper silver arsenide, Cu_2AgAs. Crystal system, cubic. Density, 8.36 g/cm^3. Occurrence: Czech Republic.

Kut Kost. Trade name of General Tool & Die Company (USA) for a series of tungsten and cobalt-tungsten high-speed steels containing up to 1.5% boron, and used for tools, dies, cutters and tipped tools.

Kutherm. Trade name of Imperial Metal Industries plc (UK) for a series of wrought copper-nickel and copper-manganese-nickel alloys. They have a service temperature range of 150-350°C (300-660°F), and are used for thermocouples, electric blankets, heating elements, heating cables and space heaters.

kutnohorite. A pink mineral of the calcite group composed of calcium manganese carbonate, $Ca(Mn,Mg)(CO_3)_2$. Crystal system, rhombohedral (hexagonal). Density, 3.12 g/cm^3; refractive index, 1.727. Occurrence: USA (New Jersey).

Kut-Steel. Trade name for steel shot used as a blasting abrasive.

Kuttwell. Trade name of Swift Levick & Sons Limited (UK) for a case-hardening steel containing 0.13-0.15% carbon, 0.9% manganese, 0.04% sulfur, 0.04% phosphorus, and the balance iron. Used for machine parts and fasteners.

kuznetsovite. A pale brown or honey-colored mineral composed of mercury arsenic chloride oxide, $Hg_6As_2Cl_2O_9$. Crystal system, cubic. Density, 8.64 g/cm^3. Occurrence: Kirgizian Republic.

Kwik Blak. Trademark of Heatbath Corporation (USA) for a room-temperature blackener used for mild steel, cast iron and malleable iron.

Kwik-Kut. Trade name of North American Steel Corporation (USA) for a fatigue-resistant steel containing 1.1% carbon, 0.5% chromium, and the balance iron. Used for hollow drills.

Kwik Seal. Trademark of DAP, Inc. (USA) for a white interior caulk for sealing tubs and tiles. *Kwik Seal Plus* is a white, crack- and mildewproof, high-gloss all-purpose caulk.

kyanite. A blue, white, gray or light-green mineral with a vitreous to pearly luster. It is composed of aluminum silicate, Al_2SiO_5, and has the same composition as the minerals *andalusite* and *sillimanite*, but differs in crystal structure and physical properties. Crystal system, triclinic. Density, 3.56-3.67 g/cm^3; hardness, 5 Mohs along the length of the crystal, and 7 Mohs at right angles to this direction; refractive index, 1.722. It decomposes to *mullite* and *cristobalite* at about 1300°C (2370°F), and has high mechanical strength and good resistance to elevated temperatures. Occurrence: Austria, France, Kenya, India, Switzerland, USA (California, Connecticut, Georgia, Massachusetts, North Carolina, South Carolina, Virginia). Used as a source of mullite in ceramics and refractories, and for precision-casting molds, brake disks, wall tile, sanitary and electrical porcelain (spark plugs), and filters. Grades with clear blue or green color are used as gemstones. Also known as *cyanite; disthene.*

Kycube. Trade name of IMI Kynoch Limited (UK) for wrought alloys containing 1.8-2% beryllium, a total of 0.2-0.6% cobalt and nickel, and the balance copper. Supplied in the form of strip and wire, they have good workability, and are used for welding electrodes, wire products and springs.

Kydene. Trademark of Lenning Chemicals (UK) for modified polyvinyl chlorides.

Kydex. Trademark of Rohm & Haas Company (USA) for a thermoformed plastic sheet made from a blend of acrylic resin and polyvinyl chloride. It has good resistance to severe service conditions, and is used for containers and housings, trays, covers, protective guards and decorative parts.

Kyloid. Trade name of Kyloid Company (USA) for casein plastics.

Kynal. Trade name of IMI Kynoch Limited (UK) for a series of wrought and cast aluminum products including several commercially pure aluminums (99+%) and numerous aluminum-copper, aluminum-magnesium, aluminum-magnesium-silicon, aluminum-manganese, aluminum-silicon and aluminum-zinc alloys. *Kynal-Core* refers to a series of aluminum-copper and aluminum-zinc alloys clad on both sides with aluminum, and used for structural members and aircraft structures.

Kynar. (1) Trademark of Atofina Chemicals Inc. (USA) for a series of hard, white, high-molecular-weight polyvinylidene fluoride (PVDF) resins supplied in unreinforced and carbon fiber-reinforced grades. They have a density of 1.75 g/cm³ (0.063 lb/in.³), excellent resistance to high humidity and to acids, alcohols, aldehydes, alkalies and hydrocarbons, poor resistance to ketones, good performance at low temperatures, good resistance to elevated temperatures up to approximately 340°C

(645°F), poor resistance to gamma radiation, and very low water absorption (less than 0.05%). Used for seals, gaskets, keyboards, keypads, microphones, laser applications, sensors for industrial and medical applications, molded and extruded parts, piping, rigid tubing, fire insulation, and for tough, flexible uniaxially or biaxially oriented films with high piezoelectric and pyroelectric activity.

(2) Trademark of Albany International, Inc. (USA) for polytetrafluoroethylene (PTFE) monofilaments.

Kynar Flex. Trademark of Atofina Chemicals Inc. (USA) for a polyvinylidene fluoride (PVDF) modified with co-monomers to increase flexibility and improve corrosion resistance and high-temperature properties. Used for sheet lining, tubing and piping.

Kynar Superflex. Trademark of Atofina Chemicals Inc. (USA) for highly flexible modified polyvinylidene fluorides (PVDFs).

Kynol. Trademark of Harbison-Carborundum Corporation (USA) for a flame-resistant *novoloid fiber* composed of a crosslinked amorphous phenolic polymer. It has outstanding high-temperature resistance, good dielectric properties, good resistance to organic solvents, and fair resistance to strong alkalies and oxidizing acids. Used as an ablative agent for spacecraft, and for flameproof apparel and protective clothing.

Kynor. Trade name of Atofina Chemicals Inc. (USA) for a vinylidene fluoride polymer supplied in the form of thin sheets. It has excellent dielectric properties and moderate thermal properties, and is used for insulating and dielectric applications.

Kyon. Trademark of Kennametal Inc. (USA) for cold-pressed advanced ceramics based on silicon aluminum oxynitride (SiAlON) providing high strength, fracture toughness and wear resistance. Supplied in the form of grooving and profiling inserts in various geometries, they are used chiefly for high-speed machining of high-temperature alloys.

Kyro-Kay. Trade name of Teledyne McKay (USA) for a series of stainless steels containing 0.02-0.03% carbon, 2.1% manganese, 0.3% silicon, 17.7% chromium, 13.6% nickel, 2.1% molybdenum, and the balance iron.

kyzylkumite. A black mineral composed of vanadium titanium oxide, $V_2Ti_3O_9$. Crystal system, monoclinic. Density, 3.75 g/cm³. Occurrence: Uzbekistan.

L

La Belle. Trade name of Crucible Materials Corporation (USA) for a series of oil-hardening shock-resisting and water-hardening plain-carbon tool steels. The shock-resisting grades contain about 0.45-0.55% carbon, and either (i) 0.8% manganese, 0.4% molybdenum and up to 2.0% silicon (AISI type S5), or (ii) 1.5% chromium, 1.4% manganese, 2-2.5% silicon, and up to 0.4% molybdenum (AISI type S6), and the balance iron. They have high toughness and good deep-hardening properties, and are used for punches, chisels, shear blades, and pneumatic tools. The water-hardening plain-carbon grades (AISI type W1) contain approximately 0.95% carbon, 0.35% manganese, 0.1-0.4% silicon, and the balance iron. They possess excellent deep-hardening properties, good impact strength, fair toughness and good cutting capacity, and are used for cold-striking and cold-heading dies.

label paper. A generic term used for paper whose composition depends largely on its use for labeling purposes. Writing and/or printing must be possible on at least one side of it.

Laboratory. Trade name of J.F. Jelenko & Company (USA) for several dental casting alloys. *Laboratory 22* is a medium-gold alloy with high strength and tarnish resistance, and outstanding workability, used for inlays, crowns and bridgework. *Laboratory 33* is a microfine, hard, medium-gold alloy (ADA type III) that provides high strength and tarnish resistance and outstanding workability, and is used for inlays, three-quarter crowns, pontics and anterior abutments. *Laboratory 44* is a microfine, heat-treatable, extra-hard medium-gold alloy (ADA type IV) that provides high strength, corrosion and tarnish resistance, and good solderability, and is used for hard inlays, quality crowns, fixed bridgework, and partial inlays.

Laboratory Plaster. Trade name of Whip Mix Corporation (USA) for a white dental plaster (ADA type II) available as a powder. Supplied in regular-set grades with a setting time of 14 minutes and fast-set grades with a setting time of 9 minutes, it is used for general-purpose laboratory work.

Lab-Putty. Trade name of Harry J. Bosworth Company (USA) for a silicone elastomer used in dentistry for laboratory applications.

Labrador feldspar. See labradorite.

labradorite. A blue, green, gray or brown plagioclase or soda-lime *feldspar* that is an isomorphous mixture of about 50-70% *anorthite* ($CaAl_2Si_2O_8$) and 30-50% *albite* ($NaAlSi_3O_8$). Crystal system, triclinic. Density, 2.70-2.72 g/cm^3; hardness, 5-6 Mohs. Occurrence: Canada. Used as a gemstone. Also known as *Labrador feldspar; Labrador spar; Labrador stone.*

Labrador spar. See labradorite.

Labrador stone. See labradorite.

Lab-Sil. Trade name of Harry J. Bosworth Company (USA) for a silicone elastomer used in dentistry for laboratory applications.

Labstone. Trade name of Heraeus Kulzer, Inc. (USA) for a die stone used for dental laboratory work.

labuntsovite. A brownish-yellow or rose-colored mineral composed of potassium titanium silicate monohydrate, $(K,Ba,Na)(Ti,Nb)(Si,Al)_2(O,OH)_7 \cdot H_2O$. Crystal system, orthorhombic. Density, 2.90 g/cm^3; refractive index, 1.702. Occurrence: Russian Federation.

Labyrinth. Austrian trade name for a patterned glass.

lac. A natural resin secreted by certain homopterous insects of southern Asia, especially the lac insect (*Coccus lacca*). Available in various forms especially as seed lac and button lac for the manufacture of shellac.

Lacanite. Trade name of Consolidated Molded Products Corporation (USA) for shellac plastics.

La Cantabra. Trade name of Vitrios y Envases SA (Mexico) for a toughened glass.

lace. (1) A fine, openwork, or mesh- or netlike fabric made by connecting base yarns with ornamental stitches, or by twisting, braiding, knotting, looping or otherwise intertwining yarns together to form an ornamental pattern or design. It can be made by hand with a needle or bobbin, or by machine. See also bobbin lace; needlepoint lace.

   (2) A fabric resembling true lace (1), but made by crocheting, embroidering, knitting, weaving or other techniques.

Lacea. Trademark of Mitsui Toatsu Chemicals (Japan) for polylactic acid (PLA) filaments and fibers.

lace leather. A strong, tough, chrome- and/or alum-tanned, oil-treated leather made from cattlehides, and used for the splicing sections of power-transmission belts.

Lacenglass. Trade name of Multiplate Glass Corporation (USA) for a laminated glass enclosing a reproduction of old laces.

lacewood. The pink to reddish-brown wood of the tree *Cardwellia sublimis*. It has a uniform texture, a beautiful figure, and an unusual, but attractive appearance marked by small silky spots. Source: Australia. Used for cabinetwork and decorative applications. Also known as *selena; silky oak.*

Lacimoid. Trade name for a building material made by treating paper, wood or textiles with a synthetic resin and compressing into dense homogeneous laminates under heat. Used for walls, ceilings, cabinets, etc.

Lacisana. Trade name for rayon fibers and yarns used for textile fabrics.

Lacolite. Trade name of Lakeside Malleable Castings Company (USA) for a cast iron used for machine-tool parts, gears, and shafts.

Lacomo. Trade name of Latrobe Steel Company (USA) for a molybdenum-type high-speed steel (AISI type M30) containing 0.8% carbon, 4% chromium, 5% cobalt, 1.5% tungsten, 1.25% vanadium, 8.5% molybdenum, and the balance iron. It has good deep-hardening properties, high hot hardness, and excellent wear resistance. Used for sawteeth, reamers, and lathe tools. Formerly known as *Electrite Lacomo.*

Lacovyl. Trade name of Atofina SA (France) for unplasticized and plasticized polyvinyl chlorides (PVCs) and vinyl chloride-vinyl acetate copolymers. The plasticized PVCs are supplied in standard and UV-stabilized grades, and three elongation ranges: up to 100%, 100-300% and above 300%. The unplasticized PVCs are stiff, strong, inherently flame-retardant thermoplastic resins supplied in rod, sheet, tube, powder and film form.

They have a density of 1.4 g/cm³ (0.05 lb/in.³), low water absorption (0.03-0.4%), good barrier properties and UV stability, an upper continuous use temperature of 50°C (148°F), and moderate chemical resistance. Used for building products, packaging (containers and bottles), and automotive components.

Lacqran. Trademark of Atofina SA (France) for acrylonitrile-bitadiene-styrene plastics.

Lacqrene. Trade name of Atofina SA (France) for high-gloss, impact-resistant polystyrene with high thermal stability, good processibility and short injection molding cycles, used for audio and video tapes, consumer electronics, and packaging. Also included under this trade name are styrene-acrylonitrile plastics.

Lacqrene Alpha. Trade name of Atofina SA (France) for crystalline polystyrene used in the manufacture of electronic components, sporting and leisure goods, and packaging.

Lacqsan. Trade name of Atofina SA (France) for styrene-acrylonitrile resins and plastics.

Lacqtene. Trademark of Atofina SA (France) for standard and UV-stabilized low-density polyethylenes available in injection molding grades. Used for tubes, hoses, pipe coatings, and packaging. *Lacqtene HD* is an unmodified and UV-stabilized high-density polyethylene supplied in rod, sheet, tube, granule and film form. It has a density of 0.95 g/cm³ (0.034 lb/in.³), very good electrical properties, relative high strength, stiffness and impact resistance, good resistance to acids, alkalies, alcohols, halogens and ketones, an upper working temperature of 55-120°C (131-248°F), and is used for containers, bottles, pipes, pipe fittings, packaging, and building products. *Lacqtene CX* is a crosslinked polyethylene, and *Lacqtene HX* and *Lacqtene LX* are linear-low-density polyethylenes.

Lacqua. Trade name of Rohm & Haas Company (USA) for a water-based organic coating.

lacquer. A quick-drying solution composed of natural or synthetic resins (e.g., shellac, latex, nitrocellulose, cellulose esters, vinyls, or acrylics) and plasticizers with or without pigments in an organic solvent. When it is applied to a substrate, the solvent evaporates, and the remainder precipitates out as a dry, hard, glossy film. Used as furniture finishes, touch-up coatings for autobodies, undercoatings, and protective and decorative coatings on metals, plastics, textiles, paper or wood.

Lacquer-Stik. Trade name of La-Co Industries Inc./Markal Company (USA) for a semi-solidified paint in stick form used for wipe-on or fill-in applications.

Lacqvyl. Trade name of Atofina SA (France) for polyvinyl chloride plastics.

Lacrinite. Trade name of Lacrinoid for phenol-formaldehyde plastics.

Lacritex. Trade name of Lati SpA (Italy) for acrylics based on polymethyl methacrylate.

lactam. A cyclic amide, such as *caprolactam*, synthesized from an amino acid by splitting off a water molecule.

Lactame 12. Trade name of Atofina SA (France) for polyamide 12 (nylon 12) resins and plastics.

Lactel. Trademark of Birmingham Polymers, Inc. (USA) for a range of biopolymers based on polycaprolactone, poly(DL-lactide), poly(L-lactide) and poly(DL-lactide-*co*-glycolide). The polycaprolactone polymers have a typical molecular weight of 100000-190000, the poly(DL-lactides) of 90000-120000 and the poly(L-lactides) of 85000-160000. The poly(DL-lactide-*co*-glycolide) copolymers are supplied with typical lactide-to-glycolide ratios ranging from 50:50 to 85:15 and molecular

weights ranging from 40000-75000 for the 50:50 composition to 75000-120000 for the 75:25 polymer. Used chiefly for drug delivery systems, and for tissue engineering applications including surgical sutures.

Lactene. Trade name of Atofina SA (France) for a series of polyethylene plastics.

lactic acid. An important hydroxy acid that exists in the form of three optical isomers: (i) D-lactic acid which is involved in muscular contraction in the body, (ii) L-lactic acid which is isolated from the fermentation products of sucrose; and (iii) DL-lactic acid which is a racemic mixture of the D- and L-isomer formed during the bacterial fermentation of certain foods (e.g. by the souring of milk and in yoghurt and cheese). *Lactic acid* can also be made synthetically from starch, milk whey, or molasses by fermentation, neutralization with calcium or zinc carbonate and subsequent concentration and decomposition of the resulting lactate mixture with sulfuric acid, and by the hydrolysis of lactonitrile. Density, 1.2 g/cm³ (0.04 lb/in.³); fusion point, above 110°C (230°F); refractive index, 1.425. Used in the manufacture of pharmaceuticals, plasticizers, plastics, adhesives and lactates, as a depressant in ore flotation, and in biology, biochemistry and biotechnology. L- and DL-lactic acids are used in the synthesis of biodegradable polymers for tissue engineering and drug delivery applications. Formula: $C_3H_6O_3$. Also known as *α-hydroxypropionic acid; milk acid.*

lactide. A compound that is the cyclic dimer of *lactic acid* and exists in two isomeric forms D-lactide and L-lactide, and as a racemic mixture of the former two known as DL-lactide. Used in biochemistry and in the synthesis of biopolymers known as "polylactides." Formula: $C_6H_8O_4$. Also known as *3,6-dimethyl-1,4-dioxane-2,5-dione.* See also DL-lactide; L-lactide; poly(lactic acid); poly(lactide).

DL-lactide. A synthetic alloy of D-lactide and L-lactide with a melting point of 96-104°C (205-219°F), used as a copolymer with L-lactide and glycolide to produce biopolymers with fast biodegradation. See also poly(DL-lactide).

L-lactide. A compound, that occurs in nature and can also be obtained by heating L-lactic acid. It has a melting point of 92-94°C (198-201°F), and is used in the synthesis of biodegradable copolymers for tissue engineering and drug delivery applications. Also known as *(3S)-cis-3,6-dimethyl-1,4-dioxane-2,5-dione.* See also poly(L-lactide).

Lactilith. Belgian trade name for casein plastics.

lactobionic acid. A monocarboxylic acid that on hydrolysis yields D-gluconic acid and D-(+)-galactose. It has a melting point of 113-118°C (236-244°F). Used in biochemistry and biotechnology. Formula: $C_{12}H_{22}O_{12}$.

Lactofil. Trade name for regenerated protein (azlon) fibers.

Lactoid. Trade name of BX Plastics (UK) for a synthetic resin made from casein and formaldehyde.

Lactoloid. Japanese trade name for casein plastics.

Lactonite. Estonian trade name for a synthetic resin made from casein and formaldehyde.

Lactophane. Trade name of British Cellophane Limited (UK) for casein plastics.

lactoprene. Any of several synthetic rubbers made by the polymerization or copolymerization of an acrylic acid ester. They have an excellent resistance to low temperatures, weathering, ozone, oxygen, and hydrocarbon oils.

lactose. A disaccharide present in milk which is synthesized by the mammary glands of animals and humans from the blood glucose. It can also be obtained commercially from milk whey.

It is yields D-glucose and D-galactose on complete hydrolysis. Used in medicine, pharmaceuticals, biochemistry and biotechnology. Formula: $C_{12}H_{22}O_{11}$. Also known as *milk sugar*.

LactoSorb. Trademark of Biomet Inc. (USA) for an essentially amorphous, bioresorbable copolymer of 82% L-lactic acid and 18% glycolic acid. A methyl group is joined to one of the carbon atoms of the lactic acid, and a hydrogen atom is bonded to the corresponding carbon atom of the glycolic acid thus making the copolymer less crystalline. Used for craniofacial fixation applications, e.g., screws, plates and mesh.

Lactovac. Trade name of Fonderie de Précision SA (France) for a water-hardening steel containing 0.2-0.5% carbon, and the balance iron. Used for machinery parts.

Lactron. Trade name Kanebo Goshen Limited (Japan) for a light, stable, wrinkle-resistant corn-based polylactic acid fiber that can be blended with cotton and wool, and is used for textile fabrics including clothing.

Lacty. Trademark of Shimadzu Corporation (Japan) for polylactic acid (PLA) filaments and fibers.

ladder polymer. A network polymer composed of two polymer chains connected or crosslinked at regular intervals. Abbreviation: LP. See also network polymers.

Ladelloy. Trademark of Foseco Minsep NV (Netherlands) for a series of alloying elements used as ladle additions in the manufacture of iron and steel.

Ladene. Trade name of SABIC (USA) for a series of flexible and rigid polyvinyl chlorides supplied in the form of powders. The flexible grades are used for film, sheeting, flexible profiles, electrical insulation, cable coverings, hoses, and artificial leather, and the rigid grades for rigid and flexible profiles, film, piping and drainage pipes, pipe fittings, flexible sheeting, bottles, injection-molded parts, and coatings. Also included under this trade name are several high-density and linear-low-density polyethylenes.

ladle brick. A refractory brick of suitable size, shape, low porosity and permanent expansion for lining ladles used to hold or transfer molten metal.

ladle coating. A coating made from water glass, iron oxide and water, and applied to metal ladles used in the melting of aluminum alloys to prevent pickup of iron.

Laestra. Trade name of Lati SpA (Italy) for syndiotactic polystyrene resins used for electronic components and devices.

Lafarge cement. Nonstaining, white or whitish cement that contains lime, calcined gypsum and marble powder, and is similar in strength to *Portland cement*. Used as a mortar and grout in the setting and hardening of granite, limestone and marble.

laffittite. A bluish white mineral composed of silver mercury arsenic sulfide, $AgHgAsS_3$. Crystal system, monoclinic. Density, 6.07-6.19 g/cm^3. Occurrence: France, USA (Nevada).

Lagal. German trade name for a zinc-free aluminum-base bearing alloy.

Lagerbronze. German trade name for a corrosion-resistant bronze composed of 85.5-89% copper, 4-5% tin, 3.5-5% zinc and 3.5-4.5% lead, and used for machine and engine bearings and parts.

lagging. Materials, such as asbestos, diatomaceous earth, or fiberglass mats, used to insulate kilns and high-temperature pipes.

Laguval. Trade name of Bayer AG (Germany) for low-cost polyester plastics with excellent electrical properties that can be formulated for room- and high-temperature applications. Used for automotive components, boat hulls and parts, furniture, and housings.

lahore. A fine, soft dyed fabric made from *cashmere* and having a small dobby-like, woven design. Used especially for dresses. See also dobby.

laid fabrics. Fabrics in which there are no weft (filling) threads and the parallel warp threads are bonded by rubber latex or other adhesive materials.

laid paper. A good-quality paper marked with closely spaced parallel lines or watermarks. The line pattern may be impressed by means of a dandy roll.

laid rope. A rope made by twisting 3 or more strands of a roping material about each other to form a helix around the same central axis. See also soft laid rope; hard laid rope; cable laid rope; hawser laid rope.

laihunite. A black mineral of the olivine group composed of iron silicate, $Fe_{1.6}SiO_4$. Crystal system, monoclinic. Density, 3.92 g/cm^3. Occurrence: China.

laitakarite. A galena-white mineral of the tetradymite group composed of bismuth selenide sulfide, Bi_4Se_2S. Crystal system, rhombohedral (hexagonal). Density, 7.93 g/cm^3. Occurrence: Finland.

lake. An organic coloring matter composed of an oil-soluble organic dye (e.g., madder, cochineal or indigo), a precipitant and an absorptive inorganic substrate (e.g., aluminum, barium, calcium or chromium). Used for interior paints, coated textiles, plastics and rubber, and as decorative coatings for metal objects. Also known as *color lake; lake pigment*.

lake asphalt. A rich asphalt obtained from Pitch Lake in Trinidad. Also known as *lake pitch*.

lake copper. A native high-purity copper (99.9%) from the Lake Superior region containing very small percentages of silver, bismuth, mercury, arsenic and antimony. It has high electrical conductivity (100% IACS), good corrosion resistance, and excellent cold and hot workability.

lake pigment. See lake.

lake pitch. See lake asphalt.

lake ore. See bog iron ore.

Lake's metal. An antifriction metal composed of 87% lead, 7% antimony and 6% tin, and used for heavy-duty bearings.

Lakritz. Trade name of former Glas- und Spiegel-Manufaktur AG (Germany) for a patterned glass.

La-Led. Trade name of LaSalle Steel Company (USA) for a free-machining carburizing steel composed of about 0.1% carbon, 1% manganese, 0.3% sulfur, 0.25% lead, 0.05% phosphorus, and the balance iron. Used for automatic screw-machine products. *La-Led-X* is a free-machining steel composed of about 0.1% carbon, 1% manganese, 0.6% phosphorus, 0.3% sulfur, 0.25% lead, traces of tellurium, and the balance iron. Used for automatic screw-machine products, hardware, and fasteners.

L-Alex. Trade name of Argonide Nanomaterials (USA) for a nano-sized aluminum powder with hydrophobic coating used in propellants.

Lall. Swiss trade name for a wrought aluminum alloy containing zinc and magnesium.

La Lucette. British trade name for antimony metal (99.35% pure).

Lambeta. Trade name for rayon fibers and yarns used for textile fabrics.

Lambeth. Trade name for polyolefin fibers and products.

Lamb & Flag. Trade name of Lamb & Flag (UK) for commercially pure tin (99.5+%) containing about 0.38% lead, 0.025% copper and 0.02% arsenic.

lambskin. (1) Leather made from the skin of lambs or sheep, the only difference being the somewhat lighter weight of the leather made from lambskin. Also known as *lambskin leather*.

(2) The skin of a lamb or sheep prepared for use as a writing material. See also parchment.

(3) Imitation lambskin made of plush.

lambskin cloth. A cotton pile fabric, usually heavily wefted and in a weft-sateen weave.

lambskin leather. See lambskin (1).

lamb's wool. The fine fleece wool obtained from an unshorn sheep up to about 8 months old. It is soft, resilient and highly valued for spinning and knitting.

lamé. (1) A satin fabric made with metallic or simulated metallic weft (transverse) threads, and used in bookbinding especially for atlases.

(2) Any rich, usually plain-weave fabric made entirely or partially of gold, silver, or other metal threads. Fibers used in combination with metal threads include acetate, triacetate, nylon, polyester, silk and viscose rayon. Used for dresses and capes.

Lamelite. Trademark of Montecatini SpA (Italy) for a series of melamine-formaldehyde resins.

lamellar clay. A plastic clay exhibiting disk- or plate-like formations.

lamellar nanocomposites. A class of organic-inorganic *nanocomposites* composed of alternate layers of a polymer and inorganic materials, such as *chalcogenides*, metal oxides and phosphates, or clays. In exfoliated lamellar nanocomposites the ratio of polymer-to-inorganic layers is fixed and the number of intralamellar polymer chains is continuously variable. In intercalated nanocomposites the compositional ratio of the polymer chains alternating with the inorganic layers is fixed, and the number of polymer layers in the intralamellar spaces is well defined. Examples of such materials include plastic-superconducting polymer-$NbSe_2$, poly(ethylene oxide)/vanadium pentoxide and polyaniline/metal phosphate composites.

Lamicoid. Trademark of Mica Insulator Company (USA) for thermosetting phenol- and urea-formaldehyde engineering plastics filled with phlogopite- or muscovite-type *mica* powder. Available in the form of tubes, sheets and molded parts, they have good dielectric properties, electrical resistance, and high stiffness and dimensional stability. Used for electrical appliances and electronic components.

Lamigadut. Trademark of Schwartz GmbH (Germany) for a series of phenol-formaldehyde plastics.

Lamigamid. Trademark of Schwartz GmbH (Germany) for a series of polyamide (nylon) resins.

Lamiglas. Trade name of PPG Industries Inc. (USA) for a sheet glass used for laminating applications.

Lamigom. Trademark of Schwartz GmbH (Germany) for thermoplastic polyester and polyether urethane elastomers.

Lamilex. Trade name of Central Glass Company Limited (Japan) for a laminating glass.

Lamilite. Trade name of Lamilite Limited (Canada) for a flat laminated safety glass.

Lamilux. Trade name of Lamilux (UK) for unsaturated polyester resins.

Laminac. Trade name of American Cyanamid Company (USA) for a series of thermosetting polyester resins often supplied reinforced with glass fibers. They have outstanding dielectric properties, excellent electrical properties, low moisture absorption, good heat resistance up to 204°C (400°F), good chemical and flame resistance, good weatherability, good color stability, and pass high-frequency radio waves. Used for reinforced structural shapes, chairs, fans, helmets, boat hulls, autobodies, radio, television and video components, aircraft glazing, and radomes.

laminar composites. Composite materials, such as sandwich laminates, that consist of two or more two-dimensional layers or plies with two of their dimensions greatly exceeding that of the third. Usually, each layer has a preferred high-strength direction.

laminate. (1) A product made by joining two or more layers or plies of similar or dissimilar materials such that the orientation of the high-strength direction either varies or not with each successive layer.

(2) A product in which thin plates or sheets, such as glass or electrical steel, are united to form a panel of greater thickness for a specific application, e.g., laminated safety glass, laminated electrical contacts or laminated transformer cores.

(3) Metal sheets or bars made up of two or more layers built up to form a structural member.

(4) A ceramic or refractory material in the form of a sheet or panel made up of two or more different layers joined by a ceramic bond, e.g., a product having a silica core and outer layers of alumina. Used for lightweight thermal insulation, kiln furniture, etc.

(5) A fibrous or fiber-reinforced composite composed of two or more layers such that the orientation of the fibers varies with each successive layer.

(6) See plastic laminate.

laminated board. A board made by bonding together two or more plies of cardboard, paper or similar sheet material, using a special adhesive. Used in building construction and for furniture, automotive head and trunk liners, door panels, and glove boxes. See also laminated paperboard.

laminated composite. See structural laminate.

laminated fabrics. (1) Fabrics composed of an outer layer of textile joined to a continuous sheet of polymeric material, such as polyurethane or polyvinyl chloride, by bonding with an adhesive, or by heating. Used especially for children's wear, jackets and pants, and low-priced coats. See also bonded fabrics.

(2) Fabrics composed of layers of similar or dissimilar textiles joined by bonding or impregnating with a polymer.

laminated glass. A safety glass made by cementing two or more sheets of plate glass together with one or more interlayers of a transparent plastic material, such as acrylic resin, cellulose acetate, polyvinyl acetal, polyvinyl butyral or silicone resin. When cracked or broken, the glass adheres to the plastic. The thickness ranges from less than 25 mm (1 in.) up to 76 mm (3 in.). Used in banks, military equipment, armored cars, airplanes, automobiles, certain pressure cookers, protective shields and covers, etc. Abbreviation: LG. Also known as *laminated safety glass; nonshattering glass; shatterproof glass.*

laminated insulating wallboard. A large, rigid sheet or panel that consists of two or more layers of material, and is used as backing or finishing material on interior walls. Depending on the particular material used, it may also provide acoustic and/ or thermal insulation. Examples of such wallboard include plywood, laminated plastics and gypsum wallboard with aluminum-foil backing on one side or strong paper on both sides.

laminated iron. A thin sheet of commercial iron (99.6+% pure) used in the construction of transformer cores with high magnetic saturation and lower eddy-current losses than solid cores.

laminated metal. See clad metal.

laminated metal composites. Composite materials consisting of two or more layers of different metals, or metals and polymers, and usually joined by press bonding at elevated temperatures. One such product consists of a layer ultrahigh carbon steel (UHCS) bonded to a layer of brass (70Cu-30Zn), combining the high strength of UHCS with the noise damping capacity of brass for aircraft and aerospace structural applications. Another example is a laminated composite of lead and polymethyl methacrylate polymer (*Plexiglass*) used for nonstructural applications.

laminated molding. See plastic laminate.

laminated paperboard. Paperboard made by cementing together two or more layers of paper using a special weather-resistant bonding agent. Available in thicknesses from 3.2-9.5 mm (0.125-0.375 in.), they are used in building construction, furniture nd automotive liners. See also laminated board.

laminated phenolics. See phenolic laminates.

laminated plastic. See plastic laminate.

laminated polyvinyl-chloride steel. A product consisting of a steel core to one or both sides of which has been applied a corrosion-resistant, electrically insulating layer of polyvinyl chloride.

laminated product. A product made from sheets of similar or dissimilar materials united by a binder.

laminated safety glass. See laminated glass.

laminated veneer lumber. An engineered wood product consisting of wood veneers (nominal thickness less than 0.250 in., or 6.4 mm), with all grains oriented in the longitudinal direction, bonded with an exterior-grade adhesive. Supplied with a total thickness of 19-178 mm (0.75-7.0 in.) and various depths and lengths, it is used for structural applications including rafters, headers, beams, joists, studs, columns, scaffold planking, and flanges for prefabricated I-beams. Abbreviation: LVL.

laminated wood. An assembly made by gluing together layers of wood (e.g., veneer or lumber) with an adhesive so that the grain of all laminations is essentially parallel, such as found in laminated structural members, e.g., decking, beams, or arches. See also built-up laminated wood; glue-laminated wood; plywood; wood laminate.

laminating paper. (1) A special brown or white-colored paper with uniform basis weight, homogeneous structure and even absorption of an impregnating resin, used in the manufacture of plastic laminates.

(2) A decorative, colored or printed paper used for bonding to cardboard or pasteboard in the manufacture of book covers, boxes, and cartons.

Laminex. (1) Trade name of Santa Lucia Cristais Ltda. (Brazil) for a laminated glass.

(2) Trademark of Schwartz GmbH (Germany) for cresol- and phenol-formaldehyde plastics.

Lamisafe. Trade name of Asahi Glass Company Limited (Japan) for a laminated safety glass.

Lamita. Trade name for rayon textile fibers, yarns and fabrics.

Lamite. Trade name of Teledyne Vasco (USA) for a series of corrosion-resistant casting alloys of cobalt, chromium, molybdenum and tungsten, used for hardfacing electrodes.

Lamitex. Trade name of Franklin Fiber–Lamitex Corporation (USA) for an extensive series of thermosetting resins and laminates including asbestos paper-filled, cotton-, nylon- or glass-fabric-reinforced or graphite-filled phenolic resins, melamine-bonded cotton- or glass-fabric laminates, glass fabric-, cotton fabric- or paper-reinforced epoxy laminates, glass mat-reinforced polyester laminates and glass-fabric reinforced polyimides. Used especially for electrical and thermal applications

and various machine parts.

lammerite. A dark green mineral composed of copper arsenate, $Cu_3(AsO_4)_2$. Crystal system, monoclinic. Density, 5.18 g/cm^3; refractive index, 1.90. Occurrence: Bolivia.

Lamo. Trade name for rayon fibers and yarns used for textile fabrics.

Lamoflex. Trade name of Fibreglass Limited (UK) for glass-fiber reinforcing materials.

Lamoltan. Trademark of LACKFA Isolierstoffe GmbH & Co. (Germany) for a series of polyurethane foams and foam products.

lampas. A heavy fabric, usually made of cotton, acrylic, rayon or a yarn blend, having a multicolored, satiny figure or pattern produced by two warp threads, one of which binds the wefts and other forms a rep ground. Originally a printed silk fabric from India. Used for drapery and upholstery.

lampblack. An almost pure, finely divided *amorphous carbon* (soot) produced by incomplete combustion of liquid or gaseous hydrocarbons, e.g., coal tars, oils, or resins, in a closed system with an insufficient supply of air. Used as a black pigment in paints, inks, rubber, carbon paper, lead pencils, crayons, ceramics, linoleum, coatings, polishes, and soaps, in the production of tungsten and titanium carbide powders, as an ingredient of insulating compositions, furnace lutes, lubricants and carbon brushes, and as a reagent in the hardening of steel by cementation processes. Also known as *hydrocarbon black*.

lamprophyllite. A brown mineral of the seidozerite group composed of sodium barium strontium titanium silicate fluoride hydroxide, $Na_2(Sr,Ba)_2Ti_3(SiO_4)_4(OH,F)_2$. Crystal system, monoclinic. Density, 3.44 g/cm^3; refractive index, 1.754. Occurrence: Russian Federation.

Lanacryl. Trade name for acrylic fibers used for the manufacture of clothing.

lana fibers. The glossy silky seed hairs obtained from silk-cotton trees (family Bombacaceae) growing especially in Venezuela and the West Indies. It is quite similar to *kapok* and used accordingly.

Lanalpha. Trade name for rayon fibers and yarns used for textile fabrics.

Lanark. Trade name of Latrobe Steel Company (USA) for a shock-resisting tool steel (AISI type S5) containing 0.5-0.6% carbon, 1.9% silicon, 0.9% manganese, 0.16-0.18% chromium, 0.28% vanadium, 1.3% molybdenum, and the balance iron. It has outstanding toughness, excellent shock resistance, high elastic limit, good deep-hardening properties and good ductility. Used for stamps, punches, chisels, shear blades, and cold-work tools.

lanarkite. A greenish, white, pale yellow, or gray mineral composed of basic lead sulfate, $PbSO_4 \cdot PbO$. It can also be made synthetically by reaction of a mixture of equal parts of lead sulfate ($PbSO_4$) and lead monoxide (PbO) at 700°C (1292°F). Crystal system, monoclinic. Density, 6.92-7.02 g/cm^3; melting point, 977°C (1791°F); hardness, 2.0-2.5 Mohs; refractive index, 2.007. Occurrence: Scotland. Used in ceramics, paints, pigments, and as a source of lead sulfate.

Lanastil. Trademark of Stilon (Poland) for nylon 6 filament yarns.

Lancastalloy. Trade name of Lancaster Steel Company (USA) for a series of pearlitic malleable cast irons.

Lan-Cer-Amp. Trade name of American Metallurgical Products Company (USA) for a foundry alloy composed of 30.0+% lanthanum, 45-60% cerium, and a total of 20-24% neodymium and praseodymium, and used in the manufacture of ferrous and nonferrous alloys.

lancewood. The tough, durable, elastic, yellowish wood of the tree *Guatteria virgata* belonging to the custard-apple family and growing in the Tropical America. It has a uniform, straight grain, a fine texture, high elasticity and hardness. Average weight, 921 kg/m³ (57 lb/ft³). Used for whip handles, fishing rods, fine cabinetwork, and as a replacement for boxwood. See also Burma lancewood; Degami lancewood.

landauite. A black mineral composed of manganese zinc iron sodium titanium oxide, $NaMnZn_2(Ti,Fe)_6Ti_{12}O_{38}$. Crystal system, rhombohedral (hexagonal). Density, 4.42 g/cm³. Occurrence: Russian Federation.

Landerig's speculum. A white, hard alloy of 70% copper and 30% tin which takes a high polish, and is used for mirrors and reflectors.

landesite. A brown mineral composed of manganese iron phosphate hydroxide trihydrate, $Mn_3Fe(PO_4)_2(OH)_3 \cdot 3H_2O$. Crystal system, orthorhombic. Density, 3.03 g/cm³; refractive index, 1.728. Occurrence: USA (Maine).

land plaster. A finely ground plaster used as a fertilizer.

Lanese. Trademark of Celanese Corporation (USA) for fine cellulose acetate staple fibers.

Langalloy. Trade name of Langley Alloys Limited (UK) for a series of cast austenitic stainless steels including several free-machining grades, and various nickel-chromium-iron, nickel-molybdenum-chromium, nickel-copper, and nickel-iron-silicon-manganese casting alloys. Used for corrosion resistant parts, and valve and pump parts.

langbanite. A black mineral composed of calcium manganese iron silicate oxide, $(Mn,Ca)_4(Mn,Fe)_9SbSi_2O_{24}$. Crystal system, hexagonal. Density, 4.70 g/cm³; refractive index, 2.36. Occurrence: Sweden, Russian Federation.

langbeinite. A colorless, yellowish, greenish, or reddish mineral with vitreous luster composed of potassium magnesium sulfate, $K_2Mg_2(SO_4)_3$. It can also be made synthetically by melting potassium sulfate (K_2SO_4) and magnesium sulfate ($MgSO_4$). Crystal system, cubic. Density, 2.83 g/cm³; refractive index, 1.536; hardness, 3.5-4 Mohs. Occurrence: Germany, India, USA (New Mexico). Used as a source of potash, e.g., for fertilizers.

langisite. A pinkish buff mineral of the nickeline group composed of cobalt nickel arsenide, $(Co,Ni)As$. Crystal system, hexagonal. Density, 8.17 g/cm³. Occurrence: Canada (Ontario).

langite. A green to blue mineral composed of copper sulfate hydroxide monohydrate, $Cu_4SO_4(OH)_6 \cdot H_2O$. Crystal system, orthorhombic. Density, 3.31 g/cm³; refractive index, 1.70. Occurrence: Germany, UK.

Langmuir-Blodgett film. A thin film with a high degree of order, produced from an amphiphilic molecular surface layer by compression into a floating monolayer and subsequent transfer to a suitable substrate by an immersion technique. A Langmuir-Blodgett film can be as thin as one monolayer or built up to considerable thickness by stacking several hundred monolayers on a substrate. It has interesting electrical and optical properties which can be tailored by changing the composition and dopant concentration. Used in biosensors and photovoltaic devices, and in catalysis. Nanowires can be produced from Langmuir-Blodgett films by microlithographic processes.

Lanital. Trade name of Snia Viscosa (Italy) for soft, silky, spun azlon (protein) fibers derived from milk casein, and used for textile fabrics.

lannonite. A chalky white mineral composed of hydrogen calcium magnesium aluminum fluoride sulfate hydrate, $HCa_4Mg_2Al_4(SO_4)_8F_9 \cdot 32H_2O$. Crystal system, tetragonal. Density, 2.22 g/cm³; refractive index, 1.478. Occurrence: USA (New Mexico).

Lanon. Trademark of Gem-O-Lite (USA) for polyester fibers used for textile fabrics.

lansfordite. A colorless to white mineral composed of magnesium carbonate pentahydrate, $MgCO_3 \cdot 5H_2O$. It can also be made synthetically from saturated magnesium bicarbonate solutions under prescribed conditions. Crystal system, monoclinic. Density, 1.70 g/cm³; refractive index, 1.469. Occurrence: Italy.

Lansil. Trade name of Lansil Limited (UK) for cellulose acetate fibers and yarns used for wearing apparel and as industrial fabrics.

Lantern. Trade name of Toledo Steel Works (UK) for a case-hardening steel containing 0.15% carbon, 0.8% manganese, 0.05% sulfur, 0.05% phosphorus, and the balance iron. Used for machine parts.

lanthana. See lanthanum oxide.

lanthanides. A series of fourteen metallic rare-earth elements, closely related in chemical and physical properties, in which the $4f$ electron orbitals are filled. In the periodic table, they follow lanthanum (atomic number 57) beginning with cerium (atomic number 58) and ending with lutetium (atomic number 71). The *lanthanides* are subdivided into two groups: (i) the light or ceric rare earths including cerium, praseodymium, neodymium and promethium, and (ii) the heavy or yttric rare earths including samarium, europium, gadolinium, terbium, dysprosium, holmium, erbium, thulium, ytterbium and lutetium.

lanthanite. A colorless, white, yellow or pale pink mineral composed of lanthanum neodymium carbonate octahydrate, $(La,Nd)_2(CO_3)_3 \cdot 8H_2O$. Crystal system, orthorhombic. Density, 2.81 g/cm³; refractive index, 1.589. Occurrence: Brazil.

lanthanum. A white, malleable, ductile, metallic element of the lanthanide series (rare-earth group) of the Periodic Table. It is available in the form of ingots, lumps, chips, rods, sheets, foil, wire and powder, and is found in nature in minerals, such as *allanite, bastnaesite, cerite* and *monazite*. Density, 6.174 g/cm³; melting point, 921°C (1690°F); boiling point, 3457°C (6255°F); hardness, 40 Vickers; atomic number, 57; atomic weight, 138.906; trivalent; superconductivity critical temperature, 4.88K; tarnishes easily; takes a high polish. Three allotropic forms are known: (i) *Alpha lanthanum* (hexagonal close-packed crystal structure) that is present at room temperature; (ii) *Beta lanthanum* (face-centered cubic crystal structure) that is stable between 310 and 868°C (590 and 1594°F); and (iii) *Gamma lanthanum* (body-centered cubic crystal structure) that is stable between 868 and 921°C (1594 and 1690°F). *Lanthanum* is used for catalytic converters, as a reducing agent, for camara lenses, and in pyrophoric alloys, cigarette lighter flints, electronic devices, battery electrodes, phosphors for X-ray screens, rocket propellants and superconductors, and in the manufacture of lanthanum compounds. Symbol: La.

lanthanum aluminate. An oxide of lanthanum and aluminum that is available as a white powder and pure or doped single crystals. Crystal system, rhombohedral. Crystal structure, perovskite. Density, 6.51 g/cm³; melting point, 2180°C (3956°F); dielectric constant, 24.5-25. Used as a substrate for the deposition of high-temperature superconductor thin films, and for microwave and high-frequency applications. Formula: $LaAlO_3$. Also known as *lanthanum aluminum oxide (LAO)*. See also thorium-doped lanthanum oxide.

lanthanum aluminum oxide. See lanthanum aluminate.

lanthanum antimonide. A compound of lanthanum and antimony supplied in the form of a high-purity powder, and used as a semiconductor. Formula: LaSb.

lanthanum arsenide. A compound of lanthanum and arsenic supplied in the form of a high-purity powder, and used as a semiconductor. Formula: LaAs.

lanthanum barium copper oxide. A ceramic, high-purity superconductor powder, crystal or thin film similar to *yttrium barium copper oxide* and related rare-earth barium cuprates. Superconductivity critical temperature, approximately 35K. Used as a raw material for the manufacture of superconducting components. Formula: $LaBa_2Cu_3O_x$. Abbreviation: LBCO; LaBCO.

lanthanum boride. Any of the following compounds of lanthanum and boron: (i) *Lanthanum hexaboride.* Black, refractory crystals or fine, black, crystalline powder (99+% pure). Crystal system, cubic. Density, 2.61 g/cm³; melting point, 2210°C (4010°F); hardness, 2770 Vickers; high electrical conductivity. Used as an electron emitter in filaments for electron microscopes. Formula: LaB_6; (ii) *Lanthanum tetraboride.* Density, 5.44 g/cm³. Used in ceramics and materials research. Formula: LaB_4; and (iii) *Lanthanum triboride.* Density, 4.92 g/cm³. Used in ceramics and materials research. Formula: LaB_3.

lanthanum bromide. White, hygroscopic crystals (99.9% pure) with a density of 5.1 g/cm³ and a melting point of 788°C (1450°F) used in ceramics, chemistry, and materials research. Formula: $LaBr_3$. Also known as *lanthanum tribromide.*

lanthanum calcium manganites. A class of compounds with perovskite crystal structures that consist of *lanthanum manganite* doped with calcium ions ($La_{0.67}Ca_{0.33}MnO_3$). The lanthanum can also be partially substituted by erbium, ytterbium or bismuth ions ($La_{0.67-x}R_xCa_{0.33}MnO_3$). They exhibit a variety of phases depending on dopant concentration and temperature. Both substituted and unsubstituted *lanthanum calcium manganites* exhibit strong magnetism over a broad range of carrier concentrations and temperatures and colossal magnetoresistance. They are extremely useful for electronic and magnetic storage and high-sensitivity sensor applications. Also known as *lanthanum calcium manganese oxides (LCMO); calcium-doped lanthanum manganites.* See also manganites.

lanthanum calcium manganese oxides. See lanthanum calcium manganites.

lanthanum carbide. Any of the following compounds of lanthanum and carbon: (i) *Lanthanum dicarbide.* Yellow crystals. Density, 5.00-5.35 g/cm³; melting point, 2360°C (4280°F); exists in three forms: tetragonal at room temperature, hexagonal at intermediate temperatures and cubic at temperatures above 1750°C (3182°F). Used in ceramics and materials research. Formula: LaC_2; and (ii) *Lanthanum sesquicarbide.* Density, 6.08 g/cm³; melting point, 2021°C (3670°F). Used in ceramics and materials research. Formula: La_2C_3.

lanthanum carbonate octahydrate. A white, hygroscopic powder (99.9% pure) with a density of 2.6-2.7 g/cm³. Formula: $La_2(CO_3)_3 \cdot 8H_2O$.

lanthanum chloride. White, hygroscopic crystals or powder (98+% pure). Crystal system, hexagonal. Density, 3.84 g/cm³; melting point, 860°C (1580°F); boiling point, 1000°C (1830°F). Used in the preparation of lanthanum metal and rare-earth transition metal borosilicates. Formula: $LaCl_3$. Also known as *lanthanum trichloride.*

lanthanum chloride heptahydrate. White, transparent, hygroscopic crystals (99.9+% pure) that decompose at 91°C (196°F), and are used in the preparation of lanthanum metals and compounds, in superconductivity studies, as analytical standard, and in biological and biochemical stain technology as a marker

for intracellular junctions. Formula: $LaCl_3 \cdot 7H_2O$.

lanthanum dicarbide. See lanthanum carbide (i).

lanthanum disilicide. See lanthanum silicide (i).

lanthanum disulfide. See lanthanum sulfide (ii).

lanthanum-doped lead zirconate-lead titanate. See lead lanthanum zirconate titanate.

lanthanum-doped lead zirconate titanate. See lead lanthanum zirconate titanate.

lanthanum ferrite. A compound of lanthanum oxide and ferric oxide. Melting point, approximately 1870°C (3400°F). Used in ceramics, electronics, and materials research. Formula: $La_2Fe_2O_6$.

lanthanum fluoride. White or off-white, hygroscopic lumps, crystals or powder (99+% pure). Crystal system, hexagonal. Density, 5.94 g/cm³; melting point, 1493°C (2719°F). Used in phosphor lamp coatings (e.g., gallium arsenide solid-state lamps), in carbon arc electrodes, in the preparation of fluoride glasses, with neodymium oxide or doped with europium or neodymium in laser systems, and in fluoride-doped superconductors with high transition temperatures. Formula: LaF_3. Also known as *lanthanum trifluoride.*

lanthanum hafnate. A compound of lanthanum oxide and hafnium oxide used in ceramics and materials research. Formula: $La_2Hf_2O_7$.

lanthanum hexaboride. See lanthanum boride (i).

lanthanum indium. An intermetallic compound of lanthanum and indium supplied as a powder or crystalline substrate. Superconductivity critical temperature, 10.4K. Used for superconductors. Formula: La_3In.

lanthanum iodide. White, hygroscopic crystals or grayish white, hygroscopic powder (99.9% pure). Density, 5.63 g/cm³; melting point of 772°C (1421°F). Used in the preparation of lanthanum metal, compounds and complexes. Formula: LaI_3. Also known as *lanthanum triiodide.*

lanthanum manganites. A class of cubic compounds with perovskite crystal structures, such as undoped lanthanum manganite ($LaMnO_3$) and various doped manganites, such as *lanthanum calcium manganite* ($La_{1-x}Ca_xMnO_3$) and *lanthanum strontium manganite* ($La_{1-x}Sr_xMnO_3$). They exhibit giant or colossal magnetoresistance and are useful for electrical, electronic and magnetic storage applications and high-sensitivity sensors. Also known as *lanthanum manganese oxides (LMO).* See also manganites; lanthanum calcium manganites; lanthanum strontium manganites.

lanthanum manganese oxides. See lanthanum manganites.

lanthanum monosulfide. See lanthanum sulfide (i).

lanthanum nickelide. An intermetallic compound of lanthanum and nickel with a density of 8.3 g/cm³ and a hydrogen-storage capacity of about 1.4 wt%. Used for the negative negative electrode of nickel-metal hydride (Ni-MH) batteries. Formula: $LaNi_5$.

lanthanum nitrate. White, hygroscopic crystals (99.9+% pure) that contain about 31.7-32.7% lanthanum. Melting point, 40°C (104°F); boiling point, 126°C (259°F) (decomposes). Used in the production of gas mantles, in superconductivity studies, as a precursor to lanthanum-based superconductors, and in biology and biochemistry as a metal stain and as an intracellular tracer that specifically displaces calcium ions. Formula: $La(NO_3)_3 \cdot 6H_2O$. Also known as *lanthanum nitrate hexahydrate.*

lanthanum nitrate hexahydrate. See lanthanum nitrate.

lanthanum nitrate pentahydrate. White chunks or crystals that contain about 33.0-33.9% lanthanum. It has a melting point of

40°C (104°F) and is used in superconductivity studies. Formula: $La(NO_3)_3 \cdot 5H_2O$.

lanthanum nitride. A compound of lanthanum and nitrogen used in ceramics, electronics and materials research. Crystal system, cubic. Density, 6.73-6.85 g/cm^3. Formula: LaN.

lanthanum oxide. A white or buff, hygroscopic powder (99.9+% pure). Crystal system, rhombohedral, or amorphous. Density, 6.51 g/cm^3; melting point, 2315°C (4200°F); boiling point, 4200°C (7590°F). Used as a substitute for lime in calcium lights, in op-tical glass, technical ceramics, incandescent gas mantles, cores for carbon-arc electrodes, in fluorescent phosphors, in refractories, as atomic absorption standard, in the solid-phase synthesis of lanthanum-copper superconductors, and in biology and biochemistry as a metal stain. Formula: La_2O_3. Also known as *lanthana; lanthanum sesquioxide; lanthanum trioxide.*

lanthanum phosphide. A compound of lanthanum and phosphorus supplied in the form of high-purity crystals or powder. Density, 5.22 g/cm^3. Used for semiconductor applications. Formula: LaP.

lanthanum-ruthenium. An intermetallic compound of lanthanum and ruthenium used in ceramics and materials research. Melting point 1427°C (2601°F). Formula: $LaRu_2$.

lanthanum sesquicarbide. See lanthanum carbide (ii).

lanthanum sesquioxide. See lanthanum oxide.

lanthanum sesquisulfide. See lanthanum sulfide (iii).

lanthanum silicate. Any of the following compounds of lanthanum oxide and silicon dioxide used in ceramics and materials research: (i) *Lanthanum silicate.* Density, 5.72 g/cm^3; melting point, 1929°C (3504°F); hardness, 5-7 Mohs. Formula: La_2SiO_5; and (ii) *Dilanthanum trisilicate.* Density, 4.85 g/cm^3; melting point, 1749°C (3180°F). Formula: $La_4Si_3O_{12}$.

lanthanum silicide. Any of the following compounds of lanthanum and silicon: (i) *Lanthanum disilicide.* Gray crystals. Crystal system, tetragonal. Density, 5.0 g/cm^3. Used in ceramics and materials research. Formula: $LaSi_2$; and (ii) *Dilanthanum silicide.* Density, 5.14 g/cm^3; melting point, approximately 1520°C (2768°F). Used in ceramics and materials research. Formula: La_2Si.

lanthanum strontium manganites. A class of compounds with perovskite crystal structures that consist of *lanthanum manganite* doped with strontium ions ($La_{1-x}Sr_xMnO_3$). The lanthanum can also be partially substituted by other ions, such as erbium (Er^{3+}), ytterbium (Yb^{3+}), etc. They are analogous to the *lanthanum calcium manganites* in that they exhibit a variety of phases when the dopant concentration and temperature are varied, strong magnetism over a broad range of carrier concentrations and temperatures, and giant magnetoresistance. They are useful for electronic and magnetic storage and high-sensitivity sensor applications. Also known as *lanthanum strontium manganese oxides (LSMO); strontium-doped lanthanum manganites.* See also manganites.

lanthanum strontium manganese oxides. See lanthanum strontium manganites.

lanthanum sulfate. White, hygroscopic crystals (99.9+% pure). Density, 2.821 g/cm^3; refractive index, 1.564. Used in the atomic weight determination of lanthanum. Formula: $La_2(SO_4)_3 \cdot 9H_2O$.

lanthanum sulfide. Any of the following compounds of lanthanum and sulfur: (i) *Lanthanum monosulfide.* Yellow crystals. Crystal system, cubic. Density, 5.61-5.86 g/cm^3; melting point, 2300°C (4170°F). Used in ceramics and materials research. Formula: LaS; (ii) *Lanthanum disulfide.* Density, 4.90 g/cm^3;

melting point, 1650°C (3000°F). Used in ceramics and materials research. Formula: LaS_2; (iii) *Lanthanum sesquisulfide.* Red, or light yellow crystals or high-purity powder (99.9% pure). Crystal system, cubic or orthorhombic. Density, 4.911 g/cm^3; melting point, 2100-2150°C (3810-3900°F); band gap, 2.7 eV. Used in ceramics and as a semiconductor. Formula: La_2S_3; and (iv) *Lanthanum tetrasulfide.* Melting point, 2099°C (3648°F). Used in ceramics and materials research. Formula: La_3S_4.

lanthanum tetraboride. See lanthanum boride (ii).

lanthanum tetrasulfide. See lanthanum sulfide.

lanthanum titanate. Any of the following compounds used in ceramics and materials research: (i) *Lanthanum titanate.* A compound with perovskite structure made by heating a mixture of La_2O_3 and Ti_2O_3 to 1200°C (2190°F) in a vacuum. Formula: $LaTiO_3$; and (ii) *Lanthanum trititanate.* A compound made from lanthanum oxide and titanium oxide. Formula: $La_2Ti_3O_9$.

lanthanum triboride. See lanthamum boride (iii).

lanthanum tribromide. See lanthanum bromide.

lanthanum trichloride. See lanthanum chloride.

lanthanum trifluoride. See lanthanum fluoride.

lanthanum triiodide. See lanthanum iodide.

lanthanum trioxide. See lanthanum oxide.

lanthanum trititanate. See lanthanum titanate (ii).

lanthanum zirconate. A compound of lanthanum oxide and zirconium oxide that is available as a white powder for use in ceramics and materials research. Formula: $La_2(ZrO_3)_3$.

Lanusa. Trade name for rayon fibers and yarns used for textile fabrics.

Lanz. Trade name of Heinrich Lanz AG (Germany) for a high-strength cast iron containing 3.4% carbon, 2.8% silicon, 0.7% manganese, and the balance iron. Used for housings, frames, gears, and castings.

lap. (1) A single layer or sheet of fibers or fabric supplied in roll form wrapped on a core.

(2) See lapping abrasives.

lapacho. The wood of any of about 20 species of hardwood trees of the genus *Tabebuia.* It has high strength and weight, high durability, good insect and decay resistance, and moderate to poor resistance to marine borers. When air-dried, it is similar to *greenwood.* Only the sapwood can be treated with preservative. Source: Latin America except Chile. Used as a heavy-duty construction wood.

LAPCalloy. Trade name of Copper Development Association (USA) for a series lead-free brass casting alloys containing small additions of bismuth.

lap cement. A cementitious building product used to seal lateral and end laps or corrugated roofing sheets.

Lapco. Trade name of Light Alloy Products Company Limited (UK) for an aluminum alloy used for light-alloy parts.

lapis lazuli. A deep blue to greenish blue crystalline rock composed chiefly of the mineral *lazurite* with small percentages of *calcite* and *sodalite.* Used in powdered form as a blue pigment (*ultramarine blue*) in paints and inks, and as a semiprecious stone.

Lapox. Trade name of Atul Limited (India) for epoxy resins and hardeners, and urea- and phenol-formaldehyde resins for wood glues.

Lapelloy. Trade name of Carpenter Technology Corporation (USA) for a series of hardened and tempered stainless and heat-resistant steels containing 0.2-0.35% carbon, 11-12% chromium, 2.5-3% molybdenum, up to 0.3% vanadium, up to 0.5% nickel, up to 2.25% copper, and the balance iron. They have good heat, scale and oxidation resistance up to 760°C (1400°F). Used for high-temperature bolts, valve stems, and turbine blades and shafts.

Lapex. Trade name of Lati SpA (Italy) for a series of polyethylene sulfide resins.

laplandite. A light gray to yellowish mineral composed of sodium cerium titanium phosphate silicate pentahydrate, $Na_4CeTiPSi_7O_{22}\cdot5H_2O$. Crystal system, orthorhombic. Density, 2.83 g/cm^3; refractive index, 1.584. Occurrence: Russian Federation.

lapping abrasives. Abrasives, such as boron carbide and silicon carbide powder in fine grain sizes (200-400 mesh) or diamond dust, used in the preparation of lapping compounds for finish grinding.

lapping compounds. Mixtures of a loose lapping abrasive and a lapping lubricant, such as alcohol, gasoline, kerosene, lard, machine oil, soda water or turpentine, that is applied to the workpiece surfaces during the lapping process (a finish grinding or ultrafinishing process). Used in the production of smooth, accurate surfaces on soft and hardened steel, glass, cemented carbides and ceramics. They are often used on parts, such as pump housings, pistons of injection pumps, injection nozzles, machinery, machine-tool parts, engine components, measuring instruments, gages, tool bits, and inserts.

Laramid. Trade name of Lati SpA (Italy) for several polyphthalamide resins.

LARC. A series of thermoplastic and thermoset polyimide resins developed by NASA at the Langley Research Center (LARC), and available in liquid and powder form for use in the manufacture of polymer-matrix engineering composites.

larch. The relatively hard and strong wood from any of a genus of trees *(Larix)* of the pine family, especially the eastern larch *(L. laricina)* and the western larch *(L. occidentalis)* of North America, and the European larch *(L. decidua)*. Used as construction lumber, for railroad ties, poles and medium-quality furniture, and as a pulpwood for paper and fiberboard manufacture. See also American larch; eastern larch; western larch; European larch.

lardarellite. A white, or yellowish mineral composed of ammonium borate hydroxide, $NH_4B_3O_6(OH)_4$. Crystal system, monoclinic. Density, 1.90 g/cm^3; refractive index, 1.509. Occurrence: Italy.

Larflex. Trade name of Lati SpA (Italy) for olefinic-based thermoplastic elastomers including ethylene-propylene-diene-monomers.

large nine-inch brick. A rectangular brick measuring approximately 229 × 171 × 64 mm (9.00 × 6.75 × 2.50 in.) that is about 50% wider than a standard nine-inch brick measuring 229 × 113 × 64 mm (9.00 × 4.4375 × 2.50 in.).

Laril. Trade name of Lati SpA (Italy) for polyphenylene oxides supplied in unmodified, glass fiber-reinforced and fire-retardant grades.

Laripur. Trade name of COIM for thermoplastic polyurethanes.

larnite. See belite.

Larodui. Trademark of BASF AG (Germany) for a series of acrylic polymers.

larosite. A mineral composed of copper silver lead bismuth sulfide, $(Cu,Ag)_{21}(Pb,Bi)_2S_{13}$. Crystal system, orthorhombic. Occurrence: Canada (Ontario).

Larpeek. Trade name of Lati SpA (Italy) for several polyether ether ketone resins.

Larport. Trade name of Osborn Steels Limited (UK) for a tough, oil- or air-hardening steel containing 0.3% carbon, 1.25% chro-

mium, 0.3% molybdenum, 4.25% nickel, and the balance iron. Used for plastic mold dies.

Larsenite. Trademark of Blasch Precision Ceramics (USA) for a ceramic composite composed of an alumina (Al_2O_3) matrix with silicon carbide (SiC) reinforcement, used for injection-molded parts. It has good thermal-shock resistance and high-temperature properties to above 1650°C (3000°F).

larsenite. A colorless, or white mineral composed of lead zinc silicate, $PbZnSiO_4$. Crystal system, orthorhombic. It can also be made synthetically. Density, 5.90 g/cm^3; refractive index, 1.95. Occurrence: USA (New Jersey).

Larton. Trade name of Lati SpA (Italy) for glass fiber- and/or glass bead-reinforced polyphenylene sulfides.

LaSalle. Trade name of LaSalle Steel Company (USA) for a series of steels supplied in the form of cold-finished bars and including various plain-carbon grades (AISI-SAE 10xx series), resulfurized free-machining grades (AISI-SAE 11xx series), resulfurized and rephosphorized free-machining grades (AISI-SAE 12xx series), standard and leaded chromium-molybdenum alloy grades (AISI-SAE 41xx series), and standard and leaded nickel-chromium-molybdenum alloy grades (AISI-SAE 86xx series).

Laselon. Trade name for polyester fibers and yarns used for textile fabrics.

laser crystals. High-purity single crystals used to generate laser beams. The original material for this purpose was a single-crystal ruby rod, i.e., a rod composed of aluminum oxide (Al_2O_3) doped with chromium (Cr^{3+}) ions. Modern laser crystals include titanium-doped sapphire, doped gadolium gallium garnet and yttrium vanadate, yttrium aluminum garnet, and yttrium lithium fluoride doped with neodymium (Nd^{3+}) ions.

laser-cut fabrics. Textile fabrics that have a design or pattern cut-in by selectively vaporizing the material with a narrow-beam laser.

laser dyes. Organic dyes, such as phthalocyanine, *p*-terphenyl, coumarin, or rhodamine, used as active elements in liquid lasers.

LaserForm. Trade name of DTM Corporation (USA) for a powder-metallurgy martensitic stainless steel (AISI type 420) alloy used for rapid prototyping of metals and plastics.

laser glass. A fluorescent glass compounded such that it can amplify electromagnetic radiation (e.g. visible radiation) upon proper electromagnetic stimulation. Typically, it is doped or mixed with several wt% neodymia (Nd_2O_3), and is used for high-powered laser applications. Abbreviation: LG.

laser materials. A group of materials used in lasers including high-purity single crystals, such as ruby, yttrium garnets and metallic molybdates or tungstates doped with rare-earth ions, several gases including argon, helium, neon and carbon dioxide, semiconductors (e.g., gallium arsenide), neodymium-doped glass, and plasmas. Abbreviation: LM.

Laserite. Trade name of DTM Corporation (USA) for several glass-filled nylons with high stiffness and heat resistance used in rapid prototyping.

Laserskin. Trademark for a biocompatible and biodegradable material made with total esterified *hyaluronan*, and supplied in the form of a 20 μm (0.0008 in.) thick membrane with regular, laser-produced microperforations. Used as artificial skin, and as a delivery system for cultured keratinocytes.

Last-A-Foam. Trademark of General Plastics Manufacturing Company (USA) for a series of structural composite cores designed for aircraft interior panel cores and sandwich panels.

Last-A-Foam FR-3700 is a rigid, high-density, flame-resistant polyurethane foam product used for aircraft components, such as edge closeouts and hard points in honeycomb cored interior laminated panels. *Last-A-Foam FR-10100* is a rigid, high-density polyisocyanurate foam product having good flame resistance and high-temperature strength. Flexible polyurethane foam grades are also available.

Lastane. Trade name of Lati SpA (Italy) for several urethane-based thermoplastic elastomers.

LasTex. Trademark of The Rubber Group (USA) for natural rubber supplied in the form of surgical-grade tubing.

Lastex. (1) Trade name of American Rubber Company (USA) for rubber fibers around which silk, rayon, cotton or synthetic fibers have been wound to make elastic yarn. *Lastex S is a spandex* (segmented polyurethane) fiber used for the manufacture of textile fabrics with excellent stretch and recovery properties.

(2) Trade name for the elastic textile fabrics obtained from the fibers described in (1), used especially for sportswear and swimsuits.

Lastic. Trade name of Roydent Dental Products (USA) for a silicone elastomer used for dental impressions.

Lasticomp. Trade name of Roydent Dental Products (USA) for a silicone elastomer used for dental impressions.

Lastil. Trade name of Lati SpA (Italy) for styrene-acrylonitriles supplied in unmodified, glass fiber-reinforced, fire-retardant and UV-stabilized grades.

Lastilac. Trade name of Lati SpA (Italy) for a series of acrylonitrile-butadiene-styrene (ABS) resins supplied in glass fiber-reinforced, medium- and high-impact, fire-retardant, high-heat, UV-stabilized, structural foam and ABS/polycarbonate alloy grades.

lasting. A soft, closely-woven textile fabric made from cotton or worsted yarn in a satin or twill weave and having a smooth, glossy surface. Used for upholstery, bag linings, and protective clothing.

Lastirol. Trade name of Lati SpA (Italy) for high-impact polystyrenes (HIPS) supplied in unmodified, UV-stabilized and fire-retardant grades. HIPS film products are also available under this trade name.

Last-O-Flash. Trade name of Manville Corporation (USA) for fiberglass scrim used to weatherproof roof flashing.

Last-O-Roof. Trade name of Manville Corporation (USA) for a one-ply roofing product composed of a principal membrane faced with polyisobutylene and welded to an asbestos support. The latter is replaced by woven glass scrim for flashings.

lastrile fibers. A generic term for synthetic fibers made from resins obtained as the copolymerization products of acrylonitrile with a diolefin (e.g., butadiene). The content of the acrylonitrile units is typically between 10 and 50 wt%.

Lasulf. Trade name of Lati SpA (Italy) for polysulfone resins supplied in unreinforced and 10 or 30% glass fiber-reinforced grades.

Latacalk. Trademark of Chemtron Manufacturing Limited (USA) for latex rubber caulking compounds.

Latamastic. Trademark of Laticrete International, Inc. (USA) for latex resin adhesives used for installing ceramic tile on interior floors and walls.

Latamid. Trade name of Lati SpA (Italy) for a series of polyamide (nylon) resins. *Latamid 6* refers to series of resins based of polyamide 6 (nylon 6) marketed in various grades: unmodified, glass or carbon fiber-reinforced, glass bead- or mineral-filled, silicone-, molybdenum disulfide- or polytetrafluoroethylene-lubricated, fire-retardant, high-impact and UV-stabilized.

Latamid 12 includes various resins based on polyamide 12 (nylon 12) and supplied in several grades: unmodified, flexible, semi-flexible, glass fiber-reinforced and UV-stabilized. *Latamid 66* refers to polyamide 6,6 (nylon 6,6) resins available in unmodified, glass or carbon fiber-reinforced, glass bead- or mineral-filled, silicone-, molybdenum disulfide- or polytetrafluoroethylene-lubricated, fire-retardant, high-impact, super-tough and UV-stabilized grades. *Latamid 66* includes various resins based of polyamide 6,6 (nylon 6,6).

Latan. Trade name of Lati SpA (Italy) for a series of acetal homopolymers and copolymers based on polyoxymethylene available in various grades: glass or carbon fiber-reinforced, silicone- or polytetrafluoroethylene-lubricated and UV-stabilized.

Latapanel. Trademark of Laticrete International Inc. (USA) for building boards and panels and prefabricated floor assemblies made from fiber-reinforced cement.

Latapoxy. Trademark of Laticrete International, Inc. (USA) for cements, grouts and mortars with or without epoxy additives, used for plasters and stucco, and for installing brick, stone or tile.

Latene. Trade name of Lati SpA (Italy) for a series of polyethylene and polypropylene homopolymers available in calcium carbonate- or talc-filled, glass-reinforced and fire-retardant grades. *Latene EP* refers to a series of polyethylene and polypropylene copolymers.

Later. Trade name of Lati SpA (Italy) for a series of polybutylene terephthalate resins available in various grades: unmodified, glass fiber- or glass bead-reinforced, mineral-filled, fire-retardant and UV-stabilized.

laterite. A red, weathered, hard, dense argillaceous rock or porous, earthy soil. It is essentially a mixture of the oxides of iron, aluminum, titanium and manganese, and may also contain bauxite.

latex. (1) The milky, viscous substance obtained from certain plants, such as rubber plants, milkweed, or poppies. It is a mixture of water, proteins, gums and carbohydrates, and can also be made synthetically. Used in the manufacture of rubber, balata and gutta-percha. Also known as *rubber latex*. See also hevea rubber.

(2) A white, viscous suspension of a natural or synthetic rubber or plastic in water. Used in adhesives, paints, and sealants.

latex cement. (1) A cementing material made by blending *Portland cement* with latex, a surfactant, and water. Upon setting it provides a bond that is stronger, tougher and more durable than that obtained with standard Portland cement.

(2) A tacky solvent solution of rubber latex used to join leather, paper, textiles, wood, etc.

latex coating. See latex paint.

latex foam. A foam rubber made from rubber latex and used for transport seating. Also known as *latex foam rubber*.

latex foam rubber. See latex foam.

latex-modified portland cement concrete. Concrete to which a polymer emulsion has been added resulting in a dense hardened cementitious product resisting the movement of moisture and chloride ions. It is usually used as a relatively thin overlay (25-38 mm, or 1-1.5 in.) on conventional concrete bridge decks.

latex paint. Paint composed of an aqueous dispersion of a synthetic resin (e.g., an acrylic, polyvinyl acetate, or styrene-butadiene) and produced by emulsion polymerization. Surfactants and protective colloids are necessary to stabilize the product. *Latex paint* is available in various colors. Interior latex paints are fast-drying finishes that impart satin or semi-gloss textures on interior walls, door and trim work, cabinets, other woodwork, and ceilings. Exterior latex paints are fast-drying finishes that are resistant to blistering, cracking, peeling and weathering. They can be used on wood siding, prefinished aluminum, vinyl and galvanized siding, and masonry, in automotive topcoats, and for coating appliances. Also known as *latex coating; latex water paint.*

latex patching compound. An alkali-, mildew-, and moisture-resistant cementing material made from *Portland cement*, synthetic rubber latex (e.g., styrene-butadiene rubber) and one or more aggregates. Used in construction as patching and for leveling applications.

latex sealant. (1) A rubber latex compound that cures or hardens to a resilient solid by water evaporation, and is used for sealing joints to prevent the entry of gases and/or liquids.

(2) An emulsion or suspension of a synthetic rubber (e.g., an acrylic or polysulfide rubber) in water. Used in building and construction as a joint sealant for interior and exterior applications.

Latexsol. Trademark of Sico Inc. (Canada) for latex floor paint used on concrete and wood surfaces.

latex water paint. See latex paint.

lath. A material in sheet form used as a base for plaster or tile on walls and ceilings. The most common types are: (i) *Wood lath* comprising a slim, narrow strip of lumber and formerly widely used to form a support for the plaster on a wall or ceiling, or make a lattice. Now, it is largely replaced by gypsum or expanded metal laths; (ii) *Gypsum lath* composed of a flat sheet or slab consisting of a rigid gypsum core with a specially treated paper cover. A standard size lath is 410 mm (16 in.) wide, 1220 mm (48 in.) long and 9.5-12.7 mm (0.375-0.500 in.) thick. It is often made with an aluminum foil vapor barrier; and (iii) *Expanded metal lath* usually composed of a copper-alloy sheet steel, slit and expanded to form openings for keying plaster. Common types of this lath include diamond mesh and flat rib. *Expanded metal lath* is usually made rust-resistant by dipping in black asphaltum paint or by galvanizing. A standard lath of this type is 690 mm (27 in.) wide and 2450 mm (96 in.) long.

lath brick. A rectangular brick whose length is much greater than its width and height.

Laticel. Trademark of Lati SpA (Italy) for a series of cellulose acetate plastics.

Laticrete. Trademark of Laticrete International, Inc. (USA) for waterproofing membranes for concrete and masonry, and rubber cement used for laying floor covering. Also included under this trademark are liquid latex compositions used with (i) cement and additives and/or fillers as tile and wall grouts and underlayments for patching and leveling walls and floors; (ii) Portland cement and sand to make mortars; (iii) Portland cement, mortars, dry grouts and sand to make grouts and mortars; and (iv) latex polymer admixtures for bonding agents for brick, concrete, grout, marble, mortar, plaster, stone and tile.

latigo leather. A tough, alum- or vegetable-tanned leather used in saddlery.

Latilon. Trade name of Lati SpA (Italy) for a series of polycarbonate resins available in various grades including unmodified, glass or carbon fiber-reinforced, fire-retardant and UV-stabilized.

latiumite. A colorless mineral composed of calcium potassium aluminum silicate sulfate, $(Ca,K)_4(Si,Al)_5O_{11}(SO_4,CO_3)$. Crystal system, monoclinic. Density, 2.93 g/cm³; refractive index,

1.606. Occurrence: Italy.

Laton. Trade name for natural rubber fibers used in the manufacture of elastic yarn for clothing, elastic bands and tapes, etc.

latrappite. A black mineral of the perovskite group composed of calcium niobium titanium oxide, $(Ca,Na)(Nb,Ti,Fe)O_3$. Crystal system, orthorhombic. Density, 4.40. Occurrence: Canada (Quebec).

Latrobe. Trade name of Latrobe Steel Company (USA) for an extensive series of machinery steels, plain-carbon, cold-work, hot-work and high-speed tool steels, and various nickel-base and cobalt-base superalloys.

latten. A metal, such as brass, steel or iron, supplied in the form of thin sheets, and used for architectural and ornamental applications.

latten brass. A copper-zinc alloy (usually *yellow brass*) rolled into thin sheets, and used for architectural and ornamental applications, hardware, and fixtures.

lattice block materials. Honeycomb-shaped structures consisting of lattice networks of fine metal or alloy wires. They are formed by first producing single flat lattices from three racks of fine wires (e.g., aluminum, magnesium, titanium, or steel wires) stacked and welded together at angles of 60°, incorporating alternate corrugated wire layers, welding the flat and corrugated layers together, and then filling the matrix with a suitable material, such as aluminum, ceramics, epoxy, or thermoplastic resins. The formed structures have high strength, low weight, good machinability and weldability, and are suitable for structural applications in the aerospace and automotive industries. Abbreviation: LBMs.

lattice brick. Hollow building blocks or perforated bricks (e.g., with diamond-shaped openings) that have very low thermal conductance, and are used for thermal insulation applications.

lauan. The dark red to light tan wood of any of several genera of trees (*Shorea, Parashorea* and *Pentacme*). It resembles genuine (American) mahogany, but is coarser in texture and less decay-resistant. *Lauan* wood is grouped for marketing purposes as *Dark-red Philippine mahogany* and *Light-red Philippine mahogany*. It has medium hardness and strength, moderate shrinkage, an open grain, and is quite easy to work. Average weight, 593 kg/m³ (37 lb/ft³). Source: Philippines, Malaya, Sarawak. Used in medium-quality furniture, cabinets, fixtures, interior trim, flush doors, wall paneling, siding, boatbuilding, and as corestock in plywood. Also known as *Philippine mahogany.* See also mahogany; tangile.

laubmannite. A yellow green mineral composed of iron phosphate hydroxide, $Fe_9(PO_4)_4(OH)_{15}$. Crystal system, orthorhombic. Density, 3.41 g/cm³. Occurrence: USA (Arkansas).

laueite. A brown mineral of the paravauxite group composed of manganese iron phosphate hydroxide octahydrate, $MnFe_2(PO_4)_2(OH)_2 \cdot 8H_2O$. Crystal system, triclinic. Density, 2.46 g/cm³; refractive index, 1.658. Occurrence: Germany.

laumontite. A colorless, white, yellow, red, brown mineral of the zeolite group composed of calcium aluminum silicate tetrahydrate, $CaAl_2Si_4O_{12} \cdot 4H_2O$. It can also be made synthetically. Crystal system, monoclinic. Density, 2.31 g/cm³.

launayite. A lead gray mineral composed of lead antimony sulfide, $Pb_{22}Sb_{26}S_{61}$. Crystal system, monoclinic. Density, 5.98 g/cm³. Occurrence: Canada (Ontario).

launderproof fabrics. Textile fabrics which do not lose color or shrink when washed under normal conditions and for prescribed periods of time.

Lauramid. Trade name of Degussa-Huels AG (Germany) for

polyamide 12 (nylon 12) resins used especially for machine components and industrial equipment.

laurel. (1) The wood from any of several trees of the genus *Cordia* including *C. alliodora, C. goeldiana* and *C. trichotoma*. It saws and machines well, and has medium strength and an attractive appearance. Source: West Indies, Central and South America. Used for cabinetwork, furniture, and boatbuilding.

(2) The strong, hard, golden brown to yellowish green wood from trees of the genus *Myrtus,* found in the southwestern United States. It takes an excellent polish and is used for highly figured veneers, panels and furniture, and turnings including bowls, trays and candlesticks. Also known as *California laurel; myrtle.*

Laurentian. Trademark of Atlas Specialty Steels (Canada) for a hollow drill steel.

laurionite. A colorless mineral composed of lead chloride hydroxide, PbClOH. It can also be made synthetically by a reaction between tetragonal lead dioxide (in aqueous suspension) and calcium chloride under prescribed conditions. Crystal system, orthorhombic. Density, 6.24 g/cm^3; refractive index, 2.116. Occurrence: Greece.

laurite. An iron-black mineral of the pyrite group composed of ruthenium sulfide, RuS$_2$. It can also be made synthetically. Crystal system, cubic. Density, 6.00 g/cm^3. Occurrence: Borneo, USA (Oregon).

Laur Silicone. Trade name of Laur Silicone Rubber Compounding, Inc. (USA) for heat-curing silicone rubber materials.

lausenite. A white, fibrous mineral composed of ferric sulfate hexahydrate, Fe$_2$(SO$_4$)$_3$·6H$_2$O. Density, 2.0 g/cm^3. Occurrence: USA (Arizona).

Lautal. Trade name of Vereinigte Leichtmetallwerke GmbH (Germany) for a wrought, heat-treatable, high-strength aluminum alloy containing 4.5-5.5% copper, up to 0.75% manganese, 0.2-0.75% silicon, and traces of iron, magnesium and manganese. Used for hardware, aircraft construction, and electric cables.

lautarite. A colorless to light yellow mineral composed of calcium iodate, Ca(IO$_3$)$_2$. It can also be made synthetically by heating calcium iodate hexahydrate under prescribed conditions. Crystal system, monoclinic. Density, 4.59 g/cm^3; refractive index, 1.840. Occurrence: Chile.

lautite. A black mineral of the sphalerite group composed of copper arsenic sulfide, CuAsS. Crystal system, orthorhombic. Density, 4.90 g/cm^3. Occurrence: Germany.

Lava. Trade name of Fabbrica Pisana SpA (Italy) for a patterned glass.

lava. A ceramic material, such as calcined talc, magnesia, magnesium silicate, mullite-alumina compounds, or steatite, used in the molding of nozzles, burner tips, electrical insulators, etc.

lava rock. A red or blackish volcanic rock supplied in sized aggregate grades for landscaping applications, and gas barbecues.

lava talc. See steatite talc.

Lavelssiere bronze. A corrosion-resistant bronze composed of 61% copper, 38% zinc and 1% tin, and used for condenser and heat-exchanger tubing, marine parts, marine hardware, and fasteners.

lavendulan. A blue mineral composed of sodium calcium copper chloride arsenate hydrate, NaCaCu$_5$(AsO$_4$)$_4$Cl·5H$_2$O. Crystal system, orthorhombic. Density, 3.34 g/cm^3; refractive index, 1.748. Occurrence: Chile, Greece.

lavenite. A colorless, yellow or brown mineral of the wohlerite group composed of sodium zirconium fluoride silicate, (Na,Ca,Mn)$_3$Zr(SiO$_4$)$_2$F. Crystal system, monoclinic. Density, 3.50 g/cm^3; refractive index, 1.750. Occurrence: Norway.

Laveten. Trademark of Primo Sverige AB (Sweden) for polyethylene and polypropylene monofilaments and fibers.

Lavin. Trade name of R. Lavin & Sons, Inc. (USA) for a series of copper-base alloys including several cast aluminum and manganese bronzes and nickel silvers as well as a wide range of deoxidizers, degasifiers and densifiers for the manufacture of copper alloys.

Lavscan. Trade name for polyester fibers and yarns used for textile fabrics.

lawn. A fine, sheer cotton, cotton blend, linen or synthetic fabric in a plain weave, used for blouses, dresses and lingerie.

lawrencite. A green or brown mineral composed of iron chloride, FeCl$_2$. Crystal system, rhombohedral (hexagonal). It can also be made synthetically. Density, 3.16 g/cm^3; melting point, 677°C (1251°F); boiling point, 1026°C (1879°F); refractive index, 1.567.

lawrencium. A short-lived synthetic radioactive element produced from californium (^{98}Cf) by bombarding with boron nuclei (^{10}B or ^{11}B). Atomic number, 103; atomic weight, 262; several other isotopes are known including ^{256}Lr, ^{257}Lr and^{260}Lr; half-life (^{257}Lr), 8 seconds; emits alpha radiation. Symbol: Lr.

lawsonbauerite. A colorless to white mineral composed of magnesium manganese zinc sulfate hydroxide octahydrate, (Mn,Mg)$_9$Zn$_4$(SO$_4$)$_2$(OH)$_{22}$·8H$_2$O. Crystal system, monoclinic. Density, 2.87 g/cm^3; refractive index, 1.608. Occurrence: USA (New Jersey).

Lawson cypress. The pale yellow to pale brown, lightweight, moderately strong high-grade wood of the tree *Chamaecyparis lawsoniana.* The heartwood is highly decay-resistant and almost indistinguishable from the sapwood. *Lawson cypress* has a strong cedar scent, a straight grain, a fine texture, good hardness and toughness, good drying properties, high stability in use, high durability, good resistance to acids, works well with hand and machine tools, and takes a good finish. Average weight, 460 kg/m^3 (29 lb/ft^3). Source: Western United States from Oregon southward along Pacific Coast to California. Used for boatbuilding, lumber for construction, sashes and doors, flooring and interior finishes, and for battery separators, mothproof chests, stadium seats, archery supplies, and patterns. Also known as *ginger pine; Oregon cedar; Port Orford cedar.*

lawsonite. A colorless or white bluish mineral composed of calcium aluminum silicate hydroxide monohydrate, CaAl$_2$Si$_2$O$_7$(OH)$_2$·H$_2$O. It can also be made synthetically. Crystal system, orthorhombic. Density, 3.07 g/cm^3. Occurrence: France, Italy, Cuba, New Caledonia, USA (California).

layered ceramic composites. A class of ceramic composites consisting of *green ceramic tapes* arranged in stacks, and laminated by a technique usually involving low heat and pressure.

layered coating. See sandwich coating.

layered silicates. See phyllosilicates.

layer silicates. See phyllosilicates.

lazarenkoite. A bright orange mineral composed of calcium iron arsenic oxide trihydrate, (Ca,Fe)FeAs$_3$O$_7$·3H$_2$O. Crystal system, orthorhombic. Density, 3.45 g/cm^3; refractive index, 1.920. Occurrence: Siberia.

lazulite. A violet to azure blue mineral with a vitreous luster composed of magnesium aluminum phosphate hydrate, MgAl$_2$(PO$_4$)$_2$(OH)$_2$. Crystal system, monoclinic. Density, 3.10-3.12 g/cm^3; refractive index, 1.636; hardness, 5-6 Mohs. Occurrence: Austria. Also known as *berkeyite; blue spar.*

lazurite. A dark blue, greenish blue or violet mineral of the sodalite group with a vitreous luster composed of sodium calcium aluminum sulfosilicate, $(Na,Ca)_8(AlSiO_4)_6(S,SO_4,Cl)_x$. Crystal system, cubic. Density, 2.38-2.45 g/cm^3; hardness, 5.0-5.5 Mohs; refractive index, approximately 1.50. Occurrence: Afghanistan. It is main component of *lapis lazuli,* and used as a semiprecious stone, and as a blue paint pigment (*ultramarine*). Also known as *false lapis.*

L.B. Blend. Trade name of Aardvark Clay & Supplies (USA) for a Long Beach Blend medium-texture clay (cone 10) made brown by the addition of colorants.

LB Bond. Trade name of Kuraray Company Limited (Japan) for a light-cure dental bonding resin.

L.B. White. Trade name of Aardvark Clay & Supplies (USA) for a Long Beach Blend buff to off-white clay (cone 10).

LC Astroloy. Trade name of Cannon-Muskegon Corporation (USA) for a nickel-base powder-metallurgy superalloy containing 15.1% chromium, 17.0% cobalt, 5.2% molybdenum, 4.0% aluminum, 3.5% titanium, up to 0.01% zirconium, 0.024% boron and 0.023% carbon. It has high heat and corrosion resistance, high stress-rupture and creep strength, and is used for turbine wheels and blades, nuclear reactor components, petrochemical equipment, and aerospace and aircraft applications, e.g., jet-engine components, nozzles, and compressor disks.

LC-Monel. Trade name of Huntington Alloys Inc. (USA) for wrought nickel alloys containing 13% copper, 1.3% iron and 0.12% carbon. Used for corrosion-resistant parts.

LD steel. A basic-oxygen furnace steel made from a charge of molten pig iron, scrap steel and fluxes in a basic-refractory lined furnace by blowing high-purity oxygen at supersonic speed through a water-cooled lance onto the surface of the molten metal to accelerate the burning off of unwanted elements. The abbreviation "LD" stands for "Linz-Donawitz" referring to the two Austrian steel mills that developed the process.

Leabrament. Trade name of JacksonLea (USA) for a polishing ce-ment.

Leabrite. Trade name of JacksonLea (USA) for a series of chemical polishes, room-temperature anodizing sealants, and plating products and processes.

leachable ceramics. Engineering ceramics composed of about 50% silica (SiO_2), 40% zircon $(ZrSiO_4)$ and 10% alumina (Al_2O_3). Commercially available in the form of rods and tubes, they have a density of approximately 2.1 g/cm^3 (0.08 $lb/in.^3$), an apparent porosity of 25%, a water absorption of 14%, a low coefficient of thermal expansion (up to 1000°C or 1830°F), an upper continuous-use temperature of 1050°C (1920°F), good resistance to metals, fair resistance to acids and halogens, and poor resistance to alkalies.

leached-glass fibers. A group of continuous silica fibers (97+%) whose manufacture involves an ion-leaching process. Often sold under trademarks, such as Refrasil or Sil-Temp, they have densities of about 2.1-2.2 g/cm^3 (0.07-0.08 $lb/in.^3$), a melt temperature of above 1760°C (3200°F) and excellent high-temperature resistance, but only moderate mechanical properties. Used as high-temperature insulation for jet aircraft, missiles, rockets, and motor components, and for insulation purposes in the molten metals and heat-treating industries.

Leacril. Trademark of Montefibre Fibre SpA (Italy) for polyacrylonitrile (acrylic) polymers and copolymers, and continuous and staple fibers as well as flock and textile fabrics made of such fibers. *Leacril* acrylic yarns are made of 88% polyacrylonitrile and 12% polyamide (nylon) and used for clothing.

lead. A soft malleable, easily melted, silvery-bluish-white to bluish-gray metallic element with bright metallic luster. It belongs to Group IVA (Group 14) of the Periodic Table, and is is commercially available in the form of ingots, sheet, foil, microfoil, pipe, tube, shot, buckles or straps, lumps, grids, bars, rods, wire, paste, powder, granules, and single crystals. The single crystals are usually grown by the Bridgeman technique. Commercial lead ores include *galena, anglesite* and *cerussite.* Crystal system, cubic. Crystal structure, face-centered cubic. Density, 11.35 g/cm^3; melting point, 327.4°C (621°F); boiling point, 1755°C (3191°F); hardness, 1.5 Mohs; atomic number, 82; atomic weight, 207.2; divalent; tetravalent; superconductivity critical temperature, 7.196K. It has good corrosion and radiation resistance, high electrical resistivity (about 20.6 μΩcm at room temperature), good acoustic, vibration and thermal insulating properties, and high absorptive power for nuclear radiation. Used for storage batteries and radiation shielding, in superconductors, solders, fusible alloys, type metals, babbitts and bearing alloys, for ammunition, roofing, plumbing supplies, containers for radioactive materials, chemical reaction equipment for corrosive liquids (piping, tank linings, etc.), lead pigments, cable covering, vibration damping in heavy construction, and tetraethyllead (formerly used as a gasoline additive). Symbol: Pb.

lead acetate. (1) A colorless, white or faintly pink, moisture-sensitive crystalline powder (90+% pure) sometimes containing 10% acetic acid. Density, 2.28 g/cm^3; melting point, 175°C (347°F). Used as a selective oxidizer in organic synthesis, in the synthesis of aryl lead triacetates, and as a laboratory reagent. Formula: $Pb(O_2C_2H_3)_4$. Also known as *lead tetraacetate.*

(2) White, hygroscopic crystals, flakes, lumps, granules, or powder (99.+% pure) obtained by treating *litharge* or thin lead plates with acetic acid. Density, 2.55 g/cm^3; melting point, loses $3H_2O$ at 75°C (167°F); boiling point, decomposes at 280°C (536°F); absorbs CO_2 from air; soluble in water and ethanol. Used as an intermediate for lead titanate $(PbTiO_3)$ by sol-gel processing, and for PbZn alkoxide acetate, in varnishes, lead driers, chrome pigments and antifouling paints, in the gold cyanidation process, in dyeing and waterproofing, and in biology and biochemistry as a metal stain for lead hydroxide. Formula: $Pb(O_2C_2H_3)_2 \cdot 3H_2O$. Also known as *acetate of lead; Goulard's powder; lead acetate trihydrate; plumbous acetate; sugar of lead.*

lead acetate trihydrate. See lead acetate (2).

lead-alkali glass. See lead glass.

lead-alkali metal. See alkali lead.

lead alloys. A large group of alloys composed predominantly of lead, and including various tin-lead and lead-tin solders, fusible alloys, lead babbitts and other bearing alloys, type metals and various casting alloys.

lead antimonate. An orange-yellow powder obtained from a solution of lead nitrate and potassium antimonate. Density, 6.58 g/cm^3. Used as a yellow pigment in overglazes, porcelain enamels and glass, and in pottery manufacture. Formula: $Pb_3(SbO_4)_2$. Also known as *antimonate of lead; antimony yellow; lead orthoantimonate; Naples yellow.*

lead antimonide. A gray powder composed of lead and antimony, and used in electronics and semicon-ductor technology. Formula: PbSb.

lead-antimony alloys. A group of alloys of lead and antimony with an eutectic temperature of 251°C (484°F) at which 3.5% antimony is soluble in the lead. Antimony has a hardening and strengthening effect on lead, and an addition of 3% antimony results in maximum hardness. Used for automotive storage batteries, cable sheathing, chemical applications (e.g., fittings for lead pipelines), components of centrifugal pumps, pipelines, water pipes, lead shot, and anodes for chromium plating. See also antimonial lead; hard lead; lead shot.

lead-antimony-tin alloys. A group of lead-base bearing alloys containing about 5-15% antimony and 1-10% tin. Up to 3% copper and small amounts of other metals, such as alkaline earths, and/or arsenic, may be added to increase the hardness and prevent segregation. *Lead-antimony-tin* alloys are corrosion-resistant, moderately fatigue-resistant, and are used for light-duty bearings. See also lead babbitts; white metal alloys.

lead babbitts. See lead-base babbitts.

lead-barium crown glass. See barium flint glass.

lead barium glass. An *optical flint glass* in which lead and barium oxides are the major ingredients.

lead-base babbitts. A group of lead-base alloys containing 5-16.5% antimony, 1-10.5% tin, 0-3% copper and 0-0.3% arsenic. They have soft matrices with embedded hard, discrete particles, good antiscoring properties, good embeddability and conformability, moderate corrosion resistance, fair to poor fatigue resistance, a hardness of 20-25 Brinell, and a maximum service temperature of 149 (300°F). Used for bearings, and as lining materials for main and connecting-rod bearings. Also known as *lead babbitts*. See also alkali lead.

lead-base porcelain enamels. Porcelain enamel frits based on lead silicate. High-lead enamels have high gloss, good acid and weather resistance, good mechanical properties, and are used for cover coats on sheet steel, cast iron, and aluminum and its alloys. A typical composition is 30-40% silicon dioxide (SiO_2), 14-45% lead monoxide (PbO), 14-20% sodium oxide (Na_2O), 15-20% titanium dioxide (TiO_2), 7-12% potassium oxide (K_2O), with the balance being barium oxide (BaO), lithium oxide (Li_2O), boric oxide (B_2O_3) and phosphorus pentoxide (P_2O_5).

lead-bearing porcelain enamel. Porcelain enamel frits containing substantial amounts of lead monoxide (PbO) as a fluxing ingredient. High-lead enamels frits for cover coats on sheet steel, cast iron, and aluminum have a high gloss, good acid and weather resistance, and good mechanical properties.

lead bisilicate. A compound composed of lead monoxide and silicon dioxide, often with small amounts of aluminum dioxide. It has a density of 4.62 g/cm³ and a melting point of 790-815°C (1454-1499°F), and is used in ceramics, and inorganic, organic and organometallic synthesis. Formula: $PbSi_2O_5$.

lead borate. A white powder made by treating a solution of lead hydroxide with boric acid. Density, 5.598 g/cm³; melting point, loses H_2O at 160°C (320°F). Used as a flux in low-temperature frits and vitrified colors, in electrically conductive ceramic coatings and glass-bonded mica, in lead glass, in waterproofing paints, as a varnish and paint drier, and in galvanoplastics. Formula: $Pb(BO_2)_2 \cdot H_2O$. Also known as *lead metaborate*.

lead borosilicate. A mixture of lead silicate and lead borate, used in the manufacture of optical glass, and in glazes.

lead brass. See leaded brass.

lead bromide. White, hygroscopic crystals or powder (95+% pure). Crystal system, orthorhombic. Density, 6.66-6.69 g/cm³; melting point, 373°C (703°F); boiling point, 916°C (1681°F). Used in ceramics, organometallic research, and as an analytical reagent. Formula: $PbBr_2$.

Lead Brick. Trade name of Instruments for Research & Industry, Inc. (USA) for polyvinyl chloride-coated lead brick used for laboratory radiation-shielding applications.

Lead Bronze. Trade name of Knowsley Cast Metal Company, Limited (UK) for a series of lead bearing bronzes containing 76-85% copper, 5-15% lead and 9-10% tin. They have excellent antifriction properties, and good to fair corrosion resistance. Used for bearings for soft to medium-hard shafts and heavy loads.

lead bronze. See leaded bronze.

lead-calcium. An antifriction metal containing 97% lead, up to 0.8% barium and calcium, respectively, and sometimes traces of lithium, sodium and magnesium. It has a hardness of 30 Brinell, a melting range of 290-460°C (580-886°F), and a pouring range of 500-600°C (958-1138°F). Used for bearings of rolling stock, automobiles and machinery.

lead-calcium alloy. A lead alloy hardened by a fractional percent of calcium (usually 0.01-0.1%). It may also contain minute quantities of tin. Used as castings and sheet in automotive and standby storage batteries.

lead carbonate. Colorless or white crystals (99.9+% pure) occuring in nature as the mineral *cerussite*. Density, 6.6 g/cm³; melting point, decomposes at 315°C (600°F). Used as a paint pigment, and in glass, porcelain enamels and glazes as a flux and fining agent. Formula: $PbCO_3$.

lead casting alloys. Lead-base alloys used in die casting and permanent-mold gravity casting. The typical composition range is about 12-16% antimony, 0-6% tin, sometimes traces of arsenic and copper, and the balance lead.

lead chloride. White crystals, needles or powder (98+% pure). It occurs in nature as the mineral *cotunnite*. Crystal system, orthorhombic. Density, 5.85-5.98 g/cm³; melting point, 501°C (934°F); boiling point, 950°C (1742°F); hardness, 2.5 Mohs; refractive index, 2.19-2.21. Used in lead chromate pigments, as an analytical reagent, in organometallic research, and in the preparation of lead salts. Formula: $PbCl_2$.

lead chromate. Yellow crystals or powder (98+% pure). It occurs in nature as the mineral *crocoite*. Density, 6.3 g/cm³; melting point, 844°C (1551°F). Used as an inorganic pigment (*chrome yellow*) in ceramic coatings, paints, plastics and rubber, and as a flux and colorant in glasses, glazes and enamels. Formula: $PbCrO_4$.

lead-coated copper. Copper sheet coated on both sides with a rough or smooth layer of lead, usually deposited by electroplating using lead fluoborate, fluosilicate or sulfamate baths, or by dipping in the molten metal. Used for roofing, chemical process equipment, and acid-resistant tanks. Abbreviation: LCC.

lead-coated steel. Steel, usually with low carbon content, coated with lead by electroplating using lead fluoborate, fluosilicate or sulfamate baths, or by dipping into the molten metal. Used for building construction components, stamped and forged parts, and as a replacement for tin-lead coatings (*terneplate*).

lead coatings. Coatings of lead or lead-rich alloys deposited on steel or copper by electroplating using fluoborate, fluosilicate or sulfamate baths, by dipping into the molten metal, or by spraying. Abbreviation: LC. See also lead-coated copper; lead-coated steel.

lead columbate. See lead niobate.

lead crown glass. A special *optical glass* that is intermediate between crown and flint glass in optical dispersion, and contains considerable quantities of lead monoxide (PbO). Also known as *crown flint glass*. See also crown glass; flint glass.

lead crystal. See crystal glass.

lead cyanamide. The lead salt of cyanamide supplied as a crystalline compound (95% pure) with a density of 6.8 g/cm³ and a melting point 580°C (1076°F). Used in metallurgy, and organic and organometallic synthesis. Formula: $PbCN_2$. See also cyanamide.

lead cyanide. A white to yellowish powder made by combining solutions of potassium cyanide and lead acetate, and used chiefly in metallurgy. Formula: $Pb(CN)_2$.

lead dioxide. Red or brown crystals or red-brown powder (97+% pure), usually obtained by treating an alkaline solution of lead hydroxide with chlorinated lime. It occurs in nature as the mineral *plattnerite*. Crystal system, tetragonal. Density, 9.37-9.64 g/cm³; melting point, decomposes at 290°C (554°F). Used for pigments, as an oxidizer, in electrodes, in the positive plates of lead-acid storage batteries, as a curing agent for polysulfide elastomers, in organometallic research, in explosives, matches and textiles, and as a versatile reagent. Formula: PbO_2. Also known as *anhydrous plumbic acid; brown lead oxide; lead peroxide; lead superoxide.*

lead driers. Nearly water-white driers that contain one or more lead compounds, and work on the body of paint films. Examples include lead acetate, lead borate, lead linoleate, lead naphthenate and lead resinate. See also drier.

lead dust. A very fine lead powder with a typical particle size of less than 100 μm (0.004 in.). See also lead powder.

leaded alloys. Alloys into which lead has been introduced during manufacture to promote machinability and/or mechanical properties. Examples include leaded brass, leaded bronze, leaded monel and leaded steel.

leaded brass. A free-cutting brass containing 57-62% copper, 34-40% zinc and 1-3.25% lead. It has good machinability, antifriction and bearing properties, and hot formability. Used for screw-machine products, hardware, engraving, clock parts, fasteners, structural shapes, and bearings. Also known as *lead brass.*

leaded bronze. A copper-base alloy containing 2-15% tin, 0-6% nickel, 0-2% zinc and 8-30% lead. It has good strength and hardness, and is used for heavy-duty bearings of automotive engines, aircraft wheels, pumps, and rolling mills, and for acid-resistant fittings and castings. Also known as *lead bronze.*

leaded commercial bronze. A free-cutting copper-base alloy containing 9-9.5% zinc and 0.5-2.25% lead. Commercially available in the form of flat products and rods, it has excellent machinability and workability, and good corrosion resistance. Used for fasteners, machine parts, hardware, screw machine parts, gilding, and percussion caps. See also commercial bronze.

leaded copper. A free-machining copper alloy that contains up to 1% lead, and is commercially available in the form of rods and structural shapes. It has high electrical conductivity (98% IACS), excellent machinability, good corrosion resistance, good cold workability, and poor hot formability. Used for connectors, switch parts, machine parts, screw-machine products, fasteners, bolts, nuts, gas-welding tips, and current-carrying studs.

leaded flanging brass. A free-machining, flange-grade brass containing 62.5% copper, 35.5% zinc and up to 2%% lead. Used for screw-machine products.

leaded glass. Windows made from pieces of clear or colored glass fixed together by strips of lead having cross sections which resemble either an "H" or and "U." This term is not synonymous with "lead glass."

leaded gunmetal. Gunmetal containing 85-90% copper, 5-10% tin, 2-5% zinc, and up to 1% lead. It has improved machinability, good corrosion resistance, and good bearing properties. Used for fittings, bearings, and gears. See also gunmetal.

leaded high brass. See free-cutting brass.

leaded litharge. See lead suboxide.

leaded manganese bronze. A copper-base casting alloy composed of 40-41% zinc, 1% lead, and a trace of manganese. It has improved machinability and good strength, and is used for valve parts, marine fittings, brackets, levers, and gears. See also manganese bronze.

leaded monel. An alloy containing about 60% nickel, 32% copper, 2.2% iron, 2.2% lead, 2% manganese, 0.9% silicon and 0.2% carbon. It has improved machinability, good to excellent corrosion resistance and good strength, and is used for chemical equipment, marine parts, pumps, valves, fittings, and machine parts. See also monel.

leaded muntz metal. A free-cutting copper-base alloy containing 38.3-39.4% zinc and 0.2-0.7% lead. It has good machinability, excellent hot formability, poor cold workability, good corrosion resistance, and good strength, stiffness and elasticity. Used for condenser-tube plates, heat exchangers and baffles. See also muntz metal.

leaded naval brass. A free-cutting brass containing 60% copper, 37.5-38.5% zinc, up to 1.75% lead and up to 0.75% tin. Commercially available in the form of rods, structural shapes and flat products, it has excellent machinability and hot forgeability, good corrosion resistance, and high strength. Used for valve stems, marine hardware and machine parts, and screw-machine products. See naval brass.

leaded nickel commercial bronze. A leaded bronze containing 89.0-90.25% copper, 6.9-8.1% zinc, 1% nickel, up to 0.1% phosphorus and 1.75-1.9% lead. Commercially available in the form of flat products and rods, it has excellent machinability, good corrosion resistance, moderate strength, good elongation, good cold workability, and poor hot formability. Used for bolts, nuts, fasteners, pole-line hardware, screw-machine products, and electrical connectors.

leaded nickel copper. A free-cutting, corrosion-resistant copper alloy containing 1% nickel, 1% lead and 0.2% phosphorus. It has excellent machinability, relatively good electrical conductivity (55% IACS), and is used for electrical contacts, connectors, control elements for power tubes, fasteners, and screw-machine products. See also nickel copper.

leaded nickel silver. A free-cutting copper-base alloy containing 22-42% zinc, 8-18% nickel, 0-2% manganese and 1-2.75% lead. Commercially available in the form of flat products, it has excellent corrosion resistance, good cold formability, and good strength. Used for key blanks, watch components, screw machine products, architectural parts, hardware, and cutlery. See nickel silver.

leaded phosphor bronze. A phosphor bronze containing 80-94% copper, 5-10% tin, 1-10% lead and up to 0.1% phosphorus. It has excellent machinability, and good resistance to corrosion, fatigue, shock and vibration resistance. Used for bearings, bushings, tubes, clutch plates, perforated sheet, gears, and spindles. See phosphor bronze.

leaded red brass. A free-cutting brass of 83-93% copper, 4-7% zinc, 1.5-5% tin and 1.5-6% lead. It has good machinability, moderate strength, and is used for hardware, nuts, bolts, screws, valves and fittings, pump parts, flanges, plumbing supplies, gears, and electrical components. See red brass.

leaded semired brass. A free-cutting brass of 76-80% copper, 9-15% zinc, 3-5% tin and 2.5-7% lead. It has good machinability, moderate strength, and is used for valves and fittings, plumbing fixtures and supplies, cocks, faucets, pipe fittings and pieces, bushings, nuts, and pump parts. See also semired brass.

leaded steel. A low- or medium-carbon steel exhibiting excellent free-machining characteristics due to the addition of 0.2-0.5% lead, usually with one or more other elements, such as sulfur and/or selenium, resulting in short, broken chips, reduced power consumption, improved surface finish and extended tool life. It has a typical steel microstructure with minute lead particles and strings distributed throughout. Used for screw-machine products. See also free-machining steel.

leaded tin bronze. A bronze containing 80-90% copper, 6-16% tin, 0-5% zinc, 1-5% lead and 0-4% nickel. It has good machinability and strength, and is used for bearings, bushings, gears, cams, guides, pump impellers, piston rings, valve components, and steam fittings. See also tin bronze.

leaded tube brass. A free-cutting *yellow brass* composed of 66-66.5% copper, 32.4-33.5% zinc and 0.25-1.6% lead. It has good machinability and cold workability, good strength and elongation, and good corrosion resistance. Used for screw-machine products, plumbing supplies, pump cylinders and liners, and ammunition primers. See also tube brass.

leaded yellow brass. A copper-base casting alloy containing 24-40% zinc, 1% tin, 0-0.3% aluminum and 1-3% lead. It has excellent machinability, good corrosion resistance, and is used for bushings, hardware, fittings, ornaments, plumbing fixtures and fittings, valve parts, cocks, ship trimmings, and battery clamps. See also yellow brass.

leaded zinc oxide. A white paint pigment made by mixing basic lead sulfate ($PbSO_4$) and zinc oxide (ZnO).

lead ferrite. A hard magnetic ceramic material with hexagonal crystal structure, usually made by pressing a blend of lead monoxide and ferric oxide powders, and sintering at prescribed temperatures to obtain the desired magnetic properties, e.g., high coercive force. Used for permanent magnets in DC motors, magnetic door latches and radio loudspeakers. Formula: $PbO \cdot 6Fe_2O_3$. Also known as *lead hexaferrite*.

lead flake. See white lead.

lead fluoride. Colorless or white crystals or white powder (99+% pure). Crystal system, rhombohedral. Density, 8.24 g/cm^3; melting point, 855°C (1571°F); boiling point, 1290°C (2354°F). Used for electronic and optical applications, as a starting material for growing single-crystal solid-state lasers, and as a dry-film lubricant for high-temperature applications. Formula: PbF_2.

lead fluosilicate. Colorless crystals that decompose when heated. Used in solutions for electrorefining lead, and in fluosilicate baths for electroplating. Formula: $PbSiF_6 \cdot 2H_2O$. Also known as *lead hexafluorosilicate; lead silicofluoride*.

lead foil. A foil made by rolling lead or a lead alloy to a prescribed thickness. Lead alloys for this purpose usually contain about 86% lead with the balance either being 7% iron, 5% aluminum and 2% tin, or 13% tin and 1% copper. It is available in thicknesses ranging from 0.001 to 3.15 mm (0.00004 to 0.125 in.), and weights from about 0.0114 to 35.8 kg/m^2, in purities of 99.95-99.9995%. Lead microfoils with thicknesses from 0.001 to 1.0 μm (0.04 to 39.4 μin.), weights from 1.1 to 1047.8 μg/cm^2 and purities of 99.99+%, are usually supplied on permanent *Mylar* support. *Lead foil* is used especially for wrapping nonedible products, and in radiation shielding.

lead-free solders. A group of environmentally conscious solders that do not contain lead, but are based on alloys, such as zinc-aluminum, tin-zinc, fusible alloys, etc., providing comparable soldering properties.

lead glance. See galena.

lead glass. A heavy glass made by introducing up to 50% lead monoxide (PbO) to a melt of ordinary *soda-lime glass* as a flux to increase the coefficient of thermal expansion, refractive index, optical dispersion and surface brilliance. It has a high electrical resistivity and moderate to poor corrosion resistance, and is used as an optical glass (for lenses, prisms, etc.), and for tableware, neon-sign tubing, and light-bulb stems. Also known as *lead-alkali glass*. This term is not synonymous with "leaded glass."

lead glaze. A glaze with a significant fluxing addition of lead monoxide (PbO) in order to effect a reduction in the fusion temperature and viscosity, an improvement of the flow properties during firing, and an enhancement of the smoothness, luster, brilliance, and resistance to chipping and water solubility.

lead halide. A compound composed of lead and a halogen, such as bromine, chlorine, fluorine or iodine.

lead hexaferrite. See lead ferrite.

lead hexafluosilicate. See lead fluorosilicate.

leadhillite. A colorless, white or greenish yellow mineral composed of lead carbonate sulfate hydroxide, $Pb_4(SO_4)(CO_3)_2(OH)_2$. Crystal system, orthorhombic. Density, 6.55 g/cm^3; refractive index, 2.00. Occurrence: Scotland.

lead-indium alloys. A group of alloys of lead and indium (UNS 51535) with outstanding corrosion resistance used for chemical process equipment. Alloys containing 50-95% lead and 5-50% indium have excellent resistance to thermal fatigue and good compatibility with gold, and are suitable for soldering electronic assemblies.

lead-indium-silver solder. A lead alloy (UNS L51510) that contains 5.0% indium and 2.5% silver, has a density of 11.1 g/cm^3 (0.40 lb/in.^{3}), and is used as specialty solder.

leading stone. See lodestone.

lead iodide. Yellow to golden-yellow crystals or powder (98.5+% pure). Crystal system, hexagonal. Density, 6.16 g/cm^3; melting point, 402°C (756°F); boiling point, 954°C (1749°C). Used in bronzing, printing, photography, cloud seeding, and organometallic research. Formula: PbI_2.

lead lanthanum zirconate titanate. A ferroelectric ceramic material that is composed of a solid solution of lead zirconate ($PbZrO_3$) and lead titanate ($PbTiO_3$), and doped with lanthanum. It has electrooptical properties, and exhibits the piezoelectric effect, i.e., it can produce a measurable change in voltage as a result of an applied mechanical pressure or stress, and vice versa. Used for transducers and other electronic components. Abbreviation: PLZT. Also known as *lanthanum-doped lead zirconate-lead titanate; lanthanum-doped lead zirconate titanate; lead lanthanum zirconate titanate ceramic; PLZT ceramic*. See also lead zirconate titanate.

leadless enamel. Enamel made from leadless frits and containing no or only a negligible amount of lead compounds.

leadless glaze. A ceramic coating that contains only a very small amount of lead.

lead linoleate. A yellowish-white paste made by heating a solution of lead nitrate with sodium linoleate, or by heating a mixture of lead monoxide (litharge) and linseed oil. Used as a paint and varnish drier. Formula: $Pb(C_{18}H_{31}O_2)_2$.

lead magnesium niobate. An electrostrictive ceramic that is a solid solution of lead magnesium oxide ($PbMgO_3$) and lead niobate ($PbNbO_3$) corresponding to the formula $Pb(Mg,Nb)O_3$. It has a density, 6.1 g/cm^3, a Curie temperature of 170°C (364°F), a high dielectric constant (5500), a high charge constant, excellent coupling properties, and a low aging tendency.

Used for sensors, actuators, receivers, hydrophones, micropositioning devices, and low-power projectors. Abbreviation: PMN.

lead metaborate. See lead borate.

lead metacolumbate. See lead niobate.

lead metaniobate. See lead niobate.

lead metasilicate. See lead silicate.

lead metatantalate. See lead tantalate.

lead metatitanate. See lead titanate.

lead metavanadate. See lead vanadate.

lead molybdate. A white to pale yellow powder (99.9+% pure) obtained from a mixture of solutions of lead nitrate and ammonium molybdate. Also available as synthetically grown crystals. Density, 6.92 g/cm^3; melting point, 1060-1070°C (1940-1958°F). Used in pigments, as an ionic conductor, in acoustooptic modulators and deflectors, with antimony compounds as an adherence-promoting agent in porcelain enamels, as a low-temperature scintillator for nuclear instruments, and in ceramics. Formula: $PbMoO_4$.

lead molybdenum sulfide. A compound of lead sulfide and molybdenum disulfide with a superconductivity critical temperature of 14.4K. Used in superconductor compounds. Formula: $PbMoS_3$.

lead monosilicate. See lead silicate.

lead monoxide. Yellow crystals or powder (99+% pure) obtained by oxidizing metallic lead, and available in purities from 99.0 to 99.999%. It exists in two modifications with tetragonal or orthorhombic crystal structure. The tetragonal form is often referred to as *litharge*, and has a density of 9.36 g/cm^3, a melting point of 888°C (1630°F), a hardness of 2 Mohs, and a refractive index of 2.665. The orthorhombic form is obtained by the oxidation of a metallic lead bath at about 345°C (653°F), occurs in nature as the mineral *massicot*, and has a density of 9.64 g/cm^3 and a melting point of 489°C (912°F). *Lead monoxide* is used in storage batteries, ceramic cements and fluxes, pottery, glazes, glasses and enamels, driers for varnishes and paints, chromium pigments, acid-resisting compositions, in oil refining, in the preparation of *red lead*, in the assay of precious metal ores, in organometallic research, and as a rubber accelerator. Formula: PbO. Also known as *plumbous oxide; yellow lead oxide.*

lead naphthenate. A soft, yellow, resinous, semi-transparent solid substance, or brown 65% solution in mineral spirits. The solid compound contains about 37% lead, and the liquid solution about 16-24% lead. Density of liquid, 1.13 g/cm^3; flash point, 104°F (40°C). Used as a paint and varnish drier, as a wood preservative, and as a lube oil additive. Formula: $C_7H_{12}O_2$·xPb.

lead neodecanoate. A compound usually supplied as a 60% solution in naphtha containing about 23-24.5% lead. It has a density of 1.11 g/cm^3, a flash point of 104°F (40°C), and is used to impart luminescent properties to organic substances. Formula: $Pb(O_2C_{10}H_{19})_2$.

lead nickel niobate. A piezoelectric ceramic that is a solid solution of lead nickelate ($PbNiO_3$) and lead niobate ($PbNbO_3$) and corresponds to the formula $Pb(Ni,Nb)O_3$. It has a density of 7.8 g/cm^3, a Curie temperature of 150°C (302°F), a high dielectric constant (5500), and a high piezoelectric longitudinal charge coefficient (d_{33}). Used for ultrasound equipment, and actuators. Formula: Abbreviation: PNN.

lead niobate. A ferroelectric material that exhibits dielectric and piezoelectric properties, and has a Curie temperature of 570°C (1058°F), high longitudinal coupling compared to planar and lateral coupling, low aging properties, and a wide range of operating temperatures. Used for high-temperature transducers, flaw detectors, thickness gages, air-blast gages, underwater sonar equipment, accelerometers, sensing devices, and defense electronics. Formula: $Pb(NbO_3)_2$ [$Pb(CbO_3)_2$]. Also known as *lead columbate; lead metacolumbate; lead metaniobate.*

lead nitrate. Colorless or white crystals (99+% pure) obtained by treating lead with nitric acid. Density, 4.53; melting point, decomposes at 470°C (878°F). Used as a photographic sensitizer, in the sol-gel preparation of lead zirconate titanate (PZT) powders, as a mordant in dyeing and printing, in lead salts, and in tanning, process engraving, lithography and microscopy. Formula: $Pb(NO_3)_2$.

lead ocher. See massicot.

lead orthoantimonate. See lead antimonate.

lead orthoplumbate. See red lead oxide.

lead oxide. See lead dioxide; lead monoxide; lead sesquioxide; lead suboxide; red lead oxide.

lead paint. Paint in which *white lead* [$2PbCO_3$·$Pb(OH)_2$] is employed as a pigment. Used as a decorative and protective coating on metal, wood, etc. Its use is now extremely limited because of the toxic nature of white lead.

lead peroxide. See lead dioxide.

lead pigments. A large group of lead compounds used to impart color to paints including: (i) basic lead carbonate (white lead); (ii) basic lead chromate (chrome red); (iii) basic lead silicate (white lead silicate); (iv) basic lead sulfate (sublimed white lead); (v) blue basic lead sulfate (blue lead); (vi) lead chromate (chrome yellow); (vii) lead monoxide; (viii) lead silicochromate; (ix) lead sulfate; (x) lead tetroxide; (xi) lead titanate; (xii) lead tungstate; and (xiii) lead vanadate.

lead phthalocyanine. A phytalocyanine derivative containing a divalent, central lead ion (Pb^{2+}). It has a typical dye content of about 80%, a maximum absorption wavelength of 698 nm, and is used as a dye and pigment. Formula: $Pb(C_{32}H_{16}N_8)$. Abbreviation: PbPc.

lead plate. An electrodeposit of lead produced on a substrate, such as copper, or low-carbon steel, from a lead fluoborate, lead fluosilicate or lead sulfamate plating bath. It protects metals against corrosives, and is used to line tanks and chemical equipment, hardware, storage battery parts, electrical components, and, owing to its softness and corrosion resistance, on bearings and bushings.

lead poly(methacrylate 2-ethylhexanoate). A radiation-dense polymer obtained by the polymerization of lead methacrylate 2-ethylhexanoate, and used in biology, biochemistry, medicine and nuclear engineering for radiation shielding applications.

lead powder. Lead (99-99.999+% pure) in the form of a fine powder (typically 5-180 μm, or 0.0002-0.007 in.), used in ceramics and powder metallurgy, in pigments, solder pastes, greases and pipe-joint compounds, and as an addition to plastics and rubber for the manufacture of radiation-shielding and sound-deadening products.

lead protoxide. See lead monoxide.

lead resinate. Yellowish-white powder or paste, or brownish, shiny, transparent lumps obtained by heating a solution of rosin oil and lead acetate. Used as a drier in paints and varnishes, and as a waterproofing agent for fabrics. Formula: $Pb(C_{20}H_{29}O_2)_2$.

lead selenide. Gray or metallic blue-gray high-purity crystals or lumps (99.9+% pure) or fine, black powder (99.99+% pure available), in cubic and hexagonal crystalline form. The cubic form occurs in nature as the mineral *clausthalite*. Crystal structure, halite. Density, 8.15 g/cm³; melting point, 1067°C (1953°F). Used for semiconductors employed in electronic devices, such as infrared detectors, thermoelectric devices, and thermistors. Formula: PbSe.

lead sesquioxide. Black monoclinic crystals or reddish-yellow amorphous powder obtained by carefully heating metallic lead. Density, 10.05 g/cm³; melting point, 539°C (986°F). Used in metallurgy, and in the manufacture of glass, glazes, porcelain enamels, and ceramic cements. Formula: Pb_2O_3.

lead sheet. Lead supplied in standard widths up to 2.4 m (8 ft.) and rolled to a thickness of up to 50.8 mm (2 in.). It is often used in the manufacture of chemical process equipment (e.g., tanks and vessels), for the acoustical insulation of buildings, for nuclear shielding applications, for cutting stencils in sign work, and in stamped form to produce washers and gaskets. *Lead sheet* may also be clad with tin by rolling lead and tin together. Also known as *sheet lead*.

lead shot. Lead in the form of small spherical particles, usually 1-6 mm (0.04-0.236 in.) in size, produced by pouring molten lead through a sieve and allowing it solidify during free fall. Purities range from about 99.8 to 99.9995%, and it may contain a small amount of arsenic. Used as shotgun ammunition, and in enclosures for radiation shielding.

lead silicate. A colorless or white crystalline powder made by combining lead acetate and sodium silicate, by heating lead oxide with silicon dioxide, or by drying the reaction product of silica gel, lead monoxide (*litharge*) and acetic acid. Density, 6.49 g/cm³; melting point, 680-730°C (1256-1346°F); refractive index, 2.01. Used in glazing pottery, in lead-bearing glazes to minimize lead solubility, in glass manufacture, in lead-fluxed bodies with high dielectric strength and lead-fluxed steatite bodies with a wide firing range, in organometallic research, in fireproofing fabrics, and as a filler and pigment. Formula: $PbSiO_3$. Also known as *lead metasilicate; lead monosilicate*. See also lead bisilicate.

lead silicochromate. A group of yellow pigments based on lead silicon chromate. Ordinary lead silicon chromate pigments are used in traffic marking paints, while basic lead silicon chromate is employed in protective coatings and primers for metals, and in high-gloss enamels.

lead silicofluoride. See lead fluosilicate.

lead-silver alloys. See lead-silver solders.

lead-silver babbitts. A group of special bearing alloys composed of 10-15% antimony, 2-5% silver, up to 5% tin, up to 0.2% copper and, sometimes, small amounts of indium. They have a maximum service temperature of 260°C (500°F), high-load carrying capacity, excellent corrosion and fatigue resistance, good embeddability and conformability, and good antiscoring properties. Used as overlays for bearings.

lead-silver solders. A group of lead-base soldering alloys containing between 1.5 and 6% silver, and usually some indium (up to 5%). They have good joint strength and high melting points, e.g., 94.5Pb-5.5Ag melts at 304°C (579°F), but only fair to poor flow characteristics, wettability and corrosion resistance. Used for soldering nonferrous metals and alloys, e.g., copper and brass. Also known as *lead-silver alloys*. See also lead-silver-tin solders.

lead-silver-tin alloys. See lead-silver-tin solders.

lead-silver-tin solders. A group of lead-base soldering alloys containing between 1.5 and 10% silver and 1 to 5% tin. Their solidus temperatures range is 266-310°C (510-590°F) and, due to the addition of tin, they possess improved flow and wetting

characteristics and increased corrosion resistance. Used for soldering electronic components and cryogenic equipment. Also known as *lead-silver-tin alloys*. See also lead-silver solders.

lead-sodium alloys. See sodium-lead alloys.

lead solders. A group of solders in which lead is the predominant constituent, and including in particular: (i) lead-tin solders; (i) lead-tin-antimony; (iii) lead-silver solders; (iv) lead-silver-tin solders; (v) lead-tin-indium solders; and (vi) lead-indium solders.

lead spar. See anglesite.

lead sponge. See sponge lead.

lead stannate. Light-colored powder obtained from lead monoxide (PbO) and tin dioxide (SnO_2). Melting point, 950°C (1742°F); loses $2H_2O$ at approximately 170°C (338°F); photosensitive properties. Used as an additive (1-5%) in barium titanate capacitors to reduce the Curie point. Also used in photosensitive materials and in pyrotechnics. Formula: $PbSnO_3 \cdot 2H_2O$. Also known as *lead stannate dihydrate*.

lead stearate. A white powder formed by heating a solution of lead acetate with sodium stearate. Density, 1.4 g/cm³; melting point, 100-115°C (212-239°F). Used as a drier in varnishes and lacquers, in high-pressure lubricants, in cutting oils, as a light and heat stabilizer in vinyl plastics, and as a constituent of greases, waxes and paints. Formula: $Pb(C_{18}H_{35}O_2)$.

lead subcarbonate. See white lead.

lead suboxide. A black, amorphous solid compound. Density, 8.34 g/cm³; melting point, decomposes on heating. Used for storage batteries. Formula: Pb_2O. Also known as *black lead oxide; leaded litharge*.

lead sulfate. White crystals or powder (98+% pure). It occurs in nature as the mineral *anglesite*. Density, 6.2 g/cm³; melting point, 1170°C (2138°F). Used as a pigment in paints, and in storage batteries. Formula: $PbSO_4$. See also basic lead sulfate; blue basic lead sulfate; tribasic lead sulfate.

lead sulfide. Silvery, metallic crystals or black powder (98+% pure). It occurs in nature as the mineral *galena*, and can also be produced synthetically. Crystal system, cubic. Crystal structure, halite. Density, 7.5-7.61 g/cm³; melting point, 1117°C (2043°F); boiling point, sublimes at 1281°C (2338°F); hardness, 2.5 Mohs; refractive index, 3.921; band gap, 0.37 eV. Used for ceramics, in infrared radiation detectors, semiconductors and glazes for claywares, and as a source of lead. Formula: PbS. Also known as *plumbous sulfide*.

lead superoxide. See lead dioxide.

lead tantalate. A ferroelectric compound of lead oxide and tantalum oxide. Density, 7.9 g/cm³; melting point, above 350°C (662°F); Curie temperature, 260°C (500°F). Used for special electroceramic materials. Formula: $PbTa_2O_6$. Also known as *lead metatantalate*.

Lead Tape. Trade name for a hard lead alloy composed of 95% lead, 4.5% antimony and 0.5% tin, and used for lead tape.

lead telluride. A compound of lead and tellurium that is available as gray or tin-white crystals, or as an off-white powder (99+% pure). It occurs in nature as the mineral *altaite*. Crystal system, cubic. Crystal structure, halite. Density, 8.16 g/cm³; melting point, 907°C (1665°F); hardness, 3 Mohs; band gap, 0.25 eV. Used as a semiconductor and photoconductor. Formula: PbTe.

lead-tellurium alloys. A group of lead alloys containing 0.06-0.10% tellurium together with 0.18-0.20% arsenic, 0.13-0.14% tin, 0.06-0.08% bismuth, 0.06% copper and traces of antimony and silver. The addition of tellurium improves the strength, hardness, toughness and corrosion resistance. Used for chemical equipment, cable sheathing, and pipe with improved resistance to hydraulic pressure. See also tellurium lead.

lead tetraacetate. See lead acetate (1).

lead tetrachloride. A yellow, oily liquid with a density of 3.18 and melting point of -15°C (5°F). Used in chemistry and materials research. Formula: $PbCl_4$.

lead tetrafluoride. White, hygroscopic crystals. Crystal system, tetragonal. Density, 6.7 g/cm^3; melting point, 600°C (1112°F). Used in chemistry and materials research. Formula: PbF_4.

lead tetroxide. See red lead oxide.

Leadtex. Trademark of Revere Copper Products, Inc. (USA) for sheet copper coated with a rough layer of lead.

lead-tin alloys. A group of alloys containing varying amounts of lead and tin. The most common alloys of this type are *lead-tin solders* containing up to 50% tin. Alloys of lead containing up to 5-25% tin are used in the form of hot-dip coatings on iron and steel sheet (known as *terneplate*) to improve the corrosion resistance. Lead-base bearing alloys (known as *lead babbitts*) contain up to about 16% antimony and up to 10% tin.

lead-tin solders. A group of lead-base solders containing up to 50% tin. They have solidus temperatures ranging from 183°C (361°F) for 50Pb-50Sn to 320°C (608°F) for 95Pb-5Sn. *Lead-tin solders* have good wetting characteristics, good strength, relatively low cost, and are used for soldering can seams and electronic components, and for coating and joining metals, lead pipes, cable sheaths, automotive components, roofing, etc.

lead titanate. A yellow crystalline powder (98+% pure) made from lead oxide (PbO) and titanium dioxide (TiO_2) at high temperatures. It has a density of 7.52 g/cm^3 and is ferroelectric below 763K. Used with lead zirconate ($PbZrO_3$) in piezoelectric lead zirconate titanate (PZT) for transducers, accelerometers and hydrophones, as an additive to barium titanate to improve piezoelectric properties, and as an industrial pigment in paints for iron and steel surfaces. Formula: $PbTiO_3$. Also known as *lead metatitanate*.

lead titanate niobate. A ferroelectric and piezoelectric ceramic that is a solid solution of lead titanate ($PbTiO_3$) and lead zirconate ($PbZrO_3$) corresponding to the formula $Pb(Ti,Nb)O_3$. It has density of 12.8 g/cm^3, a Curie temperature of 240°C (464°F), high piezoelectric constants, and a dielectric constant of 270. Used for transducers, nondestructive testing, accelerometers and hydrophones. Abbreviation: PTN.

lead trifluoroacetate. The divalent lead salt of trifluoroacetic acid with a melting point of 147-153°C (296-307°F), used in the pyrolytic preparation of fluoride glasses. Formula: $Pb(O_2C_2F_3)_2$.

lead tungstate. A white powder (99.9% pure) made from a mixture of solutions of sodium tungstate and lead nitrate. It occurs in nature as the mineral *stolzite*. Density, 8.24 g/cm^3; melting point, 1130°C (2066°F). Used as a pigment, and as a scintillator in high-energy physics. Formula: $PbWO_4$. Also known as *lead wolframate*. Abbreviation: PWO.

lead vanadate. A yellow-brown powder (99.9% pure) used as a pigment and in the preparation of other vanadium compounds. Formula: $Pb(VO_3)_2$. Also known as *lead metavanadate*.

lead vitriol. See anglesite.

lead wolframate. See lead tungstate.

lead wool. A loose rope of coarse filaments or threads of metallic lead, 0.1-0.4 mm (0.004-0.016 in.) in diameter, used for caulking and packing pipe joints, and for radiation shielding applications.

lead zirconate. A white powder (95+% pure) made from lead monoxide (PbO) and zirconium dioxide (ZrO_2). It has a density of 7.0 g/cm^3, exhibits ferroelectric and piezoelectric properties, and is antiferroelectric below 503K. Used in the preparation of lead zirconate titanate (PZT) ceramics. Formula: $PbZrO_3$.

lead zirconate titanate. A ferroelectric, piezoelectric and electro-optic ceramic that is a solid solution of lead titanate ($PbTiO_3$) and lead zirconate ($PbZrO_3$) corresponding to the formula $Pb(Ti,Zr)O_3$. It has a density of 7.5-7.6 g/cm^3, a Curie temperature above 150-360°C (300-680°F), high piezoelectric constants, a high dielectric constant (about 1000-4100) and stable electromechanical properties over a wide range of temperatures. Used for electromechanical and electroacoustic transducers, sensors, ferroelectrics in computer memories, hi-fi sets, sonar equipment, depth sounders, hydrophones, ultrasonic cleaners, and high-voltage generators. Abbreviation: PZT.

Leaf. Trade name of Pittsburgh Corning Corporation (USA) for a sculptured glass unit having a leaf design pressed into the glass on both sides.

leaf fibers. See hard fibers.

Leakpruf. Trade name of Alpha Metals Inc. (USA) for an acid flux-filled solder containing 40-60% lead, with the balance being tin. It has a solidus temperature of 183°C (361°F), and is used for soldering of nickel, monel, stainless steel, etc.

lean clay. Clay with relatively low plasticity and poor green strength.

lean concrete. Concrete that contains less cement than ordinary concrete. It has a high aggregate-to-cement ratio, usually 12:1 to 15:1, and is used as a subbase under concrete roads. Also known as *lean-mix concrete*.

lean lime. Lime that contains a considerable amount of impurities, such as oxides of aluminum, iron and silicon, will not slake readily with water, and is difficult to work. It is the opposite of *fat lime*. Also known as *poor lime*.

lean-mix concrete. See lean concrete.

lean mortar. Nonplastic mortar that is low or deficient in cement and difficult to spread.

Leantin. Trade name of Lumen Bearing Company (USA) for a lead-base babbitt containing 15% antimony and 3% tin. Used for machine and low-pressure bearings.

Learok. Trade name of JacksonLea (USA) for grease buffing compositions.

leather. An animal hide or skin that has been prepared for commercial use by removing all flesh and hair and then treating with a tanning agent (e.g., tannin) to increase its strength and abrasion resistance, make it resistant to putrefactive bacteria, enzymes and hot water, and render it more pliable and durable. *Leather* is highly poromeric, and retains the microporosity of the original skin. The hides and skins come mainly from cattle (cowhides or steerhides), although those of sheep, goats, deer, pigs, reptiles (e.g., alligators, lizards and snakes) and sharks are also employed. Used for transmission belting, gaskets, linings, coverings, footwear, jackets and other wearing apparel, protective clothing, kneepads, headwear, welder's helmets, gloves, belts, handbags, and luggage.

leatherboard. A rigid fiberboard or flexible sheet made from scrap leather using bituminous or resinous binders. It has a fiber content of not less than 75% leather, and is used for clutch linings, gaskets, paneling, suitcases, and shoe heels and linings. Also known as *fiber leather*.

leather cloth. See leather fabrics.

leather dust. Light, fleecy fibers of scrap leather from tanneries and leather-processing industries, used as fillers in adhesives and caulking compounds, and in the manufacture of napped or sueded textile fabrics.

leatherette. A collective term referring to coated fabrics with colors, textures or finishes imitating real leather. This term was formerly used as a trade name for imitation leather.

leather fabrics. An embossed, coated fabric that has an appearance and/or texture resembling that of real leather. Also known as *leather cloth*.

leather flour. Leather ground and sized for use as a filler in certain fabrics.

Leatheroid. Trademark of American Finishing Company (USA) for a compound used to coat and impregnate fabrics. It dries to a finish resembling that of artificial leather.

Leavil. Trade name of Montecatini SpA (Italy) for polyvinyl chloride fibers.

Leavin. Trade name of Montecatini SpA (Italy) for polyvinyl chloride fibers.

Leavlite. Trade name of Leavlite Limited (UK) for a laminated safety glass.

Lebanon. Trade name of Lebanon Foundry & Machine Company (USA) for an extensive series of steels including numerous plain-carbon and chromium-molybdenum and nickel-chromium-molybdenum alloy steels, several low-carbon and low-carbon low-alloy cast steels, heat- and corrosion-resistant cast steels, and various austenitic, ferritic, martensitic and precipitation-hardenable stainless steels.

Lechesne. Trade name for a corrosion-resistant alloy composed of 60-90% copper, 10-40% nickel and 0.05-0.2% aluminum, used for chemical equipment.

lecithin. Any of several light brown or brown esters of phosphatidic acid and choline (*phospholipids*) that are constituents of brain and nervous tissue and egg yolk. They are composed of fatty acids, glycerol, phosphorus and choline or inositol. *Lecithin* is also obtained commercially from soybeans, and used in biotechnology, in casein paints and printing inks, as an emulsifier and wetting agent, as an accelerator in synthetic rubbers, as a mold-release agent for plastics, as a lubricant for textile fibers, and for silk-screen media.

Leco. (1) Trade name of Lehigh Steel Corporation (USA) for a shock-resistant tool steel containing 0.33% carbon, 0.75% chromium, 0.75% molybdenum, 0.75% copper, 0.5% nickel, 0.15% titanium, and the balance iron. Used for pneumatic tools, punches, chisels, dies, and shear blades.

(2) Trademark of LECO Corporation (USA) for a complete line of metallographic products and supplies including abrasives, applicators, diamond powder suspensions, aerosol diamond sprays, diamond compound extenders, bakelite powders, epoxide resins and hardeners, aluminum-oxide optical finishing powders, cutting oils and rust inhibitors.

lecontite. A colorless mineral composed of potassium sodium ammonium sulfate dihydrate, $(K,NH_4)NaSO_4 \cdot 2H_2O$. It can also be made synthetically. Crystal system, orthorhombic. Density, 1.75 g/cm³; refractive index, 1.45. Occurrence: Honduras.

Lecoset. Trademark of LECO Corporation (USA) for cold-curing synthetic resins used for metallographic applications.

Lectrocast. Trade name of Detroit Alloy Steel Company (USA) for an abrasion- and wear-resistant nickel cast iron containing about 3-3.4% carbon, 2% silicon, 1.5-2.75% nickel, up to 0.7% chromium, and the balance iron. It has good strength and hardness, and is used for draw dies, and castings.

Lectrofluor. Trademark of General Magnaplate Corporation (USA) for a series of nonstick fluoropolymer-base protective coatings that enhance the corrosion resistance of ferrous or nonferrous metals. They have outstanding abrasion and wear resistance, excellent dry lubrication, high rigidity and toughness, good creep resistance, good load-bearing capability, excellent resistance to many corrosive acids, alkalies and solvents, exceptional mold-release properties, good high- and low-temperature properties, and a service temperature range of -84 to +177°C (-120 to +350°F). Used for food, chemical and pharmaceutical-processing applications, packaging equipment, conveyor systems, marine applications, die rolls, sealing dies, stirrers, dryers, tanks, scrubbers, chemical pumps, and hoods.

Lectroless. Trade name of Enthone-OMI Inc. (USA) for electroless coatings including *Lectroless Au* electroless gold and *Lectroless Ni* electroless nickel.

Lectro-Nic. Trade name of Enthone-OMI Inc. (USA) for a nickel electroplate and plating process.

Leda. Trade name of Firth Brown Limited (UK) for a high-speed steel containing 0.75% carbon, 4% chromium, 18% tungsten, 1-2% vanadium, 0-5.2% cobalt, and the balance iron. Used for lathe and planer tools, reamers, taps, and hobs.

Ledaloyl. Trade name of Johnson Bronze Company (USA) for an oil-impregnated sintered bronze composed of 84% copper, 10% tin, 5% lead and 1% graphite, and used for self-lubricating bearings.

Ledasto. Trade name of Rapperswill (Switzerland) for cellulose nitrate film products.

Leddel alloy. A zinc die casting alloy containing 5% aluminum and 5% copper.

ledeburite. The eutectic mixture of *austenite* and *cementite* occurring on cooling in iron-carbon alloys containing more than 2.1% but less than 6.67% carbon.

Ledebur's bearing. A series of antifriction metals composed of 77-85% zinc, 10-18% antimony and 5% copper, and used for bearings.

Ledermix. Trade name of American Cyanamid Company (USA) for a dental cement used in endodontics.

ledger paper. A strong, durable, well-sized accounting paper containing 100% white rags, and used for ledgers.

Ledloy. Trade name of Inland Steel Company (USA) for a series of carbon and alloy steels including several plain-carbon, free-machining, chromium, chromium-molybdenum and nickel-chromium-molybdenum grades.

Ledo. Trade name of Colt Industries (UK) for an oil- or water-hardening steel containing 0.4% carbon, 0.2% molybdenum, and the balance iron. Used for gears, shafts, spindles, crankshafts, and bolts.

Led-O-Loy. Trade name of Bridgeport Rolling Mills Company (USA) for a free-cutting, corrosion-resistant brass containing 87% copper, 9% zinc, 2% tin, 2% lead and 0.05% phosphorus. Used for valve parts, thrust washers, bearings, and gage and meter components.

Ledrite. Trade name of Olin Brass, Indianapolis (USA) for a series of free-cutting brasses containing 60-62% copper, 35-36% zinc and 1.5-3.75% lead. Used for screw-machine parts and hardware.

Ledron. Trademark of Südplastik GmbH (Germany) for a polyvinyl chloride film.

Leduct. Trade name of Ley's Malleable Castings Company Limited (UK) for a series of spheroidal graphitic cast irons containing 3.5-3.8% carbon, 2-2.6% silicon, 0.1-0.6% manganese, 0.005-0.01% sulfur, 0-0.06% phosphorus, 0.03-0.055% magnesium, and the balance iron. Used for automotive, truck and tractor parts.

Leff. Trade name of L-S Plate & Wire Company (USA) for precious metal dental alloys.

left-handed materials. A class of conceptual materials that exhibit one or more reversals of the usually "right-handed" relationship between electric and magnetic fields and the direction of wave propagation. For example, such materials can bend electromagnetic waves (e.g., radar or microwaves) in a direction opposite to that at which they travel through conventional "right-handed" materials. A recent practical example of such a material is a composite with a negative refractive index constructed by physicists at the University of California at San Diego, USA. It consists of several thin fiberglass sheets coated with copper wires and rings with a square-like arrangement. Abbreviation: LHM. Also known as *positive index materials.*

Lega. Trade name of Driver Harris Company (USA) for a series of low-expansion alloys containing 49-58% iron and 42-51% nickel.

Legacy. Trade name of J.F. Jelenko & Company (USA) for a ceramic dental alloy of 86% palladium, 2% gold, and less than 1% silver. It provides high strength, good finishability and polishability, and produces extremely accurate castings. Used for fusing porcelain to metal. *Legacy XT* is a high-palladium ceramic dental alloy with high strength and excellent thermal stability-sag resistance. It produces extremely accurate castings, and is used for fusing porcelain to metal, particularly in long-span bridgework.

Legal. Trade name of Siemens-Schukert AG (Germany) for a corrosion-resistant, heat-treatable aluminum alloy containing 0.5-2% magnesium, 0.3-1.5% silicon, and 0-1.5% manganese. Used for light-alloy parts.

Lega-Y. Trade name of Società Alluminio Veneto per Azioni (Italy) for a heat-treatable aluminum alloy containing 3.8-4.2% copper, 1.3-1.7% magnesium and 1.8-2.3% nickel. Used for light-alloy parts.

Legend. (1) Trade name of SS White Group (UK) for a glass-ionomer dental filling cement. *Legend Silver* is a silver-filled glass-ionomer dental cement.

(2) Trade name for a silver-free, palladium-based dental bonding alloy.

legrandite. A colorless to wax-yellow mineral composed of zinc arsenate hydroxide monohydrate, $Zn_2(AsO_4)(OH)\cdot H_2O$. Crystal system, monoclinic. Density, 3.97 g/cm^3; refractive index, approximately 1.709. Occurrence: Mexico.

Legupren. Trademark of Bayer AG (Germany) for several unsaturated polyester resins.

Leguval. Trademark of Bayer AG (Germany) for unsaturated polyester resins with good flame resistance and good resistance to corrosive chemicals and solvents. Used for hard, scratch-resistant coatings, as bonding materials, and for molded products.

Lehigh. Trade name of Lehigh Steel Corporation (USA) for a series of tool steels including various cold-work, nondeforming, shock-resisting and high-speed types.

leifite. A colorless mineral composed of sodium beryllium aluminum silicate hydroxide hydrate, $Na_6Si_{16}Al_2(BeOH)_2O_{39}\cdot 1.5H_2O$. Crystal system, hexagonal. Density, 2.58 g/cm^3. Occurrence: Greenland.

leightonite. A pale blue-green mineral composed of potassium calcium copper sulfate dihydrate, $K_2Ca_2Cu(SO_4)_4\cdot 2H_2O$. Crystal system, orthorhombic. Density, 2.95 g/cm^3; refractive index, 1.587. Occurrence: Chile.

Leipzig yellow. See chrome yellow.

Leit-C-Plast. Trademark of EMS-Chemie GmbH (Germany) for a permanently plastic, highly conductive adhesive carbon cement composed of polyisobutylene filled with carbon black, and used for mounting large specimens in scanning electron microscopy. This product is sold in the USA by Structure Probe, Inc., SPI Supplies Division.

leiteite. A colorless or brown mineral composed of iron zinc arsenate, $(Zn,Fe)As_2O_4$. Crystal system, monoclinic. Density, 4.30 g/cm^3; refractive index, 1.880. Occurrence: South West Africa.

Leitpantal. Trade name of VAW Vereinigte Aluminium-Werke AG (Germany) for a heat-treatable, wrought aluminum alloy containing 0.4-0.8% magnesium and 0.3-0.7% silicon. Used for decorative and structural parts.

Lektrocast. Trade name of Detroit Alloy Steel Company (USA) for a wear-resistant nickel cast iron used for draw dies and other castings.

Lektromesh. Trade name for screen cloth composed of copper or nickel, and made in a variety of mesh sizes.

Lektroset. Trade name for rayon fibers and yarns used for textile fabrics.

Lekutherm. Trademark of Bayer Corporation (USA) for a series of epoxy resins supplied in general-purpose, flexible, high-heat, and aluminum-, glass-, mineral-, and silica-filled casting grades. They have good chemical and electrical resistance, low shrinkage, good resistance to most organic solvents, good mechanical properties and corrosion resistance, good dimensional stability and good electrical properties. Used for molded products, electrical casings and other electrical components.

Lemac. Trade name for a series of vinyl acetate molding resins.

Lemarquand's alloy. A corrosion-resistant jewelry alloy composed of 39% copper, 37% zinc, 9% tin, 8% cobalt and 7% nickel.

Lemat's metal. A jewelry alloy composed of 80% copper, 7% nickel, 5% zinc, 5% iron, 2% tin and 1% cobalt.

Lemax. Trade name of Ley's Malleable Castings Company Limited (UK) for a series of heat-treated pearlitic malleable irons containing 2.3-2.6% carbon, 1.3-1.6% silicon, 0.4-0.6% manganese, 0.15-0.25% sulfur, up to 0.08% phosphorus, and the balance iron. Used for automotive, truck and tractor parts.

lemon chrome. A yellow inorganic paint pigment composed of lead chromate ($PbCrO_4$) and lead sulfate ($PbSO_4$).

lemon yellow. Yellow overglaze colors and paint pigments based on barium chromate ($BaCrO_4$). Also known as *Steinbühl yellow; ultramarine yellow.*

lemoynite. A pale yellow mineral composed of sodium calcium zirconium silicate pentahydrate, $(Na,K)_2CaZr_2Si_{10}O_{26}\cdot 5H_2O$. Crystal system, monoclinic. Density, 2.29 g/cm^3; refractive index, 1.553. Occurrence: Canada (Quebec).

lengenbachite. A steel-gray mineral composed of silver copper lead arsenic sulfide, $(Ag,Cu)_2Pb_6As_4S_{13}$. Crystal system, triclinic. Density, 5.85 g/cm^3. Occurrence: Switzerland.

Lennalmagman. Trade name of Reynolds Aluminiumwerke GmbH (Germany) for a non-heat-treatable aluminum alloy containing 1.6-2.5% magnesium, 0.5-1.1% manganese and up to 0.3% chromium. It has excellent resistance to seawater corrosion, and is used in shipbuilding, and for aircraft components and food-processing equipment.

Lennalman. Trade name of Reynolds Aluminiumwerke GmbH (Germany) for a corrosion-resistant, non-heat-treatable alloy containing 0.9-1.4% manganese and up to 0.3% magnesium. Used for roofing, trim, etc.

Lennalsil. Trade name of Westfälische Leichtmetallwerke GmbH (Germany) for an aluminum alloy containing 0.6-1.4% magnesium, 0.6-1.6% silicon, 0.6-1% manganese and up to 0.3% chromium. It has good weldability and formability, and is used for window frames, boats, gutters, and fan blades.

Lennedur. Trade name of Westfälische Leichtmetallwerke GmbH (Germany) for a heat-treatable aluminum alloy containing 2.5-5% copper, 0.2-1.8% magnesium and 0.3-1.5% manganese. Used for fasteners, fittings, and aircraft structures.

Lennite. (1) Trade name of Westlake Plastics Company (USA) for an abrasion-resistant, water-repellent ultra-high-molecular-weight polyethylene with excellent self-lubrication properties used for bearings, gears, and machine parts.

(2) Trade name for phenol-formaldehyde resins formerly used for denture bases.

lenoblite. A blue mineral composed of vanadium oxide dihydrate, $V_2O_4 \cdot 2H_2O$. Occurrence: Gabon.

leno-weave fabrics. Woven fabrics in which two or more warp threads cross each other and intertwine with one or more weft (filling) threads. The leno-weave pattern is used for meshed or open fabrics and nets. *Leno-weave fabrics* are usually composed of cotton or cotton-polyester blends.

Lenox. Trade name for a water-hardening tool steel containing 0.9% carbon, 1.5% manganese, 0.3% molybdenum, 0.3% silicon, and the balance iron. Used for tools, punches, dies, and knives.

lens coating. A transparent surface coating applied to optical and ophthalmic lenses and mirrors, and optical instruments to optimize light transmission. See also optical coating.

Lenslite. Trade name of Corning Glass Works (USA) for a heat-resistant borosilicate glass used for incandescent lighting fixtures.

Lenticolare. Trade name of Fabbrica Pisana SpA (Italy) for a glass with lenticular pattern.

Lentilite. Czech trade name for a patterned glass with small lenticular beads worked into the surface.

Lenzella. Trade name of Lenzing AG (Austria) for rayon fibers and yarns used for textile fabrics.

Lenzesa. Trade name of Lenzing AG (Austria) for rayon fibers and yarns.

Lenzing. (1) Trade name of Lenzing AG (Austria) for an extensive series of synthetic fibers including lyocell staple fibers (*Lenzing Lyocell*), modal staple fibers (*Lenzing Modal*), polytetrafluoroethylene staple fibers and filament yarns (*Lenzing Profilen and Lenzing PTFE*) and viscose rayon staple fibers (*Lenzing Viscose*). Used for the manufacture of textile fabrics including clothing and industrial products.

(2) Trade name of Lenzing AG (Austria) for several polyimide plastics and products.

Lenzing Fibers. Trade name of Lenzing AG (Austria) for lyocell-based cellulosic textile fibers used in the manufacture of evening and casual wear, household textiles, etc. See also lyocell fibers.

Leo. Trademark of Ivoclar Vivadent AG (Liechtenstein) for an extra-hard ceramic dental alloy containing 45.0% gold, 41.0% palladium, 11.0% platinum, 5.0% silver, 3.4% indium, 2.2% tin, 1.8% gallium, and less than 1.0% ruthenium, rhenium and lithium, respectively. It has a white color, a density of 13.3 g/cm³ (0.48 lb/in.³), a melting range of 1225-1315°C (2235-2400°F), a hardness after ceramic firing of 190 Vickers, moderate elongation, and excellent biocompatibility and resistance to oral conditions. Used for crowns, onlays, posts, bridges, and model castings.

Leolene. Trademark of Drake & Company (UK) for polypropylene filament yarns.

Leona. Trademark of Asahi Chemical Industry Company (Japan) for nylon 6,6 monofilaments and filament yarns.

Leonard. Trade name of Jessop-Saville Limited (UK) for a water-hardening tool steel containing 0.9-1% carbon, and the balance iron. Used for drills, punches and other tools.

leonite. A colorless to gray or pale yellow mineral composed of potassium magnesium sulfate tetrahydrate, $K_2Mg(SO_4)_2 \cdot 4H_2O$. It can also be made synthetically. Crystal system, monoclinic. Density, 2.20 g/cm³; refractive index, 1.483. Occurrence: Germany, USA (New Mexico).

Lepage. Trade name of Lepage's Limited (Canada) for an extensive series of adhesives and glues including various contact cements, rapid-setting epoxy adhesives, various specialty glues, and adhesive sealants, and numerous interior wall finishing products, such as plaster of paris, stucco and stipple finishes, crack fillers, paint brush and roller cleaners, paint and varnish removers, and wallpaper peelers and sizes.

Lepaz. Trade name of Ley's Malleable Castings Company Limited (USA) for a series of pearlitic malleable irons containing 2.3-2.6% carbon, 1.3-1.6% silicon, 0.4-0.6% manganese, 0.15-0.25% sulfur, up to 0.08% phosphorus, and the balance iron. Used for automotive, truck and tractor parts.

lepersonnite. A yellow mineral composed of calcium rare-earth uranyl carbonate silicate hydrate, $CaLn_2(UO_2)_{24}(CO_3)_8Si_4O_{12} \cdot 60H_2O$. Crystal system, orthorhombic. Density, 3.97 g/cm³; refractive index, 1.666. Occurrence: Zaire.

lepidocrocite. A red to red brown mineral composed of iron oxide hydroxide, γ-FeO(OH). Crystal system, orthorhombic. Density, 3.85 g/cm³; hardness, 402 Vickers; refractive index, 2.20. Occurrence: Germany, Algeria.

lepidolite. A colorless, pink, purple, grayish white or yellow mineral with a pearly luster belonging to the mica group. It is composed of potassium lithium aluminum fluosilicate, $K(Li,Al)_3(Si,Al)_4O_{10}(OH,F)_2$, and may also contain some iron. Crystal system, monoclinic. Density, 2.7-3.0 g/cm³; melting point, 1170°C (2138°F); hardness, 2.5-4 Mohs; refractive index, 1.559. Occurrence: Australia, Canada, China, Czech Republic, Germany, India, Japan, Norway, Russia, Sweden, South Africa, USA (Connecticut, California, Maine, Massachusetts, New Mexico, South Dakota), Zimbabwe. Used as an important ore of lithium, and as a flux and source of alumina in opal and flint glasses, porcelain enamels, glazes and ceramic bodies. Also known as *lithia mica; lithium mica*.

Leppestahl. Trade name of Gebrüder Höver GmbH & Co. Edelstahlwerk (Germany) for a series of high-carbon high-chromium nondeforming cold-work tool steels that may also contain small amounts of tungsten or vanadium, and are used chiefly for punches, and blanking and forming dies.

Le Provinox. French trade name for semi-nonoxidizing steel used for general engineering applications, gas-turbine compressor wheels, and structural members.

Lescalloy. Trade name of Latrobe Steel Company (USA) for an extensive series of nickel-chromium-iron alloys and numerous steels including high-carbon chromium steels for bearings, nickel-chromium molybdenum steels for structural applications and machine parts, low-carbon nickel-chromium steels for carburized parts, low-carbon high-chromium steels for heat-resistant parts, high-carbon high-chromium steels for bearings, and low-carbon nickel-chromium steels and precipitation-hardening stainless steels for corrosion-resistant parts.

Lesco. Trade name of Latrobe Steel Company (USA) for a series of steels including various stainless types, hot- and cold-work tool and die types, and high-speed types.

Lesjofors. Trade name of Lesjofors AB (Sweden) for a series of water or oil-hardening steels containing 0.1-0.8% carbon, up to 2% silicon, 0.4-1.5% manganese, up to 1% chromium, up to 1.5% nickel, and the balance iron. Used for springs.

Lestem. Trade name of Lehigh Steel Corporation (USA) for a non-tempering, water-hardening tool and die steel.

Letex. Trademark of Südplastik GmbH (Germany) for a polyvinyl chloride film.

letona. A vegetable fiber obtained from the inner bark (bast) of the an ageve plant (*Agave letonae*). Used for making textile yarns and fibers.

letovicite. A colorless mineral composed of ammonium hydrogen sulfate, $(NH_4)_3H(SO_4)_2$. It occurs in nature, but can also be made by precipitation from ammonium sulfate $[(NH_4)_2SO_4]$ and ammonium bisulfate $[NH_4HSO_4]$ in aqueous solution. Crystal system, triclinic. Density, 1.82 g/cm^3. Occurrence: Czech Republic.

leucite. A white or gray feldspathoid mineral belonging to the analcime group composed of potassium aluminum silicate, $KAlSi_2O_6$. Crystal system, tetragonal. Density, 2.47 g/cm^3; refractive index, 1.5095. Occurrence: Africa, Australia, Borneo, Brazil, Italy, Germany, Siberia, USA (Arkansas, Montana, Wyoming). Used in ceramics, glass-ceramics, dental materials (e.g., porcelains and glass-ceramics), biomedical ceramics, and biotechnology.

leucite syenite. A variety of the igneous rock *syenite* containing in excess of 5% of the mineral *leucite*.

Leuco. Trade name of Ledermann GmbH & Co. (Germany) for a series of cemented and coated cemented carbides used for cutting tools.

leucophanite. A white, green or yellow mineral of the melitite group composed of sodium calcium beryllium fluoride silicate, $NaCaBeFSi_2O_6$. Crystal system, orthorhombic. Density, 2.96 g/cm^3. Occurrence: Norway. Used as a source of beryllium.

leucophoenicite. A brown, purplish red to raspberry-red mineral of the humite group composed of manganese silicate hydroxide, $Mn_7(SiO_4)_3(OH)_2$. Crystal system, monoclinic. Density, 3.85 g/cm^3; refractive index, 1.771. Occurrence: USA (New Jersey).

leucophosphate. A buff-colored mineral composed of potassium iron phosphate hydroxide dihydrate, $KFe_2(PO_4)_2OH \cdot 2H_2O$. Crystal system, monoclinic. Density, 2.95 g/cm^3; refractive index, 1.721. Occurrence: Brazil.

leucosphenite. A light blue mineral composed of sodium barium boron titanium silicate, $Na_4BaB_2Ti_2Si_{10}O_{30}$. Crystal system, monoclinic. Density, 3.07 g/cm^3; refractive index, 1.657. Occurrence: Canada (Quebec), Greenland.

Leukorit. Trade name of Raschig GmbH (Germany) for several phenolic resins and plastics.

Leuritex. Trade name of Leuze & Rilling GmbH (Germany) for woven glass fabrics.

Leutalux. Trademark of Leuchtstoffwerk GmbH (Germany) for luminous pigments and paints.

Levaflex. Trademark of Bayer Corporation (USA) for several olefinic thermoplastic elastomers.

Levapren. Trademark of Bayer Corporation (USA) for ethylene-vinyl acetate rubber supplied in various grades with vinyl acetate contents ranging from 40 to 80%. They possess good weathering, light and ozone resistance, good hot-air resistance, and low high-temperature compression set. Used for cellular rubber goods, footwear soles, waterproofing, cable insulation and sheathing, lamp seals, moldings and extrudates, and as impact modifiers in polyvinyl chlorides.

Levasil. Trademark of Bayer Corporation (USA) for colloidal silica in aqueous solution supplied in various grades for use as a binder, polishing agent and surface modifier in electronics and materials engineering.

Levelbrite. Trade name of MacDermid Inc. (USA) for a bright nickel electroplate and plating process.

Level Chuck. Trade name of American Hard Rubber Company (USA) for hard rubber and related products.

leveling agent. A substance added to a coating or paint to improve its ability to cover a dry surface easily and form a smooth, level film without sagging or running. It also retards the formation of brush marks.

Levelume. Trade name of Atotech USA Inc. (USA) for a bright nickel electroplate and plating process.

Levepox. Trademark of Bayer AG (Germany) for epoxy resins and products.

Levesque. Trademark of Levesque Plywood Limited (Canada) for plywood and particleboard.

Levkin. Trademark of Polytox (Ukraine) for a biopolymer.

levigated abrasive. See mild polish.

levigated chrome rouge. See green rouge.

levigated chromium rouge. See green rouge.

Levilene. Trademark of SAR SpA (Italy) for polypropylene filament yarns.

Levion. Trademark of Nilit Limited (Israel) for nylon 6,6 filaments and fibers.

levyne. A colorless, reddish or grayish mineral of the zeolite group composed of calcium aluminum silicate hydrate, $Ca_3Al_{6.5}Si_{11.5}O_{36} \cdot xH2O$. Crystal system, hexagonal. Density, 2.13 g/cm^3; refractive index, 1.498. Occurrence: USA (Oregon).

Lewapren. Trade name of Lewcott Corporation (USA) for ethylene vinyl acetate (EVA) copolymers with a density of 0.94 g/cm^3, high flexibility, a maximum service temperature of 80°C (176°F), low tensile strength, excellent resistance to dilute acids and alkalies, moderate resistance to gasoline and mineral oils, and relatively low resistance to heat and solvents. Used for molded parts including bumpers, bellows, seals, gaskets, and flexible tubing.

Lewcarb. Trademark of Lewcott Corporation (USA) for modified and unmodified phenolic resins supplied as liquids, solids and powders.

Lewcott. Trade name of Lewcott Corporation (USA) for a series of polyester prepreg resins supplied on a variety of fiberglass fabrics.

Lewis Iron. Trade name of Joseph T. Ryerson & Son Inc. (USA) for a wrought iron containing 0.02% carbon, 0.02% silicon, 3% slag, and the balance iron. Used for chains, hooks, staybolts, and mine-car parts. *Lewis Special Iron* also contains 0.02% carbon, but only 0.01% silicon and 2.5% slag.

lewisite. A dark brown mineral of the pyrochlore group composed of calcium antimony oxide hydroxide, $(Ca,Fe,Na)_2(Sb,Ti)_2(O,OH)_7$. Crystal system, cubic. Density, 4.95 g/cm^3; refractive index, 2.20. Occurrence: Brazil.

Lewis metal. A fusible alloy composed of equal parts of bismuth and tin. It expands on cooling, has a melting point of 138°C (240°F), and is used for sealing and holding die parts.

Lewis Special Iron. See Lewis Iron.

lewistonite. A white mineral composed of hydrous calcium potassium sodium phosphate, $(Ca,K,Na)_5(PO_4)_3(OH)$. Crystal system, hexagonal. Occurrence: USA (Utah).

Lewtex. Trade name of Shakespeare Composites & Electronics Division (USA) for several fiberglass-reinforced composites.

Lexan. Trademark of GE Plastics (USA) for a series of polycarbonate-based engineering thermoplastics available in various grades including unfilled extrusion, injection and blow molding, and glass-filled foam. Optical grades with high light transmission and high refractive index and glass fiber-reinforced grades with enhanced strength, toughness, rigidity and thermal properties are also available. *Lexan* polycarbonates have a density of 1.2 g/cm^3 (0.04 $lb/in.^3$), high transparency, good strength, ductility and rigidity, outstanding fatigue properties, excellent dimensional stability and durability, high moduli of elasticity, high impact strength and flame resistance, good load-bearing properties, good electrical properties, a service temperature range of -215 to +121°C (-355 to +250°F), good resistance to dilute acids and alkalies, and fair resistance to other chemicals. Used for safety shields, lenses, covers and helmets, sight gages, gears and cams, automotive components, electronic, electrical and telecommunication equipment and components, power-tool housings, business machines, household appliances, sporting and recreational equipment, films, industrial glazing and foamed plastics.

Lexcel. Trademark of GE Plastics (USA) for a series of polycarbonate products including *Lexcel TW* structured panels and sheets and *Lexcel C* corrugated sheet. *Lexcel TW Twinwall* sheets have low density, high strength (stronger than acrylics and glass) and high flexibility, provide high light transmission and excellent thermal insulation properties, block harmful ultraviolet rays, can be easily cold formed, and are extremely resistant to breakage, cracking or splintering. They are available in clear, bronze and opal tints and used as glazings in a wide variety of applications including canopies, domes, greenhouses, industrial roofing, office partitions, sheds, skylights and sunrooms. *Lexcel C* single-layer corrugated sheets have a co-extruded coating on one surface that provides optimum UV protection and excellent resistance to outdoor weathering. They have low density, high impact strength, provide high light transmission and thermal insulation properties, and are extremely resistant to breakage. Used as glazings for a wide range of applications including industrial and commercial buildings, stadium roofs, greenhouses, and for wall cladding and roofing purposes.

Lexgard. Trademark of GE Plastics (USA) for laminated bullet-resistant window glazing based on *Lexan* polycarbonate resins.

Leyden blue. See cobalt blue.

Lezard. Trade name of SA Glaverbel (Belgium) for a patterned, wired glass with square mesh.

liandratite. A yellow to brown mineral of the amorphous group composed of niobium tantalum uranium oxide, $U(Nb,Ta)_2O_8$. Crystal system, hexagonal. Density, 6.80 g/cm^3; refractive index, 1.83. Occurrence: Madagascar.

Liasil. Trade name FiberGlass Industries Inc. (USA) for fiberglass reinforcements and fabrics.

liberite. A pale yellow to brown mineral composed of lithium beryllium silicate, β-Li_2BeSiO_4. It can also be made from a mixture of lithium silicate (Li_2SiO_3) and beryllium oxide (BeO) under prescribed conditions. Crystal system, monoclinic. Density, 2.69 g/cm^3; refractive index, 1.633. Occurrence: South China.

Liberty. (1) Trade name of Crucible Specialty Metals (USA) for

a tough, water-hardening tool steel containing 1.2% carbon, 0.3% chromium, 1.2% tungsten, and the balance iron. Used for reamers, broaches, dies and tools.

(2) Trade name of J.F. Jelenko & Company (USA) for a silver-free palladium-base dental alloy with good castability and finishability. Used for porcelain-to-metal bonding.

Liberty Coat. Trade name of Ferro Corporation (USA) for a porcelain enamel frit.

libethenite. A light to dark olive green mineral of the adamite group composed of copper phosphate hydroxide, $Cu_2(PO_4)(OH)$. It can also be made synthetically. Crystal system, orthorhombic. Density, 3.97 g/cm^3; refractive index, 1.743. Occurrence: Rumania.

Libradur. Trade name for polyvinyl chloride.

Lichtenberg's alloy. A German fusible alloy containing 50% bismuth, 30% lead and 20% tin. It has a melting range of 96-100°C (205-212°F), and is used for boiler safety plugs and fire extinguishers.

Lico. Trade name of Logan Iron & Steel Company (USA) for a water-hardening steel containing 0.3% carbon, and the balance iron. Used for castings.

Liddel's alloy. A zinc die casting alloy containing 5-6.5% copper and 6.5% aluminum.

liddicoatite. A dark olive-brown mineral of the tourmaline group composed of lithium calcium aluminum borate silicate hydroxide, $CaLi_2Al_7B_3Si_6O_{20}(OH,F)_4$. Crystal system, rhombohedral (hexagonal). Refractive index, 1.637. Occurrence: Madagascar.

Lideox. Trade name of Wolverine Tube, Inc. (USA) for an antifouling alloy composed of lithium, beryllium, mercury and sulfur, and used for condensers.

liebenbergite. A yellowish green mineral of the olivine group composed of nickel silicate, Ni_2SiO_4. Crystal system, orthorhombic. Density, 4.60 g/cm^3; refractive index, 1.854. Occurrence: South Africa.

liebigite. An apple or yellowish green, radioactive mineral composed of calcium uranyl carbonate decahydrate, $Ca_2UO_2(CO_3)_3\cdot10H_2O$. Crystal system, orthorhombic. Density, 2.41 g/cm^3; refractive index, 1.502. Occurrence: Czech Republic, Sweden, Canada (Saskatchewan).

Liebknecht white gold. A white, corrosion-resistant jewelry alloy composed of 80% gold, 13.9% nickel, 5% copper and 0.1% palladium.

Life. Trade name of Kerr Dental (USA) for a calcium hydroxide dental cement.

light alloys. A general term used for alloys of *light metals*, such as aluminum, beryllium and magnesium. Sometimes titanium alloys are also included under this term. Also known as *light-metal alloys; low-density alloys*.

Light-Bond. Trade name of Reliance Orthodontic Products (USA) for a dental resin used for orthodontic applications.

Light-Core. Trademark of BISCO Inc. (USA) for a light-cured, translucent, no-slump dental composite reinforced with optical fibers, and used in restorative dentistry for core-build-ups.

light-emitting polymers. A class of usually *conjugated polymers*, such as poly(*p*-phenylene vinylene), poly(fluorene) and polysilicon, that have unique optical and electrical properties including the capability to convert electrical energy into visible light. They are suitable for electronic, optoelectronic and photonic applications including especially polymer light-emitting devices (P-LEDs), and high-resolution displays. Abbreviation: LEP. See also conductive polymers; semiconductive polymers.

Lightglass. Trade name of Light Glass Industries (India) for a toughened glass.

light guides. See optical fibers.

light hydrogen. See hydrogen.

light leaded brass. See low-leaded brass.

light-metal alloys. See light alloys.

light metals. A group of metals that have densities of less than 4 g/cm^3 (0.14 lb/in.^{3}) and are strong enough for engineering applications. Examples include aluminum (2.70 g/cm^3 or 0.098 lb/in.^{3}), beryllium (1.85 g/cm^3 or 0.067 lb/in.^{3}), magnesium (1.74 g/cm^3 or 0.063 lb/in.^{3}), rubidium (1.53 g/cm^3 or 0.055 lb/in.^{3}) and lithium (0.53 g/cm^3 or 0.019 lb/in.^{3}). Also known as *low-density metals*. See also heavy metals.

light-rare-earth elements. See light rare earths.

light rare earths. A group of rare-earth metals that comprises the chemical elements with atomic numbers 57 through 63, namely, lanthanum, cerium, praseodymium, neodymium, promethium, samarium and europium. The light rare-earth elements form a subgroup within the lanthanide group of elements. Also known as *light rare-earth elements; ceric rare earths; cerics*.

light-red silver ore. See proustite.

light-reducing glass. A non-specific term used for flat glass whose ability to transmit incident light is reduced as a result of absorption, reflection and/or scattering. Depending on the degree of reduction, such a glass may be translucent or opaque.

light ruby silver. See proustite.

light-sensitive glass. See photosensitive glass.

light-sensitive material. See photosensitive material.

light stabilizers. Additives, such as carbon black, dialkyldithiocarbamates, dialkyldithiophosphates, hydroxybenzophenones, hydroxyphenylbenzotriazoles, thiobisphenolates and certain pigments (e.g., phthalocyanines), incorporated into a polymer to inhibit its degradation due to ultraviolet radiation. Also known as *photostabilizers*.

Light Super-Bond. Trade name of Carlisle/Superdent (UK) for a light-cure dental resin used for orthodontic applications.

lightweight aggregate. An inert material of low bulk density, such as pumice, foamed slag, expanded or sintered clay, shale, slate, slag, fly ash, vermiculite, perlite, clinker or scoria, used in the production of *lightweight concrete*. Some of these aggregate materials may also improve the thermal and sound-insulating properties of the final product. Abbreviation: LWA. Also known as *light aggregate*.

lightweight-coated paper. Printing paper coated on both sides with the coating weight usually being 5-10 g/m^2 (0.016-0.032 oz/ft^2) on each side. It contains some wood pulp, is available in rolls, and has a basis weight usually less than 72 g/m^2 (0.234 oz/ft^2). Abbreviation: LWC paper.

lightweight concrete. A low-density concrete (density usually below 2 g/cm^3 or 0.07 lb/in^3) made with *lightweight aggregate*. Used for slabs and blocks. Abbreviation: LWC. Also known as *light concrete; low-density concrete*.

Lignaflex. Trademark of Leeuvenburgh Fineer BV (Netherlands) for flexible wood veneer sheets on paper supports, used especially in boatbuilding and interior decoration and finishing.

Lignapal. Trademark of Leeuvenburgh Fineer BV (Netherlands) for prefinished wood veneer on phenolic supports, used especially in boatbuilding and interior decoration.

lignin. The major noncarbohydrate constituent in wood that is denser than cellulose. Chemically, it is an irregular aromatic biopolymer with substituted propylphenol groups. *Lignin* is used in the stabilization of asphalt emulsions for road surfacing, in

adhesives and coatings, as a ceramic binder, as a binder for compressed-wood products (e.g., hardboard and particleboard), as an extender and filler for phenolic plastics, in special molded products (furfural plastics), as a constituent of battery expanders, a corrosion inhibitor, and in biotechnology.

lignin plastics. Plastics derived from lignin resins and used as fillers and extenders in phenolics, or as binders in ceramics.

lignin resin. A resin obtained by reacting lignin with selected chemicals or resins, or by heating lignin at a suitable temperature.

lignin sulfonates. See lignosulfonates.

lignite. See brown coal.

lignite wax. A brownish or white wax extracted from lignite. It has a melting point of about 85°C (185°F), and is used as a replacement for beeswax and carnauba wax, in paints for roofing and other water-resistant applications, in paper sizes, as a rubber ingredient to prevent sun cracking, and in adhesives, wire coatings and electrical insulation. Also known as *montan wax*.

lignocellulose. A carbohydrate found in the cell walls and middle lamellae of woody plants and composed of normal cellulose and lignin. Abbreviation: LC.

lignocellulosic materials. A group of materials based on organic matter produced by plants, such as trees, shrubs, grasses and agricultural crops. They are composed essentially of 35-45% cellulose, 25-35% hemicelluloses and 20-25% lignin, and are used as renewable biopolymers. Also known *lignocellulosics*. See also lignocellulose.

lignoconcrete. A concrete analogous to *ferroconcrete*, but strengthened by wood instead of iron or steel.

Lignofol. Trade name for a wood laminate composed of several veneers bonded together under heat and pressure. It has a density of approximately 1.4 g/cm^3 (0.05 lb/in.^{3}), low water absorption, good workability, and is used for machine parts.

Lignostone. Trade name of Röchling Haren KG (Germany) for a dense, hard, strong compressed wood.

lignosulfonates. A group of metallic sulfonate salts, such as calcium or sodium lignosulfonate, made from the *lignin* removed from paper-pulp sulfite liquors. They are usually tan to light brown powders that decompose above 200°C (390°F), and have molecular weights ranging from 1000 to 20000. Used as ceramic binders, dispersing agents in concrete and rubber compounds, and as extenders in synthetic resins, adhesives and cementitious materials. Also known as *lignin sulfonates*.

lignum vitae. The hard, heavy wood of the evergreen trees *Guaiacum officinale* and *G. sanctum*. The heartwood is dark greenish-brown to almost black, and the sapwood yellowish and sharply defined from the heartwood. *Lignum vitae* has a high resin content, a waxy feel, will not float in water, and turns well. Average weight, 1250 kg/m^3 (78 lb/ft^3). Source: Central America, Columbia, Venezuela, West Indies. Used for pulleys, bearings, casters, mallets, turnings, etc.

likasite. A sky-blue mineral composed of a copper nitrate phosphate hydroxide, $Cu_6(NO_3)_2(PO_4)(OH)_7$. Crystal system, orthorhombic. Density, 2.97 g/cm^3. Occurrence: Zaire.

Likra. Trademark of Nylon de Mexico SA (Mexico) for elastomeric fibers and filaments.

Lilion. Trade name of Snia Fibre SpA (Italy) for nylon 6 and nylon 6,6 fibers and filament yarns used for textile fabrics including apparel and carpets.

Lille lace. A bobbin lace quite similar to *Mechlin lace*, having a fine texture and featuring dotted patterns or designs outlined by heavier threads. It is originally from Lille, a city in northern France.

lillianite. A steel-gray mineral composed of lead bismuth sulfide, $Pb_3Bi_2S_6$. It can also be made synthetically from lead sulfide (PbS) and bismuth sulfide (Bi_2S_3). Crystal system, orthorhombic. Density, 7.00 g/cm^3.

Lily Brand. Trade name of Charles Carr Limited (UK) for a series of copper-base casting alloys including various aluminum, manganese and phosphor bronzes, leaded bronzes, gunmetals, leaded gunmetals and commercial brasses.

Lima. Trade name of Vetreria di Vernante SpA (Italy) for a patterned glass.

limba. See afara.

limbric. A soft, tightly woven fabric made from quality cotton yarn in a plain weave. It is available in light and medium weights and has a weft that is coarser and denser than the warp.

lime. (1) A mixture of calcium oxide (CaO) and magnesium oxide (MgO). See also hydraulic lime.

(2) See calcium oxide; calcium hydroxide; chlorinated lime; fat lime; lean lime; sulfurated lime.

(3) See American basswood; European lime.

lime acetate. See calcium acetate.

lime-base grease. See lime grease.

lime brick. A basic refractory brick containing lime that will react chemically with acidic refractories, clays or fluxes at high temperatures. Used for furnace and converter linings.

lime-cement mortar. A masonry mortar usually made of 1 part of *mortar cement*, 1 or 2 parts of *slaked lime* or *lime putty* and 5 or 6 parts of sand by volume. Also known as *cement-lime mortar; gaged mortar; gauged mortar.*

Lime-Cote. Trade name of Rockwell Lime Company (USA) for a finishing lime used for plastering and finish-coat applications.

lime crown glass. An *optical crown glass* in which calcium oxide (CaO) is used in considerable quantity as a fluxing ingredient. See also calcium glass.

limed rosin. Rosin that has been treated with a calcium compound (e.g., calcium hydroxide), or a calcium and a zinc compound (e.g., zinc oxide). Used in paints, varnishes and molded products. Also known as *lime-hardened rosin.* See also rosin.

lime feldspar. See anorthite.

lime glass. A commercial glass containing a considerable amount of lime (up to 25% CaO) as a fluxing ingredient and network modifier, usually in combination with silicon dioxide and sodium oxide. It has good chemical properties and relatively high strength, and is used in the manufacture of bottles, containers, reinforcing fibers, etc. See also soda-lime glass.

lime grease. A smooth, buttery, water-resistant *lubricating grease* consisting of a calcium soap (e.g., calcium oleate or calcium palmitate) suspended in mineral oil. The operating temperature range of standard grades is about -30 to +80°C (-22 to +175°F), but high-temperature grades are also available. Used for the general lubrication of plain bearings, sliding surfaces, line shafting, low-load bearings operating at low to moderate speeds, water pumps, and wet lubrication points. Also known as *lime-base grease; lime soap grease; calcium-base grease; calcium grease; cup grease.*

lime-hardened rosin. See lime rosin.

lime hydrate. See calcium hydroxide.

lime marl. See calcareous marl.

lime mortar. A masonry mortar made by mixing *lime putty* with sand. It has a compressive strength up to 2.76 MPa (400 psi), and is used for nonload-bearing walls in buildings. Abbreviation: LM.

lime nitrate. See calcium nitrate tetrahydrate.

lime nitrogen. See calcium cyanamide.

Limeolith. Trade name for a precipitated chalk ($CaCO_3$) used as a white pigment extender.

lime paste. A highly plastic product with varying consistency made by slaking *quicklime* (calcium oxide) with water. Used in the manufacture of lime mortar and whitewash. Also known as *lime putty.*

lime plaster. A plastic paste obtained by mixing lime with sand and water and, sometimes, fibrous materials for use in covering walls and ceilings.

lime powder. See air-slaked lime.

lime putty. See lime paste.

lime saltpeter. See calcium nitrate tetrahydrate.

lime-sand brick. See sand-lime brick.

lime-slag cement. A special type of cement made by mixing lime and granulated blast-furnace slag.

lime slurry. Slaked lime containing an excess of uncombined water.

lime soap grease. See lime grease.

lime-stabilized zirconia. An engineering ceramic composed of zirconium oxide (ZrO_2) to which 1-16 wt% lime (CaO) has been added. Additions of up to about 8 wt% CaO are used in *partially stabilized zirconia* (PSZ) which has a two-phase microstructure composed of monoclinic zirconia solid solution and cubic zirconia solid solution. Additions of up to 16 wt% CaO are used in *fully stabilized zirconia* (FSZ) which has a cubic crystal structure. Both lime-stabilized PSZ and FSZ have excellent mechanical properties including high transverse rupture strength, good fracture toughness, good hardness, high elastic modulus, low thermal conductivity, very low coefficients of expansion, good thermal-shock resistance, good electrical insulating properties, and good resistance to many acids and alkalies. The mechanical properties (including fracture toughness and thermal-shock resistance) of PSZ are superior to those of FSZ. Used for structural applications and refractories. Also known as *calcia-stabilized zirconia.*

limestone. A sedimentary rock formed mainly from organic remains, such as shells or corals, and composed predominantly of calcium carbonate ($CaCO_3$). Calcitic limestone contains more than 95% $CaCO_3$, argillaceous limestone contains some clay, dolomitic limestone contains 5% or more magnesium carbonate ($MgCO_3$), and siliceous limestone contains some quartz or sand. Chalk and marble are varieties of limestone. *Limestone* has a density of 2.1-2.9 g/cm^3, an average compressive strength of 62 MPa (9000 psi), an average tensile strength of 2.1 MPa (300 psi), and a hardness of approximately 3 Mohs. Used as a building stone, in the manufacture of quicklime and cement, as a source of calcium in glazes, as a source of carbon dioxide, as a flux in smelting of iron ore, for road ballast and tar-macadam aggregate, and as filler.

lime wash. See whitewash (1).

lime water. See whitewash (1).

limnite. See bog iron ore.

Limoges kaolin. A high-grade *china clay* with negligible amount of iron oxide found in southern France, and used in ceramics, white paper coatings, and as a paper extender.

limonite. An ore composed of hydrous ferric oxide, $2Fe_2O_3 \cdot 3H_2O$, varying in color from dark-brown to yellowish-brown, and containing approximately 20-55% iron. Density, 3.6-4.0 g/cm^3; hardness, 5.0-5.5 Mohs. Occurrence: France, Germany, UK,

USA (Alabama, Tennessee, Virginia). Used as an ore of iron, in high-density concrete for radiation shielding, and as a yellow or brown ceramic colorant. Also known as *brown hematite; brown iron ore; brown ore.*

Linabestos. Trade name of Manville Corporation (USA) for asbestos board products.

Linacoustic. Trademark of Manville Corporation (USA) for fiberglass duct insulation.

Lin-All. Trade name of Mooney Chemicals, Inc. (USA) for a series of driers for paints, varnishes and printing inks based on tallates, i.e., metallic soaps formed of tall oil and a metal, such as calcium, cobalt, copper, iron, lead, manganese and zinc. They are available in liquid, paste and flake form.

linarite. A deep blue mineral composed of copper lead sulfate hydroxide, $CuPb(SO_4)(OH)_2$. Crystal system, monoclinic. Density, 5.35 g/cm^3; refractive index, 1.838. Occurrence: USA (Arizona), Italy.

Linco. Trade name of Latrobe Steel Company (USA) for a magnetic stainless steel containing 0.3% carbon, 1.1% manganese, 0.35% silicon, 11.5% chromium, 2.75% molybdenum, 0.35% nickel, 0.25% vanadium, and the balance iron. Available in the hardened and tempered condition, it has good strength and corrosion resistance at temperatures above 650°C (1200°F), and is used for jet-engine and gas-turbine compressor wheels and parts.

Lincoln. Trade name of Ford Motor Company (USA) for a safety glass made by laminating together two 6.35 mm (0.25 in.) thick sheets of plate glass.

lindackerite. An apple to light green mineral composed of copper arsenate nonahydrate, $Cu_5As_4O_{15}\cdot 9H_2O$. Density, 3.20 g/cm^3; refractive index, 1.662. Occurrence: Czech Republic.

Linde AF. Trade name of Union Carbide Corporation (USA) for agglomerate-free alumina powders produced by controlled de-agglomeration and classification, and used for polishing and ceramic applications.

Lindemann glass. A special glass made from lithium tetraborate ($Li_2B_4O_7$) and beryllium oxide (BeO) that is highly transparent to X-rays and is used in windows of low-voltage X-ray tubes.

linden. See American basswood; European lime.

Lindenberg. Trade name of Bergische Stahl Industrie (Germany) for a series of plain-carbon, cold-work and high-speed tool steels, and several machinery steels.

Lindlar's catalyst. A catalyst consisting of 5 wt% palladium on calcium carbonate and poisoned with lead. It is available as a powder with a typical surface area of 5-10 m^2/g.

lindgrenite. A green mineral composed of a copper molybdenum oxide hydroxide, $Cu_3(MoO_4)_2(OH)_2$. Crystal system, monoclinic. Density, 4.26 g/cm^3; refractive index, 2.002. Occurrence: Chile.

Lindiste. Trademark of American Potash & Chemical Corporation (USA) for fine abrasives composed of a natural mixture of rare-earth oxides (chiefly cerium oxides) and used for polishing glass.

lindstroemite. See lindstromite.

lindstromite. A tin-white to lead-gray mineral composed of lead copper bismuth sulfide, $PbCuBi_3S_6$. Crystal system, monoclinic. Occurrence: Sweden. Also known as *lindstroemite.*

Lineal. Trade name of Boussois Souchen Neuvesel SA (France) for a glass with linear pattern.

Lineare Filettato. Trade name of Fabbrica Pisana SpA (Italy) for a glass with linear fillet-type pattern.

Lineare Inciso. Trade name of Fabbrica Pisana SpA (Italy) for a glass with engraved linear pattern.

linear high-density polyethylenes. See high-density polyethylenes.

linear low-density polyethylenes. A group of linear polyethylene plastics that contain only a few short side chains or branches, and are copolymers of ethylene and an alkene, such as butene, hexane or propylene. They have a density range of 0.919-0.925 g/cm^3 (0.0332-0.0334 $lb/in.^3$), are white in color, and have very good chemical resistance, poor resistance to strong oxidizing agents, aromatic and halogenated aliphatic hydrocarbons, liquefied petroleum gas and solvents, very good stress-crack resistance, good abrasion and impact resistance, moderate rigidity, good (gas) weldability, a maximum service temperature of 60°C (140°F), and a brittleness temperature of -70°C (-94°F). Used for storage containers for foodstuffs, beverages, liquids and chemicals, plating and water-treatment equipment, and other industrial and commercial equipment. Abbreviation: LLDPE. See also low-density polyethylenes; polyethylenes.

linear medium-density polyethylenes. A group of linear polyethylene plastics in which the number of side chains or branches is intermediate between that of linear low-density and linear high-density grades. They have a density range of 0.926-0.940 g/cm^3 (0.0335-0.0340 $lb/in.^3$). Abbreviation: LMDPE. See also medium-density polyethylenes; polyethylenes.

Linear Metal. Trade name of Sheepbridge Engineering Limited (UK) for a cast iron containing 3.3% carbon, 1% silicon, 0.2% phosphorus, and the balance iron. Used for cylinder liners, and castings.

linear polyethylenes. Polyethylene plastics having comparatively straight and closely aligned molecular chains with relatively few side chains or branches. Also known as *linear polyethylene plastics.* See also linear high-density polyethylenes; linear low-density polyethylenes; linear medium-density polyethylenes; polyethylenes.

linear polymer. A thermoplastic polymer, such as linear polyethylene, in which each molecule consists of bifunctional monomers joined end to end in a single continuous chain with little or no branches or crosslinks. Abbreviation: LP.

line compound. A compound whose composition is indendent of temperature.

lined gold. Gold leaf or foil on a permanent or temporary support or backing of another metal.

Lineltex. Trademark of Fillatice SpA (USA) for spandex filaments and fibers used for elastic textile fabrics.

linen. (1) A strong, rigid, durable, lustrous thread or yarn spun from flax fibers. It is a good conductor of heat, more absorbent than cotton, creases and burns readily, and is resistant to insects, but attacked by mildew. See also flax.

(2) A strong, durable textile fabric knitted or woven from flax thread or yarn, and used for fine *lawn,* suiting, dress fabrics, table wear, and household furnishings. See also flax.

linen canvas. A light grayish brown, closely and uniformly woven good-quality *canvas* used especially for coats.

linen twill. A lightweight *linen* fabric in a twill weave, used especially for embroidery.

Lineon. Trademark of RK Carbon Fibers Limited (UK) for carbon fibers.

linerboard. Paperboard used as a covering material on one or both sides of *fiberboard* or *hardboard* panels. It is usually made on cylinder- or Fourdrinier-type paper machines, and may or may not be textured.

Linex. Trade name of ASG Industries Inc. (USA) for a patterned glass with concave flutes.

Linien. German trade name for a glass with a narrow line pattern.

lining leather. Chrome- or vegetable-tanned, lightly dyed leather, usually obtained from sheep, lambs, goats, kids, cattle and calves, and used for shoe linings.

lining metal. A lead-base babbitt containing 5-20 antimony and 2-20% tin, and used for bearings of locomotives, railroad cars and other rolling stock. See also lead babbitts.

linnaeite. A light to steel gray mineral of the spinel group with a reddish tarnish and a metallic luster. It is composed of cobalt sulfide, Co_3S_4, and can also be made synthetically. Crystal system, cubic. Density, 4.65 g/cm³; hardness, 5.5 Mohs. Occurrence: Germany, USA (Maryland, Missouri, Nevada). Used as an ore of cobalt. Also known as *cobalt pyrites; linneite.*

linneite. See linnaeite.

Linofelt. Trade name for an acoustical and thermal insulating product consisting of interfelted flax fibers sandwiched between two layers of strong, tough, water-resistant kraft paper.

linofil. A yarn made from flax waste.

linoleum. A resilient, washable floor covering with a hard, glossy surface made by rolling a mixture of ground cork or wood flour, hot oxidized linseed oil, an oleoresinous binder (e.g., rosin or a fossil resin), and mineral fillers and pigments (e.g., lithopone), on a fabric backing of burlap, canvas or jute. It is available in sheet and tile form.

Linotile. Trade name for a resilient linoleum floor covering available in tile form.

linotype metal. A type metal alloy containing 79-86% lead, 11-16% antimony and 3-5% tin. It has a melting point of 247°C (476°F), and is used for pressure castings, and linotype typesetting machines.

linsey. (1) A coarse fabric in a twill weave, having a weft (filling) of blended cotton and waste wool and a warp of cotton yarn.

(2) A coarse fabric made of linen. Formerly known as *linsey-woolsey.*

(3) A strong, coarse fabric having a cotton or linen warp and a woolen or worsted weft. Also known as *linsey-woolsey.*

Linstat. Trade name of Solutions Globales (France) for a range of antistatic textile fabrics including *Linstat Protektor* composed of 98% *Trevira CS* polyester and 2% DuPont *Negastat* antistatic polyester fibers, and *Linstat Negastat* composed of 98% Rhône-Poulenc *Pontella* polyester fibers and 2% DuPont *Negastat* antistatic polyester fibers. Used for commercial and industrial applications.

linters. A short, fleecy, residual fibrous material that adheres to ginned cottonseed, and is used in the manufacture of rayon, as an extender in plastics, in cellulosic plastics and nitrocellulose lacquers, as a soil-cement binder in road construction, in explosives (guncotton), as a raw material for the manufacture of cellulose derivatives (e.g., cellulose esters and absorbent cotton), and in fabrics and special papers. Also known as *cotton linters.*

Lion. Trade name of Burys & Company Limited (UK) for a series of water-hardening tool steels (AISI types W1 Grades 1 & 2), oil-hardening nonshrinking tool steels, hot-work tool steels, cobalt-tungsten high-speed steels, and stainless steels.

Lionblast. Trade name of Treibacher Schleifmittel Corporation (USA) for alumina abrasives used for blasting and cleaning applications.

Lion Brand. Trade name of Blackwell's Metallurgical Works (UK) for lead-base antifriction metals that contain 10-15% antimony and 5-15% tin and are used for crusher and rolling-mill bearings, and tin-base white metals with 10-17% lead, 3-10% antimony and 1-7% copper, used for the main bearings of marine engines.

Lionite. Trade name of Treibacher Schleifmittel Corporation (USA) for alumina abrasives.

liottite. A colorless mineral of the cancrinite group composed of sodium calcium chloride carbonate silicate sulfate hydroxide hydrate, $(Ca,Na)_4(Si,Al)_6O_{12}(SO_4,OH,Cl,CO_3)_2 \cdot xH_2O$. Density, 2.56 g/cm³; refractive index, 1.530. Occurrence: Italy.

Lipatex. Trademark of Bayer Corporation (USA) for rubber latex used in building construction, as a backing for textiles, carpets and paper, and in the manufacture of molded foams.

lipid. Any of a class of water-insoluble fatty substances that can be extracted from animal cells by organic solvents with low polarity, such as ethers or chloroform, and can also be made synthetically. The main classes of lipids are triglycerides (fat and oils), phospholipids (e.g., lecithin) and sterols (e.g., cholesterol).

Liplon. Trademark of Tai-Rah Company (Japan) for polypropylene monofilament and filament yarns.

lipoid. Compound lipid or fatty substance (e.g., cholesterol) that is like a lipid.

Lipolan. Trademark of CWH (Germany) for a styrene-butadiene copolymer.

Lipowitz alloy. An eutectic fusible alloy composed of 50% bismuth, 26.7% lead, 13.3% tin and 10% cadmium. It has a melting point of 70°C (158°F), high ductility, and is used for low-melting solders, and for casting metal patterns and ornaments. Also known as *Lipowitz metal.*

lipscombite. A black mineral composed of iron phosphate hydroxide, $Fe_3(PO_4)_2(OH)_2$. It can also be made synthetically, and may contain manganese oxide. Crystal system, tetragonal. Density, 3.66-3.80 g/cm³; refractive index, 1.67-1.80. Occurrence: Brazil.

Liqua-Sheen. Trade name of JacksonLea (USA) for a liquid burnishing compound.

Liquibuff. Trade name of M.P Idling Company, Inc. (USA) for a liquid buffing compound.

liquid bright gold. A liquid mixture of varnish and gold or gold compounds, sometimes with additions of platinum or palladium, that is applied and fired to china or other ceramics to provide a decorative gold-colored film, or make printed circuits on ceramics.

liquid bright platinum. A liquid mixture of varnish and platinum with additions of palladium, gold or bismuth that is applied and fired to pottery, glass or tile to provide a silvery finish.

liquid buffing compound. A buffing compound composed of fine abrasive suspended in a lubricant binder, and used to give a high luster to metallic surfaces.

liquid crystal. Any of a large group of organic compounds, such as ammonium oleate, cholesteryl benzoate or sodium benzoate, in a mesomorphic or intermediate state between solid and liquid that exhibits a regular crystal-like structure even in the liquid state. The two basic types are *thermotropic liquid crystals* whose liquid crystalline state depends on temperature changes and *lyotropic liquid crystals* whose liquid crystalline state depends on the concentration of the solvent. Used for electric and electronic displays (e.g., liquid crystal displays), thermometric instruments, picture tubes for color televisions, electronic clocks and calculators, optical imaging devices, etc. Also known as *liquid crystal material; mesomorphic substance.* See also cholesteric liquid crystal; nematic liquid crystal; smectic liquid crystal; columnar liquid crystal; thermotropic liquid crystal; lyotropic liquid crystal.

liquid crystal glass. A relatively new class of clear, optical materials that have properties characteristic to both liquid crystals and glasses. They are solid, stable films with highly ordered molecular structure composed of stacks of molecule layers that are slightly rotated with respect to each other and contain special additives (dopants) which allow them to actually emit and modify circularly polarized light and reflect residual light when struck by UV unpolarized light. Current applications include displays for laptops, cell phones, calculators and eyewear. Proposed future applications include selective-filter laser goggles, optoelectronic devices for optical communication and storage devices. Also known as *glassy liquid crystals.*

liquid crystalline elastomer. An elastomer, such as polysiloxane, having a three-dimensional network structure produced by crosslinking liquid crystalline side-chain polymers. It exhibits both the anisotropic properties of a liquid crystal and the rubber-like elasticity and shape stability of an elastomer. Mechanical deformation results in an optical change from translucent to transparent due to a macroscopically uniform orientation of the long molecular axis of the liquid crystal units. Used for electronic, optical and optoelectronic applications, Abbreviation: LCE.

liquid crystal material. See liquid crystal.

liquid crystal polymers. A group of engineering thermoplastics, usually copolyamides, copolyesters or polyester-amides, that develop high degrees of orientation during molding, and whose backbones are composed essentially of benzene rings. Available in fiber-reinforced and unreinforced grades, they have excellent melt processability, outstanding strength, toughness and stiffness, low coefficients of thermal expansion and high flame and chemical resistance. Often sold under trade names or trademarks, such as *Vectra* or *Xydar*, they are used for automotive components including engine components, firewall insulation, under-hood connectors, windshield washer pump gears, fuel rails and cruise-control components, electrical and electronic components and devices, such as connectors, switches, printed-wiring board components, relay and capacitor housings, sockets, brackets, lamp receptacles and coil formers, industrial equipment parts including valve and belt covers, thrust washers, pump housings, pump parts and impellers, household appliances, and cookware including components of microwave ovens, clothes dryers, steam irons, kitchen appliances, power tools, etc. Abbreviation: LCP. See also polymer liquid crystal.

Liquid Envelope. Trade name of Essex Chemical Corporation (USA) for a peelable plastic coating.

liquid glass. See soda water glass.

liquid-injection-molded plastics. Thermoplastic and thermosetting products formed by directly injecting appropriate amounts of thoroughly mixed two-part liquid resins into molds and closing the latter under pressure.

liquid-metal nuclear fuel. A nuclear reactor fuel made by dissolving a fissionable material (e.g., plutonium or uranium) in a molten metal (e.g., bismuth).

liquid metals. (1) Metals, such as cesium, gallium or mercury, that are liquid at ordinary temperatures.

(2) Metals, such as sodium (melting point, 98°C or 208°F) or potassium (melting point, 64°C or 147°F), or a combination of these metals known as NaK. Used as nuclear reactor coolants.

liquid-molded plastics. Thermoplastic and thermosetting plastics processed by any of several processes in which the starting material is a liquid polymer or a polymer/reinforcement solution. The most common liquid molding processes are compression molding, injection molding, resin-transfer molding and structural reaction-injection molding.

Liquid Nails. Trademark of Macco Adhesives (USA) for construction adhesives used on wood and plastic panels, decks, subfloors, etc.

liquid nitrided steel. See salt-bath nitrided steel.

liquid resin. A liquid polymeric material in an intermediate stage of production.

liquid semiconductor. A liquid or solid amorphous material having semiconducting properties. See also amorphous semiconductor.

liquid silicone rubber. Silicone rubber of liquid consistency used for injection-molded parts, e.g., gaskets, seals, automotive parts, and electrical insulation and encapsulation. Abbreviation: LSR.

Liquid-Vinyl. Trade name of Kwal-Howells Paint (USA) for vinyl satin house paint used for interior and exterior applications.

liquid wood filler. A low-viscosity varnish that contains extender pigments, and is used as a first coating on open-grain woods for filling the pores, and provide a nonabsorbent surface preparatory to the application of finish coats.

Liquigum. Trademark of Domtar Construction Materials (Canada) for liquid adhesives.

Liquimatic. Trade name of JacksonLea (USA) for liquid buffing compounds.

Liqui-Moly. Trade name of Liqui-Moly GmbH (Germany) for molybdenum disulfide used as a solid lubricant for machine components.

Liquispray. Trade name of The Stutz Company (USA) for a spray-applied liquid buffing compound.

Liquistone. Trademark of Niagara Protective Coatings (Canada) for masonry paints.

liquor-finished wire. Wire, usually made of a ferrous alloy, drawn through a wet solution of metallic salts (e.g., copper or tin) to ease drawing and develop a smooth bright luster. Formerly, liquor obtained from fermented grain mash was used as the drawing lubricant.

Lirelle. (1) Trademark of Unifil Limited (Ireland) for wet-strong polyester fibers used for textile fabrics, either alone or in blends with other fibers, such as cotton.

(2) Trade name of Lirelle plc (Ireland) for rayon fibers and yarns used for textile fabrics including clothing.

lironconite. A sky-blue to verdigris-green mineral composed of copper aluminum arsenate hydrate tetrahydrate, $Cu_2Al(As,P)O_4(OH)_4 \cdot 4H_2O$. Crystal system, monoclinic. Density, 2.96 g/cm³; refractive index, 1.652. Occurrence: UK.

Lisco. Trade name of Logan Iron & Steel Company (USA) for a water-hardening steel used for castings.

Lisix. Trade name of Dentamerica (USA) for a light-cure dental hybrid composite.

liskeardite. A soft, greenish white mineral composed of aluminum iron arsenate hydroxide pentahydrate, $(Al,Fe)_3AsO_4(OH)_6 \cdot 5H_2O$. Crystal system, orthorhombic. Density, 3.01 g/cm³; refractive index, 1.675. Occurrence: UK.

lisle. A fine, strong, smooth, highly-twisted yarn that has been passed through or near a gas flame to remove surface fibers, and may or may not be mercerized. Formerly used for making stockings, gloves, etc. Named after Lille (formerly Lisle), a town in northern France. Also known as *lisle thread.*

Lisse. Trade name NV Durobor (Belgium) for hollow glass blocks, flat on one side and having a broad vertical reed pattern on the other.

Listral. Trade name of Saint Gobain (France) for a cast glass that may or may not be wired and/or colored.

Litaflex. Trademark of Rex Industrie-Produkte Graf von Rex GmbH & Co. KG (Germany) for a noncombustible polymer foam material used for insulating applications (e.g., jointing and sealing), and in fire protection.

Lital. Trade name of Alcan Plate (UK) for wrought, lightweight aluminum-lithium alloys for aerospace applications including *Lital 8090* with greatly improved stiffness supplied in sheet or extruded form. Used for helicopter components, e.g., door parts, cabin floors, and fuselages.

Litecast B. Trade name of Ivoclar/Williams (USA) for a corrosion-resistant nickel-chromium dental bonding alloy.

Lite-Fil II. Trade name of Shofu Dental (Japan) for a light-cure dental hybrid composite.

Lite 'N Eze. Trademark of Arndt-Palmer Canada Inc. for lightweight autobody fillers.

LiteRock. Trademark of Consolidated Concrete Limited (Canada) for lightweight expanded clay aggregate.

Lite White. Trade name of The Wilkinson Company, Inc. (USA) for gold-based dental and medical alloys.

Litex. Trademark of CWH (Germany) for a styrene-butadiene copolymer.

Lithaflux. Trademark of Foote Mineral Company (USA) for lithium mica (*lepidolite*) used in the manufacture of glass and porcelain enamel ground coats, and as a flux in ceramic bodies and glazes.

Lithafrax. Trademark of Harbison-Carborundum Corporation (USA) for ceramics made from β-spodumene and having negligible coefficients of thermal expansion. Used as fluxes in glass, glazes and ceramic bodies. See also spodumene.

Lithaloys. Trade name of Lithaloys Corporation (USA) for a series of lithium-copper master alloys used in the manufacture of sound copper castings.

litharge. See lead monoxide.

litharge glass. Soda-lime glass in which part of the calcium oxide (CaO) is replaced by litharge (PbO). See also soda-lime glass.

litharge-glycerin cement. A product obtained by mixing glycerin with water and enough *litharge* (PbO) to yield a paste of varying consistency that upon curing yields an acid-resistant cement.

Lith-Cu. Trade name of Belmont Metals, Inc. (USA) for a lithium-copper deoxidizing alloy.

lithia. See lithium oxide.

lithia mica. See lepidolite.

N-lithiotrifluoromethanesulfonamide. A crystalline compound (97% pure) with a melting range of 234-238°C (453-460°F), used in the preparation of lithium battery electrolytes and rare-earth Lewis acid catalysts. Formula: $(CF_3SO_2)_2NLi$.

lithiophilite. A clove-brown, honey-yellow or salmon-pink mineral of the olivine group that is composed of lithium manganese phosphate, $LiMnPO_4$, and may also contain iron. A white variety can be made synthetically by heating lithium carbonate (Li_2CO_3), manganese dioxide (MnO_2) and diammonium phosphate $[(NH_4)_2HPO_4]$ under prescribed conditions. Crystal system, orthorhombic. Density, 3.34-3.50 g/cm^3; refractive index, 1.66-1.67. Occurrence: USA (Arizona, Maine, South Dakota).

lithiophorite. A black mineral composed of lithium aluminum manganese oxide hydroxide, $(Li,Al)MnO_2(OH)_2$. Crystal system, monoclinic. Density, 3.40 g/cm^3. Occurrence: Russian Fed-

eration, USA (Tennessee).

lithiophosphate. A colorless mineral composed of lithium phosphate, Li_3PO_4. It can also be made synthetically. Crystal system, orthorhombic. Density, 2.48 g/cm^3. Occurrence: Russia.

lithiotantite. A colorless mineral composed of a lithium niobium tantalum oxide, $Li(Ta,Nb)_3O_8$. Crystal system, monoclinic. Density, 7.00 g/cm^3; refractive index, above 1.90. Occurrence: Kazakhstan.

lithium. A soft, silvery-white metallic element of Group IA (Group 1) of the Periodic Table. It is commercially available in the form of ingots, rods, sheets, foils, wires, ribbons, shot, granules, powder, billets and lumps. Commercial lithium ores include *spodumene, lepidolite* and *amblygonite*, and it is also found in desert lake brines. Crystal system, cubic. Crystal structure, body-centered cubic. Density, 0.534 g/cm^3; melting point, 180.5°C (356.9°F); boiling point, 1342°C (2448°F); hardness, 0.6 Mohs (below 5 Vickers); atomic number, 3; atomic weight, 6.941; monovalent. It has good electrical conductivity (about 18.6% IACS), and is the lightest and least reactive alkali metal, and the lightest metal known. Used in calcium, silver-brazing and lead-alkali alloys, as a deoxidizer in copper and its alloys, as a deoxidizer and desulfurizer in steels and cast irons, as an alloying element in aluminum, lead, magnesium and zinc alloys (often added to increase hardness, tensile strength and/or corrosion resistance), as a lubricant additive, in glass manufacture, in battery anode materials and batteries for pacemakers, as a scavenger for inert gases, in tritium breeding, as a heat-transfer agent for nuclear reactors, in underwater buoyancy devices, in rocket propellants, in Grignard reagents for organometallics, and in biochemistry and medicine. Symbol: Li.

lithium-6. A stable isotope of lithium having a mass number of 6 and high neutron absorption, and comprising 7.5% of the element lithium. Used in nuclear reactors. Symbol: ^{6}Li.

lithium acetate dihydrate. The lithium salt of acetic acid supplied as a white, crystalline powder (98+% pure) with a density of 1.3 g/cm^3 and a melting point of 53-56°C (127-132°F). Used in chemistry, materials research, biochemistry and medicine. Formula: $LiC_2H_3O_2 \cdot 2H_2O$.

lithium acid phthalate. The lithium salt of phthalic acid. It is an alkaline metal biphthalate available in the form of single crystals with high plasticity and fissionability, and used as an analyzing crystal in X-ray spectral analysis. Deuterated lithium acid phthalate (DLAP) is also available. Formula: $LiC_8H_5O_4$. Abbreviation: LAP; LiAP. Also known as *lithium biphthalate; lithium hydrogen phthalate*.

lithium alloys. Alloys composed of varying amounts of lithium with aluminum, calcium, copper, lead, magnesium and zinc. Aluminum-lithium alloys are lightweight alloys containing 2-5% lithium, 1-3% copper, and small additions of magnesium, zirconium, and sometimes silicon and iron. They possess high tensile strengths and moduli, and are used for metal-matrix composites, and various structural and weight-sensitive applications, e.g., in the aerospace industry. Lithium-calcium alloys contain up to 50% lithium and are used to treat (e.g., degasify, desulfurize and/or deoxidize) cast irons, steels and nickel and nickel alloys, and as master alloys in the manufacture of lithium copper. Lithium-copper alloys contain up to 5% lithium and sometimes up to 7% calcium, and are used as degasifiers and deoxidizers in the manufacture of nonferrous alloys. Silver-copper brazing alloys with small amounts of lithium (typically 2%) are used for brazing stainless steels, copper and copper alloys. Lead-alkali metals are strong bearing alloys of lead hardened with small amounts (usually 0.2% or less) of alkali metals, such as lithium, calcium and sodium. Magnesium-lithium alloys are lightweight alloys containing up to 15% lithium and have good strengths, and fine-grained structures, but a maximum upper continuous service temperature of only 150°C (300°F), unless suitably modified, and are used for weight-sensitive applications in the aerospace industry and for military equipment.

lithium aluminate. White crystals or powder. Density, 2.615 g/cm³; melting point, above 1625°C (2957°F). Used as a flux in high-refractory porcelain enamels, and in ceramics and materials engineering. Single crystal substrates are used for superconductor thin film and III-V (13-15) compound deposition in the manufacture of light-emitting diodes and blue lasers. Formula: $LiAlO_2$. Also known as *aluminum lithium oxide; lithium aluminum oxide (LAO); lithium metaaluminate.*

lithium aluminosilicate. A ceramic material with excellent thermal-shock resistance made from mixtures of lithium-bearing minerals and clay, sometimes with the addition of other ceramic materials. See also lithium aluminum silicate.

lithium-aluminum alloy. See aluminum-lithium alloy.

lithium aluminum chloride. A granular, off-white to yellowish substance used in high-energy storage devices and as a supporting battery electrolyte component. Formula: $LiAlCl_4$. Also known as *lithium tetrachloroaluminate.*

lithium aluminum deuteride. White or gray moisture-sensitive crystals prepared by the reaction of aluminum chloride ($AlCl_3$) with lithium deuteride (LiD). They are soluble in tetrahydrofuran and diethyl ether, have a density of 1.02 g/cm³, and decompose at 175°C (347°F). Used as a reducing agent for the introduction of deuterium atoms into molecules. Formula: $LiAlD_4$. Abbreviation: LAD.

lithium aluminum disilicate. See lithium aluminum silicate (ii).

lithium aluminum hexasilicate. See lithium aluminum silicate (iv)

lithium aluminum hydride. A moisture-sensitive, highly reactive compound prepared by reacting aluminum chloride ($AlCl_3$) with lithium hydride (LiH). It is supplied in form of white or gray pellets or powders, or as a solution in diethyl ether, ethyl glycol dimethyl ether, diglyme or tetrahydrofuran. The solid compound has a density of 0.917 g/cm³ and decomposes at 125°C (257°F). Used as a catalyst, propellant, chemical reducing agent, and in the conversion of aldehydes, esters and ketones to amines and nitriles. Formula: $LiAlH_4$. Abbreviation: LAH.

lithium aluminum octasilicate. See lithium aluminum silicate (v).

lithium aluminum oxide. See lithium aluminate.

lithium aluminum silicate. Any of the following compounds of lithium oxide, aluminum oxide and silicon dioxide used in ceramics and glass-ceramics: (i) *Lithium aluminum silicate.* A white powder. Formula: $LiAl(SiO_3)_2$; (ii) *Lithium aluminum disilicate.* Density, 2.36 g/cm³; melting point, 1398°C (2548°F); hardness, 5-7 Mohs. Formula: $Li_2Al_2Si_2O_8$; (iii) *Lithium aluminum tetrasilicate.* Melting point, 1427°C (2600°F); hardness, 5-7 Mohs. Formula: $Li_2Al_2Si_4O_{12}$; (iv) *Lithium aluminum hexasilicate.* Density, 2.41 g/cm³; melting point, 1183°C (2161°F); hardness, 5-7 Mohs. Formula: $Li_2Al_2Si_6O_{16}$; and (v) *Lithium aluminum octasilicate.* White crystals or powder. Density, 2.39-2.46 g/cm³; melting point, 1400°C (2550°F); hardness, 6-6.5 Mohs. Formula: $Li_2Al_2Si_8O_{20}$.

lithium aluminum tetrasilicate. See lithium aluminum silicate

(iii).

lithium-base grease. See lithium grease.

lithium biphthalate. See lithium acid phthalate.

lithium borate. (1) A white powder with a melting point of 917°C (1683°F) used as spectroscopic fluxing agent, doped with boron oxide, lithium fluoride, lanthanium oxide, strontium oxide or vanadium pentoxide for electronic and optoelectronic applications, and activated with manganese in thermoluminescence dosimeters. Formula: $Li_2B_4O_7$. Also known as *lithium tetraborate.*

(2) A crystalline compound of lithium, boron and oxygen used in Q-switched lasers, such as the neodium-doped yttrium aluminum garnet and titanium-doped sapphire types. It has a high laser damage threshold, wide transparency, good nonlinear coupling properties, and good chemical and mechanical properties. Formula: $LiBO_3$. Abbreviation: LBO. Also known as *lithium boron oxide (LBO).*

lithium borate pentahydrate. A white, crystalline powder that is highly soluble in water, insoluble in alcohol, and loses $5H_2O$ at 200°C (392°F). Used in the refining and degassing of metals, as a flux in glazes and porcelain enamels, and in vacuum spectroscopy. Formula: $Li_2B_4O_7 \cdot 5H_2O$. Also known as *lithium tetraborate pentahydrate.*

lithium borohydrate. A lithium compound that is available as a flammable, moisture-sensitive white to gray, crystalline powder (99+% pure) with a melting point of 279°C (534°F) (decomposes), and as a 2.0M solution in tetrahydrofuran. Used in chemistry, medicine, and organometallic research. Formula: $LiBH_4$.

lithium boron oxide. See lithium borate (2).

lithium borosilicate. A compound of lithium oxide, boric oxide and silicon dioxide used for low-temperature enamels, and in corrosion-resistant high-temperature coatings. Formula: $Li_2B_2SiO_6$.

lithium bromide. White, hygroscopic crystals, or white to pinkish white powder (99+% pure). Crystal system, cubic. Density: 3.46 g/cm³; melting point, 550°C (1022°F); boiling point, 1265°C (2309°F). Used in air conditioners to add moisture, as a desiccant, in batteries, and as a low-temperature heat-transfer medium. Formula: LiBr.

lithium-calcium. A master alloy composed of calcium and up to 50% lithium, used to treat (e.g., degasify, desulfurize and/or deoxidize) cast irons, steels and nickel and nickel alloys, and as a melt addition in the manufacture of lithium copper.

lithium carbonate. White crystals and powder (99+% pure). Density, 2.11 g/cm³; melting point, 618°C (1144°F); boiling point, decomposes at 1310°C (2390°F). Used as a source of lithium oxide in glass, glazes and enamels, as a flux in ceramic bodies, glazes and porcelain enamels, in glass-ceramics, as a coating for arc-welding electrodes, in luminescent paints, in varnishes and dyes, and in the manufacture of aluminum. Formula: Li_2CO_3.

lithium ceramics. Ceramic materials based on *lithium aluminosilicate* obtained by blending lithium-bearing minerals (e.g., *spodumene* or *lepidolite*) and clay, sometimes with the addition of other ceramic materials. They possess excellent thermal-shock resistance and are available in a wide range of negative and positive thermal expansion coefficients.

lithium chloride. White, hygroscopic crystals or powder (99+% pure). Crystal system, cubic. Density, 2.068 g/cm³; melting point, 614°C (1137°F); boiling point, 1325-1383°C (2417-2521°F). Used in the manufacture of lithium metal, in welding and soldering fluxes, in salt baths, in heat-exchange media and dry batteries, as a drying agent, and in brines for air conditioners. Formula: LiCl.

lithium chromate. Yellow crystals or powder (99+% pure) used as oxidizers in organic synthesis, as heat-transfer media, and as corrosion inhibitors in antifreezes and water-cooled reactors. Formula: Li_2CrO_4.

lithium cobaltite. See lithium cobalt oxide.

lithium cobalt oxide. Dark blue to bluish gray, hygroscopic powder (99.8% pure) with a melting point above 1000°C (1830°F) used in porcelain enamel ground coats to combine the fluxing properties of lithium oxide and the adherence-promoting properties of cobalt oxide, in certain blue enamel compositions to stabilize and intensify the blue color, and in cathodes and electrodes of advanced batteries and fuel cells. Formula: $LiCoO_2$. Also known as *lithium cobaltite*.

lithium columbate. See lithium niobate.

lithium columbium. See lithium niobium.

lithium-conductivity bronze. A bronze that is composed of 98+% copper, up to 2% cadmium and traces of silicon and tin, and has been treated with lithium to increase its electrical conductivity. Used for high-strength conductors.

lithium copper. An oxygen-free wrought copper containing trace amounts of lithium (usually less than 0.01%) and having high electrical conductivity (above 101% IACS), high density, high elongation, and good tensile strength, toughness and deep-drawing qualities. It is made by introducing a lithium-calcium master alloy into a copper melt, and used for electrical components.

lithium-copper. Copper-base alloys containing up to 5% lithium and, sometimes, up to 7% calcium, and used as degasifiers and deoxidizers in the manufacture of nonferrous alloys.

lithium deuteride. An air- and moisture-sensitive, off-white powder or gray crystalline lithium compound containing deuterium. It has a density of 0.906 g/cm³ and a melting point of approximately 680°C (1255°F). Used in thermonuclear fusion. Formula: LiD.

lithium dialkylcopper. A group of organometallic compounds obtained by reacting an alkyllithium (RLi) with a copper halide (CuX). Examples include lithium dimethylcopper and lithium diethylcopper. The general formula is R_2CuLi. Also known as *lithium dialkylcuprate*.

lithium diarylcopper. A group of organometallic compounds obtained by treating an aryl halide (ArX) with metallic lithium (Li). An examples is lithium diphenylcopper. The general formula is Ar_2CuLi. Also known as *lithium diarylcuprate*.

lithium earth-alkali fluorides. A group of compounds consisting of lithium fluoride (LiF) and one of the earth-alkali fluorides, i.e., the fluorides of the elements in Group IIA (Group 2) of the Periodic Table, e.g., beryllium fluoride (BeF_2), magnesium fluoride (MgF_2), calcium fluoride (CaF_2) or strontium fluoride (SrF_2). The general formula is $LiAEF_3$, where AE represents Be, Mg, Ca, Sr, etc. Used for laser crystals.

lithium ethyl. See ethyllithium.

lithium ferrate. An oxide of lithium and iron used in fuel cell cathodes. Formula: Li_2FeO_4.

lithium ferrite. (1) A ferrimagnetic ceramic material with a cubic spinel crystal structure (*ferrospinel*) produced by powder metallurgy techniques. It has outstanding magnetic properties at high frequencies, very high resistivity and good corrosion resistance. Used for soft magnets in transformers for electronic and communication equipment. Formula: $Li_{0.5}Fe_{2.5}O_4$.

(2) A brownish powder with a density of 1.90 g/cm³. Formula: $LiFeO_2$. Also known *lithium iron oxide*.

lithium fluophosphate. White crystals obtained by combining the phosphate and fluoride of lithium. Formula: $LiF \cdot Li_3PO_4 \cdot H_2O$. Used in chemistry and ceramics. Also known as *lithium fluorophosphate*.

lithium fluoride. A compound of lithium and fluorine available as a fine white powder (99.8% pure), obtained by reacting lithium carbonate with hydrogen fluoride, and as white high-purity crystals (99.99+%) for infrared, visible or ultraviolet transmission. Crystal system, cubic. Density, 2.635 g/cm³; melting point, 842°C (1548°F); boiling point, 1670°C (3038°F); refractive index, 1.3915. Used as a flux and minor opacifier in porcelain enamels and glazes, in the high-temperature oxidation of noble metals using fluorine in anhydrous hydrogen fluoride, as a welding and soldering flux, in heat-exchange media, in the manufacture of lithium fluophosphate, and as an ingredient in fuel for fused salt reactors. The high-purity crystalline form is used in infrared and ultraviolet instruments and wavelength-dispersive spectrometers, in the preparation of hypersensitive thermoluminescent materials, and doped with dysprosium or manganese in thermoluminescence detectors. Formula: LiF.

lithium fluorophosphate. See lithium fluophosphate.

lithium gallate. An oxide of lithium and gallium that is available as a single crystal substrate. Crystal system, orthorhombic. Density, 4.17 g/cm³; melting point, 1600°C (2912°F). Used in the deposition of III-V (13-15) compounds for blue lasers and light-emitting diodes. Formula: $LiGaO_2$. Also known as *lithium gallium oxide (LGO)*.

lithium gallium oxide. See lithium gallate.

lithium germanate. An oxide of lithium and germanium that is available as a single crystal substrate. Crystal system, monoclinic. Density, 3.53 g/cm³; melting point, 1239°C (2262°F); soluble in water; ferroelectric and pyroelectric properties. Used in electronics, ferroelectrics and pyroelectrics. Formula: Li_2GeO_3. Also known as *lithium germanium oxide (LGO); lithium metagermanate*.

lithium germanium oxide. See lithium germanate.

lithium grease. A white *lubricating grease* composed of a lithium soap (e.g., lithium stearate or lithium hydroxystearate) suspended in mineral oil. It is water-resistant, stable when heated above its melting point, has an operating temperature range of about -30 to +135°C (-22 to +275°F) and provides good corrosion resistance. Used for low-temperature, high-load and medium-speed bearing lubrication applications. Also known as *lithium-base grease; lithium soap grease*.

lithium halide. A compound of lithium and a halogen, such as bromine, chlorine, fluorine or iodine.

lithium hydride. A flammable, moisture-sensitive, white or gray crystalline powder (93+% pure) prepared by reacting molten lithium with hydrogen. It has a density of 0.79-0.82 g/cm³, a melting point of 680°C (1256°F) [also reported as 689°C or 1272°F], and is soluble in ether. Used in ceramics, materials research, nuclear shielding, as a source of hydrogen, as a desiccant and reducing agent, as a condensing agent in organic chemistry, and in the preparation of lithium amide and double hydrides. Formula: LiH.

lithium hydrogen phthalate. See lithium acid phthalate.

lithium hydroxide. Hygroscopic, colorless crystals or white powder. Density, 2.54 g/cm³; melting point, 470°C (878°F); boiling point, 925°C (1697°F). Used for ceramics, in storage battery electrolytes, in lubricating greases and photographic developers, in biochemistry, and in lithium soaps. Formula: LiOH.

lithium hydroxide monohydrate. Colorless, hygroscopic crystals or white powder (98+% pure) with a density of 1.51 g/cm³ used in chemistry, biochemistry, ceramics and materials research. Formula: LiOH·H₂O.

lithium iodate. A compound of lithium, iodine and oxygen available as a white powder or high-purity crystals. Density, 4.48-4.49 g/cm³; melting point, 50-60°C (122-140°F). The powder is used as an oxidizer, and the crystals exhibit wide transparency, high laser damage thresholds, and good chemical and mechanical properties, and are used in Q-switched lasers. Formula: LiIO₃.

lithium iodide. White, hygroscopic crystals or powder (99+% pure). Crystal system, cubic. Density, 3.494-4.06 g/cm³; melting point, 450°C (842°F) [also reported as 469°C or 876°F]; boiling point, 1180°C (2156°F) [also reported as 1171°C or 2140°F]; refractive index, 1.955. Used in the manufacture of lithium metal, as a catalyst, and in chemistry, biochemistry and materials research. Formula: LiI.

lithium iodide trihydrate. Colorless, white or yellowish hygroscopic, light-sensitive crystals. Density, 3.43 g/cm³; melting point, 73°C (163°F). Used in air conditioning and nucleonics. Formula: LiI·3H₂O.

lithium iron oxide. See lithium ferrite (2).

lithium lactate. The lithium salt of lactic acid (95% pure) supplied as a white, crystalline, nonhygroscopic, water-soluble powder with a melting point above 300°C (570°F), and used in biochemistry, bioengineering and medicine. Formula: LiC₃H₅O₃.

lithium manganese oxide. (1) A red to reddish-brown, water-soluble powder made from lithium oxide and manganese dioxide, and used as a flux in porcelain enamels, in the manufacture of ceramic-bonded grinding wheels, in the synthesis of cathode-active materials, and for energy-storage devices. Formula: Li₂MnO₃. Also known as *lithium manganite.*

(2) A powder with a melting point above 400°C (750°F) used for electrochemical applications. Formula: LiMn₂O₄.

lithium manganite. See lithium manganese oxide (1).

lithium metaaluminate. See lithium aluminate.

lithium metaborate. A white, hygroscopic powder (98+% pure) with a density of 1.40 g/cm³ and a melting point of 845°C (1553°F), used in ceramics, as a flux in cover enamels, and in welding and brazing. Formula: LiBO₂.

lithium metaborate dihydrate. A white crystalline powder with a density of 1.79 g/cm³, used in ceramics, e.g., as a flux in porcelain enamels, in other ceramics to increase torsion resistance and tensile strength, and in nucleonics. Formula: LiBO₂·2H₂O.

lithium metacolumbate. See lithium niobate.

lithium metagermanate. See lithium germanate.

lithium metaniobate. See lithium niobate.

lithium metasilicate. See lithium silicate (ii).

lithium mica. See lepidolite.

lithium molybdate. A white, crystalline, hygroscopic powder (99+% pure). Density, 2.66 g/cm³; melting point, 705°C (1301°F). Used as an adherence-promoting agent for white enamels applied directly to steel, as a petroleum cracking catalyst, and in battery systems. Formula: Li₂MoO₄.

lithium monoxide. See lithium oxide.

lithium niobate. A ferroelectric ceramic compound available in the form of a white high-purity powder (99.98+% pure) or single crystals. Crystal structure, ilmenite. Melting point, 1250°C (2282°F); good strength; photorefractive properties. Used in ceramics, electronics and piezoelectrics, in optoelectronic, pyro-

electric and surface acoustic wave (SAW) devices, in infrared detectors, and for transducers in laser technology. Formula: LiNbO₃ (LiCbO₃) Also known as *lithium columbate; lithium metacolumbate; lithium metaniobate; lithium niobium oxide (LNO).*

lithium niobium. A double metal alkoxide supplied as a liquid with lithium and niobium contents of 0.5-0.6% and 7.2-7.4%, respectively. The lithium-to-niobium metal ratio is 1:1. Used in sol-gel and coating applications. Formula: LiNb(OR)ₓ. Also known as *lithium columbium.*

lithium niobium oxide. See lithium niobate.

lithium nitrate. Colorless or white, hygroscopic crystals or powder (99.8+% pure). Density, 2.38 g/cm³; melting point, 261°C (502°F); boiling point, decomposes at 600°C (1112°F). Used as an oxidizing flux in porcelain enamels, glazes and glass, in salt baths and heat-exchange media, in refrigeration and pyrotechnics, and as a rocket propellant. Formula: LiNO₃.

lithium nitride. Red-brown crystals or fine, free-flowing, moisture-sensitive, flammable, red or red-brown powder. Crystal system, hexagonal. Density, 1.27-1.38 g/cm³; melting point, above 813°C (1495°F). Used as a nitriding agent in metallurgy, as a source of nitrogen in organic reactions, as a nucleophilic and reducing agent in organic synthesis, and in ceramics. Formula: Li₃N.

lithium organocuprate. Any of a group of organometallic compounds obtained by treating alkyl or aryl halides with metallic lithium. Examples include lithium dimethylcuprate and lithium diethylcuprate. Used in organometallic research and the manufacture of ketones. The general formula is R₂CuLi.

lithium orthophosphate. See lithium phosphate.

lithium orthosilicate. See lithium silicate (ii).

lithium oxide. White hygroscopic crystals or powder with a density of 2.013 g/cm³ and a melting point of 1427°C (2601°F) [also reported as 1570°C or 2858°F]. Used in porcelain enamels to enhance workability and decrease firing temperatures, in electrical porcelain and dinnerware to enhance strength and gloss, in ceramic bodies and special refractories to decrease the thermal expansion and increase the resistance to thermal shocks, and as a flux in glasses with high specific resistance to enhance fluidity, workability and ultraviolet transmission. Formula: Li₂O. Also known as *lithia; lithium monoxide.*

lithium-oxide-based glass. A glass that contains considerable amounts of lithium oxide (Li₂O) and has high electrical resistivity and high transparency to X-rays. Like other alkali oxides (e.g., Na₂O, K₂O) lithium oxide tends to modify (i.e., break up) the network structure in glass, making it more fluid and workable. Used for X-ray equipment and special fiberglass products.

lithium peroxide. White crystals or fine white powder. Crystal system, hexagonal. Density, 2.14 g/cm³. Used in chemistry and materials research, e.g., as a source of active oxygen.

lithium phthalocyanine. The dilithium salt of 29*H*,31*H*-phthalocyanine supplied as a blue-purple powder with a dye content of about 70%, a flash point above 230°F (110°C) and a maximum absorption wavelength of 658 nm. Used as a dye. Formula: Li₂(C₃₂H₁₆N₈). Abbreviation: LiPc. Also known as *dilithium phthalocyanine.*

lithium phosphate. A white, crystalline powder (99+% pure) with a density of 2.54 g/cm³ and a melting point of 1205°C (2201°F), used in inorganic and organic synthesis, biochemistry and materials research. Formula: Li₂PO₄. Also known as *lithium orthophosphate.*

lithium polysilicate. A compound of lithium oxide and silicon dioxide usually available as an aqueous solution, and used in chemistry, polymer engineering, and materials research. Formula: $Li_2Si_5O_{11}$.

lithium rare-earth fluorides. A group of compounds consisting of lithium fluoride (LiF) and one of the rare-earth fluorides, e.g., erbium fluoride (ErF_3), gadolinium fluoride (GdF_3) ytterbium fluoride (YbF_3) or yttrium fluoride (YF_3). The general formula is $LiREF_4$, where RE represents a rare-earth element, such Er, Gd, Y or Yb. Used for laser crystals.

lithium ribbon. Lithium metal (99.9+% pure) in the form of a continuous strip that is up to 100 mm (4 in.) wide and up to 1.6 mm (0.06 in.) thick. It is air- and moisture sensitive, and thus usually packaged under argon or petrolatum. Used for electrical and electronic applications, e.g., batteries.

lithium silicate. Any of the following compounds of lithium oxide and silicon dioxide: (i) *Lithium metasilicate.* A colorless or white powder (98+% pure). Density, 2.52 g/cm^3; melting point, 1201°C (2194°F); hardness, 5-7 Mohs. Used as a flux and opacifier in glazes and ceramic enamels, in titania-opacified enamels to reduce firing temperature and enhance surface texture, and in welding-rod coatings. Formula: Li_2SiO_3; and (ii) *Lithium orthosilicate.* A white, moisture-sensitive powder. Density, 2.39 g/cm^3; melting point, 1253°C (2287°F); hardness, 5-7 Mohs. Used as a flux to improve surface texture of glazes and porcelain enamels, and as a minor opacifier. Formula: Li_4SiO_4. Also known as *dilithium silicate.* See also lithium polysilicate.

lithium soap grease. See lithium grease.

lithium stearate. White crystals or powder obtained by treating lithium carbonate with stearic acid. Density, 1.025 g/cm^3; melting point, 220°C (428°F). Used in plastics, greases and waxes, as a high-temperature lubricant, as a lubricant in powder metallurgy, and as a flatting agent in lacquers and varnishes. Formula: $LiC_{18}H_{35}O_2$.

lithium sulfate. Colorless crystals (99+% pure). Density, 2.22 g/cm^3; melting point, 860°C (1580°F). Used in ceramics, biochemistry, and for piezoelectric transducer elements. Formula: Li_2SO_4.

lithium sulfate monohydrate. Colorless or white crystals (99+% pure) obtained by treating the mineral *spodumene*, or lithium carbonate with sulfuric acid. Density, 2.06 g/cm^3; melting point, 130°C (266°F). Used in ceramics, biochemistry, and for piezoelectric transducer elements. Formula: $Li_2SO_4 \cdot H_2O$.

lithium sulfide. A white, hygroscopic crystals or white flammable, moisture-sensitive powder (98+% pure). Crystal system, cubic. Density, 1.64-1.66 g/cm^3; melting point, above 900°C (1652°F) [also reported as 1372°C (2502°F)]. Used in ceramics, electronics and materials research. Formula: Li_2S.

lithium tantalate. A ferroelectric ceramic compound available as a white high-purity powder with a Curie temperature above 350°C (660°F) and good strength. Used for special electrical and electronic applications, and in optoelectronic, pyroelectric and surface acoustic wave (SAW) devices. Formula: $LiTaO_3$.

lithium tantalum. A double metal alkoxide supplied as a liquid with lithium and tantalum contents of 0.5-0.6% and 14.0-14.4% respectively. The lithium-to-tantalum metal ratio is 1:1. Used in sol-gel and coating applications. Formula: $LiTa(OR)_x$.

lithium-tellurium alloy. An alloy of lithium and tellurium used as a cathode material in certain storage batteries having anodes of lithium and electrolytes of molten lithium salt.

lithium tetraborate. See lithium borate.

lithium tetraborate pentahydrate. See lithium borate pentahydrate.

lithium tetrachloroaluminate. See lithium aluminum chloride.

lithium tetrafluoroborate. A lithium compound available as a white, hygroscopic powder (98+% pure) that decomposes at 293-300°C (559-572°F). Used in ceramics, electronics, in the manufacture of solid composite electrolytes for lithium batteries, as a catalyst. Formula: $LiBF_4$.

lithium titanate. A white powder with a melting range of 1520-1564°C (2768-2847°F), used as a flux in porcelain enamels, as a powerful flux in titanium-bearing enamels, as a mill addition in glazes, and in batteries. Formula: Li_2-TiO_3.

lithium-treated lead. A lead-base bearing alloy hardened with small amounts of lithium (usually less than 0.1%) often in combination with other alkalies, such as barium, calcium and sodium. See also alkali lead.

lithium-treated steel. Steel deoxidized and desulfurized with a lithium-base master alloy. Lithium is also used in some stainless steel castings to increase fluidity.

lithium tungstate. A white powder (98% pure) with a density of 3.71 g/cm^3 and a melting point of 742°C (1367°F). Used as a reagent and as an intermediate. Formula: Li_2WO_4.

lithium zirconate. A white powder made from lithium oxide and zirconium dioxide, and used as a powerful flux and opacifier in glass, glazes and porcelain enamels. Formula: Li_2ZrO_3.

lithium-zirconium silicate. Any of the following compounds of lithium oxide, zirconium dioxide and silicon dioxide: (i) *Lithium zirconium silicate.* A white powder used as a powerful flux in enamels, glazes, and porcelains, and as a substitute for lithium zirconate. Formula: $Li_4ZrSi_2O_6$; and (ii) *Tetralithium trizirconium pentasilicate.* Density, 4.02 g/cm^3; melting point 1154°C (2109°F); hardness, 5-7 Mohs. Used in ceramics. Formula: Li_8-$Zr_3Si_5O_{20}$.

Lithobraze. Trade name of Handy & Harman (USA) for a series of silver brazing alloys containing 71-93% silver, 7.3-28% copper and 0.2-0.3% lithium. Available in the form of strips and wires, they have an average brazing range of 760-890°C (1400-1634°F), and are used for brazing stainless steels.

lithocarbon. A pure bitumen obtained from limestone found in Uvalde County, Texas, USA.

Lithocote. Trademark of Lithocote Corporation (USA) for protective coatings based on phenolic and modified epoxy or phenolic resins. They cure by oven baking, or on catalyzed epoxy resins, and are used for corrosion-resistant linings on tank cars, tanker trucks and storage vessels.

lithographic limestone. See lithographic stone.

lithographic stone. A homogeneous, compact, fine-grained limestone of yellowish white to grayish color, found chiefly near Solnhofen and Pappenheim in Bavaria, Germany. It has high purity, a hardness of 3 Mohs, and is used in lithography, and as a building stone. Also known as *lithographic limestone; Solnhofen stone.*

lithopone. A white crystalline powder composed of co-precipitated zinc sulfide (ZnS) and barium sulfate ($BaSO_4$). It has a density of 4.3 g/cm^3 (0.16 $lb/in.^3$), and is used as a pigment in paints, as a nonpoisonous replacement for *white lead*, and as a filler in textiles, paper, leather, linoleum and white rubber goods. It is now largely replaced by titanium dioxide pigments. Also known as *Charlton white; Griffith's white; Orr's white; zinc baryta white; zinc sulfide white.*

Lithosperse. Trademark of J.M. Huber Corporation (USA) for a kaolin clay.

litidionite. A blue mineral composed of potassium sodium copper silicate, $KNaCuSi_4O_{10}$. Crystal system, triclinic. Density, 2.75 g/cm^3; refractive index, 1.574. Occurrence: Italy.

litinum bronze. A free-cutting bronze containing 90% copper, 4.7% tin, 3.7% zinc and 1.6% lead, used for screw-machine products, bearings, bushings, nuts, and screws.

litmus paper. A white, unsized paper saturated with an aqueous solution of litmus, a blue coloring matter derived from lichens, and used to incidate the acicity (color changes to red) or basicity (color remains blue) of chemical solutions.

Littite. Trade name of Little Brothers Foundry Company (USA) for a cast iron used for gears, shafts and machine-tool housings.

Little's speculum. An aluminum alloy containing 31% tin, 2.3% zinc and 1.7% arsenic. It takes a fine polish, and is used for mirrors and reflectors.

Liv Carbo. Trade name of GC America, Inc. (USA) for a zinc carboxylate dental cement used for cementation and luting applications.

liveingite. A lead gray mineral composed of lead arsenic sulfide, $Pb_9As_{13}S_{28}$. Crystal system, monoclinic. Density, 5.50 g/cm^3. Occurrence: Switzerland.

lively yarn. A yarn that tends to excessively twist around itself when allowed to hang loosely. Also known as *snarly yarn.*

liver ore. See cinnabar.

Livetone. Trade name of Precision Laboratories (USA) for acrylic resins.

livingstonite. A lead-gray mineral with a red streak and an adamantine to metallic luster. It is composed of mercury antimony sulfide, $HgSb_4S_8$. It can also be made synthetically from the elements below 450°C (842°F). Crystal system, monoclinic. Density, 4.48 g/cm^3. Occurrence: Mexico. Used as an ore of antimony, and as a source of mercury.

lizardite. (1) A white mineral of the kaolinite-serpentine group that is composed of magnesium silicate hydroxide, $Mg_3Si_2O_5(OH)_4$. Crystal system, monoclinic. Density, 2.55 g/cm^3. Occurrence: UK.

(2) A green, green-blue or white mineral of kaolinite-serpentine group composed of magnesium silicate hydroxide, $(Mg,Al)_3(Si,Al)_2O_5(OH)_4$. It can also be made synthetically. Crystal system, orthorhombic. Density, 2.55-2.60 g/cm^3; refractive index, 1.565. Occurrence: Scotland, USA (Michigan).

lizard leather. A term referring to tough leather with a characteristic grain and a scaly surface made from the skins of large saurians and lizards including alligators, crocodiles, salamanders, geckos, chameleons and iguanas. Used for shoes, luggage, wallets, etc.

llama fibers. Natural protein fibers obtained from the usually white or brown wool of the llama, a domesticated South American hoofed mammal (*Lama glama*) closely related to the alpaca and guanaco. Used for the manufacture of textile fabrics.

Llumar. Trademark of Solutia Inc. (USA) for a micro-thin polymeric film that is applied to automotive or architectural glass to reduce glare, visible and ultraviolet light transmission and/or energy usage, and/or add safety and security.

L-Nickel. Trade name of International Nickel Inc. (USA) for a heat- and corrosion-resistant nickel (99.4% pure) containing 0.2% manganese, 0.15% iron, 0.1% copper, 0.05% silicon, and up to 0.02% carbon. Used for immersion heaters and chemical equipment.

Loaded Centricast. Trade name of Sheepbridge Engineering Limited (UK) for a cast iron containing 3.2% carbon, 2.8% silicon, 0.8% chromium, and the balance iron. It is made by centrifu-

gal casting, and used for cylinder liners.

loaded concrete. A special concrete for radiation shielding applications in nuclear reactors that contains elements with high capture cross sections and high atomic numbers.

Loaded Iron. Trade name of Sheepbridge Engineering Limited (UK) for a cast iron containing 3.2% carbon, 2.2% silicon, 1% chromium, 0.4% phosphorus, and the balance iron. Used for cylinder liners.

loadstone. See lodestone.

Lo-Air. Trade name of Cytemp Specialty Steel (USA) for an air-hardening medium-alloy cold-work tool steel (AISI type A6) containing 0.7% carbon, 2.25% manganese, 1% chromium, 1.35% molybdenum, and the balance iron. It has high tensile strength, good deep-hardening properties, good nondeforming properties and good wear resistance. Used for blanking, embossing, forming, and piercing dies, punches, and shear blades.

Loalin. Trademark of Catalin Corporation (USA) for a polystyrene supplied in transparent and light-colored grades. It has a density of 1.06 g/cm^3 (0.038 $lb/in.^3$), high light transmission, negligible water absorption, good dielectric properties, good resistance to acids, alkalies and alcohols, and poor resistance to aromatic hydrocarbons. Used for injection-molded parts.

Loavar. Trademark of Catalin Corporation (USA) for polystyrene resins.

loam. (1) A soil composed of about 25-50% sand, 30-50% silt and 5-25% clay, and used as a molding material in the manufacture of large metal castings, and in construction for plastering walls and filling holes.

(2) An impure potter's clay containing some iron ocher or mica.

loam sand. A molding sand that is high in clay (up to 50%) and dries hard. Used as a cementing and lining material in brick molds for loam molding of large castings.

loblolly pine. The heavy, hard, strong wood from the tall coniferous tree *Pinus taeda*. The heartwood is reddish-brown, and the sapwood yellowish-white. *Loblolly pine* has a coarse grain, a soft fiber, moderate shrinkage and good stability when adequately seasoned. Average weight, 575 kg/m^3 (36 lb/ft^3). Source: Southern and southeastern United States from Virginia to northern Florida and westward to Texas. Widely used as lumber for framing, sheathing, interior finish, subflooring and joists, and for boxes, crates, pallets, cooperage, railroad ties, poles, piles, and mine timbers. Also known as *field pine; oldfield pine; sap pine.*

Locher-Paste. Trademark of VISCOTEX Locher & Co. AG (Switzerland) for a grayish-green lithium-12-dihydrostearate based pipe-sealing compound with inorganic filler, supplied in paste form.

Lochrip. Trademark of Rippenstreckmetall-Gesellschaft mbH (Germany) for black lacquered or galvanized perforated expanded metal with ribs.

Lockalloy. Trade name of NGK Metals Corporation (USA) for a low-density engineering composite containing 62% beryllium and 38% aluminum. It has a ductile aluminum matrix reinforced with beryllium particles, a high modulus of elasticity and high yield strength. Used for structural aerospace parts, missiles, satellite launch vehicles, aircraft brakes, computer memory drums and disks, and nuclear fuel canning.

Lockbond. Trademark of Chembond Inc. (USA) for water- and solvent-based adhesives used in the automotive, converting, packaging and woodworking industries.

Lockfoam. Trade name for a polyurethane foam made with a fluorocarbon-base blowing agent by pouring *in situ.*

Lockote. Trade name of Novamex Industries (USA) Inc. for iron phosphate coatings.

Lockport Special. Trade name of Wallace Murray Corporation (USA) for a tungsten-type high-speed tool steel (AISI type T2) containing 0.7% carbon, 18% tungsten, 4% chromium, 2% vanadium, and the balance iron. It has good deep-hardening properties, excellent machinability and wear resistance, high hot hardness, and good toughness. Used for lathe and planer tools, milling cutters, and other cutting and finishing tools.

Lockstone. Trademark of Tarmac America (USA) for interlocking paving stones used for commercial applications.

Loco. Trade name of Magnolia Anti Friction Metal Company (UK) for hardened lead-base bearing alloys.

Lo-Cro. Trade name of Colt Industries (UK) for a series of corrosion-resistant steels containing up to 0.1% carbon, 4-6% chromium, and the balance iron. Used for oil-refinery equipment.

Loctite. Trademark of Loctite Corporation (USA) for a series of liquid anearobic polymer adhesives that do not cure in air when applied to the surfaces of mating metal parts, but after close fitting of these parts due to the absence of air. This trade name also includes various cyanoacrylate adhesives, silicone sealants, thread lubricants, and other adhesives, lubricants and sealants.

locust. The hard, decay-resistant wood of any of various tall North American hardwood trees of the pea family especially the black locust *(Robinia pseudoacacia)* and the honeylocust *(Gleditsia triacanthos).* Used for lumber, furniture, railroad ties, wheel spokes, fence posts, and agricultural implements. See also black locust; honeylocust.

loden. A naturally water-repellent fabric woven from coarse wool and having a fleecy or felted surface texture. Originally from the Tyrolese area of Austria, it is used for coats, jackets, capes, and traditional costumes. Also known as *loden cloth.*

lodencloth. See loden.

Lodestar. Trademark of Ivoclar Vivadent AG (Liechtenstein) for a copper-, palladium- and silver-free, extra-hard ceramic dental alloy containing 51.5% gold, 38.5% platinum, 8.5% indium, 1.5% gallium, and less than 1.0% ruthenium and rhenium, respectively. It has a white color, a density of 13.7 g/cm³ (0.49 lb/in.³), a melting range of 1215-1290°C (2219-2354°F), an as-cast hardness of 330 Vickers, moderate elongation, and excellent biocompatibility and resistance to oral conditions. Used for crowns, onlays, posts, bridges, and model castings.

lodestone. A naturally magnetic variety of the mineral *magnetite* (Fe_3O_4) that strongly attracts iron objects, and tends to align its long axis in a north-south direction and thus indicates magnetic north. Used by early mariners and navigators to determine directions. Also known as *leading stone; loadstone; natural magnet.*

Lodex. Trademark of General Electric Company (USA) for a series of anisotropic or isotropic permanent magnet materials containing varying amounts of cobalt, iron, lead and antimony. They are composed of very fine particles of iron-cobalt in lead (and sometimes antimony) powder, and can be made into any desired shape by powder metallurgy techniques, such as extrusion and pressing. *Lodex* materials have uniform magnetic and physical properties, high magnetic energy products, and high remanent magnetization and coercivity.

lodgepole pine. The pale brown, fairly soft wood from the pine *Pinus contorta.* It has a straight grain, numerous resin ducts, large shrinkage, and works well. Average weight, 560 kg/m³ (35 lb/ft³). Source: Western United States (Rocky Mountains, Pacific Coast and Alaska), Western Canada (British Columbia, Yukon and Alberta). Used for general lumber, mine timbers, railroad ties, poles, framing, siding, finish and flooring lumber, furniture, and wall paneling. Also known as *knotty pine.*

Lodlan. Trade name of Unitica (Japan) for general-purpose liquid crystal polyesters that can be reinforced with glass or carbon fibers and injection molded into high-performance composites. They can also be produced in the form of thin polymeric films (0.5 mm or 0.04 in. thick). *Lodlan* polyesters have high mechanical strength, good fluidity, low shrinkage, good chemical and heat resistance, and good gas barrier properties.

loellingite. A silver-white to steel-gray opaque mineral of the marcasite group occurring natural and also made synthetically. It is composed of iron arsenide, $FeAs_2$, and may contain some nickel, cobalt, antimony and/or sulfur. Crystal system, orthorhombic. Density, 7.40 g/cm³; hardness, 824-870 Vickers. Occurrence: Canada, Spain, Austria, UK, USA (Colorado, Maine, New Jersey, New York). Used as a source of arsenic. Also known as *löllingite.*

loess. A diluvial deposit of fine, yellowish-brown loam found in Europe and Asia, and believed to have been deposited by the wind. It is porous, and largely siliceous in composition, but also contains calcareous matter, is used in the building trades, and corresponds to *adobe clay,* a similar deposit found in Mexico and the southwestern United States.

loeweite. A colorless, white or yellowish mineral composed of sodium magnesium sulfate hydrate, $Na_4Mg_2(SO_4)_4 \cdot 5H_2O$. It can also be made synthetically. Crystal system, rhombohedral (hexagonal). Density, 2.38 g/cm³; refractive index, 1.473; hardness, 3.5 Mohs.

Lo-Ex. Trade name of Birmingham Aluminium Casting Company (USA) for an aluminum casting alloy containing 11-13% silicon, 0.5-1.2% copper, 0.9-1% magnesium and 1-3% nickel. It has a low coefficient of thermal expansion, and is used for sleeves, bushings, pistons and liners.

Loflex. Trade name of Engelhard Corporation (USA) for a bimetal with a maximum service temperature of 427°C (800°F), used for thermostats.

Lo-Flo. Trade name of Fusion Inc. (USA) for a series of silver-based brazing alloys with a melting range of 621-816°C (1150-1500°F).

Loftguard. Trade name of KoSa (USA) for high-loft polyester fibers with excellent compression recovery, supplied in cut lengths in several grades including the original *Loftguard* and *Loftguard Xtra* with super-void fiber cross section and low shrinkage. Used for bed pillows and comforters, mattress pads and quilts, and furniture cushions.

Loft-set. Trade name for nylon fibers and yarns used for lofty textile fabrics including clothing.

Loftura. Trademark for cellulose acetate fibers and yarns used for lofty clothing.

log. A term referring to an unshaped length of wood obtained from a tree (usually the trunk) and having only the branches removed.

Logan. (1) Trade name of Time Steel Service Inc. (USA) for a water-hardening tool steel (AISI type W2) containing 1% carbon, 0.25% silicon, 0.25% manganese, and the balance iron. Used for tools, dies, and mandrels.

(2) Trade name of Logan Iron & Steel Company (USA) for a water-hardening steel containing 0.4% carbon, and the balance iron. Used for castings.

(3) Trade name of Cerro Metal Products Company (USA) for a forging brass containing 58.5% copper, 39.5% zinc and 2% lead. It has excellent hot workability, high elongation and good strength and machinability. Used for forgings and pressings.

Logas. Trade name of Foseco Minsep NV (Netherlands) for degassing briquettes used to remove dissolved hydrogen from copper and nickel melts and improve the overall mechanical properties.

Lohm. Trade name of Driver Harris Company (USA) for a wrought copper alloy containing 6-7.5% nickel. It has low electrical resistance, an electrical conductivity of 18% IACS, a maximum service temperature of 200°C (390°F), and is used for electrical equipment.

Lohmanit. Trade name for a cast alloy composed of tungsten carbide (chiefly ditungsten carbide, W_2C) and molybdenum carbide, and used for hard dies and cutting tools.

Lohmann. Trade name of Friedrich Lohmann GmbH (Germany) for an extensive series of steels including several austenitic, ferritic and martensitic stainless grades, and a wide variety of cold-work, hot-work, plain-carbon and high-speed tool grades.

Loire. Trade name of Compagnie Ateliers et Forges de la Loire (France) for a series of corrosion- and/or heat-resistant alloys including several austenitic, ferritic and martensitic stainless steels, and numerous nickel-chromium and nickel-iron-chromium alloys.

Lojic. Trade name of Southern Dental Industries Limited (Australia) for an atomized corrosion-resistant spherical-particle alloy composed of 46.0% silver, 29.0% tin, 24.0% copper, 0.95% indium and 0.05% platinum with an optimum alloy-to-mercury ratio of 1:0.76 (43.2% mercury). It does not contain the corrosive tin-mercury phase (Gamma 2), and has a positive dimensional change, and high initial and final compressive and tensile strengths. Used as a dental amalgam for tooth filling. *Lojic Plus* is a corrosion-resistant, homogeneous spherical particle alloy containing 60.1% silver, 29.05% tin, 11.8% copper and 0.05% platinum, and having an optimum alloy-to-mercury ratio of 1:0.73 (42.2% mercury). It does not contain the corrosive tin-mercury phase (Gamma 2), and provides optimal plasticity, minimal dimensional change, low static creep, and very high initial and final compressive and tensile strengths. Used as a dental amalgam for restorations and core build-ups.

lokkaite. A white mineral composed of calcium rare-earth carbonate hydrate, $(Ln_{1.58}Ca_{0.23})(CO_3)_3 \cdot H_2O$. Crystal system, orthorhombic. Refractive index, 1.592. Occurrence: Finland.

Loksand. Trademark of Drake Extrusion Limited (UK) for a crimped polypropylene fiber mixed with sand and having excellent durability and wear and tear resistance. Used for the manufacture of synthetic turf for sports arenas and natural grass access roads and car parks.

Loktite. Trade name of Loktite, Inc. (USA) for polyolefin fibers and products.

löllingite. See loellingite.

Lo-Luminium. British trade name for an alloy composed of 55% tin, 33% zinc, 11% aluminum and 1% copper, and used as aluminum solder.

Lomag. Trademark of Foseco Minsep NV (Netherlands) for tableted compounds added to certain aluminum-alloy melts to remove magnesium and, sometimes, gases.

Lomat. Trade name of Okuno Chemical Industries Company Limited (Japan) for a conversion coating for zinc and zinc alloys.

Lominium. Trade name of Edgar Allen Balfour Limited (UK) for

an oil-hardening die steel containing 0.47% carbon, 1.75% chromium, 0.6% manganese, 0.2% vanadium, and the balance iron. Used for zinc die-casting dies.

Lomod. Trademark of GE Plastics (USA) for a series of thermoplastic elastomers based on polyester ether copolymers. They have excellent resistance to abrasion, cut-growth, chemicals, ozone and ultraviolet light, excellent high and low-temperature flexural properties, good impact resistance, good tensile and tear strength, good dimensional stability, and a hardness of 40 to 70 Shore D. Several grades have very low flammability, custom color capability and good printability. Used for electrical and electronic components, safety helmet support straps, air hoses, hydraulic hoses, vapor fuel lines, wire and cable coatings, bumpers, and shoe plates.

lomonosovite. A dark cinnamon-brown, yellow-brown or rose-violet mineral composed of sodium titanium silicate phosphate, $Na_5Ti_2Si_2PO_{13}$. Crystal system, triclinic. Density, 2.98 g/cm^3; refractive index, 1.75-1.77. Occurrence: Russian Federation.

Lomu. Trade name of Carpenter Technology Corporation (USA) for a nonmagnetic iron alloy containing 8-12% manganese, 6-10% nickel and 0-1% chromium. Used for springs.

Lonbella. Trade name for rayon fibers and yarns used for textile fabrics.

Lonestar. Trade name of Lonestar Industries (USA) for antimony *regulus* containing 99.7% antimony.

long clay. Fat or plastic clay with high green strength.

longcloth. A soft, semi-lustrous high-quality cotton or cotton-blend fabric made in long strips.

long glass. Glass that solidifies slowly.

Longhorn Three Star. American trade name for high-purity tin (99.86%) containing about 0.04% lead, 0.03% copper and 0.024% arsenic.

Longlast. Trade name of Kemdent (UK) for a heat-cure acrylic used for dentures.

longleaf pine. The strong wood from the yellow pine species *Pinus palustris*. The heartwood is reddish-brown, and the sapwood yellowish-white. It has moderately large shrinkage and good stability when properly seasoned. Average weight, 660 kg/m^3 (41 lb/ft^3). Source: Southern and southeastern United States from southern Virginia to southeastern Texas. Used as lumber for heavy construction, e.g., as beams, joists, posts, piles and stringers, and for boxes, crates and pallets, cooperage, railroad ties, piles, poles, and mine timbers. Also known as *Georgia pine; hard pine; hill pine*.

Long Life. Trademark of Manufacturas del Sur SA (Peru) for nylon 6 fibers and filament yarns.

long-oil alkyd. An alkyd resin, produced with a large amount of unsaturation, which is modified with up to 70% oxidizing oils, and is less reactive and more durable and elastic than short-oil grades. Used for brush-on enamels. See also short-oil alkyd.

long-oil varnish. Varnish with a high concentration of unsaturated oil. Usually there are 76-370 L (20-100 gal.) of oil per 45 kg (100 lbs.) of resin. *Long-oil varnish* is more elastic and more durable than short-oil varnish. *Spar varnish* is a type of long-oil varnish. Also known as *long varnish*. See also short-oil varnish.

Long Playing. Trade name of SA Glaverbel (Belgium) for a patterned glass featuring large grooved discs.

Long Terne. Trade name of US Steel Corporation for sheet steel coated with an alloy composed of 75-98% lead and 2-25% tin, and used for roofing, flashing and siding.

long terne. Steel plate, sheet or strip with a thickness of 0.25-2 mm (0.01-0.08 in.) coated with an alloy composed of about 75-98% lead and 2-25% tin. The typical coating thickness is 60-250 g/m² (0.28-1.17 oz/ft²). Used for automotive gasoline tanks, radiator components, air-filter containers, radio and television chassis, roofing, flashing, gutters, downspouts, siding, and electrical hardware. See also short terne; terneplate.

long varnish. See long-oil varnish.

Longwair. Trade name of British Steel Corporation (UK) for an abrasion- and wear-resistant tool steel containing 2.2% carbon, 13% chromium, and the balance iron.

long wool. Wool whose staple fibers range from about 150 to over 300 mm (6 to over 12 in.) in average length. See also medium wool; short wool.

lonsdaleite. A black mineral composed of carbon (C) and found in meteorites, or synthesized from graphite at a temperature above 1000°C (1830°F) at a typical static pressure of 130 kilobars. Crystal system, hexagonal. Density, 3.30 g/cm³.

Lonzona. Trade name of Lonza-Werke GmbH (Germany) for cellulose acetate fibers used for textile fabrics.

loop yarn. A fancy yarn that is used in certain fabrics to produce a pile with a looped or curled effect. Also known as *curled yarn*. This term is not synonymous with "loopy yarn."

loop-wale yarn. A knitted yarn made from a narrow fabric. Also known as *chainette yarn*.

loopy yarn. A low-stretch textured yarn with numerous loops of different size and shape appearing at random intervals along the filaments or fibers. This term is not synonymous with "loop yarn."

loose-fill insulation. Thermal insulation made from materials, such as fiberglass, rock or slag wool, wood fibers, shredded wood bark, sawdust, granulated cork, or ground or macerated wood pulp, perlite, vermiculite or gypsum. Depending on the particular material, its texture may range from granular to fluffy. It can be blown or poured, and is used for filling irregular and inaccessible spaces in walls, floors, roofs and attics.

loose splittings. Mica splittings of heterogeneous shape and packed in bulk form.

loose splittings with powder. Loose mica splittings dusted with finely ground mica.

Lopac. Trade name of Monsanto Chemical Company (USA) for acrylic resins.

loparite. A black mineral of the perovskite group that is composed of cerium sodium calcium titanium niobium oxide, $(Ce,Na,Ca)_2(Ti,Nb)_2O_6$, and may contain strontium and lanthanides. It can also be made synthetically. Crystal system, cubic. Density, 4.01 g/cm³; refractive index, 2.33. Occurrence: USA (Arkansas), Paraguay, Russia.

lopezite. An orange-red mineral composed of potassium chromium oxide, $K_2Cr_2O_7$. It can also be made synthetically. Crystal system, triclinic. Density, 2.69 g/cm³; refractive index, 1.7380. Occurrence: Chile.

Lophos. Trade name of Ziv Steel & Wire Company (USA) for a steel used for cold-heading dies.

Lo-pic. Trade name for polyolefin fibers and yarns used for textile fabrics.

Loralin. Trade name for polystyrene resins with good stability and moldability, good thermal and electrical resistance, good strength and modulus, poor resistance to ultraviolet radiation, poor resistance to load cracking, and a maximum service temperature of 50-100°C (120-210°F). Used for automotive and appliance parts, piping, knobs, dials, packaging, battery boxes, high-frequency insulation, and dinnerware.

lorandite. A deep red or dark lead-gray mineral composed of thallium arsenic sulfide, $TlAsS_2$. Crystal system, monoclinic. Density, 5.53 g/cm³. Occurrence: Greece, USA (Wyoming).

loranskite. A black to brownish-black mineral of yttrium cerium calcium zirconium tantalum oxide, $(Y,Ce,Ca,Zr)TaO_4$. Crystal system, orthorhombic. Density, 5-6 g/cm³; hardness, 5-6 Mohs. Occurrence: Canada, Norway, USA.

lorenzenite. A brown to black mineral composed of sodium titanium silicate, $Na_2Ti_2Si_2O_9$. It can also be made synthetically. Crystal system, orthorhombic. Density, 3.43 g/cm³; refractive index, 2.01. Occurrence: Greenland. Also known as *ramsayite*.

lorettiote. A yellow to brownish-yellow mineral composed of lead oxide chloride, $\alpha\text{-}Pb_7O_6Cl_2$. It can also be made synthetically. Crystal system, orthorhombic. Density, 7.40 g/cm³; refractive index, 2.40. Occurrence: USA (Tennessee).

Lorite. Trade name for mixtures of calcium carbonate ($CaCO_3$) and diatomaceous earth used as extenders in paints, plastics and rubber compounds.

Lorival. Trade name of United Ebonite & Lorival Limited (UK) for urea- and phenol-formaldehyde resins and plastics.

Lorkacel. Trade name of Plastimer (France) for polystyrene foams and foam products.

loseyite. A bluish white mineral composed of manganese zinc carbonate hydroxide, $(Mn,Zn)_7(OH)_{10}(CO_3)_2$. Crystal system, monoclinic. Density, 3.27 g/cm³; refractive index, 1.648. Occurrence: USA (New Jersey).

Lo-Sil. Trademark of Kaiser Aluminum & Chemical Corporation (USA) for high-alumina refractory brick and castable refractory products containing only small percentages of silica. Used in the manufacture of aluminum-melting furnaces.

Losil. Trade name of British Steel plc (UK) for a series of hot-rolled electrical steels containing 0.4-2.3% silicon. Used for rotating machinery and dynamos.

Losta. Trade name of Friedrich Lohmann GmbH (Germany) for an oil-hardening tool steel (AISI type O2) containing 0.9% carbon, 1.9% manganese, 0.1% vanadium, 0.4% chromium, and the balance iron. It has good nondeforming properties, and is used for upsetters, punches, and dies.

Lotader. Trade name of Atofina SA (France) for ethylene-acrylic ester and maleic anhydride terpolymers used for packaging applications.

Lo-Temp. Trade name of John Hewson Company (USA) for a casting brass containing 60% copper and 40% zinc.

lotharmeyerite. A reddish orange mineral composed of calcium zinc manganese arsenate hydroxide dihydrate, $CaZnMn(AsO_4)_2(OH)\cdot 2H_2O$. Crystal system, monoclinic. Density, 4.20 g/cm³; refractive index, above 1.8. Occurrence: Mexico.

Lotmessing. German trade name for corrosion-resistant brasses composed of varying amounts of copper, zinc and silicon, and used especially for hardware, bolts, and fasteners.

Lotol. Trademark of Uniroyal Chemical Division of Uniroyal Inc. (USA) for a series of compounded natural and synthetic rubber latexes.

Lo-Tran. Trade name of Houze Glass Corporation (USA) for neutral-gray sheet glass with a light transmission of 12.5%.

Lotrene. Trademark of Atofina SA (France) for a series of polyethylenes.

Lotryl. Trade name of Atofina SA (France) for ethylene-butyl acrylate and ethylene-methyl acrylate copolymers used mainly in the packaging, pulp and paper, textile and leather industries.

Lotung. Trade name of Ziv Steel & Wire Company (USA) for an oil-hardening hot-work die steel containing 0.3% carbon, 3%

chromium, 9% tungsten, 0.45% vanadium, and the balance iron. Used for brass extrusion dies and aluminum die-casting dies.

Lotus. Trade name of Lumen Bearing Company (USA) for a lead-base babbitt containing 15% antimony and 10% tin. Used for medium-duty bearings.

lotus-type porous metals. See gasarite eutectics.

loudounite. A light green to pearly-white mineral composed of sodium calcium zirconium silicate hydroxide octahydrate, Na-$Ca_5Zr_4Si_{16}O_{40}(OH)_{11} \cdot 8H_2O$. Density, 2.48 g/cm³. Occurrence: USA (Virginia).

loughlinite. A white to light green mineral composed of sodium magnesium silicate octahydrate, $Na_2Mg_3Si_6O_{16} \cdot 8H_2O$. Density, 2.16 g/cm³; refractive index, 1.505. Occurrence: USA (Wyoming).

louver cloth. A net or mesh fabric woven from coated glass yarn and used in louvers.

Louvreglas. Trade name of Doane Products Corporation (USA) for cellulose acetate plastics.

Louvrelite. Trade name of Shatterprufe Safety Glass Company (Proprietary) Limited (South Africa) for a sealed glass unit having an aluminum screen of angled louvers.

Louvrex. Trade name of ASG Industries Inc. (USA) for a patterned glass with alternate angular planes resembling the louvers in Venetian blinds.

lovdarite. A white to yellowish mineral composed of potassium sodium beryllium silicate dihydrate, $(Na,K)_2BeSi_3O_8 \cdot 2H_2O$. Crystal system, orthorhombic. Density, 2.33 g/cm³; refractive index, 1.516. Occurrence: Russian Federation.

loveringite. A mineral of the crichtonite group composed of calcium magnesium chromium iron titanium oxide, $(Ca,Ce)(Ti,Fe,Cr,Mg)_{21}O_{38}$. Crystal system, rhombohedral (hexagonal). Density, 4.41 g/cm³. Occurrence: Western Australia.

lovozerite. A black mineral of the combeite group that is composed of sodium zirconium silicate hydroxide, $Na_3ZrSi_6(O,OH)_{18}$, and may also contain potassium, manganese and calcium. Crystal system, monoclinic. Density, 2.38 g/cm³; refractive index, 1.560. Occurrence: Russian Federation.

low-alkali cement. A special *Portland cement* that contains only small amounts of sodium and/or potassium.

low-alloy cast steels. A group of cast steels containing a total of less than 8% alloying elements, such as manganese, chromium, nickel, molybdenum, vanadium, titanium and/or aluminum. They are available in a wide range of tensile strengths from good to excellent, and most of their properties including corrosion, impact and wear resistance, hardenability and machinability and high-temperature strength are superior to those of plain-carbon cast steels. Also known as *low-alloy steel castings*. See also plain-carbon cast steels.

low-alloy steel castings. See low-alloy cast steels.

low-alloy steels. A group of hardenable steels for structural applications in which the total alloy content in addition to carbon is usually 8% or less. Overall, they are about 10-30% stronger than (plain) carbon steels. Typical alloying elements include silicon, manganese, chromium, nickel, molybdenum and vanadium. Also known as *alloy constructional steels*.

low-alloy tool steels. A group of oil- or water-quenched special-purpose tool steels (AISI group L) containing 0.40-1.10% carbon, 0.10-0.90% manganese, 0-0.50% silicon, 0.60-1.20% chromium, 0-2.00% nickel, 0-0.50% molybdenum, 0.10-0.30% vanadium, and the balance iron. They have good strength and toughness, good machinability and fine grain sizes. Often sold under trademarks or trade names, such as *Bethalloy, Champaloy,*

Halvan, Nicrodie or *Tioga*, they are used for machine parts, such as cams, arbors, chucks and collets.

low-alumina silica brick. See superduty silica brick.

low brass. A yellow brass composed of 80% copper and 20% zinc, and commercially available in the form of wire and flat products. It has high ductility, excellent cold workability, good hot workability, good corrosion resistance, and is used for pump lines, flexible hose, bellows, battery caps, clock dials, musical instruments, drawn and formed parts, hardware, and diaphragms.

low-carbon cast steels. See carbon-steel castings.

low-carbon chromium-molybdenum white cast irons. See low-carbon chromium-molybdenum white irons.

low-carbon chromium-molybdenum white irons. A group of highly abrasion-resistant chromium-molybdenum white cast irons which contain 2.4-3.2% carbon, 14.0-23.0% chromium, 0.5-1.5% nickel, 0.5-1.5% manganese, up to 1.0-3.0% molybdenum, 1.0% silicon, 0.1% phosphorus, 0.06% sulfur, and the balance iron. Also known as *low-carbon chromium-molybdenum white cast irons*. See also chromium-molybdenum white irons; martensitic white irons; white irons.

low-carbon electrical sheet. A group of low-carbon steels with less than 0.1% carbon, about 0.3-4.5% silicon, and low manganese, sulfur and phosphorus contents, supplied in strip or sheet form. Included in this group are magnetic lamination steels, nonoriented electrical steels and grain-oriented electrical steels used in the manufacture of armatures, dynamos, motors, and transformers. Also known as *low-carbon sheet steel*. See also electrical sheet steel.

low-carbon ferritic steels. See ferritic steels.

low-carbon ferrochrome. See low-carbon ferrochromium.

low-carbon ferrochromium. A *ferroalloy* composed of 0.02-2% carbon, 67-73% chromium, 0.2-1% silicon, and the balance iron. It is graded according to the carbon content, and used as an alloying addition to stainless steel. Also known as *low-carbon ferrochrome*.

low-carbon ferromanganese. A *ferroalloy* composed of 0.07-0.5% carbon, 85-90% manganese, and the balance iron. It is graded according to the carbon content, and is used to introduce manganese into low-carbon steels, and for other metallurgical applications.

low-carbon ferrotitanium. A *ferroalloy* composed of up to 0.1% carbon, 20-25% titanium, 0-7.5% aluminum, and the balance iron. Used as an alloying addition to stainless steels.

low-carbon nickel-chromium white irons. See nickel-chromium white irons.

Low-Carbon Nickel. Trade name of Huntington Alloys Inc. (USA) for nickel (99.5% pure) containing 0.01% carbon, 0.18-0.2% manganese, 0.15-0.2% iron, 0.005% sulfur, 0.05-0.18% silicon, 0.05-0.13% copper, and traces of cobalt. It has high heat resistance, a service temperature exceeding 316°C (600°F), excellent corrosion resistance to caustics, halogens and hydrogen halides, many salts and food acids, high thermal and electrical conductivity, good magnetic and magnetostrictive characteristics, high ductility and softness, and a low rate of work hardening. Used especially for chemical and plating equipment, food-processing equipment, beer lines and tanks, and laboratory crucibles and dishes. Now *Nickel 201*.

Low Carbon Ontario. Trade name of Allegheny Ludlum Steel (USA) for an air- or oil-hardening cold-work tool steel (AISI type D1) containing 0.87% carbon, 12% chromium, 0.8% molybdenum, 0.5% silicon, 0.5% manganese, 0.35% nickel, 0.15%

vanadium, and the balance iron. Used for dies, tools and cutlery.

low-carbon sheet steel. See low-carbon electrical sheet.

low-carbon steel castings. See carbon-steel castings.

low-carbon steels. A group of steels with less than 0.30 wt% carbon that cannot be through-hardened by heat treatment, but can be surface-hardened (e.g., by case hardening). Their microstructure consists of *ferrite* and *pearlite* constituents. Commercially available as band iron, black iron sheet, and as bars and rods, they have low to medium strength and hardness, outstanding ductility and toughness, good machinability and weldability and easy fabricability. Used for autobody parts, structural shapes, and sheets.

Low Carbon Tatmo. Trade name of Latrobe Steel Company (USA) for a tough high-speed steel containing 0.6-0.75% carbon, 0.3% silicon, 0.25% manganese, 1.5-1.7% tungsten, 3.75% chromium, 1-1.2% vanadium, 8.2-8.7% molybdenum, and the balance iron. Used for dies, punches, taps, and cold- and hot-heading tools.

Low Chrome-Nickel. Trade name of John Matthey plc (UK) for a tough steel containing 0.35-0.45% carbon, 0.25-0.5% chromium, 0.8-1% manganese, 0.5-0.75% nickel, 0.3-0.4% silicon, and the balance iron. Used for gears, pinions, and shafts.

low-carbon white irons. A group of abrasion-resistant white cast irons having coarse pearlitic microstructures as-cast. A typical composition is 2.2-2.8% total carbon, 0.2-0.6% manganese, 1.0-1.6% silicon, 1.5% nickel, 1.0% chromium, 0.5% molybdenum, 0.15% phosphorus, 0.15% sulfur, up to 1.5% copper, the balance being iron. Copper may replace some or all of the nickel. Used for wear-resistant parts. See also white irons.

low-density alloys. See light alloys.

low-density concrete. See lightweight concrete.

low-density flexible RIM polyurethane foam. A semi-rigid polyurethane foam with a low density (below 0.3 g/cm³ or 0.01 lb/in³), made by reaction-injection molding. It has a tough, flexible skin with a smooth, essentially void-free surface, and an open-cell polyurethane foam core. Used for automobile interiors, bicycle seats, and industrial equipment.

low-density metals. See light metals.

low-density particleboard. Particleboard with a density of less than 595 kg/m³ (37 lb/ft³). It has a lower modulus of rupture, modulus of elasticity and internal bond strength than *high-density particleboard*, and is generally bonded with urea-formaldehyde resin, and used exclusively for lightweight interior applications.

low-density polyethylenes. A group of branched-chain polyethylenes with a density range of 0.910-0.925 g/cm³ (0.0329-0.0334 lb/in³), a melting point of 116°C (240°F), and a crystallinity of less than 60%. They have good strength, low thermal expansivity, a service temperature range of -60 to +90°C (-76 to +194°F), good dielectric properties, fair ultraviolet light resistance, good resistance to dilute acids, alkalies, alcohols, halogens and ketones, fair resistance to concentrated acids, and poor resistance to greases, oils and aromatic hydrocarbons. Used for wire and cable coatings, electrical insulation, packaging film, refuse and waste bags, squeeze bottles, liners for shipping containers, paper coatings, cordage, and toys. Abbreviation: LDPE. See also high-density polyethylenes; polyethylenes.

low-dielectric-constant materials. See low-k materials.

low-duty fireclay brick. Fireclay brick with a pyrometric cone equivalent above 15 and below 29. Used for refractory applications, e.g., in metallurgy. Also known as *low-heat duty brick.*

low-E coatings. Ultrathin, usually metallic coatings applied to or into window glass to block heat flow and reduce sun damage and fading of furnishings, carpets, paint and artwork. Essentially, there are three types: (i) soft metallic coatings, e.g., silver coatings, which are delicate and not resistant to scratching; (ii) hard metallic coatings, e.g., tin coatings, which are relatively resistant to scratching, and (iii) multilayered coatings which incorporate metallic and ceramic layers and are very resistant to scratching. Also known as *low-emissivity coatings.* See also low-E glass.

low-E glass. A high-performance window glass coated with a thin, invisible metallic layer, or several metallic and ceramic layers to block heat flow and significantly reduce the passage of ultraviolet and infrared light. Also known as *low-emissivity glass.* See also low-E coatings.

low-emissivity coatings. See low-E coatings.

low-emissivity glass. See low-E glass.

low-enriched uranium. See enriched uranium.

lower bainite. See bainite.

low-expansion alloys. A group of alloys containing about 30-70% nickel, 30-70% iron, and sometimes small additions of manganese and silicon, and exhibiting very small temperature coefficients of expansion (less than 1.8-9.0 × 10⁻⁶/°C or 1.0-5.0 × 10⁻⁶/°F) within specific temperature ranges. Often sold under trademarks or trade names, such as *Dilver, Dumet, Invar, Elgiloy, Elinvar, Nilver, Nivar* or *Platinite,* they are used for absolute standards of lengths (e.g., measuring rods and tapes), precision instruments, weights, bimetallic strips, thermostatic strips, clock and watch components, moving parts, such as pistons, and for glass-to-metal seals, components for electronic devices and superconducting systems. See also controlled-expansion alloys.

low-expansion iron. Any cast iron having a low coefficient of thermal expansion, especially a nickel-bearing grade.

low-expansion glass. A borosilicate glass, such as *Pyrex,* containing approximately 80-81% silicon dioxide (SiO_2), 13% boron oxide (B_2O_3), 3.5-4% sodium oxide (Na_2O), 2-2.5% aluminum oxide (Al_2O_3) and 0-0.4% potassium oxide (K_2O). It has a density of 2.20-2.30 g/cm³ (0.079-0.083 lb/in³), a refractive index of 1.47, a low coefficient of thermal expansion (typically less than 5 × 10⁻⁶/K), a high softening point (above 590°C or 1094°F), high heat and thermal-shock resistance, high toughness, and excellent chemical durability. Used for heat-resistant glassware, domestic ovenware, chemical and laboratory ware, glass-to-metal seals, and in the form of glass fibers for composite reinforcement.

low-heat cement. A cement that has a considerably lower heat of hydration than ordinary *Portland cement,* achieved by increasing the percentage of dicalcium silicate (C_2S) and tetracalcium aluminoferrate (C_4AF), and decreasing the percentage of tricalcium aluminate (C_3A) and tricalcium silicate (C_3S). Also known as *type IV cement.*

low-heat duty brick. See low-duty fireclay brick.

low-hysteresis steel. A random or oriented silicon steel (about 2.5-4% silicon) that has a low magnetic hysteresis loss, high magnetic permeability and saturation induction and high electrical resistivity. Used for transformer cores, and low-frequency power applications. See also electrical steels; silicon steels.

low-index coating. An *optical coating,* such as germanium-doped silica, that has a low refractive index, and is applied to optical fibers to keep the light path away from the fiber surfaces. See also optical fibers.

low-iron magnesite brick. A refractory brick containing at least 90% magnesia, up to 2.5% iron oxide, with the balance being other oxides.

low-k dielectrics. See low-k materials.

low-k materials. A class of materials including in particular benzocyclobutane, perfluorocyclobutane and silicon-based or fluorinated network-forming polymers, e.g., nanoporous organosilicate composite structures having dielectric constants (k) of 2.0 or less. Used in the manufacture of ultrafast microprocessors for advanced microelectronic devices. Also known as *low-k dielectrics; low-dielectric-constant materials.*

Lowland. Trade name of Lenzing Fibers Corporation (USA) for rayon fibers and yarns used for textile fabrics.

low-leaded brass. A brass containing 64.5-66.5% copper, 33-35% zinc and 0.25% lead. It has excellent machinability, high ductility, good formability and good corrosion resistance. Used for watch parts, hinges, drain tubes, plumbing supplies, pump liners, hardware, and stamped and drawn parts. Also known as *light leaded brass.*

low-leaded tube brass. A free-cutting brass containing 66-66.5% copper, 33-33.5% zinc and 0.5% lead. Commercially available in the form of tubes, it has excellent cold workability, good machinability and good corrosion resistance, and is used for plumbing accessories, pump liners, and pump and power cylinders. See also tube brass.

low-melting alloys. See fusible alloys.

low-melting glass. A chemical-resistant glass made by introducing arsenic, selenium, sulfur or thallium into the melt. Depending on the particular composition, melting points range from 127 to 349°C (260 to 660°F). Used for electrical and electronic applications, e.g., electronic components, thin films, and encapsulation.

low-melting solders. See solders.

low-modulus rayon. Rayon-based carbon fibers with a tensile modulus not exceeding 50 GPa (7×10^3 ksi). See also high-tenacity rayon; rayon fibers.

Lowmoor. Trade name of Lowmoor Best Yorkshire Iron Limited (UK) for a cast iron containing 3.2% carbon, 2.6% silicon, and the balance iron.

low-orientation yarn. A yarn whose molecular orientation is low and incomplete, and which therefore can be further oriented by drawing at high draw ratios. Abbreviation: LOY.

low-performance thermoplastic resins. A group of tough thermoplastic resins that have continuous-use temperatures up to 120°C (250°F), well below those of high-performance or engineering thermoplastics (with over 200°C or 392°F), and that are thus not suitable for engineering composites, but are used for consumer products, appliance housings, machine parts (e.g., gears and bearings), fixtures, and automotive parts.

low-phosphorus copper. See deoxidized low-residual-phosphorus copper.

low-phosphorus pig iron. Pig iron with a typical composition of 3.2-3.9% carbon, 0.3-2.0% silicon, 0.3-0.4% manganese, 0.02-0.15% sulfur, less than 0.06% phosphorus, and the balance iron. Used in the manufacture of cast steel and crucible steel. See also pig iron.

low-pressure laminates. Laminated plastics molded and cured at pressures not exceeding 2.76 MPa (400 psi). Also known as *low-pressure plastic laminates.* See also plastic laminates.

low-pressure plastic laminates. See low-pressure laminates.

low-pressure resin. See contact resin.

low-profile resins. Polyester resin systems that contain 50-70 wt% thermosetting resin and 30-50% thermoplastic resin, and offer no or very low surface waviness in the molded part. Used for the manufacture of reinforced plastics.

low quartz. Ordinary *quartz* (SiO_2) formed at a temperature not exceeding 573°C (1063°F) at which the symmetrical arrangement of the SiO_4 tetrahedra is less perfect than for quartz formed above this temperature. It has a hexagonal crystal structure, a density of 2.66 g/cm³ (0.096 lb/in.³), and a hardness of 7 Mohs. Also known as *alpha quartz.* See also high quartz; beta quartz.

low-reflection film. A transparent coating or film applied to a glass surface to minimize reflection and maximize transmission of incident light.

low-residual-phosphorus copper. See deoxidized low-residual-phosphorus copper.

Lowricryl. Trademark of Chemische Werke Lowi GmbH (Germany) for nonpolar, ultraviolet-curing, low-temperature, low-viscosity embedding media used in light and electron microscopy. This product is available in the USA from Polysciences Inc.

Lowroff bronze. A heavy-duty phosphor bronze containing 70-90.5% copper, 5-16% lead, 4-13% tin, and 0.5-1% phosphorus. Used for bushings and bearings.

Lowscore. Trade name of Dunford Hadfields Limited (UK) for a corrosion-resistant stainless steel (AISI type 442) containing up to 0.2% carbon, 18-22% chromium, and the balance iron. Used for furnace parts, and heat-resisting equipment.

low-shrink resins. Polyester resin systems that contain at least 70 wt% thermosetting resin and up to 30% thermoplastic resin, and offer low surface waviness in the molded part. Used for the manufacture of reinforced plastics.

low-silicon bronze. A silicon bronze containing 98.5% copper and 1.5% silicon. Commercially available in the form of wires, tubes and rods, it has excellent cold and hot workability, good corrosion resistance, and is used for machine parts, marine and pole-line hardware, bolts, rivets, U-bolts, screws, nuts, cable clamps, welding rod, electrical conduit, heat exchanger and condenser tubing, and hydraulic pressure lines.

low-soda alumina. Aluminum oxide (Al_2O_3) that contains less than 0.15% sodium oxide (Na_2O), and is used for high-grade electrical ceramics, e.g., insulators.

low-solubility glaze. A lead-containing glaze in which 5% or less of the lead oxide (PbO) is soluble.

Low-Temp. Trade name of Versil Limited (UK) for fiberglass flexible sections, sewn blankets, and sheets.

low-temperature carbon steels. A group of low-carbon steels that have good strength and toughness at cryogenic temperatures, and are suitable for low-temperature applications including tanks, vessels and containers for storing and handling cryogenic liquids, superconductor equipment, and structural aircraft and aerospace components.

low-temperature cement. A cementitious material produced by the hydrothermal formation of the intermediate phase *hillebrandite* (calcium silicate hydroxide, $Ca_2SiO_3(OH)_2$) at relatively low temperatures (below 300°C or 570°F) and subsequent decomposition at about 600°C (1110°F). The final product contains considerable quanties of dicalcium silicate (C_2S), and is used in the manufacture of concrete and cement products.

low-temperature gallium arsenide. Gallium arsenide (GaAs) grown at low temperatures and used as a semiconductor. Abbreviation: LT-GaAs. See also gallium arsenide.

low-temperature glaze. See soft glaze.

low-temperature indium phosphide. Indium phosphide (InP) grown at low temperatures and used as a semiconductor. Abbreviation: LT-InP. See also indium phosphide.

low-temperature resins. A group of engineering thermosets including ureas, melamines, allyls, polyesters and polyurethanes, having maximum service temperatures not exceeding 120°C (250°F). See also high-temperature resins; medium-temperature resins.

low-temperature reusable surface insulation. A thermal insulation material composed of a silica tile with a thin, white, high-emittance coating of borosilicate glass. It has a maximum service-temperature of approximately 650°C (1200°F), and is used on the surfaces of the upper wings and tail and upper sidewalls of the NASA Space Shuttle. Abbreviation: LRSI.

low-temperature solders. See fusible solders.

low-temperature steels. See cryogenic steels.

low-temperature thermoset matrix composites. Thermosetting engineering composites, such as those based on unsaturated polyesters, designed for service temperatures between about 120-150°C (250-300°F) for neat resins and 120-205 °C (250-400°F) for fiber-reinforced resins. See also medium-temperature thermoset matrix composites.

low-tenacity fiber. A textile fiber with relatively low tensile (breaking) strength. Abbreviation: LTF. See also high-tenacity fiber.

low-tenacity yarns. A class of usually textile-type yarns with relatively low tensile (breaking) strengths. They can usually be easily pulled apart by hand. See also high-tenacity yarns.

low-tin commercial bronze. A special bronze containing 90% copper, 9.5% zinc and 0.5% tin. It has good corrosion resistance, and is used for bushings, bearings, bearing sleeves and thrust washers. See also commercial bronze.

Loxley. Trade name of Sanderson Kayser Limited (UK) for a water-hardening steel containing 0.7-0.8% carbon, and the balance iron. Used for tools, taps, and punches.

Loycon. Trade name of British Steel plc (UK) for a series of constructional steels containing up to 0.15% carbon, 0.25% silicon, 1.2% manganese, 0.35% molybdenum, 0.6% chromium, 1.6% nickel and 0.12% vanadium, with the balance being iron. They have good impact strength at low temperatures, good weldability, and are used for crane parts, storage tanks, large pressure vessels and boilers, and mining equipment.

LSB. Trademark of La Seda de Barcelona SA (Spain) for nylon 6 and polyester fibers and filaments.

Lubeco. Trade name of Lumen Bearing Company (USA) for a lead-base babbitt composed of 40% tin, 12% antimony and 1% copper, and used for bearings.

Lube-Lok. Trade name of E/M Engineered Coating Solutions (USA) for oven-cured solid-film lubricants.

LubMaster. Trade name of F.J. Brodmann & Company LLC (USA) for powders used for solid lubrication applications.

Lubral. Trade name of Montecatini Settore Alluminio (Italy) for an aluminum alloy containing 6% tin, 1% copper, 1% nickel, 0.15% silicon and 0.2% iron. Used for bearings, and light-alloy parts.

Lubralloy. Trademark of Microfin Corporation (USA) for coatings produced on ferrous and nonferrous substrates by a process that chemically deposits a hard nickel alloy. They have excellent abrasion and wear resistance, high hardness, excellent lubricity and good corrosion resistance. Used for valves, transportation equipment, medical applications and electronics. The coating process is also referred to as "Lubralloy."

Lubri-Bond. Trade name of E/M Engineered Coating Solutions (USA) for an air-drying solid-film lubricant.

lubricant. Any solid, liquid or semiliquid material, such as oil, grease, wax or graphite, placed between two surfaces in contact to reduce friction and wear. Such surfaces include dies, die punches, molds, plungers, workpieces, machine parts, bearings, etc. See also lubricating greases; lubricating oils; solid-film lubricant.

lubricated yarn. A natural- or synthetic-fiber yarn that has been impregnated or otherwise treated with a suitable lubricant to enhance its knittability.

lubricating greases. Solid to semiliquid mixtures of one or more mineral oils with one or more metallic soaps, such as aluminum, barium, calcium, lead, lithium, sodium or potassium soap, petrolatum, or wax. Their textures vary greatly and may be smooth, buttery, ropy, fibrous, spongy or rubbery, and they may also have other ingredients added to impart special properties. They are often sold under trade names or trademarks, and used to reduce friction and wear between moving surfaces. See also lubricant.

lubricating oils. High-boiling paraffin-base petroleum oils, often with additives, such as rust or oxidation inhibitors, detergents, defoamers, pour-point depressants or viscosity-index improvers, used between moving surfaces to reduce friction and wear. They usually have good high-temperature stability, high low-temperature fluidity, and viscosities that change only moderately over a broad temperature range. Animal and vegetable oils were formerly also used as lubricating oils. See also lubricant.

Lubrico. Trade name of Buckeye Brass & Manufacturing Company (USA) for a heavy-duty alloy composed of 70-75% copper, 20-22% lead and 5-10% tin, and used for bushings and bearings.

Lubricomp. Trade name of LNP Engineering Plastics (USA) for an extensive series of polymer-matrix composites based on, or filled/lubricated with polytetrafluoroethylene (PTFE). In some grades the PTFE is replaced by silicone. They have excellent wear resistance, excellent processibility, good mechanical properties, low coefficients of friction, and are used for automotive applications and industrial equipment. *Lubricomp* composites are available in a wide range of grades including *Lubricomp BGU* wear-resistant reinforced and PTFE-lubricated thermoplastic semi-crystalline polyphthalamides, *Lubricomp ECL* carbon fiber-reinforced, PTFE-modified high-temperature polyetherimides, *Lubricomp IL* silicone- or PTFE-lubricated polyamide 6,12 resins, *Lubricomp KFL* PTFE-lubricated, glass-reinforced acetal composites, *Lubricomp PL* silicone- or PTFE-lubricated polyamide 6 resins, and *Lubricomp QL* silicone- or PTFE-lubricated polyamide 6,10 resins.

Lubri-Die. Trade name of Ziv Steel & Wire Company (USA) for an oil-hardening cold work steel (AISI type O6) containing 1.45% carbon, 0.8% manganese, 1.15% silicon, 0.25% molybdenum, and the balance iron. It contains small particles of graphitic carbon uniformly dispersed throughout the matrix, and has good machinability and deep-hardening properties. Used for blanking, forming and piercing dies, pneumatic hammers, punches, and a variety of other tools.

LubriLAST. Trademark of AST Products Inc. (USA) for a series of lubricious, hydrophilic coatings consisting of crosslinked supporting polymer networks. They can be formulated to incorporate various bioactive agents, such as antimicrobial, antithrombogenic and antiviral agents, or peptide and protein drugs for use in controlled drug delivery. They are supplied as aqueous based solutions for application by dip coating and subsequent low-heat curing, and adhere well to many different substrates including various metals (e.g., stainless steels, aluminum, gold, Nitinol, etc.), polymers (e.g., HDPE, LDPE, silicone, PVC, FEP, parylenes, rubber latex, etc.), glass and ceramics. *LubriLAST K* polymer coatings have a controlled release mechanism that ionically bonds the incorporated antimicrobial agents (e.g., silver halides, silver oxide, antibiotic or antiseptic compounds, etc.) to the polymer backbone, and releases them in the presence of body fluids. It can be applied to virtually any biomaterial.

Lubriloy. Trademark of LNP Engineering Plastics (USA) for nylon 6,6 alloys and composites available in unreinforced, aramid- or glass fiber-reinforced and lubricated grades. They have excellent wear and friction performance, high impact strength and low density. Used for machine parts (e.g., gears, bearings, bushings and wear strips), and automotive components (e.g., steering gear assemblies, seat belt parts, and window regulators).

Lubriplas. Trademark of Bay Resins, Inc. (USA) for a series of thermoplastic resins lubricated with polytetrafluoroethylene (PTFE), molybdenum disulfide (MoS_2) or ultrahigh-molecular-weight polyethylene (UHMWPE). The resins include carbon- or glass fiber-reinforced nylons (nylon 6 and nylon 6,6), polyacetals and polycarbonates. *Lubriplas* resins have very low wear rates and low coefficients of friction.

Lubrotec. Trade name of Eutectic Corporation (USA) for a series of metal alloy powders for spraying final coatings that are machinable, but tough and wear resistant.

Lucalen. Trademark of BASF Corporation (USA) for a series of unprocessed and semi-manufactured ethylene/acrylate copolymers. The unprocessed grades are supplied as liquids, pastes, granules and powders, and the semi-manufactured products as sheets, boards, rods and tubes. Also included under this trademark are ionomers (*Lucalen I*).

Lucalor. Trademark of Atofina SA (France) for chlorinated polyvinyl chlorides (CPVC) available in pipe and plumbing grades and supplied as unprocessed resins, and semifinished products, such as sheets, slabs, rods and tubes.

Lucalox. Trademark of General Electric Company (USA) for a transparent, pure, polycrystalline alumina (Al_2O_3) ceramic fired at high temperature to remove the micropores. Available as disks, rods and tubes, it has a light transmission up to 90%, a high modulus of elasticity, high mechanical and transverse strengths, good dielectric properties, and good heat resistance up to approximately 1980°C (3600°F). Used for high-intensity lamps, missile nose cones, instrument parts, and cutting tools.

Lucanex. Trademark of Lucas-Milhaupt, Inc. (USA) for a brazing alloy paste that is formulated to facilitate the brazing of materials, such as alumina, graphite, zirconia, and silicon nitride, which are rather difficult to wet. It is designed for use in vacuum or with high-purity argon or helium atmospheres, and can be used at service temperatures up to 260°C (500°F) in air or higher temperatures under protective atmospheres. It is also used for joining nonmetallic materials to carbon and stainless steels, cast iron, molybdenum, copper, etc.

Lucas Nifal. Trade name for a permanent magnet material composed of 63% iron, 25% nickel and 12% aluminum.

Lucerno. Trade name for a heat-resistant resistance alloy containing 65-68% nickel, 27-30% copper, 2.2-5% manganese and up to 2.4% iron.

Lucero. Trade name of Driver Harris Company (USA) for a resistance alloy composed of 70% nickel and 30% copper. It has good corrosion and heat resistance up to 500°C (930°F), and is used for bolts, springs, valves, and resistance wire.

Lucisa. Trade name of Nuova Rayon SpA (Italy) for rayon fibers and yarns used for textile fabrics including clothing.

Lucite. (1) Trademark of E.I. DuPont de Nemours & Company (USA) for acrylic resins based on polymethyl methacrylate (PMMA) and supplied in various grades including cast-sheet, casting, general-purpose and high-impact. They have good optical properties (clear or colored), outstanding light transmission and resistance to weathering, moderate strength, low hardness, good electrical resistance, low heat resistance, and a maximum service temperature of 60-93°C (140-200°F). Used for lenses, transparent enclosures, outdoor signs, nameplates, decorations, display items, dials, glazing, bottles, aircraft windows and parts, ornaments, and drafting equipment.

(2) Trademark of E.I DuPont de Nemours & Company (USA) for several acrylic products including adhesives, coatings, lacquers and syrups, mixing and tinting colors, topcoat sealers, and modifiers for other resins.

Lucitone 199. Trade name of Fricke Dental International Inc. (USA) for an acrylic resin used for dentures.

Lucky. Trade name of LarSan Chemical Company (USA) for a series of thermoplastics based on polymethyl methacrylate (PMMA) and supplied in several grades including extrusion, co-extrusion, injection-molding and high-flow.

Lucolene. Trademark of Pechiney SA (France) for plasticized polyvinyl chlorides.

Lucorex. Trade name of Pechiney SA (France) for stiff, strong, inherently flame-retardant unplasticized polyvinyl chlorides supplied in rod, sheet, tube, powder and film form. These thermoplastics have a density of 1.4 g/cm³ (0.05 lb/in.³), low water absorption (0.03-0.4%), good barrier properties and UV stability, an upper continuous-use temperature of 50°C (122°F), and moderate chemical resistance. Used for building products, containers, and bottles.

Lucovyl. Trademark of Pechiney SA (France) for a polyvinyl chloride paste.

Lucryl. Trademark of BASF Corporation (USA) for a series of acrylics based on polymethyl methacrylate (PMMA) and available in regular, general-purpose, injection molding, extrusion and impact-modified grades. *Lucryl G* refers to general-purpose acrylics and *Lucryl KR* to a range of high-impact acrylics.

luddenite. A nickel-green mineral composed of copper lead silicate hydrate, $Cu_2Pb_2Si_5O_{14} \cdot 14H_2O$. Crystal system, monoclinic. Density, 4.45 g/cm³. Occurrence: USA (Arizona).

Lüdenscheidt button metal. A German zinc die-casting alloy containing 20% copper used for various die castings and ornamental and architectural parts.

Lüdenscheidt plate. A German white metal alloy containing 72% tin, 24% antimony, 4% copper, and a trace of lead. Used for utensils.

ludlamite. A light green mineral with vitreous luster composed of iron magnesium manganese phosphate tetrahydrate, $(Fe,Mg,Mn)_3(PO_4)_2 \cdot 4H_2O$. Crystal system, monoclinic. Density, 3.15 g/cm³. Occurrence: UK, USA (Idaho).

ludlockite. A red mineral composed of lead iron arsenite, $(Fe,Pb)As_2O_6$. Crystal system, triclinic. Density, 4.40 g/cm³; refractive index, 2.055. Occurrence: Southwest Africa.

Ludloy. Trade name of Ludlow Steel Corporation (USA) for a water-hardening, shock-resistant tool steel used especially for chisels, punches, and dies.

Ludopal. Trademark of BASF AG (Germany) for several unsaturated polyesters.

Ludox. Trademark of E.I. DuPont de Nemours & Company (USA) for silicas supplied as aqueous colloidal dispersions of minute spherical particles with high surface areas. They are available in monodispersed and polydispersed grades with narrow and broad particle size distributions, and are used in adhesives and paints, as binders for granular and fibrous materials, and for the enhancement of the friction properties of smooth surfaces, such as paper or flooring.

ludwigite. A dark brown to green mineral composed of magnesium iron oxide borate, $(MgFe)_2FeBO_5$. Crystal system, orthorhombic or monoclinic. Density, 3.86 g/cm^3; refractive index, 1.832. Occurrence: Hungary, USA (Montana).

lueneburgite. A colorless or white mineral composed of magnesium boron phosphate hydroxide hexahydrate, $Mg_3B_2(PO_4)_2$-$(OH)_4 \cdot 6H_2O$. Crystal system, monoclinic. Density, 2.07 g/cm^3; refractive index, 1.54. Occurrence: Germany.

lueshite. A black to reddish brown mineral of perovskite group composed of sodium niobate, $NaNbO_3$. It can also be made synthetically. Crystal system, cubic. Density, 4.44 g/cm^3; refractive index, 2.30. Occurrence: USA (Colorado), Zaire. *Note:* Synthetic *lueshite* has a monoclinic crystal structure changing to cubic at 640°C (1184°F).

Luetecin. French trade name for corrosion-resistant copper alloy containing 6-16% nickel, 5% zinc, 5% iron, 2% tin and 1% cobalt. Used for cheap jewelry.

luetheite. A blue mineral composed of copper aluminum arsenate hydroxide monohydrate, $Cu_2Al_2(AsO_4)_2OH)_4 \cdot H_2O$. Crystal system, monoclinic. Density, 4.28 g/cm^3; refractive index, 1.773. Occurrence: USA (Arizona).

Luftura. Trade name for cellulose acetate fibers used for lofty wearing apparel and industrial fabrics.

lug brick. A brick having lugs made to simplify spacing with adjacent brick.

Lukens. Trade name of Lukens Steel (USA) for an extensive series of steels including several free-machining steels, low-carbon copper steels, high-strength carbon steel plate, high-strength low-alloy constructional and structural steels and low-temperature structural steels.

Luma-Chrome. Trade name of Atotech USA Inc. (USA) for chromium electroplates and plating processes.

Lumacryl. Trade name of The Dental Manufacturing Limited (UK) for a dough (or bulk) molded acrylic resin formerly used for dentures.

Lumapane. Trade name of Celanese Corporation (USA) for cellulose acetate plastics.

Lumarith. Trademark of Celanese Corporation (USA) for thermoplastic cellulose acetate plastics available in various grades (e.g., transparent, translucent or opaque, colored or colorless) in the form of powders, flakes, chips, chunks, blanks, blocks, slabs, sheets, rods and tubes. They have good strength and toughness, high transparency and surface gloss (transparent grades), good moldability, good chemical resistance, good resistance to oils and hydrocarbons, good machinability, poor flame resistance (but slow-burning), and a maximum service temperature of 50-93°C (120-200°F). Used for piping, appliance housings, trim, glazing, packaging, eye shades, handles, and knobs.

lumber. Hardwood or softwood sawed into pieces of uniform thickness, width and length. It may be dried, and perhaps resawed and planed, or treated chemically, but not processed in any other way. The term includes the following sawmill products: (i) boards for paneling, sheathing, flooring and trim; (ii) dimension lumber for framing members, e.g., plates, studs, rafters and sills; (iii) timbers for beams, posts, heavy stringers, etc.; and (iv) numerous specialty items. See also hardwood; softwood.

lumber core construction. Plywood having a center ply or core made of lumber strips, and face and back plies made of veneer.

LumBrite. Trademark of Toray Plastics (America) Inc. for holographic film products.

Lumdie. Trade name of Latrobe Steel Company (USA) for a chromium-type hot-work tool steel (AISI type H14) containing 0.4% carbon, 1% silicon, 0.25% manganese, 4.75% tungsten, 5.25% chromium, and the balance iron. It has good deep-hardening properties, high hot hardness, high toughness and good wear resistance. Used for gripper dies, and dies for aluminum die casting.

Lumen. Trade name of Lumen Bearing Company (USA) for an extensive series of leaded and unleaded tin bronzes and several aluminum, manganese and phosphor bronzes.

lumen bronze. A bearing alloy composed of 86% zinc, 10% copper and 4% aluminum. It has good castability and machinability, and good antifriction and antiscoring properties, but poor resistance to excessive heat or live steam. Used for high-speed bearings for medium or low loads, e.g., in electric motors and machine-tools.

lumen metal. A bearing alloy composed of 85-88% tin, 4-10% copper and 5-8% aluminum.

Lumiclad. Trade name for an organic coating made with a binder, composed of a drying oil and a synthetic resin, in which are dispersed aluminum flake and powdered asbestos. Used for roofing applications.

Lumicon. Trade name of Bayer Dental (Germany) for a zinc phosphate dental cement.

Lumicon Alloy. Trade name of Bayer Dental (Germany) for a dental amalgam alloy.

Lumifor. Trade name of Heraeus Kulzer Inc. (USA) for a light-cure dental hybrid composite.

Lumiglas. Trademark of F.H. Papenmeier Lumiglas (Germany) for sight glasses.

Luminarc. Trade name of Universal Power Corporation (USA) for an aluminum alloy used for welding rods.

luminescent dye. A dye that can be made to emit light by stimulation with an external source of radiation (e.g., X-rays or ultraviolet rays). Used in luminous paints.

luminescent enamel. Porcelain enamel containing frits that make it glow in the dark.

luminescent polymers. A group of polymers including poly(p-phenylene vinylene) and polypyridine and its derivatives that exhibit good electron transport properties, and are suitable for use in light-emitting devices.

Luminex. Trademark of Morgan Advanced Ceramics (USA) for white-colored magnesia (MgO) ceramics used for high-temperature insulation applications, and for crushable brushes and tubes in ball mills and high-speed dispensers. They have a density of 2.4 g/cm^3 (0.09 $lb/in.^3$), an open porosity of 30%, and a maximum service temperature of 1200°C (2190°F).

luminophor. A general term referring to solid and liquid substances (fluophors or phosphors) that can be stimulated to emit light by absorbing incident radiation, e.g., X-rays, cathode rays, ultraviolet radiation or alpha particles. For fluophors the emission ceases almost immediately after excitation, while for phosphors it continues for some time even after the source is removed. See also fluophor; phosphor.

luminous paint. A paint that contains luminous pigments. See also luminous pigments; luminous wall paint.

luminous pigments. A group of pigments that are either of the nonradiative type, such as barium, cadmium, calcium, strontium and zinc sulfides, which after activation with ultraviolet radiation emit visible radiation (light), or of the radioactive type containing krypton, radium, strontium, thallium, tritium, etc. Used in luminous paints and plastics.

luminous plastics. Plastics coated with or containing luminous pigments.

luminous wall paint. A special wall paint containing certain sulfide-base pigments, such as cadmium or zinc sulfide, which after activation with ultraviolet radiation emit visible radiation (light) for a short period of time. Used for walls in operating and examination rooms, laboratories and industrial establishments.

Lumirror. Trademark of Toray Plastics (America) Inc. for polyethylene terephthalate (PET) films made by a multilayer coextrusion technology. They have one smooth surface and one rougher surface suitable for processing, and are also available with primed and antistatic surfaces.

Lumite. Trademark of Johnson & Johnson Corporation (USA) for screen cloth made from flexible synthetic extruded monofilaments on a plastic, such as polyvinylidene chloride. Used as upholstering cloth, and as a replacement for woven cloth based on natural fibers.

Lumitol. Trademark of BASF Corporation (USA) for unprocessed synthetic resins including vinyls, available as liquids, pastes, chips, granules and powders, and used for coatings.

Lumnite. Trademark of Universal Atlas Cement Company (USA) for high-alumina hydraulic cement.

lump lime. Lime made from limestone burned in a vertical kiln.

Lunalite. Czech trade name for a cathedral glass with conchoidal fracture pattern.

Lunar. Trade name of J.M. Ney Company (USA) for a hard, white palladium-silver dental alloy used for porcelain bonding applications.

Lunden conductive tile. A ceramic tile, usually a floor tile, in whose manufacture an electrically conducting material, such as carbon or metal powder, has been used to dissipate static electricity.

Lundie. Trade name of Latrobe Steel Company (USA) for a die steel containing 0.4% carbon, 5.25% chromium, 4.75% tungsten, and the balance iron. Used for hot-work dies.

Lungavita. Trade name of Agriplast Srl (Italy) for durable ultraviolet- and infrared-stabilized polymer films with excellent mechanical properties and chemical resistance and good photooxidation and weathering resistance. Used for protecting flowers and crops.

Lunite. Trade name for a series of high-temperature, oxidation-resistant aluminide coatings deposited on refractory base metals (e.g., niobium) by the fused slurry process.

Lunkenheimer. Trade name of Lunkenheimer Company (USA) for a series of alloys including various low- and high-carbon steels, brasses, bronzes and copper nickels.

Lunorium. Trade name for a cast nickel alloy containing 14.9% chromium, 18.5% molybdenum, 5% iron, 4% tungsten, 0.9% cobalt, 0.4% silicon, 0.2% manganese and 0.2% carbon.

Luphen. Trade name of BASF AG (Germany) for a series of phenol-formaldehyde plastics.

Lupiace. Trade name of Mitsubishi Chemical Company (Japan)

for a series of polyphenylene ether resins.

Lupilon. Trade name of Mitsubishi Engineering Plastics (Japan) for an extensive series of polycarbonate resins supplied in a wide range of grades including glass or carbon fiber-reinforced, polytetrafluoroethylene-lubricated, fire-retardant, high-flow, structural-foam, and UV-stabilized.

Lupital. Trade name of Mitsubishi Chemical Company (USA) for a series of acetal resins.

Lupolen. Trademark of BASF Corporation (USA) for general-purpose thermoplastic high-, medium- and low-density polyethylenes and ethylene-vinyl acetates available in injection-molding, blow-molding, film and impact-modified grades. They have glossy surfaces, low water absorption, a maximum service temperature of 120°C (248°F), high durability and breaking strength, good weldability and machinability, excellent resistance to dilute acids and alkalies, fair resistance to mineral oils and gasoline and poor resistance to trichloroethylene and tetrachlorocarbon. Used for containers and packaging materials, pipes and tubing, hoses, bottles, battery cases, and injection- and blow-molded parts. *Lupolen V* refers to ethylene vinyl acetate (EVA) copolymers with a vinyl acetate content of 12%. They have a density of 0.94 g/cm³ (0.034 lb/in.³), high opacity, high flexibility, a maximum-use temperature of 80°C (176°F), low tensile strength, excellent resistance to dilute acids and alkalies, fair resistance to gasoline and mineral oils, and poor resistance to heat and solvents. Used for molded parts, including bumpers, bellows, seals, gaskets, and flexible tubing.

Luprenal. Trade name of BASF Corporation (USA) for a series of acrylic polymers.

lupuna. The very light, white, pinkish or pale reddish wood from the tree *Ceiba samauma*. It has low hardness, and poor durability and insect resistance. Source: Brazil (Amazon Basin). Used for plywood veneer.

Luralite. Trade name of Kerr Dental (USA) for a pasty zinc oxide–eugenol dental impression material.

Luran. Trademark of BASF Corporation (USA) for a series of styrene-acrylonitrile (SAN) copolymers available in various grades including general-purpose and glass fiber-reinforced. They have a density of 1.08 g/cm³ (0.039 lb/in.³), good tensile strength and impact resistance, a maximum use temperature of 90-100°C (195-210°F), excellent resistance to mineral oils and dilute alkalies, good resistance to gasoline and dilute acids, and poor resistance to tetrachloroethylene and tetrachlorocarbon. Used for household appliances, bathroom fittings, automotive trim, medical equipment, housings, precision parts, and battery boxes. *Luran S* refers to acrylic ester-modified styrene-acrylonitrile (ASA) terpolymers. They are thermoplastic materials with outstanding weatherability, excellent resistance to ultraviolet radiation and high strength and impact resistance, and are suitable for automotive and outdoor applications.

Luranyl. Trademark of BASF Corporation (USA) for unreinforced and glass fiber-reinforced polyethylene ether resins and polyphenylene ether/impact-modified polystyrene (PPE/HIPS) alloys available in standard and fire-retardant, halogen-free and structural-foam grades. They have high heat resistance, high dimensional stability, low moisture absorption, good processibility and outstanding dielectric properties. The flame-retardant grades also have excellent flow properties. Used for office and communication equipment, electrical devices and equipment, and for plumbing and sanitary systems.

Lurex. Trade name of Lurex Company, Inc. (USA) for glossy yarns, and knit and woven textile fabrics containing metallized fibers. The yarns are usually made from aluminum fibers coated with a thermosetting resin, and used for fabrics, and as embroidery and sewing threads. The fabrics are used extensively for the manufacture of wearing apparel.

Lurgi cement. Special hydraulic cement made by sintering the charge on a grate.

Lurgimetall. Trade name of Metallgesellschaft Reuterweg (Germany) for a lead-alkali metal composed of 96.5% lead, 2.8% barium, 0.4% calcium and 0.3% sodium. It has excellent antifriction properties, and is used for bearings.

Lurium. Trade name of Fromson Company Inc. (USA) for a series of highly reflective aluminums and aluminum-magnesium alloys used for reflectors, costume jewelry, automotive trim, lighting fixtures, and household appliances.

Luron. Trademark of ICI Limited (UK) for polyamide (nylon) fibers and yarns.

Lusco. Trade name of Ludlow Steel Corporation (USA) for an oil-hardening, wear- and shock-resisting steel used for pneumatic tools, punches, and chisels.

Lusix. Trade name of Westover Plastics (USA) for nylon 6,6 polymers.

Luster. Trade name of J.F. Ratcliff Metals Limited (UK) for a stainless steel.

luster color. A lustrous coating made by dissolving a metallic oxide, e.g., an oxide of chromium, cobalt, iron, titanium or zinc, in an organic solvent and applying to ceramic ware.

lustered ware. Glazed, fired ceramic ware coated with metallic pigments and re-fired to produce prismatic surface effects.

luster fabric. A plain-weave fabric made with lustrous worsted or mohair yarn, often with cotton or synthetic fiber warp threads.

Luster-Fos. Trade name of Luster-On Products Inc. (USA) for several phosphate coatings.

Lusterite. Trade name of Latrobe Steel Company (USA) for a series of stainless steels containing 0.8-1.1% carbon, 16.5-18% chromium, 0.5% molybdenum, and the balance iron. Used for bearings, cutlery, blades, knives, and surgical and dental instruments.

Luster-Lac. Trade name of Luster-On Products Inc. (USA) for a water-based clear lacquer.

lusterless paint. A light-absorbing paint that produces a dull surface finish lacking any shine or brightness. Used as camouflage paint for military vehicles.

Lusterloft. Trade name for nylon fibers and yarns used for textile fabrics including clothing.

Lusteroid. Trade name of Lusteroid Container Company (USA) for cellulose nitrate plastics.

lusterware. Earthenware decorated with metallic colors, or lustrous, iridescent glazes.

Lustra. Trade name of Heatbath Corporation (USA) for cadmium and zinc electroplates and plating processes.

Lustrablu. Trade name of ASG Industries Inc. (USA) for a blue flat glass.

Lustrac. Trade name of Lustrac Plastics Limited (UK) for cellulose acetate and polyvinyl chloride plastics.

Lustracrystal. Trade name of ASG Industries Inc. (USA) for a heavy sheet glass.

Lustrafil. Trade name for lustrous rayon fibers and yarns used for textile fabrics.

Lustraglass. Trade name of ASG Industries Inc. (USA) for highly transparent single- and double-strength quartz sheet glass.

Lustragold. Trade name of ASG Industries Inc. (USA) for a golden-amber sheet glass.

Lustragray. Trade name of ASG Industries Inc. (USA) for neutral-gray sheet glass.

Lustrakool. Trade name of ASG Industries Inc. (USA) for a greenish, heat-absorbing sheet glass.

Lustraline. Trade name of ASG Industries Inc. (USA) for a lustrous sheet glass.

Lustralite. (1) Trademark of Reichhold Chemicals, Inc. (USA) for sulfonamide-formaldehyde resins used as film-formers in certain coatings.

(2) Trade name of Enequist Chemical Company Inc. (USA) for a bronze electroplate and plating process.

Lustran. Trademark of Bayer Corporation (USA) for an extensive series of acrylonitrile-butadiene-styrene (e.g., *Lustran ABS* and *Lustran Elite HH & LMG*), acrylate-styrene-acrylonitrile (e.g., *Lustran ASA*) and styrene-acrylonitrile (e.g., *Lustran SAN*) resins available in various grades including standard, injection-molding, extrusion, plating, high- and low-gloss, fire-retardant, high-heat, and high- and medium impact. Also included under this trademark are several ABS-PVC alloys. The standard grades have a density of 1.05 g/cm³ (0.038 lb/in.³), an upper service temperature of about 70-100°C (160-210°F), and usually poor fatigue, solvent and UV resistance. Used for automotive interior trim, medical equipment, chemical equipment, enclosures for business equipment, business machines, household appliances, highway safety devices, lawn and garden equipment, refrigerator linings, electrical equipment, and toys. *Lustran Sparkle* is a transparent acrylonitrile-butadiene-styrene resin.

Lustrasol. Trade name of Reichhold Chemicals, Inc. (USA) for fast-drying, acrylic-modified alkyd resins with good color retention used for coating applications.

Lustratherm. Trade name of ASG Industries Inc. (USA) for a double glazing unit consisting of two sheets of windows glass separated by a glass spacer strip around the edges.

Lustrawhite. Trade name of ASG Industries Inc. (USA) for highly transparent glass with minimum distortion and true color transmission used for picture frames.

Lustre. (1) Trade name of Vidrieria Argentina SA (Argentina) for a patterned glass with diamond design.

(2) Trade name of W. Canning Materials Limited (USA) for *tripoli* buffing compositions used in metal finishing.

Lustre-Die. Trade name Bethlehem Steel Corporation (USA) for a die steel containing 0.5% carbon, 1% manganese, 0.3% silicon, 0.25% molybdenum, 1.1% chromium, and the balance iron. It has good weldability, and is used for plastic mold dies.

Lustreguard. Trademark of Iroquois Chemicals Corporation (Canada) for polyester, urethane and specialty coatings used for furniture and wallboard.

Lustrelac. Trademark of Iroquois Chemical Corporation (Canada) for lacquers used for furniture, wallboard and other interior wood surfaces.

Lustrelle. Trademark of Stewart Group Limited (USA) for high-luster metal-reinforced yarns.

Lustrex. Trademark of Monsanto Chemical Company (USA) for polystyrenes, polyvinyls and styrene-butadiene copolymers used as molding compounds, and for calendering and extrusion applications.

Lustropak. Trade name of Bayer Corporation (USA) for a series of acrylonitrile-butadiene-styrene resins.

Lustron. Trade name of Monsanto Chemical Company (USA) for polystyrene resins with good electrical, heat and strain resistance, good hardness, stability and moldability, good tensile and impact strength, poor resistant to ultraviolet radiation, high tendency to load cracking, and a maximum service temperature of 50-100°C (120-210°F). Used for automotive and appliance parts, piping, knobs, dials, packaging, battery boxes, high-frequency insulation, and dinnerware.

Lustrone. Trademark of Rohm & Haas Company (USA) for a dye-tinted, plasticized nitrocellulose lacquer used to produce transparent color effects on leather goods.

Lustropak. Trademark of Bayer AG (Germany) for extrusion-grade acrylonitrile-butadiene-styrenes.

Lus-Trus. Trade name of Lus-Trus Corporation (USA) for a series of nylon, olefin, polyester and saran (polyvinylidene chloride) fibers used for textile fabrics including carpets and upholstery, and clothing.

lusungite. A dark brown mineral of the alunite group composed of strontium lead iron hydroxide phosphate monohydrate, $(Sr,Pb)Fe_3(PO_4)_2(OH)_5 \cdot H_2O$. Crystal system, rhombohedral (hexagonal). Density, 4.06 g/cm^3; refractive index, 1.81. Occurrence: Central Africa.

lute. (1) A clay, cement or other adhesive substance composed of oxides and silicas, and used to pack joints or applied as a coating over a porous surface to render it impervious to the entry of fluids.

    (2) A dental cement or resin cement used in the cementation of restorations, such as bridges, crowns, inlays, onlays, etc. Also known as *luting cement*.

lutecine. A nickel silver containing 71-73% copper, 14-18% nickel, 1.5-2% cobalt, 2% tin, and the balance zinc. Used for jewelry.

Lute-It. Trademark of Jeneric/Pentron Inc. (USA) for a multipurpose, dual-cured, radiopaque, fluoride-releasing, polymer-based dental resin filled with 65 wt% barium borofluorosilicate. Supplied in several shades, it is used for cementing all-ceramic and all-composite indirect restorations, such as crowns, inlays, onlays and veneers, and for intraoral porcelain repairs.

lutetia. See lutetium oxide.

lutetium. A soft, ductile, silver-white metallic element of the lanthanide series (rare-earth group) of the Periodic Table. It is commercially available in the form of ingots, lumps, foil, powder, turnings, sheet, wire and sponge. It occurs in nature in the minerals *monazite, gadolinite, polycrase* and *xenotime*. Crystal system, hexagonal. Crystal structure, hexagonal close-packed. Density, 9.842 g/cm^3; melting point, 1652°C (3006°F); boiling point, 3395°C (6143°F); atomic number, 71; atomic weight, 174.967; trivalent. Used in nuclear technology, refractory metals, ceramics and ferrite bubble devices, and in electronics and materials research. Symbol: Lu.

lutetium boride. Any of the following compounds of lutetium and boron used in ceramics and materials research: (i) *Lutetium tetraboride*. Tetragonal crystals. Density, 7.52 g/cm^3; melting point, 2600°C (4712°F). Formula: LuB_4; and (ii) *Lutetium hexaboride*. Density, 5.74 g/cm^3. Formula: LuB_6.

lutetium bromide. White, hygroscopic crystals with a melting point of 1025°C (1877°F), used in chemistry and materials research. Formula: $LuBr_3$.

lutetium carbide. Any of the following compounds of lutetium and carbon used in ceramics and materials research: (i) *Trilutetium carbide*. Density, 10.54 g/cm^3. Formula: Lu_3C; and (ii) *Lutetium dicarbide*. Density, 8.73 g/cm^3. Formula: LuC_2.

lutetium chloride. White, hygroscopic crystals or powder (99.9+% pure). Density, 3.98 g/cm^3; melting point, 905°C (1661°F). A hexahydrate (lutetium chloride hexahydrate, $LuCl \cdot 6H_2O$) in white crystalline form (99.9%) is also available. Used in the manufacture of lutetium salts, and in materials research. Formula: $LuCl_3$.

lutetium dicarbide. See lutetium carbide (ii).

lutetium disilicide. See lutetium silicide (i).

lutetium fluoride. White crystals or powder. Crystal system, orthorhombic; Density, 8.3 g/cm^3; melting point, 1182°C (2160°F); boiling point, 2200°C (3990°F). Used in the manufacture of lutetium salts, and in materials research. Formula: LuF_3.

lutetium hexaboride. See lutetium boride (ii).

lutetium iodide. Brown, hygroscopic crystals. Crystal system, hexagonal. Density, 5.6 g/cm^3; melting point, 1050°C (1922°F). Used in chemistry and materials research. Formula: LuI_3.

lutetium nitride. A compound of lutetium and nitrogen that is available as cubic crystals with a density of 11.59, and used in ceramics and materials research. Formula: LuN.

lutetium oxide. White crystals or powder (99.9+%) usually obtained from *monazite* sand. Crystal system, cubic. Density, 9.42 g/cm^3; melting point, 2490°C (4514°F). Used in ceramics and materials research. Formula: Lu_2O_3. Also known as *lutetia*.

lutetium pentasilicide. See lutetium silicide (ii).

lutetium silicide. Any of the following compounds of lutetium and silicon used in ceramics, semiconductors and refractories: (i) *Lutetium disilicide*. Formula: LuB_2; and (ii) *Lutetium pentasilicide*. Formula: Lu_3Si_5.

lutetium sulfide. A compound of lutetium and sulfur available as light tan crystals. Crystal system, rhombohedral. Crystal structure, corundum. Density, 6.26 g/cm^3; melting point, 1750°C (3182°F); band gap, 3.18 eV. Used in ceramics, electronics and materials research. Formula: Lu_2S_3.

lutetium tetraboride. See lutetium boride (i).

lutetium telluride. A compound of lutetium and tellurium available as orthorhombic crystals with a density of 7.8 g/cm^3, and used in ceramics, electronics and materials research. Formula: Lu_2Te_3.

luting cement. See lute (2)

Lutrel. Trademark of LG Chemicals (USA) for a series polybutylene terephthalates available in glass-fiber-reinforced, flame-retardant, impact-modified and UV-stabilized grades. Used in the manufacture of molding compounds.

Lutofan. Trademark of BASF AG (Germany) for a series of polyvinyl chlorides.

Lutonal. Trademark of BASF AG (Germany) for a series of polyvinyl ether resins.

Lutrigen. Trademark of BASF AG (Germany) for a series of post-chlorinated polyvinyl chlorides.

Luvican. Trademark of BASF AG (Germany) for a series of polyacrylic plastics and compounds.

Luvitherm. Trademark of BASF AG (Germany) for a polyvinyl chloride film.

Luwipal. Trademark of BASF AG (Germany) for a series of melamine-formaldehyde plastics.

LuxaCore. Trademark of DMG Hamburg (Germany) for a dual- or self-cure composite with very high compressive strength and outstanding flow properties, used in restorative dentistry for core build-ups.

Lux-a-fill. Trade name for a light-cure dental hybrid composite.

Luxal. Trade name of Vereinigte Leichtmetallwerke GmbH (Germany) for an aluminum alloy containing 2-4% magnesium, up to 0.4% manganese, and up to 0.3% chromium. It has good resistance to seawater corrosion, and is used for fuel lines, marine parts, aircraft tanks and fittings.

Luxalloy. Trade name of Degussa Dental (USA) for a silver alloy supplied in the form of castings or sheets and used in the manufacture of dental amalgams.

Luxamatic. Trade name of The Lea Manufacturing Company (USA) for a greaseless compound used in liquid satin finishing operations.

Luxat. Trade name of DMG Hamburg (Germany) for a polyacid-modified dental composite (*compomer*).

Luxatemp. Trademark of DMG Hamburg (Germany) for a highly biocompatible, self-cure bis-acrylic composite with high bending strength and abrasion resistance, used in dentistry for temporary restorations, e.g., bridges and crowns.

Luxene. (1) Trade name Luxene Inc. (USA) for vinyl acrylate copolymers and phenol-formaldehyde resins formerly used for denture bases.

(2) Trademark of Solution Fibers, Inc. (USA) for polypropylene fibers and filament yarns.

Luxfer. Polish trade name for glass blocks.

Luxlen. Trademark of Yu-Ho Fiber Industrial Corporation (Taiwan) for polyester filaments and fibers.

Luxlite. Trade name C-E Glass (USA) for a glass with a simple, cathedral-type pattern of indefinite design.

Luxolite. (1) Trade name of Safetee Glass Company Inc. (USA) for a white laminated glass.

(2) French trade name for phenol-formaldehyde resins and plastics.

Luxon. Trade name of GC Dental (Japan) for a heat-cure acrylic resin used for dentures.

Lux-Opac. Trademark of Sico Inc. (Canada) for interior high-gloss enamel paints.

Luxor. Trade name of NV Durobor (Belgium) for a solid glass lens with one plain or reeded side and a diamond pattern on the other side.

Luxsico. Trademark of Sico Inc. (Canada) for air-drying enamel paints.

Luzerne. Trade name of Luzerne Rubber (USA) for hard rubber.

luzonite. A deep pinkish brown mineral of the chalcopyrite group composed of copper arsenic sulfide, Cu_3AsS_4. It can also be made synthetically. Crystal system, tetragonal. Density, 4.38 g/cm³. Occurrence: Philippines, Peru.

LWC paper. See lightweight-coated paper.

Lycra. Trademark of E.I. DuPont de Nemours & Company (USA) for a white, flexible, abrasion-resistant, self-extinguishing *spandex* (polyurethane) fiber usually supplied in the form of continuous monofilaments. It has a density of 1.2 g/cm³ (0.04 lb/in.³), a melting point of 230°C (446°F), good tensile strength and recovery, good resistance to oils and washing chemicals, and low water absorption. Used for stretchable garments and other elastic products, such as surgical hose.

Lynite. Trade name of Alcoa-Aluminum Company of America for an aluminum alloy containing 10.5-11% copper and 0.5% magnesium. Used for pistons.

Lynn sand. A high-purity quartz sand.

lynux bronze. A British corrosion-resistant copper alloy containing 7.2% iron and 3.8% aluminum. Used for hardware and fittings.

Lyocell. Trade name of Lenzing Fibers Corporation (USA) for lyocell fibers.

lyocell. See lyocell fibers.

lyocell fibers. A generic name for a group of biodegradable, recyclable manufactured fibers made from wood pulp cellulose. They are available as short staple fibers, long filament fibers, and microfibers in fibrillating or non-fibrillating types. *Lyocell fibers* are lustrous, breathable, absorbent and durable, have high wet strength and a soft drape, blend well with other fibers including cotton, linen, nylon, polyester, rayon, silk, spandex and wool, and take a wide range of finishes and dyes. They are often sold under trademarks and trade names, such as *Tencel* or *Lenzing*, and used especially for apparel including womens fashion garments, mens shirts, denim, chino and chambray casual wear, and for houshold textiles, such as bath towels, sheets and pillowcases. Industrial uses include conveyor belts, specialty papers, carbon shields, abrasive backings, printers blankets and medical dressings. Also known as *lyocell*. See also cellulosic fibers; fibrillating fibers; non-fibrillating fibers; microfibers.

Lyon's gold. A British alloy, similar to *tombac*, composed of 72-73% copper and 27-28% zinc and used for brazing tubes and cartridge cases.

Lyonore. Trade name of Lyon, Conklin & Company (USA) for an open-hearth copper steel containing 0.2% carbon, 0.2% copper, 0.5% chromium, 0.8% nickel, and the balance iron. Used for structural parts.

lyophilized bacteriorhodopsin. Bacteriorhodopsin obtained from the purple membranes of the bacteria *Halobacterium halobium* and *H. salinarum*, and supplied as a high-purity lyophilized (freeze-dried) powder for use in biochemistry, bioengineering, electronics, and information recording and processing. See also bacteriorhodopsin.

lyotropic liquid crystal. A liquid crystal in which the phase transition is due to the presence of a specific solvent, and for which the appearance of a mesophase depends upon the concentration of this solvent and the temperature. In such a liquid crystal the rod-like molecules (or mesogens) are actually amphiphilic surfactants consisting of polar, hydrophilic heads and nonpolar, hydrophobic tails attracted to hydrocarbons. With high solvent concentrations these molecules are arranged such that the heads are in contact with the polar solvent and the tails with the nonpolar solvent. With increasing solvent concentration the molecules self-arrange into hollow spheres, disk or rods collectively referred to as "micelles" which on further increase of solvent concentration form structures similar to cubic or hexagonal crystal lattices. Finally, at very high concentrations, they form bilayers, i.e., sheet-like molecular double layers arranged with their polar heads directed inward and toward each other. Common examples of simple lyotropic liquid crystals are mixtures of a soap or detergent with water, mixtures of myelin and water, and sodium laurate mixtures. *Lyotropic liquid crystals* are suitable for various applications including drug delivery, biological phospholipid membranes, and stabilization of hydrocarbon foams. Abbreviation: LLC. See also liquid crystal; thermotropic liquid crystal.

lyotropic liquid crystal polymer. An early thermoset variant of a *liquid crystal polymer* (aromatic polyamide) composed of two or more components and processed from solution. Abbreviation: LLCP.

Lytex. Trademark of Quantum Composites, Inc. (USA) for sheet molding compounds (SMC) consisting of a composite of chopped glass or carbon fibers in an epoxy resin matrix.

Lytherm. Trade name of Lydall Technical Papers (USA) for a high-purity ceramic paper made from a blend of high-alumina ceramic fibers with organic and inorganic binders. It has good smoothness and uniform thickness, good strength and handling characteristics prior to and after firing, and is used as parting medium between ceramic thick-film circuit substrates and saggers.

M

Maas-Glas. Trade name of Machinale Glasfabriek "De Maas" NV (Netherlands) for sheet glass.

MAC+. Trade name of MacSteel Bar Group, Quanex Corporation (USA) for decarb-free, bright-finished cold-finished steel bars.

macadam. A road-surfacing material composed of several layers of small, uniformly graded crushed stone. Each layer is rolled until solid and smooth with or without the addition of water, sand, screenings or cement. The crushed stone layer may receive a coat of bitumen, asphalt or tar. It is named after the Scottish civil engineer John L. McAdam.

Macalloy. (1) Trade name of Teledyne Vasco (USA) for a rolled steel containing 0.35-0.65% carbon, 1.25% nickel, 0.75% chromium, 0.2% molybdenum, and the balance iron. Used for machine parts, such as gears, and shafts.

(2) Trade name of McCalls Special Products (USA) for a steel containing 0.6% carbon, 0.75% chromium, and the balance iron. Used for steel bars for prestressed concrete.

macaroni yarn. See hollow-filament yarn.

Macbeth. Trade name of Corning Glass Works (USA) for gage glasses.

Macclesfield silk. A rather crisp fabric with a crepey texture, either striped or with small designs, produced from twisted spun silk yarn, and used especially for shirts and neckties.

Macco. Trade name of P.F. McDonald & Company (USA) for an extensive series of tool, die and high-speed steels.

Maccomax. Trade name of P.F. McDonald & Company (USA) for a high-speed steel containing 0.7% carbon, 4% chromium, 18% tungsten, 1% vanadium, and the balance iron. Used for tools and cutters.

macdonaldite. A colorless mineral composed of barium calcium silicate decahydrate, $BaCa_4Si_{16}O_{36}(OH)_2 \cdot 10H_2O$. Crystal system, orthorhombic. Density, 2.27 g/cm^3; refractive index, 1.524. Occurrence: USA (California).

macedonite. A yellow-brown mineral of the perovskite group composed of lead titanate, $PbTiO_3$. It can also be made synthetically from lead monoxide and titanium dioxide under prescribed conditions. Crystal system, tetragonal. Density, 7.82 g/cm^3. Occurrence: UK.

macfallite. A reddish brown to maroon white mineral composed of calcium aluminum manganese silicate hydroxide, $Ca_2(Mn,Al)_3(SiO_4)(Si_2O_7)(OH)_3$. Crystal system, monoclinic. Density, 3.43 g/cm^3; refractive index, 1.795. Occurrence: USA (Michigan).

Macgold. Trademark of MacSteel Bar Group, Quanex Corporation (USA) for precision hot-rolled steel bars supplied as rounds 25-127 mm (1-5 in.) in diameter and 4.6-12.2 m (15-40 ft.) in length. Used for automotive engine parts, and machine elements.

Mach-2. Trademark of Parkell Inc. (USA) for a fast-setting die silicone material used for dental applications.

machatschkiite. A colorless mineral composed of calcium arsenate sulfate hydrate, $Ca_6(AsO_4)_2[As_2(O_3OH)_2(SO_4)] \cdot 15H_2O$. Crystal system, rhombohedral (hexagonal). Density, 2.50 g/cm^3; refractive index, 1.593. Occurrence: Germany.

machinable carbide. A metal-matrix composite consisting of very hard titanium carbide (TiC) particles dispersed in an alloy steel matrix. It possesses high toughness, high abrasion and wear resistance, high hardness, good corrosion resistance, good machinability, good thermal-shock resistance, and a low coefficient of friction. Used for die plates, pelletizer knives, and drawing rings for gas cylinders. Abbreviation: MC.

machinable ceramics. Ceramics, such as aluminum oxide, aluminum silicate, glass, etc., usually supplied in sheets, rods and bars that are readily machinable into precision parts with conventional equipment. They have excellent thermal-shock resistance, excellent high-temperature properties, and are inert to oxidizing and reducing atmospheres.

machinable glass-ceramics. Fine-grained, crystalline ceramic materials containing about 46% silicon dioxide (SiO_2), 16% aluminum oxide (Al_2O_3), 17% magnesium oxide (MgO), 10% potassium oxide (K_2O) and 7% boron oxide (B_2O_3). Commercially available in the form of sheets, rods and bars, they have density of 2.5 g/cm^3 (0.09 $lb/in.^3$), low thermal expansivity, an upper continuous-use temperature of 800-1000°C (1470-1830°F), a Vickers hardness of 400 kgf/mm^2, relatively good machinability, high compressive strength and tensile moduli, good dielectric properties, poor resistance to concentrated acids, and fair resistance to dilute acids and alkalies. Used for molded mechanical and electrical parts, missile cones, cookware and dinnerware, radomes, high-temperature bearings, and telescope mirrors. See also glass-ceramics.

Machine Bronze. Trade name of Lumen Bearing Company (USA) for a wear-resisting bearing bronze containing 50% copper, 25% tin and 25% nickel.

machine-coated paper. Printing paper, with or without wood content, coated either inside or outside the paper machine. It has a dull to lustrous surface and the coating weight typically ranges between 5 and 20 g/m^2 (0.016 and 0.065 oz/ft^2) per side.

machine-finished paper. Paper smoothened on both sides by means of a calender located in the end section of the paper machine.

machine-glazed paper. Paper that during the drying cycle in the paper machine has been smoothened on one side using a so-called "MG cylinder."

Machinery. Trade name of Atlas Specialty Steels (Canada) for a series of plain-carbon machinery steels. *Machinery No. 30* contains 0.3% carbon, 0.8% manganese, 0.2% silicon, 0.05% sulfur, 0.04% phosphorus, and the balance iron. It has relatively high strength and moderate machinability, and is used for camshafts, shafts subject to vibration, chain and drag links, rivets, swivels, and thrust washers. *Machinery No. 40* contains 0.4% carbon, 0.8% manganese, 0.2% silicon, 0.05% sulfur, 0.04% phosphorus, and the balance iron. It has relatively high strength and moderate machinability, and is used for axles, connecting rods, crankshafts, diamond drill parts, sleeves, and constructional applications.

machinery brass. A corrosion-resistant brass composed of 83% copper, 16% zinc and 1% tin, and used for machinery parts.

machinery steel. A general term referring to any low-carbon steel with less than 0.3% carbon. Abbreviation: MS. Also known as *machine steel*.

machine steel. See machinery steel.

Mach's alloy. A corrosion-resistant aluminum alloy containing 2-10% magnesium, and used for light-alloy parts.

Mach's metal. A muntz metal-type high-strength forging brass containing 57% copper and 43% zinc. Used for hardware, fasteners, and as a brazing metal.

Mach's speculum. An alloy composed of 69% aluminum and 31% magnesium, and used for light-alloy parts.

Mac-It. Trade name of Strong, Carlisle & Hammond Company (USA) for an oil-hardening steel containing 0.4% carbon, 0.8% chromium, 1.5% nickel, and the balance iron. Used for threaded fasteners, screws, bolts, etc.

Macite. Trade name of Manufacturers Chemical Corporation (USA) for cellulose acetate plastics.

mackayite. An olive-green mineral composed of iron tellurium oxide hydroxide, $FeTe_2O_5(OH)$. Crystal system, tetragonal. Density, 4.86 g/cm^3; refractive index, 2.19. Occurrence: Mexico, USA (Nevada).

Mackenzie metal. A lead-base alloy containing 68-70% lead, 16-17% antimony, and 13-16% tin. It has good antifriction properties, and is used for bearings.

Mackenzie's amalgam. A liquid composite *amalgam* made by grinding together equal parts of a solid bismuth amalgam and a solid lead amalgam in a mortar at ambient temperatures.

mackinaw. A heavily milled woolen fabric with a thick nap on both sides, usually woven with large checks. Used for thick blankets, coats, lumber jackets, etc.

mackinawite. (1) A white gray mineral tinged with bronze. It is composed of iron sulfide, $(Fe,Ni)_9S_8$, and may contain some nickel. Crystal system, tetragonal. Density, 4.29 g/cm^3. Occurrence: Finland, USA (Massachusetts, Washington).

(2) A grayish mineral composed of iron sulfide, $FeS_{0.9}$. It is usually a synthetic product, but also occurs naturally in river sediments, e.g., in the vicinity of Boston, USA. Crystal system, tetragonal. Density, 4.1 g/cm^3.

Mack's cement. A quick-setting cement composed essentially of calcium sulfate hemihydrate ($CaSO_4 \cdot 0.5H_2O$) with small additions of calcined sodium sulfate (Na_2SO_4), and potassium sulfate (K_2SO_4). It has good adhesion and hardness, and is used for plastering walls and floors.

Macloy. Trade name of Firth-Vickers Stainless Steels Limited (UK) for a series of austenitic stainless steels containing 0.25-0.5% carbon, 0.3-1.8% silicon, 0.6-1.25% manganese, 11-17% chromium, 36-37% nickel, and the balance iron. Used for heat- and corrosion-resistant parts.

Maco. Trade name of Prolamine Products, Inc. (USA) for casein plastics.

Macoloy. Trade name of McCauley Alloy Sales Company (USA) for a series of brazing filler metals composed of copper and phosphorus. They have a melting range of 621-1038°C (1150-1980°F), and are used for brazing copper and copper alloys.

Macor. Trademark of Astro-Met Inc. (USA) for a machinable glass-ceramic containing 46% silicon dioxide (SiO_2), 16% aluminum oxide (Al_2O_3), 17% magnesium oxide (MgO), 10% potassium oxide (K_2O), and 7% boron oxide (B_2O_3). It requires no postfiring, and can be machined with conventional metalworking tools. Supplied in bar, rod and sheet form, it has a density of 2.52 g/cm^3 (0.091 $lb/in.^3$), a hardness of 400 Vickers, moderate resistance to acids and alkalies, high dielectric strength, good electrical and thermal insulating properties, and a maximum service temperature of 1000°C (1830°F). Used for electrical thermal and vacuum applications, as inlay ceramic for dentistry, and for components of the NASA Space Shuttle.

Macoustic. Trade name of National Gypsum Company (USA) for a lightweight acoustical gypsum plaster.

MacPlus. Trademark of MacSteel Bar Group, Quanex Corporation (USA) for premium-quality bright cold-finished steel bars providing tight dimensional tolerances and concentricity and defect-free surfaces. Supplied as rounds 22-127 mm (0.875 to 5 in.) in diameter and 3.0-9.8 m (10-32 ft.) in length. Used for automotive engine parts, and machine elements.

macquartite. A lemon-yellow to dark orange mineral composed of copper lead chromium oxide silicate hydroxide dihydrate, $Pb_3Cu(CrO_4)SiO_3(OH)_4 \cdot 2H_2O$. Crystal system, monoclinic. Density, 5.49 g/cm^3; refractive index, 2.31. Occurrence: USA (Arizona).

macrame. A knotted lace fabric from Arabia and Italy, used especially for scarves and shawls.

Macroballoons. Trade name for lightweight, spherical high-strength glass-fiber reinforced spheres which provide good strength, durability and buoyancy properties. They have a low density 0.7-1.0 g/cm^3 (0.02-0.04 $lb/in.^3$), a high strength-to-weight ratio, and good thermal and acoustical insulating properties. Used as a bulk filler for resin matrices, and as a lightweight core material.

Macro Brite. Trade name of MacDermid Inc. (USA) for chromate conversion coatings.

macrocomposites. See macroscopic composites.

macrocrystalline material. A solid substance, e.g., a metal, alloy or ceramic, composed of large crystals that can be discerned by the unaided eye.

macrocycle. An organic compound with a ring in its structure usually containing 15 or more atoms.

macro-defect-free cement. A composite material composed of a cement (e.g., calcium aluminate) and a polymer (e.g., polyvinyl alcohol) prepared as a paste and having relatively small voids (or defects). It consists of a uniform dispersion of fine particle additions, and is processed by first mixing the cement powder, water-soluble polymer and a small amount of water under high shear, followed by calendering, low-temperature pressure curing, and drying. Used as substrates for electronic packaging, and for structural and acoustic-damping materials. Abbreviation: MDF cement.

Macrofil. Trade name of Kennametal Inc. (USA) for a series of infiltration alloys containing 45-65% copper, 20-45% zinc, 0-15% nickel and 0-1% tin. Used in the manufacture of surface-set diamond tools.

Macrolite. (1) Trademark of 3M Company (USA) for inert, lightweight, gray ceramic spheres that contain independent closed air cells surrounded by a tough outer shell impermeable to fluids. Supplied in a size range from 300 μm to 12.7 mm (0.012 to 0.500 in), they have a low density, good thermal stability to 1093°C (2000°F), and are used for controlled-size aggregates for precast concrete, insulating bricks and blocks, oil-well cements, lightweight roofing, high-volume fillers in paints, sealants, rubber and asphalt, in liquid filtration systems, abrasives, catalyst support structures, and for vacuum molds and lightweight automotive structures.

(2) Trademark of Silvaplex, Division of Forestply Limited (Canada) for high-density particleboard.

Macroloy. Trade name of MacDermid Inc. (USA) for chromate conversion coatings for zinc alloys.

Macromal. Trade name of Rheinische Rohrenwerke AG (Germany) for a series of nonmagnetic, austenitic steels containing 0.2% carbon, 12-18% manganese, and the balance iron. Used for transformer shields, deck superstructures, and compass housings.

Macromate. Trade name of MacDermid Inc. (USA) for conversion coatings.

Macromer. Trade name for a high-impact acrylic resin for dentures.

Macroplast. Trademark of Henkel KGaA (Germany) for polyvinyl chloride resins, adhesives and pastes.

macroporous gold. A porous nanostructured material produced by first depositing densely packed layers of monodisperse, negatively charged polystyrene latex microspheres (300-1000 nm or 11.8-39.4 μin. in diameter), diluted with deionized water, onto a polycarbonate membrane, and filling the cavities of the latex crystal with colloidal gold particles (15-25 nm or 0.6-1.0 μin. in size). The latex/gold layer is then dried forming a latex-gold composite that is treated with a mixture of concentrated sulfuric acid and an inorganic oxidizer yielding meso/macro porous gold. The latter is then converted to macroporous gold by a thermal treatment in which the temperature is gradually increased to 300°C (570°F). *Macroporous gold* has the typical 3-D wire mesh structure of photonic crystals, and forbidden energy bands in the gigahertz region which make it suitable for optoelectronic devices.

macroporous material. An inorganic or organic material (e.g., a ceramic, metal, polymer, composite, etc.) with a pore size of more than 50 nm (2 μin.). Used for catalysis, separation and electronic applications (e.g., sensors, batteries, nonlinear optics, etc.), and in the manufacture of photonic crystals.

macroporous silica. Silicon with a pore size of 50 nm (2 μin.) or more used in the fabrication of two- and three-dimensional photonic crystals. See also 2D photonic crystals; 3D photonic crystals.

macroscopic composites. A class of composite materials with a dispersed phase (reinforcing element) that can be observed with the unaided eye. Examples are concrete, mortar, structural laminates (plywood, sandwich panels, etc.), and steel-belted tires. Also known as *macrocomposites*.

Macrosil. Trade name of VDM Nickel-Technologie AG (Germany) for a corrosion-resistant steel containing 0-0.15% carbon, 11-13% chromium, 1.5-2.5% nickel, 0.3-0.6% molybdenum, 17-19% manganese, and the balance iron. Used for chemical equipment.

Macrosorb. Trademark of Sterling Drug, Inc. (USA) for a sarcosine-derived *polyacrylamide* within a macroporous *kieselguhr* matrix support, used in peptide synthesis.

Macrynal. Trademark of Casella AG (Germany) for a series of acrylic polymers.

Macstone. Trade name for lightweight-concrete building blocks having a facing of sandstone.

Madam Butterfly. Trade name for rayon fibers and yarns used for textile fabrics.

madapolam. A soft-finished cotton fabric in a plain weave, usually bleached or dyed, and used especially for women's fashion.

Maderón. Trademark of Maderón (Spain) for composite materials composed of almond shells and other lignocellulosic materials. Used for seating and lamps.

madocite. A gray black-mineral composed of lead antimony sulfide, $Pb_{17}Sb_{16}S_{41}$. Crystal system, orthorhombic. Density, 5.98 g/cm^3. Occurrence: Canada (Ontario).

Madras. Trade name of Vitreal Specchi (Italy) for acid-etched glass used for interior decoration and building construction applications.

madras. A medium-to-light-weight cotton fabric in a close, plain weave, and in white or with brightly colored woven checks, stripes, or plaids. It may be machine- or hand-woven, and is used for dresses, shirts, pyjamas, neckwear, hatbands, etc. Originally from Madras, India. Also known as *Madras cotton*.

Madras cotton. See madras.

Madras hemp. See sunn hemp.

madrona. See Pacific madrone.

madrone. See Pacific madrone.

Madune. Trademark of Casella AG (Germany) for melamine-formaldehyde plastics and compounds.

Maerker Irrubigo. Trade name of Schmidt & Clemens Edelstahlwerke (Germany) for an extensive series of chromium, chromium-molybdenum, chromium-nickel, chromium-nickel-molybdenum stainless steels with good to excellent corrosion resistance used for various structural and nonstructural applications.

Maestro. Trade name of J.F. Jelenko & Company (USA) for a white-colored, microfine, extra-hard dental casting alloy (ADA type IV) containing 50% silver, 30% palladium and 3% gold. It has a density of 11.3 g/cm^3 (0.41 $lb/in.^3$), a melting range of 930-1020°C (1706-1868°F), a hardness of 180-270 Vickers and good strength and tarnish resistance. It can be cast to titanium, and is used for hard inlays, thin crowns, implant frames, fixed bridgework, and partial dentures.

Maferite. Trade name of Usines de A. Manoir (France) for a series of corrosion- and/or heat-resistant low-, medium- and high-carbon steels containing 6-30% chromium, 0-4% nickel, sometimes varying amounts of nickel, molybdenum, aluminum and silicon, and the balance iron. Used for furnace and oil-refining equipment.

magadiite. A mineral composed of sodium silicate hydroxide tetrahydrate, $NaSi_7O_{13}(OH)_3 \cdot 4H_2O$. Crystal system, monoclinic. Refractive index, 1.48. Occurrence: USA (California).

Magalloy. Trade name of REM Chemicals Inc. (USA) for mass-finishing compounds used for magnetic stainless steels.

Magaloy. Trade name of Aetna Standard Engineering Company (USA) for a ductile cast iron containing 3.2% carbon, 2% silicon, 0.7% manganese, 0.5% magnesium, and the balance iron. Used for rolls.

Magaluma. Trade name of Aluminium Belge SA (France) for a heat-treatable, corrosion-resistant aluminum alloy containing 3.4% magnesium, and 0.15% manganese. Used for light-alloy parts.

Magan-Nickel. Trade name of Vereinigte Deutsche Nickel-Werke AG (Germany) for a corrosion- and heat-resistant nickel alloy containing 0-4% aluminum, 1.5% manganese and 0-2% copper. Used for spark-plug electrodes.

magbasite. A colorless to rose-violet mineral composed of potassium barium magnesium aluminum fluoride silicate, $KBa(Mg,Fe)_6(Al,Si)Si_6O_{20}F_2$. Density, 3.41 g/cm^3; refractive index, 1.609. Occurrence: Asia.

MAG-CAL. Trademark of Timminco Limited (Canada) for a magnesium-calcium alloy used in lead refining.

Magdal. Trade name of Aluminium Français (France) for a wrought aluminum alloy containing 1.25% manganese and 1% magnesium. Used for light-alloy parts.

Magdolo. Trade name for a refractory composed of 55-62% magnesia (MgO), 30-35% calcia (CaO), and small amounts of silica (SiO_2), alumina (Al_2O_3) and ferric oxide (Fe_2O_3). It is obtained as an intermediate in the production of magnesia from seawater or brine, and used for the basic linings of LD steel converters.

Magenta. Trademark of Ivoclar Vivadent AG (Liechtenstein) for an extra-hard dental casting alloy (ADA type IV) containing 50.0% gold, 21.0% silver, 19.5% copper, 6.49% palladium, 2% indium, 1% zinc, and less than 1.0% iridium. It has a yellow color, a density of 13.0 g/cm^3 (0.47 lb/in.^{3}), a melting range of 815-875°C (1500-1605°F), an as-cast hardness of 285 Vickers, low elongation and good biocompatibility and resistance to oral conditions. Used for crowns, posts, onlays, and bridges.

Magex. Trade name of Pyramid Steel Company (USA) for a series of steels used for machine parts, such as cams, chucks, reamers, pump shafts, gears, and driveshafts.

maghemite. A strongly magnetic, brown to chocolate-brown mineral of the spinel group composed of ferric oxide, γ-Fe_2O_3. It can also be made synthetically. Crystal system, cubic. Density, 4.86-4.90 g/cm^3; hardness, 5-7 Mohs; refractive index, 2.74. Occurrence: Algeria, Japan, South Africa, USA (California). Used as an ore of iron.

Magic. Trade name of Jessop Steel Company (USA) for a shock-resisting tool steel (AISI type S5) containing 0.5% carbon, 2% silicon, 0.9% manganese, 1% molybdenum, and the balance iron. It has good deep-hardening properties, high elastic limit, good ductility and toughness, moderate wear resistance and a fine-grained microstructure. Used for cement breakers, punches, chisels, and tools.

Magicap. See Oralloy Magicap.

Magic Bond. Trademark of Devcon Corporation (USA) for a series of structural adhesives.

MagicFil. Trade name of DMG Hamburg (Germany) for a dual-cured, fluoride-releasing *compomer* used in dentistry for treating deciduous teeth.

magic sand. See *hydrophobic sand*.

Magira. Trademark of Schott DESAG AG (Germany) for an antique glass with a finely structured surface and outstanding transparency and evenness. Supplied in several colors and various sizes from 1600 × 1500 mm (63 × 59 in.) to 2050 × 1500 mm (81 × 59 in.) and thicknesses from 2.5 to 8.5 mm (1.0 to 3.4 in.), it is used for windows, doors, furniture, partitions, lamps, leaded lights, laminations, in the manufacture of laminated safety glass and insulating glass, and in the restoration of historic windows.

Magister. Trade name of Fibreglass Limited (UK) for glass-fiber reinforcing materials.

Mag-li-kote. Trade name for an artificial ground *dolomite* made from magnesium limestone and clay, and used for ingot mold coatings.

Maglite. Trade name of Whittaker, Clark & Daniels, Inc. (USA) for magnesia (MgO) supplied as a fine white powder, and used as a scorching preventive in rubber.

Mag-lith. Trade name for a lightweight, structural magnesium-lithium alloy used in space vehicles.

Maglox. Trademark of St. Gobain Industrial Ceramics (France) for an electrical-grade magnesia (MgO) powder used as mineral insulation for cables, as a filler for heating elements, and for special refractories.

Mag Metal. Trade name of J.L. Snowbar (USA) for a copper-nickel alloy used for jewelry, and as a platinum replacement.

MagnaDerm. Trade name of MarkeTech International Inc. (USA) for aluminum-free, high-purity magnesium oxide crystals for microdermabrasion and exfoliation.

Magnadize. Trademark of General Magnaplate Corporation (USA) for a synergistic coating for magnesium. In the proprietary *Magnadize* process, magnesium parts are treated in an electrolytic bath to achieve a hard coat finish with a smooth porous crystalline structure on the surface. This hardens the surface and acts as a base for impregnation with fluorocarbon (e.g., *Teflon*) or other dry lubricants. *Magnadize* coatings are available in various thicknesses ranging from 0.076 to 0.508 mm (0.0003 to 0.020 in.), they impart high hardness and corrosion resistance, and eliminate oxidation and galling.

Magnadure. Trade name of Philips Electronics North America Corporation (USA) for a series of permanent magnet materials composed of sintered barium ferrite ($BaO·6Fe_2O_3$). They have high magnetic permeability, a Curie temperature of 450°C (840°F), a high magnetic energy product, high remanent magnetization and coercive force, and a maximum service temperature of 400°C (750°F). Used for magnets in metering device and electrical and electrical equipment, and for television ring magnets.

Magnaflux steel. A high-grade structural steel for critical applications (e.g., aircraft, nuclear and chemical reactors, building construction, etc.) tested by a nondestructive technique known as the "Magnaflux" process making use of a magnetic field and magnetic particles to locate flaws (e.g., cracks).

Magnafrax. Trademark of The Carborundum Company (USA) for a series of magnesia refractory products in various forms including refractory cements used for lining industrial furnaces.

Magnaglow. Trademark of General Magnaplate Corporation (USA) for a series of fluorescent coatings with good adhesion to ferrous and nonferrous metal substrates especially iron, steel, aluminum and its alloys, copper-base alloys and titanium. They provide high visibility and good corrosion protection.

Magnagold. Trademark of General Magnaplate Corporation (USA) for extremely lightweight coatings, typically up to 3 μm (118 μin.) thick, applied by physical vapor deposition (PVD) of titanium nitride (TiN) using a reactive-plasma ion-bombardment technique in a vacuum system. This produces a hard, dense, smooth, uniform gold-colored surface on metal substrates especially stainless steels and superalloys. *Magnagold* coatings provide wear and abrasion resistance over 20 times greater than that of ordinary stainless steel. Other advantages include very high dimensional accuracy, and improvement of performance of cutting and forming tools. Used as decorative finish on jewelry, pendants and flatware, and on tools.

Magnalite. (1) Trade name of Walker M. Levett (USA) for an aluminum casting alloy containing 2.5% copper, 1.5% nickel, 0.5% zinc and 1.3% magnesium. Used for airplane construction and engine pistons.

(2) Trade name of C-E Glass (USA) for a glass with cylindrical lenses on each surface running at right angles to each other.

Magnalium. British trade name for a series of aluminum alloys containing 0-2.5% copper, 1-30% magnesium, 0-1.2% nickel, 0-3% tin, 0.2-0.6% silicon, 0-0.9% iron and 0-0.3% manganese. They have a low density, high strength, and good castability, forgeability and machinability. Used for light-alloy components, and ornamental parts.

magna metal. A magnesium alloy used for light-alloy parts.

Magnamite. Trademark of Hercules Inc. (USA) for high-strength, high-temperature unidirectional graphite-fiber-reinforced epoxy prepreg tapes used for aerospace applications.

Magnaplas. Trade name of Bay Resins Inc. (USA) for a series of injection-moldable magnetizable compounds that combine a base of thermoplastic material with predetermined loadings of magnetic materials, such as alnico, ferrites or neodymium-iron-boron alloys.

Magnaplate. Trademark of General Magnaplate Corporation (USA) for a series of synergistic coatings used to apply a harder-than-steel, dry-lubricated surface on metal components. They provide excellent protection against abrasion, corrosion, sticking and wear, and excellent mold-release properties. *Magnaplate HCR* is a coating employed to provide aluminum parts with exceptional corrosion resistance and ultrahigh surface hardness. This is accomplished by introducing a series of bimetallic particles in combination with fluorocarbons during the aluminum oxide formation. *Magnaplate HCR* can be used to increase the wear life in a broad range of corrosive and/or abrasive environments and is superior to case-hardened steels and hard-chromium electroplate. *Magnaplate HTR* is a tough, wear-resistant coating applied to steel, aluminum, copper, brass and other metals to provide a nonstick, easy-to-release surface on dies and molds. *Magnaplate HMF* is a coating used for achieving optimum wear properties on stainless steel, copper, aluminum and other alloys. It improves the original surface finish and provides a surface finish with ultrahigh hardness. Other properties include outstanding chloride resistance, low friction and superior release properties. It is used for molds exposed to operating temperatures up to 316°C (600°F). *Magnaplate SNS* is a thin, ultrahard surface coating with superior resistance to "wetting" against lead or silver solders or brazing metals and *Magnaplate TNS* is a coating with excellent anti-adhesive and release properties used on metal parts in contact with adhesives and glues.

Magne. Trade name of Uddeholm Corporation (USA) for a series of specialty steels containing 0.9% carbon, 3.25-6.25% chromium, and the balance iron. Used for permanent magnets.

Magnecon. Trademark of Dresser Industries, Inc. (USA) for basic brick cement and kiln liners.

MagneDur. Trademark of Carpenter Specialty Alloys (USA) for a strong, tough, ductile, cobalt-free semihard magnetic alloy with high magnetic remanence. Its properties can be custom-designed by processing, and it is supplied in strip form. Used for hysteresis motors, magnetic clutches, electromagnetic displays, and theft detection tags.

Magnefer. Trademark of Basic, Inc. (USA) for dead-burnt dolomite refractories used for furnace linings.

Magnel. Trade name of Capitol Castings Inc. (USA) for a cast iron containing 3% carbon, 1.5% silicon, and the balance iron. It has low hysteresis loss, high magnetic permeability, and is used for electrical and electromagnetic instruments, and magnet cores.

Magnequench. Trademark of Magnequench International, Inc. (USA) for a series of iron-neodymium-boron alloys that are available as anisotropic and isotropic powders. They have very high magnetic energy products, and are used for permanent magnets.

magnesia. See magnesium oxide.

magnesia-alumina. See magnesium aluminate.

magnesia brick. See magnesite brick.

magnesia cement. See magnesium oxychloride cement.

magnesia ceramics. A group of nonsilicate ceramics composed primarily of magnesia (MgO) and made in an electric furnace.

They have excellent resistance to severe corrosion by molten steels and basic slags, a melting point of approximately 2770°C (5018°F), and an upper service temperature of 2300°C (4170°F) in oxidizing atmospheres and 1700°C (3090°F) in reducing atmospheres. Used for furnace linings, crucibles and other components and materials in the steel industry.

magnesia-chromia. See magnesium chromite.

magnesia covering. A refractory material composed of 85% hydrated light basic magnesium carbonate ($MgCO_3$) and 15% long-fibered asbestos as a binder. It has good resistance to temperatures up to about 300°C (570°F), and is used as thermal insulation.

magnesia glass. A term referring to any glass containing small additions (typically 3-4%) of magnesium oxide (MgO), and including ordinary soda-lime silica glass for windows, containers and electric light bulbs, and special glass fibers with good chemical resistance.

magnesian limestone. Any limestone containing 65-95% calcium carbonate ($CaCO_3$) and 5-35% magnesium carbonate ($MgCO_3$).

magnesian marble. See dolomitic marble (2).

magnesian matte. A matte glaze whose magnesium oxide content is above normal.

magnesia refractories. See magnesite refractories.

magnesia spinel. See magnesium aluminate.

magnesia-stabilized zirconia. An engineering ceramic composed 92-97.2% zirconia (ZrO_2) and stabilized with 2.8-8% magnesia (MgO). Available in the form of sheets, rods, powder and shapes, it has a density of 5.74 g/cm³ (0.207 lb/in.³), low thermal expansivity, an upper continuous-use temperature of 2200°C (3990°F), a hardness of about 1250 Vickers, high tensile strength, good resistance to dilute acids, good to poor resistance to alkalies and metals, and fair resistance to concentrated acids and halogens. Used for refractories and thermal insulation applications. Abbreviation: MSZ. See also fully stabilized zirconia; lime-stabilized zirconia; partially stabilized zirconia; yttria-stabilized zirconia.

Magnesil. Trade name of Spang Specialty Metals (USA) for a series of magnetically soft alloys in strip form containing 96.75-96.85% iron and 3.15-3.25% silicon. *Magnesil-N* is a nonoriented type while *Magnesil-O* is oriented and has pronounced magnetic properties in the rolling direction. Used for magnetic shielding applications.

magnesio-anthophyllite. A white, gray, green, or clove to yellow brown synthetic mineral of the amphibole group composed of magnesium silicate hydroxide, $Mg_7Si_8O_{22}(OH)_2$. Crystal system, orthorhombic. Density, 3.21 g/cm³; refractive index, 1.602.

magnesio-arfvedsonite. (1) A mineral of the amphibole group composed of sodium calcium magnesium iron aluminum silicate hydroxide, $(Na,Ca,K)(Mg,Fe,Mn,Ti,Al)_5(Si,Al)_8O_{22}OH,O)_2$. Crystal system, monoclinic. Density, 3.14-3.17 g/cm³; refractive index, 1.662. Occurrence: India, Germany.

(2) A green, greenish black or black mineral of the amphibole group composed of potassium sodium iron magnesium aluminum silicate fluoride hydroxide, $(Na,K)_3(Fe,Mg,Al)_5Si_8O_{22}(F,OH)_2$. Crystal system, monoclinic. Density, 3.17 g/cm³; refractive index, 1.644. Occurrence: Sweden.

magnesioaxinite. A pale blue mineral of the axinite group composed of calcium magnesium aluminum boron silicate hydroxide, $Ca_2MgAl_2BSi_4O_{15}(OH)$. Crystal system, triclinic. Density, 3.18 g/cm³; refractive index, 1.660. Occurrence: East Africa.

magnesiocarpholite. A green to gray mineral of the carpholite group composed of magnesium iron aluminum silicate hydroxide, $(Mg,Fe)Al_2Si_2O_6(OH)_4$. It can also be made synthetically. Crystal system, orthorhombic. Density, 2.98 g/cm^3; refractive index, 1.60. Occurrence: France.

magnesiochromite. A dark brownish green mineral of the spinel group composed magnesium chromium oxide, $MgCr_2O_4$. It can also be made synthetically. Crystal system, cubic. Density, 4.2-4.4 g/cm^3; melting point, 2250°C (4082°F). Used for chrome-magnesite refractories, ceramic magnets, and electronic applications. Also known as *picrochromite*. See also magnesia-chromia; spinel; normal spinel.

magnesio-cummingtonite. A light green mineral of the amphibole group composed of magnesium manganese iron silicate hydroxide, $(Mg,Mn,Fe)_7Si_8O_{22}(OH)_2$. Crystal system, monoclinic. Density, 3.13 g/cm^3; refractive index, 1.644. Occurrence: Canada (Labrador).

magnesioferrite. A red to brown mineral of the spinel group composed of magnesium iron oxide, $MgFe_2O_4$. It can also be made synthetically. Crystal system, cubic. Density, 4.49-4.52; melting point, 1750°C (3182°F); refractive index, 2.35 g/cm^3; hardness, 5.5-6.5 Mohs. Used for basic refractories and ceramic magnets.

magnesio-hornblende. A dark green mineral of the amphibole group that is composed of sodium calcium iron magnesium aluminum silicate hydroxide, $(Ca,Na)_{2.26}(Mg,Fe,Al)_{5.15}(Si,Al)_8O_{22}(OH)_2$, and may also contain sodium. Crystal system, monoclinic. Refractive index, 1.658. Occurrence: Norway, USA (California).

magnesio-riebeckite. A blue or black mineral of the amphibole group composed of sodium calcium magnesium iron silicate hydroxide, $(Na,Ca)_2(Mg,Fe)_5Si_8O_{22}(OH)_2$. It can also be made synthetically. Crystal system, monoclinic. Density, 3.12-3.19 g/cm^3. Occurrence: India, Madagascar, Scotland, Finland, Bolivia.

magnesite. A colorless or yellowish white mineral of the calcite group composed of magnesium carbonate, $MgCO_3$. It can also be made synthetically. Crystal system, rhombohedral (hexagonal). Density, 3.01-3.12 g/cm^3; melting point, decomposes at about 350°C (660°F); hardness, 3.5-4.5 Mohs; refractive index, 1.700. Occurrence: Austria, Greece, USA (California, Washington). Used as an ore of magnesium, as a source of magnesium oxide, in glazes, in basic refractories for industrial furnaces, as a pigment in paints, and for thermal insulation applications. See also magnesium carbonate.

magnesite brick. A refractory brick composed of 84-92% magnesium oxide (MgO) and 8-16% other oxides. It has an excellent resistance to severe corrosion by basic slags, good high-temperature resistance up to 2150°C (3900°F), moderate elevated-temperature strength and relatively high thermal expansivity. Used for lining open-hearth and basic electric steel furnaces, copper reverberatory furnaces, and soaking pits. Also known as *magnesia brick*.

magnesite cement. A term refering to dead-burnt magnesite that has been ground to various mesh sizes.

magnesite-chrome brick. A fired or chemically bonded refractory brick made from a mixture of dead-burnt magnesia (MgO) and refractory-grade chrome ore ($FeCr_2O_4$), in which magnesia, by weight, is the predominant constituent. Also known as *magnesite-chrome refractory*.

magnesite-chrome refractory. See magnesite-chrome brick.

magnesite-dolomite brick. A refractory brick composed of a mixture of dead-burnt magnesite ($MgCO_3$) and dead-burnt dolomite [$CaMg(CO_3)_2$], in which the magnesite content exceeds that of the dolomite. Also known as *magnesite-dolomite refractory*.

magnesite-dolomite refractory. See magnesite-dolomite brick.

magnesite grain. Magnesia (MgO) in the form of small granules made by calcining magnesite ($MgCO_3$), and used for refractories.

Magnesite H-W. Trademark of Harbison-Walker Refractories Company (USA) for a refractory brick composed of at least 90% burnt magnesia (MgO). It has excellent resistance to severe corrosion by basic slags and molten metals, and is used for metallurgical furnaces (e.g., open-hearth and electric steel furnaces, copper and nickel converters, etc.) and crucibles.

magnesite refractories. Refractory materials composed principally of dead-burnt, crystalline magnesium oxide (MgO) and available in brick or cement form. They have excellent resistance to heat and corrosion, a maximum service temperature of 2300°C (4170°F), and are used for lining furnaces and melting tanks, crucibles, cement kilns, mixer furnaces, and high-temperature process equipment. Also known as *magnesia refractories*. See also dead-burnt magnesite.

magnesium. A silvery-white, ductile, malleable metallic element of Group IIA (Group 2) of the Periodic Table (alkaline-earth group). It is commercially available in the form of powder, granules, chips (turnings), flakes, ribbons, ingots, bars, rods, wire, tubing, foil, microfoil and microleaf, sheet and plate, and single crystals. The single crystals are usually grown by the Bridgeman technique. The principal ores are *magnesite* and *dolomite*, but it is also obtained from seawater and brines. Crystal system, hexagonal. Crystal structure, hexagonal close-packed. Density, 1.74 g/cm^3; melting point, 649°C (1200°F); boiling point, 1090°C (1994°F); hardness, 30-45 Vickers; atomic number, 12; atomic weight, 24.305; divalent, trivalent. It has high electrical conductivity (41% IACS), a high strength-to-weight ratio, excellent machinability, and good hot and cold workability. Used for refractories, as a desulfurizer in steelmaking, in alloys with zirconium and thorium for aircraft construction, in the production of iron, nickel, zinc, titanium, zirconium and aluminum alloys for structural parts, for die-cast machine parts, in organo-metallics, pigments and fillers, in flash bulbs, batteries, airplanes, spacecraft, missiles and racing bikes, as an anode material for galvanic protection, and for optical mirrors, precision instruments and fireworks. Symbol: Mg.

magnesium acetate. The magnesium salt of acetic acid available as colorless, deliquescent, monoclinic crystals. Density, 1.42 g/cm^3; melting point, 323°C (613°F). Used in chemistry and materials research. Formula: $Mg(C_2H_3O_2)_2$.

magnesium alloys. A group of cast or wrought alloys containing 85-100% magnesium and varying amounts of aluminum, manganese, zinc and/or rare-earth elements. They have low densities, melting points of approximately 550°C (1020°F), high brittleness, moderate strength, good strength-to-weight ratios, low moduli of elasticity, high damping capacity, excellent machinability, excellent hot workability and castability (especially die casting), fair cold workability, good weldability, good fatigue resistance, and relatively high sensitivity to stress corrosion, but good corrosion and oxidation resistance under normal atmospheric conditions. Used for ladders, luggage, loading ramps, lawnmower parts, automotive components, lightweight construction, highly stressed extrusions, pressure-tight castings, housings for machinery, hand tools and office equipment, aircraft and spacecraft structures, missiles, textile spools, structural applications in materials handling, grain shovels, and gravity conveyors.

magnesium aluminate. A synthetic spinel composed of magnesium oxide (MgO) and aluminum oxide (Al_2O_3). Crystal system, cubic. Density, 3.6 g/cm³; melting point, 2135°C (3875°F); softening range, 1000-1100°C (1830-2010°F); low thermal expansion; hardness, approximately 1000 Vickers; dielectric constant, 7.45. Used for refractories having high resistance to corrosion by slags and glasses at elevated temperatures, for ceramic magnets, as substrates for superconductor and III-V (13-15) thin film deposition, and for electronic applications. Formula: $MgAl_2O_4$. Also known as *magnesia-alumina; magnesia spinel*. See also spinel; spinels; normal spinels; synthetic spinels.

magnesium aluminosilicate. See magnesium aluminum-silicate.

magnesium-aluminum alloys. A group of cast or wrought magnesium-base alloys containing a considerable quantity of aluminum (typically up to 10%) and small additions of manganese, zinc, zirconium, rare earths, etc. They have low densities, high hardnesses, good strength-to-weight ratios, and their brittleness increases with the aluminum content. Used for automotive components, lightweight construction, highly stressed extrusions, pressure-tight castings, housings for machinery, hand tools and office equipment, aircraft and spacecraft structures, missiles, machinery, and structural applications.

magnesium-aluminum garnet. See spessartite.

magnesium aluminum oxynitride ceramics. A relatively recent class of advanced ceramics based on magnesium aluminum oxynitride (MgAlON) and usually synthesized by powder technology, e.g. hot-press sintering. They have a density of about 3.5-3.6 g/cm³ (0.13 lb/in.³), excellent chemical, mechanical and optical properties including high strength, fracture toughness, durability and good oxidation resistance up to about 1500°C (2730°F). Used for structural, electronic and optical applications. Abbreviation: MgAlON ceramics.

magnesium-aluminum silicate. Any of the following compounds of magnesium oxide (MgO), aluminum oxide (Al_2O_3) and silicon dioxide (SiO_2) used as ceramic binders, flatting agents and extenders in paints, and for glass-ceramics and glass fibers. Also known as *magnesium aluminosilicate (MAS)*: (i) *Dimagnesium dialuminum pentasilicate*. Density, 2.51 g/cm³; melting point, 1471°C (2680°F); hardness, 5-7 Mohs. Formula: $Mg_2Al_4Si_5O_{18}$; and (ii) *Tetramagnesium pentaluminum disilicate*. Melting point, 1454°C (2650°F); hardness, 5-7 Mohs. Formula: $Mg_4Al_{10}Si_2O_{23}$.

magnesium antimonide. Hexagonal crystals with a density of 3.99 g/cm³ and a melting point of 1245°C (2273°F), used in ceramics, electronics and materials research. Formula: Mg_3Sb_2.

magnesium astrophyllite. A straw yellow mineral of the astrophyllite group composed of potassium sodium iron magnesium manganese titanium oxide silicate hydroxide, $(Na,K)_4(Fe,Mg,Mn)_7Ti_2Si_8O_{24}(O,OH)_7$. Crystal system, monoclinic. Density, 3.32 g/cm³; refractive index, 1.687. Occurrence: Russian Federation.

magnesium-bentonite. See hectorite.

magnesium boride. A compound of magnesium and boron that is available as hexagonal crystals with a density of 2.57 g/cm³, and a melting point of 800°C (1472°F) (decomposes). It has superconductive properties at -233°C (-388°F). Used for ceramic and electronic applications. Formula: MgB_2.

magnesium bromide. White hygroscopic crystals or off-white hygroscopic powder (98+% pure). Crystal system, hexagonal. Density, 3.72 g/cm³; melting point, 711°C (1312°F). Used in chemistry and materials research. Formula: $MgBr_2$.

magnesium bromide hexahydrate. Colorless or white, deliquescent crystals (98+% pure). Density, 2.00 g/cm³; melting point, 172°C (341°F). Used in organic synthesis, biochemistry and medicine. Formula: $MgCl_2 \cdot 6H_2O$.

magnesium carbonate. A magnesium compound that is available as colorless or white crystals, or as a light, bulky, white powder. It occurs in nature as the mineral *magnesite*. Crystal system, rhombohedral (hexagonal). Density, 2.95-3.2 g/cm³; melting point, decomposes at 350°C (662°F); hardness, 3.5-4.5 Mohs; refractive index, approximately 1.52. Used as a high-temperature flux in glass, glazes, enamels, insulator bodies and vitreous and semivitreous ware and porcelain, as a low-temperature refractory, as an auxiliary flux in ceramic bodies, as a setting-up agent in porcelain enamels and other slips, as a thermal insulator, as a rubber reinforcing agent, and in inks and filtering media. Formula: $MgCO_3$.

magnesium casting alloys. A group of magnesium-base die-casting alloys including in particular alloys of 94% magnesium and 6% tin, and 70% magnesium, 20% cadmium and 10% lead, used for bushings, liners, and inserts.

magnesium chloride. (1) Colorless or white crystals or leaflets (98+% pure). Crystal system, rhombohedral (hexagonal). Density, 2.32 g/cm³; melting point, 714°C (1317°F); boiling point, 1412°C (2574°F). Used in the production of magnesium metal by electrolysis (Dow process), in the manufacture of magnesium-oxychloride cement, in ceramics, as a coolant for drilling tools, in refrigerating brines, as a catalyst, in paper, cotton and woolen fabrics, as a fire extinguisher, and in biochemistry, biotechnology and medicine. Formula: $MgCl_2$.

magnesium chloride hexahydrate. Colorless or white crystals (98+% pure). It occurs in nature as the mineral *bischofite*. Density, 1.56 g/cm³; melting point, loses $2H_2O$ at 100°C (212°F); boiling point, decomposes to oxychloride. Used in the production of magnesium metal by electrolysis (Dow process), in the manufacture of magnesium oxychloride cement, in ceramics, as a coolant for drilling tools, in refrigerating brines, catalysts, paper, cotton and woolen fabrics, and as a fire extinguisher. Formula: $MgCl_2 \cdot 6H_2O$.

magnesium chlorophoenicite. A colorless to white mineral of the chlorophoenicite group composed of magnesium manganese zinc arsenate hydroxide, $(Mg,Mn)_3Zn_2(AsO_4)(OH,O)_6$. Crystal system, monoclinic. Density, 3.37 g/cm³; refractive index, 1.672. Occurrence: USA (New Jersey).

magnesium chromate pentahydrate. Yellow crystals (98+% pure) with a density of 1.695 g/cm³ and refractive index of 1.55 used in chemistry, and materials research. Formula: $MgCrO_4 \cdot 5H_2O$.

magnesium chromite. A synthetic spinel composed of magnesium oxide (MgO) and chromic oxide (Cr_2O_3). Density, 4.4 g/cm³; melting point, 2250°C (4082°F); low thermal expansion. Used for chrome-magnesite refractories, ceramic magnets, and for electronic applications. Formula: $MgCr_2O_4$. Also known as *magnesia-chromia*. See also magnesiochromite.

magnesium columbate. See magnesium niobate.

magnesium dititanate. See magnesium titanate (iii).

magnesium dysprosium sulfide. Lemon-yellow crystals. Crystal system, monoclinic. Melting range, 1597-1657°C (2907-3015°F); band gap, 2.38 eV. Used in ceramics, and as as a semiconductor in electronics and materials research. Formula: $MgDy_2S_4$.

magnesium erbium sulfide. Lemon-yellow crystals. Crystal system, monoclinic. Melting range, 1836-1896°C (3337-3445°F); band gap, 2.38 eV. Used in ceramics, and as a semiconductor in electronics and materials research. Formula: $MgEr_2S_4$.

magnesium ethoxide. A metal alkoxide that typically contains 21-22% magnesium and is available in the form of a white, moisture-sensitive, flammable powder with a melting point of 270°C (518°F). It is soluble in ethanol and methanol and used as a component in olefin polymerization systems, and as an intermediate for synthetic forsterite (Mg_2SiO_4). Formula: $Mg(OC_2H_5)_2$. Also known as *magnesium ethylate.*

magnesium ethylate. See magnesium ethoxide.

magnesium-ferric oxide aerogel. An aerogel prepared from a mixture of appropriate quantities of magnesium acetate tetrahydrate [$Mg(C_2H_3O_2)_2 \cdot 4H_2O$] and ferric acetylacetonate [$Fe(O_2C_5H_7)_3$] powder by dissolving in methanol hydrolyzed with an appropriate quantity of water, and heating in an autoclave under prescribed conditions of temperature, pressure and time. The resulting product is a fine powder with spinel-type structure, suitable for use in catalysis, composites, and as a starting material for ceramics.

magnesium ferrite. A ferrimagnetic ceramic material composed of magnesium oxide (MgO) and ferric oxide (Fe_2O_3) and having a cubic spinel crystal structure (ferrospinel). Density, 4.2 g/cm³; melting point, 1750°C (3182°F); excellent magnetic properties at high frequencies; very high electrical resistivity. Used as a soft magnetic material for various applications, and in basic refractories. Formula: $MgFe_2O_4$. See also magnesioferrite.

magnesium fluoride. Colorless or white, fluorescent crystals or powder (98+% pure). High-purity grades (99.99+%) are also available. It occurs in nature as the mineral *sellaite*, and can also be made synthetically. Crystal system, tetragonal. Crystal structure, rutile. Density, 2.97-3.17 g/cm³; melting point, 1248°C (2278°F) [also reported as 1312°C or 2394°F]; boiling point, 2239°C (4062°F) [also reported as 2227°C or 4041°F]; hardness, 5.0-5.5 Mohs; refractive index, 1.35-1.38. Used as a flux in various ceramic and glass compositions particularly for infrared components used under severe conditions, in the preparation of optical thin films, for vacuum deposition applications, also in single crystal form for polarizing prisms, lenses and windows and for infrared and ultraviolet applications, and for overcoatings on aluminum. Formula: MgF_2.

magnesium fluorosilicate. White crystals or powder used in ceramics and materials research. Formula: $MgSiF_6$. Also known as *magnesium silicofluoride.*

magnesium fluosilicate hexahydrate. White, efflorescent, crystalline powder. Density, 1.788 g/cm³; melting point, decomposes 120°C (248°F). Used in ceramics, ceramic coatings, concrete hardeners, waterproofing agents, and in magnesium casting. Formula: $MgSiF_6 \cdot 6H_2O$. Also known as *magnesium hexafluorosilicate hexahydrate; magnesium silicofluoride hexahydrate.*

magnesium flux. See magnesium fluoride.

magnesium germanate. Any of the following compounds of magnesium oxide and germanium oxide used in electronics and materials research: (i) *Magnesium monogermanate.* Melting point, 1700°C (3092°F). Formula: $MgGeO_3$; (ii) *Dimagnesium germanate.* Melting point, 1854°C (3369°F). Formula: Mg_2GeO_4; and (iii) *Tetramagnesium germanate.* Melting point, 1497°C (2727°F). Formula: Mg_4GeO_6.

magnesium germanide. A compound of magnesium and germanium. Crystal system, cubic. Crystal structure, antifluorite. Density, 3.08 g/cm³; melting point, 1115°C (2039°F); band gap, 0.74 eV. Used as a semiconductor. Formula: Mg_2Ge.

magnesium germanium phosphide. A compound of magnesium, germanium and phosphorus. Crystal system, cubic. Crystal structure, sphalerite. Band gap, 2 eV. Used as a semiconductor. Formula: $MgGeP_2$.

magnesium halide. A compound of magnesium and a halogen, such as bromine, chlorine, fluorine or iodine.

magnesium hexafluorosilicate hexahydrate. See magnesium fluosilicate hexahydrate.

magnesium holmium sulfide. Off-white crystals. Crystal system, monoclinic. Melting range, 1820-1880°C (3308-3416°F). Used in ceramics, electronics, and materials research. Formula: $MgHo_2S_4$.

magnesium hydroxide. A white powder (95+% pure). Density, 2.36 g/cm³; melting point, decomposes at 350°C (662°F); noncombustible. Used in the processing of uranium, in the production of magnesium metal, as an additive for fuel oils, in sulfite pulp, and in dentistry and medicine. Formula: $Mg(OH)_2$. Also known as *milk of magnesia.*

magnesium iodide. White, hygroscopic crystals (98+% pure). High-purity grades (99.99+%) are also available. Crystal system, hexagonal. Density, 4.43 g/cm³; melting point, 634°C (1173°F). Used in ceramics and materials research. Formula: MgI_2.

magnesium lead. A compound of magnesium and lead. Crystal system, cubic. Crystal structure, antifluorite. Density, 5.1 g/cm³; melting point, 550°C (1022°F); band gap, 0.1 eV. Used in materials research and as a semiconductor. Formula: Mg_2Pb.

magnesium lime. Lime containing at least 5%, but usually 20-35% magnesium oxide (MgO). It has a relatively slow slaking time, a fast setting time, and low heat evolution and expansion. Used for mortar with a strength superior to that of high-calcium limes.

magnesium-lithium alloys. Lightweight alloys of magnesium containing up to 15% lithium and having good strength and fine grained structures, but a maximum upper continuous-use temperature of only 150°C (300°F), unless suitably modified. Used in weight-sensitive applications in the aerospace industry, and for military equipment.

magnesium-matrix composites. Composite materials consisting of magnesium matrices reinforced with fibers of alumina, graphite or silicon carbide, or whiskers of silicon carbide, alumina, etc. They have low densities, excellent high-temperature performance, high stiffness and strength, high fatigue resistance, and good dimensional stability. Used especially for aerospace applications and automotive components.

magnesium metasilicate hydrate. See magnesium silicate (i).

magnesium mica. See amber mica.

magnesium molybdate. A white, crystalline powder (99% pure) with a density of 2.208 g/cm³ and a melting point of approximately 1060°C (1940°F), used for electronic and optical applications. Formula: $MgMoO_4$.

magnesium-monel. A master alloy of 50% magnesium and 50% monel metal (a nickel-copper alloy), used to introduce nickel and copper into nonferrous alloys, and as a deoxidizer for nonferrous alloys.

magnesium monocolumbate. See magnesium niobate (i).

magnesium monogermanate. See magnesium germanate (i).

magnesium mononiobate. See magnesium niobate (i).

magnesium monosulfide. See magnesium sulfide (i).

magnesium monotitanate. See magnesium titanate (i).

magnesium naphthenate. A dark, viscous liquid that typically contains 3.5-8.0% magnesium, and is used as paint and varnish drier.

magnesium-nickel. A master alloy of 50% magnesium and 50% nickel, used to introduce nickel and nickel-base alloys into ferrous and nonferrous alloys, and as a deoxidizer for nonferrous alloys.

magnesium nickelide. An intermetallic compound of magnesium and nickel with a density of 4.1 g/cm³, used in battery electrodes of hydrogen-storage systems. Formula: Mg_2Ni.

magnesium niobate. Any of the following compounds of magnesium oxide and niobium (columbium) oxide used in ceramics and materials research, and also known as *magnesium columbate*: (i) *Magnesium mononiobate*. Melting point, approximately 1538°C (2800°F). Formula: $MgNb_2O_6$. Also known as *magnesium monocolumbate*; (ii) *Dimagnesium niobate*. Melting point, approximately 1510°C (2750°F). Formula: $Mg_2Nb_2O_7$. Also known as *dimagnesium columbate*; (iii) *Trimagnesium niobate*. Melting point, approximately 1538°C (2800°F). Formula: $Mg_3Nb_2O_8$. Also known as *trimagnesium columbate*; and (iv) *Tetramagnesium niobate*. Melting point, approximately 1483°C (2701°F). Formula: $Mg_4Nb_2O_9$. Also known as *tetramagnesium columbate*.

magnesium nitrate dihydrate. White, hygroscopic, water-soluble crystals (99+% pure). Density, 1.45 g/cm³; melting point, 95-100°C (192-212°F); boiling point, decomposes at 330°C (626°F). Used in chemistry and materials research. Formula: $Mg(NO_3)_2 \cdot 2H_2O$.

magnesium nitrate hexahydrate. White, hygroscopic, water-soluble crystals (99+% pure). It occurs in nature as the mineral *nitromagnesite*. Crystal system, monoclinic. Density, 1.636 g/cm³; melting point, 89°C (192°F); boiling point, loses $5H_2O$ at 330°C (626°F). Used in chemistry, biochemistry and materials research. Formula: $Mg(NO_3)_2 \cdot 6H_2O$. See also nitromagnesite.

magnesium nitride. Yellow, greenish-yellow or greenish-black crystals. Crystal system, cubic. Density, 2.71 g/cm³; melting point, decomposes at 1500°C (2732°F). Used in ceramics and materials research. Formula: Mg_3N_2.

magnesium oleate. The magnesium salt of oleic acid available as a yellowish powder or mass obtained by introducing magnesium chloride to sodium oleate. Used as a varnish drier, and as a lubricant for plastics. Formula: $Mg(C_{18}H_{33}O_2)_2$.

magnesium orthophosphate. White, iridescent crystals with a melting point of 1184°C (2163°F), used as a substitute for tin oxide in certain sanitaryware glazes to obtain improved color, opacity, brilliance and texture. Formula: $Mg_3(PO_4)_2$.

magnesium oxide. Colorless or white crystals or powder (98+% pure). It occurs in nature as the mineral *periclase*. Crystal system, cubic. Density, 3.58 g/cm³; melting point, 2800°C (5072°F); boiling point, 3600°C (6512°F); hardness, 5-7 Mohs; refractive index, 1.736; dielectric constant, 9.8. Used for building materials, refractories (furnace linings), crucibles, thermocouple tubing, thermal insulation and infrared windows, as a viscous flux and opacifier, in the manufacture of paper, Sorel cement (magnesium oxychloride cement) and some electronic components, as a filler in rubber, in stabilized zirconia materials, in semiconductors and electrical insulation, in crystalline form for superconductor and III-V (13-15) thin film deposition, and in polycrystalline ceramic for aircraft windshields. Formula: MgO. Also known as *magnesia*.

magnesium oxychloride cement. A relatively quick-setting cement composed of a mixture of magnesium chloride ($MgCl_2$) and magnesium oxide (calcined magnesia, MgO) that reacts with water to form a hard, strong, solid mass (magnesium oxy-

chloride). Fillers, such as sawdust, wood flour, cork, talc, sand, pulverized stone, or metal powder, may also be incorporated. Used for stone floors, man-made building stone, and stucco. Also known as *magnesia cement; oxychloride cement; Sorel cement; xylolith.*

magnesium palmitate. Crystalline needles or white lumps with a melting point of 121.5°C (251°F), used as varnish drier. Formula: $Mg(C_{16}H_{31}O_2)_2$.

magnesium peroxide. White crystals or powder. Crystal system, cubic. Density, 3.0 g/cm³; melting point, 100°C (212°F) (decomposes). Used as a bleach, oxidizer and antiseptic, and in materials research. Formula: MgO_2.

magnesium phosphide. A compound of magnesium and phosphorus available in the form of yellow crystals. Crystal system, cubic. Density, 2.02-2.06 g/cm³. Used in ceramics and materials research. Formula: Mg_3P_2.

magnesium phthalocyanine. A purple phthalocyanine derivative that contains a central magnesium atom. It has a typical dye content of about 90%, and a maximum absorption wavelength of 668 nm. Used as a dye in chemistry, biochemistry and electronics. Formula: Formula: $Mg(C_{32}H_{16}N_8)$. Abbreviation: MgPc.

magnesium pyrophosphate. Colorless crystals. Density, 2.60 g/cm³; melting point, 1383°C (2521°F). Used for porcelains and enamels, and as a substitute for tin oxide in glazes for sanitary ware. Formula: $Mg_2P_2O_7$.

magnesium pyrophosphate trihydrate. A white powder. Density, 2.56 g/cm³; melting point, loses $3H_2O$ at 100°C (212°F). Used in porcelains, enamels and glazes. Formula: $Mg_2P_2O_7 \cdot 3H_2O$.

magnesium scandium sulfide. Lemon-yellow crystals. Crystal system, cubic. Crystal structure, spinel. Melting range, 1720-1780°C (3128-3236°F); solidus temperature, 1750°C (3182°F). Used in ceramics, electronics and materials research. Formula: $MgSc_2S_4$.

magnesium selenide. Brown crystals. Crystal system, cubic. Density, 4.2 g/cm³. Used in ceramics, electronics and materials research. Formula: MgSe.

magnesium silicate. Any of the following compounds of magnesium oxide and silicon dioxide: (i) *Magnesium metasilicate hydrate*. Fine, white powder. Density, 2.6-2.8 g/cm³. Used in ceramics, glasses and refractories, as a filler for paints, varnishes, rubber and paper, and as an absorbent, filtering medium, catalyst and catalyst carrier. Formula: $3MgSiO_3 \cdot 5H_2O$. Also known as *hydrated magnesium metasilicate;* (ii) *Anhydrous magnesium silicate*. It occurs in nature as the mineral *clinoenstatite*. Crystal system, monoclinic. Density, 3.2-3.3 g/cm³; melting point, 1554°C (2829°F); hardness, 5-7 Mohs. Used in ceramics and materials research. Formula: $MgSiO_3$; and (iii) *Dimagnesium silicate*. Density, 3.22 g/cm³; melting point, 1910°C (3470°F); hardness, 5-7 Mohs. Used in ceramics and materials research. Formula: Mg_2SiO_4. See also Magnesol (1).

magnesium silicide. A moisture-sensitive, gray powder (99+% pure). Crystal system, cubic. Crystal structure, antifluorite. Density, 1.88-1.94 g/cm³; melting point, 1102°C (2016°F); band gap, 0.77 eV. Used in ceramics, materials research, and as a semiconductor. Formula: Mg_2Si.

magnesium silicofluoride. See magnesium fluosilicate.

magnesium silicofluoride hexahydrate. See magnesium fluosilicate hexahydrate.

magnesium stannate. Any of the following compounds of magnesium oxide and stannic oxide: (i) *Magnesium stannate*. Used for dielectric compositions, as a phosphor base, and in materials research. Formula: $MgSnO_3$; (ii) *Dimagnesium stannate.* Density, 4.74 g/cm³; melting point, approximately 1950°C (3542°F). Used in ceramics and materials research. Formula: Mg_2SnO_4; and (iii) *Magnesium stannate trihydrate.* White, crystalline powder that decomposes at approximately 340°C (644°F). Used as an additive in ceramic capacitors. Formula: $MgSnO_3 \cdot 3H_2O$.

magnesium stannate trihydrate. See magnesium stannate (iii).

magnesium stannide. Blue-white crystals. Crystal system, cubic. Crystal structure, antifluorite. Density, 3.53 g/cm³; melting point, 778°C (1432°F); band gap, 0.36 eV; good magnetic properties. Used as a semiconductor, in materials and thermoelectric research, and in magnetochemistry. Formula: Mg_2Sn.

magnesium stearate. A soft, white powder with a density of 1.028 and a melting point of 88.5°C (192°F), used as a paint and varnish drier, as a flatting agent and as a lubricant for plastics. Formula: $Mg(C_{18}H_{35}O_2)_2$.

magnesium sulfate. Colorless, hygroscopic crystals or white powder (97+% pure). Density, 2.66 g/cm³; melting point, decomposes at 1124°C (2055°F). Used as a suspension promoting agent in slips, as a flux in glaze composition, as a setting-up agent for enamels, in fireproofing, textile processing and paper sizing, and in biology and biochemistry. Formula: $MgSO_4$.

magnesium sulfate heptahydrate. See epsomite.

magnesium sulfide. Any of the the following compounds of magnesium and sulfur: (i) *Magnesium monosulfide.* Red-brown crystals or fine, moisture-sensitive powder available in cubic and hexagonal form. The cubic form has a density of 2.66-2.68 g/cm³ and a melting point of 2227°C (4041°F) and the hexagonal form has a density of 2.84 g/cm³ and decomposes above 2000°C (3610°F). Used in ceramics, materials research and in electronics as a semiconductor. Formula: MgS. The cubic form is also known as *niningerite*; and (ii) *Dimagnesium sulfide.* A stoichiometric intermetallic compound of magnesium and sulfur, used in ceramics, materials research and for electronic applications. Formula: Mg_2S.

magnesium telluride. A crystalline compound of magnesium and tellurium. Crystal system, hexagonal. Crystal structure, wurtzite (or zincite). Density, 3.85 g/cm³; melting point, approximately 2527°C (4851°F). Used as a semiconductor. Formula: $MgTe$.

magnesium thulium sulfide. Off-white crystals. Crystal system, monoclinic. Melting range, 1587-1647°C (2889-2997°F); solidus temperature, 1617°C (2943°F). Used in ceramics, electronics, and materials research. Formula: $MgTm_2S_4$.

magnesium titanate. Any of the following compounds of magnesium oxide and titanium dioxide: (i) *Magnesium monotitanate.* White crystals or powder (85+% pure). Crystal system, rhombohedral (hexagonal). Density, 4.0 g/cm³; melting point, 1690°C (3074°F); low dielectric constant. Used in dielectric compositions, as an addition agent to barium titanate, and in materials research. Formula: $MgTiO_3$; (ii) *Dimagnesium titanate.* Crystal system, cubic. Density, 3.52 g/cm³; melting point, 1732°C (3150°F). Used in ceramics and materials research. Formula: Mg_2TiO_4; and (iii) *Magnesium dititanate.* Density, 3.66 g/cm³; melting point, 1650°C (3000°F). Used in ceramics and materials research. Formula: $MgTi_2O_3$.

magnesium titanium. A double metal alkoxide supplied as a liquid with magnesium and titanium contents of 3.8-3.9% and 7.6-7.7%, respectively. The magnesium-to-titanium metal ratio is 1:1 and the density 1.07-1.09. Used in sol-gel and coating applications. Formula: $MgTi(OR)_x$.

magnesium tungstate. White crystals or powder (99+% pure). Crystal system, monoclinic. Density, 5.66 g/cm³. Used in fluoroscopy, fluorescent screens for X-rays and as a fluorescent paint pigment. Formula: $MgWO_4$. Also known as *magnesium wolframate.*

magnesium uranate. A compound of magnesium oxide and uranium trioxide. Melting point, 1750°C (3182°F). Used in ceramics and for nuclear applications. Formula: $MgUO_4$.

magnesium wolframate. See magnesium tungstate.

magnesium ytterbium sulfide. Yellow crystals. Crystal system, cubic. Melting range, 1618-1678°C (2944-3052°F); hardness, 318-366 Vickers; band gap, 2.5 eV. Used in ceramics, and as a semiconductor in electronics and materials research. Formula: $MgYb_2S_4$.

magnesium yttrium sulfide. Lemon-yellow crystals. Crystal system, monoclinic. Melting range, 1572-1632°C (2862-2970°F); solidus temperature, 1602°C (2916°F); band gap, 2.38 eV. Used in ceramics, and as a semiconductor in electronics and materials research. Formula: MgY_2S_4.

magnesium zippeite. A yellow, radioactive mineral of the zippeite group composed of magnesium uranyl sulfate hydroxide hydrate, $Mg_2(UO_2)_6(SO_4)_3(OH)_{10} \cdot 16H_2O$. Crystal system, orthorhombic. Density, 3.30 g/cm³; refractive index, 1.75. Occurrence: USA (Utah).

magnesium zirconate. A powder composed of magnesium oxide and zirconium oxide. Density, 4.23 g/cm³; melting point, 2150°C (3900°F). Used in electronics, dielectric compositions, and as a setter for firing ferrites and titanates. Formula: $MgZrO_3$.

magnesium zirconium. A double metal alkoxide supplied as a liquid with magnesium and zirconium contents of 2.6-2.7% and 9.9-10.1% respectively. The magnesium-to-zirconium metal ratio is 1:1 and the density 1.08-1.10. Used in sol-gel and coating applications. Formula: $MgZr(OR)_x$.

magnesium-zirconium silicate. A white solid composed of magnesium oxide, zirconium oxide and silicon dioxide. Density, 1.28 g/cm³; melting point, 1760°C (3200°F). Used for electrical resistances, in ceramics, and as an opacifier in glazes. Formula: $MgZrSiO_5$.

Magnesol. (1) Trade name for a fine *magnesium silicate* powder used in ceramics and as a filler. *Note:* This term is now also used generically as a synonym for magnesium metasilicate.

(2) Trademark of Canada Wire & Cable Limited for insulated magnet wire used for high-temperature applications.

magnet. A material that has the property or power of attracting iron, steel or other ferromagnetic materials to it. It may be a temporary magnet which loses its magnetism upon removal of the magnetizing force, or a permanent magnet which retains its magnetism even after removal of the magnetizing force. See also hard magnet; permanent magnet; soft magnet; temporary magnet.

magnet alloys. A group of alloys including *Alcomax, Alnico, Cunico, Cunife, Hycomax, Lodex, Platinax, Remalloy* and *Vicalloy* that are characterized by strong magnetic properties, large hysteresis loop areas and high magnetic energy values. They are used in the manufacture of permanent magnets.

Magnetectic. Trade name of Magnetec GmbH & Co. KG (Germany) for soft magnetic alloys and materials.

Magnetherm. Trade name of Pechiney/SOFREM (France) for an alloy containing 99.8-99.85% magnesium and 0.15-0.2% other elements. Used for shell and sand castings.

magnetic alloys. See hard magnetic materials; soft magnetic materials; magnet alloys.

magnetically hard materials. See hard magnetic materials.

magnetically soft materials. See soft magnetic materials.

Magnetic Bond. Trade name for a self-cure acrylic resin used for dental applications.

magnetic bubble material. A thin-film magnetic material, usually of the *garnet* type, that in its natural state consists of magnetic domains (i.e., microscopic regions of common magnetic moment alignment), but that by the application of an external magnetic field contract into minute cylindrical regions of magnetization. The axis of the cylindrical regions lies perpendicular to the plane of the material. Used for memory devices in data-storage systems.

magnetic ceramics. See ceramic magnets.

magnetic composites. A group of composite materials consisting of metallic, polymeric or ceramic matrices with embedded magnetic particles, e.g., iron.

magnetic copper. A high-purity wire consisting of a copper matrix with evenly spaced iron filaments around the circumference. The nominal composition is 99.96% copper and 0.04% iron. *Magnetic copper* has a density of 8.96 g/cm³ (0.324 lb/in.³), an elongation of 1.1%, good electrical conductivity, good magnetic properties, low thermal conductivity, low elongation and good strength. Used for electrical and magnetic applications.

magnetic ferroelectrics. Materials, such as ferrites and garnets, that exhibit spontaneous reversible electric-dipole alignment, electric hysteresis, piezoelectricity and parallel alignment of neighboring magnetic moments.

magnetic film. See magnetic thin film.

magnetic inspection powder. See magnetic powder.

magnetic iron ore. See magnetite.

magnetic iron oxide. See black ferric oxide.

magnetic materials. A group of metallic and ceramic materials that exhibit magnetism. They can be grouped into the following categories: (i) hard magnetic materials (e.g., magnet steels, samarium-cobalt alloys, ferrites, or garnets) with high remanence and coercive forces and large hysteresis loops used for making permanent magnets, and (ii) soft magnetic materials (e.g., iron, iron-silicon alloys, or nickel-iron alloys) with low remanence and coercive forces and small hysteresis loops used for transformer sheets, magnetic amplifiers, motors, and generators. See also magnetically soft materials; magnetically hard materials.

magnetic medium. A material, such as a flexible recording tape or a computer disk, coated with a magnetizable material on which magnetic domains can be produced by the application of an external field. See also magnetic recording materials.

magnetic metals. Metallic materials, such as iron, steel, nickel or cobalt, exhibiting ferromagnetism. See also ferromagnetic materials.

magnetic multilayers. A class of engineered structures consisting of alternating ferromagnetic layers and nonferromagnetic space layers with each individual layer being only a few atomic layers thick. Typically, the ferromagnetic layers are composed of transition metals (e.g., iron, cobalt or nickel) and the nonferromagnetic spacers of metals, such as copper, indium, tin or silver. Some of these structures, such as sputtered-deposited Co(Fe)/Cu single crystalline multilayers, exhibit colossal and giant magnetoresistance effects at temperatures close to 0K with changes in electrical resistance up to several hundred percent in relatively weak applied magnetic fields. They are used in modern magnetic disk drives to read magnetic bit states, and for magnetic recording heads.

magnetic nanocomposites. See nanomagnetic composites.

magnetic nanomaterials. See nanomagnetic materials.

magnetic polymer powder. A dry powder usually composed of considerable quantities of a magnetic material, such as magnetite (Fe_3O_4), in a polymer, and having a particle size usually in the micron range. An example is a magnetic acrylic powder composed of 25% magnetite in polymethyl methacrylate (PMMA). Used chiefly in biochemistry, immunochemistry and chromatography.

magnetic powder. A dry powder (usually high-purity iron) composed of finely divided ferromagnetic particles and available in yellow, red, black and gray colors and with fluorescent coatings applied over the surface of a ferromagnetic part to be magnetically inspected in order to locate surface and subsurface flaws (cracks, inclusions, or pipe) that tend to attract the powder. Also known as *magnetic inspection powder.*

magnetic pyrites. See pyrrhotite.

magnetic recording materials. A group of materials including barium ferrite, gamma ferric oxide, chromium dioxide, cobalt-chromium, nickel-iron, iron-aluminum-silicon, cobalt nickel and cobalt nickel chromium thin films as well as certain amorphous and nanocrystalline alloys used for the mass storage of electronic information on audio tapes, computer floppy and hard disks, video-cassette recorders, credit cards, etc. See also antiferromagnetically coupled media layers; magnetic medium; thin-film recording media.

magnetic recording paper. A reusable paper containing magnetic particles that can be oriented by a recording head to produce visible or machine-readable traces.

magnetic rubber. Synthetic rubber in strip or sheet form made by compounding with magnetic powder.

magnetic semiconductors. See diluted magnetic semiconductors.

magnetic shielding alloys. A group of high-permeability magnetic alloys, such as iron-nickel and nickel-iron, that have low hysteresis loss and are used for magnetic shielding applications. They are often sold under trade names and trademarks, such as Ferroperm, Mumetal, Netic, Sanbold, or Ultraform.

magnetic superconductors. A group of materials including certain metals (e.g., niobium and its alloys) and ceramic oxides (e.g., yttrium barium copper oxide) that exhibit both magnetic and superconducting properties. They are used for high-field solenoids, high-speed computer circuitry, thin-film devices, high-sensitivity magnetic field detectors, and other magnetic applications. Also known as *superconducting magnets; superconductor magnets.*

magnetic thin film. A magnetic material applied to a substrate as a thin film and used in computer memories. Abbreviation: MTF. Also known *magnetic film.* See also thin-film material.

magnetic wood. A material that consists of a combination of wood with a magnetic powder or fluid, or a special magnetic coating. It can be prepared by (i) impregnating ordinary wood with a water-based magnetic fluid; (ii) mixing wood powder with a ferrite powder (e.g., manganese zinc ferrite) and hot pressing into boards, or (iii) coating a ferrite powder (e.g., *Sendust* powder) onto a fiberboard. *Magnetic wood* has good magnetic characteristics and good processibility, and offers a wood texture. It also has good thermal properties and can be used as a heating board.

magnetite. A black, magnetic mineral of the spinel group with a dull, submetallic or metallic luster and a black streak. It is composed of ferroferric oxide (Fe_3O_4), and contains about 72.4%

iron. Crystal structure, cubic. Density, 4.9-5.2 g/cm³; melting point, decomposes at 1538°C (2800°F) to ferric oxide; hardness, 5.5-6.5; refractive index, 2.42. Occurrence: Central Europe, Northern Europe (Norway and Sweden), Russia, and North America (USA and Canada). Used as an important ore of iron, in washing coal and ores, for magnetic applications, as an aggregate in high-density concrete, and as a ceramic semiconductor. Also known as *magnetic iron ore*.

magnetizable polymer particles. Paramagnetic microparticles of polymers, such as polystyrenes, that exhibit magnetic attraction in the presence of a magnetic field but, after removal of the field, return to their nonmagnetic, random, particulate state. They are available in latex form and may also bear functionalized groups (amino, carboxyl, etc.). Used chiefly in affinity chromatography, antibody binding, and enzyme immobilization.

magnetoelastic materials. A group of ferromagnetic materials in which an elastic strain effects a change in magnetization. See also ferromagnetic materials.

magnetoelectric materials. A group of magnetic ferroelectrics, such as barium manganese fluoride ($BaMnF_4$), that at low temperatures exhibit a coupling between polarization and magnetization.

magnetoelectronic materials. A class of materials including ferromagnetic metals and alloys, magnetic transition-metal multilayers, granular high-magnetic-anisotopy materials, magnetic tunnel junctions and ferroelectrics that exhibit magnetic and electrical properties which make them suitable for various electronic applications, e.g., magnetic recording heads, high-sensitivity, magnetoresistant write/read heads, hard disks, magnetic storage devices and magnetic sensors. See also magnetic multilayers; ferroelectrics; high K_u materials.

Magnetoflex. Trademark of Vacuumschmelze GmbH (Germany) for a series of wrought hard magnetic alloys containing varying amounts of cobalt, iron and vanadium, and sometimes additions of chromium, copper and/or nickel. They are manufactured in wires and strips by hot or cold working (i.e., cutting, stamping and bending) and heat-treated to set the magnetic properties. *Magnetoflex* alloys have a density of 8.1 g/cm³ (0.29 lb/in.³), a hardness of 480-520 Vickers (rolled), a Curie temperature of 700°C (1290°F), a maximum service temperature of 500°C (932°F), high magnetic energy products, coercivities and remanences, and good malleability. Used for permanent magnets, electrical, electronic and magnetic equipment, rings in hysteresis motors, and tachometer systems.

magnetoliposomes. A class of biomaterials composed of nanometer-sized magnetic particles coated with a phospholipid *bilayer* membrane, and used for biological and medical applications. *Note:* A "phospholipid" is an ester of phosphatidic acid with choline, ethanolamine, serine or inositol.

magnetooptical materials. Materials, such as certain ferromagnetic ceramics as well as cobalt-platinum alloys, chromium-doped manganese-bismuth, neodymium trifluoride, and gadolinium/iron-cobalt and gadolinium/iron multilayers, in which the application of an external magnetic field results in changes in optical properties. Used for magnets in memory-storage devices, and optical disk drives.

magnetoplumbite. (1) A gray-black, magnetic mineral of the hogbomite group composed lead iron oxide, $PbFe_{12}O_{19}$. It can also be made synthetically. Crystal structure, hexagonal. Density, 5.52-5.69 g/cm³. Occurrence: Sweden. Used in ceramics and magnetic applications.

(2) See hexagonal ferrite.

magnetoresistance materials. A class of materials including ferroelectric metals, various semiconductors and layered metal structures as well as materials with perovskite crystal structures that exhibit an increase or decrease of the electrical resistivity upon application of a magnetic field. Also known as *magnetoresistive materials*. See also colossal magnetoresistance materials; giant magnetoresistance materials.

magnetoresistive materials. See magnetoresistance materials.

magnetorheological fluid. A magnetic fluid composed of a high-boiling, thermally stable carrier fluid (e.g., a silicone oil or hydrocarbon fluid), a dispersed phase (usually soft magnetic particles, such as iron or iron-cobalt particles with a size of 1-10 nm (0.04-0.4 μin.), or manganese-zinc or nickel-zinc ferrite particles with a size of less than 100 nm (3.94 μin.), and selected additives (e.g., polymers, fibrous carbon, or nanostructured zinc). In an applied magnetic field, the rheological properties of such a fluid change from those of a Newtonian fluid to those of a weak viscoelastic solid. This is due to the alignment of the magnetic particles parallel to the field. With increasing magnetic field, this viscoelastic solid exhibits a rapid and almost reversible increase in shear yield stress. Used in automotive engine mounts, seat dampers and shock absorbers, electromechanical devices, vibration-control applications, such as earthquake-resistant structures, in exercise equipment and for polishing aspherical optical lenses. Abbreviation: MRF. See also ferrofluid; magnetorheological polymer gels.

magnetorheological materials. See magnetorheological fluid; magnetorheological polymer gels.

magnetorheological polymer gels. A novel class of magnetorheological materials prepared by suspending ferromagnetic particles, such as iron powder, and selected additives in a gel composed of a polyethyleneamine, polyurethane, poly(4-vinylpyridine), polypyrrolidone or a silicone polymer. The rheological properties can be controlled by the quantity and concentration of the crosslinking agents and diluents yielding more or less gel-like materials. They are currently used for damping, vibration control and clutch applications. Abbreviation: MRPG. See also magnetorheological fluid.

magnetostrictive materials. A group of ferromagnetic materials in which a change in the degree of magnetization results in a small corresponding change in length (e.g., expansion or compression). Conversely, an externally produced strain in the magnetized material causes a change in the amount of magnetization. For example, some iron-nickel alloys tend to expand in a magnetic field, while pure nickel tends to contract. Other *magnetostrictive materials* include cobalt, iron and nickel, terbium-dysprosium and terbium-dysprosium-iron alloys, ferrites of lanthanide elements (e.g., gadolinium, samarium, terbium, holmium, erbium or thulium ferrite), yttrium-iron garnets, and amorphous iron-based metallic glasses. Used as sensors, transducers and actuators.

magnet steels. A group of steels that have high remanences and coercivities, and large hysteresis loops, and are thus suitable for making permanent magnets. Examples include martensitic carbon steels with up to 1.5% carbon and about 1% manganese, tungsten steels with about 0.7% carbon, 6% tungsten and 0.5% chromium, chromium steels with 1% carbon and 3.5% chromium, and high-cobalt steels with 0.7-0.8% carbon, 17-36% cobalt, 3-9% tungsten and 2-6% chromium. Abbreviation: MS.

magnet wire. A single-conductor, insulated aluminum or copper wire with round, square or rectangular cross sections used for the windings of electromagnets, or as coils for transformers and electrical circuits.

Magnewin. Trade name of Wintershall AG (Germany) for a series of magnesium alloys containing 7-8% aluminum, 0-0.75% zinc and 0.1-0.2% manganese. Used for light-alloy parts.

Magnico. Trade name for a cast permanent magnet material containing 12-25.5% cobalt, 14-18% nickel, 8-10% aluminum, 2.8-6% copper, and the balance iron. It has high magnetic permeability, and is used for electrical and magnetic equipment.

Magnifer. Trademark of ThyssenKrupp VDM GmbH (Germany) for an extensive series of soft-magnetic iron-nickel and nickel-iron alloys (composition range, 48-80% nickel and 20-52% iron) exhibiting high magnetic permeabilities and low coercivities. They may also contain additions of niobium, titanium or molybdenum to enhance hardness or electrical resistivity. Used for relays, transformers, low-frequency transducers, screens, etc.

Magnifrit. Trademark of Dresser Industries Inc. (USA) for a series of refractory materials.

Magnigard. Trademark of Magni Industries Inc. (USA) for corrosion-resistant topcoat finishes for metals.

Magnil. Trade name of American Silver Company (USA) for a nonmagnetic austenitic stainless steel containing 0.1% carbon, 15.5% manganese, 18% chromium, 0-0.75% nickel, and the balance iron. Used for wear plates, bellows, springs, high-modulus diaphragms, instruments and controllers, and computers and electronic components.

magniotriplite. A reddish brown mineral of the triplite group composed of magnesium iron phosphate fluoride, $(Mg,Fe,Mn)_2PO_4F$. Crystal system, monoclinic. Density, 3.57 g/cm^3; refractive index, 1.649. Occurrence: Turkestan.

Magnit. Trade name of Voest AG (Austria) for refractory compositions of the magnesite-dolomite type used in the construction of LD steel converters.

Magno. (1) Trade name of Midland-Ross Corporation (USA) for a series of low-carbon steels containing 0.05% carbon and the balance iron. They possess high permeabilities and low hysteresis losses, and are used for magnet cores.

(2) Trade name of Forjas Alavesas SA (Spain) for a series of tool steels containing 0.9-0.95% carbon, 1.1-1.9% manganese, 0-0.4% silicon, 0.4-0.5% chromium, 0.15% vanadium, 0-0.5% tungsten, and the balance iron. Used for reamers, stamps, punches, gages, and dies.

(3) Trade name of British Driver Harris Company Limited (UK) for an alloy composed of about 95-95.5% nickel and 4.5-5% manganese. Used for electrical resistance wire (e.g., in incandescent lamps and radio tubes), and for electromagnetic applications.

magnocolumbite. A pale brown mineral of the columbite group composed of magnesium niobium oxide, $MgNb_2O_6$. It can also be made synthetically from magnesium oxide (MgO) and niobium (columbium) oxide (Nb_2O_5) under prescribed conditions. Crystal system, orthorhombic. Density, 4.99 g/cm^3. Occurrence: China.

Magnogel-44. Trademark of L'Industrie Biologique Française (France) for magnetic beads containing 4% acrylamide, 4% agarose and 7% magnetite. They have a swollen bead diameter of 60-140 μm (0.002-0.006 in.) and are designed for affinity reactions in the heterogeneous phase allowing easy separation from the reaction medium without centrifugation or filtration. Used in affinity chromatography, radioimmunoassays, immuno-enzymatic assays, and in the preparation of immunosorbents.

Magnolia. Trade name of Magnolia Metal Corporation (USA) for a series of alloys including several lead babbitts, leaded bronzes and nickel-cadmium alloys. Used for bearings, bushings, and liners.

magnolia. The pale to dark brown wood of any of a genus of American trees *(Magnolia)* belonging to the same family (Magnoliaceae) as the tuliptree. Important members include the southern magnolia *(M. grandiflora)*, the Sweetbay magnolia *(M. virginiana)*, and the cucumber magnolia *(M. acuminata)*. The southern magnolia grows from South Carolina to Texas, the sweetbay is found along the Atlantic and Gulf Coasts from Long Island to Texas, and the natural range of the cucumber magnolia is from the Appalachian Forests to the Ozark Mountains and northward to Ohio. *Magnolia* wood has a straight grain, close texture, good shock resistance, moderate hardness and stiffness, and low bending and compression strength. *Cucumber magnolia* is very similar to *yellow-poplar*. Used as lumber, and for furniture, boxes, pallets, sash, doors, Venetian blinds, millwork, veneer, cabinetmaking and woodenware.

magnolia metal. An antifriction metal composed of 78-84% lead, 15-16% antimony, 0-7% tin and 0.03% iron, and used in bearings.

Magno-Nickel. Trade name of International Nickel Inc. (USA) for a nickel alloy containing 2-6% manganese. Used for tools.

Magnorite. Trademark of Norton Company (USA) for fused magnesia (MgO) supplied in granular form with a melting point of 2800°C (5070°F). Used as a refractory, for ceramic components, and in sheathing of electric heating elements.

Magnowin. Trade name of Wintershall AG (Germany) for a magnesium alloy used for light-alloy parts.

Magnox. Trade name of Birmetals Limited (UK) for a series of magnesium alloys containing 0-1% aluminum, 0-0.6% zirconium, 0-0.03% beryllium, 0-0.008% calcium, and 0-0.006% iron. They have a melting point of approximately 640°C (1184°F), and are used for light-alloy parts, structural members, and as canning materials for uranium fuel elements.

Magnum. (1) Trademark of Dow Chemical Company (USA) for a series of transparent or colored acrylonitrile-butadiene-styrene (ABS) resins available in standard and specialty grades including high- and low-gloss, fire-retardant, impact, medium-impact. high-impact and very-high impact, high-heat and sheet coextrusion. They have good toughness and chemical resistance, good processibility, high flexibility and thermal stability, and are used for automotive components and trim, refrigeration equipment, and household appliances.

(2) Trade name of Enthone-OMI Inc. (USA) for a bright nickel electrodeposit and deposition process.

(3) Trade name of Nevada Magnesite Products Company (USA) for artificial stone made from diatomaceous earth and calcined magnesia. Used in the construction trades.

Magnum Gloss. Trademark of Mississippi Lime Company (USA) for precipitated calcium carbonate (PCC) used as a white filler pigment and brightener in the manufacture of high-quality paper and paper products.

Magnuminium. Trade name of Magnesium Castings & Products Limited (USA) for a series of wrought and cast magnesium alloys containing varying amounts of aluminum, zinc, manganese, silicon, copper, silver and tin. Used for light-alloy castings and parts, aircraft and automotive parts, housings, and structural parts.

magnussonite. A grass-green mineral composed of manganese copper chloride arsenite hydroxide, $(Mn,Cu,Mg)_5(OH,Cl)$-$(AsO_3)_3$. Crystal system, cubic. Density, 4.23 g/cm^3; refractive index, 1.980. Occurrence: Sweden.

Magnyl. Trade name for thin strips of vinyl with magnetic powder applied to one side, used for small, flexible magnets.

Magox. (1) Trademark of PSC Technologies, Inc. (USA) for magnesia powder (98% pure) derived from seawater, and used as a filler for rubber and in textiles.

(2) Trademark of Basic Inc. (USA) for magnesia and calcined magnesite based bricks and construction materials supplied in bulk and bags.

Magoxid-Coat. Trademark of Luke Engineering & Manufacturing Company (USA) for a hard anodic coating for magnesium die castings, extrusions, forgings and wrought magnesium products. The adherent multilayer coating of magnesium aluminate $(MgAlO_4)$ is developed by anodizing in an oxygen-rich, slightly alkaline electrolytic bath, and builds up to a thickness of 15-25 μm (600-1000 μin.). It consists of a very porous oxide ceramic outer layer, a nonporous oxide ceramic middle layer and an inner layer forming a hard, thin barrier. It has outstanding abrasion and corrosion resistance, a scratch hardness of 7-8 Mohs, and is used on superchargers and turbochargers, aerospace components, automotive engine and transmission parts, wheels, hand tools, and sporting goods.

Magrex. Trade name of Foseco Minsep NV (Netherlands) for covering and cleansing fluxes used in the manufacture of magnesium alloys.

Magsimal. Trade name of Aluminium Rheinfelden GmbH (Germany) for a series of aluminum-magnesium-silicon casting alloys used for automotive components. *Magsimal-R* is a high-performance aluminum-magnesium-silicon die-casting alloy.

Magtrieve. Trademark of E.I DuPont de Nemours & Company (USA) for chromium dioxide.

maguey fibers. The white, glossy, lightweight, stiff vegetable fibers obtained especially from the leaves of two agaves (*Agave lurida* and *A. tequilana*) native to Mexico and Central America. Used for cordage and binder twine. See also agave fibers.

mahoe. The hard, coarse-grained, aromatic, grayish blue wood of the tree *Hibiscus tiliaceus* var. *elatus* of the mallow family. Source: Tropical America. Used as a substitute for walnut, and for gunstocks, cabinetwork and furniture. Also known as *blue mahoe; majagua*.

mahogany. The hard yellowish-brown to reddish-brown wood of any of several trees of the families Meliaceae and Geraniales, highly valued for timber and fine furniture. See also African mahogany; American mahogany; khaya; lauan; tangile.

mahogany birch. See black birch.

Maillechort. Trade name used in Europe for a series of white-colored, corrosion-resistant nickel silvers containing 65-67% copper, 13.4-13.6% zinc, 13-19.5% nickel, 0.5-3.5% iron and small amounts of tin and lead. Used for ornamental parts, hardware, and fittings.

main-chain polymer liquid crystal. A *polymer liquid crystal* in which the anisotropically shaped primitive structural units responsible for the characteristic liquid-phase long-range order of the rod- or disk-like molecules are located in the polymer main chain or backbone. Abbreviation: MCPLC. See also liquid crystal; side-chain polymer liquid crystal.

Maingold. Trademark of Heraeus Edelmetalle GmbH (Germany) for a high-gold dental casting alloy.

Main Metal. Trade name of International Alloys Limited (UK) for casting alloys composed of 50% zinc, 48% aluminum and

2% copper, supplied in the sand-cast and permanent-mold cast condition, and used for bearings.

Maintenal. Trade name of Pyramid Steel Company (USA) for a fatigue-resistant, preheat-treated steel used for wrenches, drills, gears, pump shafts, and bolts.

maintenance mix. A ready-to-use roadbuilding product consisting of bituminous material mixed with mineral aggregate. Suitable for patching holes, cracks, depressions and other defects in pavements at ambient temperatures.

Maizite. Trade name of Prolamine Products Inc. (USA) for polymer coatings deived from corn proteins.

Maizolith. Trade name for early cornstalk-derived protein plastics developed at Iowa State College (USA).

majagua. See mahoe.

majakite. A grayish white mineral composed of palladium nickel arsenide, PdNiAs. Crystal system, hexagonal. Density, 9.33 g/cm^3. Occurrence: Russian Federation.

majolica. A decorated earthenware with relatively high water absorption and low mechanical strength. See also majolica glaze.

majolica glaze. A glossy, tin oxide-opacified, white or colored overglaze decoration on earthenware, fired at relatively low temperature. See also majolica.

Majestic. (1) Trade name of United Clay, Inc. (USA) for a white-firing, naturally occurring low-carbon porcelain clay material used for tableware, tile, and other ceramic applications.

(2) Trade name of Majestic Drug Company (USA) for a dental adhesive used for dentin and enamel bonding applications.

Major. Trade name of A. Milne & Company (USA) for a high-speed steel containing 0.7% carbon, 21% tungsten, 4% chromium, 1.5% vanadium, 0.5% molybdenum, 13% cobalt, and the balance iron. Used for tools and cutters.

Major C&B-V. Trade name of Major Prodotti Dentari SpA (Italy) for a dental acrylic material (powder and liquid) for permanent crown and bridge restorations. Processed with heat polymerization and supplied in several color shades, it provides high sealing power and is compatible with all commercial dental alloys.

majorite. A purple mineral of the garnet group composed of magnesium iron silicate, $Mg_3Fe_2(SiO_4)_3$. Crystal system, cubic. Density, 4.00 g/cm^3. Occurrence: USA.

Majority. Trademark of Degussa-Ney Dental (USA) for a hard dental alloy (ADA type III) containing 50% gold, 10% silver and 5% palladium. It has a rich yellow color, a hardness of 220 Vickers, and can be hardened to ADA type IV (extra-hard) for high-stress restorations and softened to ADA type II (medium) for onlays.

makatite. A white mineral composed of sodium silicate hydroxide hydrate, $NaSi_2O_3(OH)_3 \cdot H_2O$. Crystal system, orthorhombic. Density, 2.07 g/cm^3; refractive index, 1.487. Occurrence: Kenya.

Makelot. Trade name of Makelot Corporation (USA) for phenol-formaldehyde resins and plastics.

makinenite. An orange yellow mineral composed of nickel selenide, γ-NiSe. Crystal system, hexagonal. Density, 7.23 g/cm^3. Occurrence: Finland.

Maklen. Trademark of Hemteks (Yugoslavia) for polyester staple fibers and filament yarns.

makore. See African cherry.

makori. See African cherry.

Makroblend. Trademark of Bayer Corporation (USA) for a series of engineering thermoplastics based on polycarbonate/polyethylene terephthalate (PC-PET) and polycarbonate/polybutylene terephthalate (PC-PBT) blends. The PC-PET blends have good resistance to chemicals including hydrocarbons, good moisture resistance, good resistance to ultraviolet light and weathering, outstanding dimensional stability, high toughness at very low temperatures, high impact resistance, excellent abrasion resistance, and good flowability and processibility. *Makroblend* thermoplastics are used for lawnmower decks, snow blower parts, automobile bumpers, earphones, and leveler pads.

Makrofil. Trade name for rayon fibers and yarns used for textile fabrics.

Makrofol. Trademark of Bayer Corporation (USA) for polycarbonate film materials often made conductive with carbon fillers, and supplied in various thicknesses. They have a density range of 1.28-1.35 g/cm^3 (0.046-0.049 $lb/in.^3$), high impact strength, good tensile strength, a softening temperature of 165°C (329°F), good resistance to oils, greases, alcohols, dilute acids and alkalies, fair resistance to concentrated acids, poor resistance to halogens, ketones, aromatic hydrocarbons and strong alkalies, good dielectric properties, moderate UV resistance, a heat sealing temperature of 200-230°C (392-446°F), and low permeability to carbon dioxide and oxygen. Used for wrapping, sealing and electrical applications, ID cards, and signs.

Makrolan. Trademark of Bayer Corporation (USA) for acrylic fibers used for the manufacture of clothing.

Makrolen. Trademark of Bayer Corporation (USA) for polycarbonate resins with high impact strength, good tensile strength, a softening temperature of 165°C (329°F), good resistance to oils, greases, alcohols, dilute acids and alkalies, fair resistance to concentrated acids, poor resistance to halogens, ketones, aromatic hydrocarbons and strong alkalies, and good dielectric properties. Used for molded articles, household appliances and utensils, power-tool housings, business machines, and foils.

Makrolon. Trademark of Bayer Corporation (USA) for transparent, biocompatible polycarbonate resins available in general-purpose, extrusion, blow molding, high-flow, fire-retardant, UV-stabilized, structural-foam and glass or carbon fiber-reinforced grades. They have a density of 1.2 g/cm^3 (0.04 $lb/in.^3$), good tensile strength, a maximum service temperature of about 130-140°C (266-284°F), a deformation temperature under load of 143°C (289°F), low expansivity, excellent resistance to mineral oils and gasoline, good resistance to dilute acids and alkalies, moderate resistance to trichloroethylene, low water absorption, good dielectric properties, and fair resistance to ultraviolet resistance. Used for construction appliances, consumer electronics, business equipment, telecommunications applications, medical appliances, outdoor and decorative lighting, packaging material, compact disks, sports and leisure products, machine parts, and glazing. *Makrolon FCR* is a special flame-retardant fast-cycle grade with optimum melt viscosity and high heat resistance and microwave transparency, used for molded products, e.g., office and computer equipment housings, consoles, compact disks, electrical and electronic equipment, microwave cookware, glazing, and optical and medical equipment.

Makrolux. Trade name of Atotech USA Inc. (USA) for a bright nickel electroplate and plating process.

malachite. A deep emerald-green mineral of the rosasite group with a vitreous to silky luster. It is composed of basic copper carbonate, $Cu_2CO_3(OH)_2$, and contains theoretically 57.4% copper. It can also be made synthetically from basic copper carbonate [$CuCO_3 \cdot Cu(OH)_2$] and carbonic acid (H_2CO_3) at elevated temperatures. Crystal system, monoclinic. Density, 4.03-4.05

g/cm^3; hardness, 3.5-4 Mohs; refractive index, 1.875. Occurrence: Russia, Central Africa. Used as a commercial ore of copper, as an ornamental stone, for mosaics, and as a paint pigment. Also known as *green copper carbonate; green copper ore.*

malachite green. A green pigment made of ground *malachite* and used on stoneware, and as a green dye to indicate absorption characteristics of ceramic bodies. Also known as *Victoria green.*

malacolite. See diopside.

malanite. A bright white mineral composed of copper iridium platinum sulfide, $(Cu,Ir,Pt)S_2$. Crystal system, cubic. Density, 5.83 g/cm^3. Occurrence: China.

Mal-Arc. Trade name of CMW Inc. (USA) for an abrasion-resistant alloy composed of chromium, molybdenum and cobalt, and used for hardfacing and welding rods.

Malax. (1) Trade name of Allied Steel & Tractor Products Inc. (USA) for a series of high-speed steels containing 0.75-0.85% carbon, 4-4.25% chromium, 0-5% molybdenum, 6-19% tungsten, 1-2% vanadium, 0-7.5% cobalt, and the balance iron. Used for lathe and planer tools, drills, cutters, and forming tools.

(2) Trade name of Ato Findley (France) for adhesives, coatings and mastics used for building and decorative applications.

malayaite. A colorless, pale yellow to deep orange-yellow mineral of the tilasite group composed of calcium tin silicate, CaSnSiO$_5$. It can also be made synthetically by hydrothermal treatment of the precipitated gel. Crystal system, monoclinic. Density, 4.30 g/cm^3; refractive index, 1.784. Occurrence: Malaysia, Thailand.

Malcalloy. Japanese trade name for a permanent magnet alloy containing 86.5% cobalt and 13.5% aluminum.

Malchrome. Trade name of Allied Steel & Tractor Products Inc. (USA) for an oil-hardening tool and die steel that, in addition to iron and carbon, contains varying amounts of chromium, molybdenum and tungsten.

Mal Colloy. Japanese trade name for a magnet alloy containing 65.2% cobalt, 19.8% nickel and 15% aluminum. It develops good hard magnetic properties upon heat treatment including high coercivity, residual induction and magnetic energy product, and is used for electrical and magnetic instruments and equipment.

Mal-Die. Trade name of Allied Steel & Tractor Products Inc. (USA) for an air-hardening tool steel containing 1% carbon, 5% chromium, 1.1% molybdenum, and the balance iron. Used for forming and blanking dies.

maldonite. A pinkish to silver white mineral composed of bismuth gold, Au_2Bi. Crystal system, cubic. Density, 15.46 g/cm^3. Occurrence: Australia.

maleate. A salt or ester of maleic acid.

Malecca. Trade name of Denki Kagaku (Japan) for a series styrenic copolymers.

maleic acid. A dicarboxylic acid available as colorless crystals (99% pure) with a density of 1.590 g/cm^3 and a melting point of 140-142°C (284-288°F). Used in the synthesis of maleic resins, in the synthesis various organic acids (e.g., malic, succinic, aspartic, tartaric, propionic, lactic, malonic and acrylic acid), in dyeing and finishing of cotton, wool and silk, and as a preservative for oils and fats. Formula: $C_4H_4O_4$. Also known as *maleinic acid, cis-butenedioic acid.*

maleic anhydride. A crystalline α,β-unsaturated carbonyl compound that is available in several grades and forms including rods, flakes, lumps, briquettes (99% pure), powder (95% pure) and molten. It has a density of 0.934 g/cm³ (at 20°C or 68°F), a melting point of 51-56°C (124-133°F), a boiling point of 200°C (392°F), a flash point of 218°F (103°C), and an autoignition temperature of 890°F (476°C). Used in the synthesis of polyester resins, alkyd coating resins and permanent-press resins (e.g., for textiles), in the synthesis of fumaric and tartaric acid, in Diels-Alder reactions, and as a preservative for oils and fats. Formula: $C_4H_2O_3$.

maleic resins. A group of resins obtained by a reaction of maleic acid or maleic anhydride with glycerol and rosins.

maleinic acid. See maleic acid.

Malgaloy. Trade name of Allied Steel & Tractor Products Inc. (USA) for a shock-resistant tool steel used for shear blades.

malino. A strong fiber of considerable length obtained from the leaves of the Hawaiian aloe plant, and used for cords and ropes, and for the manufacture of aloe lace fabrics.

maline. (1) A thin, stiff silk net with hexagonal open mesh, used for millinery, veils, etc. Named after the town of Malines (or Mechelen) in Belgium. Also known as *malines*. See also tulle
(2) See Mechelen lace.

malines. See maline; Mechelen lace.

malladrite. A colorless mineral composed of sodium silicon fluoride, Na_2SiF_6. It can also be made synthetically. Crystal system, hexagonal. Density, 2.71 g/cm³; refractive index, 1.3125. Occurrence: Italy.

mallardite. A colorless mineral of the melanterite group composed of manganese sulfate heptahydrate, $MnSO_4·7H_2O$. Crystal system, monoclinic. Density, 1.85 g/cm³; refractive index, 1.465. Occurrence: Japan.

malleable brass. See Muntz metal.

malleable castings. Cast forms of a metal that has been heat-treated to reduce its brittleness.

malleable cast irons. A group of white cast irons that have been converted by an annealing process involving heating to 760°C (1400°F) followed by slow cooling under carefully controlled conditions of temperature and time. This heat treatment permits the carbon to form small rosettes or spheres of carbon (*temper carbon*) in a matrix of low-carbon iron. *Malleable cast irons* are relatively ductile, and stronger and tougher than *gray cast irons*. Depending on their particular composition, the density ranges from 7.15 to 7.60 g/cm³ (0.258-0.275 lb/in.³). Four types of malleable iron are classified: (i) Standard malleable iron; (ii) Special malleable iron; (iii) Cupola malleable iron; and (iv) Pearlitic malleable iron. *Malleable cast irons* are used for automotive castings (e.g., connecting rods, transmission gears and differential casings), marine castings, housings, hardware, valves, pipe fittings, fasteners, clutches, levers, flywheels, hand tools, vises, and chains. Also known as *malleable irons.* See also blackheart malleable iron; cupola malleable iron; ferritic malleable iron; pearlitic malleable iron; special malleable iron; standard malleable iron; whiteheart malleable iron.

malleable minerals. Minerals, such as native copper or gold, that can be flattened or plastically deformed by hammering without rupture.

malleable pig iron. A grade of *pig iron* used for making *white cast irons* employed as the starting material for malleable cast iron. Typically, it contains 3.2-3.9% carbon, 0.3-2% silicon, 0.3-0.4% manganese, up to 0.06% phosphorus, 0.02-0.15% sulfur, and the balance iron.

mallet alloy. A bearing alloy containing 75% zinc and 25% copper.

Mallix. Trade name of Capitol Castings Inc. (USA) for a pearlitic

malleable cast iron containing 1.8% carbon, 1% silicon, 0.5% manganese, and the balance iron.

Mallory. Trademark of CMW Inc. (USA) for a series high-conductivity bronzes, wrought and cast copper-beryllium alloys, copper-nickel-silicon alloys, and several sintered copper and silver- and tungsten-base electric contact alloys.

Mallory No-Chat. Trademark of CMW Inc. (USA) for a sintered copper-tungsten alloy with an electrical conductivity of 14% IACS used for electric contacts.

M-Alloy. (1) Trade name of George Cook & Company Limited (UK) for a tough, oil-hardening tool steel containing 0.58% carbon, 0.35% manganese, 0.95% chromium, 0.95% silicon, 0.2% molybdenum, and the balance iron. Used for form tools, plastic mold dies, and hot and cold punches.
(2) Trade name of Thomas Bolton Limited (UK) for a corrosion- and wear-resistant steel used for aircraft forgings and parts.

Malloy. Trade name of George Cook & Company Limited (UK) for a shock-resisting tool steel containing 0.6% carbon, 1.1% silicon, 1.1% chromium, 0.45% manganese, 0.25% molybdenum, and the balance iron. Used for shear blades, punches, and dies.

Malloydium. Trade name for an acid resistant nickel silver used for tableware.

malmstone. A variety of *chert* used as a foundation material in building and paving.

Malon. Trademark of OHIS (Yugoslavia) for acrylic fibers used for textile fabrics.

Malora. Trade name of Malora SA (France) for a metallic fiber.

Malotte's metal. A fusible alloy of 46% bismuth, 20% lead and 34% tin having a melting range of 96-123°C (205-253°F).

Malta. Trade name of Jessop Steel Company (USA) for a series of sintered carbides composed of tungsten carbide and a cobalt binder, and used for cutting tools, tipped tools, etc.

Maltese lace. A *bobbin lace* similar to *Cluny lace*, made of heavy yarns, and usually having regular geometric designs featuring squares and wheels. Used especially on tablecloth and handkerchiefs.

Maltex. Trade name of Stone Manganese–J. Stone & Company Limited (UK) for a series of lead-base bearing metals containing 4-40% tin and 14-17% antimony.

Maluminum. Trade name for an aluminum alloy containing 6.4% copper, 4.8% zinc, 1.4% iron, 0.2% silicon, 0.1% manganese and 0.1% lead. Used for light-alloy parts.

MA materials. See mechanically alloyed materials.

Mammoglas. Trade name of Gottlieb Sanderfeldt & Co. (Germany) for a glass veined like marble.

Man. Trade name of Boehler GmbH (Austria) for a water-hardening steel containing 0.3% carbon, 0.25% silicon, 1.35% manganese, and the balance iron. Used for gears, shafts and machine-tool parts.

manasseite. A white or pale bluish mineral of the sjogrenite group composed of magnesium aluminum carbonate hydroxide tetrahydrate, $Mg_6Al_2CO_3(OH)_{16}·4H_2O$. Crystal system, monoclinic. Density, 2.05 g/cm³; refractive index, 1.524. Occurrence: Norway.

Manaurite. Trade name of Usines de A. Manoir, Pitres (France) for several nickel-chromium-tungsten and nickel-iron-chromium alloys including various austenitic stainless and heat-resisting steels containing 15-25% chromium and 8-35% nickel.

mandarinoite. A light yellowish green mineral composed of iron selenite hexahydrate, $Fe_2(SeO_3)_3·6H_2O$. Crystal system, monoclinic. Density, 2.93 g/cm³; refractive index, 1.80. Occurrence:

Bolivia.

Mandur. Trade name of Aluminium-Zentrale e.V. (Germany) for an age-hardenable, high-strength aluminum alloy containing 3.5-5.5% copper, 0.3-1% silicon, 0.5-1% manganese and 0.5-1.2% magnesium. Used for housings, crankcases, engine cylinder heads, and oil pans.

Mandura. Trade name of KM-Kabelmetall AG (Germany) for a copper alloy that contains 3% silicon and 1% manganese, and is used for chemical equipment and components, and pressure vessels.

Manelec. Trade name of Empire Sheet & Tin Plate Company (USA) for a soft magnetic alloy containing 99% iron and 1% silicon. It has high permeability, a small hysteresis loop, and is used for motors and electrical equipments.

Mangabraze. Trade name of Baldwin Steel Company (USA) for a wear- and impact-resistant, heat-treated steel containing 0.35% carbon, 1.92% manganese, 0.42% copper, 0.32% molybdenum, 0.28% silicon, and the balance iron. Used for hoppers, chutes, screens, conveyors, scrapers, buckets, and liners.

Manga-Kote. Trade name of Grand Northern Products Limited (USA) for a coated hardfacing electrode containing 0.8% carbon, 14% manganese, 4.7% chromium, 3.5% nickel, 3.24% molybdenum, 0.8% silicon, 0.1% boron, and the balance iron. Used for overlay and buildup of crusher parts, bucket teeth, grading buckets, and tractor rollers.

Mangal. Trade name of VAW-Vereinigte Aluminium-Werke AG (Germany) for a non-heat-treatable, corrosion-resistant aluminum alloy containing 1.5% manganese. Used for roofing, trim, commercial vehicles, and structural applications.

Mangalal. Trade name of Alcan-Booth Industries, Limited (UK) for a wrought aluminum alloy containing 1.2% manganese and 0.5% iron. Usually supplied in sheet form, it is used for roofing, containers, and cooking utensils.

Mangaloy. (1) British trade name for a heat-resistant alloy of iron, nickel and manganese, used for electrical resistors.

(2) Trade name of Bergstrom Alloys Corporation (USA) for a manganese steel containing varying amounts of carbon, manganese, nickel, chromium and iron. Used for hardfacing electrodes.

Mangan. Trade name of Krupp Stahl AG (Germany) for wear-resistant alloy steel containing 1.1-1.3% carbon, 0.3-0.5% silicon, 12-13% manganese, 0-0.1% phosphorus, 0-0.04% sulfur and 0-1.5% chromium, and the balance iron.

Manganal. (1) Trade name of Stulz Sickles Steel Company (USA) for an abrasion-resistant, high-strength plate steel containing 0.6-0.9% carbon, 11-14% manganese, 2.5-3.5% nickel, and the balance iron. Used for welding rods employed in resurfacing broken and worn high-manganese steel parts.

(2) Trade name of Sovirel (France) for a special infrared signaling glass.

manganaxinite. A pale brown mineral of the axinite group composed of calcium manganese iron aluminum boron silicate hydroxide, $Ca_2(Mn,Fe)Al_2BSi_4O_{15}(OH)$. Crystal system, triclinic. Density, 3.32 g/cm³; refractive index, 1.678. Occurrence: USA (Minnesota).

manganbabingtonite. A mineral of the pyroxenoid group composed of calcium manganese iron silicate hydroxide, $Ca_2(Mn,Fe)FeSi_5O_{14}OH$. Crystal system, triclinic. Density, 3.45 g/cm³; refractive index, 1.730. Occurrence: Russian Federation.

manganberzeliite. A yellowish to brown mineral of the garnet group composed of sodium calcium manganese arsenate, $NaCa_2Mn_2As_3O_{12}$. It can also be made synthetically. Crystal system,

cubic. Density, 4.46 g/cm³; refractive index, 1.777. Occurrence: Sweden.

manganblende. See alabandite.

manganepidote. See piemontite (1).

manganese. A hard, brittle, silvery metallic element of Group VIIB (Group 7) of the Periodic Table. It is commercially available in the form of chips, flakes, powder, foil, microfoil and vacuum melted lumps. The most important ores are *manganite, psilomelane, pyrolusite* and *rhodocrosite.* Density, 7.44 g/cm³; melting point, 1244°C (2271°F); boiling point, 1962°C (3564°F); hardness, 5 Mohs; atomic number, 25; atomic weight, 54.938; monovalent, divalent, trivalent, tetravalent, hexavalent, heptavalent; nonmagnetic; high electrical resistivity (about 160 μΩcm). Four allotropic forms are known: (i) *Alpha manganese* (body-centered cubic crystal structure) which exists at room temperature; (ii) *Beta manganese* (simple cubic crystal structure) which is present at temperatures between 742 and 1095°C (1368 and 2003°F); (iii) *Gamma manganese* (face-centered cubic crystal structure) which exists at temperatures between 1095 and 1133°C (2003 and 2071°F); and (iv) *Delta manganese* (body-centered cubic crystal structure) which exists at temperatures between and 1133 and 1244°C (2071 and 2271°F). *Manganese* is used in ferroalloys (e.g., as a deoxidizer and desulfurizer in steel manufacture), as an addition to austenitic, pearlitic, low-carbon and powder-metallurgy steels, nonferrous alloys (e.g., for improving corrosion resistance and hardness in alloys with copper and nickel), nickel-iron alloys, permanent magnet materials, titanium and magnesium alloys and superalloys, as a purifier and scavenger in metal production, in the manufacture of aluminum by the Toth process, in the manufacture of glass and paint, and in batteries. Symbol: Mn.

manganese acetate. See manganese acetate tetrahydrate.

manganese acetate dihydrate. Brown, moisture-sensitive crystals or powder (97+% pure) used as catalyst, and as mild selective oxidizing agent. Formula: $Mn(C_2H_3O_2)_3 \cdot 2H_2O$. Also known as *manganic acetate dihydrate.*

manganese acetate tetrahydrate. Pink crystals or powder (99+% pure). Density, 1.589 g/cm³; melting point, 80°C (176°F). Used as paint and varnish drier, in textile dyeing and leather tanning, as a catalyst, and in biochemistry. Formula: $Mn(C_2H_3O_2)_2 \cdot 4H_2O$. Also known as *manganese acetate; manganous acetate.*

manganese-alumina pink. A pink ceramic colorant composed of a calcined mixture of manganese carbonate ($MnCO_3$), alumina trihydrate ($Al_2O_3 \cdot 3H_2O$) and borax ($Na_2B_4O_7 \cdot 10H_2O$).

manganese aluminate. A compound of manganese oxide and aluminum oxide with a density of 4.12 g/cm³ and a melting point of 1560°C (2840°F). Used in ceramics and materials research. Formula: $MnAl_2O_4$.

manganese aluminum silicate. Any of the following compounds of manganese oxide, aluminum oxide and silicon oxide used in ceramics and materials research: (i) *Dimanganese dialuminum pentasilicate.* Melting point, 1297°C (2367°F); hardness, 5-7 Mohs. Formula: $Mn_2Al_2Si_5O_{18}$; and (ii) *Trimanganese aluminum trisilicate.* Density, 4.18 g/cm³; melting point, 1198°C (2188°F); hardness, 5-7 Mohs. Formula: $Mn_3Al_2Si_3O_{12}$.

manganese-aluminum brass. A group of strong, wear- and corrosion-resistant cast and wrought copper alloys containing varying amounts of zinc, aluminum and manganese.

manganese antimonide. Any of the following compounds of manganese and antimony used in ceramics, electronics and materials research: (i) *Manganese antimonide.* Hexagonal crystals with a density of 6.9 g/cm³ and a melting point of 840°C

(1544°F). Formula: MnSb; and (ii) *Dimanganese antimonide.* Tetragonal crystals with a density of 6.99 g/cm³ and a melting point of 948°C (1738°F). Formula: Mn₂Sb.

manganese arsenide. Any of the following compounds of manganese and arsenic used in ceramics, electronics and materials research: (i) *Manganese arsenide.* Black crystals with a density of 5.55 g/cm³. Formula: MnAs; (ii) *Dimanganese arsenide.* Tetragonal crystals with a melting point of 760°C (1400°F) Formula: Mn₂As; and (iii) *Trimanganese arsenide.* Magnetic crystals. Formula: Mn₃As.

manganese binoxide. See manganese dioxide.

manganese black. See manganese dioxide.

manganese borate. A reddish-white powder obtained by treating manganese hydroxide with boric acid, and used as a varnish and oil drier. Formula: MnB₄O₇. Also known as *manganese tetraborate.*

manganese boride. Any of the following compounds of manganese and boron used in ceramics and materials research: (i) *Manganese monoboride.* Orthorhombic crystals. Density, 6.2-6.45 g/cm³; melting point, 1890°C (3434°F). Formula: MnB; (ii) *Manganese diboride.* Grayish crystals. Crystal system, hexagonal. Density of 5.3 g/cm³; melting point, 1827°C (3321°F). Formula: MnB₂; (iii) *Dimanganese boride.* Red-brown crystals. Crystal system, tetragonal. Density, 7.2 g/cm³; melting point, 1580°C (2876°F). Formula: Mn₂B; (iv) *Manganese tetraboride.* Orthorhombic crystals. Density, 6.12 g/cm³. Formula: Mn₃B₄; and (v) *Tetramanganese boride.* Orthorhombic crystals. Formula: Mn₄B.

manganese-boron. A master alloy composed of 70-80% manganese, 20-25% boron, and small amounts of iron, aluminum and silicon, and used as a hardener and deoxidizer for brasses and bronzes, and in the manufacture of other nonferrous alloys.

manganese brass. A corrosion-resistant brass containing 70% copper, 28.7-29% zinc and 1-1.3% manganese. It possesses excellent cold formability, and is used for communications equipment, welded assemblies, and for brass products joined by resistance spot, seam or butt welding.

manganese briquettes. Ferromanganese that has been crushed, bonded with a special refractory and made into block shaped forms known as "briquettes." See also ferromanganese.

manganese bronze. (1) A tough, corrosion-resistant wrought bronze containing 58.5% copper, 39.0% zinc, 1.4% iron, 1.0% tin and 0.1% manganese. It has excellent hot workability and good strength, and is used for marine parts, clutch disks, pump rods, shafts and balls, and valve stems and bodies.

(2) A high-strength, corrosion-resistant cast bronze containing 55-67% copper, 21-42% zinc, 1-3% iron, 1-6% aluminum and 3-4% manganese. It has good castability and resistance to seawater corrosion, and is used for marine castings, gears, gun mounts, engine frames, bushings and bearings, marine racing propellers, valve stems, lever arms, screwdown nuts, hydraulic cylinder parts, and machinery parts.

manganese bromide. See manganous bromide.

manganese bromide tetrahydrate. See manganous bromide tetrahydrate.

manganese carbide. A refractory, crystalline solid. Crystal system, tetragonal. Density, 6.89 g/cm³; melting point, 1520°C (2768°F). Used in ceramics and materials research. Formula: Mn₃C.

Manganese Carbon Alloy. Trade name of LaSalle Steel Company (USA) for a steel containing 0.35-0.45% carbon, 1.2% manganese, and the balance iron. Used for spindles, pins, studs, bolts, and driveshafts.

manganese carbonate. Rose-pink crystals, or light brown amorphous powder. It occurs in nature as the mineral *rhodocrosite.* Density, 3.1-3.2 g/cm³; melting point, decomposes at 350°C (662°F). Used chiefly in ceramics, e.g., as a black, brown and purple colorant in glazes, as a paint pigment, and in chemistry, materials research, and chemical engineering. Formula: MnCO₃. Also known as *manganous carbonate.*

manganese carbonyl. Yellow, air- and heat-sensitive crystals (98% pure). Density, 1.75 g/cm³; melting point, 152-155°C (306-311°F). Used in the vapor deposition of manganese coatings, in antiknock gasoline, and as a catalyst. Formula: Mn₂(CO)₁₀. Also known as *dimanganese decacarbonyl; decacarbonyl dimanganese.*

manganese casting brass. A high-strength brass containing 58.5% copper, 40% zinc, 0.85% tin, 0.5% aluminum, 0-1.5% lead and 0.15% manganese. Used as a substitute for malleable cast iron.

manganese chloride. See manganous chloride.

manganese chloride tetrahydrate. See manganous chloride tetrahydrate.

manganese copper. (1) A group of heat- and corrosion-resistant alloys containing 29.2-89.7% copper, 8.7-51.7% manganese, 0-9.7% iron, 0-6.25% aluminum, 0-3.3% carbon, 0-1.1% silicon, 0-2.1% zinc, 0-0.4% tin and 0-0.5% lead. Used for resistances, corrosion- and heat-resistant parts.

(2) See copper-manganese.

manganese diboride. See manganese boride (ii).

manganese difluoride. See manganous fluoride.

manganese dioxide. Black crystals or powder (85-99.99% pure). It occurs in nature as the minerals *polianite* and *pyrolusite.* Crystal system, tetragonal. Crystal structure, rutile. Density, 5.026 g/cm³; melting point, decomposes to manganic oxide (Mn₂O₃) and oxygen at 535°C (995°F); hardness, 2.0-2.5 Mohs. Used as a colorant, decolorizer and scavenger in glass enamels and glazes, as a colorant in brick pigments, in paint driers, in alloy steels, cast irons and wrought irons, in uranium ore processing, in ferrites, in the manufacture of electrolytic zinc, as a depolarizer for dry cell batteries (battery manganese), as an oxidizing agent, in biochemistry, and in pyrotechnics. Formula: MnO₂. Also known as *battery manganese; manganese black; manganese peroxide.*

manganese disilicide. See manganese silicide (ii).

manganese driers. Compounds of manganese, such as manganese linoleate, manganese naphthenate, manganese oleate, manganese oxalate and manganese resinate, that are used as paint and varnish driers. They are classified as "through" driers as they act on both the top and the body of the paint or varnish film.

manganese dysprosium sulfide. Monoclinic, ceramic crystals with a melting range of 1577-1637°C (2871-2979°F) used in materials research. Formula: MnDy₂S₄.

manganese erbium sulfide. Monoclinic, ceramic crystals with a melting range of 1613-1673°C (2935-3043°F) used in materials research. Formula: MnEr₂S₄.

manganese 2-ethylhexanoate. A compound made from 2-ethylhexoic acid and manganous hydroxide and available as a clear, brown, flammable solution in mineral spirits containing 6-10% manganese. Used as a drier for enamels, paints, printing inks, and varnishes. Formula: Mn[OOCC₆H₁₃(C₂H₅)]₂. Also known *manganous 2-ethylhexanoate; manganous octoate; manganese octoate.*

manganese ferrite. A ferrimagnetic compound of manganese oxide (MnO) and ferric oxide (Fe_2O_3) with an inverse spinel structure. Density, 4.74 g/cm³; melting point, 1571°C (2860°F). Used for low-conductivity ceramic magnets, and in communication equipment. Formula: $MnFe_2O_4$.

manganese fluoride. See manganic fluoride; manganous fluoride.

manganese green. See Cassel green.

manganese halide. A binary compound of manganese and a halogen, such as bromine, chlorine, fluorine or iodine.

manganese-hoernesite. A white mineral of the vivianite group composed of manganese magnesium arsenate octahydrate, $(Mn,Mg)_3(AsO_4)_2 \cdot 8H_2O$. Crystal system, monoclinic. Density, 2.64 g/cm³; refractive index, 1.589. Occurrence: Sweden.

manganese holmium sulfide. Monoclinic, ceramic crystals with a melting range of 1567-1627°C (2853-2961°F) used in materials research. Formula: $MnHo_2S_4$.

manganese hydroxide. See manganic hydroxide; manganous hydroxide.

manganese-iron ferrite. A commercially important powder product with spinel structure, relatively high magnetic saturation magnetization and low core losses. Used for soft magnets. Formula: $Mn_xFe_{3-x}O_4$ (x = 0-3).

manganese iodide. See manganous iodide.

manganese linoleate. A dark brown mass made by boiling a manganese salt, sodium linoleate and water, and used as a paint and varnish drier. Formula: $Mn(C_{18}H_{31}O_2)_2$.

manganese molybdate. An off-white powder with a melting point above 350°C (660°F) used for optical and electronic applications. Formula: $MnMoO_4$. Also known as *manganous molybdate*.

manganese monoboride. See manganese boride (i).

manganese monophosphide. See manganese phosphide (i).

manganese monosilicide. See manganese silicide (i).

manganese monoxide. See manganous oxide.

manganese naphthenate. A hard, brown, resinous mass commercially available as a viscous liquid solution of 6-10% divalent manganese in mineral spirits or other solvent. Melting point, 130-140°C (266-284°F). Used as paint and varnish drier, and as a polymerization initiator.

manganese nickel. (1) A group of heat- and corrosion-resistant nickel alloys containing 1.5-5% manganese. Used for fittings, lead wires for electrical appliances, spark-plug electrodes, and formerly for grid and support wires in radio valve lamps.

(2) An alloy of 52-85% copper, 14-31% manganese and 3-16% nickel used for heat-resistant parts.

manganese nickel silver. A corrosion-resistant *nickel silver* containing 60-73% copper, 10-17% nickel, 2.4-20% manganese, and up to 10% tin and 8.8% zinc, respectively. Used for white metal parts.

manganese nitrate. See manganous nitrate.

manganese nitrate hexahydrate. See manganous nitrate hexahydrate.

manganese nitride. A compound of manganese and nitrogen in the form fine powder used in ceramics and materials research. Formula: Mn_4N.

manganese octoate. See manganese 2-ethylhexanoate.

manganese oleate. A brown, granular mass made by boiling manganese chloride, sodium oleate and water, and used as a paint and varnish drier. Formula: $Mn(C_{18}H_{33}O_2)_2$.

manganese oxalate. A white, crystalline powder. Density, 2.453 g/cm³; melting point, loses $2H_2O$ at 100°C (212°F). Used as paint and varnish drier. Formula: $MnC_2O_4 \cdot 2H_2O$.

manganese oxide. See manganese dioxide; manganic oxide; manganous oxide; manganese tetroxide.

manganese-palladium. A compound of manganese and palladium. It has a melting point of 1516°C (2761°F), and is used in ceramics and materials research. Formula: MnPd.

manganese peroxide. See manganese dioxide.

manganese phosphate. A reddish-white powder obtained by treating manganous hydroxide with orthophosphoric acid, and used for conversion coatings on steels, aluminum and other metals. Formula: $Mn_3(PO_4)_2 \cdot 7H_2O$. Also known as *manganous phosphate; manganous orthophosphate*.

manganese phosphate coatings. Dark gray to black, usually fine-grained coatings applied by immersion to ferrous components, such as bearings, gears and internal-combustion engine parts, for break-in galling prevention and as a corrosion-resistant base for paint. See also phosphate coatings.

manganese phosphide. Any of the following compounds of manganese and phosphorus used in ceramics, materials research and electronics: (i) *Manganese monophosphide*. Dark gray crystals. Crystal system, orthorhombic. Density, 5.49 g/cm³; melting point, 1147°C (2097°F); exhibits metamagnetic properties. Formula: MnP; (ii) *Dimanganese phosphide*. Hexagonal crystals. Density, 6-6.33 g/cm³; melting point, 1315°C (2399°F) [also reported as 1327°C (2421°F). Formula: Mn_2P; (iii) *Trimanganese phosphide*. Crystals. Density, 6.70 g/cm³; melting point, 1229°C (2244°F). Formula: Mn_3P; and (iv) *Trimanganese diphosphide*. Dark gray crystals. Density, 5.12 g/cm³; melting point, 1204°C (2199°F). Formula: Mn_3P_2.

manganese phthalocyanine. A purple crystalline *phthalocyanine* derivative that contains a divalent central manganese ion (Mn^{2+}). It exhibits canted antiferromagnetic properties and has maximum absorption wavelength of 661 nm. Used for electronic, microelectronic, biochemical and medical application. Formula: $MnC_{32}H_{16}N_8$. Abbreviation: MnPc. Also known as *manganous phthalocyanine*.

manganese protoxide. See manganous oxide.

manganese red brass. A red brass composed of 85% copper, 14% zinc and 1% manganese, and used in strip form for resistance spot and seam welded products.

manganese resinate. A dark, brownish-black mass, or pinkish white to yellow powder made by boiling manganese hydroxide, rosin oil and water. Used as a varnish and oil drier. Formula: $Mn(C_{20}H_{29}O_2)_2$.

manganese screwstock. A manganese steel containing 0.15-0.25% carbon, 0.9-1.2% manganese, 0.05-0.15% sulfur, and the balance iron. Used for case-hardened gears, shafts and other parts.

manganese selenide. A compound of manganese and selenium available in the form of black, hexagonal crystals (99.9+% pure) and gray, cubic, refractory crystals. Properties of hexagonal form: Crystal structure, wurtzite (zincite). Properties of cubic form: Density, 5.45 g/cm³; melting point, 1460°C (2660°F); refractive index, 2.697; band gap, 2.3 eV. Used in ceramics, and in materials research and electronics as a semiconductor. Formula: MnSe.

manganese-sepiolite. A red mineral of the sepiolite group composed of iron manganese silicate hydroxide octahydrate, $Mn_5Fe_5Si_{12}O_{30}(OH)_6 \cdot 8H_2O$. Crystal system, monoclinic. Density, 2.36 g/cm³. Occurrence: Greenland.

manganese sesquifluoride. See manganic fluoride.

manganese sesquioxide. See manganic oxide.

manganese-shadlunite. A grayish yellow mineral of the pentlandite group composed of copper iron manganese lead sulfide,

(Cu,Fe)$_8$(Mn,Pb)S$_8$. Crystal system, cubic. Density, 4.56 g/cm^3. Occurrence: Russian Federation.

manganese silicate. Any of the following compounds of manganese oxide and silicon dioxide: (i) *Manganous silicate.* Red crystals or yellowish-red powder. Density, 3.72 g/cm^3; melting point, 1323°C (2413°F). Used as a colorant for glass and ceramic glazes, and in materials research. MnSiO$_3$; (ii) *Dimanganese silicate.* Density, 4.05 g/cm^3; melting point, 1340°C (2444°F). Used in ceramics and materials research. Formula: Mn$_2$SiO$_4$.

manganese silicate flux. An acid flux consisting of 38-45% silicon dioxide (SiO$_2$), 36-42% manganese oxide (MnO), 4-7% calcium fluoride (CaF$_2$), up to 3% magnesium oxide (MgO), and small additions of calcium oxide (CaO), barium oxide (BaO), aluminum oxide (Al$_2$O$_3$), iron oxide (FeO), and other compounds. It has excellent operating characteristics including good current/voltage stability, moderate strength, good tolerance to rust, high heat input, good storage properties, allows for fast welding speeds and produces weld deposits with excellent bead shapes. Used in submerged-arc welding.

manganese silicide. Any of the following compounds of manganese and silicon used in ceramics and electronics: (i) *Manganese monosilicide.* Density, 5.9 g/cm^3; melting point, 1280°C (2336°F). Formula: MnSi; (ii) *Manganese disilicide.* Fine powder. Density, 5.24 g/cm^3. Formula: MnSi$_2$; (iii) *Trimanganese silicide.* Density, 6.60-6.71 g/cm^3; melting point, decomposes at 1120°C (2048°F). Formula: Mn$_3$Si; and (iv) *Pentamanganese trisilicide.* Density, 6.02 g/cm^3; melting point, decomposes at 1283°C (2341°F). Formula: Mn$_5$Si$_3$.

manganese-silicon. A master alloy composed of 73-78% silicon, 20-25% manganese, up to 1.5% iron, and up to 0.25% carbon, used to introduce manganese and silicon into metals.

manganese spar. See rhodocrosite.

manganese stainless steels. Austenitic stainless steels with excellent corrosion resistance, tensile strength, wear and abrasion resistance, and improved performance at elevated temperatures. A typical composition is 0.03-0.25% carbon, 7.0-16.0% manganese, 1.0% silicon, 16.0-22.0% chromium, 1.0-9.0% nickel, traces of nitrogen, phosphorus and sulfur, and the balance iron.

manganese steels. See carbon-manganese steels; Hadfield steels.

manganese sulfate tetrahydrate. See manganous sulfate tetrahydrate.

manganese sulfide. See manganic sulfide; manganous sulfide.

manganese tallate. The manganese salts of liquid rosin (tall oil) fatty acids. Commercial manganese tallate solutions contain 6-10% manganese, and are used as paint, varnish and oil driers.

manganese telluride. A crystalline compound of manganese and tellurium. Crystal system, hexagonal. Crystal structure, wurtzite (or zincite). Density, 6.0 g/cm^3; melting point, 1150°C (2100°F); band gap, 1 eV. Used as a semiconductor, and in ceramics and materials research. Formula: MnTe.

manganese tetraborate. See manganese borate.

manganese tetraboride. See manganese boride (iv).

manganese tetroxide. Brown crystals or brownish, hygroscopic powder (98+% pure). It occurs in nature as the mineral *hausmannite*, and is also generated in the pouring and casting of molten *ferromanganese*. Crystal system, tetragonal. Density, 4.84 g/cm^3; melting point, 1705°C (3101°F); hardness, 5.0-5.5 Mohs; refractive index, 2.46. Used in ceramics and metallurgy. Formula: Mn$_3$O$_4$. Also known as *manganous-manganic oxide; trimanganese tetroxide.*

manganese titanate. Any of the following compounds of manganese oxide and titanium dioxide used in ceramics and electronics: (i) *Manganese titanate.* Density, 4.54 g/cm^3; melting point 1359°C (2478°F). Formula: MnTiO$_3$; and (ii) *Dimanganese titanate.* Density, 4.5 g/cm^3; melting point 1454°C (2649°F). Formula: Mn$_2$TiO$_4$.

manganese-titanium. A foundry alloy used as a deoxidizing agent in high-grade steel and nonferrous alloys. Typically, it contains 38% manganese, 29% titanium, 22% iron, 8% aluminum, and 3% silicon. The standard grade has a melting point of 1454°C (2649°F), and the special grade of 1332°C (2430°F).

manganese violet. A violet paint pigment composed of manganese ammonium pyrophosphate. Also known as *permanent mauvre; permanent violet.*

manganese yttrium sulfide. Monoclinic, ceramic crystals with a melting range of 1562-1622°C (2844-2952°F) used in materials research. Formula: MnY$_2$S$_4$.

manganese-zinc ferrite. A commercially important powder product with cubic spinel microstructure, high magnetic saturation magnetization, low hysteresis losses, and a relatively high Curie temperature. Used for soft magnets. Formula: Mn$_x$Zn$_{1-x}$Fe$_2$O$_4$ (x = 0-1).

manganese-zirconium. A compound of manganese and zirconium. Melting point, 1500°C (2732°F). Used in ceramics. Formula: MnZr.

manganhumite. A brownish orange mineral of the humite group composed of magnesium manganese silicate hydroxide, (Mn,Mg)$_7$(SiO$_4$)$_3$(OH)$_2$. Crystal system, orthorhombic. Density, 3.83 g/cm^3; refractive index, 1.712. Occurrence: Sweden.

manganic acetate dihydrate. See manganese acetate dihydrate.

manganic fluoride. Red, hygroscopic crystals or moisture-sensitive, red powder (98+% pure). Crystal system, monoclinic. Density, 3.54 g/cm^3; melting point, decomposes above 600°C (1112°F); high magnetic susceptibility. Used as a fluorinating agent, and in ceramics, electronics, and materials research. Formula: MnF$_3$. Also known as *manganese sesquifluoride; manganese fluoride.*

manganic hydroxide. Grayish-black crystals or brown powder. Density, 4.2-4.4 g/cm^3; melting point, decomposes on heating; rapidly loses water to form manganese oxide hydroxide, MnO(OH). Formula: Mn(OH)$_3$. Used in ceramics, and as a pigment in textiles. Also known as *manganese hydroxide.*

manganic oxide. Black crystals or lustrous powder (99+% pure). It occurs in nature as the mineral *braunite*. Crystal system, cubic. Density, 4.32-4.82 g/cm^3; melting point, loses oxygen at 1080°C (1976°F); hardness, 4.0-5.0 Mohs. Used as a source of manganese and in materials research. Formula: Mn$_2$O$_3$. Also known as *manganese sesquioxide; manganese oxide.*

manganic sulfide. Black crystals. It occurs in nature as the mineral *hauerite*. Crystal system, cubic. Density, 3.46-3.50 g/cm^3. Used in materials research. Formula: MnS$_2$.

Manganin. Trademark of Isabellenhütte Heusler GmbH KG (Germany) for a series of electrical resistance alloys containing about 75-87% copper, 10-25% manganese and up to 5% nickel. They are commercially available in the form of foils, wires, insulated wires, powders and sheets. The standard 86Cu-12Mn-2Ni alloy has a density of 8.4 g/cm^3 (0.30 lb/in.^{3}), a melting point of 960°C (1760°F), a maximum working temperature of 140°C (285°F), a maximum service temperature (in air) of 300°C (570°F), a low temperature coefficient of resistance, high electrical resistivity, and good strength. Used for precision resistors, thermocouples, coils and shunt wires in electric instruments, instrument springs, and electrical and magnetic equipment.

Manganingot.   Trade name of Specialloy Inc. (USA) for a fused alloy containing 99.5+% manganese, used to introduce elemental manganese into ferrous and nonferrous alloys.

manganite.   A gray-black mineral with a reddish-brown to dark brown streak and a submetallic luster. It is composed of manganese oxide hydroxide, MnO(OH), and contains 62.4% manganese, 27.3% oxygen and 10.3% water. It can also be made synthetically. Crystal system, orthorhombic. Density, 4.2-4.4 g/cm^3; hardness, 4-6 Mohs; refractive index, 2.25. Occurrence: Germany, Sweden, Canada, UK, USA (Colorado, Michigan). Used as an important ore of manganese, and as a source of manganese dioxide (MnO_2). Also known as *gray manganese ore.*

manganites.   A class of manganese oxides including calcium manganite ($CaMnO_3$), lanthanum manganite ($LaMnO_3$) and calcium-doped lanthanum manganite ($La_{1-x}Ca_xMnO_3$) having body-centered cubic crystal systems or *perovskite* crystal structures, and exhibiting colossal magnetoresistance, i.e., extremely large increases in electrical resistivity in applied magnetic fields. They are suitable for various applications in magnetic data processing and storage.

mangan-neptunite.   A black mineral composed of lithium potassium sodium iron manganese titanium silicate, $Na_2KLi(Mn,Fe)_2Ti_2Si_8O_{24}$. Crystal system, monoclinic. Density, 3.23 g/cm^3; refractive index, 1.700. Occurrence: Russian Federation.

Mangan-Neusilber.   German trade name for a series of corrosion-resistant alloys containing 59-73% copper, 10-18% nickel, 2.4-20% manganese, and 5-20% zinc. Used for white metal parts.

Mangannickel.   Trade name of VDM Nickel-Technologie AG (Germany) for a series of corrosion-resistant nickel alloys containing 1.5-5.5% manganese and up to 0.25% carbon. Used for water meter parts, thermocouples, and spark plugs.

Mangano.   Trade name of Latrobe Steel Company (USA) for a nondeforming, oil-hardening steel containing 0.95% carbon, 1.65% manganese, 0.25% silicon, and the balance iron. Used for tools, dies, and gages.

manganocene.   An organometallic coordination compound that is a molecular sandwich of a manganese atom (Mn^{2+}) and two five-membered cyclopenediene (C_5H_5) rings, one located above and the other below the manganese atom plane. *Manganocene* is supplied in the form of air- and moisture-sensitive brown crystals (97% pure) (usually in the sublimed state) with a melting point of 175°C (347°F) and a flash point of 125°F (51°C). Used as a precursor in superconductivity research. Formula: $(C_5H_5)_2Mn^{2+}$. Also known as *dicyclopentadienylmanganese; bis-(cyclopentadienyl)manganese.*

manganochromite.   A brownish gray mineral of the spinel group composed of iron manganese chromium vanadium oxide, (Mn,-Fe)(Cr,V)$_2O_4$. Crystal system, cubic. Density, 4.90 g/cm^3. Occurrence: Australia.

manganocolumbite.   A pale green mineral belonging to the columbite group composed of manganese niobate, $MnNb_2O_6$. It can also be made by heating manganese dioxide (MnO_2) and niobium pentoxide (Nb_2O_5) under prescribed conditions. Crystal system, orthorhombic. Density, 5.28 g/cm^3. Occurrence: Russian Federation.

manganoepidote.   See piemontite (1).

manganolangbeinite.   A pale pink mineral composed of potassium manganese sulfate, $K_2Mn_2(SO_4)_3$. It can also be made synthetically from potassium sulfate (K_2SO_4) and manganese sulfate

($MnSO_4$) under prescribed conditions. Crystal system, cubic. Density, 3.06 g/cm^3; refractive index, 1.576. Occurrence: Italy.

manganosite.   A green mineral of the halite group composed of manganese oxide, MnO. Crystal system, cubic. It can also be made synthetically. Density, 5.36 g/cm^3; hardness, 5.0-6.0 Mohs; refractive index, 2.19. Occurrence: USA (New Jersey). Used in the manufacture of magnets, ceramics, colored glass, and paints.

manganostibite.   A black mineral composed of manganese antimony arsenic oxide, Mn_7SbAsO_{12}. Crystal system, orthorhombic. Density, 4.95 g/cm^3; refractive index, 1.95. Occurrence: Sweden.

manganotantalite.   An iron black to brownish black mineral of the columbite group composed of manganese tantalate, $MnTa_2O_6$. It can also be made synthetically from manganese dioxide (MnO_2) and tantalum pentoxide (Ta_2O_5) under prescribed conditions. Crystal system, orthorhombic. Density, 6.50 g/cm^3; refractive index, 2.25. Occurrence: Brazil.

manganotapiolite.   A dark brown mineral of the rutile group composed of iron manganese niobium tantalum oxide, (Mn,Fe)(Ta,-Nb)$_2O_6$. Crystal system, tetragonal. Density, 7.72 g/cm^3. Occurrence: Finland.

manganous acetate.   See manganese acetate tetrahydrate.

manganous bromide.   Pink, hygroscopic flakes or powder (95+% pure). Crystal system, hexagonal. Density, 4.39 g/cm^3; melting point, 650°C (1202°F) [also reported as 698°C or 1288°F]; boiling point, 1190°C (2174°F). Used in chemistry and materials research. Formula: $MnBr_2$. Also known as *manganese bromide.*

manganous bromide tetrahydrate.   Pink to red, hygroscopic crystals (98+% pure). Density, 2.01 g/cm^3; melting point, decomposes at 64.3°C (147°F). Used as a drier and catalyst. Formula: $MnBr_2 \cdot 4H_2O$. Also known as *manganese bromide tetrahydrate.*

manganous carbonate.   See manganese carbonate.

manganous chloride.   Pink, hygroscopic crystals or flakes (98+% pure). Available in trigonal and rhombohedral (hexagonal) form with the latter occurring in nature as the mineral *scacchite.* Properties of trigonal form: Density, 2.98 g/cm^3; melting point, 650°C (1202°F); boiling point, 1190°C (2174°F). Used as a catalyst in organic chlorination reactions, as a paint drier, in biochemistry and medicine, and in dyeing. Formula: $MnCl_2$. Also known as *manganese chloride.*

manganous chloride tetrahydrate.   Pink, hygroscopic crystals (98+% pure). Density, 2.01 g/cm^3; melting point, 58°C (136°F). Used as a paint drier, as a catalyst for organic reactions, in biochemistry and medicine, and in dyeing. Formula: $MnCl_2 \cdot 4H_2O$. Also known as *manganese chloride tetrahydrate.*

manganous 2-ethylhexanoate.   See manganese 2-ethylhexanoate.

manganous fluoride.   Reddish, hygroscopic crystals or light brown powder (98+% pure). Crystal system, tetragonal. Density, 3.98 g/cm^3; melting point, 856°C (1573°F) [also reported as 930°C or 1706°F]; antiferromagnetic properties. Used in ceramics, electronics, and materials research. Formula: MnF_2. Also known as *manganese difluoride; manganese fluoride.*

manganous hydroxide.   White to pink crystals. It occurs in nature as the mineral *pyrochroite.* Density, 3.258 g/cm^3; hardness, 2.5 Mohs; melting point, decomposes with heat. Used in ceramics. Formula: $Mn(OH)_2$. Also known as *manganese hydroxide.*

manganous iodide.   White, hygroscopic crystals or purple, hygroscopic, water-soluble powder (99+% pure). Crystal system, hexagonal. Density, 5.01-5.04 g/cm^3; melting point, 638°C

(1180°F); boiling point, 1061°C (1941°F). Used in chemistry and materials research. Formula: MnI$_2$. Also known as *manganese iodide.*

manganous-manganic oxide. See manganese tetroxide.

manganous molybdate. See manganese molybdate.

manganous nitrate. A manganese compound available as a corrosive, oxidizing, pink 50-52 wt% solution in water, or a 45-50 wt% solution in dilute nitric acid. It has a density of 1.536 g/cm^3 and is used in ceramics, chemistry and materials research. Formula: Mn(NO$_3$)$_2$. Also known as *manganese nitrate.*

manganous nitrate hexahydrate. Colorless or pink, water-soluble crystals. Density, 1.82 g/cm^3; melting point, 26°C (79°F); boiling point, 129°C (264°F). Used in ceramics, biochemistry, and as a catalyst. Formula: Mn(NO$_3$)$_2$·6H$_2$O. Also known as *manganese nitrate hexahydrate.*

manganous octoate. See manganese 2-ethylhexanoate.

manganous orthophosphate. See manganese phosphate.

manganous oxide. Gray crystals or powder (99+% pure). It occurs in nature as the mineral *manganosite.* Density, 5.37-5.46 g/cm^3; melting point, 1650°C (3002°F) [also reported as 1840°C or 3344°F], but converts to manganese tetroxide (Mn$_3$O$_4$) if heated in air; antiferromagnetic properties; hardness, 5.0-6.0 Mohs. Used in the manufacture of magnets, ceramics, colored glass, and paints. Formula: MnO. Also known as *manganese monoxide; manganese oxide.*

manganous phosphate. See manganese phosphate.

manganous phthalocyanine. See manganese phthalocyanine.

manganous silicate. See manganese silicate (i).

manganous sulfate tetrahydrate. Translucent, slightly purplish-pink, efflorescent prisms. Density, 2.107 g/cm^3; melting point, 30°C (86°F). Used in ceramics, paints and varnishes, in ore flotation, in textile dyes, as a catalyst in the viscose (rayon) process, and for synthetic manganese dioxide. Formula: MnSO$_4$·4H$_2$O. Also known as *manganese sulfate tetrahydrate*

manganous sulfide. Crystals or powder (99.9% pure) available in green, cubic and red, hexagonal forms. The cubic form occurs in nature as the mineral *alabandite.* Properties of cubic form: Crystal structure, sphalerite; density, 3.99 g/cm^3; melting point, 1610°C (2930°F); hardness, 3.8 Mohs. Properties of hexagonal form: Crystal structure, wurtzite (zincite); density, 3.248 g/cm^3. Used as an additive in steelmaking, in ceramics, as a pigment, and as a semiconductor. Formula: MnS. Also known as *manganese sulfide.*

Manganoid. Trade name of Dresser Industries, Inc. (USA) for a manganese steel containing 1% carbon, 1.2% manganese, and the balance iron. Used for grinding balls of pulverizers, crushers, and grinders.

Mangano Special. Trade name of Latrobe Steel Company (USA) for a nondeforming tool steel containing 0.95% carbon, 1.2% manganese, 0.5% chromium, 0.5% tungsten, and the balance iron. Used for reamers, dies, and threading taps.

manganpyrosmaltite. A brown mineral of the pyrosmaltite group composed of manganese iron chloride silicate hydroxide, (Mn,-Fe)$_8$Si$_6$O$_{15}$(OH,Cl)$_{10}$. Crystal system, hexagonal. Density, 3.13 g/cm^3; refractive index, 1.669. Occurrence: USA (New Jersey).

Manganweld. Trade name of Lincoln Electric Company (USA) for austenitic manganese steels with excellent wear and abrasion resistance, used for welding and hardfacing electrodes.

Mangcraft. Trade name of Lincoln Electric Company (USA) for austenitic manganese and manganese-nickel steels with excellent abrasion and wear resistance, used for welding rods for high-manganese and manganese-nickel steels.

Mangdie. Trade name of British Steel Corporation (UK) for an oil-hardening, nondeforming tool steel containing 0.95% carbon, 1.25% manganese, 0.5% chromium, 0.5% tungsten, 0.2% vanadium, and the balance iron. Used for dies, tools, taps, and gages.

Mangear. Trade name of British Steel Corporation (UK) for a plain-carbon steels containing 0.12% carbon, 0.2% silicon, 1.6% manganese, 0.04% sulfur, 0.04% phosphorus, and the balance iron. Used for lifting chains.

manganjiroite. A dark brownish gray mineral of the rutile group composed of sodium potassium manganese oxide hydrate, (Na,-K)Mn$_8$O$_{16}$·xH$_2$O. Crystal system, tetragonal. Density, 4.29 g/cm^3. Occurrence: Japan.

Mangjet. Trade name of Lincoln Electric Company (USA) for an austenitic manganese steel containing 0.65% carbon, 14.5% manganese, 0.14% silicon, 1.15% molybdenum, and the balance iron. It has excellent abrasion and wear resistance, and good shock resistance. Used for arc-welding and hardfacing electrodes.

mangle. The durable wood of the mangle tree, an evergreen plant belonging to the mangroves (genus *Rhizophora*) and growing in marshes and along the coasts of tropical America and western Africa.

Mangnol. Trade name for a wrought alloy containing 95% manganese and 5% nickel used for electron tube grid wires.

Mangonic. Trade name of Inco Alloys International Limited (UK) for a series of heat-resistant, magnetic nickel alloys containing 2-5% manganese. Used for electrical leads and in tungsten filament lamps.

Mango-Plate. Trade name of Pyramid Steel Company (USA) for an abrasion-resistant, work-hardened steel used for coal chutes, conveyor lines, and scraper blades.

Mangrid. Trade name of Wilbur B. Driver Company (USA) for a series of magnetic nickel alloys containing 4-20% manganese. Used for grid and lead wires.

Manhardt's alloy. A non-heat-treatable aluminum alloy containing 10% tin, 6.2% copper, 0.1% magnesium, and a trace of phosphorus. Used for light-alloy parts.

Maniflex. Trade name of Carpenter Technology Corporation (USA) for a free-machining austenitic stainless steel containing 0.2% carbon, 21% chromium, 11.5% nickel, 1.25% manganese, 0.8% silicon, 0.02% phosphorus, 0.2% sulfur, and the balance iron. It has excellent hardness and high-temperature strength, and is used for shafts and bushings in manifold exhaust heat control valves and emission control devices.

manifold paper. A thin, lightweight paper, such as *onionskin*, suitable for multiple copies.

Manila copals. Natural resins obtained by tapping trees of the genus *Agathis* found in the Philippine Islands and East Indies. They have a density of about 1.06-1.07 g/cm^3 (0.038-0.039 lb/in.^{3}), a melting point of 230-250°C (446-482°F), and are used in varnishes, lacquers, paints, printing inks, and linoleum. Also known as *Manila resins.*

Manila fibers. Long, strong, lightweight vegetable fibers obtained from the leafstalks of the abaca (*Musa textilis*), a plant of the banana family native to the Philippine Islands. They have a whitish color, high stiffness and durability, good water resistance, and high surface gloss. Used for twine, marine cordage and coarse fabrics, and in the manufacture of *Manila paper.* Also known as *abaca; abaca fibers; Manila hemp.*

Manila gold. An inexpensive low-leaded brass used for costume jewelry.

Manila hemp. See Manila fibers.

Manila maguey. See cantala fibers.

Manila paper. A term used for strong, yellowish-brown to brown wrapping paper, made originally from *Manila fibers*, but now includes any strong, dull yellow paper, such as *kraft paper*, made from chemical or mixed wood pulps.

Manila resins. See Manila copals.

Manila rope. A rope, typically 6.5-22 mm (0.250-0.875 in.) in diameter, made from long, selected *Manila fibers* and specially lubricated to retain good pliability and flexibility over a long service life. It has good to moderate strength, is not as strong as synthetic ropes, such as polyester, polypropylene or nylon, but has outstanding flexibility, good water resistance and high susceptibility to rot, mildew and bacteriological damage. Used for rigging, lifting and towing applications.

Manitoba maple. See box-elder.

manjak. A deep-black, high grade of natural asphalt that is a brittle solid which softens and flows on heating. Occurrence: Barbados, Cuba, Trinidad. Used in paints, varnishes and electrical insulation.

Mankato stone. A variety of calcitic limestone containing some alumina (Al_2O_3) and silica (SiO_2). It has an average density of 2.5 g/cm^3 (0.09 $lb/in.^3$), good compressive strength, and is used for building stone.

man-made diamond. See synthetic diamond.

man-made fibers. A generic term for staple fibers and continuous filaments made either by the polymerization of organic monomers (e.g., nylon or polyethylene fibers), or by the transformation (regeneration) of natural organic polymers (e.g., casein fibers). Also known as *manufactured fibers*. See also artificial man-made fibers; synthetic man-made fibers.

Manmo. Trade name of Teledyne Vasco (USA) for a tough, oil-hardening steel containing 0.55-0.75% carbon, 0.8% manganese, 1% chromium, 0.45% chromium, and the balance iron. Used for chisels, punches, cams, brake dies, and plastic molds.

Mannheim gold. A German jewelry alloy of 80-89% copper, 7-20% zinc and 1-9% tin. It has only moderate corrosion resistance, and is used for inexpensive jewelry and ornaments.

Manofort. Trade name of Usines de A. Manoir (France) for a series of oil-hardening, shock resisting steels containing 0.3-0.4% carbon, 2-4% nickel, 1-1.5% chromium, 0.4% molybdenum, and the balance iron. Used for gears, shafts, crankshafts and countershafts.

Manoir. (1) Trade name of Usines de A. Manoir (France) for a series of steels including several corrosion-resistant low-carbon steels containing 2-7% chromium and small additions of aluminum, silicon and/or molybdenum, and a number of low- and medium-carbon alloy steels containing varying amounts of chromium, nickel, molybdenum and vanadium. Used for machinery parts, and wear-resistant manganese steels.

(2) Trademark of Domtar Construction Materials (Canada) for asphalt shingles.

Manor Hall. Trademark of PPG Industries Inc. (USA) for a complete line of architectural paints and coatings.

mansfieldite. A light gray or green mineral of the variscite group composed of aluminum arsenate dihydrate, $AlAsO_4 \cdot 2H_2O$. Crystal system, orthorhombic. Density, 3.02 g/cm^3. Occurrence: Germany, USA (Oregon).

Mansil. Trade name of Disston Inc. (USA) for an oil-hardening die steel containing 0.90% carbon, 1.15% manganese, 0.50% chromium, 0.50% tungsten, and the balance iron. It has good deep-hardening and nondeforming properties. Used for dies, broaches, reamers, and gages.

Mansiloy. Trade name of Union Carbide Corporation (USA) for a manganese alloy containing 28-31% silicon, 0-0.07% carbon and 0-0.05% phosphorus. It reduces metal oxides from the slag, and is used in the manufacture of stainless steel.

mansonia. The wood of the hardwood tree *Mansonia altissima* resembling the *American black walnut* in appearance. It works and nails well, and has good stability and durability. Average weight, 590 kg/m^3 (37 lb/ft^3). Source: Tropical Africa. Used for high-grade joinery and general interior and exterior construction applications.

Man-Ten. Trademark of US Steel Corporation for a structural steel containing up to 0.35% carbon, 0.25-1.75% manganese, 0.10-0.30% silicon, 0.01-0.25% copper, 0-0.40% molybdenum, 0-0.20% vanadium, and the balance iron. Supplied in the form of bars, plates, sheets, strip and sections, it has good corrosion resistance and high tensile strength. Used for frames and structures of industrial equipment and machinery.

manufactured abrasives. See artificial abrasives.

manufactured alumina. Alumina produced or refined by sintering or crystallization.

manufactured carbon. A bonded granular form of carbon whose matrix has been subjected to a thermal treatment at a temperature between 900 and 2400°C (1650 and 4350°F).

manufactured fibers. See man-made fibers.

manufactured graphite. A bonded granular form of carbon whose matrix has been subjected to a thermal treatment at a temperature above 2400°C (4350°F).

manufactured marble. Marble dust mixed with plastics for use in building and construction.

manufactured sand. A fine aggregate obtained by crushing, grinding and screening rock, gravel, blast-furnace slag, hydraulic-cement concrete, etc. It has a particle size of up to about 5 mm (0.2 in.), and is used in the manufacture of concrete, and for foundry applications. Also known as *artificial sand; screenings*.

Manusolite. French trade name for cellulose acetate plastics.

Man-Van. Trade name for a water-hardening tool steel containing 0.95% carbon, 0.15% molybdenum, 0.1% vanadium, and the balance iron. Used for dies and cutting tools.

Manyro. Trade name for a vinal (polyvinyl alcohol) fiber used for textile fabrics.

Mao. Trade name of Boehler GmbH (Austria) for an austenitic stainless steel containing up to 0.05% carbon, 18% chromium, 9.5% nickel, and the balance iron. It is usually supplied in the annealed condition. *Mao Superior* is an austenitic stainless steel containing up to 0.03% carbon, 18% chromium, 9.5% nickel, and the balance iron.

Maoy. Trade name of Boehler GmbH (Austria) for an austenitic stainless steel containing up to 0.06% carbon, 18% chromium, 9.5% nickel, and the balance iron. It is usually supplied in the annealed condition.

Mapico. Trademark of Rockwood Pigments N.A., Inc. (USA) for a series of synthetic iron oxide pigments used in the manufacture of paints, and including (i) *Mapico black*, a black pigment containing 76.3% ferric oxide and 22.5% ferrous oxide, supplied as a powder with cubical particle shape; (ii) *Mapico brown*, a brown pigment containing 93.1% ferric oxide and 5% ferrous oxide, supplied as a fine powder with cubical particle shape; (iii) *Mapico crimson*, a crimson-colored pigment containing 98% ferric oxide, supplied as a powder with acicular particle shape; (iv) *Mapico lemon yellow*, a lemon-yellow pigment containing 87% ferric oxide, having a loss on ignition of almost 12%, and supplied as a fine powder with acicular parti-cle shape; and (v) *Mapico red*, a red pigment containing of 98% ferric oxide, supplied as a powder with spheroidal particle shape. Also included under this trademark are various high-purity iron oxides, and several heat-stable, tan-colored zinc ferrite and magnesium ferrite pigment powders.

mapimite. A green-blue or green mineral composed of zinc iron arsenate hydroxide decahydrate, $Zn_2Fe_3(AsO_4)_3(OH)_4 \cdot 10H_2O$. Crystal system, monoclinic. Density, 2.95 g/cm^3; refractive index, 1.678. Occurrence: Mexico.

maple. The hard, light-colored, close-grained wood of any of a large genus of trees *(Acer)* including especially the sugar maple *(A. saccharum)*, the black maple *(A. nigrum)*, the silver maple *(A. saccharinum)*, the red maple *(A. rubrum)*, and the box-elder *(A. negundo)*. Maplewood has high strength and stiffness, good shock resistance, large shrinkage, and is used for constructional applications, flooring, furniture, boxes, pallets, crates, veneer, crossties, pulpwood, woodenware, handles, spools and bobbins, and foundry patterns. See also hard maple; soft maple.

Maple Leaf. Trade name of Atlas Specialty Steels (Canada) for a water-hardening tool steel (AISI type W1) containing 0.8% carbon, 0.25% silicon, 0.3% manganese, and the balance iron. It has high toughness and shock resistance, good resistance to decarburizing, low wear resistance and moderate machinability. Used for hand chisels, wrenches, crowbars, stamps, knurls and other tools.

MA powder. See mechanically alloyed powder.

Maprenal. Trademark of Casella AG (Germany) for a series of modified melamine-formaldehyde plastics and compounds.

marabout. A fine, soft, very thin fabric made from twisted raw silk yarn, and used especially for blouses and lampshades.

Maracaibo boxwood. See zapatero.

Maradamit. Trade name for an aluminum alloy used for light-alloy parts.

Marage. Trade name of Carpenter Specialty Alloys (USA) for a series of high-strength *maraging steels* containing about 0.03% carbon, 18.5% nickel, 7.5-8.75% cobalt, 4.9% molybdenum, 0.65% titanium, and the balance iron. Used for aerospace applications, e.g., landing gears and arrestor hooks.

maraging steels. A group of ultrahigh-strength steels containing very low amounts of carbon (usually 0.03% or less) and additions of nickel (7-25%), chromium (0-14%), cobalt (up to 11%), molybdenum (up to 5%) and small amounts of titanium, aluminum and niobium (columbium), with the balance being iron. *Maraging steels* develop a ductile and tough martensitic microstructure upon cooling from the austenitizing temperature. The *martensite* is then aged at 450-500°C (840-930°F) which greatly increases strength, hardness and toughness. Used for solid rocket cases, hydrofoil struts, extrusion-press rams, aluminum die-casting dies and cores, aluminum hot-forging dies, dies for molding plastics, various tools, and aluminum extrusion dies.

Maraglass. Trade name of Acme Chemicals & Insulation (USA) for epoxy casting resins used for the manufacture of lighting fixtures, transparent equipment enclosures, and decorative products.

Maranyl. Trademark of ICI Americas Inc. (USA) for unfilled, glass-filled, impact-modified, flame-retardant, lubricated and nucleated grades of nylon 6,6, used for molded parts for the automotive, appliance and electronics industries. *Maranyl A* is a strong, stiff nylon 6,6 supplied in various grades including unmodified, high-impact, UV-stabilized, glass fiber-reinforced and mineral-filled. It has good abrasion resistance, dimensional stability and heat distortion properties, an upper service temperature of 80-160°C (176-320°F), very good resistance to alcohols, alkalies, hydrocarbons and ketones, and is used for bearings, gears, fasteners, automotive under-hood parts, powder-tool housings and sports equipment. *Maranyl B* is a nylon 6 supplied in the form of film, sheeting, rods, powder, granules and monofilaments. It has high mechanical strength and creep resistance, high impact strength, good elevated temperature properties, good processibility, a service temperature range of -40 to +160°C (-40 to +320°F), good dielectric properties, good resistance to alcohols, alkalies, hydrocarbons, greases, oils and ketones, and is used for appliances, automotive components, lawn and garden equipment, fasteners, electrical switches and connectors.

Maraset. Trade name of Acme Chemicals & Insulation (USA) for epoxy resins and compounds used for tank lining and tooling applications.

Marasperse. Trademark of American Can Company (USA) for a series of lignosulfonates supplied as brown water-soluble powders, and used in gypsum board, carbon black dispersions, and ceramics.

Marathon. Trademark of Den-Mat Corporation (USA) for a dual-cure hybrid composite resin with high wear resistance and low thermal expansion and polymerization shrinkage, used in dentistry for direct posterior restorations, stress-bearing anterior restorations and for bonding crowns, inlays, onlays and bridges.

marble. A valuable, beautiful, crystallized, metamorphic limestone composed essentially of *calcite* $(CaCO_3)$ and/or *dolomite* $[CaMg(CO_3)_2]$. It may be white, gray, brown, yellow, black, blue, red, mottled or streaked, and is capable of taking a high polish. *Marble* has an average weight of 2500-2690 kg/m^3 (156-168 lb/ft^3); a hardness of 3 Mohs, good compressive strength and good resistance to temperatures of up to 650°C (1200°F). Used for exterior and interior finish of buildings, sculptures and ornaments, for electric-power panels, as a source of calcium oxide (CaO) in glazes, in dust form as a filler or abrasive, in putties, and in casting.

marble cloth. A soft, light fabric made from silk, wool or other natural or synthetic fibers, and resembling marble in appearance. Used especially for blouses and linings. See also marble silk.

marble dust. A fine powder made by crushing and grinding marble chips, and used as a filler and mild abrasive, in the manufacture of putties, and in casting. Also known as *marble flour.*

Marble Elite. Trademark of Huber Engineered Materials (USA) for an alumina trihydrate (ATH) filler containing pigmented polymer granules. It is available in a wide range of colors and shades for use in the manufacture of *cultured marble.*

marble flour. See marble dust.

marble paper. A book paper with a color resembling marble in appearance.

marble silk. A soft, light silk fabric resembling marble in appearance, used especially for blouses and linings. See also marble cloth.

Marblette. Trade name of Marblette Corporation (USA) for thermosetting cast phenol-formaldehyde plastics.

marblewood. The hard, durable wood of the ebony tree *Diospyros kurzii*. It has a firm, close texture, works well, and takes a fine polish. The heartwood is black with yellowish bands. Average weight, 1040 kg/m^3 (65 lb/ft^3). Source: India, Andaman Islands.

Used for furniture and cabinetwork. Also known as *Andaman marblewood*.

Marblit. Trade name of Marblit (Poland) for an opaque glass with hard, brilliant finish.

Marbloid. Japanese trade name for urea-formaldehyde resins.

Marbo Film. Trade name of Marbon Corporation (USA) for rubber hydrochloride films.

Marbon. (1) Trademark of Borg-Warner Chemicals (USA) for a series of acrylonitrile-butadiene-styrene plastics with outstanding strength and toughness, good physical, chemical and electrical properties, good resistance to heat distortion, and a maximum service temperature of 121°C (250°F). Used for automotive trim, cases, appliance housings, lawn and garden equipment, refrigerator linings, highway safety devices, piping, impellers, helmets, knobs, handles, and toys.

(2) Trademark of Marbon Corporation (USA) for a series of high-styrene and styrene-butadiene copolymer rubber-reinforcing resins used as molding powders, and for dry compounding and reinforcing of rubber and synthetic elastomers, plasticizers, resins, dyes and pigments.

marcasite. A bronze-yellow, creamy white, opaque mineral composed of iron disulfide, FeS_2. Crystal system, orthorhombic. Density, 4.88-4.91 g/cm^3; melting point, transition to pyrite, 450 (840°F) or higher; hardness, 6.0-6.5. Occurrence: USA (Missouri, Oklahoma). Used as a source of sulfur and iron. Also known as *white iron pyrite*. See also iron pyrites.

Mar-Con. Trade name of Bunting Bearings Corporation (USA) for a series of continuous cast bronzes composed of varying amounts of copper, tin, lead and zinc, and used for bearings, bushings and liners.

Marconite. Trade name of Marconi Limited for a concrete containing a special electri-cally conductive aggregate used for grounding electrical equipment.

Marena. Trade name of Marena Group Inc. (USA) for regenerated protein (azlon) fibers used for the manufacture of garments.

Marenka. Trade name of Enka BV (Netherlands) for regenerated protein (azlon) fibers.

Marezzato. Trade name of Vetreria di Vernante SpA (Italy) for a glass with indeterminate pattern.

Marfran. Trade name of V. Franceschetti Elastomeri (Italy) for thermoplastic resins and elastomers.

Margard. Trademark of General Electric Company (USA) for a series of synthetic resins supplied in the form of panels or sheets.

margaritasite. (1) A yellow mineral of the carnotite group composed of cesium potassium oxonium uranyl vanadium oxide monohydrate, $(Cs,K,H_3O)(UO_2)_2(VO_4)_2 \cdot H_2O$. Crystal system, monoclinic. Density, 5.40 g/cm^3. Occurrence: Mexico.

(2) A yellow mineral of the carnotite group composed of cesium uranyl vanadium oxide, $Cs_2(UO_2)_2V_2O_8$. Crystal system, monoclinic. Density, 5.51 g/cm^3.

margarite. A grayish, pink, pale yellow or green mineral of the mica group composed of calcium aluminum silicate hydroxide, $CaAl_2(Si_2Al_2)O_{10}(OH)_2$. Crystal system, monoclinic. Density, 3.05-3.08 g/cm^3; refractive index, 1.643. It is a brittle mica. Occurrence: Japan. Also known as *calcium mica*.

margarosanite. (1) A colorless mineral composed of calcium lead silicate, $Ca_2PbSi_3O_9$. Crystal system, triclinic. Density, 4.31 g/cm^3. Occurrence: USA (New Jersey).

(2) A colorless mineral composed of lead calcium manganese silicate, $Pb(Ca,Mn)_2(SiO_3)_3$. Crystal system, triclinic. Density, 4.33 g/cm^3; refractive index, 1.771. Occurrence: USA (New

Jersey), Sweden.

Margherita Grande. Trade name of Fabbrica Pisana SpA (Italy) for a glass with a Muranese-type pattern.

Marglass. Trade name of Marglass Limited (UK) for glass yarn and woven glass fabrics and tapes.

marialite. A mineral of the scapolite group that is composed of sodium aluminum chlorate silicate, $Na_3Al_3Si_9O_{24}Cl$, and may also contain some calcium. Crystal system, tetragonal. Density, 2.56 g/cm^3. Occurrence: UK.

maricite. A colorless, gray, or pale brown mineral composed of sodium iron phosphate, $NaFePO_4$. Crystal system, orthorhombic. Density, 3.66 g/cm^3; refractive index, 1.695. Occurrence: Canada (Yukon).

Marilon. Trade name of C.G.T. (Italy) for a polyamide 6 (nylon 6) resin supplied in several grades.

Marimusume. Japanese trade name for rayon fibers and yarns used for textile fabrics.

Marine. (1) Trade name of Stone Manganese–J. Stone & Company Limited (UK) for lead-base bearing metals including *Marine I* containing 14% antimony, 4% tin and 1% copper, and *Marine II* with 22.5% tin and 7.7% antimony.

(2) Trade name of NKK Corporation (Japan) for a series of high-strength, low-alloy (HSLA) steels containing up to 0.10% carbon, up to 1.5% manganese, 0.5-0.8% chromium, 0.15-0.55% aluminum, 0.2-0.35% copper, 0-0.4% nickel, traces of niobium and vanadium, and the balance iron. Used for machinery and equipment operating in seawater.

marine alloy. An alloy composed of 48% lead, 40% tin, 10% antimony and 2% copper, and used for marine bearings, especially submerged bearings.

marine babbitt. An antifriction alloy composed of 72% lead, 21% tin and 7% antimony, and used for marine bearings.

marine bronze. A corrosion-resistant bronze containing 57.5% copper, 0.8% nickel, 0.15% tin, 0.5% aluminum, and the balance zinc. Used in shipbuilding and for marine parts.

marine glue. An adhesive, usually based on rubber or shellac, that does not dissolve in water.

Marine Glyco. Trade name of Joseph T. Ryerson & Son Inc. (USA) for a babbitt containing lead, tin and antimony, used for marine bearings.

Marine Nickel. Trade name of Lewin Metals Corporation (USA) for a babbitt metal containing tin, lead and nickel, used for bearings.

Mariner. Trade name of US Steel Corporation for a carbon steel containing 0.22% carbon, 0.8% manganese, 0.1% phosphorus, and the balance iron. Used for sheet piling in marine environments.

marine rope. A term referring to rope or cordage used for marine applications. Formerly, such rope and cordage was chiefly made from hemp or sisal fibers, but now it is also manufactured from synthetic fibers, such as polypropylene or nylon.

marine varnish. A high-grade *spar varnish* especially designed to resist prolonged immersion in salt or fresh water and exposure to marine atmospheres.

Marinite. Trade name of BNZ Materials, Inc. (USA) for non-asbestos monolithic calcium silicate structural insulation with *tobermorite* crystal structure. Supplied as 1.2 m × 2.4 m (4 ft. × 8 ft.) sheets in thicknesses from 0.25 to 3 in. (6.4 to 76 mm), it has high structural strength at elevated temperatures, excellent thermal insulating properties and thermal-shock resistance, good fire resistance to almost 1100°C (2000°F), good electrical insulating properties in dry condition, and good machinability. Used for heat processing, fire protection and electrical resistance applications.

MARIOM. An acronym for "Matrix Auto-Reinforced Organic Material" referring to a class of lightweight composite materials composed of continuous, polymeric matrices, formed from converted fibers, and reinforcing fibers of the same material. The first material of this kind was produced from bovine leather fibers by *in situ* conversion into the final composite during a die-forming process at prescribed pressure and temperature.

Marispun. Trade name for cellulose acetate fibers used for textile fabrics including wearing apparel.

Maritex. Trademark of Halyard Limited (UK) for a tough, fray- and tear-resistant fiberglass fabric with a sealed metallized skin layer. It has good fire resistance, zero oil absorption, and is used in boatbuilding.

Mark. Trade name of Atofina Chemicals Inc. (USA) for semibright nickel electroplates and plating processes.

Mark II. Trade name of Dentsply Corporation (USA) or a machinable glass-ceramic used for dental applications.

Markana. Trade name of Vereinigte Deutsche Nickel-Werke AG (Germany) for a corrosion-resistant alloy containing 58-60% copper, 38-40% zinc, and a total of 1.5-2.0% manganese and iron. It has a melting point of 880°C (1616°F), and is used for sheets for wash basins and sinks.

Marker. Trade name of Schmidt & Clemens Edelstahlwerke (Germany) for an extensive series of wrought and cast steels including various austenitic, ferritic and martensitic stainless grades, cold- and hot-work tool and die grades, water-hardening carbon and high-speed tool grades, and heat-resisting and case-hardening grades. Also included under this trade name are several nickel-base superalloys.

Markerite. German trade name for cemented carbides used for tool tips.

market brass. See high brass.

Markus alloy. A corrosion-resistant alloy composed of copper, nickel and zinc, and used for decorative, ornamental and architectural parts.

marl. A smooth-textured, white, crumbling calcareous clay containing about 35-65% calcium carbonate ($CaCO_3$), some magnesium carbonate ($MgCO_3$), and sand. Used in the production of cement and building brick, and as an anticrazing ingredient in stoneware. Also known as *marl clay; marlstone*.

marl clay. See marl.

marled yarn. See mottled yarn.

marl-effect yarn. A yarn that consists of two differently colored or dyed continuous filament yarns. See also ingrain (2).

Marlex. Trademark of Chevron Phillips Chemical Company (USA) for series of high- and medium-density, high-molecular-weight and linear low-density polyethylenes, and polypropylene homopolymers and random copolymers supplied in various grades including crosslinkable, extrusion, injection molding, blow molding, sheet/thermoforming, film, impact-modified and fiber-resin. They have good mechanical properties, good resistance to ozone, ultraviolet radiation and heat, good chemical resistance, and are used for lining reservoirs, vessels and containers, for sports and recreation equipment, marine equipment, consumer electronics, and household appliances.

Marley. Trademark of Marley Roof Tile Limited (Canada) for concrete roof tile.

Marleyflor. Trademark of Marley Werke GmbH (Germany) for flexible polyvinyl chloride film materials.

Marleylex. Trademark of Marley Werke GmbH (Germany) for flexible polyvinyl chloride film materials.

Marlok. Trademark of Rauma Materials Technology Oy Corporation (USA) for an ultahigh-strength maraging steel containing up to 0.01% carbon, 18% nickel, 11% cobalt, 5% molybdenum, up to 0.3% titanium, and the balance iron. Supplied in the form of bars and billets, it is used for long-run, aluminum die-casting applications. See also maraging steels.

marlstone. See marl.

Marly. A very thin, light, gauzelike cloth, usually made of cotton. It is named after Marly-le-Roi, a city in France.

marl yarn. Two differently colored, usually woolen-spun yarns twisted together. See also ingrain (1).

Marmorglas. Austrian trade name for colored, opaque glass with hard, brilliant finish.

Marnyte. Trade name of Bamberger Polymers International Corporation (USA) for unreinforced and glass-fiber-reinforced polyethylene terephthalates processed by injection molding and/or extrusion techniques and used for a wide range of applications including office equipment, electrical and electronic components, machine parts, furniture and household appliances.

marocain. A crepe fabric woven from high-twist silk, wool or synthetic yarn and having a cross-ribbed effect in the weft (filling). Used especially for dresses and suits. Also known as *crepe marocain; marocain crepe*.

marocain crepe. See marocain.

Marog. Trade name of Teledyne Vasco (USA) for a high-speed steel containing 0.7% carbon, 4% chromium, 18% tungsten, 1% vanadium, and the balance iron. Used for tools, drills, cutters, and lathe tools.

marokite. A black mineral of the spinel group composed of calcium manganate, $CaMn_2O_4$. Crystal system, orthorhombic. Density, 4.64 g/cm^3. Occurrence: Morocco.

Marolin. Trade name for cellulose acetate plastics.

Marothaan. Trade name of Ato Findley (France) for polyurethane coatings used for flooring applications.

Marquardt porcelain. A German porcelain that consists of *mullite* in a glassy matrix and is made from a body composed of clay, feldspar and quartz in a ratio of 11:4.5:4.5 by firing at 1400-1450°C (2550-2640°F) for a prescribed period of time during which all of the quartz dissolves. It was formerly used for furnace tubes and pyrometer sheaths.

Marques. Trade name of Société Nouvelle des Acieries de Pompey (France) for a series of free-cutting steels containing 0.08-0.2% carbon, 0.08-0.18% sulfur, up to 2% manganese, and the balance iron. Used for screw-machine products.

Marquesa Lana. Trademark of American Fibers & Yarns Company (USA) for polypropylene fibers and yarns used for textile fabrics.

Marquis. Trade name of United Clays, Inc. (USA) for a high-alumina, low-silica clay with a microstructure composed of well crystallized *kaolinite*. Mined in Tennessee, USA, it possesses a fine particle size and good rheological properties, and is used as a kaolin substitute in ceramic bodies, castables, refractories and spark plugs.

marquisette. A sheer, soft leno-weave fabric with square meshes, made of cotton, nylon, rayon, silk, etc., and used for blouses, dresses, lingerie, mosquito netting, millinery, and draperies.

Mars black. A dark brownish black paint pigment made by the oxidization of ferrous hydroxide, $Fe(OH)_2$, followed by calcination.

Mars brown. A brownish-yellow pigment made from earths and brown iron oxide.

marseilles. A thick, double-faced cotton fabric in white or other colors, woven in raised figures or stripes and having a somewhat quilted appearance. Used for bedspreads, vesting, etc. Named after Marseilles, a city in southern France.

Marsenol. Trade name of Glenroe Technologies, Inc. (USA) for a nickel-titanium shape-memory wire used for dental applications.

Marsh. Trade name of Marsh Brothers & Company Limited (UK) for a series of tool steels including several hot-work die types and numerous hot- and cold-work, tungsten and cobalt-tungsten high-speed, water-hardening and shock-resisting tool types.

Marshall Crat. Trade name of Marshall Steel Company (USA) for a carbon steel containing 0.18% carbon, 0.5% manganese, 0.2% silicon, and the balance iron. Used for fixtures, jigs, patterns and machine components.

marsh cypress. See baldcypress.

marshite. A colorless, brownish or reddish mineral of the sphalerite group composed of copper iodide, γ-CuI. It can also be made synthetically. Crystal system, cubic. Density, 5.68 g/cm^3; refractive index, 2.346. Occurrence: Australia (New South Wales).

marsh ore. See bog iron ore.

Mars pigments. A group of yellow, orange, brown, red and violet pigments made by calcining to different temperatures the precipitate formed by mixing solutions of calcium hydroxide and ferrous sulfate. See also Mars black; Mars brown; Mars yellow.

Mars Superieur. Trade name of Ugine Aciers (France) for a high-speed steel containing 0.7% carbon, 4% chromium, 18% tungsten, 1% vanadium, and the balance iron. Used for tools, dies, and cutters.

marsturite. A white to light pink mineral of the pyroxenoid group composed of sodium hydrogen calcium manganese silicate, $Mn_3CaNaHSi_5O_{15}$. Crystal system, triclinic. Density, 3.46 g/cm^3. Occurrence: USA (New Jersey).

Mars yellow. A permanent yellow pigment made from yellow iron oxide ($Fe_2O_3 \cdot H_2O$) with outstanding chemical and weather resistance, excellent opacity, and low tinting strength.

Martelado. Trade name of Vidrobrás (Brazil) for a patterned glass with hammered effect.

martempered steels. Austenitized steels that have been quenched to just above the starting temperature for the martensitic transformation, and then cooled slowly through the martensitic range to reduce the quenching stresses produced during the martensitic transformation. See also austenite; martensite.

Martenite. Trade name for a synthetic refractory that typically consists of 66.5% magnesia (MgO), 10.5% ferric oxide (Fe_2O_3), 13.4% lime (CaO), 5.2% silica (SiO_2) and 2.1% alumina (Al_2O_3), and has a loss on ignition of 2.3%. It possesses excellent wear resistance, good high-temperature resistance, and was used as a fettling material for open-hearth furnace bottoms.

martensite. (1) A metastable, supersaturated *solid solution* of carbon in iron formed by a diffusionless transformation of *austenite* if the latter is cooled in a sufficiently rapid quench. It is magnetic and has a hard, highly strained, body-centered tetragonal crystal structure with a characteristic acicular (needle-like) microstructure.

(2) A metastable transitional structure formed in certain alloys by a diffusionless phase transformation. It has a microstructure with a characteristic acicular (needle-like) pattern and can be hard and highly strained (e.g., in iron-carbon alloys) or soft and ductile (e.g., in iron-nickel alloys).

(3) The beta stabilizer supersaturated alpha product occurring in a titanium alloy if it is cooled fast enough from the beta region to prevent transformation by nucleation and growth. See also alpha titanium alloys; beta titanium alloys.

martensitic precipitation-hardenable stainless steels. A group of wrought stainless steels with martensitic microstructures in which hardening is accomplished by a final aging treatment that precipitates very fine second-phase particles (usually titanium or copper) from a supersaturated *solid solution* transforming the austenitic structure. They have excellent wear, corrosion and scaling resistance at elevated temperatures, and are used chiefly for cutlery, razor blades and instruments. Also known as *precipitation-hardenable martensitic stainless steels*. See also austenite; martensite; martensitic stainless steels; stainless steels.

martensitic stainless steels. A group of hardenable, magnetic stainless steels that contain 0.06-1.25% carbon, 1-2.5% manganese, 0.5-1.0% silicon, 9-18% chromium, 0-2.5% nickel, 0-1.5% molybdenum, and the balance iron. They are capable of being heat-treated such that *martensite* is the prime microconstituent, and are usually available in the annealed or quenched-and-tempered condition. *Martensitic stainless steels* have excellent strength, good mechanical properties, good corrosion resistance and high hardness, and are used for rifle barrels, cutlery, surgical instruments, steam turbine tubing and blading, jet-engine components, hand tools, machine parts, bushings, hardware, fittings, fasteners, valves and valve parts, springs, pump shafts, nozzles, mining equipment, pulp and paper equipment, and wear-resistant parts. Abbreviation: MSS. See also creep-resistant martensitic stainless steels; martensitic precipitation hardenable stainless steels; stainless steels.

martensitic steel. A carbon steel (minimum carbon content, 0.15%) with a martensitic microstructure that results from rapid quenching in cold water from above the critical transformation temperature.

martensitic white cast irons. A group of white cast irons that are martensitic as cast and possess superior abrasion resistance and better strength and toughness than pearlitic white irons. They also have high toughness and hardness, good corrosion resistance, and good tensile and impact strength. Used for jaw crushers, hammer mills, grinding balls, rolls, ball mill balls and liners, slurry pumps, pump components, pump impellers, feeder vanes, etc. *Martensitic white irons* can be grouped into the following three general classes: (i) *Nickel-chromium white irons*; (ii) *Chromium white irons*; (iii) *Chromium-molybdenum white irons*; and (iv) *High-chromium white irons*. Also known as *martensitic white irons*. See also pearlitic white irons; white irons.

marthozite. A green mineral of the carnotite group composed of copper uranyl selenite hydroxide heptahydrate, $Cu(UO_2)_3(SeO_3)_3(OH)_2 \cdot 7H_2O$. Crystal system, orthorhombic. Density, 4.40 g/cm^3; refractive index, 1.7825. Occurrence: Zaire.

Marties alloy. A heat- and corrosion-resistant alloy composed of 35% nickel, 27% copper, 18% zinc, 10% tin and 10% iron, and used for electrical resistances, and heat- and corrosion-resistant parts.

Martifin. Trade name of Martinswerk GmbH (Germany) for a high-quality aluminum hydroxide powder used as a filler and coating pigment in papermaking and the manufacture of paints and varnishes.

Martillado. Trade name of Vidrieria Argentina SA (Argentina) for a glass with hammered-type pattern.

Martin. Trade name of Detroit Alloy Steel Company (USA) for an air-hardening steel containing 1.4-1.6% carbon, 12-14%

chromium, 0.6-0.8% cobalt, 0.8-0.9% molybdenum, 0.35-0.4% vanadium, and the balance iron. Used for castings, shears, punches, dies, and mill liners.

Martinal. Trade name of Martinswerk GmbH (Germany) for a high-quality aluminum hydroxide powder used as a nonpolluting, fire-retardant filler for plastics and rubber products, and as a cleanser in toothpastes.

Martinique abutilon. A glossy, relatively long, yellowish white fiber obtained from a shrub (*Abutilon auretum*) belonging to the mallow family. Used for cordage and textile fabrics.

Martin's cement. A white, hard *gypsum cement* that has potash and borax added to accelerate the set, and is used for flooring, and tile imitation.

Martinsite. Trade name of Inland Steel Company (USA) for a series of high-strength carbon steels containing 0.04-0.22% carbon, 0.2-0.6% manganese, and the balance iron. Used for fasteners, tubing, and automotive and appliance parts.

Martin steel. A water-hardening steel containing 0.73% carbon, 0.4% silicon, and the balance iron. Used for hammers, dies, and blacksmith tools.

Martoxid. Trade name of Martinswerk GmbH (Germany) for high-quality aluminum oxide powders supplied in several grades with varying purity and fineness. Used in the ceramic and refractory industries for the manufacture of cutting and grinding tools, grinding media, mill liners, sliding gates, electronic components, pump and valve parts, and seal rings.

Marvac. Trade name of Latrobe Steel Company (USA) for a series of vacuum-melted *maraging steels* containing 0.01-0.03% carbon, 18-19% nickel, 7.5-9% cobalt, 4.2-5% molybdenum, 0.4-0.7% titanium, 0.1% aluminum, and the balance iron. They have good weldability, high tensile and fatigue strength, high hardness and toughness, and are used for missile and aircraft components.

Marvel. Trade name of Teledyne Vasco (USA) for an air-hardening tungsten-type hot-work tool steel (AISI type H21) containing 0.30-0.35% carbon, 0.2-0.4% silicon, 0.1-0.3% manganese, 9.5-10% tungsten, 3.25-3.75% chromium, 0.35-0.55% vanadium, and the balance iron. It has good deep-hardening properties, good abrasion resistance, high toughness, and high hot hardness. Used for hot forging, pressing, swaging and trimming dies, hot punches and shear blades, forging die inserts, hot extrusion dies and tools, compression dies, and other tools and dies.

Marvelene. Trade name for a vinyl styrene resin formerly used for denture bases.

Marvess. Trademark of Amoco Fabrics & Fibers Company (USA) for polypropylene fibers and yarns.

Marvetrite. Trade name of Fabbrica Lastre di Vetro–Pietro Sciarra SpA (Italy) for an opaque glass with hard, brilliant finish.

Marvinol. Trademark of Uniroyal Chemical Company (USA) for a series of thermoplastic resins based on polyvinyl chloride and supplied in various grades in the form of white powders and pellets. Used for plastic films and coatings, and for molded and extruded products.

Marvy Foam. Trade name of All Foam Products Company, Inc. (USA) for bricks and mortar composed of polyurethane foam and used for thermal insulation applications.

Marwe. Trade name of Mannesmann-Rohrenwerke AG (Germany) for a series of low- and medium-carbon steels used for machinery parts (e.g., bolts, gears, cams, camshafts and fasteners), and oil-refinery equipment.

Marwedur. Trade name of Mannesmann-Rohrenwerke AG (Ger-

many) for a series of corrosion- and/or heat-resistant steels used for chemical and petrochemical plant equipment, furnace parts, surgical equipment, and cutlery.

Marwin. Trade name of W. Martin Winn Limited (UK) for a free-machining steel containing 0.25% carbon, 0.3% silicon, and the balance iron. Used for machinery parts.

Mas. (1) Trade name of Thyssen Edelstahlwerke (Germany) for a tough, oil-hardening tool steel containing 0.55% carbon, 0.9% silicon, 1.2% manganese, and the balance iron. Used for chisels and punches.

(2) Trade name of Société Nouvelle des Acieres de Pompey (France) for a medium-carbon alloy steel containing 0.52% carbon, 1.0% chromium, 2.0% nickel, 0.4% molybdenum, 0.2% vanadium, and the balance iron.

mascagnite. A colorless mineral of the olivine group composed of ammonium sulfate, $(NH_4)_2SO_4$. It can also be made synthetically. Crystal system, orthorhombic. Density, 1.77 g/cm^3. Occurrence: UK, Italy.

Mascot. Trade name of Eyre Smelting Company (USA) for a bearing metal that is composed of 95-100% lead and up to 5% tin and has a melting range of 240-300°C (464-572°F).

mashru. A fabric having a warp of silk and a weft (filling) with cotton, or vice versa.

Masiloy. Trade name of Olin Corporation (USA) for a series of brasses containing 60-63.5% copper, 24.5-28% zinc, 7-12% manganese, and up to 5% nickel. Used for flatware, hollow ware, and hardware.

masking tape. A paper tape coated on one side with an adhesive substance and used for the protection (or masking) of surface areas not to be finished, for temporary positioning of paper (e.g., in artwork, drawings, etc.), and for packaging purposes. It may have a crepe-paper backing, and can be peeled off without damaging the surface.

Mask Peel. Trade name of Evans Manufacturing Inc. (USA) for a hot-dip stop-off coating used in the coating and plating industries.

Maso. Trade name of Boehler GmbH (Austria) for austenitic stainless steels including *Maso* with up to 0.05% carbon, 18% chromium, 11% nickel, 2.2% molybdenum, and the balance iron, and *Maso Superior* with up to 0.03% carbon, 18% chromium, 11% nickel, 2.2% molybdenum, and the balance iron. *Maso* austenitic stainless steels are usually supplied in the annealed condition.

Masoy. Trade name of Boehler GmbH (Austria) for austenitic stainless steels including *Masoy* with up to 0.05% carbon, 18% chromium, 12% nickel, 2.6% molybdenum, and the balance iron, and *Masoy Superior* with up to 0.03% carbon, 18% chromium, 12% nickel, 2.6% molybdenum, and the balance iron. *Masoy* austenitic stainless steels are usually supplied in the annealed condition.

Masonex. Trade name of Masonite Corporation (USA) for a water-soluble *hemicellulose* used as a binder for foundry cores and coal briquettes, and as a tackifier in adhesives.

Masonite. (1) Trademark of Masonite Corporation (USA) for a composition *hardboard* made from cellulose fibers obtained by exploding wood chips with high-pressure steam. The fibers are waterproofed with a paraffin-base sizing agent, bonded with adhesive *lignin*, arranged in thick mats or blankets, and then hot-pressed into boards or panels. Used for furniture components, siding, etc.

(2) Trademark of Masonite Corporation (USA) for a series of building products including construction fiberboard, com-

posite board, insulating board, hardboard, synthetic lumber and thermal and acoustic insulating sheets and blankets.

(3) Trademark of Masonite Corporation (USA) for a general-purpose construction adhesive.

(4) Trade name of Masonite Corporation (USA) for a *synthetic mica*, also available in plastic grades, e.g., mixtures of polypropylene and mica.

masonry cement. Any of a group of hydraulic cements composed of one or more of the following materials: Portland cement, natural cement, slag cement, hydraulic lime, Portland-pozzolan cement and/or Portland blast-furnace slag cement, in combination with specially selected additions of hydraulic lime, limestone, granulated slag, chalk, talc and/or clay.

masonry mortar. A mortar that is suitable for bonding stone, brick, tile, block and other masonry units, and for pointing and repointing joints in masonry walls. Also known as *joint mortar; pointing mortar.*

masonry unit. A building unit made of clay, stone, glass, gypsum or concrete and used for masonry walls.

mass aqua. A bluish *borosilicate crown glass* with a density of 2.36 g/cm^3 (0.085 lb/in.^{3}), a refractive index of approximately 1.50, a hardness of 6 Mohs, low dispersion, and high transmission. Used for optical applications, and as an imitation *aquamarine.*

mass concrete. Concrete cast in heavy masses, e.g., large footings, dams, etc., and often containing pozzolans or large, coarse aggregate. Also known as *bulk concrete.*

mass-finishing media. Abrasive materials, such as crushed and graded stones, hardened steel shapes, corn cobs, walnut shells, sawdust, porcelain and other vitreous ceramics or resin-bonded materials, suitable for use in mass finishing operations such as barrel, centrifugal barrel/disk, spindle or vibratory finishing.

massicot. A yellow or reddish mineral form of lead monoxide (PbO) containing 92.8% lead. It is polymorphous with *litharge*, and can also be made synthetically. Crystal system, orthorhombic. Density, 9.64 g/cm^3; melting point, approximately 600°C (1110°F); melting point, 489°C (912°F). Occurrence: USA (Colorado, Idaho, Nevada, Virginia). Also known as *lead ocher; plumbic ocher.*

Massillon. Trade name of Massillon Steel Casting Company (USA) for a range of cast steels.

massive talc. See talc (2).

massive topaz. A massive crystalline aggregate variety of the mineral *topaz* [Al$_2$SiO$_4$(F,OH)$_2$] used in the manufacture of high-alumina refractories.

mass polyvinyl chloride. Polyvinyl chloride in the form of dry-blended powders or small pellets or cubes for mass (or bulk) polymerization and melt processing by blow or injection molding, calendering or extrusion. Abbreviation: M-PVC. Also known as *bulk polyvinyl chloride.* See also polyvinyl chlorides.

Mastalloy. Trade name of Magnacast Corporation (USA) for a series of bronze and brass castings containing 71-88% copper, 4-19% tin, and up to 15% lead, 12% zinc and 4% nickel, respectively.

Master. (1) Trade name of Duke Steel Company Inc. (USA) for a water-hardening tool and die steel.

(2) Name of The Feldspar Corporation (USA) for a line of coarse- and fine-grained kaolins including various air-floated, slurry, filler, shredded and semiprocessed clays.

master alloy powder. A prealloyed metal powder with a relatively high concentration of alloying elements mixed with a base powder and used in the manufacture of powder-metallurgy ma-

terials.

master alloys. A group of alloys that are usually composed of two or more chemical elements and introduced into a molten ferrous or nonferrous metal to change its chemical composition or texture, or act as a deoxidizer. For example, ferrosilicon is a *master alloy* used to introduce silicon into iron or steel, and act as a deoxidizer. Also known as *foundry alloys.*

Masterblast. Trade name of Norton Company (USA) for abrasive blasting media.

Master Bond. Trademark of Master Bond, Inc. (USA) for a line of sealants, conductive coatings, one- and two-component epoxy adhesives and several encapsulation and potting compounds.

Mastercast. Trade name of The Feldspar Corporation (USA) for a white-firing kaolin slurry containing 45.2% silicon dioxide (SiO$_2$), 38.4% aluminum oxide (Al$_2$O$_3$), 1.8% titanium dioxide (TiO$_2$), and small additions of ferric oxide (Fe$_2$O$_3$), calcium oxide (CaO), magnesium oxide (MgO), potassium oxide (K$_2$O) and sodium oxide (Na$_2$O). Their loss on ignition is about 13.5%. *Mastercast* slurry has a density of 1.76 g/cm^3 (0.063 lb/in.^{3}), a specific surface area of 10.23 m^2/g, an average particle size of 2.1 μm (79 μin.), a pyrometric cone equivalent of 34, and excellent casting properties. Used for general ceramic applications.

Masterfil. Trade name of The Feldspar Corporation (USA) for a plastic *kaolin clay* containing 45.9% silicon dioxide (SiO$_2$), 37.8% aluminum oxide (Al$_2$O$_3$), 1.8% titanium dioxide (TiO$_2$), and small additions of ferric oxide (Fe$_2$O$_3$), calcium oxide (CaO), magnesium oxide (MgO), potassium oxide (K$_2$O) and sodium oxide (Na$_2$O). Their loss on ignition is about 13.1%. *Masterfil* clay has a density of 2.62 g/cm^3 (0.095 lb/in.^{3}), a specific surface area of 17.31 m^2/g, an average particle size of 1.2 μm (47 μin), and a pyrometric cone equivalent of 34. Used in the ceramic industry for mechanical forming applications.

Masterfloat. Trade name of The Feldspar Corporation (USA) for a white-firing, air-floated *kaolin clay* containing 45.2% silicon dioxide (SiO$_2$), 38.4% aluminum oxide (Al$_2$O$_3$), 1.8% titanium dioxide (TiO$_2$), and small additions of ferric oxide (Fe$_2$O$_3$), calcium oxide (CaO), magnesium oxide (MgO), potassium oxide (K$_2$O), and sodium oxide (Na$_2$O). Their loss on ignition is about 13.5%. *Masterfloat* clay has a density of 2.62 g/cm^3 (0.095 lb/in.^{3}), a specific surface area of 10.23 m^2/g, an average particle size of 2.1 μm (79 μin.), a pyrometric cone equivalent of 34, and excellent casting properties. Used for general ceramic applications.

Masterglass. Trademark of Saint-Gobain (France) for a clear patterned glass.

Mastergold. Trade name for a high-gold dental casting alloy.

Master Metal. Trade name of United American Metals Corporation (USA) for a babbitt composed of varying amounts of tin, copper and lead, and used for machinery bearings.

Master-Tec. Trademark of Ivoclar Vivadent AG (Liechtenstein) for a nickel- and beryllium-free ceramic dental alloy containing 55.8% cobalt, 25.0% chromium, 7.5% gallium, 5.0% tungsten, 3.0% molybdenum, 3.0% niobium, and less than 1.0% silicon. It has a white color, a density of 8.1 g/cm^3 (0.29 lb/in.^{3}), a melting range of 1250-1300°C (2280-2370°F), a hardness of 390 Vickers, good high-temperature strength, low elongation and excellent biocompatibility. Used for single crowns, posts, and bridges.

mastic. (1) A solid yellowish resinous exudate from the tree *Pistacia lentiscus* growing in the Mediterranean region, especially on the Greek Islands. Density, 1.06 g/cm³; melting point, 105-120°C (221-248°F). Used in the manufacture of adhesives and varnishes.

(2) A soft, heavy, pasty type of adhesive that may or may not be waterproof. It is composed of a mixture of asphaltic material and graded aggregate, such as asbestos, crushed rock or sand. Usually packaged in metal containers or gun cartridges, it is ready for application by spreading with a knife, float or trowel or caulking gun. Used for laying wallboards, wood paneling and floor tiles, as a protective coating suitable for application as a thermal insulation and, if waterproofed, as a waterproofing agent and sealant. Also known as *asphalt mastic*.

mastic asphalt. See mastic (2).

Mastifix. Trade name of Ato Findley (France) for adhesive mastics used for building and construction applications.

masutomilite. A pale purplish pink mineral of the mica group composed of lithium potassium manganese aluminum fluoride silicate, $K(Li,Mn)_3(Si,Al)_4O_{10}F_2$. Crystal system, monoclinic. Density, 2.94 g/cm³; refractive index, 1.569. Occurrence: Japan.

masuyite. A reddish orange, radioactive mineral of the becquerel group composed of lead uranium oxide monohydrate, $PbUO_3 \cdot H_2O$. Crystal system, orthorhombic. Density, 5.08 g/cm³; refractive index, 1.895. Occurrence: Central Africa.

mat. A thin sheet manufactured from continuous-strand fibers or randomly oriented chopped fibers by interfelting or intertwining, and with or without the application of a binder. It is available in various weights, widths and lengths. Common fiber materials supplied in mat or blanket form are glass, alumina-silica, alumina-boria-silica, zirconia, and silicon carbide. Used as a reinforcement for engineering plastics, and for acoustical, thermal and/or other insulation applications.

Matador. (1) Trade name of Dohlen-Stahl Gusstahl-Handels GmbH (Germany) for a series of high-speed steels containing 0.7-1% carbon, 4-4.3% chromium, 0-0.85% molybdenum, 1-2.5% vanadium, 8-18.5% tungsten, 0-4.7% cobalt, and the balance iron. Used for lathe and planer tools, drills, taps, broaches, and other tools.

(2) Trade name of Starcke KG Schleifmittelwerke (Germany) for waterproof, resin-bonded abrasive fabrics and papers.

matai. The yellowish brown, straight-grained wood of the pine tree *Pinus spicatus*. It has an average weight of 609 kg/m³ (38 lb/ft³). Source: New Zealand. Used for lumber and construction work. Also known as *black pine*.

Matankanese. Trade name for rayon fibers and yarns used for textile fabrics.

Matapoint. Trade name for rayon fibers and yarns.

mat binder. A synthetic resin, usually of the thermoplastic type, used in the manufacture of a fiberglass mat for holding the continuous or chopped-strand fibers in place and ensuring retention of the mat shape.

matched lumber. Lumber that is edge-dressed or shaped (e.g., rabeted) on one or both ends to make a close tongue-and-groove joint.

matching-expansion alloys. A group of alloys including various iron-nickel, nickel-iron, iron-chromium and iron-nickel-chromium types that possess coefficients of thermal expansion similar to those of ceramics and glasses and thus match the rate at which these materials cool from elevated temperatures. Used for glass-to-metal and glass-to-ceramic sealing of electronic components. See also controlled-expansion alloys.

Matchmaker. Trade name of Schottlander Limited (UK) for a dental porcelain used for metal-to-ceramic bonding applications.

Matchmate. Trade name of Schottlander Limited (UK) for a series of dental bonding alloys including *Matchmate 660* and *Matchmate HFYC* high-gold ceramic alloys, *Matchmate 640*, *Matchmate HFWC* and *Match-mate SFC* medium-gold ceramic alloys and *Matchmate 610, 615 & 620* palladium-base ceramic alloys.

match wax. A low-melting *paraffin wax* used for impregnating matches.

matelassé. (1) Originally, a rather heavy double layered, usually jacquard-woven fabric consisting of a face and back fabric that develop different degrees of shrinkage during finishing resulting in a padded, puffed or quilted effect on the face fabric. Used especially for coats, padding, furnishings, etc.

(2) Now, a fabric with a raised or puckered effect produced by the contraction during finishing of the non-heat-stabilized synthetic fiber yarns (e.g., acrylic, nylon or polyester) used in its manufacture. Used as padding, draperies, upholstery, dresses, vests, etc.

mat enamel. See matte enamel.

Matenka. Trade name for rayon fibers and yarns used for textile fabrics.

Matenkona. Trade name for rayon fibers and yarns.

materials. A term referring to substances, such as metals, plastics, ceramics, glasses, semiconductors, superconductors, dielectrics, fibers, wood, stone, sand and composites, whose properties make them useful in structures, machines, devices or other products.

Matesa. Trade name of Matesa AS (Turkey) for rayon fibers and yarns used for textile fabrics.

mat-formed particleboard. Particleboard made by arranging coated wood particles in blankets or mats of the required finished board size and then compressing by means of a flat-platen press. See also particleboard.

mat glaze. See matte glaze.

Mathesius metal. A German lead-alkali metal containing 95-96% lead, 3% calcium and 1-2% alkaline earth metals. It has excellent antifriction properties, and is used for bearings.

Matident. Trade name of Johnson Matthey plc (UK) for a series of white, heat-treatable dental casting alloys (ADA type IV) containing 54% silver, 20% palladium, 12% copper, 11% gold, and small percentages of zinc and platinum.

matildite. An iron-black to gray mineral composed of silver bismuth sulfide, $AgBiS_2$. Crystal system, hexagonal. Density, 6.90 g/cm³. Occurrence: Canada (Northwest Territories).

Matinesse. Trade name of BASF Corporation (USA) for nylon 6 fibers used for textile fabrics including clothing.

mat kid. A thin calfskin with a dull mat surface finish, used for shoe uppers.

Mat Lor. Trade name of Lorcet & Compagnie (France) for glass reinforcements.

Matobar. Trade name of Reinforcement Steel Services (UK) for hard-drawn steel bar and wire.

matlockite. A yellowish gray mineral composed of lead chloride fluoride, $PbClF$. It can be made by melting lead chloride ($PbCl_2$) and lead fluoride (PbF_2) under prescribed conditions. Crystal system, tetragonal. Density, 7.12 g/cm³; refractive index, 2.145.

Mat-O-Bel. Trade name of SA Glaverbel (Belgium) for a diffusing, nonreflective glass with a special satin finish on both surfaces.

Matreloy. Trade name of Materials Research Corporation (USA) for a wrought, nonmagnetic, heat-treatable, corrosion-resistant nickel-base alloy containing 39% chromium, 4% molybdenum, 2% titanium and 1% aluminum. It has good high-temperature properties, high yield and tensile strength, and is used for electronic components, chemical process equipment, furnace parts, and springs.

Matrimid. Trademark of Ciba-Geigy Corporation (USA) for a series of engineering resins including various bismaleimides and thermoplastic polyimides. The bismaleimides contain 4,4'-bismaleimidodiphenylmethane and 0,0'-diallyl bisphenol A in various weight ratios and have excellent high-temperature performance, high strength, excellent toughness and excellent processibility. Used for advanced composites and adhesives.

Matrix. (1) Trade name for a noneutectic fusible alloy containing 48% bismuth, 28.5% lead, 14.5% tin and 9% antimony. It expands on cooling, has a yield temperature of 116°C (241°F), a melting-temperature range of 103-227°C (217-441°F), and is used for dental models, and die mounting applications.

(2) Trade name of Teledyne Allvac (USA) for a series of tool and die steels.

matrix brass. A free-cutting brass composed of 62% copper, 37% zinc and 1-1.5% lead, and used for hardware, and engraving applications.

matrix fiber. See bicomponent fiber; biconstituent fiber.

matrix-fibril fiber. A *bicomponent fiber* consisting of a fibrous matrix material into which *fibrils* (fine, short filaments) of a chemically of physically different material have been embedded.

matrix resins. A group of thermoset or thermoplastic resins, such as epoxies, phenolics, polyesters, polyphenylene sulfides, vinyl esters, or bismaleimides, that are used to form the nonfiber components in composites.

mattagamite. A violet mineral of the marcasite group composed of cobalt telluride, $CoTe_2$. It can also be made synthetically. Crystal system, orthorhombic. Density, 8.00 g/cm^3. Occurrence: Canada (Quebec).

matte blue. A blue ceramic color obtained by mixing 6 parts of aluminum oxide (Al_2O_3), 2 parts of cobaltous oxide (CoO) and 2 parts of zinc oxide (ZnO).

matte enamel. A fired porcelain enamel that has low gloss due to the elimination of lead-bearing compounds, in particular lead oxide (PbO), and the addition of certain other oxide compounds to the frit. Also known as *mat enamel; matte porcelain enamel*.

matte glaze. A colorless or colored, fired ceramic having a low gloss due to partial devitrification induced by the elimination of lead compounds and the addition of certain oxides, such as calcium oxide (CaO), titanium dioxide (TiO_2) or zinc oxide (ZnO), into the glaze batch. Used for wall tiles. Also known as *mat glaze.*

Matte Ladle. Trade name of Edgar Allen & Company Limited (UK) for plain-carbon steel castings containing 0.37% carbon, 0.3% silicon, 0.7% manganese, and the balance iron.

matte jersey. A type of *tricot* fabric knitted with fine crepe yarn and having a lusterless surface.

matte porcelain enamel. See matte enamel.

Mattesco. Trade name for rayon fibers and yarns used for textile fabrics.

Matte Touch. Trade name for polyester fibers and yarns used for textile fabrics.

matteuccite. A colorless mineral composed of sodium hydrogen sulfate monohydrate, $NaHSO_4 \cdot H_2O$. It can also be made synthe-

tically. Crystal system, monoclinic. Density, 2.12 g/cm^3; refractive index, 1.46. Occurrence: Italy.

Matthey. Trade name of Johnson Matthey plc (UK) for an extensive series of alloys including a wide range of silver-graphite, silver-cadmium and silver-nickel electrical contact alloys, numerous cast or sintered high-conductivity tungsten-copper, tungsten-silver, copper-chromium alloys for welding electrodes and electrical contacts, several cast copper-beryllium alloys, and various lead-base soft solders.

Mattibel. Trade name of Johnson Matthey plc (UK) for a series of dental casting alloys containing 10-80% gold, 9-67% silver, 6-21% copper, and small percentages of platinum.

Mattibraze. Trade name of Johnson Matthey plc (UK) for a series of silver brazing alloys containing copper, zinc, silver and cadmium.

Matticast. Trade name of Johnson Matthey plc (UK) for a series of dental casting alloys containing gold, silver, copper, platinum, palladium and zinc.

Matticraft. Trade name of Johnson Matthey plc (UK) for a series of medium and high-gold dental casting and bonding alloys containing varying amounts of palladium, silver, and/or gallium, indium, tin, copper, platinum, rhodium and zinc. *Matticraft Alpha, E, G, JPS & S* are high-gold alloys, *Matticraft M, Y & 45* medium-gold alloys, *Matticraft B* is a palladium-silver bonding alloy, and *Matticraft C, H & 80* are silver-free palladium bonding alloys.

Mattident. Trade name of Johnson Matthey plc (UK) for a series of dental casting alloys containing gold, palladium, silver, gold, copper, indium, gallium, platinum, zinc and iron. Examples include *Mattident 60*, a 60%-gold alloy, and *Mattident G & R* high-gold alloys.

Mattieco. Trade name of Johnson Matthey plc (UK) for a series of silver-palladium dental casting alloys.

Mattiflo. Trademark of Johnson Matthey plc (UK) for a dental gold solder.

Mattiflux. Trade name of Johnson Matthey plc (UK) for silver brazing and soldering fluxes.

Mattigold AE. Trade name of Johnson Matthey plc (UK) for a high-gold dental casting alloys.

Mattiloy. Trade name of Johnson Matthey plc (UK) for a series of dental amalgam alloys containing considerable amounts of silver.

Mattinax. Trade name of Johnson Matthey plc (UK) for a series of high-gold dental casting alloys containing varying amounts of silver, palladium, platinum and copper.

Mattiphos. Trademark of Johnson Matthey plc (UK) for silver brazing alloys containing additions of phosphorus.

Mattisol. Trademark of Johnson Matthey plc (UK) for silver solders.

matulaite. A colorless to white mineral composed of calcium aluminum phosphate hydroxide hydrate, $CaAl_{18}(PO_4)_{12}(OH)_{20} \cdot 28H_2O$. Crystal system, monoclinic. Density, 2.33 g/cm^3; refractive index, 1.576. Occurrence: USA (Pennsylvania).

maucherite. A reddish platinum-gray mineral composed of nickel arsenide, $Ni_{11}As_8$. It can also be made synthetically. Crystal system, tetragonal. Density, 8.00 g/cm^3. Occurrence: Germany, Canada (Ontario, Quebec).

Mauritius fiber. A long, whitish hard vegetable fiber, similar in composition but inferior in properties to *sisal*. It is obtained from the leaves of the cabuya (*Fourcraea gigantea*), an agave plant cultivated in Central America and on the island of Mauritius. Used in making sacks, ropes and twines. Also known as

Mauritius hemp. See also figue.

Mauritius hemp. See Mauritius fiber.

Maustinox. Trade name of Usines de A. Manoir (France) for a series of heat- and corrosion-resistant austenitic stainless steels.

Mavilon. Trademark of Zoltek Magyar Viscosa RT (Hungary) for acrylic fibers used for textile fabrics.

Mawus. Trade name for polyester fibers and yarns available in several grades including *Mawus-M* and *Mawus-R*. Used for textile fabrics.

mawsonite. A brownish orange mineral of the chalcopyrite group composed of copper iron tin sulfide, $Cu_6Fe_2SnS_8$. Crystal system, tetragonal. Density, 4.66 g/cm³. Occurrence: Japan, Tasmania, Canada (Ontario).

Maxal Co. Italian trade name for a permanent magnet material containing 51% iron, 24% cobalt, 14% nickel, 8% aluminum and 3% copper. It has high permeability, and is used for magnetic and electrical equipment.

Max Bond. Trade name of H.B. Fuller Company (USA) for a series of construction, foam and panel adhesives.

Maxchip. Trade name of Edgar Allen Balfour Limited (UK) for a series of tough, shock-resistant, oil-hardening tool steels containing 0.42% carbon, 1.7% silicon, 1.05% chromium, 4.1% nickel, and the balance iron. Used for chisels and punches.

Max-El. Trade name of Crucible Materials Corporation (USA) for a series of case- or oil-hardening alloy steels containing 0.2-0.5% carbon, 1-25% manganese, 0.1-0.2% molybdenum, 0-0.6% chromium, and the balance iron. Used for gears, pinions, worms, spindles, shafts, screws, studs, and springs.

Maxel. Trade name of Crucible Materials Corporation (USA) for free-machining grades of low- and medium-carbon steels.

Maxeloy. (1) Trade name of Colt Industries (UK) for a low-carbon high-strength steel containing up to 0.15% carbon, 1.2% manganese, 0.7% phosphorus, 0.7% silicon, 0.5% nickel, 0.3% copper, a trace of vanadium, and the balance iron. Used for railroad and mine cars and structures.

(2) Trade name of Colt Industries (UK) for a high-strength low-alloy steel containing up to 0.15% carbon, 0.9-1.1% manganese, 0.3-0.5% nickel, 0-0.25% chromium, 0.20% or more copper, and the balance iron. Used for truck frames, mine and railroad cars, pumps, coal chutes, and ships.

Maxergy. Trade name of 27th Century Technologies Inc. (USA) for coatings used on steel, most other alloys, glass, fiberglass, concrete, rigid plastics, etc.

Maxhesive. (1) Trademark of Sandoz Limited (Switzerland) for adhesives used for bonding fiber-glass reinforced products, such as fittings, couplings and pipes.

(2) Trade name of The Ceilcote Company (USA) for a corrosion-resistant cement.

Maxhete. Trade name of Edgar Allen Balfour Limited (UK) for a series of austenitic chromium-nickel and low-carbon high-chromium stainless steels with excellent corrosion and heat resistance. Also included under this trade name are several nickel-iron-chromium high-temperature alloys. *Maxhete* steels and alloys are used for furnace equipment, automotive engine valves, and other high-temperature applications.

Maxibond. (1) Trade name for a corrosion-resistant cobalt-chromium dental bonding alloy.

(2) Trade name for a bonding resin used in orthodontics.

MaxiChop. Trade name of PPG Industries, Inc. (USA) for chopped-strand glass-fiber reinforcements for polymers, such as polypropylenes.

Maxi-Form. Trade name of Gulf States Steel, Inc. (USA) for a series of high-strength low-alloy (HSLA) steels containing up to 0.09% carbon, 0-0.9% manganese, 0-0.015% phosphorus, 0-0.02% sulfur, 0.02% aluminum, 0.01% or more niobium (columbium), and the balance iron. They have high impact toughness and ductility, and are used for structural members and other building and construction applications.

Maxiglue. Trade name of Ato Findley (France) for alcohol-resin adhesives used for floor-covering applications.

Maxigold. Trademark of Ivoclar Vivadent AG (Liechtenstein) for a palladium-free hard dental casting alloy (ADA type III) containing 59.5% gold, 26.3% silver, 8.5% copper, 2.7% palladium, 2.7% zinc, and less than 1.0% indium and iridium, respectively. It has a deep yellow color, a density of 13.2 g/cm³ (0.48 lb/in.³), a melting range of 840-890°C (1544-1634°F), an as-cast hardness of 150 Vickers, moderate elongation, and excellent biocompatibility and resistance to oral conditions. Used for crowns, inlays, onlays, posts, and bridges.

Maxilvry. Trade name of Edgar Allen Balfour Limited (UK) for a series of austenitic stainless steels used for chemical plant equipment, fittings, cutlery, valve plates, etc.

Maxima. Trade name of Ellwood Group Inc. (USA) for a weldable, formable, high-toughness, abrasion-resistant plate steel containing 0.23% carbon, 1.50% manganese, 0.08% silicon, 0.25% nickel, 1.97% chromium, 0.36% copper, 0.01% titanium, 0.03% niobium (columbium), 0.02% aluminum, 0.008% nitrogen, 0.002% sulfur, 0.01% phosphorus, and the balance iron. Used for heavy industrial applications including mining and construction equipment.

Maximal. Trade name for a bonding resin used in orthodontics.

Maximold. Trade name of Ziv Steel & Wire Company (USA) for an air-hardening steel containing 0.4% carbon, 5.2% chromium, 1.2% molybdenum, 1% silicon, and the balance iron. It has good resistance to heat checking, and is used for aluminum and zinc die-casting dies.

Maximum. Trade name of Westa-Westdeutsche Edelstahl-Handelsgesellschaft (Germany) for high-speed steels containing 0.74% carbon, 4.1% chromium, 1.1% vanadium, 18.5% tungsten, and the balance iron. Used for lathe and planer tools, drills, reamers, and taps.

Maximum Spezial. Trade name of Westa-Westdeutsche Edelstahl-Handelsgesellschaft (Germany) for a series of high-speed steels containing 0.75-0.95% carbon, 4.0-4.3% chromium, 0.80-0.85% molybdenum, 2.8-10% cobalt, 1.5-2.5% vanadium, 8.5-12% tungsten, and the balance iron. Used for lathe and planer tools, reamers, drills, taps, and broaches.

Maxinium. Trade name of Edgar Allen Balfour Limited (UK) for an oil-hardening steel containing 0.32% carbon, 1-2% silicon, 5.25% chromium, 5% tungsten, 0.5% molybdenum, and the balance iron. Used for die-casting dies and molds.

Maxite. Trade name of Columbia Tool Steel Company (USA) for a tungsten-type high-speed tool steel (AISI type T8) containing 0.8% carbon, 14% tungsten, 5.2% cobalt, 4% chromium and 2% vanadium, 0.6% molybdenum, and the balance iron. It has good deep-hardening properties, high hot hardness and excellent wear resistance. Used for cutting tools, boring and shaping tools, drills and reamers.

Maxit. Trade name of Metaplas Ionen GmbH (Germany) for a series of surface coatings. *Maxit AlTiN* refers to aluminum titanium nitride coatings for milling cutters, end mills and cutting wheels made of high strength steel or hard metals for high-speed cutting applications, and *Maxit CrN* is a chromium nitride coating for metal-forming and processing tools, and plastics.

Maxlen. Trade name of MRC Polymers (USA) for high-density polyethylenes supplied in several grades including unreinforced, glass fiber-, talc- or graphite-reinforced and high-impact-copolymer.

Maxmith. Trade name of Edgar Allen Balfour Limited (UK) for a tough tool steel containing 0.4% carbon, up to 0.6% manganese, 3.25% nickel, and the balance iron. Used for riveting tools, chisels, and caulking and beading tools.

Maxnap. Trade name of Edgar Allen Balfour Limited (UK) for a tough, fatigue-resistant tool steel containing 0.3% carbon, 0.6% manganese, 1.05% chromium, 0.2% vanadium, and the balance iron. Used for pneumatic hammers, riveters, and nut piercers.

Maxnite. Trade name of MRC Polymers (USA) for a 35% glass/mineral-reinforced polyethylene terephthalate resin.

Maxon. Trademark of Davis & Geck, Inc. (USA) for a biodegradable copolymer of *glycolide* and *trimethylene carbonate* (TMC) used for tissue engineering applications, e.g., surgical sutures.

Maxos. Trademark of Schott Auer (Germany) for safety sight glass made from a homogeneous high-purity borosilicate glass with outstanding chemical durability. It can withstand extreme temperatures (up to about 300°C or 570°F) and pressures and, owing to thermal prestressing (tempering), has very high thermal-shock resistance. It is supplied in the form of round sight glasses and level gauge glasses in a wide range of sizes, and used in steam vessels, pressure tanks and pipeline systems in the chemical, petrochemical and power-supply industries.

Maxpro. Trade name of MRC Polymers (USA) for a series of homopolymer and copolymer propylenes reinforced with glass beads, calcium carbonate, talc or mica, and supplied as general-purpose, impact-modified and UV-stabilized grades.

Maxtack. Trade name of A. Milne & Company (USA) for an air- or oil-hardening steel containing 2.25% carbon, 10% tungsten, 2% chromium, 2.5% manganese, 1% silicon, and the balance iron. Used for cutters, tools, and dies.

Maxtensile. Trade name of Farrell Company (USA) for an alloy cast iron containing 3.3% carbon, 2.4% silicon, 1.5% nickel, 0.8% chromium, and the balance iron. Used for hydraulic castings, couplings, sprockets, and mill rods.

Maxtress. Trade name of Jonas & Culver (Nova) Limited (UK) for a machinery steel containing 0.35% carbon, 1% chromium, 3.5% nickel, 0.5% molybdenum, and the balance iron.

Maxtuff. Trade name of Ziv Steel & Wire Company (USA) for a shock-resisting tool steel (AISI type S1) containing 0.5% carbon, 2.5% tungsten, 1.15% chromium, 0.3% manganese, 0.2% vanadium, 0.75% silicon, and the balance iron. It has high hardness, strength and toughness, and is used for chisels, mandrels, and pneumatic tools.

Maxx. Trademark of Copper & Brass Sales, Inc. (USA) for premium machining stainless steel bar.

Maxxam. Trade name of M.A. Hanna Engineered Materials (USA) for a series of thermoplastic polyolefin polymers and olefin-based compounds including impact-modified, flame-retardant polypropylenes.

Maya. Trade name of SA Glaverbel (Belgium) for a patterned glass.

Mayari Iron. Trade name of Bethlehem Steel Corporation (USA) for a heat-, wear- and acid-resistant *pig iron* containing 3.8-4.5% total carbon, 0.5-3% silicon, 0.6-2% manganese, 0.7-2% nickel, 1.5-2.5% chromium, 0-0.2% vanadium, 0.1-0.2% titanium, up to 0.1% phosphorus, up to 0.05% sulfur, and the bal-

ance iron. Used for hard castings, heavy rolls, etc. Also known as *Mayari Pig Iron*.

mayenite. A colorless mineral composed of calcium aluminum oxide, $Ca_{12}Al_{14}O_{33}$. Crystal system, cubic. Density, 2.68 g/cm³. Occurrence: Germany. It is a constituent of cement clinker and is used in ceramics and the building trades.

mayflower. See apamate.

Mayor. Trade name of Vereinigte Edelstahlwerke (Austria) for a water-hardening steel containing 0.75% carbon, and the balance iron. Used for tools, drills, punches, and springs.

Maywrap. Trade name of Mayfair Plastics Limited (UK) for glass-fiber quilts used for roof insulation applications.

Mazak. Trade name of Pasminco Europe (Mazak) Limited (UK) for a series of zinc die-casting alloys containing varying amounts of aluminum, magnesium, copper and/or nickel.

mazarine blue. See royal blue.

Mazin. (1) Trademark of E.I. DuPont de Nemours & Company (USA) for colloidal silica.

(2) Trademark of Corn Card International, Inc. (USA) for biodegradable polymers and polymer compounds used in the manufacture of extruded and molded products.

mbobomkulite. A sky-blue mineral composed of copper nickel aluminum nitrate sulfate hydroxide trihydrate, $(Ni,Cu)Al_4[(NO_3)_2(SO_4)](OH)_{12}·3H_2O$. Crystal system, monoclinic. Density, 2.30 g/cm³. Occurrence: South Africa.

M-Bronze. A leaded bronze containing 86-91% copper, 6.2-7.3% tin, 1.5-5% zinc, 1-2% lead, and up to 0.25% iron. It has good strength and good resistance to continuous temperatures up to 260°C (500°F), and is used for valves and marine parts.

MBS-PVC alloys. Rubbery materials based on blends of methacrylate-butadiene-styrene (MBS) and polyvinyl chloride (PVC). The addition of PVC greatly increase the toughness of MBS resins. *MBS-PVC alloys* are usually supplied as balanced compounds ready to be processed by extrusion, injection molding and other techniques.

McAdamite. Trade name for a strong, nonhardenable aluminum alloy containing 12-18% zinc, 3.1% copper and 0.2% magnesium. Used for light-alloy parts.

McAdams. American trade name for a series of nonhardenable aluminum alloys used for light-alloy parts. *McAdams Alloy A* contains 60% aluminum, 11-55% copper, 10-43% chromium and 20% zinc, *McAdams Alloy B* has 69% aluminum, 7.7% copper, 23% zinc and 0.3% nickel, *McAdams Alloy C* is composed of 70% aluminum, 22% zinc, 5% antimony and 3% copper, *McAdams Alloy D* contains 80% aluminum, 8% cadmium, 8% tin and 4% silver, and *McAdams Alloy E* has 82% aluminum, 12% copper, 5% cadmium and 1% silver.

mcallisterite. A white mineral composed of magnesium borate hydrate, $Mg_2B_{12}O_{20}·15H_2O$. Crystal system, rhombohedral (hexagonal). Density, 1.87 g/cm³; refractive index, 1.504. Occurrence: USA (California).

MC Alloy. Trade name of Mitsubishi Metals America Corporation for a series of corrosion-resistant alloys composed of 50% chromium, 0-25% iron, 0-5% molybdenum, and the balance nickel. They have excellent resistance to hydrofluoric, nitric, phosphoric and sulfuric acids, and are used for in the coating industry for galvanizing tank electrodes and in the semiconductor industry for wafer catching trays.

M-C asphalt. See medium-curing asphalt.

MC Bond. Trade name for an acrylic cement used to produce clear joints for cast acrylic sheets.

mcconnellite. A deep red mineral of the delafossite group composed of copper chromium oxide, $CuCrO_2$. Crystal system, rhombohedral (hexagonal). Density, 5.59 g/cm^3.

McCoy. Trademark of McCoy Foundry (Canada) for gray iron castings.

McFarland & Harder. Trade name for a series of stainless alloys containing varying amounts of nickel, copper and chromium used for heat- and corrosion-resistant parts. *McFarland & Harder Alloy A* contains 48% nickel, 43% copper and 9% chromium, *McFarland & Harder Alloy B* has 55% copper, 29% nickel and 16% chromium, *McFarland & Harder Alloy C* is composed of 59% nickel, 30% chromium and 11% copper, and *McFarland & Harder Alloy D* contains 46% nickel, 43% chromium and 11% copper.

McGill metal. An aluminum bronze containing 89% copper, 9% aluminum and 2% iron. It has good wear and corrosion resistance and high tensile strength, and is used for gears, bushings, and bearings.

mcgovernite. A bronze to red-brown mineral composed of magnesium manganese zinc arsenate silicate hydroxide, $Mg_9Mn_4Zn_2As_2Si_2O_{17}(OH)_{14}$. Crystal system, rhombohedral (hexagonal). Density, 3.72 g/cm^3; refractive index, 1.75-1.76. Occurrence: USA (New Jersey).

mcguinnessite. A light blue green fibrous mineral of the rosasite group composed of copper magnesium carbonate hydroxide, $(Mg,Cu)_2(CO_3)(OH)_2$. Crystal system, monoclinic. Density, 3.11 g/cm^3; refractive index, 1.730. Occurrence: USA (California).

M-Chrome. Trade name of Latrobe Steel Company (USA) for a deep-hardening tool steel containing 1% carbon, 1.5% chromium, and the balance iron. Used for dies, tool, and gauges.

McKay. Trade name of Teledyne McKay (USA) for an extensive series of mild, low-alloy, stainless and high-temperature steels and nickel-chromium alloys.

McKechnie. Trade name of McKechnie Metals Limited (UK) for an extensive series of coppers and copper-base alloys including deoxidized nonarsenical and electrolytic tough-pitch coppers, naval brasses, leaded and unleaded brasses and bronzes, and aluminum, manganese and silicon bronzes.

mckelveyite. A dark green or black mineral composed of sodium barium calcium rare-earth carbonate hydrate, $(Na,Ca)(Ba,Y,U)_2(CO_3)_3 \cdot xH_2O$. Crystal system, hexagonal. Density, 3.14 g/cm^3; refractive index, 1.66. Occurrence: USA (Wyoming).

McKinney. British trade name for a series of aluminum alloys containing 2-3% copper and 1-2% manganese with good corrosion resistance including seawater corrosion. Used for light-alloy parts and marine parts.

mckinstryite. A steel-gray mineral composed of copper silver sulfide, $(Ag,Cu)_2S$. Crystal system, orthorhombic. Density, 6.61 g/cm^3. Occurrence: Canada (Ontario).

McLouth. Trade name of McLouth Steel Corporation (USA) for a series of high-strength low-alloy (HSLA) steels containing 0.12-0.26% carbon, 0.45-1.5% manganese, up to 0.01% phosphorus, up to 0.02% sulfur, 0.1-0.2% silicon, 0-0.03% niobium (columbium), 0.4-0.5% chromium, 0.6-0.7% nickel, 0.2-0.3% copper, up to 0.01% molybdenum, up to 0.02% vanadium, and the balance iron. It has good formability and weldability, and is used for chutes, crane booms, derricks, dump bodies, marine parts, pressure tanks, automobile parts, buckets, truck frames, wheels, and transmission towers.

McLure Alloy. British nonhardenable aluminum alloy containing 8.2% copper, 5-6% tin, 0.9% iron, 0.3% silicon, and 0.2% manganese. Used for light-alloy parts.

McMenomy's B.M. Brown. Trade name of Aardvark Clay & Supplies (USA) for a medium-brown, medium-texture clay (cone 10).

McNamee. Trade name of R.T. Vanderbilt Company, Inc. (USA) for a kaolin clay.

mcnearite. A white mineral composed of hydrogen sodium calcium arsenate tetrahydrate, $NaCa_5H_4(AsO_4)_5 \cdot 4H_2O$. Crystal system, triclinic. Density, 2.60 g/cm^3; refractive index, 1.562. Occurrence: France.

MCrAlY coatings. Corrosion-, erosion, impact- and oxidation-resistant multicomponent alloy coatings with low chemical reactivity composed of chromium, aluminum, yttrium, and a base metal or alloy (M), usually cobalt, iron or a cobalt-nickel or nickel-cobalt alloy. They are applied to substrates (e.g., superalloys) by thermal (e.g., plasma) spraying, sputtering or physical-vapor deposition techniques, such as electron-beam evaporation. The deposited films, usually in the thickness range of 0.1-0.2 mm (0.004-0.008 in.), have a typical composition of 20% chromium, 10% aluminum and 0.3% yttrium, with the balance being the base metal or alloy (M). Depending on the base metal or alloy (M) selected, MCrAlY coatings are referred to as CoCrAlY (M = cobalt), FeCrAlY (M = iron) or NiCoCrAlY (M = nickel-cobalt alloy). *MCrAlY coatings* are often used on aircraft turbine blades and components, and other high-temperature parts.

MDF cement. See macro-defect-free cement.

MDX. Trademark of Dow Corning Inc. (USA) for medical-grade liquid silicone supplied in several grades, and used in the form of microdroplets for human skin reconstruction.

meadow ore. See bog iron ore.

Meadway. Trade name of Meadway (UK) for a heat-cure denture acrylic.

Mealorub. Trademark for a powder made from easily vulcanizable rubber latex and used in civil engineering and the construction trades.

Measle Yarn. Trade name for rayon fibers and yarns used for knitting and weaving textile fabrics.

mechanical fabrics. See industrial fabrics.

mechanical glass. Any glass that is resistant to very high impact.

mechanical leather. Any leather suitable for use in machinery, e.g., as torque transmission or conveyor belts.

mechanically alloyed materials. Materials, such as oxide-dispersion-strengthened superalloys, structural aluminum-base alloys, amorphous materials and intermetallics, produced from mechanically alloyed powders by consolidation, e.g., by extrusion or hot pressing, subsequent hot and/or cold working, and final annealing treatments. Abbreviation: MA materials. See also oxide-dispersion-strengthened alloys; aluminum alloys; intermetallics; aluminides.

mechanically alloyed powder. A homogeneous composite powder obtained from elemental powders, crushed master alloys, prealloyed powders and/or refractory-oxide powders by intensive grinding in a high-energy ball mill (e.g., an attrition mill). Alloying of the powders is obtained by the alternate cold welding and fracturing (shearing) of particles of different composition by the rapidly colliding grinding balls. Abbreviation: MA powder.

mechanically foamed plastics. Plastics whose cellular structure has been produced by (i) mechanical introduction of a gas, such as air, (ii) whipping, or (iii) other methods of agitation. See also cellular plastics.

mechanical papers. A generic term for a class of papers used for purposes other than newsprint, books and writing, and including coarse papers (e.g., kraft, jute and parchment papers and glassine), paperboards, tissue papers, building papers and felts, absorbent papers, sanitary papers, insulating papers and metal-coated papers.

mechanical pulp. A generic term for fibrous materials which are made from wood by mechanical means, e.g., grinding. Mechanical pulp is classified into: (i) purely mechanical pulp, such as groundwood pulp, refiner pulp, etc.; and (ii) chemically pre-treated pulp, such as chemical pulp, chemical refiner pulp, etc. Also known as *mechanical wood pulp*. See also wood pulp.

mechanical transfer paper. See impact paper.

Mechelen lace. A fine bobbin lace, usually on a hexagonal-mesh ground, with a floral design clearly outlined by a lustrous, silky thread. Named after the town of Mechelen (or Malines) in northern Belgium, where this type of lace was originally made. Also known as *Mechelen; maline; malines*.

Meco. Trade name for a corrosion-resistant alloy containing 50% copper, 25-30% nickel and 20-25% zinc. Used for chemical equipment.

medal bronze. A corrosion-resistant bronze containing 92-97% copper, 1-8% tin, 0-2% zinc, and 0-1% lead. Used for medals and ornaments.

Medalist. Trade name of J.F. Jelenko & Company (USA) for a white-colored palladium-silver dental alloy with excellent physical properties and good solderability and workability, used for fusing porcelain to metal particularly in implant and long-span bridgework.

medal metal. A corrosion-resistant alloy composed of 84% copper and 16% zinc used for medals and ornaments.

Medical. Trade name of Inter-Africa Dental (Zaire) for a radiopaque, water-based dental filling material containing 63% calcium hydroxide.

medical ceramics. A class of ceramic materials that are suitable for use in medicine and dentistry. Included are (i) the so-called "bioceramics" (e.g., alumina, sapphire, hydroxyapatite, glass-ceramics, etc.) that are specially designed, biocompatible ceramics for the repair or reconstruction of diseased or damaged parts of the body in particular bones and musculoskeletal hard connective tissue; (ii) dental materials, such as feldspathic porcelains and ceramic masses used to make durable and esthetically pleasing dental restorations including crowns, fillings, inlays and dentures, and dental cements; (iii) ceramics and glasses for eye glasses, diagnostic instruments, medical labware, tissue culture flasks, endoscopic fiberoptics; and (iv) insoluble porous glasses for use as antibody, antigen and enzyme carriers. See also bioceramics; dental cements; dental materials.

Medicast. Trade name for a corrosion-resistant cobalt-chromium dental bonding alloy.

Medicem. Trade name of Inter-Africa Dental (Zaire) for a fluoride-releasing glass-ionomer dental luting cement with high compressive strength, high pH-value, low solubility and acidity, and excellent bonding to the tooth structure (dentin and enamel). Used in restorative dentistry for luting crowns, bridges, inlays, onlays, pins, and orthodontic bands.

Medifill. Trade name of Inter-Africa Dental (Zaire) for a biocompatible, radiopaque, fluoride-releasing glass-ionomer dental cement that chemically bonds to the tooth structure (dentin and enamel), and has low acidity and high compressive strength. Used in restorative dentistry for fillings.

Medina quartzite. A variety of *quartz* containing 97.8% silica (SiO_2), the balance being aluminum oxide (Al_2O_3), ferric oxide (Fe_2O_3) and various alkalies. It has a melting point of 1700°C

(3090°F), and is used as a raw material for silica refractories.

Medinlay. Trade name for a high-gold dental casting alloy used for inlays.

Medior. Trade name of Medior, Inc. (USA) for a medium-gold dental casting alloy.

Medium. Trade name of Engelhard Corporation (USA) for a silver brazing solder containing 70% silver, 20% copper and 10% zinc. It has a melting point of 724-754°C (1335-1390°F), and is used for precious metals, jewelry, and art objects.

medium-alloy air-hardening steels. (1) A group of ultrahigh-strength steels containing about 0.3-0.5% carbon, 0.2-0.5% manganese, 0.8-1.2% silicon, 4.75-5.5% chromium, 1.2-1.75% molybdenum, 0.4-1.2% vanadium, and the balance iron. They are usually modifications of chromium-type hot-work tool steels, such as AISI type H11 or H13 with yield strengths of 690-1720 MPa (100-250 ksi). Used for structural applications. See also high-strength steels; ultrahigh-strength steels.

(2) A group of cold-work tool steels (AISI group A) containing about 0.45-1.5% carbon, 0.4-2.5% manganese, up to 1.5% silicon, 0.9-1.8% molybdenum, up to 5.75% chromium, 0.5-5.15% vanadium, up to 2% nickel, up to 1.5% tungsten, and the balance iron. They have good nondeforming and deep-hardening properties, good machinability, good dimensional stability, good hot hardness, relatively good toughness, and low wear resistance. Often sold under trade names or trademarks, such as *Air Hard, Airvan, Apache, Jess-Air, Sagamore, Vega* or *Windsor*, they are used for blanking punches, blanking, forming, trimming and plastic-molding dies, and gauges.

medium-carbon cast steels. See carbon-steel castings.

medium-carbon low-alloy steels. A group of ultrahigh-strength steels (including especially AISI-SAE 4130, 4140, 4340, 6150 and 8640) with yield strengths of about 690-1720 MPa (100-250 ksi), used for constructional applications. See also high-strength steels.

medium-carbon steel castings. See carbon-steel castings.

medium-carbon steels. A group of steels with carbon contents between 0.25 and 0.60% that are most often used in the tempered condition, and have microstructures of *tempered martensite*. Additions of chromium, nickel and molybdenum improve their heat-treating capacity giving rise to a variety of strength-ductility combinations. In the heat-treated condition they are stronger than low-carbon grades, but less tough and ductile. Used for structural applications, railway wheels and tracks, gears, crankshafts, bolts, anbd chisels, hammers, knives, saws and other hand tools. See also high-carbon steels; low-carbon steels.

medium-curing asphalt. A liquid bituminous product made by dissolving asphalt cement in kerosine or other petroleum diluent. Also known as *M-C asphalt; medium-curing liquid asphaltic cement.*

medium-curing liquid asphaltic cement. See medium-curing asphalt.

medium-density fiberboard. An engineered composite board product made by rubbing apart bundles of wood fibers and bonding them with a synthetic resin under heat and pressure. Many of its physical properties approach those of solid wood, and it provides high dimensional stability and consistency, excellent machinability, and has an exptionally flat, smooth, uniform and dense surface and tight edges. Used for furniture, cabinetry and architectural moldings, as a substrate for wood veneer, heat-transfer foils, vinyl films and polymer-impregnated and basis papers, and as a substitute for solid wood in many interior applications. Abbreviation: MDF. See also fiberboard.

medium-density hardboard. Hardboard similar to regular hardboard, but manufactured in large panels by bonding refined wood fibers with adhesive *lignin* and pressing together under heat to a density of 500-800 kg/m³ (31-50 lb/ft³). See also hardboard.

medium-density overlay. Exterior-type plywood with an opaque resin-treated fiber overlay. It has smooth surfaces and provides an excellent base for subsequent paint coats. Abbreviation: MDO. See also exterior-type plywood; overlay.

medium-density particleboard. Particleboard that has a density of 595-800 kg/m³ (37-50 lb/ft³), a linear expansion of about 0.25-0.35%, and a modulus of rupture, modulus of elasticity and internal bond strength intermediate between those of low- and high-density particleboard. Used for interior and exterior applications. See also particleboard; high-density particleboard; medium-density particleboard.

medium-density polyethylenes. Polyethylenes with a density between 0.926 to 0.940 g/cm³ (0.033-0.034 lb/in.³) having properties intermediate between those of high-density and low-density polyethylenes. Abbreviation: MDPE. See also polyethylenes; high-density polyethylenes; low-density polyethylenes.

medium-duty fireclay brick. Fireclay brick with a pyrometric cone equivalent between cones 29 and 31.5. See also fireclay brick.

medium-grain wood. A piece of wood yielding on average at least four annual rings per inch (25 mm) on either end.

Medium Hard. Trade name of Electro-Steel Company (USA) for a water-hardening steel containing 0.9% carbon, and the balance iron. Used for fixtures, dies, and tools.

medium-hard drawn wire. A copper wire whose tensile strength, stiffness, pliability and elongation under stress are intermediate between those of soft-drawn and hard-drawn wire. See also hard-drawn wire; soft-drawn wire.

medium-high-carbon steels. A group of *medium-carbon steels* with carbon contents between 0.23% and 0.44%.

medium-leaded brass. A free-cutting brass containing 64-65% copper, 34.5-35% zinc and 1% lead. Commercially available in the form of rods, flat products, wire and shapes, it has excellent machinability, good corrosion resistance, moderate cold workability, good electrical conductivity (26% IACS), and is used for hardware, screws, nuts, bolts, rivets, hinges, plaques, wheels, gears, ratchets, pinions, channel plates, valve stems, butts, dials, engravings, and instrument plates. See also high-leaded brass; low-leaded brass.

medium-leaded naval brass. A corrosion-resistant brass containing 60.5% copper, 38% zinc, 0.8% tin and 0.7% lead. Commercially available in the form of rods, flat products and shapes, it has good machinability and hot workability, and is used for marine hardware, screw-machine products, and valve stems. See also leaded naval brass; naval brass.

medium-low-carbon steels. A group of *low-carbon steels* with carbon contents between 0.15% and 0.23%. See also low-carbon steels.

medium-manganese steels. (1) A group of steels with a manganese content of 2-9% including in particular the medium-alloy air-hardening cold-work tool steels (e.g., AISI types A4, A6 and A10) containing about 0.65-1.05% carbon, 2% manganese, up to 0.5% silicon, 0.9-2.2% chromium, 0.9-1.4% molybdenum, up to 0.3% nickel, and the balance iron. They have excellent strengths, good deep-hardening and nondeforming properties, good machinability, relatively high toughness, and compa-

ratively low wear resistance. Often sold under trademarks or trade names, such as *Apache, Jess-Air, Lo-Air, Nutherm* or *Uni-Die,* they are used for dies, forming tools, gauges, bushings, etc.

(2) A group of wear-resistant austenitic manganese steels containing about 0.9-1.3% carbon, 5.8-6.3% manganese, 0.35-0.6% silicon, 1-1.5% molybdenum, and the balance iron. See also Hadfield steels.

medium-oil alkyd. An alkyd made with a moderate amount of unsaturated oil.

medium-oil varnish. A resin, such as an alkyd, made with a moderate amount of unsaturated oil.

medium-silicon gray irons. A group of heat-resistant, ferritic gray cast irons containing about 1.6-2.5% carbon, 4-7% silicon, 0.4-0.8% manganese, up to 0.3% phosphorus, up to 0.1% sulfur, and the balance iron. Used for fire bars, stove and furnace parts and machine parts. Also known as *medium-silicon gray cast irons.*

medium-silicon ductile irons. A group of heat-resistant ferritic ductile cast irons containing about 2.8-3.8% carbon, 2.5-6.0% silicon, 0.2-0.6% manganese, 0-1.5% nickel, 0-0.08% phosphorus, 0-0.12% sulfur, and the balance iron. Also known as *medium-silicon ductile cast irons.* See also heat-resistant ductile irons.

medium silver solder. A corrosion-resistant silver solder containing 70-75% silver, 20-23% copper, and 5-7.5% zinc.

medium staple cotton. Staple cotton fibers intermediate in length between long and short staple fibers, typically in the range of 25-38 mm (1-1.5 in.). See also long staple cotton; short staple fibers.

medium-temperature resins. A group of engineering thermosets including the epoxies, phenolics and silicones with maximum service temperatures of 121-260°C (250-500°F). See also high-temperature resins; low-temperature resins.

medium-temperature thermoset matrix composites. Thermosetting engineering composites, such as epoxies or phenolics with or without glass fabric reinforcement, designed for service temperatures of about 150-315°C (300-600°F). See also low-temperature thermoset composites.

medium wool. Wool whose fibers are intermediate in average length between long and short wools, typically in the range of 125-175 mm (5-7 in.). See also long wool; short wool.

Medusa. Trade name of Gulf Steel Corporation (USA) for a shock-resistant tool steel (AISI type S1) containing 0.5% carbon, 1.5% chromium, 2.5% tungsten, and the balance iron. Used for punches, rivet sets, pneumatic tools, chisels, dies, etc.

Meechite. Trade name of Meech Foundry Inc. (USA) for a corrosion-resistant steel containing 0.2% carbon, 29% chromium, and the balance iron. Used for melting pots for nonferrous metals.

Meehanite. Trade name of Meehanite Metal Corporation (USA) for an extensive series of high-duty cast irons that have ferritic-pearlitic microstructures with extremely uniform distributions of fine flake graphite. A typical composition is about 3% carbon, 1% silicon, and the balance iron, although alloying elements, such as chromium, copper and/or manganese may also be present. *Meehanite* is made from a charge of *gray pig iron* and *scrap steel* (usually 50% or more) in a cupola furnace. The graphite precipitation is influenced by the addition (inoculation) of calcium-silicon (in the form of calcium silicide) to the ladle and a special patented melting, mixing and casting technique. The resulting iron is wear-resistant, has a higher impact strength than regular cast iron, is mainly free of internal stresses, warpage, porosity, cracks and pipes, and can be hardened and tempered. *Meehanite* cast irons are used for engine and cylinder blocks, machine and machine-tool bases and beds, machine slides, housings, brake drums, gears, pinions, bearings, journal boxes, pistons, hydraulic cylinders, camshafts, crankshafts, connecting rods, rolls and rollers, pump parts, furnace parts, mining equipment, mill liners, and dies. Alloyed grades are suitable for heat-resisting or corrosion-resisting parts.

meerschaum. See sepiolite.

Mega-Bergerac. Trade name of Permacon (Canada) for durable paving stones with the appearance of stone slabs. Supplied in rectangular and square units of varying size in red, charcoal, gray and beige colors, they are used for walkways, patios, and steps.

Megaflon. Trademark of PPG Industries, Inc. (USA) for glossy, durable, spray-applied thermosetting fluoropolymers including *Megaflon EX* glossy, durable, spray-applied thermosetting fluoropolymer extrusion coatings with excellent weatherability and good abrasion and mar resistance, used for windows, doors, sash extrusions, poles, handrails, and fences, and *Megaflon MC* glossy, durable thermosetting fluoropolymer coatings with excellent corrosion and weathering resistance used on metal substrates in particular aluminum and clad steel for architectural and automotive applications, such as building panels, roofs, doors, decks, truck trailers, pylons, and canopies.

Megallium. Trade name of Dentsply International (USA) for a corrosion-resistant cobalt-chromium dental casting alloy.

Megalloy. Trade name of REM Chemicals Inc. (USA) for chemically-accelerated mass-finishing compounds for stainless steel products.

Megalloy EZ. Trademark of Dentsply International (USA) for a spherical dental amalgam supplied in the form of self-activating capsules.

Megaperm. Trade name of Vacuumschmelze GmbH (Germany) for a series of soft magnetic alloys containing 35-65% nickel, 10% manganese and 25-45% iron. They have high permeability, and are used for magnets, relay parts, core material and electromechanical transducers.

Megapyr. German trade name for cast alloys of iron, chromium and aluminum, used for magnets and electrical resistances.

Megasox. Trademark of A&P Technology (USA) for sleevings braided with carbon fibers having a tensile modulus of 3.86 GPa (560 ksi), or with E-glass rovings. Supplied as heavy fabric grades in two tiers, they are used as reinforcements in composites.

Megol. Trademark of Bausch & Lomb for die-cast polycarbonate used for opthalmic lenses.

Meigh Metal. Trade name of Meigh Castings Company Limited (USA) for a nonmagnetic, non-sparking, corrosion-resistant bronze containing 82% copper, 9% aluminum, 5% nickel and 4% iron. Used for castings.

meionite. A colorless mineral of the scapolite group that is composed of calcium aluminum silicate sulfate carbonate, $Ca_4Al_6(SiO_4)_6(SO_4,CO_3)$, and may contain sodium. Crystal system, tetragonal. Density, 2.74 g/cm^3; refractive index, 1.599. Occurrence: USA (Massachusetts), Norway.

Mektal. Trademark of Rogers Corporation (USA) for unreinforced and fiber-reinforced phenolic-based polymeric alloys used for molded parts.

Melafilm. Trade name of Henkel KGaA (Germany) for melamine-formaldehyde film adhesives.

Melamin. Trade name for melamine-formaldehyde resins, compounds and laminates

melamine. A white crystalline compound (99+% pure) that is a cyclic trimer of cyanamide. It has a density of 1.573 g/cm^3, a melting point of 354-360°C (669-680°F) (decomposes), and is used in the manufacture of synthetic thermosetting resins (melamine resins), in organic synthesis, in leather tanning, and with formaldehyde as a fixative in microscopy. Formula: $C_3H_6N_6$. Also known as *cyanurtriamide*.

melamine-acrylic resins. A group of fast-curing synthetic resins based on melamine and acrylic polymers and used as film formers in waterborne coatings for automotive and industrial finishes.

melamine fibers. A family of flame-resistant synthetic fibers based on melamine polymers. They have low thermal conductivity, high heat dimensional stability, are processible on ordinary textile machinery, and usually supplied in white, but can also be dyed. Used for filter media (e.g., air filters), protective clothing (e.g., industrial gloves and firefighters gear), and aircraft seating and other fire-blocking applications.

melamine-formaldehyde resins. A group of thermosetting resins of the amino family that are products of condensation reactions of melamine and formaldehyde or its polymers. Also available in modified grades include cellulose-, mineral- or glass fiber-filled, they have good resistance to elevated temperatures, good electrical properties including good arc resistance and dielectric strength, high surface hardness, good color stability, good durability and structural strength, good resistance to water, weak acids and alkalies. Used as bonding agents for glass-fiber insulation and electrical insulating parts, in insulating parts, molding compounds, electrical wiring devices, consumer products, dinnerware, and in adhesives and varnishes. Abbreviation: MF. Also known as *melamine resins*.

melamine-phenol-formaldehyde resins. A group of thermosetting resins that are condensation products of melamine, phenol and formaldehyde. They have good electrical and thermal properties, relatively high rigidity and strength, and are used for electrical components, consumer and industrial goods, etc. Abbreviation: MP. See also melamine-formaldehyde resins.

melamine plastics. Thermosetting plastics based on melamine-formaldehyde or melamine-phenol-formaldehyde resins.

melamine resins. See melamine-formaldehyde resins.

Melan. Trademark of Henkel KGaA (Germany) for melamine-formaldehyde plastics and compounds.

Melana. Trademark of Uzinade Fibre Sintetice Savienest (Rumania) for acrylic fibers used for textile fabrics.

Melange. Trade name of former Glas- und Spiegel-Manufactur AG (Germany) for a patterned glass.

mélange yarn. A yarn produced by a special dyeing/printing process (mélange printing) from fibers that have dyed and undyed areas.

melanins. A class of black and brownish, light-sensitive pigments occurring in the skin, hair and eyes of animals and humans, and in plants. It is formed naturally from the amino acid *tyrosine* by the action of the enzyme tyrosinase, and can also be prepared synthetically, e.g., by the oxidation of tyrosine with hydrogen peroxide. Chemically, it is a polyacetylene derivative with fullerene-cage like structure and has a molecular weight ranging from about 500 to 30000 Da. *Melanin* was the first known high-conductivity amorphous organic semiconductor and conductive polymer, has a band gap usually below 2 eV, is electroactive, can form charge-transfer complexes and display threshold switching, and can absorb many different types of energy (including ultrasound in the 1 MHz region) and dissipate them in the form of heat. Used in biochemistry for tissue culture applications, in active electronic devices, e.g., bistable switches, and for energy storage applications, e.g., batteries. See also allomelanins; eumelanins; pheomelanins; neuromelanins; melanogens.

melanite. A black variety of *andradite* [$Ca_3Fe_2(SiO_4)_3$] garnet.

melanocerite. A brown or black mineral composed of cerium borate silicate fluoride hydroxide, $(Ca,Ce,Y)_8(BO_3)(SiO_4)_4(F,OH)_4$. Crystal system, rhombohedral.

melanogens. Colorless precursor compounds to *melanins* that in nature are organized in natural soccer- or rugby ball-shaped structures (so-called "melanosomes") similar to fullerenes. In general, they are hydroxylated compounds (orthophenols) based on aromatic compounds, such as benzene, indole, pyridine, pyrrole, and quinoline.

melanophlogite. A colorless mineral composed of carbon silicon hydrogen oxide, $C_2H_{17}O_5 \cdot Si_{46}O_{92}$. Crystal system, tetragonal. Density, 2.01 g/cm³; refractive index, 1.425; hardness, 6.5-7 Mohs. Occurrence: Czech Republic; Italy.

melanoprotein. A conjugated protein containing *melanin*.

melanostibite. A black mineral of the corundum group composed of manganese antimony iron oxide, $Mn(Sb,Fe)O_3$. Crystal system, rhombohedral (hexagonal). Density, 5.24 g/cm³; refractive index, 2.12. Occurrence: Sweden.

melanotekite. A dark brown or blackish mineral composed of lead iron silicate, $Pb_2Fe_2Si_2O_9$. It can also be made synthetically. Crystal system, orthorhombic. Density, 6.30 g/cm³; refractive index, above 1.97; hardness, 6.5 Mohs. Occurrence: Sweden.

melanothallite. A black to bluish black mineral composed of copper oxide chloride, Cu_2OCl_2. Crystal system, orthorhombic. Density, 3.81 g/cm³. Occurrence: Russian Federation.

melanovanadite. A weakly radioactive, black mineral composed of calcium vanadium oxide, $Ca_2V_{10}O_{25}$. Crystal system, monoclinic. Occurrence: Peru.

melanterite. A pale bluish green mineral composed of iron sulfate heptahydrate, $FeSO_4 \cdot 7H_2O$. Crystal system, monoclinic. Density, 1.90 g/cm³; hardness, 2 Mohs; refractive index, 1.478.

Melantine. Trade name of Ciba-Geigy Corporation (USA) for melamine-formaldehyde plastics.

Melaplast. Trade name of BKW (Germany) for melamine-formaldehyde resins and plastics.

melaphyr. An altered or amygdaloidal (cellulor or vesicular) basaltic or andesitic rock, used especially in civil engineering and construction. See also andesite; basalt.

Melaware. Trade name of Ranton & Company (UK) for melamine-formaldehyde plastics.

Melbrite. Trademark of Montecatini Edison SpA (Italy) for melamine-formaldehyde resins, plastics and laminates.

Melchior wire. A corrosion-resistant white metal wire containing 57.3-62.7% copper, 0-25.9% zinc, 10.8-41.6% nickel and 0.6-1.1% manganese.

meldable fiber. See heterofil fiber.

meldable fabrics. Fabrics made entirely or partly of fibers that consist of two chemically and/or physically different polymers (*bicomponent fibers*) and in which fiber cohesion has been obtained by selectively melting one of these polymeric components.

Meldin. Trademark of Dixon Industries Corporation (USA) for a series of polyimide materials designed for operating tempera-

tures up to 315°C (600°F), supplied in the form of molding compounds and extruded and molded rods, tubes and sheets, and used for applications requiring reliability, precision and low wear and friction. *Meldin 2211* is a polyimide resin filled with a combination of graphite and polytetrafluoroethylene (PTFE). Its most important properties include inherent lubricity, low coefficient of friction, and retention of mechanical properties over a wide range of temperatures. Used for piston rings, seals, bearings and bushings in aerospace, automotive, and industrial applications. *Meldin 3000* is an injection-moldable thermoplastic polyimide resin with an operating temperature range of -185 to +290°C (-300 to +550°F), supplied in two grades with high strength and lubricity, two grades with high wear resistance and low coefficients of friction, and one grade with high stiffness and good electrical properties. It is suitable for various applications from office equipment and appliances to automobiles and off-road vehicles. *Meldin 9000* is a high-performance polyimide used in guidance-system bearings of missiles, and *Meldin 8100* is a missile-grade through-porous polyimide bearing material with good shedding propensity, oil retention and temperature capability, and a tensile strength superior to that of *Meldin 9000*, and used for miniature-bearing retainers.

Melinar. Trademark of E.I. DuPont de Nemours & Company (USA) for polyethylene terephthalate used for packaging applications, e.g., containers, packaging tapes and bottles for chemicals, cosmetics, liquor, etc. It can be thermoformed, injection and stretch-blow molded and is supplied in a wide range of grades including amorphous, crystalline, glass fiber-reinforced, mineral-filled, UV-stabilized, fire-retardant, high-impact, and supertough.

Melinex. Trademark of E.I. DuPont de Nemours & Company (USA) for a biaxially oriented polyester film based on polyethylene terephthalate (PET) and usually supplied in thicknesses ranging from 1.5-350 μm (59-13780 μin.) and standard and special grades (e.g., specially treated, coextruded, etc.). Opacity, clarity, surface quality, heat sealability and printability can be customized. *Melinex* film has a density of 1.4 g/cm³ (0.05 lb/in.³), a service temperature range of -40 to +170°C (-40 to +338°F), a heat-sealing temperature of 218-232°C (425-450°F), low thermal expansivity, high tensile strength, good dimensional stability, good dielectric properties, low coefficient of friction, good resistance to ultraviolet radiation, low water absorption, good resistance to acids, alcohols, greases, oils, halogens and ketones, fair resistance to aromatic hydrocarbons, poor resistance to alkalies, and low permeability to oxygen and carbon dioxide. Used for motor insulation, printed circuits, membrane touch switches, energy and solar control, labels and laminates, graphic, printing and heat-sealing applications, flexible and metallized packaging materials and magnetic audio and video tapes.

Meliodent. Trade name of Bayer AG (Germany) for a heat-cure acrylic resin used for dentures.

meliphanite. A yellow, red or black mineral of the melilite group composed of calcium sodium beryllium aluminum silicate fluoride, $(Ca,Na)_2Be(Si,Al)_2(O,F)_7$. Crystal system, tetragonal. Density, 3.00 g/cm³; refractive index, 1.6126; hardness, 5 Mohs. Occurrence: Norway.

Melite. Trade name of Melbourne Glass Company Proprietary Limited (Australia) for laminated and toughened glass.

Melkhior. Russian trade name for a copper-nickel alloy.

Melkorite. Trade name for a synthetic resin material with good resistance to continuous temperatures up to 200°C (390°F).

melkovite. A yellow mineral composed of calcium iron hydrogen molybdenum oxide phosphate hexahydrate, $CaFeH_6(MoO_4)_4(PO_4)\cdot6H_2O$. Density, 2.97 g/cm³; refractive index, 1838. Occurrence: Kazakhstan.

Melkwol. Trade name for regenerated protein (azlon) fibers obtained from milk casein and used for textile fabrics.

Mellanear. Trademark of Williams Harvey & Company Limited (UK) for high-purity tin (99.9+%) containing small amounts of lead, copper and arsenic.

Mel-Lite. Trademark of Avon Aggregates Limited (Canada) for lightweight aggregate.

mellite. A golden brown mineral with resinous luster composed of aluminum mellitate hydrate, $C_6(COO)_6Al_2\cdot18H_2O$. Crystal system, tetragonal. Density, 1.64 g/cm³; refractive index, 1.540. Occurrence: Germany.

Melloid. Trade name of Bull's Metal & Marine Limited (UK) for a series of phosphor bronzes containing varying amounts of copper, tin and phosphorus. Used for bushings, bearings, gear and worm parts, valve bodies, slide valves, pumps, marine parts, hardware, and other parts.

Melmac. Trademark of American Cyanamid Company (USA) for a series of melamine-formaldehyde plastics available in unfilled, cellulose-filled, mineral-filled and glass-fiber-filled grades. They have excellent heat resistance, good dielectric properties, good strength and hardness, good colorability and color stability, low moisture absorption, and good tensile and impact strength, especially when modified with cellulose or mineral fillers. Used as bonding agents for fibrous materials (unfilled grades), and for moldings, lighting fixtures, switchgear, switch panels, kitchen and dinnerware, and buttons.

Melment. Trademark of SKT (Germany) for melamine formaldehyde resins and products.

Melmex. Trademark of BIP Plastics (UK) for cellulose-filled melamine-formaldehyde plastics.

Melocol. Trademark of Ciba-Geigy Limited (UK) for urea- and melamine-formaldehyde resins and adhesives.

Melolam. Trademark of Ciba-Geigy Limited (UK) for melamine-formaldehyde plastics and laminates.

Melon. Trade name of Central Glass Company Limited (Japan) for a figured glass.

melonite. A steel-gray to tin white mineral of the brucite group composed of nickel telluride, $NiTe_2$. Crystal system, hexagonal. Density, 7.72 g/cm³. Occurrence: Canada (Quebec, Ontario). Also known as *tellurnickel*.

melonjosephite. A dark green mineral composed of calcium iron magnesium phosphate hydroxide, $Ca(Fe,Mg)Fe(PO_4)_2(OH)$. Crystal system, orthorhombic. Density, 3.65 g/cm³; refractive index, 1.770. Occurrence: Morocco.

Melopas. Trade name of Ciba-Geigy Limited (UK) for melamine-formaldehyde plastics and compounds.

Meloplas. Trade name of Ciba Products Corporation (USA) for polyamide-formaldehyde plastics.

Melotte alloy. A British fusible alloy containing 50% bismuth, 31% tin and 19% lead. It has a melting point of 99.5°C (211°F), and is used for sprinkler systems.

Melpure. Trade name of Magnesium Elektron Limited (UK) for magnesium ingots (99.9% pure), used as primary metals.

Melram. Trade name of Magnesium Elektron Limited (UK) for composites consisting of a magnesium-base alloy (92Mg-6Zn-1.2Cu-0.8Mn) matrix reinforced with 12 vol% silicon carbide (SiC) particles. Available in cast and wrought form, they pos-

sess high specific stiffness and creep resistance, good wear resistance, and are used for automotive and aircraft parts.

melt-blown fibers. Fine, relatively short synthetic fibers produced by first extruding a molten polymer through a die into a stream of hot air moving at high velocity and then quenching in a stream of cold air.

Meltex. Trademark of DSM Resins (USA) for a cellulose-filled melamine formaldehyde.

melton. A soft textile fabric, usually of wool or a wool blend, in a twill weave and with a felted surface. Used for overcoats and apparel. It is named after Melton Mowbray, a town in central England. Also known as *beaver cloth*.

Meltopax. Trademark of National Lead Company (USA) for a double silicate of sodium and zirconium composed of 55.5-57.5% zirconia (ZrO_2), 27-29% silica (SiO_2) and 13.5-15.5% sodium oxide (Na_2O). It has melting point of 1426°C (2600°F), and is used as a constituent in enamel frits and special glass batches.

melt-processible rubber. A thermoplastic elastomer, such as a chlorinated olefin, that exhibits high viscosity and good melt flow characteristics, and can easily be processed by extrusion, injection molding, and similar processing techniques.

Meltrite. Trade name of Pickands Mather Sales Company, Inc. (USA) for a *pig iron* containing 4-4.4% carbon, and the balance iron. Used in the manufacture of steel and cast iron.

Meltron. Trade name of Magnesium Elektron Limited (UK) for a series of magnesium alloys containing 0-8.5% aluminum, 0.3-1.5% zinc, and up to 0.4% manganese, 0.6% zirconium and 2.5% thorium, respectively. They have good formability, and are used for truck bodies, aircraft and missile components, bearing housings, cylinder heads, and control levers.

melt-spun fibers. Fibers of glass or a polymer (e.g., nylon or polyester) produced by extruding the molten starting material through a spinneret and allowing the resulting filaments to cool and solidify.

Melurac. Trademark of American Cyanamid Company (USA) for a series of melamine- and urea-formaldehyde resins.

MemCor. Trade name of Corning Corporation (USA) for a homogeneous glass/ceramic material used as a substrate for hard-disk thin-film media.

Memoreg. Trade name of Heraeus Kulzer, Inc. (USA) for a silicone elastomer used in dentistry for bite registration purposes.

memory alloys. See shape-memory alloys.

Memorywire NiTi. Trademark of American Orthodontics (USA) for a shape-memory wire composed of 52% nickel and 48% titanium, and used in dentistry and orthodontics.

MEMS materials. Materials that are used in the fabrication of microelectromechanical systems (MEMS) and differ with the particular type of system or device. Mesoscale devices (or mesomachines) are formed by electrodeposition of metals, alloys, ceramics and polymers, microscale devices (micromachines) are made by intergrated-circuit technologies involving the deposition of materials, such as gold, silicon dioxide and various semiconductors (e.g., silicon, poly-silicon, silicon nitride, etc.). Nanoscale devices (or nanomachines) are made by electrodeposition, molecular engineering and other nanotechnologies from various nanostructured materials including metals, alloys, ceramics and composites. Current applications for MEMS made from these materials include air-bag and inertial sensors, optical switches, display devices, print heads, tip and tilt meters and pressure gauges.

Mende. Trademark of Flake Board Company Limited (Canada) for thin *particleboard*.

mendelevium. A rare, radioactive, synthetic chemical element obtained in a cyclotron by bombardment of einsteinium-253 (^{253}Es) with alpha particles, or as a byproduct of nuclear fission. It is named after the Russian chemist Dimitri Ivanovich Mendeleev. Atomic number, 101; atomic weight, 256 (most stable isotope); half-life, 1.5 hours (decays by spontaneous fission). The heaviest isotope of mendelevium (^{258}Md) has a half-life of 60 days. Symbol: Md.

mendipite. A colorless, white, or gray mineral composed of lead oxide chloride, $Pb_3O_2Cl_2$. It can also be made synthetically. Crystal system, orthorhombic. Density, 7.24 g/cm³; refractive index, 2.27. Occurrence: Sweden, UK.

mendozite. A colorless or white mineral composed of sodium aluminum sulfate hydrate, $NaAl(SO_4)_2 \cdot 11H_2O$. Crystal system, monoclinic. Density, 1.73 g/cm³; refractive index, 1.460.

meneghinite. A blackish lead-gray mineral composed of copper lead antimony sulfide, $CuPb_{13}Sb_7S_{24}$. Crystal system, orthorhombic. Density, 6.36 g/cm³. Occurrence: Italy.

Menzanium. Trade name of Scheu Dental (Germany) for a nickel-free stainless steel wire used for dental applications.

Menzerna. Trade name of Menzerna-Werk GmbH & Co. (Germany) for an extensive range of grinding and polishing media based on natural or synthetic abrasives.

Menzolit. Trade name of Menzolit GmbH (Germany) for a series of molded and pressed unsaturated polyester products supplied in several grades including glass fiber-reinforced.

Mepal. Trade name of Rosti (Denmark) for melamine-formaldehyde plastics.

Meraklon. Trade name of Meraklon SpA (Italy) for polypropylene fibers used for industrial fabrics.

Meral. Trade name of Alusuisse (Switzerland) for an age-hardenable aluminum alloy containing 3.2% copper, 1% nickel, 0.8% magnesium, and 0.3% manganese. Used for light-alloy parts.

Mer-ane. Trademark of Reichhold Chemicals, Inc. (USA) for polyurethane varnishes and enamels.

meranti. See red seraya; white seraya; yellow seraya.

Mercadium. Trade name of RGH Artists Oil Paints (USA) for pigments based on cadmium or mercury sulfide, and producing a wide range of colors from light reddish yellow to dark brownish red. They have good lightfastness, high permanence and tinting strength, and good resistance to chemicals and water.

mercallite. A colorless or sky-blue mineral composed of potassium hydrogen sulfate, $KHSO_4$. It can also be made synthetically. Crystal system, orthorhombic. Density, 2.32 g/cm³; refractive index, 1.460. Occurrence: Italy.

mercerized cotton. Cotton that has been treated with a heated solution of sodium hydroxide (NaOH) at controlled temperature to increase its dye absorptivity, luster and strength. It is often used for yarns with low shrinkage and silky luster.

mercerized wool. Wool that has been treated with a solution of sodium hydroxide (NaOH) and whose fibers therefore do not have the usual felting properties. This is caustic treatment is occasionally employed to enhance the luster of wool fibers.

merchant bar. (1) Iron in bar form suitable for the market. Also known as *merchant iron*.

(2) An iron or wrought-iron bar produced from a *faggot*. Also known as *merchant bar iron*.

merchant bar iron. See merchant bar (2).

merchant iron. See merchant bar (1).

merchant pig iron. Pig iron suitable for sale to foundries. See also pig iron.

Mercoloy. Trade name of Merco Nordstrom Valve Company (USA) for a white, corrosion-resistant nickel bronze composed of 60% copper, 25% nickel, 10% zinc, 2% lead, 2% iron, and 1% tin. It has good strength, moderate elongation, and relatively good machinability. Used for valves, valve seats, and pump parts.

mercurial brass. A strong, corrosion-resistant brass composed of 70% copper, 30% zinc, and a trace (up to 0.05%) of mercury. The mercury addition retards the growth of marine organisms, such as barnacles or algae, and inhibits dezincification. Used for heat-exchanger and condenser tubing.

mercurial horn ore. See calomel.

mercuric arsenate. A yellow powder used in antifouling and water-resistant coatings and paints. Formula: $HgHAsO_4$. Also known as *mercury arsenate*.

mercuric chloride. White crystals or powder (99+% pure). Crystal system, orthorhombic. Density, 5.44 g/cm³; melting point, 276°C (529°F); boiling point, 302°C (576°F); refractive index, 1.859. Used in metallurgy, process engraving and lithography, in the manufacture of calomel, in photography and electron microscopy, in textile printing, in dry cells, as a wood preservative, and in biochemistry and medicine. Formula: $HgCl_2$. Also known as *mercury bichloride; mercury chloride; corrosive sublimate*.

mercuric cuprous iodide. A dark red, crystalline powder (99.99% pure) turning dark brown or brownish black upon heating to the α-β transition temperature of about 65-70°C (149-158°F). Used for the detection of machine bearing overheating by reversible change of color. Formula: Cu_2HgI_4. Also known as *copper-mercury iodide; copper tetraiodomercurate; cuprous mercuric iodide; mercury copper iodide*.

mercuric naphthenate. The mercury salt of naphthenic acid available as a dark amber liquid containing about 29% mercury, and used as a bacteria-, mildew- and mold retarder in paints. Also known as *mercury naphthenate*.

mercuric oleate. Yellowish to red semi-solid or solid mass usually available as a mixture of yellow mercuric oxide and oleic acid, and used in antifouling paints. Formula: $Hg(C_{18}H_{33}O_2)_2$. Also known as *mercury oleate*.

mercuric oxide. See red mercuric oxide; yellow mercuric oxide.

mercuric potassium cyanide. Colorless crystals used in the manufacture of mirrors (silvering of glass). Formula: $Hg(CH)_2 \cdot 2KCN$. Also known as *mercury-potassium cyanide*.

mercuric selenide. See mercury selenide.

mercuric silver iodide. Yellow powder with a density of 6.08 g/cm³ turning red at about 45°C (113°F). Used for the detection of journal bearing overheating by reversible color change. Formula: Ag_2HgI_4. Also known as *mercury-silver iodide; silver-mercury iodide*.

mercuric sulfate. White hygroscopic, crystalline powder (98+% pure) obtained by treating mercury with sulfuric acid. Density, 6.47 g/cm³; melting point, decomposes at red heat. Used in the extraction of gold and silver from roasted *pyrites*, as an electrolyte in batteries, in *calomel*, and as a corrosive sublimate. Formula: $HgSO_4$. Also known as *mercury persulfate; mercury sulfate*.

mercuric sulfide. See black mercuric sulfide; red mercuric sulfide.

mercuric telluride. See mercury telluride.

mercurous chloride. A white, crystalline powder (99.5+%). It occurs in nature as the mineral *calomel*. Crystal system, tetragonal. Density, 7.15 g/cm³; melting point, sublimes at 400°C (752°F). Used in ceramic paints, for electrodes, in pyrotechnics, and as a fungicide. Formula: HgCl; Hg₂Cl₂. Also known as *mercury monochloride; mercury protochloride; mild mercury chloride.*

mercurous chromate. A red powder that decomposes on heating and is used as a pigment producing green colors on ceramics. Formula: Hg₂CrO₄. Also known as *mercury chromate.*

mercurous oxide. A black powder with a density of 9.8 g/cm³ that decomposes at 100°C (212°F). Formula: Hg₂O. Also known as *black mercury oxide.*

mercurous sulfate. Yellowish white crystalline powder (99% pure). Density, 7.56 g/cm³; melting point, decomposes on heating. Used in batteries (e.g., Clark and Weston cells), and as a chemical reagent. Formula: Hg₂SO₄.

Mercury. (1) Trade name of Universal Cyclops Specialty Steel (USA) for a tool steel containing 0.7-1.2% carbon, and the balance iron. Used for tools and dies.

(2) Trade name of American Hard Rubber Company (USA) for hard rubber.

mercury. A heavy, silvery, shining metallic element of Group IIB (Group 12) of the Periodic Table. The most important ore is *cinnabar*, and it also occurs as *native mercury*. Crystal system, rhombohedral (hexagonal). Density, 13.6 g/cm³; melting point, -38.87°C (-38°F) (liquid at room temperature); boiling point, 357°C (675°F); superconductivity critical temperature, 4.15K; atomic number, 80; atomic weight, 200.59; monovalent, divalent. It has an extremely high surface tension, fair electrical conductivity (1.8% IACS), moderate thermal conductivity, and a low coefficient of linear thermal expansion. Used for amalgams (alloys with other metals), as a catalyst, in batteries, electrical equipment, mercury-vapor lamps, arc lamps and lamp tubes, advertising signs, arc rectifiers, switches for instruments and control devices, for barometers, manometers and thermometers, for diffusion pumps and other laboratory instruments, as a coolant and neutron absorber in nuclear power plants, in extractive metallurgy, in semiconductors, superconductors, mirror coatings and pigments (*vermilion*), and in dentistry. Symbol: Hg. Also known as *quicksilver.*

mercury aluminum selenide. A compound of mercury, aluminum and selenium. Crystal system, tetragonal. Crystal structure, "defect" chalcopyrite. Density, 5.05 g/cm³. Used as a semiconductor. Formula: HgAl₂Se₄.

mercury aluminum sulfide. A compound of mercury, aluminum and sulfur. Crystal system, tetragonal. Crystal structure, "defect" chalcopyrite. Density, 4.11 g/cm³. Used as a semiconductor. Formula: HgAl₂S₄.

mercury arsenate. See mercuric arsenate.

mercury bichloride. See mercuric chloride.

mercury cadmium telluride. A semiconductor compound usually obtained by molecular beam epitaxy (MBE) in which a layer of mercury telluride (HgTe) is grown on a cadmium telluride (CdTe) or cadmium zinc telluride (CdZnTe) substrate. Used for electronic components, infrared detectors, and thin-film applications. Formula: HgCdTe. Abbreviation: MCT.

mercury chloride. See mercuric chloride.

mercury chromate. See mercurous chromate.

mercury copper iodide. See mercuric cuprous iodide.

mercury gallium selenide. A compound of mercury, gallium and selenium. Crystal system, tetragonal. Crystal structure, "defect" chalcopyrite. Density, 6.18 g/cm³. Used as a semiconductor. Formula: HgGaSe₄.

mercury gallium sulfide. A compound of mercury, gallium and sulfur. Crystal system, tetragonal. Crystal structure, "defect" chalcopyrite. Density, 5.0 g/cm³. Used as a semiconductor. Formula: HgGa₂S₄. Also known as *mercury thiogallate.*

mercury indium amalgam. An alloy of 71% indium and 29% mercury, supplied in the form of lumps up to 20 mm (0.8 in.) in size.

mercury indium selenide. A compound of mercury, indium and tellurium. Crystal system, tetragonal. Crystal structure, "defect" chalcopyrite. Density, 6.3 g/cm³; melting point, 827°C (1521°F). Used as a semiconductor. Formula: HgIn₂Se₄.

mercury indium telluride. A compound of mercury, indium and tellurium. Crystal system, tetragonal. Crystal structure, "defect" chalcopyrite. Used as a semiconductor. Formula: Hg₅In₂Te₈.

mercury monochloride. See mercurous chloride.

mercury oxide. See black mercurous oxide; red mercuric oxide; yellow mercuric oxide.

mercury persulfate. See mercuric sulfate.

mercury-potassium cyanide. See mercuric potassium cyanide.

mercury protochloride. See mercurous chloride.

mercury selenide. Gray crystals or powder (99.9+% pure). It occurs in nature as the mineral *tiemannite*. Crystal system, cubic. Crystal structure, sphalerite. Density, 8.25 g/cm³; hardness, 2.5 Mohs; melting point, 797°C (1467°F). Used as a semiconductor, in heterostructures, infrared detectors, solar cells, thin-film transistors, and ultrasonic amplifiers. Formula: HgSe. Also known as *mercuric selenide.*

mercury silver iodide. See mercuric silver iodide.

mercury sulfate. See mercuric sulfate.

mercury sulfide. See black mercuric sulfide; red mercuric sulfide.

mercury telluride. Gray crystals or powder (99.9% pure). It is found in nature as the mineral *coloradoite*. Crystal system, cubic. Crystal structure, sphalerite. Density, 8.17 g/cm³; melting point, 670°C (1238°F). Used as a semiconductor, in heterostructures, infrared detectors, solar cells, thin-film transistors, and ultrasonic amplifiers. Formula: HgTe. Also known as *mercuric telluride.*

mercury tetrathiocyanatocobaltate. An organometallic compound containing divalent mercury and divalent cobalt. It is supplied in high purity (99.99%) for use as a magnetic-susceptibility standard. Formula: HgCo(SCN)₄. Also known as *cobalt tetrathiocyanatomercurate.*

mercury thiogallate. See mercury-gallium sulfide.

merenskyite. A white mineral of the brucite group composed of palladium telluride, PdTe₂. It can also be made synthetically. Crystal system, hexagonal. Density, 8.29 g/cm³. Occurrence: Transvaal.

Merico. Trade name of Meridan Steel Company Inc. (USA) for nontempering, water-hardening tool steels containing 0.35-0.4% carbon, 0-0.6% silicon, 0-0.4% manganese, 0.77% chromium, up to 0.5% nickel, 0.75% molybdenum, 0.74% copper, up to 0.5% cobalt, and the balance iron. Used for punches, chisels, pneumatic tools, rivet sets, shear blades, and blacksmith tools.

merino. (1) A fine-quality worsted wool obtained from the merino, a breed of sheep originating in Spain.

(2) The soft yarn made from merino wool (1).

(3) A thin, soft fabric made from merino yarn (2), or some substitute.

Merinova. Trade name of Merinova (Italy) for soft, silky, spun azlon fibers obtained from casein (milk) protein, and used for textile fabrics, and in blends with rabbit hair for the production of felts.

Merit. Trade name of Plating Process Systems Inc. (USA) for a bright nickel electroplate and plating process.

Merit Metal. Trade name of American Smelting & Refining Company (USA) for a lead babbitt containing varying amounts of tin and antimony. Used for bearings.

Merkalon. Trademark of Moplefan SpA (Italy) for polypropylene fibers and yarns used for padding and stuffing applications, and for the manufacture of clothing.

Merkur. Trade name of Dr. Hesse & Cie. (Germany) for a brass electroplate and plating process.

MERL. Trade name for nickel-base superalloys made by powder-metallurgy techniques and having excellent high-temperature properties including high stress-rupture and creep strength.

merlinoite. A white mineral of the zeolite group composed of potassium calcium aluminum silicate hydrate, $K_5Ca_2(Al_9Si_{23}O_{64})\cdot 24H_2O$. Crystal system, orthorhombic. Density, 2.14 g/cm^3; refractive index, 1.494. Occurrence: Italy.

Merlon. Trademark of Bayer Corporation (USA) for transparent, pale straw-colored engineering thermoplastics based on polycarbonate resins. They are commercially available in the form of rods, tubes, pipes, sheets and film extrusions, but special types are also made by blow and injection molding, extrusion and thermoforming. *Merlon* thermoplastics have a density of 1.2 g/cm^3 (0.04 $lb/in.^3$), a refractive index of 1.587, excellent batch-to-batch consistency, narrow melt flow range, high strength, modulus of elasticity, toughness and impact strength, good ductility and rigidity, a service temperature range of -215 to +121°C (-355 to +250°F), good dielectric properties, good flame and ultraviolet resistance, excellent resistance to temperature changes and moisture, and fair chemical resistance and moderate heat aging. Used for electrical and electronic components, cable wrapping, safety shields, helmets, protective equipment, covers, protective overlays for other thermoplastic sheeting, lenses, prescription and protective eyewear, sight gauges, forms, gears and cams, missile applications, and photographic films.

Mermaid. Trade name of Spear & Jackson (Industrial) Limited (UK) for a high-speed steel (AISI type T1) containing 0.75% carbon, 4.25% chromium, 18% tungsten, 1.1% vanadium, and the balance iron. It has good deep-hardening properties, good hot hardness, high abrasion and wear resistance, and is used for cutting tools, especially lathe and planer tools, end mills, milling cutters, taps and chasers.

Mermex. Trade name of ASG Industries Inc. (USA) for diamond-meshed wired glass, 5.6 mm (0.22 in.) thick, with a relatively coarse stippled pattern on one face.

Meron. Trade name of Voco (USA) for a glass-ionomer dental luting cement.

Merrifield resin. A 1 or 2% crosslinked, chloromethylated styrene-divinylbenzene copolymer resin available in mesh sizes ranging from 20-400 and chlorine anion contents from 1.0 to 4.5 mM/g. It was developed by the American chemist R.B. Merrifield as a reactive polymeric substrate for the synthesis of polypeptides and proteins from amino acids by a stepwise process.

merrihueite. A greenish blue mineral of the osumilite group composed of potassium sodium iron magnesium silicate, $(K,Na)_2(Fe,Mg)_5Si_{12}O_{30}$. Crystal system, hexagonal. Density, 2.87 g/cm^3; refractive index, 1.5955. Occurrence: USA.

Merrillite. Trade name for high-purity zinc dust used to precipitate gold and silver in the Merrill-Crowe cyanide process.

merrillite. A colorless mineral composed of calcium phosphate, $Ca_3(PO_4)_2$, and found in meteorites.

Merrilux. Czech trade name for a semi-transparent patterned glass with irregularly streaked "jelly" texture.

mersawa. The Malay name for wood from various species of trees of the genus *Anisoptera*. The same wood is available as *krabak* in Thailand and as *palosapis* in the Philippines. It is a plain wood that is not decorative and very hard to work with tools (high silica content), but can be rotary-peeled. Source: Southeast Asia (from Philippines and Malaysia to Bangladesh). Used for plywood veneer.

Merten. Trade name of Meridian Steel Company Inc. (USA) for a fatigue- and wear-resistant steel, used for machine-tool parts.

mertieite. (1) A yellow mineral composed of antimony arsenic palladium, $Pd_{11}(Sb,As)_4$. Crystal system, hexagonal. Density, 10.60 g/cm^3. Occurrence: Canada, USA (Alaska).

(2) A mineral composed of antimony palladium, Pd_8Sb_3. It can also be made synthetically by reacting high-purity palladium and antimony under prescribed conditions. Crystal system, rhombohedral (hexagonal). Density, 3.72 g/cm^3; refractive index, 1.75-1.76. Occurrence: USA (New Jersey).

merwinite. A mineral composed of tricalcium magnesium orthosilicate, $Ca_3Mg(SiO_4)_2$. It can also be made synthetically from calcium carbonate ($CaCO_3$) basic magnesium carbonate [$3MgCO_3\cdot Mg(OH)_2\cdot 3H_2O$] and silicon dioxide ($SiO_2$) by thermal treatment. Crystal system, monoclinic. Density, 3.15 g/cm^3; refractive index, 1.712. Occurrence: USA (California), Mexico, Ireland. Used in ceramics, and in the building trades.

Meryl. (1) Trademark of Rhône-Poulenc SA (France) for high-tenacity rayon fibers used for industrial fabrics.

(2) Trademark of Nylstar, Inc. (USA) for a series of strong continuous filament nylon 6 and nylon 6,6 fibers and yarns supplied in several grades including *Meryl, Meryl Nylon, Meryl Nateo* and *Meryl Souple*. Used for the manufacture of textile fabrics.

MerylMicrofiber. Trademark of Nylstar, Inc. (USA) for bright or dull, extremely fine nylon 6,6 microfibers with an individual filament diameter of less than 10 μm (394 μin.). They provide breathability, a soft handle, and excellent rain and wind resistance, and are used for closely woven and knitted fabrics including stockings, socks and lingerie.

Meryl Nateo. Trademark of Nylstar, Inc. (USA) for an intensely dull multifunctional nylon 6,6 fiber that provides a soft and smooth feel, a natural appearance, good protection against UV rays and quick transport of moisture away from the skin. It can be made into flat, texturized or twisted yarns, blended with other fibers, and is used for easy-care apparel including activewear, sportswear and fashion garments.

Meryl Nexten. Trademark of Nylstar, Inc. (USA) for a hollow polyamide (nylon) fiber used for the manufacture of light, insulating and protective fabrics including winter garments (e.g., ski jackets), activewear and sports underwear.

Meryl Satiné. Trademark of Nylstar, Inc. (USA) for a nylon fiber with a lengthened rectangular section that produces diffraction effects of reflected light. It provides a soft handle, good abrasion and pilling resistance, a smooth sheen, and is used for elegant stockings.

Meryl Skinlife. Trademark of Nylstar, Inc. (USA) for a nylon fiber with permanent bacteriostatic properties. It provides breathability, a soft handle, and is used for garments including sportswear, underwear, socks and hosiery, shoe linings, and medical and industrial fabrics (e.g., filters).

Meryl Souple. Trademark of Nylstar, Inc. (USA) for a nylon fiber with permanent antistatic properties, used especially for lingerie, underwear and stockings.

Meryl Tango. Trademark of Nylstar, Inc. (USA) for a family of nylon fibers that provide different cross sections, compositions and/or colors in the same yarn. Used for the manufacture of wearing apparel with silk-like appearance.

Meryl Techno. Trademark of Nylstar, Inc. (USA) for an abrasion resistant nylon fiber with high tensile and tear strength. It provides excellent protection against weather conditions (e.g., wind and rain), and is used especially for light, comfortable, protective activewear and sportswear, and for sports shoes and bags.

Meryl UV Protection. Trademark of Nylstar, Inc. (USA) for a family of full dull *Meryl* nylon microfiber yarns that provide excellent protection against ultraviolet rays (UVA and UVB). Used for activewear, sportswear, swimwear and other outdoor clothing.

Mesa Red. Trade name of Aardvark Clay & Supplies (USA) for a deep-red earthenware clay.

mesh. (1) A net-like structure with openings formed by a number of crossed wires, threads or strings.

(2) A fabric of thread, cord, wire, etc., knitted or otherwise interlaced into an open texture with small holes.

(3) A number of lengthwise and crosswise wires, bars or rods placed at right angles to each other and joined at the intersections by welding. Used as a concrete reinforcement. Also known as *welded-wire fabric.*

mesh reinforcement. A welded-wire fabric in sheet or roll form for embedding into fresh and uncured concrete as an additional reinforcement. Also known as *welded-wire fabric reinforcement.*

Mesmeric. Trade name of Firth Brown Limited (UK) for a case-hardening steel containing 0.2% carbon, 0.3% silicon, 0.4-1% manganese, and the balance iron. Used for shafts, axles, gears, pins, journals, and levers.

mesolite. A colorless or white mineral of the zeolite group composed of sodium calcium aluminum silicate hydrate, $Na_2Ca_2Al_6Si_9O_{30}\cdot 9H_2O$. Crystal system, monoclinic. Density, 2.26 g/cm^3; refractive index, 1.5074. Occurrence: Japan.

Mesoloy. Trade name of Molecu-Wire Corporation (USA) for a resistance alloy containing 72% iron, 23% chromium and 5% aluminum. It has a density of 7.2 g/cm^3 (0.26 $lb/in.^3$), a melting point of approximately 1510°C (2750°F), high electrical resistivity, high heat resistance, and good tensile strength. Used chiefly for resistors.

mesomorphic substance. See liquid crystal.

mesophase pitch-based precursor fibers. Carbon fibers made from petroleum pitch precursors and having high orientation resulting from the liquid crystalline order ("mesophase" order). Available in low-modulus, high-modulus and very-high-modulus grades with an average filament diameter of 8-11 μm (310-430 μin.), and a density range of 1.9-2.2 g/cm^3 (0.07-0.08 lb/in^3). Used as reinforcing fibers for composites.

mesoporous aerogel. A highly porous, low-density solid having a pore size between 2 and 50 nm (0.08 and 1.96 μin.) made from a *gel* using a gaseous dispersing medium. Examples include flexible and rigid plastic foams. Used in filtration and catalysis. See also aerogel.

mesoporous materials. Inorganic or organic materials (e.g., ceramics, metals, polymers or composite) with a pore size between 2 and 50 nm (0.08 and 1.96 μin.). Used for catalysis,

separation and electronic applications, e.g., sensors, batteries, and nonlinear optics.

mesoporous silica. A group of ordered silica-based materials with nanometer-sized pores produced by the interaction of micelle-forming surfactant molecules (e.g., *n*-cetyltrimethylammonium chloride) and silica, and used as catalysts.

mesoporous zeolite. A high-porosity *zeolite* with a pore size in the nanometer range that may or may not be loaded with metals, such as cobalt, palladium or platinum, and is used as a catalyst support in chemical engineering.

mesoporphyrin IX dihydrochloride. A synthetic *porphyrin* derivative available in the form of purple crystals (95% pure) with a maximum absorption wavelength of 401 nm. Used as a pigment and dye. Formula: $C_{34}H_{38}N_4O_4\cdot 2HCl$.

mesoporphyrin IX dimethyl ester. A synthetic *porphyrin* derivative available in the form of purple crystals with a melting point of 214-215°C (417-419°F), and a maximum absorption wavelength of 400 nm. Used as a pigment and dye. Formula: $C_{36}H_{42}N_4O_4$.

mesoscopically disordered materials. A large class of complex, structurally disordered materials whose characteristic length scale of disorder is typically in the range of 1000 nm and larger.

messaline. A thin, soft, glossy, usually plain-colored silk fabric with a surface like satin. It is now also made from synthetic fibers, e.g., acetate or polyester. Named after Valeria Messalina, the wife of the Roman emperor Claudius. Used for high-priced, luxurious dresses.

messelite. A greenish white to white mineral of the fairfieldite group composed of calcium iron manganese phosphate hydrate, $(Ca,Fe,Mn)_3(PO_4)_2\cdot 2H_2O$. Density, 3.16 g/cm^3; refractive index, 1.649. Occurrence: USA (New Hampshire), Germany, Czech Republic, Kazakhstan.

Mesta Special. Trade name of Mesta Machine Company (USA) for a cast steel containing 0.4% carbon, 1.5% nickel, 0.8% chromium, and the balance iron. Used for rolls.

meta-aluminite. A white mineral composed of aluminum sulfate hydroxide pentahydrate, $Al_2SO_4(OH)_4\cdot 5H_2O$. It can also be made synthetically. Crystal system, monoclinic. Density, 2.18 g/cm^3; refractive index, 1.512.

meta-ankoleite. A yellow mineral of the meta-autunite group composed of potassium uranyl phosphate hexahydrate, $(K_{1.7}Ba_{0.2})(UO_2)_2(PO_4)_2\cdot 6H_2O$. It can also be made synthetically. Crystal system, tetragonal; density, 3.56 g/cm^3; refractive index, 1.580. Occurrence: Uganda.

meta-autunite. A dark or yellowish green mineral of the autunite group composed of calcium uranyl phosphate hydrate, $Ca(UO_2)_2(PO_4)_2\cdot xH_2O$. It can also be made synthetically. Crystal system, orthorhombic. Density, 3.33 g/cm^3; refractive index, 1.607. Occurrence: Japan, USA (Washington).

Meta-Bond. Trademark of PPG Pretreatment & Specialty Products (USA) for microcrystalline zinc phosphate coatings applied by immersion, spraying or both. Typically, immersion coatings on steel substrates range in weight from 1.61 to 43 g/m^2 (150 to 4000 mg/ft^2), and spray coatings from 1.08 to 10.8 g/m^2 (100 to 1000 mg/ft^2). Used as protective or decorative coatings on iron, steel, galvanized steel and aluminum.

Metablen. Trade name of Atofina Metablen (France) for an extensive series of acrylic copolymers and terpolymers. *Metablen C* refers to acrylic copolymers used as lubricants for polyvinyl chlorides, and *Metablen E* to methacrylate-butadiene-styrene terpolymers used as impact modifiers. *Metablen P* refers to acrylic copolymers used as processing aids for polyvinyl chlorides, and *Metablen W* to acrylic copolymers used as impact modifiers for polyvinyl chlorides.

metaborite. A colorless or whitish mineral composed of hydrogen borate, HBO_2. It can also be made synthetically. Crystal system, cubic. Density, 2.47 g/cm^3; refractive index, 1.616. Occurrence: USA (New Jersey).

metacalciouranoite. An orange mineral of the becquerel group composed of calcium sodium barium uranium oxide hydrate, $(Ca,Na,Ba)U_2O_7 \cdot xH_2O$. Density, 4.90 g/cm^3; refractive index, 1.911. Occurrence: Russian Federation.

Meta Cast. Trade name of Metachem Resins Corporation (USA) for synthetic casting resins.

MetaCeram. Trade name of Eutectic Corporation (USA) for spray coatings based on self-bonding, iron-base powder-metallurgy alloys. They can be machined to very precise tolerances, matching original parts in surface texture and reflectivity, and applied at low substrate temperatures (less than 260°C or 500°F). Used in building up machine parts, such as shafts, rolls, sleeves, journals, etc.

metacinnabar. A black mineral of the sphalerite group composed of mercury sulfide, HgS. It can also be made synthetically. Crystal system, cubic. Density, 7.65 g/cm^3. Occurrence: USA (California), Russia. Used as a source of mercury.

Metacrylene. Trade name of Plastimer (France) for a series of butadiene-styrene and polystyrene copolymers.

metadelrioite. A mineral composed of calcium strontium vanadium oxide hydroxide, $CaSrV_2O_6(OH)_2$. Crystal system, triclinic. Density, 4.30 g/cm^3. Occurrence: USA (Colorado).

Meta-Dent. Trade name of Sun Medical Company Limited (Japan) for a self-cure acrylic resin for dental applications.

metadomeykite. A tin-white to steel-gray mineral composed of copper arsenide, β-Cu_3As. It can also be made synthetically. Crystal system, hexagonal. Density, 7.20 g/cm^3. Occurrence: USA (Michigan).

Meta-Fast. Trade name Sun Medical Company Limited (Japan) for a self-cure dental bonding resin.

Metafoam. Trade name for high- and low-density polyethylene foams available in the form of thin sheets whose tensile strength markedly decreases with increasing density.

metaheinrichite. A yellow to green mineral of the meta-autunite group composed of barium uranyl arsenate octahydrate, $Ba(UO_2)_2(AsO_4)_2 \cdot H_2O$. Crystal system, tetragonal. Density, 4.04 g/cm^3; refractive index, 1.637. Occurrence: USA (Oregon), Germany.

metahewettite. A deep red mineral composed of calcium vanadium oxide trihydrate, $CaV_6O_{16} \cdot 3H_2O$. Crystal system, monoclinic. Density, 2.94 g/cm^3; refractive index, 2.10. Occurrence: USA (Arizona, Colorado, Utah).

metakahlerite. A sulfur-yellow mineral of the meta-autunite group composed of iron uranyl arsenate octahydrate, $Fe(UO_2)_2(AsO_4)_2 \cdot 8H_2O$. Crystal system, tetragonal. Density, 3.83 g/cm^3; refractive index, 1.643. Occurrence: Germany.

metakirchheimite. A colorless, greenish yellow mineral of the meta-autunite group composed of cobalt uranyl arsenate octahydrate, $Co(UO_2)_2(AsO_4)_2 \cdot 8H_2O$. Density, 3.33 g/cm^3; refractive index, 1.644. Occurrence: Germany.

metakoettigite. A bluish gray mineral of the vivianite group composed of iron zinc arsenate hydroxide hydrate, $(Zn,Fe)_3(AsO_4)_2 \cdot 8(H_2O,OH)$. Crystal system, triclinic. Density, 3.03 g/cm^3; refractive index, 1.680. Occurrence: Mexico.

Meta-Lac. Trade name of Metal Chem Inc. (USA) for lacquer sealants.

metal. See metals.

metal alcoholates. See metal alkoxides.

metal alkoxides. A group of chemical compounds in which a metal is linked by an oxygen atom to one or more alkyl groups (ethyl, methyl, propyl, etc.). The general formula is MOR in which M represents a metal, O oxygen and R an alkyl group. They are often used in chemical vapor deposition (CVD) processes for dielectric and refractory oxides, sol-gel technology, and in the production of organometallic compounds. Examples of *metal alkoxides* include niobium ethoxide, germanium methoxide and antimony propoxide. Also known as *metal alcoholates*.

metal alloys. Solid or liquid mixtures of two or more metals, such as copper and tin or aluminum and lithium, or one or more metals with certain nonmetallic elements as in carbon steels. The final products have better mechanical properties than the pure metals.

Metalbond. Trade name of Strathmore Products, Inc. (USA) for a clear bronzing agent used in metal finishing.

metal carboxylates. A group of chemical compounds obtained by adding the functional group of carboxylic acid (COOH) to a metal. Examples include acrylates, formates and methacrylates of aluminum, calcium, potassium, sodium, etc. Used in the production of organometallic compounds.

metal-cased refractory. A basic firebrick enclosed in a thin metal casing leaving only the ends exposed. Used in the manufacture of certain steelmaking furnaces.

metal-ceramics. See cermets.

metal cloth. A fabric made with cotton or silk warp threads and aluminum, copper, gold or silver weft (filling) threads. Used for millinery, trimmings, and other decorative applications.

metal coatings. Thin layers of a metal or metal alloys produced on a base material (or substrate), often another metal, by hot dipping, electroplating, spraying or sputtering. The coatings protect the underlying material against corrosion and/or oxidation and wear, or provide a more aesthetic optical appearance. Often, there is a galvanic relation between the coating metal or alloy and the base material (or substrate). Examples of *metallic coatings* include galvanized coatings (e.g., zinc coatings on steel), tin coatings, cadmium coatings and aluminum coatings. Also known as *metallic coatings*.

metal-coated fiber. A composite fiber usually consisting of a nonmetallic fiber (e.g., carbon) with a thin electroplated metal coating (e.g., nickel). Used for engineering composites. Abbreviation: MCF.

metal composites. Composite materials consisting of metal matrices to which reinforcing fibers or whiskers have been added. See also metal-matrix composites.

Metal-Cor. Trademark of Corex Inc. (USA) for flux-cored welding wires.

Metalcor. Trademark of Ralph Wilson Plastics Company (USA) for decorative metal sheets and metal laminate sheets used for surfacing various materials including plastics.

metal diketonates. A group of chemical compounds in which a metal atom is substituted for an enolic hydroxyl hydrogen of a diketone and is linked or coordinated to one or two oxygens of the resultant diketonate anion. Used in organometallic synthesis.

metal dye. Organic dyes, such as Alizarin Cyanin RR, Alizarin Green S, Nigrosine 2Y or Naphthalene Blue RS, suitable for coloring aluminum and steel.

metal evaporated tape. Magnetic recording media fabricated by

(i) evaporating a thin film of a metal or alloy, such as cobalt, cobalt-nickel, or cobalt-nickel-chromium on a thin polymeric support tape (e.g., polyester), or (ii) evaporating a thin film of an alloy, such as cobalt-chromium, on a thin layer of a suitable metal (e.g., hexagonal close-packed titanium). Abbreviation: MET. See also magnetic medium; magnetic recording materials; thin film recording materials.

metal-faced plywood. Plywood with metal foil (e.g., aluminum, copper or steel) bonded to one or both sides to improve the overall strength.

metal fibers. See metallic fibers.

Metalfil. Trade name of Sun Medical Company Limited (Japan) for a series of hybrid dental composites. *Metalfil AP* is a highly-flowable, light-cure no-slump hybrid composite consisting of the tough, hydrophilic non-BisGMA proprietary resin "RDMA," a reactive organic filler (trimethylolpropane trimethacrylate, or TMPT), and an inorganic barium-glass filler. It has high radiopacity, mechanical strength and wear resistance, excellent polishability, low water absorption, and is used for anterior and posterior dental restorations. *Metalfil CX* is a visible light-cure, tooth-shade hybrid composite consisting of a polymer matrix and a reactive organic filler (trimethylolpropane trimethacrylate, or TMPT). It has excellent resilience, wear resistance and polishability and good flow properties, high radiopacity, mechanical strength and wear resistance, low water absorption, and is used for esthetic dental restorations. *Metalfil Flo* has a composition similar to *Metalfil AP*, provides a tooth-shade color, high radiopacity and wear resistance, excellent polishability, low water absorption, and is used in restorative dentistry for cavity repair, repair of small defects in ceramic resin restorations, and as a base/liner under composite restorations.

metal film. See metal foil.

Metalflo. Trade name of Johnson Matthey plc (UK) for a silver brazing alloy containing 33% copper, 25% silver, 25% zinc, 16.8% cadmium, and 0.2% silicon. It has a melting range of 606-720°C (1123-1320°F), and is used for brazing copper and nickel and their alloys, steels, and cast irons.

metal foam. A light metal or metal alloy produced, for example, by adding titanium or zirconium hydride to the melt to evolve hydrogen and produce a blown, sponge-like material with a very low density. Used for applications requiring low weight, and absorption of mechanical shocks. Also known as *foamed metal; foam metal.*

metal foil. A very thin sheet, usually less than 0.15 mm (0.006 in.) in thickness, made from a metal, such as aluminum, copper, steel, or tin, by beating, hammering, rolling, electroplating, or other processes. Used in the construction of electronic components, such as capacitors, for insulation and other electrical applications, in wrapping and packaging, and in laminating, e.g., aluminum foil bonded to paper or plywood, or metal foil bonded to plastics. Also known as *metal film; metallic film.*

metal-foil paper. Paper or cardboard with a bright-colored metal foil backing on one or both sides.

metal hydrides. (1) Chemical compounds, usually in the form of micron-sized powders or spongy masses, formed by the reaction of hydrogen gas with metals. Examples include aluminum hydride (AlH_3) or lithium hydride (LiH). Also known as *metallic hydrides.*

(2) A class of materials based on intermetallic alloys that are capable of absorbing hydrogen at moderate pressure and temperature and high energy-storage densities forming intermetallic hydrides. Examples include rare-earth (RE) nickel-ides corresponding to the general formula $(RE)Ni_5$, which can reversibly store 6 hydrogen atoms forming the intermetallic hydride $(RE)Ni_5H_6$, and magnesium nickelides, such as dimagnesium nickelide (Mg_2Ni) which can store up to 4 hydrogen atoms. Used in battery electrodes and hydrogen storage systems. Also known as *metallic hydrides.*

Metaline. Trade name R.W. Rhodes Metaline Company (USA) for an oil-impregnated sintered bronze made from a powder containing 90% copper and 10% tin. Used for self-lubricating bearings.

Metalite. (1) Trade name of Metalite Corporation (USA) for plywood composed of a core of balsa wood that has a strong aluminum alloy foil bonded to both sides.

(2) Trademark of Dryvit System, Inc. (USA) for exterior building wall panels.

(3) Trademark of Spherical Products Corporation (USA) for metal-clad hollow microspheres used as fillers in paints, coatings, plastics, etc.

(4) Trademark of LTV Aerospace Corporation (USA) for laminated panels composed of low-density core materials with adhesively bonded metal sheets.

Metaljoiner. Trade name of Colonial Alloys Company (USA) for a lead-tin-cadmium solder with a melting point of 343°C (650°F). Used for electrical applications.

Metallage. Trademark of Nevamar Decorative Surfaces (USA) for a decorative laminate made by bonding aluminum foil to layers of phenolic resin impregnated kraft paper under pressure at 135°C (275°F). It can be adhesively bonded to core materials, such as laminate-grade plywood, particleboard and medium-density fiberboard, does not bond to drywall, concrete or plastered walls, and is designed for interior use only. It is supplied in sheets 1.2 m (48 in.) wide, 3.1 m (120 in.) long and 1 mm (0.04 in.) thick with sanded phenolic backs, and is used for wall and ceiling panels, furniture, drawer fronts, display cases, decorative columns, store fixtures, and fascias.

metal lath. See lath (iii).

metal leaf. A very thin sheet of a metal, such as aluminum, copper, gold or silver, that is usually made by beating, and being thinner than metal foil. High-purity aluminum leaf (99.999%) is available in thicknesses and weights of 0.15-1.0 μm (6-40 mm) and 41.3-269.9 μg/cm², respectively, and high-purity copper leaf (99.99+%) is supplied in thicknesses and weights of 0.25-1.0 μm (10-40 μin.) and 219.9-899 μm/cm², respectively. High-purity gold leaf (99.99+%) is available in thicknesses and weights of 0.25-1.0 μm (10-40 μin.) and 483-1932 μg/cm², respectively, and high-purity silver leaf (99.95+%) in thicknesses and weights of 0.25-1.0 μm (10-40 μin.) and 262.3-1098 μg/cm², respectively.

metallic abrasives. A group of abrasives that are available in the form of (i) grit, i.e., angular metallic particles made from crushed, hardened cast steel shot, or chilled white cast iron shot, or (ii) shot, i.e., metallic particles, usually made of the same materials as grit, but of spherical shape. Used for the abrasive blast cleaning of ferrous and nonferrous workpieces.

metallic atomic fuel. See metallic reactor fuel.

metallic brown. A reddish-brown pigment composed of an iron-rich earth (chiefly ferric oxide, Fe_2O_3), and one or more extenders.

metallic ceramics. See cermets.

metallic cloth. See metallic fabric.

metallic coatings. See metal coatings.

metallic color. A suspension of particles of gold, silver, platinum or other metal powders in a suitable medium, such as oil. Used to produce lustrous metallic decorations on ceramic ware.

metallic composites. A general term referring to composite materials with metallic matrices. See also metal composites; metal-matrix composites.

metallic conductor. An electrical conductor having only partially filled electron energy bands. Silver is the best metallic electrical conductor, used as a reference, and its relative conductivity is arbitrarily given as 100.0%. On this relative scale the conductivities for other metals are, for example: 97.6% for copper, 76.6% for gold, 63% for aluminum, 54.6% for tantalum and 39.4% for magnesium. See also conductor.

metallic fabrics. Glossy, reflective fabrics either knitted or woven with metal yarns (e.g., aluminum, gold, or silver), a combination of metal and woolen or worsted yarns, or made entirely of nonmetallic substances (natural and/or synthetic fibers) and coated with metallic material. Used as industrial fabrics, and for clothing, e.g., coats, suits, dresses, etc. Also known as *metallic cloth.*

metallic fibers. Manufactured fibers in filament or whisker form either made entirely from a metal, such as aluminum, cobalt, gold, silver, steel, tantalum or tungsten, or composed of a metal core with a polymeric or ceramic coating (e.g., an aluminum filament coated with cellulose acetate butyrate) or a ceramic or polymeric core with a metallic coating (e.g., nickel-coated graphite fibers, or nickel- or aluminum-coated glass fibers). Depending on the particular metallic fiber material, they are used as reinforcements in engineering composites, in automotive tire cords, and for antistatic devices, chemical-resistant fabrics, heated draperies, etc. Also known as *metal fibers.*

metallic filaments. Continuous monofilaments produced in the form of thin flexible strips of metal foil (usually aluminum) coated or laminated on both sides with a tarnish-resistant polymer film (e.g., polyester). They can be produced with dull and bright metallic lusters, and are used for the manufacture of clothing.

metallic film. See metal foil.

metallic fission fuel. See metallic reactor fuel.

metallic glasses. See amorphous alloys.

metallic hydrides. See metal hydrides.

metallic magnet. A ferromagnetic alloy, such as commercial ingot iron, iron-nickel, nickel-iron-chromium, nickel-iron-molybdenum, amorphous iron, or samarium-cobalt, whose engineering applications are based mainly on its hard or soft magnetic properties. See also magnetic materials.

metallic materials. See metals.

metallic mortar. A ceramic mortar consisting of a mixture of a metal powder (e.g., lead, tungsten, or depleted uranium), ceramic oxides and water. Supplied in the form of blocks, sections, and other shapes, it possesses good weatherability, good resistance to weak acids and alkalies and thermal shocks, and good resistance to radiation including gamma and neutron rays. Used as a shielding material in X-ray and nuclear installations, and in space technology.

metallic nuclear fuel. See metallic reactor fuel.

metallic overglaze. A glaze consisting of mineral pigments based on precious metals or tinctorial metal oxides dissolved in a suitable organic solvent. They are applied to and fired on previously glazed surfaces of earthenware, tile, china, and other ceramic ware.

metallic paint. (1) Paint made by dispersing a metal pigment, e.g., a metal powder, such as aluminum or gold, or a metallic oxide, such as zinc or chromium oxide, in a suitable vehicle.

(2) Paint usually made with an iron-oxide pigment, and used exclusively on metallic surfaces.

metallic paper. (1) Paper with a surface coating of fine metal flake.

(2) A specialty paper whose surface has been coated with *zinc white* (ZnO) or clay. Although surface markings can be produced on the paper by a metal point, these cannot be erased by any means.

metallic pigment. A fine, insoluble powdered metal, metal oxide or sulfide used to give color and opacity to ceramic bodies, inks, (organic) coatings, etc., by preferentially reflecting light of certain wavelengths and absorbing that of other wavelengths.

metallic powder. A metallic element or alloy in the form of a powder with a particle size typically ranging from about 0.1-1000 μm (0.000004-0.0394 in.). Used as raw material for powder-metallurgy parts, metal-spraying applications and paint pigments, as a brazing material, as a calorizing compound, in foundry alloys, and as a catalyst. Abbreviation: MP. Also known as *metal powder.*

metallic reactor fuel. The fissionable isotopes of a metal, such as uranium, plutonium, thorium or their alloys, that are capable of acting as a source of energy and neutrons for the propagation of a chain reaction in a nuclear reactor. Also known as *metallic atomic fuel; metallic fission fuel; metallic nuclear fuel.*

metallic red. A red pigment composed of ferric oxide (Fe_2O_3) and one or more extenders.

metallic soaps. Water-insoluble salts formed of metals heavier than sodium and potassium, and organic acids, such as lauric, oleic, palmitic or stearic acid. Examples include the laurates, oleates, palmitates and stearates of barium, calcium, cerium, chromium, copper, lead, lithium, manganese, nickel, strontium and zinc. Used as driers in paints, inks, lacquers and varnishes, as flatting agents, as suspending agents and thickeners, in ceramics, lubricating greases, waterproofing agents and fungicides, and in leather and textile finishing.

metallic staple. A staple yarn produced by cutting metallic yarn to short lengths. It is often blended with other fibers and spun into yarns for the manufacture of fabrics with brilliant metallic luster.

metallic whiskers. A group of very fine and short, axially oriented fibers composed of a single crystalline metal (e.g., aluminum, cobalt, iron, nickel, rhenium or tungsten). They have high length-to-diameter ratios (typically 50 to 15000), high tensile strengths and elastic moduli, and good high-temperature properties. Used as reinforcements for metals, plastics and ceramics, and as ablative materials for aircraft and aerospace vehicles.

metallic yarn. See metallized yarn.

metallic wires. Fine wires of a metal or alloy, such as beryllium, copper, tungsten, molybdenum, or stainless steel, used as continuous reinforcements in metal-matrix composites.

metallided metal. A metal or alloy on whose surface has been produced a coating of a different metal or metalloid by electrolytically dissolving ions from the metal or metalloid anode in a bath of molten fluoride salts and diffusing into the surface of the metal or alloy cathode at a temperature of 816-1093°C (1500-2000°F). Used for aerospace and automotive applications. *Note:* Metalliding is a proprietary process. See also metalloids.

Metalline. Trademark of J. Josephson, Inc. (USA) for vinyl wall coverings.

metalline. (1) A heat- and corrosion-resistant alloy containing 35% cobalt, 30% copper, 25% aluminum, and 10% iron. Used for tools.

(2) An imitation metallic fabric woven with cotton or silk warp threads and rayon weft (filling) threads that have a brilliant metallic color or luster. Used as a dress and decorative fabric.

metallized ceramics. Ceramic materials, such as alumina or silica, coated with a slurry consisting of a metal powder (e.g., molybdenum), a metal hydride (e.g., titanium or zirconium hydride) and selected organic binders. After coating, they are fired at a prescribed temperature in a specified atmosphere. Used to facilitate the joining (e.g., brazing, sealing, etc.) of ceramics to metals.

metallized fabrics. Textile fabrics to whose surface a metallic material has been applied, either by lamination or by a chemical or electric-arc deposition process.

metallized film. A plastic film having a thermally sprayed or vacuum-deposited metal coating on one or both sides. For example, polycarbonate and polypropylene film (0.002 mm or 0.08 mil thick) is available with a deposit of aluminum on one or both sides, and metallized polyethylene terephthalate film (0.0025-0.05 mm, or 0.1-2 mils thick) has a thin deposit of aluminum, stainless steel or titanium, or a thin three-layer composite deposit consisting of a layer of copper sandwiched between two layers of stainless steel. Biaxially or uniaxially oriented polyvinylidene fluoride film (usually 0.009-0.04 mm, or 0.35-1.57 mils thick) metallized on both sides with aluminum have piezoelectric properties. *Metallized plastic film* is often used in the manufacture of yarns, for packaging, stamping foil, labels, and electrical applications. Also known as *metallized plastic.*

metallized glass. Glass to which a thin layer of metal has been permanently fused. Used especially for glass-to-metal sealing, and for electrical and electronic applications.

metallized glass fibers. Glass fibers usually made from electrical-grade glass (E-glass) and coated with a metal, such as nickel or copper. They combine the benefits of glass fibers, such as low density and good mechanical properties, with those of metals, such as high electrical conductivity and metallic luster. Used in glass-reinforced composites, thermoset bulk compounds and sheet-molding compounds.

metallized paper. A strong kraft paper having a thin metal coating on one side and used for paper capacitors. Abbreviation: MP. See also capacitor paper.

metallized plastic. See metallized film.

metallized wood. A wood product made by treating or impregnating ordinary *hardwood* or *softwood* with a molten fusible alloy in a closed container under pressure. The alloy remains in the wood cells and, upon solidification, significantly increases the hardness and compressive and flexural strength of the wood, and renders it electrically conductive in a direction parallel to the grain.

metallized yarn. (1) A textile yarn that has been combined with a metal wire (e.g., aluminum, copper, gold, or silver).

(2) A *gimped yarn* made by wrapping a metallized strip around a man-made fiber yarn. See also man-made fibers.

(3) A multiple yarn consisting of man-made fiber plies and one or more metallized-strip plies. See also plied yarn.

(4) A yarn consisting of two layers of plastic film adhesively bonded to a core composed either of a metal-dust-coated plastic film or a metal foil.

metalloaromatics. A relatively recent class of cluster compounds that are copper-, lithium- or sodium-cation capped aromatic, square-planar aluminum, gallium or indium anions synthesized by laser vaporization. They correspond to the formulas MAl, MGa and MIn, respectively. *Metalloaromatics* containing other metallic elements have also been synthesized. It is believed that these compounds could be useful as building blocks in the synthesis of novel semiconductors and superconductors.

metallocarbohedrene molecules. See metallocarbohedrenes.

metallocarbohedrenes. A class of stable molecular cluster materials that are composed of carbon atoms bound to atoms of early transition metals (M), such as chromium, hafnium, molybdenum, titanium, vanadium and zirconium, and form single or multiple cage-like structures. The single-cage cluster structures correspond to the formula M_8C_{12}. *Metallocarbohedrenes* can be synthesized from mixtures of carbon and transition-metal powders by pressing and baking and subsequent processing in a reactor using arcing electrode rods. They exhibit low ionization potentials, undergo delayed ionization, have delocalized electronic characters, and are used in the manufacture of novel electronic and optical materials, e.g., in the form of new dopants for semiconductors. Abbreviation: met-cars. Also known as *metallocarbohedrene molecules; met-cars.*

metallocenes. Coordination compounds that are obtained by bonding a transition metal or metal halide to one or more cyclopentadienyl (C_5H_5) rings. There are several types of *metallocenes* including dicyclopentadienyl-metal compounds with the general formula ($C_5H_5)_2M$; and (ii) dicyclopentadienyl-metal halides with the general formula ($C_5H_5)_2MX_{1-3}$. These compounds are molecular sandwiches in which one C_5H_5) ring is located above and the other below the metal atom plane. There are also monocyclopentadienyl-metal compounds containing only one C_5H_5) rings and having the general formula $C_5H_5MR_{1-3}$, where R represents a CO, NO, alkyl, halide or other functional group. Examples of *metallocenes* include cobaltocene, ($C_5H_5)_2Co$, ferrocene, ($C_5H_5)_2Fe$, and hafnocene dichloride, ($C_5H_5)_2HfCl$. Used in organometallic research, and in the manufacture of polymers, catalysts, ultraviolet absorbers, etc.

metalloenzyme. An *enzyme* in which a protein is bonded to a metal.

Metal-Lok. Trade name of Manville Corporation (USA) for fiberglass preformed pipe insulation with metal jacket.

metalloid powder. A powder composed wholly or in part of a metalloid, such as antimony, boron, carbon or tellurium.

metalloids. A group of chemical elements intermediate in properties between true metals and true nonmetals. They are lacking metallic bonding, and thus their valence electrons are not distributed throughout as an electron cloud (or electron gas). They do not have the plasticity of metals, and have only moderate electrical and thermal conductivities. The following elements, many of which are semiconductors, are included in this group: antimony, arsenic, bismuth, boron, carbon, germanium, polonium, phosphorus, selenium, silicon and tellurium.

Metallon. (1) Trademark of Henkel Surface Technologies (USA) for an epoxy adhesive supplied in heat-curing and room-temperature curing grades. It has relatively high peel and tensile strength up to about 82°C (180°F), and is used for metal joining applications.

(2) Trademark of SLI, Inc. (USA) for decorative metal coatings and coated products used in the construction industry on structural and nonstructural nonmetallic components and fixtures.

(3) Trademark of Metallon Engineered Materials Corporation (USA) for a strip of metal laminated wholly or in part to render the metal more electrically conductive or corrosion resistant.

metalloprotein. A conjugated *protein* in which the nonprotein portion is a metal, such as copper, iron or zinc.

Metalloy. (1) Trademark of Metals & Alloys Company Limited (Canada) for a series of metal-cored welding electrodes for low- and medium-carbon steels.

(2) Trade name of US Chemical & Plastics Inc. for a metallic filler.

metallurgical cement. See supersulfated cement.

metallurgical coke. A hard coke produced by heating bituminous or coking coal in a coke oven and used as a fuel in blast furnaces, as a base for carbon refractories, and as a source of synthesis gas. It has very high compressive strength at elevated temperatures, low ash, sulfur and phosphorus contents, and a high fixed carbon content.

metallurgical-coke-base carbon refractory. A commercial refractory product based chiefly on metallurgical coke.

metallurgical lime. Limestone used in blast-furnace ironmaking for removing impurities and, in the form of *quicklime*, as a cleansing agent in the purification of molten basic-oxygen furnace, open-hearth and electric-furnace steels, and in sinter plants. See also limestone.

Metal-Lux. Trade name of Agriplast Srl (Italy) for reflective thermal film materials composed of synthetic resins with selected additives. They provide high reflection of solar radiation, uniform distribution of visible light, and are antidrop-tested to reduce the risk of fungal diseases. Supplied in two grades, they are used in greenhouses to protect vegetable crops, such as tomatoes, against thermal inversion and excessive daytime temperatures.

Metalmate. Trade name of Jones-Blair Company (USA) for powder coatings.

metal-matrix composites. A group of composites consisting of metal bases or matrices, such as aluminum, copper, magnesium or titanium, reinforced with one or more constituents, such as continuous graphite, alumina, silicon carbide or boron fibers, or discontinuous graphite or ceramic materials in particulate or whisker form. Abbreviation: MMC. See also metal composites.

metal-mesh blanket insulation. Fiberglass insulation supplied in blanket form with a flexible metal mesh on one or both sides. Used for acoustical, thermal or other insulation applications.

MetalMix. Trademark of AMETEK Specialty Metal Products (USA) for a stainless steel powder used as an antistatic, electromagnetic interference/radio-frequency interference (EMI/RFI) shielding material or as a sound-dampening filler in various plastic compounds.

metal-nitride-oxide semiconductor. A semiconductor that has two insulating layers and consists of a silicon substrate with an intermediate layer of silicon dioxide (SiO_2) and a top layer of silicon nitride (Si_3Ni_4). Used in electronics, e.g., for memory devices. Abbreviation: MNOS.

metalodevite. A yellow mineral of the meta-autunite group composed of zinc uranyl arsenate decahydrate, $Zn(UO_2)_2(AsO_4)_2 \cdot 10H_2O$. Crystal system, tetragonal. Density, 4.01 g/cm^3; refractive index, 1.635. Occurrence: France.

Metaloplast. Trademark of Norton Performance Plastics (USA) for a polytetrafluoroethylene (PTFE) bearing material manufactured by sintering a specially formulated PTFE tape into a metal fabric. This unique material combines the advantages of PTFE (e.g., good temperature resistance, self-lubrication, conformability and low coefficient of friction) with those of metals (e.g., excellent load-bearing capabilities).

Metalor XeraFit. Trade name of Metalor Technologies SA (Switzerland) for a beryllium- and nickel-free nonprecious dental alloy containing 58.3% cobalt, 25% chromium and 2% gold.

metal-organic compounds. See organometallic compounds.

metal-organics. See organometallic compounds.

metal overlay. A thin layer of metal foil bonded to one or both surfaces of wood or plastics to provide a decorative or protective surface, or a base for subsequent paint coats.

metal-oxide semiconductor. A semiconductor in which the insulating layer is composed of the oxide of the substrate material, e.g., if the substrate is silicon the layer will consist of silicon dioxide (SiO_2), if it is germanium the layer will be germanium dioxide (GeO_2). Used in electronics, e.g., for memory devices. Abbreviations: MOS.

Metaloy. Trade name of Silver Creek Precision Company (USA) for a series of cast high-leaded bronzes containing 20-28% lead, 2-10% tin, and the balance iron. Used for bushings and bearings.

metal perxenates. A group of compounds prepared by reacting a metal in alkaline solution with xenon trioxide (XeO_3). An example is sodium perxenate ($Na_4XeO_6 \cdot 8H_2O$).

metal polish. See buffing compound; polishing abrasives.

metal powder. See metallic powder.

MetalPURE. Trademark of Superior Graphite Company (USA) for high-purity graphite and carbon powders used for powder-metallurgy applications.

metal-reinforced ceramic composites. A group of ceramic materials, such as aluminum oxide or silicon nitride, reinforced with a refractory or high-temperature metal (e.g., molybdenum, niobium, tantalum, or tungsten) usually in fiber, whisker or particulate form. They have improved strengths and moduli, a maximum service temperature range of 500-800°C (930-1470°F), and are used for a wide range of high-performance applications.

metals. (1) A group of chemical elements which for the most part are crystalline solids with metallic luster and good ductility, malleability, thermal and electrical conductivity and high chemical reactivity. Many metals are relatively hard and possess high strength, others (e.g., cesium, gallium and mercury) are liquid at ambient or slightly elevated temperatures. Metals are characterized by metallic bonding that results when each of the atoms contributes its valence electrons to the formation of an electron cloud shared by the entire material. The conduction of electricity and the principal conduction of heat are produced by the free movement of these electrons through the metal. Since the negative electron cloud surrounds each of the positive metal ions that make up the orderly three-dimensional crystal structure, strong electrical attraction holds metals together. Also known as *metallic materials*. See also actinide metals; alkali metals; alkaline-earth metals; light metals; heavy metals; noble metals; precious metals; rare-earth metals; transition metals.

(2) A general term sometimes used for alloys (solutions of metals) or mixtures of alloys, such as steel, cast iron, brass, bronze, etc. Also known as *metallic materials*.

Metalseal. Trade name of Duplate Canada Limited for laminated plate glass having an interlayer extending beyond the glass edges and metal sheets embedded in this interlayer.

Metalset. Trademark of Smooth-On Manufacturing Company (USA) for durable, solvent-free aluminum-filled epoxy resin cements used in the manufacture of fixtures and jigs, for sealing leaks, filling minor flaws in metal castings, and for smoothing surfaces.

Metalsil. Trade name of Johnson Matthey plc (UK) for a cadmium-bearing silver brazing alloy containing 20% silver, 40% copper, 25% zinc and 15% cadmium. It has a melting point of 605-765°C (1120-1410°F), and is used for brazing various ferrous and nonferrous alloys.

metal sponge. See sponge metal.

metal thread fabrics. (1) Originally, textile fabrics that contained interwoven threads of gold, silver or other precious metals, and were used for luxurious clothing, official or ceremonial dresses, robes and gowns, ecclesiastical vestments, altar hangings, etc.

(2) Now, textile fabrics that contain interwoven nonmetallic threads with brilliant metallic colors.

Metal Welder. Trade name of Devcon Corporation (USA) for primerless high-performance structural adhesives including methacrylates and high-strength epoxies used for metal-to-metal bonding of aluminum, steel and stainless steel, and for joining metals to plastics (e.g., ABS, PVC, PMMA and polyesters).

metamagnets. See metamagnetic materials.

metamagnetic materials. Materials in which a transformation from an antiferromagnetic state to a high-moment state occurs in an applied magnetic field. Also known as *metamagnets*.

Metamic. Trade name of Morgan Crucible Company Limited (UK) for a series of sintered refractory materials. One such cermet is composed of 75% nickel and 25% *mullite* and is used for bearings operating at temperatures of 500-900°C (930-1650°F), while another contains 70% chromium and 30% aluminum oxide and finds application in gas-turbine blades.

Metamold. Trade name of Ziv Steel & Wire Company (USA) for an oil-hardened steel used for zinc and aluminum die-casting dies.

metamorphic rock. A rock, such as gneiss, marble, quartzite or schist, whose initial chemical composition, structure, and/or texture have been altered by extreme variations of heat, pressure and shearing stress deep within the crust of the earth.

metanovacekite. A yellow mineral of the meta-autunite group composed of magnesium uranyl arsenate octahydrate, $Mg(UO_2)_2(AsO_4)_2 \cdot 8H_2O$. Crystal system, tetragonal. Density, 3.51 g/cm^3; refractive index, 1.632. Occurrence: Germany.

Meta-Phos. Trade name of Metal Chem Inc. (USA) for iron and zinc phosphate coatings.

metaprotein. A *protein* derivative produced by the further action of acids and alkalies.

metarossite. A light yellow to greenish-yellow mineral composed of calcium vanadium oxide dihydrate, $Ca(VO_3)_2 \cdot 2H_2O$. Crystal system, triclinic. Density, 2.80 g/cm^3. Occurrence: USA (Colorado, Utah). Used as a minor ore of vanadium.

Metarsal. Trade name of Pechiney/Trefimétaux (France) for an arsenical aluminum brass containing 76-79% copper, 18-22% zinc, 2% aluminum and 0.2-1% arsenic. It has good corrosion resistance, elongation and cold workability, and is used for evaporator, heat-exchanger and distiller tubing.

metaschoderite. A yellowish orange mineral composed of aluminum vanadium oxide phosphate hexahydrate, $Al_2PO_4VO_4 \cdot 6H_2O$. Crystal system, monoclinic. Density, 1.61 g/cm^3; refractive index, 1.604. Occurrence: USA (Nevada).

metaschoepite. A yellow, radioactive mineral composed of uranium oxide dihydrate, β-$UO_3 \cdot 2H_2O$. It is an alteration product (Phase II) of the mineral *schoepite*. Crystal system, orthorhombic. Density, 5.00 g/cm^3.

Metarsic. Trade name of Pechiney/Trefimétaux (France) for a leaded arsenical brass containing 69-70% copper, 29-30% zinc, 0.5% lead, and a trace of arsenic. It has good corrosion resistance, elongation, cold workability and machinability, and is used for evaporator, heat-exchanger and distiller tubing.

Metarstan. Trade name of Pechiney/Trefimétaux (France) for an arsenical brass containing 69-70% copper, 28-29% zinc, 1% tin and 0.2-1% arsenic. It has good corrosion resistance, elongation and cold workability, and is used for evaporator, heat-exchanger and distiller tubing.

metasideronatrite. A straw-yellow mineral composed of sodium iron sulfate hydroxide hydrate, $Na_4Fe_2(SO_4)_4(OH)_2 \cdot 3H_2O$. Crystal system, orthorhombic. Density, 2.68 g/cm^3. Occurrence: Chile.

metasilicates. (1) Metal salts derived from metasilicic acid (H_2SiO_3), e.g., sodium metasilicate (Na_2SiO_3).

(2) Silica materials in which 2 of the 4 oxygen atoms of the SiO_4 tetrahedra are shared with adjacent tetrahedra, i.e., they have 2 nonbridging oxygen atoms per tetrahedron. The general formula is $XSiO_3$, where X represents one or more metals, such as barium, cadmium, calcium, lead, magnesium or sodium.

metasilicic acid. See silicic acid.

metastudtite. A pale yellow mineral composed of uranium oxide dihydrate, $UO_4 \cdot 2H_2O$. Crystal system, orthorhombic. Density, 4.67 g/cm^3; refractive index, 1.658. Occurrence: Zaire.

metaswitzerite. A pale pink to brown mineral composed of manganese iron phosphate tetrahydrate, $(Mn,Fe)_3(PO_4)_2 \cdot 4H_2O$. Crystal system, monoclinic. Density, 2.95 g/cm^3; refractive index, 1.628. Occurrence: USA (North Carolina).

metatorbernite. An emerald-green, radioactive mineral of the meta-autunite group composed of copper uranyl phosphate octahydrate, $Cu(UO_2)_2(PO_4)_2 \cdot 8H_2O$. Crystal system, tetragonal. Density, 3.60 g/cm^3; refractive index, 1.626. Occurrence: Germany.

metatyuyamunite. A canary-yellow to greenish-yellow mineral of the carnotite group composed of calcium uranyl vanadium oxide hydrate, $Ca(UO_2)_2(VO_4)_2 \cdot xH_2O$. Crystal system, orthorhombic. Density, 3.80 g/cm^3; refractive index, 1.835. Occurrence: USA (Colorado).

metauranocircite. A yellow, radioactive mineral of the meta-autunite group composed of barium uranyl phosphate hexahydrate, $Ba(UO_2)_2(PO_4)_2 \cdot 6H_2O$. It can also be made synthetically. Crystal system, tetragonal. Density, 4.00 g/cm^3; refractive index, 1.623. Occurrence: Germany.

metauranospinite. A yellow to green, radioactive mineral composed of calcium uranyl arsenate hexahydrate, $Ca(UO_2)_2(AsO_4)_2O_{17} \cdot 6H_2O$. It can also be made synthetically. Crystal system, tetragonal. Density, 3.70 g/cm^3. Occurrence: Germany.

metavanmeerssscheite. A yellow to pale yellow, radioactive mineral of the phosphuranylite group composed of uranium uranyl phosphate hydroxide dihydrate, $U(UO_2)_3(PO_4)_2(OH)_6 \cdot 2H_2O$. Crystal system, orthorhombic. Density, 4.49 g/cm^3; refractive index, 1.65. Occurrence: Zaire.

metavanuralite. A yellow, radioactive mineral of the carnotite group composed of aluminum uranyl vanadium oxide hydroxide octahydrate, $Al(UO_2)_2(VO_4)_2(OH) \cdot 8H_2O$. Crystal system, triclinic. Density, 3.66 g/cm^3. Occurrence: Gabon.

metavariscite. A pale green mineral of the variscite group composed of aluminum phosphate dihydrate, $AlPO_4 \cdot 2H_2O$. Crystal system, monoclinic. Density, 2.51 g/cm^3; refractive index, 1.588.

Occurrence: USA (Utah).

metavauxite. A white or colorless mineral composed of iron aluminum phosphate hydroxide octahydrate, $FeAl_2(PO_4)_2(OH)_2\cdot8H_2O$. Crystal system, monoclinic. Density, 2.35 g/cm^3; refractive index, 1.561. Occurrence: Bolivia.

metavoltine. A yellow to orange-brown mineral composed of potassium sodium iron sulfate hydroxide hydrate, $Na_6K_2Fe_7(SO_4)_{12}O_2\cdot18H_2O$. Crystal system, hexagonal. Density, 2.50 g/cm^3; refractive index, 1.5925. Occurrence: Chile.

metazellerite. A synthetic mineral composed of calcium uranyl carbonate trihydrate, $CaUO_2(CO_3)_2\cdot3H_2O$. Crystal system, orthorhombic. Density, 3.66 g/cm^3.

metazeunerite. A pale green mineral of the meta-autunite group composed of copper uranyl arsenate octahydrate, $Cu(UO_2)_2(AsO_4)_2\cdot8H_2O$. Crystal system, tetragonal. Density, 3.64 g/cm^3; refractive index, 1.650. Occurrence: Germany.

met-cars. See metallocarbohedrenes.

Metco. Trade name of Sulzer Metco (USA) for an extensive series of coating products including (i) aluminum-, copper-, nickel- and polyester-based abradables used to provide clearance control in high-speed applications, e.g., jet engines; (ii) aluminum-, cobalt-, copper-, iron-, nickel- and molybdenum-base powders and chromium carbide-, tungsten carbide- and cermet-based powders used to impart properties, such as enhanced corrosion, oxidation and/or abrasion and wear resistance, enhanced hardness, toughness, machinability and/or improved antifriction, electrical, magnetic and thermal properties; (iii) cobalt- and nickel-base self-fluxing powders for coating and hardfacing applications; (iv) tungsten carbide-base high-velocity oxyfuel (HVOF) powders for thermal spraying applications; (v) nickel-iron-chromium arc-spray wires; and (vi) metal and alloy combustion wires.

Metcolite. Trademark of Sulzer Metco (USA) for hard, durable aluminum oxide blasting abrasives supplied in several grades with blocky and sharp-edged particles of varying grit size.

Metcoloy. Trade name of Sulzer Metco (USA) for a series of corrosion-resistant powders based on austenitic or ferritic stainless steels, austenitic manganese steels or nickel-iron chromium alloys. Used for thermally or flame-sprayed coatings.

Metcoseal. Trademark of Sulzer Metco (USA) for a series of one- and two-component sealers for coatings. *Metcoseal ALS* is an one-component alkyd sealer with low volatile organic content, good resistance to rural and marine atmospheres and hot gasoline (up to 60°C or 140°F) used with *Metco* coatings. *Metcoseal EPS* is a two-component epoxy-based sealer with water addition. It has a low volatile organic content, good resistance to rural/marine atmospheres and hydrocarbon solvents, and an upper use temperature of 175°C (350°F). *Metcoseal ERS* refers to a two-component, volatile organic compound free epoxy-based sealer with good resistance to acids, alkalies, alcohols and most solvents, and good heat resistance to 150°C (350°F). *Metcoseal SA* is an air-drying sealer composed of silicone resin with aluminum flakes and used with *Metco* coatings for atmospheric exposure to 480°C (900°F). *Metcoseal URS* is a single-component, general-purpose urethane resin sealer/coating with low volatile organic content and good resistance to acids, alkalies, solvents, water and aromatic hydrocarbons.

Meteor. Trade name of Teledyne Firth-Sterling (USA) for an oil-hardening tool steel containing 1.25% carbon, 0.25% chromium, 0.15% vanadium, 1.5% tungsten, and the balance iron. Used for threading taps, dental burrs and punches.

meter bronze. A bronze containing 7% zinc and 0.6% tin, and the balance copper. Used for gas and water meters.

meter leather. A sheepskin leather made impermeable by a special treatment, and used for gas-meter measuring bags.

Met Flux E. Trade name of BPI Inc. (USA) for low-melting premelted fluorspar/calcium alumina products typically containing 60% calcium fluoride (CaF_2), 21% alumina (Al_2O_3), 10% lime (CaO), 4% magnesia (MgO), 3% silica (SiO_2), 1.5% titania (TiO_2) and 0.2% ferric oxide (Fe_2O_3). Used as fluxes for metallurgical and refractory applications.

Metglas. Trademark of Allied Signal, Amorphous Metals (USA) for a series of metallic glasses composed of a base metal, such as chromium, cobalt, iron or nickel, alloyed with metalloids, such as boron, carbon, phosphorus or silicon. Commercially available in ribbon, strip, foil and wire form, they generally have high mechanical strengths and fracture toughness and are more corrosion-resistant than their crystalline counterparts. Those containing iron have excellent ferromagnetic properties including very low hysteresis and power losses. Used for brazing foils and preforms, electrical equipment, structural parts, honeycomb structures, heat exchangers, aerospace components, and soft magnets. See also amorphous alloys.

methacrylate-butadiene styrene/polyvinyl chloride alloys. See MBS-PVC alloy.

methacrylate-butadiene-styrenes. A group of butadiene-styrene-modified acrylics with outstanding toughness based on methacrylate resin monomers, and used for automotive and consumer products. Abbreviation: MBS.

methacrylate resin. A polymer or copolymer of methacrylic acid ($H_6C_4O_2$). See also acrylic plastics; acrylic resins.

methacrylic acid methyl ester. See methyl methacrylate.

methanal. See formaldehyde.

Methocel. Trademark of Dow Chemical Company (USA) for methyl cellulose ethers that have exceptional thermal gelation properties and impart superior stability to ceramic bodies in the green state. Used as binders and processing aids for ceramics, in paint removers, etc.

methoxide. A compound that is the product of a reaction in which the hydroxyl hydrogen of methyl alcohol (CH_3OH) has been substituted by a metal. Lithium methoxide ($LiOCH_3$) and sodium methoxide ($NaOCH_3$) are examples. Also known as *methylate*.

N-(4-methoxybenzylidene)-4-butylaniline. A compound that exhibits the nematic liquid crystalline phase at room temperature. It has a density of 1.027, a refractive index of 1.5496 and a flash point above 110°C (230°F) (closed cup). Used in electronics and optoelectronics. Formula: $CH_3(CH_2)_3C_6H_4N=CH-C_6H_4OCH_3$. Abbreviation: MBBA.

methyl acrylate. A flammable, colorless, volatile liquid available uninhibited and inhibited with monomethyl ether hydroquinone. The inhibited, technical grade (99% pure) has a density of 0.956, a melting point of -75°C (-103°F), a boiling point of 79-80°C (174-176°F), a flash point of 44°F (6°C) and a refractive index of 1.402, and is used as a monomer for acrylic resins, in amphoteric surfactants, and as an intermediate. Formula: $H_2C=CHCO_2CH_3$.

methylate. See methoxide.

4-methylbenzhydrylamine resin. See benzhydrylamine resin.

methylcellulose. A methyl ether of cellulose obtained from alkali cellulose by reacting with methyl chloride, methanol or dimethyl sulfate. It is available as a white grayish powder containing 25-33% methoxyl, and with a viscosity ranging from 15-4000 centipoise for a 2% solution. Used for adhesives, thickening and sizing agents, and as a protective colloid in waterborne organic coatings, films and sheeting. Abbreviation: MC. Also known as *cellulose methyl ether.*

methylethylcellulose. An ether of cellulose that contains both ethyl and methyl groups. It is available as a white or yellowish white powder or fibrous solid for use as a foaming agent, stabilizer and emulsifier. Abbreviation: MEC.

methyl fluorosilicone rubber. See fluorosilicone elastomer.

methyl methacrylate. A colorless, volatile, flammable liquid (99+% pure) derived from acetone, cyanohydrin, methanol and dilute sulfuric acid. It is usually inhibited with hydroquinone to prevent polymerization, and has a density of 0.936 g/cm³, a melting point of -48°C (-54°F), a boiling point of 100°C (212°F), a flash point of 50°F (10°C), an autoignition temperature of 790°F (421°C), and a refractive index of 1.414. It can be polymerized by catalysis, heat, light and ionizing radiation, and copolymerized with a wide range of methacrylate esters and other monomers. Used as a monomer in the production of acrylic resins, e.g., polymethacrylates, and for the impregnation of concrete. Formula: $C_5H_8O_2$. Also known as *methacrylic acid methyl ester.*

Methylon. Trademark of General Electric Corporation (USA) for phenolic-type resins used as protective coatings.

methylpentene polymer. A thermoplastic polymer based on 4-methylpentene-1 (C_6H_{12}) monomer. It has a low density, outstanding electrical properties, high optical transmission (about 90%), and is used for automotive and electronic components and laboratory ware.

methyl rubber. An elastomer made by polymerization of 2,3-dimethyl-1,3-butadiene (C_6H_{10}).

methylsilsesquioxanes. A class of inorganic polymers that upon baking and curing develop crosslinked networks of ladder structure in which each silicon atom is attached to three oxygen atoms and one methyl (CH_3) group. They have very high thermal stability (typically above 500°C or 930°F), high cracking resistance, low stress and a low dielectric constant (k = 2.7-2.9 at 1 MHz). Used as interlayer dielectrics for integrated circuit devices, they are deposited in the form of thin films by spin-on coating.

3-methylthiophene. A flammable liquid (98% pure) with density of 1.016 g/cm³, a melting point of -69°C (-92°F), a boiling point of 114°C (237°F), a flash point of 52°F (11°C) and a refractive index of 1.519. Used as a precursor to conducting polymers.

methylzinc. See dimethylzinc.

Metillure. French trade name for an acid-resistant alloy containing 84.3-85.3% iron, 14-15% silicon and 0.7% manganese. Used for evaporators, drains, and anodes.

Metite. Trade name of General Electric Company (USA) for a sintered alloy containing 79.5% copper, 9% tin, 7.5% lead and 4% graphite. It has high electrical conductivity, and is used for brush and current collectors on electrical machines.

Metlon. Trade name of Metlon Corporation (USA) for an extensive series of *metallic yarns.* The man-made fiber used in their production is usually polyester and the metal aluminum, gold or silver. Used for textile fabrics, e.g., clothing, braids, trimmings, labels, etc.

Metonal. Trade name of Pechiney/Trefimétaux (France) for a series of copper-base alloys containing 5-10% aluminum, 0-2% nickel and 0-3% iron. Used for condenser tube plates and heat-exchanger parts.

Metonic. Trade name of Pechiney/Trefimétaux (France) for a se-

ries of copper-base alloys containing 8-30% nickel, and up to 1% manganese and 2% iron, respectively. Used for heat-exchanger parts.

Metorite. Trade name for a nonhardenable aluminum alloy containing 1-2% zinc and 1-4% phosphorus. Used for light-alloy parts.

Metrocryl Universal. Trade name for a heat-cure denture acrylic.

Metronite. Trade name of Metronite Quarry (USA) for a white paint extender and filler composed of calcium carbonate, magnesium carbonate and magne-sium silicate.

Metso. Trademark of National Silicates Limited (Canada) for a series of sodium silicate compounds including (i) *Metso Beads* and *Metso Granular* based on sodium metasilicate; (ii) *Metso 200* composed essentially of sodium orthosilicate; and (iii) *Metso 99* a sodium sesquisilicate compound.

Metspec. Trade name of Metal Specialties Inc. (USA) for a series of bismuth-base fusible alloys with a melting range of 47.5-170°C (117-338°F).

Metton. Trademark of Hercules Inc. (USA) for a highly crosslinked engineering thermoset based on polydicyclopentadiene. Commonly processed by reaction injection molding, it may be glass fiber-reinforced and has excellent stiffness and flexural properties, good load-bearing properties and good impact resistance and durability. Used for molded automotive components, recreational and industrial equipment. *Metton LMR* is a thermosetting liquid molding resin based on a highly crosslinked polyolefin produced at low pressure and temperature and molded in metal (e.g., aluminum or nickel) or epoxy molds. It has superior impact strength, very low viscosity and low molding pressures (usually below 172 kPa or 25 psi) and is used for automotive and recreational vehicle components. *Metton R-Rim* is a glass-reinforced dicyclopentadiene with a low coefficient of thermal expansion, used for reaction injection molding of automotive trim, vertical exterior parts and ground-effects packages.

Metylan. Trademark of Henkel KGaA (Germany) for methylcellulose plastics and compounds.

Meubalese. Trade name for cellulose acetate fibers used for wearing apparel and industrial fabrics.

Mewlon. Trademark Unitika Limited (Japan) for vinal (polyvinyl alcohol) filaments, monofilaments and fibers used for textile fabrics.

Mexacote. Trade name for colloidal graphite powder suspended in water and spray-applied as a facing to foundry sand molds.

Mexican onyx. See onyx marble.

meyerhofferite. A white mineral composed of calcium borate hydroxide pentahydrate, $Ca_2B_6O_9(OH)_4 \cdot 5H_2O$. Crystal system, triclinic. Density, 2.12 g/cm³; refractive index, 1.535. Occurrence: USA (California).

MgAlON ceramics. See magnesium aluminum oxynitride ceramics.

M-glass. A special type of glass fiber based on beryllium oxide (BeO_2) and having a high modulus of elasticity typically 110 GPa (16 × 10³ ksi). Used as reinforcement for engineering composites.

mgriite. A gray mineral composed of copper arsenic selenide, Cu_3AsSe_3. Crystal system, cubic. Density, 4.93 g/cm³. Occurrence: Czech Republic, Russian Federation.

MHC alloys. MHC is a brand name for a series of high-temperature molybdenum alloys containing 1.2% hafnium and 0.05% carbon. They are similar to *TZC alloys,* but have greatly improved mechanical properties, and recrystallization temperatures of about 1550°C (2820°F). Used for tooling employed in hot-die forging, and other applications.

Mia-Gel. Trademark of Mia Chemical, Division of Fiberglas Canada Inc. for polyester gel coats.

Miami. Trade name of AL Tech Specialty Steel Corporation (USA) for a mold steel containing 0.38% carbon, 0.8% manganese, 0.7% silicon, 13% chromium, up to 0.3% phosphorus, up to 0.008% sulfur, and the balance iron. Used for injection, impression transfer, glass and lens-quality molds.

Mia-Pol. Trademark of Mia Chemical, Division of Fiberglas Canada Inc. for polyester resins.

miargyrite. A blackish red mineral composed of silver antimony sulfide, $AgSbS_2$. It can also be made synthetically. Crystal system, monoclinic. Density, 5.25 g/cm^3; refractive index, above 2.72. Occurrence: USA (Pennsylvania, Idaho), Spain.

Miarmi Iron. Trade name of Dayton Malleable Iron Company (USA) for a cast iron containing 3% carbon, 2% silicon, 0.7% manganese, and the balance iron. Used for high-strength castings, shafts, gears, pinions, and wheels.

MIC. Trade name of Molded Insulator Company (USA) for molded phenol-formaldehyde plastics used for electrical components and other articles.

mica. A group of mineral silicates with similar physical characteristics and atomic structures, but varying chemical compositions. The general formula is $(K,Na,Ca)(Mg,Fe,LiAl)_{2-3}(Al,Si)_4O_{10}(OH,F)_2$. Prominent members of this group are biotite, lepidolite, muscovite and phlogopite. *Mica* splits into thin partly transparent layers and, depending on the particular composition, may be colorless, white, gray, yellow, green, pink, brown, red or black. Synthetic mica is also available. Crystal system, monoclinic or hexagonal. Density range, 2.6-3.3 g/cm^3; hardness range, 2.0-4.0 Mohs; refractive index range, 1.54-1.63; excellent heat resistance; low thermal conductivity; high dielectric strength. Used for electrical and thermal insulation applications, as dielectrics in capacitors, as ceramic raw materials and in ceramic fluxes, as lubricants, for flame-resistant windows, in annealing of steel, as fillers in paints, plastics and rubber, for roofing applications, and in glass manufacture. See also synthetic mica.

Micabond. Trade name of Continental Diamond Fiber Company (USA) for engineering plastics (including shellac) filled with *phlogopite* or *muscovite* mica powder. Available in the form of tubes, sheets and molded parts, they have good dielectric properties, high electrical resistivities and high stiffness and dimensional stability. Used for electrical appliances, and in electronics.

micaceous iron oxide. A paint pigment composed of black iron oxide (FeO).

mica ceramics. Molded ceramics composed of mica powder and an inorganic binder. They have excellent dielectric properties, good physical properties, and are used for electrical insulators.

mica clay. A low-grade of *kaolin* used chiefly for earthenware glazes, and as an absorbent in oil purification.

mica flake. Ground natural or synthetic mica used in the manufacture of mica ceramics, tile, shingles, and roofing.

Mica-Flex. Trademark of Amoco Chemical Corporation (USA) for flexible insulation made from mica containing selected other ingredients, and used for generator and motor coils.

Micafolium. Trade name for an electrical insulating material consisting of *mica paper.*

Mica-Kote. Trade name for roofing materials composed of heavy felt with a special weather-resistant asphalt coating and a mica flake finish.

Mica Kote. Trade name of IMS Company (USA) for a thread lubricant containing petroleum grease and oil, mica, copper dust or fume and a corrosion inhibitor.

Micalba. Trademark of Miroglio Tessile SpA (Italy) for polyester filaments and fibers.

Micalith. Trade name for mica powders available in various grades and used as paint fillers.

mica mat. A flexible mat with excellent electrical insulating properties made from mica powder and an organic binder.

Micanite. (1) Trademark for an insulating material made by cementing together small pieces of mica with a synthetic resin, such as shellac, or a similar insulating cement.

(2) Trade name for a synthetic alkyd resin.

mica paper. A thin, flexible sheet made by bonding natural or synthetic mica powder with an organic material, such as a synthetic resin. It has high dielectric strength, good high-frequency properties, and is used for capacitors in radio and telecommunication systems.

mica pellets. Vermiculite in the form of small globules or spheres available in bags for use as loose fill thermal insulation in building ceilings and walls. See also vermiculite.

mica pigment. An extender pigment made from natural micas that are split into very thin plates or sheets.

mica powder. Natural or synthetic mica ground to a wide range of mesh sizes from 20 to over 325, and used as a filler in plastics, rubber and paint, in foundry core and mold washes, in the manufacture of mica mat and paper, ceramics, shingles, etc.

Micarta. Trademark of Westinghouse Electric Corporation (USA) for a series of plastic laminates produced by bonding paper or a fabric made from asbestos, glass or synthetic fibers with a phenolic, melamine or urea resin and curing with heat and pressure. Available in the form of sheets, rods, tubes and shapes, they have high tensile and dielectric strengths, low moisture absorption, good resistance to oils and greases, good to fair machinability, high flexural strength and toughness, and good impact strength. Used for electrical and thermal insulation, rolling-mill bearings, plating barrels, pickling tanks, oil-hardening equipment, and paper-mill equipment. *Micarta Board* refers to wallboards based on *Micarta* laminates.

Micarta BRASS. Trademark of Micarta Industries Corporation (USA) for a strong, lightweight bullet-resistant fiberglass laminate. It has good flame and smoke resistance, good richochet resistance, and can be decorated with various veneers, drywall, wallpaper, etc., and bonded to plastic laminates using a contact adhesive. Used as a reinforcing substate in architecture and building construction, e.g., for government and public buildings, postal facilities, courtrooms, corporate offices, convenience stores, bank teller stations, ATMs, gas stations, hotel lobbies, doors, and armored vehicles.

mica schist. A variety of laminated mica composed of quartz, feldspar and other minerals, and used for refractories and roofing compositions.

mica splittings. Mica, in trimmed or untrimmed form, produced from blocks by splitting to a thickness of less than 3 μm (118 μin.). Book form splittings are sheets of mica supplied in the form of books from the same block, loose splittings are of heterogeneous shape and packed in bulk form, and powdered loose splittings are loose splittings dusted with finely ground mica.

mica syenite. A variety of the igneous rock *syenite* containing appreciable amounts of the mica mineral *biotite.*

Micatherm HT. Trade name of Morgan Matroc Limited (UK) for a ceramic composite made by bonding glass with mica powder in a high-temperature molding process. It has excellent electrical and physical properties, a maximum continuous-use temperature of 450°C (840°F), and is used as a thermal barrier, and for the manufacture of electrical and electronic components.

Miccromask. Trade name of Pyramid Plastics, Inc., Tolber Division (USA) for a light-red stop-off lacquer used in hard chromium plating.

Miccropatch. Trade name of Pyramid Plastics, Inc., Tolber Division (USA) for a black or green air-drying lacquer used for repairing damaged plastisol coatings.

Miccropeel. Trade name of Pyramid Plastics, Inc., Tolber Division (USA) for a green, non-peelable stop-off lacquer for chrome and nickel plating applications.

Miccroshield. Trade name of Pyramid Plastics, Inc., Tolber Division (USA) for a transparent orange-red, high-solids, air-drying stop-off lacquer used on electroplating equipment.

Miccrostop. Trade name of Pyramid Plastics, Inc., Tolber Division (USA) for a durable, transparent-red stop-off lacquer for precious metal plating applications.

Miccrotex. Trade name of Pyramid Plastics, Inc., Tolber Division (USA) for a black, air-dry thermoplastic plating rack coating applied by dipping. The resulting film has high toughness, resiliency and flexibility.

Miccrowax. Trade name of Pyramid Plastics, Inc., Tolber Division (USA) for a range of reusable coatings for hard chromium plating applications.

Michalloy. Trade name for an abrasion-resistant *Ni-Hard*-type white cast iron containing 2.5-7% nickel. It has high toughness and hardness, good corrosion resistance and good tensile and impact strength. Used for pumps and pump components, ball-mill liners, and grinding balls.

michenerite. A grayish white mineral of the pyrite group composed of bismuth palladium telluride, BiPdTe. It can also be made synthetically. Crystal system, cubic. Density, 9.50 g/cm³. Occurrence: Canada (Ontario).

Michiana. Trade name of Michiana Products Corporation (USA) for a series of steels including austenitic chromium-nickel stainless steels, heat- and corrosion-resistant low-carbon high-chromium steels and various high-carbon high-chromium steels for special chemical and mining equipment.

Michigan. Trade name of Michigan Smelting & Refining Company (USA) for aluminum and manganese casting bronzes.

Michron. Trademark of Miroglio Tessile SpA (Italy) for polyester filaments and fibers.

Micoid. Trade name of Westinghouse Mica Insulator Company (USA) for phenol-formaldehyde resins and plastics used for electrical components and other industrial or commercial products.

Micral. Trademark of Huber Engineered Materials (USA) for alumina trihydrate (ATH) pigments and fillers.

Micrell. Trademark of American Micrell Inc. (USA) for polyester filaments and fibers.

Micria. Trade name for a series of polishing powders based on various metal oxides, such as aluminum oxide (*Micria AD*), titanium dioxide (*Micria TIS*) or zirconium oxide (*Micria ZR*).

Micro. Trade name of IMI Kynoch Limited (UK) for a corrosion-resistant wrought copper alloy containing 2% nickel. Used for locomotive boiler tubes and plates.

Micro-Aire. Trademark of Manville Corporation (USA) for preformed round fiberglass ducts and ductboards.

microalloyed steels. A group of high-strength low-alloy (HSLA) steels containing approximately 0.2-0.3% carbon and 1.6% manganese, and having small amounts of alloying elements (e.g., aluminum, niobium, titanium or vanadium) added to increase their strengths and load-bearing capacities without changing the carbon and/or manganese contents, and without reducing neither their weldability nor their notch toughness. Used mainly for structural applications, such as beams and pillars. Abbreviation: MAS.

Microballoon. Trademark of General Latex & Chemical Company (USA) for high-strength, low-weight, chemically pure hollow microspheres of glass or synthetic resins supplied in average diameters down 5 μm (197 μin.). They have good acoustical, dielectric and thermal properties, and can withstand temperatures up to 900°C (1650°F). Used in electronics, aerospace and medicine, as extenders in plastics, in the separation of helium from natural gas, and for reducing the evaporation of liquids.

Micro-Bar. Trade name of Manville Corporation (USA) for fiberglass duct liners.

Microbase. Trade name of De Trey/Dentsply International (USA) for a microwave-cure dimethacrylate (DMA) resin used for dentures.

Micro-Blast. Trade name of MDC Industries, Inc. (USA) for a white alumina (Al_2O_3) used for precision blasting.

Microbond. (1) Trade name of Detrex Corporation (USA) for phosphate coating compounds used for metal finishing applications.

(2) Trade name of Austenal Dental Inc. (USA) for a line of dental bonding alloys and dental porcelains. *Microbond Hi Life* is a high-gold dental bonding alloy, *Microbond NP2* a corrosion-resistant, beryllium-free nickel-chromium dental bonding alloy and *Microbond Porcelain* a laminate veneer porcelain.

microcarrier beads. Beads composed of glass or polymers and available in sizes ranging from 90 to 210 μm (0.0035-0.0083 in.) and specific gravities from 1.02 to 1.04 for use as substrates for cell culture applications. Collagen-coated beads in the same size and specific gravity ranges are also available. They are extensively used in biology, biochemistry, biotechnology and bioengineering. See also collagen.

Microcast MC. Trade name for a palladium-silver dental bonding alloy.

Micro-Cel. Trade name of Celite Corporation (USA) for synthetic calcium silicate.

Micro-Cell. Trade name of Hexcel Corporation (USA) for strong, lightweight aluminum honeycomb materials.

microcline. A transparent, colorless, white, pale-yellow, brick-red or green mineral belonging to the potash feldspars. It is composed of potassium aluminum silicate, $KAlSi_3O_8$, sometimes containing a little sodium. Crystal system, triclinic. Density, 2.56-2.57 g/cm³; hardness, 6.0 Mohs; refractive index, 1.52. Occurrence: Europe, South America, USA (Colorado, Connecticut, Maine, Massachusetts, New Hampshire, North Carolina, Rhode Island, South Dakota, Vermont). Used in ceramics for pottery, porcelain, hard glazes, enamels, concrete and ceramic ware, in glass manufacture, and as an abrasive.

Micro-Coustic. Trade name of Manville Corporation (USA) for fiberglass duct liners.

microcracked chromium plate. Chromium plate electrodeposited over a conventional copper-nickel or all-nickel undercoating such that the chromium contains numerous fine cracks that expose the underlying nickel. This coating is applied to many softer materials to increase the wear resistance.

microcrystalline alumina. A tough alumina (Al_2O_3) recrystallized from a molten bath by a process that results in particle or crystal sizes generally smaller than those of ordinary alumina. Used as a fine abrasive and polishing agent. See also recrystallized alumina.

microcrystalline cellulose. A highly purified particulate form of cellulose with a particle size in the microcrystalline range, typically between 1 and 150 μm (0.00004 to 0.006 in.). It has been hydrolyzed to a limiting degree of polymerization.

microcrystalline limestone. Limestone composed of micro-sized crystals.

microcrystalline materials. Solid substances, e.g., metals, alloys or ceramics, composed of crystals with diameters in the micrometer range.

microcrystalline silica. Silica (SiO_2) composed of micron-sized particles or grains and obtained from silica sand by high-temperature treatment in an arc or electrical furnace. In ground form, it can be used in the manufacture of whitewares, and in ceramic body and glaze formulations.

microcrystalline wax. A high-melting petroleum wax composed of minute crystals much smaller than those of other waxes. It has a higher viscosity and molecular weight than paraffin wax, and is used in the manufacture of adhesives, leather finishes, floor waxes, heat sealing compositions, and electrical insulation.

Microdenier Sensura. Trademark of Wellman Inc. (USA) for a microdenier polyester fiber belonging to the *Sensura* product line. It has higher tenacity than the original Sensura, blends well with cotton, and is used for textile fabrics, especially clothing.

Microdip. Trade name of Atotech USA Inc. (USA) for a microporous chromium dip applied over nickel.

microdiscontinuous plate. A collective term for microcracked and microporous chromium plate. See microcracked chromium plate; microporous chromium plate.

microduplex alloy. An alloy consisting of two phases that are uniformly distributed such that the crystal boundaries are chiefly interphase interfaces.

microengineered material. A crystalline or nanocrystalline material whose microstructure (e.g., number of phases, grain size, lattice defect geometry and structure, etc.) has been purposely controlled or influenced to produce a material with improved physical, chemical and/or mechanical properties.

microfiber batting. Textiles that contain polyester, olefin or other polymer fibers with an average diameter below 10 μm (394 μin.).

Micro-Fibers. Trade name of Manville Corporation (USA) for fine borosilicate glass fiber felt, bulk and mat insulation.

microfibers. Manufactured fibers (e.g., acrylic, polyester, nylon, rayon, lyocell, glass, or graphite) that are usually less than 10 μm (394 μin.) in diameter and used for soft, light fabrics, such as those made into rainwear, outerwear, etc., industrial fabrics, and in bulk, blanket or mat form for acoustical and/or thermal insulation applications.

Microfill. Trade name of Discus Dental Inc. (USA) for dental cements including *Microfill Pontic*, a self-cure bisphenol A glycidyl methacrylate (BisGMA) based resin cement and *Microfill C*, a dual-cure luting cement.

Microfilm. Trade name of Dental Materials Group (USA) for a dental gypsum used as a pore filler.

Microflake. Trademark of Technic Inc. (USA) for silver-palladium alloy flake used for electronic applications.

Microflat. Trade name of Washington Steel Corporation (USA) for cold-rolled, stretcher-leveled sheet and coils.

Micro-Flex. Trademark of Manville Corporation (USA) for fiberglass cushioning materials and pipe and boiler insulation.

Microflex. Trade name of Washington Steel Corporation (USA) for soft-tempered austenitic stainless steel sheet used for roofing, flashing and architectural applications.

Microflo. Trademark of Eutectic Corporation (USA) for powdered alloys.

Micro-Foil. Trademark of Manville Corporation (USA) for laminated micro-fiber felt and foil insulating tape.

microfoil. A thin metal or metal alloy film with a thickness of only a few micrometers supported on permanent backing.

Microft. Trade name of Teijin Fibers Limited (Japan) for polyester microfibers and filaments used for the manufacture of textile fabrics.

Microfyne. Trademark of Dixon Ticonderoga Company (USA) for a lubricating graphite consisting of extra-fine powdered particles. It has excellent electric conductivity, and is used for the lubrication of precision equipment, in the manufacture of electrically conductive materials, and as a coating of nonconducting parts to be plated.

Microgard. Trade name of Teijin Fibers Limited (Japan) for high-density, ultrafine microfiber fabrics for anti-tick bedding.

microgel. A material, usually a polymer, in the form of finely divided micron-sized particles. See also colloidal microgel.

Microglas. (1) Trade name of Nippon Glass Fiber Company Limited (Japan) for products made of continuous glass fibers.

(2) Trademark of Nippon Sheet Glas Company Limited (Japan) for chemical-grade borosilicate glass flakes with an average particle size of 15-600 μm (0.0006-0.02 in.), a thickness of about 5 μm (0.0002 in.) and a nominal width of 10-4000 μm (0.0004-0.16 in.). They have excellent corrosion resistance, and are used as reinforcements in thermoplastics, and in chemically-resistant coatings and linings for chemical equipment, petroleum tanks, food-processing equipment, boiler and water tanks, and plating equipment.

microglass. A thin glass used in the manufacture of cover slips for use in microscopy.

Micrograin. (1) Trade name for file-hard, fine-grain nickel electrodeposits produced on metals from a sulfate plating bath. The average grain size is 0.5 μm (20 μin.).

(2) Trade name of Belmont Metals Inc. (USA) for babbitt metal.

MicroGrid. Trademark of Delker Corporation (USA) for expanded-metal foil products. The mesh-like materials are made by slitting and stretching foil to obtain openings of exact dimensions. They are available in most metals and alloys (including aluminum, copper, zinc and lead) and provide electromagnetic shielding, heat transfer and electric conductivity.

microinfiltrated microlaminated composites. Large, bulk-laminated composite structures, usually made by a process involving tape casting and sintering, and consisting of alternating layers of a relatively ductile and tough material with low tensile strength and modulus, and a hard and brittle material with high tensile strength and modulus, whereby the ductile constituent infiltrates the brittle material. Possible ductile and brittle constituents include pure metals, metal alloys and certain intermetallics, and ceramics and refractory metals, respectively. Abbreviation: MIMLCs.

Microjoin. Trade name Sci-Pharm Inc. (USA) for a self-cure dental resin cement.

microleaf. A ductile metal or alloy in the form of a sheet of a few micrometers in thickness supported on a removable backing.

Microlene. Trademark of Polymekon SpA (Italy) for polypropylene fibers.

Microlim. Trade name of English Steel Corporation (UK) for a high-carbon chromium steel containing 0.95% carbon, 0.3% chromium, and the balance iron. Used for punches and rivet sets.

Microlite. Trademark of Manville Corporation (USA) for blanket-type thermal acoustical fiberglass insulation.

microlite. A greenish to cream-colored mineral of the pyroclore group composed of calcium sodium tantalum oxide hydroxide, $(Ca,Na,Fe)_2Ta_2(O,OH,F)_7$. Crystal system, cubic. Density, 5.30 g/cm^3; refractive index, 2.08. Occurrence: Brazil.

Microlith. Trademark of Schuller International (USA) for fiberglass insulation and textile-glass products.

Microllam. Trademark of TrusJoist/Weyerhaeuser (USA) for laminated veneer lumber (LVL) with excellent strength and stiffness, and better resistance to shrinking, warping and twisting than conventional solid sawn lumber. Used for headers and beams, especially for longer spans.

Microloft. Trade name of E.I. DuPont de Nemours & Company (USA) for polyester microfibers used for textile fabrics, especially clothing.

Microlok. Trade name of Birchwood Casey (USA) for a zinc phosphate treatment for iron and steel products.

Micro-Lok. Trademark of Manville Corporation (USA) for preformed fiberglass pipe-covering insulation.

Microlon. Trademark of Coltene-Whaledent (USA) for a heat-cure denture acrylic supplied in several shades.

Microlube. Trademark of Microfin Corporation (USA) for a high-lubricity coating and coating process. The coating has a hard, fine-grained crystalline structure that is crosslinked to a friction-inhibiting organic substance. The coating decreases friction and wear of a wide variety of ferrous and nonferrous materials.

Microlux. (1) Trademark of Meller Optics Inc. (USA) for a series of metallography-grade high-purity (99.98+%) calcined alumina abrasives supplied in particle sizes from as low as 0.05 μm (2 μin.) to over 1 μm (40 μin.).

(2) Trademark of KoSa (USA) for high-loft polyester products containing durable slick fibers (Type J30) in cut lengths of 38 mm (1.5 in.) or 63 mm (2.5 in.), and with 0.95 denier. Used especially for bedding, e.g., pillows and mattress pads.

Micro Mach. Trade name for an austenitic stainless steel containing 0.08-0.12% carbon, 17.3% chromium, 6.2% nickel, and the balance iron. Used for aircraft and missile wing and skin surfaces.

Micromattique. Trademark of E.I. DuPont de Nemours & Company (USA) for a family of microfiber polyesters used for garments and wearing apparel.

Micro-Melt. Trademark of Carpenter Technology Corporation (USA) for a series of powder-metallurgy cold-work and high-speed steels. The cold-work types are available in standard (AISI type A11) grades with 9.5% vanadium, and low-vanadium (A11-LVC) grades containing 8.5% vanadium. They possess high wear resistance and good strength and toughness and are used for punches, shears, chipper and slitter knives, blanking, cold-heading and forming dies and forming rolls. The high-speed grades are available in various molybdenum and tungsten grades (AISI types M3, M4, T15, etc.) with high toughness and wear-

resistance, and are used for broaches, milling cutters, hobs, reamers, etc. Other commercial grades include *Micro-Melt CD No. 1*, *Micro-Melt Maxamet* and *Micro-Melt PD No. 1*.

MicroMicrofiber. Trade name of E.I. DuPont de Nemours & Company (USA) for strong nylon 6,6 microfibers used for the manufacture of textile fabrics.

Micromod. Trade name for a die stone used for dental applications.

MicroModal. Trade name of Lenzing Fibers Corporation (USA) for modal microfibers used for textile fabrics. See also modal fibers.

Micronew. Trademark of BISCO, Inc. (USA) for a light-cured microfill composite reinforced with 69 wt% of a special filler of submicron particle size. It provides high strength and durability and excellent polishability. Supplied in incisal, body translucent, and body opaque shades, it is used in dentistry for direct and anterior restorations.

Micronex. Trademark of Binney & Smith Company (USA) for various grades of carbon black used as pigments in rubber and plastics, and as reinforcing agents in rubber compounding.

Micronium. Trade name Schuetz Dental GmbH (Germany) for corrosion-resistant cobalt-chromium dental casting alloys. *Micronium Exclusiv* contains 62% cobalt, 30.7% chromium, 5.7% molybdenum, 0.6% silicon, 0.5% manganese and 0.5% carbon, and has a density of 8.4 g/cm^3 (0.30 $lb/in.^3$) and a melting point of 1385°C (2525°F).

micronized barite. Barium sulfate ($BaSO_4$) reduced to a fine, micron-sized white powder, (average particle size, 5-35 μm or 197-1378 μin.) Used as extender and filler in rubber and plastics.

micronized clay. A pure china clay (kaolin) reduced to a fine, micron-sized powder (average particle size, 18-37 μm or 709-1457 μin.). Used as extender in rubber, plastics and paper.

micronized mica. Mica reduced to a fine, micron-sized powder (average particle size, 5-20 μm or 197-787 μin.). The reduction is usually carried out in a special type of dry-grinding machine known as "Micronizer." Used as extender and filler in paint, rubber and plastics.

Micro-Ohm. Trade name of Micro-Ohm Inc. (USA) for metal-filled epoxy adhesives having high conductivities and constant volume resistivities. They bond to metals, plastics, glass and polycrystalline ceramics, and are used as cold solders for electrical and electronic applications.

Micropake. Trademark of Acordis Specialty Fibers for a fine, strong, flexible, X-ray detectable yarn, melt-spun from polypropylene and loaded with 60% barium sulfate. It is radiopaque, sterilizable and can be made into woven or nonwoven fabrics for surgical gauze.

Microporite. Trademark for a highly microporous lightweight concrete prepared from steam-treated, ground silica and lime.

microporous chromium plate. Chromium plate electrodeposited in such a manner that the chromium contains numerous very fine pores with an average pore density of 16000 pores/cm^2 (100000 pores/$in.^2$).

microporous materials. A class of inorganic and organic materials (e.g., ceramics, metals, polymers, metal-organic solids, composites, natural and synthetic zeolites, xerogels, etc.) with a pore size of less than 2 nm (0.08 μin.). They are used in catalysis and separation (e.g., as molecular sieves) and for electronic applications (e.g., sensors, batteries and nonlinear optics).

Microposit. Trade name of Shipley Company (USA) for photoresists.

Micropowder. Trade name of International Specialty Products (USA) for micron-size, spherical iron powders manufactured by the chemical decomposition of iron pentacarbonyl, $Fe(CO)_5$. The powders are supplied in over 50 grades with fine, uniform particle sizes, typically between 2 and 10 µm (79 and 394 µin.) and purities of 98-99.5%. Soft (R) grades are used in conventional powder metallurgy, while hard (S) grades find application in metal injection molding. *Micropowder* powders have soft-magnetic properties, low magnetic remanence, excellent saturation characteristics, excellent resistance to magnetic shock and temperature, excellent electromagnetic properties and excellent microwave absorption, and are used in high-density powder metallurgy, for metal injection molded components, and as fillers in plastic resins, metals and alloys.

Micro-ProModal. Trade name of Lenzing Fibers Corporation (USA) for modal microfibers used for textile fabrics. See also modal fibers.

Micro-Quartz. Trade name of Manville Corporation (USA) for felt made by water deposition of fine, quartz (silica) fibers (98.5+% pure). It has a density of about 50 kg/m³ (3 lb/ft³), an upper service temperature of approximately 1090°C (1994°F) and good heat resistance. Used as thermal insulation.

Microrold. Trade name of Washington Steel Corporation (USA) for products of thin sheet and strip steel made by precision rolling in a Sendzimir mill–a cold-reduction mill invented by T. Sendzimir and consisting of a set of large-diameter backup rolls, mounted on an eccentrical shaft, and a set of smaller-diameter working rolls.

MicroSafe. Trade name of Celanese Acetate Fibers (USA) for cellulose acetate fibers used for textile fabrics.

microscopic composites. Composite materials with dispersed phases (reinforcing elements) that cannot be observed with the unaided eye. Examples are dispersion-strengthened metals or metal alloys, particle-reinforced metals, polymers and ceramics, and fiber-reinforced metals, polymers and ceramics.

MicroSeal. Trade name of SynbronEndo (USA) for a gutta-percha used in endodontics.

Microseal. Trademark of Michigan Chrome & Chemical Company (USA) for an inorganic corrosion inhibitor designed for aluminum products, but also suitable for application on zinc and cadmium-plated parts. The thin protective coating (usually less than 2 µm or 0.08 mil) does not chip, crack or peel at service temperatures ranging from -60 to +400°C (-80 to +750°F). Used on auomotive engine parts, such as intake manifolds.

Microselect. Trademark of E.I. DuPont de Nemours & Company (USA) for polyester microfibers used for textile fabrics, especially clothing.

Microsere. Trademark of International Waxes Limited (Canada) for microcrystalline waxes.

Micro-Sheet. (1) Trade name of Corning Glass Works (USA) for very thin glass drawn by a special process that imparts a virtually flawless, fire-polished surface.

(2) Trade name for titanium oxide supplied in sheets as thin as 80 µm (0.003 in.). It has good dielectric properties, and is used as a mica substitute for electrical insulation applications.

Microshell. Trade name for an epanded polystyrene foam liner.

Micro-Shim. Trade name for a series of special low-carbon and stainless steels supplied in microfoil form with a thickness of less than 12 µm (0.5 mil).

Microsil. Trade name of Magnetic Metals Company (USA) for a grain-oriented soft magnetic iron alloy containing 2.9-3.3% silicon. It has a density of 7.65 g/cm³ (0.276 lb/in.³), highly directional magnetic properties, low core losses and high permeability. Used for high-performance power transformers, magnetic amplifiers, saturable reactors, and inverter transformers.

microsilica. See silica fume.

microsommite. A colorless mineral of the cancrinite group composed of sodium aluminum chloride silicate sulfate, $(Na,Ca,K)_8$-$(Si,Al)_{12}O_{24}Cl_{2.5}$. Crystal system, hexagonal. Density, 2.45 g/cm³; refractive index, 1.521. Occurrence: Italy.

microspheres. Minute spherical particles of ceramics, glass or plastics available in a wide range of sizes from submicron to several millimeters in diameter, and used as reinforcing agents and fillers in plastics and rubbers, in acoustical and thermal insulation, and as particle size standards in polymer and biomedical research.

MicroSpun. Trademark of Wellman Inc. (USA) for a soft, light, spun polyester microfiber belonging to the *Fortel* range of fibers. It is wrinkle- and shrink-resistant, drapable, and blends well with natural and synthetic fibers. Used for wearing apparel and home fashions.

Micro-Star. Trade name of J.F. Jelenko & Company (USA) for a silver-free high-strength dental casting alloy containing 79% palladium and 2% gold. Used for joining porcelain to metal.

Microstar. Trade name of Teijin Fibers Limited (Japan) for ultra-fine microfiber fabrics for the manufacture of industrial and household wiping cloth.

Microstone. Trade name of Whip Mix Corporation (USA) for a dental stone (ADA Type III) supplied as a powder with fine particle size in golden and white colors. Used for laboratory procedures, in particular for processing acrylic dentures.

microstructured optical fibers. A class of two-dimensional *optical fibers* with nonlinear properties in which light travels in a core that consists essentially of air. They are formed by drawing structured fiber preforms, and are suitable for the manufacture of nonlinear optical devices and optical networks.

microstructured silica. A relatively recent type of porous silica (pore sizes about 190-1000 nm or 7.5-39.4 µin.) prepared by adding an aqueous solution of silicon hydroxide, $Si(OH)_4$, to a colloidal crystal film assembled by slow filtration of diluted particles of a monodisperse suspension of amidine or sulfate latexes through a porous membrane (pre size, 100 nm or 3.9 µin.). This composite is then calcined at a temperature of 450°C (840°F) for a given period of time. Potential uses include solid-state devices.

MicroSupreme. Trademark of Sterling Fibers Inc. (USA) for a colorfast acrylic microfiber (0.9 denier) supplied in staple form for spinning into yarns. It provides good drape, a soft hand, excellent moisture management, and blends well with other fibers including cotton, polyester, rayon and wool. Used especially for wearing apparel including activewear, sportswear, sweaters, and household and industrial textiles.

MicroTec. (1) Trademark of Timken Company (USA) for carbon-elevated microalloy manganese steels with ferrite-pearlite microstructures containing precipitation-strengthening elements, such as vanadium. They are supplied as machining and forging bars, seamless tubing and finished parts in several standard, machinable, high-hardness and weldable grades. Machinable grades have 0.28-0.41% carbon, 1.10-1.50% manganese, 0.08-0.18% vanadium, up to 0.05% sulfur, 0.70% silicon and 0.18% nitrogen, respectively, and the balance iron. High-hardness grades (e.g., *MicroTec 5H95A*) contain 0.52-0.57% carbon, 11.15-1.35% manganese, 0.10-0.20% vanadium, 0.04-0.06% sulfur, 0.045-0.65% silicon, 0.15-0.45% chromium, 0.014% or more nitrogen, and the balance iron. Weldable grades contain 0.10-0.32% carbon, 0.90-1.70% manganese, 0.05-0.23% vanadium, up to 0.010% sulfur, and the balance iron. *MicroTec* steels have no quenching distortion, good through-hardness and attain higher strength in the as-rolled or as-forged conditions than standard carbon and many low-alloy steels. Used in applications where moderate levels of strength and ductility are required, e.g., communication towers, automotive roller clutches and crankshafts, hydraulic cylinder barrels and cylinder rods, and structural applications.

(2) Trademark of Powdermet, Inc. (USA) for cobalt-coated, submicron tungsten carbide powders used for metal cutting and forming tools, dies and other applications.

Microtex. Trade name of Manville Corporation (USA) for fiberglass-textile-strand thermal insulation blankets.

Microthene. (1) Trademark of National Distillers and Chemical Corporation (USA) for a series of solid polyolefins including in particular polypropylenes and high-density polyethylenes, supplied in finely divided powder form for molding applications.

(2) Trademark of U.S.I. (USA) for phenol-formaldehyde plastics.

Microtherm. (1) Trade name of Microtherm Europa NV (Belgium) for a thermal insulation material with an extremely low thermal conductivity, a density of 230-240 kg/m³ (14.4-15.0 lb/ft³), and a maximum continuous-use temperature of 950°C (1740°F). Used in the manufacture of petrochemical pipelines and reactors, hot-air recuperators, nuclear reactors and diesel exhaust systems, and for aerospace applications.

(2) Trademark of KoSa (USA) for a microfine polyester yarn used for warm, comfortable, breathable high-performance apparel, and linings. It has good insulating properties, is wind- and water-resistant, and wicks moisture away from the skin.

Micro Touch. Trade name of BASF Corporation (USA) for nylon 6 fibers used for textile fabrics including clothing.

Micro-Tuf. Trade name of Manville Corporation (USA) for fiberglass insulation blanket or board used for appliances.

Microvid. Trade name of Vidrobrás (Brazil) for glass wool composed of a fleecy mass of fine, high-quality fibers, and used as acoustic and thermal insulation in industrial installations.

Microvon. Trade name of Dunlop Company (USA) for a soft, microcellular polyurethane elastomer.

Microwhite. Trade name for a white die stone used for dental applications.

Microwit. Trade name of Flachglas AG (Germany) for engineering glasses and glass products.

Microwool. Trade name of Nippon Glass Fiber Company Limited (Japan) for a series of thermal insulating products made from short-glass fibers including products, such as *Microwool MLR* (Microwool Roll), *Microwool MPI* (Microwool Pipe Insulation) and *Microwool MHR* (Microwool Home Insulation).

Microzite. Trademark of Ferro Corporation (USA) for solid, fissure-free, ultra-fine-milled zirconium silicate grinding beads supplied in four standard particle size ranges: 0.4-0.6 mm (0.015-0.024 in.), 0.6-0.8 mm (0.024-0.031 in.), 0.8-1.0 mm (0.031-0.039 in.) and 1.0-1.2 mm (0.039-0.047 in.). Used especially in the agrochemical, coating, paint and mineral industries. Ceria-stabilized zirconia, yttria-stabilized zirconia, alumina and zirconia mini-grinding beads are also included under this trademark.

Midas. Trade name of J.F. Jelenko & Company (USA) for a microfine, hard, medium-gold dental casting alloy (ADA type III) with outstanding tarnish resistance, and good grindability, burnishability and polishability. Used for fixed bridgework, crowns, and inlays.

mid chrome. A yellow, inorganic paint pigment composed of lead chromate ($PbCrO_4$).

Midcyl. Trade name of Midland Motor Cylinder Company (USA) for a cast iron containing 3.2% carbon, 2.5% silicon, and the balance iron. Used for cylinder liners.

Midflex. Trade name of Engelhard Corporation (USA) for a bimetal used as thermometal and for bimetallic strips.

Midigold 50. Trademark of Ivoclar Vivadent AG (Liechtenstein) for a hard dental casting alloy (ADA type III) containing 50.0% gold, 35.0% silver, 9.5% copper, 2.7% palladium, 3.49% palladium, 2.0% indium, and less than 1.0% iridium. It has a deep yellow color, a density of 13.2 g/cm³ (0.48 lb/in.³), a melting range of 825-915°C (1517-1679°F), an as-cast hardness of 180 Vickers, low elongation, and excellent biocompatibility and resistance to oral conditions. Used for crowns, onlays, posts, and bridges.

Mid-Max. Trade name of Midvale-Heppenstall Company (USA) for a high-speed steel containing 0.74% carbon, 0.2% manganese, 0.3% silicon, 3.75% chromium, 1.65% tungsten, 8.75% molybdenum, 1.15% vanadium, and the balance iron. Used for cutters and tools.

Midohm. Trade name of Driver Harris Company (USA) for a wrought electrical resistance alloy of 77-78% copper and 22-23% nickel. It has a density of about 8.9 g/cm³ (0.32 lb/in.³), good heat resistance, good tensile strength, and a maximum service temperature of 200°C (390°F). Used for load banks, rheostats, and resistances.

Midrex. Trademark of Midrex Corporation (USA) for a direct reduced iron supplied in the form of pellets, pigs and crushed pieces for use in the manufacture of clean, high-grade electric-arc furnace (EAF) steels.

Midvaloy. Trade name of Midvale-Heppenstall Company (USA) for an extensive series of austenitic, ferritic and martensitic stainless steels, various machinery and tool steels, and several nickel-iron-chromium heat-resistant alloys.

miersite. A yellow mineral of the sphalerite group composed of copper silver iodide, (Ag,Cu)I. Crystal system, cubic. Density, 5.45-5.64 g/cm³; refractive index, 2.20. Occurrence: Australia. Used as a semiconductor.

MightytopII. Trade name of Teijin Fibers Limited (Japan) for polyester staple fibers used for the manufacture of textile fabrics.

Migra. Trade name of Friedrich Wilhelms-Hütte (Germany) for a series of cast irons containing 3.8-4.1% carbon, 1-3% silicon, 0.4-1.5% manganese, 0-2% nickel, 0-1% chromium, and the balance iron. Used for cylinders and pistons.

Miharahyo. Trade name of Mitsubishi Rayon Company Limited (Japan) for rayon fibers and yarns used for textile fabrics.

miharaite. A pale gray to grayish white mineral composed of copper iron lead bismuth sulfide, $Cu_4FePbBiS_6$. Crystal system, orthorhombic. Density, 6.06 g/cm³. Occurrence: Japan.

Mikado. (1) Trade name of SA des Verreries de Fauquez (Belgium) for a patterned glass.

(2) Trade name of Gebrüder Hover Edelstahlwerk (Germany) for a water-hardening, wear-resistant machinery steel containing 1.45% carbon, 1.4% chromium, 0.6% manganese, 0.25% silicon , and the balance iron. Used for bushings, liners,

sleeves, bearings, and inserts.

Mikafolium. Trade name of Segliwa GmbH (Germany) for an electrical insulating material consisting of *mica paper.*

Mikanit. Trade name of Segliwa GmbH (Germany) for a rigid *mica* material used for dielectric applications.

Mikolite. Trade name for a very fine expanded vermiculite used as a filler in plastics, rubber, paints, caulking compounds and lubricating oils.

Mikrodur. Trademark of Sika AG (Switzerland) for a cement based on blast-furnace slag and supplied as a finely divided powder. It is used for blending with ordinary cements to lower the porosity and enhance the overall mechanical properties of mortars.

Mikron. Trade name for vinal (vinyl alcohol) fibers used for textile fabrics.

Milano. Trade name of Temsur SA (Argentina) used for toughened glass.

milarite. A colorless to pale green mineral of the osumilite group composed of potassium calcium beryllium aluminum silicate hydrate, $K_2Ca_4Be_4Al_2Si_{24}O_{60} \cdot H_2O$. Crystal system, hexagonal. Density, 2.53 g/cm³; refractive index, 1.532. Occurrence: Switzerland, Czech Republic, Slovakia.

Milastomer. Trade name of Mitsui Kagaku Kabushiki Kaisha Corporation (Japan) for thermoplastic olefin elastomers supplied as semiprocessed products, such as blocks, rods, plates, sheets, films and pipes, and as unprocessed pellets for further manufacture.

Milbrite. Trade name of A. Milne & Company (USA) for an oil-hardening chromium-molybdenum mold steel containing 0.5% carbon, 1% manganese, 0.3% silicon, 1.1% chromium, 0.25% molybdenum, and the balance iron.

mild abrasives. Fine *abrasives,* such as chalk, diatomaceous earth, kaolin, pumice, talc, tripoli or tin oxide, that have uniform grain sizes and hardnesses usually not exceeding 2 Mohs. Used as dental and silver polishes, and in window cleaners.

mildew-resistant fabrics. Textile fabrics that have been treated with compounds, such as phenols, formaldehyde or certain solutions of copper, mercury or zinc salts, to prevent the growth of stain-producing parasitic fungi.

mildly active fluxes. Soldering fluxes containing organic compounds (e.g., organic halides) that are decomposed by the heat of soldering. The mildly corrosive constituents in the usually rosin-based flux volatilize, leaving a residue that generally does not attack the base metal and is easily removed by washing with water. These fluxes are moderately or mildly active and have intermediate level cleaning abilities. Also known as *intermediate fluxes.* See also flux (1); soldering flux.

mild mercury chloride. See mercurous chloride.

mild polish. A very fine, chemically neutral, abrasive powder (e.g., tin oxide) used as a burnishing medium for metals, and in metallographic polishing. Also known as *levigated abrasive.*

Mild Self Hardening. Trade name of Cold Industries (UK) an oil-hardening die steel containing 1.7% carbon, 3.35% chromium, 5.25% tungsten, and the balance iron.

mild silver protein. See silver protein.

mild steels. A group of low-carbon steels containing 0.05-0.25% carbon, and about 0.5% manganese, 0.03% silicon, 0.05% sulfur, 0.05% phosphorus, and the balance iron. They are comparatively ductile and soft and cannot be hardened by heat treatment, but can be surface-hardened. *Mild steels* have low strength, good weldability, and are used for general engineering applications, sheet steel, wire, fasteners, etc. Also known

as *soft steels.* See also low-carbon steels.

Mil-ene. Trademark for a polyester film.

Milipore. Trade name of Solutions Globales (France) for a textile fabric made from 80% polyester and 20% cotton fibers.

Milium. Trade name for a thermally insulating fabric made from acetate, cotton, nylon, polyester or viscose fibers and having aluminum flakes applied to its reverse side. Used as a coat and curtain lining material.

milk acid. See lactic acid.

milk glass. See cryolite glass.

milk of lime. See whitewash (1).

milk of magnesia. See magnesium hydroxide.

milk sugar. See milk sugar.

milky glass. See cryolite glass.

mill addition. A material, other than a frit, added to a mill to complete the batch formula of a vitreous enamel or other ceramic slip. Examples of such materials are bentonite, clay, fluxing agents (e.g., lithium titanate and zinc oxide), opacifiers (e.g., titanium dioxide), setting-up agents, refractory materials, color pigments, etc. Also known as *mill batch.*

Millaloy. Trade name of Doelger & Kirsten Inc. (USA) for an oil-hardening tool steel containing 0.4% carbon, 4% nickel, 1.5% chromium, and the balance iron. Used for shear blades.

Millard. Trade name of Anaconda Company (USA) for a copper alloy containing about 8.8% aluminum, 0.6% iron, 0.1% zinc, and 0.1% tin.

mill batch. See mill addition.

millboard. A strong, rigid *paperboard* product used for making furniture panels and cartons, and in fabric pressing operations.

milled asbestos. Any of various grades of *asbestos* obtained by comminution and screen classification of suitable asbestos minerals (principally of the serpentine and amphibole groups).

milled enamel. A thick liquid composed of finely ground vitreous enamel.

milled fabrics. Knit or woven fabrics, usually of wool or a wool blend, that have been compacted, shrunk and felted by the application of moisture (e.g., a soap solution), heat and pressure. Also known as *fulled fabrics.*

milled glass fibers. Chopped or sawed, continuous-strand glass fibers reduced to very short fibers by hammer milling. Typically, milled fibers are about 0.8-6.4 mm (0.03-0.25 in.) in length. They have comparatively low length-to-diameter ratios and are used as anticrazing reinforcing fillers for adhesives, in potting compounds, and in polyesters, epoxies, phenolics, fluorocarbons, urethanes and various thermoplastics to increase stiffness, dimensional stability, and/or heat resistance.

millefiori. A specialty glass consisting of a matrix of transparent glass having a decoration of multicolored glass rods, shapes or floral patterns.

Millenite. Trade name of Lake & Elliot Limited (UK) for an oil-hardening steel containing 0.3% carbon, 1-5% nickel, 0.8% chromium, and the balance iron. Used for high-duty castings.

Millenium. Trade name of Sirius Technology Inc. (USA) for electroless nickel electroplates and plating processes.

millerite. A gray, opaque mineral in powder form composed of nickel sulfide, NiS. Crystal system, rhombohedral (hexagonal). Density, 5.50 g/cm³. Occurrence: USA (Pennsylvania). Used as an ore of nickel.

milling silver. A British corrosion-resistant nickel silver composed of 56% copper, 27.5-31% zinc, 12-16% nickel, and 0.5-1% lead. It has excellent machinability, and is used for corrosion-resistant parts.

Mill-I/Nium. Trade name of I/N Tek (USA) for cold-rolled steel sheet products supplied in coil form, and made to customer specifications.

millisite. A white to light gray mineral of the wardite group composed of sodium calcium aluminum phosphate hydroxide trihydrate, $(Na,K)CaAl_6(PO_4)_4(OH)_9 \cdot 3H_2O$. Crystal system, tetragonal. Density, 2.83 g/cm^3; refractive index, 1.598. Occurrence: USA (Florida, Utah).

Mill Mates. Trademark of Zircoa, Inc. (USA) for dense, hard, wear-resistant dispersion media based on off-white, ceria-stabilized tetragonal zirconia polycrystal beads with fine, uniform grain size. They have high fracture toughness, and are used in paints, inks, dyes, pigments, frits, glazes, ceramics, magnetic materials, cosmetics, etc. A special grade, *Mill Mates Plus*, with finer grain size and better controlled microstructure than standard products is also available.

millosevichite. A colorless mineral composed of aluminum sulfate, $Al_2(SO_4)_3$. It can be made synthetically by heating aluminum sulfate octadecahydrate $[Al_2(SO_4)_3 \cdot 18H_2O]$. Crystal system, rhombohedral (hexagonal). Density, 2.86 g/cm^3.

mill products. See rolled products.

Mills. Trade name of William Mills & Company Limited for an extensive series of chill-, die- and sand-cast aluminums, and aluminum-copper and aluminum-silicon alloys.

mill scale powder. Mill scale is the ferric oxide layer that peels off from iron or steel during hot rolling. For powder-metallurgy applications, it is often reduced to a soft spongy powder with few solid impurities.

mill white. White paint used to increase the illumination on the interior wall surfaces of industrial and commercial establishments, e.g., plants, offices and school buildings.

millwork. A term used for products that are primarily fabricated to a particular shape or assembly from lumber in a planing mill or woodworking plant. Examples of millwork include moldings, cornices, door frames, blinds and shutters, sash and window units, stairwork, porch work, kitchen cabinets, mantels and cabinets. The term does not include flooring, ceiling and siding products.

Milmold. Trade name of A. Milne & Company (USA) for an oil-hardening tool steel (AISI type P20) containing 0.3% carbon, 0.8% manganese, 0.5% silicon, 1.7% chromium, 0.4% molybdenum, and the balance iron. It has high core hardness, good deep-hardening properties and good toughness and machinability. Used for molds especially for zinc and plastic articles.

Milnair. Trade name of A. Milne & Company (USA) for a series of air- or oil-hardening tool steels containing 0.95-1% carbon, 0-2% manganese, 0-0.35% silicon, 2-5.25% chromium, 1-1.1% molybdenum, and the balance iron. Used for tools and dies.

Milne. Trade name of A. Milne & Company (USA) for a series of cold-work, hot-work, plain-carbon and high-speed tool steels.

Milo. Trade name of Hidalgo Steel Company Inc. (USA) for a series of water-hardening tool steels containing 0.4-1.1% carbon, 0-1.1% manganese, and the balance iron. Used for pneumatic tools, chisels, and punches.

milori blue. An iron blue pigment based on ferric ferrocyanide, $Fe_4[Fe(CN)_6]_3$, with small amounts of alum, barite, chalk and/or gypsum. Depending on the particular composition, it may have slight bronze overtones. *Milori blue* has good permanence and lightfastness, good resistance to water, oils, alcohols, paraffin, organic solvents and dilute acids, and poor resistance to alkalies and reducing media. Used in lithographic and printing inks, lacquers, paints, soaps, and matches.

Milpa. Trade name of Teijin Fibers Limited (Japan) for polyester filaments used for the manufacture of textile fabrics.

Milrite. Trade name of NL Industries (USA) for a white-metal bearing alloy containing lead, tin, antimony, and copper.

Milrol. Trademark of AT Plastics Inc. (Canada) for polyethylene film used in construction and agriculture.

Miltite. Trademark of C-I-L Inc. (Canada) for polyethylene film used for bulk shrink and shrink bundling.

Miltuff. Trade name of A. Milne & Company (USA) for an air- or oil-hardening shock-resisting tool steel (AISI type S7) containing 0.5% carbon, 0.7% manganese, 0.25% silicon, 3.25% chromium, 1.4% molybdenum, and the balance iron. Used for punches, chisels, rivet sets, dies, shear blades, and slitters.

Milvan. Trade name of A. Milne & Company (USA) for a series of high-speed steels containing 0.6-0.8% carbon, 4-4.25% chromium, 2.25% vanadium, 19% tungsten, and the balance iron. Used for tools and cutters.

Milvex. Trademark of Henkel Corporation (USA) for a series of specialty polyamide (nylon) resins with high clarity, excellent dimensional stability, very high elongation (over 590%), good tensile strength, high water resistance, and good adhesion to smooth metallic and nonmetallic surfaces. Used chiefly in coatings.

Milwaloy. Trade name of Milwaukee Steel Foundry Company (USA) for a series of steels including various corrosion-resistant, heat-resistant, nonmagnetic, soft-magnetic and wear-resistant steels as well as several high-carbon and nitriding steels, and high-strength cast steels.

mimetite. A yellow, orange-yellow, yellowish brown, white or colorless mineral of the apatite group composed of lead chloride arsenate, $Pb_5(AsO_4)_3Cl$. It can also be made synthetically, and may contain phosphorus. Crystal system, hexagonal. Density, 7.28 g/cm^3; refractive index, 2.263; hardness, 3.5-4 Mohs. Occurrence: UK, USA (Arizona, California, Nevada). Used as a minor source of arsenic.

Mimix. Trademark of W. Lorenz Surgical, Inc. (USA) for a synthetic *hydroxyapatite* tetra-tricalcium phosphate material supplied as a sterile, hydrophobic, white powder. The quick-setting working paste is prepared by mixing with a solution of dilute citric acid. *Mimix* exhibits excellent biocompatibility, high compressive strength and radiopacity, and is used in the repair of craniofacial bone defects.

Minacryl Universal. Trade name for a heat-cure denture acrylic.

Minalon. Trade name for cellulose acetate fibers used for wearing apparel and industrial fabrics.

minamiite. A colorless mineral of the alunite group composed of sodium potassium calcium aluminum sulfate hydroxide, $(Na,Ca,K)Al_3(SO_4)_2(OH)_6$. Crystal system, rhombohedral (hexagonal). Density, 2.80-2.85 g/cm^3. Occurrence: Japan.

Minargent. (1) Trade name for an alloy composed of 46-57% copper, 32-40% nickel, 0-28% tungsten and 0.2-0.5% aluminum, and used especially as a replacement for silver in silverware, and as a silver solder.

(2) Trade name for a silver-white alloy composed of about 56% copper, 39% nickel, 3-4% antimony and 1% aluminum, and used for silverware.

(3) Trade name for a dental amalgam alloy.

minasragrite. A blue mineral composed of vanadyl sulfate pentahydrate, $VOSO_4 \cdot 5H_2O$. Crystal system, monoclinic. Density, 2.03 g/cm^3; refractive index, 1.536. Occurrence: Peru.

Mincel. Trade name of Hercules, Inc. (USA) for crosslinked phenol-formaldehydes supplied in the form of foams.

Mindel. Trade name of Amoco Performance Products Inc. (USA) for a series of amorphous polysulfone resins and amorphous-semicrystalline polysulfone resin blends. They possess good high-temperature resistance, a heat-deflection temperature of 149°C (300°F), low warpage, high dielectric and mechanical strengths, excellent surface finishes, and good platability. Various modified and fiber-reinforced grades are also available. Used for electrical and electronic components and equipment (e.g., relays, motor starters, switches and control housings), fiberoptics, and lighting fixtures.

Mineor. Trade name of Darwin & Milner Inc. (USA) for a series of air-hardening tool steels containing 1% carbon, 0-0.2% silicon, 5% chromium, 1% molybdenum, and the balance iron. Used for mandrel, punches, rolls and dies.

mineral. (1) A naturally occurring, inorganic chemical element, compound or mixture usually having a crystalline structure, and characteristic chemical composition and physical properties expressed by a chemical formula. It is commonly obtained by mining of ores. Examples of natural minerals include bauxite, chromite, clay, cryolite, feldspar, fireclay, hematite, ilmenite, limestone, quartz, and sand.

(2) A synthetic material that matches or approximates the structure, composition and physical properties of a naturally occurring chemical element, compound or mixture. Examples of artificial or synthetic minerals include artificial barite and malachite, and synthetic cryolite, diamond, diopside, hydroxyapatite, mica and sapphire. Also known as *synthetic mineral*.

mineral abrasives. A group of nonmetallic abrasives derived from materials occurring in nature. They may be of coarse or fine particle size and include materials, such as dolomite, flint, garnet, novaculite, pumice, quartz and silica sand. Used in grinding and sanding operations and in abrasive blast cleaning of ferrous and nonferrous workpieces.

mineral black. See slate black.

mineral blue. Any of a multiplicity of *iron blue pigments*.

mineral caoutchouc. See elaterite.

mineral cotton. See mineral wool.

mineral dyes. A group of natural dyes produced from minerals and including various earth colors, such as ocher, umber, red iron oxide, Prussian blue and chrome yellow.

mineral fibers. Fibers obtained from raw materials, such as glass, rock, asbestos or slag, by drawing, or by spinning with high-speed wheels. They are used in the manufacture of wall and ceiling tiles, acoustic and thermal wall insulation, fire-resistant sprays, shingles, siding and felts. See also mineral wool; engineered mineral fibers; mineral silicate fibers.

mineral fillers. Relatively inert, finely divided substances of mineral origin, such as barite, kaolin, mica, silica flour, slate flour and talc, incorporated into a polymer, rubber, paint, paper, adhesive, etc., either in relatively large proportions to increase bulk and thus decrease cost (extenders), or in moderate proportions to modify mechanical, thermal, optical, electrical or other properties (fillers).

mineral green. A blue-green paint pigment made from basic copper carbonate $[Cu_2(OH)_2CO_3]$. Also known as *Bremen green; mountain green*.

Mineralite. (1) Trade name of HM Royal Inc. (USA) for a finely ground mica used as a filler in rubber and plastics.

(2) Trademark of Mineralite Limited (UK) for nonmetallic mineral compositions used for finishing and surfacing building walls.

(3) Trademark of Mineral Mining Company, Inc. (USA) for ground ceritic ores.

mineralizer. An inorganic substance, usually a fluxing agent, added in small quantities to a refractory composition to aid the crystal growth of compound formation.

mineral paint. A paint in which mineral substances, such as iron oxide, Prussian blue, ocher and umber, are used as pigments.

mineral pigment. A mineral substance, such as ocher or umber, used in paints or ceramics to give color, opacity and/or body.

mineral pitch. See asphalt.

mineral purple. A purplish-red paint pigment based on red iron oxide (Fe_2O_3).

mineral red. See red iron oxide; rouge.

mineral rouge. See red iron oxide; rouge.

mineral rubber. A dark brown, solid compounding material composed of petroleum asphalt. Blown asphalt, gilsonite and grahamite are varieties. *Mineral rubber* has a density of 1.07 g/cm³ (0.039 lb/in.³), and is used as a tackifier, as an extender or softener in compounding protective coatings, varnishes, rubber and paints, and also in insulating, paving and waterproofing compounds. Abbreviation: MR.

mineral silicate fibers. A class of inorganic fibers based of silicates of natural origin and including *asbestos* and *mineral wool*. See also mineral fibers.

mineral stabilizer. An inorganic substance that when added to a solid or semisolid bituminous material acts as a stabilizer.

mineral-surfaced sheet. A roofing felt coated with asphalt and surfaced with mineral granules on at least one side.

mineral tallow. See hatchettine.

mineral wax. See ozocerite.

mineral white. (1) A white extender pigment made from the mineral barite $(BaSO_4)$, and used in paints, lakes, paper coatings, etc.

(2) A white pigment composed of ground gypsum.

mineral wool. A grayish-brown or grayish-yellow fibrous material similar to *asbestos* consisting of fine filaments. It is made from molten rock or blast-furnace slag by blowing a jet stream of steam or compressed air through it, by drawing, or by spinning with high-speed wheels. The upper temperature limit for mineral (silicate) cotton blanket insulation is 500°C (932°F). Used for packing applications, as acoustic and thermal insulation in walls, in fireproofing, and as a filter medium. Compressed into mats, blankets and boards, it is also used for various acoustic and/or thermal insulating applications. Also known as *mineral cotton; rock wool; silicate cotton; slag wool*.

mineral-wool board. A building product made by compressing mineral wool into a rigid, low-density, water-resistant board, usually 25-100 mm (1-4 in.) thick. Used as thermal insulation in building walls and ceilings.

mineral yellow. See yellow ocher.

Minerva. (1) Trade name of Edgar Allen Balfour Limited (UK) for a series of oil-hardening shock-resisting tool steels containing 0.43-0.53% carbon, 1.7-1.8% chromium, 1.9-2.2% tungsten, 0.2-0.25% vanadium, and the balance iron. Used for tools, chisels, shear blades, extrusion dies, drill bits, and screwdrivers.

(2) Trade name of Mitsubishi Rayon Company Limited (Japan) for rayon fibers and yarns used for textile fabrics.

Minerva 4. Trade name of Minerva Limited (USA) for a medium-gold dental casting alloy.

Minervo Metal. Trademark of Rippenstreckmetall-Gesellschaft mbH (Germany) for black lacquered or galvanized flat-ribbed expanded metal products.

minette. See yellow ocher.

minguzzite. A green mineral composed of potassium iron oxalate trihydrate, $C_6FeK_3O_{12} \cdot 3H_2O$. Crystal system, monoclinic. Density, 2.09 g/cm³; refractive index, 1.555. Occurrence: USA (Pennsylvania, Idaho), Italy, Spain.

Mini-Baroco. Trade name of Permacon (Canada) for lightweight architectural stone with one roughcut surface supplied in square, rectangular and curved shapes in charcoal, beige and gray colors, and used for retaining walls.

Minicel. Trademark of Voltek, Division of Sekisui America Corporation (USA) for a series of fine-celled, crosslinked polyethylene foams processed by vacuum forming techniques. They have good cushioning and insulating properties, low shrinkage, high tensile strength, good fire retardancy, and are used for laminating to cloth or plastic films, e.g., in automotive vinyl/foam composites, for automotive and appliance gaskets, aircraft seating, air conditioner insulation, swimming pool covers, medical devices, toys, double-coated mounting tape, and for packaging applications.

Minifil. Trade name of Avisco (USA) for rayon and nylon filaments and yarns used for textile fabrics including clothing.

Minigold. Trademark of Ivoclar Vivadent AG (Liechtenstein) for a hard dental casting alloy (ADA type III) containing 40.0% gold, 47.0% silver, 7.5% copper, 4.0% palladium, 1% zinc, and less than 1.0% indium and iridium, respectively. It has a yellow color, a density of 12.4 g/cm³ (0.45 lb/in.³), a melting range of 865-925°C (1590-1680°F), an as-cast hardness of 215 Vickers, moderate elongation and excellent biocompatibility and resistance to oral conditions. Used for crowns, onlays, posts and bridges.

Minimax. Trade name of Graham Chemical Company (USA) for a series of corrosion-resistant alloys of silver and mercury used for dental amalgams.

minimized spangle. A coating of very small grain size produced on steel sheet or strip by dipping in a bath of molten low-lead high-grade or special high-grade zinc. When subsequently painted, the characteristic crystalline surface pattern (spangle) can be reduced to a minimum.

Minimum. Trade name of Thyssen Edelstahlwerke AG (Germany) for a series of magnetically soft alloys containing 0.04% carbon, and the balance iron. Used for electrical and magnetic equipment, and vacuum equipment.

mining steels. A group of steels suitable for making solid or detachable drill bits for mining and quarrying applications, and including various grades of plain-carbon and medium- and high-carbon alloy steels with good abrasion resistance and toughness. In the broadest sense the term *mining steels* may also include steels used in the manufacture of mining tools and equipment, such as mining drill machines, tunnel boring machines, mine hoists, coal cutters, plows and loaders, mine cars, sledges, chisels, boring bars, pick hammers, and jackhammers.

minium. A bright orange red mineral composed of lead tetroxide, Pb_3O_4, and containing up to 90.6% lead. Crystal system, tetragonal. It can also be made synthetically. Density, 8.93-9.05 g/cm³; refractive index, 2.224; hardness, 2.5 Mohs. Occurrence: USA (Colorado, Idaho, Utah, Wisconsin). Used extensively in paints for protecting iron and steel against corrosion, in glass, ceramic glazes, pottery and porcelain enamels, as a flux in ceramics, in making cement for pipes, and for packing pipe joints. See also red lead.

Min-K. Trademark of Thermal Ceramics Inc. (USA) for a vibration-resistant microporous insulation products with exception-

ally low thermal conductivity suitable for service temperatures up to 980°C (1800°F). Supplied in the form of flexible sheets and pressure-molded shapes, it is used for aviation and aerospace applications including engine nacelles, thrust reversers, auxiliary power units, on-board ovens, bleed-air and deicing ducts, and data recorders. *Min-K Flexible* is a high-temperature composite consisting of a microporous ceramic core with facings of high-temperature texiles quilted into a flexible blanket. It is supplied in thicknesses from 3 to 13 mm (0.125 to 0.500 in.) with core densities of 128, 160 and 256 kg/m³ (8, 10 and 16 lb/ft³), and used for aircraft insulating applications, e.g., engine nacelles. *Min-K Molded* is a microporous ceramic made into shapes by pressure molding, and used for fire protection applica-tions, e.g., aircraft black boxes.

Minlon. Trademark of E.I. DuPont de Nemours & Company (USA) for a family of strong, stiff, creep-resistant resins based on polyamide 6 and 6,6 (nylon 6 and 6,6) resins and supplied in various grades including glass or carbon fiber-reinforced, glass bead- or mineral-filled, polytetrafluoroethylene- or molybdenum disulfide-lubricated, UV-stabilized, high-impact, supertough and fire-retardant. They possess low warpage and good surface finish and paintability, and are used for automotive applications including wheel covers, engine styling covers and other under-hood parts. *Minlon 6* is a mineral-filled polyamide 6 (nylon 6) and *Minlon 66* a mineral-filled polyamide 6,6 (nylon 6,6).

minnesotaite. A green mineral of the talc group composed of iron silicate hydroxide, $(Fe,Mg)_3Si_4O_{10}(OH)_2$. Crystal system, monoclinic. Density, 3.01 g/cm³. Occurrence: USA (Minnesota).

Minnesota River Stone. Trade name of Vetter Stone Company (USA) for building stone.

Minofor. British trade name for an antifriction metal composed of 66-69% tin, 18-20% antimony, 9-10% zinc, 3-4% copper and up to 1% iron, and used for bearings.

Minovar. Trade name of International Nickel Inc. (USA) for a nickel cast iron containing up to 2.4% total carbon, 1-2% silicon, 34-36% nickel, up to 0.1% chromium, up to 0.5% manganese, and the balance iron. It has a low coefficient of expansion, and is used for electrical plant and equipment, dies, and gages.

Minox. Trade name of Usines de A. Manoir (France) for a series of corrosion-resistant ferritic and martensitic stainless steels used for turbine blades, valves, chemical and oil-refining equipment, cutlery, and surgical instruments.

minrecordite. A white mineral composed of calcium zinc carbonate, $CaZn(CO_3)_2$. Crystal system, rhombohedral (hexagonal). Density, 3.45 g/cm³; refractive index, 1.750. Occurrence: Namibia.

Mint Die Steel. Trade name of Jessop-Saville Limited (USA) for a water-hardening steel used for coining, cold-heading and embossing dies.

Min-U-Sil. Trademark of US Silica Company (USA) for a fine-ground *colloidal silica* containing microscopic crystalline particles, and used for ceramic parts.

Minwax. Trademark of Minwax Division of Sterling Drug Limited (Canada) for an extensive series of wood finishing products including interior oil finishes, polyurethane finishes, finishing waxes, stains, varnishes, refinishers, pencil-form colorants, paint and varnish removers, and pre-stain conditioners.

minyulite. A greenish yellow mineral composed of potassium aluminum phosphate hydroxide tetrahydrate, $KAl_2(PO_4)_2 \cdot$

(OH,F)·4H$_2$O. Crystal system, orthorhombic. Density, 2.46 g/cm^3. Occurrence: France, Australia.

Mipelon. Trademark of Mitsui Chemicals, Inc. (Japan) for an ultrahigh-molecular-weight polyethylene with a molecular weight of 2000000 or more. Supplied as a fine powder with a particle size not exceeding 25-30 μm (984-1181 μin.), it is used as an additive to resins and rubbers to enhance abrasion, chemical and impact resistance, as a food wrapping material, mixed with fillers and pigments as a modifier in polymers, and in the manufacture of heat- and water-resistant products.

Mi-Phos. Trademark of Hubbard-Hall, Inc. (USA) for iron and zinc phosphate conversion coatings for ferrous and nonferrous metals. *Mi-Phos Black* is a black zinc phosphate finish for steel, cast iron, ductile iron and malleable iron.

Miplacol. Trade name of Ato Findley (France) for acrylic adhesives used for bonding tiles, and floor and wall coverings.

Miplagres. Trade name of Ato Findley (France) for mortars used on plaster substrates.

Mipolam. Trademark of Hüls AG (Germany) for a flexible polyvinyl chloride (PVC) usually supplied in the form of tubing, film, plates, panels, profiles and shapes, and used for hoses, floor, wall and ceiling coverings, furniture, and adhesives. It is also supplied in bead form for welding and soldering PVC floor covering joints, and for composites.

Mipoplast. Trademark of Hüls AG (Germany) for a flexible polyvinyl chloride (PVC) usually supplied in film form.

mirabilite. A colorless, white, or yellow mineral composed of sodium sulfate decahydrate, Na$_2$SO$_4$·10H$_2$O. Crystal system, monoclinic. Density, 1.46-1.47 g/cm^3; liquefies at 33°C (91°F), loses water of hydration at 100°C (212°F); hardness, 1.5-2.0 Mohs; refractive index, 1.396. Used in solar heat storage, air conditioning, and in the manufacture of cellulose, water glass and dyestuffs.

Mirac. Trademark of Pemco Corporation (USA) for a hydrated sodium aluminum silicate (Na$_2$O·Al$_2$O$_3$·6SiO$_2$·H$_2$O), used in the manufacture of white, direct-on porcelain enamel for steel.

Miracast. Trade name of Degussa-Ney Dental (USA) for a hard dental casting alloy (ADA type III) containing 41% gold, 9% silver, 4% palladium, and 1% platinum. It has a rich gold color, and good processibility, a hardness of 158 Vickers, and is used for inlays, crowns, and fixed restorations.

Miracle. Trade name of D & M Inc. Alloys (USA) for a series of paste solders including *Miracle Aluminum Weld* for aluminum welds, *Miracle Copper Weld* for copper welds, *Miracle Silver Weld* for silver welds, and *Miracle Plastic Weld* for plastic welds.

Miracle 50. Trade name of GC America Inc. (USA) for a vinyl acrylate copolymer used for denture bases.

Miracle by Grilon. Trademark of Ems-Chemie AG (Switzerland) for nylon 6 staple fibers.

Miracle Mix. Trade name of GC America Inc. (USA) for a self-cure, flexible, fluoride-releasing glass ionomer cement containing 100% fine silver alloy powder. It forms a strong direct bond with tooth structure (dentin and enamel) has a tooth-like thermal expansion coefficient, and is highly resistant to chipping and flaking. Used in restorative dentistry for core build-ups, block-outs, and repairs.

Miracle Weld. Trade name of Miracle Weld Inc. (USA) for an epoxy putty supplied in the form of sticks.

Miracle Wood. Trade name of H.F. Staples & Company, Inc. (USA) for wood fillers.

Miraculoy. Trade name of Sivyer Steel Corporation (USA) for a high-strength steel containing 0.35% carbon, 1.25% manga-

nese, 0.4% silicon, 0.65% chromium, 1.5% nickel, 0.3% molybdenum, and the balance iron. Used for heavy-duty castings.

Miracure. Trademark of Pierce & Stevens Chemical Corporation (USA) for thin, clear, ultraviolet-cure coatings.

Miradapt. Trade name of Johnson & Johnson (USA) for a self-cure dental hybrid composite.

Mirafi. Trademark of Dominion Textile Inc. (Canada) for geotextiles.

Miraflex. Trademark of Owens-Corning Corporation (USA) for glass fibers and glass-fiber products.

Mirafort 4. Trade name for a medium-gold dental casting alloy.

Mirage. (1) Trade name of Mirage Dental Products (USA) for a series of dental products including *Mirage* ceramic whisker-reinforced porcelains for dental veneering applications, *Mirage ABC* adhesive bonding systems for restorative dentistry, *Mirage Bond* dentin bonding agents, *Mirage FLC* dual-cure dental resin cements and *Mirage FLC Vision* dental luting composites.

(2) Trade name of Avery Dennison Corporation (USA) for pressure-sensitive adhesive film supplied in the form of rolls.

Miraglass. Trade name of American Can Company–Glass Division (USA) for a break-resistant specialty glass chemically treated to effect a polycrystalline reorientation of the surface.

Miralene. Trade name for crimped and looped yarns based on *Terylene* polyester fibers.

Miralite. (1) Trade name of Pierce & Stevens Canada Inc. (USA) for lightweight fillers used in synthetic resin systems.

(2) Trade name of Miralite Company (USA) for an aluminum alloy containing 4% nickel, 0.3% silicon, 0.4% iron, 0.04% sodium and 0.05% lead. Used for castings and wires.

(3) Trademark of Saint-Gobain (France) for silvered glass with clear or tinted body. *Miralite Antique* is a decorative silvered glass, *Miralite-Evolution* a silvered glass with high durability, and *Miralite-Evolution Safe* a silvered safety glass.

Miralloy. Trade name of Degussa-Hüls Corporation (USA) for reflective, white or yellow copper-tin-zinc coatings applied by rack or barrel plating.

Mira Metal. Trade name for an acid- and corrosion-resisting copper-base alloy containing 16.3% lead, 6.8% antimony, and a total of 2.2% tin, zinc, iron and nickel. Used for pipes and valves.

Miramid. Trade name of Leuna-Miramid GmbH (Germany) for a series of polyamides (nylons) including nylon 6 (e.g., *Miramid D, H & P*) and nylon 6,6 (*Miramid S*).

Miramint. Trade name for a sintered tungsten carbide with cobalt binder used for hard cutting tools and dies.

Mirasol. Trademark of C.J. Osborn Chemicals, Inc. (USA) for alkyd and epoxy resins used for air-drying and baking finishes.

Mirason. Trade name of Mitsui Chemicals, Inc. (Japan) for phenol-formaldehyde plastics.

Mirathen. Trade name of Buna Sow Leuna Olefinverbund GmbH (Germany) for a series of low-density polyethylenes (*Mirathen AL, BL & CL*).

Mir-Con. Trade name of Detroit Paper Products Corporation (USA) for phenol-formaldehyde plastics and laminates.

Mirenamel. Trade name of SA des Miroiteries de Charleroi (Belgium) for a toughened enamelled glass.

Mirex. Trademark of Mitsubishi Steel Manufacturing Company Limited (Japan) for an extensive range of water-atomized metal and alloy powders supplied in various particle sizes. Available powder compositions include low-alloy, maraging, bearing, stainless, high-speed and alloy-tool steels, superalloys, soft and hard magnet alloys, low-expansion alloys, heat- and/or corrosion-resistant alloys, self-fluxing alloys, and various custom-made specialty alloy compositions. Used for powder-metallurgy products.

Mirhon. Trademark of Miroglio Tessile SpA (Italy) for polyester filaments and fibers.

Mirlon. (1) Trade name of ROTERT Osnabrück (Germany) for a solvent-resistant high-quality nylon fiber-based hand pads, supplied in fine, very fine and ultra-fine grades for grinding or polishing coated and varnished surfaces, and metals, plastics and wood.

(2) Trade name for nylon fibers used for making nonwoven industrial fabrics including coated abrasives and grinding and polishing pads.

(3) Trade name of Mirka Abrasives, Inc. (USA) for coated abrasive products made by coating backings of nylon nonwovens on one or both sides with abrasive grains of synthetic alumina or silicon carbide using a phenolic resin.

Miroflex. Trade name of Miroflex Products Company (USA) for strips of silvered glass applied to cloth.

Mirogard. Trademark of Schott DESAG AG (Germany) for a line of antireflective glasses including *Mirogard Protect*, a laminated glass that is supplied in thicknesses of 4, 6 and 8 mm (0.16, 0.24 and 0.32 in.), provides outstanding protection against UV radiation and splintering, and is used for windows, doors, pictures, etc.

Mirolit. Trade name of Schott Glas AG (Germany) for a glass with broken trellis design.

Mirolite. Trademark of Miron Inc. (Canada) for ready mixed concrete and concrete blocks.

Mirotus. Trade name of SA des Miroiteries de Charleroi (Belgium) for a toughened glass.

Mirox. Trade name of SA des Miroiteries de Charleroi (Belgium) for a silvered glass.

Mirrabend. Trade name of Mirrabend Limited (UK) for a flexible mirror glass.

Mirraloy. Trade name of Associated Steel Corporation (USA) for a turned, ground and polished steel used for journals and shafting.

Mirrex. Trademark of Nixon-Baldwin Chemicals, Inc. (USA) for calendered, unplasticized polyvinyl chloride (UPVC) film and sheeting used for packaging applications.

Mirromac. Trade name of MacDermid Inc. (USA) for a bright cyanide zinc electroplate and plating process.

Mirromold. Trade name of Carpenter Technology Corporation (USA) for a mold steel (AISI type P1) containing 0.1% carbon, 0.2% manganese, 0.1% vanadium, and the balance iron. Used for plastic molds requiring easy cold hobbing and subsequent case hardening.

Mirropane. Trade name of Libbey-Owens-Ford Company (USA) for a glass having a thin transparent coating of chromium alloy applied by a thermal evaporation process. Used for transparent mirrors.

mirror coating. See reflective coating.

Mirror-Flex. Trade name of Opals Mirror-Flex Limited (UK) for a glass product consisting of pieces of silvered glass applied to cloth.

mirror stone. See muscovite.

Mirycal. Trade name of Great Western Steel Company (USA) for a series of tough, shock-resisting tool steels containing 0.45-0.5% carbon, 0.2-0.3% manganese, 0.85-0.95% chromium, 1-1.25% tungsten, 0.15-0.2% molybdenum, and the balance iron.

Used for tools, punches, shear blades, chisels, cold sets, swages, and hot dies.

mischmetal. A mixture of rare-earth metals composed of 50% cerium, 38% lanthanum, 7% neodymium, 4% palladium and 1% other rare earths, and supplied as a natural mixture or prepared by the electrolysis of a fused rare-earth chloride mixture. The name is derived from the German word *Mischmetall* meaning "mixed metal." *Mischmetal* has a density of approximately 6.67 g/cm^3 (0.241 lb/in.^{3}), a melting point of 648°C (1198°F), and is used as a pyrophoric alloy in cigarette lighters, as a getter in vacuum tubes, in magnetic alloys, and in ferrous and nonferrous alloys, e.g., steel, cast iron, and aluminum, copper, magnesium and nickel alloys.

mischmetal-cobalt. A compound containing cobalt and *mischmetal* in a ratio of 5:1 (MMCo$_5$). Available in powder form for the manufacture of permanent magnets, its properties are quite similar to those of *samarium-cobalt* and other rare-earth magnetic materials. See also rare-earth magnets.

miscible polymer blends. A group of binary polymer blends in which the amorphous phase of one component is mutually soluble in the other. The most common types are those involving (i) two amorphous resins, (ii) an amorphous and a semicrystalline resin, and (iii) two semicrystalline resins. *Miscible polymer blends* have a single glass-transition point.

Misco. (1) Trade name of Michigan Steel Casting Company (USA) for a casting alloy composed of 55% iron, 35% nickel and 10% chromium, and used for general-purpose castings.

(2) Trade name of Michigan Steel Casting Company (USA) for a series of austenitic stainless steels typically containing up to 0.07% carbon, 29% nickel, 20% chromium, 3% copper, 2% molybdenum, 1% silicon, and the balance iron. They have good corrosion resistance, good resistance to sulfuric and mixed acids, good strength, machinability and weldability, and are used for chemical equipment, acid tanks, and structural applications.

(3) Trade name of C-E Glass (USA) for a wired glass with diamond mesh.

Miscoloy. Trade name of Michigan Steel Casting Company (USA) for a nickel-chromium steel containing a small addition of molybdenum.

Miscrome. Trade name of Michigan Steel Casting Company (USA) for a series of corrosion-resistant high-chromium steel castings, corrosion- and wear-resistant steel castings, and corrosion- and heat-resistant alloy cast irons.

misenite. A white, fibrous mineral composed of acid potassium sulfate, $K_8H_6(SO_4)_7$. Crystal system, monoclinic or orthorhombic. Density, 2.24-2.61 g/cm^3; melting point, 210°C (410°F).

miserite. A red-brown mineral composed of potassium calcium silicate hydroxide, $K(Ca,Ce)_4Si_5O_{13}(OH)_3$. Crystal system, triclinic. Density, 2.93 g/cm^3; refractive index, 1.5890. Occurrence: Canada (Quebec), USA (Arkansas).

Mishima Steel. Japanese trade name for a sintered permanent magnet material containing 65% iron, 25% nickel and 10% aluminum. It has a high magnetic coercivity, and is used for electrical and magnetic devices.

mispickel. See arsenopyrite.

Misrnylon. Trademark of Misr Rayon Company (Egypt) for nylon 6 fibers and yarns used for textile fabrics including clothing.

missile ceramics. A group of ceramic materials suitable for use in the manufacture of missiles and rockets and including in particular those ceramics which are capable of dissipating high amounts of heat and are thus used for nose cones. The latter materials are characterized by low thermal conductivity, high thermal capacity, excellent high-temperature properties, and high erosion temperatures.

Mississippi clays. A group of highly plastic, high-alumina clays mined in Mississippi, USA, and having fine particle sizes, high surface areas, good firing characteristics and low fired shrinkage. Used for general ceramic applications, in the manufacture of refractories, for fire-resistant non-asbestos ceiling tile, and in the coal-tar industry.

Mistilon. Trademark of Manufacturas del Sur SA (Peru) for nylon 6 filaments and fibers.

Mistlite. Trade name of Asahi Glass Company Limited (Japan) for a patterned glass.

Mistral. Trade name of Central Glass Company Limited (Japan) for a patterned glass.

mistral. A worsted fabric made with twisted warp and weft (filling) yarns to produce a crinkled surface effect.

Mitchalloy. Trade name of Robert Mitchell Company Inc. (USA) for a series of nickel cast irons. *Mitchalloy A* is wear-resistant, contains 2.9-3.1% total carbon, 2-2.5% nickel, 0.3-0.6% chromium, and the balance iron, and is used for brake drums. *Mitchalloy B* has 2.8-3.5% total carbon, 4-5% nickel, 1.5-2% chromium, and the balance iron. It is abrasion- and wear-resistant and finds applications in mixers, grinders and pulverizers. *Mitchalloy C* is corrosion-resistant, contains 2.7-3% total carbon, 12-15% nickel, 5-7% copper, 1.5-4% chromium, and the balance iron. It is used for pumps, valves, pipes, etc. *Mitchalloy D* is composed of 3.3% total carbon, 2.6% silicon, 1.5% nickel, 0.8% chromium, and the balance iron, and is used for grate bars.

Mi-Tech. Trademark of Mi-Tech Metals Inc. (USA) for tungsten alloys supplied in standard and special shapes.

Mitex. (1) Trademark of Millipore Corporation (USA) for polytetrafluoroethylene polymers with good chemical resistance, used for laboratory ware, and membranes for laboratory filtration assemblies.

(2) Trademark of Mitex Glasfiber AB (Sweden) for glassfiber fabrics used in the textile industry.

Mitia. Trade name of Firth Brown Limited (UK) for a series of tungsten carbides with cobalt binder. Depending on the particular composition, they are used for cutting cast iron, chilled cast iron, and mild and low-alloy steels.

Mitifine. Trade name of H. Kramer & Company (USA) for a lead-free tin babbitt supplied as a free-flowing powder. The cast product has high toughness, but exhibits slight shrinkage. Used for heavy bearings for marine engines, crossheads, and for heavy machine babbitts.

Mi-Tique. Trade name of Hubbard-Hall Inc. (USA) for antique finishes applied to decorative hardware, and including brownish finishes produced on copper, brass, bronze and muntz metal, and black finishes applied to tin, pewter, nickel, lead and muntz metal. *Mi-Tique Patina* are verdigris-green antiquing finishes for copper and brass.

Mitis Iron. British trade name for an aluminum-deoxidized wrought iron used for castings, pipes, and fittings.

mitridatite. A deep-red mineral of the arseniosiderite group composed of calcium iron phosphate hydroxide trihydrate, $Ca_3Fe_4(PO_4)_4(OH)_6 \cdot 3H_2O$. Crystal system, monoclinic. Density, 3.24 g/cm^3; refractive index, 1.85. Occurrence: USA (South Dakota), Russia.

mitscherlichite. A brilliant greenish blue mineral composed of potassium copper chloride dihydrate, $K_2CuCl_4 \cdot 2H_2O$. It can also be made synthetically by slow, room-temperature evaporation of an aqueous solution with a 2:1 molar ratio of potassium chloride (KCl) and copper chloride (CuCl$_2$). Crystal system, tetragonal. Density, 2.41 g/cm^3; refractive index, 1.638. Occurrence: Italy.

Mitsubishi. Trade name of Mitsubishi Chemical Company (Japan) for polypropylene and vinal (vinyl alcohol) fibers used for textile fabrics.

Mitsubishi Pylen. Trademark of Mitsubishi Rayon Company (Japan) for polypropylene filaments and fibers.

Mitsui B-Resin. Trademark of Mitsui Chemicals, Inc. (Japan) for polyesters supplied in amorphous grades for blending with polyethylene terephthalate (PET) or for use as multilayers with PET, and in several grades that are crystallized and preblended with PET. Used as barrier resins.

Mitsui Hi-Wax. Trademark of Mitsui Chemicals, Inc. (Japan) for a low-molecular-weight polyethylene wax used as a wax and coating modifier, and as a filler.

mix. See mixture (1), (3), (5), (6).

Mixad. Trademark of Foseco Minsep NV (Netherlands) for specialty additives used to improve the flowability, pattern impression, shatter index and bonding properties of clay-bonded sands and prevent casting defects due to sand expansion.

mixed fabrics. See mixture fabrics.

mixed-in-place road mix. A mixture of mineral aggregates (e.g., open or dense-graded aggregate, sand, etc.), cutback asphalt and bituminous emulsion or tar, produced at the construction site by means of special mixing equipment. Used in road construction for bituminous base or surface courses.

mixed oxide ceramics. A group of ceramics composed of mixtures of two or more oxides. Examples include forsterite ($2MgO \cdot SiO_2$), mullite ($3Al_2O_3 \cdot 2SiO_2$), spinel ($MgO \cdot Al_2O_3$) and cordierite ($2MgO \cdot 2Al_2O_3 \cdot 5SiO_2$).

mixed oxide fuel. A mixture of uranium oxide (UO_2) and thorium oxide (ThO_2), used in the manufacture of certain nuclear fuel elements. Abbreviation: MOX fuel.

mixed polysaccharides. A group of *polysaccharides* including gums, mucilages and hemicelluloses that yield various sugar and/or nonsugar derivatives on hydrolysis.

mixed powder. A powder composed of a uniform mixture of two or more powders with different chemical composition and/or particle size or shape. For example, a bronze powder made by mixing copper and tin powders, or an iron powder consisting of spheroidal and nodular particles.

mixite. (1) A bluish green mineral composed of copper bismuth arsenate hydroxide trihydrate, $BiCu_6(AsO_4)_3(OH)_6 \cdot 3H_2O$. Crystal system, hexagonal. Density, 3.83 g/cm^3; refractive index, 1.749; hardness, 3-4 Mohs. Occurrence: Germany.

(2) A bluish green mineral composed of copper arsenate hydrate, $Cu_3(AsO_4)_2 \cdot 6H_2O$. Crystal system, hexagonal. Density, 3.73 g/cm^3; refractive index, 1.7265. Occurrence: Germany.

mixture. (1) In chemistry, a substance consisting of two or more components that may or may not be uniformly dispersed, but keep their individual chemical properties, and can be separated by mechanical or other nonchemical means. Examples include air, cement, glass, latex, paint, petroleum, marble, seawater and wood. Also known as *mix*.

(2) In metallurgy, an *alloy* composed of two or more metals.

(3) In powder metallurgy, two or more powdered substances that have been thoroughly blended. Also known as *mix*.

(4) In polymer engineering, a solid, semi-solid or liquid material that contains one or more organic polymeric substances.

(5) In the rubber industry, natural or synthetic rubber compounded with other ingredients. Also known as *mix*.

(6) In building construction, a blend or batch of concrete, mortar or plaster ready for processing. Also known as *mix*.

mixture fabrics. Textile fabrics woven, knitted or otherwise made from mixture yarns. Also known as *mixed fabrics*.

mixture yarn. A yarn spun from a blend of two or more chemically and/or physically different fibers (e.g., different composition, color, luster, etc.).

M"(M'M")X$_4$ compound. An inverse spinel-type ceramic compound of metallic element M', metallic element M" and nonmetallic element X in which the ratio of the cations of M" to the cations of M' and M" to the anions of X is 1:1:4. Examples include FeMgFeO$_4$, FeFe$_2$O$_4$ and FeNiFeO$_4$.

M'M"X$_3$ compound. A perovskite-type ceramic compound of metallic element M', metallic element M" and nonmetallic element X in which the ratio of the cations of M' to the cations of M" to the anions of X is 1:1:3. Examples include BaTiO$_3$ and CaTiO$_3$.

M'M"$_2$X$_4$ compound. A spinel-type ceramic compound of metallic element M', metallic element M" and nonmetallic element X in which the ratio of the cations of M' to the cations of M" to the anions of X is 1:2:4. Examples include MgAl$_2$O$_4$, MgFe$_2$O$_4$ and MnFe$_2$O$_4$.

MnS+. Trade name of Pyron Corporation (USA) for manganese sulfide (MnS) powder with low moisture absorption and oxidation, used in the manufacture of powder-metallurgy parts.

moa wood. A very hard wood imported from New Zealand.

Mobil. Trade name of Mobil Chemical Company (USA) for a series of synthetic resins including *Mobil HIPS* high-impact polystyrenes supplied in unmodified, fire-retardant and UV-stabilized grades, *Mobil PE* unmodified and UV-stabilized low-density polyethylenes, and *Mobil PS* polystyrenes supplied in unmodified, glass-reinforced, medium-impact, silicone-lubricated, structural foam and UV-stabilized grades.

Mobilite. Trade name of Certain-Teed Saint-Gobain Insulation Company (USA) for glass wool used for acoustic and thermal insulation applications.

Mobilon. Trademark of Nisshinbo Industries, Inc. (Japan) for spandex fibers and yarns, used for elastic textile fabrics.

Mocar. Trade name of Hewitt Metals Corporation (USA) for a babbitt composed of 75% lead, 15% antimony and 10% tin, and used for camshaft and connecting-rod bearings.

Mocarb. Trade name of CCS Braeburn Alloy Steel (USA) for a molybdenum-type high-speed steel (AISI type M2) containing 1% carbon, 5% molybdenum, 6.5% tungsten, 4.2% chromium, 1.9% vanadium, and the balance iron. It has excellent wear resistance, good deep-hardening properties, good toughness, and high hot hardness. Used for milling cutters, end mills, lathe and planer tools, reamers, and a variety of other tools.

Mocasco. Trade name of Motor Castings Company (USA) for a series of cast irons including several compacted-graphite irons, and gray, nodular, nickel and specialty irons.

mocha. A soft, fine-grained *suede leather* made from the skin of sheep, and used for gloves and garments. Also known as *mocha leather*.

Mocha Blend. Trade name of Aardvark Clay & Supplies (USA) for a low-firing, brownish earthenware clay resembling mocca.

mocha leather. See mocha.

mocha suede. A durable, washable *suede leather* with a fine finish made from chrome-tanned skins with tight fiber structure, obtained from an Arabian breed of sheep known as "blackhead mocha."

Mo-Chip. Trade name of Teledyne Firth-Sterling (USA) for an oil-hardening high-speed steel containing 0.7% carbon, 8% molybdenum, 2.5% cobalt, 0.5% chromium, 1% vanadium, and the balance iron. Used for cutters, lathe and planer tools, and other high-speed tools.

mock crepes. Textile fabrics with crinkled surfaces obtained by special designs, weaves finishes or other means in contrast to the twisted yarns used for true *crepe fabrics*.

mockernut hickory. The tough, hard, heavy wood of the hickory tree *Carya tomentosa*. The tree derives its name from the ball- or egg-shaped nuts. The heartwood is reddish and the sapwood is thick and white. *Mockernut hickory* has high strength and shock resistance. Source: Southern Canada; eastern and central United States, especially Arkansas, Louisiana, Mississippi, Tennessee and Kentucky. Used for wheel spokes and rims, tool handles, and machine parts. Also known as *ballnut hickory; hognut hickory; white hickory*.

mock gold. A corrosion-resistant alloy resembling gold in appearance, but composed of 67-80% copper, 20-29% platinum and 0-4% zinc, and used for ornaments.

mock leno. A woven fabric with an open-mesh pattern imitating a leno weave but, unlike the latter, made with unpaired warp yarns. See also leno-weave fabrics.

mock platinum. An alloy resembling platinum in appearance, but composed of 61.5% brass and 38.5% zinc, and used for ornaments.

mock silver. (1) A corrosion-resistant, silvery aluminum alloy containing 10% tin, 5.5% copper and 0.5% phosphorus, and used for ornaments, instruments and fittings.

(2) A corrosion-resistant, silvery aluminum alloy containing about 5% silver and 5% copper. Used for ornaments and jewelry.

(3) An inverse brass composed of 55% zinc and 45% copper.

(4) A white alloy similar to pewter and Britannia metal and composed of varying amounts of tin, copper, nickel and zinc.

mock-twist yarn. A fancy *single yarn* having a mottled effect and imitating a *plied yarn* consisting of two differently colored single yarns, but usually made by spinning two differently colored rovings.

mock vermillion. Basic lead chromate (PbCrO$_4$·PbO) used as a red pigment.

Moclad. Trademark of Guardsman Product Limited (Canada) for standard and pigmented polyurethane resin compositions. They adhere well to metal, wood, fabrics and synthetic materials, and are used in protective exterior coatings, high-strength bonding adhesives, and as injection molding materials.

Moco. (1) Trade name of CCS Braeburn Alloy Steel (USA) for a molybdenum-type high-speed steel (AISI type M36) containing 0.85% carbon, 0.25% manganese, 0.3% silicon, 4.1% chromium, 8.25% cobalt, 6% tungsten, 5% molybdenum, 2% vanadium, and the balance iron. It has a great depth of hardening, high hot hardness, excellent wear resistance, and is used for heavy-duty lathe and planer tools, milling cutters, cutoff tools, boring tools, drills, and tool bits.

(2) Trade name of Mosher Company, Inc. (USA) for buffing, polishing and tumbling compounds.

moctezumite. A bright to dark orange mineral composed of lead uranyl tellurite, $Pb(UO_2)(TeO_3)_2$. Crystal system, monoclinic. Density, 5.73 g/cm^3; refractive index, above 2.11. Occurrence: Mexico.

Mocut. Trade name of CCS Braeburn Alloy Steel (USA) for a molybdenum-type high-speed steel (AISI type M1) containing 0.6-0.85% carbon, 0.3% manganese, 0.4% silicon, 3.5-4% chromium, 7.75-9% molybdenum, 1.3-1.8% tungsten, 0.9-1.3% vanadium, and the balance iron. It has a great depth of hardening, high hot hardness, excellent wear resistance, and is used for twist drills, lathe and planer tools, cutters, reamers, and hobs.

modacrylic fiber. A generic term for a synthetic fiber composed of 25-85 wt% acrylonitrile and 15-75 wt% of a second comonomer, such as vinyl chloride. It is soft and resilient, has an essentially amorphous structure, low water absorption, fair tenacity, good abrasion resistance, good resistance to acids and alkalies, good resistance to combustion (self-extinguishing), and no distinct melt temperatures, but softens at 190-240°C (375-465°F). Used for deep pile and fleece fabrics, coats, trimmings and linings, home furnishings, industrial fabrics, carpets, nonwovens, industrial filters, paint rollers, and in polymer alloys with other fibers. See also modified acrylics (2).

modacrylics. See modified acrylics (2).

Modal. Trade name of Lenzing Fibers Corporation (USA) for modal (regenerated cellulose) fibers used for textile fabrics.

modal. See modal fiber.

modal fiber. A term used by the ISO (International Organization for Standardization) for a regenerated cellulose fiber with high tenacity and wet modulus and a fairly low degree of swelling in sodium hydroxide solution. This fiber is designated as "polynosic fiber" by other standardizing agencies. Also known as *modal.*

Modal Sun. Trade name of Lenzing Fibers Corporation (USA) for modal fibers.

Modanyl. Trade name for nylon fibers and yarns.

Modar. Trademark of Ashland Chemical Inc. (USA) for a series of thermosetting resins and plastics, namely, acrylic resins available in several reinforced and modified grades. Processed by resin transfer molding (RTM), they are used for exterior and interior automotive parts, and various industrial applications. Also known as *Modar RTM.*

modderite. A synthetic mineral composed of cobalt arsenide, CoAs. Crystal system, orthorhombic. Density, 8.28 g/cm^3; stable below 950°C (1740°F).

Model Silicone. Trade name for a silicone die elastomer used in dentistry.

moderate sulfate-resistant cement. A type of Portland cement containing 44% tricalcium silicate (C_3S), 31% dicalcium silicate (C_2S), 5% tricalcium aluminate (C_3A), 13% tetracalcium aluminoferrate (C_4AF), 7% primarily simple oxides (magnesia, lime, alkali oxides, etc.) and calcium sulfate. It has lower heat of hydration, but higher sulfate resistance than ordinary *Portland cement,* and is used in the manufacture of concrete. Also known as *type II cement.*

moderator. A material used in nuclear reactors to produce slow neutrons from fast neutrons, and thus increase the probability of fission. It is made of light elements to make maximum energy from a neutron on collision. Important physical characteristics of moderators include low atomic weight, high atomic density, good stability from heat and radiation, and good thermal transfer properties. Good moderators include hydrogen, graphite (carbon), deuterium, paraffin, beryllium and ordinary water.

Modern. Trade name of Becker Stahlwerk AG (Germany) for a high-speed steel containing 0.7% carbon, 4% chromium, 18% tungsten, 1% vanadium, and the balance iron. Used for tools and cutters.

Modern Tenacin. Trade name of Dentsply/LD Caulk (USA) for zinc phosphate dental cements.

modified acetate fibers. Acetate fibers that have been streched and subsequently treated with a suitable alkaline solution. See also acetate fibers.

modified acrylics. (1) Acrylic-based thermoplastic polymers modified with suitable activators or catalysts to promote rapid curing at room temperature, and used as adhesives for structural and nonstructural applications, and as coatings on metal substrates used indoors and in marine atmospheres. See also acrylics.

(2) A group of polymers made by the copolymerization of 25-85% acrylonitrile and 75-15% of a second comonomer, usually vinyl chloride. Used chiefly for *modacrylic fiber* production by melt and solution spinning. Also known as *modacrylics.*

modified alkyds. High-molecular-weight alkyd resins either modified with drying oils, such as linseed, soybean or tung, and known as "oil-modified alkyds," or with synthetic resins, such as acrylics, silicones, urethanes or vinyls, and known as "resin-modified alkyds." Coatings produced from modified alkyds have increased durability, hardness, chemical resistance and weatherability, and are used for house paints and as baking enamels for appliances, and automotive and marine applications.

modified aluminum-silicon alloys. Hypoeutectic or hypereutectic aluminum-silicon alloys with refined silicon phase structures induced by the controlled addition of small quantities of modifiers, such as calcium, strontium, phosphorus or sodium to the melt. These alloys have improved mechanical properties.

modified asphalt. (1) Asphalt modified with suitable synthetic resins for use as waterproof coatings.

(2) Asphalt whose adhesion, penetration and tackiness have been greatly increased by blending with rosin. Used as impregnating material for flooring and roofing felts, and for laminating paper.

modified cellulose. A *cellulose* derivative formed by replacing the hydroxyl (OH) groups along the carbon chain with suitable radicals, such as alkyl, carboxyl, acetate, nitrate or xanthate groups. Examples include carboxymethylcellulose, cellulose acetate and nitrocellulose.

modified epoxy adhesives. A group of thermosetting epoxy adhesives that, depending on their particular application, are modified with various additives, such as accelerators, fillers, viscosity modifiers, flow-control and toughening agents, flexibilizer, pigments, etc. They are usually available in room-temperature or heat-curing types. Depending on the curing agents, they can be used at low or elevated temperatures (up to about 250°C or 480°F), and have relatively low peel strengths, but good to excellent lap-shear strengths. Used for structural applications.

modified gunmetal. A strong, corrosion-resistant golden-colored casting alloy containing 85-90% copper, 5-10% tin, 2-5% zinc, 0.5-5% lead and up to 0.5% nickel. It has a density of 8.7-8.8 g/cm^3 (0.31-0.32 $lb/in.^3$), good castability, machinability, bearing properties and wear resistance, and is used for valves, cocks, pump parts, bearings, gears, and hydraulic castings. It was formerly also used for cannons.

Modified Hilo. Trade name of British Driver Harris Company Limited (UK) for an electrical resistance alloy containing 80% nickel and 20% cobalt. Used for electron-tube filaments.

Modified Monel. Trade name of Manning, Maxwell & Moore Company (USA) for a group of corrosion-resistant alloys containing 60-65% nickel, 24-27% copper, 9-11% tin, 1-3% iron, and traces of manganese and silicon. Used for valves operating in superheated steam. Also known as *modified Monel metal*. See also Monel.

modified Monel metal. See modified Monel.

modified ounce metal. An alloy of 84% copper, 5% tin, 5% zinc, 5% lead and 1% nickel. It has good castability and machinability, and takes a high polish. Used for pressure-tight castings, fittings, valves, and hardware. See also ounce metal.

modified polyphenylene oxide/polyamide alloys. See MPPO/PA alloys.

modified polyphenylene oxides. Engineering thermoplastics based on polyphenylene oxide (PPO) modified with other synthetic resins to improve one or more properties, e.g., the incorporation of high-impact polystyrene (HIPS) improves the extrusion properties and the addition of polyamide (nylon) results in enhanced chemical resistance and improved dimensional stability at elevated temperatures. They have a density of approximately 1.06 g/cm^3 (0.038 lb/in.^{3}), and a glass-transition temperature of about 210°C (410°F). Abbreviation: MPPO. See also polyphenylene oxides, MPPO/PA alloys.

modified rayon. Rayon filaments or fibers made from regenerated cellulose to which other fiber-forming materials (e.g., casein) have been added during manufacture to improve certain properties, such as dry and/or wet strength, and dyeability. See also rayon.

modified resin. A synthetic resin in which one or more properties have been modified by the addition of drying oils (e.g., linseed, soybean or tung oil), natural resins, or gums.

modified rosin. Rosin whose structure has been changed by treating with catalysts or heat. The change may be brought about by hydrogenation, dehydrogenation, heat treatment or polymerization. See also rosin.

modified silicones. High-temperature silicone coating resins that have been modified with other synthetic resins to lower the baking temperature. Used for roasters and stove parts. See also silicones.

modifier. (1) A chemically inert agent added to an adhesive or resin to control flow, increase tack or impart flame resistance or color. Also known as *modifying agent*.

(2) Oxides, such as those of barium, calcium, lead, magnesium, potassium, sodium and zinc, that are added to glass to break up the silica network, lower the melting point, facilitate forming at a given temperature, and increase its chemical reactivity. Also known as *modifying agent*.

(3) A small addition of elements, such as calcium, strontium, phosphorus or sodium, made to molten aluminum-silicon alloys to effect refinement of the silicon phase in the solid state for improved mechanical properties of hypoeutectic and hypereutectic alloys.

modifying agent. See modifier.

Modiglass. Trade name of Reichhold Chemicals, Inc., Modiglass Fibers Division (USA) for a line of glass fibers and glass-fiber products.

Modipon. Trademark of Modipon Limited (India) for nylon 6 filaments and fibers.

Modolit. Trade name of former Gerresheimer Glas AG (Germany) for a patterned glass with scattered, large and small circles.

modular brick. A brick of standard size that will fit into a 4-inch (101-mm) modular masonry unit including the mortar joint.

modular masonry unit. A masonry unit of nominal dimensions based on a 4-inch (101-mm) cubical module.

Modulay. Trade name of J.F. Jelenko & Company (USA) for a medium-hard dental casting alloy (ADA type II) containing 77% gold, 14% silver and 1% palladium. It provides excellent castability and workability, high tensile strength and ductility, and is used for moderate-stress inlays, medium-hard crowns, and anterior gold facings.

Modulor. Trade name for a medium-gold dental casting alloy.

Modulvar. Trade name of Creusot-Loire (France) for a precipitation-hardening alloy containing 64% iron and 36% nickel. It has a density of 8.0 g/cm^3 (0.29 lb/in.^{3}), a high positive thermoelastic coefficient, a low coefficient of thermal expansion, good tensile strength, a melting point of 1425°C (2597°F), and is used for instruments, chronometers, etc.

Mogul. Trade name of Jessop Steel Company (USA) for a molybdenum-type high-speed steel (AISI type M1) containing 0.78% carbon, 4% chromium, 1.5% tungsten, 1.15% vanadium, 8.7% molybdenum, and the balance iron. It has a great depth of hardening, high hot hardness, excellent wear resistance, and is used for lathe and planer tools, drills, reamers, and a variety of other high-speed tools.

mohair. A yarn spun from the long, white, silky hair of the Angora goat *(Capra angorensis)*. It is strong, lustrous, has good dyeing properties and may be blended with other natural fibers, such as cotton or wool. Used for knitting, and for the manufacture of clothing and upholstery.

mohair wool. A thick, spongy, hardwearing open fabric made of mohair, or mohair and cotton or wool, used for warm, lightweight fabrics, e.g., coats, jackets, shawls, stoles, etc.

mohavite. See tincalconite.

Mohawk. Trade name of AL Tech Specialty Steel Corporation (USA) for a tungsten-type hot-work tool steel (AISI type H24) containing 0.25% carbon, 4% chromium, 15% tungsten, and the balance iron. It has a great depth of hardening, high hot hardness, good toughness, and is used for hot-forming, extrusion, trimming and swaging dies. *Mohawk Hot Die* is a tungsten-type hot-work tool steel (AISI type H24) containing 0.45% carbon, 14% tungsten, 3.5% chromium, 0.6% vanadium, and the balance iron. It has high compression strength, excellent abrasion resistance, good deep-hardening properties, high hot hardness, and is used for hot punches, shear blades, forming rolls and blanking and drawing dies, and extrusion and forging dies.

Mohdi Hot Die Steel. Trade name of Eagle & Globe Steel Limited (Australia) for a hot-work tool steel containing 0.4% carbon, 4.4% chromium, 2% tungsten, 5.5% molybdenum, 1.65% vanadium, 1.2% cobalt, and the balance iron. It has high strength and wear resistance, good hot hardness, and is used for die-casting dies for brass, extrusion tools for copper and copper-base alloys, and hot-forging dies and punches.

Moheco. Trade name of Motor Castings Company (USA) for a wear-resistant cast iron containing 2.85-3.05% total carbon, 1.75-2.1% silicon, 1.75% nickel, 0.9% chromium, 0.6-0.8% manganese, and the balance iron. Used for dies and various tools.

Mohican. Trade name of Atlas Specialty Steels (Canada) for hot-work and high-speed tool steels. *Mohican-6* is a molybdenum-type hot-work steel (AISI type H41) containing 0.62% carbon,

0.25% manganese, 0.30% silicon, 8.7% molybdenum, 3.75% chromium, 1.7% tungsten, 1% vanadium, and the balance iron. It has a great depth of hardening, high hot hardness, good shock and wear resistance, and is used for hot forming, trimming, upsetting and swaging dies, punches, rivet sets, and extrusion dies for brass, bronze and steel. *Mohican-8* is a molybdenum-type high-speed steel (AISI type M1) containing 0.8% carbon, 0.25% manganese, 0.30% silicon, 9% molybdenum, 4% chromium, 1.5% tungsten, 1.2% vanadium, and the balance iron. It has a great depth of hardening, high hot hardness, excellent wear resistance, and is used for lathe and planer tools, cutters, drills, taps, reamers, saws, and a variety of other tools.

mohite. A gray mineral with greenish tint composed of copper tin sulfide, Cu_2SnS_3. Crystal system, triclinic. Density, 4.75. Occurrence: Uzbekistan.

mohrite. A very light bluish green synthetic mineral of the picromerite group composed of ammonium iron sulfate hexahydrate, $(NH_4)_2Fe(SO_4)_2 \cdot 6H_2O$. It can be made by slow evaporation from an aqueous solution of iron sulfate ($FeSO_4$) and ammonium sulfate $[(NH_4)_2SO_2]$. Crystal system, monoclinic. Density, 1.86 g/cm³; refractive index, 1.489.

Mohr's salt. See ferrous ammonium sulfate.

Moil Point. (1) Trade name of Midvale-Heppenstall Company (USA) for an oil-hardening shock-resisting tool steel containing 0.75% carbon, 0.25% molybdenum, and the balance iron.

(2) Trade name of Colt Industries (UK) for a tough, oil-hardening steel containing 0.8% carbon, 0.5% manganese, 0.35% silicon, and the balance iron. Used for moil points.

moiré fabrics. Corded or ribbed woven fabrics having surface patterns resembling water ripples, produced by a process that may involve chemicals, heat or steam, and heavy pressure by engraved rollers. Used especially for drapery and upholstery. Also known as *watered fabrics.*

moiré taffeta. A *taffeta* fabric, usually made of silk or synthetic fibers, having a temporary or permanent surface pattern resembling water ripples, produced by embossing with engraved rollers.

moirette. A fabric made of polished cotton in a plain weave and having irregular wavy lines either in warp or weft direction.

moissanite. A bluish green mineral of the wurtzite group composed of silicon carbide, α-SiC. It can be made synthetically from a mixture of trichloromethylsilane (CH_3SiCl_3) and hydrogen (H_2) under prescribed conditions. Crystal system, hexagonal. Density, 3.22 g/cm³; refractive index, 2.648. Occurrence: USA (Arizona).

Moistop. Trade name of BSK, Division of DS Smith (UK) Limited for a strong barrier foil used in the manufacture of heat-seal bags or pouches for packaging liquid and solid products.

moisture barrier. A material or coating applied to a ceiling, floor, wall, roof, etc., to retard the passage of moisture. Examples include vulcanized rubber, phenol-formaldehyde, polyvinyl chloride and polyethylene sheeting, asphalt-coated paper, aluminum foil, and various specially formulated paints.

moisture-resistant paper. A paper that has certain substances added during manufacture to increase its resistance to moisture caused by unusual ambient or climatic conditions. Used for census forms, meter-reading forms, etc.

Moldablaster. Trade name of Heraeus-Kulzer Inc. (USA) for a dental gypsum.

moldable refractories. See plastic refractories.

Moldaloy. Trade name of Trethaway Associates (USA) for an alloy of bismuth, tin, lead and antimony. It has a melting point of 221°C (430°F), and is used for forming and forging dies, and molds.

Moldano. Trade name of Heraeus-Kulzer GmbH (Germany) for a dental gypsum.

Moldarta. Trade name for a phenolic resin available in various grades and with various reinforcing fillers.

Moldasynt. Trade name of Heraeus-Kulzer Inc. (USA) for a beige-colored synthetic dental gypsum.

Moldatherm. Trademark of Koyo Thermo System Company Limited (Japan) for a high-density ceramic fiber insulation produced by vacuum forming. The high-temperature board insulation is supplied in modular form, and in formulations for temperatures ranging from 650 to 1650°C (1200 to 3000°F). Standard shapes are available in sizes of 254 × 254 mm (10 × 10 in.), 305 × 330 mm (12 × 13 in.) and 305 × 660 mm (12 × 26 in.), and thicknesses up to 76 mm (3 in.). Used for original or retrofit applications on heat processing equipment. *Moldatherm II* is a lightweight ceramic fiber insulation without organic binder used in the manufacture of high-density insulation board products.

Mold Base. Trade name of Atlas Specialty Steels (Canada) for a mold steel used for molding dies for plastics manufacture.

mold brick. Insulating brick, usually of the fireclay type, made to fit the top section of an ingot mold. See also fireclay brick.

mold coating. (1) A fine powder or wash applied to the face of a sand mold to avoid metal penetration and enhance metal finish. It is compounded from refractory fillers (e.g., chromite, silica, or zircon flour), a vehicle (e.g., alcohol or water), and a suspension agent (e.g., bentonite, or sodium alginate). Also known *mold dressing; mold facing.* See also mold wash.

(2) A coating applied to die-casting dies and permanent molds to prevent surface defects on castings.

Moldcote. Trade name of Foseco Minsep NV (Netherlands) for core and mold dressings in paste or solution form.

mold dressing. See mold coating.

molded brick. A brick shaped in molds, usually by hand, from a suitable clay. It may or may not be subsequently pressed and/or fired.

molded composite materials. (1) A general term used for composite materials processed by molding.

(2) A specific term referring to (i) bulk molding compounds processed by compression, transfer or injection molding, (ii) sheet molding compounds, (iii) injection molding compounds or (iv) resin-transfer molding materials.

molded fabrics. Flat or pile fabrics with permanent shape retention and good heat, moisture and wear resistance achieved by a molding process that involves the application of heat and pressure to change the molecular structure of the thermoplastic fibers of which they are composed. Used for upholstery, slip covers, swimwear, gloves, etc.

molded glass. A term referring to glass shaped in a mold in contrast to glass produced by other methods, such as blowing, casting, drawing or rolling.

molded graphite. A high-purity, crystalline graphite made by mixing calcined petroleum coke with a suitable binder, such as coal-tar pitch, pressing, and heating at temperatures up to 2500°C (4530°F) in an electric furnace. It has excellent resistance to chemicals, good dimensional stability over a wide range of temperatures, an upper service temperature of approximately 400°C (750°F) in air, and up to 650°C (1200°F) in nonoxidizing atmospheres, good wear resistance, and low thermal expansion. Used for sliding bearings, bushings, seals, and packing rings.

molded plastics. See moldings.

mold facing. See mold coating.

mold flux. A *flux*, usually composed of about 25-45% calcium oxide (CaO), 20-50% silicon dioxide (SiO_2), 0-10% aluminum oxide (Al_2O_3), 1-20% sodium oxide (Na_2O), 4-10% fluorine (F), 1-25% carbon (C), 0-10% barium oxide (BaO), 0-10% boron oxide (B_2O_3), 0-10% magnesium oxide (MgO), 0-10% manganese oxide (MnO), 0-5% titanium dioxide (TiO_2), 0-5% potassium oxide (K_2O), 0-5% ferric oxide (Fe_2O_3) and 0-4% lithium oxide (Li_2O). Used in the form of powders, spherical or extruded granules, or fritted or prefused substances in the (continuous) casting of iron and steel.

Moldie. Trade name of Boyd-Wagner Company (USA) for a steam-resistant mold steel containing 0.9% carbon, 2.25% manganese, 0.1% vanadium, and the balance iron. Used for molds for rubber, and phenol-formaldehyde and melamine plastics.

molding compound. A mixture of synthetic resins, usually in the form of pellets or granules, with additives, such as colorants, fillers, flame retardants, plasticizers, pigments, reinforcements, and/or stabilizers. *Molding compounds* are sold ready for use in extrusion and molding operations. Abbreviation: MC. Also known as *molding powder*.

molding loam. A *silica sand* of coarse particle size that may or may not contain small amounts of clay or feldspar. It is often mixed with fireclay and used in the manufacture of cupola and ladle linings.

molding powder. See molding compound.

moldings. Thermoplastic or thermosetting products of varying shape and size formed by forcing pelletized or granular plastic starting materials at elevated temperatures and by pressure to become viscous and conform to the shapes of mold cavities. *Moldings* can be formed by various processing techniques including blow molding, compression molding, injection molding and transfer molding.

molding sand. Sand suitable for use in the foundry for making molds. Depending on the nature of the binder, *molding sands* can be classified into (i) natural silica-based sands (so-called "bank sands") with up to 20% clay-base contaminants; (ii) semi-synthetic sands having a *bentonite* binder added; (iii) synthetic sands (so-called "lake sands") with a relatively pure *silica, zircon, olivine* or *chromite* base in a binder, such as bentonite, *montmorillonite, kaolinite* or *illite*; and (iv) loam sands that are high in clay (up to 50%) and dry hard. A good *molding sand* should have fine grain size, high dry and green strengths and good green permeability.

Moldinoc. Trade name of Pechiney Electrométallurgie (France) for a series of cast iron inoculants containing 66-77% silicon, 0.9-3% aluminum, 0.4-1% calcium, and the balance iron.

Moldite. Trade name of Moldite Technologies (USA) for a composite sandwich consisting of a thermoset resin core (e.g., epoxy, phenolic or vinyl ester) with proprietary reinforcement particles and outer fiber skins of materials, such as aramid, carbon or glass. It has a very high strength-to-weight ratio, a very low specific gravity (usually less than 1.0), very high stiffness, high mechanical strength, good strength retention at high temperatures, good fire retardancy, good resistance to chemicals, caustics, mildew and mold. Depending on the particular materials used in its manufacture, it may also have high thermal insulating properties, or good barrier properties to liquid nitrogen.

mold lubricant. See release agent (1).

Moldmax. Trade name of Brush Wellman Engineered Materials (USA) for a weldable beryllium-copper alloy with excellent wear and corrosion resistance, and high strength and thermal conductivity. Used for plastic molds.

mold release agent. See release agent.

Mold Special. Trade name of Atlas Steels Limited (Canada) for a mold steel (AISI type P20) containing 0.35% carbon, 1.25% chromium, 0.4% molybdenum, and the balance iron. It has high core hardness, excellent resistance to decarburization, a great depth of hardening, good toughness and machinability. Used for zinc die-casting dies and plastic molds for injection, compression and transfer molding.

mold steels. A group of low-carbon steels (AISI group P) typically containing 0.1-0.4% carbon, 0.1-1% manganese, 0.1-0.8% silicon, 0.2-5.3% chromium, 0.1-4% nickel, 0.1-0.5% molybdenum, and sometimes small additions of vanadium and cobalt. Nearly all *mold steels* have uniform texture, low micro-porosity, low resistance to softening at elevated temperatures, good machinability and dimensional stability. They are often sold under trademarks and trade names, such as *Almold, Cascade, Duramold, Mold Special* or *Samson Extra*, and used for low-temperature die-casting dies, and plastic molds for injection or compression molding.

Moldtem. Trade name of Heppenstall Company (USA) for a pre-hardened, hot-work tool steel, used for zinc die-casting dies and plastic mold dies.

mold wash. A coating compound composed of an emulsion or suspension of refractory fillers (e.g., chromite, silica, or zircon flour), and a suspension agent (e.g., bentonite, or sodium alginate) in alcohol or water, and applied to the cavity of a mold to facilitate the release of a casting or ware from the mold after it has been formed. See also mold coating.

molecular composites. A group of composite materials having high length-to-diameter aspect ratios and consisting of liquid crystal thermoset matrices reinforced with a molecular dispersion of rigid rods of a *liquid crystal polymer*.

molecularly doped polymers. A class of polymers that are synthesized by the dispersion of small organic pigment molecules with semiconducting properties (chromophores) into an insulating polymer matrix. See also conductive polymers; semiconductive polymers.

molecularly engineered materials. A group of novel, complex materials, such as polymers, biocomposites, synthetic hydroxyapatite, phthalocyanines, porphyrins and other organometallics, or semiconductor nanocrystals (quantum dots) that have been designed, manipulated, modified and/or synthesized on the molecular level, i.e., their structure has been controlled on an atom-by-atom or molecule-by-molecule basis. The synthesis of these materials may involve techniques such as nanotechnology, molecular manufacturing, and MEMS (microelectromechanical system) technology. See also MEMS materials; nanomaterials.

molecular sandwich. A *metallocene* obtained by bonding a transition metal, or metal halide to one or more cyclopentadienyl (C_5H_5) rings. Depending on the number and position of cyclopentadienyl rings present and the material to which they are bonded (transition metal, or metal halide), there are two basic types: (i) dicyclopentadienyl-metal compounds with the general formula $(C_5H_5)_2M$; and (ii) dicyclopentadienyl-metal halides with the general formula $(C_5H_5)_2MX_{1-3}$. In these molecular sandwiches one C_5H_5 ring is located above and the other below the metal atom plane. Examples include cobaltocene ($C_5H_5)_2Co$, and ferrocene ($C_5H_5)_2Fe$. Used in organometallic research, and in the manufacture of polymers, catalysts, ultra-violet absorbers, etc. Also known as *sandwich molecule*.

Molecule. Trade name of SA Glaverbel (Belgium) for a glass with mosaic-type pattern.

molecule-based magnets. A relatively new class of magnetic materials based on molecular building blocks and including purely organic magnets, organic-inorganic hybrid magnets and inorganic cyanide-based magnets with network structures. They can have zero-, one-, two-, or three-dimensional structures, and be prepared in the form of thin films by low-temperature chemical vapor deposition (LTCVD) and electrodeposition. In addition to the magnetic properties displayed by many conventional transition-metal-, rare-earth-, metal/intermetallic- and metal oxide-based magnetic materials (e.g., high magnetization, remanent magnetization and susceptibility, and low anisotropy), they offer several advantages including low density and high flexibility, high transparency, solubility and biocompatibility, conductive, semiconductive or insulating properties, and photoresponsiveness. While some *molecule-based magnets* exhibit certain magnetic states (e.g., ferromagnetic, antiferromagnetic, etc.) only at very low critical temperatures (typically in the range of 2-100K) others, such as the *spin glasses* vanadium $(TCNE)_x$ (where TCNE is tetracyanoethylene) and vanadium $[Cr(CN)_6]_x$, have been obtained with critical temperatures exceeding room temperature (400K and 376K, respectively). Examples of other *molecule-based magnets* include metamagnetic tanol suberate, canted antiferromagnetic 4-cyanotetrafluorophenylthiadiazoyl and manganese phthalocyanine and ferromagnetic 4-nitrophenyl nitronyl nitroxide (NPNN). Potential applications include transformer cores, and magnetic and magnetooptical storage and recording media, e.g., magnetooptical disks, magnetic disks with higher data densities.

Moleculoy. Trade name of Molecu-Wire Corporation (USA) for a series of resistance alloys containing 20% chromium, 3-4% aluminum, 0-5% manganese, 0-1% silicon and 0-0.2% cobalt. It has a maximum operating temperature of 250°C (482°F), and high electrical resistivity. Used in the form of resistance wire for winding precision resistors and potentiometers.

Molegrain. Trade name for an alloy cast iron used for gears, shafts, and housings.

Molel. Trade name of Delsteel Inc. (USA) for an oil-hardening steel used for tools, cutters, and punches.

moleskin. A strong, heavy, closely woven fabric, usually of cotton, in a satin weave and with a soft, smooth napped back. Used for pants, sportswear, workwear, and linings.

Molex. Trade name of Associated Steel Corporation (USA) for a series of tool steels.

Molfor. Trade name of Forjas Alavesas SA (Spain) for a series of hot-work tool steels containing 0.3-0.4% carbon, 0.4-0.5% manganese, 0-1% silicon, 3-5% chromium, 1.3-2.8% molybdenum, 0.4-0.5% vanadium, and the balance iron. Used for punches, extrusion cylinders, die-casting and forging dies, and pressure casting molds.

Molibloc. Trade name of Creusot-Loire (France) for an oil-hardening mold steel containing 0.35% carbon, 0.8% manganese, 0.5% silicon, 1.6% chromium, 0.4% molybdenum, and the balance iron. It has high core hardness, good deep-hardening properties, and good toughness and machinability. Used for die-casting dies and plastic molds.

Molielastic. Trade name for flexible polyether foams and foam products.

Molin. Trade name of American Art Alloys Inc. (USA) for a cast

aluminum bronze used for gears, shafts, bearings, molds, and dies.

Molite. Trade name of Columbia Tool Steel Company (USA) for a series of molybdenum-type high-speed steels (AISI types M1 through M34) containing about 0.8-1.2% carbon, 0.3% manganese, 0.3% silicon, 3.2-9.5% molybdenum, 3.5-4.8% chromium, 1.1-6.8% tungsten, 1-4.3% vanadium, up to 8% cobalt, and the balance iron. They have excellent hot hardness, good deep-hardening properties, excellent wear resistance, high toughness, moderate machinability, and are used for cutting tools, milling cutters, lathe and form tools, and shearing, blanking and trimming dies.

molleton. A heavy wool or other fabric with two napped sides.

Molloplast. Trade name of Monopol AG (Switzerland) for a heat-cure silicone used as a dental liner.

Mollosil Plus. Trade name of 3M Dental (USA) for a self-cure silicone used as a dental liner.

molluscan glue. A mollusk-derived glue used as a tissue adhesive to enhance cell adhesion.

Molmang. Trade name of Stulz Sickles Steel Company (USA) for a wear-resistant, work hardening austenitic steel containing 11-13.5% manganese. Used for welding electrodes employed in build-up work.

Mo-Lo. Trade name of American Art Alloys Inc. (USA) for a cast iron used for cylinder liners, bushings, and piston rings.

molten cast brick. See fusion-cast brick.

molten cast refractories. See fusion-cast refractories.

Moltiflex. Trade name for flexible isocyanate foams and foam products.

Molton. Trade name for soft twill-weave cotton fabrics having napped front and back surfaces.

Moltopren. Trademark of Bayer AG (Germany) for a lightweight, crosslinked polyurethane used for adhesives, elastomers, and foams with good thermal insulating properties.

Moltrop. Trade name of Moltrop Steel Products Company (USA) for a water-hardening steel containing 0.3-0.5% carbon, and the balance iron. Used for shafts and gears.

moluranite. A black mineral with a resinous luster belonging to the amorphous group and composed of uranium molybdenum oxide hydrate, $U_4Mo_7O_{32} \cdot 20H_2O$. Density, 4.00 g/cm³; refractive index, 1.975.

Molva. Trade name of Wallace Murray Corporation (USA) for molybdenum-type high-speed steels (AISI types M2 through M10) containing about 0.8-1.3% carbon, 0.3% manganese, 0.3% silicon, 3.2-8.7% molybdenum, 1.15-6% tungsten, 3.5-4.7% chromium, 0.9-4.3% vanadium, and the balance iron. It has good hot hardness and wear resistance, excellent deep-hardening properties, and is used for cutting tools, milling cutters, lathe tools, forming tools, and blanking, shearing and trimming dies.

Moly Ark. Trade name of Jessop-Saville Limited (UK) for a high-speed tool steel containing 0.7% carbon, 5.5% tungsten, 4% molybdenum, 4% chromium, 1.5% vanadium, and the balance iron. Used for cutters and tools.

Moly Ascoloy. Trade name of Usines de A. Manoir (France) for a high-strength steel containing 0.08% carbon, 13% chromium, 2% molybdenum, and the balance iron. It has high heat resistance, and is used for nozzles, afterburners, and jet-engine components.

Moly Astroloy. Trade name of Carpenter Technology Corporation (USA) for a precipitation-hardening nickel-base superalloy containing 0.06% carbon, 15% chromium, 15% cobalt,

5.25% molybdenum, 4.4% aluminum, 3.5% titanium, and traces of boron and zirconium. Used for aircraft gas-turbine engine disks.

molybdate chrome orange. See molybdenum orange.

molybdate orange. See molybdenum orange.

molybdenite. A bluish-lead gray mineral having a gray-black streak, a metallic luster, and resembling graphite in appearance. It is composed of molybdenum disulfide, MoS_2, containing about 60% molybdenum. Crystal system, hexagonal. Density, 4.6-5.1 g/cm^3; melting point, 1185°C (2165°F); hardness, 1.0-1.5 Mohs. Occurrence: Canada, Chile, Germany, Norway, USA (Colorado, Nevada, New Jersey, New Mexico, Utah). Used as an important ore of molybdenum, as a lubricant and drawing compound for iron and steel, and for porcelain enameling. Also known as *molybdenum glance.*

molybdenum. A silvery-gray metallic element of Group VIB (Group 6) of the Periodic Table. It is commercially available in the form of ingots, rods, wire, powder, tubing, high-ductility sheets, foil, microfoil, sintered, wrought bar, arc castings and single crystals. The single crystals are usually grown by the float-zone or the radio-frequency technique. The commercial ores of molybdenum are *molybdenite* and *wulfenite.* Crystal system, cubic. Crystal structure, body-centered cubic. Density, 10.22 g/cm^3; melting point, 2617°C (4743°F); boiling point, 5612°C (8334°F); atomic number, 42; atomic weight, 95.94; divalent, trivalent, tetravalent, pentavalent, hexavalent. It has good high-temperature strength, relatively good electrical conductivity (34% IACS), a superconductivity critical temperature of 0.915K, a hardness of 200-250 Vickers, and oxidizes rapidly above 538°C (1000°F) in air at sea level, but is stable in the upper atmosphere. Used as an alloying agent in steels to increase high-temperature stability, wear resistance, hardness, hardenability and strength (e.g. in quenched and tempered steels), as an alloying element in cast iron (to increase strength, toughness and wear resistance), in high-temperature alloys, heat- and corrosion-resistant nickel-base alloys, copper-base die-casting alloys, electrical contact materials, cermets and cemented carbides, as a pigment in printing inks, paints and ceramics, in missile and aircraft parts, rocket motors, reactor vessels, nuclear applications, electric heater filaments, vacuum tube filaments, screens and grids, electrodes of mercury-vapor lamps, in filament supports for light bulbs, in filaments for metal-evaporation processes, in wire for winding electric-resistance furnaces, in special batteries, glass-to-metal seals, for electrooptical applications, as a radio isotope source, as a catalyst, and in solid lubricants. Symbol: Mo.

molybdenum alloys. A group of refractory alloys in which molybdenum is the predominant component. The most important classes of molybdenum alloys include carbide-strengthened alloys, solid-solution alloys, carbide-strengthened solid-solution alloys and dispersion-strengthened powder-metallurgy alloys. The following alloys are of particular commercial importance: (i) *Molybdenum-titanium alloys* which are carbide-strengthened alloys containing about 0.5% titanium and having good creep and tensile strength and good high-temperature structural properties. They are used in tooling for hot-forging dies and for applications requiring refractory alloys, and can be modified with up to 0.08% zirconium and 0.5% tungsten. The latter alloys are known under the trade name "TZM." They are ductile at room temperature, become brittle at lower temperatures, and have a density of 10.22 g/cm^3 (0.369 $lb/in.^3$), a melting range of 2500-2600°C (4530-4710°F), and a maximum service tempera-

ture in air of approximately 400°C (750°F); (ii) *Molybdenum-tungsten alloys* which are solid-solution alloys and include 70Mo-30W with high tensile strength, excellent resistance to corrosive molten metals (e.g., zinc), and high melting point of 2830°C (5125°F). Used for high-temperature structural parts; (iii) *Molybdenum-rhenium alloys* which are soft, ductile solid-solution alloys containing varying amounts of molybdenum and rhenium, with molybdenum usually being the predominant constituent. They are used in electronics, thermocouples, and welding rods; (iv) *Zirconia-dispersion-strengthened molybdenum alloys*, such as Z-6 with 0.5% ZrO_2, which have improved high-temperature strength and creep resistance and enhanced low-temperature ductility, and are used for high-temperature structural applications; and (v) *molybdenum alloys doped with potassium and silicon* which have excellent creep resistance and high resistance to recrystallization, and are used for electrical applications, e.g., lighting products.

molybdenum alloy steels. A group of alloy steels with a nominal molybdenum content of less than 0.52%. A typical composition is 0.1-0.5% carbon, 0.45-0.9% manganese, 0.15-0.3% silicon, 0.15-0.45% molybdenum, traces of phosphorus and sulfur, and the balance iron. Usually, this term refers to the steels in the AISI-SAE 40xx and 44xx series. Supplied in the quenched and tempered condition, they have high strength and hardness and good machinability, and are used for machinery parts, carburized parts, axle shafts, spline shafts, bolts, studs, coil and leaf springs, gears, and differential gears.

molybdenum aluminide. Any of the following compounds of molybdenum and aluminum: (i) *Molybdenum monoaluminide.* Melting point, 1700°C (3092°F). Used in ceramics, materials research and refractory applications. Formula: MoAl; (ii) *Trimolybdenum aluminide.* A fine powder with a melting point of 2150°C (3902°F). Used as a major constituent in refractory crucibles for melting titanium metal, and in ceramics and cermets. Formula: Mo_3Al.

molybdenum anhydride. See molybdenum trioxide.

molybdenum beryllide. See molybdenum beryllium.

molybdenum beryllium. Any of several intermetallic compounds of molybdenum and beryllium in particular the following two materials used in ceramics and materials research: (i) *Molybdenum diberyllide.* Melting point, above 1870°C (3398°F). Formula: $MoBe_2$; and (ii) *Molybdenum dodecaberyllide.* Density, 3.03 g/cm^3; melting point, 1650-1705°C (3002-3101°F); hardness, 950 Vickers. Formula: $MoBe_{12}$.

molybdenum black. Black, lustrous, molybdenum-base decorative coatings applied to zinc and its alloys. Also known as *moly black.*

molybdenum boride. Any of the following compounds of molybdenum and boron used as a brazing alloy to join molybdenum, tungsten, tantalum and niobium parts, in electronic components, in corrosion- and abrasion-resistant parts, in cutting tools, refractories and cermets, and in ceramics and materials research: (i) *Alpha molybdenum monoboride.* Density, 8.8 g/cm^3; melting point, 2350°C (4258°F); hardness, 8.0 Mohs. Formula: α-MoB; (ii) *Beta molybdenum monoboride.* Density g/cm^3, 8.4; melting point, 2180°C (3956°F). Formula: β-MoB; (iii) *Molybdenum diboride.* A fine powder. Density, 7.12 g/cm^3, melting point, 2100°C (3812°F), low thermal expansivity. Formula: MoB_2; (iv) *Dimolybdenum boride.* Refractory crystals. Crystal system, tetragonal. Density, 9.2-9.3 g/cm^3; melting point, 1660°C (3020°F) [also reported as 2000°C or 3632°F]; hardness, 8-9 Mohs. Formula: Mo_2B; (v) *Molybdenum pentaboride.*

Refractory crystals. Crystal system, hexagonal. Density, 7.2-7.5 g/cm³; melting point, 1600°C (2912°F) (transforms to MoB₂). Formula: Mo_2B_5; and (vi) *Trimolybdenum diboride.* Melting point, 2070°C (3758°F). Formula: Mo_3B_2.

molybdenum bromide. See molybdenum dibromide; molybdenum tribromide; molybdenum tetrabromide.

molybdenum carbide. Any of the following compounds of molybdenum and carbon used in ceramics and materials research: (i) *Molybdenum monocarbide.* Dark gray crystals, or fine gray powder. Crystal system, hexagonal. Density, 8.2-8.5 g/cm³; melting point, 2690°C (4874°F); hardness, 1800 kgf/mm². Formula: MoC; and (ii) *Dimolybdenum carbide.* Gray crystals or fine white or gray powder (99.5+% pure). It is available in orthorhombic and hexagonal form. The orthorhombic form has a density of 9.18 g/cm³, and a melting point of 2687°C (4869°F), and the hexagonal form has a density of 8.2 g/cm³, a melting range of 2505-2690°C (4540-4880°F), and a hardness of 1950 Vickers. Formula: Mo_2C.

molybdenum carbonyl. White, shiny crystals (98% pure). Density, 1.96; melting point, 150°C (302°F) (decomposes); boiling point, 156°C (313°F). Used for plating molybdenum, e.g., molybdenum mirrors. Formula: $Mo(CO)_6$. Also known as *hexacarbonylmolybdenum; molybdenum hexacarbonyl.*

molybdenum cast irons. A group of cast irons which have small quantities of molybdenum (typically 0.25 to 1.0 wt%) added in the form of calcium molybdate or *ferromolybdenum* during manufacture. The molybdenum greatly increases their abrasion and wear resistance, tensile strength, toughness and hardness, and promotes a homogeneous microstructure. These properties can be further enhanced by small additions of chromium and/or nickel.

Molybdenum Chisel. Trade name of A. Milne & Company (USA) for an oil-hardening tool steel containing 0.6% carbon, 0.5% molybdenum, and the balance iron. Used for tools and chisels.

molybdenum chloride. See molybdenum dichloride; molybdenum pentachloride; molybdenum trichloride; molybdenum tetrachloride.

molybdenum-chromium. A master alloy containing 67-72% molybdenum, 28-33% chromium and up to 0.5% iron. Used in the manufacture of nonferrous alloys.

molybdenum-copper metal-matrix composites. Engineering composites made from blends of molybdenum and copper powders by wrought powder-metallurgy methods. The typical composition range is 60-90% molybdenum and 10-40% copper. Their low thermal expansion, outstanding thermal conductivity, and good machinability make them suitable for electronic applications, e.g., as heat sinks, thermal spreaders, and circuit board cores.

molybdenum diberyllide. See molybdenum beryllium (i).

molybdenum diboride. See molybdenum boride (iii).

molybdenum dibromide. Yellow-red crystals or yellow powder with a decomposition temperature of about 900°C (1652°F), used in chemistry and materials research. Formula: $MoBr_2$. Also known as *molybdenum bromide.*

molybdenum dichloride. A yellow crystals or amorphous yellow powder with a density of 3.71 g/cm³, and a melting point of 530°C (986°F) (decomposes). Used in chemistry, organometallic synthesis, and materials research. Formula: $MoCl_2$. Also known as *molybdenum chloride.*

molybdenum diiodide. Air- and moisture-sensitive, black crystals, or amorphous brown powder (99+% pure) with a density of 5.278 g/cm³, used in chemistry, organometallic synthesis

and materials research. Formula: MoI_2. Also known as *molybdenum iodide.*

molybdenum dioxide. Violet-red or brown-violet crystals or purple, nonvolatile powder (99% pure). Crystal system, tetragonal. Density, 6.47 g/cm³; melting point, decomposes at 1100°C (2012°F). Used in pigments and materials research. Formula: MoO_2.

molybdenum diphosphide. See molybdenum phosphide (ii).

molybdenum diselenide. Gray crystals or fine, gray powder (99+% pure). Crystal system, hexagonal. Density, 6.0 g/cm³; melting point, above 1200°C (2192°F). Used as a solid lubricant. Formula: $MoSe_2$. Also known as *molybdenum selenide.*

molybdenum disilicide. See molybdemum silicide (i).

molybdenum disulfide. See molybdenum sulfide (i).

molybdenum ditelluride. A gray crystals or fine gray powder. Crystal system, hexagonal. Density, 7.7 g/cm³. Used chiefly as a solid lubricant. Formula: $MoTe_2$.

molybdenum dodecaberyllide. See molybdenum beryllium (ii).

molybdenum enamel. A *porcelain enamel* to which up to about 7.5% molybdenum trioxide (MoO_3) has been added to promote adherence and enhance acid resistance. Also known as *moly enamel.*

molybdenum fluoride. See molybdenum trifluoride; molybdenum tetrafluoride; molybdenum hexafluoride; molybdenum pentafluoride.

molybdenum germanide. A semiconductive compound of molybdenum and germanium with a melting point of 1750°C (3182°F). Used in ceramics, materials research and electronics. Formula: Mo_3Ge.

molybdenum glance. See molybdenite.

molybdenum hexacarbonyl. See molybdenum carbonyl.

molybdenum hexafluoride. White, moisture-sensitive crystals (99.9+% pure), or colorless liquid. Crystal system, cubic. Density, 2.30-2.54 g/cm³; melting point, 17.5°C (63°F); boiling point, 37°C (98°F). Used in chemistry, organometallic synthesis, and materials research. Formula: MoF_6. Also known as *molybdenum fluoride.*

molybdenum high-speed steels. A group of high-speed tool steels (AISI subgroup M) containing about 0.8-1.3% carbon, 0.3% manganese, 0.3% silicon, 3.25-10% molybdenum, 1.15-7% tungsten, 3.5-4.75% chromium, 0.95-4.25% vanadium, 0-12% cobalt, and the balance iron. They have excellent hot hardness, good deep-hardening properties, good abrasion resistance, high toughness, and moderate machinability. Often sold under trademarks and trade names, such as *Braevan, Exocut, Hypercut, Mocut, Molite, Molva, Mustang, Oglala, Sixix, Van Cut* or *Vinco,* they are used especially for cutting tools like drills, taps, chasers, reamers, broaches, milling cutters, form and lathe tools, and for shearing, blanking, thread-rolling and trimming dies. Also known as *molybdenum-type high-speed tool steels.*

molybdenum hot-work steels. A group of hot-work tool steels (AISI subgroup H40 through H59) containing 0.5-0.7% carbon, 0.3% manganese, 5-8% molybdenum, 3.75-4.5% chromium, 1-6.75% tungsten, 1.75-2.2% vanadium, and the balance iron. They have excellent resistance to softening, good hot hardness, excellent abrasion and wear resistance at elevated temperatures, low ductility, and moderate toughness and wear resistance. Often sold under trademarks or trade names, such as *Braemow Special, Electrite No. 7, Molite HW42, Mustang LC, Van-Lom* or *Vasco M-Z,* they are used especially for hot headers, punch and die inserts, hot-heading and hot-nut dies, and punches. Also known as *molybdenum-type hot-work steels.*

molybdenum iodide. See molybdenum diiodide; molybdenum triiodide.

molybdenum monoaluminide. See molybdenum aluminide (i)

molybdenum monoboride. See molybdenum boride (i) and (ii).

molybdenum monocarbide. See molybdenum carbide (i).

molybdenum mononitride. See molybdenum nitride (i).

molybdenum monophosphide. See molybdenum phosphide (i).

molybdenum nitride. Any of the following compounds of molybdenum and nitrogen used in ceramics: (i) *Molybdenum mononitride.* Hexagonal crystals. Density, 9.18 g/cm^3; melting point, approximately 748°C (1378°F). Formula: MoN; (ii) *Dimolybdenum nitride.* Gray crystals or fine grayish powder. Crystal system, cubic. Density, 8.04-9.06 g/cm^3; melting point, approximately 898°C (1649°F); hardness, 650 Vickers. Formula: Mo$_2$N; and (iii) *Trimolybdenum nitride.* Melting point, above 593°C (1099°F). Formula: Mo$_3$N.

molybdenum orange. A solid solution of lead chromate (PbCrO$_4$), lead molybdate (PbMoO$_4$) and lead sulfate (PbSO$_4$) available in the form of a fine, deep orange to light red powder, and used as a pigment in paints, plastics and inks. Also known as *molybdate chrome orange; molybdate orange.*

molybdenum oxide. See molybdenum dioxide; molybdenum sesquioxide; molybdenum trioxide; trimolybdenum oxide.

molybdenum oxychloride. Moisture-sensitive, green, crystals or powder used in chemistry, organometallic synthesis and materials research. Formula: MoOCl$_4$. Also known *molybdenum tetrachloride oxide.*

molybdenum pentaboride. See molybdenum boride (v).

molybdenum pentachloride. Greenish-black or grayish-black to dark-blue, air- and moisture-sensitive crystaks or powder (98+% pure). Density, 2.928 g/cm^3; melting point, 194°C (381°F); boiling point, 268°C (514°F). Used for vapor-deposited molybdenum coatings, as a component of fire-retardant resins, in brazing and soldering fluxes, as an intermediate for organometallic compounds, e.g., molybdenum hexacarbonyl, Mo(CO)$_6$. Formula: MoCl$_5$. Also known as *molybdenum chloride.*

molybdenum pentafluoride. Yellow crystals. Crystal system, monoclinic; melting point, 67°C (153°F); boiling point, 213°C (415°F). Used in chemistry and materials research. Formula: MoF$_5$. Also known as *molybdenum fluoride.*

Molybdenum Permalloy. See Moly Permalloy.

molybdenum phosphide. Any of the following compounds of molybdenum and phosphorus used in ceramics and electronics: (i) *Molybdenum monophosphide.* Black crystals or gray crystalline powder. Crystal system, hexagonal. Density, 7.34-7.5 g/cm^3; melting point, decomposes at 1483°C (2701°F). Formula: MoP; (ii) *Molybdenum diphosphide.* Density, 5.21 g/cm^3. Formula: MoP$_2$; and (iii) *Trimolybdenum phosphide.* Tetragonal crystals. Density, 9.14 g/cm^3. Formula: Mo$_3$P.

molybdenum-rhenium. An intermetallic compound of molybdenum and rhenium with a melting point of 2500°C (4532°F). Used in ceramics and materials research. Formula: Mo$_2$Re$_3$.

molybdenum-rhenium alloys. A group of soft, ductile molybdenum-based refractory alloys containing varying additions of rhenium (usually 11-50%) to enhance strength, plasticity, weldability and, with increasing rhenium content, hardness. Used in electronics, thermocouples, reflectors, heating elements, elastic elements (torsion bars, suspensions), and welding rods.

molybdenum selenide. See molybdenum diselenide.

molybdenum sesquioxide. A grayish-black powder frequently given the formula Mo$_2$O$_3$, although it is known only as the hydrate, Mo(OH)$_3$. Used for decorative and protective coatings on metal objects, and as a catalyst. Also known as *dimolybdenum trioxide.*

molybdenum sesquisulfide. See molybdenum sulfide (iv).

molybdenum silicide. Any of the following compounds of molybdenum and silicon: (i) *Molybdenum disilicide.* Gray crystals or dark gray, crystalline powder (99+% pure), and also supplied in the form of cylinders, rods, lumps, granules and whiskers. Crystal system, tetragonal. Density, 6.24- 6.31 g/cm^3; melting point, 1870-2030°C (3398-3686°F); hardness, 1240 Knoop; high stress-rupture strength; good oxidation resistance at elevated temperatures. Used for high-temperature protective coatings supplied by vapor deposition or flame spraying, oxidation resistant coatings for refractory metals, as a reinforcement for structural ceramic composites, as a matrix material for structural silicide compounds, for engine parts in space vehicles (molybdenum coated with molybdenum disilicide), for electrical resistors, in combination with alumina (Al$_2$O$_3$) in kiln furniture, sandblast nozzles, saggers, induction brazing fixtures and hot-press and hot-draw dies, for refractory applications, for heating elements in furnaces, and sometimes to promote special porcelain-enamel adherence. Formula: MoSi$_2$; (ii) *Trimolybdenum silicide.* Density, 8.97 g/cm^3; melting point, decomposes at 2170°C (3940°F). Used in ceramics and materials research. Formula: Mo$_3$Si; (iii) *Trimolybdenum disilicide.* Density, 8.08 g/cm^3; melting point, approximately 2093°C (3800°F). Used in ceramics and materials research. Formula: Mo$_3$Si$_2$; and (iv) *Pentamolybdenum trisilicide.* Crystal system, tetragonal. Density, 8.24 g/cm^3; melting point, 2160°C (3920°F). Used in high-temperature ceramics, as a reinforcement for structural ceramic composites, and as a matrix material for structural silicide compounds. Formula: Mo$_5$Si$_3$.

Molybdenum Silicon. Trade name of Textron Inc. (USA) for a wear-resistant molybdenum-silicon steel used for pistons and brake drums.

molybdenum-silicon. A master alloy composed of 60% molybdenum, 30% silicon and 10% iron, used to introduce molybdenum to molten steel.

molybdenum silicocarbide. A compound of molybdenum, silicon and carbon available in the form of a fine powder for use in ceramics and materials research. Formula: MoSi$_3$C.

molybdenum steels. A group of steels containing molybdenum as the principal alloying element. In ordinary steels molybdenum increases the hardenability, high-temperature strength and fatigue resistance and eliminates the temper brittleness, but reduces elongation and malleability. In tool steels it also improves cutting ability and wear resistance. See also molybdenum alloy steels; molybdenum high-speed steels; molybdenum hot-work steels.

molybdenum sulfide. Any of the following compounds of molybdenum and sulfur: (i) *Molybdenum disulfide.* Dark crystals or fine, black powder (97+%). It occurs in nature as the mineral *molybdenite.* Crystal system, hexagonal. Density, 4.6-5.1 g/cm^3; melting point, 1185°C (2165°F); hardness, 1.0-1.5 Mohs; low coefficient of friction. Used as a lubricant in greases, oil dispersions, resin-bonded films and dry powders, especially at extreme pressures and high vacua, as a friction-reducing filler in nylon gears and transmissions, in plastics to enhance the flexural strength, as a heating element for electric furnaces, and as a hydrogenation and isomerization catalyst. Formula: MoS$_2$. Also known as *molybdenum sulfide; molybdic sulfide;* (ii) *Molybdenum trisulfide.* A red-brown powder used in ceramics and materials research. Formula: MoS$_3$; (iii) *Molybdenum tetrasul-*

fide. A brown powder used in ceramics and materials research. Formula: MoS_4; (iv) *Molybdenum sesquisulfide*. A steel-gray needles that has a density of 5.9 g/cm^3 and decomposes at about 1100°C (2010°F) used in ceramics and materials research. Formula: Mo_2S_3.

molybdenum telluride. See molybdenum ditelluride.

molybdenum tetrabromide. Black, moisture-sensitive crystals or needles used in chemistry, organometallic synthesis and materials research. Formula: $MoBr_4$. Also known as *molybdenum bromide*.

molybdenum tetrachloride. Brown or black, deliquescent crystals (99.5+% pure) that decompose above 170°C (338°F), and are used in chemistry, organometallic synthesis and materials research. Formula: $MoCl_4$. Also known as *molybdenum chloride*.

molybdenum tetrachloride oxide. See molybdenum oxychloride.

molybdenum tetrafluoride. Green crystals used in chemistry and materials research. Formula: MoF_4. Also known as *molybdenum fluoride*.

molybdenum tetrasulfide. See molybdenum sulfide (iii).

molybdenum-titanium alloys. A group of refractory alloys composed of molybdenum and small additions of titanium, and sometimes zirconium and tungsten. Molybdenum alloys containing about 0.5% titanium have good creep and tensile strength, good high-temperature structural properties, and are used in tooling for hot-forging dies and for applications requiring refractory alloys. They are sometimes modified with up to 0.1% zirconium and 0.02% tungsten and sold under the trade name "TZM." These alloys are ductile at room temperature, becoming brittle at lower temperatures, and have a density is 10.22 g/cm^3 (0.369 lb/in.³), a melting range of 2500-2600°C (4530-4710°F), and a maximum service temperature in air of approximately 400°C (750°F).

molybdenum tribromide. Dark green crystals or needles. Crystal system, hexagonal. Density, 4.89 g/cm^3; melting point, 977°C (1791°F). Used in chemistry, organometallic synthesis and materials research. Formula: $MoBr_3$. Also known as *molybdenum bromide*.

molybdenum trichloride. Dark-red, hygroscopic crystals, or purple air- and moisture-sensitive powder (99.5+% pure). Crystal system, monoclinic. Density, 3.58-3.74 g/cm^3; melting point, 1027°C (1881°F). Used in chemistry, organometallic synthesis, and materials research. Formula: $MoCl_3$. Also known as *molybdenum chloride*.

molybdenum trifluoride. Brown crystals. Crystal system, hexagonal. Density, 4.64 g/cm^3; melting point, above 600°C (1112°F). Used in chemistry and materials research. Formula: MoF_3. Also known as *molybdenum fluoride*.

molybdenum triiodide. A black solid compound with a melting point of 927°C (1701°F), used in chemistry and materials research. Formula: MoI_3. Also known as *molybdenum iodide*.

molybdenum trioxide. White crystals or off-white powder (99+% pure) at room temperature that turn yellow when heated. It is also available in high-purity grades (99.9-99.999+%), and occurs in nature as the mineral *molybdite*. Crystal system, orthorhombic. Density, 4.69 g/cm^3; melting point, 795°C (1463°F); sublimes starting at 700°C (1292°F); boiling point, 1155°C (2111°F). It is an electrical insulator, but becomes a conductor when molten. Used as a source of molybdenum compounds, in the manufacture of metallic molybdenum, for adding molybdenum to alloys, as a corrosion inhibitor, in ceramic glazes, as an adherence promoter in sheet-steel enamels, cast-iron enamels, jewelry enamels and high-temperature ceramic coatings, in whiteware bodies to increase strength and to lower firing temperatures, in porcelain enamels as an opacifier, in pigments, and as a catalyst. Formula: MoO_3. Also known as *molybdenum anhydride; molybdic anhydride; molybdic oxide*.

molybdenum trisulfide. See molybdenum sulfide (ii).

molybdenum-tungsten alloys. A group of refractory alloys of molybdenum and tungsten including 70Mo-30W which has a high tensile strength, excellent resistance to corrosive molten metals (e.g., zinc), and a high melting point of 2830°C (5125°F). Used for high-temperature structural parts.

molybdenum-type high-speed tool steels. See molybdenum high-speed steels

molybdenum-type hot-work tool steels. See molybdenum hot-work steels.

molybdenum-zirconium. A compound of molybdenum and zirconium with a melting point of 1822°C (3312°F), used in ceramics, materials research and for refractories. Formula: Mo_2Zr.

molybdic acid. (1) See ammonium heptamolybdate.
  (2) See sodium molybdate.

molybdic anhydride. See molybdenum trioxide.

molybdic ocher. See molybdite.

molybdic oxide. See molybdenum trioxide.

molybdic sulfide. See molybdenum sulfide (i).

Molybdie. Trade name of A. Finkl & Sons Company (USA) for a series of nickel-chromium-molybdenum alloy steels used for forging dies, sow blocks, and machine parts.

molybdite. A lead-gray mineral composed of molybdenum trioxide, MoO_3, but often containing considerable amounts of iron and referred to as *ferrimolybdite*. Crystal system, orthorhombic. It can also be made synthetically. Density, 4.71 g/cm^3; melting point, 1185°C (2165°F). Used as an ore of molybdenum. Also known as *molybdic ocher*.

molybdofornacite. A light green mineral composed of copper lead molybdenum oxide arsenate hydroxide, $Pb_2Cu(MoO_4)$-$(AsO_4)(OH)$. Crystal system, monoclinic. Density, 6.57 g/cm^3. Occurrence: Namibia.

molybdomenite. A colorless, white or yellowish white mineral composed of lead selenite, $PbSeO_3$. It can also be made synthetically by fusion of lead monoxide (PbO) and selenium dioxide (SeO_2). Crystal system, monoclinic. Density, 7.07 g/cm^3. Occurrence: Argentina.

molybdophosphoric acid. Yellowish, water-soluble crystals (99.9+% pure) with a density of 3.15 g/cm^3 and a melting point of 78-90°C (172-194°F). Used as a catalyst, as a plating additive, in pigments, as a water-resistant additive to adhesives, cement and plastics, as a fixing agent in photography, and as an electron-dense metal stain for electron microscopy. Formula: $H_3[P(Mo_3O_{10})_4] \cdot xH_2O$ (x is usually 6-8). Also known as *phosphomolybdic acid; phosphomolybdic acid*.

molybdophyllite. A white, pale green or colorless mineral composed of lead magnesium silicate hydroxide, $Pb_2Mg_2Si_2O_7(OH)$. Crystal system, rhombohedral. Density, 4.72 g/cm^3; refractive index, 1.815. Occurrence: Sweden.

molybdosilicic acid. A yellow, crystalline powder with a density of 2.82 g/cm^3, used as a catalyst, as a plating additive, as a precipitant and ion exchanger in nuclear energy applications, and as a water-resistant additive to adhesives, cement and plastics. Formula: $H_4[Si(Mo_3O_{10})_4] \cdot xH_2O$ (x is usually 6-8). Also known as *silicomolybdic acid*.

Molybesco. Trade name of Vallouec SA (France) for a series of steels containing 0.1-0.16% carbon, 0.15-0.3% chromium, 0.3-0.6% molybdenum, and the balance iron. They have moderate to high tensile strengths, and are used for heat exchangers, condensers, furnace tubes, boilers, and superheaters.

moly black. See molybdenum black.

Moly-D. Trade name of I Squared R Element Company Inc. (USA) for thermodynamically stable molybdenum disilicide powder used as matrix material for high-temperature intermetallic composites, and high-temperature coatings.

moly enamel. See molybdenum enamel.

Moly High Speed. Trade name of McInnes Steel Company (USA) for a high-speed steel (AISI type M1) containing 0.8% carbon, 4% chromium, 4.2% molybdenum, 1.5% vanadium, 5.5% tungsten, and the balance iron. It has a great depth of hardening, high hot hardness, high tensile strength, excellent shock and wear resistance, good toughness, and is used for lathe and planer tools, drills, and other cutting tools.

Molyite. Trade name of Columbia Tool Steel Company (USA) for a high-speed steel containing 0.8-1.0% carbon, 9% molybdenum, 4% chromium, 2% vanadium, and the balance iron. Used for cutters, and lathe and planer tools.

Molykote. Trade name of Dow Corning Corporation (USA) for a high-performance solid-film lubricant consisting of molybdenum disulfide (MoS_2) dispersed in oil or grease. It has an operating temperature range of -18 to +400°C (0 to 752°F), and is used in high-pressure and elevated-temperature bearings, and in cold and hot forming to reduce friction.

Moly-Mang. Trade name of Westinghouse Electric Corporation (USA) for a wear- and abrasion resistant austenitic manganese steel containing 11-13% manganese. Used as welding rod for manganese steels.

Molymet. Trade name of Wilbur B. Driver Company (USA) for an alloy composed of 45% iron, 45% nickel and 10% molybdenum, and used for electrical applications.

Molyneaux. French trade name for a fusible alloy containing 41.5% bismuth, 16.7% tin, 25% lead and 16.7% cadmium. It melts at 60°C (140°F), and is used for fire extinguishers, fuses, etc.

Moly-Nickel. Trade name of Chicago Hardware Foundry Company (USA) for a molybdenum-nickel alloy steel used for welding rods.

Moly Permalloy. Trade name of Allegheny Ludlum Steel (USA) for a series of magnetically soft materials containing 79-81% nickel, 16-17% iron, and 2-4% molybdenum. They have a density of 8.72 g/cm³ (0.369 lb/in.³), good strength, low magnetic core losses, high initial permeability and resistivity, and low magnetocrystalline and magnetostrictive anisotropies. Used for high-frequency equipment, toroids, tape and lamination cores for transformers and relays, inductance and loading coils, and electrical equipment. Also known as *Molybdenum Permalloy.*

molysite. A brownish-yellow mineral composed of iron chloride, $FeCl_3$. Crystal system, rhombohedral (hexagonal). It can also be made synthetically. Density, 2.90 g/cm³; boiling point, 319°C (606°F); refractive index, 1.6.

Molyte. Trade name for an alloying addition composed of calcium molybdate ($CaMoO_4$) and selected fluxing agents and used to introduce molybdenum into iron and steel.

Moly Telastic. Trade name of Falk Corporation (USA) for a tough, weldable low-alloy cast steel containing 0.3-0.4% carbon, 0.7-1% manganese, up to 0.6% silicon, 0.4-0.6% chromium, 0.15-0.25% molybdenum, and the balance iron. Used for construction machinery, pressure vessels, gears, and other machinery parts.

Mo-Mang. Trade name of Abex Corporation (USA) for an austenitic manganese steel containing 0.7-0.9% carbon, 12-14% manganese, up to 0.2% molybdenum, and the balance iron. Used in the form of welding electrodes for manganese steels.

Momarc. Trade name of Teledyne Vasco (USA) for an oil- or water-hardening, wear- and abrasion-resistant steel containing 1% chromium, 1.4% chromium, 1% molybdenum, and the balance iron. Used for precision gauges, arbors, dies, ball bearings, and special tools.

Mo-Max. Trade name of Cleveland Twist Drill Company (USA) for a high-speed steel containing 0.64-0.84% carbon, 3.25-4.25% chromium, 1.25-2% tungsten, 0.75-1.25% vanadium, 7.5-9.5% molybdenum, and the balance iron. Used for high-speed tools and cutters, hot-work dies, and drills.

Momiji. Trade name of Central Glass Company Limited (Japan) for a glass with leaf design.

monadic nylon. Nylon produced from a lactam (i.e., a cyclic amide) and including nylon 6 (a polymer of caprolactam) and nylon 4 (a polymer based on butyrolactam).

Monarch. (1) Trade name of Monarch Alloy Company (USA) for a series high-leaded bronzes and leaded coppers for bearings. The cast alloys have microstructures of finely dispersed lead in copper or copper-tin matrices. See also Monarch Metal.

(2) Trademark of Hamilton Engine Packing Company (Canada) for asbestos products including sheet packings, and tapes.

Monarch Metal. Trade name of Monarch Alloy Company (USA) for a high-leaded bearing bronze containing 75% copper, 20% lead, and 5% tin. See also Monarch.

Monark. Trade name of Atlas Specialty Steels (Canada) for a series of shock-resisting tool steels. *Monark-1* (AISI type S2) contains 0.5% carbon, 1.1% silicon, 0.4% manganese, 0.45% molybdenum, and the balance iron. It has outstanding toughness, high elastic limit, great depth of hardening and good ductility. Used for pneumatic tools, chisels, shear blades, punches, bending rolls, concrete breakers, wrenches, and stamps. *Monark-2* (AISI type S5) contains 0.6% carbon, 0.75% manganese, 2% silicon, 0.3% chromium, 0.2% molybdenum, and the balance iron. Its properties and applications are similar to those of Monark-1.

Monastral Blue. Trade name for blue pigments including *Monastral Blue BF*, a bright blue pigment (Color Index No. C.I. 74160) based on *copper phthalocyanine blue* and having a dye content of about 90-97%, and *Monastral Blue B*, an aqueous suspension of 3% *phthalocyanine blue* in 0.85% sodium chloride (NaCl). *Monastral Blue* pigments are used in paints, lacquers, inks, enamels, plastics, rubber and paper, in colored chalks and pencils, and in tinplate printing.

Monax. (1) Trade name of Corning, Inc. (USA) for a white-diffusing opal glass used for lamp shades, architectural glass products, etc.

(2) Trademark of Tredegar Industries Inc. (USA) for flexible film-like laminate composed of two or more layers of plastic material stretched in one direction and used for packaging applications. *Monax+* is an uniaxially oriented high-density polyethylene film.

monazite. A yellowish, reddish-orange, or brown, radioactive mineral with a white streak and a vitreous to resinous luster. It is a rare-earth phosphate, $(Ce,La,Y,Th)PO_4$, that contains considerable amounts of thorium, cerium, lanthanum and yttrium oxide, moderate amounts of neodymium, praseodymium and samarium oxide, and trace amounts of silicon dioxide, and aluminum, gadolinium, holmium and dysprosium oxide. Crystal system, monoclinic. Density, 5.06-5.45 g/cm³; hardness, 5-5.5 Mohs; refractive index, 1.79. It is the commercial material obtained from *monazite* sand (the crude natural material) by purification. Occurrence: Australia, Brazil, Canada, East Indies, India, Sri Lanka, Switzerland, USA (Arkansas, Colorado, Florida, Idaho, Montana, North Carolina, South Carolina). Used as the principal ore of rare-earth elements.

monazite sand. See monazite.

moncheite. A steel-gray mineral of the brucite group composed of palladium platinum bismuth telluride, $(Pt,Pd)(Te,Bi)_2$. Crystal system, hexagonal. Density, 7.08 g/cm³. Occurrence: Russian Federation. It can also be made synthetically, but the synthetic mineral is composed of platinum telluride ($PtTe_2$) and has a higher density (about 10.1-10.2 g/cm³).

Mond. Trade name of Mond Nickel Company Limited (UK) for a Monel-type alloy containing 70% nickel, 26% copper and 4% manganese. It has good corrosion resistanc, high electrical resistivity, and is used for turbine blades, valve seats and stems, pump rods, resistance wire, and chemical and mining machinery. Also known as *Mond Nickel*. See also Monel.

Mond Nickel. See Mond.

Mondoza. Trademark of Petroquimica Cuyo SAIC (Argentina) for polypropylene filaments and fibers.

Monel. Trademark of Inco Alloys International, Inc. (USA) extensive series of silver-colored, highly corrosion-resistant wrought or cast alloys containing 60-70% nickel, 25-38% copper, 1-7% iron, 0.25-2% manganese, 0.2-1.5% silicon and 0.5-3% carbon. Free-machining grades contain a trace of tellurium or sulfur, and some other grades also contain small percentages of aluminum (up to 3%), titanium (up to 1%) and cobalt (up to 1%). It is commercially available in the form of foil, mesh, wire, rod, tube and powder. The standard 65Ni-33Cu-2Fe alloy has a density of 8.84 g/cm³ (0.319 lb/in.³), a melting point of about 1300-1350°C (2370-2460°F), a maximum service temperature (in air) of 450°C (840°F), and a hardness of 110-370 Brinell. All *Monel* alloys possess low thermal expansion, high tensile strength, good mechanical properties, good resistance to sulfuric, hydrofluoric, phosphoric and organic acids, good resistance to alkalies, salts, seawater and chloride cracking, fair resistance to hydrochloric acid, and poor resistance to nitric acid. Uses include valves and valve parts, pumps and pump parts, valve trim, fittings, acid and petroleum-refining equipment, oil-well components, crude petroleum stills, process vessels and piping, chemical processing equipment, gasoline and freshwater tanks, chemical and marine parts and fixtures, water meter parts, pickling baskets, feedwater heaters, heat exchangers, fishing line, bellows, doctor blades and scrapers, small machine parts, springs, fasteners, impellers, bellows, electrical and electronic components, waveguides, electronic power tubes, transistor capsules, and magnetic equipment.

Monemel. Trade name of Esperance Longdez (Belgium) for a nonaging, decarburized low-carbon steel sheet used for direct-on porcelain enameling applications. Extra deep drawing grades are also available.

Monend. Trade name of Arcos Alloys (USA) for a Monel-type alloy containing 64% nickel, 28% copper, 3.7% manganese, 0.8% titanium, 0.4% aluminum, 0.3% iron and 0.8% silicon. Used in the form of welding electrodes for nickel and nickel-base alloys including *Monel*.

Moneva. Trade name of MONEVA GmbH (Germany) for aluminum die castings.

Moneval. Trade name of Arcos Alloys (USA) for a Monel-type alloy containing 64-68% nickel, 27-30% copper and 0-1% iron. Used as welding electrodes for *Monel* and other nickel-copper alloys.

Monikrom. Trade name of Midland Motor Cylinder Company (USA) for an alloy cast iron containing 3.1-3.4% carbon, 2-2.4% silicon, 0.7% manganese, 0.8-1% chromium, 0.15-0.25% nickel, 0.15-0.25% molybdenum, and the balance iron. Used for camshafts and cam sleeves.

Monimax. Trade name of Allegheny Ludlum Steel (USA) for a magnetically soft alloy containing 50% iron, 47% nickel and 3% molybdenum. It has high saturation induction and moderate initial permeability. Used for tape recording heads, vibrator cores, small motors, special transformers, and electrical equipment.

monimolite. A yellow, gray or dark brown mineral of the pyrochlore group with a greasy or adamantine luster. It is composed of lead antimony oxide, $Pb_3Sb_2O_7$. Crystal system, cubic.

Monit. Trade mark of Nyby Uddeholm AB (Sweden) for a standard ferritic stainless steel containing 0.025% carbon, 1% manganese, 0.75% silicon, 24-26% chromium, 3-5% nickel, 3-5% molybdenum, a total of 0.07% sulfur and phosphorus, traces of nitrogen, titanium and niobium, and the balance iron. Available in sheet, pipe and tube form, it has excellent corrosion resistance even in seawater environments, and is used for valve and pump parts, marine parts and equipment, and parts exposed to highly corrosive environments.

Monix. Trade name of Saarstahl AG (Germany) for several nickel-chromium-molybdenum and chromium-nickel-molybdenum alloy steels used for machine parts and automotive components. Carburizing grades are also available.

monk's cloth. A heavy, coarse cotton fabric in a loose basket weave, used especially for draperies and upholstery. Also known as *abbot*.

MONO. Trade name of Tremco Inc. (USA) for a range of caulks and sprayable insulating foams including acrylic exterior sealants and interior/exterior latex sealants.

monoamine. An amine containing one amino group and having the general formula is RNH_2.

monoazo dye. An *azo dye* that has one azo ($-N{\equiv}N-$) group in the molecule.

monobasic ammonium phosphate. See ammonium dihydrogen phosphate.

monobasic calcium phosphate. See calcium phosphate (i).

monobasic potassium phosphate. See potassium dihydrogen phosphate.

monobasic sodium phosphate. See monosodium phosphate.

Mono-Block. Trademark of Keene Corporation (USA) for rock wool available in the form of blocks and slabs, and used for thermal insulation applications (e.g., low- or high-temperature protection).

Monobond-S. Trade name of Ivoclar Vivadent AG (Liechtenstein) for a dental bonding agent.

monocalcium ferrite. See calcium ferrite (i).

monocalcium phosphate. See calcium phosphate (i).

MonoCast. (1) Trademark of The Polymer Corporation (USA) for cast *monadic nylon* in bar, rod, plate and shape form.

(2) Trademark of Quaker Alloy Casting Company (USA) for ceramic-shell investment castings.

Mono-Coat. Trademark of Chem-Trend Inc. (USA) for an extensive series of specialized release agents used for molded plastics, and in the processing of advanced composites.

Monocoustic. Trade name of Owens-Corning Fiberglas Corporation (USA) for large-module acoustical ceiling panels.

Monocryl. (1) Trade name of Ethicon, Division of Johnson & Johnson (USA) for a biodegradable, bioabsorbable block copolymer of *glycolide* and *ε-caprolactone* that has reduced stiffness and is used in the form of monofilaments for tissue engineering applications, especially surgical sutures.

(2) Trade name of Monopol AG (Switzerland) for finishing varnishes and high-solids paints containing micaceous iron oxide.

monodisperse. A substance in which there is only one size of molecules, clusters, particles, etc.

monodisperse polymer. A polymer system in which there is only one molecular weight present. See also polydisperse polymer.

Monofer. Trade name of Monopol AG (Switzerland) for one-coat paints and synthetic resin finishes.

Monofil. Trade name for rayon fibers and yarns used for textile fabrics.

monofilament. (1) An individual, continuous fiber of small cross section, usually extruded, spun or drawn from a material, such as glass, nylon, acetate, rayon, sapphire, silicon carbide, boron, or a metal.

(2) A strong, flexible individual filament or yarn that can be braided, knitted, woven or otherwise made into textile fabrics.

monofilament yarn. A yarn made up of an individual, continuous filament of a natural or synthetic fiber.

Monoflex. Trade name of Monopol AG (Switzerland) for synthetic resin finishing coats and micaceous iron oxide paints.

MonoFoam. Trademark of DAP, Inc. (USA) for several polymer foam sealants.

Monofrax. Trademark of The Carborundum Company (USA) for fusion-cast refractories composed of 98% α- and β-alumina (Al_2O_3), about 2% other oxides and selected additives, and supplied in the form of standard bricks. They have a density range of 2.8-3.2 g/cm^3 (0.10-0.11 $lb/in.^3$), outstanding resistance to abrasion, wear, impact and chemicals, excellent oxidation resistance, and operating temperatures up to 1649°C (3000°F). Used for linings for process equipment, glass furnaces and heat-treating equipment.

Monograph. (1) Trademark of Sigri Great Lakes Carbon Corporation (USA) for monolithic graphite used for composite tooling.

(2) Trademark of Airtech International, Inc. (USA) for an epoxy resin for high-temperature tooling applications.

monohalogen compound. See halogenated hydrocarbon.

monohydrated lime. Dolomitic lime that contains at least 8% unhydrated oxides, and is made by hydration at atmospheric pressure. See also dihydrated lime; dolomitic lime.

monohydrocalcite. A mineral composed of calcium carbonate monohydrate, $CaCO_3 \cdot H_2O$. Crystal system, hexagonal. Density, 2.39 g/cm^3. Occurrence: South Australia.

Monolite. Trade name of Monowatt Electric Corporation (USA) for phenol-formaldehyde plastics and laminates used for electrical applications.

monolithic concrete. Plain or reinforced concrete cast without the formation of any joints other than construction joints.

monolithic refractories. Integral structures that are formed by casting, gunning, ramming or sintering crushed refractory materials or mixtures of materials into position without the formation of any joints other than construction joints. They are highly resistant to high-temperature corrosion by slags and glasses. Also known as *monolithics*. See also castable refractory; ramming mix.

monolithics. See monolithic refractories.

Mono-Lok. Trade name of Rocky Mountain Orthodontics Inc. (USA) for bonding resins used in orthodontics. *Mono-Lok 2* is a visible-light-activated orthodontic bonding system for metal and ceramic brackets.

Mono-Loy. Trade name of Grand Northern Products Limited (USA) for a shock- and wear-resisting steel containing 1.4% carbon, 0.9% silicon, 13% manganese, 0.5% chromium, 0.5% molybdenum, 5.5% nickel, and the balance iron. Used for hard-facing electrodes employed for overlaying or buildup on heavy industrial equipment.

monomers. A group of simple, unsaturated chemical substances of relatively low molecular weight consisting of single like or unlike molecules that are capable of joining together to form polymers or copolymers. They are the smallest repeating structures of polymers. For example, ethylene (C_2H_4), propylene ($C_2H_3CH_3$), styrene ($C_2H_3C_6H_5$) and vinyl chloride (C_2H_3Cl) are the monomers used for the preparation of polyethylene, polypropylene, polystyrene and polyvinyl chloride, respectively. See also copolymers; homopolymers; polymers.

Monophase. See SternVantage Monophase.

Monoplast. Trade name of Monopol AG (Switzerland) for plastic fillers and plastic textured finishes.

Monopren Transfer. Trade name for a dental impression silicone.

Monopur. Trade name of Monopol AG (Switzerland) for a series of coating products including one-coat and micaceous iron oxide paints, and zinc-dust and penetrating primers.

Monoquick. Trade name of Monopol AG (Switzerland) for quick-drying one-coat paints.

monosaccharide. A carbohydrate that is a derivative of a straight-chain polyhydric alcohol and cannot be hydrolyzed to simpler compounds. Examples include fructose, glucose, mannose and ribose. Also known as *simple sugar.*

Monosheer. Trade name for nylon fibers and yarns used for textile fabrics including clothing.

monosodium phosphate. Any of the following compounds: (i) *Anhydrous monosodium phosphate.* A water-soluble, white, crystalline powder (99+% pure) that forms disodium pyrophosphate at 225-250°C (437-482°F) and sodium metaphosphate at 350-400°C (662-752°F). Formula: NaH_2PO4; (ii) *Monosodium phosphate monohydrate.* Transparent, water-soluble crystals (97+% pure) that have a density of 2.040 g/cm^3 and lose H_2O at 100°C (212°F). Used in electroplating, dyeing and boiler water treatment, as an acid cleanser, emulsifier, food supplement, laboratory reagent, and acidulant, and in chemistry, biology and biochemistry as a component of buffer solutions. Formula: $NaH_2PO_4 \cdot H_2O$. Monosodium phosphate is also known as *monobasic sodium phosphate; sodium dihydrogen phosphate; primary sodium orthophosphate; sodium biphosphate; sodium acid phosphate.*

Monostrip. Trade name of Monopol AG (Switzerland) for strippable coatings.

monotropic liquid crystal. A type of *thermotropic liquid crystal*, typically below the melting point of the solid compound, whose mesomorphic state is due to a change of temperature in one direction only, usually involving either an increase in the temperature of a solid, or a decrease in the temperature of a liquid. The mesomorphic phase occurs only by supercooling the substance below its melting point. See also liquid crystal; enantiotropic liquid crystal.

monotype metal. Type metal containing about 72-78% lead, 15-18% antimony and 5-10% tin. It has excellent antifriction properties, low melting point, slight expansion on solidification, and good wear resistance and compressive strength. Used for monotyping plates.

Monova. Trademark of Stelco Inc. (Canada) for barbed wire.

monovoltine silk. Silk produced by silkworms that have only one brood per season.

Monox. (1) Trade name of George W. Prentiss & Company (USA) for an alloy of 77% nickel and 23% copper, used for instruments.

(2) Trade name for silicon monoxide (SiO).

Monsanto SEF. Trademark of Monsanto Company (USA) for modacrylic fibers.

Montan. Trade name of SWB Stahlformguss Gesellschaft mbH (Germany) for a series of water-hardening steels containing 0.7% carbon, 0-0.25% silicon, 0-0.35% manganese, and the balance iron. Used for axles, springs and rails, and various tools including punches.

Montana larch. See western larch.

Montana sapphire. A blue, steel-blue or grayish-blue *sapphire* (Al_2O_3) with slight a metallic luster, mined in Montana, USA.

Montanium. British trade name for a duralumin-type alloy composed of 96-97% aluminum, 2.5-3.5% copper and 0.5% manganese, and used for light-alloy parts.

montan wax. See lignite wax.

Montasite. South African trade name for strong asbestos fibers obtained from *amosite* (grunerite) varieties.

Montax. Trade name for a mixture of silicon dioxide powder and hydrated magnesium carbonate, used as filler in rubber, plastics and paper.

montbrayite. A yellowish white mineral composed of gold antimony telluride, $(Au,Sb)_2Te_3$. It can also be made synthetically. Crystal system, triclinic. Density, 9.94 g/cm³. Occurrence: Canada (Quebec).

montdorite. A green to brownish green mineral of the mica group composed of potassium iron magnesium manganese fluoride silicate hydroxide, $K_2(Fe^{2+},Mn,Mg)_5SiO_{20}(F,OH)_4$. Crystal system, monoclinic. Density, 3.15 g/cm³; refractive index, 1.605. Occurrence: France.

montebrasite. A white to grayish white, sometimes colorless, yellowish, pinkish, tan, greenish or bluish mineral of the amblygonite group that is composed of lithium sodium aluminum phosphate hydroxide, $(Li,Na)AlPO_4(OH,F)$, and may also contain fluorine. Crystal system, triclinic. Density, 3.03 g/cm³; refractive index, 1.6093. Occurrence: Sweden.

Montegal. Trade name of Metallgesellschaft Reuterweg (Germany) for a non-heat-treatable aluminum alloy containing 0.95% magnesium, 0.8% silicon and 0.2% calcium. Used for light-alloy parts.

monteponite. A bright red to dark brown mineral of the halite group composed of cadmium oxide, CdO. It can also be made synthetically. Crystal system, cubic. Density, 8.15 g/cm³; refractive index, 2.49. Occurrence: Italy.

monteregianite. A colorless white, gray mineral composed of potassium sodium yttrium silicate decahydrate, $(Na,K)_6Y_2Si_{16}O_{38}\cdot10H_2O$. Crystal system, orthorhombic. Density, 2.42 g/cm³; refractive index, 1.513. Occurrence: Canada (Quebec).

montgomeryite. An orange-brown mineral composed of calcium magnesium aluminum phosphate hydroxide dodecahydrate, $Ca_4MgAl_4(PO_4)_6(OH)_4\cdot12H_2O$. Crystal system, monoclinic. Density, 2.52 g/cm³. Occurrence: USA (Utah, South Dakota).

monticellite. A colorless or gray mineral of the olivine group composed of calcium magnesium silicate, $CaMgSiO_4$. It can also be made synthetically from a mixture of calcium carbonate ($CaCO_3$), basic magnesium carbonate [$3MgCO_3\cdot Mg(OH)_2\cdot 3H_2O$] and silicon dioxide ($SiO_2$) by a prescribed heating regime. Crystal system, orthorhombic. Density, 3.06 g/cm³; refractive index, 1.649. Occurrence: Found embedded in limestones. Used in ceramics.

montmorillonite. (1) Any of a group of important clay minerals that are usually white, yellow or green hydrated silicates of aluminum or magnesium, $(Al,Mg)_8(Si_4O_{10})_3(OH)_{10}\cdot xH_2O$, frequently containing small amounts of iron, calcium, sodium, potassium, etc. Crystal system, usually monoclinic, but also orthorhombic. Density range, 2.3-2.5 g/cm³; refractive index, 1.5-1.64; principal constituent of *bentonite* and *fuller's earth*; great affinity for water; high swelling capacity. Occurrence: Norway, USA (Arizona, Mississippi, Wyoming). Used as a lubricant in pottery bodies, in ceramic clays, as a bonding agent for molding sand, and in ion exchange.

(2) The grayish, blue or pale red, aluminum-rich end member of the montmorillonite group of minerals corresponding to the formula $(Al,Mg)_8(Si_4O_{10})_3(OH)_{10}\cdot12H_2O$. It is the major component of *bentonite* and *fuller's earth*. Structurally, it consists of agglomerated nanoscale particles. Used in ion exchange, ceramics, materials research, and chemistry.

Montpelier yellow. See Turner's yellow.

Montrel. Trade name of Montrel Company (USA) for polyolefin fibers and yarns used for textile fabrics.

montroseite. A black mineral of the diaspore group composed of vanadyl hydroxide, VO(OH). Crystal system, orthorhombic. Density, 4.00 g/cm³. Occurrence: USA (Utah, Colorado).

montroydite. A yellow mineral composed of mercury oxide, HgO. It can also be made synthetically. Crystal system, orthorhombic. Density, 11.22 g/cm³; refractive index, 2.5. Occurrence: France.

Monumental. Trade name of Gerresheimer Glas AG (Germany) for a patterned glass.

Monumentale. Trade name of Fabbrica Pisana SpA (Italy) for a patterned glass.

mooihoekite. A metallic yellow mineral of the chalcopyrite group composed of copper iron sulfide, $Cu_9Fe_9S_{16}$. It can also be made synthetically. Crystal system, tetragonal. Density, 4.36 g/cm³. Occurrence: South Africa.

moonstone. A translucent or semitransparent potassium and/or sodium feldspar with white to bluish opaline or pearly luster. Occurrence: Australia, Brazil, Canada, Sri Lanka, Switzerland, USA (California, Pennsylvania, Virginia). Flawless species are used as gemstones. Also known as *hecatolite*.

moonstone glass. A milky white to translucent glass that has a pearly or opalescent appearance, and resembles the mineral *moonstone*.

mooreite. A glassy, white mineral composed of magnesium manganese sulfate hydroxide octahydrate, $(Mg,Mn,Zn)_{15}(SO_4)_2(OH)_{26}\cdot8H_2O$. Crystal system, monoclinic. Density, 2.47 g/cm³; refractive index, 1.545. Occurrence: USA (New Jersey).

Moor Glo. Trade name Benjamin Moore & Company Limited (Canada) for latex house paint.

moorhouseite. A pink mineral of the hexahydrite group composed of cobalt sulfate hexahydrate, $CoSO_4\cdot6H_2O$. Crystal system, monoclinic. Density, 1.97 g/cm³. Occurrence: Canada

(Nova Scotia).

Mopack. Trade name of Mopack Handelsgesellschaft (Germany) for high-quality packaging film products including polyvinyl chloride and polypropylene shrink films, polyethylene standard and shrink films, and stretch films made of various plastics.

Moplefan. Trademark of Montecatini Edison SpA (Italy) for polypropylene film.

Moplen. Trademark of Montecatini Edison SpA (Italy) for synthetic resins including chemically inert polypropylene homopolymers and copolymers with excellent electrical properties, good fatigue strength and heat-distortion resistance, but relatively poor resistance to ultraviolet light. Used for molded articles. Also included under this trademark are synthetic continuous and staple fibers used for spinning into yarn, and for synthetic rubber.

Moplex. Trade name of Montedison SA (Italy) for polyethylene plastics.

mopungite. A mineral of the sohngeite group composed of sodium antimony hydroxide, $NaSb(OH)_6$. Crystal system, tetragonal. Density, 3.22 g/cm^3.

moquette. (1) A firm fabric, usually double woven, having a cut or uncut pile. Used especially for upholstery.

(2) A heavy, firm cotton or wool fabric with a cut or uncut mohair, worsted or nylon pile, used especially for upholstery, drapes, tablecloth, and carpets.

moraesite. A white mineral composed of beryllium phosphate hydroxide tetrahydrate, $Be_2(PO_4)(OH) \cdot 4H_2O$. Crystal system, monoclinic. Density, 1.80 g/cm^3; refractive index, 1.482. Occurrence: Brazil.

Moraine. Trade name of Moraine Manufacturing Inc. (USA) for sintered tin bearing bronzes and composite bearing metals consisting of aluminum-silicon alloy applied to a backing of steel and coated with a leaded tin bronze electrodeposit.

morass ore. See bog iron ore.

mordenite. A white, pinkish or yellowish mineral of the zeolite group composed of calcium aluminum silicate heptahydrate, $(Ca,Na_2,K_2)Al_2Si_{10}O_{24} \cdot 7H_2O$. Crystal system, orthorhombic. Density, 2.10 g/cm^3; refractive index, 1.48. Occurrence: Scotland, Italy. Used as an ion exchanger. Also known as *ashtonite*.

morelandite. A gray, or light yellow mineral of the apatite group composed of barium chloride arsenate, $Ba_2(AsO_4)_3Cl$. Crystal system, hexagonal. Density, 5.33 g/cm^3; refractive index, 1.880. Occurrence: Sweden.

morenosite. A green mineral of the epsomite group composed of nickel sulfate heptahydrate, $NiSO_4 \cdot 7H_2O$. Crystal system, orthorhombic. Density, 1.98 g/cm^3; refractive index, 1.489. Also known as *nickel vitriol*.

Moresco. Trade name of Vetreria Milanese Lucchini Perego (Italy) for a patterned glass.

Morex. Trade name of Darwins Alloy Castings (UK) a molybdenum high-speed steel containing 0.7% carbon, 4% chromium, 4-6% molybdenum, 6% tungsten, 2% vanadium, and the balance iron. Used for tools and cutters.

morganite. A rose-red variety of the mineral *beryl* $(Be_3Al_2Si_6O_{18})$ containing lithium and obtained from California, USA, and Madagascar. Used as a gemstone.

moreen. A strong, heavy fabric woven from wool, cotton, or a mixture of wool and cotton fibers, and usually having a moiré-type watered finish. Used for especially for curtains and upholstery.

morinite. A colorless to pale pink mineral composed of sodium calcium aluminum fluoride phosphate hydroxide dihydrate, Na-$Ca_2Al_2F_4(PO_4)_2(OH) \cdot 2H_2O$. Crystal system, monoclinic. Density, 2.98 g/cm^3; hardness, 4 Mohs. Occurrence: USA (South Dakota).

Morin's Chinese bronze. A leaded bronze containing 83% copper, 10% lead, 5% tin and 2% zinc, used for ornaments.

Morocco. (1) A fine, vegetable-tanned leather with characteristic pebbly grain, made from *goatskin*, and usually stained red. Used in bookbinding. Also known as *Morocco leather*.

(2) Any leather with embossed pebbled grain imitating *Morocco leather*.

Morocco leather. See Morocco.

Morocco Pinhead. Trade name of Pilkington Brothers Limited (UK) for a patterned glass.

Morones. Trade name of Vitrobrás (Brazil) for a figured glass.

moroti. See pau marfim.

Morphos. Trademark of Heatbath Corporation (USA) for iron phosphate coatings for steel, zinc and aluminum. They are supplied as spray or immersion products for use as paint bases, and corrosion resistance enhancers.

Morrison's Ductile Bronze. Trade name of William Gallimore & Sons Limited (UK) for a ductile bronze containing 91% copper and 9% tin. It has good cold workability, and is used for valves, pump parts, hardware, and gears.

mortar. A bonding agent that is a mixture of cementitious materials (e.g., cement or lime), fine aggregate (e.g., sand) and water. It is plastic and trowelable when fresh, but sets to a hard infusible solid by hydraulic action. *Mortar* possesses good acid resistance, and is available in special grades with good high-temperature properties up to 870°C (1600°F). Used as a building material, in bricklaying, and for lining chemical reaction equipment.

mortar admixture. A substance incorporated into a mortar to control the rate of setting, increase water repellency and, sometimes, effect color changes.

mortar-mix clay. A finely divided clay used as a plasticizer in masonry mortars.

Morthane. Trademark of Morton International Inc. (USA) for thermoplastic urethane elastomers and polyurethane resins used for molding compounds, and in the manufacture of adhesives and extruded products.

Mosaic. Trade name of SA des Verreries de Fauquez (Belgium) for a glass with mosaic pattern.

mosaic glass. An antique glass consisting of a matrix of transparent glass with a decoration of multicolored glass shapes, floral patterns, etc.

mosaic gold. (1) A yellow brass containing 63% copper and 37% zinc, used for ornaments and architectural parts.

(2) See artificial gold.

Mosaico. Trade name of Fabbrica Pisana SpA (Italy) for a patterned glass.

mosaic silver. An amalgam of bismuth, mercury and tin used for imitating silverwork and for bronzing applications.

mosaic tile. A porcelain or natural clay tile, glazed or unglazed, and shaped to facial dimensions of less than 152 × 152 mm (6 × 6 in.). Its nominal thickness ranges from 6.4 to 9.5 mm (0.250 to 0.375 in.), and it is often mounted on paper support to facilitate setting.

Mosaik. German trade name for glass blocks with mosaic patterns.

Mosaique. Trade name of SA Glaverbel (Belgium) for a patterned glass featuring irregular shapes.

mosandrite. A yellow to brown mineral composed of calcium titanium zirconium fluoride silicate hydroxide, $(Ca,Na,Ce)_{12}(Ti,Zr)_2Si_7O_{31}H_6F_4$. Crystal system, triclinic. Density, 3.46 g/cm³; refractive index, 1.668. Occurrence: Greenland.

moschellandsbergite. A silver-white mineral composed of mercury silver, γ-Ag_2Hg_3. Crystal system, cubic. Density, 13.60 g/cm³. Occurrence: Germany.

mosesite. A yellow to black mineral composed of mercury nitrogen chloride sulfate monohydrate, $Hg_2N(Cl,SO_4,MoO_4,CO_3)\cdot H_2O$. Crystal system, cubic. Density, 7.72 g/cm³. Occurrence: Mexico.

Mosil. Trade name of Teledyne Vasco (USA) for a shock-resistant tool steel (AISI type S5) containing 0.55-0.60% carbon, 1.8-2.0% silicon, 0.8-0.9% manganese, 0.2-0.3% chromium, 0.3-0.4% molybdenum, 0.15-0.25, and the balance iron. It has outstanding toughness, high elastic limit, a great depth of hardening and good ductility. Used for hand and pneumatic chisels, punches, shear blades, rotary shears, rotary and scrap shears, and bending dies.

moss agate. A variety of *chalcedony* quartz containing brownish or black dendrites resembling moss in appearance.

moss crepe. A crepe-weave fabric having a pebbly, spongy surface, and usually made with high-twist acetate, cotton, polyester, rayon or wool yarn. Used especially for women's fashion, e.g., dresses and suits. Also known as *pebble crepe*.

mossy tin. A powder obtained by pouring molten tin into cold water. Used chiefly for alloying applications.

mossy zinc. A powder obtained by pouring molten zinc into cold water. Used in the production of color effects on face brick.

Most. Trade name of Multi-Arc Inc. (USA) for a physical vapor deposited (PVD) coating for metal-forming tools including punches, and blanking, drawing and forming tools.

Mostar. Trade name of Société Nouvelle des Acieries de Pompey (France) for an air- or oil-hardened hot-work tool steel containing 0.32% carbon, 1.2% silicon, 5% chromium, 1.15% molybdenum, 0.35% vanadium, and the balance iron. Used for upsetters, punches, and extrusion dies.

Mo-Steel. Trade name of AMAX Corporation (USA) for a series of alloy steels. *Mo-Steel No. 1* is a tough, shock-resistant case-hardening steel containing 0.15-0.23% carbon, 0.7-1% chromium, 0.25-1% molybdenum, 0.4-0.7% manganese, and the balance iron. Used for gears and shafts. *Mo-Steel No. 2* is a water-hardening machinery steel containing 0.23-0.30% carbon, 0.8-1.1% chromium, 0.25-0.4% molybdenum, 0.5-0.8% manganese, and the balance iron. Used for crankshafts, connecting rods, front axles, propeller shafts, and bolts. *Mo-Steel No. 3* is a tough, shock-resistant, oil- or water-hardening machinery steel containing 0.3-0.4% carbon, 0.8-1.1% chromium, 0.25-0.4% molybdenum, 0.5-0.8% manganese, and the balance iron. Used for gears, heavy crankshafts, driving axles, connecting rods, and piston rods.

Motal. Trade name of Aluminiumwerke Maulbronn (Germany) for a sand-cast aluminum alloy containing 5% silicon. Used for motor housings, automobile parts, and light-alloy parts.

Mo Tap. Trade name of Latrobe Steel Company (USA) for an oil- or water-hardening steel containing 1.2% carbon, 0.6% molybdenum, 0.5% chromium, 0.8% manganese, and the balance iron. Used for taps, drills, and reamers.

Motelec. Trade name of Empire Sheet & Tin Plate Company (USA) for a magnetically soft alloy of 97.5% iron and 2.5% silicon. It possesses high saturation induction and maximum permeability, low coercivity and core losses, and is used for electrical applications.

Motemp. Trade name of CCS Braeburn Alloy Steel (USA) for a molybdenum-type high-speed steel (AISI type M10) containing 0.88% carbon, 4% chromium, 8% molybdenum, 2% vanadium, and the balance iron. It has a great depth of hardening, high hot hardness, and excellent wear resistance. Used for milling cutters, end mills, drills, saws, and a variety of other tools.

mother-of-pearl. See nacre.

Mo-Tiger. Trade name of Bethlehem Steel Corporation (USA) for a high-speed steel containing 0.7% carbon, 9% molybdenum, 4% chromium, 1.5% tungsten, 4% chromium, 1% vanadium, and the balance iron. Used for drills, cutters, and various other tools.

Moto-Fiber. Trade name of Lydall, Inc. (USA) for fiberboard used for automotive applications.

Motor. Trade name of Richard W. Carr & Company Limited (UK) for a series of high-speed steels. *Motor Magnus* is a molybdenum-type high-speed steel (AISI type M2) containing 0.8% carbon, 4.25% chromium, 5% molybdenum, 6.5% tungsten, 2% vanadium, and the balance iron. It has a great depth of hardening, excellent wear resistance, good toughness, high hot hardness, and is used for gear cutters, form tools, drills, and a variety of other tools. *Motor Maximum* is a tungsten-type high-speed steel (AISI type T1) containing 0.75% carbon, 4.25% chromium, 18% tungsten, 1.2% vanadium, and the balance iron. It has great depth of hardening, high hot hardness, excellent machinability and wear resistance, and good toughness. Used for boring tools, chisels, bushings, and die-casting dies. *Motor Special* is a tungsten-type high-speed steel containing 0.7% carbon, 4% chromium, 14% tungsten, 0.25% vanadium, and the balance iron. Used for twist drills, reamers; shear blades, chisels and slitters.

Motorgrip. Trademark of Mulco Inc. (Canada) for automobile body fillers.

Motork. Trade name of Magnetic Metals Company (USA) for a corrosion-resistant, magnetically soft iron alloy containing up to 0.6% silicon. It has high saturation induction and magnetic permeability, and low coercivity. Used for motor laminations.

Motoroku. Trade name of Asahi Glass Company Limited (Japan) for a patterned glass.

Motor Vehicle Series. Trademark of AMETEK Specialty Metal Products (USA) for low-carbon, highly compressible ferritic stainless steel powders particularly suited for the manufacture of powder-metallurgy (P/M) automotive components, such as exhaust rings and support brackets, sensor rings, seals, and rearview mirror mounts. Mechanical and chemical properties of parts produced from these powders include good atmospheric and saline corrosion resistance, hot oxidation resistance and elevated-temperature yield strength. *Motor Vehicle Series* powders are currently supplied in four grades: *MVS409CB* (0.03% carbon, 11% chromium, 0.9% silicon, 0.2% manganese, 0.5% niobium, balance iron); *MVS410L* (0.03% carbon, 12% chromium, 0.6% silicon, 0.2% manganese, balance iron); *MVS430L* (0.02% carbon, 17% chromium, 0.9% silicon, 0.2% manganese, balance iron), and *MVS434L* (0.02% carbon, 17% chromium, 1% molybdenum, 0.9% silicon, 0.2% manganese, balance iron).

mottled graywear. A *porcelain enamel* coating for steel made from a frit that contains additions of cobalt monoxide (CoO) to reduce surface misting.

mottled cast irons. A group of cast irons with high carbon content (up to 4.25%) and relatively low silicon content (typically between 1.2-2.3%). They have a semi-gray color and a microstructure intermediate between that of white and gray cast iron, i.e., a pearlitic matrix with a carbon-rich phase of lamellar (flake) graphite and *cementite*. They have white fracture surfaces with clusters of lamellar (flake) graphite resulting in a mottled appearance. Also known as *mottled irons*. See also gray cast irons; white cast irons.

mottled ware. An enamel with grayish spots used to give ware a finish resembling coarse granite.

mottled yarn. A yarn made by weaving or otherwise combining two slubbings, rovings or threads of different luster or color. Also known as *marled yarn*.

mottramite. A brown or green mineral of the descloizite group composed of copper zinc lead vanadium oxide hydroxide, $(Cu,Zn)PbVO_4(OH)$. Crystal system, orthorhombic. Density, 5.90 g/cm³; refractive index, 2.26. Occurrence: UK. Used as a source of vanadium.

Motuf. Trade name of CCS Braeburn Alloy Steel (USA) for a molybdenum-type high-speed steel (AISI type M7) containing 1% carbon, 3.75% chromium, 2.1% vanadium, 1.75% tungsten, 8.75% molybdenum, and the balance iron. It has a great depth of hardening, excellent wear resistance, high hot hardness and wear resistance, and is used for reamers, broaches, milling cutters, and many other tools.

Motung. Trade name of Universal Cyclops Specialty Steels (USA) for a series of molybdenum-type high-speed steel (AISI type M1) containing 0.65-1% carbon, 3.75-4.5% chromium, 5-8.8% molybdenum, 1.5-1.8% tungsten, 1-2% vanadium, and the balance iron. They have great depths of hardening, high hot hardness, excellent wear resistance, and are used for lathe and planer tools, form cutters, milling cutters, hot punches, routers, woodworking tools, and many other tools.

moufflin. A soft, open double-faced fabric, usually made from a wool or wool-blend yarn, and used for coats and capes.

Mouldrite. Trade name of ICI Limited and BIP Plastics (UK) for molded phenol- and urea-formaldehyde plastics, compounds and adhesives.

Moulimphy. Trade name of Creusot-Loire (France) for a mold steel used for glass rolls and molds.

Moultrex. Trade name of Saarstahl AG (Germany) for an oil-hardening mold steel containing 0.37% carbon, 1.1% manganese, 2.1% chromium, 0.45% molybdenum, 0.5% nickel, 0.08% vanadium, and the balance iron. Used for plastic molds.

mounanaite. A reddish brown mineral composed of lead iron vanadium oxide hydroxide, $PbFe_2(VO_4)_2(OH)_2$. Crystal system, triclinic. Density, 4.85 g/cm³; refractive index, above 2.09. Occurrence: Gabon.

mountain blue. See copper blue.

mountain cork. A tough, fibrous, lightweight, white to grayish white variety of *asbestos*, usually *palygorskite* or some other aluminum-magnesium silicate, that resembles cork in density and texture, and is often made into lightweight insulating boards. Also known as *mountain leather; rock cork*.

mountain crystal. See quartz crystal.

mountain elm. See wych elm.

mountain green. See mineral green.

mountainite. A white mineral composed of potassium sodium calcium silicate trihydrate, $(Ca,Na_2,K_2)_2Si_4O_{10}·3H_2O$. Crystal system, monoclinic. Density, 2.36 g/cm³; refractive index, 1.510. Occurrence: South Africa.

mountain larch. See western larch.

mountain leather. See mountain cork.

mountain oak. See chestnut oak (2).

mountain wood. A brown to grayish, compact variety of *asbestos* that resembles dry or fossil wood in appearance. Also known as *rock wood*.

Mounting Plaster. Trade name of Whip Mix Corporation (USA) for a quick-setting dental plaster supplied as a white powder. When mixed with water it exhibits low compressive strength, and low setting expansion. Used especially for mounting plates.

Mounting Stone. Trade name of Whip Mix Corporation (USA) for a fast-setting dental gypsum (dental stone) supplied as a white or blue powder. When mixed with water it develops a hard, accurate surface upon setting (low setting expansion). Used especially for mounting casts to articulators.

Mount Royal. Trademark of Federated Genco Limited (Canada) for a series of white-metal bearing alloys.

Mourey white gold. A corrosion-resistant jewelry alloy of 50% gold, 35% silver, and 15% palladium.

mourite. A violet mineral composed of molybdenum uranium oxide pentahydrate, $UMo_5O_{18}·5H_2O$. Crystal system, monoclinic. Density, 4.12 g/cm³; refractive index, above 1.79. Occurrence: Russia.

mousseline. A fine, sheer, slightly crisp fabric similar to *muslin*, originally made of silk, but now also made of rayon, cotton, wool or worsted. Used especially for dresses.

Mousepad Foam Rubber. Trade name of Ludlow Composites Corporation (USA) for a composite consisting of medium open-cell styrene-butadiene rubber latex foam with fabric facing. Used for computer mouse pads, placemats, etc.

Mousse. Trade name of SA Glaverbel (Belgium) for a patterned glass.

Mousset's silver. A corrosion-resistant alloy composed of 59.5% copper, 27.5% silver, 9.5% zinc and 3.5% nickel. Used for instruments and cheap jewelry.

Mova. Trade name of Firth Brown Limited (UK) for a heat-resistant steel containing 0.2% carbon, 0.25% silicon, 0.5% manganese, 0.7% molybdenum, 0.25% vanadium, and the balance iron. Used for turbine-rotor blades and jet-engine parts.

Mo-Van. Trade name of Colt Industries (UK) for a high-speed steel containing 0.85% carbon, 4% chromium, 1.75% vanadium, 7.5% molybdenum, and the balance iron. Used for tools, cutters, and reamers.

Movar. Trademark of Benjamin Moore & Company (USA) for polyurethane finishes.

Movil. Trademark of Montecatini Edison SpA (Italy) for synthetic fibers based on polyvinyl chloride.

Movyl. Trade name for vinyon (vinyl chloride) fibers with excellent acid, alkali and mildew resistance. Used especially for textile fabrics.

Mowicoll. Trademark of Hoechst AG (Germany) for polyvinyl acetate adhesives.

Mowilith. Trademark of Hoechst AG (Germany) for an extensive series of acrylate and vinyl acetate products including various polymer and homopolymer dispersions, vinyl acetate/ethylene copolymers, vinyl acetate/ethylene/acrylate terpolymers, vinyl acetate/acrylate copolymers, acrylate polymers and acrylate/styrene copolymers, and numerous vinyl-based wood glue binders and vinyl adhesives for bonding flooring, wall and ceiling coverings, paper, packaging materials, furniture, etc.

Mowiol. Trademark of Hoechst AG (Germany) for a series of synthetic resins and adhesives including polyvinyl alcohols.

Mowital. Trademark of Hoechst AG (Germany) for polyvinyl acetal resins and products.

Mowital-Folie. Trademark of Hoechst AG, Kalle Werk (Germany) for polyvinyl butyral film products.

Mowiton. Trademark of Hoechst AG (Germany) for polyvinyl acetate resins and products.

Mowrey. Trade name of WE Mowrey Company (USA) for several dental gold alloys including wrought gold wire available in high- and low-precious metal types, and dental casting alloys, such as *Mowrey 20/46*, a medium-gold type.

Moynel. Trade name for rayon fibers and yarns used for textile fabrics.

mpororoite. A green-yellow mineral composed of aluminum iron tungsten oxide heptahydrate, $(Al,Fe)_2W_2O_9 \cdot 7H_2O$. Crystal system, monoclinic. Density, 4.59 g/cm^3. Occurrence: Uganda.

MPPO-PA alloys. Thermoplastic polymer alloys of amorphous modified polyphenylene oxide (MPPO) and semicrystalline polyamide (PA) with glass transition temperatures of about 210°C (410°F) and 260°C (500°F) respectively. They exhibit very low moisture absorption, and good chemical resistance, dimensional stability, toughness and paintability. The addition of small amounts of graphite nanotubes can render the material conductive. Used mainly for autobody panels. Also known as *MMPO-nylon alloys*.

MPPO-nylon alloys. See MPPO-PA alloys.

MPV-Delta. Trade name of Kerr Dental (USA) for a medium-gold dental bonding alloy.

MR Lock LC. Trade name of American Orthodontics (USA) for a visible, light-cure, fluoride-releasing, one-step reinforced glass-ionomer-based dental band cement.

mroseite. A colorless to white mineral composed of calcium carbonate tellurite, $CaCO_3TeO_2$. Crystal system, orthorhombic. Density, 4.35 g/cm^3; refractive index, 1.85.

mucilage. (1) A light adhesive or glue made of water-soluble vegetable gums obtained from certain plants (e.g., acacias, elms, flax, leguminous plants, or seaweeds), and used for paper glues.
     (2) A general term sometimes used for any liquid adhesive developing bonding strength.

mucopolysaccharide. See glycosaminoglycan.

Mueller. Trade name of Mueller Brass Company (USA) for an extensive series of wrought and cast coppers and copper-base alloys including oxygen-free high conductivity, phosphorus-deoxidized and tellurium- and sulfur-bearing coppers as well as red, yellow, manganese, naval and low- and high-leaded brasses, aluminum, aluminum-silicon, silicon, manganese and Tobin bronzes, and several nickel silvers, and wrought aluminum-copper alloys.

Muflex. Trade name of Engelhard Corporation (USA) for a thermostatic bimetal with high permeability. The high-expanding metal is iron, and the low-expanding metal *Invar* (64Fe-36Ni).

MUG. Trade name of Schott DESAG AG (Germany) for a special filter glass used for tanning appliances.

Mu-Guard. Trade name of Spang Specialty Metals (USA) for a series of soft magnetic alloy strips containing 45-80% nickel, 0-5% molybdenum, and the balance iron. They have medium to very high magnetic shielding values.

muga silk. See Assam silk.

Mu-Hole Punch. Trade name of Midvale-Heppenstall Company (USA) for an oil-hardening tool steel used for punches.

mull. A soft, thin, sheer fabric, usually of cotton, silk, or a cotton-polyester blend, in a plain weave.

muirite. An orange mineral of the axinite group composed of barium calcium manganese titanium silicate hydroxide, $Ba_{10}Ca_2MnTiSi_{10}O_{30}(OH,Cl,F)_{10}$. Crystal system, tetragonal. Density, 3.86 g/cm^3; refractive index, 1.697. Occurrence: USA (California).

mukhinite. A black mineral of the epidote group composed of calcium aluminum vanadium silicate hydroxide, $Ca_2Al_2VSi_3O_{12}(OH)$. Crystal system, monoclinic. Density, 3.47 g/cm^3; refractive index, 1.733. Occurrence: Siberia.

mulberry silk. Silk obtained from silkworms (*Bombyx mori*) feeding on the leaves of white mulberry trees (*Morus alba*). See also silk.

Mulcoa. Trademark of C-E Minerals (USA) for refractory alumina-silica calcines composed of 65-77% *mullite* ($Al_6Si_2O_{13}$), 13-23% glass and up to 15% *cristobalite* (SiO_2). Supplied in several grades and grain sizes, they have bulk densities of 2.60-2.88 g/cm^3 (0.09-0.10 lb/in^3), apparent porosities of 3.2-3.6%, and pyrometric cone equivalents of 35 (1785°C or 3245°F) to 39 (1866°C or 3390°F). Used as aggregates in the refractory industry. See also calcine.

mull. A very thin, light, soft fabric of cotton or cotton-polyester blend yarn. Used for dresses and underlinings.

Mullfrax. Trademark of Harbison-Carborundum Company (USA) for a series of refractory products made from synthetic mullite ($Al_6Si_2O_{13}$) in an electrical furnaces. *Mullfrax* cement is suitable for temperatures up to 1760°C (3200°F), and is used as a construction material for furnaces and kilns.

mullite. A colorless aluminum silicate mineral ($Al_6Si_2O_{13}$) rarely found in nature and usually prepared from a stoichiometric mixture of alumina (Al_2O_3) and hydrated silica ($SiO_2 \cdot xH_2O$) by repeated grinding and subsequent heating to high temperatures (typically 1725°C or 3137°F), or by the thermal decomposition of aluminum silicate minerals, such as *andalusite, kyanite, sillimanite* and various clay minerals. The approximate composition is 71-75 wt% Al_2O_3 and 25-29 wt% SiO_2 with trace amounts of iron, calcium, chromium, magnesium, manganese, nickel, titanium, and/or zirconium. Crystal system, orthorhomic. Density, 3-3.17 g/cm^3; melting point, 1810°C (3290°F); softening temperature: 1650°C (3000°F); refractive index, 1.641. Used in refractories, as strength-producing ingredient in stoneware and porcelain, and in glass manufacture. Also known as *porcelainite*.

mullite ceramics. Silicate ceramics containing substantial quantities of *mullite*. They possess low thermal expansion, good mechanical strength and excellent corrosion and heat resistance, and are used in refractories for high-temperature applications, for furnace muffles, combustion tubes, radiant furnace tubes, kiln rollers and insulating tubing for thermocouples, as strength-producing ingredients in stoneware and porcelain, and in the manufacture of glass.

mullite porcelain. A vitreous whiteware containing *mullite* as the principal crystalline phase. It has good resistance to thermal shock, chemical attack and deformation under load. Used for technical applications, such as spark plugs and laboratory ware. Also known as *mullite whiteware*.

mullite refractories. A group of high-temperature-resistant ceramics consisting predominantly of *mullite*, and usually composed of 72 wt% alumina (Al_2O_3) and 28 wt% silica (SiO_2). They have a density ranging from 3.00 to 3.17 g/cm^3 (0.108-0.115 $lb/in.^3$), a melting point of 1810°C (3290°F), and a softening temperature of 1650°C (3000°F). Used for furnace linings, and refractory cements.

mullite whiteware. See mullite porcelain.

Mulram. Trademark of Babcock & Wilcox Company (USA) for a ground refractory material based on fused *mullite*, and supplied in various grain sizes. It has an upper service temperature of 1760°C (3200°F), and good resistance to metal and slag penetration. Used in ramming mixes for lining industrial furnaces.

Multi-ABS. Trade name of Multibase Inc. (USA) for a series of acrylonitrile-butadiene-styrene (ABS) resins available in several colors (e.g., black, gray, natural and clear) and grades including utility, extrusion, high-flow, plating, high- and medium-impact modified, flame-retardant and high-heat-resistant.

Multi-Alloy. British trade name for a nickel-base superalloy containing 0.25% carbon, 20.5% chromium, 3.5% tungsten, 3.3% cobalt, 2.9% niobium (columbium), 2.7% molybdenum, 1.2% titanium, and the balance iron. It has excellent heat resistance, and is used for gas-turbine parts.

multiaxial fabrics. Essentially warp-knitted fabrics with extra threads in warp (vertical), weft (horizontal) and diagonal direction, added over the entire width and length.

Multibase. Trade name of Multibase Inc. (USA) for plastics including acrylonitrile-butadiene-styrene (ABS) terpolymers supplied as liquids, pastes, powders and granules.

Multi-Blast. Trade name of Maxi-Blast Inc. (USA) for granulated urea blasting media.

Multibond. (1) Trade name of Metalor Technologies SA (Switzerland) for a series of medium-gold dental bonding alloys including *Multibond 2000* (54% gold), *Multibond Classic* (51.5% gold) and *Multibond Special* (52.5% gold). Also included under this trade name are several corrosion-resistant nickel-chromium dental bonding alloy.

(2) Trade name of Tokuyama Dental Corporation (USA) for an enamel and dentin bonding agent.

Multi-Brade. Trade name of C&S Engineering Corporation (USA) for mass-finishing media.

Multicell. Trade name of US Filter General Filter Products (USA) for a multicellular, short-staple rayon fiber used in the manufacture of filter papers and nonwovens.

Multico. Trade name of Hall & Pickles Limited (UK) for a cobalt-tungsten high-speed steel containing 0.85% carbon, 4.5% chromium, 0.75% molybdenum, 1.5% vanadium, 22% tungsten, 11% cobalt, and the balance iron. It has a great depth of hardening, and excellent wear resistance and hot hardness. Used for heavy-duty lathe and planer tools, milling cutters, and cut-off tools.

multicomponent fabrics. Fabrics made by bonding together two or more layers, at least one of which is a textile material, by chemical, mechanical or other means.

Multicore. Trade name of Multicore Solders Limited (UK) for a series of tin, tin-antimony, tin-silver, lead-tin, lead-tin-silver, and lead-tin-cadmium solder alloys.

Multi-Cupioni. Trade name for rayon fibers and yarns used for textile fabrics.

Multi-cure. Trademark of Dymax Corporation (USA) for a series of adhesives that can be cured by activators, heat, light, or ultraviolet radiation.

Multicut. Trade name of 3M Company (USA) for coated abrasives used for finishing stainless steel and titanium.

multidirectionally reinforced carbon/graphite matrix composites. A subcategory of *carbon-carbon composites* containing carbon or graphite fibers in three or more directions. They have relatively low thermal expansion and high strength, and are used in the aerospace and aircraft industries for nose cones, rocket motor parts, gas-turbine engines and heat-resistant structural panels, and as automotive and aircraft-brake friction materials.

multidirectionally reinforced ceramics. A group of engineering composites composed of three-dimensional preforms of continuous ceramic-fiber reinforcements embedded in ceramic matrices. They possess good oxidation resistance and high-temperature stability, good dielectric properties, and improved fracture toughness. Examples include three- and four-directional silica-silica composites and three-directional alumina-alumina composites. Used for radomes. See also preform.

multidirectionally reinforced fabrics. Fabrics made by weaving fibers of carbon/graphite, alumina, aramid, glass, or silicon carbide, into a three-directional *preform*. Used as reinforcements in engineering composites.

multidirectional tape prepreg. See tape (3).

Multidisc. Trade name of Saint-Gobain (France) for a toughened plate glass made by the method of surrounding zones.

Multifil. Trademark of Heraeus Kulzer GmbH (Germany) for ceramics and anterior composite resins used for dental restorative applications.

multifilament. A fiber consisting of two or more fine, continuous filaments of natural or synthetic material twisted together. See also filaments.

multifilamentary superconducting alloy. A superconducting composite wire consisting of several filaments, e.g., a wire composed of filaments of superconducting niobium-titanium alloy embedded in a copper matrix, or of a core of tin surrounded by several superconducting niobium filaments embedded in a copper matrix. See also filamentary alloys; composite superconductors.

multifilament yarn. A yarn made by twisting together two or more fine, continuous natural or synthetic filaments. See also filament.

Multi-Flam. Trade name of Multibase Inc. (USA) for polypropylene resins.

Multi-Flex. Trademark of Multibase Inc. (USA) for an extensive series of thermoplastic elastomers available in various colors (e.g., natural, clear, black, gray or brown), and several unfilled and filled grades. *Multi-Flex TEA* refers to an extensive series of thermoplastic elastomer alloys which are easy-to-process, adhere well to many substrates, can be painted, and are used for exterior and interior automotive applications.

Multiflex. (1) Trade name of Chemtron Manufacturing Limited (Canada) for a single-component, non-sag, low-temperature-gunnable elastomeric sealant based on a high-molecular-weight ethylene copolymer. It is has good resistance to stress, shearing, tearing and weathering, and bonds quickly and extremely well to most building materials including brick, concrete, granite, limestone, marble, glass, ferrous metals, and untreated woods.

(2) German trade name for a sand-blasted or ground patterned glass with raised portions.

Multi-Form. Trade name of Appleton Mills Forming Fabrics (USA) for multiple-layer forming fabrics for fourdrinier machines.

Multiform. Trade name of Corning Glass Works (USA) for a specialty glass developed to produce difficult shapes. It is made by mixing powdered glass with a suitable binder, pressing into shapes, and heating or sintering.

Multi-Glass. Trade name of MacPherson & Company, Inc. (USA) for a penetration-resistant composite glass used for industrial equipment window units.

Multi-Hips. Trade name of Multibase Inc. (USA) for high-impact polystyrenes.

Multi-Hone. Trade name of C&S Engineering Corporation (USA) for mass-finishing media composed of *quartz* abrasives in polyester resins.

Multilead. Trade name of Corning Glass Works (USA) for technical glass products having parallel conductors inserted into the glass which hold them rigidly in a hermetic seal.

multilobal fibers. Synthetic fibers having a cross section that resembles an inwardly curved polygon with rounded lobes or vertices. For example, hexalobal fibers have inwardly curved hexagonal cross sections with 6 rounded lobes.

Multilon. Trade name of Teijin Chemicals Limited (Japan) for polymer alloys of polycarbonate and acrylonitrile-butadiene-styrene.

Multilux. Trade name of former Erste Oesterreichische Maschinglasindustrie AG (Austria) for a cast glass with a patented prismatic patterned surface.

Multimet. Trade name of Haynes International Inc. (USA) for wrought and cast heat-resistant alloys containing 20-22.5% chromium, 19-21% nickel, 18-25% cobalt, 2.5-3.5% molybdenum, 2-3% tungsten, 1-2% manganese, 0.08-0.16% carbon, 0.75-1.5% niobium (columbium), 0.1-0.2% nitrogen, and the balance iron. They have excellent high-temperature properties, good performance under stress up to 815°C (1500°F), and moderate performance under stress to 1093°C (2000°F). Used for turbine blading, jet and combustion chambers, and welding rods.

Multimold. Trade name of Bethlehem Steel Corporation (USA) for an oil-hardening low-alloy tool steel containing 0.35% carbon, 0.7% manganese, 0.8% chromium, 0.3% molybdenum, and the balance iron. Used for die-casting dies, and plastic molds.

multioriented prepreg. See tape (3).

Multipane. Trade name of Multipane Inc. (USA) for an insulating glass.

Multipass. Trade name of Haynes International Inc. (USA) for a series of impact-resistant steels containing 0.1-0.2% carbon, 0-2% chromium, 0.4-1.4% silicon, 0.9-2% manganese, 0-0.4% molybdenum, and the balance iron. Used for electrodes employed for arc welding and build-up operations.

Multi-PF. Trademark of Degussa-Ney Dental (USA) for a palladium-free, gold-colored dental alloy containing 69% gold, 14% silver and 10% platinum. It provides excellent marginal stability, high yield strength, a hardness of 230 Vickers, and is well suited for use with low-fusing, high-expansion porcelains, such as *Duceragold*. Used in restorative dentistry, especially for crowns and bridges.

Multiphase. Trade name of SPS Technologies (USA) for a series of alloys containing cobalt, nickel, chromium, molybdenum, iron, titanium, niobium (columbium) and aluminum. They possess high fatigue strength, and good high temperature properties including creep rupture. Used for high-strength bolts and fasteners.

multiphase alloys. Alloys with two or more phases (i.e., chemically and structurally homogeneous microstructural portions). Examples include cast iron, steels, and solders. See also single-phase alloys.

multiphase materials. Materials with two or more phases (i.e., chemically and structurally homogeneous microstructural portions). Examples include multiphase alloys, cemented carbides (e.g., tungsten carbide in a cobalt binder), carbon-filled rubber, Portland cement, paints, and reinforced concrete or plastics. See also single-phase materials.

Multiplate. Trade name of PPG Industries Inc. (USA) for a bulletproof glass composed of several layers of polished plate glass interleaved with sheets of transparent plastic material.

Multiple. Trade name of Degussa Ney Dental (USA) for a series of precious, semiprecious and nonprecious dental and medical alloys including various gold-based types as well as ceramic dental bonding alloys.

multiple fabrics. (1) Hybrid fabrics made by weaving together two or more different fibers. See also hybrid fabrics.

(2) Fabrics consisting of two or more fabric layers combined by individually weaving together the warp and weft (filling) threads of each layer.

multiple-layer adhesive. An adhesive product made by applying a different dry-bond adhesive composition to each side of a suitable backing. Used to bond unlike materials.

multiple-ply plate. A relatively thick composite product made by welding together sheets composed of different steels, or wrought iron.

multiple wound yarn. A yarn made by winding together two or more filaments.

multiple yarn. See plied yarn.

Multiplex. Trademark KBAlloys Inc. (USA) for a series of aluminum master alloy hardeners consisting of two or more chemical elements in an exact ratio.

multiplex yarn. A filament yarn consisting of differently textured *single yarns*.

MultiPly. Trade name of Weyerhaeuser Limited (Canada) for utility panels made from premium-quality *hardboard* and supplied with fully sanded surfaces. Used as floor underlayments.

Multi-Pro. Trade name of Multibase Inc. (USA) for thermoplastic resins including polypropylenes available in various colors (e.g., black, gray, white, natural, dark green or yellow) and unfilled and filled grades including glass-, titanium dioxide-, calcium carbonate-, mineral-, talc- and mica-filled, high- and medium impact, high-gloss, and high- and medium-impact. Used for injection-molded and extruded products.

Multi-San. Trade name of Multibase Inc. (USA) for styrene-acrylonitrile (SAN) copolymers.

Multiseal. Trademark of Chemtron Manufacturing Limited (Canada) for a silicone sealant for nonporous surfaces.

Multisheer. Trade name for nylon fibers and yarns used for textile fabrics including clothing.

Multishift. Trade name of Sandvik Steel (USA) for a high-tensile strength strip steel used for bandsaw blades with improved service life.

Multi-strata. Trade name for rayon fibers and yarns used for textile fabrics.

Multi-Tech 2000. Trade name of Copperweld Steel Company (USA) for tellurium-treated alloy steel bar products with enhanced cold formability and machinability. The addition of tellurium changes the morphology of the sulfide inclusions from long and stringy to elliptical, resulting in a product with greatly improved isotropic properties.

Multi-Triplex. Trade name of Société Industrielle Triplex SA (France) for a laminated glass composed of four sheets of glass with three plastic interlayers.

multivoltine silk. See polyvoltine silk.

multi-walled nanotubes. See carbon nanotubes.

Multole. Trade name of Boyd-Wagner Company (USA) for a tool steel containing 0.6% carbon, 0.7% manganese, 1.9% silicon, and the balance iron. Used for punches, lathe and planer tools, and hand cutting tools.

Mumetal. Trade name of Spang & Company (USA) for a soft magnetic alloy containing 75-78% nickel, 13-17% iron, 4-6%

copper, and up to 4% molybdenum, 2% chromium, 1.5% manganese, 0.5% silicon and 0.5% carbon, respectively. Supplied in the form of strip, sheet and foil, it has a density of 8.8 g/cm³ (0.32 lb/in.³), a low coefficient of expansion at ordinary temperatures, high strength, a hardness of 105-290 Brinell, high magnetic permeability, and low field strength and hysteresis loss. Used for transformers, chokes, and magnetic shielding applications.

Mundet. Trade name of former Mundet Cork Company (USA) for 85% magnesia (MgO) based insulating cements and custom-molded thermal insulation products.

mundite. A pale yellow mineral of the phoshuranylite group composed of aluminum uranyl phosphate hydroxide hydrate, $Al(UO_2)_3(PO_4)_2(OH)_3 \cdot 5.5H_2O$. Crystal system, orthorhombic. Density, 4.05 g/cm³; refractive index, 1.682. Occurrence: Zaire.

mundrabillaite. A mineral composed of ammonium calcium phosphate monohydrate, $(NH_4)_2Ca(HPO_4)_2 \cdot H_2O$. Crystal system, monoclinic. Density, 2.05 g/cm³; refractive index, 1.542. Occurrence: Western Australia.

Munester. Trade name of Tissarex SA (France) for glass fabrics.

munga silk. See Assam silk.

mungo. (1) A yarn or fabric made from shoddy cotton. See also shoddy.

(2) A fibrous material obtained from new or old woven or milled woolen fabrics or felts.

Mungoose. Trade name of Barker & Allen Limited (UK) for a nickel silver containing 12-15% nickel, and the balance being copper and zinc. Used for ornaments, and domestic utensils.

muninga. The stable, durable wood from the tree *Pterocarpus angolensis*. It has a beautiful appearance due to the curly or interlocked grain enhancing the natural figure. It has rather good workability, and good nailing and finishing qualities. Average weight, 620 kg/m³ (38.5 lb/ft³). Source: Central and southern Africa. Used for furniture, high-class joinery, paneling, and decorative veneer.

munirite. A pearly white mineral composed of sodium vanadium oxide dihydrate, $NaVO_3 \cdot 2H_2O$. Crystal system, orthorhombic. Density, 2.43 g/cm³; refractive index, 1.757. Occurrence: Pakistan.

Muntz metal. A yellow-colored, cast or wrought brass containing 58-63% copper, 37-41% zinc, and up to 1% lead. It has a density of 8.3 g/cm³ (0.30 lb/in.³), excellent hot workability, poor cold workability, high strength, good corrosion resistance, medium machinability, and is used for nuts, pins, bolts, spindles, wires, brazing rods, forgings, tubing, condenser tubes, architectural panels and sheets, valve stems and trim, hardware, shipbuilding, marine fittings, sheathing, and electric applications. Also known as *malleable brass; yellow metal*.

Munzbronze. Trade name of VDM Nickel-Technologie AG (Germany) for a corrosion-resistant alloy of 95% copper, 4% tin and 1% zinc. Used for chemical plant and equipment.

Muralex. Trade name of ASG Industries Inc. (USA) for a glass with a finely engraved, rather discrete overall pattern.

Muraloy. Trade name of LaClede Company (USA) for a casting alloy of 55% tin, 35% copper, 5% zinc and 5% lead.

Mural Rexine. Trade name of ICI Limited (UK) for cellulose nitrate plastics.

Muranese. Trade name of SA des Verreries de Fauquez (Belgium) for a patterned glass.

murataite. A black mineral composed of sodium rare-earth zinc titanium niobium oxide fluoride, $(Na,Ln)_4Zn_3(Ti,Nb)_6O_{18}F_4$. Crystal system, cubic. Density, 4.69 g/cm³; refractive index,

2.13. Occurrence: USA (Colorado).

Murawire. Trade name of Murex Limited (UK) for plain-carbon welding electrodes. *Murawire W1* has a copper coating and contains 0.06% carbon, 1.5% manganese, 0.6% silicon, 0.06% phosphorus, and the balance iron, and *Murawire W2* has the composition, but no copper coating.

Murcolor. Trade name of Saint-Gobain (France) for prefabricated cladding panels consisting of exterior sheets of ceramic-coated glass, fiberglass cores, and interior sheets of galvanized sheet steel.

murdochite. A black, lustrous mineral composed of copper lead oxide, Cu_6PbO_8. Crystal system, cubic. Density, 6.40 g/cm³. Occurrence: USA (Arizona).

murmanite. A violet mineral with a bronze tinge composed of sodium titanium niobium silicate hydrate, $Na_2(Ti,Nb)_2Si_2O_9 \cdot xH_2O$. Crystal system, triclinic. Density, 2.84 g/cm³; refractive index, 1.765. Occurrence: Russian Federation.

Murman's alloy. A non-heat-treatable aluminum alloy containing 4-15% zinc and 3-14% magnesium, used for light-alloy parts.

Muroglue. Trade name of Ato Findley (France) for vinyl adhesives, coatings and mastics used for building and wall covering applications.

murunskite. A copper-red to pinchbeck-brown mineral of the chalcopyrite group composed of potassium copper iron sulfide, $K_2Cu_3FeS_4$. Crystal system, tetragonal. Density, 3.81 g/cm³. Occurrence: Russian Federation.

Muscle-Glass. Trade name of Corning Glass Works (USA) for a glass that bends readily when cold.

MuscleSheet. Trademark of Biomimetic Products Inc. (USA) for a soft, lightweight electroactive membrane composite with high responsivity. It is an ionic actuator and sensor that can mechanically bend under an applied low voltage, and can be supplied in a wide range of compositions and sizes. Used chiefly in electronics, microrobotics, and medicine.

muscovite. A colorless, white, gray, pale yellow, green, red or brown mineral of the mica group with a vitreous to pearly luster and a layered silicate structure. It is composed of hydrous potassium aluminate silicate, $KAl_2(AlSi_3)O_{10}(OH)_2$. Crystal system, monoclinic. Density, 2.76-3.10 g/cm³; hardness, 2.0-3.0 Mohs; refractive index, 1.59-1.60; hardness, 2-2.5 Mohs; . Occurrence: Asia, Europe (Germany, Poland), Russia, USA (California, Connecticut, Marine, New Hampshire, New Jersey, North Carolina, South Dakota, Washington). Used as an important ceramic raw material, as an insulator (below 600°C or 1110°F), as a lubricant, and in flame-resistant windows. Also known as *common mica; mirror stone; Muscovy glass; potash mica; potassium mica; white mica*.

Muscovy glass. See muscovite.

Museum Glass. Trademark of Tru Vue, Inc. (USA) for a clear, antireflective glass with excellent ultraviolet filtering properties.

musgravite. A pale olive-green mineral of the hogbomite group composed of beryllium magnesium aluminum oxide, $Be(Mg,Fe)_2Al_6O_{12}$. Crystal system, rhombohedral (hexagonal). Density, 3.68 g/cm³; refractive index, 1.739. Occurrence: Australia.

Mushet steel. An air-hardening tool steel containing 1.5-2% carbon, 1.5% silicon, 2.5% manganese, 4-9% tungsten, and the balance iron. Used for dies, and cutting tools for heavy cuts on very hard materials.

music wire. A round, cold-drawn, high-quality steel wire. A typical chemical composition is 0.7-1% carbon, 0.2-0.6% manganese, 0.1-0.3% silicon, up to 0.03% sulfur, up to 0.025% phosphorus, and the balance iron. *Music wire* is available in diameters and minimum tensile strengths ranging from 0.10 mm (0.004 in.) and 3027 MPa (439 ksi) to 6.35 mm (0.250 in.) and 1586 MPa (230 ksi), and is usually made by the electric-furnace process. It has uniform mechanical properties, high tensile strength, good fatigue properties, and high toughness and resilience. Used for mechanical springs operating at temperatures below 121°C (250°F), spiral springs, and strings of musical instruments.

muskoxite. A reddish brown mineral composed of magnesium iron oxide decahydrate, $Mg_7Fe_4O_{13}\cdot10H_2O$. Crystal system, hexagonal. Density, 3.10 g/cm^3; refractive index, 1.80-1.81. Occurrence: Canada.

muslin. A cotton or cotton-blend fabric in a plain or leno weave, having a fleecy surface nap. It is made in various weights ranging from sheer to coarse, and used for dresses, shirts, linings, interfacings, sheets, furniture and mattress covers, curtains, and polishing cloth.

Mustang. Trade name of Jessop Steel Company (USA) for a molybdenum-type high-speed steel (AISI type M2) containing 0.8-0.85% carbon, 4-4.2% chromium, 6-6.5% tungsten, 5% molybdenum, 1-2% vanadium, and the balance iron. It has a great depth of hardening, excellent wear resistance, good toughness, high hot hardness, and is used for a variety of cutting tools.

Mute. Trade name of General Refractories Company (USA) for an acoustical plaster.

Mutemp. Trade name of British Steel Corporation (UK) for a thermosensitive, magnetic alloy of 70% iron and 30% nickel. It has low initial permeability, high saturation induction, and is used for compensating shunts for electrical equipment.

Muticle. Trademark of Mitsui Chemicals America, Inc. (USA) for shaped aqueous emulsion particles supplied in three types: (i) spherical particles of highly crosslinked styrene used in the manufacture of paints and papers; (ii) fine, odd-shaped, slightly porous particles of crosslinked styrene-acrylic used as paper coatings and paint thickeners, and (iii) flattened, blood-cell-type particles of crosslinked styrene used as paper and paperboard coatings. *Muticle* have low weights, high opacity and good oil absorption, and thermal insulation properties.

mutton cloth. A loose fabric, usually cotton, plain-knitted on a circular knitting machine.

Muvar. Trade name of Hamilton Technology Inc. (USA) for a magnetically soft material containing 79% nickel, 16.7% iron, 4% molybdenum and 0.3% manganese. It has high initial permeability, low saturation induction, and is used for electrical and magnetic equipment.

MX compound. A binary ceramic compound composed of metallic element M and nonmetallic element X in which the metal ion to nonmetal ion ratio is 1:1. Examples include CsCl, NaCl, CaO, MgO, ZnO and ZnS.

MX_2 compound. A binary ceramic compound composed of a metallic element M and a nonmetallic element X in which the metal ion to nonmetal ion ratio is 1:2. Examples include CaF_2, SiO_2 and UO_2.

M_2X_3 compound. A binary ceramic compound composed of metallic element M and nonmetallic element X in which the metal ion to nonmetal ion ratio is 2:3. Examples include Al_2O_3, Cr_2O_3 and Fe_2O_3.

MXD-Faser. German trade name for a polyamide (nylon) fiber

Mxsten. Trade name of Eastman Chemical Company (USA) for high-performance polyethylenes used for blown and cast film.

My-A-Chrome. Trade name of Houghton & Richards Inc. (USA) for an oil- or water-hardening tool steel containing 0.7% carbon, 1% chromium, and the balance iron. Used for drills, punches, and tools.

Mycalex. Trademark of Mycaflex Corporation of America (USA) for a ceramic composed of ground *mica* bonded with glass or lead borate, and used in the form of sheets and rods for electrical insulators and insulator components.

Mykroy. Trademark of Mykroy/Micaflex Corporation (USA) for a glass-bonded sheet *mica* used for panels and structural parts of electronic equipment.

Mylar. (1) Trademark of E.I. DuPont de Nemours & Company (USA) for a thermoplastic polyester film made from polyethylene terephthalate (PET). Available in rolls and sheets in thicknesses ranging from 0.0015 to 0.35 mm (0.00006 to 0.01438 in.), it has a density of 1.4 g/cm^3 (0.05 $lb/in.^3$), a melting point of approximately 255°C (490°F), very high toughness, excellent fatigue and tear strength, good dielectric properties, good resistance to ultraviolet radiation, low thermal expansivity, a service temperature range of -40 to +170°C (-40 to +338°F), a heat-sealing temperature of 218-232°C (425-450°F), low permeability to oxygen and carbon dioxide, a low coefficient of friction, good resistance to humidity and solvents, good resistance to acids, alcohols, greases, oils, halogens and ketones, fair resistance to aromatic hydrocarbons, poor resistance to alkalies, and low water absorption. Used for magnetic recording tapes, videotapes and audiotapes, as a base for magnetically coated or perforated information media, for printed-circuit boards, floppy disks, capacitors, membrane switches, motor and transformer insulation, automotive tire cords, electrical, industrial and packaging applications, clothing, and cordage.

(2) Trademark of E.I. DuPont de Nemours & Company (USA) for metal-coated polyester fibers and yarns with good elevated temperatures, used for clothing, and industrial fabrics.

Myoliss. Trademark of Montefibre SpA (Italy) for acrylic filaments and fibers.

myrtle. See laurel (2).

mystic metal. A solder alloy composed of about 88.9% lead, 11% tin and 0.1% bismuth.

Mytex. Trademark of Exxon Chemical Company (USA) for a series of thermoplastic polypropylene resins for automotive applications.

N

nabaphite. A colorless mineral composed of sodium barium phosphate nonahydrate, $NaBa(PO_4) \cdot 9H_2O$. Crystal system, cubic. Density, 2.30 g/cm^3; refractive index, 1.504. Occurrence: Russian Federation.

nacaphite. A colorless mineral composed of sodium calcium fluoride phosphate, $Na_2Ca(PO_4)F$. Crystal system, orthorhombic. Density, 2.85 g/cm^3; refractive index, 1.515. Occurrence: Russian Federation.

Nacconate. Trademark of Allied Chemical & Die Corporation (USA) for a series of diisocyanates, isocyanate prepolymers and partially reacted isocyanates used in the production of rigid, semi-rigid and flexible urethane foams, elastomers, adhesives, coatings, fibers, and textile finishes.

nacre. A hard, iridescent material composed chiefly of calcium carbonate ($CaCO_3$) crystals bonded by conchiolin ($C_{32}H_{98}N_2O_{11}$), and deposited by oysters and mussels. *Natural nacre* is mainly used for ornamental objects, jewelry, buttons, etc., and *synthetic nacre* in biochemistry and biotechnology. Also known as *mother-of-pearl*.

nacreous pigment. A pigment that has guanine ($C_5H_5ON_5$) crystals added to produce an iridescent luster resembling that of *nacre* (mother-of-pearl) when applied to a surface or added to plastics or rubber. See also pearl lacquer.

nacrite. A white mineral of the kaolinite-serpentine group with bluish or red brownish tint composed of aluminum silicate hydroxide, $Al_2Si_2O_5(OH)_4$. Crystal system, monoclinic. Density, 2.65 g/cm^3; refractive index, 1.5625. Occurrence: USA (Michigan, Colorado).

Nacrolaque. Trade name of Glaceries Réunies SA (Belgium) for a bulletproof glass.

Nadir. Trade name of Hoechst Celanese Corporation (USA) for a series of chemical-resistant polysulfones supplied in tube ans sheet forms.

nadorite. A yellow-brown mineral composed of lead antimony oxide chloride, $PbSbO_2Cl$. Crystal system, orthorhombic. Density, 7.02 g/cm^3; refractive index, 2.35. Occurrence: Algeria.

Naflon. Trademark of E.I DuPont de Nemours & Company (USA) for a series of perfluorinated ion-exchange membranes of varying thickness that may or may not be reinforced with polytetrafluoroethylene (*Teflon*) powder. Also included under this trademark are perfluorinated ion-exchange powders and beads.

Nafthoflex. Trademark of Metallgesellschaft AG (Germany) for a series of plastics based wholly or in part on sulfide-containing monomers and including polysulfide elastomers.

nagashimalite. A greenish black mineral composed of barium vanadium titanium chloride borate silicate hydroxide, $Ba_4(V,-Ti)_4ClSi_8B_2O_{27}(O,OH)_2$. Crystal system, orthorhombic. Density, 4.08 g/cm^3; refractive index, 1.753. Occurrence: Japan.

nagelschmidtite. A colorless mineral composed of calcium phosphate silicate, $Ca_7(PO_4)_2(SiO_4)_2$. It can also be made synthetically. Crystal system, hexagonal. Density, 3.04 g/cm^3; refractive index, 1.660; melting point, 1760°C (3226°F).

nagyagite. A blackish lead-gray mineral composed of gold lead tellurium sulfide, $Pb_5Au(Te,Sb)_4S_x$. Crystal system, hexagonal. Density, 7.49 g/cm^3. Occurrence: Rumania. Also known as *black tellurium*.

nahcolite. A colorless to white mineral composed of sodium hydrogen carbonate, $NaHCO_3$. Crystal system, monoclinic. Density, 2.24 g/cm^3; refractive index, 1.500. Occurrence: Italy, USA (California).

nahpolite. A colorless mineral composed of hydrogen sodium phosphate, Na_2HPO_4. It can also be made synthetically. Crystal system, monoclinic. Density, 2.58 g/cm^3.

nailable concrete. A *lightweight concrete* that may contain some sawdust, and is used in the construction trades to receive and hold nails. Also known as *nailing concrete*.

nail-base fiberboard. A special building product consisting of high-density *fiberboard* (about 400 kg/m^3 or 25 lb/ft^3) used in frame construction to apply exterior horizontal siding materials, such as wood or cement-asbestos shingles, by directly nailing to the sheathing.

nailing concrete. See nailable concrete.

Nailon. Trade name for nylon fibers and products used for fabrics and in the manufacture of pin-drive anchors.

nainsook. A semisheer, soft, usually mercerized cotton fabric, intermediate in weight between *batiste* and *organdy*, in a plain weave. Used for blouses, lingerie, and children's clothes.

Nakan. Trade name of Atofina SA (France) for a series of polyvinyl chloride compounds supplied in flexible and rigid grades.

nakauriite. A sky-blue mineral composed of copper carbonate sulfate hydroxide hydrate, $Cu_8(SO_4)_4(CO_3)(OH)_6 \cdot 48H_2O$. Crystal system, orthorhombic. Density, 2.39 g/cm^3; refractive index, 1.604. Occurrence: Japan.

Nalcoag. Trademark of Nalco Chemical Company (USA) for colloidal silicas and silica sols.

Naloy. Trade name of National Broach & Machine Company (USA) for a high-speed steel containing 0.7% carbon, 18% tungsten, 4% chromium, 1% vanadium, and the balance iron. Used for form tools, broaches, reamers, and hobs.

nambulite. A red-brown mineral of the pyroxenoid group composed of lithium sodium manganese silicate hydroxide, $LiNaMn_8Si_{10}O_{28}(OH)_2$. Crystal system, triclinic. Density, 3.51 g/cm^3; refractive index, 1.710. Occurrence: Japan.

Nametal. Trade name of Foseco Minsep NV (Netherlands) for foil-wrapped metallic sodium cubes and sticks, used as modifiers in the manufacture of aluminum-silicon alloys.

namibite. A dark green mineral composed of copper bismuth vanadium oxide, $CuBi_2VO_6$ Crystal system, monoclinic. Density, 6.75 g/cm^3; refractive index, above 2.10. Occurrence: Africa (Namibia).

namuwite. A mineral composed of copper zinc sulfate hydroxide tetrahydrate, $(Zn,Cu)_4SO_4(OH)_6 \cdot 4H_2O$. Crystal system, hexagonal. Density, 2.77 g/cm^3; refractive index, 1.577. Occurrence: UK (Wales).

Nandel. Trademark of E.I. DuPont de Nemours & Company (USA) for acrylic fibers, filaments and yarns.

nanlingite. A brownish red mineral composed of calcium magnesium fluoride arsenate, $CaMg_4(AsO_3)_2F_4$. Crystal system, rhombohedral (hexagonal). Density, 3.93 g/cm³; refractive index, 1.82. Occurrence: China.

Nanlon. Trademark of Tainan Spinning Company (Taiwan) for polyester staple fibers and filament yarns.

Nanmac. Trade name of Nanmac Corporation (USA) for thermocouple alloys with high strength and elevated-temperature strength up to 1650°C (3000°F), and excellent thermal shock resistance. Used for thermocouple sheaths.

Nano 1000. Trademark of Powdermet, Inc. (USA) for submicron tungsten carbide hardmetal powders coated with submicron tungsten.

nanoalloys. A group of metallic alloys with grain sizes of less than 100 nm (3.94 μin.) produced by various techniques including powder metallurgy, electrodeposition, mechanical attrition, heavy plastic deformation, inert gas condensation and crystallization of amorphous precursors. They have significantly improved chemical, physical and mechanical properties over those of their crystalline counterparts.

Nano Alumi. Trademark of Sumitomo Electric Industries, Inc. (USA) for powder-metallurgy aluminum alloys used in the manufacture of high-strength products including engine components, and for heat sinks, electronic components, electric conductors, and magnetic applications.

nanoamorphous materials. (1) Materials consisting of amorphous regions with dimensions in the nanometer range embedded in large crystals. Such materials can be produced by creating atomic displacements (e.g., Frenkel pairs) in a crystal lattice by irradiating the material with high-energy nuclear particles and subsequent quenching.

(2) Materials produced from amorphous nanoparticles.

nanobelts. Ultrathin, flat structures composed of semiconducting metal oxides (e.g., oxides of tin or zinc) with high chemical purity, structural uniformity and oxidation resistance. They are synthesized by high-temperature evaporation of the respective oxide powders and condensation on an alumina (Al_2O_3) substrate. Each nanobelt consists of a single crystal with specific surface planes and shape. Used for electronic nanodevices, components of flat-panel displays, and small sensor devices.

Nanobond. Trademark of Nanopierce Technologies, Inc. (USA) for conductive glues and adhesives used for electronic interconnection devices.

Nanocarb. Trademark of Nanodyne Corporation (USA) for nano-sized, ceramic, metallic and metal-ceramic composite powders as well as wear-resistant coatings used in the manufacture of industrial products, such as dies, tools, cutting instruments, rolling-mill and metal forming and working equipment, equipment for the mining, gas- and oil-drilling and exploration industries, and aircraft turbine blades. Also included under this trademark are metallic carbides, such as ultra-fine-grained tungsten carbide/cobalt powders used in the manufacture of cutting tools, drill bits and wear parts.

Nanocarbon. Trademark of Jerry D. Johnson (USA) for carbon and carbon products used in nanotechnology.

Nanocer. Trademark of Nanophase Technologies Corporation (USA) for nanocrystalline metal, metal oxide and ceramic particles used for forming finished products.

Nanoceram. Trade name of Argonide Nanomaterials (USA) for alumina fibers with a diameter of 2 nm (0.08 μin.), and aspect ratios of 20-100, used as reinforcements in ceramic-, metal- and polymer-matrix composites, as catalysts and catalyst supports for precious metals, as sintering aids for ceramics, and in the manufacture of ceramic membranes and membrane reac-

tors.

Nanoclad. Trademark of Nanophase Technologies Corporation (USA) for nanocrystalline metal, metal oxide and ceramic particles used for forming finished products.

Nanoclay. Trademark of Southern Clay Products, Inc. (USA) for nanocrystalline, clay-based additives and reinforcers for plastics and rubber. They are used to adjust the melt and heat-deflection temperatures, gas permeability and flame retardancy of the aforementioned materials.

nanoclays. A term referring to *montmorillonite* and other members of the *smectite* group of clay minerals having structures consisting of agglomerated nanoscale particles which can be exfoliated by surface treatments with suitable compatibilizers. See also Nanomer (2).

nanoclusters. A generic term for loosely bound agglomerates of (aggregated) nano-sized particles.

Nanocoat. Trademark of Nanosphere, Inc. (USA) for engineered ceramic, metallic and polymeric particulate surface coverings.

nanocolloid. A *colloid* made up of crystals or particles with a typical size of less than 100 nm (3.94 μin.). Human albumin-based nanocolloids are used in medicine, e.g., for cancer treatment.

Nanocomposite. Trademark of US Technology Corporation (USA) for granulated plastic media used for deburring, deflashing and depainting workpieces.

nanocomposite coatings. A relatively new class of extremely hard and wear-resistant coatings and thin films that consist of nanocrystalline grains, usually less than 100 nm (3.94 μin.) in size, of transition-metal carbides or nitrides in amorphous matrix materials (e.g., nickel) and are typically produced by the codeposition of amorphous and nanocrystalline phases.

nanocomposites. A relatively new class of materials with nanocrystalline structure containing at least two distinct phases each of which can be nanocrystalline in nature. Examples are composites with nanostructured ceramic, metallic or polymer matrices and/or nanocrystalline reinforcements, or one-, two- or three-dimensional organic-inorganic materials composed of distinctly dissimilar components mixed at the nanometer scale. Abbreviation: NC. Also known as *nanocomposite materials*. See also organic-inorganic nanocomposites; lamellar nanocomposites; nanomagnetic composites.

Nanocor. Trademark of AMCOL International Corporation (USA) for nonmetallic minerals and mineral compositions used to make plastic molding compounds.

nanocrystalline graphite. Synthetic graphite processed such that the resulting material is composed of nano-sized crystals.

nanocrystalline materials. See nanostructured materials.

nanodots. See quantum-dot particles.

Nanodur. Trademark of Nanophase Technologies Corporation (USA) for nanocrystalline metal, metal-oxide and ceramic particles used for forming finished nanocrystalline products.

nanoengineered materials. Engineered materials (e.g., metals, alloys, ceramics, polymers or semiconductors) synthesized to have nanocrystalline structures. See also nanostructured materials.

nanofibers. Metallic or nonmetallic fibers with a diameter in the nanometer-size region, typically between 2-100 nm (0.08-3.94 μin.).

nanofoils. Nanocrystalline metals and alloys with grain sizes less than 100 nm (3.94 μin.) produced into long foils of varying thickness by techniques, such as electrodeposition. They have superior hardness and wear resistance, and unique electrical and magnetic properties, and are used in applications ranging from soft magnets and catalysts to electronics components.

NanoGard. Trademark of Nanophase Technologies Corporation (USA) for nanocrystalline metal powders and ceramic oxide powders used in the manufacture of health-care products.

Nanoglass. (1) Trademark of Allied Signal, Inc. (USA) for dielectric compositions used in semiconductor devices.

(2) Trademark of Heraeus Kulzer GmbH (Germany) for an extensive series of dental materials including ceramics and plastics used to restore bridges, crowns, teeth and dentures as well as dental veneers, fillers, adhesives, impression materials, and modeling, doubling and embedding compounds.

Nanograin. Trademark of Nanodyne Corporation (USA) for nano-sized ceramic, metal and metal-ceramic composite powders as well as wear-resistant coatings used in the manufacture of industrial products, such as dies, tools, cutting instruments, rolling-mill and metal-forming and -working equipment, equipment used in the mining, gas- and oil-drilling and exploration industries, and for aircraft turbine blades. Also included under this trademark are metallic carbides, such ultra-fine-grained tungsten carbide/cobalt powders used in the manufacture of cutting tools, drill bits, and wear parts.

NanoGram. Trademark of NanoGram (UK) for nanocrystalline materials with grain sizes as small as 5 nm (0.20 μin.). They are supplied as metal, and metal-silicon composition powders for the manufacture of various metals, alloys, carbides, oxides, nitrides and sulfides.

nanograined material. See nanogranular material.

nanogranular material. A crystalline material having a grain size of less than 0.1 μm or 100 nm (3.94 μin.). Also known *nanograined material.*

nanointermetallics. A group of *intermetallic compounds* which have a nanocrystalline microstructure with a grain size of less than 100 nm (3.94 μin.). They have unique, physical, chemical and mechanical properties that are often superior to those of their crystalline counterparts. Included are nanocrystalline aluminides, beryllides, borides, hydrides, nitrides and silicides.

Nanolatex. Trademark of Rhône-Poulenc SA (France) for ultra-fine-sized aqueous latex emulsions.

nanomagnetic composites. A group of composite materials having metallic or nonmetallic matrices (e.g., aluminum, copper, zinc, aluminum oxide, or zinc oxide) with embedded nano-sized (less than 100 nm or 3.94 μin.) magnetic particles (e.g., iron, magnetite, or chromic oxide). They can be produced by high-energy ball milling, or reaction milling, and are used for electronic and magnetic applications. Also known as *magnetic nanocomposites.*

nanomagnetic materials. A group of nanostructured materials that are either magnetic by nature, e.g., nanocrystalline iron and nanocrystalline iron-cobalt alloys, or are composites with nonmetallic matrices containing nano-sized (less than 100 nm or 3.94 μin.) magnetic particles that render them magnetic, e.g., composites consisting of aluminum oxide (Al_2O_3) or zinc oxide (ZnO) matrices with embedded nano-sized magnetic iron particles. Nanomagnetic alloys and metals have greatly enhanced magnetic properties. Also known as *magnetic nanomaterials.*

NanoMaterials. Trademark of NanoMaterials, Inc. (USA) for metal, ceramic and composite nanomaterials including nanocrystals, nanoparticles and nanotubes. Used especially for biotechnology applications.

nanomaterials. See nanostructured materials.

Nanomer. (1) Trademark of Institut für Neue Materialien gem. GmbH (Germany) for paint-like compositions used in the production of scratch-resistant, corrosion-inhibiting coatings for electronic holographic elements and displays, and for impregnating textile fabrics.

(2) Trademark of Nanocor, Inc. (USA) for clay minerals of the *montmorillonite* group surface-treated with suitable compatibilizing agents, and masterbatches containing these clays. Used for the manufacture of polymer-matrix nanocomposites.

(3) Trademark of AMCOL International Corporation (USA) for nonmetallic minerals and mineral compositions used to make plastic molding compounds.

nanomers. A generic term for discrete, non-agglomerated nano-sized particles.

NanoMet. Trademark of Powdermet, Inc. (USA) for engineered nanostructured cermet hardmetal powders with ceramic surface coatings with high adhesion and wettability.

NanoMetal. Trademark of Mott Metallurgical Corporation (US) for high-retention filter media consisting entirely of fully sintered porous metals or alloys, such as nickel, stainless steel or *Hastelloy.*

nanoparticles. See nanopowders.

Nanophase. Trademark of Nanophase Technologies Corporation (USA) for nanometer-sized ceramic, metal and metal-oxide particles that can be shaped by pressing and other techniques into solid products with improved physical and mechanical properties.

nanophase ceramics. A group of engineering ceramics (e.g., aluminum oxide, titanium dioxide, or zinc oxide) produced in powder form or a film on a substrate with particle sizes less than 100 nm (3.94 μin.) by any of various techniques including vapor-phase processing, chemical processing or electrodeposition. Used for automotive engine parts, jet aircraft components, industrial equipment, artificial hip joints, and many other products.

nanophase materials. Single- or multi-phase materials in which at least one of the phases present has nanocrystalline characteristics, i.e., has a crystal size of less than 100 nm (3.94 μin.). They can be metals or alloys, ceramics, polymers, composites or semiconductors. Abbreviation: NPM. See also nanostructured materials; nanocomposites.

Nanophaze. Trademark of Nanophase Materials Corporation (USA) for fine-grained and ultrafine-grained metallic and ceramic powders used as precursors in the manufacture of solid end products.

Nanoplast. Trade name of Polysciences, Inc. (USA) for a highly water-soluble two-component melamine/formaldehyde resin system (resin plus catalyst) used as an embedding medium for biology and biochemical high-resolution light and electron microscopy.

Nanoplate. (1) Trademark of Integran Technologies Inc. (Canada) for a series of nanocrystalline metals, alloys and composites produced by electrodeposition. Their grain sizes are less than 100 nm (3.94 μin.) and can be as small as 5 nm (0.2 μin.). Used for corrosion- and wear-resistant coatings, soft magnets, catalysts, electronic components, electroformed parts, lightweight armor, and structural components. Available in various product forms, such as coatings, free-standing foils, sheets and plates, wires, mesh and powders.

(2) Trademark of Shipley LLC (USA) for nanocrystalline copper electrodeposits.

nanoplate. A generic term referring to a nanostructured material with plate-like geometry, i.e., with nanometer dimension in one direction. A recent practical example of such a material is an ultrathin polymethyl methacrylate (PMMA) film used for magnetic recording applications. See also nanostructured materials.

Nanopolish Diamond Gel. Trademark of Intersurface Dynamics, Inc. (USA) for industrial-quality diamond products supplied as suspensions in proprietary gel carriers. Used for polishing metals and other materials.

nanoporous gold. Gold containing nanometer-sized pores and having a sponge-like surface, obtained by de-alloying, i.e., by selectively dissolving the silver out of a gold-silver alloy.

nanoporous materials. A class of solid materials, such as activated carbon and porous glass, that have nanometer-sized pores. They have large surface areas and can adsorb large quantities of liquids and gases. Used as catalysts, as biological microfilters, in molecular filtration, in the separation of chemicals, and for microelectronic applications.

nanoporous silicates. A class of low-dielectric-constant materials prepared by incorporating nano-sized air bubbles with low dielectric constant into an organosilicate-based matrix material (e.g., silsesquioxane) by reactive blending to yield an organic-inorganic composite that on vitrification and subsequent degradation of the organic polymer produces a nanoporous silicate structure. Used in the manufacture of high-speed microprocessors and other microelectronic devices.

nanopowders. Nanosized particles (typically 10-500 nm or 0.4-20 μin.) of metals, alloys, ceramics and composites that are obtained by vapor condensation, mechanical attrition, electro-explosion, chemical methods, or other processes. Used in the manufacture of nanocrystalline materials, coatings, catalysts and lubricants. Also known as *nanoparticles*.

nanoribbons. Flexible ribbons, such as carbon-based structures, formed from flattened carbon nanotubes, metal oxide-based structures grown from molecular precursors, or aluminum carbide-based structures synthesized by a self-assembly process from mixtures of aluminum, carbon and lithium. See also aluminum carbide nanostructures; carbon nanotubes.

NanoShield. Trademark of Nanophase Technologies Corporation (USA) for various coatings based on nanocrystalline materials. Depending on the starting material and processing technique(s), the resulting coating may protect the substrate against wear and abrasion, corrosion and/or ultraviolet radiation, or improve its electrical activity.

Nanosil. Trademark of Loctite Corporation (USA) for a series of high-performance industrial silicone-based adhesives that bond well to many substrates including metals, ceramics, glass, plastics, wood and paper.

Nanosphere. Trademark of Duke Engineering Corporation (US) for a line of spherical particles ranging from 0.02 to 2000 μm (0.8 μin. to 0.08 in.) in diameter. Used as size standards for instruments calibration, quality control, filter checking, and in biotechnology.

Nanostructured. Trademark of Hybrid Plastics Corporation (USA) for chemical compounds used for making plastics.

nanostructured materials. A large group of relatively new materials (e.g., metals, alloys, ceramics, semiconductors or composites) with microengineered structures in which critical length scales (e.g., grain size, particle size, or film thickness) are on the order of only a few nanometers (typically less than 100 nm or 3.94 μin.). Members of this group of materials include: (i) *zero-dimensional nanostructures*, such as individual clusters or particles, having small dimensions in all three directions; (ii) *one-dimensional nanostructures*, such as nanotubes or nanowires, having small dimensions in two directions, but considerable length in the third direction; (iii) *two-dimensional nanostructures*, such as thin films, in which only the thickness is in the nanometer range; and (iv) *three-dimensional nanostructures*, such as fully-dense nanomaterials, having large external dimensions in all three directions, but crystal sizes on the nanometer scale. As a result of the unique microstructures and high concentration of defects (e.g., large surface areas for particles, or large volume fractions of grain boundaries in three-dimensional nanostructures) these materials exhibit unique chemical, physical and mechanical properties, usually not observed for their conventional polycrystalline and amorphous counterparts. *Nanostructured materials* can be produced by many different methods including inert gas condensation, chemical and physical vapor deposition, mechanical attrition, severe plastic deformation, electrodeposition and crystallization of amorphous precursors. Current and future applications of these materials include soft and hard magnets, oxide nanoparticles for ultraviolet protection in sunscreen lotion, corrosion- and wear-resistant coatings, catalysts for a variety of processes, high-strength structural materials for nuclear steam generator repair, armor materials, and nanostructured cermets for cutting tools. They are also of large importance in the general field of nanotechnology. Also known as *nanocrystalline materials; nanomaterials*.

Nano Surface. Trademark of Advanced Surface Engineering Inc. (USA) for nanocomposite ceramic coatings.

Nanotec. Trademark of Powdermet, Inc. (USA) for coated particulates based on ceramic, metallic or intermetallic compounds. The average particle size ranges from 0.5 μm (19.7 μin.) to 3.175 mm (0.125 in.).

NanoTek. Trademark of Nanophase Technologies Corporation (USA) for nanocrystalline metal powders, and ceramic oxide powders (e.g., alumina, ceria, chromia, iron oxide, titania, yttria, etc.) supplied in particle sizes of approximately 5-50 nm (0.2-2.0 μin.). Used for electric, thermal-spray and polishing applications, advanced ceramics, and for catalytic processes.

nanotube composites. Polymer-matrix composites that are reinforced with carbon *nanotubes* and exhibit remarkable electrical properties ranging from strong antistatic to strong electrostatic. Suitable matrix resins include polyolefins, polyamides, polyetherimides and polyphenylene sulfides. Abbreviation: NTCs.

nanotubes. A relatively new class of one-dimensional *nanostructured materials* including *carbon nanotubes* and *inorganic nanotubes*.

NanoTuf. Trademark of Triton Systems, Inc. (USA) for epoxy-based nanocomposite coatings with outstanding abrasion, chemical, corrosion, flame and crack resistance, and excellent anti-reflection properties. Used on polycarbonate- and acrylic-based products, such as aircraft canopies and windows, automotive windshields, laser-protection visors and spectacles, industrial, military, sports and prescription eyewear and sunglasses, and chemical-protection headgear lenses.

nanowires. Metallic or ceramic wires having quasi-one dimensional characteristics and diameters in the nanometer range. They may be single-layered or multilayered, and are used in electronic, magnetic and optical devices.

Nanox. (1) Trademark of Nanox Limited (UK) for phosphors and phosphorescent materials used in electronic equipment, such as visual displays. Also included under this trademark are small magnetic particulates for electronic applications, and various color pigments.

(2) Trademark of US Nanocorp, Inc. (USA) for inorganic oxides used as active substances in batteries, fuel cells and similar energy-storage and conversion devices.

(3) Trademark of Elements UK Limited (UK) for zinc and zinc oxide powders used for cosmetic and healthcare applications. Also included under this trademark are chemical compounds for making plastic materials and films, zinc oxide powder coatings for decorative and printing applications, and paints, varnishes and glazes.

nantokite. A colorless to light gray mineral of the sphalerite group composed of cuprous chloride, CuCl. Crystal system, cubic. Density, 4.14 g/cm^3; refractive index, 1.930.

napa leather. Very soft, delicate, black or colored leather with a glossy surface produced from the skin of sheep, lambs or goats, and made water washable by chrome or vegetable retanning. It is named after Napa, a city in California, USA. Used for gloves, handbags, and clothing.

Napco. Trade name of Napco (USA) for bauxite, fused alumina, magnesite, fused magnesia and silicon carbide.

naphthacene. An organic compound that contains four fused benzene rings and is found in *anthracene* and *coal tar*. It is commercially available in the form of orange crystals (98% pure) with a density of 1.35 g/cm^3 and a melting point of approximately 350°C (662°F). Used in organic synthesis. Formula: $C_{18}H_{12}$. Also known as *2,3-benzanthracene; rubene; tetracene*.

naphthalene. A white, crystalline hydrocarbon whose structure consists of a double benzene ring and which is usually prepared from coal tar or petroleum fractions. It has a density of 1.145 g/cm^3, a melting point of 80-82°C (176-180°F), a boiling point of 218°C (424°F), a flash point of 176°F (80°C), and exhibits semiconductive properties. Used chiefly in the manufacture of organic compounds, such as dyes, lubricants, cutting fluids and synthetic resins, as high-purity crystals in scintillation counters, and in optoelectronic devices. Formula: $C_{10}H_8$.

naphthenate driers. Solutions of metallic salts and naphthenic acid, e.g., barium, mercuric, potassium, sodium, tin or zinc naphthenate. Used as paint driers.

naphthalocyanine. An organic dye with a dye content of about 95% and a maximum absorption wavelength of 712 nm, often used in combination with a metal, such as gallium (e.g., gallium 2,3-naphthalocyanine chloride) or tin (e.g., tin 2,3-naphthalocyanine). Formula: $C_{48}H_{24}N_8$.

2-(1-naphthyl)-5-phenyloxazole. Light green crystals or fluorescent yellow needles with a melting point of 104-106°C (219-223°F). Used as a scintillation counter and wavelength shifter in liquid scintillation spectroscopy. Formula: $C_{19}H_{13}NO$. Abbreviation: α-NPO; NPO; ANPO. Also known as α-*naphthylphenyloxazole*.

Nap-Lam. Trade name of Mathison's (USA) for thermal film products consisting of polyester substrates coated with low-melt adhesives, and supplied in several sizes and surface finishes including clear, lustrous and glossy.

Naples yellow. See lead antimonate.

napped fabrics. Textile fabrics, such as fleece, having a fuzzy or furry finish consisting of protruding fibers whose surfaces have been raised by brushing or shearing. Also known as *raised fabrics*.

napped leather. See suede.

napping cotton. A rather short, wrinkled cotton fiber suitable for making *napped fabrics*.

Napraloy. Trade name of Napraloy Company (USA) for a wear-resistant alloy of 71.5% iron, 24% chromium and 4.5% carbon, used for welding and hardfacing electrodes.

Napryl. Trademark of Pechiney SA (France) for polystyrene resins.

Nap Superior. Trade name of Vulcan Steel & Tool Company Limited (UK) for a cobalt-tungsten high-speed steel containing 0.8% carbon, 4.5% chromium, 0.5% molybdenum, 20% tungsten, 1.25% vanadium, 5% cobalt, and the balance iron. Used for cutting tools.

Napur. Trademark of Elastogran GmbH (Germany) for a series of unprocessed plastics supplied as powders, granules, pastes and liquids.

Nara Porcelain. Trade name of Aardvark Clay & Supplies (USA) for a *porcelain clay* (cone 10) with a body resembling *Grolleg*.

Narco. (1) Trademark of North American Refractories Company (USA) for an extensive series of refractory products including plastic refractories for the production of metallurgical furnaces and equipment employed in steel and gray and ductile iron manufacture, as well as refractory gunning and patching mixes.

(2) Trade name of North American Rayon Corporation (USA) for rayon fibers and yarns used for textile fabrics.

Narcocast. Trademark of North American Refractories Company (USA) for a series of castable refractories.

Narcocrete. Trademark of North American Refractories Company (USA) for a series of castable and trowelable refractories.

Narcogun. Trademark of North American Refractories Company (USA) for a series of refractory gunning mixes.

Narcoline. Trademark of North American Refractories Company (USA) for plastic refractories for the production of metallurgical furnaces and equipment employed in steel and gray and ductile iron manufacture.

Narcoloy. Trade name of N.C. Ashton Limited (UK) for an alloy composed of copper, aluminum, tin and cobalt, and used for cold forging aluminum bronze rods.

Narcon. Trade name for rayon fibers and yarns used for textile fabrics.

Narcoset. Trademark of North American Refractories Company (USA) for quick-setting refractory cements and mortars.

Narco Shur-Ram. Trademark of North American Refractories Company (USA) for plastic refractory ramming mixes.

Narcospar. Trademark of North American Refractories Company (USA) for plastic refractories.

Narene. Trade name for rayon and polyester fibers used for textile fabrics.

nargusta. See verdolago.

Narite. Trade name of N.C. Ashton Limited (UK) for a series of alloys containing 11-14% aluminum, 1-5% nickel, 4.5-5% iron, 0-1% manganese, and the balance copper. Used for deep-drawing dies, and nonsparking tools.

NARloy. Trade name of Crucible Materials Corporation (USA) for a series of brazing alloys composed of 97% copper and 3% silver. *NARloy-Z* is a wrought copper alloy which contains 3% silver and 0.5% zirconium, and is used for the main combustion chamber liner of the NASA Space Shuttle.

Narmco. Trademark of Celanese Corporation (USA) for synthetic resins including epoxies for the manufacture of adhesives with good lap-shear strength.

Narrmac. Trade name of N.C. Ashton Limited (UK) for a series of wrought aluminum bronzes containing 9.5-10.5% aluminum, 0-4.8% nickel, 2.5-4% iron, 0.5% manganese, and the balance copper. They have good strength and corrosion resistance, and good hot formability. Used for fasteners, marine parts, and pump and valve parts.

narrow fabrics. Textile fabrics, such as ribbons, tapes and webbings, produced as strips that are usually not wider than 300 mm (11.8 in.), either by braiding, knitting, weaving or otherwise interlacing fibers or yarns, or by cutting wider fabrics.

Narrow Reeded. Trade name of Chance Brothers Limited (UK) for a glass with reeded pattern.

Narrow Reedlyte. Trade name of Chance Brothers Limited (UK) for a glass with lightly reeded pattern.

narsarsukite. A honey-yellow mineral composed of sodium titanium silicate, $Na_2TiSi_4O_{11}$. Crystal system, tetragonal. It can also be made synthetically. Density, 2.75 g/cm³; melting point, 929°C (1704°F); refractive index, 1.612. Occurrence: Greenland, USA (Montana).

Nartrode. Trade name of N.C. Ashton Limited (UK) for an aluminum bronze containing 9.25-9.5% aluminum, 1-3.25% iron, 0-4.5% nickel, and the balance copper. Used for welding wire employed in automatic processes.

Naruto. Trade name of Asahi Glass Company Limited (Japan) for a figured glass with looped design.

Narve. Trade name of Uddeholm Corporation (USA) for a cold-work tool steel containing 0.65% carbon, 1% silicon, 0.5% manganese, 1.1% chromium, 0.6% molybdenum, and the balance iron. Used for shafts, dies, and tools.

NAS 10. Trade name of Novacor Chemicals, Inc. (USA) for acrylic styrene copolymers.

Nasco. Trade name of North American Steel Corporation (USA) for a tough, shock-resistant, non-tempering tool steel containing 0.5% carbon, 0.8% manganese, 1% chromium, 0.4% molybdenum, and the balance iron. Used for chisels, punches, wrenches, and tools.

Nascoloy. Trade name of North American Steel Corporation (USA) for a series of tough, wear-resistant tool steels. *Nascoloy O* is an oil-hardening type containing 0.5% carbon, 0.8% manganese, 1.8% silicon, and the balance iron. It requires tempering, and is used for chisels, rivet sets and pneumatic tools. *Nascoloy W* is a water-hardening type containing 0.4% carbon, 0.7% manganese, 0.2% silicon, 0.8% chromium, 0.3% molybdenum, 0.3% copper, and the balance iron. It does not require tempering, and is used for chisels, rivet sets and pneumatic tools.

Nashiji. Trade name of Asahi Glass Company Limited (Japan) for a figured glass.

nasinite. An orange-yellow to light gray mineral composed of sodium borate pentahydrate, $Na_4B_{10}O_{17} \cdot 5H_2O$. Crystal system, orthorhombic. Density, 2.13 g/cm³; refractive index, 1.512. Occurrence: Italy.

nasledovite. A white mineral composed of lead manganese aluminum oxide carbonate sulfate pentahydrate, $PbMn_3Al_4(CO_3)_4(SO_4)O_5 \cdot 5H_2O$. Density, 3.07 g/cm³. Occurrence: Russian Federation, Central Asia.

nasonite. A white to pearl-gray mineral of the apatite group composed of a calcium lead chloride silicate, $Ca_4Pb_6Cl_2Si_6O_{21}$. Crystal system, hexagonal. Density, 5.43 g/cm³; refractive index, 1.9453. Occurrence: Sweden, USA (New Jersey).

Nat. Trade name of Eagle & Globe Steel Limited (Australia) for a chromium-type hot-work tool steel (AISI type H11) containing 0.3-0.4% carbon, 1% silicon, 5% chromium, 1.2-1.4% molybdenum, 0.4% vanadium, and the balance iron. Used for blanking, extrusion and press-forging tools, and die-casting dies for light-metal alloys.

Natalon. Trade name for silicon carbide (SiC) products available in various shapes, grades and sizes including square, rectangular, triangular or round grinding blocks, sticks, rubbing bricks, and grinding and segmental wheels.

natanite. A light green mineral of the sohngeite group composed of iron tin hydroxide, $FeSn(OH)_6$. Crystal system, cubic. Density, 3.83 g/cm³; refractive index, 1.755.

Natene. Trademark of Pechiney SA (France) for phenol-formaldehyde plastics and molding compounds.

Nateo. See Meryl Nateo.

National. (1) Trademark of Union Carbide Corporation (USA) for carbon and graphite electrodes and brushes, carbon powder and carbon or graphite fabrics.

　　(2) Trade name of Delsteel Inc. (USA) for a tungsten-type high-speed tool steel.

　　(3) Trade name of National Plastics (USA) for nylon, polyolefin and saran (polyvinylidene chloride) fibers used for carpets and rugs, draperies, upholstery, clothing, and industrial fabrics.

National Graphitic Steel. Trade name of Capitol Castings Inc. (USA) for an abrasion-resistant graphitic steel containing 0.3% carbon, 0.5% graphite, and the balance iron. Used for steel castings.

natisite. A yellow-green to greenish gray mineral composed of sodium titanium oxide silicate, $Na_2TiOSiO_4$. Crystal system, tetragonal. Density, 3.15 g/cm³; refractive index, 1.756.

native antimony. The element *antimony* as it occurs in nature. It is a tin-white, brittle semimetal. Crystal system, rhombohedral (hexagonal). Density, 6.70 g/cm³; hardness, 3.0-3.5 Mohs. Occurrence: China, Mexico, North Africa, South Africa, South America, USA (California).

native arsenic. The element *arsenic* as it occurs in nature. It is a tin-white semimetal with metallic luster. Crystal system, rhombohedral (hexagonal). Density, 5.64-5.78 g/cm³; hardness, 3-4 Mohs.

native asphalt. Asphalt, including lake and rock asphalt, as it is mined and quarried in nature. Also known as *natural asphalt*. See also asphalt; lake asphalt; rock asphalt.

native bismuth. The element *bismuth* as it occurs in nature. It is a grayish-white semimetal with a reddish tinge and a bright metallic luster. Crystal system, hexagonal. Density, 9.77-9.83 g/cm³; hardness, 2.0-2.5 Mohs.

native copper. The element *copper* as it occurs in nature. It is a reddish metal that dulls on exposure to air, and may contain some silver and bismuth. Crystal system, cubic. Density, 8.90-8.95 g/cm³; melting point, 1083°C (1981°F); hardness, 2.5-3.0 Mohs. Occurrence: Canada, Mexico, USA (Arizona, Michigan, New Mexico).

native gold. The element *gold* as it occurs in nature. It is a yellow metal that is frequently alloyed with silver. Crystal system, cubic. Density, 19.28-19.33 g/cm³; melting point, 1063°C (1945°F); hardness, 2.5-3.0 Mohs; refractive index, 0.366. Occurrence: Australia, Canada, Russia, South Africa, USA (Alaska, California, Nevada, South Dakota, Utah).

native iron. The element *iron* as it occurs in nature, e.g. in meteorites and terrestrial rocks. It is a rare, gray or silvery white metal. Crystal system, cubic. Density, 7.88 g/cm³; hardness, 2-3 Mohs.

native metal. A metal, such as copper, gold, iron, lead, palladium, platinum or silver, that occurs naturally in the elemental or metallic state.

native paraffin. See ozocerite.

native platinum. The element *platinum* as it occurs in nature. It is a ductile, malleable, silver-gray metal often alloyed with iron, iridium, palladium, osmium and rhodium. Crystal system, cubic. Density, 13.35-21.50 g/cm³; hardness, 4.0-4.5 Mohs. Occurrence: Canada, Russia, South Africa, USA (Alaska).

native protein. A *protein* that is either fibrous or globular in nature, and consists of *polypeptide* chains linked together or held in definite folded shapes by hydrogen bonds.

native silver. The element *silver* as it occurs in nature. It is a tin-white, ductile, malleable metal which turns bronze, gray or black on exposure to air, and is frequently alloyed with gold. Crystal system, cubic. Density, 10.1-11.1 g/cm³; melting point, 961°C (1762°F); hardness, 2.5-3.0 Mohs. Occurrence: Canada, Scandinavia, USA (Arizona, Colorado, Michigan, Montana, Wisconsin).

native sulfur. The element *sulfur* as it occurs in nature. It is a bright yellow nonmetal with a resinous luster. Crystal system, orthorhombic. Density, 2.05-2.08 g/cm³; hardness, 1.5-2.5 Mohs.

native uranium. See natural uranium.

Natrelle BCF. Trade name of Solutia Inc. (USA) for strong nylon 6,6 fibers used for the manufacture of textile fabrics.

natroalunite. A white, grayish or yellowish mineral of the alunite group composed of sodium aluminum sulfate hydroxide, $(Na,K)Al_3(SO_4)_2(OH)_6$. Crystal system, rhombohedral (hexagonal). Density, 2.60 g/cm³; refractive index, 1.572. Occurrence: USA (Utah). Also known as *almerite*.

natroapophyllite. A colorless to white and sometimes brownish yellow to yellowish brown mineral of the apophyllite group composed of sodium calcium fluoride silicate octahydrate, $NaCa_4Si_8O_{20}F\cdot8H_2O$. Crystal system, orthorhombic. Density, 2.50 g/cm³; refractive index, 1.538. Occurrence: Japan.

natrobistantite. A blue-green, yellow-green or colorless mineral of the pyrochlore group composed of cesium sodium bismuth antimony niobium tantalum oxide, $(Na,Cs)Bi(Ta,Nb,Sb)_4O_{12}$. Crystal system, cubic. Density, 6.10 g/cm³. Occurrence: China.

natrochalcite. A pale green mineral composed of sodium copper sulfate hydroxide monohydrate, $NaCu_2(SO_4)_2OH\cdot H_2O$. Crystal system, monoclinic. Density, 3.49 g/cm³; refractive index, 1.655. Occurrence: Chile.

natrodufrenite. A blue-green mineral of the dufrenite group composed of sodium aluminum iron phosphate hydroxide dihydrate, $Na(Fe,Al)_6(PO_4)_4(OH)_6\cdot2H_2O$. Crystal system, monoclinic. Density, 3.20 g/cm³. Occurrence: France.

natrofairchildite. A white mineral composed of sodium calcium carbonate, $Na_2Ca(CO_3)_2$. Refractive index, 1.459.

natrolite. A white mineral of the zeolite group composed of sodium aluminum silicate dihydrate, $Na_2Al_2Si_3O_{10}\cdot2H_2O$. Crystal system, orthorhombic. Density, 2.20-2.25 g/cm³; refractive index, 1.48. Occurrence: Czech Republic, Poland. Used as a natural *zeolite*.

natromontebrasite. See fremontite.

natron. A white, gray or yellow mineral composed of sodium carbonate decahydrate, $Na_2CO_3\cdot10H_2O$. Crystal system, monoclinic. Density, 1.44-1.46 g/cm³; melting point, loses $10H_2O$ at about 33°C (91°F); refractive index, 1.425. Used as an oxidizer and flux in glass and porcelain enamels, as a neutralizer in the pickling of iron and steel prior to porcelain enameling, and in washing and bleaching textiles.

Natrona. Trade name of CCS Braeburn Alloy Steel (USA) for an molybdenum-type intermediate high-speed steel (AISI type M52) containing 0.9% carbon, 4.0% molybdenum, 4.0% chromium, 2.0% vanadium, 1.2% tungsten, and the balance iron. It has a great depth of hardening, high hot hardness, excellent wear resistance, and good resistance to decarburization. Used for tools, router bits, drills, pump parts, and bearings.

natroniobite. A white or yellowish white mineral composed of sodium niobium oxide, $NaNbO_3$. Density, 4.40 g/cm³; refractive index, 2.20. Occurrence: Russian Federation.

natrophilite. A yellow mineral of the olivine group composed of sodium manganese phosphate, $NaMnPO_4$. Crystal system, orthorhombic. Density, 3.41 g/cm³; refractive index, 1.674. Occurrence: USA (Connecticut).

natrophosphate. A colorless mineral composed of sodium fluoride phosphate hydrate, $Na_7(PO_4)_2F\cdot19H_2O$. Crystal system, cubic. Density, 1.71 g/cm³; refractive index, 1.461. Occurrence: Russia.

Natro Rez. Trade name of Natrochem, Inc. (USA) for coumarone-indene resins.

natrosilite. A colorless mineral composed of sodium silicate, β-$Na_2Si_2O_5$. Crystal system, monoclinic. Density, 2.48-2.51 g/cm³; refractive index, 1.517. Occurrence: Russia.

Natsyn. Trademark of Goodyear Tire & Rubber Company (USA) for synthetic polyisoprene rubber.

natté. A delicate, glossy fabric made of wool, cotton, silk, cellulose or synthetic fibers in several weaves, and having a pattern consisting of numerous cubes. Used for linens, ladies' wear, curtains, etc.

Natur. German trade name for a glass with irregular pattern.

natural abrasive alumina. See natural alumina.

natural abrasives. A group of abrasive materials that occur in nature and include sand (silica), tripoli, garnet, corundum, emery, flint, pumice, rouge (iron oxide), feldspar and diamond. See also abrasives; artificial abrasives.

natural adhesives. A group of adhesives consisting of a natural material, such as animal bones and hides, casein, dextrin, starch, natural rubber, or rosin, dissolved in water or alkaline water. They have poor bond strength above 100°C (212°F), poor resistance to moisture and fungi, and are used chiefly for bonding paper, paperboard, wood and metal foil. See also adhesives; synthetic adhesives.

natural alumina. Alumina (Al_2O_3) as it occurs in nature, e.g., as *corundum* or *emery*. Corundum is of relatively high purity, while emery is less pure, and contains iron oxide as the major impurity. Used as abrasives. Also known as *natural abrasive alumina*.

natural asphalt. See native asphalt.

natural cement. A type of *hydraulic cement* made from soft, pulverized argillaceous limestone by calcining at temperatures below the fusion point to drive off carbon dioxide (CO_2), and subsequent grinding to a fine powder.

natural clay tile. Tile with a characteristic, somewhat textured appearance, made from clays producing dense bodies by the dust-pressing (or dry-pressing) process (i.e., by pressing slightly moistened powdered clay into a die), or the plastic (or wet-pressing) process (i.e., by blending the clay and other ingredients with water followed by direct application of pressure).

natural composites. A group of composite materials that occur in nature, e.g., wood consisting of strong cellulose fibers in a stiffer polymer matrix (primarily lignin), and bone made up of hard, brittle mineral matter contained within an organic matrix. See also composites; synthetic composites.

natural diamond. Diamond as it occurs in nature. It is a crystalline allotrope of carbon, and the hardest known natural mineral, occurring in various colors ranging from colorless, slightly yellowish and yellow to red, green, blue and black. Crystal system, cubic. Density, 3.50-3.53 g/cm³; melting point, 3700°C (6690°F); boiling point, 4200°C (7590°F); hardness, 10 Mohs; refractive index, 2.419. It is an excellent electrical insulator, transparent to infrared radiation, and has high stability, a low coefficient of friction, and the highest thermal conductivity of any natural substance. Occurrence: Brazil, Borneo, India, South Africa, USA (Arkansas), Venezuela. See also diamond; synthetic diamond.

natural fibers. A category of textile fibers including those of animal, mineral or vegetable origin, e.g., fibers of cotton, flax, silk, wool, etc. Abbreviation: NF.

natural finish tile. Glazed or unglazed tile that has the natural color of burnt clay body.

natural flax. Flax produced by separation of the seeds from the straw and subsequent scrutching (a separation process involving beating and shaking), but without retting (partial decomposition in moisture). Also known as *green flax.*

natural garnets. A large group of silicate minerals having the general formula $X_3Y_2(SiO_4)_3$, where X represents calcium, iron, magnesium or manganese, and Y aluminum, chromium, iron or titanium. Some important members of this group are almandine, andradite, calderite, goldmanite, grossularite, katoite, knorringite, majorite, pyrobe, schorlomite, spessartite, ugrandite and uvarovite. Crystal structure, cubic. Density range, 3.5-4.3 g/cm³; hardness range, 6.5-7.5 Mohs. Used for abrasives, blast cleaning of buildings, watch bearings, and gemstones. See also garnets; synthetic garnets.

natural gypsum. See gypsum.

naturally bonded sand. See natural sand.

natural magnet. See lodestone.

natural mica. A group of minerals, all of which contain hydroxyl, aluminum silicate and an alkali, have similar physical properties and atomic structures, and can be split into flexible elastic sheets, but may be of varying chemical compositions. They have a hardness range of about 2.0-2.5 Mohs, and their general formula is $(K,Na,Ca)(Mg,Fe,Li,Al)_{2-3}(Al,Si)_4O_{10}(OH,F)_2$. Examples of natural mica minerals include biotite, glauconite, illite, lepidolite, muscovite and phlogopite.

natural optical crystal. See optical crystal.

natural polyisoprene. *Cis*-1,4-polyisoprene obtained by coagulating the latex of any of various plants of the genus *Hevea,* especially the rubber tree *(H. brasiliensis).* Used in the manufacture of *natural rubber.* See also polyisoprene.

natural polymers. Polymers derived from plants and animals, e.g., wood, natural rubber, cotton, wool, leather, silk, proteins, enzymes, carbohydrates, starches and cellulose. Abbreviation: NP. See also synthetic polymers; biopolymers.

natural protein fibers. Protein fibers obtained from animal sources and including both true and tussah silk, wool from the alpaca, beaver, Cashmere goat, guanaco, llama, mohair, otter, sheep, vicuna and yak, and camel, cow, goat, hare, horse, rabbit and other hairs. See also protein; protein fibers; regenerated protein fibers.

natural resins. Flammable, nonconductive, thermoplastic solid or semisolid organic substances of vegetable (e.g. copals, dammars, etc.) or animal origin (e.g. shellac). They dissolve in some organic solvents, and unlike gums are not soluble in water. See also resin; synthetic resins.

natural rubber. The elastic substance contained in the milky juice *(latex)* of any of various plants of the genus *Hevea,* especially the rubber tree *(H. brasiliensis).* Latex consists of an aqueous dispersion of *cis*-1,4-polyisoprene, $(C_5H_8)_n$, an unsaturated, high-molecular-weight hydrocarbon. For commercial purposes, this latex is coagulated by adding acetic or formic acid or sodium hexafluorosilicate, and subsequently concentrated by evaporation or centrifugation. The processed latex is usually dried and converted into sheets of crude rubber. Unvulcanized natural rubber has poor mechanical properties and chemical and environmental resistance, but these are significantly improved by crosslinking (or vulcanizing), usually through treatment with sulfur or special chemicals. Abbreviation: NR. See also crude rubber; hevea rubber; natural cis-polyisoprene; rubber; synthetic rubber; vulcanized rubber.

natural sand. A generic name referring to clay-bonded sands, such as silica sand, containing up to 20% clay-base contaminants as bonding materials. Used in the construction trades, and in making foundry molds. Also known as *naturally bonded sand.* See also semisynthetic sand; synthetic sand.

natural seasoned lumber. See air-dried lumber.

natural spinel. A colorless, white, blue, green, red, lavender, brown or black mineral with a white streak and a vitreous luster. It is composed of magnesium aluminate, $MgAl_2O_4$, sometimes with small amounts of chromium, iron, manganese and zinc. Crystal system, cubic. Density, 3.5-4.1 g/cm³; hardness, 8.0 Mohs; refractive index, 1.718. Occurrence: Burma, Europe, Sri Lanka, Thailand. Used as a gemstone, and in materials research. See also spinel; synthetic spinel.

natural steel. (1) Steel, such as *wootz steel*, made directly from iron ore.

(2) Steel that has been hot worked and subsequently cooled in air.

(3) Steel made by refining cast iron.

natural uranium. The element *uranium* as it occurs in nature. It contains 99.275% of the isotope ^{238}U, 0.720% of the isotope ^{235}U and 0.005% of the isotope ^{234}U. Used for nuclear applications. Also known as *native uranium; normal uranium.*

natural vermilion. See cinnabar; vermilion.

NatureFlex. Trade name of UCB Films, Inc. (USA) for a range of biodegradable, flexible, regenerated cellulose film materials derived from pulp wood. Used for agricultural and packaging applications.

Naturelle Rx. Trademark of Jeneric/Pentron Corporation (USA) for a silver-free palladium-base dental bonding alloy.

Nature Tex. Trademark of Martin Color-Fi, Inc. (USA) for polyester fibers used for textile fabrics.

NatureWorks PLA. Trade name of Cargill Dow Polymers LLC (USA) for a family of melt-processible *polylactide* (PLA) plastics made entirely from sugars or starches derived from corn, wheat, beets or rice. These natural polymers are available in a wide range of molecular weights and crystallinity variations, and can be converted into fiber form. Used for packaging, clothing, home and office furnishings, etc., and for adhesive coatings and emulsions. The fiber form serves as an intermediate between natural fibers, such as cotton, silk and wool, and synthetic fibers, such as acrylics and nylons.

Naubuc. British trade name for corrosion-resistant alloy 58% copper, 16.25% zinc, 25% nickel, and 0.75% iron. Used for knives and cutlery.

Naugapol. (1) Trademark of Uniroyal Chemical Division of Uniroyal Inc. (USA) for a series of copolymers of butadiene and styrene made into resins and elastomers. The resins are used for cable and wire insulation, floor tile and shoe soles, and the elastomers are employed in the manufacture of cable and wire insulation, adhesives and mechanical rubber products.

(2) Trademark of Uniroyal Chemical Division of Uniroyal Inc. (USA) for a series of polyvinyl chlorides and othyer vinyls.

Naugatex. Trademark of Uniroyal Chemical Division of Uniroyal Inc. (USA) for a series of butadiene-styrene latexes.

naujakasite. A silver-white mineral composed of sodium iron aluminum silicate, $Na_6(Fe,Mn)Al_4Si_8O_3$. Crystal system, monoclinic. Density, 2.62 g/cm^3; refractive index, 1.550. Occurrence: Greenland.

naumannite. An iron black mineral composed of silver selenide, Ag_2Se. It can also be made synthetically. Crystal system, orthorhombic. Density, 7.00 g/cm^3. Occurrence: Germany.

Nautal. Hungarian trade name for a corrosion-resistant aluminum alloy containing 4.5% manganese and 0.4% magnesium. Used for light-alloy parts.

Navac. Trademark of Foseco Minsep NV (Netherlands) for vacuum-processed contamination-free metallic sodium supplied in hermetically sealed aluminum containers. Used in the manufacture of aluminum, magnesium and their alloys.

Navaho. Trade name of AL Tech Specialty Steel Corporation (USA) for an air-hardening cold-work tool steel (AISI type A8) containing 0.55% carbon, 1.25% tungsten, 1.3% molybdenum, 5% chromium, 0.3% manganese, 0.9% silicon, and the balance iron. It has good machinability, high toughness, and is used for punches and dies hot and cold-work applications, shear blades, knives, forming, pressing, blanking, trimming and forging dies, and plastic molds.

navajorite. A dark brown mineral composed of vanadium oxide trihydrate, $V_2O_5 \cdot 3H_2O$. Crystal system, monoclinic. Density, 2.56 g/cm^3; refractive index, 2.02. Occurrence: USA (Arizona).

Navajo White. Trade name of Aardvark Clay & Supplies (USA) for creamy white porcelaineous stoneware clay (cone 10).

Naval. Trade name for a corrosion-resistant brass of 61% copper, 37.5-38% zinc and 1-1.5% tin, used for extrusion tools and dies.

naval aluminum. A corrosion-resistant aluminum alloy containing 1.5% copper, 0.9% manganese, 0.4% nickel and traces of iron and silicon. Used for fittings and instruments.

naval aluminum bronze. A strong, tough, corrosion resistant aluminum bronze containing 85-87% copper, 7-9% aluminum and 2.5-4.5% iron. Used for propellers and marine parts.

naval brass. See naval bronze (3).

naval bronze. (1) A tough bronze containing 88% copper, 8% tin and 4% zinc, used for expansion joints, steam and structural parts, gears, and valves.

(2) An antifriction alloy of 44% lead, 36% tin, 16% antimony and 4% copper, used for bearings.

(3) A wrought brass containing 58.5-62% copper, 37.5-39.5% zinc and 0.5-1% tin. Commercially available in the form of rods, tubes, shapes and flat products, it is difficult to machine, and has relatively good hardness, excellent resistance to seawater corrosion, excellent hot forgeability and workability, and good overall corrosion resistance. Used for marine and general hardware, fasteners, bolts, nuts, rivets, machinery parts, spindles, shafts, marine components, condenser tubes, condenser plates, pump shafts, valve stems, propeller shafts, and welding rods. Also known as *naval brass*.

naval gunmetal. A corrosion-resistant, high-strength alloy composed of 88% copper, 10% tin and 2% zinc, and used for bearings, bushings, pump impellers, valve components, sleeves, pistons, and gears.

Navalium. British trade name for an aluminum alloy containing 2.3% manganese, 0.7% tin, 0.6% iron, 0.2% silicon and 0.1% copper. Used for light-alloy parts.

Naval Journal Bearing. Trade name for a copper-base bearing alloy containing 14% tin and 3-3.5% lead.

Navaloy. British trade name for a babbitt metal containing lead, tin and antimony, used for bearings.

naval phosphor bronze. (1) A cast phosphor bronze containing 88% copper, 8% tin, 4% zinc and 0.5% phosphorus. It has good seawater corrosion resistance, and is used for bearings, gears, and marine parts.

(2) A wrought phosphor bronze containing 94% copper, 3.5% tin, 2% zinc and 0.5% phosphorus. It has good seawater resistance, and is used for pump and valve parts, and bolts.

naval stores. A collective term for a group of products derived from oleoresins obtained from certain coniferous trees, chiefly pines. It includes resins, rosins, oils, tars, pitches, spirits and turpentine. Historically, these were important items in the stores of wooden sailing vessels.

naval valve bronze. A bronze containing 88% copper, 6.5% tin, 4% zinc and 1.5% lead. Used for valve seats, valve bodies, valve stems, and fittings.

Navan. Trade name of George Cook & Company, Limited (UK) for wear-resistant, case-hardening steel, used for gears and shafts.

Navy. (1) Trade name of Paul Bergsoe & Son (Denmark) for a wear-resistant, cast tin alloy containing 13% lead, 9% antimony and 4% copper. It has a melting range of 179-352°C (355-665°F), and is used for chiefly for bearings.

(2) Trade name of American Hard Rubber Company (USA) for hard rubber.

Navy Aluminum Alloy. Trade name for a non-heat treatable aluminum alloy containing 1.5% copper, 0.9% manganese, 0.4% nickel, 0.4% iron and 0.3% silicon. Used for light-alloy parts, fittings, and instruments.

Navy Antifriction Metal. Trade name of Puget Sound Metal Works (USA) for a series of antifriction metals available in seven grades containing varying amounts of tin, antimony, lead, copper and arsenic. Used for aircraft and automotive engine bearings, and electric motor bearings.

Navy Bearing. Trade name for an antifriction alloy of 80-91% tin, 3.7-5% copper and 4.5-15% antimony. Used for bearings, bushings, sleeves, and liners.

Navy Gear Bronze. Trade name for a tough bronze of 84-86% copper, 13-15% tin, 1.5% zinc and 0.5% phosphorus. Used for gears and worm wheels.

navy pitch. A pine pitch obtained by melting rosin with pine tar, and used for marine applications. Also known as *ship pitch*.

Navy Tombasil. Trade name of Illingworth Steel Company (USA) for a corrosion-resistant cast copper alloy containing 89% copper, 6% zinc and 5% silicon. Used for valve stems.

navy wool. Glass wool made into blankets covered on both sides with a layer of fabric treated with a flameproofing compound, and used for the acoustic and thermal insulation of ducts and pipes.

Nawaro. Trademark of Elastogran GmbH (Germany) for a series of unprocessed plastics supplied as powders, granules, pastes and liquids as well as semimanufactured plastic boards, rods, tubes, and sheeting and film products.

Naxaloy. Trade name of MRC Polymers (USA) for a series of thermoplastic alloys of polycarbonate (PC) with polymethylmethacrylate (PMMA), acrylonitrile-butadiene-styrene (ABS) or polyesters. The polyester alloys are often glass-fiber-reinforced.

Naxell. Trade name of MRC Polymers, Inc. (USA) for plastic molding compounds including general-purpose and glass-reinforced polycarbonates. Used for molded and extruded products, and for film and sheeting products.

Naxos. Trade name of Naxos-Schmirgelwerk Carl Wester (Germany) for corundum and emery products supplied in the form of cutoff and grinding wheels, segmental wheels, etc.

NBR rubber. See acrylonitrile-butadiene copolymer.

NCM Alpha. Trade name for a corrosion-resistant beryllium-free nickel-chromium dental bonding alloy.

NCR paper. See carbonless paper.

nealite. A bright orange mineral composed of lead iron arsenate chloride, $Pb_4Fe(AsO_4)_2Cl_4$. Crystal system, triclinic. Density, 4.27 g/cm^3; refractive index, above 2.00. Occurrence: Greece.

Nealloy. Trade name of Cambridge Wire Cloth Company (USA) for a low-carbon steel containing up to 0.12% carbon, 0.6% silicon, 0.6% chromium, up to 1% nickel, up to 0.5% copper, and the balance iron. Used for chiefly wire cloth and woven-wire conveyor belts.

NEA material. An electronic material, such as gallium phosphide (GaP), that has been given a negative electron affinity by treating with a material, such as cesium (Ce), to reduce the surface barrier, and induce band bending resulting in a vacuum level that lies below the conduction band level. Also known as *negative-electron-affinity material*.

near-alpha titanium alloys. A class of titanium alloys including Ti-8Al-1Mo-1V and Ti-6Al-2Nb-1Ta-0.8Mo that contain small percentages (usually less than 2%) of beta stabilizers (vanadium, molybdenum, etc.). Their properties are comparable to those of *alpha-titanium alloys*. Also known as *super-alpha titanium alloys*.

near-beta titanium alloys. A class of titanium alloys including Ti-10V-2Fe-3Al that contain small percentages of alpha stabilizers (aluminum, oxygen, etc.), and whose reaction kinetics and processing usually differ from those of *beta-titanium alloys*.

Neasco. Trade name of Belmont Metals Inc. (USA) for a master alloy of copper, silicon and iron used in the manufacture of silicon bronze.

neat cement. A plastic mixture of a *hydraulic cement*, such as Portland cement, and water, but without aggregate (sand, gravel, etc.).

neat cement grout. A *grout* consisting of cement, usually a hydraulic cement, and water. Admixtures may or may not be added. Also known as *neat grout*.

neat cement paste. A plastic mixture of *hydraulic cement* and water prior to setting.

neat grout. See neat cement grout.

neat gypsum plaster. A powdered cementitious product that like *gypsum plaster* is composed of calcined gypsum, but does not contain aggregates (e.g., sand) and other additives. The latter may be added at the job site. Used as a base coats for walls. Also known as *neat plaster*.

neat leather. A bark-tanned and oil-finished cowhide.

neat plaster. See neat gypsum plaster.

neat resin. A synthetic resin that has no additives or reinforcements added.

Neatro. Trade name of Teledyne Vasco (USA) for a molybdenum-type high-speed steel (AISI type M4) containing 1.25-

1.30% carbon, 0.25-0.35% silicon, 0.2-0.3% manganese, 5.25-5.75% tungsten, 4.25-4.75% chromium, 3.75-4.25% vanadium, 4.25-4.75% molybdenum, and the balance iron. It has good deep-hardening properties, excellent abrasion and wear resistance, excellent cutting ability, and high toughness and edge strength. Used for lathe and planer tools, milling cutters, forming rolls and dies, and a wide range of tools for nonferrous metal finishing.

Nebaloy. Trade name of New England Brass Company (USA) for a cartridge brass containing 35-38% zinc, 0-0.07% lead, 0-0.05% iron, and the balance copper. It has a fine-grained structure, good cold workability, and relatively good electrical conductivity (26.5% IACS). Used for electrical terminals and connectors, stampings, shells, hardware, washers, jewelry, and drawn and stamped products.

Necomicle. Japanese trade name for a stainless steel used for chemical equipment and turbine blades.

Necroni. Trade name of National Erie Corporation (USA) for a steel containing 0.3-0.35% carbon, 0-0.35% silicon, 1% chromium, 0.5% nickel, and the balance iron. Used for crankshafts and gears.

Nedox. Trademark of General Magnaplate Corporation (USA) for a series of superhard *synergistic coatings* used for application to metals and alloys that cannot be anodized, e.g., carbon and stainless steels, copper, brass and other alloys. The application process involves the electrodeposition of a coating of chromium-nickel alloy containing numerous micropores on the metal substrate. The surface is then sealed with a precisely controlled infusion of submicron-size particles of fluorocarbons, after which it is carefully heat-treated to create a new smooth surface. The mosat important properties of the coatings include improved surface hardness, good protection against chemical attack, good abrasion resistance, and permanent lubricity. They are also suitable for high-temperature applications.

needle antimony. Stibnite (Sb_2S_3) that has developed long, needle- or spear-shaped crystals upon cooling. It is used as an adherence promoter in the manufacture of certain enamels. See also stibnite.

needle-bonded fabrics. Textile fabrics made by locking the fibers of a batt together by means of barbed needles.

needled fabrics. Textile fabrics, such as needled felt, needled mats or needle-bonded fabrics, whose fibers have been locked or interlocked by means of a needle or a set of needles.

needled felt. A textile fabric made from natural or synthetic fibers, or a combination thereof by cutting to short lengths and felting together in a needle loom. Sometimes chemical treatment, heat or moisture may form part of the processing regime, but not textile processing methods, such as knitting, stitching, weaving, thermal or adhesive bonding. Also known as *needle felt*.

needled mat. A mat manufactured from natural or synthetic fibers by cutting to short lengths and felting together in a needle loom.

needle felt. See needled felt.

needle metal. A free-cutting alloy of 85% copper, 8% tin, 5.3% zinc and 1.7% lead, used for needles, valves and fittings.

needle ore. See aikinite.

needlepoint lace. A lace made with an embroidery needle and thread, using buttonhole stitches on a heavy paper pattern. Also known as *point lace*.

needle-punched batting. Textile batting whose fibers have been mechanically entangled for stabilization purposes. Used as a filling material.

needle tubing. Tubing of stainless steel with an outside diameter of 0.36-5.16 mm (0.010-0.105 in.) and a length of up to 1.8 m (6 ft) used for surgical instruments and radon implanters.

needle wire. A tough, round tool-steel wire in coil form available in various diameters ranging from 0.254 to 2.667 mm (0.010 to 0.105 in.), and used for sewing machine needles, awls, and latch pins.

Neelium. (1) Trademark of Atlas Minerals & Chemicals, Inc. (USA) for a thin neoprene rubber coating.

(2) Trade name of General Thermoelectric Corporation (USA) for a semiconductor alloy containing bismuth, tellurium, selenium and antimony. Used for thermoelectric cooling applications.

Nefalit. Trade name of MAB Beaulieu SA (France) for strong, flexible, acid-resistant ceramic boards with maximum service temperatures of 850-1200°C (1560-2190°F), used as asbestos substitutes in sheet packing and heat insulation.

Nefalon. Trade name for nylon fibers and products.

Nefa-Perlon. Trade name for nylon fibers and yarns used for textile fabrics including clothing.

Nefmac. Trade name of Shimizu Corporation (Japan) for grid-type, fiber-reinforced plastic reinforcements used for various applications in building construction and civil engineering.

Negastat. Trademark of E.I. DuPont de Nemours & Company (USA) for antistatic polyester fibers used in the manufacture of textile fabrics.

negative crystal. A uniaxial, double-refracting crystal, such as *calcite*, in which the refractive index of the ordinary ray is greater than that of the extraordinary ray.

negative-electron-affinity material. See NEA material.

negative Poisson's ratio materials. A class of relatively new solid materials including re-entrant metal (e.g., copper) and polymer (e.g., polyurethane) foams as well as laminates, microporous materials, chiral honeycombs and materials with structural hierarchy that easily undergo volume changes and, in contrast to conventional materials, exhibit negative Poisson's ratios, i.e., expand in directions perpendicular to the upon stretching axis. Present and future applications include sponges, shock-absorbing materials, seat cushions, fasteners, air filters and biomaterials. Also known as *anti-rubbers; auxetic materials; dilatational materials.* See also re-entrant foams.

negative index materials. See right-handed materials.

neighborite. An anisotropic mineral composed of sodium magnesium fluoride, $NaMgF_3$. Crystal system, orthorhombic. Density, 3.03 g/cm^3; refractive index, 1.364. Occurrence: USA (Utah).

Neillite. Trade name of Watertown Manufacturing Company (USA) for phenol-formaldehyde resins and plastics.

Neill Ware. Trade name of Watertown Manufacturing Company (USA) for phenol-formaldehyde plastics and products.

nekoite. A colorless mineral composed of calcium silicate hydroxide pentahydrate, $Ca_3Si_6O_{12}(OH)_6 \cdot 5H_2O$. Crystal system, triclinic. Density, 2.28 g/cm^3; refractive index, 1.535. Occurrence: USA (California).

Nelco. Trade name of Nelco Metal Corporation (USA) for calcium metal (99+% pure) containing 0.25% aluminum and 0.2% magnesium.

Neloy. Trade name of National Erie Corporation (USA) for a series of tough steels containing 0.3-0.4% carbon, 0.75-1.25% manganese, 0-0.6% nickel, 0-0.2% molybdenum, 0-0.4% sili-

con, and the balance iron. Used for gears, crankshafts, and castings.

Nelson-Bohnalite. Trade name of Karl Schmidt Co. Metallschmelzwerk (Germany) for an aluminum alloy containing 10% copper, 0.3% magnesium, 0.3% nickel and 0.2% silicon. It has a low coefficient of thermal expansion, and is used for automotive pistons.

Neltape. Trade name of Dielectric Polymers Inc. (USA) for masking tape used in powder coating applications.

neltnerite. A brown or rose-red mineral composed of calcium manganese silicate, $CaMn_2SiO_{12}$. Crystal system, tetragonal. Density, 4.63 g/cm^3. Occurrence: Morocco.

nematic liquid crystal. A *liquid crystal* in a state ("mesophase") intermediate between an ordinary solid and a liquid in which the rodlike organic molecules tend to be arranged parallel to a preferred common axis, i.e., they have some degree of orientational order, but lack positional order. See also cholesteric liquid crystal; smectic liquid crystal.

Nemicle. Japanese trade name for a series of stainless steels containing varying amounts of iron, carbon, nickel, chromium and molybdenum. Used for turbine blades and chemical equipment.

nenadkevichite. (1) A brown or rose-red mineral composed of sodium niobium silicate dihydrate, $(Na,Ca)(Nb,Ti)(OSi_2O_6) \cdot 2H_2O$. Crystal system, orthorhombic. Density, 2.83-2.88 g/cm^3; refractive index, 1.686. Occurrence: Morocco, Russia.

(2) A brown or rose-red mineral composed of sodium niobium titanium silicate dihydrate, $Na(Nb,Ti)Si_2O_7 \cdot 2H_2O$. Crystal system, orthorhombic. Density, 2.68 g/cm^3. Occurrence: Morocco, Canada (Quebec).

Neo-Aloy. Trade name of Neomet Corporation (USA) for rare-earth alloys, metals and powders.

Neo-Baros. Trade name of Creusot-Loire (France) for a nonmagnetic alloy of 90% nickel and 10% chromium, used for pen points, springs and weights.

Neobond. (1) Trade name of Dex-O-Tex (USA) for a *neoprene rubber* emulsion used as a waterproofer under ceramic, marble and quarry tile and composite flooring.

(2) Trademark of FiberMark (USA) for a durable, hard-wearing art paper made from synthetic fibers. It can be printed by a variety of processes and can have security features incorporated. Used in art conservation and the graphic arts.

Neobond II. Trade name of Neobond (USA) for a corrosion-resistant cobalt-chromium dental bonding alloy.

Neocast. Trade name of Cendres & Métaux SA (Switzerland) for high-gold dental casting alloys used for crowns, bridges, fillings and other dental restorations. *Neocast 3* contains at total of 75.1% gold and platinum. Also known as *CM Neocast.*

Neoceram. Trade name of Nippon Electric Glass Company Limited (Japan) for glass-ceramics.

Neochran. Trade name of Skoda Works National Corporation (Czech Republic) for a corrosion-resistant steel containing 1-1.5% carbon, 22-30% chromium, 0.5-2.5% silicon, and the balance iron. Used for stainless castings.

Neo-Chrome. Trademark of Coburn Corporation (USA) for metallized mirror films.

Neochrome. Trademark of Sasol Fibres Proprietary Limited (South Africa) for acrylic (polyacrylonitrile) fibers used for textile fabrics.

Neocoat. Trade name of Atofina Chemicals Inc. (USA) for corrosion-resistant coatings and cements.

NeoCryl. Trademark of ICI Resins USA Inc. for a series of synthetic resins including acrylic polymers and copolymers used for protective and decorative coatings, textile and leather finishing, orthodontic applications, paper coatings, floor polishes, adhesives and paints.

neodymia. See neodymium oxide.

neodymium. A soft, malleable, silvery white to yellowish, metallic element of the lanthanide series (rare-earth group) of the Periodic Table. It is commercially available in the form of ingots, rods, sheets, foil, lumps, turnings (chips), wire and powder, and single crystals. The chief ores are *allanite, bastnaesite* and *monazite.* Density, 7.0 g/cm³; melting point, 1024°C (1875°F); boiling point, above 3030°C (5486°F); hardness, 35 Vickers; atomic number, 60; atomic weight, 144.24; trivalent. It tarnishes quickly in air, has a high electrical resistivity, paramagnetic properties, and good machinability. *Neodymium* occurs in two allotropic forms: (i) *Alpha neodymium* with hexagonal close-packed crystal structure existing up to 868°C (1594°F), and (ii) *Beta neodymium* with body-centered cubic structure present from 868°C (1594°F) to 1024°C (1875°F). Used in electronics, in alloys and permanent magnets, as a glass coloring agent (e.g., for astronomical lenses and spectacles), as an alloying addition to magnesium to increase strength and heat resistance, as a gas scavenger in iron and steel manufacture, as a yttrium-garnet laser dopant, in superconductors, and for carbon-arc lights, lighter flints and ceramic capacitors. Symbol: Nd.

neodymium acetylacetonate. Hygroscopic, violet crystals with a density of 1.618 g/cm³ and a melting point of 150-152°C (302-305°F). Used as an organometallic reagent, and as a dopant source for fiberoptics derived by sol-gel techniques. Formula: $Nd(O_2C_5H_7)_3$. Also known as *neodymium 2,4-pentanedionate.*

neodymium barium copper oxide. This term refers to any of several neodymium barium cuprates with superconducting properties, usually supplied as fine powders in varying compositions, e.g., Nd-123 ($NdBa_2Cu_3O_{7-x}$) with a Nd:Ba:Cu ionic ratio of 1:2:3 or Nd-422 ($Nd_4Ba_2Cu_2O_{10}$) with a Nd:Ba:Cu ionic ratio of 4:2:2. Some are also available as thin films. Used as high-temperature superconductors, in superconductivity research, and as a raw material for the manufacture of superconducting components.

neodymium boride. Any of the following compounds of neodymium and boron used in ceramics: (i) *Neodymium tetraboride.* Density, 5.83 g/cm³. Formula: NdB_4; and (ii) *Neodymium hexaboride.* Gray powder. Density, 4.95 g/cm³; melting point, 2538°C (4600°F). Formula: NdB_6.

neodymium bromide. Violet or green hygroscopic crystals or powder (99+% pure). Crystal system, orthorhombic. Density, 5.3 g/cm³; melting point of 684°C (1263°F); boiling point, 1540°C (2804°F). Used in the preparation of neodymium metal and salts. Formula: $NdBr_3$.

neodymium carbide. Any of the following compounds of neodymium and carbon used in ceramics, electronics and materials research: (i) *Neodymium sesquicarbide.* Density, 6.90 g/cm³. Formula: Nd_2C_3; (ii) *Neodymium dicarbide.* Yellow crystals. Crystal system, tetragonal. Density, 5.2-6.0 g/cm³; melting point, above 1980°C (3596°F). Formula: NdC_2; and (iii) *Neodymium hexacarbide.* A gray powder. Formula: NdB_6.

neodymium carbonate. A light purple to pink, hygroscopic powder (99+% pure) with a melting point of 114°C (237°F). Formula: $Nd_2(CO_3)_3 \cdot xH_2O$. Used in ceramics and superconductor research. Also known as *neodymium carbonate hydrate.*

neodymium chloride. Blue-pink, hygroscopic crystals or powder (99+% pure). Crystal system, hexagonal. Density, 4.13 g/

cm³; melting point, 784°C (1443°F), boiling point, 1600°C (2910°F). Used in the preparation of neodymium metal and salts. Formula: $NdCl_3$.

neodymium chloride hexahydrate. Red or purple, hygroscopic crystals (99.99+% pure). Crystal system, orthorhombic. Density, 2.28 g/cm³; melting point, 124°C (255°F); loses $6H_2O$ at 160°C (320°F). Used in the preparation of neodymium metal and salts. Formula: $NdCl_3 \cdot 6H_2O$.

neodymium dicarbide. See neodymium carbide (ii).

neodymium disilicate. See neodymium silicate (ii).

neodymium disulfide. See neodymium sulfide (ii).

neodymium-doped yttrium vanadate. Yttrium vanadate (YVO_4) doped with neodymium (Nd^{3+}) ions, and used as a laser crystal. Crystal system, tetragonal. Density, 4.22 g/cm³; melting point, 1825°C (3317°F); hardness, 5 Mohs; transmission range, 400-4000 nm. Formula: YVO_4:Nd. See also yttrium vanadate.

neodymium fluoride. Purple crystals or pink powder (99.9+% pure). Crystal system, orthorhombic. Density, 6.51-6.65 g/cm³; melting point, 1410°C (2570°F) [also reported as 1377°C (2510°F)]; boiling point, 2300°C (4172°F). Used in the preparation of neodymium metal and salts, and extensively in the preparation of fluorinated glasses. Formula: NdF_3.

neodymium gallate. An oxide of neodymium and gallium supplied as a single crystal substrate. Crystal system, orthorhombic. Density, 7.57 g/cm³; melting point, 1600°C (2912°F); dielectric constant, 20. Used in the thin-film deposition of high-temperature superconductors, in the deposition of III-V (13-15) nitrides for lasers, and for light-emitting diodes. Formula: $NdGaO_3$. Also *neodymium gallium oxide (NGO).*

neodymium gallium oxide. See neodymium gallate.

neodymium glass. A special glass that contains a small percentage of neodymium oxide (Nd_2O_3), and transmits 90% of the incident light rays producing the three primary colors red, green and blue, and only 10% of those producing the less desirable color yellow. Used for color television filter plates. Abbreviation: NG.

neodymium hafnate. A compound of neodymium oxide and hafnium oxide used in ceramics and materials research. Density, 8.38 g/cm³; melting point, 2400°C (4352°F). Formula: $Nd_2Hf_2O_5$.

neodymium hexaboride. See neodymium boride (ii).

neodymium hexacarbide. See neodymium carbide (iii).

neodymium iodide. Green, hygroscopic crystals or powder (99.9+% pure). Crystal system, orthorhombic. Density, 5.85 g/cm³; melting point, 775°C (1427°F) [also reported as 784°C (1443°F)]. Used in the preparation of neodymium metal and salts. Formula: NdI_3.

neodymium-iron-boron magnets. Ferromagnetic ceramic materials that consist of the intermetallic phase neodymium iron boride ($Nd_2Fe_{14}B$) and have tetragonal crystal structures. They possesses better mechanical properties than many other rare-earth magnets, and are less expensive. *Neodymium-iron-boron magnets* have exceptionally high magnetic energy densities making them one of the most compact magnet materials available. Also known as *iron-neodymium-boron magnets.*

neodymium molybdate. A compound of neodymium oxide and molybdenum trioxide. Crystal system, tetragonal. Density, 5.14 g/cm³; melting point, 1176°C (2149°F). Used in ceramics and materials research. Formula: $Nd_2(MoO_4)_3$.

neodymium monosilicate. See neodymium silicate (i).

neodymium monosulfide. See neodymium sulfide (i).

neodymium monotelluride. See neodymium telluride (i).

neodymium nitrate. Purple, or pink hygroscopic crystals or powder (99+% pure) used in the preparation of neodymium metal and salts, in materials research, and in the biosciences. Formula: $Nd(NO_3)_3 \cdot 6H_2O$. Also known as *neodymium nitrate hexahydrate*.

neodymium nitrate hexahydrate. See neodymium nitrate.

neodymium nitride. A compound of neodymium and nitrogen supplied as black crystals or powder. Crystal system, 7.69. Density, 7.69 g/cm^3. Used in ceramics and materials research. Formula: NdN.

neodymium oxalate. Rose-colored, hygroscopic crystals, or pink powder (99+% pure) used in ceramics and materials research. Formula: $Nd_2(C_2O_4)_3 \cdot 10H_2O$. Also known as *neodymium oxalate decahydrate*.

neodymium oxide. Light blue to blue-gray, hygroscopic crystals, off-white high-purity powder (99.9%), or brown impure powder (65-99%). Crystal system, hexagonal. Density, 7.24-7.28 g/cm^3; melting point, 2270°C (4118°F); boiling point, 3072°C (5,562°F). Used in glasses to impart a violet color, in technical glass to suppress yellow sodium light, as a decolorizer in heat-resisting glasses of high boric oxide content, and in various specialty glasses for the production of lasers and capacitors. Formula: Nd_2O_3. Also known as *neodymia*.

neodymium 2,4-pentanedionate. See neodymium acetylacetonate.

neodymium phosphide. A compound of neodymium and nitrogen, used in ceramics and for semiconductors. Density, 5.94 g/cm^3. Formula: NdP.

neodymium selenide. A gray powder used in ceramics and for semiconductors. Formula: Nd_2Se_3.

neodymium sesquicarbide. See neodymium carbide (i).

neodymium sesquisulfide. See neodymium sulfide (iii).

neodymium sesquitelluride. See neodymium telluride (ii).

neodymium silicate. Any of the following compounds of neodymium oxide and silicon dioxide used in ceramics and materials research: (i) *Neodymium monosilicate*. Formula: Nd_2SiO_5; and (ii) *Neodymium disilicate*. Formula: $Nd_2Si_2O_7$.

neodymium silicide. A compound of neodymium and silicon. Density, 5.84 g/cm^3; melting point, approx. 1527°C (2781°F). Used for semiconductors and refractories. Formula: $NdSi_2$.

neodymium sulfate. A white powder (99.9% pure) used in small amounts as a decolorizer in glass and in larger amounts as a glass colorant in tableware and glass-blowers' and welders' goggles. Formula: $Nd_2(SO_4)_3$.

neodymium sulfate octahydrate. Red, pink or purple, hygroscopic crystals. Crystal system, monoclinic. Density, 2.85 g/cm^3; melting point, 1176°C (2149°F). Used in decolorizing glass and coloring glass. Formula: $Nd_2(SO_4)_3 \cdot 8H_2O$.

neodymium sulfide. Any of the following compounds of neodymium and sulfur used in ceramics and materials research: (i) *Neodymium monosulfide*. Density, 6.36 g/cm^3; melting point, 2140°C (3884°F). Formula: NdS; (ii) *Neodymium disulfide*. Density, 5.34 g/cm^3; melting point, 1760°C (3200°F). Formula: NdS_2; (iii) *Neodymium sesquisulfide*. Brown crystals or olive green powder. Crystal system, orthorhombic. Density, 5.50 g/cm^3; melting point, 2010°C (3650°F); hardness, 330 Vickers. Formula: Nd_2S_3; and (iv) *Neodymium tetrasulfide*. Density, 6.02 g/cm^3; melting point, 2040°C (3704°F). Formula: Nd_3S_4.

neodymium telluride. Any of the following compounds of neodymium and tellurium used in ceramics, materials research and electronics: (i) *Neodymium monotelluride*. Melting point, 2043°C (3709°F). Formula: NdTe; (ii) *Neodymium sesquisul-*

fide. Gray crystals or fine, gray powder. Crystal system, orthorhombic. Density, 7.0 g/cm^3; melting point, 1377°C (2511°F). Formula: Nd_2Te_3; and (iii) *Neodymium tetrasulfide*. Melting point, 1862°C (3384°F). Formula: Nd_3Te_4.

neodymium tetraboride. See neodymium boride (i).

neodymium tetrasulfide. See neodymium sulfide (iv).

neodymium tetratelluride. See neodymium telluride (iii).

neodymium zirconate. A compound of neodymium oxide and zirconium dioxide supplied as a light purple powder, and used in ceramics and materials research. Formula: $Nd_2(ZrO_3)_3$.

Neo-Flash. Trade name of Paul Wissmach Glass Company Inc. (USA) for a rolled flashed *opal glass*.

Neoflex. Trademark of Mitsui Chemicals, Inc. (Japan) for a polyimide laminate supplied in two series: (i) *Series I* nonthermoplastic polyimides for flexible printed circuits and tape carrier pages, and (ii) *Series II* one- or two-layer thermoplastic polyimide for hot-melt adhesives.

Neoflon. Trade name of Daikin Industries, Limited (Japan) for fluoropolymer sheeting, and various exterior, interior and industrial fluorocarbon polymer coatings.

NEO Gel Cote. Trade name of The Glidden Company (USA) for fiberglass coatings.

NeoGen. Trademark of Dry Branch Kaolin Company (USA) for a series of aluminum silicates (kaolins) used as extender pigments in paints and inks.

neogen. A corrosion-resistant alloy composed of 58% copper, 27% zinc, 12% nickel, 2% tin, 0.5% bismuth and 0.5% aluminum, and used as imitation silver for ornamental, architectural and structural parts.

Neohecolite. Trade name for a vinyl chloride resin formerly used for denture bases.

Neolin. Trade name of Monopol AG (Switzerland) for wash primers and fillers for coatings.

Neoline. Trade name of Atofina Chemicals Inc. (USA) for corrosion-resistant coatings and cements.

Neolite. Trade name of Consortium Replastic (Italy) for 100% heterogeneous recycled multi-purpose plastics.

Neolith. Trademark of National Lead Company (USA) for dust-free, free-flowing *litharge* (lead oxide) supplied in pellet form for ceramic applications.

Neolube. Trademark of Huron Industries Inc. (USA) for a lubricant consisting of graphite dissolved in alcohol, and used to produce graphite films on bearing surfaces by solvent evaporation.

Neolyn. Trademark of Hercules, Inc. (USA) for a series of rosin-derived alkyd resins with hardnesses ranging from medium-hard to soft, and used for molded products (e.g., floor tiles). *Neolyn* solutions are used as modifiers in adhesives, lacquers, organosols, or plastisols to increase the film toughness. See also alkyd resins; rosin.

Neomagnal. German trade name for a series of alloys of aluminum, magnesium and zinc used for bearings.

neon. A colorless, odorless, tasteless, nonmetallic element of the noble gas group (Group VIIIA or Group 18) of the Periodic Table. It is a monatomic gas that forms a very small part (0.0018 vol%) of the air, and does not combine chemically with any known element. Density of gas, 0.9002 g/L (at 0°C or 32°F); density of liquid, 1.204 g/cm^3 (at -245.9°C or -411°F); liquefaction temperature, -245.9°C (-411°F); freezing point, -248.6°C (-416°F); atomic number, 10; atomic weight, 20.18; zerovalent. It is noncombustible, has good electrical conductivity, and gives off a red glow when electric current is passed through it in a low-pressure tube. The gas is used in glow-discharge lamps, luminescent electric tubes and photoelectric bulbs, photoconductors, advertising signs, floodlights, lightning arrestors, wavemeter tubes, television tubes, electronic devices, voltage indicators, and lasers. The liquid is used as a refrigerant and in cryogenics. Symbol: Ne.

Neonalium. Trade name of Aluminium-Zentrale e.V. (Germany) for an aluminum alloy containing 6-14% copper and 0.4-1% other elements. It has good machinability and heat resistance, and is used for general castings, casings, housings, and light-alloy parts.

Neopal. Trade name of Schott DESAG AG (Germany) for pot opal and flashed *opal glass*.

Neophan. Trade name of Schott DESAG AG (Germany) for a spectacle glass with a neutral tint.

neophane glass. A special yellow glass tinted with neodymium oxide (Nd_2O_3) to minimize glare, and used for automobile windshields, sunglasses, etc.

Neoplex. Trade name for rayon fibers and yarns used for textile fabrics.

Neopolen. Trademark of BASF Corporation (USA) for a series of polyethylene homopolymers and copolymers supplied as liquids, pastes and powders, and as foams in the form of boards, profiles, sheets, blocks, strips, webs, plugs, rolls and strips. The homopolymers and copolymers are used in the manufacture of plastic products, while the foams are used as thermal insulation, and for furniture, padding and stuffing, sealing agents, packaging materials, life-jackets, and sport shoes. *Neopolen E* is a lightweight, heat-resistant, resilient foam material with high-energy absorption. Produced in slab form, it is made from crosslinked polyethylene, and used for packaging fragile goods, and in building products, sports equipment, and insulation pads. *Neopolen P* is an expandable polypropylene supplied in bead form for molding into lightweight, heat-resistant, resilient foam products with high energy absorption. Used for packaging materials, as heat and sound insulation, and for automotive applications.

Neopor. Trademark of BASF Corporation (USA) for a granite-colored expandable polystyrene foam with outstanding heat-insulating properties, used for facade, masonry and exterior-wall insulation, roofing systems, etc.

Neoprene. Trade name of DuPont Dow Elastomers (USA) for a synthetic rubber made from chloroprene (2-chlorobutadiene), and vulcanized with metallic oxides. Commercially available in solid form, as latex and as flexible foam, it has a density of 1.23 g/cm³ (0.044 lb/in.³), high cohesive strength, excellent weathering, heat, oil, solvent, oxygen, ozone and corona discharge resistance, good resistance to alkalies, dilute acids, petroleum oils and greases, fair resistance to water and antifreezes, poor resistance to oxidizing agents, high flame resistance (only the isocyanate-modified grades), a service temperature range of -50 to +105°C (-58 to +221°F), and electrical properties which are not as good as those of natural rubber. The solid forms are used for mechanical rubber products, conveyors belts, V-belts, hose covers, brake diaphragms, motor mounts, rolls, seals and gaskets, adhesive cements, linings of chemical reaction equipment, tanks, oil-loading hose, wire and cable products, coatings for electric wiring, roofing membranes, flashing, footwear, and as binders for rocket fuels. The foams are used for adhesive tape, automotive components, as metal fastener substitutes, for seat cushions, carpet backing, and as sealants. The latex is used for specialty items. See also chloroprene rubber.

neoprene. A term now often used generically for *chloroprene rubber.* See also Neoprene.

neoprene latex. Neoprene rubber dispersed in water for use as a paper and textile coating, and in the manufacture of specialty items.

neoprene-lead fabrics. Highly flexible fabrics composed of neoprene impregnated with lead powder, and used for radiation shielding and protective clothing.

neoprene-phenolic adhesives. Structural adhesives based on blends of neoprene rubber and a phenolic resin. They have high peel strength, poor to fair shear strength, and high flexibility.

Neoprotec. Trade name of Sovirel (France) for a transparent glass used for welders' goggles.

Neor. Trade name of Darwin & Milner Inc. (USA) for a tool steel containing 2.3% carbon, 0.4% manganese, 0.6% silicon, 13% chromium, 0.6% nickel, and the balance iron. It has exceptional abrasion and wear resistance, and is used for press tools, punches, dies, gages, and various other tools.

Neoresit. Trade name of A. Nowack (Germany) for phenol-formaldehyde resins and plastics.

Neorex. Trade name of Creusot-Loire (France) for an air- or oil-hardening cold-work tool steel containing 2-2.1% carbon, 13% chromium, and the balance iron. Used for blanking and coining dies.

NeoRez. Trademark of ICI Resins USA Inc. for waterborne polyurethane coatings and prepolymers, and polystyrene copolymer emulsions for textile and leather finishes.

Neo Sentalloy. Trademark of GAC International, Inc. (USA) for a shape-memory alloy of 52.9% nickel and 47.1% titanium, used for orthodontic wire.

Neosid. Trademark of Neosid Corporation Division of Magnetic Materials Group Limited (UK) for iron powders and ferrite cores.

neosilicates. Silicate minerals, such as olivine, in which the SiO_4 tetrahedra do not share oxygens and are joined by ionic bonds.

Neospanal. Trade name of Kreidler Werke GmbH (Germany) for an aluminum alloy containing copper, bismuth and lead, used for light-alloy parts.

Neo-Star. Trademark of Atotech USA Inc. for bright zinc electroplates and plating processes.

Neo-Temp. Trade name for a dental cement used for temporary crown and bridge restorations.

Neotex. (1) Trademark of Columbian International Chemicals Company (USA) for a carbon black.

(2) Trade name of Porritts & Spencer Canada Inc. for nylon filter fabrics.

Neotherm. Trademark of Schott DESAG AG (Germany) for a heat-resistant furnace observation glass.

neotocite. A black mineral composed of manganese magnesium iron silicate monohydrate, $(Mn,Mg,Fe)SiO_3 \cdot H_2O$. Density, 2.43 g/cm³; refractive index, 1.62. Occurrence: USA (Wisconsin).

Neotone. Trade name of Neoloy Inc. (USA) for a plasticized acrylic resin used as a denture liner.

Neoxil. Trade name for thermoplastic polyester and polyether urethane elastomers.

Neoxydin. Trade name of W. Seibel AG (Germany) for an aluminum alloy containing magnesium, manganese, chromium and copper. It has excellent resistance to atmospheric and seawater corrosion, and is used for marine parts.

Neozote. Trade name of Expanded Rubber Company (UK) for a series of expanded synthetic rubbers.

nepheline. See nephelite.

nepheline syenite. See nephelite syenite.

nephelite. A colorless, white, gray, or yellowish feldspathoid mineral with a vitreous to greasy luster. It is composed of sodium-potassium aluminum silicate, $(Na,K)AlSiO_4$. Crystal system, hexagonal. Density, 2.55-2.65 g/cm^3; hardness, 5.5-6 Mohs; refractive index, 1.534-1.538. Used in the manufacture of ceramics, glasses and enamels, as a substitute for feldspar, and as a source of potash and aluminum. Occurrence: Canada, Central Africa, Russia, USA (Arkansas, New Jersey). Also known as *nepheline.*

nephelite syenite. A coarse-grained igneous rock composed of a mixture of nephelinic minerals ($K_2O \cdot 3Na_2O \cdot 4Al_2O_3 \cdot 9SiO_2$), soda and potash feldspars, and small amounts of mica, hornblende and magnetite. It has a density of 2.61 g/cm^3 (crystalline) or 2.28 g/cm^3 (glassy), and a hardness of 6 Mohs. Used for sanitary ware, floor and wall tile, semivitreous ware, electrical porcelains, glasses, porcelain enamels and other ceramic products, as a replacement for feldspar to lower firing temperatures and shorten firing times, and in structural clay to increase mechanical strength (however, tends to increase shrinkage). Also used in pottery, porcelain, and tile to reduce warpage, expansion, crazing, and water absorption. Also known as *nepheline syenite.*

nephrite. A tough, compact, fine-grained, bluish or greenish variety of the mineral *tremolite* belonging to the amphibole group. It is composed of calcium magnesium silicate, $Ca_2Mg_5(Si_8O_{22})(OH)_2$, sometimes with a small amount of iron replacing part of the magnesium. Crystal system, monoclinic. Density, 2.96-3.1 g/cm^3; hardness, 6-6.5 Mohs. It is the less valuable variety of *jade.* Occurrence: Mexico, New Zealand, Siberia, Turkestan, USA (Alaska, Wyoming). Used as a gemstone and for ornamental pieces. Also known as *greenstone; kidney stone.* See also jadeite.

Ne Plus Ultra. Trade name of Crown Cork & Seal Company (USA) for mineral wool finishing cement.

Nepoxide. Trademark of Atlas Minerals & Chemicals, Inc. (USA) for an epoxy resin coating that has excellent adhesive properties, and good resistance to many chemicals and solvents.

Neptune. (1) Trade name of Pyramid Steel Company (USA) for an oil-hardening low-carbon mold steel (AISI type P20) containing 0.3% carbon, 0.4% molybdenum, 1.7% chromium, 0.75% manganese, 0.5% silicon, and the balance iron. It has good deep-hardening properties, and low resistance to softening at elevated temperatures. Used for plastic molds and die-casting dies.

(2) Trademark of Pentron Laboratory Technologies (USA) for a corrosion-resistant beryllium-free nonprecious dental alloy composed of 63% nickel, 22% chromium, 9% molybdenum, and the balance titanium, aluminum, niobium, iron and yttrium. Used for crown and bridge restorations.

neptunite. A black mineral composed of sodium potassium lithium iron titanium silicate, $Na_2KLi(Fe,Mn)_2Ti_2(SiO_3)_8$. Crystal system, monoclinic. Density, 3.19 g/cm^3; refractive index, 1.6927. Occurrence: Greenland, Russia, USA (California).

neptunium. A synthetic, radioactive, metallic element of the actinide series of the Periodic Table. It was first produced by bombardment of uranium with high-speed deuterons, and is now also obtained in nuclear reactors as a byproduct in the production of plutonium. Silvery white metallic neptunium is obtained from neptunium trifluoride. Density, 20.45 g/cm^3; melting point, 637°C (1179°F); atomic number, 93, atomic weight, 237.048; trivalent, tetravalent, pentavalent, hexavalent; half-life, approxi-

mately 2.1×10^6 years. Three allotropic forms of metallic neptunium are known: (i) *Alpha neptunium* (orthorhombic crystal structure) below 279°C (534°F); (ii) *Beta neptunium* (tetragonal crystal structure) from 279-574°C (534-1065°F); and (iii) *Gamma neptunium* (body-centered cubic crystal structure) from 500-637°C (1065-1179°F). It forms compounds with many other elements including aluminum ($NpAl_2$), beryllium ($NpBe_{13}$), carbon (NpC), fluorine (NpF_3, NpF_4 and NpF_6), molybdenum ($NpMo_2$), nitrogen (NpN), oxygen (NpO_2 or Np_3O_8), and silicon ($NpSi_2$). Used in neutron detection instruments. Symbol: Np.

neptunium dioxide. A dark olive, free-flowing, radioactive powder used in the production of plutonium-238. Formula: NpO_2.

nepuite. See garnierite.

Nerane. Trade name for cellulose acetate fibers used for wearing apparel and industrial fabrics.

Nergandin. Trade name of IMI Kynoch Limited (UK) for a leaded brass composed of 70% copper, 28% zinc and 2% lead. It has good resistance to seawater corrosion, and is used for condenser tubes, and heat exchangers.

Nerofil. Czech trade name for glass-fiber filtration media.

Nervo Metal. Trademark of Rippenstreckmetall-Gesellschaft mbH (Germany) for galvanized or black-coated ribbed expanded metal used in building interiors.

Nesa. (1) Trade name of PPG Industries Inc. (USA) for a transparent, electrically conductive coating.

(2) Trademark of PPG Industries Inc. (USA) for plate or window glass with transparent, electrically conductive coating used in the manufacture of laminated and unlaminated glass units.

Nesaloy. Trade name of Nesaloy Products Inc. (USA) for a lithium alloy used to disperse lead in copper, and for alloying applications.

Nesatron. Trade name of PPG Industries Inc. (USA) for a low-resistance, electrically conductive glass with a metal oxide film produced by vacuum deposition.

nesquehonite. A colorless-white mineral composed of magnesium carbonate trihydrate, $MgCO_3 \cdot 3H_2O$. Crystal system, monoclinic. Density, 1.85 g/cm^3; refractive index, 1.569. Occurrence: Italy.

Neste. Trade name of Borealis Chemicals (USA) for a series of polyethylenes including *Neste LDPE* unmodified and UV-stabilized low-density polyethylenes and *Neste LLDPE* linear-low-density polyethylenes.

Nestem. Trade name of Shimizu Corporation (Japan) for geotextiles used in building construction and civil engineering.

Nestor. (1) Trademark of Anchor Packing Division of Robco Inc. (Canada) for braided packings made from ramie fibers.

(2) Trade name of James Ferguson & Sons (UK) for urea-formaldehyde plastics.

Nestorite. Trade name of James Ferguson & Sons (UK) for phenol- and urea-formaldehyde resins and plastics.

net. An open-mesh fabric made of strings, cords, threads or wires, knitted, knotted, woven or otherwise tied together in such a way as to leave regularly arranged holes of varying size and shape. Netting for industrial applications is often made of metal threads or wires and used for window screens, sieves, strainers, etc. Netting of cotton, viscose, nylon, polyester and other fibers is used for curtains, veils, dresses, and trimmings.

Netic. Trade name of Magnetic Shield Corporation (USA) for a series of nickel-iron alloys with low magnetic retentivity and high magnetic saturation. They are supplied in sheet and foil form for magnetic shielding applications.

nettle fibers. The soft, short fibers obtained from the stems of various species of plants of the genus Urtica, especially the stinging nettle (*U. dioica*) and the burning nettle (*U. urens*). Sometimes used for weaving into textile fabrics.

nett silk. (1) Filaments or stands of raw silk twisted and/or folded into yarns.

(2) A fabric made from nett silk yarn (1).

networked polymers. See network polymers.

network formers. See glass formers.

network modifier. See modifier (2).

network polymers. Polymers within which chemical links between the molecular chains have been set up by addition of a chemical substance (crosslinking agent) and exposure of the mixture to heat, or by subjecting to high-energy radiation (e.g., electron beams or gamma rays) to produce three-dimensional structures that are thermoset in nature and offer improved tensile strength, heat and electrical resistance, stress-crack resistance and especially good resistance to solvents and other chemicals. Examples are crosslinked polyethylenes, polyimides and polystyrenes. Abbreviation: NP. Also known as *crosslinked polymers; networked polymers.*

network silicates. Silicate materials, such as silicate glasses (e.g., fused silica, SiO_2), in which each central silicon atom (Si^{4+}) is bonded to four oxygen ions (O^{2-}) located at the corners of a SiO_4^{4-} tetrahedron. These tetrahedra combine into three-dimensional arrangements known as "network structures." See also silicates; silicate glass.

Neuberg blue. A blue pigment composed of mountain blue (ground azurite) and a mixture of iron blue (Prussian blue) and iron sulfate, and used in ceramics and paints.

Neuf Eclairs. Trade name of Creusot-Loire (France) for oil-hardening tool steel containing 0.85% carbon, 19% tungsten, 9% cobalt, 1.8% vanadium, 0.6% molybdenum, and the balance iron. Used for lathe and planer tools, milling cutters, and cutting tools.

Neupora. Trademark for polystyrene foams and foam materials.

Neuralyt. Trade name of Rasselstein AG (Germany) for steel sheets with zinc-nickel composite coating.

Neuratern. Trade name of Rasselstein AG (Germany) for steel sheets and coils coated with tin-lead alloy (terne).

Neustadt. German brass composed of 71.5% copper and 28.5% zinc, and used for deep-drawn parts including condenser tubing.

Neutraleisen. German trade name for a brittle, acid-resisting alloy of 84.3-85.3% iron, 14-15% silicon and 0.7% manganese, used for evaporators, condensers, crucibles and insoluble anodes.

neutral glass. A glass that is resistant to chemicals, e.g., a high-alkali glass.

Neutralite. Trade name of Saint-Gobain (France) for a gray, heat-absorbing sheet glass.

Neutralloy. Trade name of Bethlehem Foundry & Machine Company (USA) for a corrosion- and heat-resistant nickel casting alloy containing 14-15% chromium, 10% iron, and up to 2% silicon.

neutral refractories. Refractory materials that are intermediate between acidic and basic refractories, and chemically neutral to both acidic and basic materials, slags and fluxes at high temperatures. Examples of such materials include refractories based on carbon, chrome and mullite. Also known as *neutral refractory materials.*

neutral rope. A rope that has been constructed in such a way as to give it a certain balance or set that resists the tendency to rotate and unwind or unlay. This balance or set is often ensured by twisting the various components (e.g., threads, yarns, strands, etc.) in opposite directions.

neutral-tinted glass. A glass used as a light filter to reduce transmission with the least possible selective absorption of particular wavelengths. It is often gray in color, and of the borosilicate type.

neutral verdigris. See copper acetate.

Neutroloy. Trade name of Molecu-Wire Corporation (USA) for an alloy of 55% copper and 45% nickel. It has a maximum service temperature of 500°C (932°F), high electrical resistivity, and a low temperature coefficient of resistance. Used as resistance wire.

neutron absorber. See nuclear poison.

neutron-absorbing glass. A cadmium-borate glass with high neutron absorption capacity, and improved chemical durability due to additions of titania and zirconia.

Neutropane. Trade name of Dearborn Glass Company (USA) for a heat-absorbing, glare-reducing laminated glass with a bronze-colored interlayer of vinyl butyral plastic.

NeutroSorb. Trademark of Carpenter Technology Corporation (USA) for a series of austenitic stainless steels (similar to conventional AISI type 304). The composition range is up to 0.08% carbon, 18-20% chromium, 12-15% nickel, 0-2% manganese, up to 0.045% phosphorus, 0.03% sulfur, 0-0.75% silicon, up to 0.1% nitrogen, up to 2% boron, and the balance iron. The addition of boron provides a higher thermal neutron absorption cross-section, and higher hardness, tensile strength and yield strength. Used in the nuclear industry in burnable poison, shielding, control rods, cask baskets, and spent-fuel storage racks. *NeutroSorb Plus* is a premium-quality modified austenitic stainless steel similar to *NeutroSorb*, but with more boron.

Neutrotherm. Trade name of Boehler GmbH (Austria) for a stainless steel containing about 0.2% carbon, 13% chromium, 1.0% molybdenum, and the balance iron.

Nevada. (1) Trade name Saint-Gobain (France) for square blocks of annealed glass, molded in one piece.

(2) Trade name of Specialty Steel Company of America (USA) for an air-hardening cold-work tool steel containing 1.1% carbon, 0.75% manganese, 0.5% silicon, 5.5% chromium, 0.3% vanadium, 1.3% molybdenum, 0.015% phosphorus, 0.015% sulfur, and the balance iron.

Nevada Silver. Trade name for a nickel silver containing varying amounts of copper and nickel used for electrical resistances and ornaments.

Nevanylon. Trademark of Glanzstoff Austria AG (Austria) for polyamide (nylon) fibers and filaments.

Neveroil. Trade name of The Wakefield Corporation (USA) for a corrosion-resistant alloy of 64.5% copper and 35.5% nickel.

Neviflex. Trademark of Südplastik (Germany) for flexible polyvinyl chloride film.

Nevilan. Trademark of Südplastik (Germany) for polyvinyl chloride film.

Nevilloid. Trade name of Neville Chemical Company (USA) for coumarone-indene coating resins, and blends of coumarone-indene resins with thermosetting resins, such as melamines.

Nevindene. Trademark of Neville Chemical Company (USA) for high-melting coumarone-indene resins with a density of 1.08 g/cm³ (0.039 lb/in.³) and high hardness, used for quick-drying varnishes, rotogravure inks, aluminum paints, insulating compounds, compounding of rubber and plastics, and in dental compounds.

nevskite. A mineral of the tetradymite group composed of bismuth selenide, BiSe. Crystal system, hexagonal. Density, 8.31 g/cm^3.

nevyanskite. A tin-white variety of *osmiridium* containing about 44-58% iridium, 27-50% osmium, 0-10% platinum, 0-6% ruthenium, 1.5-3% rhodium, with the balance being copper, iron and palladium. Crystal system, cubic. Density, 17.8 g/cm^3; hardness, 6-7 Mohs; good corrosion resistance. Occurrence: Found near Nevyansk in the Ural Mountains of Russia. Used as a source of iridium and osmium, and for watch pivots, compass bearings, fountain-pen points and surgical needles.

newberyite. A colorless mineral composed of magnesium hydrogen phosphate trihydrate, $MgHPO_4 \cdot 3H_2O$. It can also be made synthetically from aqueous solutions of $MgSO_4$ and Na_2HPO_4. Crystal system, orthorhombic. Density, 2.12 g/cm^3; refractive index, 1.517. Occurrence: Australia.

New Bide. Trade name of Latrobe Steel Company (USA) for a series of cemented carbides used for cutting tools and dies.

new blue. Any of various *iron blue* pigments, such as *Prussian blue*, based on ferric ferrocyanide, $Fe_4(Fe(CN)_6)_3$, and available in various shades.

Newbray. Trade name for rayon fibers and yarns.

New Capital. Trade name of Styria-Stahl Steirische Gusstahlwerke AG (Austria) for a high-speed steel containing 0.7% carbon, 3.75% chromium, 14.5% tungsten, 0.5-0.75% vanadium, and the balance iron. Used for lathe and planer tools, various other tools, punches, and dies.

New Capital Steel. Trade name of Balfour Darwins Limited (UK) for high-speed steel containing 0.6% carbon, 14% tungsten, 3.7% chromium, 0.1% vanadium, and the balance iron. Used for cutting tools, reamers, cutters, punches, and gages.

NewCer. Trade name for a precious-metal dental bonding alloy.

New Color. Trade name for rayon fibers and yarns used for textile fabrics.

Newcomer. Trade name of Newcomer Products Inc. (USA) for a series of sintered carbides used for light to heavy-duty cutting tools.

Newcor. Trade name of British Steel plc (UK) for a silicon-free electrical steel used for motors.

Newdull. Trade name for rayon fibers and yarns with dull luster used for textile fabrics.

New-Era. Trade name for a glass-ionomer dental cement.

New-Era Silver. Trade name for a silver-filled glass-ionomer dental cement.

New-Fill. Trademark of Biotech Industry SA (Luxembourg) for a biocompatible, bioabsorbable *polylactic acid* (PLA) polymer belonging to the family of aliphatic polyesters. Used for medical and surgical applications including human body and skin reconstruction.

Newhall. Trade name of Sanderson Kayser Limited (UK) for an oil-hardening cold-work tool steel (AISI type O1) containing 0.95% carbon, 1.2% manganese, 0.55% chromium, 0.7% tungsten, 0.1-0.2% vanadium, and the balance iron. Used for lathe tools, broaches, reamers, and blanking, forming and threading dies.

New Lightweld. Trade name of Lincoln Electric Company (USA) for a carbon steel used for coated welding electrodes.

Newlo. Trade name of Gerhard Menzel Glasbearbeitungswerk KG (Germany) for a cover glass consisting of one specially treated surface that is mounted on transparencies such that the processed side faces the film.

Newlow. Trade name for rayon fibers and yarns used for textile fabrics.

Newloy. Trade name of Harrison, Fisher & Company Limited (USA) for an alloy of 64% copper, 35% nickel and 1% tin. It has good resistance to corrosion and acids, and is used as a base metal for tableware.

Newman. Trade name of Newman Clays (USA) for red clays used in the manufacture of brick, roofing tile, and pottery.

Newmax. Trade name of Ziv Steel & Wire Company (USA) for an oil-hardening chromium-molybdenum steel (similar to AISI type 4150) containing 0.45-0.53% carbon, 0.75-1% manganese, 0.8-1.1% chromium, 0.15-0.25% molybdenum, and the balance iron. Used for molds.

New Met. Trade name of Newcomer Products Inc. (USA) for a series of sintered carbides composed of a mixture of titanium carbide (TiC) and molybdenum carbide (MoC) in a nickel matrix. The nominal composition is 60-65% TiC, 31-35% MoC, and up to 9% nickel. Used for cutters.

Newood. Trade name of Taby Veneers Limited (USA) for natural and dyed wood veneers.

Newport. Trade name of Newport Steel Corporation (USA) for a series of electrical steels containing varying amounts of silicon. They have high magnetic permeability, and are used for armatures, motors, and transformers.

New Process Cold Header. Trade name of Jessop Steel Company (USA) for a water-hardening tool steel (AISI type W1) containing 1% carbon, 0.25% manganese, 0.18% silicon, and the balance iron. It has excellent machinability, good to moderate deep-hardening properties, good wear resistance and toughness, and low hot hardness. Used for punches, dies, drills, cutters, and other tools.

New Rapid. Trade name of Houghton & Richards Inc. (USA) for a tungsten-type high-speed steel used for punches, dies, and cutting tools.

new sand. New or newly mixed *molding sand.*

newsprint. A soft, coarse, inexpensive paper made chiefly from *mechanical wood pulp*, sometimes with additions of *chemical pulp*, and used for newspapers, telephone books, etc.

Newton's alloy. A white metal alloy containing 50% bismuth, 31.25% lead and 18.75% tin. It has a melting point of 95°C (202°F), and is used for fire and signal alarms, and fire extinguisher plugs. Also known as *Newton's fusible alloy; Newton's metal.*

Newton's metal. See Newton's alloy.

New Tool Steel Cast. British trade name for a corrosion-resistant alloy containing 58% nickel, 20% zinc, 12% aluminum and 10% silicon. Used for dies and cutting tools.

New Zealand beech. The strong, durable, brown wood from the tall New Zealand tree *Nothofagus solandri*. It has an average weight of 705 kg/m^3 (44 lb/ft^3), and is used chiefly in construction. Also known as *red beech; tawhai.*

New Zealand flax. See phormium fibers.

Nextel. (1) Trademark of 3M Ceramic Textiles and Composites (USA) for a series of continuous and discontinuous ceramic fibers that are transparent, unless colored with modifying components, and have a glossy appearance (like glass fibers). A commercially important grade is *Nextel 312*, a continuous alumina-boria-silica fiber containing 62% alumina (Al_2O_3), 24% silica (SiO_2) and 14% boria (B_2O_3). Commercially available in the form of continuous fibers, rovings, yarns and fabrics, it has an average diameter of 11 μm (433 μin), good high-temperature performance, an upper service temperature of 1200°C (2190°F), and a low density (2.7 g/cm³, or 0.10 lb/in³). Its chemical properties are strongly influenced by the microstructure, and it is used as a reinforcing fiber, and for high-temperature protection applications, e.g., firewall composites, fire zone wiring, flexible blankets, and gaskets. *Nextel 440* and *Nextel 480* are similar in composition and commercially available in the same forms and fiber sizes as *Nextel 312*, but with even better mechanical properties and high-temperature performance, an upper service temperature of 1430°C (2605°F), and a slightly higher density. *Nextel Z-11* contains 68% microcrystalline tetragonal zirconia (ZrO_2) and 32% amorphous silica (SiO_2), has moderate reinforcement properties, a low to moderately high modulus of elasticity, an average fiber diameter of 14 µm (551 µm), good flame resistance, an upper service temperature of 1000°C (1830°F), a low density (3.7 g/cm³, or 0.12 lb/in³) and is used as as a reinforcing fiber. Also available are *Nextel 312 Ultrafiber*, *Nextel 400 Ultrafiber* and *Nextel 610 Ultrafiber* which are discontinuous grades supplied in the form of chopped fibers, mats, or blankets.

(2) Trade name of Red Spot Paint and Varnish Company Inc. (USA) for a series of pigmented ceramic spray coatings for metals and plastics.

Nextel Ultrafiber. See Nextel (1)

Nexten. See Meryl Nexten.

Nexus. Trademark of Kerr Dental (USA) for a resin-based dental bonding cement.

Ney. Trade name of Degussa-Ney Dental (USA) for a wide range of silver-palladium dental casting alloys. An extensively used alloy of this product range is *Ney 76*, a white-colored, low-density, hard dental alloy (ADA type III) for crowns, posts, long-span bridges, and high-stress restorations.

Neycast. Trade name of Degussa-Ney Dental (USA) for a series of gold-color dental casting alloys and solders.

neyite. A gray mineral composed of lead bismuth copper silver sulfide, $Pb_7Bi_6(Cu,Ag)_2S_{17}$. Crystal system, monoclinic. Density, 7.02 g/cm³. Occurrence: Canada (British Columbia).

Ney-Oro. Trade name of Degussa-Ney Dental (USA) for an extensive series of gold-base dental casting alloys including yellow golds (up to 80% gold), white golds (50-60% gold) and silver-palladium-gold and palladium-silver-gold alloys with a total precious-metal content (including silver, platinum and palladium) of at least 75%, and a hardness ranging from soft to extra-hard. Used for inlays, bridges, crowns, bars, clasps, and bridgework.

NiAg. Trade name of Schlegel BvBa (Belgium) for a silver-metallized ripstop nylon fabric that contains a conductive, pressure-sensitive acrylic adhesive and a flame retardant, and is topcoated with corrosion-resistant, conductive nickel. It has an extremely low surface resistance, and is used for electromagnetic interference (EMI) shielding.

Niag. (1) Trade name for a corrosion-resistant alloy of 46.7% copper, 40.7% zinc, 9.1% nickel, 2.8% lead, and 0.3% manganese. Used for white metal parts.

(2) Trademark of Mueller Brass Company (USA) for a wear and corrosion-resistant alloy of 46% copper, 1% lead, some nickel, and the balance zinc. Used for hardware and plated products.

Niagara. (1) Trade name of Colt Industries (UK) for a water-hardened steel containing 0.45-0.55% carbon, and the balance iron. Used for gears, shafts, and machine components.

(2) Trade name of SA Glaverbel (Belgium) for patterned glass.

(3) Trade name of Perma Paving Stone Company (Canada) for a series of paving stones supplied as Standard Niagara units 198 × 98 × 60 mm (7.875 × 3.875 × 2.375 in.) in size, and Double Niagara units 198 × 198 × 60 mm (7.875 × 7.875 × 2.375). Supplied in several colors including natural, red, brown, charcoal and various color blends, they are ideal for driveways, walkways, patios, etc.

Niagara Blast. Trade name of Washington Mills Electro Minerals Corporation (USA) for aluminum oxide blasting media.

Niagra. (1) Trade name of Wallace Murray Corporation (USA) for a high-speed steel containing 0.7% carbon, 14% tungsten, 4% chromium, 2% vanadium, and the balance iron. Used for tools, cutters, hobs, and taps.

(2) Trade name of Pittsburgh Metallurgical Company Inc. (USA) for a series of ferroalloys including (i) *ferrochrome* composed of 66-70% chromium and 30-34% iron, and used to introduce chromium into steel; (ii) *ferrosilicon* composed of 15-90% silicon and 10-85% iron, and used to introduce silicon into steel, and (iii) *silicomanganese* composed of 12-20% silicon, 65-70% manganese and 10-23% iron, and used to introduce manganese into steel.

niahite. A pale orange mineral composed of ammonium calcium magnesium manganese phosphate monohydrate, $(NH_4)(Mn,Mg,Ca)PO_4 \cdot H_2O$. Crystal system, orthorhombic. Density, 2.39 g/cm³; refractive index, 1.604. Occurrence: Malaysia.

Nial. (1) Trademark of Wilbur B. Driver Company (USA) for a magnetic nickel-base thermocouple alloy containing 2.5% manganese, 2% aluminum, 1% silicon, and traces of cobalt, iron, magnesium and zirconium. It has a density of 8.47 g/cm³ (0.306 lb/in.³), a melting point of 1400°C (2550°F), a maximum service temperature of 1260°C (2300°F), and good oxidation resistance. Used as the negative thermoelement of standard K-type thermocouples.

(2) Trade name of Edgar Allen Balfour Limited (UK) for a permanent magnet material composed of 59% iron, 24% nickel, 13% aluminum and 4% copper. It has high permeability, and is used for lighting and ignition equipment, and loudspeakers.

Nialco. Trade name of Ugine Aciers (France) for a series of permanent magnet alloys containing 12-20% nickel, 10% aluminum, 4-20% cobalt, 2-6% copper, and the balance iron. They have high magnetic permeabilities and energy products. Used for electric and magnetic equipment, and other permanent magnets.

Nialite. Trade name of Baldwin-Lima-Hamilton Corporation (USA) for a cavitation- and corrosion-resistant cast copper alloy containing 10% aluminum, 5% nickel, 5% iron and 1.5% manganese. Used for pumps, valves, and propellers.

Ni-Alloy. Trade name of Deveco Corporation (USA) for zinc-nickel electroplates.

niangon. The wood of the tree *Tarrietia utilis* that is generally similar to *African mahogany* (khaya) and has light reddish brown heartwood. It has a coarse texture, many large pores, a herringbone figure on quartered surfaces, a characteristic greasy feel, some tendency to distort in use, and moderate durability. Average weight, 620 kg/m³ (39 lb/ft³). Source: West Africa, especially from the Ivory Coast and Ghana. Used for furniture, interior paneling, veneer, and plywood. Also known as *cola mahogany; nyankom*.

Niatrax. Trade name for refractory ceramics with outstanding acid resistance based on silicon nitride (Si_3N_4) bonded silicon carbide (SiC).

Niax. (1) Trademark of Union Carbide Corporation (USA) for a special polyurethane foam.

(2) Trademark of Nissho Iwai Textiles (USA), Inc. for textile fabrics made of natural and/or synthetic materials. The natural fabrics include flax, linen, cotton, silk and wood, and the synthetics rayons, nylons, polyesters, acrylics, polyurethanes, and vinyls.

Ni-Bar Iron. Trade name of Ingersoll-Rand Company (USA) for an alloy cast iron containing 3.3% total carbon, 0.6% manganese, 1.5% silicon, 1.5% nickel, 0.6% chromium, and the balance iron.

Niborium. Trade name of Niborium Industries Inc. (USA) for electrical contact alloys.

Nibra-Glas. Trade name of Isolierglas AG (Switzerland) for double glazing units.

Ni-Bral. Trade name of Ampco Metal (USA) for a corrosion-resistant aluminum bronze containing 78.5% copper, 10% aluminum, 5% nickel, 5% iron and 1.5% manganese. Used for valves, pumps, and bearings.

Nibrel-EN. Trade name of Zinex Corporation (USA) for a smooth, bright, medium-phosphorus (5-8%) electroless nickel and autocatalytic deposition process.

Nibrel-R. Trade name of Zinex Corporation (USA) for a bright nickel electroplate and plating process for ferrous and nonferrous metals.

Nibro-Cell. Trademark of Brown Company (USA) for a kraft laminating paper.

Nibron. Trade name of Pure Coatings Inc. (USA) for a series of nickel-boron electroless coatings that have wear lifes superior to hard chromium, and provide better wear resistance than electroless nickel-phosphorus coatings. They have excellent tribological properties including excellent resistance to galling, fretting and sliding wear, high resistance to hydrogen embrittlement, good adhesion to most commercial metals and alloys, and a continuous-use temperature up to 425°C (800°F).

Nibryl. Trade name of Brush Wellman (USA) for a series of age-hardening alloys containing 1.85-2.05% beryllium, 0.4-0.6% titanium, and the balance nickel. Supplied in the form of bars and rods, they have a density of 8.28 g/cm³ (0.300 lb/in.³), a melting point of 1330°C (2426°F) and good corrosion resistance.

Nibsi. Trade name of Western Gold & Platinum Company (USA) for a brazing powder containing 94.7% nickel, 3.5% silicon and 1.8% boron. It has a melting range of 982-1066°C (1800-1950°F), and is used for brazing stainless and high-temperature alloys.

Nica. Trade name of Breda Company (Italy) for a series of austenitic stainless steels containing up to 0.25% carbon, 18% chromium, 8% nickel, and the balance iron. Used for chemical plant equipment, filters, mixers, and tanks.

Nicalloy. British trade name for a magnetically soft material of 53% iron and 47% nickel. It has high magnetic saturation and good permeability, and is used for electrical and magnetic equipment.

Nicalon. Trademark of Nippon Carbon Company, Limited (Japan) for silicon carbide fibers produced by polymer pyrolysis. They are composed of 54.3% silicon, 30.0% carbon, 11.8% oxygen and 3.9% other elements. Commercially available in the form of continuous fibers, chopped fibers, yarns, fabrics, mats, blankets and roving, they have a density of 2.54 g/cm³ (0.092 lb/in.³), an average diameter (of round fibers) of 10-15 μm (395-590 μin), high strengths and moduli even at high temperatures, high oxidation and chemical resistance, and an upper service temperature of 1000°C (1830°F). Used as reinforcements for high-performance composites with plastic, metallic or ceramic matrices, as high-temperature insulation, and for belting, curtains, and gaskets.

Nicaloy. Trade name for a resistance alloy of 51% iron and 49% nickel that has high magnetic permeability and saturation, and is used for electrical equipment and apparatus.

Nicalun. Trade name of American Abrasive Metals Company (USA) for a wear-resistant alloy composed of a nickel-copper matrix with embedded abrasive grains. Used especially for stair treads.

Nicar. Trade name of Arcos Alloys (USA) for a hardfacing alloy containing 65-75% nickel, 12-18% chromium, 4% iron, 3.5-4% carbon, 2.5-4.5% boron, 1% silicon and 0.2% cobalt. Used for hardfacing rods.

nicarbed steel. See carbonitrided steel.

Nicaron. Trademark of Schildkroet (Germany) for polyvinyl chloride polymers and copolymers.

Nicast. Trade name of Chemtron Corporation (USA) for an alloy of 97% nickel, 2% iron and 1% silicon. Used for welding rods for cast iron.

niccolite. See nickeline.

Ni-Chem. Trade name of National Chemical Company Inc. (USA) for a series of pigments.

Nichem. Trade name of Atotech USA Inc. for an electroless nickel.

Ni-Chillite. Trade name of Gulf & Western Manufacturing Company (USA) for a chilled cast iron containing considerable amounts of nickel, chromium and molybdenum. Used for rolls.

Nichire. Trade name of Unitika Berkshire Company (Japan), formerly known as Nichire Berkshire Company, for rayon fibers and yarns used for textile fabrics.

Ni-Chro. Trade name of 3M Company (USA) for nickel-chromium dental alloys for primary and permanent molar crowns.

Nichrofry. Trade name of Creusot-Loire (France) for a series of austenitic stainless steels used for heat- and corrosion-resisting applications.

Nichrome. Trademark of Harrison Alloys (USA) for a series of heat- and oxidation-resistant alloys containing 60-62% nickel, 23-24% iron, 15-16% chromium and 0.1% carbon. Special grades include *Nichrome III* with 85% nickel and 15% chromium, *Nichrome S* with 55-56% iron, 25% nickel, 17-20% chromium and 0-2.5% silicon, and *Nichrome V* with 80% nickel and 20% chromium. *Nichrome* alloys are supplied in the form strands, strip, ribbon, wire, ingots, powder, tubes, forgings and sheet castings, have high electrical resistivity, good resistance to mine water, seawater and moist sulfurous atmospheres, good oxidation resistance at red-heat, and are used for electric resistances and heating elements.

nichromite. A dark green mineral composed of nickel chromium oxide, $NiCr_2O_4$. It can also be made synthetically. Crystal system, cubic. Density, 5.24 g/cm³. Occurrence: Morocco.

Nichrotherm. Trade name of Krupp Stahl AG (Germany) for a series of heat-resisting alloys including several chromium-nickel austenitic stainless steels, and various nickel-iron-chromium and nickel-chromium-iron alloys.

Nichrothermsteel. Trade name of Westinghouse Electric Corporation (USA) for an alloy of iron and nickel with a low coefficient of expansion.

Ni-Chro-Zink. German trade name for a die-casting alloy containing nickel, chromium and zinc.

Nickahl Iron. Trade name of Aluminium Industrie AG (Switzerland) for a tough, wear-resistant cast iron containing 1.5% carbon, 2.2% silicon, 0.8% chromium, and the balance iron. Used for automotive seat sides and door panels, and for dies used for stamping fenders.

Nickahl Steel. Trade name of Jessop Steel Company (USA) for an alloy of 58-65% iron and 35-42% nickel. It has a low coefficient of expansion, and is used for valves, thermostats, and thermostatic elements.

Nickel. Trade name of Inco Alloys International Inc. (USA) for a series of commercially pure wrought nickels containing at least 99.5% nickel along with small additions/impurities of iron, manganese, silicon, copper, carbon and sulfur. Examples include Nickel 200 and Nickel 201. They have good mechanical properties, excellent resistance to corrosion, high thermal and electrical conductivities, good magnetic and magnetostrictive properties, and are used for electrical and electronic components, lead wires, base pins, anodes, cathode shields, vacuum-tube plates, transducers, magnetostrictive devices, spun and cold-formed parts, tube structural parts, aerospace and missile parts, rocket motor cases, food-processing and chemical handling equipment, and as filler metals for TIG, MIG and SAW welding.

Nickel 201. See E Nickel.

Nickel 212. See Low Carbon Nickel.

nickel. A hard, ductile, malleable, silvery-white, ferromagnetic metallic element of Group VIII (Group 10) of the Periodic Table. It is commercially available in the form of ingots, rods, tube, sheets, foil, microfoil, pellets, shot, platelets, spheres, slugs, sponge, powder, wire, mesh, high-purity strips, cathodes and single crystals. The single crystals are usually grown by the Czochralski technique. The chief ores are *pentlandite, pyrrhotite, niccolite* and *garnierite*. Crystal system, cubic. Crystal structure, face-centered cubic. Density, 8.9 g/cm^3; melting point, 1453°C (2647°F); boiling point, 2732°C (4950°F); hardness, 100-190 Brinell; atomic number, 28; atomic weight, 58.693; divalent, tetravalent; Curie point, 360°C (680°F) above which it is paramagnetic. It has good hot and cold workability, good polishability, excellent corrosion resistance, relatively good electrical conductivity (about 23-24% IACS), and moderate thermal conductivity. Used in low-alloy steels, nickel steels, armorplate, stainless steels, corrosion-resistant alloys, copper-base alloys (nickel silvers and copper nickels, white gold, etc.), permanent magnets, electrical resistance alloys, electroplating and electroless plating, electroformed products, nickel cores for magnetostrictive devices, alkaline storage batteries, ceramics, as a colorant for green glass, in fuel cell electrodes, coinage, and as a catalyst. Symbol: Ni.

nickel-57. A radioactive isotope of nickel that has a mass number of 57, decays partly by positive beta radiation and partly by electron capture, and has a half life of 36 hours. Used as a tracer. Symbol: ^{57}Ni.

nickel-63. A radioactive isotope of nickel that has a mass number of 63, decays by negative beta radiation, and has a half-life of 92 years. Used in radioactive composition studies, and as a tracer. Symbol: ^{63}Ni.

nickel acetate. The divalent nickel salt of acetic acid available in the form of green hygroscopic crystals (98+% pure). Density of 1.744 g/cm^3; melting point, decomposes on heating to 250°C (482°F). Used as a catalyst and textile mordant. Formula: Ni(C$_2$H$_3$O$_2$)$\cdot$4H$_2$O. Also known *nickel acetate tetrahydrate*.

nickel acetate tetrahydrate. See nickel acetate.

Nickelalloy. Trade name of Teledyne McKay (USA) for pure nickel used for shielded-arc-welding electrodes.

nickel alloys. A group of strong, corrosion-resistant alloys composed predominantly of nickel. The most common alloy combinations include nickel-copper, nickel-silicon, nickel-molybdenum, nickel-chromium-iron, nickel-chromium-molybdenum, and nickel-chromium-iron-molybdenum-copper. They are often sold under trademarks and trade names, such as Duranickel, Inconel, Monel, Permalloy, or Hastelloy.

nickel aluminate. A spinel-type compound of nickel oxide and aluminum oxide used as a ceramic magnet material. Density, 4.45 g/cm^3; melting point, 2020°C (3668°F). Formula: NiAl$_2$O$_4$.

nickelalumite. A sky-blue mineral of the chalcoalumite group composed of copper nickel aluminum nitrate sulfate hydroxide trihydrate, (Ni,Cu)Al$_4$[(SO$_4$)(NO$_3$)](OH)$_{12}$·3H$_2$O. Crystal system, monoclinic. Density, 2.24 g/cm^3. Occurrence: South Africa.

nickel aluminide. A compound of nickel and aluminum with a density of 5.90 g/cm^3 (0.213 lb/in.^{3}), a melting point of 1640°C (2985°F), high hardness, excellent oxidation and thermal-shock resistance, good resistance to molten glass, and a low coefficient of thermal expansion. Supplied as a powder or as single crystals. Used for turbine blades, combustion chambers, glass-processing equipment, flame-sprayed coatings, and as a matrix material for composites. The single crystals are used for III-V (13-15) compound semiconductor deposition for light-emitting diodes and blue lasers. Formula: NiAl.

nickel-aluminum. A master alloy of 80% aluminum and 20% nickel, used to introduce nickel into high-strength aluminum alloys, and in the manufacture of aluminum bronzes.

nickel-aluminum bronze. A group of corrosion-resistant alloys (UNS C63000) containing 10-88% copper, 2-30% aluminum, 3-40% nickel, and up to 20% tin, 5% iron and 1.5% manganese, respectively. They can be shaped by forging and hot forming, and have good to excellent corrosion resistance, good hot formability, and improved strength and heat resistance. Used for ornaments, dies, molds, aircraft and turbine parts, ship propellers and shafts, plunger tips, castings, valve seats and guides, nuts, bolts, pump shafts, and structural members.

nickel-aluminum superalloy. A solid-solution alloy that corresponds to the chemical composition Ni$_3$Al and has a microstructure composed of a gamma-phase matrix with disordered crystal structure and a gamma-prime phase precipitate with ordered structure. It has outstanding high-temperature strength and corrosion resistance, and is suitable for aircraft engines.

nickel-ammonium chloride hexahydrate. See ammonium-nickel chloride hexahydrate.

nickel-ammonium sulfate hexahydrate. See ammonium-nickel sulfate.

nickel antimonide. Copper-red crystals. It occurs in nature as the mineral *breithauptite*. Crystal system, hexagonal; crystal structure, niccolite. Density, 8.74-8.23 g/cm^3; melting point, 1147°C (2097°F); hardness, 5.5 Mohs. Formula: NiSb.

nickel arsenide. Any of the following compounds of nickel and arsenic: (i) *Nickel monoarsenide*. Light red crystals. It occurs in nature as the mineral *nickeline (niccolite)*. Crystal system, hexagonal. Density, 7.3-7.8 g/cm^3; melting point, 968°C (1774°F); hardness, 5-5.5. Used in ceramics and materials research. Formula: NiAs; (ii) *Nickel diarsenide*. Grayish crystals. It occurs as the mineral *rammelsbergite*. Crystal system, orthorhombic. Crystal structure, marcasite. Density, 7.10 g/cm^3; hardness, 5.5-6 Mohs. Formula: NiAs$_2$; and (iii) *Nickel triarsenide*. Crystal structure, cubic. Crystal structure, skutterudite. Density, 6.43 g/cm^3. Used for semiconductors. Formula: NiAs$_3$.

nickelates. A class of orthorhombic compounds that are oxides of nickel and another transition metal and correspond to the general formula $RNiO_3$. Some of these compounds exhibit the giant magnetoresistance effect making them useful for applications in data processing and storage, and for high-sensitivity sensors.

nickel babbitt. A special high-tin babbitt that contains some nickel, and is used for bearings.

nickel-base heat-resisting alloys. See nickel-base superalloys.

nickel-base superalloy composites. A class of engineering composites composed of heat- and oxidation-resistant nickel alloy matrices reinforced with strong, stiff, creep-resistant fibers, usually tungsten or molybdenum, and used for turbine parts, flywheels, pressure vessels and aerospace applications.

nickel-base superalloys. A group of wrought or cast heat-resisting alloys used at temperatures above 540°C (1000°F). They contain about 48-75% nickel, 10-25% chromium, 0-20% cobalt, 0-10% molybdenum, 0-10% tungsten, 0-6% aluminum, 0-5% titanium, 0-5% niobium (columbium) and 0-5% tantalum. All contain less than 0.5% carbon, and one or several other special ingredients (e.g., boron, zirconium, hafnium, and/or yttria). *Nickel-base superalloys* have excellent high-temperature and oxidation resistance, and high tensile and fatigue strengths. Used for gas turbines, nuclear reactors, aircraft and spacecraft structures, pressure vessels, chemical process equipment, and orthopedic and dental prostheses. Also known as *high-temperature nickel alloys; nickel-base heat-resisting alloys.*

Nickel Bearing. British trade name for an alloy of 50% copper, 25% tin and 25% nickel used for heavy-duty bearings.

nickel-bearing leaded commercial bronze. A free-cutting bronze containing 89-90.25% copper, 6.9-8.1% zinc, 1.75-1.9% lead, 1% nickel, and up to 0.1% phosphorus. It has good corrosion resistance and machinability, good cold workability, poor hot workability, and is used for pole-line hardware, electrical connectors, bolts, nuts, screws, screw-machine parts, and fasteners. Also known as *nickel leaded commercial bronze.*

nickel beryllide. A compound of nickel and beryllium used in ceramics and materials research. Melting point, 1470°C (2678°F). Formula: NiBe.

nickelbischofite. An emerald-green mineral of the bischofite group composed of nickel chloride hexahydrate, $NiCl_2 \cdot 6H_2O$. It can also be made by room-temperature evaporation of an aqueous solution of nickel chloride ($NiCl_2$). Crystal system, monoclinic. Density, 1.93 g/cm³; refractive index, 1.617. Occurrence: USA (Texas).

nickelbloedite. A light green mineral composed of sodium nickel sulfate tetrahydrate, $Na_2Ni(SO_4)_2 \cdot 4H_2O$, which can also contain some magnesium. It can also be made synthetically. Crystal system, monoclinic. Density, 2.60 g/cm³; refractive index, 1.518. Occurrence: Australia.

nickel bloom. See annabergite.

nickel-bonded titanium carbide. A *cemented carbide* composed of a mixture of titanium carbide (TiC) and molybdenum dicarbide (MoC_2) with a nickel binder. Typically, the composition range is 8-25% nickel, 8-15% MoC_2 and 60-80% TiC. It has excellent high-temperature properties, high strength and hardness, relatively low abrasion resistance, a low coefficient of friction, and low thermal conductivity. Used for turbine blades, high-temperature structures, tool bits, high-temperature bearings and seals, and other wear-resistant applications.

nickel boride. Any of the following compounds of nickel and boron used in ceramics and materials research: (i) *Nickel mono-boride.* Green, refractory solid or fine, silvery granules (99+% pure). Density, 7-13-7.39 g/cm³; melting point, 1035°C (1895°F). Formula: NiB; (ii) *Dinickel boride.* A refractory solid or moisture-sensitive powder (99+% pure). Density, 7.9 g/cm³; melting point, 1226°C (2239°F). Formula: Ni_2B; and (iii) *Trinickel boride.* A refractory solid. Density, 8.19 g/cm³; melting point, 1156°C (2113°F). Formula: Ni_3B.

nickelboussingaultite. A strong bluish green mineral of the picromerite group composed of ammonium nickel sulfate hexahydrate, $(NH_4)_2Ni(SO_4)_2 \cdot 6H_2O$. Crystal system, monoclinic. Density, 1.92 g/cm³.

nickel brass. (1) See nickel silvers.

(2) A corrosion-resistant brass of 50-54% copper, 35-44% zinc, 1.5% nickel, 0.5% iron and traces of aluminum. Used for hardware, condenser tubes and plates, and heat-exchanger parts.

(3) A nickel-coated brass used for drawn, formed and stamped parts.

nickel briquettes. Nickel powder compressed into small pillow-shaped blocks for use as alloying additions in ferrous and nonferrous metal production.

nickel bromide. Yellow, deliquescent crystals or brownish-yellow powder (98+% pure). Crystal system, hexagonal. Density, 5.098 g/cm³; melting point, 963°C (1765°F). Used as a reagent in electroplating. Formula: $NiBr_2$. Also known as *nickelous bromide.*

nickel bromide trihydrate. Yellowish-green needles or green, deliquescent crystals (98+% pure) that lose $3H_2O$ at 300°C (572°F). Used as a reagent in electroplating. Formula: $NiBr_2 \cdot 3H_2O$. Also known as *nickelous bromide trihydrate.*

Nickel-Bronze. Trade name of Baldwin-Lima-Hamilton Corporation (USA) for an alloy of 85-90% copper, 3-7% nickel, 3-6% tin and 1.3% zinc. Used for gears, and screwdown nuts.

nickel bronze. A tough, fine-grained, corrosion-resistant, white-colored casting bronze composed of 80-88% copper, 5-8% tin, 5-8% nickel, 2-5% zinc and 0-5% lead. It has good strength, and moderate machinability (improves with increasing lead content). Used for structural castings, valve components, superheated steam parts, gears, bearings, machinery parts, piston cylinders, nozzles, wear guides, and decorative parts. Also known as *nickel-tin bronze.*

nickel carbonate. Light green crystals or brown powder. Crystal system, orthorhombic. Density, 2.6; melting point, decomposes on heating. Used in electroplating, in nickel catalysts and as an ingredient in ceramic colors and glazes. Formula: $NiCO_3$.

nickel carbonyl. See nickel tetracarbonyl.

Nickelcast. Trade name of Hobart Welding Products (USA) for a series of alloys containing 2% carbon, 4% silicon, 1% manganese, 0-85% nickel, 0-2.5% copper, 1.0% others, and the balance iron. Used as a weld metal for welding ductile iron and cast iron.

nickel cast irons. (1) A group of abrasion- and wear-resistant alloy cast irons containing 2.2-3.6% total carbon, 1.1-7% nickel, 0.2-1.3% manganese, 0.3-2.2% silicon, 0.5-4% chromium, 0-1% molybdenum, and the balance iron. They have high strength and electrical resistivity. Often sold under trademarks and trade names, such as *Ni Tensiliron, Ni-Tensyle* or *Ni-Hard*, they are used for gears, machine components, and wear-resistant castings. Also known as *nickel irons.*

(2) A group of nonmagnetic gray and ductile high-nickel irons with austenitic microstructures containing 1.8-3.6% total carbon, 0.5-4.5% manganese, 1-6% silicon, 1.8-11% chromium, 1% molybdenum, 0-10% copper, 0.1-0.2% phosphorus,

0.1-0.2% sulfur, 5-43% nickel, and the balance iron. They have excellent corrosion and heat resistance, good toughness and thermal-shock resistance, and good abrasion resistance. Often sold under trademarks or trade names, such as *Ni-Resist* or *Pyrocast*, they are used for pumps, compressors, valves, gas turbines, turbocharger parts, exhaust manifolds, and equipment used for handling hot corrosives. See also nickel-chromium cast irons; austenitic cast irons. Also known as *nickel irons.*

nickel cerium steel. An oil-hardening steel containing 0.4-0.75% carbon, 2-3% nickel, 0.1-1% cerium, and the balance iron. Used for machinery parts.

nickel chloride. Brown, deliquescent crystals, or yellow hygroscopic powder (98+% pure). Density, 3.51-3.55 g/cm³; melting point, 1001°C (1834°F). Used in nickel plating, biochemistry, and as a reagent. Formula: $NiCl_2$. Also known as *nickelous chloride.*

nickel chloride hexahydrate. Green, monoclinic, deliquescent crystals (99+% pure) used in fluoborate, sulfamate and Watts nickel plating baths, in biochemistry, and as a reagent chemical. Formula: $NiCl_2 \cdot 6H_2O$. Also known as *nickelous chloride hexahydrate.*

Nickel Chrome Cast Iron. Trade name for an abrasion and corrosion-resistant cast iron containing 3-3.4% carbon, 1.5-1.75% nickel, 0.5-0.7% manganese, 0.6-0.8% chromium, 0.9-1.75% silicon, and the balance iron.

nickel-chromium alloys. A group of alloys in which nickel and chromium are the predominant constituents. They are used for corrosion- and heat-resistant parts, resistance wires, thermocouples, etc. Often sold under trademarks or trade names, such as *Chromel, Cooper, Gridnic, Incoloy, Inconel, Misco* or *Thermalloy.*

nickel-chromium-aluminum alloys. A group of electrical resistance alloys containing 70-75% nickel, 20% chromium, 3% aluminum, and 0-5% copper, iron and/or manganese. They have a density range of 7-8.2 g/cm³ (0.25-0.30 lb/in.³), high strength and high electrical resistivity.

nickel-chromium-aluminum-yttrium coatings. Oxidation-resistant overlay coatings comprising nickel-base matrices with a rather large amount of chromium (15-20%), and an intermediate amount of aluminum (10-15%). Small additions of yttrium (0.2-0.5%) enhance the adherence of the oxidation product. Used on superalloys for aircraft engines, power generation equipment, etc.

nickel-chromium cast irons. A group of austenitic gray and ductile irons containing about 1.8-3% total carbon, 0.4-4.5% manganese, 1-6% silicon, 1% molybdenum, 0-10% copper, 13-43% nickel, 1-6% chromium, 0.1-0.2% each phosphorus and sulfur, and the balance iron. They have high hardness, good strength, excellent corrosion and heat resistance, good high-temperature scaling and oxidation resistance, good toughness, and good thermal-shock resistance. Often sold under trademarks or trade names, such as *Pyrocast* or *Ni-Resist*, they are used for pumps, compressors, and equipment for handling hot corrosives.

nickel-chromium-cobalt alloys. A group of *nickel-base superalloys* containing relatively high percentages of chromium (up to 25%) and cobalt (up to 20%). Used for heat-resistant applications.

nickel-chromium-iron alloys. A group of wrought and cast alloys composed predominately of nickel, chromium and iron. They have excellent oxidation resistance and strength at high temperatures, good corrosion resistance, and good resistance to many acids and caustics. Often sold under trademarks or

trade names, such as *Durimet, Esco, Incoloy, Inconel* or *Isocast.* Used for chemical equipment, pickling tanks, pumps, valves, components and equipment for handling heat-exchanger tubing in nuclear powder plants, corrosion-resistant parts, fume ducts, electrical resistances, and heating elements.

nickel-chromium-iron cast steels. A group of heat-resistant, austenitic cast steels containing 0.3-0.5% carbon, 10-20% chromium, 30-70% nickel, and the balance being iron. They have good hot strength and thermal-fatigue resistance, good resistance to both oxidizing and reducing atmospheres, and fair to poor resistance to sulfurous atmospheres. Often sold under trademarks and trade names, such as *Hoskins* or *Veriloy.* Used for chemical equipment, furnace parts, etc.

nickel-chromium-molybdenum alloys. A group of wrought and cast nickel-base alloys with high contents of chromium (15-20%) and molybdenum (9-30%). They have excellent corrosion and oxidation resistance at elevated temperatures, and high strength. Often sold under trademarks or trade names, such as *Chlorimet* or *Hastelloy.* Used for pumps, valves, chemical handling equipment, etc.

nickel-chromium-molybdenum steels. A group of alloy steels containing chromium, nickel and molybdenum as the chief alloying elements, usually in the range of 0.30-1.80% nickel, 0.20-1.20% chromium and 0.10-0.35% molybdenum. It includes the steels in the AISI-SAE 43xx, 43BVxx, 47xx, 81xx, 86xx, 87xx, 88xx, 93xx, 94xx, 97xx, and 98xx series. They have high strength and toughness, high ductility, and good forgeability and machinability. Used for shafts, chain pins, fatigue-resisting parts, carburized parts, gears, etc. Also known as *chrome-nickel-molybdenum steels; chromium-nickel-molybdenum steels.*

nickel-chromium-silicon gray irons. A group of heat-resistant austenitic gray cast irons containing about 1.8-2.6% total carbon, 0.4-1.0% manganese, 5.0-6.0% silicon, 0-1.0% molybdenum, 0-10% copper, 13-43% nickel, 1.8-5.5% chromium, 0-0.1% each phosphorus and sulfur, and the balance iron. They have austenite matrices in the as-cast condition. Often sold under trademarks or trade names, such as *Nicrosilal*, they are used for high-temperature applications, such as heat-treating boxes, furnace grids, and annealing pots.

nickel-chromium steels. A group of alloy steels containing 1.25-3.50% nickel and 0.60-1.60% chromium, and including the steels in the AISI-SAE 31xx, 32xx, 33xx, and 34xx series. They possess high tensile strength, hardness and toughness, and are used for automobile and aircraft parts, axles, shafts, gears, pinions, piston-pins, bolts, nuts, studs, transmission chains, connecting rods, crankshafts, keystock, armorplate, forgings, and carburized parts.

nickel-chromium white irons. A class of highly abrasion-resistant martensitic white cast irons with a hardness of about 550 Brinell that can be of the high-carbon, low-carbon, or nickel high-chromium type. High-carbon nickel-chromium types typically contain 3-3.6% carbon, 3.3-5.0% nickel, 1.4-4.0% chromium, up to 1.3% manganese, up to 1.0% molybdenum, up to 0.8% silicon, up to 0.3% phosphorus, up to 0.15% sulfur, and the balance iron. Low-carbon nickel-chromium types contain 2.5-3.0% carbon, 3.3-5.0% nickel, 1.4-4.0% chromium, up to 1.3% manganese, up to 1.0% molybdenum, up to 0.8% silicon, up to 0.3% phosphorus, up to 0.15% sulfur, and the balance iron. Nickel high-chromium irons contain 2.5-3.6% carbon, 7.0-11.0% chromium, 5.0-7.0% nickel, and chromium-high nickel irons 2.9-3.7% carbon, 2.7-4.0% nickel, and 1.1-1.5% chromium. Also known as *nickel-chromium white cast irons.* See also martensitic white irons; white irons.

Nickel Clad Copper Wire. Trade name of Anomet Products Inc. (USA) for a composite product consisting of a copper core clad with pure nickel. Supplied in wire and rod form with an outside diameter of 0.25-25.4 mm (0.01-1.0 in.) and nickel ratios of 10-40%, it combines the high electrical conductivity of copper with the excellent corrosion resistance of nickel.

nickel-clad copper wire. A composite wire with excellent oxidation resistance made by introducing a copper rod into a nickel tube and drawing both together.

nickel-clad steels. Low-carbon steels clad on both sides with pure nickel. They have good to excellent corrosion resistance, and are used for piping, tubing, vessels, reservoirs, tanks, containers, tank cars and trucks, retorts, stills, digesters, and electron tubes.

nickel-coated graphite fibers. Composite fibers that typically consist of carbon with thin, uniform electroplated nickel coatings. They uniquely combine the advantages of carbon fibers (e.g., low weight and high strengths and moduli) with those of nickel (e.g., high electrical and thermal conductivities). Used as electrically conductive reinforcements for plastics, as general reinforcing fibers for electromagnetic interference (EMI) shielding applications and for the elimination of statics. Abbreviation: NCGF.

nickel-coated silver powder. Silver powder coated with up to 2 wt% nickel, available in various mesh sizes for the manufacture of powder-metallurgy parts, such as electrical contacts.

nickel coatings. Thin layers of metallic nickel deposited on metal substrates (e.g., steel, brass, bronze, etc.) by electrodeposition or electroless deposition to prevent corrosion, enhance appearance, or other purposes. Also known as *nickel plates.* See also electroformed nickel; electroless nickel; Watts nickel.

nickel-cobalt steels. A group of high-strength steels containing about 9-18% nickel and 4-15% cobalt used for structural applications.

nickel-cobalt sulfate. Reddish-brown crystals used for blackening of brass and zinc. Also known as *cobaltonickelous sulfate.*

nickel-columbium. See nickel-niobium.

Nickel Copper. Trade name of American Nickeloid Company (USA) for a nickel-coated copper used for formed, stamped and drawn parts.

nickel-copper. Any of several master alloys of nickel and copper used to introduce nickel into ferrous and nonferrous alloys. Nickel-copper composed of 60% nickel, 33% copper, 3.5% manganese and up to 3.5% iron is often employed in the manufacture of acid-resistant castings, high-strength bronzes, and certain bearing bronzes. An alloy of 50% nickel and 50% copper melts at about 1270°C (2318°F), and is used in the form of slabs, ingots and shot as a ladle addition in the manufacture of iron and steel.

nickel-copper alloys. A group of alloys in which nickel and copper are the principal elements. Usually, the nickel-to-copper ratio is about 2:1, and small amounts of carbon, manganese, iron, silicon, and sulfur are also present. They possess high strength and toughness over a wide temperature range, good weldability, good cold workability and fabricability, excellent corrosion resistance, good resistance to most acids except nitric acid, good to excellent resistance to alkalies, salts, seawater and chlorides. Used for pumps, pump shafts, valves, valve parts and trim, impellers, marine parts, water-meter parts, fasteners, springs, fishing line, bellows, doctor blades and scrapers, chemical processing equipment, petroleum refining and processing equipment, gasoline and freshwater tanks, oil-well drilling collars and instruments, crude petroleum stills, process vessels and piping, electrical and electronic components, waveguides, feedwater heaters, heat exchangers, magnetic components and equipment. See also Monel.

nickel-copper-gold brazing compound. A brazing alloy of nickel, copper and gold with a solidus temperature of 950°C (1742°F) used for specialized metal-joining applications.

nickel-copper shot. Shot made from an alloy composed of equal amounts of nickel and copper, and used as a master alloy in the manufacture of iron and steel.

Nickel-Copper-Titanium. Trade name of LTV Steel Corporation (USA) for a low-carbon steel containing up to 0.15% carbon, 0-1% manganese, 0-0.5% silicon, 0-0.7% nickel, 0-0.05% titanium, 0.3% or more copper, and the balance iron. It has good welda-bility, cold formability and resistance to atmospheric corrosion. Used for guardrails and automobile bumpers.

Nickel Crowns. Trade name of Falconbridge Limited (Canada) for electrolytic nickel used for electroplating anodes.

nickel cyanide. Apple-green plates or powder obtained by treating a solution of a nickel salt with potassium cyanide. Melting point, loses $4H_2O$ at 200°C (392°F); boiling point, decomposes. Used in metallurgy and electroplating. Formula: $Ni(CN)_2 \cdot 4H_2O$. Also known as *nickel cyanide tetrahydrate.*

nickel cyanide tetrahydrate. See nickel cyanide.

nickel deuteride. A compound of nickel and deuterium. Crystal system, cubic. Crystal structure, halite. Melting point, 1987°C (3609°F). Used for semiconductors. Formula NiD.

nickel diarsenide. See nickel arsenide (ii).

nickel dimethylglyoxime. A bright red to red-purple powder that sublimes at 250°C (482°F) and is used as paint pigment and as colorant for cellulosics. Formula: $Ni(HC_4H_6N_2O_2)_2$.

nickel diselenide. See nickel selenide (ii).

nickel disilicide. See nickel silicide (ii).

nickel disulfide. See nickel sulfide (ii).

nickel ditelluride. See nickel telluride.

Nickeldur. Trade name of Janney Cylinder Company (USA) for a corrosion-resistant nickel bronze of 80% copper, 10% lead, 5% nickel and 5% tin. Used for centrifugally cast liners.

Nickeleisen. Trade name of VDM Nickel-Technologie AG (Germany) for a series of magnetically soft materials containing 62-64% iron and 36-38% nickel. They have high magnetic permeabilities, and are used for electrical and magnetic equipment, and motors.

Nickelend. Trade name of Arcos Alloys (USA) for commercially pure nickel (99.5%) containing 0.5% iron. Used for welding rods.

nickelene. A *nickel silver* containing 52-80% copper, 10-35% zinc, 5-30% nickel, 0-10% lead and 0-2% tin. It has excellent corrosion resistance, high electrical resistivity, good workability, good strength, and fair machinability improving with increasing lead content. Used for thermocouples, resistance wires, and electrical instruments.

nickel etioporphyrin III. A *porphyrin* derivative of nickel that is available in the form of purple crystals with a maximum absorption wavelength of 392 nm. Used as a dye and pigment. Formula: $Ni(C_{32}H_{36}N_4)$. Also known as *etioporphyrin I nickel.*

nickel ferrite. A ferrimagnetic purple-black ceramic powder with an inverse cubic spinel crystal structure (*ferrospinel*) usually made from nickel oxide (NiO) and ferric oxide (Fe_2O_3) by powder metallurgy techniques. Nickel ferrite powders can also be produced by spray pyrolysis. *Nickel ferrite* has a density of 5.34-5.37 g/cm³, a melting point of 1660°C (3020°F), excellent magnetic properties at high frequencies, high resistivity, and high corrosion resistance. Used as a soft magnetic materials for various applications. Formula: $NiFe_2O_4$. Also known as *diiron nickel tetraoxide*. See also ferrite; inverse spinel.

nickel fluoride. Yellow crystals or light green, hygroscopic, crystalline powder (99+% pure). Crystal system, tetragonal. Density, 4.72 g/cm³; melting point, 1474°C (2685°F); canted antiferromagnetic properties. Used in ceramics, electronics, and materials research. Formula: NiF_2. Also known as *nickelous fluoride*.

nickel fluoride tetrahydrate. A green, crystalline powder (98+% pure). Density, 4.72 g/cm³; canted antiferromagnetic properties. Used in ceramics, electronics and materials research. Formula: $NiF_2 \cdot 4H_2O$. Also known as *nickelous fluoride tetrahydrate*.

nickel glance. See gersdorffite.

Nickel-Gleam. Trade name for LeaRonal Inc. (USA) for a bright nickel electroplate and plating process.

nickel hexahydrite. A green mineral of the hexahydrite group composed of nickel sulfate hexahydrate, $NiSO_4 \cdot 6H_2O$. Crystal system, monoclinic. Density, 2.75 g/cm³.

nickel high-chromium white irons. See nickel-chromium white irons.

nickel hydroxide. See nickelic hydroxide; nickelous hydroxide.

nickelic hydroxide. A black, amorphous powder used as an active material in the positive plate of nickel-cadmium batteries. Formula: $Ni(OH)_3$. Also known as *nickel hydroxide*.

nickelic oxide. See black nickel oxide.

Nickelin. Trade name of VDM Nickel-Technologie AG (Germany) for a series of nickel silvers and copper nickels used for electrical resistances and for corrosion-resistant parts.

Nickeline. Trade name of NL Industries (USA) for several strong, heat-resistant alloys of 55-75% copper, 18-32% nickel, 0-20% zinc and 0.2-0.45% iron. Used for electrical resistances.

nickeline. A light copper-red mineral with metallic luster and dark tarnish composed of nickel arsenide, NiAs, containing as much as 43.9% nickel. Crystal system, hexagonal. Density, 7.3-7.8 g/cm³; hardness, 5-5.5 Mohs. Occurrence: Canada, Europe, Africa. Used as an important ore of nickel. Also known as *arsenical nickel; copper nickel; niccolite*.

nickel iodide. Black, hygroscopic crystals or powder (98+% pure). Crystal system, hexagonal. Density, 5.834 g/cm³; melting point, sublimes at 797°C (1466°F). Used in nickel plating, and as a reagent. Formula: NiI_2. Also known as *nickelous iodide*.

nickel-iron alloys. A group of alloys composed predominantly of iron and nickel which according to their properties can be classified into the following two categories: (i) *Magnetically soft alloys* containing about 50-80% nickel, 16-50% iron and up to 4% molybdenum. High-nickel alloys have high initial permeabilities and low saturation inductions, while low-nickel alloys have low initial permeabilities and high saturation inductions; and (ii) *Low-expansion alloys* often sold under trademarks or trade names, such as *Dumet, Invar* or *Platinite*. The typical composition range is from 48-64% iron, 31-52% nickel, 0-0.4% manganese, 0-0.1% silicon and 0-5% cobalt. Their low coefficient of thermal expansion increases with increasing nickel content.

nickel-iron-chromium alloys. A group of wrought and cast alloys composed predominantly of nickel, chromium and iron. They possess excellent oxidation resistance and good strength at high temperatures, good corrosion resistance, and good resistance to many acids and caustics. Often sold under trademarks or trade names, such as *Incoloy, Inconel, Durimet, Esco* or *Isocast*, they are used for chemical equipment, pickling tanks, pumps, valves, components and equipment for handling hot boiler feedwater in nuclear powder plants, corrosion-resistant parts, fume ducts, electrical resistances, and heating elements.

nickel irons. See nickel cast irons.

nickel-iron shot. Shot made from an alloy composed of equal amounts of nickel and iron, and used as a master alloy in the manufacture of iron and steel.

nickel leaded commercial bronze. See nickel-bearing leaded commercial bronze.

Nickel-Lume. Trade name of Atotech USA Inc. (USA) for a bright nickel electroplate and plating process.

nickel magnesia cermets. A group of composite materials consisting of magnesium oxide (MgO) particles and a nickel binder, and made at high temperatures under controlled atmospheres using powder metallurgy techniques. They have good strength and wear resistance, high resistance to oxidation and intergranular corrosion at elevated temperatures, good general corrosion resistance, excellent stress-to-rupture properties, and high toughness. Used for nuclear applications.

nickel-magnesium. A master alloy of nickel and magnesium used as a *nodulizer* in the manufacture of ductile cast iron.

Nickel Malleable. Trade name of International Nickel Inc. (USA) for a corrosion-resistant nickel (99.4% pure) containing 0.15% iron, 0.15% manganese, 0.1% copper and 0.05% carbon. Used for coinage, and as an addition to other alloys.

Nickel-Manganese. Trade name of Gilby-Fodor SA (France) for a corrosion- and heat-resistant nickel alloy containing 2-5% manganese. Used for grid wires and electronic tubes.

nickel-matrix composites. Composite materials composed of matrices of nickel or nickel alloys reinforced with tungsten or molybdenum-alloy wires or silicon carbide fibers. They have high tensile strengths, good high-temperature properties and good oxidation resistance. Used for flywheels, turbine blades, aircraft engine parts pressure vessels, loaded beams, etc.

nickel-molybdenum alloys. A group of cast and wrought nickel-base alloys containing high amounts of molybdenum. They have good corrosion and creep resistance, good high-temperature properties, high tensile strength, and good weldability. Used for process equipment, chemical equipment, oilfield pumps, valves, welding electrodes, and filler metals.

nickel-molybdenum cast irons. A group of cast irons to which 0.25-2.0% nickel and 0.25-1.5% molybdenum have been added. Sometimes they also contain other additions, such as chromium. The nickel helps refine the grain size and the size of the graphite flakes, and the presence of molybdenum produces fine and highly dispersed particles of graphite and good structural uniformity. This improves toughness, shock resistance, fatigue properties, hardenability, machinability and high-temperature strength. Used for heat-resistant castings.

nickel-molybdenum-iron alloys. A group of alloys containing 20-40% molybdenum, 40-60% nickel, up to 20% iron, 0-3% manganese, and traces of carbon. They have excellent corrosion resistance, excellent resistance to most acids, high strength, and relatively good castability and machinability. Used for chemical equipment, process equipment, and pump parts.

nickel-molybdenum steels. A group of alloy steel containing 0.85-3.50% nickel, 0.20-0.25% molybdenum and up to 0.50% carbon, and including the steels in the AISI-SAE 46xx and 48xx series. They possess good strength and toughness and good formability. Used for gears, shafts, roller bearings, heavy-duty bolts and studs, cams, chain pins, steering-knuckle pins, and minimum distortion parts.

nickel monoarsenide. See nickel arsenide (i).

nickel monoboride. See nickel boride (i).

nickel monoxide. See green nickel oxide.

nickel monoselenide. See nickel selenide (i).

nickel monosulfide. See nickel sulfide (i)

nickel naphthenate. The nickel salt of naphthenic acid available as a viscous liquid containing about 6-8% nickel, and used as paint drier.

nickel-niobium. A group of refractory master alloys available in two grades: (i) *Regular*, with 0.1% carbon, 55-65% niobium (columbium), 3% iron, 0.5% manganese, 35-45% nickel and 2.5% silicon; and (ii) *Vacuum*, with 0.1% carbon, 60-65% niobium (columbium), 1% iron, 0.05% manganese, 33-38% nickel and 0.2% silicon. Also known as *nickel-columbium*.

nickel nitrate hexahydrate. Green, deliquescent crystals. Crystal system, monoclinic. Density, 2.05 g/cm^3; melting point, 56°C (133°F); boiling point, 137°C (279°F). Used in nickel plating, in the preparation of nickel catalysts, and in the manufacture of certain brown ceramic colors. Formula: $Ni(NO_3)_2 \cdot 6H_2O$. Also known as *nickel nitrate; nickelous nitrate; nickelous nitrate hexahydrate*.

nickel nitrate tetramine. See ammoniated nickel nitrate.

nickel nitride. A compound of nickel and nitrogen used as an electrical conductor. Formula: Ni_3N.

nickelocene. An organometallic coordination compound (molecular sandwich) consisting of a nickel atom (Ni^{2+}) and two five-membered cyclopenediene (C_5H_5) rings, one located above and the other below the nickel atom plane. *Nickelocene* is supplied in the form of air- and heat-sensitive dark green crystals with a melting point of 173-174°C (343-345°F), and as a 10 wt% solution in toluene. Used as a catalyst and complexing agent in organic and organometallic synthesis, and in the biosciences. Formula: $(C_5H_5)_2Ni^{2+}$. Also known as *dicyclopentadienylnickel; bis-(cyclopentadienyl)nickel*.

nickel ocher. See annabergite.

Nickeloid. (1) Trade name Barker & Allen Limited (UK) for a corrosion-resistant alloy containing 40-45% nickel and 55-60% copper.

(2) Trade name of Pacific Metal Company (USA) for a babbitt metal containing varying amounts of lead, antimony, nickel, and tin. Used for bearings.

(3) Trade name of American Nickeloid Company (USA) for a nickel-coated zinc. It has good corrosion resistance, and is used for construction applications.

(4) Trade name of William McPhail & Sons (UK) for a corrosion-resistant alloy composed of nickel, copper and tin, and used for valves and valve seats for severe steam services.

(5) Trade name of Barker & Allen Limited (UK) for nickel silvers composed of 62% copper, 20% zinc and 18% nickel. Usually supplied in foil or tube form, they have a density of 8.72 g/cm^3 (0.315 lb/in.^{3}), a melting range of 1060-1110°C (1965-2055°F), excellent weldability and cold workability, good machinability, and poor hot formability. Used for fasteners, e.g., rivets, screws, bolts, etc., costume jewelry, and optical components.

nickelonickelic sulfide. See nickel sulfide (iv).

nickelonickelic oxide. Gray crystals or grayish-black amorphous powder. Crystal system, cubic. Used in chemistry. Formula:

Ni_3O_4.

Nickel Oreide. Trade name of Vereinigte Deutsche Nickel-Werke AG (Germany) for a corrosion-resistant alloy of 87% copper, 6.5% zinc and 6.5% nickel. Used for ornamental and architectural parts, and for hardware.

nickelous bromide. See nickel bromide.

nickelous bromide trihydrate. See nickel bromide trihydrate.

nickelous chloride. See nickel chloride.

nickelous chloride hexahydrate. See nickel chloride hexahydrate.

nickelous fluoride. See nickel fluoride.

nickelous fluoride tetrahydrate. See nickel fluoride tetrahydrate.

nickelous hydroxide. Green crystals or fine, green amorphous powder. Density, 4.15 g/cm^3; melting point, decomposes at 230°C (446°F). Used in the preparation of nickel salts. Formula: $Ni(OH)_2$. Also known as *nickel hydroxide*.

nickelous iodide. See nickel iodide.

nickelous nitrate. See nickel nitrate hexahydrate.

nickelous nitrate hexahydrate. See nickel nitrate hexahydrate.

nickelous oxide. See green nickel oxide.

nickelous phosphate. See nickel phosphate.

nickelous sulfate. See nickel sulfate.

nickelous sulfate hexahydrate. See nickel sulfate hexahydrate.

nickelous sulfate heptahydrate. See nickel sulfate heptahydrate.

nickel oxide. See black nickel oxide; green nickel oxide.

Nickeloy. British trade name for an age-hardenable aluminum alloy containing 4-5% copper and 1-2% nickel.

Nickel Pentrate. Trademark of Heatbath Corporation (USA) for a black oxide protective coating used on steel, and containing selected nickel salts and addition agents. A liquid version, *Nickel Pentrate L*, is also available.

nickel peroxide. See black nickel oxide.

nickel phosphate. A light-green powder used in electroplating and for yellow nickel. Formula: $Ni_3(PO_4)_2 \cdot 7H_2O$. Also known as *nickelous phosphate; nickel phosphate heptahydrate; trinickelous orthophosphate*.

nickel phosphate heptahydrate. See nickel phosphate.

nickel phosphide. Any of the following compounds of nickel and phosphorus used in ceramics, materials research and electronics: (i) *Nickel triphosphide*. Density, 4.16 g/cm^3. Formula: NiP_3; (ii) *Dinickel phoshide*. Gray crystals. Crystal system, hexagonal. Density, 7.33 g/cm^3; melting point, 1098°C (2008°F). Formula: Ni_2P; and (iii) *Trinickel phosphide*. Density, 7.66 g/cm^3; melting point, decomposes at 1098°C (2008°F). Formula: Ni_3P.

nickel-phosphorus coatings. (1) Hard, brittle, crystalline, nanocrystalline or amorphous nickel-phosphorus electrodeposits produced on metal or nonmetal substrates from plating baths composed of nickel sulfate, nickel chloride, phosphoric acid, phosphorous acid and selected additives at elevated temperatures using phosphate or boric acid buffers. Used for protective or decorative applications.

(2) Hard, corrosion- and wear-resistant amorphous or nanocrystalline coatings which contain about 3-12% phosphorus dissolved in nickel and are produced by electroless plating from plating solutions containing nickel chloride or sulfate as the source of nickel and sodium hypophosphite as the reducing agent. See also electroless nickel.

nickel phthalocyanine. A purple phthalocyanine powder that contains a divalent central nickel ion (Ni^{2+}). It has a typical dye content of about 85%, a maximum absorption wavelength of 670 nm, and is used as a pigment and dye. Formula: $Ni(C_{32}H_{16}N_8)$. Abbreviation: NiPc.

nickel plates. See nickel coatings.

nickel-plated steel. Steel, usually of the plain-carbon or low-alloy type, on whose surface has been deposited (e.g., by electrodeposition) a thin layer of metallic nickel, typically 5-50 μm (0.2-2.0 mil) thick. The layer serves to protect the steel against corrosive attack in rural, marine and industrial environments.

Nickel Plus. Trade name of Atotech USA Inc. (USA) for a bright nickel electroplate and plating process.

nickel powder. Nickel in the form of a fine powder supplied in various purities and particle sizes. Purities range from high (99.9-99.999%) to commercial (99.5-99.9%), and the particles are commonly spherical in shape and available in sizes from submicron (0.02 μm or 0.8 μin.) to over 150 μm (0.006 in.). Used in powder metallurgy, ceramics and electronics. Abbreviation: NP.

nickel protoxide. See green nickel oxide.

nickel-rhodium alloys. A group of alloys in which nickel and rhodium are the principal constituents. The rhodium content can be as high as 80%, and small amounts of other metallic elements, such as cobalt, copper, iridium, iron, molybdenum, palladium, platinum and/or tungsten, may be also present. Used for pen nibs, electrodes, chemical process equipment, and reflectors.

nickel selenide. Any of the following compounds of nickel and selenium: (i) *Nickel monoselenide.* Yellow-green crystals or grayish powder (99.5+% pure). Crystal system, hexagonal. Density, 7.2-8.46 g/cm³; melting point, 980°C (1796°F). Used for ceramics, materials research and electronics. Formula: NiSe; and (ii) *Nickel diselenide.* Crystal system, cubic. Crystal structure, pyrite. Used in ceramics and materials research. Formula: $NiSe_2$.

nickel sesquioxide. See black nickel oxide.

Nickel Shield. Trade name of A Brite Company (USA) for bright nickel electroplate and plating process.

nickel silicide. Any of the following compounds of nickel and silicon used in ceramics and materials research: (i) *Nickel silicide.* Crystal system, orthorhombic. Density, 7.4 g/cm³; melting point, 1255°C (2291°F). Formula: Ni_2Si; and (ii) *Nickel disilicide.* Crystal system, cubic. Density, 4.83 g/cm³; melting point, 993°C (1819°F). Formula: $NiSi_2$.

nickel-silicon alloys. A group of cast nickel alloys, typically containing up to 11% silicon and some copper, and having excellent resistance to corrosion and acids including sulfuric and hydrochloric acid. Used for pump valves and chemical equipment.

nickel-silicon brass. A strong copper-zinc alloy containing small amounts of nickel and silicon.

nickel-silicon bronze. A high-strength copper alloy containing small amounts of nickel and silicon. It has good electrical properties, and is used for electrical components, and fasteners.

nickel silvers. A group of hard, tough, silvery-white copper alloys containing 52-80% copper, 10-35% zinc and 5-35% nickel, and sometimes small amounts of lead and tin. Commercially available in the form of flat products, foils, rods, wires and tubes, they have densities ranging from 8.7 to 9.0 g/cm³ (0.315 to 0.325 lb/in.³), excellent cold workability and corrosion resistance, are rather brittle and difficult to work, and cannot be hammered into shape. Used especially for hollowware, tableware, costume jewelry, etching stock, optical parts, screws, nuts, bolts, wire, zip fasteners and nameplates, and in enameling. Also known as *German silvers; nickel brasses.*

nickel-skutterudite. A tin-white mineral composed of nickel cobalt iron arsenide, $(Ni,Co,Fe)As_{3-x}$. Crystal system, cubic. Density, 6.40 g/cm³. Occurrence: Germany.

Nickel SO. Trade name of Vacuumschmelze GmbH (Germany) for a high-purity nickel (99.85+%) made by a powder metallurgy process. It exhibits a marked increase in electrical resistivity with temperature, and is used for diesel spark plugs.

Nickel Special. Trade name of Isabellenhütte Heusler GmbH (Germany) for a resistance alloy composed of 99.4% nickel and 0.2-0.6% iron. It has a maximum service temperature of 150°C (300°F), and is used for electrical instruments and equipment.

nickel stannate dihydrate. A greenish, crystalline powder that loses $2H_2O$ at 125°C (257°F) and is used as a component in ceramic capacitors, and in barium titanate ceramics to lower the Curie temperature. Formula: $NiSnO_3 \cdot 2H_2O$.

nickel stearate. A waxy green solid with a density of 1.13 g/cm³ and a melting point of 155°C (311°F), used as a welding flux for nickel, in lubricants, and in waterproofing. Formula: $Ni(O_2C_{18}H_{35})_2$.

Nickel Staybolt. Trade name of Atlas Specialty Steels (Canada) for a tough steel containing 0.08% carbon, 0.25% manganese, 2.25% nickel, and the balance iron. Used for boiler staybolts.

Nickel Steel. Trade name of American Nickeloid Company (USA) for a nickel-bonded steel with good thermal properties at elevated temperatures, and good forming, stamping and drawing properties. Used for hardware, stampings, reflectors, and floor plates.

nickel steels. (1) A group of carbon steels, usually of the AISI-SAE 23xx and 25xx series, containing about 0.1-0.3% carbon, 04-0.9% manganese, 0.2-0.4% silicon and up to 5% nickel. They have good toughness and strength, good corrosion and shock resistance, good elevated-temperature stability, and are used for axles, shafts, propeller shafts, bolts, studs, gears, pinions, piston pins, levers, carburized parts, keystock, armor plate, structural shapes, and rails.

(2) A group of high-strength steels made by powder metallurgy techniques. They are produced by mixing 1.0-8.0% nickel with iron, 0.3-0.9% carbon and 0-2% copper, pressing and sintering. They are available in the as-sintered and heat-treated condition with a density range of 6.0-7.6 g/cm³ (0.22-0.27 lb/in.³). See also sintered nickel steel; structural nickel steel.

Nickel Storm. Trade name of A Brite Company (USA) for a bright nickel electroplate and plating process.

nickel subsulfide. See nickel sulfide (iii).

nickel sulfate. Yellowish green crystals. Crystal system, cubic. Density, 3.68 g/cm³; melting point, loses SO_3 at 840°C (1544°F). Used in nickel dip solutions to improve the adherence of porcelain enamels to steel, in nickel plating, and in blackening zinc and brass. Formula: $NiSO_4$. Also known as *nickelous sulfate.*

nickel sulfate hexahydrate. A nickel salt that is commonly used in nickel-plating baths. It is made by treating nickel with sulfuric acid, and supplied in the form of blue or pea-green crystals (98+% pure). Crystal system, monoclinic. Density, 2.070 g/cm³; melting point, loses $6H_2O$ at 280°C (536°F). Used in nickel plating, as a nickel catalyst, in blackening zinc and brass, in coatings, in nickel dip solutions to improve the adherence of porcelain enamels to steels, in ceramic products, and in the biosciences. Formula: $NiSO_4 \cdot 6H_2O$. Also known as *blue salt; nickelous sulfate hexahydrate; single nickel salt.*

nickel sulfate heptahydrate. Green crystals (99.9+% pure). It occurs in nature as the mineral *morenosite.* Crystal system, orthorhombic. Density, 1.98 g/cm³; melting point, loses $7H_2O$ at 98-

100°C (208-212°F). Formula: $NiSO_4\cdot 7H_2O$. Used in nickel dip solutions to improve the adherence of porcelain enamels to steels. Also known as *nickelous sulfate heptahydrate*.

nickel sulfide. Any of the following compounds of nickel and sulfur used in ceramics, electronics and materials research: (i) *Nickel monosulfide*. Yellow crystals or black powder. It occurs in nature as the mineral *millerite*. Crystal system, rhombohederal (hexagonal). Density, 5.3-5.65 g/cm^3; melting point, 797°C (1466°F); hardness, 3-3.5 Mohs. Formula: NiS; (ii) *Nickel disulfide*. Gray crystals or dark crystalline powder. It occurs in nature as the mineral *vaesite*. Crystal system, cubic. Crystal structure, pyrite. Density, 4.46 g/cm^3. Formula: NiS_2; (iii) *Nickel subsulfide*. Yellow crystals. Density, 5.52 g/cm^3. Formula: Ni_2S; (iv) *Nickelonickelic sulfide*. Gray-black crystals. It occurs in nature as the mineral *polydymite*. Crystal system, cubic. Crystal structure, spinel. Density, 4.77-4.81 g/cm^3; melting, 995°C (1823°F). Formula: Ni_3S_4; and (v) *Trinickel disulfide*. Hexagonal crystals. Density, 5.87 g/cm^3; melting point, 787°C (1449°F). Formula: Ni_3S_2.

nickel-tantalum. A compound of nickel and tantalum with a melting point, 1543°C (2809°F). Used in ceramics and materials research. Formula: Ni_3Ta.

nickel-tantalum alloy. A hard, ductile electrical resistance alloy containing 70% nickel and 30% tantalum, used for resistance wire.

nickel telluride. Steel-gray to tin-white crystals. It occurs in nature as the mineral *melonite*. Crystal system, hexagonal. Crystal structure, cadmium iodide. Density, 7.73 g/cm^3. Used in electronics and materials research. Formula: $NiTe_2$. Also known as *nickel ditelluride*.

nickel tetracarbonyl. An air- and heat-sensitive, colorless or yellowish volatile liquid made by passing carbon monoxide gas over powdered nickel. Density, 1.32 g/cm^3; melting point, -25°C (-13°F); boiling point, 43°C (109°F); vapor decomposes at 60°C (140°F); zerovalent compound. Used in nickel plating and vapor deposition on metallic, ceramic and glass substrates, and in the production of high-purity nickel powder by the Mond process. Formula: $Ni(CO)_4$. Also known as *nickel carbonyl*.

nickel tetrafluoroborate. A nickel compound available as a 50 wt% solution in water with a density of 1.454 g/cm^3 and as the hexahydrate $(6H_2O)$ with a density of 1.47 g/cm^3. Used for in the study of nickel salts for electroless plating applications and in the electrochemical conversion of phosphorus-chlorine bonds to phosphorus-carbon bonds. Formula: $Ni(BF_4)_2$ (anhydrous); $Ni(BF_4)_2\cdot 6H_2O$ (hexahydrate).

nickel thiophosphate. (1) Crystallized nickel thiophosphate is a double chalcogenide of nickel and phosphorus with a layered, cadmium chloride $(CdCl_2)$ type structure. It can be prepared at 700°C (1290°F) from the elements (nickel, phosphorus and sulfur) in a sealed tube.

(2) Amorphous nickel thiophosphate is a highly disordered phase of nickel thiophosphate synthesized at room temperature by mixing aqueous solutions of lithium thiophosphate (Li_2PS_3) and nickel nitrate $[Ni(NO_3)_2]$ in prescribed proportions. The initial sol can be converted to a gel through a soft chemical process and, after drying under vacuum at elevated temperatures, yields a very fine amorphized powder.

nickel-thorium. A compound of nickel and thorium with a melting point of 1532°C (2790°F). Used in ceram-ics and materials research. Formula: Ni_5Th.

Nickel Tin. Trade name of American Nickeloid Company (USA) for nickel-coated tin used for fabricated parts. It has good form-ing, stamping and drawing properties.

nickel-tin bronze. See nickel bronze.

nickel titanate. A brown powder (98+% pure) with a density of 4.56 g/cm^3. Used as a lightfast pigment in plastics, rubber, paints, textiles, and printing inks. Formula: $NiTiO_3$.

nickel-titanium. (1) An alloy composed predominantly of nickel (about 50-80%) and titanium (about 10-25%) with small amounts of aluminum, iron and silicon. Used as master alloy and deoxidizer for nonferrous alloys.

(2) A paramagnetic, intermetallic compound of nickel and titanium with high strength and good hardness. Used for magnetic and electronic components. Formula: NiTi.

nickel-titanium alloys. A group of alloys composed of nickel and titanium including (i) shape-memory alloys of 55% nickel and 45% titanium with a density of 6.5 g/cm^3 (0.235 $lb/in.^3$), a melting point of 1310°C (2390°F), high tensile strength and elastic modulus, high elongation, good corrosion resistance, and good electrical and magnetic properties. Used for sensing devices, nonmagnetic instruments, orthopedic and dental wire, and spacecraft components; and (ii) alloys with 90% nickel and 10% titanium with a melting point of about 1200°C (2444°F), high strength, and excellent corrosion resistance.

Nickel Treated. Trade name of Belmont Metals Inc. (USA) for a babbitt metal used for bearing applications.

nickel triarsenide. See nickel arsenide (iii).

nickel triphosphide. See nickel phospide (i).

nickel-uranium steel. A strong corrosion-resistant steel containing 0.2-0.8% carbon, 0.3-0.4% nickel, 0.2-0.4% uranium, and the balance iron. Used for general engineering applications.

Nickelvac. Trademark of Teledyne Allvac (USA) for double vacuum-cast alloys including several corrosion- and heat-resistant cobalt-nickel alloys, various nickel-base materials including nickel-chromium-iron, nickel-iron-chromium and nickel-cobalt-chromium alloys, and several corrosion- and/or heat-resistant austenitic chromium-nickel steels. Used for afterburners, superchargers, turbine wheels and blades, parts for gas turbines and heat engines, high-temperature springs and valves, pump parts, oil-refining equipment, chemical and petrochemical equipment, equipment for pulp and paper industries, fasteners, combustion chambers, and heaters.

nickel-vanadium steel. (1) A tough, shock-resistant steel containing 0.36% carbon, 3% nickel, 0.2-0.4% silicon, 0.1-0.45% vanadium, and the balance iron. Used for shafts, crankshafts, gears, pinions, and machinery parts.

(2) A carbon steel containing 0.28% carbon, 1% manganese, 1.5% nickel and 0.10% vanadium. Used for high-strength cast parts.

nickel vitriol. See morenosite.

nickel-zippeite. An orange to tan mineral of the zippeite group composed of nickel uranyl sulfate hydroxide hydrate, $Ni_2(UO_2)_6(SO_4)_3(OH)_{10}\cdot 16H_2O$. It can also be synthesized from acid solutions of nickel sulfate $(NiSO_4)$ and uranyl sulfate (UO_2SO_4). Crystal system, orthorhombic. Density, 3.30 g/cm^3; refractive index, 1.777. Occurrence: USA (Utah).

nickel-zirconium. (1) A corrosion-resistant alloy of 86% nickel, 6% silicon, 7.9% zirconium and 0.1% carbon. Used for chemical equipment.

(2) A master alloy composed of 40-50% nickel, 25-30% zirconium, 10% aluminum, 7.5-10% silicon and 5% iron. Used to introduce nickel and zirconium into nonferrous alloys.

nickel zirconium. Any of the following compounds of nickel and zirconium used in ceramics and materials research: (i) *Trinickel* *zirconium.* Density, 8.3 g/cm³; melting range, approximately 1600-1750°C (2910-3180°F). Formula: Ni$_3$Zr; and (ii) *Tetranickel zirconium.* Density, 8.4 g/cm³; melting point, 1650°C (3000°F). Formula: Ni$_4$Zr.

nickel zirconium steel. An oil-hardening steel containing 0.4% carbon, 3% nickel, 2.4% silicon, 0.24% zirconium, and the balance iron. Used for machinery parts, axles, shafts, and crankshafts.

Nickimphy. Trade name of Creusot-Loire (France) for a heat- and corrosion-resistant commercially pure nickel (99.78%) containing 0.1% manganese, 0.1% copper and 0.02% carbon. Used for electrical and electronic applications.

Nickolite. Trade name for a nickel silver containing 60% copper, 20% nickel, 11% zinc, 6% lead and 3% tin. It has good strength, machinability and corrosion resistance, and is used for door knobs and typewriter parts.

Nickrel. Trade name of Le Bronze Industriel (France) for a work-hardenable wrought nickel alloy containing 30% copper, 1.5% iron and 1.5% manganese. It has good corrosion resistance, good resistance to alkalies, salts, seawater, chlorides and most acids, and good strength. Used for valves, pumps, and hardware.

Nickrelk. Trade name of Le Bronze Industriel (France) for a wrought nickel alloy containing 28% copper, 3% aluminum, 1.5% iron and 1.5% manganese. It has good hardness, strength and corrosion resistance, and is used for pump parts, springs, and valves.

Nickrotherm. Trade name of Krupp Stahl AG (Germany) for a series of corrosion- and heat-resistant stainless steels containing 0.2% carbon, 1.5% silicon, 18% chromium, 0-10% nickel, and the balance iron. Used for heating muffles, autoclaves, roasting furnaces, boilers, recuperators, grates, dampers, and chemical processing equipment.

Nick Solder. Trade name of J.W. Harris Company Inc. (USA) for a lead-free solder containing 93% tin, 4% copper, and up to 2% silver and 1% nickel, respectively. It has a working temperature range of 204-316°C (440-600°F), and is used for soldering drinking water systems.

Niclad. Trade name of MacDermid (USA) for a composite sheet material produced by rolling together a sheet of nickel and a sheet of steel, or by depositing a layer of nickel on sheet steel by continuous welding. It combines the corrosion resistance of steel with the good strength of steel.

Nicloy. (1) Trade name of Babcock & Wilcox Company (USA) for a series of low-carbon steels containing 0.1-0.2% carbon, 0.4% manganese, 0.2% silicon, 3-9% nickel, 0-0.3% copper, and the balance iron. Supplied in the form of tubing, they have high-strength, good low-temperature properties, and are used for refrigerators, and liquid-air handling equipment.

(2) Trade name of Babcock & Wilcox Company (USA) for corrosion-resistant, low-expansion alloys containing 64% iron and 36% nickel. They have constant moduli of thermal expansion and are used in bimetals, Bourdon tubes, precision instruments, and hairsprings.

Nico. (1) Trade name of Northfield Iron Company (USA) for a corrosion-resistant low-carbon steel containing 0.2% carbon, 0.5% copper, 0.8% nickel, and the balance iron. Used for corrugated culvert pipes.

(2) British trade name for lead-base white metal alloys containing 4-10% tin, 1-3% nickel and 0-23% antimony. Used for bearings.

Nicoat DNC. Trademark of AHC-Oberflächentechnik Friebe &

Reininghaus GmbH (Germany) for electroless nickel plating compositions.

NiCoCrAlY coatings. See MCrAlY coatings.

Nicol. (1) Trade name of Columbia Bronze Corporation (USA) for a tough, heat-treatable cast aluminum bronze containing 78.5% copper, 10% aluminum, 5% nickel, 5% iron and 1.5% manganese. Used for ship propellers, and pump parts.

(2) Trade name for an alloy composed of 40% cobalt, 20% chromium, 15% nickel, 7% molybdenum, 2% manganese, 0.04% beryllium, 0.15% carbon, and the balance iron. It has high hardness and tensile strength, high electrical resistivity, good elevated-temperature properties, and high corrosion resistance. Used for instrument springs.

Nicolet. Trade name of C.E. Thurston & Sons, Inc. (USA) for asbestos cement and paper.

Nicolloy. Trade name of Swift Levick & Sons Limited (UK) for a series of chromium-cobalt-nickel and high-manganese austenitic steels used for hot dies.

Nicolmelt. Trademark of International Waxes Limited (Canada) for hot-melt adhesives.

Nicolon. Trademark of Schildkroet (Germany) for polyvinyl chloride film.

Niconium. Trade name of LeaRonal Inc. (USA) for a bright nickel electroplate and plating process for zinc die castings.

Ni-Copper. Trade name for a heat-resistant nickel-copper alloy used for motor winding wire.

nicopyrite. See pentlandite.

Nicoro. Trade name of GTE Products Corporation (USA) for corrosion-resistant brazing alloys containing 62% copper, 35% gold and 3% nickel. Commercially available in the form of foil, powder, wire, flexibraze, extrudable paste and preform, they have a melting range of 1000-1030°C (1830-1885°F) and are used for brazing copper, nickel, iron, steel, cobalt-base alloys, etc. A special grade, *Nicoro 80,* contains 81.5% gold, 16.5% copper and 2% nickel, and has a melting range of 910-925°C (1670-1697°F).

Nicoron. Trade name of Okuno Chemical Industries Company Limited (Japan) for an electroless nickel alloy electroplate and process.

Nicorros. Trade name of ThyssenKrupp VDM GmbH AG (Germany) for a series of *Monel*-type copper-nickel alloys with high tensile strength, good mechanical properties, and good corrosion resistance. Used for petroleum refining and processing equipment, process vessels and piping, chemical processing equipment, heat exchangers, nuclear reactor equipment, and filler metals for welding and hardfacing.

Nicoseal. Trademark of Carpenter Technology Corporation (USA) for a vacuum-melted alloy containing up to 0.02% carbon, 0.3% manganese, 0.2% silicon, 29% nickel, 17% cobalt, and the balance iron. It has a low coefficient of thermal expansion, and is used for hermetic seals, glass or ceramic to metal seals, and electronic components.

Nicosel. Trade name of Firth Brown Limited (UK) for a magnetic, low-expansion alloy containing 29% nickel, 17% cobalt, 0.15% silicon, 0.30% manganese, 0.04% carbon, and the balance iron. Used for instruments.

NiCoTef. Trademark of Nimet Industries Inc. (USA) for advanced composite coatings with nickel-phosphorus matrices and uniform dispersions of polytetrafluoroethylene (*Teflon*). They have good dry lubrication properties, low coefficients of friction, and high resistance to corrosion, abrasion and galling. Used on ferrous and nonferrous metals and alloys, e.g., carbon and stainless steels, aluminum and copper alloys, etc.

Nicotherm. Trademark of Schildkroet (Germany) for polyvinyl chloride film.

Nicovar. Trade name of Carpenter Technology Corporation (USA) for an alloy of 53.5% iron, 29% nickel and 17.5% cobalt. It has a low coefficient of thermal expansion, and is used for glass and ceramic seals.

Nicrad. Trade name of Foseco Minsep NV (Netherlands) for a ladle addition composed of nickel and chromium, and used to introduce nickel and chromium into iron and steel.

Nicral. Trade name of Nicralium Company (USA) for a series of alloys including several nickel-chromium-iron and nickel-iron-chromium alloys as well as various corrosion-resistant austenitic, ferritic and martensitic steels.

Nicralloy. Trade name of Osborn Steels Limited (UK) for heat-resistant nickel-base alloys with high electrical resistivities. *Nicralloy A* contains 80% nickel and 20% chromium, has a density of 8.4 g/cm^3 (0.30 lb/in.^{3}), and an upper service temperature of 1150°C (2100°F). *Nicralloy B* is composed 65% nickel, 20% iron and 15% chromium, and has an upper service temperature of 950°C (1740°F). *Nicralloy* alloys are used for heating elements and electrical resistors.

Nicranor. Trade name of Acieries et Forges d'Anor (France) for a heat-resistant alloy composed of 45-47% nickel, 20-22% chromium, 2-2.5% manganese, 2-2.5% silicon, 1-1.4% niobium (columbium), 0.15-0.25% carbon, and the balance iron. Used for electrical equipment.

Nicrex. Trade name of Murex Limited (UK) for high-nickel alloys with 0.07-0.2% carbon, 15-18% chromium, 0-25% iron, and 0-2% niobium (columbium). Also included under this trade name are austenitic stainless steels with up to 0.015% carbon, 18-26% chromium, 8-20% nickel, and sometimes small amounts of niobium and molybdenum, with the balance being iron. Used for welding electrodes.

Nicrfe. Trade name for a nickel alloy containing 17% chromium, 8% iron, 0-2% niobium (columbium), 0.65% manganese, 0.65% silicon, 0-0.015% phosphorus, 0.08% carbon, and 0-0.1% copper. It has good oxidation resistance and strength at high temperatures, and good corrosion resistance. Used for turbine parts and furnace equipment. See also Inconel.

Nicro. Trade name of Uddeholm Corporation (USA) for a series of nickel-chromium alloy steels (including several case-hardening grades) used for machinery parts, e.g., shafts, gears, and bolts.

NicroBlast. Trademark of Wall Colmonoy Corporation (USA) for blasting grits based on nickel-chromium alloys.

Nicrobraz. Trademark of Wall Colmonoy Corporation (USA) for a series of nickel brazing filler metals that are usually of the nickel-chromium-type and contain varying amounts of additions, such as silicon, manganese, iron, tungsten, phosphorus, boron and carbon. Commercially available in the form of strip, wire, rod, powder and transfer tape, they have a brazing range of 927-1204°C (1726-2200°F), and good to excellent strength and corrosion resistance. Used for brazing stainless steels (especially those of the AISI 300 and 400 series), nickel- and cobalt-base alloys particularly for vacuum systems, food-processing machinery, aircraft propeller hubs and steam turbine stator rings. A series of brazing aids including blasting grit, cements, filler-metal binders, fluxes and stop-offs is also available under this trade name.

Nicrocoat. Trade name of Wall Colmonoy Corporation (USA) for a series of abrasion-, corrosion- and heat-resistant metallic nickel- and cermet-type protective coating alloys. *Nicrocoat* coatings impart superior service life characteristics, and can be used for rebuilding worn base metal structures, sealing porous castings and filling surface cracks, e.g., jet-engine components, heat exchangers. mufflers, glass molds, process equipment, and turbine blades and vanes.

Nicrodie. Trade name of Columbia Tool Steel Company (USA) for an oil-hardening, low-alloy cold work tool steel (AISI type L6) containing 0.72% carbon, 0.6% manganese, 0.9% chromium, 1.5% nickel, 0.35% silicon, 0.25% molybdenum, and the balance iron. It has good deep-hardening properties, high toughness, and good wear resistance. Used for forming and trimming dies, rolls, shear blades, and structural parts.

Nicroex. Trade name of Columbia Tool Steel Company (USA) for an oil-hardening tool steel containing 0.7% carbon, 0.8% chromium, 1.25% nickel, 0.35% molybdenum, and the balance iron. Used for brake dies.

Nicrofer. Trade name of ThyssenKrupp VDM GmbH (Germany) for an extensive series of heat-resistant and austenitic stainless steels, iron-nickel superalloys and corrosion- and heat-resistant nickel-chromium alloys. Typical applications include chemical and petrochemical equipment, furnace and heat-treating equipment, heat exchangers, steam generator tubing, feedwater heaters, gas-turbine parts, and equipment for the pulp and paper industry.

Nicrolyte. Trade name of Enthone-OMI Inc. (USA) for chromium and nickel electroplates and plating processes.

Nicroma. Trade name of Ambo-Stahl-Gesellschaft (Germany) for a series of corrosion- and heat-resistant austenitic stainless steels used for chemical process equipment, oil-refining equipment, valve and pump parts, and furnace equipment and fixtures.

Nicroman. Trade name of Disston Inc. (USA) for a tough, shock-resistant tool steel containing 0.7% carbon, 1.65% nickel, 1% chromium, 0.4% manganese, 0.35% copper, and the balance iron. Used for shear blades, tools, and dies.

Nicromang. Trade name of Abex Corporation (USA) for an impact-resistant manganese steel composed of 0.8% carbon, 14.5% manganese, 4% chromium, 3.5% nickel, and the balance iron. Used for joining and rebuilding manganese steels and several other steels.

Nicromaz. Trade name of Creusot-Loire (France) for a series of corrosion-resistant steel castings containing up to 0.2% carbon, 0-2% manganese, 0-1.5% silicon, 13-22% chromium, 24-42% nickel, 0-6% molybdenum, 0-3% copper, 0-1% niobium, and the balance iron. They possess good resistance to hydrochloric, hydrofluoric, sulfuric and phosphoric acids, caustic solutions, petrochemicals, and brines. Used for chemical and petrochemical equipment, heat-treatment equipment, furnace parts, and plating equipment.

Nicromol. Trade name for alloy steel castings containing varying amounts of nickel, chromium and molybdenum.

Nicrosil. Russian trade name for a corrosion-resistant nickel cast iron containing 1.7-2% carbon, 16-20% nickel, 1.8-3% chromium, 0.8-1.3% manganese, and the balance iron. Used for furnace equipment.

Nicrosilal. Trade name of Sheepbridge Alloy Castings Limited (UK) for a series of nonmagnetic, austenitic nickel cast irons developed by the British Cast Iron Research Association (BCIRA). They contain 1.8-2.1% carbon, 5-6% silicon, 17-23% nickel, 2-5% chromium, and the balance iron, and have excellent heat and corrosion resistance. Used for high-temperature applications, such as heat-treating boxes, furnace grids, and annealing pots.

Nicrotal-SS. Trade name of Vulkan Strahlverfahrenstechnik GmbG & Co. KG (Germany) for nickel-chromium stainless steel granulates used for abrasive-blast cleaning applications.

Nicrotan. Trade name of Krupp VDM GmbH (Germany) for nickel-chromium-tantalum alloys containing small additions of aluminum, carbon and yttrium. They have excellent high-temperature strength to 900°C (1652°F), and excellent corrosion and oxidation resistance. Used for combustion chambers, heat shields, and aircraft engine parts.

Nicrothal. Trademark of Kanthal Corporation (USA) for heat-resistant nickel-base alloys with high electrical resistivities containing about 75-80% nickel, 20% chromium, and the balance, if applicable, aluminum, silicon, manganese and copper. *Nikrothal* alloys have maximum service temperatures up to 1250°C (2280°F) and are supplied in wire, ribbon, strip and foil form. Used for heating elements, electrical resistors, rheostats, and potentiometers.

Nicrotung. Trade name of Sankey & Sons Limited (UK) for a nickel-base superalloy containing 11-13% chromium, 9-11% cobalt, 7-8.5% tungsten, 3.75-4.75% aluminum, 3.75-4.75% titanium, 0.02-0.08% boron, 0.02-0.08% zirconium and up to 0.1% carbon. It has good thermal resistance up to about 1100°C (2010°F), and is used for gas turbines, and rocket engine components.

Nicuage. Trade name for a weldable low-carbon nickel-copper alloy steel used for structural applications.

Nicuend. Trade name of Arcos Alloys (USA) for an alloy containing 70% copper and 30% nickel, used to make electrodes for welding copper-nickel alloys.

Nicuite. Trade name of A.W. Cadman Manufacturing Company (USA) for a heat-treatable cast copper alloy containing 10% tin, 3.5% nickel and 2.5% zinc used for bearings and worm gears.

Niculoy. (1) Trade name of Belmont Metals Inc. (USA) for a master alloy containing copper, nickel and iron used for making various alloys.

(2) Trade name of Shipley Ronal (USA) for an electroless nickel-copper-phosphorus alloy coating.

Nicuman. Trade name of GTE Products Corporation (USA) for a series of brazing alloys containing 52-68% copper, 23-38% manganese and 9-10% nickel. Available in the form of wire, powder, foil, flexibraze, extrudable paste, and preform, they have a liquidus temperature range of 925-955°C (1697-1751°F), and a solidus temperature range of 880-925°C (1616-1697°F).

Nicusil. Trade name of GTE Products Corporation (USA) for a series of brazing alloys containing 56-71.2% silver, 28-42% copper and 0.75% nickel, and available in the form of foil, powder, wire, flexibraze, extrudable paste, and preform.

Nicu Steel. British trade name for an oil-hardening steel containing 0.3% carbon, 2.2% nickel, 0.6% manganese, 0.5% copper, and the balance iron. Used for structural work and machinery parts.

Nido d'Ape. Trade name of Fabbrica Pisana SpA (Italy) for a glass with honeycomb pattern.

niello. A metallic-black mixture of sulfides of silver, lead and copper, used for decorating metal with engraved inlays.

Nife-Optik. Trade name of AB Ruda Nya Glasindustri (Sweden) for an optical glass.

Nifer. Trade name of Texas Instruments Inc. (USA) for a carbon steel with a cladding of nickel on both sides. Used for electron tubes.

Niferloy. Trade name of Plating Process Systems Inc. (USA) for a nickel-iron electroplate and plating process.

Ni-Flex. Trade name of Materials Development Corporation (USA) for a series of nickel-base brazing alloys containing 0-15% chromium, 0-4.5% silicon, 0-4% iron and 2-3.5% boron.

Niflor. Trade name of Atotech USA Inc. for a low-friction coating produced on metallic or nonmetallic substrates by combining 18-25% polytetrafluoroethylene (PTFE) particles with electroless nickel. The resulting solid-lubricant surface is highly corrosion resistant. Both the process and coating are called *Niflor.*

Niflow. Trade name of MacDermid Inc. (USA) for a semibright nickel electroplate and plating process.

nifontovite. A colorless mineral composed of calcium borate dihydrate, $Ca_3B_6O_6(OH)_{12} \cdot 2H_2O$. Crystal system, monoclinic. Density, 2.36 g/cm³; refractive index, 1.578. Occurrence: Russia.

nigerite. A honey-colored mineral of the hogbomite group composed of tin zinc aluminum iron oxide hydroxide, $(Sn,Zn)(Al,Fe)_4O_8(OH)_{0.2}$. Crystal system, hexagonal. Density, 4.08 g/cm³; refractive index, 1.791. Occurrence: Brazil, Nigeria.

Nigerlene. Trademark of Nichemtex Industries Limited (Nigeria) for polyester staple fibers and filament yarns.

niggliite. A silver-white mineral of the nickeline group composed of platinum tin, PtSn. It can also be made synthetically. Crystal system, hexagonal. Density, 13.18 g/cm³. Occurrence: South Africa.

Nigy. Trade name of Creusot-Loire (France) for a heat- and corrosion-resistant nickel alloy containing 2-3% silicon, 0.5-1% manganese and 0.1% carbon. Used for spark plugs.

Ni-Hard. Trade name of Thomas Foundries, Inc. (UK) for a series of abrasion-resistant, martensitic white cast irons containing up to 3.8% carbon, 3.5-7% nickel, 1.4-11% chromium; 0.3-1.5% manganese, 0-1% molybdenum, 0-0.8% silicon, and the balance iron. They have high toughness and hardness, good corrosion resistance, good tensile and impact strength, and are used for jaw crushers, hammer mills, grinding balls, rolls, ball mill balls and liners, slurry pumps, pump components, and feeder vanes.

Nika. Japanese trade name for a tough and corrosion-resistant bearing alloy containing copper, tin and lead.

Nikalet. Trade name of Nippon Plastics Company (Japan) for polyvinyl chlorides.

Nikalium. Trade name of Delta Metal (BW) Limited (UK) for a tough, wear- and corrosion-resistant copper alloy containing 9% aluminum, 4.5% iron and 5% nickel. Used for gears, ship propellers and marine turbines.

Nikamelamine. Trade name of Nippon Plastics Company (Japan) for a series of melamine-formaldehyde plastics and compounds.

Nikavinyl. Trade name of Nippon Plastics Company (Japan) for polyvinyl chlorides.

Nike. Trade name of Breda Company (Italy) for a series of austenitic stainless steels containing up to 0.12% carbon, 18% chromium, 8-10% nickel, 0-2.5% molybdenum, traces of titanium, and the balance iron. Used for chemical equipment, mixers, tanks, vessels, and filters.

Nikelet. Trade name of Nippon Plastics Company (Japan) for melamine-formaldehyde plastics and compounds.

Niklad. Trade name of MacDermid Inc. (USA) for electroless nickel coatings and plating solutions.

NiklTech. Trademark of Lucent Technologies (USA) for nickel electrodeposits and plating processes.

Niko. Trade name of Swift Levick & Sons Limited (UK) for a low-expansion alloy containing varying amounts of iron, nickel and cobalt.

Nikon. Trade name of Massey-Harris Limited (Canada) for a cast iron containing 3.3% carbon, 0.7% manganese, 2.2% silicon, and the balance iron. Used for machine parts.

Nikora. Trade name of Atotech USA Inc. for an electroless nickel deposit and plating process.

Nikothane. Trade name of G.J. Nikolas & Company, Inc. (USA) for a quick-drying urethane.

Nikro M. Trade name of Teledyne Vasco (USA) for a special-purpose tool steel (AISI type L6) containing 0.68-0.73% carbon, 0.2-0.4% silicon, 0.5-0.6% manganese, 1.35-1.5% nickel, 0.8-0.9% chromium, 0.2-0.3% molybdenum, and the balance iron. It has great depth of hardening, high hardness and strength, high toughness, and good shock and wear resistance. Used for brake, die-casting, plastic-molding, drop-forging and thread-rolling dies, screwdriver bits, shear blades, punches, chisels, saws, and machine parts, such as clutches, spindles, pins and gears.

Nikrome. Trade name of Joseph T. Ryerson & Son Inc. (USA) for a series of strong, nickel-chromium alloy steels used for axles, spindles, shafts, and mandrels.

Nilad. Trademark of Diamonex (USA) for a diamond-like carbon coating applied to metals to provide them with high wear performance and enhanced corrosion and friction properties.

Nilamid. Trade name of Euronil (Italy) for a series polyamide (nylon) resins including nylon 6 and nylon 6,6, supplied in several grades.

Nilamon. Trade name of Ecomid (Italy) for a several polyamides and polyamide blends including resins based on nylon 6 and nylon 6,6.

Nilcor. Trade name of National Standard Company (USA) for a nonmagnetic nickel alloy used for springs.

Nilgro. Trade name of Darwins Alloy Castings (UK) for a series of controlled low-expansion alloys containing 58-64% iron and 36-42% nickel. Used for thermostats.

Nilo. Trademark of Inco Alloys International Inc. (USA) for a series of iron-nickel and iron-nickel-chromium alloys with low thermal expansion. Used for thermostats, glass-to metal sealing, and instrument components.

Nilomag. Trademark of Inco Alloys International Inc. (USA) for a series of soft-magnetic alloys of the nickel-iron-molybdenum and nickel-iron-copper-molybdenum type. Available in the form of sheet and strip, they have high magnetic permeabilities and low core losses. Used for magnetic amplifiers, cores of telephone transformers, inductors, magnetic shields, tape recorder heads and computer memories.

Nilstain. Trade name of Wilbur B. Driver Company (USA) for an extensive series of austenitic stainless steels used for springs, bolts, screws, nuts, welding wire, furnace parts, instruments, and other corrosion-resistant parts.

Nilvar. Trade name of Harrison Alloys Inc. (USA) for a controlled expansion alloy containing 64% iron and 36% nickel (same composition as *Invar*). It has a density of 8.0 g/cm³ (0.29 lb/in.³), a melting point of 1495°C (2723°F), a hardness of 160 Brinell, an electrical conductivity of 2% IACS, a very low coefficient of expansion at room temperature, and high strength and elastic modulus. Used for instrument parts, bimetals, thermostats, measuring tapes, length standards, and glass-to-metal joining applications.

Nimac. Trade name of MacDemid Inc. (USA) for nickel electro-

deposits and plating process.

Nimar. Trade name of Arcos Alloys (USA) for a series of maraging steels containing up to 0.03% carbon, 18.5% nickel, 7.5-8.5% cobalt, 4.9% molybdenum, 0.4-0.7% titanium, 0.1% aluminum, and the balance iron.

Nimark. Trade name of Carpenter Technology Corporation (USA) for a series of low- and medium-carbon maraging steels with good strength and ductility, used for high-strength components.

Nimbus. Trade name of Horbach & Schmitz GmbH (Germany) for a series of molybdenum- and tungsten-type high-speed steels used for lathe and planer tools, milling cutters, end mills, broaches, various other tools, and dies.

Nimend. Trade name of Arcos Alloys (USA) for a nickel-base alloy containing of 0.1-0.15% carbon, 14-18% chromium, 15-18% molybdenum, 5.3% iron and 4-5% tungsten. It has good corrosion resistance, and is used for hardfacing electrodes.

Nimetic Grip. Trade name of 3M ESPE (USA) for several self-cure dental resin cements.

nimite. A yellowish green mineral of the chlorite group composed of magnesium nickel aluminum silicate hydroxide, $(Ni,Mg,Al)_6(Si,Al)_4O_{10}(OH)_8$. Crystal system, monoclinic. Density, 3.19 g/cm³; refractive index, 1.647. Occurrence: South Africa.

Nimo. Trade name of Darwin & Milner Inc. (USA) for an oil-hardening steel containing 0.5-0.6% carbon, 2.5-2.8% nickel, 0.4-0.5% molybdenum, 0.3-0.5% chromium, 0.1-0.15% vanadium, and the balance iron. Used for tools and dies.

Ni-Moc. Trade name of Société Nouvelle du Saut-du-Tarn (France) for a heat- and corrosion-resistant nickel alloy containing 16% chromium, 17% molybdenum, 7% iron and 4% tungsten. Used for pumps, valves, and pharmaceutical, chemical and food-processing equipment.

Nimocast. Trade name of Inco Alloys International Inc. (USA) for an extensive series of nickel-base heat-resistant casting alloys containing 50-75% nickel, 0-25% chromium, 0-20% cobalt, 0-18% molybdenum, 0-10% iron, 0-6.5% aluminum, 0-12% tungsten, 0-5.2% titanium, 0-6.5% niobium and up to 0.15% carbon. They have good tensile, creep and fatigue strengths, high oxidation resistance at elevated temperatures, and good corrosion resistance. Used for gas-turbine parts, jet-engine parts, aerospace equipment, turbochargers rotors, diesel-engine parts, and other high-temperature applications.

Nimofer. Trade name of ThyssenKrupp VDM GmbH (Germany) for a series of corrosion-resistant alloys containing about 67-69% nickel, 28% molybdenum, 1.8-4.0% iron, 0.7% chromium, and small amounts of other elements. They are supplied in several grades the form of semi-finished products, and used for chemical, petrochemical and water treatment equipment and other corrosion-resistant equipment and parts.

Nimol. Trade name of Mond Nickel Company Limited (UK) for nickel-chromium cast irons.

Ni-Mold. Trade name for a mold steel (AISI type P6) containing 0.08-0.13% carbon, 0.2-0.4% silicon, 0.20-0.4% manganese, 3.25-3.75% nickel, 1.4-1.75% chromium, and the balance iron. It has great depth of hardening, exceptional core strength and toughness, outstanding resistance to decarburization, and takes a smooth, high luster when polished. Used for plastic molding dies.

Nimoloy. Trademark of Inco Alloys International Inc. (USA) for nickel-chromium-cobalt superalloys with high tensile strengths and shock resistance, used for hot-working tools, shear blades, and forging tools.

Nimonic. Trademark of Inco Alloys International Inc. (USA) for an extensive series of wrought and cast nickel-base superalloys containing 50-75% nickel, 0-25% chromium, 0-20% cobalt, 0-9% molybdenum, 0-42% iron, 0-12% tungsten, 0-5% aluminum, 0-3% titanium, 0-5% niobium and up to 0.2% carbon. They have good tensile, creep and fatigue strengths, good high-temperature properties, high oxidation resistance at elevated temperatures, good corrosion resistance, and are used for gas-turbine parts, jet-engine parts, valves, combustion chambers, aerospace equipment, missile and rocket parts, welding rods, and other high-temperature applications.

Nimpkish Red Cedar. Trade name of Canadian Forest Products Limited for red cedar-based panelwood.

Nimuden. Trade name of Uyemura International Corporation (USA) for electroless nickel.

nine-inch brick. A standard, rectangular refractory brick measuring approximately $229 \times 113 \times 64$ mm ($9.000 \times 4.4375 \times 2.500$ in.).

990 gold. A jewelry alloy of 99.0 wt% gold, hardened with up to 1.0 wt% titanium. It has the impact and wear resistance of conventional 99.0 wt% gold alloys, but improved durability and hardness. Heat treatment by homogenization, followed by solution heat treating and quenching, and cold working greatly enhances its ductility, while age hardening of the wrought material significantly enhances its hardness.

ningyoite. A brownish green or yellow mineral of the rhabdophane group composed of calcium uranium phosphate monohydrate, $CaU(PO_4)_2 \cdot H_2O$. Crystal system, orthorhombic. Density, 4.63 g/cm³; refractive index, 1.64. Occurrence: Japan.

niningerite. A colorless mineral of the halite group composed of magnesium sulfide, MgS. Crystal system, cubic. Density, 2.66 g/cm³. See also magnesium sulfide (i).

ninon. A lightweight, open-mesh fabric, usually of silk or a synthetic fiber, such as nylon or rayon, woven in plain weave and having a smooth surface. Used especially for clothing (e.g., lingerie and evening wear), and curtains.

niobite. An iron-black to brownish-black mineral with submetallic luster and dark red to black streak that is composed of niobium tantalum iron manganese oxide, $(Fe,Mn)(Nb,Ta)_2O_6$, and may contain about 44-70% Nb_2O_3 (Cb_2O_3) and 0.4-7% Ta_2O_5. Crystal system, orthorhombic. Density, 5.2-7.9 g/cm³; hardness, 6 Mohs. Occurrence: Africa (Zaire, Nigeria), Brazil, Canada, Germany, Greenland, Malaysia, Russia, USA (Idaho, Maine, North Carolina, South Dakota). Used as an ore of niobium (columbium) and tantalum, and for lamp filaments. Also known as *columbite; dianite; greenlandite*.

niobium. A rare, silvery white to steel gray, ductile metallic element of Group VB (Group 5) of the Periodic Table. It is commercially available in the form of plates, rods, wire, foil, microfoil, tube, turnings, powder, and high-purity single crystals. The crystals are usually grown by the float-zone technique. The commercial niobium ores are *niobite (columbite)* and *pyrochlore*. Crystal system, cubic. Crystal structure, body-centered cubic. Density, 8.57 g/cm³; melting point, 2468°C (4474°F); boiling point, 4927°C (8901°F); hardness, 115-160 Vickers; atomic number, 41; atomic weight, 92.906; superconductivity critical temperature, 9.25 K; divalent, trivalent, tetravalent, pentavalent. It has good superconductive properties, fair electrical conductivity (about 10.5% IACS), good resistance to tarnishing or oxidizing at room temperature, and good resistance to radiation damage. Used as an alloying element in stainless steels (e.g., in austenitic stainless steels to reduce susceptibility to intercrystalline corrosion), as *ferroniobium* master alloy in al-

loy steels, in superconducting and magnetic alloys (with tin and titanium), in cermets, for nuclear reactor applications, for missiles and rockets, and for cutting tools, welding rods, magnets, pipelines, and cryogenic equipment. Symbol: Nb; Cb. Also known as *columbium*.

niobium alloys. A group of alloys based on niobium (columbium) including (i) niobium-titanium and niobium-tin alloys used for magnetic and high-temperature applications and for superconductors; (ii) niobium-tungsten, niobium-hafnium and niobium-tantalum alloys used for high-temperature applications; (iii) niobium-uranium alloys in the preparation of nuclear fuel; (iv) niobium-zirconium alloys used for high-strength and high-temperature applications, and for sodium vapor lamps; and (v) *ferroniobium* used to introduce niobium into alloy and stainless steels.

niobium aluminide. Any of the following compounds of niobium (columbium) and aluminum also known as *columbium aluminide*: (i) *Niobium trialuminide*. Density, 4.50 g/cm³; melting point, above 1755°C (3191°F); excellent high-temperature strength up to 1430°C (2605°F); superconductivity critical temperature, 17.5K. Used for refractory coatings and as a superconductor. Formula: $NbAl_3$ ($CbAl_3$). Also known as *columbium trialuminide*; (ii) *Diniobium aluminide*. Used as a superconductor. Formula: Nb_2Al ($CbAl_3$). Also known as *dicolumbium aluminide*; and (iii) *Triniobium aluminide*. Superconductivity critical temperature, 18.9K. Used as a superconductor. Formula: Nb_3Al ($CbAl_3$). Also known as *tricolumbium aluminide*.

niobium beryllide. Any of the following compounds of niobium and beryllium used in high-temperature applications, and also known as *columbium beryllide*: (i) *Niobium diberyllide*. Melting point, approximately 2080°C (3776°F); good strength at elevated temperatures. Formula: $NbBe_2$ ($CbBe_2$). Also known as *columbium diberyllide*; (ii) *Niobium pentaberyllide*. Melting point, approximately 1829°C (3324°F); good strength at elevated temperatures. Formula: $NbBe_5$ ($CbBe_5$). Also known as *columbium pentaberyllide*; and (iii) *Niobium dodecaberyllide*. Density, 2.91 g/cm³; melting point, 1690°C (3074°F); good strength at elevated temperatures. Formula: $NbBe_{12}$ ($CbBe_{12}$). Also known as *columbium dodecaberyllide*. Other less common niobium beryllides include Nb_2Be_{17} (Cb_2Be_{17}) with a melting point of 1750°C (3182°F) and good strength at elevated temperatures, and (ii) Nb_2Be_{19} (Cb_2Be_{19}) with a density of 3.15 g/cm³, a melting point of approximately 1705°C (3100°F), and good strength at elevated temperatures.

niobium boride. Any of the following compounds of niobium and boron used for high-temperature applications, and also known as *columbium boride*: (i) *Niobium monoboride*. Gray crystals. Crystal system, orthorhombic. Density, 7.2-7.6 g/cm³; melting point, above 2270°C (4118°F); hardness, 8 Mohs. Formula: NbB (CbB). Also known as *columbium monoboride*; (ii) *Niobium diboride*. Gray crystals or powder. Crystal system, hexagonal. Density, 6.8-6.97 g/cm³; melting point, 3050°C (5522°F). Formula: NbB_2 (CbB_2). Also known as *columbium diboride*; (iii) *Triniobium tetraboride*. Density, 7.3 g/cm³; melting point, 2700°C (4892°F). Formula: Nb_3B_4 (Cb_3B_4). Also known as *tricolumbium tetraboride*; and (iv) *Triniobium diboride*. Melting point, 1816°C (3300°F). Formula: Nb_3B_2 (Cb_3B_2). Also known as *tricolumbium diboride*.

niobium bromide. See niobium tribromide; niobium pentabromide.

niobium carbide. Any of the following compounds of niobium and carbon, also known as *columbium carbide*: (i) *Niobium* *monocarbide.* A lavender-gray crystalline powder (97% pure). Crystal system, cubic. Crystal structure, halite. Density, 7.6-7.82 g/cm^3; melting point, approximately 3500°C (6330°F); boiling point, 4300°C (7792°F); hardness, above 9 Mohs; high modulus of rupture; high electrical resistivity. Used for cemented carbide-tipped tools, special steels, preparation of niobium (columbium) metal, coating graphite for atomic reactors, and in materials research. Formula: NbC (CbC). Also known as *columbium monocarbide;* (ii) *Diniobium carbide.* Refractory crystals. Crystal system, hexagonal. Density, 7.85 g/cm^3; melting point, approximately 3087°C (5590°F). Used in ceramics and materials research. Formula: Nb_2C (Cb_2C). Also known as *dicolumbium carbide.*

niobium chloride. See niobium trichloride; niobium pentachloride; niobium tetrachloride.

niobium diberyllide. See niobium beryllide (i)

niobium diboride. See niobium boride (ii).

niobium dioxide. White crystals or powder (99+% pure). Crystal system, tetragonal. Density, 5.9 g/cm^3; melting point, 1902°C (3456°F). Used for ceramics, refractories and other high-temperature applications. Formula: NbO_2 (CbO_2). Also known as *columbium dioxide.*

niobium diselenide. See niobium selenide.

niobium disilicide. See niobium silicide (i).

niobium disulfide. See niobium sulfide (i).

niobium ditelluride. See niobium telluride (ii).

niobium dodecaberyllide. See niobium beryllide (iii).

niobium ethoxide. A metal alkoxide available as a pale yellow moisture-sensitive, flammable liquid (99.9+% pure). Density, 1.258 g/cm^3; melting point, 5-6°C (41-43°F); boiling point, 142°C (288°F)/0.1 mm; flash point, 74°C (165°F); refractive index, 1.516. Used as an intermediate for sol-gel perovskites and lead magnesium columbates, as an intermediate for acentric piezoelectric lithium niobium oxide ($LiNbO_3$) films, in the manufacture of dielectric films, for the impregnation of paper for dielectric purposes, and in the preparation of ferroelectric films in combination with strontium and barium alkoxides. Formula: $Nb(OC_2H_5)_5$ [$Cb(OC_2H_5)_5$]. Also known as *columbium ethylate; columbium ethoxide; niobium ethylate.*

niobium ethylate. See niobium ethoxide.

niobium fluoride. See niobium trifluoride; niobium pentafluoride; niobium tetrafluoride.

niobium foil. Niobium in foil form available as regular foil (99.8-99.9% pure) in thicknesses ranging from 0.0025-2.0 mm (0.0001-0.08 in.) and as disk-shaped microfoil (99.9% pure) on permanent *Mylar* support in thicknesses from 0.0025-0.01 μm (0.1-0.4 μin.) and weights ranging from 2.2-21 μg/cm². Also known as *columbium foil.*

niobium germanide. Any of the following compounds of niobium and germanium, also known as *columbium germanide:* (i) *Triniobium digermanide.* Melting point, 1650°C (3000°F). Used as a refractory for high-temperature applications. Formula: Nb_3Ge_2 (Cb_3Ge_2). Also known as *tricolumbium digermanide;* (ii) *Diniobium germanide.* Melting point, 1910°C (3470°F). Used for refractories and other high-temperature applications. Formula: Nb_2Ge (Cb_2Ge). Also known as *dicolumbium germanide;* and (iii) *Triniobium germanide.* Cubic crystals. Melting point, 1910°C (3470°F); superconductivity critical temperature, 23.2K. Used for refractories, high temperature applications and superconductors. Formula: Nb_3Ge (Cb_3Ge). Also known as *tricolumbium germanide.*

niobium hydride. A gray high-purity powder with a density of

6.67 g/cm^3, used in ceramics and materials research. Formula: NbH (CbH). Also known as *columbium hydride.*

niobium iodide. See niobium tetraiodide; niobium pentaiodide.

niobium methoxide. A metal alcoholate with a melting point of 53°C (127°F), used in the manufacture of thin dielectric films. Formula: $Nb(OCH_3)_5$ [$Cb(OCH_3)_5$]. Also known as *columbium methoxide; columbium methylate; niobium methylate.*

niobium methylate. See niobium methoxide.

niobium-molybdenum disilicide composite. A ceramic-matrix composite consisting of a molybdenum disilicide ($MoSi_2$) matrix reinforced with niobium fibers, and used for high-strength high-temperature applications. Also known as *columbium-molybdenum disilicide composite.*

niobium monoboride. See niobium boride (i).

niobium monocarbide. See niobium carbide (i).

niobium mononitride. See niobium nitride (i).

niobium monotelluride. See niobium telluride (i).

niobium monoxide. A compound of columbium and oxygen supplied as gray crystals or powder (99.9%). Crystal system, cubic. Density, 7.3 g/cm^3; melting point, 1937°C (3519°F). Used for ceramics, high-temperature applications, and in refractories. Formula: NbO (CbO). Also known as *columbium monoxide.*

niobium nitride. Any of the following compounds of niobium and nitrogen, also known as *columbium nitride:* (i) *Niobium mononitride.* Gray crystals or gray-black powder (99% pure). Crystal system, cubic. Density, 8.47 g/cm^3; melting point, 2300°C (4170°F). Used for superconductors, ceramics, refractories, and high-temperature applications. Formula: NbN (CbN). Also known as *columbium mononitride;* (ii) *Diniobium nitride.* Density, 8.31 g/cm^3; melting point, approximately 2316°C (4200°F). Used for ceramics, refractories, and high-temperature applications. Formula: Nb_2N (Cb_2N). Also known as *dicolumbium nitride;* and (iii) *Tetraniobium trinitride.* Used for ceramics, refractories, and for high-temperature applications. Formula: Nb_4N_3 (Cb_4N_3). Also known as *tetracolumbium trinitride.*

niobium oxide. See niobium dioxide; niobium monoxide; niobium pentoxide.

niobium pentaberyllide. See niobium beryllide (ii).

niobium pentabromide. Orange crystals or yellow, moisture-sensitive crystalline powder (98+% pure). Crystal system, orthorhombic. Density, 4.36 g/cm^3; melting point of 254°C (489°F); boiling point of 361.6°C (682°F). Used in the preparation of niobium compounds, and as an intermediate. Formula: $NbBr_5$ ($CbBr_5$). Also known as *columbium bromide; columbium pentabromide; niobium bromide.*

niobium pentachloride. Yellow, moisture-sensitive crystals (99+%). Crystal system, monoclinic. Density, 2.75-2.78 g/cm^3; melting point, 194-205°C (381-401°F); boiling point, 245-254°C (473-489°F); melting and boiling point vary greatly with purity. Used in the preparation of pure niobium (columbium) and as an intermediate. Formula: $NbCl_5$ ($CbCl_5$). Also known as *columbium chloride; columbium pentachloride; niobium chloride.*

niobium pentafluoride. Colorless, moisture-sensitive crystals or white powder (98+%). Crystal system, monoclinic. Density, 3.92 g/cm^3; melting point, 72°C (161°F) [also reported as 80°C (176°F)]; boiling point, 229°C (444°F). Used in the preparation of niobium compounds, and as an intermediate. Formula: NbF_5 (CbF_5). Also known as *columbium fluoride; columbium pentafluoride; niobium fluoride.*

niobium pentaiodide. Yellow-black crystals or moisture-sensitive, black powder. Crystal system, monoclinic. Density, 5.32 g/cm³; melting point, decomposes above 200°C (392°F). Used in the preparation of niobium compounds, and as an intermediate. Formula: NbI_5 (CbI_5). Also known as *columbium iodide; columbium pentaiodide; niobium iodide.*

niobium pentoxide. A white crystalline, ferroelectric powder (99.5% pure). Crystal system, orthorhombic. Density, 4.47-5.05 g/cm³; melting point, 1520°C (2768°F); Curie temperature, 200-275°C (392-527°F). Used as an intermediate, in electronics, materials research, and for high-temperature applications. Formula: Nb_2O_5 (Cb_2O_5). Also known as *columbium oxide; columbium pentoxide; niobium oxide.*

niobium phosphide. A compound of niobium and phosphorus supplied as a powder. Crystal system, tetragonal. Density, 6.40-6.54 g/cm³; melting point, decomposes between 1660 and 1730°C (3020 and 3146°F). Used for semiconductors, ferroelectric applications, and electronics. Formula: NbP (CbP). Also known as *columbium phosphide.*

niobium-potassium oxyfluoride. Colorless or white, lustrous, monoclinic plates or leaflets used in the separation of niobium (columbium) from tantalum, and in the preparation of niobium (columbium) metal by electrolysis. Formula: $K_2NbOF_5 \cdot H_2O$. Also known as *columbium-potassium oxyfluoride; potassium-columbium oxyfluoride; potassium-niobium oxyfluoride; potassium oxyfluocolumbate; potassium oxyfluoniobate; potassium pentafluocolumbate; potassium pentafluoniobate.*

niobium selenide. Gray crystals or gray-black powder. Crystal system, hexagonal. Density, 6.3 g/cm³; melting point, above 1316°C (2400°F); vacuum-stable from -170 to +1315°C (-430 to +2400°F); higher electrical conductivity than graphite. Used as a lubricant and conductor for high temperatures and high vacuum, and with silver, copper, or other metal powders for self-lubricating bearings, and gears. Formula: $NbSe_2$ ($CbSe_2$). Also known as *niobium diselenide; columbium selenide; columbium diselenide.*

niobium silicide. Any of the following compounds of niobium and silicon, also known as *columbium silicide:* (i) *Niobium disilicide.* Gray crystals or fine powder. Density, 5.37 g/cm³; melting point, 1950°C (3542°F). Used in refractories, in semiconductors and in materials research. Formula: $NbSi_2$ ($CbSi_2$). Also known as *columbium disilicide;* (ii) *Niobium trisilicide.* Density, 7.05 g/cm³. Used for refractories and semiconductors. Formula: $NbSi_3$ ($CbSi_3$). Also known as *columbium trisilicide;* (iii) *Triniobium disilicide.* Used for refractories, in semiconductors and in materials research. Formula: Nb_3Si_2 (Cb_3Si_2). Also known as *tricolumbium disilicide;* (iv) *Tetraniobium silicide.* Density, 8.01 g/cm³; melting point, approximately 1950°C (3540°F). Used for refractories, semiconductors, and ceramics. Formula: Nb_4Si (Cb_4Si). Also known as *tetracolumbium silicide;* and (v) *Pentaniobium trisilicide.* Density, 7.34-7.75 g/cm³; melting point, approximately 1950-2000°C (3540-3630°F). Used for refractories, ceramics, and semiconductors. Formula: Nb_5Si_3 (Cb_5Si_3). Also known as *pentacolumbium trisilicide.*

niobium-stabilized stainless steel. An austenitic stainless steel to which a small amount of niobium (typically 0.3-1.0%) has been added to reduce susceptibility to intercrystalline corrosion. Also known as *columbium-stabilized stainless steel.*

niobium stannide. See niobium-tin.

niobium steel. A carbon steel to which have been added up to 0.1% niobium (columbium) to improve weldability and formability and promote grain refinement. Used for machine parts, forgings, dies, gages, etc. Also known as *columbium steel.*

niobium sulfide. Any of the following compounds of niobium and sulfur used in ceramics and materials research, and also known as *columbium sulfide:* (i) *Niobium disulfide.* Black crystals or powder (99.% pure). Crystal system, rhombohedral. Density, 4.4 g/cm³. Formula: NbS_2 (CbS_2). Also known as *columbium disulfide;* and (ii) *Niobium trisulfide.* Formula: NbS_3 (CbS_3). Also known as *columbium trisulfide.*

niobium telluride. Any of the following compounds of niobium and tellurium used in ceramics, refractories and semiconductors: (i) *Niobium monotelluride.* Crystalline powder. Crystal system, hexagonal; melting point, 1650°C (3000°F) Formula: $NbTe$ ($CbTe$). Also known as *columbium telluride;* and (ii) *Niobium ditelluride.* Crystalline powder. Crystal system, hexagonal. Density, 7.6 g/cm³. Formula: $NbTe_2$ ($CbTe_2$). Also known as *columbium ditelluride.*

niobium tetrachloride. Violet-black crystals. Crystal system, monoclinic. Density, 3.2 g/cm³. Used in ceramics, materials research, in the preparation of niobium compounds, and as an intermediate. Formula: $NbCl_4$ ($CbCl_4$). Also known as *columbium chloride; columbium tetrachloride; niobium chloride.*

niobium tetrafluoride. Black, hygroscopic crystals. Crystal system, tetragonal. Density, 4.01 g/cm³; melting point, decomposes above 350°C (660°F). Used in ceramics, materials research, in the preparation of niobium compounds, and as an intermediate. Formula: NbF_4 (CbF_4). Also known as *columbium fluoride; columbium tetrafluoride; niobium fluoride.*

niobium tetraiodide. Gray crystals. Crystal system, orthorhombic. Density, 5.6 g/cm³; melting point, 503°C (937°F). Used ceramics, materials research, in the preparation of niobium compounds, and as an intermediate. Formula: NbI_4 (CbI_4). Also known as *columbium iodide; columbium tetraiodide; niobium iodide.*

niobium-tin. A compound of niobium (columbium) and tin with a superconductivity critical temperature of 18.3K, used for superconductor magnets in communication equipment, and in the containment of plasmas in thermonuclear fusion reactors. Formula: Nb_3Sn (Cb_3Sn). Also known as *columbium-tin; niobium stannide; columbium stannide.*

niobium-titanium. An alloy of 52-56% niobium (columbium) and 44-48% titanium with a melting point of approximately 2200°C (3990°F). It is also available as a superconducting composite wire consisting of individual superconducting filaments of niobium (columbium) and titanium in a copper matrix, or as a single filament of niobium-titanium clad with copper. Used for magnetic devices with fields up to 100 kG, and for superconductor applications. Also known as *columbium-titanium.*

niobium trialuminide. See niobium aluminide (i).

niobium tribromide. Dark brown solid used ceramics, materials research, in the preparation of niobium compounds, and as an intermediate. Formula: NbI_4 (CbI_4). Also known as *columbium bromide; columbium tribromide; niobium bromide.*

niobium trichloride. A black solid with a density of 4.2 g/cm³, used in ceramics, materials research, in the preparation of niobium compounds, and as an intermediate. Formula: $NbCl_3$ ($CbCl_3$). Also known as *columbium chloride; columbium trichloride; niobium chloride.*

niobium trifluoride. Blue crystals. Crystal system, cubic. Density, 4.2 g/cm³. Used in ceramics, materials research, in the preparation of niobium compounds, and as an intermediate. Formula: NbF_3 (CbF_3). Also known as *columbium fluoride; columbium trifluoride; niobium fluoride.*

niobium trioxide. A compound of niobium and oxygen with a melting point of 1775°C (3227°F), used in ceramics and for high-temperature applications. Formula: Nb_2O_3 (Cb_2O_3). Also known as *columbium trioxide.*

niobium trisilicide. See niobium silicide (ii).

niobium trisulfide. See niobium sulfide (ii).

niobium-tungsten. An alloy of 84% niobium, 11% tungsten, 3% molybdenum and 2% hafnium. It has high strength at temperatures of up to 1093°C (2000°F), and is used for instruments. Also known as *columbium-tungsten.*

niobium-uranium. An alloy of 80% niobium and 20% uranium with good strength retention and hardness at a temperature of 870°C (1598°F). Used as a nuclear fuel. Also known as *columbium-uranium.*

niobium-zirconium. A refractory alloy composed of 99% niobium and 1% zirconium. Density, 8.59 g/cm³; melting point, 2399°C (4350°F); hardness, 65-140 kgf/mm²; high strength at 1093°C (2000°F); good high-temperature mechanical properties. Used for high-temperature parts, and missile and aircraft components. Also known as *columbium-zirconium.*

niobo-eschynite. A red brown to dark brown mineral with a resinous luster of the columbite group. It is composed of calcium cerium rare-earth titanium niobium oxide, $(Ce,Ca,Nd,Ln)(Nb,Ti)_2O_6$. Crystal system, orthorhombic. Density, 5.04 g/cm³; refractive index, 2.32. Occurrence: USA (Alaska), Russian Federation.

niobophyllite. A chocolate brown mineral of the astrophyllite group composed of potassium iron niobium aluminum silicate, $(K,Na)_3(Fe,Mn)_6(Nb,Ti)_2(Si,Al)_8(O,OH,F)_{31}$. Crystal system, triclinic. Density, 3.42 g/cm³; refractive index, 1.760. Occurrence: Canada (Labrador).

niocalite. A lemon-yellow mineral of the wohlerite group composed of calcium niobium silicate, $(Ca,Nb)_4Si_2(O,OH,F)_9$. Crystal system, monoclinic. Density, 3.32 g/cm³; refractive index, 1.714. Occurrence: Canada (Quebec).

Nioloy. Trade name of Teledyne Ohiocast (USA) for a chilled cast iron containing 3% carbon, some nickel, and the balance iron. Used for rolls for steel mills.

Ni-O-Nel. Trade name of Inco Alloys International Inc. (USA) for highly corrosion-resistant wrought alloys containing 42% nickel, 19-21.5% chromium, 3-5.5% molybdenum, 1.7-2.25% copper, 0-1% titanium, 0.5% manganese, 0.1% aluminum, 0.03-0.05% carbon, 0.015% sulfur, and the balance iron. They have good resistance to strong chemicals and hot corrosive gases including hot sulfuric, nitric and phosphoric acids, and ammonium hydroxide. Used for chemical processing equipment, pollution control equipment, acid evaporators, pickling equipment, nuclear-fuel reprocessing equipment, piping for oil and gas wells, tank trucks, and welding electrodes for arc and gas welding.

Nioro. Trademark of GTE Products Corporation (USA) for a brazing alloy containing 82% gold and 18% nickel. Commercially available in the form of foil, flexibraze, powder, wire, extrudable paste, and preform, it has an eutectic temperature of 950°C (1740°F), a brazing temperature range of 950-1004°C (1742-1840°F), and excellent flow and wetting properties. Used for brazing iron, nickel and cobalt-base alloys, tungsten, molybdenum, copper, and Kovar.

Nioroni. Trade name of GTE Products Corporation (USA) for a brazing alloy containing 73.8% gold and 26.2% niobium. Commercially available in the form of foil, flexibraze, powder, wire, extrudable paste, and preform, it has a liquidus temperature of 1010°C (1850°F), and a solidus temperature of 980°C (1795°F). Used for brazing iron, nickel and cobalt-base alloys, tungsten, molybdenum, copper, and Kovar.

Niostan. Trade name for a superconductor alloy composed of a niobium (columbium) ribbon with a thin niobium (columbium) stannide coating, and used for computer equipment operating in subzero environments.

Nipeon. Trade name of Japanese Geon (Japan) for polyvinyl chlorides.

Niperm. Trade name of Vereinigte Edelstahlwerke (Austria) for a series of soft-magnetic high-permeability iron-nickel alloys.

Nipermag. Trade name of Cinaudagraph Corporation (USA) for a cast permanent magnet alloy containing 32% nickel, 12% aluminum, 0.4% titanium, and the balance iron. It has high permeability, and is used for loudspeakers, motors and generators.

Niphoplate. Trademark for an electroplated coating composed of a nickel-phosphorus alloy (87Ni-13P) with a corrosion resistance superior to that of electroless-nickel deposits and a wear resistance equal to that of hard-chromium coatings. The coating can be overcoated with gold for electrical-contact applications.

Nipigon. Trade name of Atlas Specialty Steels (Canada) for a cobalt-tungsten type high-speed steel (AISI type T5) containing 0.78% carbon, 0.3% silicon, 0.25% manganese, 18.5% tungsten, 8% cobalt, 4.25% chromium, 1.9% vanadium, 0.85% molybdenum, and the balance iron. It has good deep-hardening properties, excellent red hardness and wear resistance, low toughness and poor machinability. Used for heavy-duty lathe and planer tools, boring tools, form cutters, and dies.

Niplon. Trade name for nylon fibers and resins.

Nipol. Trade name of Japanese Geon (Japan) for butadiene rubber.

Nipolon. Trade name of Nippon Plastics Company (Japan) for phenol-formaldehyde and high-density polyethylene plastics.

Niposit. Trade name of Shipley Ronal (USA) for electroless nickel.

nipple brass. A leaded brass containing 63% copper, 35% zinc and 2% lead, usually supplied in wire form.

Nippon. Trade name of Nippon Stainless Steel Company Limited (Japan) for a series of austenitic, ferritic and martensitic stainless steels.

Nippon Evershining Steel. Trade name of Nihon Jyokiko Seikosho Goshi (Japan) for austenitic stainless steel containing 0.15% carbon, 18% chromium, 8% nickel, and the balance iron. Used for stainless parts.

Niproteq. Trade name of Atotech USA Inc. for bright nickel electrodeposits and plating processes.

Nipure. Trade name of Wilbur B. Driver Company (USA) for a vacuum-melted high-purity nickel (99.77+%) containing up to 0.05% iron, 0.05% chromium, 0.04% copper, 0.02% manganese, 0.01% silicon, 0.01% magnesium, 0.01% aluminum, 0.01% titanium, 0.01% lead, and 0.02% carbon. Used in electronics.

Nirabond. Trade name of Niranium, Division of CMP Industries, LLC(USA) for a corrosion-resistant beryllium-free nickel-chromium dental bonding alloy.

Niranium. Trademark of CMP Industries, Inc. (USA) for a heat-resistant alloy of 64.2% cobalt, 28.8% chromium, 4.3% nickel, 2% tungsten, 0.7% aluminum, 0.2% carbon and 0.1% silicon. Used for high-temperature applications. Also included under this trademark are corrosion-resistant chromium-cobalt and nickel-chromium dental casting alloys.

Nircord. Trademark of Nirlon Limited (India) for nylon 6 filament yarns used for tire cords and other industrial applications.

Ni-Resist. Trade name of Fahralloy Company (USA) for a series of austenitic high-nickel cast irons containing 2.7-3.1% carbon, 12-36% nickel, 1-1.5% manganese, 1-3% silicon, 4-8% copper, 1.25-4% chromium, and the balance iron. They have excellent heat and corrosion resistance, good resistance to high-temperature scaling and growth, high toughness and thermal shock resistance, an upper service temperature of 950°C (1740°F), and low coefficients of expansion. Used for chemical plant equipment, food-processing equipment, pumps, filter presses, valves, pipes, fittings, automotive engine pistons and sleeves, bearings, hydraulic turbines, turbocharger casings, manifolds, impellers, propellers, nozzles, oil burners, gages, glass and ingot molds, paper rolls, switchgear, furnace parts, stove tops, and cookware.

Nirester. Trademark of Nirlon Limited (India) for polyester filament yarns.

Niresult. Trade name of Michigan Steel Casting Company (USA) for heat- and corrosion-resistant cast irons containing about 1.1% carbon, 24-30% chromium, 12-15% nickel, 5-7% copper, and the balance iron.

Nirex. Trade name of Harrison Alloys Inc. (USA) for a strong, ductile, stainless, heat-resistant nickel-base alloy containing 11-15% chromium and 0-15% iron.

Nirlon. Trademark of Nirlon Limited (India) for nylon 6 filament yarns.

Niromet. Trade name of Wilbur B. Driver Company (USA) for a series of controlled expansion alloy containing 52-64% iron and 36-48% nickel. They have high electrical resistivity, and are used for bimetals, precision springs, time devices, glass-to-metal seals, leads and terminals for resistors, terminal bands in vitreous enameled resistors, cores and armature for relays, motors and transformers, and fiberoptics.

Niron. (1) Trade name of Wilbur B. Driver Company (USA) for a series of controlled expansion alloy containing 51-54% iron and 46-49% nickel. Used chiefly for glass-to-metal seals.

(2) Trade name of Enthone-OMI Inc. (USA) for a nickel-iron electrodeposits and plating processes.

Nirosad. Trade name of Metalltechnik Schmidt GmbH & Co. (Germany) for an austenitic stainless steel having a microstructure with controlled amounts of *martensite*. It is composed of 0.1% carbon, 18% chromium, 10.5% nickel, 1.1% manganese, 1% silicon, 0.3% molybdenum, and the balance iron. Used as abrasive-blast cleaning shot.

Nirosta. Trade name of Krupp Stahl AG (Germany) for an extensive series of austenitic, ferritic, martensitic and precipitation-hardenable stainless steels used for cutlery, household appliances and articles, residential and laboratory sinks, chemical equipment, food-processing equipment, surgical instruments, and other corrosion-resistant items.

Nisat. Trade name of Plating Process Systems Inc. (USA) for a satin nickel electroplate and plating process.

nisbite. A white mineral of the marcasite group composed of antimony nickel, $NiSb_2$. It can also be made synthetically. Crystal system, orthorhombic. Density, 8.01 g/cm³. Occurrence: Canada (Ontario).

Nishikalon. Trade name for vinyon (vinyl chloride) fibers with excellent acid, alkali and mildew resistance. Used especially for textile fabrics.

Nishimura catalyst. A black powder composed of rhodium sesquioxide (Rh_2O_3) and platinum dioxide (PtO_2) with a rhodium-to-platinum ratio of 3:1, and used as a catalyst. Also known as *rhodium-platinum oxide*.

Nisiloy. Trade name of International Nickel Inc. (USA) for a nickel alloy containing 30% silicon and 10% iron. It has a melting point of 982°C (1800°F), and is used as an iron inoculant, and in the manufacture of gray iron castings.

Ni-Span C. Trademark of Inco International Alloys Inc. (USA) for an age-hardening, wrought nickel alloy containing 42-42.25% nickel, 48.5-50% iron, 2.4-2.6% titanium, 5.3-5.5% chromium, 0.5-0.8% silicon, 0-0.7% aluminum and small additions of silicon, copper, carbon and sulfur. It has a low temperature coefficient (zero in the temperature range between 0 and 100°C, or 32 and 212°F), a constant modulus of elasticity, relatively good weldability, and high strength. It is ferromagnetic up to 204°C (400°F). Used for Bourdon tubes, precision springs, diaphragms, thermostats, bimetals, tuning forks, and resonators.

Nistan. Trade name of Atotech USA Inc. for a tin-nickel alloy electrodeposit and plating process.

Nital. Trade name of CCS Braeburn Alloy Steel (USA) for an oil-hardening hot-work steel containing 0.27% carbon, 3% chromium, 0.2% vanadium, 10.5% tungsten, 0.2% molybdenum, 1.5% nickel, and the balance iron. Used for brass extrusion dies.

Nitec. Trade name of Heatbath Corporation (USA) for low-, medium- and high-phosphorus electroless nickel.

Ni-Tek. Trade name of Ultimate Wireforms (USA) for a nickel-titanium shape-memory wire used for dental applications.

Ni-Tensyl. Trade name of Sheepbridge Engineering Limited (UK) for a nickel cast iron containing 2.8% carbon, 1.5-2% nickel, 1.25-1.75% silicon, 0-0.35% chromium, and the balance iron. It has high strength, good hardness, and good abrasion and wear resistance. Used for flywheels, gears, or wheels.

niter. A colorless mineral of the aragonite group composed of potassium nitrate, KNO_3. Crystal system, orthorhombic. Density, 2.10; hardness, 2 Mohs; refractive index, 1.505. See also potassium nitrate.

niter cake. See sodium bisulfate.

NiTerne. Trade name for a duplex-coated, cold-rolled sheet with a highly corrosion-resistant flash coat of nickel electrodeposited prior to hot dipping in a *terne* bath of 85-97% lead and 3-15% tin. It has improved resistance to pinhole rusting, and enhanced surface wettability. Used for automobile fuel tanks, fuel filter cans, and electrical condenser cans.

NiTi. Trademark of ORMCO Corporation (USA) for a shape-memory wire composed of 52.4% nickel and 47.6% titanium, and used for dental and orthodontic applications.

Nitinol. (1) Trade name of TiNi Alloy Company (USA) for non-magnetic, corrosion-resistant alloys composed of 53-60% nickel and 40-47% titanium, and exhibiting the shape-memory effect. They have good tensile strength, good electrical and mechanical properties, a long fatigue life, and can be hardened by heat treatment to high hardnesses. Used for chemical plant equipment, spacecraft components, sensing devices, nonmagnetic tools and instruments, and biomedical and dental applications, e.g., orthodontic and orthopedic wire. See also shape-memory alloys.

(2) Trademark of 3M Unitek (USA) for a shape memory alloy composed of 52% nickel, 45% titanium and 3.0% chromium, and used for orthodontic wire. *Nitinol SE* orthodontic wire contains 52.8% nickel and 47.2% titanium.

Nitinol-60 Balls. Trade name of Abbott Ball Company (USA) for balls made of *Nitinol* (60Ni-40Ti). They have high hardness, strength, impact strength and corrosion resistance, and are used for ball bearings and races.

Nitiray. Trade name for nylon fibers and yarns used for textile fabrics including clothing.

Niti-Vilon. Trademark of Nitivy Company (Japan) for vinal (polyvinyl alcohol) fibers used for textile fabrics.

Nitlon. Trade name for acrylic fibers used for the manufacture of textile fabrics.

Nitral. Trade name of Acieries Nouvelle de Pompey (France) for low- and medium-carbon chromium-molybdenum alloy steels.

Nitralloy. Trade name of Nitralloy Corporation (USA) for a nitriding steel containing 0.2-0.5% carbon, 0.4-0.7% manganese, 0.2-0.4% silicon, 0.7-1.5% aluminum, 0-1.8% chromium, 0.15-1.0% molybdenum, 0-1.2% copper, 0-3.5% nickel, and the balance iron. After nitriding, it has a hard case and a tough core. Used for tools, gages, gears, and shafts.

Nitrard. Trade name for a nitriding steel.

Nitrasil. Trademark of Tenmat Inc. (USA) for a range of silicon nitride (Si_3N_4) products. *Nitrasil R* is a reaction-bonded silicon nitride with a density of 2.4 g/cm^3 (0.09 $lb/in.^3$), an upper continuous-use temperature of 1200-1500°C (2190-2730°F), good strength and oxidation resistance, good wear properties, and good resistance to various chemicals and molten metals. Used for gas-turbine parts, supports, and automotive applications.

nitratine. A colorless mineral of the calcite group composed of sodium nitrate, $NaNO_3$. Crystal system, rhombohedral (hexagonal). Density, 2.26 g/cm^3; refractive index, 1.608.

Nitray. Trademark of Winter & Co. GmbH (Germany) for artificial leather and leather substitutes.

Nitrelmang. Trademark of Foote Mineral Company (USA) for a high-purity nitrided manganese used in the manufacture of free-cutting steels, certain high-temperature alloy steels, and high-nitrogen tinplate.

Nitrex. Trade name of Carpenter Technology Corporation (USA) for a series of nitriding steels used for camshafts, cams, gears, cylinder liners, and thread guides. *Nitrex I* is a wear- and abrasion-resistant nitriding steel containing 0.38-0.45% carbon, 1.4-1.8% chromium, 0.6% manganese, 0.3% silicon, 0.85-1.2% aluminum, 0.3-0.45% molybdenum, and the balance iron. *Nitrex II* is a precipitation-hardening nitriding steels containing 0.2-0.27% carbon, 0.6% manganese, 1-1.5% chromium, 0.8-1.2% aluminum, 3.25-3.75% nickel, 0.3% silicon, 0.2-0.3% molybdenum, and the balance iron.

Nitricast. Trade name of Certified Alloy Products Inc. (USA) for a nitrided cast iron containing 2.5% carbon, 2.5% silicon, 1% chromium, 1% aluminum, 0.2% molybdenum, and the balance iron. Used for oilwell tooling, sleeves, liners, and cams.

Nitricastiron. Trade name of Nitralloy Corporation (USA) for a wear-resistant nitrided cast iron containing 2.75% total carbon, 0.6-0.9% combined carbon, 1.9% graphitic carbon, 2.6-2.7% silicon, 1.2-1.3% chromium, 0.16% vanadium, 0.25% molybdenum, 1% aluminum, traces of sulfur and phosphorus, and the balance iron. It has a high case hardness, used for valves, cylinders, and cams.

nitrided cast iron. A cast iron on whose surface a very hard, wear-resistant nitride-containing case has been produced by heating in a nitrogen atmosphere (e.g., ammonia gas). A typical composition is 2.5-2.75% carbon, 2.5-2.7% silicon, 1-1.3% chromium, 1% aluminum, 0.2-0.3% molybdenum, and the balance iron. Used for oilwell tooling, sleeves, liners, valves, cylinders, and cams.

nitrided steels. Strong, tough low-carbon steels on whose sur-

faces a very hard nitride-containing case (about 0.4 mm, or 0.16 in. thick) has been produced by heating at 500-540°C (930-1000°F) in a nitrogen atmosphere (e.g., ammonia gas, molten cyanides) for about 50 hours. Used for valve stems, piston rods, worms, engine cylinders, crankshafts, camshafts, cams, piston pins, cylinder liners, gears, bushings, rolls, thread guides, gages, fittings, and wear-resistant parts. Also known as *nitriding steel.*

nitride fibers. A group of continuous or discontinuous ceramic fibers that are prepared by pyrolysis of polymers or chemical vapor deposition (CVD). Typical nitride fibers include continuous boron nitride fibers, and continuous and discontinuous silicon nitride fibers. *Nitride fibers* have high tensile strengths and moduli of elasticity, and are used as reinforcing fibers for engineering composites. Abbreviation: NF.

nitride fuel. A fissionable nuclear fuel based on uranium nitride (UN_2).

nitrides. Binary compounds of nitrogen and a more electropositive metal, such as aluminum, gallium, iron or titanium, that correspond to the general formula Me_xN_y.

nitriding steels. See nitrided steels.

nitrile. An organic cyanide containing a carbon-nitrogen triple bond represented by the general formula $RC \equiv N$. Examples include acrylonitrile ($H_2C=CHC \equiv N$) and acetonitrile ($CH_3C \equiv N$).

nitrile-butadiene latex. A viscous suspension of *acrylonitrile-butadiene rubber* in water. Abbreviation: NBL.

nitrile-butadiene rubber. See acrylonitrile-butadiene copolymer.

nitrile resins. A group of amorphous copolymers based on 70% acrylonitrile, 20-30% styrene or methyl methacrylate, and 0-10% butadiene. They have high toughness, excellent barrier properties, good thermal stability, and form viscous liquids above 200°C (390°F). Abbreviation: NR.

nitrile rubber. See acrylonitrile-butadiene copolymer.

nitrile-silicone rubber. A copolymer of acrylonitrile and silicone obtained when one or more of the side-bonded atoms or atom groups (e.g., hydrogen or methyl) in some molecular groups are replaced by one or more nitrile group ($-C \equiv N-$). It has high flexibility, good resistance to jet fuels, solvents, hot and cold oils, high resistance to degradation and swelling in oils and solvents, a service temperature range of -73 to +260°C (-100°F to +500°F). Used for gaskets, seals and O-rings, diaphragms, and chemical, gasoline and oil hose and tubing. Abbreviation: NSR.

Nitrilon. Trade name of Changshu Chunwei Textile Limited (China) for acrylic (polyacrylonitrile) fibers used for the manufacture of textile fabrics.

Nitrix. Trademark of Uniroyal Chemical Division of Uniroyal Inc. (USA) for nitrile rubber latexes used in leather and paper finishing.

Nitro. Trade name of Carpenter Technology Corporation (USA) for a water-hardening tool steel (AISI type W2) containing 1% carbon, 0.35% manganese, 0.25% silicon, 0.2% vanadium, and the balance iron. Used for drills, taps, cutters, and springs.

nitrobarite. A colorless mineral composed of barium nitrate, $Ba(NO_3)_2$. Crystal system, cubic. Density, 3.24 g/cm^3; refractive index, 1.571. Used in ceramics and electronics. See also barium nitrate.

Nitroblack. Trade name of Lexington Coating Technology Inc. (USA) for a nitrocarburizing/oxidizing corrosion- and wear-resistant coating for ferrous metals that produces a nitrided layer under a porous black coating of pure ferroferric oxide (Fe_3O_4).

nitrocalcite. A colorless mineral composed of calcium nitrate tetrahydrate, $Ca(NO_3)_2 \cdot 4H_2O$. Crystal system, monoclinic. Density, 1.90 g/cm³; refractive index, 1.498. Used in ceramics.

nitrocellulose. A nitric ester of cellulose made by treating cellulose with a mixture of nitric and sulfuric acid. It contains between 10 and 14% nitrogen, and only the high-nitrogen form (*guncotton*) is explosive. The low-nitrogen form (*pyroxylin*) is used for collodion, lacquers, varnishes, and plastics. The form used for plastics usually contains about 11% nitrogen and has a viscosity of about 20-40 centipoise. Density, 1.66 g/cm³; flash point, 12.8°C (55°F). Used as a binder in conductive and other coatings, in the manufacture of plastics and photographic film, in lacquers and varnishes, for high explosives and rocket propellant, and leather finishing compounds. Also known as *cellulose nitrate; nitrocotton*.

nitrocellulose lacquer. An extremely fast-drying lacquer comprised of high- or low-viscosity nitrocellulose resin (nitrogen content, 11-13.5%), a plasticizer (e.g., dibutyl phthalate or blown castor oil), and a solvent (e.g., ethanol, toluene or xylene). It is often modified with other resins, such as alkyds, ester gum, or rosin. Alkyd-resin-modified nitrocellulose lacquer has improved durability. *Nitrocellulose lacquer* has good hardness and abrasion resistance, good to fair exterior durability, and is used for protective coatings on metallic parts (e.g., autobodies), paper products, plastics and textiles, as touch-up and aerosol lacquers, and as furniture finishes.

nitrocellulose rayon. Rayon made by denitration of cellulose nitrate fibers. See cuprammonium rayon; rayon fiber; viscose rayon.

Nitrochrome. Trade name of Shieldalloy Metallurgical Corporation (USA) for a series of master alloys containing 64-65% chromium, 3-6% nickel, 0-1.5% silicon, up to 0.1% carbon, and the balance iron. Used as ladle additions for ferrous alloys.

nitrocotton. See nitrocellulose.

Nitrodur. Trade name of Thyssen Edelstahlwerke AG (Germany) for a series of nitriding steels typically containing 0.27-0.33% carbon, 1.1-2.3% chromium, 0-1% nickel, 0-1.1% aluminum, 0-0.2% molybdenum, 0-0.1% vanadium, and the balance iron. Used for valve stems, worms, gears, bushings, crankshafts, camshafts, cams, and wear-resistant parts.

Nitrofil. Trade name of IMI Rod & Wire (UK) for a copper alloy containing 0.2-0.3% titanium and 0.2-0.3% aluminum. It has a melting point of 1080°C (1976°F), and used for filler wires for nitrogen-arc and inert-gas shielded arc welding of copper.

nitrogen. A colorless, tasteless, odorless, diatomic, gaseous element of Group VA (Group 15) of the Periodic Table. It occurs in the earth's atmosphere to the extent of 78.0 vol% or 75.5 wt%, and is also a constituent of nitrates, ammonium salts, soil and all living matter. Density, 1.251 g/L (0°C or 32°F); freezing point, -210°C (-346°F); boiling point, -195.5°C (-320°F); atomic number, 7; atomic weight, 14.007; monovalent, divalent, trivalent, tetravalent, pentavalent. Used in gas-filled electric lamps, in the electric and electronic industries, as a refrigerant, for cryogenic applications, in rocket fuels, in gas thermometry, as a diluent in combustion processes, for chilling in aluminum foundries, in bright annealing of steel, for inflating tires, in the manufacture of various chemicals (e.g., ammonia, acrylonitrile, cyanamide, cyanides, nitrides, etc), in explosives, as an inert gas for purging, blanketing and exerting pressure, and in biochemistry, biotechnology and medicine. Symbol: N.

nitrogen-15. A stable isotope of nitrogen with a mass number of 15.00011 that occurs in natural nitrogen to the extent of 0.365%.

Used as a tracer. Symbol: ^{15}N.

nitrogen-strengthened stainless steels. A group of special austenitic stainless steels in which manganese has been substituted for all or part of the nickel to allow greater amounts of nitrogen to be dissolved in the matrix. Nitrogen increases the annealed strength at cryogenic and elevated temperatures and considerably improves the resistance to pitting corrosion.

nitromagnesite. A colorless to white mineral composed of magnesium nitrate hexahydrate, $Mg(NO_3)_2 \cdot 6H_2O$. Crystal system, monoclinic. Density, 1.58 g/cm³; refractive index, 1.506. Occurrence: USA.

Nitron. (1) Trade name of Monsanto Chemical Company (USA) for cellulose nitrate plastics.

(2) Trademark of Kaneka Corporation (Japan) for acrylic fibers used for the manufacture of textile fabrics.

Nitronic. Trade name of Armco International (USA) for a series of austenitic stainless steels containing 0.06-0.40% carbon, 15.0-24.0% chromium, 0.5-14% nickel, 1.5-14.0% manganese, 0-1.0% silicon, 0.15-0.40% nitrogen, traces of molybdenum, phosphorus, sulfur and vanadium, and the balance iron. Commercially available in the form of bars, rods, billets, wires, sheets, strip, pipes and tubing, they have excellent mechanical strength and corrosion resistance, good high-temperature properties, good resistance to oxidation and chemicals, good wear and galling resistance, and very low magnetic permeability. Used for pole-line hardware, springs, cold-headed parts, heat exchangers, chemical equipment, aircraft exhaust systems, cryogenic applications, and for equipment for the petrochemical, food-processing, and pulp and paper industries.

NitroSil. Trademark of Pyromatics, Inc. (USA) for high-purity, high-density opaque *quartz* (SiO_2) products used for chemical vapor deposition (CVD), epitaxy and heat shielding applications.

nitroso ester terpolymer. A flame- and heat-resistant terpolymer of trifluoronitrosomethane, tetrafluoroethylene and methyl-4-nitrosoperfluorobutyrate crosslinked by peroxides. It has a glass-transition temperature of approximately -50°C (-58°F), a decomposition temperature of 275°C (527°F), high tensile strength, a very high elongation of 500% or more, and a durometer hardness of 78 Shore. Used for fireproof interiors of aircraft and space vehicles. Also known as *nitroso rubber*. Abbreviation: NET; AFMU.

nitroso polymer. A copolymer of trifluoronitrosomethane (CF_3-NO) and tetrafluoroethylene ($F_2C=CF_2$) having excellent flame and heat resistance. See also nitroso ester terpolymer.

nitroso rubber. See nitroso ester terpolymer.

Nitro Special Vanadium. Trade name of Carpenter Technology Corporation (USA) for a water-hardening tool steel (AISI type W2) containing 0.60-1.40% carbon, 0.1-0.4% manganese, 0.1-0.4% silicon, up to 0.15% chromium, up to 0.20% nickel, 0.25-0.35% vanadium, and the balance iron. It has a good machinability, good to moderate deep-hardening properties, good wear resistance and toughness, and low hot hardness. Used for cutting tools, such as milling cutters, planer tools, drills, and other tools.

NitroSteel. Trademark of MacSteel Bar Group, Quanex Corporation (USA) for carbon and alloy steel bars and tubes surface hardened to 64-71 Rockwell C by a proprietary, combined nitriding/oxidizing treatment known as "Nitrotec". Supplied in several diameters and lengths, *NitroSteel* possesses excellent corrosion resistance, good wear- and dent-resistance, good resistance to surface pitting, flaking and hydrogen embrittlement.

Used for automotive shafts, hydraulic and piston rods.

2-nitrothiophene. A light-sensitive organic compound available in technical grades (85% pure) with a melting point of 43-45°C (109-113°F), a boiling point of 224-225°C (435-437°F), and a flash point of 201°F (93°C). Used in organic synthesis, e.g., for organic conductors.

Nittany. Trade name of Cerro Metal Products Company (USA) for a series of low-, medium- and high-leaded and free-cutting brasses typically containing 61-64% copper, 0-3.5% lead, and the balance being zinc. Used especially for high-speed screw-machine products, hardware, fasteners, and bolts.

Nitto. Trade name of NITTO Deutschland GmbH (Germany) for epoxy potting and encapsulating compounds for electronic components and epoxy casting resins for the manufacture of printed circuits.

Nittobo. Trade name of Mitsubishi Rayon Company Limited (Japan) for rayon fibers and yarns used for textile fabrics.

Nituf. Trade name of CCS Braeburn Alloy Steel (USA) for an air-hardening die and tool steel (AISI type A9) containing 0.5% carbon, 0.4% manganese, 1% silicon, 5.25% chromium, 1.5% nickel, 1% vanadium, 1.35% molybdenum, and the balance iron. Used for dies, die inserts, punches, mandrels, and hammers.

Nituff. Trademark of Nimet Industries Inc. (USA) for a polytetrafluoroethylene (*Teflon*) penetrated aluminum hardcoat produced by anodizing. It firmly bonds to the aluminum substrate, and thus resists peeling and chipping. The coating provides a clean, file-hard, dry-lubricating surface with low friction and high resistance to corrosion, abrasion and galling.

Nitung. Trade name of Pennsylvania Steel Corporation (USA) for an oil-hardening hot work tool steel containing 0.3% carbon, 2.75% chromium, 0.3% molybdenum, 9.5% tungsten, 1.6% nickel, and the balance iron. Used for hot-work and extrusion tools.

Nivac. Trade name of Crucible Materials Corporation (USA) for a series of nickel products including several high-purity nickels containing 0.007% carbon, and several high-permeability magnet alloys containing 50-80% nickel and 20-50% iron.

Nivaflex. Trade name of Vacuumschmelze GmbH (Germany) for a heat- and corrosion-resistant alloy of 45% cobalt, 21% nickel, 18% chromium, 6.7% iron, 4% molybdenum, 4% tungsten, 1% titanium, 0.3% beryllium and 0.03% carbon. Used for high-temperature springs.

Nivar. Trade name for a low-expansion alloy containing varying amounts of nickel and iron. Used for watch parts, instruments, and machine parts requiring low coefficients of thermal expansion.

Nivarox. Trade name of Vacuumschmelze GmbH (Germany) for a precipitation-hardening constant-modulus alloy containing 54% iron, 37% nickel, 8% chromium, 0.85% manganese, 0.90% beryllium, 1% titanium, 0.20% silicon and 0.10% carbon. Used for instrument springs, high-accuracy watch and clock springs, and diaphragms.

Nivco. Trademark of Westinghouse Electric Corporation (USA) for a precipitation-hardening cobalt-base alloy containing 22.5% nickel, 1.8% titanium, 1-1.1% zirconium, 0-1% iron, 0-0.35% manganese, 0-0.2% aluminum, 0-0.15% silicon, and 0.02-0.05% carbon. It has good strength-to-weight ratio, and good heat resistance and strength up to 650°C (1202°F). Used as a damping material for steam turbine blade applications.

Ni-Vee. Trade name of Olds Alloy Company (USA) for a series of copper-base alloys containing 5% nickel, 5% tin, 0-5% zinc

and 1-20% lead. They have good corrosion resistance, and good to moderate strength. Used for gears, cams, rollers, guides, valves, pump castings, fittings, plumbing parts, bearings and bushings.

Nivion. Trademark of LIBA SpA (Italy) for nylon 6 fibers and yarns used for textile fabrics.

Nivonplast. Trade name of Enichem SpA (Italy) for polyamide (nylon) resins. *Nivonplast A* is a polyamide 6,6 (nylon 6,6) supplied in unmodified, glass-fiber-reinforced and mineral-filled grades, and *Nivonplast B* is a polyamide 6 (nylon 6) supplied in unmodified, glass-fiber-reinforced, elastomer-copolymer, and mineral-filled grades.

Ni-Weld. Trade name of Crucible Materials Corporation (USA) for a pure nickel used in the form of welding rods for cast irons.

Nixonite. Trade name of Nixon Nitration Works (USA) for a cellulose acetate materials.

Nixonoid. Trade name of Nixon Nitration Works (USA) for celluloid-type cellulose nitrate plastics.

Nizec. Trade name of British Steel plc (UK) for a low-carbon steel electrolytically coated with a zinc-nickel alloy.

NiZn-Cote. Trade name of Thomas Steel Strip Corporation (USA) for an electrolytic nickel-zinc coating for steel.

NK Hiten. Trade name of NKK Corporation (Japan) for a series of structural steel plate products with excellent weldability and cold-cracking resistance, and good low-temperature properties.

NK Super BR-Core. Trade name of NKK Corporation (Japan) for high-silicon magnetic sheet products. They have ultralow core losses, low residual magnetic flux density, and are used for electrical and magnetic applications.

NLO crystal. See nonlinear optical crystal.

NLO materials. See nonlinear optical materials.

no-bake binder. A liquid sand binder composed of a phenol-formaldehyde and a polyurethane resin (e.g., methylene diphenyl isocyanate) and a special liquid catalyst. Curing is achieved at room temperature (i.e., without baking), upon addition of the catalyst to the sand-resin mixture. Used in foundries for core and molding making.

Nobelex. Trademark of Istrochem (Slovak Republic) for polypropylene fibers and yarns used for textile fabrics.

nobelium. A synthetic, radioactive element of the actinide series of the Periodic Table. Atomic number, 102, atomic weight, 259. Isotopes with mass numbers of 254-258 have also been prepared. Symbol: No.

Nobellon. Trade name of Nobell Plastics Company (USA) for phenol-formaldehyde resins and plastics.

Nobeloy. Trade name for a corrosion-resistant gold-base jewelry alloy.

Nobestos. Trademark of Lydall, Inc. (USA) for a series of fiber-elastomer, non-asbestos gasket materials and millboards containing inorganic and organic binders for optimum cutting and sealing characteristics and temperature resistance. The gasket materials have good temperature resistance up to 400°C (750°F), and the millboards up to 1260°C (2300°F). *Nobestos* is available in the form of sheets and rolls, and used in the manufacture of exhaust and fuel system gaskets, engine gaskets, and jacketed heat exchanger gaskets.

Nobil-Ceram. Trademark of Jeneric/Pentron Inc. (USA) for a corrosion-resistant beryllium-free nickel-chromium dental bonding alloy.

Nobilium. Trade name of Nobilium Company (USA) for a corrosion-resistant cobalt-base dental casting alloy containing 28%

chromium, 5% molybdenum, 1% vanadium, 0.5% iron, 0.4% carbon, 0.1% nickel, 0.1% manganese, and 0.05% silicon.

Nobil Solder. Trade name of Nobilium Company (USA) for a dental gold solder.

Nobipor. Trademark of Nobis GmbH (Germany) for polystyrene foam and foam products.

noble fir. The light, moderately strong wood from the fir tree *Abies procera.* Source: Western United States. Used for lumber for interior finish, siding, moldings, aircraft construction and plywood veneer.

Noble J. Trade name for a high-gold dental bonding alloy.

noble metals. A term referring to the metals osmium, iridium, platinum and gold with atomic numbers 76, 77, 78 and 79, respectively, all of which are located in Period 6 of the Periodic Table, and the metals ruthenium, rhodium, palladium and silver with atomic numbers 44, 45, 46 and 47, respectively, which are in Period 5. Their standard electrode potentials are more positive (or noble) than that of hydrogen, and thus they are highly resistant to corrosion and oxidation. Used for electric contacts, electroplating, jewelry, thin-film circuits, metal-film resistors, in the synthesis of noble-gas fullerenes, etc. The terms *precious metals* and *noble metals* are not synonymous, although some precious metals may be noble as well. For example, gold has the characteristics of both noble and precious metals, whereas osmium is considered noble, but not precious. Abbreviation: NM.

nobleite. A colorless mineral composed of calcium borate tetrahydrate, $CaB_6O_{10} \cdot 4H_2O$. Crystal system, monoclinic. Density, 2.09 g/cm^3; refractive index, 1.520. Occurrence: USA (California).

Noblen. Trademark of Sumitomo Corporation (Japan) for polystyrene resins and plastics.

Noblen PP. Trademark of Sumitomo Corporation (Japan) for polypropylene homopolymers and copolymers supplied in standard, talc-filled, or UV-stabilized grades.

Nobricella. Trade name for rayon fibers and yarns used for textile fabrics.

no-carbon-required paper. See carbonless paper.

nocerite. See fluoborite.

No-Chat. Trade name of CMW Inc. (USA) for a sintered material of 90% tungsten, 6% nickel and 4% copper. It has density of 16.96 g/cm^3 (0.613 $lb/in.^3$), a high modulus of rigidity, an electrical conductivity of 14% IACS, and is used for cutoff tools, boring bars, arbors, and other tools.

Nocolok. Trademark of Solvay Fluor GmbH (Germany) for a noncorrosive torch and furnace aluminum brazing paste for temperatures over 560°C (1040°F). It contains a nonhygroscopic, nonreactive flux, which imparts a protective coating eliminating cleaning after brazing.

No-Clamp. Trademark of Dural Products Limited (Canada) for contact cement.

Nodular. Trademark for ductile cast iron.

nodular cast irons. See ductile cast irons.

nodular graphite. A form of graphite found in malleable irons and consisting of rounded temper carbon clusters.

nodular fireclay. A rock composed of aluminous and/or ferruginous nodules bonded by *fireclay.* Also known as *nodule clay.*

nodular irons. See ductile cast irons.

nodular powder. A metal powder composed of irregular, knotted or rounded particles.

nodule clay. See nodular clay.

Nodulite. Trade name of Hamilton Foundry (USA) for a series of strong, tough, wear-resistant ductile cast irons typically con-

taining 3.3% carbon, 2.5% silicon, 0.7% manganese, 0.05% magnesium, and the balance iron. Used for gears, cams, bearings, crankshafts, pistons, pump bodies, valves, dies, pipe fittings, and brake drums.

nodulizer. An agent or foundry alloy, such as unalloyed magnesium, a nickel-magnesium or magnesium-ferrosilicon alloy, or a cerium-bearing alloy, added to molten iron just before pouring to effect the formation of ball-shaped (nodular) or spheroidal graphite during solidification.

nodulizing alloy. See nodulizer.

Noduloy. Trade name of Cyprus Foote Mineral Company (USA) for a series of foundry alloys of the magnesium-ferrosilicon type, with or without cerium, added to molten cast iron to effect the formation of nodular or spheroidal graphite.

No-Du-Mag. Trade name of Ferranti Limited (UK) for an austenitic ductile (nodular) cast iron containing 3.3% carbon, 2.5% silicon, 5.6% manganese, 10-11% nickel, 0.17% magnesium, and the balance iron. It has good strength and corrosion resistance, and is used for switchgear; magnetic chucks, and corrosion-resistant grids.

no-fines concrete. A concrete containing little or no aggregate of less than 0.187 in. (4.76 mm) in diameter, and usually has a high proportion of communicating pores.

nogal. See tropical walnut.

Nogroth. Trade name of Empire Steel Castings Company (USA) for a high-nickel cast iron.

Noheet. Trade name of Ardal Limited (UK) for a lead-alkali metal of 98% lead, 1.4% sodium, with the balance being tin and antimony. It has good antifriction properties, and is used for bearings.

Noil. Trade name of Baker, Perkins & Company Limited (UK) for a bronze containing 80% copper and 20% tin. Used for piston rings.

noils. The short, broken fibers combed out during the manufacture of cotton, wool, or silk yarn. They can be carded and spun into coarse yarns, used in blends for the manufacture of yarns and fabrics, or made into felts. Also known as *comber noils.*

no-iron fabrics. Smooth-textured fabrics made from synthetic or synthetic/natural fiber blends with or without the use of resin finishes, and requiring little or no ironing or pressing after laundering.

Noise-Stop. Trade name of Owens-Corning Fiberglass Company (USA) for glass-fiber boards used for noise control and baffles.

nolanite. A black mineral composed of iron vanadium oxide hydroxide, $(V,Fe,Ti)_{10}O_{14}(OH)_2$. Crystal system, hexagonal. Density, 4.69 g/cm^3. Occurrence: Australia, Canada (Saskatchewan).

Nolex. Trade name of Nolex Engineering for insulating material made from finely divided synthetic *fluorine mica* using a hot-molding process. It has good dielectric properties, high dimensional accuracy, and good machinability. Used for molded parts for insulating appli-cations.

Nolibond. Trademark of Rhône-Poulenc SA (France) for a powdered adhesive resin based on reactive, low-molecular weight thermoplastic polyamide-imide (PAI). It has excellent high-temperature toughness, and is used for aerospace, automotive, industrial, electrical and electronic applications.

Nolicoat. Trademark of Rhône-Poulenc SA (France) for a liquid coating resin based on reactive, low-molecular weight thermoplastic polyamide-imide (PAI). It has excellent high-temperature toughness, and is used for aerospace, automotive, industrial, electrical and electronic applications.

Nolimold. Trademark of Rhône-Poulenc SA (France) for a liquid molding resin based on reactive, low-molecular weight thermoplastic polyamide-imide (PAI).

Noll Special. Trade name of Carpenter Technology Corporation (USA) for a series of water-hardening tool steels (AISI type W1) containing 1.05% carbon, 0.2% manganese, 0.2% silicon, and the balance iron. Used for hand tools, cutting tools, and dies.

Nomag. Trade name of Mond Nickel Company Limited (UK) for a nonmagnetic, austenitic cast iron containing 3% carbon, 5-6% manganese, 10-12% nickel, 2-3% silicon, and the balance iron. It has high electrical resistance, and is used for switch covers and resistance grids.

Nomar. Trademark of Nomar Canvas Products (USA) for a series of coated polyester and polycarbonate films providing good abrasion and chemical resistance.

Nomelle. Trade name for acrylic fibers used for the manufacture of textile fabrics.

Nomex. (1) Trademark of E.I. DuPont de Nemours & Company (USA) for polyaramid fibers, yarns and fabrics with low density (only about 0.9 g/cm^3 or 0.03 lb./in.^{3}), good tensile strength, high toughness, self-extinguishing properties, good dielectric properties, low coefficients of friction, low thermal expansivities, an upper service temperature of 180-250°C (355-480°F), good resistance to dilute acids, alcohols, alkalies, aromatic hydrocarbons, greases, oils and ketones, and fair resistance to concentrated acids. Used for bag filters for industrial dust control, for pressboard, insulation of aerospace vehicles, and in the manufacture of core materials for advanced honeycomb structures. *Nomex Pressboard* is an insulation material with a continuous-use temperature of 220°C (430°F), made of 100% aramid fibers. Used for mold, platen, and wire insulation for vacuum forming and injection molding equipment, pelletizer heat shields, and fire barriers for aircrafts.

(2) Trademark of E.I DuPont de Nemours & Company (USA) for a series of nylon fibers.

Nomichi. Trade name of Asahi Glass Company Limited (Japan) for a glass with irregular quadrangular shapes.

nonacarbonyliron. See iron nonacarbonyl.

Non-Actinic. Trade name of Pilkington Brothers Limited (UK) for a rolled glass of soft greenish tint. It is opaque to ultraviolet light, and greatly reduces fading.

nonactive fluxes. A group of rosin-type soldering fluxes dissolved in alcohol or turpentine. They are only used on highly solderable surfaces. See also rosin.

Non-Chrome Blue. Trade name of Pavco Inc. (USA) for a bright blue chromate coating and process that uses no trivalent or hexavalent chromium.

noncoated paper. A paper whose surface is not coated but may be treated and/or pigmented.

noncombustible. A solid, liquid or gaseous material that will neither ignite nor actively support combustion. Examples include silicon dioxide, polyimide, water, and carbon dioxide. Also known as *noncombustible material.*

noncombustible fabrics. Textile fabrics treated to prevent both ignition (combustion) and emission of combustible vapors when exposed to external ignition sources.

noncombustible material. See noncombustible.

nonconductor. See insulator.

nonconiferous wood. See hardwood.

Noncorrodible Aluminum. Trade name of Haywoods NCA Metal Limited (UK) for wrought and cast aluminum alloys containing 3.0% copper, 2.2% nickel, 0.2% manganese and 0.2% nitrogen. They have excellent seawater corrosion resistance, and are used for marine parts, and corrosion-resistant parts. Abbreviation: NCA.

noncorrosive fluxes. See activated rosin fluxes.

noncrystalline material. A solid or liquid that has no long-range atomic order. Examples include silicate glasses, amorphous metals, and most plastics. It is sometimes also referred to as an *amorphous material, glassy material,* or *vitreous material.*

Non Curl. Trademark of Form-Mate Carbon Products Limited (Canada) for carbon paper.

nondeforming steels. A group of tool steels, usually of the oil hardening-type, which greatly resist deformation when heat-treated. The typical composition range is 0.5-1.6% carbon, 1-1.4% manganese, 0.5% silicon, 0-0.5% chromium, 0-0.6% tungsten, and the balance iron. They possess good machinability, but only fair to poor shock resistance. Used for dies, gages, and precision tools. Also known as *nonshrinking steels.*

nonductile material. See brittle material.

Non Dulling. Trade name of L.B. Allen Corporation (USA) for stainless steel solders used for decorative sanitary work.

nonenyl succinic anhydride. A yellowish liquid (85+% pure) used as an epoxy hardener and in electron microscopy. Formula: $C_{13}H_{20}O_3$. Abbreviation: NSA. Also known as *(2-nonen-1-yl)succinic anhydride.*

Nonerode. Trade name of Coulter Steel & Forge Company (USA) for an oil-hardening hot-work tool steel (AISI type H11) containing 0.35% carbon, 1.5% molybdenum, 5% chromium, 0.4% vanadium, and the balance iron. It has excellent ductility, good machinability, and good high-temperature strength up to 540°C (1004°F). Used for hot-work tools, punches and die-casting dies.

Nonferal. Trade name for a finely divided precipitated calcium carbonate used as a pigment extender.

nonferromagnetic materials. A group of materials which exhibit no permanent magnetism. Included are paramagnetic materials (e.g., aluminum, magnesium, platinum, and tin) and diamagnetic materials (e.g., copper, gold, silver, and zinc). Abbreviation: NFMM.

nonferrous alloys. A generic term referring to (i) alloys neither containing nor derived from iron or iron-base alloys, and (ii) alloys that have as their predominant base a metal other than iron. For example, an alloy containing 55% nickel and 45% iron is a nonferrous alloy, while an alloy of 60% iron and 40% nickel is a ferrous alloy. Abbreviation: NFA. See also nonferrous metals.

nonferrous metals. A generic term for metals that contain no iron, e.g., aluminum, copper, zinc or lead. They are subdivided according to density into *light metals* and *heavy metals.* Light metals (e.g., aluminum, beryllium or magnesium) have densities of less than about 4-5 g/cm^3 (0.14-0.18 lb/in^3), while heavy metals (e.g., chromium, cobalt, copper, lead, nickel or tungsten) have higher densities. Abbreviation: NFM. See also nonferrous alloys.

nonfibrillating fibers. Textile fibers that have been treated with enzymes or special chemicals to prevent their ends from peeling back, splintering and protruding as minute fibrils from the surface. In addition to enzyme and chemical treatments fibrillation can also be prevented by reducing moisture and agitation during dry cleaning. See also fibrils; fibrillated fibers; fibrillating fibers.

Non-Flam. Trademark of Coburn Corporation (USA) for nonflammable films.

Non-Flame. Trade name of Flame Control Coatings, Inc. (USA) for fire- and flame-retardant paints.

nonflammable fabrics. See flameproof fabrics.

Non Gamma 2. See Cavex Non Gamma 2.

Nongram. Trade name for an alloy composed of 87% copper, 11% tin and 2% zinc, used for valves, bearings, and bushings.

Non-Gran Bronze. Trade name of American Non-Gran Bronze Company (USA) for a sintered tin bronze with a nongranular microstructure, used for pulleys, and machine parts.

nonlustrous glaze. A dull, matte, or low-gloss inseparable ceramic glaze or enamel that has been fire-bonded to the surface of a product.

nonlinear dielectrics. Dielectric materials, such as certain ceramics and polymers, which when under the influence of an electric field exhibit nonlinear polarizations, i.e., polarizations that do not vary in proportion to that field.

nonlinear materials. Materials in which certain influences, e.g., electrical or magnetic fields, mechanical stresses, etc., produce responses, e.g., electric polarizations, magnetizations, or mechanical strains, that do not change linearly in proportion to those influences.

nonlinear optical crystal. A synthetic optical crystal, e.g., a lithium niobium oxide ($LiNbO_3$) crystal, for which the dielectric response function to optical radiation is strongly nonlinear, i.e., does not vary in proportion with that radiation. Abbreviation: NLO crystal.

nonlinear optical materials. A broad class of materials in which interactions with electromagnetic radiation (light) from lasers produce nonlinear optical responses, i.e., responses that do not vary in proportion (linearly) with that radiation. For example, they may produce blue light from incident red light. Examples of such materials include ammonium dihydrogen phosphate (ADP), β-barium borate (BBO), lithium niobium oxide (LNO), intercalated lamellar nanocomposites, microstructured optical fibers, photorefractive materials, polymer-liquid crystals and certain polar polymers, e.g., dipolar polyamide, and hyperpolarizable molecules, such as julodinyl-6-isoxazolone and julolidinyl-6-*N,N*-diethylthiobarbituric acid. They are suitable for photonic and optoelectronic devices, such as optical sensors and optoelectronic switches. Abbreviation: NLO materials.

nonlustrous tile. A tile coated with a nonlustrous glaze.

nonmagnetic materials. A group of nonmagnetizable materials, such as glass, paper, and polymers, that are not affected by magnetic fields. They have relative permeabilities of or close to unity. Abbreviation: NMM.

nonmagnetic steel. A steel that has a very low relative permeability and, at normal temperatures, is not affected by magnetic fields. Examples include (i) manganese steels containing more than 12% manganese and smaller amounts of nickel and sometimes chromium; (ii) nickel steels with about 18-30% nickel; and (iii) austenitic stainless steels (18-8 type).

nonmetallic abrasives. Abrasives, such as dolomite, flint, garnet, glass beads, ground glass, plastics (e.g., nylon and polycarbonate), pumice, quartz, and various agricultural products (sawdust, crushed walnut or pecan shells, or rice hulls) used in abrasive blast cleaning of metals, plastics, ceramics, glass, and wood.

nonmetallic materials. See nonmetals.

nonmetallic powder. A powder that is not produced from a metal, e.g., a powder made of boron, carbon, selenium, or silicon.

nonmetals. (1) The chemical elements arsenic, astatine, bromine, carbon, chlorine, fluorine, hydrogen, iodine, nitrogen, oxygen, phosphorus, polonium, selenium, silicon, sulfur and tellurium as well as the noble gases (argon, helium, krypton, neon, radon and xenon), that accept or share electrons in their valence shells and whose electronic structure, bonding characteristics, and resulting physical and chemical properties differ from those of metals. In general such elements lack characteristic metallic properties, such as good electrical and thermal conductivity, low electronegativity, high luster, etc. Also known as *nonmetallic materials*.

(2) Natural or synthetic materials that are not metals. Natural nonmetals include amber, asbestos, leather, mica, petroleum, waxes, water, and wood, and synthetic nonmetals ceramics, glasses, plastics, and rubber. Also known as *nonmetallic materials*.

nonoriented electrical steel. A special, flat-rolled, low-carbon steel with silicon (0.3-4.3%) and/or aluminum whose grains do not exhibit a preferred magnetization in any particular direction. Depending on the silicon content, it possesses good stamping properties, good permeability and low core loss, and is available as strip or sheet for use in electrical applications, such as motors, dynamos and generators, rotating machinery, transformers, ballasts, and relays. Also known as *nonoriented silicon steel*.

nonoriented silicon steel. See nonoriented electrical steel; silicon steels (1).

nonoxide fibers. A group of ceramic fibers that are not based on oxides, but on carbides or nitrides, e.g., fibers of silicon carbide or boron nitride. See also oxide fibers.

Non-Oxidizable. Trade name for an abrasion-, corrosion-, and heat-resistant alloy containing 1.1% carbon, 10% manganese, 25% chromium, 0.95% silicon, and the balance iron.

Nonpareil. Trade name for a porous brick with low heat transmission composed of diatomaceous earth. The air pockets are due to the incorporation of ground cork into the raw batch that burns out during molding. Used for thermal insulation.

Non-Pareil. Trade name of Theodore Hiertz Metal Company (USA) for an antifriction metal of 78% lead, 17% antimony and 5% tin. It has a melting point of about 300°C (570°F), and is used for bearings and solders.

nonplastic ceramics. Ceramic materials other than the plastic clays. Also known as *nonplastics*.

nonpolar material. A material that has neither positive or negative electric polarization, i.e., no electric dipole moment.

nonpolar polymers. Synthetic polymers, such polytetrafluoroethylene and other fluorinated polymers, that have low dielectric constants and losses, and no permanent electric dipole moment. They have potential applications in electronics, optoelectronics and photonics, e.g., for packaging, and charge-electret and photonic devices. See also polar polymers.

nonpolar resin. A synthetic resin, such as polyethylene or polystyrene, which has no concentration of electrical charges on a molecular scale.

nonreinforced concrete. See plain concrete.

nonrigid plastics. Plastics that have flexural or tensile moduli of elasticity not exceeding 70 MPa (10 ksi) at a temperature of 23°C (70°F), and a relative humidity of 50%. See also rigid plastics; rigid resin.

Non-Scuff. Trade name of Coulter Steel & Forge Company (USA) for a tough hot-work tool steel (AISI type H43) containing 0.6% carbon, 3.7% chromium, 1.8% vanadium, 8.5% molybdenum, and the balance iron. It has excellent abrasion and wear resistance, and is used for hot headers, punches and shear blades, die inserts, and other hot-work tools.

nonshattering glass. See laminated glass.

Nonshock Tungsten. Trade name of Coulter Steel & Forge Company (USA) for a shock-resistant tool steel (AISI type S1) containing 0.55% carbon, 2.75% tungsten, 1.25% chromium, 0.2% vanadium, and the balance iron. It has high tensile strength, good fatigue resistance and good toughness. Used for dies, chisels, and other tough, shock-resistant tools.

Non-Shrink. (1) Trade name of Bisset Steel Company (USA) for a nonshrinking, oil-hardening steel containing 1.5% carbon, 13% chromium, 1.5% tungsten, and the balance iron. Used for dies, tools, and production dies.

(2) Trade name of Osborn Steels Limited (UK) for a nondeforming steel containing 0.95% carbon, 1.25% manganese, 0.5% chromium, 0.5% tungsten, and the balance iron. Used for tools and dies.

Non-Shrinkable. Trade name of Teledyne Vasco (USA) for oil-hardening cold-work tool steels (AISI type O1) containing 0.9-1% carbon, 0.2-0.4% silicon, 1.15-1.25% manganese, 0.5% chromium, 0.4-0.6% tungsten, 0.15-0.25% vanadium, and the balance iron. They have low tendency to warping or shrinking, and are used for lathe and cutting tools, blanking, forming, stamping, trimming and various other dies, plastic molds, punches, shears, bushings, gages, and various other tools.

nonsilicate glasses. Glasses, such as those based on boron oxide (B_2O_3), phosphorus pentoxide (P_2O_5), germanium disulfide (GeS_2), beryllium tetrafluoride (BeF_4) or zirconium tetrafluoride (ZrF_4), whose silica (SiO_2) content is low or nil.

nonsilicate oxide ceramics. Ceramic oxide materials, such as alumina (Al_2O_3), beryllia (BeO), magnesia (MgO), spinel ($MgAl_2O_4$) or zirconia (ZrO_2), whose composition includes no or very little silica (SiO_2).

nonshrinking steels. See nondeforming steels.

nonslip concrete. A concrete with a mechanically roughened surface or a sand-like surface made by additions of fine alumina (Al_2O_3) or silica (SiO_2) grains to the surface prior to hardening. Used for steps and other areas of pedestrian traffic to prevent slipping.

nonspinning asbestos. Fibrous asbestos, usually of the amphibole group, that cannot be spun. Used in the manufacture of asbestos paper and shingles, and as a filler or reinforcement in plastics, rubber, or paint.

Non-Split. Trade name of Salzburger Glasveredelungsindustrie (Austria) for a multilayered laminated glass.

Non Stain. Trade name of Sanderson Kayser Limited (UK) for a stainless steel containing 0.3% carbon, 1% manganese, 13% chromium, 0.5% nickel, and the balance iron. Used for corrosion-resistant parts.

nonstoichiometric compound. A compound, such as *wuestite*-type iron oxide ($Fe_{1-x}O$ with x = 0.05), that does not have an exact or fixed ratio of chemical elements. See also stoichiometric compound.

nonstoichiometric intermetallic compound. An intermetallic compound formed of two or more components (metals), and having a composition, structure and properties different from either component (metal), and in which the ratio of the components present is not exact, but variable in certain ranges. See also stoichiometric intermetallic compound.

non-stress-graded lumber. Softwood construction lumber that is not intended for structural applications, and therefore does not need to have specified requirements with respect to strength, stiffness and defects. Examples include boards (commons), battens, crossarms and planks. Also known as *yard lumber*. See also stress-graded lumber.

nonstripping agent. See adhesion promoter (2).

Nonsulite. Trade name of Michiana Products Corporation (USA) for a heat- and corrosion-resistant cast steel containing up to 0.5% carbon, 28% chromium, 8% nickel, and the balance iron. It has good resistance to sulfurous atmospheres at elevated temperatures, and is used for chemical plant and equipment, furnace equipment, etc.

Non-Tarnishable. Trade name for a leaded brass of 63.6% copper, 31% zinc, 3.25% tin, and 2% lead. It has excellent resistance to corrosion and tarnishing.

non-tarnish paper. Paper, such as metallic interleaving paper, which is free from corrosion-promoting substances which may attack metals.

Non-Tempering. (1) Trade name of Allied Steel & Tractor Products Inc. (USA) for an oil- or water-hardening, shock-resistant tool steel containing 0.35% carbon, 0.75% manganese, 0.5% molybdenum, 0.45% silicon, 0.8% chromium, 0.25% tungsten, and the balance iron. Used for punches, chisels, hand tools, and blacksmith tools.

(2) Trade name of Bethlehem Steel Corporation (USA) for a shock-resistant tool steel containing 0.35% carbon, 0.8% chromium, 0.7% manganese, 0.3% copper, 0.3% molybdenum, and the balance iron. Used for blacksmith tools, chisels, punches, and stamps.

non-torque yarn. A yarn that does not tend to rotate or twist when hanging freely. The opposite of a *torque yarn*.

non-transition elements. (1) A term now used for the metallic elements in Group 1 (lithium through francium), Group 2 (beryllium through radium), Group 13 (aluminum through thallium), Group 14 (tin and lead), Group 15 (antimony and bismuth) and Group 16 (polonium) of the Periodic Table. Abbreviation: NTM. Also known as *non-transition metals*. See also transition metals.

(2) A term formerly used to refer to the metallic elements in the A Groups of the Periodic Table, i.e., Groups IA, IIA, IIIA, IVA, VA, and VIA. Abbreviation: NTM. Also known as *non-transition metals*. See also transition elements.

non-transition metals. See non-transition elements.

nontronite. A green or yellowish green mineral of the smectite group that is composed of calcium magnesium aluminum iron silicate hydroxide hydrate, $(Ca,Mg)_{0.5}Fe_2(Si,Al)_4O_{10}(OH)_2 \cdot xH_2O$, but in which calcium, magnesium and aluminum may be replaced by sodium. Crystal system, hexagonal. Density, 3.42 g/cm³; refractive index, 1.760. Occurrence: USA, Brazil.

No-Nubbin. Trade name of Schaffner Manufacturing Company Inc. (USA) for nubless buffing compositions.

nonuniform sand. Sand with nonuniform particle size, i.e., containing sand particles of various sizes. See also uniform sand.

Nonvar. Trade name of Firth Brown Limited (UK) for a nondeforming, oil-hardening plain-carbon steel containing 0.92% carbon, 1.75% manganese, 0.3% silicon, and the balance iron. Used for tools and dies.

nonveneered panel. A wood panel made without veneer plies, but consisting of reconstituted wood. Examples include particleboard, waferboard, and oriented strandboard.

nonvitreous ceramics. Ceramics, such as brick or tile, that have not been vitrified. They absorb relatively high amounts of water (more than about 3% for brick and more than 7% for floor and wall tile). Also known as *nonvitrified ceramics*.

nonvitrified ceramics. See nonvitreous ceramics.

Non-Wair. Trade name of British Steel Corporation (UK) for an air- or oil-hardening nondistorting alloy tool steel containing 2.2% carbon, 13% chromium, and the balance iron. Used for tire and brick molds, various other tools, dies, and gages.

nonwoven composites. Composite materials, usually in sheet or board form, made from natural and/or synthetic fibers (e.g., lignocellulosic, textile or waste-paper fibers) by first producing a continuous, loosely consolidated mat followed by mechanical interlocking or another bonding method with or without the application of heat. Used for building products, underlayment, automotive interiors, and absorbent products for personal hygiene. See also woven composites.

nonwoven fabrics. Loose, porous textile fabrics made from natural or synthetic fibers, yarns, rovings, etc., with or without a binder (e.g., a synthetic resin or rubber latex) by pressing or interlocking with or without the application of heat. Used for filtration products, carpet backing, support for plastic film, napkins, drapes, etc. Also known as *nonwovens; nonwoven textiles*. See also continuous nonwoven fabrics; dry-laid nonwoven fabrics; wet-laid nonwoven fabrics; woven fabrics.

nonwovens. See nonwoven fabrics.

nonwoven textiles. See nonwoven fabrics.

No-Ox. Trade name Boller Development Corporation (USA) for an aluminized high-temperature iron alloy containing 6% aluminum, small amounts of carbon, manganese and silicon, and the balance iron. It has good oxidation resistance up to 1200°C (2190°F), and is used for pack carburizing containers, baffles and flame deflectors, burners, and furnace parts.

Nopalon. Trademark of Rhodiatoce (France) for polyamide (nylon) fibers used for wearing apparel and industrial fabrics.

Nopalplast. Trademark of Rhodiatoce (France) for a series of polyamide (nylon) resins.

Noral. Trade name of Alcan-Booth Industries, Limited (UK) for a series of wrought and cast aluminum products including commercially pure aluminums (99+%) and various aluminum-copper, aluminum-magnesium, aluminum-magnesium-silicon, aluminum-manganese, aluminum-silicon, aluminum-silicon-magnesium and aluminum-zinc alloys.

Noralide. Trademark of Norton Company (USA) for silicon nitride (Si_3N_4) that is hot-pressed at temperatures exceeding 1700°C (3090°F). It has high tensile strength and modulus, good compressive strength, high hardness (1700-2200 kgf/mm^2), good fatigue and wear resistance, very low porosity, a low coefficient of friction, good resistance to halogens and dilute acids and fair resistance to concentrated acids. Used for ball and roller bearings, tool bits and cutting components.

Noranda. Trade name of Noranda Inc. (Canada) for an extensive series of wrought coppers and copper alloys supplied in sheet, strip, tube, rod, bar and wire form. The coppers include electrolytic tough-pitch, silver-bearing tough-pitch, phosphorus-deoxidized and tellurium-bearing phosphorus-deoxidized grades, and the copper alloys include various admiralty, forging, gilding, cartridge, naval, low, red, yellow, manganese and aluminum brasses, unleaded and medium- and high-leaded brasses and free-cutting brasses, unleaded and leaded commercial bronzes, aluminum, manganese, low-silicon, phosphor and low-fuming bronzes, and several copper nickels, nickel silvers, and leaded nickel silvers. *Noranda Al* is a magnesium-aluminum die-casting alloy with improved elevated temperature creep resistance.

norbergite. A tan-colored mineral of the humite group composed of magnesium fluoride silicate, $Mg_3SiO_4F_2$. It can also be made synthetically from magnesium oxide, magnesium fluoride and silicic acid at 25°C (77°F). Crystal system, orthorhombic. Density, 3.15 g/cm^3; refractive index, 1.573. Occurrence: Sweden, USA (New Jersey).

Norbide. Trade name of Norton Company (USA) for a cemented boron carbide (B_4C) containing 78% boron, 21% carbon and 0.14% iron, and made in an electric furnace by heating coke and boric acid. *Norbide* has a density of 2.52 g/cm^3 (0.09 lb./in.^{3}), excellent wear resistance, very high hardness (above 9 Mohs), good compressive strength, low thermal expansivity, melting point of 2455°C (4450°F), an upper continuous-use temperature of 980°C (1795°F), and fair resistance to acids, alkalies and halogens. Used for wire-drawing dies, nozzles and nozzle linings for abrasive blasting operations, drills for tungsten carbide, die nibs, abrasives for grinding and lapping hard materials (e.g., tungsten carbide or tantalum carbide), and as a deoxidizer for steels.

Norbond. Trademark of Norton Performance Plastics (USA) for a line of hot-melt adhesive products.

Norchem. Trademark of Quantum Chemical Corporation (USA) for a series of low-density polyethylenes with good electrical properties used for cable and wire insulation. Some grades are available with good environmental and or low-temperature resistance.

Norcast. Trademark of Norton Pakco Industrial Ceramics (USA) for a series of monolithic castables.

NorCor. Trademark of Norfield Corporation (USA) for plastic honeycomb materials that can be easily formed without loss of strength.

Norcore. Trademark of Norfield Corporation (USA) for strong, lightweight high-impact styrene honeycomb panels with excellent dimensional stability and resistance to warping, and good workability.

Nordel. Trademark of DuPont Dow Elastomers (USA) for a rubberlike polymeric material based on a sulfur-curable ethylene-propylene-diene terpolymer. It has good processibility, a low yellowness index, low odor, and is used for automotive components (e.g., hose and brake components), electrical and appliance components, cable and wire insulation, belts, hose, mechanical rubber goods, air ducts, and window profiles.

nordite. A light or dark brown to black mineral composed of sodium manganese strontium lanthanum silicate, $Na_3MnSrLaSi_6O_{17}$. Crystal system, orthorhombic. Density, 3.48 g/cm^3. Occurrence: Russian Federation.

Nordot. Trademark of Synthetic Surfaces Inc. (USA) for a series of epoxy, urethane and other adhesives that bond well to metals, wood, concrete, asphalt, vinyls, urethanes, rubber, foams and carpeting. Used especially for bonding outdoor and indoor sport and recreational surfaces (e.g., synthetic turf, playground and tennis court surfaces, etc.) and for various industrial and constructional bonding applications.

nordstrandite. A gray to yellow mineral composed of aluminum hydroxide, $Al(OH)_3$. Crystal system, triclinic. It can also be made synthetically. Density, 2.44 g/cm^3; refractive index, 1.580. Occurrence: Guam, Malaysia.

nordstromite. A lead-gray mineral composed a bismuth copper lead selenide sulfide, $CuPb_3Bi_7S_{10}Se_4$. Crystal system, monoclinic. Density, 7.13 g/cm^3. Occurrence: Sweden.

Norem. Trade name of Amax R&D Center (USA) for carbide-strengthened, wear-resistant iron-chromium-base hardfacing alloys containing relatively high amounts of manganese (about 4-5%) and nickel (about 4-5%). They possess exceptional cavitation-erosion wear resistance, high corrosion and galling-wear resistance, and good arc weldability. Supplied in powder, wire and rod form, they are used for nuclear and hydroelectric plant components, e.g., valve seats and disks, and pump parts.

Norepol. Trade name of US Department of Agriculture for vulcanized vegetable oil based polymers.

Noresco. Trade name of Boehler GmbH (Austria) for various types of steels including several low- and medium-carbon alloy steels containing varying amounts of chromium, tungsten, vanadium, molybdenum, nickel, and silicon, and several high-speed tool steels. *Noresco Extra Tough Hard* is a tough tool steel containing 0.9% carbon, 0.15% vanadium, and the balance iron. *Noresco Dominator* are tool steels containing 1.5-1.9% carbon, 11.5-12% chromium, 0-2% tungsten, 0-0.8% molybdenum, 1% vanadium, and the balance iron. Used for cutting tools, reamers, and dies. *Noresco Favorit* is a high-carbon alloy steel containing 0.9% carbon, 0.2% chromium, 0.15% vanadium, and the balance iron. *Noresco Parforce* are case-hardening steels containing about 0.15% carbon, 0.75% chromium, 3-4% nickel, and the balance iron. Used for machine parts and case-hardened parts. *Noresco Tyrant Extra* is an alloy steel containing 0.45% carbon, 1% silicon, 1.05% chromium, 2% tungsten, and the balance iron.

Norflex. Trademark of Norton Performance Plastics (USA) for clear polystyrene film products, and adhesives for bonding grinding wheels.

Norian SRS. Trade name of Stratec Medical (Switzerland) for an injectable calcium phosphate bone cement. It has high compressive strength and, after implantation, gradually transforms into new bone. Used in orthopedics and maxillofacial surgery, especially for the treatment of cancellous bone.

Norit. (1) Trademark of American Norit Company, Inc. (USA) for activated carbon available in various grades and particle sizes including fine mesh (100 mesh or finer) and pellets. It has excellent adsorptive properties, and is used in liquid-phase adsorption systems.

(2) Trade name of Christian Geyer Co. (Germany) for phenol-formaldehyde resins and plastics.

Norlig. Trademark of American Can Company (USA) for a series of unmodified or partially modified *lignosulfonates* manufactured from waste sulfite liquor. Commercially available in the form of powders and brownish liquids, they are used as road binders, as binders in foundry products, in linoleum paste and gypsum board, and in leather tanning.

Norma. Trade name of Circeo Filati S.r.l. (Italy) for a yarn blend composed of 50% polyester and 50% cotton, and supplied in several counts. Used for textile fabrics.

Normag. Trade name of Norsk Hydro (Norway) for a series of strong, ductile, tough magnesium alloys used for pressure-die casting applications.

normalized steels. Steels heated to about 55°C (100°F) above the upper critical temperature (i.e., into the *austenite* region), held there until the temperature is uniform throughout, and then cooled in still air to room temperature. They have homogeneous microstructures consisting of relatively fine pearlite, and uniform mechanical properties, good ductility and toughness by grain size refinement, and are free from stresses that may have developed during machining, welding or forming operations.

Normalloy. (1) Trade name of US Steel Corporation for a normalized steel containing 0.4-0.5% carbon, 0.9-1.3% manganese, 0.3-0.6% chromium, a trace of vanadium, and the balance iron. Used for heavy-duty automotive machinery, and machine elements.

(2) Trade name of Foote Mineral Company (USA) for a normalized alloy steel containing 0.4% carbon, 0.8% manganese, 1.5% chromium, 0.2% vanadium, and the balance iron.

normal metallic elements. See normal metals.

normal metals. Metals, such as copper, gold or silver, with completed inner shells and three or fewer valence electrons. Also known as *normal metallic elements*.

Normal Ni. Trade name of British Driver-Harris Company Limited (UK) for a high-purity nickel whose purity is intermediate between that of *Active Ni* and *Passive Ni*. Used for electrical and electronic applications.

normal spinels. Cubic spinels, such as magnesium aluminate ($MgAl_2O_4$), nickel aluminate ($NiAl_2O_4$), zinc aluminate ($ZnAl_2O_4$) or zinc ferrite ($ZnFe_2O_4$), having all divalent cations in 4-fold (tetrahedral) interstitial sites, and all trivalent cations in 6-fold (octahedral) sites. Many important magnetic ceramics are normal spinels. See also cubic spinels; inverse spinels; spinels; ferrites.

normal uranium. See natural uranium.

normal-weight concrete. Concrete with a unit weight of about 2400 kg/m³ (150 lb/ft³).

Norman brick. A special building brick that is 68 × 102 × 305 mm (2.5 × 4.0 × 12.0 in.) in size.

Normandie. Trademark of Société Asbestos Limitée (Canada) for crude asbestos and asbestos fibers.

Normant reagents. Vinylmagnesium halides that are similar in chemical properties to *Grignard reagents*, and used in organometallic synthesis.

Normar. Trade name of Ziv Steel & Wire Company (USA) for a nondeforming, shock-resistant steel containing 0.9% carbon, 1.5% manganese, 0.5% chromium, 0.25% molybdenum, and the balance iron. Used for punches, crimpers, and dies.

Normount. Trademark of Norton Performance Plastics (USA) for self-adhesive polyurethane sealing and mounting tape.

Noro. Trade name of Hoffmann & Co. KG (Germany) for a series of cold-work and high-speed tool steels.

Norpex. Trademark of Chevron Phillips Chemical Company (USA) for a series of polyphenylene ethers/oxides available in unreinforced, glass-filled, mineral-reinforced, high-heat and high-impact grades for the manufacture of molded and extruded products.

Norplant. Trademark of Wyeth-Ayerst Pharmaceuticals (USA) for a nondegradable silicone rubber used in subdermal implants for the controlled release of drugs.

Norplex. Trade name of Norplex Oak Inc. (USA) for an extensive series of plastic laminates including fiberglass-fabric-reinforced copper-clad or unclad polyimide and paper-based or nylon-reinforced phenolic laminates, various copper-clad and unclad unreinforced or aramic-reinforced epoxy laminates and prepregs, copper-clad or glass-fabric-reinforced polytetrafluoroethylene laminates, and several polyester film laminates with copper foils or polyester adhesives and polyimide films with polyester adhesives. Used especially for electrical and electronic applications including flexible circuits and printed wiring boards.

Norply. Trademark of Northwood Pulp & Timber Limited (Canada) for plywood.

Norpol 2000. Trademark of Norton Pakco Industrial Ceramics (USA) for an alumina-based polishing slurry supplied as a concentrate in aqueous solution, and used in the optical and glass industries.

Norprene. Trademark of Norton Performance Plastics (USA) for thermoplastic elastomers available in standard and special grades (including food grades). They have good electrical and thermal stability, a service temperature range of -59 to +135°C (-75 to +275°F), high flexibility, a durometer hardness of 61 Shore A, good heat resealability, excellent resistance to dilute and weak acids and bases, good resistance to strong and concentrated acids, fair resistance to strong and concentrated bases, excellent radiation resistance, good ultraviolet radiation resistance, and good nonaging and nonoxidizing properties. Used for flexible tubing, and various film products.

Norseal. Trademark of Norton Performance Plastics (USA) for self-adhesive, polyvinyl chloride (PVC) foamed fluoropolymer sealing tape.

norsethite. A clear to white mineral of the calcite group composed of a barium magnesium carbonate, $BaMg(CO_3)_2$. Crystal system, rhombohedral (hexagonal). Density, 3.84 g/cm^3; refractive index, 1.694. Occurrence: USA (Wyoming).

Norskalloy. Trade name for a special *pig iron* made from Norwegian ores (magnetic iron ores). A typical composition is 4.00-4.50% total carbon, 0.50-1.50% silicon, 0.20% manganese, 0.20-0.25% phosphorus, 0.30-0.40% vanadium, 0.40-0.80 titanium, and the balance iron.

Norsodyne. Trademark of Atofina SA (France) for unsaturated polyester resins.

Norsolene. Trademark of Atofina SA (France) for a series of coumarone-indene resins.

Norsomix. Trademark of Atofina SA (France) for glass-fiber-reinforced unsaturated polyesters.

Norsoran. Trademark of Atofina SA (France) for acrylonitrile-butadiene-styrene resins.

Norsorex. Trademark of Atofina SA (France) for a polynorbornene elastomer supplied as powders and masterbatches.

North Carolina pine. See shortleaf pine.

Northern Forest Stone. Trade name of Vetter Stone Company (USA) for building stone and marble.

Northern Gold. Trade name of MDF Temple (USA) for lumber and lumber products including hardboard, fiberboard, and wood trim.

northern pine. The hard, yellowish wood from the pine tree *Pinus sylvestris*. Average weight, 510 kg/m^3 (32 lb/ft^3). Source: Northern Europe and British Isles; now also planted in Canada and the United States. Used in building construction, carpentry and joinery, for railway sleepers, telegraph, telephone and transmission poles, scaffolding, and as pulpwood. Also known as *Baltic pine; Baltic redwood; Danzig pine; red deal; Scotch fir; Scotch pine; Scots fir; Scots pine; yellow deal*.

northern white cedar. See eastern white cedar.

northern white pine. See eastern white pine.

Northlite. Trade name Shatterprufe Safety Glass Company (Proprietary) Limited (South Africa) for a sealed glass unit with a diffusing interlayer of fiberglass.

North Star. (1) Trade name of Swedish American Steel Corporation (USA) for a nondeforming tool and die steel.

(2) Trademark of North Star Cement Limited (Canada) for Portland cement.

northupite. A yellow, white or gray mineral of the tychite group composed of sodium magnesium chloride carbonate, $Na_3Mg(CO_3)_2Cl$. Crystal system, cubic. Density, 2.38 g/cm^3; refractive index, 2.38. Occurrence: Canada (Labrador), USA (California).

North Wood Plastics. Trade name of North Wood Plastics (USA) for wood fiber-filled polypropylene and high-density polyethylene resins, some of which are supplied in molding and extrusion grades.

Norton. Trademark of Norton Company (USA) for an extensive range of refractories and refractory products.

Norton Abrasives. Trade name of Norton Company (USA) for an extensive range of abrasive products.

Norton Boron Master Alloy. Trade name of Norton Company (USA) for a ferroalloy containing 81-84% iron, 8.5% boron, 2.7-3.3% silicon, 1-1.7% carbon, and a total of 2.5-4.3% aluminum and titanium. Used as a steel inoculant for the improvement of the deep-hardening properties.

Norton Crystolon. See Crystolon.

Nortuff. Trade name of Quantum Chemical Company (USA) for a series of impact-resistant polyolefins including various fiber- or mineral-reinforced polypropylenes and high-density polyethylenes. Used for automotive and electrical components, machine parts, power tool housings, outdoor equipment and furniture, etc.

Norway iron. A charcoal pig iron made from high-purity iron ore (*magnetite*) mined in northern Sweden and Norway. It has very high ductility and toughness, good soft magnetic properties including high permeability, and is suitable for use in transformer cores.

Norway maple. The hard, close-grained, yellowish white wood of medium-sized tree *Acer platanoides*. Source: Central and Eastern Europe and Scandinavia, and now also grown in North America. Used for woodworking, lumber, furniture, flooring, and plywood.

Norway pine. See red pine (1).

Norway spruce. The tough, white wood of the spruce tree *Picea abies*. It has a straight, even grain, high elasticity, moderate durability, and is rather difficult to work. Average weight, 460 kg/m^3 (29 lb/ft^3). Source: Central and Northern Europe, and British Isles. Used for building construction, flooring and general joinery work, pulpwood, and packaging applications. Also known as *Baltic whitewood; European spruce; spruce fir; white deal; white fir*.

Norwegian saltpeter. See calcium nitrate tetrahydrate.

Noryl. Trademark of GE Plastics (USA) for thermoplastic resins based on polyphenylene oxide (PPO) or nylon 6,6. Supplied in unfilled, carbon-, glass- or mineral-filled, injection or blow molding, extrusion, and foam grades. *Noryl* has high strength and modulus, excellent corrosion resistance, good abrasion and impact resistance, high creep and chemical resistance, high heat deflection temperatures, good dimensional stability, low density, very low flammability and water absorption, outstanding processibility, and good sound-dampening properties. Used for household appliances, pumps, office equipment, industrial equipment, automotive components, medical equipment, and construction equipment. *Noryl GTX* refers for polymer blends of amorphous PPO and semicrystalline nylon 6,6. They have enhanced chemical resistance, improved dimensional stability at elevated temperatures, low moisture sensitivity and mold shrinkage, very good paint adhesion and good resistance to chipping. Used for automobile doors, hatchbacks, wheel panel, and other exterior body parts. *Noryl CT, GFN, HB & V* refers modified polyphenylene ether (PPE) plastics and blends.

NorZon. Trademark of Norton Company (USA) for fused zirconia-alumina abrasives supplied in grit sizes from 20 to 220, and used for organically bonded grinding wheels and coated abrasives.

nosean. A colorless, gray, blue or brown mineral of the sodalite group composed of sodium aluminum sulfate silicate, Na_8Al_6-

$Si_6O_{24}SO_4$. It can also be made synthetically. Crystal system, triclinic. Density, 3.42 g/cm^3; refractive index, 1.760; hardness, 5.5 Mohs. Also known as *noselite*.

noselite. See nosean.

No Shock. Trade name of E.I. DuPont de Nemours & Company (USA) for an antistatic, carbon-containing, continuous-filament nylon 6,6 fiber used for carpets and rugs, and antistatic floor coverings.

NoSilver. Trade name for a silver-free medium-gold dental bonding alloy.

no-slump dental composite. A ceramic or synthetic resin based dental composite that experiences no or only negligible slump (shrinkage) upon setting.

no-slump concrete. Concrete with a slump of 25 mm (1 in.) or less. For concrete the term *slump* refers to the perpendicular distance through which the top of a molded mass of fresh concrete drops on removal of the mold.

notch-ductile steel. A steel, usually with relatively low carbon content, that exhibits ductile fracture at stress concentration points, i.e., experiences considerable plastic deformation upon fracture.

Nottingham lace. A flat, wide coarse lace made by machines. It is named after Nottingham, England, where it was originally made. Used for curtains, bedspreads and tablecloth.

noumeite. See garnierite.

Nous. Trade name for a plasticized acrylic resin used as a denture liner.

Nouvelle. Trademark of Hercules, Inc. (USA) for polypropylene fibers and yarns.

Nova. (1) Trade name of Jonas & Colver (Nova) Limited (UK) for a series of molybdenum- and tungsten-type tool steels.

(2) Trade name of Nova Chemicals Inc. (USA) for high-impact polystyrenes.

(3) Trade name of Atotech USA Inc. for a bright nickel electrodeposit and plating process.

(4) Trade name of Keramos (USA) for a series of modified lead titanates belonging to the *Kézite* family of advanced piezoelectric ceramics. They provide low planar coupling and high ring-around sensitivity for use in medical, imaging and nondestructive testing (NDT) transducers.

Novabat. Trade name of Chemtech Finishing Systems, Inc. (USA) for a bright acid tin electrodeposit and plating process.

Novabestos. Trademark of Raybestos-Manhattan, Inc. (USA) for sheet and tape materials with good dielectric properties made by bonding asbestos fibers with silicone resin. Used for electrical insulation applications.

Novablend. Trademark of Novatec Plastics & Chemical Company, Inc. (USA) for plastic compounds in particular polyvinyl chlorides, supplied as molding and extrusion liquids, pellets, and powders.

Novabond. Trade name of The Argen Corporation (USA) for a medium-gold dental bonding alloy. *Novabond 2* is a silver-free palladium dental bonding alloy.

NovaBone. Trademark of USBiomaterials Corporation for an osteoproductive particulate *Bioglass* bone-grafting material used in orthopedic surgery for spine fusion, revision arthroplasty and general defect filling, and in oral and dental surgery for implanted graft cases. See also NovaBone-CM.

NovaBone-C/M. Trademark of USBiomaterials Corporation for a particulate *Bioglass* bone-grafting materials used in craniofacial and maxillofacial surgery.

novacekite. A yellow, radioactive mineral of the autunite group composed of magnesium uranyl arsenate hydrate, $Mg(UO_2)_2$-$(AsO_4)_2 \cdot xH_2O$. It can also be made synthetically. Crystal system, tetragonal. Density, 3.60 g/cm^3; refractive index, 1.637. Occurrence: Germany.

Novacite. Trade name of Malvern Minerals Company (USA) for novaculitic tripoli and modified novaculite.

novaculite. A hard, fine-grained, bluish-white or opaque white sedimentary rock composed predominantly of microcrystalline *quartz* with some *chalcedony*. It has a silica content of 99.5%, and a hardness 6-6.5 Mohs. Source: USA (Arkansas). Used in the manufacture of silica refractories, and as an abrasive for blast cleaning and whetstones. Also known as *razor stone*.

Novacor. Trade name of Nova Polymers (USA) for high-impact polystyrenes.

Novadip. Trade name of Chemtech Finishing Systems Inc. (USA) for dip coatings applied on zinc plate.

Novafix. Trade name of Ato Findley (France) for alcohol-resin adhesives used for floor covering applications.

Novaflex. Trade name of Bayer Corporation (USA) for polyurethane adhesives for laminating and structural bonding, and polyurethane and epoxy coatings for floors.

Novakap. Trade name of Malvern Minerals Company (USA) for treated *quartz* microform abrasives.

novakite. A steel gray mineral composed of copper silver arsenide, $(Cu,Ag)_4As_3$. Crystal system, tetragonal. Density, 6.70 g/cm^3. Occurrence: Czech Republic.

Noval. Trade name of Novalis Fibres GmbH (Germany) for nylon 6 and nylon 6,6 fibers and yarns.

Novalast. Trade name of Nova Polymers (USA) for a series of thermoplastic elastomers.

Novalastik. Trademark of Norton Performance Plastics (USA) for foam materials with polyvinyl chloride cores and butyl coverings, used for sealing applications.

Novalene. (1) Trade name of Nova Polymers, Inc. (USA) for a series of thermoplastic elastomers.

(2) Trade name of Novaceta SpA (Italy) for cellulose acetate fibers used for the manufacture of wearing apparel.

Novalite. (1) Trade name of Novalite (UK) for an aluminum casting alloy containing 12.5% copper, 1.4% nickel, 0.5% silicon, 0.5% iron and 0.3% magnesium. Used for pistons.

(2) Trademark of Selby, Battersby & Company (USA) for marble products made by bonding marble chips and formulated powders with a synthetic resin. Used for flooring applications.

(3) Trademark of Polistone SpA (Italy) for precast interior and exterior building panels made of various stone materials.

Novalon. (1) Trade name of Tenax Finishing Products Company (USA) for textured finishes.

(2) Trademark of Indopco, Inc. (USA) for starch-based adhesives used in the manufacture of corrugated board.

Novamat. Trade name for rayon fibers and fabrics.

Novamet. Trademark of Novamet Specialty Products/Inco Specialty Powder Products (USA) for a series of specialty metal particulates including the following: (i) *Silver Coated Nickel Spheres* (15% silver) with a particle size of 10 μm (390 μin.) and an apparent density of 2.5 g/cm^3 (0.09 lb/in.³); (ii) *Nickel Coated Graphite* (60% fully encapsulated nickel) having a particle size of 100 μm (3900 μin.), and an apparent density of 1.6 g/cm^3 (0.06 lb/in.³); (iii) *Conductive Nickel Spheres* with a particle size of 8-9 μm (310-350 μin.), an apparent density of 3.2-3.5 g/cm^3 (0.12-0.13 lb/in.³); (iv) *Conductive Nickel Pigment* having a particle size of 3.2 μm (125 μin.); and (v) *HCA-1 High-*

Conductivity Flake with a screen mesh of -400 (99%), an apparent density of 0.90 g/cm³ (0.03 lb/in.³), and a thickness of 1.0-1.1 µm (39-43 µin.). Used for electronic applications, e.g., EMI/RFI shielding, conductive films, coatings, adhesives, sealants, and gaskets.

Novamid. Trademark of Mitsubishi Chemical Corporation (Japan) for polyamide 6 (nylon 6) polymers used for making electrical, mechanical, automotive and railway parts, building products, fishing nets and lines, and plastic films. They are supplied in a wide range of grades including casting, elastomer-copolymer, stampable sheet, unmodified, carbon- or glass fiber-reinforced, silicone-, polytetrafluoroethylene- or molybdenum disulfide-lubricated, glass bead- or mineral-filled, fire-retardant, high-impact, and UV-stabilized.

Novantiox. Trade name of Chiers-Chatillon (France) for an austenitic stainless steel containing about 0.06% carbon, 2% manganese, 1% silicon, 19% chromium, 10% nickel, and the balance iron. Used for chemical and food-processing equipment, and architectural molding and trim.

Novapol. Trademark of Nova Chemicals Inc. (USA) for a series of linear polyethylene resins including low-density, linear-low-density, high-density and medium-density grades. Depending on the grade, they are supplied as films, masterbatches, or blow molding, extrusion, concentrate and rotational molding resins.

Novarex. (1) Trade name of Pentron Laboratory Technologies (USA) for a corrosion-resistant dental ceramic alloy composed of 55% cobalt, 25% chromium, 10% tungsten, and the balance aluminum, ruthenium, niobium, yttrium and zirconium. Used for crown and bridge restorations.

(2) Trademark of Mitsubishi Kasei Corporation (Japan) for polycarbonate resins used for molded products, and in the recording media industries.

Novastan. Trade name of Chemtech Finishing Systems Inc. (USA) for an immersion tin and related coating process.

Novatec. Trade name of Mitsubishi Chemical Company (Japan) for phenol-formaldehyde plastics.

Novatect. Trade name of Atotech USA Inc. for electroless nickel.

Novatex. Trade name of Novatex International (USA) for rayon fibers and yarns used for textile fabrics.

Nova Thene. Trademark of Polymer International (NS) Limited (Canada) for coated industrial and tarpaulin fabrics.

Novatherm. (1) Trademark of Norton Performance Plastics (USA) for hot-melt butyl polymer sealing compounds.

(2) Trademark of Gebrüder Ditzel GmbH (Germany) for heat-radiating plastic, paper and textile materials in bulk form used for thermal insulation applications.

Novatron. Trademark PFE Limited (UK) for polypropylene staple fibers.

Novellon. Trade name of British Celanese (UK) for cellulose acetate plastics.

novelty yarn. See fancy yarn.

Novex. Trade name of BP Chemicals (UK) for several unmodified and UV-stabilized low-density polyethylenes.

Novite. Trade name of American Marsh Pumps Inc. (USA) for a nickel cast iron containing 3% carbon, 1.5% nickel, 0.5% chromium, and the balance iron. It has good abrasion resistance, and high strength and hardness, and is used for water pumps, impellers, and cylinder heads.

Novitex. Trade name of J.H. Benecke GmbH (Germany) for a flexible polyvinyl chloride film.

Novo. Trade name of H. Boker & Company (USA) for a series of tungsten-type high-speed steels, and several corrosion-resis-

tant steels including ferritic and martensitic (high-chromium), and austenitic (chromium-nickel) stainless grades.

Novocel. Trademark of Unnafibras Textil Ltda. (Brazil) for viscose rayon filament yarns.

Novocoat. Trademark of Superior Environmental Products, Inc. (USA) for a wide range of multifunctional novolac coatings for various applications, such as concrete pipes, steel tanks, chutes, hoppers, floors, foundations, and wastewater-management and chemical equipment.

Novodur. Trademark of Bayer Corporation (USA) for a series of acrylonitrile-butadiene-styrene (ABS) copolymers supplied in extrusion, injection molding, glass fiber-reinforced, fire-retardant, high-heat, high-impact, medium-impact, and UV-stabilized grades. They have a density of approximately 1.1 g/cm³ (0.04 lb/in.³), good tensile strength and impact resistance, an upper service temperature of about 90-100°C (194-212°F), excellent resistance to mineral oils and dilute alkalies, good resistance to gasoline and dilute acids, and poor resistance to trichloroethylene and tetrachlorocarbon. Used for automotive components, appliance housings, business machines, enclosures, shielding, covers, etc. *Novodur KU* refers to modified ABS resins and *Novodur W* to styrene-acrylonitrile (SAN) resins.

Novo Enormous. Trade name of Jonas & Colver (Nova) Limited (UK) for a series of cobalt-tungsten high-speed steels containing about 0.75% carbon, 4.5% chromium, 18% tungsten, 1.25% vanadium, 10% cobalt, and the balance iron. Used for cutting tools.

Novokonstant. Trade name of VDM Nickel-Technologie AG (Germany) for a resistance alloy containing 82.5% copper, 12-13.5% manganese, 3-4% aluminum and 1-1.5% iron. It has good high-temperature resistance, high electrical resistivity, and a low coefficient of expansion. Used for precision resistors.

novolac epoxies. See epoxy novolacs.

novolac phenolics. See novolacs.

novolacs. A group of brittle, linear, thermoplastic phenol-formaldehyde resins made under acidic conditions (i.e., with acid catalysts and excess phenol). They can be cured by addition of a hardener (e.g., hexamethylenetetramine, or *p*-formaldehyde) and heating at 93-204°C (200-400°F). The resulting crosslinked thermoset phenolics are used as bonding and molding materials, as additives in nitrile rubber, in brake linings and clutch facings, in grinding wheels and electrical insulation, and in air-drying varnishes. Also known as *novolac phenolics; two-stage phenolic resins.*

Novolen. Trademark of BASF Corporation (USA) for a series of polypropylene block copolymers supplied in standard, talc-filled, elastomer-modified, UV-stabilized and injection molding grades. They have a density of 0.91 g/cm³ (0.033 lb/in.³), good tensile strength, good resistance to elevated temperatures, high toughness and rigidity, excellent resistance to dilute acids and alkalies, fair resistance to mineral oils, gasoline and trichloroethylene, and poor resistance to carbon tetrachloride. Used for fittings, protective covers, and carrying cases. Also included under this trademark are a series of polystyrene resins

novoloid. See novoloid fiber.

novoloid fiber. A man-made fiber composed of 85 wt% or more crosslinked novolac. It has good high-temperature and dielectric properties, and good resistance to organic solvents and non-oxidizing acids. Also known as *novoloid.* See also novolacs; epoxy novolacs.

Novomax. Trade name of Jonas & Colver (Nova) Limited (UK) for a series of cobalt-tungsten high-speed steels containing about 0.75% carbon, 4.5% chromium, 18% tungsten, 1.25% vanadium, 5% cobalt and the balance iron. Used for cutting tools.

Novon. (1) Trade name of Novon Company (USA) for starch-based polymers.

(2) Trademark of Novamont SpA (Italy) for a biodegradable vegetable starch supplied in the form of pellets and granules for use as a polymer replacement for making containers, sheets, and packaging materials.

Novonit. Trade name of Krupp Stahl AG (Germany) for an extensive series of austenitic, ferritic and martensitic stainless steels used as filler metals for welding applications.

Novonox. Trade name of Krupp Stahl AG (Germany) for a series of austenitic, ferritic, martensitic and precipitation-hardening stainless steels used for chemical and pharmaceutical equipment, textile machinery, tanks, agitators, valves, valve trim, food-processing equipment, aircraft parts, household appliances, automotive, structural and architectural parts, dental and surgical instruments, cutlery, knives, ball bearings, hardware, fittings, and fasteners.

Novopan. Trademark of Novopan urea-formaldehyde plastics.

Novoply. Trademark of US Plywood Corporation for a strong panel material supplied in thicknesses up to 0.75 in. (19 mm) and composed of a core of wood chips bonded with synthetic resin, and overlaid with hardwood veneers. Used for cabinetwork, furniture, and paneling.

NovoRez. Trademark of PolySpec (US) for a series of concrete coatings and tank linings which, depending on the particular grade, may be resistant to general chemicals, acids, and/or industrial solvents.

Novoston. Trade name of Stone Manganese–J. Stone & Company Limited (UK) for several corrosion-resistant aluminum bronzes containing 69-78.5% copper, 11-13% manganese, 7-9% aluminum, 2-4% iron, and 1.5-5% nickel. Available in the as-cast and as-forged conditions, they are used for ship propellers, impellers and stators.

Novo Superb. Trade name Jonas & Colver (Nova) Limited (UK) for a series of cobalt-tungsten high-speed steels containing about 1.25% carbon, 4.25% chromium, 10.5% tungsten, 3.75% molybdenum, 3.2% vanadium, 10% cobalt and the balance iron. Used for cutting tools.

Novo Superior. Trade name of H. Boker & Company (USA) for a series of tungsten-type high-speed steels containing about 0.75-0.85% carbon, 18-21% tungsten, 4-4.5% chromium, 0.5-2% vanadium, and the balance iron. They have good machinability and wear resistance, high hot hardness, and good toughness. Used for cutting tools, form tools, drills, and reamers. *Novo Superior Vanadium* tungsten-type high-speed steels have similar chromium and tungsten contents, but 1% carbon and 3% vanadium. They have excellent wear resistance and cutting ability, high hot hardness, and good toughness and machinability. Used for lathe and planer tools, cutters, drills, dies, and punches.

Novotekt. Trade name of Braun-Lötfolien (Germany) for a series of brazing fluxes.

Novotex. (1) Trademark for laminated plastics composed of cotton fabrics bonded with phenolic resin. Used for gears, bearings, and for insulation applications.

(2) Trademark of Champion International Corporation (USA) for lumber products including plywood, paneling, siding, wood roofing, and wood-base acoustical tile.

(3) Trademark of New World Textiles, LLC (USA) for metallic yarns and threads used in the textile industry.

(4) Trademark of Novocon International Inc. (USA) for reinforcing fibers for use in refractories and concrete.

Novotherm. (1) Trademark of Stahlwerke Südwestfalen AG (Germany) for a series of austenitic, ferritic and martensitic stainless steels, and several creep- and heat-resistant stainless steels. The stainless grades are used for chemical equipment, agitators, mixers, filters, tanks, food handling equipment, oil-refining equipment, household appliances, pump and valve parts, and turbine parts, and the heat-resisting grades are used for heat-treating equipment, furnace parts, and oxidation-resistant components.

(2) Trademark of FMC Corporation (USA) for a high-temperature insulating specialty coatings used on underwater drilling equipment.

Novul Crete. Trade name of E-Poxy Industries, Inc. (USA) for an elastomeric concrete.

Novus. Trade name of The Hygiene Corporation (USA) for a fluoroelastomer used as a denture liner.

nowackiite. A lead-gray mineral composed of copper zinc arsenic sulfide, $Cu_6Zn_3As_4S_{12}$. Crystal system, hexagonal. Density, 4.38 g/cm^3. Occurrence: Switzerland.

No-Wear. Trade name of GTE Sylvania (USA) for an abrasion- and wear-resistant alloy containing tungsten carbide, chromium, cobalt and iron. Used for hardfacing welding electrodes.

Nowofol. Trademark of Nowofol Kunststoffprodukte GmbH & Co. KG (Germany) for stretched and unstretched, foamed and unfoamed film products of polypropylene, high-density polyethylene and polytetrafluoroethylene polymers.

Noxide. Trade name of Pratt & Lambert, Inc. (USA) for enamels and primers.

Noxyde. Trademark of Rust-Oleum Corporation (USA) for a water-based, one-part coating system providing excellent corrosion, ultraviolet and weather resistance, and good resistance to abrasion, cracking and peeling.

Noyo River Redwood. Trademark of Georgia-Pacific Corporation (USA) for redwood lumber and lumber products.

nozzle brick. A refractory brick of hollow tubular shape used in a ladle for teeming steel.

npn semiconductor. A semiconductor junction consisting of two layers of an n-type semiconducting material and one thin interlayer of p-type semiconducting material. See also n-type semiconductor; p-type semiconductor.

np semiconductor. A semiconductor with an n-type and a p-type region. See also n-type semiconductor; p-type semiconductor.

NRC. Trademark of National Rayon Corporation (India) for nylon 6 filament yarns.

nsutite. A black mineral composed of manganese oxide hydroxide, $Mn(O,OH)_2$. Crystal system, hexagonal. Density, 4.55 g/cm^3. Occurrence: West Africa.

n-type germanium. A type of *germanium* usually grown by the Czochralski technique and predominantly doped with donor-type impurity atoms of pentavalent Group VA (Group 15) elements, such as arsenic or antimony, and in which the negative charge carriers (conduction electrons) dominate over the positive charge carriers (holes). Therefore the conduction electrons are primarily responsible for electrical conduction. Abbreviation: n-Ge.

n-type oxide. A transition-metal oxide semiconductor, such as zinc oxide, that has a stoichiometric $M_{1+x}O$ defect structure, and whose metal ions (M) donate electrons to the conduction band for n-type semiconduction.

n-type semiconductor. An *extrinsic semiconductor* predominantly doped with donor-type impurity atoms of pentavalent Group VA (Group 15) elements, such as phosphorus, bismuth, antimony or arsenic, and in which the negative charge carriers (conduction electrons) dominate over the positive charge carriers (holes) and are responsible for electrical conductivity.

n-type superconductor. A high-temperature superconductor in which superconduction occurs by means of an n-type or electron carrier. For example, layered strontium barium copper oxide [(Sr,Ba)CuO] compounds, composed of laminations of alternate strontium barium [Sr(Ba)] and copper oxide (CuO_2) layers and produced by high-pressure synthesis, exhibit superconductivity critical temperatures exceeding 60K. See also high-temperature superconductors.

n-type silicon. A type of *silicon* usually grown by the Czochralski or the float-zone technique and predominantly doped with donor-type impurity atoms of pentavalent Group VA (Group 15) elements, such as arsenic, antimony or phosphorus, and in which the negative charge carriers (conduction electrons) dominate over the positive charge carriers (holes). Abbreviation: n-Si.

Nual. Trademark of Alcan Wire and Cable (Canada) for aluminum conductors and rods.

Nuall. Trade name of Flockton, Tomkin & Company Limited (USA) for an oil-hardening tool steel used for chuck jaws, reamers, and drills.

NuAlloy DP. Trade name of New Stetic (Brazil) for a dental amalgam alloy.

Nubelar. Trade name of Glidden Industrial Coatings (USA) for a mono-bake primer used for aluminum and aluminum alloys.

Nublack. Trade name of Chas. F. L'Hommedieu & Sons Company (USA) for buffing compositions used on copper and aluminum parts.

Nublite. Trade name for rayon fibers and yarns used for textile fabrics.

Nu-Block. Trademark of Norton Company (USA) for fully-cured, preformed, polymer-bonded refractory shapes used for metal melting applications.

Nu-Bronze. British trade name for a corrosion-resistant bronze of 95.4% copper, 3.25% nickel, 0.25% manganese and 0.7-1.1% silicon.

Nubuck. A white or cream-colored, proprietary buck leather.

nubuk. Suede leather obtained from chrome-tanned cow or calf leather ground on the grain side only.

Nubun. Trade name for a synthetic latex based on styrene-butadiene rubber.

nub yarn. A *fancy yarn* with prominent knots (nubs) of different color or different material arranged at regular or irregular intervals. Also known as *knop yarn; knotted yarn.*

Nucalloy. Trade name of Stoody Company (USA) for a series of cast nickel alloys containing 12-14% chromium, 2-3% boron, 4-5% silicon, and 0.5-0.7% carbon. They have good high-temperature properties, and good corrosion and wear resistance. Used for hardfacing applications.

NuCap. Trade name of Suarez Import (Brazil) for a calcium hydroxide dental cement.

Nucast. Trade name of Apothecaries Hall Company (USA) for a commercially pure nickel (99.0+%) used for anodes.

Nucerite. Trademark of Pfaudler Company (USA) for a series of glass-ceramics usually bonded to steel substrates. They have good abrasion and impact resistance, good thermal-shock resistance, good heat resistance to about 650°C (1200°F), and good resistance to corrosive chemicals and atmospheres. Used for linings of reaction equipment, and tanks, pipes, flanges, valves and nozzles.

Nuchar. Trademark of Westvaco (USA) for a series of activated carbon products used in decolorizing and deodorizing.

Nucite. Trade name of PPG Industries Inc. (USA) for a white chalkboard made of heat-strengthened glass (6.4 mm or 0.25 in. thick) with a permanently fused-on porcelain enamel containing fine abrasive particles.

Nucleant. Trade name of Foseco Minsep NV (Netherlands) for a series of grain refiners supplied in tablet form, and used as ladle additions in the manufacture of light metals (e.g., aluminum and magnesium) and their alloys.

nuclear ceramics. See reactor ceramics.

nuclear fuel. See reactor fuel.

nuclear-grade materials. See reactor-grade materials.

nuclear materials. Materials used in nuclear reactors, e.g., as fuels, shielding materials, neutron absorbers, or moderators.

nuclear poison. A material used in a nuclear reactor to absorb neutrons without propagating them. Elements with high neutron absorption cross-sections include gadolinium (49000 barns), samarium (5820 barns), europium (4100 barns), cadmium (2450 barns), dysprosium (930 barns), boron (672 barns), indium (194 barns), iridium (425 barns), mercury (375 barns), rhodium (150 barns), erbium (160 barns), thulium (115 barns) and hafnium (103 barns). "Barn" is a unit used to express nuclear cross sections. One barn is equal to 10^{-24} cm^2 per nucleus. Also known as *neutron absorber; poison.*

nuclear-reactor ceramics. See reactor ceramics.

nucleated glasses. See glass-ceramics.

nucleated resins. Thermoplastic resins, such as polyethylene or polypropylene, to which a *nucleating agent (2)* has been added to accelerate the formation of crystallized regions, and thus improve its mechanical properties.

nucleating agent. (1) A substance added to molten metal to promote nucleation, i.e., the start of the growth of many grains or a new phase to produce fine grain structures.

(2) A substance, such as a mineral additive (talc, kaolin, silica), organic salt or polymer, added in the form of fine solid particles (1-10 μm or 39-390 μin.) in concentrations of up to 0.5% to a molten thermoplastic polymer, such as polyethylene terephthalate, polyamide or polypropylene, to accelerate the formation of crystallized regions, and thus improve its mechanical properties including tensile strength, yield point, modulus of elasticity, hardness and impact strength.

(3) A substance, such as titanium dioxide (TiO_2), phosphorus pentoxide (P_2O_5) or zirconium oxide (ZrO_2), that when added to ordinary glass as a fine dispersion of particles significantly increases the nuclei density resulting in the formation of a hard, strong, nonporous, fine-grained, partly crystalline and partly glassy ceramic materials commonly known as *nucleated glasses* or *glass-ceramics.*

nucleic acid. An organic acid found in the nucleus and protoplasm of all cells and yielding phosphoric acid, sugar, pyrimidines and purines on hydrolysis. The two principal types are *deoxyribonucleic acid* (DNA) and *ribonucleic acid* (RNA).

Nuclon. Trademark of PPG Industries, Inc. (USA) for a pale yellow polycarbonate-based engineering thermoplastic with good dimensional stability, good impact, heat and thermal-shock resistance, good electrical properties, and high clarity. Used for machine components, and housings.

Nu-Coat Poly. Trade name of McGean-Rohco Inc. (USA) for several non-chromium, corrosion-protective topcoat finishes for cadmium, zinc, zinc-alloy and chromate electrodeposits and autobody parts. *Nu-Coat Poly B* produces a dark black film on most metallic surfaces, and *Nu-Coat Poly C* forms a semiclear polymer film.

Nucon. Trademark of J. King Company (USA) for a refractory made from magnesia (MgO) and chrome ore (FeCr$_2$O$_4$) by burning at high temperatures, and used for copper and lead furnace roofs, sidewalls and roofs of steel furnaces, and linings of rotary kilns for lime, dolomite and magnesia manufacture.

Nucrel. Trade name of E.I. DuPont de Nemours & Company (USA) for ethylene-maleic anhydride (EMA) copolymer resins supplied in blown-film, cast-film and extrusion coating grades. They provide excellent hot-tack strength, excellent adhesion to nylon, paper, aluminum foil, etc., good toughness and delamination resistance, and low seal-initiation temperature. Used as seal layers in coextruded or laminated structures.

Nucut. Trade name of Wallace Murray Corporation (USA) for an oil-hardening die steel.

Nu-Die. Trade name of Crucible Materials Corporation (USA) for a chromium-type hot-work steel (AISI H11) containing 0.35-0.4% carbon, 1.2% silicon, 5-5.2% chromium, 1.5% molybdenum, 0.4% vanadium, and the balance iron. It has excellent ductility, good deep-hardening properties, high hot hardness, good toughness and machinability, and good wear resistance. Used for die-casting, plastic-mold, extrusion and other dies. *Nu-Die V* is a chromium-type hot work steel (AISI type H13) containing 0.4% carbon, 1.05% silicon, 0.35% manganese, 5.25% chromium, 1.35% molybdenum, 1.05% vanadium, and the balance iron. It has properties and applications similar to those of *Nu-Die*. *Nu-Die XL* is a premium-quality chromium-type hot-work steel (AISI type H13) containing 0.4% carbon, 0.35% manganese, 5.2% chromium, 1.3% molybdenum, 1% silicon, 0.95% vanadium, up to 0.005% sulfur, and the balance iron. It has high hot hardness and abrasion resistance, a uniform microstructure, no segregation and microbanding, and high cleanliness, and is used for long-run aluminum die-casting dies, extrusion and forging dies, plastic molds, die inserts, and mandrels.

Nueral. Trade name of Aluminiumwerke Nürnberg GmbH (Germany) for a heat-treatable alloy of aluminum, silicon, copper and nickel. Used for cylinder heads.

Nufaply. Trademark of Newfoundland Hardwoods Limited (Canada) for hardwood plywoods.

nuffieldite. A lead-gray mineral composed of lead bismuth copper sulfide, Pb$_2$Cu(Pb,Bi)Bi$_2$S$_7$. Crystal system, orthorhombic. Density, 7.01 g/cm^3. Occurrence: Canada (British Columbia).

Nu-Gild. Trade name of Century Brass Products Inc. (USA) for a brass composed of 87-87.5% copper and 12.5-13% zinc. It has good corrosion resistance, excellent cold workability, good hot workability, good deep-drawing qualities, and is used for costume jewelry, ornaments, fasteners, and hardware.

Nu-Gold. British trade name for a red brass containing 87.7% copper and 12.3% zinc. It has good corrosion resistance, excellent cold workability, good deep drawing qualities, good hot workability, and a melting temperature of about 927°C (1700°F). Used for art objects, ornaments and jewelry. See also jewelry bronze.

Nu-Iron. Trade name for an iron powder made in a two-stage process by first reducing ferric oxide (Fe$_2$O$_3$) in a furnace under a hydrogen atmosphere at approximately 700°C (1290°F) to ferrous oxide (FeO), and then at approximately 590°C (1095°F) to iron. Used for powder-metallurgy parts.

NU-Klad. Trademark of Ameron International Corporation (US) for a series of solvent-free acid-, alkali- and solvent-resistant monolithic epoxy surfacers for application to concrete surfaces exposed to heavy traffic and abrasion.

nukundamite. A copper-colored mineral composed of copper iron sulfide, (Cu,Fe)$_4$S$_4$. It can also be made synthetically. Crystal system, hexagonal. Density, 4.30 g/cm^3. Occurrence: Fiji Islands.

Nulite F. Trade name of NSI Dental Pty. Limited (Australia) for a light-cure high-strength fiber-reinforced dental hybrid composite.

nullaginite. A bright green mineral of the rosalite group composed of nickel carbonate hydroxide, Ni$_2$(CO$_3$)(OH)$_2$. Crystal system, monoclinic. Density, 3.56 g/cm^3. Occurrence: Western Australia.

Nuloy. Trade name of Sivyer Steel Corporation (USA) for a tough steel containing 0.3% carbon, 1.5% nickel, 0.8% chromium, and the balance iron. Used for steel castings.

Numa. Trade name for a *spandex* (segmented polyurethane) fiber used for the manufacture of textile fabrics with excellent stretch and recovery properties.

Numax. Trade name of RFD–GQ Limited (UK) for polyvinyl chlorides and vinyl plastics.

Numet. Trademark of National Lead Company (USA) for depleted uranium–a grade of metallic uranium containing less than 0.720 wt% of the fissile isotope uranium-235 (^{235}U). It has a density of 18.9 g/cm^3 (0.68 lb/in.^{3}), good strength, and is used as a structural material for radiation shielding applications, casks for fuel elements and radioisotopes, balance weights for airplanes, and high-speed rotors for gyrocompasses.

Numetal. Trade name of Allegheny Ludlum Steel (USA) for an alloy containing 77% nickel, 17% iron, 4.5% copper and 1.5% chromium. It has high magnetic permeability, and is used for sensitive relays and audio transformers.

nun's veiling. (1) A thin, lightweight plain-weave cotton, silk, wool or synthetic fabric, usually black, brown or white, and used for religious and other ceremonial dresses.

(2) A soft, thin, lightweight, plain-weave silk or worsted fabric similar to that described under (1), but made in a wide range of colors, and used for blouses, dresses and nightwear.

Nupol. Trade name of Cook Composites and Polymers (USA) for high-temperature vinyl ester resins.

Nu-Poly. Trade name of Nudo Products, Inc. (USA) for fiberglass and vinyl panels used for walls, ceilings and signs.

Nupron. Trade name for rayon fibers.

Nupronium. Trade name for rayon fibers and yarns used for textile fabrics.

Nu-Pyr-Loy. Trade name of Pyramid Steel Company (USA) for a tough, oil-hardening, shock-resistant tool steel used for pneumatic and shock tools, and shear blades.

Nural. Trade name of Aluminium-Zentrale e.V. (Germany) for a series of wrought and cast aluminum-copper, aluminum-silicon, aluminum-silicon-magnesium and aluminum-zinc alloys. Used for aircraft parts, engine pistons, housings, cylinder heads, fittings, forgings, and castings.

Nurel. Trademark of Nurel SA (Spain) for nylon 6 fibers and yarns used for textile fabrics.

Nürnberger Gold. German trade name for a gold-colored, corrosion-resistant copper alloy containing 2-7.5% aluminum and 0.2-2.5% gold. Used as a gold substitute, and for jewelry. In English-speaking countries it is also referred to as "Nuremberg gold."

Nurox. Trade name of Capitol Castings Inc. (USA) for a cast iron used for gears, shafts, and housings.

Nuroz. Trademark of Reichhold Chemicals, Inc. (USA) for polymerized wood rosin with a melting point of 76°C (169°F) and high oxidation resistance, used as a soldering flux, and in adhesives, hot-melt mixes, varnishes, waxed paper, paper coatings, and synthetic resins.

Nusat. Trade name of Atotech USA Inc. for a satin-finish nickel electroplate and plating process.

Nushank. Trade name of Atlas Specialty Steels (Canada) for a wear-resistant hollow drill steel containing 0.43% carbon, 0.6% manganese, 0.25% silicon, 3% nickel, 0.4% chromium, 0.25% molybdenum, and the balance iron. It has high strength, stiffness and toughness, and is used for tools, shanks, and detachable bit shanks for rock drill.

Nusite. Trade name of Nusite Steel Process Company (USA) for a heat-treated cobalt steel containing 0.9% carbon, 13% cobalt, and the balance iron. Used for permanent magnets.

Nutherm. Trade name of Atlas Specialty Steels (Canada) for a medium-alloy, air-hardening cold-work tool steel (AISI type A6) containing 0.7% carbon, 2% manganese, 0.3% silicon, 1.35% molybdenum, 1% chromium, and the balance iron. It has good nondeforming and deep-hardening properties, and good wear resistance. Used for plastic molds and inserts, blanking, coining, forming and other dies, bending tools, punches, shear blades, and other tools.

Nutmeg Chromium Plus. Trademark of Nutmeg Chrome Corporation (USA) for chrome/*Teflon* coatings.

nuttall oak. The hard, reddish-brown wood from the hardwood tree *Quercus nuttallii* belonging to the American red oaks. It has a coarse texture, and moderate strength and durability. Source: Lower Mississippi Valley (Mississippi, Missouri and Louisiana) and western Alabama. Used as lumber for constructional purposes, veneer, furniture, boxes, interior trim, agricultural implements, and handles. See also American red oak.

nutty putty. A name given to a highly viscoelastic silicone polymer that upon application of a gradually increasing tensile stress flows (elongates) like a highly viscous fluid, while bouncing elastically when made into a ball and dropped onto a flat surface. This product is sold by Binney & Smith Inc., USA, under the trademark *Silly Putty*.

Nuva-Sil. Trademark of Henkel Loctite Corporation (USA) for UV-curable electronic silicone-based potting and sealing compounds used for circuit board components, sensors, etc.

Nu-Vu. Trade name of PPG Industries Inc. (USA) for a 6.4 mm (0.25 in.) thick glass sheet that has microscopic ceramic particles permanently fused into the front surface. Used for rear-projection screens.

Nuweld. Trade name of ASG Industries Inc. (USA) for a wired glass with a diamond mesh.

Nu-Weld. Trade name of Dural Products Limited (Canada) for wood glues.

nyankom. See niangon.

Nyblade. Trade name of Sanderson Kayser Limited (UK) for a cold-work tool steel containing 0.5% carbon, 3% nickel, 1.1% silicon, 0.65% chromium, and the balance iron. Used for punches, chisels, shear blades, and picks.

Nybrad. Trade name of Glassmaster Monofilament Division (USA) for a nylon monofilament abrasive.

Nyby. Trade name of Nyby, Granges AB (Sweden) for an extensive series of austenitic, ferritic and martensitic stainless steels.

Nycast. Trademark of Cast Nylons, Limited (USA) for nylon-6 castings supplied as blocks, sheets, rods and tubes in unmodified, plasticized, heat-stabilized, molybdenum disulfide-filled and food grades.

Nycel. Trademark of Celanese Mexicana SA (Mexico) for nylon 6 fibers and yarns used for textile fabrics.

Nydur. Trademark of Bayer AG (Germany) for a polyamide (nylon 6) resin with good chemical resistance, durability and thermal stability. Glass-reinforced and impact-modified grades are also available. Used for housings for appliances, pumps, etc.

Nykon. Trade name of LNP Engineering Plastics (USA) for a series of polyamides (nylons). *Nykon P* is a molybdenum disulfide-lubricated polyamide 6 (nylon 6) and *Nykon R* a molybdenum disulfide-lubricated polyamide 6,6 (nylon 6,6).

Nykrom. Trade name for a nickel-chromium steel.

Nylacast. Trade name of Nylacast Corporation (USA) for cast polyamide 6 (nylon 6). It has high mechanical strength and creep resistance, high stiffness, good elevated temperature properties, a service temperature range of -40 to +160°C (-40 to +320°F), good resistance to alcohols, alkalies, hydrocarbons, greases, oils and ketones. Used for appliances, industrial equipment, automotive components, and lawn and garden equipment.

Nylafil. Trademark of DSM Engineering Plastics (USA) for impact-resistant glass-fiber-reinforced nylon 6, 6 and nylon 6,10 resins used for automotive, electrical and industrial applications.

Nylaflow. Trade name of The Polymer Corporation (USA) for low-friction nylon 6,6 resins used for low- and high-pressure tubing.

Nylamid. Trade name of Polymer Service Corporation (USA) for polyamide 6 and 6,6 (nylon 6 and 6,6) resins supplied in unreinforced and glass- and/or mineral-reinforced grades used for injection-molded and extruded products.

Nylasint. Trademark of Polymer Service Corporation (USA) for filled and unfilled polyamide (nylon) powder used for sintered bearings.

NylaSteel. Trademark of DSM Engineering Plastics (USA) for a composite material consisting of a core of stainless steel (AISI type 303) or cold-rolled steel (AISI-SAE type 1117) covered with *Nylatron* cast nylon, e.g., *Nylatron GSM* for impact resistance, *Nylatron NSM* for wear resistance, *Nylatron MC-907* for food-grade applications, or *Nylatron MC-901* for heat resistance to 175°C (260°F). Supplied as stock shapes in billet and blank form for use in machining gears, pulleys, rollers, sprockets, or wheels.

Nylatron. Trademark of DSM Engineering Plastics (USA) for an extensive series of modified polyamide thermoplastics based on nylon 6 (*Nylatron Nylon 6*) or nylon 6,6 (*Nylatron Nylon 6,6*). They are supplied as powders, granules, pellets and emulsions, often reinforced with carbon or glass fibers, lubricated with molybdenum disulfide, polytetrafluoroethylene or silicone, or filled with glass beads or minerals, and supplied in a wide range of grades including casting, elastomer-copolymer, fire-retardant, high-impact, supertough, stampable-sheet, and UV-stabilized.

Nylawear. Trademark of A.L. Hyde Company (USA) for wear-resistant low-friction lubricated cast nylon used for bushings, gears, rollers, pulleys, and timing screws.

Nylco. Trade name of Worthen Industries Inc., Nylco Division (USA) for metal-coated nylon fibers and the textile fabrics made from these fibers.

Nylene. Trademark of Custom Resins (USA) for an extensive series of polyamide 6 (nylon 6) homopolymers and copolymers available in extrusion, film, glass-reinforced, mineral- or molybdenum disulfide-filled, impact-modified, and nucleated grades.

Nylenka. Trade name of Enka BV (Netherlands) for nylon fibers and yarns used for textile fabrics including clothing.

Nylex. Trade name for lightweight, resilient, durable nylon fibers used for textile fabrics and brooms.

Nylfil. Trademark of Nylon de Mexico SA (Mexico) for nylon 6 fibers and yarns used for textile fabrics.

Nylfrance. Trade name of Rhône-Poulenc SA (France) for nylon fibers and products.

Nylhair. Trademark of Inquitex SA (Spain) for nylon 6 filament yarns used for textile fabrics.

Nylocel. Trademark of Quintex SA (Colombia) for nylon 6 filament yarns used for textile fabrics.

Nyloft. Trade name for nylon fibers and yarns used for textile fabrics including clothing.

Nyloil. Trademark of Cast Nylon, Limited (USA) for lubricated nylon castings supplied as blocks, sheets, rods and tubes.

Nylon. Trademark of E.I. DuPont de Nemours & Company (USA) for an extensive range of plastics and synthetic fibers based on polyamide polymers. The term "nylon" is now often used generically for thermoplastic polyamides. See also nylons.

nylon. See nylons.

nylon 4. A polyamide polymer based on butyrolactam (2-pyrrolidone). It has high tenacity, good abrasion resistance, high moisture absorption, an upper service temperature of about 100-150°C (210-300°F), and is commonly used in the form of textile fibers for fabrics. Formula: $[-C_4H_9N-]_n$. Also known as *poly(pyrrolidone)*.

nylon 4,6. A polyamide polymer that is the condensation product of adipic acid and tetramethylenediamine made with or without the addition of a heat stabilizer. Often supplied in pellet, film, sheet or rod form, it has a density of 1.18 g/cm³ (0.043 lb/in.³), a melting point of 295°C (563°F), good strength, low dielectric strength, good heat resistance, high stiffness, low creep at elevated temperatures, good processibility, a service temperature range of -40 to +155°C (-40 to +311°F), low thermal expansivity, and self-extinguishing properties. Formula: $[-NH(CH_2)_4NHCO(CH_2)_4CO-]_n$. Used for molded parts, packaging films, and electrical and electronic components. Also known as *poly(tetramethylene adipamide)*.

nylon 6. A polyamide polymer made from caprolactam with or without the addition of a heat stabilizer. It has a density of 1.08-1.13 g/cm³ (0.039-0.041 lb/in.³), good strength, high elonga-tion, good resistance to dilute acids, aromatic hydrocarbon, greases, oils, halogens and ketones, fair resistance to alcohols and alkalies, poor resistance to concentrated acids, fair resistance to ultraviolet radiation, good dielectric properties, a melting temperature of 228.5°C (443°F), a continuous service temperature range of -40 to +160°C (-40 to +320°F), a glass-transition temperature of 62.5°C (144°F), low thermal expansivity, a low coefficient of friction, and very good self-extinguishing properties. Used for molded parts, sheet, monofilament, rods and film. Formula: $[-NH(CH_2)_5CO-]_n$. Also known as *poly(caprolactam)*.

nylon 6,6. A polyamide polymer that is the condensation product of adipic acid (obtained by catalytic oxidation of cyclohexane) and hexamethylenediamine. Usually supplied in pellet or granule form, it has a density of 1.09-1.14 g/cm³ (0.039-0.041 lb/in.³), good strength, melting point of 267°C (513°F), a continuous service temperature range of -30 to +180°C (-22 to +356°F),

a glass-transition temperature of 45°C (113°F), a refractive index of 1.565, a low thermal expansivity, a low coefficient of friction, good resistance to dilute acids, alkalies, aromatic hydrocarbons, greases, oils, halogens and ketones, fair resistance to alcohols, poor resistance to concentrated acids, fair resistance to ultraviolet radiation, good dielectric and self-extinguishing properties. It is available in neat, fiber-reinforced and lubricated grades. Used for synthetic fibers, engineering resins, films and sheets, gears and mechanical parts, in biochemistry, chemistry, biotechnology and medicine. Formula: $[-NH(CH_2)_6NHCO(CH_2)_4CO-]_n$. Also known as *poly(hexamethylene adipamide)*.

nylon 6,66. A polyamide that is a copolymer of polyhexamethylene adipamide and caprolactam made with the addition of a heat stabilizer, a nucleating agent and a lubricant. Usually supplied in pellet form, it has a density of 1.30 g/cm³ (0.047 lb/in.³), a melting point of 250-260°C (480-500°F). Used as a general-purpose thermoplastic resin, and as a heat stabilizing and nucleating agent in other polymers. Formula: $[-NH(CH_2)_6NHCO(CH_2)_4CO-]_x[-NH(CH_2)_5CO-]_y$. See *poly(hexamethylene adipamide-co-caprolactam)*.

nylon 6,9. A polyamide polymer that is a condensation product of undecanedioic acid and hexamethylendiamine. Often supplied in pellet form, it has a density of 1.08 g/cm³ (0.039 lb/in.³), a melting point of 210°C (410°F), and a glass-transition temperature of 58°C (136°F). Used for molded parts, and for sheets and films. Formula: $[-NH(CH_2)_6NHCO(CH_2)_7CO-]_n$. Also known as *poly(hexamethylene nonanediamide)*.

nylon 6,10. A polyamide polymer made from sebacic acid and hexamethylendiamine, and is often supplied in pellet or monofilament form. It has a density of 1.04 g/cm³ (0.038 lb/in.³), high toughness, good resistance to elevated temperatures, a melting point of 240°C (464°F), a glass-transition temperature of 40°C (104°F), low water absorption, and a refractive index of 1.565. Used for paint and other brush bristles, fishing line, sewing thread, in dentistry for denture bases, and in biochemistry and biotechnology. Formula: $[-NH(CH_2)_6NHCO(CH_2)_8CO-]_n$. Also known as *poly(hexamethylene sebacamide)*.

nylon 6,12. A polyamide polymer made from hexamethylenediamine and 1,12-dodecanedioic acid with or without the addition of a lubricant. Often supplied in pellet form, it has a melting point of 218°C (425°F), and a glass-transition temperature of 46°C (115°F). Used as a general-purpose thermoplastic resin. Formula: $[-NH(CH_2)_6NHCO(CH_2)_{10}CO-]_n$. Also known as *poly(hexamethylene dodecanediamide)*.

nylon 9. A polyamide polymer obtained from soybean oil (9-aminononanoic acid) by reacting with ozone. It has a low moisture absorption, good dimensional stability, and good electrical resistance (better than nylon 6 and nylon 6,6). Used for coatings on metals, and electrical and electronic components. Also known as *poly(aminononanone)*.

nylon 11. A polyamide polymer made from castor bean oil (ω-amino acids). Often supplied in pellet or rod form, it has a density of 1.02-1.04 g/cm³ (0.037-0.038 lb/in.³), good strength, good dimensional stability, a melting point of 198°C (388°F), a glass-transition temperature of 46°C (115°F), a service temperature of -50 to +130°C (-58 to +266°F), low thermal expansivity, a low coefficient of friction, low water absorption, good resistance to dilute acids, alkalies, aromatic hydrocarbons, greases, oils, halogens and ketones, fair resistance to concentrated acids and alcohols, fair resistance to ultraviolet radiation, good dielectric properties, and good self-extinguishing properties. Used for powder coatings on metals applied by electrostatic-spray or fluidized-bed techniques, in the biosciences, and for injection-molded parts. Formula: $[-NH(CH_2)_{10}CO-]_n$. Also known as *poly(undecanolactam); poly(undecanoamide)*.

nylon 12. A polyamide polymer made from butadiene. It has a density of 1.01 g/cm³ (0.036 lb/in.³), good strength, good dimensional stability, a melting temperature of 178°C (352°F), a glass-transition temperature of 37°C (99°F), a low coefficient of friction, low thermal expansivity, low water absorption, good resistance to dilute acids, alkalies, aromatic hydrocarbons, greases, oils, halogens and ketones, fair resistance to concentrated acids and alcohols, fair resistance to ultraviolet radiation, and good dielectric and self-extinguishing properties. Used for powder coatings on metals applied by electrostatic-spray or fluidized-bed techniques, in dentistry for denture bases, in biochemistry and medicine, for molded parts, and in film form for packaging and wrapping applications. Formula: $[-NH-(CH_2)_{11}CO-]_n$. Also known as *poly(lauryllactam)*.

nylon-ABS alloys. Thermoplastic blends of crystalline nylon and amorphous acrylonitrile-butadiene-styrene (ABS) with high toughness and impact resistance, outstanding processibility, excellent flow characteristics, good abrasion, chemical and heat resistance, good surface finish, and reduced moisture sensitivity. Used for lawn and garden equipment, power-tool housings, sporting goods, and automotive components.

nylon block copolymer. A copolymer of nylon 6 and rubber corresponding to the specific A-B-A block structure. Its properties can be greatly modified by varying the quantity and type of rubber. It combines the excellent strength, stiffness and elevated-temperature properties of nylon 6 with the toughness and flexibility of rubber. Abbreviation: NBC. See also block copolymer; nylon 6; nylon-EPDM alloys.

Nylon Ducilo. Trademark of DuPont SA (Argentina) for nylon 6 and nylon 6,6 filament yarns used for textile fabrics.

nylon-elastomer alloys. Blends of nylon and elastomers, such as ethylene-propylene terpolymer (EPDM), having good resiliency, elongation and impact resistance, and good strength, stiffness and other structural properties. Used for automotive bumper components, power-tool housings, and sporting goods.

nylon-EPDM alloys. Blends of crystalline nylon and elastomeric ethylene-propylene-diene monomer (EPDM) rubber having good resiliency, elongation and impact resistance, high strength and stiffness, and other structural properties. Used for automotive bumper components, power-tool housings, and sporting goods. See also nylon block copolymer.

nylon fibers. Generic term for a group of synthetic long-chain polyamides (including polyamide 6 and 6,6) with high abrasion resistance, strength and toughness, good elasticity, low permanent elongation, high gloss, low water absorption, good mildew resistance, poor resistance to mineral acids, and good resistance to alkalies. Used for clothing, draperies, curtains, floor coverings including carpets, upholstery, yarns, tire cords, etc. See also nylon; polyamide fibers (2).

nylon film. Any of several nylon polymers supplied in the form of a film of varying thickness. *Nylon 6 film* is available in thicknesses of 0.05-0.50 mm (0.002-0.02 in.) and *Nylon 6,6 film* in thicknesses of 0.15-0.50 mm (0.006-0.02 in.). They have a heat-sealing temperature of 0°C (32°F), low permeability to carbon dioxide and oxygen, and good dielectric properties. Other nylons are also available in film form. Used for heat-sealing applications, wrapping of food products, etc.

nylon foam. A flexible or rigid foam made from *nylon 6* and containing numerous bubbles or cells resulting from the action of a blowing agent. Used for buoys, safety products, flotation devices, and light construction. Also known as *cellular nylon; foamed nylon*.

Nylon Industrial Fiber. Trade name of Solutia Inc. (USA) for a heavy-denier nylon 6,6 fiber used exclusively for industrial applications including tire cords, hoses, ropes, conveyor belts, and airbags.

nylon monofilament. A single-strand or thread of nylon, typically 0.15-1.0 mm (0.006-0.04 in) in diameter and 10-1000 m (33-3280 ft) in length, with good strength, elasticity and toughness, high gloss, low permanent elongation, low water absorption, and good resistance to alkalies. Used for brush bristles, strings for tennis rackets, fishing lines, and netting.

nylon-PAR alloys. Thermoplastic blends of crystalline nylon and amorphous polyarylate (PAR), which have good dimensional stability, high impact strength, freedom from warpage, good chemical properties, and are used for automotive body panels.

nylon plastics. A family of plastics based on any of a group of thermoplastic polyamides (*nylons*) that are condensation products of aliphatic diamines and aliphatic dicarboxylic acids.

nylon-PPE alloys. Polymer blends of a crystalline nylon and amorphous polyphenylene ether (PPE) with good thermal properties and chemical resistance used for automotive components including body panels.

nylon-polyphenylene oxide alloys. See nylon-PPO alloys.

nylon-PPO alloys. Immiscible blends of crystalline nylon 6,6 and amorphous polyphenylene oxide with enhanced chemical resistance, improved dimensional stability at elevated temperatures, low moisture sensitivity and mold shrinkage, very good paint adhesion, and good resistance to chipping. Used for consumer and industrial products, and automobile components including doors, hatchbacks and wheel panels.

nylon rope. Rope made from nylon filament fibers typically 5-102 mm (0.2-4 in.) in diameter. It has excellent elastic and tensile properties, very high dry and wet breaking strength, high creep resistance, outstanding impact resistance, excellent abrasion and wear resistance (much better than manila rope), good ability to withstand severe and sudden shocks, negligible softening with increasing temperature up to a maximum service temperature of approximately 150°C (300°F), good resistance to oils, solvents and alkalies, poor resistance to chemicals, most acids, paint and linseed oil, good dielectric properties (when dry and clean), and good resistance to rot, mildew and fungi. Used for various lifting and towing applications.

nylons. Generic term for a group of thermoplastic polyamides that are condensation products of aliphatic diamines and aliphatic dicarboxylic acids. They have high strengths and moduli, high rigidity and impact resistance, good thermal, electrical and chemical resistance, excellent resistance to dilute and weak acids and bases, good resistance to most organic solvents, fair resistance to strong and concentrated bases, fair ozone resistance, poor resistance to alcohols, glycols, strong and concentrated acids, and ultraviolet radiation, good processibility, self-extinguishing properties, a typical service temperature range of -51 to +150°C (-60 to +200°F). Used for bearings, bushings, gears, cams, fasteners, tubing, hose, automotive fuel tanks, electrical and thermal insulation, gaskets, tire cord, cordage, netting, bristles for brushes, carpets and upholstery, sporting goods, parachutes, film products, fibers, yarns and fabrics, coatings for metallic substrates, molding powders, molded products (e.g., rods, bars, and sheets), etc. See also nylon 4; nylon 4,6, nylon

6; nylon 6,6; nylon 6,9; nylon 6,10; nylon 6,12; nylon 6,66; nylon 6,12; nylon 9; nylon 11; nylon 12.

Nylon Salt. Trade name of Solutia Inc. (USA) for an aqueous solution of nylon 6,6 monomers produced from adipic acid and hexamethylene diamine by condensation polymerization. Used as a polymerization intermediate for the manufacture of nylon fibers and resins.

nylon-spandex fibers. Bicomponent fibers composed of alternating nylon and spandex (polyurethane) segments and combining the unique properties of both polymers. Used for textile fabrics including clothing.

Nylopak. Trademark of Bengal Fiber Industries (Pakistan) for nylon 6 filament yarns.

Nylotex. Trademark of Aberton Textiles Limited (Canada) for nylon knitting yarns.

Nylsuisse. Swiss trade name for nylon fibers and yarns used for cordage and textile fabrics.

Nyma. Trade name for rayon fibers and yarns used for textile fabrics.

Nymarilon. Trade name for acrylic fibers used for the manufacture of textile fabrics.

Nymata. Trade name for rayon fibers and yarns used for textile fabrics.

Nymatco. Trade name for rayon fibers and yarns.

Nymax. Trade name of M.A. Hanna Engineered Materials (USA) for mineral- and/or glass-fiber-filled nylon 6,6 compounds used for automotive applications, and for the manufacture of crystalline polymers.

Nymcel. Trade name for rayon fibers and yarns used for textile fabrics.

Nymcord. Trade name for high-tenacity rayon fibers used for industrial fabrics including tire cords.

Nymcrylon. Trade name for acrylic fibers used for the manufacture of textile fabrics.

Nymella. Trade name for rayon fibers and yarns used for textile fabrics.

Nymphwrap. Trade name of Sylvania Industrial Corporation (USA) for cellophane-type regenerated cellulose used for wrapping purposes.

Nypel. Trademark of AlliedSignal Corporation (USA) for a series of polyamides based on nylon 6,6, and supplied in unreinforced and glass- or carbon-fiber-reinforced, high-impact, supertough, fire-retardant, UV-stabilized, molybdenum disulfide- or polytetrafluoroethylene-lubricated, and glass bead- or mineral-filled grades. *Nypel Polypro* refers to a series of polyolefin fibers, resins and products.

Nyrim. Trademark of DSM Engineering Plastics (USA) for engineering thermoplastics based on a nylon-6 block copolymer. They are usually processed by reaction injection molding, or reactive casting methods, and have very high damage tolerance, excellent toughness, good high and low-temperature impact resistance, good resistance to extremely high and low temperatures, good abrasion resistance, good strength and durability, good corrosion resistance, and good surface finish. Used for casings and housings for industrial equipment, automotive components, such as cooling fans, sheaves, rollers, pulleys, gears, and wheel chocks.

Nytelle. Trade name for nylon fibers and yarns used for textile fabrics.

Nytex. Trademark of Shreeji Screens & Filters (India) for a hybrid fabric woven from nylon and rayon fibers, and used for tire cord.

nytril fibers. A generic term for man-made textile fibers containing 85% or more of a long-chain vinylidene dinitrile ($-CH_2-C-(CN)_2-$) polymer, where the vinylidene dinitrile content is at least every other unit in the polymer chain. They have excellent thermal resistance at temperatures exceeding 150°C (300°F), good to excellent resistance to acids, fair resistance to cold dilute alkalies, poor resistance to strong alkalies, outstanding resistance to outdoor weathering in sunlight, and good mildew resistance. Used for apparel, sweaters, knitting yarns, and pile and industrial fabrics.

NZPlus. Trademark of Norton Company (USA) for fused zirconia-alumina abrasives supplied in grit sizes of 20-220, and used for coated abrasive applications.

O

oak. The tough, hard, durable wood of any of a genus *(Quercus)* of deciduous trees and shrubs. North American oaks are divided into *red* and *white oaks* with each group containing several species. Common European oaks include the English oak *(Q. robur)* and the chestnut oak *(Q. petraea)*. The Japanese oak *(Q. mongolica* var. *grosseserrata)* is a familiar Asian species. Certain Australian and Tasmanian eucalyptus trees, such as *Eucalyptus obliqua, E. regnans* and *E. gigantea,* are also referred to as "oaks." Oak is used in the construction of buildings, and for furniture, millwork, shipbuilding, veneer, etc. See also American red oak; American white oak; European oak; Japanese oak; Tasmanian oak; red oaks; white oaks.

oakum. Loose hemp fibers obtained by untwisting and picking apart old ropes. They are usually impregnated with tar or pitch and may contain considerable quantities of jute fibers and new tow fibers. Formerly used for stopping up the seams or cracks in ship decks and sides, and for general caulking applications and for packing caisson and pipe joints.

Oamaru stone. A white limestone with granular texture quarried at Oamaru, New Zealand, and used as a valuable building stone.

Oasis. (1) Trademark of E.I. DuPont de Nemours & Company (USA) for polyimide/fluoropolymer composite films.

(2) Trade name of Technical Absorbents/Acordis Group (UK) for a superabsorbent fiber that can be spun into yarns and made into woven and nonwoven fabrics.

oatmeal cloth. A soft, usually heavy fabric having a specked, uneven surface. Used especially for drapery and upholstery.

oatmeal paper. A paper with a coarse, flaky surface texture made by adding fine sawdust or wood flour during the forming of the sheet on the wire screen of a paper machine. Used for wallpaper and sketching paper.

obeche. The light, soft, almost white wood from the tree *Triplochiton scleroxylon*. It has poor decay resistance, low strength, works and machines well, takes a fine finish, and has good staining and painting qualities. Average weight, 380 kg/m^3 (24 lb/ft^3). Source: West and Central Africa. Used for veneer and corestock. Also known as *samba; wawa.*

Oberlander's Resin. Trade name for a phenol-formaldehyde resin formerly used for denture bases.

oboyerite. A milk-white mineral composed of hydrogen lead tellurium oxide dihydrate, $H_6Pb_6(TeO_3)_3(TeO_6)\cdot 2H_2O$. Crystal system, triclinic. Density, 6.40 g/cm^3. Occurrence: USA (Arizona).

obscure glass. See translucent glass.

obsidian. A hard, dark-colored glassy rock formed by the rapid cooling of molten lava. It is high in siliceous materials, has a relatively low water content, and resembles *granite* in composi-

tion. Occurrence: Iceland, Mexico, Italy. Used for ornamental and decorative applications. Also known as *Iceland agate.*

obsidianite brick. A highly siliceous fireclay used to make lightweight, acid-resisting brick that is burnt to a glassy mass and somewhat resembles *obsidian* in appearance.

Occidental Furane. Trade name of Occidental Chemical Corporation (USA) for a series of furane resins.

Occlusin. Trade name of GC America, Inc. (USA) for a posterior dental composite.

Ocean. Trademark of Ocean Construction Supplies Limited (Canada) for packaged concrete dry-mixes, and concrete blocks, brick and pipe.

Oceane. Trade name for cellulose acetate fibers used for wearing apparel and industrial fabrics.

ocher. A naturally occurring yellow, red or brown powder composed of hydrated ferric oxides, clay and siliceous matter (sand), and used as a paint pigment, in papermaking, and in engobe slips, underglaze colors and overglaze decorations. See also burnt ocher.

ochoo pine. The coarse-textured wood of the tall coniferous tree *Hura crepitans* of the Euphorbiaceae family. It has pale yellow heartwood, low durability, low shrinkage, moderate permeability, high susceptibility to blue stain, good machinability and finishability. Average weight, 440 kg/m^3 (27 lb/ft^3). Source: Humid tropical and transitional subtropical forests. Used for carpentry work, doors, furniture, moldings, plywood, fiberboard and particleboard, boxes, coffins, toys, etc. Also known as *ochoo; possumwood; assacu.*

ocote. The strong, hard, heavy, yellowish to reddish wood from the pine tree *Pinus oocarpa*. It has good workability and machinability and moderate shrinkage. Source: Northwestern Mexico; Guatemala, Nicaragua. Used in building construction, boxes, pallets, crates, cooperage, poles, and piles. Also known as *okote.*

Ocrate. Trade name for a cement treated with gaseous silicon tetrafluoride (SiF_4) that reacts with any free lime (CaO) to produce calcium fluoride (CaF_2). It has increased density, and enhanced resistance to abrasion, wear, acids and chemicals. The treatment is also referred to as "Ocrate."

octaethylporphine. Purple crystals (97% pure) used in the synthesis of organometallic complexes. Formula: $C_{36}H_{46}N_4$. Abbreviation: OEP. See also porphine.

octahedrite. See anatase.

Octanium. Trade name of Parker Pen Company (USA) for a nonmagnetic, noncorrosive alloy of 40% cobalt, 20% chromium, 15.5% nickel, 15% iron, 7% molybdenum, 2% manganese, 0.15% carbon and 0.03% beryllium. Used for fountain pen nibs.

Octight. See Cavex Octight.

octithiophene. An organic semiconductor that can be synthesized by oxidative dimerization of quaterthiophene (4T) and purified by vacuum sublimation. The resulting material is a crystalline, bright-red powder with a melting point of 370°C (698°F). Solid *octithiophene* is an elongated planar molecule having photovoltaic and p-type electron transport properties. It is useful for Schottky and p-n junction devices. Abbreviation: 8T.

Octiva. Trade name of Mathison's (USA) for thermal film products for the graphic arts and allied trades. They consist of polyester substrates coated with copolymer adhesives, and are supplied in several sizes and surface finishes including Clear Gloss, Luster and Matte and Textured Satin. *Octiva Write Erase* is a special heat-activated film product with clear surface finish, used especially for applications involving dry erase markers.

Octiva Lo-Melt. Trade name of Mathison's (USA) for thermal

807

film products consisting of polyester or polyvinyl chloride substrates coated with low-melt adhesives, and supplied in several sizes and surface finishes including Clear Gloss, Heat Set Gloss, Luster, Matte and Textured Satin.

octylthiophene. A liquid organic compound (97% pure) with a density of 0.920 g/cm^3, a boiling point of 106-107°C (223-224°F)/3 mm, a flash point above 230°F (110°C) and a refractive index of 1.492. Used as a precursor for conducing polymers. See also thiophene; poly(thiophene).

o'danielite. A pale violet mineral composed of hydrogen sodium zinc arsenate, $NaZn_3H_2(AsO_4)_3$. Crystal system, monoclinic. Density, 4.24 g/cm^3; refractive index, 1.753. Occurrence: Namibia.

Odessa. Trade name for a corrosion-resistant alloy composed of 42.5% copper, 33.25% silver, 15.75% zinc and 8.5% nickel. Used as a silver solder.

Odinit. Trade name of Heinrich Knoell Co. (Germany) for phenol-formaldehyde resins and plastics.

ODM alloys. A family of oxide-dispersion strengthened alloys developed by Dour Metal SA (Belgium) and produced from mechanically alloyed starting powders by consolidation into billets, followed by two-stage hot working at high and low temperature, and two final heat treatments. The alloys contain 13-20% chromium, 3-6% aluminum, 1.5% molybdenum, 0.6% titanium, 0.5% yttria, and the balance iron. They have high strength and oxidation resistance at temperatures up to 1200°C (2190°F), and are suitable for high-temperature applications, e.g., heat-exchanger components.

Oerstit. Trade name of Thyssen Edelstahlwerke AG (Germany) for an extensive series of cast or sintered isotropic and anisotropic permanent magnet alloys of the iron-cobalt-nickel-aluminum-copper type, sometimes containing titanium. Used for magnetic and electrical applications.

OFE-HIT. Trade name of Copper & Brass Sales, Inc. (USA) for oxygen-free electronic copper.

off-axis laminate. A laminate in which the principal axis and direction of loading or stress is oriented at an angle other than 0° or 90° with respect to a reference direction, e.g., the fiber direction in a fiber composite.

offhand glass. Glassware formed by a craftsman, working without the benefit of molds.

office paper. A collective term for paper used in business and administrative offices, and including typewriter, writing, xerographic, facsimile, onionskin, accounting, chart, airmail, carbon and carbonless paper.

offretite. A colorless mineral of the zeolite group composed of potassium calcium magnesium aluminum silicate hydrate, $(K,Ca,Mg)_3Al_5Si_{13}O_{36}\cdot14H_2O$. Crystal system, hexagonal. Density, 2.13 g/cm^3; refractive index, 1.489. Occurrence: France.

offset paper. A coated or uncoated, somewhat porous paper that meets the requirements of offset printing in terms of material composition and surface properties.

offshore steels. A group of normalized or quenched-and-tempered steels including medium-strength structural carbon steels, such as ASTM A36, and high-strength low-alloy structural steels, such as ASTM A633, that owing to their good weldability and fabricability, and high toughness are particularly suitable for offshore structures, such as drilling rigs and platforms, and shipbuilding applications.

OFHC copper. See oxygen-free high-conductivity copper.

OFHC sulfur copper. See oxygen-free high-conductivity sulfur copper.

oganwo. The wood of the African mahogany *Khaya senegalensis.* The heartwood is dark reddish brown, and the sapwood grayish to pinkish red. *Oganwo* machines and finishes well and has moderate stability. Source: West and East Africa from Senegal to Uganda and southward to Angola. Used for fine furniture, interior paneling, boat construction, veneer and plywood. Also known as *dry-zone mahogany; kail.*

ogdensburgite. A dark reddish brown mineral composed of calcium zinc iron arsenate hydroxide pentahydrate, $Ca_2ZnFe_6(AsO_4)_5(OH)_{11}\cdot5H_2O$. Density, 2.92 g/cm^3; refractive index, 1.775. Occurrence: USA (New Jersey).

ogeche tupelo. The moderately heavy, strong, stiff, light-colored wood of the hardwood tree *Nyssa ogeche.* It is very similar to water tupelo, but less common. *Ogeche tupelo* has a close grain, a uniform texture, high hardness and moderate shock resistance. Source: Southeastern United States (Coastal Plain swamps of northern Florida and southeastern Georgia). Used for boxes, crates, pallets, baskets, veneer, etc. Also known as *gopher plum; ogeechee tupelo; sour tupelo.* See also water tupelo.

Oglala. Trade name of AL Tech Specialty Steel Corporation (USA) for a molybdenum-type intermediate high-speed steel (AISI type M52) containing about 0.9% carbon, 0.1-0.4% manganese, 0.1-0.4% silicon, 4% molybdenum, 4% chromium, 2% vanadium, 1.25% tungsten, and the balance iron. It has good deep-hardening properties, high hot hardness, and excellent wear resistance. Used for woodworking tools, hydraulic pump parts, bearings, router bits, taps, chasers and drills.

Ohio. Trade name of Teledyne Ohiocast (USA) for a series of cast corrosion- and/or heat-resistant steels, alloy machinery steels, and high-strength structural steels.

Ohio buckeye. See American horse chestnut.

Ohio Die. Trade name of Teledyne Vasco (USA) for an air-hardening high-carbon, high-chromium cold-work tool steel (AISI type D2) containing about 1.5-1.6% carbon, 0.3-0.45% silicon, 0.2-0.3% manganese, 11.5-12.5% chromium, 0.75-0.85% molybdenum, 0.7-0.9% vanadium, and the balance iron. Free-machining and precision-ground grades are also available. *Ohio Die* has high abrasion and wear resistance, good deep-hardening properties, high hardness, outstanding toughness, and good machinability and nondeforming properties. Used for a variety of dies, and for shear blades, punches, lathe tools, bending, burnishing, seaming and spinning tools, bushings, rolls, knurls, and gages.

Ohioloy. Trade name of Teledyne Ohiocast (USA) for a series of cast steel and nickel products including various plain-carbon steels, low-alloy machinery and structural steels, chromium-molybdenum alloy steels, heat-resistant and stainless steels as well as several nickel and nickel-copper alloys.

Ohmal. Trade name for a heat-resistant alloy composed of 87.5% copper, 9% manganese and 3.5% nickel, and used for electrical resistors.

Ohmaloy. (1) Trade name of Wilbur B. Driver Company (USA) for a steel containing 0.6% carbon, 13% chromium, 4% aluminum, and the balance iron. It loses ductility above 870°C (1600°F), and is used for resistance wire, rheostats, and resistors.

(2) Trade name of Allegheny Ludlum Steel (USA) for a heat-resistant stainless steel containing 0.2% carbon, 13% chromium, 4% aluminum, and the balance iron. Used for rheostats, and heating elements.

Ohmalloy. Trade name of Gilby-Fodor SA (France) for a heat-resistant alloy containing 80% iron, 15% chromium and 5% aluminum. Used for electrical and electronic applications.

Ohmalon. Trade name of Wilbur B. Driver Company (USA) for a heat-resistant alloy of 74% iron, 14% aluminum and 12% chromium.

Ohmax. Trade name of Driver Harris Company (USA) for an alloy of 71.5% iron, 20% chromium and 8.5% aluminum. It has excellent heat resistance to 1200°C (2190°F), and is used for electrical resistances.

Ohmi. Trade name for rayon fibers and yarns used for textile fabrics.

ohmilite. A light pink mineral composed of strontium titanium silicate hydroxide dihydrate, $Sr_3TiSi_4O_{12}(OH) \cdot 2H_2O$. Crystal system, monoclinic. Density, 3.38 g/cm³. Occurrence: Japan.

oil asphalt. See asphalt oil.

oil black. A finely divided form of *carbon black* that is made by burning vaporized heavy oil fractions (e.g., mineral oils) in a furnace under conditions where combustion is incomplete. Used as a black pigment, and as filler in rubber and molding compounds. Incorporated into polymers, it enhances the protection against ultraviolet radiation and improves weatherability.

oil blue. A bluish-purple pigment based on copper sulfide used in varnishes.

oilcloth. (1) A fabric of cotton, linen, jute or hemp made glossy and waterproof on one side by coating with a composition obtained by mixing oil with chalk, colorants and turpentine, or synthetic resins. Upon drying it may be painted and/or decorated with printed designs. It has now been replaced by synthetic fabrics, but was formerly widely used for protective coverings, floor covering and tablecloth.

(2) A fabric made waterproof by treating with an oil, such as linseed. Used for protective clothing and coverings. Also known as *oilskin*.

oil-coated fabrics. Fabrics sealed and waterproofed by the application of an oil, such as linseed.

Oilcrat. Trade name of Marshall Steel Company (USA) for an oil-hardening nondeforming steel containing 0.95% carbon, 1.25% manganese, 0.15% vanadium, 0.5% chromium, 0.5% tungsten, and the balance iron. Used for punches, taps, crimpers and spindles.

Oil Cups. Trade name for an alloy composed of 88% copper, 7% zinc and 5% tin and used for oil cups and fittings.

Oildag. Trademark of Acheson Colloids Company (USA) for a suspension of graphite in oil used as a lubricant.

Oil Die Smoothcut. Trade name of Columbia Tool Steel Company (USA) for an oil-hardening cold-work tool steel (AISI type O3) containing 1.05% carbon, 0.8% manganese, 1.6% chromium, 0.5% tungsten, and the balance iron. It has good free-cutting properties, and is used for punches, hobs, and dies.

oiled silk. A soft silk fabric waterproofed by treatment with a drying oil, such as boiled linseed.

oiled viscose. A viscose fabric waterproofed by treating with a drying oil.

oiled paper. Wood-free paper, frequently orange in color, whose permanent transparency and imperviousness to water is obtained by treating with drying oils. Abbreviation: OP. Also known as *oil paper*.

oiled wool. (1) Dyed knitting wool that has not been scoured.

(2) Dyed spinning wool that still contains oil additions.

Oil Engine Babbitt. Trade name of Hoyt Metal Company of London Limited (UK) for a babbitt composed of tin, lead and antimony. It has a melting point of 239°C (462°F), and is used for oil-engine bearings.

oil-extended rubber. (1) A synthetic rubber with an addition of up to 50% of a processing oil, such as petroleum in emulsion to extend or increase its volume and thus decrease cost, and promote low-temperature properties. Abbreviation: OER.

(2) A synthetic rubber, such as nitrile or styrene-butadiene, whose volume has been extended or increased by the addition of plasticizers (usually aromatic or cyclic hydrocarbons) without a considerable decrease in its engineering properties. Abbreviation: OER.

oil-gas tar. A tar produced during the manufacture of oil gas by high-temperature cracking of oil vapors.

oil grain. Heavy, vegetable-tanned, oil-finished side leather with a pebbly effect on the grain side of the skin. Used for shoes

Oilgraph. Trade name of AL Tech Specialty Steel Corporation (USA) for a graphitic oil-hardening cold-work tool steel (AISI type O6) containing 1.45% carbon, 1.15% silicon, 0.8% magnesium, 0.2% chromium, 0.25% molybdenum, and the balance iron. It has good deep-hardening properties, good machinability, good wear and abrasion resistance, and good toughness. Used for structural parts, such as bushings, arbors, bodies and shanks for cutting tools and gages, and dies.

oil-hardening cold-work tool steels. A group of cold-work tool steels (AISI group O) containing about 0.85-1.55% carbon, 0.30-1.80% manganese, up to 1.50% silicon, 0.30-0.90% chromium, up to 0.30% molybdenum, up to 2.00% tungsten, 0.40% vanadium, and the balance iron. They attain their good deep-hardening properties by quenching in oil, and have good machinability, reduced resistance to deformation, moderate wear resistance and low hot hardness. Often sold under trade names and trademarks, such as *Badger, Col-Graph, Halgraph, Keewatin, Oilgraph, Saratoga, Stentor, Teenax, Trueglide* or *Wando*, they are used for structural parts, arbors, bushings, gages, machine parts, jigs, shanks for cutting tools, forming tools, medium-duty cutting tools, taps, dies, reamers, broaches and press tools.

oil-hardening steel. A carbon or alloy steel hardened by quenching in an oil, such as prepared mineral, vegetable, animal or fish oil. It exhibits improved hardness and toughness, and very low warpage.

oil-impregnated powder-metallurgy parts. Powder-metallurgy parts, such as self-lubricating bronze bearings, that have been finished after sintering and sizing by saturation with oil, either by immersion or by drawing the latter through the oil by external pressure, or vacuum.

Oilite. Trademark of Chrysler Corporation (USA) for an extensive series of powder metallurgy (sintered) materials used for self-lubricating bearings. *Oilite Aluminum* contains 85-87% aluminum, 4-5% copper, 3-4% tin, 3-4% lead, and up to 3% other elements. It has an as-sintered density of 2.2-2.4 g/cm³ (0.08-0.09 lb/in.³). *Oilite Brass* contains 77-80% copper, 1-2% lead, 0-0.25% iron, 0-0.1% tin, the balance being zinc. Its as-sintered density is 7.2-7.7 g/cm³ (0.26-0.28 lb/in.³). *Oilite Copper* contains up to 98% copper, and has an as-sintered density of 8.0 g/cm³ (0.29 lb/in.³) or more. *Oilite Lead Bronze* is composed of 76-78.5% copper, 14-16% lead, 6.5-8.5% tin, 0-1% iron, and 0-1% carbon. Its as-sintered density is 6.8-7.2 g/cm³ (0.25-0.26 lb/in.³). *Oilite Nickel Silver* contains 63.0-66.5% copper, 16.5-19.5% nickel and 16.5-19.5% zinc. It has an as-sintered density of 7.5-8.8 g/cm³ (0.27-0.32 lb/in.³). *Oilite Stainless Steel* contains 0.08-0.2% carbon, 70.5-73% iron, 17-19% chromium and 8-10% nickel, and has an as-sintered density of 6.0-6.4 g/cm³ (0.22-0.23 lb/in.³). *Oilite Steel* has up to 1% carbon, 2% nickel, 0.5% molybdenum, 0.4% manganese, the balance being iron. Its as-sintered density is 6.0-7.2 g/cm³ (0.22-0.26 lb/in.³). *Oilite Stressite Bronze* has a density of 7.2-7.8 g/cm³ (0.26-0.28 lb/in.³), and is composed of 93-96% copper, 4-6% tin and up 2.5% other elements.

oil lacquers. Air-drying dispersions of resins in oil varnish and volatile solvents that weather like oil paints.

oil-modified alkyd coatings. High-molecular-weight *alkyd resins* modified with drying oils, such as linseed, soybean or tung, and employed to produce surface finishes with increased hardness, durability, chemical and weather resistance. Used as house paints, baking enamels for appliances, and for automotive and marine applications. See also alkyd coatings; long-oil alkyd; short-oil alkyd.

oilnut. See butternut.

oil paint. A paint made by mixing pigments, such as red lead, white lead or zinc oxide, with oil varnish. Drying oils (e.g., linseed, tung, etc.) and siccatives can be added as required. *Oil paint* coats are soft and elastic, and are used to protect buildings, vehicles, bridges and other structures against corrosion. Abbreviation: OP.

oil paper. See oiled paper.

Oil Pump. British trade name for a leaded brass composed of 85% copper, 9% zinc, 3% tin and 3% lead. Used for oil pump parts.

oil-repellent textiles. Textile fibers, yarns or fabrics that have been specially treated with chemicals to enhance their resistance to wetting by oils and oily liquids.

oil sand. A porous sandstone rock or free-flowing sand bonded or impregnated with crude oil or other hydrocarbons. Occurrence: Canada (Alberta), USA (California, Utah), Trinidad, Venezuela. Used in oil production, and for high-strength foundry cores. Abbreviation: OS.

oil shale. A hard, dark brown, fine-grained, sedimentary carbonaceous rock that yields substantial amounts of oil when heated in a closed retort to 430-540°C (800-1000°F) or by direct combustion *in situ* in interior excavations. It contains a high percentage of kerogen which is the precursor for the formation of oil. Occurrence: Australia, USA (Colorado, Utah, Wyoming). Used chiefly in oil production. Abbreviation: OS.

oilskin. See oilcloth (2).

oil-soluble resin. A synthetic resin that can be modified by dissolving or dispersing in, or reacting with, a drying oil, such as linseed, soybean or tung at moderate temperatures, and that thereafter forms a uniform film when applied to a substrate.

oil stain. A wood finish that contains little pigment and unlike paint does not totally mask the wood, but changes its color while preserving its distinguishing characteristics. *Pigmented (opaque) oil stain* contains finely ground color pigments in solution with linseed oil. It covers the wood's natural color and obscures the grain pattern while revealing its texture. *Penetrating oil stain* adds color, while revealing both the grain pattern and the texture. *Oil stain* is supplied in various tones including cedar, cherry, ebony, mahogany, maple, oak, redwood, rosewood and walnut. Abbreviation: OS. See also wood stain; stain.

oilstone. A natural or synthetic, fine-grained, abrasive stone whose rubbing surface is oiled. Used as a whetstone for sharpening tools, and in finely ground form for grinding and polishing applications.

oilstone powder. A finely ground oilstone used for grinding and polishing metals.

Oil Stop. Trade name of Irvington Varnish and Insulator Company (USA) for protein plastics derived from cashew nut shells.

Oiltemp. Trade name of Bethlehem Steel Corporation (USA) for an oil-hardening steel containing 0.9% carbon, 1.1% manganese, 0.5% tungsten, 0.5% chromium, 0.2% vanadium, and the balance iron. Used for tools and dies.

oil-tempered steel. Steel heated above the transformation temperature into the *austenite* range, quenched in oil to form *martensite*, tempered (i.e., reheated to a temperature below the transformation range to increase toughness and ductility), and then cooled in air. It has a microstructure of *tempered martensite* composed of *alpha ferrite* and *cementite*.

oil varnish. A liquid made from resinous substances dissolved in a drying oil of vegetable origin (linseed, tung, etc.), and used to produce a glossy finish on wood, metal and other substrates. The formation of the glossy finish is due to polymerization of the oil. See also long-oil varnish; short-oil varnish.

Oilway. Trade name of H. Boker & Company (USA) for an oil-hardening steel containing 0.94% carbon, 1.2% manganese, 0.5% tungsten, 0.44% chromium, and the balance iron. Used for tools and dies.

oil-well cement. A slow-setting hydraulic cement suitable for use in oil wells under conditions of high pressures and temperatures, e.g., to set well casings, and support pipe. Abbreviation: OWC.

oil white. A mixture of lithopone and white lead ($2PbCO_3 \cdot Pb(OH)_2$) or zinc white (ZnO), and sometimes additions of gypsum ($CaSO_4 \cdot 2H_2O$), magnesia (MgO), silica (SiO_2) and whiting ($CaCO_3$). Used as a nontoxic replacement for white lead in house paint.

Ojibway. Trade name of Atlas Specialty Steels (Canada) for a water-hardening die steel containing 0.95% carbon, 0.3% manganese, 0.3% silicon, and the balance iron.

ojuelaite. A light, yellowish green mineral of the arthurite group composed of zinc iron arsenate hydroxide tetrahydrate, $ZnFe_2(AsO_4)_2(OH)_2 \cdot 4H_2O$. Crystal system, monoclinic. Density, 3.39 g/cm³; refractive index, 1.730. Occurrence: Mexico.

Okadur. Trade name of KM-Kabelmetall AG (Germany) for a series of age-hardenable aluminum alloys containing 2.5-5% copper, 0.2-1.8% magnesium and 0.3-1.5% manganese. Used for structural parts. A free-cutting grade containing 0.5-2.5% lead, and small additions of bismuth, cadmium and tin is also available for screw-machine products.

Okadurplat. Trade name of KM-Kabelmetall AG (Germany) for an age-hardenable aluminum alloy containing 2.5-5% copper, 0.2-1.8% magnesium and 0.3-1.5% manganese. Used for aircraft structures.

Okalux. Trademark of Okalux Kapillarglas GmbH (Germany) for opaque and transparent insulating and glass/plastic laminated glass.

okanoganite. A pale pink mineral composed of sodium calcium rare-earth fluoride borate silicate, $(Na,Ca)_3Ln_{12}Si_6B_2O_{27}F_{14}$. Crystal system, rhombohedral (hexagonal). Density, 4.35 g/cm³; refractive index, 1.753. Occurrence: USA (Washington).

OkemCoat. Trade name of Chemetal, Oakite Products Inc. (USA) for a conversion coating for zinc.

okenite. A colorless mineral composed of calcium silicate dihydrate, $CaSi_2O_5 \cdot 2H_2O$. Crystal system, triclinic. Density, 2.30 g/cm³. Occurrence: India, USA (California).

Oker. British trade name for a free-cutting brass containing 54-69% copper, 30-45% zinc, up to 0.5% tin and up to 1% lead. It has good strength and corrosion resistance, and good workability and machinability. Used for hardware, fittings, tubes, sheets, trim, architectural and ornamental applications.

Oker-Cast. British trade name for a free-cutting casting brass containing 72% copper, 24% zinc, 2.3% iron and 1.7% lead used for turbine parts.

okote. See ocote.

okoume. The pale-pink wood of the tree *Aucoumea klaineana*. It is lighter and softer than true mahoganies, but widely used as a substitute. *Okoume* has a uniform texture, a high luster, and saws poorly because of its high silica content, but can be peeled for veneer. Average weight, 430 kg/m³ (27 lb/ft³). Source: West Central Africa, Equatorial Guinea, Congo, Gabon. Used for decorative plywood paneling, furniture, doors, chests, boxes and boats. Also known as *gaboon; Gaboon mahogany*. See also mahogany.

Okulit. German trade name for a wired cast glass. Also known as *Drahtokulit*.

Olane. Trade name for polyolefin fibers and products.

Olasal. French trade name for phenol-formaldehyde resins and plastics.

Olde Rosedale. Trade name of Perma Paving Stone Company (Canada) for paving stone supplied in rectangular units, 225 × 150 × 60 mm (9 × 6 × 2.375 in.), square units, 150 × 150 × 60 mm (6 × 6 × 2.375 in.), and half squares, 75 × 150 × 60 mm (3 × 6 × 2.375 in.). They are available in several colors including terracotta, slate, and gray-blend. Used for courtyards, driveways, walkways, and patios.

Olde World Cobble. Trade name of Perma Paving Stone Company (Canada) for strong, durable paving stone that resembles antique European cobblestone and is supplied in random shapes in terracotta and gray colors.

oldfield pine. See loblolly pine.

Old Forge. Trade name of Permacon (Canada) for terracotta, gray or beige paving and wall stone with the rustic, time-worn appearance of old natural stone.

Old Genuine Babbitt. Trade name of Lumen Bearing Company (USA) for a *babbitt* containing 89% tin, 7.5% antimony and 3.5% copper, which corresponds closely to the composition of the original babbitt metal. Used for heavy-duty bearings. See also tin babbitt.

oldhamite. A tan-colored mineral of the halite group composed of calcium sulfide, CaS. It can also be made synthetically. Crystal system, cubic. Refractive index, 2.137.

Old Hickory. Trade name of Old Hickory Clay Company (USA) for a ball clay.

Old Mill Cobble. Trade name of Perma Paving Stone Company (Canada) for paving stone with a rough texture and cobble effect supplied in full, square, and half shapes in terracotta, granite, and antique-buff colors.

Oldopal. Trademark of Buesing & Fasch KG (Germany) for a series of unsaturated polyesters.

Olds Bearing Bronze. Trade name of Olds Alloys Company (USA) for a corrosion-resistant bronze containing 60-65% copper, 17-19% lead, 9-11% zinc, 4.5-6% nickel and 4.5-6.5% tin. Used for formed parts, bushings and bearings.

Olds Leaded Bronze. Trade name of Olds Alloys Company (USA) for a high-leaded tin bronze containing 70% copper, 20% lead and 10% tin. It is cast on steel backs, and used as a bimetal surface layer for bushings and bearings.

Oldsmoloy. Trade name of Olds Alloys Company (USA) for an alloy of 45% copper, 35% zinc, 15% nickel, and a total of 5% chromium, tin and manganese. It has good resistance to salt water and food acids, and is used for hardware, gears, bear-

ings, and food processing equipment.

Olefane. Trade name for a flexible, transparent polypropylene film with excellent resistance to oils, solvents and water, and a maximum service temperature of 121°C (250°F).

olefin copolymers. See polyolefin copolymers.

olefin elastomer. An elastomer that is the reaction product of the copolymerization or terpolymerization of olefins, such as butylene, ethylene, propylene or isoprene.

Olefinesse. Trademark of Billermann KG (Germany) for polypropylene staple fibers and filament yarns.

olefin fabrics. See polyolefin fabrics.

olefin fibers. See polyolefin fibers.

olefinics. See polyolefin plastics.

olefin-modified styrene-acrylonitriles. A class of opaque engineering thermoplastics made by combining styrene-acrylonitrile (SAN) with a grafted, saturated olefinic elastomer, such as ethylene-propylene rubber. They have a density of about 1.02 g/cm³ (0.037 lb/in.³), high toughness and ductility, excellent weather resistance, good processibility, and are used for automotive components, such as bumpers, exterior and interior trim, body moldings, for construction and building products, such as window frame components, shutters, fencing, downspouts, gutters, siding, trim, swimming pool components, boat hulls, outdoor furniture, etc. Abbreviation: OSA. See also styrene-acrylonitriles.

olefin plastics. See polyolefin plastics.

olefin resins. See polyolefin resins.

olefin rubber. See olefin elastomer.

olefins. (1) A class of highly reactive unsaturated hydrocarbons, including ethylene, propylene, butene and pentene, with one or more carbon-carbon double bond. The general formula is C_nH_{2n}.
(2) See olefin fibers.

Olehard. Trade name of Chisso America (USA) for filled polypropylenes.

oleoresin. A mixture of an essential oil and one or more plant resins.

oleoresinous coating. A slow-curing, oil-drying coating produced from oleoresinous varnish. It has excellent wetting properties and is used on poorly prepared surfaces.

oleoresinous paint. A paint based on an unsaturated oil, such as linseed, and a drier.

oleoresinous varnish. A varnish prepared from drying oils (e.g., tung, linseed, or fish oil) and natural or synthetic resins, such as rosins, gum congo or rosin-modified or oil-soluble phenolics. It produces pale-colored coatings (finishes) that turn yellow on exposure.

Oleplate. Trade name of Amoco Performance Products, Inc. (USA) for polypropylene resins.

Oletex. Trade name of RBX Corporation (USA) for polyolefin fibers and yarns used for textile fabrics.

Olevac. Trade name of Amoco Performance Products, Inc. (USA) for polypropylene resins.

olgite. A bright blue, or bluish green mineral composed of sodium barium strontium phosphate, Na(Sr,Ba)PO₄. Crystal system, hexagonal. Density, 3.94 g/cm³; refractive index, 1.623. Occurrence: Russian Federation.

oligomer. A low-molecular-weight polymer molecule composed of only a few monomer units, e.g., a dimer, trimer or tetramer.

Olin. Trade name of Olin Brass (USA) for an extensive series of coppers and copper-base alloys including various electrolytic, oxygen-free, fire-refined and beryllium coppers, aluminum, phosphorus and silicon bronzes, architectural, cartridge and jewelry bronzes, red and yellow brasses, low-, medium- and high-leaded brasses, copper nickels, nickel silvers, gilding metals, copper-iron-cobalt-tin-phosphorus alloys and copper-manganese brazing alloys.

olive drab. A dull, dark greenish-yellow dyed wool fabric used especially for army uniforms.

olivenite. A olive-green mineral of the adamite group with an adamantine to vitreous luster. It is composed of copper arsenate hydroxide, $Cu_2AsO_4(OH)$. Crystal system, orthorhombic. Density, 4.38 g/cm^3; refractive index, 1.810; hardness, 3 Mohs. Occurrence: Chile, UK, USA (Utah).

olive wood. The attractive wood of a family of evergreen, subtropical trees (genus *Olea*). It is brownish yellow in color, very hard, and has dark streaks. Used for ornamental carving and novelty items.

olivine. A term referring to a group of olive-green or grayish green to yellowish brown minerals including *chrysolite, forsterite, fayalite, peridot, monticellite* and *tephroite*. They are composed of magnesium-iron silicate $(Mg,Fe)_2SiO_4$. Crystal system, orthorhombic. Density, 3.2-3.6 g/cm^3; hardness, 6.5-7 Mohs. Used for refractories, cement, foundry sand, in the manufacture of electronic components, and ceramic-to-metal seals. Some varieties are used as gemstones.

olmsteadite. A deep brown or red-brown mineral composed of potassium iron niobium tantalum oxide phosphate dihydrate, $KFe_2(Nb,Ta)(PO_4)_2O_2 \cdot 2H_2O$. Crystal system, orthorhombic. Density, 3.31 g/cm^3; refractive index, 1.755. Occurrence: USA (South Dakota).

Olopol. Trade name of Monopol AG (Switzerland) for a protective interior tank coating.

olsacherite. A colorless mineral composed of lead selenate sulfate, $Pb_2(SO_4)(SeO_4)$. Crystal system, orthorhombic. Density, 6.72 g/cm^3; refractive index, 1.966. Occurrence: Bolivia.

olshanskyite. A colorless mineral composed of calcium borate nonahydrate, $Ca_3B_4O_9 \cdot 9H_2O$. Crystal system, monoclinic. Density, 2.23 g/cm^3; refractive index, 1.568.

Olympia. Trade name of J.F. Jelenko & Company (USA) for a high-strength ceramic dental alloy containing 51.5-57.0% gold and 35.0-38.5% palladium. Used for fusing metal to porcelain. *Olympia II* is a high-palladium ceramic dental alloy that provides outstanding castability and porcelain adherence, and is used for fusing dental porcelain to metal.

Olympic. (1) Trade name of William Oxley & Company (UK) for a high-speed steel containing 0.7% carbon, 18% tungsten, 4% chromium, 1% niobium, and the balance iron. Used for cutting tools, such as drills, lathe tools, milling cutters, reamers, and taps.

(2) Trade name of Latrobe Steel Company (USA) for an air-hardening, high-carbon, high-chromium-type cold-work tool steel (AISI D2) containing 1.5% carbon, 0.5% manganese, 0.3% silicon, 12% chromium, 0.9-1% vanadium, 0.75% molybdenum, and the balance iron. It has high hardness, good abrasion resistance, and good nondeforming properties. Used for tools and dies.

(3) Trade name of Chase Brass & Copper Company (USA) for a series of silicon bronzes and brasses containing about 77-96% copper, 1-4.25% silicon and 1-22% zinc. They have good strength and corrosion resistance, and are used for hardware, welding rod, cables, springs, and fasteners.

(4) Trade name of PPG Industries, Inc. (USA) for a complete line of exterior architectural coatings and wood stains used for siding, decks and outdoor furniture.

OMAg. Trade name of OMG Americas Metal Powders Division (USA) for silver containing a uniform dispersion of 0.5 wt% (*OMAg5*) or 1.0 wt% (*OMAg10*) nanometer-sized aluminum oxide (Al_2O_3) particles. Produced by a special powder metallurgy technique involving an internal oxidation process, they are supplied in powder, strip, bar, and wire form, can be processed by various techniques including hot extrusion and press-sinter repressing, and can also be combined with various other metal oxides (e.g., cadmium, tin, zinc and cobaltocobaltic oxide) to produce high-strength and high-conductivity products. Used for electrical contact buttons, fuel cells, etc.

olympite. A colorless mineral composed of sodium phosphate, Na_3PO_4. Crystal system, orthorhombic. Density, 2.80 g/cm^3; refractive index, 1.510. Occurrence: Russian Federation.

Omanitbronze. Trade name of Ostermann GmbH & Co. (Germany) for a corrosion-resistant bronze containing 81% copper, 10% aluminum, and a total of 9% iron and nickel. Used for fittings.

Oman Metal. Trade name of Oman Non-Friction Metal Company (USA) for a self-lubricating bearing alloy composed of 75% copper and 25% lead. Used for heavy-duty bearings.

Ombral. Trade name of Vereinigte Glaswerke (Germany) for a heat-absorbing, glare-reducing polished plate glass with a grayish to greenish tint.

ombré. Fabrics that either have shaded colors, or dyed, printed or woven designs whose colors change gradually from light to dark tones.

Omega. (1) Trade name of Bethlehem Steel Corporation (USA) for a shock-resisting tool steel (AISI type S5) containing 0.6% carbon, 0.7% manganese, 1.85% silicon, 0.25% vanadium, 0.5% molybdenum, and the balance iron. It has outstanding toughness, excellent shock resistance, high elastic limit, and good ductility and fatigue resistance. Used for pneumatic chisels, shear blades, rivet sets, punches, and blacksmith tools.

(2) Trade name of Stapleton Technologies (USA) for an electroless nickel coating.

Omega Alpha. Trade name of OmegaDent (South Korea) for corrosion-resistant, beryllium-free nickel-chromium dental bonding alloys.

Omega Extra. Trade name of Remystahl (Germany) for nickel-molybdenum machinery steels.

Omega Super. Trade name of Dr. Hesse & Cie. (Germany) for a semibright nickel electroplate and plating process.

Omega VMK. Trade name of OmegaDent (South Korea) for dental porcelains used for metal-to-ceramic bonding applications.

omeiite. A steel-gray mineral composed of osmium arsenide, $OsAs_2$. Crystal system, orthorhombic. Density, 11.21 g/cm^3. Occurrence: China.

Ommet Iron. Trade name of Molybdenum Co., N.O. (Austria) for a pure, sintered iron containing 0.01% phosphorus, 0.01% manganese, 0.001% carbon, and the balance iron.

Omnifit. Trademark of OmniTECHNIK GmbH (Germany) for anaerobic one-component adhesives.

Omniflex. Trade name of GC America, Inc. (USA) for a silicone elastomer used for dental impressions.

Omni-Nylon. Trademark of Fibrasomni SA (Mexico) for nylon 6 monofilaments.

Omni Saran. Trademark of Fibrasomni SA (Mexico) for saran (polyvinylidene chloride) monofilaments and fibers with good chemical resistance, good weatherability, and good resistance to mildew and insects. Used for carpets and rugs, draperies, upholstery, clothing, industrial fabrics, etc.

Omnisil. Trademark of Coe Laboratories, Inc. (USA) for a series of silicone elastomer-based dental impression materials, bonding agents and putties.

Omnite. Trade name of Omnite Company (Netherlands) for phenol-formaldehyde resins and plastics.

omphacite. A dark green mineral of the pyroxene group composed of calcium sodium magnesium iron aluminum silicate, $(Ca,Na)(Mg,Fe,Al)Si_2O_6$. Crystal system, monoclinic. Density, $3.39 \ g/cm^3$; refractive index, 1.693. Occurrence: Norway.

Omyacarb. Trade name of OMYA GmbH (Austria) for a range of carbon-derived fillers for paper, paints, lacquers, varnishes, plastics and adhesives.

Onanth. Trade name for nylon fibers and products.

Onazote. Trade name of Expanded Rubber Company (UK) for hard rubber.

O.N.C. Trade name of General Electric Company (USA) for asbestos insulating cable products.

Ondine. Trade name of SA Glaverbel (Belgium) for a figured glass.

Ondule. Trade name for rayon fibers and yarns.

ondule. A plain-weave fabric of cotton, silk or synthetic fibers, having an irregularly woven warp creating a wavy surface effect.

Ondulette. French trade name for rayon fibers and yarns used for textile fabrics.

Ondulex. Trade name of Verreries de la Gare et A. Belotte Réunies SARL (France) for a reinforced, corrugated glass.

One Coat Bond. Trade name of Coltene-Whaledent (USA) for a single-component, light-cure dental adhesive system with gel-like consistency. It provides high bond strength, excellent wetting characteristics, and can be used for bonding tooth structure (i.e., dentin and enamel), and metals, ceramics, glass ionomers and compomers.

one-coat ware. (1) Ceramic ware to which only one coat of porcelain enamel has been applied, i.e., the ground and cover coat have been combined into one coat to serve protective, decorative, and adherence-promoting functions.

(2) Ceramic ware to which one coat of porcelain enamel has been applied over a ground coat.

one-component adhesive. A film or paste adhesive containing a latent curing agent or catalyst which is activated by heat, e.g., an adhesive consisting of a low-melting epoxy resin and a curing agent that is activated at relatively high temperature. Also known as *one-part adhesive*.

one-dimensional nanostructures. See nanostructured materials.

one-dimensional photonic crystals. See 1D photonic crystals.

one-directional fabrics. See unidirectional fabrics.

one-face fabrics. Textile fabrics that have only one side with an appearance suitable for use as the face side.

One-Five-One. Trade name of George Cook & Company Limited (UK) for an air-hardening cold-work tool steel (AISI type A2) containing 1% carbon, 5% chromium, 1% molybdenum, 0.5% manganese, 0.3% vanadium, and the balance iron. Used for dies, cutters, and various tools.

Oneida. Trademark of Associated Steel Corporation (USA) for an oil-hardening tool steel (AISI type O1) containing 0.9% carbon, 1.2% manganese, 0.5% tungsten, 0.5% chromium, 0.2% vanadium, and the balance iron. Used for press tools, taps, reamers, and broaches.

one-part adhesive. See one-component adhesive.

Oneral. British trade name for cobalt-base superalloys containing 27-28% chromium, 6-17% nickel, 5-10% molybdenum, 0-4% iron, 0.3-1% carbon, 0.03% titanium and 0.03% zirconium. They have high heat resistance, and is used for high-temperature components, and cast turbine blades.

One-Step. Trade name of BISCO Inc. (USA) for a one-step dental bonding system.

one-step resin. See A-stage resin.

One Ton Brass. British trade name for a corrosion-resistant brass of 61% copper, 38% zinc and 1% tin. Used for marine parts, condenser tubes and plates, and heat exchangers.

1-2-3 superconductors. Ceramic high-temperature superconductors, such as yttrium barium copper oxide ($YBa_2Cu_3O_x$), having a metal ion ratio (e.g., Y:Ba:Cu) of 1:2:3. Other examples of such superconductors include samarium barium copper oxide ($SmBa_2Cu_3O_x$) and gadolinium barium copper oxide ($GdBa_2Cu_3O_x$).

One-Up Bond F. Trade name of J. Morita Company (Japan) for a self-etching, fluoride-releasing, light-cure bisphenol A glycidyl-ether dimethacrylate (BisGMA) based dental resin that contains the proprietary MAC-10 adhesive monomer. It provides strong adhesion to the tooth structure (i.e., dentin and enamel), and is used in dentistry as a bonding agent for restorations, and for sealing exposed cervical and root surfaces.

one-way fabrics. Textile fabrics with a unidirectional nap or pile, i.e., whose fuzzy, furry or looped surface fibers have been laid in one direction.

one-way slab. A slab of reinforced concrete in which the steel rods are vertical to the supporting beam.

on-fire glaze. A durable overglaze applied and fired on the glazed surface of ceramic ware. Used as a decoration. Also known as *on-glaze*.

on-glaze. See on-fire glaze.

O-Nickel. Trade name of Henry Wiggin & Company Limited (UK) for a heat- and corrosion-resistant commercially pure nickel (99.5+% including some cobalt) with a controlled amount of magnesium. Used for valve components and tube cathodes.

Onion fusible alloy. A white metal alloy of 50% bismuth, 30% lead and 20% tin. It has a melting point of 92°C (198°F), and is used for fuses and safety plugs.

onioncloth. A strong large-mesh netting, usually in a leno weave, used in the form of bags for holding onions.

onionskin paper. A thin, glossy, translucent bond paper usually made from sulfite pulp. It is lightweight and durable, often supplied in white and canary, and resembles the dry skin of an onion in appearance. Used for multiple typewriter copies, air mail or branch office correspondence, interleaving order books, etc.

Onnex-HT. Trade name of BFD Inc., Advanced Composite Materials (USA) for a wear-resistant composite consisting of a ceramic material, such as aluminum oxide (Al_2O_3), interpenetrated by a continuous high-temperature metallic phase (e.g., aluminum bronze, nickel alloy, nickel-chromium-aluminum alloy or intermetallics). It can withstand high temperatures in erosive and corrosive environments, and its properties can be tailored choosing different ceramic or metallic phases. Applications include thermal exhaust parts, thermal-reactor environments, chemical processing equipment, and internal combustion engine parts.

Onnex-W. Trade name of BFD Inc., Advanced Composite Materials (USA) for a family of wear-resistant carbide or boride based composites interspersed with a tailorable hard material such as alumina, silicon carbide, boron carbide, or titanium diboride. It is machinable, and can be made to near-net shape.

Applications include bearing slides, valve parts, apex liners, nozzles, and mining equipment.

onoratoite. A mineral composed of antimony oxide chloride, $Sb_8O_{11}Cl_2$. Crystal system, triclinic. Density, 5.30 g/cm^3; refractive index, 2.205. Occurrence: Italy.

On-Plus. Trade name of Peninsular Steel Company (USA) for a high-speed steel containing 0.7% carbon, 4% chromium, 9.5% molybdenum, 1.5% tungsten, 1% vanadium, and the balance iron. Used for tools and high-speed cutters.

Onplus. Trade name for a high-speed steel containing 0.7% carbon, 18% tungsten, 4% chromium, 2% vanadium, and the balance iron. Used for cutting tools.

Ontario. Trade name of AL Tech Specialty Steel Corporation (USA) for a high-carbon, high-chromium cold-work tool steel (AISI type D2) containing 1.5% carbon, 12% chromium, 0.8-1% molybdenum, 0.9% vanadium, and the balance iron. It has high hardness, good abrasion resistance, good nondeforming properties, and high hardness. Used for long-run forming and blanking dies, punches, and shear blades. A sulfurized grade (0.12% sulfur) called *"Ontario-EZ"* with improved machinability, and a grade with lower carbon content, *Ontario Low Carbon* (AISI type D1) are also available.

Ontex. Trade name Research Polymers Inc. (USA) for a series of thermoplastic elastomers.

Ontop. Trade name for a high-speed steel containing 0.7% carbon, 18% tungsten, 4% chromium, 1% vanadium, 5% cobalt, and the balance iron. Used for cutters and tools.

On-X. Trademark of the Medical Carbon Research Institute, LLC (USA) for a strong, tough, silicon-free, pure *pyrolytic carbon* with a density of 1.9 g/cm^3 (0.07 lb/in.^{3}), high hardness, low friction, and excellent wear resistance. It is highly thrombo-resistant, and is suitable as a coating on long-term implants, such as artificial heart valves.

Onyx. Trade name of Premier Refractories and Chemicals Inc. (USA) for silicon carbide plastic refractory mixes.

onyx. A translucent variety of quartz made up of straight, parallel layers of different colored *chalcedony* and *opal*. The alternating layers are usually white and black, white and brown, or white and red. It can also be artificially colored. Used as a semiprecious stone, and as a cameo and ornamental building stone.

Onyx Elite. Trademark of Huber Engineered Materials (USA) for an extensive series of alumina trihydrate (ATH) pigments and fillers with high brightness and whiteness used primarily for cast polymer applications especially *cultured onyx.*

onyx marble. A translucent, layered, crystalline form of *calcite* ($CaCO_3$) in delicate colors deposited in the form of stalactites in caves. Occurrence: Algeria, Egypt, Gibraltar, USA (Arizona, California, Montana, New Mexico). Used for ornaments and lamp stands. It is not related to true onyx, which is a variety of quartz. Also known as *Algerian onyx; Gibraltar stone, Mexican onyx; oriental alabaster.*

Onyx Spring Steel. Trade name of Crucible Materials Corporation (USA) for a water-hardening carbon steel containing 0.7% carbon, and the balance iron. Used for springs, spring washers, etc.

oolitic limestone. A limestone with an even, granular texture composed essentially of small spherical particles of calcium carbonate.

oosterboschite. A mineral composed of copper palladium selenide, $(Pd,Cu)_7Se_5$. Crystal system, orthorhombic. Density, 8.48 g/cm^3. Occurrence: Zaire.

Ooze Calf. Trade name for a proprietary, velvet- or suede-finished calfskin.

Opacarb. Trademark of Minerals Technologies, Inc. (USA) for a family of coating materials based on precipitated calcium carbonate.

Opaceta. Trade name for cellulose acetate fibers used for textile fabrics including wearing apparel.

opacifiers. Materials used in glasses, glazes and porcelain enamels to increase diffuse reflection, refraction and diffraction, and produce an opaque appearance. *Opacifiers* are usually second-phase particles of antimony, tin oxide, titanium compounds or zircon with refractive indices greater than that of glass (about 1.5). The degree of opacification depends on the average particle size and concentration, and the mismatch of refractive indices.

Opal. (1) Trade name of Dentsply Ceramco (USA) for opalescent dental porcelain used for metal-to-ceramic bonding applications.

(2) Trade name of Dentsply Corporation (USA) for a light-cure dental resin cement used for luting restorations.

opal. A noncrystalline mineral occurring in many varieties and colors (white, yellow, red, brown, green, gray and blue), and composed of amorphous hydrated silica ($SiO_2 \cdot xH_2O$). Some varieties have a peculiar iridescent play of colors, and are valued as gems. Density, 2.1-2.3 g/cm^3; hardness, 5.5-6.5 Mohs. Used as a precious stone and in ceramics. Crystalline opals have spectral properties that make them suitable for photonic applications. See also fire opal.

Opal-Bauelemente. Trade name of Glasbauelemente GmbH (Germany) for composite cladding panels having centers of foamed glass, inner facings of plasterboard and outer facings of a toughened, ceramic-coated glass, such as *Polycolor* or *Delogcolor*.

OpalDam. Trademark of Ultradent Products, Inc. (USA) for a light cure methacrylate-base resin that is passively adhesive and light reflective. Used in dentistry as a reflective resin barrier.

opalescent glass. See opal glass.

opalescent glaze. A ceramic glaze that due to the addition of fluoride compounds to the batch fires to a milky or iridescent appearance.

opal glass. A translucent or opaque glass of white or milky appearance due to small light-dispersing particles or bubbles in the body of the glass. The bubbles are caused by the addition of fluorine compounds (usually fluorite or cryolite) to the melt. Also known as *opalescent glass; white glass.*

Opalhur. Trade name of Hurlingham SA (Argentina) for internal cladding materials made from opaline glass.

Opalika. Trademark of Schott DESAG AG (Germany) for a white flashed opal glass with outstanding UV-protection and light diffusion and showing negligible yellowing. Used for glazing applications, e.g., in public buildings, museums, etc.

Opalin. Trade name of Vereinigte Glaswerke (Germany) for a colored opaque glass ground and polished on one side.

Opalina. Trade name of Hurlingham SA (Argentina) for a colored, opaque glass.

Opaline. Trade name of Boussois Souchon Neuvesel SA (France) for an opaque glass colored throughout.

opaline chert. Chert composed essentially of opal ($SiO_2 \cdot xH_2O$). See also chert.

Opalit. Trademark of Saint-Gobain (France) for a translucent enameled glass.

Opalite. Trade name of Catalin Corporation and Monsanto Chemical Company (USA) for phenol-formaldehyde resins, compounds and plastics.

opalized wood. See silicified wood.

opalizer. Any fluoride-based compound, such as cryolite, fluorspar, or sodium fluoride, that produces an opalescent effect in glasses and glazes.

Opal Luting Cement. Trade name of 3M Dental (USA) for an opalescent resin-based dental luting cement for use with 3M Dental's bonding agents for porcelain veneers.

Opalon. Trademark of Monsanto Chemical Company (USA) for synthetic resins and resinous compositions including polyvinyl chloride and phenol-formaldehyde resins.

Opal Ronobil. Trade name of Bruestungselemente GmbH (Germany) for glass cladding panels.

Opalspun. Trade name of Plyglass Limited (UK) for plastic-bonded, pigmented glass gauze used in the lettering of glass, and as an interlayer in composite glass.

Opalux. Trade name of GC America, Inc. (USA) for a light-cure dental hybrid composite.

Opalwax. Trade name of Parchem (USA) for a solvent-resistant wax composed of hydrogenated castor oil, and supplied in the form of pearl-white flakes. It has a melting point of 85-88°C (185-190°C). Used as electrical insulation, for candles, in carbon paper, and as a substitute for carnauba wax. See also Castorwax.

opaque enamel. Enamel having a high degree of opacity due to purposely incorporated small second-phase pores or particles.

opaque glaze. A nontransparent, colored or colorless glaze of bright satin or glossy finish applied to the surface of a ceramic product.

opaque material. A material that does not transmit electromagnetic radiation.

Opax. Trademark of National Lead Company (USA) for a zirconium oxide (88+% pure) containing up to a total of 12% silicon dioxide, sodium oxide and aluminum oxide. It has a melting point of 2480°C (4495°F), and is used for ceramic enamels and glazes on dinnerware and wall tiles.

Opaxit. Trade name of Walbrzyh (Poland) for a colored cladding glass with uniquely smooth surface.

Opelon. Trademark of Toray–DuPont Company (Japan) for spandex fibers and filament yarns used for elastic fabrics.

open-burning clay. Clay whose porosity is increased by firing.

open-cell cellular polymer. See open-cell foam.

open-cell foam. A thermoplastic or thermosetting plastic or elastomer in the form of a flexible or rigid foam with numerous small, predominately open (or interconnected), spherical, polyhedral or honeycomb-shaped cells. Also known as *open-cell cellular polymer; open-cell foamed polymer.* See also closed-cell foam.

open-cell foamed polymer. See open-cell foam.

open-cell material. An organic or inorganic material containing numerous small cells, most of which are open (or interconnected). See also closed-cell material.

open clay. A highly porous clay with a sandy texture.

open-end yarn. A yarn made by a high-speed spinning process in which twist is transferred from formed yarns to continously fed fibers or highly drafted slivers by mechanical or other techniques. The name comes from the "break" or "open end" in the fiber flow created by the sliver feedstock.

open-face fabrics. Face fabrics that when joined to another material reveal selected areas of the substrate. Opposite of closed-face fabrics.

open-graded aggregate. Graded mineral aggregate that when com-

pacted produces relatively large voids or spaces between the aggregate particles. Abbreviation: OGA. See also dense-graded aggregate; aggregate.

open-grained wood. Wood having rather large, open pores that usually require filling before finishing. Examples of such woods include ash, chestnut, hickory, mahogany, oak, poplar and walnut. Also known as *coarse-textured wood.* See also close-grained wood.

open-hearth steel. A carbon or low-alloy steel made by a process that involves melting selected pig iron and scrap steel with or without the addition of pure iron ore in a reverberatory furnace. The preferred term in Europe is *Siemens-Martin steel.* Abbreviation: OHS.

opening material. A material, such as chamotte, flint, grog, or sand, added to a plastic clay to increase its porosity, minimize shrinkage and accelerate drying.

open sand. Foundry sand with high permeability or porosity that give passage for air, gases and steam to escape when the molten metal is poured.

opepe. See bilinga.

Ophtasan. Trade name of Schott DESAG AG (Germany) for a light pink, tinted ophthalmic glass.

ophthalmic glass. A high-grade glass having great uniformity and selected optical and physical properties that make it suitable for prescription eyeglasses.

Oplen. Trademark of Jeil Synthetic Fibers Company (South Korea) for polyester filament yarns.

Oplexmatt. Trade name for rayon fibers and yarns used for textile fabrics.

Opotow. Trade name of Teledyne Water Pik Technologies, Inc. (USA) for zinc oxide/eugenol dental cements, impression materials and bite registration pastes. *Opotow Alumina EBA* is an alumina and ethoxybenzoic acid (EBA) reinforced grade.

Oppanol. Trademark of BASF AG (Germany) for thermoplastic polyisobutylene polymers with excellent resistance to water, acids and alkalies. Supplied in several grades, they are used in waterproofing textiles.

opsin. A colorless, light-sensitive pigment that is the protein constituent of *rhodopsin.*

Opsite. Trade name for a thin polyethylene sheet material suitable for use in wound care applications.

OPTA. Trade name of E.I. DuPont de Nemours & Company (USA) for nylon 6,6 fibers.

Optal. Trade name of Wieland-Werke AG (Germany) for a series of aluminum alloys containing 0.5-7% magnesium, up to 1% manganese and up to 0.5% chromium. Used for structural parts, and in building construction.

Optalloy. Trademark of Dentsply International (USA) for dental amalgam alloys supplied in pellet form for restorative work.

Opt-E-Bond. Trademark of H.B. Fuller Company (USA) for hot-melt adhesives used for bonding board, film, foil and paper. They were formerly sold by Eastman Adhesives (USA) under the trade name *Eastobond.*

Optec. Trademark of American Thermocraft Corporation (USA) for dental porcelains and bonding agents. *Optec HSP* are leucite-reinforced high-ceramic dental porcelains used for core-build-ups. *Optec Universal* is a multi-purpose dental bonding agent.

Optema. Trade name of Exxon Chemical Company (USA) for ethylene-maleic anhydride copolymers used for compounding, blown film, extrusion coating, injection molding and adhesives and sealants.

Optene. Trade name of Borealis Chemicals (USA) for ethylene-vinyl acetates supplied with vinyl acetate contents of 12, 25 and 33%.

Optibent. Trademark of Süd-Chemie, Inc. (USA) for a range of additives based on *smectite* or *bentonite* clays, and used in cement and other building products.

Optibloc. Trademark of Minerals Technologies, Inc. (USA) for a talc-based high-clarity antiblocking agent used in the manufacture of polyethylene and other film materials for packaging applications.

Optibond. Trade name of Kerr Dental (USA) for fluoride-releasing bisphenol A glycidylether dimethacrylate (BisGMA) based dental adhesive resins and bonding system. *Optibond FL* is a fluoride-releasing multi-purpose dental bonding system. *Optibond Solo* is a single-component (primer plus adhesive) light-cure dentin/enamel adhesive resin system with low polymerization shrinkage and microleakage, and high bond strength. *Optibond* products are used for bonding composite, compomer and porcelain restorations, and composite and porcelain inlays and onlays.

optical brightener. See brightener (3).

optical calcite. See Iceland spar.

optical cement. A transparent synthetic adhesive (e.g., an acrylic resin or compound) that has good high-temperature properties, and is used in cementing (bonding) optical elements.

optical coating. (1) A coating of a highly reflective material applied to a prepared glass surface. See also mirror coating.

  (2) A coating applied to a glass surface to reduce the amount of light reflected at the surface. See also antireflection coating.

  (3) See low-index coating.

optical crown glass. See crown glass.

optical crystal. A crystal of natural or synthetic origin used for infrared and ultraviolet optics, in the detection of short-wave radiation, and for piezoelectric applications. Natural optical crystals include calcium chloride, sodium chloride, potassium iodide and silver chloride. Anthracene and stilbene are examples of synthetic optical crystals.

optical fibers. Long, thin, fine strands of highly transparent materials, usually quartz glass or plastics, suitable for transmitting light over great distances. They are of extremely high purity, have controlled profiles for the refractive index, and exceptionally low light absorption. Low-index coatings (e.g., germanium-doped silica) are employed to keep the light path away from the fiber surfaces. Used for long-distance communication transmission, remote-sensing devices, analytical instrumentation, radiation dosimeters, and high-temperature thermometers. There are also novel microstructured two-dimensional optical fibers suitable for applications in nonlinear optics and photonics. Abbreviation: OF. Also known as *fiber-optics; light guides*. See also low-index coating; microstructured optical fibers.

optical flint glass. (1) An optical glass with a high index of refraction and high dispersion.

  (2) Any optical glass with a reciprocal dispersive power (nu value) of less than 50.0.

  (3) Any optical glass having a reciprocal dispersive power (nu value) of 50.0 to 55.0 and a refractive index of less than 1.60.

  (4) See flint glass (1).

optical glass. A glass of extremely high quality and uniformity that meets stringent requirements for composition, homogeneity, transmission, refractive index and light dispersion. It may be a *crown glass* (containing lime) or a *flint glass* (containing lead oxide). Used for vision correction lenses, lenses for binoculars, cameras, microscopes and other instruments, prims, mirrors, vacuum tubes, and electrical equipment.

optical-grade silicon. A high-purity single- or polycrystalline silicon (99.999%) with high electrical resistance and an infrared transparency in the wavelength range of about 500 to 15000 nm.

optically active material. A material, such as *quartz crystal*, that rotates the plane of linearly polarized light.

optically anisotropic materials. A group of materials that have non-cubic crystal structures and exhibit different optical properties in different crystallographic directions.

optically biaxial crystal. An anisotropic monoclinic, orthorhombic or triclinic crystal that exhibits double refraction, but has two optic axes and three refractive indices.

optically inactive material. A material that does not rotate the plane of linearly polarized light.

optically isotropic materials. A group of materials that have cubic crystal structures and refractive indices which are identical for all propagation directions and therefore do not respond to cross-polarized light.

optically uniaxial crystal. A hexagonal, rhombohedral or tetragonal crystal, such as *calcite* or *quartz*, that is double-refracting in all light-propagation directions except for the direction of the principal crystallographic axis in which it is single refracting, i.e., it has only one axis (the "optic axis") along which monochromatic light can travel without exhibiting double refraction.

optical materials. Materials that are suitable for optical applications, and may or may not be transparent to visible, infrared, ultraviolet or X-ray radiation. Examples include optical glass, fused silica, quartz, calcite, polytetrafluoroethylene and polymethyl methacrylate.

optical plastics. Plastics, such as acrylics, methyl methacrylates, polycarbonates, or polystyrenes, that are transparent to visible radiation, and can thus be used for the manufacture of opti-cal lenses, mirrors and prisms.

optical sapphire. A high-purity *sapphire* (99.9+%) with high hardness, high refractive index (above 1.76), and low optical dispersion. Used for the manufacture of optical lenses.

optical vacuum coatings. Transparent or semitransparent metal films of aluminum, chromium, copper, gold, platinum, rhodium, silver, or titanium, deposited from a source in a high-vacuum environment onto a glass, ceramic or plastic substrate. Transparent vacuum coatings are useful for first-surface mirror applications, e.g., on automotive rearview mirrors and reflectors for sealed-beam lamps, and on scientific instruments, such as microscopes, monochromators and astronomical telescopes. Semitransparent vacuum coatings are used on beamsplitters, light-attenuators, sunglasses, and neutral-density filters.

optical whitener. See brightener (3).

optical wire alloy. A nickel silver containing 54% copper, 28% zinc and 18% nickel, used for optical instruments.

Opticast. Trade name of Cendres & Métaux SA (Switzerland) for high-gold dental casting alloys used for crowns and other restorations. Also known as *CM Opticast*.

Opticolor. Trade name of Chicago Dial Company (USA) for a glass filter plate containing neodymium oxide (Nd_2O_3). It transmits 90% of the light rays producing the three primary colors red, green and blue, and only 10% of those producing the less desirable color yellow. Used for color television tubes.

Opticure. Trade name of Glidden Industrial Coatings (USA) for high-speed, ultraviolet-arc-curing coatings.

Optifloat. Trade name of Flachglas AG (Germany) for a float glass.

Optigel. Trademark of Süd-Chemie, Inc. (USA) for a range of gelling agents.

Optilon 399. Trademark of Coltene-Whaledent (USA) for a heat-cured rubber/methyl methacrylate resin having high impact strength and resilience, and used for denture bases.

Optimax. Trade name of Uddeholm Corporation (USA) for a specialty stainless mold steel (AISI type 420) with a high level of microcleanliness. It contains 0.38% carbon, 0.5% manganese, 0.9% silicon, 13.6% chromium, 0.3 vanadium, and the balance iron.

Optimum. Trade name of Otto Wolff Handelsgesellschaft (Germany) for high-carbon, high-chromium tool steels used for forming and blanking dies.

Option. Trademark of Degussa-Ney Dental (USA) for a silver-free dental ceramic alloy containing 78.8% palladium and 2% gold. It has a melting range of 1100-1190°C (2012-2174°F), very high yield strength, and a hardness of 425 Vickers. Used for high-stress restorations and implant cases.

Optital. Trade name of Vulkan Strahlverfahrenstechnik GmbH & Co. KG (Germany) for cast-steel granulates available in several grades including *Optital-S* for abrasive-blast cleaning and *Optital-SP* for shot peening.

Optix. Trade name of Plaskolite Inc. (USA) for clear and colored acrylic flat sheeting based on polymethyl methacrylate, and used for awnings, displays, sunrooms, signage, and skylights. *Optix TemperElite* refers to green, acrylic flat sheeting.

optoelectronic materials. A class of materials including *semiconductors*, such as gallium arsenide (GaAs), gallium indium arsenide (GaInAs), aluminium gallium arsenide (AlGaAs), indium phosphide (InP) and indium gallium phosphide (InGaP), and various *heterostructures* and *quantum-well materials*, that are suitable for use in the manufacture of solid-state and other electronic devices, transmission systems, communications equipment, optical sensors and storage devices, waveguides, optoelectronic packaging, and photonic devices. See also electro-optic materials.

Optosil P. Trade name Bayer Corporation (USA) for silicone elastomers used for dental impressions.

Optum. Trade name of Ferro Corporation (USA) for a range of polyolefin alloys supplied in thermoforming and extrusion grades. They provide fast cycling times, high heat-deflection temperatures, and good chemical resistance and barrier properties.

Opus. Trade name Kerr Dental (USA) for an extensive range of dental cements including *OpusCem* glass-ionomer cement, *OpusFil W* glass-iono-mer filling cement, *Opus PCF* zinc polycarboxylate cement and *Opus Silver* silver-filled glass-ionomer cement.

Ora-Crylic. Trade name of Henry P. Boos Dental Laboratories (UK) for an acrylic resin used for dental applications.

Oraglas. Trade name of Lenning Chemicals (UK) for a series of plexiglas-type polymethyl methacrylate resins.

Oralite. Trade name of Oralite Company (UK) for a cellulose nitrate formerly used for denture bases.

Oralloy. Trade name of Coltene-Whaledent (USA) for a zinc-free, high-silver dental amalgam alloy that does not contain the corrosive tin-mercury phase (Gamma 2 phase). Supplied in powder and tablet form, it has a low, well balanced mercury content, outstanding corrosion resistance and good compressive strength. Used in restorative dentistry. Also available are *Oralloy Magicap* dental amalgam capsules.

oralloy. A term sometimes used for *depleted uranium*. See depleted uranium.

orange cadmium. See cadmium sulfide.

orange chrome. An inorganic pigment composed of basic lead chromate ($PbCrO_4 \cdot PbO$).

Orange Label. (1) Trade name of A. Milne & Company (USA) for a precision-cast tool steel containing 1-1.05% carbon, and the balance iron.

(2) Trade name of Wallace Murray Corporation (USA) for a fatigue-resistant tool steel containing 0.45-0.55% carbon, 1.25-1.75% chromium, 2-3% tungsten, 0.2-0.3% vanadium, and the balance iron. Used for chisels, punches, and rivet sets.

orange lead. A fine, bright orange-red powder prepared by roasting basic lead carbonate [$2PbCO_3 \cdot Pb(OH)_2$]. It contains 95.5+% red lead oxide (Pb_3O_4), has a density of 9.0 g/cm^3 (0.32 $lb/in.^3$), low tinting strength, and high brilliance. Used as an inorganic pigment in printing inks and primers. Also known as *orange mineral*.

orange mineral. See orange lead.

orange oxide. See uranium trioxide.

orange uranium oxide. See uranium trioxide.

Oranium Bronze. Trade name of Olin Brass, Indianapolis (USA) for a series of aluminum bronzes containing 88-97% copper and 3-12% aluminum. Used for castings, gears, propellers, hardware, and bearings.

Orbis. Trade name of Uddeholm Corporation (USA) for a cold-work steel containing 0.95% carbon, 1.5% silicon, 0.75% manganese, 1% chromium, 0.1% vanadium, and the balance iron. Used for dies.

Orbit. Trade name of Crucible Materials Corporation (USA) for a tough, air-hardening, free-machining steel containing 0.7% carbon, 2% manganese, 0.15% sulfur, 1% chromium, 1.35% molybdenum, and the balance iron. Used for blanking and forming dies, hobs, and rolls.

orcelite. A white mineral composed of copper iron nickel arsenide sulfide, $(Ni,Fe,Cu)_{4.2}(As,S)_2$. Crystal system, hexagonal. Density, 7.25 g/cm^3. Occurrence: Spain.

Orchid. Trade name of SA Glaverbel (Belgium) for a glass with an orchid-type floral design.

ordinary steels. See plain-carbon steels.

Ordix. Trade name of Saarstahl AG (Germany) for a cold-work tool steel containing 0.58% carbon, 1.8% manganese, 0.75% manganese, 0.4% chromium, and the balance iron. *Ordix Extra* is an oil-hardening, shock-resistant tool steel that contains 0.67% carbon, 1.3% silicon, 0.5% manganese, 0.5% chromium, and the balance iron, and is used for springs and punches. *Ordix Special* contains 0.56% carbon, 1.05% manganese, 1.15% chromium, 0.15% nickel, 0.12% vanadium, and the balance iron, and is used for tools, dies, crimpers, and punches.

ordnance bronze. A dense, soft phosphorus-deoxidized tin bronze composed of 80% copper, 10% tin and 10% lead. It has excellent machinability, good corrosion resistance, moderate strength, and is used for bearings operating at heavy pressures and high speeds, pressure-tight castings, and pumps. See also high-leaded tin bronze.

ordnance steels. Steels used in the manufacture of military materiel including vehicles, machinery and tools. Such steels are usually made by rigidly controlled methods of manufacture to ensure the highest level of quality and meet the most demanding requirements [e.g., MIL-SPEC (Military Specifications) or MIL-STD (Military Standards)] regarding cleanliness and ultrasonic and magnaflux magnetic testing.

ordonezite. A light to dark brown mineral of the rutile group composed of zinc antimony oxide, $ZnSb_2O_6$. Crystal system, tetragonal. Density, 6.64 g/cm^3; refractive index, above 1.95. Occurrence: Mexico.

or doublé. A copper alloy with a thin cladding of gold used for costume jewelry. See also doublé; duplex metal.

ore. A naturally occurring mineral that contains a valuable substance, such as a metal, which can be commercially extracted. For example, cassiterite, galena and hematite are ores of tin, lead and iron, respectively.

Oregon ash. See western ash.

Oregon cedar. See Lawson cypress.

oregonite. A white mineral composed of iron nickel arsenide, Ni_2FeAs_2. Crystal system, hexagonal. Density, 7.04 g/cm^3. Occurrence: USA (Oregon).

Oregon maple. See bigleaf maple.

Oregon pine. See Douglas fir.

Oregon white pine. See ponderosa pine.

oreide bronze. A group of brass compositions containing 68-92% copper, 10-32% zinc, 0-4% tin and 0-1% lead. They develop a golden color when polished, and are used for hardware, carriage and harness hardware, plumbing, ornamental purposes, and jewelry.

Orel. Trademark of E.I. DuPont de Nemours & Company (USA) for polyester fibers and filaments.

Orelloy. Trade name of Oregon Steel Mills (USA) for a series of steel products including various abrasion-resistant quenched- and tempered grades as well as various high-strength low-alloy structural grades.

Oremet. Trademark of Oregon Metallurgical Corporation (USA) for high-quality titanium supplied in bar, billet and plate form.

Orevac. Trade name of Atofina SA (France) for ethylene-vinyl acetate/maleic anhydride terpolymers used for packaging and automotive applications. *Orevac Grafted* refers to grafted polyethylenes and polypropylenes used for packaging applications.

Orfan. Trade name of Th. Goldschmidt AG (Germany) for a series of thermosetting resins supplied as decorative films for laminating wood-based boards.

Orgalloy. Trade name of Atofina SA (France) for a series of polyamide (nylon) alloys including *Orgalloy LE* containing nylon 6 and *Orgalloy R* containing nylon 6,6.

Orgalon. Trade name of Atofina SA (France) for a series of polycarbonate resins.

Orgamide. Trade name of Atofina SA (France) for polyamide 6 (nylon 6) supplied in unmodified, glass fiber- or glass bead-reinforced, mineral-filled, fire-retardant, high-impact and UV-stabilized grades. *Orgamide* nylon has high mechanical strength and creep resistance, high impact strength, good elevated temperature properties, good processibility, a service temperature range of -40°C to +160°C (-40 to +320°F), good dielectric properties, and good resistance to alcohols, alkalies, hydrocarbons, greases, oils and ketones. Used for appliances, automotive components, lawn and garden equipment, fasteners, and electrical switches and connectors.

organdy. A light, sheer or semi-sheer, plain-woven cotton or cotton-blend fabric, usually with a relatively crisp finish, used for blouses, dresses, evening wear, children's clothes, curtains, bedspreads, and apparel trim, e.g., collars and cuffs.

organic charge-transfer salt. See electron-transfer salts.

organic coatings. A group of surface finishes including paints, varnishes, lacquers, enamels, asphaltic materials and rubbers, that essentially consist of a volatile vehicle, such as a ketone, ester alcohol, petroleum solvent, or water, a nonvolatile vehicle, such as an acrylic, alkyd, epoxy or polyurethane resin, and pigments, such as drying oils (e.g., linseed or tung oil), as well as selected plasticizers, ultraviolet absorbers, emulsifiers and dispersing agents. Used to protect metallic and nonmetallic structures against marine, rural, industrial and chemical exposure. Abbreviation: OC.

organic cold-molded plastics. A group of plastic materials consisting of asbestos fibers bonded with organic materials, such as asphalt, pitch, oil, or phenolic or melamine resins, and used for electrical components. See also cold-molded plastics; inorganic cold-molded plastics.

organic composites. Mixtures or mechanical combinations of two or more distinct materials, at least one of which is an organic material. They are solid in the finished state, and mutually insoluble. Examples include (i) laminates of paper, fabric or wood and a thermosetting material (resin, rubber or adhesive), commonly used for tire carcasses, plywood and electrical insulating structures, and (ii) reinforced plastics composed principally of organic or inorganic fibers in thermosetting or thermoplastic matrices.

organic compounds. A large group of compounds that contain at least the element carbon, but may also have other elements, such as hydrogen, oxygen, nitrogen, phosphorus, sulfur, or a halogen. Examples of such compounds include carbohydrates, hydrocarbons, alcohols, ethers, ketones, carboxylic acids, organometallic compounds and synthetic resins.

organic conductors. (1) A class of electrically conductive polymers, such as polyacetylene, polyaniline, poly(p-phenylene), poly(p-phenylene sulfide) and poly(p-phenylene vinylene), whose conduction properties are either inherent, or due to the addition of several percent of dopants such electron acceptors (e.g., arsenic pentafluoride) or electron donors (e.g., alkali metal ions, or iodine).

(2) A fairly recent class of complex organic materials, such as the *electron transfer salts*, that are essentially composed of relatively large organic molecules containing approximately 20 atoms per molecule. They exhibit a wide range of interesting electronic properties including conductivity or superconductivity, and low Fermi energies as well as other intriguing electronic states, such as spin density waves, charge density waves and the quantum Hall effect, and greatly enhanced Shubnikov-de Haas and de Haas-van Alphen effects. See also organic superconductors.

organic electroluminescent materials. A class of organic materials, such as anthracene, poly(p-phenylene vinylene) and doped poly(vinyl carbazole), that emit light in response to an applied electric field, and are therefore suitable for the manufacture of electroluminescent displays and devices. Abbreviation: OEL materials. See also electroluminescent materials.

organic fibers. Natural or synthetic fibers made from organic materials and having length-to-diameter ratios of 100:1 or above. Examples include cotton, hemp, aramid and polyolefin fibers. Used for the manufacture of textile fabrics, cordage, brooms, etc., and as reinforcements in advanced composites.

organic fluxes. A group of moderately active soldering fluxes that consist of organic acids and bases, and have moderate cleaning abilities. They are corrosive during the soldering operation, but generally become non-corrosive and/or inert thereafter. *Organic fluxes* are most efficient in the temperature range of 93-316°C (200-600°F).

organic glass. An amorphous, transparent, glass-like plastic, such as polymethyl methacrylate. Abbreviation: OG. See also polymethyl methacrylates; Plexiglass

organic-inorganic hybrid magnets. See molecule-based magnets.

organic-inorganic nanocomposites. A class of nanocomposites with inorganic matrices (e.g., metal or ceramic) into which inorganic components have been incorporated. These components can be zero- or one-dimensional materials, such as molybdenum selenide chains and clusters, two-dimensional layered materials, such as chalcogenides, metal oxides and phosphates or clay-based materials, and three-dimensional systems, such as zeolites. Potential applications include nonlinear optics, battery cathodes and ionics, sensors, nanowires, and lightweight mechanically reinforced structures, bioceramics, and biomineralization. See also nanocomposites; nanostructured materials; lamellar nanocomposites.

organic magnets. See molecule-based magnets.

organic materials. Materials based on carbon chains or rings, often in combination with hydrogen and/or other elements. Examples are hydrocarbons, petroleum and petroleum derivatives, esters, polymers, foams, resins, detergents, dyes, organic composites, coal and coal derivatives, wood products, and organometallics.

organic-matrix composites. A group of composite materials consisting of matrices based on an organic material, such as polyester, polyphenylene sulfide, or vinyl ester, reinforced with inorganic or organic fibers or particles. Examples of *inorganic-organic composites* are fiberglass-reinforced plastics and graphite-epoxy laminates or prepregs, and *organic-organic composites* include aramid-epoxy composites, and plywood.

organic membrane. A microporous thin-film structure made of organic materials (e.g., cellophane, collodion, or other polymers), and used as permeable or semipermeable membrane in osmosis, and in liquid and gas separation.

organic-organic composites. Composite materials whose matrices and reinforcements consist of organic material. Examples include aramid-epoxy, aramid-polyphenylenesulfide and aramid-vinylester composites, laminates of paper, fabric or wood and thermosetting materials (e.g., resins, elastomers or adhesives) and mixed fabrics, i.e., woven combinations of wool or cotton and synthetic fibers. See also organic composites.

organic pigment. A pigment based on one or more organic compounds of animal, vegetable or synthetic origin. Animal-based pigments include *rhodopsin* and *melanin*, vegetable-derived pigments *chlorophyll* and *indigo*, and synthetic pigments *phthalocyanine*, *toluidine* and *lakes*.

organic polymer. A commercial polymer in which the main chain is made up of carbon atoms. Most polymers are organic.

organic semiconductors. Organic materials, such as metal *phthalocyanines*, metal *tetracyanoquinodimethanes*, *dihexylhexathiophene*, *pentacene* or *melanin*, exhibiting conductive or semiconductive properties which make them useful for the manufacture of electronic, optoelectronic, photonic and photovoltaic devices and components.

organic superconductors. A class of organic compounds that are essentially crystals consisting of donor molecules or cations, such tetramethyltetraselenathiafulvalene (TMTSF), or bisethylenedithiotetrathiofulvalene (BEDT-TTF), and acceptor anions, such as PF$_6$, or ClO$_4$. In general, they can be synthesized by

electrocrystallization, precipitation from mixed donor-anion solutions, or direct donor-anion reactions. Examples of *organic superconductors* include the *Bechgaard salts* (TMTSF)$_2$PF$_6$ and (TMTSF)$_2$ClO$_4$, other Bechgaard-type salts with the general formula (TMTSF)$_2$X, and bisethylenedithiotetrathiafulvalene salts with the general formula (BEDT-TTF)$_2$X. Owing to their similar electronic properties and electron-transfer mechanisms, *alkali-doped fullerenes* are generally included in this class of superconductors. See also organic conductors (2).

organic zinc-rich paint. A two-component paint containing finely divided zinc dust in an organic vehicle, such as an epoxy resin. Used as a primer on ferrous substrates.

organisol. (1) A suspension of finely divided particles of a resin, e.g., a vinyl, dispersed in a liquid organic mixture. See also organosol.

(2) Abrasion- and chemical-resistant vinyl coatings containing considerable amounts of solvents.

organoceramics. A group of high-strength composites composed of ceramic matrices toughened by the interspersion of molecular-level polymers, usually polyvinyl alcohol, polydimethyldiallyl ammonium chloride or polydimethylbutyl ammonium iodide in the form of layered or threaded polymer chains. They possess high strength, and are used for structural applications.

organoclay. A kaolin- or montmorillonite-type clay to which an organic structure has been bonded by chemical means. Also known as *organopolysilicate*.

organofunctional silanes. A group of silane-derived compounds corresponding to the general structural formula R$_a$Si(OR)$_{4-a}$, where R is an alkyl, aryl, or other organic functional group, and OR is a methoxy, ethoxy, or acetoxy group). Examples of such compounds include methylaminopropyltrimethoxysilane and *n*-octyltrimethoxysilane. Used in the manufacture of silicones, silanes, siloxanes, acrylics, epoxies, phenolics, melamines, nylons, polyvinyl chlorides, polyolefins, polyurethanes and nitrile rubbers, for free-radical, cross-linked acrylics, polyester, styrenics, and various other polymers, in the production of adhesives, sealants, paints and coatings, pigments and fillers, glass fibers and fabrics, and in electrical engineering and electronics, foundry moldmaking, and for many other applications.

organomagnesium compounds. A group of organometallic compound formed by reacting an alkyl or aryl halide with metallic magnesium in an inert solvent, such as ether or tetrahydrofuran. Examples include ethyl magnesium iodide (C$_2$H$_5$MgI) and phenyl magnesium chloride (C$_6$H$_5$MgCl). The general formula is RMgX. See also Grignard reagent.

organometallic compounds. A group of organic compounds that have one or more metal-carbon bonds. The general formula is MR$_x$, where M is a metal and R an organic radical. Examples include butyllithium [C$_4$H$_9$Li], diethylzinc [Zn(C$_2$H$_5$)$_2$], tetraethylgermane [Ge(C$_2$H$_5$)$_4$], tetraphenyllead [Pb(C$_6$H$_5$)$_4$], tributyl aluminum [Al(C$_4$H$_9$)$_3$], copper phthalocyanine [Cu(C$_{32}$H$_{16}$N$_8$)] and the metallocenes. Also known as *metal-organic compounds; metalorganics; organometallics*.

organometallics. See organometallic compounds.

organophilic material. An essentially water-immiscible material with strong dipole moments that is attracted to hydrocarbons and hydrocarbon-miscible substances.

organopolysilicate. See organoclay.

organosilane. An organic chemical compound made by bonding silicon to carbon, and used as a coupling agent in the manufacture of organic coatings. Also known as *organosilicon*.

organosilicon.   See organosilane.

organosilicon polymers.   Polymers, such as *polysilanes* or *polysilazanes*, containing silicon-carbon bonds.

organosiloxanes.   See silicones.

organosol.   A colloidal dispersion of finely divided particles of a synthetic resin, such as polyvinyl chloride, in an organic liquid (e.g., a plasticizer) that is not a solvent. It is used to apply tough, hard, thin coatings, for molded products, and in the manufacture of plastic films. See also organisol; plastisol.

organza.   A thin, smooth fabric woven from silk, rayon, nylon, or polyester yarn, usually in a plain weave, and being more crisp and sheer than *organdy*. Used especially for bridal and evening wear, curtains, and millinery.

organzine yarn.   A yarn produced by twisting together two or more single raw silk yarns in opposite directions.

Orgater.   Trade name of Atofina SA (France) for a series of polybutylene terepthalates supplied in glass-fiber-reinforced, fire-retardant and UV-stabilized grades. They have a density of about 1.3 g/cm^3 (0.05 lb/in.^{3}), good impact strength, high strength and dimensional stability, high toughness, good durability, good electrical properties, high dielectric strength, good chemical resistance, and low water absorption. Used for automotive and electrical components, power-tool housings, etc.

Orgues.   Trade name of SA Glaverbel (Belgium) for a patterned glass with broad vertical reeds.

Orient.   Trade name of SA Glaverbel (Belgium) for a patterned glass with oriental design.

oriental alabaster.   See onyx marble.

oriental amethyst.   A purple gem variety of the mineral *corundum* (Al_2O_3).

oriental topaz.   A yellow gem variety of the mineral *corundum* (Al_2O_3).

Oriented.   (1) Trade name of Armco International (USA) for a series of oriented electrical steels.

(2) Trade name of Oriented Plastics (USA) for oriented saran (polyvinylidene chloride) fibers with good chemical resistance, good weatherability, and good resistance to mildew and insects. Used for carpets and rugs, draperies, upholstery, clothing, industrial fabrics, etc.

oriented composites.   Composite materials whose macroconstituents are aligned in one or several directions. Examples include *oriented strand board*, which is a composite of wood strands oriented at right angles to each other in a phenolic matrix, and oriented fiber-reinforced plastics which are polymer-matrix composites containing directionally aligned fibers. See also oriented materials; oriented polymers.

oriented crystal.   (1) A crystal whose grains have been aligned in a particular direction.

(2) A crystal whose crystallographic axes have been aligned with respect to an electric or magnetic field.

oriented electrical steel.   A low-carbon sheet steel with up to 3.15% silicon that contains large grains, and has highly directional magnetic properties with low core losses and high magnetic permeability when the flux path is parallel to the rolling direction. Used in sheet form for power and distribution transformers. Also known as *grain-oriented electrical steel; grain-oriented sheet; grain-oriented silicon steel; oriented silicon steel*.

oriented fibers.   (1) Natural or synthetic fibers that are, or have been aligned in a preferred direction.

(2) Natural fibers, such as cotton or wool, that are preferably aligned in a direction determined during growth.

(3) Synthetic fibers in which the linear molecules are aligned in a preferred direction, usually the fiber axis. This alignment is usually produced during the fiber extrusion or drawing process.

oriented flakeboard.   See oriented strand board.

oriented materials.   Materials, such as polymers or composites, in which the micro- or macroconstituents are directionally aligned. One-directionally aligned materials are referred to as "unidirectional," and two-directionally aligned materials as "bidirectional." See also oriented composites; oriented fibers; oriented polymers.

oriented polymers.   Polymers in which the molecules are directionally aligned. See also oriented materials.

oriented silicon steel.   See oriented electrical steel; silicon steels (1).

oriented strand board.   A manufactured panel of reconstituted wood strands that are mechanically oriented at right angles to each other, coated with wax, bonded with thermosetting resin (e.g., phenolic) and then compressed under heat. The resulting structural panel has excellent strength, stiffness and stability. The most common panel size is 1.2 × 2.4 m (4 × 8 ft.) with thicknesses ranging from 6.4 to 19 mm (0.25 to 0.75 in.). The material was originally designed specifically for construction applications (e.g., roof and wall sheathing, subflooring, siding, etc.), but its good appearance on both sides has resulted in many other applications, such as wall paneling, shelving, center layers in composite panels, and I-beam webs. Abbreviation: OSB. Also known as *oriented flakeboard*.

oriented strand lumber.   Structural composite lumber made by bonding flaked wood strands of high length-to-thickness ratios with a special synthetic adhesive, and then orienting and forming the bonded strands into a mat followed by pressing. Used for millwork, studs and other structural building products. Abbreviation: OSL.

oriented tape.   A polymer tape with high uniaxial (longitudinal) strength, usually composed of a thermoplastic olefin, and produced by first extruding as a film or sheet, cutting into tape, and then hot stretching the tape along one axis to induce preferred molecular orientation. See also thermoplastic olefins; tape.

orientite.   A red-brown to black mineral composed of calcium manganese silicate heptahydrate, $Ca_2Mn_3(SiO_4)_3OH$. Crystal system, orthorhombic. Density, 3.05 g/cm^3; refractive index, 1.78. Occurrence: Cuba, USA.

orient yellow.   A cadmium yellow pigment made by coprecipitating cadmium sulfide (CdS) with barium sulfate ($BaSO_4$).

Oriex.   Trademark of Huels America Inc. (USA) for biaxially oriented vinyl sheeting and film products used for shrink packaging applications.

original monel metal.   A corrosion-resistant nickel casting alloy containing 23% copper and 2.5% silicon, and used for pump parts and chemical equipment. See also monel metal.

Orion.   (1) Trade name of Degussa-Ney Dental (USA) for several dental bonding alloys. *Orion* is a copper- and silver-free dental bonding alloy containing 52% gold, 39% palladium, and additions of indium and gallium to improve fluidity. It has a very light oxide layer, high yield strength and ductility, and a hardness of 220 Vickers. Used in restorative dentistry especially for fixed restorations. *Orion Argos* is a palladium-silver alloy, *Orion Delphi* a medium-gold alloy and *Orion Libra* a silver-free palladium alloy.

(2) Trade name of National Alloys Limited (USA) for several plain-carbon tool steels.

(3) Trade name of Creusot-Loire (France) for a series of cast, corrosion-resistant ferritic stainless steels.

Orkan. Trade name of Hufnagel GmbH (Germany) for a series of cobalt-tungsten-type high-speed steels containing 1.2-1.4% carbon, 4-4.2% chromium, 1-4% molybdenum, 10-12% tungsten, 3-4% vanadium, 5-10% cobalt, and the balance iron. Used for finishing and lathe tools.

Orkot. Trade name of Orkot Limited (UK) for several polymers including polyamides, supplied as blocks, rods, tubes, sheets and sections, and used in the manufacture of bearings, bushings, journals, washers, thrust collars, wear pads and rings, and guide strips.

Orleans. Trade name of Wallace Murray Corporation (USA) for an oil-hardening, shock-resisting tool steel (AISI type S5) containing 0.55% carbon, 2% silicon, 0.8% manganese, 0.25% chromium, 0.35-0.45% molybdenum, 0.2% vanadium, and the balance iron. It has outstanding toughness, high elastic limit, good deep-hardening properties and good ductility. Used for pneumatic chisels, punches, concrete breakers, and cutters.

Orlon. Trademark of E.I. DuPont de Nemours & Company (USA) for an acrylic (polyacrylonitrile) fiber. It has a density of 1.14-1.17 g/cm^3 (0.041-0.042 $lb/in.^3$), good tensile strength, high elasticity, a softening temperature of 235°C (455°F), good resistance to mildew, weathering, water and ultraviolet light, good resistance to mineral acids, alcohols, acetone, benzene, carbon tetrachloride and petroleum ether, good to fair resistance to weak alkalies, and blends well with wool or other fibers. Used for knitted clothing, rugs, draperies, outdoor fabrics, and filter fabrics.

Orm. Trade name of Delta Metal Company (UK) for a brass containing 40% zinc, varying amounts of tin, lead, iron, aluminum, manganese and nickel, the balance copper. Used for extrusions.

Ormecon. Trademark of Zipperling Kessler & Co. (Germany) for dispersible, intrinsically conductive *polyaniline*-base polymer powders, varnishes and lacquers. The powders are used in the manufacture of films, housings, electrical and electronic components, optical switches, gas-separation membranes, and the varnishes and lacquers for the production of rust preventive and electrically conductive coatings, and for electromagnetic shielding applications.

ormocers. Organic-inorganic hybrid materials in which the organic component is a synthetic resin and the inorganic component a ceramic material. Used for porous membranes, porous insulators for noise reduction applications, in the synthesis of biomaterials, as biosensors, in restorative dentistry, and for the control of electrical, optical, mechanical and/or thermal properties of devices.

Ormogold. Trademark of Degussa-Hüls AG (Germany) for dental veneer used for restorative applications.

ormolu. A gold-colored alloy composed of copper and zinc, and used as gold imitation, and in the decoration of furniture, clocks, etc.

ormosils. Organic-inorganic hybrid materials in which the inorganic substance is a silicone or siloxane compound. Used for porous membranes, porous insulators for noise reduction applications, in the synthesis of biomaterials, as biosensors, and for the control of electrical, optical, mechanical and/or thermal properties of devices.

Ormspun. Trade name of Zephyr Inc. (Canada) for spun synthetic yarns.

Ormulu. Trade name of Lumen Bearing Company (USA) for an alloy composed of 58-94% copper, 0-25.3% zinc and 6-16.7% tin, and used for instruments, utensils, hardware, bearings, bushings, springs, and electrical contacts and switches.

Ornal. Trade name of Kreidler Werke GmbH (Germany) for a series of wrought aluminum alloys containing 0.4-1.2% magnesium and 0.4-1.2% silicon, and used for instrument components and optical equipment.

ornamental brick. Face brick of standard rectangular or other shape whose surface has been decorated with a design or pattern. Used chiefly for ornamental or decorative applications.

ornamental concrete. A concrete that, unlike ordinary concrete, is used primarily for decorative applications.

ornamental tile. A conventional tile or a tile of special size or shape usually with a design or pattern, and used for decorative applications.

Orobraze. Trade name of Johnson Matthey plc (UK) for a series of copper and gold brazing alloys. The copper-base alloys contain gold or silver (e.g., 70Cu-30Au, 62.5Cu-37.5Ag, etc.), and the gold-base alloys copper and iron, nickel, or silver (e.g., 80Au-20Cu+Fe, 75Au-25Ni, 70Au-30Ag, etc.).

Oro-Caps. Trade name for an encapsulated low-copper dental amalgam alloy.

Orocast. Trade name of Engelhard Industries (USA) for a series of medium-gold dental casting alloys containing varying amounts of platinum. Available in the annealed or age-hardened condition, they have a melting range of 850-910°C (1560-1670°F).

Oroglas. Trade name of Atoglas, Atofina Chemicals Inc. (USA) for transparent polymethyl methacrylates supplied as granules, beads, resins, and cast sheets. They have a density of 1.2 g/cm^3 (0.04 $lb/in.^3$), outstanding light transmission and resistance to weathering, moderate strength, good ultraviolet resistance, low heat and flame resistance, low water absorption, good resistance to dilute acids and alkalies, fair resistance to concentrated acids and poor resistance to aromatic hydrocarbons, greases, oils, halogens and ketones. Used for signs, nameplates, decorations, display items, and glazing.

Oromerse. Trade name of Technic Inc. (USA) for immersion gold coatings supplied in various grades for nickel, electroless nickel, silver and silver alloys.

Orosene. Trade name of Technic Inc. (USA) for cobalt-brightened gold finishes supplied in various grades including lead-resistant, heat-resistant, etc.

Orosphere. Trade name of Engelhard Industries (USA) for dental amalgam alloys of the spherical-particle type. *Orosphere II* is a low-copper dental amalgam alloy, and *Orosphere Plus* a high-copper dental amalgam alloy.

Orotemp. Trade name of Technic Inc. (USA) for ultrapure, ductile, heat-resistant neutral gold used for plating applications.

orpheite. A colorless to gray, pale blue, pale green or yellow green mineral of the zeolite group composed of lead aluminum hydrogen phosphate sulfate hydroxide hydrate, $H_6Pb_{10}Al_{20}(PO_4)_{12}(SO_4)_5(OH)_{40} \cdot 11H_2O$. Crystal system, rhombohedral (hexagonal). Density, 3.75 g/cm^3; melting point, approximately 1200°C (2190°F). Occurrence: Bulgaria.

orpiment. Lemon to golden yellow mineral with resinous luster composed of arsenic trisulfide, As_2S_3, and containing 61% arsenic and 39% sulfur. Crystal system, monoclinic. Density, 3.4-3.5 g/cm^3; melting point, 300°C (572°F); boiling point, 707°C (1305°F); hardness, 1.5-2.0 Mohs; refractive index, 2.81. Occurrence: Central Europe, Peru, USA (Nevada, Utah). Used as a source of arsenic, and as a pigment (king's yellow).

Orplid. Trade name of C. Hafner Gold- und Silberscheideanstalt (Germany) for high-gold dental alloys including *Orplid LFC* high-gold alloy with rich yellow color used with low-fusing ceramics and *Orplid Keramik* high-gold porcelain-bonding alloy.

Orr's white. See lithopone.

Ortalion. Trademark of Bemberg SpA (Italy) for nylon 6 fibers and filament yarns used for textile fabrics.

Ortho Band. Trademark of Pulpdent Corporation (USA) for a radiopaque, fluoride-releasing glass-ionomer cement that forms a strong chemical bond with metals and dental enamel, and is used for the cementation of orthodontic bands.

orthobrannerite. A black mineral of the amorphous group composed of uranyl titanium oxide hydroxide, $UTi_2O_6(OH)$. Crystal system, orthorhombic. Density, 5.46 g/cm³; refractive index, 2.328. Occurrence: China.

Orthocem B. Trade name for a glass-ionomer dental cement.

Orthochrom. Trade name of Rohm & Haas Company (USA) for a pigmented, plasticized nitrocellulose lacquer used for leather and textile finishing.

orthoclase. A colorless, white, gray, pale yellow or reddish mineral of the (potash) feldspar group composed of potassium aluminum silicate, $KAlSi_3O_8$. Crystal system, monoclinic. Density, 2.57 g/cm³; refractive index, 1.520. Occurrence: Australia, Canada, Central Europe, Scandinavia, USA (California, Colorado, Connecticut, Maine, Massachusetts, New Hampshire, North Carolina, Rhode Island, South Carolina, South Dakota, Texas, Vermont, Virginia). Used for pottery, enamel and ceramic ware, glass, abrasives, cements and concretes, insulating compositions, electrical and other porcelains, and roofing materials. Also known as *common feldspar; orthoclase feldspar; orthose.*

Orthocryl. Trade name of Otto Bock Health Care (USA) for a self-cure dental acrylic for orthodontic applications.

Orthodontic Plaster. Trade name of Whip Mix Corporation (USA) for a super-white, dental plaster (ADA Type II) supplied as a powder for mixing with water. Upon setting it develops a very hard, accurate surface, and has a high setting expansion (0.20%). Used especially for orthodontic applications.

Orthodontic Stone. Trade name of Whip Mix Corporation for a super-white dental stone (ADA Type III) supplied as a powder for mixing with water (28 mL water per 100 g powder). The mixture has a smooth, easy-flowing consistency, and upon setting develops a hard, accurate surface, and has a low setting expansion (0.09%). Used especially for orthodontic applications.

Orthodur. Trade name of Vereinigte Metallwerke Ranshofen-Berndorf AG (Austria) for a wrought, heat-treatable aluminum alloy containing 3.5-4.5% copper, 0.4-1.0% magnesium and 0.3-1% manganese. It has high strength, poor to moderate corrosion resistance, and is used for aircraft and automotive components and structural parts.

orthoericssonite. A brownish black mineral of the seidozerite group composed of barium manganese iron silicate hydroxide, $BaMn_2FeSi_2O_8(OH)$. Crystal system, orthorhombic. Density, 4.22 g/cm³; refractive index, 1.840. Occurrence: Japan.

Orthofan. Trade name of Schott DESAG AG (Germany) for a brown-tinted spectacle glass.

orthoferrites. A group of ferrimagnetic materials with orthorhombic crystal structure. See also ferrites; ferrimagnetic materials.

orthoferrosilicite. (1) A green or dark brown synthetic mineral of the pyroxenoid group composed of iron silicate, $FeSiO_3$. It can be made from a stoichiometric mixture of metallic iron, hematite and silica gel under prescribed conditions. Crystal system, orthorhombic. Density, 3.88 g/cm³; refractive index, 1.780.

(2) A brown mineral of the pyroxenoid group composed of iron magnesium silicate, $(Fe,Mg)SiO_3$. Crystal system, orthorhombic. Density, 3.83 g/cm³. Occurrence: South Africa.

Orthofoam. Trade name for high- and low-density polyethylene foams supplied in sheet form.

Ortho Gold. Trade name for a zinc phosphate dental cement.

Orthogonal. Trade name of Spang Specialty Metals (USA) for a grain-oriented, soft magnetic alloy composed of 50% iron and 50% nickel, and supplied in the form of thin strips. It has high saturation induction and relatively low initial permeability, and is used for flux counters, flux switching devices, and magnetic amplifiers.

Orthometal. Trade name of Telcon Metals Limited (UK) for a soft magnetic alloy with a square hysteresis loop used for magnetic amplifiers.

Orthomite Super-Bond. Trade name for a bonding resin used for orthodontics.

Orthomumetal. Trade name of Telcon Metals Limited (UK) for a soft magnetic alloy.

Orthonik. Trade name of Armco International (USA) for a magnetically soft alloy containing 45-50% nickel and 50-55% iron. It has a grainy structure, a rectangular hysteresis loop and high permeability. Used for magnetic and electrical equipment, and magnetic tape and foil.

Orthonol. Trademark of Rock Mountain RMO (USA) for a shape-memory alloy composed of 51.8% nickel and 48.2% titanium, used for dental and orthodontic wire.

orthophthalic polyesters. See orthophthalic resins.

orthophthalic resins. A group of general-purpose polyester resins made from alloys of phthalic anhydride and fumaric acid or maleic anhydride. They have moderate thermal stability, poor high-temperature performance, good chemical resistance and good processibility. Also known as *ortho resins; orthophthalics; orthophthalic polyesters.*

orthopinakiolite. A black mineral of the ludwigite group composed of magnesium manganese oxide borate, $(Mg,Mn)_2MnBO_5$. Crystal system, orthorhombic. Density, 4.03 g/cm³. Occurrence: Sweden.

Orthoplus. Trade name of Orthoplus (Israel) for a silicone elastomer used for dental impressions.

ortho resins. See orthophthalic resins.

orthose. See orthoclase.

Orthosil. Trade name of Thomas & Skinner Inc. (USA) for a thin silicon steel sheet containing up to 5% silicon, and the balance iron. It has good soft magnetic properties, high magnetic saturation, and is used for transformer laminations, chokes, filters, and reactors.

orthosilicates. (1) Metal salts derived from orthosilicic acid (H_2SiO_4), e.g., magnesium orthosilicate (Mg_2SiO_4).

(2) Silica materials in which none of the four oxygen atoms of the SiO_4 tetrahedra are shared with adjacent tetrahedra. The general formula is $XSiO_4$, where X represents one or more metals, such as calcium, lead, magnesium, manganese, sodium or zinc.

orthosilicic acid. See silicic acid.

orthotropic material. A material, such as timber and certain reinforced plastics, whose elastic properties vary considerably in different (e.g., horizontal or vertical) planes.

orthotungstic acid. See tungstic acid.

Orthovisc. Trademark of Antika Therapeutics (USA) for a viscous hyaluronan (hyaluronic acid) product used in the treatment of osteoarthritis as an intra-articular injection to improve joint mobility and relieve pain.

Ortiflor. Trade name of Vetreria di Vernante SpA (Italy) for a patterned glass.

Orvar. Trade name of Uddeholm Corporation (USA) for a series of abrasion- and shock-resistant chromium-type hot-work tool steels (AISI type H13) containing 0.35-0.4% carbon, 0-1.2% silicon, 5-5.5% chromium, 1.2-1.5% molybdenum, 0.9-1.1% vanadium, and the balance iron. Used for forging, extrusion and die-casting dies, punches, and shears. *Orvar Supreme* are premium-grade chromium-type hot-work tool steels (AISI type H13) used for molds.

osarizawaite. (1) A green mineral of the alunite group composed of lead copper aluminum sulfate hydroxide, $Pb(Al,Cu)_3(SO_4)_2$-$(OH)_6$. Crystal system, rhombohedral (hexagonal). Density, 4.04 g/cm^3. Occurrence: Australia.

(2) A pale yellow green mineral of the alunite group composed of lead copper aluminum iron sulfate hydroxide, $PbCu$-$(Al,Fe)_2(SO_4)_2(OH)_6$. Crystal system, rhombohedral (hexagonal). Density, 4.20 g/cm^3. Occurrence: Japan.

osarsite. A gray mineral of the marcasite group composed of osmium ruthenium arsenide sulfide, $(Os,Ru)AsS$. Crystal system, monoclinic. Density, 8.44 g/cm^3. Occurrence: USA (California).

Osborn. Trade name of Osborn Steels Limited (UK) for an extensive series of steels including numerous low-carbon chromium-molybdenum alloy grades, several austenitic, ferritic and martensitic stainless grades (AISI types 303 through 431), various cold-work tool grades (AISI types D, A and O), chromium- and tungsten-type hot-work tool grades (AISI types H10 through H21), and molybdenum- and tungsten-type high-speed tool grades (AISI types M1 through M42, and T1 through T6).

osbornite. A yellow mineral of the halite group composed of titanium nitride, TiN. It can be made synthetically. Crystal system, cubic. Density, 5.25 g/cm^3.

Oscillumin. German trade name for an aluminum alloy containing 12.6-13.2% silicon, 0.8% copper, and 0.4-0.6% iron. It possesses high fluidity, and is used for light-alloy castings.

Osemund. Trade name of Stahlwerke R. & H. Plate (Germany) for an oil-hardening, wear-resistant steel containing 1.4% carbon, 0.3% chromium, 0.1% vanadium, and the balance iron. Used for cutter, bearings, and liners.

Osmagal. Trade name of Kabel- und Metallwerke AG (Germany) for a non-heat-treatable aluminum alloy containing 1.8% manganese. It has good corrosion resistance, and is used for heat exchangers, truck panels, fixtures, and ductwork.

osmic acid. See osmium tetroxide.

osmiridium. A tin-white to light steel-gray natural alloy with metallic luster whose composition varies greatly from 17-80% osmium, 10-77% iridium, 0-18% ruthenium, 0-17% rhodium, 0-10% platinum, 0-3% iron, 0-1% copper, and traces of gold and palladium. Alloys containing more iridium than osmium are usually referred to as *iridosmine*. Crystal system, hexagonal. Density, 18.8-21.1 g/cm^3; hardness, 6-7 Mohs; good corrosion resistance; good resistance to acids including aqua regia. Used as a source of iridium and osmium, for hardening of platinum, as primary standards of weight and length, and for jewelry, watch pivots, compass bearings, fountain-pen points and surgical needles.

osmium. A white to grayish blue metallic element of Group VIII (Group 8) of the Periodic Table. It is commercially available in the form of sponge, powder and arc-melted buttons, and occurs as native osmium, in native platinum and in *osmiridium*. Crystal system, hexagonal. Crystal structure, hexagonal close-packed. Density, 22.5 g/cm^3; melting point, 3045°C (5513°F); boiling point, 5027°C (9081°F); hardness, 300-1000 Vickers; superconductivity temperature, 0.66K; atomic number, 76; atomic weight, 190.23; monovalent, divalent, trivalent, tetravalent, pentavalent, hexavalent, heptavalent, octavalent. It belongs to the noble metal group, has relatively good electrical conductivity (about 19.5% IACS), good corrosion and wear resistance, and cannot be worked by standard metallurgical processes. Used as a hardener for iridium and platinum, for fountain-pen points, clock bearings, instrument pivots, compass needles, lamp filaments, electrical contacts, decorative applications, in coatings, and as a catalyst. Symbol: Os.

osmium boride. Any of the following compounds of osmium and boron used in ceramics and materials research: (i) *Osmium diboride*. Density, 12.8-14.8 g/cm^3. Formula: OsB_2; and (ii) *Osmium pentaboride*. Density, 15.2 g/cm^3. Formula: Os_2B_5.

osmium bromide. Dark gray crystals that decompose at 340°C (644°F) used in chemistry and materials research. Formula: $OsBr_3$. Also known as *osmium tribromide*.

osmium diboride. See osmium boride.

osmium carbonyl. See triosmium dodecacarbonyl.

osmium chloride. See osmium dichloride; osmium trichloride; osmium tetrachloride.

osmium dichloride. Dark brown, hygroscopic crystals used in chemistry and materials research. Formula: $OsCl_2$. Also known as *osmium chloride*.

osmium diboride. See osmium boride (i).

osmium dioxide. Yellow-brown crystals. Crystal system, tetragonal. Density, 11.4 g/cm^3; melting point, decomposes at 650°C (1200°F). Used in chemistry and materials research. Formula: OsO_2. Also known as *osmium oxide*.

osmium disulfide. See osmium sulfide (i).

osmium fluoride. See osmium tetrafluoride; osmium pentafluoride; osmium hexafluoride; osmium octafluoride.

osmium hexafluoride. Yellow or greenish yellow crystals. Crystal system, cubic. Density, 4.1 g/cm^3; melting point, above 33°C (91°F); boiling point, 46°C (115°F). Used in chemistry and materials research. Formula: OsF_6. Also known as *osmium fluoride*.

osmium monoxide. Black crystals used in chemistry and materials research. Formula: OsO. Also known as *osmium oxide*.

osmium octafluoride. Lemon-yellow crystals with a melting point of 34°C (93°F) and a boiling point of 47°C (117°F). Used in chemistry and materials research. Formula: OsF_8. Also known as *osmium fluoride*.

osmium oxide. See osmium monoxide; osmium dioxide; osmium sesquioxide; osmium tetroxide.

osmium pentaboride. See osmium boride (ii).

osmium pentafluoride. Blue crystals with a melting point of 70°C (158°F), used in chemistry and materials research. Formula: OsF_5. Also known as *osmium fluoride*.

osmium-potassium chloride. Red-purple, hygroscopic crystals. Crystal system, cubic. Density, 3.42 g/cm^3; melting point, 600°C (1112°F) (decomposes). Used as a reagent. Formula: K_2OsCl_6. Also known as *potassium chloroosmate; potassium hexachloroosmate*.

osmium plutonium. An intermetallic compound of osmium and plutonium used in ceramics and nuclear engineering. Melting point, 1500°C (2732°F). Formula: OsPu.

osmium sesquioxide. Dark brown crystals used in chemistry and materials research. Formula: Os_2O_3. Also known as *osmium oxide.*

osmium silicide. A compound of osmium and silicon used in ceramics and materials research. Formula: OsSi.

osmium-sodium chloride. A crystalline compound usually supplied in anhydrous form as orange crystals containing approximately 40.3% osmium, or in hydrated form $(OsNa_2Cl_6 \cdot xH_2O)$ as an orange-red, hygroscopic powder used as a reagent. The anhydrous form is also employed as a catalysts in oxidation reactions. Formula: $OsNa_2Cl_6$. Also known as *sodium-osmium chloride; sodium chloroosmate; sodium hexachloroosmate.*

osmium sulfide. Any of the following compounds of osmium and sulfur used in chemistry and materials research: (i) *Osmium disulfide.* Black crystals. Crystal system, cubic. Formula: OsS_2; and (ii) *Osmium tetrasulfide.* Brown-black crystals or powder used in chemistry and materials research. Formula: OsS_4.

osmium tetrachloride. Red-brown needles or or red-black crystals. Crystal system, orthorhombic. Density, 4.38 g/cm³; boiling point, 450°C (842°F). Used in chemistry and materials research. Formula: $OsCl_4$. Also known as *osmium chloride.*

osmium tetrafluoride. Yellow crystals or brown powder used in chemistry and materials research. Formula: OsF_4. Also known as *osmium fluoride.*

osmium tetrasulfide. See osmium sulfide (ii).

osmium tetroxide. Pale yellow crystals or yellow, amorphous powder. Crystal system, monoclinic. Density, 4.9-5.1 g/cm³; melting point, 39.5-42°C (103-108°F); boiling point, 135°C (275°F). Used in microscopic staining, photography, organic synthesis, and as a catalyst. Formula: OsO_4. Also known as *osmic acid; osmium oxide; perosmic oxide.*

osmium tribromide. See osmium bromide.

osmium trichloride. Brown, moisture-sensitive, alcohol- and water-soluble crystals (99.9% pure). Crystal system, cubic. Melting point, decomposes above 450°C (842°F). It is also available in hydrated form $(OsCl_3 \cdot xH_2O)$ as dark green or black crystals that decompose above 500°C (932°F). Used in chemistry and materials research. Formula: $OsCl_3$. Also known as *osmium chloride.*

osmocene. An organometallic coordination compound (stable molecular sandwich) consisting of an osmium atom (Co^{2+}) and two five-membered cyclopenediene (C_5H_5) rings, one located above and the other below the osmium atom plane. *Osmocene* is supplied in the form of white crystals with a melting point of 226-230°C (439-446°F), and used as a catalyst, intermediate and UV-absorbers. Formula: $(C_5H_5)_2Os^{2+}$. Also known as *dicyclopentadienylosmium; bis(cyclopentadienyl)osmium.*

osmosis kaolin. An electroosmotically deposited *kaolin* made into a fine powder, and used in the manufacture of electrical insulators and synthetic mica. Electroosmosis is the movement of a liquid absorbed by a porous substance to an electrode by the application of an electric field.

Osna. Trade name of KM-Kabelmetall AG (Germany) for sulfur-bearing and leaded coppers with free-machining properties.

osnaburg. A strong, heavy or medium-weight, semifinished fabric, usually cotton, in a plain weave, used for curtains, bedspreads, upholstery, sacks, and industrial applications.

Osnalium. Trade name of KM-Kabelmetall AG (Germany) for a series of corrosion-resistant wrought aluminum alloys containing 2-7.5% magnesium, 0-1% manganese and 0-0.3% chro-

mium. Used for light-alloy parts, marine hardware, aircraft structures, and light-alloy parts.

Osnalen. Trademark of Hagedorn–Plastic GmbH (Germany) for phenol-formaldehyde and polypropylene plastics including several film products.

Osnisil. Trade name of KM-Kabelmetall AG (Germany) for a series of copper alloys containing 1.3-3.5% nickel, 0.5-1% silicon and 0.2-0.3% chromium. Available in the form of rods, profiles, tubes, wires, finished parts and strip, they are used for electrical components, fasteners, and mechanical components.

Ospelon. Trademark of Hagedorn–Plastic GmbH (Germany) for a series of modified polyvinyl chloride (PVC) plastics and PVC copolymers supplied in film and several other forms.

Osprey CE. Trade name of Osprey Metals Limited (UK) for a group of lightweight silicon-aluminum alloys with a typical aluminum content of 30-60%, produced by a proprietary spray forming technique. They are characterized by low and controlled, mainly temperature-independent coefficients of expansion that, depending on the composition, vary between those of aluminum and silicon. Other important properties include high thermal conductivity and low density. They can be easily machined and electroplated with nickel, gold and silver, and are used for electronic packaging applications.

ossein. The organic basis of bone tissue that can be obtained by dissolving bone in hydrochloric acid. It is closely related to gelatin, and used in the manufacture of certain glues and fibers.

ossein fibers. Rather brittle continuous fibers obtained from *ossein.*

osseoalbuminoid. A water-insoluble protein that forms part of the organic matrix of bone.

Osstyrol. Trademark of Hagedorn–Plastic GmbH (Germany) for acrylonitrile-butadiene-styrenes and styrene-butadiene copolymers supplied in film and several other forms.

Ostermann. Trade name of Ostermann GmbH & Co. (Germany) for a series of cast and wrought copper-base alloys including various copper nickels, phosphorus coppers, aluminum, nickel and phosphor bronzes, leaded and unleaded tin bronzes, and red brasses. Used for machinery parts, bearings, fittings, valve and pump parts, ship propellers, and electrical parts.

Ostron 100. Trade name of GC Dental (Japan) for a self-cure dental acrylic resin supplied in clear, blue, pink and white colors. Used for impression tray and baseplate applications.

osumilite. (1) A black mineral composed of potassium sodium magnesium iron aluminum silicate monohydrate, $(K,Na,Ca)(Mg,Fe)_2(Al,Fe)_2(Si,Al)_{12}O_{30} \cdot H_2O$. Crystal system, hexagonal. Density, 2.64 g/cm³; refractive index, 1.546. Occurrence: Japan.

(2) A black mineral composed of potassium iron magnesium aluminum silicate monohydrate, $(K,Na)(Mg,Fe)_2(Al,Mg)_3(Si,Al)_{12}O_{30} \cdot H_2O$. It can also be made synthetically. Crystal system, hexagonal. Density, 2.63 g/cm³; refractive index, 1.541. Occurrence: Canada.

otavite. A white or yellow mineral of the calcite group composed of cadmium carbonate, $CdCO_3$. It can also be made synthetically. Crystal system, rhombohedral (hexagonal). Density, 5.03 g/cm³; refractive index, 1.842. Occurrence: Southwest Africa.

O-Ten. Trade name of USX Corporation/USS Division (USA) for a low-carbon high-strength structural steel (API 2Y-50) with enhanced toughness and good weldability. Usually supplied in plate form, it is used especially for offshore drilling platforms, pressure vessels, penstocks, and construction and transportation equipment components.

Otiscoloy. Trade name of LTV Steel (USA) for a steel containing 0.08-0.12% carbon, 0.9-1.25% manganese, 0.35% copper, 0.1% nickel, 0.05% chromium, and the balance iron. Used for deep-drawn parts.

Otisel. Trade name of Otis Elevator Company (USA) for a series of wrought and cast plain-carbon, heat-resistant and stainless steels.

otjisumeite. A white or colorless mineral composed of lead germanium oxide, $PbGe_4O_9$. Crystal system, triclinic. Density, 6.09 g/cm^3. Occurrence: Namibia.

Otsoeristeet. Trade name of A. Ahlström Osakeyhtiö (Finland) for insulation products composed of coarse glass fibers. *Otso* is the general group name for these products.

Ottawa. (1) Trade name of Allegheny Ludlum Steel (USA) for a heat-treatable, galling-resistant iron containing 3.25% carbon, 1% chromium, 1% molybdenum, 12% vanadium, and the balance iron.

(2) Trade name of Atlas Specialty Steels (Canada) for a high-quality carbon hollow drill steel containing 0.8% carbon, 0.25% manganese, 0.15% silicon, 0.018% sulfur, 0.018% phosphorus, and the balance iron. Used for hollow rock drills, mining drills and rods, and detachable bit shanks.

Ottawa sand. A relatively pure silica sand with naturally rounded quartz grains found near Ottawa, Illinois, USA. Used for testing hydraulic cement.

ottemannite. A gray mineral composed of tin sulfide, Sn_2S_3. It can also be made synthetically. Crystal system, orthorhombic. Density, 4.76 g/cm^3. Occurrence: Bolivia.

ottoman. A medium- to heavyweight knit or woven fabric, often of silk or rayon with wool or cotton, having wide crosswise ribs or cords. Used especially for women's apparel, drapery and upholstery.

Otto's speculum. An alloy containing 69% copper and 31% tin used for mirrors and telescope reflectors.

ottrelite. A gray to black monoclinic variety of the mineral *chloritoid* composed of iron manganese aluminum silicate hydrate, $(Fe,Mn)(Al,Fe)_2Si_3O_{10}·H_2O$. It is a brittle mica. Occurrence: USA.

otwayite. A bright green mineral composed of nickel carbonate hydroxide monohydrate, $Ni_2CO_3(OH)_2·H_2O$. Crystal system, orthorhombic. Density, 3.41 g/cm^3. Occurrence: Western Australia.

Ouatisol. Trade name of Saint-Gobain (France) for flexible felt of superfine glass fibers impregnated with silicone resins.

ounce metal. A composition metal originally made by introducing 1 oz (28.35 g) each of lead, tin and zinc to 1 lb (453.6 g) of copper. The nominal composition is given as 5% lead, 5% tin, 5% zinc and 85% copper. It has good castability and machinability, and takes a high polish. Used for bearings, hardware, screws, nuts, hydraulic castings, such as pump parts, carburetors, and valves. See also composition metal.

ourayite. A mineral of the lillianite group composed of silver lead bismuth sulfide, $Ag_{25}Pb_{30}Bi_{41}S_{104}$. Crystal system, orthorhombic. Density, 7.11 g/cm^3. Occurrence: USA (Colorado).

ouricury wax. A hard, brown vegetable wax obtained from the leaves of the South American ouricury palm *(Cocos coronata)*. It has a density of 0.97 g/cm^3 (0.035 $lb/in.^3$), a melting point of 85°C (185°F), and is used as a substitute for carnauba wax, and for blending with carnauba and other waxes.

oursinite. A mineral composed of oxonium cobalt magnesium uranyl silicate trihydrate, $(H_3O)_2(Co,Mg)(UO_2)_2(SiO_4)_2·3H_2O$. Crystal system, orthorhombic. Density, 3.67 g/cm^3; refractive index, 1.640. Occurrence: Zaire.

Outokumpu. (1) Trade name of Outokumpu Metals (USA) Inc. for a series of commercially pure coppers (UNS Nos. C10100 through C13000) including various silver-bearing and silver- and oxygen-free high-conductivity, oxygen-free electronic and phosphorus-deoxidized grades. Also included under this trade name are various zirconium and chromium-zirconium coppers and tellurium-bearing coppers as well as a wide range of wrought brasses (cartridge, and yellow brasses, low brasses, red brasses, etc.), wrought bronzes (commercial bronzes, aluminum bronzes, high and low-silicon bronzes, phosphor bronzes, etc.), wrought copper nickels, and nickel silvers. Usually supplied as sheets, strips, tubes, rounds, rods, squares, semifinished products, or structural shapes.

(2) Trade name of Outokumpu Metals (USA) Inc. for a series of zinc die-casting alloys.

Ovako. Trade name of Ovako Steel Hellefors AB (Sweden) for a series of steel products including various air-hardening, case-hardening and quenched-and-tempered grades, plain-carbon grades, boron alloy grades, nitrided/nitrocarburized grades, chromium roller bearing grades, constructional and machinery grades and high-strength microalloyed structural grades.

Ovation. Trademark of Degussa-Ney Dental (USA) for a copper-free dental alloy containing 69.8% palladium, 6.5% silver, and 3% gold. It has a light oxide layer, high yield strength, a hardness of 255 Vickers, and is used in restorative dentistry especially for long-span bridges and implant cases.

oven-dried lumber. See kiln-dried lumber.

oven-dried material. A material dried to virtually constant mass in an oven or similar heated enclosure under prescribed conditions of humidity and temperature.

oven glass. Glass having a low coefficient of thermal expansion, and exhibiting high thermal-shock resistance. Used for baking dishes, casseroles and other cookware.

ovenware. Ceramic whiteware or glass (e.g., casseroles, ramekins, etc.) that is able to withstand the heat of an oven without cracking.

overcup oak. The hard, strong, durable, light brown wood of the medium-sized tree *Quercus lyrata* belonging to the American white oaks. It has a close, coarse grain, good decay resistance, works well with power tools, and has good gluing and nailing qualities. Source: USA (Southern and south Atlantic States from New Jersey to Texas). Used for lumber, millwork, furniture, carvings, ship and boat structures, barrels and kegs, railroad ties, fencing, posts, and veneer. See also American white oak.

overglaze. A second glaze applied over a previously glazed surface of ceramic ware.

overglaze colors. A mixture of finely divided low-melting glasses and pigments fired as a decorative design on glazed ceramic surfaces (china, pottery, or tile) usually at a temperature of 705-815°C (1300-1500°F).

overite. A light green to colorless mineral composed of calcium magnesium aluminum phosphate hydroxide tetrahydrate, $CaMgAl(PO_4)_2(OH)·4H_2O$. Crystal system, orthorhombic. Density, 2.53 g/cm^3; refractive index, 1.574. Occurrence: USA (Utah).

overlaid particleboard. A special type of *particleboard* veneered with fiber sheets or materials, such as *hardboard* or *plastic laminates*, and used for paneling, cabinetwork, furniture, doors, and countertops.

overlay. (1) A thin layer of paper, metal foil, plastic film or other material bonded to one or both surfaces of panel materials or lumber to provide a decorative or protective face, or a base for subsequent paint coats. See also high-density overlay; medium-density overlay; overlaid particleboard.

(2) A mat of glass or synthetic fibers employed as a surface layer to give a smooth surface finish, reduce the appearance of the fiber pattern, or facilitate precision grinding or machining. Also known as *overlay sheet; surfacing mat.*

(3) A concrete topping used to repair worn concrete surfaces.

overlay coating. A corrosion- and/or oxidation-protective coating composed of multicomponent alloys, such as MCrAlY (where the base metal (M) is cobalt, iron or a cobalt-nickel alloy), applied to a substrate, usually a superalloy, by plasma spraying, sputtering, physical-vapor deposition or other techniques. See also MCrAlY.

overlay paper. A non-filled, lightfast paper manufactured from bleached high-purity cellulose. It forms the transparent top layer of laminating papers, and has a basis weight of about 20-50 g/m^2 (0.065-0.163 oz/ft^2).

overlay plywood. Plywood having a resin-impregnated fiber surface. See high-density overlay; medium-density overlay; overlay; plywood.

overlay sheet. See overlay (2).

Ovignose. Trade name for vinyon (vinyl chloride) fibers with excellent acid, alkali and mildew resistance. Used especially for textile fabrics.

Owens Corning. Trademark of Owens-Corning (USA) for an extensive series of building materials and composite systems based on asbestos, fiberglass, or other fibrous materials.

Owens Corning Fiberglas. General trade name of Owens-Corning Fiberglas (USA) for an extensive series of fiberglass products. Abbreviation: OCF.

owyheeite. A silver-gray mineral composed of silver lead antimony sulfide, $Ag_2Pb_5Sb_6S_{15}$. Crystal system, orthorhombic. Density, 6.25 g/cm^3. Occurrence: USA (Idaho).

oxalaldehyde. See glyoxal.

oxalic acid. An organic acid that occurs in wood sorrel, rhubarb and various other plants, and can also be prepared synthetically. It is available in anhydrous and hydrated form. The anhydrous form is available as colorless, transparent water-soluble orthorhombic crystals (99+% pure) with a melting point of 187°C (decomposes). The dihydrate is supplied as monoclinic tablets or prisms with a melting point of 101-102°C (214-216°F). Used in textile bleaching, stain removal, dye manufacture, in printing, as a precipitating agent for rare earths, in metal cleaning, as an automobile radiator cleanser, as a catalyst, as a stripper for permanent-press resins, as an intermediate and purifier, in biochemistry, biotechnology and biology, and in medicine. Formula: $C_2H_2O_4$; (anhydrous); $C_2H_2O_4 \cdot 2H_2O$ (dihydrate).

oxammite. A colorless mineral composed of ammonium oxalate monohydrate, $C_2H_8N_2O_4 \cdot H_2O$. Crystal system, orthorhombic. Density, 1.50 g/cm^3; refractive index, 1.549.

oxford. (1) A soft fabric, usually made of cotton or a cotton/synthetic fiber blend in a plain or basket weave. Used for shirts, sportswear and other apparel. Also known as *oxford cloth.*

(2) A fabric produced from a blend of black and white yarns.

oxford cloth. See oxford.

oxidation-protective coatings. A group of coatings used to protect superalloy and refractory-metal substrates against corrosive attack by oxidizing media at high temperatures.

oxide-base cermets. A group of composite materials consisting of ceramic oxide particles, such as aluminum oxide, chromic oxide, magnesium oxide, silicon dioxide or uranium dioxide, and a metallic binder, such as aluminum, cobalt, chromium, iron, nickel, silicon, steel, tantalum, titanium or zirconium. They are made at high temperatures under controlled atmospheres using powder-metallurgy techniques involving pressing and sintering. *Oxide-base cermets* have outstanding high-temperature strength and wear resistance, high resistance to oxidation and intergranular corrosion at elevated temperatures, good corrosion resistance, excellent stress-to-rupture properties, and high toughness. Used for super-high-speed cutting tools, turbojet engines, gas turbines, rocket motors, nuclear reactor components, chemical equipment, pumps for severe service, high-temperature-resistant coatings, electrical components, seals, and bearings.

oxide ceramic coatings. A group of coatings based on oxide materials, such as aluminum oxide, barium oxide, beryllium oxide, boron oxide, calcium oxide, cobalt oxide, silicon dioxide, or zirconium oxide. Used to provide metal substrates (except refractory metals) with protection against oxidation at elevated temperatures, and with a high degree of thermal insulation.

oxide ceramics. A group of ceramics made by dry pressing or slip casting of essentially pure oxides, such as aluminum oxide, beryllium oxide, magnesium oxide, thorium dioxide, and zirconium oxide, and sintering at high temperatures. They have relatively low densities, high brittleness, high hardness, good wear resistance, low thermal conductivities, low coefficients of friction, low mechanical and thermal-shock resistance, low rupture strength, good hardness and strength retention up to about 1093°C (2000°F), and a softening temperature of 1400°C (2550°F). Used for cutting tools and high-temperature applications. Also known as *cemented oxides; sintered oxides.*

oxide coatings. (1) Surface coatings produced on finish-ground tools by steam oxidation or immersion into an alkali-nitrate bath usually to increase the service life and prevent adhesion to workpieces.

(2) High-temperature, oxidation-resistant coatings, usually of refractory oxides, such as aluminum oxide, zirconium oxide, or thorium dioxide, applied to metal substrates.

oxide conversion coatings. Protective and/or decorative coatings of aluminum oxide produced on the surface of aluminum or aluminum alloys by immersion in a bath of suitable composition, e.g., a mixture of sodium chromate and sodium carbonate, sodium chromate, sodium carbonate and sodium silicate, or sodium carbonate and sodium dichromate. The aluminum oxide surface coating is formed by a chemical reaction of the aluminum or aluminum-alloy surface with the reagents in the bath. See also chemical conversion coatings.

oxide-dispersion-strengthened alloys. A group of alloys hardened and strengthened by dispersion of small concentrations (typically 0-15 vol%) of small-diameter oxide particles (alumina, beryllia, silica, thoria, or yttria). Examples include thoria-dispersed nickel, sintered aluminum powder (SAP) alloys, yttria-dispersed iron alloys, and yttria-dispersed nickel alloys. Used for high-temperature turbine-engine nozzles and seals, thermal protection systems for spacecraft, aircraft and spacecraft control surfaces, and heat shields. Abbreviation: ODS alloys.

oxide fibers. A group of continuous or discontinuous ceramic fibers that, in general, are based on oxides of aluminum, silicon or zirconium. See also nonoxide fibers.

oxide glasses. A group of noncrystalline solids in which one or more oxides form the predominant components. They are further subdivided into (i) silicate oxide glasses (e.g., vitreous silica, borosilicate and window glass, or fiberglass), and (ii) nonsilicate oxide glasses (e.g., boron oxide, germanium dioxide or phosphorus pentoxide glass).

oxide minerals. Naturally occurring minerals in which the constituents are essentially in oxide form. Examples include cassiterite (SnO_2), chromite ($FeCr_2O_4$), chrysoberyl ($BeAl_2O_4$), corundum (Al_2O_3), cuprite (Cu_2O), hematite (Fe_2O_3), ilmenite ($FeTiO_3$), magnetite (Fe_3O_4), rutile (TiO_2) and zincite (ZnO).

oxide nuclear fuel. A fissionable nuclear fuel based on uranium oxide (UO_2) or plutonium oxide (PuO_2).

oxidized asphalt. See blown asphalt.

oxidized cellulose. See oxycellulose.

oxidized polyacrylonitrile. A black material with moderate mechanical properties obtained from *polyacrylonitrile* by thermal rearrangement and oxidization. Used as a precursor for certain carbon fibers, and for glowproof cotton wool, protective clothing, aircraft, race-car and high-speed train brakes, and in carbon-carbon composites. See also carbon fibers; polyacrylonitrile carbon fibers.

oxidizer. A compound that readily evolves oxygen, attracts electrons, and removes or replaces hydrogen in another compound. Also known as *oxidizing agent.*

oxidizing agent. See oxidizer.

Oxiron. Trademark of FMC Corporation (USA) for an epoxidized polyolefin for casting and laminating resins.

2-oxohexamethyleneimine. See ε-caprolactam.

oxomethane. See formaldehyde.

oxosilanes. See siloxanes.

Oxy. Trademark of Occidental Chemical Corporation (USA) for vinyl resins.

Oxyblend. Trademark of Occidental Chemical Corporation (USA) for dry-blend compounds of polyvinyl chloride resins.

oxycellulose. Cellulose or regenetated cellulose that has been treated with an oxidizing agent, such as nitrogen dioxide (NO_2). It is usually supplied in the form of gauze, lint, pellets, or powder. *Oxycellulose* is insoluble in water and common organic solvents, soluble in alkaline solution and many amines, and slowly degrades at room temperature. Used chiefly as a cation exchanger in the life sciences, and as an bioadhesive. Also known as *oxidized cellulose.* See also cellulose; regenerated cellulose.

oxychloride cement. See magnesium oxychloride cement.

Oxyclear. Trade name of Occidental Chemical Corporation (USA) for unplasticized polyvinyl chloride resins and compounds, and structural foams used in the manufacture of construction materials, and for packaging applications.

oxydiethanol. See diethylene glycol.

oxygen. A colorless, odorless, tasteless, nonflammable, diatomic gaseous element of Group VIA (Group 16) of the Periodic Table. It occurs combined with hydrogen as water (H_2O), is a constituent of most common rocks and minerals, and occurs in a many organic compounds, and uncombined in the earth's atmosphere to the extent of 20.95 vol% or 23.16 wt%. Density of gas, 1.429 g/L (at 0°C or 32°F); density of liquid, 1.149 g/cm³ (at -183°C or -297°F); density of solid, 1.426 g/cm³ (at -252.5°C or -422°F); melting point, -218.8°C (-362°F); boiling point, -183.0°C (-297°F); atomic number, 8; atomic weight, 15.999; divalent; actively supports combustion. Used in oxyacetylene and oxyhydrogen welding and cutting, in steelmaking, in various other metallurgical processes, in glassmaking, in water purification, in the chemical industry, in biochemistry, medicine, and for many other applications. Symbol: O.

oxygen-18. A stable isotope of oxygen having a mass number of 18 and occurring in a proportion of 8 parts to 10000 of ordinary oxygen (oxygen-16) in air, water, rocks, etc. Used as a tracer. Also known as *heavy oxygen.* Symbol: ^{18}O.

oxygen-free copper. A high-purity copper (99.9+%) made by induction melting of high-grade cathode copper under nonoxidizing conditions. It is commercially available in the form of rods, wires, tubes, pipes, structural shapes and flat products, and in special grades (e.g., cryogenic). It has a high conductivity (101% IACS), exceptional ductility, excellent hot and cold workability, good forgeability, low gas permeability, low outgassing tendency, and excellent corrosion resistance. Used for high-power electron tubes, anodes for vacuum tubes, vacuum seals, klystrons, microwave tubes, rectifiers, and other electrinic devices and components. Abbreviation: OF Cu. See also oxygen-free high-conductivity copper.

oxygen-free electronic copper. A commercially pure copper (99.99%) with good to excellent corrosion resistance, excellent cold and hot workability, good forgeability, high electrical and thermal conductivity, and excellent corrosion resistance. It is commercially available in the form of rods, tubes, pipes, flat products, and shapes. Used for electrical conductors, busbars, waveguides, lead-in wires, anodes, vacuum seals, rectifiers, klystrons, transistors and microwave components, and other electrinic devices and components. Abbreviation: OFE Cu.

oxygen-free extra-low-phosphorus copper. A commercially pure copper (99.95%) containing no more than 0.003% phosphorus. It has high electrical conductivity, excellent hot and cold workability, good forgeability, good corrosion resistance, and good weldability and brazeability. It is commercially available in the form of rods, tubes, pipes, flat products and shapes. Used for electrical conductors, and busbars. Abbreviation: OFXLP Cu.

oxygen-free high-conductivity copper. A commercially pure copper (99.9-99.98%) with high electrical conductivity (101% IACS), excellent hot and cold workability, good forgeability and good corrosion resistance. It is commercially available in the form of rods, wires, tubes, pipes, flat products and shapes. Used for electrical conductors, busbars, waveguides, lead-in wires, Dumet wire, cables, and anodes, transistors, radar and microwave components, and other electronic components and devices. Abbreviation: OFHC Cu.

oxygen-free high-conductivity sulfur copper. A commercially pure copper (99.90% min) containing 0.2-0.6% sulfur. It has excellent corrosion resistance, excellent cold and hot workability, good machinability and electrical and thermal conductivity. Used for screw-machine products, high-conductivity parts, contact pins, inserts, cables, electrical connectors, motor and switch components, and electronic components.

oxygen furnace steel. See basic-oxygen-furnace steel.

oxygenized iron. A high-strength cast iron produced by first melting in a cupola, subjecting to a blast of air, and then returning to the cupola.

oxymethylene. See formaldehyde.

Ozalid. Trade name of Hoechst AG (Germany) for paper, fabrics and plastic films having a light-sensitive emulsion coating applied to the surface. Developing with ammonia produces positive prints. Used for making photocopies, blueprints and other photoprints.

ozocerite. Yellowish brown, green or blackish, relatively impure *paraffin wax* occurring in nature. It is an amorphous, water-soluble mixture of hydrocarbons. Density, 0.85-0.95 g/cm³; melting point, 55-110°C (131-230°F). Occurrence: Australia, Central Europe, Caspian Sea region. Used for electric insulation, electrotypers' wax, paints, wood fillers, floor, furniture and leather polishes, waxed paper and fabrics, paper and textile sizes, carbon paper, rubber goods, lubricants, greases and crayons, as a source of ceresin wax, and as a substitute for beeswax and carnauba wax. Also known as *earth wax; fossil wax; mineral wax; native paraffin; ozokerite.*

ozokerite. See ozocerite.

ozone. An unstable, pale blue, triatomic gas, O_3, with characteristic odor produced by electric discharge in oxygen, and naturally in the atmosphere by lightning. It occurs as a natural constituent in the earth's atmosphere in concentrations of about 0.01 ppm. Density of gas, 2.144 g/L (at 0°C or 32°F); density of liquid, 1.614 g/L (at -195.4°C or -319°F); melting point, -192°C (-314°F); boiling point, -112°C (-170°F); much more reactive than oxygen. Used in the purification of air and water, in air conditioning, as a powerful oxidizer, as a bleaching agent, in various chemical processes, and in biochemistry and medicine.

P

PA-ABS alloys. Thermoplastic polymer blends of about 50% crystalline polyamide (PA) and 50% amorphous acrylonitrile-butadiene-styrene (ABS), often supplied in injection molding grades. They have high toughness and impact resistance, outstanding processibility, excellent flow characteristics, good abrasion, chemical and heat resistance, good surface finish, and reduced moisture sensitivity. Used for lawn and garden equipment, power-tool housings, sporting goods and automotive components.

PAA-CORE. Trademark of Alcore, Inc. (USA) for an aluminum honeycomb material used for airframe primary structures.

paakkonenite. A dark gray mineral composed of antimony arsenic sulfide, Sb_2AsS_2. Crystal system, hexagonal. Density, 5.21 g/cm^3. Occurrence: Finland.

PAAMA 2. Trade name of Southern Dental Industries Limited (Australia) for a two-component (primer plus adhesive) resin system based on an aromatic dimethacrylate (DMA) monomer that chemically and mechanically bonds to the tooth structure. It provides high bonding strength, very low polymerization shrinkage and high facture resistance. Used in restorative dentistry for dentin- and/or enamel-to-composite bonding.

Pabco. Tradename of former Plant Rubber & Asbestos Works (USA) for a series of fibrous asbestos and non-asbestos products including magnesia-based high-temperature blocks and pipe coverings as well as fibered gypsum wall plasters.

Pabco Super Caltemp Gold. Trade name of Johns Manville Company (USA) for high-temperature fiberglass block and pipe insulation.

pabstite. A colorless to white mineral of the benitoite group composed of barium tin silicate, $BaSnSi_3O_9$. It can also be made synthetically. Crystal system, hexagonal. Density, 4.03 g/cm^3; refractive index, 1.685. Occurrence: USA (California).

Pace. Trade name of Pace Technologies (USA) for a high-chromium cast iron containing 1.6% carbon, up to 1.5% manganese, up to 2.0% silicon, 28% chromium, 2% nickel, 2% molybdenum, and the balance iron. It outperforms many premium stainless steels and abrasion-resistant alloys in abrasive and highly corrosive environments, and is used for fluid-handling systems, pumps, valves, impellers, casings, and wear plates. "PACE" stands for "Protection Against Corrosion/Erosion."

Pac-Grade. Trademark of Orpac Inc. (USA) for boron nitride coatings that can be applied by dipping, painting and spraying and are used as high-temperature lubricants on metals and plastics, in the extrusion and superplastic forming of superalloys, and as a nonstick surface for molds employed for pressing and forming ceramics and composites.

pachnolite. A colorless or whitish mineral composed of sodium calcium aluminum fluoride monohydrate, $NaCaAlF_6 \cdot H_2O$. Crystal system, monoclinic. Density, 3.00 g/cm^3; refractive index, 1.413; hardness, 3 Mohs. Occurrence: Greenland.

Pacific. Trade name of Pilkington Brothers Limited (UK) for patterned glass.

Pacific madrone. The reddish-brown, hard, heavy, brittle wood of the large, evergreen tree *Arbutus menziesii*. It has large shrinkage, good workability and large warpage. Source: Pacific Coast of North America from British Columbia to California. Used for fencing, furniture, as fuelwood (charcoal), and for tanning leather. Also known as *madrone; madrona*.

Pacific red cedar. See western red cedar.

Pacific silver fir. See silver fir (2).

Pacific yew. The dark reddish-brown, heavy, strong wood of the coniferous tree *Taxus brevifolia*. It has a fine grain, exceptional durability and very high hardness, but is very difficult to work. Average weight, 670 kg/m^3 (42 lb/ft^3). Source: Western Canada and United States from British Columbia to California. Used for archery bows, fenceposts, and cabinetwork.

packaged concrete. A packaged mixture of dry ingredients, i.e., cement and admixtures, and accelerators, gas formers, retarders, hardeners, etc., that requires only water to yield concrete.

pack-carburized steel. A low-carbon or low-alloy steel with an initial carbon content of 0.06-0.30% into whose surface layer carbon has been introduced by heating above the transformation temperature range, typically between 850 and 1050°C (1560 and 1920°F), while packed in a closed container in intimate contact with a solid carbonaceous material, such as hardwood charcoal with barium carbonate additions. The resulting steel has a ductile and shock-resistant core and a hard, abrasion- and wear-resistant surface.

packfong. An alloy of copper, zinc and nickel with a silvery appearance used for jewelry and cutlery. It was introduced from China in the 18th Century. See also nickel silver.

Packing. (1) Trade name for an alloy of 82% lead, 13% antimony and 5% tin with a pouring temperature of 324°C (615°F), used for metallic packing

(2) Trade name for an antifriction alloy composed of 71% tin, 24% antimony and 5% copper, and used for valve packing.

packing. (1) A material, such as asbestos, rubber, fiber or plastic sheet, or a lead-antimony, tin-antimony or copper-lead alloy, used to pack or make airtight, watertight, steamtight, oiltight, etc.

(2) Any soft and pliable material, such as rubber, that can be used to restrict or prevent the passage of matter between surfaces that move relative to each other.

(3) A material used in powder metallurgy to embed compacts during presintering and sintering to protect them against contamination. Also known as *packing material*.

(4) A material, such as gravel, sand or mill scale, that is used in annealing boxes or pots as a casting support to prevent warpage. Also known as *packing material*.

packing leather. A highly stuffed leather, usually a chrome- and/or vegetable-tanned cattlehide, used as piston packing and for gaskets used in pump valves.

packing material. See packing (3), (4).

packing metal. An alloy composed of 51.8% copper, 17% lead, 15% nickel, 14.3% zinc and 1.9% tin, and used for metallic packing.

PacMaster. Trade name of F.J. Brodmann & Company LLC (USA) for pack diffusion powders.

padding. A usually soft material, such as polymer foam, rubber, cotton or synthetic fibers, used for cushioning or stuffing applications.

PAD lumber. See partially air-dried lumber.

padouk. The hard wood from trees of the genus *Pterocarpus*. The heartwood is light to dark brownish red with black stripes. *Padouk* often has a brilliant red color and takes a fine polish. Source: Africa; Asia (Burma, India). Used for furniture, cabinetwood, veneer, turnings, and musical instruments.

PA-EPDM alloys. Blends of crystalline polyamide (PA) and elastomeric ethylene-propylene-diene monomer (EPDM). They possess good resiliency, elongation and impact resistance, good strength, stiffness and other structural properties, and are used for automotive bumper components, power-tool housings and sporting goods.

Pagalin 2. Trade name of Metalor Technologies SA (Switzerland) for a hard, silver-based dental casting alloy.

Pagalinor. Trade name of Metalor Technologies SA (Switzerland) for several dental alloys. *Pagalinor 2* is a hard, silver-based casting alloy and *Pagalinor 4* an extra-hard, silver-palladium-based universal alloy for low-fusing ceramic restorations.

Page-Allegheny. Trade name of American Chain & Cable (USA) for an extensive series of corrosion- and/or heat-resistant austenitic, ferritic and martensitic stainless steel welding electrodes.

Pageant. Trade name for a palladium-silver dental bonding alloy.

Paglia. Trade name of Vetreria di Vernante SpA (Italy) for a patterned glass.

Pagwood. Trademark of PAG Presswerk AG (Germany) for cellulose-filled phenol-formaldehyde plastics and products.

painite. A deep red, transparent mineral composed of calcium aluminum boron zirconium oxide, $CaZrBAl_9O_{18}$. Crystal system, hexagonal. Density, 4.01 g/cm^3; refractive index, 1.8159. Occurrence: Burma.

paint. (1) A general term used for an organic coating that consists of a uniformly dispersed mixture of a vehicle of film-forming materials, e.g., synthetic resins in a solvent, oil or water, and one or more organic or inorganic pigments with or without drying agents. The viscosity may range from a thin liquid to a semisolid paste. *Paint* is applied as a thin colored film to various surfaces for decoration, protection against corrosion, oxidation or other type of deterioration, illumination, sanitation, aid to morale, safety, fire retardation, or other applications.

(2) A term sometimes used to refer to the pigment in paint (1) alone.

Paintbond. Trade name of Detrex Corporation (USA) for a phosphate coating for paint bonding applications.

paint clay. A ferruginous or manganiferous clay of light-yellow to dark reddish brown color that readily mixes with linseed oil.

paint drier. A metallic compound that promotes oxidation or drying of oils when added to a paint. Most are solutions of acetates, linoleates, naphthenates, octoates and resinates of metals, such as barium, cobalt, iron, lead, potassium, sodium, tin or zinc, in oils or volatile solvents.

paint filler. A white insoluble substance, such as barium sulfate, calcined gypsum, clay, kieselguhr or whiting, used as an extender for paint pigments.

paint pigment. A fine, insoluble powder composed of organic or inorganic substances and used to give color and opacity to organic coatings by preferably reflecting light of certain wavelengths and absorbing light of other wavelengths. Examples include black rouge, copper blue, pearl white, vermilion red, yellow ocher, etc.

Paintstik. Trade name of La-Co Industries Inc. (USA) for touch-up paint used for metal parts, and supplied in the form of sticks.

Paintwell. Trade name of US Steel Corporation for a zinc-coated low-carbon steel that is chemically treated to enhance its paint-adhesion characteristics.

Pair-A-Pane. Trade name of Nudor of Indiana Inc. (USA) for insulating glass.

Pair Glass. Trade name of Asahi Glass Company Limited (Japan) for double glazing units.

Paklon. Trademark of 3M Company (USA) for film backing used for pressure-sensitive adhesive tapes.

Paladon 65. Trade name for a heat-cure acrylic for dentures.

Palakav. Trade name for a self-cure acrylic composite used in restorative dentistry.

Palapress. Trade name of Heraeus-Kulzer Inc. (USA) for an acrylic resin used for denture lining and relining applications.

palarstanide. A steel-gray mineral composed of palladium tin arsenide, $Pd_8(Sn,As)_3$. Crystal system, hexagonal. Occurrence: Russian Federation.

Palasiv 62. Trade name for a plasticized acrylic resin used as a denture liner.

Palatal. Trademark BASF Corporation (USA) for unsaturated polyester resins supplied in flexible and rigid casting grades. They have good flame resistance and good resistance to corrosive chemicals and solvents. Used for hard, scratch-resistant coatings.

Palatemp. Trademark of Heraeus Kulzer GmbH (Germany) for dental resins used for crowns, bridges, palatal plates, dentures, fillings, etc.

Palatex. Trade name of Rockland Dental Company (USA) for acrylic resins used for dental applications.

Palatone. Trade name of Schwab & Frank, Inc. (USA) for acrylic resins and products.

Palatray LC. Trade name of Heraeus-Kulzer Inc. (USA) for a light-cure dental acrylic resin used for impression trays.

palau. (1) A white-colored, corrosion-resistant alloy of 80% gold and 20% palladium, used as a platinum substitute, and for jewelry, chemical equipment, and laboratory ware.

(2) A white-colored, corrosion-resistant alloy of 60% nickel, 20% platinum, 10% palladium and 10% vanadium used as a platinum substitute, and for jewelry, chemical equipment, and laboratory ware.

Palaural. Trade name of Johnson Matthey plc (UK) for a white dental casting alloy containing 46% gold, 30% palladium, 19% copper, and the balance zinc. Used for posts and pins.

Palavit. Trade name of Heraeus-Kulzer Inc. (USA) for a series of acrylic resins including *Palavit L*, a self-cure resin for dental trays and *Palavit 55* an acrylic for dental bonding and lining applications.

Pala Xpress. Trade name of Heraeus-Kulzer Inc. (USA) for an injection-molded acrylic resin used for denture applications.

Pal-Bond Extra. Trade name for a silver-free palladium-base dental bonding alloy.

Palco. (1) Trade name of GTE Products Corporation (USA) for a brazing alloy of 65% palladium and 35% cobalt. Commercially available in the form of foil, wire, flexibraze, powder, extrudable pastes and preforms, it has a liquidus temperature of 1235°C (2255°F) and a solidus temperature of 1230°C (2246°F). Used for brazing molybdenum and tungsten, and for cathodes.

(2) Trademark of Pacific Lumber Company (USA) for various lumber products made from redwood bark fibers including wood wool, interior and exterior siding, thermal insulating

boards, moldings and trim.

(3) Trademark of Pacific Lumber Company (USA) for a series of concrete products including building blocks, bricks, pavers, stepping stones, pier blocks, and wall ends and caps.

Palcusil. Trade name of GTE Products Corporation (USA) for a series of silver brazing alloys containing 54-70% silver, 20-32% copper and 4-25% palladium. Commercially available in the form of foil, flexibraze, wire, powder, extrudable paste and preforms, they have good to excellent flowability and wettability, and a melting range of 807-905°C (1485-1742°F). Used for brazing ferrous and nonferrous metals to ceramics, and for high-temperature applications.

paldao. A quite hard, exotic wood from the Philippines having a very attractive grain pattern and large, partially plugged pores. Used for architectural fixtures, and for built-ins for public or institutional buildings.

pale glass. A term referring to any pale green glass.

Paleo. Trade name of Permacon (Canada) for a series of durable paving stones with the appearance of ancient stone. They are supplied in several shapes and sizes (rectangular, square, and curved) in various colors, and used for walkways, patios, and steps.

palermoite. A white mineral composed of lithium sodium strontium calcium aluminum phosphate hydroxide, $(Li,Na)_2(Sr,Ca)Al_4(PO_4)_4(OH)$. Crystal system, orthorhombic. Density, 3.22 g/cm^3; refractive index, 1.642. Occurrence: USA (New Hampshire).

pale yellow gold. A corrosion-resistant gold-base jewelry alloy containing a total of 8% silver and iron.

Palfique Estelite. Trade name of J. Morita Company (Japan) for a 100% spherically filled radiopaque polymer composite with high abrasion resistance, used for anterior and posterior dental restorations.

Palgard. Trademark of Pratt & Lambert Inc. (USA) for epoxy coatings.

Palid. German trade name for an antifriction alloy composed of 82% lead, 11% antimony and 7% arsenic, and used for bearings.

Paliney. Trade name of J.M. Ney Company (USA) for durable, heat-treatable alloys containing varying amounts of silver and palladium. They have high yield strength, good resistance to corrosion, heat and wear, and a service temperature range above 150°C (300°F). Used for low-voltage sliding contacts, and in electronics for miniaturization.

Palissade. Trade name of SA Glaverbel (Belgium) for a patterned glass.

palissander. The hard, durable wood of any of various trees of the genus *Dalbergia*, especially *D. latifolia* and *D. sissoo*. The heartwood is dark purplish-brown with blackish stripes. *Palissander* has a beautiful grain figure on flat-sawn surfaces, resembles *Brazilian rosewood* in appearance and, owing to its calcareous deposits, is difficult to work with hand tools, but turns well. Average weight, 850 kg/m^3 (53 lb/ft^3). Source: India except northwest. Used for high-quality furniture, fine cabinetwork, veneer, parquet floors, turnery, decorative purposes, ornaments, and carving. Also known as *Indian rosewood*.

Palitinho. Trade name of Vidrobrás (Brazil) for patterned glass.

Pallabraze. Trade name of Johnson Matthey plc (UK) for a series of high-temperature brazing metals including various alloys of silver and palladium, copper, silver and palladium, copper and palladium, and palladium and nickel. Used chiefly for electronic brazing applications.

Pallacast. Trade name of Engelhard Industries (USA) for a series of low-gold palladium-base dental casting alloys containing varying amounts of gold and silver. Available in the annealed or age hardened condition, they have a melting range of 920-1025°C (1688-1877°F).

Pallacast Wire. Trade name of Engelhard Industries (USA) for a wrought gold alloy wire used for dental applications.

Pallacon. Trade name for a low-gold palladium-base dental casting alloy.

Palladent. Trade name for an alloy of 60% aluminum and 40% palladium, used in dentistry.

Palladex. Trade name of Enthone-OMI Inc. (USA) for low-stress palladium plating products.

palladium. A silver-white, ductile metallic element of Group VIII (Group 10) of the Periodic Table. It is commercially available in the form of sponge, shot, granules, powder, rods, ingots, sheets, foil, microfoil, leaf, tube, wire and single crystals. The single crystals are usually grown by the Czochralski technique. It occurs as native palladium and in ores with gold, platinum, etc. Crystal system, cubic. Crystal structure, face-centered cubic. Density, 12.0 g/cm^3; melting point, 1554°C (2829°F); boiling point, 3140°C (5684°F); hardness, 4.8 Mohs (40-100 Vickers); atomic number, 46; atomic weight, 106.42; divalent, trivalent, tetravalent. It is a precious metal with good resistance to tarnishing in air at normal temperatures, good general corrosion resistance and poor resistance to highly oxidizing environments, moderate electrical conductivity (16% IACS), low linear thermal expansion, and good room-temperature fabricability. Used in alloys for electrical contacts, telephone relays and telecommunication switching systems, in silver brazing alloys, in the manufacture of aircraft spark plugs, catalytic converters, nonmagnetic watches and precision balances, in metallizing ceramics, in dentistry, in hydrogen separation, in protective and conductive coatings, and for jewelry (white gold alloys). Symbol: Pd.

palladium acetylacetonate. Yellow to orange, crystalline compound (99% pure) that typically contains 34.8-35.0% divalent palladium. Melting point, decomposes above 205°C (401°F); soluble in toluene and pentanedione. Used to prepare palladium oxide-silicon dioxide $(PdO-SiO_2)$ composite xerogels by sol-gel reactions with tetraethoxysilane $[Si(C_2H_5O)_4]$, and as a versatile organometallic reagent. Formula: $Pd(O_2C_5H_7)_2$. Also known as *palladium 2,4-pentanedionate*.

palladium alloys. A group of alloys containing palladium as the principal element. Common palladium alloys include: (i) alloys of palladium and ruthenium, palladium and silver, palladium, silver and nickel and palladium and copper used for low-voltage electrical contact applications; (ii) alloys of palladium and platinum used in electrochemistry as insoluble anodes for electrolytic protection; (iii) high-melting alloys of silver and palladium, silver, copper and palladium, palladium and nickel and palladium and cobalt with good corrosion resistance, flowability and wettability, used for brazing ferrous and nonferrous metals to ceramics; (iv) alloys of palladium, silver and gold used as casting alloys in dentistry; and (v) corrosion resistant alloys of palladium and rhodium, or palladium and silver used for jewelry.

palladium aluminide. A compound of palladium and aluminum with a melting point of 1642°C (2988°F). Used in ceramics and materials research. Formula: PdAl.

palladium beryllide. Any of the following compounds of palladium and beryllium used in ceramics and materials research: (i)

Palladium monoberyllide. Melting point, 1465°C (2669°F). Formula: PdBe; and (ii) *Palladium dodecaberyllide.* Density, 3.18 g/cm^3. Formula: PdBe$_{12}$.

palladium black. A finely divided form of metallic palladium available as a flammable, black powder (99.8% pure) with an apparent density of approximately 8.8-9.8 g/cm^3 (0.32-0.35 lb/in.^{3}), and a typical surface area of 23 to over 40 m^2/g. Used as a catalyst and as an adsorbent for gases. Also known as *palladium mohr.*

palladium bromide. Red-black, hygroscopic crystals (99+% pure). Crystal system, monoclinic. Density, 5.173 g/cm^3; melting point, decomposes at 250°C (482°F). Used as a catalyst and in electroless coating. Formula: PdBr$_2$. Also known as *palladium dibromide; palladous bromide.*

palladium chloride. Rust-colored, hygroscopic crystals (99+% pure). Crystal system, rhombohedral. Density, 4.0 g/cm^3; melting point, 678-680°C (1252-1256°F). It may contain 59-60% palladium. Used for electroless coating on metals, sometimes in porcelain compositions, and also in photographic chemicals, for leak detection in gas lines, in indelible inks, as a catalyst, and in microscopy as a metal stain and for the demonstration of unsaturated hydrophilic lipids. Formula: PdCl$_2$. Also known as *palladium dichloride; palladous chloride.*

palladium chloride dihydrate. Dark brown crystals or powder used for electroless coating on metals, sometimes in porcelain compositions, in photographic chemicals, indelible inks, and as a catalyst. Formula: PdCl$_2$·2H$_2$O. Also known as *palladous chloride dihydrate.*

palladium-copper alloys. A group of electrical contact alloys composed of varying amounts of palladium and copper. A 60Pd-40Cu alloy has a density of 10.67 g/cm^3 (0.385 lb/in.^{3}), a solidus temperature of 1200°C (2190°F), an electrical conductivity of 8% IACS, and good tensile strength. Used for relays and sliding contacts.

palladium dibromide. See palladium bromide.

palladium dichloride. See palladium chloride.

palladium difluoride. See palladium fluoride.

palladium diiodide. See palladium iodide.

palladium disulfide. See palladium sulfide (ii).

palladium dodecaberyllide. See palladium beryllide (ii).

palladium flake. Flat or scalelike palladium powder particles of relatively small thickness (less than 1 μm or 40 μin.) Used in coatings.

palladium fluoride. Violet, hygroscopic crystals. Crystal system, tetragonal. Density, 5.76 g/cm^3; melting point, 952°C (1746°F). Used for electroless coating on metals, as a catalyst. Formula: PdF$_2$. Also known as *palladous fluoride.*

palladium gold. (1) A native alloy of 90-95% gold and 5-10% palladium, used as a platinum substitute, and in jewelry and white gold. Also known as *porpezite.*

(2) An alloy of 40% gold, 40% copper, 10% silver and 10% palladium, used as a platinum substitute, and in jewelry and white gold alloys.

palladium iodide. Black crystals or powder (99+% pure). Density, 6.0 g/cm^3; melting point, decomposes at 350°C (660°F). Used for electroless coating, and as a catalyst. Formula: PdI$_2$. Also known as *palladous iodide.*

palladium leaf. Palladium made into extremely thin foils or leaves (usually less than 1.0 μm or 40 μin. thick) by hammering or rolling, and often supplied on plastic backing. Used for architectural and ornamental applications, and for embossing leather and fabrics.

palladium mohr. See palladium black.

palladium monoberyllide. See palladium beryllide (i).

palladium monosulfide. See palladium sulfide (i).

palladium monoxide. See palladium oxide.

palladium-nickel alloys. A group of brazing alloys composed of varying amounts of palladium and nickel with high melting points (e.g., 1238°C or 2260°F for 60Pd-40Ni), good corrosion resistance, and good flowability and wettability. Used for brazing ferrous and nonferrous metals, and for brazing ceramics to metals.

palladium nitrate. A compound that is available as brown, deliquescent crystals (99.9+% pure). Used as a catalyst, and as an analytical reagent. Formula: Pd(NO$_3$)$_2$. Also known as *palladous nitrate.*

palladium oxide. Black powder or black-green to amber crystals (98+% pure). Crystal system, tetragonal. Density, 8.3-8.7 g/cm^3; melting point, 870°C (1598°F) [also reported as 750°C or 1382°F]. Used as a catalyst in organic and inorganic synthesis, and as a metal stain in microscopy. Formula: PdO. Also known as *palladium monoxide.*

palladium 2,4-pentanedionate. See palladium acetylacetonate.

palladium-polyethylenimine. See Royer catalysts.

palladium-potassium chloride. A red, hygroscopic powder with a density of 2.738 g/cm^3, used as a reagent. Formula: K$_2$PdCl$_6$. Also known as *potassium chloropalladate; potassium hexachloropalladate.*

palladium powder. A finely divided form of platinum made by atomization processes or chemical reduction. The free-flowing, spherical atomized powder has a purity of 99.9+%, and is supplied in particle sizes of 740-297 μm (0.0029-0.0117 in.). The chemically reduced amorphous powder has a purity of 99.9+%, and is available in fine particle sizes from submicron to 3.5 μm (less than 40 to 138 μin.). Used for coatings, and in the manufacture of parts for high-temperature applications.

palladium-rhodium alloys. A group of corrosion-resistant jewelry alloys containing varying amounts of palladium and rhodium, usually in a ratio of about 8:2 or 9:1.

palladium-ruthenium alloys. A group of corrosion-resistant alloys containing varying amounts of palladium and ruthenium, and used for precious-metal applications. A common alloy composition is Pd-4.5Ru.

palladium-silver alloys. A group of alloys containing varying amounts of palladium and silver. The 60Pd-40Ag alloy has a density of 11.3 g/cm^3 (0.408 lb/in.^{3}), a solidus temperature of 1338°C (2440°F), and an electrical conductivity of 4% IACS. *Palladium-silver alloys* have relatively good tensile strength, and are used for telephone-type relay contacts and similar applications. The 67Pd-33Ag alloy has high corrosion resistance and is used for jewelry, while the 75Pd-25Ag alloy is employed as a catalyst.

palladium-sodium chloride. See sodium tetrachloropalladate.

palladium sponge. See sponge palladium.

palladium subsulfide. See palladium sulfide (iii).

palladium sulfide. Any of the following compounds of palladium and sulfur used in materials research and chemical synthesis: (i) *Palladium monosulfide.* Gray crystals or brown-black powder. Crystal system, tetragonal. Density, 6.6-6.7 g/cm^3; melting point, 950°C (1742°F). Formula: PdS; (ii) *Palladium disulfide.* Dark brown crystals. Formula: PdS$_2$; and (iii) *Palladium subsulfide.* Greenish-gray crystals with a density of 7.3 g/cm^3 and a melting point of 800°C (1472°F) (decomposes). Formula: Pd$_2$S.

palladium-uranium. A compound of palladium and uranium with a melting point of 1638°C (2980°F) used in ceramics and materials research. Formula: Pd_3U.

Palladius-6. Trade name of Vident (USA) for a palladium-based semi-precious dental alloy.

palladoarsenide. A mineral composed of palladium arsenide, Pd_2As. It can also be made synthetically. Crystal system, monoclinic. Density, 10.13 g/cm^3. Occurrence: Russian Federation.

palladobismutharsenide. A cream-colored mineral composed of palladium bismuth arsenide, $Pd_{10}As_4Bi$. It can also be made synthetically. Crystal system, orthorhombic. Density, 10.86 g/cm^3.

Palladon. Trade name for a special plastic used for dentures.

palladous bromide. See palladium bromide.

palladous chloride. See palladium chloride (1).

palladous chloride dihydrate. See palladium chloride (2).

palladous dichloride. See palladium chloride (2).

palladous fluoride. See palladium fluoride.

palladous iodide. See palladium iodide.

palladous nitrate. See palladium nitrate.

palladseite. A mineral composed of palladium selenide, $Pd_{17}Se_{15}$. Crystal system, cubic. Density, 8.33 g/cm^3.

Palladure. Trade name of Shipley Ronal (USA) for palladium finishes for electronic connectors.

Pallament. Trade name of Shipley Ronal (USA) for palladium-nickel finishes for electronic connectors.

Pallamerse. Trade name of Technic Inc. (USA) for solderable, tarnish-resistant, pore-free immersion palladium deposits up to 0.5 µm. (19.7 µin.) thick, applied over nickel or over copper alloys.

Pallas. (1) Trade name of Gebrüder Hover Edelstahlwerk (Germany) for a water-hardening, wear-resistant steel used for bearings, cutters, liners, and sleeves.

(2) Trade name for low-gold palladium-base dental casting alloys.

PallaTech. Trademark of Lucent Technologies (USA) for palladium and palladium-alloy deposits and plating processes.

Palliag. Trademark of Degussa-Ney Dental (USA) for a series of low-gold silver-palladium dental alloys and solders. *Palliag LS* is a white, low-density dental casting alloy containing 56% silver and 36.9% palladium. Supplied in nugget form, it has high yield strength, and a hardness of 230 Vickers. Used in restorative dentistry especially for crowns and bridges. *Palliag M* is a white, low-density, extra-hard dental casting alloy (ADA type IV) containing 58.5% silver, 27.4% palladium and 2% gold. It has a fine grain structure, very high mechanical strength, and a hardness of 310 Vickers. Used in restorative dentistry for crowns, bridges and implant superstructures.

Pallium 3. Trade name for a low-gold dental casting alloy.

Pallnic. Trade name of Engelhard Electro Metallics Division (USA) for a palladium-nickel alloy electroplate and plating process.

Pallorag 35. Trade name of Cendres & Métaux SA (Switzerland) for a low-gold dental casting alloy containing a total of 41% gold and platinum.

Palloro. Trade name for a low-gold dental casting alloy.

Palma. Trademark of J.L. & P. Weidner KG (Germany) for powders and pastes of aluminum and bronze.

Palmansil. Trade name of GTE Products Corporation (USA) for a brazing alloy containing 75% silver, 20% palladium and 5% manganese. It has a liquidus temperature of 1072°C (1962°F), and a solidus temperature of 1008°C (1846°F). Commercially available in the form of foil, flexibraze, wire, powder, extrudable paste and preforms, it has good flowability and wettability, and is used for high-temperature brazing, and for brazing metals to ceramics, etc.

palm fibers. (1) Cellulose fibers obtained from the fruit husks of palm trees, especially the coconut (*Cocos nucifera*) and the raffia (*Raphia raffus*). Used for ropes, fish nets, brooms, brushes, mats and baskets. See also coir; raffia.

(2) A term sometimes used for cellulose fibers obtained from any chiefly tropical or subtropical tree or shrub of the palm family (Palmae) including the coconut (*Cocos nucifera*), the raffia (*Raphia raffus*), the palmetto (genus *Sabal*) and the palmyra (genus *Borassus*).

palmierite. A colorless mineral composed of potassium lead sulfate, $K_2Pb(SO_4)_2$. It can also be made from a 1:1 molar mixture of potassium sulfate (K_2SO_4) and lead sulfate ($PbSO_4$) under prescribed conditions. Crystal system, rhombohedral (hexagonal). Density, 4.33 g/cm^3; refractive index, 1.712. Occurrence: Italy.

palm wax. A yellowish amorphous wax from a South American palm tree *(Ceroxylon andicola)*. It has a density of 0.990-0.995 g/cm^3, a melting point of 102-105°C (216-221°F), and is used as a substitute for beeswax, and in the manufacture of candles and floor polishes.

palmyra. A reddish-brown, water-resistant, durable, moderately stiff vegetable fiber obtained from the bases of fanlike leaves of the Indian palm tree *Borassus flabellifer*. Used in broom manufacture.

Palni. Trade name of Western Gold & Platinum Company (USA) for a brazing alloy of 60% palladium and 40% nickel. Commercially available in the form of foil, wire, flexibraze, powder, extrudable paste and preforms, it has a liquidus temperature of 1238°C (2260°F), a solidus temperature of 1238°C (2260°F) and good flowability and wettability. Used for high-temperature brazing, for brazing ferrous and nonferrous metals, and for brazing ceramics to metals.

Palnicusil. Trademark of Morgan Crucible Company plc (UK) for a brazing alloy of 48.6% silver, 22.5% palladium, 10% nickel and 18.9% copper. Commercially available in the form of foil, wire, flexibraze, powder, extrudable paste and preforms, it has a liquidus temperature of 1179°C (2155°F), and a solidus temperature of 910°C (1670°F). Used for high-temperature brazing, and for brazing ferrous and nonferrous metals.

Palniro. Trademark of Western Gold & Platinum Company (USA) for a series of brazing alloys containing 30-70% gold, 8-34% palladium and 22-36% nickel. It is commercially available in the form of foil, wire, flexibraze, powder, extrudable pastes and performs, and is used for high-temperature brazing.

Paloro. Trademark of Morgan Crucible Company (UK) for a brazing alloy composed of 92% gold and 8% palladium. Commercially available in the form of foil, wire, flexibraze, powder, extrudable pastes and preforms, it has a liquidus temperature of 1240°C (2264°F), a solidus temperature of 1200°C (2192°F). Used for brazing tungsten, molybdenum and their alloys.

palosapis. The Philippine name for wood from various species of trees of the genus *Anisoptera*. The same wood is sold as *krabak* in Thailand and as *mersawa* in Malaysia. It is a plain wood that is not decorative and very hard to work (high silica content), but can be rotary-peeled. Source: Southeast Asia (from Philippines and Malaysia to Bangladesh). Used for veneer for plywood.

Paloy. Trade name for a low-gold dental casting alloy.

Palruf. Trademark of Suntuf Inc. (Canada) for tough, durable, corrugated polyvinyl chloride roofing panels that are 10 times stronger than fiberglass panels and have excellent environmental resistance. They are supplied in clear, white and green colors with standard panel widths of 660 mm (26 in.) and lengths of 2.4 m (8 ft.) and 3.6 m (12 ft.).

Palsil. Trade name of GTE Products Corporation (USA) for a brazing alloy composed of 90% silver and 10% palladium. Commercially available in the form of foil, wire, flexibraze, powder, extrudable pastes and preforms, it has a melting range of 1002-1066°C (1835-1950°F), good flowability and wettability, and is used for high-temperature brazing.

Palusol. Trademark of BASF Corporation (USA) for intumescent fiberboard made chiefly of sodium silicate. It is easy to handle, machine and work, offers superior resistance to fire and smoke, and possesses excellent noise- and heat-insulating properties. Used for fire-resistant building components and sound-insulating doors.

palygorskite. A group of tough, fibrous, lightweight, white to grayish white clays related to *attapulgite*, a hydrated aluminum magnesium silicate in which an extensive amount of magnesium is replaced by aluminum. *Palygorskite* found in monoclinic crystalline form composed of $MgAlSi_4O_{10}(OH) \cdot 4H_2O$ or $(Mg,Al)_5(Si,Al)_8O_{20}(OH)_2 \cdot 8H_2O$, and in orthorhombic crystalline form composed of $Mg_5(Si,Al)_8O_{20}(OH)_2 \cdot 8H_2O$. Density, 2.36-2.40 g/cm^3; refractive index, 1.52. Occurrence: UK (Leicestershire and Scotland), USA (Alaska, Florida, Georgia, New Mexico, Washington). Used as a source of alumina and magnesia.

Pam-O-Pack. Trademark of Norton Performance Plastics (USA) for sealing strips of polytetrafluoroethylene polymers.

Pamylon. Trademark of Kohap Limited (South Korea) for polyester and nylon 6 staple fibers and filament yarns.

Pan. Trade name for acrylic (polyacrylonitrile) fibers used for the manufacture of textile fabrics.

Panabond. Trade name of Panadent Limited (UK) for a series of dental bonding alloys including *Panabond 2* silver-free palladium-base alloy, *Panabond 45* medium-gold alloy, and *Panabond Yellow* high-gold alloy.

Panacast. Trade name of Panadent Limited (UK) for a series of dental casting alloys including Panacast 5 low-gold alloy and Panacast 60 medium-gold alloy.

Panachrome. Trade name of Panadent Limited (UK) for a corrosion-resistant cobalt-chromium dental casting alloy.

Panadura. Trade name of Paramet Chemical Corporation (USA) for phenol-formaldehyde plastics.

PAN-Alu-Quarzal. Trade name of PAN-Metallgesellschaft (Germany) for an aluminum-base bearing alloy containing 2-5% copper.

panama. A crisp, lightweight, one- or two-color fabric, usually of cotton or wool, in a plain weave, having a coarse, basketweave effect resembling that of panama hats. Used especially for summer dresses and suits.

Panapren. Trademark of Wacker Chemie (Germany) for a silicone elastomer used for dental impression.

Panaroc. Trade name of Westroc Industries Limited (Canada) for extremely durable and water-resistant cement board panels made of glass fiber-wrapped Portland cement and used inside shower stalls or around bathtubs.

Panasil. Trademark of Kettenbach GmbH & Co. KG (Germany) for a silicone elastomer used for dental impression.

panasqueiraite. A pink mineral composed of calcium magnesium fluoride phosphate hydroxide, $CaMgPO_4(OH,F)$. Crystal system, monoclinic. Density, 3.27 g/cm^3; refractive index, 1.596. Occurrence: Portugal.

Panavia. Trademark of J. Morita Company (Japan) for self-cure two-component adhesive resin cements used in dentistry for luting restorations and cementing veneers and inlays. *Panavia* consists of a powder composed of a bisphenol A diglycidylether dimethacrylate (BisGMA) based resin filled up to 76 wt% with quartz particles and containing a benzoyl peroxide initiator, and a liquid composed of aliphatic and aromatic methacrylates, phosphate monomers, a 10-methacryloyloxydecyl dihydrogen phosphate (MDP) adhesive monomer, and selected activators and stabilizers. *Panavia 21* is a modified version of *Panavia* supplied in the form of two pastes and having a dentin/enamel primer incorporated that contains 10-methacryloyloxydecyl dihydrogen phosphate (MDP), hydroxyethyl methacrylate (HEMA) and *N*-methacryloyl 5-aminosalicylic acid (5-NMSA).

Pan-Brick. Trademark of Pan-Brick Inc. (Canada) for plywood wall panels with urethane foam insulation and brick facing.

pancake. An artificial leather made by cementing skived leather scraps together under heavy pressure. Used for shoe heels.

Pandaloy. Trade name of Allegheny Ludlum Steel (USA) for a stainless chromium-nickel steel used especially for papermaking equipment.

Pandex. Trade name of Latrobe Steel Company (USA) for an age-hardenable, heat-resistant austenitic steel containing up to 0.08% carbon, 1.5% manganese, 0.75% silicon, 24-28% nickel, 13-16% chromium, 1.75-2.25% titanium, 1-1.5% molybdenum, 0.1-0.5% vanadium, 0-0.2% aluminum, and the balance iron. It has an excellent high-temperature properties at over 700°C (1292°F), and is used for jet-turbine wheels and blades, superchargers, jet-engine components and fuel nozzles.

Pandura. Trade name of Paramet Chemical Corporation (USA) for phenol-formaldehyde resins and plastics.

Panelag. Trademark of Quigley Company Inc. (USA) for high-temperature refractory cements.

panelboard. A strong, stiff *paperboard* suitable for use as a paneling material in buildings and autobodies. It may be treated for water resistance, and the surface may be painted, grained, textured or otherwise decorated.

Panelbond. Trademark of Quigley Company Inc. (USA) for high-temperature refractory cements.

panel brick. An elongated refractory silica brick used for lining coke ovens.

Panelglas. Trade name of Manville Corporation (USA) for fiberglass acoustical ceiling lay-in panels.

paneling. Wood panels used in architecture, construction and interior decoration, and usually supplied as (i) plywood composed of various thin sheets, (ii) composite plywood consisting of veneer faces bonded to different types of wood cores, or (iii) nonveneered panels, such as hardboard, waferboard, particleboard, oriented strand board, etc.

paneling board. Any material, such as plywood, waferboard, particleboard, hardboard or oriented strand board, made into a rigid board or sheet for use as structural and/or decorative panel for doors, countertops, cabinets, siding, sheathing and floor underlayment.

Panelstone. Trademark of Ruberoid Company (USA) for asbestos and asbestos-cement sheets used for roofing and siding applications.

Panelyte. Trademark of Laminated Plastics, Inc. (USA) for a series of plastic laminates in sheet or block form made by impregnating strong paper (e.g., kraft) or wood with a thermosetting resin, such as a phenolic, melamine or silicone. Supplied in a wide range of colors, wood grains, pattern and surface designs, they are used for decorative applications.

panethite. A pale amber mineral composed of sodium calcium magnesium iron phosphate, $(Na,Ca)_2(Mg,Fe)_2(PO_4)_2$. Crystal system, monoclinic. Density, 2.90 g/cm^3; refractive index, 1.576. Occurrence: USA.

Panex. Trademark of Zoltek Corporation (USA) for high-strength, high-modulus continuous carbon fibers made from a polyacrylonitrile (PAN) precursor. They have large filament counts (45700), a fiber diameter of 7.2 μm (283 μin.), a density of 1.81 g/cm^3 (0.065 $lb/in.^3$), and good and consistent mechanical properties. Used in the manufacture of fabrics, tow and yarn, and as a reinforcement for engineering composites, laminates, molding compounds, ablative phenolics, rocket motors, and brake disks.

PAN fibers. See polyacrylonitrile fibers.

Pangel. Trade name of J.M. Huber Corporation (USA) for a micronized *sepiolite* mineral thickener.

Pangold. Trade name of Degussa Dental (USA) for a low-gold dental casting alloy.

Pangold Keramik. Trade name of Degussa Dental (USA) for a palladium-silver ceramic bonding alloy used in dentistry.

PAni. Trade name of Ormecon Chemie GmbH (Germany) for a conductive, protective coating for application to metallic substrates. It is based on *polyaniline* powder blended with a thermoplastic resin, such as polymethyl methacrylate (PMMA).

Panilax. Trade name of Micanite & Insulators Company (UK) for phenol-formaldehyde plastics and laminates.

Panlite. Trade name of Teijin Chemicals Limited (Japan) for polycarbonate resins supplied in the form of pellets for making compact audio disks, in film form for packaging and other applications, and in fiber-reinforced grades for engineering applications. *Panlite XM* refers to polycarbonate/polybutylene terephthalate (PC/PBT) alloys.

Pannakril. Trade name for acrylic fibers used for the manufacture of textile fabrics.

panne. A knit satin or pile fabric that has been given a lustrous surface by flattening with a roller.

Panneau-Sol. Trade name of SA Isoverbel NV (Belgium) for glass-wool floor insulation materials.

Panneau-Toiture. Trade name of SA Isoverbel NV (Belgium) for board-type glass-wool roof insulation materials.

panne velvet. A lightweight fabric whose thick, short-cut nap or pile has been flattened in one direction. See also velvet.

Panobrick. Trade name of Hill Brothers Glass Company Limited (UK) for glass blocks of varying size and design supplied in a wide range of colors. They have been combined into panels by dry-fixing, precasting or built-in-place processes.

Panoclad. Trade name of Hill Brothers Glass Company Limited (UK) for color-fired, toughened infill glass panels.

Panodur. Trade name of Unalit (France) for dark brown hardboard made from virgin and recycled wood fibers by bonding with a natural glue of wood. Supplied in various panel grades with one smooth and one rough side in sizes up to 1.7 × 5.5 m (5.5 × 18 ft) and thicknesses from 2 to 8 mm (0.08 to 0.31 in.).

Panofin. Trademark of Forex Leroy Inc. (Canada) for thin particleboard.

Panofor. Trademark of Forex Leroy Inc. (Canada) for waferboard.

Pan-O-Glass. Trade name of Splintex Belge SA (Belgium) for a toughened, colored cladding glass.

Pan-O-Pac. Trade name of Splintex Belge SA (Belgium) for a cladding panel consisting of an outer face of *Pan-O-Glass* as insulating core, and a backing sheet composed of a metal or other material.

Panorama. Trade name for a high-copper dental amalgam alloy.

Panoroc. Trade name of Glaceries de Saint-Roch SA (Belgium) for a ceramic-enameled, toughened glass.

Panoscreen. Trade name of Hill Brothers Glass Company Limited (UK) for a toughened laminated glass.

Panotuff. Trade name of Hill Brothers Glass Company Limited (UK) for a toughened glass.

Panox. Trademark of SGL Carbon Group (USA) for polyacrylonitrile-based carbon fibers.

Panseri. British trade name for a high-temperature-resistant aluminum alloy containing 11.5% silicon, 4.5% nickel, 1% copper, 0.5% iron and 0.4% manganese. Used for pistons.

Pansil. Trade name of J.M. Huber Corporation (USA) for a micronized *sepiolite* mineral thickener.

Pantal. Trade name of VDM Aluminium GmbH (Germany) for a series of wrought, corrosion-resistant aluminum-magnesium and aluminum-silicon alloys used for structural applications, welding rods, etc.

Pantanax. Trade name of Thyssen Edelstahlwerke AG (Germany) for a series of highly wear-resistant austenitic manganese steels.

Pantaplast. Trade name of Koepp AG (Germany) for polyurethane foams and foam products.

Pantarin. Trade name of Koepp AG (Germany) for flexible polyurethane foam materials including *Pantarin S* polyester foam and *Pantarin T* polyether foam.

Panteg. Trade name of British Steel Corporation (UK) for a series of wrought austenitic and ferritic stainless steels, usually supplied in sheet form, and used for chemical and food-processing equipment, tanks, brewing equipment, welded structures, household appliances, decorative trim, etc.

Panther. (1) Trade name of McGean Inc. (USA) for black oxide coatings.

(2) Trade name of AL Tech Specialty Steel Corporation (USA) for a series of cobalt-tungsten high-speed steels containing about 1.5% carbon, 4% chromium, 12% tungsten, 5% vanadium, 5% cobalt, and the balance iron. They have high wear resistance and red-hardness, good deep-hardening properties, good machinability, and are used for lathe and planer tools, milling cutters, broaches, form tools, punches, and blanking dies.

(3) Trade name of Ernst Haiss Eisen- und Metallwerk KG (Germany) for steel wool and abrasive bands and cloth.

(4) Trademark of Engineered Carbons, Inc. (USA) for a series of oil-pelletted furnace-grade carbon blacks (92-95% pure) with an iodine content of about 24-95 mg/g, an average particle size of 25-90 nm (1.0-3.5 μin.), and a density 1.7-1.9 g/cm^3 (0.06-0.07 $lb/in.^3$).

Panther Black. Trade name of McGean Inc. (USA) for black oxide coatings.

Panther Special. Trade name of AL Tech Specialty Steel Corporation (USA) for a tungsten-type high-speed steel (AISI type T4) containing 0.75% carbon, 4% chromium, 18-19% tungsten, 1% vanadium, 5% cobalt, and the balance iron. It has good deep-hardening properties, high hot hardness, excellent wear resistance, and good machinability. Used for cutting tools.

pan tile. A roofing tile whose cross section resembles the letter "S." It is installed such that the up curve of one tile overlaps the down curve of an adjoining tile.

Panzer. Trade name of Rudolf Schmidt Stahlwerke (Austria) for a high-speed steel containing 0.7% carbon, 19% tungsten, 4% chromium, 2% vanadium, and the balance iron. Used for tools and cutters.

Panzerholz. Trademark of Blomberger Holzindustrie B. Hausmann GmbH & Co. KG (Germany) for bulletproof plywood.

Pao. Trademark Jang Dah Nylon Corporation (Taiwan) for nylon 6 fibers and filament yarns.

paolovite. A white mineral composed of palladium tin, Pd_2Sn. Crystal system, orthorhombic. Density, 11.32 g/cm^3. Occurrence: Russian Federation.

papagoite. A cerulean blue mineral of the axinite group composed of calcium copper aluminum silicate hydroxide, $CaCuAl(SiO_3)_2(OH)_3$. Crystal system, monoclinic. Density, 3.25 g/cm^3; refractive index, 1.641. Occurrence: USA (Arizona).

PA-PAR alloys. Thermoplastic blends of a crystalline polyamide (PA) and an amorphous polyarylate (PAR). They possess good dimensional stability, high impact strength, freedom from warpage, good chemical properties, and are used for automotive applications, e.g., body panels.

paper. A sheet-like, essentially fibrous material that is formed on a fine-wire screen by removing the water from a dilute fiber suspension. The resulting matted or felted sheet is subsequently pressed and dried. The raw materials for its manufacture are wood, rags (cotton or linen), other fibers (e.g., hemp, bagasse, flax, jute, etc.), cellulose, waste paper, fillers, such as kaolin, barium sulfate, gypsum, talc, etc., and starches or other sizing materials. Synthetic paper made from man-made fibers is also available. *Paper* can be classified in the following groups according to composition: (i) *First-class paper* made wholly from cotton or linen rags; (ii) *Second-class paper* made partly from rags (also from cotton, linen or hemp) with no more than 50% cellulose; (iii) *Third-class paper* containing varying amounts of rags and cellulose; and (iv) *Fourth-class paper* made wholly from woodpulp. Used for writing, printing, drawing, packing and wrapping, for sanitary applications, for the manufacture of paperboard and paper laminates, in photography, for electrical applications (e.g., capacitor paper), etc. See also synthetic paper.

paper birch. See American white birch.

paperboard. A heavy, strong, rigid, stiff grade of paper with an approximate sheet thickness of more than 0.25 mm (0.01 in.), used for cardboard for making cartons, boxes, etc.

paper clay. A white, fine-grained, high-grade clay, such as *kaolin*, that is added to paper pulp as a filler and extender, and as a whiteness and surface finish improver.

paper coating. A coating composition made from a suspension of high-grade clays, sizing materials, starches, wax, casein, rosin and/or polymers (e.g., urea or melamine resins). It is used to add strength to paper, improve its finish and degree of whiteness and give special characteristics to its surface. For example, sizing agents and waxes add protection against running and penetration of ink to writing paper, and urea and melamine resins add moisture resistance to the surface of packing paper.

paper fibers. Fibers suitable for use as additions to wood pulp in the manufacture of paper. Examples include cotton, flax, linen, silk and certain synthetic fibers, such as nylon, rayon and vinyl fibers.

paper laminate. A composite consisting of any of several types of thermosetting material (phenolic, polyester or epoxy resin, silicone rubber, adhesives, etc.) bonded to paper. It has high tensile strength, excellent dielectric strength and low moisture absorption. Used for electrical insulating applications. Abbreviation: PL.

papermaker's alum. See aluminum sulfate octadecahydrate.

paper-plastic composites. Strong, stiff composite materials produced by a thermokinetic mixing process in which a high-speed impeller disperses reinforcing wood fibers, obtained from waste paper or wood, with the aid of polar waxes into thermoplastic polymers, such as polyolefins or polystyrenes, followed by solid-phase extrusion and drawing to align the polymer chains with the wood fibers. Used in building construction for load-bearing applications, such as beams and joists.

paper pulp. See wood pulp.

paper yarn. (1) A yarn made by twisting or rolling one or more strips of moist paper in the longitudinal direction. Used for knitted and woven fabrics.

(2) A yarn that essentially consists of one or more strips of paper.

PAPI. Trademark of Dow Chemical Company (USA) for polymeric methylene diisocyanate (MDI) with delayed gel time and fast rise profile, used for semiflexible foam applications in the automotive industry, e.g., instrument panels and interior trim.

Papi. Trademark of Upjohn Company (USA) for a series of urethane polymers including various methylene diphenyl types, and copolymers of phenyl isocyanate and formaldehyde. Used in the manufacture of rigid urethane foams, adhesives, and foam sealants.

papier-mâché. A lightweight, moldable mass that is made from comminuted waste paper pulped with glue, starch, gypsum and clay, and upon drying becomes hard and strong, and can be sanded, drilled, painted, and coated with varnish. The production of molded articles is facilitated by impregnation with linseed oil. Used for modelling, toys, and in the textile industries for dress forms.

PA-PPE alloys. Polymer blends of a crystalline polyamide (PA) and an amorphous polyphenylene ether (PPE) with good thermal properties and chemical resistance, used for automotive applications, e.g., body panels.

PA-PPO alloys. Immiscible polymer blends of crystalline polyamide (PA) and amorphous polyphenylene oxide (PPO) with enhanced chemical resistance, improved dimensional stability at elevated temperatures, low moisture sensitivity and mold shrinkage, very good paint adhesion and good resistance to chipping. Used for automotive applications, e.g., doors, hatchbacks and wheel panels.

papreg. A group of plastic laminates produced by bonding together, under heat and pressure, sheets of paper impregnated with a thermosetting resin, such as a phenolic or melamine. Used for decorative and insulating applications (usually in the form of panels or boards), and for cores employed as carriers of flexible materials.

papyrus. A paper-like writing material obtained from the water plant *Cyperus papyrus* of Northern Africa by slicing the pith and pressing the wet strips together. It was already used by the ancient Egyptians, Greeks and Romans. Also known as *Egyptian paper*.

papyrus fibers. The cellulose fibers obtained from the stems of the water plant *Cyperus papyrus* of Northern Africa. It was already used by the ancient Egyptians, Greeks and Romans for cordage, mats, sails and clothing.

Pa-Qel. Trade name for acrylic fibers used for the manufacture of textile fabrics.

Para. Trade name of Cytemp Specialty Steel (USA) for a deep-hardening steel containing 1.25% carbon, 1.6% tungsten, 0.4% chromium, 0.2% vanadium, and the balance iron. Used for cutting tools and dies.

para-alumohydrocalcite. A white mineral composed of calcium aluminum carbonate hydroxide hexahydrate, $CaAl_2(CO_3)_2$-$(OH)_4 \cdot 6H_2O$. Crystal system, triclinic. Density, 2.00 g/cm³. Occurrence: Russian Federation.

para-aramid fiber. See aramid.

parabutlerite. A light orange mineral composed of iron sulfate hydroxide dihydrate, $FeSO_4(OH) \cdot 2H_2O$. Crystal system, orthorhombic. Density, 2.55 g/cm³; refractive index, 1.663. Occurrence: Chile.

paracasein. A compound obtained by the conversion of the protein *casein* in milk by the enzyme rennin. It combines with calcium to form insoluble calcium paracaseinate.

paracelsian. A colorless or pale yellow mineral of the feldspar group composed of barium aluminum silicate, $BaAl_2Si_2O_8$. Crystal system, monoclinic. Density, 3.32 g/cm³; refractive index, 1.5824. Occurrence: UK (Wales), Italy.

parachute cloth. A strong, tight, dense, yet lightweight fabric, usually of nylon or silk, used for parachutes, luggage, outerwear, etc.

Paracon. Trade name of Bell Telephone Laboratories (USA) for polyester resins.

paracoquimbite. A pale violet mineral composed of iron sulfate nonahydrate, $Fe_2(SO_4)_3 \cdot 9H_2O$. Crystal system, rhombohedral (hexagonal). Density, 2.11 g/cm³. Occurrence: Chile.

paracostibite. A white mineral composed of antimony cobalt sulfide, CoSbS. Crystal system, orthorhombic. It can also be made synthetically. Density, 6.90 g/cm³. Occurrence: Canada (Ontario).

paracoumarone. See cumar gum.

Para Cover. Trade name of Paramount Glass Manufacturing Company Limited (Japan) for glass-wool pipe covers.

Paracril. Trademark of Uniroyal Chemical Company (USA) for a group of acrylonitrile-butadiene rubbers with excellent resistance to aliphatic hydrocarbons, gasoline and petrochemicals, mineral and vegetable oils and fats, fair mechanical properties, low elasticity, and poor low-temperature properties. Used for carburetor and gasoline tanks, pump parts and gaskets, and as a plasticizer for thermosets.

paracrystalline materials. Materials, such as certain semicrystalline high polymers, with pseudocrystalline structures, i.e.., structures which lack long-range order, although there is some local highly disordered structure.

paradamite. A yellow mineral composed of zinc arsenate hydroxide, $Zn_2(AsO_4)OH$. Crystal system, triclinic. Density, 4.55 g/cm³; refractive index, 1.771. Occurrence: Mexico.

paradocrasite. A silver-white mineral composed of antimony arsenic, $Sb_2(Sb,As)_2$. Crystal system, monoclinic. Density, 6.50 g/cm³. Occurrence: Australia.

paraelectric materials. Materials, usually with symmetrical unit cells, that contain permanent dipoles (e.g., dipole molecules) and exhibit only small polarizations with applied electric fields.

paraffin duck. A firm, heavy *duck* waterproofed with paraffin, and used for outerwear.

paraffin wax. A flammable, white wax composed of a mixture of high-molecular-weight hydrocarbons (saturated, straight-chain *n*-paraffins and branched isoparaffins) obtained predominantly from petroleum or shale by refining. It has a density of 0.880-0.915 g/cm³, a melting range of 47-65°C (117-149°F), and a flash point of 198°C (390°F). Used in paper coatings and sealants, in the manufacture of candles and tapers, in carpet backing, food cartons, electrical insulation, polishes and lubricants, and as a waterproofing agent. Abbreviation: PW.

Parafilm. Trademark of American National Can Company (USA) for a water-resistant, self-sealing, translucent thermoplastic film used for laboratory applications.

Paraglass. (1) Trade name of Semi-Elements Inc. (USA) for a water-white, transparent, magnetically susceptible high-purity glass.

(2) Trade name of Degussa AG (Germany) for *plexiglas*-type acrylic plastics.

Paragon. (1) Trademark of Coltene-Whaledent (USA) for a heat-cured, crosslinked acrylic denture resin with good working properties supplied in several shades.

(2) Trademark of Paragon, Inc. (USA) for a palladium-based dental bonding alloy.

Paragon Cast. Trade name of Paragon Die Casting Company (USA) for high-quality aluminum and zinc die castings.

Paragon clay. A white powder of hydrous aluminum silicate with a density of 2.5-2.7 g/cm³ (0.09-0.10 lb/in.³), used as a filler and extender in rubber and paper.

paragonite. A colorless or yellow mineral of the mica group composed of sodium aluminum silicate hydroxide, $NaAl_2(AlSi_3O_{10})$-$(OH)_2$. It can also be made synthetically. Crystal system, monoclinic. Density, 2.85 g/cm³; refractive index, 1.602. Occurrence: Switzerland, Germany.

paraguanajuatite. A dark gray mineral of the tetradymite group composed of bismuth selenide, Bi_2Se_3. Crystal system, rhombohedral (hexagonal). Density, 7.68 g/cm³. Occurrence: Mexico.

parahopeite. A colorless mineral composed of zinc phosphate tetrahydrate, $Zn_3(PO_4)_2 \cdot 4H_2O$. Crystal system, triclinic. Density, 3.31 g/cm³; refractive index, 1.625. Occurrence: Zimbabwe, Canada (British Columbia).

paraindene. See cumar gum

parakeldyshite. A white mineral with a slight bluish tint composed of sodium zirconium silicate, $Na_2ZrSi_2O_7$. Crystal system, triclinic. Density, 3.39 g/cm³; refractive index, 1.692. Occurrence: Norway.

parakhinite. A dark green mineral composed of copper lead tellurate hydroxide, $Cu_3PbTeO_4(OH)_6$. Crystal system, hexagonal. Density, 6.69 g/cm³; refractive index, 2.155. Occurrence: USA (Arizona).

paralaurionite. A colorless or whitish mineral composed of lead chloride hydroxide, PbCl(OH). Crystal system, monoclinic. Density, 6.15 g/cm³; refractive index, 2.15. Occurrence: Greece.

Paraline. Trade name of Central Glass Company Limited (Japan) for a glass comprising straight parallel wires at intervals of about 50 mm (2 in.).

Parallam. Trademark of Weyerhaeuser Inc. (USA) for *parallel strand lumber* (PSL) manufactured from long, slender wood strands bonded together by microwave curing. It is ideal for long-span situations that require beams with large cross sections, and for columns, posts, flooring and other heavy-load applications.

parallel laminate. A laminate in which all material plies or layers are oriented essentially parallel with respect to the grain or high-strength direction in tension.

Parallel-O-Bronze. Trade name of Libbey-Owens-Ford Company (USA) for a twin-ground, polished plate glass supplied in bronze shades.

Parallel-O-Grey. Trade name of Libbey-Owens-Ford Company (USA) for a twin-ground, polished plate glass with neutral-gray color.

Parallel-O-Plate. Trade name of Libbey-Owens-Ford Company (USA) for a twin-ground, polished plate glass.

parallel strand lumber. Structural composite lumber with high bending strength, made of parallel-arranged long wood veneer strands bonded with a special adhesive, and used especially for beams, headers, and load-bearing columns. Abbreviation: PSL.

Paralloy. Trade name of Youngstown Foundry & Machine Company (USA) for a nickel cast iron containing 3% carbon, 2.5-3% nickel, 2% silicon, 1% manganese, 0.5-0.75% chromium, 0.2-0.25% molybdenum, and the balance iron. It has high hardness, good tensile strength, and is used for cylinders and drawing dies.

Paraloid. Trade name of Rohm & Haas Company (USA) for acrylonitrile-butadiene-styrene plastics.

paralstonite. A colorless to smoky white mineral of the aragonite group composed of barium strontium calcium carbonate, $(Ba,Sr)Ca(CO_3)_2$. Crystal system, hexagonal. Density, 3.60 g/cm^3; refractive index, 1.672. Occurrence: USA (Illinois).

Paramafil. Trade name for rayon fibers and yarns used for textile fabrics.

paramagnetic iron. Ferromagnetic iron made paramagnetic by the application of a high pressure, usually just above 10 GPa (1.45×10^3 ksi).

paramagnetic materials. Materials that have relative magnetic permeabilities slightly greater than 1 (typically between 1.00 and 1.01), and for whose atoms the total sum of the magnetic moments of all orbital electrons is not equal to zero. They are magnetized parallel to an applied magnetic field to an extent usually proportional to the strength of the field. Examples of such materials include aluminum, gadolinium, iridium, lithium, magnesium, molybdenum, platinum, tin and tungsten.

paramelaconite. A mineral composed of copper oxide, Cu_4O_3. Crystal system, tetragonal. Density, 6.04 g/cm^3. Occurrence: USA (Arizona).

paramontroseite. A grayish black mineral of the diaspore group composed of vanadium oxide, VO_2. Crystal system, orthorhombic. Density, 4.00 g/cm^3. Occurrence: USA (Colorado).

p-aramid fibers. See aramid.

Paramount. (1) Trademark of Bramco Division of Savolite Chemical Company Limited (Canada) for caulking compounds.

(2) Trade name for rayon fibers and yarns used for textile fabrics.

Parana pine. See araucarian pine.

paranatrolite. A colorless mineral of the zeolite group composed of sodium aluminum silicate trihydrate, $Na_3Al_2Si_3O_{10}\cdot3H_2O$. Crystal system, orthorhombic. Density, 2.21 g/cm^3. Occurrence: Canada (Quebec).

Parapet. Trade name of Kuraray Company (Japan) for polymethyl methacrylate resins supplied several grades.

parapierrotite. A black mineral composed of thallium antimony sulfide, $TlSb_5S_8$. It can also be made synthetically. Crystal system, monoclinic. Density, 5.05-5.07 g/cm^3. Occurrence: Yugoslavia.

Paraplast. Trademark of Fisher Scientific (USA) for mixtures of highly purified paraffin and regulated molecular-weight polymers. *Paraplast Plus*, a special grade, contains dimethyl sulfoxide (DMSO). Used in biology and biochemistry as tissue embedding media for light microscopy.

Paraplex. Trademark of C.P. Hall Company (USA) for low-cost polyester resins with excellent electrical properties. They can be formulated for room- or high-temperature use, and are often

fiber-reinforced. Used for fiber-glass boats, autobody components, chairs, fans and helmets, coatings for wood, metals, fabrics and paper, calendered sheet and film, extruded and molded items, electrical wire insulation, and for blending with other resins.

ParaPost Cement. Trademark of Coltene-Whaledent (USA) for a chemically curing, fluoride-releasing thixotropic dental resin cement with high retention, and excellent handling properties. Used for the cementation of posts, and for fixed prostheses.

Parapremium. Trademark of Timken Company (USA) for a series air-melted fatigue-resistant steels with very high cleanness used for pinions, gears, crankshafts, bearings, aircraft engine components, downhole oil tools, etc.

Paraprene. Trademark of Para Paints Canada Inc. for exterior and interior latex floor paints.

Parapro. Trade name for polyolefin fibers and products.

pararammelsbergite. A gray mineral composed of nickel arsenide, $NiAs_2$. Crystal system, orthorhombic. Density, 7.25 g/cm^3; hardness, 762-792 Vickers. Occurrence: Canada (Ontario).

pararealgar. A bright yellow to orange-yellow to orange-brown mineral composed of arsenic sulfide, AsS. Crystal system, monoclinic. Density, 3.52 g/cm^3; refractive index, above 2.02. Occurrence: Canada (British Columbia).

Para rubber. A collective term applied to *natural rubber* obtained from the tree *Hevea brasiliensis* originally native to South America, especially the Brazilian state Para, but now also grown in Southeast Asia (Sri Lanka, Indonesia, Malaysia, etc.). See also hevea rubber.

paraschachnerite. A gray mineral composed of silver mercury, Ag_3Hg_2. It can also be made synthetically. Crystal system, orthorhombic. Density, 12.98 g/cm^3. Occurrence: Germany.

paraschoepite. A yellow, radioactive mineral composed of uranium oxide hydrate, $UO_{2.86}\cdot1.5H_2O$. It is an alteration product (Phase III) of the mineral *schoepite*. Crystal system, orthorhombic.

parascholzite. A white to colorless mineral composed of calcium zinc phosphate dihydrate, $CaZn_2(PO_4)_2\cdot2H_2O$. Crystal system, monoclinic. Density, 3.12 g/cm^3; refractive index, 1.588. Occurrence: Germany. See also scholzite.

Paraseal. Trademark of Savolite Chemical Company Limited (USA) for caulking and sealing compounds.

Parasol. Trade name of Bruin Plastics (USA) for fade-resistant, dimensionally stable acrylic fabrics with fluorocarbon finishes, supplied in wide range of colors. They have good dirt, spot, water and oil resistance, good crack and tear resistance and excellent transparency. Used for awnings, terrace canopies, garden and pool products, furniture, marine covers, etc.

paraspurrite. A colorless mineral composed of calcium carbonate silicate, $Ca_5(SiO_4)_2CO_3$. Crystal system, monoclinic. Density, 3.00 g/cm^3; refractive index, 1.672. Occurrence: USA (California).

parastibite. A white mineral composed of antimony cobalt sulfide, CoSbS. It can also be made synthetically. Crystal system, orthorhombic. Density, 6.90 g/cm^3. Occurrence: Canada (Ontario).

parasymplesite. A light greenish blue mineral of the vivianite group composed of iron arsenate octahydrate, $Fe_3(AsO_4)_2\cdot8H_2O$. Crystal system, monoclinic. Density, 3.07 g/cm^3; refractive index, 1.660. Occurrence: Morocco, Japan.

paratacamite. A green to greenish black mineral that is composed of copper chloride hydroxide, $Cu_2(OH)_3Cl$, can also contain zinc, and can be made synthetically. Crystal system, monoclinic. Density, 3.74 g/cm³; refractive index, 1.843. Occurrence: Chile, Iran.

Paratect. Trade name for a series of sealants and paints based on bituminous materials.

paratellurite. A grayish white mineral composed of tellurium oxide, TeO_2. Crystal system, tetragonal. Density, 5.60 g/cm³. Occurrence: Mexico.

Paratex. (1) Trade name of Bayer AG (Germany) for coating materials based on chlorinated rubber.

(2) Trademark of Blocksom & Company (USA) for latex-bonded natural fibers used for packing and insulating applications, and for stuffing mattresses.

Parathane. Trademark of Para Paints Canada Inc. for exterior and interior polyurethane floor enamel paints.

Paratherm. Trade name of American Optical Corporation (USA) for a spectacle glass with green tint.

Para-Tissue. Trade name of J.H. McNairn Limited (Canada) for waxed paper.

Paratite. (1) Trademark of J.H. McNairn Limited (Canada) for waxed paper.

(2) Trademark of Paramount Packaging Corporation (USA) for plastic film used for securing items on pallets.

Paratone. Trade name of Paramount Glass Manufacturing Company Limited (Japan) for glass-fiber acoustical ceiling board.

paraumbite. A colorless to white mineral composed of potassium zirconium hydrogen silicate hydrate, $K_3Zr_2H(Si_3O_9)_2 \cdot H_2O$. Crystal system, orthorhombic. Density, 2.59 g/cm³; refractive index, 1.601. Occurrence: Russian Federation.

paravauxite. A colorless to pale greenish white mineral of the paravauxite group composed of iron aluminum phosphate hydroxide octahydrate, $FeAl_2(PO_4)_2(OH)_2 \cdot 8H_2O$. Crystal system, triclinic. Density, 2.36 g/cm³; refractive index, 1.559. Occurrence: Bolivia.

Para-Wrap. Trademark of J.H. McNairn Limited (Canada) for waxed paper used for wrapping applications.

Parbond. Trademark of Parr Industries Limited (Canada) for liquid rubber sealer.

Par-Cast. Trademark Para Paints Canada Inc. for exterior stucco-texture coatings for use on masonry and wood.

parchment. The dried and tanned skin of sheep, goats, etc. prepared for use as a smooth, hard-finished, high-grade writing material (e.g., for diplomas, records, legal documents and maps). It is also used for banjo and drum heads, lampshades, etc. Vegetable parchment is made by treating paper in a bath of hot sulfuric acid. See also parchment paper; vellum.

parchment paper. A tough, grease- and oil-resistant wrapping paper that resembles *parchment* in appearance and is made by treating unsized base paper with hot sulfuric acid and subsequent rinsing with water and ammonia. The sulfuric acid effects a swelling of the cellulose in the paper.

Parcolac. Trademark of Parker Chemical Company (USA) for heavy zinc phosphate coatings used as rustproofing on metals.

Parcolene Z. Trademark of Parker & Amchem (USA) for a titanium phosphate-based activating agent used in zinc phosphate coating of iron, steel, zinc and aluminum substrates. It is usually applied by immersion or spraying.

Par-Ex. Trademark of Para Paints Canada Inc. for exterior alkyd house paints.

Par-Exc. Trade name of Teledyne Vasco (USA) for a shock-resisting tool steel (AISI type S1) containing 0.50-0.55% carbon, 0.2-0.35% silicon, 0.1-0.3% manganese, 1.75-2.25% tungsten, 1.5-1.8% chromium, 0.2-0.3% vanadium, and the balance

iron. It has a great depth of hardening, good toughness and hardness, good nondeforming properties, and is used for hot-forming dies, heading and upsetting dies, hubbing dies, die-casting dies, hand and pneumatic chisels, punches, shear blades, pipe cutters, and wear parts.

Parez. Trademark of American Cyanamid Company (USA) for a series of melamine- and urea-formaldehyde resins.

Parfait. Trade name for a cellulose nitrate formerly used for denture bases.

Parafe. Trade name for rayon fibers and yarns used for textile fabrics.

Par-Flat. Trademark of Para Paints Canada Inc. for interior latex flat paints.

Par-Flex. Trademark of Para Paints Canada Inc. for interior latex house paints.

Parforce Special. (1) Trade name of Boehler GmbH (Austria) for case-hardening steels containing about 0.15% carbon, 0.75% chromium, 3-4% nickel, and the balance iron. Used for machine parts, case-hardened parts, etc.

(2) Trade name of Boehler GmbH (Austria) for medium-carbon steels containing about 0.4% carbon, 1.4% chromium, 4-5% nickel, 1% tungsten, and the balance iron. Used for plastic dies.

pargasite. A brown mineral of the amphibole group that is composed of sodium calcium magnesium aluminum silicate hydroxide, $NaCa_2Mg_4Al_3Si_6O_{12}(OH)_2$, and may also contain iron. Crystal system, monoclinic. Density, 3.05-3.21 g/cm³; refractive index, 1.59-1.61. Occurrence: Canada (Quebec), Finland.

Parian cement. A *gypsum cement* made by mixing completely or partially dehydrated gypsum (calcium sulfate) with borax and enough water to produce a trowelable consistency. Used primarily for finish coats.

Parian paste. A ceramic body consisting of 67 wt% feldspar and 33 wt% china clay fired at approximately 1200°C (2190°F).

parianite. A purified *lake asphalt* imported from Trinidad.

Pariflux. Trademark of Foote Mineral Company (USA) for fluorspar (fluorite) used for welding electrode coatings to improve arc stabilization, aid fluxing of weld metal and form protective slags.

Paris blue. See Prussian blue.

Paris green. See copper acetoarsenite.

parisite. A brown or yellow mineral of the bastnaesite group composed of calcium cerium lanthanum fluoride carbonate, $(Ce,La)_2Ca(CO_3)_3F_2$. Crystal system, rhombohedral (hexagonal). Density, 4.32 g/cm³; refractive index, 1.676. Occurrence: USA (California, Massachusetts).

Paris metal. A nickel silver containing 71-73% copper, 14-18% nickel, 1.5-2% cobalt, 2% tin, and the balance zinc. Used for jewelry.

parison alloy. An alloy composed of 69% copper, 19.5% nickel, 6.5% zinc and 5% cadmium. Used for cheap jewelry.

Paristone. Trademark of Domtar Construction Materials (Canada) for a series of plasters.

Paris white. A white, finely ground, washed chalk ($CaCO_3$) obtained from France. Used as an extender in paper, rubber and plastics, and for *whiting*.

Paris yellow. See chrome yellow.

Parkaloy. Trade name of Parker-Kalon Corporation (USA) for a cold-forging alloy composed of 0.3% carbon, 7% chromium, 1.2% nickel, and the balance iron. Used for hardware and fasteners.

parkerite. A bronze mineral composed of nickel bismuth sulfide,

$Ni_3Bi_2S_2$. The natural, orthorhombic form is found in Ontario, Canada, and the synthetic, monoclinic form can be obtained by crystallization from the melt. Density, 8.49-8.53 g/cm^3.

parkerized steel. A steel treated in a hot solution of manganese dihydrogen phosphate to produce a dark, fine-grained, corrosion-resistant coating of ferric phosphate. The steel is subsequently coated with paraffin oil. *Note:* Parkerizing is a proprietary process.

Parkerizing. Trade name of Henkel Surface Technologies (USA) for dark, corrosion-resistant phosphate coatings and coating processes for ferrous metals.

Parker's chrome alloy. A corrosion-resistant alloy composed of 60% copper, 20% zinc, 10% chromium and 10% nickel, and used for fountain-pen points.

Parklane. Trade name of Bondex International Inc. (USA) for latex paints.

Parklite. Trade name of Parkwood Corporation (USA) for cellulose acetate plastics.

Parkwood. Trade name of Parkwood Corporation (USA) for cellulose acetate plastics.

Par-Lac. Trademark of Para Paints Canada Inc. for interior alkyd gloss and semi-gloss paints.

Parlodion. Trademark of Mallinckrodt Inc. (USA) for a purified grade of pyroxylin usually supplied in the form of chips for biological, biochemical and medical applications.

Parlon. Trademark of Hercules, Inc. (USA) for chlorinated rubber in the form of a white, nonflammable powder. Available in several viscosities, it has a density of 1.6 g/cm^3 (0.057 $lb/in.^3$), good film-forming properties, good corrosion resistance, good resistance to acids, alkalies, solvents and alcohols, and good compatibility with most natural and synthetic resins. Used for marine, swimming-pool, traffic and masonry paints, protective coatings, varnishes, adhesives, inks, plastics, and concrete treatment.

Par-Luminum. Trademark of Para Paints Canada Inc. for aluminum paint.

Par-Lux. Trademark of Para Paints Canada Inc. for exterior and interior high-gloss enamel paint.

Parlux. Trade name of Vereinigte Glaswerke (Germany) for bronze-colored, heat-reflecting glass blocks with fused-on coatings of transparent metal oxides.

Parmastic. Trademark of Parr Industries Limited (UK) for a polysulfide sealing compound.

Parmolite. Trade name of Window Glass Limited (Italy) for a patterned glass with lozenge-type design.

parnauite. A pale blue, blue green, yellow green or green mineral composed of copper arsenate sulfate hydroxide heptahydrate, $Cu_9(AsO_4)_2(SO_4)(OH)_{10} \cdot 7H_2O$. Crystal system, orthorhombic. Density, 3.09 g/cm^3; refractive index, 1.704. Occurrence: USA (Nevada).

PAR-PA alloys. Thermoplastic blends of an amorphous polyarylate (PAR) and a crystalline polyamide (PA). They possess good dimensional stability, high impact strength, very low warpage, and good chemical properties. Used for automotive applications, e.g., body panels.

Parr. (1) British trade name for a corrosion-resistant nickel-base alloy containing 18% chromium, 8.5% copper, 3.3% tungsten, 2% aluminum, 1% manganese, 0.2% titanium and 0.2% boron. Used for machinery parts, and chemical apparatus and equipment.

(2) British trade name for a corrosion-resistant alloy composed of 80% nickel, 15% chromium and 5% copper, and used

for machinery parts, and chemical equipment.

parsettensite. A brown mineral composed of manganese silicate hydroxide, $Mn_5Si_6O_{13}(OH)_8$. Crystal system, orthorhombic. Density, 3.78 g/cm^3. Occurrence: Switzerland, Italy.

Parsol. Trade name of Saint-Gobain (France) for a glare-reducing, heat-absorbing plate glass available in bronze and gray colors.

parsonite. A pale yellow, radioactive mineral composed of lead uranyl phosphate hydrate, $Pb_2UO_2(PO_4)_2 \cdot xH_2O$. Crystal system, triclinic. Density, 5.72 g/cm^3. Occurrence: France, Zaire.

Parsons. Trade name of Manganese Bronze Company (UK) for a series of copper alloys including several copper-zinc-nickel alloys, manganese bronzes and high-tensile brasses as well as a number of tin-base bearing alloys.

Parson's alloy. A tough, corrosion-resistant manganese bronze containing 55-60% copper, 39-42% zinc, 0-1.4% iron, 0.7-1.0% tin, 0.5-1.0% aluminum and 0-3.5% manganese. It has good resistance to mild acids, good tensile strength, and is used for valves, pump and sluice valve spindles, valve stems, marine engine parts, condenser tubes, machine parts, engine frames, propellers, and propeller blades. Also known as *Parson's manganese bronze.*

Parson's manganese bronze. See Parson's alloy.

Parson's white brass. (1) A hard, tough cast brass composed of 62% tin, 35% zinc and 3% copper, and used for marine and automobile bearings.

(2) A tin babbitt containing 14% lead, 7% antimony and 5% copper. It has good antifriction properties, and is used for steam pumps, reciprocating engines, thrust blocks, marine main bearings, and automobile bearings.

Parsteel. Trade name of Paragon Steel Company (USA) for a tough gear steel containing 0.5% carbon, 3% nickel, 1% chromium, and the balance iron.

Par-Ten. Trade name of US Steel Corporation (USA) for a high-strength, low-alloy (HSLA) constructional steel containing 0.12% carbon, 0.75% manganese, 0.1% silicon, 0.04% vanadium, and the balance iron. Used for mine cars, autobodies, and agricultural and railroad equipment.

Par-Tex. Trademark of Para Paints Canada Inc. for interior alkyd flat paints.

partially acetylated cotton. Cotton, usually in the form of fibers or fabrics, with good heat, rot and mildew resistance, but lower tensile strength and abrasion than fully acetylated cotton. It obtained by treatment with acetic acid, acetic anhydride or perchloric acid in the presence of a suitable catalyst. This treatment partially converts the raw cotton fiber to cellulose acetate. See also acetylated cotton; cellulose acetate; cotton.

partially air-dried lumber. Green, unseasoned lumber dried for a short period of time by exposure to outside air. It has an average moisture content of 19% or more. Abbreviation: PAD lumber. See also air-dried lumber; kiln-dried lumber.

partially oriented yarn. A synthetic filament yarn with considerable, but not complete molecular orientation obtained by controlled extrusion of the liquid or semiliquid starting polymer through a die or spinneret. Abbreviation: POY.

partially stabilized zirconia. An engineering ceramic composed of zirconium oxide (ZrO_2) to which some (typically 1 to 8 wt%) calcia (CaO) or magnesia (MgO) has been added to produce a tetragonal phase which transforms under stress. Unlike *fully stabilized zirconia*, that has a cubic microstructure, it has a two-phase microstructure containing monoclinic zirconia solid solution and cubic zirconia solid solution. Its mechanical properties including fracture toughness and thermal-shock resistance are superior to those of the fully stabilized material. Used for structural applications, and in the manufacture of bioceramics (e.g., for orthopedics). Abbreviation: PSZ. See also lime-stabilized zirconia; magnesia-stabilized zirconia; transformation-toughened zirconia; yttria-stabilized zirconia.

particleboard. A hard-surfaced, man-made panel composed of wood chips, flakes, shavings, slivers, particles or sawdust bonded together with a synthetic resin (urea-formaldehyde resins for interior types, or phenolic resins for interior and exterior types) or another binder under heat and pressure. The bonding agent used and the particular manufacturing process determine the suitability of the panel for exterior applications. *Particleboard* is classified into low-, medium- and high-density grades according to the panel weight in kg/m³ (or lb/ft³) which ranges from as low as 595 kg/m³ (37 lb/ft³) to over 800 kg/m³ (50 lb/ft³). It is available in 1.2 × 2.4 m (4 × 8 ft.) sheets in various thicknesses, typically between 6.3 and 19.0 mm (0.25 and 0.75 in.). Used as a building board, for furniture components, as a substructure for plastic laminates, as floor underlayment, and as a core material for surface veneers. See also high-density particleboard; medium-density particleboard; low-density particleboard.

particleboard corestock. Particleboard used as a core material for overlaying with wood veneer, fiber sheets, hardboard, plastic laminates, etc. Typical applications include cabinetwork, paneling, furniture, countertops, doors, dividers, and wainscots. Also known as *core-type particleboard.*

particleboard panelstock. Particleboard used predominantly for paneling applications. Its surfaces may be specially treated to obtain decorative effects, e.g., overlaid with wood veneers, or plastic or fiber sheets.

particleboard underlayment. Particleboard made to close thickness tolerances and specifically engineered as a smooth substrate or underlayment for carpeting and other finish flooring.

particle-oriented paper. A chart or recording paper with a surface emulsion coating consisting of microscopic magnetic flakes and oil. The flakes are oriented by the magnetic field produced by the recording head and, depending on the orientation, absorb or scatter indicident light producing a dark, visible image that can be read by magnetic means.

particle-reinforced composite. See particulate composite.

particulate. A solid or liquid substance in particle form, e.g., the dispersed particles of alumina in dispersion-strengthened aluminum, copper or iron.

particulate composites. Microscopic composites consisting of metallic or nonmetallic particles suspended or embedded in the matrices of other materials. The dispersed phase particles are relatively large, roughly equal in all dimensions, and often bear some fraction of an applied load. Examples include tungsten or molybdenum particles in copper matrices, tungsten carbide or titanium carbide particles embedded in cobalt or nickel matrices, silicon carbide in aluminum matrices, rubber reinforced with carbon black particles or finely ground silica, thermoplastic polymers, such as polystyrene, polyvinyl chloride or polymethyl methacrylate, reinforced with rubber particles, etc. Also known as *particle-reinforced composite.*

particulate recording media. Ferromagnetic or ferrimagnetic recording media consisting of discrete, submicron-sized particles of iron oxides (typically γ-Fe_2O_3), chromium dioxide (CrO_2), cobalt (Co^{2+})-doped iron oxides, metals (especially iron) or hexagonal ferrites (especially $BaFe_{12}O_{19}$, $PbFe_{12}O_{19}$ and

$SrFe_{12}O_{19}$) dispersed in a polymeric or resinous binder and applied to a nonmagnetic support (e.g., a plastic film or aluminum plate). See also magnetic medium; magnetic recording materials.

particulate-reinforced metal-matrix composites. A group of composites consisting of metallic or nonmetallic particles embedded or suspended in metal or metal alloy matrices. Examples include silicon carbide particulate-reinforced aluminum-matrix composites, molybdenum or tungsten particulate-reinforced copper-matrix composites, and tungsten carbide or titanium carbide particulate reinforced cobalt- or nickel matrix composites. Abbreviation: PRMMC. See also particulate composites.

parting agent. See release agent.

parting compound. A waterproof material, such as graphite or silica, in powder form that prevents moist foundry sand from adhering to the pattern, and allows the sand on the parting surfaces of cope and drag to separate without sticking.

parting powder. A waterproof, nonsiliceous powder, usually chalk or bone meal, applied to a foundry pattern to facilitate removal from the molding sand without sticking or clinging.

parting sand. A fine, round-grained sand, often with tripolite or bentonite additions, dusted over the drag of a foundry mold to prevent the cope from sticking to it. Also known as *foundry parting sand.*

Partinium. British trade name for a group of aluminum alloys for lightweight aircraft and automobile parts. Typical compositions are: (i) 88.5% aluminum, 7.4% chromium, 1.7% zinc, 1.3% iron and 1.1% silicon; and (ii) 96.0% aluminum, 2.4% antimony, 0.8% tungsten, 0.6% copper and 0.2% tin.

Partitionlite. Trade name of Australian Window Glass Proprietary Limited for 8.7 mm (0.34 in.) thick, gray, decorative figured glass sheets.

partition tile. A tile used in the construction of nonload-bearing partitions and interior walls.

partridgeite. See bixbyite.

partridgewood. See acapau.

Par-Trim. Trademark of Para Paints Canada Inc. for exterior and interior alkyd trim enamel paints.

ParVAR. Trade name of Ellwood City Forge (USA) for a series of superclean carbon, alloy and tool steels and several ferritic and martensitic stainless steels made by a process that combines air melting, ladle refining and vacuum degassing. They have less than 0.001 wt% sulfur, 0.008 wt% phosphorus, 10 ppm oxygen and 1.5 ppm hydrogen, respectively, and are usually supplied as forgings.

Par-Vin. Trademark of Para Paints Canada Inc. for interior latex flat paint.

partzite. An olive-green mineral with black tarnish. It belongs to the pyrochlore group and is composed of copper antimony oxide hydroxide, $Cu_2Sb_2(O,OH)_7$. Crystal system, cubic. Density, 3.80 g/cm³; refractive index, 1.70. Occurrence: USA (California).

parwelite. A pale brown to orange mineral of the fluorite group composed of calcium magnesium manganese antimony arsenic silicate, $(Mn,Mg,Ca)_5SbAsSiO_{12}$. Crystal system, monoclinic. Density, 4.62 g/cm³; refractive index, 1.85. Occurrence: Sweden.

Par-Wood. Trade name of Para Paints Canada Inc. for a Danish oil used to protect wood and provide it with a beautiful, lustrous finish.

ParyLAST. Trademark of AST Products Inc. (USA) for parylene coatings with good electrical insulating properties, high dielectric strengths, excellent to good substrate adhesion and excellent barrier properties. They are applied by vapor deposition/plasma molecular activation, and used on pacemaker leads, and on medical implants as a protection against corrosive body fluids.

Parylene. Trade name of Carbide Corporation (USA) for poly(*p*-xylylene) polymers. See also parylenes.

parylene-F. A thermoplastic thin-film polymer based on poly(tetrafluoro-*p*-xylylene). It has a dissociation temperature of 520°C (968°F), a low dielectric constant (k ~ 2.3), high resistivity and dielectric strength and high chemical and thermal stability. Usually produced by vapor deposition techniques, it is used as an interlayer dielectric for ultra-large-scale-integrated circuits. Also known as *poly(tetrafluoro-p-xylylene)*.

parylene-N. A thermoplastic thin-film polymer based on poly(*p*-xylylene). It has a dielectric constant of about 2.6, and produces pure, uniform, pore-free vapor-deposited coatings or thin films on various substrates including metals, ceramics, paper and plastics. Used as dielectric in capacitors, thin-film circuits, etc., and as a chemical and electrical barrier on sensitive electrical, electronic or medical components and equipment. Also known as *poly(p-xylylene)*.

parylenes. A group of thermoplastic polymers based on poly(*p*-xylylene) or poly(tetrafluoro-*p*-xylylene) that produce pure, uniform, pore-free coatings or thin films on various substrates including metals, ceramics, paper and plastics. Copolymers formed by mixing *parylene-N* with co-monomers, such as perfluorooctylmethacrylate or tetravinyltetramethylcyclotetrasiloxane, can also be prepared. Used as dielectrics in capacitors, thin-film circuits and electronic miniaturization, and as chemical and electrical barriers on sensitive electrical, electronic or medical components and equipment. See also parylene-F; parylene-N.

pascoite. A yellow-orange mineral composed of calcium vanadium oxide hydrate, $Ca_3V_{10}O_{28} \cdot 17H_2O$. It can also be made synthetically. Crystal system, monoclinic. Density, 1.87 g/cm³; refractive index, 1.824. Occurrence: Peru. Used as an ore of vanadium.

pashmina. The very fine hair of the inner coat of the pashmina goat *(Capra hircus)* native to the Himalayas. Used for making yarns and fabrics.

passive iron. Iron on whose surface a thin oxide has been formed by immersion in a suitable passivating solution (e.g., concentrated nitric acid) to prevent further corrosive attack.

passively smart material. An engineered material that is considered to be "passively smart" because of its ability to respond to an external change without assistance. See also actively smart material; smart material.

passive metal. A metal, such as aluminum, bismuth, chromium, cobalt, iron, nickel, tin or titanium, with a highly adherent and very thin surface oxide layer. This layer may either have formed naturally or be produced by treatment with suitable passivating agents inhibiting or preventing further corrosive attack.

Passive Ni. Trade name of British Driver-Harris Company Limited (UK) for a high-purity nickel used for thermionic tubes.

paste. (1) A term referring to an adhesive, semisolid composition, such as a mixture of starch or dextrin and water, or latex and water, used to bond paper, cardboard, etc., to various substrates. Also known as *paste adhesive*.

(2) A mixture of cement and water used as a cementing ingredient in concrete.

(3) The material used to form a porcelain body. See also

hard paste; soft paste.

(4) A ferromagnetic material made into finely divided particles and suspended in a liquid, such as water or a light petroleum distillate. Used in the wet-particle method of magnetic particle inspection.

(5) A sheet molding compound consisting of a filled thermosetting resin, a chopped or continuous glass-fiber reinforcement and certain additives (e.g., catalysts, fillers, pigments, flame retardants, etc.). See also sheet molding compound.

(6) An intimate mixture of a finely powdered brazing filler metal and an inorganic or organic flux, or neutral vehicle in the proper proportions to form a paste.

paste adhesive. See paste (1).

pasteboard. A relatively thin, hard and stiff cardboard-type material composed either of several sheets of paper pasted together or pressed and dried paper pulp. Used for the manufacture of boxes and cartons. See also cardboard.

paste brazing filler metal. See paste compound.

paste compound. An intimate mixture of finely powdered brazing filler metal and an inorganic or organic flux, or neutral vehicle in the proper proportions to form a paste. Also known as *paste brazing filler metal*.

paste epoxies. A class of solid epoxy structural adhesives supplied in one- and two component systems in a wide range of viscosities from flowable to thixotropic pastes. Used to seal metal components, bond nylon to metals, join composite substrates, etc.

paste paint. An organic coating whose pigment is concentrated enough to allow for a significant reduction with the vehicle prior to application.

paste resin. A soft, viscous, usually semisolid mixture of finely divided resin powder mixed with a plasticizer, but without a solvent.

paste solder. An intimate mixture of finely powdered soldering metal, a suitable inorganic or organic or flux, and a tinning agent. It is usually applied to the soldered joint by syringe or special applicator gun. Also known as *paste soldering filler metal*.

paste soldering filler metal. See paste solder.

paste wood filler. An inert compound supplied in the form of a stiff paste for filling the pores in open-grain woods like ash, hickory, mahogany, oak, poplar or walnut.

patching cement. (1) A mixture of Portland cement, fine aggregate and sometimes other components, such as organic binders, that becomes plastic and trowelable when tempered with water. Used for repairing concrete and mortar, repairing and patching wallboard, plaster, etc.

(2) A special, fireclay-type, cementitious material used in repairing and patching furnace walls, glass mold bottoms, etc. See also fireclay.

pâte dure. See hard paste porcelain.

Patent Cobalt-Chrome Steel. Trade name of Darwin & Milner Inc. (USA) for an air-hardening, nondeforming, wear-resistant tool steel containing about 1.3-1.5% carbon, 0.3% manganese, up to 0.6% silicon, 11.8-13.8% chromium, 2.7-3.3% cobalt, up to 0.6% nickel, 0.5-0.9% molybdenum, and the balance iron. Used for metal cutting and forming tools.

patent alum. See aluminum sulfate octadecahydrate.

patented steel wire. High-strength, medium- or high-carbon steel wire that has been heated above the upper critical temperature, and then cooled in air, or molten lead or salt.

patent leather. A black leather with smooth, high-gloss surface that is usually finished with a clear plastic coating of high solids content, or by film lamination. This leather manufacturing technique was previously patented. Used especial for shoes and handbags. Also known as *japanned leather.*

patent plate. Plate glass ground and polished on both sides.

patent yellow. See Turner's yellow.

pâte tendre. See soft paste porcelain.

PATH. Trademark of Huber Engineered Materials (USA) for alumina trihydrate (ATH) pigments and fillers.

Patina. Trademark of Paris Brick Company Limited (Canada) for calcite bricks.

patina. (1) The green or greenish coating or film of copper carbonate that forms naturally over time on the surface of copper and copper alloys, such as bronze, exposed to the atmosphere.

(2) A green or greenish coating or film analogous to that in (1), but produced on copper and copper alloys by a suitable chemical treatment (e.g., with an aqueous solution of copper nitrate, ammonium chloride and calcium chloride, or an aqueous solution of acetic acid, ammonium chloride, sodium chloride, cream of tartar and copper acetate) as a corrosion-resistant and/or antique, ornamental finish.

Patina Line. Trade name of Toppan Printing Company Limited (Japan) for embossed and printed polyvinyl chloride laminates.

Patlon. Trademark of Amoco Fabrics & Fibers Company (USA) for polypropylene fibers, yarns and monofilaments.

patronite. A black mineral composed of vanadium sulfide, VS_4. Crystal system, monoclinic. Density, 2.83 g/cm³. Occurrence: Peru. Used as a source of vanadium.

pattern alloy. (1) A British aluminum alloy containing aluminum, 8% copper and 2% tin used for pistons.

(2) A British bearing alloy containing 87% lead and 13% antimony.

pattern coatings. Special materials applied to wooden foundry patterns as protective coatings against moisture and abrasion by molding sand. See also pattern varnishes.

patterned hardwood. A *hardboard* panel into whose surface special patterns have been machined or pressed to provide a decorative touch. Used as siding material for houses, and for wainscoting and dividers.

patterned lumber. Lumber that in addition to being dressed, matched and/or shiplapped is also shaped to a pattern.

pattern metal. An antifriction alloy of 30-40% zinc, 20-42% lead, 15-40% tin and 0-3% copper, used for bearings.

Pattern Resin. Trade name of GC Dental (Japan) for a brush-applied acrylic resin with very low shrinkage upon curing. Supplied as a two-component powder-liquid product, it is used for producing stable dental patterns and models.

pattern wood. A kiln-dried hardwood or softwood that is suitable for foundry patterns. Common woods used for this purpose are birch, cherry, fir, mahogany, maple, poplar, sugar pine, walnut, white pine and whitewood. The most important characteristics of *pattern wood* are good workability and machinability, high durability, low warpage, shrinkage and expansion, good resistance to checking and cracking, good gluing qualities, good surface finish, relative low cost, and low moisture content (typically less than 6%).

pattern varnishes. A group of varnishes used for coating foundry patterns in order to protect them against moisture. Examples include black, red and yellow shellac varnishes.

Pattex. Trademark of Henkel KGaA (Germany) for a series of general-purpose adhesives based on chloroprene rubber.

Pattrex. Trade name of Foseco Minsep NV (Netherlands) for

pattern stone powders that set with a satin smooth finish. Used in casting shops in the preparation of pattern plates, Keller models, etc., to produce hard, durable surfaces.

Patty Paper. Trademark of J.H. McNairn Limited (Canada) for waxed paper.

paulingite. A colorless mineral of the zeolite group composed of potassium calcium aluminum silicate hydrate, $K_2(Ca,Ba)_{1.3}(Si,Al)_{12}O_{24} \cdot 14H_2O$. Crystal system, cubic. Density, 1.473 g/cm³; refractive index, 1.473. Occurrence: USA (Washington).

Paulite. Trade name of Simpson Brothers Machine Works (USA) for a steel containing 3.2% carbon, 2.5% nickel, 1.5% nickel, 0.8% chromium, and the balance iron.

paulmooreite. A mineral composed of lead arsenite, $Pb_2As_2O_5$. Crystal system, monoclinic. Density, 6.91-6.95 g/cm³; refractive index, above 1.9. Occurrence: Sweden.

paulownia. The strong wood of the so-called empress or princess tree *(Paulownia tomentosa)*, originally a medium-sized oriental tree, but now also grown in the United States. It has a coarse grain, good resistance to warping, an average weight of 264 kg/m³ (16 lb/ft³), and is used for lumber. It is named after the Russian princess Anna Paulowna. Also known as *empress*. See also kiri.

pau marfim. The tough, impact-resistant wood of the tree *Balfourodendron riedelianum*. It resembles birch, and hard maple in appearance. Source: Southern Brazil, Paraguay, northern Argentina. Used for turned items, furniture, cabinetmaking, veneer, etc. Also known as *moroti*.

Pavchrome. Trade name of Pavco Inc. (USA) for an extensive series of greenish and yellow conversion coatings for zinc and cadmium.

paving asphalt. See asphalt cement.

paving brick. High-strength vitrified clay brick with high abrasion resistance and low water absorption, that is often provided with spacing lugs and made with smooth or wire-cut surfaces. Used in the construction of roads, sidewalks, driveways and other pavements. Also known as *road brick.*

paving-brick clay. Impure refractory clay or shale with high strength and plasticity used to make paving brick.

paving concrete. Concrete suitable for use in road construction, and including coarse-aggregate bituminous concrete, fine-aggregate bituminous concrete, hot-laid coarse and fine tar concrete and cold-laid coarse and fine tar concrete.

paving sand. Sand with a particle size ranging from 0.25 to 200 mesh depending upon application. It is suitable for use in the construction of asphaltic and concrete pavements, and for grouting applications.

pavonite. A grayish white mineral composed of silver bismuth sulfide, $AgBi_3S_5$. Crystal system, monoclinic. Density, 6.54 g/cm³. Occurrence: USA (Colorado), Bolivia.

PavPhos. Trade name of Pavco Inc. (USA) for an extensive series of iron and zinc phosphate coatings for ferrous metals and alloys.

Pax. Trade name of Sanderson Kayser Limited (UK) for a shock-resisting tool steel (AISI type S1) containing 0.5% carbon, 0.8% silicon, 1.5% chromium, 2.2% tungsten, 0.2% vanadium, and the balance iron. It has excellent toughness, good wear and abrasion resistance, and good hot hardness. Used for shear blades, punches, chisels, etc.

paxite. A steel-gray mineral composed of copper arsenide, Cu_2As_3. Crystal system, orthorhombic. Density, 5.30 g/cm³. Occurrence: Czech Republic, Slovakia.

Pax Non-Break. Trade name of Sanderson Kayser Limited (UK) for a shock-resisting tool steel containing 0.4% carbon, 1.5% chromium, 2.2% tungsten, 0.2% vanadium, and the balance iron. It has excellent toughness, good wear and abrasion resistance, and good hot hardness. Used for chisels, punches, dies, and hot-work tools.

Paxolin. Trade name of Micanite & Insulators Company (UK) for phenol-formaldehyde resins, plastics and laminates.

Paxon. Trade name of Paxon Polymer Company LP (USA) for a series of thermoplastic compounds based on polyolefins and including high- and medium-density polyethylene flake, powder and granules used for making extruded and injection-, blow- or rotational-molded products including packaging films, and bottles and other containers.

PBI fibers. See polybenzimidazole fibers.

PBT-elastomer alloys. Thermoplastic blends of a crystalline polybutylene terephthalate (PBT) and an elastomer. They have excellent resistance to abrasion, chemicals, ozone and ultraviolet light, excellent high- and low-temperature flexural properties, good impact resistance, good tensile and tear strengths, good dimensional stability and good chemical resistance. Used for electrical and electronic components, pneumatic and hydraulic hoses, vapor fuel lines, wire and cable coatings, automotive bumpers and shoe plates.

PBT-LDPE alloys. Polymer blends of polybutylene terephthalate (PBT) and low-density polyethylene (LPDE). The addition of 15-25% LDPE to PBT combines two crystalline polymers resulting in an alloy of enhanced mechanical properties and processibility, and decreased moisture absorption.

PBT-PAR alloys. Thermoplastic polymer blends of crystalline polybutylene terephthalate (PBT) and amorphous polyarylate (PAR). These blends have good chemical and impact resistance and good dimensional stability, and are used for automotive applications, e.g., body panels.

PBT-PC alloys. Polymer blends of crystalline polybutylene terephthalate (PBT) and amorphous polycarbonate (PC). They combine the good chemical resistance of PBT with the high toughness, good creep and impact resistance and good dimensional stability of PC. Used for automotive exterior components, e.g., bumpers, bumper covers, etc.

PBT-PET alloys. Polymer blends of crystalline polybutylene terephthalate (PBT) and crystalline polyethylene terephthalate (PET). The addition of PET enhances surface gloss and improves dimensional stability and rigidity while the PBT facilitates processing and increases toughness, fatigue endurance, and impact resistance. Other important properties include excellent durability, high heat resistance, good chemical resistance, good ultraviolet stability, good electrical properties, low thermal expansion, water absorption and flammability. Used for appliance housings, business machine cabinets and consoles, and exterior parts for automobiles.

PBT-PPE alloys. Polymer blends of crystalline polybutylene terephthalate (PBT) and amorphous polyphenylene ether (PPE). They combine the good chemical resistance, high stiffness and low melt viscosity of PBT with the high toughness, good impact resistance and dimensional stability of PPE.

PC-ABS alloys. Tough, ductile polymer blends containing more than 50% polycarbonate (PC) with the balance being acrylonitrile-butadiene-styrene (ABS). Commonly supplied in injection molding and plating grades, they have good mechanical and thermal properties, good impact resistance, low surface tension, good interfacial adhesion and relatively low viscosity for easier processing. Used for business machine and appli-

ance housings, automotive interior and exterior applications (e.g., instrument panels, grilles and wheel covers), power tools, electrical components, such as terminal blocks and switches, food trays, etc.

PC-Crete. Trademark of American Polymer Corporation (USA) for epoxy putties used for patching concrete cracks.

PC-Kleber. Trademark of BASF AG (Germany) for a polyvinyl chloride adhesive.

PC-Lumber. Trademark of Protective Coating Company (USA) for epoxy putties used for filling wood imperfections.

PC-Marine. Trademark of Protective Coating Company (USA) for epoxy putties for marine applications.

PC-Metal. Trademark of Protective Coating Company (USA) for epoxy putties used for filling imperfections on metallic surfaces.

PC-PA alloys. Blends of amorphous polycarbonate (PC) and crystalline polyamide (PA). They combine the high toughness, impact resistance and good electrical and thermal properties of PC with the outstanding chemical resistance and low friction propeties of PA.

PC-PBT alloys. Polymer blends of crystalline polybutylene terephthalate (PBT) and amorphous polycarbonate (PC). They combine the good chemical resistance of PBT with the high toughness, good creep and impact resistance, and good dimensional stability of PC. Used for automotive exterior components, e.g., bumpers, bumper covers, etc.

PC-PET alloys. Polymer blends of amorphous polycarbonate (PC) and crystalline polyethylene terephthalate (PET) with excellent mechanical compatibility. They combine the good chemical resistance of PET with the high toughness, good creep resistance and excellent dimensional stability of PC. Used for small mechanical components, lawnmower parts, helmets, etc.

PCS fiber. See plastic-clad silica fiber.

PC-SMA alloys. Immiscible polymer blends of amorphous polycarbonate (PC) and amorphous rubber-modified styrene-maleic anhydride (SMA) copolymer. They combine the good strength, toughness, ductility and heat resistance of PC with the good melt flow characteristics and moldability and low cost of SMA. Also, the heat deflection temperature is raised by about 10-15°C (18-30°F) as compared with the base copolymers. Used for automotive interior trim (e.g., instrument panels), power-tool housings, and electric components.

PC-SuperEpoxy. Trademark of Protective Coating Company (USA) for translucent, pasty industrial epoxy adhesives.

PCT fibers. See poly(1,4-cyclohexanedimethylene)terephthalate fibers.

PC-TPUR alloys. Polymer blends of amorphous polycarbonate (PC) and elastomeric thermoplastic polyurethane (TPUR). The TPUR elastomer adds resiliency and impact resistance, while the PC improves the rigidity and high-temperature resistance. These blends are used for automotive exterior components, e.g., bumpers.

PCU. Trade name for vinyon (vinyl chloride) fibers with excellent acid, alkali and mildew resistance. Used especially for textile fabrics.

PC-Woody. Trademark of Protective Coating Company (USA) for epoxy pastes.

PdPro. Trademark of Degussa-Ney Dental (USA) for a dental ceramic alloy containing 54.9% palladium and 37.5% gold. It has a very light oxide layer, a hardness of 250 Vickers, and is used for esthetic restorations.

PDS. Trademark of Ethicon Inc., Division of Johnson & Johnson

(USA) for a biodegradable polymer synthesized by the ring-opening polymerization of *p*-dioxanone, and used in the form of monofilaments for surgical sutures.

peachblossom ore. See erythrite.

peached fabrics. Fabrics with soft, sueded surface finishes suggestive of the fuzzy skins of peaches, and usually produced by sanding or chemical treatment.

peacock blue. A ceramic color typically composed of 18 parts of *china stone*, 13 parts of *cobalt oxide*, 6 parts of *flint* and 3 parts of *standard black*.

peacock ore. See bornite.

pea gravel. Gravel screened and sized to a particle size range between 6.3 and 12.5 mm (0.250 and 0.500 in.). Used for asphaltic surfaces and roofing materials.

peanut fibers. Soft, fairly bulky regenerated textile fibers produced from peanut protein.

pear. The pale pinkish-brown to rosy-red wood from the deciduous tree *Pyrus communis*. It resembles apple in appearance, and has a fine, uniform texture and a straight or irregular grain (depending on the trunk shape). Pearwood is relatively strong and tough, may blunt saws, but turns well, and takes a fine finish. Source: Europe, North America. Used for veneer in cabinetry and inlays for reproducing flesh colors, and for bowls, handles and carvings. Also known as *common pear; pearwood*.

pearceite. A mineral composed of silver copper arsenic sulfide, $(Ag,Cu)_{16}As_2S_{11}$. Crystal system, monoclinic. Density, 6.13 g/cm³. Occurrence: USA (Colorado, Montana, Utah), Mexico.

Pearl. Trade name of UCB Films, Inc. (USA) for a pearlized cellulose film supplied in a wide range of colors for food packaging applications.

Pearlalith. Trade name of B. Schwanda & Sons (USA) for casein plastics.

pearl alum. See aluminum sulfate octadecahydrate.

pearl ash. Commercial potassium carbonate (K_2CO_3) usually made by refining potash. It has a density of 2.3 g/cm³, a melting point of 909°C (1668°F), and is used as a flux in glass, glazes and porcelain enamels. Also known as *purified potash*. See also potassium carbonate.

Pearlbrite. Trademark of Enthone-OMI Inc. (USA) for a uniform satin nickel finish for metals and plastics that can be overplated with black nickel, gold, silver and other metals.

Pearl Brite. Trade name of Laminated Glass Corporation (USA) for a white, translucent laminated glass used for partitions and room dividers.

Pearl Crown. Trade name of Superior Threads Company (USA) a colorfast, machine washable continuous filament rayon sewing thread supplied in a wide range of colors.

pearl filler. Anhydrous calcium sulfate ($CaSO_4$) in the form of white crystals or powder for use as a paper filler.

Pearlilux. Czech trade name for a glass with a pattern consisting of regular chains of very small pearls.

pearlite. A mixture of alternate lamellae of *ferrite* and *cementite* formed in slowly cooled iron-carbon alloys (steels and cast irons) during the eutectoid reaction. It is the result of the transformation of *austenite* at temperatures above the *bainite* range.

pearlitic cast irons. A group of cast irons containing about 2.2-3.5% carbon and about 1.5-3.5% silicon. They have microstructures consisting of pearlitic matrices and interspersed graphite lamellae, and excellent abrasion and wear resistance and high toughness. Used for automotive engine blocks, bearings, brake drums, chains and chain links, clutch parts, crankshafts, cylinder housings, gears, and pistons. Also known as *pearlitic irons*.

pearlitic malleable irons. A group of cast irons produced either by controlled heat treatment of the same base white irons used to produce ferritic malleable cast iron or by alloying to prevent decomposition of carbides. Depending on heat treatment and desired hardness, they have microstructures of *pearlite*, spheroidite or *tempered martensite*. A typical composition is 2.00-2.70% total carbon, 0.25-1.25% manganese, 1.00-1.75% silicon, 0-0.05% phosphorus, 0.03-0.18% sulfur, the balance being iron. *Pearlitic malleable irons* have a density of 7.35-7.44 g/cm³ (0.266-0.269 lb/in.³), excellent wear resistance, high strength, good to fair machinability, high hardenability, and moderate ductility and shock resistance. Used for axle housings, differential housings, universal-joint yokes, gears, camshafts and crankshafts for automobiles, machine parts, ordnance equipment and tools. See also malleable irons.

pearlitic manganese steels. See carbon-manganese steels.

pearlitic steel. See eutectoid steel.

pearlized coating. A coating applied to fabrics to produce a lustrous, pearly surface. Often applied to fabrics used for making outerwear.

pearl lacquer. A lacquer that contains suspended guanine crystals (obtained from fish scales), and is used as a surface coating to produce a brilliant, pearly luster.

Pearls. Trade name of Gerresheimer Glas AG (Germany) for a patterned glass.

pearls. Smooth, rounded, white, cream, pink, bluish-gray or black-colored nacreous concretions of calcium carbonate (known as "nacre" or "mother-of-pearl") formed as deposits around a foreign body in the shells of mollusks, such as oysters, clams and mussels, and chiefly used as gems. See also nacre; synthetic nacre.

pearl sinter. See geyserite.

pearl spar. A variety of the mineral *dolomite* with a rhombohedral crystal structure and a vitreous pearly luster.

Pearlthane. Trademark of Merquinsa (USA) for thermoplastic polyurethanes based on polycaprolactone-copolyesters. They have very high rates of crystallization, a density of about 1.19 g/cm³ (0.043 lb/in.³), a melting point of 185°C (365°F), high elongations at break (730%), and are supplied as white or pale yellow pellets for extrusion into heat-resistant blown or flat polyurethane films for sealing applications. They adhere well to rubber, polyurethanes, polyvinyl chlorides, and fabrics.

Pearlveil. Trade name of Schmelzer Industries Inc. (USA) for fiberglass surfacing veils.

pearl white. A white paint pigment composed of bismuth oxychloride (BiOCl).

PearlyLene. Trademark of Tung Ho Spinning Company (Taiwan) polyester staple fibers.

pearwood. See pear.

peau de soie. A soft, fine, slightly lustrous fabric made from silk or synthetic yarn (e.g., acetate or polyester) in a modified satin weave, and having a slightly corded or grained appearance. Used for dresses.

Pebax. Trademark of Atofina SA (France) for thermoplastic engineering elastomers based on polyether block amides (PEBA) that combine the resiliency and flexibility of rubber with the good processibility of thermoplastics. Supplied as granules or fine powders, they have outstanding mechanical properties, very high elastic memory, excellent impact resistance, good chemical resistance, and good resistance to low temperatures (-40°C or -40°F). Used for sports and leisure products, electrical and electronic equipment, tubes, bellows, seals, diaphragms, and protective coatings.

pebble. (1) A small stone, usually worn round and smooth by being rolled about by water, and ranging from about 2-64 mm (0.08-2.5 in.) in diameter.

(2) Hard flint, hard-burnt porcelain or other heavy, abrasion-resistant material resembling a pebble (1) used as a grinding medium in ball mills or pebble mills.

pebble crepe. See moss crepe.

pebbled goat. A tanned goatskin with a pebbly surface finish on the grain side produced by passing between rollers under pressure.

pebble lime. See calcium oxide.

Pebblex. Trade name of ASG Industries Inc. (USA) for glass with a deeply imprinted irregular pattern of pebbles.

pecan. The hard, heavy wood from the large hickory tree *Carya pecan* (also called *C. illinoensis*). It is considered inferior to true hickories, reddish brown in color with occasional dark streaks, and has large shrinkage. *Pecan* is more popular for its edible nuts than its wood. Average weight, 720 kg/m^3 (45 lb/ft^3). Source: Central and southern United States; Mexico. Used as veneer for furniture and paneling, and for tool and implements, handles, flooring and pallets. See also hickory.

pecan shell media. Brown to tan-colored, crushed and ground shells of *pecan* nuts available in coarse, medium and fine grades for use as soft-grit abrasive media in blast cleaning, removal of rust, scale, paint, lacquer, carbon deposits, etc., and for deflashing of molded plastic parts.

peccary. A chrome-tanned leather made from the skin of the peccary, a wild boar native to Central and South America. Used for gloves.

Pe-Ce. German trade name for very durable polyvinyl chloride fibers for textile fabrics, decorative and insulating applications, cordage, netting, etc.

Pechiney. Trade name of Pechiney Electrométallurgie (France) for a series of cast iron nodularizers containing 3.5-10% magnesium, 1-3% calcium, 45-50% silicon, 0-1% aluminum, 0-2% rare earths, and the balance iron.

Pechiney Strontium. Trade name of Pechiney Electrométallurgie (France) for an alloy containing 99+% strontium, 0.4% barium, 0.1% calcium, 0.1% sodium, 0.05% magnesium, 0.05% nitrogen, 0.01% iron, and 0.0005% iron. Used to introduce strontium into aluminum-silicon alloys in order to modify the eutectic structure and improve mechanical properties and machinability.

Pechko white gold. A corrosion-resistant alloy composed of 60% gold, 30% palladium, 8-10% platinum and 0.1-2% iridium, used for ornaments and jewelry.

Peckrite. Trade name of Peckovers, Limited (Canada) for a cast iron containing 2.8-3% total carbon, 1.6-1.8% silicon, 1.2-1.5% nickel, 0.75% manganese, and the balance iron. Used for cams, gears, bushings, and piston rings.

pecoraite. A green mineral of kaolinite-serpentine group composed of nickel silicate hydroxide, Ni$_3$Si$_2$O$_5$(OH)$_4$. Crystal system, monoclinic. Density, 3.08 g/cm^3; refractive index, 1.65. Occurrence: Australia.

pectolite. A white, grayish, or colorless mineral of the pyroxenoid group composed of sodium calcium hydrogen silicate, NaCa$_2$HSi$_3$O$_9$. Crystal system, triclinic. Density, 2.85 g/cm^3; hardness, 5; refractive index, 1.606. Occurrence: Scotland, USA (New Jersey), Greenland.

Pedilen. Trade name of Otto Bock Company (USA) for flexible (soft) and rigid foams based on liquid polyol mixtures containing tertiary aliphatic amines. Used in the medical and health care industries.

PEEK-Optima. Trade name of Victrex USA Inc. for a semicrystalline polyetheretherketone (PEEK) thermoplastic with exceptional chemical resistance and strength, excellent radiation resistance and tribological properties, high rigidity and toughness, a melting temperature of 340°C (644°F), a glass-transition temperature of 145°C (293°F), and an elastic modulus that can be matched to that of human bone. Supplied in three unfilled grades with low-, medium- or high- viscosity, it is used for implants, e.g., human fingers and vertebrae for spinal surgery. *PEEK-Optima LT* is an inherently lubricated grade used for dental, hip and spinal implants, heart valve components, etc.

PEEK-PESV alloys. Binary blends of crystalline polyetheretherketone (PEEK) and amorphous polyether sulfone (PESV) which combine the toughness, abrasion resistance, and good load-bearing properties of PEEK with the excellent dimensional stability of PESV. Used for electronic packaging and consumer items.

peelable lacquer. A special coating product consisting of a film of lacquer on a flexible support or backing that can be peeled off for application to a substrate.

Peel Coat. Trade name of Evans Manufacturing, Inc. (USA) for a strippable, hot-melt protective plastic coating for metals.

Peelcote. Trade name of ACI Chemicals Inc. (USA) for temporary barrier coatings for paint spray booth applications.

peen coating. A coating of zinc, usually 25-75 μm (1-3 mils) thick, produced on ferrous parts of suitable size and shape by tumbling in a rotary barrel holding zinc powder and numerous glass beads. Also known as *peen-galvanized coating*.

peen-galvanized coating. See peen coating.

Peerafilter. Trademark of Peerless Coatings Limited (UK) for an impact-resistant precoated acrylic or polycarbonate sheet material used for optoelectronic applications. It is supplied in sizes of 850 mm × 850 mm (33.5 in. × 33.5 in.) and 1220 mm × 900 mm (48.0 in. × 35.4 in.) with thicknesses of 0.5-6.0 mm (0.02-0.24 in.). The applied coating has excellent abrasion resistance and minimizes glare.

Peeraguard. Trademark of Peerless Coatings Limited (UK) for a high-quality, UV-cured antireflective hard coating with good optical properties (from clear to glossy) for application to plastic substrates, such as acrylics and polycarbonates. It has excellent abrasion and chemical resistance, and optical clarity. Typical applications include automotive interiors and instrumentation, architectural glazing, instrument and LCD displays, safety visors, machine guards, printed and machined parts, etc. *Peeraguard Automotive* refers to a range of automotive finishes including hardcoat lacquer systems for plastic headlamps and rear lamps, anti-reflective, textured and patterned finishes for plastic automotive interior components, and smooth and high-gloss base coats. *Peeraguard Exterior* is an exterior-grade hardcoat for polycarbonate substrates. It has excellent resistance to abrasion, weathering and yellowing, good optical clarity, good resistance to many chemicals including basic fluids to aggressive solvents, and is used on architectural and industrial glazing, e.g., bus shelters, road barriers, phone booths, and advertising displays.

Peeramist. Trademark of Peerless Coatings Limited (UK) for a semihard, antimist coating for acrylic or polycarbonate substrates that provides excellent optical properties and good abrasion resistance.

Peerastat. Trademark of Peerless Coatings Limited (UK) for a soft-coat finish with excellent antistatic properties.

Peerglass. Trade name of Pilkington Perkin-Elmer (UK) for two-way mirrors.

Peerless. Trade name of Crucible Materials Corporation (USA) for a series of hot-work steels used for extrusion, forging and gripping dies, swages, etc.

Pegamoid. Trade name for an artificial leather.

Pegase. Trade name of Société Nouvelle des Acieries de Pompey (France) for free-machining medium-carbon steels used for screw machine products, and fasteners. A typical composition is 0.5% carbon, 0.08% silicon, 0.5% manganese, and the balance iron.

Pegasus. Trade name for denture and tray acrylics.

pegboard. Hardboard panel with uniform, closely spaced punched or drilled holes that is used with special hardware (pegs, hooks, etc.) to form storage walls, or hold tools, displays, etc. Also known as *perforated hardboard*.

pegmatite. A coarse-grained igneous rock with large interlocking crystals that consists essentially of feldspar, quartz and mica. Used for building construction, and as a source of lithia, zircon, tin, tungsten, tantalum or uranium.

Peguform. Trademark of Pegulan-Werke AG (Germany) for flexible polyvinyl chlorides.

pehrmanite. A light green mineral of the hogbomite group composed of beryllium iron aluminum oxide, $BeFe_2Al_6O_{12}$. Crystal system, rhombohedral (hexagonal). Density, 4.08 g/cm^3; refractive index, 1.79. Occurrence: Finland.

peisleyite. A white mineral composed of sodium aluminum sulfate phosphate hydroxide hydrate, $Na_3Al_{16}(SO_4)_2(PO_4)_{10}(OH)_{17}\cdot20H_2O$. Crystal system, monoclinic. Density, 2.12 g/cm^3; refractive index, 1.510. Occurrence: South Australia.

Peka. German trade name for a laminated safety glass similar to *Sekurit*.

Pekafill. Trade name for a light-cure dental hybrid composite.

Pekalux. Trade name for a light-cure dental hybrid composite.

pekin. A high-quality fabric in a novelty weave, having vertical stripes spaced at regular intervals. It can be made of cotton or silk, or alternating velvet and satin stripes.

pekoite. A lead gray mineral of the stibnite group composed of copper lead bismuth selenide sulfide, $CuPbBi_{11}(S,Se)_{18}$. Crystal system, orthorhombic. Density, 6.80 g/cm^3. Occurrence: Australia.

Pelaspan. Trademark of Dow Chemical Company (USA) for expandable polystyrene supplied in the form of heat-activated beads and pellets.

Pelco. Trade name for synthetic fibers based on a polyester-caprolactone block copolymer.

Pelcoloy. Trade name of Molecu-Wire Corporation (USA) for a resistance alloy of 70% nickel and 30% iron. It has high electrical resistivity, and a maximum service temperature of 590°C (1094°F). Used for resistance wire.

Pelifix. Trademark of Pelikan AG (Germany) for a blue glue stick used to bond paper. The glue goes on blue to simplify gluing, but the color disappears in a few minutes.

Pelikanol. Trademark of Pelikan AG (Germany) for a white-colored paper adhesives.

pellet. A small rounded, spherical or rectangular body made by pressing a light bulky material, such as carbon black, plastic molding powder, metal powder, ores, etc.

Pelletex. Trade name for a carbon black used as a reinforcing agent in rubber products.

Pellethane. Trademark of Dow Chemical Company (USA) for thermoplastic polyurethane elastomers based on polyester polycaprolactone. They possess high clarity, good resiliency and impact resistance, excellent low-temperature flexibility, excellent heat and fungus resistance, and excellent chemical, fuel and oil resistance. Some grades also provide enhanced abrasion and mar resistance and/or paintability. Used for exterior and interior automotive trim, single-sided window encapsulation, protective covers for wire and cable, tubing and hose covers, etc. *Pellethane 2363* is a series of biocompatible polyetherurethanes (PEU) used as biomaterials for the manufacture of implantable devices including pacemaker leads and cardiac prosthesis devices.

Pellon. Trademark of Pellon Chemotextiles Inc. (Canada) for non-woven fabrics.

pellyite. A colorless to pale yellow mineral composed of barium calcium iron magnesium silicate, $Ba_2Ca(Fe,Mg)_2Si_6O_{17}$. Crystal system, orthorhombic. Density, 3.51 g/cm^3; refractive index, 1.645. Occurrence: Canada (Yukon).

Pelseal. Trademark of Pelseal Technologies LLC (USA) for a series of acid-resistant *Viton*-type fluoroelastomer based on one- and two-component caulks and sealants supplied in various grades for horizontal and/or vertical applications, such as sealing cracks in concrete, concrete and steel joints and industrial flooring joints, bonding rubber gaskets to metal, and bonding and splicing O-rings and fluoroelastomers. They have high resistance to chemicals including sulfuric, nitric and hydrochloric acids, excellent flexibility, good resistance to salt spray and fungal growth, good resistance to ozone, UV-radiation and sunlight, good resistance to temperatures ranging from -40 to +204°C (-40 to +400°F), and good adherence to various metallic and nonmetallic materials. Also included under this trademark are various liquid fluoroelastomer adhesives and coatings.

pelt. See hide.

Pemco. Trademark of Glidden-Durkee Division of SCM Corporation (USA) for coloring oxides and ceramic frits.

Penacolite. Trademark of Glidden-Durkee Division of SCM Corporation (USA) for a series of thermosetting resins and adhesives based on resorcinol-formaldehyde or resorcinol-phenol-formaldehyde.

Penchlor. Trademark of Atofina Chemicals Inc. (USA) for an acidproof, sodium silicate-type cement used for lining chemical containers and process equipment.

pencil cedar. See eastern red cedar.

pencil metal. A metal composed of about 53% bismuth, 42% lead and 5% mercury, and used for pencil cores.

pencil stone. See pyrophyllite.

Pencoyd. Trade name of US Steel Corporation for a structural steel containing 0.3% carbon, 1% nickel, and the balance iron. Used in bridge construction.

penetrating stain. Wood stain that penetrates below the surface into the wood fibers. It is made by dissolving an oil-soluble dye in alcohol or oil. Penetrating oil stains add color, while revealing both the grain pattern and the texture. They are supplied in various tones including cedar, cherry, ebony, mahogany, maple, oak, redwood and walnut. See also oil stain; wood stain; stain.

penetration macadam. A paving material composed of screened gravel or crushed stone aggregate in two sizes bonded together by a bituminous material, such as asphalt, tar or a cement grout. Used in road construction. Also known as *grouted macadam*.

penfieldite. A white mineral composed of lead chloride hydroxide, $Pb_2Cl_3(OH)$. Crystal system, hexagonal. Density, 5.82 g/cm^3. Occurrence: Greece, Chile.

Pen Hob. Trade name for a case-hardened steel containing up to 0.1% carbon, 1.25% nickel, 0.6% chromium, and the balance iron. Used for hobbed cavity molds.

penikisite. A blue to green mineral of the bjarebyite group composed of barium iron magnesium aluminum phosphate hydroxide, $Ba(Mg,Fe)_2Al_2(PO_4)_3(OH)_3$. Crystal system, triclinic. Density, 3.79 g/cm^3; refractive index, 1.688. Occurrence: Canada (Yukon).

penkvilksite. A white mineral composed of sodium titanium silicate pentahydrate, $Na_4Ti_2Si_8O_{22}\cdot5H_2O$. Density, 2.58 g/cm^3; refractive index, 1.640. Occurrence: Russian Federation.

pen metal. (1) A corrosion-resistant alloy composed of 67% gold, 25% copper and 8% silver, and used for pen points.

(2) A corrosion-resistant alloy composed of 85% copper, 13% zinc and 2% tin, and used for pen points.

Penn-Air. Trade name of Pennsylvania Steel Corporation (USA) for a medium-alloy air-hardening tool steel (AISI type A2) containing 1% carbon, 2% manganese, 5% chromium, 1% molybdenum, and the balance iron. It has good deep-hardening properties, and good wear resistance and toughness. Used for blanking and forming tools, punches, and shear blades.

Pennant. Trade name of Delsteel Inc. (USA) for a nondeforming steel containing 0.9% carbon, 1.6% manganese, 0.2% vanadium, and the balance iron.

pennantite. A brown mineral of the chlorite group composed of manganese aluminum silicate hydroxide, $(Mn_5Al)(Si_3Al)O_{10}(OH)_8$. Crystal system, monoclinic. Density, 3.15 g/cm^3; refractive index, 1.667. Occurrence: UK (Wales). Used as a source of manganese.

Pennchem. Trade name of Atofina Chemicals, Inc. (USA) for vinyl ester resin-based adhesive mortars and polymer concretes.

Penn Crete. Trade name of IPA Systems, Inc. (USA) for stucco.

Penn-Cut. Trade name of Pennsylvania Steel Corporation (USA) for a series of high-speed steels. *Penn-Cut* is a standard tungsten-type high-speed steel (AISI type T1) containing 0.7% carbon, 18% tungsten, 4% chromium, 1% vanadium, and the balance iron. It has high hot hardness, excellent machinability and wear resistance and good toughness. *Penn-Cut 5* is a tungsten-type high-speed steel (AISI type T5) containing 0.8% carbon, 18% tungsten, 8% cobalt, 4% chromium, 1-2% vanadium, and the balance iron. It has excellent red hardness and wear resistance, and good resistance to decarburization. *Penn-Cut Moly* is a molybdenum-type high-speed steel (AISI type M2) containing 0.8% carbon, 6.3-6.5% tungsten, 5% molybdenum, 4% chromium, 2% vanadium, and the balance iron. It has excellent wear resistance, good toughness, and high hot hardness. *Penn-Cut* steels are used for lathe and planer tools, cutters, drills, taps, reamers, broaches, hobs, chasers, and form and cutoff tools.

Penn-Flex. Trade name of Pennsylvania Steel Corporation (USA) for an oil- or water-hardening tool steel containing 0.33% carbon, 0.72% manganese, 0.25% silicon, 0.85% chromium, 0.42% tungsten, 0.45% molybdenum, and the balance iron. Used for plastic molds.

Pennguard. Trade name of Atofina Chemicals Inc. (USA) for borosilicate glass blocks used for building and construction applications.

Pennrold. Trade name of Brush Wellman Corporation (USA) for a series of age-hardenable, fatigue- and corrosion-resistant copper alloys containing 0.4-2.1% beryllium, 0-2.6% cobalt and

0-2.6% nickel. Used for current-carrying springs, mechanical parts, switch parts, switch blades, circuit breaker parts, contacts, and bellows.

Penntrowel. Trade name Atofina Chemicals Inc. (USA) for epoxy-based floor surfacing compounds, adhesives mortars, monolithic floor toppings, etc.

Pennvernon. Trade name of PPG Industries Inc. (USA) for a sheet glass made by the Pennvernon or Pittsburgh glass-drawing process

Pennverto. Trade name of Pennitalia SpA (Italy) for sheet glass made by the Pennvernon or Pittsburgh glass-drawing process.

Pen-O-Four. Trade name of Peninsular Steel Company (USA) for an oil-hardening steel containing 0.75% carbon, 1.75% nickel, 0.9% chromium, 0.7% manganese, 0.35% molybdenum, and the balance iron. Used for tools and dies.

Penros. Trademark of Reichhold Chemicals, Inc. (USA) for a polymerized wood rosin used as a soldering flux, and in oleoresinous and spirit varnishes, hot-melt compounds, adhesives, waxed paper and paper coatings.

penroseite. A steel-gray mineral of the pyrite group that is composed of nickel copper selenide, $(Ni,Cu)Se_2$, and may also contain some sulfur. Crystal system, cubic. Density, 6.51-6.58 g/cm^3. Occurrence: Bolivia, Zaire.

Penstock. Trade name of Lukens Steel (USA) for a series of steels containing up to 0.5% carbon, 0.9-1.35% manganese, 0.15-0.3% silicon, and the balance iron. Used for penstock, welded structures, etc.

pentacarbonyliron. See iron pentacarbonyl.

pentacene. A highly reactive aromatic compound consisting of 5 fused benzene rings and corresponding to the formula $C_{22}H_{14}$. It is supplied in the form of deep-blue crystals that sublime at 290-300°C (554-572°F) and decompose in air above 300°C (572°F). *Pentacene* is an organic semiconductor, has a typical room-temperature electron mobility range 10^{-1} to 10^0 cm^2/V-s, and can be grown as a thin film on glass or silicon (001) substrates for use in the manufacture of organic thin-film transistor (OTFT) displays. It is also used as a photoconductor, e.g., in copiers and in high-efficiency organic photocells.

Pentacite. (1) Trademark of Reichhold Chemicals, Inc. (USA) for light-colored modified rosin esters and rosin ester gums used in coatings and rubber adhesives.

(2) Trademark of Reichhold Chemicals, Inc. (USA) for alkyd resins made with pentaerythritol (PETN), and used in coatings and printing inks.

Pentaclear. Trade name of Kloeckner Pentaplast GmbH (Germany) for clear rigid polyvinyl chloride sheeting used for packaging applications.

pentacolumbium trisilicide. See niobium silicide (v).

pentaerythritol. White, combustible crystals or powder (98+% pure) with a density of 1.40 g/cm^3 (at 25°C or 77°F), a melting point of 255-259°C (491-498°F), a boiling point of 276°C (529°F), and a refractive index of 1.55. Used as wavelength-dispersive spectrometer crystals, and in the manufacture of alkyd resins, rosin and tall oil esters, synthetic lubricants, paint swelling agents, special varnishes, plasticizers and explosives. Formula: $C_5H_{12}O_4$. Abbreviation: PET; PETN.

Pentaform. Trade name of Kloeckner Pentaplast GmbH (Germany) for rigid polyvinyl chloride sheeting and film used for deep-drawn parts.

Pentaftal. Trade name of Rheinpreussen AG (Germany) for a series of phthalate resins

pentagonite. A blue mineral composed of calcium vanadyl silicate tetrahydrate, $Ca(VO)Si_4O_{10}·4H_2O$. Crystal system, orthorhombic. Density, 2.34 g/cm^3; refractive index, 1.544. Occurrence: USA (Oregon).

pentahafnium trisilicide. See hafnium silicide (iii).

pentahydrite. A colorless mineral of the chalcanthite group composed of magnesium sulfate pentahydrate, $MgSO_4·5H_2O$. It can also be made synthetically from an aqueous solution of magnesium chloride, magnesium sulfate and sodium chloride under prescribed conditions. Crystal system, triclinic. Density, 1.90 g/cm^3; refractive index, 1.492.

pentahydroborite. A colorless mineral composed of calcium borate pentahydrate, $CaB_2O_4·5H_2O$. Density, 2.00 g/cm^3; refractive index, 1.536. Occurrence: Russia.

pentalobal fibers. Synthetic fibers with enhanced luster and cross sections resembling inwardly curved pentagons with rounded lobes or vertices.

Pentalyn. Trademark of Hercules Inc. (USA) for modified and unmodified pentaerythritol (PETN) esters of rosin used in adhesives, varnishes, protective coatings, paints and printing inks.

pentamanganese trisilicide. See manganese silicide (iv).

pentaniobium trisilicide. See niobium silicide (v).

pentapeptide. A *peptide* that contains 5 amino acid residues.

Pentapharm. Trade name of Kloeckner Pentaplast GmbH (Germany) for rigid polyvinyl chloride film and sheeting used in the pharmaceutical industries as packaging materials.

Pentaplus. Trade name of Kloeckner Pentaplast GmbH (Germany) for unplasticized polyvinyl chloride sheeting used as waterproofing material for foundation walls.

Pentaprint. Trade name of Kloeckner Pentaplast GmbH (Germany) for printable rigid polyvinyl chloride film and sheeting.

penta resin. A hard semisynthetic compound made by the esterification of *rosin* with *pentaerythritol* (PETN), and used in varnishes, cellulosic lacquers, paints, etc.

Pentasound. Trade name of Kloeckner Pentaplast GmbH (Germany) for rigid polyvinyl chloride supplied in board form for use as acoustical insulation.

pentatantalum silicide. See tantalum silicide (iii).

pentatantalum trisilicide. See tantalum silicide (iv).

Pentathane. Trade name of Molded Dimensions, Inc. (USA) for a series of polyurethane formulations and molded products for applications requiring high abrasion and impact resistance.

Pentatherm. Trade name of Kloeckner Pentaplast GmbH (Germany) for rigid polyvinyl chloride film and sheeting used as thermal insulation for pipes.

pentatitanium trisilicide. See titanium silicide (iv).

pentatungsten trisilicide. See tungsten silicide (ii).

pentavanadium trisilicide. See vanadium silicide (iii).

Pentawax. Trade name for a series of waxes made from vegetable oils and pentaerythritol (PETN) and used as a paper coating, in printing inks, etc.

pentazirconium trisilicide. See zirconium silicide (vii).

Pent Core. Trade name for self-cure dental composite resins for core-build-ups.

Pent-Caps. Trade name for a high-copper dental amalgam alloy.

Pentecor. Trade name of C-E Glass (USA) for glass with a pattern of parallel prisms.

pentetic acid. See diethylenetriaminepentacetic acid.

Pentex. (1) Trademark of Honeywell Performance Fibers (USA) for a high-performance polyester fiber based on polyethylene naphthalate (PEN). It has high tenacity, high elastic modulus, excellent dimensional stability, and low creep properties. Used as a reinforcement in high-performance tires and tire carcasses,

and mechanical rubber goods as well as for cordage and narrow- and broad-woven industrial fabrics.

(2) Trade name of UK Plastics Limited (UK) for cellulose nitrate plastics.

pentlandite. A light bronze-yellow mineral with metallic luster that is composed of iron-nickel sulfide, $(Fe,Ni)_9S_8$, and contains of 35.4% nickel. Crystal system, cubic. Density, 4.6-5.1 g/cm^3; hardness, 3.5-4.0 Mohs. Occurrence: Canada (Ontario), Norway, UK. Used as an important ore of nickel. Also known as *nicopyrite*.

Penton. Trademark of Hercules, Inc. (USA) for a thermoplastic resin based on chlorinated polyether. Usually supplied as a fine, natural, black or olive-green molding powder, it has a density of 1.4 g/cm^3 (0.05 $lb/in.^3$), good tensile and impact strength, good dimensional stability, very low water absorption, excellent chemical resistance, a maximum service temperature of 120°C (250°F), good electrical properties, good moldability and processibility, and poor flame resistance (but self-extinguishing). Used for chemical processing equipment, laboratory equipment, tank and valve linings, valve bodies, tanks, meters, pumps and pump parts, piping, fittings, bearings, monofilament filter supports, and column packings.

pentosan. A *polysaccharide* found in the woody tissues of plants such as oat hulls, corncobs and rice hulls, which yields pentose on hydrolysis.

pentose. A polypentoside sugar whose molecules contain five carbon atoms and three asymmetric carbon atoms. It is obtained from *pentosan* by hydrolysis. Formula: $C_5H_{10}O_5$.

Pentrate Ultra. Trade name of Heatbath Corporation (USA) for black oxide protective coatings for application to steel. A liquid grade (*Pentrate Ultra Liquid*) is also supplied.

pentylcyanophenyl. A liquid crystal material used for optical and optoelectronic display applications. Formula: C_5H_{11}-Ph-Ph-CN. Abbreviation: PCB; 5CB; CB5.

4'-pentyl-4-biphenylcarbonitrile. A nematic liquid crystal compound with a density of 1.008 g/cm^3, a boiling point of 140-150°C (284-302°F)/5mm, a refractive index of 1.532, and a flash point above 110°C (230°F). Used in electronics and optoelectronics. Formula: $CH_3(CH_2)_4C_6H_4C_6H_4CN$.

4-(trans-4-pentylcyclohexyl)benzonitrile. A nematic liquid crystal compound with a melting point of 30-55°C (86-131°F) and a flash point above 110°C (230°F). Used in electronics and optoelectronics. Formula: $CH_3(CH_2)_4C_6H_{10}C_6H_4CN$.

pentyl esters. See amyl esters.

4'-(pentyloxy)-4-biphenylcarbonitrile. A nematic liquid crystal compound with a flash point above 110°C (230°F). Used in electronics and optoelectronics. Formula: $CH_3(CH_2)_4OC_6H_4C_6H_4CN$.

Peonia. Trade name of Vetreria di Vernante SpA (Italy) for glass with an unusual floral pattern.

PEP. Trade name of O. Hommel Company (USA) for a porcelain enamel powder.

Pepcoat. (1) Trademark for a coating composed of a combination of resins and solid lubricants. When applied to steel components and force cured, it quickly forms a hard, protective surface layer providing excellent resistance to abrasion and galling, and a long service life over a wide range of service temperatures from -220 to +310°C (-364 to +590°F). Particularly suited for coating pulp and paper mill and power-plant equipment, and hydrocarbon-processing plants.

(2) Trade name of Pep & Joss Metal Finishers for a corrosion- and wear-resistant amorphous phosphorus-nickel alloy

deposit with an as-deposited hardness of 500-550 Vickers, and outstanding adhesion to various metallic substrates including carbon, alloy and stainless steels, cast irons, aluminum, copper and copper alloys, nickel and titanium. It is deposited by an autocatalytic reduction process (electroless), and is used for a wide range of applications in the chemical, food-processing, petroleum-refining, medical, pharmaceutical and many other industries.

pepper-and-salt fabrics. Textile fabrics with a speckled appearance produced by using either twisted yarns of two colors, e.g., black and white, or intricate weave patterns.

peptide. A compound formed by the combination of 2 or more amino acids chemically bound together with amide linkages (CONH) with the splitting out of water. See also dipeptide; pentapeptide; tripeptide; polypeptide.

Peptizoid. Trade name of Frederick Gumm Chemical Company Inc. (USA) for cleaning compounds, surface treatments, and burnishing and deburring compounds used in the metal finishing industries.

Pera. Trade name of Ramge Chemie GmbH (Germany) for specialty adhesives used for attaching flooring materials.

Peraluman. Trade name of Alusuisse (Switzerland) for an extensive series of wrought and cast aluminum alloys containing 0.5-10% magnesium, 0.2-1.5% manganese, 0-0.4% iron, 0-0.4% silicon, 0-0.25% zinc, 0-0.1% titanium, 0-0.2% chromium, and 0-0.01% copper. Used for architectural and decorative applications, shipbuilding, boat hulls, marine parts, aircraft parts, containers, fittings, automotive components, machine parts, roofing, paneling, fittings, trim, fasteners, hollowware, extrusions, and chemical and food processing equipment.

Perax. Trade name of Marsh Brothers & Company Limited (UK) for a tough, oil-hardening tool steel used for punches, dies and general tools.

Perbuna C. Trademark of Bayer Corporation (USA) for a chloroprene rubber with excellent resistance to weathering, heat, flames and corona discharge, good resistance to oils, solvents, oxygen and ozone, a service temperature range of -20 to +100°C (-4 to +212°F), and a hardness of 30-90 Shore. Used for mechanical rubber goods, in lining chemical tanks, and for belts, hoses, seals and gaskets.

Perbunan. Trademark of Bayer Corporation (USA) for acrylonitrile-butadiene rubber (NBR) with excellent resistance to swelling in organic solvents, excellent resistance to vegetable, animal and petroleum oils and greases, good resistance to gasoline, fair mechanical and poor low-temperature properties, a service temperature range of -25 to +100°C (-13 to +212°F), and a hardness of 20-92 Shore. Used for gasoline, chemical and oil hoses, seals, O-rings, gaskets, and pump parts. *Perbunan NT* is an abrasion- and wear-resistant acrylonitrile-butadiene rubber (NBR) supplied in several grades including standard, fast-vulcanizing, and blended with polyvinyl chloride (NBR content, 70%). It has excellent resistance to liquid fuels, mineral oils and greases, low permeability to gases, good aging resistance, and is used for technical moldings, seals, membranes, sleeves, bellows, buffers, vibration dampers, hose, conveyor belting, brake and clutch linings, sponge rubber, footwear soles, gloves, cable sheathing, and rubberized fabrics.

Perc. Trade name of Ferro Corporation (USA) for inorganic powder coatings and porcelain enamels used on metallic materials.

percale. A firm, smooth, medium-weight cotton fabric in a plain weave that may be printed, and is made in different grades. High-quality percale is closely woven and has a lustrous surface finish. Used for dresses, shirts, sportswear, sheets and book covers.

percaline. A fine, light cotton fabric with a glossy finish, used as a garment lining material.

percussion cap brass. A gilding metal containing 90% copper, 9.6% zinc and 0.4% lead. Used for percussion caps, gilding, and electrical parts.

Percy Aluminum. British trade name for wrought and cast heavy-duty copper alloys containing 7.5-13.0% aluminum, 0-2% lead and 0-1.5% manganese. Used for stripper nuts, and bearings.

percylite. A light blue mineral composed of lead copper chloride hydroxide, $PbCuCl_2(OH)_2$. Crystal system, cubic. Density, 5.25 g/cm³; hardness, 2-2.5 Mohs.

Perdonal. Trade name of Wieland Werke AG Metallwerke (Germany) for corrosion-resistant, non-heat-treatable, wrought aluminum alloys containing 1.5-3% magnesium, 0.5-1.5% manganese and 0-0.3% chromium. They possess excellent resistance to seawater corrosion, and good weldability and formability, and are used for shipbuilding, automobile bodies, window frames, roofing, architectural trim, and hydraulic tubing.

Perduro. Trade name of Dresser Industries (USA) for a malleable cast iron containing 1.6-1.8% total carbon, 0.4-0.6% combined carbon, 0.5-0.7% manganese, 1.5% silicon, 0.9-1.1% copper, and the balance iron. It has good resistance to elevated temperatures up to 538°C (1000°F), and is used for chains, sprockets, chain links, rolls, levers, etc.

peretaite. A colorless mineral composed of calcium antimony oxide sulfate hydroxide dihydrate, $CaSb_4O_4(SO_4)_2(OH)_2 \cdot 2H_2O$. Crystal system, monoclinic. Density, 3.80 g/cm³; refractive index, 1.841. Occurrence: Italy.

Perfan. Trade name of Fidenza Vetraria SpA (Italy) for glass diffusers used in building construction.

Perfect. Trade name for a resin-modified glass-ionomer dental luting cement.

perfect crystal. See perfect crystalline substance.

perfect crystalline substance. A crystalline substance that has a completely ordered array, i.e., in which each atom, ion, or molecule has a specific site to occupy and a specific orientation. It is of high purity and has a minimum number of vacancies, dislocations and other defects. Also known as *perfect crystal*.

Perfection. Trademark of Den-Mat Corporation (USA) for a homogeneous, non-agglomerated, self-polishing, microfine composite resin with excellent color stability and stain and microcrack resistance, and high tensile strength and wear resistance. Used in dentistry for the restoration of anterior teeth, the resurfacing of composite restorations, and the veneering of posterior composite resins and worn acrylic facings.

perfluoroethylene. See tetrafluoroethylene.

Perf Foam. Trade name of General Foam Corporation (USA) for perforated, flexible polyurethane foam supplied in slit rolls for packaging applications.

Perfil. (1) Trademark of DuPont Canada Inc. for fibrillated polyolefin tape.

(2) Trade name for vinyon (vinyl chloride) fibers with excellent acid, alkali and mildew resistance. Used especially for textile fabrics.

Perfilon. Trade name of E.I DuPont de Nemours & Company (USA) for nylon fibers and yarns used for textile fabrics.

Per-Fit. Trade name for a self-cure silicone used for denture liners.

Perflex. Trademark of Union Carbide Corporation (USA) for elastic yarns and spandex fibers.

perfluoroalkoxy alkanes A class of translucent thermoplastic fluoropolymer resins obtained by adding an alkoxy side chain onto a base of tetrafluoroethylene. They have a density of 2.13-2.16 g/cm^3 (0.078 lb/in.^{3}), a melting point of 300-310°C (572-590°F), a durometer hardness of 60 Shore D, excellent thermal resistance, negligible water absorption, excellent resistance to acids, bases, ozone and ultraviolet radiation, good abrasion resistance, good dielectric properties, low coefficients of friction, and a service temperature range of -195 to +260°C (-320 to +500°F). Used for piping, flexible tubing, valves, pumps, filter cartridges, etc. Abbreviation: PFA.

perfluorocarbon. A fluorocarbon compound that contains only fluorine-carbon bonds, and no hydrogen-carbon bonds. Abbreviation: PFC.

perfluorocyclobutane. A thermosetting aromatic ether oligomer with three reactive functional groups that is a derivative of cyclobutane. It has good thermal stability to about 350°C (660°F), a glass-transition temperature of 380°C (716°F), low moisture absorption, and a low dielectric constant (k = 2.24). Used as a heat-transfer agent and refrigerant, and in the form of thin films as an interlayer material for integrated circuit devices. Formula: C_4F_8. Also known as *octafluorocyclobutane.*

perfluoro elastomer. A chlorinated fluoroelastomer with excellent resistance to elevated temperatures (up to 300°C or 570°F) and good resistance to extreme environments including solvents, alkalies, concentrated acids and aromatic fuels.

perforated brick. Building brick whose weight has been reduced by several holes symmetrically arranged parallel with the face.

perforated fabrics. Patterned textile fabrics produced by punching out small holes or motifs with a metal punch or roller. Also known as *punched fabrics.*

perforated hardboard. See pegboard.

perforated metal. Sheet metal (e.g., carbon steel, stainless steel, aluminum, brass, monel, etc.) having round, square, rectangular, triangular, diamond or other perforations blanked, punched or pierced out. It is available in various designs, such as checkerboard, criss-cross, diamond, cane and tire track, and used as a partitioning material for tool cribs, warehouses, etc., for decorative and safety applications, such as cashier windows, lounges, lobbies and cafeterias, for grillework, light diffusers, ventilating screens, heat transfer baffles, dividers, shelving, furniture, machine guards, heater guards and dipping and pickling baskets, and in screening and sizing crushed stone, sand, coal, milled ores, etc.

Perform. (1) Trade name of Krupp Stahl AG (Germany) for an exten-sive series of high-grade carbon and alloy steels used for structural applications.

(2) Trade name of Coltene-Whaledent (USA) for an acrylic resin used for denture lining and relining applications.

PerformancePBN. Trade name of Morgan Advanced Ceramics (USA) for pyrolytic boron nitride used in the manufacture of electrical, microwave and semiconductor components, and crucibles for the production of gallium arsenide crystals.

PerforMax. Trademark of Owens Corning (USA) for long glass reinforcing fibers.

PerformMAX. Trademark of Huber Engineered Woods (USA) for strong, stiff, water-resistant four-layer structural panels made by mixing chopped wood strands with a urethane-phenolic resin blend, orienting the strands in the two core layers normal to the two surface layers, applying a resin-impregnated overlay, compressing under high heat, and then applying a final non-

skid surface coating. Used for subflooring applications.

Perform-Soft. Trade name of Coletene-Whaledent (USA) for an acrylic resin used for denture lining and relining applications.

Pergo. Trademark for laminate flooring with thin aluminum oxide coating, and good resistance to stains, scratches, fading and wear. Also included under this trademark are *Pergo Moisturbloc* moisture-barrier underlayments and *Pergo Soundbloc* sound-barrier underlayments.

Pergopak. Trademark for white titanium-oxide extender pigments for the paper and cardboard industries.

Pergut. Trademark of Bayer AG (Germany) for a chemical, water-chlorinated cold rubber used in corrosion-protective paints, for the formulation of adhesives and printing inks, and in the form of films for various industrial applications.

perhamite. A light brown to white mineral composed of calcium aluminum phosphate silicate hydroxide hydrate, $Ca_3Al_7(SiO_4)_3$-$(PO_4)_4(OH)_3 \cdot 16.5H_2O$. Crystal system, hexagonal. Density, 2.64 g/cm^3; refractive index, 1.564. Occurrence: USA (Maine).

periclase. A colorless, transparent, grayish white, yellow, brown, greenish or black mineral with a vitreous luster. It is a natural magnesium oxide, MgO, occurring in granules, and can also be made synthetically by calcining magnesia at high temperatures. Crystal structure, cubic. Density, 3.56-3.70 g/cm^3; hardness, 5.5-6.0 Mohs; refractive index, 1.732. Occurrence: Europe, USA (California, New Mexico). Used in refractories for lining steelmaking furnaces, and in glass manufacture.

periclase brick. Magnesite brick containing 90% or more magnesium oxide (MgO). See periclase refractory.

periclase grain. Granular crystalline *magnesite* containing 85% or more magnesium oxide (MgO).

periclase refractory. A nonsilicate oxide ceramic composed of 90.0% magnesia (MgO), 3.0% silica (SiO_2), 3.0% ferric oxide (Fe_2O_3), 2.5% calcia (CaO), 0.5% chromia (Cr_2O_3) and 1.0% alumina (Al_2O_3). It has an apparent porosity of 22%, and is used as refractory material for lining steelmaking furnaces, and in glass manufacture.

peridot. A clear, yellow-green variety of the mineral *olivine* from St. John's Island (Red Sea), used chiefly as a gemstone.

Perilon. Trademark of SASA (Turkey) for polyester staple fibers and filament yarns.

periodic acid. White, water-soluble crystals (98+% pure) with a melting point of 122°C (252°F). It loses $2H_2O$ at 100°C (212°F) and decomposes at 130°C (266°F), and is used as a chemical reagent, as an oxidizer, in photographic paper, enhancing the wet strength of paper, and in electrophoresis. Formula: H_5IO_6.

periodic copolymer. A *copolymer* in which the monomeric repeating units (A, B and C) are arranged as ordered sequences. It is usually referred to as poly(A-*per*-B-*per*-C).

PerioGlass. Trademark of USBiomaterials Corporation (USA) for a surface-active, osteostimulative, particulate *Bioglass* bone-grafting material used in dentistry to repair and regenerate alveolar bones and tissue lost to periodontal disease, and repair extraction sockets.

Peripor. Trademark of BASF AG (Germany) for a flame-retardant expandable polystyrene with excellent resistance to pressure and moisture used in building construction, e.g., for perimeter insulation, flat-roof insulation, and frost blankets in road and railroad construction.

perite. A sulfur-yellow mineral composed of lead bismuth oxide chloride, $PbBiO_2Cl$. Crystal system, orthorhombic. Density, 8.16 g/cm^3; refractive index, above 2.4. Occurrence: Sweden.

Perking brass. British brass containing 76-80% copper, 20-24%

and a trace of zinc used for reflectors and ornaments.

Perlapoint. Trade name for rayon fibers and yarns used for textile fabrics.

Perlargon. Trade name for nylon fibers and yarns used for textile fabrics.

Perlatex. Trademark of Sico Inc. (Canada) for an interior latex semi-gloss enamel paint.

Perlglo. Trade name for rayon fibers and yarns with pearly luster used for textile fabrics.

Perlinato. Trade name of Fabbrica Pisana SpA (Italy) for a patterned glass.

Perlit. German trade name for a tough, wear-resistant nickel cast iron containing 1.7-3.5% carbon, 0.5-1.5% silicon, 0.2-0.6% phosphorus, 0.6-1% manganese, varying small amounts of nickel and chromium, and the balance iron. Used for brake drums, automotive engine blocks, frames and housings.

perlite. A *volcanic glass* composed of small spheroids usually less than 10 mm (0.4 in.) in diameter. It is commonly grayish or green, has a pearly luster, and contains 65-75% silica (SiO_2), 10-20% alumina (Al_2O_3), 2-5% water, and small percentages of lime (CaO), potash (K_2O) and soda (Na_2O). When heated to a suitable temperature, it will expand to more than ten times its original size and form a light, fluffy material with a cellular structure. Occurrence: USA (California, Colorado, Nevada, New Mexico, Oregon). Used as an acoustic and thermal insulating material, for lightweight wallboard, preformed perlite insulating board, and as a lightweight aggregate in concrete, mortar and plaster. See also obsidian.

Perlitguss. German trade name for a tough, high-grade cast iron containing about 2.2-3.5% carbon and 1.5-3.5% silicon and having a microstructure comprising a purely pearlitic matrix with interspersed graphite lamellae. It has excellent wear resistance, and is used for automotive engine blocks, brake drums, cylinder housings, gears, bearings, pistons, crankshafts, chains and chain links, and clutch parts.

perloffite. A dark brown to greenish brown or black mineral of the bjarebyite group composed of barium manganese iron phosphate hydroxide, $BaMn_2Fe_2(PO_4)_3(OH)_3$. Crystal system, monoclinic. Density, 3.32 g/cm³; refractive index, 1.803. Occurrence: USA (South Dakota).

Perlofil. Trade name for nylon fibers and yarns used for textile fabrics including clothing.

Perlon. Trade name of Bayer AG (Germany) for a polyamide 6 (nylon 6) fiber made from caprolactam and having high strength and toughness, good elasticity, low permanent elongation, high gloss, low water absorption, good resistance to dilute acids, aromatic hydrocarbon, greases, oils, halogens and ketones, fair resistance to alcohols and alkalies, poor resistance to concentrated acids, and good self-extinguishing properties. Used as a textile fiber, and for fishing lines, ships' ropes, and wires for paper machines. *Perlon-L* refers to a series of polyamide (nylon 6) fibers for clothing and industrial fabrics, and *Perlon-U* to spandex (segmented polyurethane) fibers used for the manufacture of textile fabrics with excellent stretch and recovery properties.

Perlon-Draht. Trademark of Bayer Faser GmbH (Germany) for nylon 6 monofilaments.

Perlon-Hoechst. Trade name of Hoechst AG (Germany) for nylon 6 fibers used for textile fabrics including clothing.

Perltex. Trade name of W.R. Grace & Company (USA) for textured asbestos- or perlite-based plasters and related building products.

Perlux. Trademark of Plastica Moderna SA de CV (Mexico) for nylon 6 monofilaments.

Perm. Trade name for a self-cure acrylic resin used for dentures.

Perma-Blast. Trade name of MDC Industries, Inc. (USA) for steel-based abrasive blasting media used in metal finishing.

Permabrasive. Trade name of MDC Industries Inc. (USA) for malleable iron shot used in abrasive blast cleaning of gray iron parts.

Permabright. Trademark for luminous pigments and paints.

Perma Brush Crete. Trade name of CGM Inc. (USA) for brush-on cementitious waterproof coatings.

Perma-Cap. Trade name of Owens-Corning-Ford Company (USA) for a relatively thick mat of fine glass fibers, laid down in an irregular pattern, reinforced with glass yarns, and then bonded together into a tough sheet using a resinous binder.

PermaCem. Trademark of DMG Hamburg (Germany) for a self- and light-curing, biocompatible dental cement used for permanent bridge and crown restorations.

Perma-Chrome. Trademark of Jonergin Company Inc. (USA) for metallized paper and paper products.

Permaclad. (1) Trade name of Permali Limited (UK) for copper-clad epoxy glass laminates used in printed circuit manufacture.

(2) Trade name of Sherwin-Williams Company (USA) for a polyester coating for aluminum extrusions.

(3) Trade name RDM International Corporation (USA) for a carbon steel that has a layer of stainless steel bonded to one side.

PermaClear. Trademark of Wellman Inc. (USA) for a series of clear polyethylene terephthalate (PET) resins supplied in food, UV-stable and other grades. Used for the manufacture blow-molded beverage bottles and in other packaging applications.

Permacoat. Trade name of Iroquois Chemicals Corporation (Canada) for paper coatings based on thermosetting resins.

Perma-crete. Trademark of Porter Paints (USA) for texture coatings.

Permadyne. Trade name of 3M Espe Dental (USA) for a polyether elastomer used for dental impressions.

Permaflex. (1) Trade name of Arndt-Palmer Canada Inc. (USA) for flexible autobody filler.

(2) Trade name for a heat-cure silicone used in restorative dentistry as a liner.

PermaFlow. Trademark of Ultradent Products, Inc. (USA) for a flowable, radiopaque, light-cure methacrylate-based composite with an average particle size of 0.07 μm (2.8 μin.). *PermaFlow DC* is a 70% filled, fluoride-releasing, dual-cured methacrylate-based composite resin that provides high compressive strength. *PermaFlow* resins are used for luting and cementing dental work.

Permafly. Trade name of Creusot-Loire (France) for a soft magnetic alloy of 80% nickel and 20% iron. It has high initial permeability, low saturation induction, and is used for magnetic and electrical equipment.

Perma-Foam. Trade name of Insulfoam (USA) for expanded polystyrene insulation products.

Permafoam. Trademark of Hudson Foam Plastics Corporation (USA) for a lightweight, foamed polyester resin with good strength, good oxidation resistance, high flame retardancy and good resistance to many oils and solvents. Used for thermal insulation, upholstery, carpet padding, and cushioning applications.

Permafresh. (1) Trademark of Sun Chemical Corporation (USA) for an extensive range of synthetic resins including thermosetting urea-formaldehydes, modified carbamide resins and melamine or cellulose derivatives. Used in the textile industries to impart moisture- and wrinkle resistance, reduced shrinkage, enhanced surface finish or other properties, and/or add body to fabrics.

(2) Trademark of Drake Extrusions (UK) for an extruded polypropylene fiber that has a chemical compound (Ultrafresh) incorporated providing protection against bacteria, fungi, molds and dust mites. Used in especially for floor coverings and soft furnishings used in hospitals, sports facilities, etc.

Permag. Trademark of Alloy Metal Products Inc. (USA) for a permanent magnet alloy containing 60% copper, 20% nickel, and 20% iron. Its magnetic orientation is developed by rolling or other mechanical working. *Permag* has a density of 8.6 g/cm³ (0.31 lb/in.³), a maximum service temperature of 350°C (660°F), a Curie temperature of 410°C (770°F), high coercive force and remanent magnetization, and a high magnetic energy product. Used for magnetic and electrical equipment.

Permagel. Trade name for a fine, yellowish white powder composed of a hydrated double silicate of magnesium and aluminum derived from the mineral *attapulgite*, and used as a flatting agent and extender in paints, as a suspension agent, as an emulsifier.

Permaglas Epoxy. Trade name of BTR Permali RP (UK) for nonwoven glass-fabric or glass-prepreg epoxy laminates.

Permaglas Polyester. Trade name of BTR Permali RP (UK) for chopped-glass and woven-glass roving epoxy laminates.

Permaglass. Trade name of Guardian Industries Corporation (USA) for a toughened and laminated glass.

Permagrid. Trade name of Driver-Harris Company (USA) for a nickel alloy containing 4% titanium and 0.3% magnesium. Used for grid wire.

Permagum. Trade name for a silicone elastomer used in restorative dentistry as a liner.

Permal. Trade name of Perry Barr Metal Company Limited (UK) for an aluminum alloy containing 5% silicon. It has good fluidity, and is used for die castings.

Permalife. Trade name of Standard for Electric Company Inc. (USA) for glass used for battery separators, retainer and surfacing mats, air filter media, and milled fibers.

Permalight. Trademark of Permalight GmbH (Germany) for highly reflecting polymer film.

Permalloy. Trademark of B & D Industrial & Mining Services, Inc. (USA) for a series of soft magnetic alloys containing typically about 36-81% nickel, 19-64% iron, and traces of carbon, cobalt, chromium and vanadium. A special grade with up to 5% molybdenum is sold under the trade name *Moly Permalloy*. *Permalloy* alloys have very high magnetic permeability, saturation flux density, low coercive force, low hysteresis loss, high electrical resistivity, and are used for electrical parts subject to alternating magnetic fields, magnetic amplifiers and controllers, switches, and relay armatures.

Permalok. Trademark of Aluminum Company of Canada, Limited for aluminum extrusions.

Permalon. Trade name Visking Corporation (USA) for saran (polyvinylidene chloride) fibers with good chemical resistance, good weatherability, and good resistance to mildew and insects. Used for carpets and rugs, draperies, upholstery, clothing, industrial fabrics, etc.

Permalume. Trademark of Atotech USA Inc. for an electroplating bath containing nickel sulfate, nickel chloride, boric acid, and organic addition agents, and the resulting semi-bright nickel coating.

permanent dye. A lightfast pigment or coloring substance which when applied to or incorporated into a material (e.g., a textile fabric) brings about a permanent change in the original color of that material.

permanent green. See Guignet's green.

Permanente. Trade name of Permanente Magnesium Inc. (USA) for a heat-treatable magnesium casting alloy containing 6% aluminum, 2.9% zinc, and 0.25% manganese.

Permanente 165 AF. Trade name of National Refractories & Minerals Corporation (USA) for refractory ram-cast mixes.

permanent mauvre. See manganese violet.

permanent magnet materials. A group of magnetic materials that due to their domain alignments retain a high degree of magnetization virtually unchanged for a long period of time even after removal of the external magnetic field. They have constant magnetic fields, high coercive forces, large hysteresis loop areas and large energy products. Examples of such materials include certain steels (e.g., martensitic carbon steels and high-cobalt steels), Alnico, Cunife, Cunico, Vicalloy and Remalloy alloys, ferrites, neodymium-iron-boron alloys, cobalt/rare earth alloys, chromium cobalt-iron alloys, manganese-bismuth alloys, and platinum-cobalt, platinum-iron and platinum-nickel alloys. They are used for hard magnetic applications. See also hard magnetic materials.

permanent-mold castings. Castings with good surface finish and high dimensional accuracy produced by pouring a liquid metal into a permanent (reusable) metal, ceramic or graphite mold by gravity or low pressure, and allowing the poured metal to solidify. Alloys of aluminum, bronze, lead, zinc and magnesium as well as cast iron are commonly supplied in the form of permanent-mold castings.

permanent-press fabrics. (1) Textile fabrics that have received a finishing treatment to impart crease resistance and permanent hot creasing or pleating. The finishing treatment may consist of the application of a synthetic resin and curing before or after the fabrics have been fabricated into garments or, if the fabrics are made of heat-settable fibers, high-temperature pressing. Also known as *durable-press fabrics*. See also permanent-press resins.

(2) Textile fabrics, such as certain blends of polyester with cotton and rayon, that have inherent crease resistance and do not required a finishing treatment. Also known as *durable-press fabrics*.

permanent-press resins. Synthetic resins, usually formaldehyde or maleic anhydride based thermosets, that are applied to textile fabrics or fibers to impart crease resistance and permanent hot creasing or pleating. They may be cured before or after the textiles have been fabricated into garments. Also known as *durable-press resins*. See also permanent-press fabrics.

permanent violet. See manganese violet.

permanent white. See artificial barite.

Permanickel. Trademark of Inco Alloys International, Inc. (USA) for age-hardening alloys containing 98.5% nickel, 0.40-0.50% titanium, 0.35% magnesium, 0.10-0.35% iron, 0.25% manganese, 0.10-0.25% carbon, and up to 0.18% silicon, 0.05% sulfur and 0.13% copper, respectively. They have good mechanical properties, good corrosion resistance, good thermal and electrical properties, and are used for high-temperature springs, diaphragms and springs, wire, thermostat contacts, magnetostriction devices, solid-state capacitors, and fuel cells.

Permanit. Trade name of Vereinigte Edelstahlwerke (Austria) for a series of cast permanent magnet materials including several iron-nickel-aluminum, iron-nickel-cobalt, iron-cobalt-chromium, iron-cobalt-nickel-aluminum and platinum-cobalt alloys.

Permanite. (1) Trademark of Foote Mineral Company (USA) for a purified magnetic iron ore.

(2) Trade name for a magnetically hard alloy containing iron, chromium, cobalt and tungsten. It has high coercive force, and is used for permanent magnets.

(3) Trademark of Koch Engineering Company, Inc. (USA) for acid- and alkali-resisting furfural alcohol resin coatings used for tanks, acid vessels and other vessels and containers.

(4) Trademark of TBA Industrial Products Limited (UK) for asbestos, balata, flax, gutta percha, hemp, rubber and metal jointing and packing materials.

(5) Trade name of Parker Pen Company (USA) for cellulose nitrate plastics.

Permant. Trade name of Firth Brown Limited (UK) for a low-expansion alloy containing 36% nickel, 0.2% manganese, 0.05% silicon, 0.08% carbon, and the balance iron. It has a very low coefficient of thermal expansion, and is used for instruments, clocks, watches and scales.

Perma-Pane. Trade name of Whizzer Industries Inc. (USA) for a hermetically sealed glass unit with aluminum channel spacer.

Permaplate. Trade name of Bradford Glass Company Limited (UK) for a toughened polished plate glass.

Perma-Ply-R. Trademark of Fiberglas Canada Inc. for fiberglass used for thermal insulation applications.

Permapol. Trademark of Products Research & Chemical Corporation (USA) for a series of polymeric materials including polythioether polymers, polyurethanes and flexibilized and elastomeric adhesives, silicones and syntactic foams.

PermaQuick. Trademark of Ultradent Products, Inc. (USA) for a 40-45% filled, fluorine-releasing, radiopaque, light-cure acrylate-based resin used for dental enamel bonding applications.

Permaquik. Trademark of Permaquik (Canada) Limited for asphalt membranes.

PermaRez. Trademark of PolySpec (US) for an extensive series of filled or unfilled reinforced linings based on polymers, such as vinyl ester, and epoxy/novolac systems.

Perma-Rez. Trademark of C-K Composites, Inc. (USA) for cast epoxy products.

Permas. Trade name of Fisher Scientific Company (USA) for an austenitic stainless steel containing up to 0.25% carbon, 2% manganese, 1.5% silicon, 24-26% chromium, 19-22% nickel, traces of phosphorus and sulfur, and the balance iron.

PermaSeal. Trademark of Ultradent Products, Inc. (USA) for a light-cure, unfilled methacrylate-based resin used in dentistry as a composite sealer and bonding agent.

Permaseal. (1) Trade name of Chemtron Limited (Canada) for concrete floor sealants.

(2) Trade name of Kelley Technical Coatings, Inc. (USA) for nitrile rubber based sealing compounds.

Permaslik. Trade name of E/M Engineered Coating Solutions (USA) for solid-film lubricants.

PermaStat. Trade name of RTP Corporation (USA) for antistatic, carbon black-free acrylonitrile-butadiene styrene materials for application to the surfaces of molded articles, such as household products, calculators, chip carriers and matrix trays.

Permat. Trade name for an alloy containing 45% copper, 30% cobalt and 25% nickel, used for magnetic and electrical equipment.

Permatex. Trademark of Loctite Corporation (USA) for anaerobic flange sealants.

Permatone. Trademark of Perma-Flex Inc. (Canada) for urethane-based products.

Permatreat. Trade name of BetzDearborn, Division of Hercules Inc. (USA) for conversion coatings.

Permatuff. Trade name of Bradford Glass Company Limited (UK) for a toughened, polished black glass and toughened sheet and cast glass.

Perm-Au-Tone. Trade name of Advanced Chemical Company (USA) for gold electroplates and plating processes for jewelry.

Permax. Trademark of Vacuumschmelze GmbH (Germany) for crystalline soft magnetic alloys containing 54-68% nickel and 32-46% iron. They have high initial and maximum permeability, and relatively high saturation flux density. Used for circuit breakers, instrument transformers, cores, chokes, magnetic amplifiers, leakage-current protection switches and power electronics.

Permcol. Trade name of British Hard Rubber Company (UK) for hard rubber (ebonite).

Perm-Cote. Trademark of Novamax Technologies Holdings, Inc. (USA) for an anticorrosive and rust-inhibiting phosphate coating used for rustproofing ferrous and nonmetal metals.

permeability alloys. A group of soft magnetic nickel-iron alloys containing about 50-80% nickel. The high-nickel alloys have low saturation induction, but high initial permeability and low magnetocrystalline and magnetostrictive anisotropy, while the low-nickel alloys have higher saturation induction, but lower initial permeability. Molybdenum additions (typically 4%) increase the resistivity and reduce both the magnetocrystalline and magnetostrictive anisotropy. Often sold under trade names or trademarks, such as *Alfenol, Duraperm, Hipernik, Permenorm, Permalloy, Perminvar, Supermalloy, Supermendur* or *Vicalloy*, they are used for transformers, magnetic cores and amplifiers, low-frequency amplifiers, chokes, relays, rectifiers, measurement instruments, and magnetic shielding systems.

Permelastic. Trade name of Kerr Dental (USA) for a resilient polysulfide-based dental impression material.

Permendur. Trade name of Telcon Metals Limited (UK) for a series of soft magnetic alloys composed of 48.8-76% iron, 24-50% cobalt, 0-0.4% manganese and 0-2% vanadium, and usually supplied in the form of sheets and strips. The commercially most important alloy in this series is 49Co-49Fe-2V with a density of 8.15 g/cm³ (0.294 lb/in.³), high saturation induction, a Curie temperature of 940°C (1724°F), high permeability and good mechanical properties. *Permendur* alloys are used for diaphragms, stators, magnetostriction devices, and magnetic and electrical equipment.

Permenorm. Trademark of Vacuumschmelze GmbH (Germany) for a series of crystalline soft-magnetic alloys containing about 35-50% nickel and 50-65% iron, and made by magnetic annealing and extreme cold reduction (rolling) in one direction. They have high saturation flux densities, relatively low permeabilities, low static coercivities, square hysteresis loops, and are used for transformers, magnetic cores and amplifiers, low-frequency amplifiers, chokes, relays, mechanical rectifiers, measurement instruments, magnetic shielding, and magnetic systems.

Permet. (1) Trade name of Colt Industries (UK) for a magnetically soft high-purity iron with high saturation induction, low remanent magnetization and coercive force, used for magnetic equipment.

(2) Trade name of Crucible Materials Corporation (USA) for a magnetically soft material composed of 70% iron and 30% cobalt. It has very high saturation induction, a high magnetic energy product, low coercive force, and is used for magnetic strips.

(3) Trade name for a magnet alloy containing 50% copper, 29% cobalt and 21% nickel.

permingeatite. A brown-rose mineral of the chalcopyrite group composed of copper antimony selenide, Cu_3SbSe_4. Crystal system, tetragonal. Density, 5.86 g/cm^3. Occurrence: Czech Republic.

Perminvar. Trade name of Western Electric Company (USA) for a series of soft magnetic alloys composed of 43-70% nickel, 22-34% iron, 7-25% cobalt, 0-0.6% manganese. They have high, constant magnetic permeabilities for small magnetic field strengths, and are used for electrical communication equipment, magnetic circuits, magnetic coils, transformers, and magnets.

Permite. (1) Trade name of Aluminum Industries Inc. (USA) for an extensive series of cast aluminum-copper, aluminum-magnesium, aluminum-silicon, aluminum-silicon-copper and aluminum-tin alloys.

(2) Trade name of Southern Dental Industries Limited (Australia) for a biocompatible, corrosion-resistant admix of spherical and lathe-cut alloy particles. It contains 56% silver, 27.9% tin, 15.4% copper, 0.5% indium and 0.2% zinc, has an optimum alloy-to-mercury ratio of 1:0.92 (47.9% mercury), does not contain the corrosive tin-mercury phase (Gamma 2), and has controlled dimensional expansion and low static creep. Used in restorative dentistry for retrograde amalgam fillings.

Permlastic. Trademark of Chemical Bank Corporation (USA) for a polysulfide elastomer used for dental impressions.

Permo. Trade name of Permo Inc. (USA) for a corrosion- and wear-resistant alloy of osmium, rhodium and rubidium, used for instrument bearings and fountain-pen tips.

Permold. Trade name of Permold Inc. (USA) for a series of strong sand- and permanent mold-cast aluminum-silicon and aluminum-silicon-copper alloys used for automotive components, home appliance parts, cooking utensils, aircraft parts, machinery parts, typewriter parts, architectural applications, etc.

Permutit. Trademark of The Permutit Company (USA) for a synthetic zeolite (sodium aluminum silicate) used as an ion exchanger in the softening of water to replace calcium, manganese and iron by sodium. It is a granular powder made from sodium carbonate, aluminum silicate and silicon dioxide by a special melting process.

Pernambuco wood. The bright-red wood of the tree *Caesalpinia brasiliensis* growing in Brazil and the tropical Americas. It takes a fine, brilliant polish, and is used for cabinetwork, fine furniture, violins, and as source of red or purple dye (brazilwood extract). See also brazilwood.

Pernervo Metal. Trademark of Rippenstreckmetall-Gesellschaft mbH (Germany) for galvanized or black-coated perforated expanded metal with ribs, used in building interiors.

Pernifer. Trademark of ThyssenKrupp VDM GmbH (Germany) for a series of alloys containing about 64% iron and 36% nickel. *Pernifer 2918* is a special grade containing 53% iron, 29% nickel and 17.5% cobalt. The iron-nickel-grades have very low thermal expansion in the temperature range between 20 and 200°C (68 and 390°F), and are used for magnetic and electrical equipment.

Pernima 72. Trademark of ThyssenKrupp VDM GmbH (Germany)

for a controlled-expansion alloy containing 72% manganese, 18% copper and 10% nickel. Used for electrical resistances, heating elements, and glass-sealing applications

peroba de campos. The strong, heavy wood from the tree *Paratecoma peroba*. The heartwood is brown and the sapwood yellowish gray. *Peroba de campos* has a fine texture, and machines very well. Source: Eastern Brazil. Used for lumber, high-quality millwork, furniture and veneer.

perosmic acid anhydride. See osmium tetroxide.

perosmic oxide. See osmium tetroxide.

perovskite. A yellow, yellowish-white, reddish brown or grayish-black mineral with a submetallic to subadamantine luster. It is composed of calcium titanate, $CaTiO_3$, and can also be made synthetically, e.g., by mixing equimolar amounts of calcium oxide (CaO) and titanium dioxide (TiO_2), pelletizing, and subsequent heating to 1000-1200°C (1830-2190°F) for a prescribed period of time in an oxidizing atmosphere. Crystal system, cubic. Density, 4.0 g/cm^3; melting point, 1915°C (3479°F); hardness, 5.5 Mohs; refractive index, 2.38. Occurrence: Russia, USA. Used in stannate, titanate and zirconate dielectric bodies. The synthetic material is also used in emission-control devices (catalyzers).

perovskites. A group of ceramic oxides corresponding to the *perovskite* crystal structure, ABO_3, where A denotes a divalent metal, such as barium, calcium, magnesium or strontium, B a tetravalent metal, such as tin, titanium or zirconium, and O is oxygen. Examples include barium titanate ($BaTiO_3$), strontium zirconate ($SrZrO_3$) and strontium stannate ($SrSnO_3$). Many *perovskites* are useful as magnetic and electronic materials. Also known as *perovskite-type oxides*.

Perplex. Trade name of SWB Stahlformguss Gesellschaft mbH (Germany) for a water-hardening, wear-resistant steel containing 1.05% carbon, 1% chromium, 1.15% tungsten, 0.9% manganese, and the balance iron. Used for cutters, liners, sleeves and bearings.

perrhenic acid. Strong, stable, water-soluble monobasic acid (99.9+% pure) supplied as a 65-80 wt% solution in water. It has a density of 2.16 g/cm^3, and is used in the preparation of highly conductive and superconductive synthetic metals. Formula: $HReO_4$.

perrierite. A brownish black mineral composed of cerium titanium silicate, $Ce_2Ti_2Si_2O_{11}$. Crystal system, monoclinic. Density, 4.30 g/cm^3; refractive index, 2.01. Occurrence: USA (Virginia).

Persian red. A fine paint pigment composed of red iron oxide (Fe_2O_3) and formerly made from *hematite* imported from Asia.

persimmon. The strong, hard, heavy, close-grained wood from the common persimmon tree *(Diospyros virginiana)* belonging to the ebony family. The narrow heartwood is dark brown to blackish, and the wide sapwood is light brown. *Persimmon* takes fine polish and has an average weight of 785 kg/m^3 (49 lb/ft^3). Source: Southeastern and central United States from Rhode Island to Florida and eastward to Texas. Used for shoe lasts, weaving shuttles, golf-stick heads, and tools.

Perspex. Trademark of ICI Acrylics (UK) for thermoplastic polymethyl methacrylate (PMMA) cast into sheets, rods or tubes, and available in colorless transparent (92% transmission) and colored grades. It has a density of 1.2 g/cm^3 (0.04 $lb/in.^3$), good electrical resistance, low hardness and heat resistance, low thermal expansivity, good ultraviolet resistance, low flame resistance, a maximum service temperature of 60-93°C (140-200°F), good machinability and moldability, low water absorption, good resistance to dilute acids and alkalies, fair resistance to concentrated acids, and poor resistance to aromatic hydrocarbons, greases, oils, halogens and ketones. Used for lenses, signs, lighting fixtures, nameplates, aircraft domes, decorations, display items, dials, glazing and bottles. *Perspex CQ* is a polymethyl methacrylate (PMMA) thermoplastics supplied as monomer cast sheeting for glazing, covers, and guards.

Perstorp. Trade name of Perstorp AB (Sweden) for a series of polymeric materials including *Perstorp Melamine* cellulose-filled melamine formaldehydes, *Perstorp Phenolic* phenolic resins supplied in wood- or natural fiber-filled general-purpose, mineral-filled high-heat, mica-filled electrical, glass-reinforced high-impact, cotton-filled medium-shock, cellulose-filled shock-resistant and other grades, *Perstorp Polyester DMC* bulk (or dough) molding compounds supplied in unmodified, fire-retardant, high-heat and low-profile grades, and *Perstorp Urea* cellulose-filled urea-formaldehyde resins.

Perstrip. Trade name of Composition Materials Company Inc. (USA) for plastic media used for dry stripping applications.

Pertac. Trademark of Espe Dental-Medizin GmbH & Co. KG (Germany) for dental cements, lacquers, composites and bonding agents. *Pertac II* is a hybrid dental restorative composite, *Pertac Hybrid* refers to a light-cure all-purpose composite resin for dental restorative applications including inlays, and *Pertac Universal Bond* is a dental adhesive for dentin bonding applications.

Pertinax. Trademark of Felten GmbH (Germany) for a laminate composed of paper impregnated with phenol-formaldehyde resin.

Perunal. Trade name of Alusuisse (Switzerland) for a series of wrought aluminum alloys containing 5-6.2% zinc, 2-2.9% magnesium, 1.2-2% copper, 0-0.5% iron, 0-0.3% magnesium, 0-0.3% chromium and 0-0.05% titanium. Used for light-alloy parts, vehicle parts, and machinery.

Pervenac. Trade name of Peterborough Paper Converters Inc. (Canada) for heat sealing paper.

perylene. A crystalline aromatic compound (99+% pure) that has semiconducting properties and decomposes at 278-280°C (532-536°F). Used as a laser crystal and as an organic semiconductor for electronic and microelectronic applications. Formula: $C_{20}H_{12}OH$. Also known as *dibenz[de,kl]anthracene*.

Petaflex. Trademark of National Starch & Chemical Investment Corporation (USA) for a series of two-component polyester resins supplied as solids, in solution, and in several urethane-modified polyester resin grades. The solid grades have excellent heat resistance to 200°C (390°F), and are used for adhesives, coatings, flexible packaging and industrial laminates, flexible circuits, microwavable, outer-ply retorts, sterilizable medical packaging, and cable wrap.

petalite. A colorless, white, gray or pinkish mineral with a white streak and a vitreous luster. It is composed of lithium aluminum silicate, α-$LiAlSi_4O_{10}$, and usually contains about 4.5-11.8% Li_2O. Crystal system, monoclinic. Density, 2.35-2.47 g/cm³; melting point, 1400°C (2552°F); hardness, 6.0-6.5 Mohs; refractive index, 1.510; low thermal expansion coefficient. Occurrence: Africa, Sweden, USA (New England states). Used as a source of lithia in porcelain enamels, glazes, glass and specialty bodies, as a flux in ceramics to promote fusion, reduce thermal expansion and improve thermal-shock resistance, as a filler, and as a source of lithium salts.

petarasite. A light green to yellow mineral of the combeite group composed of sodium zirconium chloride silicate hydroxide dihydrate, $Na_5Zr_2Si_6O_{18}(Cl,OH)\cdot2H_2O$. Crystal system, orthorhom-

bic. Density, 2.88 g/cm³; refractive index, 1.598. Occurrence: Canada (Quebec).

petersite. A bright yellowish green mineral of the mixite group composed of calcium copper rare-earth yttrium phosphate hydroxide trihydrate, $(Y,Ln,Ca)Cu_6(PO_4)_3(OH)_6\cdot3H_2O$. Crystal system, hexagonal. Density, 3.41 g/cm³; refractive index, 1.666. Occurrence: USA (New Jersey).

Pethapane. Trade name for chlorinated polyether products.

Petite Marquerite. Trade name of Boussois Souchon Neuvesel SA (France) for a patterned glass featuring flowers.

Petlon. Trade name of Albis North America (USA) for a series of engineering thermoplastics based on modified semicrystalline polyethylene terephthalate (PET). Commercially available in regular, flame-retarding, and mineral- and/or glass-reinforced grades, they have good dimensional stability, high stiffness, easy flowability in thin sections, good chemical resistance, outstanding electrical properties, good thermal resistance, low moisture absorption, a service temperature range of -40 to +224°C (-40 to +435°F) and high heat-deflection temperature of 224°C (435°F). Used for automotive components, machine parts, electronic components, etc.

Peton. Trade name of United Insulator Company (USA) for phenolic plastics.

petong. A Chinese copper alloy with high hardness.

PET-PC alloys. Polymer blends of crystalline polyethylene terephthalate (PET) and amorphous polycarbonate (PC) with excellent mechanical compatibility. They combine the good chemical resistance of PET with the high toughness, good creep resistance and excellent dimensional stability of PC. Used for mechanical components, lawnmower parts, helmets, etc.

Petra. Trade name of Allied Signal Corporation (USA) for a series of engineering thermoplastics based on polyethylene terephthalate (PET) and supplied in a wide range of grades including amorphous, crystalline, glass-fiber-reinforced, fire-retardant, mineral-filled and UV-stabilized. Used for electrical and office equipment, and electronic components.

Petralit. Trade name of Associated Dental Products Limited, Kemdent Works (UK) for a zinc silicophosphate dental cement.

Petrarch. (1) Trademark of United Chemical Technologies, Inc. (USA) for silicone resins and elastomers used as adhesives, sealants and encapsulants, for gaskets, protective coatings for electronics and medical devices, fiberoptics, etc.

(2) Trademark of Petrarch Claddings Limited (UK) for stone composition cladding sheets.

Petrex. Trademark of Hercules Inc. (USA) for a series of alkyd resins used for adhesives, coatings, varnishes, lacquers an inks, and in the manufacture of molded articles.

Petrofibe. Trade name of International Waxes Limited (Canada) for industrial-grade petrolatum.

petrolatum. A yellowish to light amber semisolid mixture of mineral oil and certain hydrocarbon waxes, better known by the trade name *Vaseline*. It is used in ointments, cosmetic preparations, leather greases, lubricating greases, and rust-preventive compounds for iron and steel parts. A liquid form, known as *white mineral oil*, and used as a lubricant and dispersing agent is also available. Also known as *petroleum jelly*.

petroleum asphalt. See asphalt oil.

petroleum coke. A carbonaceous residue remaining after the complete distillation of oil. It is a fine granular form of *coke* containing about 90-95% fixed carbon, 5-10% volatile and combustible matter, up to 0.3% ash, and up to 1% sulfur. Used for refractory furnace linings, molded carbon products, battery carbons and carbon pencils (electric carbons), electrodes for the electrolytic recovery of aluminum, and in the electrothermal manufacture of calcium carbide, phosphorus, silicon carbide and calcium carbide.

petroleum coke-base refractories. Commercial carbon refractory composed principally of calcined petroleum coke. See also carbon refractories.

petroleum jelly. See petrolatum; Vaseline.

petroleum pitch. See asphalt oil.

petroleum waxes. A group of high-molecular-weight hydrocarbon waxes obtained from petroleum and including paraffin, microcrystalline and petrolatum waxes.

Petrolite. Trade name of St. Lawrence Resin Products Limited (Canada) for petroleum-based microcrystalline waxes.

Petromat. Trade name for polyolefin fibers and fiber products.

Petrothene. Trademark of Quantum Chemical Corporation (USA) for a series of polypropylenes, ethylene-vinyl acetates and high- and low-density polyethylenes used for molded or extruded articles, blown and cast films, and coatings for wire and cable. *Petrothene XL* is a crosslinked polyethylene grade.

petrovicite. A cream-colored mineral composed of copper lead mercury bismuth selenide, $Cu_3HgPbBiSe_5$. Crystal system, orthorhombic. Density, 7.76 g/cm³. Occurrence: Czech Republic.

petscheckite. A black mineral of the amorphous group composed of iron uranium niobium tantalum oxide, $UFe(Nb,Ta)_2O_8$. Crystal system, hexagonal. Density, 5.00 g/cm³. Occurrence: Madagascar.

petzite. A steel-gray mineral composed of gold silver telluride, Ag_3AuTe_2. Crystal system, cubic. Density, 8.70 g/cm³. Occurrence: Rumania.

Pewlon. Trademark of Asahi Chemical Industry Company (Japan) for polyacrylonitrile fibers and filament yarns.

pewter. (1) A term formerly used for a dull, soft, ductile alloy of 80-90% tin and 10-20% lead with good workability used for ornamental parts, domestic utensils and dishes.

(2) A term now used for a silvery-white, corrosion-resistant alloy of 75-92% tin, 6-7.5% antimony and 1.5-3.5% copper. It is a *Britannia metal*-type alloy used for eating and cooking utensils, cast decorative objects and ornaments, and printing plates.

Pex. Trade name for polyethylene fibers and products.

Pfinodal. Trademark of Ametek, Inc. (USA) for a roll-compacted sintered spinodal copper alloy made from a powder composed of 77% copper, 15% nickel and 8% tin. After sintering, it is subjected to a thermomechanical treatment resulting in spinodal hardening and a high-tensile-strength microstructure. Used for electrical and electronic components.

Pfizer. Trade name of Pfizer Inc. (USA) for high-purity cobalt and nickel, nickel-iron controlled expansion alloys, corrosion-resistant nickel-copper alloys, iron-nickel-cobalt glass-sealing alloys, high-cobalt alloys, and several clad metals, such as nickel-clad carbon steel, stainless steel-clad carbon steel, stainless steel clad aluminum, and titanium-clad aluminum.

PGA. Trademark of Lukens Corporation (USA) for a *polyglycolide*-based biopolymer used for tissue engineering applications, e.g., sutures.

Phanocel. Trade name of Sylvania Industrial Corporation (USA) for cellophane-type regenerated cellulose.

pharmacolite. A colorless mineral of the gypsum group composed of calcium hydrogen arsenate dihydrate, $CuHAsO_4 \cdot 2H_2O$. Crystal system, monoclinic. Density, 2.68 g/cm³; refractive index, 1.589. Occurrence: France.

pharmacosiderite. A green mineral composed of potassium iron arsenate hydroxide hydrate, $K_2Fe_4(AsO_4)_3(OH)_5 \cdot 6.3H_2O$, and sometimes small amounts of barium and aluminum. Crystal system, cubic. Density, 2.80 g/cm³; refractive index, 1.693. Occurrence: UK, Germany. Also known as *cube ore; iron sinter.*

Phasealloy. Trade name of Phase Change Corporation (USA) for regular- and fast-set dental amalgam alloy capsules.

Phase Alpha. Trademark of Composite Polymers Division of Ashland Chemical Inc. (USA) for a sheet molding compound used for automotive and industrial applications.

phase-change material. A material that changes from the solid to the liquid to the gaseous state and can store immense quantities of thermal energy.

Phase Epsilon. Trademark of Composite Polymers Division of Ashland Chemical Inc. (USA) for a sheet molding compound used for automotive and industrial applications.

Phenac. Trade name of American Cyanamid Company (USA) for *bakelite*-type phenolic resins and plastics.

phenacite. See phenakite.

phenakite. A colorless, yellowish or brownish mineral with white streak and vitreous luster composed of beryllium silicate, Be_2SiO_4. Crystal system, rhombohedral (hexagonal). Density, 2.98 g/cm³; refractive index, 1.653; hardness, 5-6.5 Mohs. Occurrence: Brazil, France, Norway, Mexico, USA (Colorado, Montana, New Hampshire). Lustrous species are used as gemstones. Also spelled *phenacite.*

Phenalin. Trade name for molded plastic articles made from phenolic resins.

phenanthrene. A colorless crystalline compound (90+%) obtained from coal tars by fractional distillation followed by recrystallization from alcohol. It is supplied in technical grades with a purity of 90%, high-purity grades (96+%), and zone-refined grades (99.5+%). Density, 1.063 g/cm³; melting point, 99-101°C (210-214°F); boiling point, 340°C (644°F). Used in biochemistry, biotechnology, organic synthesis of phenanthrene derivatives, manufacture of pharmaceuticals and explosives, as a dyestuff, in liquid chromatography, and in electronics and optoelectronics. Formula: $C_{14}H_{10}$.

1,10-phenanthroline. A heterotricyclic compound available as a hygroscopic, white, crystalline powder with a melting point of 114-117°C (237-242°F) used as a building block for metallomacrocycles, as a ligand in the spectrophotometric determination of metals, in the photocatalytic reduction of carbon dioxide, in the preparation of indicators, as a starting material for the formation of complex compound with ferrous ions, and as a drier in coatings. Formula: $C_{12}H_8N_2$.

phenate. See phenoxide.

Phenix. (1) Trade name of International Nickel Inc. (USA) for a resistance alloy composed of 75% iron and 25% nickel.

(2) Trade name of Vereinigte Edelstahlwerke (Austria) for a series of hardenable, corrosion-resistant high-chromium-type steels containing 0.15-0.2% carbon, 0-0.3% manganese, 0-0.4% silicon, 0-0.3% vanadium, 0-0.5% tungsten, 12-13% chromium, 1.15% molybdenum, 0.4-0.7% nickel, and the balance iron. Used for oil refinery and chemical plant equipment, table flatware, knives, cutlery, surgical instruments, bearings, hardware, and valves.

Phenocarb. Trademark of Lewcott Corporation (USA) for solid, modified and unmodified phenolic resins, and phenolic-based composites supplied on substrates, such as fabrics, fibers and particles of carbon, silica and glass. They have good high-temperature stability, and good char, erosion and corrosion resistance.

Phenofoam. Trademark of Lewcott Corporation (USA) for foamed, solid, fire-retardant phenolics used for thermal insulation applications.

Phenoid. Trade name of Mica Manufacturing Company (UK) for unfilled and mica-filled bakelite-type phenolic plastics.

phenolate. See phenoxide.

phenol-formaldehyde resins. A family of thermosetting resins made by a condensation reaction between phenol (C_6H_5OH) and formaldehyde (CH_2O). They are grayish to black when cured and cannot be effectively colored. *Phenol-formaldehydes* have good thermal resistance, good electrical and flame resistance, good rigidity and dimensional stability, good strength and durability, high hardness, low moisture absorption, good sound and noise-absorption properties, good processibility by molding, casting and laminating processes, good resistance to organic solvents, fair resistance to alkalies, poor resistance to strong acids and alkalies, moderate resistance to weathering and moisture, and excellent thermal stability to over 150°C (300°F). Often sold under trade names and trademarks, such as *Bakelite, Durez, Nipolon, Novatec, Resine, Resinox* or *Textolite,* they are used for handles, pulleys, wheels, electrical equipment, fuse blocks, plugs, insulators, electrical fixtures, television and radio cabinets, computer components, machine, instrument, motor and appliance housings, containers, buttons, toilet seats, laminates, shell mold binders, coil forms, piping, conduits, ducts, thermal and acoustic insulation, brake linings and clutch facings, machine parts, chemical equipment, chemical-resistant mortars, laminating and impregnating plywood and glass-fiber composites, ablative coatings for aerospace applications, paint and baked enamel coatings, and bonding powders. Abbreviation: PF. *Note:* The term "bakelites" is now also used generically as a synonym for phenol-formaldehyde resins and products.

phenol-furfural resins. A family of thermosetting resins that are somewhat similar to phenol-formaldehydes, but made by a condensation reaction of phenol (C_6H_5OH) and furfural ($C_5H_4O_2$). They are suitable for use in the manufacture of injection-molded parts. Abbreviation: PFF.

phenolic aldehyde. A compound, such as salicylaldehyde, produced by introducing an aldehyde group (–CHO) into the aromatic ring of a phenol.

phenolic coatings. Hard, air-drying coatings prepared with resin binders that incorporate phenol-formaldehyde resins in drying oils (linseed, tung, etc.). They have good abrasion resistance, good water resistance, good resistance to alkalies and solvents, and excellent resistance to corrosion including seawater corrosion. Used for can and tank linings, ship bottoms, marine structures, coatings and linings for chemical process equipment, coatings for equipment used in the pulp and paper industry, and as general maintenance paints for metallic substrates.

phenolic ester. An ester produced by reacting a carboxylic acid, acid chloride, or anhydride with a phenol or phenol derivative, usually with the elimination of water. The general formula is RCOOPh.

phenolic ether. An ether produced from a phenol by reacting in alkaline solution with alkyl halides, or methyl sulfate. The general formula is PhOR.

phenolic foam. A phenolic resin processed into a foam with numerous small cells by the action of a blowing or foaming agent, such as sodium bicarbonate.

phenolic foam boards. Rigid insulating boards consisting of closed- or open-cell phenol-formaldehyde foam with or without water-repellent exterior surface skins. They have a high thermal insulating value, but must be protected from exposure to sunlight and in some cases water. See also rigid board insulation.

phenolic laminates. Composite materials consisting of fabrics (e.g. canvas, linen or cotton), kraft paper, asbestos or glass fibers bonded with phenolic resin and compressed at high temperatures. They possess good impact and shock resistance, good electrical insulating properties, relatively high strength, high flexural strength and toughness, low thermal conductivity, good moisture resistance, and good resistance to oil, grease and various fluids. Used for gears, low- and medium-speed bearings, detachable facings for vise jaws, mallets, electrical components, etc. Also known as *laminated phenolics.*

phenolic microspheres. Minute spherical particles of phenolic plastics filled with nitrogen and used for making foams of thermosetting plastics, such as epoxies or unsaturated polyesters.

phenolic plastics. A family of thermosetting engineering plastics obtained by compounding phenolic resins with reinforcing agents (e.g., cellulose, minerals or glass fibers), fillers, colorants, lubricants, etc. They have excellent dimensional and thermal stability, good creep resistance, high hardness and compressive strength, good load-bearing properties at elevated temperatures, good electrical insulating properties, good moldability, good resistance to solvents, weak acids, oils, greases, petrochemicals and automotive fluids, poor resistance to strong acids and alkalies, and fair resistance to ultraviolet radiation. Used for automotive components, electrical connectors, circuit breakers, switchgear, brushholders, wiring devices, handles and knobs for household appliances, motor housings, pulleys, brake pistons, pump housings, and foundry cores and molds. Also known as *phenolics.* See also phenol-formaldehyde resins; phenol-furfural resins.

phenolic resins. A group of thermosetting resins that are condensation products of phenol or substituted phenols and an aldehyde, such as acetaldehyde, formaldehyde or furfural. They are soluble in oils (including drying oils) and may be compounded with a large number of reinforcements (e.g., cellulose, minerals or glass fibers), resins, fillers, etc. Used in the manufacture of molded products, as binders in foundry cores and molds, in plywood and particleboard manufacture, in coated abrasives and grinding wheels, and in adhesives, varnishes, paints, coatings, rubber tackifiers, concrete, etc. Also known as *phenolics.*

phenolic-resin primer-sealer. A phenolic resin-based wood finish that is well suited for softwoods, especially fir, and penetrates deep into the wood pores, and when dried and balances the density of soft and hard grains.

phenolics. See phenolic plastics; phenolic resins.

phenolic urethane no-bake binders. See phenolic urethane resin binders (i).

phenolic urethane resin binders. Liquid resin systems used as binders to hold sand grains together in foundry molds or cores. Included are: (i) *No-bake binders* composed of a liquid phenol-formaldehyde resin, a liquid polyurethane resin (e.g., methylene diphenyl isocyanate), and a special liquid catalyst. Curing is achieved at room temperature (i.e., without baking) upon addition of the catalyst to the sand-resin mixture; and (ii) *Two-part cold-box binders* composed of a mixture of liquid phenol-formaldehyde resin and a liquid polyurethane resin. They cure upon blowing a vaporized amine catalyst through the sand-resin mixture in a vented core box.

Phenolite. Trademark of NVF Company (USA) for a plastic laminate composed of sheets of paper impregnated with phenolic resin, and used for decorative applications.

phenol-phosphor resins. Engineering thermoplastics based on phenolic resins modified with hexachlorocyclotriphosphazene (phosphonitrilic chloride), and used for elevated temperature applications.

Phenolyte. Trade name for laminated phenolic plastics and phenolic-impregnated paper.

Phenopreg. Trade name of Detroit Wax Paper Company (USA) for phenolic resins and laminates.

phenothiazine. Greenish gray or greenish yellow, water-insoluble flakes, granules or powder (98+% pure) with a melting point of 182-187°C (359-368°F), and a boiling point of 371°C (699°F) (sublimes at 130°C/266°F/1 mm). Used in the manufacture of dyes, in the preparation of charge-transfer semiconducting complexes, as an antioxidant, as a inhibitor in polymers, in the manufacture of pharmaceuticals, and in biochemistry. Formula: $C_{12}H_9NS$. Also known as *thiodiphenylamine*.

phenoxide. A compound formed by substitution of the hydrogen of a phenolic hydroxyl group by a metal, e.g., sodium phenolate. Also known as *phenate; phenolate*.

phenoxy resins. Clear, thermoplastic polyester resins of high molecular weight which are copolymers of *bisphenol A* and *epichlorohydrin*. They are available in many grades, can be cured by crosslinking with anhydrides, polyisocyanates, etc., and are used for adhesives and coatings, and for blow-molded, injection-molded, or extruded parts (containers, piping, etc.). Molded parts have good dimensional stability, low mold shrinkage, and fair heat and corrosion resistance. Abbreviation: PO.

Phenrok. Trade name of Detroit Wax Paper Company (USA) for phenolic plastics and laminates.

phenylamine. See aniline.

phenylethylene. See styrene.

Phenylon. Trade name for an aramid fiber.

phenylsilane resins. Thermosetting resins formed by the polymerization of silicone and phenolic resins. They are usually supplied as solutions for compounding with reinforcing agents, such as glass fibers or mineral fillers, and molding into parts with excellent heat resistance up to 290°C (554°F). Used for applications similat to those of phenolic plastics.

3-phenylthiophene. An organic compound (95% pure) with a melting point of 90-93°C (194-199°F) used as a precursor for conducting polymers. See also thiophene; poly(thiophene).

phenylzinc. See diphenylzinc.

pheochromes. A group of yellow, reddish or violet, crystallizable pigments related to *pheomelanins* and found in animal feathers, and animal and human hair. They are soluble in dilute hydrochloric acid, and are considered for biological electric conductor applications. Also known as *trichosiderins; trichochromes*.

pheomelanins. A group of *melanins* found in animal feathers, and animal and human hair, and produced by the radical-type polymerization of a sulfurated melanogen, possibly from a cysteinyldopa precursor. They are reddish brown to brown, amorphous pigments that are insoluble in dilute hydrochloric acid, and show great potential for biological electric conductor applications. See also melanogens.

Philadelphia Bronze. Trade name of Ampco Metals (USA) for a series of strong, corrosion-resistant cast aluminum and nickel bronzes used for gears, worm wheels, impellers, cams, valves and valve seats, pump parts, fittings, bearings, bushings, bolts, hardware, gun mounts, drawing dies, etc.

Philblack. Trademark of Phillips Chemical Company (USA) for a series of carbon blacks supplied in various grades and used for specialty applications, e.g., as fillers for natural and synthetic rubber.

Philippine mahogany. See lauan.

philipsbornite. A colorless mineral of the alunite group composed of lead aluminum arsenate hydroxide monohydrate, $PbAl_3(AsO_4)_2(OH)_5 \cdot H_2O$. Crystal system, rhombohedral (hexagonal). Density, 4.35 g/cm³; refractive index, 1.79. Occurrence: Namibia.

Philite. Trade name of Philips Lamps (UK) for phenol-formaldehyde plastics used for lighting and other electrical components.

Phillips 66. Trademark of Phillips Chemical Company (USA) for nylon 6,6 fibers.

phillipsite. A colorless, white, yellowish mineral of the zeolite group composed of potassium calcium aluminum silicate hydrate, $K_2Ca_2(Al,Si)_{16}O_{32} \cdot 13.5H_2O$. Crystal system, monoclinic. Density, 2.20 g/cm³; refractive index, 1.5042. Occurrence: Italy.

Philo. Trade name of SiMETCO (USA) for a series of ferroalloys including: (i) *High-carbon type ferrochrome* containing 4-6% carbon, 1-2% silicon, 65-70% chromium, and the balance iron. Used in the production of chromium steels and cast irons, and as silicon addition to steel and iron, and (ii) *Low-carbon type ferromanganese* composed of 0.7% carbon, 78-82% manganese, 0-1% silicon, and the balance iron. Used as high-manganese addition to low carbon steels and for metallurgical applications. In addition to these ferroalloys various grades of ferrosilicon are also sold under this trade name. They have typical compositions of 50-75% silicon and 48.5-22% iron, 75% silicon and 22.5% iron, 85% silicon and 12.5% iron, and 91% silicon and 6.5% iron. Used for metallurgical applications, as a graphitizer and deoxidizer, as silicon addition, in high-silicon alloys, and as an inoculant for cast iron.

Philprene. Trademark of Phillips Petroleum Company (USA) for a series of styrene-butadiene rubbers produced by the cold rubber process (i.e., polymerization at 4°C or 40°F) or the hot-rubber process (i.e., polymerization at 50°C or 120°F). Oil-extended (i.e., compounded with plasticizers based on aromatic or cyclic hydrocarbons), pigmented and nonpigmented grades also are available. *Philprene* rubber has good abrasion and heat resistance, and is used for tire carcasses and treads, wire and cable coverings, insulation, coated textiles, floor tile, and for molded and extruded products.

Philtread. Trademark of Phillips Paint Products Limited (Canada) for protective coatings based on chlorinated rubber.

phlogopite. See amber mica.

Phöbus. Trademark of Wieland Dental + Technik GmbH & Co. KG (Germany) for an extra-hard, white silver-palladium dental casting alloy.

phoenicochroite. A dark red mineral composed of lead chromium oxide, $Pb_2(CrO_4)O$. It can also be made synthetically by heating lead monoxide (massicot) and chromic oxide together under prescribed conditions. Crystal system, monoclinic. Density, 7.01 g/cm³; refractive index, 2.44; hardness, 3-3.5 Mohs. Occurrence: USA (Arizona).

Phoenix. (1) Trade name of British Steel Corporation (UK) for a series of plain-carbon steels containing up to 0.2% carbon, 0.7% manganese, 0.06% sulfur, 0.06% phosphorus, and the balance iron. Free-machining grades with 0.25% sulfur and 0.1% phosphorus are also available. Used for machine parts.

(2) Trade name of Phoenix Steel Corporation (USA) for corrosion-resistant, ferritic stainless steels (AISI type 430) used for chemical equipment, heat-resisting parts, etc.

(3) Trade name of NL Industries (USA) for a copper-base bearing alloy containing varying amounts of tin and antimony. Used for heavy-duty bearings.

(4) Trademark of Eckart America LP (USA) for pearlescent pigments.

Phoenixite. Japanese trade name for cellulose nitrate plastics and products.

Phoenixloy. Trade name of Duraloy Blaw-Knox (USA) for a series of superhard alloy cast irons containing 2.5-3.5% carbon, a total of 5% nickel and chromium, and the balance iron. Used for mill rolls.

Phoenix Mangear. Trade name of United Engineering Steels Limited (UK) for a plain-carbon steel containing 0.12% carbon, 0.2% silicon, 1.6% manganese, 0.04% sulfur, 0.04% phosphorus, and the balance iron. Used for lifting chains.

Phoenolan. Trademark of Phoenix AG (Germany) for thermoplastic polyester and polyether urethane elastomers.

Phoenolit. Trademark of Phoenix AG (Germany) for modified polyvinyl chlorides.

Phoenopren. Trademark of Phoenix AG (Germany) for polyurethane foams and foam products.

Phonitherm. Trade name of Ato Findley (France) for heat- and soundproofing products used in the building and construction industries.

Phono. Trade name of Olin Brass, Indianapolis (USA) for a series of high-copper alloys containing small amounts of tin, cadmium or aluminum. They possess relatively high electrical conductivity, and good corrosion resistance and ductility, and are used for trolley wire, electric wire and cable, electrical conductors, and hardware.

phonolite. A grayish, greenish or brownish igneous rock composed mainly of potash and soda feldspars and related minerals. It derives its name from the fact that it is sonorous when struck with a hammer. Used in the manufacture of ceramics, glass and aluminum, as a concrete aggregate, and as a building stone. Also known as *clinkstone.*

Phonstop. Trade name of Flachglas AG (Germany) for an acoustic insulating glass.

phormium fibers. The glossy, durable, relatively strong cellulose fibers obtained from the leaves of a perennial herb (*Phormium tenax*) of the lily family, native to New Zealand and the surrounding islands, but also cultivated in Argentina, Brazil, South Africa, Japan, and the western United States. They are white or reddish- to brownish-yellow, up to 3.6 m (12 ft.) long, and softer than other hard vegetable fibers, but not resistant to moisture and water. Used for the manufacture of cordage, twine, mats and textiles, and as a replacement for sisal. Also known as *New Zealand flax; phormium.*

Phosbrite. Trade name of Harstan Division, Chemtech Industries, Inc. (USA) for phosphating compounds used in metal finishing.

Phosco. Trade name of United Wire & Supply Company (USA) for a self-fluxing brazing alloy containing 92.5% copper and 7.5% phosphorus. It has a melting range of 716-788°C (1320-1450°F), and is used for brazing nonferrous metals.

Phos-Coat. Trade name of Dober Chemical Corporation (USA) for iron phosphate coatings.

Phos-Copper. Trade name of Westinghouse Electric Corporation (USA) for a series of copper-base brazing alloys containing 4-

8% phosphorus. They have a melting range of 707-893°C (1305-1640°F), high electrical conductivity, and are used for brazing brass, copper alloys and electrical connections.

Phos-Dip. Trademark of Heatbath Corporation (USA) for zinc and manganese phosphate coatings used as paint bases or for corrosion resistance. The manganese coatings also provide excellent lubrication.

PhosGuard. Trademark of Rockwood Pigments N.A., Inc. (USA) for anticorrosive pigments based on zinc phosphate.

phosgenite. A yellowish white to yellowish brown, colorless, white, gray or greenish mineral composed of lead chloride carbonate, $Pb_2(CO_3)Cl_2$. Crystal system, tetragonal. Density, 6.15 g/cm³; refractive index, 2.15. Occurrence: UK. Also known as *crom-fordite; horn lead.*

Phosguard. Trade name of Tronex Chemical Corporation (USA) for a series of phosphate coatings for use on metallic substrates.

phosinaite. A brownish rose mineral composed of sodium calcium cerium silicon phosphate monohydrate, $Na_3(Ca,Ce)SiPO_7 \cdot H_2O$. Crystal system, orthorhombic. Density, 3.00 g/cm³; refractive index, 1.572. Occurrence: Russian Federation.

Phosnic. Trade name of Chase Brass & Copper Company (USA) for an age-hardenable, corrosion-resistant bronze containing 0.85-1.35% nickel, 0.18-0.3% phosphorus, and the balance copper. Used for electrical conductors, marine hardware, and spring clips.

Phoson. Trademark of J.W. Harris Company (USA) for a series of copper-phosphorus and copper-silver-phosphorus brazing filler metals used for joining copper and copper-base alloys, and other nonferrous alloys.

Phosphacap. Trademark of Vivadent AG (Liechtenstein) for an encapsulated zinc phosphate dental cement.

Phosphalloy. Trade name of Sheffield Smelting Company (UK) for a series of brazing alloys containing 5-6.5% phosphorus, 2-15% silver, and the balance copper. The 80Cu-15Ag-5P alloy has a melting range of 625-780°C (1157-1436°F), and the 91.5Cu-6.5P-2Ag alloy of 690-835°C (1274 to 1535°F). Used for brazing copper alloys.

phosphammite. A colorless mineral composed of ammonium hydrogen phosphate, $(NH_4)_2HPO_4$. It can also be made synthetically. Crystal system, monoclinic. Density, 1.62 g/cm³; refractive index, 1.518.

phosphate. (1) A salt or ester of phosphoric acid (H_3PO_4), e.g., copper phosphate or zinc phosphate.

(2) A general term used in the field of ceramics to designate phosphorus-bearing compounds for the manufacture of ceramic bodies, glass and glazes, and bioceramics. Examples include bone ash, calcium phosphate, potassium phosphate, sodium phosphate, and minerals, such as amblygonite and apatite.

phosphate-bonded ceramic coatings. Ceramic coatings, up to 50 mm (2 in.) thick, formed by treating a metal oxide, such as aluminum, chromium or zinc oxide, with phosphoric acid (H_3PO_4). They possess low density, low thermal conductivity, high refractoriness after curing, and are used for the protection of metallic substrates against high temperatures (over 2400°C or 4350°F), and as binders in thin ceramic paint films.

Phosphate Cement. Trade name of Heraeus Kulzer, Inc. (USA) for a zinc phosphate dental cement.

phosphate coatings. Conversion coatings, usually 3-50 μm (0.1-2 mils) thick, produced on iron, steel, galvanized steel, aluminum, cadmium, tin, titanium and zinc by treating (e.g., immersing, brushing or spraying) with suitable phosphate solutions (usually aqueous solutions of phosphoric acid) to resist corrosion, e.g., as bases for paint or protective oils, improve lubrication or paint adhesion, and/or resist abrasion and wear. Four principal types are in general use: (i) *Iron phosphate coatings;* (ii) *Lead phosphate coatings;* (iii) *Manganese phosphate coatings;* and (iv) *Zinc phosphate coatings.* Examples of trade names for phosphate coating processes include Bonderizing, Granodizing and Parkerizing. See also conversion coatings.

phosphate-coated steel. See phosphatized steel.

phosphate crown glass. A *crown glass* containing a considerable amount of phosphorus pentoxide (P_2O_5) as a network former, and used for optical applications. Abbreviation: PCG. See also phosphate glass.

phosphated steel. See phosphatized steel.

phosphate glass. A clear glass in which the essential network former is phosphorus pentoxide (P_2O_5) as a partial replacement for silica (SiO_2), and which also contains significant amounts of alumina (Al_2O_3). It is resistant to hydrofluoric acid (HF) and other fluorine chemicals. Also known as *phosphorus glass.*

phosphate porcelain enamels. A group of low-melting porcelain enamels based on high-phosphate enamel frits. A typical melted-oxide composition (in wt%) is 40% phosphorus pentoxide (P_2O_5), 23% alumina (Al_2O_3), 20% sodium oxide (Na_2O), 8% boron oxide (B_2O_3), 5% fluorine (F_2), 4% lithia (Li_2O) and some titania (TiO_2). *Phosphate porcelain enamels* have good resistance to acids, but only fair to poor resistance to alkalies and water, and are used on aluminum and aluminum alloys.

phosphate rock. A rock that is rich in phosphates, especially calcium orthophosphate, $Ca_3(PO_4)_2$, and is used as a source of phosphorus and phosphates. Also known as *phosphorite.*

phosphate slag. See slag (4).

phosphatic bauxite. A porous, yellowish white aluminous rock containing approximately 32% aluminum oxide (Al_2O_3), 25% phosphorus pentoxide (P_2O_5), 8% ferric oxide (Fe_2O_3), 7% silicon dioxide (SiO_2), 1% titanium dioxide (TiO_2), and the balance water. Used as a source of aluminum.

phosphatic slag. See basic slag.

Phosphatine. Trade name of Svedia Dental for a zinc phosphate dental cement.

phosphatized steel. A steel on whose surface a phosphate conversion coating, typically 5-10 μm (0.2-0.4 in.) thick, has been produced by chemical treatment with aqueous solutions containing metallic phosphates using immersion, spraying or brushing techniques. The coating provides the steel with an inexpensive protection against filiform corrosion and is an adherent primer for subsequent paint coats. Used especially for autobodies. Also known as *phosphate-coated steel; phosphated steel.*

phosphazene fluorocarbon elastomer. A phosphazene polymer in which the substituents are fluorocarbon groups. Also known as *phosphonitrilic fluorocarbon elastomer; phosphorus nitrile fluoroelastomer.* Abbreviation: PNF. See also fluorocarbon elastomers; phosphazene polymers.

phosphazene polymers. A group of inorganic chain or ring polymers based on phosphonitrilic repeating units, i.e., repeating units containing alternating phosphorus and nitrogen atoms with two substituents, usually organic groups (e.g., halogens or aminos), on each phosphorus atom. Used as elastomers for low- and high-temperature applications. Also known as *phosphazenes; phosphonitriles; phosphonitrilic plastics.*

phosphides. Compounds of phosphorus and a metal or semimetal, such as aluminum, boron, barium, gallium, iron, tantalum, vanadium or zinc. Metal phosphides have good thermal stability, and are often used in electronics and electrical engineering as

semiconductors, and for ferroelectric applications.

phosphine. A pyrophoric, colorless gas with a density of 1.185 g/cm³, a melting point of -133.5°C (-208°F), a boiling point of -87°C (-124°F), and an autoignition temperature of 100°F (37°C). Available in high-purity electronic grades (99.999+%), it is used as a dopant for amorphous and crystalline silicon and other n-type semiconductors, as as a precursor to III-V (13-15) semiconductor thin films, as a condensation catalyst in organic reactions, and as a polymerization initiator. Formula: PH_3. Also known as *hydrogen phosphide.*

phosphoferrite. A white, yellow or pale green mineral composed of iron manganese phosphate trihydrate, $(Fe,Mn)_3(PO_4)_2 \cdot 3H_2O$. Crystal system, orthorhombic. Density, 3.29 g/cm³; refractive index, 1.674. Occurrence: Germany.

phosphomolybdic acid. See molybdophosphoric acid.

phosphonitriles. See phosphazene polymers.

phosphonitrilic fluorocarbon elastomer. See phosphazene fluorocarbon elastomer.

phosphonitrilic plastics. See phosphazene polymers.

phosphophyllite. A colorless to pale greenish blue mineral that is composed of zinc iron manganese phosphate tetrahydrate, $Zn_2Fe(PO_4)_2 \cdot 4H_2O$, and may also contain manganese. Crystal system, monoclinic. Density, 3.11 g/cm³. Occurrence: Germany, Bolivia.

phosphoprotein. Any conjugated protein that is linked with phosphoric acid, e.g., casein in milk and vitellin in egg yolk.

phosphor. A solid or liquid substance belonging to the *luminophors,* i.e., materials that can be stimulated to emit light by absorbing incident radiation (e.g., X-rays, cathode rays, ultraviolet radiation, alpha particles etc.). For *fluophors* the emission ceases almost immediately after excitation, while for *phosphors* it continues for some time even after the source is removed.

phosphor bronzes. A group of hard, strong, corrosion-resistant tin bronzes composed of 79.2-98.75% copper, 1.25-10% tin, 0-1.0% lead and traces of zinc and iron, and deoxidized with 0.1-2.5% phosphorus. They are commercially available as rods, flats, sheets, foil, wire and powder in four standard grades: (i) *Grade A* (UNS No. C51000) with 5.0% tin; (ii) *Grade C* (UNS No. C52100) with 8.0% tin; (iii) *Grade D* (UNS No. C52400) with 10.0% tin, and (iv) *Grade E* (UNS No. C50500) with 1.25% tin. *Phosphor bronzes* possess good to excellent cold workability, good hot workability, medium to excellent machinability, good tensile strength, ductility and shock resistance, low coefficients of friction, and are used for electrical contacts, flexible hose, pole-line hardware, valve parts, thrust washers, shafts, gears, pinions, bearings, fittings, hardware, bellows, clutch disks, pins, springs, hairsprings, switch parts, wires, wire brushes, welding rods, and machinery components. Abbreviation: P-bronzes.

phosphor copper. (1) A series of corrosion-resistant alloys available in two grades: (i) *Grade A* with 0-0.15% iron, 0.14+% phosphorus, and the balance copper; and (ii) *Grade B* with 0-0.15% iron, 0.10+% phosphorus, and the balance copper. Used for bearings and machine castings.

(2) A master alloy composed of phosphorus and copper and available in the form of slabs and shot in three grades with 5%, 10% and 15% phosphorus, respectively. Used for deoxidizing brasses and bronzes, for enhancing the hardness and fatigue strength of bronzes, and in the manufacture of phosphor bronze. Also known as *phosphorus copper.*

phosphorescent paint. A luminous paint that gives off light as a result of the absorption of certain radiation, such as X- or ultraviolet rays, even after the exposure to this radiation has ceased. It is made phosphorescent by using pigments, usually metal sulfides or combinations thereof, such as the sulfides of cadmium, calcium, strontium and zinc. Used to make objects visible in the dark, e.g., for airfield marking, signals and signs.

phosphorescent pigments. Pigments that are suitable for use in phosphorescent paints. They are usually metal sulfides or combinations thereof, such as sulfides of cadmium, calcium, strontium and zinc. See also phosphorescent paint.

phosphoric anydride. See phosphorus pentoxide.

phosphoric bromide. See phosphorus pentabromide.

phosphoric chloride. See phosphorus pentachloride.

phosphoric fluoride. See phosphorus pentafluoride.

phosphoric oxide. See phosphorus pentoxide.

phosphoric perbromide. See phosphorus pentabromide.

phosphoric perchloride. See phosphorus pentachloride.

phosphorite. See phosphate rock.

phosphorized admiralty metal. A copper alloy (CDA 445) containing 28-29% zinc, 0.9-1.2% tin and up to 0.1% phosphorus. It has good corrosion resistance and strength, good cold workability, and is used for tubing in marine equipment, condenser tubes, evaporator and heat exchanger tubing, ferrules, etc. Also known as *phosphorized admiralty*.

phosphorized arsenical copper. A copper (99.68% pure) containing 0.3% arsenic, and deoxidized with 0.02% phosphorus. Commercially available in the form of rods, tubes and flat products, it has good resistance to corrosion including pitting, high elongation, excellent hot and cold workability, and an electrical conductivity of 45% IACS. Used for heat exchangers, condensers, evaporators, distiller tubing, and staybolts. Also known as *phosphorus-deoxidized arsenical copper*.

phosphorized copper. Any of various commercially pure coppers deoxidized with phosphorus and including *high-residual-phosphorus copper* with 99.90% copper and 0.02-0.4% phosphorus, and *low-residual-phosphorus copper* with 99.90-99.99% copper and 0.004-0.12% phosphorus. Also known as *phosphorus deoxidized copper*.

phosphorogen. A substance, such as a sulfide of cadmium, copper, manganese or silver, used to induce phosphorescence in another material or compound.

phosphorroesslerite. A mineral composed of magnesium hydrogen phosphate heptahydrate, $MgHPO_4 \cdot 7H_2O$. Crystal system, monoclinic. Density, 1.73 g/cm^3. Occurrence: Austria, USA (Nevada).

phosphor tin. A foundry alloy containing up to 5% phosphorus, used in the manufacture of phosphor bronze to introduce phosphorus.

phosphorus. A nonmetallic element of Group VA (Group 15) of the Periodic Table. It is commercially available in the form of lumps, sticks and powder, and occurs in nature the the form of phosphates in minerals, such as *amblygonite, apatite, pyromorphite* and *wavellite*, in *phosphate rock* and in all living matter. Atomic number, 15; atomic weight, 30.974; monovalent, trivalent, tetravalent, pentavalent. There are three allotropic forms: (i) *White* (or *yellow*) *phosphorus*. A waxy, transparent solid. Crystal system, cubic. Density, 2.34 g/cm^3, melting point, 44°C (111°F); boiling point, 280°C (536°F); hardness, 0.5 Mohs; high electrical resistivity; good phosphorescence at room temperature. Used for smoke screens, analytical chemistry, etc.; (ii) *Red phosphorus*. A violet-red, amorphous powder. Density, 2.34 g/cm^3; autoignition temperature, 260°C (500°F); high electrical resistivity. Used in phosphors, electroluminescent coatings,

phosphor bronzes, metallic phosphides, compound semiconductors, safety matches, and in the manufacture of phosphorus compounds; and (iii) *Black phosphorus*. A black, electrically conducting solid made by heating white phosphorus under pressure, and resembling graphite in appearance. Crystal system, monoclinic. Density, 2.25-2.69 g/cm^3. Used in chemistry and materials research. Symbol: P.

phosphorus-32. A radioactive isotope of phosphorus with a mass number of 32 that is made by irradiation of phosphorus and phosphorus compounds and potassium dihydrogen phosphate, has a half-life of 14.3 days and emits beta rays. Used in chemical, biological and medical research (e.g. as a tracer), in medicine as a treatment for leukemia, skin lesions, etc., in the study of bone and teeth formation and deposition, in thickness and wear studies, and in lead detection. Symbol: ^{32}P. Also known as *radiophosphorus*.

phosphorus copper. See phosphor copper (2).

phosphorus-deoxidized arsenical copper. See phosphorized arsenical copper.

phosphorus-deoxidized copper. See phosphorized copper.

phosphorus glass. See phosphate glass.

phosphorus nitride. An amorphous, nonhygroscopic white solid which decomposes at 800°C (1472°F) and is used for doping semiconductors. Formula: P_3N_5.

phosphorus nitrile fluoroelastomer. See phosphazene fluorocarbon elastomer.

phosphorus oxychloride. A colorless, fuming liquid (99+% pure) with a density of 1.675 g/cm^3, a melting point of 2°C (35°F), a boiling point of 105.8°C (222°F) and a refractive index of 1.460. It is also available in high-purity (99.999%) electronic grades. Used as a catalyst, chlorinating agent and gasoline additive, in the manufacture of tricresyl phosphate, cyclic and acyclic esters for plasticizers, in organophosphorus compounds, and in hydraulic fluids and fire-retarding agents. The high-purity grades are used as dopants for semiconductor-grade silicon. Formula: $POCl_3$. Also known as *phosphoryl chloride*.

phosphorus pentoxide. A soft, white hygroscopic powder (97+% pure) that has a density of 2.39 g/cm^3, a melting point of 580-585°C (1076-1085°F), and sublimes at 300°C (572°F). High-purity grades (99.998-99.9999%) are also available. Used as an opalizer and network former in glass, in asphalt coatings, fire extinguishing compounds, in the manufacture of phosphorus oxychloride and metaphosphoric acid, and as a reagent in chemistry and biochemistry. Formula: P_2O_5. Also known as *anhydrous phosphoric acid; phosphoric anhydride; phosphoric oxide*.

phosphorus steel. A low-carbon steel with up to 0.3% phosphorus added to increase hardness, strength, corrosion resistance, and machinability.

phosphosiderite. A light purple pink mineral of the variscite group composed of iron phosphate dihydrate, $Fe(PO_4) \cdot 2H_2O$. Crystal system, monoclinic. Density, 2.74 g/cm^3; refractive index, 1.728; hardness, 3.5 Mohs. Occurrence: Germany.

phosphosilicate glass. A silicate glass containing a considerable amount of phosphorus pentoxide (P_2O_5) as a network former. See also phosphate glass.

phosphotungstic acid. See tungstophosphoric acid.

phosphotungstic pigments. A group of blue or green pigments made by precipitating *phosphomolybdic acid* or *phosphotungstic acid* with malachite green [$Cu_2CO_3(OH)_2$], Victoria blue ($C_{33}H_{31}N_3 \cdot HCl$) or other basic dye. Used for paints, enamels, paper, inks, etc.

phosphowolframic acid. See tungstophosphoric acid.

phosphuranylite. A yellow, radioactive mineral of the phospho-ferrite group composed of calcium uranyl phosphate hydroxide hexahydrate, $Ca(UO_2)_3(PO_4)_2(OH)_2 \cdot 6H_2O$. Crystal system, orthorhombic. Density, 3.50 g/cm^3; refractive index, 1.710. Occurrence: Australia.

Phos-Silver. Trade name of American Brazing Alloys Company (USA) for a series of brazing alloys containing 1.8-18.5% silver, 4.9-8.0% phosphorus, and the balance copper. They have a melting range of 618-869°C (1145-1595°F), and are used for brazing copper alloys including brass and bronze.

Phos-Trode. Trade name of Ampco Metals (USA) for a phosphor bronze (Grade C) containing 7-9% tin, 0.35% phosphorus, up to 0.5% other elements, and the balance copper. Used for shielded-arc welding rods.

Photac-Fil Aplicap. Trade name for an encapsulated light-cure glass-ionomer used for restorative dentistry.

Photac-Fil Quick. Trade name for a fast-setting resin-modified glass-ionomer cement used for restorative dentistry.

Photact. Trademark of Keuffel & Esser Canada Inc. for photographic films and papers.

Photo Bond. Trade name of Espe (USA) for a dual-cure bonding agent used in restorative dentistry for dentin bonding.

Photo-Bond Aplicap. Trade name of Espe (USA) for an encapsulated light-cure glass-ionomer used for restorative dentistry.

photocatalyst. See photoinitiator; photocatalytic material.

photocatalytic material. Any of a group of materials that exhibit catalytic abilities upon exposure to light, and include photoinitiators such as those used in the photopolymerization of polymers, minerals such as the anatase form of titanium dioxide, certain aluminosilicates and titanates, and pollutants, such as atmospheric aerosols. Also known as *photocatalyst.*

photoceramics. Earthenware or whiteware, such as pottery or china, with a photographed surface image.

photochemical glass. A photosensitive glass exposed in a camera to an image of the desired design or pattern, and developed to expose the lines or areas to be etched or eaten away in a subsequent acid bath. The resulting design or picture is three-dimensional.

photochromic glass. Glass that reversibly changes color or optical density on exposure to light, but which returns to its original color and clarity when the light source is removed. This color change is due to the action of light on minute silver chloride or silver bromide crystals distributed throughout the glass matrix. Used for sunglasses, automobile windshields, electronic applications, etc.

Photo Clearfil. Trademark of J. Morita Company (Japan) for a self-curing dental material based on bisphenol glycidylmethacrylate (BisGMA) that provides excellent bonding to dental enamel and dentin, and is used for filling tooth cavities, and for tooth surface restoration. It is supplied in a kit including adhesives, bonding and etching agents, fillings and accessories.

photoconductive film. A material in film form, e.g., a semiconductor, such as cadmium selenide or cadmium sulfide, whose electrical conductivity varies with the absorbed amount of electromagnetic radiation (light).

photoconductive materials. See photoconductors.

photoconductors. Semiconductor materials, such as cadmium selenide and cadmium sulfide, that change their electrical resistivity when illuminated with light energy. They have insulator properties when in the dark, but when exposed to light charge carriers are produced by exciting electrons into the conduction band. The quantity of illumination on the surface determines the number of electron-hole pairs generated, and consequently the conductivity of photoconductors. Also known as *photoconductive materials; photoresistors; photoresistive materials.*

photocopying paper. A generic term or various types of paper used in photocoping and including: (i) *Silver-sensitized paper;* (ii) *Diazo-type paper;* (iii) *Standard photocopying paper* composed of combinations of coatings with and without silver sensitization; (iv) *Direct* and *indirect electrostatic copying paper;* and (v) *Direct* and *indirect thermocopying paper.*

photodegradable plastics. Plastics in which the bonds between the carbon and hydrogen atoms are broken down upon absorption of ultraviolet light. Most plastics are photodegradable with the rate of degradation ranging from a few days to several years. Some additives (e.g., carbon black), known as *ultraviolet stabilizers,* are used to screen out UV rays altogether, while other additives (e.g., hydroxybenzophenones), known as *ultraviolet absorbers,* are employed to absorb and convert harmful UV rays into thermal radiation. In some plastics photodegradation is welcome and promoted by the use of ultraviolet accelerators.

photodichroic material. A material in which the optical properties of dichroism and double refraction are induced by electromagnetic radiation (light).

photoelastic material. An isotropic, transparent dielectric material, such as a plastic, which exhibits changes in optical properties (e.g., double refraction) when subjected to mechanical pressure or stress.

photoelectric material. See photoemissive material.

photoelectromagnetic material. A material, such as certain photoconductors and intermetallic semiconductors, in which a potential difference is produced when placed in a magnetic field and exposed to electromagnetic radiation (light).

photoemissive material. Any of a class of materials including III-V (13-15) semiconductors and rare-earth doped glasses that emit electrons when struck by electromagnetic radiation (light). Used in photocells and phototubes, relays, photonic crystals for narrow-linewidth lasers, etc. Also known as *photoelectric material.*

photoengraving magnesium. A commercially pure magnesium containing small additions of other elements, and used for lightweight, wear-resistant photoengraving plates.

photoengraving zinc. A pure, fine-grain zinc containing small additions of iron, cadmium, magnesium and manganese, and used for photoengraving plates.

photoferroelectric material. Any of a class of materials including especially ferroelectric ceramics, such lead lanthanum zirconate titanate, in which incident electromagnetic radiation (light) at or near the energy band gap effects changes in the electric field created by an applied voltage, and may also effect their ferroelectric remanent polarization properties.

Photoglaze. Trademark of Lord Techmark, Inc. (USA) for a series of ultraviolet- and electron beam-curing coatings.

photographic base paper. A wood-free paper of uniform transparency and finish that has a photosensitive surface coating. It is usually moisture-resistant, dimensionally stable, chemically neutral, and free of any impurities.

photographic film. A thin, flexible strip or sheet of cellulose acetate or other suitable plastic (e.g., nitrocellulose, polyvinyl chloride, etc.) coated on one side with a photosensitive emulsion (usually a silver halide suspended in gelatin), and used to make photographic negatives, transparencies, etc.

photographic paper. A specially prepared, high-grade paper having a photosensitive surface coating, usually composed of a silver halide suspended in gelatin. Used as a material to make photographic positives by exposure of negatives during development.

Photogray. Trade name of Corning Glass Works (USA) for a photochromic glass used for ophthalmic lenses.

photoinitiator. A chemical compound, such as benzoin ethyl ether, DL-camphorquinone or methyl benzoyl formate, that is added to a prepolymer or mixture of monomers to start photopolymerization upon exposure to visible or ultraviolet light. Also known as *photocatalyst.*

photolithographic materials. A class of materials including silicon, silicon dioxide, and polymeric photoresists suitable for use in the lithography process for producing integrated circuit element patterns on wafers. See also photoresist.

photomagnetic material. Any of a group of materials whose magnetic susceptibility is changed by light. Examples include inorganic solids, such as iron borate ($FeBO_3$), and inorganic-organic hybrids, such as hydous potassium cobalt ferrocyanide ($K_{0.4}Co_{1.3}[Fe(CN)_6] \cdot xH_2O$). Used in electronics, optoelectronics and photonics.

Photo Mount. Trade name of 3M Company (USA) for a permanent, clear-color aerosol adhesive containing volatile petroleum distillate and acetone, and used for mounting photos, art prints, illustrations, maps, etc.

photon-gated material. An electronic material in which the presence of an initiating light beam is required for persistent spectral holeburning with a narrow-band laser to occur.

photonic band-gap materials. See photonic crystals.

photonic crystals. A class of artificial crystals with highly periodic structures operating in the near-infrared (780-3000 nm) and visible regions of the electromagnetic spectrum. Structurally, they are transparent dielectrics with high refractive indices (about 3.0) containing numerous minute air holes with lower refractive indices arranged in a lattice pattern. Light (photons) passing through these crystal is confined to either within the dielectric or the air holes, resulting in the formation of a photonic band gap, i.e., allowed and forbidden photonic energy regions. The dielectric blocks light with wavelengths in the photonic band gap, while allowing others to pass. *Photonic crystals* can be one-dimensional (1D), two-dimensional (2D) or three-dimensional (3D). The ideal photonic crystal has a three-dimensional lattice structure and can be produced, for example, by the accurate arrangement of micromachined silicon wafers in successive layers, the self-arrangement of submicron-sized silica spheres in colloidal suspension and filling of the air voids with a high-refractive-index titanium dioxide solution, or from a suspension of polymer spheres in a hydrogel film. *Photonic crystals* for laser applications can also be made from photoemissive materials, such as III-V (13-15) semiconductors and rare earth-doped glasses, by growing thick layers of silicon dioxide on a silicon substrate and subsequently adding a layer of silicon nitride, or by integrating synthetically grown opal spheres into silicon. *Photonic crystals* with band gaps at microwave and radio frequencies are currently being used for mobile-phone antennas and optical-communications applications, e.g., high-Q filters, waveguides, channel-drop filters, etc. Also known as *photonic band-gap materials; photonic materials.* See also 1D photonic crystals; 2D photonic crystals; 3D photonic crystals; colloidal photonic crystals; Yablonovite.

photonic materials. See photonic crystals.

photophor. See calcium phosphide.

photopigment. A pigment, such as a *chromophore*, that changes chemically or physically when exposed to light.

photopolymers. A class of photosensitive polymers, such as polyvinyl acetate esters, that undergo a spontaneous and irreversible chemical or physical change on exposure to visible radiation.

photopolymerizable liquid crystal polymers. A class of polymers produced by photopolymerization of oriented liquid crystalline monomers, such as acrylate derivatives and styrenes. Polymeric films so produced have unique optical properties making them suitable for use in electronics, optoelectronics and photonics. See also liquid crystal polymers.

photoprotecting paper. An opaque, chemically neutral, black-colored, machine-finished cellulose paper that can be folded without breaking. The average basis weight is about 80-90 g/m^2 (0.26-30 oz/ft^2). Used for wrapping photographic films and plates.

photorefractive materials. A class of nonlinear optical materials that experience appreciable refractive index changes under the effect of inhomogeneous illumination at low power densities. Examples include barium titanate, bismuth germanate, bismuth titanate, lithium niobate, certain semiconductor materials (e.g., gallium arsenide, cadmium telluride and indium phosphide), and several organic crystals and polymers. Used in lasers, coherent sources, optoelectronic devices, optical processors, computers, etc. Abbreviation: PR materials.

photoresist. A photosensitive polymeric material used in the integrated-circuit lithography process for producing metal or vitreous silica patterns on silicon wafers in the fabrication of photonic crystals, etc. Metal patterns require a *negative photoresist* in which exposure to ultraviolet radiation results in cross-linking of the polymer and subsequent removal of unexposed material by a suitable solvent. For silica patterns a *positive photoresist* is used in which the polymer is depolymerized by ultraviolet radiation and subsequently removed. Conventional technologies for the manufacture of integrated circuits use optical lithography in which the laser energy of the imaging radiation is in the visible (436 nm) or ultraviolet (below 400 nm) wavelength range. More recently developed photolithography techniques utilize chemically-applied photoresists which are more practicable for modern excimer lasers having typical imaging radiation wavelengths in the deep ultraviolet range (below 250 nm). For argon fluoride (ArF) and fluorine (F_2) excimer laser sources these wavelengths can be as low as 193 nm and 157 nm, respectively. See also photolithographic materials.

photoresistive materials. See photoconductors.

photoresistors. See photoconductors.

photosensitive glass. Glass containing small quantities of submicroscopic particles of metals, such as copper, gold or silver, that make it sensitive to visible radiation in a way analogous to photographic films. The photograph is usually developed by heat treatment. Also known as *light-sensitive glass.*

photosensitive material. A material that emits electrons from its surface when energized by light or other radiant energy. It may have photoconductive, photoemissive or photovoltaic properties. Also known as *light-sensitive material.*

photosensitizer. A substance that is added to a material to enhance its sensitivity to electromagnetic radiation (light).

photostabilizers. See light stabilizers.

photothermoelastic material. An isotropic, transparent dielectric material that exhibits changes in optical properties when subjected to temperature gradient-induced mechanical stress.

photoviscoelastic material. A transparent, viscoelastic material, such as a plastic, that exhibits changes in optical properties when subjected to mechanical stress.

photovoltaic materials. A group of electronic materials which exhibit the photovoltaic effect, i.e., which are capable of producing a voltage by the absorption of an electromagnetic radiation, such as light. Examples include single-crystalline, polycrystalline and amorphous silicon as well as cadmium telluride, copper indium selenide, gallium arsenide, iron sulfide, nanocrystalline thin films, and several conjugated conductive polymers. *Photovoltaic materials* are particularly useful in solar energy technology and energy conversion, and optoelectronics. Abbreviation: PV materials.

photovoltaic polymers. Conjugated conductive polymers, such as *polythiophene*, that have band gaps in the 1.5-3.0 eV range making them suitable for photovoltaic devices and optoelectronic devices operating in the visible light range. See also conductive polymers; conjugated polymers

Phrikolat. Trade name of Phrikolat-Chemische Erzeugnisse GmbH (Germany) for carboxymethyl celluloses and carboxymethylhydroxyethyl celluloses.

Phrilon. Trademark for a nylon fiber used for the manufacture of textile fabrics.

Phrix. Trade name for rayon fibers and yarns used for textile fabrics.

Phrix Perlon. Trade name for nylon fibers and yarns used for textile fabrics including clothing.

m-phthalic acid. See isophthalic acid.

p-phthalic acid. See terephthalic acid.

phthalocyanine blue. See copper phthalocyanine

phthalocyanine liquid crystals. A class of synthetic phthalocyanines that form columnar liquid crystals capable of conducting light energy or electricity along their columnar axes. These materials are currently under investigation for use in electronic, optoelectronic and photonic applications.

phthalocyanine materials. See phthalocyanines (1).

phthalocyanine pigments. See phthalocyanines (2).

phthalocyanines. (1) A group of materials that are organic molecules (e.g., liquid crystals, dendrimers, ladder polymers, etc.) with large aromatic pyrrole ring structures which can contain a central metal ion (e.g., cobalt, copper, iron, lithium, magnesium, nickel, etc.). The general formula for metal-free phthalocyanine is $C_{32}H_{16}N_8$. Formerly used chiefly as pigments and dyes (see phthalocyanines (2)), they are now produced by molecular epitaxial deposition, Langmuir-Blodgett, self-assembly and other advanced techniques and supplied as crystals, thin films, composites, etc. *Phthalocyanines* are often used for their light-absorbing and photoconductive properties, and their electron mobilities in the range of about 10^{-1} to 10^{-2} cm²/V-s, e.g., as laser dyes, photoconductors for laser printers, in nonlinear optics, optical data storage, recordable CDs, electronic sensors, electrochromic displays, xerography, solar energy conversion, industrial heterogeneous catalysis, and as organic semiconductors for various microelectronic applications. Also known *phthalocyanine materials.*

(2) A group of green or blue organic pigments including (i) *Phthalocyanine* (bluish green); (ii) *Copper phthalocyanine* (bright blue); (iii) *Chlorinated copper phthalocyanine* (green), and (iv) *Sulfonated phthalocyanine* (green). They exhibit good lightfastness and chemical stability, high tinting strength, and good resistance to elevated temperatures. Used in baking enamels, automobile finishes, and plastics and inks. Also known as *phthalocyanine pigments.*

Phthalopal. Trademark of BASF AG (Germany) for a series of maleic resins.

phuralumite. A lemon-yellow mineral of the phosphuranylite group composed of aluminum uranyl phosphate hydroxide decahydrate, $Al_2(UO_2)_3(PO_4)_2(OH)_6 \cdot 10H_2O$. Crystal system, monoclinic. Density, 3.50 g/cm³; refractive index, 1.616. Occurrence: Zaire.

phurcalite. A yellow mineral of the phosphouranylite group composed of calcium uranyl phosphate hydroxide tetrahydrate, $Ca_2(UO_2)_3(PO_4)_2(OH)_4 \cdot 4H_2O$. Crystal system, orthorhombic. Density, 4.03 g/cm³; refractive index, 1.730. Occurrence: Germany.

phyllosilicates. A group of polymerized silicates with a two-dimensional sheetlike or layered structure, produced by the sharing of 3 of the 4 oxygen ions in each SiO_4 tetrahedron with adjacent tetrahedra. The repeating unit formula may be represented by $(Si_2O_5)_2$. Examples of *phyllosilicates* include kaolinite clay [$Al_2(Si_2O_5)(OH)_4$], talc [$Mg_3(Si_2O_5)_2(OH)_2$] and the micas. Also known as *layered silicates; layer silicates; sheet minerals; sheet silicates.*

Piacryl. Trademark for polymethyl methacrylate resins and products.

Piadural. Trademark for modified urea-formaldehyde plastics and products.

Piaflor. Trademark for urea-formaldehyde foams and foam products.

piano wire. A round, cold-drawn steel wire with high tensile strength containing 0.70-1.00% carbon. It is available in diameters and tensile strengths ranging from 0.76 mm (0.03 in.) and 2.76 GPa (400 ksi) to 1.65 mm (0.065 in.) and 2.41 GPa (350 ksi), has uniform mechanical properties, good fatigue properties, high toughness and resilience, and is used for mechanical and piano springs, and wire screens.

pianlinite. A pure, white and grayish white mineral composed of aluminum silicate hydroxide, $Al_2Si_2O_6(OH)_2$. Crystal system, orthorhombic. Density, 2.45 g/cm³; refractive index, 1.539. Occurrence: China.

Piastrelle. Trade name of Vetreria di Vernante SpA (Italy) for a patterned glass.

Piatherm. Trademark for urea-formaldehyde foams and foam products.

Pibiflex. Trade name of Enichem SpA/P Group (Italy) for a flexible polyester copolymer.

Pibiter. Trade name of Enichem SpA/P Group (Italy) for a polybutylene terephthalate supplied in unmodified, carbon- or glass-fiber-reinforced, fire-retardant, glass bead-filled, silicone- or polytetrafluoroethylene-lubricated, UV-stabilized and structural foam grades.

pickeringite. A colorless or whitish mineral of the halotrichite group composed of magnesium aluminum sulfate hydrate, $MgAl_2(SO_4)_4 \cdot 22H_2O$. Crystal system, monoclinic. Density, 1.79 g/cm³; refractive index, 1.79. Occurrence: USA (New Mexico).

Piccolyte. Trademark of Hercules, Inc. (USA) for a thermoplastic terpene resin used in varnishes.

Piccoumaron. Trademark of Hercules, Inc. (USA) for thermoplastic coumarone-indene resins used in adhesives, paints, etc.

Piccovol. A coal-tar derivative used as a rubber softener.

picker leather. Leather, usually rawhide or belting leather, used for pickers in textile machinery.

pickle alum. See aluminum sulfate octadecahydrate.

picotite. A yellowish to brown variety of the mineral *hercynite* composed of oxides of magnesium, aluminum, and chromium, $(Mg,Fe)O·(Al,Cr)_2O_3$. It occurs frequently in basic refractory slag, and can also be made synthetically. Crystal system, cubic. Crystal structure, spinel. Density, 4.08 g/cm³; hardness, 7.8-8.2 Mohs. Used as a refractory and in materials research. Also known as *chrome spinel.*

picotpaulite. A cream-white mineral of the wurtzite group composed of thallium iron sulfide, $TlFe_2S_3$. Crystal system, orthorhombic. Density, 5.23 g/cm³. Occurrence: Yugoslavia, Croatia, Serbia.

picrochromite. See magnesiochromite; magnesia-chromia; spinel; normal spinel.

picromerite. A colorless mineral composed of potassium magnesium sulfate hexahydrate, $K_2Mg(SO_4)_2·6H_2O$. It can also be made synthetically by slow, room-temperature evaporation of a 1:3 aqueous solution of potassium sulfate (K_2SO_4) and magnesium sulfate $(MgSO_4)$. Crystal system, monoclinic. Density, 2.03 g/cm³; refractive index, 1.462. Occurrence: Germany.

picropharmacolite. A white mineral composed of calcium magnesium arsenate hexahydrate, $(Ca,Mg)_3(AsO_4)_2·6H_2O$. Crystal system, monoclinic. Density, 2.62 g/cm³; refractive index, 1.571. Occurrence: Germany.

piemontite. (1) A brown to black mineral of the epidote group composed of calcium aluminum iron manganese silicate hydroxide, $Ca_2(Al,Fe,Mn)_3Si_3O_{12}OH$. Crystal system, monoclinic. Density, 3.47 g/cm³; refractive index, 1.7876. Occurrence: Sweden. Also known as *manganepidote.*

(2) A purplish red mineral of the epidote group composed of calcium aluminum silicate hydroxide, $Ca_2Al_3(SiO_4)_3(OH)$. Crystal system, monoclinic. Density, 3.49 g/cm³; refractive index, 1.766. Occurrence: New Zealand.

pierrotite. A gray black mineral composed of thallium antimony arsenic sulfide, $Tl(Sb,As)_5S_8$. Crystal system, orthorhombic. Density, 4.97 g/cm³. Occurrence: France.

Pierrot's metal. An antifriction metal containing 83-84% zinc, 2-3% copper, 7.5% tin, 3.5% antimony and 3% lead. It has only moderate heat resistance, and is used for medium and light-duty bearings and bushings.

Piertex. Trademark of Crown Diamond Paints Limited (Canada) for acrylic latex-base texture finishes.

pietra serena. A gray sandstone used as a building stone in Italy.

piezoelectric ceramics. Ceramic materials, such as barium titanate, lead metaniobate, lead zirconate-titanate, etc., that exhibit the piezoelectric effect. Abbreviation: PEC. See also piezoelectric materials.

piezoelectric crystals. Crystals composed of a material, such as quartz or barium titanate, that exhibit the piezoelectric effect. See also piezoelectric materials.

piezoelectric foams. Electrically charged foams produced from polymers, such as polytetrafluoroethylene, polypropylene and other low-dielectric-constant polymers, that exhibit nonsymmetric charge distributions, and very high quasistatic piezoelectric coefficients (or piezoelectric longitudinal charge coefficients, d_{33}). They have potential applications in electronics and optoelectronics. See also piezoelectric polymers.

piezoelectric materials. A group of materials that produce a voltage after experiencing mechanical pressure or stress, and vice versa. Examples of such materials include quartz, Rochelle salt, lithium sulfate, piezoelectric ceramics, and certain plastics. Used as transducers, acoustic sensors, sonar and ultrasonic devices, etc. Abbreviation: PEM.

piezoelectric polymers. Polymer, usually in film form, that have piezoelectriic properties and are therefore capable of converting thermal and mechanical energy into electricity. Abbreviation: PEP. Also known as *piezoelectropolymer; piezopolymer.* See also piezoelectric materials.

piezoelectric semiconductor. A semiconductor, such as Rochelle salt, barium titanate or quartz, having piezoelectric properties. Abbreviation: PES. See also piezoelectric materials.

piezoelectropolymer. See piezoelectric polymer.

piezomagnetic material. A solid material in which a magnetic moment is induced by a mechanical stress, e.g., by collision with another solid.

piezooptic material. A crystalline material whose degree of birefrigence changes due the application of a mechanical stress.

piezopolymer. See piezoelectric polymer.

piezoresistive material. A material, such as a semiconductor, whose electrical resistivity changes due to the application of a mechanical stress.

pig aluminum. Aluminum as it comes from the melting furnace, cast into solid forms.

pigeonite. A light purplish brown mineral of the pyroxene group composed of iron magnesium calcium silicate, $(Fe,Mg,Ca)SiO_3$. Crystal system, monoclinic. Density, 3.30 g/cm³; refractive index, 1.7066. Occurrence: Japan, USA (Minnesota).

pig iron. Crude iron made in a blast furnace or electric furnace by reduction of iron ore in the presence of coke and limestone. A typical composition is 3.2-5.5% carbon, 0.3-2.5% silicon, 0.3-5.0% manganese, 0.07-2.2% phosphorus, and 0.005-0.15% sulfur, and the balance iron. *Pig iron* contains many impurities and, owing to its high carbon content, is rather brittle and only a small fraction is used directly in the manufacture of mechanical parts. Most pig iron is further refined, and used as a raw material in the production of cast irons and steels. Pig iron coming from the furnace is usually cast into uniform, oblong shapes called "pigs." Abbreviation: PI.

pig lead. Commercial-grade lead cast into blocks or bars.

Pigment Blue 15. See copper phthalocyanine blue.

Pigment Blue 27. See indigo (2).

Pigment Blue 66. See Prussian blue.

pigment-dyed fabrics. Fabrics having an insoluble, more or less permanent colorant applied, often in the form of a paste or emulsion, which is subsequently heat cured and bonded with synthetic resins or other agents.

pigmented oil stain. A wood stain containing finely divided, insoluble color pigments in solution with drying oil, varnish, turpentine, mineral spirits, etc. It covers the wood's natural color, and obscures the grain pattern, while revealing the texture. The depth of color can be controlled by the amount of stain wiped off.

pigment-finished leather. Leather finished with compositions containing opaque pigments to hide the original grain pattern to varying degrees.

Pigment Green 7. See copper phthalocyanine green.

pigment-printed fabrics. Fabrics that have been printed with a paste or emulsion colorant, composed of an insoluble pigment, a binder and a thickner, which is subsequently heat cured and bonded with synthetic resins or other agents. Also known as *pigment prints.*

pigment prints. See pigment-printed fabrics.

pigments. Fine, insoluble powders used to give color and opacity to ceramic bodies, inks, textiles, (organic) coatings, etc., by preferably reflecting light of certain wavelengths and absorbing light of other wavelengths. There are two basic categories of pigments: (i) *Inorganic pigments* including oxides and sulfides of metals, such as antimony, cobalt, iron, lead, manganese, titanium, zinc, etc., metal powder suspensions (e.g., aluminum or gold), earth colors, and carbon blacks; and (ii) *Organic pigments*, such as rhodopsin, chlorophyll, indigo, phthalocyanine, toluidine, lakes, etc., which are based on animal, vegetable or synthetic compounds.

pig metal. A metal, such as aluminum, iron or lead, cast in an oblong shape for remelting, storage and transportation.

pignut hickory. (1) A collective term referring to the heavy, hard wood of any of a class of hickory trees including the sand hickory *(Carya pallida)*, pignut hickory *(C. glabra)*, black hickory *(C. texana)*, and scrub hickory *(C. floridana)*. The heartwood is reddish and the sapwood white. *Pignut hickory* has high toughness and strength, high shock resistance, and large shrinkage. Source: Eastern and central United States. Used for tool handles, ladder rungs, agricultural implements, furniture and pallets. See also hickory.

(2) A term referring specifically to the heavy, hard wood from the pignut hickory tree *(Carya glabra)*. See also hickory; pignut hickory (1).

pigskin. A term referring to leather made from the hide of hogs and pigs, and sometimes including leather made from peccaries and carpinchos. Used for gloves, belts, handbags, and clothing.

pig tin. Tin (99.80+% pure) as it comes from the melting furnace, cast into solid forms.

Piladuc. Trade name of Youngstown Steel (USA) for a rolled carbon steel containing 0.2% carbon, and the balance iron. Used for tinplate.

pile fabrics. Textile fabrics, such as velvet, plush or corduroy, with raised surfaces woven with loops of yarn which may be uncut or cut. The weave is made by using two warp yarns and one weft (or filling) yarn, or one warp yarn and two filling yarns. Woven towels, rugs and carpets are pile fabrics. Also known as *pile-weave fabrics.*

Pilger Roll Steel. Trade name of Firth Brown Limited (UK) for hardened and tempered steel castings containing 1.1% carbon, varying amounts of chromium and molybdenum, and the balance iron. Used for Pilger rolls in tube-reducing mills.

Pillarcote. Trademark of Pillar-Greenwood Limited (Canada) for high-gloss lacquers.

pilling-resistant fabrics. Textile fabrics that do not, or only to a very small extent form small fuzzy balls of loose fibers on the surface due to wear. They are either treated with antipilling chemicals, or knitted or woven from special non-pilling yarns.

Pillnay. Trade name of W. R. Grace & Company (USA) for special coating lacquers used for sheetmetal and glass packaging applications.

pillow lace. See bobbin lace.

Pilray. Trade name for rayon fibers and yarns used for textile fabrics.

Pil-Trol. Trademark of Solutia Inc. (USA) for a durable, color-fast low-pill acrylic fiber with excellent shape retention and stain resistance. Used particularly for uniform and school sweaters.

Pima cotton. A fine, long-staple hybrid cotton used especially in shirt and dress fabrics. Named for Pima County in southeastern Arizona, USA.

piña. See pineapple fiber.

Pinacryptol Yellow. A hygroscopic crystalline powder (98% pure)

with a melting point of 260-262°C (471-504°F) and a maximum absorption wavelength of 385 nm used as a photosensitizing dye. Formula: $C_{21}H_{22}N_2O_7S$.

pinacyanol bromide. A green powder with a dye content of about 90-95%, a melting point of 286°C (547°F) (decomposes) and a maximum absorption wavelength of 607 nm used as a photosensitizing dye. Formula: $C_{25}H_{25}N_2Br$.

pinacyanol chloride. A green powder with a melting point of 270°C (518°F) (decomposes) and a maximum absorption wavelength of 604 nm used as a photosensitizing dye. Formula: $C_{25}H_{25}N_2Cl$.

pinacyanol iodide. A dark green powder (97% pure) with a melting point of 295°C (563°F) (decomposes) and a maximum absorption wavelength of 614 nm used as a photosensitizing dye. Formula: $C_{25}H_{25}N_2I$.

pinchbeck. A red brass composed of 88-94% copper and 6-12% zinc, and used for cheap jewelry to imitate gold. Also known as *pinchbeck metal.*

pinchite. A dark brown to black mineral composed of mercury oxide chloride, $Hg_5O_4Cl_2$. Crystal system, orthorhombic. Density, 9.17 g/cm³; refractive index, above 2.00. Occurrence: USA (Texas).

pincord. A general term referring to fabrics, such as certain corduroys, that have narrow ribs, ridges or wales. Also known as *baby cord; pinwale.*

pine. The wood of any of a large genus *(Pinus)* of conifers found throughout North and Central America, Asia and Europe. Pine is one of the most important timber trees in the world, yielding not only lumber, but also turpentine, tar, pitch and medicinal oil. The most important species in North America are the jack pine *(P. banksiana)*, loblolly pine *(P. taeda)*, lodgepole pine *(P. contorta)*, longleaf pine *(P. palustris)*, pitch pine *(P. rigida)*, ponderosa pine *(P. poderosa)*, red pine *(P. resinosa)*, sugar pine *(P. lambertiana)*, eastern white pine *(P. strobus)* and western white pine *(P. monticola)*. Also known as *pinewood. Note:* For properties and uses, see particular species.

pineapple fibers. The fine, white, glossy cellulose fibers obtained from the leaves of the pineapple plant (genus *Ananas*) native to tropical and subtropical America, but also cultivated in other warm regions. Used for the manufacture of fine fabrics. Also known as *piña.*

pine gum. The amorphous solid residue left after the distillation of crude turpentine obtained from the oleoresinous substance tapped from certain living European and North American softwood trees (chiefly pine). It has a density of 1.08 g/cm³ (0.039 lb/in.³), a melting range of about 100-140°C (210-285°F), and is used in the manufacture of varnish and paint driers, printing inks, lubricants, glues, etc., for synthetic amber, and in hardening steel. Also known as *gum rosin; pine resin; yellow resin.*

pine pitch. A dark-colored amorphous solid substance obtained as a byproduct in the distillation of pinewood. Used for waxes and as a wood preservative.

pine resin. See pine gum.

pine tar. A brownish-black, viscous, liquid or semisolid obtained by the destructive distillation of pinewood or other wood from coniferous trees. It has a density of 1.03-1.07 g/cm³ (0.037-0.039 lb/in.³), a boiling range of about 240-400°C (465-750°F). Used for roofing materials, asphaltic compositions, paints and varnishes, plastics and rubber processing, tar soaps, road paving, as a general preservative, as an adherent lubricant, and in ore flotation. Also known as *wood tar.*

Pine Tree Castings. Trade name of Pine Tree Castings, Ruger

Advanced Materials Group (USA) for ferrous investment castings.

pinewood. See pine.

pinpoint oxford. A fabric made from fine yarn, usually cotton, in a weave that resembles, but is smaller than that of true oxford. See also oxford (1).

Pinhead. Trade name of Window Glass Limited (India) for a glass with regular pinhead pattern.

pinheiro do Parana. See araucarian pine.

pinho do Parana. See araucarian pine.

pinite. A compact, brown, gray or green form of *mica*, essentially of the muscovite type. It is a hydrous silicate of alumina and potash that inverts to *mullite* at 1125°C (2057°F). Used in ceramics, and in the manufacture of dense, abrasion-resistant refractories, e.g., for kiln linings.

Pink Label. Trade name of T. Turton & Sons Limited (UK) for a tool steel containing 0.9% carbon, 0.2% vanadium, and the balance iron.

pink salts. See rare-earth sodium sulfates.

pink topaz. (1) Pink-colored *topaz* produced from yellow or brown varieties by heating under prescribed conditions.

(2) A naturally, pink-colored variety of *topaz*.

Pinkus brass. A free-cutting brass containing 88.1% copper, 6.9% zinc, 2.5% tin, 1.8% lead, 0.3% nickel, and 0.4% antimony. Used for hardware and fittings.

Pinkus bronze. A bronze containing 72.5% copper, 14.7% tin, 8.8% lead, 2.5% antimony, and 1.5% zinc. Used for heavy-duty bearings and bushings.

Pink Vosges. Trade name of Carriere de Rothbach (France) for a pinkish sandstone from the Vosges Mountains in northeastern France.

PinkWrap. Trademark of Owens Corning Fiberglass (USA) for pink-colored, tear-resistant housewrap for use with *Fiberglas Pink* wall and attic insulation.

pin metal. Wrought brass, usually in the form of cold-drawn wire, containing about 62-63% copper and 37-38% zinc. Used for pins.

pinnoite. A yellow mineral composed of magnesium borate trihydrate, $Mg(BO_2)_2 \cdot 3H_2O$. Crystal system, tetragonal. Density, 2.27 g/cm^3; refractive index, 1.565; hardness, 3-4 Mohs. Occurrence: Germany.

Pinsbac. Trade name of Aktiebolaget Svenska Metallverken (Sweden) for a water-hardening steel containing 0.3-0.5% carbon, and the balance iron. Used for gears and machinery parts.

Pinstripe. Trade name of ASG Industries Inc. (USA) for a glass with single-strand wire insert.

pinwale. See pincord.

Pioloform. Trademark of Wacker-Chemie GmbH (Germany) for a series of polyvinylacetal resins.

Pioneer. (1) Trade name of Time Steel Service Inc. (USA) for an air-hardening, medium-alloy cold-work tool steel (AISI type A8) containing 0.55% carbon, 0.3% manganese, 0.9% silicon, 5.1% chromium, 1.45% molybdenum, 1.25% tungsten, and the balance iron. It has high hot hardness, and good wear resistance and toughness. Used for dies, shear blades, punches, etc.

(2) Trade name of The Feldspar Corporation (USA) for a ground hydrous magnesium silicate (talc) mined in Texas. Supplied in several grades with particle sizes of 4.2-14.2 μm (165-560 μin.), it is a soft, inert white powder with excellent uniformity and consistency containing 63.6% silicon dioxide (SiO_2), 25.9% magnesium oxide (MgO), 6.5% aluminum oxide (Al_2O_3), 1.5% calcium oxide (CaO), and small additions of ferric oxide (Fe_2O_3), potassium oxide (K_2O) and sodium oxide (Na_2O). The loss on ignition is about 8.7%, and it has a density of 2.85 g/cm^3 (0.103 lb/in.^{3}), excellent abrasion, weathering and chemical resistance, and is used as a functional filler in exterior paints and coatings, adhesives, sealants and caulks.

pipe. A tubular product of concrete, plastic or metal used to convey fluids including liquids, gases and finely divided solids.

pipe clay. A fine-grained, white-burning clay, marl or fireclay of high plasticity containing little or no iron.

Pipeglass. Trade name of Termac SA (Argentina) for glass-fiber tissue used to protect pipelines against corrosion.

Pipe-Line. Trade name of SA Glaverbel (Belgium) for patterned glass.

pipe steel. A carbon or low-alloy steel used in the manufacture of standard pipe (for the residential and industrial transmission of fluids), transmission or line pipe, oil-country tubular goods (e.g., casing, tubing and drill pipe), water-well pipe, pressure pipe, and special pipe (e.g., conduit, piling and nipples).

pipestone. A fine-grained, bright-red clay composed primarily of hydrous aluminum silicates. It is readily compressible, and has a high surface friction. Formerly used for tobacco pipes, it is now used for gaskets for high-pressure equipment. Also known as *catlinite*.

Pipewrap. Trade name of Fibreglass Limited (UK) for glass-fiber insulation supplied in the form of strips, and used for the protection of cold water pipes against frost.

piqué. A medium-weight fabric, usually cotton or a cotton blend, woven with narrow, longitudinal or vertical cords or ribs, or a honeycomb, bullseye or bird's-eye texture. Used for dresses, shirts, and sportswear.

Pireks. Trade name of Edgar Allen Balfour Steels Limited (UK) for a series of wrought or cast heat-resisting iron-nickel-chromium and nickel-chromium-iron alloys. Depending on the particular composition, they may also have good resistance to atmospheric corrosion. Used for resistors, heat-treating equipment, furnace parts, and pyrometer tubes.

pirquitasite. A brownish gray mineral of the chalcopyrite group composed of silver zinc tin sulfide, Ag_2ZnSnS_4. Crystal system, tetragonal. Density, 4.84 g/cm^3. Occurrence: Argentina.

Pirsch's German silver. A corrosion-resistant nickel silver containing 71-80% copper, 16-17% nickel, 1-7.5% zinc, 1-2.5% tin, 1-2.8% antimony, 1-2% cobalt, 1-1.5% iron and 0-0.5% aluminum. Used for tableware and ornaments.

pirssonite. A colorless to white mineral composed of sodium calcium carbonate dihydrate, $Na_2Ca(CO_3)_2 \cdot 2H_2O$. It can also be made synthetically. Crystal system, orthorhombic. Density, 2.38 g/cm^3; refractive index, 1.509. Occurrence: USA (California, Wyoming).

Pisopal. Trade name of Hurlingham SA (Argentina) for floor covering made from toughened opaline glass.

Piston. Trade name of Uddeholm Corporation (USA) for a water-hardening steel containing 1.05% carbon, 0.1% vanadium, and the balance iron. Used for pistons, drills, and taps.

pita. A strong fiber obtained from the thick, fleshy leaves of agaves (genus *Agave*), especially the century plant (*A. americana*) growing in the southwestern United States, and Central and South America. Used chiefly for the manufacture of sacks, ropes and twine. Also known as *pita fiber; pita hemp*. See also agave fibers.

Pitaloy. Trade name of Pittsburgh Steel Foundry Corporation (USA) for a series of tough cast steels containing 0.3-0.4% carbon, 0.9% manganese, 0.35% silicon, 0-1.6% nickel, 0-0.4%

molybdenum, 0.1-1% vanadium, and the balance iron. Used for general castings, locomotive frames, rolling mill machinery, crossheads, etc.

pitch. (1) The black or dark brown, viscous carbonaceous substance obtained at about 177°C (350°F) as a residue in distilling coal tar, pine tar, petroleum, etc. It liquefies upon heating and has a melting range of 140-157°C (285-315°F). *Pitch* also occurs in natural form as asphalt or glance pitch. Used in waterproofing compounds for ships, in roofing compounds, sealants and pavements, as a wood preservative, and in the manufacture of high-modulus carbon fibers.

(2) A resin obtained from the wood of certain coniferous trees, e.g., pines. See also pine pitch.

pitch-bearing basic ramming mix. A refractory ramming mix composed of basic grains to which pitch has been added.

pitch-bearing ramming mix. A refractory ramming mix that contains pitch, and is used to form monolithic furnace linings.

Pitchblack. Trade name of A Brite Company (USA) for blackening products for metals and alloys.

pitchblende. A massive variety of *uraninite* containing 55-75% uranium dioxide (UO_2), up to 30% uranium trioxide (UO_3), a small amount of water, and often other elements. It occurs in greenish-brown to black, pitchlike masses, has a dull to pitchy luster and is radioactive. Density, 6.5-8.5 g/cm³; hardness, 5.5. Occurrence: Australia, Canada, Europe, South Africa, USA (Colorado, North Carolina, Utah), Zaire. Used as a high-grade ore of uranium, and as a source of radium and actinium.

pitch-bonded basic brick. Unburned basic refractory brick bonded with pitch. Also known as *pitch-bonded basic refractory*.

pitch-bonded basic refractory. See pitch-bonded basic brick.

pitch coke. A coke obtained by destructive distillation of coal-tar pitch. It has both low sulfur and ash content, and very high carbon content, and is used in the manufacture of electrode carbon.

pitch-impregnated basic brick. Burnt basic refractory brick impregnated with pitch after firing. Also known as *pitch-impregnated refractory*.

pitch-impregnated refractory. See pitch-impregnated basic brick.

pitch oil. See creosote.

pitch pine. (1) The relatively hard, heavy, resinous wood of the medium-sized tree *Pinus rigida*. Source: Eastern United States from Maine to Georgia. Used for lumber, as fuel, as pulpwood, and as a source of pine oil.

(2) The strong, heavy wood of any of various species of pine including the North American longleaf pine *(Pinus palustris)* and slash pine *(P. elliottii)*. A pitch pine imported from the Caribbean region is *P. caribaea*. Used principally for constructional work, and to some extent for interior fittings.

Pitho. Trade name of Sanderson Kayser Limited (USA) for a nonshrinking tool steel containing 0.9-0.95% carbon, 1.2% manganese, 0.5% chromium, 0.5-0.7% tungsten, 0-0.15% vanadium, and the balance iron. Used for reamers, broaches, centers, collets, gages, and other tools.

Pitt-Char XP. Trademark of PPG Industries Inc. (USA) for a flexible, intumescent epoxy coating used for fire and corrosion protection applications.

Pittsburgh. (1) Trade name of Teledyne Pittsburgh Tool Steel (USA) for an air-hardening tool steel (AISI type A2) containing 1% carbon, 5.25% chromium, 1.1% molybdenum, 0.5% manganese, 0.25% vanadium, and the balance iron. Used for tools, dies, jigs, and precision parts.

(2) Trade name of Wheeling-Pittsburgh Steel Corporation

(USA) for a carbon steel containing 0.1% carbon, and the balance iron. Used for welding rods.

(3) Trade name of PPG Industries, Inc. (USA) for an extensive range of glass fibers and glass-fiber products.

Pitt-Ten. Trade name of Wheeling-Pittsburgh Steel Corporation (USA) for a series of high-strength low-alloy (HSLA) structural steels, usually of the niobium-vanadium-type, and supplied in the form of plate, bar, shapes and sheet piling for welded, bolted and riveted bridges and buildings, bus and truck bodies, mine cars, and industrial equipment.

Pitt-Therm. Trademark of PPG Industries, Inc. (USA) for air-dry silicone coatings used to protect carbon and stainless steels from chloride attack and stress corrosion cracking at dry heat and temperatures to 455°C (850°F).

Piuma. Trade name of Circeo Filati S.r.l. (Italy) for a yarn blend composed of 50% polyester microfibers and 50% viscose modal microfibers, and supplied in several counts. Used for textile fabrics, especially clothing.

Pivot Drill Rod. Trade name of Colt Industries (UK) for a water-hardening steel containing 1.25% carbon, and the balance iron. Used for pivots and drills.

PK-Tex. Trademark for melamine-formaldehyde plastics and products.

Placet. British trade name for an electrical resistance alloy containing 60% nickel, 20% iron, 15% chromium and 5% manganese. Used for applications requiring high heat resistance.

placing sand. A sand, usually of the silica type, that is free of iron and fluxing ingredients, and is used in the placement of ware in kilns to prevent sticking to shelves and setter plates during the firing operation.

Placovar. Trade name of Hamilton Technology Inc. (USA) for a ductile permanent magnet material that is an intermetallic compound of 76.8 wt% platinum and 23.2% cobalt. It has a density of 15.5 g/cm³ (0.56 lb/in.³), high coercive force and residual induction, a high magnetic energy product, no magnetic orientation, and a Curie temperature of approximately 480°C (896°F). Used for small magnets for electric instruments and wrist watches, and for small relays.

plagioclases. A collective name for a group of sodium and calcium feldspars including *albite* ($NaAlSi_3O_8$) and *anorthite* ($CaAl_2Si_2O_8$). Also known as *soda-lime feldspars; sodium-calcium feldspars*.

plagionite. A blackish gray mineral with metallic luster composed of lead antimony sulfide, $Pb_5Sb_8S_{17}$. Crystal system, monoclinic. Density, 5.54 g/cm³. Occurrence: Germany.

plaided fabrics. Fabrics with a pattern consisting of a repeated design of broad and narrow stripes crossing each other at right angles.

plain-carbon steels. A group of steels in which carbon is the principal alloying element, the amount of manganese does not exceed 1.65%, and the copper and silicon contents are less than 0.60%, respectively. The three main types are: (i) *Low-carbon steels* with 0.08-0.35% carbon; (ii) *medium-carbon steels* with 0.35-0.50% carbon, and (iii) *high-carbon steels* with 0.50-2.0%. Also known as *carbon steels; ordinary steels; straight carbon steels; wrought carbon steels*.

plain-carbon tool steels. A group of tool steel that contain carbon as the principal alloying element (0.60-1.50%), and only minimal amounts of other elements including manganese and silicon, chromium and vanadium. They have very high resistance to decarburization, good machinability, high toughness, good resistance to softening at elevated temperatures, low to medium wear resistance, and medium resistance to cracking. Used for cutting tools, such as reamers, taps, milling cutters and planer tools, and for chisels, screwdriver blades, cold punches, nail sets, vise jaws, anvil faces, chuck jaws, wood augers, threading dies, and die parts. Also known as *carbon tool steels*. See also water-hardening tool steels.

plain concrete. (1) Ordinary, unreinforced *concrete* that may or may not contain a fibrous admixture to keep shrinkage and temperature cracking at a minimum. Also known as *nonreinforced concrete; unreinforced concrete*. See also reinforced concrete.

(2) Ordinary *concrete* containing only aggregate, cement and water, but no admixtures, e.g., accelerators, retarders, air entrainers, etc.

plainsawn lumber. Hardwood lumber sawn lengthwise, tangent to the growth rings. The grain pattern on the surface of the sawn material is of the common "U" or "V" shape, and the annual rings are very apparent. *Note:* Softwood lumber so sawn is referred to as "flat-grained lumber."

plain-weave fabrics. Woven fabrics in which the warp threads or yarns pass over and under alternate weft (filling) threads or yarns, and vice versa.

Planatex. Trade name of Dunloplan, Division der Dunlop AG (Germany) for a flexible polyvinyl chloride film.

plancheite. A greenish blue mineral composed of copper silicate hydroxide hydrate, $Cu_8(Si_4O_{11})_2(OH)_4 \cdot xH_2O$. Crystal system, orthorhombic. Density, 3.65 g/cm^3; refractive index, 1.718. Occurrence: Argentina.

Plancher. Trade name of Ziv Steel & Wire Company (USA) for an oil-hardening, shock-resisting tool steel containing 0.5-0.7% carbon, 0.5% molybdenum, 0.8% manganese, 2% silicon, and the balance iron. Used for especially for punches, chisels and shear blades.

planed lumber. See surfaced lumber.

planed-to-caliper hardboard. Special *hardboard* machined to very close thickness tolerances.

Planemel. Trade name of Esperance Longdoz (Belgium) for a nonaging low-carbon steel sheet for porcelain enameling applications.

Planet. Trade name of A.R. Purdy Company Inc. (USA) for a series of tool steels including water-hardening plain-carbon, high-speed tool and drill-rod grades.

Planetaloy. Trade name of Kobe Steel Limited (Japan) for a high-strength, low-alloy (HSLA) steel powder made by high-pressure water atomization. It contains 3.4% nickel, 0.9% manganese, 0.9% chromium, 0.4% molybdenum, 0.4% silicon, and the balance iron. Sintered parts made from this powder have a homogeneous transformed structure, excellent mechanical properties including high tensile strength, and a small dimensional change.

Planeweld. Trade name of Lincoln Electric Company (USA) for a series of chromium-molybdenum steels containing 0.1-0.15% carbon, 0.8% chromium, 0.2% molybdenum, and the balance iron. Used for shielded-arc welding electrodes.

planewood. The moderately hard and strong wood of any of several North American, European and Asian hardwood trees of the genus *Platanus* making up the plane tree family (Platana-ceae). Important European and Asian plane trees species include the London plane (*P. acerifolia*) and the Oriental plane (*P. orientalis*). The very large American sycamore or buttonwood (*P. occidentalis*) is an American species. Also known as *sycamore*. See also American sycamore.

Planilux. Trade name of Fabbrica Pisana SpA (Italy) for tough-

ened glass blocks.

plank. (1) In softwood terminology, a square-sawn timber 50-100 mm (2-4 in.) in thickness, and 279 mm (11 in.) or more in width. See also board.

(2) In hardwood terminology, a square-sawn or unedged timber 50 mm (2 in.) or more in thickness, and of varying width. See also board.

planing mill products. Wood products, such as ceilings, flooring and siding, which have been worked (planed) to pattern.

plantain fibers. See banana fibers.

plant mix. See cold-laid plant mix; hot-laid plant mix.

plant-mixed concrete. Concrete made in a central mixing plant and transported to the job site in special equipment.

PlasGlas. Trademark of Plastics Engineering Company (USA) for a series of thermoset polyester bulk molding compounds that are available in general-purpose, high- and medium-strength, electrical, low-shrinkage and heat-resistant grades.

Plasite. Trade name of Wisconsin Protective Coating Corporation (USA) for a corrosion-resistant plastic coating used in metal finishing.

Plaskon. (1) Trademark of Amoco Chemical Company (USA) for a series of alkyd, melamine, urea and nylon molding resins, polyester and polyethylene resins for coating, bonding and impregnating, polytetrafluoroethylene resins for molded products, and several chlorotrifluoroethylene resins.

(2) Trademark of Plaskon Materials, Inc. (USA) for thermosetting epoxy and melamine- or urea-formaldehyde molding resins. They have good heat and electrical resistance, high rigidity, strength, hardness and stability, good moldability, good resistance to organic solvents, hydrocarbons and oils, poor resistance to strong acids and alkalies, and fair resistance to weathering and moisture. Used for electric devices, containers, cabinets, buttons, toilet seats, etc.

Plaslok. Trade name of Plaslok Corporation (USA) for a series of phenolic molding materials.

Plaslube. Trademark of DSM Engineering Plastics (USA) for a series of lubricated acetal, nylon and polycarbonate thermoplastics with improved friction and wear properties, used for machine parts, such as drive and pinion gears, bearings and cams.

plasma. A bright grass-green variety of *chalcedony* quartz used as a gemstone.

plasma coatings. See plasma-sprayed coatings.

Plasmadize. Trademark of General Magnaplate Corporation (USA) for composite thermal-spray coatings composed of a matrix of hard, ultrafine particles of ceramics, such as aluminum oxide, chromic oxide or tungsten carbide, infused with selected polymers and/or metals to enhance the structural integrity of the coating and impart lubricity. *Plasmadize* coatings provide outstanding wear resistance, corrosion resistance and release properties, and can be applied to many metals and alloys.

plasma-nitrided steel. Steel, usually in the form of a finished part, such as a gear, crankshaft, tool spindle, cutting tool or forging die, that has been case-hardened by using high-voltage electrical energy to ionize a low-pressure nitrogen gas and thus form a plasma through which the nitrogen ions produced are accelerated and impinged upon the part. Plasma nitriding increases the temperature of the part, and induces diffusion of the ions into the surface layer to produce a hard nitrided case. Also known as *ion-nitrided steel*. See also case-hardened steel.

Plasmaphan. Trademark of Enka AG (Germany) for a transparent cellulose used in medicine for plasmapheresis membranes.

Plasmaplate. Trade name of General Magnaplate Corporation (USA) for protective coatings of refractory metals, such as tungsten, tantalum or molybdenum, applied by a plasma spraying torch.

plasma-sprayed coatings. Coatings produced on substrates, such as ferrous metals and alloys, hafnium, niobium, tantalum, tungsten, ceramics (e.g., alumina and zirconia), tungsten and vanadium carbides, and zirconium diboride, by utilizing an electric arc formed between the electrode and the nozzle of a spray gun to heat a gas, such as argon, nitrogen or hydrogen, or a mixture of gases to such high temperatures (often exceeding 8000°C or 14430°F) that it becomes a plasma or ionized gas. The powdered coating materials are melted, atomized by the plasma and propelled onto the substrates at high speed. *Plasma-sprayed coatings* typically have a thickness of about 1.25-5.0 mm (0.05-0.2 in.) on metallic substrates and up to 0.4 mm (0.015 in.) on ceramics. They are used for improved wear and corrosion resistance, or as thermal barriers. Also known as *plasma coatings*.

Plasmold. Trade name of Firth Brown Limited (UK) for an oil-hardening, shock-resisting steel containing 0.35% carbon, 4.3% nickel, 1.3% chromium, 0.3% molybdenum, 0.5% manganese, 0.25% silicon, and the balance iron. Used for plastic dies and molds, forging dies, crankshafts, bolts and gears.

Plaspreg. Trademark of Furane Plastics Inc. (USA) for a special *plaster of Paris* impregnated with low-viscosity furan resin.

Plastacele. Trademark of E.I. DuPont de Nemours & Company (USA) for thermoplastic cellulose acetate plastics that are available in transparent, translucent, opaque, colored and colorless grades. They have good strength and toughness, high surface gloss, good chemical resistance, good moldability and machinability, a maximum service temperature of 50-93°C (120-200°F), and good resistance to hydrocarbons and oils. Used for piping, appliance housings and trim, glazing, packaging, eye shades, handles, and knobs.

Plastalloy. Trade name of Disston Inc. (USA) for a water-hardening low-alloy mold steel containing 0.08% carbon, 0.43% manganese, 1.3% nickel, 0.56% chromium, and the balance iron. It has a fine grain size, and is used for hobbed plastic molds, and for dies.

Plastapak. Trade name of H. Brand GmbH (Germany) for flexible polyvinyl chlorides.

Plastapan F. Trade name of Plasta (Germany) for flexible polyvinyl chloride plastics and products.

Plastazote. Trademark of BP Chemicals (UK) for a series of closed-cell, crosslinked, conductive polyethylene foams used in the packaging industry.

Plasted. Trade name of Stedfast Inc. (Canada) for artificial leather.

plaster. (1) A generic term for any of a group of plastic pastes obtained by mixing gypsum or lime with sand and water and, sometimes, fibrous materials. Used in covering walls, ceilings, etc.

(2) See gypsum plaster; plaster of Paris.

plaster-base finish. A material, such as wood, gypsum, fiberboard or expanded-metal lath, used as a base for plaster finishes.

plasterboard. See gypsum wallboard.

plaster-mold castings. Castings, often of intricate design, that have smooth surface finishes and good dimensional accuracy and are produced by pouring molten metal into a mold made of plaster or a combination of plaster and sand, and allowing them solidify. Metals and alloys commonly supplied as plaster-mold

castings include aluminum and manganese bronzes and yellow brass as well as aluminum and magnesium and their alloys.

plaster of Paris. A dry, white powder made by grinding and calcining (heating) gypsum. It is partially dehydrated gypsum composed chiefly of calcium sulfate hemihydrate, $CaSO_4·0.5H_2O$. When mixed with water it forms a paste which hardens quickly. Originally from Paris, France, it is used for gypsum plaster, molds, models, inexpensive statuaries, casts, in building construction, as a glass batch material, as an additive to certain glazes, as a resorbable ceramic in bone repair, and as a mounting medium for optical glasses. Also known as *calcined gypsum*. See also calcium sulfate.

plaster retarder. A substance, such as dextrin, glue, hair or animal blood, that slows or delays the setting of plaster.

Plastex. Trade name for a heat-cure acrylic used for dentures.

Plast-Grit. Trademark of Composition Materials Company Inc. (USA) for a reusable, dry stripping plastic abrasive used in the aircraft and automotive industries for surface cleaning and paint removal.

Plastibrade. Trade name of JacksonLea (USA) for a polishing cement.

Plastic. Trade name of Plasticwerk Scherer & Trier oHG (Germany) for profiles and products made of unreinforced and glass fiber-reinforced thermoplastics, such as polyamides (nylons), polyethylenes, polypropylenes or acetals, and profiles made of plastic/metal, plastic/textile or plastic/wood composites.

plastic abrasives. Mild abrasives made from plastics, such as polycarbonate or nylon, and used for abrasive blast cleaning of plastic parts, and for deburring metal parts.

Plasticalk. Trade name for nonoxidizing plastic resins with good resistance to cracking, drying, and shrinking used as replacement for putty, in joining glass to glass and metal to wood, and for caulking purposes.

Plasticarve. Trade name of Schwab & Frank, Inc. (USA) for acrylic resins.

plastic alloys. See alloys (2).

Plastic-Armor. Trade name of Permagile Epoxies, Division of American Safety Technologies, Inc. (USA) for corrosion-resistant epoxy-base surface coatings used on wood, metals, concrete, and masonry surfaces.

plastic bearing materials. Plastics, such as acetal, phenolic, polyamide (nylon) and tetrafluoroethylene resins, or plastic laminates, that are used in the manufacture of sleeve and roller bearings, e.g., for bearing balls, races and rollers. They have excellent corrosion resistance, quiet operation, good moldability and processibility, and outstanding compatibility.

plastic bronze. A bronze containing about 66-68% copper, 26-30% lead, 5% tin and 1% nickel. It has good strength, good to fair fatigue resistance, high plasticity and low hardness, and good antiscoring properties. Used for heavy-duty bearings. See also semiplastic bronze.

Plastic-Cell. Trade name for a lightweight expanded polyvinyl chloride used for safety devices (buoys, floats, etc.), and thermal insulation applications.

plastic cement. (1) A cement used to seals openings in concrete.

(2) A hydraulic cement that has one or more plasticizers added to improve its workability and moldability.

(3) A mixture of trowelable consistency composed of cutback asphalt stabilized with asbestos or other mineral fibers. It is sometimes known as *flashing cement*.

plastic-clad silica fiber. An optical-fiber cable consisting of a core fiber of silica (SiO_2) and an outer covering (cladding) of a suitable plastic material. Abbreviation: PCS fiber.

plastic clay. A clay that when blended with water will produce a moldable mass.

plastic coatings. A class of organic coatings based on thermoplastic or thermosetting resins and deposited on substrates, such as metals, plastics, ceramics, concrete, masonry, glass, wood, paper, cardboard, fabrics, leather or combinations of these, by processes including extrusion, immersion, brushing, flame spraying, knife coating (spreading), roller coating, calendering or fluidized-bed techniques. Plastic resins commonly used for coating applications include acrylics, epoxies, polyamides (nylons), polyethylenes, polypropylenes, polyvinyl chlorides and polytetrafluoroethylenes. *Plastic coatings* serve any of the following functions: (i) corrosion and weather protection of metallic parts and equipment (e.g., coatings on chemical-process or outdoor equipment); (ii) electrical insulation (e.g., dielectric coatings on electrical and electronic components); (iii) lubricity (e.g., nonstick surfaces on cooking utensils); (iv) water and moisture protection (e.g., damp- and waterproof coatings on concrete or masonry); (v) strengthening and stiffening (e.g., enhancement of strength and/or reduction of moisture absorption of paper, cardboard or textiles); and (vi) wear and abrasion resistance (e.g., nylon coatings on soft metallic parts).

plastic composites. A family of composites containing two or more distinct materials, at least one of which is a plastic. Examples of such composites include (i) *Fiber-reinforced plastics,* i.e., thermosetting or thermoplastic materials reinforced with glass, metal or ceramic fabrics, mats, or strands, whiskers or any other fiber form, and (ii) *Plastic laminates,* i.e., paper, cloth or wood bonded with thermosetting resin, rubber or adhe-sives. *Plastic composites* find application in autobody components, ablative coatings, domestic and industrial appliances and equipment, pressure vessels, tire carcasses, plywood, electrical insulation, etc. See also polymer-matrix composites.

plastic dielectrics. Thermoplastic and thermosetting materials that are used in capacitors, cable and conductor insulation, etc., for their electrically insulating or dielectric properties including high dielectric strengths, high electrical resistivities, high arc resistances, etc. Typical plastic dielectrics include phenolics, polyamides (nylons), polybutadienes, polycarbonates, polychloroprenes, polyesters, polyethylenes, polypropylenes, polystyrenes, polysulfones, polytetrafluoroethylenes, polyvinyl chlorides, polyvinylidenes, and some liquid crystal polymers.

Plasticell. Trade name of BTR Industries (UK) for polyvinyl chlorides including various cellular grades.

plastic fibers. (1) Single component fibers consisting entirely of one particular plastic material, e.g., nylon fibers.

(2) Multi-component fibers in which the core fibers and claddings consist of similar or dissimilar plastic materials.

plastic film. A polymeric material, such as cellophane, polyester, polyethylene, polypropylene, polystyrene, polyvinyl chloride or polyvinylidene, in film form, usually less than 1.3 mm (0.05 in.) in thickness, produced by blow molding, extrusion or coextrusion. It has a low density, high degree of flexibility, high tensile and tear strength, and good resistance to chemicals and moisture. Used for wrapping and packaging food products, textiles and other products, in heat sealing, for photographic film, for lining wood, fabric, paper and masonry, on slip surfaces, and in waterproofing garments.

plastic firebrick. High-duty fireclay brick made from plastic fireclay by molding, extruding or tamping. See also firebrick.

plastic fireclay. (1) Fireclay that is naturally plastic and thus capable of bonding nonplastic materials. Also known as *bond fireclay.* See also fireclay.

(2) A fireclay that is suitable for use as a plasticizing agent in mortar. Also known as *bond fireclay.* See also fireclay.

plastic foams. See cellular plastics.

Plastic Hobbing. Trade name of Edgar Allen Balfour Limited (UK) for a plain-carbon steel containing 0-0.1% carbon, 1% silicon, 0.4% manganese, and the balance iron. Used for plastic molds, and dies.

Plastic Insulcrete. Trademark of Quigley Company, Inc. (USA) for plastic refractories.

plasticized polymer. See plastisol.

Plasticlad. Trade name of JacksonLea (USA) for a greaseless satin-finishing compound.

plastic laminate. (1) A plastic product in the form of a thin sheet made by bonding together, usually in a mold by a procedure involving heat and pressure, layers or plies of a reinforcing material, such as fiberglass, asbestos, textile fabric or wood impregnated with a synthetic, often thermosetting resin. Also known as *laminated plastic; laminated molding.*

(2) A composite in the form of a sheet, rod or tube consisting a thermosetting plastic (e.g., epoxy, phenolic, polyester or silicone) bonded to asbestos, fiberglass, paper, textile, wood or other materials by adhesives, impregnation, or heat and pressure. It exhibits high tensile and dielectric strengths, and low moisture absorption. Also known as *laminated plastic.*

plastic materials. See plastics (2).

plastic metal. (1) A lead-base bearing metal containing varying amounts of antimony and tin.

(2) A tin-base babbitt containing 9.5% chromium, 8.6% antimony and 1.4% iron, used for bearings and bushings.

plastic-metal laminate. A durable plastic laminate consisting a thermosetting plastic (e.g., polyvinyl chloride or polyester) bonded to a metal, usually by means of adhesives. Used mainly for decorative applications. See also plastic laminate.

plastic microspheres. Micron-sized spheres made of plastics, such as polyesters, polystyrenes, epoxies or phenolics, used in the manufacture of plastic foams, as resin extenders, as fillers, in sandwich construction, and as acoustic and/or thermal insulation.

plastic mortar. A *mortar* of high plasticity and trowelable consistency.

Plasticoil. Trade name of Schwab & Frank Inc. (USA) for cellulose acetate plastics.

plastic paint. An organic coating in which the film-forming material is a synthetic resin or polymer, such as an acrylic, alkyd, chlorinated rubber, epoxy, fluorocarbon, nitrocellulose, phenolic, polyester, polyurethane, silicone or vinyl.

plastic pipe. A strong, lightweight, corrosion-resistant pipe made from thermoplastic or thermosetting plastics with or without glass-fiber or other reinforcements, and used for the distrubution and transmission of water, gas, oil, etc. Common plastics used for these applications include acrylonitrile-butadiene-styrene (ABS), polyvinyl chloride (PVC), high-density polyethylene (HDPE), polyesters, and phenolics.

plastic powder coatings. Coatings deposited on metallic or nonmetallic substrates from thermoplastic or thermosetting materials, such as acetals, acrylics, cellulosics, epoxies, fluorocarbons, polyamides (nylons), polycarbonates, polyethylenes or polyvinyl chlorides, in powder or granule form using processes, such as fluidized bed, flame spraying, plasma spraying, electrostatic spraying, flow coating, etc. See also plastic coatings.

plastic refractories. Graded, water-tempered refractory materials or mixtures with a consistency that they can be extruded, and of adequate workability that they can be rammed into position to form monolithic structures (e.g., furnace linings). They may or may not be air-hardening. Also known as *moldable refractories*.

Plasticresin. Trade name of JacksonLea (USA) for a polishing cement used in metal finishing.

plastics. (1) A term referring to materials, such as glass, vulcanite, phenol-formaldehyde, nylon, etc., which harden and retain their shape after being molded or shaped when subjected to heat or pressure.

(2) A term referring to a group of soft or rigid organic materials that are composed predominantly of nonmetallic elements or compounds, and produced synthetically from raw materials, such as petroleum, urea, phenol, glycerin, etc. They can be readily molded and shaped into desired form. Examples of widely used plastics include polyesters, polyethylenes, polystyrenes and polyvinyls. Usually, elastomers and synthetic fibers are not classified as plastics. Also known as *plastic materials*.

plastic semiconductor. An organic semiconductor either based on a synthetic resin, such as polyacetylene, polyparaphenylene or polyparaphenylene sulfide, having an inherently conductive or semiconductive backbone, or on an initially nonconductive resin made conductive or semiconductive by incorporating a conductive additive. In the latter, the level of semiconductivity can be varied according to requirements.

plastic sheeting. A continuously cast, extruded or molded layer of plastic whose length and width are large as compared to its uniform thickness that usually exceeds 1.3 mm (0.05 in.). Also known as *sheet plastic*.

Plastic Welder. Trademark of Devcon Corporation (USA) for a line of structural adhesives including methacrylates and high-strength epoxies used for bonding plastics, elastomers, ceramics, concrete and brick.

Plastic Wood. Trademark of Bondex International, Inc. (USA) for synthetic wood fillers used for repairing surface imperfections.

plastic wood. A composite made by compounding wood flour or wood cellulose with a suitable high-molecular-weight synthetic resin. Used as a filler for holes and seams in wood products.

PlastiDip. Trademark of PDI Inc. (USA) for tough, flexible air-dry multi-purpose rubber coatings.

Plastiform. (1) Trade name of British Steel Corporation (UK) for an air- or oil-hardening alloy tool steel containing 0.3% carbon, 4% nickel, 1.3% chromium, 0.25% molybdenum, and the balance iron. Used for stamping dies, plastic molds, and drop stamping dies.

(2) Trade name of 3M Company (USA) for flexible and easily cut permanent magnet materials in sheet and strip form composed of barium ferrite or strontium ferrite bonded with nitrile rubber, polyamide, polyethylene or polyphenylene. Supplied in rigid or flexible forms, they have high coercive force and residual inductance, and are used for instruments, and electrical systems and controls.

plastigel. A *plastisol* that has a thixotropic or gelling agent (e.g., bentonite clay) incorporated to increase viscosity.

Plastiglue. Trade name of Ato Findley (France) for neoprene adhesives used for building and floor-covering applications.

Plasti-Grit. Trade name of Composition Materials Company, Inc. (USA) for recyclable abrasive plastic media used for dry stripping and cleaning of paints and coatings from vehicles, boats, machine parts and turbine blades.

Plasti-Grit Clear Cut. Trade name of Composition Materials Company, Inc. (USA) for abrasive plastic media used for blast cleaning of composites and soft metals.

Plastikon. Trademark of Plastikon Industries (USA) for a flexible rubber sealing putty with good resistance to cracking upon long standing, and good adhesion on wood and concrete substrates.

Plastiktrim. Trade name of R.D. Werner Company (USA) for cellulose acetate plastics.

Plastilock. Trademark of Sovereign Engineered Adhesives, LLC (USA) for liquid spray adhesives used for bonding plastics, rubber, metal, glass, wood, paper, cardboard, cork, and fabrics.

Plastilux. Trade name of Plastilux GmbH (Germany) for thermoplastic, elastomeric and co-extruded profiles and shapes available in a wide range of cross sections.

Plastin. Trademark of Forchheim GmbH (Germany) for polyethylene film materials.

Plastine. (1) Trade name of Sillcocks-Miller Company (USA) for cellulose nitrate plastics.

(2) French trade name for a black-colored cellulose acetate powder.

Plastiplate. Trade name for a strong, lightweight panel material made from wood flour by combining with a phenolic bonding agent and curing with a special accelerator. Used in the construction industry.

Plastirol. Trade name of Zotefoames (France) for a polyethylene foam materials.

Plastiron. Trade name of Disston Inc. (USA) for a carburized-grade of low-carbon steel containing up to 0.12% carbon, and the balance iron. Used for plastic molds, and dies.

Plast-Iron. Trade name of National Radiator Company (USA) for high-purity electrolytic iron powder (99.98% pure) and reduced iron oxide powders. They are used for powder-metallurgy parts, pole pieces, magnets, radio cores, welding-rod coatings, self-lubricating bearings, etc.

Plastisol. (1) Trade name of ND Industries (USA) for a polyvinyl chloride supplied as a liquid dispersion for use in a proprietary open-molding process producing soft foams and hard, dense solid shapes. It is self-extinguishing and has excellent weatherability, good resistance to chemicals, and high dielectric strength.

(2) Trade name of Roesler GmbH (Germany) for polystyrene foams and foam products.

plastisol. A finely divided polymer or copolymer, such as polyvinyl acetate or polyvinyl chloride, suspended in a liquid plasticizer. Stabilizers, pigments and other additives may also be incorporated. It can be molded, cast and converted to a continuous film by curing at elevated temperatures. The film has high toughness, and excellent abrasion and corrosion resistance. Also known as *plasticized polymer*. See also organisol; plastigel.

plastisol coatings. A family of glossy, highly stable vinyl dispersion coatings applied to heated metal parts by dipping into a *plastisol* and subsequent baking in an oven. *Plastisol coatings* have high chemical resistance and dielectric strength and good resilience and flexibility. They are often applied as protective coatings to tanks, pipes, ductwork, and plating racks and baskets.

PlastiSpan. Trademark of Plasti-Fab Limited (Canada) for a polystyrene foam used for thermal insulation applications.

Plastite. Trade name of Wisconsin Protective Coating Corporation (USA) for a corrosion-resistant coating.

Plastithane. Trademark of Plasti-Fab Limited (Canada) for rigid polyurethane insulation.

Plastitube. Trade name of Schwab & Frank Inc. (USA) for cellulose acetate plastics supplied in the form of tubes and pipes.

Plastivin. Trademark of Actol Chemicals Limited (Canada) for polyvinyl acetate emulsions.

Plastizote. Trademark of Zotefoams plc (UK) for block-type polyolefin foams and foam products.

Plast-Manganese. Trade name of SCM Corporation (USA) for electrolytic manganese powder used for powder-metallurgy parts, welding rod coatings, and fuses.

Plast-Nickel. Trade name of SCM Corporation (USA) for nickel powder used for the manufacture of powder-metallurgy parts and magnets, welding rod coatings, and filters.

Plasto. Trade name of SWB Stahlformguss Gesellschaft mbH (Germany) for a series of case-hardening steels (including chromium, chromium-manganese and nickel-chromium-molybdenum grades) and several medium-carbon high-chromium stainless grades.

Plastocor. Trademark of Dipl.-Ing. Ernst Kreiselmaier GmbH & Co., Wasser- und Metall-Chemie KG (Germany) for plastic coatings used for general applications, and on containers, tubes, pipes, tube sheets, heat exchangers, condensers, waterboxes, etc.

Plastodur. Trade name of VAW Vereinigte Aluminium-Werke AG (Germany) for a heat-treatable, wrought aluminum alloy containing 2.5-5.0% copper, 0.2-1.8% magnesium and 0.3-1.5% manganese. It has low corrosion resistance, high strength, and is used for structural, automotive and aircraft parts.

Plastoflex. (1) Trade name of Advance Solvents & Chemical Corporation (USA) for cellulose acetate plastics.

(2) Trade name of Elkoflex (Germany) for flexible polyvinyl chlorides.

Plastone. Trade name National Plastics Inc. (USA) for inorganic- and cottonseed hull-filled phenolics.

Plastothen. Trademark of Forchheim GmbH (Germany) for polyethylene film materials.

Plastotrans. Trademark of Forchheim GmbH (Germany) for polyethylene film materials.

Plastrim. Trade name of Michigan Molded Plastics Inc. (USA) or cellulose acetate plastics.

Plastshine. Trade name of JacksonLea (USA) for buffing compounds used for plastics.

Plast-Silicon. Trade name of SCM Corporation (USA) for silicon powder used for fuses and pyrotechnics.

Plast-Sponge. Trade name of SCM Corporation (USA) for a high-grade, finely divided porous form of iron made by the reduction of iron oxide, and used in the manufacture of powder-metallurgy parts.

Plastupalat. Trademark of Bayer AG (Germany) for a series of acrylate polymers.

Plastylene. Trade name for a series of high-density polyethylenes.

Platabond. Trade name of Atofina SA (France) for ethylene-vinyl acetate (EVA) copolymers supplied as hot-melt powders and used for bonding textiles and leather.

Platalargan. Trade name for a corrosion-resistant alloy of platinum, aluminum and silver, used for pen points.

Platamid. Trade name of Atofina SA (France) for co-polyamides supplied as hot-melt granules and powders, and used for bonding textiles and leather.

Platanex. Trade name of Enthone-OMI Inc. (USA) for a platinum electroplate and plating process.

Platano. Trade name of Fabbrica Pisana SpA (Italy) for a patterned glass featuring leaves.

platarsite. A gray mineral of the pyrite group composed of platinum rhodium ruthenium arsenide sulfide, $(Pt,Rh,Ru)(As,S)_2$. Crystal system, cubic. Density, 8.00 g/cm^3. Occurrence: Southern Africa.

Plate. Trade name of Stahlwerke R. & H. Plate (Germany) for an extensive series of tool steels (including plain-carbon, cold-work, hot-work, nondeforming and high-speed grades), mold steels, case-hardening steels (including chromium, chromium-manganese, chromium-molybdenum, nickel-chromium, and nickel-chromium-molybdenum grades.), corrosion and/or heat-resistant austenitic, ferritic and martensitic stainless steels and several high-temperature steels.

plate. A flat-rolled metal product whose length and width are much greater than its thickness. The minimum dimensions depend on the type of metal. For example, steel plate has a thickness of 4.6 mm (0.180 in.) or greater, and a width of 203 mm (8.0 in.) or greater, and aluminum plate has a thickness greater than 6.3 mm (0.200 in.).

plate crystal. A crystal that has a high aspect ratio, i.e., whose diameter or thickness greatly exceeds its length.

plate glass. A high-quality flat glass with plane, parallel surfaces formed by a casting or rolling process. Both sides are ground and polished to provide high clarity and smoothness. Used for storefronts, large windows in industrial, commercial and residential buildings, office partitions, and mirrors.

Plate-Loy. Trade name for hot-dip terne-coated steel plate with good solderability and corrosion resistance. See also terne-coated steel.

Platemex. Trade name of Cristales Inastillables de México SA (Mexico) for laminated plate glass.

Platergal. Trade name of Lavorazione Leghe Leggere SpA (Italy) for a heat-treatable, wrought aluminum alloy containing 0.5-1.5% zinc. Used for finstock and cladding.

plater's brass. A brass alloy typically composed of 70-90% copper and 10-30% zinc, and used for plating anodes.

Plater's Metal. Trade name of Century Brass Products Inc. (USA) for a high-strength alloy containing 85-88% copper, 1.75-2.5% tin, and the balance zinc. Used for diaphragms and springs.

plate steels. A group of plain-carbon, alloy and high-strength low-alloy (HSLA) steels grades with carbon contents below 0.40% that possess good weldability, formability and machinability and high tensile strength. Used in the manufacture of steel plate for the construction of buildings, bridges, railroad cars, pressure vessels, etc.

Platherm. Trade name of Atofina SA (France) for co-polyesters supplied as hot-melt granules and powders, and used for bonding textiles and leather.

Plathuran. Trade name of Atofina SA (France) for polyurethanes supplied as hot-melt powders, and used for surface treatment applications.

Platigo. Trade name for dental alloys including *Platigo G* medium-gold platinum dental casting alloy, and *Platigo J* high-gold dental casting alloy.

Platikut. Trade name of Disston Inc. (USA) for a steel containing 0.2% carbon, 0.8% manganese, 0.6% nickel, 0.5% chromium, 0.2% molybdenum, and the balance iron. Used for molds and dies for the plastic industries.

Platilon. Trade name of Atofina SA (France) for co-polyamides, co-polyesters and polyurethanes supplied as hot-melt films, and thermoplastic polyurethane elastomers. Used mainly in the automotive, leather and textile industries.

platina. (1) A British corrosion-resistant zinc alloy containing 20.2-46.6% copper and 0.2-0.4% iron. Used for hardware, ornaments and buttons. Also known as *Birmingham platina*.

(2) A British white-colored, corrosion-resistant, brittle zinc alloy containing 20-25% copper and small amounts of iron. Used for hardware, ornaments and jewelry. Also known as *plating platina*.

PlatinaTech. Trademark of Lucent Technologies (USA) for platinum electrodeposits and deposition processes.

Platinax. Trade name of Johnson Matthey plc (UK) for a permanent magnet alloy composed of 76.7% platinum and 23.3% cobalt. It has a density of 15.5 g/cm³ (0.56 lb/in.³), very high coercivity and magnetic energy product, no magnetic orientation, high tensile strength, excellent formability, a Curie temperature of 480°C (895°F), and a maximum service temperature of 350°C (660°F). Used for small and extremely powerful magnets.

Platinel. Trade name of Engelhard Corporation (USA) for a series of alloys for high-temperature thermocouple alloys including 65Au-35Pd (for the negative leg), 83Pd-14Pt-3Au (for the positive leg), and 55Pd-31Pt-14Au.

plating platina. See platina (2).

platinic bromide. See platinum tetrabromide.

platinic chloride. See platinum tetrachloride.

platinic fluoride. See platinum tetrafluoride.

platinic iodide. See platinum tetraiodide.

platinic gold. A native alloy composed of 84.6% gold, 11.4% platinum, 2.9% silver, 0.9% copper and 0.2% iron.

platinic oxide. See platinum dioxide.

platinic sodium chloride. See sodium hexachloroplatinate.

platiniridium. A silvery, granular native alloy composed essentially of platinum and iridium with small additions of osmium, palladium, ruthenium, etc. Crystal system, cubic. Density, approximately 22.7 g/cm³ (0.82 lb/in.³); hardness, 6-7 Mohs. Used for pen nibs and jewelry.

Platinite. Trade name of Creusot-Loire (France) for a low-expansion alloy composed of 48-58% iron and 42-52% nickel. It has density of 8.2 g/cm³ (0.30 lb/in.³), a melting point of approximately 1470°C (2680°F), high heat resistance, and a low coefficient of linear expansion (same as that of platinum and glass). Used as platinum substitute in electric light bulbs, and for glass-to-metal seals.

platinized asbestos. An asbestos product of high chemical activity made by soaking asbestos fibers of suitable quality in an aqueous solution of platinum chloride and subsequent heating.

Platinlloyd. Trade name of BEGO Bremer Goldschlaegerei (Germany) for platinum-based dental alloys.

platino. A precious metal alloy composed of 89% gold and 11% platinum.

platinoid. (1) A nickel silver containing 54-60% copper, 20-24% zinc, 14-25% nickel, 0.3-2.0% tungsten, 0-0.5% iron and 0.2% manganese. It has high electrical resistance, and is used for thermocouples and heating elements.

(2) A corrosion-resistant alloy containing 50-90% copper, 3-40% nickel, 0.1% aluminum and 0.4% zinc. Used for chemical equipment and cheap jewelry.

platinous bromide. See platinum dibromide.

platinous chloride. See platinum dichloride.

platinous iodide. See platinum diiodide.

platinous oxide. See platinum monoxide.

platinous sulfide. See platinum sulfide (i)

Platinore. Trade name for a corrosion-resistant cobalt-chromium dental casting alloy.

Platinox. Trade name of Kloeckner Stahl GmbH (Germany) for a plated, extra-deep-drawing stainless steel supplied in cold- and hot-rolled sheet grades.

Platinum. Trade name of Technic Inc. (USA) for platinum plating finishes, electrodeposits and deposition processes including *Platinum AP*, a dark hard platinum finish and plating process for nickel, and *Platinum TP*, a hard, pore-free platinum electrodeposit for electronic applications.

platinum. A silvery white, malleable, ductile metallic element of Group VIII (Group 10) of the Periodic Table. It is commercially available in the form of powder (platinum black), sponge, tube, rod, wire, sheet, foil, microfoil, gauze, special compositions for electronics, metallizing and ceramic and metal decoration, and as single crystals. The single crystals are usually grown by the Czochralski technique. *Platinum* occurs native, and is often mixed with ores of copper, nickel, etc. Crystal system, cubic. Crystal structure, face-centered cubic. Density, 21.45 g/cm³; melting point, 1772°C (3222°F); boiling point, 3827°C (6921°F); hardness, 40-100 Vickers; atomic number, 78, atomic weight, 195.078; monovalent, divalent, trivalent, tetravalent. It is a noble and precious metal, and has excellent corrosion resistance, good resistance to mineral and organic acids, nitric, sulfuric and hydrofluoric acid, poor resistance to aqua regia and fused alkalies, moderate electrical conductivity (16% IACS), moderate thermal conductivity (approximately 71-72 W/m·K), and a low coefficient of thermal expansion (about 9.0×10^{-6}/K). Used for electroplating, crucibles, laboratory ware, chemical process and reaction vessels, containers, electrical contacts, thermocouples, electric furnace windings, bushings, permanent magnets, platinum resistance thermometers, spinnerets for rayon and glass fiber manufacture, automotive catalyzers, in the manufacture of electrical/electronic devices used in space, chemical, and laboratory equipment, in the production of metallic colors, as a general catalyst, for jewelry, for surgical wire, in dentistry, for coating missile and rocket nose cones, and for aircraft fuel nozzles. Symbol: Pt.

platinum alloys. A group of alloys having platinum as the principal element. The most common platinum alloys are: (i) *Platinum-rhodium alloys* used for automobile pollution-control equipment, in thermocouples, in the chemical industry as catalysts, for high temperature vessels, furnace resistors, thermocouples and resistance thermometers, in components of gas-turbine aircraft engines, and for spinnerets employed in manufacture of glass fibers, rayon fibers, etc.; (ii) *Platinum-iridium alloys* used as insoluble anodes for the production of persulfates and perchlorates and in electroplating, for jewelry and electrical contact materials, and in fuse wire and hypodermic needles; (iii) *Platinum-palladium alloys* used as insoluble anodes for electrolytic protection in electrochemistry; (iv) *Platinum-gold alloys* used for spinnerets employed in manufacture of glass, rayon and other fibers, and in jewelry; (v) *Thoriated platinum-tungsten alloys* used for spark-plug electrodes; (vi) *Platinum-cobalt alloys* which exhibit very high coercivities, and are used for permanent magnet applications; (vii) *Platinum-ruthenium alloys* used for electrical contacts and in jewelry; and (viii) *Platinum-nickel alloys* used for the manufacture of jewelry, and in solid-state electronics.

platinum arsenide. Tin-white crystals that occur in nature as the mineral *sperrylite*. Crystal system, cubic. Density, 10.58-10.60 g/cm^3; hardness, 6.5 Mohs. Used in chemistry and materials research. Formula: $PtAs_2$.

platinum-barium cyanide. See barium-platinum cyanide.

platinum beryllide. A compound of platinum and beryllium used in ceramics and materials research. Density, 4.53 g/cm^3. Formula: $PtBe_{12}$.

platinum black. A finely divided form of metallic platinum (99.9+% pure) of deep black color obtained by the reduction of a solution of a platinum salt with magnesium or zinc. It has an apparent density of 15.8-17.6 g/cm^3 (0.57-0.64 $lb/in.^3$), is available in several grades (fuel-cell, low-surface-area, low-bulk-density, etc.) and is used as a catalyst, in fuel cells, gas absorption (hydrogen, oxygen, etc.), and in gas ignition apparatus. Also known as *platinum mohr*.

platinum bromide. See platinum dibromide; platinum tribromide; platinum tetrabromide.

platinum chloride. See platinum dichloride; platinum trichloride; platinum tetrachloride.

platinum coatings. A group of highly corrosion-resistant coatings based on platinum, and produced by various processes including vapor deposition and electroplating using platinum dichloride or tetrachloride as the bath electrolyte. Used for springs, mechanical parts in electrical and electronic instruments and devices, and corrosion-resistant parts.

platinum-cobalt alloys. A group of extremely expensive permanent magnet alloys of up to 80% platinum and 20% or more cobalt. Often sold under trade names, such as *Platinax* or *Ultramag*, they have very high coercivities and magnetic energy products, no magnetic orientation, high tensile strength, excellent formability and machinability, and melting points above 1420°C (2590°F). Used for small magnets.

Platinum Colour Solder. Trade name of Engelhard Industries (USA) for a gold-base solder containing varying amounts of palladium and platinum. It has a melting range of 832-860°C (1530-1580°F), and is used in dentistry.

platinum dibromide. A red-brown powder (98% pure). Density, 6.65 g/cm^3; melting point, decomposes at 250°C (482°F). Used for platinum salts and in electroplating. Formula: $PtBr_2$. Also known as *platinous bromide; platinum bromide*.

platinum dichloride. Green crystals or olive-brown powder (98+% pure). Crystal system, hexagonal. Density, 6.05 g/cm^3; melting point, decomposes at 581°C (1077°F). Used for platinum salts and in electroplating. Formula: $PtCl_2$. Also known as *platinous chloride; platinum chloride*. See also chloroplatinic acid.

platinum diiodide. A black powder (98% pure). Density, 6.40 g/cm^3; melting point of 360°C (680°F) (decomposes). Used for platinum salts, and in electroplating. Formula: PtI_2. Also known as *platinous iodide; platinum iodide*.

platinum dioxide. Dark brown or black crystals or powder with a typical platinum content of 80-85%. Crystal system, hexagonal; melting point, 450°C (842°F); high surface area. Used in ceramics, and in the form of Adams' catalysts for platinum hydrogenation. Formula: PtO_2. Also known as *platinic oxide*. See also Adams' catalyst.

platinum disulfide. See platinum sulfide (ii).

platinum flake. Laminar or scalelike platinum powder particles of relatively small thickness (less than 1 μm or 40 μin.). Used for coating applications.

platinum fluoride. See platinum tetrafluoride.

platinum foil. Platinum in the form of foil or microfoil. Platinum microfoil (99.99% pure) on permanent plastic backing is avail-able with a thickness of 0.01-0.25 μm (0.4-10 μin.). Regular platinum foil (99.85-99.95% pure) is usually supplied in thicknesses ranging from 0.0005-0.125 mm (0.00002-0.005 in.). Used especially for electronics, dentistry, and jewelry.

platinum gauze. High-purity platinum (99.9%) in the form of gauze woven from wire with a diameter of 0.06-0.2 mm (0.002-0.008 in.). It is available in various sizes from 45 to 80 mesh, and used as a catalyst.

platinum gold. (1) A white, corrosion-resistant alloy composed of 60-70% gold and 30-40% platinum, and used for ornaments and jewelry.

(2) A corrosion-resistant alloy composed of 58% platinum, 25% silver and 17% gold, and used for jewelry.

platinum-gold alloys. Alloys that contain platinum and gold as the major constituents, and are used especially for spinnerets employed in the manufacture of glass, rayon and other fibers, and in jewelry.

platinum-group metals. See platinum metals.

platinum iodide. See platinum diiodide; platinum tetraiodide.

platinum iridium. (1) A corrosion-resistant alloy with a hardness of 400 Brinell (about 6 Mohs) composed of 70% platinum and 30% iridium, and used for surgical instruments and hypodermic needles.

(2) A corrosion-resistant alloy with a hardness of 170 Brinell (about 4 Mohs) composed of 95% platinum and 5% iridium, and used for jewelry.

platinum-iridium alloys. A group of alloys of platinum and up to 30% iridium for which the hardness, chemical resistance and melting point increases with the iridium content. The medium-hard corrosion-resistant 95Pt-5Ir alloy is used for jewelry, and the hard 90Pt-10Ir alloy has a density of 21.56 g/cm^3 (0.780 $lb/in.^3$), a melting point of 1800°C (3270°F), and is used for high-temperature thermocouples, electrical contacts, fuse wire, jewelry, and formerly as the metric standard of length. The hard 80Pt-20Ir is used for magneto contact points, fuse wire and hypodermic needles, and the corrosion-resistant 70Pt-30Ir alloy has a very high hardness (400 Brinell), and is used surgical instruments and hypodermic needles.

platinum lead. See Birmingham platinum.

platinum-lithium. A brittle solid made directly from the elements at a temperature of about 200°C (392°F), and used in the preparation of active platinum catalysts. Formula: $LiPt_2$.

platinum metals. A group of 6 transition metals (ruthenium, rhodium, palladium, osmium, iridium and platinum) that are in Group VIII (Groups 8 to 10) of the Periodic Table. It is often subdivided into two triads: (i) the Period 5 triad containing ruthenium, rhodium, palladium, and (ii) the Period 6 triad with osmium, iridium and platinum. The members of the former triad all have densities of about 12-12.5 g/cm^3 (0.43-0.45 $lb/in.^3$), while those of the latter triad all have densities above 21.4 g/cm^3 (0.77 $lb/in.^3$),. Also known as *platinum-group metals*.

platinum mohr. See platinum black.

platinum monoxide. Black crystals or violet-black powder. Crystal system, tetragonal. Density, 14.1 g/cm^3; melting point, decomposes at 325°C (617°F). Formula: PtO. Used in chemistry and materials. Also known as *platinous oxide*.

platinum monosulfide. See platinum sulfide (i).

platinum-nickel alloys. A group of corrosion-resistant alloys composed of platinum and nickel, and used in the manufacture of jewelry and ornaments, and in solid-state electronics. A typical composition is about 90% platinum and 10% nickel.

platinum-osmium alloys. A group of corrosion-resistant electrical contact alloys composed of platinum and osmium with the osmium content being usually 35% or more.

platinum oxide. See platinum dioxide; platinum monoxide.

platinum-palladium alloys. Alloys composed predominantly of platinum and palladium, and used for relays, electrical contact (often with ruthenium, or gold and silver additions), and in insoluble anodes for electrochemistry.

platinum-polyethylenimine. See Royer catalysts.

platinum powder. A finely divided form of platinum made by atomization processes or chemical reduction. The free-flowing, spherical atomized powder has a purity of 99.9%, and is supplied in particle sizes of 35-700 µm (0.0014-0.276 in.). The chemically reduced amorphous powder has a purity of 99.9+%, and is available in fine particle sizes from 0.25-3.5 µm (10-138 µin.). Used chiefly as a catalyst, and in powder metallurgy.

platinum-rhenium. A compound of platinum and rhenium used in ceramics and materials research. Melting point, 2450°C (4440°F). Formula: PtRe.

Platinum Rhodium. Trade name for a heat-resistant alloy of 80-100% platinum and 0-20% rhodium. Used for thermocouples.

platinum-rhodium alloys. A group of alloys composed of platinum and up to 40% rhodium and having higher hardnesses, melting points and high-temperature strengths and better acid resistance particularly to aqua regia than unalloyed platinum. The 87Pt-13Rh alloy has a density of 19.64 g/cm³ (0.710 lb/in.³), a melting point of 1860°C (3380°F), and is used for standard thermocouples in oxidizing or inert atmospheres at up to 1480°C (2700°F). The 90Pt-10Rh alloy is a standard thermocouple alloy with a density of 19.97 g/cm³ (0.721 lb/in.³), a melting point of 1850°C (3360°F), a hardness of 90-165 Brinell, and is employed in oxidizing or inert atmospheres at up to 1480°C (2700°F). The 70Pt-30Rh alloy has a melting point of 1927°C (3501°F), and is used for standard thermocouples employed in oxidizing and inert atmospheres, or in vacuum at up to 1700°C (3100°F). *Platinum-rhodium alloys* are also used in automobile pollution-control equipment, in the glass industry particularly as glass-fiber extrusion bushings, in the chemical industry as catalysts, in high-temperature vessels, furnace resistors and resistance thermometers, in components of gas-turbine aircraft engines, and in spinnerets for the manufacture of glass, rayon and other fibers.

platinum-ruthenium alloys. A group of relatively hard and wear-resistant alloys composed of platinum and up to about 15% ruthenium. Alloys with 86-95% platinum and 5-14% ruthenium are often used for electrical contacts, and jet-engine glow plugs. The 95Pt-5Ru alloy has a solidus temperature of 1775°C (3230°F), an electrical conductivity of 5% IACS, a density of 20.57 g/cm³ (0.743 lb/in.³), and is used for electrical contacts, and jewelry, and the 90Pt-10Ru alloy has a melting point of 1800°C (3272°F), a density of 20.10 g/cm³ (0.726 lb/in.³), and an electric conductivity of 4% IACS.

platinum sesquisulfide. See platinum sulfide (iii).

platinum silicide. Orthorhombic crystals with a density of 12.4 g/cm³ and a melting point of 1229°C (2244°F). Used as a semiconductor with infrared sensitivity up to 5000 nm. Formula: PtSi.

platinum silver. A corrosion-resistant alloy of 66.7% silver and 33.3% platinum, used for jewelry and ornaments.

platinum-sodium chloride. See sodium hexachloroplatinate; sodium tetrachloroplatinate.

platinum solder. A corrosion-resistant alloy composed of 73% silver and 27% platinum, and used as a solder for platinum alloys.

platinum sponge. See sponge platinum.

platinum substitute. Any of a group of silvery white gold-, silver-, nickel- and zirconium-base alloys including (i) corrosion-resistant alloys composed of about 72% nickel, 24% aluminum, 3.5% bismuth and 0.5% gold used for ornamental parts; (ii) corrosion-resistant jewelry alloys composed of 70% silver, 25% platinum and 5% cobalt or nickel; and (iii) heat-resistant alloys composed of about 70% gold, 25% silver, 5-7% nickel or platinum, and used for electrical contacts.

platinum sulfide. Any of the following compounds of platinum and sulfur used chiefly in chemistry, materials science and electronics: (i) *Platinum monosulfide.* Gray, tetragonal crystals or black powder. It occurs in nature as the minerals *cooperite* and *braggite.* Density, 8.85 g/cm³ (braggite); 10.26 g/cm³ (cooperite). Formula: PtS. Also known as *platinous sulfide;* (ii) *Platinum disulfide.* Hexagonal crystals or black powder. Density, 7.22 g/cm³. Formula: PtS_2; and (iii) *Platinum sesquisulfide.* A gray powder. Density. 5.52 g/cm³. Formula: Pt_2S_3.

platinum tetrabromide. Brown-black crystals or brown powder (99.9% pure) with a density of 5.69 g/cm³ that decomposes at 180°C (356°F), and is used for platinum salts and in electroplating. Formula: $PtBr_4$. Also known as *platinic bromide.*

platinum tetrachloride. Brown-red, moisture-sensitive crystals (98+% pure). Crystal system, cubic. Density, 4.30 g/cm³; melting point, decomposes at 370°C (698°F). Used for platinum salts and in electroplating. Formula: $PtCl_4$. Also known as *platinic chloride.*

platinum tetrafluoride. Red crystals with a melting point of 600°C (1112°F) used in chemistry and materials research. Formula: PtF_4. Also known as *platinic fluoride; platinum fluoride.*

platinum tetraiodide. An amorphous brown-black powder (99.9% pure) that decomposes at 130°C (266°F), and is used for platinum salts and in electroplating. Formula: PtI_4. Also known as *platinic iodide.*

platinum tribromide. Green-black crystals that decompose at 200°C (392°F), and is used in chemistry and materials research. Formula: $PtBr_3$. Also known as *platinum bromide.*

platinum trichloride. Green-black crystals. Density, 5.25 g/cm³; melting point, decomposes at 435°C (815°F). Used for platinum salts and in electroplating. Formula: $PtCl_3$. Also known as *platinum chloride.* See also chloroplatinic acid.

platinum wire. Platinum in wire form supplied in purities from 99.9 to 99.99%, and diameters of 0.2 µm to 1.5 mm (8 µin. to 0.06 in.). The high-purity grades are often used for thermocouples. Fine platinum wire (typically 0.2-10 µm or 8-394 µin. in diameter) is produced by the Wollaston process, and used for scientific instruments. *Platinum wire* with insulating coatings of polyurethane, polyamide (nylon), polyimide, polyester and polytetrafluoroethylene is also available for electrical applications. See also Wollaston wire.

Platnam. Trade name of Hopkinsons, Britannia Works (UK) for a nickel alloy containing 33% copper, 13% tin, 0.5% iron and 0.3% aluminum. It has good elevated temperature properties up to 600°C (1110°F), and is used for valve disks and seats.

Platnic. Trade name for a corrosion-resistant alloy composed of platinum and nickel, and used for jewelry and ornaments.

Platogum. Trade name for dental impression plaster.

plattnerite. An iron-black mineral of the rutile group composed of lead oxide, PbO_2. Crystal system, tetragonal. Density, 9.56 g/cm³; refractive index, 2.3. Occurrence: Mexico.

Plattzink. Trademark of Platt Brothers & Company (USA) for zinc and zinc-alloy rod, strip and metallizing wire.

Plavia. Trade name for rayon fibers and yarns used for textile fabrics.

Plax Methacrylate. Trade name of Plax Corporation (USA) for a series synthetic resins including polymethyl methacrylates (*Plax Methacrylate*) and polystyrenes (*Plax Polystyrene*).

playfairite. A lead-gray mineral composed of lead antimony arsenic sulfide, $Pb_{16}(Sb,As)_{18}S_{43}$. Crystal system, monoclinic. Density, 5.64 g/cm^3. Occurrence: Canada (Ontario).

pleated fabrics. Fabrics that have one or more flat, relatively narrow folds produced by doubling the material on itself and pressing into place.

Plei-tech. Trademark of Pleiger Plastics Company (USA) for a series of polyurethane elastomers that can be processed by various techniques including compression and open-cast molding, and spin casting to produce low-density cellular foam. *Plei-tech* elastomers have good wear resistance and shock absorption properties, and are used for diaphragms, seals, wipers, couplings, gears, rolls, wheels, etc.

Plenco. Trademark of Plastics Engineering Company (USA) for a series of synthetic resins and molding compounds including *Plenco Alkyd* alkyd resins supplied in long- and short-glass-fiber reinforced, mineral-filled, fire-retardant and high-impact grades for the manufacture of molded products, *Plenco GMC* general-purpose polyester molding compounds, *Plenco MF* cellulose-filled melamine formaldehyde molding compounds, *Plenco P'Est* flexible and rigid cast polyesters, and *Plenco PF* phenolic molding materials, cotton fabrics, glass fabrics, and paper laminates.

pleochromic dyes. A class of synthetic organic dyes used in certain colored liquid crystal displays and having different light absorptions and thus colors along different liquid crystal alignment axes.

pleonaste. See ceylonite.

plessite. An intimate mixture of *kamacite* (a low-nickel iron phase) and *taenite* (a high-nickel iron phase), found between pure kamacite areas in some meteorites.

Plessy's green. A deep-green, impure pigment of chromium phosphate ($CrPO_4$) and 2-6 water molecules (H_2O).

Plestar. Trade name for a polymer film based on polycarbonate.

Plettenberg. Trade name of Plettenberger Gusstahlfabrik (Germany) for an extensive series of steels including various alloy machinery grades (chromium-manganese, chromium-molybdenum, nickel-chromium-molybdenum, nickel-chromium-molybdenum-vanadium and silicon-manganese types including several carburizing grades), numerous cold-work, hot-work, and water-hardening plain-carbon tool grades, and martensitic stainless grades.

Plex. Trademark of Roehm GmbH (Germany) for a series of polymethyl methacrylate resins.

Plexalkyd. Trademark of Roehm GmbH (Germany) for alkyd resin/methyl methacrylate resins.

Plexalloy. Trademark of Roehm GmbH (Germany) for modified polymethyl methacrylate polymers.

Plexar. Trademark of Quantum Chemical Corporation (USA) for a series of polyolefin-based materials including polyethylene coating resins for bonding paper, paperboard, wood and various plastics (nylon, polyethylene, etc.), and extrudable polyolefin adhesives for use in multilayer barrier packaging, and polar-to-nonpolar plastics bonding.

Plexene. Trademark of AtoHaas America Inc. (USA) for a series of styrene-acrylic resins used for molded parts with good dielectric properties.

Plexidur. Trademark of Roehm GmbH (Germany) for polmethyl methacrylate and polacrylonitrile resins.

Plexiglas. Trademark of Atoglas, Atofina Chemicals Inc. (USA) for transparent thermoplastic polymethyl methacrylate (PMMA) polymers supplied in the form of beads, granules, mold and extrusion resins, and cast sheets. They are available in numerous grades including general-purpose, high-heat, high-flow, optical, impact-modified and gamma ray-resistant, and have a density of 1.19 g/cm^3 (0.043 $lb/in.^3$), a refractive index of 1.49, outstanding light transmission and resistance to weathering, fair mechanical properties, good electrical resistance, low hardness and heat resistance, a maximum service temperature of 60-93°C (140-200°F), excellent resistance to shattering, relatively high thermal expansion, good ultraviolet resistance, good resistance to dilute acids and alkalies, fair resistance to concentrated acids, poor resistance to aromatic hydrocarbons, greases, oils, halogens, and ketones, and poor flame resistance (but are slow burning or self-extinguishing). Used as glass substitutes, for windows, lenses, contact lenses, decorative illuminated signs, letters for signs, nameplates, decorations, ornaments, lighting fixtures, drafting equipment, furniture components, transparent enclosures, display items, bottles, dials, light diffusers, industrial and architectural glazing, aircraft canopies and windows, and windshields for boats.

Plexigum. Trademark of AtoHaas America Inc. (USA) for a synthetic acrylate resin that has elastomeric properties and is supplied in powder, film or liquid form. It is used for injection molding compounds, and for bonding glass or plastic sheets.

Plexileim. Trade name of Rohm & Haas Company (USA) for polyacrylic resins and compounds used for the manufacture of adhesives.

Plexlith. Trade name of Rohm & Haas Company (USA) for acrylate resins.

Pleximon. Trade name of Rohm & Haas Company (USA) for acrylate resins.

Plexine. Trade name of Rohm & Haas Company (USA) for polystyrene resins and products.

Plexisol. Trade name of Rohm & Haas Company (USA) for polymethyl methacrylate resins.

Plexit. Trade name of Rohm & Haas Company for polymethyl methacrylate plastics.

Plexite. Trade name of ASG Industries Inc. (USA) for a glazing material composed two sheets of glass, one of double and the other of single strength, bonded with *Plexigum*. See also double-strength glass; single-strength glass.

Plextol. Trade name of Rohm & Haas Company (USA) for acrylate resins.

Plexon. (1) Trademark of Sandstrom Products Company (USA) for high-gloss, corrosion-resistant acrylic urethane enamel coatings used on concrete, steel, and other materials.

(2) Trade name of Experimental Fabrics, Inc./Plexon (USA) for vinyl-coated fabrics.

Plextol. Trademark of Bayer Corporation (USA) for rubber latex used in building construction, as a backing for textiles, carpets and paper, and in the manufacture of molded foams. Also included under this trademark are plexiglas-type acrylic resins.

Plexus. Trade name of ITW Devcon (USA) for a series of two-component structural methacrylate adhesives that possess high tensile and peel strength, and bond well to many substrates including aluminum, steel, ferrites, ceramics, composites and many thermoplastics and thermosets. Some grades have high chemical and fuel resistance, and are stable to ultraviolet light. Used in the automotive, chemical, transportation and marine industries.

Plex-Weld. Trade name of Bell Chemical Company (USA) for acrylic cement.

Pliaglas. Trade name of Pioneer Suspender Company (USA) for polyvinyl chlorides and other vinyl plastics.

Pliana. Trademark of Industrias Polifil SA (Mexico) for polypropylene staple fibers and filament yarns.

Plibrico. Trademark of Plibrico Company (USA) for plastic refractories.

Plicast. Trademark of Plibrico Company (USA) for hydraulic-setting refractories.

plied yarn. A yarn made by twisting together two or more individual yarns. Also known as *doubled yarn; folded yarn; formed yarn; multiple yarn*. See also single yarn.

Pligun. Trade name of Plibrico Company (USA) for refractory gunning mixes.

Pliobond. Trade name of W.J. Ruscoe Company (USA) for nitrile rubber construction adhesives.

Pliochlor. Trade name of Goodyear Tire & Rubber Company (USA) for chlorinated synthetic rubber.

Pliofilm. Trade name of Goodyear Tire & Rubber Company (USA) for rubber hydrochloride available as a coating material for producing tough, water-resistant films on paper or textiles, and as a transparent film or sheet with a heat-sealing temperature of 105-130°C (221-266°F) for wrapping and heat-sealing applications.

Plioflex. Trademark of Goodyear Tire & Rubber Company (USA) for a series general-purpose extended styrene-butadiene rubbers.

Plioform. Trade name of Goodyear Tire & Rubber Company (USA) for a thermoplastic material obtained by polymerizing rubber and supplied in transparent, translucent, opaque and colored grades. It possesses good processibility and machinability and good resistance to acids, alkalies, alcohols, ketones and esters. Used in lacquers, as a molding powder, and for molded products.

Pliolite. Trademark of Goodyear Tire & Rubber Company (USA) for a series of polymerized, cyclized or isomerized rubber products including various highly chlorinated rubbers used in insulating compounds, adhesives and protective paints, styrene-acrylate copolymers used in protective coatings, synthetic and natural rubber latexes, styrene-butadiene copolymers, and rubber reinforcing resins.

Pliolithe. Trade name of Goodyear Tire & Rubber Company (USA) for styrene-butadiene rubber.

Pliosheen. Trade name of Goodyear Tire & Rubber Company (USA) for rayon fabrics with elastomeric coatings.

Pliotec. Trade name of Goodyear Tire & Rubber Company (USA) for a family of waterborne resins used as industrial coatings on metals, plastics, ceramics, etc. They are available in general-purpose and high-performance grades.

Pliovic. Trademark of Goodyear Tire & Rubber Company (USA) for thermoplastic resins that are copolymers of vinyl chloride and vinylidene chloride. They are available in the form of fine white powders for compounding, and can be processed by standard methods including blow, compression and injection molding, calendering and extruding. *Pliovic* resins are usually rigid, but can be made flexible with plasticizers. They have poor resistance to heat distortion, and are used for piping, garden hose,

electrical wire insulation, floor coverings, tile and coatings, and formerly for phonograph records.

Plioway. Trademark of Goodyear Tire & Rubber Company (USA) for vinyl-acrylic resins used for low-odor, solvent-based coatings.

Plisulate. Trade name of Plibrico Company (USA) for refractory insulating cements.

Pliseal. Trade name of Plibrico Company (USA) for refractory boiler/wall coatings.

plissé. A plain-colored or printed cotton or cotton-polyester fabric in a plain weave having a pleated or puckered surface produced either by chemical treatment (e.g., printing with caustic soda), or by the use of yarns with different degrees of shrinkage. It does not need ironing or pressing, and is used especially for blouses, skirts, housecoats, swimwear, and underwear.

Plitex. Trade name of Hood Rubber Company (USA) for phenol-formaldehyde resins, plastics and laminates.

plombierite. A pink to red brown mineral composed of calcium hydrogen silicate hexahydrate, $Ca_5H_2Si_6O_{18}·6H_2O$. Density, 2.02 g/cm³; refractive index, 1.495. Occurrence: Ireland.

Plowface. Trade name of Champion Rivet Company (USA) for an abrasion- and impact-resistant cast iron containing 4.5% carbon, 6.5% manganese, 2% silicon, 30% chromium, and the balance iron. Used for hard-surfacing electrodes.

plow steel. A high-grade steel containing 0.5-0.9% carbon, and having high tensile strength. It is used for wire rope, in the manufacture of wire for prestressed concrete, and for plow shares and blacksmith tools.

plugging compound. A putty-like mixture of inorganic materials, such as powdered clay or frit, with water, used to plug holes and provide a smooth, uniform surface on cast iron prior to porcelain enameling.

plumalsite. A green or yellow, and sometimes black or colorless mineral composed of lead aluminum silicate, $Pb_4Al_2(SiO_3)_7$. Crystal system, orthorhombic. Density, 4.38 g/cm³; refractive index, above 1.782. Occurrence: Ukraine.

plumbago. See graphite.

plumbago bricks. Refractory bricks made from a mixture of about equal parts of graphite and refractory clay.

plumber's white. A group of free-cutting, corrosion-resistant alloys of 54-58% copper, 25-27% zinc, 13-17% nickel, 1-7% lead, 0-1% iron, and 0-1% tin. Used for plumbing fixtures, cocks, faucets, etc.

plumber's wiping solder. See wiping solder.

plumbic bronze. A bronze containing 69.6% copper, 26% lead, 1.7% manganese, 1.5% tin and 1.2% iron. It has good machinability, and is used for heavy-duty bearings and bushings.

plumbic ocher. See massicot.

plumboferrite. A black mineral composed of lead iron oxide, $PbFe_4O_7$. It can also be made synthetically. Crystal system, hexagonal. Density, 6.07 g/cm³. Occurrence: Sweden. Used in ceramics, materials research and electronics.

plumbogummite. A brown mineral of the alunite group composed of lead aluminum phosphate hydroxide monohydrate, $PbAl_3(PO_4)_2(OH)_5·H_2O$. Crystal system, rhombohedral (hexagonal). Density, 4.01 g/cm³; refractive index, 1.652. Occurrence: France.

plumbojarosite. A brown mineral of the alunite group composed of lead iron sulfate hydroxide, $PbFe_6(SO_4)_4(OH)_{12}$. It can also be made synthetically. Crystal system, rhombohedral (hexagonal). Density, 3.65 g/cm³; refractive index, 1.875. Occurrence: Greece, USA (New Mexico).

plumbonacrite. A synthetic mineral composed of lead oxide carbonate hydroxide, $Pb_{10}(CO_3)_6(OH)_6O$. Crystal system, hexagonal. Density, 7.07 g/cm^3.

plumbopalladinite. A white or grayish white mineral with a rose tint. It belongs to the nickeline group and is composed of lead palladium, Pb_2Pd_3. It can also be made synthetically. Crystal system, hexagonal. Density, 12.40 g/cm^3; hardness, 400-440 Vickers. Occurrence: Russian Federation.

plumboplumbic oxide. See red lead oxide.

plumbopyrochlore. A red mineral of the pyrochlore group composed of lead rare-earth niobium oxide hydroxide, $(Pb,Ln)_{2-x}(Nb,Ta)_2O_6(OH)$. Crystal system, cubic. Density, 6.75 g/cm^3; refractive index, 2.08. Occurrence: Russia.

Plumb-O-Sil. Trademark of National Lead Company (USA) for soft, white powders that are coprecipitates of lead orthosilicate and silica gel. Used in the manufacture of translucent or colored vinyl film, and as stabilizers in vinyl plastics.

plumbotellurite. A gray, or yellow-gray to brown mineral composed of lead tellurite, α-$PbTeO_3$. Crystal system, orthorhombic. Density, 7.20 g/cm^3; refractive index, 2.23. Occurrence: Kazakhstan.

plumbotsumite. A colorless mineral composed of lead silicate hydroxide, $Pb_5Si_4O_8(OH)_{10}$. Crystal system, orthorhombic. Density, 5.60 g/cm^3; refractive index, 1.933. Occurrence: Namibia.

plumbous acetate. See lead acetate (2).

plumbous oxide. See lead monoxide.

plumbous sulfide. See lead sulfide.

Plumbral. Trade name of Foseco Minsep NV (Netherlands) for covering and cleansing fluxes for high-leaded copper alloys. They limit oxide formation, prevent pick-up of gases during melting, and aid in the formation of a microstructure with uniform lead distribution.

Plumbsol. Trade name of Johnson Matthey plc (UK) for soldering alloys containing 95-100% tin and 0-5% silver. The 95Sn-5Ag alloy has melting range of 221-241°C (430-465°F). *Plumbsol* solders have good wettability, high joint strength, and are used as soft solders for plumbing installations.

Plumrite. Trade name of Olin Brass, Indianapolis (USA) for a series of leaded and unleaded brasses used for water pipes, plumbing, flexible hose, and deep-drawn parts. The average composition range is 60-85% copper, 15-40% zinc and 0-1% lead. The leaded grades have free-cutting properties. Also included under this trade name are several high-purity coppers.

Pluralite. Trade name of C-E Glass (USA) for glass with a shallow, fluted pattern.

Pluralloy. Trade name of Teledyne McKay (USA) for a series of covered welding electrodes based on carbon or low-alloy steel, and used for shielded metal-arc welding.

Pluramelt. Trade name for engineering composites made by bonding various types of stainless steel (austenitic, ferritic, etc.) by intermelting.

Plus Castables. Trade name of A.P. Green Industries Inc. (USA) for castable refractories.

plush. A fabric, usually knitted or woven from cotton, silk, wool or synthetic yarn and having a cut pile on one side which is softer and longer than that of *velvet*. Used for clothing, draperies, upholstery and stuffed toys.

Pluswood. Trademark of Pluswood Inc. (USA) for particleboard.

Pluto. Trade name of former Vereinigte Edelstahlwerke (Austria) for a tungsten-type high-speed steel containing 0.7% carbon, 4% chromium, 19% tungsten, 2% vanadium, and the balance iron. Used for lathe and planer tools, cutters, drills, broaches, reamers, dies, and punches.

Pluto 2P. Trade name for a high-gold dental casting alloy with rich yellow color.

Plutocrat. Trade name of Richard W. Carr & Company Limited (UK) for a high-speed steel containing 0.8% carbon, 22% tungsten, 4.5% chromium, 1% vanadium, 0.5% molybdenum, and the balance iron. Used for lathe, cutoff and boring tools, milling cutters, reamers, and heavy-duty drills.

plutonium. A silvery white, synthetic, radioactive, metallic element of the actinide series of the Periodic Table. It was first produced by bombardment of uranium-238 with deuterons in a cyclotron. Trace amounts occur naturally in uranium ores (e.g., pitchblende). The most important isotopes are ^{238}Pu, ^{239}Pu, ^{240}Pu, ^{241}Pu and ^{244}Pu. Atomic number, 94; atomic weight, 244; trivalent, tetravalent, pentavalent, hexavalent. Six allotropic forms are known: (i) *Alpha plutonium* (monoclinic crystal structure) that is stable below 118°C (244°F) and has a density of 19.86 g/cm^3; (ii) *Beta plutonium* (monoclinic crystal structure) that is stable between 118 and 200°C (244 and 392°F) and has a density of 17.70 g/cm^3; (iii) *Gamma plutonium* (orthorhombic crystal structure) that is stable between 200 and 312°C (392 and 594°F) and has a density of 17.14 g/cm^3; (iv) *Delta plutonium* (face-centered cubic crystal structure) that is stable between 312 and 458°C (594 and 856°F), and has a density of 15.92 g/cm^3; (v) *Delta-prime plutonium* (tetragonal crystal structure) that is stable between 458 and 480°C (856 and 896°F) and has a density of 16.00 g/cm^3; and (vi) *Epsilon plutonium* (body-centered cubic crystal structure) that is stable between 480 and the melting point of 640°C (896 and 1184°F), and has a density of 16.51 g/cm^3. *Plutonium* is used as a nuclear reactor fuel, in nuclear batteries, pacemakers, nuclear weapons, production of radioisotopes, and in film cleaners.

plutonium-238. The first synthetic radioisotope of plutonium having a mass number of 238. It can be produced by bombardment of uranium-238 with deuterons in a cyclotron, decays by alpha-particle emission and has a half-life of approximately 89.6 years. Symbol: ^{238}Pu.

plutonium-239. A synthetic radioisotope of plutonium having a mass number of 239. It can be produced by bombardment of nonfissionable uranium-238 with slow electrons in a nuclear reactor, decays by alpha-particle emission, and has a half-life of 24360 years. Used as a nuclear reactor fuel and in nuclear weapons. Symbol: ^{239}Pu.

plutonium-240. A synthetic radioisotope of plutonium having a mass number of 240. It decays by alpha-particle emission, and has a half-life of 6580 years. Used for nuclear applications. Symbol: ^{240}Pu.

plutonium-241. A synthetic radioisotope of plutonium having a mass number of 241. It emits beta and gamma radiation, and has a half-life of 14 years. Used for nuclear applications. Symbol: ^{241}Pu.

plutonium-aluminum alloy. Plutonium to which a small amount of aluminum (typically 0.1 wt%) has been added as a delta-phase stabilizer resulting in the retention of the delta phase at room temperature. Used for nuclear applications.

plutonium beryllide. A compound of plutonium and beryllium. Density, 4.36 g/cm^3; melting point, 1700°C (3092°F). Used for ceramic and nuclear applications. Formula: $PuBe_{13}$.

plutonium boride. Any of the following compounds of plutonium and boron used for ceramic and nuclear applications: (i) *Plutonium monoboride*. Density, 14.10 g/cm^3. Formula: PuB; (ii) *Plutonium diboride*. Density, 12.81 g/cm^3. Formula: PuB_2;

(iii) *Plutonium tetraboride.* Tetragonal crystals. Density, 9.36. Formula: PuB_4; and (iv) *Plutonium hexaboride.* Density, approximately 7.25 g/cm^3. Formula: PuB_6.

plutonium carbide. Any of the following compounds of plutonium and carbon used in pellet form in nuclear fuel elements: (i) *Plutonium monocarbide.* Density, 13.5-14.0 g/cm^3; melting point, approximately 1655°C (3011°F). Formula: PuC; and (ii) *Plutonium sesquicarbide.* Cubic crystals. Density, 12.7 g/cm^3. Formula: Pu_2C_3.

plutonium diboride. See plutonium boride (ii).

plutonium dioxide. See plutonium oxide (iii).

plutonium disilicide. See plutonium silicide (i).

plutonium hexaboride. See plutonium boride (iv).

plutonium-iron alloy. A low-melting alloy of plutonium and iron used as a liquid reactor fuel.

plutonium monoboride. See plutonium boride (i).

plutonium monocarbide. See plutonium carbide (i).

plutonium monosulfide. See plutonium sulfide (i).

plutonium monoxide. See plutonium oxide (i).

plutonium oxide. Any of the following compounds of plutonium and oxygen used for nuclear applications: (i) *Plutonium monoxide.* Density, 13.9 g/cm^3; Formula: PuO; (ii) *Plutonium sesquioxide.* Hexagonal crystals. Density, 10.2-11.2 g/cm^3; melting point, 2216°C (4021°F). Formula: Pu_2O_3; and (iii) *Plutonium dioxide.* Density, 11.46 g/cm^3; melting point, 2241°C (4066°F). Formula: PuO_2.

plutonium phosphide. A compound of plutonium and phosphorus with a density of 10.88 g/cm^3. Used for ceramic and nuclear applications. Formula: PuP.

plutonium sesquicarbide. See plutonium carbide (ii).

plutonium sesquioxide. See plutonium oxide (ii).

plutonium sesquisulfide. See plutonium sulfide (ii).

plutonium silicide. Any of the following compounds of plutonium and silicon used for ceramic and nuclear applications: (i) *Plutonium disilicide.* Density, 9.12-9.18 g/cm^3. Formula: $PuSi_2$; (ii) *Plutonium trisilicide.* Formula: $PuSi_3$; and (iii) *Triplutonium disilicide.* Density, 11.98 g/cm^3. Formula: Pu_3Si_2.

plutonium sulfide. Any of the following compounds of plutonium and sulfur used for ceramic and nuclear applications: (i) *Plutonium monosulfide.* Density, 10.60 g/cm^3; and (ii) *Plutonium sesquisulfide.* Density, 8.41 g/cm^3; melting point, approximately 1720°C (3128°F). Formula: Pu_2S_3.

plutonium tetraboride. See plutonium boride (iii).

plutonium trisilicide. See plutonium silicide (ii).

Pluto Paramount. Trade name of Richard W. Carr & Company Limited (UK) for a cobalt-tungsten high-speed steel containing 0.8% carbon, 4.5% chromium, 0.7% molybdenum, 18% tungsten, 1.25% vanadium, 10% cobalt, and the balance iron. Used for cutting and lathe tools, and broaches.

Pluto Perfectum. Trade name of Richard W. Carr & Company Limited (UK) for a cobalt-tungsten high-speed steel containing 0.8% carbon, 4.8% chromium, 0.8% molybdenum, 18% tungsten, 1.2% vanadium, 5% cobalt, and the balance iron. Used for cutting tools, milling cutters, reamers, and lathe and planer tools.

Pluto Premium. Trade name of Richard W. Carr & Company Limited (UK) for a high-speed steel containing 1.55 % carbon, 4.5% chromium, 3% molybdenum, 6.3% tungsten, 5% vanadium, 5% cobalt, and the balance iron. Used for cutting tools and broaches.

ply. (1) A general term for a single layer in an assembly composed of laminated pieces.

(2) A single layer of metal in a laminate.

(3) A strand or twist of yarn or rope, e.g., a four-ply yarn consists of four strands of yarn twisted together.

(4) A single layer of laminated fabric or felt.

(5) Any of the single layers of fabric used in the construction of an automotive tire.

(6) Any of the thin layers or sheets of veneer that make up a plywood panel.

Plyamine. Trade name of Reichhold Chemicals, Inc. (USA) for amino and urea resins and laminates.

Plyamul. (1) Trademark of Reichhold Chemicals, Inc. (USA) for vinyl acetate homopolymers used in the manufacture of polyvinyl acetate adhesives.

(2) Trademark of Reichhold Chemicals, Inc. (USA) for polyvinyl acetate adhesives.

Plyform. Trademark of APA-The Engineered Wood Association (USA) for a construction plywood used for concrete forms.

Plyglass. Trade name of Plyglass Limited (UK) for units composed of two or more layers of glass having permanently sealed spaces between them.

plyglass. (1) Generic term for a colored, sandwich-like structure consisting of a layer of glass fibers between two layers of sheet glass. Used for decorative applications, e.g., light fixtures.

(2) A glass product made by covering one or both sides of opal glass with a clear glass of corresponding coefficient of thermal expansion. Used for lamp shades, and globes.

plying cement. A bituminous composition suitable for bonding plies (sheets) of felts, fabrics, or glass mats to each other or other substrates.

Plykrome. Trade name of Krupp AG (Germany) for a composite steel made by welding a sheet of high-chromium stainless steel onto the surface of a non-stainless steel. Used for corrosion-resistant tanks, vessels, containers, and reservoirs.

Plyloc. Trade name for nylon fibers and products.

Ply-Mat. Trade name of Manville Corporation (USA) for a plied fiberglass mat used as a reinforcement for plastics.

Plymax. Trade name for a strong laminate consisting of a plywood core with aluminum foil bonded to it.

plymetal. (1) A composite consisting of two or more layers of dissimilar metals bonded together, e.g. nickel bonded to steel, or copper bonded to aluminum.

(2) A strong laminate consisting of a plywood core with aluminum or steel foil bonded to one or both sides.

Plymite. Trade name of P.L. & M. Company (USA) for a tungsten alloy used for wear-resistant parts.

Plymouth. Trade name for polyolefin fibers and products.

PlyoForm. Trade name of Plyonex Corporation (USA) for prepregs consisting of high-strength fibers impregnated with *PolyPly* aqueous resin emulsions, and used in the manufacture of structural composites.

Plypac. Trademark of Montecatini (Italy) for a flexible polyvinyl chloride packaging film.

Plypane. Trade name of Plypane Inc. (Canada) for sheet glass used for insulating applications.

Plypanel. Trademark of Douglas Fir Plywood Association (USA) for a construction plywood.

Plyron. Trademark of American Plywood Association (USA) for a hardwood-plywood laminate bonded with a water-resistant glue.

Plyscord. Trademark of Douglas Fir Plywood Association (USA) for unsanded exterior or interior sheathing-grade plywood panels with plugged veneer on one side. Used for various utility applications in building construction.

Plytanium. Trademark of Georgia-Pacific (USA) for strong, stiff, durable *southern pine* plywood produced by alternating thin wood plies (or veneers) at 90° angles without altering the wood grain. They have good resistance to shrinkage, swelling and sagging, and are supplied for different building applications including radiant barrier roof sheathing, wall sheathing, siding panels and *Sturd-I-Floor* subflooring.

Plyvit. Trademark of Montecatini (Italy) for polyvinyl chlorides.

Plywall. Trademark of Robinson Plywood & Timber Company (USA) for plywood used in the construction of interior walls.

plywood. A wood panel constructed by gluing together with a synthetic resin (usually a phenol- or resorcinol-formaldehyde) a number of layers (veneers or plies) of wood with the grain direction turned at right angles in each successive layer. An odd number of veneers (3, 5, 7, 9, etc.) is used so they will be balanced on either side of a center core, and the grain of the outside layers will run in the same direction. *Plywood* can also be made by bonding surface veneers or plies to a solid core (lumber core plywood). The standard panel size is 1.2 × 2.4 m (4 × 8 ft.), and the thickness ranges from 3.2 to over 25 mm (0.125 to over 1 in.). The two basic types of plywood are interior- and exterior-type plywood. *Plywood* has a high strength-to-weight ratio, high heat capacity, low thermal expansion, low water absorption, and is used for furniture, cabinets, structural subfloors, walls, partitions, building interiors and exteriors, shelving, shipping containers, ducts, linings of vehicles, etc. See also exterior-type plywood; interior-type plywood; structural plywood.

plywood block. A dense, hard block made by first bonding together a number of layers (plies) of hardwood or softwood with the grain of adjoining plies running at right angles, impregnating with a synthetic resin, such as a phenolic or urea-formaldehyde, and subjecting to a heat and pressure treatment. It has good resistance to water and oils, and is used for machine parts, such as bearings, gears, pulleys and rollers.

ply yarn. A yarn consisting of several strands twisted together, e.g., four-ply yarn consists of four strands. The strength of the yarn increases with the number of plies.

PLZT. See lead lanthanum zirconate titanate.

PM 2000. Trade name of Plansee GmbH (Austria) for an iron-based oxide-dispersion-strengthened (ODS) superalloy with body-centered-cubic microstructure. It has a nominal composition of about 20% chromium, 5.5% aluminum, 0.5% yttria (Y_2O_3), and the balance iron. Used for high-strength, high-temperature applications.

P/M brass. See sintered brass.

P/M bronze. See sintered bronze.

P/M copper. See sintered copper.

P/M copper iron. See sintered copper iron.

PMC UF. Trade name of Plastic Manufacturing Company (USA) for cellulose-filled urea-formaldehydes.

P/M copper steel. See sintered copper steel.

P/M forgings. See powder-metallurgy forgings.

PM γ-MET. Trade name of Plansee Metallwerk-Gesellschaft (Austria) for a gamma titanium aluminide powder-metallurgy alloy with a nominal composition of 46.5 at% aluminum, a total of 4 at% chromium, niobium, tantalum and boron, and the balance titanium. Used for high-temperature aerospace applications.

P/M infiltrated steel. See powder-metallurgy infiltrated steel.

P/M iron. See sintered iron.

P/M iron-copper. See sintered iron-copper.

P/M iron-nickel. See sintered iron-nickel.

P/M material. See sintered material.

P/M nickel silver. See sintered nickel silver.

P/M nickel steel. See sintered nickel steel.

P/M part. See powder-metallurgy part.

PMR polyimides. A group of crosslinked, high-molecular-weight polyimides having excellent elevated-temperature, physical and mechanical properties, and high thermooxidative stability. Used as high-temperature matrix resins for fiber-reinforced plastics for aircraft and spacecraft engines. "PMR" stands for "*in situ* polymerization of monomer reactants" referring to the fact that the monomers contained in them are latent, but react at high temperatures by addition reaction. The glass-, graphite- or carbon-fiber-reinforced polyimide plastics are also known as "PMR polyimides."

P/M stainless steels. See sintered stainless steels.

P/M steels. See sintered steels.

P/M titanium. See sintered titanium.

Pneu-Die. Trade name of Disston Inc. (USA) for a tough tool steel containing 0.48% carbon, 2.15% chromium, 0.3% molybdenum, and the balance iron. Used for punches and chisels.

pneumatically placed concrete. See dry-mix shotcrete.

Pneumo. Trade name of George Cook & Company Limited (UK) for a tough, oil-hardening tool steel containing 0.45% carbon, 1.25% chromium, 0.75% silicon, 0.35% manganese, 0.2% tungsten, and the balance iron. Used for rivet sets and upsetters.

Pneutough. Trade name of British Steel Corporation (UK) for an oil-hardening, shock-resisting tool steel containing 0.5% carbon, 0.7% silicon, 1.5% chromium, 0.25% vanadium, 2.25% tungsten, and the balance iron. Used for pneumatic chisels, die casting dies, pneumatic tools, and guillotine blades.

pnictides. Compounds of a pnictogen, i.e., an element of Group V (Group 15) of the Periodic Table and one or more electropositive elements. Examples include aluminum antimonide (AlSb), cadmium tin phosphide ($CdSnP_2$), gallium nitride (GaN) and yttrium arsenide (YAs).

pnictogens. The elements of Group V (Group 15) of the Periodic Table including nitrogen, phosphorus, arsenic and antimony.

Pnusnap. Trade name of Firth Brown Limited (UK) for oil- or air-hardening hot-work steels containing 0.35-0.43% carbon, 0.6-1% silicon, 0.3-0.5% manganese, 1-1.1% chromium, 1.8-1.9% tungsten, and the balance iron. Used for tools, dies, punches, shears, and shear blades.

Pobedit. Russian trade name for a cobalt-bonded cemented carbide with a nominal composition of 90.0% tungsten carbide and 10.0% cobalt. Used for cutting tools.

Pocan. Trademark of Albis North America (USA) for a series of thermoplastic polybutylene terephthalate (PBT) resins and PBT/polyethylene terephthalate (PET) blends that can be readily processed by injection molding, and are available in unreinforced, glass- and mineral-filled, flame-retarded and impact-modified grades. They possess exceptional chemical resistance, high rigidity and hardness, high heat deflection temperatures up to 210°C (410°F), good friction and abrasion properties, easy processibility, excellent electrical properties, and low water absorption. Used for pump housings, electrical connectors, electrical and electronic equipment, automotive components, household appliances, business machines, and lighting equipment.

Pocan B. Trademark of Albis North America (USA) for a series of thermoplastic polybutylene terephthalate (PBT) resins that can be readily processed by injection molding, and are available in unreinforced, glass- and mineral-filled, UV-stabilized, flame-retarded and impact-modified grades.

Poco. Trademark of Poco Graphite, Inc. (USA) for fine-grained, isotropic, formed graphite materials supplied in several density grades ranging from about 1.5 to 1.9 g/cm^3 (0.05-0.07 lb/in.^{3}). They have good flexural strength, high tensile strength, and are used for jigs and fixtures for electronic components, electrodischarge machining, and for several applications in the aerospace and nuclear industries. *Poco XT* is a graphite material with a particle size of 20 μm (780 μin.) supplied in block sizes up to 150 × 610 × 1015 mm (6 × 24 × 40 in.), precision-machined shapes and custom-cut sizes, and used for furnace fixturing, continuous casting and heater elements.

PocoFoam. Trademark of Poco Graphite Inc. (USA) for low-density, open-celled graphite foam with high thermal conductivity, and a maximum service temperature of 3000°C (5430°F). Used for heat sinks, heat exchangers, and thermal transfer applications.

podo. The hard, light-colored wood of several species of trees of the genus *Podocarpus*. It has a fine, uniform grain, an even texture, good workability, low nailing qualities, and poor decay resistance. Average weight, 510 kg/m^3 (32 lb/ft^3). Source: East Africa. Used for timber, joinery, and interior work.

Point 4. Trade name of Kerr Dental (USA) for a light-cure universal dental composite.

pointelle. A knitted fabric having a pattern of holes produced by transfer stitches.

Pointex. Trade name of ASG Industries Inc. (USA) for an almost transparent glass with an overall pattern of very small bosses of varying size.

pointing mortar. See masonry mortar.

point lace. See needlepoint lace.

poison. See nuclear poison.

poitevinite. A salmon-colored mineral of the kieserite group composed of copper iron sulfate monohydrate, $(Cu,Fe)SO_4 \cdot H_2O$. Crystal system, monoclinic. Density, 3.30 g/cm^3. Occurrence: Canada (British Columbia).

Pokalon. Trademark of Lonza-Werke GmbH (Germany) for polycarbonate film materials including *Pokalon C* is a polycarbonate film that is often made conductive with carbon fillers and supplied in a wide range of thicknesses. It has a density range of 1.28-1.35 g/cm^3 (0.046-0.049 lb/in.^{3}), good tensile strength, good resistance to oils, greases, alcohols, dilute acids and alkalies, fair resistance to concentrated acids, poor resistance to halogens, ketones, aromatic hydrocarbons and strong alkalies, good dielectric properties, moderate UV resistance, and low permeability to carbon dioxide and oxygen. Used for electrical applications, ID cards, etc.

Polacryl. Trade name of Polacryl, Inc. (USA) for a series of polymer coatings, paints, rheology modifiers and dispersants.

Polan. Trade name for nylon fibers and products.

Polana. Trademark Stilon (Poland) for nylon 6 staple fibers used for textile fabrics.

Polane. Trade name of Sherwin-Williams Company (USA) for polyurethane coatings.

Polar. (1) Trade name of Teledyne Allvac (USA) for a carbon steel containing 0.2% carbon, and the balance iron. It has high permeability, and is used for pole pieces for magnetos.

(2) Trade name of SA Glaverbel (Belgium) for patterned glass.

Polar Chem. Trade name of Mead Corporation (USA) for paperboard used for packaging applications.

Polar Crete. Trade name of Consolidated Coatings Corporation (USA) for a patching cement used for freezers and coolers.

polar crystal. See ferroelectric crystal.

Polarflex. Trade name of Unionglas AG (Germany) for a figured glass processed on the patterned side.

Polarguard. Trademark of KoSa (USA) for strong, durable high-performance continuous-filament polyester insulation for outdoor apparel and sleeping bags. It provides excellent loft retention, very low moisture absorption, and hypoallergenic properties including insect, mildew and fungus resistance. *Polarguard* is supplied in various grades including the original *Polarguard Classic* as well as *Polarguard HV* with high-void triangular fiber cross section, *Polarguard 3D* with high-void triangular fiber cross section and very fine denier, *Polarguard Delta* with very-high-void cross section and modified batt formation, and *Polarguard Home* for bedding.

Polaris. (1) Trade name of Time Steel Service Inc. (USA) for a series of mold steels. *Polaris 2* is a low-carbon mold steel (AISI type P2) containing 0.06% carbon, 0.3% manganese, 0.15% silicon, 0.95% chromium, 0.25% molybdenum, and the balance iron. It has good core strength, good wear properties at elevated temperatures, and good dimensional stability. Used for plastic molding dies. *Polaris 20* is a medium-carbon mold steel (AISI type P20) containing 0.37% carbon, 1% manganese, 0.3% silicon, 1.25% chromium, 0.35% molybdenum, 0.15% vanadium, and the balance iron. It has high core hardness, and good toughness and machinability, and is used for zinc die-casting dies, plastic molds for injection, compression and transfer molding.

(2) Trade name of Atotech USA Inc. (USA) for bright cyanide copper and plating process.

Polarit. Trade name of Outokumpo Metals (USA) Inc. for a series of wrought, corrosion-resistant austenitic chromium-nickel stainless steels.

polarite. A grayish mineral composed of palladium lead bismuth, $Pd(Pb,Bi)$. Crystal system, orthorhombic. Density, 12.51. Occurrence: Russian Federation.

polarized ceramics. A group of ceramics, such as lead zirconate titanate, barium titanate and lead metaniobate, that can be polarized to obtain piezoelectric properties. They have high electromechanical conversion efficiency, and are used in transducers for ultrasonic inspection and cleaning, in underwater sonar applications, accelerometers, and sensing devices.

polarizing coating. A coating that produces polarized electromagnetic radiation (light) with a frequency above about 3×10^{11} Hz. It is usually composed of one or more double-refracting materials, and used on certain optical instruments and devices.

polarizing materials. Material, such as calcite crystals, polarized glass, certain polymer films (polyvinyl alcohol, etc.) and sandwich-type laminates, that consist of a rigid polarizer between two sheets of plastic, or a polymer film between two sections of ground and polished glass. Abbreviation: PM.

Polar Light. Trademark of Gestenco International AB (Sweden) for a light-cure dental composite for bonding applications.

Polarlite. Czech trade name for a smooth cathedral glass.

polar material. A material that has either positive or negative electric polarization.

polar polymers. Synthetic polymers, usually with amorphous structures, that have a permanent electric dipole moment, which may be due to the incorporation of certain additives or functional groups, e.g., *chromophores*. They show great potential for applications in photonics and nonlinear optics. See also nonpolar polymers.

Polar Star. Trademark of Celanese Canada Inc. for a water-repellant polyester fabric.

Polartec. Trademark of Malden Mills Industries, Inc. (USA) for polyester fibers and yarns used for the manufacture of warm, breathable, windproof outdoor clothing.

PolarTherm. Trademark of Advanced Ceramics Corporation (USA) for a series of chemically inert, highly thermally conductive boron nitride fillers. They have theoretical density of 2.25 g/cm^3 (0.08 lb/in.^{3}), and a dielectric constant of 3.9, good flow properties at high loadings, high volume resistivity, and high moisture resistance. Supplied as fine hexagonal powders and low- and high-density agglomerates, they are used in polymers for the manufacture of electronic components and assemblies to increase thermal dissipation properties and improve electronic and electrical (dielectric) properties.

Polastic. Trade name of W. Canning Materials Limited (USA) for a buffing composition for plastics.

Polastor. Trade name of Boisellier (France) for unsaturated polyester.

Polathane. Trademark of Polaroid Corporation (USA) for a series of thermosetting urethane elastomers. Some are polyester-based and possess excellent chemical stability, while others are polyether-based and offer good hardness, abrasion resistance and shock absorption. The properties and uses vary greatly with the particular formulation. Typical applications include conveyor systems, wheels, tires, belts, shock pads, and abrasion-resistant shields.

Polathene. Trade name for polyolefin fibers and products.

Polectron. Trademark of GAF Corporation (USA) for modified vinylpyrrolidone resins used as adhesives on various substrates (e.g., metal, paper, wood, glass, etc.), as binders for wood, paper and fiberglass, and as precoats on photosensitive paper.

Polestron. Trade name of Tenax Finishing Products Company (USA) for polyester coatings.

polhemusite. A black mineral of the sphalerite group composed of mercury zinc sulfide, $(Zn,Hg)S$. Crystal system, tetragonal. Density, 4.2-5.2 g/cm^3. Occurrence: USA (Idaho).

Polibrid. Trademark of Polibrid Coatings, Inc. (USA) for a wide range of chemical and/or wear-resistant solventless, elastomeric polyurethane coatings and linings.

Poliafil. Trade name for polyolefin and nylon fibers.

Polidro. Trademark of LMP Corporation (USA) for polyvinyl chlorides.

Polifil. Trade name of Plastics Group of America (USA) for unreinforced and glass-reinforced polyamides (nylon 6 and 6,6) and calcium carbonate-filled polypropylenes with good strength and rigidity used for machine parts, such as gears, brackets, valves, bearings, etc. Also included under this trade name are *Polifil HM* talc- or calcium carbonate-filled polypropylenes as well as several glass-reinforced acrylonitrile-butadiene-styrene compounds for the manufacture of automotive trim, business machines, sporting goods, power-tool housings, and household appliances.

Poligal. Trademark of Sumar SA (Chile) for polyester staple fibers and filament yarns.

Polilac. Trade name of CHI MEI (Italy) for an acrylonitrile-butadiene-styrene resin available in various grades.

polish. (1) A short-term decorative coating that may or may not provide protection to the substrate.

(2) Any solid, semisolid or liquid mixture applied to a substrate to impart surface protection, decorative finish or smoothness. Examples include solid substances, such as finely divided red ferric oxide (rouge), for polishing glass and mirror backs,

and semisolid (pasty) or liquid substances, such as carnauba or candellia wax in a suitable solvent, for the protection and enhancement of wood and leather.

polished cotton. A plain-woven cotton fabric having a glazed, polished, more or less lustrous finish produced either by using a particular type of weave (e.g., satin), by calendering, or by the application of a synthetic resin coating.

polished plate glass. A high-quality plate glass with plane-parallel surfaces formed by a casting or rolling process. Both sides are ground and polished to provide high clarity and smoothness. Used for storefronts, large windows in industrial, commercial and residential buildings, office partitions, and mirrors.

polished wire glass. A sheet of glass that has a layer of meshed wire incorporated to resist shattering when broken, and is ground and polished on one or both sides.

polished yarn. A cotton yarn, twine or sewing thread having a smooth, polished, lustrous finish produced by treating with a suitable chemical, e.g., gelatin or starch, and then passing over rollers.

polishing abrasives. Fine abrasives, such as emery, corundum, polishing rouge, chromium oxide, rottenstone or pumice, usually attached to a cloth, wheel or belt, and used to put a smooth, bright finish on metal surfaces by rubbing action.

polishing crocus. A fine, soft natural or synthetic ferric oxide powder of bright red color used in cleaning and polishing for minimum stock removal, e.g., in buffing cutlery and some nonferrous metals.

polishing lubricants. Lubricants that are usually mixtures of fatty acids, animal tallow and waxes, and used to extend the life and improve the surface finish of polishing wheels and belts.

polishing paste. A compound consisting of fine abrasive particles (aluminum oxide, silicon carbide, etc.) immersed in a grease or wax binder, and used for polishing metal surfaces.

polishing powder. A very fine solid powder, such as red iron oxide (rouge), pumice, rottenstone, alumina, titania, zirconia, etc., that imparts smoothness or decorative finish to the surface of metals, glass and plastics.

polishing rouge. See rouge.

Polital. Trade name of Dürener Metallwerke (Germany) for a corrosion-resistant, heat-treatable aluminum alloy containing 0.5-4% magnesium, 0-1.5% silicon, and 0-1% manganese. Used for light-alloy parts, general structures, scaffolds, booms, and transmission towers.

Politen-Omni. Trademark of Fibrasomni SA (Mexico) for polyethylene monofilaments.

Politit. Trade name of Hoffmann & Co. KG (Germany) for a series of austenitic, ferritic and martensitic stainless steels.

Polivit. Trade name of Dürener Metallwerke (Germany) for an aluminum alloy containing 1.8% manganese, 0.6% copper and 0.2% silver, used for light-alloy parts.

Poli-Wood. Trademark or Morval-Durofoam Limited (Canada) for plastic moldings.

polka dots. A fabric having a repetitive, regular pattern of embroidered, flocked or printed dots or round spots.

Pollopas. Trade name of Hüls AG (Germany) for colored, cellulose-filled urea-formaldehyde resins and plastics.

pollucite. A colorless transparent, white or grayish-white mineral of the analcime group composed of cesium sodium aluminum silicate hydrate, $(Cs,Na)_2Al_2Si_4O_{12} \cdot xH_2O$. It can also be made synthetically in anhydrous form $(CsAlSi_2O_6)$ from a powder prepared by heating cesium nitrate, aluminum nitrate and silicon dioxide under prescribed conditions, followed by pressing and firing at high temperatures. Crystal system, cubic. Density, 2.80-2.95 g/cm³; hardness, 6.5 Mohs; refractive index, 1.520. Occurrence: Canada, Japan, Southwest Africa, USA (Maine). Used as a source of cesium, in fluxes, in welding materials, as a catalyst, in ion propulsion, in thermocouples, and as a gemstone. Also known as *pollux*.

pollux. See pollucite.

polo cloth. A knitted or woven wool fabric having a pronounced nap, used especially for sportswear.

Polofil. Trade name Vivadent AG (Liechtenstein) for light-cure hybrid composites used for restorative dentistry.

polonides. Compounds of polonium (Po) and a metal (M), such as beryllium, cadmium or zinc. The general formula is MPo.

polonium. A radioactive element of Group VIA (Group 16) of the Periodic Table. It occurs naturally in trace amounts in uranium ores (e.g., pitchblende), and can also be produced synthetically in nuclear reactors by bombarding bismuth with neutrons. The most commonly produced isotope is polonium-210 (half-life, approximately 138 days). Polonium-209 is the most stable isotope (half-life, approximately 103 years). Important properties of polonium: Density, 9.4 g/cm³; melting point, 254°C (489°F); boiling point, 962°C (1764°F); atomic number, 84; atomic weight, 210; divalent, tetravalent, hexavalent. It has poor resistance to concentrated sulfuric and nitric acid, aqua regia, dilute hydrochloric acid, good to fair resistance to alkalies, and is chemically similar to bismuth and tellurium. Two allotropic forms are known: (i) *Alpha polonium* (simple cubic crystal structure) with a density of 9.32 g/cm³, and (ii) *Beta polonium* (rhombohedral crystal structure) with a density of 9.41 g/cm³. *Polonium* is used as a source of alpha radiation, as a neutron source, in nuclear batteries, as a film cleaner, in instrument calibration, oil-well logging, moisture determination, in intermetallic and semiconductive compounds, and as a power source. Abbreviation: Po.

polonium chloride. Yellow, hygroscopic crystals with a molecular weight of 351 g/mol, a melting point of 300°C (572°F) and a boiling point of 390°C (734°F). Used in chemistry and materials research. Formula: $PoCl_4$.

polonium oxide. Yellow, cubic crystals with a density of 8.9 and a decomposition temperature of 500°C (932°F). Used in chemistry and materials research. Formula: PoO_2.

Polsilon. Trademark of Chemitex-Anilana (Poland) for glass fibers.

Polvin. Trademark of Monsanto Chemical Company (USA) for polyvinyl chloride resins.

Polwinit. Polish trade name for polyvinyl chlorides.

Poly. Trade name of Lepage's Limited (Canada) for an extensive series of home restoration products including wallpaper pastes and sizes, wallpaper removers, super-strength cements, stipple-finish ceiling and wall textures, stucco-type white interior textures, patching compounds and crack fillers, subfloor leveling compounds, tile grouts, paint and varnish removers, etc.

polyacenaphthylene. A powder obtained by the polymerization of acenaphthylene ($C_{12}H_8$). It has an average molecular weight of approximately 5000-100000 and a glass-transition temperature of 214°C (417°F). Used in biochemistry and biotechnology.

polyacetal-elastomer alloys. Alloys (or blends) of crystalline polyacetal polymers and elastomers, such as butadiene rubber (BR) or ethylene-propylene terpolymer (EPDM). They combine the high resiliency, elongation, and impact resistance of elastomers with the strength, stiffness, and chemical resistance of polyacetals. Used for sporting goods, automotive and industrial components, housings, and shoe soles. See also acetal resins.

polyacetal resins. See acetal resins.

polyacetals. See acetal resins.

polyacetylene. See acetylene polymer.

polyacrylamide. A high polymer ($CH_2CHOONH_2)_n$ produced by the polymerization of acrylamide with *N,N'*-methylene bis-acrylamide. It is available as a white, water-soluble powder, and as a 10-50 wt% solution in water. Used as a suspension agent or thickener for ceramic slips and slurries, as a binder, in adhesives, in gel form in electrophoresis, as a molecular-weight standard, and in the heat treating industries for the formulation of polymer quenchants. Abbreviation: PAA; PAAM; PAAm.

poly(acrylamide-*co*-acrylic acid). A copolymer of acrylamide and acrylic acid available in powder and granule form with a typical weight-average molecular weight ranging between 5000000 and 15000000. Used in the biosciences.

poly(acrylamide-*co*-acrylic acid) sodium salt. The sodium salt of a copolymer of acrylamide and acrylic acid available as a powder in low carboxyl (10%), medium carboxyl (40%) and high carboxyl (70%) contents, and used chiefly as a polyelectrolyte.

polyacrylate elastomers. A group of elastomers based on polymers of butyl or ethyl acrylate. They have outstanding resistance to oils including sulfur-bearing grades, fair mechanical properties, fair to good dielectric properties, and a service temperature range of -40 to +204°C (-40 to +400°F). Used for seals, O rings, gaskets, etc.

polyacrylates. A group of thermoplastic resins that are polymerization products of an ester or salt of acrylic acid and methacrylic acid. They typically have great optical clarity, possess a high degree of light transmission, and are used in emulsion paints, surface coatings, paper and leather finishes, elastomers, etc. See also acrylates.

polyacrylic acid copolymer. A copolymer of acrylic acid and another monomer, such as acrylamide or maleic acid.

polyacrylic fibers. Continuous fibers made from acrylate resins by extrusion through a spinneret. See also acrylic fibers.

polyacrylonitrile. A polymer obtained by the polymerization of acrylonitrile (C_3H_3N) and supplied as a powder in molecular weights ranging from 25000 to over 1000000. Copolymers with butadiene, and terpolymers with butadiene and styrene are also available. Used as a base material or precursor in the production of certain carbon fibers for engineering applications, in the manufacture of various synthetic textile fibers (e.g., *Orlon*) and, when blended with other materials, in the manufacture of hard, heat-resistant resins for wall panels, molded articles, and medical and pharmaceutical packaging. Abbreviation: PAN.

poly(acrylonitrile-*co*-butadiene). See acrylonitrile-butadiene copolymer.

polyacrylonitrile-butadiene-styrene plastics. See acrylonitrile-butadiene-styrene resins.

poly(acrylonitrile-*co*-butadiene-co-styrenes). See acrylonitrile-butadiene styrenes.

polyacrylonitrile carbon fibers. Carbon fibers based on a polyacrylonitrile precursor. They have high tensile strength and elastic modulus, high conductivity, and are used as reinforcing fiber in composites, and as conductive reinforcements in plastic materials. Also known as *carbon-polyacrylonitrile fibers*. See also polyacrylonitrile; carbon fibers.

polyacrylonitrile fibers. A generic term for man-made fibers composed of a linear synthetic polymer having 85 wt% or more recurring acrylonitrile units (–CH₂–CHCN–) in the chain. Abbreviation: PAN fibers. See also polyacrylonitrile; polyacrylonitrile carbon fibers.

poly(acrylonitrile-*co*-methyl acrylate). See acrylonitrile-methyl acrylate copolymer.

poly(acrylonitrile-*co*-vinyl chloride). See acrylonitrile-vinyl acrylate copolymer.

poly(acrylonitrile-*co*-4-vinylpyridine). See acrylonitrile-vinylpyridine copolymer.

Polyactive. Trademark of Osteotech Limited for a biocompatible, biodegradable, re-sorbable synthetic copolymer of polyethylene oxide (PEO) and polybutylene terephthalate (PBT) used in tissue engineering and for medical implants.

Poly-Ac-Zen. Trade name of Hentzen Coatings, Inc. (USA) for air-dry acrylic enamel paints and coatings.

Polyaire. Trade name for polyurethane foams and foam products.

poly-L-alanine. A polymer of the amino acid L-alanine with a typical molecular weight of 3000-25000. It is available as a white, crystalline powder with a softening point of 95-97°C (203-207°F), and used in biochemistry, bioengineering, and in the manufacture of biopolymers.

poly(alkylene oxide). A water-soluble, nonionic copolymer synthesized from ethylene oxide (C_2H_4O) and propylene oxide (C_3H_6O). Used in the manufacture of plastic products, and in the heat treating industry for formulating polymer quenchants. Abbreviation: PAO.

Polyall. Trademark of Allied Chemical Corporation (USA) for a series of fast-curing, thermosetting compounds based on alkyd resins. They have a density, 2.1 g/cm³ (0.08 lb/in.³), high strength, good dimensional stability, good electrical and flame resistance, a heat distortion temperature of 204°C (400°F), good ultraviolet resistance, good fungus resistance, and good resistance to solvents and dilute acids.

polyallomers. A group of white-colored, highly crystalline copolymers of propylene and another olefin, such as ethylene. They can be processed by blow molding, injection molding, thermoforming, vacuum forming and extrusion, and are available in standard and modified grades. *Polyallomers* are thermoplastics with high resistance to flexing fatigue, fair to good stiffness, excellent to fair impact strength, good abrasion resistance, and fair hardness. Used for molded products, such as food containers, typewriter cases, threaded container closures, snap clasps, luggage shells, film, sheeting, and wire and cable. Abbreviation: PA.

poly(allylamine). A polymer of allylamine (C_3H_7N) that is usually supplied as a 20 wt% solution in water with an average molecular weight ranging from 17000-65000, a room-temperature viscosity of about 50-250 centipoise, a density of 1.020 g/cm³ (0.037 lb/in.³), and a refractive index of 1.382-1.383. Used chiefly in biochemistry and biotechnology.

Polyamid. Trade name of H. Roemmler GmbH (Germany) for a series of polyamide (nylon) fibers and plastics supplied in several grades.

polyamide–acrylonitrile-butadiene-styrene alloys. See PA-ABS alloys.

polyamide–ethylene-propylene-diene-monomer alloys. See PA-EPDM alloys.

polyamide fibers. (1) Natural fibers, such as silk, wool and animal hair, composed of polymers that contain recurring amide groups (–CONH–) in the chain.

(2) Synthetic fibers whose fiber-forming substance is any

long-chain synthetic polymer having amide (–CONH–) repeating groups in the chain of which 85 wt% or more are linked directly to aliphatic or cycloaliphatic groups. See also nylon fibers.

polyamide–polyarylene alloys. See PA-PAR alloys.

polyamide–polyphenylene ether alloys. See PA-PPE alloys.

polyamide–polyphenylene oxide alloys. See PA-PPO alloys.

polyamide-imides. A group of thermoplastic high-performance polymers which are condensation products of trimellitic anhydride and aromatic diamines containing alternating amide and imide linkages. They have a density of 1.4 g/cm³ (0.05 lb/in.³), exceptional toughness and dimensional stability, exceptional mechanical strength, high resistance to fatigue and stress rupture, outstanding wear resistance, low coefficients of friction, low coefficients of thermal expansion, excellent thermal stability, a wide service temperature range of -200 to +260°C (-328 to +500°F), good electrical properties, low water absorption, good resistance to acids, alcohols, aromatic hydrocarbons, greases, oils and ketones, poor resistance to alkalies, and good resistance to alpha, beta, gamma and ultraviolet radiation. Often sold under trade names or trademarks, such as *Ryton, Torlon, Ultem, Vespel* or *Victrex*, they are used for electrical and electronic connectors, burn-in sockets, photocopier parts, machine elements, such as gears, bearings, bushings, liners, etc., automotive engine components, fasteners, washers, seal rings, wear rings, shrouds, housings, valve parts, engineering composites, laminates, and prepregs. Abbreviation: PAI. Also known as *polyamideimide resins*.

polyamide plastics. Plastics based on any of a group of thermoplastic polyamides that are condensation products of aliphatic diamines and aliphatic dicarboxylic acids. See also nylon plastics; nylons.

polyamide resins. See polyamides (2); nylons.

polyamides. (1) A group of natural or synthetic polymers that contain an amide group (–CONH–), as a recurring part of the molecular chain. Natural polyamides include casein, soybean and zein, and synthetic polyamides are exemplified by the numerous grades of nylon. Abbreviation: PA.

(2) A group of thermoplastic polymers made by the condensation of polyamines or diamines with dibasic acids (e.g., polycarboxylic acid), or by the polymerization of amino acid. Many polyamides are fiber-forming. Abbreviation: PA. Also known as *polyamide resins*. See also nylon; polyamino acids.

polyamino acids. A group of synthetic biopolymers, such as poly-L-alanine, poly-D-lysine and poly-L-tyrosine, that are based on natural amino acids coupled with nonamide linkages. They exhibit considerable mechanical strength and good biocompatibility but, in pure form, have a high degree of crystallinity and fairly poor processability. Thus, they are usually copolymerized or modified with tyrosine derivatives. Used for tissue engineering and biomedical device applications. See also amino acid; pseudo-polyamino acids.

poly(aminononanone). See nylon-9.

poly(4-aminostyrene). See aminopolystyrene resin.

polyampholytes. Polymeric solutions of amphoteric electrolytes used chiefly in electrophoresis and in isoelectric focusing for the separation of proteins. They are available with a wide range of pH values and contain specific functional groups that can react with specific compounds. See also ampholyte.

Polyan. Trade name for an injection molding acrylic used for dental applications.

Polyane. Trade name for phenol-formaldehyde film materials.

polyanhydrides. A class of hydrophobic synthetic biopolymers, such as poly(SA-HAD anhydride) which contains sebacic acid (SA) and hexadecanoic acid (HAD) in its molecular structure. They are usually synthesized by melt polycondensation involving the dehydration of diacid molecules. *Polyanhydrides* degrade primarily by surface erosion, possess excellent *in vivo* compatibility, and biostability, and are currently used for drug delivery applications.

polyaniline. An inherently conductive polymer that can be produced electrochemically in the form of a thin film directly on a conductive substrate, such as indium tin oxide (ITO), and chemically in de-doped or doped emeraldine base (EB) form, that is soluble in *N*-methylpyrrolidinone (NMP), and is complexed (doped) with an organic sulfonic surfactant, such as *p*-toluenesulfonic acid or dodecylbenzenesulfonic acid. Used in the manufacture of electronic, microelectronic, optoelectronic, photonic devices, e.g., Schottky barrier devices and photoelectrochemical cells, as coatings or blends for electrostatic dissipation and EMI shielding applications, etc. Abbreviation: PAn; PAni.

poly(anilinesulfonic acid). A water-soluble, self-doped conducting polymer usually supplied as aqueous solution with a degree of sulfonation of about 100%, a number-average molecular weight of 10000, a density of 1.0 g/cm^3 (0.036 $lb/in.^3$), and a boiling point of about 100°C (210°F). Used for electronic and optoelectronic applications.

Polyanyl. Trade name for polyamide (nylon) film materials.

polyaramid. See aramid.

Polyarmor. Trade name of Plastic Flamecoat Systems (USA) for a polyethylene copolymer based powder coating.

Polyarns. Trade name for olefin, vinyon and nylon fibers and yarns used for textile fabrics.

polyarylate–polyamide alloys. See PAR-PA alloys.

polyarylates. A group of tough, heat-resistant, thermoplastic, aromatic polyesters made from aromatic dicarboxylic acids and diphenols. They have excellent resistance to ultraviolet radiation, excellent dimensional stability, low flammability, low warp resistance, high durability, good electrical and mechanical properties, and good creep and warpage resistance. Often sold under trade names or trademarks, such as *Ardel, Durel* or *Xydar*, they are used for electrical connectors, relay housings, switch and fuse covers, helmets, face shields, automotive headlight housings, brake light reflectors, exterior trim on cars and trucks, and traffic signal lenses. Abbreviation: PAR.

polyarylene ethers. A class of thermoplastic polymers produced from an activated difunctionalized aromatic precursor and bisphenol by a reaction involving heat and a basic catalyst. They have high thermal stability, long shelf lives, good moisture and solvent resistance, low dielectric constants (k < 3.0) and glass-transition temperatures ranging from as low as 290°C (555°F) to over 450°C (840°F). Used for engineering applications and as thin films for integrated-circuit devices. Abbreviation: PAE.

polyaryletherketones. A group of linear, aromatic, semicrystalline thermoplastic engineering materials with high tensile strength, good elongation and impact strength, good chemical resistance, good hydrolytic and thermal stability, good integral lubricity, good overall electrical properties, good retention of mechanical properties at elevated temperatures, high temperature stability, high heat-distortion temperature exceeding 300°C (570°F), high glass-transition temperatures, and low flammability and smoke generation. Often sold under trade names or

trademarks, such as *Kadel* or *Ultrapek*, they are used for engine components, air ducts, interior cabin material, nonstructural components for aircraft and spacecraft, wire and cable, pump components, bearing surfaces, backup seals, and monofilament. Abbreviation: PAEK.

polyarylsulfones. A group of engineering thermoplastics produced from aromatic polyether sulfone and readily processed by compression and injection molding and extrusion, machining and ultrasonic welding. They have outstanding resistance to high and low temperatures (from -240 to +260°C or -400 to +500°F), a glass-transition temperature of about 275°C (527°F), high tensile and impact strength, good electrical insulating properties, high transparency, good processibility, high hydrolytic stability, and good resistance to oils, many chemicals and most solvents. Often sold under trade names or trademarks, such as *Radel* or *Ryton PAS*, they are used for electronic and electrical applications. Abbreviation: PAS; PASU.

Polyäthylen-Draht. German trade name for polyethylene fibers, yarns and related products.

poly(azelaic anhydride). A polymer of azelaic acid ($C_9H_{14}O_3$) that is usually supplied as a moisture-sensitive solid with less than 7 wt% free acid. It has an average molecular weight of approximately 1800, a density of 1.07 g/cm^3 (0.039 $lb/in.^3$), and a melting point of 61°C (142°F). It is also available in 7 wt% carboxy-terminated grades with a melting point of 52-54°C (126-129°F). Used chiefly in biochemistry and biotechnology. Abbreviation: PAPA. Also known as *polyazelaic polyanhydride*.

Poly BD. Trademark of Atofina Chemicals, Inc. (USA) for an extensive series of epoxidized and hydroxylated polybutadienes.

Polybead. Trademark of Polysciences, Inc. (USA) for monodisperse microparticles based on various polymers, e.g., polystyrenes.

polybenzimidazole fibers. Synthetic fibers whose fiber-forming substance is a long-chain aromatic polymer that has repeating benzimidazole groups as an essential part of its backbone. They possess excellent resistance to most chemicals and solvents, and excellent flame and high-temperature resistance. Used as reinforcement fibers for advanced composites and for high-performance protective clothing, e.g., firefighters coats and space suits. Abbreviation: PBI fibers. See also polybenzimidazoles.

polybenzimidazoles. A group of high-molecular weight engineering thermoplastics produced by the condensation of diphenyl isophthalate and 3,3'-diaminobenzidine. They have a density of 1.3 g/cm^3 (0.05 $lb/in.^3$), high tensile, compressive and flexural strengths, high elastic moduli, good dimensional stability, low flammability, good dielectric properties, excellent thermal properties, low thermal expansivity, an upper service temperature of 260-400°C (500-750°F), low coefficients of friction, excellent ablation resistance, good resistance to alcohols, aromatic hydrocarbons, greases, oils and ketones, good resistance to dilute acids, good to fair resistance to alkalies, and fair resistance to concentrated acids. Often sold under trade names and trademarks, such as *Celazole*, they are used for fibers, in composites for aircraft and aerospace applications (e.g., missile nose cones, rocket nozzles and hypersonic systems), for bearings, seals, gaskets, valve seats and backup rings, in adhesives with high adhesion to steel, titanium, beryllium and aluminum alloys, and in coatings and ablative materials. Abbreviation: PBI.

polybenzocyclobutanes. A class of homopolymers produced by heating benzocyclobutane to a temperature above 200°C (390°F). They have low dielectric constants and dissipation factors, excellent thermal stability at over 300°C (570°F), and low water absorption. Uses include adhesives for aerospace and electronic applications, matrix resins for composites, potting compounds, and coatings. Also known as *benzocyclobutane polymers*.

poly(γ-benzyl-L-glutamate). A polymer obtained by the polymerization of the γ-benzyl ester of L-glutamic acid ($C_{12}H_{15}NO_4$), and available in molecular weights ranging from 15000 to over 1000000. Used in biochemistry and biotechnology.

poly(benzyl methacrylate). A polymer of benzyl methacrylate ($C_{11}H_{12}O_2$) available as a powder with an average molecular weight of about 70000, a melting point of 54°C (129°F) and a refractive index of 1.568. Used in chemistry, biochemistry and biotechnology.

polyblend. A mixture made from polymerized components, such as (i) two natural or synthetic homopolymers, e.g., a rubber-polystyrene alloy, (ii) a homopolymer and a copolymer, e.g., a butadiene-styrene rubber, or (iii) two copolymers, e.g., an alloy of isobutylene-isoprene and butadiene-acrylonitrile.

Poly-Bond. Trade name of Bridgeport Insulated Wire Company (USA) for polymer-coated solderable enamel wire.

Polybond. (1) Trademark of Schlegel Corporation (USA) for urea-formaldehyde resins.

(2) Trademark of Flexible Products Company (USA) for polyurethane adhesives used to join the skin of prefab constructional wall or roofing panels to the core.

(3) Trademark of Polyglass SpA (Italy) for waterproof roofing membranes.

(4) Trademark of Moore Diversified Products, Inc. (USA) for plastic building products, such as moldings, raceway ducts, and riser guards.

PolyBolt. Trade name of Reichhold Chemicals, Inc. (USA) for structural adhesives including epoxies, urethanes, methacrylates and hybrids used in building construction, for automotive and marine applications, and in the manufacture of showers and tubs.

Polybolta. Trade name for polyolefin fibers and products.

polyborosilanes. A class of organosilicon polymers, analogous to polysilanes, $(R_2Si)_n$, but containing boron atoms. The general formula is $(R_2SiBR)_n$ in which R usually represents hydrogen, or an alkyl, such as ethyl. Used in the manufacture of silicones, and as preceramic precursors for the preparation of silicon carbide and other engineering ceramics. See also polysilanes; polycarbosilanes.

polyborosilazanes. A class of organosilicon polymers, analogous to polysilazanes [(RR'SiNR")$_n$], but with a main chain containing silicon, nitrogen and boron atoms and, usually, hydrogen and/or organic side groups (e.g., imides). While several general formulas are possible, two of the most common ones are $(N(BR_2)SiR_2)_n$ and $(BRNHSiRNH)_n$, in which in which R represents the organic side groups. Used as preceramic precursors for the preparation of $Si_xB_xN_y$, $SiBN_xC$ and other engineering ceramics. See also polysilazanes; polysilsesquiazanes.

polyborosiloxanes. A class of organosilicon polymers in which the main chain consists of alternating silicon and oxygen atoms with side groups of alkyls, aryls, arenyls, etc. (e.g., methyl or phenyl), and boron. Used as preceramic precursor materials for the manufacture of $SiBO_xN_y$ engineering ceramics. See also polysiloxanes; siloxanes.

polybutadiene. See butadiene rubber.

polybutadiene elastomer. See butadiene rubber.

poly(butadiene oxide). A polymer obtained from polybutadiene by epoxidation (i.e., attachment of epoxide groups). It typically has a molecular weight of 700, and an oxirane content of 7-8%. Also known as *epoxidized polybutadiene*.

polybutadiene rubber. See butadiene rubber.

polybutenes. A group of thermoplastic, isotactic, stereoregular polyolefins made by the polymerization of 1-butene (C_4H_8) and characterized by high crystallinity, excellent toughness, flexibility and creep resistance, and good chemical and electrical properties. Used in the manufacture of adhesives, sealants, films, coatings and synthetic rubber. Abbreviation: PB. Also known as *polybutylenes*. See also polybutylene terephthalates.

poly(butyl acrylate). A polymer of butyl acrylate ($C_7H_{12}O_2$) with a typical average molecular weight of 60000-100000. It is usually supplied as a solution in toluene, and used in the manufacture of acrylic plastic products, in the preparation of copolymers (e.g., with ethyl acrylate and methacrylic acid), in adhesives and coatings, and in the biosciences. Abbreviation: PBA.

poly(1,4-butylene adipate-*co*-1,4-butylene succinate). A biodegradable thermoplastic polymer of 1,4-butylene adipate and 1,4-butylene succinate that is usually extended with 1,6-diisocyanatohexane, and has a density of 1.2 g/cm³ (0.04 lb/in.³), and a melting point of 98°C (208°F).

poly(1,4-butylene adipate-*co*-polycaprolactam). A semicrystalline biodegradable thermoplastic polymer of 1,4-butylene adipate and polycaprolactam with a density of 1.07 g/cm³ (0.039 lb/in.³) and a melting point of 125°C (257°F).

polybutylenes. See polybutenes.

poly(1,4-butylene succinate). A biodegradable thermoplastic polymer of 1,4-butylene succinate that is usually extended with 1,6-diisocyanatohexane, and has a density of 1.3 g/cm³ (0.05 lb/in.³) and a melting point of 120°C (248°F).

polybutylene terephthalate–elastomer alloys. PBT-elastomer alloys.

polybutylene terephthalate–linear-density polyethylene alloys. See PBT-LDPE alloys.

polybutylene terephthalate–polyarylate alloys. See PBT-PAR alloys.

polybutylene terephthalate–polycarbonate alloys. See PBT-PC alloys.

polybutylene terephthalate–polyethylene terephthalate alloys. See PBT-PET alloys.

polybutylene terephthalate–polyphenylene ether alloys. See PBT-PPE alloys.

polybutylene terephthalates. A group of semicrystalline engineering thermoplastics made by polycondensation of 1,4-butanediol and dimethyl terephthalate (DMT), and supplied in unmodified and glass- or mineral-filled grades. Unmodified grades have a density of 1.31 g/cm³ (0.047 lb/in.³), and glass-filled grades of 1.63 g/cm³ (0.059 lb/in.³). *Polybutylene terephthalates* have high strength and dimensional stability, high toughness, good durability, good electrical properties, high dielectric strength, low coefficients of friction, low moisture absorption, good thermal resistance, an upper service temperature of 120-200°C (250-390°F), good resistance to acids, alcohols, aromatic hydrocarbons, greases and oils, fair resistance to alkalies, and good ultraviolet stability. Often sold under trade names and trademarks, such as *Arnite T, Celanex, Crastin, Pibiter, Pocan, Thermofil PBT, Ultradur B, Valox, Vandar* or *Vestodur*, they are used for electrical and electronic components and devices (e.g., cable protection, chip carriers, connectors, fuse cases, junction boxes, motor housings, network interfaces, relays and switches), automotive appli-cations (e.g., body panels, brake systems, door handles, fenders, headlamp systems, water pumps, windshield wiper assemblies and wheel covers), gear wheels, bearings, pipes, conduits, fittings, pump components, household appliance parts, outboard motor propellers, protective helmets, and power-tool housings. Abbreviation: PBT.

poly(n-butyl methacrylate). A polymer of n-butyl methacrylate $(C_8H_{14}O_2)$ usually supplied in the form of crystals or powder with a typical weight-average molecular weight of 330000 and a glass-transition temperature of 50°C (122°F). Used in the manufacture of acrylic plastic products, in adhesives and coatings, etc. Abbreviation: PBMA.

poly(3-butylthiophene-2,5-diyls). A class of conductive polymers that are available in regioregular and regiospecific conformations. The former are black solids which have a head-to-tail regiospecific conformation of 97% and average molecular weights of about 142000, while the latter are red solids in which the ratio of the head-to-head/head-to-tail linkages of regioisomers is about 1:1. Used for electronic and optoelectronic applications.

polycaprolactam. See nylon 6.

poly(ε-caprolactone). A biodegradable, semicrystalline homopolymer synthesized by ring-opening polymerization of ε-caprolactone $(C_6H_{10}O_2)$. It has a melting point of 59-64°C (138-147°F), a glass-transition temperature of -60°C (-76°F), an average molecular weight ranging from about 14000 to over 65000, good biocompatibility, fairly slow degradation properties (up to 2 years). It is available in flake, pellet and other physical forms, and is used in tissue engineering, e.g., for biodegradable sutures. In addition to the homopolymers there are several copolymers that have higher rates biodegradability and bioabsorbability, e.g., copolymers of ε-caprolactone and DL-lactide, and block copolymers of ε-caprolactone and glycolide. The latter are typically available as monofilaments and are suitable for surgical sutures. Abbreviation: PCL.

poly(ε-caprolactone)-poly(DL-lactic-co-glycolic acid) composite. A porous, biodegradable polymer-matrix composite consisting of a blend of poly(ε-caprolactone) and poly(DL-lactic-co-glycolic acid) with *hydroxyapatite* granules incorporated. It develops a three-dimensional osteoconductive scaffold that supports bone cell growth both on the surface and throughout, and shows great potential for applications in human bone-tissue engineering.

Polycarbafil. Trade name of Fiberfil (USA) for glass-fiber-reinforced polycarbonate plastics.

polycarbamide. See polyurea fibers.

polycarbamide fibers. See polyurea fibers.

Polycarb Waterset. Trademark of Pulpdent Corporation (USA) for a water-activated polycarboxylate dental cement that contains polyacrylic acid, and is used for the permanent and temporary cementation of crowns, bridges, inlays and orthodontic brackets, and as hard restoration base.

polycarbomethylsilane. A polycarbosilane polymer with an average molecular weight of 800-3500 supplied in various grades, and used as a polymeric precursor for silicon carbide ceramics. Formula: $[-Si(CH_3)HCH_2-]_n$. Also known as *poly(methylsilylene)methylene*. Abbreviation: PCMS. See also polycarbosilanes.

polycarbonate-acrylonitrile-butadiene-styrene alloys. See PC-ABS alloys.

polycarbonate plastics. Thermoplastic polyesters based on polycarbonate resins. See also polycarbonates.

polycarbonate-polyamide alloys. See PC-PA alloys.

polycarbonate-polybutylene terephthalate alloys. See PC-PBT alloys.

polycarbonate-polyethylene terephthalate alloys. See PC-PET.

polycarbonate resins. See polycarbonates.

polycarbonates. A group of transparent, amorphous engineering thermoplastic derived from bisphenol A and phosgene. They are linear, high-molecular-weight, low-crystalline polyesters of carbonic acid. They are commercially available in the form of films, metallized films, sheets, rods, fibers, granules and pellets, and can be easily processed by blow, injection and rotational molding, thermoforming, vacuum forming, extrusion and fluidized-bed techniques. They have a density of 1.2 g/cm³ (0.04 lb/in.³), high light transmission, excellent colorability, high gloss, high dimensional stability, good strength, high modulus of elasticity, high creep resistance, outstanding impact resistance and ductility, high hardness, negligible moisture absorption, good dielectric properties, good electrical resistivity, low mold shrinkage, low thermal expansivity, high heat resistance, a service temperature range of -170 to +125°C (-274 to +257°F), low coefficients of friction, good resistance to weathering and ozone, fair resistance to ultraviolet radiation, high flame retardancy, good biocompatibility, good resistance to dilute acids, mineral acids, alcohols, greases, oils, fair resistance to concentrated acids, and poor resistance to alkalies, aromatic hydrocarbons, solvents, halogens and ketones. Often sold under trade names or trademarks, such as *Apec, Calibre, Lexan, Makrolon, Makrolen, Merlon, Panlite, Plestar* or *Xantar*, they are used for molded products, safety shields, helmets, ophthalmic lenses, light globes, CD disks, glazing materials, unbreakable windows, meter face plates, household appliances (e.g., food processors, refrigerator drawers, air-conditioner housings, vacuum cleaner parts, microwave cookware and pitchers), automotive components (e.g., headlights, taillights, spoilers and instrument panels), mailboxes, filter bowls, power-tool housings, computer components and peripherals, prosthetic devices, business machine housings, connectors, terminal blocks, runway markers, marine propellers, baby bottles, bases for photographic films, solution-cast or extruded film, tubes and piping, and emulsion coatings. Abbreviation: PC. Also known as *polycarbonate resins*.

polycarbonate-styrene-maleic-anhydride alloys. See PC-SMA alloys.

polycarbonate-thermoplastic polyurethane alloys. See PC-TPUR alloys.

polycarbosilanes. A group of organosilicon polymers that can be prepared from polysilanes $[-(CH_3)_2Si]_n$ by Kumada rearrangement reactions at elevated temperatures. The general formula is $(R_2SiR')_n$, in which R is usually hydrogen or methyl (CH_3), and R' a methylene (CH_2) or ethynylene $(C≡C)$ group. Used as preceramic precursors for the preparation of silicon carbide and other engineering ceramics. See also polycarbomethylsilane; polysilanes.

polycarbosiloxanes. A class of organosilicon polymer in which the main chain consists of alternating silicon and oxygen atoms with side groups of hydrogen, or alkyls, aryls, arenyls, etc. (e.g., methyl or phenyl), and carbon. Used as preceramic precursor materials for the preparation of Si-O-C engineering ceramics. See also polysiloxanes; siloxanes.

Polycast. (1) Trade name of Polycast Technology Corporation

(USA) for cell-cast acrylic sheeting.

    (2) Trademark of Austenal, Inc. (USA) for resinous material used for dentures.

    (3) Trademark of Morrison Molded Fiber Glass Company (USA) for polymer concrete products including floor tiles and blocks, drainage systems, structural beams, and slabs.

Polyceraguard. Trade name of BetzDearborn, Division of Hercules Inc. (USA) for a spray-booth coating used for grates.

Polychem. Trademark of Budd Company (USA) for a series of phenolic and melamine resins and cotton fabric, glass fabric and paper laminates. The phenolic resins are supplied in numerous grades including wood or natural fiber-filled general-purpose, mineral-filled high-heat, mica-filled electrical, glass-reinforced high-impact, cotton-filled medium-shock, chopped-fabric-filled medium-impact and cellulose-filled shock-resistant foam. *Polychem MF* refers to cellulose-filled melamine formaldehydes.

polychlal fibers. Bicomponent fibers produced by grafting a number of vinyl chloride units to polyvinyl alcohol units, and emulsion spinning into staple and tow. Used for carpets, upholstery, curtains, blankets, etc.

polychloroprene rubber. See chloroprene rubber.

polychlorotrifluoroethylenes. See chlorotrifluoroethylenes.

Polychromatic. Trade name of SA Glaverbel (Belgium) for an extremely thick cast glass with variations in color tint in each sheet.

polychromic glass. A term referring to any multi-colored glass. The variety of colors is derived from colorants added to the glass batch during manufacture. Used for optical applications, and in building construction.

Polyclear. Trademark of Hoechst AG (Germany) for amorphous polyethylene terephthalates.

Polyco. Trademark of Borden Chemical, Division of The Borden Company (Canada) for an extensive series of thermoplastic polymers including acrylic copolymers, butadiene-styrene copolymers, water-soluble polyacrylates, polystyrenes, and vinyl acetate, vinyl chloride and vinylidene chloride copolymers and polymers. Supplied in the form of solvent solutions or aqueous emulsions, they are used in adhesives and coatings, and for textile and leather finishing.

PolyColl. Trade name of Polysciences, Inc. (USA) for collagen-coated microcarriers supplied in sizes ranging from 90 to 210 µm and specific gravities from 1.02 to 1.04 for use as substrates in cell culture applications. They can be easily trypsinized, aid in rapid cell attachment, provide strong adhesion, but can also be easily cleaned and re-used. Used extensively in biology, biochemistry, biotechnology, and bioengineering.

Polycolor. Trademark of former Spiegelglaswerke Germania AG (Germany) for toughened, ceramic-coated glass.

Polycor. (1) Trademark of Carlew Chemicals Limited (Canada) for vinyl compounds.

    (2) Trademark of Cook Composites & Polymers Company (USA) for polyester molding and laminating resins, gel coats and enamels.

Polycore. (1) Trademark of Pre Finish Metals, Inc. for a lightweight laminate consisting of a polypropylene-film material sandwiched between two outer skins of aluminum-killed, drawing-quality cold-rolled steel. The laminates can be slit, drawn, roll-formed, punched and projection-welded, and have good sound-deadening properties and good thermal and electromagnetic shielding properties. Used for motor and engine housings, thermal barriers, and electromagnetic (EMI) shielding applications.

    (2) Trademark of Cansew Inc. (Canada) for sewing thread composed of a polyester filament core with cotton or rayon wrapping.

    (3) Trademark of Jiffy Foam, Inc. (USA) for a rigid urethane foam plastic used for boat hulls, honeycomb aircraft structures, and as a core material for paneling

    (4) Trademark of MSC Laminates & Composites Inc. (USA) for a series of constrained-layer composites.

    (5) Trade name of Industrial Dielectrics Inc. (USA) for several chemical-resistant glass-fiber-reinforced vinyl ester resins used for compression-molded pumps, fittings and other industrial components.

Poly-Cote. Trade name of General Polymers Corporation (USA) for 100% solids urethane enamel paint for ceilings, walls and floors.

polycoumarone resins. See coumarone-indene resins.

Poly-Cover. Trademark of Flex-O-Glass, Inc. (USA) for polyethylene film, 25-250 µm (1-10 mil) thick, supplied in clear and black opaque colors in width up to 12.2 m (40 ft.). Used for agricultural applications.

Polycrest. Trade name for polyolefin fibers and products.

Polycron. Trademark of Quimica Industrial SA SA (Chile) for polyester staple fibers.

Polycron III. Trademark of PPG Industries, Inc. (USA) for spray-applied high-solids polyester extrusion coatings with good weatherability and impact and mar resistance for finishing aluminum substrates, e.g., RV components, high-rise curtainwalls, window and door frames, railings, trim, etc.

Poly-Cryl. Trade name of XIM Products, Inc. (USA) for two-part polyurethane finishes.

Polycrylic. Trademark of Minwax Division of of Sterling Drug Limited (Canada) for water-based, low-odor protective urethane finishes for interior wood surfaces (e.g., furniture, floors, railings, etc.) supplied in gloss, semi-gloss and satin.

polycrystalline boron nitride. See cubic boron nitride.

polycrystalline diamond. An extremely hard and wear-resistant, colorless diamond material with polycrystalline structure obtained from natural or synthetic diamond powders by compaction and sintering. Crystal system, cubic. Density, 3.5 g/cm^3. Used for indexable inserts for high-speed cutting tools. Abbreviation: PCD.

polycrystalline material. A material that is made up of a multitude of small crystals (also called crystallites) or grains, and therefore contains numerous grain boundaries. Most engineering metals and alloys and many ceramics are polycrystalline.

polycrystalline silicon. See polysilicon.

Polycure. Trade name of BP Chemicals International (USA) for weather-resistant polyethylene resins for electrical applications including power cable insulation and lighting fixtures.

Polycut. Trademark of 3M Company (USA) for coated abrasives.

polycyclic compound. An organic compound, such as anthracene or naphthalene, that contains 3 or more identical or different rings in its structures.

polycyclic hydrocarbon. See polynuclear hydrocarbon.

poly(1,4-cyclohexylenedimethylene)terephthalate. A linear polyester prepared by a condensation reaction of terephthalic acid with 1,4-cyclohexanedimethanol and available in fiber form. Abbreviation: PCT.

poly(1,4-cyclohexylenedimethylene)terephthalate fibers. Linear polyester fibers based on poly(1,4-cyclohexanedimethylene)terephthalate (PCT) and having a density of 1.23 g/cm^3

(0.044 lb/in.^{3}) a melting point of 290°C (555°F), low moisture absorption, good chemical resistance to many acids, alkalies and solvents, and excellent long-term heat resistance. Used for carpets, chemically resistant film and circuit board lamination. Abbreviation: PCT fibers.

poly(1,3-cyclopentylenevinylenes). See polynorbornenes.

Poly-Dap. Trade name of Industrial Dielectrics Inc. (USA) for a series of mineral-filled, or short or long glass-fiber-filled diallyl phthalate (DAP) resins and compounds with excellent electrical and high-temperature properties. Used for electrical and electronic applications including terminal, insulators and switches. Some grades are suitable for the encapsulation of electronic and electrical components.

Polydamp. Trade name of Polymer Technologies Inc. (USA) for a series of acoustical, thermal and vibration-damping composites including viscoelastic-foil damping composites, non-burning melamine-foam acoustical and thermal insulators, and reinforced acoustical barrier composites. Used for aerospace and aircraft applications.

poly(3-decylthiophene-2,5-diyls). A class of conductive polymers that are available in regioregular and regiospecific conformations. The former are black solids that have a head-to-tail regiospecific conformation of more than 98.5%, and average molecular weights of about 42000, while the latter are red solids in which the ratio of the head-to-head/head-to-tail linkages of the regioisomers is about 1:1. Used for electronic and optoelectronic applications.

Polydet. Trade name of Flachglas AG (Germany) for glass-fiber-reinforced sheets and foils based on unsaturated polyesters. Used in the form of profiles on balconies and building exteriors.

poly(dibutyltitanate). A polymeric metal alkoxide supplied as an ethylene glycol stabilized liquid with a titanium content of 22.0-23.0%. It has a unit molecular weight of 210.10, a density of 1.07-1.10 g/cm^3 (0.039-0.040 lb/in.^{3}), and a viscosity of 3200-3500 centistokes. Used in sol-gel and coating applications. Formula: $[(C_4H_9O)_2TiO]_n$.

poly(diethoxysiloxane). A polymeric metal alkoxide supplied as a liquid with a silicon content of 20.5-21.5%. It has a unit molecular weight of 134.20, a density of 1.05-1.07 g/cm^3 (0.038-0.039 lb/in.^{3}), and a viscosity, 4-5 centistokes. Used in sol-gel and coating applications. Formula: $[(C_2H_5O)_2SiO]_n$.

poly(diethoxysiloxane-co-ethylphosphate). A metal alkoxide supplied as a hygroscopic liquid consisting of a copolymer of diethoxysiloxane and ethylphosphate. It has silicon and phosphorus contents of 19.1-19.6% and 1.4-1.5%, respectively. Used in sol-gel and coating applications. Formula: $[(C_2H_5O)_2SiO]$-$[(C_2H_5O)OPO]_n$.

poly(diethoxysiloxane-co-ethyltitanate). A metal alkoxide supplied as a liquid consisting of a copolymer of diethoxysiloxane and ethyltitanate. It has silicon and titanium contents of 19.1-19.6% and 2.1-2.3%, respectively. The viscosity is 15-25 centistokes. Used in sol-gel and coating applications. Formula: $[(C_2H_5O)_2SiO][(C_2H_5O)_2TiO]$.

poly-1,1-dihydroperfluorobutyl acrylate. A white, nonflammable elastomeric polymer with a density of 1.5 g/cm^3 (0.05 lb/in.^{3}), good mechanical properties, good resistance to synthetic lubricants, solvents, hydraulic fluids, oils, etc., and low flexibility below -17°C (1.4°F). Used for hose and tubing, diaphragms, O-rings, gaskets, seals, sheeting, coatings for metallic and nonmetallic substrates, and textile coating compositions.

poly(dimethoxysiloxane). A polymeric metal alkoxide supplied as a liquid with a silicon content of 26.0-27.0%. It has a unit molecular weight of 106.15, a density of 1.14-1.16 g/cm^3 (0.041-0.042 lb/in.^{3}), and a viscosity of 6-9 centistokes. Used in sol-gel and coating applications. Formula: $[(CH_3O)_2SiO]_n$.

poly(dioxanone). A biodegradable and bioabsorbable polyether-ester based biopolymer produced by ring-opening polymerization of p-dioxanone at elevated temperatures in the presence of a suitable catalyst. It has a crystallinity of about 55%, a glass-transition temperature of -10 to 0°C (14 to 32°F), and good biocompatibility. Used in monofilament form for tissue engineering applications, such as sutures. Abbreviation: PDO.

polydisperse. A material consisting of molecules of different molecular weights.

polydisperse polymer. A polymer system in which there are several molecular weights presents. See also monodisperse polymer.

poly(3-dodecylthiophene-2,5-diyls). A class of conductive polymers that are available in regioregular and regiospecific conformations. The former are black solids that have head-to-tail regiospecific conformations of more than 98.5%, and average molecular weights of about 162000, while the latter are red solids in which the ratio of the head-to-head/head-to-tail linkages of the regioisomers is about 1:1. Used for electronic and optoelectronic applications.

Polydros. Trade name of Polydros SA (Spain) for cellulated glass having one surface coated with colored glass.

Polydur. (1) Trade name of Hüls AG (Germany) for polyester bulk molding compounds supplied in unmodified, electrical, fire-retardant, high-heat and low-profile grades.

(2) Trade name of Denaco (UK) for a series of thermoplastic polyurethanes with good dimensional stability and rigidity, good moldability and processibility, low coefficients of friction, and good paint-ability even without primers. Used for automotive body moldings.

polydymite. A light- to steel-gray mineral of the spinel group composed of nickel sulfide, Ni_3S_4. Crystal system, cubic. Density, 4.81 g/cm^3; melting point, 995°C (1823°F). Occurrence: Russian Federation.

polyelectrolytes. A class of natural or synthetic high polymers that contain anionic or cationic ions (e.g., in the form of metallic fillers) and, in aqueous solution, exhibit electrical conduction. Natural polyelectrolytes include proteins and certain gums, and synthetic polyelectrolytes salts of polyacrylic acid and polyethylenimine.

polyelectrolite cement. Any of a class of composite materials made by mixing a polymer (e.g., polyacrylic acid) and an inorganic filler, such as zinc oxide powder. They have excellent mechanical properties, very rapid rates of setting, good low-temperature processibility, adhere well to many engineering substrates, and are used chiefly in dentistry and biomedicine. Also known as ionomer cement.

Polyen. Trademark of Degussa AG (Germany) for a polyethylene film.

polyepichlorohydrin rubber. An elastomer resulting from the homopolymerization of epichlorohydrin, or the copolymerization of epichlorohydrin with ethylene oxide. See also epichlorohydrin elastomer.

Poly-Epoxy. Trade name of Pratt & Lambert, Inc. (USA) for an epoxy resin coating.

polyester coatings. Tough, relatively thick coatings based on a polyester resin, most frequently with glass fiber or flake reinforcement. They cure by chemical reaction, and have good resistance to abrasion and water, high film hardness, good to fair chemical resistance, high durability, and high gloss. Used on metal and wood substrates, for coil coatings, and for specialty bake coats.

polyester composites. A group of polymer-matrix engineering composites in which the reinforcing phase, usually fibers of glass, aramid or carbon in the form of rovings or mats, is dispersed in a thermosetting or thermoplastic polyester resin. Also known as *polyester-matrix composites*.

polyester–copolyester elastomer alloys. Polymer blends composed of a crystalline polyester (e.g., polybutylene terephthalate or polyethylene terephthalate) and a copolyester elastomer (COPE). They combine the good structural properties (e.g., stiffness and strength) of crystalline polyester with the resilience, impact resistance and elongation of copolyester elastomers. Used for automotive applications, housings, etc.

polyester fibers. A generic name for nonflammable, synthetic fibers made from a substance containing at least 85 wt% thermoplastic polyester resin. Available as staple fibers and continuous filaments, they have high tensile strength, a melting point of 264°C (507°F), and low water absorption. Often sold under trade names and trademarks, such as *Celanar, Dacron, Fortel, Terylene* or *Vycron*, they are used in fabric casing plies for tires, seat belts, paint roller covering, rubber hose reinforcement, fire hose, and clothing fabrics.

polyester film. A thin sheet made of thermoplastic polyester resin, usually by continuous extrusion processes. It is commercially supplied with a standard thickness of 0.015-0.35 mm (0.6-14 mils), and has very high toughness, excellent fatigue and tear strength, high dielectric strength, high electrical resistivity, good resistance to ultraviolet radiation, low thermal expansion, low coefficient of friction, good resistance to humidity and solvents, good resistance acids, alcohols, greases, oils, halogens and ketones, fair resistance to aromatic hydrocarbons, and poor resistance to alkalies and low water absorption. Often sold under trade names, such as *Melinex, Mylar, Scotchpar* or *Terefilm*, it is used for motor and transformer insulation, automotive tire cords, electrical, industrial and packaging applications, and cordage. Sensitized polyester film is used in magnetic recording tapes and in reprography.

polyester laminates. High-strength composites that are commercially available in the form of bars, sheets, rods, tubes and structural shapes made by impregnating glass-fiber mats or glass fabrics with polyester-resin solution, sometimes using high proportions of a filler, such as calcium carbonate or calcium sulfate, followed by final curing.

polyester-matrix composites. See polyester composites.

polyester–polyarylate alloys. Thermoplastic polymer blends of a crystalline polyester (e.g., polybutylene terephthalate or polyethylene terephthalate) and an amorphous polyarylate. They have good chemical and impact resistance, good dimensional stability and stiffness, and are used for automotive exterior parts, e.g., body panels.

polyester–polycarbonate alloys. Polymer blends of a crystalline polyester (e.g., polybutylene terephthalate or polyethylene terephthalate) and an amorphous polycarbonate. They combine the good chemical resistance, freedom from warpage and high stiffness of polyesters with the high toughness, good creep resistance and excellent dimensional stability of polycarbonates. Used for automobile bumpers, small mechanical components, lawnmower parts, and helmets.

polyester–polyphenylene oxide alloys. Engineering thermoplastics based on a blend of a crystalline polyester (e.g., polybutylene terephthalate or polyethylene terephthalate) and an amorphous polyphenylene oxide. They combine the good chemical resistance, freedom from warpage, and high stiffness of polyesters with the good impact resistance, high toughness, and outstanding energy absorption characteristics of polyphenylene oxide. Used for exterior automotive components, e.g., body panels.

polyester-polysulfone alloys. Polymer blends composed of a crystalline polyester (e.g., polybutylene terephthalate or polyethylene terephthalate) and an amorphous polysulfone. They combine the high stiffness, good chemical resistance and good melt flow characteristics of polyesters with the good dimensional properties, high-temperature resistance and impact resistance of polysulfones. Used for automotive applications, and electrical and electronic components.

polyester resins. A group of thermosetting or thermoplastic resins prepared by the condensation of polybasic and monobasic acids with polyhydric alcohols. They have high strengths and good resistance to chemicals and moisture. See also thermoplastic polyesters; unsaturated polyesters.

polyester rope. A rope that is made from polyester fibers and is stronger than polypropylene or nylon, but heavier in weight. It has good strength retention when wet, excellent creep and abrasion resistance, good shock-absorbing ability (about two-thirds that of nylon), good resistance to ultraviolet radiation (does not degrade in sunlight), no softening with increasing temperature, excellent dielectric properties, good resistance to acids and alkalies, fair to poor resistance to chemicals, and good resistance to rot, mildew and fungi. Used for towing and lifting purposes, and as barrier rope.

polyester rubber. See polyurethane elastomer.

polyesters. A group of synthetic resins or plastics that contain the ester group –CO–O–, and are made by polycondensation of polybasic and monobasic acids with polyhydric alcohols. See also polybutylene terephthalates; polyethylene terephthalates; polyester fibers; polyester film; polyester resins.

poly(ester-urethane). A polyurethane in which the alcoholic hydroxyl group is a polyester derived from adipic acid. Abbreviation: PAUR. See also polyurethanes.

polyetheretherketone–polyethersulfone alloys. See PEEK-PESV alloys.

polyetheretherketones. A group of linear, aromatic, crystalline engineering thermoplastics often supplied in glass fiber- or carbon fiber-reinforced grades. They have a density of 1.32 g/cm³ (0.048 lb/in.³), good resistance to alpha radiation, fair resistance to ultraviolet radiation, high degrees of oxidative and hydrolytic stability, low thermal expansion, good heat resistance, low glass transition temperatures, a continuous use-temperature of 250°C (480°F), good processibility, low coefficients of friction, good resistance to dilute acids, alcohols, alkalies, solvents, aromatic hydrocarbons, greases, oils, halogens and ketones and poor resistance to concentrated acids and chlorine. Often sold under trade names or trademarks, such as *Ketron, Larpeek* or *Victrex*, they are used as a matrix resin for fiber-reinforced composites for aerospace applications. Abbreviation: PEEK.

polyether foam. A flexible or rigid *polyurethane foam* made by the use of a polyether, such as polypropylene glycol.

polyether-imides. A group of linear, amorphous, high-performance thermoplastic polymers with repeating aromatic imide and ether units. Available in standard and various modified and reinforced grades, they have a density (unreinforced) of 1.27 g/cm³ (0.046 lb/in.³), high strength, creep resistance and rigidity at room and elevated temperatures, high moduli of elasticity, excellent dimensional stability, low shrinkage, low water absorption, a glass transition temperature of 217°C (423°F), a maximum service temperature of 170-180°C (338-356°F), good dielectric properties, good chemical resistance, good hydrolytic stability, good ultraviolet radiation resistance, good flame resistance, and good melt processibility. Used for interior aircraft parts, aircraft panels and seat component parts, pump housings, pump parts, oil chambers, automotive engine parts, electrical parts, printed-circuit boards, chip carriers, burn-in sockets, fluid and air handling components, fasteners, reflectors, temperature sensors, packaging materials, medical appliances, and as a matrix material for engineering composites. Abbreviation: PEI.

polyether resins. A group of polymers in which the repeating unit in the chain contains ether groups. Also known as *polyethers*.

polyethers. See polyether resins.

polyether sulfones. A group of melt-processible, amorphous, high-temperature engineering thermoplastics consisting of repeating phenyl groups with ether and sulfone linkages. Supplied in standard and reinforced grades, they are usually processed by blow molding, injection molding, vacuum forming, and extrusion. *Polyether sulfones* have a density of 1.37 g/cm³ (0.049 lb/in.³), high transparency, good mechanical strength, good creep resistance and dimensional stability, outstanding load-bearing properties, high toughness, good resistance to alpha radiation, fair resistance to ultraviolet radiation, good flame retardancy, low smoke emission, good dielectric properties, low thermal expansion, high thermal stability, a service temperature range of -110 to +220°C (-165 to +430°F), good resistance to dilute acids, alcohols, alkalies, greases, oils and halogens, fair resistance to concentrated acids and aromatic hydrocarbons, and poor resistance to ketones. Often sold under trade names and trademarks, such as *Radel, Ultrason* or *Victrex*, they are used for pumps, valve bodies and seats, seals, meter components, fluid-handling equipment, electrical and electronic components (e.g., battery parts, circuit connectors, fuse housings, lighting fixtures, terminal blocks, switches, printed-circuit components and thyristors), automotive components (e.g., heater fans, bearing cages and gears), ducts, and cookware. Abbreviation: PES, PESV.

polyethylene fibers. A group of high-strength high-modulus fibers based on linear high-density polyethylene, usually of high or ultrahigh molecular weight. They have a density of 0.95-0.97 g/cm³ (0.034-0.035 lb/in.³), high elongation, excellent mechanical properties at ambient temperatures, good abrasion resistance, low water absorption, and a softening point of 123°C (253°F). Often sold under trade names or trademarks such as *Dylan, Dyneema* or *Spectra*, they are used in the manufacture of rope and cordage, as reinforcing fibers, for paper and cloth laminates, nonwoven fabrics, piping, film packaging, wire insulation, and industrial gloves.

polyethylene film. High- or low-density polyethylene in the form of a blow-molded or extruded sheet, typically 0.01-0.5 mm (0.4-20 mils) thick, in standard and biaxially oriented grades. *Polyethylene film* has good tear strength, high elongation at break (600-700%), good permeability to carbon dioxide and oxygen, good dielectric properties, good resistance to alkalies, alcohols,

halogens, ketones, greases, oils and dilute acids, fair resistance to concentrated acids, aromatic hydrocarbons, greases and oils, an upper service temperature of 50-90°C (120-195°F) and fair resistance to ultraviolet resistance. Often sold under trade names and trademarks, such as *Hicor, Miltite, Poly-Cover, Saratran* or *Zendel*, it is used for packaging applications including food items, sealing tape, etc.

polyethylene foam. A lightweight, flexible cellular plastic film or sheeting made from high- or low-density polyethylene. It has very low water absorption, and is used for electrical and thermal insulation applications.

polyethylene glycol terephthalates. See polyethylene terephthalates.

polyethylene naphthalates. A group of colorless, inherently UV-stable semicrystalline polyester resins supplied in granule form in several resin grades, and in fiber and film form in biaxially oriented and heat-stabilized grades. They have a density of 1.36 g/cm³ (0.049 lb/in.³), good electrical properties, good chemical and temperature resistance, good strength and toughness, good dimensional stability, an upper service temperature of 155°C (311°F), excellent barrier properties to moisture, oxygen and carbon dioxide, and a high glass-transition temperature. Used for injection-molded products, such as food and beverage containers, in the manufacture of film and sheeting for plastic bags, bubble packs, labels, wrapping and packaging applications, for flexible printed circuitry, electrical insulation, etc., and for compounding and blending applications. Abbreviation: PEN.

polyethylene oxides. A group of polymers made by the polymerization of ethylene oxide (C_2H_4O) and available with viscosity-average molecular weights ranging from about 100000 to over 7000000. They have a density of 1.13 g/cm³ (0.041 lb/in.³), a melting temperature of about 65°C (149°F), good dimensional stability, and good resistance to oils, greases and hydrocarbons, used as replacements for phenolic resins. Abbreviation: PEO.

polyethylene plastics. Plastics based on reaction products of the polymerization of ethylene, or the copolymerization of ethylene with other monomers. Also known as *ethylene plastics*. See also ethylene copolymers.

polyethylene-polyethylene composites. Engineering composites consisting of polyethylene matrices reinforced with polyethylene fibers.

polyethylene resins. See polyethylenes.

polyethylenes. A group of strong, lightweight, thermoplastic synthetic materials made by the polymerization of ethylene (C_2H_4). They usually contain 85% or more ethylene and 95% or more total olefins, and are available in four basic grades: *high-density polyethylene* (HDPE), *low-density polyethylene* (LDPE), *linear low-density polyethylene* (LLDPE), and *ultrahigh-molecular-weight polyethylene* (UHMWPE). *Polyethylenes* with molecular weights of more than 4000 are white, tough, translucent, waxy solids, and those with less than 4000 are greasy liquids. Solid polyethylenes have high toughness, a Shore D durometer hardness of 50, melting point of 100-112°C (210-234°F), a maximum service temperature, -73 to +80°C (-100 to +175°F), relatively low coefficients of friction, good electrical insulating properties, good moisture resistance, excellent resistance to dilute and weak acids, good resistance to weak and strong bases, poor resistance to most strong and concentrated acids, excellent ultraviolet radiation resistance, and good ozone resistance. Often sold under trade names or trademarks, such as *Alathon, Bapolene PE, Dowlex, Eltex, Empee PE,*

Flexirene, Hostalen G, Lacqrene or *Marlex.* Used for molded household and shipping containers, refuse containers, bulk storage containers, spray tanks, automotive gas tanks, battery parts, truck-bed liners, tool boxes, flexible bottles, pipes, tubing, hose, toys, can liners, insulation, fiberoptic cable jackets, thin films or sheets for packaging, and high-strength high-modulus fibers for tensile applications, such as rope and cord. Abbreviation: PE. Also known as *ethylene resins; polyethylene resins.*

polyethylene terephthalate–polycarbonate alloys. See PET-PA alloys.

polyethylene terephthalates. A group of engineering thermoplastics that are condensation products of terephthalic acid or dimethyl terephthalate and ethylene glycol. They are commercially available in the form of oriented films, metallized films, laminates, sheets, fabrics, fibers, rods, granules, etc. Glass fiber- and mineral-filled, and glycol-modified grades are also available. They have a density of 1.4 g/cm³ (0.05 lb/in.³), high strength, toughness and dimensional stability, high hardness and wear resistance, fatigue and tear strength, a melting point of 245°C (473°F), a service temperature range of -40 to +170°C (-40 to +340°F), low coefficients of friction, low moisture absorption, high flame retardancy, good dielectric properties, good electrical resistance, good resistance to ultraviolet radiation, good resistance to acids, alcohols, greases, oils, halogens, ketones and solvents, fair resistance to aromatic hydrocarbons, and poor resistance to alkalies. Often sold under trade names and trademarks, such as *Arnite G, Celanar, Dacron, Hostadur, Impet, Melinex, Mylar, Petra, Rynite* or *Vestan,* they are used for magnetic recording tape, packaging films, clothing (often blended with cotton, wool or other fibers), soft-drink bottles, automotive tire cords, injection-molded parts, hand tools, pump and motor housings, furniture, electrical components (e.g., connectors, relays and switches), household appliance components, automotive components (e.g., alternator housings, headlamp reflectors, lamp sockets and mirror backs), and luggage racks. Abbreviation: PET. Also known as *polyethylene glycol terephthalates.*

polyethylene wax. A low-molecular-weight polyethylene used as a filler and modifier in various waxes, and as a modifier in certain coatings to promote gloss and toughness.

poly(ethyl oxazoline). A water-soluble, nonionic homopolymer synthesized from ethyl oxazoline (C_5H_9NO). Used in the manufacture of plastic products, and in the heat treating industry for formulating polymer quenchants. Abbreviation: PEOX.

Polyextra. Trade name for polyester fibers and products.

Polyfabs. Trade name of A. Schulman Inc. (USA) for acrylonitrile-butadiene-styrene terpolymers.

Polyfibre. Trade name of Dow Chemical Company (USA) for polystyrene resins and fibers.

Poly-Fil. Trade name of Fairfield Processing Corporation (USA) for polyester fiberfill.

Polyfil. Trademark of J.M. Huber Corporation (USA) for a kaolin clay used as a filler and/or extender in plastics, elastomers and other materials.

Polyfin. Trademark of Polysciences, Inc. (USA) for a mixture of fine paraffin waxes and copolymer alloys supplied in the form of opaque pellets with a melting point of 55°C (131°F). Used in biology and biochemistry as an embedding and infiltration medium for light microscopy.

Polyflex. (1) Trademark of LMP Corporation (USA) for flexible polyvinyl chlorides.

(2) Trade name of Plax Corporation (USA) for polysty-

rene resins and products.

Polyflon. Trade name of Daikin Kogyo Company, Limited (Japan) for a series of fluoropolymers supplied in the form of powders, granules, pastes, dispersions, and enamels.

Poly Flow. Trade name of Hedeman Manufacturing Corporation (USA) for extruded plastics supplied in various grades.

Polyfluoron. Trademark for a fluorocarbon polymer with outstanding inertness, good resistance to high temperature, good electrical resistance, low coefficient of friction, relatively low strength (but reinforceable), and a maximum service temperature of 204-260°C (400-500°F). Used for bearings, filters, piping, valves, seals, hose, electrical insulation, electronic components, enamels, and high-temperature coatings.

Polyfluron. Trademark of Schnabel & Co. KG (Germany) for polytetrafluoroethylene (PTFE) used for flexible tubing and hoses.

polyfluorochloroethylene resins. See chlorotrifluoroethylene polymers.

Polyfoam. Trademark of Foamex LP (USA) for rigid polyurethane foam sheets made from a polyether. Used for thermal insulation applications.

Polyform. Trade name Foamex LP (USA) for a high-strength, stretch-resistant thermoplastic based on polyformaldehyde, and used in the manufacture of orthopedic devices (e.g., protective, supportive and corrective splints), and for heart valve stents. *Polyform 2000* refers to polyurethane foam sheeting used for cushioning, bedding, packaging and medical applications, for furniture, and as a carpet underpadding.

Polyfort. Trade name of A. Schulman Inc. (USA) for polyethylene and polypropylene resins supplied in powder, granule, pellet and bead form for further processing by molding and extrusion techniques.

Poly F Plus. Trade name for a zinc polycarboxylate dental cement.

polyfunctional polymer. A network polymer with two or more reaction sites for each mer. See also network polymers.

Polygalva. Trade name for a zinc alloy containing controlled amounts of aluminum, magnesium, tin and zinc, and used as a master alloy for galvanizing silicon-killed steels with 0.05-0.20% silicon.

PolyGlas. Trade name of Polysciences, Inc. (USA) for glass microcarrier beads supplied in sizes ranging from 90 to 210 μm (0.003 to 0.008 in.) and densities from 1.02 to 1.04 g/cm³(0.037-0.038 lb/in.³). Used as substrates for cell culture applications in biology and biochemistry, and in biotechnology and bioengineering.

Polyglas. (1) Trade name of Westinghouse Electric Corporation (USA) for glass fiber-reinforced plastic shapes made by a continuous molding process.

(2) Trade name of Goodyear Tire & Rubber Company (USA) for glass-fiber tire cords.

Polyglass. (1) Trade name of Glaceries de Saint-Roch SA (Belgium) for an insulating plate glass.

(2) Trade name for light-cure, glass-filled hybrid composite used in dentistry for inlays.

Poly-Glaze. Trade name of Schnee-Morehead Inc. (USA) for silicone sealants.

Poly Glide. Trade name of Ludlow Steel Company (USA) for ultrahigh-molecular-weight polyethylene with good antifriction properties.

Polygloss. Trademark of J.M. Huber Corporation (USA) for a glossy kaolin clay powder used as a filler in plastics, elastomers and other materials.

poly(glycolic acid). A synthetic, biodegradable, linear aliphatic polyester prepared by first synthesizing a glycolide monomer by the dimerization of glycolic acid ($HOCH_2COOH$) and subsequent ring-opening polymerization of this monomer. It has a density of 1.53 g/cm^3 (0.055 $lb/in.^3$), a melting point of 220-230°C (428-446°F), a glass-transition temperature of 35-40°C (95-104°F), a high molecular weight, a high degree of crystallinity (typically 45-55%), and is insoluble in most solvents. It can be produced in the form of high-strength and modulus monofilaments and fibers, and copolymerized with poly(lactide), trimethylene carbonate (TMC) or *p*-dioxanone. Used as a biomaterial in tissue engineering, e.g., for absorbable sutures. Abbreviation: PGA. Also known as *polyglycolide*. See also glycolide.

polyglycolide. See also polyglycolic acid.

poly(glycolide-*co*-trimethylene carbonate). See polyglyconate.

polyglyconate. A flexible, biodegradable A-B-A block copolymer synthesized from glycolide and trimethylene carbonate (TMC) at elevated temperatures in the presence of a special catalyst, and consisting of a central glycolide-TMC block (B) and glycolide terminal blocks (A) with a typical glycolide-to-TMC ratio of 2:1. It is used for surgical sutures and for medical fixation devices, such as tacks and screws. Abbreviation: PGA-TMC. Also known *poly(glycolide-co-trimethylene carbonate)*.

Poly-Grit. Trade name of MDC Industries, Inc. (USA) for abrasive blasting media used in metal finishing.

Poly-Guard. Trade name of Symplastics Limited (Canada) for high-density and ultrahigh-molecular-weight polyethylene sheets and rods.

Polyguard. Trade name of KoSa (USA) for polyester fibers supplied in several grades *Polyguard 3D, Polyguard HV, Polyguard Classic* and *Polyguard Home*, and used for textile fabrics including industrial products and home furnishings.

polyhalite. A colorless, white or gray mineral composed of potassium calcium magnesium sulfate dihydrate, $K_2Ca_2Mg(SO_4)_4 \cdot 2H_2O$. Crystal system, triclinic. Density 2.78 g/cm^3; refractive index, 1560. Occurrence: Germany, Netherlands, USA (New Mexico, Texas). Used as a source of potash for fertilizers.

polyhalogen compound. See halogenated hydrocarbon.

polyhedral oligomeric silsesquioxanes. Cage-like hybrid polymer molecules of silicon and oxygen ($RSiO_{1.5}$) that are chemically intermediate between silica and silicone. Supplied as monomers in liquid and solid form, they are used as additives to enhance the heat and/or abrasion resistance of paints and coatings, in polymers to improve the mechanical properties especially the elastic modulus and hardness, and the thermal resistance and upper service temperatures, fire retardation and viscosity. They are also used as ablative materials and as precursors to glassy and ceramic matrices. Abbreviation: POSS.

poly(hexafluoropropylene-*co*-vinylidene fluoride)s. See hexafluoropropylene vinylidene fluorides.

poly(hexamethylene adipamide). See nylon 6,6.

poly(hexamethylene adipamide-*co*-caprolactam). See nylon 6,6.

poly(hexamethylene dodecanediamide). See nylon 6,12.

poly(hexamethylene nonanediamide). See nylon 6,9.

poly(hexamethylene sebacamide). See nylon 6,10.

poly(3-hexylthiopene-2-diyl). A conducting polymer that is supplied in regiospecific and regiorandom conformations. The regiospecific grade has an average molecular weight of about 87000 with typical head-to-tail conformations of more than 98.5%. The regiorandom grade is supplied as a red solid with a head-to-head/head-to-tail ratio of regioisomer linkages of 1:1. Used as an organic semiconductor with a typical electron mobility of about 10^{-1} cm^2/V-s for various microelectronic and optoelectronic applications.

Polyhipe. Trademark of Joseph Crosfield & Sons Limited (UK) for emulsion-derived porous polymeric foams including polystyrene.

Poly-Hone. Trade name of Automated Finishing Inc. (USA) for abrasive plastic media used for mass finishing applications.

polyhydroxyalkanoates. A family of polyesters based on hydroxyalkanoic acids, such as 3-hydroxybutyric acid, 3-hydroxypropionic acid or 3-hydroxyvaleric acid. Depending on their composition and processing (e.g., extrusion, molding, etc.), they may be hard crystalline thermoplastics with melting temperatures of approximately 50-180°C (122-356°F), or highly elastic rubbers. Abbreviation: PHA. See also poly(hydroxybutyrates); poly(3-hydroxybutrate-valerates).

poly(4-hydroxybenzoates). A family of thermoplastic polyesters or copolyesters based on 4-hydroxybenzoic acid (HOC_6H_4-$COOH$). They are supplied in unreinforced and reinforced grades and used chiefly in biochemistry and biotechnology. Abbreviation: PHB; POB.

poly(4-hydroxybenzoic acid-*co*-ethylene terephthalate). A liquid crystalline copolyester of 4-hydroxybenzoic acid and ethylene terephthalate. Formula: $(-OC_6H_4CO-)_x(-OCH_2CH_2O_2C-C_6H_4-4-CO-)_y$.

poly(4-hydroxybenzoic acid-*co*-6-hydroxy-2-naphthoic acid). A random thermoplastic liquid crystalline copolymer that can be unreinforced or reinforced with glass fibers. It has an average molecular weight of more than 20000, a density of 1.5 g/cm^3 (0.05 $lb/in.^3$), and a melting point of 280°C (536°F). Formula: $(-OC_6H_4CO-)_x(-OC_{10}H_6CO-)_y$.

poly(hydroxybutyrates). A group of brittle, biodegradable, highly-crystalline natural thermoplastic homopolymers derived from the dry matter of the aseptically grown bacterium *Alcaligenes eutrophus*. Available in technical and research grades in the form of films, sheets, nonwoven fabrics, staple fiber wools, flocculent and granules, they have a density of 1.5g/cm^3 (0.05 $lb/in.^3$), a melting point of 176°C (349°F), an upper service temperature of 95°C (203°F), good dielectric properties, fair resistance to ultraviolet radiation, high tensile moduli, good resistance to greases and oils, fair resistance to alcohols and weak acids and poor resistance to alkalies. Used for orthopedic and surgical devices (e.g., bone plates), packaging, personal hygiene products, slow-release systems for herbicides, drugs, etc. Abbreviation: PHB. Also known as *poly(3-hydroxybutyric acid)s*.

poly(3-hydroxybutyrate-valerates). A group of random, biodegradable, natural copolymers derived from the dry matter of the aseptically grown bacterium *Alcaligenes eutrophus*. 3-hydroxyvalerate can be incorporated into the polymer by influencing the nutrition of the bacteria culture. Depending on the conditions of the culture, the copolymers may contain 80-93% poly(3-hydroxybutyrate), and 5-20% poly(3-hydroxyvalerate). Available in technical and research grades in the form of flocculent and granules, they have a density of 1.5 g/cm^3 (0.05 $lb/in.^3$), an upper service temperature of 95°C (203°F), good dielectric properties, fair resistance to ultraviolet radiation, high tensile moduli, good resistance to greases and oils, fair resistance to alcohols and weak acids, and poor resistance to alkalies. Used for orthopedic and surgical devices, tissue engineering, packaging, personal hygiene products, slow-release systems for herbicides, drugs, etc. Abbreviation: PHBV. Also known as *poly(3-hydroxybutyric acid-co-3-hydroxyvaleric acid)*.

polyimide film. Polyimide in the form of a thin film, typically 0.008-0.125 mm (0.3-5.0 mils) thick, with good tear strength, high elongation at break (70+%), good permeability to carbon dioxide and oxygen, good dielectric properties including a dissipation factor of approximately 0.01 at 1 MHz and a typical dielectric strength of 220-280 kV/mm (560-710 V/mil). Used for capacitors, printed-circuit boards, electric motors, insulation of aircraft, and missile wire cables.

polyimide foam. A lightweight, flame-retardant, cellular polyimide used for floor panels, wallboard, and thermal insulation.

polyimide resins. See polyimides.

polyimides. A group of high-temperature engineering polymers made by reacting aromatic dianhydrides with aromatic diamines and have the imide group –CONHCO– in the main chain. They are available in thermoplastic and thermoset grades, with a density of 1.42 g/cm³ (0.051 lb/in.³), good tensile strength, high modulus of elasticity, high room-temperature rigidity and stiffness, good high-temperature stability, good wear resistance, a heat-distortion point above 260°C (500°F), a glass-transition temperature above 427°C (800°F), low thermal expansion, good dielectric properties, low water absorption, good flame retardancy, excellent resistance to nuclear radiation and thermal oxidative degradation, fair chemical resistance, good resistance to organic substances, poor resistance to strong alkalies, low outgassing in high vacuum, and low coefficients of friction. Often solder under trade names and trademarks, such as *Gemon, Kapton, Kinel, Polymer SP* or *Vespel*, they are used for high-temperature coatings, ablative agents, adhesives, as binders in abrasive wheels, in magnetic recording tapes, wire and cable insulation, lightweight foams and fibers, in matrix resins for engineering laminates and composites for aerospace applications, semiconductor applications, molded parts (e.g., valve seats, retainers, bearings, printed circuits), and low-dielectric-constant thin films for integrated-circuit devices. Abbreviation: PI. Also known as *polyimide resins*. See also thermoplastic polyimides; thermoset polyimides.

polyindene resins. See coumarone-indene resins.

polyisobutenes. Thermoplastic, isotactic polymers made by the polymerization of isobutene (C_4H_8). They are available in molecular weights ranging from 500 to over 1000000. Their consistency varies with molecular weight from elastomeric solids (typical molecular weight, 75000 or more) to viscous liquids (typical molecular weight, 500-2700). Used in the manufacture of adhesives, sealants, films, coatings, and synthetic rubber. Abbreviation: PIB. Also known as *polyisobutylenes*.

polyisobutylenes. See polyisobutenes.

polyisocyanate resins. See isocyanate resins.

polyisocyanates. See isocyanate resins.

polyisocyanurate foam boards. See isocyanurate foam boards.

polyisocyanurate foam insulation. See isocyanurate foam insulation.

polyisocyanurate resins. See isocyanurate resins.

polyisocyanurates. See isocyanurate resins.

Polyisoplast. Trademark of Chemtron Manufacturing Limited (Canada) for an insulating glass sealant.

polyisoprene. (1) The main ingredient of rubber-like substances, such as natural rubber, gutta percha, balata, etc. There are two stereospecific polyisoprene isomers: (i) *cis*-1,4-polyisoprene

(natural rubber) and (ii) *trans*-1,4-polyisoprene (gutta percha). Unvulcanized polyisoprene is thermoplastic and combustible. Used in the manufacture of rubber goods. Abbreviation: PI. See also natural polyisoprene.

    (2) *Cis*-1,4-polyisoprene and *trans*-1,4-polyisoprene produced synthetically by the effect of heat and pressure on isoprene in the presence of stereospecific catalysts (usually organometallic compounds). Used in the manufacture of synthetic rubber goods. Abbreviation: PI. Also known as *synthetic polyisoprene*.

polyisoprene fibers. See elastodiene fibers; rubber fibers.

Polyite. Trade name of Minerva Dental Laboratories (UK) for polyester resins.

Polyjel. Trade name of Dentsply/Caulk (USA) for a gel-like polyether-based dental impression material.

Polykent. Trade name of Kent Dental/Hejco (UK) for a polycarboxylate dental cement.

Polyklon. Trademark of Polymekon SpA (Italy) for polypropylene staple fibers used for textile fabrics.

Poly-Kore. Trade name of EDCO Products, Inc. (USA) for polystyrene insulating boards used for sheathing applications.

Poly-kote. Trade name of Polysciences, Inc. (USA) for a 1% solution of dichlorodimethylsilane in benzene used to render electrophoresis gel tubes nonwettable.

Polylac. Trademark of Chi Mei Industrial Company, Limited (Taiwan) for acrylonitrile-butadiene-styrene resins supplied in general-purpose, extrusion, sheet extrusion, high-flow, flame-retardant and high-heat grades.

poly(L-lactic acid). A synthetic, biodegradable poly(hydroxy ester) polymer that is based on L-lactic acid and releases it during *in vivo* degradation. It is widely used as a biomaterial in tissue engineering, e.g., for sutures. Abbreviation: PLA; PLLA. See also lactic acid; poly(L-lactide).

polylactic acid fiber. A relatively new biodegradable synthetic fiber made from celluloses and cornstarch. It can be blended with cotton, rayon and wool, and is used for yarns, knitted and woven fabrics, nonwovens, etc. Abbreviation: PLA fiber.

poly(DL-lactic-co-glycolic acid). A synthetic, biodegradable copolymer of L-lactic acid and glycolic acid. It releases lactic acid and glycolic acid during *in vivo* degradation, and is widely used as a biomaterial in tissue engineering, e.g., for sutures. Abbreviation: PLGA. See also lactic acid; glycolic acid.

polylactide. A biopolymer synthesized from L-lactide, D-lactide or DL-lactide. Abbreviation: PLA. See also poly(lactic acid); poly(L-lactide); poly(DL-lactide).

poly(L-lactide). A semicrystalline, biodegradable homopolymer synthesized by the polymerization of L-lactide, the naturally occurring optical isomer of lactide. It has a degree of crystallinity of about 37%, a melting point of 175-178°C (347-352°F), a glass-transition temperature of 60-65°C (140-149°F), fairly slow biodegradation compared to poly(DL-lactide), high tensile strength, high modulus, and low elongation. It can be copolymerized with DL-lactide or glycolide to produce materials with faster biodegradation. Used for orthopedic fixation devices, and in tissue engineering for sutures. Abbreviation: LPLA. See also L-lactide.

poly(DL-lactide). An amorphous, biodegradable polymer synthesized by the polymerization of DL-lactide, a synthetic alloy of D-lactide and L-lactide. It has a lower tensile strength and modulus, higher elongation and faster biodegradation than poly(L-lactide). It can be copolymerized DL-lactide or glycolide, and is suitable for drug delivery systems. Abbreviation: DLPLA.

See also DL-lactide.

poly(DL-lactide-*co*-caprolactone). A synthetic, biodegradable copolymer of lactide (3,6-dimethyl-1,4-dioxane-2,5-dione) and caprolactone. Available in serveral lactide to caprolactone monomer ratios ranging from 86:14 to 40:60, it is used as a biomaterial in tissue engineering. Abbreviation: DLPLA-PCL. See also poly(ε-caprolactone); poly(DL-lactide).

poly(DL-lactide-*co*-glycolide). A synthetic, biodegradable amorphous copolymer of DL-lactide and glycolide. It is available in several lactide to glycolide monomer ratios ranging from 85:15 to 30:70. Depending on the monomer ratio the average molecular weight ranges from 50000 to over 120000. The glass-transition temperature for the 85:15 and 75:25 copolymers is about 50-55°C (122-131°F), and for the 65:35 and 50:50 copolymers 45-50°C (113-122°F), respectively. Used as a biomaterial in tissue engineering and for drug delivery applications. Abbreviation: DLPLG; PGA-DLPLA. See also poly(DL-lactide); glycolide.

poly(L-lactide-*co*-glycolide). A synthetic, biodegradable amorphous copolymer of L-lactide and glycolide. It is available in serveral lactide to glycolide monomer ratios ranging from 75:25 to 30:70. Used as a biomaterial in tissue engineering, e.g. for absorbable sutures. Abbreviation: LPLG; PGA-LPLA. See poly(L-lactide); polyglycolide.

poly(L-lactide-*co*-DL-lactide). A synthetic, biodegradable copolymer of poly(L-lactide) (LPLA) and poly(DL-lactide) (DLPLA) that has faster biodegradation and bioabsorption than LPLA alone, and is used for angioplastic plugs, surgical pins and other surgical devices, and for dental guided-tissue-regener-ation (GTR) membranes. Abbreviation: LPLA-DLPLA. See also poly(L-lactide); poly(DL-lactide).

polylactic resins. Soft flexible resins prepared by the reaction of lactic acid with a fatty oil, such as castor oil at elevated temperatures. Used in the production of durable coatings.

poly(lauryllactam). See nylon 12.

Polylene. Trademark of Polylen AS (Turkey) for polyester filament yarns used for textile fabrics.

Polyliner 40. Trade name for a heat-cure silicone used as a denture liner.

Polylite. (1) Trade name of Reichhold Chemicals, Inc. (USA) for an extensive series of synthetic resins including various unsaturated polyesters used for laminating, tooling, casting, furniture, boatbuild-ing, and for cultured granite, marble and onyx.

(2) Trade name of Minerva Dental Laboratories (UK) for acrylics used for dentures.

polylithionite. A colorless mineral of the mica group composed of potassium lithium aluminum fluoride silicate, $KLi_2AlSi_4O_{10}F_2$. It can be made synthetically. Crystal system, monoclinic. Density, 2.83 g/cm^3; refractive index, 1.552.

Polyloom. Trade name for polyolefin fibers and yarns used for woven and knitted fabrics including clothing.

Polylure. Trademark of Lilly Industries Inc. (USA) for spray-applied coatings with high-solids content used on aluminum extrusions.

poly-L-lysine. A polymer of the amino acid *L-lysine* available in molecular weights ranging from 1000 to over 300000. *Polylysines* with molecular weights exceeding 70000 are used in cell biology to improve the adhesion of cells to solid substrates.

Polyman. Trade name of A. Schulman, Inc. (USA) for neat and olefin-modified styrene-acrylonitriles and acrylonitrile-butadi-ene-sty-rene/polyvinyl chloride alloys. *Polyman ABS* refers to acrylonitrile-butadiene-styrene resins supplied in transparent, low-gloss, high-heat, UV-stabilized, fire-retardant, glass fiber-rein-forced, high- and medium-impact, plating, and structural foam grades.

Poly Mask. Trademark of Sealed Air Corporation (USA) for protective plastic tape coated with adhesive.

PolyMat. German trade name for glass-fiber mats suitable for use as prepregs in the manufacture of glass-fiber-reinforced thermoplastics using matrix resins such as polyamide, polyester and polypropylene. Supplied in weights from 200 to 1500 g/m^2 (5.9 to 44.3 oz/yd^2) and widths up to 2.5 m (8.2 ft), they have excellent flow properties and impregnability, good dimensional stability, good thermal stability, and good thermal and acoustical insulating properties. Used for engineering composites for automotive interiors.

polymer. See polymers.

polymer alloys. See alloys (2).

polymer blends. See alloys (2).

polymer concrete. A strong, corrosion- and impact-resistant, crosslinked polymer composite made by first mixing selectively-graded aggregates with a polymer resin, and then molding and curing under prescribed conditions. The precast concrete is available in a variety of colors, and can be reinforced with glass fibers to further enhance rigidity and strength. Used as a substitute for Portland cement concrete products including floor tiles and blocks, drainage systems, and structural beams and slabs.

polymer-derived engineering ceramics. A class of engineering ceramics including silicon carbides, silicon oxycarbides, silicon boronitrides and silicon nitrides prepared from silicon-containing polymer precursors, such as silanes, silizanes and siloxanes. They are produced by first synthesizing the polymer from monomer or oligomer precursors and then forming into the desired (near-net) shape (e.g., fibers, coatings, tapes, foams or bulk components), curing to a thermoset at elevated temperatures, and thermally induced decomposition (pyrolysis) in an inert or reactive atmosphere at temperatures between 500 to 1500°C (930 to 2730°F) to produce an organic/inorganic hybrid. Inert or reactive fillers (e.g., powder or fibers) may also be incorporated. Many polymer-derived ceramics exhibit properties intermediate between those of polymers and ceramics, and are used as reinforcements in composites, and in the manufacture of various structural and nonstructural components for aircraft, spacecraft, transportation systems, communication equipment, etc. See also preceramic polymers.

polymer-dispersed liquid crystal. A nematic or chiral nematic liquid crystal prepared by dispersing micron-sized droplets of liquid crystal material in a solid polymer matrix with a typical polymer concentration of more than 20 wt%. It is usually made by microencapsulation or phase separation techniques. Nematic types are usually prepared by microencapsulation which involves mixing of a liquid crystal with an aqueous polymer solution. Both nematic and chiral nematic types can be made by polymerization-, temperature- or solvent-induced phase separation of a homogeneous mixture of a polymer or prepolymer and a liquid crystal material. The orientation of the molecules in a *polymer-dispersed liquid crystal* can be changed by the application of an electric field resulting in a variation of the transmitted light intensity. The configurations of the liquid crystal drops can be axial, bipolar or radial. They are extremely suitable for optoelectronic applications including switchable windows, light shutters. projection devices, etc. Abbreviation: PDLC. See also liquid crystal; polymer-stabilized liquid crystal; polymer liquid crystal.

polymer electrets. A class of low-dielectric-constant polymers, such as certain fluoropolymers and fluorinated polymers, either with oriented dipole moments, or with the ability to quasipermanently store electric charges in the bulk or on the surface. They have potential applications in electronic packaging. See also electret.

polymer gel. An insoluble solid polymer beyond the gel point, i.e., the point at which the extent of crosslinking exceeds a critical value. It is produced from a liquid, soluble polymer (sol), and combines liquid and solid properties. See also biomimetic gels; gel; sol.

polymer glasses. See amorphous plastics.

polymeric composites. See fiber-reinforced plastics.

polymeric liquid. A term used to designate a solution of polymers, or a molten polymer. Such a liquid exhibits a non-Newtonian behavior below a given critical molecular-weight value (the so-called "critical molecular weight for entanglement").

polymeric materials. See polymers.

polymer-impregnated concrete. A concrete into whose pore spaces a synthetic resin has been introduced. See also polymer concrete.

polymer liquid crystal. A polymer whose liquid crystal properties are due to the presence of structural units consisting of rigid disk- or rod-like molecules (known as "mesogens") in their main chains (backbones) or side chains (branches). It is suitable for use in various applications ranging from high-strength fibers to optical applications including waveguides and electrooptical modulators, various other nonlinear optical devices, spatial light modulators, optical amplifiers, tunable notch filters, and laser-beam defectors. Abbreviation: PLC. See also liquid crystal; main-chain polymer liquid crystal; side-chain polymer liquid crystal; liquid crystal polymers.

polymer-matrix composites. See fiber-reinforced plastics.

polymer-modified bitumen. A dark-colored bituminous material, such as asphalt, pitch or tar, whose properties have been modified by a polymer dispersion.

polymer-modified cement. A mixture of a cement paste and small amounts of a rubber latex, such as styrene-butadiene. Abbreviation: PMC.

polymer paint. A water-base organic coating composed of acrylic or vinyl resin. Upon surface application the water evaporates leaving a continuous, uniform, flexible, water-resistant polymer film.

polymer plastics. Synthetic high polymers unmodified or with additions of modifying agents, such as fillers, colorants, plasticizers, etc. Based on their thermal behavior, they can be classified as either thermoplastic or thermosetting. See also high polymers; plastics; polymer.

polymer quenchants. A group of synthetic quenching media including polyvinyl alcohol, polyalkylene oxide, polyethyl oxazoline and polyvinyl pyrrolidone, used (usually in aqueous solution) in the ferrous and nonferrous heat treating industry as substitutes for traditional water, brine and mineral-oil quenchants.

Polymers. Trade name for polyolefin and nylon fibers and products.

polymers. A group of plastic and elastomeric engineering materials in which large molecules (macromolecules) composed of many repeating units (mers) are arranged in long chains or networks. *Polymers* are usually addition or condensation products of polymerization reactions. For example, polyethylene is an engineering polymer formed by the polymerization of the mer ethylene. Also known as *polymeric materials*.

polymer sol. A soluble, liquid polymer at a stage prior to reaching the gel point, i.e., the point at which the extent of crosslinking exceeds a critical value. The sol stage is often followed by the gel stage. See polymer gel.

Polymer SP. Trade name of E.I. DuPont de Nemours & Company (USA) for polyimide resins.

polymer-stabilized liquid crystal. A cholesteric or nematic liquid crystal to which up to 10 wt% of a monomer (e.g., a diacrylate or a photopolymerizing monomer) have been added to form a polymer network, stabilize its texture and improve its electro-optical properties. It has various applications in electronics and optoelectronics especially the cholesteric types which are used in bistable reflective displays, photo-tunable devices and light shutters. Abbreviation: PSLC. See also liquid crystal; polymer-dispersed liquid crystal; polymer liquid crystal.

polymet. A composite material composed of a metal matrix (usually aluminum) and a high-melting-temperature liquid crystalline polymer, and made by hot extrusion of an metal/polymer powder mixture. The extrusion process elongates the polymer into aligned filaments and bonds the metal powder particles into a solid matrix.

polymethacrylates. A group of thermoplastic resins that are polymerization products of an ester or salt of methacrylic acid and acrylic acid. They are used in emulsion paints, surface coatings, paper and leather finishes, elastomers, resins, etc. See also polymethacrylic acid.

polymethacrylic acid. A water-soluble methacrylic or methacrylate polymer obtained by the polymerization of methacrylic acid ($C_4H_6O_2$). It is often supplied as aqueous solution, e.g., in form of the sodium or potassium salt. Used especially in the manufacture of acrylic products and copolymers, in adhesives and coatings, as a suspending agent, and as a textile size.

polymethylene wax. A white, lustrous, highly durable microcrystalline wax based on polymethylene.

polymethyl methacrylates. A group of thermoplastic polymers obtained from methyl methacrylate ($C_5H_8O_2$). Commercially available in the form of sheets, rods, tube, powder, granules, solutions, and emulsions, they have a density of 1.2 g/cm³ (0.04 lb/in.³), a refractive index of 1.49, high transparency (above 90%), excellent optical properties, moderate hardness and strength, good resistance to ultraviolet radiation, good dielectric properties, low heat resistance, poor flame resistance, an upper service temperature of 90°C (194°F), low water absorption, good weatherability, good machinability and moldability, good resistance to dilute acids and alkalies, fair resistance to concentrated acids, and poor resistance to aromatic hydrocarbons, greases, oils, halogens and ketones. Often sold under trade names and trademarks, such as *Altuglas, Baycryl, Degulan, Elvacite, Lucite, Lucryl, Perspex, Plexiglas* or *Vedril*, they are used for lenses, optical instruments, transparent enclosures, furniture components, outdoor signs, lighting fixtures, light diffusers, nameplates, decorations, display items, dials, bottles, industrial and architectural glazing, aircraft canopies and windows, windshields for boats, etc., aircraft parts, ornaments, drafting equipment, surgical appliances, and as facing materials for composites. Abbreviation: PMMA.

polymethylpentenes. A group of transparent, semicrystalline polyolefin resins with a density of 0.84 g/cm³ (0.030 lb/in.³), high clarity and light transmission (above 90%), high gloss, high hardness, good impact strength, fair resistance to ultraviolet light, good dielectric properties, excellent heat resistance, poor flame resistance, a service temperature range of -30 to +115°C (-22 to +239°F), good resistance to acids, alcohols, alkalies, greases, oils and ketones, fair resistance to aromatic hydrocarbons, and poor resistance to carbon tetrachloride and cyclohexane. Often sold under trade names and trademarks, such as *Crystalor, TPX* or *Zeonex*, they are used for electronic equipment, light reflectors, hospital equipment, laboratory ware (beakers, etc.), packaging of frozen food, and food containers, e.g. trays for TV dinners. Abbreviation: PMP.

poly(methylsilylene). See polycarbomethylsilane.

poly(α-methyl styrene). A low-molecular-weight polymer of α-methyl styrene (C_9H_{10}) usually supplied in the form of beads or powder. Used chiefly in the manufacture of polyester and other plastic products. Abbreviation: PAMS.

poly(3-methylthiophene). A conductive conjugated polymer of 3-methylthiophene (C_5H_6S) used in the manufacture of photo-electrochemical cells. Abbreviation: P3MTh.

Polymid. Trademark of Lawter International Inc. (USA) for polyamide-based synthetic resins for printing inks, and the like.

Polymisr. Trademark of Misr Rayon Company (Egypt) for polyester filament yarns used for textile fabrics.

Polymist. Trademark of Ausimont USA for white, free-flowing, micronized polytetrafluoroethylene-based lubricant powders used as additives in thermoplastics, thermosets, elastomers, coatings, paints, inks, oils and greases to enhance wear, antiscuff, slip, and/or release properties.

Poly-mount. Trade name for crystal clear, quick-drying, permanent acrylic mounting media for cover glasses used in microscopy.

Polynak. Russian trade name for acrylic (polyacrylonitrile) textile fibers.

Polynam-A. Trade name of Duralac Inc. (USA) for a lacquer used on brass and silver.

polynaphthalenes. A class of double-ring polymers produced in the form of highly adhesive, transparent thin films from fluoro-1,2-diethynylbenzene (polynaphthalene-F) or 1,2-diethynylbenzene (polynaphthalene-N) by vapor deposition polymerization, or in the form of brown, insoluble, low-molecular-weight powders from solution thermolysis of 1,2-diethynylbenzene. The thin films have relatively high hardness and dissociation temperatures and are used as low-dielectric-constant materials for integrated circuit applications.

Polynil. Trademark of Nilit Limited (USA) for polyamide 6,6 supplied in high-viscosity and lubricated grades. They have a density of 1.14 g/cm³ (0.041 lb/in.³), and a melting point of 257-258°C (495-498°F). The high-viscosity grades have a maximum service temperature of 69°C (156°F), and the lubricated grades of 80°C (176°F).

polynorbornene rubber. A synthetic rubber based on norbornene (C_7H_{10}) or a norbornene derivative. Abbreviation: PNR.

polynorbornenes. A class of isotropic, aliphatic hydrocarbon polymers with rigid polycyclic backbones. They have a weight-average molecular weight of 2000000, a density of 0.96 g/cm³ (0.035 lb/in.³), good thermal stability up to 400°C (750°F), a glass-transition temperature of approximately 350°C (660°F) and low dielectric constants (k ~ 2.4-2.6). Supplied as powders and thin films, they are and used as interlayer dielectrics for integrated circuits and for other electronic applications. Formula: $(-C_5H_8CH=CH-)_x$. Also known as *poly(1,3-cyclopentylenevinylene.*

polynosic fibers. See modal fibers.

polynuclear aromatic compounds. An aromatic compounds containing two or more closed benzenoid rings, e.g., anthracene ($C_{14}H_{10}$), naphthalene ($C_{10}H_8$) and phenanthrene ($C_{14}H_{10}$). Also known as *polynuclear aromatics (PNA).* See also aromatic compound.

polynuclear aromatic hydrocarbon. A hydrocarbon containing 3 or more closed benzene rings. Abbreviation: PNAH. See also aromatic hydrocarbon

polynuclear aromatics. See polynuclear aromatic compounds.

poly(octafluoro-*bis*-benzocyclobutene). A vapor-deposited thin film polymer produced by deep ultraviolet photolysis of the monomer octafluoro-*bis*-benzocyclobutene. It has high thermal stability and is used as a low-dielectric-constant material in microelectronics.

poly(3-octathiophene-2,5-diyls). A class of conductive polymers available in regioregular and regiospecific conformations. The former are black solids with a head-to-tail regiospecific conformation of more than 98.5%, and average molecular weights of about 142000, while the latter are red solids in which the ratio of the head-to-head/head-to-tail linkages of regioisomers is about 1:1. Used for electronic and optoelectronic applications.

polyolefin copolymers. Thermoplastic reaction products of the copolymerization of two or more olefin monomers, e.g., propylene and ethylene, butylene and propylene, ethylene and butylene, ethylene and hexylene, etc. Also known as *olefin copolymers.*

polyolefin fabrics. Woven or nonwoven fabrics made from fibers or filaments based on polyolefins, e.g., butylene, ethylene or propylene. Also known as *olefin fabrics.*

polyolefin fibers. Synthetic fibers in which the fiber-forming substance is a long-chain polymer containing 85 wt% or more of a simple olefin, such as butylene, ethylene or propylene, but excluding amorphous polyolefins that normally qualify as rubber. Also known as *olefin fibers.*

polyolefin plastics. A group of plastics based on polyolefins, such as polybutylene, polyethylene, polypropylene, etc. Also known as *olefinics; olefin plastics.*

polyolefin resins. A group of resins made by the polymerization of olefins, e.g., polyethylene from ethylene or polypropylene from propylene, or by the copolymerization of olefins (at least 50 wt%) with other monomers, e.g., propylene with ethylene, butylene with propylene, ethylene with butylene, ethylene with hexylene, etc. Also known as *olefin resins.*

polyolefins. A group of thermoplastic hydrocarbon polymers prepared by the polymerization of a simple olefin, such as butylene, ethylene or propylene. Abbreviation: PO.

Poly-Optic. Trade name of Polytek Development Corporation (USA) for clear casting resins used for optical applications.

polyorthoesters. A class of hydrophobic synthetic biopolymers that contain hydrolytic linkages which are sensitive to acids, but stable to bases. They degrade by surface erosion and are used for drug delivery applications, e.g., in implants.

Polyost PE. Trademark of Polyost NV (Belgium) for polyethylene staple fibers used for textile fabrics.

Polyost PP. Trademark of Polyost NV (Belgium) for polypropylene staple fibers used for textile fabrics.

Polyox. Trademark of Union Carbide Corporation (USA) for a series of water-soluble ethylene oxide polymers supplied in the form of white granules for the production of films, fibers and molded goods. They have molecular weights ranging from about 100000 to several millions. The water-soluble, grease-resistant films are used for packaging detergents, soaps, etc. Other uses include water-soluble adhesives, paper coatings, textile warp sizes, and in latex paints.

polyoxadiazole. The reaction product of the polymerization of oxadiazole (C_2N_2O), usually made into high-temperature-resistant fibers.

polyoxamide. A polyamide-type polymer made from oxalic acid and diamines.

polyoxybenzoates. See aromatic polyesters.

poly(3,4-oxyethyleneoxythiophene)/poly(styrene sulfonate). A conductice polymer alloy usually supplied as a 1.3 wt% dispersion in water with 0.8% polystyrene and 0.5 wt% polydioxin, and used for electronic and optoelectronic applications, and in the synthesis of novel polymers.

polyoxymethylenes. See acetal resins.

Poly-Pack. Trademark of Poly-Pack Verpackungs-GmbH (Germany) for a range of polyethylene packaging films.

Poly-Pad. (1) Trade name of MBI Products Company (USA) for plastic-encapsulated acoustical insulation.

(2) Trade name of Polyurethane Products Corporation (USA) for polyurethanes used for industrial applications.

Poly-Pale. Trademark of Hercules, Inc. (USA) for water-soluble pale-colored polymerized rosin.

Polypenco. (1) Trademark of The Polymer Corporation (USA) for polyamide, polyamide-imide, polycarbonate and acetal resins supplied in rods, strips and tubing forms for electrical and machine parts.

(2) Trademark of DSM Engineering Plastics (USA) for a series of industrial thermoplastics including polycarbonates possessing high heat-distortion temperature, good low-frequency and high voltage insulation, high impact resistance and strength. Also included under this trademark are various modified and unmodified polyamide-imide products with good high-temperature and wear properties.

polypeptide. A peptide with a molecular weight lower than that of a protein (usually 10000), and in which two or more amino acids are linked by peptide bonds. See also peptide.

Polyphane. Trademark of W. Ralston (Canada) Inc. for polyethylene film, tubing and sheeting.

polyphenylcarbynes. A group of carbon-base polymers in which the carbon atoms are arranged in a tetrahedral diamond-like structure with atoms rings (R groups) dangling from the carbon base. Upon heating to 200-400°C (390-750°F) the bonds between the carbon base and the R groups are broken and the polymer changes into diamond or diamond-like carbon. Used as precursors for diamond and diamond-like coatings.

polyphenylene ether–high-impact polystyrene alloys. See PPE-HIPS alloys.

polyphenylene ether–polyamide alloys. See PPE-PA alloys.

polyphenylene ethers. See polyphenylene oxides.

polyphenylene oxide–nylon alloys. See PPO-PA alloys.

polyphenylene oxide–polyamide alloys. See PPO-PA alloys.

polyphenylene oxide–polybutylene alloys. See PPO-PBT alloys.

polyphenylene oxide–polyester alloys. See PPO-polyester alloys.

polyphenylene oxide–polystyrene alloys. See PPO-PS alloys.

polyphenylene oxides. A group of engineering thermoplastics based on linear, noncrystalline polyethers derived by the oxidative polycondensation of 2,6-dimethylphenol in the presence of a copper-amine complex catalyst. They can be modified by the inclusion of high-impact polystyrene to improve the extrusion properties, and may also be filled with glass to improve the tensile strength and modulus. *Polyphenylene oxides* can be processed on conventional extrusion and injection molding equipment, and have a density of 1.06 g/cm³ (0.038 lb/in.³), excellent mechanical properties, good resistance to ultraviolet radiation, excellent electrical properties, good self-extinguishing properties, a service temperature range of -170 to +190°C (-274 to +374°F), low thermal expansivity, good hydrolytic stability, good resistance to dilute acids and alkalies, fair resistance to alcohols, greases, oils, detergents, ketones and concentrated acids, and poor resistance to aromatic and chlorinated hydrocarbons and halogens. Often sold under trade names and trademarks, such as *Iupiace, Laril, Norpex* or *Vestoran*, they are used for office furniture, business machines, automotive components (e.g., steering columns, instrument panels, wiper blades and interior trim), chemical-process and fluid-handling equipment, hospital and laboratory equipment, nose cones for space vehicles, dielectric components, fiberoptic connectors, ceiling boxes, control housings, pumps, impellers, pipe, valves, and fittings. Abbreviations: PPO; PPE. Also known as *polyphenylene ethers.*

polyphenylene sulfides. A group of crystalline, aromatic engineering thermoplastics produced commercially by the reaction of *p*-dichlorobenzene with sulfur compounds, such as sodium sulfide. Commercial engineering grades are often fiber-reinforced, but there are also glass-filled grades containing 40-60 wt% glass, and glass/mineral-filled grades are also available. They have good tensile strength, high moduli of elasticity, moderate flexural properties, low elongation and impact strength, good thermal stability, an upper service temperature of 220-260°C (428-500°F), inherent flame resistance, very low water absorption, good resistance to alcohols, alkalies, greases, oils, ketones, aromatic hydrocarbons and dilute acids, and fair resistance to halogens and concentrated acids. Often sold under trade names or trademarks, such as *Bayfide, Craston, Fortron, Primef, Ryton, Supec, Techtron* or *Tedur*, they are used for automotive components (e.g., components of air-conditioning, brake, cooling and emission-control, and fuel and ignition systems including generator and alternator parts), appliance housings, handles, disk drives, microwave ovens, electrical and electronic components (e.g., connectors, light sockets, relay parts and switches), heat shields, pump housings, valve parts, pump vanes and impellers, and as matrix resins of engineering composites. Abbreviation: PPS.

polyphenylene sulfones. See polysulfones.

poly(p-phenylene vinylenes). A class of conjugated conductive polymers, either unfunctionalized or functionalized by the addition of cyano (–CN) groups to a dialkoxy derivative. Unfunctionalized poly(p-phenylene vinylene) is a good electron-hole transporter, while cyano-functionalized poly(p-phenylene) is a good electron transporter. *Poly(p-phenylene vinylenes)* exhibit electroluminecence, and are used for Schottky-barrier and photovoltaic devices. Abbreviation: PPV.

polyphenylsulfones. A group of transparent thermoplastic engineering resins with a density of 1.29 g/cm³ (0.047 lb/in.³), a refractive index (at 20°C/68°F) of 1.6720, and a glass-transition temperature of 208°C (406°F). Used for structural and nonstructural applications. Abbreviation: PPSU.

polyphosphazene rubber. An elastomeric or thermoplastic chain or ring polymer having alternating phosphorus and nitrogen atoms in the main chain and two organic side groups (e.g., amino or halogen) on each phosphorus atom. It can be modified with 55 wt% fluorine, has a service temperature range of

-55 to +175°C (-67 to +347°F), and is used mainly for compounding applications, and for tires.

Polyplank. Trademark of Poly Foam Products Limited (USA) for low-density polyethylene foamboard.

PolyPlas. Trade name of Polysciences, Inc. (USA) for plastic microcarrier beads available in sizes ranging from 90 to 210 μm (0.003 to 0.008 μin.) and densities from 1.02 to 1.04 g/cm³ (0.037-0.038 lb/in.³). Used as substrates for cell culture applications, and extensively for various other applications in biology, biochemistry, biotechnology and bioengineering.

Polyplate. Trademark of J.M. Huber Corporation (USA) for a kaolin clay.

Polyplus. (1) Trade name of US Plastic & Chemical Corporation for polymeric blast cleaning media used in metal finishing operations.

    (2) Trade name of Yale Cordage, Inc. (USA) for braided synthetic ropes.

Poly-Ply. Trade name of Electro Chemical Engineering & Manufacturing Company (USA) for rigid polyvinyl chloride used for sheet linings.

Polyply. (1) Trademark of Polyply Inc. (USA) for fiberglass-fabric-reinforced, copper-clad or unclad epoxy laminates, copper-clad or unclad glass-mat-reinforced polyester laminates, and polytetrafluorethylene-modified copper-clad polyester laminates. Used for electrical and electronic applications including circuit boards.

    (2) Trademark of Plyotherm, Inc. (USA) for a range of synthetic resins supplied in several grades, and used as matrix resins for fiber-reinforced composites and laminates.

    (3) Trade name of Plyonex Corporation (USA) for aqueous emulsions of modified acrylic, styrenic and urethane polymers used in the manufacture of *PlyoForm* prepregs.

Poly Pore. Trade name of General Polymeric Corporation, Gen-Pore Division (USA) for porous elastomers supplied in the form of precision-molded components for filtration and fluid-handling applications.

Polypreg. Trade name of Polyméric NV (Netherlands) for polymer prepreg glass mats.

Polypren. Trademark of Polymer-Synthese Werk (Germany) for a series of styrene-butadiene and nitrile rubbers.

Poly-Prim. Trademark of Dominion Textile Inc. (Canada) for industrial fabrics.

Polyprime. (1) Trademark of American Polymer Corporation (USA) for polymer undercoatings for concrete, metal and wood.

    (2) Trademark of Ludlow Corporation (USA) for polymer-coated, glueable and printable paperboard.

Poly-Pro. (1) Trade name of Polyurethane Products Corporation (USA) for polyurethanes used for industrial products.

    (2) Trade name of Gehr Plastics, Inc. (USA) for high-performance thermoplastics supplied in the form of extruded slabs and rods.

Polypro. (1) Trademark of Mitsui Plastics (Japan) for polypropylene homopolymers and copolymers with excellent fatigue strength and stress-crack resistance, good flexural and electrical properties and thermal resistance. Supplied in unmodified and UV-stabilized grades, they are used for molded articles and packaging film.

    (2) Trademark of Columbian Rope Company (USA) for polypropylene monofilaments used fore cordage.

Poly-pro. Trademark of TAMKO Roofing Products, Inc. (USA) for asphalt-based roll roofing.

Polyprop-Omni. Trademark of Fibrasomni SA (Mexico) for polypropylene monofilaments and filament yarns.

polypropylene–ethylene-propylene-diene-monomer alloys. See PP-EPDM alloys.

polypropylene fibers. Strong, lightweight fibers based on polymers or copolymers of polypropylene, and sometimes modified or blended with other polymers and/or additives. They have high strength, high abrasion resistance, low weight, low thermal conductivity, very low moisture absorption, good resistance to moisture and weak acids, alkalies and solvents, a melting point above 150°C (300°F), and blend well with other fibers including cotton and wool. They are available as monofilament for filter cloth and as multifilament yarns for fabrics, e.g., sportswear, workwear, thermal clothing and blankets, and for rope and cordage.

polypropylene film. Polypropylene in the form of a thin, flexible film, typically 0.004-0.02 mm (0.2 to 0.8 mil) thick, in standard and biaxially oriented grades. *Polypropylene films* have high clarity, good tear strength, high elongation at break (550-1000%), good permeability to carbon dioxide and oxygen, good dielectric properties, a heat-sealing temperature of approximately 140-305°C (285-580°F), an upper service temperature of 120°C (248°F), good resistance to acids, alkalies, alcohols, ketones, solvents and moisture, fair resistance to greases, oils and aromatic hydrocarbons, poor resistance to halogens, and fair resistance to ultraviolet resistance. Often sold under trade names or trademarks, such as Bicor, Dynafilm or Olefane, they are used for electrical, packaging and heat-sealing applications.

polypropylene gycol terephthalates. See polypropylene terephthalates.

polypropylene oxide. A polymer based on propylene oxide (C_3H_6O) and used in the manufacture of polypropylene oxide rubber and as an intermediate for urethane foams. Abbreviation: PPOX.

polypropylene oxide rubber. An elastomer based on polypropylene oxide $[(C_3H_6O)_n]$, and having rather poor mechanical properties without reinforcing fillers, good resistance to ozone, heat aging and low temperatures, good flexural properties, and fair oil resistance. Used for insulation, seals, and moderately stressed rubber goods.

polypropylene rope. A rope made from multifilament polypropylene yarns commonly in diameters of 4.76-12.7 mm (0.187-0.500 in.). It is not as strong as polyester or nylon rope, but lighter in weight, and will float on water. It has good strength retention when wet, low elasticity, good low-temperature flexibility, a shock-absorbing ability about half that of nylon, high susceptibility to softening with increasing temperature, fair to poor ultraviolet resistance (degrades in sunlight), good moisture resistance, excellent dielectric properties, excellent resistance to concentrated and dilute acids and alkalies, fair to poor resistance to solvents, and good resistance to rot and mildew. Used for marine applications, and as a towing and barrier rope.

polypropylenes. A group of lightweight, crystalline thermoplastics made by the polymerization of high-purity propylene gas with a stereospecific, organometallic catalyst. They are translucent, off-white solids with molecular weights of more than 40000, and commercially available in the form of molding powders, sheets, rods, film, metallized film, staple and continuous filament yarn, fibers, granules and low-density foam, and can be processed by blow and injection molding, extrusion and other processes. *Polypropylenes* have a density of 0.90 g/cm³ (0.033 lb/in.³), a melting range of 168-171°C (334-340°F), excellent fatigue strength and stress-crack resistance, good flexural characteristics, good to fair abrasion resistance, good hardness and toughness, fair to poor resistance to ultraviolet radiation, fair ozone resistance, good dielectric properties, good heat resistance, a service temperature range of -60 to +120°C (-76 to +248°F), fair impact resistance, poor impact strength below -10°C (14°F), low thermal expansivity, low coefficients of friction, poor flame resistance (but are low-burning), good (gas) weldability, very low water absorption (less than 0.02%) and moisture permeability, excellent resistance to alcohols, alkalies and weak and strong acids, good resistance to aldehydes, esters, aliphatic hydrocarbons and ketones, fair resistance to aromatic, chlorinated and halogenated hydrocarbons, greases, oils and strong oxidizing agents, poor resistance to halogens, broad chemical resistance to pickling and plating solutions, and good resistance to organic chemicals and fungi and bacteria. Often sold under trade names or trademarks, such as *Acclear, Appryl, Dexflex, Cefor, Edistir, Eltex P, Escon, Fortilene, Meraklon, Moplen, Sequel, Tenite* or *Typar*, they are used for packaging film, fibers, cordage, wire and cable coatings, coated and laminated products, reinforced plastics, tanks, bottles, pipe and tubing, plating and pickling lines, etch tanks for processing silicon wafers, food containers, luggage, cabinets, printing plates, fittings, toys, wearing apparel, synthetic paper, automotive components, appliance parts, artificial grass and turfs, fish nets, surgical casts, strapping, disposable filters, and heat- and chemical-resistant parts. Abbreviation: PP.

polypropylene terephthalates. A group of engineering thermoplastics that are condensation products of terephthalic acid or dimethyl terephthalate and propylene glycol. They are commercially available in the form of films, sheets rods, granules, fibers, etc., and in reinforced and mineral-filled grades. *Polypropylene terephthalates* have high strength, toughness and dimensional stability, good electrical resistance, and good to moderate resistance to most ordinary chemicals. Used chiefly for electrical components, industrial and consumer products. Abbreviation: PPT. Also known as *polypropylene glycol terephthalates*.

polypyrrole. A conductive conjugated polymer doped with an organic acid, such toluenesulfonic acid ($CH_3C_6H_4SO_3H$). It has a density of 1.0 g/cm³ (0.036 lb/in.³), a melting point of -2°C (28°F), a boiling point of 100°C (212°F), and is used in the manufacture electronic and optoelectronic devices, e.g., Schottky devices and photoelectrochemical cells. Polymer-supported polypyrroles are also available. The latter are often supplied as (i) water-dispersible, undoped polypyrrole shells on polyurethane cores, (ii) water-dispersible, p-doped conductive polypyrrole shells on polyurethane core binder resins, or (iii) p-doped conductive polypyrrole shells on waterborne polyurethane core resin binders. Abbreviation: PPy.

poly(pyrrolidone). See nylon 4.

Poly-R. Trade name of Seaman Corporation (USA) for polymer-coated high-performance fabrics.

polyrayon gaberdine. A lightweight, twill-textured *gaberdine* fabric made from a blend of 65% polyester and 35% rayon fibers. Used especially for easy-care clothing.

Poly Repair-it. Trademark of Kwik Mix Materials Limited (Canada) for polymer cement patch.

Poly-Rib. Trade name of Flex-O-Glass, Inc. (USA) for ribbed sheeting of plastics, such as polyethylenes, acrylics, etc.

Polyrite GMC. Trade name of Polyply Inc. (USA) for general-purpose molding compounds available in unmodified, high-impact and mineral-filled grades.

Polyrite PP. Trade name of Polyply Inc. (USA) for unreinforced

and glass-fiber-reinforced polyesters for electrical applications.

Polyrite SMC. Trade name of Polyply Inc. (USA) for polyester sheet-molding compounds supplied in unmodified and low-profile grades.

Polyroqq. Trade name of Dental Ventures of America (USA) for an urethane-based dental cement.

Polyron. Trade name for thermoplastic polyester and polyether urethane elastomers.

polysaccharide. A complex, high-molecular-weight carbohydrate, such as glycogen, starch, dextrin or cellulose, consisting of many *monosaccharide* molecules.

Polysar. Trademark of Bayer Corporation (USA) for polymer latexes, resins, rubbers and plastics based on butadiene, butyl, styrene, ethylene-propylene and other monomers and copolymers, and supplied as blow, injection molding or extrusion compounds. *Polysar ABS* is an acrylonitrile-butadiene-styrene resin. *Polysar S* is an emulsion styrene-butadiene rubber pre-cross-linked with divinylbenzene, and used as partial replacement for general-purpose rubbers, and as a processing aid to improve the dimensional stability and surface smoothness of extruded and calendered products.

Polyseal. Trademark of Chemtron Manufacturing Limited (Canada) for a polyvinyl acetate caulking compound.

Poly/seal. Tradename of Polysciences, Inc. (USA) for a flexible, biologically compatible sealing material used for electrical implants.

Polysil. (1) Trademark of Polydyne, Inc. (USA) for silicone-based roof coatings.

(2) Trademark of Ernst Muhlbauer KG (Germany) for dental filling and impression materials.

(3) Trademark of Scican, Division of Lux & Zwingenberger Limited (Canada) for dental filling materials.

(4) Trade name for a polymer-impregnated silica with good mechanical strength and dielectric properties.

(5) Trademark for silver-based paste used in the manufacture of solid-state devices.

polysilanes. A class of organosilicon polymers obtained by the polymerization of organosilicon compounds, such as dimethyl-dichlorosilane ($C_2H_6SiCl_2$), at elevated temperatures and, for halosilanes, usually involving catalytic dehalogenation. The general formula is $(R_2Si)_n$ in which R usually represents an alkyl, such as ethyl, methyl, phenyl, etc. Used in the manufacture of silicones, and as preceramic precursors for the preparation of silicon carbide and other engineering ceramics. See also organofunctional silanes; polyborosilazanes; polysilsesquiazanes.

polysilazanes. A class of organosilicon polymers whose main chains consist of alternating silicon and nitrogen atoms with side groups of hydrogen, or alkyls, aryls, arenyls, etc. (e.g., methyl or vinyl). The general formula is $(RR'SiNR'')_n$ in which R can represent hydrogen, and R' and R'' identical or different organic groups. Used as preceramic precursors for the preparation of silicon nitride, silicon carbonitrides and other engineering ceramics. See also polyborosilazanes; polysilsesquiazanes.

polysilicon. A contraction of the term "polycrystalline silicon" referring to a form of silicon in which the atoms have the diamond lattice arrangement. See also amorphous silicon; crystalline silicon.

polysiloxanes. A group of straight-chain polymers in which the main chain consists of alternating silicon and oxygen atoms with side groups (R) of hydrogen, or alkyls, aryls, arenyls, etc. (e.g., methyl or phenyl). The general formula is $(R_2SiO)_n$. Depending upon their molecular weight, they can exist as liquids, greases and solid rubbers, resins or plastics. Used as a preceramic precursor material for the manufacture of silicon oxycarbide engineering ceramics. See also siloxanes; polyborosiloxanes; polycarbosiloxanes.

polysilsesquiazanes. See silsesquiazanes.

polysilsesquicarbodiimides. See silsesquicarbodiimides.

polysilsesquioxanes. See silsesquioxanes.

polysilylcarbodiimide. An organosilicon polymer obtained by the polymerization of silyl (SiH_3) and carbodiimide ($H_2N–C\equiv N$). The general formula is $[R_2Si(N=C=N)]_n$, in which R usually represents hydrogen, or an alkyl (e.g., methyl). Used in the synthesis of novel polymers and polymer composites, and as preceramic precursors for the manufacture of engineering ceramics.

Poly Slik. Trademark of Rexam Release (USA) for polyolefin-coated release paper.

poly(sodium acrylate). A water-soluble, ionic homopolymer synthesized from sodium acrylate ($C_3H_3O_2Na$) monomer in the presence of a catalyst. Used in the heat treating industry for formulating polymer quenchants. Abbreviation: PSA.

Polysolit. Trademark of LMP Corporation (USA) for polyvinyl chlorides.

Polysorb. Trademark of US Surgical Corporation (USA) for a biodegrable, bioabsorbable polymer based on poly(L-lactide-co-glycolide), and used for surgical plates, screws and mesh, and for tissue engineering applications. See also poly(L-lactide-co-glycolide).

PolySpec. Trademark of PolySpec (US) for an extensive series of primers, sealants, bonding agents and non-sag gels. The primer grades are available for epoxies on concrete and steel, and vinyl esters and polyesters on concrete and steel. The sealants are based on fluoroelastomers, and the non-sag gels are used with vinyl esters and polyesters.

Polyspur. Trade name of A. Schulman Inc. (USA) for alloyed thermoplastic elastomers.

Polystar. Trademark of Fibres South, Inc. (USA) for polypropylene filament yarns.

Polysteen. Trademark of Steen & Co. GmbH (Germany) for polyethylene, polypropylene and polyester staple fibers.

Poly-Stik. Trade name for multifilament polyesters for screens, wire cloth, strainers, filters, and screen printing.

Polystone. Trademark of Röchling Haren AG (Germany) for high-molecular-weight polypropylene and low-density polyethylene, high-pressure polyethylene and various polyamides used in the manufacture chemical tanks, processing equipment, and ventilation equipment for the semiconductor and electroplating industries.

Polystruc. Trade name of Industrial Dielectrics Inc. (USA) for chemical- and corrosion-resistant glass-fiber-reinforced melamine resins for structural, electrical and various other industrial applications.

polystyrene dielectrics. Engineering thermoplastics with outstanding electrical properties including high electrical resistivity, high dielectric strength, a dielectric constant of 2.4-3.1 at 1MHz, and a dissipation factor of 4×10^{-4} at 1 MHz. Owing to these properties, they find application as dielectric materials in certain film capacitors, and as electrical insulators in battery cases, lighting panels, electrical equipment, appliance housings, and consumer electronics.

polystyrene fibers. Synthetic fibers in which the fiber-forming substance is a long-chain styrene polymer. They have low thermal conductivity and good resistance to acids, alcohols, alkalies and many solvents.

polystyrene foam. A lightweight, expanded, cellular polystyrene, usually of the rigid type, supplied in the form of prefoamed blocks, boards or sheets with a density range of 16-80 kg/m³ (1-5 lb/ft³), and used for acoustic and thermal insulation, cold storage, light construction as in boats, flotation devices, airport runways, highway construction, ice buckets, water coolers, drinking cups, expendable casting patterns, fillers and partitions in shipping containers, furniture construction, and toys. Also known as *cellular polystyrene; foamed polystyrene; styrene foam.*

polystyrene–polyphenylene oxide. See PS-PPO alloys.

polystyrene resins. See polystyrenes.

polystyrenes. A group of transparent homopolymer thermoplastics obtained by the polymerization of styrene by free radicals using a peroxide initiator, and often copolymerized or blended with other materials. They are commercially available in the form of films, sheets, rods, rigid foams, powders, granules, flocculent, fibers, liquid polymer and expandable beads or spheres, and can be processed by blow, compression and injection molding, thermoforming and extrusion. *Polystyrenes* have a density of 0.9 g/cm³ (0.03 lb/in.³), high optical clarity, excellent electrical properties, good thermal and dimensional stability, high rigidity, high hardness, rather high brittleness when unmodified, low impact strength unless reinforced, poor light fastness when unmodified, fair resistance to ultraviolet radiation, good staining resistance, excellent dielectric properties, an upper service temperature of 95°C (203°F), low thermal expansivity, good thermal insulating properties, poor flame resistance (but are slow-burning), good processibility, good resistance to acids, alcohols, alkalies and many solvents, good to poor resistance to greases and oils, and poor resistance to ketones and aromatic hydrocarbons. Often sold under trade names and trademarks, such as *Aim, Gulf PS, Empee PS, Lacqrene, Lustrex, Norflex, Rexolite, Styrolux, Styron* or *Vestyron,* they are used for electrical insulation, capacitors, battery cases, lighting panels, machine housings, electrical equipment, compact cases, container lids, knobs, wall tiles, appliance housings, refrigerator doors, clock and radio cabinets, containers and molded household wares, dishes, brush backs, toys, packaging, as film formers in paints, as foams in thermal insulations, and light construction and packaging. Abbreviation: PS. Also known as *polystyrene resins.* See also high-impact polystyrenes; medium-impact polystyrenes.

polystyrene-silica hybrid. An organic-inorganic hybrid material prepared by covalently incorporating an organic glass, usually trialkoxysilyl-functionalized polymethyl methacrylate, into a silica glass network by co-condensing with tetraethylorthosilicate (TEOS) in a sol-gel process.

Polystyrol. Trademark of BASF Corporation (USA) for a series of polystyrenes that are available clear and colored in regular, fire-retardant, glass-reinforced and impact-modified grades. They have a density of 1.05 g/cm³ (0.038 lb/in.³), high gloss, high hardness and brittleness, a softening temperature of 80°C (176°F), and excellent electrical properties. Most properties including brittleness can be modified by copolymerizing with other monomers. Used for electrical components, e.g., housings, coils, battery cases, lighting panels, etc., machine and appliance housings, refrigerator doors, air conditioner cases, containers and molded household wares, toys, and packaging materials.

polysulfide. See polysulfide rubber.

polysulfide adhesives. A group of crosslinked adhesives based on polysulfide rubber and having excellent environmental durability in contact with water and fuel, but only fair peel and shear strengths. Used for structural applications involving moderate stresses, and as sealants for aircraft and marine applications.

polysulfide coatings. Protective coating for various substrates (e.g., metal, masonry and concrete) produced by the polymerization of an inorganic polysulfide with a chlorinated alkyl polyether. They have exceptional resistance to ozone, oxidation, weathering and ultraviolet light.

polysulfide elastomer. See polysulfide rubber.

polysulfide rubber. A synthetic polymer that is a condensate of sodium polysulfides with organic dihalides, and is supplied in solid and liquid form. It has poor mechanical properties (including strength, compression set and abrasion resistance), poor heat and flame resistance, high resiliency, very good low-temperature flexibility, good aging characteristics, an upper service temperature of 121°C (250°F), excellent resistance to oils, fuels, solvents, oxygen and ozone, good resistance to ultraviolet light, and high impermeability to gases. Used for oil and gasoline hoses, gaskets, washers, diaphragms, adhesive compositions, binders, coatings, and in sealants for aircraft, building and marine applications. Also known as *polysulfide; polysulfide elastomer.*

polysulfide sealants. Liquid or semiliquid compositions made from polysulfide rubber and commonly applied with spray or caulking guns. They are supplied in several colors, produce flexible seals upon curing, and have excellent durability. Used especially for aircraft, building and marine applications.

polysulfone–acrylonitrile-butadiene-styrene alloys. See PSU-ABS alloys.

polysulfone–polybutylene terepthalate. See PSU-PBT alloys.

polysulfone–polyethylene terephthalate. See PSU-PET alloys.

polysulfone–polyester alloys. See PSU-polyester alloys.

polysulfone resins. See polysulfones.

polysulfones. A group of transparent, melt-processible engineering thermoplastics produced by polycondensation of bisphenol A and dichlorodiphenylsulfone. They can be processed by blow and injection molding, extrusion or other processes, and have a density of 1.24 g/cm³ (0.045 lb/in.³), high strength, impact resistance and ductility, high rigidity and toughness, good dimensional stability, high flexural moduli, good creep resistance, high hardness, low shear sensitivity, good high-temperature resistance, low coefficient of expansion, a service temperature range of -100 to +160°C (-148 to +320°F), a glass-transition temperature of 185°C (365°F), good dielectric properties, high electrical resistivity, good self-extinguishing properties, high resistance to burning, good processibility and fabricability, excellent hydrolytic stability, fair to poor moisture resistance, good oxidation resistance, good resistance to corrosive acids, alkalies, alcohols, detergents, greases, oils and many solvents and chemicals, and poor resistance to aromatic hydrocarbons, ketones and chlorinated hydrocarbons. Often sold under trade names and trademarks, such as *Ensifone, Lasulf, Mindel, Udel* or *Ultrason S,* they are used for automotive components, power-tool housings, electrical equipment, printed-circuit boards, electronic connectors, computer components, medical instrumentation, food trays, microwave cookware, appliances, extruded pipe and sheet, as base matrices for stereotype printing plates, in orthopedics, and as matrix resins for engineering composites. Ab-

breviation: PSU. Also known as *polyphenylene sulfones; polysulfone resins.*

polysynthetically twinned titanium aluminide. A titanium alloy with 48 wt% aluminum that has improved creep resistance as a result of having a refined lamellar structure by controlled twinning.

polyterephthalate. A thermoplastic polyester having the terephthalate $[C_6H_4(COOR)_2]$ group as a repeating unit in the polymer chain.

polyterpene resins. A class of thermosetting resins or amber, viscous liquids prepared by the polymerization of turpentine using mineral acids, aluminum chloride, etc., as catalysts, and used for adhesives and hot-melt coatings, as paper impregnants, and as rubber curing agents and plasticizers.

polytetrafluoroethylene fibers. Synthetic fibers that have been spun from polytetrafluoroethylene polymers and possess good flame resistance and good resistance to acids, alcohols, alkalies, greases, oils, and solvents. Abbreviation: PTFE fibers. See also fluoro fibers; polytetrafluoroethylenes.

polytetrafluoroethylenes. A group of fluorocarbon polymers made by the polymerization of tetrafluoroethylene, and commercially available in rod, tube, film, sheet, insulated wire, monofilament, multifilament fiber, granule, fine powder, extrusion and molding powder form, and as water-base dispersions and glass-filled grades. They have a density of 2.13-2.19 g/cm³ (0.77-0.79 lb/in.³), high toughness, good strength when reinforced, high melt viscosity, good resistance to nuclear radiation, good resistance to ultraviolet radiation and ozone, good weatherability, good dielectric properties, excellent electrical resistance, relatively high thermal expansivity, very high thermal stability, a service temperature range of -260 to +260°C (-436 to +500°F), very low coefficients of friction, excellent antistick properties, excellent resistance to oxidation, good flame retardancy, and good resistance to acids, alcohols, alkalies, aromatic hydrocarbons, greases, oils, solvents, halogens, ketones and oxidizing agents. Often sold under trade names and trademarks, such as *Algoflon, Chemfluor, Fluon, Fluorosint, Hostaflon, Rulon, Teflon, Voltalef* or *Zitex,* they are used for fuel hoses, flexible hose, gaskets, seals, sealing rings, liners and tapes, packings, cylinder-head gasket coatings, chemical process equipment, bearings, piston rings, linings for fittings, pipes, pumps, tanks, tubes, valves, coatings for metals and fabrics, coaxial spacers, wire coating and tape, pressure-sensitive tape, insulators, printed-circuit board laminates, vascular grafts and artificial ligaments, ablative coatings for aerospace applications, lubricants, and nonstick cookware. Abbreviation: PTFE.

polytetrafluoronaphthalenes. See poly(naphthalenes).

poly(tetrafluoro-p-xylylene). See parylenes; parylene-F.

poly(tetramethylene adipamide). See nylon 4,6.

Polytex. Trade name of Tenax Finishing Products Company (USA) for one-coat textured coatings.

polythene. A term used in the UK for *polyethylenes.*

Polytherm. Trademark of Forchheim GmbH (Germany) for polyvinyl chloride film materials.

polythiophene. A conducting polymer that exhibits the photovoltaic effect. It is usually supplied as a powder and used as an organic semiconductor in the manufacture of solid-state and microelectronic devices, e.g., Schottky devices and field-effect transistors, and as a photoconductor in photoelectrochemical cells. Abbreviation: PTh.

Polytie. Trademark of Poly-Twine Corporation (Canada) for polypropylene monofilaments and fibers.

Poly-tite. (1) Trademark of Hanson Inc. (Canada) for a compressible urethane joint sealant.

(2) Trademark of Sandell Manufacturing Company, Inc. (USA) for caulking and sealing compounds.

Poly-Tone. Trademark of General Dental Products, Inc. (USA) for an acrylic resin used for denture bases.

polytrifluorochloroethylene resins. See chlorotrifluoroethylenes.

Polytron. (1) Trade name of Industrial Dielectrics Inc. (USA) for high-performance polyester compounds used for electrical and electronic applications.

(2) Trade name of B.F Goodrich Company (USA) for chemical-resistant polyvinyl chlorides for electronic packaging applications, and office equipment.

Polytrope. Trademark of A. Schulman Inc. (USA) for olefinic-based thermoplastic elastomers supplied as bulk powders, pellets, granules and beads, and used for further processing by extrusion and molding processes.

Poly-Twine. Trademark of Poly-Twine Corporation (Canada) for polyethylene and polypropylene monofilaments and fibers.

poly(undecanoamide). See nylon 11.

poly(undecanolactam). See nylon 11.

Polyunion. Trademark of Polyunion Kunststoffwerk GmbH (Germany) for polyethylene fibers, yarns and products.

polyurea. See polyurea fibers.

polyurea fibers. A generic term for man-made fibers composed of long-chain polymers having in the chain aliphatic repeating units linked to ureylene (–HNCONH–) units, and in which the sum of the aliphatic and ureylene units is 85 wt% or more of the chain. Also known as *polycarbamide; polycarbamide fibers.*

polyurethane adhesives. A group of rubber-base one- or two-part adhesives that are the products of reactions of polyisocyanate and hydroxy-terminated polyethers or polyesters. They cure at room or elevated temperatures, and have high toughness and flexibility, high peel strength at low and high temperatures, good shear strength at low and moderately high temperatures, good chemical and heat resistance, high cohesive strength, a service temperature range of approximately -253 to +149°C (-423 to +300°F), high toughness and moisture sensitivity. Used for structural and nonstructural bonding applications.

polyurethane coatings. Plastic coatings derived from thermosetting polyurethane resins and applied by brush, spray or dipping. They are usually chemically cured (air-dried or heat cured), but catalyst-cured grades are also available. *Polyurethane coatings* have high toughness, good abrasion and impact resistance, good resistance to chemicals, acids, alkalies, most solvents and water, poor resistance to aromatic and chlorinated solvents, good weatherability, high flexibility, high hardness, good durability, and high gloss, and are difficult to topcoat. Used for aircraft finishes, marine equipment, chemical-process equipment, metal and plastic coatings, coatings for structural steel, baked coatings, wire coatings, maintenance paints, masonry coatings, linings for storage tanks, coatings for elastomers and leather, etc. Also known as *urethane coatings.*

polyurethane elastomer. A transparent polyurethane-resin elastomer obtained by the reaction of polyisocyanates with linear polyesters or polyethers containing hydroxyl groups. It has a durometer hardness of 85 Shore, good abrasion resistance, high tensile and tear strength, excellent ozone resistance, good weatherability, good resistance to oils, organic solvents and oxygen, good resistance to weak and dilute acids, good resistance strong and weak bases, poor resistance to strong or concentrated ac-

ids, good high-temperature performance, increased susceptibility to hardening and low-temperature embrittlement, and a service temperature of -56 to +85°C (-70 to +185°F). Used for sealants, adhesives, caulking agents, films and linings, binders, abrasive wheels, in electronics for encapsulation, automotive components (e.g., bumpers, fenders, etc.), and shoe heels. Abbreviation: UE; UR. Also known as *polyester rubber; polyurethane rubber; urethane elastomer.*

polyurethane fibers. A generic term for man-made fibers composed of long-chain polymers having in the chain aliphatic repeating units linked to urethane (–NHCOO–) units, and in which the sum of the aliphatic and urethane units is 85 wt% or more of the chain. They are crystalline fibers usually obtained by the reaction of hexamethylene diisocyanate with 1,4-butanediol, and have high moduli of elasticity, good electrical resistance, poor flame resistance and high moisture resistance. Used for textiles, bristles for brushes, etc. See also elastane fibers; spandex fibers.

polyurethane foam. A flexible or rigid cellular material made by treating a polyol, such as polyester, polyether, polypropylene glycol or polytetramethylene glycol, with a diisocyanate, such as hexamethylene diisocyanate or toluene diisocyanate, in the presence of water and a suitable catalyst, such as amines, organotin compounds, or tin soaps. The cellular structure is produced by carbon dioxide formed by a reaction of the water with the isocyanate groups. Flexible foams are used for cushions, padding, mattresses, furniture, upholstery, automotive accessories, carpet underlayment, and packaging. Rigid foams are employed for acoustic and thermal insulation (e.g., in buildings, vehicles, refrigerators and pipelines), transport containers, furniture, packaging and packing, marine flotation devices, automotive components, boat hulls, and in shipbuilding (for buoyancy). Also known as *diisocyanate foam; urethane foam.* See also cellular polyurethane; flexible polyurethane foam; polyurethane foam insulation; rigid polyurethane foam.

polyurethane foam boards. Rigid insulating boards consisting of closed-cell polyurethane foam filled with gases (e.g., fluorocarbons). They are available in a wide range of sizes and usually come double-faced with foil, and sometimes bonded with interior or exterior finishes. They have a high thermal insulating value, but must be protected from exposure to sunlight and water and covered with a fire-resistant material. Polyurethane foam boards are used on buildings and structures for acoustical, air and vapor barrier applications. See also rigid board insulation.

polyurethane foam insulation. A type of foamed-in place thermal insulation based on polyurethane. The foam is supplied in liquid form and sprayed directly onto building surfaces in layers typically less than 50 mm (2 in.). It expands in place up to 28 times its original size and sets instantly to a pale yellow, closed-cell rigid material. Polyurethane foam insulation can also be used for air barrier applications, but must be protected from prolonged exposure to sunlight and, when used indoors, must be covered with a fire-resistant material.

polyurethane foam sealants. Liquid or semiliquid polyurethanes supplied in dispensing systems with spray nozzles or in individual spray cans. Upon application they expand and cure to very durable foams. Depending on the particular ingredients and the amount of precuring, they are available in slow- and fast-expansion foam grades. They bond well to many materials except polyethylenes, polytetrafluoroethylenes and silicones, and are widely used for a wide range of sealing applications in the building and construction industries.

polyurethane–acrylonitrile-butadiene-styrene alloys. See PUR-ABS alloys.

polyurethane resins. See polyurethanes.

polyurethane rubber. See polyurethane elastomer.

polyurethanes. A group of polymers that may be of the thermoplastic type (linear polymers) or thermosetting type (network polymers), and are reaction products of organic diisocyanates (e.g., toluene diisocyanate, diphenylmethane diisocyanate, etc.) with any of various compounds containing hydroxyl groups (e.g., diols, polyols, polyethers, polyesters, etc.). The thermosetting polymers are available in cellular, flexible, rigid or solid form. Often sold under trade names and trademarks, such as *Baydur, Bayflex, Ecolan or Superflex, Vitel* or *Vistram,* they are used in paints, varnishes and adhesives, for elastomers and foams, in protective coatings, as binders, sizing and sealants, fibers, and as potting or casting resins. Abbreviation: PUR. Also known as *polyurethane resins; urethane resins; urethanes.* See also thermoplastic polyurethanes; thermosetting polyurethanes.

Poly V. Trade name of Accurate Set, Inc. (USA) for polyvinylsiloxane-based dental impression materials.

Poly-Vent. Trade name CRP (USA) for waterproof, microporous nonwoven polyethylene films used for cleanroom applications.

Polyverbel. Trade name of SA Glaverbel (Belgium) for double or multiple glazing units.

Polyvin. Trade name of A. Schulman Inc. (USA) for elastomers and plastics including plasticized polyvinyl chlorides (PVCs) supplied as bulk powders, granules, pellets and beads for further processing by extrusion and extrusion. The plasticized PVCs are supplied in three elongation ranges: 0-100%, 100-300% and above 300%.

polyvinyl acetals. A group of thermoplastic vinyl resins that are products of a condensation reaction of a polyvinyl alcohol with acetaldehyde. They are colorless and lightfast, and have good resistance to gasoline, oils, chlorinated hydrocarbons, and moisture. Used in paints, lacquers, adhesive and films, and in the manufacture of cast, extruded, molded or coated products. Also known as *vinyl acetal resins.*

polyvinyl acetate emulsion adhesive. An adhesive whose polymeric portion contains polyvinyl acetate, polyvinyl acetate copolymers or combinations thereof, and may or may not have binders and modifiers incorporated. It is suitable for bonding metal, wood, glass and ceramic substrates, and has a maximum service temperature of 74°C (165°F). Also known as *polyvinyl resin emulsion adhesive; polyvinyl resin emulsion glue.*

polyvinyl acetate resins. See polyvinyl acetates.

polyvinyl acetates. A group of colorless, transparent, thermoplastic polymers produced by the polymerization of vinyl acetate with peroxide catalysts. Commercially available in the form of granules, solution, latices and pastes, they have good weatherability and light stability, transparency to ultraviolet radiation, poor flame retardancy, good flexibility, toughness and hardness, high dielectric strength, good moisture resistance, good resistance to gasoline, oils and fats, and fair to poor resistance to alcohols, esters, benzene and chlorinated hydrocarbons. Used as binders in sizing compounds for glass-fiber textiles, in bookbinding and textile finishing, as adhesives for paper, metals, wood, glass and ceramics, as strengthening agents for cements and latex water paints, for paper and fabric coatings, in paperboard, as bases for inks and lacquers, and for sealants. Abbreviation: PVAc or PVAC. Also known as *polyvinyl acetate resins; vinyl acetate resins.*

polyvinyl alcohol fibers. See vinal fibers; vinylal fibers.

polyvinyl alcohols. A group of colorless, water-soluble thermoplastic polymers made by hydrolysis of polyvinyl esters (e.g., polyvinyl acetate), and commercially available in form of white to yellowish-white granules or powder in low- and high-molecular-weight grades. Their viscosity, strength, elongation, tear resistance and flexibility increase with increasing molecular weight. *Polyvinyl alcohols* have a density of 1.29 g/cm³ (0.046 lb/in.³), a refractive index of 1.51, poor permeability to gases, good resistance to solvents, greases, oils and petroleum hydrocarbons, and poor resistance to acids and alkalies. The particular properties of these polymers depend on the degree of polymerization, and they are used for general adhesives, paper coatings, binders for ceramics, fabrics, leather and paper, molding powders, cements and mortars, as additions to glazes and bodies to enhance dry strength prior to firing, as sizings and adhesives for glass fibers, as thickening and suspension agents for ceramic slurries, textile finishes and sizes, as an emulsifying agent, in grease-proofing paper, in photosensitive films, in the heat treating industry for the formulation of polymer quenchants, in the *in vivo* controlled release of proteins, and in histochemistry. Abbreviation: PVA, PVAl or PVAL.

polyvinyl butyrals. A group of colorless, thermoplastic vinyl resins that are products of a condensation reaction of a polyvinyl alcohol with butyraldehyde. They have high flexibility and toughness, and are used as interlayers in safety glass and in shatter-resistant protection in aircraft, and in adhesives, films, lacquers and paints. Abbreviation: PVB.

poly(N-vinylcarbazoles). A group of brown thermoplastic resins prepared by the reaction of acetylene (C_2H_2) with carbazole ($C_6H_4NHC_6H_4$). They have a softening temperature of 150°C (302°F), excellent electrical properties including electrical conductivity, good chemical and heat resistance, and poor mechanical strength. Doped poly(N-vinylcarbazole) exhibits electroluminescence. Used as an impregnant for paper capacitors, in Schottky-barrier and electroluminescent devices, and as replacement for mica in electrical appliances and devices. Abbreviation: PVCZ; PVCz; PVK.

polyvinyl chloride acetates. A group of colorless thermoplastics that are copolymers of vinyl chloride ($CH_2=CHCl$) and vinyl acetate ($CH_3COOCH=CH_2$) with a typical chloride content of 85-97%. When compounded with plasticizers, they can be made into polymers that are superior to polyvinyl chloride in flexibility and rubber in aging properties. They have good resistance to water and concentrated acids and alkalies, and are used for cable and wire coatings and jackets, protective clothing, and for lining chemical equipment.

polyvinyl chloride–acrylonitrile-butadiene-styrene alloys. See PVC-ABS alloys.

polyvinyl chloride–acrylic alloys. See PVC-acrylic alloys.

polyvinyl chloride–chlorinated polyethylene elastomer alloys. See PVC-CPE alloys.

polyvinyl chlorides–chlorinated polyvinyl chloride alloys. See PVC-CPVC alloys.

polyvinyl chloride–ethylene-propylene-diene-monomer alloys. See PVC-EPDM alloys.

polyvinyl chloride/ethylene-vinyl acetate alloys. See PVC-EVA alloys.

polyvinyl chloride fibers. Spun thermoplastic fibers whose fiber-forming substance is any long-chain synthetic polymer made up essentially of vinyl chloride ($-CH_2-CHCl-$) units. Abbreviation: PVC fibers. See also chloro fibers; saran fibers.

polyvinyl chloride–methacrylate-butadiene-styrene alloys. See PVC-MBS alloys.

polyvinyl chloride–nitrile rubber alloys. See PVC-nitrile rubber alloys.

polyvinyl chloride–polymethyl methacrylate alloys. See PVC-PMMA alloys.

polyvinyl chlorides. A family of engineering thermoplastics produced by the polymerization of vinyl chloride (CH_2=$CHCl$) by free radicals with a peroxide initiator. They are easily compounded into flexible forms by the use of plasticizers, stabilizers and fillers, and readily processed by blow molding, extrusion, calendering, fluid-bed coating and other processes. Reinforced grades are also available, e.g., polyester-cord reinforcement for flexible tubing. *Polyvinyl chlorides* have excellent dimensional stability, outstanding resistance to weathering, fungi, water and alcohols, good resistance to acids and alkalies, fats and petroleum hydrocarbons, good dielectric properties, poor resistance to gamma radiation, and melting point of 148°C (298°F). They are often sold under trade names and trademarks, such as *Benvic, Dalvin, Hostalit, Lacqvyl, Leavil, Nakan, Marvinol, Vestolit, Vinidur* or *Unichem*. Rigid polyvinyl chlorides are used for piping, conduits, plumbing, house siding, gutters, window and door frames, decorative profiles and trim, containers, lining of reservoirs and chemical tanks, flooring, packaging, toys, outdoor furniture, consoles and cabinets of business machines and consumer electronics. Flexible polyvinyl chlorides are used for coating glass bot-tles, paper and textiles, for strip curtains, glass-fiber fabrics, as adhesives and bonding agents, as plastisols and organosols, for cable and wire coatings, clothing, gaskets and seals, flooring, tennis court playing surfaces, flexible tubing, garden hose, shoes, magnetic tape, film and sheeting and fibers, and formerly also for phonograph records. Abbreviation: PVC. Also known as *vinyl chlorides; vinyl chloride resins.* See also unplasticized polyvinyl chloride.

polyvinyl dichlorides. A group of polymers of chlorinated polyvinyl chloride with high strength, outstanding chemical resistance over wide temperature range, good resistance to inorganic reagents, aliphatic hydrocarbons and alcohols, and good self-extinguishing properties. Used for fittings and pipes employed in handling hot corrosives (not exceeding 100°C or 212°F).

polyvinyl ethers. See polyvinyl ethyl ethers.

polyvinyl ethyl ethers. Rubbery solids or viscous gums prepared by polymerization of vinyl ethyl ether (CH_2=$CHOC_2H_5$). They have good resistance to weak and strong alkalies and weak acids, and are used for pressure-sensitive tape. Abbreviation: PVE. Also known as *polyvinyl ethers.*

polyvinyl fluorides. A group of highly crystalline polymers of vinyl fluoride (CH_2=CHF), often supplied as thin films. They have high strength and flexibility, good resistance to weather, good dielectric properties, low permeability to air and water, and good resistance to oils, solvents and many chemicals. Used for electrical equipment, protective material for outdoor use, packaging of nonfood items, and surface coatings. Abbreviation: PVF.

polyvinyl formals. A group of colorless, thermoplastic vinyl resins which are products of a condensation reaction of a polyvinyl alcohol with formaldehyde. They have density of 1.23 g/cm^3 (0.044 lb/in.^{3}), a glass-transition temperature of 108°C (226°F), a refractive index of 1.502 (at 20°C/68°F), high flexibility and toughness, and good resistance to greases and oils. Used in adhesives, films, lacquers, paints, wire coatings, as mounting media in metallography and electron microscopy, and for cast, extruded, or molded objects. Abbreviation: PVF; PVFM.

polyvinyl formal solution. A 0.25 or 0.50 wt% solution of polyvinylformal in ethylene dichloride used in electron microscopy.

polyvinyl formate resins. Transparent resins made by polymerizing vinyl formate (CH_2=$CHOOCH$), and used in the manufacture of hard, solvent-resistant plastics.

polyvinylidene chloride fibers. See chloro fibers; saran fibers.

polyvinylidene chloride plastics. See polyvinylidene chlorides.

polyvinylidene chlorides. A group of thermoplastic materials usually made by polymerization of vinylidene chloride (CH_2=CCl_2) and available in the form of powder, films, oriented fibers, extruded and molded products, and as copolymer with acrylonitrile or vinyl chloride. They have a density of 1.63 g/cm^3 (0.059 lb/in.^{3}), good abrasion resistance, good self-extinguishing properties, fair resistance to ultraviolet radiation, good dielectric properties, an upper service temperature of 80-100°C (176-212°F), low coefficients of friction, low vapor transmission, good resistance to chemicals, good resistance to acids, alcohols, alkalies, greases, oils and halogens, fair resistance to aromatic hydrocarbons and ketones, and poor resistance to chlorinated hydrocarbons. Often sold under trade names and trademarks, such as *Diofan D, Duran, Geon, Ixan, Saran* or *Tygan*, they are used for packaging of food products, in latex coatings, piping for chemical equipment, upholstery, fibers, and bristles. Abbreviation: PVDC. Also known as *polyvinylidene chloride plastics.*

polyvinylidene fluorides. A group of a melt-processible, thermoplastic fluorocarbon polymers made by emulsion or suspension polymerization of vinylidene fluoride (CH_2=CF_2). They are available in the form of powder, pellets, films, metallized films, sheets, tubes, solutions and dispersions, and can be processed by compression and injection molding, extrusion and other processes. *Polyvinylidene fluorides* have a density of 1.76 g/cm^3 (0.064 lb/in.^{3}), a melting point of 155-170°C (311-338°F), a refractive index of 1.42, good tensile and compressive strength, high impact strength, excellent rigidity and abrasion resistance, good (gas) weldability, good electrical properties, good self-extinguishing properties, excellent resistance to nuclear and ultraviolet radiation, excellent ozone resistance, excellent stress-crack resistance, good thermal stability, a service temperature range of -40 to +220°C (-40 to +428°F), good processibility, low coefficients of friction, very low water absorption, good resistance to oxidative degradation, good weatherability, good resistance to acids, alcohols, alkalies, greases, oils and halogens, fair resistance to aromatic hydrocarbons, and poor resistance to ketones. Often sold under trade names and trademarks such as *Dyflor, Ensikem, Floraflon, Kynar* or *Solef*, they are used for the insulation of high-temperature wire, for tank and drum linings, chemical tanks and tubing, piping, industrial battery casings, pump, valve and impeller parts, in precious metal recovery, protective paints and coatings, in weatherable coatings on steel and aluminum, for shrinkage tubing for resistors, diodes, and in sealants and filter cartridges. Abbreviation: PVDF.

polyvinylidenes. A group of polymers composed of CH_2=CR_2 mers, where R is a chlorine (Cl) radical as in polyvinylidene chloride, or a fluorine (F) radical as in polyvinylidene fluoride. Also known as *vinylidene resins.*

polyvinyl isobutyl ether. A polymer produced by the polymerization of vinyl isobutyl ether [CH_2=$CHOCH_2CH(CH_3)_2$] with peroxides or acid catalysts, and supplied as a white elastomer and as an opaque or viscous liquid in various grades and molecular weights. It has a density of 0.92 g/cm³ (0.033 lb/in.³), a refractive index of 1.45-1.46, good water resistance, good resistance to ethanol, and acetone, dilute and concentrated alkalies and dilute acids, and poor resistance to most organic solvents. Used for adhesives, surface coatings, plasticizers, tackifiers, waxes, lubricating oils, laminating agents, and cable filling. Abbreviation: PVI; PVIE.

polyvinyl methyl ether. A colorless, viscous, liquid high-molecular-weight polymer prepared by the polymerization of vinyl methyl ether (CH₂=CHOCH₃) with acid catalysts or peroxides, and used for hot-melt adhesives, pressure-sensitive tape, plasticizer, and tackifier. Abbreviation: PVM; PVME.

poly(2-vinylpyridine). A vinyl-based conductive polymer (–CH-(C₅H₄N)-CH₂–)ₙ used in photographic dyes, tablet coatings, anstatic agents, and in the manufacture of Schottky devices. Abbreviation: P2VP.

polyvinyl pyrrolidone. A water-soluble, nonionic homopolymer synthesized from vinyl pyrrolidone (C₆H₇NO) and used in the manufacture of elastomeric and plastic products, in the synthesis of biomaterials, and in the heat treating industry for formulating polymer quenchants. Abbreviation: PVP. Also known as *poly(N-vinylpyrrolidone)*.

polyvinyl resin. See polyvinyls.

polyvinyl resin emulsion glue. See polyvinyl acetate emulsion adhesive.

polyvinyls. A group of polymers derived from vinyl monomers and including polyvinyl chloride (PVC), polyvinylidene chloride (PVDC), chlorinated polyvinyl chloride, copolymers of vinyl and vinylidene chloride (VC-VDC), vinyl chloride-vinyl acetate copolymers (VC-VA), polyvinyl formal (PVFM), polyvinyl butyral (PVB), etc. Also known as *polyvinyl resins; vinyl polymers; vinyl resins*.

Polyviol. Trademark of Bayer Corporation (USA) for polyvinyl alcohol resins.

Polyvoltac. Trademark of Polyvoltac Canada Inc. (USA) for flame-retardant silicon rubber foams that maintain their elastomeric properties between -60 and +230°C (-76 and +446°F) in continuous service. They have good dielectric properties, good permanent set resistance, good resistance to ozone and ultraviolet radiation, and good corona resistance. Used for aircraft, bus and subway seating, ship and hospital mattresses, fire and welding blankets, insulation in electrical and electronic products, and gaskets and seals.

polyvoltine silk. Silk produced by silkworms that have several broods per season. Also known as *multivoltine silk*.

Polyweld. Trade name of American Phenolic Corporation (USA) for weldable polystyrene plastics.

Poly Wound. Trade name of Polygon Company (USA) for filament-wound polymeric composites supplied in the form of tubes and shapes.

Poly-Wrap. Trade name of Flex-O-Glass, Inc. (USA) for extruded plastic sheeting and film used for packaging applications. It is available in several plastic grades include polybutyrate, acrylic, linear-density polyethylene, and ionomer.

Polywrap. Trade name for polyolefin fibers, films and other products for wrapping and packaging applications

poly(p-xylylene). See parylenes; parylene-N.

Poly-Zen. Trade name of Hentzen Coatings, Inc. (USA) for air-drying polymeric coatings.

Polyzinc. Trade name of The Ceilcote Company (USA) for an organic zinc-rich coating used as a primer on ferrous substrates.

Pomet. Trade name of Powder Metals Inc. (USA) for a series of ferrous and nonferrous powder-metallurgy materials including several sintered high-purity irons, low-carbon steels, nickel-chromium stainless steels, bronzes, and aluminum alloys.

Pomoloy. Trade name of Fairbanks, Morse & Company (USA) for a series of gray cast irons containing 3.2% carbon, 2-2.2% silicon, and the balance iron. Used for pump casings.

Pomona. Trade name of Aardvark Clay & Supplies (USA) for medium toasty orange clay (cone 10).

Pompey. Trade name of Société Nouvelle des Aciers de Pompey (France) for an extensive series of case-hardening, carburizing and nitriding steels, numerous carbon and low-alloy structural steels, carbon and alloy machinery steels, free-machining steels, tool steels (e.g., plain-carbon, hot-work, shock-resisting, and nondeforming types), mold steels, spring steels, austenitic manganese steels, and stainless steels.

Pompton. Trade name of AL Tech Specialty Steel Corporation (USA) for a series of water-hardening steels (AISI type W1) containing 0.5-1.5% carbon, 0.25% manganese, 0.25% silicon, and the balance iron. Available in various carbon grades, they have excellent machinability, good to moderate deep-hardening properties, wear resistance and toughness, and low hot hardness. Used for pneumatic tools, chisels, shear knives, threading dies, cold-heading dies, trimming and blanking dies, punches, and drills.

Ponal. Trademark of Henkel KGaA (Germany) for polyvinyl acetates used for the manufacture of adhesives.

pondcypress. The soft, light, durable wood of a distinctive variety of the baldcypress (*Taxodium distichum* var. *nutans*) native to the swamps and marshes of Southeastern United States and Mexico. It has essentially the same properties and uses as its parent species, e.g., construction work, paneling, cooperage, etc. See also baldcypress.

ponderosa pine. The soft, reddish-brown wood of the large pine tree *Pinus ponderosa*. It has a fine, straight grained, uniform texture, low strength, low tendency to twist or warp, works very well. Average weight, 450 kg/m³ (28 lb/ft³). Source: Western North America from southern British Columbia to California and eastward to South Dakota and Wyoming. Used for window and door frames, sash, moldings, paneling, cabinetmaking, veneer, particleboard, boxes, crates, other millwork, toys, models, railroad ties, piles, posts, mine timbers, paper pulp, and as fuel. Also known as *Oregon white pine; western yellow pine*. See also yellow pine.

pond pine. The wood from the softwood tree *Pinus serotina* which is similar to pitch pine. Source: Atlantic coastal region from New Jersey to Florida. Used for general construction, railroad ties, posts and poles. Also known as *swamp pine*. See also pitch pine.

pongee. A soft, light-to-medium-weight, washable fabric made of cotton, silk, nylon or other fibers in a plain weave with irregular yarns. Silk pongee has a characteristic tan color and excellent draping qualities. Used for blouses, dresses, nightwear, curtains, and linings.

Ponolith. Trademark of E.I. DuPont de Nemours & Company (USA) for a white, *lithopone*-type pigment consisting of about 2 parts of barium sulfate (BaSO₄) and 1 part of zinc sulfide (ZnS). Used chiefly as a paper coloring agent.

Ponsard's high manganese brass. A British corrosion-resistant manganese brass containing 50-75% copper, 20-25% manganese, 2-15% zinc, and 0-16% iron. Used for marine parts and corrosion-resisting parts.

Pontallor. Trade name of Degussa Dental (USA) for low-gold precision-metal casting alloys used in dentistry and dental engineering.

Pontalite. Trade name of E.I. DuPont de Nemours & Company (USA) for a methyl methacrylate supplied in the form of a molding powder. It is molded under heat (about 140-165°C or 285-330°F) and pressure (about 17-29 MPa or 2.5-4.2 ksi) and must be cooled to ambient temperature under maximum pressure. The resin is tough, hard, water white to clear, has a heat distortion temperature of about 65°C (150°F), is resistant to etchants and adheres well to specimens used for metallurgical mounting applications.

Pontan. Trade name for a water-repellent, insect-resistant fabric which is surface-coated or impregnated with rubber.

Pontella. Trademark of Rhône-Poulenc–Setila SA (France) for polyester fibers and yarns.

pontiac. A strong, dark gray, waterproof fabric knitted from wool yarn and used for coats and skirts.

pontianak. The grayish white gum obtained from *jelutong* trees (*Dyera costulata* and *D. laxifolia*) native to Borneo and Malaya. It is similar to *gutta-percha* and used as electrical insulation, as a coating for transmission belts, and in adhesives, lacquers, paints, and varnishes. Also known as *pontianak gum*.

Pontilhado. Trade name of Vidrobrás (Brazil) for a patterned glass.

Pontor. Trade name of Metalor Technologies SA (Switzerland) for an extensive series of dental casting and universal alloys. *Pontor 2* is an extra-hard casting alloy containing 63% gold, 3.0% palladium and 0.5% platinum, and *Pontor MPF* an extra-hard casting alloy containing 72% gold and 3.6% platinum. *Pontor 4CF* is an extra-hard universal alloy containing 65.1% gold and 10.0% platinum, and *Pontor LFC* an extra-hard universal alloy containing 69.4% gold and 9.4% platinum. Both are used for low-fusing ceramic restorations.

Pontova. Trade name for rayon fibers and yarns used for textile fabrics.

poodle cloth. (1) See astrakhan.

(2) A soft, medium- to heavyweight knitted or woven fabric in plain colors or with random effects having a distinctive curled or looped finish made to imitate the coat of a poodle. It is made from wool, acrylic or other yarns including yarn blends. Used for coats, dresses and jackets.

pool paint. See swimming-pool paint.

poor lime. See lean lime.

poor pine. See spruce pine.

popcorn. A knitted or woven fabric made with a yarn that has thick spots or specks resembling popcorn in appearance.

Pope's Island. A term referring to a group of French white metal alloys composed of 67-70% copper, 13-15% zinc, 14-20% nickel and 0-1% tin, and used for jewelry, tableware, and as a base for plated ware.

poplar. The soft, light wood from any of various trees of the genus *Populus*. Commercially important species in North America include the balsam poplar (*P. balsamifera*), the bigtooth aspen (*P. grandidentata*) and the quaking or trembling aspen (*P. tremuloides*). The black poplar (*P. nigra*), gray poplar (*P. canescens*) and white poplar (*P. alba*) are commercially important European species which, however, are now also widely grown in North America. See also aspen.

poplin. A strong fabric, usually of cotton, wool, silk, nylon, polyester or yarn blends, in a plain weave and having a fine, crosswise rib. Used for sportswear, raincoats, shirts, blouses, dresses, pyjamas, etc.

Popril. Trademark of Zoltek Magyar Viscosa RT (Hungary) for polypropylene staple fibers.

Poprolin. Trademark for polypropylene resins and products.

Porasil. Trademark of Waters Associates, Inc. (USA) for spherical porous silica available in various mesh sizes, and used as an adsorbent in gas chromatography.

Porcast. Trade name for a high-gold dental bonding alloy.

Porcedent. Trade name for a vinyl ester resin formerly used for denture bases.

Porcelain. Trade name of Aardvark Clay & Supplies (USA) for a smooth, white porcelain clay (cone 10).

porcelain. A generic term for a glazed or unglazed high-quality ceramic whiteware fired at very high temperatures (typically 1300-1450°C or 2370-2640°F). It consists of varying amounts of kaolin, clay, quartz and feldspar, and is characterized by a vitreous fracture. *Porcelain* is available in domestic, chemical, electrical, mechanical, structural and thermal grades. Ordinary porcelain has relatively good impact resistance, good to fair compression strength, and low tensile and flexural strengths, while technical grades have improved tensile, compressive and/or other properties. Other important properties include a density of 2.3-2.6 g/cm^3 (0.08-0.09 lb/in.^{3}), an upper service temperature of about 1090°C (2000°F), high translucency, high brittleness (unless modified), good scratch resistance, low water absorption, high impermeability to gases, poor electrical and thermal conductivity, good resistance to acids (except hydrofluoric acid), poor resistance to hot concentrated alkalies, and high refractoriness. Used for dinnerware, laboratory ware, reaction vessels, piping, valves, pumps, corrosion-resistant equipment, food-processing equipment, grinding media for balls mills, high-voltage insulators, electrical resistors, spark plugs, electron tubes, and bioceramics. See also European porcelain.

porcelain balls. Hard, dense, uniform, abrasion-resistant spheres of porcelain used as crushing and grinding media in ball mills. They are low in porosity and high in toughness. Also known as *porcelain grinding balls.*

porcelain brick. A white, brick of porcelain made by molding and hard firing, and used in the construction of enamel mill interiors.

porcelain cement. A cement that is often mixed with gutta-percha and shellac, and used for porcelain-to-porcelain bonding applications.

porcelain clay. Any high-quality refractory clay, such as *kaolin*, that fires to a white or almost white color, and can thus be used in the manufacture of porcelain. Also known as *porcelain earth.*

porcelain color. A pigment suitable for decorating porcelain and chinaware.

porcelain earth. See porcelain clay.

porcelain enamel. A thin, glassy or vitreous, clear or colored inorganic coating consisting of intimately ground blends of low-sodium frit, clay, feldspar, applied and fired on a metal, such as sheet steel, cast iron or aluminum, at temperatures above 425°C (800°F). The resulting surface is smooth, hard, glossy or semi-glossy, and has good abrasion, corrosion, and thermal-shock resistance. Used for lining chemical and high-temperature process equipment, tanks, pipes, pumps, water treatment equipment, light reflectors, laboratory bench tops, storage containers, marine parts, appliances, and cooking utensils. Abbreviation: PE. Also known as *enamel; vitreous enamel.*

porcelain enameled iron. A premium grade of commercially pure iron that may contain as much as 0.20% manganese, and is used as a base metal for porcelain enamel. Also known as *enam-*

eling iron.

porcelain enameled sheet steels. A group of low-carbon sheet steels used as bases for porcelain enamels or other vitreous coatings. Steels suitable for this application include cold-rolled aluminum-killed steel, cold-rolled rimmed steel, decarburized enameling steels, enameling iron, and interstitial-free enameling steel. *Porcelain-enameled sheet steels* require only one cover coat, have good sag resistance, and exhibit no carbon boiling or fishscale formation. Also known as *enameled sheet steels.*

porcelain enamel frit. A glassy material composed of a mixture of ceramic materials and made by melting and quenching in air or water to form small friable particles that are subsequently ground for use in the manufacture of porcelain enamels. Also known as *enamel frit.*

porcelain enamel oxides. A wide range of inorganic oxides or mixtures of calcined oxides, such as lead oxide (PbO), zinc oxide (ZnO), or zirconia (ZrO_2), used on sheet steel and cast iron in ground coats to promote adherence, or in cover coats to impart color and improve the appearance. Also known as *enamel oxides.*

porcelain enamel sanitary ware. Porcelain-enameled ware used for hygienic and sanitary applications, e.g., bathtubs, lavatories, sinks, bidets, and bathroom fixtures.

porcelain grinding balls. See porcelain balls.

porcelain insulator. An electrical insulator of porcelain made by mixing the starting materials–kaolin, clay, quartz and feldspar–and molding into the desired shape. After drying and coating with a glazing compound, they are fired at high temperatures.

porcelainite. See mullite.

porcelain paste. A plastic porcelain body in the unfired state.

porcelain tile. A dense, fine-grained, smooth ceramic mosaic tile that has a sharply formed face, and is usually made by the dry-pressing (dust pressing) process.

Porceleen. Trade name of G.J. Nikolas & Company Inc. (USA) for clear lacquer.

Porcelfina. Trademark of Degussa-Ney Dental (USA) for a fine-grained dental porcelain with excellent handling characteristics, good resistance to checking, cracking, crazing and tearing, and a thermal expansion coefficient that is compatible with most dental alloys.

Porcelguard. Trade name of Akzo Porcelain America Inc. (USA) for an epoxy finish used for aluminum windows.

Porcelite. Trade name of Kerr Dental (USA) for a light-cure dental resin cement.

Porcelite Dual Cure. Trade name of Kerr Dental (USA) for a dual-cure dental resin cement.

Porcerax II. Trade name of International Mold Steel Inc. (USA) for a special permeable stainless steel with an average pore size is 7 μm (275 μin.), and high porosity (25 vol% air). It is used for plastic injection molds required to vent gas.

pore-forming material. A substance that is incorporated into a metal-powder blend and volatilizes during the sintering process yielding a specific type and degree of porosity in the finished part.

Porex. Trade name of Moraine Manufacturing Inc. (USA) for a powder-metallurgy bronze composed of 89-95% copper and 5-11% tin, and used for diffusers, filters, and porous membranes.

Poro Bronze. Trade name of Poro Metals Limited (USA) for a babbitt composed of 80% tin, 13% antimony and 17% copper, and used for bushings, sleeves and bearings.

Porocel. Trademark of Engelhard Corporation (USA) for an activated bauxite available in various regular, low-iron and low-silica grades (20-60 mesh) for use as an absorbent in catalysis, and as a filtering medium.

Porocell. Trade name for polystyrene foams and foam products.

Porodur. Trademark of Sieg GmbH (Germany) for phenolic resins and foams.

poromer. See poromeric material.

poromeric material. A pliant synthetic material, similar to leather in appearance, that is resistant to water penetration, scuffing and abrasion, but is permeable to both air and water vapor. An example of such a material is a polyurethane fabric treated with polyester. Also known as *poromer.*

Poron. Trademark of The Rogers Corporation (USA) for a high-density cellular urethane foam that exhibits excellent physical and mechanical properties. Available in regular, flame-retardant, adhesive and foil-clad grades, it has exceelent energy absorption and vibration dampening characteristics, low compression set, good shock resistance, good high and low temperature strength and impact resistance, good hydrolytic stability, internal strength and dimensional stability, very good fabricability, good resistance to ozone and chemicals, and good corrosion resistance. Used for vibration mounts, bumper pads, noise absorption pads, automotive trim tape, pads in seat belt retractors, shock pads, mounting tape substrates, tote box liners, camera case liners, gaskets, seals, spacers in liquid crystal displays, athletic padding, various cushioning and padding applications, and electronic and general industrial applications.

Poroplast. Trade name of Orbitaplast for flexible polyvinyl chloride foams.

Porosil. Trade name of Atofina Ceca (USA) for diatomites and kieselgurs used as fillers for plastics, rubber, paints, varnishes, inks, and paper.

Porosint. Trade name of Sintered Products Limited (UK) for a series of porous, sintered products including various bronzes, copper-nickels, and stainless steels used for filtering applications.

porous carbon. A product, usually in disk, plate or tube form, made from uniform carbon particles pressed together without the use of a binder. Its microstructure consists of small interconnecting pores, and it has higher tensile and compressive strengths, but lower oxidation resistance than porous graphite. Used in the filtration of corrosive fluids.

porous graphite. A product, usually in disk, plate or tube form, made from uniform graphitic particles pressed together without the use of a binder. Its microstructure consists of small interconnecting pores, and most of its properties are quite similar to those of porous carbon, but it has higher oxidation resistance and lower tensile and compressive strengths.

porous materials. Materials, such as stone, concrete, ceramic materials, powdered metals or cellular plastics, that contain open or closed voids, pores, cells, interstices or similar openings. Depending on the pore size, they are commonly classified as: (i) *Macroporous* (pore diameters greater than 5000 nm or 197 μin.), (ii) *Microporous* (pore diameters less than 200 nm or 8 μin.), or (iii) *Mesoporous* (pore diameters of 200-5000 nm or 8-197 μin.) materials. Used for membranes employed in the separation of chemicals and hot gases, for catalytic chemical processing, as substrates for electronics, optical precursors, thermal insulation, electrodes for batteries and fuel cells, hydrogen storage, and reusable particulates.

porous metals. A group of metals, such as steel, aluminum, brass, bronze, nickel, silver and tungsten, made by any of several methods including powder metallurgy (e.g., powder sintering, slip casting, slip foaming, etc.), casting (e.g., lost-foam, infiltration, foaming, etc.), chemical, electrochemical or electroless deposition, physical vapor deposition and gas-eutectic transformation. Depending on the method of manufacture they may be permeable or impermeable, but in general have microstructures consisting of uniformly distributed pores of controlled size and shape. Compared to nonporous metals they possess better acoustic damping and shock absorption properties, but poorer strength, plasticity, electrical conductivity and corrosion resistance. Typical applications of permeable porous metals include catalysts, mufflers, electrolytic and fuel cells, flame arresters, filters (e.g., for hot and corrosive fluids), self-lubricating bearings, thermal screens, vibration dampers, and metal-matrix composites. Impermeable porous metals are used for thermal insulation, seals for turbojet engines, and hydrogen storage. See also powdered metals; sintered materials.

Porous Poly. Trade name of General Polymeric Corporation, Gen-Pore Division (USA) for porous plastics supplied in the form of precision-molded components used for fluid filtration and handling applications.

porous wood. Hardwood with pores or vessels large enough to be observed by the unaided eye.

Porox. (1) Trademark of Ferro Corporation (USA) for vitrified silica media with a density of 2.4 g/cm³ (0.09 lb/in.³) and a hardness of 7.5 Mohs, used for grinding media for fine or micromilling.

(2) Trademark of Ferro Corporation (USA) for high-density refractory porcelain grinding balls and grinding-mill linings.

(3) Trademark of Ferro Corporation (USA) for acidproof porcelain blocks and cement used on building walls and floors, and for lining containers.

porpezite. See palladium gold (1).

porphyrins. A class of biologically active nitrogenous compounds with parent structures consisting essentially of four pyrrole (C_4H_5N) rings joined through their α-carbon atoms by four methene (=C–) groups. They also contain two hydrogen atoms that can be replaced by various central atoms. Examples of naturally occurring porphyrin derivatives include (i) *chlorophyll*, a plant pigment which catalyzes photosynthesis and contains a central magnesium atom, (ii) *heme*, a constituent of hemoglobin and contains a central iron atom, and (iii) the *cytochrome* and *phthalocyanine pigments*. Many *porphyrins* with central metal atoms can also be made synthetically, and have optical and electrically conductive properties that make them suitable for use in optoelectronics, photonics and bioengineering.

porphyry. A general term for any igneous rock consisting of a relatively fine-grained matrix material (known as the groundmass) and well-defined embedded crystals (known as phenocrysts). For example, porphyry copper consists of a groundmass of porphyry rock with copper minerals distributed throughout.

porpoise leather. A tough, strong, flexible, alum-tanned, oil-impregnated leather made from the hides of dolphins and small toothed whales. It is sometimes used as lace leather for splicing sections of power-transmission belts.

Pors-on. Trademark of Degussa-Ney Dental (USA) for a series of palladium-silver dental alloys including (i) *Pors-on 4*, a microfine-grained dental ceramic alloy with 58% palladium and 30% silver, (ii) *Pors-on Lite,* a microfine-grained dental ceramic alloy with 61.4% palladium and 26% silver and (iii) *Pors-on Plus,*

a microfine-grained dental ceramic alloy with 47.5% palladium, 27.5% silver and 15.5% silver. *Pors-on* alloys have excellent physical properties, high corrosion and tarnish resistance, high yield strength, hardnesses ranging between 245 and 260 Vickers, and good compatibility with most dental porcelains. Used in restorative dentistry for porcelain bonding applications and long- and short-span bridges.

Porta. Trademark of Wieland Dental + Technik GmbH & Co. KG (Germany) for an extensive series of dental alloys including (i) gold-platinum ceramic alloys, such as the yellow-colored, hard *Porta Geo* and *Porta Geo 2*, the yellow-colored, extra-hard *Porta Geo Ti* and the pale yellow, extra-hard *Porta Impulse*; (ii) gold-platinum-palladium ceramic alloys, such as the yellow-colored, hard *Porta KL90*, the pale yellow, hard *Porta AB76* and the pale-yellow, extra-hard *Porta Top H* as well as the white-colored extra-hard *Porta PK* and *Porta K6*; (iii) gold-palladium ceramic alloys, such as the white-colored, extra-hard *Porta SMK 80* and *Porta SMK 82*; and (iv) several universal alloys, such as the yellow-colored, extra-hard *Porta Aurium, Porta Maximum* and *Porta Optimum* for low-fusing, high-expansion ceramic restorations.

Portadur. Trademark of Wieland Dental + Technik GmbH & Co. KG (Germany) for an extensive series of yellow, gold-based dental casting alloys including several medium-hard types, such as *Portadur IN* and *Portadur T*, and extra-hard types, such as *Portadur 2* and *Portadur P4*.

Portagold. Trademark of Wieland Dental + Technik GmbH & Co. KG (Germany) for a yellow, medium-hard gold-based dental casting alloy.

Portalloy 54. Trademark of Wieland Dental + Technik GmbH & Co. KG (Germany) for a white, extra-hard gold-palladium dental ceramic alloy.

Porta Plast. Trade name of Porta Plast GmbH (Germany) for plastic profiles made from rigid polyvinyl chloride.

PortaSmart. Trademark of Wieland Dental + Technik GmbH & Co. KG (Germany) for a white, extra-hard semiprecious dental ceramic alloy.

Portex. Trade name of Portland Plastics (UK) for dough (or bulk) molded acrylic resins for dentures, and several vinyl-based resins.

Portland blast-furnace cement. A hydraulic cement made by intimately grinding together granulated basic blast-furnace slag and Portland cement clinker. The slag content is usually between 30 and 85%. Also known as *Portland blast-furnace slag cement*.

Portland cement. A *hydraulic cement* produced by fine grinding and intimate mixing of clay, lime, silica, alumina, magnesia and ferric oxide in the proper proportions, and then heating (calcining) the mixture to about 1400°C (2550°F) in a rotary or shaft kiln. The resulting bluish gray "clinker" product is then ground into a very fine powder to which a small amount of gypsum is added to retard the setting process. The principal constituents of Portland cement are tricalcium silicate (C_3S) and dicalcium silicate (C_2S). The composition range is 28-45 wt% C_3S, 19-49 wt% C_2S, 4-11 wt% C_3A, 8-13 wt% C_4AF, and 6-9 wt% other compounds including simple oxides, such as lime (CaO) and magnesia (MgO), alkaline oxides and calcium sulfate ($CaSO_4$). The name "Portland cement" originates from the fact that the fully hardened cement resembles a natural limestone quarried on the Isle of Portland, a peninsula in southern England. Used for mortar, concrete, plaster and clinkers. Abbreviation: PC. Also known as *type I cement. Note:* In cement terminology C_3S, C_2S, C_3A and C_4AF refer to tricalcium silicate ($3CaO \cdot SiO_2$), dicalcium silicate ($2CaO \cdot SiO_2$), tricalcium aluminate ($3CaO \cdot Al_2O_3$) and tetracalcium aluminoferrate ($4CaO \cdot Al_2O_3 \cdot Fe_2O_3$), respectively.

Portland cement clinker. See clinker (2).

Portland cement concrete. A macroscopic engineering composite consisting of Portland cement, a fine aggregate (usually sand), a coarse aggregate (usually gravel) and water. Used as a construction material. Abbreviation: PCC. See also concrete.

Portland cement plaster. A *plaster* made by mixing together an aggregate, water and a cementitious material based on Portland cement, a Portland cement/masonry cement blend or a Portland cement/lime blend. Used for finishing walls.

portlandite. A white mineral of the brucite group composed of calcium hydroxide, $Ca(OH)_2$. It can also be made synthetically, and occurs in Portland cement. Crystal system, hexagonal. Density, 2.24 g/cm^3; refractive index, 1.704. Occurrence: Ireland.

Portland pozzolan cement. A *hydraulic cement* made by intimately and uniformly blending or grinding together Portland cement, Portland-cement clinker, Portland blast-furnace slag cement and *pozzolan*.

Portland stone. A yellowish white natural limestone quarried on the Isle of Portland, a peninsula in southern England. Used as building stone.

Port Orford cedar. See Lawson cypress.

Poscal. Trade name of Demedis Dental (Netherlands) for a zinc phosphate dental cement.

positive crystal. An anisotropic, optically uniaxial crystal, such as quartz, in which the refractive index of the ordinary ray is smaller than that of the extraordinary ray.

positive index materials. See left-handed materials.

positive photoresist. A *photoresist* in which the photosensitive polymeric material is depolymerized by ultraviolet radiation and subsequently removed. Used in the integrated-circuit lithography process for producing vitreous silica patterns on silicon wafers.

posnjakite. A light blue mineral composed of copper sulfate hydroxide monohydrate, $Cu_4SO_4(OH)_6 \cdot H_2O$. It can also be made synthetically. Crystal system, monoclinic. Density, 3.35 g/cm^3; refractive index, 1.680.

POSS. See polyhedral oligomeric silsesquioxanes.

possumwood. See ochoo pine.

Postaer Sandstein. Trademark of Sächsische Sandsteinwerke GmbH (Germany) for a high-grade sandstone from Saxony.

Postalloy. Trademark of Postle Industries, Inc. (USA) for welding and hardfacing wires and rods based on austenitic manganese steels, low-alloy materials, and chromium and tungsten carbides.

postcard paper. A fine, lightweight cardboard, similar to *Bristol board*, made by the soda, or the sulfite pulping process. It has a smooth, firm surface suitable for writing, typing and printing, and is used for picture postcards, greeting cards, etc.

Post Cement Hi-X. Trademark of BISCO, Inc. (USA) for a fast-setting, self-cured dental cement with high radiopacity used for cementing posts.

Post Com II. Trade name for self and light-cure hybrid composite resins used as posterior dental restorative materials.

Post Comp. Trade name of Generic Pentron (USA) for posterior composite resins used as dental restorative materials.

poster board. A stiff, compact, white or colored *pasteboard* composed of two or more webs of paper available in various qualities and thicknesses, and used for show cards, signs, display advertising, etc.

poster cloth. A smooth-faced drill fabric that has been heavily finished with starch and may or may not be treated with suitable chemicals, such as pyroxylin, to enhance weather- and termite resistance. Used for outdoor advertisements, placards, etc.

poster paper. A strong, white or colored paper that does not curl upon application of a paste or glue. It is treated with a waterproofing compound, and colored grades also contain nonfading pigments. Used for billboard posters for outdoor advertisements and public announcements.

poster paint. A glossy, non-transparent, quick-drying water paint with gum or resin binder, used for posters and artwork.

post locust. See black locust.

post oak. The strong, hard, durable wood of the small broadleaved tree *Quercus stellata* belonging to the American white oak group. The heartwood is grayish-brown, and the sapwood nearly white. *Post oak* has a close, coarse grain and good decay resistance. Source: Southeastern and central United States from Illinois to New Jersey and from Texas to Florida. Used for lumber, millwork, furniture, carvings, bent parts, barrels, fencing, and posts.

post-tensioned concrete. A type of reinforced concrete in which stresses are applied after the concrete has attained sufficient or maximum strength. Sheet metal or rubber tubes (so-called "tendons") are located inside and pass through the forms around which the concrete is poured. After the concrete has hardened, steel wires are fed through the resulting holes and tensile stress is applied to the wires imposing compressive stresses on the concrete. Used for buildings and structures with high load-bearing capacity. Abbreviation: PTC.

potarite. A silver-white mineral composed of mercury palladium, HgPd. It can also be made synthetically. Crystal system, tetragonal. Density, 14.88 g/cm^3. Occurrence: Guinea.

potash. A term usually referring to potassium (K), potassium carbonate (K_2CO_3), potassium hydroxide (KOH) or potassium oxide (K_2O).

potash alum. See potassium-aluminum sulfate.

potash blue. A blue pigment prepared by precipitating ferrous ferrocyanide from a solution of ferrocyanide and ferrous sulfate, followed by oxidation with potassium dichromate. Used in the manufacture of carbon paper.

potash chrome alum. See chromium potassium sulfate.

potash feldspars. See potassium feldspars.

potash mica. See muscovite.

potash water glass. Colorless lumps or granules, or colorless liquid composed of varying amounts of silicon dioxide (SiO_2) and potassium oxide (K_2O). It has a density range of 1.25-1.39 g/cm^3 (0.045-0.050 $lb/in.^3$), and is used in ceramics as a source of potassium and silica and as an antiblooming agent, for high-temperature mortars, in the manufacture of glass and refractories, in welding rods, as a binder in carbon arc-light electrodes, and as an adhesive. Formula: K_2SiO_3. Also known as *potassium silicate; silicate of potash*.

potassium. A soft, silvery, metallic element of Group IA (Group 1) of the Periodic Table. It is commercial available in the form of ingots, rods sealed under argon in glass, and sticks in mineral oil. Important potassium ores are *carnallite, polyhalite* and *sylvite*, and it also occurs in soils. Crystal system, cubic. Crystal structure, body-centered cubic. Density, 0.86 g/cm^3; melting point, 63.25°C (146°F); boiling point, 760°C (1400°F); hardness, 0.5 Mohs; atomic number, 19; atomic weight, 39.098;

monovalent. It is an alkali metal with fair electrical conductivity (about 25% IACS) and poor resistance to moist air. Used in the manufacture of glass and lenses, in heat-exchange alloys, in ceramics, for seeding combustion gases in magnetohydrodynamic generators, in matches and gunpowder, as a substitute for sodium, in the preparation of potassium peroxide, and in the biosciences. Symbol: K.

potassium-40. A natural radioisotope of potassium with a mass number of 40. It is present in potassium to the extent of 0.01%, has a half-life of 1.31×10^9 years and decays by a process involving beta particle emission and electron capture. Symbol: ^{40}K. Also known as *radiopotassium*.

potassium acetate. The potassium salt of acetic acid that is available in the form of a white, hygroscopic, crystalline powder (99+% pure). Density, 1.57 g/cm^3; melting point, 292°C (558°F). Used as a flux in the manufacture of crystal glass, as a dehydrating agent, and as a textile conditioner. Formula: $KC_2H_3O_2$.

potassium acid fluoride. See potassium bifluoride.

potassium acid phosphate. See potassium dihydrogen phosphate.

potassium acid phthalate. The potassium salt of phthalic acid. It is an alkaline metal biphthalate available in the form of single crystals with high plasticity and used as an analyzing crystal in X-ray spectral analysis. Formula: $KC_8H_5O_4$. Abbreviation: KAP. Also known as *potassium biphthalate; potassium hydrogen phthalate*.

potassium acid tartrate. See potassium bitartrate.

potassium alum. See potassium-aluminum sulfate.

potassium aluminate. A compound of potassium oxide and aluminum oxide that is available as colorless crystals with a melting point above 1650°C (3000°F), and used in ceramics and as a paper sizing. Formula: $K_2Al_2O_4$.

potassium aluminosilicate. See potassium-aluminum silicate.

potassium-aluminum disilicate. See potassium-aluminum silicate (ii).

potassium-aluminum hexasilicate. See potassium-aluminum silicate (iv).

potassium-aluminum monosilicate. See potassium-aluminum silicate (i).

potassium-aluminum silicate. Any of the following compounds containing potassium oxide, aluminum oxide and silicon dioxide, used in ceramics and materials research, and also known as *potassium aluminosilicate*: (i) *Potassium-aluminum monosilicate*. Formula: $K_2Al_2SiO_6$; (ii) *Potassium-aluminum disilicate*. Density, 2.6 g/cm^3; melting point, 1750°C (3180°F); hardness, 5-7 Mohs. Formula: $K_2Al_2Si_2O_8$; (iii) *Potassium-aluminum tetrasilicate*. Density, 2.47 g/cm^3; melting point, 1688°C (3070°F); hardness, 5-7 Mohs. Formula: $K_2Al_2Si_4O_{12}$; and (iv) *Potassium-aluminum hexasilicate*. Melting point, 1150°C (3000°F); hardness, 5-7 Mohs. Formula: $K_2Al_2Si_6O_{16}$.

potassium-aluminum sulfate. A colorless or white crystalline substance which is soluble in water. It occurs in nature as the mineral *kalinite*. Density, 1.757 g/cm^3; melting point, 92°C (198°F); boiling point, loses $9H_2O$ at 64.5°C (148°F); hardness, 2-2.5 Mohs. Used as a cement hardener, as a dyeing mordant, in papermaking, matches, paints, tanning agents, waterproofing agents, and for water purification applications. Formula: $AlK(SO_4)_2 \cdot 12H_2O$. Also known as *alum; aluminum-potassium sulfate; potash alum; potassium alum*.

potassium-aluminum tetrasilicate. See potassium-aluminum silicate (iii).

potassium amalgam. A mixture of sodium amalgam and potash used in combination with other amalgams, and as a chemical reagent. See also amalgam.

potassium antimonate. White crystalline granules or powder used in starch sizes, in flame-retarding compounds, and in cytochemistry and histochemistry. Formula: $K_2H_2SbO_7 \cdot 4H_2O$. Also known as *potassium pyroantimonate*.

potassium argentocyanide. See silver-potassium cyanide.

potassium aurichloride. See gold-potassium chloride.

potassium bichromate. See potassium dichromate.

potassium bifluoride. Corrosive, colorless crystals or white powder. Crystal system, cubic. Density, 2.37 g/cm^3; melting point, 239°C (462°F). Used as an etchant for glass, as a flux in silver solders and as an electrolyte in fluorine manufacture. Formula: KHF_2. Also known as *potassium acid fluoride; potassium-hydrogen fluoride*.

potassium biphthalate. See potassium acid phthalate.

potassium bitartrate. The monopotassium salt of tartaric acid available as white crystals or powder with a density of 1.954 g/cm^3, and used in galvanic tinning of metals. Formula: $KHC_4H_4O_6$. Also known as *cream of tartar; potassium acid tartrate; potassium-hydrogen tartrate*.

potassium borate. A compound of potassium, boron and oxygen available as nonlinear high-purity crystals for single and third harmonic generation applications. Formula: KBO_3.

potassium borofluoride. Colorless crystals obtained by treating potassium hydroxide with boric acid and hydrogen fluoride. Crystal system, cubic or orthorhombic. Density, 2.50 g/cm^3; melting point, decomposes at 350°C (662°F). Used for soldering and brazing fluxes, as a grinding aid in resinoid grinding wheels, in sand casting of aluminum and magnesium, and in electrochemical processes. Formula: KBF_4. Also known as *potassium fluoborate*.

potassium borohydride. A moisture-sensitive, white, crystalline powder (98+% pure) obtained by reacting potassium hydroxide with sodium borohydride. Density, 1.178 g/cm^3; melting point, decomposes above 400°C (752°F), refractive index, 1.494. Used as a reducing agent for aldehydes, ketones and acid chlorides, as a foaming agent for plastics, and as a source of hydrogen. Formula: KBH_4.

potassium bromide. Colorless, hygroscopic crystals or white, hygroscopic granules or powder (99+% pure). Crystal system, cubic. Density, 2.75 g/cm^3; melting point, 730°C (1346°F); boiling point, 1435°C (2615°F); refractive index, 1.559. Used in photography (i.e., in the preparation of silver bromide emulsions for photographic films, plates and papers), process engraving and lithography, spectroscopy, infrared transmission, and as a reagent. Formula: KBr.

potassium carbonate. A white, hygroscopic granular powder (99+% pure). Crystal system, monoclinic. Density, 2.43 g/cm^3; melting point, 891°C (1636°F); boiling point, decomposes. Used as an addition of K_2O in glasses, glazes and enamels, as an opacifier, fluxing agent and glass former, in special glasses (e.g., optical and color television tubes), as a dehydrating agent, and in pigments, printing inks and explosives. Formula: K_2CO_3. Also known as *potash; salt of tartar*. See also pearl ash.

potassium chloride. Colorless or white, hygroscopic granules or powder (99+% pure). It occurs in nature as the mineral *sylvite*. Crystal system, cubic. Crystal structure, halite. Density, 1.98 g/cm^3; melting point, 776°C (1429°F); boiling point, 1500°C (2730°F); hardness, 2 Mohs; refractive index, 1.490. Used as a set-up agent in porcelain enamel slips, in molten salt baths for heat treatment of steels, in acid chloride zinc plating baths, as a source of potassium salts, in photography and spectroscopy, and in biology, molecular biology and biochemistry. Formula: KCl. Also known as *potassium muriate.*

potassium chloroaurate.   See gold-potassium chloride

potassium chloroiridate.   See iridium-potassium chloride.

potassium chloroosmate.   See osmium-potassium chloride.

potassium chloropalladate.   See palladium-potassium chloride.

potassium chlororhodite.   See rhodium-potassium chloride.

potassium chromate.   Yellow crystals (99% pure). It occurs in nature as the mineral *tarapacaite.* Crystal system, orthorhombic. Density, 2.732 g/cm³; melting point, 971°C (1780°F). Used as a yellow or orange pigment in porcelain enamels and glazes, as an analytical reagent, and in aniline black. Formula: K_2CrO_4.

potassium-chromium sulfate.   See chromium-potassium sulfate.

potassium cobaltinitrite.   See cobalt-potassium nitrite.

potassium columbate.   See potassium niobate.

potassium-columbium oxyfluoride.   See niobium-potassium oxyfluoride.

potassium-copper cyanide.   White crystals obtained by evaporating a water-soluble solution of cuprous cyanide and potassium cyanide. Density, 2.38 g/cm³; decomposed by heating in water. Used in copper electroplating baths. Formula: $KCu(CN)_2$. Also known as *copper-potassium cyanide; cuprous potassium cyanide; potassium cuprocyanide.*

potassium cuprocyanide.   See potassium copper cyanide.

potassium cyanide.   White crystals or powder (97+% pure). Crystal system, cubic. Density, 1.52 g/cm³; melting point, 634.5°C (1174°F). Used in the extraction of gold and silver from ores, in copper and gold plating baths, as a neutralizer in the pickling of metals, and in the porcelain enameling and case hardening of steels. Formula: KCN.

potassium cyanoargentate.   See silver potassium cyanide.

potassium cyanoaurite.   See gold potassium cyanide.

potassium dichromate.   Bright orange, transparent crystalline substance (99+% pure). Density, 2.692 g/cm³; melting point, 396°C (745°F); boiling point, decomposes at 500°C (932°F). Used in electroplating and brass pickling compositions, for pigments and ceramic colors, in the production of pink, red, green and purplish-red colors in glazes, in ceramics, alloys, chrome glues and adhesives, in leather tanning, in wood stains, in microscopy as a metal stain, as a battery depolarizer, and in pyrotechnics. Formula: $K_2Cr_2O_7$. Also known as *red potassium chromate.*

potassium dicyanoargentate.   See silver-potassium cyanide.

potassium dicyanoaurate.   See gold-potassium cyanide.

potassium dihydrogen phosphate.   White powder or colorless crystals (99+% pure). Crystal system, tetragonal. Density, 2.338 g/cm³; melting point, 253°C (487°F); ferroelectric below 123K; good chemical and mechanical properties; wide transparency; high laser damage threshold. Potassium dihydrogen phosphate for electrooptical applications often has a deuteration of over 98 at% (DKDP). Used in spectrometry, as a laboratory reagent, as a volumetric standard, as ferroelectric crystals in scientific equipment, and in Q-switched lasers of the titanium-doped sapphire and neodymium-doped yttrium aluminum garnet types. Formula: KH_2PO_4. Abbreviation: KDP. Also known as *potassium acid phosphate; monobasic potassium phosphate.*

potassium feldspars.   A group of colorless, white, pale-yellow or reddish potassium-bearing feldspars that have the general formula $KAlSi_3O_8$, and may also contain small amounts of sodium. It includes in particular the minerals *microcline* and *orthoclase.* Used in ceramics for pottery, electrical and other porcelains, hard glazes, enamels, cements, concretes, ceramic ware, in glass manufacture, as an abrasives, as a bond for abrasive wheels, and in insulating compositions and roofing materials. Also known as *potash feldspars.*

potassium ferric oxalate.   Green, light-sensitive crystals. Crystal system, monoclinic. Melting point, loses 3H₂O at 100°C (212°F); decomposes at 230°C (446°F). Used in photography and blueprinting. Formula: $K_3Fe(C_2O_4)_3\cdot3H_2O$.

potassium ferricyanide.   Bright red, lustrous, light-sensitive crystals or granular powder (99+% pure). Density, 1.89 g/cm³; melting point, decomposes. Used in tempering steel, in the manufacture of pigments, as an etchant, in electroplating, for silvering mirrors, as a sensitive coating on blueprint paper, and as a cytochemical stain. Formula: $K_3Fe(CN)_6$. Also known as *potassium hexacyanoferrate; red potassium prussiate; red prussiate; red prussiate of potash.*

potassium ferrocyanide.   Lemon-yellow crystals or powder (99+% pure). Crystal system, monoclinic. Density, 1.85 g/cm³; melting point, loses 3H₂O at 70°C (158°F); boiling point, decomposes. Used for case-hardening of steel, in dry colors, and in process engraving and lithography. Formula: $K_4Fe(CN)_6\cdot3H_2O$. Also known as *potassium hexacyanoferrate; yellow potassium prussiate; yellow prussiate of potash.*

potassium fluoborate.   See potassium borofluoride.

potassium fluoride.   Colorless, hygroscopic crystals or white powder (99+% pure). Crystal system, cubic. Density, 2.48 g/cm³; melting point, 846°C (1555°F); boiling point, 1505°C (2741°F); refractive index, 1.363. Used as a glass etchant, as a soldering flux, as a flux in the preparation of ferroelectric crystals of barium titanate, as a fluoride ion source, as a fluorinating agent, in the measurement of electronic polarizabilities of ions in alkali halide based polymers, and in the room-temperature oxidation of noble metals in hydrogen fluoride. Formula: KF.

potassium fluoride dihydrate.   Colorless or white hygroscopic crystals (98% pure). Crystal system, monoclinic. Density, 2.454 g/cm³; melting point, 41°C (106°F); boiling point, 156°C (313°F); refractive index, 1.352. Used as a glass etchant, as a flux in the preparation of ferroelectric crystals of barium titanate, as a stationary phase in gas chromatography, and as a soldering flux. Formula: $KF\cdot2H_2O$.

potassium fluorophosphate.   White crystals (98+% pure). Density, 2.75 g/cm³; melting point, 575°C (1067°F); boiling point, decomposes. Used as a reagent, in biochemistry and medicine, and in ceramics. Formula: KPF_6. Also known as *potassium hexafluorophosphate.*

potassium fluosilicate.   White crystalline powder. Crystal system, cubic. Density, 2.66 g/cm³; melting point, decomposes on heating. Used as a fluxing ingredient in porcelain enamels, in the metallurgy of aluminum and magnesium, in ceramics, for synthetic mica, and in opalescent glass. Formula: K_2SiF_6. Also known as *potassium silicofluoride; potassium hexafluorosilicate.*

potassium fluotantalate.   See tantalum-potassium fluoride.

potassium fluozirconate.   See zirconium-potassium fluoride.

potassium-gold chloride.   See gold-potassium chloride.

potassium-gold cyanide.   See gold-potassium cyanide.

potassium hexachloroiridate.   See iridium-potassium chloride.

potassium hexachloroosmate.   See osmium-potassium chloride.

potassium hexachloropalladate.   See palladium-potassium chloride.

potassium hexachlororhenate.   See rhenium-potassium chloride.

potassium hexacyanocobaltate.   See cobalt-potassium cyanide.

potassium hexacyanoferrate. See potassium ferricyanide; potassium ferrocyanide.

potassium hexafluorophosphate. See potassium fluorophosphate.

potassium hexafluorosilicate. See potassium fluosilicate.

potassium hydrate. See potassium hydroxide.

potassium hydride. A moisture-sensitive compound of potassium and hydrogen supplied as white needles, gray powder or a slurry in the form of a flammable, moisture-sensitive 20-35 wt% dispersion in mineral oil. Used in photosensitive coatings for photoelectric cells. Formula: KH.

potassium-hydrogen fluoride. See potassium bifluoride.

potassium-hydrogen phthalate. See potassium acid phthalate.

potassium-hydrogen phosphate. See dibasic potassium phosphate.

potassium-hydrogen tartrate. See potassium bitartrate.

potassium hydroxide. White, hygroscopic crystalline solid (85+% pure) with corrosive properties usually obtained electrolytically from a concentrated potassium chloride solution. It is commercially available in the form of broken pieces, granules, lumps, sticks, pellets and flakes in standard and high-purity (99.99%) grades, and as a 40-50% liquid solutions. Crystal system, orthorhombic. Density, 2.044 g/cm^3; melting point, 360°C (680°F); boiling point, 1324°C (2415°F); very strong alkaline properties. Used in electroplating, in cleansing baths for scouring metals, in quenching baths for steel, as an electrolyte in alkaline batteries and certain fuel cells, as an absorbent for carbon dioxide, as a drying agent for gases and dyestuffs, in paint removers and ceramics, and in biology, biochemistry, biotechnology and medicine. The high-purity grade is also used for semiconductor applications. Formula: KOH. Also known as *caustic potash; potassium hydrate.*

potassium iodide. Colorless or white, hygroscopic crystals, granules or powder (99+% pure). Crystal system, cubic. Density, 3.13 g/cm^3; melting point, 686°C (1267°F); boiling point, 1330°C (2426°F); refractive index, 1.677. It can be doped with ions of halogens other than iodine or alkali metals, silver, etc., and is used for infrared transmission and scintillation, in the precipitation of silver in photographic emulsions, in spectroscopy, and in biochemistry and biotechnology. Formula: KI.

potassium-iridium chloride. See iridium-potassium chloride.

potassium manganate. Dark green powder or crystals. Crystal system, orthorhombic. Melting point, decomposes at 190°C (374°F). Used in batteries, water purification and photography, and as an oxidizer. Formula: K$_2$MnO$_4$.

potassium metatantalate. See potassium tantalate.

potassium mica. See muscovite.

potassium monosulfide. See potassium sulfide (i).

potassium muriate. See potassium chloride.

potassium naphthenate. The potassium salt of naphthenic acid available as a powder containing about 13.1% potassium, and used as a drier.

potassium niobate. (1) A ferroelectric compound available in the form of a white powder, also in high-purity grades (99.999%). It has a Curie temperature of 435°C (814°F) and is used for ferroelectric applications. Formula: KNbO$_3$. Also known *potassium columbate.*

(2) A ferroelectric compound made from potassium oxide and niobium trioxide. Curie temperature: 420°C (788°F). Used for ferroelectric applications. Formula: K$_8$Nb$_6$O$_{13}$. Also known as *potassium columbate.*

potassium-niobium oxyfluoride. See niobium-potassium oxyfluoride.

potassium nitrate. Transparent, colorless or white crystals or powder (99+% pure). It occurs in nature as the mineral *niter.* Crystal system, orthorhombic. Density, 2.10 g/cm^3; melting point, 334°C (633°F); boiling point, decomposes at 400°C (752°F); hardness, 2 Mohs. Used as an oxidizer and fluxing agent for glass, glazes and porcelain enamels, in metallurgy (tempering of steel), in pyrotechnics, for matches, and as an oxidant in solid rocket propellants. Formula: KNO$_3$. Also known as *saltpeter.*

potassium nitrite. White or slightly yellowish, deliquescent prisms or sticks. Density, 1.915 g/cm^3; melting point, decomposes at 350°C (662°F); boiling point, explodes at 537°C (999°F). Used as a color stabilizer, anti-tearing agent and set-up addition in porcelain enamels, in the regeneration of heat-transfer salts, as a rust inhibitor, in manufacture of dyes, and as an oxidizer. Formula: KNO$_2$.

potassium oxide. Colorless or gray crystals. Crystal system, cubic. Density, 2.32-2.35 g/cm^3; melting point, 350°C (662°F). Used as a deflocculating agent in engobes and in casting and glaze slips, as a color stabilizer and fluxing agent in glass, glazes and porcelain enamels. Formula: K$_2$O. Also known as *potash.*

potassium oxycolumbate. See niobium-potassium oxyfluoride.

potassium oxyniobate. See niobium-potassium oxyfluoride.

potassium oxyfluocolumbate. See niobium-potassium oxyfluoride.

potassium oxyfluoniobate. See niobium-potassium oxyfluoride.

potassium pentafluocolumbate. See niobium-potassium oxyfluoride.

potassium pentafluoniobate. See niobium-potassium oxyfluoride.

potassium pentasulfide. See potassium sulfide (iv).

potassium permanganate. Dark purple crystals with bluish metallic luster. Crystal system, orthorhombic. Density: 2.703 g/cm^3; melting point, decomposes at 240°C (464°F). Used as an oxidizer and bleaching agent, in air and water purification, and in microscopy as a metal stain. Formula: KMnO$_4$.

potassium perrhenate. White crystals (99+% pure). Density, 4.887 g/cm^3; melting point, 550°C (1022°F); boiling point, approximately 1365°C (2489°F); refractive index, 1.643. Used as an oxidizer and in the manufacture of rhenium metal. Formula: KReO$_4$.

potassium phosphate. See dibasic potassium phosphate; monobasic potassium phosphate; tripotassium phosphate.

potassium selenide. Red, hygroscopic crystals. Crystal system, cubic. Density, 2.29 g/cm^3; melting point, 800°C (1472°F). Used in ceramics, materials research, and electronics. Formula: K$_2$Se.

potassium silicate. See potash water glass.

potassium silicofluoride. See potassium fluosilicate.

potassium-silver cyanide. See silver-potassium cyanide.

potassium-sodium alloys. See sodium-potassium alloys.

potassium-sodium tartrate. Colorless crystals or white powder (99+% pure). Crystal system, orthorhombic. Density, 1.78 g/cm^3; melting point, 70-80°C (158-176°F); boiling point, loses 4H$_2$O at 215°C (419°F); ferroelectric and piezoelectric properties. Used for ferroelectric and piezoelectric applications, as an addition to copper and bronze plating baths, in silvering mirrors, and as a cement for joining two metal surfaces together by applying in the hot plastic condition. Formula: KNaC$_4$H$_4$O$_6$·4H$_2$O. Also known as *potassium-sodium tartrate tetrahydrate; Rochelle salt; Seignette salt; sodium-potassium tartrate; sodium-potassium tartrate tetrahydrate.*

potassium stannate. White to light tan crystals (99.9% pure).

Crystal system, trigonal. Density, 3.20 g/cm³; melting point, decomposes at 140°C (284°F). Used in alkaline tin plating electrolytes, in organic and organometallic synthesis, in immersion tinning of aluminum, and in textile dyeing and printing. Formula: $K_2SnO_3 \cdot 3H_2O$. Also known as *potassium stannate trihydrate*.

potassium sulfate. Colorless or white crystals, granules, or powder (99+% pure). It occurs in nature as the mineral *arcanite*. Crystal system, orthorhombic. Density, 2.66 g/cm³; melting point, 1072°C (1962°F). Used in alum manufacture, as a raw material in glassmaking and gypsum cements. Formula: K_2SO_4.

potassium sulfide. Any of the following compounds of potassium and sulfur: (ii) *Potassium monosulfide*. Yellow-red or red, deliquescent crystals or powder. Crystal system, cubic. Density 1.75 g/cm³; melting point, above 910°C (1670°F). Used as an analytical reagent, and in biochemistry and medicine. Formula: K_2S; (ii) *Potassium trisulfide*. Brownish-yellow crystals. Melting point, 252°C (486°F). Formula: K_2S_3; (iii) *Potassium tetrasulfide*. Red-brown crystals. Melting point, 145°C; boiling point, decomposes at 850°C (1562°F). Formula: K_2S_4; and (iv) *Potassium pentasulfide*. Orange crystals. Melting point, 206°C (403°F). Formula: K_2S_5.

potassium tantalate. A ferroelectric compound with a high dielectric constant (above 4000) at the Curie temperature of -260°C (-436°F). Used for special ferroelectric applications. Formula: $KTaO_3$. Also known as *potassium metatantalate*.

potassium-tantalum fluoride. See tantalum-potassium fluoride.

potassium tetrachloroaurate. See gold-potassium chloride.

potassium tetrasulfide. See potassium sulfide (iii).

potassium titanate. A white powder (95+% pure) obtained by treating potassium hydroxide with titanic acid. Density, 3.2 g/cm³; melting point, 1370°C (2498°F). Used in the production of thermal insulating fibers, and for nuclear-power applications and rockets and missiles. Formula: K_2TiO_3.

potassium titanate fibers. Thermal insulating fibers made from potassium oxide and titanium dioxide. The approximate fiber composition is $K_2O \cdot (TiO_2)_x$, where x is 4-7. Available in the form of continuous and staple fibers, mats, fabrics, and other products, it has a density of 3.2 g/cm³ (0.12 lb/in.³), a melting point of 1370°C (2500°F), a high refractive index, an upper continuous-use temperature of 1150°C (2100°F), and can diffuse and reflect infrared radiation. Used for nuclear-power applications, rockets and missiles, and as a thermal insulator.

potassium-titanium fluoride. See titanium-potassium fluoride.

potassium titanyl phosphate. A nonhygroscopic compound of potassium, titanium, phosphor and oxygen supplied as high-purity crystals with high nonlinear coefficient and high damage threshold, and used as high-frequency doubler in Nd:YAG lasers, and in optoelectronics as a modulator in waveguides. Formula: $KTiOPO_4$. Abbreviation: KTP.

potassium trisulfide. See potassium sulfide (ii).

potassium uranate. A compound of potassium oxide and uranium trioxide that is available as orange-yellow crystals. Crystal system, orthorhombic. Melting point, 1620°C (2948°F). Used for ceramic and nuclear applications. Formula: K_2UO_4.

potassium-yttrium halide. Any of several compounds of potassium and yttrium with any of the halogens. Examples are potassium yttrium chloride, potassium yttrium fluoride, etc. Used for laser crystals, and in electronics.

potassium-zinc silicate. A compound of potassium oxide, zinc oxide and silicon dioxide with a melting point of 1297°C (2367°F). Used in ceramics. Formula: K_2ZnSiO_4.

potassium zirconifluoride. See zirconium-potassium fluoride.

potassium zirconium chloride. See zirconium-potassium chloride.

pot clay. A refractory clay used in the production of melting pots and crucibles for the glass industry.

pot earth. See potter's clay.

pot glass. See pot metal (3).

Potingris. French trade name for an alloy made from *Potinjaune* by adding more lead and tin. Used for hardware.

Potinjaune. French trade name for a yellow brass composed of 72% copper, 24.8% zinc, 2% lead and 1.2% tin, and used for hardware and in the manufacture of other alloys, such as Potingris.

pot metal. (1) A bearing alloy containing 67-80% copper, and 20-33% lead, and small amounts of tin or zinc.

(2) Cast iron used in the manufacture of pots, and other hollow ware.

(3) A glass melted in a pot or crucible. Also known as *pot glass*.

(4) A glass that is colored while in the melting pot.

Potomac. Trade name of AL Tech Specialty Steel Corporation (USA) for a series of chromium-type hot-work tool steels including *Potomac* ((AISI types H12), *Potomac A* (AISI types H11 and *Potomac M* (AISI types H13) containing about 0.32-0.40% carbon, up to 1.00% silicon, up to 0.30% manganese, 5.00-5.25% chromium, 1.00-1.75% molybdenum, up to 1.50% tungsten, 0.20-1.00% vanadium, and the balance iron. They have good deep-hardening properties, high toughness, good machinability, high hot hardness, good abrasion resistance, and are used for hot-forging dies, hot-extrusion tooling, hot-nut forming tools, and for aircraft, missile and rocket cases.

potosiite. A mineral of the cylindrite group composed of lead tin iron antimony sulfide, $Pb_{24}Sn_9Fe_4Sb_8S_{56}$. Crystal system, triclinic. Density, 6.20 g/cm³. Occurrence: Bolivia.

potosi silver. A highly corrosion resistant nickel silver containing varying amounts of copper and nickel used for ornamental and architectural applications.

potter's clay. A pure, highly plastic, non-fissile, iron-free clay, such as ball clay, that fires to a clean, white, ivory or buff color, and is suitable for making pottery on a potter's wheel. Also known as *argil; pot earth; potter's earth*.

potter's flint. A finely ground flint used as an ingredient in ceramic products (e.g., earthenware or pottery) to minimize shrinkage due to drying and firing.

potter's red cement. An intimately ground mixture of Portland cement and crushed and sintered red clay used for decorative applications.

pottery. A generic term for all burnt clayware including vases, dishes, bowls, plates and pots formed from moist clay and hardened by firing. It usually does not include technical, structural and refractory clay products, porcelain and stoneware.

pottery body stains. Finely divided pigments usually composed of calcined oxides used for coloring ceramic bodies, e.g., chinaware, terra cotta and tile.

pottery plaster. A plaster suitable for making molds for pottery and ceramic wares.

potting compounds. See potting materials.

potting materials. A group of materials including thermosetting insulating polymers (e.g., alkyds, epoxies, urethanes and silicones) and ceramic compounds (e.g., alumina and magnesia) used to protect potted electrical or electronic components from shock, vibration, air, moisture, etc. Also known as *potting com-*

pounds.

poubaite. A silver white mineral of the tetradymite group composed of lead bismuth selenide telluride, $PbBi_2(Se,Te)_4$. Crystal system, rhombohedral (hexagonal). Density, 7.88 g/cm^3. Occurrence: Czech Republic.

poughite. A yellow mineral composed of iron tellurite sulfate trihydrate, $Fe_2(TeO_3)_2SO_4 \cdot 3H_2O$. Crystal system, orthorhombic. Density, 3.76 g/cm^3; refractive index, 1.985. Occurrence: Mexico.

pounce. Pumice in the form of a very fine powder used to stop ink from spreading on a writing surface, prepare tracing cloth and parchment for writing, and in the graphic arts to transfer designs through stencils. See also *pumice.*

pouncing paper. Paper coated with *pumice* powder, and used for fine polishing.

pouring-pit refractory. (1) Any of a group of refractory products (e.g., nozzles, sleeves, stoppers, ladle brick, etc.) employed in the transfer and flow control of steel from the melting furnace to the mold or ingot.

(2) Any refractory product used in the casting of molten metal.

pour-in-place concrete. See cast-in-place concrete.

Pour-N-Cure. Trade name for a self-cure dental acrylic.

Povorex. Trademark of Low-Density Products Limited (Canada) for cellular concrete blocks.

powder. (1) A general term for dry, finely divided particles obtained from a solid substance either mechanically by comminution (e.g., pounding, crushing or grinding), chemically by precipitation or combustion, or by other processes. Examples include inorganic pigments, ceramic and metal powders and plastic molding powders.

(2) A metallic element or alloy in the form of a powder with a particle size range of about 0.1 to 1000 µm (4 µin. to 0.0394 in.). Used as a raw material for powder-metallurgy parts, in metal-spraying, for paint pigments, for brazing materials, in calorizing compounds and foundry alloys, and as a catalyst.

powder blend. A dry, free-flowing plastic molding compound containing all required ingredients, and ready for use in the molding operation. Also known as *dry blend.*

powder blue. A mixture of cobalt oxide, silicon dioxide and a fluxing agent, such as potassium carbonate, employed as a blue colorant in the manufacture of glass, glazes and vitreous enamels.

powder-bonded nonwoven fabrics. Nonwoven fabrics produced by a process in which thermal bonding is achieved by dispersing a heat-sensitive powder in the fiber batt or web. Also known as *powder-bonded nonowovens.* See also thermally bonded nonwoven fabrics.

powder coatings. A group of protective and/or electrically insulating organic coatings produced on suitable items (e.g., wire goods, metal parts, appliance housings, or chemical and laboratory equipment) by first applying a fine, dry polymer or resin powder (e.g., acrylic, epoxy, fluorocarbon, polyamide, polyester, polyethylene, polyurethane or polyvinyl chloride) by processes, such as electrostatic spraying, fluidized-bed or electrostatic fluidized-bed techniques, and then fusing the powder particles at temperatures above the melting point of the powder.

powder compact. See compact.

powdered activated carbon. Activated carbon in particle sizes primarily 80 mesh or smaller. See also activated carbon.

powdered coal. A highly volatile, intimately ground coal used as a fuel for heating industrial furnaces, ovens and kilns.

powdered emery. Dark, crystalline aluminum oxide (Al_2O_3) in the form of a fine powder used as a grinding and polishing abrasive.

powdered flux. See welding powder.

powdered glass. Ground glass used a filler for plastics and coatings.

Powdered Iron. Trade name of Connelly-GPM, Inc. (USA) for *sponge iron* supplied in the form of aggregate or filings for use as a catalyst and purifier.

powdered lime. See air-slaked lime.

powdered metals. Metals made into powdered form by any of various processes including chemical precipitation, comminution, condensation, electrodeposition, hydrometallurgy, gas or water atomization, granulation, machining, milling, reduction and thermal decomposition. Used as raw materials for the manufacture powder-metallurgy parts, in metal spraying, for hardfacing and brazing, as master alloys and alloying additions, as catalysts, as paint and ink pigments, in pyrotechnics, etc. Also known as *powder metals.*

powdered polymers. Thermoplastic and thermosetting polymers made in powdered form. Powdered thermoplastics are used in powder molding, a technique that produces articles by melting a plastic powder against the inside of a mold, and powdered thermosets are used in the production of sprayed coatings on automobiles, and industrial equipment.

powdered rubber. See rubber powder.

powdered talc. See French chalk.

powder-forged products. Fully dense ferrous or nonferrous powder-metallurgy products formed by first pressing loose, blended or unblended powders into preforms of the desired shapes and contours, sintering at high temperatures under prescribed conditions, and then forging in a standard forging press.

powder insulation. A finely divided material, such as asbestos, porcelain, or a glass or plastic, held between a hot and a cold surface to serve as thermal insulation in decreasing the transfer of heat between the surfaces by convection and radiation.

powder lubricant. A substance blended with or otherwise included into a ceramic or metal powder to reduce friction between the particles and between the particles and a mold, die, punch, core rod, etc. and thus simplify pressing and subsequent ejection of the compact.

powder-metallurgy brass. See sintered brass.

powder-metallurgy bronze. See sintered bronze.

powder-metallurgy copper. See sintered copper.

powder-metallurgy copper iron. See sintered copper iron.

powder-metallurgy copper steel. See sintered copper steel.

powder-metallurgy forgings. (1) Powder-metallurgy parts produced from green, presintered or fully sintered compacts by additional working, usually involving hot or cold forming methods. Abbreviation: P/M forgings.

(2) Powder-metallurgy parts processed to the desired shape by conventional forging of powders enclosed in hermetically sealed containers. Abbreviation: P/M forgings.

powder-metallurgy infiltrated steel. A high-density steel containing about 0.3-1.0% carbon and 8-25% copper, prepared by infiltrating the copper alloy into the porous steel matrix during sintering for a specified period of time at about 1120°C (2048°F) in an endothermic gas atmosphere. The infiltrated steel has a final density of about 7.3 g/cm^3 (0.26 $lb/in.^3$), is nearly fully dense, and has high strength, enhanced mechanical properties and good machinability. Abbreviation: P/M infiltrated steel. Also known as *infiltrated steel; P/M infiltrated steel.*

powder-metallurgy iron. See sintered iron.

powder-metallurgy iron-copper. See sintered iron-copper.

powder-metallurgy iron-nickel. See sintered iron-nickel.

powder-metallurgy material. See sintered material.

powder-metallurgy nickel silver. See sintered nickel silver.

powder-metallurgy nickel steel. See sintered nickel steel.

powder-metallurgy part. See sintered part.

powder-metallurgy stainless steels. See sintered stainless steels.

powder-metallurgy steels. See sintered steels.

powder-metallurgy titanium. See sintered titanium.

powder metals. See powdered metals.

powder-molded plastics. See powder moldings.

powder moldings. Plastic parts of varying shapes and sizes usually made by techniques in which powdered starting materials are essentially melted against the cavities of stationary or rotational molds. Also known as *powder-molded plastics*.

powder paint. A type of organic coating composed of plastic resins, pigments and additives and made by blending the aforementioned ingredients together into a uniform mixture, heating to the melting point, extruding into thin sheets, cooling, and finally grinding into a fine powder of controlled particle size and optimum flowability.

Powdiron. Trade name of GKN Powder Met Inc. (USA) for a series of powder-metallurgy parts made from iron powder or a blend of iron powder, carbon (or graphite) and/or copper by pressing and sintering, sometimes followed by oil infiltration. Used for bearings, structural parts, wheels, toys, automotive components, and machine-tools.

powellite. A colorless mineral of the scheelite group composed of calcium molybdenum oxide, $CaMoO_4$. It can also be made from calcium chloride ($CaCl_2$) and sodium molybdate (Na_2MoO_4) under prescribed conditions. Crystal system, tetragonal. Density, 4.23 g/cm^3; refractive index, 1.974. Occurrence: Siberia, USA (California, Idaho, Michigan, Nevada, Texas). Used as a ore of molybdenum.

Power. (1) Trade name of Magnolia Metal Corporation (USA) for a cast alloy composed of tin, copper and antimony, and used for bearings for heavy loads.

(2) Trade name for a soft magnetic alloy containing 0-5% silicon, and the balance iron. It has high magnetic permeability, and is used for laminations in power units.

Power Bond. Trademark of Industrial Formulators of Canada Limited for a polysulfide bonding agent for concrete.

Powerclad. Trade name of Sherwin Williams Company (USA) for cathodic and anodic electrodeposition coatings used in metal finishing.

Powercron. Trade name of PPG Industries Inc. (USA) for electrocoatings.

Power-Prime. Trade name of PPG Industries, Inc. (USA) for primers for electrocoatings used on metal substrates.

Powersilk. Trade name of BASF Corporation (USA) for nylon 6 fibers used for textile fabrics including clothing.

Powersteel. Trade name of Crobalt Inc. (USA) for a cast steel containing 0.9% carbon, 4% cobalt, and the balance iron. Used for cutters, and tools.

Power-Tread. Trade name of Tnemec Company Inc. (USA) for industrial polymer floor topping systems.

Power-Ty. Trademark of Poli-Twine Division of Niagara Structural Steel Limited (Canada) for baler twine.

Power-Weld. Trade name of Magnolia Metal Corporation (USA) for nickel genuine babbitt.

Powhatan. Trade name of Atlas Specialty Steels (Canada) for a tungsten-type high-speed steel (AISI type T4) containing 0.83% carbon, 18% tungsten, 5-10.5% cobalt, 4% chromium, 1-1.7% vanadium, 1% molybdenum, and the balance iron. It has high wear resistance and red hardness, good deep-hardening properties, low toughness, relatively poor machinability, and is used for lathe, planer and various other cutting tools.

Pow'r Pump. Trademark of Bondfast Inc. (USA) for clear and white adhesive sealants supplied in a cartridge with incorporated application system for interior and exterior applications.

Poxycol. Trademark of Mulco Inc. (USA) for an epoxy-polymer adhesive used for concrete and metal protection applications.

Poxy-gard. Trade name of Pratt & Lambert, Inc. (USA) for epoxy-based rust inhibitive primers and maintenance coatings.

poyarkovite. A deep raspberry to cherry-red mineral that turns black on exposure to air. It is composed of mercury chloride oxide, Hg_3ClO. Crystal system, monoclinic. Density, 9.50 g/cm^3. Occurrence: Kirgizian Republic.

pozzolan. (1) A special ingredient, usually a finely divided burnt clay or shale, that is added to certain cements to facilitate hardening underwater.

(2) A finely divided material, such as certain blast-furnace slags, fly ashes, etc., that will exhibit cementitious properties when combined with lime in the presence of water. Used in cements.

(3) A natural cement of volcanic origin that occurs in powdered form near Pozzuoli in Italy, on the Grecian island of Santorin, and in the Rhine valley in Germany and many other locations in Central Europe. It is composed essentially of silica and alumina, and used in cement manufacture. Also known as *pozzolan cement; pozzoulana cement*.

pozzolan aggregate mixture. A mixture of pozzolan and other lightweight aggregates used in the manufacture of certain concretes and mortars. Abbreviation: PAM.

pozzolan cement. See pozzolan (3).

Pozzolith. Trade name of Master Builders, Inc. (USA) for an air-entraining agent consisting of a mixture of calcium lignosulfonate, sodium lauryl sulfate, calcium chloride and fly ash. Used as an admixture in concrete. See also air-entraining agent; air-entrained concrete.

pozzoulana cement. See pozzolan (3).

PPE-HIPS alloys. Alloys of about 50-80 wt% polyphenylene ether (PPE) and high-impact polystyrene (HIPS) with good melt mixability in all proportions, good processibility, low viscosity, low coefficients of thermal expansion, outstanding heat and chemical resistance and improved toughness. Used for automotive, equipment and other applications. See also HIPS-PPE alloys.

PPE-PA alloys. Polymer blends of an amorphous polyphenylene ether (PPE) and a crystalline polyamide (PA) having good thermal properties and chemical resistance. Used for automotive applications, e.g., body panels.

PP-EPDM alloys. Polymer blends of crystalline polypropylene (PP) and elastomeric ethylene-propylene-diene monomer (EPDM). They combine the impact resistance, elongation and resiliency of EPDM elastomers with the chemical resistance, toughness, strength and other structural properties of PP resins. Used for automobile bumper parts, power-tool housings, sporting goods, etc.

P Pfersee. Trademark of Chemische Fabrik Pfersee GmbH (Germany) for natural and synthetic resins supplied in paste, powder and liquid form.

PPG. Trade name of PPG Industries, Inc. (USA) for an extensive series of glass fibers and glass-fiber products.

PPO-nylon alloys. See PPO-PA alloys.

PPO-PA alloys. Immiscible blends of amorphous polyphenylene oxide (PPO) and crystalline polyamide (PA). They possess enhanced chemical resistance, improved dimensional stability at elevated temperatures, low moisture sensitivity and mold shrinkage, very good paint adhesion and good resistance to chipping. Used for automobile doors, hatchbacks and wheel panels. Also known as *PPO-nylon alloys*.

PPO-PBT alloys. Engineering thermoplastics based on blends of amorphous polyphenylene oxide (PPO) and semicrystalline polybutylene terephthalate (PBT). See also PPO-polyester alloys.

PPO-polyester alloys. Engineering thermoplastics based on polymer blends of amorphous polyphenylene oxide (PPO) and crystalline polyester (e.g., polybutylene terephthalate or polyethylene terephthalate). They combine the good chemical resistance, low warpage and high stiffness of polyesters with the good impact resistance, high toughness and outstanding energy absorption characteristics of polyphenylene oxides. Used for exterior automotive parts, e.g., body panels.

PPO-PS alloys. Engineering thermoplastics based on miscible polymer blends of amorphous polyphenylene oxide (PPO) and amorphous polystyrene (PS). They combine the high-temperature resistance and good toughness and chemical resistance of PPO with the good mechanical and thermal properties (including impact strength) of PS. Their glass-transition temperature is about 155°C (311°F), and they are used for automobile interiors (e.g., instrument panels and seat backs) and exteriors (e.g., wheel covers and mirror housings), business machines (e.g., keyboard and printer bases and terminal housings), household appliances, and power tools.

PQ Hollow Spheres. Trade name of PQ Corporation (USA) for lightweight hollow *Q-cel* glass spheres used as additives for plastic compounds, ceramics, refractories, etc.

Prague red. A red pigment consisting essentially of ferric oxide (Fe_2O_3).

prase. A translucent, dull-green *chalcedony* quartz used as a gemstone.

praseodymia. See praseodymium oxide.

praseodymium. A malleable, ductile yellowish metallic element of the lanthanide series (rare-earth group) of the Periodic Table. It is commercially available in the form of ingots, lumps, sheets, rods, foil, turnings (chips), powder and wire. The commercial praseodymium ores are allanite, bastnaesite, cerite and monazite, and it is also obtained as a fission product. Density, 6.782 g/cm³; melting point, 931°C (1708°F); boiling point, 3512°C (6354°F); hardness, 40 Vickers; atomic number, 59; atomic weight, 140.908; trivalent, tetravalent. It has high susceptibility to tarnishing, exhibits paramagnetic properties, and has low electrical conductivity (about 2.5% IACS). There are two allotropic forms: (i) *Alpha praseodymium* (hexagonal closed-packed crystal structure). Density, 6.782 g/cm³; stable up to 798°C (1468°F); and (ii) *Beta praseodymium* (body-centered cubic crystal structure). Density, 6.64 g/cm³; stable from the transformation temperature of 798°C (1468°F) to the melting point of 931°C (1708°F). Used as a scavenger (oxygen, sulfur, etc.) for ferrous alloys, as an ingredient of mischmetal (e.g., for lighter flints), for permanent magnets (e.g., praseodymium-cobalt, PrCo₅), phosphors, superconductors and lasers, as a core material for carbon arc-lights, in searchlights, as a coloring agent (greenish-yellow) for glazes, glasses and ceramics, and as a petroleum cracking catalyst. Symbol: Pr.

praseodymium barium copper oxide. A fine powder that unlike other rare-earth barium cuprates (e.g., yttrium barium copper oxide) is nonconducting and nonmetallic. Used as a thin-film material in high-temperature superconducting devices, and as a barrier layer in tunnel junction devices. Formula: $PrBa_2Cu_3O_7$. Abbreviation: PrBCO, Pr-123.

praseodymium boride. A compound of praseodymium and boron used in ceramics and materials research: (i) *Praseodymium tetraboride*. Density, 5.20 g/cm³. Formula: PrB_4; and (ii) *Praseodymium hexaboride*. Black crystals. Crystal system, cubic. Density, 4.84-4.86 g/cm³; melting point, 2610°C (4730°F). Formula: PrB_6.

praseodymium bromide. Green, hygroscopic crystals or powder. Density, 5.28 g/cm³; melting point, 693°C (1279°F). Used in ceramics and materials research. Formula: $PrBr_3$.

praseodymium carbide. Any of the following compounds of praseodymium and carbon used in ceramics and materials research: (i) *Praseodymium dicarbide*. Yellow crystals. Density, 5.73 g/cm³; melting point, 2535°C (4595°F). Formula: PrC_2; and (ii) *Praseodymium sesquicarbide*. Cubic crystals. Density, 6.62 g/cm³. Formula: Pr_2C_3.

praseodymium chloride. Blue-green, hygroscopic needles or powder (99.9+% pure). Crystal system, hexagonal. Density, 4.02 g/cm³; melting point, 786°C (1447°F); boiling point, 1700°C (3092°F). Used in ceramics and materials research. Formula: $PrCl_3$.

praseodymium chloride heptahydrate. Green, hygroscopic crystals (99.9% pure). Crystal system, triclinic. Density, 2.250 g/cm³; melting point, 115°C (239°F). Used in ceramics and materials research. Formula: $PrCl_3 \cdot 7H_2O$.

praseodymium dicarbide. See praseodymium carbide (i).

praseodymium dioxide. A brown-black powder used in chemistry, ceramics (e.g., ceramic colors) and materials research. Formula: PrO_2.

praseodymium disulfide. See praseodymium sulfide (ii).

praseodymium fluoride. Green crystals or light blue powder (99.9+% pure). Crystal system, hexagonal. Density, 6.3 g/cm³; melting point, 1395°C (2543°F). Used in ceramics and materials research. Formula: PrF_3.

praseodymium hexaboride. See praseodymium boride (ii).

praseodymium iodide. Orthorhombic crystal. Density, 5.8 g/cm³; melting point, 737°C (1359°F). Used in ceramics and materials research. Formula: PrI_3.

praseodymium monosulfide. See praseodymium sulfide (i).

praseodymium nitride. A compound of praseodymium and nitrogen available as cubic crystals with a density of 7.46-7.49 g/cm³. Used in ceramics and materials research. Formula: PrN.

praseodymium oxalate. Light-green, hygroscopic crystals or powder (99.9+% pure) used in ceramics and materials research. Formula: $Pr_2(C_2O_4)_3 \cdot 10H_2O$.

praseodymium oxide. (1) See praseodymium dioxide; praseodymium trioxide.

(2) A black, hygroscopic powder (99.9+% pure) used in the production of yellow and green ceramic colors. Formula: Pr_6O_{11}.

praseodymium phosphide. A compound of praseodymium and phosphorus used in ceramics, electronics and materials research. Density, 5.72 g/cm³. Formula: PrP.

praseodymium-ruthenium. A compound of praseodymium and ruthenium used in ceramics and materials research. Melting point, 1680°C (3056°F). Formula: $PrRu_2$

praseodymium sesquicarbide. See praseodymium carbide (ii).

praseodymium sesquisulfide. See praseodymium sulfide (iii).

praseodymium silicate. A compound of praseodymium oxide and silicon dioxide used in ceramics and materials research. Melting point, 1398°C (2548°F). Formula: Pr_2SiO_5.

praseodymium silicide. A compound of praseodymium and silicon available as tetragonal crystals with a density of 5.64 and a melting point of 1712°C (3114°F). Used in ceramics and materials. Formula: $PrSi_2$.

praseodymium sulfide. Any of the following compounds of praseodymium and silicon used in ceramics, electronics and materials research: (i) *Praseodymium monosulfide.* Density, 6.03 g/cm³; melting point, 2230°C (4046°F). Formula: PrS; (ii) *Praseodymium disulfide.* Density, 5.16 g/cm³; melting point, 1782°C (3240°F). Formula: PrS_2; (iii) *Praseodymium sesquisulfide.* Brown crystals or powder. Crystal system, orthorhombic. Density, 5.04-5.27 g/cm³; melting point, 1765°C (3209°F); band gap, 1.68 eV. Formula: Pr_2S_3; and (iv) *Praseodymium tetrasulfide.* Density, 5.77 g/cm³; melting point, 2100°C (3812°F). Formula: Pr_3S_4.

praseodymium telluride. A compound of praseodymium and tellurium available as cubic crystals with a density of 7.0 g/cm³ and a melting point of 1500°C (2732°F). Used in ceramics, electronics and materials. Formula: $PrTe_2$.

praseodymium tetraboride. See praseodymium boride (ii).

praseodymium tetrasulfide. See praseodymium sulfide (iv).

praseodymium trioxide. A white, hygroscopic crystals or yellow-green, hygroscopic powder (99.9+% pure). Crystal system, hexagonal. Density, 6.9-7.07 g/cm³; melting point, 2200°C (3992°F); absorbs carbon dioxide from air. Used as a glass and ceramic pigment, in materials research, and as a laboratory reagent. Formula: Pr_2O_3. Also known as *praseodymia; praseodymium oxide.*

praseodymium yellow. A ceramic colorant composed of a mixture of silica, zirconia and approximately 5% of praseodymium oxide. Used in the production of yellow colors in glazes.

prealloyed powder. A metal powder made up of at least two different elements and alloyed during powder manufacture. Each particle is of the same chemical composition as the entire powder batch. Examples include steel powders containing iron, carbon, silicon, manganese, nickel and chromium, blended as required, and brass powders containing copper and zinc and, if required, lead in the proper proportions. Abbreviation: PAP.

preblended rubber. Rubber blended with bitumen prior to use in building and road construction.

precast concrete. Concrete cast in some shape in forms or molds in a factory or at the job site prior to fixing into final position in a structure. It can be ordinary concrete, prestressed concrete or reinforced concrete, and is available in various forms, such as building units (blocks), slabs, or sandwich panels. Also known as *prefabricated concrete.*

precast gypsum block. Gypsum cast into building blocks in forms or molds in a factory prior to fixing into its final position in a building. Also known as *gypsum block.*

Pre-Cat. Trademark of Ciba-Geigy Corporation (USA) for a graphite-fabric-reinforced epoxy resin used for tooling applications.

Precedent. Trade name for a glass-ionomer dental cement.

preceramic polymers. A relatively recent class of chain polymers synthesized from suitable organometallic precursor materials (e.g., siloxanes, silazenes, etc.). Crosslinking at moderately elevated temperatures (typically 100-250°C or 210-480°F) in controlled atmospheres results in the conversion of these polymers into preceramic networks which, after pyrolytic transformation, i.e., heating to high temperatures (typically in the range of 800-1000°C or 1470-1830°F), form amorphous metastable ceramic materials. These can then be converted (crystallized) into stable, homogeneous ceramics. Silicon carbide, silicon nitride, silicon oxynitride and many non-silicon ceramics can be produced from preceramic precursors. See also polymer-derived ceramics.

precious metals. A group of relatively scarce and thus valuable metals, such as gold, silver and platinum, used for coinage, ornaments, jewelry, and similar applications. The terms *precious metals* and *noble metals* are not synonymous, although some precious metals may be noble as well. For example, gold has the characteristics of both noble and precious metals, whereas osmium is considered noble, but not precious. Abbreviation: PM.

precipitated calcium carbonate. A white filler pigment synthesized from *calcite* ($CaCO_3$) by precipitation. It has an average particle size of approximately 0.3-1.2 µm (12-47 µin.), a specific surface area of approximately 8-21 m²/g, and a refractive index of 1.59. Generally, it is available in three morphologies-prismatic, scalenohedral and rhombohedral-with the latter having the highest pigment scattering factor. Used as a brightener in the manufacture of high-quality paper and paper products. Abbreviation: PCC. See also calcium carbonate; chalk (2).

precipitated chalk. See chalk (2).

precipitated iron carbonate. See brown iron oxide.

precipitation-hardenable alloys. See age-hardenable alloys.

precipitation-hardenable austenitic stainless steels. See austenitic precipitation-hardenable stainless steels.

precipitation-hardenable martensitic stainless steels. See martensitic precipitation-hardenable stainless steels.

precipitation-hardenable semiaustenitic stainless steels. See semiaustenitic precipitation-hardenable stainless steels.

precipitation-hardenable stainless steels. A group of wrought *stainless steels* containing about 0.05-0.2% carbon, 11.0-18.0% chromium, 3.0-10.0% nickel, 0.5-1.3% manganese, 0.1-1.0% silicon, additions of copper, aluminum, molybdenum, niobium and titanium, and the balance iron. Available in austenitic, semiaustenitic and martensitic grades, they are hardened by a final aging treatment that precipitates very fine second-phase particles from a supersaturated solid solution. See also austentic precipitation-hardenable stainless steels; semiaustentic precipitation-hardenable stainless steels; martensitic precipitation-hardenable stainless steels.

Precipitator Spin-Glas. Trade name of Johns Manville Company (USA) for fiberglass board insulation used for building and construction applications.

Precise. Trade name for a silicone elastomer used for dental impressions.

Precision. (1) Trade name of Cold Metal Products Company, Inc. (USA) for a plain-carbon structural steel containing 0.3% carbon and the balance iron.

(2) Trade name of Precision Casting Company (USA) for a series of die-cast aluminum-silicon, aluminum-silicon-copper, zinc-aluminum-copper, and magnesium-aluminum alloys, and various die-cast silicon and aluminum brasses.

precision resistance alloys. A group of alloys including nickel-chromium, nickel-chromium-aluminum and nickel-iron that exhibit low electrical resistance changes with temperature making them useful for precision-wound resistors. See also resistance alloys.

precision sheet. Flat-rolled metal of short length, usually at least 610 mm (24 in.) wide and between 0.127 and 0.381 mm (0.005 and 0.015 in.) thick.

precision strip. Flat-rolled metal longer than a precision strip and, usually less than 610 mm (24 in.) wide and between 0.127 mm and 0.381 (0.005 and 0.015 in.) thick.

precoated metal products. Rolled products that have metallic, organic or conversion coatings (phosphate, chromate, oxide, etc.) applied at the mill prior to their manufacture into finished parts.

PreCoCast. Trade name of Alu-Team Form- und Schmiedetechnik GmbH (Germany) for forging-grade aluminum and aluminum alloy castings.

Precolite. Trade name of Precision Convex Manufacturing Company (Philippines) for laminated safety glass.

precursor-derived ceramics. A class of high-purity, high-temperature ceramics prepared from low-molecular-weight inorganic, organometallic or organic precursors, such as polysiloxanes or polysilazanes. See also preceramic polymers; polymer-derived ceramics.

Pre-Fax. Trade name of Unifrax Corporation (USA) for ceramic-fiber reinforcements supplied in the form of preforms for metal-matrix composites.

prefabricated concrete. See precast concrete.

prefabricated masonry. Masonry products made in a factory or other location and shipped to the building site for rapid final assembly.

prefinished metals. Metal products, usually in sheet form, that have been prepainted, preplated, clad or bonded with metal or other materials at the mill to produce decorative and/or protective finishes, and reduce or eliminate the need of subsequent finishing.

preform. (1) The initially pressed powder-metallurgy, green, presintered or fully sintered compact to be subjected to cold or hot repressing.

(2) A preshaped reinforcement of continuous aramid, carbon, glass or other fibers arranged in the form of a cloth or mat and injected or impregnated with hot-melt matrix resin (epoxy, bismaleimide, polyimide, etc.). Also known as *fiber preform*.

(3) A compact pill or tablet used in the molding of thermosets. It is formed into a compact by pressing premixed material to achieve efficiency and accuracy in handling and control of uniformity of charge for mold loading operations. Also known as *biscuit*.

(4) A sintered or prefired compact of glass powder used in the manufacture of glass-to-metal seals.

(5) A filler metal for brazing or soldering, made into a special shape or form for particular applications. Also known as *preformed filler metal*.

preform binder. The hot-melt matrix resin (e.g., epoxy, bismaleimide, polyimide, etc.) used to impregnate or inject the preform described under "preform" (2). It is usually cured to provide shape retention and facilitate handling.

preformed filler metal. See preform (5).

preformed foam. A foam produced in a special foam generation equipment, and already in a foamed state when added to the mixer together with other admixtures in the production of cellular concrete. This is in sharp contrast to conventional gas-forming chemicals or foaming agents that are introduced into concrete to produce lightweight cellular structures by chemical action.

prefused flux. A dry, granular blend of minerals, such as fluorspar and minerals containing certain oxides (e.g., oxides of aluminum, calcium, magnesium, manganese, silicon and titanium), melted and poured at very high temperatures (usually below 3100°C or 5610°F), and then crushed and sized. Used in submerged-arc welding.

Prega. Trade name of Uddeholm Corporation (USA) for an oil-hardening hot-work die steel containing 0.45% carbon, 3.0% chromium, 0.6% manganese, 0.45% molybdenum, and the balance iron.

Preglas. Trade name of Herberts GmbH (Germany) for unsaturated polyesters supplied in several grades including glass-fiber-reinforced.

Preglasit. Trade name of Berliner Glas KG (Germany) for a reflection-reducing glass used for picture glazing applications.

Pregwood. Trade name Formica Corporation (USA) for phenolic-impregnated wood laminates used for building and decorative applications.

prehnite. A colorless, white or pale green mineral with a vitreous luster related to the zeolites. It is composed of calcium aluminum silicate hydroxide, $Ca_2Al_2Si_3O_{10}(OH)_2$. Crystal system, orthorhombic. Density, 2.8-2.95 g/cm^3; hardness, 6.0-6.5 Mohs; pyroelectric properties. Occurrence: Australia (New South Wales), USA (New Jersey). Used in ceramics, electronics, and materials research. The translucent green species are used as gemstones.

preimpregnated material. See prepreg.

preisingerite. A white to grayish white mineral composed of bismuth oxide arsenate hydroxide, $Bi_3O(OH)(AsO_4)_2$. Crystal system, triclinic. Density, 6.00 g/cm^3; refractive index, 2.16. Occurrence: Argentina.

preiswerkite. A pale greenish mineral of the mica group composed of sodium magnesium aluminum silicate hydroxide, $NaMg_2Al(Al_2Si_2O_{10})(OH)_2$. Crystal system, monoclinic. Density, 2.96 g/cm^3; refractive index, 1.614. Occurrence: Switzerland.

Prelana. Trademark of Premnitz (Germany) for acrylic fibers used for the manufacture of textile fabrics.

Preline. Trade name for a calcium hydroxide dental lining cement.

Prelude. Trade name for a light-cure composite used for dental inlays.

Premabraze. Trade name of Handy & Harman (USA) for a series of high-purity specialty brazing filler metals containing varying amounts of either silver, copper and indium, gold and nickel, or gold, silver and palladium. They are used in the average brazing range of 600-1000°C (1110-1830°F).

Premag. Trade name of Murex Limited (UK) for an alloy composed of 99.2% copper, and 0.8% silver. Used for welding rods for welding copper and its alloys.

Premalox. Trademark of Alcoa Inc. (USA) for calcined aluminas (99.8% pure) with a fired density of about 3.76 g/cm^3 (0.136 lb/in.³) used for engineering ceramics.

Premat. Trade name of Turner Brothers Asbestos Company Limited (UK) for preimpregnated glass-fiber mats.

Pre-Max. Trade name of Unifrax Corporation (USA) for ceramic-fiber reinforcements supplied in the form of preforms for metal-matrix composites.

Premet. Trade name of Allegheny Ludlum Steel (USA) for a cast alloy composed of 0.12% carbon, 30% cobalt, 19% tungsten, 5% molybdenum, 2% vanadium, 0.2-0.5% manganese, and the balance iron, used for cutting tools.

Premier. (1) Trademark of Dofasco Inc. (Canada) for continuous galvanized steel sheets and coils.

(2) Trade name of United Clays, Inc. (USA) for a highly plastic, white-fired premium clay mined in Tennessee, USA. It has a low carbon content, good pressing and predictable shrinking properties. Used for wall tile, and for various fast-firing applications.

(3) Trade name for rayon fibers and yarns used for textile fabrics.

Premier Nickel Chrome. Trade name of H.K. Porter Company, Inc. (USA) for a corrosion-resistant nickel alloy containing 25% iron, 14-16% chromium and 0.01% carbon. Used for wire cloth, heating units, dipping baskets, and rheostats.

Premi-Glas. Trademark of Premix, Inc. (USA) for a series of bulk molding compounds (BMC) and sheet molding compounds (SMC). Examples include *Premi-Glas 1100 BMC*, a glass-reinforced isopolyester bulk molding compound (BMC) composed essentially of isophthalic acid, maleic anhydride and styrene monomer, and providing high color stability, impact resistance and stain resistance, and high microwave transparency, *Premi-Glas 1200 VE*, a strong, tough, resilient, moisture-resistant vinyl ester sheet molding compound (SMC) which retains its mechanical properties at high temperatures, and *Premi-Glas SMC*, a compression-molded, glass-reinforced sheet molding compound (SMC) providing excellent resistance to rust and corrosion from hydrocarbons, solvents and detergents, outstanding hydrolytic stability, high impact strength and dimensional stability, very low flammability, and good processibility.

Premium. (1) See Anso Premium.

(2) Trade name for a high-gold dental bonding alloy.

Premium Alkyl. Trademark of Polifarb Becker Debisa SA (Poland) for a high-quality flat, interior acrylic paint for brush, roller or spray application to concrete, plaster and gypsum walls and ceilings.

premium-grade titanium alloys. Titanium alloys of aircraft quality used frequently for the manufacture of jet engines. Also known as *premium-quality titanium alloys*.

premium quality titanium alloys. See premium-grade titanium alloys.

Premium Pennvernon. Trade name of PPG Industries Inc. (USA) for a sheet glass made on an improved Pennvernon drawing machine.

Premix. Trademark of Premix Inc. (USA) for a series of polyester bulk molding compounds (BMC) and sheet molding compounds (SMC).

premix. (1) A uniform, thorough mixture or blend of two or more powders of different composition, ready-to-use in the production of powder-metallurgy compacts. Also known as *premixed powder*.

(2) See prepreg (1).

premixed powder. See premix (1).

Premo. Trade name of Uddeholm Corporation (USA) for an oil-, water- or case-hardening mold steel (AISI type P4) containing 0.04-0.05% carbon, 3.9-4% chromium, 0.5% molybdenum, 0.1% manganese, 0.1% silicon, and the balance iron. It has high core hardness, high strength at elevated temperatures and good toughness. Used especially for plastic mold dies.

Prenite. Trademark of Chimiplast for flexible polyvinyl chlorides.

Prenylon. Trade name for nylon fibers and products.

preobrazhenskite. A colorless or lemon-yellow mineral composed of magnesium borate hydroxide, $Mg_3B_{11}O_{15}(OH)_9$. Crystal system, orthorhombic. Density, 2.45 g/cm^3; refractive index, 1.570. Occurrence: Kazakhstan.

prepacked concrete. Concrete made by placing clean, graded coarse aggregate in a form, compacting and injecting a grout composed of Portland cement and sand into the form to fill the voids. Also known as *preplaced-aggregate concrete*.

prepainted metals. Metal products, usually in sheet form, that have organic surface coatings applied at the mill as decorative and/or protective finishes, and/or to reduce or eliminate the need for subsequent finishing.

Preparakote. Trademark of C-I-L Paints (Canada) for a high-solids primer-surfacer.

prepared calcium carbonate. See precipitated calcium carbonate; chalk (2).

prepared chalk. See chalk (2).

preplaced-aggregate concrete. See prepacked concrete.

preplated metals. Metal products, usually in sheet form, that have been plated at the mill to produce decorative and/or protective finishes, and/or to reduce or eliminate the need of subsequent finishing.

preplied prepregs. Quasi-isotropic multidirectional prepregs fabricated with multiple plies of unidirectional tape oriented in any of the following four directions with respect to the longitudinal fiber axis: (i) parallel (0°); (ii) at a right angle (90°); (iii) diagonal in positive direction (+45°), or (iv) diagonal in negative direction (-45°). See also multidirectional prepregs.

Prep-N-Cote. Trade name of Henkel Surface Technologies (USA) for an iron phosphate coating for steel products.

prepolymer. (1) A polymer with a degree of polymerization intermediate between that of the original monomer reactant or reactants and the final, fully cured product.

(2) The resin precursor of a thermoset.

(3) A reaction intermediate or product of a hydroxyl-containing substance (e.g., a polyol or drying oil) and an isocyanate (e.g., toluene diisocyanate or diphenylmethane diisocyanate) in which the isocyanate is in considerable excess. Used in the production of polyurethane coatings and foams.

pre-post-tensioned concrete. Prestressed concrete in which a portion of the tendons are pretensioned and some are post-tensioned. See also post-tensioned concrete; prestressed concrete; pretensioned concrete.

prepreg. (1) A reinforcing material in sheet form (e.g., cloth, mat or paper), usually containing fillers and other additives, impregnated with thermosetting resin, ready for molding or winding, and stored for use. Also known as *premix; preimpregnated material*.

(2) Fibers preimpregnated with matrix resin in the uncured state. Also known as *preimpregnated material*.

prepreg resin. A thermosetting resin, such as an acrylamate, epoxy, polyimide or vinyl ester, used as a matrix material for preimpregnating continuous or chopped fibers in the manufacture of prepreg and prepreg tow.

prepreg tow. An untwisted bundle of continuous filaments or a strand of fibers (e.g., aramid, carbon and fiberglass) impregnated with thermosetting matrix resin and wound on a core. Also known as *towpreg*.

prereduced iron-ore pellets. Semi-metallized pellets of iron ore produced from low-grade *taconite* ore concentrates by rolling to shape, drying, calcining and roasting in a reducing atmosphere. Used in the manufacture of pig iron. Also known as *semimetallized pellets*.

Prescoloy. Trade name of Pressed Steel Car Company (USA) for a high-strength steel containing 0.2% carbon, 0.9% nickel, and the balance iron. Used for freight car castings and truck side frames.

Prescollan. Trade name for thermoplastic polyester and polyether urethane elastomers supplied in the form of films and foams.

preservative. (1) A substance that extends the service life of a material.

(2) A chemical substance that will prevent or inhibit the development and action of wood-destroying fungi, marine borers other harmful insects, and wood-deteriorating weather conditions.

(3) A substance that when added to uncompounded rubber latex slows down putrefaction and accompanying premature coagulation.

preserved rubber latex. Uncompounded rubber latex that has a preservative added to slow down putrefaction and accompanying premature coagulation.

Pres Glas. Trade name of Birma Products Corporation (USA) for glass insulation and acoustical products supplied in the form of rigid, semi-rigid or flexible boards, and/or molded shapes.

PresMaster. Trademark of F.J. Brodmann & Co. LLC (USA) for pressing powders supplied in pure metal, metal alloy (e.g., steel, superalloys, amorphous or dispersion-strengthened) and intermetallic grades.

preshrunk concrete. (1) Concrete mixed about 1 to 3 hours prior to placement to minimize shrinkage during setting.

(2) Concrete mixed in a stationary mixer before being conveyed to a truck mixer.

preshrunk fabrics. Textile fabrics shrunk during manufacture to reduce or eliminate shrinkage during use.

preshimmed sealant. A sealant containing encapsulated solids or discrete particles that minimize its deformation within a joint when subjected to compressive forces. Used in building construction.

President. Trademark of Coltene-Whaledent (USA) for a series of addition-polymerizing silicone elastomers and vinyl polysiloxane (VPS) dental impression materials. Included are *President Jet-Bite*, a fast-setting, yellow-colored no-slump vinyl polysiloxane (VPS) dental impression material of creamy, thixotropic consistency having high accuracy and end hardness, outstanding dimensional stability, and suitable for use in bite registration, *President MicroSystem*, a silicone elastomer used for dental impressions, *President Putty*, a yellow-colored vinyl polysiloxane (VPS) dental putty with high rigidity, and *President Putty Soft*, a soft, creamy, beige-colored vinyl polysiloxane (VPS) dental putty with excellent handling properties and accuracy.

Pressal-Leim. Trademark of Henkel KGaA (Germany) for melamine-formaldehyde adhesives.

Pressamine. Trade name of Atofina SA (France) for urea-formaldehyde adhesives and impregnating resins, melamine-urea-formaldehyde and melamine-urea-phenol-formaldehyde resins and binders, and phenol-formal adhesives.

Pressant. Trade name of Otto Wolff Handelsgesellschaft (Germany) for a series of case-hardening steels containing 0.15-0.2% carbon, up to 0.25% silicon, 0.4-1.25% manganese, up to 1.15% chromium, and the balance iron. Used for gears, pinions, cams and camshafts.

Pressco. Trade name of Pressco Casting & Manufacturing Corporation (USA) for a series of die-cast leaded brasses, yellow brasses, aluminum bronzes, manganese bronzes and leaded manganese bronzes, and silicon brasses and bronzes.

Press-Die. Trade name of A. Finkl & Sons Company (USA) for a precipitation-hardened die steel containing 0.18-0.23% carbon,

3.25-3.45% molybdenum, 0.60-0.80% manganese, 3.00-3.50% nickel, 0.20% chromium, 0.05-0.08% vanadium, and the balance iron. Used for forging and upsetter dies and inserts, and piercing and punching dies.

pressed bar. A green powder-metallurgy compact in the form of a bar.

pressed brick. Brick usually made from a clay with low moisture content (5-7%) by densifying in molds under high pressure to eliminate imperfections of shape and texture prior to firing. It has improved homogeneity and strength.

pressed brick clay. A stiff clay of uniform color and low moisture content (5-7%) suitable for pressed brick.

pressed fuel. A fuel, such as coal dust or charcoal, often combined with other ingredients (e.g., tar) and pressed into a small cake or block. See also briquette.

pressed glass. Glass formed between a plunger and a mold by the application of pressure while in the plastic (molten) state.

pressings. (1) Metal or metal alloy products of varying shape and size formed by press forging, i.e., by shaping the preformed starting materials in forging-press dies by the gradual application of pressure to allow them to fully fill the die cavities.

(2) Metal or metal alloy sheets and plates produced by shallow drawing.

Press'n Sand. Trademark of 3M Company (USA) for adhesive-backed coated abrasives.

Pressurdie. Trade name of CCS Braeburn Alloy Steel (USA) for a series of chromium-type hot-work tool steels (AISI types H10 to H19) containing about 0.35-0.40% carbon, 3.25-5% chromium, 0.4-2% vanadium, 0-4.25% tungsten, up to 2.5% molybdenum, 0-4.25% cobalt, and the balance iron. They possess good deep-hardening properties, high hot hardness, high toughness, good wear resistance, and are used for dummy blocks, dies, mandrels, die inserts, punches, and hot shears.

pressure bag moldings. Reinforced plastic moldings produced by first securing suitable flexible bags over resin-impregnated reinforcements placed on molds, applying fluid pressure (e.g., air or water pressure) to the bags, and then curing the pressed materials.

pressure castings. Metal or metal-alloy castings made by a process, such as centrifugal, cold-chamber pressure, die or squeeze casting or injection molding, that involves the application of pressure to the molten or plastic materials.

pressure-hydrated lime. See autoclaved lime.

pressure-sensitive adhesives. A group of adhesives that develop maximum bond strength at room temperature with the application of very light pressure. They are usually viscous materials, such as natural rubber, synthetic rubber (e.g., styrene-butadiene, butyl, butadiene-acrylonitrile or polyacrylate rubber), thermoplastic elastomers, etc., supplied as water-base, hot-melt or organic-solvent-base systems, either unsupported or supported on a substrate (e.g., paper, plastic films, fabrics or metallic foils). *Pressure-sensitive adhesives* have permanent room-temperature tack, usually fair to poor heat resistance, and are used for labels, and in tape form in the construction, electrical, automotive and appliance industries.

pressure-sensitive adhesive paper. (1) Paper that has a coating on one side which makes it permanently adhesive. The adhesive coating is usually protected until usage by a sheet of anti-adhesive paper or, when supplied in the form of rolls, with a special anti-adhesive treatment of the noncoated surface.

(2) Paper coated with a dry adhesive that becomes adhesive by the application of pressure.

pressure-sensitive tape. A backing material in the form of a tape coated on one side with an adhesive that with the application of very light pressure sticks instantaneously to many surfaces. For example, a polytetrafluoroethylene film with a silicone adhesive backing.

pressure-sintered materials. Ceramic, metal or metal-alloy products usually formed from powdered starting materials by the application of relatively low pressure, but high continuous or discontinuous sintering temperatures.

Pressure Tite Iron. Trade name of Jenney Cylinder Company (USA) for a centrifugally cast iron containing 2.85% total carbon, 0.65% combined carbon, 0.8% manganese, 1.6% silicon, 1.25% nickel, 0.4% chromium, and the balance iron. Used for liners, engine sleeves and oil pumps.

pressure-treated lumber. Lumber subjected to a special process that forces preservatives deep into the cells of the wood. It has good resistance to rot, insects and weathering, and is well suited for exterior applications even in exposed situations, e.g., decks and porches, and for applications requiring soil contact.

Prestem. (1) Trade name of Heppenstall Company (USA) for a prehardened hot-work tool steel containing 0.3% carbon, 3-3.2% nickel, 3.4% molybdenum, and the balance iron. Used for tools and dies.

(2) Trade name of Creusot-Loire (France) for hot-work die steel containing 0.2% carbon, 0.65% manganese, 0.30% silicon, 3.15% nickel, 3.4% molybdenum, and the balance iron.

Prestige. Trademark of Domtar Inc. (Canada) for bond and offset paper.

Prestite. Trademark of Westinghouse Electric & Manufacturing Company (USA) for unglazed, nonporous porcelain-type ceramic products made from blends of ball clay, feldspar, flint and kaolin. They have good dielectric properties, zero moisture absorption, good compressive strength, good resistance to many acids, poor resistance to hot concentrated alkalies, and are used for molded parts, electrical insulators, and electrical parts.

Pres-Tite. Trademark of Lepage's Limited (Canada) for a rubber-base contact cement.

Presto. (1) Trade name of Carpenter Technology Corporation (USA) for a low-alloy tool steel containing 1.0% carbon, 1.4% chromium, and the balance iron. Used for machine parts and tools.

(2) Trade name of Carpenter Technology Corporation (USA) for a chromium alloy steel containing 1.4% carbon, 1% chromium, and the balance iron. Used for ball and roller bearings.

(3) Trademark of Remystahl (Germany) for a series of tungsten and cobalt-tungsten high-speed steels.

prestressed concrete. Concrete into which compressive stresses of such magnitude have been introduced by pretensioning or post-tensioning steel bars, strands, wires or rods that the tensile stresses resulting from the service loads are offset to a desired degree. Used for highways and railway bridges. Abbreviation: PC.

prestressing steel. Any high-strength steel suitable for use in bar, strand, wire or rod form in prestressed concrete.

pretensioned concrete. Concrete cast around steel bars, rods, strands or wires that are under tension. This tension compresses the concrete increasing its strength accordingly. The bars, rods, strands, or wires can also be bent to exert force in any direction in order to offset the effects of the pressure of a load (e.g., flexure, cracking, etc.) on the concrete. Used in the manufacture of buildings requiring high load-carrying capacity.

Preussag Anthrazit. Trade name of Preussag AG (Germany) for a high-grade carbon concentrate used in heat generation and water filtration.

Preuss alloy. A magnetic alloy composed of 0.04-0.06% carbon, 0.2-1.5% silicon, 30% cobalt, and the balance iron. It has high magnetic permeability, and is used for magnetic circuits.

Prevail. Trade name of Dow Chemical Company (USA) for engineering thermoplastic blends of acrylonitrile-butadiene-styrene and polyurethane, used for a wide variety of injection-molded products including automotive components.

Prevex. Trade name of GE Plastics (USA) for polyphenylene-oxides and polyphenylene ether/polystyrene alloys with good creep resistance and chemical resistance, processibility and weatherability, supplied in unmodified, glass fiber-reinforced, fire-retardant, and structural-foam grades. Used for office office equipment and automotive components including instrument panels and trim.

PreVISION. Trade name for a dental cement used for temporary crown and bridge restorations.

prevulcanized rubber latex. Rubber latex whose particles have been partially vulcanized. Rubber articles and films can be produced from this latex by simple drying. See also rubber latex.

Prewesta. Trade name of Stahlwerk Stahlschmidt GmbH & Co. (Germany) for a series of tough, oil-hardening tools steels containing about 0.35-0.55% carbon, 1.8% tungsten, 0-1% chromium, 0-1% nickel, 0.18% vanadium, and the balance iron. Used for cold punches, crimpers, and upsetters.

Prexi. Trade name of Uddeholm Corporation (USA) for an oil- or case-hardening mold steel containing 0.15% carbon, 1.1-1.2% chromium, 1% manganese, 0.25% silicon, 0.25% molybdenum, and the balance iron. Used for plastic mold dies and die-casting dies.

Prezenta. Trade name for rayon fibers and yarns used for textile fabrics.

priceite. A white mineral composed of calcium borate heptahydrate, $Ca_4B_{10}O_{19} \cdot 7H_2O$. Density, 2.41-2.42 g/cm³; refractive index, 1.59. Occurrence: USA (California), Turkey.

priderite. A black mineral of the rutile group composed of potassium titanium iron oxide, $(K,Ba)(Ti,Fe)_8O_{16}$. Crystal system, tetragonal. Density, 3.86 g/cm³; refractive index, above 2.10. Occurrence: Western Australia.

Prima. German trade name for a series of water-hardened plain-carbon tool steels used for hand tools, drills, taps, reamers, dies, springs, rails, gears, shafts and bolts.

Primablend. Trade name of Prima Plastics (USA) for a series of blends of polycarbonate and acrylonitrile butadiene-styrene, polycarbonate and polymethyl methacrylate, or polycarbonate and polyester. They are available in unreinforced and glass-fiber-reinforced grades.

Primacor. Trademark of Dow Chemical Company (USA) for a series of water-dispersible ethylene-acrylic acid copolymers supplied in blown film, cast film, extrusion and dispersible resin grades. They have high melt indices, good dispersibility, high optical clarity, low heat-sealing temperature, good adhesion to glass, metal foil and many other substrates, good resistance to tears, punctures, moisture, air, grease and chemicals. Used as sealing films, e.g., for packaging fatty and greasy products. Extrusion coating grades are used in the manufacture of drink cartons, toothpaste tubes, and wire and cable.

Primalit. Trade name of Esperanza SA (Spain) for hollow glass blocks.

Primalith. Trade name of Saint-Gobain (France) for glass blocks consisting of two hollow half-blocks welded together.

Primallor G. Trademark of Degussa-Ney Dental (USA) for an extra-hard, fine-grained dental casting alloy (ADA type IV) containing 67.5% gold, 13.5% silver, 3% palladium and 1% platinum. It has very high mechanical strength, a hardness of 260 Vickers, and is used in restorative dentistry for full cast crowns, short- and long-span bridges, cast partial dentures and implants.

Primaloft. Trade name for polyester fibers used for textile fabrics.

Primamid. Trade name of Prima Plastics (USA) for a series of thermoplastics based on nylon 6,6, and available in unreinforced, glass fiber-reinforced and/or mineral- or polytetrafluoroethylene-reinforced and impact-modified grades.

Primanate. Trade name of Prima Plastics (USA) for a series of polycarbonate resins available in unreinforced, general-purpose, economy, glass-fiber-reinforced, structural-foam, flame-retardant, impact-modified and wear-resistant grades.

Primanex. Trade name of Prima Plastics (USA) for polyacetals supplied in unreinforced and glass-fiber-reinforced grades.

Primanite. Trade name of Prima Plastics (USA) for polyethylene butylene and polyethylene terephthalate resins supplied in unreinforced and glass fiber-, mineral-, polytetrafluoroethylene-, and/or silicone-reinforced grades.

Primapro. Trade name of Prima Plastics (USA) for a series of polypropylene homopolymers and copolymers reinforced with barium sulfate, calcium carbonate, talc, mica, or glass beads or fibers. They are available in extrusion, impact-modified and UV-stabilized grades.

Primapron. Trademark of Prima Plastics (USA) for a series of thermoplastics based on polyamide (nylon) 6 and available in unreinforced and impact, high-impact, glass bead- or glass fiber-reinforced, or and mineral-filled grades.

Prima-Rock. Trade name of Whip Mix Corporation (USA) for a dental gypsum (ADA Type V) supplied as a powder in violet and yellow colors. It is compatible with all types of impression materials.

primary alloy. An alloy whose principal element is produced directly by refining from ore, as contrasted with a *secondary alloy* in which it is produced from recycled scrap.

primary aluminum. Virgin or new aluminum metal produced from alumina (Al_2O_3) in an electric furnace, as compared with *secondary aluminum* that is produced from aluminum scrap. Also known as *virgin aluminum.*

primary calcium phosphate. See calcium phosphate (i).

primary cellulose acetate. Cellulose acetate that contains 60% combined acetic acid and is formed by acetylating purified cellulose with acetic anhydride in a suitable solvent, e.g., acetic acid or methylene chloride, in the presence of a catalyst, e.g., perchloric or sulfuric acid. See also cellulose acetate; secondary cellulose acetate.

primary derived protein. Any of several protein derivatives, such as proteans, metaproteins and coagulated proteins, in which the relative size of the protein molecules has essentially not been modified.

primary lead. See primary metal.

primary metal. A metal obtained from minerals, natural brines or ocean water, as contrasted to a *secondary metal* that is made from recycled scrap metal. Examples include primary aluminum, primary lead and primary zinc. Also known as *virgin metal.*

primary sodium phosphate. See monosodium phosphate.

primary zinc. See primary metal.

Primatel. Trade name of Prima Plastics (USA) for a series of thermoplastic polyolefin alloys.

Primathon. Trade name of Prima Plastics (USA) for high-density polyethylenes supplied in unreinforced, and glass fiber-, graphite-, or talc-reinforced grades.

Primatran. Trade name of Prima Plastics (USA) for a series of impact-modified polystyrenes available in calcium carbonate- and glass fiber-reinforced grades.

primavera. The durable wood from the tree *Cybistax donnellsmithii.* It has good machinability, beautifully figured patterns, high dimensional stability, and is used for furniture, paneling and trim.

Primax. Trade name of Air Products & Chemicals Inc. (USA) for polyethylene particles used in the reinforcement of engineering plastics.

Prime & Bond. Trade name of Dentsply Caulk (UK) for a light-cure, fluorine-releasing, filled dental bonding system based on urethane dimethacrylate (UDMA), and used for tooth structure (dentin and enamel) bonding.

Primeau Argo. Trade name of The Premier Group (Canada) for autoclaved concrete blocks.

primed hardboard. Hardboard panel with a special base coat applied to improve paintability. Also known as *undercoated hardboard.* See also hardboard.

primed particleboard. Exterior- and interior-type *particleboard* having a factory applied painted base coat. Used for products to be painted. Also known as *undercoated particleboard.*

Primef. Trade name of Solvay Polymer Inc. (USA) for 40% glass-fiber-reinforced polyphenylene sulfide resins.

Primel. Trademark of Polmeros Colombianos SA (Colombia) for polyester staple fibers and filament yarns.

primer. (1) A paint, varnish or lacquer that cures by air drying or oven baking, and is suitable for direct application to a substrate (e.g., a metal, plastic, etc.) as the first of two or more coats. Primers have good adhesion properties, wetting characteristics and corrosion protection, and often contain pigments, such as ammonium ferrous phosphate, barium potassium chromate, manganese phosphate, red lead, zinc chromate, zinc phosphate, or zinc yellow.

(2) A highly viscous, liquid bituminous material that when applied to a surface penetrates, bonds and stabilizes it and improves its adhesion prior to the heavy application of bitumen.

(3) See adhesion promoter (1).

primer brass. A brass composed of 70% copper and 30% zinc, and used for cartridge primers.

primer gilding. Gilding metal composed of 97% copper, 3% zinc, and traces of lead and iron. Used as a base for fire-enameled parts and prime coats.

primes. Metal products that are usually in plate or sheet form, of prime quality, and free of visible imperfections.

prime sheet. A high-quality metal sheet whose entire surface is free of visible imperfections.

prime western zinc. A less-pure grade of *slab zinc* containing up to 1.4% lead, 0.05% iron, 0.20% cadmium and 0.05% aluminum. Used for galvanizing applications.

Primopol. Trade name of Monopol AG (Switzerland) for a series of cellulose lacquers.

primrose chrome. A pale yellow paint pigment composed of lead chromate, lead sulfate and aluminum oxide.

Prince. Trade name of A. Milne & Company (USA) for a water-hardening chromium steel used for ball bearings.

Prince Albert fir. See Alaskan pine.

Prince's metal. Brass containing 61-85% copper and 15-39% zinc, used for hardware and flexible hose.

princess tree. See kiri; paulownia.

Principle. Trade name of Dentsply/Caulk (USA) for a dental compomer cement.

print. Textile fabrics that have patterns printed on them by means of dyes applied with blocks, rollers, screens, etc.

print-bonded nonwoven fabrics. Nonwoven fabrics produced by a process in which bonding is achieved by the application of controlled amounts of adhesive to selected areas of the fiber batt or web. Also known as *print-bonded nonwovens.*

printed-circuit laminates. Fabric- or paper-based laminates having a thin face sheet of copper foil for subsequent etching to produce a circuit pattern. Used in the manufacture of lightweight printed circuits by photofabrication processes.

printed foil. A thin metal film or foil having a pattern or design printed on one side, and used as a decorative covering on metals, plastics and composites.

printed yarns. Yarns that have patterns or designs printed on them prior to knitting or weaving into fabrics.

printing paper. A generic name for high-quality coated or non-coated rag paper that has a high degree of whiteness and is suitable for printing. Coated printing paper is usually supercalendered. Abbreviation: PP. See supercalendered paper.

Priphane. Trade name for transparent, flexible sheeting produced in a multistage process (viscose process) by treating cellulose with sodium hydroxide and carbon disulfide. It has high impermeability to nonaqueous substances, and may have a moisture-proof surface coating.

Prism. Trade name of Bayer Corporation (USA) for solid reaction-injection-molded polyurethane supplied as two-component liquid systems based on methylene diisocyanate (MDI).

Prisma AP.H. Trade name of Dentsply Corporation (USA) for a light-cure all-purpose hybrid composite resin in restorative dentistry.

Prisma Universal Bond. Trade name of Dentsply Corporation (USA) for a dental bonding agent used for dentin bonding applications.

Prisma VLC Dycal. Trade name of Dentsply Corporation (USA) for a light-cure calcium hydroxide dental cement.

Prisma-Fil. Trade name of Dentsply Corporation (USA) for a light-cure anterior composite resin used in restorative dentistry as a fissure sealant.

Prisma Microfine. Trade name of Dentsply Corporation (USA) for anterior composite resins used for dental restorative applications.

Prismaonda. Trade name of Fabbrica Pisana SpA (Italy) for a prismatic glass having one ribbed and one wavy side.

Prismasol. Trade name of Fabbrica Pisana SpA (Italy) for a glass with fluted pattern.

prismatic glass. A special translucent glass made up of parallel prisms that produce an iridescent and sometimes multicolored appearance.

Prize Ribbon. Trade name of Hewitt Metals Corporation (USA) for a tin babbitt containing varying amounts of copper and antimony. Used for bearings.

Prizrenit. Trademark of Progres (Yugoslavia) for polyester filament yarns.

ProBalsa. Trade name of Diab Group (USA) for end-grain *balsa* wood core materials.

Probedit. Russian trade name for sintered tungsten carbide used for cutting tools and dies.

probertite. A colorless mineral composed of sodium calcium borate hydroxide trihydrate, $NaCaB_5O_7(OH)_4 \cdot 3H_2O$. Crystal system, monoclinic. Density, 2.14 g/cm^3; refractive index, 1.524. Occurrence: USA (California).

ProBond. (1) Trade name of Dentsply Caulk (USA) for a multipurpose dental bonding agent.

(2) Trade name of Borden Company (USA) for a range of adhesives including professional wet-tack wood glues and waterproof, super-strong polyurethane glues that bond well to a wide range of building materials.

(3) Trade name of Silicon Carbide Products Inc. (USA) for abrasion- and wear-resistant nitride- and composite-bonded silicon carbide materials used for power-plant burner parts, wear nozzles, immersion tubes, kiln furniture, spray nozzles, and thermocouple protection tubes.

Procal. (1) Trademark of Foseco International Limited (UK) for ceramic-fiber insulation products with outstanding stability at high temperatures (above 1650°C or 3000°F), superior resistance to chemical attack and optimum performance in high-velocity gaseous environments. It is available in various shapes and sizes including boards, tubes, pipes, sleeves, rings, nozzles, etc., and used in the metallurgical and ceramics industries, and for domestic cookers and water heaters.

(2) Trademark of Stauffer Chemical Company (USA) for a pressure-sensitive adhesive backed vinyl film used for decorative and protective applications, and in printing and die-cutting.

(3) Trade name for a calcium hydroxide dental cement.

Procom. (1) Trademark of BASF Corporation (USA) for a series of polypropylenes.

(2) Trade name for a self-cure dental hybrid composite.

Procon. Trademark of Toyoba Company Limited (Japan) for polyphenylene sulfide staple fibers and filament yarns.

Prodag. Trademark of Acheson Industries, Inc. (USA) for solutions of *colloidal graphite* in water used for foundry facings, mold washes, ingot molds, parting compounds for molds, molds for mechanical rubber products, and stop-off coatings and lubricants.

Prodec. Trademark of Avesta Sheffield Inc. (USA) for a weldable, machinable and electro-polishable austenitic stainless steel bar and plate supplied in AISI types 304 (a low-carbon modification of type 302), 304L (an extra-low-carbon modification of type 304), 316 (a highly corrosion-resistant type) and 316L (an extra-low-carbon modification of type 316).

Prodigy. Trade name for a light-cure dental hybrid composite.

Pro-Fax. Trademark of Himont USA Inc. for polypropylene homopolymers, copolymers and elastomers available in various grades including mica- or calcium carbonate-filled, glass fiber-reinforced, UV-stabilized, fire-retardant, elastomer-modified and structural-foam. The resins are suitable for food packaging applications, and the elastomers for the manufacture of automotive components requiring high impact, e.g., bumpers.

Proferall. Trade name of Textron Inc. (USA) for wear-resistant, high-strength cast irons containing 3.1-3.3% total carbon, 2.2-2.4% silicon, 0.5-0.6% manganese, 0.8-1.0% chromium, 0.2-0.5% molybdenum, 0.4-1.5% nickel, and the balance iron. Used for crankshafts, gears, pistons, cylinders, diesel-engine parts, and refrigeration parts.

Profex. Trade name of PROFEX Kunststoffe GmbH (Germany) for special plastic profiles and shapes including pipes with square or rectangular cross sections.

Profilcolor. Trade name of Vetreria di Vernante SpA (Italy) for colored, formed glass building elements.

Profile. Trade name for a self-cure dental hybrid composite.

profile fibers. Textile fibers whose filaments have cross sections of particular shape or profile, e.g., tri-, penta- or octalobal, or dogbone. See also multilobal fibers.

Profilglas. Trade name of former Glas- und Spiegel-Manufaktur AG (Germany) for formed glass building elements.

Profilit. Trade name of Moosbrunner Glasfabrik AG (Austria) for formed glass building elements.

Profilite. Trade name of C-E Glass (USA) for formed glass building elements.

Profilm. (1) Trademark for a waxed paper with a thin shellac coating used in ceramics for silk screen work.

(2) Trade name of Protective Lining Corporation (USA) for strong, chemical-resistant plastic (e.g., polyethylene) films for bags, covers, etc.

Profilor. Trade name of Boussois Souchon Neuvesel SA (France) for translucent U-shaped glass channels that have coatings deposited on their interior surfaces which give them a golden-brown tinge.

Profilver. Trade name of Vetreria di Vernante SpA (Italy) for formed glass building elements.

Progen. Trade name of Seaboard Steel Company of America (USA) for a tough, shock-resisting, self-tempering steel containing 0.33-0.4% carbon, 0.55-0.75% chromium, 0.5-0.7% molybdenum, 0.45-0.65% copper, 0.5-0.7% silicon, 0.1% titanium, and the balance iron. Used for die-casting dies, tools, chisels, punches, and picks.

Progilite. French trade name for bakelite-type phenol-formaldehyde resins and plastics.

Progress. Trade name of former Glas- und Spiegel-Manufaktur AG (Germany) for glass blocks.

Project 7000. Trademark of Carpenter Technology Corporation (USA) for a series of stainless steels. *Project 7000 Stainless Type 203* is an austenitic stainless steel (AISI type 203) containing 0.07% carbon, 3.5% silicon, 1% manganese, 14.75% chromium, 4.5% nickel, 0.4% niobium, 0.5% molybdenum, and the balance iron. Supplied in solution-treated and aged conditions in the form of bars, extrusions, forgings, rings and wires, they have outstanding machinability, excellent corrosion resistance, high strength and fracture toughness, and is used for making machine parts, fasteners and fittings, airframes, etc. *Project 7000 Stainless Type 416* is a machinable martensitic stainless steel (AISI type 416) containing up to 0.15% carbon, 1.0% silicon, 1.25% manganese, and 0.6% phosphorus, at least 0.15% sulfur, 12-14% chromium, and the balance iron.

prolamine. Any of a group of simple proteins, such as gliadin in wheat and zein in corn, that are soluble in strong ethyl alcohol, insoluble in water, and not coagulated by heat.

Prolastic. Trade name of Discas Recycled Products (USA) for olefinic-based thermoplastic elastomers.

Prolen. Trademark of Chemosvit (Slovak Republic) for polypropylene filament yarns.

Prolex. Trade name of BSK, Division of D.S. Smith (UK) Limited for a highly water-resistant roofing membrane for residential and industrial buildings. It also provides protection against poisonous radon or methane gas ingress.

Prolite. Trade name of Murex Limited (UK) for tungsten carbide materials used for tools.

Proloft. Trade name for polyolefin fibers and yarns used for textile fabrics.

Prolon. Trademark of Productos Plasticos de Puebla SA (Mexico)

for nylon 6 monofilaments.

Proloy. Trademark of GE Plastics (USA) for a family of thermoplastic acrylonitrile-butadiene-styrene/polycarbonate blends available in various grades including several injection-molding and automotive grades. *Proloy* blends have good toughness and impact strength, high heat resistance and good processibility. Used for molded parts, such as home appliance parts, power tools, vacuum cleaner parts, lawn and garden products, telecommunications products, and office machines.

Promal. Trade name of FMC Corporation (USA) for a malleable cast iron containing 1.8% total carbon, 1% silicon, 0.3% manganese, and the balance iron. It has good strength, fatigue resistance and toughness, and is used for sprockets, gears, brake drums, chains, chain links, and valve parts.

Promat. Trademark of Owens-Corning (USA) for combination fabrics made by assembling a 0° unidirectional warp, a 90° unidirectional weft, and a chopped strand mat using the proprietary *Knytex* stitch-bonding process. Used for hand lay-up and resin-transfer molding applications.

Promet. Trade name of American Crucible Products Company (USA) for an extensive series of cast copper alloys including various manganese bronzes, leaded and unleaded brasses, leaded and unleaded tin bronzes, and lead- and tin-base babbitts.

Prometal. Trade name of Ste. de Produits Métallurgiques (France) for corrosion- and heat-resistant stainless steels.

Promethium. Trade name for a high-ductility alloy composed of 67% copper, 30% zinc and 3% aluminum, and used for condenser tubes.

promethium. A silvery-white, radioactive, metallic rare-earth element of the lanthanide series of the Periodic Table. The most abundant isotope is *promethium-147* which is prepared in a cyclotron by bombarding neodymium with protons, recovered from spent uranium fission products, or obtained by reduction of the chloride or fluoride with an alkali metal. Density, 7.2 g/cm^3; melting point, 1160°C (2120°F); boiling point, 2730°C (4946°F); atomic number, 61; atomic weight, 145 (most stable isotope); trivalent; half-life, 18 years. It exists in two allotropic forms: (i) *Alpha promethium* (hexagonal crystal structure) stable at room temperature; and (ii) *Beta promethium* (body-centered cubic crystal structure) present at high temperatures. Used for auxiliary nuclear power generators, nuclear batteries and luminescent paint, as a source of beta rays for thickness gages, as an X-ray source, and in certain tungsten cermets. Symbol: Pm. Formerly also known as *illinium*.

promethium oxide. A compound of promethium and oxygen used for nuclear applications and in ceramics. Formula: Pm_2O_3.

Promilan. Trademark of Toray Industries, Inc. (Japan) for nylon 6,6 filament yarns.

ProModal. Trade name of Lenzing Fibers Corporation (USA) for modal fibers.

Promogan. Trade name for a wound dressing material based on a sterile, freeze-dried matrix made of collagen and oxidized regenerated cellulose, and supplied in 3 mm (0.1 in) thick hexagonal sheets of varying size.

promoted resin. A synthetic resin to which one or more accelerators, but no catalyst have been added.

promoter. See accelerating agent (2).

Promount. Trademark of 3M Canada Inc. (USA) for heat-activated film-mounting adhesives.

Pro-Mount Foam. Trade name of Pro Tapes & Specialties, Inc. (USA) for double-coated foam tapes.

Prompt. Trade name ESPE Dental (Germany) for a dental bonding agent for tooth structure (dentin/enamel) bonding. *Prompt L-POP* is a fluorine-releasing methacrylate-type bonding system for orthodontic brackets.

ProNectin F. Trademark of Protein Polymer Technologies, Inc. (USA) for a protein-based biopolymer designed using two oligopeptide blocks–a silk fibroin-like structural block and a fibronectin-like functional block. It is produced by a technique involving genetic engineering and culture media (microorganisms) and the final crystallized material has high thermal stability, good human fibronectin cell attachment, good adhesion to plastic surfaces and high insolubility in aqueous solutions at body temperature. Used chiefly as a cell attachment coating. See also fibroin; fibronectin.

Pronto. Trademark of 3M Company (USA) for a series of transparent, ready-to-use one-part cyanoacrylate adhesives. They have 100% solids (solvent-free), high bond strengths, short setting times, and are supplied in several formulations with: (i) high-, low- and elevated-temperature resistance to fuels, lubricating oils and others chemicals; (ii) gap filling properties; (iii) longer curing times for repositioning; and (iv) high peel and impact strengths. Used for bonding metals, plastics, rubber and other materials, and for electronic applications.

proof gold. A commercial fine gold having a purity of 99.99+%, a density of 19.32 g/cm³ (0.698 lb/in.³), a hardness of 33 Brinell, and a melting point of 1064°C (1947°F). Used in electronics. See also commercial fine gold; fine gold.

proof silver. A commercially pure silver (99.9%) with a density of 10.49 g/cm³ (0.379 lb/in.³), and a melting point of 960°C (1760°F). Used for electronic applications, fuses and vacuum brazing.

propadiene. See allene (2).

Propak. Trade name of PolyPacific (USA) for polypropylenes.

Propathene. Trademark of ICI Americas Inc. (USA) for a series of polypropylene plastics available as molding powders, sheets, tubes, rods and special shapes. They have a density of 0.91 g/cm³ (0.033 lb/in.³), excellent resistance to dilute acids and alkalies, fair resistance to mineral oils, gasoline and trichloroethylene, and poor resistance to tetrachlorocarbon. Used for fittings, and for parts with elevated-temperature and chemical resistance.

Propel HD. Trade name of Albany International (USA) for thermosetting polyimide with high dimensional stability at temperatures above 250°C (482°F), and good chemical resistance.

propeller brass. A strong, tough, cast copper alloy containing 40% zinc, 1% tin, 1% iron, 0.2% manganese and 0.2% aluminum. Used for marine propellers.

Propex. Trademark of Amoco Fabrics Limited (Canada) for polypropylene and polyethylene industrial fabrics.

Propilan. Trademark of Propilan SA (Spain) for polypropylene staple fibers.

ProPink. Trademark of Owens-Corning Fiberglass (USA) for loosefill fiberglass insulation used for the thermal insulation of attics.

Pro Plast. Trade name of American Pro-Fol (USA) for cast polypropylene film.

Proplatinum. Trade name of a corrosion-resistant white alloy composed of 72.0% nickel, 23.6% silver, 3.7% bismuth and 0.7% gold, and used for jewelry, corrosion-resistant parts, and as a replacement for platinum.

Propocon-Isocon. Trade name of Lankro Chemicals Urethane Division (USA) for semirigid polyurethane foams.

propoxide. A compound that is the product of a reaction involving the substitution of the hydroxyl hydrogen of *n*-propyl alcohol (C_3H_7OH) by a metal. Titanium propoxide, $Ti(OC_3H_7)_4$, and magnesium propoxide, $Mg(OC_3H_7)_2$, are examples. Also known as *propylate*.

propylate. See propoxide.

propylene oxide rubber. A group of thermosetting elastomers that are copolymers of allyl glycidyl ether ($C_6H_{10}O_2$) and propylene oxide (C_3H_6O). They have a service temperature range of -55 to +130°C (-67 to +266°F), and good low-temperature properties. Both the resistance to oxidation and ultraviolet light are good, while the resistance to oil is poor. Used for mechanical rubber products.

propylene plastics. A group of tough, hard olefin plastics made from polypropylene or copolymers of propylene with other monomers.

Propylex. Trade name of Courtaulds Advanced Materials (UK) for a chemical- and weather-resistant polypropylene copolymer used for food-processing and industrial equipment.

Propylon. Trade name for polyolefin fibers and yarns used for textile fabrics and plastic products.

Propylux. (1) Trademark of Westlake Plastics Company (USA) for polypropylene resins supplied in the form of rods, tubes and rigid sheets. They have high rigidity, good thermal properties, a maximum service temperature of 121°C (250°F), good resistance to chemicals, moisture and electricity, good processibility and good strength and toughness. Used for piping, tubing, packaging, television cabinets, automotive trim, hinges, and other hardware.

(2) Trademark of Jackstadt GmbH (Germany) for self-adhesive and self-fixing paper and plastic foils used for making stickers and label.

Prosital. Trade name of Vetreria Italiana Balzaretti Modigliani SpA (Italy) for rigid glass-fiber panels used for thermal insulation applications.

prosopite. A colorless mineral composed of calcium aluminum fluoride hydroxide, $CaAl_2(F,OH)_8$. Crystal system, monoclinic. Density, 2.89 g/cm³; refractive index, 1.503; hardness, 4.5 Mohs. Occurrence: USA (Colorado).

Prospector. Trade name of J.F. Jelenko & Company (USA) for a hard dental casting alloy (ADA type III) containing 38% silver, 20% gold and 20% palladium. It provides a brilliant gold color, a high lustrous polish, high tarnish resistance, and is used for crowns, fixed bridgework and hard inlays. Also available is *Prospector Plus,* a hard low-gold dental casting alloy (ADA type III) with brilliant gold color, and properties and applications similar to those of Prospector.

prosperite. A white to colorless mineral composed of hydrogen calcium zinc arsenate hydroxide, $HCaZn_2(AsO_4)_2(OH)$. Crystal system, monoclinic. Density, 4.31 g/cm³; refractive index, 1.748. Occurrence: Namibia.

Prostat-H. Trade name for an antistatic fractional melt propylene copolymer with high toughness, good processibility and formability and good surface resistivity. Used in electronic packaging (e.g., corrugated profiles, shipping tubes for integrated circuits, etc.), blister packs, blow-molded containers, thermoformed trays and boxes, and profile extrusions.

Prostran. Trade name for polyolefin fibers and yarns used for textile fabrics.

protactinium. A rare, silver-gray, radioactive metallic element of the actinide series of the Periodic Table. It is a constituent of all uranium ores, and can be produced by irradiation of thorium-230. The longest lived isotope is *protactinium-231* which decays by alpha emission into actinium-227, and has a half-life of about 33000 years. Density, 15.37 g/cm³; melting point, 1575°C (2867°F); atomic number, 91; atomic weight, 231.036; tetravalent, pentavalent. Two allotropic forms are known: (i) *Alpha protactinium* (body-centered tetragonal crystal structure) that has a density of 15.37 g/cm³ and is stable below 1165°C (2129 C); and (ii) *Beta protactinium* (body-centered cubic crystal structure) that has a density of 13.87 and is stable between 1165 and 1575°C (2129 and 2867°F). Used in chemical and other research. Symbol: Pa.

Protal. Trade name of Siemens AG (Germany) for wrought, age-hardenable high-strength aluminum alloy containing 2.5-5% copper, 0.2-1.8% magnesium and 0.3-1.5% manganese. Used for structural parts for the aircraft and automotive industries. A free-cutting grade containing some lead is also available.

Protal 7200. Trade name of Denso North America Inc. (USA) for a 100% solids epoxy resin used as a protective coating on pipes and pipelines.

protamine sulfate. Any of several proteins obtained from salmon as salmine or from herring as clupeine. They are available as white or yellowish white amorphous powders or as 1% aqueous solutions for use in biochemistry, biology and medicine. The solutions are also employed as adhesives in electron microscopy.

Protane. Trade name of Sovirel (France) for glass with maximum transmission in the greenish/yellow zone. Used for oxyacetylene welding goggles.

Protean. Trade name of Reichhold Chemicals, Inc. (USA) for fast-drying acrylic emulsion/alkyd hybrid resins with good adhesion properties. Used for coating applications.

ProTec CEM. Trade name of Ivoclar Vidadent (USA) for a hybrid resin/glass-ionomer dental cement

ProTech. Trade name for acrylic resin for relining denture.

Pro-Tech. Trade name of International Polarizer, Inc. (USA) for polarized sheets used for sunglass lenses.

Protech. Trade name of Progress Paint Manufacturing Company (USA) for an epoxy catalyzed polyester coating.

ProTechtor. Trademark of National Nonwovens (USA) for moldable, lightweight, high-performance, mechanically interlocked nonwovens. They provide high abrasion resistance, excellent flame retardancy, good acoustical and thermal insulating properties, and are used for fire protection apparel, protective gloves and vests, helmets, body and vehicle armor, and blast containment devices.

Proteclair. Trade name of Sovirel (France) for glass with maximum transmission in the greenish/yellow zone. Used for welders' goggles.

Protectatin. Trade name for a thin clear oxide coating developed by the US Tin Research Institute and produced on tinplate by immersion in a solution of trisodium phosphate, sodium dichromate and sodium hydroxide. It increases protection against sulfur stains and promotes paint adhesion.

protective coating. A film of metal, ceramic, glass, plastic, rubber or paint applied to a substrate to inhibit corrosion and reduce wear, and sometimes provide decorative effects.

Protective Deck Plate. Trade name for a case-hardened steel containing 0.2-0.3% carbon, 3.5% nickel, 1.5% chromium, and the balance iron. Used for armor plate.

protective fabrics. Fabrics coated or impregnated on one or both sides with an organic coating, such as a lacquer or varnish, or a polymer, such as natural or synthetic rubber, vinyl, etc., and used for tarpaulins, protective clothing, outerwear, linings, book covers, wall facings, gasketing, and upholstery. See also coated fabrics.

Protect Liner. Trade name for light-cure composite resins supplied in standard and microfine grades used in restorative dentistry as lining materials.

Protecto. (1) Trademark of Dudick Corrosion Proof Inc. (USA) for acid-resistant coatings and linings for metallic substrates.
(2) Trade name of Celluloid Corporation (USA) for *celluloid*-type cellulose nitrate plastics.

Protectocote. Trademark of Alcan Canada Foils, Division of Aluminum Company of Canada Limited for transparent plastic film products.

Protectopane. Trade name of Dearborn Glass Company (USA) for a burglar-resisting laminated glass having a thin vinyl interlayer.

Protectopure. Trademark of Alcan Canada Foils, Division of Aluminum Company of Canada Limited for wax-coated cellophane and foil sheets.

Protector. Trade name of Vidrobrás (Brazil) for a toughened and laminated glass.

Protectoseal. Trade of Alcan Canada Foils, Division of Aluminum Company of Canada Limited used for foil and wax laminates.

Protect-O-Sheet. Trademark of Flex-O-Glass, Inc. (USA) for a specially processed, peelable polyethylene masking film with excellent adhesion to most flat, nonporous surfaces.

Protectox. Trade name of Technic, Inc. (USA) for anti-tarnish for silver, brass and gold.

protein. Any of a group of high polymers found in all living cells and consisting of alpha-amino acids joined together by peptide linkages. Proteins can be grouped into three classes: (i) *simple proteins*, such as albumins, globulins and prolamines; (ii) *conjugated proteins*, such as glycoproteins, phosphoproteins and lipoproteins; and (iii) *derived proteins*, such as polypeptides, peptones and proteoses. Proteins from animal and vegetable sources are used in the manufacture of fibers, polymers and plastics. See also protein fibers; protein plastics; protein polymers.

protein fibers. Natural or regenerated fibers based on proteins. Natural protein fibers include both true and tussah silk, wool from the alpaca, beaver, Cashmere goat, guanaco, llama, mohair, otter, sheep, vicuna and yak, and camel, cow, goat, hare, horse and rabbit hairs. Regenerated protein fibers are made by the chemical regeneration of natural protein fibers and include fibers based on casein, peanut, soybean or zein. Used for the manufacture of textile fabrics. See also natural fibers; regenerated fibers; protein.

protein plastics. Plastics produced from proteins, i.e., from high-molecular-weight polymers found in animal and vegetable products. Examples include casein, the chief protein in milk, as well as the proteins occurring in fish, peanuts, soybean, etc. Used for fibers and molded articles. See also casein plastics; protein.

protein polymers. Protein-based natural or synthetic biopolymers. Examples of natural protein polymers include silk fibroin, collagen and elastin, and synthetic protein polymers include *ProNectin F*, which is bioengineered from amino acid blocks involving microorganisms. Natural protein polymers in fiber form are widely used as textile fibers and in the life sciences, and synthetic protein polymers in the form of fibers, films, gels and coatings have found various applications in biomedicine and surgery, e.g., as coatings for providing good human cell attachment. See also protein.

ProTek-R. Trade name of Foam Plastics of New England (USA) for plastic foams for building insulation and packaging applications.

Protek-Coat. Trade name for cutback asphalt.

Protektor 220. Trade name of Solutions Globales (France) for a flame-resistant textile fabric made entirely from *Trevira CS* polyester fibers. Used in protective and industrial applications.

Protek-Wrap. Trade name of Daubert Industries, Inc. (USA) for coated paper.

Protel. Trademark of Amoco Fabrics & Fibers Company (Canada) for polypropylene staple fibers and filament yarns.

Protemp 3 Garant. Trade name of 3M ESPE (USA) for a chemical-cure bismethylacrylate (BMA) type dental composite resin used for temporary crown and bridge restorations.

Protex. (1) Trade name of Cristales y Vidrios SA (Chile) for a toughened glass.

(2) Trade name of Tolber Division of Pyramid Plastics (USA) for an transparent, air-drying, corrosion protective coating for ferrous metals.

(3) Trade name of Mask-Off Company, Inc. (USA) for protective pressure-sensitive papers.

Protex-A-Crete. Trade name of Protex-A-Cote, Inc. (USA) for concrete sealers and hardeners.

Protex-A-Wood. Trade name of Protex-A-Cote, Inc. (USA) for a linseed oil-based emulsion paint used for protective coatings on wood surfaces.

Protex AEA. Trade name of Protex-A-Cote, Inc. (USA) for an air-entraining agent used in concrete manufacture.

Protexit. Trade name of Erste Oesterreichische Maschinglasindustrie AG (Austria) for a toughened glass.

Protexo-Cote. Trade name of Thermo-Cote Inc. (USA) for a room-temperature strippable protective coating for metal substrates.

Proth-Auro 22. Trade name of Metalor Technologies SA (Switzerland) for an extra-hard dental casting alloy containing 68.5% gold, 3.8% palladium and 3.5% platinum.

Protherm. Trade name of Brush Wellman Engineered Materials (USA) for a weldable beryllium-copper alloy with excellent wear resistance and corrosion resistance, and good thermal conductivity, strength and hardness. Used for plastic molds.

Protite. Trade name of Progress Paint Manufacturing Company (USA) for a two-component polyurethane coating.

protium. See hydrogen.

Protochrome. Trade name for an abrasion- and heat-resistant cast iron containing 2-2.5% carbon, 24-27% chromium, and the balance iron. Used for hardfacing electrodes.

Protocol. Trademark of Ivoclar Vivadent AG (Liechtenstein) for an extra-hard ceramic dental alloy containing 75.2% palladium, 6.5% silver, 6.0% gold, 6.0% gallium, 6.0% indium, and less than 1.0% ruthenium and lithium, respectively. It has a white color, a density of 11.0 g/cm³ (0.397 lb/in.³), a melting range of 1270-1310°C (2320-2390°F), a hardness of 235 Vickers, moderate elongation, and excellent biocompatibility. Used for crowns, onlays, posts and bridges.

Protofer DS. Trade name of Monopol AG (Switzerland) for an anticorrosive paint used on ferrous substrates.

Protoflex. Trade name of Glyco Products Company (USA) for flexible casein plastics.

ProtoFunctional. Trademark of DSM Somos (USA) for a range of resin-based solid imaging materials in liquid or powder form that experience a phase change to a solid when exposed to intense radiation such as that of a laser beam, and can replicate the performance parameters of production materials. *Somos*

7100 refers to a series of resin-based general-purpose materials with high heat deflection temperatures and high humidity tolerances. *Somos 8100* and *Somos 9100* are materials that can imitate the properties of polyethylene and polypropylene, respectively.

Protoloy. Trade name of Molecu-Wire Corporation (USA) for a resistance alloy composed of 80% nickel and 20% chromium. It has high heat and corrosion resistance, high electrical resistivity, a maximum service temperature of 1100°C (2010°F). Used for heating elements, and resistors.

Protonex. Trademark of Enthone-OMI, Inc. (USA) for gold electroplates and plating processes.

Protopol. Trade name of Monopol AG (Switzerland) for anticorrosive and chemically curing primers.

protoporphyrin. (1) A violet porphyrin acid derived by removing the bound iron from heme or hemin. Formula: $C_{34}H_{34}N_4O_4$. See also heme; hemin.

(2) A porphyrin acid derivative containing a central metal atom, and corresponding to the general formula $(C_{34}H_{34}N_4O_4)$Me where Me is a metal, such as zinc or sodium.

Protor. Trade name of Cendres & Métaux SA (Switzerland) for high-gold dental casting alloys used for crowns, bridges, fillings and other dental restorations. *Protor 3* contains a total of 75.05% gold and platinum. Also known as *CM Protor.*

Pro-Tuft. Trade name for polyolefin fibers and yarns used for textile fabrics.

Pro 2000. Trade name of Canadian Adhesives, Division of Rexnord Canada Limited for a professional elestomeric interior/exterior door and window sealant belonging to the *Bulldog Grip* product line. It has twice the stretch of premium acrylic caulking.

proudite. A silver-gray mineral composed of copper lead bismuth selenide sulfide, $Cu_2Pb_{15}Bi_{19}(S,Se)_{44}$. Crystal system, monoclinic. Density, 7.08 g/cm³. Occurrence: Australia.

proustite. A light ruby red, scarlet or vermilion mineral with an adamantine luster. It belongs to the pyrargyrite group and is composed of silver sulfarsenite, Ag_3AsS_3, and contains theoretically about 65.4% silver. Crystal system, rhombohedral (hexagonal). Density, 5.51-5.66 g/cm³; melting point, 480°C (896°F); hardness, 2.0-2.5; refractive index, 2.9789. Occurrence: Canada (Ontario), Chile, Germany, Mexico, Peru, USA (Colorado, Nevada, New Mexico). Used as an ore of silver. Also known as *light-red silver ore; light ruby silver.*

ProviCem. Trade name for an eugenol-free dental cement used for temporary bridge and crown restorations.

Provifil. Trade name for Inter-Africa Dental (Zaire) for a light-cure, one-part dental filling material with short setting time and low discoloration, used in restorative dentistry for temporary inlay/onlay filling.

Provil. Trademark of Bayer Corporation (USA) for dental impression materials with silicone additions.

Provil Novo. Trademark of Heraeus Kulzer Inc. (USA) for silicone elastomer-based dental impression materials.

Provista. Trade name of Eastman Chemical Company (USA) for a series of tough, clear, easy-to-process copolymers supplied as extruded profiles for use in the manufacture of pen barrels, store displays, pricing channels, etc.

Pro-Wood. Trademark of Universal Forest Products, Inc. (USA) for lumber used for manufactured housing applications.

Proxcote. Trademark of Pierce & Stevens Canada Inc. for gloss paper coatings.

Proxseal. Trademark of Pierce & Stevens Canada Inc. for heat-seal paper coatings.

Proxyl. Trade name of Lee S. Smith & Son (USA) for cellulose nitrate plastics.

Pruf-Stok. Trade name of General Plastics Manufacturing Company (USA) for a flame-resistant rigid polyurethane foam tape material.

Prussian blue. An intensely blue pigment (Color Index C.I. 77510) composed of ferric ferrocyanide, $Fe_4[Fe(CN)_6]_3$. It is obtained by the action of potassium ferrocyanide on a ferric salt, and is used in paints, printing inks and plastics, in paper dyeing, as a layout ink in machine shops, dissolved in oil to indicate high spots on bearings, and in baking enamels and industrial finishes. Also known as *Berlin blue; Paris blue; Pigment Blue 27*.

Prussian red. A group of red pigments obtained from ferric oxide or potassium ferrocyanide, and used in paints, inks and plastics.

Prym. Trade name of W. Prym-Werke GmbH & Co. KG (Germany) for an extensive series of coppers and copper-base alloys including oxygen-free high-conductivity, electrolytic tough-pitch, phosphorus-deoxidized, silver-bearing and iron-bearing coppers, gilding, red, low, cartridge, high-leaded and yellow brasses, Muntz metal and leaded Muntz metal, forging brasses, free-cutting brasses, architectural, commercial and phosphor bronzes, copper nickels, nickel bronzes, and nickel silvers.

Prystal. Trade name of Catalin Corporation (USA) for transparent phenolic resins and plastics.

Prystaline. French trade name for urea-formaldehyde resins.

przhevalskite. A bright yellow mineral of the meta-autunite group composed of lead uranyl phosphate dihydrate, $Pb(UO_2)_2(PO_4)_2 \cdot 2H_2O$. It can also be made synthetically. Crystal system, orthorhombic. Refractive index, 1.749. Occurrence: Russia.

pseudoalloy. An alloy that is not produced by melting, but by sintering.

pseudo-autunite. A pale yellow to white mineral composed of calcium oxonium uranyl phosphate pentahydrate, $(H_3O)_4Ca_2(UO_2)_2(PO_4)_4 \cdot 5H_2O$. Crystal system, orthorhombic. Density 3.28 g/cm³; refractive index, 1.568. Occurrence: Russian Federation.

pseudoboleite. A blue mineral composed of lead copper chloride hydroxide dihydrate, $Pb_5Cu_4Cl_{10}(OH)_8 \cdot 2H_2O$. Crystal system, tetragonal. Density, 4.85 g/cm³; refractive index, 2.03. Occurrence: USA (California), Mexico, Peru.

pseudobrookite. A dark brown mineral composed of iron titanium oxide, Fe_2TiO_5. It can also be made synthetically by reacting equimolar quantities of ferric oxide (Fe_2O_3) and titanium dioxide (TiO_2) under prescribed conditions. Crystal system, orthorhombic. Density, 4.39 g/cm³; refractive index, 2.39.

pseudo-ixiolite. A black mineral composed of manganese tantalum niobium oxide, $(Mn,Ta,Nb)O_2$. Crystal system, orthorhombic. Density, 6.95 g/cm³. Occurrence: Canada (Manitoba).

pseudolaueite. An orange-yellow mineral composed of manganese iron phosphate hydroxide hydrate, $MnFe_2(PO_4)_2(OH)_2 \cdot xH_2O$. Crystal system, monoclinic. Density, 2.46 g/cm³; refractive index, 1.650. Occurrence: Germany.

pseudo-poly(amino acids). A group of biopolymers synthesized by modifying *poly(amino acids)* with tyrosine derivatives or copolymerizing with other biopolymers. See also tyrosine-derived polymers.

pseudorutile. A mineral composed of iron titanium oxide, $Fe_2Ti_3O_9$. Crystal system, hexagonal. Density, 2.46 g/cm³; refractive index, 1.650. Occurrence: Germany.

Pseudotique. Trade name of Safetee Glass Company Inc. (USA) for a reproduction of antique window glass.

pseudowollastonite. A white mineral composed of calcium silicate, α-$CaSiO_3$. It can be made by heating calcium oxide (CaO) and silicon dioxide (SiO_2) under prescribed conditions. Crystal system, triclinic. Density, 2.91 g/cm³; refractive index, 1.611.

psilomelane. A black mineral with a brownish-black streak, and a submetallic luster composed of hydrous barium manganese oxide, $BaMn_9O_{16}(OH)_4$, containing varying amounts of calcium, cobalt, copper, magnesium and nickel. Crystal system, orthorhombic. Density, 3.7-4.7 g/cm³; hardness, 5-6 Mohs. Occurrence: Cuba, India, Russia, South Africa, USA (Arkansas, Georgia, Virginia). Used as an important ore of manganese.

PS-PPO alloys. Engineering thermoplastics based on miscible polymer blends of amorphous polystyrene (PS) and amorphous polyphenylene oxide (PPO). The addition of PS to PPO reduces the tensile strength and heat deflection temperature somewhat and increases the thermal expansion. *PS-PPO alloys* combine the high-temperature resistance, and good toughness and chemical resistance of PPO with the good mechanical and thermal properties (including impact strength) of PS. Their glass transition temperature is about 155°C (311°F), and is used for automobile interiors (e.g., instrument panels and seat backs, etc.) and exterior (wheel covers and mirror housings, etc.), business machines (keyboard and printer bases, and terminal housings), household appliances, and power tools.

PSU-ABS alloys. Immiscible blends of an amorphous polysulfone (PSU) and a amorphous acrylonitrile-butadiene-styrene (ABS). The addition of PSU increases the creep resistance and/or improves the hydrolytic stability of ABS. Used for food-processing equipment and food service trays.

PSU-PBT alloys. Polymer blends of amorphous polysulfone (PSU) and semicrystalline polybutylene terephthalate (PBT). See also PSU-polyester alloys.

PSU-PET alloys. Polymer blends of amorphous polysulfone (PSU) and semicrystalline polyethylene terephthalate (PET). See also PSU-polyester alloys.

PSU-polyester alloys. Polymer blends composed of amorphous polysulfone (PSU) and a crystalline polyester (e.g., polybutylene terephthalate or polyethylene terephthalate). They combine the high stiffness, good chemical resistance and good melt flow characteristics of polyesters with the good dimensional properties, high-temperature resistance and impact resistance of PSU. Used for automotive applications and electrical and electronic components.

PSX. Trademark of Ameron International Corporation (US) for a line of engineered siloxane coatings.

PTFE fibers. See polytetrafluoroethylene fibers.

P-Tin Alloy. Trade name of Capper Pass & Son Limited (UK) for a soft solder containing 55.5% tin, 41.1% lead and 3.4% antimony.

p-type germanium. A type of *germanium* usually grown by the Czochralski or zone-leveling technique and doped predominantly with acceptor-type impurity atoms of trivalent Group IIIA (Group 13) elements, such as aluminum, gallium or indium, and in which the positive charge carrier concentration (holes) exceed the negative charge carrier concentration (conduction electrons). Therefore the holes are primarily responsible for electrical conduction. Symbol: p-Ge.

p-type oxide. A transition-metal oxide semiconductor, such as ferrous or nickel oxide, having a stoichimetric $M_{1-x}O$ defect structure. The trivalent metal ions of the oxide accept electrons from the bivalent metal ions resulting in the formation of holes in the valence band for p-type semiconduction. See also defect semiconductor.

p-type semiconductor. An *extrinsic semiconductor* doped predominantly with acceptor-type impurity atoms of trivalent Group IIIA (Group 13) elements, such as aluminum, boron, gallium or indium, and in which the positive charge carrier concentration (holes) exceed the negative charge carrier concentration (conduction electrons) and dominate electrical conductivity.

p+-type semiconductor. A *p-type semiconductor* in which the concentration of positive charge carriers (holes) greatly exceeds the negative charge carriers (conduction electrons) and thus dominates electrical conductivity.

p-type silicon. A type of *silicon* usually grown by the Czochralski or float-zone technique and doped predominantly with acceptor-type impurity atoms of trivalent Group IIIA (Group 13) elements, such as boron, aluminum or gallium, and in which the positive charge carrier concentration (holes) exceed the negative charge carrier concentration (conduction electrons). Symbol: p-Si.

Pucaro-Triflexil. Trade name of Christian Authenrieth GmbH & Co./Pucaro-Werk (Germany) for plastic laminated pressboard.

pucherite. A reddish amber mineral composed of bismuth vanadium oxide, $BiVO_4$. Crystal system, orthorhombic. Density, 6.25 g/cm^3; refractive index, 2.50; hardness, 4 Mohs. Occurrence: Germany. Used as a vanadium ore.

pucker embroidery. An embroidered fabric with a crinkle or pucker on the surface, intentionally produced by the embroidery stitches.

puckered fabrics. Textile fabrics, such as *plissé* and *seersucker*, having curly, crimpy or crinkled surfaces, produced either by weaving with yarns of different sizes, tensions or degrees of shrinkage, or by special chemical or other treatments.

puddled steel. A steel made in a reverberatory-type furnace of special design known as a "puddling furnace."

puddling iron. A grade of pig iron having an average composition of 3.5-4.5% carbon, 3-5% manganese, and up to 1% silicon, 0.3% phosphorus and 0.04% sulfur, with the balance being iron. Used in the production puddled or wrought iron.

Pueblo White. Trade name of Aardvark Clay & Supplies (USA) for white, superplastic earthenware clay.

puffing agent. A synthetic organic substance used to enhance the viscosity in varnishes and paints.

Puff Stuff. Trade name for polyester fibers and yarns used for textile fabrics.

pulp. See cotton pulp; wood pulp.

pulpboard. A relatively inexpensive *chipboard* made with a considerable quantity of *mechanical wood pulp*.

Pulpdent OBA. Trademark of Pulpdent Corporation (USA) for a visible light-cure, fluoride-releasing dental adhesive of thick consistency used for bonding metal, ceramic and clear sapphire orthodontic brackets.

pulpstone. Sandstone cut into wheels of specified diameter and width and used for crushing and grinding woodpulp and other grinding and polishing operations. Now alumina or silicon carbide products of such shape are also referred to as pulpstone.

pulpwood. Any wood, such as aspen, balsam fir, Douglas fir, hemlock, pine or spruce, used primarily for the production of *wood pulp* for making paper.

Pulse. Trademark of Dow Chemical Company (USA) for a series of engineering thermoplastics based on amorphous blends of polycarbonate and acrylonitrile-butadiene-styrene. Supplied in

various grades including unreinforced, glass fiber-reinforced, injection molding and plating. They have good mechanical and thermal properties, good impact resistance, low surface tension, good interfacial adhesion, and relatively low viscosity for easier processing, and are used for home appliance parts, power tools, lawn and garden products, automotive interior and exterior applications, telecommunications, and office machines.

Pulsus. Trade name of Saarstahl AG (Germany) for tool steel containing 0.35-0.42% carbon, 4.7-5% chromium, 1.5% nickel, 0.5% molybdenum, 0.2% vanadium, and the balance iron. Used for hot-work tools and dies.

Pultrex. Trademark of Creative Pultrusions, Inc. (USA) for lightweight high-strength pultruded composite structural shapes including angles, bars, beams, channels, rods, sheets, strips and tubes. They have excellent dimensional stability, excellent dielectric properties, good corrosion resistance, and good thermal insulating properties.

pultrusions. Composite products of constant cross-sectional shape (e.g., solid rod and bar stock, or structural profiles) formed by first pulling reinforcing fibers through impregnation baths, containing formulated, usually thermoplastic resins, preshaping the impregnated fibers to the desired profile, and then continuously pulling them through heated shaping dies where the final curing takes place.

Pulva-Lure. Trade name of Glidden Industrial Coatings (USA) for powder coatings used in metal finishing.

Pulvatex. Trademark for a powder composed of 60% rubber and 40% diatomaceous earth, and used in the construction trades.

pulverized lime. Burnt lime that will pass a 0.250-in (6.3-mm) US standard sieve.

pulverized fuel ash. See fly ash.

pulverized silica. Silicon dioxide obtained from crushed and finely ground quartz. See also silica powder.

Puma. Trade name of Hufnagel GmbH (Germany) for a cobalt-tungsten high-speed steel containing 0.65% carbon, 18% tungsten, 15% cobalt, 4.25% chromium, 1.6% vanadium, 0.8% molybdenum, and the balance iron. Used for cutting tools for rough machining.

pumice. A spongy volcanic rock or stone with glassy texture and relatively high silica content (up to 75% SiO_2) that is usually ground into a fine powder. Occurrence: Italy, New Zealand, USA (California, Idaho, Nebraska, New Mexico and Oregon). It has a density range of 0.37-0.90 g/cm^3 (0.013-0.033 $lb/in.^3$), and is used as a cleaning and polishing abrasive, a lightweight aggregate in concrete, for acoustic and thermal insulation, as a raw material in the manufacture of brick, in filtration and road-building, and in rubber and plastics. Also known as *pumice stone*.

pumice concrete. A lightweight concrete made with *pumice* powder as the essential aggregate. See also lightweight concrete.

pumice slag. See artificial pumice.

pumicite. A volcanic ash similar in composition to *pumice*, but occurring naturally in fine-grained, finely divided form. Used as a cementing ingredient with Portland cement in concrete, as a fine abrasive, and as a scouring agent.

pumpcrete. See pumped concrete.

pumped concrete. Any concrete that is conveyed through a hose or pipe by the use of a pump. Also known as *pumpcrete*.

pumpellyite. A bluish green mineral that is composed of calcium aluminum iron silicate hydroxide monohydrate, $Ca_2(Al,Fe)_3Si_3O_{11}(OH)_2 \cdot H_2O$, and also contain manganese or magnesium. Crystal system, monoclinic. Density, 3.20 g/cm^3; refractive index, 1.706. Occurrence: USA (California, Michigan), Japan, New Zealand.

pumpkin ash. The strong wood from the small to medium-sized tree *Fraxinus profunda*. Its properties are similar to those of *water ash*. Source: Eastern and central United States (especially Mississippi Valley, northern Florida and coastal regions of Virginia and the Carolinas). Used for furniture, handles, baskets and crates. See also ash.

punched fabrics. See perforated fabrics.

Punktalglas. German trade name for a specially ground spectacle glass used to correct astigmatism.

PUR-ABS alloys. Polymer blends which uniquely combine the outstanding toughness and abrasion resistance of polyurethane (PUR) with the lower cost and higher stiffness and dimensional stability of the acrylonitrile-butadiene-styrene (ABS).

Purac. Trademark of Gerardus Kaalverink (Germany) for polyurethane integral skin foams used in the manufacture of products for the automotive, building, furniture, toy and sporting goods industries.

Pural KR. Trademark of Sasol North America Inc. (USA) for high-purity aluminum oxide.

Puralloy. Trade name of Lobdell Company (USA) for a cast iron used in the manufacture of rolls for the pulp and paper industry.

Puralox. Trademark of Sasol North America Inc. (USA) for high-purity aluminum oxide.

Purco Chisel. Trade name of A.R. Purdy Company Inc. (USA) for a tough tool steel containing 0.7% carbon, 2% silicon, 0.9% manganese, and the balance iron. Used for punches and chisels.

Purebide. Trade name of Pure Carbon, Division of Morgan Advanced Materials & Technology (USA) for a series of silicon carbides. *Purebide R* is a reaction-bonded alpha silicon carbide material with high modulus of elasticity, high compressive strength, and excellent wear and corrosion resistance up to 1370°C (2500°F). It is used for bearings, seal rings, liners, and pressure plates. *Purebide S* is a molded and self-sintered material containing at least 95% pure silicon carbide and no free silicon or graphite. It is characterized by excellent abrasion, wear and corrosion resistance, good high-temperature resistance to 1370°C (2500°F) and good chemical resistance to strong acids and alkali solutions.

Purebrite. Trademark of Alcoa-Aluminum Company of America (USA) for anodized aluminum alloy sheeting.

Purecel. Trade name for precipitated calcium carbonate.

pure clay. A *clay* consisting theoretically of 39.5% alumina (Al_2O_3), 46.6% silica (SiO_2) and 13.9% water (H_2O).

pure gum. A nonpigmented, translucent, basic polymer.

pure iron. Iron (99.9+% pure) produced by electrolytic deposition from solutions of a ferrous salt (electrolytic iron), thermal decomposition of iron pentacarbonyl, $Fe(CO)_5$, or high-temperature hydrogen annealing of soft iron for prolonged periods of time. In spite of its purity, it contains traces of carbon, manganese, silicon, phosphorus, sulfur, copper and nickel. *Pure iron* has a density of 7.87 g/cm³ (0.284 lb/in.³), high magnetic permeability, high electrical resistivity, low coercivity, and is used for electrical and magnetic equipment.

Pure Metals. Trademark of Nevamar Decorative Surfaces (USA) for anodized aluminum sheets supplied with brushed and polished finishes in sizes of 600 × 2400 mm (24 × 96 in.), 600 × 3070 mm (24 × 120 in.), 1200 × 2400 mm (48 × 96 in.) and 1200 × 3070 mm (48 × 120). Used for wall and ceiling panels,

signage, furniture, fascias, columns, and store fixtures.

pure oxides. Oxide ceramics with very low impurity levels (typically less than 1 wt%) made by special chemical separation and subsequent processing. See also oxide ceramics.

pure silver solder. A corrosion-resistant, eutectic solder of 72% silver and 28% copper. Supplied in strip, wire and powder form, it has a melting temperature of 779°C (1435°F) and a brazing range of 779-899°C (1435-1650°F). Used for joining ferrous and nonferrous metals.

Puretung. Trade name of GTE Products Corporation (USA) for a commercially pure tungsten (99.9%) used for welding electrodes.

purified cellulose. See alpha cellulose.

purified cotton cellulose. Chemical cellulose made from cotton fibers and linters. See also alpha cellulose.

purified ozocerite. See ceresin.

purified potash. See pearl ash.

purified wood cellulose. Chemical cellulose made from wood. See also alpha cellulose.

purifier. A substance, usually a fluxing agent, added to a molten metal or alloy to absorb, reduce, oxidize or decompose impurities, and remove them in the form of gases or slags. See also flux (2).

Purilon. Trade name for rayon fibers and yarns used for textile fabrics.

Purite. Trademark of Olin Corporation (USA) for fused soda ash containing about 98% sodium carbonate (Na_2CO_3) used as a desulfurizing, fluxing and refining agent for iron, steel, and nonferrous metals.

Purlboard. Trademark of Guildfords Limited (Canada) for polyurethane insulation.

purl fabrics. (1) Textile fabrics knitted with inverted stitches to produce horizontal ridges and therefore a ribbed appearance.

(2) Textile fabrics, such as certain laces, braids or ribbons, that have chains of small loops along the edges.

Puro. Trademark of Remystahl (Germany) for an extensive series of austenitic, ferritic and martensitic stainless steels including various heat-resisting, free-machining and stabilized grades. The latter have improved weldability for structural applications. Used for chemical equipment, tanks, textile machinery, springs, ball and roller bearings, machine parts, surgical instruments, knives, cutlery, gages, pump and valve parts, trim, turbine wheels and blades, hardware, aircraft parts, vehicles, fittings, and household appliances.

Purocel. Trade name of Lawrence Industries (UK) for a bleached, purified wood cellulose used in the manufacture of high-quality writing paper.

Puron. Trade name of Westinghouse Electric Corporation (USA) for a high-purity iron (99.95%) used for spectroscopic and magnetic standards. See also pure iron.

Puros Allograft. Trademark of Sulzer Medica Limited (Switzerland) for a biocompatible, mineralized bone allograft material used in surgery to stimulate natural bone growth.

purple brown. A paint pigment based on iron oxide.

purple copper ore. See bornite.

Purple Cut Steel. Trade name of Allegheny Ludlum Steel (USA) for a tool steel containing 1.2% carbon, 1.2% silicon, 1.2% molybdenum, 1.2% chromium, 0.4% tungsten, 0.3% manganese, 0.25% vanadium, and the balance iron. It retains a keen cutting edge at purple heat, and is used for cutting tools, and dies.

purpleheart. See amaranth.

Purple Label. Trade name of Jessop Steel Company (USA) for a tungsten-type high-speed steel (AISI type T4) containing 0.74-0.80% carbon, 0-0.8% molybdenum, 14-18.5% tungsten, 5-8% cobalt, 1.5-2% vanadium, 4-4.5% chromium, and the balance iron. It has good deep-hardening properties, high hot hardness, excellent wear resistance and good machinability. Used for high-speed cutting tools, lathe and planer tools, drills and broaches. Also available is *Purple Label Extra*, a tungsten-type high-speed steel (AISI type T5) with similar properties and applications, but containing 0.78% carbon, 0.8% molybdenum, 18-18.5% tungsten, 4% chromium, 8% cobalt, 2% vanadium, and the balance iron.

purple of Cassius. See gold-tin purple.

purpurite. A rose to reddish purple mineral of the olivine group composed of manganese iron phosphate, $(Mn,Fe)PO_4$. Crystal system, orthorhombic. Density, 3.20 g/cm^3; refractive index, 1.86. Occurrence: USA (California, Dakota, Maine), Southwest Africa, Portugal.

PUR-SCHAUM. Trademark of Hanno-Werk GmbH & Co KG (Austria) for polyurethane foams.

PurSil. Trademark of Polymer Technology Group, Inc. (USA) for a series of hydrophobic, optically clear, biocompatible thermoplastic segmented silicone polyether urethane copolymers. They have high tensile strength, outstanding hydrolytic stability and nonadhesive silicone surfaces, and are supplied in the form of granules and free-flowing pellets. Used for medical implants including artificial heart components and ventricular assist devices.

pussy willow. (1) The fairly soft and light wood of a North American willow (*Salix discolor*). Pussy willow trees and shubs are more popular for their soft, furry, silvery gray catkins than their wood.

(2) A fabric made from net silk yarn in a plain weave and having delicate lines in horizontal direction.

putty. (1) A dough-like mixture of pigment and oil (usually whiting and linseed oil, sometimes mixed with white or red lead). Used for setting glass in window frames, filling imperfections (e.g., small holes and cracks) in wood or metal surfaces, as a filler for patterns, and for general sealing and caulking applications. See also caulking compound (2).

(2) A mild abrasive composed of crude tin oxide, a mixture of tin and lead oxides. Used for polishing glass, stone, ceramics, metals and jewelry, for dental polishes, and on enamels for giving opaque whiteness. Also known as *jeweler's putty; putty powder*.

Puzzle. Trade name of SA Glaverbel (Belgium) for a patterned glass featuring jig-saw shapes.

PVA fibers. See vinal fibers; vinylal fibers.

PVC-ABS alloys. Blends of polyvinyl chloride (PVC) and acrylonitrile-butadiene-styrene (ABS) resins that are commercially available in several grades including self-extinguishing for power-tool housings, and high-impact, high- and low-gloss for various injection-molded and extruded products. Plasticized blends of ABS and PVC are used for automobile dashboard assemblies.

PVC-acrylic alloys. Blends of polyvinyl chloride (PVC) and either polymethyl methacrylate (PMMA), acrylonitrile-butadiene-styrene (ABS), methacrylate-butadiene-styrene (MBS) or butylacrylate in which the PVC content is about 50-80%. They combine the toughness, chemical resistance and flame retardancy of PVC with the rigidity, and resistance to impact and heat distortion of acrylics, and are used for power-tool hous-

ings, automobile components and trim, computer and office equipment housings, modems, etc. See also PVC-ABS alloys.

PVC-CPE alloys. Blends of polyvinyl chloride (PVC) and chlorinated polyethylene elastomer (CPE) combining the toughness, flame retardancy, weatherability of PVC with the low-temperature flexibility of CPE. Commonly processed by injection molding or extrusion, they are used for electrical wire and cable coatings and jackets, extruded and molded shapes and film and sheet products.

PVC-CPVC alloys. Blends of regular-grade polyvinyl chloride (PVC) and chlorinated polyvinyl chloride (CPVC). The incorporation of CPVC significantly increases the chemical resistance and service temperature range (softening temperature range) of PVC. Used for piping and fittings for handling hot corrosives, and laboratory ware.

PVC-EPDM alloys. Blends of polyvinyl chloride (PVC) and ethylene-propylene diene monomer (EPDM). The incorporation of the elastomeric EPDM significantly enhances the toughness (impact strength) of PVC. Used for various molded products.

PVC-EVA alloys. Blends of polyvinyl chloride (PVC) and ethylene-vinyl acetate (EVA) elastomer combining the toughness, flame retardancy and weatherability of PVC with the flexibility, resistance to environmental stress-cracking and ultraviolet radiation of EVA. Used for molded articles including automotive components, equipment housings, and electrical products.

PVC fibers. See polyvinyl chloride fibers.

PVC-MBS alloys. Blends of polyvinyl chloride (PVC) and methacrylate-butadiene-styrene (MBS). They combine the toughness, chemical resistance and flame retardancy of PVC with the rigidity, and resistance to impact and heat distortion of MBS. Used for equipment housings and automobile components.

PVC-nitrile rubber alloys. Polymer blends which combine the toughness, chemical resistance and flame retardancy of polyvinyl chloride (PVC) with the high elasticity and flexibility and exceptional wear resistance of nitrile rubber (NR).

PVC-PMMA alloys. See PVC-acrylic alloys.

PVDC fibers. Polyvinylidene chloride fibers. See chloro fibers; saran fibers.

pyinkado. The very hard, durable wood of trees of the species *Xylia xylocarpa* and *X. dolabriformis*. Average weight, 980 kg/m^3 (61 lb/ft^3). Source: Burma. Used for heavy constructional work, marine structures, and flooring.

Pylon. (1) Trademark of Pylon Industries Limited (Bangladesh) for nylon 6 fibers and filament yarns used for textile fabrics.

(2) Trademark of Kohap Company (South Korea) for polypropylene staple fibers.

Pyral. Trade name of Creusot-Loire (France) for a corrosion-, creep- and heat-resistant stainless steel containing 0.11% carbon, 1% manganese, 1% silicon, 6% chromium, 0.5% molybdenum, and the balance iron. It has good mechanical properties at moderately elevated temperatures, and is used for heat exchangers and oil-refinery equipment.

Pyralin. Trade name of E.I DuPont de Nemours & Company (USA) for polyimide and cellulose nitrate plastics.

Pyralloy. Trade name of Dunford Hadfields Limited (UK) for a corrosion- and heat-resistant steel containing 0.3% carbon, 12-16% chromium, and the balance iron. It has good heat resistance to 800°C (1470°F), and is used for valves for internal combustion engines, and furnace parts and equipment.

Pyralux. Trademark of E.I. DuPont de Nemours & Company (USA) for flexible, solderable copper-clad laminates, bond plies, coverlays and modified acrylic bonding adhesives for printed circuit applications. They are available in four grades: *Pyralux AP, Pyralux FR, Pyralux LF* and *Pyralux PC.*

Pyram. Trade name of Corning Glass Works (USA) for flat sheets of *Pyroceram* glass-ceramic.

Pyramid. (1) Trade name of St. Marys Cement Company (Canada) for Portland and masonry cements.

(2) Trademark of BISCO, Inc. (USA) for stratified aggregate composites used for dental restorations. *Pyramid Dentin* handles like amalgam and is supplied in six shades, and *Pyramid Enamel* can easily be sculpted, takes a high polish, and is available in three shades.

Pyramidal. (1) Trade name of Windows Glass Limited (India) for a figured glass, with pyramid-shaped indentations.

(2) Trade name of SA des Verreries de Fauquez (Belgium) for a glass featuring pyramids in squares.

Pyramid Babbitt. Trade name of Magnolia Metal Corporation (USA) for a lead babbitt containing varying amounts of tin and antimony. It has a pouring temperature range of about 470-540°C (880-1005°F), and is used for bearings.

Pyran. Trademark of Schott Glas Jena (Germany) for a fire-resistant glass.

Pyranova. Trademark of Schott Glas Jena (Germany) for a fire-resistant glass.

pyrargyrite. A reddish, gray or black mineral with a red streak and an adamantine luster. It is composed of silver antimony sulfide, Ag_3SbS_3, and contains 22.3% antimony and 59.8% silver. Crystal system, rhombohedral (hexagonal). Density, 5.77-5.86 g/cm³; melting point, 486°C (907°F); hardness, 2.5 Mohs; refractive index, 3.084. Occurrence: Canada, Europe, Mexico, South America, USA (Colorado, Nevada, New Mexico). Used as an ore of silver. Also known as *dark-red silver ore; dark ruby silver.*

Pyra-Shell. Trade name of Shoeform Company (USA) for cellulose nitrate plastics.

Pyrasteel. Trade name of Chicago Steel Foundry (USA) for a series of corrosion- and heat-resistant chromium-nickel and nickel-chromium stainless steels used for heat-treating equipment, petroleum refining equipment, furnace parts and fixtures.

Pyr-Au-Bond. Trade name of Enthone-OMI Inc. (USA) for gold electroplates and plating processes used for wire bonding.

Pyrax. Trademark of R.T. Vanderbilt Company, Inc. (USA) for finely divided, white mineral powder obtained from *pyrophyllite* or *talc.* Used as filler in rubber, paints, and paper.

Pyre-ML. Trade name for polyimide resins.

pyrene. A polynuclear aromatic compound that is available as a colorless to yellow-orange powder (98+% pure) obtained from coal tar. It has a density of 1.27, a melting point of 149-156°C (300-313°F) and a boiling point of 404°C (759°F). Used in biochemistry and for optical applications. High-purity, zone-refined grades (99.9+%) are also used as aromatic reference standards. Formula: $C_{16}H_{10}$.

Pyrex. Trademark of Corning Inc. (USA) for a borosilicate glass containing 81% silica (SiO_2), 13% boria (B_2O_3), 3.5% sodium oxide (Na_2O), and 2.5% alumina (Al_2O_3). Commercially available in flat and hollow forms, it has a density of 2.25 g/cm³ (0.081 lb/in.³), a refractive index of 1.474, excellent thermal and thermal-shock resistance, a low coefficient of expansion (about 4.6×10^{-6}/K), high toughness, excellent chemical durability, and a softening temperature of 820°C (1508°F). Used for domestic ovenware, other heat-resistant glassware, chemical and laboratory ware and in fiber manufacture.

pyridine. A heterocyclic compound that is obtained from coal tar and has a six-membered ring containing one atom of nitrogen. It is available as a flammable, colorless or pale yellow liquid (99+% pure) with a density of 0.978 g/cm³, a melting point of -42°C (-44°F), a boiling point of 115°C (239°F), a flash point of 68°F (20°C), an autoignition temperature of 900°F (482°C) and a refractive index of 1.510. Used in the manufacture of pharmaceuticals (e.g. sulfa drugs, antihistamines and steroids), as a denaturant for ethyl alcohol, in the production of textile waterproofing agents, as a rubber accelerator, in high-performance liquid chromatography, spectrophotometry, chemistry, biochemistry, biotechnology and medicine. Formula: Formula: C_5H_5N.

pyridine polymer. A polymer or copolymer of 4-vinylpyridine (C_7H_7N) or methylvinylpyridine (C_8H_9N), such as *poly(4-vinylpyridine)* or *poly(4-vinylpyridine-co-styrene).*

Pyrista. Trade name of Firth-Vickers Stainless Steels Limited (UK) for a ferritic stainless steel containing 0.09% carbon, 29% chromium, 1.8% nickel, and the balance iron. It is usually supplied in bar form.

pyrite. A bronze to brass-yellow, or brown mineral with a brownish-black or greenish streak and metallic luster. It is composed of iron sulfide, FeS_2, contains theoretically 53.3% sulfur and 46.7% iron, but is frequently mixed with small amounts of arsenic, cobalt, copper, gold, nickel, selenium or other elements. Crystal system, cubic. Density, 4.8-5.2 g/cm³; melting point, 642°C (1188°F); hardness, 6.0-6.5 Mohs. Occurrence: Canada, France, Germany, Hungary, Italy, North Africa, Norway, Portugal, Spain, Sweden, UK, USA (Colorado, Connecticut, Massachusetts, New Jersey, New York, Utah, Virginia). Used as an ore of iron and sulfur, in the manufacture of sulfuric acid, sulfur dioxide and ferrous sulfate, in copperas, in the recovery of copper, gold and silver, in amber glass, as a filler in resin-bonded abrasives and brake linings, and for cheap jewelry. Also known as *fool's gold; iron pyrite.*

Pyrmant. Brand name for a commercially pure tin (99.85%) containing 0.048% lead, 0.022% copper, and 0.043% arsenic.

pyroaurite. A yellowish white mineral of the sjogrenite group with pearly or greasy luster. It is composed of magnesium iron carbonate hydroxide hydrate, $Mg_6Fe_2CO_3(OH)_{16} \cdot 4H_2O$. Crystal system, rhombohedral (hexagonal). Density, 2.12 g/cm³; refractive index, 1.564; hardness, 2-3 Mohs. Occurrence: Sweden.

pyrobelonite. A red mineral of the descloizite group composed of lead manganese vanadium oxide hydroxide, $PbMn(VO_4)(OH)$. Crystal system, orthorhombic. Density, 5.82 g/cm³; refractive index, 2.36. Occurrence: Sweden. Used as vanadium ore.

PyroCarbon. Trademark of Ascension Orthopedics, Inc. (USA) for a durable, stable, high-strength bioceramic coating over graphite substrate used in the manufacture of orthopedic implants (e.g., hands, feet, upper extremities, etc.) and heart valves.

Pyrocast. (1) Trade name of Pacific Foundry Company (USA) for a heat-resistant cast steel containing 1.75-2% carbon, 24-28% chromium, and the balance iron. Used for furnace parts and equipment, and heat-treating equipment.

(2) Trade name for a high-strength chromium-nickel cast iron.

pyrocellulose. See guncotton.

Pyroceram. Trademark of Corning Inc. (USA) for opaque-white *glass-ceramics* with nonporous structures of silicate grains. They have a density of 2.5 g/cm³ (0.09 lb/in.³), a hardness of about 7 Mohs, outstanding strength, high thermal-shock resistance and flexural strength, good corrosion resistance, and a softening temperature of 1350°C (2460°F). Used in cements for sealing inorganic materials, in coatings, for balls of valves and bearings, for heat-exchanger tubes, molded electrical and mechanical parts, telescope mirrors, and in special-purpose ceramic products.

pyrochlore. A reddish brown to red or yellow mineral of the pyrochlore group composed of calcium niobium titanium oxide fluoride, $(Ca,Na)_2(Nb,Ti)_2O_6F$. Crystal system, cubic. Density, 4.15 g/cm³; refractive index, 2-2.2; hardness, 5-5.5 Mohs. Occurrence: Canada (Quebec, British Columbia), Malawi, Russia. Used as an ore of niobium.

pyrochlore oxides. A family of refractory oxides composed of two cations of metallic element A, two cations of metallic element B and seven oxygen (O) anions corresponding to the general formula $A_2B_2O_7$. The A cation can be monovalent, divalent or trivalent, and the B cation tetravalent or pentavalent. *Pyrochlore oxides* are synthesized by co-precipitation of the oxides from solution, freeze-drying or sol-gel techniques. Many of these oxides are characterized, among other things, by their piezoelectric, ferrimagnetic or ferromagnetic, or semiconducting properties. Examples of such materials include neodymium zirconate, samarium zirconate, gadolinium zirconate and yttrium tin oxide.

pyrochroite. White to pale green or blue mineral of the brucite group composed of manganese hydroxide, $Mn(OH)_2$. It can also be made synthetically. Crystal system, hexagonal. Density, 3.25-3.27 g/cm³; hardness, 2.5 Mohs; refractive index, 1.724. Used in ceramics and materials research.

Pyrochrom. Trade name of Pose-Marre Edelstahlwerk GmbH (Germany) for a series of alloys containing up to 0.05% carbon, 15-25% chromium, 20-80% nickel, and the balance iron. Used for electric heating elements.

Pyro Die. Trade name of Crucible Materials Corporation (USA) for a tool steel containing 0.4% carbon, 1% chromium, 0.25% vanadium, and the balance iron. Used for hand tools and dies.

Pyrodie. Trade name of British Steel plc (UK) for an alloy steel containing 0.4% carbon, 1% silicon, 5% chromium, 1.3% molybdenum, 1% vanadium, and the balance iron. Used for dies.

Pyrodur. Trade name of Bergische Stahl Industrie (Germany) for a series of cast heat-resistant chromium-silicon, chromium-nickel-silicon and nickel-chromium-silicon steels, and several cast cobalt-chromium-iron and nickel-iron-chromium superalloys.

pyroelectric materials. A group of polar crystals, such as lithium sulfate monohydrate, tourmaline and barium titanate, in which a change in temperature simultaneously produces positive and negative electrical charges on the crystal surface.

Pyroferal. Czech trade name for a heat-resistant steel used for furnace equipment.

Pyrofil. Trademark of Grafil Inc. (USA) for pyrolytic carbon fibers.

Pyrofine. Trade name of Atofina SA (France) for a series of aluminum nitride powders used in the manufacture of engineering ceramics, and as reinforcements for ceramic-, polymer- and metal-matrix composites.

Pyrofix. Trade name of TradeARBED Inc. (USA) for a series austenitic stainless steel containing up to 0.25% carbon, 0-2% manganese, 0-1.5% silicon, 17-26% chromium, 19-37% nickel, and the balance iron. Used for high-temperature applications.

Pyroflex. Trade name of Spoldzielnia Pracy (Poland) for a toughened glass.

PyroFoam HP. Trade name of Rex Roto Corporation (USA) for a

refractory composite with excellent thermal insulating properties and good resistance to thermal cracking. It is composed of over 50% aluminosilicate fibers embedded in an amorphous silicon dioxide matrix, and its surface is infiltrated with a special ceramic to increase abrasion and erosion resistance, and strength. Used for nonferrous casting applications.

Pyro-Glass. Trade name of Precision Fiberglass Products, Inc. (USA) for high-temperature fiberglass tubing supplied in lengths up to 812 mm (32 in.).

Pyrograf. Trade name of Applied Sciences Inc. (USA) for vapor-grown high-purity carbon fibers made by gas-phase deposition of a hydrocarbon in the presence of a suitable catalyst. They have high stiffness, and electrical and thermal conductivity.

pyrographite. See pyrolytic graphite.

pyrolusite. An iron-black or dark steel-gray mineral with a black or bluish-black streak and a dull or metallic luster. It is composed of manganese dioxide, β-MnO_2, and can also be made synthetically. Crystal system, tetragonal. Density, 4.73-4.86 g/cm³; hardness, 2-2.5 Mohs. Occurrence: Argentina, Australia, Brazil, Canada, Cuba, Europe, Ghana, India, USA (Arkansas, Georgia, Lake Superior region, Massachusetts, New Mexico, Vermont, Virginia). Used as the most important ore of manganese, as an oxidizer, as an adherence-promoting agent for porcelain enamels on sheet iron and steel, and as a purple or red colorant in glazes, glass and porcelain enamels.

pyrolytic boron nitride. A high-temperature protective boron nitride (BN) that is an electrical insulator and good thermal conductor, resistant to corrosive atmospheres (acids, hot ammonia and most molten metals), has a density of 2.1 g/cm³ (0.07 lb/in.³), decomposes at 2500°C (4530°F), and has a maximum service temperature (in air) of 800°C (1470°F). It is applied as a coating to protect and seal graphite, and for compound semiconductor crystal growth and wafer-processing applications.

pyrolytic carbon. Carbon formed by low-pressure sublimation of pyrolytic graphite at approximately 2277°C (4130°F), and now considered an allotrope of carbon. The original pyrolytic carbon had about 3-8% silicon added as an alloying element, but more recently silicon-free pure carbon varieties have been produced. *Pyrolytic carbon* has a two-dimensional or turbo-stratic crystal structure with densities ranging from 1.4-2.1 g/cm³ (0.05-0.07 lb/in.³), and is used as a thin coating on artificial heart valves. See also highly oriented pyrolytic carbon; pyrolytic graphite.

pyrolytic coatings. Thin coatings of a material, such as boron nitride, carbon, silica, silicon carbide or silicon nitride, deposited on a suitable substrate by thermal decomposition of a volatile compound, usually in a vacuum.

pyrolytic graphite. A dense, nonporous, ultrapure graphite made by the thermal decomposition of a carbonaceous gas, such as methane. It is usually grown with a closely controlled crystal orientation, and the direction of applied stress is parallel to the direction of the greatest modulus and strength. Available in coating form, it is impervious to fluids, and has high flexural and tensile strengths, good thermal resistance and strength retention even at 3300°C (5970°F) and high flame resistance. Used for rocket nozzles, parts for missiles, spacecraft and reentry vehicles, nuclear reactors, as an anisotropic electrical and thermal conductor, for other high-temperature applications, in the redensification of carbon-carbon composites, and on silicon-carbide fibers to improve the conductivity. Abbreviation: PG. Also known as *pyrographite*.

Pyromark. Trade name of Tempil, Inc. (USA) for an organic paint based on silicone that transforms to organic silica at temperatures above 1090°C (1994°F). Used as an ablative paint for temperatures up to 1370°C (2500°F).

pyromelane. See brookite.

Pyromet. Trade name of Carpenter Technology Corporation (USA) for a series of austenitic stainless steels, hot-work and high-speed tool steels, and high-strength precipitation-hardenable nickel-base superalloys.

pyromets. A group of high-strength crystalline alloys based on chromium, cobalt, iron, nickel and/or other transition metals strengthened with a metalloid, such as boron. *Pyromet* alloys are manufactured by first producing a melt-spun amorphous metal ribbon or tape that is then reduced to a powder and subsequently hot isostatically pressed or hot-extruded and crystallized.

Pyromic. Trade name of Telcon Metals Limited (UK) for a series of electrical resistance alloys containing nickel, chromium and sometimes iron. The 80Ni-20Cr alloy can be used up to 1150°C (2100°F).

pyromorphite. A yellow, green or brown mineral of the apatite group composed of lead chloride phosphate, $Pb_5(PO_4)_3Cl$. It can also be made synthetically. Crystal system, hexagonal. Density, 6.5-7.04 g/cm^3; hardness, 3.5-4 Mohs; refractive index, 2.058. Used as an ore of lead. Also known as *green lead ore*.

Pyron. (1) Trade name of SWB Stahlformguss Gesellschaft mbH (Germany) for an extensive series of corrosion- and heat-resistant chromium-silicon and chromium-nickel steels used for furnace equipment, heat-treating equipment, valves, pumps, turbine parts, petroleum-refining equipment, and chemical-process equipment.

(2) Trade name of Pyron Corporation (USA) for a series of low-density iron powders used in friction applications.

(3) Trade name of Zoltek Corporation (USA) for oxidized, polyacrylonitrile precursor-derived carbon fibers with an average diameter 13 μm (512 μin.). They have excellent resistance to organic solvents and weak acids, good resistance to strong acids and weak bases, poor resistance to strong bases, good electrical properties, and good thermal stability.

Pyroneal. Trade name of Heppenstall Company (USA) for an oil or air-hardening hot-work tool steel containing 0.55% carbon, 2.2% nickel, 0.9% chromium, 0.7% molybdenum, 0.6% silicon, and the balance iron. Used for press and upsetter dies, hot-work dies, and punches.

Pyro-Paint. Trademark of Aremco Products, Inc. (USA) for a series of ceramic coatings based on boron nitride and zirconium oxide.

pyrope. A colorless mineral of the garnet group composed of magnesium aluminum silicate, $Mg_3Al_2(SiO_4)_3$. It can also be made synthetically. Crystal system, cubic. Density, 3.58; refractive index, 1.702. Used in ceramics, and as a gemstone. Also known as *Bohemian garnet*.

Pyropel. Trademark of Albany International, Inc. (USA) for a lightweight, nonwoven high-temperature insulation material based on polyimide. It has a fibrous, three-dimensionally reinforced construction, exceptional rigidity and strength, extended performance at cryogenic temperatures (-149°C or -236°F) and elevated temperatures (316°C or 600°F), good thermal and dimensional stability, low heat release, low smoke evolution, good resistance to most chemicals, and very good thermal insulation and acoustic damping properties. Used as a core material for thermal and acoustic insulation, interior panels and ducts for aircraft and ships, and composite fire stops in aircraft, ships,

and vehicles. *Pyropel HD* is a thermoset polyamide plate machining stock with good to excellent chemical resistance and a maximum service temperature of 285°C (545°F). Used for valve parts, seals and manifolds in chemical-processing equipment, exhaust and clamp rings, thermal spacers, and soldering fixtures.

pyrophanite. A gray, or yellowish brown mineral of the corundum group composed of manganese titanium oxide, $MnTiO_3$. It can be made from equimolar amounts of manganese carbonate ($MnCO_3$) and potassium titanate (K_2TiO_3). Crystal system, rhombohedral (hexagonal). Density, 4.54 g/cm^3; refractive index, 2.48; hardness, 5 Mohs.

pyrophoric alloys. Alloys that spark when slight friction is applied by another metal (e.g., by striking with a hammer at a certain angle). Examples include mischmetal, ferrocerium, zirconium-lead alloys, and alloys of 45% iron, 20% chromium, 10% manganese, 10% antimony and 15% titanium. Used in lighters for gas stoves and for cigarette lighters. See also Auer metal; Kunheim metal; sparking metal.

pyrophoric material. A solid (e.g., titanium dichloride or phosphorus) or liquid (e.g., tributylaluminum, triethylgallium or trimethylindium) that will self-ignite spontaneously in air at elevated temperatures. Used for the tips on pocket lighters and similar devices.

pyrophoric powder. A finely divided metal or metal-alloy powder whose particles will auto-ignite and burn upon exposure to oxygen or air.

pyrophyllite. A white, gray, pale blue, green or brown phyllosilicate mineral of the talc group with a pearly to greasy luster. It is composed of hydrated aluminum silicate, $Al_2Si_4O_{10}(OH)_2$. Crystal system, monoclinic. Density, 2.8-2.9 g/cm^3; melting point, 1800°C (3270°F); hardness, 1-2 Mohs; refractive index, 1.58-1.59. Occurrence: Canada (Newfoundland), Japan, New Zealand, USA (California, North Carolina). Used in ceramics, refractories, castables, plastic and gunning mixes, as paint extender and paper filler, also in wallboard, tile, slate pencils and insulators, as replacement for talc, e.g., in dusting rubber, as buffer in high-pressure equipment, and as sealer in pressure forming of synthetic diamonds at elevated temperatures. Also known as *pencil stone*.

Pyro-Putty. Trademark of Aremco Products Inc. (USA) for trowelable putties made by blending stainless steel powder with an aqueous silicate ceramic binder. They can be used for continuous temperatures up to 1090°C (1994°F), and have thermal expansion coefficients matching those of iron and steel. *Pyro-Putty 2400* is a relatively thin room-temperature-curing formulation for filling small cross sections (pinholes), and *Pyro-Putty 653* is a viscous formulation that cures in 1 hour at 120°C (250°F) and is used for larger cross section up to 20 mm (0.75 in.) in diameter. *Pyro-Putty* is used in the repair of iron and steel components operating at high temperatures, such as automotive gaskets and manifolds, carbon and stainless structures, industrial furnaces and ovens, exhaust systems, and in foundries for plugging defects in castings.

PyroQuartz. Trade name of Pyromatics, Inc. (USA) for high-purity (99.998+%) fused quartz and seamless, precision-molded fused quartz products used in the chemical industries for vessels, tanks, boats, beakers and dishes.

Pyros. Trade name of Creusot-Loire (France) for a heat-resisting nickel alloy containing 7% chromium, 5% tungsten, 3% manganese, and 3% iron. Used in dilatometric temperature indicating or control devices.

Pyrosil SNM. Trade name of Pyro Shield Inc. (USA) for ceramic-free, nonrespirable needled silica insulation with a service temperature up to 1540°C (2800°F) used for furnace linings, and various applications in the heat-treating industries.

pyrosilicates. Silica materials in which only one of the four oxygen atoms of the SiO_4 tetrahedra is shared with adjacent tetrahedra, i.e., they have three nonbridging oxygen atoms per tetrahedron. The general formula is XSi_2O_7, where X represents one or more metals, such as calcium, sodium, samarium or ytterbium.

pyrosmaltite. A green or brown mineral composed of manganese iron silicate chloride hydroxide, $(Mn,Fe)_8Si_6O_{15}(OH,Cl)_{10}$. Crystal system, hexagonal. Density, 3.06 g/cm³; refractive index, 1.6785; hardness, 4 Mohs. Occurrence: Australia.

pyrostibite. See kermesite.

pyrostilpnite. A red mineral composed of silver antimony sulfide, Ag_3SbS_3. Crystal system, monoclinic. Density, 5.94-5.98 g/cm³. Occurrence: Czech Republic, Germany.

Pyrostop. Trade name of Flachglas AG (Germany) for a fire-resistant glass.

Pyrotem. Trade name of Heppenstall Company (USA) for an oil-hardening, shock-resistant tool steel containing 0.55% carbon, 2.1% nickel, 0.87% chromium, 0.73% molybdenum, and the balance iron. Used for hot-work dies and cold shear knives.

Pyrotex. (1) Trademark of Raymark Industries, Inc. (USA) for a felt composed of a mixture of ceramic fibers and long-fiber *chrysotile* asbestos supplied in the form of rolls, sheets and tapes. Used as thermal insulation, for electrical insulating tape, and for protective applications.

(2) Trademark of Raymark Industries, Inc. (USA) for strong, asbestos-laminated phenolics with good electrical and high-temperature insulating properties. Used for aerospace, electrical and structural applications.

Pyrotherm. Trade name of Pose-Marre Edelstahlwerk GmbH (Germany) for an extensive series of wrought and cast corrosion- and/or heat-resistant steels including various austenitic, ferritic and martensitic stainless grades. Used for the manufacture of furnaces for the metallurgical, ceramics and glass industries, and for furnace parts and fixtures, chemical plant equipment, heat-treating equipment, heat exchangers, petroleum refining equipment, combustion chambers, gas-turbine parts, etc.

Pyrotough. Trade name of British Steel Corporation (UK) for a hot-work tool steel containing 0.35% carbon, 2.75% chromium, 0.5% tungsten, 0.25% vanadium, and the balance iron. Used for hot-work tools and dies.

Pyrovan. Trade name of Latrobe Steel Company (USA) for an air-hardening hot-work tool steel containing 0.75% carbon, 1% silicon, 1.1% molybdenum, 5.2% chromium, 2.5% vanadium, and the balance iron. It has good wear and high-temperature resistance, high hardness, and is used for forging and forming dies and hot-press tools and dies.

Pyrowear. Trademark of Carpenter Technology Corporation (USA) for special corrosion-resistant chromium-cobalt-nickel carburizing steels with high impact strength and fracture toughness, and good corrosion and fatigue resistance. Used for high-temperature applications including gears operating in high temperature environments, such as gear boxes and airframe parts for aircraft and helicopters, and for bearings.

pyroxenes. A group of rock-forming silicate minerals corresponding to the general formula $ABSi_2O_6$, where A usually represents magnesium, calcium, lithium, sodium or divalent iron, and B magnesium, aluminum or divalent iron. Prominent members of this group include *augite, diopside, enstatite, jadeite* and *spodumene*. Density range, 3.2-4.0 g/cm³; hardness range, 5.5-6.0 Mohs.

pyroxferroite. A yellow mineral of the pyroxenoid group composed of iron calcium silicate, $(Fe_{0.86}Ca_{0.14})SiO_3$. Crystal system, triclinic. Density, 3.80 g/cm³; refractive index, 1.755. Occurrence: Moon (collected during NASA Apollo 11 mission).

pyroxmangite. A rose-pink mineral of the pyroxenoid group that are composed of manganese silicate, $MnSiO_3$, and may also contain iron. It can also be made synthetically. Crystal system, triclinic. Density, 3.69 g/cm³; refractive index, 1.736. Occurrence: Japan, Scotland, Sweden, USA (South Carolina, Idaho).

pyroxylin. A low-nitrogen form of *nitrocellulose* (usually 10-11% nitrogen) that is soluble in ether-alcohol mixes and acetone, and is used in the manufacture of collodion, lacquers, varnishes and plastics, and in biology and biochemistry. Also known as *collodion cotton; soluble guncotton; soluble nitrocellulose.*

pyroxylin cement. A common cement made by dissolving *nitrocellulose* in a chemical solvent with or without compounding ingredients, such as synthetic resins or plasticizers. It dries by solvent evaporation, adheres to many substrates, and is used for various cementing applications.

pyrrhotite. A bronze-yellow to brownish red, weakly magnetic mineral of the nickeline group with black streak and metallic luster. It is composed of iron sulfide, $Fe_{1-x}S$ (x = 0-0.2), also contains small amounts of nickel, copper, cobalt and manganese, and exists in several polytypes including 1C through 7C and 11C. The 3C and 4C polytypes have a slightly different iron-to-sulfur ratio represented by the formula Fe_7S_8. They occur naturally, e.g., in India and Rumania, and can also be made synthetically. Crystal system, hexagonal. Density, 4.58-4.67 g/cm³; hardness, 4 Mohs. Occurrence: Canada (Manitoba, Ontario), India, Japan, Mexico, Rumania, Switzerland, USA (Pennsylvania, Tennessee). Used as an ore of nickel, an minor ore of iron and in the manufacture of sulfuric acid. Also known as *magnetic pyrites; pyrrhotine.*

pyrrhotine. See pyrrhotite.

pyrrole. A heterocyclic compound that contains four carbon atoms and one nitrogen atom in its ring structure, and is the parent compound of several biological substances including chlorophyll and the porphyrins. It is available as a brownish to yellow oil (98+% pure) with a density of 0.967 g/cm³, a melting point of -23°C (-9°F), a boiling point of 131°C (268°F), a flash point above 92°F (33°C) and a refractive index of 1.509. Used as a chemical reagent in chemistry, biochemistry, biotechnology, materials research, electronics, medicine, and pharmacology. Formula: C_4H_5N.

pyrrole black. A black pigment based on pyrrole (C_4H_5N) that on oxidation yields tetracarboxylic acid, and is an amorphous semiconductor with a small band gap.

pyrrolidine. A flammable, colorless to light yellow liquid (95+% pure) with a density of 0.866 g/cm³, a melting point of -60°C (-76°F); a boiling point of 87-88°C (187-190°F), a flash point of 37°F (2°C) and a refractive index of 1.443. Used as a curing agent for epoxies, in rubber accelerators, as an inhibitor, as an intermediate for several chemicals, and in biochemistry and biotechnology. Formula: C_4H_9N.

2-pyrrolidone. A pale yellow liquid (99% pure) with a density of 1.12 g/cm³, a melting point of 23-25°C (73-77°F), a boiling point of 245°C (473°F), a flash point of 265°F (129°C) and a refractive index of 1.487. Used as a monomer for polybutyrolactam (nylon 4), as as a solvent for polymers, sugars and polyhydroxylic alcohols, and as a plasticizer for acrylic floor polish latexes. Formula: $(CH_2)_3C(O)NH$. Also known as *2-pyrrolidinone*.

Pytex. Trademark of Shreeji Screens & Filters (India) for polyester monofilament fabrics.

Pythagoras Ware. Trade name of W. Haldenwanger Technische Keramik GmbH & Co. KG (Germany) for a mullite-type porcelain with good crushing strength, fair tensile strength, low thermal expansion, and an upper service temperature above 1400°C (2550°F). Used for gastight ceramic whiteware.

Python. Trade name of AL Tech Specialty Steel Corporation (USA) for a water-hardening tool steel (AISI type W2) containing 0.8-1.2% carbon, 0.25% manganese, 0.25% silicon, 0.2% vanadium, and the balance iron. It has excellent machinability, good to fair deep-hardening properties, good to fair wear resistance and toughness, low hot hardness, and is used for machine parts, dies, and hand and cutting tools.

PZT. See lead zirconate titanate.

Q

Q-Alloys. Trade name of Alloy Engineering & Casting Company (USA) for an extensive series of corrosion- and/or heat-resistant cast alloys including various iron-chromium, iron-chromium-nickel, iron-nickel-chromium and nickel-chromium-iron alloys as well as several austenitic stainless steels. Depending on the particular composition and heat treatment, they are used in plant and equipment for severe heat and corrosion conditions, such as industrial furnaces and furnace parts, heat-treating equipment, equipment for the chemical and pulp and paper industries, rolling-mill equipment, and for high-temperature applications.

QA Wire. Trade name of Engelhard Industries (USA) for a gold-based alloy containing platinum and palladium, and supplied in wire form.

Qazul. Trade name of ICI Limited (UK) for nylon fibers and yarns used for the manufacture of luggage.

Qbeads. Trademark of Quantum Dot Corporation (USA) for encoded nano-sized beads used for genotyping, gene expression and proteomics.

Q Brand. Trade name of R.W. Carr & Company Limited (UK) for plain-carbon steels containing 0.4% carbon, 0.8% manganese, and the balance iron. Used for die inserts.

Q-Cel. Trademark of PQ Corporation (USA) for sodium silicate supplied as hollow microspheres, and used as low-density filler in materials, such as plastics, ceramics and composites.

Qdot. Trademark of Quantum Dot Corporation (USA) for highly fluorescent semiconductor nanocrystals (about 2-10 nm) with highly crystalline structures used for biotechnology and biochemistry applications including biological labeling, detection reagents, DNA chips, immunology and flow cytometry.

Q-Fiber. Trademark of Schuller International (USA) for glass fibers.

Q-Fiber-Felt. Trade name of Manville Corporation (USA) for silica fiber insulation products with a heat resistance up to 980°C (1795°F), supplied in felt form.

Qiana. Trademark E.I. DuPont de Nemours & Company (USA) for a synthetic fiber based on polyamide (nylon), and used for flexible fabrics with a feel and drape similar to silk.

qiviut. A soft, durable natural textile fiber obtained from the Arctic musk ox (*Ovibos moschatus*).

Q-ROK. Trademark of US Silica Company for unground silica (quartz) supplied in several grades.

QST steel. See quenched and self-tempered steel.

Q-Temp. Trade name of US Steel Corporation for a series of alloy steels including several low-carbon boron and low-carbon chromium and chromium-vanadium grades used for ma-

chinery parts, and fasteners.

Q-thane. (1) Trademark of K.J. Quinn & Company, Inc. (USA) for thermoplastic polyurethane elastomers.

(2) Trademark of K.J. Quinn & Company, Inc. (USA) for adhesives used for bonding leather, leather substitutes and leather products (e.g., shoes), plastic film, fabrics, floors, wood, paper, and foil.

(3) Trademark of K.J. Quinn & Company, Inc. (USA) for oil-free, moisture-curing protective polyurethane floor and wall coatings that provide resistance to abrasion and chemicals.

Quacorr. Trade name of QO Chemicals (USA) for several furane resins.

Quadrant. Trade name of Cavex (Netherlands) for dental resin composites and bonding systems including *Quadrant Core*, a light-cure dental composite for core-build-ups, *Quadrant Uni-Bond*, a light-cure, fluorine-releasing bonding system containing bisphenol A glycidylether dimethacrylate (BisGMA) and hydroxyethyl methacrylate (HEMA), and *Quadrant Unifix*, a multi-purpose dental adhesive agent.

Quadrax Biaxial Tape. Trademark of Quadrax Corporation (USA) for biaxial tape preforms.

Quadrello. Trade name of Vetreria di Vernante SpA (Italy) for a patterned glass.

Quadrettato. Trade name of Fabbrica Pisana SpA (Italy) for a patterned glass.

Quadrionda. Trade name of Fabbrica Pisana SpA (Italy) for patterned glass. A small version of this pattern is known as *Quadrionda Piccolo*.

Quaker. Trade name of Quaker Alloy Casting Company (USA) for a series of plain-carbon and alloy steel castings including various heat-resisting and low-temperature-resisting chromium-molybdenum, nickel-chromium-molybdenum and copper-nickel grades. Also included under this trade are several casting alloys of the nickel-molybdenum-iron and nickel-copper type.

Quality Steel. Trade name of Engelhard Corporation (USA) for a series of air-hardening steels containing 0.3-0.35% carbon, 1.2-1.5% chromium, 4.5-5% nickel, 0.3-0.5% manganese, and the balance iron. Used for gears and pinions.

quality wools. A term used for wools with a quality grading between 28 and 100, and including superior-quality *merino* wools (above 60) and coarse wools (28-40).

Quallofil. Trademark of E.I. DuPont de Nemours & Company (USA) for high-loft thermal insulation made of a proprietary blend of seven-hole fibers (including polyester), and used in high-performance outerwear and sleeping bags. An antimicrobial grade, known as *Quallofil with Allerban*, is also available.

Qualloform. Trade name of E.I. DuPont de Nemours & Company (USA) for polyester fibers used for textile fabrics.

quandilite. A colorless synthetic mineral of the spinel group composed of magnesium titanium oxide, Mg_2TiO_4. It is made by heating titanium dioxide (TiO_2) and magnesium carbonate ($MgCO_3$) under prescribed conditions. Crystal system, cubic. Density, 3.55 g/cm^3.

quantum dots. See quantum-dot particles.

quantum-dot colloidal particles. See quantum-dot particles.

quantum-dot molecules. See quantum-dot particles.

quantum-dot particles. A nanocrystal composed of a semiconductor material, such as cadmium selenide, with a typical particle size range of 2-10 nm (0.08-0.4 μin.) for effective three-dimensional electron confinement. Used chiefly in electronics, for high-density and high-speed devices, and in biological labeling and medical diagnostics. Also known as *nanodots; quantum-dot colloidal particles; quantum-dot molecules; quantum dots*.

quantum well. A potential well of semiconducting material in which the excitons or carriers are confined in one dimension within a layer that has a thickness in the nanometer range and may be as small as only 2 nm (0.08 μin). Used in the manufacture of miniaturized electronic and optoelectronic components.

quantum well materials. A group of semiconducting materials, such as gallium arsenide (GaAs), indium phosphide (InP), aluminum gallium arsenide (AlGaAs), indium gallium arsenide (InGaAs), indium gallium phosphide (InGaP) and indium gallium arsenide (InGaAs), that have thicknesses in the nanometer range and are suitable for use in the manufacture of electronic components and vertical-cavity surface emitting lasers (VCSELs).

quantum wire. A two-dimensional *quantum well.*, i.e., a quantum well in which the electrons are confined in two dimensions.

Quar-A-Poxy. Trademark of H.B. Fuller Company (USA) for epoxy-base concrete tile adhesives.

quarry tile. A glazed or unglazed floor tile whose top surface area and thickness are usually 39 cm² (6 in.²), or more and 13 to 19 mm (0.50 to 0.75 in.), respectively. It is highly resistant to abrasion and many liquids, and is made by the extrusion process from natural clays or shales.

quartered oak. Oak sawed or sliced in a radial direction resulting in a striking grain pattern. Used where spectacular wood grain effects are desired, e.g., for cabinets.

quarter-hard wire. An aluminum wire processed to produce a tensile strength intermediate between that of soft wire and half-hard wire. See also half-hard wire; hard-drawn wire; soft wire.

quarter-sawn lumber. See edge-grained lumber.

Quarto. Trade name of Libbey-Owens-Ford Company (USA) for a laminated glass composed of two sheets of glass, one of which has a ground and polished exterior surface.

quarto plate. Steel plate, typically ranging from 5 to 100 mm (0.2 to 4.0 in.) in thickness, up to 3.2 m (10.5 ft.) in width and up to 14 m (46 ft.) in length, produced in a four-high mill, i.e., a rolling mill with two small-diameter working rolls and two large-diameter backup rolls. It is a heavy engineering product used in the manufacture of chemical tanker ships, industrial process equipment, and other heavy-duty applications. Also known as *reversing mill plate*.

Quartz. Trade name of SA Glaverbel (Belgium) for a patterned glass.

quartz. A colorless, white to reddish polymorphic crystalline mineral with a vitreous luster composed of silicon dioxide, SiO_2. It occurs in many varieties, such as *agate, chalcedony, chert, flint, or opal*, and can also be made synthetically by crystal-growing techniques. Crystal system, hexagonal. Density, 2.66 g/cm³; melting point, 1710°C (3110°F); hardness, 7 Mohs; refractive index, 1.54. It has piezoelectric and pyroelectric properties, exhibits double refraction, and is transparent to ultraviolet radiation. Occurrence: Found abundantly on every continent. Used as a glass former, as a vitrification aid in ceramic composition, as a abrasive in barrel-finishing, polarizing prisms, spectrographs, prisms and lenses for infrared systems, lamps, resonators, piezoelectric controls in filters, radio and television components, instrument bearings, as a substrate for electronic applications, in electronic components including tubing, and as a gemstone. See also fused quartz; synthetic quartz; high quartz;

low quartz; beta quartz.

Quartz-coat. Trademark of Aremco Products, Inc. (USA) for a white, quartz-base glass-ceramic filler with a density of 1.52 g/cm³ (0.055 lb/in.³), and excellent adhesive strength and color stability. Used for reflective coatings for heater and quartz lamp applications.

quartz crystal. A pure, colorless, transparent variety of *quartz* with low brilliance, found in Brazil, Madagascar, Japan and the USA (Arkansas). Used for lenses, wedges and prism components in optical instruments, and for jewelry, and ornaments. It was formerly widely used for radio oscillators. Also known as *mountain crystal; rock crystal*. See also crystal (1).

Quartzel. Trademark of Quartz & Silice SA (France) for high-purity (99.99%) fused silica fibers with an average diameter of 14 μm (551 μin.). They have high tensile strength, excellent resistance to thermal shock, a very low coefficient of thermal expansion, a very low dielectric strength, and negligible water absorption. Used for structural and electronics composites.

quartz fibers. Fibers drawn from the mineral *quartz* and having a density of 2.20-2.25 g/cm³ (0.079-0.081 lb/in.³), an average filament diameter of 1-15 μm (39-591 μin), moderate tensile strength, an upper service temperature of 1050°C (1920°F), and outstanding temperature resistance and electrical signal transparency. Used for reinforcing fibers in ablatives, thermal barriers, antenna windows, and radomes, for torsion threads in delicate instruments, and for yarns and fabrics.

quartz glass. See fused quartz.

quartzite. See ganister.

quartz paper. A ceramic paper made by mixing *quartz fibers* with a bonding agent (e.g., bentonite) and forming into mats or sheets on a paper machine. Used for electrical insulation. See also quartz.

quartz pebbles. Hard, tough, rounded stones of quartz used for grinding media in ball, pebble and tube mills for grinding cement, minerals, ores, and other materials.

quartz topaz. See citrine.

quartz yarn. A highly durable yarn spun from quartz fiber and discontinuous textile waste fibers whereby the former forms the backbone of the yarn. See also quartz; quartz fibers.

Quarzal. Trade name of PAN-Metallgesellschaft (Germany) for a series of aluminum-copper alloys, used for bearings and wear-resistant machine parts.

Quasar. Trademark of Rocky Mountain Orthodontics Inc. (USA) for a composite resin used in orthodontics for bonding ceramic brackets.

Quash. Trademark of Dow Chemical Company (USA) for foams made from Dow *Index* Interpolymers, and used for sound management applications.

Quasiceram. Polish trade name for glass-ceramics.

quasicrystalline materials. Compounds formed from metals, such as aluminum, copper, iron, magnesium or manganese, that have quasiperiodic atomic structure sequences, i.e., atomic structure sequences that are not strictly repetitive. Quasicrystals are composed of three-dimensional icosahedra arranged in a way similar to two-dimensional Penrose tilings. They exhibit 5-fold symmetry and have properties that are not characteristic of other metallic compounds. Typical commercial quasicrystalline materials include aluminum-copper-iron, aluminum-palladium-manganese, titanium-zirconium-nickel and aluminum-manganese lanthanide. Used for surgical and acupuncture needles, dental reamers, as selective solar absorbers, in surface coatings, in photomagnetic recording, as heat barriers, and for hydrogen storage applications. Also known as *icosahedral phases; quasi-crystals.*

quasi-isotropic laminate. A laminate whose layers or plies have been oriented such that the high-strength direction varies. It thus approximates isotropic behavior. See also isotropic laminate.

quasi-optic composites. Composite materials usually manufactured from rare-earth materials and ferrites (e.g., gadolinium/ferrite composites) that change the plane of polarization of incident microwave and millimeter range waves, and are used in radar-deception equipment.

quaternary alloy. An alloy composed of four principal element. For example, a free-cutting phosphor bronze is a quaternary alloy of about 88% copper, 4% lead, 4% zinc and 4% tin.

quaternary steel. A steel that contains two alloying elements in addition to carbon and iron, e.g., a steel composed of carbon, iron, chromium and nickel.

p-quaterphenyl. A crystalline compound (99.5+% pure) with a melting point of 316-319°C (601-606°F), a boiling point of 428°C (802°F)/18 mm, a minimum absorption wavelength of 350 nm, and a peak lasing wavelength of 374 nm. Used as a laser dye, and as a wavelength shifter in soluble scintillators. Formula: $C_6H_5C_6H_4C_6H_4C_6H_5$.

quaterpolymer. A polymer consisting of four repeating units or monomers, e.g., a polymer made of acrylonitrile, ethylene, propylene and styrene monomers.

Quatre Eclairs. Trade name of Creusot-Loire (France) for an oil-hardening tool steel containing 0.85% carbon, 0.6% molybdenum, 1.6% vanadium, 18% tungsten, and the balance iron. Used for lathe and planer tools.

Quatrex. Trademark of Ciba-Geigy Corporation (USA) for several *bisphenol A* epoxy resins.

Quattro Punte. Trade name of Fabbrica Pisana SpA (Italy) for a figured glass.

Quazite. Trademark of Strongwell Company (USA) for strong, durable, corrosion-resistant, lightweight, precast polymer concrete products used for underground construction applications, e.g., enclosures, service boxes and panel vaults.

Quebec. Trademark of Ciment Quebec Inc. (Canada) for Portland-type standard and masonry cements.

quebracho. The wood of any of several South American hardwood trees of the genera *Aspidosperma, Quebracho* and *Schinopsis* in the order of Sapindales. *Red quebracho* is the hard, durable, brownish-red wood of the tree *A. quebracho* belonging to the dogbane family and *white quebracho* the hard, durable, light wood of any of several trees of the genus *Schinopsis* belonging to the cashew family, especially *S. lorentzii* and *S. balansae*. Source: Argentina, Brazil, Paraguay. Used as a source of tannin, and for medicine, timber, crossties, posts and firewood. See also red quebracho; white quebracho.

Queen. Trademark of China Man-Made Fiber Corporation (Taiwan) for viscose rayon staple fibers and filament yarns.

Queensland walnut. The hard, strong, brown wood from the hardwood tree *Endiandra palmerstonii*. It resembles true walnut in appearance, and has an average weight of 640 kg/m³ (40 lb/ft³). Source: Australia. Used for ornamental applications, veneer and furniture.

queen's metal. A silvery-white casting alloy containing 50.5-88.5% tin, 7-16.5% antimony, 1-16.5% zinc, 0-16.5% lead, up to 3.5% copper and 0-1% bismuth. Used for utensils and type metal.

queenwood. See angico.

queitite. A colorless to pale yellow mineral composed of lead zinc sulfate silicate, $Pb_4Zn_2(SO_4)(SiO_4)(Si_2O_7)$. Crystal system, monoclinic. Density, 6.08 g/cm³; refractive index, 1.901. Occurrence: Namibia.

quenched and self-tempered steels. High-strength low-alloy steel shapes of structural quality produced by an in-line process that combines the heat treatment and quenching process and ensures a fine-grained microstructure and controlled chemical composition, and eliminates subsequent tempering after rolling. The heat treatment consists of water cooling of the surface of the shapes after hot rolling, interrupting the cooling process before the core is affected, and then allowing the flow of heat from the core to the surface to self temper the outer layers of the shape. QST steel shapes have high toughness and can be readily arc-welded. Used for structural applications including bolted, riveted and welded bridges and buildings. Abbreviation: QST steel.

quenched-and-tempered steel. A low-carbon constructional alloy steel that contains 0.25% or less carbon, and has been uniformly heated into the *austenite* region, quenched in a suitable medium under uniform conditions to form *martensite*, uniformly reheated to the proper tempering temperature, usually between 204 and 704°C (400 and 1300°F), and allowed to cool in air to room temperature. It may contain alloying elements, such as boron, vanadium and molybdenum, all of which contribute to hardenability, and vanadium, molybdenum and titanium which form persistent carbides to resist softening upon tempering. It has good hardenability, high strength, good toughness and good heat and corrosion resistance. Used in the form of plate for the construction of welded pressure vessels and structural members for large steel structures, mining equipment, earthmovers, etc. See also heat-treatable steel.

quenching media. Chemical media used for quenching and thus hardening ferrous and nonferrous alloys. The most common media include water, brine solutions, caustic solutions, molten salts, molten metals, animal, vegetable and mineral oils, polymer solutions, and gases. Also known as *hardening media.*

quenselite. A pitch-black mineral composed of lead manganese oxide hydroxide, $PbMnO_2(OH)$. Crystal system, monoclinic. Density, 6.84 g/cm³. Occurrence: Sweden.

quenstedtite. A mineral composed of iron sulfate hydrate, $Fe_2(SO_4)_3 \cdot 11H_2O$. Crystal system, triclinic. Density, 2.15 g/cm³. Occurrence: Chile.

Questra. Trademark of Dow Chemical Company (USA) for a family of high-heat, glass-filled semicrystalline engineering resins derived from syndiotactic aromatic polymers using a proprietary catalyst. They are supplied in several grades including *Questra 2730* that is impact-modified for medical applications, *Questra EA* with high strength or high toughness for circuit breakers, connectors, fuse holders and electronic applications, and *Questra WA* with excellent to good heat and chemical resistance for automotive applications including chassis/powertrains, and lighting, electronic and coolant system components.

quetzalcoatlite. A capri-blue mineral composed of copper zinc tellurate hydroxide, $Cu_4Zn_8(TeO_3)_3(OH)_{18}$. Crystal system, hexagonal. Density, 6.05 g/cm³; refractive index, 1.802. Occurrence: Mex-ico.

Quick Cast. Trade name of Koronis Parts, Inc. (USA) for ultra-high-molecular-weight polyethylene supplied in extruded profiles.

Quickcast. Trade name for a kaolin with a mesh size of 200 used for whiteware, tile and other ceramic products chiefly to increase the degree of whiteness.

Quick Gel. Trademark of 3M Company (USA) for an industrial-grade, non-running instant adhesive with gel-like consistency that will bond metals, ceramics, plastics, rubber, cork and foam. It has good adhesion to vertical surfaces, short curing times (2-120 seconds) and good resistance to temperature between -54 and +82°C (-65 and +180°F), and bonds well to poorly matched substrates.

quicklime. See calcium oxide.

quick malleable iron. A malleable cast iron composed of approximately 2.2% carbon, 0.3-0.6% manganese, 1.5% silicon, 0.75-1% copper, and the balance iron. It has high tensile strength, and is used for automotive engine castings.

QuickPac. Trademark of Dexter Corporation (USA) for a *Hysol* type ethylene-vinyl-acetate hot-melt adhesives used for sealing cartons.

Quickram. Trade name of Premier Refractories & Chemicals Inc. (USA) for non-phosphate bonded plastic refractory mixes that set in air, or with heat and/or chemicals.

Quick Set. Trademark of Kwik Mix Materials Limited (Canada) for several hydraulic cement patching compounds.

quick-setting cement. (1) A cement, usually of the alumina-calcium silicate or calcium sulfate type that, due to its special composition and fineness of grind, sets in a shorter time than conventional Portland types. See also Portland cement.

(2) A *dental cement* having relatively short setting time. Many glass-ionomer cements are of the quick-setting type.

quicksilver. See mercury.

quicksilver solder. A corrosion-resistant solder alloy composed of 57-63% silver, 21-25% copper, 3.8-6.2% tin and 10-12% zinc.

quicksilver vermilion. See red mercuric sulfide.

Quickstone. Trade name of Whip Mix Corporation (USA) for a dental stone (ADA Type III) supplied as a powder for mixing with water. Supplied in buff and blue colors, it exhibits low setting expansion (0.16%) upon hardening, and is used for laboratory procedures.

Quicktemp. Trade name of Schottlander Limited (UK) for a self-cure, two-component bis-acryl dental resin used for temporary crown and bridge restorations.

Quiet Steel. Trade name of MSC Laminates & Composites Inc. (USA) for a composite material made by laminating two steel sheets to a custom-selected thin viscoelastic polymer sheet. Supplied in coils up to 183 cm (72 in.) wide, it reduces noise and vibration, can be stamped, punched, rolled and welded, and is used for automotive components, such as dash panels, floor pans and valve covers, and for lawnmower decks.

QuietZone. Trademark of Owens Corning Fiberglass (USA) for an acoustical insulation material supplied in the form of pink-colored fiberglass batts and blankets for use in floors and walls.

Quik Cote. Trade name of PK Insulation Manufacturing Company, Inc.(USA) for thermal insulating and finishing cements.

Quikcrete. Trade name for a range of concrete and mortar repair products including (i) concrete resurfacers supplied as blends of Portland cement, sand, polymer modifiers and other additives, (ii) anchoring cements, (iii) quick-setting cements, (iv) hydraulic water-stop cements, (v) crack sealers, (vi) vinyl patching cements, and (vii) mortar repair masses supplied in gun cartridges.

quilot. The fine, lustrous, white inner fiber obtained from the abaca (*Musa textilis*), a plant of the banana family native to the Philippine Islands. Used for fine fabrics. See also Manila fibers.

Quilticel. Trade name for cellulose acetate fibers used for quilting.

Quina. Trade name for a nylon fiber with silk-like properties, supplied in filament form, and used especially for the manufacture of nonyellowing, crease-resistant knitted fabrics.

Quin-Coat. Trade name of Chemtech Finishing Systems Inc. (USA) for a post-chromate coating.

Quinel. German trademark of Winter & Co. GmbH (Germany) for an artificial leather.

quinhydrone. Dark green crystals obtained by the oxidation of hydroquinone with sodium dichromate. It has a density of 1.40 g/cm^3 and a melting point of 171-174°C (340-345°F). Used as an electrode for measuring the pH in neutral and acid solutions. Formula: $C_6H_4O_2C_6H_4(OH)_2$.

Quin-Lac. Trade name of Chemtech Finishing Systems Inc. (USA) for acrylic-based lacquers.

quinol. See hydroquinone.

quinoline black. A black pigment based on quinoline (C_9H_7N) and related to melanin. See also melanin.

Quintana. Trademark Moplefan SpA (Italy) for polypropylene fibers and filament yarns.

Quintess. Trade name of Quintess (France) for polyester fibers and products.

Quintesse. Trade name of ICI Limited (UK) and DuPont (UK) Limited for nylon 6,6 staple fibers used for upholstery.

Quinze Eclairs. Trade name of Creusot-Eclairs (France) for an oil-hardening tool steel containing 1% carbon, 20% tungsten, 15% cobalt, 1.1% vanadium, 1% molybdenum, and the balance iron. Used for cutting and boring tools.

Quso. Trademark of Philadelphia Quartz Company (USA) for a series of soft, white precipitated amorphous silicas with extremely fine particle sizes (10-20 nm or 0.4-0.8 μin.). Used for paper coatings, as fillers in plastics, and as thickening, flatting, anticaking and reinforcing agents.

Qwik-Set. Trade name of Foster Wheeler (USA) for quick-setting thermal insulating and fitting cements.

R

rabbit hair. The soft protein fibers obtained from the protective hairy coat of wild rabbits (family Leporidae, especially genera *Oryctolagus* and *Sylvilagus*). They are sometimes spun as yarn, but more often used for the production of felt.

rabbittite. A pale green, radioactive mineral composed of calcium magnesium uranyl carbonate hydroxide hydrate, $Ca_3Mg_3(UO_2)_2(CO_3)_6(OH)_4 \cdot 18H_2O$. Crystal system, monoclinic. Density, 2.60 g/cm^3; refractive index, 1.508. Occurrence: USA (Utah).

Rabi. Trade name of Osborn Steels Limited (UK) for a hardened and tempered mold steel containing 0.3% carbon, 4.1% nickel, 1.3% chromium, 0.3% molybdenum, and the balance iron.

Rabifix. Trade name of Wilhelm vom Hofe GmbH (Germany) for black-lacquered, galvanized ribbed expanded metal.

Racalloy Manganese. Trade name of Reid-Avery Company (USA) for an austenitic manganese-nickel steel used for coated electrodes for hardfacing manganese steels.

racemic mixture. An optically inactive mixture of equal parts of dextro- and levorotatory isomers that does not rotate the plane of polarized light, e.g., (±)-lactic acid. Symbol: ±.

racemized protein. An equimolar mixture of dextro- and levorotatory forms of a protein.

racing sand. Special molding sand with improved bond properties. See also molding sand; foundry sand.

Racofix. Trademark of R. Avenarius GmbH & Co. KG Chemische Fabriken (Germany) for a quick-setting assembly cement.

Radeco. Trade name of Westa-Westdeutsche Edelstahlhandelsgesellschaft (Germany) for high-speed steel containing 1.3% carbon, 4.3% chromium, 3.8% vanadium, 12% tungsten, 0.9% molybdenum, and the balance iron. Used for punches and forming and blanking dies.

Radel. (1) Trademark of BP Amoco (USA) for polyarylsulfones and polyphenylene sulfones that are available in unfilled and glass fiber-reinforced extruded grades with high strength and impact resistance. Used for electronic, electrical and automotive components, and industrial equipment. *Radel A* refers to polyethersulfones supplied in unreinforced and 30% glass fiber-reinforced grades with good mechanical strength, good creep resistance and dimensional stability, and high thermal stability. *Radel R* are glass fiber-reinforced polyethersulfones with very good mechanical strength, outstanding load-bearing properties, high toughness, and good creep resistance and dimensional stability.

(2) Trademark of Union Carbide Corporation (USA) for an oriented polyethylene oxide polymer used to make molded articles, and biaxially oriented film.

Radial. Trade name of NV Durobor (Belgium) for hollow glass blocks, plain on one side and reeded on the other.

radial brick. A brick with a curved face used in circular construction (e.g., concentric cylinders).

Radiametal. Trade name of Allegheny Ludlum Corporation (USA) for an alloy composed of 53% iron and 47% nickel, and used for electrical equipment.

Radianite. Trade name of Latrobe Steel Company (USA) for a stainless steel containing 0.7-0.8% carbon, 0.5% manganese, 0.5% silicon, 16.5-18% chromium, and the balance iron. It takes a high polish, keeps a keen cutting edge, and is used for stainless products and cutlery.

Radiant. Trade name of NV Durobor (Belgium) for hollow glass blocks with one side plain and the other cross-reeded.

radiation-cure coating. A coating whose resin binder is composed of an organic monomer or polymer that cures to a continuous film when irradiated with an electron beam or ultraviolet radiation at or slightly above room temperature.

Radiaver. Trade name of Glaceries Réunies SA (Belgium) for an electrically conductive glass.

Radiflam. Trade name of Radici Plastics (USA) for polyamide (nylon) and polybutylene terephthalate resins supplied in unreinforced and fiber-reinforced grades, and including *Radiflam A* nylon 6,6, *Radiflam B* polybutylene terephthalate and *Radiflam S* nylon 6.

Radikal. Trade name of J.C. Soding & Halbach (Germany) for a series of high-speed steels containing 0.8-1.25% carbon, 4.0-4.5% chromium, 0.8-9.2% molybdenum, 1.3-3.4% vanadium, 2-8.7% tungsten, 0-8% cobalt, and the balance iron. Used for cutters, drills, taps, broaches, reamers, and various other machining and finishing tools.

Radilene. Trademark of Deufil GmbH (Germany) for polypropylene fibers and filaments.

Radilon. Trademark of Radici Plastics (USA) for polyamide (nylon) resins used for blowing, extrusion and pressing applications. Examples include *Radilon A* nylon 6,6, *Radilon CA* nylon 6,66, *Radilon CS* nylon 6,6/6T and *Radilon S* nylon 6.

radioactive carbon. See carbon-14.

radioactive cobalt. Any of the radioactive isotopes of cobalt with mass numbers 55 through 62. See also cobalt-60.

radioactive elements. Metallic elements, such as promethium, radium, thorium and uranium, that emit radiant energy in the form of alpha, beta or gamma rays as a result of the spontaneous transitions in their atoms. Also known as *radioactive metals*.

radioactive lead. Any of the radioactive isotopes of lead with the following mass numbers: 194, 195, 196, 197 (isomer), 198, 199 and isomer, 200, 201 and isomer, 202 and isomer, 203 and isomer, 204 and isomer, 205 and isomer, 207 (isomer), 209, 210, 211, 212 and 214. Also known as *radiolead*.

radioactive materials. Materials, such as metallic elements (e.g., radium and uranium) or minerals (e.g., carnotite and pitchblende) that emit radiant energy in the form of alpha, beta or gamma rays as a result of the spontaneous transitions in their atoms.

radioactive metals. See radioactive elements.

radioactive paint. A luminous paint that contains permanent pigments based on radioactive elements or isotopes of radium, strontium, krypton, promethium, thallium or tritium that emit alpha, beta or gamma rays. Used for watch hands, clock and instrument dials, etc.

radioactive salt. A salt, such as radium chloride or uranium hexafluoride, that contains radioactive atoms and is thus suitable for use in certain luminous paints

radio alloys. A group of electrical resistance alloys containing 78-98% copper and 2-22% nickel and characterized by uniform electrical resistivities, stable resistances, reproducible temperature coefficients of resistance, relatively high tensile strengths, good ductility and corrosion resistance, and a service temperature range of 350-400°C (660-750°F). Used for radio components, such as wires and resistors, and for rheostats, voltage control relays and other electrical instruments.

radiobromine. See bromine-82.

radiocalcium. See calcium-45.

radiocarbon. See carbon-14.

radiocesium. See cesium-137.

radiochemical. A chemical substance that contains radioactive elements and is used as a tracer, e.g., in biology, chemistry and medicine.

radiochromium. See chromium-51.

radiocobalt. See cobalt-60.

radiogold. See gold-198.

radiographic paper. A sheet of opaque paper that on one side has a coating composed of gelatin and silver salts and a developer. The developing action is due to an activator, usually an alkaline solution, upon exposure. Used in radiographic inspection to produce permanent images.

Radiohm. Trade name of Driver-Harris Company (USA) for an electrical resistance alloy composed of 78-85% iron, 10-17% chromium and 5% aluminum. It has good heat resistance to 1000°C (1830°F), and is used for resistors.

radioiodine. See iodine-131.

radioiron. See iron-59.

radiolarian earth. A porous sediment formed from the siliceous ooze containing the skeletal remains of radiolaria (microscopic aquatic organisms). Its uses are similar to those of diatomaceous earth, e.g., as filter medium, catalyst and absorbent.

radiolarite. A red or brown, argillaceous, very hard substance formed from the skeletal remains of radiolaria (microscopic aquatic organisms).

radiolead. See radioactive lead.

radio metal. A group of soft magnetic, *Permalloy*-type alloys containing 50-64% iron, 36-50% nickel and up to 5% copper. They have high initial magnetic permeability, high magnetic saturation, low hysteresis loss and high electrical resistivity, and are used for transformer and transductor cores, relays, etc.

Radionix. Trade name of Krupp Thyssen Nirosta GmbH (Germany) for radiation-absorbing stainless steel supplied as cold- and hot-rolled strip, sheet and plate.

Radiopaque. Trademark of Norton Performance Plastics (USA) for fluoropolymer tubing used with X-ray equipment.

radiopaque substance. A substance that is not transparent to radiation, especially X-rays. Many materials used in medicine and dentistry are radiopaque.

radiophosphorus. See phosphorus-32.

Radior. Trade name of NV Durobor (Belgium) for glass blocks with a reeded pattern, vertical on one side and horizontal on the other.

radiosodium. See sodium-24.

radiostrontium. See strontium-90.

radiosulfur. See sulfur-35.

Radipol. Trade name of Polymers International (USA) for polyamide 6,6 (nylon 6,6) polymers.

Radite. Trade name of Shaeffer Pen Company (USA) for cellulose nitrate plastics.

Raditer. Trade name of Radici Plastics (USA) for a series of polybutylene terephthalates (*Raditer B*) and unmodified and glycol-modified polyethylene terephthalates (*Raditer E*).

radium. (1) A rare, silvery-white, luminescent, radioactive metallic element of Group IIA (Group 2) of the Periodic Table (alkaline earth group). It occurs naturally in uranium ores, especially *pitchblende* and *carnotite*. Crystal system, cubic. Crystal structure, body-centered cubic. Density, 5.0 g/cm³; melting point, 700°C (1292°F); boiling point, 1140°C (2084°F); atomic number, 88; atomic weight, 226.025; divalent. It emits radiant energy in the form of alpha, beta and gamma rays, and turns dark on exposure to air. Used in the manufacture of luminous and luminescent paints, as a neutron source, in industrial radiography, and as a source of radon. Symbol: Ra.

(2) A brilliant, glossy fabric, usually of acetate, triacetate, silk or rayon yarn, in a plain weave and with a *taffeta*-like crispness and a crepe-like drape, used especially for blouses, robes, and lingerie.

radium bromide. White, radioactive crystals turning yellow or rose-red on standing. Crystal system, monoclinic. Density, 5.79 g/cm³; melting point, 728°C (1342°F); boiling point, sublimes at 900°C (1652°F). Used in luminous paint, and in physical research and medicine. Formula: $RaBr_2$.

radium chloride. Pale yellow, radioactive crystals turning yellow or rose-red on standing. Crystal system, monoclinic. Density, 4.91 g/cm³; melting point, above 1000°C (1832°F). Used in luminous paint, and in physical research and medicine. Formula: $RaCl_2$.

radium emanation. See radon.

radon. A rare, colorless, radioactive, gaseous element belonging to the noble gas group (Group VIIIA, or Group 18) of the Periodic Table. It is produced by the disintegration of radium, and can form yellow to orange-red, phosporescent, opaque crystals. Radon-222 has a half-life of 3.823 days and decays by alpha particle emission. Density of gas, 9.72 g/L; density of liquid, 4.4 g/cm³ (at -61.8°C/-79°F); density of solid, 4.0 g/cm³; melting point, -71°C (-96°F); boiling point, -61.8°C (-79°F); atomic number, 86; atomic weight, 222; divalent, tetravalent, hexavalent; heaviest known gas. Used as a tracer in leak detection, and in flow-rate measurement, radiography, and earthquake prediction. Symbol: Rn. Also known as *radium emanation*.

Rafaga. Trade name of Vidrieria Argentina SA (Argentina) for a glass with linear pattern.

raffia. The soft cellulose fibers obtained from the leafstalks of the Madagascan raffia palm (*Raphia ruffia*). They can be readily woven into baskets, mats and hats.

Raffinal. Trade name of Alusuisse (Switzerland) for corrosion-resistant, high-purity aluminum (99.99%). It has a high electrical conductivity, and is used for electrical conductors, chemical plant equipment, and window frames.

rag-containing paper. Paper that contains at least 10 wt% cotton or linen rags.

rag-content paper. Paper made from materials containing cotton or linen rags.

Raggiante. Trade name of Vetreria di Vernante SpA (Italy) for a sandblasted glass featuring irregular geometric shapes.

rag paper. An expensive, high-quality paper composed of at least 25% cotton or linen rags, but also available in grades with 50, 75 and 100%. Its strength, resistance to aging and discoloration, and degree of whiteness increase with the rag content.

See also rags.

Ra-Grid. Trade name of ASG Industries Inc. (USA) for a tempered glass with fused-on aluminum grid with electrical connectors.

rag roofing felt. Roofing felt made by interfelting absorbent animal and/or vegetable fibers and, usually, saturating with bituminous material. Used for shingles, siding, roll roofing, etc.

rags. The waste fabrics including garment cutouts, piece ends and rejected or discarded patterns. Cotton and linen rags are sometimes used in the manufacture of paper. See also rag paper.

raguinite. (1) A bronze-colored mineral composed of iron thallium sulfide, $TlFeS_2$. Crystal system, orthorhombic. Density, 6.40 g/cm³. Occurrence: Greece, Macedonia.

(2) A violet black synthetic mineral composed of iron thallium sulfide, $FeTlS_2$. Crystal system, monoclinic. Density, 5.66 g/cm³.

Raiado. Trade name of Vidrobrás (Brazil) for a glass with fluted pattern.

Railan. Trade name for rayon fibers and yarns used for textile fabrics.

Rail-Arc. Trade name of Chemetron Corporation (USA) for a manganese steel containing 0.06% carbon, 3.3% manganese, 0.5% silicon, 17.8% chromium, 8.1% nickel, and the balance iron. Used for arc-welding electrodes for multi-pass buildup on carbon and austenitic manganese steel rails.

Railender. Trade name of Champion Rivet Company (USA) for a steel containing 0.35% carbon, 1.8% chromium, 0.12% vanadium, and the balance iron. Used chiefly for hardfacing electrodes.

Railite. Trade name of Chrysler Corporation (USA) for an alloy composed of 75% iron and 25% copper, and used for sintered, self-lubricating bearings and oil-cushion bearings.

Railko. Trade name of Railko (USA) for a cotton-fabric phenolic laminate.

railroad ballast. Crushed stone or coarse gravel used in railroad beds.

railroad bronze. A group of strong, tough, heavy-duty bronzes containing 73.5-89% copper, 0-20% tin, 0-22.2% zinc, 0-20% lead and up to 0.5% phosphorus. Used for locomotive bearings, axle bearings, gears, injectors, slide valves and piston rods.

railroad thermit. A thermit composed of a finely divided mixture of aluminum, red iron oxide, nickel, manganese and steel. See also thermit.

Rails. Trade name of SA Glaverbel (Belgium) for a figured glass with pinstripe design.

rail steel. A wear-resistant, deoxidized steel, usually of the low- or high-carbon, or chromium- or manganese-alloy type, used for making rails.

rail steel reinforcement. Concrete reinforcement in the form of bars made by hot rolling standard T-section rails.

Railway. Trade name of Stone Manganese–J. Stone & Company Limited (UK) for bearing alloys. *Railway A* are lead-base alloys containing 40% tin, 16% antimony and 1% copper, and *Railway C* are tin-base alloy containing 11% antimony, 5% lead and 4% copper.

Railway Axle Box. Trade name of A. Cohn Limited (UK) for a tough alloy containing 83% copper, 7% tin, 6% zinc and 4% lead. Used for castings and bearings.

Railwear. Trade name of Hollup Corporation (USA) for a tough, wear-resistant manganese steel containing 0.7% carbon, 13% manganese, and the balance iron. Used for coated electrodes for hardfacing rail joints.

Rainbow. (1) Trade name of Stratec Medical (Switzerland) for a biomimetic calcium phosphate coating that has outstanding mechanical properties, good biocompatibility, and closely mimics natural bone. It can be used on porous and nonporous materials including metal implants, such as hip replacements.

(2) Trade name for rayon fibers and yarns used for textile fabrics.

rainbow granite. A *granite* either pink to pale purple with dark spots, or black to dark green with pink or pale red or yellow spots, used for ornamental and decorative applications, and in building construction.

rainbow yarns. Novelty yarns with a shaded or graduated color effect.

Raindrops. Trade name of SA des Verreries de Fauquez (Belgium) for a patterned glass that features raindrops.

Raion FTC 3000. Trademark of Nuova Rayon Italia SpA (Italy) for viscose rayon fibers and yarns.

raised fabrics. See napped fabrics.

raite. A gold to brown mineral composed of sodium manganese silicate nonahydrate, $Na_4Mn_3Si_8(O,OH)_{24} \cdot 9H_2O$. Crystal system, orthorhombic. Density, 2.39 g/cm³; refractive index, 1.542. Occurrence: Russian Federation.

Rajah. Trade name of J.F. Jelenko & Company (USA) for a microfine, hard, dense dental casting alloy (ADA type III) containing 58% gold, 27% silver and 3.5% palladium. It provides good strength, takes a lustrous polish, and is used for veneer crowns, fixed bridgework and hard inlays.

rajite. A bright green mineral composed of copper tellurium oxide, $CuTe_2O_5$. Crystal system, monoclinic. Density, 5.76 g/cm³; refractive index, 2.135. Occurrence: USA (New Mexico).

Rakel's metal. A corrosion-resistant alloy composed of 88% copper, 10% nickel, 1% manganese and 1% zinc.

Raku Clay. Trade name of Aardvark Clay & Supplies (USA) for a buff-colored earthenware clay with 30 and 60 mesh sands.

Raku Glaze. Trade name of Aardvark Clay & Supplies (USA) for ceramic glazes supplied in red, white, blue, green shell and copper colors.

ralstonite. A colorless to white mineral composed of sodium magnesium aluminum fluoride hydroxide monohydrate, $NaMgAlF_6 \cdot H_2O$. Crystal system, cubic. Density, 2.67 g/cm³; refractive index, 1.37; hardness, 4.5. Occurrence: Greenland.

Ramaboard. Trade name for rigid asbestos-based board products for high-temperature applications, such as furnace tops and end boards for induction coils.

Ramak. Trade name of North American Refractories Company (USA) for refractory ramming mixes.

Ramal. Trade name North American Refractories Company (USA) for refractory ramming mixes.

Ramax. Trade name of Uddeholm Corporation (USA) for a free-machining stainless mold steel (AISI type 420F) containing 0.3% carbon, 1.35% manganese, 0.35% silicon, 0.12% sulfur, 16.7% chromium, and the balance iron. Used plastic and rubber mold components.

ramdohrite. A dark gray mineral of the lillianite group composed of silver lead antimony sulfide, $Pb_6Ag_3Sb_{11}S_{24}$. Crystal system, orthorhombic. Density, 5.43 g/cm³. Occurrence: Bolivia.

rameauite. An orange, radioactive mineral of the becquerel group composed of potassium calcium uranyl oxide nonahydrate, $K_2Ca(UO_2)_6O_8 \cdot 9H_2O$. Crystal system, monoclinic. Density, 5.60

g/cm³; refractive index, 2.0. Occurrence: France.

Ramelon. Trade name for *vinyon fibers* with excellent acid, alkali and mildew resistance. Used especially for textile fabrics.

Ramet. Trade name of Fansteel Metals (USA) for micrograin tantalum carbide (TaC) with excellent shock and impact resistance, good wear resistance, and a strength similar to high-speed steel. Used as a cutting-tool material for heavy roughing at low speeds, for machining high-temperature/high-strength alloys, work-hardening stainless steels and hard metals, and for cut-off, screw-machine and milling applications.

Ramfrax. Trademark of The Carborundum Company (USA) for a refractory ramming cement.

ramie. A fine, white, lustrous fiber obtained from the stalks of the tall perennial plant *Boehmeria nivea* of the nettle family. It can be spun and woven, but is combustible and not self-extinguishing. *Ramie* has high wet strength, good resistance to rot and mildew, a tensile strength four times that of flax and almost three times that of hemp, an elasticity greater than that of flax, good absorbency, and good wearing qualities. Source: Brazil, China, Egypt, India, Taiwan, USA (Florida). Used for coarse fabrics, cordage, mats, high-grade paper, patching water mains, marine gland packings, stern-tube packing in ships, and twine and yarn. Also known as *China grass.*

ramin. The light-colored, moderately fine-textured, straight-grained wood of several species of trees of the genus *Gonystylus,* especially *G. bancanus* and *G. macrophyllum.* It has a beautiful appearance, fair workability, low nailing qualities, and poor decay resistance, but can be treated with preservatives. Average weight 660 kg/m³ (41 lb/ft³). Source: Southeast Asia from Malay Peninsula to Sumatra and Borneo. Used for furniture, joinery, doors, handles, plywood for doors, and interior trim.

Ramix. Trademark of Basic Refractories, Inc. (USA) for high-magnesia brick and refractories used for furnace lining and repair.

Ramman. Trade name of Central Glass Company Limited (Japan) for a patterned glass with floral design.

rammelsbergite. A gray opaque mineral of the marcasite group composed of nickel arsenide, $NiAs_2$. It can also be made synthetically. Crystal system, orthorhombic. Density, 7.10 g/cm³. Occurrence: Morocco. A tin-white, cobalt-bearing variety with red tinge is found in northern Australia.

ramming material. A coarsely ground refractory material or mixture of refractory materials used in making ramming mixes.

ramming mix. A refractory product made by tempering a batch of ground refractory materials with water. It is suitable for ramming into place to form monolithic furnace linings, or to patch special shapes.

Ram Press. Trade name of Aardvark Clay & Supplies (USA) for an earthenware clay with a basic pressing body, used for ram-press applications.

ramsayite. See lorenzenite.

ramsdellite. A steel-gray to iron-black mineral of the rutile group composed of manganese oxide, MnO_2. Crystal system, orthorhombic. Density, 4.83 g/cm³. Occurrence: USA (New Mexico).

Ramsos. Trade name of Robert-Leyer-Pritzkow & Co. (Germany) for a series of high-speed, hot-work and water-hardening plain-carbon tool steels.

Ramtite. Trade name of C-E Refractories (USA) for several refractory ramming mixes.

Ranalloy. Trade name of US Steel Corporation for a series of wear-resistant steels supplied in the form of hardfacing electrodes.

rancieite. A black or brownish violet mineral composed of calcium manganese oxide trihydrate, $(Ca,Mn)Mn_4O_9·3H_2O$. Crystal system, hexagonal. Density, 3.20 g/cm³. Occurrence: USA (Virginia).

Randex. Trade name of ASG Industries Inc. (USA) for a patterned glass with a mat finish on one side and straight, parallel randomly spaced ribs on the other.

Randolf metal. A corrosion-resistant dental alloy.

Random Clear. Trade name of Owens-Illinois, Inc. (USA) for a glass block consisting of two halves made with slightly dissimilar molds effecting subtle irregularities in the block's face contours.

random copolymer. A statistical copolymer containing disordered sequences of monomeric repeating units. It is usually referred to as poly(A-*ran*-B), e.g., a random copolymer of ethylene glycol and propylene glycol may be referred to as poly(ethylene glycol-*ran*-propylene glycol). Also known as *random polymer.* Abbreviation: RC. See also copolymer; statistical copolymer.

Random-set. Trade name for nylon fibers and yarns used for textile fabrics including clothing.

Random-tone. Trade name for nylon fibers and yarns used for textile fabrics including clothing.

random yarn. A novelty yarn, either of multiple color, or made by twisting together differently colored strands.

Raney cobalt. Trademark of W.R. Grace & Company (USA) for a cobalt powder supplied as slurry in water, and used as an active catalyst.

Raney copper. Trademark of W.R. Grace & Company (USA) for a fine, air-sensitive copper powder supplied as 50% aqueous solution with high surface adsorption, and used as an active catalyst. Abbreviation: RC.

Raney nickel. Trademark of W.R. Grace & Company (USA) for gray, spongy, pyrophoric powders and crystals obtained by leaching the aluminum from an 50Al-50Ni alloy with a 25% sodium hydroxide solution. Usually supplied as aqueous slurry in a solution with a pH value greater than 9, and a surface area of about 80-100 m²/g. Molybdenum-promoted grades are also available containing about 8-12% aluminum, 1-3% molybdenum and 85-93% nickel. *Raney nickel* is used as an active hydrogenation catalyst. Abbreviation: RN.

Ranite. Trade name of Rankin Manufacturing Company (USA) for wear-resistant cobalt- and iron-base hardfacing electrodes.

rankamaite. A white mineral composed of potassium sodium tantalum niobium oxide hydroxide, $(Na,K,Pb)_3(Ta,Nb)_{11}(O,OH)_{30}$. Crystal system, orthorhombic. Density, 5.50 g/cm³. Occurrence: Zaire, Congo.

rankinite. A colorless mineral composed of calcium silicate, $Ca_3Si_2O_7$. It can also be made synthetically. Crystal system, monoclinic. Density, 2.96-3.00 g/cm³; refractive index, 1.644. Occurrence: New Zealand.

Ransco Alloy. Trade name of Rasmussen Manufacturing Company (USA) for a centrifugally-cast iron containing 2.5-3.25% carbon, 1-2% silicon, 12-15% nickel, 5-7% copper, and the balance iron. Used for cylinder and diesel-engine liners.

ranunculite. A gold-yellow mineral of the meta-autunite group composed of oxonium aluminum uranyl phosphate hydroxide hydrate, $Al(H_3O)(UO_2)(PO_4)(OH)_3·3H_2O$. Crystal system, monoclinic. Density, 3.40 g/cm³; refractive index, 1.664. Occurrence: Zaire.

Raonel. Trade name of Rolled Alloys (USA) for a heat-resisting nickel alloy containing 14-17% chromium, 6-10% iron, 0-1%

manganese and up to 0.15% carbon. It has good hot and cold workability, and is used for high-temperature applications, and regenerators.

Rapco Foam. Trademark of Jesco, Inc. (USA) for foamed-in-place acoustical and thermal insulation.

Rapid. Trade name of Coltene-Whaledent (USA) for a hydrophilic condensation-type silicone dental impression material.

rapid-curing asphalt. Liquid asphalt made by diluting asphalt cement with gasoline, naphtha or similar substance. Also known as *R-C asphalt; rapid-curing liquid asphaltic cement.*

Rapide. Trade name of Aubert & Duval (France) for a series of oil-hardening high-speed steels containing 0.6-0.9% carbon, 4-5% chromium, 5.7-19% tungsten, 1-2.2% vanadium, 0-9.5% cobalt, and the balance iron. Used for machining and finishing tools including milling cutters, reamers, broaches and drills, and punches.

Rapidex. (1) Trade name of Swiss Aluminium Limited (Switzerland) for a series of age-hardenable aluminum extrusion alloys composed of 0.3-0.6% silicon, 0.1-0.3% iron, 0.02-0.1% copper, 0-0.1% manganese, 0.3-0.6% magnesium, 0-0.05% chromium, 0-0.15% zinc and 0-0.15% titanium. Available in billet form in various strength grades, they are used for electrical conductors and automotive applications.

(2) Trademark of H.B. Fuller Company (USA) for reactive hot-melt adhesives.

Rapid-Fire. Trademark of Babcock & Wilcox Company (USA) for castable refractories.

Rapidflex. Trade name of Vetreria Italiana Balzaretti Modigliani SpA (Italy) for a pipe insulation material composed of a cylinder of kraft-aluminum paper containing glass fibers treated with thermosetting resin.

rapid-hardening cement. A finely ground special-grade of *Portland cement* that at almost identical setting time develops its strength more rapidly than ordinary Portland cement. Used for cold-weather construction. See also high-early-strength cement.

rapidly solidified alloys. Amorphous or metastable crystalline metal alloys made by cooling (undercooling, or chilling) a melt below its melting point at a high quench rate (typically 10^5 to 10^6 °C/s). Depending on the particular processing technique various commercial forms may be produced. Techniques, such as melt spinning, planar flow casting and melt extraction, produce thin (approximately 25-100 μm or 0.001-0.004 in.) ribbons, tapes, sheets or fibers. Resolidification and surface melting produce thin surface layers, and atomization produces powders with particle size in the range of 10-1000 μm (0.0004-0.04 in.). Abbreviation: RSA.

rapidly solidified powder. A ferrous or nonferrous powder (particle size range approximately 10-1000 μm, or 0.0004-0.04 in.) quenched at rates greater than 10^4 °C/s. It may be produced by any of various processes including water and gas atomization, rotating electrode techniques, melt spinning, melt extraction and plasma deposition. Used as a starting material in powder metallurgy. Abbreviation: RSP.

Rapid Panther. Trade name of Styria-Stahl Steirische Gusstahlwerke AG (Austria) for a tungsten-type high-speed steel containing 0.7% carbon, 4% chromium, 19% tungsten, 2.25% vanadium, and the balance iron. It has high wear resistance and hot hardness, and good toughness. Used for lathe tools, milling cutters, reamers, broaches, drills, taps, and hobs

RapidPost. Trade name of King Packaged Products Company (Canada) for a concrete mix that reaches initial set within one hour and is used especially for setting posts and poles.

Rapid Spezial. Trade name of Thyssen Edelstahlwerke AG (Germany) for a tungsten-type high-speed steel containing 0.7% carbon, 4% chromium, 19% tungsten, 2.2% vanadium, and the balance iron. It has high wear resistance and hot hardness, and good toughness. Used for cutting tools.

Rapid Stone. See Hydrock/Rapid Stone.

Rapitex. Trade name of Fibreglass Limited (UK) for reinforced glass-fiber materials.

rare-earth alloys. A group of alloys containing one or more of the following rare-earth elements as their principal constituent: lanthanum (La), cerium (Ce), praeseodymium (Pr), neodymium (Nd), promethium (Pm), samarium (Sm), europium (Eu), gadolinium (Gd), terbium (Tb), dysprosium (Dy), holmium (Ho), erbium (Er), thulium (Tm), ytterbium (Yb), yttrium (Y) and/or lutetium (Lu). Examples of rare-earth alloys include mischmetal, neodymium-iron-boron magnet alloys and samarium-cobalt magnet alloys. Abbreviation: REA.

rare-earth aluminum garnets. A group of powder-based ferrimagnetic ceramic materials, such as yttrium aluminum garnet (YAG), represented by the general formula $M_3Al_5O_{12}$, where M represents a rare-earth element, such as dysprosium, erbium, europium, gadolinium, holmium, samarium, terbium, thulium, ytterbium, or yttrium.

rare-earth–cobalt magnet. Any of various types of permanent magnets made with a plastic or metallic binder from a rare-earth element and cobalt. They have high magnetic induction, high coercive force, high permanency, and a high energy product. Examples include cerium-cobalt ($CeCo_5$), holmium-cobalt ($HoCo_5$), praseodymium-cobalt ($PrCo_5$) and samarium-cobalt (Sm_2Co_7 and $SmCo_5$). See also rare-earth magnets; permanent magnets.

rare-earth elements. A group of 15 metallic elements with atomic numbers between 57 and 71 having similar chemical and physical properties. They form Group IIIB (the lanthanide series) of the Periodic Table and are: lanthanum (57), cerium (58), praseodymium (59), neodymium (60), promethium (61), samarium (62), europium (63), gadolinium (64), terbium (65), dysprosium (66), holmium (67), erbium (68), thulium (69), ytterbium (70) and lutetium (71). This group is subdivided into the cerium subgroup containing elements 57 through 62, and the yttrium subgroup containing elements 63 through 71. Yttrium, *didymium* and thorium, although not rare earths, are closely associated. Also known as *rare earths; rare-earth metals.* See also lanthanide; light rare earths.

rare-earth garnets. A group of powder-based ferrimagnetic ceramic materials that have very complex crystal structures represented by the general formulas $M_3Fe_5O_{12}$ and $M_3Al_5O_{12}$, where M represents a rare-earth element, such as dysprosium, erbium, europium, gadolinium, holmium, samarium, terbium, thulium, ytterbium or yttrium. Specific types are referred to as yttrium-aluminum garnet (YAG), yttrium-iron garnet (YIG), gadolinium-iron garnet (GdIG), etc. Used in solid-state electronics, lasers, and microwave devices. See also garnets.

rare-earth iron garnets. A group of powder-based ferrimagnetic ceramic materials, such as yttrium-iron garnet (YIG), that correspond to the general formula $M_3Fe_5O_{12}$, where M represents a rare-earth element, such as dysprosium, erbium, europium, gadolinium, holmium, samarium, terbium, thulium, ytterbium or yttrium. See also rare-earth garnets.

rare-earth magnets. Permanent magnet materials composed of one or more rare-earth elements with or without the addition of other elements. They have exceptionally high coercive forces and energy densities. Examples include magnets of samarium-cobalt (Sm_2Co_7 and $SmCo_5$) and neodymium-iron-boron ($Nd_2Fe_{14}B$). Used for electronic and electrical applications, computers, signaling devices, and communications equipment. Abbreviation: REM. Also known as *rare-earth magnetic materials*. See also rare-earth cobalt magnet.

rare-earth metals. See rare-earth elements.

rare-earth minerals. A group of relatively rare minerals composed principally of rare-earth oxides, e.g., allanite, bastnaesite, cerite, gadolinite and monazite.

rare-earth oxides. A group of oxide compounds corresponding to the general formula $(RE)_2O_3$, where RE represents a rare-earth element, such as dysprosium, erbium, yttrium, etc.

rare-earth perovskites. A group of ferrimagnetic, orthorhombic oxides corresponding to the general formula $(RE)MO_3$, where RE represents a rare-earth element, such as yttrium, europium, dysprosium, erbium, etc., and M a trivalent metal, such as aluminum.

rare earths. See rare-earth elements.

rare-earth salts. A group of crystalline compounds derived from an acid by replacing the hydrogen by a rare-earth metal. Included are various rare-earth acetates, carbonates, bromides, chlorides, fluorides, iodides, nitrates, phosphates and sulfates. Rare-earth salts are commonly used for ceramic applications as coloring agents, glass decolorizers, cores for arc carbons, electronic components, fiberoptic and laser glasses, incandescent gas mantles, magnetic compositions, phosphors, polishing compounds, and ultraviolet absorbers.

Rare Earth Silicide. (1) Trade name of Globe Metallurgical Inc. (USA) for a master alloy containing 30-35% silicon, 30-35% iron, and a total of 30-35% rare-earth elements. Used for neutralizing sulfur, in steels and cast irons for overall quality improvement and inclusion minimization, and in the manufacture of compacted-graphite and ductile cast irons.

(2) Trade name Cyprus Foote Mineral Company (USA) for a ferroalloy containing 28-33% silicon, 15-18% cerium, a total of 30-35% other rare earths, and the balance iron. Used as deoxidizers, desulfurizers and sulfide-shape controllers in steels, and as a source of cerium and rare earths for cast irons.

rare-earth-sodium sulfates. A group of compounds that are composed of sodium sulfate, a rare-earth sulfate and, usually, water, and correspond to the general formula $(RE)_2(SO_4)_3 \cdot Na_2SO_4 \cdot 2H_2O$, where RE represents a rare-earth element, such as erbium, gadolinium or ytterbium. They are available in the form of pink crystals, and used as intermediates in the manufacture of rare earths, and as ingredients in ultraviolet-absorbing glass. Also known as *pink salts*.

rare metals. A group of metallic elements, such as astatine and francium (both of which are seldom found in nature) as well as technetium and the transuranic elements which, due to their difficulty of production and/or extraction from ores and resulting high price, were previously not often used.

raschel. Any fabric made with a warp-knitting machine known as "Raschel machine" which can produce double-rib effects and laces (raschel laces) with a wide range of patterns and textures.

rasorite. See kernite.

raspite. A yellowish brown mineral of the monazite group composed of lead tungstate, $PbWO_4$. Crystal system, monoclinic. Density, 8.47 g/cm³; refractive index, 2.27. Occurrence: Australia (New South Wales).

Raster. Trade name of Glasindustrie Pieterman NV (Netherlands) for a patterned glass.

rasvumite. A steel gray mineral composed of potassium iron sulfide, KFe_2S_3. Crystal system, orthorhombic. Density, 3.10 g/cm³. Occurrence: USA (California), Russian Federation.

Ratacanda. Trademark of Ratcliffs (Great Bridge) Limited (UK) for copper and brass strip.

rathite. A lead gray mineral composed of lead arsenic sulfide, $(Pb,Tl)_3As_5S_{10}$. Crystal system, monoclinic. Density, 5.37 g/cm³. Occurrence: Switzerland.

ratine. A loose, plain-woven fabric with a coarse irregular texture made with *ratine yarn*. Used especially for dresses, suits, overcoats, and drapery.

ratine yarn. A plied yarn with planned knots and curls produced by twisting heavy and fine yarns together under uneven tension. Used to make fabrics with irregular textures.

rattail cord. A cord, usually of satin construction, that has been woven in the form of a tube. Also known as *American cord*.

Rattan. Trade name of Pilkington Brothers Limited (UK) for a patterned glass.

rattan. The long, pliable, jointed stems of any of several climbing palms (genera *Calamus* and *Daemonorops*) found in Sri Lanka, Malaya and Laos. *Rattan* has high strength and durability, and is used for wickerwork, baskets, canes, furniture, and heavy cordage.

Ratujal. Trade name for rayon fibers and yarns used for textile fabrics.

Rauchberg. German trade name for a corrosion-resistant high-leaded bearing bronze containing 66-75% copper, 15-19% lead, 5-10% tin and 1-5% antimony.

rauenthalite. A colorless mineral composed of calcium arsenate decahydrate, $Ca_3(AsO_4)_2 \cdot 10H_2O$. Crystal system, triclinic. Density, 2.38 g/cm³; refractive index, 1.552. Occurrence: France.

rauli. See Antarctic beech.

rauvite. A purplish or bluish black, radioactive mineral composed of calcium uranyl vanadium oxide hydrate, $Ca(UO_2)_2V_{10}O_{28} \cdot 16H_2O$. Density, 2.92 g/cm³; refractive index, 1.88-1.89. Occurrence: USA (Colorado, Utah).

Rauxite. Trade name of US Industrial Alcohol Company (USA) for urea-formaldehyde plastics.

Rauzene. Trade name of US Industrial Alcohol Company (USA) for phenol-formaldehyde resins and plastics.

Raven. Trademark of Columbian International Chemicals Company (USA) for an extensive range of carbon blacks.

Ravinil. Trade name of EVC (USA) for several unplasticized polyvinyl chlorides.

raw clay. An impure clay as it is obtained from weathering rocks, prior to processing.

raw cotton. Ginned cotton lint suitable for use in the manufacture of textiles.

raw dolomite. Dolomite that is crushed, but not yet calcined. Used in construction of metallurgical furnaces. See also dolomite.

raw glaze. A ceramic glaze that is made entirely from raw materials, i.e., materials which have not been prefused.

raw gypsum. See gypsum.

rawhide. Untanned hide of cattle or other animals that has been dehaired, dried and treated with an oil, grease or preservative. Used in the manufacture of mechanical goods, such as gears, pinions, gaskets, transmission belt lacings, luggage, etc.

raw material. (1) A term used for any unprocessed or partially processed material from which one or several useful products can be derived. For example, bauxite is the raw material for aluminum, coal for coal tar, cotton and wool for fabrics, ores for metals, polyisoprene for natural rubber, and wood for paper.

(2) A general term used for any constituent or ingredient used in the preparation of a mixture or product. For example, woodpulp, cellulose and cotton rags are raw materials used in the manufacture of paper, and polymer resins, pigments, antioxidants and fillers are raw materials for plastics.

raw mix. A ceramic blend composed of finely ground raw materials in the proper proportion and prior to calcining.

raw refractory dolomite. A natural *limestone* essentially composed of equal quantities of calcium carbonate ($CaCO_3$) and magnesium carbonate ($MgCO_3$) and used as a refractory material, or as a component in refractory materials.

raw rubber. See crude rubber.

raw sienna. A brownish yellow pigment obtained from the earth. It contains hydrated ferric oxide, manganese oxide and siliceous matter, and is used in the manufacture of *burnt sienna*, as a colorant in oil paints, etc. See also sienna.

raw silk. A single thread of *silk* consisting of several fine filaments. It is made by winding the silk filaments produced by the silkworm (the larva of the Asiatic moth *Bombyx mori*) up onto a reel. Each filament is composed of 80% *fibroin* and 20% *sericin*.

raw umber. A brown pigment obtained from the earth and usually ground and levigated. It consists of hydrated ferric oxide, manganese oxide and siliceous matter. See also umber.

Raxa. Trade name of Heinrich Reining GmbH (Germany) for an extensive series of steels including various case-hardening types, cast and wrought corrosion- and/or heat-resisting stainless types, cold- and hot-work tool types, and plain-carbon-, tungsten- and molybdenum-type high-speed grades.

Raxit. Trade name of Heinrich Reining GmbH (Germany) for a series of corrosion-resistant steel castings of the iron-chromium-nickel, or iron-nickel-chromium type.

Raxotherm. Trade name of Heinrich Reining GmbH (Germany) for heat-resistant steel castings of the iron-chromium, iron-chromium-nickel and iron-nickel-chromium type, used in the manufacture of metallurgical and other industrial furnaces, and heat-treating equipment.

Rayado. Trade name of Vidrieria Argentina SA (Argentina) for a patterned glass featuring a fluted design.

Rayban. Trade name of Bausch & Lomb Inc. (USA) for a green-tinted spectacle glass.

Raybel. Trade name of SA Glaverbel (Belgium) for an electrically conductive glass.

Raybestos. Trademark of Raybestos Products Company (USA) for thermal insulating materials based on asbestos, and used for asbestos boards, heating panels, packings, and automotive brake linings.

Raybrite. Trade name for a series of metallic fibers supplied in several grades including *Raybrite MF* and *Raybrite MM*.

Raycron. Trade name of PPG Industries, Inc. (USA) for radiation-curable organic coatings.

Raydoon. Trademark of Canada Cordage Inc. for rayon twine.

Rayflex. Trade name of Rayflex Limited (UK) for rayon fibers and yarns used for textile fabrics including clothing.

rayite. A lead-gray mineral composed of lead silver thallium antimony sulfide, $Pb_8(Ag,Tl)Sb_8S_{21}$. Crystal system, monoclinic. Density, 6.13 g/cm³. Occurrence: India.

Raynile. Trade name for strong, flexible fabrics with rayon warp (longitudinal) threads and nylon weft (transverse) threads, and used for conveyor belts.

Rayo. Trade name for a resistance alloy composed of 85% nickel and 15% chromium. It has high electrical resistivity, high heat resistance, and a maximum service temperature of 1150°C (2100°F). Used for resistances and heating elements.

Rayocord. Trademark of ITT Rayonier Inc. (USA) for a high-tenac-ity rayon used in the manufacture of tire cords, and in the form of reinforcing fabrics in polymer-matrix composites.

Rayoface. Trade name of UCB Films, Inc. (USA) for an extensive series of biaxially oriented polypropylene (BOPP) film materials supplied in several grades for self-adhesive applications. *Rayoface C* and *Rayoface W* are uncoated, glossy films for facestock applications. The former is a clear and the latter a white-colored film. *Rayoface CA* and *Rayoface WA* are top-coated, glossy films for facestock applications. The former is a clear and the latter a white-colored film. *Rayoface WI* is a white, top-coated, computer-imprintable film for facestock applications. *Rayoface CL* is a clear, glossy film for overlamination applications, and Rayoface CR a clear film suitable for siliconization and use as a release liner.

Rayolanda. Trade name for rayon fibers and yarns used for textile fabrics.

Rayolit. German trade name for a patterned glass.

rayon. See cuprammonium rayon; nitrocellulose rayon; rayon fiber; viscose rayon.

rayon fiber. Any of a group of fibers composed of regenerated cellulose and made from wood pulp by the viscose process, from cotton linters by the cuprammonium process, or from nitrocellulose fibers. They can be modified and blended with other cellulose-forming materials (e.g., cotton) and synthetic fibers, and have a density of about 1.5 g/cm³ (0.08 lb/in.³), high, low, or medium tenacity, poor resistance to moisture, and are not self-extinguishing. Used for woven and nonwoven fabrics (e.g., clothing, home furnishings, industrial fabrics, etc.). See also cuprammonium rayon; nitrocellulose rayon; viscose rayon.

rayon fabrics. Any of various textile fabrics made from rayon fiber, usually by spinning.

rayon-HP. High-performance grades of rayon with increased tenacity, wet modulus and/or other properties. Used for industrial applications, e.g., tire cords.

rayon staple. Rayon fibers produced in short, spinnable lengths (typically 25-125 mm or 1-5 in.) either directly, or by chopping continuous fibers (*rayon tow*).

rayon tow. A loose, untwisted rope consisting of numerous parallel, continuous rayon filaments and used in the manufacture of rayon staple, and for flock.

Rayophane. Trade name of UCB Films (formerly British Rayophane, UK) for a regenerated cellulose supplied in the form of a thin, strong, flexible film having a density of 1.44 g/cm³ (0.05 lb/in.³), high water absorption, moderate mechanical properties, excellent resistance to alkalies, dilute acids, greases, oils and air. Untreated film softens on exposure to heat at about 150°C (300°F) and has very low permeability to permanent gases. Used for wrapping and packaging of food and goods.

Rayoweb CR. Trade name of UCB Films, Inc. (USA) for a clear biaxially oriented polypropylene (BOPP) film suitable for siliconization and use as a release liner.

Rayox. Trademark of R.T. Vanderbilt Company, Inc. (USA) for titanium dioxide used in paint and paper manufacture.

Raysorb. Trade name for absorbent rayon fibers.

razor stone. See novaculite.

R-C asphalt. See rapid-curing asphalt.

RCI Reinforced. Trade name for a glass-ionomer/cermet dental cement.

Reactal. Trade name of J.J. Rieter & Company (USA) for a corrosion- and heat-resisting cast iron containing 3.2% carbon, 2.2% silicon, 1.5% nickel, 0.8% chromium, and the balance iron.

Reaction. Trademark of Ludlow Composites Corporation (USA) for a composite with rubbery feel, consisting of open-cell styrene rubber latex foam with sheet or fabric facings.

reaction-bonded silicon carbide. A fine-grained, low-porosity silicon carbide (SiC) made by infiltrating silicon into a pre-shaped green body of silicon carbide and carbon powder, and subsequent firing. It has a density of 3.10 g/cm^3 (0.112 lb/in.^{3}), high strength up to 1350°C (2460°F), good high-temperature and oxidation resistance, high hardness (2500-3000 Vickers), good thermal-shock resistance, good resistance to acids, and poor as-fired machinability. Used for gas-turbine parts, furnace applications, bearings, and seals. Abbreviation: RBSC; RBSiC.

reaction-bonded silicon nitride. Silicon nitride (Si_3N_4) made by nitriding a compacted silicon powder and subsequent firing. It has a density of 2.4-2.5 g/cm^3 (0.08-0.09 lb/in.^{3}), a hardness of about 3000 Vickers, an upper continuous-use temperature of 1200-1500°C (2190-2730°F), good strength and oxidation resistance, good wear properties, and good resistance to various chemicals and molten metals. Used for gas-turbine parts, supports, and automotive applications. Abbreviation: RBSN.

reaction-cured glass. A black coating with high optical emittance composed of borosilicate glass and silicon tetraboride, and used in the manufacture of high-temperature reusable surface insulation for the NASA Space Shuttle. Abbreviation: RCG.

reaction flux. A soldering flux containing at least one component that upon heating reacts chemically with the base metal resulting in fast fluxing action.

reaction-injection-molded plastics. Solid or microcellular plastics, such as epoxies, polyureas or polyurethanes, formed by mixing the liquid reactants in the appropriate amounts and filling the mixture into a mold where it reacts and cures. Abbreviation: RIM plastics. See also reinforced-reaction-injection-molded composites.

reaction-spun fibers. Synthetic fibers produced by the polymerization of reactants while being forced through a multi-hole extrusion die.

reactive aggregate. An aggregate composed of siliceous minerals or rocks that when added to cement will react chemically with the alkalies (e.g., calcium, magnesium, potassium and sodium), and produce a destructive internal expansion in concrete or mortar after hardening. Also known as *alkali-reactive aggregate*.

reactive alloys. A group of alloys based on beryllium, hafnium, molybdenum, niobium (columbium), tantalum, titanium, tungsten, zirconium, etc., that readily combine with oxygen at elevated temperatures.

reactive metals. A group of metals including beryllium, hafnium, molybdenum, niobium (columbium), tantalum, titanium, tungsten and zirconium, that have a great affinity for oxygen at elevated temperatures.

reactive powder concrete. A cementitious material that obtains its ultrahigh strength through a post-set heat treatment at temperatures between 200-400°C (390-750°F). Its average compressive strength ranges from 200 to 800 MPa (29 to 116 ksi). Used in the building and construction trades. Abbreviation: RPC.

reactive rubber adhesives. A class of adhesives, such as polysulfide, polyurethane and silicone, based on monomers that polymerize *in situ* to produce synthetic rubber. Used for structural applications.

reactor ceramics. A group of special ceramic materials used in nuclear reactors, e.g., for control and fuel elements, moderators, and shielding. Ceramics for control elements include several rare-earth oxides and boron tetracarbide, ceramics for fuel elements include uranium dioxide and uranium mono- and dicarbide, and ceramis for moderators and shielding include concrete, silicon carbide, beryllia and carbon. Also known as *nuclear ceramics; nuclear-reactor ceramics*.

reactor fuel. Any of the fissionable isotopes of uranium, plutonium and thorium that are capable of acting as sources of energy and as source of neutrons for the propagation of a chain reaction in a nuclear reactor. Also known as *atomic fuel; fission fuel; nuclear fuel*. See also fissionable material.

reactor-grade materials. A group of materials that are free of neutron-absorbing contaminants and can be used in nuclear reactor construction. Examples of materials produced in reactor grades are beryllium oxide, boron tetracarbide, silicon carbide, zirconium chloride and zirconium hydride. Also known as *nuclear-grade materials*.

Reading. Trade name of Reading Alloys, Inc. (USA) for a series of refractory and reactive master alloys of the chromium-aluminum, aluminum-vanadium, vanadium-aluminum, zirconium-aluminum, molybdenum-nickel and chromium-nickel type.

Ready-Flow. Trade name of Engelhard Minerals & Chemicals Corporation (USA) for a brazing alloy composed of 56% silver and 44% copper. It has a melting point of 629°C (1165°F), and is used for brazing most ferrous and nonferrous metals.

Ready-Mark. Trade name of Brown & Sharpe Manufacturing Company (USA) for an oil-hardening tool steel (AISI type O1) containing 0.9% carbon, 1.2% manganese, 0.5% chromium, 0.5% tungsten, 0.2% vanadium, and the balance iron. It has a low tendency to warping and shrinking, and is used for dies, punches, reamers, tools, and shear blades.

ready-mixed concrete. Concrete mixed by any means and delivered to the construction site in the plastic state ready for placement. Abbreviation: RMC.

Readyweld. Trade name of Lincoln Electric Company (USA) for a low-carbon steel used for welding rods.

realgar. A red to aurora-red mineral with an orange-red streak and a resinous luster. It is composed of arsenic monosulfide, AsS, and contains 70.1% elemental arsenic. Crystal system, monoclinic. Density, 3.55-3.57 g/cm^3; melting point, 307°C (585°F); hardness, 1.5-2.0 Mohs; refractive index, 2.684. Occurrence: Chile, USA (Utah, Nevada, Wyoming). Used as a pigment, and as a source of arsenic. Also known as *red arsenic; red orpiment; sandarac*.

Realor. Trade name for a low-gold dental casting alloy.

Realox. Trademark of Alcoa–Aluminum Company of America (USA) for a series of low-soda calcined aluminum oxide materials with crystal sizes from 0.5-4.0 µm (20-160 µin.). Used for the manufacture of ceramic products.

Réaumur porcelain. A special type of porcelain composed essentially of devitrified or fritted glass.

reavy yarn. A plied yarn made by twisting together a single and a two-ply yarn.

rebar. See reinforcing bar.

Rebaron. Trade name of GC America, Inc. (USA) for self-cure dental acrylic resins used for acrylic denture rebasing/relining. *Rebaron LC* is a pink-colored, visible-light-cured acrylic resin supplied as a two-component powder-liquid product.

Rebilda. Trade name for a self-cure dental composite used for core-build-ups.

rebonded fused grain refractory. A fired refractory, usually a brick or special shape, composed essentially or entirely of fused refractory grain. See also fused grain refractory.

Recal. Trade name for a calcium hydroxide cement used for dental applications.

recent resin. A natural resin, such as a *dammar*, obtained from the exudate of a living tree. See also copal.

Recidal. Italian trade name for a series of corrosion-resistant, free-cutting aluminum alloys containing 5-6% copper, 0.1-0.3% lead, and up to 0.3% cadmium, 0.6% zinc, 0.3% iron and 0.25 silicon, respectively. Used for screw-machine products and fasteners.

reclaimed enamel. Porcelain-enamel overspray and glaze scrapings from dip tanks, spray booths or washed ware that are reconditioned for use.

reclaimed rubber. Rubber prepared partially or entirely from natural or synthetic scrap rubber (e.g. old tires or processing scrap). Blends of virgin and scrap rubber are used for high and medium-grade and scrap rubber alone for low-grade rubber goods. Rubber reclamation processes may entail grinding, digesting in suitable caustic solutions, and forming into powder, pellets, sheets or slabs.

reclaimed sand. Burnt or partially burnt foundry sand that has been treated for reuse.

reclaimed wool. Wool fibers that have been obtained by the reclamation of either unused felted or woven fabrics, or wool rags and manufactured waste. See also reprocessed wool; reused wool.

Reco. Trade name of Usines Emile Henricott SA (Belgium) for a series of permanent magnet alloys containing 19-25% nickel, 0-20% cobalt, 5-13% aluminum, 4-7% copper, 0-7% titanium, and the balance iron. They have high permeability, and are used for electrical and magnetic equipment.

Recolglass. Trade name of Floreal Balzaretti (Argentina) for glass tiles supplied in various colors.

reconditioned sand. Burnt or partially burnt foundry sand that has been treated for reuse (*reclaimed sand*) or is reused directly after being reconditioned (*system sand*).

reconstituted fibers. A term referring to synthetic fibers made from recovered waste polymers, or blends of new and recovered waste polymers.

reconstituted mica. Mica made by grinding natural or synthetic mica scrap into flake or powder, combining with a suitable binder and pressing into sheets or variously shaped objects for use as electrical insulation.

reconstructive biomaterials. A class of biodegradable or biostable materials that are suitable for the repair and reconstruction of damaged or diseased human organs and body parts. Examples of include alumina ceramics for load-bearing prostheses (e.g., hip and knee joint parts, and bone screws) and dental applications (e.g., jaw-bone reconstruction), bioactive glass-ceramics for orthopedic and dental implants, silicone rubber for the reconstruction of soft tissue, expanded polytetrafluoroethylene for blood-vessel grafts, and self-reinforced polyglycolic acid or polylactic acid for implants.

Re'Cord. Trade name of Harry J. Bosworth Company (USA) for a blue-colored polyvinylsiloxane-based dental elastomer for bite registration purposes.

Record. Trade name of American Smelting & Refining Company (USA) for lead-base bearing alloys.

Recovac. Trademark of Vacuumschmelze GmbH (Germany) for

a series of soft magnetic alloys containing 77% nickel and 23% iron. They have good wear resistance, high initial permeability, low coercivity and saturation induction, a Curie temperature of 280-440°C (535-825°F), and a hardness of 160-240 Vickers. Used for magnetic heads, magnetic head shielding, and relay components.

recovered wool. See reused wool.

recrystallized alumina. A tough, abrasion-resistant, high-purity aluminum oxide (Al_2O_3) produced by recrystallization from a molten bath and composed essentially of single crystals. It has a density of 3.9 g/cm³ (0.14 lb/in.³), excellent high-temperature properties, good compressive strength, high hardness (above 1500 Vickers), an upper continuous-use temperature of 1800°C (3270°F), exceptional resistance to mechanical damage, and good resistance to acids, alkalies and halogens. Used in the manufacture of bearings for watches and scientific instruments, lasers, fiber-forming dies, and as a fine abrasive and polishing agent. Also known as *single crystal alumina*. See also microcrystalline alumina.

recrystallized graphite. Graphite that has been purified by repeated recrystallization at high temperatures.

recrystallized silicon carbide. Silicon carbide (98+%) that has been purified by one or more recrystallization treatments.

Rectiprene. Trade name of Ato Findley (France) for neoprene adhesives used for bonding foams and furniture.

rectorite. (1) A white to yellowish gray or yellow mineral of the smectite group composed of sodium aluminum silicate hydroxide dihydrate, $(Na,Ca)Al_4(Al_{1.7}Si_{6.3})O_2(OH)_4 \cdot 2H_2O$. Crystal system, monoclinic. Density, 2.39 g/cm³. Occurrence: Pakistan, South Africa.

(2) A white or yellow mineral of the mixed-layer group composed of potassium aluminum silicate hydroxide hydrate, $K_{0.6}Al_2Si_4O_{10}(OH)_2 \cdot xH_2O$. Crystal system, monoclinic. Density, 2.92 g/cm³. Occurrence: South Africa.

Recupex. Trade name of Foseco Minsep NV (Netherlands) for a series of covering and refining fluxes used in the melting of copper and nickel alloy scrap (e.g., turnings or chips) in particular to cleanse the melt and decrease oxidation losses.

recycled plastics. Plastics prepared from scrap (e.g., chips and discarded articles) by direct processes (i.e., cleaning, regrinding, reprocessing and subsequent addition to other plastic material), hydrolytic processes (i.e., chemical decomposition effected by reaction with water at high temperature and pressure) or pyrolytic processes (i.e., thermal decomposition without oxidation). Principally, direct processes are suitable for thermoplastics (e.g., polyethylene, polypropylene, polystyrene and polyvinyl chloride), hydrolysis for polyurethanes, polyamides and polyesters, and pyrolysis for nearly all thermosets, thermoplastics and elastomers.

recycled polyethylene terephthalate. Polyethylene terephthalate (PET) plastics prepared from discarded food and non-food packaging containers (e.g., water and soft drink bottles, salad dressing and cooking oil containers, and household cleaner bottles) by regrinding and reprocessing, or pyrolysis. It can be used for the manufacture of a wide range of new PET materials including bottles and containers for food and non-food packaging, sheet and film products, automotive components including bumpers, door panels and luggage racks, carpets, fabrics for underwear and shirts, shoes, upholstery, and fiberfill for coats and sleeping bags. Abbreviation: RPET.

recycled wool. Wool fibers reclaimed from felted, knitted, spun or woven wool products that have either been used or not used by the ultimate consumer.

Recyclospheres. Trade name of Sphere Services Inc. (USA) for strong alumina-silica spherical fillers and extenders supplied in particle sizes of 150-300 μm (6-12 mils). See also Cenospheres.

Red Action. Trademark of Superex Canada Limited for autobody fillers.

red alder. The tough, moderately heavy and strong wood of the hardwood tree *Alnus rubra*. Freshly cut red alder wood is nearly white to light pinkish brown and turns reddish-brown after cutting. *Red alder* has a fine, even grain, high resiliency, low shock resistance and shrinkage, good workability, and takes a fine polish. Average weight, 450 kg/m³ (28 lb/ft³). Source: Pacific Coast of North America from Alaska to California. Used for furniture, cabinets, interior finish, sash, doors, millwork, panel stock, as pulpwood and fuelwood, and as a substitute for walnut and mahogany.

Redalloy. Trade name of Chase Brass & Copper Company (USA) for a corrosion-resistant alloy composed of 85% copper, 14% zinc and 1% tin. It has high elongation, excellent cold workability, good strength, and is used for condenser and heat-exchanger tubing.

Red Anchor Drill Rod. Trade name of Teledyne Allvac (USA) for a water-hardening tool steel (AISI type W1) containing 0.95-1.1% carbon, and the balance iron. It is made with a bright, lustrous and smooth finish from high-grade stock, and has uniform composition, structure and tolerances, excellent resistance to decarburization, and good hardenability and machinability. Used for drills, threading taps, pins, punches, motor shafts, spindles, mandrels, wrenches, anvils, punches, cold-heading dies, and dental tools.

red antimony. See kermesite.

red arsenic. See realgar.

red arsenic glass. See arsenic sulfide (i).

red arsenic sulfide. See arsenic sulfide (i).

Redart. Trade name of Resco Products, Cedar Heights Clay Division (USA) for a red clay.

red ash. The strong, hard wood of the tree *Fraxinus pennsylvanica*. It is quite similar to black ash. Green ash was formerly categorized as a hairless variety of red ash, but now the two are combined. Average weight, 625 kg/m³ (39 lb/ft³). Source: Eastern and central United States and southern Canada (from Nova Scotia to Alberta). Used for furniture, veneer, containers and cooperage. See also black ash; green ash.

red balsa. Balsawood obtained from the tree *Ochroma velutina* growing in Central America. It is used as a source of a dark-colored fiber suitable for thermal insulation and padding applications. See also balsa.

red beech. See European beech; New Zealand beech.

red brass. See hydraulic bronze; wrought red brass.

red brick. A red building brick, usually of standard size, made of ferruginous clay (red clay).

red casting brass. See hydraulic bronze.

red cedar. See eastern red cedar; western red cedar.

red cement mortar. A cement mortar to which a predetermined percentage of fired clay has been added.

red chalk. Red ocher mixed with some clay. See also red ocher.

Red Chip. Trade name of Firth Sterling Inc. (USA) for a tungsten-type high-speed steel (AISI type T4) containing 0.75% carbon, 4% chromium, 18% tungsten, 1% vanadium, 5% cobalt, and the balance iron. It has high hot hardness, excellent wear resistance, and good machinability. Used for lathe and

planer tools, milling cutters, and various other cutting and finishing tools.

red clay. A reddish-brown, fine-grained *ferruginous clay* that produces a red color when fired, and is used in the manufacture of brick, roofing tile and pottery. Also known as *brown clay*.

red cobalt. See erythrite.

red copper ore. See cuprite.

red copper oxide. Reddish-brown, moisture-sensitive crystals or red powder (97+% pure). It occurs in nature as the mineral *cuprite*. Crystal system, cubic. Density, 5.75-6.09 g/cm³; melting point, 1235°C (2255°F); boiling point, 1800°C (3270°F). Used in glass, glazes, porcelain enamels and other ceramics primarily for the production of red colors, in copper salts, electroplating, antifouling paints and brazing preparations, and in photocells. Formula: Cu_2O. Also known as *copper hemioxide; copper oxide; copper protoxide; copper suboxide; cuprous oxide*.

Red Cut Cobalt. Trade name of Teledyne Vasco (USA) for a tungsten-type high-speed steel (AISI type T4) containing 0.70-0.75% carbon, 0.2-0.4% silicon, 0.1-0.3% manganese, 17.75-18.75% tungsten, 4-4.5% chromium, 0.4-0.9% molybdenum, 4.5-5 cobalt, 1.0-1.15% vanadium, and the balance iron. It has good deep-hardening properties, high hot hardness, excellent wear resistance, and good machinability. Used for lathe and planer tools, milling cutters, and various boring, finishing and shaping tools.

Red Cut Superior. Trade name of Teledyne Vasco (USA) for a tungsten-type high-speed steel (AISI type T1) containing 0.5-0.8% carbon, 0.25-0.4% silicon, 0.1-0.3% manganese, 17.5-18.5% tungsten, 3.75-4.25% chromium, 0.95-1.10% vanadium, and the balance iron. It has good deep-hardening properties, high hot hardness, high cutting ability, excellent abrasion and wear resistance, and high toughness. Used for cutting, machining, finishing and spinning tools, woodworking and machine knives, punches, lamination and forming dies, and springs for elevated temperatures.

red cypress. See baldcypress.

red deal. See northern pine.

Red Devil. Trade name of Champion Rivet Company (USA) for a series of flux-coated welding rods made from steel containing 0.05-0.1% carbon, up to 0.9% molybdenum, and the balance iron.

Red Diamond. Trade name of Spartan Redheugh Limited (UK) for wear-resistant steel plate available in numerous grades including carbon-manganese with 0.45% carbon and 0.75% manganese, high-carbon chromium with 1.15% carbon, 0.60% manganese and 1.50% chromium, austenitic manganese with 1.15% carbon and 12.5% manganese, wear-resistant alloy with 0.25% carbon, 0.75% manganese, 1.00% chromium and 0.20% molybdenum, hardened and tempered alloy with 0.40% carbon, 0.60-0.75% manganese, 1.10-1.25% chromium, 0.25-0.30% molybdenum and up to 1.50% chromium, and corrosion-resistant high-chromium stainless containing up to 0.03% carbon and 11.5% chromium.

reddingite. A pinkish white to pale rose-pink, or yellowish white to colorless mineral of the phosphoferrite group composed of iron manganese phosphate trihydrate, $(Mn,Fe)_3(PO_4)_2 \cdot 3H_2O$. Density, 3.23 g/cm³; refractive index, 1.664. Occurrence: Germany, USA (Connecticut).

Redel. Trademark of Winter & Co. GmbH (Germany) for artificial leather and leather substitutes. *Redel-Nappa* refers to nappa leather substitutes.

red elm. See slippery elm.

red ferric oxide. See red iron oxide.

red fir. (1) The light, moderately strong wood from the large coniferous tree *Abies magnifica*. Source: Western United States (mainly California and Oregon). Used as lumber for building construction, millwork, doors, crates, boxes, veneers and plywood, and as pulpwood. Also known as *California red fir; golden fir.*

(2) A term sometimes used to refer to the reddish-brown wood of young *Douglas fir* trees.

red glass. (1) Any glass that has been made red by the addition of cuprous oxide, or gold chloride (e.g. as purple of Cassius).

(2) A soda-zinc glass with small additions of cadmium and selenium.

red gold. Reddish yellow, corrosion-resistant alloys composed of up to 75% gold and 25% copper, with little or no silver. Used for jewelry and ornaments.

red gum. (1) A term used to refer to the lustrous, reddish-brown, usually highly figured heartwood of the broad-leaved tree *Liquidambar styraciflua* belonging to the sweet gums, and found throughout Mexico and the southeastern United States. See also sweet gum.

(2) The strong, tough, orange red wood of the red gum tree *Eucalyptus calophylla* growing in Australia. It has good durability, a fine grain with numerous gum veins, and is used for lumber and structural applications.

red hematite. See hematite.

Redifast. Trade name for a fast-setting, self-cure dental acrylic.

Redi-Kote. Trade name of US Steel Corporation for a low-carbon steel used to make sheet products with remelted zinc coating.

Redilon. Trade name for a heat-curing acrylic resin for dentures. *Redilon DB* refers to a self-curing denture acrylic.

Redi-Mold. Trademark of Teledyne Vasco (USA) for a chromium-type hot-work tool steel (AISI type H13) containing 0.35-0.37% carbon, 0.9-1.1% silicon, 0.25-0.45% manganese, 5-5.5% chromium, 1.1-1.4% molybdenum, 0.9-1.2% vanadium, and the balance iron. Usually supplied as flats and squares with a smooth surface finish, it has good wear resistance and red hardness, excellent resistance to thermal fatigue, heat checking, pitting and oxidation, and a low coefficient of thermal expansion. Used for die-casting cores, and plastic molding and aluminum extrusion dies and inserts.

Red Indian. Trade name of Atlas Specialty Steels (Canada) for a chromium-type hot-work tool steel (AISI type H14) containing 0.35% carbon, 0.3% manganese, 1% silicon, 5% chromium, 4.5% tungsten, 0.3% molybdenum, 0.3% vanadium, 0.5% cobalt, and the balance iron. It has good deep-hardening properties, moderate hot wear resistance and red hardness, low cold wear resistance, high toughness, good resistance to heat checking and moderate machinability. Used for die-casting and extrusion dies, and mandrels.

red iron ore. See hematite.

red iron oxide. Red-brown to black crystals or red powder supplied in technical (99.5% pure) and electronic (99.999% pure) grades. It occurs in nature as the minerals *hematite* and *maghemite*. Crystal system, cubic. Density, 5.12-5.24 g/cm³; melting point, 1565°C (2849°F); hardness, 5.5-6.5 Mohs. The alpha form exhibits canted antiferromagnetic properties. Used in metallurgy as a source of iron, as a component of thermit, in foundry cores and molds to improve hot compressive strength, as a paint and rubber pigment, as a pigment to produce various colors in glazes and glasses, in electronic pigments for televi-

sions, in the manufacture of ferrites, permanent magnets, memory cores for computers and magnetic recording tapes, in polishing compounds for glass and other substances, and as a catalyst. Submicron-sized particles are also used as particle size standards in polymer and biomedical research. Formula: Fe_2O_3. Also known as *ferric oxide; ferric trioxide; iron sesquioxide; red ferric oxide; red iron trioxide.* See also jeweler's rouge; mineral red; mineral rouge; red ocher; rouge.

red ironstone clay. See argillaceous hematite.

red iron trioxide. See red iron oxide.

Redi-Scribe. Trademark of Teledyne Vasco (USA) for *Ohio-Die* cold-work tool steels (AISI type D2), *Colonial No. 6* cold-work tool steel (AISI type O1) and *Air-Hard* cold-work tool steels (AISI type A2) supplied with smooth, shiny uniform finish provided by horizontal spindle grinding. Used in tool and die making.

redistilled zinc. High-purity zinc (99.99+%) made from zinc containing impurities, such as cadmium and lead, by a selective distillation process.

Reditray. Trade name for a self-cure acrylic for dental trays.

red juniper. See eastern red cedar.

red khaya. The light pinkish brown to dark reddish brown wood of the hardwood tree *Khaya ivorensis*. It closely resembles American mahogany in appearance, but has a slightly coarser texture, a more pronounced grain pattern and greater shrinkage. *Red khaya* is highly figured, with interlocking grain, easy to season, machines and finishes well, and has moderate decay resistance. Average weight, 530 kg/m³ (33 lb/ft³). Source: Tropical West Africa (Ivory Coast, Ghana and Nigeria). Used for fine furniture, interior paneling, store fixtures, art objects, boat construction, veneer, and plywood. Also known as *red mahogany; sassandra mahogany.* See also khaya.

Red Label. Trade name of Simonds Worden White Company (USA) for a water-hardening tool steel (AISI type W1) containing 0.9-1.2% carbon, 0.25% silicon, 0.25% manganese, and the balance iron. It has excellent machinability, good wear resistance and toughness, low hot hardness, and is used for hand and cutting tools.

Red Label Extra. Trade name of Wallace Murray Corporation (USA) for a water-hardening tool steel (AISI type W2) containing 0.25% manganese, 0.25% silicon, up to 0.25% vanadium. The carbon content can be varied between 0.6 and 1.4%. It has excellent machinability, good wear resistance and toughness, low hot hardness, and is used for hand and cutting tools.

red lead. See red lead oxide.

red lead ore. See crocoite.

red lead oxide. Bright red to dark brown powder or crystals (98+% pure) obtained by heating lead monoxide (*litharge*) in air. It occurs in nature as the mineral *minium*. Crystal system, tetragonal. Density, 8.9-9.1 g/cm³; melting point, decomposes above 800°C (1500°F); hardness, 2.5 Mohs. Used extensively in paints for the protection of iron and steel against corrosion, in varnishes, storage batteries, glasses, ceramic glazes, pottery and enamels, as a flux in ceramics, in making cement for pipes, and in packing pipe joints. Formula: Pb_3O_4. Also known as *lead orthoplumbate; lead tetroxide; plumboplumbic oxide; red lead.*

redledgeite. A black mineral of the rutile group composed of magnesium chromium titanium silicate hydroxide, $Mg_4Cr_6Ti_{23}Si_2O_{61}(OH)_4$. Density, 3.72 g/cm³. Occurrence: USA (California, Colorado).

red locust. See black locust.

red mahogany. See red khaya.

Redmanol. Trade name Union Carbide Corporation (USA) for phenol-formaldehyde resins and plastics.

red maple. The soft, white wood of the medium-sized hardwood tree *Acer rubrum*. The heartwood is pale reddish-brown, and the sapwood white with reddish-brown tinge. *Red maple* has good strength, hardness and shock resistance. Average weight, 608 kg/m³ (38 lb/ft³). Source: Eastern United States (from Maine to Florida) and southeastern Canada (from Ontario to Nova Scotia). Used for furniture, boxes, pallets, crates, and woodenware.

red meranti. See red seraya.

red mercuric oxide. Red or orange-red crystals, or red powder (98+% pure). It occurs in nature as the mineral *montroydite*. Crystal system, orthorhombic. Density, 11.1 g/cm³; melting point, decomposes at 500°C (930°F). Used as a paint pigment, as a fungicide and antifouling agent in paints, in ceramic colors, in dry batteries, in polishing compounds, as an oxidizer, in the manufacture of mercury salts, and in biochemistry and medicine. Formula: HgO. Also known as *red mercury oxide; red precipitate.*

red mercuric sulfide. Red crystals or fine, brilliant scarlet-red powder (99% pure). It occurs in nature as the mineral *cinnabar*. Crystal system, hexagonal. Density, 8.10 g/cm³; melting point, sublimes at 583.5°C (1082°F); hardness, 2.0-2.5 Mohs. Used as a pigment. Formula: HgS. Also known as *Chinese vermilion; mercuric sulfide; mercury sulfide; quicksilver vermilion; red mercury sulfide.*

red mercury oxide. See red mercuric oxide.

red mercury sulfide. See red mercuric sulfide.

red metal. (1) An impure intermediate product containing about 48% copper obtained from roasted copper sulfide ores (e.g., chalcocite or covellite) by melting.

(2) A free-cutting copper alloy containing 20% zinc, 6% lead and 4% tin, and used for hardware.

red mud. A reddish byproduct sludge obtained in the Bayer process by the extraction of pure alumina from bauxite. It contains 30-60% ferric oxide (Fe_2O_3), the balance being chiefly silicon dioxide (SiO_2) and titanium dioxide (TiO_2). Used in steelmaking. See also Bayer alumina.

red oak. See American red oak.

red ocher. See red iron oxide; rouge.

Re-Dolite. Trade name of L.C. Chase & Company (USA) for polyvinyl chlorides and other plastics.

Redon. German trade name for acrylic monofilaments and fibers used for textile fabrics.

red orpiment. See realgar.

Redox. Trademark of Grothe Keramik (Germany) for red ceramic colorants.

red oxide of copper. See cuprite.

red oxide of zinc. See zincite.

Redphase P. Trade name for a silicone elastomer used for dental impressions.

red phosphorus. See amorphous phosphorus.

red pine. (1) The moderately heavy, strong, stiff, soft and shock-resistant wood of the medium-tall pine *Pinus resinosa*. The heartwood is light red to reddish-brown, and the sapwood almost white. It has a straight grain, and is nonporous and very resinous. Average weight, 540 kg/m³ (34 lb/ft³). Source: Eastern Canada (Maritime provinces) and United States (New England, New York, Pennsylvania, Lake states). Used chiefly for lumber, poles, piles, posts, railway ties, cabin logs, and as a pulpwood and fuelwood. Also known as *Norway pine*.

(2) The straight-grained, reddish-brown wood of the tall tree *Dacrydium cupressinum*. It has good workability, and an average weight of 590 kg/m³ (37 lb/ft³). Source: New Zealand. Used for furniture, millwork, and as a pulpwood. Also known as *rimu*.

red potassium chromate. See potassium dichromate.

red potassium prussiate. See potassium ferricyanide.

red precipitate. See red mercuric oxide.

red prussiate. See potassium ferricyanide.

red prussiate of potash. See potassium ferricyanide.

red prussiate of soda. See sodium ferricyanide.

red quebracho. The hard, durable, brownish-red wood of the hardwood tree *Aspidosperma quebracho* of the dogbane family in the order Sapindales. It is obtained frequently with black spots, has high brittleness, and takes a high polish. Average weight, 1250 kg/m³ (78 lb/ft³). Source: Argentina, Paraguay. Used for crossties, posts, firewood, medicine, and tanning. See also quebracho; white quebracho; yellow quebracho.

Red Ray. Trade name for an electrical resistance alloy composed of 85% nickel and 15% chromium. It has high heat resistance, and is used for heating elements.

red rubber. (1) Rubber either made with pigments producing red color, or colored red by other means (e.g., coatings).

(2) Rubber whose reddish color is produced by vulcanizing with antimony pentasulfide. This production method is now virtually obsolete.

redruthite. See chalcocite.

Red Sabre. Trade name of Bethlehem Steel Corporation (USA) for a tungsten-type high-speed steel (AISI type T15) containing 1.5% carbon, 4-4.75% chromium, 12% tungsten, 5% vanadium, 5% cobalt, and the balance iron. It has good deep-hardening properties, excellent wear resistance and hot hardness, and is used for tool bits, cutters, dies, and lathe and planer tools, broaches.

RedSeal. Trademark of Pratt & Lambert, Inc. (USA) for a line of latex flat and satin paints used for exterior and interior finishes.

red seraya. The Sabah name for wood of various hardwood species of the genus *Shorea* also known by the Malaysian and Sarawak name of "red meranti." It is supplied as light-red meranti, a lightweight utility wood, and dark-red meranti, a heavier wood used for more exacting applications. *Red seraya* has good working and nailing qualities, and resembles *lauan* and *mahogany* in appearance. Average weight, 550 kg/m³ (34 lb/ft³). Source: Indonesia, Malaysia, Sarawak, Sumatra and surrounding area. Used as a replacement for mahogany, for lumber, high-class furniture, plywood and superior joinery, and extensively for interior construction work. Also known as *red meranti*. See also white seraya; yellow seraya.

Red Shadow. (1) Trade name of Ziv Steel & Wire Company (USA) for an oil-hardening high-speed steel containing 0.8% carbon, 4% chromium, 5% molybdenum, 6.5% tungsten, 2% vana-dium, and the balance iron. Used for milling cutters, drills, and various other machining and finishing tools.

(2) Trade name of Agawam Tool Company (USA) for a hot-work tool steel containing 0.4% carbon, 3.1% chromium, 10% tungsten, 0.25% manganese, 0.25% vanadium, and the balance iron. Used for tools and dies.

red shellac varnish. A red solution made by first dissolving *shellac* in grain or wood alcohol and then adding finely ground Chinese vermilion (red mercuric sulfide). Used to coat patterns for metal castings.

red spruce. The wood from the medium-sized softwood tree *Picea rubens* belonging to the Eastern spruces. No distinction is made between black, red and white spruce in marketing. *Red spruce* is light-colored, has only slightly different heartwood and sapwood, moderate strength, dries well, and is stable after drying. Average weight, 448 kg/m³ (28 lb/ft³). Source: Canada (Maritime provinces and southern Quebec), USA (New England states and Appalachian Mountain region). Used chiefly for pulpwood, millwork, framing lumber, boxes and crates, and sounding boards for pianos. Also known as *Canadian spruce; West Virginia spruce.*

Red Star. Trade name of Teledyne Vasco (USA) for a water-hardening tool steel (AISI type W1) with excellent machinability and good to moderate deep-hardening properties. Used for hand tools, knives, punches, chisels, shear blades, dies, and gages. *Red Star Vanadium* is a water-hardening carbon-vanadium tool steel (AISI type W1) with excellent machinability, good to moderate deep-hardening properties, and a fine-grained microstructure. *Red Star Tungsten* is an oil- or water-hardening cold-work tool steel (AISI type O7) containing 1.15-1.25% carbon, 0.2-0.4% silicon, 0.2-0.3% manganese, 1.5-1.7% tungsten, 0.6-0.8% chromium, 0.15-0.25% vanadium, 0.2-0.3% molybdenum, and the balance iron. It has a great depth of hardening, good toughness and machinability, high hardness and good wear resistance. Used for cutting and finishing tools, paper and woodworking knives, dies, and gages.

Red Streak. Trade name of Wallace Murray Corporation (USA) for a tungsten-type high-speed steel (AISI type T1) containing 0.75% carbon, 4% chromium, 18% tungsten, 1% vanadium, and the balance iron. It has a great depth of hardening, high hot hardness, excellent machinability and wear resistance, good toughness, and is used for cutting tools, drills, and various machining and finishing tools.

red thermit. A *thermit* consisting of a mixture of finely divided aluminum and red ferric oxide (Fe_2O_3). Used for welding iron and steel.

Red Tiger. Trade name of Bethlehem Steel Corporation (USA) for a high-speed steel containing 1% carbon, 18.5% tungsten, 4.5% chromium, 2.6% vanadium, 0.65% molybdenum, and the balance iron. Used for taps, reamers, cutters and tools.

Red Tip. Trade name of Mueller Brass Company (USA) for a series of free-cutting brasses containing varying amounts of copper, zinc and lead. Used for screw-machine products.

Red Top. Trademark of US Gypsum Company (USA) for acoustical and patching plasters, gypsum wall plasters, wood fiber plasters, fireproofing compounds, and thermal insulating wool.

reduced iron. Finely divided iron powder, usually with a particle size of 0.02-800 µm (0.8 µin to 0.03 in.) and a purity of 99.0+%, produced by reduction of high-purity ferric oxide or oxalate in a stream of hydrogen. Also known as *spongy iron.*

reduced metal powder. A metal powder made by the chemical reduction of a metal compound (e.g., a metal oxide, hydroxide, carbonate or oxalate).

reduced nickel. Nickel produced by high-temperature hydrogen reduction of the product obtained from the precipitation of nickel carbonate or hydroxide onto kieselguhr. Used as a catalyst.

reduced powder. A metal or nonmetal powder made by chemical reduction of a compound without melting.

reducing agent. A substance that effects the reduction or removal of the oxygen in a compound. It supplies electrons to the compound and, during the reduction process, becomes oxidized.

reductant. A reducing agent or substance, such as coal or coke,

used in the smelting of ores or concentrates for removing oxygen.

Redux. (1) Trademark of Hexcel Corporation (USA) for a heat-curing epoxy-phenolic adhesive. Supplied in powder, film and foam form, it has low peel strength and good lap shear strength at low and high temperatures (-73°C to +260°C, or -100 to +500°F). Used as a structural adhesive for joining metals, plastics, wood and composites.

(2) Trade name of Ciba-Geigy Composites Inc. (USA) for a bismaleimide (BMI) film adhesive used for bonding composites to composites. It is compatible to BMI prepregs, such as *Fibredux.*

red vitriol. See bieberite.

redwood. (1) The soft, light, nonresinous, brownish-red wood of the giant coniferous tree *Sequoia sempervirens*. The highly decay-resistant heartwood ranges from light cherry to dark mahogany in color, and the narrow sapwood is nearly white. *Redwood* has a fine and even texture, a straight grain, outstanding durability, good water and moisture resistance, splits easily, and is stable in use when dry. Average weight, 420 kg/m³ (26 lb/ft³). Source: Pacific coast of North America from California to Oregon. Used for outdoor constructional work, outside finish, outdoor furniture, interior paneling and fittings, siding, doors, sash, blinds, fencing, plywood, lumber for vats, tanks, silos, wood-stave pipes, etc. Also known as *California redwood; coast redwood; Humboldt redwood; sequoia.* See also sequoia.

(2) See northern pine.

redwood bark fiber. Short, twisted fibers obtained by shredding the bark of North American redwood (*Sequoia sempervirens*). They possess good fire and water resistance, and can be readily interfelted and spun. Used as thermal insulation for refrigerators, house walls and, blended with wool or other fibers, for blankets.

red zinc oxide. See zincite.

red zinc ore. See zincite.

reed brass. A corrosion-resistant brass composed of 67-69% copper, 30-32% zinc and 1% tin. Used for condenser and heat-exchanger tubing, fittings, hardware, pipes and tubes.

Reeded. Trade name of Chance Brothers Limited (UK) for a patterned glass featuring narrow or broad vertical reeds. See also Broad Reeded; Narrow Reeded.

Reedlyte. Trade name of Chance Brothers Limited (UK) for patterned glass featuring narrow or broad vertical reeds. See also Broad Reedlyte; Narrow Reedlyte.

reedmergnerite. A colorless mineral of the feldspar group composed of sodium boron silicate, $NaBSi_3O_8$. Crystal system, triclinic. Density, 2.78 g/cm³; refractive index, 1.565. Occurrence: USA (Utah).

Reedrop. Trade name of Chance Brothers Limited (UK) for patterned glass featuring broad vertical reeds and irregularly spaced raindrops.

reeled silk. Untwisted and unthrown *raw silk* as wound from the cocoons of silk-producing insects. See also raw silk; thrown silk.

Reel-Satin. Trade name of OMG Fidelity (USA) for a pure, matte tin electrodeposit and deposition process.

Reemay. Trade name of DHJ Industries (USA) for two-dimensional nonwoven fabrics made from nondirectional continuous polyester fibers. Supplied in weights from 17 to 200 g/m² (0.5 to 5.9 oz/yd²) and widths up to 3.75 m (12.3 ft.). They can be processed by all conventional processes including sewing, needling, pleating, calendering, creping and welding, and have high tensile strength, good dimensional stability and chemical resistance, and an upper use temperature of 175°C (347°F). Used as backing materials for polymer films, and in filtering equipment, cable coverings, coating bases, glass-fiber-reinforced plastics, furniture, upholstery, and shoes.

re-entrant foams. Metallic and polymeric foam materials specially processed, usually in a processing regime involving compression and heat, to effect an inward protrusion of the cell ribs in their cellular architectures. They exhibit a negative Poisson's ratio. See also *negative Poisson's ratio materials.*

Reevecote. Trademark of Reeves Brothers, Inc. (USA) for a polyethylene terephthalate (*Dacron*) fabric coated with fluorocarbon resin (*Kel-F*). Used for gaskets, seals, and diaphragms.

reevesite. A green to yellow mineral of the sjogrenite group composed of nickel iron carbonate hydroxide tetrahydrate, $Ni_6Fe_2(CO_3)(OH)_{16} \cdot 4H_2O$. Crystal system, hexagonal. Density, 2.87 g/cm^3; refractive index, 1.72. Occurrence: South Africa.

Reevon. Trademark of Reeves Brothers, Inc. (USA) for polyethylene fibers and fabrics used for upholstery.

Reex. Italian trade name for heat- and corrosion-resistant steel containing 0.45-0.9% carbon, 14% nickel, 11% chromium, 2% tungsten, 1%, and the balance iron. Used for valves, and valve and pump components.

Refax Metal. Trade name of Watsontown Foundry (USA) for a cast iron containing 3.2% carbon, 2% silicon, 1% nickel, 0.8% chromium, and the balance iron. Used for valve and pump bodies.

Refel. Trademark of Tenmat Inc. (USA) for a range of silicon carbide products. *Refel F* is a reaction-bonded silicon carbide with a density of 3.10 g/cm^3 (0.112 $lb/in.^3$), high strength up to 1350°C (2460°F), good high-temperature and oxidation resistance, high hardness (2500-3000 Vickers), good thermal-shock resistance, and good resistance to acids. Used for gas-turbine parts, furnace applications, bearings, and seals.

reference material. A material that has a well defined chemical composition and one or more properties that are within established ranges, making it useful for calibration, standardization and or control of an apparatus, process or substance.

Ref Grog 64-A. Trade name of BPI Inc. (USA) for a crushed and sized refractory grog with a typical chemical composition of 64% alumina (Al_2O_3), 30% silica (SiO_2), 2.7% titania (TiO_2), 1.7% ferric oxide (Fe_2O_3), the balance being potassium oxide (K_2O), sodium oxide (Na_2O) and moisture.

refikite. A white mineral composed of dihydrodextropimaric acid, $C_{20}H_{32}O_2$. Density, 1.09 g/cm^3. Occurrence: Germany, Italy.

Refined. See Atlas Refined.

refined shellac. A purified grade of orange, yellow or white *shellac* that has the wax removed.

refined tar. A thick, sticky, brownish-black liquid obtained by removing the water from tar. Used in waterproofing compositions, and road surfacing. Also known as *dehydrated tar.*

Refinite. Trademark of Refinite Corporation (USA) for bentonite.

Reflect. Trade name for a silicone elastomer used for dental impressions.

Reflectal. Trade name of Alusuisse (Switzerland) for a series of age-hardenable, wrought aluminum-magnesium and aluminum-magnesium-silicon alloys with high to medium strength and good corrosion resistance. Used for chemical equipment, decorative applications, automotive parts, ship- and boatbuilding, aircraft parts, building construction, machine parts, window

frames, fan blades, hardware, door fittings, ornamental and architectural trim and parts, mirrors, watch cases, costume jewelry, electrical conductors, and extrusions.

Reflectalloy. Trade name of McGean Inc. (USA) for a zinc alloy electroplate and plating process.

Reflectasol. Trade name of Saint-Gobain (France) for a solar control glass.

reflective coating. A coating of highly reflective material, usually 0.03-0.1 µm (0.001-0.004 mil) thick, applied to a prepared glass surface. The reflective material is often aluminum with a thin abrasion-resistant overcoat of silicon oxide deposited by a vacuum coating process. Used for mirror applications, e.g., microscopes, telescopes and automotive rearview mirrors. Also known as *mirror coating.*

reflective insulation. Insulation employed to resist the flow or transmission of heat radiation. It is usually in the form of a metal foil or foil-faced material with shiny surfaces, e.g., aluminum foil in sheets or corrugations supported on paper, steel sheets with special coatings, aluminized paper, gold and silver foil or coatings, polished white metals, or fiberglass blankets or batts with aluminum foil facings. Foil and foil-faced materials differ from other insulating materials in that the number of reflecting surfaces, not the thickness of the material, determines their insulation value. Also, in order to be effective, the foil must be exposed to an air space. Also known as *reflective insulator.*

Reflector. Trade name of SA des Verreries de Fauquez (Belgium) for a glass with curvilinear pattern.

reflector sheet. A laminated metal in sheet form composed of a base of strong, ductile commercially pure aluminum or an aluminum-manganese alloy and a facing or cladding of high-polish, heat- and light-reflective high-purity aluminum on one or both sides.

Reflectovue. Trade name of New York Air Brake Company (USA) for a coated glass that reduces solar heat transmission by reflection.

Reflektal. Trade name for a corrosion-resistant, high-purity aluminum.

Reflelite. Czech trade name for a patterned glass with flaked surface.

Reflex. Trade name for a nickel-titanium wire used for dental applications.

Reflexite. Trademark of Permalight GmbH (Germany) for highly reflective, flexible polyvinyl chloride film materials supplied in various colors for sewing, gluing, heat-sealing and welding applications.

Reflite. Italian trade name for phenol-formaldehyde resins and plastics.

Reflo. Trade name of Reflo AG (Switzerland) for a reflection-free glass used in picture glazing.

Reformend. Trade name of Arcos Alloys (USA) for a low-carbon steel used for welding electrodes for cast iron.

Refractaloy. Trademark of Westinghouse Electric Corporation (USA) for a series of precipitation-hardenable superalloys containing up to 0.08% carbon, 20-40% nickel, 16-22% cobalt, 16-20% chromium, 3-8% molybdenum, 0.5-2.0% silicon, 0.4-1.0% manganese, up to 4% tungsten, up to 3% titanium, up to 0.3% aluminum, and the balance iron. They have high tensile, yield and creep strength, high ductility, and good corrosion resistance up to 815°C (1500°F). Used for gas-turbine blades and disks, and springs.

refractories. (1) Ores, cements, sand and metallic or ceramic materials that have relatively high melting temperatures.

(2) A group of inorganic, nonmetallic natural or synthetic materials that have a melting point of at least 1580°C (2875°F) and will withstand continuous temperatures above 540°C (1000°F). They are usually also resistant to abrasion, pressure, thermal shock and corrosion by molten metals, slags and hot dusty gases, and have low thermal conductivities and at least moderate load-bearing capacities. *Refractories* can be classified according to composition into: (i) *Acidic refractories* (e.g., silica and fireclay); (ii) *Basic refractories* (e.g., magnesite and dolomite); and (iii) *Neutral refractories* (e.g., carbon, chrome and mullite). Other important refractory materials include alumina, bauxite, chromite, forsterite, graphite, kaolin, silicon carbide, sillimanite, zirconia and zirconium silicate. Used for lining steel and glass-melting furnaces and coke ovens, for metal-melting crucibles and pots, in kiln construction, and for gas-turbine blades and other continuous high-temperature applications. Also known as *refractory materials*.

Refractory. British trade name for a superalloy containing 30% cobalt, 21% nickel, 20% chromium, 14% iron, 8% molybdenum, 4% tungsten, 0.05% carbon, 3% other elements. It has high tensile strength and heat resistance, and is used for jet-engine components.

refractory aggregate. Refractory materials made into a conglomerate mass with a suitable matrix binder.

refractory alloys. See refractory metals.

refractory block. A refractory material that is usually rectangular in shape and of large size.

refractory bonding mortar. A high-temperature mortar used in the construction of industrial furnaces to provide a structural bond between individual refractory units.

refractory brick. Brick of varying shape and size that is capable of resisting high temperatures and, frequently, abrasion, corrosion and thermal shocks during use as a furnace, kiln or oven lining. Common sizes are 23 × 15 × 6 cm (9.0 × 6.0 × 2.5 in.), and 34 × 15 × 6 cm (13.5 × 6.0 × 2.5 in.). Important types include silica brick, fireclay brick, high-alumina brick (e.g., corundum, mullite and sillimanite), dolomite brick, chromite brick, and carbon and graphite brick.

refractory cement. A finely ground mixture of refractory ingredients, such as fireclay, ganister, silica sand, water glass, chrome ore, magnesite, zircon or crushed brick, that when tempered with water becomes plastic and trowelable and can be used in laying and bonding refractory bricks and linings, as a patching material, or as a form filler for cracks and holes. Often sold under trade names and trademark, such as *Alfrax, Basifrit, Carbofrax, Firefrax, Hadesite* or *Mullfrax*. Also known as *refractory mortar*.

refractory ceramics. A class of ceramics including fireclay (e.g., high-purity and high-alumina fireclay), acid refractories (e.g., high-silica types), basic refractories (e.g., periclase, magnesia and magnesite) and special refractories (e.g., alumina, beryllia, zirconia, carbon, graphite, chrome ore, mullite and silicon carbide). *Refractory ceramics* are characterized by excellent high-temperature stability, chemical inertness when exposed to severe environments and the ability to provide thermal insulation. Typical applications include furnace linings in metal refining, glass manufacturing, metallurgical heat treatment, and power generation and aerospace applications. Abbreviation: RC. Also known as *high-temperature ceramics*.

refractory chrome ore. A refractory ore composed chiefly of chrome-bearing spinels and used in the manufacture of refractory products. See also chrome ore; spinel.

refractory clay. A clay that has a melting point above 1600°C (2910°F) and strong bonding power and thus is suitable for use in firebrick and linings for furnaces, kilns, ovens, reactors, etc. See also fireclay.

refractory coatings. A group of inorganic, protective, high-temperature coatings based on refractory enamels or intermetallic compounds (e.g., metal borides, carbides, nitrides, oxides or silicides), and used on brickwork, metals and other materials that are exposed to elevated temperatures. They are applied by troweling, dipping, flame spraying, plasma-arc deposition, electrophoresis or other techniques. Typical applications include coatings for industrial furnace brickwork, aircraft and spacecraft exhaust systems, pyrometer sheaths, furnace components, heat-treating and chemical-processing equipment, nuclear power plant equipment, and heat exchangers.

refractory composite coating. See composite refractory coating.

refractory concrete. A concrete that is suitable for use in high-temperature applications, and is usually made by bonding refractory aggregate together using a heat-resistant aluminous cement.

refractory enamel. A porcelain enamel that is usually of special composition, contains additions of silica, alumina and/or zircon, and is used as a protective coating on metals and superalloys in contact with hot and corrosive gases.

refractory fibers. A class of inorganic, nonmetallic, continuous or discontinuous carbide, nitride or oxide fibers (e.g., alumina, alumina-silica, alumina-boria-silica, zirconia-silica, zirconia-yttria, silicon carbide or silicon nitride) used in applications requiring resistance to temperatures above 540°C (1000°F). In addition to their heat resistance, *refractory fibers* usually have one or several of the following properties: excellent mechanical properties, good electrical and/or acoustic insulation properties and good chemical and oxidation resistance. Used for thermal and acoustic insulation applications, as reinforcements for metallic, plastic and ceramic materials, in filtration and packing, etc. Abbreviation: RF. See also ceramic fibers; carbide fibers; oxide fibers; continuous fibers; discontinuous fibers.

refractory hard metals. Machinable metal-matrix composites consisting of very hard particles of refractory metal carbides, borides, beryllides or silicides dispersed in metallic matrices (e.g., alloy steel). They are produced from crystalline powders by pressing and sintering, and have high hardness, toughness, abrasion and wear resistance, high melting points, good corrosion resistance, low coefficients of friction, good machinability and hot workability, and good thermal shock resistance. Used for dies and die plates, pelletizer knives, wire-drawing dies, drawing rings for gas cylinders, and in the manufacture of gages, metal-working tools and spray nozzles.

refractory insulating concrete. A type of refractory concrete with low thermal conductivity.

refractory lime. Limestone, usually of the dolomitic type, that is fired at very high temperatures, and resistant to atmospheric hydration and recombination with carbon dioxide (CO_2). Used as a refractory material, as a component in refractories, and as an ingredient in certain cements.

refractory lining. A lining composed essentially of high-grade refractory clay, fireclay or ganister that is highly resistant to high temperatures, and is used in the construction of industrial furnaces and boiler foundations.

refractory magnesia. See dead-burnt magnesite.

refractory materials. See refractories (2).

refractory metals. A group of nonferrous metallic materials with extremely high melting points (about 1875-3410°C, or 3405-6170°F) including chromium, vanadium, niobium (columbium), molybdenum, rhenium, tantalum, tungsten and their alloys. They possess strong interatomic bonding, high moduli of elasticity, and high strengths and hardnesses at ambient as well as elevated temperatures. However, due to their high reactivity with oxygen, they require protective coatings and controlled-atmosphere conditions. Used as alloying elements in stainless steels, and in the manufacture of dies and structural parts, incandescent light filaments, X-ray tubes, and welding electrodes.

refractory mortar. See refractory cement.

refractory ore. An ore containing a valuable substance (e.g., a metal) whose commercial extraction or recovery is rather difficult.

refractory oxides. Oxides such as those of aluminum, calcium, magnesium, silicon, titanium, yttrium or zirconium, that have high melting points, and can be used singly or in combination for high-temperature applications, e.g. as coatings for refractory metals.

refractory patching cement. A refractory cement used to patch or fill damaged areas, cracks, holes and other discontinuities in kilns, furnaces, glass tanks, refractory molds, etc. See also refractory cement.

refractory porcelain. Ceramic whiteware that is highly resistant to heat- and/or chemical attack, e.g., porcelain based on alumina, cordierite, magnesia, steatite, titania or zircon. It has good mechanical and electrical properties and thermal-shock resistance, and is used for applications, such as electrical insulators, spark plugs, furnace and kiln furniture, combustion boats, and laboratory and chemical process equipment.

refractory sand. Sand based on high-melting materials, such as chromite, olivine, silica or zircon. Used in the manufacture of refractories, foundry cores, molds and mold facings, as an opacifier in ceramics, as a source of refractory metals and nonmetals, and in glassmaking.

refractory solder. An alloy composed of 50% copper and 50% zinc. It has a liquidus temperature of 871°C (1600°F), and is used for special soldering and brazing applications.

refractory stone. Any stone or rock, such as sandstone, soapstone or quartzite, that is resistant to high temperatures and used for solid blocks, or in crushed form in the manufacture of bricks using a suitable binder.

refractory ware. Ware that will not fuse, crack or disintegrate when subjected to relatively high temperatures. It includes saggers, crucibles, pyrometer tubes, and refractory brick and shapes.

Refrasil. Trademark of Hitco Carbon Composites Inc. (USA) for white, vitreous silica fibers (97.9+% pure) and woven textile products. They are commercially available in the form of bulk fibers, batts, yarns, cordage, fabrics, tapes, sleevings and flakes. *Refrasil* fibers have a density of 2.1 g/cm³ (0.08 lb/in.³), a liquidus temperature above 1760°C (3200°F), an average fiber diameter of 6 μm (236 μin.), excellent high-temperature resistance, an upper service temperature of 1095°C (2003°F) and fair mechanical properties. Used as high-temperature insulation for missile, jet aircraft and industrial applications, and for insulation applications in the molten metals and heat-treating industries. *Refrasil* textile fabrics resist oxidation and corrosive chemicals, and are used for welding blankets, furnace curtains, and as ceramic fiber replacements. *Refrasil B200* refers

to a series of asbestos-free, silica-based fiber insulating products supplied in blanket form.

Refrax. Trademark of Harbison-Carborundum Corporation (USA) for silicon carbide refractories with silicon nitride binders. Commercially available in the form of bricks, shapes and finished parts, they have good high-temperature and thermal-shock resistance, a service temperature up to 1230°C (2245°F), good crushing strength, good dimensional stability and good abrasion and corrosion resistance. Used for pumps and pump parts, valve parts, lining of electrolytic cells for smelting aluminum, burners, pyrometer protection tubes, brazing and furnace fixtures, conveyor equipment, hot-spray nozzles, rocket motor parts, and nuts, bolts and other fasteners.

refrigeration-grade high-impact polystyrene. A special grade of high-impact polystyrene with enhanced chemical resistance to chlorofluorocarbon blowing agents and oily foodstuffs found in refrigerators and industrial cooling equipment. See also high-impact polystyrenes.

Regal. (1) Trademark of 3M Company (USA) for coated abrasives.

(2) Trademark of Benjamin Moore & Company Limited (Canada) for a range of paints including *Regal Aquaglo* semi-gloss latex paints, *Regal Aquavelvet* eggshell, flat latex enamel paints and *Regal Wall Satin* interior latex flat paints.

(3) Trademark of Cabot Corporation (USA) for anh extensive series of carbon blacks supplied in a wide range of grades for various applications.

Regalite. (1) Trade name of 3M Company (USA) for resin-bonded abrasives composed of the tough and hard patented *Cubitron* mineral on durable polyester cloth, and used for dry and wet grinding applications.

(2) Trademark of O'Sullivan Corporation (USA) for polyvinyl sheeting.

(3) Trademark of Hercules Inc. (USA) for hydrocarbon resins used in the manufacture of adhesives, coatings, sealants and printing inks.

(4) Trademark of Lushan Company (USA) for plastic sheeting.

Regel Star. Trade name for a dental amalgam alloy.

Regenal. Trademark of Wuppermetall GmbH (Germany) for pure and high-grade aluminum, remelted aluminum, and various aluminum alloys.

regenerated anhydrite. Gypsum from which the water of hydration has been removed. See also anhydrite; gypsum.

regenerated cellulose. Cellulose fibers or sheets obtained by first dissolving a cellulosic substance, such as wood or cotton, and then regenerating it by precipitation (e.g., treatment with acid as in the cuprammonium process) or extrusion (e.g., spinning through minute openings of a spinneret as in the viscose process). See also cuprammonium rayon; rayon; viscose rayon.

regenerated cellulose fibers. (1) A term used for man-made fibers manufactured from *regenetated cellulose.*

(2) Man-made fibers manufactured from *regenetated cellulose* in which 15 or less of the hydrogen atoms of the hydroxyl groups have been replaced by other functional groups. See also cuprammonium rayon; modal fibers; rayon; viscose rayon.

regenerated fibers. Synthetic man-made fibers, such as rayon, made by a process that usually involves the dissolution of natural fiber-forming substances, such as wood or cotton, and subsequent regeneration of the original polymer structure, e.g., by extrusion or precipitation.

regenerated protein fibers. Fibers obtained by the chemical regeneration of natural protein substances, e.g., ardein, casein soybean or zein. Used especially in blends for the manufacture of textile fabrics.

Regenex. Trade name of Foseco Minsep NV (Netherlands) for powdered cleansing and degassing fluxes used for melts of copper- and nickel-alloy scrap containing considerable amounts of nonmetallic contaminants.

Regent. (1) Trade name of A. Heckford Limited, Birmingham Metal Works (UK) for a wrought copper alloy containing 20% nickel. It has an electrical conductivity of 6.5% IACS, and is used for electrical resistances.

(2) Trade name of United Clays, Inc. (USA) for a highly plastic, low-carbon clay used for sanitaryware, tableware and other ceramic applications.

(3) Trade name for carbon blacks.

Regent Vac-Arc. See Vac-Arc Regent.

Regin. (1) Trade name of Uddeholm Corporation (USA) for a series of hot-work tool steels containing 0.4-0.5% carbon, 1-1.2% chromium, 2.25-2.5% tungsten, up to 0.25% molybdenum, 0.9% silicon and 0.2% vanadium, respectively, and the balance iron. Used for hot-work tools, chisels, shear blades, and dies.

(2) Trade name of Uddeholm Corporation (USA) for a shock-resisting tool steels (AISI type S1) containing 0.5% carbon, 0.9% silicon, 1.2% chromium, 2.5% tungsten, 0.2% vanadium, and the balance iron. Used for chisels, shear blades, and dies.

Regina. Trade name of Regina Glass Fibre Limited (UK) for glass tissue.

Regis. Brand name for a high-purity tin containing about 0.011% lead, 0.001% copper and 0.004% arsenic.

Registrado. Trade name of Voco Australia Proprietary Limited for a silicone paste used in dentistry for bite registration purposes.

Reglex. Trade name of ASG Industries Inc. (USA) for a patterned glass with 64 pyramidal indentations per square inch (about 10 indentations per square centimeter).

Reglit. Trade name of Bauglasindustries AG (Germany) for shaped glass building elements.

Regular. Trade name of Latrobe Steel Company (USA) for a steel containing 0.3-0.4% carbon, 13-14% chromium, up to 0.5% manganese, and the balance iron. It is stainless when hardened and polished, and used for cutlery, knives, swords, needles, clippers, bayonets, gages, dies, pump rods, pistons, and cold sets.

regular alumina. Alumina (Al_2O_3) made by recrystallization and composed of relatively large crystals. It has an average purity of about 95-99%. See also alumina; recrystallized alumina.

Regular Straight Carbon. Trade name of Jessop-Saville Limited (UK) for a water-hardening tool steel containing 0.7-1.3% carbon, and the balance iron. Used for hand tools, such as chisels and punches, and for various cutting and finishing tools, such as drills, taps, reamers, or chasers.

Regulus. Trade name for a hard lead containing about 25% antimony. Used for castings, acid valves, cocks, flanges, and chemical apparatus.

regulus. (1) A commercially pure form of antimony.

(2) See regulus metal.

regulus metal. Hard lead composed of 88-94% lead and 6-12% antimony, and used for chemical plant and equipment, bearings, valves and cocks, flanges, and tank linings. Also known as *regulus*.

regulus of Venus. A purple alloy composed of 50% copper and 50% antimony.

Reichhardtsdorfer Sandstein. Trademark of Sächsische Sandsteinwerke GmbH (Germany) for a high-grade sandstone from Saxony.

Reichhold UF. Trade name of Reichhold Chemical Company (USA) for cellulose-filled and foam-grade urea-formaldehydes.

Reich-O-Bond. Trade name of Reichhold Chemical Company (USA) for a general-purpose, pressure-sensitive adhesive with good adhesion to vinyl, low tack, and high shear strength. Used for bonding films, labels, and specialty tapes.

Reich-O-Melt. Trade name of Reichhold Chemical Company (USA) for a multi-purpose, pressure-sensitive adhesive with good adhesion to polyolefin and good heat resistance used for bonding laser labels, pharmaceutical labels, etc.

Reich's bronze. A German bronze composed of 85% copper, 7.5% iron, 6.4% tin, 0.6% aluminum and 0.5% manganese. It has good strength and corrosion resistance.

Reilen PP. Trademark of Reinhold KG (Germany) for polypropylene staple fibers.

Reilon N. Trademark of Reinhold KG (Germany) for nylon 6 staple fibers.

Reinaluminium. German trade name for wrought, commercially pure aluminum (98-99.9%) used for extrusions, electrical parts, electrical conductors, household appliances and utensils, packaging (e.g., foils, cans and tubes), building materials (e.g., roofing, gutters, trim and paneling), containers and vessels for the chemical industry, and cladding.

reinerite. A light yellow-green mineral composed of zinc arsenate, $Zn_3(AsO_3)_2$. Crystal system, orthorhombic. Density, 4.27 g/cm³; refractive index, 1.790. Occurrence: Southwest Africa.

reinforced asbestos roving. A bundle of usually untwisted, continuous asbestos fibers containing a core composed of other fibers.

reinforced carbon-carbon. A lightweight thermal insulation material composed a pyrolyzed *carbon-carbon composite* consisting of a carbon matrix reinforced with graphite fibers and coated with a thin, oxidation-resistant layer of silicon carbide (SiC). It has high-strength and rigidity, high resistance to fatigue and impact, low thermal expansion, a maximum service temperature of approximately 1650°C (3000°F), and is used for the nose cap and leading wing edges of the NASA Space Shuttle. Abbreviation: RCC. See also carbon-matrix composite.

reinforced concrete. A composite material consisting of concrete reinforced or strengthened in tension while in the plastic state by embedding steel bars, rods, wires or mesh. Theoretically, the steel reinforcement takes all the tensile forces, while the concrete takes all the compressive forces. Since steel and concrete have almost the same coefficient of thermal expansion (about 11.5×10^{-6}/°C for steel and 13×10^{-6}/°C for concrete, respectively), the relatively strong adhesive bond is not affected by changes in temperature. The introduction of high-modulus reinforcing fibers, such as asbestos, glass, nylon, or polyethylene into concrete is a relatively recent technique. *Reinforced concrete* is used for building construction (e.g., beams, trussed girders, slabs, walls, floors and stairs), civil engineering (e.g., bridges, overpasses, roads, harbors, tunnels and runways), fences, telegraph poles, pipes for pressure lines, reactor shielding, etc. Abbreviation: RC. See also fiber-reinforced concrete.

Reinforced Coverall. Trademark of Flex-O-Glass, Inc. (USA) for a high-strength cord-reinforced polyethylene film.

reinforced foam. A polymer foam, such as polyurethane or polystyrene foam, filled with glass fibers or flakes, or ground minerals (e.g., mica) to enhance its stiffness, impact resistance or other mechanical properties.

reinforced gummed tape. A special tape product coated with adhesive, and composed of glass, rayon or sisal fibers bonded with rubber latex, synthetic resins or bituminous materials, and sandwiched between two sheets of strong kraft paper.

reinforced materials. Composite materials composed of two or more distinct materials in which one of the materials serves as the matrix and the other material or materials as reinforcements to improve the mechanical properties. Examples include reinforced concrete and reinforced plastics.

reinforced molding compounds. A ready-to-use plastic molding compound consisting of a polymer or synthetic resin and a fibrous reinforcing agent (e.g., glass or synthetic fibers), or a reinforcing filler (e.g., cotton or minerals).

reinforced phenolics. Phenolic resins whose mechanical properties including toughness and dimensional stability, and thermal stability have been improved by adding up to 45-65 wt% reinforcements and fillers (e.g., wood flour, cellulose, carbon, glass or polyamide fibers, glass or graphite fabrics, aluminum powder, rubber, or minerals). Depending on the particular reinforcement used, the upper service temperature ranges from 160 to 180°C (320 to 355°F). Used as matrix materials for medium-temperature thermoset composites, and for motor housings, automotive components (e.g., distributors, drive pulleys and pump housings), electrical components (e.g., fixtures and circuit breakers), telephones, and plywood and particleboard.

reinforced plastic mortar. A composite material manufactured from a thermosetting resin (e.g., a polyester or epoxy), a glass-fiber reinforcement, and sand or other aggregates. Used for the manufacture of pipes and manhole covers. Abbreviation: RPM.

reinforced plastics. A group of thermosetting and thermoplastic polymers whose strength and stiffness have been greatly enhanced by the incorporation of reinforcements (e.g., aramid, carbon, glass, metal, boron or graphite), usually in the form of continuous or chopped strands, whiskers, mats or woven fabrics. Abbreviation: RP. See also fiber-reinforced plastics; polymeric composites; reinforced thermoplastics; reinforced thermosets.

reinforced polypropylene. Polypropylene reinforced with up to 50 wt% mineral fillers (e.g., talc, mica, or calcium carbonate) or fibrous reinforcements (e.g., glass or carbon fibers). It has improved dimensional stability, tensile strength and stiffness, and is used for household appliances, garden equipment, and automotive components, e.g., seat backs, fans and dashboard parts.

reinforced polyurethane foam. Polyurethane foam filled with glass fibers or flakes, mica or other ground minerals. It has higher stiffness and impact resistance and lower thermal expansion and heat sag than regular polyurethane foam. The reinforcing filler is usually premixed with the liquid foam components.

reinforced product. A product, such as a plastic article or concrete slab, whose strength has been increased by the incorporation of mechanical reinforcements (e.g., fibers, whiskers, fabrics, wires, or particulates).

reinforced reaction-injection-molded composites. Solid or microcellular composites, often with polyurethane matrices, formed by: (i) mixing the liquid reactants in the appropriate amounts, adding suitable reinforcements (e.g., chopped or milled glass fibers), and filling the mixture into a mold where it reacts and cures; or (ii) placing reinforcing fiber mats (e.g., glass) into a mold and filling it with the liquid reactive mixture which then penetrates the mat and solidifies. Composites produced using reinforcing fiber mats are known as "structural reaction-injection-molded composites" (or SRIM composites). Abbreviation: RRIM composites. See also reaction-injection molded plastics.

reinforced silicone. A silicone resin reinforced with carbon, glass or other fibers. Abbreviation: RSIL, or R-SIL.

reinforced thermoplastics. Engineering plastics composed of thermoplastic matrix resins (e.g., acrylics, polyamides, polycarbonates, polystyrenes, or polysulfones) reinforced with continuous or chopped fibers or woven fabrics (e.g., aramid, carbon, or glass). Abbreviation: RTP.

reinforced thermosets. Engineering plastics composed of thermosetting matrix resins (e.g., bismaleimides, polyimides, phenolics, epoxies, silicones, or unsaturated polyesters) reinforced with continuous or chopped fibers or woven fabrics (e.g., carbon, graphite, or glass). Abbreviation: RTS.

reinforcement. (1) Any inert material, such as certain continuous and chopped fibers, whiskers or particulates (metallic or nonmetallic particles) embedded in a metallic or nonmetallic matrix to improve its strength and stiffness.

(2) Bars, rods, wires or mesh, usually of steel, but also of asbestos, glass, nylon or polyethylene, embedded in a concrete, mortar or plaster to reinforce (or strengthen) it. Theoretically, the reinforcement takes all the tensile forces, while the concrete takes all the compressive forces.

(3) Chopped or continuous fibers of boron, carbon, glass, aramid, silicon carbide or aluminum oxide, or metal wires embedded in a metallic or nonmetallic matrix to provide additional strengthening including improved tensile and/or impact strength or toughness. Also known as *reinforcement fiber; reinforcing agent.*

reinforcement fiber. See reinforcement (2).

reinforcing agent. (1) A material, such as carbon black, kaolin, or zinc oxide, in finely divided form, added to elastomers or plastics, usually in relatively high percentages, to increase strength, hardness and/or abrasion resistance. Although a reinforcing agent may increase bulk, the term should not be confused with the terms "filler" and "extender."

(2) See reinforcement (3).

reinforcing bar. A steel bar or rod of varying shape embedded in concrete as a reinforcement. Also known as *concrete reinforcing bar; rebar; reinforcing rod; steel reinforcing bar.* See also reinforcing steel; reinforced concrete.

reinforcing steel. Steel bars, rods, wires or mesh embedded into a concrete, mortar or plaster to reinforce (or strengthen) it.

reinhardbraunsite. A colorless mineral composed of calcium silicate hydroxide, $Ca_5(SiO_4)_2(OH)_2$. Density, 2.83 g/cm^3.

Reinkupfer. German trade name for commercially pure copper (99.9%) with good corrosion resistance and high electrical conductivity, used in resistance alloys for electrical equipment and instruments.

Reinnickel. German trade name for commercially pure nickel (99+%) with good corrosion resistance and a maximum service temperature of 700°C (1290°F), used in the manufacture of resistance alloys for electrical equipment and instruments, and for food processing and handling equipment, and chemical equipment.

Reinstaluminium. German trade name for high-purity, wrought aluminum (99.98%) with a density of 2.71 g/cm³ (0.098 lb/in.³), high corrosion resistance, and good processibility. Used for reflectors, lighting fixtures, automotive components (e.g., trim and fittings), art objects, high-quality tableware, and equipment for the chemical and food-processing industries.

Reisholz. Trade name of Mannesmann-Rohrenwerke AG (Germany) for an extensive series of case-hardening and heat-treatable steels, plain-carbon tool steels, and stainless steels.

Reith's alloy. A German high-leaded bronze containing about 10-12% tin, 9-10% lead, 5% antimony, and the balance copper. Used for heavy-duty bearings.

Rekford Eminent. Trade name of Boyd-Wagner Company (USA) for a series of high-speed tool steels.

Rekord. Trade name of Idealstahl Breidenbach KG (Germany) for a series of high-speed steels containing 0.7-1.35% carbon, 0-1.2% molybdenum, 12-18.5% tungsten, 0-10% cobalt, 0-3.8% vanadium, 4-4.75% chromium, and the balance iron. Used for lathe and planer tools, milling cutters, drills, taps, reamers, and forming dies.

relaxed yarn. A nylon or polyester filament yarn that has been specially treated to reduce and even out shrinkage and twist and provide it with uniform dyeing qualities.

Relay. (1) Trade name of AL Tech Specialty Steel Corporation (USA) for a series of soft magnetic alloys containing 97.5-99% iron and 0.5-2.5% silicon. They have high magnetic permeability, and are used for armatures, relays, solenoid switches, and magnets.

(2) Trade name of AL Tech Specialty Steel Corporation (USA) for an oil-hardening, shock-resistant tool steel containing 0.5% carbon, 0.7% chromium, 0.5% molybdenum, and the balance iron.

(3) Trade name of TISSI Dental (Italy) for an addition-curing silicone material used for dental impressions.

release agent. (1) A lubricant, such as graphite, soap, wax or silicone oil, applied over the work surfaces of a mold, or added to the material to be molded to prevent adhesion, reduce friction and facilitate removal of molded plastic or ceramic products from the mold. Also known as *mold lubricant; mold release agent; parting agent.*

(2) A material, such as a lubricant, oil or wax, applied to the surface of concrete to prevent it from sticking to another material.

release film. An impermeable film or layer that does not stick or adhere to other materials. For example, in certain plastic molding processes release films are used to prevent the mold from sticking to the resins being cured.

release paper. See anti-adhesive paper.

Reliakote. Trade name of Reliable Coatings, Inc. (USA) for powder coating materials used in metal finishing.

Reliance. (1) Trade name of Reliance Steel Casting Company (USA) for a water-hardening steel containing 0.25-0.4% carbon, and the balance iron. Used for steel castings.

(2) Trade name of Emsco Derrick & Equipment Company (USA) for a steel containing 0.07-0.13% carbon, 0.2-0.5% manganese, 3-3.5% nickel, 1.4-1.6% copper, and the balance iron. Used for oil-well equipment.

(3) Trade name of Teledyne Pittsburgh Tool Steel (USA) for a water-hardening tool steel (AISI type W1) containing 0.95% carbon, 0.35% manganese, 0.2% silicon, and the balance iron.

(4) Trademark of Metal Distributors Limited (Canada) for several lead products (including pipes and sheets), type metals

and solder alloys.

Relianite. Trade name of American Meter Company (USA) for a nodular cast iron containing 3.2% carbon, with the balance being iron. Used for gears, shafts and housings.

Reliapreg. Trade name of Ciba-Geigy Corporation (USA) for a series of phenolic laminates and resins. The laminates are composed of cotton, fabrics, or paper bonded with phenolic resin. The resins are supplied in wood or natural fiber-filled general-purpose, mineral-filled high-heat, mica-filled electrical, glass-reinforced high-impact, cotton-filled medium-shock, chopped fabric-filled medium-impact, cellulose-filled shock-resistant foam and other grades.

relief fabric. See cloque.

Reliner. Trade name of Voco-Chemie GmbH (Germany) for soft and extra-soft acrylic resins used for denture relining applications.

Relit. Russian trade name for a hard sintered tungsten carbide with cobalt binder used for cutting tools.

Relleum. Trade name of Mueller Brass Company (USA) for a free-cutting brass containing 67% copper, 31% zinc and 2% lead. Used for screw-machine products, and hardware.

Relon. (1) Trade name for polyamide 6 (nylon 6) fibers and yarns used for textile fabrics including clothing.

(1) Trade name for polyamide 6 (nylon 6) plastics and products supplied in unmodified, glass- or mineral-reinforced, graphite-, molybdenum disulfide- or carbon black-filled, heat-stabilized, UV-resistant, injection-molding, extrusion and several other grades.

Rely. Trade name of Rely Metal Works (South Africa) for lead- and tin-base bearing alloys.

RelyX. Trademark of 3M ESPE (USA) for a line of dental restorative resin and composite cements and adhesives. *RelyX ARC* is an adhesive resin cement composed of bisphenol A glycidyl-ether methacrylate (BisGMA) and triethylene glycol dimethacrylate (TEGDMA) polymers with radiopaque, wear-resistant zirconia/silica filler. It adheres strongly to metals, ceramics, porcelains, precured composites and tooth structure (dentin and enamel), and is used for indirect restorations, especially for the cementation of crowns, bridges, onlays, inlays, endodontic posts, and adhesively bonded amalgams. *RelyX Luting* is a fast-setting, fluoride-releasing, resin-modified glass-ionomer cement with mousse-like consistency. It has high fracture toughness, and is used for luting metal crowns, crowns with alumina or zirconia, bridges, inlays, onlays, orthodontic appliances, and endodontic posts. *RelyX Unicem* is a dual (light and chemical) curing, self-adhesive universal dental resin cement containing phosphoralated methacrylates. Used for composite, porcelain, metal and metal-ceramic restorations. *RelyX Veneer Cement* is a methacrylate-based resin cement used as a veneering material.

Rema. (1) Trade name of Great Western Steel Company (USA) for a case-hardening steel containing 0.05% carbon, 0.2% manganese, 0.1% silicon, and the balance iron. Used for molds for the plastic industry.

(2) Trade name of Fagersta Bruks AB (Sweden) for an oil-hardening steel containing 1% carbon, 1.2% manganese, 0.5% chromium, and the balance iron. Used for molds and dies for plastic industry.

Remalloy. (1) Trade name of Arnold Engineering Company (USA) for a series of molybdenum-iron-cobalt permanent magnet materials. *Remalloy 1* contains 71% iron, 17% molybdenum and 12% cobalt and *Remalloy 2* has 68% iron, 20% molybdenum and 12% cobalt. *Remalloy* materials have high coercivities, high magnetic energy products, high residual flux densities, a Curie temperature of 900°C (1650°F), and no magnetic orientation. Used for receiver magnets, and electrical and magnetic equipment.

(2) Trade name of Rennie Tool Company Limited (UK) for an oil-hardening high-speed steel containing 0.7% carbon, 4% chromium, 18% tungsten, 12% cobalt, 2% vanadium, and the balance iron. Used for lathe and planer tools, and various other machining and finishing tools.

(3) Trade name of Hoytland Steel Company (USA) for a case-hardening steel containing 0.1% carbon, 0.5% manganese, 0.6% chromium, 1.5% nickel, and the balance iron. Used for plastic dies and molds.

Remanit. Trade name of Thyssen Edelstahlwerke AG (Germany) for an extensive series of wrought and cast austenitic, ferritic and martensitic stainless steels and numerous heat-resisting steels. The austenitic grades contain about 0.03-0.1% carbon, 18% chromium, 8% nickel and varying amounts of molybdenum, copper, titanium and niobium. They are highly corrosion-resistant, readily weldable, and have good strength, but are usually difficult to machine. The martensitic grades contain about 0.15-0.9% carbon, 13-18% chromium, and sometimes 1-2% molybdenum or nickel. They have high strength, are less resistant to corrosion than the austenitic grades, and difficult to machine and weld. The low-carbon ferritic grades have carbon and chromium contents of up to 0.1% and 12-18%, respectively, and are easy to machine, but difficult to weld and less corrosion resistant than the austenitic grades. Heat-resisting grades usually contain additions of chromium and aluminum, and are highly oxidation-resistant.

Remanium. Trade name of Dentaurium J.P. Winkelstroeter (Germany) for several corrosion-resistant cobalt-chromium dental bonding and casting alloys including *Remanium CS*, a beryllium-free grade.

Rem Arms. Trade name of Remington Arms Company Inc. (USA) for an extensive series of ferrous powder-metallurgy materials including various sintered high-purity irons, iron-nickels, iron-silicons, medium- and high-carbon steels, copper steels, high-, medium- and low-nickel steels, and austenitic, ferritic and martensitic stainless steels.

remanufactured lumber. Softwood lumber whose size and shape has been changed by further processing after grading. The term includes factory and shop lumber used by millwork plants in the fabrication of windows, doors, moldings and other trim items. Also known as *remanufacture lumber.*

remanufactured wool. Wool fibers reclaimed from previously used or processed wool and including short mungo and shoddy fibers obtained from woven, knitted or felted products as well as the short waste fibers (noils) obtained during cotton processing. Used in blends with new wool or other textile materials for the manufacture of fabrics.

Rematitan. Trademark of Dentaurum–J.P Winkelstroeter KG (Germany) for a shape-memory alloy composed of 53.2% nickel and 46.8% titanium, and used for dental and orthodontic wire. *Rematitan Lite* contains 53.5% nickel and 46.5% titanium.

Remember. Trademark of Solutia Inc. (USA) for soft, colorfast acrylic fibers used for the manufacture of craft yarns for hand-knit and crocheted fabrics with good shape retention after laundering.

Remendur. Trade name of Wilbur B. Driver Company (USA) for a series of reactive, refractory, magnetically semihard materi-

als containing about 48-49% cobalt, 2-5% vanadium, and the balance iron. They possess high remanence, high residual induction, square hysteresis loops, nondirectional properties, high Curie temperatures, and are used for magnetic devices, telephone and reed switches, and relay components.

Remic. Trade name of Detco–Helmut Detter & Co. (Germany) for micaceous iron oxide products.

Remiral. German trade name several for aluminum-magnesium alloys.

remoistening adhesives. Adhesives, such as animal glue, dextrin, or gum arabic, that are applied to surfaces in the activated state, but upon drying become nonadhesive and remain latent until reactivated by the application of water. Used for postage, revenue and trading stamps, etc.

Remount. Trade name of George Cook & Company Limited (UK) for a water-hardening tool steel (AISI type W2) containing 1.3% carbon, 0.4% manganese, 0.2% silicon, 0.1% vanadium, and the balance iron. Used for cold forming tools and wire drawing dies.

Remy. Trade name of Remystahl (Germany) for an extensive series of alloys including various corrosion-resistant and heat-resistant steels, nickel-chromium-molybdenum alloy steels, case-hardening steels, plain-carbon, cold-work and hot-work tool steels, commercially-pure nickels, and various nickel alloys including nickel-copper, nickel-chromium-iron, nickel-molybdenum-iron, nickel-chromium-molybdenum, nickel-silicon-copper and nickel-chromium-cobalt alloys.

Renaissance. Trademark of Arriscraft Corporation (Canada) for masonry units.

Renal. Trade name of J. & A. Erbsloh Aluminium (Germany) for a high-purity aluminum (99.99%) with high electrical conductivity and excellent corrosion resistance, used for electrical conductors and capacitors.

Renamel. Trade name of Cosmedent, Inc. (USA) for a light-cure, microfill hybrid composite used in restorative dentistry.

Renault alloy. A French nonhardenable aluminum alloy containing 10% zinc and 2% copper, and used for light-alloy parts.

Rene. Trade name for metallic fibers.

René. Trademark of Cannon-Muskegon Corporation (USA) for a series of corrosion-, heat- and oxidation-resistant nickel-base superalloys containing 8-16% cobalt, 9-20% chromium, 0-24% iron, 0-7% tungsten, 0.5-6% aluminum, 0-5.5% tantalum, 2.1-3.5% niobium (columbium), 2.3-5% titanium, 2-10.5% molybdenum, 0-1.6% vanadium, 0-0.25% manganese, 0.02-0.3% carbon, 0.03-0.09% zirconium and 0.01-0.02% boron. They have a maximum service temperature of 1038°C (1900°F), good long-time stability and durability, and are used for turbine blades, jet-engine and gas-turbine components, afterburners, high-temperature springs and bolts, turbine frames, and compressor disks.

Renew. Trademark of BISCO, Inc. (USA) for a hydrophobic, non-sticky, light-cured hybrid composite, used for dental restorations and core build-ups. Supplied in incisal and body translucent and opaque shades, it has high durability, high stain and wear resistance, and low shrinkage.

renierite. An orange-bronze mineral of the tetrahedrite group composed of copper iron germanium arsenic sulfide, $Cu_3(Fe,Ge)(S,As)_4$. Crystal system, cubic. Density, 4.30 g/cm³. Used as a source of germanium. Occurrence: Zaire.

rennet. A dried extract containing *rennin*.

rennet casein. Casein precipitated from skimmed milk by the enzyme rennin, and used especially in the manufacture of industrial plastics, and biopolymers. See also casein.

Rennite. Trade name of B.A. Field Company (USA) for an oil-hardening high-speed tool steel containing 0.7% carbon, 4% chromium, 18% tungsten, 2% vanadium, 12% cobalt, and the balance iron. It has excellent wear resistance, high hot hardness, and is used for lathe and planer tools, cutters, broaches, reamers, hobs, taps, and tool bits.

Renown. (1) Trade name of Latrobe Steel Company (USA) for a water-hardening tool steel containing 1% carbon, 0.1% chromium, 0.2% vanadium, and the balance iron. Used for stamping dies and fine-edged tools.

(2) Trade name of Renown Steel, Division of Fobasco Limited (Canada) for a range of steel products including sheets, blanks, slit coils and round strip as well as galvanized, electrogalvanized and aluminized tin plate and tin-mill black plate.

RenShape. Trade name of Ciba, Polymers Division (USA) for epoxy and polyurethane resins used for rapid tooling systems.

Reny. Trade name of Mitsubishi Chemical Company (Japan) for nylon 6,6 polymers.

Renyl C. Trade name of Montedison SA (Italy) for a series of polyamide (nylon) resins.

Renyx. (1) Trade name of Allied Die Casting Corporation (USA) for a corrosion-resistant aluminum die-casting alloy containing 4% nickel, 4% copper and 0.5% silicon.

(2) Trade name of Allied Die Casting Corporation(USA) for a corrosion-resistant zinc die-casting alloy containing 4% aluminum, 3% copper and 1% manganese.

Reo. Trade name of Time Steel Service Inc. (USA) for a series of water-hardening tool steel (AISI type W2) containing 1% carbon, 0.35% manganese, 0.2% silicon, up to 0.2% vanadium, and the balance iron.

Reostene. British trade name for a heat-resistant alloy composed of nickel and iron, and used for electrical resistances.

rep. A heavy, plain-weave fabric, often of cotton, silk or wool, having characteristic round, padded ribs in the weft (filling). Used especially for robes, neckties, upholstery and drapery. Also known as *repp*.

Repelit. Trade name of Siemens-Schuckert AG (Germany) for a laminate composed of phenolic-impregnated paper.

Replay. Trade name of Huntsman Chemical Company (USA) for a series of polystyrenes with good impact resistance containing selected quantities of post-consumer recycled polystyrene.

Replicast. Trademark of Cotronics Corporation (USA) for electrically nonconductive, room-temperature-curing epoxies that produce accurate, durable molds, patterns, models and prototypes, and possess good machinability, castability and resistance to solvents and most chemicals.

Replikote. Trade name of Richter Precision Inc. (USA) for a family of carbon- and metal-base coatings.

Repolem. Trade name of Atofina SA (France) for vinyl copolymers, emulsions and dispersions, polyvinyl acetates and acrylic dispersions, used chiefly in the paint, varnish, paper, textile and leather industries.

repp. See rep.

Reprean. Trade name of Discas Inc. (USA) for several thermoplastic compounds and polymers including ethylene copolymers made from recycled scrap. Used for footwear, containers, lawn furniture and automotive components.

repressed brick. Brick shaped by repressing blanks cut from a column of soft or stiff clay in a mechanical press.

reprocessed fibers. Reinforcing fibers recovered from cut and trimmed scrap accumulating during fiber-reinforced composite processing. After removal of the matrix material and/or sizing, they are either reused immediately, or cut into short fibers.

reprocessed plastics. Thermoplastics, such as polyethylene, polypropylene, polystyrene or polyvinyl chloride, prepared from scrap (e.g., chips, discarded articles or rejected parts) or nonuniform or nonstandard virgin materials by direct processes that usually involve cleaning, regrinding, reprocessing and subsequent addition to other thermoplastic materials.

reprocessed wool. Wool fibers that have been obtained from unused felted, knitted or woven fabrics by any of several reclamation processes. See also remanufactured wool; reused wool.

Reprodent Hard. Trade name for acrylic resins including *Reprodent Hard*, a self-cure acrylic for dental applications, and *Reprodent Soft*, a plasticized acrylic used as a denture liner.

Reprorubber. Trademark of Flexbar (USA) for a metrology casting material available in two viscosity grades: (i) a quick-setting putty consisting of a base and a catalyst putty that are kneaded together prior to use. Its final color on setting is light blue; and (ii) a two-component thin pour for internal-shape applications that sets at room temperature in about 10-15 min. and has a final light green color. *Reprorubber* is used to make impressions of components made of metals, plastics, rubber, ceramics, glass, wood, paper or cardboard.

Reprosil. Trade name of Dentsply/Caulk (USA) for addition-polymerizing silicone elastomers used for dental impressions.

Republic. Trade name of Republic Steel Corporation (USA) for an extensive series of steels including various machinery steels, high-strength low-alloy structural steels, and several readily weldable ultrahigh-strength structural steels with 0.25-0.45% carbon, and about 9% nickel and 4% cobalt, respectively. Also included under this trade name are numerous tool materials including water-hardening plain-carbon steels, molybdenum and tungsten-type high-speed steels, chromium-, molybdenum- and tungsten-type hot-work steels, high-carbon high-chromium, medium-alloy air-hardening and oil-hardening cold-work steels, shock-resisting tool steels, low-alloy special-purpose tool steels, and mold steels. A wide range of corrosion- and heat-resistant alloys, and titanium alloys for aircraft and jet-engine parts are also sold under this trade name.

Resarit. Trade name of Resart-IHM AG (Germany) for acrylic injection molding compounds.

Resartglas. Trade name of Resart-IHM AG (Germany) for a transparent thermoplastic polymethyl methacrylate polymer supplied in the form of sheets, rods, tubes and molding compounds. Cast, flame-retardant grades for applications in building construction and interior decorating are also available.

resawn lumber. Dressed and graded lumber reduced in thickness by sawing.

Resbond. Trademark of Cotronics Corporation (USA) for a series of room-temperature curing high-temperature ceramic and epoxy adhesives and sealants supplied in paste form in several grades including ultrahigh-temperature, electrically-resistant, one-component and thermally-conducting. They have good electrical properties, good resistance to chemicals, solvents, oxidizing and reducing atmospheres and molten metals, continuous service temperatures (depending on grade) of 1260-2200°C (2300-4000°F), and bond well to standard and stainless steels, ceramics, iron, and most metals. Used for sealing and bonding firebrick, brazing fixtures, exhaust systems and gas-turbine components, for electrical and electronic applications, e.g., potting, sealing and coating, and for forming electrical seals.

Rescor. Trademark of Cotronics Corporation (USA) for a series of engineering ceramics including machinable alumina silicates, fused silica foams and glass-ceramics supplied as prefabricated products, and in the form of plates and rods for refabrication with conventional equipment. They have continuous service temperatures of 1150°C (2100°F) for the alumina silicate grades, 540°C (1000°F) for the glass-ceramic types, and 1650°C (3000°F) for the fused silica foams. Castable ceramics of alumina, ceramic foam, silica, silicon carbide and zirconia are also available. They have high strength, thermal-shock resistance at temperatures to 1650°C (3000°F) in oxidizing, reducing or vacuum atmospheres and high corrosion resistance, good electrical properties. *Rescor* materials are used for electrical insulators, furnace parts, vacuum feedthroughs, welding and soldering fixtures, and many other products.

resenes. The unsaponifiable constituents of *rosin* and other natural resins.

Resi-Bond. Trademark of Georgia-Pacific Resins, Inc. (USA) for liquid phenol-formaldehyde resins used to dry- or wet-process hardboard. The resin imparts outstanding strength, surface hardness and water resistance.

residual asphalt. Untreated bitumen obtained by the steam distillation of asphaltic and semi-asphaltic crude oil.

residual ochers. Mineral pigments consisting of mixtures of limonite or hematite and varying amounts of clay. Examples include *sienna* (yellow), *umber* (brown) and *Indian red* (dark red).

Resiform. Trade name for phenolic molding compounds.

Resilacrete. Trademark of Lepage's Limited (Canada) for an oil- and grease-resistant urethane floor coating for interior and exterior use on concrete and wood. It is available in silver-gray, medium-gray, terra-cotta and sand colors.

Resi-Lam. Trademark of Georgia-Pacific Resins, Inc. (USA) for high-temperature-resistant phenol-formaldehyde laminating resins used in the manufacture of aerospace structural components and insulation, circuit-board components, and molded electrical and electronic components for automotive and other applications.

Resilia. Trade name of Bethlehem Steel Corporation (USA) for an alloy steel containing 0.7% carbon, 2% silicon, 1.1% manganese, and the balance iron. Used for springs.

Resilient. Trade name for a self-cure dental composite used for core-build-ups.

Resilute. Trademark of Pulpdent Corporation (USA) for a strong, dual-cure, fluoride releasing 68.2% filled dental resin luting cement with a very low film thickness (22 μm or 0.9 mil) and high radiopacity. Supplied in universal and clear shades, it is used for the permanent cementation of metal castings, the indirect restoration of ceramic and composite inlays, onlays and laminate veneers, and for bonded amalgam restorations.

Resi-Mat. Trademark of Georgia-Pacific Resins, Inc. (USA) for a liquid modified phenol-formaldehyde resin for glass matting, used in the production of building products, such as flooring, roofing and insulation. Modification with latex significantly enhances its strength.

Resimelt. Trademark for a series of hot-melt adhesives used for structural applications.

Resimene. Trademark of Monsanto Chemical Company (USA) for a series of melamine- and urea-formaldehyde resins usually supplied as liquid solutions. They can be butylated, isobutylated or methylated, and are used in paints, varnishes and lacquers for automobiles, appliances and machinery, and for laminating and impregnating paper and fabrics.

Resiment. Trade name of Septodont, Inc. (USA) for a bisphenol A glycidyl dimethacrylate (BisGMA) resin cement for permanent dental restorations.

Resi-Mix. Trademark of Georgia-Pacific Resins, Inc. (USA) for liquid, thermosetting phenol-formaldehyde adhesives used for bonding plywood veneers.

resin. See resins.

resinate. A salt that is a mixture of resin acids (e.g., abietic and pimaric acid) found in *rosin*, and is used in lubricating greases, as a gel thickener, and in paints as a drier and flatting agent.

resin binder. (1) A synthetic resin forming the nonvolatile portion (or film former) of a paint and serving to bind the pigment particles.

(2) A thermosetting resin, such as a phenolic, urea, melamine or furane resin, used as a foundry binder in the manufacture of cores and shell molds.

resin cement. A *dental cement* composed of a synthetic resin, such as bisphenol A glycidylether methacrylate (BisGMA) or urethane dimethacrylate (UDMA), filled with 30-80 wt% micron- and/or submicron-sized mineral or other particles. It can be of the visible-light, chemical or dual (light and chemical) activated (cured) type. Used especially for luting porcelain and cast ceramic and composite resin restorations. Also known as *dental resin cement.*

resin-coated sand. Foundry sand of core or molding quality that has a synthetic resin applied by a cold- or hot-coating process to serve as a binder. Abbreviation: RCS.

resin coatings. A group of organic coatings based on polymer resins, such as alkyds, acrylics, cellulosics, epoxies, melamines, phenolics, polyesters, silicones, ureas, urethanes or vinyls, and dissolved or dispersed in a suitable solvent, or water. Abbreviation: RC.

resin concrete. See polymer concrete.

Resine. Trade name for a series of phenolics including phenol furfurals and phenol formaldehydes with good heat and electrical resistance, high rigidity, strength, hardness and stability, low moisture absorption, and good moldability. They can be made fairly resistant to chemicals, but are severely attacked by all strong acids and alkalies. Used for laminates, shell forms, handles, pulleys, wheels, television and radio cabinets, plugs, fuse blocks, and coil forms.

Resinform. Trade name of Resinform SRL (Italy) for a cellular rubber.

Resinite. (1) Trade name of Borden, Inc. (USA) for a phenol-formaldehyde resin made by condensation polymerization.

(2) Trademark of Borden, Inc. (USA) for plastic packaging film.

(3) Trademark of 3M Company (USA) for resin-coated abrasives.

resin-matrix composites. Composite materials consisting of matrices based on a thermoset or thermoplastic resin filled or reinforced with other materials, such as glass or carbon fibers, or wood flour. Examples include epoxy-matrix composites and phenolic-matrix composites.

resin-modified alkyds. High-molecular-weight *alkyd resins* that have been modified with a drying oil, such as linseed, soybean or tung oil, and a synthetic resin, such as an acrylic, silicone, urethane or vinyl resin. Coatings produced from these alkyds have increased hardness, durability, chemical and weather resistance, and are used for house paints and baking enamels for automotive, marine and appliance applications.

Resinoid. Trade name of Resinoid Engineering Company (USA) for a series of resins, molding compounds, laminates and other plastic materials. *Resinoid GMC* are general-purpose plastic molding compounds supplied in unmodified, fire-retardant, high-impact and mineral-filled grades. *Resinoid PF* refers to a series of phenol-formaldehyde resins and laminates. The laminates consist of cotton, glass fabrics or paper, bonded with phenolic resin. The resins are supplied in numerous grades including wood and natural fiber-filled general-purpose, mineral-filled high-heat, mica-filled electrical, glass-reinforced high-impact, cotton-filled medium-shock, chopped fabric-filled medium-impact and cellulose-filled shock-resistant foam grades.

resinoid. A thermosetting resin, such as a phenolic, that is either in the soluble and fusible initial stage (A-stage) or in the insoluble and infusible final stage (C-stage). See also A-stage resin; C-stage resin.

Resinomer. Trade name BISCO, Inc. (USA) for a dual-cure dental resin cement used for lining and luting restorations.

resinous cement. A type of cement that is based on a synthetic resin and upon setting becomes hard and impervious to acids and moisture. Used to join or place chemical stoneware and acid-resisting brick.

Resinox. Trademark of Monsanto Chemical Company (USA) for a series of phenolic resins that are available in a variety of forms (e.g., powders, flakes, granules, pellets, and liquids) and can be compounded with a large number of resins, glass fibers, or mineral or cellulosic fillers. They possess excellent thermal stability to over 150°C (300°F), good chemical stability, good moldability, high hardness and dimensional stability, and are used as bonding agents for grinding wheels, foundry cores and molds, brake and clutch linings, as special coatings, in cellulose, fiberglass and foam insulation, as impregnating and bonding agents for paper, fibers and fabrics, in linings for chemical containers, cans, drums and tank cars, and for motor housings, distributors for cars and trucks, and electrical fixtures.

resin paste. A paste based on an unsaturated polyester resin and containing various other materials, such as pigments, flame retardants, ultraviolet absorbers, mold-release agents, fillers, glass fiber reinforcements, or catalysts, in exact proportions to obtain optimum processibility, moldability, and mechanical and physical properties. Used as a starting material in the sheet-molding compound process. See also sheet-molding compound.

ResinRock. Trade name of Whip Mix Corporation (USA) for dental gypsum composed of alpha gypsum and a synthetic resin, and supplied as a powder for mixing with water. When cast from such a mixture, upon setting the die stone develops good surface hardness and abrasion resistance, high dimensional stability and low setting expansion (about 0.08%). Supplied in ivory, gray, peach and blue colors, it is suitable for high-accuracy implants and complex restorations.

resins. A class of natural and synthetic materials of organic composition that can be solid, semisolid or pseudosolid, usually have high molecular weights, wide softening or melting ranges, and tend to flow when subjected to stress. They are highly reactive materials that in their initial state are pourable liquids, but upon activation are transformed into solids. See also ion-exchange resins; natural resins; synthetic resins.

resin solder. See rosin solder.

resin-transfer-molded plastics. Plastics and reinforced plastics produced by slowly pumping (injecting) premixed liquid resins under low pressure into two-piece, matched metal molds, which may or may not contain reinforcements (e.g., glass fibers, or knit, woven or nonwoven fabrics), and then closing the molds. The molds and resins may or may not be heated. Abbreviation: RTM plastics.

resin-treated fabrics. Fabrics treated with polymer resins to impart or improve firmness, stability, weight and/or wrinkle resistance, reduce shrinkage, and/or add special surface effects, such as embossing or glazing.

Resisco. (1) Trade name of IMI Kynoch Limited (UK) for a corrosion-resistant alloy of 91% copper, 7% aluminum and 2% nickel. Used for coolers and condensers.

(2) Trade name of IMI Yorkshire Alloys Limited (UK) for an aluminum bronze containing 92.5-93.5% copper and 6.5-7.5% aluminum. Used for condenser tubes, pump parts and hardware.

Resi-Set. Trademark of Georgia-Pacific Resins, Inc. (USA) for phenol-formaldehyde *resole* resins supplied in one- and two-component (with catalyst) grades. Used in the manufacture of fiberglass-reinforced plastic products by filament winding, bulk and sheet compound molding, rapid-transfer molding, pultrusion, or hand layup. The resulting products exhibit excellent flame resistance and low smoke generation, and are suitable for use in the aerospace, construction and mass transportation industries.

Resist. Trade name of NSI Dental Pty. Limited (Australia) for a light-cure, low-viscosity bisphenol glycidylether methacrylate (BisGMA) modified urethane dimethacrylate (UDMA) resin used as a dental pit and fissure sealant.

resist. (1) A protective film or covering, such as wax, paper, metal, foil or plastic, that is placed over an area or surface to protect it from subsequent applications of colors and glazes or from decorative treatments, such as etching or sandblasting.

(2) An acid-resistant, nonconducting coating material used to shield or protect parts of an integrated circuit during etching.

(3) A patternable material of high radiation sensitivity used in semiconductor imaging. It is applied to all or part of the substrate surface to shield or protect it from being altered during processing.

(4) A material applied to certain portions of the work to prevent them from being deteriorated or contaminated with brazing filler metal.

(5) A nonconductive material, such as a lacquer, wax or tape, applied to a portion of the surface of the work to prevent chemical or electrochemical action during electroplating or coating. Also known as *stop-off*.

Resista. Trade name of Glass Tougheners Proprietary Limited (Australia) for a toughened glass.

Resistab. Trademark of Allis-Chalmers Corporation (USA) for several urethane castings.

Resistac. Trade name of American Manganese Bronze Company (USA) for a series of strong, corrosion-resistant, wrought and cast aluminum bronzes containing 88-90% copper, 8-11% aluminum, 1-5% iron, up to 5% nickel, and the balance copper. Used for chemical apparatus, and gears, pumps and impellers.

Resista-Cote. Trade name of The Demp-Nock Company (USA) for anti-stick, plastic surfacing materials.

Resistal. (1) Trade name of Degussa Dental (USA) for a series of nonprecious metal alloys used in dentistry and dental engineering. *Resistal P* is a corrosion-resistant nickel-chromium dental bonding alloy.

(2) British trade name for a corrosion-resistant aluminum bronze composed of 90% copper, 9% aluminum and 1% iron, and used for gears and slides.

(3) Trademark of Didier Refractories Corporation (USA) for high-alumina refractory bricks.

(4) Trademark of Ambler Asbestos Shingle & Sheathing Company (USA) for asbestos composition board.

Resistall. Trademark of L.L. Brown Paper Company (USA) for a paper impregnated with a thermosetting resin. It possesses outstanding water resistance and wet strength, and is used in the manufacture of maps, writing and packaging paper, moisture-resistant documents, etc.

Resistaloy. Trade name of Cerro Metal Products Company (USA) for a nickel brass containing 59% copper, 38% zinc, 2% aluminum and 1% nickel. It has good strength and resistance against seawater and other corrosion, and is used for propeller shafts, shafts, bearings, hardware, fasteners, boat hardware, marine applications, and high-tensile forgings.

resistance alloys. A group of alloys including radio, resistor and heating alloys as well as thermostat metals whose primary properties include uniform electrical resistivity and stable resistance, and which are used in instruments and control equipment to measure and regulate electrical characteristics (resistor and radio alloys), in furnaces and appliances to generate heat (heating alloys) or in applications where heat generated in metal resistors is converted to mechanical energy (thermostat alloys). Examples include alloys of nickel and chromium, nickel and manganese, nickel, chromium and iron, chromium, aluminum and iron, and copper and manganese. Many resistance alloys are sold under trade names and trademarks, such as Alumel, Calorite, Cromel, Excello, Fecraloy, Hoskins, Hytemco, Incoloy, Manganin, Nicrothal, Nichrome, Therlo or Tophet. Also known as *electrical resistance alloys*.

resistance materials. A group of materials including carbon, graphite, silicon carbide and the *resistance alloys* which have sufficiently high electrical resistivity to be useful in the manufacture of fixed and variable linear resistors. Also known as *electrical resistance materials*.

resistance wire. Wire made from an electrical resistance alloy or metal having high and uniform resistivity and good corrosion resistance, e.g., copper-manganese, nickel-manganese, nickel-chromium, nickel-chromium-iron or chromium-aluminum-iron alloys. They are often sold under trade names and trademarks, such as Alumel, Chromel, Kanthal, Manganin, Nichrome, or Therlo.

Resistat. (1) Trademark of Composition Materials Company, Inc. (USA) for a nontoxic, dry stripping plastic abrasive with antistatic additive, used for deflashing, blast cleaning and deburring.

(2) Trademark of BASF Corporation (USA) for nylon and polyester fibers and yarns.

(3) Trademark of Shakespeare Monofilaments & Specialty Polymers (USA) for strong, flexible nylon fibers whose outer skins are saturated (suffused) with electrically conductive carbon particles. Supplied as monofilaments, multifilament yarns, tow, staple and supported yarns, they are used for static control applications, e.g., carpets, brushes, knitted, woven and non-woven industrial textiles.

(4) Trademark of Polydon Associates (USA) for antistatic polyethylene packaging film used for electronic applications.

Resistco. Trade name of STD Services Limited (UK) for plain-carbon steel tubing containing 0.11% carbon, up to 0.3% silicon, 0.4% manganese, 0.3% copper, 0.05% sulfur, 0.05% phosphorus, and the balance iron. Used for boiler feed tubes.

Resistherm. Trade name of Isabellenhütte Heusler GmbH KG (Germany) for an electrical resistance alloy composed of 70% nickel, 29% iron and 1% chromium, and having good heat resistance and a maximum service temperature of 800°C (1470°F).

Used for electrical equipment, and instruments for temperature measurements and control functions.

Resistin. British trade name for a tough, strong, corrosion- and heat-resistant electrical resistance alloy composed of 85-86.5% copper, 11.7-12% manganese and 1.8-3% iron.

Resisto. Trade name of Uddeholm Corporation (USA) for an oil-hardening, shock-resistant tool steel containing 0.6% carbon, 1.85% silicon, 0.7% manganese, 0.45% molybdenum, 0.2% vanadium, and the balance iron. Used for shear blades, pneumatic tools, and chisels and punches.

Resisto-Cast. Trade name of Grand Northern Products Limited (USA) for a cast iron containing 3.5% carbon, 3.2% silicon, 0.6% manganese, 1.3% nickel, 0.2% molybdenum, 0.2% copper, 0.4% phosphorus, 0.07% sulfur, 0.3% chromium, and the balance iron. Used for oxyacetylene repairs of cast-iron castings.

Resistoflex. Trademark of Unidynamics Corporation (USA) for a series of vinyls supplied in flexible and rigid grades for extrusion and molding applications. They possess high stiffness at low temperatures, good abrasion resistance, good flame and electrical resistance, good resistance chemicals and oils, good weatherability, good processibility, and a maximum service temperature of 138°C (280°F). Used for floor and wall coverings, upholstery, tubing, sheeting, insulation, safety glass interlayers, rainwear, and toys. They were formerly also used for phonographic records.

Resisto-Loy. Trade name of Grand Northern Products Limited (USA) for a cast iron containing 3.5% carbon, 1% manganese, 0.46% silicon, 29.3% chromium, 4% molybdenum, 0.3% copper, and the balance iron. It has good resistance to abrasion and corrosion, and is used for coated, hardfacing electrodes for overlay and buildup applications on plows, shovel teeth and dredging tools, and for valves in contact with corrosive chemicals.

Resiston. Trade name of American Hard Rubber Company (USA) for a hard rubber.

resistor composition. A composition made by mixing a finely divided semiconductor metal with a temporary organic carrier and glass binder, applying to a glass or ceramic material by brushing, dipping, or spraying, and then firing into the material at about 505-760°C (940-1400°F). It has good abrasion and moisture resistance, good resistance to elevated temperatures, good reproducibility, and low temperature coefficients of resistance. Used for fired-on resistors employed for electronic applications.

Resistox. Trademark of Glidden-Durkee Division of SCM Corporation (USA) for stabilized copper powder (99+% pure) supplied in a wide range of particle sizes. It has a density of 8.9 g/cm³ (0.32 lb/in.³), an apparent density range of 2.0-3.5 g/cm³ (0.07-0.13 lb/in.³), and is used in the manufacture powder-metallurgy bearings, sintered machine parts, electrical contacts and brushes, metallic paints, *Sorel cement*, and catalysts.

Resi-Stran. Trademark of Georgia-Pacific Resins, Inc. (USA) for a liquid phenol-formaldehyde resin used as a binder in the manufacture of waferboard and oriented strand board (OSB).

Resistvar. Trade name of Hamilton Technology Inc. (USA) for an electrical resistance alloy containing 55% copper and 45% nickel. It has a density of 8.89 g/cm³ (0.321 lb/in.³), a melting point of 1276°C (2329°F), a boiling point above 2400°C (4350°F), a maximum operating temperature of 600°C (1110°F), high electrical resistivity, a negative temperature coefficient of resistance, and good corrosion resistance. Used for electrical instruments, precision resistors, reducing rheostats, and shunts.

Resital. Trade name of Gioca Binder Systems Limited (Canada) for a series of synthetic resins, such as phenol-formaldehyde, used as binders in foundry applications.

resite. See C-stage resin.

Resitex. Trade name for laminated phenolics.

resitol. See B-stage resin.

Resiweld. Trademark of H.B. Fuller Company (USA) for thermosetting, reactive adhesives including epoxies, phenolics, polyesters, polysulfides and urethanes, and epoxy-based hot melts. Used for structural applications.

Resmetal. Trademark of Borden Chemical (USA) for a special composition of resin and metal that when catalyzed produces solids similar to metals. Used for moldmaking, and in the repair of metal surfaces.

Resocel. Trade name of Micafil (UK) for mica-filled bakelite-type phenolics.

Resoform. Trade name of Micafil (UK) for unfilled and mica-filled bakelite-type phenolic plastics.

Resogil. Trade name of Resopal SA (France) for melamine-formaldehyde plastics.

Resoglaz. Trade name of Advance Solvents & Chemical Company (USA) for several polystyrenes.

resole. See A-stage resin.

Resolite. Trade name of Oralite Limited (UK) for a vinyl ester resin formerly used for denture bases.

Resolut. Trademark of W.L. Gore Company (USA) for a biodegradable polymer based on poly(DL-lactide-*co*-glycolide), and used for tissue engineering applications, and dental guided-tissue-regeneration (GTR) membranes.

Resopal. Trademark of Resopal Werk–H. Roemmler GmbH (Germany) for light-colored, highly durable high-pressure laminates composed of several sheets of paper bonded with amino, melamine or phenolic resin, and several resin-bonded chipboard products. Used for wall and ceiling panels, tabletops, and other building and interior decoration products.

Resopalit. Trade name of Resopal Werk–H. Roemmler GmbH (Germany) for melamine-formaldehyde plastics and laminates used for building products.

Resophene. French trade name for phenol-formaldehyde resins and plastics.

resorbable bioceramics. A group of dense, nonporous or porous ceramics including calcium sulfate, tricalcium phosphate and calcium phosphate salts, that degrade biologically over time and are gradually replaced by tissue. Used for surgical sutures, and in hard tissue replacement. See also bioresorbable polymers.

resorcinol. A white, light-sensitive crystalline compound (99% pure) with a density of 1.272 g/cm^3, a melting point of 109-111°C (228-232°F), a boiling point of 281°C (538°F), a flash point of 261°F (127°C), and an autoignition temperature of 1126°F (607°C). Used for the preparation of resorcinol-formaldehyde adhesives, aerogels and resins, as a crosslinker for neoprene, as a rubber tackifier, in dyes, cosmetics and pharmaceuticals, and as a chemical reagent. Formula: $C_6H_4(OH)_2$. Also known as *1,3-benzenediol; m-dihydroxybenzene; resorcin.*

resorcinol adhesives. A group of strong, chemically resistant adhesives that are based on resorcinol-formaldehyde resins dissolved in water, ketones or alcohol. They are either of the fast- or room-temperature-curing type, and are used for gluing wood and plywood, and for joining rubber to textiles.

resorcinol-formaldehyde aerogel. A porous, three-dimensional,

low-density resorcinol-formaldehyde particulate system with fractal microstructure obtained by sol-gel polymerization. It is an organic, electrically insulating material that is used as a precursor for carbon aerosols, and can be made into composite sheets by impregnating nonwoven carbon paper with resorcinol-formaldehyde gel. Abbreviation: RF aerogel.

resorcinol-formaldehyde resins. Room-temperature-curing resins made by the polymerization of *resorcinol* with formaldehyde, and used in the manufacture of resorcinol adhesives and aerogels, and acid-resistant plastics. Abbreviation: RF resins.

resorcinolphthalein. See fluorescein.

Resorit. Trade name Resopal Werk–H. Roemmler GmbH (Germany) for an engineering thermoplastic based on polymethyl methacrylate polymers and available in various grades and forms. It has a density of 1.2 g/cm^3 (0.04 lb/in.^{3}), good to moderate mechanical properties, a maximum service temperature of 60-93°C (140-200°F), good optical properties, excellent resistance to gasoline and mineral oils, and moderate resistance to dilute acids and alkalies, and trichloroethylene. Used for lenses, glazing, signs, enclosures, and molded articles.

Resorsabond. Trademark of Georgia-Pacific Resins, Inc. (USA) for a liquid phenol-resorcinol formaldehyde resin binder supplied as a room temperature-curing two-component (resin plus hardener) product for use in the manufacture of softwood I-beams.

Resource. Trade name of Osborn Steels Limited (UK) for a corrosion-resistant steel containing 0.3% carbon, 13% chromium, and the balance iron. Used for plastic molds and dies.

Resovin. Trade name of S.S. White Dental Manufacturing Company (USA) for a vinyl ester resin formerly used for denture bases.

Resproid. Trade name of Respro Inc. (USA) for polyvinyl chlorides and other plastics.

resteel. Any form of steel suitable for use in the reinforcement of concrete in a construction. See also reinforcement (2); reinforcing bar.

Restex. Czech trade name for a toughened glass.

Restil. Trade name for styrene-acrylonitrile (SAN) resins.

Restocem-PL. Trade name of Lee Pharmaceuticals (USA) for a glass-ionomer dental luting cement.

Restocore-PL. Trade name Lee Pharmaceuticals (USA) for a glass-ionomer dental cement used for core-build-ups.

Resto-Crete. Trade name of Western Group (USA) for exterior wall coatings.

Restofil-PL. Trade name Lee Pharmaceuticals (USA) for a glass-ionomer dental cement used for fillings.

Restolux. Trade name Lee Pharmaceuticals (USA) for a light-cure hybrid composite used in restorative dentistry.

Restomer. Trade name Lee Pharmaceuticals (USA) for resin-modified glass-ionomer dental cements including *Restomer-Base* used as a base liner, *Restomer-Cem* used for luting applications and *Restomer-Fil* used as a filler.

Restover. Trademark of Schott DESAG AG (Germany) for a machine-drawn antique glass with irregular surface structure supplied in thicknesses from 2.5 to 3.0 mm (0.10 to 0.12 in.), and used in the restoration of historic window glazing.

resulfurized carbon steels. A group of free-machining carbon steels (AISI-SAE series 11xx) containing about 0.1-0.6% carbon, 0.3-1.65% manganese, 0.08-0.35% sulfur, 0.04% phosphorus, and the balance iron. Some resulfurized grades are also rephosphorized (AISI-SAE series 12xx) and contain 0.04-0.12% phosphorus. The sulfur and, if applicable, phosphorus content of these steels are intentionally increased to improve machinability. Used for adapters, shafts, crankshafts, spindles, studs, pins, handles, yokes, washers, screwstock, screws, nuts, and carburized parts.

Result. Trade name for a dental amalgam alloy.

Resurfo. Trademark of The Reardon Company (USA) for a patching paste used on interior wallboard, and for plaster repairs.

Resystretched. Trademark of Rubafilm (France) for a prestretched polyolefin film with a thicknesses of 8-10 μm (0.3-0.4 mil) used for food packaging and palletizing applications.

Retain. Trademark of Dow Chemical Company (USA) for a series of acrylonitrile-butadiene-styrene (ABS) resins and ABS/polycarbonate alloys with 25% post-consumer recycled content. Available in standard, paintable, medium-heat and high-flexural-modulus grades, they are used for automotive applications, e.g., exterior and interior trim, and instrument panels.

retarded gypsum plaster. A *gypsum plaster* whose setting time has been prolonged by the introduction of a *retarder*, such as dextrin, glue or blood.

retarder. (1) A compounding ingredient added to rubber to minimize premature vulcanization.

(2) A substance introduced into cement, mortar, plaster or stucco to lengthen the setting time. After the initial set it does not affect the properties in any way.

(3) A liquid incorporated into paints or lacquers to delay the evaporation of their solvents.

(4) See inhibitor (3).

retgersite. A pale blue-green mineral composed of nickel sulfate hexahydrate, $NiSO_4 \cdot 6H_2O$. It can also be made synthetically. Crystal system, tetragonal. Density, 2.04 g/cm³; refractive index, 1.513.

reticulated glass. An ornamental glass containing an interlaced network of lines and used for decorative applications.

retinal. A yellowish to orange, unsaturated aldehyde derived from vitamin A (retinol). It forms the prosthetic group of the protein *rhodopsin* in the rod cells of mammalian retinas, and is transformed to yellow-colored 11-*trans*-retinal by the impingement of light. Also known as *retinene; retinene-1*. See also rhodopsin.

retinene. See retinal.

Retino. Trade name of Fabbrica Pisana SpA (Italy) for a patterned glass.

Retinonda. Trade name of Fabbrica Pisana SpA (Italy) for a glass with wavy pattern.

Retort. Trade name of Edgar Allen Balfour Limited (UK) for plain-carbon steel castings containing 0.3-0.35% carbon, 0.4-0.8% silicon, 0.6-0.8% manganese, and the balance iron. Used for machinery parts, gears, and castings.

retort carbon. A dense carbon or graphite formed during coal-gas manufacture in the upper sections of the retort. It is sometimes added to glazes or similar ceramics to produce localized reduction during firing.

retort clay. A plastic, semirefractory clay that is suitable for use in the manufacture of gas and zinc retorts.

retouch enamel. A brushed-on coating or fine overspray of vitreous enamel, usually applied for touch-up purposes.

Retractyl. Trademark of Rhovyl SA (France) for vinyon (vinyl chloride) fibers with excellent acid, alkali and mildew resistance. Used especially for textile fabrics.

retzian. (1) A chocolate brown mineral composed of manganese cerium arsenate hydroxide, $Mn_2CeAsO_4(OH)_4$. Crystal system, orthorhombic. Density, 4.57 g/cm³; refractive index, 1.788. Oc-

currence: Sweden.

(2) A pinkish brown to reddish brown mineral composed of manganese neodymium arsenate hydroxide, $Mn_2Nd(AsO_4)(OH)_4$. Crystal system, orthorhombic. Density, 4.20 g/cm³; refractive index, 1.782. Occurrence: USA (New Jersey).

reused wool. Wool fibers that have been reclaimed from wool *rags* and manufactured waste. Used especially for the manufacture of *shoddy* and *mungo* yarns and fibers. Also known as *recovered wool*.

Revalon. Trade name of Revere Copper Products, Inc. (USA) for wrought aluminum brasses containing 76% copper, 22% zinc, 2% aluminum and 0.05% arsenic. They have high to medium strength, high toughness, and good resistance to seawater and impingement corrosion. Used for condenser and heat-exchanger tubes, hardware, and marine parts.

revdite. A colorless or white mineral composed of sodium silicate pentahydrate, $Na_2Si_2O_5 \cdot 5H_2O$. Crystal system, triclinic. Density, 1.94 g/cm³; refractive index, 1.482. Occurrence: Russian Federation.

Revere. Trade name of Revere Copper Products, Inc. (USA) for an extensive series of alloys including several aluminum and magnesium alloys, coppers and high-copper alloys, wrought and cast bronzes and brasses, cupronickels and nickel silvers.

(2) Trade name of Revertex Limited (UK) for rubber fibers used in the manufacture of elastic yarn for clothing, and elastic bands and tapes.

Revers-A-Blok. Trade name of Perma Paving Stone Company (Canada) for architectural stone with textures on one or two surfaces resembling natural stone. Two stone systems are available: (i) *Revers-A-Blok I* with regular, corner, coping and capping units, 75 mm (3 in.) high; and (i) *Revers-A-Blok II* with regular, tapered and corner units, 150 mm (6 in.) high, and coping and capping units, 75 mm (3 in.) high. They are supplied in several colors.

reverse bronze. A group of corrosion-resistant bronzes, usually containing between 50 and 90% tin, and the balance copper. Used especially for bearings.

reversible bonded fabrics. A term used for rather stiff two-ply fabrics made by bonding together two face fabrics. Either side can be used as the face side.

reversible colloid. A colloid, such as a gel, that after being coagulated can be redispersed, e.g., by the addition of water, or by heating.

reversible fabrics. Double-faced textile fabrics either side of which can be used as the face side.

reversing mill plate. See quarto plate.

Revertex. Trademark of Revertex Limited (UK) for rubber latex supplied natural or concentrated by evaporation. The rubber suspension also contains protective reversible colloids.

Revetor II. Trade name of Metalor Technologies SA (Switzerland) for a palladium-free dental solder containing 75% gold and having a melting range of 750-800°C (1380-1470°F).

Revicol. Trade name of Cristales y Vidrios SA-Cristavid (Chile) for painted and fired semi-toughened sheet glass used chiefly for cladding.

Revlite. Trade name for a lightweight metal-matrix composite (MMC) consisting of a matrix of AA-6061 aluminum (a wrought heat-treatable high-aluminum alloy) reinforced with 25 vol% silicon carbide (SiC) particulate, and used for aerospace and aircraft applications.

Revolene. Trademark of Anchor Packing Division Robco Inc. (Canada) for asbestos braided packings.

Revolite. Trade name of Revolite Corporation (USA) for phenolic resins and plastics.

Revolution. Trademark of Sybron Corporation (USA) Inc. for a light-cured, radiopaque, fluoride-releasing bisphenol A-glycidylether methacrylate (BisGMA) based hybrid composite resin that is 63% filled with ceramic particles having an average size of 1.7 μm (67 μin.). It has flowable consistency, good compressive and tensile strengths and good wear resistance. Used in dentistry for restorations, enamel and porcelain repairs, small core-build-ups, for veneering, as a pit and fissure sealant, and as a liner.

Revotek LC. Trade name of GC America, Inc. (USA) for a light-curing, two-component bismethacrylate (BMA) type dental composite resin used for temporary crown and bridge restorations.

Revy. Trade name of Revelstoke Companies Limited (Canada) for softwood lumber.

reworked plastics. Thermoplastics that after having been processed by any of a number of commercial techniques (e.g., molding or extrusion) are reground, pelletized or solvated for reuse in the manufacture of new products.

Rex. (1) Trade name of Crucible Materials Corporation (USA) for an extensive series of air-melted tungsten- and molybdenum-type high-speed tool steels for cutting tools and cutoff blades. Powder-metallurgy grades, e.g., *CPM Rex 121,* are also available. See also CPM Steels.

(2) Trade name of United Clays, Inc. (USA) for a white-firing, coarse, premium *ball clay* mined in Tennessee, USA. It has high alumina and potassium contents, superior defloculation properties, moderate strength, and is used for sanitaryware, tableware and tile, for fast-firing applications, and in the manufacture of fire-resistant ceiling tile.

(3) Trademark of Pentron Laboratory Technologies (USA) for several dental ceramic alloys for crown and bridge restorations. *Rex 4* is a beryllium-free alloy composed of 62% nickel, 22% chromium, and the balance tungsten, aluminum, silicon and lanthanum. *Rex CC* is a nickel- and beryllium-free alloy composed of 62% cobalt, 26% chromium, 9% tantalum, 5% molybdenum, 3% aluminum, and the balance manganese and hafnium.

Rexal. Trademark of Didier Refractories Corporation (USA) for refractory bricks.

Rexalloy. (1) Trade name of Alloy Cast Products, Inc. (USA) for cast alloys containing 44-64% cobalt, 28-35.5% chromium, 4.5-17% tungsten, 1.2-2.3% carbon, 0-2% iron, 0-1% manganese, 0.5% silicon, up to 0.5% molybdenum and 0.2-0.22% boron. They have good wear resistance and high tensile strength, and are used for cutting tools.

(2) Trademark of Pentron Laboratory Technologies (USA) for a beryllium-free dental alloy composed of 67% nickel, 14% chromium, 8% gallium, 8% molybdenum, and the balance cobalt, aluminum, iron, silicon, manganese and zinc. Used for crown and bridge restorations.

Rexarc. Trade name of Sight Feed Generator Company (USA) for a series of arc-welding electrodes used to apply steel hardfacing materials.

Rex Bronze. Trade name of Whipple & Choate Company (USA) for an alloy composed of 80% copper, 10% zinc and 10% lead.

Rex Champion. Trade name of Colt Industries (UK) for a tungsten-type high-speed steel containing 0.7% carbon, 4% chromium, 14% tungsten, 2% vanadium, and the balance iron. Used for cutting and finishing tools, and twist drills.

Rex Composition. Trade name of Whipple & Choate Company (USA) for a free-cutting brass composed of 76-86% copper, 3-6% tin, 5-14% zinc and 2-6% lead, and used for castings.

REXE. Trademark of Teijin Fibers Limited (Japan) for a polyetherester elastic yarn with excellent resistance to heat, moisture and chlorine bleach. Used for the manufacture of women's apparel, and sportswear.

Rexell. Trademark of Huntsman Company (USA) for several linear-low-density polyethylenes supplied as cast film extrusions in co-extruded grades with good mechanical and sealing properties, in general-purpose grades and in grades suitable for food packaging and shippings sacks.

Rexene. Trade name of Rexene Corporation (USA) for a series of thermoplastic polyolefin resins including polyethylenes and polypropylenes used for molded products, fibers and films. *Rexene PE* is a low-density polyethylene supplied in standard and UV-stabilized grades, and *Rexene PP* refers to polypropylene homo-polymers and copolymers supplied in standard and UV-stabilized grades.

Rexenite. Trade name of Rexenite Corporation (USA) for cellulose acetate and cellulose acetate butyrate plastics.

Rexflex. Trade name of Rexene Corporation (USA) for a series of flexible polyolefin resins including polypropylenes.

Rexillium. Trademark of Pentron Laboratory Technologies (USA) for a range of nonprecious dental ceramic alloys for crown and bridge restorations. *Rexillium III* is a corrosion-resistant alloy composed of 76% nickel, 14% chromium, 6% molybdenum, 2.5% aluminum, 1.99% beryllium, and the balance titanium and cobalt. *Rexillium V* is composed of 74% nickel, 14% chromium, 9% molybdenum, 2.5% aluminum, 1.99% beryllium, and the balance titanium and cobalt. *Rexillium NBF* is a nickel- and beryllium-free alloy composed of 52% cobalt, 25% chromium, 14% tungsten, 8% gallium, and the balance aluminum, yttrium and rhenium.

Rexine. Trade name of ICI (Rexine) Limited (UK) for cellulose nitrate plastics including *celluloid*-type and film materials.

Rexite. Trade name of Colt Industries (UK) for a series of tough, wear-resistant sintered tungsten carbides with cobalt binder, used for cutting tools.

Rexol. French trade name for polystyrene plastics.

Rexolite. Trade name of C-Lec Plastics, Inc. (USA) for a cross-linked polystyrene resin with a density of 1.05 g/cm³ (0.038 lb/in.³), good tensile strength, high thermal and electrical resistance, good stress-crack resistance, good resistance to alpha, beta, gamma, X-ray and ultraviolet radiation, an upper service temperature of 93°C (200°F), good resistance to dilute acids and alcohols, fair resistance to concentrated acids, alkalies, greases and oils, and poor resistance to aromatic hydrocarbons, halogens and ketones. Used as a dielectric in X-ray equipment, and radiation detectors.

Rexophen. Trademark of Forex Leroy Inc. (Canada) for exterior-type particleboard.

Rexor. Trade name of Allied Steel & Tractor Products Inc. (USA) for a tough, wear-resistant tool steel containing 0.35% carbon, 0.7% manganese, 0.6% chromium, 0.35% molybdenum, 0.15% silicon, 0.2% nickel, and the balance iron. Used for pneumatic tools, chisels, and punches.

Rexprint. Trade name for a silicone elastomer used for dental impressions.

Rex-Sil. Trade name for a silicone elastomer used for dental impressions.

Rex Super Cut. Trade name of Colt Industries (UK) for a cobalt-tungsten high-speed steel (AISI type T5) containing 0.77% carbon, 18-18.5% tungsten, 4% chromium, 2% vanadium, 8% cobalt, and the balance iron. It has excellent machinability and wear resistance, high hot hardness, good toughness, and is used for lathe tools, cutters, drills, and various other cutting or finishing tools.

Rex Super Van. Trade name of Colt Industries (UK) for a tungsten-type high-speed steel (AISI type T2) containing 0.7% carbon, 18% tungsten, 4% chromium, 2% vanadium, 0.65% molybdenum, and the balance iron. It has excellent machinability and wear resistance, high hot hardness, good toughness, and is used for milling cutters, drills, and various other cutting or finishing tools.

Rextac. Trademark of Rexene Products Company (USA) for amorphous polyalphaolefins (PAOs) with good low-temperature flexibility, water resistance and chemical inertness, high softening points, consistent tack and compatibility, and constant processibility. Used for laminates, adhesives and sealants.

Rextil. Trademark of REX Industrieprodukte Graf von Rex GmbH & Co. KG (Germany) for asbestos fabrics, ropes and yarns.

Rexweld. Trade name of Colt Industries (UK) for a series of hardfacing and welding electrodes composed of corrosion- and wear-resistant cobalt-chromium-tungsten-carbon alloys, corrosion-resistant nickel-molybdenum-chromium-iron or nickel-chromium-silicon alloys, or high-carbon iron-nickel-chromium alloys.

Rexwood. Trademark of Rexwood Products Limited (Canada) for flakeboard.

reyerite. A white mineral composed of potassium sodium calcium aluminum silicate hydroxide monohydrate, $(Na,K)Ca_7Si_{11}AlO_{29}(OH)_4 \cdot H_2O$. Crystal system, hexagonal. Density, 2.51 g/cm³; refractive index, 1.563. Occurrence: Greenland.

Reymet. Trade name of Reynolds Company (USA) for metallic fibers.

Reynbow. Trade name of Reynbow Products (USA) for metal/polymer bicomponent fibers.

Reynobond. Trademark of Reynolds Metals Company (USA) for composite building panels composed of two sheets of corrosion-resistant aluminum bonded to an extruded thermoplastic core. Available in regular and fire-resistant grades, they have high durability and formability, and high strength-to-weight ratios and rigidity. Used for architectural and building construction applications, e.g., fascias and canopies, exterior cladding, column covers, beam wraps, equipment enclosures, furniture, kiosks, and ATM machines.

Reynolds. (1) Trade name of Reynolds Metals Company (USA) for an extensive range of aluminum products including several grades of commercially pure aluminum, and numerous cast and wrought aluminum-copper, aluminum-magnesium, aluminum-magnesium-silicon, aluminum-manganese, aluminum-silicon and aluminum-zinc alloys. They are supplied in various forms including sheets, foils and fibers. *Reynolds Wrap* refers to aluminum foil used for food wrapping and flexible packaging applications. It is made from an aluminum alloy containing 98.5% aluminum, with the balance being chiefly iron and silicon. *Reynolds Wrap* has high strength, good heat, cold and puncture resistance, and is available in several sizes.

(2) Trade name of Reynolds Company (USA) for readily brazeable low-carbon molybde-num steel tubing used for boiler tubes, and pipelines.

Reynolite. Trademark of Reynolds Metals Company (USA) for a strong, flat, lightweight aluminum composite material made by bonding two sheets of aluminum with polyester finish to a thermoplastic core. The composite panel has excellent workability and formability, high rigidity and durability, and good weather and corrosion resistance. Used for commercial and industrial signage.

Reynolon. Trademark of Reynolds Metals Company/Reynolon (USA) for a premium aluminum shrink film used for industrial packaging applications.

Reyspun. Trade name for spun metal/polymer fibers.

Rez. Trade name of Lepage's Limited (Canada) for wood finishing products including semi-transparent and solid-color wood stains, and various deck and furniture stains.

Rezcore. Trademark of Flake Board Company Limited (Canada) for particleboard corestock.

Rezcote. Trademark of Flake Board Company Limited (Canada) for thin, coated particleboard.

Rezfil. Trademark of Flake Board Company Limited (Canada) for surface-filled and sanded particleboard.

Rezflake. Trademark of Flake Board Company Limited (Canada) for particleboard made of wood flakes bonded together under pressure.

Rezinwood. Trade name of L.F. Laucks Inc. (USA) for phenolic-impregnated wood laminates.

Rezistal. Trade name of Colt Industries (UK) for an extensive series of stainless steels and nickel-chromium alloys used for applications requiring high strength and durability, and good resistance to acids, corrosion and elevated temperatures.

Rezistan. Trade name of Uniworld Corporation of America (USA) for a corrosion-resistant alloy cast iron used for pumps, valves and combustion chambers.

Rezite. Trademark of Flake Board Company Limited (Canada) for coated particleboard products.

Rezi-Weld. Trade name of W.R. Meadows, Inc. (USA) for epoxy resins used for sealants and waterproofing agents.

Rezklad. Trademark of Atlas Minerals & Chemicals, Inc. (USA) for acid-resistant epoxy and acrylic latex compositions used as modifiers in Portland cement concrete, and for topping, coating or pouring floors. *Rezklad Bonding Cement* refers to acid- and alkali-resistant adhesives.

Rez-N-Bond. Trade name of Schwartz Chemical Company, Inc. (USA) for a synthetic resin cement for bonding polymethyl methacrylate (e.g., Plexiglass, or Lucite) and polystyrene.

Rezprime. Trademark of Flake Board Company Limited (Canada) for primed particleboard products.

Rezyl. Trademark of Koppers Company, Inc. (USA) for an extensive series of modified alkyd resins including numerous long- and short-oil alkyds, used for baking and brushing enamels and coatings.

RGI. Trade name for several glass-ionomer dental cements including *RGI Fil* filling cement, *RGI Fil* luting cement and *RGI Reinforced* cermet-reinforced cement.

R-Glass. Trademark of Vetrotex Italia Nuova Italtess (Italy) for R-glass fibers.

R-glass. Fiberglass with a density of about 2.55 g/cm³ (0.092 lb/in.³), high tensile strength (up to 4.33 GPa), high modulus of elasticity (up to 85 GPa) and good resistance to corrosion by most acids. Used as reinforcing fiber in engineering composites. See also fiberglass.

rhabdite. See schreibersite.

rhabdophane. (1) A brown, pinkish or yellow white mineral composed of cerium phosphate monohydrate, $CePO_4 \cdot H_2O$. Crystal system, hexagonal. Density, 3.97 g/cm³; refractive index, 1.654.

Occurrence: UK.

(2) A brown or pinkish mineral composed of yttrium lanthanum phosphate monohydrate, $(La,Y)PO_4·H_2O$. Crystal system, hexagonal. Density, 3.94 g/cm^3; refractive index, 1.654. Occurrence: USA (Connecticut).

Rheindur. German trade name for an age-hardenable high-strength aluminum alloy containing 3.5-5% copper, 0.5-1.2% magnesium, 0.5-1% manganese and 0.5-1% silicon. Used for aircraft and jet-engine components, hardware and fasteners.

Rheinrohr. Trade name of Rheinische Rohrenwerke AG (Germany) for a series of steels including various stainless, heat-resisting, low-temperature-resistant, case-hardening, machinery and boilerplate grades.

rhenia. See rhenium trioxide.

rhenium. A rare, silvery white metallic element of the Group VIIB (Group 7) of the Periodic Table. It is commercially available in the form of powder, rods, wires, strips, ribbons, sheets, foils, mi-crofoils, sintered bar, tubing, arc-melted buttons, pellets and single crystals. The single crystals are usually grown by the float-zone or the radio-frequency technique. It occurs in nature in the mineral *columbite* and in platinum and molybdenum ores (e.g., *molybdenite*). Crystal system, hexagonal. Crystal structure, hexagonal close-packed. Density, 21.02 g/cm^3; melting point, 3180°C (5756°F); boiling point, 5627°C (10161°F); hardness, 280-700 Vickers; superconductivity critical temperature, 1.70K; atomic number, 75, atomic weight, 186.207; monovalent, divalent, trivalent, tetravalent, pentavalent, hexavalent, heptavalent. It has high tensile strength and elastic modulus, an electrical conductivity of about 9% IACS, poor resistance to strong oxidizing agents (nitric and sulfuric acids) and good resistance to hydrochloric acid and seawater. Used to improve the ductility of molybdenum- and tungsten-based alloys, in welding rods, electronic filaments, electrical contacts, high-temperature thermocouples, igniters for flash bulbs and refractory metal parts, as a catalyst, in jewelry, and in electroplating and vapor-phase deposition of metals. Symbol: Re.

rhenium boride. Any of the following compounds of rhenium and boron used for ceramics and materials research: (i) *Trirhenium boride*. Density, 19.4 g/cm^3. Formula: Re_3B; and (ii) *Rhenium triboride*. Formula: Re_7B_3.

rhenium bromide. See rhenium tribromide; rhenium pentabromide.

rhenium carbonyl. White or yellowish crystals (98% pure) that decompose at 170°C (338°F). Used as catalyst and reagent. Formula: $Re_2(CO)_{10}$. Also known as *dirhenium decacarbonyl*.

rhenium chloride. See rhenium trichloride; rhenium tetrachloride; rhenium pentachloride; rhenium hexachloride.

rhenium dioxide. Gray orthorhombic crystals. Density, 11.4 g/cm^3; decomposition temperature, 900°C (1652°F). Used in chemistry and materials research. Formula: ReO_2.

rhenium diphosphide. See rhenium phosphide (ii).

rhenium disilicide. See rhenium silicide (ii).

rhenium disulfide. See rhenium sulfide (i).

rhenium fluoride. See rhenium tetrafluoride; rhenium pentafluoride; rhenium hexafluoride; rhenium heptafluoride.

rhenium halide. A compound of rhenium and a halogen, such as bromine, chlorine, fluorine or iodine.

rhenium heptafluoride. Yellow, cubic crystals with a melting point of 48.3°C (119°F) and a boiling point of 73.7°C (165°F), used in chemistry and materials research. Formula: ReF_7.

rhenium heptasulfide. See rhenium sulfide (ii).

rhenium heptoxide. A brownish-yellow, hygroscopic powder or yellow crystals. Density, 6.10 g/cm^3; melting point, approximately 297°C (567°F); sublimes above 250°C (480°F). Used in ceramics, materials research, and in the production of rhenium compounds. Formula: Re_2O_7.

rhenium hexachloride. A red-green solid with a melting point of 29°C (84°F), used in chemistry and materials research. Formula: $ReCl_6$.

rhenium hexafluoride. Yellow liquid or cubic crystals. Density, 6.0 g/cm^3; melting point, 18.5°C (65°F); boiling point, 47.6°C (118°F). Used in chemistry and materials research. Formula: $ReCl_6$.

rhenium-molybdenum alloys. See molybdenum-rhenium alloys.

rhenium monophosphide. See rhenium phosphide (i).

rhenium monosilicide. See rhenium silicide (i).

rhenium oxide. See rhenium dioxide; rhenium heptoxide; rhenium pentoxide; rhenium trioxide.

rhenium pentabromide. A brown solid that decomposes at 110°C (230°F). Used in chemistry and materials research. Formula: $ReBr_5$.

rhenium pentachloride. Brown-black solid, or green-black crystals. Density, 4.9 g/cm^3; melting point, 220°C (428°F). Used in chemistry and materials research. Formula: $ReCl_5$.

rhenium pentafluoride. A yellow-green solid. Melting point, 48°C (118°F); boiling point, 220°C (428°F). Used in chemistry and materials research. Formula: ReF_5.

rhenium pentoxide. Blue-black, tetragonal crystals with a density of 7.0 g/cm^3, used in chemistry and materials research. Formula: Re_2O_5.

rhenium phosphide. Any of the following compounds of rhenium and phosphorus used in ceramics, electronics and materials research: (i) *Rhenium monophosphide*. Density, 12.0 g/cm^3; melting point, 1204°C (2200°F). Formula: ReP; (ii) *Rhenium diphosphide*. Density, 8.33. Formula: ReP_2; and (iii) *Dirhenium phosphide*. Density, 16.4 g/cm^3. Formula: Re_2P.

rhenium-potassium chloride. A green powder (99.9+% pure) with a density of 3.34 g/cm^3, used as a reagent and in materials research. Formula: K_2ReCl_6. Also known as *potassium hexachlororhenate*.

rhenium powder. Rhenium in powder form obtained by reducing ammonium perrhenate (NH_4ReO_4) in hydrogen. It is available in typical purities from 99.9 to 99.995% and particle sizes ranging from 3 to over 150 μm (0.0001 to 0.006 in.). Used as a starting material for the production of wrought rhenium and rhenium-alloy products by powder metallurgy, and as a catalyst.

rhenium silicide. Any of the following compounds of rhenium and silicon used in ceramics and materials research: (i) *Rhenium monosilicide*. Density, 13.04 g/cm^3; melting point, approximately 1900°C (3450°F). Formula: ReSi; (ii) *Rhenium disilicide*. Density, 10.71 g/cm^3; melting point, approximately 1927°C (3500°F). Formula: $ReSi_2$; and (iii) *Rhenium trisilicide*. Density, 15.44 g/cm^3; decomposes above 1921°C (3490°F). Formula: Re_5Si_3.

rhenium sulfide. Any of the following compounds of rhenium and sulfur: (i) *Rhenium disulfide*. Triclinic crystals. Density, 7.6 g/cm^3. Used in chemistry and materials research. Formula: ReS_2; and (ii) *Rhenium heptasulfide*. Brown-black tetragonal crystals. Density, 4.866 g/cm^3 (85% pure). Used as a catalyst. Formula: Re_2S_7.

rhenium tetrachloride. A purple-black, hygroscopic crystals. Density, 4.9 g/cm^3; decomposes at 300°C (570°F). Used in chemistry and materials research. Formula: $ReCl_4$.

rhenium tetrafluoride. Blue, tetragonal crystals. Density, 7.49 g/cm^3; sublimes above 300°C (570°F). Used in chemistry and materials research. Formula: ReF_4.

rhenium triboride. See rhenium boride (ii).

rhenium tribromide. Red-brown crystals. Crystal system, monoclinic. Density, 6.1 g/cm^3; sublimes at 500°C (930°F). Used in chemistry and materials research. Formula: $ReBr_3$.

rhenium trichloride. A red-black, hygroscopic crystals or powder. Crystal system, hexagonal. Density, 4.81 g/cm^3; decomposes at 500°C (930°F). Used in chemistry and materials research. Formula: $ReCl_3$.

rhenium trioxide. Red powder or red-purple crystals. Crystal system, cubic. Density, 6.9 g/cm^3; decomposes at 400°C (750°F); high surface adsorption. Used for alloying purposes. Formula: ReO_3. Also known as *rhenia*.

rhenium trisilicide. See rhenium silicide (iii).

rhenium-tungsten. An intermetallic compound of rhenium and tungsten with a melting point above 2998°C (5428°F), used for refractory metals and various high-temperature applications. Formula: Re_3W_2.

Rhenoblend. Trademark of Bayer Corporation (USA) for crosslinkable reclaims and polymer blends.

Rhenoflex. Trade name of Hüls AG (Germany) for chlorinated polyvinyl chlorides.

Rhenosin. Trademark of Bayer Corporation (USA) for several hydrocarbon and coumarone-indene resins, plasticizers, tackifiers and homogenizing agents.

Rheocin. Trademark of Süd-Chemie, Inc. (USA) for a range of thixotropic materials based on castor oil, or castor oil derivatives.

Rheotan. British trade name for a series of copper alloys containing 12% manganese, 0-25% nickel, 2-12% iron and 2-18% zinc. Used for electrical resistances.

Rheotanium. Trade name for a corrosion-resistant jewelry alloy composed of 60-90% gold and 10-40% palladium.

Rhino. Trade name of Commentryenne (France) for a series of molybdenum-cobalt, molybdenum-tungsten, molybdenum-cobalt-vanadium and molybdenum-vanadium tool steels used for cutters, drills, bits, saws, punches and dies

Rhino-Hyde. Trademark of Kunststoffe Arthur Krüger (Germany) for several high-performance plastics.

Rhino Linings. Trademark of Rhino Linings USA, Inc. for flexible 100% solids sprayed polyurethane linings used to provide abrasion-, corrosion-, impact- and chemical-resistant surfaces on a wide range of substrates.

Rhodamine dyes. A large group of synthetic dyes usually supplied as crystals and/or powders, and used as laser dyes, as fluorescent dyes, e.g., in the life sciences, and in colorimetry. Examples include Rhodamine B ($C_{28}H_{31}ClN_2O_3$), Rhodamine 6G perchlorate ($C_{28}H_{31}ClN_2O_7$) and Rhodamine 123 ($C_{21}H_{17}ClN_2O_3$).

Rhodesian teak. The durable, reddish-brown wood from the tree *Baikiaea plurijuga*. It has a very attractive appearance and a smooth hard surface, but is not a true *teak*. Average weight, 900 kg/m^3 (56 lb/ft^3). Source: Southern Africa. Used for chiefly for flooring.

rhodesite. A white mineral of gold group composed of calcium potassium silicate hydrate, $(Ca,K,Na)_8Si_{16}O_{40} \cdot 11H_2O$. Density, 2.36 g/cm^3; refractive index, 1.505. Occurrence: USA (California), South Africa.

Rhodex. Trade name of Enthone-OMI Inc. (USA) for a rhodium electrodeposit and plating process.

Rhodia. Trade name of Rhône-Poulenc SA (France) for cellulose acetate fibers used for the manufacture of lingerie, pyjamas, shirts, ties and swimwear, and in staple form in blends for suitings, sportswear, knitting yarns, household textiles, carpets, and cable insulations.

Rhodiaceta. Trade name of Rhône-Poulenc SA (France) for nylon fibers and yarns used for textile fabrics including clothing.

Rhodiafil. Trade name of Rhône-Poulenc SA (France) for rayon fibers and yarns used for textile fabrics.

Rhodialin. Trade name of Rhône-Poulenc SA (France) for cellulose acetate fibers used for wearing apparel and industrial fabrics.

Rhodianyl. Trade name of Fairway Filamentos SA (Brazil) for nylon 6,6 staple fibers used for textile fabrics including clothing.

Rhodipor. Trade name of Rhodius Chemie-Systeme GmbH (Germany) for packaging, shipping containers, and molded products based on rigid polystyrene foams.

rhodium. A silvery-white metallic element of Group VIII (Group 9) of the Periodic Table. It is commercially available in the form of powder, sponge, wire, foils, microfoils, sheets, rods and single crystals, and occurs naturally in platinum ores. Crystal system, cubic. Crystal structure, face-centered cubic. Density, 12.41 g/cm^3; melting point, 1965°C (3569°F); boiling point, 3727°C (6741°F); hardness, 120-300 Vickers; atomic number, 45; atomic weight, 102.906; divalent, trivalent, tetravalent, pentavalent, hexavalent. It is a noble and precious metal, and has good resistance to dilute acids (including aqua regia), poor resistance to concentrated sulfuric acid, high tensile strength, high surface reflectivity, and good electrical conductivity (about 37% IACS). Used as alloying element with platinum, for nontarnishing electrodeposits on jewelry and ornaments of Sterling silver or white gold, in electrodeposited coatings on metals, in mirror coatings, in coatings on glass, alloyed with platinum for high-temperature thermocouples (positive leg), for rayon spinnerets, in headlight reflectors, furnace windings, laboratory crucibles, fountain-pen points, mirrors for optical instruments, electrical contacts, telephone relays, spark plugs and surgical instruments, and as a catalyst. Symbol: Rh.

rhodium acetate dimer. Black-green-red crystals (99.9+% pure) used as homogeneous catalysts. Formula: $Rh(O_2C_2H_3)_4$.

rhodium antimonide. A compound of rhodium and antimony. Crystal system, cubic. Crystal structure, skutterudite. Melting point, 897°C (1647°F); band gap, 0.8 eV. Used as a semiconductor, and in ceramics. Formula: $RhSb_3$.

rhodium arsenide. A compound of rhodium and arsenic. Crystal system, cubic. Crystal structure, skutterudite. Melting point, 997°C (1827°F); band gap, 0.85 eV. Used as a semiconductor, and in ceramics. Formula: $RhAs_3$.

rhodium black. A finely divided black-colored form of metallic rhodium obtained by reduction of a solution of a rhodium salt with a suitable material. Used as a catalyst.

rhodium carbonyl. An organometallic compound that contains rhodium, is available in the form of black crystals, and is used as an organometallic catalyst, and for rhodium coatings. Formula: $Rh_6(CO)_{16}$. Also known as *hexarhodium hexadecacarbonyl*.

rhodium chloride. Red crystals or brown-red, moisture-sensitive powder (98+% pure). Crystal system, monoclinic. Density, 5.38 g/cm^3; decomposes at 450°C (840°F); boiling point, 717°C (1323°F). Used in chemistry, biology, biochemistry, biotechnology, and materials research. Formula: $RhCl_3$. Also known as *rhodium trichloride.*

rhodium chloride hydrate. A dark red, hygroscopic powder or crystals (99.9+% pure) decomposing at 100°C (212°F). It typically contains 38-43.5% rhodium, and is used in chemistry, biochemistry and materials research. Formula: $RhCl_3 \cdot xH_2O$.

rhodium coatings. A group of decorative and engineering coatings based on rhodium. The decorative coatings (typically less than 5 µm or 0.2 mil thick) are pinkish white, nontarnishing electrodeposits produced on jewelry and silver or platinum products using solutions of rhodium and phosphoric or sulfuric acid. The engineering coatings are electrodeposits (typically 25-200 µm or 1-8 mils thick) produced from solutions of rhodium and sulfuric acid with the addition of selenic acid or magnesium sulfate and sulfamate on various metal substrates, or by barrel plating. Rhodium engineering coatings are hard, wear- and corrosion-resistant, highly reflective, provide low electrical resistivity, and are used on electronic components and devices, electrical contacts, reflectors, surgical instruments, missiles, and optical equipment.

rhodium dioxide. Brown-black crystals or powder (99.9% pure). Crystal system, tetragonal. Density, 7.2 g/cm³. It is also available as the dihydrate ($RhO_2 \cdot 2H_2O$). Used chemistry and materials research. Formula: RhO_2. Also known as *rhodium oxide.*

rhodium fluoride. See also rhodium trifluoride; rhodium hexafluoride.

rhodium hexafluoride. Black cubic crystals. Density, 3.1 g/cm³; melting point, 70°C (158°F). Used in chemistry and materials research. Formula: RhF_6. Also known as *rhodium fluoride.*

rhodium iodide. Black, hygroscopic, crystals or powder. Crystal system, monoclinic. Density, 6.4 g/cm³. Used in chemistry and materials research. Formula: RhI_3. Also known as *rhodium triiodide.*

rhodium monosulfide. See rhodium sulfide (i).

rhodium oxide. See rhodium dioxide; rhodium sesquioxide.

rhodium phosphide. A compound of rhodium and phosphor. Crystal system, cubic. Crystal structure, skutterudite. Melting point, 1197°C (2187°F). Used as a semiconductor, and in ceramics. Formula: RhP_3.

rhodium-platinum oxide. See Nishimura catalyst.

rhodium-polyethylenimine. See Royer catalysts.

rhodium sesquioxide. Gray crystals or powder (99.8+% pure). Crystal system, hexagonal. Density, 8.20 g/cm³; melting point, 1100°C (2012°F) (decomposes). It is also available in hydrated form, e.g., as the pentahydrate ($Rh_2O_3 \cdot 5H_2O$), a yellow powder that decomposes on heating. Used in chemistry, biochemistry and materials research. Formula: Rh_2O_3. Also known as *rhodium oxide.*

rhodium sesquisulfide. See rhodium sulfide (ii).

rhodium silicide. A compound of rhodium and silicon used as a semiconductor. Formula: $RhSi$.

rhodium sulfate. A liquid compound formed by dissolving a prescribed amount of rhodium metal in specified amount of concentrated sulfuric acid. Used as an electrolyte in rhodium plating. Formula: $Rh_2(SO_4)_3$.

rhodium sulfide. Any of the following compounds of rhodium and sulfur used in ceramics and materials research: (i) *Rhodium monosulfide.* Gray-black crystals. Formula: RhS; and (ii) *Rhodium sesquisulfide.* Black crystals. Formula: Rh_2S_3.

rhodium trichloride. See rhodium chloride.

rhodium trifluoride. Red, hexagonal crystals. Density, 5.4 g/cm³; boiling point, above 600°C (1110°F). Used in chemistry and materials research. Formula: RhF_3. Also known as *rhodium fluoride.*

rhodium triiodide. See rhodium iodide.

rhodizite. A colorless mineral composed of cesium beryllium aluminum borate hydroxide, $CsAl_2Be_4B_{11}O_{25}(OH)_4$. Crystal system, cubic. Density, 3.44 g/cm³; refractive index, 1.693. Occurrence: Madagascar.

rhodocrosite. A pale pink to rose-red or brownish mineral of the calcite group with a white streak and a vitreous to pearly luster. It is composed of manganese carbonate, $MnCO_3$, and may contain some iron, calcium, magnesium and/or zinc. It can also be made synthetically. Crystal system, rhombohedral (hexagonal). Density, 3.3-3.7 g/cm³; hardness, 3.0-4.5 Mohs; refractive index, 1.816; photoluminescent properties. Occurrence: Europe, USA (Colorado, Connecticut, Montana, Nevada, New Jersey). Used as an ore of manganese, and in electronic and materials research. Also known as *manganese spar.*

Rhodoid. Trade name of Rhône-Poulenc SA (France) for cellulose acetate plastics.

rhodolite. A rose or violet-red variety of the mineral *garnet* with a brilliant luster. It is composed essentially of a mineral mixture of 2 parts of *almandite* [$Fe_3Al_2(SiO_4)_3$] and 1 part of *pyrope* [$Mg_3Al_2(SiO_4)_3$]. Occurrence USA (North Carolina). Used chiefly as a gemstone.

Rhodomerse. Trade name of Technic Inc. (USA) for an immersion rhodium deposit and related deposition process.

rhodonite. A pale pink, rose-red or brown mineral of the pyroxenoid group composed of manganese silicate, $MnSiO_3$. Crystal system, triclinic. Density, 3.4-3.7 g/cm³; hardness, 5.5-6 Mohs. Occurrence: Russia, USA (New Jersey). Used for ornamental applications.

rhodopsin. A light-sensitive, purple pigment found in the rod cells of vertebrate retinas and composed of *retinal* and *opsin.* It makes vision in dim light possible, and is regenerated in the dark. Used in biochemistry and biotechnology. Also known as *visual purple.* See also bacteriorhodopsin.

Rhodorsil. Trademark of Rhône-Poulenc SA (France) for a series of two-part silicone room-temperature vulcanisates (RTVs) used for molding, tooling and fabricating composites. *Rhodorsil 585* has low viscosity and high tear resistance, and is formulated for polyester, decorative concrete polymer blends and flame-inhibition and heat-resistant applications. *Rhodorsil 1547* is a heat-resistant, high-strength, high-durometer, room-temperature molding and casting vulcanisate for rigid urethanes and epoxies, and metal casting. *Rhodorsil 1556* offers heat resistance, low durometer hardness and high strength for metal forming, vacuum blankets, bladder and plaster molding applications. It has also good resistance to cuts, nicks and tears.

rhodostannite. A reddish mineral of chalcopyrite group composed of copper iron tin sulfide, $Cu_2FeSn_3S_8$. It can also be made synthetically. Crystal system, hexagonal. Density, 4.79 g/cm³.

rhodplumsite. A pale to dark gray mineral composed of rhodium lead sulfide, $Rh_3Pb_2S_2$. Crystal system, rhombohedral (hexagonal). Density, 9.85 g/cm³. Occurrence: Russia.

rhoenite. A brown mineral of the aenigmatite group composed of calcium iron aluminum silicate, $(Ca,Na)_2(Fe,Mg,Ti)_6(Si,Al)_6O_{20}$. Crystal system, triclinic. Density, 3.64 g/cm³; refractive index, 1.806. Occurrence: USA (Texas), Germany.

Rhofibre. Trade name of Rhovyl Corporation (France) for *vinyon fibers* used especially for textile fabrics.

Rhofil. Trade name of Rhovyl Corporation (France) for *vinyon fibers* with excellent acid, alkali and mildew resistance. Used especially for textile fabrics.

rhomboclase. A colorless, white, gray or yellow mineral composed of iron hydrogen sulfate tetrahydrate, $FeH(SO_4)_2 \cdot 4H_2O$. It can also be made synthetically. Crystal system, orthorhombic. Density, 2.23 g/cm^3; refractive index, 1.555. Occurrence: USA (California), Hungary, Czech Republic.

rhombohedral iron ore. See hematite.

Rhombolux. Czech trade name for a glass with diamond patterns.

Rhonel. Trade name of Rhovyl Corporation (France) for polyamide (nylon) fibers.

Rhonite. Trademark of Rohm & Haas Company (USA) for urea-formaldehyde resins usually supplied as aqueous solutions.

Rhoplex. Trademark of Rohm & Haas Company (USA) for white, opaque acrylic latexes (essentially water dispersions of acrylic copolymers) and acrylic resin emulsions. Their general properties including tackiness, hardness and flexibility differ greatly with composition. Used for coating and finishing paper and textiles, in emulsion paints, as bonding agents for fibers and pigments, and in transparent or colored coatings on various substrates including metals and wood. *Rhoplex* coatings are highly permanent and durable.

rhotanium. An alloy of rhodium and gold used for jewelry and ornamental applications.

Rhovyl. Trademark of Rhovyl SA (France) for *vinyon fibers* used for industrial fabrics, and clothing.

RhTech. Trademark of Lucent Technologies (USA) for a rhodium electrodeposit and plating process.

rhyolite. A fine-grained, light-colored *volcanic rock* composed largely of alkali feldspar and quartz, and used as a concrete aggregate.

rhyolitic tuff. Tuff composed of fragments of highly viscous, silica-rich lava. See also rhyolite; tuff.

rib. (1) A general term for textile fabrics having one or more straight, raised cords or ridges.

(2) A stretchable fabric knit with alternating purl (inverted) and plain stitches.

Ribbed. Trade name of ASG Industries Inc. (USA) for a glass with linear pattern.

ribbon. (1) A synthetic or natural fiber, usually rectangular in cross section, whose width is at least 4 times its thickness.

(2) A narrow textile fabric of fine texture that has a weight not exceeding 15 oz/yd^2 (510 g/m^2), and is used for trimming and decorative applications.

(3) See amorphous ribbon.

(4) See nanoribbon.

Ribbon Straw. Trade name for cellulose acetate fibers.

Riblene. Trade name of Enichem SpA/Polimeri Europa (Italy) for high- and low-density polyethylenes supplied in unmodified and UV-stabilized grades.

Ribmet. Trademark for ribbed industrial steel mesh.

Ribolux. Czech trade name for a patterned glass having parallel rows of prominent ribs or flutes with wavy cross section.

ribonucleic acid. A nucleic acid present in all living cells, and involved in the coding of genetic information with DNA, transferring it from the nucleus into the cytoplasm and translating it into specific enzymes and proteins. Abbreviation: RNA. See also deoxyribonucleic acid.

ribose. A pentose sugar that exists in D- and L-form. D-ribose is available as a crystalline compound (98+% pure) with a melting point of 88-92°C (190-198°F) and a specific optical rotation of -19.7° (at 20°C/68°F), and is an essential component of many nucleotides including RNA. L-ribose is available as a crystalline compound (98% pure) with a melting point of 81-82°C (178-180°F) and a specific optical rotation of +19° (20°C/68°F). Used in biochemistry, bioengineering, DNA research, and medicine. Formula: $C_5H_{10}O_5$.

ribosomal RNA. Ribosomal ribonucleic acid that is a RNA molecule associated with the structure of the ribosomes (cytoplasmic organelles) and serves as a template (or replica) for cytoplasmic protein synthesis. Abbreviation: rRNA.

rice-hull media. A finely ground flour obtained from rice hulls, and used as a soft-grit abrasive for blast cleaning, deburring and finishing plastics and delicate items.

Ricem. Trademark of Montefibre SpA (Italy) for acrylic (polyacrylonitrile) staple fibers.

rice paper. (1) A thin paper made from the stem pith of the rice-paper tree *(Tetrapanax papyriferum)* growing in Taiwan.

(2) A thin, high-quality paper obtained from the straw of rice by pulping in an alkaline solution.

(3) A cigarette paper made from flax.

Richard's alloy. A British zinc die-casting alloy containing 4% aluminum.

Richard's bronze. A British corrosion-resistant, high-strength bronze containing 55-56% copper, 42% zinc, 1-2% aluminum and 1% iron. Used for marine hardware.

Richardson's speculum. A British bronze containing 65% copper, 30% tin, 2% arsenic, 2% silicon and 0.7-1% zinc. It has high hardness and toughness, takes a high polish, and is used for reflectors.

Richard's plastic babbitt. A babbitt composed of 82.13% tin, 9.77% antimony and 8.1% copper, and used for antifriction alloys and bearings.

Richard's solder. (1) An alloy composed of 71.5% tin, 25% zinc and 3.5% aluminum, and used as an aluminum solder.

(2) A solder containing 61-65% copper, 29-33% zinc, 3% aluminum and 3% tin.

rich clay. A plastic clay with high green strength and good workability.

rich concrete. A concrete mixture that contains a high proportion of cement.

richellite. A reddish to yellow brown mineral of the amorphous group composed of calcium iron aluminum phosphate hydroxide, $(Ca,Fe)(Fe,Al)_2(PO_4)_2(OH,F)_2$. Crystal system, tetragonal. Density, 3.66 g/cm^3. Occurrence: Belgium.

richelsdorfite. A turquoise to sky blue mineral composed of calcium copper antimony chloride arsenate hydroxide hexahydrate, $Ca_2Cu_5SbCl(OH)_6(AsO_4)_4 \cdot 6H_2O$. Crystal system, monoclinic. Density, 3.20 g/cm^3; refractive index, 1.765. Occurrence: Germany.

richetite. A black, radioactive mineral of the becquerel group composed of lead uranium oxide monohydrate, $U-Pb-O \cdot H_2O$. Crystal system, monoclinic. Refractive index, 1.98. Occurrence: Zaire.

rich gold metal. See copper-rich brass.

rich lime. See fat lime.

rich low brass. See wrought red brass.

Richloy. Trade name of National Cable & Metal Company (USA) for a series of soft solders containing 20-65% tin and 35-80% lead. Their melting points range from 183 to 277°C (361 to 531°F).

Richmond. Trade name for saran (polyvinylidene chloride) fibers with good chemical resistance, good weatherability, and good resistance to mildew and insects. Used for carpets and rugs, draperies, upholstery, clothing, and industrial fabrics.

Richply. Trademark of Richmond Plywood Corporation Limited (Canada) for plywood.

richterite. (1) A brown, yellow or rose-red mineral of the amphibole group composed of sodium calcium magnesium silicate hydroxide, $Na_2CaMg_5Si_8O_{22}(OH)_2$. It can be made synthetically. Crystal system, monoclinic. Density, 2.99 g/cm^3.

(2) A pale reddish brown mineral of the amphibole group composed of potassium sodium calcium magnesium silicate hydroxide, $KNaCaMg_5Si_8O_{22}(OH)_2$. Crystal system, monoclinic. Density, 3.12 g/cm^3. Occurrence: Australia.

Richware. Trade name of Makalot Corporation (USA) for phenol-formaldehyde plastics.

rickardite. A deep purple mineral composed of copper telluride, Cu_7Te_5. It can also be made synthetically. Crystal system, orthorhombic. Density, 7.54 g/cm^3. Occurrence: El Salvador, Brazil, USA (Arizona, Colorado).

ridge tile. A roofing tile of special shape used to form the junction of two sloping sides of a roof. See also hip tile.

Ridover. Trade name of Saint-Gobain (France) for formed glass building elements having one patterned surface and a wire reinforcement running in longitudinal direction.

riebeckite. A dark blue asbestos mineral of the amphibole group composed of sodium iron silicate hydroxide, $(Na,Ca)_2(Fe,Mn)_3$-$Fe_2(Si,Al)_8O_{22}(OH,F)_2$. Crystal system, monoclinic. Density, 3.38-3.40 g/cm^3; refractive index, 1.701. Occurrence: Russian Federation, Rumania, South Africa. Used for fireproof fabrics, and for thermal and electrical insulation applications. See also amphibole asbestos; crocidolite.

Rieke zinc. Trademark of Rieke Metals, Inc. (USA) for highly reactive zinc metal supplied as a suspension in tetrahydrofuran (THF) and having a density of 0.949 g/cm^3 (0.034 $lb/in.^3$) and a flash point of 1°F (-17°C).

rift-sawn lumber. See edge-grained lumber.

Rigel Star. Trade name for a low-copper silver-tin dental amalgam alloy.

rigging leather. A flexible, heavy-duty leather tanned in water extracts of wood or bark, and used for rigging applications.

right-handed materials. A class of conventional materials in which the relationship between electric and magnetic fields and the direction of wave propagation is "right-handed." Abbreviation: RHM. Also known as *negative index materials*. See also left-handed materials.

Rigid-Board. Trade name of Manville Corporation (USA) for semi-rigid fiberglass boards used for metal building insulation.

rigid board insulation. Lightweight insulation usually manufactured from closed-cell plastic-foam materials (e.g., expanded or extruded polystyrenes, phenolics, polyisocyanurates or polyurethanes) in the form of rigid boards and in a wide range of sizes. It has a high thermal insulating value per unit thickness (also known as R-value or heat-flow resistance value) and can be used for the acoustical and thermal insulation roof decks, ceilings, basement floors and walls, as exterior sheathing, and as an air-barrier material. See also board insulation; semi-rigid board insulation.

rigid cellular core. See rigid foam.

Rigidex. (1) Trademark of BP Amoco plc (UK) for tough, electrically insulating, chemically resistant medium- and high-density polyethylenes used for molded articles.

(2) Trademark of Fibergrate Composite Structures Inc. (USA) for fiberglass-reinforced plastic tiles and panels.

rigid foam. Lightweight polystyrene or polyurethane foam available in the form of prefoamed blocks, boards or sheets. It has excellent resistance to heat and permeation by air and water, and is used for the thermal insulation of boxcars, tank and hopper cars, refrigerated cars, containers for aircraft, ships, trucks and trailers; storage vessels, pipelines, building blocks, windows and doors; in the manufacture of coolers, display cases, refrigerators, and freezers; as filling and/or acoustic insulation in automobile engines and trunk hoods, boat hulls, surfboards, and skis; and for flotation devices, automotive components (e.g., bumpers), furniture and packaging. Also known as *rigid cellular core.*

rigid insulation. A thermal insulation material of varying composition supplied in the form of blocks, boards, bricks, sheets or slabs.

rigid integral skin foam. Polyurethane foam consisting of a porous core sandwiched between solid skin layers with a transition zone between the core and each layer. It has a low weight and high stiffness and rigidity, and is used for appliance housings (e.g., radios, televisions and computers), window-frame profiles, skylight frames, and furniture, such as tables, desks, cabinets and book shelves.

Rigidite. Trademark of Narmco Materials (USA) for several bismaleimide resins.

(2) Trademark of BASF Structural Materials, Inc. (USA) for composite materials composed of reinforced resins in any of various matrix materials.

rigidized metal. See textured metal.

rigid plastics. Plastics made from rigid resins.

rigid resin. A synthetic resin having a flexural or tensile modulus exceeding 700 MPa (100 ksi) at a temperature of 23°C (73°F) and a relative humidity of 50%.

rigidsol. A special type of *plastisol* usually made with a cross-linking plasticizer and having a high modulus of elasticity.

Rigidtex. Trade name of BP Chemicals (UK) for polyethylene plastics.

Rigid-Tex. Trade name of Rigid-Tex Corporation (USA) for structurally reinforced ferrous and nonferrous metal sheets.

Rigiduct. Trade name of PPG Industries Inc. (USA) for glass-fiber boards used in the manufacture of air-handling systems.

rigid urethane foam. Polyurethane foam obtained by treating polyethers made from methyl glucoside, sorbitol or sucrose with a diisocyanate in the presence of some water and a catalyst (e.g., an organotin compound, or an amine). It has excellent thermal and acoustic insulating properties, high impermeability to air and water, and fair to poor flame resistance. Used for thermal insulation applications (e.g., coolers, display cases, refrigerators, deep freezers, containers for aircraft, ships, trucks, trailers, boxcars, refrigerated cars, tank and hopper cars, and for storage vessels, pipelines, building blocks, windows, doors, filling and acoustic insulation, boat hulls, surfboards, skis, packaging applications, furniture, buoyancy and flotation devices, and automobile bumpers.

Rigipore. Trade name of BP Chemicals (UK) for polstyrene structural foams.

Rigips. Trademark of Rigips GmbH (Germany) for gypsum wallboards.

Rigor. Trade name of Uddeholm Corporation (USA) for an air-hardening cold-work tool steel (AISI type A2) containing 1% carbon, 0.6% manganese, 0.2% silicon, 5.3% chromium, 2% vanadium, 1.1% molybdenum, and the balance iron. It has good nondeforming properties, wear resistance and toughness, and is used for press tools, and dies.

Rigortex. Trademark of Koppers Company, Inc. (USA) for a highly corrosion-resistant vinyl copolymer coating.

Riken. Trade name of Riken Metal Manufacturing Company (USA) for a series of wrought or sand- or die-cast magnesium alloys containing up to 11% aluminum, 0.1-2.5% manganese and 0-3.5% zinc.

Rilsan. (1) Trademark of Atofina Chemicals Inc. (USA) for nylon 11 and nylon 12 resins supplied in unmodified, flexible, semiflexible, carbon and glass fiber-reinforced, fire-retardant, UV-stabilized and coating grades. They have a density range of 1.01-1.04 g/cm³ (56 lb/in.³), good strength, good dimensional stability, a melting point of 198°C (388°F), a glass-transition temperature range of of about 37-46°C (99-115°F), a service temperature of -50 to +130°C (-58 to +266°F), low thermal expansivity, good dimensional stability, a low coefficient of friction, low water absorption, good resistance to dilute acids, alkalies, aromatic hydrocarbons, greases, oils, halogens and ketones, fair resistance to concentrated acids and alcohols, fair resistance to ultraviolet radiation, and good dielectric and self-extinguishing properties. Used for powder coatings on metals, in coatings for glass fibers for fiberoptic cables, for hoses, and and for molded, precision-engineering and low-temperature-tough parts. Common grades include *Rilsan A* and *Rilsan B* as well as *Rilsan N*, a transparent grade.

(2) Trademark of Atofina Chemicals Inc. (USA) for polyamide 11 (nylon 11) fibers used for industrial and other textile fabrics.

Rimcast. Trade name of Nippon Steel Corporation (Japan) for continuously cast high-strength, bake-hardenable hot-rolled steel sheets with 0.1% carbon, about 70 ppm aluminum, and nitrogen. They have greatly improved fatigue resistance and are used especially for automotive applications.

RIMline. Trademark of ICI Americas Inc. (USA) for a series of polyurethane-based systems used for reaction-injection molding (RIM) including engineering thermoplastic and rigid RIM systems for both industrial and consumer applications. The *RIMline RS* series contains several rigid structural systems with high stiffness, good surface characteristics and good mold flow. The *RIMline E* series consists of elastomeric systems featuring a range of flexural moduli and related properties as well as high impact strength and temperature performance. This series also offers very fast process times, and can be modified with conventional fillers and reinforcements.

rimmed steel. A steel that essentially has not been deoxidized and therefore contains enough iron oxide to effect a continuous evolution of carbon monoxide during solidification. The resulting ingot has a relatively soft, easily machinable rim that is free of voids, and can contain blowholes, but no pipe. The core is high in strength, low in toughness, and contains numerous entrapped gas bubbles. Sheet and strip products produced from this ingot have good cold workability and excellent surface quality. Also known as *rimming steel; unkilled steel.*

Rimplast. Trademark of Hüls AG (Germany) for a series of thermoplastics modified by the addition of reactive *polysiloxanes* to form an interpenetrating polymer network during processing. This modification results in better moldability and reduced mold shrinkage, higher chemical resistance, good mechanical properties, and improved thermal properties, such as lower glass-transition temperatures and higher thermal stability. Supplied in pellet form for extrusion, blow forming and injection molding, they are used in the electrical and electronic industries, for protective and peel-off films for the packaging indus-

try, for medical equipment and instruments, piston rings, bushings, and valves.

RIM plastics. See reaction-injection-molded plastics.

Rimtec. Trade name of Rimtec (USA) for several vinyl resins.

rimu. See red pine (2).

ringas. The dark red, fine-textured wood of several species of trees of the genus *Melanorrhoea*. Source: Borneo. Used for chiefly carving. Also known as *Borneo rosewood.*

ring silicates. See cyclosilicates.

ring-spun yarn. A relatively strong yarn produced from roving on a spinning machine (ring spinner) consisting of a fast-rotating bobbin contained within a slower moving ring and a traveler (or slide) which rotates around the edge of the ring. The roving is fed to the traveler, twisted into yarn and wound onto the bobbin.

ringwoodite. A purple or bluish-gray mineral of the spinel group composed of magnesium silicate, Mg_2SiO_4. It can also be made synthetically, and may contain some iron. Crystal system, cubic. Density, 3.90 g/cm³; refractive index, 1.768. Occurrence: USA.

Rinmann's green. See cobalt green.

rinneite. A colorless, rose, purple or yellow mineral composed of potassium sodium iron chloride, $K_3NaFeCl_6$. Crystal system, rhombohedral (hexagonal). Density, 2.55 g/cm³; refractive index, 1.588. Occurrence: Italy, Germany.

Rio Red. Trade name of Aardvark Clay & Supplies (USA) for a dark brown to red clay (cone 10).

Riotglas. Trade name of Safetee Glass Company Inc. (USA) for a thick laminated safety glass.

Riotshield. Trade name of Laminated Glass Corporation (USA) for a burglarproof laminated safety glass used for storefront windows and interior display cases.

Rip. Trademark of Rippenstreckmetall-Gesellschaft mbH (Germany) for a ribbed expanded metal.

ripped lumber. Dressed and graded lumber sawed into narrower pieces by ripping, i.e., by sawing along the grain.

Rippling. Trade name of SA des Verreries de Fauquez (Belgium) for a patterned glass.

ripstop. A woven fabric with squares on the surface formed by ribbed yarns evenly spaced in the warp and weft (filling). Used especially for outerwear, leisure wear and sports wear.

Risomur. Trademark of Richard Sommer GmbH (Germany) for industrial lacquers and varnishes.

Rita. Trade name of Cannon-Stein Steel Company (USA) for a series of steels including various structural, case-hardening and machinery grades, high-speed, plain-carbon and hot-work tool grades, and several die grades.

Riteflex. Trademark of Hoechst Celanese Corporation (USA) for thermoplastic polyester elastomers with outstanding fatigue resistance, excellent chemical resistance, good low-temperature impact resistance, and a wide service temperature range of -40 to +121°C (-40 to +250°F). Used for applications requiring elastomers that are tougher than rubber and related elastomers.

Ritex. Trademark of General Refractories Company (USA) for several refractory products including magnesite brick.

Ritox. Trade name of Oxelösunds Järnverk (Sweden) for blackboards produced by blasting and painting glass surfaces.

rivadavite. A colorless mineral composed of sodium magnesium borate hydrate, $Na_6MgB_{24}O_{40}\cdot22H_2O$. Crystal system, monoclinic. Density, 1.90 g/cm³; refractive index, 1.481. Occurrence: Argentina.

River Ace. Trade name of Kawasaki Steel Corporation (Japan) for a series of quenched-and-tempered high-strength low-alloy (HSLA) steels containing 0.1-0.16% carbon, 1.04-1.37% manganese, 0.009-0.016% phosphorus, 0.006-0.016% sulfur, 0.23-0.36% silicon, 0.21-0.47% copper, 0.08-1.54% nickel, 0.43-0.51% chromium, 0.05-0.66% molybdenum, 0.01-0.05% vanadium, 0.0018-0.0035% boron, and the balance iron. Used for weldable structures and construction equipment.

river cottonwood. See swamp cottonwood.

River Flex. Trade name of Kawasaki Steel Corporation (Japan) for steel plate products that have high yield strengths, and contain titanium and niobium additions.

River Hi-Zinc. Trademark of Kawasaki Steel Corporation (Japan) for an electrogalvanized sheet steel rustproofed on one or both sides with a thin electrodeposit of a zinc-nickel alloy (about 12% nickel). It has excellent coatability, weldability and corrosion resistance, and is used for automotive body panels.

River Hi-Zinc Super. Trademark of Kawasaki Steel Corporation (Japan) for an electrogalvanized sheet steel rustproofed on one or both sides with two layers of protective coatings. The thin inner zinc-nickel alloy electrodeposit provides good formability and weldability, while the outer iron-phosphorus layer imparts good paint adhesion and thus exceptional surface appearance. Used in the automotive industry for body panels, doors, fenders, hoods and trunk lids.

River Lite. Trade name of Kawasaki Steel Corporation (Japan) for high-strength chromium-aluminum steel containing small additions of zirconium (0.05%) and lanthanum (up to 0.08%). It has good high-temperature oxidation resistance, and is used for automotive catalytic converters.

river maple. See silver maple.

Riverside. Trade name of Riverside Metals Corporation (USA) for an extensive series of copper-base alloys including various phosphor and jewelry bronzes, cupronickels, low brasses, nickel silvers, and copper-base bearing alloys.

riversideite. A white mineral composed of calcium silicate hydroxide, $Ca_5Si_6O_{16}(OH)_2$. Crystal system, orthorhombic. Density, 2.38 g/cm^3; refractive index, 1.601. Occurrence: Ireland, USA (California).

River Ten. Trade name of Kawasaki Steel Corporation (Japan) for a series of highly corrosion-resistant and readily weldable high-strength low-alloy (HSLA) structural steels.

rivet steel. A soft, tough low-carbon steel containing approximately 0.1-0.25% carbon. It has good shearing strength and is used for making rivets.

Rivitex. Trade name of Hood Rubber Company (USA) for *bakelite*-type phenolic plastics and laminates.

R-Max. Trademark of Partek Insulations Limited (Canada) for cryogenic wool.

R-Monel. Trade name Huntington Alloys Inc. (USA) for a free-machining alloy containing 66.0-67.0% nickel, 30.0-31.5% copper, 1.2-1.4% iron, 0.9-1.0% manganese, 0.15-0.20% carbon, 0.1-0.2% silicon and 0.04-0.05% sulfur. It has high tensile strength, excellent corrosion resistance over wide range of temperatures and conditions, good weldability, and improved machining characteristics due to the controlled sulfur additions. Used for screw-machine products, fasteners, bolts, screws, precision parts, valve components, regulators, fire extinguishers, water-meter components, and instruments. See also Monel.

road asphalt. See asphalt cement.

road ballast. A natural mixture of small stones and fine gravel that may or may not be bonded with clay. Used in road construction.

road brick. See paving brick.

road-marking paint. See traffic paint.

road materials. Materials used in the construction of roads and highways and including broken stone, stone chips and flour, gravel, sand, cement, concrete, asphalt, bitumen, pitch, tar, wood, steel, polymeric materials, paving bricks and stones, and joint sealing compounds. Also known as *highway materials*.

road mesh. Steel reinforcing mesh that after being embedded in concrete forms part of the base course of a concrete road.

road tar. A bituminous product obtained from coal tar and made suitable for road paving by various treatments. Abbreviation: RT.

roaldite. A synthetic mineral composed of iron nitride, γ'-Fe_4N. It can also be made synthetically. Crystal system, cubic. Density, 7.21 g/cm^3. Occurrence: USA (California).

Robax. Trademark of Schott Glas AG (Germany) for a high-quality protective and diffusing sheet glass.

Roberts. Trade name of JacksonLea (USA) for a greaseless buffing compound used for satin finishing applications.

robertsite. A shiny black mineral of the arseniosiderite group composed of calcium manganese phosphate hydroxide trihydrate, $Ca_3Mn_4(PO_4)_4(OH)_6 \cdot 3H_2O$. Crystal system, monoclinic. Density, 3.17 g/cm^3; refractive index, 1.82. Occurrence: USA (South Dakota).

Robins. Trade name of Robins Engineers & Constructors Inc. (USA) for a series of gray and nickel cast irons.

robinsonite. A black mineral of composed of lead antimony sulfide, $Pb_4Sb_6O_{13}$. It can also be made synthetically, and may contain bismuth. Crystal system, triclinic. Density, 5.20-5.63 g/cm^3. Occurrence: Spain, Canada (British Columbia), USA (Nevada).

roble. See apamate.

Roburit. Trade name of Emmabody Glasverk AB (Sweden) for a toughened sheet glass.

Robust. Trade name of Saarstahl AG (Germany) for a series of tool steels including several hot-work and high-speed grades.

Roc. Trade name of Forjas Alavesas SA (Spain) for a high-carbon, high-chromium air-hardening tool steel containing 1.9-2.1% carbon, 0.4% manganese, up to 0.5% silicon, 12% chromium, 0.7% tungsten, and the balance iron. Used for stamping and thread-rolling dies, sand-blasting nozzles, and finishing tools.

Rocan. Trade name of Revere Copper Products, Inc. (USA) for a corrosion-resistant, high-strength sheet copper used for roofing and conduits.

ROCC. Trade name of PPG Pretreatment & Specialty Products (USA) for a reactive, organic conversion coating.

Roccia. Trade name of Vetreria di Vernante SpA (Italy) for a glass having an indeterminate pattern.

Roc Extra. Trade name of Forjas Alavesas SA (Spain) for a high-carbon, high-chromium air-hardening tool steel containing 1.55-1.65% carbon, 0.4% manganese, 0-0.5% silicon, 12% chromium, 0.6-0.8% molybdenum, 0.5% tungsten, 0.1-0.25% vanadium, and the balance iron. Used for cold-work dies and tools, forming tools, and metal saws.

Röch. Trade name of Röchling Burbach GmbH (Germany) for a series of machinery and tool steels. Also known in English-speaking countries as "Roch" or "Roech."

Rochelle salt. See potassium-sodium tartrate.

Rochet. Trade name of Teijin Fibers Limited (Japan) for polyester filaments used for the manufacture of textile fabrics.

Rochflex. Trademark of Rochevert Inc. (Canada) for polyvinyl chloride and thermoplastic rubber compounds.

Röchling. Trade name of Röchling Burbach GmbH (Germany) for an extensive series of steels including various machinery and case-hardening grades, various wrought and cast corrosion-, wear- and/or heat-resistant grades, and numerous plain-carbon and cold- and hot-work die and tool grades. Also known in English-speaking countries as "Rochling" or "Roechling."

Röchlingstahl. Trade name of Röchling Burbach GmbH (Germany) for an extensive series of steels including various machinery and case-hardening grades, plain-carbon, cold-work and hot-work die and tool grades and a wide range of high-speed tool grades. Also known in English-speaking countries as "Rochlingstahl" or "Roechlingstahl."

Rochrome. Trade name for a liquid crystal tape material used for thermographic testing applications.

rock. A naturally formed aggregate of grains composed of one or several mineral species. Rocks can be generally classed into three major groups according to their origin: (i) igneous rocks (e.g., basalt or granite); (ii) sedimentary rocks (e.g., shale or sandstone), and (iii) metamorphic rocks (e.g., gneiss or schist). Used in civil engineering, building and road construction, and in broken or ground form in the manufacture of various products including asphalt, cement and concrete.

Rockaloy. Trade name of Industries Trading Company (USA) for a series of cemented tungsten carbides with cobalt binders, used for cutting tools.

rock asphalt. Asphalt extracted from a porous limestone or sandstone impregnated or saturated by natural processes, and used as a source of asphalt for paving and construction, and in the manufacture of mastic.

rockbridgeite. A greenish black mineral composed of iron phosphate hydroxide, $Fe_5(PO_4)_3(OH)_5$. Crystal system, orthorhombic. Density, 3.39 g/cm³; refractive index, 1.880. Occurrence: USA (Virginia).

rock cement. A relatively soft and weak, rapid-hardening natural cement that is composed of *argillaceous limestone*, and has the capability of setting under water. Also known as *Roman cement*.

rock cork. See mountain cork.

rock crystal. (1) Blown glassware that has been highly polished, and either handcut or engraved.

(2) See quartz crystal; crystal (1).

rock elm. The tough, strong, dense wood of the elm tree *Ulmus thomasii*. The heartwood is light brown, and the sapwood almost white. *Rock elm* has a fine texture, high hardness, excellent bending qualities and good abrasion resistance. Average weight, 700 kg/m³ (44 lb/ft³). Source: Eastern United States and Canada (Southern Ontario). Used for dock and wharf construction, fenders, belting for ships, rowing boats and other small craft, bent work, e.g., railcar roofs, boxes and barrels. Also known as *cork elm; hickory elm*.

Rocket. Trade name of Lehigh Steel Corporation (USA) for an air- or oil-hardening, wear- and abrasion-resistant tool and die steel.

rock gypsum. A massive variety of *gypsum* with coarsely crystalline to finely granular microstructure.

Rockingham ware. A decorated earthenware or semivitreous ware coated with a brown or mottled manganese glaze.

Rockite. Trade name for a concrete anchoring and patching cement supplied as a powder blend for mixing with water. It dries in 15 minutes and becomes stronger than concrete in 1 hour.

Rockland. Trade name of Rolland Inc. (Canada) for bond papers.

Rocklath. Trademark of Canadian Gypsum Company Limited for *lath*-type plaster bases.

Rocklite. Trademark of Rocklite, Inc. (USA) for lightweight aggregate used in the manufacture of concrete.

rock maple. See sugar maple.

Rockoustile. Trade name for a acoustic building tile composed of exfoliated mica.

Rockrite. (1) Trade name of Tube Reducing Corporation (USA) for tubing composed of chromium steel (AISI-SAE 52100) and used for bearings, bearing races, and liners.

(2) Trade name of F.A. Hughes & Company (USA) for phenol-formaldehyde resins and plastics.

rock salt. See halite.

Rocktex. Trade name for *mineral wool* formed by blowing air or steam through molten rock, and used as thermal insulation.

Rockwall. Trade name of National Gypsum Company (USA) for an acoustic plaster belonging to the *Gold Bond* line of products.

Rockwire. Trade name of Central Glass Company Limited (Japan) for a wired glass with hammered-type pattern.

rock wood. See mountain wood.

Rockwool. Trade name of National Gypsum Company (USA) for a thermal insulating cement belonging to the *Gold Bond* line of products.

rock wool. See mineral wool.

Rocky Mountain juniper. The soft, fragrant wood of the small juniper tree *Juniperus scopularum*. Source: USA (Western states), Canada (southern British Columbia and Alberta). Used for lumber, fence posts, and poles.

Rocogips. Trade name of Rocogips GmbH (Germany) for an extensive series of gypsum building products including natural gypsum (*anhydrite*), adhesive gypsum plaster, building and stucco plaster, casting and molding plaster, and spackling compounds.

Rocoloy. Trade name for a high-strength, corrosion-resistant steel containing 0.4% carbon, 1.35% chromium, 1.35% cobalt, 1.3% silicon, 0.8% nickel, 0.5% molybdenum, 0.3% tungsten, 0.15% vanadium, and the balance iron. Used especially for fasteners.

Roctec. Trade name of Boride Products, Inc. (USA) for extremely hard, homogeneous and void-free composite carbides produced by reacting tungsten carbide, vanadium carbide or molybdenum carbide using a proprietary rapid omnidirectional compaction (ROC) process. Two important grades are: *Roctec 100*, a molybdenum dicarbide–tungsten carbide composition, and *Roctec 500*, a tungsten carbide composition. *Roctec* carbides are used for cutting tools, blasting and spray nozzles, wear parts, bearings, and wire-drawing dies.

rod. A slender, straight bar of metal, plastic or wood, usually with circular cross section.

rodalquilarite. An emerald-green mineral composed of iron hydrogen chloride tellurate, $Fe_2H_3O(TeO_3)_4Cl$. Crystal system, triclinic. Density, 5.10 g/cm³; refractive index, approximately 2.2. Occurrence: Spain.

Rodar. Trademark of Carpenter Technology Corporation (USA) for an iron-base superalloy containing 29% nickel, 17% cobalt and 0.3% manganese. Supplied in the form of bars, wires and strips, it has good heat and thermal-shock resistance, good machinability, weldability and brazeability. Used for metal-to-glass seals.

Rodfor. Trade name of Forjas Alavesas SA (Spain) for oil-hardening tool steels containing 1% carbon, 1.5% chromium, up to 0.4% silicon, 0.3% manganese, and the balance iron. Used for drills, pins, shafts, dies, and gages.

Rodierglas. Trade name of Leonard Rodier Company (USA) for a jewel-fused glass composed of 3 or more sheets with permanent colors and interspersed colored glass chips.

Rodinox. Trade name of Teledyne Rodney Metals (USA) for a series of austenitic, ferritic and martensitic stainless steels used mainly for valves, springs, fasteners, couplings, diaphragms, and bellows.

Rodip. Trade name of McGean Inc. (USA) for chromate conversion coatings and topcoats.

rod-shaped low-molecular-mass liquid crystal. A *liquid crystal* consisting of elongated, anisotropic, rod-shaped organic molecules of low molecular mass that arrange preferentially along a particular axis in space.

Rod's Bod. Trade name of Aardvark Clay & Supplies (USA) for a light- to medium-tan clay (cone 10) with speckles.

Rodseal. Trade name of Teledyne Rodney Metals (USA) for a series of iron-nickel and iron-nickel-cobalt alloys used for glass-to-metal seals, bimetallic strips, and electronic components.

roeblingite. A white mineral composed of calcium lead manganese silicate sulfate hydroxide tetrahydrate, $Ca_6Pb_2Mn(SO_4)_2Si_6O_{18}(OH)_2 \cdot 4H_2O$. Crystal system, monoclinic. Density, 3.43 g/cm³; refractive index, 1.64. Occurrence: Sweden, USA (New Jersey).

roedderite. A colorless mineral of the osumilite group composed of sodium potassium magnesium aluminum silicate, $(Na,K)_2(Mg,Fe)_5(Al,Si)_{12}O_{30}$. Crystal system, hexagonal. Density, 2.60-2.65 g/cm³; refractive index, approximately 1.536. Occurrence: USA (Kansas).

roemerite. A clove to honey brown mineral composed of iron sulfate hydrate, $(Ca,K,Na)Si_{16})_2O_{40} \cdot 11H_2O$. Density, 2.18 g/cm³; refractive index, 1.570; hardness, 3-3.5 Mohs. Occurrence: USA (California, Utah).

roentgenite. See rontgenite.

Roesch. Trade name for an alloy composed of 50% zinc, 49% tin, 0.7% antimony and up to 0.2% copper, and used as an aluminum solder.

roesslerite. A colorless mineral composed of magnesium hydrogen arsenate heptahydrate, $MgHAsO_4 \cdot 7H_2O$. It can also be made synthetically. Crystal system, monoclinic. Density, 1.95 g/cm³; refractive index, 1.507.

Rogard Prime. Trade name of McGean Inc. (USA) for a protective coating used for zinc-plated items.

Rogers. Trademark of Rogers Corporation (USA) for a series of polymeric materials including *Rogers* phenolic molding materials, *Rogers DAP* diallyl phthalate resins, supplied in long and short glass fiber-reinforced, mineral and/or synthetic fiber filled, fire-retardant and other grades, and *Rogers Envex* polyimides, supplied in unmodified, glass fiber-reinforced, and graphite-, molybdenum disulfide- or polytetrafluoroethylene-lubricated grades.

roggianite. A white yellow mineral composed of calcium aluminum silicate hydrate, $Ca_8Al_8Si_{16}O_{40}(OH)_{16} \cdot 13H_2O$. Crystal system, tetragonal. Density, 2.02 g/cm³; refractive index, 1.527. Occurrence: Italy.

Rohacell. Trademark of Rohm GmbH (Germany) for a range of rigid, closed-cell plastic foams based on polymethacrylamide (PMA) and available in the form of blocks, sheets and cast pieces. It has outstanding resistance to most technical solvents, can be thermoformed, machined and bonded by reaction adhesives, and is used as a floating material, and for architectural modeling applications. *Rohacell A* is an autoclavable, thermoformable PMA-based foam with a heat-distortion temperature of 180°C (356°F), used for aerospace applications (e.g., aircraft manufacture, antennas and radomes). *Rohacell HF* is a thermoformable PMA foam with very fine cell structure that provides very low dielectric values, high radiation transmission and outstanding mechanical properties, and is used for radomes, antennas and X-ray equipment. *Rohacell IG* is a thermoformable PMA-based foam used as a core material for structural composites in model building, shipbuilding, automobile construction, sporting goods and medical technology. *Rohacell P* is a PMA foam with a heat-distortion temperature of 130°C (266°F) and good anisotropic properties, used as a core material for high-quality skis. *Rohacell PMI* is a lightweight, stiff, high-strength polymethacrylimide foam with isotropic structure, excellent processibility and good resistance to organic solvents, used as a core material for advanced aerospace sandwich constructions. *Rohacell S* is a self-extinguishing PMA foam used in shipbuilding, and vehicle and aircraft construction. *Rohacell WF* is a thermoformable PMA foam that can be produced by co-curing and processed by autoclave and RTM methods, and is used as a core material for engineering composites for the aircraft, aerospace and missile industries.

Rohn. Trade name for a heat- and scale-resistant alloy composed of 50% nickel, 30% chromium, 17% iron and 3% silicon, and used for electrical resistances.

Rohtal-HK. Trade name of Vulkan Strahlverfahrenstechnik GmbH Co. KG (Germany) for chilled white cast iron shot used for abrasive-blast cleaning applications.

Roica. Trademark of Asahi Chemical Industry Company (Japan) for *spandex* fibers and yarns used for elastic textile fabrics.

Rokide. Trade name of Norton Pakco Industrial Ceramics (USA) for a system of corrosion- and wear-resistant refractory coatings based on aluminum oxide (*Rokide A, MBA, MBAT 97/3, SA* and *UPHA*), zirconium oxide (*Rokide Z* and *EZ*), zirconium silicate (*Rokide ZS*), alumina-titania (*Rokide MBAT 87/13*), chrome oxide (*Rokide C, MBC* and *TC*) or magnesium aluminate (*Rokide MA*). They are produced by melting the end of a rod composed of the refractory coating material and projecting the small, molten particles through a rapidly moving stream of air onto the cool ceramic or metal substrate surface.

rokuknite. A light yellowish brown mineral composed of iron chloride dihydrate, $FeCl_2 \cdot 2H_2O$. It can also be made synthetically. Crystal system, monoclinic. Density, 2.35 g/cm³; refractive index, 1.633. Occurrence: Germany.

Rokusho. A coloring solution commercially produced in Japan for coloring art and jewelry metals and alloys. It produces red-brown colors on copper, purplish to purplish-black colors on *Shaku-do* alloy (95% copper and 5% gold) and a light gray color on *Shibu-ichi* alloy (75% copper and 25% silver). It is not available outside Japan, but a substitute can be prepared by mixing 6 g cupric acetate, 1.5 g copper sulfate and 1.5% table salt with 1 L distilled water.

Rokwal. Trade name of Perma Paving Stone Company (Canada) for rectangular architectural stone made with one surface texture resembling natural stone. It is supplied with a size of 300 × 190 × 125 mm (12 × 7.5 × 5 in.) in several colors ranging from charcoal to brown and reddish brown. Used mainly for retaining walls, straight patio steps, and planters.

Rolan. Trademark of Rolan Company for acrylic fibers used for the manufacture of textile fabrics.

Rolands' cement. General trade name for relatively slow-setting, rapid-hardening aluminate cements. See also aluminate cement.

Roley. Trade name of Robert-Leyer-Pritzkow & Co. (Germany) for an extensive series of steels including various high-speed, plain-carbon and cold-work tool grades, austenitic, ferritic and martensitic stainless grades, and machinery and case-hardening grades.

Roll-A-Glass. Trademark of O'Sullivan Corporation (USA) for transparent, flexible polyvinyl chloride sheeting.

Roll-A-Tex. Trade name of Bondex International Inc. (USA) for an interior/exterior texture paint supplied in fine, medium and coarse grades, and used to produce roll-on decorative effects on walls.

Roll-Bonded Clad. Trade name of Lukens Steel Company (USA) for a composite material consisting of carbon steel clad with a high-nickel alloy, such as *Inconel* or *Inco Alloy C-276* (a nickel-molybdenum-chromium alloy). The cladding process involves hot rolling of the steel and high-nickel alloy to produce a single plate that is metallurgically and integrally bonded across the entire steel-nickel interface. It has excellent corrosion resistance including crevice corrosion, high surface smoothness, and is used for vessels, reservoirs, tanks, containers, and chemical and petrochemical equipment.

Rollbryt. Trade name of Pax Surface Chemicals, Inc. (USA) for a burnishing compound.

rolled compact. A powder-metallurgy compact made into a relatively long sheet, strip or rod by passing the starting powder through a continuous rolling mill. Many metals and alloys including copper, nickel, titanium, brass, bronze and steel are available in this form.

rolled glass. (1) A thick flat glass formed by passing a roller over the molten or plastic glass. A design may also be worked into the surface by a roller with patterned or textured surface.

(2) An optical glass rolled into plates instead of being cooled in a melting pot and then processed.

rolled gold. A relatively inexpensive base metal, such as brass or cupronickel, clad or plated on one or both surfaces with a thin layer of gold alloy (10 karat or more) in which the amount of gold alloy is less 5% of the total weight. Used for jewelry and ornaments. Also known as *rolled plate*. See also gold-filled metal.

rolled materials. See rolled products.

rolled metal. A relatively inexpensive nonprecious base metal clad or plated on one or both surfaces with a thin precious-metal layer, and used for jewelry and ornaments.

rolled plate. See rolled gold.

rolled products. The commercial products of a rolling mill including structural shapes (e.g., angles, beams, channels, piling, pipes and tubes, rails, and tees), wire and wire products, bars (e.g., flats, hexagonals, octagonals, rounds and squares), plates, sheets, strips and coils. Also known as *mill products; rolled materials.*

rolled steel. Any product of a steel rolling mill including structural shapes, bars, rods and wires, pipes and tubes, plates, sheets, strips and coils. Abbreviation: RS. See also cold-rolled steel; hot-rolled steel; rolled products.

roller-compacted concrete. A stiff, no-slump concrete mixture that can be placed and compacted using conventional asphalt pavement rollers. Used for highway and off-highway pavements. Abbreviations: RCC.

Rollisol. Trade name of Saint-Gobain (France) for glass wool-based double-foil building insulation materials.

Rollo. Trade name of British Steel Corporation (UK) for a carbon steel containing 0.9% carbon, 0.2% vanadium, and the balance iron. Used for turning tools.

Rolloy. Trade name of Dresser Industries (USA) for a wear-resistant cast iron containing 2.5-3.5% total carbon, 4.5% nickel, 1.5% chromium, and the balance iron.

roll roofing. Roofing that consists of mineral granules on asphalt saturated felt or fiberglass sheets, and is supplied in the form of rolls. It is usually applied in overlapping strips parallel to the eaves, and can be used as a main roof covering, or as a flashing material.

Roll Steel. Trademark of Atlas Specialty Steels (Canada) for several machinery steels.

Roll-X. Trade name for a beryllium-free, corrosion-resistant nickel-chromium dental bonding alloy.

Rol-Man. Trade name of Manganese Steel Forge Company (USA) for a nonmagnetic, austenitic manganese steel containing 1-1.4% carbon, 11-14% manganese, and the balance iron. It has high strength and elongation, good abrasion resistance, fair to poor machinability, and is used for bushings, wear plates, and woven wire screens.

romaine. A light, lustrous plain-weave fabric made of acetate, rayon, silk, wool or other fibers. It is produced with a low thread count and has an uneven texture.

Roman. Trade name of Permacon (Canada) for paving stone supplied in rectangular and square formats in various colors including beige, brown, charcoal, gray and terracotta. Used for walkways, steps, patios, etc.

Roman brick. A building brick that is $51 \times 102 \times 305$ mm ($2 \times 4 \times 12$ in.) in size, and used chiefly on the exterior or facing of a structure or wall. See also building brick; facing brick.

Roman bronze. A high-strength *naval bronze* containing 39% zinc, 0.75% tin, and the balance copper. It has good hot workability and corrosion resistance, and is used for welding rod, hardware, fasteners, and fittings.

Roman cement. See rock cement.

romanechite. An iron-black mineral composed of barium manganese oxide hydroxide, $BaMn_9O_{16}(OH)_4$. Crystal system, monoclinic. Density, 4.71 g/cm³; refractive index, 1.505. Occurrence: USA (California), Germany.

romanium. A high-aluminum alloy containing 1.75% nickel, 0.25% copper, 0.15% tin and 0.15% tungsten. Used for light-alloy parts.

Roman lime. A relatively impure, highly hydraulic lime.

Roman ocher. A naturally occurring, dark orange *ocher* used as a pigment.

Roman pewter. A dull, soft, ductile alloy composed of approximately 70% tin and 30% lead, and used for ornamental parts, domestic utensils, and dishes.

Romany-Spartan. Trade name of US Ceramic Tile Company (USA) for glazed ceramic floor and wall tile.

romarchite. A black mineral composed of tin oxide, SnO. It can also be made synthetically. Crystal system, tetragonal. Density, 6.40 g/cm³. Occurrence: USA (California).

romeite. A reddish brown mineral of the pyrochlore group composed of calcium antimony oxide fluoride hydroxide, $CaSb_2O_6(F,O,OH)$. Crystal system, cubic. Density, 4.70 g/cm³; refractive index, 1.817; hardness, 5.5-6.5 Mohs. Occurrence: Sweden.

Romilly brass. A corrosion-resistant free-cutting brass containing 70.5% copper, 29% zinc, 0.2% tin and 0.3% lead. Used for hardware and fittings.

Rondalite. Trade name of Kokomo Opalescent Glass Company Inc. (USA) for a rough-rolled sheet glass with bull's-eye pattern.

Rondelite. Trade name of Australian Window Glass Proprietary Limited (Australia) for a prismatic patterned glass.

Rondolit. Trade name of Gerresheimer Glas AG (Germany) for a patterned glass with large and small circles.

Ronensil. Trade name of Saarstahl AG (Germany) for a nonmagnetic, high-manganese steel with good abrasion and corrosion resistance, used for stainless parts, and tableware.

Ronfalin. Trade name of DSM Engineering Plastics (USA) for acrylonitrile-butadiene-styrenes supplied in high- and medium-impact, high-heat, UV-stabilized and fire-retardant grades. They have a density of approximately 1.05 g/cm³ (0.038 lb/in.³), excellent tensile strength, an upper service temperature about 70-100°C (158-212°F), good electrical properties, excellent resistance to mineral oils and dilute alkalies, good resistance to gasoline and dilute acids, poor resistance to trichloroethylene, tetrachlorocarbon and ketones, and poor fatigue, solvent and UV resistance (unless stabilized). Used for automotive components and interior trim, boat hulls, appliance housings, business machines, enclosures, piping, and casings.

Ronilla. Trade name of J.M. Steel & Company (UK) for polystyrene plastics.

Ronoval. Trade name of Shipley Ronal (USA) for gold finishes used for electronic connector applications.

rontgenite. A wax-yellow to brown mineral composed of calcium cerium fluoride carbonate, $Ca_2Ce_3(CO_3)_5F_3$. Density, 4.20 g/cm³; refractive index, 1.662. Occurrence: Greenland. Also known as *roentgenite*.

roof cement. See roofing cement.

Roofchrome. Trademark of Quigley Company, Inc. (USA) for chromia-based gunning cements.

roofing. See roofing materials.

roofing cement. A black mixture of trowelable consistency composed of cutback asphalt stabilized with glass, asbestos or mineral fibers, and used for caulking roofs, and repairing seams, cracks, blisters and separations around chimneys, stacks, vents, flashings, etc. Also known as *roof cement; roofing putty; roof putty; flashing cement*. See also slater's cement.

roofing copper. Soft, hot-rolled electrolytic tough-pitch copper in sheet form used for gutters, flashing and other roofing applications.

roofing felt. A dry felt soaked or saturated with asphalt or coal tar and used under shingles as a sheathing paper, and as a lamination in constructing built-up roofs. It is supplied in rolls of different lengths, widths and weights. Also known as *slater's felt*.

roofing granules. See granules (3).

roofing materials. Materials, such as asphalt, wood, metal and mineral shingles, slate, tile, roll roofing, galvanized iron, aluminum and copper, applied to the structural parts of a roof to protect it against sun, rain, snow, wind and dust. Also known as *roofing*.

roofing putty. See roofing cement.

roofing sand. A white silica sand suitable for roofing applications, e.g., for coating roll roofing, roofing felt, or shingles.

roofing slate. A hard, fine-grained, foliated variety of *slate*, usually of gray or black color, obtained from coal beds and made into sheets about 300 × 150 mm (12 × 6 in.) to 600 × 350 mm (24 × 14 in.) in size, and 3-19 mm (0.125-0.750 in.) in thickness. Used as a roofing material.

roofing tile. A roofing product available in the form of overlapping or interlocking structural units composed of molded, hard-burnt shale, mixtures of shale or clay, or concrete. Clay roofing tile is hard, fairly dense, durable and usually unglazed. *Roof-*

ing tile is supplied in a wide variety of shapes, colors and textures.

Roofinsul. Trademark of Nordfibre Company (Canada) for a lightweight fiberboard used as a roof sheathing material.

roof insulation materials. (1) A term referring to a wide range of different materials used in the acoustical and thermal insulation, and moistureproofing of roof decks and attics. Materials, such as thermally insulating and weather- and moisture-resistant polystyrene foam boards with high heat flow resistance (R-values) are installed to the outside and/or inside of roof decks. Attic insulation materials also have high R-values and include fiberglass batts and blankets, rafter vents, and fiberglass or mineral-wool loosefill. Also included in this term are roofing materials, such as shingles, roll roofing, and roofing tile and concrete.

(2) An insulating material, usually a lightweight concrete, used over structural roofing.

roof insulation board. A lightweight board made from a pulp of wood, cane or other cellulosic fibers, and combining strength with acoustical and thermal insulating properties. It is primarily used as a sheathing material for roofs and as such is often impregnated with asphalt to improve moisture resistance.

Roofloy. Trade name of American Smelting & Refining Company (USA) for sheet lead containing 0.20% tin, 0.02% calcium and 0.01% magnesium. It is usually supplied in the form of rolls and has good strength, high stiffness and rigidity, good creep resistance and good weatherability. Used for roofing applications, e.g., flashing and gutters.

roof putty. See roofing cement.

Roofrap. Trade name of Versil Limited (UK) for a fiberglass insulation material used for domestic roofs.

room-temperature-curing adhesive. An adhesive that sets to handling strength at a temperature between 20 and 30°C (68 and 86°F) within 60 minutes, and thereafter obtains full strength without the application of heat. Also known as *room-temperature setting adhesive*.

room-temperature vulcanisate. A crosslinked polymeric product, usually a silicone or other elastomer, vulcanized (cured) at about room temperature (20-30°C or 68-86°F) by chemical reaction. Used for gaskets, molds, electronic encapsulation, conformal coatings, and in bonding metals and nonmetals. Abbreviation: RTV.

rooseveltite. A white mineral of the monazite group composed of bismuth arsenate, $BiAsO_4$. It can also be made synthetically. Crystal system, monoclinic. Density, 7.01 g/cm³; refractive index, 2.2. Occurrence: Bolivia.

rope. A long, strong flexible line or cord, usually more than 4 mm (0.16 in.) in diameter, consisting of three or more laid, twisted or braided strands of natural or synthetic fibers. See also fiber rope; braided rope; laid rope; twisted rope.

rope belting. Strong, flexible cotton rope belting made by twisting four cotton strands, each with a protective covering consisting of ten spirally twisted cotton yarn cords, around a central core.

rope materials. Materials used in the manufacture of fiber and wire ropes. The most common natural fibers utilized in rope-making are asbestos, cotton, manila and sisal, and the most popular synthetic materials are nylon, polyester and polypropylene. Materials used in the manufacture of wire rope include iron, phosphor bronze, traction steel, plow steel and bridge-rope steel.

roping yarn. A cotton or other yarn used in rope manufacture to form a strand.

roquesite. A gray or slightly bluish mineral of the chalcopyrite group composed of copper indium sulfide, $CuInS_2$. It can also be made synthetically. Crystal system, tetragonal. Density, 4.75 g/cm^3. Occurrence: France.

Rorschach. Trade name of Rorschach Aluminiumwerke (Switzerland) for a series of aluminum-copper-magnesium, aluminum-magnesium, aluminum-magnesium-manganese, aluminum-manganese and aluminum-silicon-magnesium alloys.

rosaniline. A synthetic triphenylmethane dye (Color Index C.I. 42510) available in the form of reddish-brown crystals with a melting point of 186°C (367°F) (decomposes), and a maximum absorption wavelength of 545-550 nm. Used as a dye, as a biological stain, and as a fungicide. Formula: $C_{20}H_{21}N_3O$.

rosasite. A greenish blue mineral composed of copper zinc carbonate hydroxide, $(Cu,Zn)_2(CO_3)(OH)_2$. Density, 4.10 g/cm^3; refractive index, above 1.780. Occurrence: Bulgaria, Italy.

Rosal. Czech trade name for a brownish-pink sun-protection glass.

Rosalin. Trade name of Schott DESAG AG (Germany) for a pink-tinted spectacle glass.

roscherite. (1) A dark to reddish brown mineral composed of calcium beryllium phosphate hydroxide dihydrate, $(Ca,Mn)_3Be_3(PO_4)_3OH_3·2H_2O$. Crystal system, monoclinic. Density, 2.93 g/cm^3; refractive index, 1.641. Occurrence: Brazil, Germany.

(2) A dark brown mineral composed of beryllium calcium iron magnesium aluminum phosphate hydroxide dihydrate, $Be_2Ca(Fe,Mg)_2Al(PO_4)_3(OH)_3·2H_2O$. Crystal system, monoclinic. Density, 2.77 g/cm^3. Occurrence: Brazil.

roscoelite. A colorless, green, red, tan or brown mineral of the mica group with a pearly luster. It is essentially a modified *muscovite* mica composed of potassium aluminum vanadium silicate hydroxide, $KAlV_2Si_3O_{10}(OH)_2$, and contains up to 28% V_2O_3 and 1.5-3.5% vanadium. Crystal system, monoclinic. Density, 2.8-3.0 g/cm^3; hardness, 2.5 Mohs; refractive index, 1.63. Occurrence: Australia, USA (Arizona, California, Colorado, Utah). Used as an ore of vanadium. See also vanadium mica.

Rose. (1) Trade name of Sovirel (France) for a pink-tinted spectacle glass.

(2) Brand name for a high-purity tin (99.95%) containing very small amounts of lead, copper and arsenic.

Rosein. British corrosion-resistant white metal alloy composed of 40% nickel, 30% aluminum, 20% tin and 10% silver, and used for ornaments and jewelry.

Rose Label. Trade name of A. Allan & Son (USA) for a plain-carbon tool steel.

roselle. A relatively strong cellulose fiber obtained from the leaves of a plant (*Hibiscus lobata*) cultivated in temperate climates. It is virtually indistinguishable from commercial *jute* and thus used as a substitute, and in the manufacture of textile fabrics.

roselite. A dark rose-pink mineral composed of calcium cobalt magnesium arsenate dihydrate, $Ca_2(Co,Mg)(AsO_4)_2·2H_2O$. Crystal system, monoclinic. Density, 3.69 g/cm^3; refractive index, 1.704; hardness, 3.5 Mohs. Occurrence: Germany.

rose maple. The fragrant, brownish pink wood of the tree *Cryptocarya ethyroxylon*. It has high hardness and a wavy grain, and is not a true maple. Average weight, 720 kg/m^3 (45 lb/ft^3). Source: Australia. Used for cabinets, furniture and paneling.

rosemary pine. See shortleaf pine.

rosenbuschite. An orange to gray mineral of the seidozerite group composed of sodium calcium iron titanium zirconium fluoride silicate, $(Na,Ca)_3(Fe,Ti,Zr)(SiO_4)_2F$. Crystal system, triclinic.

Density, 3.30 g/cm^3; refractive index, 1.687. Occurrence: Norway.

rosenhahnite. A colorless to buff mineral composed of calcium silicate hydroxide, $Ca_3(Si_3O_8(OH)_2)$. Crystal system, triclinic. Density, 2.89 g/cm^3; refractive index, 1.640. Occurrence: USA (California).

Rosenmuster. Trade name of former Glas- und Spiegel-Manufactur AG (Germany) for a glass with floral pattern suggesting roses.

Rosenstiehl's green. See Cassel green.

rose point. See grospoint (2).

rose porcelain. A hard porcelain with a bright, shining decoration of red or reddish enamel. See also hard paste porcelain.

rose quartz. A pale to deep pink crystalline variety of *quartz* used for ornamental applications.

Rose's alloy. See Rose's metal.

Rose's fusible alloy. See Rose's metal.

Rose's metal. (1) A white-metal alloy containing 50% bismuth, 22% tin and 28% lead. It has a density of 9.85 g/cm^3 (0.356 lb/in.^{3}), a solidus temperature of 96°C (204°F), and a liquidus temperature of 107°C (225°F). Used for fuses, safety plugs, for soldering safety equipment and heat-sensitive devices, and for mold applications. Also known as *Rose's alloy; Rose's fusible alloy*.

(2) A white metal alloy containing 35% bismuth, 35% lead and 30% tin. It has a melting point of 98°C (208°F), and is used for fire extinguishers, and safety plugs. Also known as *Rose's alloy; Rose's fusible alloy*.

rosette graphite. A graphite found in gray cast iron and consisting of flakes arranged in whorls or radiating from the centers of crystallization.

rose vitriol. See bieberite.

rosewood. The very hard, durable, dark-reddish wood from any of several tropical trees (genus *Dalbergia*) of the pea family including Brazilian rosewood (*D. nigra*) and Indian rosewood (*D. latifolia* and *D. sissoo*). See Brazilian rosewood; Indian rosewood; jacaranda.

Rosin. British trade name for an alloy composed of 40% nickel, 30% aluminum, 20% tin and 10% silver, and used for ornamental articles and jewelry.

rosin. A translucent, amber-colored solid resin. *Gum rosin* is obtained in the distillation of crude turpentine from the sap of pine trees. *Wood rosin* is obtained from stumps or dead wood by steam distillation. *Tall oil rosin* is obtained from tall oil (a mixture of rosin acids formed as residues in papermaking by the sulfate pulping process). *Rosin* is hard and friable at room temperature, and soft and sticky at elevated temperatures. It has a density of 1.08 g/cm^3 (0.039 $lb/in.^3$), a melting range of 100-150°C (210-300°F), and a flash point of 187°C (370°F). Used as soldering flux for copper and tin, in paper sizing, as an ingredient in paints, varnishes, plastics and synthetic rubber, and in hot-melt adhesives, mastics, sealants, ester gums and insulating compounds. See also colophony.

rosin-alkyd resin. An alkyd resin whose properties have been modified by the addition of *rosin*. See also alkyd resins.

rosin-core solder. See rosin solder.

rosin ester. A hard, light-colored semisynthetic resin made by heating abietic acid (obtained from rosin, or rosin acids) and a polyhydric alcohol, such as glycerin or pentaerythritol. It has a flash point of 190°C (375°F), and is used in paints, varnishes, cellulosic lacquers, enamels, and adhesives. Also known as *ester gum*.

rosin-extended rubber. A *cold rubber* extended with up to 50% rosin.

rosin fluxes. A group of low-activity soldering fluxes that are based on rosin, and are the least effective fluxes in cleaning off metal oxides or tarnishes. They are classified by their activity as nonactive, mildly active and fully active, and are used for electrical and electronic soldering applications.

rosin solder. A solder in the form of a tube whose interior is filled with an acid-free, noncorrosive flux based on rosin or *colophony*. Also known as *resin solder; rosin-core solder.*

Rosite. (1) Trademark of Allen-Bradley Company (USA) for ceramic products composed of calcium-aluminum silicate mixed with asbestos. They have excellent thermal resistance, high compressive strength, an upper service temperature of 480°C (896°F), good resistance to alkalies, and good electrical insulating properties. Used for molded electrical parts and panels.

(2) Trademark of Rostone Corporation (USA) for thermoset polyester bulk and sheet molding compounds available in glass-reinforced, conductive and other grades. They have good creep and wear resistance, good dimensional stability, and high flexural modulus, stiffness and impact strength. Used for molding high-precision parts, business equipment, and machine parts.

Ross alloy. A British bronze composed of 68% copper and 32% tin, and used for reflectors.

rossite. A colorless mineral composed of calcium vanadium oxide tetrahydrate, $CaV_2O_6 \cdot 4H_2O$. It can be made synthetically. Crystal system, triclinic. Density, 2.43 g/cm^3; refractive index, 1.641. Occurrence: Brazil, USA (Colorado).

Rosslyn. Trade name of American Clad Metals Inc. (USA) for a sheet of copper having a cladding of stainless steel on both sides. It has good thermal conductivity, good corrosion resistance and formability, and is used for cooking utensils, kettles, food-processing equipment, and heat exchangers.

rostite. A mineral composed of aluminum sulfate hydroxide pentahydrate, $Al(SO_4)(OH) \cdot 5H_2O$. Crystal system, orthorhombic. Density, 1.92 g/cm^3; refractive index, 1.460. Occurrence: Iran.

Rostodur. Trade name of Hoffmann Elektrogusstahlwerk (Germany) for corrosion-resistant steels containing 0.15-0.20% carbon, 13% chromium, 1% molybdenum, and the balance iron. Used for chemical-process and oil-refining equipment, cutlery, knives, surgical instruments, valve and pump parts, springs, and bearings.

rotary-cut veneer. Wood veneer formed in a special lathe by rotating a log against the edge of a broad cutting knife. It is cut in a continuous strip or sheet much as paper is unwound from a roll.

rotary kiln block. A special circular brick or curved refractory shape that usually has an outside chord of 9 in. (230 mm) and an inside chord of 6-9 in. (150-230 mm) in radial length, and a thickness of 4 in. (100 mm). Used for lining circular and rotary kilns.

Rotas. Trade name of Worthington Steel & Annealing Company (UK) for a case-hardening carbon steel containing 0.15-0.25% carbon, and the balance iron. Used for case-hardened parts, such as gears, axles and rollers.

rotating electrode powder. A metal powder produced by rotating a consumable electrode at high speeds about its longitudinal axis, while melting one end by an electric arc, plasma arc, electron beam, or laser. The droplets flying off the molten end form solid spherical or near-spherical powder particles of varying size (typically 50-400 μm or 0.002-0.016 μin.). Abbreviation: REP.

rotational castings. Hollow, cast thermoplastic products formed by placing the starting materials, usually in liquid or paste form, in hollow molds, heating and rotating the molds until the molten plastics coat the mold surfaces, and then cooling in the still rotating molds. Also known as *rotocast plastics.*

rotational moldings. Hollow, molded thermoplastic products (e.g., polycarbonates, polyethylenes, polyvinyl chlorides, or nylons 6 and 11) formed by charging the starting materials, usually in the form of dry, finely divided powders, in hollow molds, heating and rotating the mold until the molten plastics fuse to the mold surfaces, and then cooling in the still rotating molds. Also known as *rotomolded plastics.*

Rotelloy. Trade name of Telcon Metals Limited (UK) for a series of soft magnetic alloys composed of varying amounts of iron, cobalt and vanadium. They have high saturation induction, high strength, relative good machinability and workability, and are used for speed generators.

Rotguss. German trade name for *red casting brass* composed of 82-93% copper, 6-18% tin, 1-15% zinc and 0-15% lead. Used for fittings, valves, hardware, worms, tubing, bearings, bushings, liners, sleeves, and pump casings.

Rotisol. Trade name of Vetreria Italiana Balzaretti Modigliani SpA (Italy) for an insulating felt made from special glass fibers, treated with a thermosetting resin, and having tarred paper on one surface and perforated kraft paper on the other.

Roto-Brite. Trade name Roto-Finish Company Inc. (USA) for barrel and vibratory finishing compounds.

rotocast plastics. See rotational castings.

Roto-Forms. Trade name of Roto-Finish Company Inc. (USA) for ceramic and plastic-bonded finishing media.

rotogravure paper. A coated or uncoated paper that meets the requirements of rotogravure printing regarding composition and surface properties. Rotogravure is a printing process employing an engraved copper cylinder.

rotomolded plastics. See rotational moldings.

Roto Nickel. Trade name of RotoMetals, Inc. (USA) for babbitt bearing metals.

Rotorenaluminium. Trade name of Alusuisse (Switzerland) for primary aluminum foundry alloys containing 0.5% iron, 0.4% silicon, 0.02% copper, 0.01% magnesium and 0.07% zinc. Used for machinery and rotors.

rotor-grade titanium. Titanium alloys, such as certain alpha and near-alpha alloys (e.g., Ti-8Al-1Mo-1V) and alpha-beta alloys (e.g., Ti-6Al-4V), suitable for exacting rotating-part applications, such as gas-turbine engine blades, disks and wheels.

Rotosil. Trademark of Heraeus Amersil Inc. (USA) for a series of high-purity, nonporous, opaque *fused quartz* products supplied in the form of tubes, drums, crucibles, bell jars, dishes and pots. They have excellent resistance to thermal shocks, corrosion and electrical influences, high chemical inertness at various temperatures, pressures and pH levels, and low coefficients of thermal expansion, and are used for high-temperature processing of metallic and nonmetallic materials, and for electrical insulators.

Rotothene. Trade name for linear-low-density polyethylene used for molded and rotomolded products.

rotovinyl. Vinyl flooring sheets having surface patterns produced by rotogravure, a printing process employing an engraved copper cylinder.

Rotoxit. British trade name for a corrosion-resistant alloy composed of 96% copper and 4% silicon, and used for bearings, pump and valve parts, marine parts, bells, hardware, fasteners, and fittings.

Rotpunkt. Trade name of SWB Stahlformguss Gesellschaft mbH (Germany) for a series of high-speed steels containing 0.7-0.9% carbon, 4% chromium, 1% vanadium, varying amounts of tungsten and molybdenum, and the balance iron. Used for lathe and planer tools, drills, and various other metal-finishing tools.

rottenstone. See tripoli.

Rotuba. Trade name of Rotuba Plastics (USA) for a series thermoplastics based on cellulose acetate.

Rotung. Trade name of Plansee Metallwerk-Gesellschaft (Austria) for an electrical contact material consisting of tungsten impregnated with copper.

Rotwyla. Trade name for rayon fibers and yarns used for textile fabrics.

roubaultite. A green mineral composed of copper uranyl hydroxide pentahydrate, $Cu_2(UO_2)_3(OH)_{10}\cdot 5H_2O$. Crystal system, triclinic. Density, 4.81 g/cm^3; refractive index, 1.800. Occurrence: Zaire.

rouge. A fine, high-grade, red or reddish powder consisting of hydrated *ferric oxide* with an average particle size of 1-10 μm (39-394 μin.), a density of 4.20-5.25 g/cm^3 (0.150-0.190 lb/in.^{3}), and a Mohs hardness of 5.5-6.5. Used in metal buffing and polishing, especially for producing high polishes on aluminum, brass, gold and silver, and as a polishing agent for glass and jewelry, as a pigment, and in bearing lubricants. Also known as *jeweler's rouge; mineral red; mineral rouge; polishing rouge; red ocher.*

rouge flambé. A decorative glaze containing colloidal copper, that when flowed over pottery products and fired in a reducing atmosphere produces a characteristic red color.

roughcast glass. Flat glass that has one surface textured using a special roller with a patterned or textured face.

roughened finish tile. See rough-finish tile.

rough-finish tile. A tile whose back surface has been roughened before firing by mechanical means, e.g., wire cutting or wire brushing, to promote bonding to substrates, such as mortar or plaster. Also known as *roughened finish tile.*

rough glass. Rolled sheets of glass cut into workable size.

roughleaf dogwood. The hard, heavy wood of the medium-sized dogwood tree *Cornus drummondii*. It has a close grain, and an uniform texture. Source: Central and midwestern United States. Used for weaving shuttles, bobbins, pulleys, jewelers' blocks, skate rollers, tool handles, mallets, and golf-club heads. See also dogwood.

Roughlite. Trade name of Asahi Glass Company Limited (Japan) for a glass with rough surface pattern.

Rough Sawn. Trademark of Canadian Forest Products Limited for a plywood used as siding.

rough-sawn lumber. Unsurfaced and undressed lumber cut to rough size with saws.

Roundel. Trade name of Saint-Gobain (France) for a cast glass with bull's-eye pattern.

roundwood. Timber that is not squared or cut into lumber, but used in the round for logs, poles and pulpwood.

routhierite. A violet-red mineral of the chalcopyrite group composed of thallium mercury arsenic sulfide, $TlHgAsS_3$. Crystal system, tetragonal. Density, 6.81 g/cm^3. Occurrence: France.

Rovana. Trademark of Badische Corporation (USA) for thermoplastic saran (vinylidene chloride) copolymer fibers and filaments with good chemical resistance, good weatherability, and good resistance to mildew and insects. Used for carpets and rugs, draperies, upholstery, clothing, and industrial fabrics.

Rovcloth. Trademark of Fiber Glass Industries Inc. (USA) for fiberglass fabrics used as reinforcements.

Rovel. Trademark of Dow Chemical Company (USA) for a series of weatherable, rubber- and olefin-modified styrene-acrylonitriles used for exterior and interior automotive applications, and sporting and recreational equipment.

Rover. Trade name of Porcher-Soieries SARL (France) for a glass cloth used as a reinforcement, and for electrical and sealing applications. *Rover-Heliover* glass cloth is used for curtains and wall coverings.

Roviex. Trade name of Schott DESAG AG (Germany) for a pinkish-brown spectacle glass.

Rovicella. Trade name for rayon fibers and yarns used for textile fabrics.

Roviglas. Trade name of NV Syncoglas SA (Belgium) for a twistless glass fiber roving cloth used as reinforcement for plastics.

roving. A bundle of continuous, normally untwisted fibers, often wound on a roll and called a "roving package." The bundle usually contains less than 10000 filaments.

roving asbestos. This term usually refers to *chrysotile* asbestos in the form of a strand, but is sometimes used loosely to denote any other mineral or organic fiber in strand form, e.g., cotton.

roving ball. A bundle of continuous, usually untwisted fibers wound on a roll or cardboard tube.

roving cloth. A coarse textile fabric woven from a bundle of continuous, usually untwisted fibers.

Rovmat. Trade name of Fiber Glass Industries Inc. (USA) for glass-fiber reinforcement mats.

Rovtex. Trade name of Fibres de Verre SA (Switzerland) for woven glass-fiber rovings.

roweite. An amber to brownish mineral composed of calcium manganese borate hydroxide, $Ca_2Mn_2B_4O_7(OH)_6$. Crystal system, orthorhombic. Density, 2.93 g/cm^3; refractive index, 1.658. Occurrence: USA (New Jersey).

Rowela. Trade name of Royalin GmbH (Germany) for high-quality nonwovens based on acrylic, cellulose, glass, polyester or polyamide fibers, and used in the plastic, electrical, building, chemical, filtration and sporting-goods industries.

rowlandite. A drab-green mineral of the amorphous group composed of yttrium iron calcium cerium silicate fluoride hydroxide, $(Y,Fe,Ca,Ce)_3(SiO_4)_2(F,OH)$. It may be red in color when altered. Density, 4.39 g/cm^3. Occurrence: USA (Texas).

Roxaprene. Trade name for chlorinated rubber used for coating applications.

Roxo. Trade name of Marsh Brothers & Company Limited (UK) for a silicon cast iron.

Roxol. Trade name of Ato Findley (France) for floor finishing coatings.

Roxul. Trademark of Roxul Company (Canada) for an extensive series of lightweight, noncombustible, water-repellent or -resistant insulation products containing 94-99% mineral fibers, obtained from basalt rock and slag, in 1-6% cured, urea-extended phenol-formaldehyde binder. They are supplied in high-melting grayish green fibrous batts, blankets or boards for commercial, industrial and residential thermal, acoustical and fire protection applications. *Roxul AFB* is a chemically inert batt insulation product with excellent dimensional stability, fire resistance and acoustical damping properties, used in wall partitions. *Roxul RHF* refers to insulation products supplied in the form of boards with high thermal resistance and an upper service temperature of 650°C (1200°F), used as pipe and tank wrap. *Roxul RHT* are rigid insulation products supplied in the form of boards in thicknesses from 25 to 76 mm (1 to 3 in.), and used for the insulation of storage tanks, drying equipment, ovens, petrochemical and power plants, and for commercial construction applications. *Roxul RXL Safe* is a semi-rigid material supplied in batt form with excellent fire performance and acoustical insulation properties, low thermal conductivity, and a melting temperature of 1177°C (2150°F), used for commercial, industrial and residential applications. *Roxul RW* flexible thermal insulation products supplied in the form of blankets with good fire resistance and sound absorption properties, good resistance to fungi, mildew and vermin growth, and an upper service temperature of about 650°C (1200°F), used for the insulation of boilers, furnaces, towers, ovens, drying equipment, and petrochemical and power plants.

Royal. (1) Trade name of Pilkington Brothers Limited (UK) for a glass with a large formal pattern.

(2) Trade name for a precision-ground low-carbon steel containing 0.18% carbon, and the balance iron. It is usually supplied as flat stock in standard lengths of 610 mm (24 in.), ground on all four sides to a satin-smooth finish. A case-hardenable grade is also available.

royal blue. A rich, dark blue ceramic color containing approximately 40-60% cobalt oxide, and used as an overglaze or underglaze. Also known as *mazarine blue*. See also smalt.

Royal Bride. Trade name of Chance Brothers Limited (UK) for decorative table glassware with a pattern of small roses around the edges.

Royalcast. Trade name of Uniroyal Chemical Company, Inc. (USA) for a series of castable polyurethane plastics for general-purpose and engineering applications.

Royal Coat. Trade name of Aervoe-Pacific Company, Inc. (USA) for white aerosol spray paints.

Royal Crest. Trade name of Hurlock Brothers Company, Inc. (USA) for cardboard products.

Royal Crete. Trade name of Royal Oil Company (USA) for clear and colored masonry coatings.

Royalene. Trademark of Uniroyal Chemical Company (USA) for a high-performance ethylene-propylene-diene monomer (EPDM) rubber with good colorability, low-temperature resiliency, flexing characteristics, thermal resistance, weatherability, electrical properties and resistance to ozone and ultraviolet radiation and many chemicals. Used for automotive components, tires, tank cars, hoses, and electrical insulation.

Royalex. Trademark of Royalite Thermoplastics Division of Uniroyal Chemical Company (USA) for polyvinyl chloride/acrylonitrile-butadiene-styrene foam core sheets.

Royalglas. Trade name of Uniglass Industries (USA) for yarn dye glass fabrics.

royal green. See copper acetoarsenite.

Royalite. Trademark of Royalite Thermoplastics Division of Uniroyal Chemical Company (USA) for rigid thermoplastic sheets based on acrylonitrile-butadiene-styrene (ABS) and ABS alloys. They have good fire resistance and impact strength, excellent formability and colorability, and are used for aircraft interior applications, such as galley and lavatory structures, seat trays, armrests, window surrounds, and side walls.

Royal Marine. Trademark of Georgia-Pacific Corporation (USA) for wood and lumber products.

Royal Nickel Genuine. Trademark of Federated Genco Company (Canada) for white-metal bearing alloys.

Royalstat. Trademark of Royalite Thermoplastics Division of Uniroyal Chemical Company (USA) for permanently static-conductive thermoplastic acrylonitrile-butadiene-styrene (ABS) resins and ABS alloy sheets available in various sizes. They have high surface resistivity, impact strength and flexibility, and are used in the manufacture of circuit boards.

Roydazide. Trade name for silicon nitride used for coatings and molded parts. It has good corrosion resistance, high hardness and good high-temperature properties.

Royer catalysts. A group of catalysts composed of polyethyleneimine (PEI) with 1-3 wt% of a noble metal, such as palladium, platinum, rhodium or ruthenium, and supplied as 40-200 mesh powders or 20-40 mesh beads on a silica gel support.

Roylar. Trade name of Royalite Thermoplastics Division of Uniroyal Chemical Company (USA) for urethane-based thermoplastic elastomers.

Roylon. Trade name for nylon fibers and yarns used for textile fabrics.

rozenite. A white mineral of the starkeyite group composed of iron sulfate tetrahydrate, $FeSO_4 \cdot 4H_2O$. It can also be made synthetically. Crystal system, monoclinic. Density, 2.19 g/cm³; refractive index, 1.536. Occurrence: Poland.

RPET. See recycled polyethylene terephthalate.

RRIM composites. See reinforced-reaction-injection molded composites.

R-Rounds. Trademark of Inco Limited (Canada) for electrolytic nickel (99.99+%) supplied in the form of rounds, and used as a nonactivated anode material for general-purpose electroplating processes.

RTM plastics. See resin-transfer-molded plastics.

RTV silicone rubber. A group of room-temperature vulcanizing silicone rubber compounds ranging in consistency from flowable and pourable to thixotropic and pasty. Supplied in adhesive, sealant, encapsulation, construction, glass fiber-filled molded grades, they have good physical and dielectric properties, low compression set, excellent chemical resistance, and good heat resistance. Used as adhesives and sealants, agents for bonding metals, polymers and ceramic substrates to each other, as protective barriers against liquids and certain vapors, for gaskets for automobiles, airplanes, spacecraft and appliances, for molds for plastics and low-melting metals, for flexible molds for electronic applications, as dielectrics, potting and encapsulation compounds, for dielectric gels and oils, conformal coatings, and insulation products.

ruarsite. A mineral of the marcasite group composed of ruthenium arsenide sulfide, RuAsS. Crystal system, monoclinic; density, 7.08 g/cm³; refractive index, 1.641.

Ruba. Trade name of Synthetic Latex Corporation (USA) for synthetic rubber.

Rubadur. Trade name of Ruberoidwerke AG (Germany) for building facing tiles.

Rubatex. Trademark of Rubatex Corporation (USA) for sponge and synthetic rubber supplied in the form of sheets, boards, tubes and mats for weatherstripping, seals, gaskets, vibration dampeners, sound-deadening parts, shock-absorbing parts, insulation for refrigerators, ice boxes, freezers, athletic paddings, and wet suits. Also included under this trademark are adhesives for bonding sponge rubber insulation materials as well as synthetic rubber adhesives for joining asbestos, cork, fiberglass and sheet metals.

rubellite. A pink to red, lithium-containing variety of the mineral *tourmaline* found in Brazil, Madagascar and California (USA). Crystal system, orthorhombic (hexagonal). Density, 3.04 g/cm³; hardness, 7-7.5 Mohs. Used chiefly as a gemstone.

rubber. (1) A natural, synthetic or modified elastomer with a high degree of elasticity, i.e., which can be greatly deformed or stretched and returns substantially to its original shape when the deforming or streching force is removed. See also elastomers; crude rubber; hevea rubber; natural rubber; synthetic rubber; vulcanized rubber.

(2) See rubber fiber.

rubber adhesives. (1) High-molecular-weight natural or synthetic rubber dissolved in suitable evaporative organic solvents, e.g., hydrocarbon or chlorinated hydrocarbons, with or without tack improvers and antioxidants. They usually have high peel strength, low shear strength, and good environmental properties. Used for laying floor covering, and other bonding applications. Also known as *rubber-base adhesives; rubber cements.*

(2) Natural or synthetic virgin or scrap rubber compounds containing tack improvers, fillers, antioxidants, or other additives and coated onto fabric supports to produce pressure-sensitive tapes.

(3) Room-temperature-curing, two-component blends of rubber, a catalyst and a solvent that cure *in situ* and are mixed shortly before use.

rubber-base adhesives. See rubber adhesives.

rubber-base paint. An organic coating in which the film-forming material or vehicle is composed of chlorinated rubber or synthetic rubber latex, and which cures by solvent evaporation. It is suitable for chemical and marine environments and applications requiring water immersion and alkali or acid resistance (e.g., ship bottoms and superstructures, marine equipment, shore and offshore installations, swimming pools, locks, or chemical-processing equipment). They are also used under conditions of high humidity (e.g., in chemical plants, pulp and paper equipment, refrigerators, washers, and air conditioners). Its resistance to solvents, ultraviolet light and heat is usually rather poor.

rubber-bonded cork. A solid rubber to which granules of cork have been added to introduce air into the material. The modulus of elasticity depends on the density of the rubber, the volume fraction and density of the cork (including air bubbles) and the dimensions of the cork granules.

Rubber Calk. Trademark of Products Research & Chemical Corporation (USA) for rubber-base caulking compounds.

rubber cements. See rubber adhesives (1).

rubber clay. A light colored, fine-grained clay, such as kaolin, used as a filler in rubber.

rubber compound. An intimate mixture of a synthetic or natural rubber and one or more compounding ingredients (e.g., pigments, crosslinking agents or ultraviolet stabilizers) present in proportions adequate for forming desired finished products.

rubber-core yarn. An elastic yarn consisting of a core of round or square rubber filaments with wrappings of one or more natural or man-made fiber yarns.

rubber fibers. Man-made fibers whose fiber-forming substance consists of natural or synthetic rubber, and which are used in the manufacture of elastic yarn for clothing, and elastic bands and tapes. Also known as *rubber.*

rubber foam. See sponge rubber.

rubber hydrochloride. The thermoplastic reaction product of natural rubber and hydrochloric acid obtained at low temperatures under pressure. It contains about 30% chlorine, and is available as a white powder or transparent film. The film is very resistant to water, acids and oils, but not resistant to ultraviolet radiation, unless modified with suitable stabilizers. It has good flame resistance and a softening temperature of 110-120°C (230-250°F), and is used for protective coverings, films for food packaging, and protective clothing, e.g., rainwear.

rubberized bitumen. A bituminous material, such as asphalt, pitch or tar, modified with a natural or synthetic rubber, and used for building and construction applications.

rubberized fabrics. Flexible, water-resistant textile fabrics, such as cotton, coated or impregnated with liquid rubber and pressed between rollers. Used for clothing, and industrial applications.

rubber latex. See latex (1).

rubber-modified polymer. a polymer-matrix composite whose toughness has been significantly increased by a discontinuous rubber phase that is more flexible and ductile than the polymer. An example is an acrylic resin modified with styrene-butadiene rubber (SBR).

rubber powder. A fine powder that is composed of raw or unvulcanized rubber particles with an average diameter of 5 mm (0.2 in.) or less, and may or may not contain compounding ingredients. It is made by mechanical reduction of bales or sheets of rubber, e.g., by means of rotary or slab cutters, and used in the manufacture of molded rubber goods, and adhesives. Also known as *granulated rubber; powdered rubber.*

rubber products. A term referring to commercial articles or goods manufactured essentially from rubber, e.g., hose, tires, bands, balls, tapes, electrical insulation, footwear, or mechanical items.

rubber sheeting. A simple cotton fabric heavily coated with cured rubber on one or both sides.

rubber sponge. See sponge rubber.

rubbing brick. A block of bonded abrasive, such as silicon carbide, used for rubbing down castings, dressing and polishing marble and granite, smoothing concrete, or scouring chilled-iron rolls. Also known as *rub brick.*

rubbing stone. (1) Any stone suitable for use in removing surface imperfections by smoothing, sharpening, or rubbing.

(2) Any fine-grained abrasive stone suitable for removing imperfections from vitreous-enameled and glazed surfaces by rubbing.

rubbing varnish. A hard-drying varnish that can be rubbed to a suitable abrasive, mixed with oil or water, and applied to a uniform leveled surface.

rubble. Rough, irregular stones broken from larger masses of rock, either naturally by geological action, or artificially in quarrying, cutting, or blasting. Used for coarse masonry, as a filler between facing courses of house walls, and as a roadbuilding and concrete aggregate.

rubble aggregate. Aggregate consisting of rough, broken stones or bricks from collapsed or demolished buildings. The fragments are usually 150 mm (6 in.) or more in size, but do not weigh more than 45 kg (100 lb). Used in roadbuilding and concrete manufacture.

rubble concrete. (1) Concrete made with rubble aggregate. See rubble aggregate.

(2) A mass concrete of the cylopean type, but with aggregate consisting of smaller stones. See also cyclopean concrete.

rubble masonry. A large block of irregular or broken stone placed and arranged roughly in courses and used in the construction of foundations, for coarse masonrywork, or as a backing material.

Rubbon. Trade name of Chemetron Corporation (USA) for a lead-free, self-fluxing aluminum solder with a working range of 374-382°C (705-720°F).

rub brick. See rubbing brick.

Rubcofil. Trade name of Kloeckner-Schott Glasfaser GmbH (Germany) for glass-fiber impregnated fabrics.

Rubel Bronze. Trade name of Allgemeine Deutsche Metallwerk GmbH (Germany) for a series of strong, corrosion-resistant bronzes composed of 40-55% copper, 22-54.5% zinc, 3-15% nickel, 1-3% manganese, 1-2% iron and 0.5-3% aluminum, and used for propellers, rotor covers, centrifugal drums, bushings, bearings, steam and safety jackets, piston and rings, gears, and valve parts.

Rubelit. Trade name of Allgemeine Deutsche Metallwerk GmbH (Germany) for a corrosion-resistant alloy composed of 40-55% copper, 22-54.5% zinc, 3-15% nickel, 1-3% manganese, 1-2% iron and 0.5-3% aluminum, and used for bushings, bearings, valve guides, and piston bolts.

Rubel metal. A tough, ductile alloy composed of 50-60% copper, 35-40% zinc and a total of 4-5% iron and manganese, and used for fittings and hardware.

rubene. See naphthacene.

Rubens brown. See Vandyke brown.

Ruberstein. Trade name of Ruberoidwerke AG (Germany) for building wall and facing tiles.

Rubide. Trade name of Firestone Tire & Rubber Company (USA) for rubber hydrochloride.

rubidium. A rare, soft, silvery-white, metallic element of Group IA (Group 1) of the Periodic Table. It is commercially available in the form of ingots, rods, and lumps usually packaged under argon or kerosene. The principal source is the mineral *carnallite*. Crystal structure, cubic. Crystal structure, body-centered cubic. Density, 1.532 g/cm^3; melting point, 38.89°C (102°F); boiling point, 686°C (1267°F); atomic number, 37; atomic weight, 85.468; monovalent, divalent, trivalent, tetravalent. It is weakly radioactive, and has high heat capacity, a high heat transfer coefficient, high chemical reactivity, and an electrical conductivity of about 14% IACS. Used as a scavenger in vacuum tubes, in photocells, as a heat transfer medium, as a catalyst or catalyst promoter, and the manufacture of ceramics and glasses. Symbol: Rb.

rubidium-87. A radioactive isotope of rubidium with a mass number of 87 and a half-life of 6.0×10^{10} years. It decays by negative beta-particle (negatron) emission, and is used in nuclear magnetic resonance (NMR) spectroscopy. Symbol: ^{87}Rb.

rubidium acetate. The rubidium salt of acetic acid available in the form of white, hygroscopic crystals (98+% pure) with a melting point 246°C (474°F). Formula: $RbC_2H_3O_2$.

rubidium acid phthalate. The rubidium salt of phthalic acid available in the form of single crystals with high plasticity and fissionability. Used as an analyzing crystal in wavelength-dispersive X-ray spectroscopy. Formula: $RbC_8H_5O_4$. Abbreviation: RAP; RbAP. Also known as *rubidium biphthalate; rubidium hydrogen phthalate.*

rubidium biphthalate. See rubidium acid phthalate.

rubidium bromide. White, hygroscopic crystals (99+% pure). Crystal system, cubic. Density, 3.35 g/cm^3; melting point, 682°C (1260°F); boiling point, 1340°C (2444°F); refractive index, 1.553. Used as a source of rubidium metal. Formula: RbBr.

rubidium carbonate. Colorless or white, extremely hygroscopic crystals or powder (99+% pure). Melting point, 837°C (1539°F); dissociates above 900°C (1650°F). Used for special glass formulations. Formula: Rb_2CO_3.

rubidium chloride. White, hygroscopic crystals or lustrous, crystalline powder (99+% pure). Crystal system, cubic. Density, 2.76-2.80 g/cm^3; melting point, 715°C (1319°F); boiling point,

1390°C (2534°F); refractive index, 1.493. Used as a source of rubidium metal, and in ceramics and materials research. Formula: RbCl.

rubidium chromate. Yellow, orthorhombic crystals (99+% pure) with a density of 3.518 g/cm^3. Formula: Rb_2CrO_4.

rubidium disulfide. See rubidium sulfide (ii).

rubidium fluoride. White, hygroscopic crystals or powder (99+% pure). Crystal system, cubic. Density, 3.2-3.56 g/cm^3; melting point, 775°C (1427°F); boiling point, 1410°C (2570°F); refractive index, 1.398. Used for glass formulations, in the preparation of rubidium metal and rubidium compounds. Formula: RbF.

rubidium-germanium fluoride. See rubidium hexafluorogermanate.

rubidium halide. A compound of rubidium and a halogen, such as bromine, chlorine, fluorine, or iodine.

rubidium hexafluorogermanate. A compound of rubidium fluoride and germanium fluoride available as a white crystalline powder with a melting point of 696°C (1284°F) that is readily soluble in hot water. Used in electronics and materials research. Formula: Rb_2GeF_6. Also known as *rubidium-germanium fluoride.*

rubidium hexasulfide. See rubidium sulfide (v).

rubidium hydrate. See rubidium hydroxide.

rubidium hydrogen phthalate. See rubidium acid phthalate.

rubidium hydroxide. A fused solid supplied in the form of a white, hygroscopic powder. Density, 3.203 g/cm^3; melting point, 300-302°C (572-576°F); attacks glass at room temperature. It is also available as a corrosive, colorless 50 wt% solution in water (99+% pure) with a density of 1.740 g/cm^3. Used as an electrolyte in low-temperature storage batteries. Formula: RbOH. Also known as *rubidium hydrate.*

rubidium iodide. White, hygroscopic, light-sensitive crystals (99+% pure). Crystal system, cubic. Density, 3.55 g/cm^3; melting point, 642°C (1188°F); boiling point, 1300°C (2372°F); refractive index, 1.6474. Used as a source of rubidium metal. Formula: RbI.

rubidium monosulfide. See rubidium sulfide (i).

rubidium monoxide. Yellow-brown, hygroscopic crystals or pieces of varying size. Crystal system, cubic. Density, 3.7-4.0 g/cm^3; decomposes at 400°C (750°F). Used for chemistry and materials research. Formula: Rb_2O.

rubidium nitrate. White crystals (99+% pure) with a density of 3.11 g/cm^3 and a refractive index of 1.52. Formula: $RbNO_3$.

rubidium oxide. See rubidium monoxide; rubidium peroxide; rubidium superoxide; rubidium tetroxide; rubidium trioxide.

rubidium pentasulfide. See rubidium sulfide (iv).

rubidium peroxide. White, orthorhombic crystals. Density, 3.8 g/cm^3; decomposition temperature, 600°C (1112°F). Used in chemistry and materials research. Formula: Rb_2O_2.

rubidium selenide. A compound of rubidium and selenium available as white, cubic crystals. Density, 3.22 g/cm^3; melting point, 733°C (1351°F). Formula: Rb_2Se.

rubidium sulfate. White crystals (99+% pure). Density, 3.613 g/cm^3; melting point, 1050°C (1922°F); boiling point, approximately 1700°C (3090°F); refractive index, 1.5130. Formula: Rb_2SO_4.

rubidium sulfide. Any of the following compounds of rubidium and sulfur used in electronics and materials research: (i) *Rubidium monosulfide.* White crystals or powder. Density, 2.91 g/cm^3; melting point, 425°C (797°F). Formula: Rb_2S; (ii) *Rubidium disulfide.* Dark red crystals. Melting point, 420°C (788°F). Formula: Rb_2S_2; (iii) *Rubidium trisulfide.* Red-yellow crystals.

Melting point, 213°C (415°F). Formula: Rb_2S_3; (iv) *Rubidium pentasulfide*. Red, orthorhombic, deliquescent crystals. Density, 2.61 g/cm³; melting point, 225°C (437°F). Formula: Rb_2S_5; and (v) *Rubidium hexasulfide*. Brownish red crystals. Melting point, 201°C (393°F). Formula: Rb_2S_6.

rubidium superoxide. Off-white, tetragonal crystals. Density, 3.0 g/cm³; melting point, 412°C (774°F). Used for chemistry and ma-terials research. Formula: RbO_2.

rubidium tetroxide. Yellow crystals with a density of 3.1 g/cm³ and a melting point of 280°C (536°F). Formula: Rb_2O_4.

rubidium trioxide. Black crystals with a density of 3.53 g/cm³ and a melting point below 500°C (930°F). Formula: Rb_2O_3.

rubidium trisulfide. See rubidium sulfide (iii).

Rubinal. Trademark of Didier Refractories Corporation (USA) for refractory bricks.

Rubprene. Trade name of Firestone Tire & Rubber Company (USA) for neoprene rubber products.

Rub-Tex. Trade name of Richardson Company (USA) for *hard rubber* products.

ruby. A deep red, transparent variety of the mineral *corundum* (Al_2O_3) whose color is due to the presence of small amounts of chromic oxide. Synthetic ruby can be made in the form of single-crystal sheets or rods from aluminum oxide by adding small amounts of chromic oxide. Crystal system, hexagonal. Density, 4 g/cm³; hardness, 9 Mohs. Occurrence: Burma, Cambodia, East Africa, India, Sri Lanka, Thailand, USA (Montana, North Carolina). Natural ruby is used as a gemstone and abrasive, and synthetic ruby as an abrasive and, in the form of high-purity crystals, in masers and lasers.

ruby alumina. An abrasive grade of the mineral *corundum* having a ruby-red color due to the presence of small amounts of chromic oxide. See also ruby.

ruby arsenic. See arsenic sulfide (i).

ruby copper ore. See cuprite.

ruby glass. A deep-red glass produced by adding finely divided selenium or cadmium sulfide, copper oxide, or gold chloride to the melt. Used for decorative applications. See also copper ruby; gold ruby; selenium ruby.

ruby mica. See Indian mica.

ruby silver. See proustite; pyrargyrite.

ruby spinel. A clear-red variety of *magnesium spinel* ($MgAl_2O_4$) that resembles true *spinel* in color only, and is chiefly used as a gemstone.

ruby zinc. A term referring to deep-red, transparent varieties of *sphalerite* (ZnS) and *zincite* (ZnO).

ruche. A gauze- or lace-type fabric with a surface pattern of crimps or flutes. Used especially for trimmings.

rucklidgeite. A silver-white mineral of the tetradymite group composed of bismuth lead telluride, $(Bi,Pb)_3Te_4$. Crystal system, rhombohedral (hexagonal). Density, 7.74 g/cm³. Occurrence: Russia, Armenia.

Ruda-Glas. Trade name of AB Ruda Nya Glasindustri (Sweden) for a glass used for molded lenses, glassware, and technical applications.

Ruflux. Trademark of National Lead Company (USA) for a series of metallurgical and welding fluxes based on rutile (TiO_2), zircon ($ZrSiO_4$), zirconia (ZrO_2), potassium titanate ($K2TiO_3$) or sodium titanate ($Na_2Ti_3O_7$).

rugby ball. See buckminsterfullerene.

ruizite. An orange to brown mineral composed of calcium manganese silicate hydroxide dihydrate, $CaMn(SiO_3)_2(OH)\cdot2H_2O$. Crystal system, monoclinic. Density, 2.90 g/cm³; refractive in-dex, 1.715. Occurrence: USA (Arizona).

Rulli. Trade name of Vetreria Milanese Luccini Perego (Italy) for a glass with bull's eye pattern.

Rulon. Trademark of Furon Company (USA) for a series of engineering thermoplastics based on polytetrafluoroethylene (PTFE). *Rulon II* is an injection-moldable PTFE material with self-lubricating properties, a low coefficient of friction, good wear, corrosion and elevated-temperature resistance, good dielectric properties, and good resistance to most acids, bases, hydrocarbons, hot water, etc. Used for bearings, microwave ovens, office equipment, clothes dryers, and linear, oscillatory and rotary motion applications, e.g. in machine components. *Rulon J* is a reinforced PTFE compound with self-lubricating properties and a very low coefficient of friction, which finds application in bearings, e.g., for start-stop applications where stick-slip must be eliminated. *Rulon M 2000* is an injection-moldable PTFE with high strength and rigidity, inherent lubricity, and a wide range of operating temperatures. It is employed for bearings, thrust washers, valve seats, wear rings, compressor vanes, piston rings, and seals.

Rune. Trade name of Uddeholm Corporation (USA) for a water-hardening steel containing 1.4% carbon, 0.45% chromium, and the balance iron. Used for razors.

Ruolz silver. A corrosion-resistant nickel silver composed of 50-60% copper, 15-25% nickel, 5-15% zinc, and about 20% silver, used for ornamental parts and silverware.

rupee silver. A corrosion-resistant wrought silver alloy composed of 91.6% silver and 8.4% copper, and used as coinage in India, Pakistan, Mauritius, Sri Lanka and Nepal.

rusakovite. A yellow or orange mineral composed of iron vanadium phosphorus oxide hydroxide trihydrate, $(Fe,Al)_5(VO_4,PO_4)_2(OH)_9\cdot3H_2O$. Density, 2.73 g/cm³; refractive index, 1.833. Occurrence: Russian Federation.

Ruscar. Trade name of Charles Carr Limited (UK) for a babbitt composed of varying amounts of lead, tin and antimony, and used for antifriction bearings.

Russell Cast Stone. Trade name of W.N. Russell & Company (USA) for cast stone used for architectural applications.

russellite. A yellow to green mineral composed of bismuth tungsten oxide, Bi_2WO_6. Crystal system, tetragonal. Density, 7.44 g/cm³. Occurrence: Western Australia, UK.

Russia. See Russian leather (1).

Russian leather. (1) Fine, smooth leather, commonly dyed red and obtained from calfskins by treating with birch oil. Used for bookbinding and shoes. Also known as *Russia*.

(2) A fancy leather made from calfskins.

Russian River. Trade name of Aardvark Clay & Supplies (USA) for an orange-brown clay with smooth texture.

Rust-Ban. Trademark of Exxon Corporation (USA) for an extensive series of inorganic and organic coatings used for the protection of steel structures in highly corrosive environments, e.g., industrial, marine, shore and offshore equipment, ships, boats and other marine vessels, chemical plants, and pulp and paper mills. The inorganic coatings are based on zinc or zinc alloys, and the organic coatings on resins, such as alkyds, epoxies, phenolics, silicones or vinyls.

rustenburgite. A light cream mineral composed of palladium platinum tin, $(Pt,Pd)_3Sn$. Crystal system, cubic. Density, 15.08 g/cm³. Occurrence: South Africa.

Rustic. Trade name of Permacon (Canada) for paving slabs with rustic surface textures supplied in square units, 295 × 295 mm (11.75 × 11.75 in.), and rectangular units, 295 × 595 mm (11.75

× 23.5 in.) in several colors including gray, charcoal, beige, and red. Used for walkways, back yards, and patios.

Rustic Stone. Trademark of Canadian Forest Products Limited for panelwood.

Rustic Wall. Trade name of Perma Paving Stone Company (Canada) for architectural stone with rough surface texture supplied as regular, tapered, corner and coping units in natural, sandstone and walnut colors.

rust-inhibiting paint. See antirust paint.

rust-inhibitive washes. Special solutions used to etch iron or steel thereby forming a dull gray rust-inhibitive coating of uniformly fine texture receptive to priming coats.

rust inhibitor. An organic or inorganic substance, such as a phosphate or chromate, that dissolves in a corroding medium, but is capable of forming a protective layer at either the anodic or cathodic areas, and thus reduces the rate of a chemical or electro-chemical reaction.

Rustique. Trademark of Graybec Inc. (Canada) for rough-textured building bricks.

Rus-Tique Brik. Trade name of Graybec Inc. (Canada) for building bricks.

Rustless. Trade name of Armco Steel Corporation (USA) for a series of austenitic, ferritic and martensitic steels including various free-machining and heat-resisting grades used for valves, fasteners, shafts, turbine blades, dental and surgical instruments, cutlery, automotive exhaust systems, furnace equipment, automotive trim, and fittings.

Rust-Oleum. Trade name of Rust-Oleum Corporation (USA) for rust preventive and protective paints and coatings with excellent weather, fade and chip resistance, used on metal, wood and concrete surfaces.

Rustop. Trade name of BSK, Division of DS Smith (UK) Limited for a series of vapor corrosion inhibitor (VCI) papers supplied in standard VCI coating grades and in specialty grades with additional strong, waterproof layers. Used for packaging ferrous metal products.

rust preventives. Temporary coatings, usually blends of petroleum, petrolatum or oils with selected inhibitors and additives of varying properties and effectiveness, applied to the surfaces of iron and steel parts to protect them against corrosion during manufacture, storage and use. Also known as *rust-preventive compounds.*

rustproof material. A material that resists rust either owing to its composition (e.g., a ceramic, plastic, or nonferrous metal) or a protective, rust-resisting surface coating produced by electroplating, galvanizing, aluminizing, or other techniques.

Rust Shield. Trademark of Armor Paint Limited (Canada) for masonry paint.

rustumite. A colorless mineral composed of calcium chloride silicate hydroxide, $Ca_{10}(Si_2O_7)_2(SiO_4)Cl_2(OH)_2$. Crystal system, monoclinic. Density, 2.85 g/cm³. Occurrence: Scotland, Afghanistan.

RuTech. Trademark of Lucent Technologies (USA) for a ruthenium electrodeposit and plating process.

ruthenarsenite. A mineral composed of ruthenium arsenide, (Ru,-Ni)As. Crystal system, orthorhombic. Density, 9.81 g/cm³. Occurrence: Papua and New Guinea.

ruthenates. A class of layered perovskite-type compounds that contain atomic planes of ruthenium and oxygen, or ruthenium, copper and oxygen with strontium or strontium and rare-earth metal ions (Gd, Eu, Sm) located between them. They exhibit superconductivity transition temperatures ranging from as low

as 1.5K for strontium ruthenium oxide (Sr_2RuO_4) to as high as 58K for the ruthenium-cuprate superconducting magnet rubidium strontium rare-earth copper oxide ($CuSr_2(Gd,Eu,Sm)Cu_2O_8$. See also perovskites.

Ruthenex. Trade name of Enthone-OMI Inc. (USA) for a ruthenium electrodeposit and plating process.

ruthenic bromide. See ruthenium bromide.

ruthenic chloride. See ruthenium chloride.

ruthenic chloride hydrate. See ruthenium chloride hydrate.

ruthenic fluoride. See ruthenium fluoride.

ruthenic iodide. See ruthenium iodide.

ruthenium. A brittle, grayish white metallic element Group VIII (Group 8) of the Periodic Table. It is commercially available in the form of rods, disks, microfoils, sponge, powder and single crystals. The single crystals are usually grown by the float-zone or the radio-frequency technique. Ruthenium occurs in platinum ores (e.g., iridosmine). Crystal system (α-ruthenium below 1035°C or 1895°F), hexagonal. Crystal structure (α-ruthenium), hexagonal close-packed. Several other allotropic crystal forms of ruthenium have also been reported including: (i) *Beta ruthenium* that is stable between 1035 and 1200°C (1895 and 2192°F); (ii) *Gamma ruthenium* that is stable between 1200 and 1500°C (2192 and 2732°F); and (iii) *Delta ruthenium* that is stable between 1500 and 2310°C (2732 and 4190°F). Other important properties of ruthenium include: Density, 12.2 g/cm³; melting point, 2310°C (4190°F); boiling point, 3900°C (7052°F); hardness, 350-750 Vickers; atomic number, 44; atomic weight, 101.07; superconductivity critical temperature, 0.49K; zerovalent, monovalent, divalent, trivalent, tetravalent, pentavalent, hexavalent, heptavalent, octavalent. It is both a noble and precious metal, and has relatively good electrical conductivity (about 22.5% IACS), poor workability at any temperature, good corrosion resistance, good resistance to water, acids (including aqua regia) and alcohol, and poor resistance to fused alkalies. *Ruthenium* is used as a hardener for platinum and palladium alloys, in corrosion-resistant alloys, high-temperature supercon-ductors, electrical contacts, aircraft magneto contacts, solar cells, printed circuits, thickness gages, medical instruments, electroplating, thick-film resistor pastes, catalysts, fountain-pen points, jewelry, and ruthenium/titanium oxide-coated electrodes. Symbol: Ru.

Ruthenium V. Trade name of Technic Inc. (USA) for a mirror-bright ruthenium electrodeposit for decorative and industrial applications.

ruthenium boride. Any of the following compounds of ruthenium and boron used in ceramics and materials research: (i) *Ruthenium diboride.* Density, 7.6-10.1 g/cm³. Formula: RuB_2; (ii) *Ruthenium pentaboride.* Density, 9.2 g/cm³. Formula: Ru_2B_5; and (iii) *Ruthenium triboride.* Formula: Ru_7B_3.

ruthenium black. A black, finely divided form of metallic ruthenium (99.9+% pure) with a typical apparent density of 9-10 g/cm³ (0.32-0.36 lb/in.³), used chiefly as a catalyst.

ruthenium bromide. Brown crystals or blackish, crystalline powder. Crystal system, hexagonal. Density, 5.30 g/cm³; melting point, decomposes above 400°C (750°F). Used as a reagent. Formula: $RuBr_3$. Also known as *ruthenic bromide; ruthenium tribromide.*

ruthenium carbonyl. See triruthenium dodecacarbonyl.

ruthenium chloride. Brown crystals or black, hygroscopic powder. Crystal system, hexagonal. Density, 3.11 g/cm³; decomposes above 500°C (930°F). Used as a reagent. Formula: $RuCl_3$. Also known as *ruthenic chloride; ruthenium trichloride.*

ruthenium chloride hydrate. A black, hygroscopic powder with a density of 1.48 g/cm³. Used as a reagent. Formula: $RuCl_3 \cdot xH_2O$. Also known as *ruthenic chloride hydrate.*

ruthenium diboride. See ruthenium boride (i).

ruthenium dioxide. Dark black, tetragonal crystals or black, hygroscopic powder (99.9% pure) with a density of 6.97-7.05 g/cm³. It is also available in hydrated form ($RuO_2 \cdot xH_2O$) as a black powder with high surface area. Used for thick film resistor pastes, ruthenium/titanium oxide coatings for dimensionally stable titanium anodes for the electrolysis of chlorine or caustic soda, in vacuum evaporation, and in ceramics and materials research. Formula: RuO_2. Also known as *ruthenium oxide.*

ruthenium fluoride. See ruthenium trifluoride; ruthenium pentafluoride; ruthenium hexafluoride.

ruthenium halide. A compound of ruthenium and a halogen, such as bromine, chlorine or iodine.

ruthenium hexafluoride. Dark brown crystals. Crystal system, orthorhombic. Density, 3.54 g/cm³; melting point, 54°C (129°F). Used in chemistry and materials research. Formula: RuF_6. Also known as *ruthenium fluoride.*

ruthenium iodide. Black, hexagonal crystals. Density, 6.0 g/cm³; melting point, 590°C (1094°F) (decomposes). Used as a chemical reagent. Formula: RuI_3. Also known as *ruthenic iodide; ruthenium triiodide.*

ruthenium monosilicide. See ruthenium silicide (i).

ruthenium oxide. See ruthenium dioxide; ruthenium tetroxide.

ruthenium pentaboride. See ruthenium boride (ii).

ruthenium pentafluoride. Green crystals or dark green powder. Crystal system, monoclinic. Density, 3.9 g/cm³; melting point, 86.5°C (188°F); boiling point, 227°C (441°F). Used as a chemical reagent. Formula: RuF_5. Also known as *ruthenium fluoride.*

ruthenium-polyethylenimine. See Royer catalysts.

ruthenium sesquisilicide. See ruthenium silicide (ii).

ruthenium silicide. Any of the following compounds of ruthenium and silicon used in ceramics, materials research and electronics: (i) *Ruthenium monosilicide.* Metallic prisms. Formula: RuSi; and (ii) *Ruthenium sesquisilicide.* Formula: Ru_2Si_3.

ruthenium sulfide. Gray-black, cubic crystals with a density of 6.0-6.9 g/cm³. It occurs in nature as the mineral *laurite.* Used in ceramics and materials research. Formula: RuS_2.

ruthenium tetroxide. Yellow, monoclinic crystals. Density, 3.29 g/cm³; melting point, 25.5°C (78°F); boiling point, approximately 40°C (104°F) (decomposes). It is also available as a yellow, stabilized liquid in the form of a 0.5 wt% solution in water. Used as a strong oxidizer for biological and polymeric materials. Formula: RuO_4. Also known as *ruthenium oxide.*

ruthenium triboride. See ruthenium boride (iii).

ruthenium tribromide. See ruthenium bromide.

ruthenium trichloride. See ruthenium chloride.

ruthenium trifluoride. Dark brown crystals. Crystal system, rhombohedral. Density, 5.36 g/cm³; melting point, decomposes above 600°C (1110°F). Used in chemistry and materials research. Formula: RuF_3. Also known as *ruthenic fluoride; ruthenium fluoride.*

ruthenium triiodide. See ruthenium iodide.

ruthenocene. An organometallic coordination compound (molecular sandwich) consisting of a ruthenium atom (Ru^{2+}) and two five-membered cyclopenediene (C_5H_5) rings, one located above and the other below the ruthenium atom plane. *Ruthenocene* is supplied in the form of light yellow crystals (97% pure) with a melting point of 194-201°C (381-394°F) and used in organo-metallic research, in biochemistry, and in medicine as a tumor affinitive compound. Formula: $(C_5H_5)_2Ru^{2+}$. Also known as *dicyclopentadienylruthenium; bis(cyclopentadienyl)ruthenium.*

rutherfordine. A white to pale yellow, radioactive mineral composed of uranyl carbonate, UO_2CO_3. Crystal system, orthorhombic. Density, 4.80 g/cm³; refractive index, 1.730. Occurrence: East and Central Africa.

rutherfordium. A very unstable, artificial radioactive transuranium element with an atomic number of 104, a mass number of 261 (longest-lived isotope ^{261}Rf or ^{261}Ku), and a half-life of approximately 65 seconds (^{261}Rf). The name "kurchatovium" was suggested by researchers in the Russian Federation and the name "rutherfordium" by scientists in the United States. Symbols: Ku; Rf; Unq. Also known as *element 104; kurchatovium; unnilquadium.*

rutile. A reddish-brown, dark red or black mineral with a light brown streak and a submetallic to adamantine luster which can also be made synthetically, e.g., by the Verneuil crystal growing technique. It is one of the three polymorphic forms of titanium dioxide (TiO_2) the other two being *anatase* and *brookite.* Rutile contains theoretically 60% titanium, and may also have varying amounts of iron, niobium and tantalum. Crystal system, tetragonal. Density, 4.18-4.25 g/cm³; melting point, 1640°C (2984°F); hardness, 6-7 Mohs; refractive index, 2.65; dielectric constant, 86-200; transmission wavelength range, 450-500 nm. Occurrence: Australia, Brazil, India, USA (Arkansas, Florida, Massachusetts, Virginia). Used as an important ore of titanium, as a source of titanium compounds, in ceramics, as a steel deoxidizer, in welding-rod coatings, as an opacifier and colorant in porcelain enamels, glazes and tile, in the manufacture of tan-colored glass, as a paint pigment, as a component in titanate dielectrics, and as a synthetic crystal for fiber-optic prisms and substrates.

Ruvea. Trade name for nylon fibers and yarns used for textile fabrics.

Rx Bond. Trade name for a dual-cure multi-purpose dental bonding cement.

Rx Imperial 2. Trade name for a high-gold dental bonding alloy.

Rx Naturelle. Trade name of Jeneric/Pentron Corporation (USA) for a silver-free, palladium-base dental bonding alloy.

Ryanite. Trademark of Allyne-Ryan Foundry Company (USA) for a nickel cast iron containing 3.2-3.4% carbon, 2.4% silicon, 1.5% nickel, 0.5-4% chromium, and the balance iron. It has good hardness and strength, and is used for cylinder blocks, brake drums, crankcases, cylinders, and dies.

Rycase. Trade name of Joseph T. Ryerson & Son Inc. (USA) for a free-cutting carburizing steel (AISI-SAE 1117) containing 0.14-0.20% carbon, 1.00-1.30% manganese, up to 0.10% silicon, 0.08-0.13% sulfur, up to 0.04% phosphorus, and the balance iron. Used for carburized parts and screwstock.

Ryco. Trade name for a dough or bulk-molded acrylic resin used for dentures.

Ryco-Sep. Trade name for a dental gypsum used as a pore filler.

Rycrome. Trade name of Joseph T. Ryerson & Son Inc. (USA) for a chromium-molybdenum steel containing 0.4% carbon, 0.9% chromium, 0.2% molybdenum, 0.9% manganese, 0.2% silicon, and the balance iron. Used for pinions, crankshafts, bolts, studs, machinery parts, and forgings.

Rycut. Trade name Joseph T. Ryerson & Son Inc. (USA) for a series of free-cutting steels containing varying amounts of lead (typically 0.15-0.35%). Resulfurized grades containing up to 0.1% sulfur are also supplied. Used for machine parts, such as gears, spindles, shafts, axles, cams, collets, bushings, studs, and brake dies.

Ry-Die. Trade name Joseph T. Ryerson & Son Inc. (USA) for a nonshrinking, air-hardening tool steel (AISI type A2) containing 1% carbon, 5.2% chromium, 1.1% molybdenum, 0.6% manganese, 0.25% vanadium, and the balance iron. It has good deep-hardening properties, good toughness, fair wear resistance, and is used for shear blades, and blanking and forming dies.

Ryercite. Trade name of Joseph T. Ryerson & Son Inc. (USA) for phenol-formaldehyde resins and plastics.

Rynalloy. Trade name of Teledyne Ryan Aeronautical (USA) for an alloy steel composed of up to 2.5% carbon, up to 6% silicon, up to 1% manganese, 20% nickel, 1.8% chromium, and the balance iron. It has good antigalling properties, a maximum service temperature of 982°C (1800°F), and is used for ball and socket joints in aircraft exhaust systems.

rynersonite. (1) A reddish pink mineral of the columbite group composed of calcium niobium tantalum oxide, $Ca(Ta,Nb)_2O_6$. Crystal system, orthorhombic. Density, 6.40 g/cm^3; refractive index, above 2.05. Occurrence: USA (California).

(2) A pale yellow mineral of the columbite group composed of calcium tantalum oxide, $CaTa_2O_6$. It can be made synthetically by heating stoichiometric quantities of tantalum pentoxide (Ta_2O_5) and calcium carbonate ($CaCO_3$) in air under prescribed conditions. Crystal system, orthorhombic. Density, 7.02 g/cm^3.

Rynite. Trademark of E.I. Du Pont de Nemours & Company for a series of tough, chemical-resistant thermoplastic engineering polyesters including polybutylene terephthalates and polyethylene terephthalates with good elevated temperature properties up to 150°C (300°F). They are supplied in neat, amorphous, crystalline, mica- and glass-filled, glass fiber reinforced, high-impact, supertough, fire-retardant and UV-stabilized grades, and in special electrical grades with high thermal and electrical conductivities. *Rynite* polyesters have high stiffness, good high-temperature resistance, high tensile and impact strength, good electrical properties, and are used for automotive components and trim, electrical and electronic components and equipment, shop and garden tools and equipment, appliance and equipment housings, sporting goods, furniture, pump housings, and hardware.

ryolex. A *volcanic rock* composed of about 7 parts of silica (SiO_2) and 3 parts of alumina (Al_2O_3), and used for the manufacture of lightweight heat insulation.

Ryton. Trademark of Chevron Phillips Chemical Company (USA) for semicrystalline engineering thermoplastics based on polyphenylene sulfide (PPS) supplied in unmodified, carbon or glass fiber-reinforced, glass bead-filled and polytetrafluoroethylene-lubricated grades. *Ryton* thermoplastics have good mechanical properties, excellent dimensional stability, good moldability and rigidity, good corrosion resistance, outstanding resistance to chemicals including strong acids and solvents and highly caustic or acidic chemicals, good resistance to lubricants, fuels and hydraulic fuels, good flame resistance, good creep resistance, good thermal stability up to 260°C (500°F), a glass-transition temperature of 85°C (185°F), a melting point of 285°C (545°F), good vibrational resistance, and good electrical insulating properties. Used for electronic and electrical components, containers and casings, couplings and pump housings, and as matrix resins for engineering composites. Also included under this trademark are polyphenylene sulfide fibers. *Ryton PAS* are high-temperature thermoplastics based on polyarylene sulfide (PAS) with a glass transition temperature of 145°C (293°F), and a melting point of 170°C (345°F). Used for electrical and electronic applications.

RZ powder. A German reduced iron powder made from the scale of pig iron. RZ stands for the German word "Roheisenzunder," meaning "pig iron scale."

S

Saarstahl. Trade name of Saarstahl AG (Germany) for an extensive series of steels and superalloys including various carbon, cold-work, hot-work and high-speed tool steels, corrosion- and/ or heat-resistant steels, case-hardening and carburizing steels, high-strength and structural steels, free-cutting steels, bearing and spring steels, and cobalt- and nickel-base superalloys.

sabatierite. A bluish-gray mineral composed of copper thallium selenide, Cu_6TlSe_4. Crystal system, orthorhombic. Density, 6.83 g/cm^3. Occurrence: Czech Republic.

Sabeco. Trade name of Manco Products, Inc. (USA) for a series of bronzes composed of 69-71% copper, 18-26% lead and 4.5-17% tin, and used for bearings and bushings.

Saben. Trade name of Sanderson Kayser Limited (UK) for a tungsten-type high-speed steel containing 0.7% carbon, 4% chromium, 14% tungsten, 1% vanadium, and the balance iron. *Saben Extra* is a tungsten-type high-speed steel (AISI type T1), and *Saben Tenco* (AISI type T5) and *Saben Wunda* (AISI type T4) are cobalt-tungsten-type high-speed steels. *Saben 6-5-2* is a molybdenum-type high-speed steel (AISI type M2) and *Saben HC* a tungsten-vanadium type high-speed steel. *Saben* tool steels are used for cutting tools, lathe and planer tools, milling cutters, drills, threading taps, reamers, hobs and broaches

Sabex. Trade name of Sanderson Kayser Limited (UK) for a hot-work tool steel containing 0.28% carbon, 0.3% silicon, 0.3% manganese, 2.25% nickel, 2.5% chromium, 9.5% tungsten, 0.15% vanadium, and the balance iron. Used for dies and tools.

Sabinite. Trade name of US Gypsum Company (USA) for an acoustical plaster.

Sable Lite. Trade name of Safetee Glass Company Inc. (USA) for a black laminated glass.

SABRE. Trademark of Dow Chemical Company (USA) for a series of chemical-resistant polycarbonate/polyester alloys with excellent high- and low-temperature properties used for injection-molded automotive, electrical and electronic components, various consumer products, and industrial equipment.

Sabre. (1) Trademark of Chemetall, Oakite Products, Inc. (USA) for abrasive compounds and pastes used in metal finishing.

(2) Trade name of Atlas Specialty Steels (Canada) for a tungsten-type high-speed steel (AISI type T15) containing 1.6% carbon, 0.3% silicon, 0.25% manganese, 4.25% chromium, 12.5% tungsten, 5% vanadium, 5.5% cobalt, and the balance iron. It has excellent abrasion and wear resistance, high hot hardness, good deep-hardening properties, good machinability, and is used for tool bits, milling cutters, form tools, cold-forming and blanking dies, and cutting tools for automatic screw machines.

(3) Trade name of Carborundum Abrasive Company (USA) for bonded abrasives.

sabugalite. A bright to lemon-yellow mineral of the autunite group composed of aluminum uranyl hydrogen phosphate hydrate, $HAl(UO_2)_4(PO_4)_4 \cdot 16H_2O$. Crystal system, tetragonal. Density, 3.20 g/cm^3; refractive index, 1.582. Occurrence: Portugal.

saccharide. See monosaccharide; disaccharide; polysaccharide.

saccharin. A white, sweet-tasting, crystalline powder (98+% pure) with a melting point of 226-230°C (439-446°F) that is the anhydride of *o*-sulfimide benzoic acid. It is also available as the calcium, hemicalcium and sodium salt. Used in electroplating as a grain refiner, and in beverages, foods and medicines as a low-calorie artificial sweetener. Formula: $C_7H_5NO_3S$. Also known as o-*benzosulfimide;* o-*sulfobenzimide.*

Sachtolith. Trademark of Sachtleben Corporation (USA) for zinc sulfide (ZnS) used as a white pigment in paints, and in lithopone pigments.

Sachtoperse. Trademark of Sachtleben Corporation (USA) for white, ultrafine titanium oxide (TiO_2) powders used as pigments in ceramics, paints, coatings, paper, plastics, rubber, and other materials.

sacking. Coarse fabrics, usually made of burlap, flax, hemp, jute or polyolefin, used especially for making bags and sacks.

sacrificial coating. An active metal or metal-alloy coating, e.g., a zinc coating on iron or steel that corrodes preferentially for the sake of protecting the substrate metal or alloy.

sacrofanite. A colorless mineral of the cancrinite group composed of sodium calcium aluminum silicate carbonate sulfate hydrate, $(Na,Ca,K)_9(Si,Al)_{12}O_{24}(SO_4,CO_3,Cl,OH)_4 \cdot 0.3H_2O$. Crystal system, hexagonal. Density, 2.42 g/cm^3; refractive index, 1.505. Occurrence: Italy.

saddle clay. A clay of fine particle size, usually containing considerable quantities of fluxing agents, that can be fused at relatively low temperatures. Used for natural stoneware or electrical porcelain glazes.

saddle leather. (1) A relatively flexible, vegetable-tanned leather of natural tan color made from cattlehide and used for saddles, harnesses and other leather furnishings for horses.

(2) Any leather similar in appearance and flexibility to the leather described in (1), but used for clothing, handbags, suitcases, etc.

Sader. Trade name of Ato Findley (France) for an extensive series of adhesives, mastics, coatings, sealants and waterproofing agents. The adhesives are based on acrylics, styrene-butadiene rubber, cellulosics, cyanoacrylates or vinyls and are used mainly in electronics, building and construction, and in the manufacture of plastic and rubber products.

Saderfix. Trade name of Ato Findley (France) for acrylic adhesives used for floor covering applications.

Saderprene. Trade name of Ato Findley (France) for neoprene adhesives used in woodworking and carpentry.

Sadertac. Trade name of Ato Findley (France) for acrylic adhesives used for floor covering applications.

Sadrifill. Trademark of DS Fibres NV (Belgium) for polyester and polypropylene staple fibers.

Sadriloft. Trademark of DS Fibres NV (Belgium) for polyester staple fibers used for textile fabrics including clothing.

Sadrilux. Trademark of DS Fibres NV (Belgium) for polyester and polyester staple fibers.

SAE babbitts. White metal bearing and bushing alloys as classified by the Society of Automotive Engineers International (SAE International). The classification contains several alloy groupings and is based on a SAE number designation, e.g., SAE 11, SAE 16, SAE 795, etc. The SAE alloy groupings are: (i) tin-base alloys; (ii) lead-base alloys; (iii) lead-tin overlays; (iv) copper-lead alloys; (v) copper-lead-tin alloys; (vi) aluminum-base alloys, and (vii) other copper-base alloys.

SAE bronzes. Bearing bronzes as classified by the Society of Automotive Engineers International (SAE International). The classification contains several alloy groupings and is based on a SAE number designation, e.g., SAE 790, SAE 795, etc.

SAE steels. See AISI-SAE steels.

SAFA. Trademark of Nylstar SA (Spain) for nylon 6 and nylon nylon 6,6 fibers and filament yarns.

SAFASIL. Trademark of SAFAS Corporation (USA) for a flexible, weather- and waterproof, nonshrinkable, air-curing silicone caulking/sealing compound supplied in gun cartridges for interior and exterior joining and assembling applications.

Safe'n'Sound. Trademark of Roxul Inc. (Canada) for noncombustible, water-resistant mineral wool insulation used as noise and fire protection in interior building walls and ceilings. See also Roxul.

Safecap. Trade name for a series of corrosion-resistant dental amalgam alloys that do not contain the corrosive tin-mercury phase.

Safeflex. Trade name of Monsanto Chemical Company (USA) for polyvinyl butyral sheeting used as interlayer in laminated safety glass.

Safegard. Trade name of Sanchem Inc. (USA) for a conversion coating for aluminum products.

Safeglaze. Trade name of Guardian Industries Corporation (USA) for a toughened glass.

Safeguard. Trade name of Chrysler Corporation (USA) for toughened and laminated automotive replacement glass products.

Safe-Lite. Trade name of Duplate Canada Limited for a thick, laminated sheet glass.

Safelite. Trade name of Royal Industries Inc., Safelite Division (USA) for a laminated glass.

Saferlite. Trade name of Safety Glass Industria e Comércio de Vidros Ltda. (Brazil) for a laminated glass.

Safeshield. Trade name of Australian Window Glass Proprietary Limited (Australia) for a toughened sheet glass.

Safetyclad. Trade name of Safeguard Safety Glass Manufacturers Proprietary Limited (Australia) for a ceramic-coated, toughened glass used for cladding applications.

safety glass. See laminated glass; tempered glass; wired glass.

safety paper. A special type of paper with or without genuine watermark that is made reactive by the introduction of certain chemicals into the liquid stock.

Safety Samson. Trade name of Celanese Corporation (USA) for cellulose acetate plastics.

Safety-Silv. Trade name of J.W. Harris Company, Inc. (USA) for a series of cadmium-free silver brazing alloys.

Safetyware. Trade name of Bryant Electric Company (USA) for bakelite-type phenol-formaldehyde plastics used for electrical components and parts.

SafeSorb. Trade name of National Nonwovens (USA) for a chemical-free nonwoven composite that can absorb 20 times its weight of water.

Safe-T-Duct. Trademark of Schott Process Systems, Inc. (USA) for borosilicate glass ductwork.

Safety-Walk. Trademark of 3M Company (USA) for antislip floor surfacing materials.

Safevue. Trade name of Shatterprufe Safety Glass Company Proprietary Limited (South Africa) for a laminated glass.

Safeweld. Trade name of Standard Safety Glass Company (USA) for a polished wired glass.

Safex. Trade name of Hindustan Safety Glass Company Limited (India) for a toughened and laminated safety glass.

SAFF. Trademark of Steen & Co. GmbH (Germany) for polypropylene staple fibers.

Saffalloy. See Cerrosafe.

Saffil. Trademark of Imperial Chemical Industries, plc (UK) for a series of engineering ceramics composed of 95-97% alumina (Al_2O_3) and 3-5% silica (SiO_2) and available in various grades in the form of short (discontinuous), chopped and bulk fibers, nonwoven fabrics, mats and blankets. They have a density range of 3.3-3.5 g/cm³ (0.12-0.13 lb/in.³), an average fiber diameter of 3 μm (118 μin.), very high tensile strengths and moduli, an upper continuous-use temperature of 1600°C (2910°F), and good resistance to acids, alkalies, halogens and metals. Used as reinforcements in metal and ceramic composites.

safflorite. A tin-white mineral of the marcasite group composed of cobalt iron arsenide, $(Co,Fe)As_2$. It can also be made synthetically. Crystal system, orthorhombic. Density, 7.20 g/cm³. Occurrence: Canada (Ontario).

Saffron. Trade name for a medium-gold dental casting alloy with orange-yellow color.

Safir. Trademark of Renker GmbH & Co. KG (Germany) for a wide variety of paper and plastic film products including thermal printing paper, photoprinting and blueprint paper, carbon paper, label, graph and recording paper, lampshade paper and parchment, metallized paper, plastic film products and fabric-lined paper.

Safist. Trade name Cheapa Glass Company Limited (Australia) for a laminated glass.

Saflex. (1) Trademark of Solutia Inc. (USA) for flexible polyvinyl butyral (PVB) supplied in clear, graduated, translucent, transparent and other grades, and used as interlayer in annealed, tempered or heat-stabilized safety glass for residential windows and doors, and automotive windshields.

(2) Trade name of Engelhard Corporation (USA) for thermostatic bimetals that are active between 260 and 427°C (500 and 800°F).

Safplate. Trademark of Strongwell Company (USA) for tough, corrosion- and slip-resistant, gray fiberglass-gritted floor plate with antiskid surface made by coating proprietary *Extren* fire-retardant pultruded fiberglass-reinforced polyester plate with epoxy resin containing 35-50 mesh silica grit. Supplied in 1.2 × 2.4 m (4 × 8 ft.) panels in thicknesses ranging from 3 to 19 mm (0.125 to 0.750 in.) as solid plate or bonded with proprietary *Duratek* or *Duragrid* fiberglass grating. Used for trench covers, pedestrian bridge walkways and overpasses, bridge and pool decks, splash walls, and odor control covers.

Saf-Pro-Tek. Trademark of Pokonobe Industries Inc. (Canada) for slip-resistant floor coatings.

Saf-T. Trademark of Elmer's Products Inc. (USA) for a nonflammable, solvent-free, fast-setting latex contact cement.

Saf-T-Bond. Trade name of Capitol USA, LLC (USA) for solvent-free adhesives.

Sagamore. Trade name of AL Tech Specialty Steel Corporation (USA) for an air-hardening medium-alloy cold-work steel (AISI type A2) containing 1% carbon, 5% chromium, 1% molybdenum, 0.25% vanadium, and the balance iron. It has good deep-hardening and nondeforming properties, good wear resistance and toughness, and is used for press tools, punches, and blanking, forming and trimming and thread-rolling dies. A free-machining grade (*Sagamore-EZ*) and a grade (*Sagamore V* – AISI type A7) with higher carbon (about 2.00-2.85%) and chromium (about 5.00-5.75%) contents and additions of 0.50-1.50% tungsten are also available.

sagger clay. A rather uniform clay, either open-firing, semi-open-firing or tight-firing, that is relatively resistant to repeated heating and cooling, and is thus used in the manufacture of saggers (i.e., containers for firing porcelain and pottery).

Sagittarius. Trademark of Ivoclar Vivadent AG (Liechtenstein) for a ceramic dental alloy containing 75.0% gold, 16.8% palladium, 2.0% platinum, 2.0% silver, 2.0% indium, and less than 1.0% copper and rhenium, respectively. It has a white color, a density of 16.4 g/cm³ (0.59 lb/in.³), a melting range of 1130-1255°C (2065-2290°F), an as-cast hardness of 280 Vickers, moderate elongation, and excellent biocompatibility and resistance to oral conditions. Used for crowns, onlays, posts and bridges.

Sago Grob. Trade name of former Gerresheimer Glas AG (Germany) for a patterned glass.

sahamalite. A colorless mineral composed of magnesium cerium lanthanum carbonate, (Mg,Fe)(Ce,La,Nd,Pr)$_2$(CO$_3$)$_4$. Crystal system, monoclinic. Density, 4.30 g/cm³; refractive index, 1.776. Occurrence: USA (California).

Sahara. (1) Trade name of SA Glaverbel (Belgium) for a patterned glass.

(2) Trade name of Société Industrielle Triplex SA (France) for a laminated glass with tinted interlayer.

Sahis. Trademark of Filament Fiber Technology Corporation (USA) for polypropylene fibers and filament yarns.

sahlinite. A light yellow mineral composed of lead oxide chloride arsenate, Pb$_{14}$(AsO$_4$)$_2$O$_9$Cl$_2$. Crystal system, monoclinic. Density, 7.95 g/cm³. Occurrence: Sweden.

sailcloth. (1) Strong, heavy, closely woven cotton fabrics, such as *canvas* or *duck*, used for making sails, sailor's clothes, etc.

(2) A term used for fabrics made of aramid, nylon or polyester, and used almost exclusively in the manufacture of sails.

(3) A stiff, ribbed, plain-colored fabric, usually made of rayon, cotton or a polyester-cotton blend in a plain or basket weave. It is often finished with resin and has good pleating and draping qualities. Used for blouses, dresses, pants, summer jackets and leisure wear.

Sailsagar. Trade name of Ranchi, Steel Authority of India for a quenched-and-tempered copper-bearing, high-strength low-alloy (HSLA) structural steel supplied in plate form. It has an acicular ferritic microstructure, good low-temperature toughness, good weldability, and is used in shipbuilding.

sainfeldite. A colorless mineral composed of calcium hydrogen arsenate tetrahydrate, Ca$_5$H$_2$(AsO$_4$)·4H$_2$O. Crystal system, monoclinic. Density, 3.04 g/cm³; refractive index, 1.610. Occurrence: France.

Saint Anne marble. A dark bluish black *marble* with white veins imported from Belgium.

Saint Baume marble. A yellowish *marble* with brownish or reddish veins imported from France.

Saint Juery Extra. Trade name of Société Nouvelle du Saut-du-Tarn (France) for a high-speed steel containing 0.7% carbon, 4% chromium, 18% tungsten, 1% vanadium, and the balance iron. Used for tools, dies and cutters.

Saint-Just. Trademark of Saint-Gobain (France) for blown and drawn glass products.

Saint Peter's sandstone. A sandstone from Illinois, USA, valued as a building stone.

sakhaite. A gray-white mineral composed of calcium magnesium carbonate borate hydrate, Ca$_3$Mg(PO$_3$)$_2$(CO$_3$)·xH$_2$O. Crystal system, cubic. Density, 2.78 g/cm³; refractive index, 1.640. Occurrence: Siberia.

sakharovite. A gray mineral composed of lead antimony bismuth sulfide, Pb(Bi,Sb)$_2$S$_4$. Crystal system, monoclinic. Occurrence: Russia.

Sak-Mix. Trade name of Ato Findley (France) for an extensive series of adhesive cements and mortars, plasters, mastics, additives and coatings for building and construction applications, e.g., tile fixing, floor finishing and home decoration.

Sakorn. (1) Trademark of Oriental Fibre (Thailand) for polyester fibers and filament yarns.

(2) Trademark of Hantex Corporation (Thailand) for nylon 6 fibers and filament yarns.

sakuraiite. A greenish or steel-gray mineral of the chalcopyrite group composed of copper zinc iron silver indium tin sulfide, (Cu,Zn,Fe,Ag)$_3$(In,Sn)S$_4$. Crystal system, tetragonal. Density, 4.45 g/cm³. Occurrence: Japan.

Salamander. Trade name of General Electric Company (USA) for asbestos-insulated electrical conducting wire.

sal ammoniac. A colorless mineral consisting of ammonium chloride, NH$_4$Cl. Crystal system, cubic. Density, 1.53 g/cm³; melting point, sublimes at 340°C (644°F); boiling point, 520°C (968°F); refractive index, 1.642. It can also be made synthetically. Used as an electrolyte in dry cells, as a soldering flux, as a pickling agent in zinc coating and tinning, and in electroplating and textile printing. See also ammonium chloride.

Salar. Trade name of St. Lawrence Resin Products Limited (USA) for a series of hydrocarbon resins.

sal chalybis. See ferrous sulfate.

saleeite. A yellow to pale canary-yellow mineral of the autunite group composed of magnesium uranyl phosphate hydrate, Mg(UO$_2$)$_2$(PO$_4$)$_2$·xH$_2$O. It can also be made synthetically. Crystal system, tetragonal. Density, 3.20 g/cm³; refractive index, 1.57-1.58. Occurrence: Germany, Portugal, Central Africa.

salesite. A bluish green mineral composed of copper iodate hydroxide, CuIO$_3$(OH). Crystal system, orthorhombic. Density, 4.77 g/cm³; refractive index, 2.070. Occurrence: Chile.

Salge's alloy. A zinc alloy containing 10% tin, 4% copper, 1% antimony and 1% lead. Used for novelty items.

Salgemma. Trade name of Vetreria di Vernante SpA (Italy) for a patterned glass.

salicide. A self-aligned *silicide* made by depositing a thin film of a material, such as titanium, on a silicon wafer by a physical vapor deposition technique, followed by short rapid thermal annealing at about 600°C (1110°F) and 800° (1470°F), respectively. Used in integrated-circuit manufacture.

Salivan. Trade name for a silver-indium dental casting alloy.

Sallit's speculum. A corrosion-resistant alloy containing 65% copper, 31% tin and 4% nickel. Used for metallic mirrors.

Salmax. Trade name of Salmax Gesellschaft für oberflächenveredelte Feinbleche mbH (Germany) for hot-dip galvanized sheet steel, plastic-coated hot-dip galvanized sheet steel and trapezoidal profiles of hot-dip galvanized sheet steel.

salmon brick. A relatively soft, underfired building brick of salmon color used for applications that require only moderate strength.

salmon gum. The dense, hard, yellowish-pink wood from the tree *Eucalyptus salmonophloria*. It has a fine, open grain, and an average weight of 960 kg/m³ (60 lb/ft³). Source: Southern

United States. Used for fine furniture, cabinetmaking, veneer, moldings, inside trim, plywood, cooperage, baskets, woodenware, boxes, pallets and crates.

sal soda. See sodium carbonate decahydrate.

salt. (1) A solid ionic compound made up of a cation other than H^+ and an anion other than OH^- or O^{2-}. It is usually derived from an acid by replacing all or part of the hydrogen by a metal or electropositive radical. The reaction of an acid with a base yields a salt and water. Sodium chloride (NaCl), potassium bromide (KBr) and copper sulfate ($CuSO_4$) are examples of salts.

(2) A term referring to common salt, i.e., sodium chloride (NaCl). See also sodium chloride; halite.

Salta. Trade name of Westerwald AG (Germany) for patterned glass blocks featuring rectangles.

salt-bath nitrided steel. A carbon or alloy steel that has been nitrided in a molten nitrogen-containing salt bath (usually composed of a mixture of sodium and potassium cyanides) at a temperature between 510 and 565°C (950 and 1050°F). It has a surface layer (or case) consisting of hard metallic carbides, improved wear resistance, and increased fatigue endurance. Also known as *liquid nitrided steel*. See also nitrided steel

salt cake. Impure *sodium sulfate* (90-99% Na_2SO_4) with a density of 2.67 g/cm³ and a melting point of 888°C (1630°F), used as a source of sodium, as an antiscumming agent for glazes and glass, and in the manufacture of plate and window glass, paper pulp and sodium salts.

salt glaze. A lustrous glaze formed on the surface of a ceramic body toward the end of the firing cycle by putting salt into the firebox. The salt volatilizes and the fumes enter into a thermochemical reaction with the silicates and other ingredients of the ceramic.

salt-glazed tile. A facing tile with a salt-glaze finish. See also salt glaze.

salt of tartar. See potassium carbonate.

salt of tin. See stannous chloride.

saltpeter. See potassium nitrate.

Salus. Trademark of Filament Fiber Technology, Inc. (USA) for polypropylene fibers and filament yarns.

Salvid. Trade name of Cristales y Vidrios SA-Cristavid (Spain) for a laminated glass.

Salvo. Trade name of Midvale-Heppenstall Company (USA) for a tough tool steel containing 0.6% carbon, 1% chromium, 2% nickel, 0.2% vanadium, and the balance iron. It has good deep-hardening properties, and is used for tools and rivet sets.

samaria. See samarium oxide.

samarium. A pale gray or silvery, brittle, metallic element of the lanthanide series (rare-earth group) of the Periodic Table. It is commercially available in the form of ingots, lumps, rods, filings (chips), sheets, foils, wires and powder, and occurs naturally associated especially with cerium, yttrium and neodymium in rare-earth minerals. Melting point, 1077°C (1971°F); boiling point, 1791°C (3256°F); hardness, 39 Vickers; atomic number, 62; atomic weight, 150.36; divalent, trivalent. It ignites in air at about 500°C (930°F), develops an oxide film in air, and has high electrical resistivity (about 90 μΩcm), and a very high neutron absorption capacity. Two allotropic forms are known: (i) *Alpha samarium* (rhombohedral crystal structure) that has a density, 7.54 g/cm³ and is stable up to 917°C (1683°F), and (ii) *Beta samarium* (body-centered cubic crystal structure) that has a density of 7.40 g/cm³ and is stable between 917 and 1077°C (1971°F). *Samarium* is used as a neutron absorber, as a dopant for laser crystals, in masers, phosphors and ceramic condens-

ers, in metallurgical and materials research, in superconductors and permanent magnets, and as an reducing agent. Symbol: Sm.

samarium aluminate. A compound of samarium oxide and aluminum oxide with a melting point of 1982°C (3600°F) used in ceramics and materials research. Formula: $Sm_2Al_2O_6$.

samarium barium copper oxide. A high-temperature superconductor that is available as a fine, high-purity powder and as a thin film material. It similar to *yttrium barium copper oxide*, but has a samarium-barium solid solution. Used as a raw material for manufacture of superconducting components. Formula: $SmBa_2Cu_3O_{7-x}$. Abbreviation: SBCO; SmBCO; Sm-123.

samarium boride. Any of the following compounds of samarium and boron used in ceramics and materials research: (i) *Samarium tetraboride*. Density, 6.18 g/cm³. Formula: SmB_4; and (ii) *Samarium hexaboride*. A refractory solid. Density, 5.07 g/cm³; melting point, approximately 2538°C (4600°F); hardness, 2500 Vickers. Formula: SmB_6.

samarium bromide. See samarium dibromide; samarium tribromide.

samarium carbide. Any of the following compounds of samarium and carbon used in ceramics and materials research: (i) *Samarium dicarbide*. Yellow crystal. Crystal system, hexagonal. Density, 6.50 g/cm³; melting point, above 2204°C (4000°F). Formula: SmC_2; and (ii) *Samarium sesquicarbide*. Density, 7.47 g/cm³. Formula: Sm_2C_3; and (iii) *Trisamarium carbide*. Density, 8.14 g/cm³. Formula: Sm_3C.

samarium carbonate. Yellow, hygroscopic crystals (99+% pure) with a melting point above 500°C (930°F). It is also available in hydrated form [$Sm_2(CO_3)_3 \cdot xH_2O$]. Formula: $Sm_2(CO_3)_3$.

samarium chloride. See samarium dichloride; samarium trichloride.

samarium chloride hexahydrate. Light yellow, hygroscopic crystals (99+% pure) obtained by treating samarium carbonate or oxide with hydrochloric acid. Density, 2.383 g/cm³; melting point, loses $5H_2O$ at 110°C (230°F). Used as a catalyst. Formula: $SmCl_3 \cdot 6H_2O$.

samarium cobalt. A group of ferromagnetic compounds containing cobalt and samarium in ratios of 5:1 ($SmCo_5$) and 7:2 (Sm_2Co_7). They are available as fine powders for the manufacture of samarium-cobalt permanent magnets.

samarium-cobalt magnet. A permanent magnet made from *samarium cobalt* (Sm_2Co_7 or $SmCo_5$). It has exceptionally high magnetic induction, high coercive force and magnetic energy density, and is extremely resistant to demagnetization. See also rare-earth magnets.

samarium dibromide. Brown crystals with a melting point of 669°C (1236°F), used in chemistry and materials research. Formula: $SmBr_2$. Also known as *samarium bromide*.

samarium dicarbide. See samarium carbide (i).

samarium dichloride. Brown to reddish-brown crystals or powder (99.9+% pure) with a density of 3.68 g/cm³ and a melting point of 855°C (1571°F). Used in chemistry and materials research. Formula: $SmCl_2$. Also known as *samarium chloride*.

samarium disilicate. See samarium silicate (ii).

samarium disulfide. See samarium sulfide (ii).

samarium fluoride. White crystals or off-white powder (99.9+%) with a density of 6.928 g/cm³, a melting point of 1306°C (2382°F), a boiling point of 2323°C (4213°F) and a refractive index of 1.396. Used as a reagent and catalyst. Formula: SmF_3. Also known as *samarium trifluoride*.

samarium hexaboride. See samarium boride (ii).

samarium iodide. See samarium triiodide.

samarium monosilicate. See samarium silicate (i).

samarium monosulfide. See samarium sulfide (i).

samarium monotelluride. See samarium telluride (i).

samarium naphthenate. A viscous liquid typically containing 12-15% samarium, and used chiefly as a paint and varnish drier.

samarium nitride. A compound of samarium and nitrogen with a density of 8.50 g/cm^3, used in ceramics and materials research. Formula: SmN.

samarium oxide. Yellow-white crystals, or white- or cream-colored, hygroscopic powder (99.9+% pure). Crystal system, cubic. Density, 7.6-8.35 g/cm^3; melting point, 2335°C (4235°F); refractive index, 1.97; high thermal neutron cross section. Used in infrared-absorbing glass and luminescent glass, as a phosphor activator and neutron absorber, in nuclear control rods, in the preparation of samarium salts, and as a catalyst. Formula: Sm_2O_3. Also known as *samaria*.

samarium phosphide. A compound of samarium and phosphorus with a density of 6.34 g/cm^3, used in ceramics and materials research. Formula: SmP.

samarium selenide. A compound of samarium and selenium with a melting point of 2093°C (3800°F), used in ceramics and materials research. Formula: SmSe.

samarium sesquicarbide. See samarium carbide (ii).

samarium sesquisulfide. See samarium sulfide (iii).

samarium sesquitelluride. See samarium telluride (ii).

samarium silicate. Any of the following compounds of samarium oxide and silicon dioxide used in ceramics: (i) *Samarium monosilicate.* Density, 6.36 g/cm^3; melting point, 1940°C (3524°F). Formula: Sm_2SiO_5;(ii) *Samarium disilicate.* Density, 5.20; melting point, 1777°C (3231°F). Formula: $Sm_2Si_2O_7$; and (iii) *Disamarium trisilicate.* Density, 5.76 g/cm^3; melting point, 1920°C (3488°F). Formula: $Sm_4Si_3O_{12}$.

samarium silicide. A compound of samarium and silicon with a density of 6.26 g/cm^3. Used in ceramics. Formula: $SmSi_2$.

samarium sulfate. White or pale yellow, hygroscopic crystals (99.9+% pure). Crystal system, monoclinic. Density, 2.93 g/cm^3; melting point, loses $8H_2O$ at 450°C (842°F). Used for red and infrared phosphors. Formula: $Sm_2(SO_4)_3 \cdot 8H_2O$. Also known as *samarium sulfate octahydrate*.

samarium sulfide. Any of the following compounds of samarium and sulfur exhibiting thermoelectric properties and used for ceramic, electronic and nuclear applications: (i) *Samarium monosulfide.* Density, 6.01 g/cm^3; melting point, 1940°C (3524°F). Formula: SmS; (ii) *Samarium disulfide.* Density, 5.66 g/cm^3; melting point, 1730°C (3146°F). Formula: SmS_2; (iii) *Samarium sesquisulfide.* Red-brown crystals or red powder. Crystal system, orthorhombic. Density, 5.73-5.87 g/cm^3; melting point, 1720°C (3128°F); band gap, 1.71 eV. Formula: Sm_2S_3; and (iv) *Samarium tetrasulfide.* Density, 6.14 g/cm^3; melting point, 1799°C (3270°F). Formula: Sm_3S_4.

samarium telluride. Any of the following compounds of samarium and tellurium used in ceramics, electronics and materials research: (i) *Samarium monotelluride.* Crystalline compound. Melting point, 1915°C (3479°F). Formula: SmTe; and (ii) *Samarium sesquitelluride.* Orthorhombic crystals. Density, 7.31 g/cm^3. Formula: Sm_2Te_3.

samarium tetraboride. See samarium boride (i).

samarium tetrasulfide. See samarium sulfide (iv).

samarium tribromide. Yellow crystals with a melting point of 640°C (1184°F) used in chemistry and materials research. Formula: $SmBr_3$. Also known as *samarium bromide*.

samarium trichloride. White to yellowish, hygroscopic crystals, powder or beads (99.9+% pure) with a density of 4.465 g/cm^3 and melting point of 682°C (1260°F). Formula: $SmCl_3$. Also known as *samarium chloride*.

samarium triiodide. Orange-yellow crystals or powder (99.9+% pure) with a melting point of 850°C (1562°F). Used in chemistry and materials research. Formula: SmI_3. Also known as *samarium iodide*.

samarium trifluoride. See samarium fluoride.

samarskite. A velvet-black or brown, radioactive mineral of the columbite group composed of rare-earth niobium oxide, $(Y,Ce,Er,U,Fe,Pb,Ca,Th)(Nb,Ta,Ti,Sn)_2O_6$. Crystal system, orthorhombic. Density, 5.15-5.70 g/cm^3; refractive index, 2.20. Occurrence: USA (California), Europe (Norway). Used as a source of rare earths. Also known as *uranotantalite*.

Samba. Trade name of Circeo Filati S.r.l. (Italy) for 100% pure polyester yarn based on cationic *Dacron* fibers and used for textile fabrics.

samba. See obeche.

sampleite. A light blue-green mineral composed of sodium calcium copper chloride phosphate pentahydrate, $NaCaCu_5(PO_4)_4Cl \cdot 5H_2O$. Crystal system, orthorhombic. Density, 3.20 g/cm^3; refractive index, 1.677. Occurrence: Chile.

Sampson. (1) British trade name for an oil-hardening steel containing 0.43% carbon, 1.22% nickel, 0.43% chromium, 0.43% manganese, and the balance iron. Used for gears and tools.

(2) Trade name for a zinc-base antifriction metal containing 8% aluminum and 4% copper. Used for bearings.

Sam-Sol. Trade name of Multiplate Glass Company (USA) for aluminum screening embedded in plain or tinted vinyl resin and sandwiched between two sheets of one-way glass.

Samson. (1) Trade name of Carpenter Technology Corporation (USA) for a series of chromium and nickel-chromium steels used for tool shanks, shafts, gears, bolts, clutches, and coil and leaf springs.

(2) Trade name for nylon fibers and yarns used for textile fabrics including clothing.

Samson Extra. Trade name of Carpenter Technology Corporation (USA) for a mold steel (AISI type P5) containing 0.1% carbon, 0.2% silicon, 0.3% manganese, 2.3% chromium, and the balance iron. It has a great depth of hardening, high core strength, good hobbability and good toughness and machinability, and is used for plastic molding dies.

samsonite. A steel-black mineral composed of silver manganese antimony sulfide, Ag_4MnSb_2S. Crystal system, monoclinic. Density, 5.51 g/cm^3. Occurrence: Germany.

samuelsonite. A colorless mineral of the apatite group composed of barium calcium iron manganese aluminum phosphate hydroxide, $(Ca,Ba)_9Fe_2Mn_2Al_2(PO_4)_{10}(OH)_2$. Crystal system, monoclinic. Density, 3.35 g/cm^3; refractive index, 1.650. Occurrence: USA (New Hampshire).

Sanalux. Trade name of Schott DESAG AG (Germany) for a hand-blown glass with high transparency in the ultraviolet range.

SAN-ABS alloys. See ABS-SAN alloys.

SanBlend. Trade name of United Clays, Inc. (USA) for a series of refined ball clays supplied in various grades in the form of slurries and noodles. Depending on the particular grade, they are used for a multitude of ceramic applications including sanitaryware and tableware.

Sanbold. (1) Trade name of Sanderson Kayser Limited (UK) for a series of high-permeability magnet alloys composed of 35-77% nickel and 23-65% iron. Used for transformers, relays, chokes, and magnetic shielding applications.

(2) Trade name of Sanderson Kayser Limited (UK) for a series of steels including several plain-carbon grades containing between 0.2 and 1.0% carbon, medium-carbon chromium grades containing 0.4-0.6% carbon and 0.5-1% chromium, high-carbon chromium grades with 1.1% carbon and 1.3% chromium, low- and medium-carbon nickel grades containing 0.25-0.40% carbon, 1.5% manganese and 0.4-1.0% nickel, low-carbon molybdenum grades having 0.35% carbon, 1.5% manganese and 0.3-0.4% molybdenum, and numerous low- and medium-carbon alloy grades containing 0.1-0.5% carbon and varying amounts of one or more of the following alloying elements: molybdenum, chromium, or nickel.

sanbornite. A colorless or white mineral composed of barium silicate, β-$BaSi_2O_5$. It can also be made synthetically from a 1:2 molar mixture of barium carbonate ($BaCO_3$) and silica gel under prescribed conditions. Crystal system, orthorhombic. Density, 3.70 g/cm³; refractive index, 1.617. Occurrence: Germany, USA (California).

Sanbron. Trade name of Sanderson Kayser Limited (UK) for a series of high-speed and hot-work tool steels, and several austenitic stainless steels and corrosion-resistant low-carbon high chromium steels.

SAN-butyl acrylate. A polymer blend made by adding butyl acrylate to styrene-acrylonitrile (SAN) in order to improve outdoor stability.

Sanco. Trade name of Isolierglaswerk Langgöns Wolff + Meier GmbH & Co. KG (Germany) for a multi-pane insulating glass.

Sancy. Trade name of Aubert & Duval (France) for a nondeforming, oil- and air-hardening tool steel containing 1.7% carbon, 13% chromium, 0.5% tungsten, 0.5% molybdenum, and the balance iron. Used for forming and blanking dies.

sand. Angular or spherical grains of worn-down or disintegrated rock composed essentially of quartz grains ranging from 0.05 to 5 mm (0.002 to 0.2 in.) in size, sometimes in combination with particles of feldspar, mica, or other mineral materials. Used in the manufacture of glass, glazes, porcelain enamels, ceramic bodies, artificial stone, Portland cement, in building and construction work, as an abrasive, as a setting medium for the firing of ceramic ware, and as a core material in foundry molds.

sandalwood. The light-colored, fragrant heartwood of any of several evergreen trees of the genus *Santalum* especially white sandalwood *(S. album)*. It has a close grain and high hardness. Source: Southern Asia. Used for cabinetwork, ornamental carved objects, and chests, boxes and fans.

sandarac. (1) The durable, highly balsamic, mahogany-colored wood of the coniferous sandarac tree *Callitris quadrivalvis* growing in Northwest Africa (e.g., Morocco) used chiefly in cabinetmaking. See also sandarac gum.

(2) See realgar.

sandarac gum. A soft, yellow, solid, resinous exudate obtained from the coniferous sandarac tree *Callitris quadrivalvis* growing in Northwest Africa (e.g., Morocco). Available in the form of an amorphous powder or lumps, it has a density of 1.05-1.09 g/cm³ (0.038-0.039 lb/in.³), a melting point of 135-145°C (275-293°F), and is used in the manufacture of paints, varnishes, lacquers, putties, dental cements, incense, erasers, linoleum and oilcloth, and as a fumigant. Also known as *juniper gum.* See also sandarac (1).

sand asphalt. A natural blend of asphalt and varying amounts of loose sand grains. Also known as *asphaltic sand.*

(2) A hot-laid mixture composed of local sand and asphalt without special grading, and used in road construction.

(3) A mixture of local sand and asphalt, with or without mineral filler, laid by mixing in place, or by mixing in a traveling plant. Used in road construction. See also sheet asphalt.

sandblasted glass. Fine glassware that has been ornamented by projecting fine sand grains, crushed flint, or iron powder at high speed onto its surface to produce a design or pattern in mat finish. Used especially for architectural applications and in the arts.

sandblast sand. A coarse or fine mineral abrasive that, driven by compressed air, is sprayed over surfaces of metal, stone or glass for cleaning, grinding, cutting or decorating purposes. It may be ordinary silica sand or finely divided quartz, flint, garnet, dolomite, pumice, novaculite or slag. Its particle size typically ranges from about 5 mm (0.20 in.) to less than 350 μm (0.014 in.)

sand castings. Ferrous or nonferrous *castings* formed by pouring liquid or viscous metals or alloys into cavities, formed in molds made of sand, and allowing them to solidify.

sand-cement grout. See sanded grout.

sand crepe. An irregularly textured crepe fabric woven from silk or synthetic yarns and having a finish resembling sand in appearance. See also crepe.

sande. The yellowish-white to light brown wood obtained from several trees of the genus *Brosimum*. It has poor resistance to decay, insects and staining. Source: Pacific Ecuador and Columbia. Used for lumber, plywood and moldings. Also known as *cocal.*

sanded fabrics. Fabrics that have been finished with sandpaper or other abrasives to create special surface effects, raise surface fibers, or produce a sueded or peached quality or feel.

sanded grout. A Portland cement *grout* containing a fine aggregate, such as silica sand. Also known as *cement-sand grout; sand-cement grout; sand grout.*

sanded plaster. A plaster containing a considerable quantity of sand. For finishing exterior walls and waterproofing foundation walls, it is usually a mixture of cement, sand and water. Sanded plasters for finish coats on interior walls usually consist of a mixture of a special sand, gypsum or lime, cement and water. The smoothness of the finish depends chiefly on the coarseness of the sand.

Sanderite. Trade name for nylon fibers and products.

Sanderson. Trade name of Sanderson & Kayser Limited (UK) for an extensive series of steels including various nickel and nickel-chromium alloy machinery and case-hardening grades, corrosion-resistant medium-carbon high-chromium grades, high-carbon high-chromium cold-work tool and die grades, water hardening plain-carbon tool and die grades, and tungsten magnet steel grades.

Sanderson Extra. Trade name of Crucible Materials Corporation (USA) for a water-hardening tool steel (AISI type W1) containing 1-1.1% carbon, 0.25% manganese, and the balance iron. It has excellent machinability, good to moderate wear resistance and toughness, low hot hardness, and is used for hand tools, cutting tools, punches and dies.

sand finish. Structural clay products, such as brick or tile, with sand-covered surfaces resulting from their manufacture by the soft- or stiff-mud process.

sand grout. See sanded grout.

sanding sealer. A lacquer used as a seal coat over a wood filler and under finishing lacquers.

Sand-Light. Trademark of United Abrasives Inc. (USA) for abrasive pads consisting of abrasive grain-impregnated nonwoven synthetic fibers. Supplied in several grades from heavy-duty for removing rust and scale and cleaning welds on steels and nonferrous metals, and general-duty for cleaning and finishing ferrous and nonferrous metals to light-duty for cleaning and polishing plastics, fiberglass and ceramics.

sand-lime brick. A refractory brick, usually white in color, made by mixing silica sand with hydrated lime and water and molding under prescribed conditions. Also known as *lime-sand brick*.

Sandow. Trade name of LTV Steel (USA) for a tough, oil-hardening steel containing 0.4% carbon, 0.9% chromium, 0.2% molybdenum, and the balance iron. Used for axles, gears, crane hooks, and mandrels.

sandpaper. (1) A strong, heavy paper coated on one side with sharp particles of sand, i.e., particles of the quartz variety *flint*. It is a common abrasive in the woodworking industries, often used for smoothing surfaces and removing existing paint finishes, and is also used for sanding, cleaning or polishing metals, plastics, fiberglass, etc. The name was probably derived from the off-white color and somewhat sandy texture of flint. See also flint paper; quartz paper.

(2) A term now used for any paper coated on one side with an abrasive material, such as flint, garnet, emery, aluminum oxide or silicon carbide.

Sandro Plast. Trade name of Sandrock Bautenschutz GmbH (Germany) for water-repellent bituminous paints used as protective coatings on buildings, and for waterproofing concrete.

Sandstone. Trade name of Aardvark Clay & Supplies (USA) for an earthenware clay with a tan body resembling sandstone.

sandstone. A sedimentary rock formed predominantly by fine grains of quartz sand, typically 0.05-5 mm or 0.002-0.2 in. in size, cemented together with clay, silica, iron oxide or calcium carbonate. Depending on the particular cementing material, sandstone is classified into the following varieties: (i) *Argillaceous sandstone* (cemented with clay minerals); (ii) *Calcareous sandstone* (cemented with calcium carbonate); (iii) *Ferruginous sandstone* (cemented with iron oxide minerals); and (iv) *Siliceous sandstone* (cemented with silica). *Sandstone* has an average weight of 2290 kg/m³ (143 lb/ft³) and typical ultimate tensile and compressive strengths of 2.07 MPa (300 psi) and 62.05 MPa (9000 psi), respectively. Used as a building stone, and in the manufacture of abrasives, ceramics and glass.

sand-struck brick. A brick made by molding, usually by hand, a rather wet clay (typical moisture content between 20 and 30%) in a mold whose inside has been coated with sand to prevent sticking of the brick to the mold.

Sandvik. Trade name of Sandvik Steel (Sweden) for an extensive series of alloys including several corrosion- and/or heat-resistant austenitic, ferritic and martensitic and duplex stainless steels (also supplied as welding electrodes and wires), numerous high-speed steels, a wide range of carbon and low-alloy machinery and structural steels, various nickel-chromium, nickel-chromium-iron, nickel-chromium-molybdenum and nickel-iron alloys, several high-titanium and high-zirconium alloys, and a wide range of sintered tungsten carbides.

Sandvik SAF. Trade name of Sandvik Steel (Sweden) for a series of *superduplex stainless steels* with excellent pitting and crevice and stress-corrosion corrosion resistance, high mechanical strength, and good forgeability and weldability. Used for boat- and shipbuilding, and other equipment operating in highly corrosive environments, e.g., in the chemical and petrochemical industries.

Sandwich. Trade name of Importglas GmbH & Co. (Germany) for a multi-sheet insulating glass.

sandwich coating. A type of mechanical coating consisting of two or more layers of different metals obtained by introducing the metals in powder form individually to a coating barrel. Also known as *layered coating*.

sandwich constructions. See sandwich laminates.

sandwich laminates. A class of structural laminates composed of two strong outer sheets separated by a thin layer of a less dense material. Such an arrangement provides for a light, relatively strong and stiff structure. The lightweight core between the two outer sheets is usually composed of *foamed plastics* (e.g., epoxies, phenolics, silicones or polyurethanes), *honeycomb*, etc., and the outer sheets are made of aluminum, fiberglass, graphite or aramid. The core is usually joined to the facings by brazing, diffusion-bonding or gluing using a structural adhesive in liquid, film or paste form. Also known as *sandwich constructions; sandwich materials; sandwich panels*. See also honeycomb structure; hybrid composite; hybrid laminate.

sandwich moldings. Thermoplastic moldings usually consisting of foamed or reinforced core materials, and unfilled skin materials having good surface appearance. They are produced from polymer melts by a process in which the polymer forming the skin is injected first into the mold, promptly followed by injection of the polymer forming the core, whereby the latter pushes the former to the extremities of the mold and, in passing, solidifies the skin layer adhering to the cooled mold surfaces.

sandwich molecule. See molecular sandwich.

sandwich panels. See sandwich laminates.

Sandwiform. Trade name of Venture Industries (USA) for a strong, stiff, lightweight sandwich panel composed of a honeycomb core based on a polymer, such as cellular polypropylene, and skin layers based on a thermoplastic, such as glass fiber-reinforced polypropylene. It has high toughness and good formability, and is used for automotive applications.

saneroite. A bright orange mineral composed of sodium magnesium vanadium silicate hydroxide, $Na_2Mn_{10}VSi_{11}O_{34}(OH)_4$. Crystal system, triclinic. Density, 3.47 g/cm³; refractive index, 1.742. Occurrence: Italy.

Sanforized fabrics. Fabrics, such as cotton or linen, that have been compressively shrunk by a proprietary process before they are made into clothing (e.g., shirts) to limit subsequent shrinkage during laundering. The residual shrinkage of Sanforized fabrics is usually less than 1%.

sanglier. A plain-weave fabric, often made of mohair and worsted fibers and having a compact, wiry texture and a rough overall surface finish.

Sanicro. (1) Trade name of Sandvik Steel Company (USA) for a series of austenitic stainless steels composed of 0.02-0.07% carbon, 0.5-2.5% manganese, 0.5-1.0% silicon, 21.0-28.0% chromium, 31.0-34.0% nickel, 3.0-4.0% molybdenum, 0.6-1.4% copper, 0-0.5% titanium, 0-0.35% aluminum, 0.030% phosphorus, 0.030% sulfur, and the balance iron. They have excellent corrosion and oxidation resistance, and are used for seamless tubing and piping and wire products for the oil and gas production industries.

(2) Trade name of Sandvik Steel Company (USA) for austenitic nickel-base alloys containing 16-20% chromium, 0-9% iron, 0.3-0.4% titanium, 0-0.25% aluminum, 0.03-0.05% carbon, 0.8-3% manganese, and 0-0.4% silicon. They have excellent corrosion and oxidation resistance, and are used for weld-

ing wire, corrosion and oxidation resistant parts, and tubing.

sanidine. (1) A colorless, transparent mineral of the feldspar group composed of potassium sodium aluminum silicate, $(K,Na)AlSi_3O_8$. Crystal system, monoclinic. Density, 2.54 g/cm^3; refractive index, 1.520. Occurrence: Germany.

(2) A colorless or white, sometimes pink, yellow or red mineral of the feldspar group composed of potassium aluminum silicate, $KAlSi_3O_8$. It can be made synthetically. Crystal system, monoclinic. Density, 2.56 g/cm^3; refractive index, 1.5275. Occurrence: Germany, Italy.

Saniro. Trade name for a modacrylic fiber with excellent moth and mildew resistance, relatively high tenacity and abrasion resistance, good resistance to acids and alkalies, and fair to poor resistance to ketones. Used chiefly for clothing and industrial fabrics.

Sanitary. Trademark of GE Silicones (USA) for silicone-base construction sealants.

sanitary ware. Glazed, vitrified whiteware or porcelain-enameled fixtures, such as bathtubs, sinks, lavatories and bidets.

sanjuanite. A white mineral composed of aluminum phosphate sulfate hydroxide nonahydrate, $Al_2PO_4SO_4OH\cdot9H_2O$. Crystal system, monoclinic. Density, 1.94 g/cm^3. Occurrence: Argentina.

Sanmac. Trade name of Sandvik Steel Company (USA) for hollow steel bar supplied in austenitic stainless grades AISI 304, 304L, 316 and 316L.

sanmartinite. A dark brown mineral of the wolframite group composed of zinc iron tungsten oxide, $(Zn,Fe,Ca,Mn)WO_4$. It can also be made synthetically, but the synthetic form does not contain iron, calcium and manganese. Crystal system, monoclinic. Density, 6.70 g/cm^3. Occurrence: Argentina.

SanMix SB. Trade name of United Clays, Inc. (USA) for a special blend of plastic clays with moderately high *kaolinite* and *mica* contents. It has superior casting and deflocculation characteristics, and is used for sanitaryware and general ceramic casting applications.

SAN-PC alloys. Polymer blends of amorphous styrene-acrylonitrile (SAN) and amorphous polycarbonate (PC). They uniquely combine the outstanding processibility of SAN with the many excellent physical and mechanical properties of PC.

SAN-PVC alloys. Polymer blends of styrene-acrylonitrile (SAN) and polyvinyl chloride (PC). They uniquely combine the outstanding processibility of SAN with the excellent outdoor weatherability and flame retardancy of PVC.

sansevieria. See bowstring hemp.

Sans Rival. Trademark of Société Nouvelle du Saut-du-Tarn (France) for a high-speed steel containing 0.7% carbon, 18% tungsten, 4% chromium, 1% vanadium, and the balance iron. Used for tools, dies, and cutters.

San-Splint. Trademark of Smith & Nephew Associated Companies Limited (UK) for a stretch-resistant thermoplastic material used in the manufacture of protective, supportive and corrective splints.

Sanstat. Trademark of Shakespeare Monofilaments & Specialty Polymers (USA) for strong, flexible nylon fibers whose outer skins are saturated (suffused) with electrically conductive carbon particles. Supplied as monofilaments, multifilament and supported yarns, tow, and staple fibers, they are used for static control applications, e.g., carpets, brushes, knitted, woven and nonwoven industrial textiles.

santaclaraite. A pink or tan mineral of the pyroxenoid group composed of calcium manganese silicate hydroxide monohy-

drate, $CaMn_2Si_5O_{14}(OH)_2\cdot H_2O$. Crystal system, triclinic. Density, 3.31 g/cm^3; refractive index, 1.696. Occurrence: USA (California).

santafeite. A black mineral composed of sodium manganese calcium strontium vanadium oxide tetrahydrate, $NaMn_3(Ca,Sr)(V,As)_3O_{13}\cdot4H_2O$. Crystal system, orthorhombic. Density, 3.38 g/cm^3. Occurrence: USA (New Mexico).

Santa Maria. The hard, strong wood of the tree *Calophyllum brasiliense*. It has good workability, and pinkish to red heartwood and light-colored sapwood. Source: Southern Mexico, Central America; northern South America; West Indies. Used as veneer for plywood and in boat construction.

santanaite. A yellow mineral composed of lead chromium oxide, $Pb_{11}CrO_{16}$. Crystal system, hexagonal. Density, 9.16 g/cm^3. Occurrence: Chile.

santite. A colorless mineral composed of potassium borate tetrahydrate, $KB_5O_8\cdot4H_2O$. Crystal system, orthorhombic. Density, 1.74 g/cm^3; refractive index, 1.436.

Santocel. Trademark of Monsanto Chemical Company (USA) for a light, porous silica aerogel used in thermal insulation, and as a thickener in volatile and nonvolatile liquids.

Santolite. Trademark of Monsanto Chemical Company (USA) for a series of synthetic resins formed by the condensation of toluol sulfonamides and formaldehyde. Depending on the particular composition, they are supplied as soft, viscous liquids or brittle lumps, and used as plasticizers and in special lacquers.

Santoprene. Trademark of Monsanto Chemical Company (USA) for black, vulcanized, olefinic-based thermoplastic rubber supplied in various grades. It can be processed by blow and injection molding, extrusion and other melt-processing techniques. *Santoprene* has excellent ozone and flex-fatigue resistance, good fluid resistance, poor compatibility to acetals and polyvinyl chlorides, and is used for automotive parts, electrical products, mechanical rubber goods, sheets for gaskets and other goods, hose and tubing, and as a replacement for thermosetting elastomers, such as chlorosulfonated polyethylene, polychloroprene and ethylene-propylene-diene-monomer.

Santorin earth. A light-colored *volcanic ash* from the Grecian island of Santorin. It contains 60-65% silica (SiO_2), 13% alumina (Al_2O_3), 5-6% ferric oxide (Fe_2O_3), 3-4% calcia (CaO), 2% magnesia (MgO), and varying amounts of alkalies and titania (TiO_2). Used in cement and concrete. See also pozzulana; trass.

Santotac MRS. Trademark of Solutia Inc. (USA) for a thermoplastic resin produced from recycled polyvinyl butyral (PVB), and used as a heat, light and impact modifier in the manufacture of resilient flooring.

Santowax. Trademark of Monsanto Chemical Company (USA) for a series of microcrystalline waxes.

SAP alloys. See sintered aluminum powder alloys.

sapele. The dark brown wood of the very large tree *Entandrophragma cylindricum* belonging to the mahogany family. It has an attractive appearance with a regular stripe or roe figure resembling mahogany, but unlike the latter is close-grained. *Sapele* has good strength and durability, good machining, finishing, nailing and gluing qualities, and a high tendency to warp in use. The heartwood is moderately durable and resists preservative treatment. Average weight, 620 kg/m^3 (39 lb/ft^3). Source: Tropical Africa (Sierra Leone to Angola and east through the Congo to Uganda). Used for decorative flooring, joinery, furniture and general construction (e.g., window frames, stairways, doors and pianos), and rotary-peeled for veneer for decorative plywood. Also known as *sapele mahogany; scented mahogany; West African cedar.*

Saphir. Trade name of Ugine Aciers (France) for a water-hardening tool steel used for drills, taps, reamers, chasers, planer tools, shears, punches, chisels, etc.

saponified cellulose acetate. A cellulose acetate filament yarn whose dimensional stability and strength-to-volume ratio have been significantly enhanced by steam heating, streching by about 4-10 times its original length, winding onto bobbins and saponifying the streched and softened yarn by treating with a sodium hydroxide solution, followed by washing, oiling, drying and rewinding.

saponite. A soft, white, yellowish, grayish, bluish or reddish mineral with a soapy feel belonging to the same group as *montmorillonite* (smectite group). It is a hydrous silicate of magnesium and aluminum, $(Mg_2Al)(Si_3Al)O_{10}(OH)_2 \cdot 5H_2O$, that may also contain calcium and iron. It can also be made synthetically. Crystal system, monoclinic. Density, 2.16-2.50 g/cm^3; hardness, 1.0-1.5 Mohs; refractive index, 1.53. Occurrence: Europe (Poland, UK), USA (California). Used in ceramics.

sappanwood. The brilliant red wood from the tree *Caesalpinia sappan* of the pea family. It takes a bright, beautiful polish. Source: India, Sri Lanka, Malay peninsula. Used for fine furniture, cabinetwork and musical instruments, e.g., violin bows.

sapphire. A bright-blue, transparent variety of the mineral *corundum* composed of aluminum oxide (Al_2O_3) and containing traces of cobalt, chromium and titanium. It can also be made synthetically. Synthetic sapphire is grown directly from the melt into rods, spheres, disks, squares, rectangles, whiskers and single crystals. Crystal system, rhombohedral (hexagonal). Density, 3.98 g/cm^3; melting point, 2040°C (3704°F); hardness, 9.0 Mohs; refractive index, 1.765; good infrared and ultraviolet transmission; high tensile strength; excellent high-temperature stability; upper continuous-use temperature, 1800-1950°C (3270-3540°F); good electrical properties including a dielectric constant of 7-10.5 and a dielectric strength of 17 kV/mm, and good resistance to acids, alkalies, halogens and metals. Occurrence: Australia, Sri Lanka, Burma, India, Thailand, USA (California, Montana, North Carolina). Used as a gemstone, as an abrasive and polishing material, for pointers, wearing points of instruments and formerly for phonograph needles, for high-precision bearings, valve and roller-bearing balls, ring and plug gages, thread guides of textile machines, for lenses, prisms, optical windows, electron and microwave tubes and optoelectronic components, as a substrate for thin-film components, integrated circuits and superconductors, for infrared detectors, for optical elements in radiation detectors, in dental implants, as whiskers and fibers for composite reinforcement, and for the manufacture of aluminum composites.

Sapphire Blue. Trade name of Hermes Abrasives Limited (USA) for ceramic alumina abrasive grain used in metalworking operations.

sapphire glass. Glass with a bright blue, sapphire-like color.

sapphire whiskers. Axially-oriented, single-crystalline filaments of sapphire (Al_2O_3) with high tensile strengths and moduli of elasticity. They have an average fiber diameter of 1 μm (39 μin.), an average fiber length of 3 mm (0.125 in.), good dielectric properties, excellent high-temperature and corrosion resistance, and an upper service temperature of 1300-1320°C (2370-2410°F). Used as reinforcements in metal-matrix composites, as fillers in plastics to increase heat resistance and dielectric properties, and for electrical and thermal insulating applica-

tions.

sapphirine. A blue, bluish-gray, or green mineral of the aenigmatite group composed of magnesium aluminum silicate, $Mg_{3.5}Al_9Si_{1.5}O_{20}$. Crystal system, monoclinic. Density, 3.40 g/cm^3; refractive index, 1.719. Occurrence: Greenland.

sap pine. See loblolly pine.

sapwood. The layers of wood closest to the bark that are actively involved in the life processes of a tree. *Sapwood* is usually lighter in color than *heartwood*, but less durable and more susceptible to decay. It is not necessarily weaker than heartwood of the same species.

sarabauite. A carmine-red mineral composed of calcium antimony oxide sulfide, $CaSb_{10}O_{10}S_6$. Crystal system, monoclinic. Density, 4.80 g/cm^3. Occurrence: Malaysia.

Saran. Trademark of Dow Chemical Company (USA) for a series of vinyls (e.g., polyvinyl acetate, polyvinyl alcohol, polyvinyl chloride and polyvinylidene chloride) that are originally rigid, but can be made flexible with plasticizers. They are often manufactured as thin, flexible sheets and films, or textile fibers and available in various transparent and colored grades. *Saran* vinyls have excellent resistance to rotting, soiling and damage, good weatherability and flame resistance, good resistance to solvents, oils, brines and many chemicals, good abrasion and electrical resistance, good low-temperature stiffness, good processibility, and a maximum service temperature of 138°C (280°F). Used for pipe and tubing, garden hose, electrical wire insulation, floor and wall coverings, safety glass interlayers, toys, upholstery, automobile seat covers, insect screens, clothing, rainwear, bristles for brushes and brooms, packaging film, coatings, and formerly for phonographic records.

saran. A generic term referring to a group of thermoplastics made by the polymerization of vinylidene chloride ($CH_2=CCl_2$) and available in the form of powders, films, oriented fibers, extruded and molded products, and as copolymers with acrylonitrile or vinyl chloride. They have a density of 1.63 g/cm^3 (0.059 lb/in.^{3}), good abrasion resistance and self-extinguishing properties, good dielectric properties, an upper service temperature of 80-100°C (176-212°F), fair resistance to ultraviolet radiation, low coefficients of friction, low vapor transmission, good resistance to acids, alcohols, alkalies, greases, chemicals, oils and halogens, fair resistance to aromatic hydrocarbons and ketones, and poor resistance to chlorinated hydrocarbons. Used for packaging food products, in latex coatings, for piping for chemical equipment, for upholstery, fibers, bristles, etc. Also known as *polyvinylidene chlorides; polyvinylidene chloride plastics.*

Saranac. Trade name of Colt Industries (UK) for a tool steel containing 0.9% carbon, 0.9% chromium, and the balance iron.

Saranex. Trademark of Dow Chemical Company (USA) for a polyvinylidene film laminated with polyethylene to give heat-sealing qualities, and used for wrapping and packaging applications, and for freezer and food storage bags.

saran fibers. Synthetic fibers whose fiber-forming substance is a long-chain polymer containing at least 80 wt% vinylidene chloride repeating units ($-CH_2=CCl_2-$). They have a softening temperature of 115-137°C (239-279°F), good resistance to acids, alcohols, alkalies, greases, oils and halogens, fair resistance to aromatic hydrocarbons and ketones, poor resistance to chlorinated hydrocarbons, good weatherability, good resistance to mildew and insects, and poor flame resistance (but self-extinguishing). Used for carpets and rugs, curtains and drapes, filter cloth, upholstery and screens.

saran latex. An aqueous dispersion of polyvinylidene chloride plastic that produces glossy, greaseproof and waterproof coatings on paper for food packaging applications. It is also used as an impregnant for textile fabrics.

saran rope. A synthetic rope made by twisting a number of polyvinylidene chloride monofilaments into a strand and three strands into a rope. It is flexible, resistant to rotting, soiling and many chemicals, and about as strong as sisal, but softens greatly above 75°C (167°F). Used for general-purpose and light service applications.

Saranspun. Trade name of National Plastic Products Company (USA) for spun saran (polyvinylidene chloride) fibers.

Saratan. Trademark of Karl Dickel & Co. KG (Germany) for polyethylene, polypropylene, polyvinyl chloride and polyvinylidene chloride shrink and stretch films.

Saratoga. Trade name of AL Tech Specialty Steel Corporation (USA) for an oil-hardening cold-work tool steel (AISI type O1) containing 0.9% carbon, 1.2% manganese, 0.5% chromium, 0.5% tungsten, 0.38% silicon, and the balance iron. It has good nonshrinking properties, a great depth of hardening, good machinability, and is used for blanking, forming and trimming dies, threading and drawing dies, plug gages, and punches.

sarcolite. A flesh-red mineral composed of sodium calcium aluminum silicate, $(Ca,Na)_4Al_3(Al,Si)_3Si_6O_{24}$. Crystal system, tetragonal. Density, 2.92 g/cm^3; refractive index, 1.6039. Occurrence: Italy.

sarcopside. A colorless, or gray to brown mineral of the olivine group that is composed of iron manganese phosphate, $(Fe,Mn)_3(PO_4)_2$, and may also contain magnesium. Crystal system, monoclinic. Density, 3.95 g/cm^3. Occurrence: USA (South Dakota, New Hampshire).

sard. A brownish variety of *chalcedony* quartz used as a semiprecious stone.

Sardo. Trade name of Hidalgo Steel Company Inc. (USA) for a nonshrinking, air-hardening steel containing 1.5% carbon, 13% chromium, and the balance iron. Used for dies and tools.

sardonyx. A mineral made up of brown layers of *sard* and white or black layers of *chalcedony* quartz. Used for cameos.

Sarelon. Trade name for regenerated protein (azlon) fibers.

Sarfa. Trade name for rayon fibers and yarns used for textile fabrics.

Sarille. Trademark of Courtaulds Limited (UK) for a crimped viscose rayon staple fiber that blends well with other fibers, and is used for the manufacture of fabrics with wool-like textures and warm, bulky feel, e.g., clothing and blankets.

sarkinite. A red or reddish yellow mineral of the triplite group composed of manganese arsenate hydroxide, $Mn_2AsO_4(OH)$. Crystal system, monoclinic. Density, 4.13 g/cm^3; refractive index, 1.807. Occurrence: Sweden.

Sarlink. Trade name of Novacor Chemicals Inc. (USA) for a series of high-performance thermoplastic elastomers. The *Sarlink 1000 Series* has excellent oil and fuel resistance, the *Sarlink 2000 Series* offers outstanding compression-set and damping characteristics, excellent low-temperature flexibility and low permeability, and the *Sarlink 3000 Series* has excellent moldability and compression set characteistics, high resiliency, and good weatherability.

Sarlon. (1) Trademark for saran (polyvinylidene chloride) fibers with good chemical resistance, good weatherability, and good resistance to mildew and insects. Used for carpets and rugs, draperies, upholstery, clothing, industrial fabrics, etc.

(2) Trademark of Sarlon Industries Proprietary Limited (Australia) for polypropylene fibers and filament yarns.

sarmientite. A pale yellow-orange mineral composed of iron arsenate sulfate hydroxide hydrate, $Fe_2(AsO_4)(SO_4)OH \cdot 5H_2O$. Crystal system, monoclinic. Density, 2.58 g/cm^3; refractive index, 1.635. Occurrence: Argentina.

sartorite. A lead-gray mineral with metallic luster composed of lead arsenic sulfide, $PbAs_2S_4$. Crystal system, orthorhombic. Density, 5.10 g/cm^3. Occurrence: Switzerland.

saryarkite. A white mineral composed of aluminum calcium yttrium phosphate silicate hydroxide nonahydrate, $(Ca,Y,Th)_2Al_4(SiO_4,PO_4)_4(OH)_6 \cdot 9H_2O$. Crystal system, tetragonal. Density, 3.07 g/cm^3.

Sasa. Trade name of Nippon Sheet Glass Company Limited (Japan) for a patterned glass.

sasaite. A white mineral composed of aluminum iron phosphate sulfate hydroxide hydrate, $(Al,Fe)_{14}(PO_4)_{11}(SO_4)(OH)_7 \cdot 84H_2O$. Crystal system, orthorhombic. Density, 1.75 g/cm^3; refractive index, 1.473. Occurrence: South Africa.

Saskfor. Trademark of Saskatchewan Forest Products Corporation (Canada) for lumber.

Saskply. Trademark of Saskatchewan Forest Products Corporation (Canada) for plywood and plywood products.

sassafras. The durable, moderately hard and heavy wood of the medium-sized tree *Sassafras albidum* belonging to the laurel family. The heartwood is dull grayish-brown, dark brown or reddish brown, and the sapwood light yellow. *Sassafras* has good shock and decay resistance, and poor bending and compression qualities. Source: Eastern United States. It is of limited commercial importance, but sometimes used as lumber for small boats, fenceposts, rails, containers, millwork, barrels and buckets, and as a fuelwood.

sassandra mahogany. See red khaya.

sassoline. See sassolite.

sassolite. A white or colorless mineral composed of hydrogen borate, H_3BO_3. It can also be made synthetically. Crystal system, triclinic. Density, 1.46 g/cm^3; refractive index, 1.456. Also known as *sassoline*.

Sastiga. Trade name for rayon fibers and yarns used for textile fabrics.

Satan. Trade name of A. Milne & Company (USA) for a chromium-type hot-work tool steel containing 0.23% carbon, 0.6% manganese, 1.25% silicon, 10% chromium, 1% vanadium, 0.75% nickel, 0.45% tungsten, 0.45% molybdenum, and the balance iron. Used for press and forging dies, extrusion dies, and tools.

Satco. Trade name of NL Industries (USA) for a white metal alloy containing 97.5% lead and a total of 2.5% tin and calcium. It has a melting point of 420°C (788°F), and is used for high-speed bearings, trailer and driving boxes, etc.

sateen. A woven fabric, usually of mercerized cotton or other spun yarn, in a satin weave and having a smooth, lustrous, water-repellent surface. Used for draperies, linings, bedspreads, and fancy dresses and costumes.

Satene. Trade name of W. Canning Materials Limited (USA) for greaseless satin finishing compositions used in metal finishing.

satimolite. A white mineral composed of potassium sodium aluminum borate chloride hydrate, $KNa_2Al_4(B_2O_3)_3Cl_3 \cdot 13H_2O$. Crystal system, orthorhombic. Density, 2.10 g/cm^3; refractive index, 1.552. Occurrence: Russia.

satin. A soft, closely-woven fabric, usually of wool, silk, rayon or synthetic fibers, with a smooth, lustrous face and a dull-finish back, used for clothing and footwear.

Satin Blend. Trade name of F.L & J.C. Codman Company (USA) for nonwoven abrasives.

satin-back crepe. See crepe-back satin.

satin calf. Leather made from the grain side of cattlehide by splitting, dressing with oil and smooth finishing.

Satincoat. Trademark of Dofasco Inc. (USA) for continuous galvanized steel sheets and coils.

Satiné. See Meryl Satiné.

satinee. The moderately hard, reddish brown wood of the tree *Ferolia guianensis* belonging to the Rosaceae family in the order of Rosales. It has a fine grain, beautiful appearance, and takes a high polish. Average weight, 865 kg/m³ (54 lb/ft³). Source: Guianas, Central America. Used for fine furniture and cabinet-making.

satinella. A lustrous, satin-weave cotton fabric used for clothing and linings.

satinet. A thin silk satin or imitation satin fabric used for clothing.

Satinex. Trade name of Glaceries de Saint-Roch SA (Belgium) for a glass with acid-treated surface.

Satin Finish. Trademark for polyvinyl chloride compounds with gloss levels ranging from basic matte to high satin. Some grades provide controlled gloss. Used for automotive components, weatherstripping, and wire and cable.

Satinflo. Trademark of Porritts & Spencer Canada Inc. for papermaker's felts.

Satin-Glide. Trade name of National Standard Company (USA) for stainless-steel welding wire.

satin glaze. A semimatte glaze, frequently of the tin-zinc-titanium type, having a characteristic satin-like appearance. Used on wall tiles. Also known as *satin-vellum glaze; vellum glaze.*

Satin-Grey. Trade name of Australian Window Glass Proprietary Limited (Australia) for a gray figured glass.

Satinlite. Trade name of Australian Window Glass Proprietary Limited (Australia) for a patterned glass.

Satinol. Trade name of ASG Industries Inc. (USA) for a glass treated on one or both surfaces to give an appearance like satin.

Satinovo. Trademark of Saint-Gobain (France) for an acid-etched glass.

satin rouge. Lampblack in brick form used for polishing silverware, celluloid and bone. See also lampblack.

Satin Shield. Trade name of Birchwood Laboratories Inc. (USA) for a water-emulsion blend of waxes.

satin spar. See fibrous gypsum.

satin stone. See fibrous gypsum.

satin-vellum glaze. See satin glaze.

Satin Vinyl. Trade name of Frazee Industries (USA) for vinyl paints used for exterior and interior applications.

satin-weave fabrics. Woven fabrics, usually with smooth, glossy surfaces, having the weft (transverse) threads or yarns alternately cross over several warp (longitudinal) threads or yarns and under a single thread or yarn. Also known as *harness-weave fabrics.*

satin white. A paint pigment and extender composed of calcium sulfate and calcium aluminate. The particle size ranges from 0.2 to 2.0 μm (8 to 78 μin.) and it is also used as a high-bulking filler in paper coating formulations.

satinwood. The hard, lightweight, yellowish-brown wood of the tree *Chloroxylon swietenia* belonging to the mahogany family. It has a satiny luster, a close grain, and high durability. Source: India, Sri Lanka. Used for fine furniture, and farm tools.

Satmumetal. Trade name of Telcon Metals Limited (UK) for a soft magnetic alloy composed of 53-58% nickel and 42-47% iron. It has high saturation induction and low coercivity, and is used for toroids, small distribution transformers, instrument transformers, and ground leakage protective devices.

satpaevite. A canary to saffron-yellow mineral composed of aluminum vanadium oxide hydrate, $Al_{12}V_8O_{37} \cdot 30H_2O$. Density, 2.40 g/cm³. Occurrence: Russia, Kazakhstan.

Satran. Trade name of MRC Polymers (USA) for a series of acrylonitrile-butadiene-styrene, styrene-acrylonitrile and polystyrene resins supplied in reinforced and unreinforced grades with special impact and heat resistance.

satterlyite. A brownish yellow mineral composed of iron magnesium phosphate hydroxide, $(Fe,Mg)_2(PO_4)(OH)$. Crystal system, hexagonal. Density, 3.68 g/cm³; refractive index, 1.7195. Occurrence: Canada (Yukon).

saturated felt. Felt saturated with hot asphalt or bitumen for use in the manufacture of roll roofing, and shingles.

saturation-bonded nonwoven fabrics. Nonwoven fabrics produced by a process in which bonding is achieved by the application of a liquid adhesive to the entire fiber batt or web. Also known as *saturation-bonded nonwovens.*

Saturn. (1) Trade name of Cytemp Specialty Steel Division (USA) for a fast-finishing tool steel (AISI type F2) containing 1.25% carbon, 3.5% tungsten, and the balance iron. Used for cutting tools, drills, chasers, dies, and wire-drawing plates.

(2) Trade name for a dual-cure dental composite.

Satylite. Trade name of Enthone-OMI Inc. (USA) for a satin nickel electrodeposit and plating process.

sauconite. A brown to pinkish cinnamon mineral of the smectite group composed of zinc magnesium aluminum silicate hydrate, $(Zn,Mg)_3(Si,Al)_4O_{10}(OH)_2 \cdot xH_2O$. Crystal system, monoclinic. Refractive index, 1.592. Occurrence: USA (Arkansas, Colorado, Pennsylvania).

Sauder. Trademark of Sauder Industries Limited (Canada) for prefinished plywood.

Sava. Trade name for rayon fibers and yarns used for textile fabrics.

Savbit. Trade name of Multicore Corporation (USA) for a solder composed of 50% tin, 48.5% lead and 1.5% copper. It has a liquidus temperature of 183°C (361°F) and a solidus temperature of 215°C (419°F).

Saville. Trade name of J.J. Saville & Company Limited (UK) for a series of alloy steels (e.g., nickel-chromium, nickel-chromium-molybdenum and chromium-molybdenum-vanadium types including carburizing and nitriding types), stainless steels (e.g., austenitic chromium-nickel types), tool steels (e.g., plain-carbon, cold-work, hot-work and high-speed types) and mold steels.

sawdust. The particles of wood obtained in sawing, and used in the manufacture of particleboard, as an aggregate for concrete, in kraft paper pulp, as a mild abrasive for tumbling and vibratory finishing of metals, as a packaging material, as a fuel, and in meat smoking.

sawdust concrete. A lightweight concrete of relatively low strength in which sawdust is used as the main aggregate. Used as a *nailable concrete* in construction applications.

sawn lumber. Lumber cut from a log by means of a saw using any of various methods. See also lumber; plain-sawn lumber; resawn lumber; rough-sawn lumber; quarter-sawn lumber.

sawn veneer. Veneer produced by simple sawing as compared to rotary-cut or sliced veneer.

saw-sized lumber. Rough lumber sawn to almost uniform and exact size within a specified positive or negative sawing tolerance.

SAX. Trademark of Alcan Wire and Cable (Canada) for non-dusting, free-flowing aluminum grains.

Saxon. Trade name of Pittsburgh Corning Corporation (USA) for glass blocks with linear patterns.

Saxonia. German trade name for metallurgical lead (99.9% pure) used in the manufacture of lead alloys.

Saxonia metal. British zinc alloy containing 6% copper, 5.3% tin, 3% lead and 0.9% aluminum. It has poor resistance to heat and live steam, and is used for bearings.

saxony. (1) A high-grade wool used to make suits and coats in twill or herringbone weaves. Originally from Saxony, Germany.

(2) A high-grade, finely twisted wool or worsted yarn used to make good-quality fabrics.

(3) A soft, heavy fabric with a slightly fuzzy or furry surface, made from saxony (1) or other high-quality wool, and used especially for making suits and coats.

(4) A soft *tweed* fabric of fine-quality wool, used especially for sport coats.

Saxony blue. See smalt.

Sayelle. Trade name of Monsanto Chemical Company (USA) for acrylic yarns used for the manufacture of knitted and woven clothing.

sayrite. A yellow-orange or red-orange mineral composed of lead uranyl oxide hydroxide tetrahydrate, $Pb_2(UO_2)_5O_6(OH)_2 \cdot 4H_2O$. Crystal system, monoclinic. Properties Density, 6.76 g/cm³. Occurrence: Zaire.

Sazanami. Trade name of Asahi Glass Company Limited (Japan) for a figured glass with hexagonal wire mesh.

sazhinite. A white-, gray- or cream-colored mineral composed of sodium cerium silicate hexahydrate, $Na_3CeSi_6O_{15} \cdot 6H_2O$. Crystal system, orthorhombic. Density, 2.61 g/cm³; refractive index, 1.528. Occurrence: Russian Federation.

S-Babbitt. Trade name of American Smelting & Refining Company (USA) for a lead babbitt containing 15% antimony, 1% tin, 1% arsenic and 0.5% copper. Used for bearings, liners and sleeves.

SB Boron. Trade name of SB Boron Corporation (USA) for elemental amorphous boron powders including *SB Boron 92* (90-92% pure) and *SB Boron 95* (95+% pure) with a melting point of 2167°C (3935°F). The former has a maximum particle size of 1 μm (39 μin.) and is used in welding fluxes, coatings, linings and in various automotive, aerospace and industrial applications, and the latter has a maximum particle size of 0.8 μm (31 μin.) and is used in solid fuels and slurries, explosive primers, ceramic formulations, and for nuclear absorption applications.

SB-Mercier. Trade name of Ato Findley (France) for an extensive series of adhesive cements and mortars, plasters, concretes, mastics, additives, primers and coatings used for building, construction and architectural applications, e.g., tile fixing, floor finishing, and home decoration.

sborgite. A colorless mineral composed of sodium borate pentahydrate, $NaB_5O_8 \cdot 5H_2O$. It can also be made synthetically. Crystal system, monoclinic. Density, 1.71 g/cm³; refractive index, 1.44. Occurrence: Italy.

SBQ steels. See special-quality steel bars.

scacchite. A pink mineral composed of manganese chloride, $MnCl_2$. It can also be made synthetically from manganese chloride tetrahydrate, $MnCl_2 \cdot 4H_2O$. Crystal system, rhombohedral (hexagonal). Density, 3.00 g/cm³; refractive index, 1.62. Occurrence: Volcanic regions, such as the vicinity of Vesuvius, Italy.

S-C asphalt. See slow-curing asphalt.

Scaldura. Trade name for rayon fibers and yarns used for textile fabrics.

Scaldyna. Trade name for rayon fibers used for textile fabrics.

scaleboard. Wood made into a relatively thin sheet and used as veneer.

scandia. See scandium oxide.

scandium. A soft, silvery-white, very light metallic element of Group IIIB (Group 3) of the Periodic Table. It is commercially available in the form of chips, ingots, lumps, sheets, foils, wire and powder. The principal scandium ores are *wolframite* and *thortveitite*. Density, 2.99 g/cm³; melting point, 1541°C (2806°F); boiling point, 2831°C (5128°F); hardness, 78-136 Brinell; atomic number, 21; atomic weight, 44.956; trivalent. *Scandium* has high electrical resistivity (about 63 μΩcm), good resistance to tarnishing in air, is chemically similar to rare-earth elements. There are two allotropic forms (i) *Alpha scandium* (hexagonal crystal structure) that is stable at room temperature, and (ii) *Beta scandium* (body-centered cubic crystal structure) that is stable at high temperatures. *Scandium* is used in semiconductors, superconductors, leak detectors (scandium isotopes), ceramics, and for aerospace applications. Symbol: Sc.

scandium boride. A refractory, solid compound of scandium and boron used in ceramics and materials research. Density, 3.17-3.67 g/cm³; melting point, 2249°C (4080°F). Formula: ScB_2.

scandium bromide. White, hygroscopic crystals. Melting point, 969°C (1776°F) [also reported as 967°C (1772°F)]. Used in the preparation of scandium metal and compounds. Formula: $ScCl_3$.

scandium carbide. A compound of scandium and carbon with a density of 3.59 g/cm³, used for high-temperature semiconductors, and in ceramics. Formula: ScC.

scandium chloride. White, hygroscopic crystals or powder (99.9+% pure). Density, 2.39 g/cm³; melting point, 939°C (1722°F). Used in the preparation of scandium metal and compounds. Formula: $ScCl_3$.

scandium disilicate. See scandium silicate (ii).

scandium fluoride. A white powder (99.9+% pure) with a melting point of 1515°C (2759°F), used in the preparation of scandium metal, and in special glasses. Formula: ScF_3.

scandium halide. A compound of scandium and a halogen, such as bromine, chlorine, fluorine or iodine.

scandium monosilicate. See scandium silicate (i).

scandium nitride. A compound of scandium and nitrogen. Density, 3.6 g/cm³; melting point, 2700°C (4890°F). Used for aerospace applications, and as a crucible material for the preparation of high-purity single crystals of gallium. Formula: ScN.

scandium oxide. White crystals or powder (99.9+% pure). Crystal system, cubic. Density, 3.864 g/cm³; melting point, 2485°C (4505°F); refractive index, 1.964. Used in ceramics, as a network former in glass, in high-temperature systems, for electronic applications, in scandium-doped yttrium-based superconductors, and in the preparation of scandium fluoride. Formula: Sc_2O_3. Also known as *scandia*.

scandium phosphide. A compound of scandium and phosphorus with a density of 3.28 g/cm³, used as a semiconductor. Formula: ScP.

scandium silicate. Any of the following compounds of scandium oxide and silicon dioxide used in ceramics and materials research: (i) *Scandium monosilicate*. Density, 3.49 g/cm³; melting point, 1950°C (3540°F). Formula: Sc_2SiO_5; and (ii) *Scandium disilicate*. Density, 3.39 g/cm³; melting point, 1860°C

(3380°F). Formula: $Sc_2Si_2O_7$.

scandium sulfide. A compound of scandium and sulfur available as yellow crystals. Crystal system, orthorhombic. Density, 2.91 g/cm³; melting point, 1775°C (3227°F). Used in ceramics. Formula: Sc_2S_3.

scandium telluride. A compound of scandium and tellurium available as black crystals. Crystal system, hexagonal. Density, 5.29 g/cm³. Used in ceramics. Formula: Sc_2Te_3.

Scandruden. Trade name of Scanglas A/S (Sweden) for an insulating glass.

Scanex. Trade name of Scanglas A/S (Sweden) for a laminated glass.

scantling. (1) Square-sawn timber that is 50 to less than 100 mm (2 to less than 4 in.) thick and 50 to less than 125 mm (2 to less than 5 in.) wide. Used for studding.

(2) A British term for a building stone with a length of 1.8 m (6 ft.) or more.

scapolite. A group of tectosilicate minerals including *marialite, meionite* and *sarcolite,* and varying in composition from sodium aluminum silicate with sodium chloride to calcium aluminum silicate with calcium carbonate. It is usually white to gray in color with a vitreous to pearly luster, but also occurs in bluish, greenish, yellowish and reddish varieties. Crystal system, tetragonal. Density, 2.65-2.74 g/cm³; hardness, 5-6 Mohs. Used as a gemstone. Also known as *wernerite.*

Scarab. Trade name of Beetle Products Company (UK) for urea-formaldehyde resins.

scarbroite. (1) A white mineral composed of aluminum hydroxide monohydrate, $Al_2(OH)_6 \cdot H_2O$. Crystal system, monoclinic. Density, 2.21 g/cm³. Occurrence: Montenegro.

(2) A white mineral composed of aluminum carbonate hydroxide, $Al_5(OH)_{13}(CO_3) \cdot 5H_2O$. Crystal system, triclinic. Density, 2.17 g/cm³; refractive index, 1.509. Occurrence: UK.

scarlet chrome. See Chinese scarlet (2).

scarlet lake. A pigment made by precipitation of aniline color upon a base of alumina trihydrate and barium sulfate.

scarlet oak. The hard, moderately strong and durable wood of the medium-sized tree *Quercus coccinea* belonging to the red oak family. The heartwood is reddish-brown and the sapwood white. It has a coarse texture and a beautiful grain. Source: Eastern and central United States (from southern Maine to southern Alabama, and from Missouri to Maryland); southeastern Canada. Used for lumber, veneer, furniture, boxes, agricultural implements, handles, etc. See also American red oak; oak; red oak.

Scason. Trade name of Pechiney/Eurotungstène (France) for a wear-resistant tungsten carbide with cobalt binder, used for anti-skid studs for tires, and studded straps for snow and ice chains.

scavenger. (1) A reactive metal introduced into a molten metal or alloy to combine with and remove dissolved gases, such as oxygen and/or nitrogen.

(2) A chemically active substance, such as barium or barium alloy, calcium, magnesium, niobium, tantalum, thorium, thorium-mischmetal, zirconium, palladium black, charcoal or silica gel, used in a vacuum tube to remove residual gases by adsorption or combination. Also known as *degasser; getter.*

scawtite. A colorless mineral composed of calcium carbonate silicate dihydrate, $Ca_7(Si_6O_{18})(CO_3) \cdot 2H_2O$. Crystal system, monoclinic. Density, 2.77 g/cm³; refractive index, 1.605. Occurrence: USA (California), Northern Ireland.

scented mahogany. See sapele.

sceptre brass. A high-strength brass composed of 62% copper, 36% zinc, 0.7-1.4% iron, 0.5-1.1% aluminum and 0.07% lead. Used for ornamental and architectural parts, and hardware.

schachnerite. A gray mineral composed of silver mercury, $Ag_{1.1}Hg_{0.9}$. It can also be made synthetically. Crystal system, hexagonal. Density, 13.38 g/cm³. Occurrence: Germany.

schafarzikite. A mineral composed of iron antimony oxide, $FeSb_2O_4$. Crystal system, tetragonal. Density, 4.30 g/cm³; refractive index, above 1.74. Occurrence: Czech Republic.

schairerite. A colorless mineral composed of sodium chloride fluoride sulfate, $Na_3SO_4(F,Cl)$. Crystal system, hexagonal. Density, 2.61 g/cm³. Occurrence: USA (California).

schallerite. A light brown mineral of the pyrosmaltite group composed of manganese silicate hydroxide, $(Mn,Mg,Fe)_8(Si,As)_6O_{15}(OH)_{10}$. Crystal system, hexagonal. Density, 3.34-3.37 g/cm³. Occurrence: USA (New Jersey).

schapbachite. A reddish-black or lead-gray mineral composed of silver bismuth sulfide, α-$AgBiS_4$. It can also be made synthetically. Crystal system, cubic.

schappe silk. (1) A term formerly used for a yarn spun from partially degummed silk. Also known as *schappe. Note:* The fermentation degumming process was known as "Schapping".

(2) A term now used for yarns or fabrics of spun silk or silky synthetics. Also known as *schappe.*

Scharr-Stat. Trade name of Scharr Industries Inc. (USA) for a wear-resistant polyethylene film for heat-sealing applications.

schaurteite. A white mineral of the fleischerite group composed of calcium germanium sulfate hydroxide tetrahydrate, $Ca_3Ge(SO_4)_2(OH)_4 \cdot 4H_2O$. Crystal system, hexagonal. Density, 2.65 g/cm³; refractive index, 1.569. Occurrence: Southwest Africa.

Scheele's green. A fine, light green or yellowish-green powder composed of copper arsenite ($CuHAsO_3$) and used as a pigment.

scheelite. A colorless, yellowish white, gray, brown or green mineral with a vitreous luster. It is composed of calcium tungstate, $CaWO_4$, and contains theoretically 80.6% WO_3 and 19.4% CaO. Crystal system, tetragonal. Density, 5.9-6.1 g/cm³; hardness, 4.5-5.0 Mohs; refractive index, 1.918. Occurrence: Asia (China), Australia, Brazil, Europe (Czech Republic, Germany, Italy, UK), USA (Arizona, California, Colorado, Nevada, Utah). Used as an ore of tungsten, as a phosphor, and as an alloying addition to steel.

Schein 20/20. Trade name of Schein Rexodent (UK) for a light-cure hybrid composite used in dentistry.

Schenk. Trade name of Schenk Leichtgusswerke (Germany) for a series of corrosion-resistant die- and sand-cast aluminum, magnesium and zinc alloys including various aluminum-copper-silicon, aluminum-magnesium, aluminum-silicon, aluminum-silicon-copper, aluminum-zinc-copper, magnesium-aluminum, magnesium-aluminum-zinc and zinc-aluminum types.

schertelite. A mineral composed of ammonium magnesium hydrogen phosphate tetrahydrate, $Mg(NH_2)_2H_2(PO_4)_6 \cdot 4H_2O$. Crystal system, orthorhombic. Density, 1.82 g/cm³.

schieffelinite. A white mineral composed of lead tellurate sulfate monohydrate, $Pb(Te,S)O_4 \cdot H_2O$. Crystal system, orthorhombic. Density, 4.98 g/cm³. Occurrence: USA (Arizona).

Schiff bases. A class of colorless, weakly basic solid compounds obtained by the condensation of aldehydes or ketones with primary amines. They are used in liquid crystals for electronic applications (e.g., displays), as intermediates, as rubber accelerators, and in organic dyes, such as naphthol and phenylene blue. The general formula is RR'X=NR".

Schirm. Trade name of Dörrenberg Edelstahl GmbH (Germany) for a series of water-hardening tool steels containing 0.85-1% carbon, up to 0.25% silicon and manganese, respectively, and the balance iron. Used for springs, lathe tools, taps, reamers, broaches and drills.

schirmerite. A lead-gray mineral composed of silver lead bismuth sulfide, $AgPb_2Bi_3S_7$. Crystal system, orthorhombic. Density, 6.74 g/cm^3. Occurrence: USA (Colorado).

schist. Any of various crystalline metamorphic rocks, such as feldspar, hornblende, mica and quartz, that due to their foliated structures can readily be split into thin plates or slabs. Used in building construction, e.g., as concrete aggregate.

schlossmacherite. A light green mineral of the alunite group composed of oxonium calcium aluminum arsenate sulfate hydroxide, $(H_3O,Ca)Al_3(SO_4,AsO_4)_2(OH)_6$. Crystal system, rhombohedral (hexagonal). Density, 3.00 g/cm^3; refractive index, 1.597. Occurrence: Chile.

schmitterite. A straw-yellow mineral composed of uranyl tellurate, UO_2TeO_3. Crystal system, orthorhombic. Density, 6.88 g/cm^3; refractive index, above 2.05. Occurrence: Mexico.

Schneider. Trade name of Creusot-Loire (France) for an extensive series of austenitic, ferritic and martensitic stainless steels, various plain-carbon, cold-work, hot-work and high-speed tool steels, and a wide range of alloy steels including several case-hardening grades.

schneiderhoehnite. A deep brown mineral composed of iron arsenic oxide, $Fe_8As_{10}O_{23}$. Crystal system, triclinic. Density, 4.30 g/cm^3. Occurrence: Morocco, Southwest Africa.

schoderite. A yellow mineral composed of aluminum vanadium oxide phosphate octahydrate, $Al_2(PO_4)(VO_4)\cdot9H_2O$. Crystal system, monoclinic. Density, 1.92 g/cm^3; refractive index, 1.563. Occurrence: USA (Arkansas).

schoenfliesite. A reddish brown mineral of the sohngeite group composed of magnesium tin hydroxide, $MgSn(OH)_6$. It can also be made synthetically. Crystal system, cubic. Density, 4.30 g/cm^3; refractive index, 1.590. Occurrence: USA (Alaska).

schoepite. A radioactive mineral composed of uranium oxide dihydrate, $UO_3\cdot2H_2O$. It occurs in three phases: Phase I is amber to brown in color and alters to Phases II and III in air. Phase II and Phase III are yellow in color and are known as *metaschoepite* and *paraschoepite,* respectively. *Schoepite* can also be made synthetically. Crystal system, orthorhombic. Density, 4.49-4.92 g/cm^3; refractive index, 1.720. Occurrence: Congo.

scholzite. A white or colorless mineral composed of calcium zinc phosphate dihydrate, $CaZn_2(PO_4)_2\cdot2H_2O$. It can also be made synthetically by crystallization from hot aqueous solutions of calcium chloride, zinc chloride and phosphoric acid. Crystal system, orthorhombic. Density, 3.11 g/cm^3; refractive index, 1.586. Occurrence: Germany, Australia.

Schomberg alloy. An antifriction alloy composed of 87% zinc, 10% tin and 3% copper, and used for bearings.

Schönox. Trade name of SCHÖNOX GmbH (Austria) for an extensive range of building products including adhesives for bonding flooring, parquet flooring and wall coverings, several jointing, wall and floor compounds, various tile and stone cements, and several other cements and sealants.

schoonerite. A brown to copper-red mineral composed of iron manganese zinc phosphate hydroxide nonahydrate, $ZnMnFe_3(PO_4)_2(OH)_2\cdot9H_2O$. Crystal system, orthorhombic. Density, 2.89 g/cm^3; refractive index, 1.652. Occurrence: USA (New Jersey).

schorl. A black mineral of the tourmaline group composed of sodium iron aluminum boron silicate hydroxide, $NaFe_3Al_6(BO_3)_3Si_6O_{18}(OH)_4$. It can also be made synthetically. Crystal

system, hexagonal. Density, 3.13 g/cm^3; refractive index, 1.658. Occurrence: Norway. Also known as *schorlite.*

schorlomite. A black mineral with a vitreous luster belonging to the garnet group. It is composed of calcium iron titanium silicate, $Ca_3(Fe,Ti)_2(Si,Ti)_3O_{12}$. Crystal system, cubic. Density 3.81-3.88 g/cm^3; hardness, 7-7.5 Mohs; refractive index, 1.94. Occurrence: USA (California), Canada (Quebec).

Schottglas. Trade name of Schott Glas AG (Germany) for an extensive series of of glass products.

schreibersite. A silver- to tin-white mineral composed of iron nickel phosphide, $(Fe,Ni)_3P$. It can also be made synthetically. Crystal system, tetragonal. Density, 7.00 g/cm^3. Occurrence: Philippines. Also known as *rhabdite.*

schreyerite. A black or reddish brown mineral composed of vanadium titanium oxide, $V_2Ti_3O_9$. Crystal system, monoclinic. Density, 4.48 g/cm^3; refractive index, 2.7. Occurrence: Kenya.

schroeckingerite. A greenish yellow, radioactive mineral with a vitreous luster. It is composed of sodium calcium uranyl carbonate fluoride sulfate decahydrate, $NaCa_3(UO_2)(CO_3)_3(SO_4)F\cdot10H_2O$, and can also be made synthetically. Crystal system, orthorhombic. Density, 2.54 g/cm^3; hardness, 2.5 Mohs; refractive index, 1.543; fluorescent in ultraviolet light. Occurrence: Argentina, USA (Arizona, Utah, Wyoming), Europe.

schubnelite. A black mineral composed of iron vanadium oxide dihydrate, $Fe_2(V_2O_8)\cdot2H_2O$. Crystal system, triclinic. Density, 3.28 g/cm^3. Occurrence: Gabon.

schuetteite. A canary-yellow mineral composed of mercury oxide sulfate, $Hg_3(SO_4)O_2$. It can also be made synthetically by treating mercury sulfate $(HgSO_4)$ with water. Crystal system, hexagonal. Density, 8.18 g/cm^3. Occurrence: USA (California, Idaho, Nevada, Oregon).

schuilingite. A blue mineral composed of copper lead rare-earth carbonate hydroxide hydrate, $PbCuLn(CO_3)_3(OH)\cdot1.5H_2O$. Crystal system, orthorhombic. Density, 5.20 g/cm^3. Occurrence: Zaire.

Schulaflex. Trademark of A. Schulman Inc. (USA) for a series of flexible elastomers.

Schulamid. Trademark of A. Schulman Inc. (USA) for a series of polyamide (nylon) polymers and formulated, compounded or alloyed ionomers supplied as beads, pellets, granules or bulk powders for the manufacture of extruded and molded articles.

Schulamid Nylon 6. Trade name of A. Schulman Inc. (USA) for a series of resins based on polyamide 6 (nylon 6) and available in various grades including unmodified, casting, elastomer-copolymer, stampable-sheet, glass or carbon fiber-reinforced, glass bead- and/or mineral-filled, silicone-, molybdenum disulfide- or polytetrafluoroethylene-lubricated, fire-retardant, high-impact and UV-stabilized.

Schulatec. Trademark of A. Schulman Inc. (USA) for unprocessed or semiworked thermoplastics supplied as beads, pellets and powders for the plastics and rubber industries.

Schulink. Trademark of A. Schulman Inc. (USA) for a series of crosslinkable high-density polyethylenes supplied as bulk powders or pellets.

Schulman FPP. Trade name of A. Schulman Inc. (USA) for elastomer-modified and calcium carbonate-filled polypropylenes.

schultenite. A colorless mineral composed of lead hydrogen arsenate, $PbHAsO_4$. It can also be made synthetically. Crystal system, monoclinic. Density, 6.04 g/cm^3; refractive index, 1.910. Occurrence: Germany, Southwest Africa.

Schulz. German trade name for a heat- and corrosion-resistant alloy composed of 35% tungsten, 31% cobalt, 15% chromium,

10% molybdenum, 6.3% nickel, 2% iron and 0.7% carbon. Used for tools.

Schulz alloy. A free-cutting alloy composed of 91% zinc, 6% copper and 3% aluminum, and used for ornamental parts, die castings, and type metals.

schumacherite. A yellow mineral composed of bismuth hydroxide vanadium oxide, $Bi_3O(OH)(VO_4)_2$. Crystal system, triclinic. Density, 6.86 g/cm^3. Occurrence: Germany.

Schwabenstahl. Trade name of Schwäbische Hüttenwerke GmbH (Germany) for a cold-drawn, free-cutting steel containing up to 0.15% carbon, 0.7-1.2% manganese, 0.24-0.35% sulfur, 0.04-0.12% phosphorus, 0.15-0.35% lead, and the balance iron. Used for screw-machine products, handles, disks, and pins.

schwartzembergite. A brown-yellow mineral composed of lead chlorate iodate hydroxide, $Pb_6(IO_3)_2Cl_2O_2(OH)$. Crystal system, orthorhombic. Density, 7.09 g/cm^3. Occurrence: Chile.

Schwarza. Trade name for nylon and rayon fibers and yarns used for textile fabrics.

Schwarzpunkt. Trade name of SWB Stahlformguss Gesellschaft mbH (Germany) for an oil-hardening tool steel used for cutters, reamers and dies.

Schweinfurt green. See copper acetoarsenite.

Schwermetall. Trade name of Plansee Metallwerk-Gesellschaft (Austria) for a heavy, sintered tungsten alloy used for radioactive shields, counterweights and gyroscopes.

Scimitars. Trade name of Barker & Allen Limited (UK) for a nickel silver containing 30% zinc, 18% nickel, and the balance copper. Used for ornaments and household utensils.

Scintibloc. French trade name for a thallium-doped sodium iodide [NaI(Tl)] used for scintillators in gamma spectrometry and X-ray analysis. See also sodium iodide thallide.

scintillator. A crystalline material that converts incident ionizing radiation into visible light. The light (optical photon) is emitted in the form of rapid flashes. Used in light-sensitive detectors to measure the intensity of X-ray beams, and as part of nuclear radiation counters. See also scintillator crystals.

scintillator crystals. Transparent phosphor crystals that emit minute flashes of visible light when struck (excited) by ionizing radiation (e.g., alpha, beta, gamma or X-rays). Examples of such crystals include thallium-activated sodium iodide (NaI:Tl), thallium-activated cesium iodide (CsI:Tl), cerium-activated yttrium aluminum garnet (YAG:Ce), bismuth germanium oxide (BGO) and gadolinium silicon oxide (GSO). See also scintillator.

Sclair. Trademark of Nova Chemicals Inc. (USA) for a series of polyolefin resins including various very-low-density, linear-low-density, linear-medium-density, medium-density and high-density polyethylenes supplied in film, extrusion, blow molding, injection molding and rotational molding grades.

Sclairfilm. Trademark of E.I. DuPont de Nemours & Company (USA) for clear, tough, durable polyolefin films supplied in several grades. They offer excellent hot-tack and heat-seal strength, and are used as laminating films, sealant layers in multilayer structures, monolayer bag films, and for vacuum packaging applications.

Sclairlink. Trademark of Novacor Chemicals Inc. (USA) for cross-linkable polyethylene resins used in the manufacture of molded articles.

Scleron. Trade name of Vereinigte Leichtmetallwerke GmbH (Germany) for a series of age-hardened aluminum alloys containing (i) 4% silicon and 1.5% copper, (ii) 4% copper and 0.1% lithium, (iii) 12% zinc, or (iv) 12% zinc, 3% copper, 0.6%

manganese, 0.5% silicon, 0.4% iron and 0.1% lithium. Used for machinery and structural parts, connecting rods, etc.

scleroprotein. See albuminoid.

Scobalit. Trade name for unsaturated polyester resins used for panels.

scolectite. A white or colorless mineral of the zeolite group composed of calcium aluminum silicate trihydrate, $CaAl_2Si_3O_{10}\cdot 3H_2O$. Crystal system, monoclinic. Density, 2.24 g/cm^3; refractive index, 1.517. Occurrence: Thailand.

Scolefin. Trade name of Buna Sow Leuna Olefinverbund GmbH (Germany) for several polyolefin plastics including *Scolefin PE* high-density polyethylenes, *Scolefin-UHMW* ultrahigh-molecular-weight polyethylenes and *Scolefin PP* unmodified, modified or reinforced polypropylenes.

Sconablend. Trade name of Buna Sow Leuna Olefinverbund GmbH (Germany) for thermoplastic polyolefins based on polyolefin/polyamide blends.

Scooter. Trade name of British Rolling Mills Limited (UK) for a general-purpose free-cutting steel containing 0.15% carbon, 0.3% sulfur, and the balance iron. Used for screw-machine parts.

scored-finish tile. Structural tile that has been grooved, notched or scratched to promote its bond with mortar, plaster or stucco.

scorodite. A grayish green mineral of the variscite group composed of iron arsenate dihydrate, $FeAsO_4\cdot 2H_2O$. It can also be made synthetically, but this form is usually colorless, greenish brown, bluish, violet or yellow. Crystal system, orthorhombic. Density, 3.3 g/cm^3; refractive index, 1.79. Occurrence: India, USA (Utah, Washington).

Scorpion. Trade name of T. Turton & Sons Limited (UK) for a low-carbon alloy steel containing 0.3% carbon, 1.3% chromium, 3.4% nickel, and the balance iron.

scorzalite. A sky-blue mineral composed of iron magnesium aluminum phosphate hydroxide, $(Fe,Mg)Al_2(PO_4)_2(OH)_2$. Crystal system, monoclinic. Density, 3.30 g/cm^3; refractive index, 1.667. Occurrence: Brazil, USA (South Dakota).

Scotch. Trademark of 3M Company (USA) for an extensive series of epoxy adhesives including adhesive tapes (e.g., transparent, clear, removable, mounting, strapping, filament and other types), hot-melt adhesives, structural adhesives, sealants, plastic film and sheeting, adhesive tape dispensers, and removable self-sticking notes.

Scotchbond. Trademark of 3M Dental (USA) for a multi-purpose dental adhesive system consisting of an adhesive, an etchant and a primer. It is light-curing (but has dual-cure capabilities), fast-setting, has excellent bonding strength and bonds well to various dental surfaces including fresh amalgam, indirect restorations, self-cure composites, etc.

Scotchcal. Trademark of 3M Company (USA) for a blackout film used in the painting of automobiles. It consists of a release liner, a pressure-sensitive acrylic adhesive and a vinyl film layer.

ScotchCap. Trademark of 3M Company (USA) for laminating films.

Scotch-core. Trademark of 3M Company (USA) for syntactic cores employed in multiple-ply laminate structures to provide increased stiffness, impact resistance, greater machinability, and low volatile losses. Used for various engineering applications, such as engine nacelles, wing and fuselage skins, etc.

Scotchdamp. Trademark of 3M Company (USA) for sound and vibration dampening materials.

Scotch elm. See wych elm.

Scotchfil. Trademark of 3M Company (USA) for putties used for electrical applications.

Scotch fir. See northern pine.

Scotchfoam. Trademark of 3M Company (USA) for single-coated foam tapes.

Scotchguard. Trademark of 3M Company (USA) for oil- and water-repellent finishes for textiles and other materials.

Scotch-Grip. Trademark of 3M Company (USA) for a series of elastomer-based adhesives including contact, rubber-and-gasket, industrial and plastic grades. Used to bond metal, wood, rubber, plastics, etc.

Scotch Kote. Trademark of 3M Company (USA) for powdered epoxy coatings for pipelines and concrete reinforcing bars.

Scotchlite. Trademark of 3M Company (USA) for bubbles composed of soda-borosilicate glass and supplied in various particle sizes. Used in molded plastics, cast syntactic foams, building and construction products, putties, caulking and spackling compounds, tile grouts, cultured marble, plywood patches, synthetic wood, coring materials, thermal insulation, and as a sensitizers in industrial explosives.

Scotch MagicPlus. Trade name of 3M Company (USA) for removable transparent tape that is invisible and has a clean-release adhesive. It removes without surface damage, and is used for releasable envelopes, copywork, graphic arts, and similar applications.

Scotch-melt. Trademark of 3M Company (USA) for hot-melt adhesives.

Scotch-Mount. Trademark of 3M Company (USA) for double-coated foam tapes.

Scotchpak. Trademark of 3M Company (USA) for heat-sealable plastic film.

Scotchpar. Trademark of 3M Company (USA) for polyester films.

Scotch pine. See northern pine.

Scotchply. Trademark of 3M Company (USA) for epoxy and carbon/epoxy prepregs with very low moisture absorption, good hot/wet temperature performance to 165°C (325°F) and high toughness. They are made by impregnating carbon, glass or *Kevlar* (aramid) fibers with epoxy resin.

Scotchpro. Trademark of 3M Company (USA) for a film backing material used for pressure-sensitive adhesive tapes.

Scotch-Seal. Trademark of 3M Company (USA) for a series of sealants.

Scotchtint. Trademark of 3M Company (USA) for a plastic film with improved solar-control properties.

Scotch-Weld. Trademark of 3M Company (USA) for one- and two-part epoxy and urethane structural adhesives available in various grades (including electric grades), and in a wide range of colors including translucent, green, gray, tan, brown, white, off-white, cream and yellow. Also included under this trademark are heat-curing thermosetting adhesive films used for bonding metals to metals, ceramics, glasses or honeycomb materials.

Scots fir. See northern pine.

Scots pine. See northern pine.

Scott. Trade name of Time Steel Service Inc. (USA) for an air- or oil-hardening high-carbon, high-chromium tool and die steel (AISI type D4) containing 2.25% carbon, 11.5% chromium, 0.5% silicon, 0.35% manganese, 0.2% vanadium, 0.8% molybdenum, and the balance iron. It has good deep-hardening properties, good wear resistance, moderate hot hardness, and is used for blanking and trimming dies, deep-drawing and wire-drawing dies, swaging and forming dies, thread-rolling dies, and punches.

Scottfelt. Trademark of Scott Paper Company (USA) for cellular polyurethane sheeting used for foam filters.

Scott's cement. A quick-setting cement made by grinding lime with calcined gypsum.

scoured wool. Wool from which dirt, natural oils and fats and other impurities have been removed by washing with aqueous or solvent media.

scoured textiles. Textile fibers, yarns and fabrics from which impurities, such as dirt particles, oils, fats, waxes or lubricants, have been removed by treating with detergents, solvents or other chemicals (scouring agents).

scouring abrasives. Relatively fine abrasives, such as silica sand or quartz powder, used in the wet or dry finishing of metals, and in the manufacture of certain soaps.

scouring block. A ceramically bonded abrasive block of alumina, silicon carbide or other abrasive material used in the grinding and polishing of metals and ceramics.

scove brick. Refractory brick that in the unfired condition is suitable for use in the construction of scove kilns (updraft kilns for firing brick and similar materials).

Scovil. Trade name of JacksonLea (USA) for a polishing cement.

Scovill. Trade name of Century Brass Products Inc. (USA) for a series of coppers and copper-base alloys including various sulfur-bearing coppers, low brasses, low- and high-leaded brasses, free-cutting brasses, forging, cartridge, admiralty and naval brasses, yellow brasses, leaded red brasses, tin and aluminum brasses, Muntz metals, commercial bronzes, gilding and spring bronzes, low-silicon bronzes, phosphor and manganese bronzes, nickel silvers and copper nickels.

ScratchGuard. Trademark of E.I. DuPont de Nemours & Company (USA) for scratch-resistant non-stick coatings.

scratch-resistant coatings. Coatings applied to glass to reduce the effects of marking or cutting when in frictive contact with rough or sharp objects, e.g., special coatings applied to optical or ophthalmic lenses and mirrors.

SCR brick. A brick that was developed by the Structural Clay Products Research Foundation in the United States and is 2.66 × 6.00 × 12.00 in. (68 × 152 × 305 mm) in size.

Screen. Trade name of SA Glaverbel (Belgium) for a patterned glass.

screen-back hardboard. Hardboard made into panels that have a mesh or screen pattern on the back produced during manufacture by hot-pressing of the wet or damp wood-fiber mat and subsequent press drying. It has one patterned and one smooth surface. See also hardboard.

screen cloth. A woven tissue, usually with screen openings of 12, 14, 16 and 18 mesh (1.70, 1.40, 1.18 and 1.00 mm), used in screen decks. Also known as *screen fabric; screen mesh.*

screen fabric. See screen cloth.

Screenglas. Trade name of Colcombet Fois & Compagnie SA (France) for glass threads coated with vinyl chloride.

screening paste. An oil suspension of finely milled color oxides used to decorate porcelain enamel, glazes and other ceramic ware in the silk-screen printing process.

screenings. See manufactured sand.

Screenlok. Trade name of Safetee Glass Company Inc. (USA) for a shade-screen laminated glass.

screen mesh. See screen cloth.

Screen-Tex. Trademark of Bruin Plastics (USA) for durable, flame-retardant textile fabrics with outstanding printability including screen printing used for signs, banners, and flexible billboards.

screw bronze. A free-cutting bronze composed of 93.5% copper, 5% zinc, 1% tin and 0.5% lead, and used for hardware, screws, nuts, bolts, rivets, pins and studs.

screw metal. A free-cutting alloy composed of 60% copper, 38.5% zinc and 1.5% lead, and used for hardware, studs, screws, and bolts.

screw nut bronze. A high-strength bronze composed of 86% copper, 11% tin and 3% zinc, and used for screws, nuts, and bolts.

screwstock. Metals or alloys, such as a free-cutting steel, cast iron, brass or aluminum, supplied in the form of bars, rods or wires for the manufacture of screw-machine products, i.e., screws, bolts and other threaded fasteners.

Scribe-It. Trade name of Marshall Steel Company (USA) for an oil-hardening tool steel (AISI type O1) containing 0.95% carbon, 1.2% manganese, 0.25% silicon, 0.5% chromium, 0.2% vanadium, 0.5% tungsten, and the balance iron. It has good deep-hardening properties, good machinability, low tendency to shrinking or warping, moderate hot hardness, and is used for blanking, forming and trimming dies, shear knives, punches, reamers, taps, etc.

scrim. (1) An inexpensive, open nonwoven fabric made from continuous-filament yarn, and used as a reinforcement or support in the manufacture of certain composites.

(2) A light, loose plain-weave fabric, usually of cotton, used as a support for needlework or carpets, as a backing for fabrics, and in curtains.

scrim cloth. A fabric embedded in film adhesives to aid in handling and bond-thickness control. Also known as *carrier cloth*.

SC-Special. Trade name of Teledyne Vasco (USA) for a tungsten-type hot-work tool and die steel (AISI type H24) containing 0.47-0.52% carbon, 0.2-0.4% silicon, 0.2-0.4% manganese, 14-15% tungsten, 2.75-3.25% chromium, 0.4-0.6% vanadium, and the balance iron. They have high heat resistance, good nondeforming properties, and are used for extrusion dies, hot-forging dies and punches, and hot-forming rolls and piercers.

scrub pine. See Virginia pine.

Sculpture. Trade name of Jeneric Pentron (USA) for a *ceromer*-type dental composite for esthetic crown and bridge restorations.

sculpture alloys. A group of alloys including silicon bronze, beryllium copper, white bronze, pewter and jewelers' manganese bronze used for casting sculptures, art objects, etc.

Sculptured Chromspun. Trade name of Eastman Chemical Company (USA) for cellulose acetate fibers used for wearing apparel and industrial fabrics.

Sculptured Estron. Trade name of Eastman Chemical Company (USA) for cellulose acetate fibers used for wearing apparel and industrial fabrics.

Scutabond. Trade name for a dental bonding cement used for temporary crown and bridge restorations.

scutched fibers. (1) Flax fibers obtained from deseeded or retted flax straw by breaking up and removing the woody portions.

(2) Cotton that has been mechanically opened, cleaned and formed into a continuous lap or sheet.

SD-Nickel. Trade name of Metal & Thermit Corporation (USA) for a commercially pure nickel (99.9%) used for anodes with improved plating properties.

Seabeads. Trade name of Pittsburgh Corning Corporation (USA) for small glass nodules.

SeaCru. Trademark of Crucible Materials Corporation (USA) for a series of steels including nitrogen-strengthened stainless steels (modified AISI 304 and 316 types) and precipitation-strengthened martensitic stainless steels. Used for marine applications, e.g., boat shafts.

Sea-Cure. Trademark of Trent Tube Division of Crucible Materials Corporation (USA) for a series of highly alloyed ferritic stainless steels (UNS type S44660) containing 0.025% carbon, 1.00% manganese, 1.00% silicon, 25.0-27.0% chromium, 1.5-3.5% nickel, 0.4% phosphorus, 0.03% sulfur, 2.5-3.5% molybdenum, a total of up to 0.8% titanium and niobium, 0.035% nitrogen, and the balance iron. Commercially available in the form of sheet, strip and welded tubing, they have excellent resistance to seawater, brines, brackish water, most pollutants, sour gases and many organic and inorganic chemicals, outstanding resistance to chloride pitting and crevice corrosion, high resistance to stress corrosion, good resistance to scaling, and good mechanical properties. Used for marine structures and equipment, feedwater heaters, condensers, heat exchangers, etc.

Seadrift. Trade name of Australian Window Glass Proprietary Limited (Australia) for a patterned glass.

Seafoam. Trademark of Woodbridge Foam Corporation (USA) for polyurethane foams.

sea foam. See sepiolite.

SeaGard. Trademark of Honeywell International (USA) for polymeric marine overlay finishes used on cordage to enhance the wet performance.

Sea Island cotton. Soft, smooth, lustrous high-grade staple cotton fibers, typically about 51 mm (2 in.) long, used to make high-quality fabrics. Sea Island cotton comes from Florida, Mexico and Central America.

seal. See sealskin leather.

Sea-Lac. Trade name of Duralac Inc. (USA) for baking enamels used in metal finishing.

sealants. Organic substances of liquid or semisolid consistency which, after being applied to a joint or spread over a surface, cure or harden to resilient elastomeric solids. Examples of such materials include synthetic polymers, such as acrylic, polychloroprene, polyurethane, rubber or silicone (often modified with mineral fillers and pigments), as well as bituminous materials, waxes, etc. Used widely in building construction to prevent gas or liquid entry (e.g., on foundation walls, concrete surfaces, windows and doors), and for automotive, aerospace and various industrial applications.

Sealcrete. Trademark of Heidelberger Zement (Germany) for concrete sealing products.

Sealed Brilliance. Trade name of Safetee Glass Company Inc. (USA) for single- and double-face laminated mirrors having standard density.

sealed coating. An anodic oxide coating with reduced porosity produced on aluminum (especially castings) that has previously been treated in boiling deionized water or other aqueous media, e.g., slightly acidified hot water.

sealed hardboard. See filled hardboard.

sealed particleboard. See filled particleboard.

sealer. (1) A liquid coating composition, such as a thin varnish or clear lacquer, that also contains pigments, and is used for sealing porous surfaces of wood, plaster or metal, often preparatory to the application of finish coats.

(2) A liquid polymeric substance, e.g., acrylic- or urethane-based, used to seal and protect concrete surfaces, paving and patio stones, etc. It is usually resistant to gasoline, oil, and road salts.

(3) A black liquid substance, e.g., acrylic-, coal-tar emulsion- or bitumen-based, used to seal and protect driveways, airport runways, etc., against the elements.

Seal-Glaz. Trade name of Seal Glaz Corporation (USA) for sealed glass units.

seal grain. Split side leather with an artificial grain produced by passing through rollers under pessure to imitate tanned genuine sealskin.

sealing alloys. A group of special metals and alloy that have the same or nearly the same coefficient of expansion as glass, and are thus suitable for glass-to-metal sealing applications. Glass-to-metal seals are made by first heating the glass-metal interface into the melting range of the glass such that the molten glass wets the oxide layer on the metal or alloy surface and upon cooling sets forming a hermetic seal. Materials suitable for glass sealing include platinum, copper, tungsten, molybdenum, and iron-chromium, nickel-iron and nickel-iron-cobalt alloys. Also known as *ceramic sealing alloy; glass-metal seal alloys; glass sealing alloys; glass-to-metal seal alloys.*

sealing compound. (1) A chemical compound in paste or tape form used to make airtight seals on threaded pipe joints.

(2) A bituminous material used for filling and sealing joints and cracks in concrete or mortar.

(3) A special curing compound used for concrete.

(4) A compound used to prevent air and moisture from entering capacitor blocks, dry batteries, transformers and other electrical components and devices.

sealing glass. A glass possessing special thermal expansion and flow characteristics that enable it to bond to another glass or solid.

sealing glass solder. A *sealing glass* whose softening temperature is low enough to make it suitable for use as an intermediate bonding material.

sealing tape. (1) A long-wearing adhesive tape that offers instant adhesion and maximum holding power, is stronger and heavier than masking and transparent tape, and often waterproof. It is usually a gummed tape based on a plastic, such as polypropylene, or kraft paper, and supplied in the form of rolls on cardboard or plastic cores. Used for sealing parcels, cartons, envelopes, etc.

(2) A sealant of preformed shape used in building construction, usually in a joint with initial application of compressive force. Also known as *tape sealant.*

sealing wax. A substance composed of a mixture of resins, shellac, turpentine and pigments that is hard at room temperature, but becomes soft and moldable when heated. Used for sealing containers, documents, etc.

Sealithor. Belgian trademark for a sulfate-resisting cement.

Sealkote. Trademark of Chemtron Manufacturing Limited (USA) for asphalt cements, undercoatings and sound-deadening materials.

seal leather. See sealskin leather.

Sealmet. (1) Trademark of Allegheny Ludlum Steel (USA) for a series of controlled-expansion alloys containing 6-28% chromium, 0-42% nickel, and the balance iron. Supplied in wire, strip and sheet form, they are used for glass-to-metal seals, and in instruments.

(2) Trade name of ZYP Coatings, Inc. (USA) for a thin, paintable coating that provides metallic substrates with long-term protection against chemical attack and high-temperature oxidation at temperatures to 1300°C (2370°F). The standard coating thickness is 0.1-0.2 mm (5-8 mils).

Seal N'Sound. Trademark of Witco Chemical Canada Limited for automobile underbody coatings.

Seal-Rite. Trademark of Pulpdent Corporation (USA) for a series of light-cure, fluoride-releasing dental pit and fissure sealants including *Seal-Rite*, a flowable, 34.4% filled grade with pearlescent color, *Seal-Rite Low Viscosity*, a highly flowable, 7.7% filled grade with off-white color, and *Seal-Rite UDMA*, a free-flowing, wear-resistant, 23.8% filled urethane dimethacrylate (UDMA) resin with off-white color.

sealskin leather. A tough, tanned leather with an attractive grain, made from the hides of any of several families of carnivorous aquatic mammals including true seals, eared and fur seals and sea lions. Split *walrus leather* is often used as a substitute. Used for clothing, handbags, suitcases, etc. Also known as *seal; seal leather.*

Seal-Stat. Trade name of Scharr Industries Inc. (USA) for a polyester polyethylene laminate film for heat-sealing applications.

Seal Tex. Trade name of Texo Corporation (USA) for phosphate coatings for zinc and iron products.

Seal-Therm. Trade name for silicone rubber supplied in the form of sheets and tapes with or without backing.

Sealvac. Trade name of Vacuum Metals Corporation (USA) for a controlled (thermal) expansion alloy composed of 54% iron, 29% nickel and 17% cobalt, and used for glass-to-metal seals, and for sealing leads in light bulbs and radio tubes.

Sealvar. Trademark of Pfizer Inc. (USA) for a controlled (thermal) expansion alloy composed of 54% iron, 29% nickel and 17% cobalt, and used for glass-to-metal seals, and for hermetic seals with the harder glasses and ceramics used in the electronic industries.

seamanite. A transparent, pale- to wine-yellow mineral composed of manganese borate phosphate hydroxide, $Mn_3(PO_4)B(OH)_6$. Crystal system, orthorhombic. Density, 3.13 g/cm³; refractive index, 1.663. Occurrence: USA (Michigan).

seamed tubing. Tubing of steel, cast iron, aluminum, brass, copper, or other metals, usually made by rolling flat stock (i.e., skelp, strip, sheet or plate) into a cylinder and joining the seam by butt, electric-resistance or fusion welding, or by rolling thin flat stock into a two-layer cylinder and then furnace-brazing the lapped surfaces. In contrast to *seamless tubing*, the resulting product has a longitudinal seam. Also known as *welded tubing.*

Seamfree. Trademark of MacSteel Bar Group, Quanex Corporation (USA) for bright, cold-finished steel bar products.

seamless tubing. Tubing of steel, cast iron, aluminum, brass, copper, or other metals, usually made by piercing a solid heated billet passed between two heavy, tapered rollers, mounted at an angle and over a mandrel (Mannesmann process), by forcing a pierced, heated solid rod over a mandrel by means of rollers (Pilger process), or by hot extrusion. In contrast to *seamed tubing*, the resulting product does not have a longitudinal seam.

Searcher. Trade name of Osborn Steels Limited (UK) for a shock-resisting tool steel containing 0.45-0.63% carbon, 0.4-1% chromium, 1.6-2% tungsten, and the balance iron. Used for shear blades, chisels, punches, crimpers and wedges.

searlesite. A white mineral composed of sodium boron silicate hydroxide, $NaBSi_2O_5(OH)_2$. It can also be made synthetically. Crystal system, monoclinic. Density, 2.46 g/cm³; refractive index, 1.531. Occurrence: USA (California, Wyoming).

seasoned lumber. Lumber whose moisture content has been decreased by air or kiln drying to the level appropriate for its grade and use. Seasoning makes lumber more durable and serviceable, and reduces its weight. See also air-dried lumber; kiln-dried lumber.

seawater bronze. A corrosion-resistant bronze containing 44.5% copper, 33% nickel, 16% tin, 5.5% zinc and 1% bismuth. It has excellent seawater corrosion resistance, and is used for marine parts.

seawater magnesia. Magnesia (MgO) precipitated from seawater by treatment with slaked lime or lightly calcined dolomite. See also magnesia.

seawater magnesite. See synthetic magnesite.

seaweed fibers. See alginate fibers.

Seaworthy. Trade name of Canada Cordage Inc. (Canada) for Manila rope.

SeBiLoy. See EnviroBrass.

Secaero. Trade name of Seaboard Steel Company of America (USA) for an air-hardening cold-work tool steel (AISI type A2) containing 1% carbon, 5% chromium, 1.1% molybdenum, 0.65% manganese, 0.3% vanadium, and the balance iron. It has good deep-hardening and nondeforming properties, good toughness, moderate hot hardness and wear resistance, and is used for punches, trimming, blanking, beading, extrusion and stamping dies, forming rolls, mandrels, etc.

Secar. Trademark of Lafarge Company (USA) for pure calcium aluminate cement used in the manufacture of special refractory shapes and castables for industrial furnaces and fire chambers.

Secco Clay. Trade name of Southeastern Clay Company (USA) for hard rubber clays and adhesives.

Secobalt. Trade name of Seaboard Steel Company of America (USA) for a high-speed steel containing 0.75% carbon, 18% tungsten, 11% cobalt, 5% chromium, 1.5% vanadium, 0.8% molybdenum, and the balance iron. Used for lathe and planer tools, tool bits, form cutters, milling cutters, drills, cutoff tools, hobs, etc.

Secodie. Trade name of Seaboard Steel Company of America (USA) for a nondeforming die steel containing 1.35% carbon, 1.5% molybdenum, 1.2% silicon, 0.6% chromium, 0.35% vanadium, and the balance iron.

Secoleo. Trade name of Seaboard Steel Company of America (USA) for an oil-hardening tool and die steel containing 0.75% carbon, 1.65% nickel, 1.1% chromium, 0.45% manganese, 0.2% silicon, and the balance iron.

Secon. Trade name of Secon Metals Corporation (USA) for a series of electrical resistance alloys based on molybdenum-palladium, rhodium-platinum, tungsten-platinum or platinum-rhodium-ruthenium, and used for potentiometers and resistance wire.

secondary alloy. An alloy whose principal element is produced from recycled scrap, in contrast to a *primary alloy* in which it is produced directly by refining from an ore.

secondary barium phosphate. See barium hydrogen phosphate.

secondary calcium phosphate. See calcium phosphate (ii).

secondary cellulose acetate. Cellulose acetate formed from *primary cellulose acetate* by partial hydrolysis, i.e., by the addition of excess water to the residual acetic anhydride.

secondary metal. A metal made from recycled scrap, in contrast to a *primary* or *virgin metal* obtained from minerals (ores), natural brines, or ocean water. Examples include secondary aluminum, copper, lead and zinc.

secondary sodium orthophosphate. See disodium phosphate.

second-split leather. A low-quality leather obtained from the inner layer of cattlehide and sometimes used for inexpensive leather goods.

seconite. A fine plastic clay that is often mixed with molding sand for use as a binder.

Secovan. Trade name of Seaboard Steel Company of America (USA) for a high-speed tool steel containing 0.7% carbon, 18%

tungsten, 4.5% chromium, 2.25% vanadium, 0.65% molybdenum, and the balance iron. Used for tools and cutters.

Secretan. British trade name for a strong, corrosion-resistant copper alloy containing 3-7% aluminum, 1.5% magnesium and 0.5% phosphorus.

secunda kraft paper. A paper made from kraft pulp containing no more than 50% (waste) kraft paper.

Secura. Trade name of Spraylat Corporation (USA) for powder coating materials.

Securacoat GPX. Trade name of General Plasma (USA) for a series of thermal spray coating systems used on severe-service ball valves made of stainless steel, titanium, nickel-base alloys, etc. Included are coatings of tungsten carbide/cobalt (*Securacoat GPX 9660*), tungsten carbide/nickel (*Securacoat GPX 9657*), chromium carbide/nickel-chromium (*Securacoat GPX 9176*), chromium oxide/ceramic (*Securacoat GPX 9160*) and iron-chromium-nickel-molybdenum alloy (*Securacoat GPX 9346*). *Securacoat* coatings are applied by the high-velocity oxyfuel (HVOF) process and have excellent wear resistance, moderate to excellent corrosion and impact resistance, smooth surfaces, high densities, and hardness values from 600 to 1200 Vickers.

Secure. Trade name of DuBois Chemicals, Division of Diversey Lever (USA) for a phosphate coating used in metal finishing.

SecurFilm. Trade name of Mathison's (USA) for a laminating product used in the graphic arts and printing trades. It consists of a printable polyester film with a proprietary, ink-jet receptive coating that can also act as and adhesive for mounting applications.

Securipoint. Trademark of Saint-Gobain (France) for a glass with high mechanical performance.

Securistretch. Trade name of Rubafilm (France) for stretch film with personalized printed identification used for wrapping palets for shipping.

Sécurit. Trademark of Saint-Gobain (France) for a toughened plate glass suitable for automotive windshields.

Sécurit Contact. Trademark of Saint-Gobain (France) for a slip-retardant toughened plate glass.

Securiton. Trade name of Securitron (Switzerland) for a vacuum alarm glass composed of an exterior pane of plate or sheet glass, a wafer-thin empty space and an interior barrier of laminated safety glass. Breaking of the vacuum triggers an alarm.

Secur-Lite. Trade name of Amerada Glass Company (USA) for a laminated safety glass with a tough, high-tensile plastic interlayer. Used for shop windows.

Secur-Tem & Poly. Trade name of Globe Amerada Architectural Glass (USA) for a chemically-strengthened glass used for correc-tional facilities and mental institutions.

Securus. Trademark of Honeywell Performance Fibers (USA) for energy-absorbing synthetic fibers based on *Pelco* polyester-caprolactone block copolymers and used for safety belt webbing, e.g., in automotive seat belts and child safety restrains.

Sedanol. Trade name for a zinc oxide–eugenol dental cement.

sederholmite. A yellow to orange-yellow mineral of the nickeline group composed of nickel selenide, $\beta\text{-Ni}_{0.85}\text{Se}$. Crystal system, hexagonal. Density, 7.11 g/cm^3. Occurrence: Finland.

sedimentary rock. A rock, such as shale, sandstone, limestone or gypsum, formed by the accumulation and consolidation of mineral and particulate matter (sediment) deposited by the action of water (aqueous sediment), air (eolian sediment) or glacial ice (glacial sediment).

Sedo. Trade name of Flachglas AG (Germany) for all-glass double glazing units.

Sedona Red. Trade name of Aardvark Clay & Supplies (USA) for a clay (cone 5) with a buff-red smooth throwing body.

sedovite. A reddish brown mineral composed of uranium molybdenum oxide, $U(MoO_4)_2$. Crystal system, orthorhombic. Density, 4.20 g/cm^3.

Sedura. Trade name for rayon fibers and yarns used for textile fabrics.

seed crystal. A small piece of *single crystal* that serves as a nucleus for growing larger crystals.

seed fibers. The silky fibers and hairs obtained from the seed pods of plants, such as cotton, cottongrass or kapok, and made into yarns and fabrics, or used as stuffing.

Seekure. Trade name of BSK, Division of DS Smith (UK) Limited for a strong, tough laminated product supplied in standard and fire-retardant grades. Used as a case liner, lidding or wrapping paper for commercial and industrial export packaging applications, and for the temporary protection of industrial plants and buildings.

seeligerite. A yellow mineral composed of lead oxide chloride iodide, $Pb_3O_4Cl_3I$. It can also be made synthetically. Crystal system, orthorhombic. Density, 6.83 g/cm^3; refractive index, 2.32. Occurrence: Chile.

seersucker. A lightweight fabric, usually of cotton, a cotton blend, rayon or nylon, and generally produced with alternate plain and crinkled stripes. Used for summer suits and sportswear.

SEF. Trademark of Solutia Inc. (USA) for modacrylic fibers.

segelerite. A yellow-green mineral composed of calcium magnesium iron phosphate hydroxide tetrahydrate, $CaMgFe(PO_4)_2(OH)\cdot4H_2O$. Crystal system, orthorhombic. Density, 2.67 g/cm^3; refractive index, 1.635. Occurrence: USA (South Dakota).

Seger's porcelain. A German porcelain made from a batch of feldspar, quartz and kaolin, and covered with a glaze composed of feldspar, flint, whiting and kaolin. It is first kiln-fired at low temperatures prior to glazing, and then subjected to a second kiln fire at Seger cone 9.

Segliwa. Trade name of Segliwa GmbH (Germany) for muscovite and phlogopite mica products.

segmented polyurethanes. A class of polyurethanes containing hard segments (e.g., diphenylmethane diisocyanate) and soft segments (e.g., polytetramethylene oxide or polycarbonate). They have properties quite similar to the *Lycra Spandex* textile fiber including outstanding flex-fatigue resistance, high strength, excellent wear and fatigue resistance, excellent biocompatibility, biostability and thromboresistance and low water absorption. Used as biomaterials for ventricular-assist devices (VADs), vascular grafts, prostheses and total artificial hearts (TAHs), and for textile fabrics.

Seguridad. Trade name of Vidrios y Envases SA (Mexico) for a toughened glass.

Seifert. British trade name for an antifriction alloy composed of 73% tin, 21% zinc, 5% lead, 0.5% antimony and 0.5% phosphorus, and used for bearings and ornaments.

Seidlitz salt. See epsomite.

seidozerite. A brownish red mineral composed of sodium manganese titanium zirconium silicate fluoride, $Na_4MnTi(Zr_{1.5}Ti_{0.5})O_2(F,OH)(Si_2O_7)_2$. Crystal system, monoclinic. Density, 3.47 g/cm^3; refractive index, 1.758. Occurrence: Russian Federation.

seignette-electric materials. See ferroelectric materials.

Seignette salt. See potassium-sodium tartrate.

Seilon. Trade name for acrylonitrile-butadiene-styrene plastics with good physical, chemical and electrical properties, high tensile strength and toughness, high modulus of elasticity, good impact resistance and good dimensional stability to 121°C (250°F). Used for automotive trim, cases, appliance housings, piping, impellers, helmets, knobs and handles.

seinajokite. A rose-gray mineral of the marcasite group composed of antimony iron, $FeSb_2$. It can also be made synthetically. Crystal system, orthorhombic. Density, 8.27 g/cm^3. Occurrence: Finland.

sekaninaite. A blue or violet-blue mineral of the cordierite group composed of iron aluminum silicate, $Fe_2Al_4Si_5O_{18}$. It can also be made synthetically. Crystal system, orthorhombic. Density, 2.77 g/cm^3; refractive index, 1.57. Occurrence: Czech Republic.

Sekril. Trademark of Courtaulds Limited (UK) for a polyacrylonitrile (PAN) staple fiber.

Sekurit. Trademark of Sekurit-Glas Union GmbH (Germany) for a nonshattering, toughened safety plate glass used for vehicles, industrial and commercial buildings.

Sekurit S. Trademark of Sekurit-Glas Union GmbH (Germany) for a toughened glass with two fracture zones.

Sekurit-Verbund. Trademark of Sekurit-Glas Union GmbH (Germany) for a multilayer safety glass.

Selar. Trademark of E.I. DuPont de Nemours & Company (USA) for a series of synthetic resins including polyesters, nylons and polyamide/polyethylene alloys. *Selar PA* refers to impact-resistant, crystal-clear, amorphous polyamide (nylon) barrier resins for flavor-sensitive refrigerated foods and beverages. They can be blended with polyamide 6 (nylon 6) and ethylene-vinyl alcohol, and are supplied in several grades. *Selar PT* refers to tough, heat-resistant specialty polyester resins possessing excellent aroma and flavor protection, good chemical resistance, high melt strength and low coefficients of friction. They are suitable for blow molding, sheet and foam extrusion, cast and blown film and extrusion coating applications for the food, personal care and cosmetics industries including films for snacks and lidding, bakery and deli trays, cosmetic, juice and condiment bottles and coated board for juice cartons and frozen entrees. *Selar RB* refers to high-density polyethylene laminar barrier resins supplied in several grades for automotive fuel tanks.

Selee Ceramic Foam. Trade name of Selee Corporation (USA) for an alumina-reticulated open-cell ceramic foam.

Selciato. Trade name of Vetreria di Vernante SpA (Italy) for a patterned glass with an hammered effect.

Select Coat. Trade name of Nordson Corporation (USA) for a conformal coating for application to printed circuit boards.

select lumber. An appearance grade of softwood lumber that is available in the following four grades: (i) *Grade A,* that is almost free of defects and suitable for natural finishes and staining; (ii) *Grade B,* that contains a few, small, visible defects and is suitable for natural finishes; (iii) *Grade C,* that contains some defects which, however, will be hidden if it is painted; and (iv) *Grade D,* that contains several defects similar to those in "Grade C" and is suitable for painted projects.

(2) A good grade of hardwood lumber with one ungraded side and the other at least 90% clear. It is suitable for cabinetwork, but some culling may have to be done.

Selectron. Trade name of PPG Industries Inc. (USA) for low-cost polyester resins supplied in neat and fiber-reinforced grades. They have excellent electrical properties, can be formulated for room or high-temperature applications, and are used for fiberglass boats, autobody components, chairs, fans, helmets, etc.

Select-Sil. Trademark of Freudenberg-NOK (USA) for a synthetic compound consisting of ground cork in a silicone matrix in which the cork functions as a lubricant greatly improving the wear characteristics. It has good dimensional stability at elevated temperatures, good low- and high-temperature characteristics, high tear strength, and good abrasion resistance. Used for sealing applications.

selena. See lacewood.

Selene. Trade name of Vetreria di Vernante SpA (Italy) for a glass with indeterminate pattern.

selenic acid. A white, water-soluble, hygroscopic solid compound (94+% pure) with a density of 2.51 g/cm^3, a melting point of 58°C (136°F) and a refractive index of 1.5174. Used as a reagent. Formula: H_2SeO_4.

selenious acid. White or colorless transparent, moisture-sensitive crystals (98+% pure) with a density of 3.01 g/cm^3 and a melting point of 70°C (158°F) (decomposes). Used as a reagent, e.g., for the preparation of amorphous selenium. Formula: H_2SeO_3. Also known as *selenous acid*.

selenite. A finely crystallized variety of the mineral *gypsum* ($CaSO_4 \cdot 2H_2O$) found in transparent, colorless, monoclinic crystals and foliated masses, somewhat resembling mica in appearance. Also known as *gypsum spar; sparry gypsum*.

selenium. A nonmetallic chemical element of Group VIA (Group 16) of the Periodic Table resembling sulfur in chemical properties. It is commercially available in the form of foils, pellets, lumps and powder, can be made to high-purity (99.999%), and is obtained from sulfide ores (e.g., *copper pyrites, iron pyrites* and *zinc blende*) and in copper refining. Atomic number, 34; atomic weight, 78.96; divalent, tetravalent, hexavalent. It forms binary alloys with silver, copper, zinc, lead, etc., and exists in a gray, crystalline form, and as a fine, amorphous red powder. *Crystalline selenium:* There are three room-temperature forms of crystalline selenium: *alpha selenium, beta selenium and gamma selenium*. The former two have monoclinic crystal structures, and the latter a hexagonal crystal structure. Density, 4.8 g/cm^3; melting point, 217°C (423°F); boiling point, 685°C (1265°F); hardness, 2.0 Mohs; refractive index, 3.0. It is a p-type semiconductor, has photoconductive and photovoltaic properties, and its electrical conductivity (about 14.5% IACS) varies with light irradiation. Used as a decolorizer in glass, in the production of rose and ruby colors in glass, glazes and porcelain enamels, in steel, copper and Invar (for degasification and enhancement of machinability), as an additive to lead-antimony battery grid metal, as a rubber accelerator, as a catalyst, in photocells, solar cells, solar batteries, metallic rectifiers, electronic components, magnetic computer cores, photocopier plates, xerographic drums and television cameras, as a pigment in paints, plastics and ceramics, and in the manufacture of photonic crystals. *Amorphous selenium* is obtained by reduction from selenious acid, has a density of 4.26 g/cm^3 and a softening temperature of 40°C (104°F), and becomes black on standing and crystalline on heating. Symbol: Se.

selenium copper. An alloy of 99.5% copper and 0.5% selenium. It has good machinability and hot workability, high electrical conductivity, and is used for electrical contacts.

selenium dioxide. A white, yellowish-white or pink crystalline powder (98+% pure) with a density of 3.95 g/cm^3 and a melting point of 315°C (599°F) (sublimes). Used in analytical chemistry, as a catalyst, as an antioxidant for lubricating oils, and as a mold oxidizing reagent for allylic olefins and acetylenes. Formula: SeO_2. Also known as *selenium oxide*.

selenium disulfide. See selenium sulfide (ii).

selenium halide. A compound of selenium and a halogen, such as bromine, chlorine, fluorine or iodine.

selenium monosulfide. See selenium sulfide (i).

selenium oxide. See selenium dioxide; selenium trioxide.

selenium ruby. A ruby-red glass that contains selenium oxide, cadmium sulfide, arsenic oxide and carbon, and is made in a reducing atmosphere. Used for signal lenses.

selenium stainless steel. A stainless steel containing about 0.05-0.15% selenium. The selenium greatly improves machinability, and has a lesser effect on the corrosion resistance than sulfur.

selenium steel. A low-carbon or stainless steel to which small amounts of selenium (usually 0.05 to 0.35%) have been added to promote the formation of short chips and thus improve machinability.

selenium sulfide. Any of the following compounds of selenium and sulfur: (i) *Selenium monosulfide*. Black, or orange-yellow crystals or powder. Density, 3.056 g/cm^3; decomposes at about 118-119°C (244-246°F). Used as a semiconductor. Formula: SeS; and (ii) *Selenium disulfide*. Bright orange to red powder. Melting point, below 100°C (210°F); boiling point, decomposes. Used as a dry-film lubricant. Formula: SeS_2.

selenium trioxide. An amorphous, pale yellow powder with a density of 3.6 g/cm^3 and a melting point of 120°C (248°F) (decomposes). Used as a chemical reagent. Formula: SeO_2. Also known as *selenium oxide*.

selenourea. Off-white, light- and moisture-sensitive crystals (99.9+% pure) that decompose at 200°C (392°F) and are commercially available in practical and optical grades. Formula: NH_2CSeNH_2.

selenous acid. See selenious acid.

self-blended yarn. A single yarn spun from a blend of fibers of different lengths, counts, etc., but of the same species.

self-bonding material. A material that when applied to a substrate by thermal spraying has the tendency to form a metallurgical bond.

self-cleaning enamel. An acid- and thermal shock-resistant porcelain enamel that contains selected mill additions and when applied to the inside of culinary ovens will continuously provide oxidation of grease and oven spills during use.

self-contained chemical paper. A sheet of paper that has two materials added during manufacture or coated onto the formed sheet, which will react upon application of pressure to form a visible image.

self-contained mechanical paper. A sheet of paper with a pigmented substrate that becomes visible when the surface coating is rendered transparent by the application of pressure.

self-copying paper. Generic name for several papers including (i) *intermediate transfer paper*, (ii) *single-sided self-copying paper* and (iii) *double-sided self-copying paper*. Also known as *copying carbon paper*.

self-curing adhesive. An adhesive that undergoes vulcanization (curing) without application of heat. Also known as *self-vulcanizing adhesive*.

self-curing concrete. Concrete that hardens without aftertreatment, i.e., without periodical wetting during the initial stages of curing.

self-extinguishing plastics. Thermoplastics or thermosets that will burn, but will extinguish themselves within a prescribed period of time after removal of the flame. Examples include hexafluoropropylene vinylidene fluoride copolymer, polyamide (nylon), polyaramid, polyethylene terephthalate, polyphenylene oxide, polysulfone, polyvinylidene chloride, polyvinylidene fluoride and unplasticized polyvinyl chloride.

self-fluxing alloy. A soldering alloy that can be used without additional fluxing agents, since it wets the substrate when heated to its melting point. For example, soft solders are often supplied in the form of wires containing acid or rosin cores for fluxing, or as prealloyed powders suspended in fluxing agents that constantly wet the surface during soldering.

self-fluxing iron ore. Pellets composed of iron ore concentrates and the proper proportion of limestone for use as combined blast-furnace charge and flux in the manufacture of pig iron.

self-glazed porcelain. An oriental porcelain with a glaze of only one tint.

self-hardening cement. A cement that sets at or slightly above room temperature.

self-hardening mortar. A mortar that sets at or slightly above room temperature.

self-hardening steel. See air-hardening steel (1).

Selfoc. Trade name of Nippon Sheet Glass Company Limited (Japan) for a special optical glass fiber whose refraction becomes steadily larger from its exterior to the center.

self-reinforcing rubber. A synthetic rubber that has attained the desired strength, hardness, abrasion resistance, etc., without the addition of reinforcing agents.

self-skinning foam. A polyurethane foam that produces a tough, solid skin over a porous core upon curing. It is widely used in the automotive industry.

self-stressed concrete. An expansive-cement concrete in which the increase in volume during setting and initial hardening has been restrained thereby inducing persistent compressive stresses.

self-vulcanizing adhesive. See self-curing adhesive.

seligmannite. A lead-gray to black mineral of the bournonite group composed of copper lead arsenic sulfide, $CuPbAsS_3$. It can be made synthetically. Crystal system, orthorhombic. Density, 5.41 g/cm^3. Occurrence: Switzerland, USA (Utah).

sellaite. A colorless mineral of the rutile group composed of magnesium fluoride, MgF_2. Crystal system, tetragonal. Density, 2.97-3.17 g/cm^3; hardness, 5.0 Mohs; refractive index 1.378. It can also be made synthetically. Used as a flux in various ceramic and glass compositions, particularly in infrared components used under severe conditions, and as single crystals for polarizing prisms, lenses and windows.

Sellotape. Trade name of Sellotape Products (UK) for polyethylene resins and products.

Semalloy. Trade name of Semi-Alloys (USA) for an extensive series of fusible (low-melting) alloys used for solders, fusible elements in sprinklers, safety plugs, etc., encapsulation of electronic components, patterns and models, low-temperature castings and dental applications. Also included under this trade name are various tin-lead and lead-tin solders as well as brazing filler metals based on aluminum, silver, copper, nickel, brass, gold or cobalt, and several other precious or nonprecious metals and alloys.

Sembilan. Trademark of Chempaka Negri Lakshmi Textiles Snn. Bhd. (Malaysia) for a polyester fibers and filament yarns.

Semdex. Trade name of Heppenstall Company (USA) for a water-hardening tool steel containing 0.6% carbon, 0.25% silicon, 0.8% manganese, 0.25% chromium, and the balance iron. Used for hand tools, chisels, cold punches, nail sets, and forg-

ing and trimming dies.

semenovite. A brown or gray mineral composed of rare-earth sodium calcium beryllium silicate fluoride hydroxide monohydrate, $(Ln,Na,Ca)_{12}(Si,Be)_{20}O_{40},(F,OH)_8·H_2O$. Crystal system, tetragonal. Density, 3.14 g/cm^3; refractive index, 1.614. Occurrence: Greenland.

semi-alloyed powder. A metal powder made by partial alloying of two or more elements during manufacture.

semiaustenitic precipitation-hardenable stainless steels. A group of wrought stainless steels, usually modifications of standard 18-8 stainless steels with lower nickel contents, that have austenitic microstructures in the solution-annealed condition, and can be cold-worked like other austenitic stainless steels, and transformed to martensite by thermal and/or thermomechanical treatments that precipitate very fine second-phase particles (e.g., aluminum or copper) from a supersaturated solid solution. Also known as *precipitation-hardenable semiaustenitic stainless steels*. See also austenitic precipitation-hardenable stainless steels; precipitation-hardenable stainless steels.

semicompreg. A plywood material that has been impregnated with phenolic resin (e.g., phenol-formaldehyde), compressed to a density of less than 1.25 g/cm^3 (0.045 $lb/in.^3$) and then cured. See also compreg.

semiconducting compounds. See semiconductor compounds.

semiconducting crystal. A crystal, such as germanium, silicon or tin, that exhibits an electrical conductivity intermediate between that of a conductor and an insulator.

semiconducting glaze. A ceramic glaze containing metal oxides, such as cobaltocobaltic oxide (Co_3O_4), chromic oxide (Cr_2O_3), cupric oxide (CuO), ferric oxide (Fe_2O_3), ferroferric oxide (Fe_3O_4), manganese dioxide (MnO_2) or titanium dioxide (TiO_2), in sufficient quantities to promote a degree of electrical conductivity to prevent surface discharge or flashover in electrical ceramics (e.g., porcelain insulators).

semiconducting materials. See semiconductors.

semiconducting polymers. A class of molecularly doped polymers including *polyaniline* and *polythiophene* that have been synthesized by the incorporation of small semiconducting organic pigment molecules (known as *chromophores*), either as side chains or directly into the main chain or backbone. Used in electroluminescent devices, as photoreceptors in electrophotographic devices like photocopiers and printers, and as semiconductive layers in optoelectronic devices. See also conductive polymers; molecularly doped polymers.

semiconductor compounds. Compounds whose electrical resistivity varies in the range from about 10^{-4} to 10^7 $(\Omega\text{-m})^{-1}$. They consist mainly of combinations of ions from Groups III and V (Groups 13 and 15) and Groups II and VI (Groups 12 and 16) of the Periodic Table, but many other covalently bonded compounds also have desirable semiconductive properties. The *III-V (13-15) compounds* are MX compositions where M is a trivalent (3^+) element and X a pentavalent (5^+) element, while the *II-VI (12-16) compounds* combine a divalent (2^+) element with a hexavalent (6^+) element. These materials are intrinsic semiconductors in the "pure" state, but can be made into extrinsic semiconductors by doping. They have negative temperature coefficients of electrical resistivity. Examples of semiconductor compounds include aluminum antimonide (AlSb), boron phosphide (BP), cadmium selenide (CdSe), gallium arsenide (GaAs), indium antimonide (InSb), mercury telluride (HgTe), zinc sulfide (ZnSe) and zinc telluride (ZnTe). Also known as *compound semiconductors; semiconducting compounds*. See also extrinsic semiconductor; intrinsic semiconductor.

semiconductors. A group of elements and compounds including germanium, silicon, tin, cadmium sulfide, gallium arsenide, gallium phosphide, indium antimonide and zinc selenide, that have filled valence bands at 0K and relatively narrow energy band gaps. Their electrical conductivity is intermediate between that of a conductor (e.g., a metal or alloy) and a nonconductor or insulator (e.g., glass or mica). Typically, the room-temperature electrical conductivity ranges between 10^{-6} and 10^{+4} (Ω-m)$^{-1}$. Abbreviation: SC. Also known as *semiconducting materials*. See also compound semiconductor; extrinsic semiconductor; intrinsic semiconductor; n-type semiconductor; p-type semiconductor.

Semicosil. Trademark of Wacker Silicones Corporation (USA) for silicone fluids and elastomers for electronic packaging applications.

semicrystalline plastics. Plastics, such as acetals, fluoropolymers, polyamides, polyethylenes, polybutylene terephthalates, polyethylene terephthalates, polypropylenes, polyphenylene sulfides, polyaryletherketones and polyetheretherketones, in which some of the molecular chains have arranged into a regular crystalline array through chain folding. For practical purposes, plastics are considered "crystalline" if the crystallinity (i.e., crystalline or ordered regions) amounts to at least 5 wt% of the total plastic. Also known as *crystalline plastics; crystalline polymers; semicrystalline polymers.*

semicrystalline polymers. See semicrystalline polymers.

Semi-dul. Trade name for rayon fibers and yarns.

semifinished products. (1) A generic term used in the steel industry for the blooms, billets and slabs supplied to the rolling mill or extrusion plant for the manufacture of structural shapes, bars, plates, pipes, etc.

(2) A term used in steel manufacture to refer to the finished products of rolling mills and extrusion plants including structural shapes (e.g., beams, angles, tees, zees, channels or piling), bars (e.g., rounds, squares, half rounds, hexagons or octagons), plates, sheets and strips, pipes and tubes, wire, wire rope, etc.

semifriable abrasives. Abrasives whose friability (brittleness) is intermediate between that of friable and tough abrasives. They exhibit a considerable resistance to grain breakage, relatively slow formation of new sharp edges, and are therefore used for intermediate grinding operations only. They are not suitable for finish and precision grinding. See also friable abrasives.

semifriable alumina. A hard, abrasive grade of recrystallized alumina containing about 96-98% alumina. Abbreviation: SFA. See also friable alumina.

semihard rubber. A relatively hard rubber containing about 25-30 wt% sulfur. See also hard rubber.

Semi-High. Trade name of Disston Inc. (USA) for a semi-high-speed tool steel containing 1.1% carbon, 1.1% tungsten, 0.3% chromium, 0.2% vanadium, and the balance iron.

semihydraulic lime. Lime whose properties and composition are intermediate between that of *hydraulic lime* and high-calcium or rich lime. See also rich lime.

semi-insulating polycrystalline oxygen-doped silicon. See SIPOS.

semi-interpenetrating network. A mixture of polymers in which a crosslinked polymer partially penetrates another polymer. It results when two separate polymer networks or structures are partially intertwined. Abbreviation: SIPN; S-IPN. Also known as *semi-interpenetrating polymer network.* See also interpenetrating network.

semi-interpenetrating polymer network. See semi-interpenetrating network.

semikilled steel. Steel that owing to the addition of appreciable quantities of aluminum, ferrosilicon or manganese to the ladle has only been partially deoxidized to obtain an almost uniform distribution of phosphorus and sulfur segregations across the entire cross section. In a semikilled steel solidification shrinkage is offset. See also killed steel.

semimagnetic semiconductors. See diluted magnetic semiconductors.

semimat glaze. A colorless or colored ceramic glaze that exhibits only a moderate degree of gloss and is thus considered to be intermediate between a high-gloss and a mat glaze in appearance.

semimetal. A chemical element, such as boron, arsenic or selenium, that depending on particular conditions can react as either a metal or a nonmetal. Abbreviation: SM. Also known as *semimetallic element.*

semimetallized pellets. See prereduced iron-ore pellets.

Seminole. Trade name of AL Tech Specialty Steel Corporation (USA) for a shock-resisting tool steel (AISI type S1) containing about 0.5% carbon, 2.0-2.5% tungsten, 1.3% chromium, 0.2% vanadium, and the balance iron. It has a great depth of hardening, high hardness and strength, excellent toughness, good wear and abrasion resistance, and good hot hardness. Used for chisels, punches, headers, shear blades, forming tools, dies, and pneumatic tools.

semiopaque enamel. See semitransparent enamel.

semiplastic bronze. A bearing bronze with moderate plasticity containing about 75-79% copper, 13.5-16.5% lead and 7-9% tin. See also plastic bronze.

semiporcelain. A generic term for dinnerware with low translucency and a low to moderate degree of water absorption (typically between 0.3 and 3.0%).

Semi Rigid-Board. Trade name of Manville Corporation (USA) for semi-rigid fiberglass boards used for metal building insulation.

semi-rigid board insulation. Lightweight thermal insulation usually manufactured from glass-fiber materials and supplied in the form of semirigid boards in a wide range of sizes. It may or may not have a water-repellent covering, has a high thermal insulating value per unit thickness (R-value or heat-flow resistance value) and, owing to a certain degree of flexibility (semi-rigidity), can be easily put into place. Used for the thermal insulation of buildings and structures. See also board insulation; rigid board insulation.

semi-rigid cast elastomer. A cast polyurethane elastomer, either solid or containing numerous microscopic cells, that possesses excellent abrasion and wear resistance, and relatively high stiffness. Used for shock absorbers, autobody parts, cyclone separators, and load-transmission elements.

semi-rigid molded foam. A polyurethane foam having numerous small, predominately open (or interconnected) cells and exhibiting considerable elastic hysteresis. Used for applications requiring mechanical damping, such as crash pads, head rests, etc.

semi-rigid plastics. Plastics made from semirigid resins.

semi-rigid resin. A synthetic resin having a flexural or tensile modulus between 69 and 690 MPa (10 and 100 ksi) at a temperature of 23°C (73°F) and a relative humidity of 50%. See also rigid resin.

semisilica brick. Firebrick containing 72-85% silica (SiO_2), with the balance being chiefly alumina (Al_2O_3) and titania (TiO_2). It has a maximum service temperature of about 1480°C (2700°F), high-load-bearing capacity, excellent resistance to spalling and fluxing, and good dimensional stability. Used for furnace applications. Also known as *semisilica firebrick; semisilica fireclay brick*.

semisilica refractories. Refractory materials containing at least 72% of silica (SiO_2), with the balance being primarily alumina (Al_2O_3). Also known as *semisiliceous refractories*.

semisiliceous refractories. See semisilica refractories.

semi-solid processed metals. Metals and metal alloys produced by first heating the starting material to a liquid state, and then cooling it into the solid-liquid range and mechanically breaking it up (e.g., by continuous stirring) to obtain a viscous, semi-solid material with uniform microstructure and predominantly rounded grains. This semi-solid is then forced into a die and cooled to room temperature. Semi-solid processed metals and alloys have better mechanical properties and structural integrity than traditional liquid-cast metals and alloys, and can be further processed by forging, rolling, extrusion and other processes.

Semi-Steel. British trade name for a cast iron containing 3-3.5% total carbon, 1.4-1.8% silicon, 0.5-0.8% combined carbon, and the balance iron. Used for frames, housings and cast gears.

semisteel. A high-strength cast iron made by adding considerable quantities (typically 25-60%) of low-carbon steel scrap to molten pig iron in a cupola. Abbreviation: SS.

semisynthetic fibers. Fibrous materials, such as rayon, glass, boron, boron nitride, carbon, graphite or fused silica, that are manufactured from natural materials, but that do not occur in the free state. See also natural fibers; synthetic fibers.

semisynthetic sand. Natural foundry sand whose properties have been changed by adding a clay binder, such as *bentonite*. See also natural sand; synthetic sand.

Semi-Thik. Trade name of Manville Corporation (USA) for fiberglass batts used as thermal home insulation.

semitransparent enamel. An enamel glaze that is intermediate in light transmission between an opaque and a transparent glaze. Also known as *semiopaque enamel*. See also opaque glaze; transparent glaze.

Semitron. Trademark of DSM Engineered Plastics (USA) for a series of static dissipative thermoplastics based on copolymer acetals, polyether-imides, polyether sulfones and polytetrafluoroethylenes. They have an upper service temperature of about 260°C (500°F), and are used for integrated circuits, circuit boards, disk drives, wafer combs and various other electrical and electronic components, such as sockets and contactors.

Semivac 90. Trade name of Vacuumschmelze GmbH (Germany) for a permanent magnet material composed of iron, chromium, cobalt, nickel and molybdenum. Supplied in the form of wire and strip with bright metallic surface, it has a density of 7.85 g/cm^3 (0.284 lb/in.^{3}), a hardness of 700 Vickers, a Curie temperature of 700°C (1290°F), a maximum service temperature of 450°C (840°F), and high magnetic remanence and squareness. Used in the deactivation of anti-theft devices.

semivitreous ceramics. (1) Ceramics whose degree of vitrification is indicated by a moderate water absorption of 0.2-3.0%

(2) Ceramic floor and wall tile whose degree of vitrification is indicated by a water absorption of 3.0-7.0%.

semivitreous china. White or ivory-colored dinnerware or other ceramic product exhibiting moderate water absorption (usu-

ally 4-10%) and intermediate strength. See also vitreous china.

semi-worsted spun yarn. A worsted yarn spun from roving or carded and gilled slivers.

Semmum. Trade name for rayon fibers and yarns used for textile fabrics.

Semperal. Trade name of Metallwerke AG (Austria) for a heat-treatable aluminum alloy containing 4% copper, 1% zinc, 0.7% iron, 0.6% manganese and 0.3% magnesium. Used for light-alloy parts.

semseyite. A gray to black mineral of the lillianite group composed of lead antimony sulfide, $Pb_9Sb_8S_{21}$. Crystal system, monoclinic. Density, 6.03 g/cm^3; hardness, 2.5 Mohs. Occurrence: Hungary.

senaite. A black mineral of the crichtonite group composed of lead titanium iron oxide, $Pb(Ti,Fe,Mn)_{21}O_{38}$. Crystal system, rhombohedral (hexagonal); density, 5.20 g/cm^3. Occurrence: Brazil.

senarmontite. A colorless, white or grayish mineral with a white streak and a resinous luster. It is composed of antimony trioxide, Sb_2O_3, and contains theoretically 83.3% antimony. It can also be made synthetically. Crystal system, cubic. Density, 5.2-5.6 g/cm^3, hardness, 2.5 Mohs; refractive index, 2.087; turns yellow on melting. Occurrence: Northern Africa, Europe, Mexico, USA (Montana, Nevada). Used as an ore of antimony, as a paint pigment, and as a flameproofing agent. See also valentinite.

Sendust. Trade name of Siemens & Halske AG (Germany) for a brittle cast alloy of 85% iron, 9.6% silicon and 5.4% aluminum. It is a low-cost soft magnetic material used for cores in electromagnetic equipment.

Seneca. (1) Trade name of Senaca Wire & Manufacturing Company (USA) for a carbon steel containing 0.7% carbon, and the balance iron. Used for wire products.

(2) Trade name of Atlas Specialty Steels (Canada) for a tungsten-type hot-work tool steel (AISI type H21) containing 0.35% carbon, 0.3% manganese, 0.3% silicon, 9.5% tungsten, 3.25% chromium, 0.4% vanadium, and the balance iron. It has a great depth of hardening, high hot hardness and hot wear resistance, low cold-wear resistance, high to moderate toughness, and moderate machinability. Used for extrusion molds, tools and dies, dummy blocks, die-casting dies for copper alloys, hot trimming, swaging and blanking dies, forging die inserts, hot-nut tools, hot punches, and hot bolt and rivet dies.

Senegal gum. See acacia gum.

sengierite. A green, radioactive mineral of the carnotite group composed of copper uranyl vanadium oxide hydroxide hydrate, $Cu_2(UO_2)_2V_2O_8(OH)_2\cdot xH_2O$. Crystal system, monoclinic. Density, 4.41 g/cm^3; refractive index, 1.92-1.94. Occurrence: Congo, Zaire. Used as an ore of vanadium.

Senotex. Trade name of H.B. Fuller Company (USA) for two-part polyurethane coatings with a content of 100% solids.

sensitizing compounds. Metallic salts, such as the chlorides of copper, gold, palladium and tin, or certain salts of aluminum, barium, cadmium, iridium, palladium or silver, dissolved in water or an organic solvent, that when applied to glass and ceramic ware produce an invisible film and initiate or accelerate subsequent surface treatments, such as plating or silvering.

sensitive metal. The common name of a pyrophoric material composed of 75% cerium and 25% platinum.

Sensorvac. Trade name of Vacuumschmelze GmbH (Germany) for a permanent magnet material composed of iron, nickel, aluminum and titanium. Supplied in the form of strip with bright metallic surface, it has a density of 7.65 g/cm³ (0.284 lb/in.³), a hardness of 600 Vickers, a Curie temperature of 630°C (1165°F), a maximum service temperature of 300°C (570°F), and high magnetic remanence and squareness. Used in the deactivation of anti-theft devices.

Sensura. Trade name of Wellman Inc. (USA) for a soft, resilient, breathable polyester fiber. It has excellent durability, drapability and crease retention, and is used for apparel including sportswear and casual wear. See also Microdenier Sensura.

Sentalloy. Trademark of GAC International, Inc. (USA) for a shape-memory alloy composed of 52.6% nickel and 47.4% titanium, used for dental and orthodontic wire.

Sentinol. Trade name for a nickel-titanium shape-memory dental wire.

Sentoku. Japanese trade name for bronze alloys.

SentryGlas. Trademark of E.I. DuPont de Nemours & Company (USA) for composite glass composed a sheet of glass with a *Butacite* polyvinyl butyral interlayer that has an exterior layer of clear strong polyester film with abrasion-resistant coating bonded to it. It is extremely resistant to breakage and spalling, and used in window systems.

Seomag. German trade name for hard, corrosion-resistant protective coatings of magnesium oxide (MgO) produced on magnesium and aluminum alloys by anodic oxidation, i.e., by converting the magnesium surface to magnesium oxide by making the metal or alloy the anode in an electrolytic cell.

separate-application adhesive. An adhesive supplied in two components or parts. One part or component is applied to the first substrate and the other to the second substrate. Uniting of the two substrates results in an adhesive joint. See also two-component adhesive.

separated aggregate. Concrete aggregate that has been classified into coarse and fine components.

separating powder. A powder applied to a mold surface to prevent adhesion and facilitate removal of ceramic ware after molding.

Separit. Trade name of Foseco Minsep NV (Netherlands) for superfine, nonsiliceous parting powders that are fully waterproofed and applied to foundry patterns to ensure clean lifts of castings from molding sands.

Sephacryl. Trademark of Pharmacia Biotech AB (Sweden) for a series of crosslinked copolymers of allyl dextran and *N,N'*-methylenebis(acrylamide) used in gel-filtration chromatography of proteins.

Sephadex. Trademark of Pharmacia Biotech AB (Sweden) for a series of synthetic, organic compounds derived from the polysaccharide dextran by crosslinking (e.g., with epichlorohydrin) to produce a three-dimensional network. Various ionic functional groups (carboxymethyl, quaternary aminoethyl, etc.) can be attached by ether linkages to the glucose units of the main polymer chain. *Sephadex* products are supplied in a wide range of forms and grades for use as anion and cation exchangers for the chromatographic purification of proteins, gel filtration of biological materials, and as biomaterials.

Sepharose. Trademark of Pharmacia Biotech AB (Sweden) for a series of synthetic, organic compounds derived from the polysaccharide agarose, and used as anion and cation exchangers in ion exchange chromatography of biological materials, and as biomaterials.

Sephacel. Trademark of Pharmacia Biotech AB (Sweden) for a series of synthetic organic compounds derived from cellulose, usually diethylaminoethyl (DEAE) cellulose, and used as an-

ion and cation exchangers in ion exchange chromatography of biological materials.

Sepia. Trade name of Société Industrielle Triplex SA (France) for a laminated glass with tinted interlayer.

sepia. A powder made from the calcified portion of the backs of cuttlefish (order Sepioidea, genus *Sepia*), and used as a fine polishing powder for jewelry, ornaments, etc. Also known as *cuttlefish bone.*

sepiolite. A white, gray, pink, greenish, pale yellow mineral composed of magnesium silicate hydrate, $Mg_4Si_6O_{15}(OH)_2 \cdot 6H_2O$. Crystal system, orthorhombic. Density, 2.00-2.26 g/cm³; hardness, 2.0-2.5 Mohs. Occurrence: Japan, Spain, Turkey, USA (Utah). Used as a soft lightweight, absorbent clay in the manufacture of tobacco pipes and ornamental carvings. Also known as *meerschaum; sea-foam.*

sepiomelanin. A black *melanin* found in the ink sacs of cuttlefish, especially the genus *Sepia officinalis,* and certain squids. It is a 5,6-dihydroxyindole pigment, and a linear polymer stabilized with hydroxyl groups. Used in biochemistry and biotechnology.

Seprafilm. Trademark of Genzyme for a bioresorbable sodium hyaluronate-based membrane used for the prevention or reduction of postoperative adhesions after abdominal surgery or uterine myomectomy.

Septalloy. Trade name for dental amalgam alloys.

Septon. Trademark of Kuraray Company Limited (Japan) for diblock (SEP) and triblock (SEPS) copolymers conisting of polystyrene blocks and soft polyolefin (e.g., polyethylene and polypropylene) rubber blocks. They are thermoplastics with flowlike consistency at elevated temperatures, while exhibiting rubberlike properties within a wide temperature range. *Septon* copolymers have low toxicity, excellent elasticity and tensile strength, excellent low temperature properties and heat-aging resistance, high electrical insulating properties, and good resistance to acids, alkalies and alcohols. The diblock (SEP) copolymers can be blended with process oil and tackifiers to produce heat- and weather-resistant adhesives and with olefinic acids to enhance physical properties, and can also be used as compatibilizers between olefinic and styrenic plastics. The triblock (SEPS) copolymers can be blended with process oil and polyolefins to produce soft compounds for use as vulcanized rubber substitutes.

Sequel. (1) Trademark of Solvay Engineered Polymers (USA) for a series of engineering polypropylenes.

(2) Trademark of Degussa-Ney Dental (USA) for copperfree dental palladium-silver-gold ceramic alloys with light oxide layers and good yield strength. *Sequel Plus* is a copper-free grade containing 74.4% palladium, 6.5% silver and 5.3% gold. It has very high ductility and a hardness of 220 Vickers. Used for metal-to-porcelain restorations.

sequoia. A term referring to the soft, light, nonresinous, brownish-red wood of coniferous trees of the genus *Sequoia,* especially the redwood (*S. sempervirens*) and the giant sequoia (*S. gigantea*). The latter grows in the Sierra Nevada, USA, and is now protected by law and therefore no longer harvested commercially. See also redwood.

Seracelle. Trade name of Courtaulds Limited (UK) for cellulose acetate plastics and fibers.

Seraceta. Trade name of Courtaulds Limited (UK) for a continuous-filament cellulose acetate used for wearing apparel and as industrial fabrics.

Seraglass. Trade name of Seratuff Proprietary Limited (Australia) for a repetitively ceramic-printed toughened glass used for industrial applications.

Séralit. Trademark of Saint-Gobain (France) for a screen-printed toughened glass.

serandite. A rose-red mineral of the pyroxenoid group composed of sodium manganese calcium hydrogen silicate, $Na(Mn,Ca)_2$-$Si_3O_8(OH)$. Crystal system, triclinic. Density, 3.41 g/cm³; refractive index, 1.64-1.67. Occurrence: Canada (Quebec), Guinea.

Seratelle. Trade name for cellulose acetate fibers used for wearing apparel and as industrial fabrics.

Seratuff. Trade name of Seratuff Proprietary Limited (Australia) for a repetitively ceramic-printed and toughened glass used for industrial applications.

seraya. See red seraya; white seraya; yellow seraya.

Serelle. Trademark of KoSa (USA) for durable high-loft continuous filament polyester fiberfill with excellent compression recovery and insulation performance, used for bedding, e.g., pillows and mattress pads.

serendibite. A dark blue mineral of the aenigmatite group composed of calcium magnesium aluminum boron silicate, Ca_2-$(Mg,Al)_6(Si,Al,B)_6O_{20}$. Crystal system, triclinic. Density, 3.38 g/cm³; refractive index, 1.703. Occurrence: Canada (Northwest Territories), USA (New York), Sri Lanka.

Serene. Trademark of KoSa (USA) for durable high-loft continuous filament polyester fiberfill with excellent compression recovery and insulation performance, used for bedding, e.g., pillows and mattress pads.

serge. Generic term for smooth-faced fabrics, usually made from wool, worsted or silk yarn, in a twill weave and having slanting lines or ridges on their surfaces. Used primarily for suits and pants, uniforms, and as a lining material for clothing.

sericin. A gummy, proteinaceous substance comprising about 20-30 wt% of *raw silk*. It is usually removed during preparation by a boil-off process. Also known as *silk gum*.

sericite. A soft, white, fine-grained potash mica, $K_2O \cdot 3Al_2O_3 \cdot 6SiO_2 \cdot 2H_2O$, that is a hydrous variety of *muscovite*. It is available as a fine powder for use in aluminum paint, as a foundry core and mold wash, and as a filler for plastics and rubber. Also known as *damourite; sericite mica*.

Sermabond. Trademark of Sermatech International (USA) for an electrically conductive coating for composites, ceramics and glasses. It provides a bond strength after curing of above 35 MPa (5 ksi), and has good resistance to temperatures up to 650°C (1200°F).

Sermalloy. Trade name of SERMAG–Société d'Etudes et de Recherches Magnetique (France) for a series of *alnico*-type permanent magnet alloys.

SermaGard. Trademark of Sermagard, Division of Telefix Inc. (USA) for a protective coating composed of a sacrificial base coat with a topcoat or sealer. It provides protection from corrosion, stress-corrosion cracking and hydrogen embrittlement for fasteners in contact with dissimilar metal parts.

Sermagard. Trade name of MetoKote Corporation (USA) for a water-based aluminized coating with ceramic binder.

SermaLon. Trademark of Sermatech International Inc. (USA) for a three-part coating system consisting of an aluminum-ceramic base coat, an inhibitor and a non-stick sealcoat. It provides outstanding corrosion resistance for parts operating in severe environments, e.g., compressor rotors, turbine parts, etc.

Sermaloy. Trademark of Sermatech International Inc. (USA) for advanced high-temperature coatings that are equally effective against both high- and low-temperature corrosion and oxida-

tion. Used for gas-turbine engines, die-casting molds, radiant heat tubes, and diesel turbochargers.

SermaTel. Trademark of Sermatech International Inc. (USA) for a series of advanced coatings including (i) several abrasion- and corrosion-resistant high-temperature polymeric coatings for light metals, such as aluminum and magnesium; and (ii) several composite coatings consisting of aluminum-filled chromate-phosphate bond coats and either organic barrier film topcoats, chemically inert chromate-phosphate topcoats or aluminum ceramic overlay coatings. They are designed for heat-sensitive alloys, heat-treated aluminum and titanium alloys and high-strength steels. *SermaTel* coatings are often used on turbine-engine components and other high-temperature components.

Sernshi. Japanese trade name regenerated cellulose supplied in film form.

serotonin black. A black pigment based on the neurotransmitter *serotonin* ($C_{10}H_{12}N_2O$) and related to *melamin*. Used in biochemistry and biotechnology.

serpentine. A group of green, greenish-yellow, yellowish-brown, greenish-gray or greenish-black *asbestos* minerals including *antigorite* and *chrysotile*. Chemically, they are hydrous magnesium silicates, $Mg_3Si_2O_5(OH)_4$, and also contain 2-8% iron oxide, moderate amounts of aluminum and small amounts of manganese monoxide, calcium oxide, potassium oxide and sodium oxide. *Serpentine* minerals have a greasy to silky luster and a somewhat soapy feel. Crystal system, monoclinic, or orthorhombic. Density, 2.50-2.65 g/cm³; hardness 2.5-4.0 Mohs. Occurrence: Canada (Quebec), New Zealand, South Africa, USA (California, Vermont, Arizona). Used in forsterite refractories, in building trim, for roofing granules and terrazzo, and in the manufacture of ornaments and novelties. The *chrysotile* variety is used for fireproofing applications. Also known as *serpentine asbestos*.

serpentine crepe. A plain-weave fabric made from natural or synthetic fibers and having a twisted filling thread producing a crepy surface effect.

serpierite. A green mineral composed of calcium copper zinc sulfate hydroxide trihydrate, $Ca(Cu,Zn)_4(SO_4)_2(OH)_6 \cdot 3H_2O$. Crystal system, monoclinic. Density, 3.07 g/cm³; refractive index, 1.641. Occurrence: Greece.

service hardboard. Hardboard with a density of about 880 kg/m³ (55 lb/ft³) made without additives (e.g., moisture inhibitors). It has a lower tensile strength and resistance to moisture than other types, but has improved dimensional stability. Available in thicknesses from 0.125 to 0.500 in. (3.2 to 12.7 mm), it can have one or two smooth sides (S1S or S2S). The water absorption of the S1S type increases from 25 to 30% with decreasing thickness, while that of the S2S type increases from 18 to 30%.

service-tempered hardboard. See tempered service hardboard.

Servofrax. Trade name of Servo Corporation of America (USA) for an infrared transmitting glass.

Setila. Trademark of Rhône-Poulenc–Setila SA (France) for polyester fibers and filament yarns.

Setilmat. Trade name for cellulose acetate fibers and products used for textile fabrics.

Setilose. Trade name for cellulose acetate fibers used for wearing apparel and as industrial fabrics.

Setina. Trade name for cellulose acetate fibers used for wearing apparel and as industrial fabrics.

Setit. Trademark of National Lead Company (USA) for a colloidal aluminum oxide used as a suspension aid in the ceramic industry.

sett. British name for a small, dressed, wear-resistant, usually rectangular paving stone composed of basalt, granite or quartzite. It is usually heat-treated by burning, and used for heavy traffic areas.

setting-up agent. An electrolyte, such as magnesium carbonate, potassium carbonate, magnesium sulfate or sodium oxide, that is introduced into a glaze, porcelain enamel, clay or other slurry to promote flocculation and increase the suspension properties. Also known as *set-up agent.*

set yarn. See stabilized yarn.

Seva. Trade name of IMI Yorkshire Alloys Limited (UK) for a corrosion-resistant arsenical brass containing 69-71% copper, 29-31% zinc and 0.02-0.04% arsenic. Usually supplied in tube form, it has good malleability and ductility, and is used for condenser and evaporator tubing, coolers, juice heaters, and vacuum pan tubes.

Sevacarb. Trade name for carbon blacks.

Seven Star. Trade name of Carpenter Technology Corporation (USA) for a molybdenum-type high-speed tool steel (AISI type M7) containing 1.00% carbon, 8.75% molybdenum, 4.00% chromium, 2.00% vanadium, 1.75% tungsten, and the balance iron. It has a great depth of hardening, excellent wear resistance and high hot hardness, and is used for lathe and planer tools, cutters, reamers, etc.

seventy-thirty. A corrosion-resistant alloy containing 70% copper and 30% nickel. Used for heat-exchanger and condenser tubes.

Severn. Trade name of Pasminco Europe (Mazak) Limited (UK) for high-purity zinc (99.5%) supplied in the form of ingots.

Seville. Trade name of Aardvark Clay & Supplies (USA) for an orange-brown medium-coarse clay (cone 10).

Sevritron. Trade name for self-cure acrylic cements and resins used in dentistry for crown and bridge restorations and tooth filling.

sewer brick. An abrasion-resistant brick of low water absorption used in drainage systems and similar structures.

sewer tile. A hollow, impervious structural clay product, usually of circular cross section, that is used in drainage systems.

sewing thread. A flexible, cabled or multiple yarn that has been twisted onto a reel, tube or other support in counterclockwise direction (Z-twist) and given a special finishing treatment. Used for needlework and stitching. See also cabled yarn; multiple yarn.

Seycast. Trade name of Seymour Products Company (USA) for a controlled-grain nickel (99% pure) used for anodes.

Seymour. Trade name of Seymour Products Company (USA) for a series of coppers and copper alloys including various grades of commercially pure copper, and several high and low brasses, red and yellow brasses, high and low leaded brasses, cartridge brasses, commercial and gilding bronzes, copper nickels, and leaded and unleaded nickel silvers.

Seymourco. Trade name of Seymour Products Company (USA) for a nickel silver containing 61% copper, 34% zinc and 5% nickel. Used for costume jewelry.

Seymourite. Trade name of Seymour Products Company (USA) for a white-colored, highly corrosion-resistant copper-base alloy containing 18% nickel and 18% zinc. Used for ornaments and electrical equipment.

SF-Extra. Trade name for a high-purity grade of zirconia powder used in the manufacture of moderate-strength structural ceramics.

SG irons. See ductile cast irons.

S-Glass. Trademark of Owens-Corning Fiberglas (USA) for gigh-strength glass fibers. See also S-glass.

S-glass. Magnesium aluminosilicate glass used in the form of high-strength filaments for reinforcing metal, ceramic and organic-matrix composites. A typical composition is 64% silica (SiO_2), 25% alumina (Al_2O_3), 10% magnesia (MgO), 0.3% soda (Na_2O) and a total of 0.7% other oxides. *S-glass* has a density of 2.48-2.50 g/cm³ (0.089-0.090 lb/in.³), a melting point of about 1450°C (2640°F), a high elastic modulus (86 GPa or 12.5 × 10³ ksi, or more), very high tensile strength (4.6 GPa or 665 ksi, or more), and good elevated temperature properties. It is available in two grades almost identical in composition, but having different surface coatings: *S-2 glass* for commercial applications and *S-glass* for more exacting applications.

S-2 glass. A high-performance *S-glass* developed by Owens-Corning Fiberglass (USA). S-2 glass fibers have a density of 2.49 g/cm³ (0.090 lb/in.³), an ultimate tensile strength of about 4.6 GPa (665 ksi) and an elastic modulus of 87 GPa (12.5 × 10³ ksi). They are supplied in fiber, yarn and roving form for commercial composite reinforcement applications.

shadecloth. (1) A stiff, plain-woven fabric treated with oil, starch or special chemicals to enhance its opacity. Used for window and lamp shades.

(2) See holland cloth (1).

Shade Green. Trade name of Owens-Illinois Inc. (USA) for glass blocks made from a bluish-green glass.

Shadelite. Trade name of Duplate Canada Limited for a graduated, tinted glass used for motor buses.

shading lacquer. A transparent colored spray-applied lacquer used for shading applications.

shadlunite. A grayish yellow mineral of the pentlandite group composed of copper iron lead cadmium sulfide, $(Cu,Fe)_8(Pb,Cd)S_8$. Crystal system, cubic. Density, 4.72 g/cm³. Occurrence: Russian Federation.

shandite. A mineral composed of nickel lead sulfide, $Ni_3Pb_2S_2$. It also be made synthetically. Crystal system, rhombohedral (hexagonal). Density, 8.65 g/cm³. Occurrence: Australia, Tasmania.

Shadowlite. Trade name of Triplex Safety Glass Company Limited (UK) for a laminated glass with gray tinted interlayer.

shadowy organdy. A light, crisp, sheer *organdy* fabric with a shadowy surface effect produced by repeatedly printing the same color onto itself.

Shagbark. Trade name for a combed, textured cotton having a fine pile and a crisp surface finish.

shagbark hickory. The hard, tough, elastic wood from the tall hickory tree *Carya ovata.* The heartwood is reddish brown and the sapwood white. It has a fine, even, straight grain, high strength (stronger than oak), high resiliency and shock resistance, fair durability, high shrinkage and fair workability. Average weight, 720-820 kg/m³ (45-51 lb/ft³). Source: Southern Canada; eastern United States (Atlantic, central and Lake states from Maine to Illinois and from Florida to Texas). Used for axe, pick, shovel and tool handles, for ladder rungs, athletic equipment, agricultural implements, furniture, pallets and blocking, as sawdust and chips for panel materials, and for meat smoking.

shagreen leather. A hard, strong, untanned leather made from the skin of horses, donkeys, seals and other animals. It has a characteristic surface finish obtained by embossing grains into the skin to imitate shark leather.

shahtoosh. The hair obtained from the chiru (*Pantholops hodgsoni*), a Tibetan goat antelope. It was formerly made into fabrics. *Note:* This animal is now listed as an endangered species.

shake. A hand-split wood shingle with at least one rough surface due to the natural split of the wood along the grain.

Shaker. Trade name of SA Glaverbel (Belgium) for patterned glass.

Shaku-do. Japanese corrosion-resistant jewelry alloy composed of 94-96% copper, 3.7-4.2% gold, 0.1-1.6% silver, and traces of arsenic, iron and lead. Used for art objects and ornamental parts.

shale. A fine-grained sedimentary rock formed from clay that has been subjected to great pressure. It contains considerable quantities of quartz, mica and other minerals, has a specific gravity of 2.6-2.9 g/cm³ (natural) or 1.5 g/cm³ (as-quarried), and splits easily into thin layers. Used in building construction, and ceramics. See also oil shale.

shale clay. Finely divided *shale* used in ceramics as a clay replacement.

shalloon. A twilled woolen or worsted cloth, used chiefly as a lining for coats, special uniforms, etc. Named after Chalons-sur-Marne, a city in northeastern France where it was originally made.

Shalltray. Trade name of Teijin Fibers Limited (Japan) for polyester filaments used for the manufacture of textile fabrics.

Shamra. Trade name for mullite refractories.

Shanton. Trade name for polyester fibers and yarns used for textile fabrics.

shantung. (1) A heavy, soft (tussah) silk or rayon fabric, somewhat like *pongee* in appearance and texture, made in a plain weave with irregular nub yarns, and used for suits and dresses. It is named after the province of Shantung in northeastern China. See also tussah.

(2) A cotton fabric similar to silk or rayon shantung (1) and *broadcloth*, and used for dresses and childrens' wear.

shantung-type yarn. An irregular nub yarn made to imitate the silk yarn used for making true shantung. See also shantung (1).

Shapal-M. Trademark of Tokuyama Corporation (Japan) for a machinable ceramic of high-purity aluminum nitride (AlN) supplied in the form of sheets and rods. It has a density of 2.9 g/cm³ (0.10 lb/in.³), high strength and thermal conductivity, a moderate resistance to acids and alkalies, an upper continuous-use temperature above 1000°C (1830°F), and can be machined with carbide tools. Used for metal-melting crucibles, heat sinks, and in the electronics industry for insulating components.

shape-memory alloys. A group of alloys that after being deformed can return to their original shape upon heating, and thus are said to have "remembered" their previous shape. The reversible change in shape is due to a diffusionless martensitic phase transformation that is essentially time independent. The shape memory effect is common to many nonferrous alloys with internally twinned or faulted martensitic structures including nickel-titanium, cobalt-iron, copper-aluminum, gold-cadmium, indium-thallium, and several titanium-base alloys. An example of a commercial shape-memory alloy is an alloy of 53-60% nickel and 40-47% titanium, sold under trade name *Nitinol*. Abbreviation: SMA. Also known as *memory alloys*.

shape-memory ceramics. A class of antiferroelectric ceramics, such as lead lanthanum zirconate titanate (PLZT) and barium stannate titanate (BST), in which the application of an electric field results in deformation (i.e., an expansion or contraction) by a reversible process that transforms the antiferroelectric phase of the initial state to a ferroelectric phase. Used in the manufacture of actuators and transducers.

shape-memory plastics. Crystalline, radiation-crosslinked plastic materials, such as certain fluoropolymers, polyolefins and elastomers, which when heated to high enough temperatures become rubbery, shrink (typical shrink ratio up to 4:1) and temporarily lose their crystalline structures (but not their random crosslinks) and, upon cooling, regain their crystallinity and expand returning to their original shape. They are widely used in the form of molded parts and tubing as wire and cable insulation, sleeves for soldering devices and crimp splices, abrasion-, corrosion-, thermal- and chemical-resistant protection for mechanical components, etc. Also known as *heat-shrinkable plastics*.

shape-memory stainless steels. A group of stainless steels typically containing 13-21% manganese, 4.9-5.5% silicon, 8-10% chromium and 4.4-6.3% nickel, and the balance iron, that experience a reversible shape-memory effect owing to a stress-induced transformation of face-centered cubic (fcc) austenite to hexagonal close-packed (hcp) ε-martensite during cooling. They have low stacking fault energies, and are presently used chiefly for pipe couplings. See also shape-memory alloys.

shape-selective catalyst. A solid catalyst that discriminates among molecules of different shape and size. It is produced by introducing a transition metal into a crystalline aluminosilicate (or *zeolite*). Used in catalytic reactions, e.g., for cracking straight-chain hydrocarbons.

Sharalloy. Trade name of Sharon Steel Corporation (USA) for a series of high-strength low-alloy (HSLA) steels containing 0.1-0.16% carbon, 0.4-0.5% manganese, 0.1-0.3% silicon, 0.01% niobium, and the balance iron. Used in the form of hot-rolled sheet and strip for automotive and structural applications.

Shareen. Trade name for nylon fibers and yarns used for textile fabrics.

shark leather. A durable, nonscuffing leather made from the top grain of the skins of sharks. It is usually tanned and has a relatively hard, but pliable surface. Used for bookbinding, handbags, and shoes.

sharkskin. (1) A heavy, two-color woven fabric, somewhat resembling *taffeta*, made from fine, white- and black-colored yarns of wool, worsted, rayon or cotton, and used for dresses, suits and sportswear.

(2) A crisp, usually white fabric, often of acetate, triacetate or rayon filament yarn, having a grainy, crepey texture with reduced luster. Used for tennis clothes and uniforms.

Sharon. Trade name of Sharon Steel Corporation (USA) for an extensive series of alloy steels, austenitic stainless steels, Hadfield manganese steels, case-hardening steels, structural steels and tool steels.

Sharpealoy. Trade name of Brown & Sharpe Manufacturing Company (USA) for a high-speed tool steel containing 1% carbon, 8.35% molybdenum, 3.85% chromium, 3.85% cobalt, 1.75% tungsten, 1.1% vanadium, and the balance iron. Used for cutting tools, such as drills, taps, milling cutters, end mills, lathe tools, chasers, etc.

sharpite. An olive-green, radioactive mineral composed of calcium uranyl carbonate hydroxide hexahydrate, $Ca(UO_2)_6(CO_3)_5(OH)_4 \cdot 6H_2O$. Crystal system, orthorhombic. Density, 3.33 g/cm³. Occurrence: Congo.

sharp sand. (1) Coarse sand with angular particle shape.

(2) Foundry sand essentially free of bonding agents, such

as clay.

Shat-R-Proof. Trade name of Shatterproof Glass Corporation (USA) for a laminated safety glass.

shatterproof glass. See laminated glass.

shattuckite. A deep blue mineral composed of copper silicate hydroxide, $Cu_5(SiO_3)_4(OH)_2$. Crystal system, orthorhombic. Density, 4.01 g/cm³; refractive index, 1.782. Occurrence: USA (Arizona).

Shawinigan. (1) Trade name of Shawinigan Chemicals Limited (Canada) for a series of heat- and corrosion-resistant austenitic, ferritic and martensitic stainless steels.

(2) Trade name of Chevron Phillips Chemical Company (USA) for various grades of high-purity *acetylene black* with high electrical conductivity and liquid absorption capacity. Used as conductive fillers in plastics and rubber.

Shawnee. Trade name of AL Tech Specialty Steel Corporation (USA) for a heavy-duty hot-work tool steel (AISI type H19) containing 0.4% carbon, 4.25% cobalt, 4.25% chromium, 4.25% tungsten, 2% vanadium, and the balance iron. It has excellent resistance to abrasion, shock and wear at elevated temperatures, good deep-hardening properties, and superior hot hardness and resistance to heat checking. Used for press forging dies, brass extrusion dies, valve extrusion dies, hot punches, dummy blocks, die inserts, mandrels, etc.

shcherbakovite. A yellow or reddish brown mineral composed of sodium potassium barium titanium silicate, $Na(K,Ba)_2(Ti,Nb)_2(Si_2O_7)_2$. Crystal system, orthorhombic. Density, 2.97-3.34 g/cm³; refractive index, 1.745. Occurrence: Russian Federation, Australia.

shcherbinaite. A yellowish green mineral composed of vanadium oxide, V_2O_5. Crystal system, orthorhombic. It can also be made synthetically. Density, 3.32 g/cm³; refractive index, 2.42. Occurrence: Russian Federation.

Shearcut. Trade name of Gulf Steel Corporation (USA) for an oil-hardening shock-resisting tool steel (AISI type S5) containing 0.55% carbon, 2% silicon, 0.8% manganese, 0.4% molybdenum, 0.3% vanadium, 0.25% chromium, and the balance iron. It has good deep-hardening properties, high toughness, good ductility, a high elastic limit, and a fine-grained structure. Used for chisels, punches, jaws, shear blades, forming tools, stamps, dies, etc.

shear steel. A steel made by shearing *blister steel* to short lengths, heating to a high temperature, joining by hammering and/or rolling, and finishing by hammering. Used for making cutlery.

Sheartough. Trade name of British Steel Corporation (UK) for an oil-hardening alloy tool steel containing 0.45% carbon, 0.3% manganese, 0.25% silicon, 2.25% nickel, 1.15% chromium, 0.2% vanadium, and the balance iron. Used for shear blades, guillotine blades, shafts, etc.

sheath-core. A multicomponent synthetic fiber comprising a continuous central core enclosed in a continuous outer sheath or envelope.

sheathing. An underlayment fastened to the framework of walls and roofs before finishing material (e.g., brick, wood or vinyl siding, roofing, etc.) is added. It can be made of plywood, composite board, fiberboard, gypsum board, or foamed polystyrene or polyurethane plastics. Also known as *sheathing board*.

sheathing bronze. Trade name for an alloy of 44.5% copper, 32.5% nickel, 16% tin, 5.5% zinc and 1.5% bismuth. It has good resistance to seawater corrosion, and is used for sheathing and in marine construction.

sheathing lead. A special antimonial lead containing about 1%

antimony and traces of bismuth, copper and silver. It has good corrosion and fatigue resistance and good extrusion characteristics. Used as sheathing over rubber, plastic or paper-insulated copper power and telephone cables. Also known as *cable lead*.

sheathing paper. A heavy, high-quality building paper applied over *sheathing* in wall, floor and roof construction to resist passage of air and moisture.

Shedisol. Trade name of Saint-Gobain (France) for glass wool ceiling boards laminated with aluminum foil. Used in industrial buildings.

sheepskin. Soft, fine leather obtained from the skin of sheep, and used for polishing and buffing wheel disks, parchment, shoe linings, and clothing.

sheer fabrics. A category of very thin, transparent or translucent fabrics made from fine yarns in open weaves and including chiffon, gauze, organdie and voile.

sheet. (1) A material, usually a metal or plastic, of uniform thickness and considerable length and width as compared to its thickness. The minimum and/or maximum dimensions depend on the particular type of material, e.g., flat-rolled metal sheet has a maximum thickness of 6.5 mm (0.25 in.), with thicker metal products being usually referred to as "plates," and thinner ones as "foil," and plastic sheeting has a minimum thickness of 0.5 mm (0.02 in.), with thinner products being usually referred to as "film." See also plate; foil, plastic film; plastic sheeting.

(2) A large rectangular piece of textile fabric, usually cotton, linen or a cotton blend, used to cover a mattress on a bed.

sheet asphalt. A plant-mix of asphalt cement composed of graded sand (particle size less than 2 mm or 0.08 in.) and a mineral filler. Used as a road-paving material that yields a smooth, continuous surface. See also sand asphalt.

sheet copper. A corrosion-resistant copper, usually of commercial purity, hot-rolled into sheets, and either used in the as-hot rolled condition or subsequently cold-rolled to increase strength and hardness. It is often used for roofing applications (e.g., gutters, flashing, etc.), and chemical process equipment.

sheet glass. (1) Flat glass produced by floating, i.e., by continuously flowing melted soda-lime glass onto the flat surface of molten tin contained in a vat. As it flows over the tin, a ribbon of glass is formed that has smooth, parallel surfaces. The glass cools and becomes a rigid continuous sheet that is then annealed, inspected and cut to size. Nominal thicknesses range from about 2.5 to 22 mm (0.100 to 0.875 in.). It is inferior to ground and polished *plate glass*, and used in regular windows.

(2) A generic term for all forms of glass that are flat and including sheet, plate, rolled and float glass.

sheet iron. Flat rolled iron product in the form of sheets or thin plates.

sheeting. (1) A continuous film of plastic whose length and width are large as compared to its uniform thickness. See also plastic sheeting.

(2) Soft, plain or printed textile fabrics of varying composition, count and size, often cotton or a cotton blend, used for sheets, curtains, valances, and pillowcases.

sheet lead. See lead sheet.

sheet metal. Flat-rolled sheets of metal with a maximum thickness of 6.5 mm (0.25 in.). Abbreviation: SM.

sheet mica. A relatively flat *mica* that is essentially free of defects in structure and thus suitable for punching or stamping into shapes and products for insulating applications in electronics and electrical engineering. *Sheet mica* is classified into block mica, film mica and mica splittings. Also known as *book* *mica.* See also block mica; film mica; mica splittings.

sheet molding compound. A composite material consisting of a mixture of a glass-fiber reinforcement, a thermosetting resin (e.g., unsaturated polyester) and all necessary additives (fillers, pigments, etc.), processed into sheet form to simplify molding. A *sheet molding compound* has excellent flow properties and can be compression molded into articles with intricate shapes and excellent surface finish. Abbreviation: SMC.

sheet minerals. See phyllosilicates.

sheet plastic. See sheeting; plastic sheeting.

Sheetrock. Trademark of Canadian Gypsum Company (Canada) for *gypsum wallboard* supplied in standard, fire-retardant, moisture-resistant and lightweight grades. Also included under this trademark are various drywall compounds.

sheetrock. A term often used generically in Canada and the United States for large sheets of *gypsum wallboard* sandwiched between two layers of a special paper.

sheet rubber. Crude rubber made by coagulating latex with acetic or formic acid or sodium hexafluorosilicate, smoking and drying, and subsequent conversion into smooth or ribbed sheets. See also natural rubber; rubber latex.

sheet silicates. See phyllosilicates.

sheet steel. See steel sheet.

sheet-steel enamel. A porcelain enamel well suited for application onto sheet iron and steel.

sheet tin. A thin sheet of iron or steel with a thin surface coating of tin applied by hot dipping, electroplating or immersion. The coating provides a nontoxic, protective, decorative surface, facilitates soldering and/or assists in bonding to other metals.

Sheetweld. Trade name of Westinghouse Electric Corporation (USA) for a carbon steel used for welding electrodes for light sheet metals.

Sheffalloy. Trade name of New England Collapsible Tube Company (USA) for an alloy of lead and tin used for collapsible tubes.

Sheffield. Trade name of Jessop-Saville Limited (UK) for a water-hardening die steel containing 0.82-0.85% carbon, 0.3% manganese, 0.5-0.7% chromium, 0.18-0.22% molybdenum, and the balance iron. Used for coining and cutlery dies.

Sheffield hard alloy. A British corrosion-resistant copper-base alloy containing 31% tin and 20% zinc. Used for ornamental parts.

Sheffield Hi-Strength. Trade name of Armco Steel Company (USA) for a series of readily weldable high-strength low-alloy (HSLA) steels used for mine cars, bus and truck bodies, derricks, cranes, booms, and structural applications.

Sheffield nickel silver. British nickel silver containing 55-63% copper, 17-37% zinc, 11-19% nickel and 0.3% lead. Used as a base for plated tableware.

Shef-Lo-Temp. Trade name of Armco Steel Company (USA) for a tough, shock-resistant hardened and tempered steel containing 0.18% carbon, 0.7-1.35% manganese, 0.15-0.3% silicon, 0.2% chromium, 0.2% nickel, 0.07% molybdenum, 0.2% copper, and the balance iron. Used for low-temperature applications, structural members, pressure vessels, derricks, booms and bridges.

Shef-Super-Lo-Temp. Trade name of Armco Steel Company (USA) for a wear- and shock-resistant steel containing 0.16% carbon, 1.2-1.45% manganese, 0.2-0.5% silicon, 0.2% chromium, 0.2% nickel, 0.07% molybdenum, 0.2% copper, and the balance iron. Used for low-temperature applications, pressure vessels, derricks, booms and bridges.

Shef-Ten. Trade name of Armco International (USA) for a low-carbon alloy steel containing 0-0.28% carbon, 1.4% manganese, 0.4% nickel, 0.2% copper, and the balance iron. It is usually supplied in the as-rolled condition, has good formability, and is used for structural applications, shovels, mining equipment, and railroad cars.

Shelblast. Trade name of Agrashell, Inc. (USA) for soft-grit agricultural blasting media (e.g., nut shells) used in metal finishing.

Sheldrite. Trade name of B.F. Goodrich Company (USA) for conductive, water-based coatings used in metal finishing.

Shell. Trade name of Shell Chemical Company (USA) for a series of polyolefins including *Shell PB* polybutylene plastics and *Shell PP* polypropylene homopolymers and copolymers supplied in unmodified, elastomer-modified and UV-stabilized grades.

shellac. An alcohol-soluble natural resin secreted by the lac insect *(Laccifer lacca)* and deposited on the twigs of certain trees in southern Asia. After purification it is formed into thin sheets that are subsequently fragmented into thin orange, yellow or bleached white flakes. Used in the manufacture of lacquers and varnishes for wood or metal, in floor, furniture and leather polishes, in textile finishes, in the manufacture of plastics, in dielectric coatings and deKhotinsky cement, and formerly as a binder for grinding wheels.

shellac plastics. Plastics made from *shellac* and having poor heat resistance and high water absorption.

shellac varnish. Shellac dissolved in denatured alcohol and used in wood finishing. It produces a hard, durable, quick-drying, light-colored coating, but the finish is not resistant to alcohol and water.

shellac wax. A hard wax containing 3% shellac used in polishes and insulating materials.

shellbark hickory. The heavy, tough reddish wood of the tall hardwood tree *Carya laciniosa* belonging to the true hickories. Commercially almost no distinction is made between the wood from the shagbark, shellbark and mockernut hickory. *Shellbark hickory* has high strength, hardness and shock resistance, high shrinkage, and is difficult to work. Source: Southern Canada and eastern United States. Used for tool handles, ladder rungs, agricultural implements, and furniture. Also known as *kingnut hickory.* See also shagbark hickory; mockernut hickory.

shell castings. See shell moldings.

shell-crosslinked knedels. A relatively new class of nano-sized polymers consisting of hydrophobic, solid polymer spheres (cores) contained within hydrophilic, hollow polymer spheres (shells). They can be synthesized by first forming a block copolymer, quaternizing with a hydrophilic polymer, organizing into polymer micelles, and finally crosslinking the shell and/or core. An example is a polymer consisting of a core of hydrophobic polystyrene and a shell of quaternized polyvinylpyridine crosslinked by irradiation. *Shell-crosslinked knedels* are thought to be suitable for drug delivery and encapsulation applications. Abbreviation: SCK. *Note:* The word "knedel" has been adopted from a Polish term referring to a food product consisting of meat-filled dough.

Shelldie. Trade name of A. Finkl & Sons Company (USA) for a chromium-type hot work steel (AISI type H11) containing about 0.36% carbon, 0.6% manganese, 0.9% silicon, 5% chromium, 1.8% molybdenum, 0.27% vanadium, and the balance iron. It has excellent deep-hardening properties, ductility and toughness, good resistance to heat checking, good machinability, and fairly good shock and abrasion resistance. Used for mandrels, punches, heading dies, forging die inserts, and extrusion tools.

Shellex. Trade name of A. Finkl & Sons Company (USA) for a tough, oil-hardening tool steel containing 0.4% carbon, 1% silicon, 3.3% chromium, 2.4% molybdenum, 0.5% manganese, 0.4% vanadium, and the balance iron. Used for extrusion tools, forging dies, inserts, and punches.

shell moldings. Castings with excellent surface finish and good dimensional accuracy produced by pouring a molten metal or alloy into a thin, shell-shaped mold produced from fine silica sand and a thermosetting (e.g., phenolic) resin binder. Also known as *shell castings*.

shell molding resin. A thermosetting resin powder (e.g., phenol-formaldehyde resin, novolac, etc.) used in shell molding of metals and alloys as a sand binder.

sherardized steel. A steel on whose surface a thin, tightly adhering, corrosion-resistant zinc-rich coating (average coverage, 0.4-1.8 g/cm^2 or 0.1-0.4 oz/ft^2) has been produced by a diffusion process involving heating in a closed drum, with or without tumbling action, for a specified period of time at about 300-420°C (570-790°F) in contact with zinc dust, or a mixture of zinc dust and silica sand.

sherwoodite. A blue-black mineral composed of calcium aluminum vanadium oxide hydrate, $Ca_9(Al_2V_{28}O_{80})\cdot56H_2O$. Crystal system, tetragonal. Density, 2.80 g/cm^3; refractive index, 1.765. Occurrence: USA (Colorado).

Shetland. (1) A fine, hairy, strong worsted spun from the wool of a breed of sheep raised in the Shetland Islands, Scotland. Widely used for knitting or weaving sweaters, shawls, coats, etc. Also known as *Shetland wool*.

(2) A soft woven or knit wool *tweed* fabric of Shetland wool (1) or wool of similar characteristics, with a raised texture, used for sweaters, coats, suits, sportswear, etc.

Shetland wool. See shetland (1).

Shibu-ichi. Japanese jewelry alloy containing 51.1-67.3% copper, 32.1-48.9% silver, 0-0.5% iron, and 0-0.12% gold. Used for art objects and ornamental parts.

Shield. Trade name of E.L. Yencken & Company Proprietary Limited (Australia) for mirror wall tiles.

shielding cement. A ceramic mortar consisting of a mixture of lead powder, ceramic oxides and water. It has good weatherability, good resistance to weak acids and alkalies and good resistance to radiation including gamma and neutron radiation, and to thermal shocks. Used in the form of blocks, sections, etc., in the shielding of X-ray and nuclear installations.

shielding glass. A transparent protective glass containing considerable quantities of the oxides of heavy elements, such as lead, niobium, tantalum and tungsten. It absorbs high-energy electromagnetic radiation, and is thus suitable for nuclear applications, such as shielding one region of space from ionizing radiation emanating from another.

shielding paint. A radiation-protective paint that may contain lead, and is used in nuclear installations, and on X-ray equipment and rooms.

Shieldrite. Trade name of B.F. Goodrich Company (USA) for a series of water-based nickel-filled acrylic conductive coatings providing effective EMI/RFI shielding. They provide good adhesion to various plastic substrates, good conductivity, smooth surface finishes, low film thickness and good aging characteristics. Used for business machines, computers and other electronic equipment. See also EMI/RFI shielding materials.

shiga. A Japanese corrosion-resistant white gold alloy composed of 60-90% gold, 5-20% nickel and 5-20% chromium, and is used for jewelry and ornamental parts.

Shimmereen. Trade name of BASF Corporation (USA) for nylon-6 fibers and yarns used for textile fabrics including clothing.

shim steel. Steel, usually a bright-finished tempered carbon or stainless grade, made into flat, thin pieces of varying thickness (typically between 0.025 and 0.508 mm, or 0.001 and 0.020 in.) and extreme close tolerances. Steel shims are used between mating parts to provide proper clearance (e.g., for adjusting the fit of bearings and machined parts), in the manufacture of thickness gauges, and for packing applications.

shim stock. Materials, such as hard-temper, mirror-bright finish cold-rolled brass or steel, manufactured to extreme close tolerances and within a standard thickness range of 0.025 to 0.508 mm (0.001 to 0.020 in.). Used in the adjustment or alignment of surfaces, for gaskets, etc.

shingle. A thin, oblong sheet of building material, such as asbestos, wood, mineral fiber, metal, etc., laid in overlapping rows as a roof covering or siding of a building.

shingle spruce. See white spruce.

Shinglewall. Trademark of Canadian Forest Products Limited for panelwood.

shinglewood. See western red cedar.

Shinko. Trade name for rayon fibers and yarns used for textile fabrics.

Shinlon. Trademark of Shinkong Synthetic Fibers Corporation (Taiwan) for polyester staple fibers and filament yarns.

ship-and-galley tile. Quarry tile whose face contains a pattern of indentations to prevent slipping when walked upon, even when wet.

Ship Cote. Trade name of Ellis Paint Company (USA) for marine paints.

ship nail brass. A free-cutting brass containing 64% copper, 25% zinc, 8.5% lead, and 2.5% tin. Used for hardware and ship nails.

ship pitch. See navy pitch.

Shirley Cloth. British trademark for a very closely woven cotton fabric resembling *oxford cloth*, and used for military uniforms.

Shiro Diafil. Trade name for rayon fibers.

shirred fabrics. A term referring to fabrics that are usually printed and made from cotton or polyester, and have one elasticized edge. Used for skirts, dresses, and casual clothing.

Shock-Die. Trade name of Columbia Tool Steel Company (USA) for an air-hardening, shock-resistant steel (AISI type S7) containing 0.5% carbon, 0.75% manganese, 1.4% molybdenum, 3.2% chromium, and the balance iron. It has a great depth of hardening, excellent toughness, and is used for punches, chisels, slitters, hobs, and blanking, gripper, forming and swaging dies.

Shock Proof. Trade name of Wheeling-Pittsburgh Steel Corporation (USA) for a malleable cast iron containing 3% carbon, 1.5% silicon, and the balance iron. Used for general castings, gears, and machine parts.

shock-resisting tool steels. A group of tools steels (AISI group S) with low carbon content (typically 0.40-0.65%) and a combination of high strength and toughness and low to medium wear resistance. The principal alloying elements are manganese (0.10-1.50%), silicon (0.15-2.50%), chromium (0.30-3.50%), tungsten (0-3.00%), molybdenum (0.20-1.80%) and vanadium (0-0.25%). They are often sold under trademarks or trade names, such as *Atsil, Bearcut, Halvan, Hy-Tuf, La Belle, Lanark, Omega, Orleans, Seminole, Simoch* or *Vibro*. The hardenability of *shock-resisting tool steels* varies from shallow to deep hardening, and their nonwarping properties are only fair. Used for pneumatic and hand chisels, rivet sets, punches, headers, piercers, driver bits, forming tools, bending rolls, and shear tools. Also known as *shock-resistant tool steels*.

Shock-Rite. Trade name of St. Lawrence Steel Company (USA) for an oil- or water-hardening, shock-resistant tool steel (AISI type S5) containing 0.65% carbon, 2% silicon, 0.85% manganese, 0.25% chromium, 0.25% molybdenum, 0.2% vanadium, and the balance iron. It has good deep-hardening properties, good ductility and toughness, improved machinability, and is used for pneumatic tools, shear blades, hand tools, and forming tools.

Shocktough. Trade name of Osborn Steels Limited (UK) for a tough high-carbon tool steel containing 1.2% chromium and nickel, 2% tungsten, and the balance iron. It has good shock and abrasion resistance, and is used for pneumatic chisels, rock drills, etc.

shoddy. (1) An inferior grade of wool made from wool waste, old rags, cotton yarn, etc.

(2) A fabric made from wool waste.

Shoeflex. Trade name for nylon fibers and fabrics.

shoe leather. Leather suitable for the manufacture of shoes and boots and obtained from the hides of cows, calves, goats, sheep, etc., and including upper leather, sole leather and insole leather.

shoe nail brass. A yellow brass containing 63% copper and 37% zinc, used for shoe nails, hardware and fasteners.

shoe tip metal. A red brass containing 88% copper and 12% zinc used for shoe tips.

Shofu. Trade name of Shofu Dental Corporation (Japan) for a dental amalgam alloy.

shop lumber. Softwood lumber produced or selected primarily for remanufacturing purposes. It is a low-grade of *factory lumber* used by millwork plants for the fabrication of windows, doors, moldings and other trim items.

short-cycle malleable iron. Malleable iron, usually with high silicon content, that can be thoroughly annealed in a relatively short period of time. See also high-silicon malleable iron; malleable iron.

short-fiber composites. See discontinuous fiber composites.

short-fiber-reinforced thermoplastics. See discontinuous fiber reinforced thermoplastics.

short fibers. See discontinuous fibers.

short glass. A quick-setting glass.

shortite. A colorless to light yellow mineral composed of sodium calcium carbonate, $Na_2Ca_2(CO_3)_3$. Crystal system, orthorhombic. Density, 2.63 g/cm³; refractive index, 1.555. Occurrence: USA (Wyoming).

shortleaf pine. The wood from the tall pine tree *Pinus echinata* belonging to the Southern pines. The heartwood is reddish-brown and the sapwood yellowish-white. It has moderate strength and shrinkage, and high stability, when properly seasoned. Average weight, 575 kg/m³ (36 lb/ft³) Source: Southeastern United States (especially Georgia, Alabama, North Carolina, Arkansas and Louisiana) and westward to Texas and Oklahoma. Used as lumber for building, interior finish, joists, sheathing, subflooring, boxes, pallets, crates, cooperage, structural plywood, and for preservative-treated railroad ties, piles, poles and mine timbers. Also known as *North Carolina pine; rosemary pine*. See also Southern pine.

short-oil alkyd. An alkyd resin, produced with a small amount of unsaturation, which is modified with up to 30-45% nonoxidiz-

ing oils, is more reactive and brittle and dries more quickly than a *long-oil alkyd*. Used for baking enamels.

short-oil varnish. A varnish made with a relatively low proportion of oil to resin. It is less elastic and less durable than a *long-oil varnish*. Also known as *short varnish*.

short terne. Light-gage steel sheet with a thickness of less than 0.25 mm (0.01 in.), coated with a corrosion-resistant lead-tin alloy (usually 75-98% lead and 2-25% tin). It is similar in thickness to *tinplate*, but no longer of great commercial importance. See also long terne.

short varnish. See short-oil varnish.

short wool. Wool whose fibers range from about 25 to 150 mm (1 to 6 in.) in average length. See also medium wool; long wool.

shot. Metal in the form of small spherical or elongated particles. It can be obtained by dropping molten metal from a certain height into a suitable quenchant (e.g., water). Most metals are commercially available in this form. Common examples of shot include lead shot used as shotgun ammunition and in radiation shielding, and cast-steel and cast-iron shot used in abrasive blasting and shot peening operations. Also known as *shot metal*.

shotcrete. See air-blown mortar.

shot metal. See shot.

showcase metal. A corrosion-resistant copper alloy containing 22.5-24% zinc and 8-18% nickel. Used for architectural parts and showcases.

showerproof fabrics. Soft cotton or polyester-cotton fabrics that have been made resistant to light rain by spraying a thin layer rubber latex on one side. The right side of these fabrics is somewhat wrinkled resembling cheesecloth in appearance. Used especially for raincoats.

Shreelon. Trademark of Shree Synthetics Limited (India) for nylon 6 fibers and filament yarns.

Shreester. Trademark of Shree Synthetics Limited (India) for polyester fibers and filament yarns.

shrink-mixed concrete. Concrete whose ingredients are first partially blended in a stationary mixer and then transferred to a truck mixer where mixing is completed in transit to the building site.

shrinkproof fabrics. Textile fabrics impregnated with a thermosetting resin to increase stability, and impart crease and shrink resistance, while maintaining softness and texture.

shrink tape. A polymer tape (e.g., polyester) used in vacuum molding to consolidate plastic parts during elevated temperature cures. They usually come with release coatings (e.g., *Teflon*) to prevent bonding to the parts being cured.

shroud laid rope. A rope consisting of four strands either helically twisted or laid around each other or around a central core.

shuiskite. A dark brown mineral of the pumpellyite group composed of calcium magnesium chromium silicate hydroxide hydrate, $Ca_2MgCr_2(SiO_4)(Si_2O_7)(OH)_2 \cdot H_2O$. Crystal system, monoclinic. Density, 3.24 g/cm³. Occurrence: Russia.

Shuman. Trade name of Shuman Plastics (USA) for a series of synthetic resins including *Shuman HIPS* high-impact polyethylenes, available in unmodified, fire-retardant and UV-stabilized grades, *Shuman PE* low-density polyethylenes, supplied in unmodified and UV-stabilized grades, and *Shuman PS* polystyrene resins available in unmodified, medium-impact, glass-fiber-reinforced, fire-retardant, silicone-lubricated, structural-foam and UV-stabilized grades.

shumard oak. The wood from the medium-sized tree, *Quercus shumardii*, resembling *scarlet oak* and belonging to the southern red oak group. The heartwood is reddish-brown and the sapwood white. *Shumard oak* has a coarse texture and a beautiful grain. Source: Southern, southeastern and central United States from Texas to Florida and from North Carolina to Missouri. Used for lumber, veneer, furniture and cabinetmaking. See also American red oak; oak; red oak.

Shunt Steel. Trade name of Japan Special Steel Company (Japan) for a magnetic, temperature-sensitive alloy composed of 70% iron and 30% nickel, and used for compensating shunts for electrical equipment.

Shurbond. Trade name of Anchor Alloys Inc. (USA) for a series of copper- and silver-brazing filler metals including copper-phosphorus, copper-silver-phosphorus, copper-zinc-silver cadmium, silver-copper-zinc, silver-zinc-copper and silver-copper-zinc-cadmium. Used for joining carbon, low-alloy and stainless steels, and copper and its alloys.

Shur-Ram. Trademark of North American Refractories Company (USA) for plastic refractories.

Shurti. Trade name for polyolefin fibers and products.

Shuswap. Trade name of Federated Cooperatives Limited (Canada) for lumber and plywood.

sial. An aluminoborosilicate glass having high thermal and chemical resistance.

Sialfrax-HDAR. Trademark of The Carborundum Company (USA) for silica-alumina refractory bricks used for blast furnace linings.

siallites. A generic term for ceramic materials and clays composed essentially of silicon dioxide and aluminum oxide.

Sialon. Trademark for a hard, chemically inert silicon aluminum oxynitride (SiAlON) ceramic, usually made from silicon nitride, aluminum nitride and aluminum oxide by pressureless sintering. It has good high-temperature and oxidation resistance and is used for high-tech applications. See also silicon aluminum oxynitride ceramics.

Sialon II. Trademark for a chemically inert, isostatically pressed high-temperature silicon aluminum oxynitride (SiAlON) ceramic similar to *Sialon*, but with significantly enhanced impact fracture toughness, higher tensile strength and thermal conductivity, and excellent nonwetting properties and surface smoothness. It is designed for severe industrial applications including those in the nonferrous molten metal industries including protection tubes, and heater tubes for degassing systems. See also silicon aluminum oxynitride ceramics.

SiAlON ceramics. See silicon aluminum oxynitride ceramics.

sibirskite. A gray mineral composed of calcium hydrogen borate, $CaHBO_3$. Crystal system, monoclinic. Refractive index, 1.643. Occurrence: Russian Federation.

Sibitle. Italian trade name for urea-formaldehyde resins.

Sibley alloy. A British, strong, nonhardenable alloy composed of 67-90% aluminum and 10-33% zinc, and used for light-alloy parts and castings.

Sicalite. French trade name for casein plastics.

siccative. See desiccant.

Siccoform. Trade name for a silicone elastomer used for dental restorations.

sicilienne. A fabric originally made in Sicily from mohair and cotton, but now also produced from cotton, silk and wool blends. Used especially for dresses.

sicklerite. A dark brown mineral of the olivine group composed of lithium manganese iron phosphate, $Li(Mn,Fe)PO_4$. Crystal system, orthorhombic. Density, 3.36 g/cm^3; refractive index, 1.735. Occurrence: Canada (Manitoba), USA (California).

Sicoester. Trademark of Sico Inc. (Canada) for epoxy ester enamel

paints.

Sicofill. Trademark of Sico Inc. (Canada) for wood fillers.

Sicogard. Trademark of Sico Inc. (Canada) for lead silicochromate primers used for enamel paints.

Sicoid. French trade name for cellulose acetate plastics.

Sicolac. Trademark of Sico Inc. (Canada) for pure shellac varnishes.

Sicolum. Trademark of Sico Inc. (Canada) for interior and exterior aluminum paints.

Sicolux. Trademark of Sico Inc. (Canada) for interior and exterior alkyd enamel paints.

Sicomet. Trademark of Henkel Surface Technologies (USA) for an extensive series of ethyl-, methyl- and methoxyethyl-based cyanoacrylate adhesives supplied in a wide range of viscosities and set times for bonding ceramics, ferrites, metals, plastics and elastomers.

Sicomill. Trademark of Superior Graphite Company (USA) for a series of processebd silicon grains and powders available in various grades for use in the manufacture of engineering ceramics.

Sicon. Trademark of Dexter Corporation (USA) for silicone-based coatings used in metal finishing.

Siconaval. Trademark of Sico Inc. (Canada) for an exterior alkyd marine enamel.

Siconex. Trade name of 3M Company (USA) for a ceramic-matrix composite consisting of *Nextel* fibers in a silicon carbide matrix. It is manufactured by chemical vapor deposition of silicon carbide on a woven or molded fiber mat preform. The resulting composite is a lightweight, shatter-resistant product with 70% dense silicon carbide and 30% fiber. *Siconex* has outstanding thermal shock resistance up to 1204°C (2200°F), and can be used for producing complex, thin-walled shapes.

Siconide. Trademark of Superior Graphite Company (USA) for silicon nitride powders for high-performance ceramics. *Siconide S* and *Siconide P* are high-purity grades, and *Siconide U* is an ultrahigh-purity grade.

Siconium. Trademark of Sico Inc. (Canada) for an aluminum paint.

Sicopoxy. Trademark of Sico Inc. (Canada) for epoxy coatings.

Sicopure. Trademark of Sico Inc. (Canada) for an exterior alkyd gloss paint.

Sicorama. Trademark of Sico Inc. (Canada) for an interior alkyd flat enamel paints.

Sicorethane. Trademark of Sico Inc. (Canada) for polyurethane varnishes.

Sico-Satin. Trademark of Sico Inc. (Canada) for interior acrylic latex paint with satin finish.

Sicosto. Trademark of Sico Inc. (Canada) for vinyl sealers for wood.

Sicotemp. Trademark of Sico Inc. (Canada) for an heat-resistant aluminum paint.

Sico-Tex. Trademark of Sico Inc. (Canada) for an exterior acrylic latex paint.

Sicovel. Trademark of Sico Inc. (Canada) for an interior alkyd semi-gloss velvet enamel paint.

Sicromal. Trade name of Rheinische Rohrenwerke AG (Germany) for a series of heat- and corrosion-resistant high chromium and chromium-nickel steels including various stainless grades. Used for furnace equipment, petroleum-refining equipment, heat-treating equipment, textile machinery, chemical process equipment, acid storage vessels, mixers, tanks, pharmaceutical equipment, superheaters and recuperators, valve and pump parts, and heating elements and resistances.

Sicromo. Trade name of Timken Company (USA) for a series of heat- and corrosion-resistant alloy steels of the chromium-silicon-molybdenum type used for high-temperature and refinery tubing.

Sicron. Trade name of Himont USA for plasticized polyvinyl chlorides supplied in three elongation ranges: 0-100%, 100-300% and above 300%. Sicron UPVC are unplasticized polyvinyl chlorides supplied in standard, crosslinked, high-impact, structural-foam and UV-stabilized grades.

Sicursiv. Trade name of Società Italiana Vetro SpA (Italy) for a safety glass.

side-arch brick. A wedge-shaped brick having its faces inclined toward each other.

side-chain polymer liquid crystal. A *polymer liquid crystal* in which the mesogenic units consisting of rigid disk- or rod-like molecules are not located in the main chain or backbone, but are found in the side chains or branches. Abbreviation: SCPLC. See also main-chain polymer liquid crystal.

side-construction tile. A structural clay product designed to receive its principal stress approximately at right angles to the axes of the cells.

side-cut brick. A brick shaped by wire cutting along the sides.

side leather. Leather manufactured from the grain side of cattlehides split along the backbone prior to tanning. Used for shoe uppers, insoles and bags.

Sideraphite. British trade name for a stainless and corrosion-resistant alloy composed of 62% iron, 23% nickel, 5% aluminum, 5% copper and 5% tungsten, and used for acid-resisting vessels, apparatus and equipment, and for chemical process equipment.

siderazot. A grayish synthetic mineral composed of iron nitride, $Fe_{2.5}N$. Crystal system, hexagonal. Density, 3.08 g/cm^3.

siderite. (1) A gray, green, light yellowish brown or brownish-red mineral of the calcite group with a white streak and a vitreous to pearly luster. It is composed of iron carbonate, $FeCO_3$, containing 48.2% iron, and sometimes calcium, magnesium, manganese and/or zinc. Crystal system, rhombohedral (hexagonal). Density, 3.83-3.93 g/cm^3; hardness, 3.5-4 Mohs; refractive index, 1.8728. Occurrence: Europe (Austria, Germany, Hungary, UK), Greenland, USA (Connecticut, Massachusetts, New York, North Carolina, Ohio, Pennsylvania, Vermont). Used as an ore of iron, and a yellow to red colorant in ceramic bodies and glazes. High-manganese *siderite* is used in the manufacture of *spiegeleisen*. Also known as *chalybite; sparry iron ore; spathic iron; spathic iron ore; spathic ore.*

(2) A term referring to iron occurring in meteorites.

sideronatrite. A lemon-yellow mineral composed of sodium iron sulfate hydroxide trihydrate, $Na_2Fe(SO_4)_2(OH)\cdot3H_2O$. Crystal system, orthorhombic. Density, 2.28 g/cm^3; refractive index, 1.525. Occurrence: Chile.

siderophyllite. A blue-green mineral of the mica group composed of potassium iron aluminum silicate fluoride hydroxide, $KFe_2Al_3Si_2O_{10}(F,OH)_2$. It can also be made synthetically. Crystal system, monoclinic. Density, 3.04 g/cm^3; refractive index, 1.625. Occurrence: Germany.

siderotil. A pale green mineral of the chalcanthite group composed of iron sulfate pentahydrate, $FeSO_4\cdot5H_2O$. Crystal system, triclinic. Density, 2.15 g/cm^3; refractive index, 1.525. Occurrence: USA (Nevada), Italy.

sidorenkite. A pale rose mineral of the bradleyite group composed of sodium manganese carbonate phosphate, Na_3Mn_4-

$(PO_4)(CO_3)$. Crystal system, monoclinic. Density, 2.90 g/cm^3; refractive index, 1.563. Occurrence: Russian Federation.

siegenite. A light gray mineral of the spinel group composed of nickel cobalt sulfide, $NiCo_2S_4$. Crystal system, cubic. Density, 4.50 g/cm^3. Occurrence: Germany.

Siemens-Halske. Trade name of Siemens & Halske AG (Germany) for a stainless, corrosion-resisting nickel alloy containing 5-30% tantalum. Used for corrosion- and heat-resistant parts and spark-plug electrodes.

siemensite. A *superrefractory* made by melting bauxite, chromite and magnesite in an open electric-arc furnace.

Siemens-Martin steel. See open-hearth steel.

sienna. A pigment obtained from the earth and composed chiefly of hydrated ferric oxide (Fe_2O_3) and manganic oxide (Mn_2O_3). Raw sienna is yellowish brown and has a density of 3.27 g/cm^3 (0.118 $lb/in.^3$), and burnt (roasted) sienna is orange-red or reddish brown and has a density of 3.95 g/cm^3 (0.142 $lb/in.^3$). Used as a paint pigment, and as a colorant for slips, bodies, and glazes especially celadons. See also umber.

Sierra. (1) Trade name of Atlas Specialty Steel (Canada) for a hollow drill steel.

(2) Trade name of The Wilkinson Company, Inc. (USA) for a series of gold-based dental and medical alloys.

Siesta. Trade name of Pilkington Brothers Limited (UK) for a patterned glass.

Sifalbronze. Trade name of Sifbronze (UK) for aluminum-bronze welding rods.

Sifalumin. Trade name of Sifbronze (UK) for aluminum-copper, aluminum-magnesium and aluminum-silicon brazing and welding rods.

Sifas. Trademark of SIFASA (Turkey) for nylon 6 staple fibers and filament yarns.

Sifbrass. Trade name of Sifbronze (UK) for copper-zinc alloys used for brazing cast iron, copper and copper alloys.

Sifbronze. Trade name of Sifbronze (UK) for a series of copper-zinc based brazing and braze-welding filler metals, sometimes containing small additions of tin, iron, silicon, aluminum and manganese. Filler metals containing 10% nickel and having greatly improved tensile strength are also available. Used for brazing copper and its alloys, cast iron, malleable steel, and other alloys.

Sifcoloy. Trade name of Spuck Iron & Foundry Company (USA) for an erosion-resistant cast iron containing about 3.1% total carbon, 1.4% silicon, 0.8% molybdenum, 0.6% copper, and the balance iron. Used for lock gate valves and nozzle castings.

Sifcoro. Trade name for of Sifbronze (UK) for a series of gold-based brazing filler metals containing 20% copper and 5-20% silver. Available in various forms including wire, paste, powder, foil and preform, it is used in high-temperature brazing operations.

Sifcupron. Trade name of Sifbronze (UK) for copper-phosphorus and copper-phosphorus-silver brazing alloys for joining copper and copper alloys.

Sifmig. Trade name of Sifbronze (UK) for aluminum, aluminum-magnesium and aluminum-silicon welding rods for joining aluminum alloys.

Sifsteel. Trade name of Sifbronze (UK) for carbon, low-alloy and stainless-steel hardfacing and welding filler rods and welding electrodes.

Sigal. Trademark of Siemens AG (Germany) for an aluminum casting alloy containing 13% silicon. It has good resistance to atmospheric and seawater corrosion, and high strength. Used for castings and lightweight construction applications.

Sigelit. Trade name used in the former Czechoslovakia for phenol-formaldehyde resins and plastics.

Sigeron. Trade name of Prescott Company (USA) for a nickel cast iron containing 3.2% carbon, 2.2% silicon, 1.5% nickel, and the balance iron.

Sigit. Trade name used in the former Czechoslovakia for phenol-formaldehyde resins and plastics.

Sigla. Trade name of Flachglas AG (Germany) for a range of laminated safety glass products made by cementing two or more sheets of plate glass together with one or more sheets of transparent plastic material. When cracked or broken, the glass will adhere to the plastic rather than fly. It is available in various grades including bronze-colored *Sigla-Bronze*, smoke-colored *Sigla-Rauch*, and decorative *Sigla-Edelit, Sigla-Exquisit, Sigla-Print, Sigla-Rondo* and *Sigla-Strip*. Used for banks, armored cars, airplanes, automobiles, safety glasses and televisions.

sigloite. A straw-yellow mineral of the paravauxite group composed of iron aluminum phosphate octahydrate, $FeAl_2(PO_4)_2(O,-OH)_2 \cdot 8H_2O$. Crystal system, triclinic. Density, 2.35; refractive index, 1.586. Occurrence: Bolivia.

Sigma cement. Trade name for a special hydraulic cement obtained by blending ordinary Portland cement with a particle size of less than 30 μm (0.001 in.) with 15-50% of an inert filler, such as basalt, flint, limestone, etc. with a particle size between 30 to 200 μm (0.001 and 0.008 in.).

Sigmacover. Trade name of Sigma Coatings (Netherlands) for epoxy coatings for industrial applications.

Sigmadur. Trade name of Sigma Coatings (Netherlands) for polyurethane topcoats.

Sigmalumin. Trade name of Aluminium Belge SA (Belgium) for an age-hardenable aluminum alloy containing 3.8% copper, 0.8% silicon and 0.7% manganese. Used for light-alloy parts.

Sigma Monofil. Trade name of BP Metal Composites Limited (UK) for a strong, stiff, lightweight silicon-carbide fiber made by chemical-vapor deposition. Supplied in the form of monofilaments, it is used as reinforcement in metals and intermetallics in particular for aerospace applications.

signal glass. Glass in various colors used in signal devices.

Signature Hardwoods. Trademark of Georgia-Pacific Corporation (USA) for hardwood lumber.

sign enamel. A brilliant, glossy porcelain enamel suitable for use in sign work.

Sigrafil. Trademark of SGL Carbon GmbH (Gerrmany) for a carbon fibers and yarns. *Sigrafil O* refers to *Panox*-type polyacrylonitrile-based carbon fibers.

Sigratherm. Trademark of SGL Carbon GmbH (Germany) for graphite and carbon felt products for high-temperature insulation applications.

Sigribond. Trademark of Sigri Great Lakes Carbon Corporation (USA) for composite material composed of polyacrylonitrile fibers bonded with pitch and synthetic resin.

SIKA. Trade name of St. Gobain Industrial Ceramics (France) for silicon carbide abrasive grains for use in the manufacture of bonded abrasives (e.g., honing stones, grinding wheels), coated abrasives (e.g., grinding cloth and paper), as wiresawing and polishing abrasives for granite and marble, and as lapping and polishing powders.

Sikadur. Trademark of Sika Canada Inc. for concrete bonding compounds.

Sikaflex. Trademark of Sika-Plastiment GmbH (Austria) for a range of construction and industrial adhesives.

SIKA MET. Trade name of St. Gobain Industrial Ceramics (France) for a metallurgical-grade silicon carbide supplied in the form of briquettes and loose grains for use as alloying and deoxidizing agents and inoculants in the foundry and steel industries.

Sikastix. Trade name of Sika Canada Inc. for structural adhesives.

SIKA REF. Trade name of St. Gobain Industrial Ceramics (France) for silicon carbide with high abrasion and thermal-shock resistance, high chemical inertness at elevated temperatures. Used in the manufacture of refractory products, such as crucibles, monolithics, kiln furniture, incinerators, furnace bricks, and tapholes and runners for metallurgical furnaces.

SIKA TECH. Trade name of St. Gobain Industrial Ceramics (France) for submicron silicon carbide powders used in the manufacture of reaction-bonded, recrystallized and sintered silicon carbide products, such as kiln furniture, burner nozzles, diffusion parts, seal rings, slurry pump parts, etc.

SikaTop. Trade name of Sika Canada Inc. for ready-to-use mortars.

Sicomet. Trade name of Sichel-Werke GmbH (Germany) for a line of synthetic adhesives for bonding metals, plastics and elastomers.

Sikron 600. German trade name for crystalline quartz flour with a typical particle size of 2.5-6.3 μm (98-248 μin.). Used in the manufacture of ceramic glazes.

Silafont. Trade name of Swiss Aluminium Limited (Switzerland) for a range of aluminum products including several cast and wrought aluminum-silicon, aluminum-silicon-magnesium and aluminum-silicon-manganese alloys.

Silagum. Trademark of DMG Hamburg (Germany) for addition and condensation silicones used for dental impressions and bite registration.

Sil-Aid. Trade name of United Wire & Supply Company (USA) for an alloy of 35% silver, 26% copper, 21% zinc and 18% cadmium. It has a melting range of 607-704°C (1125-1300°F), and is commercially available in the form of strips, wire, powder, and clad sheet or strip. Used in silver brazing of metals.

Silal. (1) Trade name of Eduard Hueck Metallwalz- und Presswerk (Germany) for a series of aluminum alloys containing 1-4% magnesium, 0-0.8% manganese, 0-0.5% silicon, 0-0.3% titanium and 0-0.3% chromium. Used for light-alloy parts.

(2) Trade name of Sheepbridge Engineering Limited (UK) for a scale-resistant cast iron containing about 2.5% carbon, 5% silicon, 0-1% aluminum, 1% manganese, 0.2% phosphorus, 0-0.5% chromium, and the balance iron. Used for fire bars, and stove and furnace parts.

Silanca. British corrosion-resistant silver alloy containing 4-4.5% antimony, 1-3% cadmium and 0-2.5% zinc. Used for silverware.

silane. A pyrophoric, moisture-sensitive gas with a density of 1.114 g/cm^3, a melting point of -185°C (-301°F), and a boiling point of -112°C (-169°F). It is supplied in various grades including standard (99.9+%), electronic (99.998% pure), semiconductor (99.99+% pure) and deuterated electronic grades [e.g., silane-d_4 (SiD_4)]. Used as a dopant for solid-state devices, in the preparation of amorphous silicon, as a precursor to plasma and hot-wire deposition of silicon thin films and in the manufacture of silane derivatives and organofunctional silanes. Formula: SiH_4. Also known as *silicon tetrahydride*. See also polysilanes.

Silaplast. Trade name for a silicone elastomer used for dental impressions.

Silaprene. Trademark of Uniroyal Adhesives & Sealants Company Inc. (USA) for an extensive series of adhesives and sealants including several one and two-part polyurethane sealants used for marine applications, e.g., for bonding metal, fiberglass and wood above and below the water line, and for waterproofing windshields and hatches. Also included under this trademark are a number of flexible, two-part room-temperature-curing epoxy adhesives for aerospace and aircraft applications.

Silar. (1) Trademark of Arco Metals Company (USA) for discontinuous silicon carbide reinforcement fibers.

(2) Trade name for a self-cure, microfine, silicone-based dental composite.

Silasoft. Trade name for a silicone elastomer used for dental impressions.

Silasox. Trademark of A&P Technology (USA) for sleevings braided with *E-glass* roving. Supplied in light-, medium- and heavy-fabric grades, they are used as reinforcements for composites.

Silastene. Trade name for silicone elastomers.

Silastic. (1) Trademark of Dow Corning Corporation (USA) for a line of high-performance silicone rubber products including *Silastic* High Consistency Rubber (HCR) compounds and bases, *Silastic* Fluorosilicone Rubber (FSR) compounds and bases, *Silastic* Liquid Silicone Rubber (LSR), and rubber additives, adhesion promoters, pigments and primers.

(2) Trademark of Dow Corning Corporation (USA) for a white silicone elastomer containing organosilicon polymers. It has a density of 0.5 g/cm^3 (0.02 $lb/in.^3$), an elongation of 70-150%, high tensile strength, excellent resistance to compression set, good adhesion properties, excellent resistance to high and low temperatures (-90 to +250°C or -130 to +480°F), high thermal conductivity, excellent electrical properties, good corona resistance and weatherability, low water absorption, good resistance to dilute acids, greases and lubricating greases and oils, fair resistance to concentrated acids, alcohols, alkalies, ketones, and poor resistance to aromatic hydrocarbons and halogens. Used for electrical insulation, insulating components of electronic parts, wire and cable, gaskets, miscellaneous mechanical products, textile coatings, gaskets, diaphragms, seals, O-rings, and hose and tubing.

Silatec. Trade name for a silicone resin used in dentistry for laboratory duplicating purposes.

Silback. Trade name of Teijin Fibers Limited (Japan) for polyester filaments used for the manufacture of textile fabrics.

Sil-Bar. Trademark of Dow Corning Corporation (USA) for a corrosion-resistant thermosetting room-temperature-curing silicone coating. It has a service temperature range of -40 to +200°C (-40 to +390°F) and good resistance to moisture and corrosive gases. Used for corrosion protection of ferrous and nonferrous parts.

Silber 70. German trade name for a high-copper dental amalgam alloy.

Silbereisen. Trade name of Friedrich Wilhelms-Hütte (Germany) for a series of strong cast irons containing 2.8% carbon, 1-2.5% silicon, 0-2% nickel, 0.4-1.5% manganese, up to 0.8% phosphorus, up to 0.04% sulfur, and the balance iron. Used for cylinders, pump bodies, high-pressure valves, turbine housings, and steamship and locomotive cylinders.

Silberkreuz. Trade name of Plettenberger Gusstahlfabrik (Germany) for a series of high-speed steels containing 0.74-0.86% carbon, 4.1% chromium, 0.85% molybdenum, 1.1-2.5% vanadium, 8.7-18.5% tungsten, and the balance iron. Used for lathe and planer tools, drills, taps and reamers.

Silberlot. German trade name for brazing alloys containing mainly copper, zinc and silver, and sometimes additions of cadmium, manganese, phosphorus and nickel. They produce strong, clean joints and have a brazing range of 610-860°C (1130-1580°F). Used for brazing steel, copper and copper alloys, and noble metals.

Silberpunkt. Trade name of SWB Stahlformguss Gesellschaft mbH (Germany) for high-speed steel containing 0.74% carbon, 4.1% chromium, 18% tungsten, 1.1% vanadium, and the balance iron. Used for lathe and planer tools, taps, broaches, reamers, drills, hobs, chasers, and milling cutters.

Silbione. Trademark of Rhône-Poulenc SA (France) for medical-grade silicone polymers.

Silblock. Trade name of Foote Mineral Company (USA) for an alloy of 50-75% iron and 25-50% silicon. Used for blocking open-hearth and electric-furnace steel.

Sil Bond. Trade name for sodium silicate foundry resins used as no-bake core binders.

Sil-Bond. Trade name of United Wire & Supply Company (USA) for a series of brazing alloys containing 30-50% silver, 15-34% copper, 15.5-33% zinc, 18-24% cadmium and 0-3% nickel. They have a melting range of 607-754°C (1125-1390°F), and are used for general-purpose brazing.

Silbralloy. Trade name of Johnson Matthey plc (UK) for a silver brazing alloy containing 91.5% copper, 6.5% phosphorus and 2% silver. It has a melting range of 644-825°C (1190-1515°F).

Silbrass. Trade name for a corrosion-resistant alloy containing 83-86% copper, 10-12% zinc, and 4-5% silicon.

Silbrax. Trade name of Arcos Alloys (USA) for a self-fluxing welding rod used for deoxidized copper.

Silbraze. Trade name of Sheffield Smelting Company Limited (UK) for brazing alloys containing about 41% zinc, 1% silver and 0.3% silicon. They have a melting range of 886-893°C (1625-1640°F).

Silcalfa. Trade name of Alfa Romeo (Italy) for a corrosion-resistant aluminum alloy containing 11-14% silicon. Used for automotive castings.

Silcaride. Trade name of Washington Mills Ceramics Corporation (USA) for silicon carbide blasting media used in metal finishing.

Sil-Co. Trade name of United Wire & Supply Company (USA) for a silver solder containing 70% silver, 20% copper and 10% zinc. It has a melting range of 668-738°C (1235-1360°F).

Sil-Con. Trade name of United Wire & Supply Company (USA) for a silver solder composed of 50% silver, 34% copper and 16% zinc. It has a melting range of 682-766°C (1260-1410°F).

Silcoro. Trade name of GTE Products Corporation (USA) for a series of gold-based brazing filler metals containing 20% copper and 5-20% silver. Available in several forms including wire, paste, powder, foil, preform, etc., it is used for high-temperature brazing.

Sil-Co-Sil. Trademark of US Silica Company (USA) for ground and fine-ground silica supplied in several grades and particle sizes.

Silcrome. Trademark of Allegheny Ludlum Steel (USA) for a series of heat-, oxidation- and corrosion-resistant steels used for automotive and aircraft valves, valve seats, and supercharger wheels for jet engines.

Silcut. Trade name for a free-machining steel used for screwstock.

Sildul. Trade name of Novaceta S.p.a. (Italy) for dull cellulose acetate fibers used for the manufacture of wearing apparel.

Silectron. (1) Trade name of Allegheny Ludlum Steel (USA) for grain-oriented, soft magnetic material containing 97% iron and 3% silicon. It has high permeability and low core loss at high induction. Used for magnetic cores and power transformers.

(2) Trade name of George W. Prentiss & Company (USA) for an alloy containing 95% nickel, 4% manganese and 1% silicon. Used for spark plugs.

Sil-Een. Trade name of United Wire & Supply Company (USA) for an alloy of 40% silver, 36% copper and 24% zinc. It has a melting range of 721-785°C (1330-1445°F), and is used as a silver solder.

Silenco. Trade name of Ato Findley (France) for bituminous adhesives and seal coats for building and construction.

Silene. (1) Trademark of PPG Industries Inc. (USA) for calcium silicate powders used as reinforcing agent in rubber.

(2) Trademark of Harry J. Bosworth Company (USA) for a putty-like silicone material for making dental inlay, crown and bridge impressions.

(3) Trademark of Novaceta SpA (Italy) for glossy and delustered cellulose acetate filament yarns used for wearing apparel.

Silenka. Trade name of NV Silenka AKU-Pittsburgh (Netherlands) for textile glass fibers including yarns, rovings, chopped strands and reinforcing mats.

Silentus. Trade name of Safetee Glass Company Inc. (USA) for acoustical insulating glass.

Silesia. (1) Trade name of Vidriobrás (Brazil) for glass with a chequerboard design.

(2) Trade name of Georgsmarienwerke Selesiastahl GmbH (Germany) for a series of alloy steels including several case-hardening grades, water-hardening plain-carbon and hot-work tool and die steels, mold steels, etc.

silesia. A fine, light cotton fabric in a twill weave, having a smooth, glazed surface produced by calendering. Used especially for linings. Named after Silesia, a region in eastern Europe where it was originally made.

Silesit. Polish trade name for bakelite-type phenolic resins and plastics.

Silesitol. Polish trade name for bakelite-type phenolic resins and plastics.

Sil-Ex. Trade name of United Wire & Supply Company (USA) for an alloy composed of 50% silver, 15.5% copper, 16.5% zinc and 18% cadmium. It has melting range of 627-635°C (1160-1175°F), and is used as a silver solder.

Silex. Trade name for a strong, heat and shock-resistant glass containing 98+% quartz.

silex. (1) A chemically inert form of silica that does not shrink or absorb moisture, and is used widely in making paste wood fillers.

(2) A paint filler made of finely ground quartz.

Silferal. Trade name of Aluminiumwerke Maulbronn (Germany) a nonhardenable, corrosion-resistant aluminum-silicon casting alloy. It has good castability, a dense, fine-grained structure, good machinability, weldability and grindability and good to fair corrosion resistance, and is used for cylinder heads and motorcycle cylinders.

Sil-Fil. Trade name of United Wire & Supply Company (USA) for an alloy containing 50% silver, 15.5% copper, 15.5% zinc, 16% cadmium and 3% nickel. It has a melting range of 632-688°C (1170-1270°F), and is used as a silver solder.

Silfirtex. Trade name of Compagnie Française des Isolants SA (France) for glass cloth.

Silflake. (1) Trademark of Handy & Harman (USA) for a commercially pure silver containing about 1% of an organic lubricant and trace amounts of iron. Available in various grades with different conductivities and covering powers, and with particle sizes ranging from 2-10 μm (79-394 μin.). Used for conductive coatings and adhesives.

(2) Trademark of Technic Inc., Engineered Powders Division (USA) for silver flake used in conductive adhesives and coatings.

Sil-Fos. Trade name of Handy & Harman (USA) for a series of brazing filler metals containing 74-93% copper, 2-18% silver and 5-7.25% phosphorus. They have a brazing range of 625-810°C (1157-1490°F), and are used for joining copper, nickel and their alloys.

Silfram. Trade name of Stoody Company (USA) for an abrasion-corrosion- and wear-resistant cast iron containing 2% carbon, 2% silicon, 31% chromium, 10% nickel, 2% tungsten, and the balance iron. Used for welding and hardfacing rod for agricultural equipment.

Silfrax. Trademark of The Carborundum Company (USA) for bonded refractories available in various grades with silicon carbide contents ranging from 40-78%, and porosities between 9-18%. They have high refractoriness, good resistance to wear and abrasion, good resistance to flame abrasion, good resistance to spalling and clinker adhesion, high strength, high thermal conductivity, and good thermal-shock resistance. Used for kiln furniture in ceramic kilns, shapes for boiler furnaces, air-cooled furnace linings, glass lehrs, pit furnaces, enameling-furnace ware supports, and bricks for boiler and furnace installations.

Silfresh. Trade name of Novaceta S.p.a. (Italy) for cellulose acetate fibers used for the manufacture of wearing apparel.

Silglaze. Trademark of GE Silicones (USA) for silicone-base construction and weatherproofing sealants.

Sil-Gon. Trade name of United Wire & Supply Company (USA) for a solder containing varying amounts of silver and copper. It has a melting range of 691-774°C (1275-1425°F), and is used for soldering most ferrous and nonferrous alloys.

Silgrip. Trade name of GE Silicones (USA) for silicone sealants and adhesives.

silhydrite. A white mineral composed of silicon oxide monohydrate, $Si_3O_6 \cdot H_2O$. Crystal system, orthorhombic. Density, 2.14 g/cm³; refractive index, 1.466. Occurrence: USA (California).

silica. A silicate material occurring in crystalline form as *quartz, tridymite, cristobalite, coesite* and *stishovite,* in crypocrystalline form as *chert, flint* and *chalcedony* and in amorphous form as *opal.* It is also the chief component of *sand* and *diatomite,* and can be made synthetically as a noncrystalline solid or glass called *fused silica* (vitreous silica). All crystalline and noncrystalline polymorphs have a basic tetrahedron structure. *Silica* is commercially available in the form of colorless crystals, as a white powder (99.5+% pure), and in high-purity electronic grades (99.999%). Crystal system, cubic. Density, 2.2-2.6 g/cm³; melting point, 1710°C (3110°F); boiling point, 2330°C (4225°F); hardness, 7 Mohs; refractive index, 1.5. It has excellent resistance to acids (except hydrofluoric acid) and alkalies, a thermal conductivity about half that of glass, a dielectric constant of about 3.2-4.2, and high heat and thermal-shock resistance. Used in the manufacture of glass, refractories, abrasives, numerous whiteware bodies and glazes, porcelain enamels, foundry molds, carborundum, ferrosilicon, concrete and mortars and water glass, electric and electronic products, in the manufacture of optoelectronic and photonic devices, in semiconductor technology, in materials research, chemistry, biochemistry, bioengineering and medicine, in water filtration and thermal insulation, as an inert pigment, as a filler in paper, as flattening agent in paints, and as a source of silicon. The amorphous form is used in silica gels, and hydrated and precipitated grades are used as reinforcing agents in rubber. Formula: SiO_2. Also known as *silicon dioxide*. See also amorphous silica.

silica abrasives. A group silica (SiO_2) based natural abrasives which are chemically similar, but differ in physical make-up. Examples include chert, diatomite, diatomaceous earth, flint, infusorial earth, quartzite, sand, sandstone, tripoli and tripolite. Depending on their hardness, they are used for applications ranging from buffing and polishing to grinding and sandblasting. Also known as *silicon dioxide abrasives*.

silica aerogel. See synthetic silica.

silica brick. An acid refractory brick, usually made from *ganister* and containing at least 90-92% silica (SiO_2) bonded with hydrated lime, and fired at a high temperature. It has high strength at elevated temperatures, high thermal conductivity, good abrasion resistance and poor resistance to molten basic slags. Used for lining furnace and kiln roofs. *Note:* This term is not synonymous with *siliceous brick*. See also acid brick; acid refractory; siliceous brick; Dinas brick.

silica cement. A refractory mortar or cement consisting of a finely divided mixture of quartzite, silica and fireclay. Used in kilns and furnaces. Also known as *silica fireclay*.

silica fibers. Continuous or discontinuous ceramic oxide fibers containing up to 99.95% silica (SiO_2). Commercially available in the form of whiskers, continuous and discontinuous filaments, yarns and fabrics, they have high to moderate strength, moderate moduli of elasticity, good high-temperature resistance, and a continuous-use temperature of 1000-1100°C (1830-2010°F). Used as reinforcement fibers for composites, in rocket and missile construction, as high-temperature insulation fibers for motor components, and as thermal insulating fibers in various industries. For constructional purposes, they are used in laminate form impregnated and bonded with high-temperature resins. Also known as *silicon dioxide fibers*.

silica fireclay. See silica cement.

silica flour. A finely ground quartz (usually from 80-325 mesh) that contains about 99.5% silica (SiO_2) and is used as an additive in casting slips and molding sands, in paints and refractory ramming mixes, for facing molds for sand casting, and in the manufacture of flooring bricks.

silica fume. A byproduct of silica production valued as a cementing material in concrete manufacture, and also used in the chemical industry. It is essentially composed of micron-sized particles or grains. Also known as *microsilica*.

silica gel. An amorphous, highly absorbent form of silica (SiO_2) obtained by treating sodium silicate with sulfuric acid and washing and grinding the resulting jelly-like product. It is commercially available in various sizes ranging from 3-425 mesh, and can withstand temperatures up to 260-315°C (500-600°F). Used as a drying and deodorizing agent in air conditioners, etc., as a dehumidifying and dehydrating agent, as a catalyst and catalyst carrier, as a drying agent for compressed gases, in refrigerants and oils, in the production of gasoline and oil, and in chromatography, biochemistry, bioengineering and medicine. Abbreviation: SG. See also silicic acid.

silica glass. See fused silica.

silica glass paper. A porous, nonhydrating paper made from high-purity silica-glass fibers with or without binder. Used in filtration, electrical and thermal insulation, and for high-temperature applications.

silica hydrogel. A colorless, jelly-like substance composed of hydrated silica ($SiO_2 \cdot xH_2O$) and formed by coagulation from an aqueous suspension of *colloidal silica*. Used for paper and textile coatings.

Silical. (1) Swedish trade name for an aluminum alloy containing 11-14% silicon. It has a dense, fine-grained structure, good castability, machinability, weldability and grindability, and good to fair corrosion resistance. Used for exhaust valves.

(2) Trade name for a finely ground powder consisting of the mineral *pyrophyllite* (hydrated aluminum silicate) obtained from Newfoundland, Canada. Used as a replacement for talc in dusting rubber.

Silicap. Trade name for an encapsulated silicate cement used in dentistry.

silica powder. Silicon dioxide (SiO_2) in the form of a fine white amorphous or crystalline powder. Crystalline powders are usually made from crushed quartz. *Silica powder* is available in particle sizes ranging from less than 1 to over 150 μm (less than 0.00004 to over 0.006 in.), and purities from 99.5% for regular grades to 99.999% for electronic grades. It has a density of 2.6 g/cm³ (0.09 lb/in.³), a melting point of 1500°C (2730°F), and is used as an abrasive, in refractories and ceramics, in glass manufacture, in rubber compounding, as flatting agent for paints, in paper coatings, for electric and electronic applications, etc. See also silica flour.

silica ramming mix. A refractory product made from a batch of ground *ganister* and/or sand, silica fluor, fireclay and bentonite. Used for placing or patching acid refractories and furnace linings.

Silicarb. Trade name of Bethlehem Steel Corporation (USA) for a water or oil-hardening, shock-resistant tool steel containing 0.55% carbon, 2% silicon, 0.75% manganese, 0.2% molybdenum, and the balance iron. Used for impact tools, hand tools, stamps, and punches.

Silicarbide. Trade name of Washington Mills Electro Minerals Corporation (USA) for silicon carbide blasting media.

silica refractories. Refractory materials containing 96.6% silica (SiO_2), 2.4% calcia (CaO), 0.7% magnesia (MgO) and 0.3% alumina (Al_2O_3), and having an apparent porosity of 25%, excellent high-temperature load-bearing capacities, and good resistance to silica-rich (acid) slags. Used for the roofs of steel- and glassmaking furnaces, and for acid-slag containment vessels. See also acid refractories; basic refractories; neutral refractories; siliceous refractory.

silica rock. A natural rock, such as *Dinas rock*, containing a very high amount of silica or quartz.

silica sand. A natural sand that contains 98+% free silica (SiO_2) and is used as a source of silicon, in the manufacture of glass, glazes, porcelain enamels, ceramic bodies, artificial stone and Portland cement, as a setting medium in the firing of ceramic ware, in building and construction work, as an abrasive and blasting sand, for the initial grinding or surfacing of plate glass, as a cutting medium for gang saws on stone, and as a core material in foundry molds

silica-silica composites. A group of unidirectional or multidirectional engineering composites consisting of silica matrices (usually high-purity colloidal silica) with silica fiber reinforcements (usually continuous fibers of fused quartz). They have good mechanical properties, and a continuous-use temperature range of 900-950°C (1650-1740°F). Used for structural and nonstructural components, e,g., for aircraft and spacecraft equipment.

silica sol. A colloidal suspension of silica powder particles in water used in ceramic engineering as a binder.

silica sonogel. A *gel* produced from a *silica sonosol* by the sol-gel technique, and used for optical and telecommunication applications.

silica sonosol. A *sol* produced from tetramethoxysilane, formamide and acidic water by using ultrasound as a reaction promoter. It is a precursor in the preparation of *silica sonogel* and *silica xerogel* by the sol-gel technique.

silica stone. A hard, durable sedimentary rock consisting predominantly of siliceous minerals. It is difficult to work, and used in the construction trades.

silicate abrasives. See silicates (2).

silicate brick. An imprecise term usually referring to *forsterite* firebrick.

silicate cement. (1) An acidproof cement made by mixing an inert powder with a sodium silicate solution. Used for lining and jointing chemical and laboratory stoneware and acid resisting bricks, and in restorative dentistry.

(2) The silicate of *soda glue* used for bonding cardboard and plywood boxes.

silicate coatings. Brittle ceramic coatings with a standard thickness of 25-50 μm (1-2 mils) prepared from silicate powders (frits) with or without mill-added refractories. The typical composition of a melted frit is 37.5-57.0% silicon dioxide (SiO_2), 6.5-6.0% boron oxide (B_2O_3), up to 20.0% aluminum oxide (Al_2O_3), up to 18.0% sodium monoxide (Na_2O) as well as varying amounts of the oxides of barium, beryllium, bismuth, calcium, cerium, cobalt, manganese, nickel, potassium, titanium, vanadium, zinc and zirconium. *Silicate coatings* have good chemical stability at elevated temperature and good resistance to wear, abrasion and general abuse. Used for aircraft combustion chambers, turbines, exhaust manifolds, and heat exchangers.

silicate cotton. See mineral wool.

silicated gamma alumina. An aluminum oxide compound containing 97.5% *gamma alumina* (γ-alumina) and 2.5% silica. It is available in the form of white pellets with a surface area of 246 m²/g and pore volume of 0.88 cm³/g. Used as catalyst and bed supports in chromatography and filtration.

silicate garnets. A class of natural garnets corresponding to the general formula $A_3B_2Si_3O_{12}$, where A is a calcium, magnesium, iron or manganese cation, and B an aluminum, iron, manganese, chromium or vanadium cation. They have body-centered cubic (bcc) structures consisting of alternating SiO_4-tetrahedra and BO_6-octahedra. Examples include almandine ($Fe_3Al_2Si_3O_{12}$), andraite ($Ca_3Fe_2Si_3O_{12}$) grossularite ($Ca_3Al_2Si_3O_{12}$), pyrope ($Mg_3Al_2Si_3O_{12}$), spessartine ($Mn_3Al_2Si_3O_{12}$) and uvarovite ($Ca_3Cr_2Si_3O_{12}$). Many silicate garnets are useful as gemstones, abrasives, and recyclable water-filter materials.

silicate of potash. See potash water glass.

silicate of soda. See soda water glass.

silicate paint. An organic coating in which the binder is *soda water glass* (sodium silicate) dissolved in water. Used on mortar, masonry, concrete, etc.

silicates. (1) Compounds composed of silicon, oxygen and one or more metals, and sometimes hydrogen. Structurally, they consist of central silicon atoms each of which is bonded to four oxygen atoms situated at the corners of an SiO_4^{4-} tetrahedron. Silicates are found widely in rocks, and are used in building materials, such as glass, bricks or cement, and in ceramics.

(2) A group of natural abrasives consisting of silica combined chemically with metal oxides. Examples include fused glass, pumice and pumicite. Used especially for metal polishing, glass cutting, and scouring and cleaning applications. Also known as *silica abrasives*.

silica wash. Finely divided silica mixed with water and selected ingredients for brushing or spraying on foundry core or mold faces.

silica white. See carbon white.

silica xerogel. A homogeneous *xerogel* produced from tetramethoxysilane, formamide and acidic water by the sol-gel technique. Used for optical and telecommunication applications. See also silica sonogel; silica sonosol.

siliceous brick. A type of acid refractory brick containing 78-92% silica (SiO_2), with the balance being chiefly alumina (Al_2O_3). Used for furnace linings. *Note:* This term is not synonymous with *silica brick*. See also acid brick; acid refractory; silica brick.

siliceous calamine. See willemite.

siliceous clay. A clay containing considerable quantities of uncombined silica (SiO_2) of microscopic and/or macroscopic particle size.

siliceous earth. A general term for any of several white, yellowish or light-gray, friable, porous, highly siliceous materials derived from the remains of microorganisms, such as diatoms and radiolaria, and including both diatomaceous and radiolarian earth. See also diatomaceous earth; radiolarian earth.

siliceous fireclay. A low-fluxing fireclay containing considerable quantities of uncombined silica.

siliceous fireclay brick. A fireclay brick containing considerable amounts of free silica, and is usually low in fluxing ingredients. See also acid brick; acid refractory; silica brick.

siliceous material. (1) A material that is low in metallic oxides and alkalies, and composed chiefly of silica.

(2) A material containing appreciable quantities of silica or silicates.

siliceous refractories. Refractory materials that contain between 78 and 92% silica (SiO_2), with the balance being essentially alumina (Al_2O_3). See also acid refractories; basic refractories; neutral refractories; silica refractories.

siliceous sinter. See geyserite.

silicic acid. White, amorphous powder or lumps obtained by gradual drying and igniting of the jelly-like precipitate obtained by the acidification of a sodium silicate solution. It corresponds to the formula $SiO_2 \cdot xH_2O$, where x is usually 1 or 2. *Metasilicic acid* (H_2SiO_3 or $SiO_2 \cdot H_2O$) has a density of 2.1-2.3 g/cm³ and is not soluble in water, and *orthosilicic acid* (H_4SiO_4 or $SiO_2 \cdot 2H_2O$) has a density of 1.5-1.6 g/cm³ and is soluble in water. Used in chromatography, as a laboratory reagent and as a rubber reinforcer. See also silica gel.

silicide coatings. Thin, brittle protective coatings composed of silicon and one or more electropositive elements, such as aluminum, chromium, molybdenum, niobium or titanium, and formed on substrates (usually refractory metals, such as tantalum, molybdenum, rhenium, niobium, tungsten and their alloys) when heated in an oxygen atmosphere. They have good self-healing properties and chemical stability, high oxidation resistance, good to fair emittance and adherence, and high susceptibility to crack formation. Used for the protection of refractory metals against oxidation, e.g., on gas and rocket engine components, rocket nozzles, and turbine blades and vanes.

silicides. Binary compounds of silicon and a more electropositive metal, such as aluminum, chromium, molybdenum, niobium, titanium or tungsten. Many silicides are suitable for high-temperature structural applications under oxidizing conditions, typically in the range of 1200-1600°C (2190-2910°F), e.g., as oxidation-resistant coatings for refractory metals, as reinforcements for structural ceramic composites, and as matrix materials for structural silicide compounds. Some silicides are used as solid-film lubricants, electrical resistors, in hydrogen storage for fuel cells, and in superconductor research. See also high-temperature structural silicides.

silicified wood. A material in which quartz, jasper, opal and/or chalcedony have replaced the wood fibers of a tree. Depending on the particular replacing mineral, it may be known as *agatized wood, jasperized wood, opalized wood* or *petrified wood.* Used in the manufacture of tiles, can and umbrella handles, paperweights, tabletops, etc.

Silico. Trade name of Krupp Stahl AG (Germany) for a series of water-hardening steels used for shafts, gears and fasteners and bolts.

Silico Alloy. Trade name of Columbia Tool Steel Company (USA) for an oil- or water-hardening shock-resisting tool steel (AISI type S5) containing 0.55-0.58% carbon, 1.95-2% silicon, 0.8% manganese, 0.4% molybdenum, 0.3% chromium, 0.5% molybdenum, and the balance iron. It has outstanding toughness, excellent shock resistance, a high elastic limit, a great depth of hardening and good ductility. Used for pneumatic tools, shear blades, mandrels, bending dies, pipe cutters, chisels, punches, shears and stamps.

silicofluorides. See fluorosilicates.

silicomanganese. An alloy composed of 1-3% carbon, 65-70% manganese, with the balance being silicon. Used in the manufacture of low-carbon and silicon-manganese steels for springs and high-strength structural parts.

silicomanganese brass. A strong, corrosion-resistant brass containing 40% zinc, 1-2% manganese, 0.2-0.3% iron, 0.05-1% silicon, and the balance copper. Used for architectural and ornamental applications, and for marine parts.

silicomanganese steel. A spring steel containing 0.5-0.6% carbon, 0.4-1.9% silicon, 0.55-1.5% manganese, and the balance iron.

silicomolybdic acid. See molybdosilicic acid.

silicon. A nonmetallic element of Group IVA (Group 14) of the Periodic Table, usually prepared by reducing silica with carbon in an electric furnace, and commercially available in the form of lumps, chunks, random pieces, sheets, boules, powder, deposited rod, epitaxial, float-zone and Czochralski single crystal granules or slices. It is not found free in nature, but is a major component of silica and silicates (rocks, quartz, sand, clays, etc.). The two important forms are: (i) *Crystalline silicon,* that is brittle, hard and steel-gray and has a diamond crystal structure, and (ii) *Amorphous silicon* that is a dark brown powder. A *graphitic silicon* in the form of lustrous, black plates has also been reported. Density (crystalline form), 2.34 g/cm³; melting point, 1410°C (2570°F); boiling point, 2355°C (4271°F); hardness, 7 Mohs; atomic number, 14; atomic weight, 28.086; tetravalent. It is an intrinsic semiconductor, and has a high electrical resistivity (23×10^{10} μΩcm). Used as an alloying element in steels (increases hardenability, elasticity, deoxidation, hot hardness, corrosion resistance) usually in the form of ferrosilicon, as an addition to cast iron to promote the decomposition of cementite into ferrite and carbon, as an addition to aluminum, copper and other metals, as a n- or p-type semiconductor for solid-state devices (e.g., transistors, diodes, photovoltaic cells, solar cells, computer circuitry, rectifiers, etc.), in the manufacture of optoelectronic devices and 2D and 3D photonic structures, in the manufacture of silicon carbide and silicon nitride, cermets, refractories, etc., in the manufacture of organosilicon compounds, lubricating greases and oils, and in the manufacture of cement and glass. Symbol: Si.

silicon alloys. A group of alloys in which silicon is a major alloying element, and including in particular alloys of silicon with aluminum, copper, iron or nickel.

silicon-aluminum. A master alloy composed of 50% silicon and 50% aluminum supplied in the form of small lumps. It has a melting point of 1050°C (1920°F), and is used to introduce silicon into aluminum-alloy melts, and as an intermediate and hardener alloy.

silicon-aluminum bronze. A corrosion-resistant aluminum bronze containing varying amounts copper, aluminum and silicon. See also aluminum bronze; aluminum-silicon bronze; silicon bronze.

silicon aluminum oxynitride ceramics. A group of hard, chemically inert ceramics based silicon aluminum oxynitride (SiAlON) and usually made from silicon nitride, aluminum nitride and aluminum oxide by pressureless sintering. They have a density of 3.2-3.3 g/cm³ (0.11-0.12 lb/in.³), high strength, low thermal expansivity below 1000°C (1830°F), an upper continuous-use temperature of about 1500°C (2730°F), good high-temperature and oxidation resistance, a hardness of 8-8.5 Mohs, good thermal-shock resistance, low thermal expansion, high fracture toughness, high wear resistance and toughness, exceptional room-temperature flexural strength, and high corrosion resistance and chemical inertness. Used for engine components, dies, bearings, wear pads, kiln furniture, boats, crucibles, retorts, refractories, thermocouple sheaths, extrusion tools, drawing dies and plugs, as inserts for high-speed machining of high-temperature alloys, and in the form of coatings. Abbreviation: SiAlON ceramics; Sialon ceramics. See also Sialon.

silicon-aluminum-vanadium. A master alloy of 80% silicon, 10% aluminum and 10% vanadium, used in the manufacture of nonferrous alloys.

silicon boride. Any of the following compounds of silicon and boron: (i) *Silicon monoboride.* Stable in air to 1370°C (2498°F). Used in ceramics. Formula: SiB; (ii) *Silicon tetraboride.* Gray-black powder (98% pure). Melting point, decomposes at 1870°C (3400°F). Used for refractories and ceramics. Formula: SiB_4; (iii) *Silicon hexaboride.* Melting point, 1950°C (3540°F). Used for refractories and ceramics. Formula: SiB_6; and (iv) *Silicon tetraboride/silicon* (SiB_4+Si). A borosilicate glass made by reacting silicon and boron in air. It is stable in air to 1550°C (2820°F), has good thermal-shock resistance, and is used as a refractory.

silicon borocarbide. A compound of silicon, boron and carbon supplied in the form of a fine powder for ceramic applications. Formula: SiB_4C.

silicon brass. See silicon casting brass; silicon red brass.

silicon bromide. See silicon tetrabromide.

silicon bronze. (1) A wrought or cast copper-base alloy containing up to 4% silicon, used for electrical applications, welding rods for welding copper to steel, and for fasteners, hardware, castings, bearings, and pumps.

(2) A term used loosely for wrought and cast alloys of copper, zinc and silicon, preferably referred to as "silicon brasses," and for aluminum bronzes containing silicon.

(3) See high-silicon bronze; low-silicon bronze.

silicon carbide. A ceramic material that consists of silicon and carbon and is commercially available in the form of powder, filaments, whiskers, sheets, rods, tubes and single crystals. The regular grade is made in an electric furnace by heating carbon (usually anthracite or coke) with silica (silica sand) at 2000°C (3630°F) forming bluish-black crystals. Hot-pressed and reaction-bonded grades are also available. Density, 3.1-3.2 g/cm³; melting point, 2700°C (4890°F); hardness, 9+ Mohs. It has high toughness, good resistance to acids, good to poor resistance to alkalies and halogens, high thermal-shock resistance, a low coefficient of thermal expansion, excellent thermal conductivity and good electrical conductivity. *Silicon carbide* is often sold under trademarks or trade names, such as *Carborundum, Carbosilite, Carbowalt, Crystolon* or *Silundum.* Used as an abrasive also for grinding wheels, as a refractory for kiln furniture, retorts, nozzles, combustion chambers and nuclear reactors, as a semiconductor, in light-emitting diodes to produce green or yellow light, for high-strength, high-modulus continuous fibers and whiskers, and as particulate for engineering composites. Formula: SiC. See also amorphous silicon carbide.

silicon carbide abrasives. Synthetic abrasives composed of hard, sharp grains of silicon carbide (Mohs hardness, 9+) that fracture easily providing new cutting surfaces. Used extensively in grinding, polishing, honing, and sand blasting.

silicon carbide-alumina composites. Ceramic composites consisting of alumina (aluminum oxide) matrices reinforced with silicon carbide fibers or whiskers. They provide improved impact resistance (toughness), good high-temperature creep resistance, good thermal-shock resistance, and are used as cutting-tool inserts for machining hard metal alloys.

silicon carbide-aluminum composites. Composites consisting of aluminum or aluminum-alloy matrices reinforced with silicon carbide particulate or continuous fibers. They have high tensile strength, high stiffness, low density, and are used for aerospace applications, e.g., structural elements for wings, compression tubes, fins for projectiles, pressure vessels, casings for missile bodies, and for cutting tools, and electronic packaging.

silicon carbide briquettes. Silicon carbide crushed and made into block-shaped forms known as "briquettes." Used to deoxidize and inoculate gray cast iron melted in cupolas.

silicon carbide cement. An acidproof refractory cement based on silicon carbide and having a low coefficient of thermal expansion, excellent thermal conductivity and high electrical conductivity. Used for crucible and furnace linings.

silicon carbide-copper composites. Composites consisting of copper or copper-alloy (usually bronze or brass) matrices reinforced with continuous silicon carbide fibers. They have good tensile strength and low density, and are used for aerospace applications, e.g., missile components, propellers, etc.

silicon carbide fibers. Ceramic fibers based on silicon carbide and commercially available in continuous, yarn, chopped fiber, fabric, mat, roving and discontinuous (whisker) form. They have a density range of density of 2.3-3.2 g/cm³ (0.08-0.12 lb/in.³), high tensile strength and modulus of elasticity, high stiffness, and a maximum service temperature range of 1200-1600°C (2190-2910°F). Used as reinforcement in metal-, ceramic- and organic-matrix composites.

silicon carbide firebrick. A refractory brick consisting predominantly of silicon carbide, and characterized by good thermal shock resistance, good abrasion and wear resistance, and good resistance to many chemicals.

silicon carbide foam. A lightweight ceramic consisting of silicon carbide foamed into shapes, and available in (i) low-density grades (0.27 g/cm³ or 0.01 lb/in.³) with a porosity of 85-95% and good compressive and tensile strength; and (ii) high-density grades (0.53 g/cm³ or 0.02 lb/in.³) with a porosity of 80% and excellent compressive and tensile strengths. It has good resistance to hot chemicals and good machinability, and is used in ceramics and as thermal insulation.

silicon carbide–glass-ceramic composites. Composites consisting of glass-ceramic matrices (e.g., lithium aluminosilicate) reinforced with continuous silicon carbide fibers. They have high toughness, good tensile strength, low density and a maximum service temperature of 1000°C (1830°F). Used chiefly for aerospace applications.

silicon carbide-magnesium composites. Composites consisting of magnesium or magnesium-alloy matrices reinforced with continuous silicon carbide fibers. They have high tensile strength, high stiffness and low density, and are used chiefly for aerospace applications.

silicon carbide refractories. Refractory products in which silicon carbide is the principal constituent. They possess high thermal-shock resistance, low coefficients of thermal expansion, excellent thermal conductivity, high electrical conductivity, high toughness, good resistance to acids, and good to poor resistance to alkalies and halogens. Used for kiln furniture, retorts, nozzles, combustion chambers and nuclear reactors.

silicon carbide-reinforced lithium aluminosilicate. A composite consisting of a glass-ceramic matrix based on lithium aluminosilicate ($Li_2O–Al_2O_3–SiO_2$ system) reinforced with continuous silicon carbide fibers. It has high toughness, good tensile strength, low density, a low thermal expansion coefficient, and a maximum service temperature of 1000°C (1830°F). Used for high-temperature applications, and for aerospace applications.

silicon carbide-reinforced magnesium aluminosilicate. A composite consisting of a glass-ceramic matrix based on magnesium aluminosilicate ($MgO–Al_2O_3–SiO_2$ system) reinforced with continuous silicon carbide fibers. It has high toughness, good tensile strength, low density, a low thermal expansion coefficient, and a maximum service temperature of 1200°C (2190°F). Used for high-temperature applications.

silicon carbide-reinforced silicon carbide. A high-temperature, ceramic composite consisting of a silicon carbide matrix reinforced with silicon carbide fibers or whiskers. It is produced by chemical vapor infiltration and deposition. The introduction of silicon carbide fibers into the silicon carbide matrix greatly improves the overall strength and fracture toughness.

silicon carbide-reinforced silicon nitride. A high-temperature, ceramic composite consisting of a silicon nitride matrix reinforced with silicon carbide fibers or whiskers, and produced by chemical vapor infiltration and deposition. The introduction of silicon carbide fibers into the silicon nitride matrix greatly improves the overall strength and fracture toughness.

silicon carbide–titanium composites. Composites consisting of titanium or titanium-alloy matrices reinforced with continuous silicon carbide fibers or particulate. They have high tensile strength, high stiffness and low density, and are used for aerospace and automotive applications, e.g., driveshafts, turbine-engine parts, turbine disks, fan blades, etc.

silicon carbide whiskers. Axially-oriented, single crystalline filaments of silicon carbide made by chemical reaction or rice-hull pyrolysis. They have a predominant fiber length of 10-200 μm (390-7900 μin.), an average fiber diameter of 0.1-10 μm (4-390 μin.), very high tensile strength and elastic moduli, high heat capacity, and an upper service temperature of 1600-1800°C (2910-3270°F). Used in the manufacture of composite structures with metal, ceramic or glass matrices for aircraft and space vehicle applications, e.g., as ablative agents.

silicon casting brass. A casting brass containing 65-83% copper, 14-34% zinc and 1-4% silicon. It has high strength, fair to good machinability and good corrosion resistance, and is used for bearings, gears, valve stems, fasteners, and die castings.

silicon cast irons. See acid-resistant cast irons.

silicon chloride. See silicon tetrachloride.

silicon-copper. (1) A group of alloys composed of 80-90% copper and 10-20% silicon, and used to introduce silicon into copper and copper-alloy melts, in the manufacture of deoxidized copper, and as a deoxidizer and hardener in copper-base alloys.

(2) A hard, tough alloy of 70-90% copper and 10-30% silicon obtained from silica and copper by electrolysis, and used in the manufacture of silicon bronze. Also known as *copper silicide.*

(3) A group of copper alloys containing 0.25-3.25% silicon used for electrical applications and in the form of welding rods for welding copper to steel.

silicon dioxide. See silica.

silicon dioxide abrasives. See silica abrasives.

silicon dioxide fibers. See silica fibers.

silicon dioxide glass. See fused silica.

Silicone II. Trade name of GE Silicones (USA) for low-odor silicone sealants supplied in different colors (clear, almond, white, gray, brown, etc.) for use in interior and exterior applications, e.g., sealing of windows, doors, roofs, concrete, masonry, wood, tiles, bathtubs, etc.

silicone adhesives. A group of one- and two-part adhesives based on polyfunctional siloxanes with or without mineral fillers. They are soft, flexible elastomers supplied as moisture and UV-curing systems, and possess excellent peel and impact strength, good heat and moisture resistance, good processibility, good resistance to polar solvents, good adhesion to many substrates, poor shear strength, low cohesive strength, a cure temperature range of 20-260°C (68-500°F), and a maximum service temperature of 260°C (500°F). Used for bonding metals, ceramics and glass. Also known as *silicone cements.*

silicone-alkyd coatings. Protective coatings whose binder is obtained by reacting silicone with alkyd resins. They provide high resistance to extreme sunlight, good durability and gloss and good chemical resistance (except alkalies and solvents), and are used on ferrous and nonferrous metal parts, e.g., process equipment for chemical and petroleum industries.

silicone-alkyd resins. High-molecular-weight alkyd resins modified with a drying oil, such as linseed, soybean or tung, and a silicone resin. They possess increased hardness, durability, chemical and weather resistance, and are used as film formers in protective coatings, house paint, baking enamels, etc.

silicone cements. See silicone adhesives.

silicone coatings. Protective coatings based on silicone resins cured by air drying, or baking. They have good exterior durability, gloss and color retention and high heat resistance. Used for exterior coatings, and on stove parts, roasters, etc.

siliconed film. A polyester, polyethylene or metallic film treated with silicone on one or both sides, and used for temporary protection applications.

silicone elastomer. See silicone rubber.

silicone enamel. A heat- and chemical-resistant paint made by grinding or mixing mineral pigments with silicone resin. It dries by oven baking forming a smooth, hard surface on metals or ceramics. See also baking enamel.

silicone foam. A flexible, spongy cellular silicone rubber made by foaming, and used as a cushioning material, for vibration dampening and shock insulation, and for thermal and acoustic insulation applications.

silicone grease. A water-resistant, dark blue lubricating grease consisting of silicone oil with an organic thickener. It has an operating temperature range of about -5 to +300°C (-23 to +570°F), and is used for the lubrication of plain and antifriction bearings.

silicone plastics. Plastics based on *silicone resins.*

silicone resins. Thermosetting resins that are usually heat curing and based on inorganic three-dimensional network polymers with siloxane (–Si–O–Si–) repeat units having side groups, such as methyl, phenyl, vinyl, amino, hydroxy, ethoxy or fluoro, attached to the silicon atom. They are commercially available in various forms as liquid or solid resins, neat or compounded with plasticizers, pigments, fillers (e.g., mica or silica), or reinforcements (e.g., carbon or glass fibers). The mechanical properties of silicone resins are relatively stable between -100 and +300°C (-148 and +570°F). They also possess good thermal stability, moisture resistance and electrical (dielectric) properties, and are used for laminates (with glass cloth), patterns, molding compounds, filament winding, electrical insulation, high-temperature insulation, electronic applications (e.g., encapsulation of electronic components), for vibration-damping devices, for biomedical applications (e.g., catheters), in the manufacture of high-temperature components for the aircraft and aerospace industries, for sealants, adhesives and protective coatings, as a modifier for alkyd resins, for impregnating electric coils, and as bonding agents. Abbreviation: Si; SR. See also silicones.

silicone rubber. Any of a group of cross-linked elastomeric materials having backbone molecule chains that alternate silicon and oxygen atoms with side-bonded atoms or atom groups (e.g., hydrogen, methyl, etc.). They are available in various grades including standard, high-temperature (over 300°C or 570°F), low compression-set, high tensile-strength and room-temperature-vulcanizing. They have a density of 0.5 g/cm³ (0.02 lb/in.³), moderate physical properties and relatively low strength (unless reinforced), a service temperature range of -90 to +250°C (-130 to +482°F), excellent electrical properties, good weatherability, good resistance to dilute acids, greases and oils, fair resistance to concentrated acids, alcohols, alkalies and ketones, and poor resistance to aromatic hydrocarbons and halogens. Used for high- and low-temperature insulation and electrical insulation, for the encapsulation of electronic parts, for automotive engine components, for diaphragms, gaskets, seals, tubing, surgical membranes and implants, in soft tissue reconstruction, for air locks, flexible windows for face masks, various mechanical products, and silicone foam products. Abbreviation: SR. Also known as *silicone elastomer.*

silicones. A large group of polymers containing alternate silicon and oxygen atoms whose properties are determined by the organic radicals attached to the silicon atoms. They are derived from silica (sand) and methyl chloride, or by a reaction of silicon tetrachloride and a Grignard (organometallic) reagent. Depending on the degree of polymerization and molecular weight, they can be liquid, semiliquid or solid, and are commercially available in the form of fluids, emulsions, solutions, resins, powders, pastes and elastomers. Their wide service temperature range, excellent water repellency, high lubricity, good thermal and electrical properties, good resistance to oxidation and weathering and relatively low chemical activity make them suitable as adhesives, lubricants, sealants, waxes, polishes, coatings and paints, dielectric fluids, coolants, synthetic rubber, and the like. Also known as *organosiloxanes*. See also silicone resins; silicone elastomer.

silicone sealants. Liquid or semiliquid compositions made from silicone rubber and commonly applied with spray or caulking guns. They are supplied in clear/transparent and colored (black, brown, white, etc.) grades with excellent water repellency, high lubricity, good resistance to oxidation and weathering, good adhesion to most metallic and nonmetallic surfaces, relatively low chemical activity and good resistance to dilute acids, greases and oils. Used as flexible gasket materials in sealing machine housings (engines, gear boxes, pumps, etc.), and in building construction as a flexible compound for sealing concrete, masonry and wood, roofs, windows and doors, bathtubs and tiles, gutters, etc.

silicone tape. A flat, self-bonding tape that is made of silicone and does not contain adhesives. It has good resistance to moisture, oxygen, ozone and electrical voltages, a service temperature range of -65 to +257°C (-150 to +500°F), and fuses to itself in about 24 hours forming an inseparable bond. Supplied in rolls, it is used as a cohesive insulating barrier for cleanroom applications.

silicon-ferrochrome. A master alloy composed of 45-55% chromium, 30-50% iron, 1-17% silicon and 2.5-7% carbon, and used to introduce chromium and silicon into steel.

silicon fluoride. See silicon tetrafluoride.

silicon-gold alloy. See gold-silicon alloy.

silicon halide. A compound of silicon and a halogen, such as bromine, chlorine, fluorine or iodine.

silicon hexaboride. See silicon boride (iii).

Silico Nickel. Trade name of Gilby-Fodor SA (France) for a heat-resistant alloy composed of 96% copper, 2.5% silicon and 1.5% manganese. Used for cathodes and filaments in electronic tubes.

silicon iodide. See silicon tetraiodide.

siliconized graphite. Graphite products whose surface has been chemically converted to silicon carbide. They combine the outstanding lubricity of graphite with the excellent abrasion and wear resistance of silicon carbide, and have good high-temperature resistance up to 1370°C (2500°F), and good chemical and corrosion resistance. Used for bearings, seal rings, liners, etc.

siliconized iron and steel. Iron and steel products (e.g., screws, bolts, fittings, small forgings and castings) on whose surface an acid- and corrosion-resistant silicon-rich (up to 14%) coating has been produced by treating with silicon carbide or ferrosilicon powder in the presence of chlorine gas at temperatures of 900-1000°C (1650-1830°F) for several hours.

silicon irons. See acid-resistant cast irons.

silicon-iron electrical steel. See electrical steel; silicon steels.

silicon-manganese. A master alloy composed of 65-70% manganese, 1-3% carbon, and the balance being silicon. Available in

four grades with 1.0%, 2.0%, 2.5% and 3.0% carbon, respectively, it is used to introduce manganese into steel, and as a deoxidizer and scavenger in steel.

silicon-manganese steels. See structural silicon steels.

silicon metal. Foundry alloys with high silicon contents usually between 50 and 95%. See also ferrosilicon.

silicon monoboride. See silicon boride (i).

silicon monoxide. An amorphous, black or dark brown powder (99.9+% pure) made by reducing silica with carbon in an electric furnace, and available in the form of lumps, granules, powders and tablets. Density, 2.15-2.24 g/cm³; melting point, above 1700°C (3095°F); boiling point, 1880°C (3415°F); high hardness and abrasiveness. Used for thin protective overcoatings on aluminum and its alloys, in optical equipment and mirrors, for replicas in electron microscopy, and as a brown pigment. Formula: SiO.

silicon nanoparticles. A relatively new class of nano-sized particles produced by an electrochemical process in which the surface layer of a silicon wafer is eroded by etching in a bath of hydrofluoric acid and hydrogen peroxide with an applied electric current. The resulting intermediate consists of a network of weakly interconnected nanostructures. Ultrasound treatment of the wafer intermediate then produces individual nano-sized particles which can be separated by size. The smallest four sizes are blue, green, yellow and red photostable, luminescent particles that fluoresce in ultraviolet light and two-photon infrared light. It is currently believed that these particles could be useful for flash memories, electronic displays and other microelectronic and optoelectronics applications, and as fluorescent markers for biological and biomedical applications.

silicon-nickel. A master alloy composed of 40-80% nickel, 16-18% silicon and 2.5-30% iron, and used to introduce nickel into steel.

silicon nitride. A compound of silicon and nitrogen commercially available in the form of grayish amorphous or crystalline shapes, tubes, sheets, rods, powder, and continuous and discontinuous fibers. The regular grade is made by a reaction of silicon and nitrogen powders in an electric furnace at 1300°C (2370°F). Hot-pressed and reaction-bonded grades are also available. It has a density of 3.44 g/cm³ (0.124 lb/in.³), a melting point of 1900°C (3450°F), a hardness above 9 Mohs, high compressive strength, low thermal expansion, high resistance to thermal shock and chemicals, an upper continuous-use temperature of 1100-1650°C (2010-3000°F), good resistance to oxidation and various corrosive media, good resistance to dilute acids and halogens, fair resistance to concentrated acids, good to poor resistance to alkalies, good resistance to molten aluminum, lead, tin and zinc, and poor resistance to hydrofluoric acid. Used as an abrasive, in refractories, refractory coatings and mortars, in crucibles for zone-refining germanium, for bonding silicon carbide, as a thermal insulator, in thermocouple tubes, in stator blades for high-temperature gas turbines, in rocket nozzles, as a passivating agent in transistors and other solid-state devices, as a catalyst support, and for high-strength fibers and whiskers. Formula: Si_3N_4.

silicon nitride fibers. Ceramic fibers based on silicon nitride and available in both continuous and discontinuous (whisker) form. They have an upper service temperature of 1500°C (2730°F), high tensile strength and elastic moduli, low coefficients of thermal expansion, good wear resistance, good dielectric properties, and are used as reinforcements for composites with ceramic, metal, glass or glass-ceramic matrices.

silicon oxide. See silica; silicon monoxide.

silicon oxycarbide. A compound of silicon carbide and silicon dioxide with good high-temperature properties used as a high-temperature ceramic material, in the manufacture of fibers, sheets and other products, and for thermal insulation applications. Abbreviation: SiOC.

silicon oxycarbide glass. A dark, hard, amorphous solid that contains only silicon, oxygen and carbon, and can be prepared from silicone-based starting polymers (e.g., methyl trichlorosilane with or without dimethyl dichlorosilane) by treating in a solvent of water, toluene and isopropyl alcohol. The resulting *silanol* intermediate forms a siloxane ring-containing silicon-oxygen chain polymer by the elimination of water. This polymer is then pyrolized to form silicon oxycarbide glass, and is available in powder form and as a dense product (sheet, fiber, etc.). It has excellent high-temperature strength and chemical stability, good resistance to crystallization and oxidation above 1000°C (1830°F), and higher elastic modulus and hardness than *fused silica*.

silicon oxynitride. A compound of silicon nitride and silicon dioxide with good high-temperature properties used as a stable refractory in the manufacture of plates, crucibles and tubes for fusing salts and nonferrous metals, for ceramic composites, and as reinforcing fiber. Formula: Si_2ON_2. Abbreviation: SiON.

silicon oxynitride-reinforced silicon carbide. A ceramic composite consisting of a silicon carbide matrix reinforced with silicon oxynitride fibers. It has a density of about 2.7 g/cm^3 (0.10 lb/in.³), and is used for high-temperature applications.

silicon phosphide. A ceramic compound of silicon and phosphorus with a band gap of 2 eV, used as a semiconductor. Formula: SiP_2.

silicon red brass. A wrought red brass containing 81-82% copper, 14-17% zinc and 1-4% silicon. It has high tensile strength, excellent hot workability, good corrosion resistance, and is used for valve stems, and in resistance welding. Also known as *silicon brass*.

silicon spiegel. See silicospiegel.

silicon steels. (1) A group of low-carbon steels (less than 0.1%) that contain between 0.3 and 4.3% silicon, and have good soft magnetic properties, good electrical resistivity, high hardness and brittleness and poor to fair workability. They are often sold under trademarks or trade names, such as *Cubex, Hipersil* or *Orthosil*. Depending to the magnetic orientation of the grains, it can be classified into: (i) *Oriented silicon steels* containing up to 3.15% silicon. They have highly directional magnetic properties with very low core losses and very high magnetic permeability when the flux path is parallel to the rolling direction, and are used in the form of sheets for power and distribution transformers; and (ii) *Nonoriented silicon steels* which are a special, flat-rolled, low-carbon or alloy steels that contain 0.3-4.3% silicon and sometimes additions of aluminum, and whose grains do not exhibit preferred magnetization in any particular direction. Depending on the silicon content, they possesses good stamping properties, good permeability, low core losses, and are available as strip and sheet for use in electrical applications, such as motors, dynamos and generators, rotating machinery, transformers, ballasts, relays, etc. See also electrical steel.

(2) A group of strong, tough, shock-resisting tool steel containing about 0.3-1.0% carbon, 0.9-2.5% silicon, 0.3-1.5% manganese, 0.3-1.4% molybdenum, up to 0.5% vanadium, and the balance iron. Often sold under trademarks or trade names, such as *Black Giant, Mosil, Silico Alloy, Silman, Simoch* or *Solar*, they are used for forming tools, thread-rolling dies, chisels, pneumatic tools, shear blades, and punches. See also shock-resisting tool steels.

(3) Nonweldable structural steels containing 0.25% carbon, 0.5% silicon, and the balance iron. Used for bridges, towers and structures.

(4) Spring steels containing about 0.45-0.55% carbon and 1-3% silicon. They are available in wire form in the annealed and hardened-and-tempered condition, and have a high elastic limit, and high yield and fatigue strength. Used for coil springs and automotive leaf springs.

silicon tetraboride. See silicon boride (ii).

silicon tetrabromide. A colorless, moisture-sensitive liquid (99+% pure) with a density of 2.772 g/cm^3, a melting point of 5°C (41°F) and a boiling point of 154°C (309°F). Used as a reagent. Formula: $SiBr_4$. Also known as *silicon bromide; tetrabromosilane*.

silicon tetrachloride. A colorless, moisture-sensitive liquid (98+% pure) with a density of 1.483 g/cm^3, a melting point of -70°C (-94°F), a boiling point of 58°C (136°F), and a refractive index of 1.412. It is also available in high-purity semiconductor grades (99.998+%) and fiber-optic grades (99.9999%). Used in the manufacture of high-purity silica and fused silica, in the preparation of silicones, ethyl silicate and similar compounds, in fiberoptics, as a reagent, as a source of silica, silicon and hydrogen chloride, in smoke screens, and in biochemistry and medicine. Formula: $SiCl_4$. Also known as *silicon chloride; tetrachlorosilane*.

silicon tetraethoxide. See tetraethyl orthosilicate.

silicon tetrafluoride. A colorless gas (99.5+% pure) with a density of 3.570 g/cm^3, a melting point of -90°C (-130°F), and a boiling point of -86°C (-122°F). It is also available in high-purity electronic grades (99.99+%). Used in the preparation of low-temperature silicon films by chemical vapor deposition (CVD), in the manufacture fluosilicic acid, in the manufacture of pure grades of silicon, as an intermediate and reagent, etc. Formula: SiF_4. Also known as *silicon fluoride; tetrafluorosilane*.

silicon tetrahydride. See silane.

silicon tetraiodide. White crystals, or off-white, moisture-sensitive powder (99.9+% pure) with a density of 4.198 g/cm^3, a melting point of 120.5°C (249°F) and a boiling point of 287.5°C (549°F). Used as a reagent. Formula: SiI_4. Also known as *silicon iodide; tetraiodosilane*.

silicon tin bronze. A wrought, corrosion-resistant alloy of 97.5% copper, 1.75% tin and 0.75% silicon. Used for electrical equipment.

silicon-vanadium. A master alloy of 60% silicon and 40% vanadium used in the manufacture of nonferrous alloys.

silicophosphate gel. A homogeneous, transparent gel that can be prepared by a sol-gel technique by first mixing appropriate quantities of ethanol, tetraethyl orthosilicate (TEOS), water and phosphoric acid. The resulting mixture is then allowed to gel in an air oven at elevated temperatures.

silicospiegel. A *spiegeleisen* containing 15-30% manganese, 7-15% silicon, 0.15% phosphorus, 0-5% carbon, and the balance iron. Used as a deoxidizer in steels, as an alloying addition for open-hearth steels, and for the introduction of manganese into cast iron. Also known as *silicon spiegel*.

silicotungstic acid. See tungstosilicic acid.

silicowolframic acid. See tungstosilicic acid.

Siliglass. Trade name of Nippon Mineral Fiber Manufacturing

Company Limited (Japan) for a thermally resistant, high-silicic acid glass-fiber product composed of more than 96% silicon dioxide.

Silika. Trade name of RLB Silika (Russia) fore silica-based nonwovens made by needle punching and used for acoustical, electrical and thermal insulation applications.

Silikon. Trademark of Richard James Inc. (USA) for liquid injectable silicone supplied in several grades including *Silikon 1000* and *Silikon 5000* for human skin reconstruction.

SILIKON SR. Trademark of Hanno-Werk GmbH & Co KG (Austria) for silicone rubber-based adhesives and sealants suitable for a wide range of construction applications.

Silimaz. Trade name of Krupp Stahl AG (Germany) for acid-resistant alloy steel of 0.6% carbon, 15.5% silicon, 2.6% manganese, and the balance iron. It is brittle in the as-cast condition, has poor thermal-shock resistance, and is used for fittings, valves, pumps, drains, agitators and castings.

Silimite. Trademark of Allis-Chalmers Manufacturing Company (USA) for a *dolomitic lime* with high magnesium content used in the reduction of silicon dioxide in hot-process water softening equipment.

Silimo. Trade name of Bethlehem Steel Corporation (USA) for a water- or oil-hardening, shock-resistant tool steel containing 0.5% carbon, 1.1% silicon, 0.4% manganese, 0.5% molybdenum, 0.2% vanadium, and the balance iron. Used for impact tools and pneumatic tools.

Silione. Trademark of Vetrotex Deutschland GmbH (Germany) for continuous glass fibers.

Silionne. Trade name of Société du Verre Textile SA (France) for glass yarn spun from continuous glass filaments.

Silit. Trademark of Saint-Gobain Industrial Ceramics, Inc. (USA) for a material that consists essentially of silicon carbide, and is usually supplied in the form of preshaped rods. It has a density of 3.2 g/cm³ (0.12 lb/in.³), high electrical resistivity, a maximum service temperature of 1400-1600°C (2550-2910°F), and is used for heating elements, and refractory furnace parts, such as crucibles, boats, saggars, nozzles, rods and rollers.

Silit-SK. Trademark of Saint-Gobain Industrial Ceramics, Inc. (USA) for reaction-sintered, silicon-infiltrated silicon carbide.

Silitef. Trademark of Nimet Industries Inc. (USA) for an autocatalytic coating made by combining nickel and phosphorus with a fluoropolymer and silicon carbide. Used on ferrous and nonferrous metal substrates.

silk. (1) A fine, soft, tough protein-based (fibroin) textile fiber obtained in filament form from the cocoons of the silkworm, the larva of the moth *Bombyx mori*. Source: China, India, Italy, Japan and France. Used for the preparation of raw and thrown silk. See also raw silk; thrown silk.

(2) The continuous protein filaments (threads) spun out from the abdomens of various species of insects and spiders, or produced by their larvae. They are usually not used for textiles but sometimes as crosshairs for certain precision optical instruments. See also spider silk.

silk cotton. See kapok.

silk gum. See sericin.

Silkiss. Trademark of Miroglio Tessile SpA (Italy) for polyester fibers and filament yarns.

silk lace. A delicate silk or rayon fabric hand- or machine-made in various pattern, and used especially for veils and evening gowns.

silk noil yarn. A yarn spun from the short waste fibers removed from silk during dressing.

Silkon. Trade name for regenerated protein (azlon) fibers.

Silkorrit. Trade name of Georgsmarienwerke Selesiastahl GmbH (Germany) for a series of austenitic stainless steels containing up to 0.12% carbon, 18% chromium, 8.5-10.5% nickel, 0-2.2% molybdenum, 0-0.5% titanium, and the balance iron. Used for welded structures, chemical plant and equipment, tanks, mixers and agitators.

Silksite. Trade name of Nippon Sheet Glass Company Limited (Japan) for a diffuse reflection glass.

Silk-Skin. Trademark of Winter & Co. GmbH (Germany) for an artificial leather imitating silk.

silk-spun yarn. A staple yarn other than silk produced on silk finishing and manufacturing machinery.

Silkwool. Trade name for regenerated protein (azlon) fibers.

silky oak. See lacewood.

Silky-Rock. Trade name of Whip Mix Corporation (USA) for a dental gypsum (ADA type IV) supplied as a powder. When cast from a mixture of 23 mL water per 100 g powder, it has a smooth, silky pouring quality. Upon setting the die stone develops outstanding surface hardness and density, and low setting expansion (0.09%). Supplied in yellow, white and violet colors, it is compatible with all types of dental impression materials.

Silky Touch. Trade name of BASF Corporation (USA) for nylon-6 and polyester fibers and yarns used for textile fabrics, especially clothing.

sillenides. A group of bismuth compounds with body-centered cubic *sillenite*-type structure including bismuth germanate, bismuth silicate and bismuth titanate.

sillenite. A greenish mineral composed of bismuth silicon oxide, $Bi_{12}SiO_{20}$. Crystal system, cubic. Density, 9.30 g/cm³. Occurrence: Mozambique, Mexico.

sillimanite. A white, yellow, brown, grayish or bluish green mineral with a vitreous luster. It is composed of aluminum silicate, has the same formula as the minerals *andalusite* and *cyanite*, Al_2SiO_5, and contains theoretically 35-37% SiO_2 and 60-63% Al_2O_3, with the balance, if applicable, being Fe_2O_3 and MnO. Crystal system, orthorhombic. Density, 3.23-3.27 g/cm³; hardness, 6.0-7.5 Mohs; refractive index, 1.66. It has high heat resistance, decomposes at 1545°C (2813°F) forming mullite and silica and, on further heating to 1810°C (3290°F), *corundum* and glass. Occurrence: Australia, Brazil, India, Kenya, USA (California; Connecticut, Georgia, Massachusetts, New Hampshire, Pennsylvania, South Carolina, South Dakota). Used for special porcelains and refractories for spark plugs, pyrometric tubes, chemical laboratory ware, special porcelain shapes, and as a patching compound for furnaces. Also known as *fibrolite*.

sillimanite refractories. Refractories composed essentially of sillimanite, Al_2SiO_5 (55-64% Al_2O_3 and 25-36% SiO_2), with the balance being ferric oxide (Fe_2O_3) and manganese oxide (MnO). They have good spalling resistance, high refractoriness under load, and are used for refractory bricks, glass-tank furnaces and kiln furniture, and as patching compounds for furnaces.

Sillitin 85. Trademark of W. Haldenwanger Technische Keramik GmbH & Co. KG (Germany) for a mixture of 75-85% *quartz* and 15-25% *kaolinite*, also known as "Neuburg chalk." It has a fineness of 20 µm (788 µin.) or less. Used in ceramic glazes.

Sillman bronze. A corrosion-resistant aluminum bronze containing 86.4% copper, 9.7% aluminum, and 3.9% iron. Used for bullet shells and jewelry.

Sil-Lo. Trade name of United Wire & Supply Company (USA) for a brazing alloy containing 15% silver and 5% phosphorus. Commercially available in the form of strips, wires, rods and powder, it has a melting range of 641-704°C (1185-1300°F), and is used for silver brazing nonferrous alloys.

Sil-Lon. Trade name of United Wire & Supply Company (USA) for a brazing alloy containing 40% silver, 30.5% copper and 29.5% zinc. It has a melting range of 677-732°C (1250-1350°F), and is used for silver brazing metals.

Sil-Loy. Trade name of United Wire & Supply Company (USA) for a brazing alloy containing 65% silver, 20% copper and 15% zinc. It has a melting range of 685-718°C (1265-1325°F), and is used for silver brazing.

Silly Putty. Trademark of Binney & Smith Inc. (USA) for an abrasive-filled, flowable plastic medium used in metal finishing. See also nutty putty.

Silma. Trade name of Degussa AG (Germany) for a silver brazing alloy containing 85% silver and 15% manganese. Commercially available in the form of strips, wire, powder and sheet, it has a melting point of 977°C (1790°F), and is used for high-temperature brazing applications.

Sil-Mag. Trade name of SiMETCO (USA) for a series of cast-iron inoculants containing 42-65% silicon, 7.5-20% magnesium, and the balance iron. A special grade containing 1% mischmetal is also available. Used in the manufacture of nodular cast iron.

Silmal. French trade name for an aluminum alloy containing 1.75% silicon and 1% magnesium. Used for light-alloy parts.

Silmalar. Trade name of British Alcan Wire Limited (UK) for an aluminum alloy containing 0.5% silicon and 0.5% magnesium. It has good electrical conductivity (about 54% IACS), and is used for conductors.

Silmalec. Trade name of Alcan-Booth Industries Limited (UK) for an age-hardening aluminum alloy containing 0.5% silicon and 0.5% magnesium. Used for light-alloy parts.

Silman. Trade name of Teledyne Vasco (USA) for a shock-resisting tool steel (AISI type S4) containing 0.52-0.57% carbon, 1.8-2.2% silicon, 0.8-0.9% manganese, 0.2-0.3% chromium, 0.15-0.25% vanadium, and the balance iron. It has good abrasion and wear resistance, good fatigue resistance, high toughness, high elastic limit, and good resistance to high-temperature grain growth. Used for punches, hand and pneumatic chisels, track chisels, shear blades, rivet sets, tools shanks, and coal cutters.

Silmanal. Trade name of GE Company (USA) for a silver-based permanent magnet material containing 8.5-9% manganese and 4-4.5% aluminum. It possesses high coercive force and low flux density, and is usually supplied in the cold rolled condition. Used for magnetic and electrical equipment.

Silmar. Trademark of BP Chemicals (UK) for unsaturated polyester resins used for molding compounds, electrical components, boat decks, surfboards, truck cabs, fiberglass-reinforced composites, etc.

Silmate. Trademark of GE Silicones (USA) for silicone adhesives.

Silmelec. French trade name for aluminum alloy containing 1% silicon and 0.6% manganese. It has high electrical conductivity, and is used for electrical conductor wires.

Silmet. (1) Trademark of Barker & Allen Limited (UK) for a nickel silver composed of 62% copper, 20% zinc and 18% nickel. Usually supplied in foil or tube form, it has a density of 8.72 g/cm³ (0.315 lb/in.³), a melting range of 1060-1110°C (1940-2030°F), excellent weldability and cold workability, good

machinability and poor hot formability. Used for deep stamping and spinning applications, and in manufacture of electrotype.

(2) Trade name for a dental amalgam alloy.

Silmicro. Trade name of Novaceta S.p.a. (Italy) for cellulose acetate microfibers used for the manufacture of wearing apparel.

Silmo. Trade name of Timken Company (USA) for a steel containing 0-0.15% carbon, 1.15-1.65% silicon, 0.45-0.65% molybdenum, and the balance iron. It has good heat resistance to 538°C (1000°F), and is used for high-temperature tubing.

Silnic. Trade name of Chase Brass & Copper Company Inc. (USA) for a corrosion-resistant, high-strength bronze containing 1.6-2.2% nickel, 0.45-0.75% silicon, and the balance copper. Commercially available in the form of rod and wire, it is used for structural components, fasteners and bolts.

Sil-Nik. Trade name of United Wire & Supply Company (USA) for an alloy of 40% silver, 30% copper, 28% zinc and 2% tin. It has a melting range of 670-780°C (1240-1435°F), and is used as a silver solder.

Silnova. Trade name of Novaceta S.p.a. (Italy) for cellulose acetate fibers used for the manufacture of wearing apparel.

Silo. Trademark of Rockwood Pigments N.A., Inc. (USA) for high-purity iron oxides and various black-, brown-, red- or yellow-colored iron oxide paint pigments.

Sil-O-Cel. Trademark of Manville Corporation (USA) for a diatomaceous earth with good heat resistance to 870°C (1600°F). Supplied in the form of blocks or powder, it is used for thermal insulation applications.

Sil-O-Crete. Trademark of M.A. Bruder & Sons, Inc. (USA) for a clear, water-repellent, nonfilm-forming silicone resin-based coating used on above-grade exterior masonry.

Sil-Oid. Trade name of United Wire & Supply Company (USA) for a brazing alloy of 34% copper, 31.5% silver, 19% cadmium and 15.5% zinc. It has a melting range of 629-754°C (1165-1390°F).

Silon. (1) Trademark of Silon AS (Czech Republic) for nylon 6 monofilaments and filament yarns.

(2) Trademark of Cheil Synthetics Company (South Korea) for polyester fibers and filament yarns.

(3) Trademark of Bio Med Sciences, Inc. (USA) for polymeric film materials used for biomedical applications.

Siloprene. Trademark of GE Silicone (USA) or a silicone elastomer with excellent resistance to high and low temperatures, an service temperature range of -90 to +250°C (-130 to +482°F), excellent electrical properties, high tensile strength and elasticity, and good resistance to weathering and lubricating greases and oils. Used for electronic parts, electrical insulation, gaskets, and various mechanical products.

Silor. Trade name of Westerwald AG (Germany) for sun-filtering glass blocks with a warm tone.

Siloset. Trade name for silicone polymers.

Sil-O-Spar. Trade name of The Feldspar Corporation (USA) for a high-quality blend of silica and feldspar (silicate of sodium, potassium, calcium and aluminum) used for general ceramic applications including sanitaryware.

Silotex. Trademark of Anaconda Wire & Cable Company (USA) for magnet wire having a uniform covering consisting of a continuous-filament glass yarn wrapping treated with silicone resin.

siloxanes. A group of straight-chain polymers in which the main chain consists of alternating single-bonded silicon and oxygen atoms with alkyl side groups (e.g., methyl). The general formula is R_2SiO in which R represents the alkyl group. *Siloxanes* are available as liquids, greases, solid rubbers, resins and plastics. Also known as *oxosilanes*. See also polysiloxanes.

Siloxicon. American trade name for silicon carbide refractories.

Silpalon. Trademark of Mitsubishi Rayon Company (Japan) for acrylic fibers and filament yarns used for the manufacture of textile fabrics.

Silpowder. (1) Trademark of Handy & Harman (USA) for a commercially pure silver powder (99.9+%) obtained by precipitation from silver salt solutions or by atomization, and supplied in various grades with particle sizes ranging from 5 to 125 μm (0.0002 to 0.005 in.). Used for powder-metallurgy parts (e.g., electrical contacts, brushes, batteries, etc.), as pigments for conductive coatings, and for adhesives.

(2) Trademark of Technic Inc., Engineered Powders Division (USA) for precipitated amorphous silver powders used in conductive coatings, adhesives, electrical contacts, etc.

Silpruf. Trademark of GE Silicones (USA) for silicone base construction sealants.

Silres. Trademark of Wacker Silicones Corporation (USA) for liquid silicone resins used for paints and coatings.

silsesquiazanes. A class of thermosetting organosilicate polymers analogous to *silsesquioxanes*, but with NH– functional groups replacing all of the oxygens. They are used as preceramic precursor powders for the manufacture of engineering ceramics. The general formula is $(RSi(NH)_{3/2})_n$ in which R represents hydrogen, or an alkyl or aryl, e.g., methyl or vinyl. Also known as *polysilsesquiazanes*.

silsesquicarbodiimides. A class of organosilicon polymers analogous to *silsesquioxanes*, but with carbodiimide (–N=C=N–) groups replacing all of the oxygens. The general formula is $(RSi(N=C=N)_{3/2})_n$ in which R represents hydrogen, or an alkyl or aryl, e.g., methyl. Used as preceramic precursors for the preparation of engineering ceramics. Also known as *polysilsesquicarbodiimides*. See also silsesquioxanes.

silsesquioxanes. A class of thermosetting organosilicate polymers, i.e., partially condensed organically modified silicates produced from functionalized derivatives, such as $RSiX_3$ (in which X represents OR, halogen, etc.), by acid- or base-catalyzed hydrolysis. They may have random or ladder structures and upon curing exhibit good electrical properties including low dielectric constants and losses and very good thermal stability. They are supplied as thin films for microelectronic applications (e.g., low-dielectric-constant applications), and as preceramic polymer precursor powders for the manufacture of engineering ceramics. The general formula is $(RSiO_{3/2})_n$ in which R represents hydrogen, or an alkyl or aryl, e.g., methyl or phenyl. Abbreviation: SSQ. Also known as *polysilsesquioxanes*.

SilSkin. Trademark of Richard James Inc. (USA) for a medical-grade polydimethylsiloxane (PDMS) based liquid injectable silicone for medical implants and reconstructive plastic surgery applications.

SilSkin II. Trademark of Richard James Inc. (USA) for a soft silicone polymer for maxillofacial restorations.

Silsoft. Trade name of Novaceta S.p.a. (Italy) for supple cellulose acetate fibers used for the manufacture of wearing apparel.

Silsphere. Trademark of Technic Inc., Engineered Powders Division (USA) for precipitated spherical silver powders used for electronic applications.

silt. Very fine-grained particles with a particle size of 0.05-0.005 mm (0.002-0.0002 in.) composed of earth, sand, etc., and resulting from the disintegration of rock, carried by moving water and deposited as sediment.

Sil-Temp. Trademark of Ametek, Inc. (USA) for continuous leached-glass fibers containing 98+% silica. They have a density of 2.2 g/cm³ (0.08 lb/in.³), moderate mechanical properties, a melt temperature above 1760°C (3200°F), and are often used in laminate form impregnated and bonded with high-temperature resins. They are also used as high-temperature insulation for motor components and similar applications, in high-temperature and flame-resistant hose, and in the manufacture of missile and rocket parts.

Sil-Tex. Trade name of United Wire & Supply Company (USA) for a brazing alloy compsed of 60% silver, 25% copper and 15% zinc. It has a melting range of 666-720°C (1230-1330°F).

Sil-Tin. Trade name of United Wire & Supply Company (USA) for a brazing alloy of 56% silver, 22% copper, 17% zinc and 5% cadmium. It has melting range of 618-652°C (1145-1205°F).

Sil-Tite. Trade name of United Wire & Supply Company (USA) for an alloy of 45% silver, 30% copper and 25% zinc. It has melting range of 670-743°C (1240-1370°F), and is used for silver brazing metals.

Sil-Tron. Trade name of United Wire & Supply Company (USA) for an alloy composed of 80% silver, 16% copper and 4% zinc. It has a melting range of 746-804°C (1375-1480°F), and is used for silver brazing.

siltstone. A fine-grained compacted rock chiefly composed of silt particles and intermediate in texture between shale and sandstone. Used as concrete aggregate.

Silumin. Trade name of Vereinigte Deutsche Metallwerke AG (Germany) for a series of wrought and cast aluminum alloys containing 11-14% silicon. They have a density range of 2.63-2.65 g/cm³, high strength and good corrosion resistance. The casting alloys are supplied as sand, chill, die and permanent-mold castings. *Silumin-Beta* refers to sand-cast aluminum alloys containing 12% silicon, 0.5% manganese and 0.3% magnesium, and *Silumin-Gamma* to wrought and cast aluminum alloys containing 12% silicon, 0.5% manganese and 0.3% magnesium. Used for cylinders, automotive parts, instruments and housings.

Siluminite. Trade name of Franklin Fiber–Lamitex Corporation (USA) for phenolic or melamine resin impregnated or bonded asbestos papers and fabrics used for the manufacture of electrical and mechanical goods, such as gaskets, washers and commutator parts.

Silundum. Trade name for a silicon carbide made in an electric furnace by heating carbon (usually anthracite or coke) with silica (silica sand) at high temperature. It has high hardness, high electrical resistance and good resistance to oxidation to 2912°C (5274°F). Used for refractory products.

Sil-Utec. Trade name of United Wire & Supply Company (USA) for a silver solder containing 72% silver and 28% copper. Commercially available in the form of strips, wire, and powder, it has a melting point of 779°C (1435°F).

Silux. Trademark of 3M Dental (USA) for dental restoratives and enamel bonding materials.

Silux Enamel Bond. Trademark of 3M Dental (USA) for dental adhesives used in restorative dentistry for enamel bonding applications.

Silux Plus. Trademark of 3M Dental (USA) for a light-cure methacrylate-based composite 40 vol% filled with silica particles (0.04 μm or 1.5 μin.). It is non-sticky and nonslumping, has low to moderate mechanical properties, good luster retention, and can take a high polish. Supplied in a wide range of shades and opaque (dentin) and translucent (enamel) opacities, it is used in dentistry for anterior restorations.

Silva Bronze. Trade name of Imperial Metal Industries (UK) for a wrought aluminum bronze used for fire back boilers.

Silvacel. Trademark of Weyerhaeuser Company (USA) for wood fibers used as thermal and acoustical insulation, in the manufacture of special papers and boards, as filter media, etc.

Silvacon. Trademark of Weyerhaeuser Company (USA) for a series of wood products made from the bark of Douglas fir *(Pseudotsuga menziesii)*. Supplied in the form of granules, powder, fibers and finished products, they are used for acoustical and floor tile, as additives in foundry sands, in dusting powders and paints, as fillers for plastics, in asphalt and fibrous paints, as extenders for phenolic adhesives, in the manufacture of sponge rubber, and as substitutes for cork.

Silvaise. Trade name for a silicon alloy containing 10% titanium, 10% vanadium, 6% aluminum, 6% zirconium and 0.5% boron.

Silvaloy. Trade name of Engelhard Corporation (USA) for an extensive series of silver-base brazing filler metals containing varying amounts of silver, copper and/or zinc, cadmium, nickel, tin and manganese. They are suitable for joining most ferrous and nonferrous alloys.

Silvanite. Trade name of Columbia Tool Steel Company (USA) for a special-purpose high-speed tool steel (AISI type F8) containing 1.3% carbon, 8% tungsten, 4% chromium, 0.5% vanadium, and the balance iron. Used for paper knives, scrapers and woodworking tools.

Silvan Star. Trade name of Teledyne Firth Sterling (USA) for a carbon tool steel containing about 1% carbon, 0.25% manganese, 0.25% silicon, 0.20% vanadium, and the balance iron. Used for dies and tools.

Silvar. Trademark of Texas Instruments Inc. (USA) for an advanced composite made by infiltration of molten silver into a porous nickel-iron alloy preform. The resulting powder-metallurgy matrix has numerous interconnected paths of silver. *Silvar* composites have high thermal conductivity and good dimensional stability, good heat-sink capabilities, high durability, excellent machinability (mirror finishes possible), good processibility by lapping, threading, drilling, etc., and good workability by rolling, coining, stamping and forging. Used for electronic applications, e.g., power hybrid circuits, laser diode mounts, operational amplifiers and leadframes.

SilvaTech. Trademark of Lucent Technologies (USA) for silver and silver-alloy electrodeposits and plating processes.

Silvaz. Trade name of Union Carbide Corporation (USA) for a series of alloys containing 35-50% silicon, 10% titanium, 6-10% vanadium, 6-7% aluminum, 6-7% zirconium, 0.5-0.6% boron, and the balance iron. Used in steelmaking as deoxidizers and grain size controllers, in hardening of steel, and for the introduction of boron into steel.

Silvel. Trade name for a nickel silver containing 67.5% copper, 16% zinc, 6.5% nickel and 2.2% iron. Used for electrical contact springs, tableware and plumbing fixtures.

Silvelmet. Trade name of Metro Cutanit Limited (UK) for a series of sintered materials consisting of varying amounts of silver and tungsten, or silver and tungsten carbide.

silver. A white, lustrous, ductile, precious metallic element of Group IB (Group 11) of the Periodic Table. It is commercially available in the form of rods, ingots, bullions, foils, microfoils, microleaves, plates, sheets, moss, lumps, shot, powder, tubing, wire, needles, whiskers and single crystals. The single crystals are usually grown by the Bridgeman technique. The most important silver ores are native silver, *argentite* and *cerargyrite*.

Crystal system, cubic; crystal structure, face-centered cubic; density, 10.5 g/cm³; melting point, 961.9°C (1763°F); boiling point, 2212°C (4014°F); hardness, 25-95 Vickers; atomic number, 47; atomic weight, 107.868; monovalent, divalent. It has good cold workability, excellent electrical conductivity (up to 106% IACS), excellent thermal conductivity, high light reflectivity, good resistance to alkalies and many chemicals, and poor resistance to nitric acid, hot sulfuric acid and alkali cyanide solutions. Used for electrical contacts, electrical and electronic devices, magnet windings, corrosion-resistant equipment, chemical reaction equipment, silver plating, mirror coatings, solar-tower reflectors, jewelry, coinage, cutlery, dishes, special batteries, solar cells, brazing alloys, bearing metals, photographic film paper and chemicals, and photosensitive glass. It also used in precipitated, powdered, fluxed or paste form as a decoration for pottery, glass and porcelain-enameled ware, for dental, medical and scientific equipment, and in dental alloys and amalgams. Symbol: Ag.

silver acetylacetonate. A light- and moisture-sensitive metal alkoxide (98%) that has a melting point of 138-143°C (280-289°F), but decomposes slowly above 70°C (158°F). Used as a component in low-cost glass-frit extended silver conductive pastes. Formula: $Ag(O_2C_5H_7)$.

silver alloys. A group of alloys in which silver is the principal constituent. Examples include silver-copper alloys for jewelry, silver-cadmium, silver-copper, silver-copper-nickel and silver-palladium alloys used as electrical contact materials, and silver-lead alloys used for bearings.

silver aluminum selenide. A compound of silver, aluminum and selenium. Crystal system, tetragonal. Crystal structure, chalcopyrite. Density, 5.07 g/cm³; melting point, 947°C (1737°F). Used as a semiconductor. Formula: $AgAlSe_2$.

silver aluminum sulfide. A compound of silver, aluminum and sulfur. Crystal system, tetragonal. Crystal structure, chalcopyrite. Density, 3.94 g/cm³. Used as a semiconductor. Formula: $AgAlS_2$.

silver aluminum telluride. A compound of silver, aluminum and tellurium. Crystal system, tetragonal. Crystal structure, chalcopyrite. Density, 5.07 g/cm³; melting point, 727°C (1341°F). Used as a semiconductor. Formula: $AgAlTe_2$.

silver amalgam. A solid solution of mercury and silver found in nature, e.g., in the mineral *arquerite*, containing between 26 and 96% silver, and also made synthetically.

silver antimony. A silver-white crystalline compound of silver and antimony. It occurs in nature as the mineral *dyscrasite*. Crystal system, rhombohedral. Density, 9.74 g/cm³; hardness, 3.5-4.0 Mohs. Used as a semiconductor. Formula: Ag_3Sb.

silver antimony selenide. A compound of silver, antimony and selenium. Crystal system, rhombohedral. Density, 6.6 g/cm³; melting point, 637°C (1179°F). Used as a semiconductor. Formula: $AgSbSe_2$.

silver antimony telluride. A compound of silver, antimony and tellurium. Crystal system, rhombohedral. Density, 7.12 g/cm³; melting point, 557°C (1035°F); very low thermal conductivity. Used as a p-type semiconductor, and in thermoelectric applications. Formula: $AgSbTe_2$.

Silver Babbitt. Trade name of Magnolia Anti-Friction Metal Company (UK) for tin-base bearing metals containing varying amounts of silver, cadmium and nickel. Used for heavy-duty applications.

silver-base powder-metallurgy materials. A group of materials made by blending silver powder with selected metallic and/or nonmetallic powders, compacting (pressing) at ambient or elevated temperatures and sintering the resulting shapes at specified temperatures. Some of the suitable powders for blending include graphite, tungsten, tungsten carbide, cadmium oxide, nickel and molybdenum. They have excellent electrical and thermal conductivity, and outstanding oxidation resistance. Used for electrical contacts, and electrical and electronic components.

silver-bearing copper. An oxygen-free, high-purity copper (99.95%) containing 8-25 oz/ton (227-851 g/metric ton) of silver. It possesses good corrosion resistance, excellent properties at elevated temperatures, high electrical conductivity (100.5% IACS), good hot and cold workability, and is used for electrical and electronic components, conductivity wires, commutators, high-speed motors, radiators, gaskets, busbars, printing rolls, printed-circuit foil, and chemical process equipment.

silver-bearing fire-refined tough-pitch copper. A fire-refined tough-pitch copper (99.88%) containing up to about 25 oz/ton (78 g/metric ton) of silver. It has high ductility, good hot and cold workability, good corrosion resistance, excellent electrical conductivity, and is used for electrical conductors, busbar, electronic components, waveguides, klystrons, etc.

silver-bearing solders. A group of tin, tin-lead and lead solders containing 2-5% silver. Silver has a considerable hardening effect on tin and tin-lead alloys. *Binary tin-silver alloys* with an eutectic point of 221°C (430°F) at 3.5% silver, have good wetting characteristics and high joint strengths. *Binary lead-silver alloys* have good joint strength and high melting points (e.g., the 94.5Pb-5.5Ag alloy melts at 304°C or 579°F), but have fair to poor flow characteristics, wettability and corrosion resistance. The addition of up to 5% tin to lead-silver solders greatly improves the flow and wetting characteristics and increases the corrosion resistance. See also lead-silver solders; tin-lead solders; tin-silver solders.

silver-bearing tough-pitch copper. Tough-pitch copper (99.90%) with 0.04% oxygen and 8-25 oz/ton (227-851 g/metric ton) of silver. See also silver-bearing copper; tough-pitch copper.

silver beech. The strong, light-brown wood from the tall tree *Nothofagus menziesii*. It has a straight grain, and an average weight of 545 kg/m³ (34 lb/ft³). Source: New Zealand. Used for furniture, barrels, casks, implements, etc.

silver bell metal. A white metal alloy composed of 60% tin and 40% copper, used for bells.

silver birch. (1) The tough, white to yellowish-white wood of the hardwood tree *Betula pendula* with white to silvery white bark. It has a fairly straight grain, a fine texture, and an average weight of 600 kg/m³ (37 lb/ft³). Source: Europe including Scandinavia, Eastern Europe and British Isles. Used for turnery, furniture, house construction, wagons, and for the manufacture of plywood. Also known as *white birch*.

(2) See yellow birch.

silver bismuth selenide. A compound of silver, bismuth and selenium. Crystal system, rhombohedral. Used as a semiconductor. Formula: $AgBiSe_2$.

silver bismuth sulfide. A compound of silver, bismuth and sulfur. Crystal system, rhombohedral. Used as a semiconductor. Formula: $AgBiS_2$.

silver bismuth telluride. A compound of silver, bismuth and tellurium. Crystal system, rhombohedral. Used as a semiconductor. Formula: $AgBiTe_2$.

Silverbond. Trade name of Jessop Steel Company (USA) for a stainless clad steel used for structural applications.

Silver Bond B. Trade name of Unimin Specialty Minerals Inc. (USA) for ground crystalline and microcrystalline silica used as paint extender.

silver brazing alloys. A group of hard solders that melt at much higher temperatures (610-900°C or 1130-1650°F) than soft solders, and produce much stronger joints. The typical composition range is 20-92.5% silver, 0-40% copper, 0-35% zinc, 0-4.5% nickel, 0-24% cadmium, 0-15% manganese and 0-10% tin. Commercially available in the form of strips, wire, powder, clad sheets or strips, and rods. They are often sold under trademarks or trade names, such as *Easy-Flow, Sil-Fil, Sil-Fos* or *Sil-Utec*, and used for joining most ferrous and nonferrous metals, except magnesium and aluminum, and also for making ceramic-to-metal seals. Also known as *silver solders*.

silver bromide. Pale yellow, light-sensitive crystals or powder (99+% pure). It occurs in nature as the mineral *bromargyrite*. Crystal system, cubic. Density, 6.473 g/cm³; melting point, 432°C (810°F); boiling point, decomposes at 700°C (1292°F); darkens on exposure to light. Used as a semiconductor, for photographic film and plates, and in photochromic glass. Formula: AgBr.

silver bronze. A silvery-white corrosion-resistant bronze containing 54.7-70.7% copper, 13-23% zinc, 16-18% nickel, 0-2% aluminum, 0-2% lead and 0.3% silicon. Used for hardware, bearings and bushings.

silver-cadmium alloys. A group of alloys in which silver and cadmium are the principal constituents. Common examples include alloys containing silver, cadmium, copper and zinc used for brazing ferrous and nonferrous metals, and alloys of silver and cadmium, sometimes with nickel, and/or copper. Used as electrical contact materials providing improved arc-quenching characteristics.

silver carbonate. Yellow to yellowish-gray, light-sensitive crystals or powder (98+% pure). Density, 6.077 g/cm³; melting point, 218°C (424°F). Used in the production of iridescent stains or sheens on glazes and as an ingredient in some glass stains. Formula: Ag_2CO_3.

silver cedar. See Atlas cedar.

silver chloride. White, granular, light-sensitive crystals or powder (99+% pure). It occurs in nature as the mineral *cerargyrite* (horn silver). Crystal system, cubic. Density, 5.56 g/cm³; melting point, 455°C (851°F); boiling point, 1550°C (2822°F); refractive index, 2.071; darkens on exposure to light; crystals transmit more than 80% of the wavelengths from 50 to 200 μm. Used in ceramics for yellow glazes, purple of Cassius and silver lusters, in photochromic glass, in optics, photography and photometry, in silver plating, in the preparation of pure silver, and in batteries. Single crystals are used for infrared absorption cells and lens elements. Formula: AgCl.

silver chromate. A dark brownish-red crystalline powder (99+% pure). Crystal system, monoclinic. Density, 5.625 g/cm³. Used as a laboratory reagent. Formula: Ag_2CrO_4.

silver-clad powder. A copper powder coated with a thin layer of silver, and used in the manufacture of electrical contacts.

silver-clad sheet. A sheet of ferrous or nonferrous alloy that has a thin, reflective layer of pure silver (99.9+%) applied by rolling. For general properties and applications, see silver-clad steel.

silver-clad steel. Steel or stainless steel that has a thin, electrically conductive layer of silver applied to one side by rolling. It is highly resistant to food products, hot concentrated solutions of many organic acids, and hot caustic solutions. Used for chemical-process equipment, food-processing equipment, water distillation and purification systems, medical and scientific equipment, machinery bearings, shims, and optical components, mirrors and reflectors.

silver coatings. Coatings of metallic silver (typically 2.5-10 μm or 0.1-0.4 mil thick) produced on other metals or metal objects especially by electroplating. The coatings may serve to enhance the corrosion resistance of chemical equipment and storage containers of iron or steel, improve the appearance of stainless steel or pewter for tableware and flatware, or enhance the reflectivity of optical systems or the electrical and thermal conductivity of materials for electrical contacts. See also silver plate.

silver copper. A copper (99.75-99.8%) containing 0-0.11% magnesium, 0-0.06% phosphorus and 0-0.034% silver. It has good corrosion resistance, good hot and cold workability, good conductivity, and good strength and softening resistance. Used for electronic components, electrical contacts and connectors, springs, diaphragms, fittings, clamps, and resistance welding electrodes.

silver-copper alloys. Alloys composed of silver and copper, with silver being the principal constituent. Common examples include alloys containing silver, copper, cadmium and zinc used for brazing ferrous and nonferrous metals, alloys of silver and copper, sometimes with cadmium and/or nickel, used as electrical contact materials combining good electrical properties with a higher hardness than unalloyed silver, and alloys containing about 92.5% silver and 7.5% copper, used for tableware, jewelry, and electric contacts.

silver-copper brazing alloy. An eutectic brazing alloy composed of 72% silver and 28% copper, and having a melting temperature of 780°C (1435°F). Commercially available in the the form of strips, powder and wire, it has a brazing range of 780-900°C (1435-1650°F). Used for brazing most ferrous and nonferrous metals, except aluminum and magnesium.

silver-copper-tin brazing alloy. A brazing alloy composed of 60% silver, 30% copper and 10% tin and having a solidus temperature of 602°C (1115°F) and a liquidus temperature of 718°C (1325°F). Commercially available in the form of strips and wire, it has a brazing range of 718-843°C (1325-1550°F), and good ductility. Used for brazing most ferrous and nonferrous metals, except aluminum and magnesium.

Silvercote. Trade name of Little Falls Alloy Inc. (USA) for age-hardenable beryllium copper (Cu-0.5Be-0.25Co) with a light coating of silver. It has good electrical conductivity (65-70% IACS), high strength and fatigue resistance, and good corrosion resistance. Used for electrical conducting wires, lead wires, connectors, springs, and pins.

silver cyanide. White, light-sensitive crystals or powder (99+% pure). Crystal system, hexagonal. Density, 3.95 g/cm³; melting point, decomposes at 320°C (608°F); refractive index, 1.685; darkens on exposure to light. Used in silver plating. Formula: AgCN.

silver difluoride. A hygroscopic, white or gray crystals or hydroscopic, light-sensitive, brown powder (98+% pure). Density, 4.580 g/cm³; melting point, 690°C (1274°F); boiling point, 700°C (1292°F). Used as a laboratory reagent. Formula: AgF$_2$.

silver-exchanged zeolites. A class of *zeolites* containing various amounts of silver and supplied in various compositions including (i) granules (+20 mesh) corresponding to the formula Ag$_{34}$-Na$_2$[(AlO$_2$)$_{86}$(SiO$_2$)$_{106}$]·xH$_2$O, containing about 35% silver, having 0.74 nm pore openings, and providing excellent adsorption up to about 500°C (930°F) and an equilibrium water capacity of 20 wt%; and (ii) moisture-sensitive pellets (1.6 mm) with

mordenite crystal structure corresponding to the formula Ag$_{7.6}$-Na$_{0.4}$[(AlO$_2$)$_8$(SiO$_2$)$_{40}$]·xH$_2$O, containing about 35-40% silver, having a density of 1.07 g/cm³, a melting point above 500°C (930°F), and 0.74 nm pore openings, and providing excellent adsorption up to about 500°C (930°F), an equilibrium water capacity of 7 wt%, and self-binding properties with high acid stability. The granules and pellets are used for the selective removal of halogens from gas streams, and the granules are also used as catalysts and as trace hydrogen scavengers.

silver fir. (1) The tough, white wood of the fir tree *Abies alba*. It often combined with European spruce *(Picea abies)* and sold as whitewood. It is similar to *European spruce*, and has high elasticity, moderate durability, and an average weight of 480 kg/m³ (30 lb/ft³). Source: Europe including British Isles. Used for building construction, flooring and general joinery work, and as pulpwood. Also known as *European silver fir.*

(2) The creamy white to pale brown, lightweight wood from the softwood tree *Abies amabilis*. It is typically straight-grained, and has slightly higher strength than the firs in eastern North America. Source: Pacific Coast of United States (Washington, Oregon, California). Used for building construction, millwork, sash, doors, interior finish, boxes, crates, and veneer. Also known as *Pacific silver fir.*

silver flake. Silver (99.9+%) in the form of small platelets having a particle size range of about 3-20 μm (118-787 μin.). Used in the production of conductive and reflective coatings.

Silver-Flo. Trade name of Johnson Matthey plc (UK) for a series of cadmium-free silver brazing alloys containing 12-56% silver, 21-55% copper, 17-40% zinc and 0-5% tin. They have a melting range of 618-830°C (1144-1526°F), and are used for general brazing of steel, malleable iron, copper, nickel and their alloys.

silver fluoride. Hygroscopic, light-sensitive, water-soluble, yellow-brown crystals or crystalline mass, or hygroscopic, yellow powder (98+% pure). Crystal system, cubic. Density, 5.852 g/cm³; melting point, 435°C (415°F); boiling point, 1159°C (2118°F). Used in materials science, biochemistry and medicine, and as a replacement of fluorine for bromine and chlorine in organic synthesis. Formula: AgF.

silver foil. (1) A thin sheet of silver available in several grades: (i) *Microfoil* (99.95+% silver) supplied in thicknesses and weights ranging from 0.001 to 1.0 μm (0.04 to 40 μin.) and 1.1 to 1098 μg/cm², respectively; (ii) *Microleaves* (99.95+% silver) available on removable support in thicknesses and weights ranging from 0.25 to 1.0 μm (10 to 40 μin.) and 262.3 to 1098 μg/cm² respectively; and (iii) *Standard foils* (99.9+% silver) supplied in thicknesses from 1.0 μm to 6.0 mm (40 μin. to 0.24 in.) and various shapes and sizes (e.g., disks, squares, rounds, etc.). Used for electrical and optical applications.

(2) A term used for any thin sheet of silver-colored metal, e.g., aluminum foil, tin foil, etc. used for wrapping, insulation, etc.

Silver Fox. Trade name of British Steel Corporation (UK) for a series of corrosion- and/or heat-resistant austenitic, ferritic and martensitic stainless steels including several free-cutting grades.

silver gallium selenide. A compound of silver, gallium and selenium. Crystal system, tetragonal. Crystal structure, chalcopyrite. Density, 5.84 g/cm³; hardness, 4400 Knoop; melting point, 847°C (1557°F); low optical absorption, wavefront distortion and scattering. Used as a semiconductor, for infrared wave plates, differential absorption LIDAR applications, nonlinear optics, and in surgery. Formula: AgGaSe$_2$.

silver gallium sulfide. A compound of silver, gallium and sulfur. Crystal system, tetragonal. Crystal structure, chalcopyrite. Density, 4.72 g/cm³. Used as a semiconductor and in nonlinear optics. Formula: $AgGaS_2$. Also known as *silver thiogallate*.

silver gallium telluride. A compound of silver, gallium and tellurium. Crystal system, tetragonal. Crystal structure, chalcopyrite. Density, 6.05 g/cm³; hardness, 1800 Knoop; melting point, 717°C (1323°F). Used as a semiconductor. Formula: $AgGaSe_2$.

silver germanium phosphide. A compound of silver, germanium and phosphorus. Crystal system, tetragonal. Hardness, 6150 Knoop; melting point, 742°C (1368°F). Used as a semiconductor. Formula: $AgGe_2P_3$.

silver germanium selenide. A compound of silver, germanium and selenium. Crystal system, tetragonal. Used as a semiconductor. Formula: Ag_2GeSe_3.

silver germanium telluride. A compound of silver, germanium and tellurium. Crystal system, tetragonal. Used as a semiconductor. Formula: Ag_2GeTe_3.

silver glance. See argentite.

Silvergleam. Trade name of Zinex Corporation (USA) for a bright silver electrodeposit and noncyanide plating process.

silver-graphite. A powder-metallurgy material made by blending silver powder with graphite, compacting at ambient or elevated temperatures and sintering the resulting shapes. A typical composition is 1-15% graphite, with the balance being silver. Used for electrical contacts especially sliding and brush contacts.

silver halide. A binary compound of silver and a halogen, such as bromine, chlorine, fluorine or iodine.

silver indium selenide. A compound of silver, indium and selenium. Crystal system, tetragonal. Crystal structure, chalcopyrite. Density, 5.81 g/cm³; hardness, 1850 Knoop; melting point, 780°C (1436°F). Used as a semiconductor. Formula: $AgInSe_2$.

silver indium sulfide. A compound of silver, indium and sulfur. Crystal system, tetragonal. Crystal structure, chalcopyrite. Density, 5.0 g/cm³; hardness, 2250 Knoop. Used as a semiconductor. Formula: $AgInS_2$.

silver indium telluride. A compound of silver, indium and tellurium. Crystal system, tetragonal. Crystal structure, chalcopyrite. Density, 6.12 g/cm³; melting point, 602°C (1278°F). Used as semiconductor. Formula: $AgIn_2Te_2$.

Silverine. Trade name of Stone Manganese–J. Stone & Company Limited (UK) for a corrosion-resistant alloy composed of 71-80% copper, 16-17% nickel, 1-8% zinc, 1-2.8% tin, 1-2% cobalt and 1-1.5% iron. Used for ornamental and sanitary fittings.

silvering. (1) A coating of metallic silver produced on another metal or metal object especially by electroplating. See also silver coatings.

(2) A thin film of a metal, such as silver, deposited on a glass or other substrate by chemical deposition, burning-on, sputtering or evaporation.

(3) A silver-base reflection coating with an amalgam containing tin, lead and bismuth.

silver iodide. Pale yellow, light-sensitive powder (99+% pure). It occurs in nature as the mineral *iodargyrite*. Crystal system, hexagonal. Crystal structure, wurtzite (zincite). Density, 5.67-5.68 g/cm³; melting point, 558°C (1036°F); boiling point, 1506°C (2743°F); hardness, 1.5-2.5; refractive index, 2.22; darkens on exposure to light. Used in photography, for cloud seeding, and as a semiconductor. Formula: AgI.

silver iron sulfide. A compound of silver, iron, and sulfur. Crystal system, tetragonal. Crystal structure, chalcopyrite. Density, 4.53 g/cm³. Used as a semiconductor. Formula: $AgFeS_2$.

Silverite. British trade name for a nickel silver containing varying amounts of copper and nickel. It has excellent corrosion resistance, and good strength. Used for condensers and heat exchangers, evaporators, pump parts, and hardware.

Silverjet. Trade name of Shipley Ronal (USA) for silver finishes used for semiconductor applications.

Silver Knight. Trademark of U.S.E. Hickson Products Limited (Canada) for a reflective aluminum roof coating used for sealing cracks, nail holes and other discontinuities.

Silver Label. (1) Trade name of Wallace Murray Corporation (USA) for a nondeforming tool steel containing 0.9% carbon, 1.1% manganese, 0.5% chromium, 0.5% tungsten, and the balance iron. Used for hobs, taps, reamers and dies.

(2) Trade name of Peninsular Steel Company (USA) for an oil- or water-hardening shock-resisting tool steel (AISI type S5) containing 0.6% carbon, 1.85% silicon, 0.8% manganese, 0.45% molybdenum, 0.2% vanadium, and the balance iron. It has a high elastic limit, good ductility, and high toughness. Used for punching and shearing dies.

silver leaf. (1) Silver beaten into very thin sheets or leaves and used mostly for lettering on glass.

(2) Tinfoil (Sn-8.25Zn-0.75Pb) resembling silver leaf in appearance, and used for wrapping purposes.

Silver-Lite. Trademark of Soft-Lite LLC (USA) for a multilayered low-emissivity coating for application to insulated glass by a vacuum-coating process. The coating consists of a thin layer of titanium dioxide (TiO_2) directly applied to the glass, intermediate layers of nickel-chromium, silver and again nickel-chromium, and a top layer of silicon nitride (Si_3N_4). The double- and triple-glass windows containing this coating are also sold under the trademark "Silver-Lite". See also low-E coatings; low-E glass.

Silver-Lume. Trademark of Atotech USA Inc. for bright silver electrodeposits and plating processes. The electrodeposits are produced from plating solutions composed of silver cyanide ($AgCN$), potassium cyanide (KCN), potassium carbonate (K_2CO_3), and selected addition agents. Used on electronic components, tableware, flatware, jewelry, watch cases, etc.

silver manganate. See silver permanganate.

silver maple. The wood of the maple tree *Acer saccharinum*. The heartwood is light reddish-brown and the sapwood white with slight reddish-brown tinge. *Silver maple* is softer and less heavy than *sugar maple*. Average weight, 529 kg/m³ (33 lb/ft³). Source: Southern, central and eastern Canada; northeastern and central United States. Used for furniture, crates, pallets, boxes and woodenware. Also known as *river maple; swamp maple; white maple*.

silver-matrix composite. An electrically conductive material comprising a silver or silver-alloy matrix with continuous nickel fibers.

silver-mercury iodide. See mercuric silver iodide.

silver metal. (1) A British alloy composed of 2 parts of zinc and 1 part of silver. Used for ornaments.

(2) A British corrosion-resistant nickel silver composed of 52-54% copper, 7-8.5% zinc, 31-32% nickel, 3-3.6% lead, 1-1.4% tin, 0.8% iron, and 0.3% manganese.

silver-molybdenum alloys. A group of alloys of silver containing up to 50% molybdenum and made by powder metallurgy techniques, usually by pressing and sintering molybdenum powder into a compact and subsequent infiltration with silver, or by pressing the alloyed powder to shape and then sintering. The addition of the refractory metal molybdenum increases the hardness and wear resistance of the silver. Used as electrical contact materials for switchgear contacts, circuit breakers, etc.

silver-nickel alloys. Alloys of silver containing up to 50% nickel and made by powder metallurgy techniques usually involving pressing, sintering and repressing, and sometimes rolling. Used for electrical contacts.

silver nitrate. A white, orthorhombic crystalline salt (99+% pure) obtained by treating silver with dilute nitric acid. It turns gray on exposure to light. Density, 4.352 g/cm³; melting point, 212°C (414°F); boiling point, decomposes at 444°C (831°F); strong oxidizer and corrosive. Used in glass manufacture to introduce silver into glass, as a yellow colorant in glazes, as a silvering compound for mirrors, for photographic film, in silver plating and ink manufacture, and in biological and biochemical electrophoresis and electron microscopy. Formula: $AgNO_3$. Also known as *caustic silver.*

silver nitrite. Yellow or grayish-yellow needles or powder (99% pure). Crystal system, orthorhombic. Density, 4.453 g/cm³; melting point, 140°C (284°F). Used in the preparation of aliphatic nitrogen compounds, as a reagent for alcohols, in standard solutions for water analysis, in biochemistry, and in analytical chemistry. Formula: $AgNO_2$.

Silveroid. Trade name of Henry Wiggin & Company Limited (UK) for a corrosion-resistant alloy composed of 54% copper, 45% nickel and 1% manganese. Used for cutlery.

silver orthophosphate. See silver phosphate.

silver oxide. Brown-black crystals or dark brown to black, light-sensitive powder (99+% pure). Crystal system, cubic. Density, 7.143 g/cm³; melting point, decomposes above 300°C (570°F). Used as a yellow colorant in glass and glazes, in glass stains, as a glass polishing material, as a strong oxidizing agent, and as a catalyst in water purification. Formula: Ag_2O. Also known as *argentous oxide.*

Silver Pal-Bond. Trade name for a palladium-silver dental bonding alloy.

silver-palladium alloys. A group of alloys in which silver and palladium are the predominant constituents. They are used for electrical contacts and in high-temperature brazing. A 90Ag-10Pd contact alloy has a solidus temperature of 999°C (1830°F), an electrical conductivity of 27% IACS, and a density of 10.57 g/cm³ (0.382 lb/in.³). The palladium increases the hardness, but reduces the conductivity of the silver. *Silver-palladium alloys* have excellent flow and wetting characteristics, and are used for brazing various metals, for ceramic-to-metal joints, and for electronic and aerospace applications.

silver paper. Paper coated on one side with a layer of metal resembling silver. Used for the decoration of greeting cards, for wrapping chocolate bars, etc.

silver permanganate. Violet, light-sensitive, crystalline powder. Crystal system, monoclinic. Density, 4.27 g/cm³; melting point, decomposes. Used in gas masks. High-purity grades are used in electronics and chemistry. Formula: $AgMnO_4$. Also known as *silver manganate.*

silver peroxide. A grayish or black powder (99.9% pure). Crystal system, monoclinic or cubic. Density, 7.44-7.50 g/cm³; melting point, decomposes above 100°C (212°F). Used in the manufacture of silver-zinc batteries, as a surface catalyst in the epoxidation of alkenes, and in the direct conversion of benzyl halides into benzyl ethers. Formula: AgO or Ag_2O_2. Also known as *argentic oxide.*

silver phosphate. A yellow, light-sensitive powder (99+% pure). Density, 6.37 g/cm³; melting point, 849°C (1560°F). It turns brown on exposure to light or heat, and is used as a catalyst, in photographic emulsions and in pharmaceuticals. Formula: Ag_3PO_4. Also known as *silver orthophosphate.*

silver plate. See silver coatings; silvering (1).

Silver-Ply. Trade name for a composite plate or sheet consisting of a relatively soft low-carbon steel that has a hard stainless steel layer bonded to one or both sides by hot rolling. The layer may be 10-20% of the total thickness of the composite.

silver-potassium cyanide. White light-sensitive crystals or powder (99.9+% pure) obtained by treating a solution of potassium cyanide with silver chloride. Density, 2.36 g/cm³ (25°C/77°F). Used in silver-plating solutions. Formula: $KAg(CN)_2$. Also known as *potassium argentocyanide; potassium cyanoargentate; potassium-silver cyanide; potassium dicyanoargentate.*

silver powder. Silver in the form of a crystalline or amorphous powder with a purity of 99.9 to 99.999%, obtained by chemical reduction, atomization or electrolysis. Particle sizes range from as low as 0.07 μm (2.8 μin.) to over 250 μm (0.01 in.), and particle shapes from dendritic to spherical. Amorphous powders are usually chemically reduced. Used in electroplating, and for electrical and electronic applications, e.g., integrated circuits, and electrical contacts.

silver protein. A colloidal, light-sensitive preparation of either 7.5-8.5% silver plus protein (known as "strong silver protein") or 19-23% silver plus protein (known as "mild silver protein"). It is usually made by reacting a silver compound with gelatin in the presence of an alkali, and used in aqueous solutions as an antibacterial, e.g., on mucous membranes, and in biological and biochemical microscopy.

silver selenide. Gray needles, or moisture-sensitive crystalline lumps. It occurs in nature as the mineral *naumannite.* Crystal system, orthorhombic. Density, 8.0-8.2 g/cm³; melting point, approximately 880°C (1616°F). It is also available in high-purity grades (99.999%). Used in materials research and electronics. Formula: Ag_2Se.

silver-sensitized photocopying paper. Paper coated with a silver halide, e.g., silver bromide (AgBr) or silver chloride (AgCl). A latent image is produced on this paper by electromagnetic and especially visible radiation which can then be made visible and permanent (durable) by developing and fixing, or electrolysis.

silver solders. See silver-brazing alloys.

silver spruce. See Sitka spruce.

silver steel. A bright-drawn high-carbon steel containing about 1-1.3% carbon, 0-0.3% silicon, 0.4-0.5% manganese, 0.04% sulfur, 0.04% phosphorus, 0.3-0.5% chromium (optional), and the balance iron. Commonly supplied centerless-ground in the as-rolled condition for the manufacture of dowel and location pins. Abbreviation: SS.

SilverStone. Trademark of E.I. DuPont de Nemours & Company (USA) for fluoropolymer coatings used on cookware.

Silver Stripe High Speed. Trade name of Great Western Steel Company (USA) for a tungsten-type high-speed tool steel containing 0.7% carbon, 18% tungsten, 4% chromium, 1% vanadium, and the balance iron. Used for lathe and planer tools, milling cutters, and boring and chasing tools.

silver sulfide. Lead-gray to black or grayish-black crystals, or grayish or black, light-sensitive powder (99+% pure). It occurs in nature as the minerals *argentite* and *acanthite.* Crystal system, monoclinic (argentite); orthorhombic (acanthite). Density,

7.32 g/cm³; melting point, 825°C (1517°F); boiling point, decomposes; hardness, 2-2.5 Mohs. Used for inlaying in niello metalwork, and in ceramics and electronics. Formula: Ag_2S.

silver telluride. Black crystals (99+% pure). It occurs in nature as the mineral *hessite*. Crystal system, monoclinic. Density, 8.4-8.5 g/cm³; melting point, 955°C (1751°F); hardness, 2.5 Mohs. Formula: Ag_2Te.

Silvertem. Trade name of Heppenstall Company (USA) for a prehardened hot-work tool steel containing 0.35% carbon, 4.75% chromium, 1.25% tungsten, 1.5% molybdenum, 1% silicon, 0.2% silver, and the balance iron. Used for hot-work dies.

Silvertex. Trade name of Colcombet Fois & Cie. SA (France) for glass-fiber cloth and tape used for fiber-reinforced plastics.

silver thiogallate. See silver gallium sulfide.

silver tin selenide. A compound of silver, tin and selenium. Crystal system, tetragonal. Used as a semiconductor. Formula: $AgSnSe_3$.

silver tin telluride. A compound of silver, tin and tellurium. Crystal system, tetragonal. Used as a semiconductor. Formula: Ag_2SnTe_3.

silver tissue paper. A wood-free tissue paper from which all silver-attacking substances have been removed, and which may or may not have chemicals added to protect the silver. It is machine-finished, and has a basis weight of 18 g/m² (0.003 oz./ft.²).

silver tungstate. A white powder (99% pure) made from tungstic acid and silver oxide, and used for electronic and chemical applications. Formula: Ag_2WO_4.

silver-tungsten alloys. A group of silver alloys containing up to 50% tungsten and made by powder-metallurgy techniques either by pressing and sintering tungsten powder into a compact and subsequent infiltration with silver, or by pressing alloyed tungsten powder to shape and sintering. The addition of tungsten increases the hardness and resistance to arcing and wear of the silver. Used for electrical contacts of switching devices.

silvery iron. A high-silicon *pig iron* with a light-gray color, a fine grain and a silvery fracture. It contains about 0.25-1.5% carbon, 15-20% silicon, 1% manganese, up to 0.15% phosphorus, and the balance iron. Used as a ferrosilicon alloy for cast iron and some steels and, in pulverized form, in the beneficiation of certain ores and minerals. Also known as *silvery pig iron*.

silver-zinc alloys. A group of alloys composed of varying amounts of silver and zinc. Alloys of about 75% silver and 25% zinc are commonly used as contact alloys for telephone jacks, and alloys of silver, zinc, copper and sometimes nickel are used as brazing filler metals for ferrous and nonferrous metals and alloys.

Silvit. Trade name of Saint-Gobain (France) for a patterned glass imitating tree bark.

Silvium. Trade name of ESB Inc. (USA) for a corrosion-resistant alloy of lead and antimony used for battery grids.

Silvrex. Trade name of Enthone-OMI Inc. (USA) for bright silver electrodeposits and plating processes.

Silvung. Trade name of Plansee Metallwerk Gesellschaft (Austria) of powder-metallurgy materials composed of tungsten impregnated with silver, and used in electrical contacts for interrupters and voltage regulators, and for facing contacts.

Sil-X. Trademark of Williams Advanced Materials (US) for a silver-based alloy with an engineered microstructure consisting of small, uniform grains. It has excellent optical reflectivity and sunlight resistance, high aging resistance, and stable jitter. Used for the reflective (L1) and semireflective (L0) layers in DVD disks.

Simaf. Trade name of Pechiney Electrométallurgie (France) for a cast iron nodulizer containing 45-50% silicon, 5-6.5% magnesium, 0.4-0.6% calcium, 0-0.4% lanthanum, 0.8-1.3% aluminum, and the balance iron.

Simagal. Trade name of Kreidler Werke GmbH (Germany) for aluminum alloy containing 0.6-1.4% magnesium, 0.6-1.6% silicon, 0.4-1% manganese and 0-0.3% chromium. It has good formability and weldability, and is used structural parts, window frames and profiles, fan blades and gutters and boats.

Simalec. Trade name of Aluminum Wire & Cable Company Limited (UK) for a wrought aluminum alloy containing 0.5% magnesium, 0.5% silicon and 0.04% copper. Used for electrical applications.

Simalloy. Trade name of Simonds Worden White Company (USA) for several *Inconel*-type nickel-chromium-iron alloys and various modified austenitic stainless steels.

Simanal. Trade name of Ohio Ferro Alloys Corporation (USA) for an alloy of 40% iron, 20% silicon, 20% aluminum and 20% manganese. Used as a deoxidizer in steelmaking.

Simancon. Trade name of British Steel Corporation (UK) for normalized, weldable plain-carbon steel plate containing 0.22-0.24% carbon, 0.6% silicon, 1.6% manganese, 0.045% sulfur, 0.045% manganese, 0.04% niobium, and the balance iron. Used for engineering and constructional applications.

Simbrax. Trade name for a corrosion-resistant alloy composed of 85-97% copper, 2-7% zinc, 1-3.5% silicon and 1.5% iron. Used for pressure gages.

Simco. Trade name of Sillcocks-Miller Company (USA) for cellulose nitrate plastics.

Simcronite. Trade name of Foseco Minsep NV (Netherlands) for a combined graphitizing and stabilizing inoculant for cast iron.

Simeteora. British trade name for high-speed tool steel containing 0.76% carbon, 16.9% tungsten, 2.9% chromium, 0.2% manganese, and the balance iron. Used for tools, cutters, drills, taps, reamers, hobs, chasers, etc.

Simgal. Trade name of Alcan-Booth Industries Limited (UK) for a heat-treatable aluminum alloy containing 0.6-0.75% magnesium and 0.4-0.5% silicon. Used for architectural sections, window frames, and wires.

Simichrome. Trade name of Competition Chemicals (USA) for a chromium metal polish.

Simidur. Trade name of Wieland Dental + Technik GmbH & Co. KG (Germany) for several dental alloys including (i) white-colored, extra-hard palladium-base ceramic alloys, such as *Simidur KF plus*, *Simidur S1S* and *Simidur S2*; and (ii) white-extra-hard universal alloys, such as *Simidur E*, for low-fusing high-expansion ceramic restorations.

Simo. (1) Trade name of Acciaierie Valbruna SpA (Italy) for a shock-resisting tool steel containing 0.6% carbon, 1.9% silicon, 0.8% manganese, 0.35% molybdenum, up to 0.3% chromium, and the balance iron. Used for hand and pneumatic tools, forming tools, and punches.

(2) Trade name of Aubert & Duval (France) for a corrosion- and heat-resistant steel containing 0.4% carbon, 2.5% silicon, 10% chromium, 0.9% molybdenum, and the balance iron. Used for high-temperature valves.

Simoch. Trade name of Teledyne Vasco (USA) for an air-hardening, shock-resisting tool steel (AISI type S7) containing 0.46-0.5% carbon, 0.8-1% silicon, 0.2-0.4% manganese, 3-3.5% chromium, 1.3-1.5% molybdenum, 0.2-0.3% vanadium, and the balance iron. It has good deep-hardening properties, very high impact strength, excellent toughness, and moderate wear resistance. Used for shear knives, shear blades, punches, rivet sets, chisels, slitting cutters, and gripper dies.

Simonds. Trade name of Wallace Murray Corporation (USA) for series of alloys including several corrosion- and heat-resistant nickel-iron alloys, various iron-nickel bimetals, iron-cobalt magnet alloys, and a number of steels including cobalt and chromium magnet grades, high-nickel grades, cold- and hot-work tool steels, water-hardening tool steels, and tungsten and cobalt-tungsten high-speed steels.

simonellite. A white yellowish mineral composed of 1,1-dimethyl-7-isopropyl-1,2,3,4-tetrahydrophenanthrene, $C_{19}H_{24}$. Crystal system, orthorhombic. Density, 1.08 g/cm³. Occurrence: Italy.

simonkollerite. A synthetic mineral composed of zinc chloride hydroxide monohydrate, $Zn_5(OH)_8Cl_2 \cdot H_2O$, made from zinc chloride ($ZnCl_2$) and zinc hydroxide [$Zn(OH)_2$] under prescribed conditions. Crystal system, rhombohedral (hexagonal). Density, 3.34 g/cm³.

Simpa. Trade name for a self-cure silicone used for dental lining purposes.

simple alloy steel. See ternary steel.

simple carbohydrate. A simple form of sugar, e.g., glucose or lactose. See also monosaccharide.

simple protein. A protein, such as a protamine, albumin, globulin or prolamine, that on hydrolysis yields only amino acids or their derivatives.

simple sugar. See monosaccharide.

Simplex. (1) Trade name of Vidrios Planos Lirquen SA (Chile) for a patterned glass featuring narrow vertical reeds.

(2) Trade name of Union Carbide Corporation (USA) for a ferrochrome composed of 0-0.1% carbon, 5-7% silicon, 63-66% chromium, and the balance iron. Available in the form of quick-dissolving pellets, it is used for the manufacture of stainless steel.

(3) Trade name of Colt Industries (UK) for a carburizing steel composed of 0.2% carbon, 1.25% nickel, 0.6% chromium, and the balance iron. Used for forgings, gears, pinions, shafts, etc.

(4) Trade name for a series dental acrylics used for restorations.

Simplicity. Trade name of Apex Materials (USA) for a light-curing, self-etching, two-step, universal dental bonding/adhesive system used for direct restorations, composite restorations, and for bonding bridges, crowns, inlays, onlays and posts.

simplotite. A dark green mineral composed of calcium vanadium oxide pentahydrate, $CaV_4O_9 \cdot 5H_2O$. Crystal system, monoclinic. Density, 2.64 g/cm³; refractive index, 1.767. Occurrence: USA (Colorado, Utah).

simpsonite. A yellowish brown mineral composed of aluminum tantalum oxide hydroxide, $Al_4Ta_3O_{13}OH$. Crystal system, hexagonal. Density, 6.70 g/cm³; refractive index, 2.040. Occurrence: Brazil, Zimbabwe, Western Australia.

sin-chu. A Japanese brass composed of 66.5% copper, 33.4% zinc and 0.1% iron, and used for hardware.

Sinfonie. (1) Trade name of Gerresheimer Glas AG (Germany) for a patterned hollow glass blocks.

(2) Trade name of ESPE GmbH & Co. KG (Germany) for dental hybrid composites.

Singapore gum. See East India gum (ii).

Single Add Premix. Trademark of Asbury Wilkinson Graphite Foundry Supply (Canada) for graphite-base foundry sand binders and facing additives.

Single Bond. Trademark of 3M Dental (USA) for a single-component (primer plus adhesive), fluoride-releasing, light-cure dentin/enamel adhesive resin system based on bisphenol A glycidyl methacrylate (BisGMA) and containing hydroxyethyl methacrylate (HEMA). It has very low polymerization shrinkage and microleakage, and very high bond strength. Used in dentistry for bonding metal, composite, compomer and porcelain restorations, and composite and porcelain inlays and onlays.

single-burnt dolomite. See calcined dolomite.

single crystal alumina. See recrystallized alumina.

single crystals. Crystalline solids (elements or compounds) for which the regular and repeated atomic order extends throughout their entirety. They may be cut from natural crystals (e.g., fluorite, garnet, ruby, sapphire, etc.) or grown artificially by electrodeposition, physical or chemical vapor deposition, saturated solution or melt cooling, growth in gel media or solution evaporation. Common crystal-growing techniques used in electronics and semiconductor technology include Czochralski (for silicon, germanium, gallium arsenide, indium antimonide, etc.), Bridgman (for metals, certain II-VI (12-16) semiconductors), float-zone (for silicon), zone leveling (for germanium, gallium arsenide, indium arsenide, indium antimonide, etc.) and Verneuil (for refractory oxides). *Single crystals* are used in electronics and semiconductor technology (e.g., semiconductors, miniaturized components, computer memory systems, lasers, masers, whiskers, etc.), in materials science and engineering, mineralogy, and in the life sciences.

single-end rovings. Continuous rovings of glass fibers made by impregnating with synthetic resin and winding onto a mandrel, or pulling through a hot die.

single-hydrated lime. See monohydrated lime.

single nickel salt. See nickel sulfate hexahydrate.

single oxide ceramics. A group of ceramics that are composed of one metal or nonmetal oxide only, e.g., alumina (Al_2O_3), magnesia (MgO), beryllia (BeO), silica (SiO_2), thoria (ThO_2), uranium dioxide (UO_2) and zirconia (ZrO_2).

single-phase alloys. Amorphous, single crystalline or polycrystalline alloys that contain only one phase. Polycrystalline alloys contain many grains of the same phase, each grain differing only in crystal orientation, but not in chemical composition. These alloys are solid solutions of two or more components. See also multiphase alloys.

single-phase materials. Amorphous, single crystalline or polycrystalline materials that contain only one phase, e.g., single-phase alloys, window glass, polystyrene, ruby gems, etc. See also multiphase materials.

single-screened refractory. A refractory material from which particles exceeding a specified size range have been eliminated by screening, but which apart from that contains essentially the same particle size distribution as processed by crushing and/or grinding.

single-sized aggregate. Aggregate in which the majority of the particles are of sizes lying between narrow limits. See also aggregate.

single-stage phenolic resin. A thermosetting resin produced by a condensation-type reaction of phenol with formaldehyde and an alkaline or basic catalyst in which water is formed as a by-product. Owing to its reactive groups it can be subsequently cured by the application of heat to form an infusible, insoluble, crosslinked material (C-stage resin). Also known as *single-stage resole*. See also phenolic resin; A-stage resin; B-stage resin; C-stage resin.

single-stage resole. See single-stage phenolic resin.

single-strength glass. Clear sheet glass with a nominal thickness of about 2-2.5 mm (0.08-0.10 in.) and smooth, truly parallel surfaces produced by floating a ribbon of glass on molten tin. Used for small window panes and picture frames. Abbreviation: SS glass.

single-wall carbon nanohorn. See carbon nanohorn.

single-walled nanotubes. See carbon nanotubes.

single yarn. (1) A bundle or assembly of untwisted natural or man-made fibers or filaments suitable for use in knitting and weaving of textile fabrics.

(2) A bundle or assembly of natural or man-made fibers or filaments made by spinning or twisting several staple fibers or single filaments together to form a strand, suitable for use in knitting and weaving of textile fabrics.

sinhalite. A mineral of the olivine group composed of aluminum magnesium borate, $AlMgBO_4$. It can also be made synthetically. Crystal system, orthorhombic. Density, 3.49 g/cm³. Occurrence: Germany, Sri Lanka.

Sinimax. Trade name of Allegheny Ludlum Steel (USA) for a magnetically soft material containing 55% iron, 42% nickel and 3% silicon. Used for autotransformers, electrical equipment, and tape recording heads.

Sinitex. Trade name for saran (polyvinylidene chloride) fibers with good chemical resistance, good weatherability, and good resistance to mildew and insects. Used for carpets and rugs, draperies, upholstery, clothing, industrial fabrics, etc.

Sinkral. Trade name of Enichem SpA (Italy) for acrylonitrile-butadiene-styrene resins available in various grades including unreinforced, glass fiber-reinforced, high-heat, UV-stabilized, medium-, high- and very high-impact, low-gloss and plating. They are supplied as sheeting for the manufacture of molded articles, laminated glass products and for aircraft insulation, and also as film, blocks, rods and shapes for the manufacture of household appliances, kitchen and bathroom accessories, industrial parts, and automotive components.

sinnerite. A steel-gray mineral of the sphalerite group composed of copper arsenic sulfide, $Cu_6As_4S_9$. It can also be made synthetically. Crystal system, triclinic. Density, 4.47 g/cm³. Occurrence: Switzerland.

sinoite. A colorless to light gray synthetic mineral composed of silicon oxide nitride, Si_2N_2O. Crystal system, orthorhombic. Density, 2.83 g/cm³.

sinopis. A variety of the mineral *hematite* used as a red paint pigment.

Sinsen. Trade name for acrylic fibers used for the manufacture of textile fabrics.

Sintag. Trade name of Handy & Harman (USA) for a corrosion-resistant sintered material composed of 85-90% silver and 10-15% cadmium oxide (CdO), and used for electrical contacts.

Sintaloy. Trade name of Dixon Sintaloy Inc. (USA) for a series of ferrous and nonferrous powder-metallurgy materials including high-purity iron and high-conductivity coppers, and various carbon, copper and nickel steels, high-strength low-alloy steels and copper-infiltrated steels as well as several nickel silvers and leaded and unleaded brasses and bronzes. Used for sintered structural and machinery parts.

Sinteel. Trade name of Electro Metal Corporation (USA) for a series of copper-impregnated sintered steels that have a density ranging from 7.2-7.6 g/cm³ (0.26-0.27 lb/in.³), and are used for machinery parts.

sinter. (1) An agglomerate in the form of a medium-sized clinker made by mixing fine-particle iron ore, high-ash coke and fine particles of limestone flux followed by controlled sintering (burning). Used as iron-blast furnace charge.

(2) See sintered part.

Sinterbronze. Trade name of Plansee Metallwerk Gesellschaft (Austria) for a sintered bronze made from an elemental powder blend composed of 90% copper and 10% tin, and used for self-lubricating porous bearings.

Sintercast. Trade name for a noble-metal foil used for dental crown applications.

sintered alumina. A refractory produced by heating natural aluminum oxide (e.g., corundum) to temperatures just below the melting point to effect a microstructure with grains ranging in size from microcrystalline to coarsely crystalline. It has high density, abrasion resistance, physical strength and dielectric strength and a low-power factor. Used in the manufacture of abrasives, high-temperature coatings, ceramic-metal seals, machine tools, automotive and aircraft spark plugs, thread guides, and electrical components. Abbreviation: SA.

sintered aluminum powder. See sintered aluminum powder alloys.

sintered aluminum powder alloys. A group of dispersion-strengthened aluminum alloys containing fairly small concentrations of small-diameter aluminum oxide (Al_2O_3) particles (typically 0.01-0.1 μm in diameter). A dispersion of only 10 vol% Al_2O_3 in aluminum increases its tensile strength by a factor of more than four. *Sintered aluminum powder (SAP) alloys* are produced by causing a very thin and adherent Al_2O_3 coating to form on the surface of extremely small flakes of aluminum dispersed within the aluminum-metal matrix. The material is then processed by powder metallurgy involving compaction and sintering. Commercially available in various forms including sheet, rods and tubes, SAP alloys have good corrosion and oxidation resistance, good high-temperature stability and high strength, and are used for pistons, combustion chambers, cylinder heads, jet-engine parts, heat exchangers, etc. Abbreviation: SAP alloys. Also known as *sintered aluminum powder.* See also dispersion-strengthened alloys; dispersion-strengthened materials.

sintered beryllia. Refractory ceramic produced by heating beryllia to temperatures just below the melting point. The resulting compacted product has high electrical and thermal conductivity, good wetting resistance, good physical and mechanical properties, and high heat-stress resistance. See also beryllia; beryllia ceramics.

sintered brass. Brass produced from prealloyed powder of 63-90% copper, 10-35% zinc and 0-2% lead by pressing and sintering. It has a density range of 7.2-8.4 g/cm³ (0.26-0.30 lb/in.³), good corrosion resistance and free-machining properties, considerably high strength and good ductility. Used for bearings, gears, cams, latch bolts, actuators, trim, fittings, cylinder locks, etc. Also known as *P/M brass; powder-metallurgy brass.*

sintered bronze. A solid mass of bronze produced from elemental premixes or prealloyed powders of 90% copper and 10% tin by pressing and sintering. It has a density range of 5.6-7.2 g/cm³, and is used for bearings, hardware, bolts, nuts, screws, rivets and pins. Also known as *P/M bronze; powder-metallurgy bronze.*

sintered carbides. See cemented carbides.

sintered copper. Copper produced from atomized, oxide-reduced or electrolytic copper powder by pressing and sintering. Also known as *P/M copper; powder-metallurgy copper.*

sintered copper iron. An iron-copper alloy produced by blending iron powder with up to 4.0% copper and 0.3% carbon, and pressing and sintering. In the as-sintered condition, it has a density range of 6.0-7.2 g/cm³ (0.22-0.26 lb/in.³). Also known as *P/M copper iron; powder-metallurgy copper iron.*

sintered copper steel. A steel produced by blending iron powder with 1.5-22.0% copper and up to 1.0% carbon, and pressing and sintering. It is supplied in the as-sintered and heat-treated condition, and has a density range of 5.6-7.2 g/cm³ (0.20-0.26 lb/in.³). Also known as *P/M copper steel; powder-metallurgy copper steel.*

sintered glass. Bonded, but unsealed glassware of controlled porosity and desired shape and strength, made by pressing and sintering particles of glass. Used in aeration, filtration, etc.

sintered iron. A medium- to high-density iron produced by blending iron powder with up to 0.3% carbon, and pressing and sintering. It is supplied in the as-sintered condition with a density range of 5.8-7.6 g/cm³ (0.21-0.27 lb/in.³), and used for magnetic and electrical applications. Also known as *P/M iron; powder-metallurgy iron.*

sintered iron-copper. An alloy produced by blending iron powder with 9.5-10.5% copper and up to 0.3% carbon, and pressing and sintering. It is supplied in the as-sintered condition with a density of 5.6-6.0 g/cm³ (0.20-0.22 lb/in.³). Also known as *P/M iron-copper; powder-metallurgy iron-copper.*

sintered iron-nickel. A powder-metallurgy alloy made by blending iron powder with 1.0-8.0% nickel, up to 0.3% carbon and 2.5% copper, and pressing and sintering. It is available in the as sintered condition with a density range of about 6.4-7.6 g/cm³ (0.23-0.27 lb/in.³). Also known as *P/M iron-nickel; powder-metallurgy iron-nickel.*

sintered magnesia. Any of a group of special ceramics made by dry-pressing or slip-casting of essentially pure magnesia (MgO) followed by sintering at high temperatures. They have high hardness, good strength, thermal properties and wear resistance, and are used for high-temperature components, tools, electrical components, etc.

sintered material. A low-, medium- or high-density material made by blending selected metallic and/or nonmetallic powders, compacting (pressing) at ambient or elevated temperatures and sintering (heating) the resulting shapes to complete the metallurgical bond between the powder particles. Examples include sintered iron, sintered steel, sintered superalloys, sintered bronze and sintered brass. Also known as *P/M material; powder metallurgy material.*

sintered nickel silver. A powder-metallurgy alloy produced by pressing and sintering from a prealloyed nickel powder consisting of a blend of 65% copper, 18% zinc, 17-18% nickel and usually up to 1.5% lead (to improve machinability) and mixtures of the stearates of zinc and lithium. It has a density range of 7.6-8.4 g/cm³ (0.27-0.30 lb/in.³). Also known as *P/M nickel silver; powder metallurgy nickel silver.*

sintered nickel steel. A high-strength steel made by powder-metallurgy techniques, usually involving mixing of 1.0-8.0% nickel with iron, 0.3-0.9% carbon and 0-2.0% copper, and subsequent pressing and sintering. Available in the as-sintered and heat-treated condition, it has a density range of 6.0-7.6 g/cm³ (0.22-0.27 lb/in.³). Also known as *P/M nickel steel; powder-metallurgy nickel steel.*

sintered oxides. See oxide ceramics.

sintered part. A mechanical or structural part of definite shape made from powders by pressing and subsequent sintering, i.e.,

heating below the melting point of the major powder component. Also known as *P/M part; powder-metallurgy part; sinter.*

sintered refractories. Refractories produced from starting materials (usually in powder form), such as alumina, silica, magnesia, dolomite, mullite, zirconia, silicon carbide, or blends thereof by firing at high temperatures. They are usually highly resistant to abrasion, pressure, thermal shock and corrosion by molten metals, slags, etc., and have low thermal conductivities and at least moderate load-bearing capacities.

sintered silicon carbide. Fully dense silicon carbide (SiC) produced by hot pressing, i.e., by compacting the starting powder at an elevated temperature. Usually supplied in sheet form, it has a density of 3.05-3.15 g/cm³ (0.110-0.114 lb/in.³), high tensile strength and modulus, high compressive strength, a hardness of 2400-2800 Vickers, and an upper continuous use temperature of 1500-1700°C (2730-3090°F). Used for high-strength and high-temperature applications. Abbreviation: SSiC or SSC. Also known as *hot-pressed silicon carbide.*

sintered silicon nitride. Fully dense silicon nitride (Si_3N_4) usually produced by first cold isostatic pressing of the starting material (silicon nitride powder plus sintering aids) and subsequent pressureless or pressure sintering of the resulting green body at temperatures exceeding 1750°C (3180°F). Usually supplied in sheet form, it has a density of 3.10-3.20 g/cm³ (0.112-0.116 lb/in.³), high tensile strength and elastic modulus, high compressive strength, a hardness of 1500-2200 Vickers, and an upper continuous use temperature of 1100-1650°C (2010-3000°F). Used for high-strength and high-temperature applications. Abbreviations: SSN.

sintered stainless steel. A group of ferrous powder-metallurgy materials made in austenitic, ferritic and martensitic grades from fully alloyed powder blends by high-pressure compaction (pressing) and sintering at high temperatures (usually 1120-1315°C or 2050-2400°F). Available in the as-sintered and heat-treated condition with a density range of 6.2-7.6 g/cm³ (0.22-0.27 lb/in.³), they have good corrosion resistance and high tensile strength. Also known as *P/M stainless steels; powder metallurgy stainless steels.*

sintered steels. A group of ferrous powder-metallurgy materials made by blending iron powder with 0.3-1% carbon, and pressing and sintering for a specified period of time at high temperatures. The carbon (usually in the form of graphite) then dissolves and forms steel. They are available in the as-sintered and heat-treated condition with a density range of 6.4-7.2 g/cm³ (0.23-0.26 lb/in.³), good tensile and yield strength, good hardness and very low elongation. Also known as *P/M steel; powder-metallurgy steels.*

sintered titanium. Titanium or a titanium-base alloy made from blended elemental or prealloyed powder by pressing or forging and sintering. It has outstanding corrosion resistance and a high strength-to-weight ratio. Also known as *P/M titanium; powder-metallurgy titanium.*

sintered zirconia. Zirconia produced from a starting powder by a pressing and sintering (heating). Used for refractories and high-temperature components.

Sinterfil. Trade name of Klöckner-Draht GmbH (Germany) for plastic-coated and sintered barb wire of iron or steel.

sinter-HIPed products. Dense powder-metallurgy products produced in a single-stage sintering and hot-isostatic pressing process in which the powder is heated in an evacuated vessel at sintering temperature, and consolidated by means of argon gas as the medium for exerting equal from all directions. The process is also known as "pressure-assisted sintering."

Sinterloy. Trade name of Charles Hardy & Company (USA) for powder-metallurgy materials made from steel powders containing 0.15%, carbon, and the balance iron. Used for machine parts, such as gears, cams, fasteners and washers.

sinter magnesia. See dead-burnt magnesite.

Sintox. Trade name of Sintox Corporation of America (USA) for sintered alumina (Al_2O_3) used for cutting-tool inserts.

Sintramant. Trade name of Plansee Metallwerk-Gesellschaft (Austria) for sintered tungsten carbide used for hardfacing electrodes.

Sintrex. Trade name of Easton Metal Powder Inc. (USA) for a series of electrolytic iron powders containing 99.0+% iron and up to 0.12% other nonvolatile elements. They are available in various grades including 100-mesh annealed standard powder, 100-mesh annealed sponge powder and 200-mesh annealed powder, and used for sintered soft magnetic iron parts.

Sinvet. Trade name of Enichem SpA (Italy) for polycarbonate resins available in unmodified, high-flow, food and drug, UV-stabilized and glass-fiber-reinforced grades.

Sioplas. Trade name of AEI Compounds (USA) for crosslinked polyethylene.

Sioux. Trade name of Atlas Specialty Steels (Canada) for an oil-hardening high-carbon alloy machinery steel containing 1.1% carbon, 0.4% manganese, 0.2% silicon, 1.4% chromium, 0.4% molybdenum, 0.03% sulfur, 0.03% phosphorus, and the balance iron. It has high tensile strength, wear resistance and surface hardness (file-hard) and moderate toughness. Used for conveyor pins, dowel pins, lock washers, nail sets, pushers, wearing pins and plates, tong and vise jaws, guides, hammers, ratchets, screwdrivers, lathe centers, straightening and swaging rolls, forming and ring dies, roller bearings and roller bearing races. Also known as *Atlas KK*.

Siplex. Trade name of Glaswerke Haller GmbH (Germany) for a laminated glass having an interlayer of thin wire.

Sipo mahogany. See Assie mahogany.

Sipomer. Trademark of Alcolac Inc. (USA) for acrylate, methacrylate and specialty monomers, such as allyl glycolate, diethyl and dimethyl maleate and hydroxyethyl methacrylate.

Siporex. Trademark of American Siporex Corporation (USA) for an aerated concrete used in the manufacture of lightweight structural building materials including slabs, columns, beams, lintels, blocks, etc.

siporex. A building material made from a slurry of sand, lime or cement, and aluminum powder by first casting into molds and then steaming for a prescribed period of time. Used in the manufacture of roofing slabs, wall blocks, door lintels and other building materials with high acoustical and thermal insulation properties.

SIPOS. A semi-insulating polycrystalline oxygen-doped silicon obtained as a thin film deposited on silicon. Depending on the oxygen concentration (typically 15-50 at%), its as-deposited microstructure is either polycrystalline, nanocrystalline or amorphous. It is used as a resistor in CMOS circuits, as an emitter in heterojunction transistors, in solar cells, and as a surface passivation layer for high-voltage power devices.

Sirfene. Trade name for phenol-formaldehyde resins and laminates.

Sirius. Trade name for rayon fibers and yarns used for textile fabrics.

Sironze. Trade name of Wilbur B. Driver Company (USA) for a corrosion-resistant silicon bronze containing 96% copper, 3% silicon and 1% iron. Used for hardware, bolts, nuts, screws, rivets and pins.

Siroplast. Trade name of Hegler Plastik GmbH (Germany) for plastic drainage pipes used in road construction and civil engineering.

Sirtene. Trade name for high-density polyethylene plastics.

sisal fibers. The strong, hard, yellowish white to reddish fibers obtained from the leaves of several varieties of agave plants especially *Agave sisalina* of the amaryllis family. *Sisal* is cultivated in the Bahamas, Central America, East Africa, Haiti, Indonesia and Mexico, and used in the manufacture of rope, cordage, sacking, binder twine and brush bristles, and for carpets, upholstery, mattress liners, sailcloth, cement plasters for walls, and laminated plastics. Also known as *sisal; sisal hemp.*

Sisalkraft. Trade name of Fortfiber Corporation (USA) for a series of heavy, waterproof kraft papers made from sisal fibers and used as water and vapor control layers in building construction, especially for timber frame construction applications.

sisal rope. A sisal fiber-based rope with a typical diameter of 6-19 mm (0.25-0.75 in.). It has moderate strength (only about 75-80% that of Manila rope), is coarser than Manila, and has a safe working load about 20% of its tensile strength, poor moisture resistance and high susceptibility to rot, mildew and bacteriological damage. Used for general-purpose medium-duty applications, and for tiedown, guard and light-hauling purposes. See also sisal fibers; Manila fibers.

siserskite. A light steel-gray variety of the natural alloy *osmiridium* composed of 57% osmium, 8% ruthenium, and the balance rhodium and iridium. Crystal system, hexagonal. Density, 19-21 g/cm^3; hardness, 6-7 Mohs; good corrosion resistance; good resistance to acids including aqua regia. Occurrence: Russia. Used as a source of iridium and osmium, for hardening platinum, and for fountain-pen points.

sissoo. The hard, durable wood of the tree *Dalbergia sissoo* belonging to the Indian rosewoods. The heartwood is dark purplish-brown with blackish stripes, and has a beautiful grain figure on flat-sawn surfaces. *Sissoo* resembles Brazilian rosewood in appearance, and turns well. Average weight, 850 kg/m^3 (53 lb/ft^3). Source: India. Used for high-quality furniture, fine cabinetwork, veneer, parquet floors, turnery, decorative purposes, ornaments, and carving. See also Brazilian rosewood.

Sistal. Trade name of Teledyne Vasco (USA) for a hot-work tool steel containing 0.47% carbon, 8.25% molybdenum, 1.35% molybdenum, 1.2% tungsten, 0.9% silicon, 0.3% manganese, 0.3% vanadium, and the balance iron. Used for extrusion presses, upsetting rams, and pressure dies.

Sitka cypress. See Alaska cedar.

Sitka spruce. The soft, lightweight, high-quality wood from the very tall coniferous tree *Picea sitchensis*. The heartwood is light pinkish-brown, and the sapwood creamy white. *Sitka spruce* has a close, straight grain, a fine, uniform texture, is free from knots and resin ducts and possesses good strength, toughness, elastic properties, and stiffness. Average weight, 430 kg/m^3 (27 lb/ft^3). Source: Pacific coast of North America from California to Alaska. Used chiefly for lumber, as a pulpwood, and for cooperage, furniture, millwork, boxes, crates, piano-sounding boards and violins, masts, spars, oars, and paddles. Also known as *silver spruce; western spruce; yellow spruce.*

Sitkin. Trade name of Sitkin Smelting & Refining Company (USA) for an extensive series of cast leaded and unleaded brasses and bronzes (red and yellow brasses, aluminum, manganese and silicon bronzes, nickel silvers, etc.), and numerous copper-base bearing alloys containing varying amounts of lead, tin, and/or zinc and nickel.

Situssa. Trade name of Novaceta S.p.a. (Italy) for cellulose acetate fibers used for the manufacture of wearing apparel.

Siveras. Trademark of Toray Industries, Inc. (Japan) for liquid crystal polymer resins.

Sivovinyl. Trademark of Sico Inc. (Canada) for vinyl coatings.

Sivyer. Trade name of Sivyer Steel Corporation (USA) for a extensive series of plain-carbon steels (AISI-SAE 10xx series), low-alloy steels (AISI-SAE 43xx and 86xx series) including several boron, nitriding and carburizing grades, various chromium and chromium-nickel electric grades, austenitic manganese grades, and heat- and corrosion-resistant nickel-chromium stainless grades.

Six Eighty Alloy. Trade name of Johnson Matthey plc (UK) for a 65% silver alloy, used for dental amalgams.

Sixix. Trade name of Atlas Specialty Steels (Canada) for a molybdenum-type high-speed tool steel (AISI type M2) containing 0.84% carbon, 0.3% silicon, 0.25% manganese, 6.5% tungsten, 5% molybdenum, 4% chromium, 1.9% vanadium, and the balance iron. It has a great depth of hardening, good toughness, high hot hardness, and is used for twist drills, threading taps, broaches, reamers, hobs, counterbores, countersinks, form cutters, end mills, milling cutters, gear cutters, etc. A free-machining quality, *Sixix-fm*, with similar composition, but increased manganese (0.3%) and additions of 0.12% sulfur is also available.

6 Tile. Trademark of Dry Branch Kaolin Company (USA) for a translucent, bright-firing air-floated *kaolin* with a high degree of whiteness, and high plasticity and green strength. Used for whitewares and ceramic tiles.

size. (1) A material, such as rosin, wax or asphalt, added to fiberboard or particleboard during manufacture to improve its moisture resistance. This term does not include binders, such as phenol-formaldehyde resins, etc.

(2) A temporary coating of gelatin, starch, wax, oil, rosin, silicone resin, etc., applied to fibers and yarns during manufacture to provide surface protection and increased handleability. It is usually removed prior to use in composites.

(3) A temporary or permanent coating of gelatin, starch, oil, wax, casein, asphalt, rosin or synthetic resin applied to leather, fabrics, paper, etc., to promote handleability, processibility, smoothness, stiffness and strength. Also known as *sizing compound*.

(4) A liquid, water-soluble solution of starch, glue or soap used to treat pottery, plaster and similar ceramics, usually with the purpose to seal or fill porous surfaces, and prevent excessive absorption of finishing materials.

(5) A thin, liquid composition or varnish used as an initial coat on metal to promote adhesion of subsequent coats.

sized compact. A sintered powder-metallurgy compact that has been pressed to the desired size in a die by means of a punch.

sized gypsum. Crushed gypsum graded according to particle size.

sizing compound. See size.

sjogrenite. A white to yellowish mineral composed of magnesium iron carbonate hydroxide tetrahydrate, $Mg_6Fe_2CO_3(OH)_{16} \cdot 4H_2O$. Crystal system, hexagonal. Density, 2.11 g/cm^3; refractive index, 1.573. Occurrence: Sweden.

Skagit. Trademark of Clayburn Refractories Limited (Canada) for fireclay and special-duty fireclay brick.

Skai. (1) Trademark of Konrad Hornschuh AG (Germany) for artificial leather used for clothing, gloves, etc.

(2) Trademark of Konrad Hornschuh AG (Germany) for plastics available as sheets and foils, and as laminates with fabrics, felt, paper, foamed materials, plastics, etc., and used for shoe soles and heels, roofing, table top coverings, and the like.

Skamex. Trademark of E.I. DuPont de Nemours & Company (USA) for fluorocarbon plastics used as ladle additives in the manufacture of ferrous and nonferrous metals.

Skanopal. Trade name of Perstorp AB (Sweden) for urea-formaldehyde resins.

Skapalon. Trade name for nylon fibers and yarns used for textile fabrics including clothing.

Skefko. Trade name of Hofors Steel Works (Sweden) for water-hardening steel containing 0.95-1.10% carbon, 0.25-0.35% manganese, 0.40-1.65% chromium, up to 0.40% molybdenum, and the balance iron. Used for ball bearings.

skein. A continuous strand, filament, yarn or roving in the form of a loose, flexible coil.

Skeleton. Trade name for a medium-gold dental casting alloy.

skelp. Flat strip, sheet or plate of iron, steel or nonferrous material used for making welded tubing or pipe.

skelp iron. Wrought iron in the form of flat-rolled bars used for making welded tubing or pipe.

Skenandoa. Trade name for rayon fibers and fabrics.

Skendo. Trade name for rayon fibers and yarns used for textile fabrics.

skiagite. A mineral of the garnet group composed of manganese silicate, $FeMnSi_3O_{12}$. Crystal system, cubic. Occurrence: Scotland, India. Used in ceramics.

Skillcast 60. Trade name for a 60%-gold dental casting alloy.

skin. See hide (2).

Skinlife. See Meryl Skinlife.

skinnerite. A gray, synthetic mineral composed of copper antimony sulfide, Cu_3SbS_3, grown in the form of crystals from a sulfide vapor phase at 400°C (752°F). Crystal system, monoclinic. Density, 5.07 g/cm^3.

Skinoid. Trademark for polymer-base leather substitute used for book covers.

skiver. The thin, soft grain split of a sheepskin used for small leather goods, and sweatbands for hats.

Skivertex. Trademark of Winter & Co. GmbH (Germany) for a textile-based leather substitute imitating skiver, and used for book covers.

sklodowskite. A yellow, radioactive mineral of the uranophane group composed of magnesium uranyl silicate hydroxide pentahydrate, $Mg(UO_2)_2(SiO_3OH)_2 \cdot 5H_2O$. Crystal system, monoclinic. Density, 3.54 g/cm^3; refractive index, 1.635. Occurrence: Zaire.

skutterudite. A tin-white to silver-gray mineral with a metallic luster. It is composed of cobalt arsenide, $CoAs_3$, and can also be made synthetically in powder form. Crystal system, cubic. Density, 6.50 g/cm^3; hardness, 5.5-6 Mohs. Occurrence: Rumania, Norway, Canada, USA (Colorado). Used as minor cobalt ore.

skutterudites. A group of materials with the general formula RE-TM$_4$Pn$_{12}$, where RE is a rare-earth element, such as cerium, TM a transition metal, such as iron, and Pn a pnictogen, such as antimony, arsenic or phosphorus. Structurally, they consist of pnictides (i.e., pnictogen compounds) arranged in tilted octahedra with the rare-earth element located in the simple cubic environment of the transition metals, each of which is in turn located in the octahedra. Used in ceramics.

Skybloom. Trade name for rayon fibers and fabrics.

Skybond. Trade name for polyimide resins.

Skyloft. Trade name for rayon fibers and yarns used for textile fabrics.

Skylon. Trademark of Sunkyong Industries Limited (South Korea) for polyester and cellulose acetate fibers and filament yarns.

Skyrock. Trade name a self-cure dental composite used for core-build-ups.

Skytex. Trade name of ASG Industries Inc. (USA) for a patterned glass having eight parallel ribs per inch (2.5 cm).

Skytrol. Trade name of Pittsburgh Corning Corporation (USA) for a glass block having a fibrous glass screen cast into a rigid, steel-reinforced concrete grid.

slab. (1) The outside pieces of wood and bark cut from logs during the sawing of lumber.

(2) A rectangular semifinished metal or alloy, either hot-rolled or strand-cast, and ready for rolling into plates. It usually has a minimum width of 25 cm (10 in.) and a minimum cross-sectional area of 105 cm^2 (16 in.^{2}).

(3) A flat surfaced stone or marble.

(4) A flat, horizontal section of concrete of uniform thickness laid as a single unjointed unit. See also concrete slab.

slab glass. A block of optical glass obtained by forming or cutting chunk glass into slabs or plates of suitable size for subsequent processing. See also chunk glass.

slab zinc. A commercial zinc cast into various shapes and sizes. The following six grades are available: special high purity (99.99%), high purity (99.90%), continuous galvanizing (99.5% pure), controlled lead grade, prime western (98.0% pure) and remelt (99.0% pure).

slag. (1) A nonmetallic substance formed either in metal smelting by the combination of a flux (e.g., limestone) with the gangue of an ore, or in metal refining by the combination of the impurities with the additive or additives introduced for the purpose of facilitating the refining. *Slags* can also be formed by chemical reaction between refractories (furnace linings) and fluxing agents, e.g., coal ash, or between dissimilar refractories. Slags from smelting and refining operations are usually high in silicates, and are used as railroad ballast, in highway construction, as cement and concrete aggregate, and for mineral wool and cinder block. Also known as *cinder.*

(2) A nonmetallic byproduct obtained in the molten condition simultaneously with iron in a blast furnace. It consists primarily of silicates and aluminosilicates of calcium and other alkaline materials. Once solidified, it has a porous structure, and is usually crushed and used in making concrete aggregate, Portland cement, railroad ballast, slag wool and slag bricks. Also known as *blast-furnace slag.*

(3) A light, spongy, highly siliceous pozzolanic material used in the production of Portland cement.

(4) A calcium silicate (CaSiO$_3$) slag obtained as a byproduct of electric furnaces and used in the production of phosphorus from phosphate rock and in the manufacture of glass. Also known as *phosphate slag.*

slag brick. A brick made from blast-furnace slag.

slag cement. A synthetic cement made by intimately grinding and blending blast-furnace slag and hydrated lime or Portland cement. Also known as *cold-process cement.*

slag concrete. A concrete made with slag cement and/or slag sand as fine aggregate.

slag flour. See basic phosphate slag.

slag former. Materials, such as asbestos, feldspar, magnesium

carbonate, silicon dioxide or titanium dioxide, that are used as ingredients in the welding fluxes coated onto electrodes. The slag shields the molten metal against oxygen from the atmosphere and removes impurities from the molten metal.

slag sand. Blast-furnace slag that has been finely crushed and sized for use as aggregate in mortar and concrete, and in the manufacture of slag bricks.

slag shingle. Crushed slag sometimes used in road construction.

slag wool. See mineral wool.

slaked lime. See calcium hydroxide.

slash pine. The hard, strong wood of the coniferous tree *Pinus elliottii* belonging to the southern pine group. The heartwood is reddish-brown and the sapwood yellowish-white. It has moderate shrinkage, and somewhat resembles *loblolly pine.* Average weight, 605 kg/m^3 (38 lb/ft^3). Source: Swampy areas (slashes) of southeastern United States (Georgia, Alabama, Florida, Mississippi and Louisiana). Used in construction, cooperage, kraft paper pulp, etc.

slate. A dense, fine-grained, usually grayish black or bluish-gray rock derived from shale and composed chiefly of micas, chlorite, quartz, hematite and clays. It easily splits into thin, smooth, tough sheets or slabs, has an average weight of 2800 kg/m^3 (175 lb/ft^3), and an average ultimate tensile and compressive strength of 3.4 MPa (0.5 ksi) and 96.5 MPa (14 ksi), respectively. Occurrence: Europe, USA (California, Colorado, Maine, New York, Pennsylvania, Vermont, Virginia). Used as a roofing and flooring material, for decorative stone, and blackboards. Slate powder and flour are used as an abrasive, as a pigment and filler in putties, paint and rubber.

slate black. Inorganic black pigments made by grinding black slate, shale, carbonaceous rock, slaty coal, graphite, etc., and used in coatings, inks, plastics, etc. Also known as *mineral black.*

slate cement. (1) Crushed slate mixed with asphalt or tar for use in roofing.

(2) A hydraulic cement containing crushed slate as aggregate and used in building construction.

slate flour. Slate in the form of a finely divided grayish powder used in the manufacture of Portland cement, ceramics, glass, linoleum and metal polishes, as a building material, as a filler for putties, paint, rubber and plastics, in caulking compounds, asphalt surfacing mixtures, and roofing mastics.

slate granules. Particles of crushed slate, approximately 8 mesh (2.36 mm) in size, used as surfacing material on asphalt roofing and shingles. See also roofing granules.

Slatekote. Trade name for a heavy, weather- and fire-resistant roofing felt saturated with asphalt and coated with slate granules of varying color.

slate lime. A finely divided mixture of approximately 60 wt% quicklime and 40 wt% calcined slate, used in the manufacture of porous insulating concrete.

Slater. Trade name of Slater Steels (USA) for an extensive series of corrosion- and heat-resistant austenitic, ferritic, martensitic and precipitation-hardening stainless steels including hardenable, welding and free-machining grades. Used for turbine and jet-engine parts, burner nozzles, heat-treating equipment, industrial furnaces, furnace parts and fixtures, heat exchangers, aircraft and missile components, high-temperature bolting, pump and valve parts, valve trim, bearings, cutlery, knives, dental and surgical instruments, machine parts, fasteners, fittings, kitchen utensils, food-processing equipment, chemical, petrochemical and pharmaceutical equipment, textile and pulp and paper machinery, welded structures, gaskets, automotive and aircraft parts, afterburners, and combustion chambers.

slater's cement. A water-resistant, putty-like roofing material, usually gray in color, used as a caulking compound for covering exposed bolt heads, end and side laps of corrugated roofing, and other exposed areas. See also *roofing cement*.

slater's felt. See *roofing felt*.

slavikite. A greenish yellow mineral composed of magnesium iron sulfate hydroxide octadecahydrate, $MgFe_3(SO_4)_4(OH)_3 \cdot 18H_2O$. Crystal system, hexagonal. Density, 1.90 g/cm^3. Occurrence: Argentina, Czech Republic.

slawsonite. A mineral of the feldspar group composed of strontium aluminum silicate, $SrAl_2Si_2O_8$. It can also be made synthetically from aqueous gels of strontium oxide, aluminum oxide and silicon dioxide, or by firing a strontium sulfate-kaolin mixture under prescribed conditions. Crystal system, monoclinic. Density, 3.12 g/cm^3. Occurrence: USA (Oregon).

Slax. (1) Trademark of Foseco Minsep NV (Netherlands) for a slag-coagulating ladle flux used in the manufacture of gray, white and malleable iron.

(2) Trademark of Foseco Minsep NV (Netherlands) for slag coagulants used in the manufacture copper and/or nickel and their alloys for slag control in the ladle and reduction of inclusions in the castings.

sleeve bearing materials. A class of comparatively soft bearing alloys suitable for bonding in one or more layers to relatively thick steel, bronze or brass backings. Examples of such materials include tin and lead-base alloys, copper, copper-lead and copper-lead-tin alloys, aluminum-base alloys, silver-base alloys, sintered metals and lead-tin overlays.

sleeve brick. A tubular firebrick used as a slag vent lining in ladles for molten metals.

Sleipner. Trade name of Uddeholm Steel Company (USA) for a wear- and chipping-resistant chromium-molybdenum-vanadium tool steel that can be treated by a high-temperature tempering process to develop high surface hardness. Used for drawing, forming, blanking, coining, cold-extrusion and cold-forging tools.

sliced veneer. Veneer made by moving a log or flitch sideways against a knife. See also *veneer*.

slicker fabrics. Textile fabrics, such as cotton, rayon or silk, that have been coated with a waterproof film. Used for raincoats and other waterproof clothing.

slicker solder. A solder composed of 60-66% tin and 34-40% lead, having a melting point of 183-190°C (361-374°F), and used as a commercial solder especially in the electronics industry.

slide valve bronze. A high-strength, corrosion-resistant bronze for slide valves containing 88.5% copper, 9% zinc, and 2.5% tin.

Slidex. Trade name of Jones & Rooke Limited (UK) for a wrought copper nickel containing 10-18% nickel, 17-25% zinc, and the balance copper. Supplied in strip and wire form, it has excellent cold workability and is used for slide fasteners.

Slim-Thinline. Trade name of Precision Specialty Metals Inc. (USA) for stainless steel supplied in strip, sheet and coil form in thicknesses ranging from 0.1 to 2 mm (0.005 to 0.080 in.) and widths from 10 to 914 mm (0.500 to 36 in.).

slip. A thin, watery suspension of finely divided ceramic material (clay, cement, etc.) in a liquid. Also known as *slurry*.

slip-cast materials. Ceramic, powder-metallurgy or refractory products, often of large size and complex shape, formed by first pouring a stiff slip or slurry, composed of a liquid mixture of the ceramic, metal or refractory starting powder, into a plaster mold, allowing a solid layer or shell of the desired thickness to deposit on the mold wall, and then drying the shell in the mold, followed by firing or sintering of the shell at high temperatures.

slip clay. Fine-grained clay with low firing shrinkage that contains a high proportion of fluxing impurities and fuses readily at a relatively low temperatures (705°C or 1300°F) producing a natural glaze on the resulting clayware.

slip coating. A ceramic coating applied to a body or shape in the form of a slip (or slurry) and subsequently fired to the required maturity. See also *engobe*.

slip glaze. A glaze composed essentially of easily fusible clay or silt and other ingredients blended with water to develop a creamy consistency.

slippery elm. The soft, coarse-textured wood of the red elm (*Ulmus fulva* or *U. rubra*), a small- to medium-sized forest tree, once found throughout most of eastern North America. The stand is now severely threatened by Dutch elm disease (a fungal disease) and phloem necrosis. The name "slippery elm" refers to the slimy, slippery inner bark of the tree used for making medicine, and the alternate name "red elm" to the red-hairy buds. The heartwood is light brown, and the sapwood is almost white. *Slippery elm* has an average weight of 593 kg/m^3 (37 lb/ft^3), good bending qualities, and is used for lumber, furniture, and veneer for decorative paneling. Also known as *red elm*.

slip stains. Ceramic pigments calcined and added to a slip rather than a body. They are intended to decrease the amount of colorant necessary to produce a desired color.

slit film. A yarn made from extruded film by cutting lengthwise into strips.

sliver. A loose assembly of untwisted staple or filament fibers in continuous form.

Slocrode. Trade name of Sheepbridge Alloy Castings Limited (UK) for a nickel cast iron containing 3.2% carbon, 0.75-1.25% nickel, 0.5-0.75% chromium, and the balance iron. Used for high-duty castings.

slop. A homogeneous slurry composed of glaze ingredients blended with water for application to ceramic ware by brushing, dipping or spraying. Also known as *slop glaze*.

slop glaze. See *slop*.

Slotera. Trademark of Slovensky Hodvab (Slovak Republic) for polyester fibers and filament yarns.

Slotoloy. Trade name of Dr. Ing. Max Schloetter (Germany) for a zinc-nickel alloy coating and coating process.

Slotoposit. Trade name of Dr. Ing. Max Schloetter (Germany) for electroless copper.

Slotozid. Trade name of Dr. Ing. Max Schloetter (Germany) for bright zinc electroplate and plating process.

Slovina. Trade name for rayon fibers and yarns used for textile fabrics.

slow-curing asphalt. A liquid asphalt made by diluting asphalt cement with a gas oil of slow volatilization. Also known as *S-C asphalt; slow-curing liquid asphaltic cement*.

Slow Recovery. Trademark of Ludlow Composites Corporation (USA) for a closed-cell polyvinyl chloride latex foam.

slubbed fabrics. Fabrics made with slubbed yarns.

slubbed yarn. A yarn that has uneven thick and thin sections at irregular intervals either resulting from manufacturing defects, or purposely produced by changing the degree of tension during spinning.

Sluggos. Trade name of Hudson Industries (USA) for mass finishing media used in metal finishing.

slurry. See slip.

slurry grout. A mixture of materials, such as cement, clay or sand, in water.

slurry-infiltration coatings. Ceramic-matrix materials applied to man-made fibers by passing the fiber tows through slurries composed of mixtures of ceramic powders, organic binders and carrier liquids. The ceramic coatings infiltrate the fiber surfaces forming strong, durable bonds upon setting.

slurry-sinter coatings. Coatings produced on refractory metal substrates by spray or dip application of metal particles plus binder, followed by solid or liquid-phase sintering in a vacuum or inert environment.

slush. A relatively thin slurry made by mixing Portland cement and sand with water, and applied to a ceramic surface by pouring, slushing or spreading.

slush castings. Hollow castings with thin shells consisting of low-melting alloys of lead, zinc, tin, antimony and bismuth. They are made by first pouring the molten alloy into a coreless metal mold and then, after the skin has frozen, turning the mold upside down to remove the still liquid excess metal. Used for ornaments, toys, and hollow parts.

slushing agent. A temporary coating composed of a nondrying oil, low-melting grease or similar compound applied to metals, machine parts, etc., to prevent or reduce corrosion during fabrication, shipping or storage. Although it adheres well to the metal or part, it is nevertheless easily removable. Depending on the particular composition and consistency, it may either be a "slushing grease" or a "slushing oil." Also known as *slushing compound.*

slushing compound. See slushing agent.

slushing grease. See slushing agent.

slushing oil. See slushing agent.

slush moldings. Thin-walled thermoplastic products formed by first pouring liquid resin dispersions (slushes) into heated molds, allowing viscous skins of desired thickness to form inside the molds, draining off excess liquids, and then cooling the molds to room temperature.

SMA. Trade name of Atofina SA (France) for styrene-maleic anhydride resins and copolymers used in paints and varnishes.

smalt. A deep blue pigment prepared by fusing together silica (SiO_2), potash (Na_2O) and cobalt oxide (CoO), and reducing to a powder the glass thus formed. It contains about 65-72% SiO_2, 16-21% Na_2O, 6-7% CoO, and sometimes a small amount of alumina (Al_2O_3). *Smalt* has moderate covering power, and is used as a paint and varnish pigment applied to freshly coated surfaces to provide unusual decorating effect, as a ceramic pigment, for coloring glass, glazes and porcelain enamels, as a bluing agent for paper and textiles, and as a rubber colorant. Also known as *azure blue; Dumont's blue; Saxony blue.* See also royal blue.

smaltite. A silver-white to metallic-gray mineral that is composed of cobalt nickel arsenide, $(Co,Ni)As_2$, and sometimes contains sulfur. Crystal system, cubic. Density, 6.5 g/cm³; hardness, 5.5-6 Mohs.

SMA-PC alloys. Immiscible polymer blends of an amorphous rubber-modified styrene maleic anhydride (SMA) copolymer and an amorphous polycarbonate (PC). They combine the good strength, toughness, ductility and heat resistance of PC with the good melt flow characteristics and moldability and low cost of SMA. Also, blending raises the heat deflection temperature

by about 10-15°C (18-30°F) as compared with the base copolymers. Used especially for automotive components and interior trim, and consumer goods.

Smartan. Trade name of Atofina SA (France) for styrene-maleic anhydride (SMA) resins and copolymers used mainly in the paint, varnish, textile, leather and pulp and paper industries.

Smart Bond. Tradenark of Gestenco International AB (Sweden) for a chemical-curing, ethylcyanoacrylate-based composite resin for orthodontic bonding applications.

SmartCoat. Trademark of Apache Products Company (USA) for exterior urethane coatings.

Smartcoat. Trade name for a multi-layer coating developed in the United Kingdom. It is composed of an underlayer of chromium oxide and a top layer of aluminum oxide. *Smartcoat* provides excellent protection to corrosion and pitting corrosion for industrial and marine gas turbines at high and low operating temperatures.

SmarTemp. Trademark of Parkell, Inc. (USA) for a durable, tough, nonsticky dental composite resin that is 64% filled with radiopaque barium glass and fumed silica, and has very low shrinkage and high polishability. Supplied in four shades from ultra-light to dark, it is used for temporary crown and bridge restorations.

smart coatings. A group of multilayer films with a total thickness of less than 10 μm (0.4 mil) and good high-temperature properties that comprise several planar sensor-circuitry layers between tough, electrically insulating layers. Used for *in situ* monitoring of aircraft engine parts.

smart glass. A glass that changes its degree of transparency in an applied electrical field. For example, a glass containing a liquid crystal thin film between transparent electrodes which, upon application of an electric field, changes in appearance from frosted and opaque to transparent.

smart materials. Engineering and/or engineered materials that are considered to be "smart" because of their ability to adapt to changes in the surroundings. Materials that respond to external change without assistance are referred to as "passively smart," while materials that recognize external changes and respond accordingly are known as "actively smart" or "intelligent." Four widely used smart materials are electrostrictive lead magnesium niobate, piezoelectric lead zirconate titanate, magnetostrictive terbium-dysprosium-iron alloy (*Terfenol*) and nickel-titanium shape-memory alloy (*Nitinol*). Used for actuators, transducers, sensors, etc. See also intelligent materials; shape-memory alloys.

smectic liquid crystal. A liquid crystal in a mesomorphic state ("mesophase") composed of rodlike organic molecules arranged in more or less organized layers. Currently there are more than a dozen identified smectic phases known as smectic-A (Sm-A), smectic-B (Sm-B), smectic-C (Sm-C), etc. In the Sm-A phase, the molecules in each layer are fluid, can glide easily over each other, and are oriented perpendicular to the layers. In the Sm-B phase, they have hexagonal ordering and strong interlayer correlations, and in the Sm-C phase, the molecules in each layer are fluid, can glide over each other, and are oriented with their axes tilted with respect to the layer normal. See also liquid crystal; cholesteric liquid crystal; nematic liquid crystal.

smectite. A term referring to kaolinite minerals of the mixed-layer group composed of aluminum silicate hydroxide monohydrate, Al-Si-O-OH-H_2O. Crystal system, monoclinic. Occurrence: Poland.

smelter-grade alumina. Aluminum oxide produced by the Bayer process from bauxite containing about 30-60% alumina. A typical composition (in wt%) is 99.3-99.7% aluminum oxide (Al_2O_3), 0.20-0.50% sodium oxide (Na_2O), 0.005-0.025% silicon dioxide (SiO_2), 0.005-0.020% ferric oxide (Fe_2O_3), 0.001-0.008% titanium dioxide (TiO_2), 0.005-0.015% gallium oxide (Ga_2O_3), less than 0.005-0.040% calcium oxide (CaO), less than 0.001-0.010% zinc oxide (ZnO), less than 0.0001-0.0015% phosphorus pentoxide (P_2O_5), less than 0.001-0.003% vanadium pentoxide (V_2O_5) and 0.05-0.20% sulfur trioxide (SO_3). Used as a starting material for the production of aluminum.

smirnite. A mineral composed of bismuth tellurium oxide, Bi_2TeO_5. It can also be made synthetically from bismuth oxide and tellurium dioxide. Crystal system, orthorhombic. Density, 7.83 g/cm^3.

smithite. A brown-orange mineral composed of silver arsenic sulfide, Ag_2AsS_2. It can also be made synthetically. Crystal system, monoclinic. Density, 4.88 g/cm^3. Occurrence: Switzerland.

smithsonite. A colorless, white, gray, yellow, brown or green mineral with a vitreous luster. It belongs to the calcite group and is composed of zinc carbonate, $ZnCO_3$. Crystal system, rhombohedral (hexagonal). Density, 4.30-4.45 g/cm^3; hardness, 4.5-5.0; refractive index, 1.841. Occurrence: Greece, Southern Africa, USA (Arkansas, Colorado, Missouri, New Mexico, Wisconsin). Used as an ore of zinc. The translucent green or bluish varieties are used as gemstones. Also known as *zinc spar*.

smoked elkskin. A soft, tough, smoked chrome-tanned cow or steer hide resembling *elkskin*, but also made in black and other colors. Used for shoes.

smoked glass. Gray or smoky-brown commercial glassware made by addition of certain chemicals to the melt, or by exposure to a reducing atmosphere during melting and cooling.

smoked sheet rubber. Natural rubber (latex) coagulated with acetic or formic acid, or sodium hexafluorosilicate, formed into sheets and exposed to wood smoke to kill bacteria. See also natural rubber; rubber latex.

smoky quartz. A smoky brownish-gray variety of *quartz* found in Switzerland, Spain and the USA (Colorado), and used chiefly as a gemstone.

S-Monel. Trade name of Henry Wiggin & Company Limited (UK) for a series of extra-hard nonmagnetic casting alloys containing 61-65% nickel, 30-32% copper, 3.5-4.0% silicon, 1-3% iron, up to 1% manganese and 0.12% carbon. They possess good corrosion, abrasion and wear resistance, good nongalling properties, high strength and hardness and low sparking properties. Used for valves and valve parts, pumps and pump parts, chemical and marine parts, disks and bushings for high temperature steam environments, and pump liners and sleeves for severe conditions. See also Monel.

Smooth-Cut. Trade name of Columbia Tool Steel Company (USA) for an air- or oil-hardening, nondeforming cold-work tool steel (AISI type A6) containing 0.7% carbon, 2% manganese, 1.1% chromium, small amounts of sulfur, and the balance iron. It has excellent deep-hardening properties, fair wear resistance and machinability, and is used for blanking and forming dies, coining and trimming dies, shear blades, punches, and mandrels.

smooth-finish tile. Unglazed tile with a smooth surface, left as removed from the forming die.

smooth glass. Glass finely ground on one or two surfaces and ready for polishing.

smooth one-side hardboard. Hardboard that has one smooth surface (or face) and a mesh pattern or other rough texture on the opposite side (or back). Abbreviation: S1S hardboard. See also hardboard.

Smooth Rough. Trade name of C-E Glass (USA) for a patterned glass.

smooth two-side hardboard. Hardboard with two smooth surfaces (face and back) produced from a dry mat by pressing between two smooth, heated platens. Abbreviation S2S hardboard. See also hardboard.

smythite. A mineral of the nickeline group composed of iron sulfide, Fe_9S_{11}. Crystal system, rhombohedral (hexagonal). Density, 3.95 g/cm^3. Occurrence: USA (Indiana).

snakeskin. The thin, durable, vegetable-tanned leather obtained from the skin of large snakes and used for fancy leather goods, novelty items and shoe uppers mainly because of its beautiful, conspicuous markings.

snakeskin glaze. A glaze of very low expansion or high surface tension that has a tendency to crawl during firing producing a decorative effect resembling snakeskin on pottery and other whiteware ceramics.

Snap. Trademark of Parkell Inc. (USA) for a color-stable vinyl ethyl methacrylate-based powder-liquid dental resin with very low polymerization shrinkage used for temporary crown and bridge restorations.

Snap-Stone. Trade name of Whip Mix Corporation (USA) for a fast-setting dental stone supplied as a light pink powder. When cast from a mixture of 23 mL water per 100 g powder, it develops high early compressive strength. It is compatible with all impression materials and suitable for many applications from mouth guards and orthodontic appliances to temporary restorations and denture repairs.

snarl yarn. A compound yarn made from a lively, highly twisted yarn so as to have snarls or kinks of varying size and frequency protruding from the core.

snarly yarn. See lively yarn.

Sniaform. Trade name of Nyltech (USA) for a series of polyamide (nylon) resins.

Snialon. Trade name of SNIAFA (Argentina) for nylon 6 fibers and filament yarns.

Sniamid. Trade name of Nyltech (USA) for a series of polyamide (nylon) resins including *Sniamid 6* based on polyamide 6 (nylon 6) and supplied in unmodified, glass fiber-reinforced, glass bead- or mineral-filled, fire-retardant, high-impact and UV-stabilized grades, *Sniamid 66* based on polyamide 6,6 (nylon 6,6) and supplied in unmodified, glass fiber- and bead-reinforced, mineral-filled and fire-retardant grades, and *Sniamid ASN,* a transparent polyamide.

Sniater. Trade name of Baroda Rayon Corporation (India) for polyester fibers and filament yarns.

Sniavitrid. Trade name of Nyltech (USA) for polyamide 6 (nylon 6) filled with 25% glass beads.

S-Nickel. Trademark of Inco Limited (Canada) for a sulfur-activated anode material containing 99.96% nickel. Supplied in the form of small spherical pellets, it is used in nickel plating with baskets.

Sniol. Trade name for vinyon (vinyl chloride) fibers with excellent acid, alkali and mildew resistance. Used especially for textile fabrics.

Sno-Flake. Trade name of Henkel Surface Technologies (USA) for water-wash spray booth compounds, booth coatings, paint strippers and polishing compounds.

Snowcrete. Trademark of Associated Portland Cement Manufacturers Limited (UK) for a white Portland cement.

Snow*Tex. Trademark of US Silica Company (USA) for a white calcined kaolin supplied as a powder with an average particle size of 1.5 μm (59 μin.) and a density of 2.58 g/cm³. It is composed of 57.0% silicon dioxide (SiO_2), 39.0% aluminum oxide (Al_2O_3), 1.00% ferric oxide(Fe_2O_3), 0.70% titanium dioxide (TiO_2), 1.70% potassium oxide (Na_2O) plus sodium oxide (Na_2O), and 0.35% calcium oxide (CaO) plus magnesium oxide (MgO).

SnTech. Trademark of Lucent Technologies (USA) for tin electrodeposits and deposition processes.

Soalon. Trademark of Mitsubishi Rayon Company (Japan) for a strong cellulose triacetate fiber with good resistance to dilute solutions of weak acids, excellent mildew and sunlight resistance, good dimensional stability, and good resistance to temperatures up to about 250°C (480°F). Used for textile fabrics including woven and knitted clothing.

soap brick. Brick whose width is one-half the usual dimension, i.e., about 2 in. (50 mm).

soap-drawn wire. A wire produced from wire rod by first removing the scale and then coating with a soap solution and drawing through the tapered hole of a die, or through a series of dies.

soaprock. See talc (2).

soapstone. See talc (2).

Soarnol. Trademark of Atofina SA (France) for ethylene-vinyl alcohol (EVA) copolymers used for packaging applications.

sobelevskite. A grayish white mineral of the nickeline group composed of bismuth palladium, PdBi. Crystal system, hexagonal. Density, 11.88 g/cm³. Occurrence: Russian Federation.

soda. Any of several sodium-containing compounds, such as sodium carbonate (calcined soda), sodium hydroxide (caustic soda), sodium bicarbonate (baking soda) and sodium oxide. See also sodium bicarbonate; sodium carbonate; sodium carbonate decahydrate; sodium carbonate monohydrate; sodium hydroxide; sodium oxide.

soda alum. See sodium-aluminum sulfate.

soda aluminosilicate glass. A chemically strengthened glass composed predominantly of silicon dioxide, aluminum oxide and sodium monoxide, and used for special applications, e.g., aircraft windshields.

soda ash. See sodium carbonate.

soda-ash briquettes. Pure soda ash in the form of blocks, often with a hydrocarbon binder, used as a ladle or furnace addition for desulfurization of ferrous melts (iron and steel). See also sodium carbonate.

soda-ash pellets. Soda ash in the form of pellets used as a ladle or furnace addition for desulfurization of ferrous melts (iron and steel). It may also contain other purifying ingredients. See also sodium carbonate.

soda ball. A black product composed of sodium carbonate, sodium sulfide, carbon and mineral matter, and obtained by concentrating and heating black liquor in rotary furnaces. Also known as *black ash.*

soda crystals. See sodium carbonate monohydrate.

soda feldspar. See albite.

soda-lime-borosilicate glass. A special type of soda-lime glass containing about 64-68% silica (SiO_2), 11-15% lime (CaO), a total of 7-10% soda and potash (Na_2O and K_2O), 4-6% boria (B_2O_3), 3-5% alumina (Al_2O_3), 2-4% magnesia (MgO), 0-1% barium oxide (BaO) and 0-0.8% iron oxide (Fe_2O_3). It has good chemical stability in corrosive environments, and is used in the form of reinforcing fibers for engineering composites.

soda-lime feldspars. See plagioclases.

soda-lime glass. A commercial glass with a relatively low melting temperature consisting of 72-76% silicon dioxide (SiO_2), 14-16% soda (Na_2O), 5-10% lime (CaO), 0-4% magnesia (MgO) and 1-2% alumina (Al_2O_3). It has high durability, good workability, and is used for window and plate glass, containers, art objects, light bulbs, and industrial products. Also known as *soda-lime-silicate glass.*

sodalite. (1) A feldspathoid mineral with an opaque to vitreous luster that is usually blue, but also occurs in white, yellow, gray or green colors. It is composed of sodium aluminum chloride silicate, $Na_4Al_3Si_3O_{12}Cl$, and, structurally, is a tectosilicate with aluminosilicate tetragonal framework. Crystal system, cubic. Density, 2.3 g/cm³; hardness, 5.5-6 Mohs; refractive index, 1.488. Occurrence: Canada (British Columbia, Ontario), USA (Maine), Italy. Used for ornamental purposes, and as a catalyst.

(2) See cage zeolite.

soda microcline. See anothoclase.

soda monohydrate. See sodium carbonate monohydrate.

soda niter. See sodium nitrate.

soda nitrate. See sodium nitrate.

soda pulp. A chemical woodpulp obtained from the digestion of wood chips (chiefly poplar) by caustic soda.

soda spar. A feldspar that contains 7% or more soda (Na_2O), and is used for ceramic enamels.

soda water glass. Greenish, glassy lumps, grayish white powder or a viscous, turbid or transparent liquid varying in composition from $Na_2O \cdot 3.75SiO_2$ to $2Na_2O \cdot SiO_2$ and having varying proportions of water. It is obtained by fusing sand and soda ash, and is used as an adhesive, in concrete hardeners, as a binder and deflocculating agent in the manufacture of cements, mortars, paints, abrasive wheels, foundry cores and molds, as a glass foamer, in the manufacture of silica gel, protective coatings and pigments, for linings of converters, digesters and acid concentrators and other chemical or metallurgical equipment, in ore flotation, and as a flame retardant. Also known as *liquid glass; silicate of soda; sodium silicate; sodium silicate glass; soluble glass.*

soddyite. A yellow, radioactive mineral composed of uranyl silicate dihydrate, $(UO_2)_2(SiO_4) \cdot 2H_2O$. It can also be made synthetically by precipitating from an aqueous solution containing sodium silicate and uranyl acetate. Crystal system, orthorhombic. Density, 4.70 g/cm³; refractive index, 1.685. Occurrence: Zaire.

Soderfors. Trademark of Soderfors AB (Sweden) for an extensive series of steels including various water-hardening plain-carbon, carbon-tungsten special-purpose, cold-work, hot-work and high-speed tool steels, numerous austenitic, ferritic and martensitic stainless steels, and several alloy steels including case-hardening grades.

Soding. Trade name of J.C. Soding & Halbach (Germany) for an extensive series of steels including numerous water-hardening plain-carbon, cold-work, hot-work and high-speed tool steels, and various austenitic, ferritic and martensitic stainless steels.

sodium. A soft, silver-white metallic element of Group IA (Group 1) of the Periodic Table. It is commercially available in the form of ingots, lumps, bricks, spheres and sticks, usually packaged under vacuum, nitrogen or argon in ampules or bottles, or in kerosene or mineral spirits, and as 40 or 50 wt% dispersion in light oil, paraffin or mineral spirits. It occurs in nature only in compounds especially as the chloride (*halite*) and nitrate (*soda niter*). Density, 0.968 g/cm³; melting point, 97.8°C (208°F); boiling point, 892°C (1638°F); atomic number, 11;

atomic weight, 22.99; monovalent. It is waxy at room temperature and brittle at low temperatures, and has excellent electrical conductivity (about 41% IACS), high heat-absorbing capacity, very high chemical reactivity and sensitivity to air and moisture, high flammability and a moderately low neutron-capture cross section. There are two forms of sodium: (i) *Alpha sodium* that has a body-centered cubic crystal structure and exists at room temperature; and (ii) *Beta sodium* that has a hexagonal crystal structure and exists at low temperatures (below 36K or -237.15°C). *Sodium* is used as an alloying addition for aluminum, lead and zinc, as a reactant for the deoxidation of metals, as a coolant in nuclear reactors, as an electrical conductor in homopolar generators, as a heat-transfer agent in solar-powered electric generators, in electric power cables (encased in polyethylene), as a working fluid for evaporative heat pipes, in street lights and nonglare lighting for highways, in sodium-sulfur batteries, fuel cells and scintillators, in titanium reduction, in glass manufacture, in the manufacture of sodium compounds and tetraethyl and tetramethyl lead, and as a polymerization catalyst for synthetic rubber. The radioactive forms (e.g., sodium-24) are used in tracer studies.

sodium-24. A radioactive isotope of sodium with a mass number of 24 produced by deuteron bombardment of sodium. It emits beta and gamma rays, decays to magnesium-24 and has a half-life of 15.1 hours. Used in chemical, biological and medical research (e.g., in the study of metabolic pathways in animals and plants) and for tracer studies. Symbol: ^{24}Na. Also known as *radiosodium.*

sodium acetate. The sodium salt of acetic acid available as colorless, water-soluble crystals or white powder (99+% pure). Density, 1.528 g/cm³; melting point, 324°C (615°F). Used in electroplating, photography and tanning, as a dye and color intermediate, as a dehydrating agent, as laboratory reagent, in biochemistry and medicine, in microscopy, and in food production. Formula: $NaC_2H_3O_2$.

sodium acetate trihydrate. Colorless, water-soluble crystals (99+% pure). Crystal system, monoclinic. Density, 1.45 g/cm³; melting point, 58°C (136°F). Used in electroplating, photography and tanning, as a dye and color intermediate, as a dehydrating agent, as laboratory reagent, in biochemistry and medicine, and in food production. Formula: $NaC_2H_3O_2 \cdot 3HO$.

sodium acid carbonate. See sodium bicarbonate.

sodium acid fluoride. See sodium bifluoride.

sodium acid phosphate. See monosodium phosphate.

sodium acid pyrophosphate. White, crystalline powder. Crystal system, monoclinic. Density, 1.862 g/cm³; melting point, decomposes at 220°C (428°F). Used in electroplating, metal cleaning and phosphatizing. Formula: $Na_2H_2P_2O_7 \cdot 6H_2O$. Abbreviation: SAPP. Also known as *dibasic sodium pyrophosphate; disodium dihydrogen pyrophosphate; disodium pyrophosphate; disodium diphosphate.*

sodium acid sulfate. See sodium bisulfate.

sodium acid sulfite. See sodium bisulfite.

sodium alginate. The sodium salt of alginic acid available in the form of a colorless to pale yellow solid obtained from brown seaweed by extraction and purification. It forms aqueous colloidal solutions and is used in water-base and latex paints, in paper coatings, in films, in cement compositions, as a suspension agent and thickener, and in biochemistry and biotechnology. Formula: $NaO_6C_6H_7$.

sodium aluminate. White, hygroscopic powder (99.9% pure). Melting point, 1800°C (3270°F). Used in the manufacture of

milk glass because of its opacifying or obscuration properties, in porcelain-enamels and glaze slips to improve the suspension and working properties, in hardening building stones, in zeolites, water purification and paper sizing. Formula: $Na_2Al_2O_4$. Also known as *sodium metaaluminate.*

sodium-aluminum sulfate. A colorless crystalline substance obtained by adding sodium chloride to a hot solution of aluminum sulfate. Density, 1.675 g/cm³; melting point, 61°C (142°F). Used in waterproofing of fabrics, in dry colors and ceramics, in tanning and paper sizing, in matches, in engraving, and in water purification. Abbreviation: SAS. Formula: $AlNa-(SO_4)_2 \cdot 12H_2O$. Also known as *alum; aluminum-sodium sulfate; soda alum.*

sodium amalgam. A silvery white, porous, highly flammable, crystalline mass that contains 2-20% sodium metal, and is used in the reduction of metal halides, in metal separation, hydrogen manufacture, fuel cells, and analytical chemistry. Formula: Na_xHg_y.

sodium antimonate. See antimony sodiate.

sodium aurichloride. See gold-sodium chloride.

sodium aurocyanide. See gold-sodium cyanide.

sodium-base grease. See sodium grease.

sodium benzoate. The sodium salt of benzoic acid available as a hygroscopic, white, crystalline compound with a melting point above 300°C (570°F), formed by a reaction of aqueous sodium hydroxide and benzonitrile at high temperatures, or by a reaction of benzoic acid and sodium bicarbonate. Used as a rust inhibitor for steel and iron, etc., as an antiseptic, as a food preservative, as an intermediate for dyes and liquid crystals. Formula: $NaC_7H_5O_2$.

sodium-beryllium fluoride. See beryllium-sodium fluoride.

sodium bicarbonate. White crystals or powder. Crystal system, monoclinic. Density, 2.16-2.20 g/cm³; melting point, loses CO_2 at 270°C (518°F). Used as a deflocculant in special casting slips, in the preparation of cobalt body stains, as a body wash to improve body-glaze reactions, in cleaning solutions for enameler's steel, in gold and platinum plating, in sponge rubber, as a fire extinguisher, and in chemistry and biochemistry as buffer component. Formula: $NaHCO_3$. Also known as *baking soda; sodium acid carbonate; sodium-hydrogen carbonate.*

sodium bichromate. See sodium dichromate.

sodium bifluoride. White, crystalline powder, granules (98% pure), or small pellets (99% pure). Density, 2.08 g/cm³; melting point, decomposes on heating. Used in the room-temperature synthesis of fluoride glasses, in etching glass, and in the manufacture of tinplate. Formula: $NaHF_2$. Also known as *sodium acid fluoride; sodium-hydrogen fluoride.*

sodium biphosphate. See monosodium phosphate.

sodium bismuth tungstate. A double tungstate of sodium and bismuth available as an optically dense single crystal for use as a scintillator, and as a Cerenkov radiator. Formula: $NaBi-(WO_4)_2$. Abbreviation: NBWO.

sodium bisulfate. Colorless crystals or white fused lumps. Density, 2.44 g/cm³; melting point, decomposes above 315°C (599°F). Used in the manufacture of brick and magnesia cements, in the manufacture of paper, soap, industrial cleaners and metal pickling compounds, and as a flux for decomposing minerals. Formula: $NaHSO_4$. Also known as *niter cake; sodium acid sulfate; sodium-hydrogen sulfate.*

sodium bisulfite. White crystals or crystalline powder. Density: 1.48 g/cm³; melting point, decomposes. Used as a polymerization initiator, in copper and brass plating, as a metallographic etchant, as a mordant for textiles, as a bleaching agent for groundwood, wool, etc. Formula: NaHSO$_3$. Also known as *sodium acid sulfite; sodium-hydrogen sulfite*.

sodium boltwoodite. A pale yellow to white mineral of the uranophane group composed of potassium sodium oxonium uranyl silicate monohydrate, (Na,K)(H$_3$O)UO$_2$SiO$_4$·H$_2$O. Crystal system, tetragonal. Density, 4.10 g/cm^3. Occurrence: Russian Federation.

sodium borate. See borax.

sodium borate decahydrate. See borax.

sodium borohydride. A white, moisture-sensitive crystalline powder, granules or pellets (99+% pure). Density, 1.07 g/cm^3; melting point, approximately 400°C (752°F) (decomposes). It is also available in solutions, and in cobalt-doped grades. Used in the recycling of gold and platinum group metals, in organometallics, in biochemistry and biotechnology, in electroless nickel plating, in bleaching wood pulp and as a blowing agent for plastics. The cobalt-doped grades are used as a solid source of hydrogen gas and in the production of anaerobic reducing atmospheres in biology and biochemistry. Formula: NaBH$_4$. Also known as *sodium tetrahydridoborate*.

sodium bromide. White, hygroscopic crystals or crystalline powder (99+% pure). Crystal system, cubic. Density, 3.203 g/cm^3; melting point, 755°C (1391°F) [also reported as 747°C or 1377°F]; boiling point, 1390°C (2534°F); refractive index, 1.641. Used in the preparation of bromine compounds, in photography and in biochemistry and medicine. Formula: NaBr.

sodium-calcium feldspars. See plagioclases.

sodium-calcium grease. A water-resistant lubricating grease consisting of a complex sodium-calcium soap dissolved in a synthetic oil. It is suitable for lubricating bearings operating at high speeds and has a standard operating temperature range of -60 to +120°C (-76 to +250°F).

sodium carbonate. A white, hygroscopic powder (99.5+% pure). Density, 2.532 g/cm^3; melting point, 851°C (1563°F); boiling point, decomposes; refractive index, 1.535. Used as a flux in glass, glazes and enamels, as an acid neutralizer in the treatment of metals for porcelain enamels, in biochemistry and medicine, in ceramics and in petroleum refining. Formula: Na$_2$CO$_3$. Also known as *calcined soda; soda ash*.

sodium carbonate decahydrate. White, monoclinic crystals. Density, 1.44 g/cm^3; melting point, loses 10H$_2$O at 32.5-34.5°C (90.5-94°F). Used as an oxidizer and flux in glasses and porcelain enamels, as a neutralizer in the pickling of iron and steel prior to porcelain enameling, in biochemistry and medicine, and in washing and bleaching textiles. Formula: Na$_2$CO$_3$·10H$_2$O. Also known as *sal soda; washing soda*.

sodium carbonate monohydrate. White crystals or crystalline powder (99.5+% pure). Density, 2.25 g/cm^3; melting point, loses H$_2$O at 109°C (228°F). Used in glass manufacture, in photography, as an intermediate in thermochemical reactions, as an analytical reagent, in biochemistry and medicine, and in cleaning and bleaching compounds. Formula: Na$_2$CO$_3$·H$_2$O. Also known as *crystal carbonate; soda crystals; soda monohydrate*.

sodium carboxymethyl cellulose. See carboxymethylcellulose.

sodium chlorate. Colorless crystals or white powder (99+% pure). Crystal system, cubic or trigonal. Density, 2.490 g/cm^3; melting point, 248-261°C (479-502°F); boiling point, decomposes. Used as an electrolyte in the chemical machining, as an oxidizer, in biochemistry, in bleaching textiles and paper pulp, in ore processing, and in pyrotechnics. Formula: NaClO$_3$.

sodium chloride. Colorless hygroscopic crystals or white, crystalline powder (99+% pure). It occurs in nature as the mineral *halite*. Crystal system, cubic. Crystal structure, halite. Density, 2.165 g/cm^3; melting point, 801°C (1474°F); boiling point, 1413°C (2575°F); hardness, 2 Mohs; refractive index, 1.542. Used for ceramic glazes, in glass, soda-ash and soap, in metallic sodium production, in metallurgy, in paper pulp and water softeners, in deicing of roads and sidewalks, in ion exchange, in nuclear reactors, in fire extinguishers, in supercooled solutions, in biochemistry and medicine, and in tanning and spectroscopy. Single crystals are used for ultraviolet and infrared transmission, spectroscopy (e.g. in wavelength-dispersive spectrometers), optical lenses and windows, and as selective reflectors. Formula: NaCl.

sodium chloroaurate. See gold-sodium chloride.

sodium chloroiridate. See iridium-sodium chloride.

sodium chloroosmate. See osmium-sodium chloride.

sodium chloroplatinate. See sodium tetrachloroplatinate; sodium hexachloroplatinate.

sodium chromate. Yellow, translucent, deliquescent crystals (98% pure). Crystal system, monoclinic. Density, 1.483 g/cm^3; melting point, 19.9°C (68°F). Used as a corrosion inhibitor for ferrous materials, as a paint pigment, in inks, and as a wood preservative. Formula: Na$_2$CrO$_4$.

sodium chromate tetrahydrate. Yellow, deliquescent, monoclinic crystals (99% pure) used in the manufacture of pigments, in chromium compounds, in leather tanning, and as corrosion preventives. Formula: Na$_2$CrO$_4$·4H$_2$O.

sodium citrate. The trisodium salt dihydrate of citric acid available as white crystals or granular powder. Melting point, loses 2H$_2$O at 150°C (302°F) and melts above 300°C (570°F); boiling point, decomposes at red heat. Used in electroplating baths, in photography, and in biochemical and biological electrophoresis and microscopy. Formula: C$_6$H$_5$O$_7$Na$_3$·2H$_2$O. Also known as *trisodium citrate; trisodium citrate dihydrate*.

sodium cobaltinitrite. See sodium hexanitritocobaltate.

sodium columbate. See sodium niobate.

sodium-copper alloys. Alloys composed of 85-95% copper and 5-15% sodium and used for desulfurizing copper-base alloys, especially bronzes and brasses.

sodium-copper cyanide. White, crystalline, double salt of sodium cyanide and copper cyanide with a density of 1.01 g/cm^3, a melting point of 100°C (210°F) (decomposes) used in the preparation and maintenance of sodium cyanide base copper plating baths. Formula: NaCu(CN)$_2$. Also known as *copper-sodium cyanide; sodium cyanocuprate*.

sodium cyanate. White, crystalline powder (90+% pure) with a density of 1.937 g/cm^3 and a melting point of 700°C (1970°F). Used in the heat treating of steel, in organic synthesis, and in the manufacture of medical preparations. Formula: NaOCN.

sodium cyanide. Colorless or white, deliquescent crystals or crystalline powder (95+% pure). Crystal system, cubic. Melting point, 563.7°C (1047°F); boiling point, 1496°C (2725°F). Used in electroplating and metal cleaning, as an addition to improve the performance of neutralizer baths in preparing steels for porcelain enameling, in the case hardening of steel, in various heat-treating baths, in the extraction of gold and silver from ores, in ore flotation, in biochemistry and biology, in the preparation of hydrogen cyanide, in chelating compounds, and in the manufacture of pigments and nylon. Formula: NaCN.

sodium cyanoaurite. See gold-sodium cyanide.

sodium cyanocuprate. See sodium-copper cyanide.

sodium dachiardite. A colorless mineral of the zeolite group composed of sodium aluminum silicate hydrate, $Na_4(Al_4Si_{20})$-$O_{48} \cdot 13H_2O$. Crystal system, monoclinic. Density, 2.16 g/cm^3. Occurrence: Japan.

sodium deuteride. A moisture-sensitive, gray powder usually supplied as a 20 wt% slurry in oil with 98 at% deuterium and used as a reducing agent and in the preparation of deuterated sodium compounds. Formula: NaD.

sodium dichromate. Red or orange, deliquescent crystals (99+% pure). Crystal system, monoclinic. Density, 2.52 g/cm^3; melting point of 357°C (675°F); decomposes at 400°C (752°F) and loses $2H_2O$ on prolonged heating at 100°C (212°F). Used in colorimetry, in biology, biochemistry and medicine, as an orange-yellow colorant for glazes and porcelain, and as a pigment for coloring metals. Formula: $Na_2Cr_2O_7 \cdot 2H_2O$. Also known as *sodium dichromate dihydrate; sodium bichromate.*

sodium dihydrogen phosphate. See monosodium phosphate.

sodium diuranate. A yellow-orange solid used as a pigment to produce yellow or ivory shades in bodies, glazes and porcelain enamels, and in the manufacture of opalescent, fluorescent yellow uranium glass. Formula: $Na_2U_2O_7 \cdot 6H_2O$. Also known as *uranium yellow; yellow oxide.*

sodium edetate. See monosodium EDTA.

sodium feldspar. See albite.

sodium ferricyanide. Ruby-red, deliquescent crystals obtained by introducing chlorine into sodium ferrocyanide solution, and used in the manufacture of pigments, and in printing. Formula: $Na_3Fe(CN)_6 \cdot H_2O$. Also known as *red prussiate of soda; sodium hexacyanoferrate hydrate.*

sodium ferrocyanide. Lemon-yellow, semitransparent, monoclinic crystals with a density of 1.458 g/cm^3 used in case hardening of steel, as a blue pigment in paints and printing inks and blueprint paper, in metal pickling, as a catalyst, and as a photographic fixing agent. Formula: $Na_4Fe(CN)_6 \cdot 10H_2O$. Also known as *yellow prussiate of soda; sodium hexacyanoferrate decahydrate.*

sodium fluoaluminate. Colorless or white crystals obtained from the mineral *cryolite,* or made synthetically from fluorite (CaF_2), sodium carbonate, sulfuric acid and hydrated aluminum oxide. Crystal system, hexagonal. Density, 2.90 g/cm^3; melting point, above 1000°C (1830°F); refractive index, 1.338. Used as a flux and opacifier in opal glass and porcelain enamels, as a flux in whiteware bodies, in light bulbs and welding-rod fluxes, as a flux in the production of aluminum from bauxite (electrolysis process), as a filler in grinding wheels, as constituent in dental cements, as a binder for abrasives, in the manufacture of soda, and in electric insulation. Formula: Na_3AlF_6. Also known as *sodium hexafluoroaluminate; synthetic cryolite.*

sodium fluoborate. A white, moisture-sensitive powder (98+% pure) with a density of 2.47 g/cm^3 and a melting point of 384°C (723°F). It is slowly decomposed by heat, and is used in fluxes for nonferrous metals, in electrochemical processes, in sand casting of aluminum and magnesium, as an oxidation inhibitor, and as a fluorinating agent. Formula: $NaBF_4$. Also known as *sodium tetrafluoroborate.*

sodium fluoride. Colorless or white crystals or powder (99+% pure). It occurs in nature as the mineral *villiaumite,* and is also available in synthetic, high-purity optical grades. Crystal system, cubic or tetragonal. Crystal structure, halite. Density, 2.56-2.78 g/cm^3; melting point, 988°C (1810°F); boiling point, 1695°C (3083°F); hardness, 2-2.5 Mohs; refractive index, 1.336. Used in the manufacture of synthetic cryolite and glass, as a flux and gas- or bubble-type opacifier in porcelain enamels, in

electroplating baths, in biochemistry and medicine, and in the degassing of steel. Single crystals are used in windows for ultraviolet- and infrared-radiation detection systems. Formula: NaF. Also known as *fluorol.*

sodium fluosilicate. Colorless crystals or white, free-flowing powder (99+% pure). Density, 2.679 g/cm^3; melting point, decomposes at red heat, refractive index, 1.310. Used as a flux and opacifier in porcelain enamels, as an opacifier in glass, in opalescent glass, in metallurgy (aluminum and beryllium), in the manufacture of pure silicon, and in organic synthesis. Formula: Na_2SiF_6. Also known as *sodium hexafluorosilicate; sodium silicofluoride.*

sodium germanate. A compound of sodium monoxide and germanium oxide supplied in the form of white, deliquescent crystals. Crystal system, monoclinic. Density, 3.31 g/cm^3; melting point, 1083°C (1981°F); refractive index, 1.59; soluble in water. Used in special glasses, in electronic devices, such as diode rectifiers and transistors, and as a catalyst for the preparation of polyesters. Formula: Na_2GeO_3. Also known as *sodium metagermanate; sodium germanium oxide (NGO).*

sodium germanium oxide. See sodium germanate.

sodium glucoheptonate. A light tan, crystalline powder used as a sequestering agent for polyvalent metals, in metal cleaning, in paint stripping and as an etchant for aluminum. Formula: $NaC_7H_{13}O_8$.

sodium gluconate. The sodium salt of gluconic acid available as a white to yellowish crystalline powder, and used as a sequestering agent, in metal cleaning, in paint stripping, as an aluminum deoxidizer, in metal plating, and in rust removal. Formula: $NaC_6H_{11}O_7$.

sodium glycolate. A white powder used as a buffer in electroless plating. Formula: $NaH_3C_2O_3$. Also known as *sodium hydroxyacetate.*

sodium-gold chloride. See gold-sodium chloride.

sodium-gold cyanide. See gold-sodium cyanide.

sodium grease. A lubricating grease composed of a sodium soap suspended in mineral oil. It is not resistant to water, but completely dissolved in it, has a texture varying from fibrous to smooth and an operating temperature range for standard grades of about -20 to +90°C (-8 to +195°F), but high-temperature grades are also available. Used for the lubrication of wheel bearings, high-temperature components, transmissions, etc. Also known as *sodium-base grease.*

sodium halide. A compound of sodium with a halogen, such as bromine, chlorine, fluorine, or iodine.

sodium hexachloroiridate. See iridium-sodium chloride.

sodium hexachloroosmate. See osmium-sodium chloride.

sodium hexachloroplatinate. A compound usually supplied as the tetrahydrate ($Na_2PtCl_6 \cdot 4H_2O$) in the form of a yellow, water-soluble powder, or as the hexahydrate ($Na_2PtCl_6 \cdot 6H_2O$) as an orange, hygroscopic powder with a density of 2.5 g/cm^3. Used as a reagent, in etching on zinc, in plating, microscopy and photography, and for mirrors. Also known as *platinic sodium chloride; platinum-sodium chloride; sodium platinichloride.*

sodium hexacyanoferrate. See sodium ferricyanide; sodium ferrocyanide.

sodium hexafluoroaluminate. See sodium fluoaluminate.

sodium hexafluorophosphate. White, hygroscopic crystals (98+% pure) with a density of 3.369 g/cm^3 and a melting point above 200°C (390°F). Formula: $NaPF_6$.

sodium hexafluorosilicate. See sodium fluosilicate.

sodium hexametaphosphate. Hygroscopic, water-soluble, white crystals or powder (96+% pure) that are water miscible and have dispersing, deflocculating and sesquesting properties. Used in water softening, as a corrosion inhibitor for steels, in deicing salt mixtures, as a dispersant in clays and pigments, in scale removal, etc. Formula: $(NaPO_3)_6$.

sodium hexanitritocobaltate. A compound containing trivalent cobalt, and available as an orange powder that decomposes at 220°C or 428°F). It is usually obtained by adding sodium nitrite and acetic acid to cobalt salt solution. Used as a pigment, and in biochemistry and medicine. Formula: $Na_3Co(NO_2)_6$. Also known as *sodium cobaltinitrite; cobalt-sodium nitrite.*

sodium hydrate. See sodium hydroxide.

sodium hydride. A silver-gray, moisture-sensitive, flammable powder, or a gray moisture-sensitive, microcrystalline 25-80 wt% dispersion in mineral oil. Properties of powder: Density, 1.39 g/cm³; melting point, decomposes at 425°C (797°F); particle size range, 5-50 μm (0.0002-0.002 in.). Used as a descaling agent for hot-rolled strip. Formula: NaH.

sodium-hydrogen carbonate. See sodium bicarbonate.

sodium-hydrogen fluoride. See sodium bifluoride.

sodium hydrogen phosphate. See disodium phosphate.

sodium-hydrogen sulfate. See sodium bisulfate.

sodium-hydrogen sulfite. See sodium bisulfite.

sodium hydroxide. A brittle, hygroscopic, white solid obtained by treating calcium oxide with a hot sodium carbonate solution, or by the interaction of calcium hydroxide and sodium carbonate. It is commercially available in the form of beads or pellets, and as aqueous solutions. Density, 2.130 g/cm³; melting point, 318.4°C (605°F); boiling point, 1390°C (2534°F); strong corrosive alkali. Used in ceramics as a de-enameling agent, as a solvent for glass, in metal-cleaning compounds, as an etchant, in ion exchange, in electrodeposition, in biochemistry, biology and medicine, in the manufacture of cellophane, paper and rayon, and in rubber reclamation. Formula: NaOH. Also known as *caustic soda; sodium hydrate; white caustic.*

sodium hydroxyacetate. See sodium glycolate.

sodium hydroxystannate. See sodium stannate.

sodium hypophosphite. Colorless, pearly, crystalline plates or white granular powder used as a reducing agent in electroless nickel plating solutions for metals and plastics and as a laboratory reagent in chemistry, biochemistry, medicine and materials research. Formula: $NaH_2PO_2 \cdot H_2O$. Also known as *sodium hypophosphite monohydrate.*

sodium hyposulfite. See sodium thiosulfate.

sodium iodide. White, cubic crystals or white, hygroscopic powder (99.5+% pure). Crystal system, cubic. Density, 3.667 g/cm³; melting point, 651°C (1204°F); boiling point, 1304°C (2379°F); refractive index, 1.7745. Used in scintillators (thallium-activated sodium iodide crystals) e.g., for analysis and detection of nuclear energies, and in cloud seeding. Formula: NaI.

sodium iodide thallide. A hygroscopic, high-purity, thallium-activated sodium iodide crystal with high luminescence and negligible self-absorption. Crystal system, cubic. Density, 3.67 g/cm³; hardness, 2 Mohs, melting point, 651°C (1204°F). Used in scintillators, e.g., for the analysis and detection of nuclear energies. Formula: NaI(Th); NaI:Tl. Also known as *thallium-activated sodium iodide; thallium-doped sodium iodide.*

sodium-lead alloys. A group of alloys of sodium and lead including: (i) 90Pb-10Na, a flammable, corrosive solid alloy used in the manufacture of lead tetraethyl, and for the introduction of sodium into certain alloys; (ii) 98Pb-2Na, used as a deoxidizer and homogenizer in nonferrous lead-bearing metals; and (iii)

lead alloys with small quantities of sodium, used as deoxidizers and stabilizers for lead in cable sheathing. Also known *lead-sodium alloys.*

sodium metaaluminate. See sodium aluminate.

sodium meta-autunite. A yellow mineral of the meta-autunite group composed of sodium uranyl phosphate octahydrate, $Na_2(UO_2)_2(PO_4)_6 \cdot 8H_2O$. It can also be made synthetically. Crystal system, tetragonal. Density, 3.58 g/cm³. Occurrence: Russian Federation.

sodium metagermanate. See sodium germanate.

sodium metasilicate. A compound of sodium monoxide and about 44-47% silicon dioxide, available as a grayish white, hygroscopic, crystalline powder or white granules. Crystal system, monoclinic. Density, 2.4 g/cm³; melting point, 1088°C (1990°F). Used in metal cleaning, in cleaning drawing compounds from metals for porcelain enameling, and in de-inking paper. Formula: Na_2SiO_3. Also known as *sodium silicate.* See also soda water glass.

sodium metatantalate. A white, ferroelectric powder (99.9% pure) that has a perovskite structure at room temperature, a melting point of 630°C (1166°F) and Curie temperature of 475°C (887°F). Used for electroceramic applications. Formula: $NaTaO_3$. Also known as *sodium tantalate.*

sodium metavanadate. Colorless crystals, or white to pale green, monoclinic, crystalline powder with a melting point of 630°C (1166°F) used as a corrosion inhibitor in gas scrubbers, and in ferroelectric ceramics. Formula: $NaVO_3$. Also known as *sodium vanadate.*

sodium monoxide. See sodium oxide.

sodium molybdate. The sodium salt of molybdic acid available as a white powder (98+% pure). Density, 3.78 g/cm³; melting point, 687°C (1269°F); refractive index, 1.714. Used as a deflocculant, adherence promoter and rust inhibitor in porcelain enameling, as a paint pigment, in metal finishing, as a brightener in zinc plating, as a catalyst in the preparation of dyes and pigments, and in biology, biochemistry and medicine. Formula: Na_2MoO_4

sodium molybdate dihydrate. White, rhombohedral crystals or crystalline powder (99+% pure) with a density of 3.28 g/cm³ used as a paint pigment, in metal finishing and as a brightener in zinc plating, in the preparation of dyes and pigments, and in biology, biochemistry and medicine. Formula: $Na_2MoO_4 \cdot 2H_2O$. Also known as *sodium molybdate crystals.*

sodium molybdophosphate. Yellow, water-soluble crystals with a density of 2.83 g/cm³ used in photography, as a waterproofing agent for adhesives, cements and plastics, in pigments and in biochemistry and biology as a metal stain for microscopy. Formula: $Na_3[P(Mo_3O_{10})_4]$. Also known as *sodium phospho-12-molybdate.*

sodium molybdosilicate. Yellow crystals with a density of 3.44 g/cm³ used as a waterproofing agent for adhesives, cements and plastics, in pigments, as a fixing and oxidizing agent in photography, as a catalyst and reagent, in plating, in nuclear energy production (precipitant and ion exchanger). Formula: $Na_4[Si(Mo_3O_{10})_4] \cdot xH_2O$. Also known as *sodium silico-12-molybdate.*

sodium naphthenate. The sodium salt of naphthenic acid available as a powder containing about 8.5% sodium, and used in paints as bodying agent, emulsifier and drier.

sodium niobate. (1) A ferroelectric compound available as a white powder (99.9+% pure). Curie temperature, 360°C (680°F); Néel temperature, 627K. Used for electroceramic applications. Formula: $NaNbO_3$ ($NaCbO_3$) Also known as *sodium columbate.*

(2) A crystalline compound obtained by treating a niobium (columbium) compound with hot concentrated sodium hydroxide (NaOH). It consists of sodium monoxide and niobium pentoxide, and is used in the purification of niobium (columbium) materials. Formula: $Na_2Nb_2O_6 \cdot 7H_2O$ ($Na_2Cb_2O_6 \cdot 7H_2O$). Also known as *sodium columbate.*

sodium nitrate. Colorless or white, hygroscopic crystals (99+%). Crystal system, trigonal. Density, 2.267 g/cm³; melting point, 308°C (586°F). Used as an oxidizer and flux in glass, porcelain enamels and glazes, as a general oxidizing agent, in solid rocket propellants, in pyrotechnics, etc. Formula: $NaNO_3$. Also known as *soda niter; soda nitrate.*

sodium nitrite. Pale yellowish or white, hygroscopic crystals, pellets, sticks or powder (97+% pure) made by heating sodium nitrate. Crystal system, orthorhombic. Density, 2.16 g/cm³; melting point, 271°C (520°F); boiling point, decomposes at 320°C (608°F); ferroelectric properties below 437K. Used as a metal cleaner, acid neutralizer and rust inhibitor, as a tear-resistant additive in porcelain-enamel slips, in rubber accelerators, as a photographic reagent, and in dye manufacture. Formula: $NaNO_2$.

sodium orthophosphate. See disodium phosphate; monosodium phosphate; trisodium phosphate.

sodium-osmium chloride. See osmium-sodium chloride.

sodium oxide. Gray, deliquescent crystals, or white, amorphous powder (97% pure). Density, 2.27 g/cm³; melting point, sublimes at 1275°C (2327°F). Used as a network modifier in glass formation, in the manufacture of soda-lime and other glasses, and in enamels and glazes. Formula: Na_2O. Also known as *sodium monoxide.*

sodium-palladium chloride. See sodium tetrachloropalladate.

sodium pentaborate. A white, crystalline, free-flowing powder with a density of 1.72 g/cm³ used as a flux in glass manufacture and in fireproofing compositions. Formula: $Na_2B_{10}O_{16} \cdot 10H_2O$. Also known as *sodium pentaborate decahydrate.*

sodium peroxide. A yellowish-white, hygroscopic powder (93+% pure) with a density of 2.805 g/cm³ and a melting point of 675°C (1247°F) (decomposes). Used as an oxidizing agent, as a bleaching agent for paper, textiles, etc., in the manufacture of organic chemicals and pharmaceuticals, in water purification, in calorimetry, in the oxygen generation for undersea vessels, in ore processing, in analytical chemistry, and in biochemistry and medicine. Formula: Na_2O_2.

sodium persulfate. White crystalline powder with a density of 2.40 g/cm³ used as battery depolarizer, in emulsion polymerization and in bleaching. Formula: $Na_2S_2O_8$. Also known as *sodium peroxydisulfate.*

sodium phosphate. See disodium phosphate; monosodium phosphate; trisodium phosphate.

sodium phospho-12-molybdate. See sodium molybdophosphate.

sodium phosphotungstate. See sodium tungstophosphate.

sodium platinichloride. See sodium hexachloroplatinate.

sodium polysulfide. Yellow-brown, granular, free-flowing powder used in electroplating, and in oil-resistant synthetic rubber. Formula: Na_2S_n.

sodium-potassium alloys. A group of alloys available as silvery, air- and moisture-sensitive mobile liquids or solids. They are commercially supplied in three grades: (i) 78K-22Na with a density of 0.847 g/cm³ (0.030 lb/in.³) at 100°C (212°F), a melting point of -11°C (12°F) and a boiling point of 784°C (1443°F); (ii) 56K-44Na with a density of 0.886 g/cm³ (0.032 lb/in.³) at 100°C (212°F), a melting point of 19°C (66°F) and a boiling point of 825°C (1517°F); and (iii) 85.3Na-14.7K. *Sodium-potassium alloys* are used as heat exchanger fluids, as coolants in certain types of nuclear reactor, and as electric conductors. Also known as NaK. Also known as *potassium-sodium alloys.*

sodium-potassium tartrate. See potassium-sodium tartrate.

sodium pyroborate. See borax.

sodium pyrophosphate. See tetrasodium pyrophosphate.

sodium rhodanate. See sodium thiocyanate.

sodium selenide. Hygroscopic crystals or amorphous powder with a density 2.62 g/cm³ and a melting point above 875°C (1607°F). Used in chemistry and materials research. Formula: Na_2Se.

sodium selenite. White, moisture-sensitive crystals or powder (99+% pure). Density, 3.0 g/cm³; melting point, above 350°C (662°F). Both the anhydrous compound and the pentahydrate ($Na_2SeO_3 \cdot 5H_2O$), supplied as a white powder, are used as decolorizers in glass, and in the production of rose and ruby colors in glass, porcelain enamels and glazes. Formula: Na_2SeO_3.

sodium silicate. See soda water glass; sodium metasilicate.

sodium silicate glass. See soda water glass.

sodium silicofluoride. See sodium fluosilicate.

sodium silico-12-molybdate. See sodium molybdosilicate.

sodium silicotungstate. See sodium tungstosilicate.

sodium stannate. White to light tan, hexagonal crystals or powder (95% pure). Melting point, loses $3H_2O$ at 140°C (284°F); soluble in water. Used as a source of tin oxide, as an opacifier in glass, porcelain enamels and glazes, as a source of tin for electroplating and immersion plating, in blueprint paper, and in textile fireproofing. Formula: $Na_2SnO_3 \cdot 3H_2O$. Also known as *sodium hydroxystannate; sodium stannate trihydrate.*

sodium stannate trihydrate. See sodium stannate.

sodium subsulfite. See sodium thiosulfate.

sodium sulfate. White crystals or powder (99+% pure). It occurs in nature as the mineral *thenardite.* Crystal system, orthorhombic. Density, 2.68 g/cm³; melting point, 888°C (1630°F); refractive index, 1.484. Used in glazes and glass, as a source of sodium oxide, as an antiscumming agent, in manufacture of kraft paper and paperboard, in textile fiber processing, in biochemistry and biology, and in freezing mixtures. Formula: Na_2SO_4.

sodium sulfate decahydrate. The sodium salt of sulfuric acid available as large, transparent crystals, small needles or as a granular powder. It occurs in nature as the mineral *mirabilite.* Crystal system, monoclinic. Density, 1.464 g/cm³ (crystals); melting point, liquefies at 33°C (91°F); boiling point, loses $10H_2O$ at 100°C (212°F). Used in solar heat storage, air conditioning, textile dyeing, in the manufacture of cellulose, water glass and dyestuffs, and in biochemistry, biology and medicine. Formula: $Na_2SO_4 \cdot 10H_2O$. Also known as *Glauber's salt.*

sodium sulfide. White, hygroscopic, cubic crystals or off-white, flammable, hygroscopic powder (95+% pure) with a density of 1.86 g/cm³, a melting point of 950°C (1742°F). It is also available in electronic grades. Used in organic chemicals, in sulfur dyes, in the manufacture of viscose rayon, leather and paper pulp, in electronics and optoelectronics, in hydrometallurgy and extractive metallurgy (e.g., for gold ores and for sulfiding oxidized lead and copper ores), in photography, engraving and lithography, and as an analytical reagent. Formula: Na_2S.

sodium sulfide nonahydrate. A white, flammable crystalline compound (98+% pure). Crystal system, tetragonal. Density, 1.427 g/cm³; melting point, 920°C (1742°F) (decomposes). Used in organic chemicals, in sulfur dyes in the manufacture of viscose rayon, leather and paper pulp, in hydrometallurgy and extractive metallurgy (e.g., for gold ores and for sulfiding oxidized lead and copper ores), in photography, engraving and lithography, in electronics, biochemistry and medicine, and as an analytical reagent. Formula: $Na_2S\cdot9H_2O$.

sodium sulfocyanate. See sodium thiocyanate.

sodium superoxide. Yellow, cubic crystals with a density of 2.2 g/cm^3 and melting point of 552°C (1026°F). Used in chemistry and materials research. Formula: NaO_2.

sodium tannate. The sodium salt of tannic acid used as a deflocculating agent for clay slips.

sodium tantalate. See sodium metatantalate.

sodium tetraborate. See anhydrous borax.

sodium tetraborate decahydrate. See borax.

sodium tetrachloropalladate. A crystalline double salt of palladium chloride and sodium chloride used in electroplating. Formula: Na_2PdCl_4. Also known as *palladium-sodium chloride; sodium-palladium chloride.*

sodium tetrafluoroborate. See sodium fluoborate.

sodium tetrahydridoborate. See sodium borohydride.

sodium thiocyanate. Colorless, hygroscopic, light-sensitive, orthorhombic crystals or white powder with a melting point of 287°C (549°F) used in black nickel plating. Formula: NaSCN. Also known as *sodium rhodanate; sodium sulfocyanate.*

sodium thiosulfate. White, hygroscopic, translucent, monoclinic crystals or powder (99+% pure) obtained by heating a solution of sodium sulfite with powdered sulfur. Density, 1.729 g/cm^3; melting point, 40-45°C (104-113°F); boiling point, decomposes at 48°C (118°F); and loses $5H_2O$ at 100°C (212°F). The anhydrous salt (99% pure) is a white hygroscopic, crystalline compound with a density of 1.667 g/cm^3. Used in the extraction of silver from its ores, as a fixing agent in photography, in chrome tanning, in bleaching and papermaking, as a reducing agent, in the dechlorination of water, and in biology, biochemistry and medicine. Formula: $Na_2S_2O_3$ (anhydrous); $Na_2S_2O_3\cdot5H_2O$ (pentahydrate). Also known as *sodium hyposulfite; sodium sub-sulfite; sodium thiosulfate pentahydrate.*

sodium titanate. White crystals or powder (95+% pure) with a density of 3.35-3.50 g/cm^3 and a melting point of 1128°C (2062°F). Used chiefly in welding. Formula: $Na_2Ti_3O_7$.

sodium tungstate dihydrate. White, orthorhombic crystals (99+% pure) with a density of 3.24 g/cm^3, that lose $2H_2O$ at 100°C (212°F), and have a melting point of 692°C (1277°F) and a refractive index of 1.553. The anhydrous salt is commercially available as a white powder. Used as a chemical reagent, in combination with hydrogen peroxide for the oxidation of secondary amines to nitrones, in the preparation of tungsten and phosphotungstate compounds, as a fireproofing agent for cellulose and textiles, in biochemistry, e.g., for the preparation of protein-free filtrates and as an alkaloid precipitant. Formula: $Na_2WO_4\cdot2H_2O$. Also known as *sodium wolframate dihydrate.*

sodium tungstophosphate. A yellowish-white, water-soluble powder used in the manufacture of organic pigments, as a waterproofing agent for adhesives, cements and plastics, as a reagent, and in biochemistry as a metal stain. Formula: $Na_3PO_4\cdot12WO_3\cdot xH_2O$. Also known as *sodium phosphotungstate.*

sodium tungstosilicate. White, water-soluble, crystalline powder usually supplied as the pentahydrate and used chiefly as an additive in plating processes, as a catalyst in organic reactions, as a precipitant and inorganic ion exchanger, and as a metal stain for biological and biochemical microscopy. Formula: $Na_4SiO_4\cdot12WO_3\cdot5H_2O$. Also known as *sodium silicotungstate; sodium tungstosilicate; sodium tungstosilicate pentahydrate.*

sodium uranospinite. A yellow, radioactive mineral of meta-autunite group composed of sodium uranyl arsenate hydrate, $Na(UO_2)(AsO_4)\cdot3.5H_2O$. It can also be made synthetically. Crystal system, tetragonal. Density, 3.68 g/cm^3.

sodium vanadate. See sodium metavanadate.

sodium wolframate dihydrate. See sodium tungstate dihydrate.

sodium-zinc alloy. A brittle alloy composed of 98% zinc and 2% sodium, and used as a deoxidizer for nonferrous alloys.

sodium zincate. A compound of sodium oxide and zinc oxide used for waterproofing asbestos-cement shingles. Formula: Na_2ZnO_2.

sodium zippeite. A yellow synthetic mineral of the zippeite group composed of sodium uranyl sulfate hydroxide tetrahydrate, $Na_4(UO_2)_6(SO_4)_3(OH)_{10}\cdot4H_2O$. Crystal system, orthorhombic. Density, 3.30 g/cm^3; refractive index, 1.690.

Soflen. Trademark of SAR SpA (Italy) for polypropylene fibers and filament yarns.

Soft. Trade name of Electro-Steel Company (USA) for a water-hardening carbon steel used for tools, shafts, and gears.

Softalon. Trade name for nylon fibers and yarns used for textile fabrics including clothing.

soft brass. A ductile brass annealed after drawing and rolling to remove the effects of cold working.

soft-burnt lime. A highly porous and reactive quicklime calcined at relatively low temperature.

Soft Cell. Trademark of Ludlow Composites Corporation (USA) for a soft, lightweight closed-cell polyvinyl chloride foam.

Soft-Clad. Trade name of Reichhold Chemicals, Inc. (USA) for polyurethane and polyester polyol resins used for coating applications.

soft clay. A hand-moldable clay that is fusible at relatively low temperatures.

soft coal. See bituminous coal.

soft-drawn wire. (1) A wire drawn or rolled to final size and then heated to increase its ductility. Also known as *soft wire.*

(2) An annealed aluminum or copper wire with a rather low tensile strength, but high elongation. It is very pliable and easily bent. Also known as *soft wire.* See also hard-drawn wire.

soft enamel. A porcelain enamel that is fusible at a relatively low temperature.

Softerex. Trade name for a plasticized acrylic resin used for denture lining.

soft fibers. See soft vegetable fibers.

soft-fired clay. Clay products having only moderate compressive strengths and relatively high water absorption due to firing at relatively low temperatures, well below the vitrification point. Also known as *soft-fired wire.*

soft glass. An easily fusible glass with a relatively low viscosity at elevated temperatures, low softening point, or low resistance to abrasion, scratching and other mechanical damage, or any combination thereof.

soft glaze. A glaze that fires at a relatively low temperature of 1050°C (1920°F) or below. It is softer and less chemically resistant than hard glazes. Also known as *low-temperature glaze.* See also hard glaze.

Softglo. Trade name of BASF Corporation (USA) for nylon-6 fibers and yarns used for textile fabrics including clothing.

soft iron. Iron that unlike the harder cast iron can be readily machined with cutting tools and worked with hand tools. It has a density of 7.86 g/cm^3 (0.284 lb/in.³), a hardness of 150-350 Brinell, a tensile strength of 180-260 MPa (26-38 ksi), an elongation of 45-50% in 50 mm (2 in.), high magnetic permeability and low coercivity. Used widely for electrical and magnetic applications. Abbreviation: SI.

soft laid rope. A rope that consists of loosely twisted or laid strands. It is flexible and deformable, but usually stronger than *hard-laid rope.*

Softlex. Trade name for an elastomer used for dental impressions.

Soft-Liner. Trade name of GC Dental (Japan) for a dental acrylic resin used for the temporary rebasing/relining of dentures.

Softliner. Trade name of Inter-Africa Dental (Zaire) for a biocompatible addition-cure silicone denture relining material with very high adhesion.

Softlite. Trade name of Bausch & Lomb Inc. (USA) for pink-tinted protective spectacle glass.

Soft-Lite. (1) Trade name of Pittsburgh Corning Corporation (USA) for a white opal glass insert fused to the inside edge seal of two glass block halves.

(2) Trademark of Soft-Lite LLC (USA) for a ultraviolet- and weather-resistant polyvinyl chloride used for windows and doors.

soft magnetic ceramics. A group of ceramic mate-rials that possess spinel crystal structures ($MO \cdot Fe_2O_3$, where M represents a metal), and are composed of fixed mixtures of ferric oxide and appropriate compounds of divalent metals, such as barium, cobalt, copper, lead, magnesium, manganese, nickel, strontium or zinc. They exhibit ferromagnetic, ferrimagnetic, antiferromagnetic, magneto-optical and magnetostrictive properties, and are used in antennas, computer memory cores, ferrite cores for coils, computer disks, recording tapes, telecommunications systems, etc. Also known as *soft magnetic ferrites.*

soft magnetic ferrites. See soft magnetic ceramics.

soft magnetic materials. A group of ferromagnetic or ferrimagnetic materials that are easily demagnetized after removal of the magnetizing field. They are characterized by high saturation flux density, large relative permeability, small hysteresis loops, and small coercive forces. Examples include silicon steel, ingot iron, *Puron, Supermalloy, Hypernik* and *Perminvar. Soft-magnetic materials* are used for AC and DC motors and generators, transformers, adjustable cores in radio-frequency communication coils, high-voltage electrical lines, communication equipment, headphones, microphones, loudspeakers, magnetic and magnetooptical disks, fax machines, magnetic resonance imaging devices, magnetic separators, couplings, frictionless bearings, sensors, switches, medical implants, etc. Also known as *magnetically soft materials; soft magnets.*

soft maple. A collective term for the soft, strong wood of the red maple (*Acer rubrum*), silver maple (*A. saccarinum*) and box-elder (*A. negundo*). The heartwood is light reddish-brown, and the sapwood white with slight reddish-brown tinge. It has an average weight of 569 kg/m³ (35 lb/ft³) ranging from 529 kg/m³ (33 lb/ft³) for *silver maple* to 609 kg/m³ (38 lb/ft³) for *red maple*. Source: Eastern and central United States; southern Canada. Used for lumber, veneer, crossties, furniture, woodenware, boxes, pallets, crates, etc., and as a pulpwood.

soft mica. Mica that has a tendency to split into thin flexible sheets when bent.

soft-mud brick. Brick molded by hand or, more frequently, by machine from wet soft clay bodies containing 20-35% water.

Softpane. Trade name of Nippon Sheet Glass Company Limited (Japan) for patterned glass.

soft paste porcelain. A ceramic whiteware made from a body containing a glassy frit and a high content of fluorspar by fir-

ing at relatively low temperatures, usually 1100-1300°C (2010-2370°F). Also known by the French term "pâte tendre." It is the opposite of hard porcelain (or pâte dure) referring to porcelain fired at relatively high temperatures, usually 1300-1445°C (2370-2635°F). Also known as *fritted porcelain; soft porcelain.*

soft porcelain. See soft paste porcelain.

Softral. Trade name of Central Glass Company Limited (Japan) for a figured glass.

Soft Ray. Trade name of General Motors Corporation (USA) for tinted windshield glass.

soft rubber. A soft substance made by first compounding crude rubber with about 3-8 wt% sulfur, and frequently softeners, fillers, antioxidants, etc., followed by a relatively short vulcanization treatment. The elasticity and ductility increases with decreasing sulfur content. It is very susceptible to loss of elasticity by natural aging (e.g., absorption of oxygen from the air) usually accelerated by heat, light or cold. The aged rubber is either a hard and brittle or a soft and tacky substance. Used for water hose, seals, vee belts, shock absorbers, couplings and clutches, rollers, cable sheathing, conveyor belts, protective clothing, automotive tires, and shoe soles.

soft silicas. A generic term used in the abrasives industry for fine-grained, porous siliceous substances, especially microcrystalline silica, rottenstone and tripoli.

soft solders. See solders.

soft steels. See mild steels.

SoftStrand. Trademark of Owens-Corning (USA) for a flexible, high-strength single-end wound glass-fiber reinforcement impregnated with a uniform, abrasion-resistant polyurethane-based coating, and used as a nonconductive, protective central member for optical telecommunication cables.

soft superconductor. A superconductor material, such as aluminum, lead, mercury or tin, that is completely diamagnetic, i.e., rejects virtually all magnetic flux from its interior while in the superconducting state. At field strengths below the critical magnetic field (H_c) it remains superconducting, and upon reaching the critical value, conduction becomes normal resulting in complete flux penetration. Abbreviation: SSC. Also known as *ideal superconductor; type I superconductor.*

soft vegetable fibers. A class of vegetable fibers including cotton, flax, jute and ramie, that are obtained from the seeds or the basts of plants. They are used extensively for the manufacture of textiles. See also bast fibers; vegetable fibers.

Softwire. Trade name of Central Glass Company Limited (Japan) for a wired glass.

soft wire. See soft-drawn wire.

softwood. The wood produced by cone-bearing or coniferous trees, such as cedars, cypresses, firs, hemlocks, larches, pines, redwoods and spruces, belonging to the botanical class Gymnospermae. The term should not always be taken literally, e.g., the wood of some softwoods (e.g., Douglas fir, longleaf pine and yew) is harder than that of hardwoods, such as aspen, balsa and yellow poplar. Also known as *coniferous wood*. See also hardwood.

softwood lumber. Softwood timber, logs, beams, boards, etc. roughly sawn and prepared for use. The following grading system for softwood lumber has been established by the American Lumber Standards Committee (ALSC): (i) *Select grades:* (a) Grade A that is almost free of defects and suitable for natural finishes and staining, (b) Grade B that contains a few, small, visible defects and suitable for natural finishes, (c) Grade C that contains some defects which however will be hidden if it is to be painted, and (d) Grade D that contains several defects similar to those in "Grade C" and is suitable for painted projects; (ii) *Common grades:* (a) Grade 1 that refers to a sound material that will contain tight knots and a limited number of blemishes, (b) Grade 2 that should not have splits or distortions like warp, but can have defects, such as loose knots, end checks and some blemishes and discolorations, (c) Grade 3 that can contain various types of defects some of which should be removed before the material is used as medium-quality construction lumber, (d) Grade 4 that is a low-quality construction lumber that may contain many defects, even open knot holes, and (e) Grade 5 that is the lowest quality material that can be used as a filler; and (iii) *Structural grades:* (a) Construction grade that is a high-quality structural lumber, (b) Standard grade that is a good-quality lumber similar to the construction grade, but with some defects, (c) Utility grade that is a poor-quality structural lumber which must always be reinforced with addition components, and (d) Economy grade that is a structural lumber of very low quality.

sogdianite. A violet mineral of the osumilite group composed of lithium potassium zirconium silicate, $(K,Na)_2Li_2(Li,Fe,Al)_2Zr-Si_{12}O_{30}$. Crystal system, hexagonal. Density, 2.90 g/cm^3; refractive index, 1.606. Occurrence: Tadzhik Republic.

sohngeite. A light brown mineral composed of gallium hydroxide, $Ga(OH)_3$. Crystal system, cubic. Density, 3.84 g/cm^3; refractive index, 1.736. Occurrence: Southwest Africa.

Soho Self-Hardening. Trade name of Hobson, Houghton & Company (USA) for an oil-hardening tool and die steel containing 0.9% carbon, 4% chromium, 0.2% vanadium, and the balance iron.

Soibell. Trademark of Kenebo Limited (Japan) for bicomponent staple fibers.

soil aggregate. A natural or synthetic mixture of aggregates containing sand, gravel, stone and substantial quantities of silt-clay materials having a particle size smaller than 75 μm (0.003 in.). See also aggregate.

soil cement. A mixture consisting of *Portland cement* bonded with soil, that when blended with water forms a concrete with low to moderate engineering properties. The density of compacted soil-cement mixtures ranges from about 1440 kg/m^3 (90 lb/ft^3) for clayey and silty soils to 2160 kg/m^3 (135 lb/ft^3) for gravelly soils. Used as a base course for road pavements.

soisette. An extremely fine and soft fabric made from mercerized cotton yarns, and used for nightwear and negligees.

Sokalan. Trademark of BASF Corporation (USA) for a series of copolymers of acrylic acid and maleic acid.

sol. A colloidal suspension of particles in a liquid or gas. If the liquid is water the suspension is called a "hydrosol." A suspension in a gas is referred to as an "aerosol." The sol state is often followed by the gel state. See also gel; sol-gel glass.

Solablue. Trade name of Chance-Pilkington Limited (UK) for blue-tinted sunglass with a high absorption of ultraviolet radiation and over 50% daylight transmission.

Solabraze. Trade name of International Harvester Company (USA) for a series of brazing alloys 43-47% zirconium, 43-48% titanium, 0-12% nickel, 0-5% aluminum, and 2-5% beryllium. They have a melting range of 800-925°C (1470-1700°F), a brazing range of 900-955°C (1650-1750°F). Used for brazing titanium to itself and to ceramics.

Soladur. Trademark of Schott DESAG AG (Germany) for a special filter glass.

Solane. Trade name of Sovirel (France) for brownish-pink spectacle glass.

Solar. (1) Trade name of Carpenter Technology Corporation (USA) for a water-hardening shock-resisting tool steel (AISI type S2) containing 0.5% carbon, 1% silicon, 0.4% manganese, 0.5% molybdenum, and the balance iron. It has outstanding toughness, a great depth of hardening, high strength, moderate wear resistance, and moderate machinability. Used for pneumatic tools, rivet busters, chisels, screwdrivers, stamps and shear blades.

(2) Trade name of Westa-Westdeutsche Edelstahlhandelsgesellschaft (Germany) for an oil- or water-hardening tool steel containing 1.05% carbon, 1.15% tungsten, 1% chromium, 0.9% manganese, and the balance iron. Used for forming dies, reamers, cutters, milling cutters, taps, molds, stamping and punching tools, knives, etc.

So-Lara. Trademark of Solutia Inc. (USA) for acrylic fibers used for the manufacture of textile fabrics.

Solarban. Trade name of PPG Industries Inc. (USA) for a double glazing unit with a selective-reflective coating on the inside surface.

Solarbronze. Trade name of PPG Industries Inc. (USA) for heat-absorbing, glare-reducing plate glass in brownish yellow shade with slight rose tint.

Solarcast 20. Trademark of Ivoclar Vivadent AG (Liechtenstein) for a hard dental casting alloy (ADA type III) containing 40.0% silver, 20.0% gold, 15.0% palladium, 11.0% indium, 10.0% copper, 2% tin and 2% zinc. It has a yellow color, a density of 10.8 g/cm^3 (0.39 lb/in.^{3}), a melting range of 815-875°C (1500-1605°F), an as-cast hardness of 240 Vickers, low elongation, and good biocompatibility. Used for crowns, posts, onlays, and bridges.

Solarex. Trade name of Pilkington Brothers Inc. (UK) for a heat-absorbing glass.

Solargray. Trade name of PPG Industries Inc. (USA) for neutral-gray, heat-absorbing polished plate glass.

Solaris. Trade name of Westerwald AG (Germany) for glass blocks.

Solarmatic. Trade name of Bachmann Brothers Inc. (USA) for a self-adjusting glass lens whose optical density and color are increasing and decreasing with glare.

SolarMax. Trademark of E.I. DuPont de Nemours & Company (USA) for nylon 6,6 supplied in 30-210 denier high-tenacity, and 200-400 denier bright and semidull luster yarns. They provide superior breaking strength and ultraviolet resistance, and are used for outdoor fabric applications, such as flags, hot-air balloons, parachutes, life jackets, and tents.

Solaro. Trade name of Metalor Technologies SA (Switzerland) for several dental casting alloys including *Solaro 3*, an extra-hard alloy containing 56% gold, 5% palladium and 0.4% platinum, and *Solaro Special*, a hard alloy containing 50% gold and 5% palladium.

Solarpane. (1) Trade name of Oxelösungsa Järnverk (Sweden) for heat-absorbing, sun-reflecting glass having a thin nickel coating on one side which gives a neutral gray shade to reflected and transmitted light.

(2) Trade name of Solarpane Manufacturing Company (Canada) for double glazing units.

Solar Seal. Trade name of Solar Seal Corporation (USA) for insulating glass units.

Solar Shield. Trade name of Corning Glass Works (USA) for a heat-shielding window glass consisting of borosilicate glass coated with a thin ceramic layer.

Solarshield. Trade name of Shatterprufe Safety Glass Company Proprietary Limited (South Africa) for a heat-reflecting, transparent laminated glass that comprises two sheets of glass, one of which has a metallic film on the inside deposited by a thermal evaporation process.

Solbisky's alloy. A German nonhardenable aluminum alloy containing 1.4% zinc, 0-3% cadmium, 0.3% tin and 0.5-1% nickel. Used for light alloy parts.

Solbor. Trade name of Sollac-Usinor (France) for hot-rolled quench-hardening manganese-boron steels with excellent wear resistance and good hot and cold formability. They are supplied in a wide range of widths and thicknesses, and in several grades including *Solbor 1500* and *Solbor 1700,* in which the number indicates the ultimate tensile strength in MPa. Used for cutting tools, parts of earthworking and quarrying equipment.

Solbond. Trade name of Sollac-Usinor (France) for corrosion-resistant coil-coated steel sheets that have adhesive layers (5-17 µm or 200-680 µin. thick) applied to their surfaces which can be thermally reactivated. They are supplied in a wide range of widths and thicknesses, and in several grades including *Solbond G, Solbond P, Solbond R* and *Solbond V.* Used for bonding to wood, paper, cardboard, mineral wool, thermoplastic or thermosetting polymers, elastomers and foams, and/or polymer alloys, e.g., in automotive applications.

Solbright. Trade name of Sollac-Usinor (France) for a range of corrosion-resistant steel sheets with a coil-coated layer of polyester lacquer with a standard thickness of 25 µm (1.0 mil). They provide high surface brightness, good coating adherence, flexibility, UV-stability and resistance to stains, heat and detergents, and are used in the manufacture of domestic appliances. They are supplied in a wide range of widths and thicknesses, and in several grades including *Solbright 1, Solbright 2, Solbright 3* and *Solbright HT* (high-temperature grade).

Solchrom-Pal. Trade name of Metalor Technologies SA (Switzerland) for a nonprecious dental solder solid in tube and wire form. It contains 30% palladium and has a melting range of 865-940°C (1590-1725°F).

Solclean. Trade name of Sollac-Usinor (France) for a dry-film temporary protection for pickled hot-rolled steels that provides good cleanness and adherence and fairly good corrosion resistance. Supplied on pickled coils, slit strips, or cut-to-length sheets.

Solcolor. Trade name of Sollac-Usinor (France) for corrosion-resistant, cold-rolled galvanized or electrogalvanized steel sheets with a thin, coil-coated layer of polyester or polyurethane lacquer. They have good resistance to acids, aromatic, aliphatic and chloride-containing solvents, household products and mineral oils, and moderate resistance to alkalies. They are supplied in a wide range of widths and thicknesses, and in several qualities including *Standard, Flexible, Lum, Mat* and *Mono* with different coating thicknesses and levels of surface brightness.

Solconfort. Trade name of Sollac-Usinor (France) for sandwich sheets consisting of two bare or metallic or organic coated cold-rolled steel facings with a 25-45 µm (1-1.8 mils) thick core of polyester or acrylic resin. They are supplied in a wide range of widths and thicknesses, and provide good noise reduction by damping vibrations, and possess good stampability and oil resistance. Used in the manufacture of household appliances and air-conditioners for casings and motor housings and supports; also for automotive components, staircases, sliding doors, curtain rails, etc.

Soldamoll. Trademark for a series of special solders including various tin-lead, tin-cadmium, tin-silver, tin-copper, tin-lead-silver, tin-antimony, cadmium-zinc, cadmium-zinc-silver, lead-silver and cadmium-silver alloys. Depending on the particular composition, they have a general melting range of 145-400°C (295-750°F), a density range of 7.2-11.2 g/cm³ (0.26-0.40 lb/in.³), and a hardness range of 9-50 Brinell. Used for soldering commutators, refrigeration equipment, food processing equipment, capacitors, printed circuit boards, ceramics, etc.

Soldate. Trade name of Aardvark Clay & Supplies (USA) for clay (cone 10) with medium toasty all-round body supplied in several grades.

Solderal. Trade name of Non-Corrodal Alloys Limited (USA) for a corrosion-resistant aluminum alloy containing 5% silicon. Used for castings and aluminum solder.

solder alloys. See solders.

Solder Auro. Trade name for a dental gold solder.

Solderex. Trade name of Enthone-OMI Inc. (USA) for tin and tin-lead electroplates and plating processes.

Solderfast. Trade name of Uyemura International Corporation (USA) for a tin-lead electrodeposit and deposition process.

solder glass. A glass that has a relatively low softening point, and is suitable for sealing or bonding pieces of higher-melting glass without the need for deforming or softening them. Also known as *solder sealing glass.*

soldering alloys. See solders.

soldering fluxes. Cleaning agents that are applied to surfaces to be soldered to remove oxides and other surface compounds prior to soldering. They may also keep the solder and base metal from oxidizing during soldering, should have good wetting action and promote better fusion by increasing fusibility and reducing melting temperature. *Soldering fluxes* are divided into three basic groups: (i) inorganic, (ii) organic, and (iii) rosin-based fluxes. See also inorganic fluxes; organic fluxes; rosin-based fluxes.

soldering paste. A thin paste or solution containing the soldering fluxes in the proper proportions.

soldering wire. See solder wire.

Solderon. Trade name of Shipley Ronal (USA) for a wide range of surface finishes based on pure tin and tin-copper, tin-bismuth, and tin-lead alloys.

solders. A group of alloys that melt below 450°C (840°F) and below the melting point of the base metals to be joined, and are often referred to as "soft solders," since they are composed of relatively soft metals (e.g., tin, lead, etc.). Soft-soldered joints are flexible, but have low strength. The most commonly used solders are based on lead and tin. Other commercial solders include tin-antimony, tin-antimony-lead, tin-silver, tin-silver-lead, tin-zinc, cadmium-silver, cadmium-zinc and zinc-aluminum alloys as well as fusible alloys. Used for electrical and electronic applications, in roofing and plumbing, for joining automobile bodies, refrigeration and air-conditioning systems and piping, and for joining like or unlike metals. Also known as *low-melting solders; soft solders; solder alloys; soldering alloys.*

solder sealing glass. See solder glass.

solder wire. Solder in the form of a solid or flux-cored wire. Also known as *soldering wire.*

Solderzit. Trade name of L & R Manufacturing Company (USA) for a lead-tin solder supplied in the form of a prealloyed powder with the required fluxing medium added.

Soldier. Trade name of Burys & Company Limited (UK) for a series of hot-work die and tool steels.

Soldur. (1) Trade name of Sollac-Usinor (France) for an extensive series of hot-rolled high-strength sheet steels with a typical composition range (in wt%) of up to 0.10% carbon, 0.5-2.0% manganese, up to 0.025% phosphorus, 0.010-0.020% sulfur, 0.03-0.40% silicon, and additions of aluminum, either alone or with niobium and/or titanium, vanadium and molybdenum, with the balance being iron. They are supplied in several grades, and have excellent cold formability, good mechanical properties including high resistance to low-temperature brittle fracture, and good corrosion resistance and weldability. Used in the manufacture of trucks, trailers, chassis, agricultural and construction machinery, cranes and jibs, automobile wheels, railroad equipment, beams, lamp posts and masts, industrial shelving, etc.

(2) Trade name of Sollac-Usinor (France) for a series of cold-rolled high-strength steels with a typical composition range (in wt%) of 0.08% carbon, 0.6-0.9% manganese, 0.025-0.030% phosphorus, 0.04-0.35% silicon, 0.02% or more aluminum, 0-0.1% niobium, and the balance being iron. They are supplied in several grades with elongations of 21% or more, and good cold formability, limited drawability and good weldability, fatigue strength and shock resistance. Used for structural applications.

Solef. (1) Trademark of Solvay Polymers, Inc. (USA) for pure, chemically inert polyvinylidene fluoride (PVDF) that can be processed by injection molding, roto-molding, machining and welding. It has excellent creep resistance, outstanding chemical resistance even to harsh chemicals, such as bromic, chromic, hydrochloric, hydrofluoric, nitric, phosphoric and sulfuric acids, good low and high-temperature performance from -40 to +150°C (-40 to +300°F), high mechanical strength, and excellent durability. Used for chemical process equipment, solid piping systems, scrubbers, heat exchangers, vessels, pipes, valves, pumps, fittings, filter housings, and for coating and lining chemical tanks, containers and process equipment.

(2) Trademark of Solvay Polymers, Inc. (USA) for mono-oriented and bi-oriented transparent or metallized piezoelectric and pyroelectric films based on polyvinylidene fluoride (PVDF). Supplied in standard thicknesses of 25 μm (1 mil), they possess high longitudinal deflection, and are used for mechanical and optical switches, vibrators, and damping control systems.

Soleil. Trade name of Creusot-Loire (France) for a series of cast and wrought austenitic, ferritic and martensitic stainless steels used for turbine blades and wheels, cutlery, knives, surgical instruments, bearings, machinery parts, pump and valve parts, household appliances, chemical process equipment, petrochemical and oil refining equipment, and fixtures and fittings.

sole leather. A heavy, dry-finished leather made from the hides of cattle and horses, and used in the manufacture of outer soles, insoles, heels, and toecaps.

Soléma. Trade name of Sollac-Usinor (France) for enameling steel sheet with an average carbon content of 0.025 wt% designed for conventional and two-coat/one-firing processes. It has a good enamelled surface appearance, good weldability, improved mechanical properties, and good formability by bending and stamping, good tensile and yield strengths, and a very high elongation. Supplied in various widths and thicknesses, it is used for domestic appliance casings, cooking hobs, etc., and for architectural panels.

Soléma Plus. Trade name of Sollac-Usinor (France) for decarburiz-ed enameling steel sheet with an average carbon content of less than 0.01 wt% that is designed for conventional and two-coat/one-firing processes, but cannot be used for direct enameling applications. It has a good enamelled surface appearance, good weldability, good mechanical properties (but somewhat lower tensile and yield strength than *Soléma*), good formability by bending and stamping, and very high elongation. It is supplied in several widths and thicknesses, and used for domestic appliance casings, cooking hobs, etc., and for architectural panels.

Solemail. Trade name of Sollac-Usinor (France) for hot-rolled enameling steel sheets with about 0.14-0.17% carbon for one-side enameling by powder and liquid application processes. They are supplied in several width and thicknesses, and in several grades including *Solemail 250* has *Solemail 310*. Used for enameled pressure tanks, panels and structural parts, water heater bottoms and lids, etc.

Solex. Trade name of PPG Industries Inc. (USA) for a blue-green, heat-absorbing plate glass.

Solfer. Trade name of Sollac-Usinor (France) for a direct enameling sheet steel with a very low carbon content (less than 0.004 wt%) for single- and multi-coat coverage by any conventional enameling process. The enamel layer is usually 13 μm (0.5 mil) thick, and the enameled sheet has excellent surface cleanness and appearance, good formability by bending and stamping, good weldability, good shock resistance, good high-temperature creep resistance, and good resistance to nail marks. It is supplied in several widths and thicknesses, and used for domestic appliances, especially casings, cooking hobs, side panels, pans, covers, oven muffles, sinks and washbasins, and for architectural paneling.

Solfer Plus. Trade name of Sollac-Usinor (France) for a direct enameling sheet steel similar to *Solfer*, but with slightly lower yield and tensile strengths and higher elongation.

Solfilm. Trade name of Sollac-Usinor (France) for a range of coil-coated steel sheets, either galvanized or plain cold-rolled, with flexible, low-friction thermolaminated functional polymer topcoat. They have excellent hardness, good formability, bendability and fabricability, good shock and scratch resistance, good chemical and heat resistance, good corrosion resistance, and good foodstuff compatibility. *Solfilm* sheets are supplied in several grades with different coatings including *Solfilm A & B* with polyethylene terephthalate (PET) topcoats and *Solfilm C* with a thick complex film with PET topcoat. Used as inside facings of sandwich panels for coldrooms, cold-storage buildings, refrigerated trucks, etc.

Solfilm Brilliant. Trade name of Sollac-Usinor (France) for a range of coil-coated galvanized steel sheets with white, flexible, low-friction thermolaminated polyethylene terephthalate (PET) topcoats with smooth, granular or orange-peel appearance. They have very high gloss and brightness, good formability, bendability and fabricability, good shock resistance, good chemical and heat resistance, and are used in the manufacture of domestic appliances.

Solflex. Trademark of Goodyear Tire & Rubber Company (USA) for styrene-butadiene rubber made by solution polymerization, and used for making tires and other rubber goods.

sol-gel glass. A glass produced by first hydrolyzing silica gel, colloidal silicates, amorphous silica and selected metal salts. By removal of the solvent, the resulting homogeneous colloidal solution (or sol) is then transformed to a rigid multicomponent noncrystalline gel. This gel then undergoes thermal treatments involving the removal of the volatiles and initial densification. Finally, the resulting reactive noncrystalline gel is densified by sintering. *Sol-gel glass* is optically transparent, has good mechanical properties, and about one-half the density of ordinary glass. It is supplied in various forms including powders, coatings, fibers, thin films, composites, monoliths and porous membranes. See also sol; gel.

solid. An amorphous or crystalline substance that has a definite shape and volume, and can be compressed only slightly as compared to a liquid or gas.

solid asphalt. Asphalt in the form of granules or powder having a kinematic viscosity of at least 10^3 St (cm²/s) at a temperature of 40°C (104°F). Also known as *solid bituminous material*.

solid bituminous material. See solid asphalt.

Solid Bond. Trade name of Heraeus Kulzer Inc. (USA) for dental adhesive for dentin and enamel bonding.

solid-drawn tube. A seamless metal tube drawn from a hollow ingot on mandrels of successively decreasing diameters.

Solid Drill. Trade name of Agawam Tool Company (USA) for a water-hardening tool steel containing 1.05% carbon, 0.35% manganese, and the balance iron. Used for granite drills.

Solidek. Trade name of Dri-Dek Corporation (USA) for interlocking acid-resistant plastic floor tiles.

solid electrolytes. Electrolytes in solid form, such as salt mixtures, halide salts, oxide mixtures and glasses, used in batteries and fuel cells.

Solidex. Trade name of Shofu Dental Corporation (Japan) for a light-cured porcelain-filled dental hybrid composite that provides excellent abrasion resistance and elasticity, outstanding handling characteristics, exhibits a natural porcelain-like appearance, and is used as a veneering material for restorations including implants, inlays and onlays.

solid fiberboard. A solid board made by laminating two or more layers of paperboard, and used for shipping containers. See also containerboard; fiberboard; paperboard.

solid-film lubricant. Solid materials, such as graphite, molybdenum disulfide, tantalum disulfide, tantalum diselenide, titanium diselenide, titanium ditelluride, zirconium diselenide, boron nitride, talc, mica, babbitt, lead, silver, metallic oxides, or polytetrafluoroethylene (PTFE), that are used in conjunction with fluid or grease lubricants to provide solid-film lubrication under high-load, slow-speed or oscillating load conditions. Commercially available as pastes and sprays, they are used for bearings, machine beds, guideways, screw threads, gears, chains, shafts, machine components, and in hot- and cold-forming operations. Also known as *dry-film lubricant; dry lubricant; solid lubricant*.

SOLID-H. Trademark of Hydrogen Components Inc. (USA) for *metal hydrides* that absorb large amounts of hydrogen and are used in batteries and hydrogen-storage devices.

solid insulator. An insulator composed of a solid, dielectric material, such as amber, barium titanate, glass, mica, polystyrene, polyvinyl chloride, porcelain, quartz, rubber, wood, etc.

solid lubricant. See solid-film lubricant.

solid masonry unit. A masonry unit, such as a building block or brick, in which the net cross-sectional area in any plane parallel to the bearing surface is at least 75% of its total cross-sectional area measured in the same plane. See also hollow masonry unit.

solid solution. A homogeneous mixture of two or more solids (e.g., two metals, a metal and a nonmetal, or two nonmetals) in which (i) the atoms or ions of the solute replace the atoms or ions of the solvent (substitutional solid solution), or (ii) the atoms or ions of the solute take positions interstitially in the crystal lattice of the solvent (interstitial solid solution).

Solid Zinc Strip. Trade name of Alltrista Zinc Products Company (USA) for solid, rolled strip of approximately 99% pure zinc. It is produced in coils and supplied in various widths, gauges and finishes with good electrical conductivity (26-28% IACS), a density of 7.14-7.18 g/cm³, good corrosion resistance, formability, platability, solderability and resistance weldability, and good nonsparking properties. Used for coinage, plumbing hardware, roofing, flashing, terrazzo strip, weath-erstripping, low-voltage electrical parts, automotive leaf spring liners, decorative trim, and EMI/RFI shielding applications.

Solifarm. Trade name of Sollac-Usinor (France) for a range of steel sheets with a coil-coated layer of flexible polyester topcoat applied on a crack-resistant galvanized coating (*Galflex B*). They have excellent shock resistance, very good drawability, good bendability, a maximum continuous-use temperature of 90°C (194°F), excellent resistance to outdoor corrosion and aggressive livestock atmospheres, very good resistance to mineral oils, and are used for roofing and exterior wall paneling.

Solissime. Trade name of Sollac-Usinor (France) for a range of steel sheets with coil-coated layers of flexible polyesters or polyurethanes. Prior to coil coating, the sheets receive thin, crack-resistant galvanized coatings (e.g., *Galfan* or *Galflex B*). The coil-coated sheets have excellent shock resistance, good drawability, a maximum continuous-use temperature of 90°C (194°F), good photochemical stability, very good resistance to mineral oils, good to very good resistance to acids and alkalies, and good resistance to industrial, seacoast, and dust-laden atmospheres. Used for architectural and building applications, such as roll-formed roofing, wall facing panels, fascias, awnings, and interior and exterior architectural products.

Solist. Trademark of DMG Hamburg (Germany) for a light-cured, one-component dentin/enamel bonding agent used in restorative dentistry for compomers, ormocers and composites.

Solitaire. Trade name of Heraeus Kulzer Inc. (USA) for a light-cure, packabe, high-viscosity polyglass dental resin composite for posterior restorations.

Solite. (1) Trademark for lightweight aggregate used in the manufacture of concrete.

   (2) Trade name of AFG Industries, Inc. (USA) for a flat glass.

   (3) Trademark of Monarch Rubber Company (USA) for lightweight rubber goods.

Solform 800. Trade name of Sollac-Usinor Sacilor (France) for an ultralow-carbon bainitic (ULCB) steel containing 0.04% carbon, 1.8% manganese, 0.25% silicon, 0.03% aluminum, 0.055% niobium, small additions of molybdenum, titanium and boron, and the balance iron. It has good formability, high tensile and yield strengths, and an elongation of 17%. Used for automotive wheels.

Sollaser. Trade name of Sollac-Usinor (France) for hot-rolled steel sheets with carbon contents ranging from about 0.08-0.17%, which are available with guaranteed flatness for automatic thermal and mechanical cutting machines. They have high surface cleanness, excellent weldability, good formability and stampability, and are supplied in several qualities (including *Sollaser 220*, *Sollaser 260* and *Sollaser 380*) as edge-sheared, cut-to-length sheets in the pickled or unpickled (black) condition with various thicknesses, widths and lengths.

Sollight. Trade name of Sollac-Usinor (France) for lightweight sandwich sheets consisting of two sheet-steel facings with a thermoplastic polymer core. The steel facing sheets have a crack-resistant galvanized coating (*Galflex B*) with an organic topcoat (e.g., polyester, polyvinylidene fluoride, etc.). The sandwich sheets have good surface quality especially good flatness, and high stiffness, excellent drawability, good bendability and formability, good indentation and puncture resistance, good resistance to atmospheric corrosion, good UV stability, a service temperature range of -20 to +80°C (-4 to +176°F), and can be joined by adhesive bonding, riveting or lockseaming. They are supplied in several widths, thicknesses and grades including *Sollight Industrie, Sollight Bâtiment HR* and *Sollight Bâtiment ST*. Used for building ceiling panels, fascias, interior cladding, covers and casings of industrial machinery, ceiling panels for railroad vehicles, etc. See also galvanized steel.

Solnhofen stone. See lithographic stone.

Solobond Plus. Trade name for a light-cure dental bonding agent used for tooth structure (i.e., dentin and enamel) bonding.

Solocoat. Trade name of Wall Colmonoy Corporation (USA) for a series of one-step, self-bonding powder alloys including stainless steel, aluminum bronze and nickel-aluminum-molybdenum alloys used for thermal-spray metallizing applications.

solongoite. A colorless mineral composed of calcium borate hydroxide chloride, $Ca_2B_3O_4(OH)_4Cl$. Crystal system, monoclinic. Density, 2.51 g/cm³; refractive index, 1.510. Occurrence: Russian Federation.

Solpanel. Trade name of Sollac-Usinor (France) for composite panels consisting of a core of cellular plastic (e.g., polypropylene), 5-150 mm (0.2-5.9 in.) thick, adhesively bonded two 0.5-1.5 mm (0.02-0.06 in.) thick facings of metallic or organic coated sheetmetal. Depending on the particular coating, the panels may be provide corrosion and/or crack resistance. Supplied as standard flat panels in varying lengths and widths, and as custom-made angles, they are used in the building construction, transport and office furniture industries for partitions, floors, doors and facings.

Solphor. Trade name of Sollac-Usinor (France) for a series of cold-rolled, rephosphorized high-strength steels with a typical composition range (in wt%) of up to 0.10% carbon, up to 0.70% manganese, up to 0.12% phosphorus, up to 0.5% silicon, 0.02% or more aluminum, with the balance being iron. They have good tensile and yield strengths, high elongation, good drawability and stampability, good weldability, and high strain hardening coefficients. Supplied in several grades, they are used for structural applications.

Solplus. Trade name of Sollac-Usinor (France) for cold-rolled, bake hardenable steels with a typical composition range (in wt%) of from less than 0.02 to 0.10% carbon, up to 0.70% manganese, from less than 0.04 to 0.12% phosphorus, up to 0.50% silicon, 0.02% or more aluminum, with the balance being iron. Supplied in several grades good tensile and yield strengths, high elongation and good indentation resistance, they are used for structural applications, and in the automotive and domestic appliance industries for parts requiring good appearance.

Solprene. Trademark of Phillips Chemical Company (USA) for a thermoplastic elastomer that is a copolymerization product of butadiene and styrene, and is used for footwear, wire and cable, sponge rubber, floor tile and miscellaneous molded and extruded products.

Solready. Trade name of Sollac-Usinor (France) for galvanized or *Galfan*-coated sheet steels with a thin organic coating based on an acrylic thermoplastic resin. Supplied in three variants in the form of coils, blanks, sheets and slit strips, they are used for casings of domestic appliances, such as washers and dryers, covers and structural components, and hi-fi equipment and air-conditioner parts. See also galvanized steel.

Solrec. Trade name of Sollac-Usinor (France) for continuously annealed, cold-rolled sheet steels supplied in three grades with a composition range (in wt%) of up to 0.085% carbon, up to 0.55% manganese, up to 0.030% phosphorus, 0.030% sulfur, 0.020% or more aluminum, and the balance iron. Made by continuous casting, they have good weldability, flatness, surface cleanness, uniform mechanical properties. Supplied in several grades including *Solrec 230, Solrec 270* and *Solrec 290*, they are used drum bottoms and panel radiators, refrigerator casings and doors and metal furniture, shelving and desktops.

Solstamp. (1) Trade name of Sollac-Usinor (France) for hot-rolled, high-formability drawing steels with a typical composition range (in wt%) of 0.05-0.10% carbon, 0.3-0.5% manganese, up to 0.025% phosphorus, 0.025% sulfur, 0.03% silicon, 0.01% or more aluminum, sometimes small additions of titanium, and the balance iron. Supplied in several grades including *Solstamp 25, Solstamp 30, Solstamp 33* and *Solstamp 37* with good mechanical properties and arc weldability, they are used for drawn and/or deep-drawn parts.

(2) Trade name of Sollac-Usinor (France) for cold-rolled carbon steels with from less than 0.01 to 0.10% carbon. They are supplied in several grades with good mechanical properties including high elongation and stiffness. Some grades are made by interstitial-free (IF) refining. Used for deep- and very deep-drawing applications in the manufacture of domestic appliances, heating and ventilation equipment, metal furniture, profiled tubes, etc.

Solstrip. Trade name of Sollac-Usinor (France) for a series of cold-rolled high-carbon strip steels supplied in several grades with high, as-annealed ultimate tensile strengths, minimum elongations of 20%, and hardnesses of 85-91 Rockwell B. The final hardness is achieved after custom forming and cutting by a quenching treatment. Used in the manufacture of precision mechanical parts, such as elastic fasteners, clutch parts, chain links, blades, piston segments, etc.

soluble glass. See soda water glass.

soluble guncotton. See pyroxylin.

soluble nitrocellulose. See pyroxylin.

soluble starch. A starch usually obtained from potatoes and available as a white, water-soluble powder formed by very slight hydrolysis. It decomposes at 256-258°C (493-496°F), and is used as an emulsifier, as a paper coating, as a textile size, and in gel electrophoresis.

Solugold. Trade name of Working Solutions Inc. (USA) for protective coatings used on aluminum, magnesium and zinc.

So-Luminum. Trade name for an alloy of 55% tin, 33% zinc, 11% aluminum and 1% copper used as an aluminum solder.

Soluna. Trademark of Mitsubishi Rayon Company (Japan) for polyester fibers and filament yarns used for textile fabrics.

Soluphene. Trade name of Atofina SA (France) for phenol-formaldehyde adhesives used in the building and construction industries.

Solus. Trademark of Taconic (USA) for strong, durable, moisture-resistant architectural glass fabrics coated on both sides with polytetrafluoroethylene (PTFE). Used for glazing applications.

Solusal. Trade name of Electro-Steel Company (USA) for a lightweight, corrosion-resistant alloy composed of aluminum, copper and silicon. Used for airplane, automobile and marine parts.

solution. A homogeneous mixture of two or more components in a single phase.

solution ceramics. Solutions consisting of a decomposable metal salt and thermoplastic resin or porcelain enamel applied and matured on the hot substrate surface and converting to a ceramic or glassy coating exhibiting high resistance to thermal shock.

solution-hardened alloys. Metal alloys, such as nickel-copper or copper-zinc, that have been heated to, and held at an appropriate temperature for a sufficient period of time to allow one or more constituents to enter into solid solution, and then cooled rapidly to prevent the constituent from precipitating. The increased hardness and mechanical strength of the cooled solid solution is due to the pinning of dislocations by solute atoms. Also known as *solution-heat-treated alloys*.

solution-heat-treated alloys. See solution-hardened alloys

Solvar. (1) Trade name of Solvay Polymers, Inc. (USA) for polyvinyl alcohol (PVAL) and polyvinyl acetate (PVAC) resins.

(2) Trade name of Rea Magnet Wire Company (USA) for aluminum and copper magnet wire.

solvent-activated adhesive. An adhesive that is essentially dry to the touch, but can be made tacky by means of a solvent just prior to use.

solvent adhesive. An adhesive, other than a water-base one, whose vehicle is a volatile organic liquid.

solvent-bonded nonwoven fabrics. Nonwoven fabrics produced by a process in which bonding is achieved by the softening of the fiber surfaces in a batt or web with a suitable solvent. Also known as *solvent-bonded nonwovens*.

solvent cement. An adhesive composed of a solution consisting of a synthetic resin or resin compound dissolved in a suitable organic liquid.

solvent moldings. Thermoplastic moldings formed either by applying a plastic resin dispersion to a mold and drawing off the solvent, leaving behind an adhering plastic film of varying thickness, or by first immersing a male mold into a dissolved resin and then removing it after an adhering layer of desired thickness has formed.

solvent-release sealant. A sealant that cures essentially by evaporation of the solvent.

Solvic. Trademark of Solvay Polymers, Inc. (USA) for unmodified and modified polyvinyl chlorides (PVCs) including *Solvic S* impact modified PVCs, and *Solvic-Premix* ready-to-process PVC premixes supplied in powder form.

Solvron. Trademark of Nitivy Company (Japan) for polyvinyl alcohol fibers and filament yarns.

Somdie. Trademark Walter Somers Limited (UK) for a high-carbon die-block steel containing 1% chromium, 1% nickel, 0.5% molybdenum, and the balance iron.

Somel. Trademark of E.I. DuPont de Nemours & Company (USA) for thermoplastic polyolefins.

Sona-Metal. Trade name of Manos Limited (Switzerland) for heat-treatable aluminum alloy composed of 3-4% copper, 1% iron, 0.7% tin, 0.2-0.7% silicon, 0.5% magnesium, and 0.2% nickel. Used for light-alloy parts.

SonaSpray. Trade name of International Cellulose Corporation (USA) for a cellulose-based acoustic ceiling finish supplied in spray cans.

Sonderbronze. German trade name for a series of alloys containing at least 78% copper and significant additions of aluminum, beryllium, iron, nickel, manganese, lead, silicon and/or tin. They often derive their names from major alloying elements, e.g., aluminum bronze, beryllium bronze, leaded bronze, silicon bronze, etc. In general, *Sonderbronze* alloys possess high corrosion resistance, good antifriction properties, high electrical conductivity and often good strength properties. Used for fittings, valves, pump housings, gears, worms and worm wheels, bearings, shafts, springs, fasteners, piston pins, chemical equipment, and wear parts.

Sonderlegierung. Trade name of Vereinigte Leichtmetallwerke GmbH (Germany) for corrosion-resistant aluminum alloy of 1% manganese, 0.5% iron and 0.2% silicon. Used for deep-drawn parts.

Sondermessing. German trade name for a series of copper-zinc alloys to which small amounts of other metals, such as aluminum, iron, lead, manganese, nickel, silicon or tin, have been added in order to improve certain properties, e.g., hardness and strength, corrosion resistance and wear resistance. Used for hardware, nuts, bolts, screws, fittings, tubes, bushings, guides, sleeves, electrical and electronic applications, architectural purposes, and appliances.

Sonderstahl. Tradename of Stahlwerk Kabel, C. Pouplier (Germany) for a series of die and tool steels.

Sonderstähle. Trade name of Kind & Co. Edelstahlwerk (Germany) for a series of carbon tool steels.

Sonex. Trademark of Illbruck-Sonex (Germany) for a spongy, open-cell foam used to absorb and muffle acoustic energy, and supplied in various colors (blue, brown, beige, charcoal, etc.) and thicknesses (50-100 mm, or 2-4 in.). Used for industrial and pro-audio applications.

Sonobatts. Trade name of Owens-Corning-Ford Company (USA) for glass-fiber pads.

Sonoboard. Trademark of Owens-Corning-Ford Company (USA) for acoustical glass-fiber ceiling boards having white film facings.

Sono-Cem. Trade name for a dual-cure dental resin cement.

Sonocor. Trademark of Owens-Corning-Ford Company (USA) for acoustical glass-fiber ceiling tiles with white, stippled film surfaces.

Sonoface. Trade name of Owens-Corning-Ford Company (USA) for acoustical glass-fiber tile and ceiling boards having film facings with salt and pepper design.

Sonoflex. Trademark of Owens-Corning-Ford Company (USA) for acoustical glass-fiber ceiling panels having white, woven-textured patterns.

Sonoform. Trade name of Owens-Corning-Ford Company (USA) for acoustical glass-fiber ceiling tiles with vacuum-formed films deposited over standard surface finishes.

sonogel. A *gel* produced from a *sonosol* by the sol-gel technique.

Sonoglas. Trademark of Fiberglas Canada Inc. (USA) for acoustical glass-fiber insulation products.

Sonolite. Trade name Noise Control Products Limited (UK) for acoustical glass-fiber ceiling panels with polyvinyl chloride coatings.

sonolite. A pink brown to brown mineral of the humite group composed of manganese silicate hydroxide, $Mn_9Si_4O_{16}(OH)_2$, and sometimes also containing zinc. Crystal system, monoclinic. Density, 3.87 g/cm^3; refractive index, 1.778. Occurrence: USA (New Jersey), Japan.

Sonoma. Trademark of Degussa-Ney Dental (USA) for a low-density dental alloy containing 56% silver and 37% palladium. It has excellent sag resistance, and is designed for matching with the *Duceragold* porcelain system. Used in restorative dentistry, especially for long-span bridges.

Sonora. Trademark of E.I. DuPont de Nemours & Company (USA) for a group of polymers used in apparel, upholstery fabrics, resins and monofilaments.

sonoraite. A yellowish green mineral composed of iron tellurate hydroxide monohydrate, $FeTeO_3(OH)\cdot H_2O$. Crystal system, monoclinic. Density, 3.95 g/cm³; refractive index, 2.023. Occurrence: Mexico.

Sonora White. Trade name of Aardvark Clay & Supplies (USA) for a smooth off-white clay (cone 10).

sonosol. A *sol* produced by using ultrasound as a reaction promoter. It is a precursor for the preparation of a *sonogel*.

Sonoston. Trade name of Stone Manganese–J. Stone & Company Limited (UK) for a manganese casting alloy with high damping capacity composed of 30% copper, 4% aluminum and 2.5% nickel.

Sonotherm. Trademark of Fiberglas Canada Inc. (USA) for glass-fiber insulation with excellent acoustical and thermal properties.

Sonowax. Trademark of Witco Chemical Company, Inc. (USA) for microcrystalline wax emulsions used in the manufacture of resins and as textile top finishes.

Sontara. Trademark of E.I. DuPont de Nemours & Company (USA) for spunlaced nonwoven fabrics

Sooflex. Trade name for nylon fibers and products.

sopcheite. A gray mineral with brownish tint composed of silver palladium telluride, $Ag_4Pd_3Te_4$. Crystal system, orthorhombic. Density, 9.95 g/cm³. Occurrence: Russian Federation.

Sopilen. Trademark of Sopron (Hungary) for polypropylene fibers and filament yarns.

Soprasolin. Trademark of Soprema (France) for bituminous roof sealing courses and dampproof roofing membranes.

Soprefelex. Trademark of Soprema (France) for bituminous roofing felt, roof sealing courses, and dampproof membranes.

Sorane 500. Trade name of ICI Polyurethane (UK) for a soft microcellular polyurethane elastomer.

Sorba-Glas. Trademark of Industrial Noise Control Inc. (USA) for noise-absorbing quilted fiberglass blankets.

sorbite. (1) A now obsolete term formerly used for a fine aggregate of *ferrite* and *cementite* that is softer than *troostite* and obtained in a steel either by tempering after hardening, or by regulating the cooling rate. The former is now known as *tempered martensite*, and the latter as very fine *pearlite*.

(2) In a tool steel, a mixture of *ferrite* and *cementite* obtained by regulating the cooling rate such that it is too fast to produce *pearlite* and too slow to produce *martensite*.

Sorbitex. Trademark of Kaltwalzwerk Brockhaus GmbH (Germany) for texture-rolled steel strip used for spings.

Sorbo-Cel. Trade name of Manville Corporation (USA) for a diatomaceous silica used for filtration applications.

sorbyite. A black mineral composed of lead antimony arsenic sulfide, $Pb_{17}(Sb,As)_{22}S_{50}$. Crystal system, monoclinic. Density, 5.52 g/cm³. Occurrence: Canada (Ontario).

Soreblister. Trade name of Atofina SA (France) for highly transparent, impact-resistant, deep-drawing grade rigid polyvinyl chloride sheeting for blister packs.

Soreflon. Trade name of Atofina SA (France) for polytetrafluoroethylenes, either unfilled or filled with graphite, glass fibers or bronze.

Sorel cement. See magnesium oxychloride cement.

Sorelmetal. Trademark of QIT-Fer et Titane Inc. (Canada) for high-purity pig irons with high carbon (as high as 4.3%) or low

carbon (as low as 2.85%) contents. A typical composition range is 2.85-4.3% carbon, 0.18-1.1% silicon, 0.009% manganese, 0.027% phosphorus, 0.02% sulfur, and the balance iron. Used as raw materials in the production of all grades of ductile irons and some special grades of gray and malleable irons, as preferential charge materials for melting high-grade steels, and for recarburizing applications.

Sorel's alloy. A zinc-base bearing alloy containing 1-10% copper and 1-10% iron, used for ornamental applications.

Sorelslag. Trademark of QIT-Fer et Titane Inc. (Canada) for a slag rich in titanium dioxide (about 70%) made by electric-furnace smelting of iron-titanium ores. In its final form, the slag is granular with a maximum particle size of 16 mm (0.625 in.), and is used primarily for the production of white pigments for the paint, ceramic, plastic, rubber and automobile industries, and in the manufacture of welding-electrode coatings and ferrotitanium.

sorensenite. A colorless, pinkish or white mineral composed of sodium beryllium tin silicate dihydrate, $Na_2SnBe_2(Si_3O_9)_2\cdot 2H_2O$. It can also be made synthetically. Crystal system, monoclinic. Density, 2.90 g/cm³; refractive index, 1.585. Occurrence: Greenland.

Soretherm. Trade name of Atofina SA (France) for highly transparent polyvinyl chloride shrink-on film used for packaging applications.

Sormite. (1) Russian trade name for an air-hardening, nondeforming tool steel containing 1.75% carbon, 15.5% chromium, 2% nickel, 2% silicon, and the balance iron. Used for blanking and forming dies, punches, gages, etc.

(2) Russian trade name for hard, wear- and corrosion-resistant alloy containing 3.2% carbon, 30% chromium, 4.5% silicon, 5.3% nickel, 1.1% manganese, and the balance iron. Used for hardfacing applications and as a substitute for *Stellite*-type alloys.

sorosilicates. Silicate minerals, such as *akermanite* ($Ca_2MgSi_2O_7$), in which two SiO_4 tetrahedra share one oxygen atom.

sosedkoite. A colorless mineral composed of potassium sodium aluminum antimony niobium tantalum oxide, $(K,Na)_5Al_2(Ta,Nb,Sb)_{22}O_{60}$. Crystal system, orthorhombic. Density, 6.90 g/cm³. Occurrence: Russian Federation.

soucekite. A lead-gray mineral of the bournonite group composed of bismuth copper lead sulfide selenide, $CuPbBi(S,Se)_3$. Crystal system, orthorhombic. Density, 7.60 g/cm³. Occurrence: Czech Republic.

Soudinox. Trade name of Creusot-Loire (France) for a series of austenitic stainless steels containing 0-0.15% carbon, 1-7% manganese, 0-3.7% silicon, 13-30% chromium, 0-25.5% nickel, 2.5-4.3% molybdenum, and the balance iron.

Sounda. Trade name of Owens-Corning-Ford Company (USA) for acoustical glass-fiber ceiling tiles used for direct application to existing ceiling.

sound barrier materials. See acoustical materials.

sound-damping alloy. An alloy, such as copper-manganese or cast iron, that absorbs sound and sound vibrations, and is used for machinery housings, jackhammers, power tools, etc.

sound-deadening board. An acoustic insulating product based on cellulosic fiberboard, and used for soundproofing ceiling, floor and wall assemblies.

Sound-Gard. Trade name of ASG Industries Inc. (USA) for a sound-control glass consisting of one or more interlayers with a total thickness of 1.14 mm (0.045 in.) sandwiched between two layers of transparent, bronze, gray or green plate or sheet glass. It is particularly effective in the frequency range of 0.25-4.00 kHz.

sound insulators. See acoustical materials.

Sound Sealant. Trade name of Acoustical Solutions Inc. (US) for a non-flammable, non-staining, permanently flexible latex-base acoustical sealant that bonds well to concrete, drywall, metal, wood and many other substrates. It is supplied in cartridges for gun application, and used in building and construction in and around joints, cutouts and other openings of wall partition systems to reduce sound transmission.

Soundtropane. Trade name of Dearborn Glass Company (USA) for a laminated glass with a viscoelastic interlayer that is 7-19 mm (0.28-0.75 in.) thick. Used in sound transmission control.

Souple. See Meryl Souple.

source material. Any material capable of rendering a fissile material upon extraction.

sourgum. The hard, tough wood of the tree *Nyssa sylvatica*. The heartwood is brownish gray and the sapwood grayish white. It has a fine, uniform texture, high to medium strength and stiffness and moderate shock resistance. Average weight, 561 kg/m³ (35 lb/ft³). Source: Eastern and southeastern United States from southern Maine to Florida and from Texas to Michigan and Missouri. Used for lumber, furniture, boxes, crates, veneer, and paper pulp. Also known as *black gum; black tupelo.*

sour tupelo. See ogeche tupelo.

southern cypress. See baldcypress.

Southern Asbestos. Trade name of C.E. Thurston & Sons, Inc. (USA) for asbestos products including fabrics, sewing thread, and *amosite* felt.

Southern Gold. Trademark of Georgia-Pacific Corporation (USA) for plywood and plywood products.

southern pine. A collective term for the wood of several species of pine trees sold together. The four most important of these species are longleaf *(Pinus palustris),* shortleaf *(P. echi-nata),* loblolly *(P. taeda)* and slash pine *(P. elliottii).* The heartwood is reddish-brown and the sapwood yellowish-white. It has moderate shrinkage and an average weight of 575-660 kg/m³ (36-41 lb/ft³). *Longleaf* and *slash pine* are somewhat heavier than *shortleaf* and *loblolly pine.* Source: Southern United States (from North and South Carolina and Georgia to Texas, Arkansas and Oklahoma). Used for interior and exterior construction work (e.g., stringers, joists, beams, posts and piles, subflooring, interior finish and sheathing), boxes and crates, cooperage, plywood, railroad piles, poles and ties, and mine timbers. Also known as *southern yellow pine.*

southern white cedar. See Atlantic white cedar.

southern yellow pine. See southern pine.

souzalite. A blue green mineral composed of magnesium iron aluminum phosphate hydroxide dihydrate, $(Mg,Fe)_3(Al,Fe)_4(PO_4)_4(OH)_6 \cdot 2H_2O$. Crystal system, triclinic. Density, 3.09 g/cm³; refractive index, 1.642. Occurrence: Brazil.

sovereign gold. A standard gold alloy (22-karat) containing 91.7% gold and 8.3% copper.

Soviden. Trade name for saran (polyvinylidene chloride) fibers with good chemical resistance, good weatherability, and good resistance to mildew and insects. Used for carpets and rugs, draperies, upholstery, clothing, industrial fabrics, etc.

Sovirel. Trade name of Sovirel (France) for a range of laboratory glassware and optical glass products.

soybean fiber protein. A relatively new type of protein fiber obtained from soybean and made by wet spinning involving the addition of a functional agent. It is soft, lustrous, highly absor-

bent and breathable, resembles silk and cashmere in feel and appearance, and can be blended with flax, wool, silk, cashmere and spandex for the manufacture of apparel and underwear. Abbreviation: SFP.

soybean fibers. Soft, white to light tan, spun protein fibers with wool-like appearance and a natural crimp, obtained by chemical regeneration from soybean. Used occasionally in textile blends.

soybean glue. A relatively strong, moderately water-resistant vegetable glue made from soybean meal. It is usually supplied as a dry powder, and used for joining panel materials, such as plywood.

Soylon. Trade name of Ford Motor Company (USA) for protein fibers derived from soybeans.

Spa Alloys. Trade name of Abex Corporation (USA) for a series of stainless cast steels containing 1.6% carbon, 28% chromium, 2% nickel, 2% molybdenum, 1% copper, 0-2% silicon, 0-1.5% manganese, and the balance iron. They have excellent corrosion and erosion resistance, very good abrasion resistance, and are used for chemical process equipment, metallurgical equipment, etc.

spaceboard. A composite material developed by the US Forest Products Laboratory. It is a three-dimensional board with numerous small cells ("spaces") made of old cardboard and newspapers or mixed office waste. It is manufactured by first producing a slurry on conventional papermaking equipment, dewatering and then press-drying the wet fibers against patterned rubber molds. The resulting boards are subsequently bonded together with a suitable adhesive, and are stronger than ordinary corrugated fiberboard.

SpaceMaster. Trademark of F.J. Brodmann & Co. LLC (USA) for various powders of materials, such as sapphire, barium titanate, ferrites, magnetite, synthetic diamond, boron nitride and cubic boron nitride, lanthanum chromite, molybdenum disilicide, aluminum nitride, porcelain, etc. Depending on the particular powder composition, they are used as fillers for flat screen spacers, radar absorbers, superabrasives, and high dielectric, electrical or thermal conduction applications.

Space Shield. Trade name of Republic Glass Company Inc. (USA) for insulating glass.

spackling compound. A type of plastering material that is used to repair surface irregularities and cracks in plaster.

Spallshield. Trademark of E.I. DuPont de Nemours & Company (USA) for anti-lacerative plastic sheeting for lamination to glass surfaces for use in commercial windows, window walls, etc.

Spanacoustic. Trade name of Manville Corporation (USA) for acoustical fiber-glass panels.

Spanasil. Trade name of Kreidler Werke GmbH (Germany) for a heat-treatable aluminum alloy containing 1% silicon, 1% lead and 0.8% magnesium. It has improved free-cutting properties, and is used for light-alloy parts.

Spancolor. Trade name of Central Glass Company Limited (Japan) for toughened spandrel glass, glazed on one side with a pigmented ceramic material.

Spancore. (1) Trade name of Manville Corporation (USA) for acoustical fiberglass panels.

(2) Trademark of Rubyco Inc. (Canada) for elastic yarns and spandex fibers.

Spandavan. Trademark of Gomelast CA (Venezuela) for spandex fibers and filament yarns used for elastic textile fabrics.

Spandelle. Trade name for a spandex (segmented polyurethane) fiber used for the manufacture of textile fabrics with excellent stretch and recovery properties.

spandex fiber. A generic name used in North America for a very light, elastic synthetic fiber in which the fiber-forming substance is composed of 85% or more of a segmented polyurethane. Often the polyurethane is shaped like a core around which cotton or other fibers are wound to make elastic yarn. Used to impart elasticity to garments, e.g., girdles, hosiery, socks, pants, etc. The ISO (International Organization for Standardization) designates this fiber as "elastane." Also known as *spandex*. See also elastane fiber.

Spandrel Ceramiclite. Trade name of C-E Glass (USA) for a fully toughened, permanently ceramic-coated glass.

spandrel glass. A special architectural glass that is used in curtain walls, usually in nonvision areas, and in the cladding of buildings.

Spandreline. Trade name of PPG Industries Inc. (USA) for a window material consisting of a heat-strengthened glass pane that combines a clear-vision area with a ceramic-colored spandrel area.

Spandrelite. Trade name of PPG Industries Inc. (USA) for a ceramic-colored structural plate glass used for cladding buildings.

Spang Chalfant. Trade name of National Supply Company (USA) for a case-hardening steel containing 0.1-0.2% carbon, 0.5% manganese, 0-2% chromium, 0-0.65% molybdenum, and the balance iron. Used for gears, shafts, and rolls.

spangle sheet. A hot-dip galvanized sheet steel with a characteristic crystalline surface pattern, formed by the solidified zinc.

spangolite. A dark or bluish green mineral with vitreous luster composed of copper aluminum chloride sulfate hydroxide trihydrate, $Cu_6Al(SO_4)Cl(OH)_{12} \cdot 3H_2O$. Crystal system, hexagonal. Density, 3.14 g/cm^3. Occurrence: USA (Arizona, Utah).

Spanish blonde lace. A large, floral, hand-made *bobbin lace*, usually of raw silk and in white, yellowish-white or black colors, stitched to a mesh ground having a heavy yarn outlining.

Spanish cedar. The soft, reddish, fragrant wood of any of various large trees of the genus *Cedrela*. It has a light to dark reddish brown heartwood, medium strength, good resistance to decay and insect, works and seasons very well, splits readily, takes a beautiful polish and has good gluing qualities. *Spanish cedar* somewhat resembles the lighter grades of true mahogany. Average weight, 480 kg/m^3 (30 lb/ft^3). Source: Central and South America, and West Indies. Used for construction applications, furniture, cabinetwork, joinery, interior trim, boatbuilding, cigar boxes, foundry patterns, and as a mahogany replacement. Also known as *cedro*.

Spanish chestnut. See sweet chestnut.

Spanish leather. Leather having a somewhat creased or wrinkled surface featuring elongated grains produced by tanning in a strong liquor based on *quebracho*.

Spanish mahogany. The hard, tough wood of the mahogany tree *Swietenia mahagoni*. It has high strength, very high stability, an attractive appearance, and is denser and finer textured than other mahoganies, but now largely replaced by Honduras and African mahoganies. Average weight, 720 kg/m^3 (45 lb/ft^3). Source: West Indies. Used for fine, high-quality furniture and interior decorative work. Also known as *Cuban mahogany*.

Spanish white. (1) A white, ground and washed chalk ($CaCO_3$), similar to *Paris white*, obtained from Spain, and used as an extender in paper, rubber, plastics, etc., and in *whiting*.

(2) See bismuth oxynitrate.

Silkiss. Trademark of Kuraray Company (Japan) for spandex fibers and filament yarns used for elastic textile fabrics.

Spanzelle. Trade name for a spandex (segmented polyurethane) fiber used for the manufacture of textile fabrics with excellent stretch and recovery properties.

spar. (1) A collective name used for light-colored, readily cleavable, crystalline minerals, such as feldspar, fluorspar (fluorite), calcspar (calcite), Iceland spar and dogtooth spar.

(2) See spar varnish.

Sparkaloy. Trade name of Wilbur B. Driver Company (USA) for a wear-resistant iron alloy containing 4% manganese and 1% silicon. Used for spark-plug wire.

Sparkalure. Trade name for polyolefin fibers and yarns used for textile fabrics.

Sparkel. Trade name of Pilkington Brothers Limited (UK) for a patterned glass.

sparking metal. An archaic term for a pyrophoric alloy, i.e., an alloy that sparks when slight friction is applied by another metal (e.g., by striking with a hammer at a certain angle). It often refers to *Auer metal* that is composed of 65% mischmetal (chiefly cerium and lanthanum) and 35% iron, and was formerly used for gas and cigarette lighters. See also pyrophoric alloys.

Sparkling. Trade name for nylon fibers and yarns used for the manufacture of glossy fabrics including clothing.

Spark-L-lite. Trade name for saran (polyvinylidene chloride) fibers with good chemical resistance, good weatherability, and good resistance to mildew and insects. Used for carpets and rugs, draperies, upholstery, clothing, industrial fabrics, etc.

Spark-O-Lite. Trade name of The Paul Wissmach Glass Company Inc. (USA) for a special color-reflection-transmitting glass.

Sparkonite. Trade name of CMW Inc. (USA) for a series of sintered materials of varying compositions used for EDM (electrical discharge machining) electrodes. Some typical compositions include: (i) 95% copper and 5% $BaCO_3$; (ii) 97.2% copper and 2.8% ZrO_2; (iii) 68-70% tungsten and 30-32% copper, and (iv) graphite-base materials.

sparry gypsum. See selenite.

sparry iron ore. See siderite (1).

Sparta. Trade name of Cytemp Specialty Steel Division (USA) for an air-hardening medium-alloy cold-work tool steel (AISI type A2) containing 1.05% carbon, 5.00% chromium and 0.90-1.25% molybdenum, 0.25% vanadium, and the balance iron. It has a great depth of hardening, good nondeforming properties, and good wear resistance and toughness. Used for press tools, punches, cutting tools, and blanking, forming and trimming dies.

spartalite. See zincite.

Spartan. (1) Trademark of Lukens Steel Company (USA) for a very-low-carbon high-strength steel containing up to 1.75% copper as the primary precipitation-strengthening element. It has good low-temperature properties, high strength, toughness and ductility, good formability and weldability, good resistance to atmospheric corrosion, and requires little or no preheat before welding. Used for dredging and mining equipment, ship hulls, offshore drilling rigs, cold-climate machinery, truck frames, and large valves.

(2) Trademark of Atlas Specialty Steels (Canada) for a series of hot-work and high-speed tool steels. *Spartan-5* is a hot-work tool steel (AISI type H26) containing 0.5% carbon, 0.3% manganese, 0.3% silicon, 18% tungsten, 4% chromium, 1% vanadium, and the balance iron. It has excellent wear resistance and high hot hardness, and is used for extrusion dies for

brass, bronze and steel, hot-forging dies and die inserts, hot trimming and blanking dies, coining and tieplate dies, glass-forming dies, punches and grippers, hot nut piercers, and shear blades. *Spartan-7* is a tungsten-type high-speed tool steel (AISI type T1) containing 0.73% carbon, 0.25% manganese, 0.3% silicon, 18% tungsten, 4% chromium, 1% vanadium, and the balance iron. It has high hot hardness, excellent machinability and wear resistance, good toughness, and is used for flat and twist drills, threading taps, thread chasers, broaches, reamers, countersinks, lathe and planer tools, boring tools, slotters, slitting cutters, woodworking knives, chisels and punches, mandrels, forming rolls, nail and lamination dies, and knurling tools.

(3) Trademark of Ivoclar Vivadent AG (Liechtenstein) for palladium-based dental alloys.

(4) Trade name of J.H. France Refractories Company (USA) for high-alumina brick.

Spartan Plus. Trademark of Ivoclar Vivadent AG (Liechtenstein) for a silver-free ceramic dental alloy containing 78.8% palladium, 10.0% copper, 9.0% gallium, 2.0% gold, and less than 1.0% indium, germanium and iridium, respectively. It has a white color, a density of 10.7 g/cm^3 (0.387 lb/in.^{3}), a melting range of 1180-1210°C (2155-2210°F), a hardness after ceramic firing of 310 Vickers, moderate elongation, excellent biocompatibility, and prevents ceramic discoloration. Used for crowns, onlays, posts, and bridges.

Spartan Redheugh. Trade name of Spartan Redheugh Limited (UK) for a series of structural steels and austenitic, ferritic and martensitic stainless steels.

Spartex. Trade name of Egyptian Lacquer Manufacturing Company Inc. (USA) for one-coat texture enamels.

spar varnish. A very durable, weather-resistant *long-oil varnish* consisting of one or more drying oils, one or more natural or synthetic resins (e.g., rosin, phenolics, etc.), and one or more volatile thinners and driers (usually linoleates, naphthenates or resinates of cobalt, lead or manganese). It has excellent resistance to heat, sunlight and water, and is used for on exterior surfaces in severe service applications. Spar varnish was named from its suitability for the wooden decks and spars of ships. Also known as *spar*.

spathic iron. See siderite (2).

spathic ore. See siderite (2).

Spaulding. Trademark of Spaulding Composites Company (USA) for an phenolic-based electrical insulating paper.

Spauldite. Trademark of Spaulding Composites Company (USA) for phenolic, melamine and epoxy resins and laminates. The phenolic laminates may be reinforced with aramid, canvas, cotton, glass or linen fabrics, or paper, the melamine laminates with glass or linen fabrics, and the epoxy laminates with glass fabrics.

Spaulrad. Trademark of Spaulding Composites Company (USA) for a radiation-resistant woven-glass-fiber-reinforced polyimide laminate.

Spear. Trade name of Spear & Jackson (Industrial) Limited (UK) for a series of steels including several plain-carbon grades containing 0.2-1.0% carbon, low-carbon nickel grades containing 0.1-0.3% carbon and 3-3.2% nickel, low-carbon alloy grades with 0.1-0.3% carbon, 0.8-1.2% chromium, 0.75-4.1% nickel and sometimes up to 0.3% molybdenum, numerous tool steel grades of the carbon-tungsten, cold-work and hot-work types, and various stainless steel grades.

Special. (1) Trade name of Jessop Steel Company (USA) for a high-speed steel containing 0.7% carbon, 18% tungsten, 4%

chromium, 1-2% vanadium, and the balance iron. Used for reamers, cutters, broaches, lathe and planers tools, and drilling tools.

(2) Trade name of Electro-Steel Company (USA) for an oil-hardening steel containing 0.8% carbon, 0.8% chromium, 0.2% vanadium, and the balance iron. Used for tools and cutters.

(3) Trade name of CCS Braeburn Alloy Steel (USA) for a plain-carbon tool steel (AISI type W1) containing 0.5-0.8% carbon, 0.25% manganese, 0-0.2% vanadium, and the balance iron.

Special Advance. Trade name of British Driver-Harris Company Limited (UK) for a thermocouple alloy of 58% copper and 42% nickel.

Special Alloy. (1) Trade name of Disston Inc. (USA) for a series of water-hardening tool steels containing 0.45% carbon, 0.8% chromium, and the balance iron. Used for tools, dies, punches, and chisels.

(2) Trade name of Edgar T. Ward's Sons Company (USA) for a water-hardening tool steel containing 0.7% carbon, 1% manganese, 0.4% chromium, and the balance iron. Used for tools and dies.

(3) Trade name of Wallace Murray Corporation (USA) for a series of water-hardening tool steels (AISI type W2). *Special Alloy-8* contains 0.8% carbon, 0.2% manganese, 0.2% silicon, 0.2% vanadium, and the balance iron. It has high toughness, low wear resistance and red hardness and moderate machinability. Used for rivet sets, clamping dies and pneumatic chisels. *Special Alloy-10* contains 1.05% carbon, 0.2% manganese, 0.2% silicon, 0.2% vanadium, and the balance iron. It has good toughness, moderate wear resistance and machinability, low red hardness, and is used for embossing dies, engravers rolls, die headers, striking dies, and rock-drill parts.

Special Aluminum. Trade name for a heat-treatable copper alloy containing 10-11% aluminum, 3.6% nickel, 3.1-4.25% iron and 0-1.6% manganese. Used for worm wheels, gears, and trolley shoes.

Special Ardho. Trade name of Spencer Clark Metal Industries Limited (UK) for a tool steel containing 0.4% carbon, 0.8% chromium, 3% nickel, and the balance iron. Used for chisels, punches, and blacksmithing tools.

special bar quality steel. See special quality bar steel.

Special Bloc. French trade name for an oil-hardening hot-work die steel containing 0.55% carbon, 1% chromium, 0.75% manganese, 0.55% nickel, 0.45% molybdenum, 0.05% vanadium, and the balance iron.

Special Chrome Vanadium. Trade name of Edgar Allen Balfour Limited (UK) for an oil-hardening tool steel containing 0.6% carbon, 0.9-1% chromium, 0.2% vanadium, and the balance iron. Used for punches and tools.

Special Conqueror. Trade name of Joseph Beardshaw & Son Limited (UK) for a water-hardened tool steel (AISI type W1) containing 0.85-1.1% carbon, and the balance iron. It has good machinability, fair deep-hardening properties and toughness, and is used for shear blades, punches, rock drills, taps, knives, and chisels.

Special Echo. Trade name of Hall & Pickles Limited (UK) for a high-speed tool steel containing 0.73% carbon, 4.3% chromium, 18.2% tungsten, 1.3% vanadium, and the balance iron.

Special Inlay. Trade name of J.F. Jelenko & Company (USA) for a soft, microfine dental casting alloy (ADA type I) containing 83% gold, 10% silver, and 1% palladium. It has a density of

16.6 g/cm³ (0.600 lb/in.³), a melting range of 940-960°C (1725-1760°F), and is used for soft, low-stress inlays.

Special Lohys. Trade name of Sankey & Sons Limited (UK) for a soft magnetic material containing 99.25% iron and 0.75% silicon. It has high magnetic permeability, low coercive force, low hysteresis loss, and is used for electrical equipment and magnetic instruments.

Specialloy. (1) Trade name of Specialloy Inc. (USA) for an extensive series of copper, manganese, nickel and niobium-base master alloys used in the manufacture of ferrous and nonferrous alloys.

(2) Trade name of Specialty Steel Company of America (USA) for an oil hardening medium-carbon low-alloy tool steel.

special malleable iron. Malleable iron produced to meet specialized requirements imposed by alloying, special heat-treatment or both. It is completely annealed and derives its particular properties from the compositional effect on the ferritic matrix. Two types of special malleable iron are classified: (i) *high-silicon malleable iron*; and (ii) *alloyed malleable iron*.

Special Manganese Nickel. Trade name of American Chain & Cable (USA) for a manganese steel containing 0.4% carbon, a total of 12-14% manganese and nickel, and the balance iron. Used for shielded-arc electrodes for welding and hardfacing.

special nuclear materials. A term referring to nuclear materials such as *plutonium-239, uranium-233, enriched uranium* and any other *enriched material* suitable for use as a nuclear fuel. Abbreviation: SNM. See also nuclear material.

Special Oil Hardening. Trade name of Jessop Steel Company (USA) for an oil-hardening, cold-work tool steel (AISI type O2) containing 0.9% carbon, 1.2-1.8% manganese, 0-0.2% chromium, and the balance iron. It has a great depth of hardening, good nondeforming properties, good machinability, good to moderate wear resistance, and is used for spindles, hobs, press tools, punches, and crimpers.

Special Oilway. Trade name of H. Boker & Company (USA) for an oil-hardening cold-work tool steel (AISI type O2) containing 0.9% carbon, 1.6% manganese, and the balance iron. It has good nondeforming properties, machinability and wear resistance, and is used for tools, dies, cutters, punches, reamers, broaches, saws, etc.

Special Presse. Trade name of Creusot-Loire (France) for hot-work tool steel containing 0.2% carbon, 3.4% molybdenum, 3.15% nickel, and the balance iron. Used for forging and forming dies.

Special Purpose Alloy. Trade name of Sharon Steel Corporation (USA) for a steel containing about 1.2-1.3% carbon, 0.1-0.4% manganese, 0.3% silicon, 0.1-0.3% chromium, and the balance iron. Used for hacksaw blades, razor blades, and cutlery.

special-purpose tile. Glazed or unglazed floor or wall tile made to specifications of shape, size, thickness, color, surface decoration or design, electrical properties, resistance to chemicals, mechanical and thermal shocks, abrasion and wear, and/or frosting and staining that normally do not apply to standard tile.

special-purpose tool steels. A group of oil-hardening tool steels subdivided into two classes: (i) *Low-alloy types* (AISI type L) and (ii) *Carbon-tungsten types* (AISI type F). Low-alloy types contain about 0.45-1.20% carbon, 0.20-0.75% manganese, 0.70-1.50% chromium, 0-2.00% nickel, 0-0.35% molybdenum, 0.10-0.30% vanadium, and the balance iron. They have fine grain sizes, good deep-hardening properties and tensile strength, fair machinability and good to fair toughness. Often sold under trademarks or trade names, such as *Halvan, Champalloy, Crown*

Superb, Metamold, Nicroman, Presto or *Titon*, they are used for arbors, chucks, collets, cams, knuckle pins, clutch parts, dies, etc. Carbon-tungsten types contain about 1.00-1.25% carbon, 0.2-0.4% manganese, 0.30-3.50% tungsten, and the balance iron. Other alloying elements may be present in small amounts. They possess fair deep-hardening properties and machinability, good to fair wear resistance and poor toughness. Often sold under trademarks or trade names, such as *Alloy Finishing, Colonial-4, Saturn* or *Silvanite,* they are used for special applications including tools, and machine parts.

special-quality steel bars. Steel bars, free of visible pipe and excessive chemical segregation and with minimum surface imperfections, rolled from specially conditioned and/or inspected billets and blooms. They have characteristics that make them suitable for structural applications that necessitate heat treatment, hot forging, cold drawing and forming, machining, etc. Also known as *special bar quality steels; SBQ steels.*

special refractories. A group of ceramic materials, such as aluminum oxide, magnesium oxide, beryllium oxide, zirconium oxide, silicon dioxide, silicon nitride, thorium oxide, uranium dioxide, carbon, graphite, mullite and spinel, used for highly specialized refractory applications.

specialty cast iron. (1) An alloy cast iron, i.e., a cast iron into which alloying elements, such as nickel, chromium, molybdenum or vanadium, have been introduced to improve properties, such as acid-, corrosion, heat- and/or wear resistance.

(2) A cast iron, such as *Meehanite*, produced by a special process. This term excludes standard gray, white, ductile and malleable cast irons.

specialty steel. A steel whose special properties are due to the addition of alloying elements, or special processing. Examples of such steels include high-strength steels, wear-resistant steels, acid- and corrosion-resistant steels, heat-resistant steels, creep-resistant steels, bearing steels, valve steels, nonmagnetizable steels, and permanent magnet steels. Abbreviation: SS.

specialty yarns. A general term referring to yarns with special effects and including fancy and novelty yarns, metallic and tinsel yarns, and crepe, snarl, nubbed, knobbed and tufted yarns.

Special Vanadium. Trade name of A. Milne & Company (USA) for a water-hardening tool steel containing 0.8-1.1% carbon, 0.1-0.2% vanadium, and the balance iron. Used for hand and cutting tools.

Special White. Trademark of Degussa-Ney Dental (USA) for an extra-hard dental alloy containing 45% gold, 40% palladium and 6.5% silver. It has excellent physical properties, high yield strength, a hardness of 260 Vickers, and is used in restorative dentistry, especially for short- and long-span bridges and single crowns.

Special Wolfram. Trade name of Swift Levick & Sons Limited (UK) for a hot-work die steel containing 0.5% carbon, 1.2% chromium, 2.2% tungsten, and the balance iron.

speckled ware. Ceramic ware with decorative surface finishes in which spots of one color appear on a relatively uniform background of another color.

specification cement. A cement prepared in accordance with a prescribed specification or standard. For example, in the United States a specification cement may be prepared in accordance with the specifications set forth by the American Society for Testing and Materials (ASTM), the Portland Cement Association (PCA) or the American Concrete Institute(ACI). Also known as *standard cement.*

Speckelon. Trade name for nylon fibers and yarns used for textile fabrics including clothing.

Speclar. Trade name of Pax Surface Chemicals Inc. (USA) for a bright electroless nickel coating.

Speclar+. Trade name of MSC Laminates & Composites Inc. (USA) for a coating of durable, and highly scratch-resistant and reflective material produced on aluminum or steel substrates by sputter coating. Used on lighting fixtures, light reflectors, flash-bake ovens, etc.

SpecTape. Trade name of SpecTape Inc. (USA) for a series of EMI/RFI shielding materials based on polyester laminates containing rubber and aluminum, or on polyvinyl chlorides or polyethylene terepthalates.

Spectar. Trade name of Eastman Chemical Company (USA) for tough, high-clarity polyester copolymer resins used for thick-sheet applications.

Spectra. (1) Trademark of AlliedSignal Corporation (USA) for a series of high-strength, high-modulus ultrahigh-molecular-weight extended-chain polyethylene fibers and filaments produced by a gel-spinning process. They have a density of 0.97 g/cm^3 (0.035 lb/in.^{3}), a low melting point, excellent mechanical properties at ambient temperatures, and very high tensile strength (10 times stronger than ordinary steel). Two common grades are *Spectra 900* with an ultimate tensile strength of 2.68 GPa (388 ksi) and an elastic modulus of 117 GPa (17 × 10^3 ksi), and *Spectra 1000* with an ultimate tensile strength of 3.12 GPa (452 ksi) and an elastic modulus of 173 GPa (25 × 10^3 ksi), respectively. Other grades include *Spectra Fusion, Spectra Guard, Spectra Shield* and *Spectra Shield Plus.* Used for tensile applications, such as rope and cord, reinforcing fibers, and industrial gloves.

(2) Trade name of NV Hardmaas (Netherlands) for a toughened sheet glass painted and fired in various colors.

SpectraBond. Trade name for a nickel-titanium wire used for dental applications.

Spectra-Cel. Trade name of American Fillers & Abrasives, Inc. (USA) for specially coated wood fibers used for applications requiring high-intensity colored texture effects.

SpectraFlex. Trademark of Honeywell/AlliedSignal Corporation (USA) for strong, yet flexible polyethylene fibers used for the manufacture of industrial products.

Spectrafloat. Trade name of Pilkington Brothers Limited (UK) for heat-rejecting float glass. The heat-rejecting effect is produced by concentrating closely packed, extremely fine metal particles immediately below the surface of the glass to form a layer.

Spectra-Glaze. Trademark of Genstar Corporation (USA) for a series of glaze-faced lightweight concrete blocks and masonry units.

Spectra Guard. Trademark of Spectra Technologies (USA) for an engineered multicomponent yarn consisting of a core, made up of a *Spectra* polyethylene fiber and a glass fiber, and a wrapping of *Spectra* polyethylene fiber. It has several times the cut resistance of aramid fibers, and provides excellent tear resistance, good launderability, good resistance to bleaches and other chemicals, and blends well with other fibers including nylons and polyesters. Used for protective gloves and sleeves, and protective and safety fabrics. *Spectra Guard CX* is astrong, lightweight engineered yarn consisting of a core, made up of a *Spectra* polyethylene fiber and a glass fiber, and wrapping of spun cotton. It provides nearly twice the cut resistance of aramid, a comfort and suppleness like cotton, excellent resistance to bleaches and other chemicals and can be coated, laminated or

dotted for improved grip. Used for protective gloves, and for broad-woven, coated industrial fabrics including tarpaulins, cargo curtains, etc.

Spectraguard. Trademark of AlliedSignal Corporation (USA) for a series of conductive, one- and two-part, silver-filled acrylic, epoxy and polyurethane coatings providing good chemical and abrasion resistance and EMI/RFI shielding effectiveness of about 70-80 dB (at 1-3000 MHz), tack-free times between 15 and 60 minutes depending on grade, flash points of 2°C (35°F) for epoxies, 10°C (50°F) for polyurethanes and 13°C (55°F) for acrylics, and service temperatures of -51 to +204°C (-60 to +400°F) for epoxies, -51 to +177°C (-60 to +350°F) for polyurethanes, and -51 to +135°C (-60 to +275°F) for acrylics. They can be applied to plastics, composites and primed or unprimed metals, and are used for ground-plane applications requiring high conductivity across a nonconductive seam. The acrylic coatings are also suitable for surfaces exposed to water, solvent, chemical and salt-spray corrosion.

Spectran. Trademark of Honeywell International (USA) for a series of acrylic and polyester fibers.

Spectra Shield. Trademark of Honeywell/AlliedSignal Corporation (USA) for an engineering composite in which nonwoven fibers of *Spectra* ultra-high-molecular-weight polyethylene fibers are laid up in a cross-ply (0°/90°) in a (usually thermoset) polymer matrix. A modified grade, *Spectra Shield Plus,* is also available. Used for armor applications, e.g., military equipment, shields, helmets, etc.

Spectra-Seal. Trademark of 3M Unitek (USA) for gold-based dental alloys.

Spectrim. Trademark of Dow Plastics (USA) for a series of polyurethane-base resins including regular *Spectrim,* which is a reaction-moldable polycarbamate resin available in several grades including glass-reinforced. It has excellent physical and mechanical properties, and is used for automotive applications, such as bumpers, window encapsulation, fascia, claddings, interior and exterior trim, and body, door and instruments panels. Other important products in this series are: (i) *Spectrim BP,* an impact-modified polyurethane for reaction-injection molding (RIM) of automotive components, e.g., body panels, fenders and claddings; (ii) *Spectrim BST,* a tough polyurethane for reaction-injection molding (RIM) that exhibits a low-coefficient linear thermal expansion, good paintability and gravel resistance, and is used for automotive applications, e.g., exterior trim and claddings; (iii) *Spectrim HH,* an impact-resistant polyurethane for reaction-injection molding (RIM) of exterior automotive components, such as wheel covers, body panels, fenders, filler panels and other exterior trim. It is designed to withstand a bake heat of 175 to 200°C (345 to 390°F) and provides high rigidity, good processibility and paintability, and excellent surface finish; (iv) *Spectrim MM,* a polyurethane with high energy ab-sorption for reaction-injection molding (RIM) of automotive front and rear systems including bumper beams; and (v) *Spectrim RL* polyurethane resins for reaction-injection molding (RIM) of automotive components, e.g., interior door panels, quarter-trim panels and polyvinyl chloride/thermoplastic olefin (PVC/TPO) covered interior trim.

Spectrum. (1) Trademark of RoyaliteThermoplastics Division of Uniroyal Inc. (USA) for plastic sheeting based on thermoplastic olefin, or high- or medium-density polyethylene.

(2) Trade name of American Orthodontics and Dentsply/Caulk (USA) for light-cure dental hybrid resin composites used for bonding orthodontic brackets and other dental applications.

specular hematite. See specularite.

specular iron ore. See specularite.

specularite. A brilliant gray or grayish black variety of the mineral *hematite* with a metallic luster. It is composed of ferric oxide, Fe_2O_3, and contains up to 70% iron. It occurs in both foliated micaceous masses and tabular or disklike crystals. Density, 4.8-5.3 g/cm^3; hardness, 5.5-6.5 Mohs. Used as an ore of iron. Also known as *gray hematite; iron glance; specular hematite; specular iron; specular iron ore.*

SpecularPlus. Trade name of MSC Laminates & Composites Inc. (USA) for durable high-reflectance materials made by bonding scratch-resistant, mirror-like finishes to aluminum or steel sheet substrates using magnetron-sputtered vacuum deposition techniques.

Specular+SR. Trade name of MSC Laminates & Composites Inc. (USA) for a laminate with mirror-like appearance and high reflectance composed of a steel substrate coated with adhesive, metallized with silver and covered with a thin polyester film with a scratch-resistant coating.

speculum. A hard, brittle, brilliant silvery white bronze of varying compositions including: (i) 66-68.25% copper and 31.75-34% tin; (ii) 66% copper, 34% tin and traces of arsenic; (iii) 64% copper, 32% tin and 4% nickel; and (iv) 50-60% copper and 40-50% tin. Microstructurally, in speculum alloys copper and tin form the intermetallic compounds Cu_4Sn and Cu_3Sn. A 67Cu-33Sn speculum alloy has a density of 8.6 g/cm^3 (0.31 lb/in.^{3}) and a melting point of 745°C (1373°F). *Speculum* takes a high polish, has high toughness and good resistance to tarnishing, and is used in the manufacture of metal mirrors, reflection-diffraction gratings, reflecting telescopes and other optical instruments, and for electroplating cutlery and household goods. Also known as *speculum alloy; speculum metal.*

speculum plate. A hard, corrosion-resistant, silvery white electrodeposited coating composed of an alloy of 50-60% copper and 40-50% tin. The commonest coating is composed of 60% copper and 40% tin, and used for decorative applications requiring a silver-like appearance, in the form of protective coatings for food-processing and chemical process equipment, and as reflective coatings for mirrors and optical equipment. See also speculum.

Spedex. Trademark of Barker & Allen Limited (UK) for a free-machining nickel silver containing 53-65% copper, 8-20% nickel, 25% zinc and 2% lead. It is usually supplied in foil or tube form, and has a density of 8.72 g/cm^3 (0.315 lb/in.^{3}), a melting range of 1060-1110°C (1940-2030°F), excellent weldability and cold workability, good machinability, and poor hot formability. Used for hardware, screws, nuts, bolts, keys, carburetors, gas and electric meters, and sundries.

Speed. Trade name for a nickel-titanium wire used in dentistry.

Speed-Alloy. Trade name of Teledyne McKay (USA) for carbon and low-alloy flux-cored gas-shielded welding electrodes and wires.

Speedaloy. Trade name of Hoytland Steel Company (USA) for silvery white, cast cobalt-base alloy containing 32% chromium, 18% tungsten and 2% carbon. It has high tensile strength, good high-temperature properties, and is used for dies, cutters and tool bits.

Speedbonder. Trademark of Loctite Americas (USA) for a series of anaerobic adhesives for high-speed, high-strength bonding of metals, plastics, glass and ceramics at room temperature.

Speed Case. Trade name of Jones & Laughlin Steel Company (USA) for a free-machining, case-hardening steel containing 0.2% carbon, 1.25% manganese, 0.25% silicon, 0.02% phosphorus, and the balance iron. Used for carburized parts, such as gears, cams, shafts, jigs, camshafts and fixtures.

Speed Crete. Trade name of TAMMS (USA) for concrete repair and patching materials.

Speed-Cut. Trade name of Teledyne Vasco (USA) for a free-machining mold steel (AISI type P20) containing 0.38-0.48% carbon, 0.2-0.4% silicon, 0.7-1% manganese, 1-1.3% chromium, 0.4-0.6% molybdenum, and the balance iron. Available in the annealed and heat-treated conditions, it has excellent machinability, outstanding wear resistance and nongalling properties, and a file-hard surface obtainable by proper heat treatment. Used for plastic molding dies, die-casting dies, press-brake dies, cavity and backing plates, ejector plates, spacer blocks, milling cutter and reamer bodies, collets, arbors, boring bars, bushings, gears, pinions, spindles, shafts, and axles.

Speedex. Trade name of Coltene-Whaledent (USA) for a biocompatible condensation-curing silicone-based dental impression material providing a snap set after about 2 minutes.

Speedicut. Trade name of Firth Brown Limited (UK) for a series of tool steels including *Speedicut, Speedicut Leda, Speedicut Maximum* and *Speedicut Vanleda* tungsten high-speed steels, and *Speedicut Sixleda* and *Speedicut Superleda* cobalt-tungsten high-speed steels. Used for cutting tools, e.g., lathe and planer tools, and form and milling cutters.

Speedie. Trade name of The Buckeye Products Company (USA) for buffing and polishing compositions used in metal finishing.

Speediset. Trademark of Westroc Industries Limited (Canada) for quick-setting spackling compounds.

Speedmask. Trademark of Dymax Corporation (USA) for a liquid, fast UV-curing, 100% solids temporary masking resin used to protect components during blasting, shot peening, machining, plasma spraying and plating operations.

Speed Set. Trade name of Georgia Pacific Corporation (USA) for a quick-setting insulating cement.

Speed Star. Trade name of Carpenter Technology Corporation (USA) for a molybdenum-type high-speed tool steel (AISI type M2) containing 0.8% carbon, 5.5-6% tungsten, 4.25-5% molybdenum, 4% chromium, 1.5-2% vanadium, and the balance iron. It has a fine grain size, a great depth of hardening, excellent wear resistance, good toughness and high hot hardness. Used for lathe and planer tools, milling cutters, drills, taps, reamers, broaches, etc.

Speedstitch. Trademark of Stelco Inc. (Canada) for metal stitching wire.

Speed Stone. Trade name for a dental stone used particularly for modelling applications.

Speedtex. Trademark of Technical Coatings Limited (Canada) for quick-drying spray enamel paints.

Speed Treat. (1) Trade name of LTV Steel (USA) for a free-machining structural steel containing 0.42-0.53% carbon, 1-1.25% manganese, 0.2-0.3% sulfur, up to 0.045% phosphorus, and the balance iron.

(2) Trade name of Sainte d'Escaut & Meuse (France) for steel containing 0.45% carbon, 1.25% manganese, and the balance iron. Used for gears, pinions, shafts, and leadscrews.

Spelloy. Trademark of Kolon Company (USA) for a series of heat-resistant, thermoplastic olefin alloy resins with a density of 1.2 g/cm^3 (0.04 lb/in.^{3}), low water absorption (0.3%), high low-temperature impact resistance, an elongation at break of 90%, an upper continuous-use temperature of 110°C (230°F), a dielectric constant of 3 (at 1 MHz) and good dimensional stability. Used for industrial and electrical parts.

Speltafast. Trade name of Richard, Thomas & Baldwins Limited (UK) for galvanized mild steel sheet.

spelter. A commercial *slab zinc* with a purity of up to 99.6% and containing lead and/or iron impurities. Used for galvanizing and coating applications.

spelter solder. See brazing brass.

Spenard. Trade name of Spencer Clark Metal Industries Limited (UK) for a nickel-chromium tool steel containing 0.32% carbon, 0.3% silicon, 0.5% manganese, 4.1% nickel, 1.3% chromium, 0.3% molybdenum, and the balance iron. Used for punches, plastic molds, and trimming dies.

spencerite. A white mineral composed of zinc phosphate hydroxide trihydrate, $Zn_4(PO_4)_2(OH)_2 \cdot 3H_2O$. Crystal system, monoclinic. Density, 3.24 g/cm^3; refractive index, 1.602. Occurrence: Canada (British Columbia).

Spence's metal. A mixture obtained by adding iron disulfide (FeS_2), zinc sulfide (ZnS), usually in the form of the mineral *sphalerite*, and lead sulfide (PbS), usually in the form of the mineral *galena*, to molten sulfur and metallic oxides. It has a melting point of 160°C (320°F), expands on cooling, and is used as corrosion-resistant lute for joining pipes.

Spenkel. Trademark of Reichhold Chemicals, Inc. (USA) for oil-modified uralkyd, moisture-curing polyurethane resins and polyurethane prepolymers with good color retention and abrasion resistance. Used for coating applications.

Spenlite. Trademark of Reichhold Chemicals, Inc. (USA) for aliphatic urethane lacquers, moisture-curing urethane resins and aliphatic urethane prepolymers.

Spensol. Trademark of Reichhold Chemicals, Inc. (USA) for various polyurethanes including oil-modified polyurethane resins and polyurethane elastomer dispersions for coating applications.

spergenite. See Indiana limestone.

sperrylite. A tin-white mineral of the pyrite group with a black streak and a metallic luster. It is composed of platinum arsenide, $PtAs_2$. Crystal system, cubic. Density, 10.6-10.8 g/cm^3; hardness, 1080-1145 Vickers. Occurrence: Canada (Ontario), USA (Nevada, North Carolina, Wyoming), South Africa.

spertiniite. A blue or blue-green mineral composed of copper hydroxide, $Cu(OH)_2$. Crystal system, orthorhombic. Density, 3.93 g/cm^3. Occurrence: Canada (Quebec).

Spesin. Trademark of Kolon Company (USA) for a series of heat-stable, abrasion-resistant thermoplastic polybutylene terephthalate resins with a density of 1.3 g/cm^3 (0.05 lb/in.³), a melting point of 224°C (435°F), an upper service temperature of 155°C (311°F), low water absorption (0.08%), excellent chemical and mechanical properties, excellent electrical properties (including a dielectric constant of 3.1), high gloss, and good processibility. Used for industrial parts, and for electrical and electronic components.

spessartine. A brownish-red or hyacinth-red mineral of the garnet group composed of manganese aluminate, $Mn_3Al_2(SiO_4)_3$. It can also be made synthetically. Crystal system, cubic. Density, 4.00 g/cm^3. Used in ceramics. The transparent red varieties are used as gemstones.

Sphäroguss. German trade name for spheroidal graphite cast iron.

sphalerite. A white, yellow, brown, black or red mineral with a resinous luster. It is composed of zinc sulfide, ZnS, and contains theoretically 67% zinc, usually some cadmium, iron and manganese, and sometimes mercury. It can also be made syn-

thetically. Crystal structure, cubic. Density, 3.9-4.1 g/cm^3; hardness, 3.5-4.0 Mohs. Occurrence: Australia, Canada, Europe, Mexico, USA (Colorado, Idaho, Kansas, Missouri, Montana, Oklahoma, Wisconsin). Used as the most important ore of zinc, as a source of cadmium, as a source of sulfur dioxide for sulfuric acid manufacture, and as a phosphor. The clear species are used as gemstones. Also known as *black jack; blende; false galena; zinc blende.*

sphalerites. A group of ceramic materials corresponding to the sphalerite (ZnS) crystal structure, AX, where A denotes a metal, such as zinc, beryllium or silicon, and X usually either carbon or a chalcogen, such as oxygen, sulfur, selenium or tellurium. Examples include beta silicon carbide (β-SiC), beryllium oxide (BeO), zinc sulfide (ZnS) and cadmium telluride (CdTe).

sphene. See titanite.

Spheralloy. Trade name for a spherical-particle dental amalgam alloy.

spherical-particle alloy. (1) A *dental amalgam* made by mixing finely divided spherical particles of a silvery-white alloy, usually composed of about 20-70% silver, 12-28% tin, 2-15% copper and 0-1% zinc, with liquid mercury.

(2) A *dental amalgam* made by mixing finely divided spherical particles of a silver-based alloy (typically 25-28% tin, 11-15% copper, 0-9% palladium, 0-0.3% zinc, 0-0.05% platinum, balance silver) with a liquid gallium-base alloy containing varying amounts of indium, tin and sometimes a trace of bismuth.

spherical powder. A metal or nonmetal powder consisting of rounded particles of globular or ball shape. Abbreviation: SP.

Sphericel. Trade name for glass spheres used as reinforcement in engineering thermoplastics.

Spheriglass. Trademark of Potters Industries Inc. (USA) for silica glass spheres used as fillers in composites.

Spherix. Trademark of Pechiney Electrométallurgie (France) for a cast iron inoculant composed of 72% silicon, 1.5% calcium, 1% bismuth, 0.9% aluminum, 0.5% rhenium, and the balance iron.

spherocobaltite. A pink to purple mineral of the calcite group composed of cobalt carbonate, $CoCO_3$. It can also be made synthetically by hydrothermal synthesis from cobalt chloride, carbon dioxide and sodium bicarbonate. Crystal system, rhombohedral (hexagonal). Density, 4.13 g/cm^3; refractive index, 1.855.

spheroidal graphite. Graphite that is shaped somewhat like a sphere and has a polycrystalline radial structure. It can be obtained in a cast iron by suitable treatment.

spheroidal graphite cast irons. See ductile cast irons.

spheroidal irons. See ductile cast irons.

spheroidal powder. A powder consisting of rounded particles of nearly globular or oval shape.

spheroidized steel. A steel subjected to a heat treatment that usually involves heating to a temperature just below the Ae_1 critical temperature (about 723-738°C or 1333-1360°F) and holding at this temperature for a prolonged period of time to allow the *cementite* to form hard, small, separate globules or spheres in the predominantly soft *ferrite* matrix, followed by slow cooling to room temperature. The resulting steel has good ductility and can be formed or machined. *Note:* The Ae_1 critical temperature is the temperature at which the formation of *austenite* starts under equilibrium conditions.

Spheron. Trade name for a series of carbon blacks.

Spherosil. Trademark of IBF Corporation (USA) for spherical porous silica available in various mesh sizes. Used as an adsorbent in gas chromatography.

Spherulite. Trade name of Vulcan Foundry Company (USA) for a nodular cast iron containing 3% carbon, 2.5% silicon, and the balance iron. Used for gears, shafts, and housings.

spherulitic cast irons. See ductile cast irons.

spherulitic graphite irons. See ductile cast irons.

spherulitic irons. See ductile cast iron.

Sphinx. Trade name of Ugine Aciers (France) for oil-hardening tool steel containing 0.8% carbon, 2% manganese, 0.2% vanadium, and the balance iron. Used for punches, headers, crimpers, cutters, and dies.

spianter. A British hard zinc-base bearing alloy containing 8% antimony and 2% copper.

spider silk. (1) Silk that is secreted in the form of a liquid by glands located in the posterior portion of the abdomens of spiders and hardens (polymerizes) before it is released from small finger-like appendages known as "spinnerets." The released silk is a fibrous protein substance composed of more than 50% fibroin, a polymerized protein essentially made up of the amino acids glycine and alanine, with a molecular weight of approximately 200000-300000. It is waterproof, finer than human hair, lighter than cotton, stronger than aramid (*Kevlar*) and 5 times stronger than steel (on a weight for weight basis), and can be stretched to 2-4 times its length without breakage. Natural spider silk was formerly used for the cross hairs of optical instruments, such as levels, transits and astronomical telescopes, and in gun sights. Also known as *natural spider silk.* See also dragline silk.

(2) Spider silk (1) replicated or synthesized in the laboratory by genetic engineering techniques, such as recombinant DNA. Several synthesis routes are or have been pursued. In one such technique cloned portions of the genes for the silk proteins of spiders, such as the golden orb weaver (*Nephila clavipes*), were implanted in yeast or bacteria (e.g., *E. coli*) which then produced silk protein in solution that, once forced through fine tubes, yielded synthetic silk fibers. In another technique spider genes are first implanted into mammal cells (e.g., from cows or hamsters), and then water-soluble proteins are produced from the resulting intermediate substance, which are forced through minute holes, similar to those in the spinnerets of spiders, and issue in the form of fine, synthetic silk fibers. The silk thus produced is said to be lustrous, flexible and strong, yet not as strong as the natural substance, has the feel of silkworm silk, and, once perfected, could be used for the manufacture of protective clothing and bulletproof vests, biodegradable fishing lines, improved medical sutures and bandages, artificial tendons and ligaments, etc. Also known as *artificial spider silk; bioengineered spider silk; synthetic spider silk.*

spiegeleisen. A *pig iron* containing 4.5-6.5% carbon, 10-30% manganese, up to 1% silicon, up to 0.1% phosphorus, up to 0.04% sulfur, with the balance being iron. It is made from high- and low-grade manganese ores, and graded according to manganese content. The name is derived from the German word *Spiegeleisen* meaning "mirror iron" and referring to its shiny, mirror-like fracture surface. It has a melting range of 1065-1240°C (1950-2265°F), and is used in steelmaking as a deoxidizer and to increase the manganese content, in the manufacture of Bessemer and open-hearth iron and steel, and for other metallurgical applications. Also known as *spiegel; spiegel iron.*

spiegel iron. See spiegeleisen.

Spimalit. Trade name for fiber-reinforced unsaturated polyester resins.

spin-bonded nonwoven fabrics. Nonwoven fabrics composed of thermoplastic synthetics, such as polyamides, polyesters or polyolefins, in which the continuous filaments exiting the spinning extruder form a random web that is subsequently thermally bonded. They have outstanding strength, and are used as a backing materials, and for industrial applications. Also known as *spin-bonded nonwovens.*

spin-drawn filaments. Partially or highly oriented filaments which receive most of their orientation by a process which involves spinning and subsequent drawing.

spinel. A colorless, white, blue, green, red, lavender, brown or black mineral with a white streak and a vitreous luster. It is composed of magnesium aluminate, $MgAl_2O_4$, and sometimes contains small amounts of chromium, iron, manganese and zinc. It can also be made synthetically. Crystal system, cubic. Density, 3.5-4.1 g/cm^3; hardness, 8.0 Mohs; refractive index, 1.718. Occurrence: Burma, Europe, Sri Lanka, Thailand. Used as a gemstone, and in ceramics and crystallography. See also natural spinel; synthetic spinel.

spinel ruby. See balas ruby.

spinels. A group of ceramic compounds consisting of two oxides and identified by the general formula AB_2O_4, where A is a divalent metal, such as magnesium, iron, nickel or zinc, and B a trivalent metal, such as aluminum or iron. Examples include magnesium aluminate($MgAl_2O_4$), nickel aluminate($NiAl_2O_4$), zinc aluminate ($ZnAl_2O_4$) and zinc ferrite($ZnFe_2O_4$). See also spinel; natural spinels; normal spinels; synthetic spinels.

Spinel 25. Trade name of CE-Minerals (USA) for a fused aggregate containing 74.3% alumina (Al_2O_3), 25.0% magnesia (MgO), and small additions of silica (SiO_2), calcia (CaO) and sodium oxide (Na_2O). Supplied in a range of grain sizes, it has a bulk density of 3.3 g/cm^3 (0.12 $lb/in.^3$), an apparent porosity of 7.5%, and is used chiefly to enhance the resistance of pure alumina to hot metals, slag, erosion and corrosion.

Spin-Glas. Trademark of Manville Corporation (USA) for thermal and acoustical fiberglass boards and blankets supplied in various grades, and used as equipment and duct insulation.

spin glasses. A large group of different materials for which long-range atomic ordering does not exist, but which exhibit local spatial correlations of adjacent atomic spins. They have paramagnetic spin alignments, but in contrast to paramagnets their spin orientations remain fixed or vary only slightly with time, and the variation with temperature of their magnetic susceptibility changes abruptly at the so-called "freezing temperature."

spinning brass. A highly ductile brass of yellow color composed of two parts of copper and one part of zinc and supplied in bar and sheet form for the manufacture of deep-drawn or spun objects, such as lamp and lighting fixtures, hardware, and novelty items.

spinning paper. Paper made by cutting small strips from sodium cellulose or sulfite cellulose paper. It has exceptionally high tensile strength in machine direction, and is used in the manufacture of paper twine and yarn.

spinning silver. A British nickel silver composed of 67% copper, 17% zinc and 16% nickel, and used for spun and drawn articles, spoons, forks and knives.

spinning solution. A solution of a fiber-forming polymer for extrusion through a spinneret. In the textile industries the terms "spinning solution" and "dope" are often used synonymously.

spinodal hardening alloys. Alloys such as those of copper-nickel with chromium or tin additions that harden by means of a homogeneous two-phase decomposition reaction (known as "spinodal decomposition") initiated by solution treatment at high temperatures and subsequent quenching. They have soft, ductile microstructures, can be cold-worked or cold-formed and retain their dimensional stability during the hardening process.

Spinsulation. Trade name of Manville Corporation (USA) for a glass-fiber material with kraft paper-faced vapor barrier.

Spinzwel. Trade name of Henry Wiggin & Company Limited (UK) for a nickel silver containing 64% copper, 26% zinc and 10% nickel. Used for drawn and spun articles.

Spiralloy. Trademark of Hercules, Inc. (USA) for filament-wound composite structures consisting of glass, aramid or carbon filaments impregnated with matrix resins, such as epoxies or unsaturated polyesters. Used for pressure vessels, radomes, cases for rocket motors and torpedoes, aerospace structures, and underwater equipment.

Spiralok. Trademark of Poli-Twine Corporation (Canada) for polypropylene monofilaments.

Spirit. Trade name for a silver-free, palladium-based dental bonding alloy.

spiroffite. A red-purple mineral that is composed of manganese tellurate, $Mn_2Te_3O_8$, and may also contain zinc. Crystal system, monoclinic. Density, 5.01 g/cm³; refractive index, 1.91. Occurrence: Mexico.

spiropolymers. A subclass of uniform, insoluble and soluble ladder polymers in which adjacent rings have one atom in common. Incorporation of compounds containing six hydroxyl (OH) groups results in hyperbranched ladder polymers. See also hyperbranched polymer; ladder polymer.

splat powder. A powder consisting of flat, flaky particles produced by rapid cooling or quenching of liquid metal.

Splintex. Trade name of Splintex Limited (UK) for a laminated and toughened glass.

split film. A network of interconnected synthetic fibers produced by mechanically cracking or splitting a polymer film or tape that has been oriented in the direction of extrusion.

splittings. Mica, in trimmed or untrimmed form, produced from blocks, thins and splitting blocks to a thickness of less than 0.03 mm (0.0012 in.) and a minimum usable area of 484 mm² (0.75 in.²). Bookform splittings are sheets of mica supplied in form of books from the same block, loose splittings are of heterogeneous shape and packed in bulk form, and powdered loose splittings are loose splittings dusted with finely ground mica.

SPM Sil-Bond. Trade name of SPM International Corporation (USA) for silicone coatings used for gaskets.

spodumene. A white, gray, pale green to bright green, pink or purple mineral of the pyroxene group with a white streak and a vitreous luster. It is composed of lithium aluminosilicate, α-$LiAlSi_2O_6$, and contains 8.4% lithium oxide (Li_2O), some of which may be replaced by sodium. Alpha spodumene inverts to beta spodumene on heating. Crystal system, monoclinic. Density, 3.12-3.20 g/cm³; hardness, 6.5-7.0 Mohs; refractive index, 1.665; very low thermal expansion. Occurrence: Brazil, Mozambique, USA (California, Massachusetts, North Carolina, South Dakota). Used as an ore of lithium, as a flux, and to improve thermal-shock resistance in glass, porcelain enamels, glazes and ceramic bodies. Also known as *triphane*.

Sponge. Trade name of SA Glaverbel (Belgium) for a patterned glass.

sponge. See cellular plastic; metal foam; sponge metal; sponge rubber.

sponge gold. A spongy, compact mass of gold made by distillation of gold amalgam. Also known as *cake of gold*.

sponge iron. Iron in porous or finely powdered form made by reducing an iron oxide with charcoal or coke, or in a coke-oven gas or natural-gas atmosphere at temperatures below the melting point. Used in the precipitation of copper or lead from solutions of their salts, in electric-furnace steelmaking, in powder metallurgy, and as a catalyst. Also known as *iron sponge*.

sponge iron powder. A porous, chemically reduced powder that is made by grinding and sizing of sponge iron which may be purified and/or annealed.

sponge lead. A pure, porous lead used as the negative plates of lead-acid storage batteries. Also known as *lead sponge*.

sponge metal. Metal, such as iron, nickel, titanium or zirconium, produced in porous or finely powdered form. Also known as *metal sponge*. See also sponge iron; sponge lead; sponge nickel; sponge titanium; sponge zirconium.

sponge nickel. A gray, spongy, pyrophoric powder obtained by leaching the aluminum from an 50Al-50Ni alloy with a 25% sodium hydroxide solution. See also Raney nickel.

sponge palladium. Metallic platinum (99.9% pure) in a blue-gray, finely-divided and spongy form suitable for use in gas lighters, and as a catalyst. Also known as *palladium sponge*.

sponge platinum. A finely divided, porous form of metallic platinum (99.9+% pure) of grayish-black to bluish-gray color obtained by ignition of ammonium hexachloroplatinate or other salts. It has a typical particle size of less than 20 mesh (850 μm). Used as a catalyst, in gas absorption (hydrogen, oxygen, etc.), and in gas ignition. Also known as *platinum sponge*.

sponge rubber. A flexible, spongy foam of natural or synthetic rubber containing numerous uniform pores produced by beating air into latex and vulcanizing, or by incorporating a blowing or foaming agent (e.g., ammonium carbonate, sodium bicarbonate, etc.). Used as a cushioning material for seats, as underlayment for carpets, rugs and business machines, for mattresses and upholstery, for vibration dampening and shock insulation, and for orthopedic insoles. Also known as *cellular rubber; foam rubber; rubber foam; rubber sponge*.

sponge titanium. Crude titanium of sponge-like appearance produced by reduction of titanium tetrachloride with molten magnesium (Kroll process) or sodium (Hunter process) in an argon or helium atmosphere, and subsequent consolidation by heating. It is available in a wide range of particle sizes ranging from fine powders of less than 650 μm (0.0256 in.) to pieces of 0.08-0.5 in. (2-12 mm). Also known as *titanium sponge*.

sponge titanium powder. A porous, spongy powder which is made by grinding and sizing of sponge titanium.

sponge zirconium. An air-sensitive, porous form of zirconium which results from decomposition or reduction without fusion. It usually contains up to 2.5% hafnium, but can be refined to purities of 99+% by special processes. The refined material has a particle size of about 30 mesh (600 μm). Used as a catalyst, and in powder metallurgy. Also known as *zirconium sponge*.

Spongex. Trademark of Sponge Rubber Products Company (USA) for a silicone foam used as a cushioning material, and for vibration damping applications.

spongy iron. See reduced iron.

Spontex. Trade name for rayon fibers and yarns used for textile fabrics.

Sportouch. Trade name of BASF Corporation (USA) for nylon 6 fibers used for textile fabrics.

Spot-Lite Glo. Trade name for a phosphorescent frit that contains zinc sulfide (ZnS), and is used in the manufacture of luminous ceramic ware, and signs.

Spotlyte. Trade name of Pilkington Brothers Limited (UK) for a patterned glass.

Spotswood. Trade name of Australian Window Glass Proprietary Limited (Australia) for a patterned glass giving a stippled effect.

Spot-Weld. Trademark of Sealed Air Corporation (USA) for laminated polyethylene foam.

Sprababbitt. Trademark of Sulzer Metco (USA) for lead and tin-base alloys. *Sprababbitt A* is a tin alloy containing 7.5% antimony, 3.5% copper and 0-0.25% lead, and *Sprababbitt L* contains 76.75% lead, 13% antimony, 10% tin and 0.25% copper. *Sprababbitt* is supplied in wire form for metal spraying applications, e.g., on high-speed and heavy duty bearings.

Sprabond. Trademark of Sulzer Metco (USA) for refractory metals, such as molybdenum, usually in wire or electrode form suitable for the production of corrosion-resistant, arc-sprayed coatings on metallic substrates.

Sprabrass. Trade name of Sulzer Metco (USA) for a wrought alloy composed of 66% copper and 34% zinc, and supplied in wire form for metal spraying applications.

Sprabronze. Trade name of Sulzer Metco (USA) for a series of wrought bronzes containing varying amounts of copper and zinc or copper and tin, frequently with the addition of aluminum, tin, manganese, iron and/or phosphorus. Supplied in wire form, they are used for metal spraying applications.

Spra-Gard. Trade name of OMG Fidelity (USA) for spray-booth coatings.

Sprairon. Trade name of Sulzer Metco (USA) for iron supplied in wire form for metal spraying applications.

Sprasteel. Trade name of Sulzer Metco (USA) for several plain-carbon steels (0.1-1.2% carbon) and nickel-manganese alloy steels supplied in wire form in several grades for metal spraying applications.

Spraybond. Trade name Sulzer Metco (USA) for refractory metals, such as molybdenum, usually in wire or electrode form suitable to produce corrosion-resistant, arc-sprayed coatings on metallic substrates.

spray-bonded nonwoven fabrics. Nonwoven fabrics produced by a process in which bonding is achieved by spraying droplets of adhesive into the fiber batt or web. Also known as *spray-bonded nonwovens*.

spray coating. A fine, uniform metallic or nonmetallic coating applied by any of several thermal spraying technique including flame, electric-arc and plasma-arc spraying. Also known as *spray deposit*. See also thermal spray coatings.

spray deposit. See spray coating; thermal spray coating.

sprayed asbestos. Hydrated asbestos applied with a spray gun and containing cement that is about 55-65% asbestos fiber.

sprayed mortar. See air-blown mortar.

Sprayflon. Trademark for polytetrafluoroethylene sprays used as parting agents and lubricants.

spray-foam insulation. A type of foamed-in place thermal insulation based on polyurethane or isocyanurate foams. The foam is supplied in liquid form and sprayed directly onto building surfaces or poured into enclosed building cavities by means of a pump-driven spray gun. It expands in place and sets instantly. Also known as *foamed-in place insulation*. See also polyurethane foam insulation; isocyanurate foam insulation.

spray-formed alloys. Ferrous and nonferrous alloys formed by atomizing inductively melted starting alloys into fine droplets and spraying them onto substrates were they undergo rapid cooling. Starting alloys suitable for spray forming include high-chromium cast irons, stainless and other high-alloy steels, high-carbon high-speed tool steels, aluminum-silicon alloys, certain copper-nickel alloys, and some superalloys. Spray-formed alloys are supplied in many different product shapes including billets, disks, rings, rolls and tubing.

spray-formed silicon-aluminum. A group of silicon-aluminum alloys with a fine, isotropic microstructure produced by spray forming, i.e., by atomizing the inductively melted starting alloy into fine droplets and spraying them onto a rotating plate were they undergo rapid cooling. They are characterized by high thermal conductivity, low density and low thermal expansivity, and can be easily machined and electroplated with nickel, gold and silver. Supplied in billet form, they are used for electronic packaging applications, and in the aerospace and automotive industries.

Spray Guard. Trademark for water-based masking agents for application by brushing, dipping or thermal spraying.

Spray-It. Trade name of The Buckeye Products Company (USA) for a liquid buffing compound used in metal finishing.

spray lime. Slaked lime, $Ca(OH)_2$, of such extremely fine particle size that almost all of the particles will pass the No. 325 (45 μm) US standard sieve.

Spray Mix. Trade name of National Refractories & Minerals Corporation (USA) for wet refractory mortars.

Spray-Mount. Trade name of 3M Company (USA) for aerosol adhesives used for mounting photos, etc.

Spraypac. Trademark of Dexter Corporation (USA) for a *Hysol*-type ethylene-vinyl-acetate hot-melt adhesive spray for bonding paper, styrofoam, plastics and woods, and for closing cartons.

Spraytex. Trademark of Sico Inc. (Canada) for semi-gloss latex finishes for concrete surfaces.

Spred Gel Flo. Trademark of Glidden Company (USA) for a gelled exterior alkyd paint.

Spred Latex. Trademark of The Glidden Company (USA) for a latex semi-gloss enamel paint.

Spreemetall. Trade name of Allgemeine Elektrizitäts-Gesellschaft (Germany) for a corrosion-resistant alloy of 55% copper, 43% zinc, 1.5% manganese and 0.5% lead. Supplied as sheets, bars and forgings.

S-Prime. Trademark for a paintable coating used to protect metallic materials against liquid metals, fluxes, gases, slags, etc. It has good thermodynamic stability, excellent corrosion resistance, and a maximum service temperature of 1500°C (2730°F). Used for furnace linings, retorts, trays, etc.

S-Prime-Mod. Trademark for modified *S-Prime* paintable coatings.

spring brass. A hard-drawn wrought brass containing 66-72% copper and 28-34% zinc, and usually used in the "spring hard" temper. Commercially available in the form of round sections and flat strips, it has good electrical conductivity, cold workability, fair tensile strength, fair spring qualities, an upper service temperature of 65°C (149°F) and good low-temperature properties. Used for springs, condenser tubes, water boiler tubes, and deep-drawn shells.

spring bronze. A wrought bronze containing 85-87.5% copper, 11.5-14.5% zinc and 0.8-1% tin, and usually used in the "spring hard" or "extra-hard" temper. It has good electrical conductivity and corrosion resistance, good strength and hardness, excellent cold workability, an upper service temperature of 100°C (212°F) and good low-temperature properties. Used for rolled sheet and strip for relay springs, electrical contacts and switch parts, and flat springs.

spring gold. A corrosion-resistant alloy composed of 50% copper, 25% gold and 25% silver, and used for dental applications and jewelry.

spring lay rope. A rope consisting of 6 strands around a central core, each strand being composed of a fiber core around which alternating fiber or wire components have been twisted or laid.

spring materials. High-resilience materials, such as certain high-carbon steels, alloy and stainless steels and copper- and nickel-base alloys, suitable for making springs.

spring silver. (1) A nickel silver containing 54.6% copper, 27.3% zinc and 18% nickel. It has excellent corrosion resistance, good tensile strength and cold workability, and is used for springs, resistance wire, electrical contacts, and optical parts.

(2) Silver (99.4+% pure) containing small amounts of magnesium and nickel. It has a density of 10.34 g/cm^3 (0.374 lb/in.^{3}), air hardens by internal oxidation, and has good electrical conductivity (70% IACS) and good spring properties. Used for relays, switches, springs, and spring contacts.

spring steels. A group of alloy and high-carbon steels with low phosphorus and sulfur content and excellent elastic and fatigue properties and high tensile and yield strengths. Common spring steels for general engineering applications (e.g., machinery, industrial equipment, automobiles, etc.) include quenched-and-tempered steels, silicon, silicon-chromium and silicon-manganese steels, and chromium-vanadium steels. High-temperature spring steels for valve springs in combustion engines are often of the tungsten-chromium-vanadium type and retain good mechanical properties at temperatures exceeding 500°C (930°F). Several austenitic and martensitic stainless steels (e.g., AISI 302, 304, 316, 414, 420 and 431) are also used for springs.

spring wire. High-quality wire made from alloy, high-carbon or stainless steels and used for springs. Examples include music spring steel wire, piano wire and alloy steel spring wire. This term also includes wire made from nonferrous alloys, such as brass, bronze, nickel-base alloys or cobalt-chromium alloys.

spruce. The soft, light, resinous, moderately strong wood of any of a genus *(Picea)* of coniferous trees belonging to the pine family, and found in North America, Europe and Asia. Important North American spruces include the black spruce *(P. mariana),* the Engelmann spruce *(P. engelmannii),* the red spruce *(P. rubens),* the sitka spruce *(P. sitchensis)* and the white spruce *(P. glauca).* The Norway spruce *(P. abies)* is common to Northern and Central Europe including the British Isles, but is now also found in North America.

spruce fir. See Norway spruce.

spruce pine. The moderately strong wood of the medium-sized coniferous tree *Pinus glabra.* It has light brown heartwood and almost white sapwood. Source: USA (Coastal southeastern South Carolina, Georgia, northwestern Florida, Alabama, Mississippi and Louisiana). Used for lumber and plywood, as a pulpwood and a as fuel. Also known as *cedar pine; poor pine; Walter pine.*

Spuma. Trade name of Vetreria di Vernante SpA (Italy) for a patterned glass.

Spun-black. Trade name for rayon fibers and yarns used for textile fabrics.

spun-bonded olefins. High-strength nonwovens produced by the continuous spinning and mechanical or thermal bonding of endless polyolefin fibers, e.g., polyethylene or polypropylene. A popular spun-bonded polyolefin product is DuPont's *Tyvek* used especially as housewrap in building and construction, and in the manufacture of packaging and clothing. See also spun-bonded nonwovens.

spun-bonded nonwoven fabrics. Nonwoven fabrics produced in the form of sheets, tapes and laminates by first extruding a polymer through a multi-hole extrusion die, cooling the emerging filaments and laying them down in form of a web on a continuous-belt conveyor, and then bonding the web by the application of heat, pressure and/or adhesives. Also known as *spun-bonded nonwovens.*

Spuncast. Trademark of Esco Corporation (USA) for spun-cast heat- and corrosion-resistant stainless steel.

spun concrete. A concrete product, such as a pipe, that has been compacted or densified by centrifugal action. Also known as *centrifugally cast concrete.*

Spunenka. Trade name of Enka BV (Netherlands) for rayon fibers and yarns used for textile fabrics.

spun glass. An individual continuous filament, staple fiber or a mass of fine threads of attenuated glass with average diameters of less than 25 μm (0.001 in.). Abbreviation: SG. See also fiber; filament; glass fiber; glass wool.

Spungo. Trade name for spun rayon fibers and yarns used for textile fabrics.

spunlaced aramid sheet. A low-density, lightweight nonwoven sheet in which webs of short *aramid fibers* are tangled without binder resin by jets of high-pressure water. It has excellent drapability, good impregnating properties, and is used for surfacing veil, fire-blocking layers in aircraft seating, and for circuit boards.

spunlaced nonwoven fabrics. Nonwoven fabrics produced by mechanically entangling polymer fibers without binder resins by jets of high-pressure water. Also known as *hydroentangled nonwoven fabrics; hydroentanged nonwovens; spunlaced nonwovens.*

spunlaid nonwoven fabrics. Nonwoven fabrics produced by first extruding a polymer through a spinneret, laying down the emerging filaments in web form and then bonding the web by any of several techniques. Also known as *spunlaid nonwovens.*

Spunlo. Trade name for spun rayon fibers and yarns used for textile fabrics.

Spunloc. Trade name for spun polyester yarns used for knitted and woven fabrics.

Spunnaire. Trademark of Wellman Inc. (USA) for a soft, white, lightweight, air-spun polyester staple fiber that belong to the *Fortrel* range of fibers. It is optically bright, colorfast and wrinkle- and pilling-resistant, and is used for sportswear, sleepwear, underwear, thermal wear, knitwear, socks, blankets, and throws and other home fashions.

Spunnesse. Trade name of Wellman Inc. (USA) for spun polyester fibers that belong to the *Fortrel* range of fibers and are used for textile fabrics including clothing.

spun polyester. A soft, light fabric knitted or woven from spun polyester yarn. Used especially for sportswear, dresses and nightwear.

spun rayon. (1) A yarn spun from short *rayon* filaments.
(2) A fabric made from spun rayon yarn (1).

spun roving. A strand of aramid or glass fibers made by doubling continuous filaments back on each other.

spun silk. (1) Yarn spun from short filaments obtained from silk wastes.
(2) Fabrics made from spun silk yarn (1).

spun viscose. (1) A yarn spun from viscose rayon fibers.
(2) Soft, drapey, plain or printed fabrics made from spun

viscose (1) in a plain weave. Used especially for blouses, dresses, shirts, and nightwear.

spun yarn. A continuous strand of natural or synthetic fibers or filaments made from staple fibers or continuous filaments, usually by drawing and twisting.

spurrite. A pale gray mineral composed of calcium carbonate silicate, $Ca_5(SiO_4)_2CO_3$. Crystal system, monoclinic. Density, 3.03 g/cm³. Occurrence: USA (California, New Mexico), Mexico.

sputtered coating. A fine-grained metal or nonmetal coating, usually a thin film, applied in a vacuum chamber onto a substrate, such as a glass, plastic, metal, or paper. Ions of the material to be deposited are sputtered from a cathode as a result of heavy ion (e.g., argon ions) impact. Sputtered coatings are used to improve the surface properties of materials.

Squamettato. Trade name of Fabbrica Pisana SpA (Italy) for a patterned glass with hammered effect.

Square. (1) Trade name of SA Glaverbel (Belgium) for a patterned glass featuring horizontal and vertical reeds.

(2) Trade name of Magnetic Metals Company (USA) for a series of soft magnetic alloys containing 45-80% nickel, 20-50% iron and 0-4% molybdenum. They possess medium to high initial permeability, high magnetic saturation, low coercive force and square hysteresis loop. Used for magnetic amplifiers, controllers, high-frequency equipment, cores for transformers and relays, and electrical equipment.

square-cut glass. Optical-grade glass with ground and polished faces, cut into small squares, graded according to weight, and used, e.g., as optical windows, and in optical equipment.

Squarelite. Trade name of Australian Window Glass Proprietary Limited (Australia) for wired glass with a square mesh.

Square-tex. Trademark of Canadian Forest Products Limited for panelwood.

SRC-Glass. Trade name of Taconic (USA) for fiberglass fabrics coated with silicone rubber.

srilankite. A black-brown mineral composed of titanium zirconium oxide, $(Ti,Zr)O_2$. Crystal system, orthorhombic. Density, 4.77 g/cm³. Occurrence: Sri Lanka.

SRIM composites. See reinforced-reaction-injection-molded composites.

SR Ivocap. Trade name of Ivoclar Vivadent AG (Liechtenstein) for a heat-cure acrylic resin used for dentures.

SR Ivocon. Trade name Ivoclar Vivadent AG (Liechtenstein) for a dental acrylic used for veneering applications.

SR-Nickel. Trade name of Vereinigte Deutsche Nickel-Werke AG (Germany) a pure nickel (99.8%) used for chemical equipment and apparatus, and for electroplating anodes.

S-Rounds. Trademark of Inco Limited (Canada) for electrolytic nickel (99.915% pure) supplied in the form of rounds as an activated anode material for nickel plating.

SSCC. Trade name of Aerocote Corporation (USA) for a fine, dense, nonmetallic crystalline conversion coating produced on the surface of stainless steels, chromium-iron, and high-nickel alloys.

Staalglas. Trade name of Staalglas NV (Netherlands) for a toughened glass.

Staballoy. Trademark of Jessop Saville Limited (UK) for steel bars used in the manufacture of downhole tools and other equipment for the oil and gas exploration industries.

Stabar. Trademark of Imperial Chemical Industries plc (UK) for a series of thin, high-performance thermoplastic films produced by combining *Victrex* polyethersulfone (PES) and polyether-

etherketone (PEEK) polymers. Common grades include *Stabar K* noncrystallized PEEK, *Stabar S* PES, and *Stabar XK* crystallized PEEK. Supplied in thicknesses ranging from 0.005 to 0.025 mm (0.0002 to 0.001 in.), *Stabar* films have good chemical and hydrolytic resistance, good heat resistance, good mechanical and electrical properties, good dimensional stability, low flammability, low smoke and toxic-gas emission on burning, and good heat sealability and thermoformability. Used in the aerospace, electronic and electrical industries, e.g., for flexible printed circuits, and masking tapes of circuit boards.

Stabilenka. Trade name of Enka BV (Netherlands) for nylon fibers and products.

Stabilit. (1) German trade name for hard rubber used as electric insulating material.

(2) Trademark of Henkel KGaA (Germany) for two-part resin adhesives used for bonding ceramics, glass, metals, and other materials.

(3) Trademark of Stabilit, SA de CV (Mexico) for fiberglass structural materials used for liner panels.

Stabilite. (1) Trade name for laminated wood in which the individual layers of veneer have been impregnated with phenolic resin.

(2) Trademark of Hill & Griffith Company (USA) for a pulverized furfural residue used in the manufacture of steel castings.

(3) Trademark of Eagle Forest Products Limited (Canada) for wood panels used for flooring and subflooring.

stabilized fabrics. Textile fabrics impregnated or coated with thermosetting resin to impart crease and shrink resistance, and promote greater resiliency.

stabilized ferritic stainless steels. See superferritic stainless steels.

stabilized rubber latex. Rubber latex treated with certain stabilizing agents to reduce premature coagulation.

stabilized yarn. A yarn whose tendency to shrinkage, stretch, twist and/or snarl has been minimized by a special setting or stabilizing treatment which may involve a heating and cooling process. Also known as set yarn.

stabilized zirconia. See fully stabilized zirconia; lime-stabilized zirconia; magnesia-stabilized zirconia; partially stabilized zirconia; yttria-stabilized zirconia.

stabilizer. (1) An oxide, such as calcium oxide (lime), aluminum oxide or titanium dioxide that when introduced into a frit, glaze, or color oxide stabilizes the coloring during the firing process.

(2) Additives used to reduce polymer degradation by oxygen, ozone, heat, ultraviolet radiation, microbial action, etc., and promote physical and chemical properties during processing and use.

(3) A substance added to a solution or suspension of a cementitious material to increase its stability by preventing precipitation.

stabilizer alloy. A master alloy composed of magnesium and varying amounts of aluminum, copper, nickel, Monel metal or zinc. Used as a deoxidizer for nonferrous alloys.

Stabilo Temp. Trade name for a self-cure acrylic resin used in dentistry for temporary restorations.

Stabilor. Trademark of Degussa-Ney Dental (USA) for a series of medium-gold dental casting alloys containing varying amounts of other precious metals. Included are: (i) *Stabilor G*, a fine grained, extra-hard dental casting alloy (ADA type IV) containing 58% gold, 23.3% silver and 5.5% palladium; (ii) *Stabilor LS*, a white, microfine grained, extra-hard dental casting alloy (ADA type IV) containing 58% gold, 25% silver and

12.95% palladium; and (iii) *Stabilor H*, a microfine grained, hard dental casting alloy (ADA type III) containing 56% gold, 32% silver and 4% palladium. *Stabilor* alloys have good mechanical properties, and are used in restorative dentistry for crowns, bridges, and implants. *Stabilor LS* was developed for use with *Duceragold* porcelain in the restoration of single crowns, long-span bridges and implant superstructures.

Stabilothermo. Trade name of Isometal International Isolation SA (France) for a thermal insulation material composed of a sheet of polyethylene bubbles filled with dry air and having a 30 μm (1.2 mil) thick aluminum foil bonded to both sides. Used as insulation for water tanks, homes, boats, recreational vehicles, campers, etc.

stable-base film. A high-stability polymeric material in film form having good resistance to stretching and shrinkage.

Stable-Clad. Trade name of Norplex Oak Inc., Division of Allied-Signal (USA) for epoxy laminates used for printed-circuit applications.

stable fabrics. Dimensionally stable textile fabrics that exhibit high resistance to slippage of yarn segments in different directions.

Stabuff. Trade name of MacDermid Inc. (USA) for an electroless nickel coating.

Stackpole. Trade name of Stackpole Corporation (USA) for a series of permanent magnet materials.

StadiaTurf. Trademark of Southwest Recreational Industries, Inc. (USA) for tufted polypropylene fabrics for sand-filled turf applications in sports arenas.

Stadip. Trademark of Saint-Gobain (France) for a series of laminated safety glass products including colored *Stadip Color*, *Stadip Protect* with polyvinyl butyrate interlayer, and *Stadip silence* with polyvinyl butyrate interlayer and enhanced acoustic insulating properties.

Stae. Trade name of Southern Dental Industries Limited (Australia) for a single-component (primer plus adhesive), fluoride-releasing, light-cured dentin/enamel adhesive resin system based on urethane dimethacrylate (UDMA). It has very low polymerization shrinkage and microleakage and very high bond strength. Used in dentistry for bonding composite, compomer and porcelain restorations, and composite and porce-lain inlays and onlays.

Sta-Fit. Trade name of Manville Corporation (USA) for glass-fiber batting for thermal insulation applications.

Stag. Trade name of Edgar Allen Balfour Limited (UK) for a series of tool steels included *Stag Extra Special* cobalt-tungsten high-speed tool steel for drills, hobs, lathe and planer tools, reamers and threading taps, *Stag Major* cobalt-tungsten high-speed steel for drills, hobs, lathe and planer tools and reamers, *Stag Special* tungsten high-speed steel for drills, hobs, lathe tools and shaper and planer tools, and *Stag Vanco* cobalt-tungsten high-speed steel with excellent cutting ability, high wear resistance and high hot hardness for high-speed cutting tools.

Sta-Gloss. Trademark of Jessop Steel Company (USA) for a series of stainless steels containing 0.2-0.6% carbon, 12-17% chromium, and the balance iron. Used for valves, airplane parts, arch supports, ball bearings, cutlery, surgical instruments, rolls, piston rods, dies, and gages.

Stahlrump. Trade name of J.H. Arnold Rump Sohn GmbH & Co. (Germany) for strip steels supplied in standard and special grades.

Stahlschmidt. Trade name of Stahlwerk Stahlschmidt GmbH & Co. (Germany) for an extensive series of plain-carbon and al-loy machinery steels including several case-hardening grades, and various plain-carbon, cold-work, hot-work and high-speed tool steels.

stain. See ceramic stain; glass stain; wood stain.

stained glass. Glass colored by any of various means including immersion in a solution of a color-forming metal salt and subsequent firing to a temperature at which the color is developed and absorbed into the glass surface, exposure to the vapors of a color-forming salt at elevated temperatures in a closed furnace, or incorporation of colorants (e.g., metallic oxides) in the glass batch. Used in the production of variable-color mosaics, church windows, etc.

Stainless. Trade name of Uddeholm Corporation (USA) for a series of corrosion-resistant, low- and medium-carbon high-chromium steels containing about 0.1-0.35% carbon, 13-15% chromium, and low-carbon chromium-nickel steels containing up to 0.15% carbon, 16-18% chromium, 6-8% nickel, and small additions of other elements, such as titanium and aluminum. Used for valve and pump parts, cutlery, knives, kitchen utensils, turbine parts, and chemical equipment.

stainless alloy. Any ferrous or nonferrous alloy that is highly resistant to corrosion.

stainless-clad aluminum. A metal product, usually in sheet or strip form, consisting of a core of aluminum or aluminum alloy that has a layer of stainless steel bonded to one or both sides. It combines the low weight and good formability and machinability of aluminum with the superior corrosion resistance of stainless steel. Used for automotive trim, aircraft parts, marine equipment, and machinery parts and equipment.

stainless-clad copper. A metal product consisting of a copper or copper-alloy sheet that has a layer of stainless steel bonded to both sides. It combines the high thermal conductivity and good formability of copper with the outstanding corrosion resistance of stainless steel. Used for cooking utensils, kettles, food processing equipment, and heat exchangers.

stainless-clad steel. A metal product consisting of a relatively soft low-carbon steel core sandwiched between hard, corrosion-resistant stainless steel surface layers.

Stainless Invar. (1) Japanese trade name for a corrosion-resistant alloy composed of 63.5% iron, 31.5% nickel and 5% cobalt, and used for stainless parts.

(2) Trade name for a low-expansion alloy composed of 54% cobalt, 36.5% iron and 9.5% chromium, and used for instruments.

stainless irons. See ferritic stainless steels.

stainless steel casting alloys. A group of cast corrosion-resistant steels containing about 0.01-0.20% carbon, more than 11% chromium and up to 30% nickel. They are extensively used for applications requiring resistance against corrosion by aqueous solutions at or near room temperature and hot corrosive gases and liquids at temperatures to 650°C (1200°F).

stainless steel fabrics. Fabrics made wholly or in part from heat- and corrosion-resistant stainless steel yarn. They may be blended with textile fibers (cotton, wool, etc.), and are used for coverings, special carpeting, etc.

stainless steel flake. Flat or scalelike stainless steel powder particles of relatively small thickness, used for powder-metallurgy parts and as paint pigments.

stainless steel foil. A flat-rolled stainless steel product that is available in coils with a thickness of 0.005-0.13 mm (0.0002-0.005 in.), a width of less than 610 mm (24 in.) wide, and a length of up to 100 m (328 ft.). Used for laminating metals, ceramics, plastics, wood, etc., and for pressure-sensing bellows and diaphragms.

stainless steel plate. A flat-rolled or forged stainless steel product that is more than 4.76 mm (0.1875 in.) thick and 254 mm (10 in.) or more wide, and supplied in the annealed condition. Used for machinery, equipment, and appliances.

Stainless Steel Plus. Trade name of Ametek (USA) for a series of corrosion-resistant powders produced by blending conventional austenitic stainless steel powder with a 10% copper-nickel-tin (Cu-15Ni-8Sn) powder. Used for powder-metallurgy parts.

stainless steel powder. Stainless steel in the form of a fully alloyed powder usually produced by an atomization process. The composition of austenitic grades ranges from 0.01-0.03% carbon, 0.7% silicon, 0.2-0.4% manganese, 16.5-20% chromium, 8-14% nickel, 2.5-3.5% molybdenum, and the balance iron. A typical analysis for ferritic grades is 0.02% carbon, 0.25% manganese, 0.7% silicon, 17% chromium, 1.0% molybdenum, and the balance iron. Martensitic grades typically have 0.02-0.15% carbon, 11.5-13.5% chromium, 0.4% manganese, 0.7% silicon, and the balance iron. The particle sizes of *stainless steel powder* ranges from about 325 to over 100 mesh (45 to over 150 μm), and it is used for powder-metallurgy parts, and as a paint pigment.

stainless steels. A group of alloy steels that are highly resistant to corrosion and staining in a wide range of environments. Their predominant alloying element is chromium (10% or more). Other corrosion resistance enhancing additions include nickel and molybdenum. Stainless steels can be divided into the following five classes: (i) *Austenitic stainless steels;* (ii) *Ferritic stainless steel;* (iii) *Martensitic stainless steels;* (iv) *Precipitation-hardening stainless steels;* and (v) *Duplex stainless steels.*

stainless steel sheet. A flat-rolled stainless steel product that is less than 4.76 mm (0.1875 in.) thick and 610 mm (24 in.) or more wide, and supplied in the form of coils or cut lengths. Used for machinery, vehicles, equipment, appliances, etc.

stainless steel strip. A flat-rolled stainless steel product that is 0.13-4.76 mm (0.005-0.1875 in.) thick and 610 mm (24 in.) or less wide, and made by cold rolling, or initial hot rolling followed by cold rolling.

stainless steel wire. A stainless steel product with high dimensional accuracy and excellent surface finish obtained from hot-rolled and annealed rod by cold drawing or finishing. It is available in a wide range of shapes including (i) round, hexagonal, octagonal, square or shaped with a diameter or size of 12.70 mm (0.500 in.) or less, and (ii) flat with a thickness from less than 4.76 mm (0.1875 in.) to 0.25 mm (0.010 in.) and a width of 1.59-9.52 mm (0.0625-0.375 in.). Used in the manufacture of cable, cord and rope for aircraft, ships, and elevators, for springs, cold heading and cold forging applications, screens in industrial plants, etc. Insulated stainless steel wire with epoxy, polyester or tetrafluoroethylene insulation for electrical applications is also available.

stainless steel wool. Long, fine threads or shavings of stainless steel usually supplied in the form of batts, pads or ribbons. See also steel wool.

Stainmaster. Trademark of E.I. DuPont de Nemours & Company for nylon 6,6 fibers available in several grades including *Stainmaster, Stainmaster Luxra* and *Stainmaster XTRA*. Used for the manufacture of wear- and stain-resistant residential carpets.

stain wax. A special wood finish that produces a color resembling penetrating oil stain, but with a waxy luster.

Stakusit. Trade name of SKS-Stakusit Kunststoff GmbH (Germany) for profiles made of a wide range of different plastics.

Staline. Trade name for an ethoxybenzoic acid (EBA) dental lining cement.

Stalinite. Russian trade name for a sintered alloy containing 11.5% manganese, 9.5% chromium, and the balance iron. Used for machine components, agricultural and related equipment, and dredging and oil-drilling equipment.

Stalloy. British trade name for a soft magnetic steel containing 3-4% silicon, and sometimes small additions of aluminum. It has high magnetic permeability, low coercivity, low energy losses and high magnetic saturation. Used for telephone receiver and loudspeaker diaphragms, electrical equipment, instruments, and transformer cores.

Staloy N. Trade name of DSM Engineering Plastics (USA) for a polyamide/acrylonitrile-butadiene-styrene alloy.

Stamark. Trademark of 3M Company (USA) for plastic film used for pavement marking.

Stamax P. Trademark of Sabic BV (Netherlands) for a long-glass-fiber-reinforced polypropylene composite with excellent mechanical and thermal properties including high tensile strength, high creep and impact resistance, high heat-deflection temperatures and low thermal expansion. Used as a thermoplastic molding compound in the manufacture of automotiove components, such as dashboard carriers, door modules, front-end modules and underbody shielding.

Staminal. Trade name of Latrobe Steel Company (USA) for a tough, air- or oil-hardening tool steel containing 0.55% carbon, 0.9% manganese, 1% silicon, 0.4% chromium, 2.7% nickel, 0.13% vanadium, 0.45% molybdenum, and the balance iron. Used for chisels, punches, stamps, cold cutters, dies, and die blocks.

Stamina Wood. Trade name of Fibron Products, Inc. (USA) for impregnated wood laminated with phenolic resin.

Stamylan. Trade name of DSM Engineering Resins (USA) for an extensive series of polyethylene and polypropylene resins including: (i) *Stamylan HD*, an unmodified and UV-stabilized high-density polyethylene supplied in rod, sheet, tube, granule and film form with a density of 0.95 g/cm³ (0.034 lb/in.³), very good electrical properties, relative high strength, stiffness and impact resistance, good resistance to acids, alkalies, alcohols, halogens and ketones, an upper service temperature of 55-120°C (130-248°F), used for containers, bottles, pipes, pipe fittings, and packaging; (ii) *Stamylan LD*, an unmodified and UV-stabilized low-density polyethylene supplied in the form of sheets, rods, tubes, granules, powders or fibers. It has high toughness, good electrical insulation properties, good chemical resistance, and is used for electrical applications, containers, linings, and packaging film; (iii) *Stamylan MD*, a medium-density polyethylene; (iv) *Stamylan P* polypropylene homopolymers and copolymers supplied in unmodified, talc-filled, elastomer-modified and UV-stabilized grades; and (v) *Stamylan UH*, an chemically inert high-molecular weight polyethylene. It has a density of 0.94 g/cm³ (0.034 lb/in.³), excellent wear and abrasion resistance, high impact resistance, a low coefficient of friction, high melt viscosity, good resistance to most chemicals except halogens and aromatic hydrocarbons, and is used for industrial equipment and machine parts.

Stamylex. Trade name of DSM Engineering Resins for a series of polyethylenes including *Stamylex PE* linear-low-density polyethylenes, and *Stamylex XL* crosslinked polyethylenes.

Standalloy. Trade name for a dental amalgam alloy.

Standard. (1) Trade name of Standard Safety Glass Company

(USA) for a wired glass.

(2) Trade name of Boyd-Wagner Company (USA) for a series of water-hardening tool steels containing varying amounts of carbon (typically between 0.7 and 1.4%) and 0.25-0.35% manganese, with the balance being iron.

(3) Trade name of Metalor Technologies SA (Switzerland) for several gold-based dental solders including *Standard LFC 1*, a palladium-free solder containing 75% gold and having a melting range of 855-880°C (1570-1615°F), *Standard LFC 11*, a palladium- and copper-free solder containing 69.7% gold and 0.5% platinum, and having a melting range of 875-970°C (1605-1780°F) and *Standard LFC 12*, a palladium-free solder containing 90.6% gold and having a melting range of 940-970°C (1725-1780°F). Used for soldering universal dental alloys.

(4) Trade name for rayon fibers and yarns used for textile fabrics.

standard admiralty bronze. A tough, corrosion- and wear-resistant alloy containing 88% copper, 10% tin and 2% zinc. Used for trolley wheels, worm wheels, gears and bronze castings. See also admiralty bronze.

Standard Alloy. Trade name of Michigan Standard Alloy Inc. (USA) for an extensive series of heat- and corrosion-resistant steels containing varying amounts of nickel, chromium, manganese, silicon and carbon. Used for chemical and food-processing equipment, furnace parts and equipment, carburizing boxes, retorts, and melting pots.

Standard Alnico. Trade name of Harsco Corporation (USA) for a permanent magnet material containing 11-13% aluminum, 19-23% nickel, 4.5-5.5% cobalt, and the balance iron. See also alnico.

standard black. A black ceramic color of varying composition.

standard brick. Brick or firebrick made to a standard nominal size, e.g., 57 × 102 × 203 mm (2.25 × 4 × 8 in.) for building brick. See also brick; firebrick.

Standard Cadmium. Trade name for a brazing alloy containing 92.5% silver, 5.75% copper and 1.75% cadmium. Used for joining most ferrous and nonferrous metals.

standard cement. See specification cement.

Standard Glyco. Trade name of Joseph T. Ryerson & Son Inc. (USA) for a babbitt containing varying amounts of lead, tin and antimony. Used for bearings.

standard gold. See coinage gold (1).

standard hardboard. Hardboard manufactured from wood fibers by bonding with adhesive *lignin* and compressing under heat. It is brownish in color, and has a density of about 0.5-1.3 g/cm³ (0.02-0.05 lb/in.³), and often one textured surface. Since it has no inhibitors added, its moisture resistance is only moderate with the water absorption increasing from 12 to 40% with decreasing thickness. It is available in thicknesses from 0.083 to 0.375 in. (2.1 to 9.5 mm), a standard width of 4 ft. (1.2 m) and a standard length of 8 ft. (2.4 m). *Standard hardboard* can have one or two smooth sides (S1S or S2S), and is used for furniture components, siding, and inside wall coverings.

standard malleable iron. A term sometimes used for regular *malleable cast iron* to distinguish it from other malleable irons, such as cupola or special malleable iron. The average chemical composition of white iron for the manufacture of standard malleable iron ranges from 2.00-2.70% carbon, 0.80-1.20% silicon, less than 0.55% manganese, 0.20% phosphorus, 0.18% sulfur, with the balance being iron. It is available in general-purpose and high-strength, high-elongation grades with both strength and ductility increasing with decreasing carbon con-

tent.

Standard Misco. Trade name of Michigan Steel Casting Company (USA) for a wear-, heat- and corrosion-resisting steel containing 0.6% carbon, 35% nickel, 15% chromium, 0.5% manganese, and the balance iron. Used for heat treating boxes, furnace parts, valves and stills.

standard nickel bronze. A strong, tough, fine-grained, corrosion-resistant nickel casting bronze containing 88% copper, 5% tin, 5% nickel and 2% zinc. Used for valve parts, marine parts, gears, bearings, nozzles, machinery parts, corrosion resistant parts, and wear guides. See also nickel bronze.

standard phosphor bronze. A hard, strong, corrosion-resistant casting bronze containing 80% copper, 10% tin, 10% lead and up to 0.25% phosphorus. It has a density of 8.95 g/cm³ (0.323 lb/in.³), good machinability, and is used for heavy-duty, high-speed bearings, pressure-tight castings, pumps and valves, and corrosion-resistant parts. See also phosphor bronze.

standard plywood. Plywood that is usually composed of an odd number of veneers (3, 5, 7, etc.) placed such that the grain in each ply is at right angles to the adjoining one, and the grain direction of the back and front surface veneers follows the long dimension of the panel. The standard panel size is 1.2 × 2.4 m (4 × 8 ft.), and the thickness ranges from about 3 to over 25 mm (0.125 to over 1 in.). See also plywood.

standard polyolefin monofilament. A flat strand of a polyolefin, such as polyethylene or polypropylene, having an approximate width of 2.54 mm (100 mils) and a thickness of 50 μm (2 mils). See also polyolefins.

standard reference material. A reference material whose composition and physical and chemical properties have been certified by a recognized testing laboratory or standardizing authority, e.g., the American Society for Testing and Materials (ASTM) in the USA. Abbreviation: SRM.

standard sand. A relatively pure silica sand, such as *Ottawa sand*, that contains naturally rounded grains, is essentially free from organic matter, and will pass the No. 20 (850 μm), but be retained on the No. 30 (600 μm) US standard sieve.

standard section. Any of several sections of steel or light-metal alloy, such as equal or unequal angles, channels, I-beams, wide-flange sections, piling, tees, zees, etc., made by rolling to standard dimensions. See also structural shapes.

standard silver. See sterling silver.

standard steel. A steel of selected chemical composition and proven quality that is extensively used for a wide range of applications. Its particular composition and quality is specified by a recognized standardizing body, e.g., the American Iron and Steel Institute (AISI) in the USA. In most cases standard grades of steel can be used to replace other equivalent grades.

standard sterling silver. See sterling silver.

standard tin. A commercial tin with a purity of 99.75+%.

standard zinc dust. A finely divided powder made by blending metallic zinc powder with zinc oxide powder, and used in protective coatings (primers and topcoats) for metals.

standard zinc-lead white. A white pigment composed of a mixture of *basic white lead* ($PbSO_4 \cdot PbO$) and *zinc white* (ZnO).

Standart. Trademark of Eckart-Werke Carl Eckart GmbH & Co (Germany) for aluminum powders supplied in flake or micro-ground form for various chemical and general engineering applications, and gold bronze powders used for embossing and offset printing and in the manufacture of spray paints, varnishes, and wallpaper.

Standox. Trademark of Herberts GmbH Lackfabriken (Germany) for automotive color-match lacquers.

Stanelec. Trade name of Empire Sheet & Tin Plate Company (USA) for a commercial iron with a purity of 99.75% containing 0.25% silicon. It has a density of 7.85 g/cm³ (0.284 lb/in.³), high magnetic saturation and permeability, high electrical resistivity, and is used for electrical equipment, e.g., armatures motors, etc.

stanfieldite. A clear reddish to amber mineral composed of calcium iron magnesium phosphate, $Ca_4(Mg,Fe)_5(PO_4)_6$. Crystal system, monoclinic. Density, 3.15 g/cm³; refractive index, 1.622. Occurrence: USA.

Stanfire. Trade name of Standard Brake Shoe & Foundry Company (USA) for a heat-resisting cast iron containing 3.2% carbon, 1.5% nickel, 0.8% chromium, and the balance iron. Used for fire-box grates.

Stan-Fos. Trade name of Johnson-Matthey plc (UK) for a phosphorus-bearing silver brazing alloy containing 86.25% copper, 6.75% phosphorus and 7% tin. It has a melting range of 640-680°C (1185-1255°F).

stanleyite. An aquamarine to deep cobalt-blue mineral composed of vanadyl oxide sulfate hexahydrate, $VOSO_4 \cdot 6H_2O$. Crystal system, orthorhombic. Density, 1.95 g/cm³; refractive index, 1.519. Occurrence: Peru.

stannic acid. See stannic oxide.

stannic anhydride. See stannic oxide.

stannic bromide. Moisture-sensitive, white crystals (99% pure). Crystal system, orthorhombic. Density, 3.34 g/cm³; melting point, 31°C (87°F); boiling point, 203-204°C (397-399°F). Used in organic and organometallic synthesis, and for mineral separations. Formula: $SnBr_4$. Also known as *tin bromide; tin tetrabromide.*

stannic chloride. A colorless, corrosive, moisture sensitive liquid (99+% pure). Density, 2.226 g/cm³; melting point, -33°C (-27°F); boiling point, 114.1°C (237°F). Used in the production of abrasion-resistant coatings on glass, as an electrically conducting film on glass and ceramics, in electroconductive and electroluminescent coatings, in ceramic coatings, in the manufacture of sensitized paper (e.g., blueprint paper), and for tin salts. Formula: $SnCl_4$. Also known as *tin chloride; tin perchloride; tin tetrachloride.*

stannic chloride pentahydrate. Monoclinic crystals or moisture-sensitive, off-white lumps with a melting point of 56°C (133°F). Formula: $SnCl_4 \cdot 5H_2O$. Used as a substitute for stannic chloride ($SnCl_4$). Also known as *tin chloride pentahydrate.*

stannic chromate. A yellowish brown, crystalline powder obtained by treating stannic hydroxide with chromic acid. Melting point, decomposes on heating. Used in coloring porcelain and china rose or violet. Formula: $Sn(CrO_4)_2$. Also known as *tin chromate.*

stannic fluoride. Moisture-sensitive, white crystals or powder (95+% pure). Crystal system, tetragonal. Density, 4.780 g/cm³; melting point, sublimes at 705°C (1301°F). Formula: SnF_4. Also known as *tin fluoride; tin tetrafluoride.*

stannic dioxide. See stannic oxide.

stannic iodide. Moisture-sensitive, orange crystals (95+% pure). Crystal system, cubic. Density, 4.473 g/cm³; melting point, 144.5°C (292°F); boiling point, 364.5°C (688°F); refractive index, 2.106. Formula: SnI_4. Used in organic, inorganic and organometallic synthesis. Also known as *tin iodide; tin tetraiodide.*

stannic oxide. A white or gray crystals or powder (99.9+% pure) precipitated from a solution of stannic chloride by ammonium hydroxide. It occurs in nature as the mineral *cassiterite*. Crystal system, tetragonal. Crystal structure, rutile. Density, 6.85-7.0 g/cm³; melting point, 1127°C (2061°F); boiling point, sublimes at 1800-1900°C (3272-3452°F); refractive index, 1.9968; hardness, 6-7 Mohs; low thermal expansion. Used as an opacifier in white and colored porcelain enamels, glazes, refractory glazes and alabaster, milk and opaque glasses, in putties, for tin salts, as a catalyst, in the low-pressure chemical vapor deposition (LPCVD) of tin oxide surfaces, and as a polishing and grinding powder for steel, glass, etc. Formula: SnO_2. Also known as *flowers of tin; stannic acid; stannic anhydride; stannic dioxide; tin anhydride; tin ash; tin dioxide; tin oxide; tin peroxide.*

stannic selenide. A red-brown, crystalline compound of tin and selenium. Density, 5-5.1 g/cm³; melting point, 650°C (1202°F). Used as a semiconductor. Formula: $SnSe_2$. Also known as *tin selenide.*

stannic sulfide. See artificial gold.

stanniol. (1) A very thin, silvery foil composed of 96-99% tin, 0.7-2.5% lead, 0.3-1% copper, 0.4% nickel and 0.1% iron. Used for wrapping products, and for electrical applications, e.g., electrodes, capacitors, and sheathing. See also tinfoil.

(2) A silvery foil of aluminum with a thickness of 0.01 mm (0.0004 in.) or less, used for wrapping food products and cigarettes, and as thermal insulation.

stannite. Gray to black mineral with a metallic luster. It belongs to the chalcopyrite group and is composed of copper iron tin sulfide, Cu_2FeSnS_4, containing about 28.5% tin. Crystal system, tetragonal. Density, 4.40-4.47 g/cm³; hardness, 4.0 Mohs. Occurrence: Bolivia, France, Southeast Asia. Used as an ore of tin. Also known as *bell-metal ore; tin pyrites.*

stannoidite. A grayish mineral of the chalcopyrite group composed of copper iron tin sulfide, $Cu_8(Fe,Zn)_3Sn_2S_{12}$. Crystal system, orthorhombic. Density, 4.29 g/cm³. Occurrence: Japan, France.

Stannol. Trademark of Stannol Lötmittelfabrik Wilhelm Paff GmbH & Co. KG (Germany) for tin solders (in bar, block, wire, ribbon and powder form), extruded tin solders, silver brazing filler metals and various soft solders, bismuth solders, brazing and soldering pastes, soldering fluids and resins, solders for electronic applications, rosin-core solders, rosin-based soldering fluxes, welding fluxes for aluminum, copper, brass, zinc, cast iron and tool steel, various types of lead, hard-lead, zinc and zinc-alloy anodes, lead, tin, tin-lead and bismuth powders and granules, and several special preparations for galvanizing, tinning and leading applications.

Stannolume. Trade name of Atotech USA Inc. (USA) for a bright acid tin electrodeposit and deposition process.

stannomicrolite. A yellowish brown mineral of the pyrochlore group composed of tin niobium tantalum oxide, $Sn_2(Ta,Nb)_2O_7$. Crystal system, cubic. Density, 8.34 g/cm³. Occurrence: Finland.

stannopalladinite. A brown-rose mineral of the nickeline group composed of palladium tin, Pd_3Sn_2. Crystal system, hexagonal. Density, 10.20 g/cm³.

Stannostar. Trade name of Enthone-OMI Inc. (USA) for a bright acid tin electrodeposit and deposition process.

stannous acetate. Moisture-sensitive, gray or yellow crystals or off-white powder (95+% pure). Density, 2.30 g/cm³; melting point, 182°C (359°F); boiling point, sublimes at 155°C (311°F)/ 0.1 mm. Used as a reducing agent, in the preparation of tin (stannic) oxide thin films by photochemical vapor deposition, as a catalyst with palladium diacetate, and in fabrics to promote dye uptake. Formula: $Sn(C_2H_3O_2)_2$. Also known as *tin acetate; diacetoxytin.*

stannous bromide. An air- and moisture-sensitive yellow powder. Crystal system, orthorhombic. Density, 5.12 g/cm³; melting point, 232°C (449°F); boiling point, 620°C (1148°F). Used as a catalyst and reducing agent. Formula: $SnBr_2$. Also known as *tin bromide; tin dibromide.*

stannous chloride. White, air and moisture-sensitive crystals (98+% pure) obtained by dissolving tin in hydrochloric acid. Crystal system, orthorhombic. Density, 3.950 g/cm³; melting point, 246.8°C (476°F); boiling point, 623°C (1153°F). Used as a sensitizing agent for silvering mirrors, in the manufacture of glass and the plating of plastics, in tin galvanizing, in immersion tinning of metals, as a soldering agent, in conductor and resistor coating on glass, porcelain enamels and ceramics for surface heating, in the manufacture of phosphors, as a catalyst, as a reducing agent, and in biochemistry. Formula: $SnCl_2$. Also known as *salt of tin; tin chloride; tin crystals; tin dichloride; tin salts; tin protochloride.*

stannous chloride dihydrate. White, monoclinic crystals (98+% pure). Density, 2.710 g/cm³; melting point, 37.7°C (100°F); boiling point, decomposes. Used as a chemical reagent, catalyst, and reducing agent. Formula: $SnCl_2 \cdot 2H_2O$. Also known as *tin chloride dihydrate; tin dichloride dihydrate.*

stannous chromate. A brown powder obtained by action of sodium chromate on stannous chloride. Used as a colorant in the decoration of porcelain and pottery. Formula: $SnCrO_4$. Also known as *tin chromate.*

stannous fluoborate. A tin compound available as a 47% solution in water. Density, 1.60 g/cm³; melting point, decomposes above 130°C (266°F). Used in tin and tin-lead plating baths. Formula: $Sn(BF_4)_2$. Also known as *stannous fluoroborate; tin fluoborate; tin fluoroborate.*

stannous fluoride. White, air and moisture-sensitive crystalline powder (97+% pure). Crystal system, monoclinic. Density, 4.57 g/cm³; melting point, 219°C (426°F); boiling point, 850°C (1526°F). Used in ceramics, in organic and organometallic synthesis, and in dental preparations as a prevention of tooth demineralization. Formula: SnF_2. Also known as *tin difluoride; tin fluoride.*

stannous fluoroborate. See stannous fluoborate.

stannous iodide. Yellow-red crystals or red-orange, air and moisture-sensitive powder (90+% pure). Crystal system, monoclinic. Density, 5.280 g/cm³; melting point, 320°C (608°F); boiling point, 720°C (1328°F). Used in ceramics, and in inorganic, organic and organometallic synthesis. Formula: SnI_2. Also known as *tin diiodide; tin iodide.*

stannous oxalate. White crystals or crystalline powder (98% pure) obtained by treating stannous oxide with oxalic acid. Density, 3.56 g/cm³; melting point, decomposes at 280°C (536°F). Used in the formation of transparent conductive films on hot glass, as a transesterification catalyst in the manufacture of phthalate esters, and in textile dyeing and printing. Formula: SnC_2O_4. Also known as *tin oxalate.*

stannous oxide. Black, brownish-black or gray crystalline powder (99+% pure). Crystal system, cubic or tetragonal. Density, 6.45 g/cm³; melting point, decomposes at 1080°C (1976°F). Used as a ceramic color, as an opacifier in enamels and refractory glass, in the manufacture of ruby glass, as a coating for conductive glass, as a soft abrasive (putty powder) for polishing and grinding, in plating, as a reducing agent, as an intermediate in preparation of stannous salts, and in the preparation of

indium tin oxide (ITO). Formula: SnO. Also known as *tin monoxide; tin oxide; tin protoxide.*

stannous pyrophosphate. A white, crystalline powder (98+% pure). Density, 4.01 g/cm³; melting point, decomposes above 400°C (752°F); soluble in hydrochloric acid. Used as a diagnostic aid in radioactive bone scanning, as an additive in toothpastes, and in medical, biochemical and dental research. Formula: $Sn_2P_2O_7$. Also known as *tin pyrophosphate.*

stannous selenide. A gray, crystalline compound of tin and selenium. Crystal system, cubic. Crystal structure, halite. Density, 6.18 g/cm³; melting point, 860°C (1580°F). Used as a semiconductor. Formula: $SnSe$. Also known as *tin selenide.*

stannous sulfate. Moisture-sensitive, white powder, or white to yellowish crystals (95+%) obtained by treating stannous oxide with sulfuric acid. Density, 1.35 g/cm³; melting point, decomposes above 360°C (680°F) to SnO_2 and SO_2. Used in drawing of steel wire, and for tin plating electrolytes. Formula: $SnSO_4$. Also known as *tin sulfate.*

stannous sulfide. Dark gray or black crystalline powder. It occurs in nature as the mineral *herzenbergite*. Crystal system, orthorhombic. Density, 5.08 g/cm³; melting point, 880°C (1616°F); boiling point, 1230°C (2246°F). Used in the manufacture of bearing materials, as a polymerization catalyst, and as a reagent. Formula: SnS. Also known as *tin monosulfide; tin protosulfide; tin sulfide.*

stannous telluride. A gray, crystalline compound of tin and tellurium. Crystal system, cubic. Crystal structure, halite. Density, 6.45-6.50 g/cm³; melting point, 790-807°C (1454-1485°F); band gap, 0.5 eV. Used as a semiconductor. Formula: $SnSe$. Also known as *tin selenide.*

Stannum Metal. Trade name of Lumen Bearing Company (USA) for a high-tin babbitt containing 6% antimony and 4% copper. It has a maximum operating temperature of 149°C (300°F), excellent embeddability and antiscoring properties, and good corrosion resistance. Used for machine bearings.

Stanobond. Trade name of Henkel Surface Technologies (USA) for a bronze immersion coating used on metals.

Stanomerse. Trade name of Technic Inc. (USA) for bright immersion tin deposits used for decorative and electronic applications.

Stantufont. Trade name of Società Alluminio Veneto per Azioni (Italy) for an age-hardening aluminum alloy containing 12.4-13% silicon, 0.5-1% copper, 0.7-1% magnesium and 2-2.4% nickel. Used for light-alloy parts.

Stanuloy. Trade name MRC Polymers, Inc. (USA) for recycled polyesters supplied in unfilled and glass-filled grades, and recycled polycarbonate/polyethylene terephthalate alloys.

Stanyl. Trade name of DSM Engineering Plastics (USA) for high-temperature engineering plastics based on nylon 4,6 with a processing cycle 30% faster than nylon 6,6 and 50% faster than polyphenylene sulfide. It has a density of 1.18 g/cm³ (0.043 lb/in.³), a melting point of 295°C (563°F), good heat resistance, high stiffness, good strength, low creep at elevated temperatures, low dielectric properties, good processibility, a service temperature range of -40 to +155°C (-40 to +311°F), low thermal expansivity, good self-extinguishing properties, and is used for molded parts and packaging films.

StanylEnka. Trade name of DSM Engineering Plastics (USA) for engineering fibers based on nylon 4,6. See also Stanyl.

Stanyl High Flow. Trade name of DSM Engineering Plastics (USA) for reinforced flame-retardant nylon 4,6 resins with good flow and processing properties.

Stapa. Trade name of Eckart-Werke Carl Eckart GmbH & Co (Germany) for aluminum paste supplied in many special types for various industrial applications.

Staple Fiber. Trade name of Nichias Corporation (USA) for discontinuous reinforcing fibers composed of 60-68% alumina (Al_2O_3), 23-32% silica (SiO_2) and 4-9% boria (B_2O_3).

staple fiber. A synthetic or natural fiber in short lengths (typically 25-150 mm or 1-6 in.), made either directly, or by chopping continuous fibers. See also continuous fiber; discontinuous fiber; short fiber.

staple fiber fabrics. Fabrics made from staple natural mor synthetic fibers by any of various manufacturing processes including weaving, knitting or pressing.

staple glass fiber. A relatively short individual glass fiber less than 432 mm (17 in.) long and 7-10 μm (280-380 μin.) in diameter, usually made by attenuating molten glass.

staple glass yarn. A yarn made from individual glass filaments having a nominal length of 200-380 mm (8-15 in.).

staple yarn. Yarn made from staple natural or synthetic fibers.

Stapron. Trade name of DSM Engineering Plastics (USA) for styrene-maleic anhydride (SMA) copolymers and acrylonitrile-butadiene-styrene (ABS) alloys including *Stapron S* unreinforced and glass-fiber-reinforced SMA copolymers, *Stapron N* acrylonitrile-butadiene-styrene (ABS) terpolymers and ABS/nylon alloys, and *Stapron C* polycarbonate/ABS alloys.

Star. Trade name of A. Cohn Limited (UK) for a leaded brass containing 10% zinc, 6% tin, 6% lead, and the balance copper. It has good corrosion resistance, excellent machinability, and is used for castings.

Staramide 6. Trade name of Ferro Corporation (USA) for mineral-filled or 30% glass fiber-reinforced polyamide 6 (nylon 6).

Staramide 66. Trade name of Ferro Corporation (USA) for polyamide 6,6 (nylon 6,6) 33% reinforced with carbon fibers, or 40% filled with minerals.

star antimony. A highly refined grade of antimony metal with a characteristic sparkling surface pattern resembling stars or fern leaves. Also known as *star metal*.

Starblast. Trademark of E.I DuPont de Nemours & Company (USA) for blasting abrasives made from sand based on the mineral *staurolite* (iron magnesium aluminum silicate hydroxide) mined in Florida, and composed of dense, rounded, subangular grains. They are supplied in several grades for a wide range of metal-blasting applications.

Star Blue Chip. Trade name of Teledyne Firth-Sterling (USA) for a tungsten-type high-speed tool steel (AISI type T7) containing 0.7% carbon, 4% tungsten, 14% chromium, 2% vanadium, and the balance iron. Used for milling cutters, reamers, hobs, drills, and lathe tools.

Star Boron. Trade name of Carpenter Technology Corporation (USA) for a molybdenum-type high-speed tool steel (AISI type M40) containing 0.5-0.6% carbon, 8% molybdenum, 2% tungsten, 1.5% vanadium, 8% cobalt, 0.5% boron, and the balance iron.

star-branched polymer. A *branched polymer* produced by a polymerization technique in which the formed side chains or branches emanate from a single point. See also branched polymer.

Starbrite. Trademark of Star Fibers, Inc. (USA) for nylon 6 and nylon 6,6 staple fibers used for textile fabrics.

Starburst. (1) Trademark of Dendritech, Inc. (USA) for an extensive series of PAMAM-based dendrimers with different surface hydroxyl groups.

(2) Trade name for high- and low-copper silver-tin dental amalgam alloy.

Starburst Non-Gamma II. Trade name for a high-copper silver-tin dental amalgam alloy that does not contain the corrosive tin-mercury (Gamma 2) phase.

Starcast. Trade name of United Clays, Inc. (USA) for a quick-firing, low-carbon clay with good casting and forming properties, superior suspension characteristics, and moderate to high plasticity. Used in the manufacture of ceramic bodies and glazes.

starch. A polysaccharide composed of about 20-25% water-soluble *amylose* and 75-80% water-insoluble *amylopectin*. It is the storage form of carbohydrates in most green plants including cereal grains, peas, beans, potatoes, and green fruits, and is commercially available as white, amorphous powder or granules, or in crystalline form. It forms an irreversible gel in hot water and swells at room temperature in contact with formic acid, formamide, metallic salt and strong bases. Used in the manufacture of polymers, as an adhesive, in papermaking, in food products, as an indicator in analytical chemistry, as an abherent and mold-release agent, as a fabric stiffener, in oil-well drilling fluids, in electrophoresis gels, etc. See also hydrolyzed starch; soluble starch.

starch-based polymer. A crosslinkable, low-viscosity polyol polymer obtained by subjecting a slurry consisting of an aqueous mixture of starch with dibasic acids, hydrogen donors and selected catalysts to high temperatures and pressures yielding a low-viscosity polymer in a 50% solids aqueous solution. Used as water-resistant adhesive, as a moisture barrier material, as a paper-coating binder, and in high-wet-strength paper.

starch coating. A protective coating made by first soaking a starch in a small amount of cold water to break up lumps and then adding boiling water. The resulting transparent preparation is then mixed to a creamy consistency, cooled to room temperature and applied to surfaces previously coated with flat paint, or to colored wallpapers with a large paint or calcimine brush. The coating is stippled while wet to remove brush marks, and can be removed by means of water.

starch gum. See dextrin.

starch-plastic blends. A blend of any of various starches, such as corn, wheat or potato starch, with a polymer, such as low-density polyethylene.

Star Columbium. Trade name of Carpenter Technology Corporation (USA) for a molybdenum-type high speed tool steel (AISI type M8) containing 0.8% carbon, 4% chromium, 5% molybdenum, 5% tungsten, 1.5% vanadium, 1.2% niobium (columbium), and the balance iron. Used for dies.

Star-C. Trade name of Ferro Corporation (USA) for a series of reinforced plastics including *Star-C PA6*, a 30% carbon fiber-reinforced polyamide 6 (nylon 6), *Star-C PA66*, a polyamide 6,6 (nylon 6,6) reinforced with 10 or 30% carbon fibers, *Star-C PBT*, a 30% carbon fiber-reinforced polybutylene terephthalate, *Star-C PC*, a 30% carbon fiber-reinforced polycarbonate, and *Star-C PPS*, a 30% carbon fiber-reinforced polyphenylene sulfide.

Starelio. Trademark of Saint-Gobain (France) for a solar control glass.

Starflam. Trade name of Ferro Corporation (USA) for a series of reinforced plastics including *Starflam ABS* fire-retardant acrylonitrile-butadiene-styrenes, *Starflam PA6* 30% glass fiber-reinforced, fire-retardant polyamide 6 (nylon 6) resins, and *Starflam PBT* fire-retardant polybutylene terephthalates reinforced with 20 or 30% glass fibers.

Starglas. Trade name of Ferro Corporation (USA) for a series of reinforced plastics including *Starglas PBT* 30% glass-fiber-reinforced polybutylene terephthalates, *Starglas PC* 30% glass fiber-reinforced polycarbonates, and *Starglas PSU* 30% glass fiber-reinforced polysulfones.

staringite. A gray mineral of the rutile group composed of iron tin tantalum oxide, $(Sn,Fe)(Sn,Ta,Nb)_2O_6$. Crystal system, tetragonal. Density, 7.17 g/cm^3. Occurrence: Brazil.

Star J-Metal. Trade name of Haynes International, Inc. (USA) for a wear-resistant sintered cobalt alloy containing 31-33% chromium, 16-18% tungsten, 0-3% iron, 0-3% nickel, 0-1% manganese, 0-1% silicon, 0-1% boron and 2.2-2.7% carbon. It has a density of 8.58 g/cm^3 (0.310 $lb/in.^3$), high transverse strength, and is used for cutting tools, wear-resistant parts and high-temperature applications. Also known as *Stellite Star-J Metal.*

Stark. Trade name of Latrobe Steel Company (USA) for a tough, abrasion-resistant molybdenum-type high-speed steel (AISI type M4) containing about 1.3% carbon, 5.5% tungsten, 4.5% chromium, 4.5% molybdenum, 4% vanadium, and the balance iron. It has good hot hardness, wear resistance and deep-hardening properties, and is used for lathe and planer tools, chisels, reamers, drills, chasers, plastic core pins, broaches, and milling cutters. Formerly known as *Electrite Stark.*

starkeyite. A colorless mineral composed of magnesium sulfate tetrahydrate, $MgSO_4 \cdot 4H_2O$. It can also be recrystallized from an aqueous solution of magnesium sulfate. Crystal system, monoclinic. Density, 2.01 g/cm^3; refractive index, 1.491. Occurrence: USA (Missouri).

Star-L. Trade name of Ferro Corporation (USA) for a series of lubricated polyamide (nylon) resins including *Star-L PA6*, a polyamide 6 (nylon 6) lubricated with 20% polytetrafluoroethylene, and *Star-L PA66*, a polyamide 6,6 (nylon 6,6) lubricated with 20% polytetrafluoroethylene.

Starli. Trade name of Paul Bergsoe & Son (Denmark) for a shock-resistant tin babbitt containing 10% antimony and 5% copper. It has a melting range of 227-332°C (440-630°F), a maximum operating temperature of 150°C (300°F), good embeddability and antiscoring properties, and good corrosion resistance. Used for engine bearings.

Starlight. Trade name for a light-cure hybrid composite.

starlite. A blue variety of the colorless mineral *zircon* produced by artificially coloring by heat treatment. Used as a gemstone.

Starlux. Trade name of ASG Industries Inc. (USA) for twin-ground polished plate glass.

Starmax. Trade name of Carpenter Technology Corporation (USA) for a molybdenum-type high-speed steel (AISI type M1) containing about 0.85% carbon, 8.7% molybdenum, 4% chromium, 1.7% tungsten 1.2% vanadium, and the balance iron. It has good deep-hardening properties, high hot hardness, and ex-cellent wear resistance. Used for lathe and planer tools, milling cutters, drills, taps, reamers, and broaches. A free-machining grade, *Starmax FM*, is also available.

star metal. See star antimony.

star mica. A brown variety of *phlogopite* mica found in Canada and New York State and exhibiting a six-rayed star (asterism) when a point source of light is viewed through it.

Staro. Trade name of Stahlwerk Stahlschmidt GmbH & Co. (Germany) for a martensitic stainless steel containing 0.4% carbon, 0.3% manganese, 0.4% silicon, 13% chromium, and the balance iron. Used for cutlery, valves, surgical instruments, and dental equipment.

Starpylen. Trade name of Ferro Corporation (USA) for a 30% glass fiber-coupled polypropylene resin.

Starrlum. Trade name for preformed alumina balls used in abrasive tumbling.

star quartz. A variety of *vitreous quartz* from Canada that exhibits a star-like effect (asterism) in transverse crystal sections when cut en cabochon (in convex form).

star ruby. A ruby-red variety of the mineral *corundum* that has a stellate opalescence when observed by reflected light in the direction of the crystallographic c-axis. It can also be made synthetically from corundum by the addition of small amounts of impurities. Used as a gemstone.

Starry. Trade name of SA des Verreries de Fauquez (Belgium) for a glass with star design.

star sapphire. A variety of the mineral *corundum* from Sri Lanka that shows a six-rayed star (asterism) in transverse crystal sections when cut in en cabochon (in convex form). It can also be made synthetically. Used as a gemstone.

Star-Twist. Trade name of UCB Films, Inc. (USA) for a high-performance cellulose film for confectionary twist-wrapping applications.

Star VPS. Trade name for an addition cured vinyl polysiloxane (VPS) resin used for dental impressions.

Star Zenith. Trade name of Carpenter Technology Corporation (USA) for a tungsten high-speed tool steel (AISI type T1) containing 0.5% carbon, 18% tungsten, 4% chromium, 1% vanadium, and the balance iron. It has good deep-hardening properties, good wear resistance and high hot hardness. Used for drills, cutting tools, and punches.

Stat-BR. Trade name for a butadiene rubber used in dentistry for bite registrations.

Statex. Trademark of Columbian International Chemicals Company (USA) for a series of carbon blacks.

Staticguard. Trade name for antistatic nylon fibers and yarns used in the manufacture of carpets, upholstery and clothing.

Static Intercept. Trade name of Orbis-Menasha Corporation (USA) for a statically dissipative polypropylene copolymer resin made by injection molding. It has a density of 1.22 g/cm^3 (0.044 $lb/in.^3$), high impact resistance, a service temperature range of 4-107°C (39-225°F), and is used for containers.

statistical copolymer. A *copolymer* in which the monomer unit sequences are distributed according to well-known statistical laws, such as the Markovian statistics, etc. A copolymer consisting of statistically distributed sequences of monomeric units A and B is referred to as poly(A-*stat*-B), e.g., poly(ethylene-*stat*-propylene) is a statistical copolymer of ethylene and propylene.

Stat-Kon. Trademark of LNP Engineering Plastics (USA) for an extensive series of conductive and statically dissipative thermoplastic composites including noncarbon antistatic polystyrenes, stainless steel fiber-reinforced acrylonitrile-butadiene styrene, fiber-reinforced polycarbonate and carbon powder-filled polypropylene. Available in various forms including sheet and film, they are used for automotive components (fuel filters, etc.), electronic packaging, components for electrical and electronic devices, electronics handling devices, business machines, electronic component bins, and surface mount applications for tapes and reels.

Stat-Loy. Trademark of LNP Engineering Plastics (USA) for several grades of inherently antistatic extrusion composites available in various thermoplastic base resins.

StatNot. Trademark of Micarta Industries Corporation (USA) for a statically dissipative plastic laminate.

Statoil. Trade name of Borealis Chemicals (USA) for a series of unmodified and UV-stabilized polyolefins including *Statoil HDPE* high-density polyethylenes, *Statoil LDPE* low-density polyethylenes and *Statoil PP* polypropylene homopolymers and copolymers.

Stat-Rite. Trade name of BF Goodrich Static Control Polymers (USA) for polymer alloys composed of base polymers, such as acrylonitrile-butadiene styrene, acetal, acrylic, polyethylene terephthalate glycol, polypropylene and soft polyurethane, blended with suitable high-molecular-weight dissipative polymers. Used in the manufacture of permanently dissipative thermoformed, injection-molded and extruded parts for electrical and electronic applications.

statuary bronzes. A group of reddish casting bronzes containing 89.2-94.8% copper, 1.4-10.2% tin, 0.4-9.5% zinc, 0-6.25% lead, 0-0.71% nickel and up to 0.35% phosphorus. They possess good castability, good to medium corrosion resistance, and are used for statues, ornaments, plaques, etc.

statuary marble. A high-grade marble that is free from defects, inclusions or other discontinuities and thus suitable for use in ornaments and statues.

StatusBlue. Trade name of DMG Hamburg (Germany) for an addition-cured silicone dental impression material with high dimensional stability.

staurolite. A brown, reddish-brown or black mineral with a resinous to vitreous luster. It is composed of iron magnesium aluminum silicate hydroxide, $(Fe,Mg)_4Al_{18}H_2Si_3O_{48}$. Crystal system, orthorhombic. Density, 3.75 g/cm³; hardness, 7-7.5 Mohs; refractive index, 1.745. Occurrence: Switzerland. Used as a gemstone, and as a blasting abrasive.

Stavanger. Trade name of Stavanger Company (Norway) for a series of corrosion- and heat-resistant austenitic, ferritic and martensitic stainless steels.

Stavax ESR. Trade name of Uddeholm Corporation (USA) for an electroslag remelted, deep-hardening, oil-hardened martensitic stainless steel (AISI type 420) containing 0.35% carbon, 0.45% manganese, 0.45% silicon, 13.6% chromium, and the balance iron. It has high strength, good nondeforming properties, excellent polishability, good corrosion resistance, and is used for plastic molds.

Staybelite. Trademark of Hercules, Inc. (USA) for a pale, thermoplastic resin obtained from the hydrogenated constituents of *rosin*. It has a softening point of 75°C (167°F), and is used for adhesives, paper sizes, protective coatings, and in dentistry as an endodontic sealant.

Stayblade Max. Trade name of Firth-Vickers Stainless Steels Limited (UK) for an austenitic stainless steel containing 0.12% carbon, 18.5% chromium, 8.5% nickel, 1.4% aluminum, 0.8% titanium, and the balance iron. It has good heat resistance to 900°C (1650°F), and is used for turbine blades, boiler drums, etc.

Stayblade Steel. Trade name of Firth-Vickers Stainless Steels Limited (UK) for a heat- and corrosion-resisting steel containing 0.22% carbon, 20% chromium, 8.5% nickel, 1.2% titanium, 1% silicon, 0.6% manganese, and the balance iron.

staybolt iron. A wrought iron used for staybolts.

Stay-Brite. Trade name of J.W. Harris Company, Inc. (USA) for a series of low-temperature solders containing 0-6% silver, and the balance tin. They have a melting range of 221-279°C (430-535°F), and are used for joining all metals except aluminum, and for refrigeration joints.

Staybrite. Trade name of Firth-Vickers Stainless Steels Limited (UK) for a series of wrought and cast austenitic stainless steels containing about 0.03-0.1% carbon, 18% chromium, 8% nickel, varying amounts of molybdenum, copper, titanium and niobium, and the balance iron. Supplied in standard and free-machining grades, they have good corrosion resistance, weldability and strength. Used for aircraft and transport equipment, chemical and food-processing equipment, and for cutlery, surgical instruments, household appliances, filters, and pump parts.

Stay-Dry. Trade name of Pittsburgh Corning Corporation (USA) for pipe insulation made from cellular glass.

Stayer's. Trademark of Fuji Spinning Company (Japan) for modal staple fibers used for textile fabrics.

StayGard. Trademark of Honeywell Performance Fibers (USA) for fine-denier nylon-6 fibers used in the manufacture of airbags.

Stay-Silv. Trade name of J.W. Harris Company, Inc. (USA) for a series of brazing alloys containing 80-93% copper, 0-15% silver and 5-7.2% phosphorus. They have a melting range of 643-816°C (1190-1500°F), and are used for brazing ferrous and nonferrous alloys including those of copper.

steacyite. A brown mineral of the ekanite group composed of potassium sodium calcium thorium silicate, $(Na,Ca)_2KThSi_8O_{20}$. Crystal system, tetragonal. Density, 2.95 g/cm³; refractive index, 1.573. Occurrence: Canada (Quebec).

steam bronze. See hydraulic bronze.

steam-cured concrete. Concrete that has been rapidly cured (hardened) in water vapor either at atmospheric pressure and ambient temperatures of 40-95°C (100-200°F), or at high pressure and temperatures of 170-215 °C (340-420°F) in an autoclave.

steam metal. (1) A general term for any copper-base alloy suitable for exposure to steam environments.

(2) A free-cutting bronze containing 6% tin, 3% lead, 3% zinc, and the balance copper. Used for injectors, valves, steam valves, fittings, and other steam equipment. See also hydraulic bronze.

steam-refined asphalt. Asphalt refined in the presence of water vapor during the fractional distillation of petroleum.

steatite. (1) A ceramic material composed of 80-85% talc, 5-15% clay and up to 5% feldspar, and fired at a temperature of about 1400°C (2550°F). See also talc; steatite porcelain; steatite talc.

(2) See talc (2).

steatite ceramics. Silicate ceramics in which steatite, $Mg_3Si_4O_{10}(OH)_2$, is the principal crystalline phase. See also talc (1) and (2).

steatite porcelain. A vitreous ceramic whiteware containing about 64% silica (SiO_2), 30% magnesia (MgO), 5% alumina (Al_2O_3), and 1% other ingredients. It has good dielectric properties, and is used for electrical and industrial applications. Also known as *steatite whiteware*.

steatite talc. A high-grade massive talc, or ground talc powder composed essentially of steatite, $Mg_3Si_4O_{10}(OH)_2$, and suitable for use in electrical insulators. Also known as *lava talc*. See also talc (1) and (2).

steatite whiteware. See steatite porcelain.

Steb Metal. Trade name of Samuel Taylor Limited (UK) for a special alloy produced by rolling silver into copper, and used for enameling applications, and electrical contacts.

Stedmetal. Trade name of Stedman Foundry & Machine Company (USA) for a nickel cast iron containing 3.2% carbon, 0.7% manganese, 2.5% silicon, 1.5% nickel, and the balance iron. Used for crushers, grinders, and pulverizers.

steel. See steels.

steel aggregate. Concrete aggregate made from scrap steel. See also concrete aggregate.

steel bars. Steel products of round, square, hexagonal, octagonal, flat, triangular or half-round cross section made from billets by hot rolling on a bar mill. For flats the minimum thickness is 5.16 mm (0.203 in.) and the minimum width 203 mm (8 in.), and for other cross sections the minimum size is 9.52 mm (0.375 in.).

Steelblast. Trade name for steel grit used in sandblasting and tumbling operations.

Steel Bond. Trade name of Hy-Poxy Systems, Inc. (USA) for steel-filled epoxy resin repair kits.

steel-bonded carbides. A group of composites composed of ultrahard, rounded and smooth carbide micrograins (e.g., titanium carbide, TiC) distributed throughout a tool-steel matrix. They combine the superior wear resistance and lubricity of carbides with the good hardenability and machinability of steel. Abbreviation: SBC. See also cemented carbides.

steel bronze. A hard, strong, corrosion-resistant tin bronze containing 92% copper and 8% tin, and deoxidized with 0.1-2.5% phosphorus. It was formerly cast in iron molds and used in the manufacture of guns. See also phosphor bronze.

Steelcast. Trade name of Steele's (Contractors) Limited (UK) for a toughened rough-cast glass.

steel castings. Castings made by pouring molten steel into molds of the desired configuration and allowing it to solidify. They can be classified by composition into the following general groups: (i) *Low-carbon steel castings* containing less than 0.20% carbon; (ii) *Medium-carbon steel castings* containing 0.20-0.50% carbon; (iii) *High-carbon steel castings* containing more than 0.50% carbon; (iv) *Low-alloy steel castings* containing less than 8% alloying elements; and (v) *High-alloy steel castings* containing 8% or more alloying elements. *Steel castings* are used for engine and pump casings and parts, turbine casings and wheels, valve bodies, wheel rims, spokes and sets, fittings, forging presses, impellers, machine parts, shipbuilding and marine equipment, agricultural, construction, mining and other industrial machinery, railroad car frames, and a variety of other applications. Also known as *cast steels.*

steel cord. A string of several steel strands or filaments twisted together.

Steelcote. Trade name of ACI Chemicals, Inc. (USA) for a phosphate coating for steel products.

steel emery. An abrasive composed of chilled white cast iron shot and used for barrel finishing and stone grinding. See also chilled cast iron.

steel fiber-reinforced cement. A durable, high-strength composite having a cement-based matrix reinforced with continuous steel fibers. The matrix may be modified with chemical admixtures, polymer additives, etc. Used in building and civil-engineering construction. Abbreviation: SFRC. See also reinforced concrete.

steel filament. An individual continuous steel fiber used in the manufacture of strands, cords, ropes, etc.

steel foil. A thin sheet of carbon or alloy steel, usually less than 0.15 mm (0.006 in.) thick, and often coated with aluminum, tin, zinc, or a synthetic resin. Used in laminating, packaging, etc.

steel grit. Strong, crushed and graded steel shot in the form of angular particles with many sharp corners. It is available in particle sizes from 10 mesh (2 mm) down to 80 mesh (180 μm), and in three general Rockwell hardness ranges: 45 HRC,

56 HRC and 65 HRC. Used in abrasive blast cleaning to produce bright finishes, in the cleaning of castings to remove heavy scale, rust or welding flux and spatter, and in blast cleaning of nonferrous metals. See also grit.

Steel Master. Trade name for steel-filled epoxy resins which bond to plastics, ceramics, concrete and many metals including aluminum. Used to repair and rebuild worn or otherwise damaged metal parts.

Steelmet. Trade name of CMW Inc. (USA) for a series of prealloyed steel powders used to produce strong steel parts by powder metallurgy techniques.

Steelon. Trade name for nylon fibers.

steel pipe. A cylindrical tubular steel product made to standard combinations of outside diameter and wall thickness, and commonly used for pipelines and piping systems. The most common types of steel pipe are: (i) *Standard pipe* for industrial and domestic water, steam, oil or gas transmission, and distribution or service lines; (ii) *Line pipe* for transmission of water, oil and gas; (iii) *Pressure pipe* for elevated-temperature and/or pressure service, e.g., water-tube and fire-tube boilers; (iv) *Water-well pipe* for casings or drive pipes; (v) oil-country tubular goods including drill pipe, casings and tubing; and (vi) *Special pipe* for conduits, piling, pipe nipples, etc.

Steelplate. Trade name of Steele's (Contractors) Limited (UK) for a toughened polished plate glass.

steel plate. A carbon or alloy steel product obtained by reducing the thickness of a *slab* by hot rolling. The minimum thickness and width are 4.60 mm (0.180 in.) and 203 mm (8 in.), respectively. Practically, thicknesses and widths may range from 5 to 200 mm (0.188 to 8 in.) and 1220 to 3050 mm (48 to 120 in.), respectively. *Steel plate* has good strength and toughness, and good formability, weldability and machinability, and is used in the construction of bridges, buildings, ships, railroad cars, military equipment and vehicles, storage tanks, pressure vessels, large industrial machinery, etc.

steel powder. Steel in the form of a fully alloyed powder, usually produced by an atomization process. It may also consist of iron powder blended with carbon. Prealloyed steel powder contains iron, carbon, silicon, manganese, nickel, chromium, and/or other alloying elements. Available in a wide range of particle sizes, it is used chiefly for powder-metallurgy parts, in composites, and in paints. See also prealloyed steel powder; stainless steel powder.

steel reinforcing bar. See reinforcing bar.

steels. Iron-base alloys containing between 0.008 and 2.11 wt% carbon, and often appreciable amounts of alloying elements, such as manganese, silicon, vanadium, chromium, nickel, molybdenum, tungsten and/or cobalt. Their microstructures usually consist of alpha *ferrite* and iron carbide (*cementite*) phases, and are available in a wide range of forms including structural shapes (angles, beams, tees, zees, etc.), rails, plates, sheets and strips, coils, bars (rounds, squares, flats, hexagons, octagons, etc.), rods, wires, wire ropes, wire fabrics, wool, and pipes and tubes. *Steels* can be classified by composition into the following general groups: (i) *Plain carbon steels* including low-carbon steels with less than 0.2% carbon, medium-carbon steels with 0.2-0.5% carbon and high-carbon steels with more than 0.5% carbon; (ii) *Low-alloy steels* with up to 8% alloying elements; and (iii) *High-alloy steel* containing more than 8% alloying elements and including stainless and tool steels. *Steel* is used in the construction of bridges, buildings, industrial plant, machinery and machine parts, in shipbuilding, for automobile bodies, tools, dies, industrial and medical equipment, reinforced concrete, castings, hardware, turbines, engines, aerospace equipment, and a variety of other products.

steel scrap. Steels or steel products that owing to their composition, size, grade, etc., have been discarded, but are suitable for use as charge materials for certain steelmaking furnaces, and as aggregates in concrete.

steel sheet. A cold- or hot-rolled carbon or alloy steel product intermediate in thickness between steel foil and plate. Usually, the minimum and maximum thickness is 0.15 mm (0.006 in.) and 4.57 mm (0.180 in.), respectively, and the minimum width is 305 mm (12 in.). It has good strength and excellent finishing characteristics, good general appearance, good processibility and deep-drawing properties, and good compatibility with other materials and coatings. Used in the construction of buildings, industrial equipment, household appliances, automobiles, truck and trailer bodies, chemical equipment and tanks, electrical and magnetic equipment, etc. Also known as *sheet steel*.

Steelshine. Trade name of Matchless Metal Polish Company (USA) for buffing compounds for stainless steel.

steel shot. Spherical and nearly spherical particles of cast steel obtained by blasting a stream of molten steel with water. Typically, the Rockwell hardness ranges from 40 to 50 HRC. It is available in sizes from No. 70 to No. 930 with the shot number being the nominal diameter of an individual pellet in 0.0001 in. (2.5 μm). Used in abrasive blast cleaning, for removing scale, sand, etc., in the cleaning of castings, and for shot peening. See also shot; peening shot.

steel slab. A rectangular semifinished steel product hot-rolled down from an ingot in a primary rolling mill, or strand-cast in a continuous casting machine. Its minimum width is usually 25 cm (10 in.), and its minimum cross-sectional area 105 cm^2 (16 in.^{2}). A steel slab can be more than 100 m (328 ft.) long, weigh more than 8 tons, is usually deseamed to provide a smooth surface, and ready for rolling into plates, skelps or strips.

steel slag. A slag obtained during the smelting and refining of iron and steel and including blast-furnace and steel-furnace slags. See also slag.

steel strand. A number of steel filaments or wires twisted together, e.g., for making steel wire rope.

steel strip. A flat-rolled steel product that is 5.8 mm (0.23 in.) or less thick and 305 mm (12 in.) or less wide, i.e., it is narrower than, but otherwise similar to a *steel sheet*. Also known as *strip steel*.

Steeltec. Trademark of Pratt & Lambert, Inc. (USA) for an aluminum silicone resin coatings for thermal protection applications up to 649°C (1200°F). *Steeltec Universal RIP* is a durable, rust-inhibiting phenolic-modified alkyd primer for the protection of iron and steel.

steel tubing. Hollow steel products of symmetrical shape (usually square, round, oval or rectangular) and including (i) *Pressure tubing* for elevated internal and/or external temperature or pressure applications; (ii) *Structural tubing* for welded, riveted or bolted constructions; and (iii) *Mechanical tubing* for mechanical applications. Also known as *tube steel*.

steel tubular products. A term referring to hollow symmetrical steel products of cylindrical, square, oval, rectangular or any other shape, and including both *steel pipe* and *steel tubing*.

Steeltuff. Trade name of Steele's (Contractors) Limited (UK) for a toughened sheet glass.

steel wire. A wire of round, half-round, oval, square, flat, triangular, hexagonal or octagonal cross section cold-drawn from carbon or alloy steel wire rod. Round wire is usually available in sizes from about 0.127 mm (0.005 in.) to 25.4 mm (1 in.). Steel wire is used for baling and strapping, as reinforcement in concrete and other composites, as telephone and telegraph wire, for fasteners, springs and wire rope, for structural applications, for musical instruments, etc.

steel wire rod. A semifinished steel product made from a billet by hot rolling on a rod mill. It is used primarily for the manufacture of wire. Round steel rod is usually produced in nominal diameters from 5.6 mm (0.22 in.) to 18.7 mm (0.73 in.).

steel wool. Long, fine threads or shavings of steel usually in the form of batts, pads or ribbons. It is available in grades 3, 2, 1, 0, 2/0, 3/0 and 4/0 (finest). Used for cleaning and polishing metals or wood.

steerhide. Leather made from the hides of steer or oxen and used for protective clothing, linings, suitcases, traveling bags, shoe soles, uppers and insoles, upholstery, etc. See also cattle hides.

Stefanite. Trade name of Usines Emile Henricott SA (Belgium) for a series of cemented carbides used to make cutting tools for rough, medium and finish machining.

steigerite. A yellow to green mineral composed of aluminum chromium vanadium oxide trihydrate, $(Al,Cr)VO_4 \cdot 3H_2O$. Crystal system, monoclinic. Density, 2.52 g/cm^3. Occurrence: Kazakhstan, USA (Colorado, Utah).

Steinbühl yellow. See lemon yellow.

Steinit. Trademark of Steinhaus GmbH (Germany) for wear-resistant ceramics composed of 90% alumina (Al_2O_3). They have good antifriction properties, good high temperature and chemical resistance and good noise dampening properties. Supplied as shaped parts for lining hoppers and chutes, and as rings for lining pipes, pipe elbows, etc.

Steklofon. Russian trade name for a differentially toughened glass for automobile windshields.

Stelanneal. Trademark of Stelco Inc. (Canada) for annealed and galvanized steel sheets and coils.

Stelclad. Trademark of Stelco Inc. (Canada) for precoated steel culvert stock.

Stelco. Trademark of Stelco Inc. (Canada) for a wide range of steel products including semifinished billets, blooms, slabs and rods, tinplate and hot- and cold-rolled strip and sheet, bar-mill products including flats, rounds, bars, bands, angles, structural shapes and sections, steel plate, drop forgings, machinery steels, and welding rods.

Stelcoat. Trademark of Stelco Inc. (Canada) for galvanized sheet steel and steel culvert stock.

Stelcolour. Trademark of Stelco Inc. (Canada) for prefinished steel tin-mill products and galvanized and cold-rolled steel products.

Stelcoloy. Trade name of Stelco Inc. (Canada) for a series of high-strength low-alloy (HSLA) steels containing up to 0.25% carbon, up to 1.5% manganese, 0.15-0.3% silicon, 0.2-0.5% copper, 0.2-0.5% nickel, up to 0.5% chromium, 0.1% vanadium and 0.05% niobium, respectively, and the balance iron. Used for welded, riveted and bolted constructions, bridges, buildings, etc.

Stelform. Trademark of Stelco Inc. (Canada) for spiral-welded steel pipe.

Stella. Trademark of Stella Werner Deussen (Germany) for hollow glass products.

Stellar. Trade name of Specialty Steel Company of America (USA) for an oil-hardening low-alloy tool steel containing 0.34% carbon, 0.9% chromium, 0.8% manganese, 0.5% silicon, 0.35%

molybdenum, 0.33% copper, and the balance iron. Used for dies and hand tools.

stellerite. A white mineral of the zeolite group composed of calcium aluminum silicate hydrate, $Ca_2Al_4Si_{14}O_{36} \cdot 14H_2O$. Crystal system, orthorhombic. Density, 2.13 g/cm^3; refractive index, 1.496. Occurrence: Italy.

Stellite. Trademark of Deloro Stellite Limited (UK) for an extensive series of nonferrous alloys containing 20-80% cobalt, 10-40% chromium, 0-25% tungsten, 0-10% molybdenum, 0-2.5% carbon, and sometimes additions of iron, nickel, manganese, silicon, niobium, vanadium, copper and/or boron. They have very high hardness, excellent abrasion, corrosion and wear resistance, good thermal-shock resistance, good high-temperature properties, good oxidation resistance, high stress rupture, low creep, and are used for high-speed cutting tools, special castings, powders, electrodes and rods for hardfacing applications, jet-engine turbine blades and vanes, and firing racks and tools.

Stellite Star J-Metal. See Star-J Metal.

Stellon. British trade name for a dough (or bulk) molded acrylic resin formerly used for denture bases.

Stellram. Trade name of Stellram Limited (Switzerland) for a sintered tungsten carbide used for cutting tools.

Stelmax. Trade name of Stelco Inc. (Canada) for a series of high-strength low-alloy (HSLA) steels containing 0.15% carbon, 1.5% manganese, up to 0.02% nitrogen, 0.005% or more niobium, 0.01% or more vanadium, and the balance iron. Used for automotive bumpers, reinforcements, and high-strength sheet.

Steltex. Trademark of Stelco Inc. (Canada) for textured cold-rolled steel products.

Stelvetite. Trade name of British Steel plc (UK) for a plastic-coated low-carbon steel.

Stelweld. Trademark of Stelco Inc. (Canada) for weldable concrete reinforcing bars made of high-strength steel.

Stemallit. Russian trade name for a sheet glass coated with ceramic paints or enamels, and used as a facing material for office building exteriors.

Stemalloy. Trade name of Lunkenheimer Company (USA) for a corrosion- and wear-resistant alloy composed of copper, tin, zinc and lead, and used for valve stems and seats, and globe valves.

stencil metal. Sheet lead and zinc used to make stencils for enameling operations.

stencil base paper. Paper coated or impregnated such that it becomes permeable for printing ink in those areas in which it is subjected to pressure by stencils, writing utensils, typewriters, or other office equipment.

stenhuggarite. An orange mineral composed of calcium iron antimony oxide arsenate, $CaFe(AsO_2)(AsSbO_5)$. Crystal system, tetragonal. Density, 4.63 g/cm^3. Occurrence: Sweden.

stenonite. A colorless mineral composed of strontium aluminum fluoride carbonate, $Sr_2AlF_5CO_3$. Crystal system, monoclinic. Density, 3.86 g/cm^3; refractive index, 1.527. Occurrence: Greenland.

Stentor. Trade name of Carpenter Technology Corporation (USA) for an oil-hardening cold-work tool steel (AISI type O2) containing 0.9% carbon, 1.6% manganese, 0.2% silicon, and the balance iron. It has good nondeforming properties and machinability, and good to moderate wear resistance. Used for tools, gages, dies and cutters.

step-growth polymer. See condensation polymer.

stephanite. An iron-black mineral with metallic luster composed of silver antimony sulfide, Ag_5SbS_4. Crystal system, orthorhombic. Density, 6.25 g/cm^3. Occurrence: Germany, Mexico. Used as an ore of silver. Also known as *black silver; brittle silver ore; goldschmidtine.*

Stephenson. (1) A tough British copper alloy containing 8.3% zinc, 2.9% lead, 4.3% lead and 0.4% iron. Used for piston rings.

(2) A British casting alloy containing 31% tin, 31% iron, 19% zinc and 19% copper.

(3) A British bearing metal containing 7.5% tin, 8% lead, 5% zinc, and the balance copper. Used for locomotive bearings.

step-reaction polymer. See condensation polymer.

STERalloy. Trade name of Hapco Inc. (USA) for food- and drug-grade liquid molding polymer alloys including Series 2000 elastomeric alloys and Series 5000 rigid alloys. They have good chemical and physical properties, good processibility, and high impact resistance.

Sterbon. Trade name for silicon carbide.

stercorite. A colorless mineral composed of sodium ammonium hydrogen phosphate hydroxide tetrahydrate, $Na(NH_4)(PO_3)OH \cdot 4H_2O$. It can also be made synthetically. Crystal system, triclinic. Density, 1.57 g/cm^3; refractive index, 1.442.

Stereo. Trade name of Blackwells Metallurgical Limited (UK) for a brass containing 38% zinc, 2% iron, and the balance copper.

stereoblock polymer. A polymer in which rather long sections having a particular stereospecific (three-dimensional) structure are interrupted by shorter segments of different stereospecific structure. For example, it may consist of long sections with isotactic structure separated by short segments with syndiotactic structure. See also block copolymer.

stereoisomer. (1) A chemical compound that has the same molecular formula as another compound, but differs from it in the way its atoms are arranged in three dimensions. See also optical isomer; geometric isomer.

(2) A polymeric isomer in which the order of the branches or side groups within a mer unit along the backbone is identical, but its three-dimensional arrangement is different.

Stereon. Trademark of Firestone Polymers (USA) for a series of styrene-butadiene multiblock thermoplastic elastomers used for blending with polystyrenes, polyolefins and high-impact polystyrenes, in adhesives, and as primary polymers for modified asphalt.

stereoregular polymers. Polymers in which a specific three-dimensional order or regularity of molecular arrangement has been produced by the use of certain organometallic catalysts. Such polymers are said to have an ordered crystalline or semicrystalline structure (e.g., polyethylene) that differs from that of amorphous polymers (e.g., polyvinyl chloride) in which the molecules are randomly arranged. Depending on the position of the atoms or atom groups (radicals) attached to the carbon backbone of the molecular chain, stereoregular polymers are classified as: (i) *isotactic polymers* (all atoms or groups on one side of the chain), (ii) *syndiotactic polymers* (atoms or groups alternately on one side of the chain and then on the other); and (iii) *atactic polymers* (atoms or groups randomly oriented on the chain). Also known as *stereospecific polymers.*

stereorubber. (1) An isotactic synthetic rubber made from *cis*-1,4-polyisoprene (C_5H_8) using a stereospecific organometallic catalyst. See also isoprene rubber.

(2) An isotactic synthetic rubber made from 1,3-butadiene (C_4H_6) by polymerization with stereospecific organometallic

catalysts in the presence of a solvent (e.g., benzene). It consists of butadiene molecules arranged in the 1,4-position (*cis*-1,4-polybutadiene), and is superior to natural rubber in both elasticity and abrasion resistance. See also butadiene rubber.

stereo SBR. A stereospecific *styrene-butadiene rubber* (SBR), i.e., a SBR having a specific three-dimensional molecular order or regularity and produced by polymerization in the presence of an organometallic catalyst, usually an organolithium compound, such as butyllithium.

stereospecific polymers. See stereoregular polymers.

stereospecific styrene-butadiene rubber. See stereo SBR.

stereotype metal. A *type metal* alloy usually containing 13-15% antimony, 3-7% tin, and the balance lead. It has a melting point of 242°C (468°F), and is used for stereotype printing plates, and for castings.

Sterifood. Trade name of Silicofab, Division of Robco Inc. (Canada) for food-grade silicone elastomer tubing.

Sterling. (1) British trade name for a solder of 62% tin, 15% zinc, 11% aluminum, 8.3% lead, 2.5% copper and 1.2% antimony.

(2) Trade name of Castings Corporation (USA) for an austenitic stainless steel (AISI type 302) containing 0.1% carbon, 18% chromium, 8% nickel, and the balance iron. Used for chemical plant equipment.

(3) Trade name of Teledyne Firth-Sterling (USA) for a water-hardening plain-carbon tool steel (AISI type W1) containing 0.9-1.1% carbon, 0.3-0.4% manganese, 0.1-0.5% chromium, 0.15-0.25% vanadium, and the balance iron. Used for drills, taps, shear blades, knives, etc.

(4) Trade name of Sterling Die Casting Company, Inc. (USA) for aluminum and zinc die castings.

(5) Trade name of Taracorp IMACO, Inc. (USA) for lead-free solders.

(6) Trademark of Cabot Corporation (USA) for an extensive series of carbon blacks.

(7) Trade name of Sterling-Clark-Lurton Corporation (USA) for caulking compounds.

sterlinghillite. A white to light pink mineral composed of manganese arsenate tetrahydrate, $Mn_3(AsO_4)_2 \cdot 4H_2O$. Occurrence: USA (New Jersey).

Sterling Metal. Trade name for a free-cutting alloy of 66% copper, 27-33% zinc, 0.7% iron, 0-2% lead and 0-6.3% tin. Used for fittings and hardware.

sterling silver. A corrosion-resistant, high-grade alloy containing at least 92.5% pure silver, with the balance usually being copper as a hardener. Available in 10-70% cold-worked grades, it has a density of 10.4 g/cm³ (0.38 lb/in.³), and is used for electrical contacts, coins, medallions, cutlery, silverware, jewelry, ornaments, and utensils. Also known as *coin silver; standard silver; standard sterling silver.*

sterling silver solder. (1) A corrosion-resistant silver solder composed of 80% silver, 18% zinc and 2% copper.

(2) A corrosion-resistant brazing filler metal composed of 92.8% silver, 7% copper and 0.2% lithium. It has a liquidus temperature of 900°C (1650°F), and is used for brazing sterling silver.

Sterlite. Trade name of Sterlite Foundry & Manufacturing Company (USA) for a tough, corrosion-resistant alloy composed of 53% copper, 25% nickel, 20% zinc and 2% manganese. Used for surgical instruments, laundry equipment, diary machines, ship fittings, etc.

Stermet. Trade name of Sterling Alloy Casting Corporation (USA) for a series of cast heat- and corrosion-resistant iron-chromium-

nickel and iron-nickel-chromium alloys used for furnace parts and fixtures, heat-treating equipment, transportation equipment, and chemical and food processing equipment.

sternbergite. A pinchbeck-brown mineral of the cubanite group composed of silver iron sulfide, $AgFe_2S_3$. Crystal system, orthorhombic. Density, 4.16 g/cm³. Occurrence: Czech Republic.

stern tube metal. A white metal composed of 68.5% tin, 30% zinc and 1.5% copper, and used in shipbuilding for propeller shaft bushings.

Sterngold 66. Trade name for a dental casting alloy containing 66% gold.

Sternite. Trade name of Sterling Moulding Materials (UK) for phenol-formaldehyde resins, plastics, and molding compounds.

SternVantage. Trademark of Omega Handelsgesellschaft mbH & Co. KG (Germany) for a line of dental impression materials including *SternVantage Putty* and *SternVantage Putty Soft* dough-like impression materials based on addition-curing vinyl silicones, *SternVantage Light Body* low-viscosity polyvinyl siloxane (PVS) impression material for crowns, bridges, inlays, onlays, implants, prostheses, *SternVantage Quick Light Body* quick-setting VPS impression material, *SternVantage Quick Bite* thixotropic bite registration material based on addition curing vinyl silicones, *SternVantage Monophase* medium-viscosity polyvinyl siloxane impression material based on addition-curing vinyl silicones, and *SternVantage Adhesive* tray adhesive for silicone impression materials.

Steroxy. Trademark of Sterling Limited (Canada) for epoxy powder coatings.

sterro metal. A strong, hard casting brass composed of 55-60% copper, 38-42% zinc, 1.8-4.7% iron, and sometimes small amounts of tin and/or manganese. Used for pressure-tight castings, marine castings, and hydraulic cylinders subject to high pressure.

sterryite. A black mineral composed of silver lead antimony arsenic sulfide, $Ag_2Pb_{10}(Sb,As)_{12}S_{29}$. Crystal system, orthorhombic. Density, 5.91 g/cm³. Occurrence: Canada (Ontario).

Stervac. Trade name of Teledyne Firth-Sterling (USA) for a series of cast heat-resistant nickel-chromium-cobalt and nickel-iron-chromium alloys used for gas-turbine and jet-engine parts, valves, bolts, etc.

stetefeldtite. A yellow mineral of the pyrochlore group composed of silver antimony oxide hydrate, $AgSb_2(O,OH,H_2O)_6$. Crystal system, cubic. Density, 4.60 g/cm³; refractive index, 1.95. Occurrence: USA (Nevada).

stevensite. A mineral of the smectite group composed of sodium calcium magnesium iron manganese silicate hydroxide, $(Ca,Na)_xMg_3Si_4O_{10}(OH)_2 \cdot zH_2O$. Crystal system, orthorhombic. Occurrence: Japan, USA (New Jersey).

Steve's White. Trade name of Aardvark Clay & Supplies (USA) for a bright white earthenware clay used for throwing and hand building applications.

stewartite. A brownish yellow mineral composed of manganese iron phosphate hydroxide octahydrate, $MnFe_2(PO_4)_2(OH)_2 \cdot 8H_2O$. Crystal system, monoclinic. Density, 2.94 g/cm³; refractive index, 1.658. Occurrence: Brazil, USA (California).

stibarsen. A tin-white or reddish gray mineral of the tetradymite group composed of antimony arsenide, (As,Sb). Crystal system, rhombohedral (hexagonal). Density, 6.00 g/cm³. Occurrence: Sweden.

stibic anhydride. See antimony pentoxide.

stibiconite. A white to pale yellow, orange or brown mineral of the pyrochlore group composed of antimony oxide hydroxide,

$Sb_3O_6(OH)$. Crystal system, cubic. Density, 4.40 g/cm³.

stibiobetafite. A brownish black mineral of the pyrochlore group composed of calcium antimony niobium titanium tantalum oxide hydroxide, $(Ca,Sb)_2(Ti,Nb,Ta)_2(O,OH)_7$. Crystal system, cubic. Density, 5.30 g/cm³. Occurrence: Czech Republic.

stibiocolumbite. A brown to yellowish brown mineral of the columbite group composed of antimony niobium tantalum oxide, $Sb(Nb,Ta)O_4$. It can also be made synthetically. Crystal system, orthorhombic. Density, 5.73-6.30 g/cm³; refractive index, 2.42. Occurrence: USA (California).

stibiopalladinite. A silvery white to steel gray mineral composed of antimony palladium, Pd_5Sb_2. Crystal system, orthorhombic. Density, 9.42-9.50 g/cm³. Occurrence: South Africa.

stibiotantalite. A honey- to greenish yellow mineral of the columbite group composed of antimony tantalum oxide, $Sb(Ta,Nb)O_4$. Crystal system, orthorhombic. Density, 7.30-7.59 g/cm³; refractive index, above 2.11.

stibivanite. A yellow-green mineral composed of antimony vanadium oxide, Sb_2VO_5. Crystal system, monoclinic. Density, 5.12 g/cm³. Occurrence: Canada (New Brunswick).

stibnite. A lead- or steel-gray mineral with a metallic luster and a blackish tarnish. It is composed of antimony sulfide, Sb_2S_3, and contains up to 71.8% antimony, and sometimes small amounts of silver, gold and/or copper. The average antimony content of the ore is 45-60%. It can also be made synthetically. Crystal system, orthorhombic. Density, 4.52-4.63 g/cm³; hardness, 2 Mohs. Occurrence: Bolivia, Chile, China, France, Germany, Hungary, Japan, Mexico, Peru, South Africa, USA (Alaska, Arkansas, California, Idaho, Nevada). Used as a principal ore of antimony. Also known as *antimonite; antimony glance; gray antimony.*

stichtite. A lilac to pink mineral of the sjogrenite group composed of magnesium chromium carbonate hydroxide tetrahydrate, $Mg_6Cr_3CO_3(OH)_{16} \cdot 4H_2O$. Crystal system, rhombohedral (hexagonal). Density, 2.16 g/cm³; refractive index, 1.516. Occurrence: Tasmania, Scotland, Canada (Quebec), southern Africa.

Stick. Trade name for an adhesive used for dental impression trays.

stick. Abrasive grains held together in stick form usually by a vitrified, resinoid or metal bond, and used for dressing abrasive wheels and sharpening tools, and in precision honing.

stick rouge. A stick made from *rouge,* a fine, high-grade, reddish powder composed of hydrated iron oxide, and used for buffing and polishing applications.

stick shellac. Shellac available in a wide range of colors in the form of a stick, and used in the repair of furniture, and in refinishing to fill cracks and scratches in wood.

Sticky Post. Trade name for a self-cure dimethacrylate (DMA) dental cement used for the cementation of posts.

stiff clay. See stiff mud.

stiff mud. A stiff, but plastic clay with a moisture content of approximately 12-15%. Also known as *stiff clay.*

stiff-mud brick. Brick made by extruding a stiff, but plastic clay with a moisture content of approximately 12-15% through a die.

Stiktit. Trademark of 3M Company (USA) for pressure-sensitive coated abrasives.

trans-stilbene. The *trans*-form of stilbene available as colorless to pale yellow crystals (96+% pure) with a density of 0.970 g/cm³, a melting point of 122-124°C (252-255°F), a boiling point of 305-307°C (581-585°F)/744 mm. Used in transition-metal catalyzed asymmetric epoxidation and dihydroxylation, in the manufacture of dyes and optical bleaches, and in biochemistry. High-purity, zone-refined crystals are used as phosphors and scintillators. Formula: $C_{14}H_{12}$. Also known as *trans-1,2-diphenylethylene; toluelene.*

stilbite. A creamy yellow or colorless mineral of the zeolite group composed of sodium calcium aluminum silicate hydrate, $NaCa_2\text{-}Al_5Si_{13}O_{36} \cdot 14H_2O$. Crystal system, monoclinic. Density, 2.16 g/cm³; refractive index, 1.5. Occurrence: Iceland, Canada (Nova Scotia).

Stilla. Trade name of Vetreria di Vernante SpA (Italy) for glass with a cross-hatch pattern of droplets formings one set of diagonal lines.

stilleite. A gray mineral of the sphalerite group composed of zinc selenide, ZnSe. It can also be made synthetically. Crystal system, cubic. Density, 5.42 g/cm³; refractive index, 2.89. Occurrence: Zaire.

stillwaterite. A creamy-gray mineral composed of palladium arsenide, Pd_8As_3. It can also be made synthetically. Crystal system, hexagonal. Density, 10.40 g/cm³. Occurrence: USA (Montana).

stillwellite. A colorless mineral composed of cerium boron silicate, $CeBSO_5$. Can be made synthetically. Crystal system, monoclinic. Density, 4.61 g/cm³; refractive index, 1.765. Occurrence: Australia.

Stilon. Trademark of Chemitex-Stilon (Poland) for nylon 6 fibers and filament yarns used for textile fabrics.

stilpnomelane. A dark brown mineral composed of iron magnesium aluminum silicate hydroxide, $(Fe,Mg)_6(Si,Al)_8O_{19}(OH)_9$, and sometimes also containing manganese. Crystal system, triclinic. Density, 2.83 g/cm³; refractive index, 1.735. Occurrence: USA (Michigan), New Zealand.

Stimulan. Trademark of Biocomposites Limited (UK) for a biocompatible, fully resorbable, high-purity synthetic calcium sulfate used in bone repair.

Stipple. Trade name of PPG Industries Inc. (USA) for a patterned rough-cast plate glass.

Stippletone. Trademark of Trent Coatings Limited (Canada) for ready-to-use, mildew-resistant interior and exterior vinyl textured stucco finishes for ceilings and walls, available in adobe, gray, peach, topaz and white colors.

Stippolyte. Trade name of Pilkington Brothers Limited (UK) for a patterned glass.

stishovite. A colorless mineral of the rutile group composed of silicon dioxide, SiO_2. Crystal system, tetragonal. Density, 4.35 g/cm³; refractive index, 1.799. Occurrence: USA (Arizona).

stitch-bonded fabrics. Fabrics consisting either of yarns stitched together or stitched webs of fibers.

stitching wire. (1) An alloy of 74.5% nickel, 18% chromium and 7.5% manganese, used for electrical resistances.

(2) A light-drawn steel wire used in wire-stitching machines.

Stitchmat. Trademark of Vetrotex CertainTeed Corporation (USA) for a woven glass mat composed of E-glass fibers and polyester threads.

stittite. A brown mineral of the sohngeite group composed of iron germanium hydroxide, $FeGe(OH)_6$. Crystal system, tetragonal. Density, 3.60 g/cm³; refractive index, 1.737. Occurrence: Southwest Africa.

St. Joe Chemical Lead. A high-purity pig lead (99.94+%) containing about 0.05% copper, 0.005% silver, 0.0003% cadmium and 0.0003% nickel. It has a density of 11.35 g/cm³ (0.410 lb/in.³), a melting point of 326°C (618°F), and is used for acid containers and chemical equipment.

St. Lawrence. Trademark of St. Lawrence Cement Inc. (Canada) for standard, high-early-strength, sulfate-resistant and block-type Portland and masonry cements, and ready-mix concrete.

Stockade. Trademark of Canadian Forest Products Limited for panelwood.

Stockholm tar. A term formerly used for a tar obtained in the distillation of pinewood and used in wooden boatbuilding. It originally referred to the kiln-burned tar obtained in Scandinavia from the wood of the Baltic pine (*Pinus sylvestris*).

stockinet. A soft, elastic, machine-knitted fabric made from cotton, wool, or synthetic yarn with plain stitches, and used especially for making underwear and apparel.

stock steel. A term referring to standard steels including regular and stiff stock grades. Regular stock steel is bright, nonhardened carbon steel, usually of the low-carbon (0.08-0.17%) or medium-low-carbon (0.17-0.26%) type. Stiff stock steel is usually made from medium-low-carbon (0.17-0.26%) or medium-high-carbon (1.00-1.50) grades, and is also bright and nonhardened, but stiffer, tougher and harder than regular stock steel.

stoiberite. A black mineral composed of copper vanadium oxide, $Cu_5V_2O_{10}$. Crystal system, monoclinic. Density, 5.00 g/cm³. Occurrence: Central America.

stoichiometric compound. A compound, such as magnesium oxide (MgO) or aluminum oxide (Al_2O_3A), having an exact or fixed ratio of chemical elements. See also nonstoichiometric compound.

stoichiometric intermetallic compound. A compound formed of two or more components (metals) and having a composition, structure and properties different from either component (metal), and in which the ratio of the components present is exact or fixed. Examples include aluminium antimonide (AlSb), nickel aluminide (Ni_3Al) and magnesium sulfide (Mg_2S). See also nonstoichiometric intermetallic compound.

stoichiometric semiconductor. A stoichiometric intermetallic compound, such as aluminum antimonide (AlSb) or gallium arsenide (GaAs), having semiconductive properties. See also semiconductor compounds; stoichiometric intermetallic compound.

stokesite. A colorless mineral of the pyroxenoid group composed of calcium tin silicate dihydrate, $CaSnSi_3O_9 \cdot 2H_2O$. Crystal system, orthorhombic. Density, 3.21 g/cm³; refractive index, 1.6125. Occurrence: UK.

Stoklosar. Trademark of Stoklosar Marble Quarries Limited (Canada) for terrazzo chips.

Stol. Trade name of KM-Kabelmetall AG (Germany) for a series of copper alloys containing 0-2.3% iron, 0.15% tin, 0.1-0.6% magnesium, 0-0.1% zinc, 0.05% silver and 0.03-0.05% phosphorus. Used in electronics for lead frames and other applications.

Stollberg brass. A free-cutting brass containing 32.8% zinc, 0.4% tin, 2% lead, and the balance copper.

stolzite. A pale yellow white mineral of the scheelite group composed of lead tungstate, $PbWO_4$. It can be made synthetically by precipitation from solutions of lead nitrate and sodium tungstate. Crystal system, tetragonal. Density, 8.34 g/cm³; refractive index, 2.27.

stone. An individual block, mass or fragment of consolidated rock used widely in building and road construction.

Stoneblast A. Trade name for a 25% zirconia-alumina abrasive for special applications including frosting and lettering on granite and marble monuments and ornaments.

Stone Bronze. Trade name of Stone Manganese–J. Stone & Company Limited for a corrosion-resistant bronze containing 58% copper, 39% zinc, 1.5% iron, 0.8% aluminum, 0.5% manganese and 0.2% tin. Used for hardware, nuts, bolts, screws, and marine propellers.

stone china. A glazed or unglazed, nonporous dinnerware made from clay that is capable of vitrification. See also ironstone china.

stone-cutting sand. A screened, but ungraded abrasive sand with tough grains of uniform size used for cutting and sawing stone and marble.

Stone Mason. Trademark of U.S.E. Hickson Products Limited (Canada) for an extensive series of products including concrete and masonry crack fillers, instant anchoring cements for setting posts, poles, railings, etc., in concrete and masonry, moisture sealants for interior and exterior walls, fast-setting mortars for patching purposes, water-based acrylic concrete, paving stone sealers, etc.

Stone's Gear Bronze. Trade name of Belmont Metals Inc. (USA) for a wear-resistant bronze containing 88.5% copper, 11% tin, 0.25% lead and 0.25% phosphorus. Used for gears and worm wheels.

stoneware. A hard, nonporous, opaque vitreous or semivitreous ceramic ware fired at a temperature that is higher than that for *earthenware*, but lower than that for *porcelain*. It is made from a blend containing chiefly clay, silica, feldspar and kaolin, and may be gray, yellow or brown in color. It has a fine texture, and excellent chemical resistance, but usually low thermal-shock resistance and tensile strength. Used for laboratory, industrial and residential appliances and equipment, such as tanks, reaction chambers, sinks, fittings, pump parts, chemical containers, etc. See also chemical stoneware.

stoneware clay. A semirefractory plastic clay with long firing range and low fire shrinkage used in the manufacture of stoneware.

stone-washed fabrics. Textile fabrics, such as cotton or jeans, with a soft hand and fading colors purposely produced by washing with stone pebbles.

Stonewell Babbitt. Trade name of United American Metals Corporation (USA) for a babbitt containing varying amounts of lead, tin and antimony. Used for heavy-duty bearings and machinery bearings.

Stoodite. (1) Trademark of Stoody Company (USA) for cast tungsten carbide.

(2) Trade name of Stoody Company (USA) for an abrasion-resistant material containing 4% carbon, 4% manganese, 1.5% silicon, 31% chromium, and the balance iron. Used for hardfacing rods.

Stoody. Trade name of Stoody Company (USA) for an extensive series of hardfacing alloys supplied in the form of powders, wires and electrodes and including various nickel- and cobalt-base alloys, tungsten carbides in steel binders, austenitic manganese steels, and austenitic, ferritic and martensitic stainless steels.

stop-off. See resist (5).

Stopox. Trade name of SB Boron Corporation (USA) for a refractive additive composed of a high-quality metal alloy blended with a special boron-magnesium enhancer.

Stoppinosiv. Trade name of Società Italiana Vetro (Italy) for staple glass fibers used in the manufacture of tapes.

Stopray. Trade name of SA Glaverbel (Belgium) for a heat-reflective glass having one surface coated with a thin metallic film.

Storalon. Trade name for silicon carbide.

Storm King. Trademark of U.S.E. Hickson Products Limited (Canada) for a series of waterproof and sun-reflecting coatings for recreational-vehicle exteriors (walls, roofs, etc.) that help reduce interior temperatures.

Stovall. Trade name for a corrosion-resistant dental alloy containing 71-73% gold, 15-17% copper, 13-14% nickel and 2% platinum.

stove lacquer. A lacquer that hardens by the application of heat, e.g., from an infrared lamp or an oven.

straetlingite. A mineral composed of calcium aluminum silicate octahydrate, $Ca_2Al_2SiO_7 \cdot 8H_2O$. It can also be made synthetically. Crystal system, monoclinic. Density, 1.90 g/cm³. Occurrence: Germany.

straight brasses. A group of wrought binary alloys containing about 60-95% copper and 5-40% zinc. It includes gilding metal (95% copper), commercial bronze (90% copper), jewelry bronze (87.5% copper), red brass (85% copper), low brass (80% copper), cartridge brass (70% copper), yellow brass (65% copper) and Muntz metal (60% copper). For straight brasses there is a direct relationship between the copper content and the density, melting point, Young's modulus and electrical and thermal conductivities, and an inverse relationship between the copper content and the tensile strength, hardness, work-hardening tendency, and coefficient of thermal expansion. They have good elongation, fair to excellent corrosion resistance, fair to poor machinability, excellent cold workability (except Muntz metal), and good hot workability.

straight brick. A rectangular brick that is up to 343 mm (13.5 in.) long and has a width greater than its thickness.

straight carbon steels. See plain-carbon steels.

straight chromium steels. A group of low-alloy steels (AISI-SAE 50xx, 51xx and 52xx) containing chromium as the only alloying element other than carbon, manganese and silicon, as contrasted with other chromium steels containing additional alloying elements, such as molybdenum (AISI-SAE 41xx) or vanadium (AISI-SAE 61xx). See also chromium steels.

straight-grained wood. Wood having the fibers running parallel to the axis of the piece.

strained silicon. A relatively recent form of silicon produced from ordinary silicon by deposition onto a substrate whose atoms are spaced farther apart than those of the silicon. Thus the atoms of the deposited silicon lattice are stretched to line up with the substrate atoms. This form of silicon offers less resistance to electron flow, and is used for the manufacture of ultra-fast microchips.

straited hardboard. A special *hardboard* panel with machined-in shallow, randomly-spaced surface grooves running parallel to the long dimension of the panel.

straited plywood. A special *plywood* panel with machined-in shallow, randomly-spaced surface grooves running parallel to the long dimension of the panel.

Straits tin. A high-purity commercial tin (99.895%) used for tinplate, foil, pewter, etc.

strand. (1) A general term for any untwisted bundle of continuous filaments, slivers, tows, yarns, etc.

(2) In the steel industries, a number of steel filaments or wires twisted together, e.g., for making steel wire rope.

(3) In the glass industries, a number of continuous glass filaments made into an untwisted bundle or assembly.

(4) In the textile industries, a general term for a linear textile material.

strand castings. See continuous castings.

Strandofoam. Trademark of Dow Chemical Company (USA) for a series of high-performance foam materials.

stranskiite. A blue mineral composed of copper zinc arsenate, $(Zn,Cu)_3(AsO_4)_2$. It can also be made synthetically. Crystal system, triclinic. Density, 5.23 g/cm³; refractive index, 1.842. Occurrence: Southwest Africa.

strashimirite. A white to pale green mineral composed of copper arsenate hydroxide pentahydrate, $Cu_8(AsO_4)_4(OH)_4 \cdot 5H_2O$. Crystal system, monoclinic. Density, 3.81 g/cm³. Occurrence: Bulgaria.

Strass. Trade name of SA Glaverbel (Belgium) for a patterned glass.

Strata. Trade name for rayon fibers and yarns used for textile fabrics.

Stratabond. (1) Trademark of Lubrizol Corporation (USA) for a series of organic phosphate coatings for spray, immersion, flow-coat or roller-coat application to aluminum, steel and galvanized steel in preparation for subsequent painting operations.

(2) Trademark of Technisand, Inc. (USA) for chemically coated sand used in oil and gas production.

(3) Trade name of Stratford-Cookson (UK) for a dental bonding agent.

StrataShield. Trade name of Tnemec Company Inc. (USA) for a line of polymer floor coatings.

Strata-Slub. Trade name for rayon fibers and slub yarns used for textile fabrics.

Stratella. Trade name for rayon fibers and yarns.

Stratford. Trade name of Perma Paving Stone Company (Canada) for paving stones supplied in regular, double regular, tapered, circle and square units in several colors including salmon, red brown, gray and terra cotta. Used for driveways, walkways, etc.

Stratifil. Trade name of Société du Verre Textile SA (France) for glass-fiber roving made from *Silionne* continuous filament yarns.

Stratimat. Trade name of Société du Verre Textile SA (France) for chopped strand mats made from *Silionne* continuous filament glass yarns.

Stratipreg. Trade name of Société du Verre Textile SA (France) for textile glass fiber prepregs.

Strativerre. Trade name of Cotton Frères & Compagnie (France) for glass fabrics used for fiber-reinforced plastics.

Stratolux. (1) Trade name of Plyglass Limited (UK) for a composite glass consisting of an upper leaf of rough, cathedral-type glass and a lower leaf of transparent glass.

(2) Trade name of Atotech USA Inc. for a bright nickel electrodeposit and plating process.

Strator. Trade name for a low-gold dental casting alloy.

Strat-O-Sheen. Trade name of LeaRonal Inc. (USA) for burnishing compounds used in metal finishing.

Stratosphere. Trade name for a dental amalgam alloy.

Stratton. Trade name of United Clays, Inc. (USA) for a strong, plastic clay with low organic content used in the ceramic industries for various extrusion and pressing applications.

Strauss Metal. Trade name of Ziv Steel & Wire Company (USA) for a cemented tungsten carbide with cobalt binder used for high-speed cutting tools.

strawboard. A low-grade *pasteboard* made from straw or *straw fibers* and used as rigid thermal insulation.

straw cotton. A stiff, bristly, heavily starch-sized cotton thread sometimes used in the textile and allied industries as a substitute for straw.

straw fibers. Pale-yellow, relatively short cellulose fibers obtained from the stalks of stems of cereal plants, such as wheat and rice. They can be pulped and made into coarse paper and blended with other fibers to yield high-quality paper. Straw and straw fibers are also used for the manufacture of low-cost paperboard (strawboard), can be woven into hats, baskets and bottle covers, and are sometimes used for thermal insulation applications.

Strawn. Trade name for rayon fibers and yarns used for textile fabrics.

straw yarn. A monofilament yarn that is usually made by an extrusion process and resembles natural straw in cross section and appearance.

stream tin. Tin ore obtained by washing sand and gravel containing the mineral *cassiterite*.

strelkinite. A yellow mineral of the carnotite group composed of sodium uranyl vanadium oxide hexahydrate, $Na_2(UO_2)_2V_2O_8 \cdot 6H_2O$. Crystal system, orthorhombic. Density, 3.70 g/cm³; refractive index, 1.881. Occurrence: Russian Federation.

Strenes. Trade name of Advance Foundry Company (USA) for a series of cast irons containing about 2.8-3.% carbon, 1.5-2.2% silicon, up to 1% manganese, varying amounts of nickel, chromium and molybdenum, and the balance iron. Used for tools, dies, fixtures and bushings.

strengite. A lavender-puple to pale red mineral of the variscite group that is composed of iron phosphate dihydrate, $Fe(PO_4) \cdot 2H_2O$, and sometimes also contains aluminum. Crystal system, orthorhombic. Density, 2.84 g/cm³; refractive index, 1.720. Occurrence: Germany.

Strenlite. Trade name of Stelco Inc. (Canada) for a high-tensile steel containing up to 0.28% carbon, 1.1-1.6% manganese, up to 0.31% silicon, 0.2% or more copper, and the balance iron. Used for structural applications, e.g., riveted or bolted bridges.

stress-graded lumber. Softwood construction lumber of any width with a minimum thickness of 5 cm (2 in.) that is intended for structural applications, and therefore must have specified requirements with respect to its mechanical properties including strength and stiffness. Other critical considerations include the number, size and locations of defects (e.g., knots, splits, shakes, etc.), and the slope of the grain. Examples of stress-graded lumber products include beams, decking, posts, stringers, timbers and structural boards. See also non-stress-graded lumber.

Stressproof. Trademark of Stelco Inc. (Canada) for a free-machining steel containing 0.4-0.48% carbon, 1.35-1.65% manganese, up to 0.04% phosphorus, 0.24-0.33% sulfur, 0.15-0.3% silicon, up to 0.25% copper, and the balance iron. Available in the form of high-strength cold-drawn steel bars, it has high consistency and wear resistance, low warpage and excellent machinability. Used for bolts, shafts, spindles, studs and pins.

stress-relief-annealed steel. A steel heated to a temperature below the eutectoid temperature, held there long enough to attain a uniform temperature, and then slowly air-cooled to room temperature to remove residual stresses (e.g., from quenching or machining operations) and avoid distortion and possible cracking.

Stretch-aire. Trademark of KoSa (USA) for an atmospherically dyeable single polyester stretch yarn with cotton feel that belnds well with other yarns. Used for sportswear including socks, thermal wear, casual wear, lingerie, T-shirts and other clothing.

Stetchcoat. Trade name of Atofina Chemicals Inc. (USA) for in-

mold powder coatings.

Stretch ever. Trade name for a spandex (segmented polyurethane) fiber used for the manufacture of textile fabrics with excellent stretch and recovery properties.

stretch fabrics. Knit or woven fabrics, usually made with stretch or elastomeric yarns, such as *spandex*, having a greater degree of and recovery from stretch than ordinary fabrics.

stretch-spun rayon. Rayon filaments with increased strength produced from moistened, stretched filaments by spinning prior to the final coagulation process.

stretch-spun yarn. A high-tenacity spun yarn produced by a process involving substantial stretching after extrusion.

stretch two-way. Woven or knit fabrics, usually made with an elastic yarn, such as *spandex*, that have excellent elastic properties in both the longitudinal and transverse direction.

stretch yarn. A spun yarn or thermoplastic filament characterized by a high degree of curl and elastic stretch under stress, and fast recovery upon relaxation. For example, *spandex* is a stretch yarn.

Stria. Trade name of Owens-Corning-Ford Company (USA) for acoustical fiberglass tiles having painted surfaces with decorative parallel striations.

Strialine. Trade name for polyester fibers and yarns used for textile fabrics.

Striato. Trade name of Vetreria di Vernante SpA (Italy) for a patterned glass available in narrow- and broad-fluted designs.

Strimek. Trade name for a high-strength low-alloy (HSLA) for precision cold-rolled strip steel.

string. A fine cord intermediate in diameter between a thread and rope, and used for tying and fastening applications

stringhamite. An azure-blue mineral composed of calcium copper silicate dihydrate, $CaCuSiO_4 \cdot 2H_2O$. Crystal system, monoclinic. Density, 3.17 g/cm³; refractive index, 1.717. Occurrence: USA (Utah).

strip. (1) A long, narrow, flat piece of metal, wood, paper or other material.

(2) A flat-rolled metal product, somewhat narrower than a sheet, whose length is many times its width.

(3) A relatively long, narrow metal powder product compacted in a rolling mill.

(4) A thin, narrow board or thin batten of wood.

Stripcoat. Trademark of Bartlett Services, Inc. (USA) for paint-like, strippable coatings based on cellulosics. Available for spray and roll-on applications to control surface contaminations, and in decontamination and masking operations.

strippable coatings. See temporary coatings.

strip steel. See steel strip.

Strofil. Trademark of Vinisa (Argentina) for polypropylene monofilaments.

Stroh. Trade name of SPS Industries Inc. (USA) for a wear- and shock-resistant steel containing 0.4% carbon, and the balance iron. Used for steel castings.

stromeyerite. A steel-gray or blue mineral composed of silver copper sulfide, AgCuS. Crystal system, orthorhombic. Density, 6.20-6.25 g/cm³. Occurrence: Siberia, Canada (Ontario), USA (Arizona).

Strong Fibro. Trade name for high-tenacity rayon fibers.

strong silver protein. See silver protein.

strontia. See strontium oxide.

strontianite. A colorless, white, gray, yellowish or greenish mineral with a vitreous luster. It belongs to the aragonite group and is composed of strontium carbonate, $SrCO_3$. Crystal system, orthorhombic. Density, 3.68-3.79 g/cm³; hardness, 3.5-4.0 Mohs; refractive index, 1.663. Occurrence: Germany, Mexico, Scotland, USA (California, New York, Washington). Used as a source of strontium and strontium compounds.

strontioborite. A colorless mineral composed of strontium calcium magnesium borate dihydrate, $(Sr,Ca,Mg)B_8O_{13} \cdot 2H_2O$. It can also be made synthetically. Crystal system, monoclinic. Density, 2.81 g/cm³; refractive index, 1.510. Occurrence: Russian Federation, Ukraine.

strontiodressite. A white mineral of the dundasite group composed of strontium aluminum carbonate hydroxide monohydrate, $(Sr,Ca)Al_2(CO_3)_2(OH)_4 \cdot H_2O$. Crystal system, orthorhombic. Density, 2.71 g/cm³; refractive index, 1.583. Occurrence: Canada (Quebec).

strontioginorite. A colorless mineral composed of strontium calcium borate octahydrate, $(Sr,Ca)_2B_{14}O_{23} \cdot 8H_2O$. Crystal system, monoclinic. Density, 2.25 g/cm³; refractive index, 1.524. Occurrence: Germany.

strontium. A silvery white to pale yellow, soft metallic element of Group IIA (Group 2) of the Periodic Table (alkaline-earth group). It is commercially available in the form of rods, lumps, pieces, turnings and wire. The most important strontium ores are *celestite* and *strontianite*, and it also occurs in mineral springs. Density, 2.6 g/cm³; melting point, 770°C (1418°F); boiling point, 1375°C (2507°F); hardness, 1.8 Mohs; atomic number, 38; atomic weight, 87.62; divalent; oxidizes quickly in air. Three allotropic forms are known: (i) *Alpha strontium* that has a face-centered cubic crystal structure and exists below 235°C (455°F); (ii) *Beta strontium* that has a hexagonal close-packed crystal structure and is stable between 235 and 540°C (455 and 1004°F); and (iii) *Gamma strontium* that has a body-centered cubic crystal structure and is stable between 540 and 770°C (1004 and 1418°F). *Strontium* is used in alloys, as a getter in electron tubes, in nuclear batteries, phosphorescent paints, ceramics, superconductors and fireworks, and as a beta-radiation source. Symbol: Sr.

strontium-90. A radioactive isotope of strontium with a mass number of 90, usually obtained from the fission products of nuclear reactor fuels, but also occurring in the fallout from nuclear explosions. It has a half-life of 28 years, emits beta radiation, and is used as a beta-radiation source in thickness gages, in electronics for studying strontia in vacuum tubes, for the elimination of static charges, as a phosphor activator, as a source of ionizing radiation in luminous paint, and in nuclear batteries. Also known as *radiostrontium*.

strontium acetate. A white, hygroscopic, crystalline powder (99.9+% pure) obtained by a reaction between strontium hydroxide and acetic acid and subsequent crystallization. Density, 2.099 g/cm³; melting point, decomposes. Used as an intermediate for strontium compounds, in the preparation of catalysts, and in organometallic and superconductor research. Formula: $Sr(C_2H_3O_2)_2$.

strontium aluminate. Any of the following compounds of strontium oxide and aluminum oxide used in ceramics, electronics and materials research: (i) *Strontium monoaluminate*. Melting point, 2010°C (3650°F). Formula: $SrAl_2O_4$; and (ii) *Strontium dialuminate*. Density, 3.03 g/cm³; melting point, 1770°C (3218°F). Formula: $SrAl_4O_7$.

strontium-aluminum. A series of master alloys of strontium and aluminum, or strontium, aluminum and magnesium used chiefly to efficiently and effectively introduce strontium into aluminum alloys. Typical compositions include 90% strontium and 10% aluminum, and 45% strontium, 35% aluminum and 25% magnesium.

strontium-aluminum silicate. A compound of strontium oxide, aluminum oxide and silicon dioxide used in ceramics and materials research. Density, 3.12 g/cm³; melting point, 1660°C (3020°F); hardness, 5-7 Mohs. Formula: $SrAl_2Si_2O_8$.

strontium-apatite. A pale green to yellowish green or colorless mineral of the apatite group composed of strontium hydroxide phosphate, $Sr_5(PO_4)_3OH$. Crystal system, hexagonal. Density, 3.84 g/cm³; refractive index, 1.651. Occurrence: Russian Federation.

strontium arsenate. A compound of strontium oxide and arsenic pentoxide used in ceramics and electronics. Density, 4.60 g/cm³; melting point, 1637°C (2979°F). Formula: $3SrO \cdot As_2O_5$.

strontium boride. A compound of strontium and boron available as black crystals or dark crystalline powder (99+% pure). Density, 3.39-3.42 g/cm³; melting point, 2235°C (4055°F). Used in nuclear absorption control rods, for high-temperature insulation, as a nuclear control additive, and in energy sources when using the radioisotope (strontium-90). Formula: SrB_6. Also known as *strontium hexaboride*.

strontium bromide. White, hygroscopic, crystals or powder (99+% pure). Crystal system, tetragonal. Density, 4.21 g/cm³; melting point, 643°C (1189°F). Used for strontium salts, as a laboratory reagent, and in chemistry, materials research and medicine. Formula: $SrBr_2$. Also known as *anhydrous strontium bromide*.

strontium bromide hexahydrate. White crystals (99+% pure). Crystal system, hexagonal. Density, 2.35 g/cm³; melting point, loses $4H_2O$ at 89°C (192°F) and remaining $2H_2O$ at 180°C (356°F). Used for strontium salts, as a laboratory reagent, and in medicine. Formula: $SrBr_2 \cdot 6H_2O$.

strontium carbide. A compound of strontium and carbon. Crystal system, tetragonal. Density, 3.19 g/cm³; melting point, about 2000°C (3630°F). Used in chemistry, ceramics and materials research. Formula: SrC_2.

strontium carbonate. A white powder (98+% pure) obtained from the mineral *celestite* by fusing with sodium carbonate. It is also available in high-purity grades (99.995+%). Crystal system, orthorhombic. Density, 3.70 g/cm³; melting point, decomposes at 1100-1340°C (2012-2444°F). Used in the manufacture of iridescent glass, in radiation-resistant glass for color television tubes, for introducing strontium oxide into glass, glazes and enamels, in low-temperature leadless glazes, in ceramic ferrites, and in the preparation of strontium-containing superconductors. Formula: $SrCO_3$.

strontium chloride. White, hygroscopic, crystalline needles or powder (99+% pure). Crystal system, cubic. Density, 3.05 g/cm³; melting point, 873°C (1603°F); boiling point, 1250°C (2280°F); refractive index, 1.650. Used for strontium salts, in electron tubes, and in pyrotechnics. Formula: $SrCl_2$. Also known as *anhydrous strontium chloride*.

strontium chloride hexahydrate. White crystals or needles (99+% pure). Crystal system, trigonal. Density, 1.93 g/cm³; melting point, loses $6H_2O$ at 115°C (239°F). Used for strontium salts, as a chemical reagent, and in biochemistry. Formula: $SrCl_2 \cdot 6H_2O$.

strontium chromate. A light yellow pigment with corrosion-resistant and rust-inhibiting characteristics. Crystal system, monoclinic. Density, 3.84 g/cm³; good heat resistance and light stability; low reactivity in highly acidic media. Used in protective coatings for metals, as a colorant in polyvinyl chloride resins, and in electroplating baths. Formula: $SrCrO_4$.

strontium dialuminate. See strontium aluminate (ii).

strontium dioxide. See strontium peroxide.

strontium dioxide octahydrate. See strontium peroxide octahydrate.

strontium-doped lanthanum manganites. See lanthanum strontium manganites.

strontium dysprosium sulfide. A light ivory-colored crystalline compound of strontium sulfide and dysprosium sulfide. Crystal system, orthorhombic. Melting point, 1850°C±30°C (3362°F ±54°F). Used in ceramics and materials research. Formula: $SrDy_2S_4$.

strontium erbium sulfide. A pale rose crystalline compound of strontium sulfide and erbium sulfide. Crystal system, orthorhombic. Density, 5.47 g/cm³; melting point, 1891°C±30°C (3436°F±54°F); hardness, 315-385 Vickers; band gap, 3.54 eV. Used in ceramics, electronics, and materials research. Formula: $SrEr_2S_4$.

strontium 2-ethylhexanoate. An off-white solid compound available in superconductor grades for the preparation of superconducting thin films. Formula: $Sr(O_2C_8H_{15})_2$. See *strontium octoate*.

strontium ferrite. An anisotropic, magnetically hard ceramic material usually made by pressing micro-sized blended strontium ferrite powder and sintering at prescribed temperatures to obtain desired magnetic properties. It has a hexagonal crystal structure, a Curie temperature of approximately 460°C (860°F), high coercive force, a high magnetic energy product, and a maximum service temperature of 400°C (750°F). Used for permanent magnets. Formula: $SrO\cdot6Fe_2O_3$ ($SrFe_{12}O_{19}$). Also known as *strontium hexaferrite*.

strontium fluoride. White, hygroscopic crystals or powder (99+% pure). Crystal system, cubic. Density, 4.24 g/cm³; melting point, above 1450 (2640°F); boiling point, 2460°C (4510°F); refractive index, 1.442. Used for single-crystal components in lasers, often doped with samarium, for electronic and optical applications, as a substitute for other fluorides, and in high-temperature dry-film lubricants. Formula: SrF_2.

strontium gadolinium sulfide. A light ivory compound of strontium sulfide and gadolinium sulfide. Crystal system, cubic. Density, 5.63 g/cm³ (theoretical); melting point, 1980°C±30°C (3596°F±54°F); band gap, 2.7 eV. Used as a semiconductor, and in ceramics and materials research. Formula: $SrGd_2S_4$.

strontium hexaboride. See strontium boride.

strontium hexaferrite. See strontium ferrite.

strontium hexafluoroactylacetonate. A metal alkoxide available in the form of white crystals for use in the chemical vapor deposition of superconductors. Formula: $Sr(O_2C_5HF_6)_2$. Also known as *strontium hexafluoro-2,4-pentanedionate*.

strontium holmium sulfide. A crystalline compound of strontium sulfide and holmium sulfide. Crystal system, orthorhombic. Density, 5.37 g/cm³; melting point, 1930°C±30°C (3506°F±54°F); hardness, 357-510 Vickers. Used as a semiconductor, and in ceramics and materials research. Formula: $SrHo_2S_4$.

strontium hydrate. See strontium hydroxide.

strontium hydride. Orthorhombic crystals with a density of 3.26 g/cm³ and a melting point of 1050°C (1922°F). Used in chemistry and materials research. Formula: SrH_2.

strontium hydroxide. White, deliquescent crystals (95+% pure) with a density of 3.625 g/cm³ and a melting point, 375°C (707°F). Used chiefly as a reagent, as a stabilizer in adhesives, plastics and glasses, and in lubricants, soaps and greases. Formula: $Sr(OH)_2$. Also known as *strontium hydrate*.

strontium hydroxide octahydrate. White, or colorless, hygroscopic, tetragonal crystals (95+% pure) with a density of 1.90 g/cm³. Used chiefly as a reagent, as a stabilizer in adhesives, plastics and glasses, and in lubricants, soaps and greases. Formula: $Sr(OH)_2\cdot8H_2O$.

strontium iodide. White hygroscopic crystals, or off-white powder (99+% pure). Density, 4.55 g/cm³; melting point, 538°C (1000°F); boiling point, decomposes at 1773°C (3223°F). Used chiefly as a chemical intermediate and in biochemistry and medicine, e.g., as a source of iodine. Formula: SrI_2.

strontium iodide hexahydrate. White to yellowish white, deliquescent crystals that decompose in moist air. Crystal system, hexagonal. Density, 4.42 g/cm³; melting point, decomposes at 90°C (195°F). Used chiefly as a chemical intermediate, and in biochemistry and medicine, e.g., as a source of iodine. Formula: $SrI_2\cdot6H_2O$.

strontium lanthanum aluminate. An oxide of strontium and aluminum available as a single crystalline substrate. Crystal system, tetragonal. Melting point, 1348°C (2458°F); dielectric constant, 16.8. Used as substrate for the deposition of high-temperature superconductors. Formula: $SrLaAlO_3$. Also known as *strontium lanthanum aluminum oxide (SLAO)*.

strontium lanthanum gallate. An oxide of strontium, lanthanum and gallium available as a single crystalline substrate. Crystal system, tetragonal. Melting point, 1238°C (2260°F); dielectric constant, 300. Used as substrate for deposition of high-temperature superconductors. Formula: $SrLaGaO_3$. Also known as *strontium lanthanum gallium oxide (SLGO)*.

strontium lanthanum sulfide. A compound of strontium sulfide and lanthanum sulfide available as ivory-colored crystals. Crystal system, cubic. Density, 4.82 g/cm³; melting point, 1860°C±30°C (3380°F±54°F); hardness, 380-600 Vickers. Used as a semiconductor, and in ceramics and materials research. Formula: $SrLa_2S_4$.

strontium molybdate. A crystalline powder. Density, 4.0 g/cm³; melting point, 1600°C (2912°F). Used for electronic and optical applications, and corrosion-protective pigments. The single crystal grades are used for solid-state laser applications. Formula: $SrMoO_4$. Also known as *strontium molybdenum oxide (SMO)*.

strontium molybdenum oxide. See strontium molybdate.

strontium monoaluminate. See strontium aluminate (i).

strontium monosilicate. See strontium silicate (i).

strontium monosulfide. Gray, moisture-sensitive crystals or powder (99.9+% pure). Crystal system, cubic. Crystal structure, halite. Density, 3.70 g/cm³; melting point, above 2000°C (3630°F); band gap, 4.1 eV. Used as a semiconductor, in luminous paints with a bluish green glow, and in the manufacture of strontium chemicals. Formula: SrS. Also known as *strontium sulfide*.

strontium neodecanoate. An off-white solid compound available in superconductor grades containing about 16-21% strontium. Used in the preparation of high-temperature superconductors. Formula: $Sr(O_2C_{10}H_{19})_2$.

strontium neodymium sulfide. A light-green compound of strontium sulfide and neodymium sulfide. Crystal system, cubic. Density, 5.18 g/cm³; melting point, 1830°C±30°C (3326°F±54°F); hardness, 455-650 Vickers; band gap, 2.58 eV. Formula: $SrNd_2S_4$. Used as a semiconductor, and in ceramics and materials research.

strontium nitrate. A white or yellowish white, hygroscopic crystalline powder (99+% pure). Crystal system, cubic. Density, 2.986 g/cm³; melting point, 570°C (1058°F); boiling point, 645°C (1193°F). Used in the preparation of lanthanum strontium copper oxide superconductors, in marine and railroad-signal lights, in matches and military flares, as an oxidizer, and in biology, biochemistry and medicine. Formula: $Sr(NO_3)_2$.

strontium nitride. A compound of strontium and nitrogen. Melting point, 1483°C (2700°F). Used as a semiconductor. Formula: Sr_3N_2.

strontium octoate. See strontium 2-ethylhexanoate.

strontium oxide. An oxide of strontium available in technical grades as a brown moisture-sensitive powder, or porous lumps, and in high-purity grades (99.9+%) as a white, moisture-sensitive powder. Crystal system, cubic. Density, 4.7-5.1 g/cm³; melting point, 2430°C (4405°F); boiling point, about 3000°C (5430°F). Used in pigments, as a colorant in glass, in the manufacture of strontium salts, in materials research, and as a desiccant. Formula: SrO. Also known as *strontia*.

strontium peroxide. White, moisture-sensitive crystals or powder obtained by passing oxygen over heated strontium oxide (SrO). Crystal system, tetragonal. Density, 4.56-4.78 g/cm³; melting point, decomposes at 215°C (419°F). Used chiefly in fireworks, bleaching, and as an antiseptic. Formula: SrO_2. Also known as *strontium dioxide*.

strontium peroxide octahydrate. Colorless crystals or crystalline powder. Density, 1.95 g/cm³; melting point, loses $8H_2O$ at 100°C (212°F). Used chiefly in fireworks, bleaching, and as an antiseptic. Formula: $SrO_2·8H_2O$. Also known as *strontium dioxide octahydrate*.

strontium phosphate. A compound of strontium oxide and phosphorus pentoxide used in ceramics and electronics. Density, 4.53 g/cm³; melting point, 1766°C (3211°F). Formula: $Sr_3P_2O_8$.

strontium praseodymium sulfide. A light-green compound of strontium sulfide and praseodymium sulfide. Crystal system, cubic. Density, 5.05-5.18 g/cm³; melting point, 1890±30°C (3434±54°F); hardness, 455-650 Vickers; band gap, 2.58 eV. Formula: $SrPr_2S_4$. Used as a semiconductor, and in ceramics and materials research.

strontium samarium sulfide. A bright-yellow compound of strontium sulfide and samarium sulfide. Crystal system, cubic. Density, 5.4 g/cm³; melting point, 1880°C±30°C (3416°F±54°F); band gap, 2.32 eV. Formula: $SrNd_2S_4$. Used as a semiconductor, and in ceramics and materials research.

strontium scandate. A compound of strontium oxide and scandium oxide with a density of 4.59 g/cm³, used in ceramics. Formula: $SrSc_2O_4$.

strontium selenide. White, cubic crystals with a density of 4.54 g/cm³ and a melting point of 1600°C (2910°F), used in chemistry and materials research. Formula: SrSe.

strontium silicate. Any of the following compounds of strontium oxide and silicon dioxide used in ceramics and materials research: (i) *Strontium monosilicate.* Colorless prisms. Density, 3.65 g/cm³; melting point, 1580°C (2876°F). Formula: $SrSiO_3$; and (ii) *Distrontium silicate.* Density, 3.84 g/cm³; melting point, above 1705°C (3100°F); hardness, 5-7 Mohs. Formula: Sr_2SiO_4.

strontium silicide. Silver-gray, cubic crystals with a density of 3.35 g/cm³ and a melting point of 1100°C (2012°F), used in chemistry and materials research. Formula: $SrSi_2$.

strontium stannate. A compound of strontium oxide and stannic oxide. Melting point, above 1400°C (2550°F). Used as an ingredient in titanate bodies to reduce the Curie temperature. Formula: $SrSnO_3$.

strontium sulfate. Hygroscopic, colorless or white crystals, or powder (99+% pure) usually obtained by grinding the mineral *celestite*. Crystal system, orthorhombic. Density, 3.96 g/cm³; melting point, 1605°C (2920°F); refractive index, 1.622. Used as a fining agent in the production of crystal glasses, in the production of iridescence on the surfaces of glass and pottery glazes, in pigments, as a brightening agent in paints, and in paper manufacture. Formula: $SrSO_4$.

strontium sulfide. See strontium monosulfide; strontium tetrasulfide.

strontium terbium sulfide. A light gray, crystalline compound of strontium sulfide and terbium sulfide. Crystal system, orthorhombic. Melting point, 1859°C±30°C (3378°F±54°F); band gap, 2.88 eV. Used as a semiconductor, and in electronics, ceramics and materials research. Formula: $SrTb_2S_4$.

strontium tetrasulfide. Reddish, water-soluble crystals with a melting point of 25°C (77°F), available in anhydrous form, but usually supplied as the hexahydrate. Formula: SrS_4 (anhydrous); $SrS_4·6H_2O$ (hexahydrate). Also known as *strontium sulfide*.

strontium thulium sulfide. A crystalline compound of strontium sulfide and thulium sulfide. Crystal system, orthorhombic. Melting point, 1780°C±30°C (3236°F±54°F). Used in ceramics and materials research. Formula: $SrTm_2S_4$.

strontium titanate. Off-white powder (99+% pure) or single crystals (99.5+% pure). Crystal system, cubic. Crystal structure, perovskite. Density, 4.81-5.12 g/cm³; melting point, 2060°C (3740°F); dielectric constant, 300; negative temperature coefficient. Used as a dielectric material in electronics and electrical insulation applications, as substrates for superconductor applications, in low-melting glazes, as an additive to barium titanate to decrease the Curie temperature, in other titanates, in temperature-compensating materials, and in ceramics. Formula: $SrTiO_3$. Also known as *strontium titanium oxide (STO)*.

strontium titanium. A double metal alkoxide supplied as a liquid with strontium and titanium contents of 6.4-6.6% and 3.5-3.6%, respectively. It has a strontium-to-titanium metal ratio of 1:1 and a density of 0.95-1.00 g/cm³. Used in sol-gel and coating applications. Formula: $SrTi(OR)_x$.

strontium titanium oxide. See strontium titanate.

strontium uranate. A compound of strontium oxide and uranium trioxide with a melting point of about 1800°C (3270°F) used in ceramics. Formula: $SrUO_4$.

strontium white. A white paint pigment composed of strontium sulfate ($SrSO_4$).

strontium ytterbium sulfide. A crystalline compound of strontium sulfide and ytterbium sulfide. Crystal system, orthorhombic. Melting point, 1856°C±30°C (3373°F±54°F). Used in ceramics and materials research. Formula: $SrYb_2S_4$.

strontium yttrium sulfide. A light gray crystalline compound of strontium sulfide and yttrium sulfide. Crystal system, orthorhombic. Melting point, 1836°C±30°C (3437°F±54°F); band gap, 3.1 eV. Used as a semiconductor, and in ceramics and materials research. Formula: SrY_2S_4.

strontium zirconate. A white powder (95+% pure). Density, 5.48 g/cm³. melting point, above 2600°C (4710°F). Used in ceramics, electronics, and in small amounts in dielectric compositions to reduce the Curie temperature. Formula: $SrZrO_3$.

strontium zirconium. A double metal alkoxide supplied as a liquid with strontium and zirconium contents of 5.9-6.1% and 6.2-6.4%, respectively. It has a strontium-to-zirconium metal ratio of 1:1 and a density of 0.96-1.01 g/cm³. Used in sol-gel and coating applications. Formula: $SrZr(OR)_x$.

Structo-Foam. Trademark of Stauffer Chemical Company (USA) for rigid polystyrene foam used in the thermal insulation of buildings, refrigerator cars and trucks, in flotation devices and moorings, and as fillers in shipping containers.

Structolite. Trademark of Canadian Gypsum Company Limited for gypsum plaster.

Structur. Trade name for dental resins used for temporary bridge and crown restorations.

structural adhesives. A group of one- or two-component engineering adhesives, usually of the polyimide, bismaleimide, modified epoxy, neoprene rubber, acrylic or silicone-type, that are designed to transfer loads between the members making up a joint (the so-called "adherends"). In contrast to ordinary adhesives, they possess excellent bond strength, are resistant to humidity, creep, solvents, and high and/or low temperatures. The peel and shear strengths vary greatly with composition, temperature and application. Used for bonding metals to each other, or to ceramics, wood and plastics, and in assembling components of composites into larger structures. Abbreviation: SA.

structural board. A stress-graded board of softwood lumber used for structural applications.

structural building products. Building units that are either of the load-bearing type (i.e., capable of carrying loads in addition to their own weight) or the non-load-bearing type (i.e., capable of carrying only their own weight). Also known as *structural products*.

structural ceramic composites. A group of composite materials with ceramic matrices, that may contain one or more discrete crystalline or amorphous phases, reinforced with carbon, ceramic, glass or metal fibers or whiskers. Used for high-performance structural engineering applications.

structural-clay facing tile. A structural clay tile with good appearance and structural integrity suitable for use on interior or exterior walls, column, partitions, etc.

structural-clay products. Conventional engineering ceramics, such as bricks, tiles, and sewer pipes, composed essentially of clay, and used in applications requiring structural integrity.

structural-clay tile. A hollow masonry building unit of good structural integrity made of burnt clay and having parallel cells. Used as a facing, furring, floor, fireproofing and/or header tile, or as a load-bearing or partition tile.

structural composite lumber. A group of engineered wood products made by bonding wood veneer or strands with a special adhesive, and including *laminated veneer lumber* (LVL), *oriented strand lumber* (OSL) and *parallel strand lumber* (PSL).

structural composite panel. An engineered wood product made in three- or five-layer arrangements bonding veneer face layers to a core composed of *oriented strand board* or other wood-base material, followed by pressing. A three-layer panel contains one wood-fiber core layer sandwiched between two veneer faces, and a five-layer panel has a central veneer crossband. Used for sheathing, siding, etc.

structural composites. A class of composite materials used chiefly for their mechanical properties (e.g., tensile strength and elastic modulus) relevant for structural applications, e.g., laminar composites or sandwich panels. Abbreviation: SC.

structural concrete. Concrete exhibiting characteristics that make it suitable for bearing a structural load, or forming an integral part of a structure.

structural facing unit. A building unit of good overall appearance including color and surface finish that make it suitable for use in those areas of a wall requiring one or more faces to be exposed in the finished state.

structural foams. Expanded plastics of high rigidity based on polyurethane or a thermoplastic resin, such as acrylonitrile-butadiene-styrene, polycarbonate, polyethylene, polyester or a polyphenylene ether blend. They have dense, solid, integral outer skins and porous, foamed cores, and are used in acoustic and thermal insulation, and for exterior and interior automotive components, boat hulls, furniture, etc. Abbreviation: SF. See also cellular plastics; integral skin foam.

structural glass. (1) Glass in the form of hollow blocks often with patterned faces, used as structural building units in walls, windows, partitions, etc.

(2) Clear, opaque or colored flat glass, often ground and polished, used for structural applications.

structural insulating board. See insulation board.

structural laminate. Two or more two-dimensional sheets or panels, each having a preferred high-strength direction, stacked and bonded together such that the orientation of the high-strength direction is different for each successive layer. Such a laminate has high strength in many directions. Examples include plywood, and aligned or continuous fiber-reinforced plastics. Also known as *laminated composite*.

structural lightweight concrete. A *structural concrete* with a low density of about 1.44-1.85 g/cm^3 (0.052-0.067 lb/in.^{3}) made with pumice, foamed slag, expanded clay, or other lightweight aggregates. See also lightweight concrete.

structural materials. Engineering materials, such as ferrous and nonferrous metals, ceramics, glasses, glass-ceramics, composites and thermosetting and thermoplastic polymers, that due to their superior mechanical properties, are used primarily in structural applications.

structural nickel steel. A structural steel (AISI-SAE 2345) containing 0.45% carbon, 0.70% manganese, 3.25% nickel, and the balance iron. It has high tensile strength (up to 690 MPa or 100 ksi), and a minimum elongation of 18%. Used in building construction.

structural plastics. A group of thermosetting and thermoplastic materials used in building and construction for their good mechanical properties. Examples of structural thermosets include aminos, epoxies, phenolics and unsaturated polyesters. Structural thermoplastics include acetals, acrylics, polycarbonates, polyphenylene ethers, polyurethanes, thermoplastic polyesters and nylons. Abbreviation: SP.

structural plywood. A strong, stiff structural wood panel manufactured by bonding together with a synthetic resin (usually a phenol- or resorcinol-formaldehyde) a number of layers (veneers or plies) of wood with the grain direction turned at right angles in each successive layer. Used especially for sheathing, siding, and sanded and concrete forms. See also exterior-type plywood; interior-type plywood; plywood.

structural reaction-injection-molded composites. See reinforced-reaction-injection-molded composites.

structural RIM composites. See reinforced-reaction-injection-molded composites.

structural products. See structural building products.

structural sandwich. An engineering composite made up of two thin, strong outer sheets composed of aluminum, aramid, fiberglass or graphite separated by a layer of a lightweight material, such as a foamed plastics, honeycomb, etc. It has high strength and stiffness, and is used for load-bearing applications especially in the aircraft, spacecraft and shipbuilding industries.

structural sealants. A class of sealants that are capable of transferring dynamic and/or static loads across joint members and are therefore suitable for structural applications,

structural shapes. Metal sections of standard design including angles, beams, channels, beams, tees, zees and piling used in building and construction.

structural silicon steels. (1) A group of high-strength steels containing up to 0.4% carbon, 0-0.05% sulfur, 0-0.04% phosphorus, 0.6% or more manganese and 0.2% silicon, with the balance being iron. Additions of up to 0.2% copper may be introduced to increase corrosion resistance. Used for structural applications, such as bridges, derricks, plates and shapes.

(2) A group of alloy steels containing 0.5-0.65% carbon, 1.40-2.20% silicon, 0.65-1.00% manganese, up to 0.65% chromium, 0.035% phosphorus and 0.040% sulfur, respectively, and the balance iron. Usually, this term refers to the high-strength steels in the AISI-SAE 92xx series used for structural applications and automotive leaf springs. Also known as *silicon-manganese steels*.

structural steels. A group of steels usually made by the basic-oxygen-furnace or electric-furnace process, and used as constructional materials in shipbuilding, bridge and building construction, bolted, riveted or welded structures, for reinforcing bars in concrete, in structural engineering, mechanical engineering (e.g., axles, shafts, pins, keys, gears, normally or highly stressed parts, etc.), and for automobile bodies, truck frames, brackets, derricks, crane booms, dock and sea walls, bulkheads, rail cars, and mining equipment. Characteristic properties of structural steels include high tensile and yield strengths, and good formability and weldability. They are supplied in the form of structural shapes (angles, beams, channels, piling, tees, zees, etc.), bars (e.g., rounds, squares, hexagons, octagons, flats, triangulars, etc.), wire and wire products, pipes and tubes, plates, sheets and strips. Also known as *constructional steels*. See also high-strength structural steels; high-strength low-alloy steels.

structural thermoplastic composites. Fiber-reinforced thermoplastic composites (e.g., acetals, acrylics, polycarbonates, polyesters or nylons) that have mechanical performance characteristics making them suitable for structural applications in building and construction. Abbreviation: STC.

structural wood panels. A group of engineered wood products used especially for structural applications, such as flooring, siding, wall and roof sheathing, stair treads and risers, concrete forms, furniture, boats, crates, pallets, etc. The most common types of structural wood panels are *composite panels*, *plywood* and *oriented strand board*.

Strudon. Trademark of Strudex Fibres Limited (USA) for polypropylene yarns used for upholstery, carpets, braids, etc.

strueverite. A mineral of the rutile group composed of iron tantalum titanium oxide, $(Ti,Ta,Fe)O_2$. Crystal system, tetragonal. Density, 5.25 g/cm^3; refractive index, 2.50. Occurrence: USA (South Dakota).

strunzite. A straw-yellow to brownish yellow mineral composed of manganese iron phosphate hydroxide octahydrate, $MnFe_2(PO_4)_2(OH)_2 \cdot 8H_2O$. Crystal system, monoclinic. Density, 2.52 g/cm^3; refractive index, 1.670. Occurrence: Germany, USA.

struvite. A colorless to yellow mineral composed of ammonium magnesium phosphate hexahydrate, $NH_4MgPO_4 \cdot 6H_2O$. Crystal system, orthorhombic. Density, 1.71 g/cm^3; refractive index, 1.496; hardness, 2 Mohs. Occurrence: Iceland.

Strux. Trade name of Strux Corporation (USA) for a lightweight plastic foam based on cellulose acetate and made by extrusion.

STRUXural Foam. Trade name of Strux Corporation (USA) for rigid polyurethane foam supplied in the form of blocks, boards and sheets.

Stryton. Trade name for nylon fibers and yarns.

Stubs. Trade name of Peter Stubs (UK) for silver steel containing about 1.1% carbon, 0.2% silicon, 0.35% manganese, 0.045% sulfur, 0.045% phosphorus, and the balance iron. Commonly supplied centerless ground in the as rolled condition for use in the manufacture of gages and pins.

Stuccato. Trademark of Canexel Hardboard Inc. (Canada) for hardboard used as siding.

stucco. (1) A mixture of trowelable, plasterlike consistency made from Portland cement, sand and a small percentage of lime by blending with an adequate amount of water. It is applied to exterior walls or other exterior surfaces on buildings or structures, and upon setting forms a hard, smooth or rough finish. The finish coat may be tinted by adding coloring, or the surface may be painted with a suitable material.

(2) A fine plaster used on ceilings and walls and for architectural decorations. It is made of white marble powder, fine sand, gypsum and water, or gypsum, glue and water.

Stuco. Trademark for a liquid synthetic resin mixture used as a water substitute in the mixing of gypsum mortar or cement.

Stuc-O-Kote. Trade name of Contact Paint & Chemical Corporation (USA) for masonry and swimming pool paints.

Studal. French trade name for an aluminum alloy containing 1.3% manganese and 1% magnesium, used for light-alloy parts.

studtite. A colorless mineral composed of uranium oxide tetrahydrate, $UO_4 \cdot 4H_2O$. It can also be made synthetically. Crystal system, monoclinic. Density, 3.58 g/cm^3; refractive index, 1.555. Occurrence: Germany.

stuetzite. A lead-gray synthetic mineral composed of silver telluride, Ag_5Te_3. Crystal system, hexagonal. Density, 7.61 g/cm^3.

stuffing box alloy. A corrosion-resistant casting alloy composed of 61.5% copper, 15.5% nickel, 11% zinc, 10% lead and 2% tin. Used for stuffing boxes.

stumpflite. A creamy white mineral of the nickeline group composed of antimony bismuth platinum, Pt(Sb,Bi). Crystal system, hexagonal. Density, 13.52 g/cm^3. Occurrence: South Africa.

Stupalith. Trade name for lithium aluminosilicate ceramics made from a mixture of lithium-bearing minerals and clay, sometimes with the addition of other ceramic materials. They possess excellent thermal-shock resistance, and have a wide range of negative and positive thermal expansion coefficients.

Stupalox. Trade name for a tough, wear-resistant, hot-pressed aluminum oxide. It has a density of 3.95 g/cm^3 (0.143 $lb/in.^3$), very high compressive and transverse rupture strength, and is used for plug and ring gages.

Sturdicast. Trade name of J.F. Jelenko & Company (USA) for an extra-hard, heat-treated dental casting alloy (ADA type IV) containing 60% gold, 22% silver and 4% palladium. It provides high strength, low weight, good resiliency, and a lustrous polish. Used for hard inlays, partial dentures, thin crowns, and fixed bridgeworks.

Sturd-I-Floor. Trademark of Georgia-Pacific (USA) for *Plytanium* southern pine plywood used as a quite, stiff, single-layer subflooring material under hardwood flooring, ceramic or vinyl tile, or carpeting.

sturmanite. A bright yellow mineral of the ettringite group composed of calcium iron boron sulfate hydroxide hydrate, $Ca_6Fe_2(SO_4)_2[B(OH)_9](OH)_{12} \cdot 26H_2O$. Crystal system, hexagonal. Density, 1.85 g/cm³; refractive index, 1.5. Occurrence: South Africa.

Stycast. Trademark of Emerson & Cuming (USA) for epoxy resins and molding compounds. The casting resins are unfilled or filled with aluminum, glass, minerals or silica, and supplied in general-purpose, lightweight, fire-retardant and thermally conductive grades. The molding compounds are filled with minerals and/or glass fibers, and supplied in standard, high-heat or fire-retardant grades.

Stycond-X. Trade name for a black, extrusion-grade carbon-conductive amorphous thermoplastic based on a styrenic polymer blend. It has low shrinkage, and is used for profile extrusion, sheet extrusion/thermoforming, and for thin-gage packaging roll stock.

Stylefluid CF. Trade name of Metalor Technologies SA (Switzerland) for a hard universal dental alloy containing 58% gold, 8% palladium and 2% platinum. Used for low-fusing ceramic restorations.

Stylon. Trade name for nylon fibers and yarns.

stylus metal. A lead babbitt containing 10% antimony and 5% tin, and used for bearings.

Styraclear. Trade name of Westlake Plastics Company (USA) for a transparent polystyrene with excellent dielectric properties and good chemical resistance used for electrical components and for tubing and protective equipment for handling corrosive liquids.

Styrafil. Trademark of Wilson-Fiberfill International (USA) for glass-fiber-reinforced polystyrene.

Styraloy. Trademark of Dow Chemical Company (USA) for polystyrene molding resins.

Styramic. Trade name of Monsanto Chemical Company (USA) for polystyrene resins and products.

styrenated alkyd. An alkyd resin reacted with styrene monomer (C_8H_8) and used for protective coatings.

styrenated polyester. An unsaturated polyester reacted with styrene monomer (C_8H_8).

styrene. An aromatic hydrocarbon available as a flammable, colorless, oily liquid (99+% pure) that readily undergoes polymerization when exposed to light, heat or a peroxide catalyst. It has a density of 0.909 g/cm³, a melting point of -31°C (-24°F), a boiling point of 145-146°C (293-295°F), a flash point of 88°F (31°C), an autoignition temperature of 914°F (490°C) and a refractive index of 1.547. Used as a monomer in the manufacture of acrylonitrile-butadiene-styrene, styrene-acrylonitrile and polystyrene resins and styrene-butadiene rubber, in rubber-modified polysty-rene, protective coatings based on alkyds or styrene-butadiene latex, as a copolymer in various synthetic resins, as an interme-diate, and in biotechnology. Formula: C_8H_8. Also known as *phenylethylene; styrene monomer; vinylbenzene.*

styrene-acrylonitriles. A family of random, linear, transparent thermoplastic copolymers of styrene and acrylonitrile. Rubber-modified versions are also available, e.g., olefin-modified styrene-acrylonitriles (OSAs) and acrylic-styrene-acrylonitriles (ASAs). They have a density of about 1.08 g/cm³ (0.039 lb/in.³), outstanding dimensional stability, high hardness and rigidity, excellent load-bearing properties, high gloss, good environmental stress-cracking resistance, good heat-deflection properties, excellent thermal properties, glass-transition temperatures of 109-116°C (228-241°F), and excellent resistance to acids, alkalies, salts and some solvents. They are often sold under trade names and trademarks, such as *Lacqsan, Luran, Lustran SAN, Tyril* or *Vestoran.* Used for consumer electronics, terminal

boxes, automotive instrument lenses, battery caps, housewares, household appliances, boat hulls, display racks, etc. Abbreviation: SAN.

styrene-acrylonitrile/polycarbonate alloys. See SAN-PC alloys.

styrene-acrylonitrile/polyvinyl chloride alloys. See SAN-PVC alloys.

styrene block copolymer. See styrenic block copolymer.

styrene-butadiene-acrylonitrile. A terpolymer of styrene and variable amounts of butadiene and acrylonitrile that has exceptional rigidity, excellent thermal stability, high impact strength (toughness), and extremely high gloss. Abbreviation: SBA; S/BA.

styrene-butadiene copolymer. See styrene-butadiene rubber.

styrene-butadiene elastomer. See styrene-butadiene rubber.

styrene-butadiene latex. A white, viscous dispersion or suspension of styrene-butadiene rubber in water, used for latex paints, styrene-butadiene rubber, etc. Abbreviation: SBL.

styrene-butadiene resin. A copolymer of butadiene and at least 50 wt% styrene with good chemical resistance and excellent film-forming properties.

styrene-butadiene rubber. A synthetic rubber that is a random copolymer of styrene and butadiene made either by polymerization in emulsion in the presence of peroxide-type catalysts at elevated temperatures, by polymerization in the presence of metal peroxides at relatively low temperature (cold rubber) or by polymerization in solution with stereospecific organolithium compounds. It has good physical properties (when reinforced with carbon black, etc.), excellent abrasion and impact resistance, good electrical and thermal properties, moderate fatigue resistance, tensile strength, resilience and hysteresis, good water resistance, poor oil, ozone and weather resistance and poor general chemical resistance. It is often sold under trade names and trademarks, such as *Buna CB and Buna S, Budene, Dyradene, Intene, Krylene* or *Polysar,* and used for automotive tires and tubes, gaskets, shoe soles and heels, mechanical goods, coatings, adhesives and sealants. Abbreviation: SBR. Also known as *styrene-butadiene copolymer; styrene-butadiene elastomer.*

styrene-butadiene-styrene block copolymer. A thermoplastic elastomer that is a *block copolymer* composed of hard, stiff thermoplastic styrene (S) mers and soft, elastic butadiene (B) mers arranged in alternating positions along the molecular chain. Abbreviation: S-B-S; SBS.

styrene-butylene resin. A thermoplastic copolymer of styrene and butylene with good moldability and electrical properties, and low water absorption.

styrene-chloroprene copolymer. See styrene-chloroprene rubber.

styrene-chloroprene rubber. A synthetic rubber that is a copolymer of styrene and chloroprene (2-chlorobutadiene). Used for a wide range of mechanical rubber products. Abbreviation: SCR. Also known as *styrene-chloroprene copolymer.*

styrene copolymers. A group of engineering thermoplastics produced by combining styrene with one or more monomers. For example, copolymers of styrene with butadiene (SBR), copolymers of styrene with acrylonitrile (SAN), copolymers of styrene, butadiene and acrylonitrile (S/BA or ABS), etc.

styrene-ethylene-butadiene-styrene block copolymer. A *block copolymer* composed of styrene (S), ethylene (E) and butadiene (B) mer units clustered in blocks along the molecular chain. Abbreviation: S-E-B-S; SEBS.

styrene-isoprene rubber. A synthetic rubber that is a copolymer of styrene and isoprene (3-methyl-1,3-butadiene) used for a wide range of mechanical rubber products. Abbreviation: SIR.

styrene-isoprene-styrene block copolymer. A thermoplastic elastomer that is a *block copolymer* composed of hard, stiff thermoplastic styrene (S) mers and soft, elastic isoprene (I) mers arranged in alternating positions along the molecular chain. Abbreviation: S-I-S; SIS.

styrene foam. See polystyrene foam.

styrene-maleic anhydrides. Copolymers of styrene monomer and maleic anhydride excellent colorability and heat resistance, good toughness and impact strength, good dimensional stability, and good resistance to aging and chemicals. They are often sold under trade names or trademarks, such as *Arloy, Artem, Dylark* or *Smartan*, and supplied in standard and (glass) reinforced high-temperature, high-impact or high-modulus grades. The unreinforced grades have a density range of about 1.06-1.08 g/cm^3 (0.038-0.039 lb/in.^{3}) and the glass-reinforced grades of 1.13-1.23 g/cm^3 (0.041-0.044 lb/in.^{3}). *Styrene-maleic anhydrides* are used for automotive instrument panels and exterior and interior trim, consoles, dishwasher trays, electrical components, consumer and industrial products, and in paints and coatings. Abbreviation: SMA; S/MA.

styrene-methylstyrene copolymer. A copolymer resin produced by combining styrene monomer with α-methylstyrene monomer. Abbreviation: SMS.

styrene monomer. See styrene.

styrene plastics. A group of thermoplastics based on polymers resulting from the polymerization of styrene monomer, or the copolymerization of styrene monomer with other monomers, such as acrylonitrile, alkyd, butylene, maleic anhydride or unsaturated polyester. See also styrene-acrylonitriles; styrene maleic anhydrides, styrenated alkyd; styrenated polyester; acrylonitrile-butadiene styrenes; styrene butylene resins.

styrene-rubber plastics. Plastics with a styrene content of 50 wt% or more, made by mixing a styrene plastic with a rubber and several compounding materials. Abbreviation: SRP.

styrenic block copolymer. A copolymer composed of block segments of a hard, stiff thermoplastic styrene mer and a soft, elastic mer, such as butadiene or isoprene, alternating positions along the molecular chain. Also known as *styrene block copolymer.* See also styrene-butadiene-styrene block copolymer; styrene-ethylene-butadiene-styrene block copolymer; styrene-isoprene-styrene block copolymer.

styrenics. A group of thermoplastic elastomers based on polymers made by the copolymerization of styrene monomer with butadiene or isoprene. They have generally good physical properties (especially when reinforced), excellent abrasion and impact resistance, good electrical and thermal properties, good water resistance, fair mechanical properties, poor chemical, oil, ozone and weather resistance. Used for automotive tires and tubes, sheets, tubes, gaskets, adhesives, sealants, coatings, shoe soles and heels, mechanical goods, wire and cable, floor tile, and miscellaneous molded and extruded products.

Styresol. Trademark of Reichhold Chemicals, Inc. (USA) for a series of styrenated alkyd coating resins with good resistance to acids, alkalies and gasoline, low water absorption and excellent air-drying and baking properties.

Styrex. (1) Trademark of Nova Group, Inc. (USA) for extruded and intruded pellets of plastics including polystyrene and polyacrylonitrile.

(2) Trade name of Dow Chemical Company (USA) for polystyrenes.

Styria. Trade name of former Vereinigte Edelstahlwerke (Austria) for an extensive series of carbon and alloy steels, tool steels (e.g., plain-carbon, hot-work, cold-work, high-speed, etc.), and several iron-nickel low-expansion alloys.

Styrite. Trade name of Dow Chemical Company (USA) for polystyrene products.

Styroblend. Trademark of BASF AG (Germany) for polystyrenes and polystyrene alloys.

Styrocel. Trade name for expandable polystyrenes.

Styrodur. Trademark of BASF Corporation (USA) for a green-colored, chlorofluorocarbon-free, extruded rigid polystyrene foam with high compressive strength, good thermal insulating properties and low water absorption. Supplied in a variety of grades, it is used in building construction or civil engineering for the thermal insulation of roofs, ceiling, window and door lintels, above and below-grade insulation, and as antifrost layer in road and railroad construction.

Styro-Ex. Trademark of Huntsman Chemical Company (USA) for expandable polystyrene beads.

Styroflex. Trademark of BASF Corporation (USA) for polystyrene materials supplied in film form, and used for radio-frequency cables, underwater cables, etc. Also included under this trademark are various other unprocessed plastics supplied as powders, liquids, granules, or pastes.

Styrofoam. Trademark of Dow Chemical Company (USA) for a line of lightweight rigid multicellular polystyrene foams with good resistance to moisture and water, excellent thermal insulation properties, an upper service temperature of 77°C (171°F) and good resistance to mold and mildew. Used for thermal insulation (cold storage), boats, battery cases, etc., as a packaging material, as a separator in packing containers, for expendable foundry patterns, for airport runway and highway construction, for rigid home insulation, for toys, drinking cups, etc.

Styrofoam SM. Trademark of Dow Chemical Company (USA) for polystyrene foam with high toughness and compressive strength, excellent moisture resistance, and excellent thermal insulation properties. Used in the insulation of roof decks, interior ceilings, basement floors, and below-grade exterior walls.

Styrofoam Cladmate. Trademark of Dow Chemical Company (USA) for blue-colored, rigid polystyrene home insulation with excellent air-barrier and thermal insulation properties. It is supplied in 0.6 × 2.4 m (2 × 8 ft) sheets of varying thickness, and used as exterior wall sheathing.

Styrofoam Wallmate. Trademark of Dow Chemical Company (USA) for moisture-resistant, rigid *Styrofoam* polystyrene insulation. It has a high R-value (heat flow resistance), and is supplied in slotted 0.6 × 2.4 m (2 × 8 ft) sheets for the insulation of basement walls and crawlspaces.

Styrolux. Trade name of BASF Corporation (USA) for polystyrenes, and styrene-butadiene (S/B) and styrene-butadiene-styrene triblock (S/B/S) polymers.

Styron. Trademark of Dow Chemical Company (USA) for a series of polystyrene resins available in various grades including general-purpose, medium-impact, high-impact, heat-resistant, impact-heat resistant, glass-reinforced, light-stabilized and structural foam. They can be processed by extrusion, injection molding and thermoforming, and have good electrical, heat and stain resistance, good hardness, stability and moldability, poor resistance to ultraviolet radiation, high tendency to load cracking, and a maximum service temperature of 40-100°C (120-210°F). Used for automotive components, appliance parts, machine housings, piping, bottle closures and containers, battery cases, lighting equipment, computer electronics, high-frequency insulation, dials, knobs, medical and pharmaceutical applications, dinnerware, drinking cups, packaging, toys, and foam sheets.

Styron A-Tech. Trademark of Dow Chemical Company (USA) for high-impact polystyrene resins made by a patented proprietary technology. They have high gloss, stiffness, toughness and good flow characteristics and processibility, and are used for packaging applications (containers, disposable cups, preformed dairy cups, form-fill seal dairy sheets, etc.), large and small appliances (e.g., refrigerator and freezer parts), lawn and garden equipment, business equipment, toys, etc.

Styron HIPS. Trade name of Dow Chemical Company (USA) for unmodified, fire-retardant and UV-stabilized high-impact polystyrenes.

Styroplas. Trade name of F.A. Hughes Company (UK) for polystyrene plastics.

Styropor. Trademark of BASF Corporation (USA) for a usually white, lightweight polystyrene foam used for acoustic and thermal insulation, lightweight construction (boats, etc.), flotation devices, ice buckets, water coolers, fillers in shipping containers, as a packaging material and in furniture construction.

Styrospan. Trademark of Dow Chemical Company (USA) for *Styrofoam* rigid home insulation with high R-value supplied in 0.6 × 2.4 m (2 × 8 ft) sheets for above and below grade applications.

suanite. A white mineral composed of magnesium borate, Mg_2-B_2O_5. Crystal system, monoclinic. Density, 2.91 g/cm^3; refractive index, 1.639. Occurrence: Korea.

Subdoo. Trade name of Laminated Glass Corporation (USA) for glare-reducing, heat-absorbing laminated glass.

subfoundation board. A thin, highly flexible fibrous-felted board or liner with good dimensional stability and ply-bond used in building construction in combination with a *foundation board* as a carrier for subsequent trim. It may be treated for water resistance and have a natural or coated surface.

sublimed blue lead. See blue lead (1).

sublimed white lead. See basic white lead.

submicron-grained material. A crystalline material having a grain size of less than 1 μm (39.4 μin.), but more than about 0.1 μm or 100 nm (3.94 μin.).

submicron powder. A finely divided powder whose particles are smaller than 1 μm (39.4 μin.).

Subo. Trade name of Ato Findley (France) for self-adhesive acrylic coatings and acrylic adhesives used for packaging applications.

Substraight. Trademark of Honeywell Performance Fibers (USA) for a line of ultra-low-shrink high-performance polyester fibers with high strength, excellent durability and good processibility. Used for outdoor furniture, upholstery, umbrellas, backlit signs and awnings, lightweight truck covers, tarpaulins, vinyl-coated mesh, architectural purposes, single-ply roofing fabrics, marine covers, and various other applications requiring coated or laminated fabrics.

Success. Trade name of Inter-Africa Dental (Zaire) for a cold-polymerizing dental material with high modulus of elasticity and transverse strength, short working time and low discolorization. Used in restorative dentistry for temporary crowns, bridges, inlays, and onlays.

succinylated amylose. An *amylose* (linear starch polymer) that has been treated with a succinic or succinyl compound to produce a biocompatible adhesive for medical and biotechnology applications.

Suda. Trade name for rayon fibers and yarns used for textile fabrics.

Sudalon. Trademark of Inversiones Aragua (Venezuela) for nylon 6 fibers and filament yarns used for textile fabrics.

sudburyite. A white mineral with yellow tint composed of palladium antimony, PdSb. It can also be made synthetically. Crystal system, monoclinic. Density, 9.37 g/cm^3. Occurrence: Canada (Ontario).

sudoite. A grayish white mineral of the chlorite group composed of magnesium aluminum silicate hydroxide, $Mg_2Al_3O(Si_3Al)$-$O_{10}(OH)_8$. Crystal system, monoclinic. Density, 2.42 g/cm^3. Occurrence: Japan.

suede. A soft, chrome- or alum-tanned leather, usually made from calfskin, whose surface has been run on a carborundum or emery wheel to separate the fibers and give the leather a velvetry nap. Used for pocketbooks, coats, hats, gloves, shoe uppers, etc. Also known as *napped leather; suede leather.*

suede calf. Calfskin finished on the flesh side by holding it against an abrasive wheel to give it a nap or velvety surface. Used for shoes.

sueded fabrics. Textile fabrics with a thin covering of leather dust made to imitate real suede leather.

suede fabrics. Woven or knitted textile fabrics with a short nap that has the look and feel of real suede leather and is produced by a special surface finishing process. Used for dresses, skirts, jackets, sportswear, etc.

suede leather. See suede.

Suede-skin. Trade name for rayon fibers and fabrics.

Sue-Del. Trademark of Winter & Co. GmbH (Germany) for an artificial leather imitating suede.

Südglas. Trade name of Südglas Klumpp & Arretz GmbH (Germany) for a laminated safety glass.

Suedois. Trade name of Compagnie Ateliers et Forges de la Loire (France) for water-hardening, wear-resistant steel containing 0.8-1% carbon, 0.4-0.7% manganese, 0.2% silicon, and the balance iron. Used for chisels, punches, taps, cutters, drills, precision tools, springs, bearings, and bumper bars.

suessite. A cream-white mineral of the iron group composed of iron silicide, Fe_3Si. Crystal system, monoclinic. Density, 7.08 g/cm^3. Occurrence: Australia.

sugar. See sugars.

sugar-cane wax. A hard wax obtained from sugar cane (genus *Saccharum*) by solvent extraction. It is tan to brownish in color, melts below 80°C (176°F), and is used as a substitute for carnauba wax, in floor and furniture polishes, and in the manufacture of carbon paper.

sugar maple. The hard, strong, light-tan wood of the large maple tree *Acer saccharum*. The heartwood is light to medium reddish-brown and the sapwood white with reddish-brown tinge. *Sugar maple* has a fine, even texture, a beautiful grain pattern, high hardness and stiffness, good abrasion and shock resistance, large shrinkage, medium gluing qualities, low nail qualities, good machinability, and is difficult to work with hand tools. Average weight, 705 kg/m^3 (44 lb/ft^3). Source: Eastern North America (Lake, central and middle Atlantic states and southern Ontario). Used for flooring, bowling alleys, paneling, interior decoration, quality furniture, cabinetwork, veneer, crossties, woodenware, boxes, crates, pallets, spools and bobbins, handles, and as a pulpwood. Also known as *rock maple*. See also hard maple.

sugar of lead. See lead acetate.

sugar pine. The soft, durable, moderately strong wood of the tall coniferous tree *Pinus lambertiana.* The heartwood is dull yellow or light brown, occasionally tinged with red, and the sapwood yellowish white. *Sugar pine* has a straight grain, an uniform texture, numerous, large resin ducts, good durability and workability, small shrinkage and warpage, and good nailing properties. Average weight, 415 kg/m³ (26 lb/ft³). Source: USA (California and southwestern Oregon). Used for sash and door construction, interior house finish, quality millwork, shingles, boxes, woodenware, and foundry patterns.

sugars. A class of carbohydrates, including glucose, fructose and sucrose, that are aldehydic or ketonic derivatives of the higher alcohols. Sugars are colorless, usually more or less sweet-tasting, readily water-soluble and essentially crystallizable. See also disaccharide; monosaccharide; trisaccharide.

sugilite. A light brownish yellow mineral of the osumilite group composed of lithium potassium sodium iron silicate monohydrate, $(K,Na)(Na,Fe)_2(Li,Fe)_3Si_{12}O_{30} \cdot H_2O$. Crystal system, monoclinic. Density, 2.74 g/cm³; refractive index, 1.610. Occurrence: Japan.

Suiko. Trade name for rayon fibers and yarns used for textile fabrics.

SULC steels. See super-ultralow-carbon steels.

Suleine. Trade name for rayon fibers and yarns.

sulfar fiber. A man-made fiber composed of long-chain synthetic polysulfides with 85 wt% or more of the sulfide linkages joined directly to two six-carbon rings. Also known as *sulfar.*

sulfate paper. A strong kraft-type paper made by the sulfate pulping process in which sodium sulfate is added to the caustic liquor. It has a lesser degree of whiteness than soda or sulfite papers, and is used for wrapping, packaging, containerboard, etc. See also kraft paper; soda paper; sulfite paper.

sulfate-resistant cement. Portland cement that is low in tricalcium aluminate (C_3A) and has excellent resistance to dissolved sulfates. The composition in weight percent is: 38% tricalcium silicate (C_3S), 43% dicalcium silicate (C_2S), 4% tricalcium aluminate (C_3A), 9% tetracalcium aluminoferrate (C_4AF), and a total of 6% simple oxides (lime, magnesia, alkali oxides, etc.) and calcium sulfate. Used in the manufacture of sulfate-resistant concrete, e.g., for marine structures. Also known as *type V cement.*

sulfite paper. A paper made by the sulfite pulping process in which wood chips (mostly from spruce and other coniferous woods) are digested with a solution of ammonium, calcium or magnesium disulfite in the presence of free sulfur dioxide. It has a higher degree of whiteness than *sulfate papers,* and is used for writing, printing, drawing, etc.

sulfoaluminate cement. An *expansive cement* that is essentially composed of anhydrous calcium aluminosulfate ($4CaO \cdot 3Al_2O_3 \cdot SO_3$) and made by grinding a mixture of gypsum and high-alumina cement.

o-sulfobenzimide. See saccharin.

sulfoborite. A colorless mineral composed of magnesium borate sulfate nonahydrate, $Mg_6B_4O_{10}(SO_4)_2 \cdot 9H_2O$. Crystal system, orthorhombic. Density, 2.38 g/cm³; refractive index, 1.541. Occurrence: Russian Federation.

sulfohalite. A mineral composed of sodium chloride fluoride sulfate, $Na_6(SO_4)_2ClF$. Crystal system, cubic. Density, 2.51 g/cm³; refractive index, 1.455. Occurrence: USA (California).

sulfolane. See tetramethylene sulfone.

Sulfor. Trade name of Forjas Alavesas SA (Spain) for free-machining, low-carbon steel containing 0.1% carbon, 1.1% manganese, 0-0.06% silicon, 0-0.05% phosphorus, 0.3% sulfur, 0.2%

lead, and the balance iron.

sulfur. A light-yellow, nonmetallic element belonging to Group VIA (Group 16) of Periodic Table. It is commercially available in the form of pieces, flake and powder, occurs in nature as native sulfur, and is also obtained from iron pyrites, natural gas, crude oil, etc. Atomic number, 16; atomic weight, 32.066; divalent, tetravalent, hexavalent; high room-temperature electrical resistivity (about 2×10^{23} μΩcm); low thermal conductivity. Two stable crystalline forms are known: (i) *Alpha sulfur.* Orthorhombic yellow crystals. Density, 2.07 g/cm³; reversible transformation to beta sulfur at 94.5°C (202°F); melting point, about 113°C (235°F); refractive index, 2.0377; and (ii) *Beta sulfur.* Monoclinic, slightly yellow crystals. Density, 1.96 g/cm³; melting point, 119°C (246°F); boiling point, 444.6°C (832°F); flash point, 207°F (405°F); autoignition temperature, 232°C (450°F). Two amorphous (liquid) forms are also known. *Sulfur* is used for the vulcanization of rubber, in sulfur compounds (e.g., sulfur dioxide and sulfuric acid), in the manufacture of free-machining alloys, in the manufacture of gunpowder, matches and explosives, in pulp and paper manufacture, for electrical insulators, in cement sealants, in low-temperature mortars, as a binder and asphalt extender in road paving, in food processing, as a colorant in glass to produce yellows and ambers, and with cadmium sulfide in selenium ruby glass. Symbol: S.

sulfur-35. Radioactive and radiotoxic isotope of sulfur with a mass number of 35 usually made by pile irradiation of elemental sulfur and sulfur-containing chlorides. It has a half-life of 87.1 days, emits beta radiation, and is used in biological, chemical and medical research, e.g., as a tracer in protein metabolism studies and in the determination of the turnover rate of plasma proteins, in the study of rubber vulcanization, in the polymerization of synthetic rubber, in the study of surface phenomena, in the study of the effect of sulfur in steel and in plating solutions, in the study of the relationship between sulfur and engine wear, etc. Symbol: ^{35}S. Also known as *radiosulfur.*

sulfurated lime. A grayish yellow powder composed of calcium sulfide and calcium sulfate obtained by roasting calcium sulfate with coke, and used for luminous paints.

sulfurated potash. See sulfurated potassium.

sulfurated potassium. A dipotassium thiosulfate mixture with potassium sulfide that contains 12.8% or more sulfur, and is available in the form of hygroscopic lumps. It is initially liver brown, but turns greenish-yellow. Used in the production of decorative color effects on nonferrous metals, such as brass, bronze and nickel. Also known as *sulfurated potash.*

sulfur cement. A cement made by mixing pitch and sulfur in equal proportions, and used for joining iron components.

sulfur-concrete. A mixture of powdered stoneware, glass and sulfur that has been melted and poured into molds.

sulfur-bearing copper. A free-cutting copper that contains up to 0.40% sulfur and is usually available in rod or wire form. It has good machinability and corrosion resistance, high electrical conductivity (approximately 96% IACS), high elongation, and excellent cold and hot workability. Used for screw-machine products, soldering coppers, fasteners, welding-torch tips, electrical connectors and components, etc. Also known as *sulfur copper.*

sulfur copper. See sulfur-bearing copper.

sulfuric anodized aluminum. Aluminum or aluminum alloy on whose surface a hard, colorless, transparent, protective coating of aluminum oxide (Al_2O_3) has been produced by making the aluminum or aluminum alloy the anode in an electrolytic cell and using a sulfuric acid electrolyte. Abbreviation: SAA. See also anodized aluminum.

sulfur-impregnated abrasive. A bonded abrasive whose connected pores have been filled with sulfur. Also known as *sulfur-impregnated abrasive product*.

sulfur mortar. A material made by dispersing carbon black, silica flour or a similar filler in sulfur, usually with the addition of small percentages of other additives.

sulfur powder. Ultra-high-purity sulfur (99.999+%) in the form of a fine powder for use in the manufacture of semiconductors.

sulvanite. A bronze-gold-colored mineral of the sphalerite group composed of copper vanadium sulfide, Cu_3VS_4. Crystal system, cubic. Density, 3.86 g/cm³. Occurrence: USA (Utah), Australia.

Sulzer Metco. Trade name of Sulzer Metco (USA) for a series of coating products including (i) nickel-, iron-, chromium- and tungsten carbide-base powders for corrosion and/or wear-resistant arc and thermal sprayed coatings; (ii) cobalt-base abradable coatings; and (iii) monoaluminide-chromium-aluminum-yttrium overlay coatings.

sumac wax. A yellowish solid obtained from plants of the genus *Rhus* growing primarily in Japan. It has a density of 0.97-0.98 g/cm³ (0.035-0.036 lb/in.³), a melting point of 53°C (127°F), and is used as a replacement and extender for beeswax, in floor waxes, in polishes, candles, etc. Also known as *Japanese tallow; Japan tallow; Japan wax*.

Sumet Bronze. Trade name of Sumet Corporation (USA) for a series of high-leaded bronzes containing 10-32% lead, varying amounts of copper and tin, and sometimes nickel. Used for high-speed bearings, and packing rings.

Sume. Trademark of Sulzer Metco (USA) for a series of metal, alloy and ceramic coating systems developed for various industrial applications. *Sume Cal* and *Sume Roll* coating systems are designed for paper machinery, *Sume Plant* coating systems are used for medical applications (e.g., coating of orthopedic implants), *Sume Print* and *Sume Cer* coatings systems for applications in the printing industry, *Sume Sol* coating systems for applications under starved lubrication conditions, *Sume Shield* coating systems for applications in the nuclear reactor technology and *Sume Tex* coating systems are especially suitable for textile machinery.

Sumika. Trademark of Sumitomo Limited (Japan) for continuous fibers composed of 85% alumina (Al_2O_3) and 15% silica (SiO_2), and used for metal-matrix composite reinforcement.

Sumikaflex. Trademark of Occidental Chemical Corporation (USA) for polyvinyl acetate resins.

Sumikon. Trademark of Occidental Chemical Corporation (USA) for strong, dimensionally stable glass-fiber and mineral-filled phenolic molding compounds available in injection and compression molding grades. They have a density of 1.64 g/cm³ (0.059 lb/in.³), low water absorption (0.15%), excellent heat resistance, good dielectric properties, and a maximum service temperature in air of 200°C (390°F). Used for machine parts, such as pulleys and clutch disks.

Sumistrong. Trade name of Sumitomo Metal America Inc. for a series of high-strength low-alloy (HSLA) structural steels supplied in pipe or plate form for various structural applications, e.g., marine structures, bridges, buildings and pressure vessels.

Sumiten. Trade name of Sumitomo Metal America Inc. for a series of high-strength low-alloy (HSLA) structural steels supplied in plate form for various structural applications including marine structures, bridges, buildings, and pressure vessels.

Sumizinc V. Trade name of Sumitomo Metal Industries Limited (Japan) for an electrogalvanized steel sheet with a thin coating (30 g/m² or 0.1 oz/ft²). It has improved formability, paintability, weldability and higher corrosion resistance than standard electrogalvanized sheet. Used for autobody parts.

Summer. Trade name of Chance Brothers Limited (UK) for decorative table glassware with a pattern depicting various summer flowers.

Summit. Trade name of Aurident, Inc. (USA) for a dental ceramic alloy containing 74.2% gold, 7.3% palladium and 4.1% platinum. It has a light yellow color, is compatible with most porcelains, and used for porcelain-to-gold restorations.

Summum. Trade name for rayon fibers and yarns used for textile fabrics.

Sumo. Trade name for a glass-ionomer dental cement.

sun-baked brick. An unburned brick baked by exposure to the sun. See also adobe.

Sun Bronze. (1) British trade name for a heat-resistant alloy composed of 40-60% cobalt, 30-50% copper and 10% aluminum, and used for high-temperature fittings.

(2) British trade name for a strong, corrosion-resistant aluminum bronze composed of 95% copper and 5% aluminum, and used for bushings, liners, sleeves, pump rods, and propeller blade bolts.

Sunburst. Trademark of World Alloys & Refining, Inc. (USA) for a copper-aluminum dental ceramic alloy with a density of 7.8 g/cm³ (0.28 lb/in.³), a melting range of 970-1040°C (1780-1905°F), a hardness of 138 Vickers, and moderate elongation. Used for crown and bridge restorations, as an implant material, and as a post and core material.

Suncast. (1) Trade name of J.F. Jelenko & Company (USA) for a deep yellow colored, hard porcelain-to-metal dental casting alloy (ADA type III) containing 50-55% gold, 28.5-35% silver and 4-5% palladium. Used for crowns, hard inlays, and fixed bridgework.

(2) Trade name of SA Glaverbal (Belgium) for heat-absorbing cast glass with bluish tint.

Suncast-D. Trade name of J.F. Jelenko & Company (USA) for a hard dental casting gold (ADA type III) with high tarnish resistance used for crowns, hard inlays and fixed bridgework.

Suncast-DKF. Trade name of J.F. Jelenko & Company for a hard, non-heat-treatable cast dental gold with a rich yellow color. Used for crowns, hard inlays, and fixed bridgework.

Suncut. Trade name of Asahi Glass Company Limited (Japan) for heat-reflecting glass.

Sundance. (1) Trademark of Celanese Chemical Company (USA) for linings made from *Arnel* cellulose triacetate fibers.

(2) Trademark of Georgia-Pacific Corporation (USA) for lumber and wood products (e.g., siding).

(3) Trade name of Porcelanite, Inc. (USA) for glazed floor tile.

Sundeel. Trade name for a corrosion-resistant dental alloy containing 80% gold, 10% tungsten, 7% nickel, and small additions of other metals..

sundiusite. A white or colorless mineral composed of lead sulfate oxide chloride, $Pb_{10}(SO_4)Cl_2O_8$. Crystal system, monoclinic. Density, 7.00 g/cm³; refractive index, above 2.10. Occurrence: Sweden.

Sundora. Trade name of E.I. Du Pont de Nemours & Company (USA) for cellulose acetate plastics.

sun-dried brick. See adobe.

Sunfix. Trade name of Vereinigte Glaswerke (Germany) for hollow glass blocks.

Sunglas. Trademark of Ford Glass Limited (Canada) for coated and low-emissivity glass products.

Sunigum. Trademark of Goodyear Tire & Rubber Company (USA) for acrylate terpolymers.

Sunkong. Trademark of Sunkong Industries Limited (South Korea) for polyester and cellulose acetate fibers and filaments.

sunlight-resistant textiles. Textile fibers and fabrics, such as acrylics, modacrylics and nytrils, that have excellent resistance to damage (e.g., discoloration, loss of strength, etc.) caused by the ultraviolet rays of the sun.

Sun Line. Trademark of Ube Kasei Company (Japan) for polyethylene monofilaments.

Sunlit. Trade name of Semon Bache & Company (USA) for a glass that transmits a high percentage of the incident ultraviolet radiation.

sunn hemp. The flexible, durable vegetable fibers obtained from a shrub (*Crotalaria juncea*) native to India. These cellulose fibers are bast fibers growing between the bark and the stem of the shrub. Used in the manufacture of ropes, cords, twine, oakum, paper, fabrics, and fish nets. Also known as *Bombay hemp; Indian hemp; Madras hemp.*

Sunplate. Trade name of SA Glaverbel (Belgium) for heat-absorbing plate glass.

Sunprene. Trade name of Résinoplast (France) for polyvinyl chloride elastomers used for automotive and building products, and sporting and leisure goods.

Sunray Gold. Trade name of Jones & Rooke Limited (UK) for a corrosion-resistant alloy composed of copper and nickel, and used for jewelry and imitation gold.

Sunrise. Trade name for a medium-gold dental casting alloy.

Sunscreen. Trade name of Colcombet Fois & Compagnie SA (France) for a fabric woven from vinyl chloride-coated glass fibers. It eliminates up to 84% of the solar radiation, while remaining crystal clear, and is used for window blinds.

sunshine carbon. Molded coal-tar carbon with a core of metallic cerium used in electric-light carbons to produce a spectral wavelength about that of sunlight. The cerium generates bluish an effect in the light.

Sunshade. (1) Trade name of PPG Industries Inc. (USA) for shaded, glare-reducing glass.

(2) Trade name of Parlee Company (USA) for a tinted plastic film with metallic particles that is applied to sheet glass to reflect thermal radiation.

Sunshine. Trade name for polyolefin fibers and products.

Sunspun. Trade name for spun rayon fibers and yarns used for textile fabrics.

sunstone. See aventurine (2).

Suntex. (1) Trade name of Esperanza SA (Spain) for lightly patterned, rectangular glass lenses used for vertical glazing.

(2) Trade name of Suntex Safety Glass Industries Limited (UK) for laminated and toughened glass.

Suntrol. Trade name of Pittsburgh Corning Corporation (USA) for green fibrous insert for use in glass blocks.

Suntuf. Trademark of Suntuf Inc. (Canada) for tough, durable, corrugated polycarbonate roofing panels that are 20% stronger than fiberglass panels, resist yellowing, block ultraviolet rays, and have excellent environmental resistance. They are supplied in clear, white, gray and green colors with standard panel widths of 660 mm (26 in.) and lengths of 2.4 m (8 ft.) and 3.6 m (12 ft.).

Sunt-X. Trade name of Atul Glass Company Limited (USA) for a laminated glass.

Sunwax. Trade name of Sunwax International Company (Hong Kong) for a microcrystalline wax.

Sunylon. Trademark of Formosa Chemicals & Fiber Corporation (Taiwan) for nylon 6 fibers and filament yarns.

suolunite. A colorless to honey-yellow mineral composed of calcium silicate hydroxide monohydrate, $Ca_2Si_2O_5(OH)_2 \cdot H_2O$. Crystal system, orthorhombic. Density, 2.63 g/cm^3; refractive index, 1.6199. Occurrence: Bosnia.

Supac. Trade name for polyolefin fibers and products.

Supalex. Trade name of Rhodia (USA) for bright nickel electrodeposit and deposition process.

Supalux. Trade name of Fulgurit Baustoffe GmbH (Germany) for asbestos-cement fire-protection panels.

Supamuff. Trade name of Fibreglass Limited (UK) for a glass-fiber-filled cylinder jacket used for hot-water storage systems.

Supard Chrome. Trade name of Jenney Cylinder Company (USA) for a steel containing 0.4% carbon, 11.5-14% chromium, 0-1% nickel, 0-0.5% molybdenum, and the balance iron. It has good resistance to sulfuric acid, and is used for bushings, plunger sleeves, and pump liners.

Supavit. Trade name of British Steel Corporation (UK) for decarburized mild steel sheet used for porcelain enameling applications.

Supawrap. Trade name of Fibreglass Limited (UK) for a fiberglass thermal insulation material supplied in batts or blankets, typically 51 mm (2 in.) thick, for residential attic insulation.

Supazote. Trademark of Zotefoams plc (UK) for block-type polyolefin foams and foam products.

Supec. Trademark of GE Plastics (USA) for a polyphenylene sulfide (PPS) resin reinforced with 40% glass fibers. It has a density of 1.66 g/cm^3 (0.060 $lb/in.^3$), good mechanical properties, excellent dimensional stability, good moldability and rigidity, good corrosion resistance, good resistance to chemicals including acids, solvents, lubricants, fuels and hydraulic fuels, good flame resistance, good creep resistance, good thermal stability up to 260°C (500°F), melting point of 285°C (545°F), and good electrical insulating properties. Used for automotive components, aircraft parts, electrical and electronic equipment and components, and various other industrial applications.

Super. (1) Trade name of Agawam Tool Company (USA) for a high-speed steel containing 1.1% carbon, 0.3% manganese, 3.75% chromium, 17.5% tungsten, 0.8% molybdenum, 3.25% vanadium, and the balance iron. Used for lathe and planer tools.

(2) Trade name for autobody and premixed fiberglass fillers.

(3) Trade name for castable and plastic refractories.

superabrasives. A group of abrasive materials including cubic boron nitride (CBN), hexagonal boron nitride (HBN) and natural and synthetic diamond that have very high hardness and toughness together with thermal and wear characteristics that make them suitable for use in grinding and machining of advanced materials, such as superalloys, titanium alloys, nickel and titanium aluminides, etc. See also cubic boron nitride; hexagonal boron nitride; natural diamond; synthetic diamond.

Super-Alloy. Trade name of Kloster Steel Corporation (USA) for a shock-resisting tool steel (AISI type S1) containing 0.5% carbon, 0.75% silicon, 1.15% chromium, 2.5% tungsten, 0.2% vanadium, and the balance iron. It has good deep-hardening properties, good toughness, and high hardness and strength. Used for dies and hand tools.

superalloys. Nickel-, iron-nickel and cobalt-base alloys suitable for use at temperatures above 540°C (1000°F). They have excellent resistance to tensile and vibratory stresses and thermal shocks, excellent high-temperature mechanical properties, optimal oxidation and creep resistance, and can be processed by melting, forging, sheet rolling, powder processing, investment casting and joining. *Nickel-base superalloys* typically contain the following alloying additions: 5-25% chromium, 0-19% cobalt, 0-10% molybdenum, 0-10% tungsten, 0-6% aluminum, 0-5% titanium, 0-3% iron and 0-0.2% carbon. *Iron-nickel-base superalloys* contain: 10-55% nickel, 5-25% chromium, 1.3-9% molybdenum, 0-3% titanium, 0-2% aluminum, 0-0.3% carbon, and the balance iron. *Cobalt-base superalloys* contain: 0-25% nickel, 0-30% chromium, 0-15% tungsten, 0-9% tantalum, 0-4.5% aluminum, 0-4% titanium, 0-4% niobium (columbium) and 0-0.9% carbon. *Superalloys* are often sold under trademarks or trade names, such as *Astroloy, Discaloy, Elgiloy, Hastelloy, Inconel, Pyromet, René, Udimet* or *Waspaloy*. Used in wrought, cast or powder-processed forms for turbines, nuclear reactors, aircraft parts, high-temperature structural applications (e.g., furnaces and heat-treating equipment), etc. Abbreviation: SA. Also known as *heat-resistant alloys; heat-resisting alloys; high-temperature high-strength alloys.*

Super-Allstep. Trade name of Ludlow Composites Corporation (USA) for a closed-cell polyvinyl chloride latex foam.

super-alpha titanium alloys. See near-alpha titanium alloys.

Superalumag. Trade name of Tefileries & Laminoirs du Havre (France) for a series of corrosion-resistant aluminum alloys containing 3-8% zinc, 1.5-2% magnesium, 0-1% copper, 0.4-0.5% manganese and 0.2-0.3% chromium. Used for engine parts, and aircraft parts and structures.

Superarnum. Trade name for rayon fibers and yarns used for textile fabrics.

superaustenitic stainless steels. A group of highly alloyed *austenitic stainless steels* typically containing up to 0.02% carbon, 20-25% chromium, 17-22% nickel, 4.5-7.3% molybdenum, 0-0.5% nitrogen, 0-0.6% copper, 0-3% manganese, and the balance iron. Compared to conventional austenitic stainless steels, they have greatly improved resistance to both localized and stress-corrosion cracking in oxidizing chloride and sulfide-containing solutions and various process chemicals, and are well suited for chemical and seawater applications.

Super Austenite. Trade name of T. Inman & Company Limited (USA) for a high-speed tool steel containing 0.75% carbon, 4.5% chromium, 18% tungsten, 0.5% molybdenum, 1% vanadium, and the balance iron. Used for lathe and planer tools, reamers, broaches, drills, and cores for pressure-die castings.

Super Avional. Trade name of Alusuisse (Switzerland) for age-hardenable aluminum alloy containing 4.5% copper, 1.2% manganese, 1% magnesium and 0.8% silicon. Used for structural members, aircraft parts, and automotive parts.

Superb. Trade name of Latrobe Steel Company (USA) for an oil and water-hardening special-purpose tool steel containing 0.7% carbon, 0.25% silicon, 0.25% manganese, 0.95% chromium, 0.18% vanadium, and the balance iron. It has a great depth of hardening, excellent toughness, good to moderate machinability and moderate wear resistance. Used for hand tools, cutting tools, and dies.

Super-Beckacite. Trademark of Reichhold Chemicals, Inc. (USA) for phenolic resins used in coatings.

Super-Beckamine. Trademark of Reichhold Chemicals, Inc. (USA) for melamine-formaldehyde resins used in coatings.

Superbend. Trade name for a low-strength magnesium-silicon alloy used for bendable extruded shapes with minimum orange peel.

Super-Bioney. Trademark of Degussa-Ney Dental (USA) for a copper-, silver- and palladium-free dental alloy containing 82% gold and 15% platinum. It has a yellow color, a light oxide layer, high yield and bond strengths, a hardness of 220 Vickers, and a thermal coefficient compatible with most porcelains. Used in dentistry for porcelain restorations.

Superblack. Trade name of Pavco Inc. (USA) for a noniridescent black coating.

Superbolt. Trade name of Timken Company (USA) for a tough steel containing 0.35-0.45% carbon, 0.5-0.8% manganese, 1.5-2% nickel, 0.2-0.3% molybdenum, 0.25% or more chromium, and the balance iron. Used for crosshead bolts.

Super Bond. Trade name of Kwik Mix Minerals Limited (Canada) for topping cement.

Superbond. Trade name of Carlisle Dental for a bonding resin used in orthodontics.

Super-Bond C&B. Trade name of Sun Medical Company Limited (Japan) for a self-curing polymer-based dental adhesive resin cement that forms a strong bond between tooth structure (i.e., dentin and enamel) and metal, porcelain or composite resin restorations. Used for the cementation of crowns, bridges, inlays, onlays and posts, for bonding splints, and as a protective lining of vital teeth.

Super-Bond D-Liner II Plus. Trade name of Sun Medical Company Limited (Japan) for a self-curing methacrylate-based dental bonding agent with thin to medium film thickness that strongly bonds tooth structure (i.e., dentin and enamel) to composite-resin or amalgam restorations.

Super Breda. Trade name for rayon fibers.

superbronze. Any of several strong, corrosion-resistant brasses modified with aluminum and magnesium, and sometimes iron. The typical composition range is 57-69% copper, 21-38% zinc, 2-3.5% manganese, 1.2-6% aluminum and 0-2% iron. They have high strength and hardness, poor workability and machinability, and are used for marine parts, propeller blades, bearings, and bushings.

supercalendered paper. A highly rolled and coated, usually wood-free *printing paper* with a shiny and smooth surface produced by supercalendering. It is often colored, and used for printing fine-screen halftones. The coating composition is such that optimum print results are obtained. Also known as *art paper.*

Super Capital. Trade name of Eagle & Globe Steel Limited (Australia) for a high-performance, high-speed tool steel containing 1.3% carbon, 4.25% chromium, 3.1% molybdenum, 9% tungsten, 3.5% vanadium, 8.5-9.5% cobalt, and the balance iron. Used for tools for abrasive and heavy cutting spplications, and for reamers and hobs.

Super-Carb. Trade name of BF Goodrich Aerospace (USA) for a carbon/carbon composite material consisting of a vapor-infiltrated carbon matrix reinforced with polyacrylonitrile fibers. Used for aerospace components.

Super Carbon Patch. Trade name of Wellsville Fire Brick Company (USA) for plastic refractories.

Supercase. Trade name of Supersteels Inc. (USA) for a free-machining case-hardening steel containing 0.08-0.15% carbon, and the balance iron. Used for camshafts, transmission shafts, gears, bolts, worms, and sprockets.

Super Cast. Trademark of Pentron Laboratory Technologies (USA) for a corrosion-resistant nonprecious dental casting alloy composed of 64% cobalt, 30% chromium, 6% molybdenum, and the balance manganese, silicon, carbon and nitrogen. Used for partial denture applications.

Super Cello-Bond. Trade name of Cello-Foil Products, Inc. (USA) for *cellophane* and polyethylene laminations used for flexible packaging applications.

Super Cesco. Trade name of Crucible Electric Steel Company (USA) for a shock-resisting tool steel containing 0.55% carbon, 2% silicon, 1% manganese, 1.3% molybdenum, 0.35% vanadium, and the balance iron. Used for pneumatic tools, shear blades, punches, etc.

Super Chlor. Trade name of Duriron Company Inc. (USA) for a cast alloy steel containing 0.95-1.1% carbon, 14.2-14.7% silicon, 4-5% chromium, and the balance iron. It has good resistance to sulfuric, nitric and hydrochloric acid, and is used for chemical and industrial equipment.

Super Chrome. Trade name of Time Steel Service Inc. (USA) for an air- or oil-hardening high-carbon, high-chromium type cold-work tool steel (AISI type D5) containing 1.4% carbon, 0.3% manganese, 0.6% silicon, 13% chromium, 0.5% nickel, 0.6% molybdenum, 3.3% cobalt, and the balance iron. It has excellent deep-hardening properties, and good wear resistance and hot hardness, and is used for trimming, coining, blanking and forming dies, shear blades, and punches.

superclean steels. A class of steels with extremely low contents of inclusion-forming elements, such as oxygen and sulfur, and impurities, such as hydrogen and phosphorus. The high cleanness is usually achieved by vacuum-arc remelting or electroslag remelting, or by a combination of air melting, ladle refining and vacuum degassing treatments.

Super Coat. Trade name of Yawata Iron & Steel Company Limited (Japan) for sheet metal coated with thin layers of chromium and chromium trioxide. It has good resistance to staining and corrosion, and is used for especially beverage cans, motor oil cans, and dry cells.

Super Cobalt. Trade name of Latrobe Steel Company (USA) for tungsten-type high-speed tool steels (AISI type T5) containing 0.85% carbon, 18.75% tungsten, 8.25% cobalt, 4.1% chromium, 1.9% vanadium, and the balance iron. They have a great depth of hardening, excellent red hardness, excellent wear resistance, and are used for high-speed cutting tools, such as lathe and planer tools, cutoff tools, etc. Formerly known as *Electrite Super Cobalt.*

Supercon. (1) Trademark of Supercon Limited (USA) for an extensive series of superconducting ceramic powders with varying metal-ion ratios, wet or dry-processed from carbonates, nitrates or oxides. Included are several yttrium barium copper oxides (e.g., YBCO-123 and YBCO-124), bismuth strontium calcium copper oxides (e.g., BSCCO-1112, BSCCO-2212 and BSCCO-2223), thallium barium calcium copper oxides (e.g., TBCCO-2212, TBCCO-2223 and TBCCO-4334) and several other high-temperature superconductor powders.

(2) Trademark of Supercon Limited (USA) for drawn refractory metal and alloy wire products used as superconductor materials for cryogenic applications.

(3) Trademark of A. Schulman, Inc. (USA) for compounded, formulated or alloyed polymers used in the manufacture of molded and extruded products.

superconcrete. An engineering composite developed at Oak Ridge National Laboratory, Tennessee, USA. It is made by packing the pores of standard concrete with recycled polystyrene or similar polymer using a special process. It is superior to standard concrete in strength, and more resistant to erosive chemicals.

superconducting alloys. See superconductor alloys.

superconducting composite wire. A conductor consisting of a superconductor material, e.g. a multifilamentary wire of superconducting niobium-titanium alloy filaments in a copper matrix, a core of tin surrounded by superconducting niobium filaments in a copper matrix, or an individual filament of superconducting niobium-titanium alloy clad with copper.

superconducting compounds. See superconductor compounds.

superconducting gallium arsenide. See gallium arsenide semiconductor.

superconducting materials. See superconductors.

superconducting magnets. See magnetic superconductors.

superconducting metals. See superconductor metals.

superconductive alloys. See superconductor alloys.

superconductive compounds. See superconductor compounds.

superconductive metals. See superconductor metals.

superconductor alloys. A group of alloys that at a temperature near absolute zero lose both their electrical resistance and magnetic permeability, i.e., have almost infinite electrical conductivity. Examples of such alloys include those of niobium and zirconium, niobium and titanium, and bismuth and lead. Abbreviation: SCA. Also known as *superconducting alloys; superconductive alloys.*

superconductor compounds. A group of intermetallic compounds, such as niobium-tin (Nb_3Sn), niobium-germanium (Nb_3Ge), vanadium-gallium (V_3Ga), titanium-cobalt (Ti_2Co) and lanthanum-indium (La_2In), that at a temperature near absolute zero lose both their electrical resistance and magnetic permeability, i.e., have almost infinite electrical conductivity. Also known as *superconducting compounds; superconductive compounds.*

superconductor magnets. See magnetic superconductors.

superconductor metals. A group of metallic elements that lose virtually all their electrical resistance at temperatures approaching 0K (-273.15°C). Examples of such metals include aluminum (1.175K), beryllium (0.026K), gallium (1.08K), hafnium (0.128K), indium (3.41K), lanthanum (4.88K), lead (7.196K), mercury (4.15K), niobium (9.25K), rhenium (1.70K), ruthenium (0.49K), tantalum (4.47K), thallium (2.38K), tin (3.722K), vanadium (5.40K), zinc (0.85K) and zirconium (0.61K). Superconductivity does not occur in alkali metals, noble metals and ferromagnetic and antiferromagnetic metals. Also known as *superconducting metals; superconductive metals.*

superconductors. A group of materials including metallic elements, alloys, ceramics, intermetallic compounds and semiconductors that lose their resistance to the flow of electric current when cooled to low (cryogenic) temperatures. They are used for magnets and memory elements in high-speed computers, electromagnets, magnetic levitation devices, ultrafast switches, and laboratory and research equipment. Also known as *cryogenic conductors; superconducting materials.* See also high-temperature superconductors; hard superconductors; soft superconductors.

superconductor thin film. See thin film superconductor.

Supercontryx. Trade name of Saint-Gobain (France) for a glass with high-lead content, used for X-ray protection applications.

Super Cordura. Trade name of E.I. DuPont de Nemours & Company (USA) for a high-tenacity rayon used for tire cords.

Supercote. Trademark of Imperial Optical Company Limited (Canada) for antireflection coatings used for optical lenses.

Super-Cure. Trade name for a heat-cure acrylic used for dentures.

Super Cyclone. Trade name of British Steel Corporation (UK) for a high-speed steel containing 0.8% carbon, 4.5% chromium, 21.5% tungsten, 1.5% vanadium, 12% cobalt, and the balance iron. Used for cutting tools.

Super-D. Trademark of Clayburn Refractories Limited (Canada) for fireclay bricks.

Super-Die. Trade name of Whip Mix Corporation (USA) for a high-accuracy, fast-setting dental gypsum (ADA type IV) supplied as a powder. When cast from a mixture of powder and water of specified proportions, it develops an extremely smooth and dense surface with a glazed, low-friction finish. It has a low setting expansion, and is suitable for use as dye stone with hydrocolloid impressions.

Superdie. Trade name of Belmont Metals Inc. (USA) for a high-carbon, high-chromium cold-work tool steel (AISI type D3) containing 2.2% carbon, 12% chromium, 0.8% tungsten, 1% silicon, and the balance iron. It has excellent abrasion and wear resistance, excellent nondeforming properties, a great depth of hardening, good compressive strength, and is used for blanking and trimming dies, extrusion and drawing dies and burnishing tools.

Superdraw. Trade name of Edgar Allen Balfour Limited (UK) for a water-hardening die steel containing 1.3% carbon, 0.3% manganese, 2.25% tungsten, and the balance iron. Used for cold-drawing dies.

SuperDrop. Trade name for a series of instant-binding cyanoacrylate adhesives used for general joining applications.

superduplex stainless steels. A group of highly alloyed duplex (ferritic/austenitic) stainless steels typical containing up to 0.04% carbon, 18-25% chromium, 4-7.5% nickel, 0-4% molybdenum, 0-0.3% nitrogen, 0-2% copper, and the balance iron. Compared to conventional *duplex stainless steels*, they have greatly improved resistance to pitting corrosion (due to their chromium and molybdenum content) and stress-corrosion cracking. Abbreviation: SDSS.

superduty fireclay brick. Fireclay brick having a pyrometric cone equivalent (PCE) greater than or equal to Cone 33, a linear shrinkage of less than 1% when reheated to 1598°C (2908°F), and a maximum of 4% weight loss in the panel-spalling test (preheated to 1649°C or 3000°F). See also fireclay brick.

superduty fireclay plastic refractory. A special fireclay plastic refractory product having a higher pyrometric cone equivalent (PCE), a lower shrinkage, and a lower spalling loss than an ordinary *fireclay plastic refractory*.

superduty silica brick. A special *silica refractory brick* in which the total content of alumina (Al_2O_3), titania (TiO_2) and alkalies is much lower (usually less than 1%) than normal. Also known as *low-alumina silica brick*.

Superdux 64. Trademark of Nippon Yakin Kogyo Company Limited (Japan) for a superplastic low-carbon microduplex stainless steel containing 25% chromium, 6.5% nickel, 3.2% molybdenum, 0.1% nitrogen, and the balance iron. Used for commercial aircraft sinks.

Super Dylan. (1) Trademark of Koppers Company, Inc. (USA) for high-density polyethylene used for molded products, extruded film, sheeting, pipe, filaments, wire insulation, etc.

(2) Trade name of Arco Chemical Company (USA) for linear-low-density polyethylene.

Super EBA. Trade name of Harry J. Bosworth Company (USA) for an ethoxybenzoic acid (EBA) reinforced zinc oxide–eugenol dental cement.

Super-E-Core. Trade name of NKK Corporation (Japan) for electrical steel sheet with a silicon content of 6.5% providing higher magnetic permeability, lower magnetostriction and reduced energy losses than conventional silicon sheet steels. It is produced by coating conventional cold-rolled silicon steel sheet (containing approximately 3% silicon) with a uniform layer of silicon in a continuous furnace using a chemical-vapor deposition (CVD) technique. The silicon addition is then diffused into the steel at 1200°C (2190°F). Supplied in widths of 400-600 mm (16-24 in.) and thicknesses of 0.2-0.3 mm (0.008-0.012 in.), it is used for uninterruptible power supplies, high-frequency transformer cores, high-speed generators and motors, and electric reactors for train and induction-heater applications.

Superex. Trade name of Thermal Ceramics, Inc. (USA) for a series of thermal insulation products composed of asbestos or glass fibers bonded with calcined or uncalcined powdered *diatomaceous earth* and molded into blocks. They can resist temperatures of up to 1040°C (1904°F), and are used for boiler walls, industrial furnaces, etc.

Superez. Trade name for a cellulose acetate formerly used for denture bases.

Superface. Trade name of PPG Industries Inc. (USA) for a glass-fiber duct liner in the form of a semi-rigid, dual-density blanket having a relatively low-density body, and a thin, high-density surface layer.

SuperFast. Trade name of Reichhold Chemical Company (USA) for an abrasion-resistant synthetic rubber latex with a soft hand and excellent fastness, used for textile fabrics.

superferritic stainless steels. A group of highly alloyed *ferritic stainless steels* (typically up to 29% chromium, 4% molybdenum and 0-4% nickel) with extra-low carbon (up to 0.01%) and nitrogen (up to 0.025%) contents, and additions of titanium and niobium. They have improved toughness, weldability and reduced weld cracking, excellent resistance to seawater, brines, brackish water, most pollutants, sour gases and many organic and inorganic chemicals, outstanding resistance to chloride pitting and crevice corrosion, high resistance to stress corrosion, good resistance to scaling and good mechanical properties. Often sold under trade names and trademarks, such as *Monit* or *Sea-Cure*, they are used for marine structures and equipment, feedwater heaters, condensers, heat exchangers, chemical and oil processing equipment, etc. Also known as *stabilized ferritic stainless steels*.

Superfine. (1) Trademark of Stauffer Chemical Company (USA) for fine, air-classified sulfur flour used in casting aluminum and magnesium, in glassmaking, sulfur cements, chrome oxide pigments, etc.

(2) Trade name of PPG Industries Inc. (USA) for fine glass fibers bonded with synthetic resin into light blankets of varying densities and thicknesses.

(3) Trade name of Fibreglass Limited (UK) for an exceptionally strong, resilient and easy-to-handle thermal insulation made from extremely fine, long glass fibers.

(4) Trade name of Nippon Mineral Fiber Manufacturing Company Limited (Japan) for acoustical and thermal insulation materials based on mineral fibers.

superfine red lead. Fume *red lead* in the form of a very fine powder having an average particle size in the submicron to 1 μm (39.4 μin. or less) range. Used as a pigment.

superfines. Metal powder particles having a size of less than 10 μm (394 μin.).

superfine wool. A high-quality wool with an average fiber diameter of approximately 15-18 μm (590-710 μin.).

SuperFlex. Trade name of Arnco (USA) for polyurethanes.

Super-Flex. Trade name of Proko Industries, Inc. (USA) for elastomeric interior and exterior wall coatings.

Superflex. Trade name for rayon fibers and yarns used for textile fabrics.

Superflex Ultra-Blue. Trade name of Loctite Americas (USA) for a patented silicone sealant that provides controlled swell when in contact with oil to create an even tighter seal. It has high flexibility, is noncorrosive to aluminum, steel and iron, and possesses a maximum operating temperature of 260°C (500°F) and low susceptibility to hardening, crumbling, cracking and drying. Used for automotive applications including oil pans, camshaft covers, timing chain covers, valve covers and differential housings, and as a general industrial sealant against hydrocarbons, fluids and greases.

Super-Genuine Babbitt. Trade name of Magnolia Metal Corporation (USA) for a tough babbitt composed of tin, copper and antimony, and used for marine, airplane and internal-combustion engine bearings.

Super Glas. Trade name of Oatey Company (USA) for fiberglass supplied in the form of a gel for building and roofing applications.

Super Glass. Trade name of Newalls Insulation & Chemical Company Limited (USA) for glass-fiber insulation.

SuperGlaze. Trade name of Lincoln Electric Company (USA) for aluminum metal-inert-gas (MIG) welding wires supplied in several diameters. They provide smooth surfaces and stable arcs, and virtually eliminate burnback, tangling and birdsnesting.

Super-Glue. Trademark of Laporte Construction Chemicals North America, Inc. (USA) for an epoxy adhesive that dries to a clear, permanent bond in just a few seconds.

Supergraphite. Trade name for a recrystallized graphite with an upper service temperature of 3040°C (5504°F) used for heat-resisting parts.

Superguard. Trademark of Iroquois Chemicals Corporation (Canada) for interior wood coatings.

Super Hallamite. Trade name of Hallamshire Steel Company (USA) for an oil-hardening, high-speed tool steel containing 0.7% carbon, 18% tungsten, 4% chromium, 1% vanadium, 5% cobalt, and the balance iron. Used for saws, saw blades, cutting tools, etc.

superhard coating materials. A class of carbon- or noncarbon-based materials that exhibit hardness values in excess of about 40 GPa (5.8×10^3 ksi) and include diamond-like carbon, cubic boron nitride (CBN), nanocomposite thin films, fullerene-like carbon nitrides, and multilayered nanostructured materials. Used for coating cutting tools, hard-disk components, wear parts, etc.

SuperHardSteel. Trade name for a tough steel coating applied by thermal spraying to various substrates. It has very high hardness (up to 1500 Vickers), high abrasion, corrosion, impact and wear resistance, and low coefficient of friction.

Super Hardtem. Trade name of Heppenstall Company (USA) for an oil-hardening, shock-resistant die steel containing 0.43% carbon, 2.6% nickel, 1.35% chromium, 0.55% molybdenum, 0.18% vanadium, and the balance iron.

Super-High Speed. Trade name for a high-speed steel containing 0.75-0.9% carbon, 17-22% tungsten, 4.5-6% chromium, 1-2.5% vanadium, 0.75-1.5% molybdenum, 8-15% cobalt, and the balance iron. It has high red hardness, and is used for tools, such as lathe and planer tools, cutoff tools, drills, milling cutters, etc.

Super Hi-Mo. Trade name of Teledyne Firth-Sterling (USA) for a high-speed steel (AISI type M30) containing 0.8% carbon, 8.5% molybdenum, 4% chromium, 5% cobalt, 1.8% tungsten, 1.2% vanadium, and the balance iron. It has good deep-hardening properties, high hot hardness, and good wear resistance. Used for cutting tools.

Super Hi-Naraco. Trade name formerly used for rayon fibers and yarns.

Super Hi-Talc. Trademark of Bakertalc Inc. (Canada) for micronized talc.

Super-Holfos (WW). Trade name of Holcroft Castings & Forgings Limited (UK) for an abrasion- and wear-resistant centrifugally cast bronze containing 11.5% tin, 0.2% phosphorus, and the balance copper. Used for helical and worm gears, couplings, driveshafts, etc.

Super Hydra. Trade name of Osborn Steels Limited (UK) for a tungsten high-speed steel containing 0.75% carbon, 4.25% chromium, 18% tungsten, 1.1% vanadium, and the balance iron. Used for lathe and planer tools, cutters, threading taps, drills, hobs and reamers.

Super Hy-Tuf. Trade name of Colt Industries (UK) for a tough, oil-hardening steel containing 0.32-0.43% carbon, 1.3% manganese, 2.3% silicon, 1.4% chromium, 0.35% molybdenum, 0.2% vanadium, and the balance iron. It has high tensile strength, good shock resistance, and is used for aircraft structures and landing gears.

Super-Impacto. Trademark of Atlas Specialty Steels (Canada) for a tough, fatigue-resistant machinery steel containing 0.12% carbon, 3.25% nickel, 1.5% chromium, 0.5% manganese, and the balance iron. Used for transmission shafts, clutch dogs, heavy-duty gears, bearings, and plastic molds.

Super Impacto PQ. Trademark of Altlas Specialty Steels (Canada) for a mold steel (AISI type P6) containing 0.1% carbon, 3.5% nickel, 1.5% chromium, and the balance iron. It has a great depth of hardening, exceptional core strength, outstanding resistance to decarburization, good toughness, and a high luster. Used for plastic molding dies.

Super Inmanite. Trade name of T. Inman & Company Limited (USA) for a cobalt-tungsten type high-speed tool steel containing 0.8% carbon, 22% tungsten, 4.5% chromium, 1.5% vanadium, 10% cobalt, 1% molybdenum, and the balance iron. Used for heavy-duty applications.

Super Invincible. Trade name of Joseph Beardshaw & Son Limited (UK) for a series of high-speed tool steels containing 0.75% carbon, 14-19% tungsten, 4-4.25% chromium, 1.5-2% vanadium, 5-10% cobalt, and the balance iron. Used for lathe and planer tools, drills, broaches, cutters, reamers, hobs, chasers, shear blades, and punches.

Super Invar. Trademark of Carpenter Technology Corporation (USA) for a low-expansion alloy containing 31-32% nickel, 5-5.25% cobalt, 0.35% manganese, 0.3% silicon, 0.05% carbon, and the balance iron. It has a density of 8.1 g/cm³ (0.29 lb/in.³), a hardness of 160 Brinell, high electrical resistivity, a Curie temperature of 260°C (500°F), high tensile strength and modulus of elasticity, a coefficient of thermal expansion about one-half that of ordinary *Invar* with 36% nickel. Used for supports for laser and optical instruments, and for structural components.

Superior. (1) Trade name of Bethlehem Steel Corporation (USA) for a water-hardening tool steel (AISI type W2) containing 0.7-1.3% carbon, 0.15-0.25% vanadium, and the balance iron. It has excellent machinability, good to moderate depth of hardening, wear resistance and toughness, and low hot hardness. Used for cutting tools, such as milling cutters, taps, broaches and reamers, hand tools, such as knives, chisels and punches, and dies.

(2) Trade name of CCS Braeburn Alloy Steel (USA) for a series of high-carbon high-chromium type cold-work tool steels (AISI types D2 to D5) containing about 1.4-2.2% carbon, 12-12.5% chromium, 0-0.5% nickel, 0-0.8% molybdenum, 0-0.8% vanadium, 3.5% cobalt, and the balance iron. Used for drawing and forming tools, dies, shear blades, and punches.

(3) Trade name of Superior Steel Corporation (USA) for austenitic, ferritic and martensitic stainless steels.

Superior Casting Alloy. Trademark of Federated Genco Limited (Canada) for babbitt metals used for bearings.

Superior Shafting. Trademark of Atlas Specialty Steels (Canada) for a high-quality electric-furnace machinery steel containing 0.4% carbon, 1.1% manganese, 0.2% silicon, 0.15% molybdenum, 0.08% sulfur, 0.03% phosphorus, and the balance iron. Supplied as centerless ground stock, it has relatively high strength and good machinability. Used for vehicle axles, connecting rods, motor and engine shafts, machine-tool spindles, etc.

Superior Spur. Trade name of T. Turton & Sons Limited (UK) for a tungsten-type high-speed tool steel containing 0.72% carbon, 4% chromium, 18% tungsten, 0.3% molybdenum, 1.1% vanadium, and the balance iron.

Super-Job. Trademark of CC Chemicals Limited (Canada) for a polymer concrete resurfacer mix.

Superkleen. Trade name of Dexter Plastics (USA) for a series of flexible polyvinyl chloride resins that can be processed by injection molding or extrusion into medical tubing and various health care products.

Super-Kut. Trade name of Great Western Steel Company (USA) for a high-speed tool steel containing 0.7% carbon, 18% tungsten, 4% chromium, 5% cobalt, 0.5% molybdenum, 1% vanadium, and the balance iron. Used for broaches, reamers, chasers, milling cutters and lathe and planer tools.

Super-L. Trade name for rayon fibers.

Superla. Trade name of Boyd-Wagner Company (USA) for a water-hardening tool steel (AISI type W2) containing 1.3% carbon, 0.2-0.3% vanadium, and the balance iron. It has good machinability and toughness, and is used for lathe and planer tools, taps, drills, and reamers.

super lacquer. A solution of blends of polyurethane and ethylene-vinyl acetate, or blends of nitrocellulose and polyurethane.

Super La-Led. Trade name of LaSalle Steel Company (USA) for a free-machining steel containing 0-0.13% carbon, 0.85-1.35% manganese, 0.5% sulfur, 0.15-0.35% lead, and the balance iron. Used for screw-machine products, bolts, fasteners, and shafts.

Super Latex. Trade name of Induron Coatings, Inc. (USA) for latex house paint.

Superleda. Trade name of Firth Brown Limited (UK) for a cobalt-tungsten high-speed steel containing 0.78% carbon, 4.8% chromium, 1% molybdenum, 18.8% tungsten, 2% vanadium, 9.3% cobalt, and the balance iron. Used for cutting tools, reamers, broaches, hobs, etc.

Super-Light. Trade name of Philip Carey Manufacturing Company (USA) for 85% magnesia high-temperature insulating blocks and pipe coverings.

Super Lion. Trade name of Burys & Company Limited (UK) for a series of high-speed steels containing 0.75% carbon, 0.75% molybdenum, 19-20% tungsten, 4-4.5% chromium, 1-1.25% vanadium, 10-15% cobalt, and the balance iron. Used for lathe and planer tools, reamers, drills, and hobs.

Super-Loy. Trade name of General Steel Industries (USA) for a carbon steel containing 0.2% carbon, and the balance iron. Used for fasteners.

Superloy. (1) Trade name of Resisto-Loy Company, Inc. (USA) for hardfacing electrodes used to produce hard, wear-resistant deposits of the iron-chromium-cobalt-molybdenum-tungsten-type. They consist of a tube of soft steel containing the hardfacing material in granule form.

(2) Trade name of Acme Electric Welder Company (USA) for copper alloys used for resistance and spot-welding electrodes.

(3) Trade name of Washington Iron Works, Inc. (USA) for a series of shock-resistant alloy steels used for machine parts (e.g., gears, shafts, crankshafts), castings, machine-tool parts, housings, etc.

Superlume. Trademark of Atotech USA Inc. for a super-leveling electroplating process and the resulting bright deposits produced from a bath containing nickel sulfate, nickel chloride, boric acid and selected additives. Used on steel, copper, brass, zinc, etc.

Superlux. (1) Trademark of Sico Inc. (Canada) for alkyd enamel coatings used for automotive applications.

(2) Trademark of DMG Hamburg (Germany) for a range of composites and bonding agents used in restorative dentistry. Examples include *Superlux Dual* fast-setting, dual-cured universal dentin/enamel bonding agent, *Superlux P/P Anterior* self-cured, micro-filled composite, *Superlux P/P Posterior* self-cured microfine hybrid composite, *Superlux Solar* light-cure microfine hybrid composite and *Superlux Universalhybrid* light-cure, fine hybrid composite.

Supermagaluma. Trade name of Aluminium Belge SA (Belgium) for heat-treatable, corrosion-resistant aluminum alloy containing 5.4% magnesium, 0.15% manganese, and the balance iron. Used for light-alloy parts.

Supermal. Trade name of Dresser Industries (USA) for a wear-resistant pearlitic malleable iron containing 1.7% total carbon, 0.3% manganese, 1.2% silicon, and the balance iron. It has high fatigue strength, and is used for conveyor equipment, buckets, and chain links.

Supermalloy. Trade name of Magnetic Metals Company (USA) for a vacuum-melted soft-magnetic material containing 79-80% nickel, 15-16% iron, 4-5% molybdenum, 0.3-1% manganese, and traces of carbon, silicon and sulfur. Usually supplied in the form of sheets, foils and microfoils, it has extremely low coercive force, very high permeability, low hysteresis loss and low electrical resistivity. Used for magnets, transformer and relay cores, and communication and radar equipment.

Super Manganese Bronze. Trade name of American Manganese Bronze Company (USA) for a strong, corrosion-resistant manganese bronze containing 69% copper, 20% zinc, 6.5% aluminum, 2.5% iron and 2% manganese. Used for marine parts, aircraft engine components, propeller parts, gears, and bushings.

Super Max. Trade name of Atotech USA Inc. for a bright nickel electrodeposit and deposition process.

Supermendur. Trade name of Westinghouse Electric Corporation (USA) for a soft, malleable magnetic material containing 49% iron, 49% cobalt and 2% vanadium. It has high magnetic permeability, high magnetic saturation, low hysteresis loss, and a rectangular hysteresis loop. Used for pulse and power transformers, and amplifiers.

Supermetonic. Trade name of Pechiney/Trefimétaux (France) for a corrosion-resistant alloy containing 66% copper, 30% nickel, 2% iron and 2% manganese. Used in the form of tubing for heat exchangers and condensers.

Supermex. Trade name of Cristales Inastillables de México SA (Mexico) for laminated sheet glass.

Super Mo Chip. Trade name of Teledyne Firth-Sterling (USA) for a molybdenum-type high-speed steel (AISI type M40) containing 0.7% carbon, 8% molybdenum, 4% chromium, 1.5% vanadium, 8% cobalt, and the balance iron. Used for cutting tools.

Supermold. Trade name of Latrobe Steel Company (USA) for a general-type mold steel (AISI type P20) containing 0.3% carbon, 1.7% chromium, 0.4% molybdenum, and the balance iron. Used for die casting dies, and plastic injection molds.

Super Motung. Trade name of Cytemp Specialty Steel (USA) for a series of molybdenum-type high-speed steels (AISI type M30) containing 0.8-0.9% carbon, 3.75-4% chromium, 5-8.25% cobalt, 1.25-2% vanadium, 1.5-1.75% tungsten, 8.5-9.5% molybdenum, and the balance iron. They have good red-hardness and wear resistance, a great depth of hardening, and are used for cutting tools, form tools, broaches, drills, taps, chasers, end mills, milling cutters, and lathe tools.

Super Motung Special. Trade name of Cytemp Specialty Steel (USA) for a series of molybdenum-type high-speed steels (AISI type M34) containing 0.9% carbon, 8% molybdenum, 4% chromium, 2% tungsten, 2% vanadium, 8% cobalt, and the balance iron. They have good red-hardness and wear resistance, and are used for cutting tools, broaches, drills, taps, chasers, cutters, etc.

Super Mumetal. Trade name of Telcon Metals Limited (UK) for vacuum-melted soft-magnetic alloys containing 77% nickel, 14% iron, 5% copper and 4% molybdenum. They have high magnetic permeability, low losses, and are used for electrical and magnetic equipment.

Super-Narco. Trade name for rayon fibers and yarns used for textile fabrics.

Super Nerva. Trade name of Compagnie Français de Métaux (France) for high-speed steel containing 0.7% carbon, 18% tungsten, 4% chromium, 1% vanadium, and the balance iron. Used for tools and cutters.

Supernickel. Trade name for a corrosion-resistant alloy of 70% copper and 30% nickel. It has a density of 8.94 g/cm³ (0.323 lb/in.³), good tensile strength, good cold and hot workability, and is used for tubes and plates for condensers and heat exchangers.

Super Nitane. Trademark of ORMCO Corporation (USA) for a 51.3%-nickel, 48.7%-titanium wire used in dentistry and orthodontics.

Super Nitralloy. Trade name of Anaconda Company (USA) for a nitriding steel containing 0.23% carbon, 0.5% chromium, 2% aluminum, 0.25% molybdenum, 5% nickel, 0.1% vanadium, and the balance iron. Used for machine parts, and wear-resistant parts.

Super Novex. Trade name of Jonas & Culver Limited (UK) for a low-carbon alloy steel containing 0.3% carbon, 0.8% chromium, 3% nickel, 0.25% molybdenum, and the balance iron.

Supernylon Cydsa. Trademark of Fibras Quimicas SA (Mexico) for nylon 6 monofilaments and filament yarns.

Superock. Trade name of Superock Block Company (USA) for expanded slag used as a lightweight aggregate.

Super Olefin Polymer. Trade name of Toyota Motor Corporation (Japan) for an engineering polymer made by introducing hard polypropylene segments into an elastomeric matrix consisting of amorphous ethylene propylene rubber plus noncrystalline polypropylene. The resulting polymer alloy has an interpenetrating network (IPN) structure. *Super Olefin Polymer* has good impact properties, toughness and stiffness, a low coefficient of thermal expansion, and is used for automotive components, e.g., bumper parts. Abbreviation: SOP.

Superon. Trade name for a self-cure dental cement for crown and bridge restorations.

superopaque enamel. A porcelain enamel characterized by an extremely high opacity produced from a frit high in opacifiers, such as titanium oxide.

Super Oralium. Trade name of Engelhard Industries (USA) for white-colored low-gold dental casting alloys containing varying amounts of palladium and platinum. Supplied in the annealed and age hardened condition, they have a melting range of 923-975°C (1693-1787°F).

Superox. Trademark of Ferox Coatings Inc. (Canada) for polyurethane coatings.

SuperPac. Trademark of Dexter Corporation (USA) for *Hysol*-type ethylene-vinyl-acetate hot-melt adhesives used for sealing cartons.

Super Panther. Trade name of AL Tech Specialty Steel Corporation (USA) for a tungsten-type high-speed steel (AISI type T5) containing 0.8% carbon, 19% tungsten, 8% cobalt, 4% chromium, 2% vanadium, 0.7% molybdenum, and the balance iron. It has a great depth of hardening, excellent red hardness and wear resistance. Used for lathe and planer tools, reamers, and form cutters.

superparamagnetic materials. A group of materials whose magnetic susceptibility is intermediate between that of ferromagnetic and paramagnetic materials. They are composed of particles of ferromagnetic or ferrimagnetic materials, such as iron, that have ferromagnetic bulk properties which are sub-domain in size and separated by a nonmagnetic matrix (e.g., cobalt particles in a copper matrix) or aqueous or organic carrier phases to form liquid magnets or ferrofluids.

Superpax. Trademark of Upjohn Company (USA) for a fine zirconium silicate powder (92-94.5% pure) with an average particle size of 5 µm (0.0002 in.) or less. Used as an opacifier in white ceramic glazes, and as a filler for plastics and rubber.

superplastic aluminum alloys. Alloys of aluminum that develop extremely high tensile elongations at temperatures above room temperature. Supplied in sheet thicknesses from 3-15 mm (0.125 to 0.6 in.), they can be thermoformed into complex shapes.

superplastic ceramics. A group of ceramics with grain sizes in the micron range that, unlike ordinary ceramics, can be shaped (e.g., stretched or bent) at high temperatures without breaking.

superplastic steel. (1) A general term referring to a fine-grained steel with about 1.3-1.9% carbon that at elevated temperatures can develop high elongations of more than 400% at relatively small strain rates, and at ordinary temperatures has both high yield and tensile strengths. The superplasticity is produced by thermomechanical treatment.

(2) A superplastic *duplex stainless steel* that contains about 24-30% chromium and 4-9% nickel, and has a very fine-grained microduplex structure in which *ferrite* and *austenite* are distributed uniformly. It has high strength and toughness, and good

hot workability and corrosion resistance. See also superplastic steel (1).

superplastic zinc alloys. A group of fine-grained wrought alloys composed of about 76-80 wt% zinc and 20-24 wt% aluminum, sometimes with small amounts of copper and/or magnesium, that at elevated temperatures and low strain rates can undergo large amounts of plastic deformation (elongations of more than 2000%) without necking. They are usually supplied in the as-rolled condition with properties varying with the rolling direction, and can be easily formed into complex shapes by deep drawing, stretch forming, compression molding, vacuum forming or thermoforming processes. The copper and/or magnesium additions result in increased strength, but reduce ductility. An increase in strength can also be obtained by annealing and subsequent air cooling.

Super Poly. Trade name of Astrup (USA) for polyester-laminated sheeting.

superpolymers. See engineering plastics.

Super Purity. Trade name of British Aluminium Company, Limited (UK) for a corrosion-resistant, high-purity aluminum (99.99%). Available in various tempers, it is used for reflectors, roofing flashings, fittings, trim, lighting fixtures, etc.

Super Pyrobrite. Trade name of Rhodia (USA) for a bright copper electrodeposit and deposition process.

Super Pyroneal. Trade name of Heppenstall Company (USA) for an oil-hardening die steel containing 0.42-0.47% carbon, 0.3-0.6% manganese, 0.5-0.7% silicon, 4.1-4.5% nickel, 1.25-1.65% chromium, 0.7-0.8% molybdenum, 0.12-0.16% vanadium, and the balance iron. It has high compressive strength, and good resistance to softening at high temperatures.

Super Pyrotem. Trade name of Heppenstall Company (USA) for an oil-hardening die steel containing 0.42-0.47% carbon, 0.3-0.6% manganese, 0.5-0.7% silicon, 4.1-4.5% nickel, 1.25-1.65% chromium, 0.7-0.8% molybdenum, 0.12-0.16% vanadium, and the balance iron. It has high compressive strength and good resistance to softening at high temperatures.

Super Radiometal. Trade name of Telcon Metals Limited (UK) for a vacuum-melted soft magnet material composed of 50% iron and 50% nickel. It has high magnetic permeability, low hysteresis loss, and is used for electrical and magnetic equipment, as a core material for transformers, and for magnetic shielding applications.

Super-Rapid Extra. Trade name of Houghton & Richards Inc. (USA) for a high-speed steel containing 0.7% carbon, 18% tungsten, 4% chromium, 1% vanadium, and the balance iron. Used for cutting tools, dies and punches.

Super-Rayflex. Trade name for rayon fibers and yarns used for textile fabrics.

superrefractories. A group of refractories, such as zirconium diboride and green silicon carbide, that can be subjected to exceptionally high temperatures and/or severe operating conditions including highly corrosive atmospheres, and extreme mechanical abuse.

Super Rose. Trademark of Fiberglas Canada Limited for pink fiberglass insulation with good thermal properties used for commercial, industrial and residential applications.

Super Samson. Trade name of Carpenter Technology Corporation (USA) for an air-hardening low-carbon mold steel (AISI type P4) containing 0.1% carbon, 0.3% manganese, 0.2% silicon, 5% chromium, 0.9% molybdenum, 0.25% vanadium, and the balance iron. It has high core hardness, high strength at elevated temperatures, a great depth of hardening, good wear

resistance, good toughness, and good dimensional stability. Used for plastic molds and die-casting dies.

Super Satin. Trade name of Induron Coatings, Inc. (USA) for semi-gloss enamel paints.

Superseal. Trade name of Pavco Inc. (USA) for a conversion coating.

Superseas. Trade name for rayon fibers and yarns used for textile fabrics.

Super-Sep. Trade name for a gypsum used in dentistry as a pore filler.

Super-Shedisol. Trade name of Saint-Gobain (France) for glass-fiber thermal insulation used for factory roofs.

Super-Shock. Trade name of Allied Steel & Tractor Products Inc. (USA) for a shock- and wear-resistant steel containing 0.5% carbon, 0.75% silicon, 1.15% chromium, 2.5% tungsten, 0.2% vanadium, and the balance iron. Used for cold-battering and chipping tools, and caulking and beading tools.

Super Siri. Trade name for cellulose acetate fibers and yarns used for textile fabrics.

Superskin. Trademark of Otto Bock Health Care (USA) for a liquid thermoplastic polyurethane-based film containing volatile tetrahydrofuran (THF) and alcohol solvent additives. Used in medicine and surgery for coating or laminating applications.

SuperSonex. Trademark of Illbruck, Inc. (USA) for a lightweight, porous melamine material with excellent resistance to constant temperatures up to 150°C (300°F) and intermittent temperatures up to 250°C (480°F), excellent noise control properties in the frequency range between 0.1-1.0 KHz, elimination of echoes and feedback, and low flame spread and smoke emission. Used in acoustic insulation applications, e.g., dead rooms.

Super Special Echo. Trade name of Hall & Pickles Limited (UK) for a tungsten high-speed steel containing 0.7% carbon, 4.4% chromium, 21% tungsten, 1.4% vanadium, and the balance iron. It has excellent red hardness and wear resistance, and is used for single-point cutting tools and inserts.

Super Special Express. Trade name of Leadbeater & Scott Limited (UK) for a high-speed steel containing 0.7% carbon, 18% tungsten, 4% chromium, 1% vanadium, and the balance iron. Used for cutting tools.

Super Speed Star. Trade name of Carpenter Technology Corporation (USA) for a molybdenum-type high-speed steel (AISI type M3) containing about 1% carbon, 6% molybdenum, 6% tungsten, 4% chromium, 2.4% vanadium, and the balance iron. It has good wear resistance and hot hardness, and is used for cutting tools.

superstainless steels. A group of special *stainless steels* including especially superaustenitic, superferritic and superduplex grades that due to their high alloy contents and low impurity levels possess extreme resistance to pitting, crevice corrosion and stress-corrosion cracking in highly corrosive environments, e.g., chlorine, seawater, acidic solutions, corrosive chemicals, etc. See also superaustenitic stainless steels; superferritic stainless steels; superduplex stainless steels.

Super Star. (1) Trade name of Carpenter Technology Corporation (USA) for a high-carbon, molybdenum-cobalt high-speed steel (AISI type M42) containing about 1.1% carbon, 0.25% silicon, 0.25% manganese, 3.75% chromium, 9.5% molybdenum, 1.5% tungsten, 1.15% vanadium, 8% cobalt, and the balance iron. It has a great depth of hardening, high hot hardness, excellent wear resistance, and is used for tools bits, form tools, drills, milling cutters, drills, broaches, hobs, lathe tools, taps, end mills, and gear cutters.

(2) Trade name of J.F. Jelenko & Company (USA) for microfine palladium-silver dental alloy with high strength and excellent castability used for fusing porcelain to metal particularly implants, long-span bridges and single crowns.

Super Star Zenith. Trade name of Carpenter Technology Corporation (USA) for a high-speed steel (AISI type T2) containing 0.8% carbon, 18% tungsten, 4% chromium, 2% vanadium, and the balance iron. It has excellent deep-hardening properties, good hot hardness, wear resistance and machinability. Used for cutting tools.

Superston. Trademark of Stone Manganese–J. Stone & Company Limited (UK) for a series of corrosion-resistant cast aluminum bronzes containing up to 10% aluminum, and sometimes high amounts of manganese (11-15%) together with small amounts (0-3%) of both nickel and iron. Used for marine parts and hardware, shafts, propellers, gears, guides, valve and pump parts, engine components, etc.

Superston Seventy. Trade name of Stone Manganese–J. Stone & Company Limited (UK) for aluminum bronze castings containing 72% copper, 15% manganese, 7.5% aluminum, 3% iron, and 2.5% nickel. Used for marine applications, such as propellers, and fittings.

supersulfated cement. A sulfate-resistant *hydraulic cement* made by mixing ground blast-furnace slag, calcium sulfate and a small quantity of lime. Also known as *metallurgical cement.*

Super-Suprenka. Trade name of Enka BV (Netherlands) for high-tenacity rayon fibers.

Super-Temp. Trade name of BF Goodrich Aerospace (USA) for a family of ceramic-matrix composites and oxidation-resistant carbon-carbon composites for high-temperature applications, e.g., in reentry vessels, rocket motors, and turbine engines.

Supertemp. (1) Trade name of Bethlehem Steel Corporation (USA) for an oil-hardening, creep-resistant steel containing 0.35% carbon, 0.8% manganese, 0.45% chromium, 1% tungsten, 0.6% molybdenum, and the balance iron. Used especially for fasteners and forgings.

(2) Trade name of Eagle Picher Industries (USA) for insulating blocks made by combining granulated mineral wool with a mineral binder and pressing. They have an upper service temperature of 925°C (1697°F), and are used for thermal insulation applications.

(3) Trademark of RJF International Corporation (USA) for flexible bonded magnetic strips and sheets that contain barium and/or strontium ferrite, and are used in electronics.

Super Tenax. Trade name for rayon fibers and yarns used for textile fabrics.

Supertensile. Trade name for a round ultrahigh-strength high-carbon steel wire used for strings of musical instruments.

Super Terrific. Trade name of Wardlows Limited (UK) for a high-speed die and tool steel containing 0.7-0.8% carbon, 18% tungsten, 5% cobalt, 4% chromium, 1.3% vanadium, and the balance iron.

Supertherm. Trade name of Abex Corporation (USA) for a stainless steel containing 0.5% carbon, 35% nickel, 26% chromium, 15% cobalt, 5% tungsten, 1.6% silicon, and the balance iron. It has good hot ductility, and high heat resistance to 1260°C (2300°F), and is used for high-temperature applications, e.g., furnaces and kilns.

Super-Thik. Trade name of Manville Corporation (USA) for fiber-glass batts used as thermal home insulation.

Super Thoroseal. Trademark of Thoro System Products (USA) for a hard, wear-resistant, cement-base, 100% waterproof coating for use on concrete and masonry surfaces.

Super Tiger. Trade name of Bethlehem Steel Corporation (USA) for a high-speed steel containing 0.8% carbon, 18.5% tungsten, 4.5% chromium, 0.8% molybdenum, 1.7% vanadium, and the balance iron. Used for lathe and planer tools, drills, and milling cutters.

Supertough. Trade name of British Steel Corporation (UK) for a series of steels including several plain carbon grades containing 0.2-0.35% carbon and 1.5% manganese, low-carbon nickel grades with 0.25% carbon, 1.5% manganese and 0.4% nickel, low-carbon molybdenum grades having 0.35% carbon, 1.6% manganese and 0.28-0.45% molybdenum, and several low- and medium-carbon alloy steels containing 0.18-0.40% carbon, 1.3-1.6% manganese, 0-0.5% chromium, 0-0.75% nickel, and 0.20-0.45% molybdenum.

Super Tufcor. Trade name of Jonas & Culver Limited (UK) for a plain carbon steel with high core toughness containing 0.15% carbon, 0.9% manganese, and the balance iron. Used for machinery parts.

Super Tuff. (1) Trade name of Time Steel Service Inc. (USA) for an air- or oil-hardening, shock-resistant tool steel (AISI type S7) containing 0.5% carbon, 0.7% manganese, 3.25% chromium, 1.4% molybdenum, and the balance iron. It has good deep-hardening properties, high toughness, good hot hardness, and is used for punches, chisels, shear blades, and dies.

(2) Trade name for polyolefin fibers and products.

Super Ugimax. Trade name of Ugine Aciers (France) for a permanent magnet material containing 40-51% iron, 24-35% cobalt, 14% nickel, 8% aluminum and 3% copper. It has a high energy product, and high residual induction, permeability and coercive force, and is used for magnetic and electrical applications.

Super Unicut. Trade name of Cytemp Specialty Steel (USA) for a molybdenum-type high-speed steel (AISI type M15) containing 1.6% carbon, 4% chromium, 3.5% molybdenum, 6.5% tungsten, 5% vanadium, 5% cobalt, and the balance iron. Used for cutting tools.

super-ultralow-carbon steels. A group of high-strength sheet steels with extremely low carbon contents usually less than 250 ppm. A typical composition (in wt%) is 0.020-0.025% carbon, 0.15% manganese, 0.01% silicon, 0.006% phosphorus, 0.002% nitrogen, 0.03% aluminum, 0.03 titanium, and the balance iron. These steels have high tensile and yield strengths, high elongation, good aging resistance, excellent to good cold formability, and are used for automotive and structural applications. Also known as *SULC steels.*

Superwear. (1) Trade name of Hall & Pickles Limited (UK) for a wear-resistant tool steel containing 2% carbon, 13% chromium, 0.5% cobalt, and the balance iron.

(2) Trademark of Alloy Technology International Inc. (USA) for metal-matrix composites (MMC) formerly sold under the trade name *Ferro-Tic.* They consist of very hard, rounded titanium carbide (TiC) particles uniformly distributed throughout alloy steel or superalloy matrices. The matrix alloy can be a tool steel, a martensitic stainless steel, a maraging steel or an age-hardenable nickel-base or nickel-iron alloy. *Superwear* alloys have high toughness, high abrasion and wear resistance, high hardness, good corrosion resistance, low coefficients of friction, good machinability, good thermal shock resistance, and are used for die plates, pelletizer knives, drawing rings for gas cylinders, high-production tooling, etc.

Superwhite. Trademark of Schott DESAG AG (Germany) for a high-transparency glass used for electronic applications.

(2) Trade name for white rayon and cellulose acetate fibers used for textile fabrics.

Superwind. Trade name for rayon fibers and yarns used for textile fabrics.

Superwool. Trademark of Thermal Ceramics Inc. (USA) for high-temperature amorphous ceramic wool with low biopersistence, used as fibrous insulation for applications in the nonferrous molten-metals industry and for automotive components. *Superwool 607 Max* is a silica-calcia-magnesia ceramic wool with low linear shrinkage and a continuous service temperature of 1200°C (2190°F) used for automotive brake-pad compositions.

Super-Y. Trade name of Chicago Malleable Casting Company (USA) for a malleable cast iron containing 3.3% carbon, 1.5% nickel, 0.5% chromium, and the balance iron.

Super-Z. Trade name of Toyo Soda Manufacturing Company Limited (Japan) for a partially stabilized zirconia/alumina hybrid powder composed of 75.7 wt% zirconia (ZrO_2), 20.0 wt% alumina (Al_2O_3) and 3 mol% yttria (Y_2O_3). It has a bulk density of 5.50 g/cm³ (0.199 lb/in.³), can be processed by hot isostatic pressing, and is used for producing ceramic components with room-temperature bending strengths of up to 2.4 GPa (348 ksi).

Superzinc. Trade name of MacDermid Inc. (USA) for a zinc alloy electrodeposit and deposition process.

Super Zorite. Trade name of Michiana Products Corporation (USA) for an iron-nickel-base superalloy containing 39-46% iron, 37-40% nickel and 17-21% chromium. It has high heat resistance to 982°C (1800°F), and is used for high-temperature applications, e.g., furnace parts.

SuPima. Trade name for fine, silky cotton fabrics made from long-staple *Pima cotton*.

Supplex. Trademark of E.I. DuPont de Nemours & Company (USA) for strong, durable, colorfast nylon 6,6 staple fibers and filament yarns with the soft, supple touch of cotton. Used for casualwear, outer-wear, sportswear, and swimwear.

supported needled felt. Needled felt that is supported (backed) with a knitted, stitched or woven textile fabric. See also needled felt.

Supra. (1) British trade name for an aluminum alloy containing 20% silicon, 5% copper, 2% manganese and 0.7% iron. Used for lightweight pistons.

(2) Trade name of Glidden Industrial Coatings (USA) for hybrid high-solids coatings for metals.

Supral. (1) Trade name of Superform Metals Limited (USA) for a series of corrosion-resistant aluminum alloys with good superplastic forming properties.

(2) Trademark of BASF Corporation (USA) for rayon fibers and yarns used for textile fabrics.

Supralen. Trademark of BASF Corporation (USA) for a series of high- and low-density polyethylenes.

Supralon. Trademark of Progres (Yugoslavia) for nylon 6 fibers and filament yarns used for textile fabrics.

Supramid. Trademark of BASF Corporation (USA) for a polymer with a protein-like structure used in surgery for bone replacements and as suture material.

Supramid Extra. Trademark of BASF Corporation (USA) for synthetic polyfilament or monofilament surgical sutures.

supramolecular materials. A class of self-assemblying and self-organizing materials in which molecular arrangements have been brought about by manipulating noncovalent interactions. In their synthesis they have been structurally controlled on the nanoscale, and their overall characteristics are completely different from those of any of their components. *Supramolecular materials* include a wide range of materials, such as molecularly engineered liquid crystals, polymers with controlled crystallinity and viscosity-modified coatings.

Supranium. Trade name for a corrosion-chromium dental bonding alloy.

Supraplast. Trade name of Süd-West-Chemie GmbH (Germany) for synthetic resins and molding compounds.

Supra Rekord. Trade name of Stahlwerke Südwestfalen (Germany) for a series of high-speed steels containing 0.76-0.86% carbon, 12-18% tungsten, 2.8-10% cobalt, 4.2-4.3% chromium, 0.8-0.9% molybdenum, 1.5-2.1% vanadium, and the balance iron. Used for cutting tools.

Suprasca. Trade name for rayon fibers and yarns used for textile fabrics.

Supra-Stone. Trade name of Kerr Dental (USA) for a dense, hard die stone used for dental applications.

Supratest. Trade name of ASG Industries Inc. (USA) for laminated safety glass.

Suprax. Trade name of Schott Glas AG (Germany) for gage glass having a very low thermal expansion.

Supr-Clean. Trademark of Montgomery Welding & Manufacturing (USA) for electropolished stainless steel used for cleanroom furniture and fixtures.

Suprel. Trademark of Vista Chemical Company (USA) for injection-moldable styrene-vinyl-acrylonitrile (SVA) thermoplastics and acrylonitrile-butadiene-styrene/polyvinyl chloride (ABS/PVC) blends. They have high-performance properties, good heat-deflection temperature and impact strength, and are used for business machines and telecommunications equipment.

Suprema. Trade name for rayon fibers and yarns used for textile fabrics.

Supremat. Trade name of Fibreglass Limited (UK) for hand lay-up low-pressure glass-fiber mold mats.

Supreme. (1) Trademark of LTV Steel (USA) for precision-ground steel shafting.

(2) Trade name of Atotech USA Inc. (USA) for a bright nickel electrodeposit and deposition process.

Supreme Seal. Trademark of Canadian Adhesives Division of Rexnord Canada Limited for a high-quality acrylic latex sealant supplied in cartridge form for interior and exterior use.

Supremus. Trade name of Jessop Steel Company (USA) for a tungsten-type high-speed steel (AISI type T1) containing 0.73% carbon, 18% tungsten, 4% chromium, 1% vanadium, and the balance iron. It has a great depth of hardening, high hot hardness, excellent machinability and wear resistance, good toughness, and is used for cutting tools.

Supremus Extra. Trade name for Jessop Steel Company (USA) for a tungsten-type high-speed steel (AISI type T2) containing 0.85% carbon, 18.5% tungsten, 4% chromium, 2% vanadium, 0.7% molybdenum, and the balance iron. It has a great depth of hardening, excellent machinability and wear resistance, high hot hardness, good toughness, and is used for cutting tools.

Suprenka. Trade name of Enka BV (Netherlands) for rayon fibers and yarns used for textile fabrics.

Suprima. Trademark of E.I. DuPont de Nemours & Company (USA) for fine dual-denier synthetic fabrics with superior drape, feel and wrinkle resistance produced by a proprietary spinning technology. They can be blended with natural fibers including wool, and are used for tailored apparel.

Suprimpacto. Trade name of Atlas Specialty Steels (Canada) for a deep-hardening carburizing steel containing 0.12% carbon, 0.5% manganese, 0.2% silicon, 1.5% chromium, 3.75% nickel, up to 0.03% phosphorus, 0.03% sulfur, and the balance iron. It has very high core strength, good toughness, wear and fatigue resistance, high tensile strength, and moderate machinability. Used for carburized parts subject to severe operating conditions, anvils, shafts subject to vibration, transmission shafts, bushings, heavy-duty bearings and gears, truck and tractor gears, chuck jaws, clutch dogs, collets, straightening rolls, diesel engine injectors, plastic molds, and pneumatic tools.

Supron. Trade name for nylon fibers and yarns used for textile fabrics.

Sura. Trade name of Surahammars Bruk (Sweden) for a series of nickel-chromium, nickel-chromium-molybdenum, chromium-molybdenum machinery steels, and various nonoriented and grain-oriented silicon steels for transformers, motors, dynamos, generators, armatures, etc.

Suracel. Trade name for a cellulose acetate fibers and yarns.

surah. A soft, relatively heavy fabric of silk, acetate, triacetate, polyester or rayon yarn in a twill weave used for blouses, dresses, suits, scarves, and ties. It is named after Surat, a city in western India.

Sur Coat. Trade name of Coral International (USA) for a conversion coating and coating process.

Surcoat. Trade name of Sur-Fin Chemical Corporation (USA) for a chromate coating for aluminum.

Sure Cure. Trade name of Dymax Corporation (USA) for tough, glossy, two-component conformal coatings of the ultraviolet or chemical-curing type.

SureFil. Trade name of Dentsply/Caulk (USA) for a light-cure, packable, high-viscosity dental resin composite for posterior restorations.

surface coatings. Coatings produced on the surface of a material for corrosion and/or wear resistance, decoration, texture, electrical insulation, lubricity, and/or protection against high temperatures. Typical examples are chromate and phosphate coatings, black oxide coatings, anodic coatings, organic coatings, nanocrystalline coatings, inorganic and vitreous coatings, and electrodeposits. See also coatings.

surfaced lumber. Lumber which is surfaced or finished by running through a planer. It may be surfaced on one side (S1S), two sides (S2S), one edge (S1E), two edges (S2E), one side and one edge (S1S1E), two sides and one edge (S2S1E), one side and two edges (S1S2E), or four sides (S4S). Also known as *dressed lumber; planed lumber.*

surface-engineered material. A material having a surface purposely designed, treated and/or finished to impart certain well-defined, desirable properties, such as hardness, appearance, finish, corrosion, heat and/or wear resistance, chemical activity or other properties.

surface glaze. A thin, clear glaze applied over both the ceramic body and the decoration.

Surface Modifying End Groups. Trademark of Polymer Technology Group, Inc. (USA) for surface-active oligomers, such as silicone, fluorocarbon, sulfonate, hydrocarbon and polyethylene oxide groups, covalently bonded to the backbone of a base polymer, such as polyurethane, to control the surface chemistry. For biopolymers the attachment of these end groups often results in increased biostability, thromboresistance and abrasion resistance. Abbreviation: SMEs.

surfacer. A pigmented composition used to smoothen or improve a substrate or primer prior to the application of finish coats.

surface-stabilized ferroelectric liquid crystal. A *ferroelectric liquid crystal* in which the helical structure is constrained. It

has a lower switching time from the white (or relaxed) state to the black state and vice versa than other liquid crystals, and is ideally suited for various optoelectronic applications especially high-definition televisions. Abbreviation: SSFLC. See also liquid crystal.

surfacing mat. See overlay.

Surfacite. Trade name of The Shwayder Company (USA) for wear-resistant hardfacing electrodes and preforms composed of tungsten carbide (WC) particles in a bronze matrix enclosed in a shell of low-carbon steel. They possess good weldability and impact strength, and can be heated and bent to conform to any surface.

Surfastone. Trademark of Sternson Limited (Canada) for an exposed aggregate wall coating.

Surfcote. Trademark of Imperial Optical Company Limited (Canada) for tinting and antireflection coatings used for optical lenses.

Surf-Hard. Trade name of Rio Algom Corporation (Canada) for an abrasion- and wear-resistant nitriding steel containing 0.4% carbon, 0.6% manganese, 1.6% chromium, 0.35% molybdenum, 1.05% aluminum, and the balance iron. Used for gears, shafts, bushings, sleeves, ejectors pins, etc.

Surflac. Trade name of Pax Surface Chemicals Inc. (USA) for water-based coatings.

Sur-Flex. Trademark of Flex-O-Glass, Inc. (USA) for ionomer-based plastic films including glossy, strong, puncture-resistant *Surlyn* skin packaging films and glossy, strong, cold-crack resistant stretch pack films.

Surflex F. Trade name for a polysulfide elastomer used for dental impressions.

Surgaloy. Trade name of Davis & Geck Inc. (USA) for an austenitic stainless steel containing 0-0.02% carbon, 18% chromium, 8% nickel, and the balance iron. Used for medical applications.

surgical alloys. A group of alloys including in particular several cobalt-base superalloys and stainless steels used for surgical and dental instruments. They are very resistant to corrosive attack by body acids and fluids, and have no harmful effects on human tissue.

Surgical Simplex. Trade name for a self-cure acrylic dental cement.

Surinam greenheart. See greenheart (1).

surinamite. A blue mineral composed of iron magnesium aluminum silicate, $(Mg,Fe)_3Al_4BeSi_3O_{16}$. Crystal system, monoclinic. Density, 3.30 g/cm^3; refractive index, 1.743. Occurrence: Surinam.

Surinam teak. The hard, heavy, dark brown wood from the tree *Hymenea courbaril.* It has fair to poor workability and moderate durability, and is used for flooring, paneling, and general construction. Also known as *West Indian locust. Note:* Surinam teak is not a true teak.

surite. A white mineral composed of lead aluminum carbonate silicate hydroxide, $PbAlSi_2O_5CO_3OH$. Crystal system, monoclinic. Density, 4.00 g/cm^3. Occurrence: Argentina.

Surmat. Trademark of Nicofibers (USA) for glass fiber filaments, yarns.

Surlyn. Trade name of E.I. Du Pont de Nemours & Company, Inc. (USA) for a family of ionomer (polyolefin) resins that can be processed by blow or injection molding or thermoforming. They have high transparency, toughness and impact resistance, good dura-bility and tensile strength, a softening point of 71°C (160°F), high resiliency and flexibility, good resistance to oils, greases and solvents, and poor resistance to acids. Used for golf balls, coatings, and packaging films.

Surprex W2010X. Trade name of Fujimi Inc. (Japan) for a composite cermet powder made from tungsten carbide, with a metal or alloy matrix binder. It has enhanced durability and impact resistance, and strong adhesion to many substrates. Used in form of a coating applied by high-velocity oxyfuel (HVOF) and high-velocity air-fuel (HVAF) processes.

Surprise. Trade name for a dental porcelain for metal-to-ceramic restorations.

sursassite. A copper-colored mineral of the pumpellyite group composed of manganese aluminum silicate trihydrate, Mn_5-$Al_4Si_5O_{21}\cdot3H_2O$. Crystal system, monoclinic. Density, 3.26 g/cm^3; refractive index, 1.755. Occurrence: Switzerland.

sursulfate cement. A cement composed of 70% slag and 30% calcium sulfate.

Survival. Trade name of Sylvania Industrial Corporation (USA) for regenerated cellulose.

Susini. Japanese trade name for an aluminum alloy containing 1.5-4.5% copper, 1.5-2% manganese and 0.5-1.5% zinc. Used for light-alloy parts.

suspender. See suspension agent (2).

suspension. A system in which minute solid, semisolid or liquid particles are dispersed in a fluid suspending medium (liquid or gas). The particles are referred to as the "dispersed phase" and the suspending medium as the "continuous phase." Suspensions in which the particles are solid and the suspending medium is a liquid include wood pulp suspended in water and chalk suspended in water. Aerosol and fog are examples of suspensions in which the particles are liquids and the suspending medium is a gas.

suspension agent. (1) An agent added to a solid-in-liquid suspension to decrease the rate of settlement of the particles.

(2) A material, such as clay, bentonite, borax or sodium aluminate, added to a porcelain-enamel or glaze slip to promote or stabilize suspension of the solid particles in the liquid suspending medium. Also known as *suspending agent; suspender.*

suspension polymer. A polymer, usually in the form of powder, spheres or beads, that is the end product of a polymerization reaction carried out in suspension at standard pressure (101.325 kPa or 14.696 psi) and at temperatures between 60 and 80°C (140 and 176°F) with the reacting species in emulsified form. Abbreviation: SP.

suspension polyvinyl chloride. Polyvinyl chloride, usually in the form of a dry-blended powder, small pellets or cubes, used for suspension polymerization and melt processing by blow or injection molding, calendering or extrusion. Abbreviation: S-PVC. See also suspension polymer.

sussexite. A white to buff or straw yellow colorless mineral that is composed of manganese borate hydroxide, $MnBO_2(OH)$, and may also contain magnesium. Crystal system, orthorhombic. Density, 3.30 g/cm^3; refractive index, 1.728. Occurrence: Switzerland, USA (New Jersey).

Sustaglide. Trade name of Sustaplast Inc. (USA) for a cast polyamide (nylon) containing a special thermoplastic lubricant. The unique combination of the excellent sliding properties of the lubricant with the good mechanical and thermal properties of the polyamide provides a material suitable for applications requiring wear resistance and a wide service temperature range from -40 to +105°C (-40 to +220°F).

Suwefa. Trade name of Stahlwerke Südwestfalen (Germany) for an extensive series of austenitic, ferritic and martensitic stain-

less steels, austenitic manganese steels, numerous alloy steels including several case-hardening and nitriding grades, and an extensive range of tool steels, e.g., plain-carbon, hot-work, cold-work and high-speed grades.

suzukiite. A bright green mineral of the pyroxenoid group composed of barium vanadium silicate, $BaVSi_2O_7$. Crystal system, orthorhombic. Density, 4.00 g/cm^3; refractive index, 1.739. Occurrence: Japan.

svabite. A colorless or pale lilac, transparent mineral of the apatite group composed of calcium chloride fluoride arsenate hydroxide, $Ca_5(AsO_4)_3(OH,Cl,F)$. Crystal system, hexagonal. Density, 3.53 g/cm^3; refractive index, 1.716. Occurrence: Siberia.

svanbergite. A colorless, yellow or reddish mineral of the alunite group that is composed of strontium aluminum phosphate sulfate hydroxide, $SrAl_3(PO_4)(SO_4)(OH)_6$, and may also contain some calcium. Crystal system, rhombohedral (hexagonal). Density, 3.22 g/cm^3; refractive index, 1.633. Occurrence: Sweden.

Svedion. Trade name for a corrosion-resistant cobalt-chromium dental casting alloy.

sveite. A white mineral composed of potassium aluminum chloride nitrate hydroxide octahydrate, $KAl_7(NO_3)_4Cl_2(OH)_{16}\cdot8H_2O$. Crystal system, monoclinic. Density, 2.00 g/cm^3. Occurrence: Venezuela.

Svenska. Trade name of Svenska Metallverken AB (Sweden) for a series of wrought coppers and copper alloys including several high-conductivity coppers and various leaded and unleaded brasses (e.g., high and low-leaded brasses, aluminum brasses, admiralty brasses and muntz metals) and nickel silvers.

Sverker. Trade name of Uddeholm Corporation (USA) for a series of high-carbon high-chromium cold-work tool steels (AISI types D2 and D6) containing 1.5-2.1% carbon, 11.8-13% chromium, 0.2-0.8% vanadium, 0-1.3% tungsten, 0-1% molybdenum, and the balance iron. Used for dies, plastic molds, hobs and punches.

Svitontex. Trademark of Chemosvit (Slovak Republic) for nylon 6 staple fibers and filament yarns.

Svorit. Czech trade name for a smoke-colored, anti-sun spectacle glass.

Swallow. Trade name of Central Glass Company Limited (Japan) for a glass with swallow pattern.

swamboite. A pale yellow mineral of the uranophane group composed of hydrogen uranyl silicate hydrate, $UH_6(UO_2SiO_4)_6\cdot30H_2O$. Crystal system, monoclinic. Density, 4.03 g/cm^3; refractive index, 1.661. Occurrence: Zaire.

swamp cedar. See eastern white cedar.

swamp cottonwood. The moderately strong wood of the broad-leaved tree *Populus heterophylla*. The heartwood is grayish-white to light-brown and the sapwood whitish. *Swamp cottonwood* has a fine, open grain, a uniform texture, and works fairly well. Other properties include low to moderate strength, excellent gluing and nailing qualities and high warping tendency. Average weight, 450 kg/m^3 (28 lb/ft^3). Source: Eastern United States (especially along the Mississippi and Ohio, and along the Atlantic coast from Maryland to South Carolina). Used for lumber and veneer used crates, pallets, and boxes, paneling, and as a pulpwood and fuelwood. Also known as *river cottonwood; swamp poplar.*

swamp gum. See water tupelo.

swamp maple. See silver maple.

swamp ore. See bog iron ore.

swamp pine. See pond pine.

swamp poplar. See swamp cottonwood.

swamp tupelo. See water tupelo.

swan's-down. A very soft, heavily napped fabric made of cotton, or a blend of cotton or silk with wool.

Swansea Vale. Trade name of Pasminco Europe Limited (UK) for a commercially pure zinc (98.5%) supplied in ingot form.

swartzite. A green, radioactive mineral composed of calcium magnesium uranyl carbonate dodecahydrate, $CaMg(UO_2)(CO_3)_3 \cdot 12H_2O$. Crystal system, monoclinic. Density, 2.30 g/cm³; refractive index, 1.51. Occurrence: USA (Arizona).

Sweat-On Paste. Trade name of Wall Colmonoy Corporation (USA) for an abrasion- and wear-resisting alloy composed of 82% chromium and 18% boron. Used as an alloy paste for welding and hardfacing applications.

swedenborgite. A colorless or wine-yellow mineral composed of sodium beryllium antimony oxide, $NaBe_4SbO_7$. Crystal system, hexagonal. Density, 4.28 g/cm³. Occurrence: Sweden.

Swedish green. See cobalt green.

Swedish iron. A charcoal pig iron made from high-purity iron ore (*magnetite*) mined in the vicinity of Dannemora, Sweden, and in northern Norway. Also known as *Dannemora iron.*

Swedish putty. A spackling compound mixed with a paste paint.

sweet birch. See black birch.

sweet buckeye. See yellow buckeye.

sweet chestnut. The durable wood of the hardwood tree *Castanea sativa*. It is similar in appearance to oak, but is lighter and not as strong. Average weight, 550 kg/m³ (34 lb/ft³). Source: Europe including British Isles. Used for furniture, building construction purposes, and in marine construction. Also known as *European chestnut; Italian chestnut; Spanish chestnut.*

sweetgum. The moderately strong, hard and stiff wood of the hardwood tree *Liquidambar styraciflua*. It varies in color from pinkish to brown and even gray. The heartwood, also known as "red gum," is lustrous and reddish-brown, may be highly figured and has a ribbon-stripe pattern due to interlocked grain, and the sapwood is light-colored. *Sweetgum* has a fine, close grain, an even fine texture, good working and machining properties, high tendency to warping, good gluing, staining and nailing qualities and high shock resistance. Average weight, 575-640 kg/m³ (36-40 lb/ft³). Source: Mexico, USA (from southwestern Connecticut westward into Missouri and southward to the Gulf states). Used for fine furniture, cabinetmaking, veneer; moldings, inside trim, plywood, cooperage, woodenware, boxes, toys, railroad ties, and as a pulpwood and fuel. It is sometimes used as a substitute for *satinwood*. See also red gum.

Sweet Pea. Trade name of Nippon Sheet Glass Company Limited (Japan) for a patterned glass.

Swelan. Swedish trade name for viscose rayon fibers.

swelling clay. A clay, such as *sodium bentonite*, that has a high swelling capacity in water.

Swelan. Trademark of Svenska Rayon AB (Sweden) for viscose rayon staple fibers.

Swift. Trade name of Reichhold Chemical Company (USA) for a pressure-sensitive adhesive with high tack and peel strength that bonds well to polyethylene, and is used for bag sealing applications and various applications in the beverage, packaging, pulp and paper, woodworking and textile industries.

Swift Set. Trade name of King Packaged Products Company (Canada) for a fast-setting cement for concrete repairs.

Swiftweld. Trade name of Lincoln Electric Company (USA) for a carbon steel containing 0.2% carbon, and the balance iron. Used for welding electrodes for mild steel.

swimming-pool paint. An air-drying organic coating for swimming pools usually based on a synthetic rubber (e.g., chlorinated rubber), and having good resistance to water, alkalies and acids. Also known as *pool paint.*

Swina. Trade name for rayon fibers and yarns used for textile fabrics.

swinefordite. A greenish gray mineral of the smectite group composed of lithium magnesium aluminum silicate fluoride hydroxide, $Li(Al,Mg)_4Si_8O_{20}(OH)_4 \cdot xH_2O$. Crystal system, orthorhombic. Refractive index, 1.524. Occurrence: USA (North Carolina).

Swirl. Trade name of PPG Industries Inc. (USA) for a patterned, rough-cast plate glass.

swiss. A fine, shear, crisp cotton fabric frequently having a pattern of woven or chemically applied dots or figures. Used for aprons, dresses and curtains. Originally from Switzerland. See also dotted swiss.

Swissflam. Trademark of Saint-Gobain (France) for an insulated, fire-resistant glass.

Swissflam Lite. Trademark of Saint-Gobain (France) for a non-insulated, fire-resistant glass.

switch copper. A strong, high-conductivity electrolytic copper (99.9+% pure) with excellent cold and hot workability used for switch blades and electrical components.

Sybraloy. Trade name of Kerr Dental (USA) for a regular- and fast-set dental amalgam alloy capsules.

sycamore. See American sycamore; planewood.

Sycob. Trade name of Aubert & Duval (France) for oil- or air-hardening, nondeforming tool and die steel containing 2% carbon, 12.5% chromium, 0.5% cobalt, 0.4% molybdenum, 0.2% vanadium, and the balance iron.

Syenite. Trade name of C-E Glass (USA) for a glass with irregular pattern.

syenite. A plutonic (coarse-grained), light colored (typically reddish or grayish-white) igneous rock composed chiefly of *orthoclase* feldspar with small amounts of hornblende, biotite, pyroxene, zircon, apatite and magnetite. There are several different types including *nepheline syenite* that contains the mineral nephelite in excess of 5%, *mica syenite* that contains appreciable amounts of biotite, and *leucite syenite* with more than 5% of the mineral leucite.

syeporite. A brown mineral composed of cobaltous sulfide, CoS. Crystal system, hexagonal. Density, 5.45 g/cm³; melting point, above 1100°C (2010°F).

Syl-Carb. Trade name of GTE Products Corporation (USA) for a series of tungsten carbide powders used in the manufacture of cemented carbide products, such as metal cutting tools, dies, and drilling tools.

Sylek. Trade name of Surface Engineered Consulting Limited (UK) for electroless nickel-boron coatings and plating solutions.

Sylgard. Trademark of Dow Corning Corporation (USA) for silicone resins and elastomers used as encapsulants in electronics.

Syloid. Trademark of W.R. Grace & Company (USA) for micronsized silica gel used in adhesives, lacquers, baking finishes, insulation, etc.

Sylomer. Trademark of Getzner Werkstoffe GmbH (Austria) for polymer-based acoustic insulating materials.

Sylphrap. Trade name of Sylvania Industrial Corporation (USA) for regenerated cellulose.

Sylramic. Trade name of Dow Corning Corporation (USA) for strong, high-temperature silicon carbide fibers used as reinforcements in ceramic, polymer and metal-matrix composites, in woven tapes and braids, and in high-temperature insulation products.

Sylvacote. Trademark of MacMillan Bloedel Limited (Canada) for plywood used for concrete forms, and specialty plywood boards for siding, soffits, etc.

Sylvaloy. Trade name of Wilbur B. Driver Company (USA) for a magnetic alloy of 97% nickel and 3% silicon.

Sylvamix. Trademark of Glidden-Durkee Division of SCM Corporation (USA) for tall oil resin derivatives used as additives in concrete masonry and mortar joints to reduce efflorescence, enhance texture and decrease water absorption.

Sylvania. Trade name of GTE Products Corporation (USA) for a series of refractory alloys based on tungsten or molybdenum, made by powder-metallurgy techniques, and used for electrical contacts, flywheels, counterweights, radioactive shielding, high-temperature filaments, gyroscopes, heat exchangers, heat engines, etc. Also included under this trade name are iron-nickel alloys supplied in wire form for controlled expansion glass-to-metal seals, high-purity iron for filaments, and nickel-base alloys for nuclear reactor equipment.

Sylvania Cellophane. Trade name of Sylvania Industrial Corporation (USA) for cellophane-type regenerated cellulose.

sylvanite. A silver-white, steel-gray or brass-yellow mineral with a brilliant metallic luster. It belongs to the calaverite group, is composed of gold silver telluride, $(Au,Ag)Te_2$, and contains about 24.5% gold and 13.3% silver. Crystal system, monoclinic. Density, 7.9-8.3 g/cm^3; hardness, 1.5-2 Mohs; darkens on exposure. Occurrence: Australia, Rumania, USA (California, Colorado). Used as a source of gold. Also known as *goldschmidtite; graphic tellurium; white tellurium; yellow tellurium.*

Sylvaply. Trademark of MacMillan Bloedel Limited (Canada) for plywood.

sylvine. See sylvite.

sylvite. A colorless, white, reddish or bluish mineral with a white streak and a vitreous luster. It belongs to the halite group, is composed of potassium chloride, KCl, and contains 43% potassium chloride and 57% sodium chloride. Crystal system, cubic. Density, 1.99 g/cm^3; melting point, approximately 790°C (1454°F); hardness, 2.0 Mohs; refractive index, 1.490. Occurrence: Europe, USA (New Mexico, Texas). Used as an important source of potassium, and in fertilizers. Also known as *sylvine.*

Symbioceram. Trademark of Degussa-Ney Dental (USA) for a hydrothermal porcelain that is compatible to most ceramic alloys with thermal expansion coefficients in the range of 14.2-15.1 × 10^{-6}/K. It has excellent wear properties, and good polishability and handling characteristics, and is used for dental restorations.

symmetrical compound. A compound prepared from *benzene* by substituting the hydrogen atom on three alternate carbon atoms (e.g., C-1, C-3 and C-5) with another chemical element or functional group.

symmetrical laminate. A laminate in which the sequence of layers (or plies) above and below the midsurface or midplane is symmetrical. *Note:* The midsurface is located at half the total laminate thickness.

Symphony. Trade name of J.F. Jelenko & Company (USA) for a hard dental casting alloy (ADA type III) containing 61% silver, 27% palladium, and 2% gold. It provides excellent castability, adheres well to dental porcelain, and is used for crowns, inlays and fixed bridgework.

symplesite. A light green mineral of the vivianite group composed of iron arsenate octahydrate, $Fe_3(AsO_4)_2·8H_2O$. Crystal system, triclinic. Density, 3.01 g/cm^3; refractive index, 1.668; hardness, 2.5 Mohs. Occurrence: Rumania.

Simplicity. Trade name of Gervase Structures Limited (UK) for a double glazing unit with an aluminum U-section channel.

synadelphite. A red mineral composed of manganese arsenate hydroxide dihydrate, $Mn_9As_3O_{11}(OH)_9·2H_2O$. Crystal system, orthorhombic. Density, 3.57 g/cm^3; refractive index, 1.751. Occurrence: Sweden.

synchysite. (1) A brownish red mineral of the bastnaesite group composed of calcium cerium fluoride carbonate, $CaCe(CO_3)_2F$. Crystal system, orthorhombic. Density, 3.92 g/cm^3; refractive index, 1.648. Occurrence: Italy.

(2) A light grayish blue or light violet mineral of the bastnaesite group composed of calcium neodymium fluoride carbonate, $CaNd(CO_3)_2F$. Crystal system, orthorhombic. Density, 4.21 g/cm^3; refractive index, below 1.66. Occurrence: Czech Republic.

(3) A brownish red, yellow or gray mineral of the bastnaesite group composed of calcium yttrium fluoride carbonate, $CaY(CO_3)_2F$. Crystal system, orthorhombic. Density, 3.40-3.94 g/cm^3. Occurrence: USA (New Jersey, Colorado).

Synco-Komplex. Trade name of NV Syncoglas SA (Belgium) for a combination of woven glass rovings and chopped-strand or continuous-strand mat, or surfacing tissue.

Syncomat. Trade name of NV Syncoglas SA (Belgium) for a surfacing tissue of chopped glass strands or continuous glass, acrylic or polyester fibers.

Syncopreg. Trade name of NV Syncoglas SA (Belgium) for a pre-impregnated glass mats and woven rovings.

SynCore. Trademark of Dexter Corporation (USA) for low-density, syntactic-film core materials based on toughened bismaleimide epoxy resins. Used for engine nacelles and radomes on commercial jet aircraft.

Syndecrete. Trademark of Syndesis Inc. (USA) for a cement-based composite made from recycled materials.

Syndetikon. German trademark for a viscous bone glue.

syndiotactic polymer. A polymer with a regular, alternate arrangement of side groups (R) on each side of the polymer chain. Examples of such polymers are syndiotactic polypropylene $(R=CH_3)$, and syndiotactic polystyrene $(R=C_6H_5)$.

synergistic coatings. A term used for coatings produced during multi-step processes combining the advantages of anodizing on hard-coat plating with the controlled infusion of low friction polymers and/or dry lubricants. These coatings become an integral part of the top layers of the base metal.

Synergy. (1) Trademark of Dow Chemical Company (USA) for strong, tough, resilient soft-touch performance foams made from a proprietary blend of custom-designed low-density polyethylenes and Dow *Index Interpolymers*. They have excellent shock-absorbing, vibration dampening and insulating properties. Supplied in black or natural colors in plank form, 50 × 600 × 2700 mm (2 × 24 × 108 in.), they are available in three grades with different degrees of softness: *Synergy 1000* is a soft foam with a density of 29 kg/m^3 (1.8 $lb/ft.^3$), *Synergy 3000* an even softer foam with a density of 26 kg/m^3 (1.6 $lb/ft.^3$) and *Synergy 5000* is the softest grade with a density of 29 kg/m^3 (1.8 $lb/ft.^3$). *Synergy* foams are used for barrier and buoyancy components, dunnage and cushioning components for packaging applications, automotive materials handling, padding, protection of electronic equipment, and touch-sensitive applications.

(2) Trademark of Coltene-Whaledent (USA) for light-cure, no-slump micro-hybrid composite that polish to a high ceramic-like luster. Used in restorative dentistry.

Synflex-N. Trade name for nylon fibers, yarns and products.

SynFoam. Trademark of Powdermet, Inc. (USA) for a family of lightweight closed-cell structural metallic syntactic foams composed of hollow ceramic microballons (e.g., alumina, carbon, glass, or mullite) in a high-strength alloy matrix. Typical matrix materials include magnesium, aluminum, rhenium, titanium, tantalum, iron/steel, nickel and cobalt-based alloys. *SynFoam* composites are made by an activated sintering process, and can be formed into complex shapes by compressive molding, isostatic pressing, powder injection molding, slip or slurry casting or vacuum sintering. They have high strength, stiffness and dimensional stability, high shear strength, high heat resistance, and very low moisture and water absorption. Used as core materials, for lightweight structures, and in high-strength tooling.

syngenite. A white mineral composed of potassium calcium sulfate monohydrate, $K_2Ca(SO_4)_2 \cdot H_2O$. It can also be made synthetically. Crystal system, monoclinic. Density, 2.56 g/cm^3.

Synolite. Trade name of DSM Resins (USA) for polyester casting resins available in flexible and rigid grades.

Synopal. Trade name of Karl Kroyer (Denmark) for a crystallized glass substance that is used to increase the skid resistance of roads.

Synres. Trade name for unsaturated polyesters.

Synsil. Trademark of Minerals Technologies, Inc. (USA) for a family of synthetic silicate minerals used in glass manufacture, especially for lowering the melting temperatures.

Syn-Slag 79. Trade name of BPI Inc. (USA) for a blend of lime and premelted calcium aluminate with fluorspar. A typical composition is 79% lime (CaO), 12% calcium fluoride (CaF_2), 5% alumina (Al_2O_3), 1.5% magnesia (MgO), 1.5% silica (SiO_2), and a total of 1% titania (TiO_2), ferric oxide (Fe_2O_3), sodium oxide (Na_2O), phosphorus pentoxide (P_2O_5) and sulfur. Used as a synthetic slag for metallurgical applications.

Syntac. (1) Trademark of W.R. Grace & Company (USA) for a series of syntactic cellular plastics.

(2) Trade name of Ivoclar Vivadent AG (Liechtenstein) for a selective-etch, light-cure acrylate-base dental enamel bonding system. *Syntac Single Component* is a light-cure single-component acrylate-based dental bonding system, and *Syntac Sprint* a fluorine-releasing, single-step dental adhesive for dentin bonding applications.

syntactic cellular plastic. See syntactic foam.

syntactic foam. A lightweight cellular plastic made by mixing preformed cells (hollow microspheres of glass, phenolic, epoxy, etc.) with liquid synthetic resin and the necessary additives. Also known as *syntactic cellular plastic*.

syntan. (1) A synthetic organic tanning agent, such as a sulfonated phenol or a naphthol condensed with formaldehyde. Used for tanning various leathers.

(2) Leather that has been tanned with synthetic organic tanning agents. Combined syntan-vegetable and syntan-chrome tanned leather is often used for shoe soles and uppers.

Syntex. Trade name of US Polymers, Inc. (USA) for a wide range of synthetic resins.

synthacyl. A copolymer of dimethyl acrylamide and acrylosarcosine methyl ester used as a substrate resin in the synthesis of polypeptides and oligonucleotides.

Synthane. Trademark of Synthane Corporation (USA) for laminated phenolic condensation products.

Synthemul. (1) Trademark of Reichhold Chemicals, Inc. (USA) for an abrasion-resistant synthetic rubber latex with a soft hand and excellent washfastness. Used for textile fabrics.

(2) Trademark of Reichhold Chemicals, Inc. (USA) for acrylic emulsions used in the manufacture of durable exterior paints, ceramic tile mastics, and paints for metals and wood.

Synthemul CPS. Trade name of Reichhold Chemicals, Inc. (USA) for acrylic and styrene-acrylic copolymers used as hydraulic cement additives, and in decorative interior paints.

synthetic abrasives. See artificial abrasives.

synthetic adhesives. A group of organic adhesives including those manufactured from thermoplastic and thermosetting resins (e.g., polyethylene, polyamide, polyimide, epoxy, phenol-formaldehyde, etc.) as well as and elastomer-solvent cements, silicone and chlorinated rubber adhesives, etc. See also adhesives; natural adhesives.

synthetically bonded sand. See synthetic sand.

synthetic cast iron. A cast iron made from a charge of scrap iron and carbon by melting in an electric induction furnace.

synthetic cold-rolled sheet. A steel sheet that, after being hot-rolled and pickled, has received a final pass in a temper mill to impart a surface finish approaching that of a cold-rolled sheet. See also hot-rolled sheet; hot-rolled and pickled sheet.

synthetic composites. Composite materials made up of at least two synthetically produced phases, e.g., fiberglass and a synthetic resin, bonded to produce a material with properties superior to those of either phase. See also composites; natural composites.

synthetic cryolite. See sodium fluoaluminate.

synthetic diamond. A usually colorless *diamond* made by heating a carbonaceous material (e.g., carbon or graphite) and a metallic catalyst (e.g., chromium, cobalt, nickel, etc.) in an electric furnace at about 1650°C (3000°F) under high pressure. Crystal structure, cubic. Density, 3.36 g/cm^3; hardness, above 6900 Knoop. Used as an abrasive (in the form of polishing powders, grinding wheels, etc.) for grinding or polishing hard materials, in drill bits, bonded tips on cutting tools, wire-drawing dies, and in glass and metal cutting. Also known as *manmade diamond*. See also natural diamond.

synthetic diopside. A white synthetic form of the mineral *diopside* obtained from sand, lime and a special binder and catalyst. It is composed of calcium magnesium silicate corresponding to the formula $CaMg(SiO_3)_2$, and usually supplied as a fine powder. Crystal system, monoclinic. Density, 3.2-3.3 g/cm^3; hardness, 5-6 Mohs. Used for refractories, in welding-rod coatings, and as a component in whiteware bodies, glazes and glass.

synthetic fiber felt. A felt-like nonwoven structure made from synthetic fibers, such as acrylic, nylon, polyester, polypropylene, polytetrafluoroethylene, etc. Used for fancy dresses, costumes, fabric decorations and ornamentations.

synthetic fibers. A collective term for man-made fibers including (i) thermoplastic organic fibers, such as nylon, polyester and polypropylene, (ii) thermosetting organic fibers, such as aramid, and (iii) artificial inorganic fibers, such as glass, carbon, silicon carbide, alumina, and boron. Used for textile fabrics and in advanced composites. Also known as *synthetics*. See also artificial fibers; synthetic man-made fibers.

synthetic fluoroapatite. See fluoroapatite.

synthetic garnets. Ferrimagnetic ceramic materials having very complicated crystal structures that can be represented by the general formulas $M_3Fe_5O_{12}$ and $M_3Al_5O_{12}$, respectively, where M usually represents a rare-earth element, such as dysprosium, erbium, europium, gadolinium, holmium, samarium, terbium, thulium, ytterbium or yttrium. Used in solid-state electronics, lasers, and microwave devices. See also natural garnets.

synthetic gem. A man-made precious and semiprecious stone, such as diamond, ruby or sapphire, used principally for industrial applications, e.g., in bearings for clocks, in scales and precision instruments, in laser and maser equipment, in optical lenses and windows, as abrasives, for tool tips, etc. See also synthetic diamond; synthetic ruby; synthetic sapphire.

synthetic graphite. A high-purity, crystalline *graphite* made by heating petroleum coke to about 3000°C (5430°F) in an electric furnace. Also known as *artificial graphite*.

synthetic gypsum. Gypsum ($CaSO_4 \cdot 2H_2O$) obtained as a byproduct from chemical or industrial operations, as compared to that found in nature. See also calcium sulfate dihydrate; gypsum.

synthetic high polymers. See synthetic resins.

synthetic hydroxyapatite. A usually green or bluish-green surface-reactive bioceramic based on calcium phosphate hydroxide with outstanding biocompatibility and available in regular and microcrystalline grades. It is a finely divided, nonstoichiometric powder with a hexagonal or pseudohexagonal crystal structure, and a density of 3.08-3.15 g/cm³ (0.111-0.114 lb/in.³). Used in chromatography, as a coating material on metallic prostheses and implants (e.g., steel or titanium-alloy based) in bone reconstruction and regeneration, in dental implants, and for biocomposites. See also hydroxyapatite.

synthetic leather. A leather substitute made either from scrap leather, or by coating a textile fabric with a mixture of pyroxylin, castor oil, pigments and solvent. Also known as *artificial leather*. See also leather; leatherboard.

synthetic magnesite. Dead-burnt magnesia (97-99% pure) obtained chemically from seawater or brines. Also known as *seawater magnesite*. See also magnesite; dead-burnt magnesia.

synthetic man-made fibers. Staple fibers and filaments made by the polymerization of organic monomers. Examples include acrylic, modacrylic, nylon, polyethylene, polypropylene, polyester, polyurethane, polyvinyl alcohol and trivinyl fibers. See also synthetic fibers; man-made fibers.

synthetic mica. A soft, translucent noncombustible solid made from *muscovite* (ruby mica), *phlogopite* (amber mica) or *pegmatite* usually in a crucible or electric furnace using a crystal-growing technique involving heating and subsequent controlled cooling. Its color ranges from colorless to slight red for muscovite to brown or greenish yellow for phlogopite. It is available in the form of blocks, sheets, powder and single crystals. Density, 2.6-3.2 g/cm³; hardness, 2.8-3.2 Mohs; refractive index, 1.56-1.60. It has excellent mechanical properties, good dielectric properties, good heat resistance to 600°C (1110°F), very low water absorption; good resistance to alkalies and dilute acids, and good to fair machinability. Used for electrical equipment, incandescent lamps, vacuum tubes, specialty paper for insulation and filtration, as a flux in glass and ceramic manufacture, for windows in high-temperature equipment, in cement for wallpaper and wallboard, as a dusting agent, and as a filler in exterior paints and rubber. See also mica.

synthetic mineral. See mineral (2).

synthetic mullite. A colorless ceramic aluminum silicate material ($Al_6Si_2O_{13}$) usually prepared from a stoichiometric mixture of alumina (Al_2O_3) and hydrous silica ($SiO_2 \cdot xH_2O$) by repeated grinding and subsequent heating to high temperatures (typi-

cally 1725°C or 3135°F) in an electric furnace. The approximate composition is 71-75 wt% Al_2O_3, 25-29 wt% SiO_2 and trace amounts of iron, calcium, chromium, magnesium, manganese, nickel, titanium and zirconium. Crystal system, orthorhombic. Density, 3.00-3.17 g/cm³; melting point, 1810°C (3290°F); softening temperature, 1650°C (3000°F); refractive index, 1.641; excellent resistance to corrosion and heat; good spalling resistance; low thermal expansion; good flame resistance. Used as refractories for high-temperature applications, as a strength-producing ingredient in stoneware and porcelain, and in glass manufacture. Also known as *artificial mullite*. See also mullite.

synthetic nacre. A material that is analogous to natural *nacre*, but synthesized in the laboratory from calcium carbonate and conchiolin ($C_{32}H_{98}N_2O_{11}$) and used chiefly in biochemistry and biotechnology.

synthetic optical crystal. See optical crystal.

synthetic paper. A paper made from synthetic fibers, such as polyester or polyacrylonitrile or blends of synthetic fibers and wood pulp, on conventional papermaking machinery. Depending on the production process, they can either resemble ordinary paper or fabrics in appearance and properties. Used in filtration, and for electrical insulation, interlinings in clothing, and shoe fabrics.

synthetic cis-polyisoprene. See isoprene rubber; polyisoprene (2).

synthetic polymers. Man-made polymers, i.e., polymer that are not derived from plants and animals, but manufactured from natural, inorganic or organic materials. Examples include various thermoplastic and thermosetting resins (e.g., nylon, polyvinyl chloride, polyethylene, polystyrene, polyurethane, phenolics and polyesters) as well as vulcanized and unvulcanized elastomers. Abbreviation: SP. See also natural polymers.

synthetic quartz. A large, uniform, high-purity *quartz* crystal grown at high temperature and pressure around a seed of quartz that is suspended in an alkaline solution containing natural quartz crystals. Used in precision oscillators, piezoelectric controls in filters, selective wave filters, spectrographic equipment, prisms and lenses for infrared analysis, and radio and television components.

synthetic resins. A group of complex, man-made high polymers produced by a chemical reaction between two or more relatively simple substances. Examples include acrylic resins, alkyd resins, phenolic resins, urethane resins, melamine resins, epoxy resins, polyester resins, synthetic rubbers and elastomers, etc. Also known as *synthetic high polymer*. See also resin; natural resins; high polymers.

synthetic rubber. Any of a group of elastomers that resemble *natural rubber* in physical and chemical properties, and have its unique properties of deformation (high elastic yield strain) and elastic recovery after vulcanization with sulfur or other crosslinking agents. *Synthetic rubber* can be produced from various starting materials, in particular by polymerization of unsaturated compounds, but also by addition and condensation polymerization of suitable starting products. It is often distinguished from natural rubber by its improved properties, such as higher resistance to abrasion, gasoline, oil, oxygen and heat, and lower gas permeability. Examples of include butadiene and styrene-butadiene rubber, nitrile rubber, isoprene rubber, chloroprene rubber and silicone rubber. Abbreviation: SR.

synthetic ruby. Ruby made in the form of single-crystal sheets or rods from aluminum oxide by adding small amounts of chromic oxide. Used especially in maser and laser equipment, and as an abrasive. See also ruby.

synthetics. (1) Textile fabrics made of synthetic fibers.
(2) See synthetic fibers.

synthetic sapphire. Aluminum oxide, Al_2O_3, containing traces of cobalt, chromium and titanium, and made by crystal-growing techniques. It is available in the form of rods, spheres, disks, squares, rectangles, whiskers and single crystals, and used as an abrasive, for pointers, wearing points of instruments, high-precision bearings, valves rings and plug gages, thread guides on textile machines, lenses, prisms, optical windows, electron and microwave tubes, as a substrate for thin-film components and integrated circuits, in infrared detectors, for optical elements in radiation detectors, and as whiskers and fibers in aluminum composites. It was formerly also widely used for phonograph needles. See also sapphire.

synthetic sand. A *foundry sand* with a base of *natural sand* (e.g., silica, chromite, olivine, zircon, etc.) from which all or most of the natural bonding materials have been removed, and to which a binder, such as bentonite, and water have been added. It has several advantages over natural foundry sands including a more uniform grain size, lower moisture content, better moldability, and a higher refractoriness. Also known as *synthetically bonded sand.* See also semisynthetic sand.

synthetic silica. An extremely light, highly porous, open-cell, semitransparent *silica* (SiO_2) material available in the form of a silica powder made from a silica gel using a gaseous dispersing medium, or as an open-cell silica foam produced by the sol-gel process and subsequently dried by supercritical extraction. Used as a reinforcement in rubber, as a filler in plastics, as a gloss reducer in paints, in atomic particle detectors, for aerospace applications, and as a low-temperature insulating materials, e.g., for refrigerator walls. Also known as *silica aerogel.* See silica.

synthetic spider silk. See spider silk (2).

synthetic spinels. A group of magnetic ceramics with the general formula AB_2O_4, where A is a divalent metal, such as barium, calcium, cobalt, copper, magnesium, manganese, nickel, strontium or zinc, and B is a trivalent metal, such as aluminum, chromium or iron. *Synthetic spinels* can be classified into the following groups: (i) *Inverse spinels* that are cubic spinels, such as magnesium ferrite ($MgFe_2O_4$) and nickel ferrite ($NiFe_2O_4$) with the divalent cations (e.g., Mg^{2+}, Ni^{2+}) in 6-fold (octahedral) interstitial sites, and the trivalent cations (e.g., Fe^{3+}) equally divided between 6-fold (octahedral) and 4-fold (tetrahedral) sites; and (ii) *Normal spinels* that are cubic spinels, such as nickel aluminate ($NiAl_2O_4$) or zinc ferrite ($ZnFe_2O_4$) with all divalent cations in 4-fold (tetrahedral) interstitial sites and all trivalent cations in 6-fold (octahedral) sites. *Synthetic spinels* are used as refractories, in electronic applications, as substrates for Group III-V (Group 13-15) thin film deposition and superconductor applications, and in the manufacture of ceramic colors. See also spinels.

synthetic zeolite. A hydrous sodium aluminosilicate compound made by first reacting bauxite with sodium hydroxide, and then adding sodium silicate to the intermediate product. The approximate general formula is $Na_2O \cdot Al_2O_3 \cdot xSiO_2 \cdot yH_2O$. Used in ion-exchange reactions (e.g., in water softeners), as adsorbents and desiccants, and as heating and cooling agents in solar collectors.

Synthofil. Trade name for a vinal (polyvinyl alcohol) fiber formerly made in Germany. Used for textile fabrics.

Syntho-Glass. Trademark of Neptune Research, Inc. (USA) for a fiberglass fabric supplied with water-activated resin, and used for pipe repairs.

Syntholvar. Trade name of Varflex Company (USA) for polyvinyl chlorides and other vinyl plastics.

Synvarite. Trademark of Synvar Corporation (USA) for solid phenol-formaldehyde and urea-formaldehyde resins supplied in powder, granule and lump form.

SynVisc. Trademark of Biomatrix Inc. (USA) for a viscous hyaluronan (hyaluronic acid) product used in the treatment of osteoarthritis as an intra-articular injection to improve joint mobility and relieve pain.

Syracuse Genuine Babbitt. Trade name of United American Metals Corporation (USA) for a shock-resistant tin babbitt containing varying amounts of copper, antimony and lead. Used for bearings resistant to mechanical shocks and high temperatures.

Sysorb. Trade name of Synos (USA) for a biodegradable, bioabsorbable polymer based on poly(DL-lactide) and used in tissue engineering.

system sand. Burnt or partially burnt *foundry sand* that is directly re-used after being reconditioned.

Sytex. Trademark for acoustic, electrical and thermal insulating mats and sheets.

Syton. Trademark of E.I. DuPont de Nemours & Company (USA) for colloidal silica.

SY-Ultra 5.2. Trade name of Z-Tech Corporation (USA) for yttria-stabilized zirconia powders that do not contain binders. They can be isostatically pressed into machinable parts of the desired shape, and then finished before firing. The fired products have high sintered density and low porosity.

Szabo YPG. Trade name for a high-gold dental bonding alloy.

szaibelyite. A white-yellow mineral composed of magnesium borate hydroxide, $MgBO_2(OH)$. It can also be made synthetically Crystal system, orthorhombic. Density, 2.60 g/cm^3; refractive index, 1.649. Occurrence: Canada (British Columbia), Germany.

szmikite. A colorless mineral of the kieserite group composed of manganese sulfate monohydrate, $MnSO_4 \cdot H_2O$. It can also be made synthetically. Crystal system, monoclinic. Density, 3.15 g/cm^3; refractive index, 1.595.

szomolnokite. A yellow red mineral of the kieserite group composed of iron sulfate monohydrate, $FeSO_4 \cdot H_2O$. It can also be made synthetically. Crystal system, monoclinic. Density, 3.05 g/cm^3; refractive index, 1.623. Occurrence: Hungary, Czech Republic, Slovakia.

T

taaffeite. A red mineral of the hogbomite group composed of beryllium magnesium aluminum oxide, $BeMg_3Al_8O_{16}$. Crystal system, hexagonal. Density, 3.61 g/cm³; refractive index, 1.721. Occurrence: Sri Lanka.

Tab 2000. Trade name for a dental resin used for temporary crown and bridge restorations.

table salt. See halite.

Taboren. Trade name of Silon A/S (Czech Republic) for modified polypropylene plastics.

Tabs. Trade name of Tekno Polimer AS for acrylonitrile-butadiene-styrene plastics.

tabular alumina. Corundum (99.5+% pure) prepared by heating alumina (Al_2O_3) to temperatures above 1980°C (3595°F) until near total conversion to alpha alumina (α-Al_2O_3) is obtained. It is composed of tabular crystals, contains little soda (usually less than 0.03%) and is available as white pellets, coarse granules and fine powder. *Tabular alumina* has a density of 3.4-4.0 g/cm³ (0.12-0.14 lb/in.³), a melting point of 2040°C (3704°F), a hardness of 9 Mohs, a porosity of approximately 5%, excellent dielectric properties, high strength and good volume stability at elevated temperatures. Used for high-grade refractories for lining ceramic and metallurgical furnaces, for kiln furniture, for electrical insulators, electroceramics and high-quality porcelains, in investment casting, for abrasive products, for catalyst supports, and as a filler for plastics. See also corundum.

tabular spar. See wollastonite.

tacamahac. See balsam poplar.

tacharanite. A white red mineral composed of calcium aluminum silicate hydrate, $Ca_{12}Al_2Si_{18}O_{51}\cdot18H_2O$. Crystal system, monoclinic. Density, 2.33 g/cm³; refractive index, 1.54. Occurrence: Germany.

tachyhydrite. A yellow mineral composed of calcium magnesium chloride dodecahydrate, $CaMg_2Cl_6\cdot12H_2O$. Crystal system, hexagonal. Density, 1.67 g/cm³; refractive index, 1.520. Occurrence: Germany.

tachylite. See basalt glass.

tackifier. An agent used to impart or increase adhesive stickiness of a material, such as rubber.

Tackle Twill. Trade name for rayon and cotton fabrics in a twill weave used for sportswear and uniforms.

tack-spun fabrics. Textile fabrics manufactured by melting a polymer film with backing substrate by means of a heated roller. The film adheres to the roller and upon separation a fibrous pile is produced.

taconite. A sedimentary rock composed of a mixture of fine-grained silica, hematite and magnetite, and used as a low-grade iron ore containing about 25-30% iron. Occurrence: USA (Taconic Mountains, Lake Superior region, Minnesota).

Tacryl. Trademark for acrylic fibers used for the manufacture of textile fabrics.

Tactel. Trademark of E.I. DuPont de Nemours & Company (USA) for specialty nylon 6,6 filaments and fibers including *Tactel, Tactel Aquator, Tactel Ispira* and *Tactel Strata* as well as *Tactel Micro* microfibers. Used for apparel, hosiery, and other fabrics.

Tactesse. Trademark of DuPont (UK) Limited for nylon 6,6 staple fibers and filament yarns used for "wool-like" textile fabrics including carpets.

Tactic. Trademark of Dow Chemical Company (USA) for a series of epoxy resins.

tactic polymer. A *stereoregular polymer* whose molecular arrangement or structure is regular or symmetric. The three main types of tactic polymers are: (i) *Atactic polymers* having side groups randomly arranged on one side of the polymer chain or backbone; (ii) *Isotactic polymers* having all side groups located on the same side of the chain; and (iii) *Syndiotactic polymers* with a regular, alternate arrangement of side groups on each side of the chain.

Tactix. Trademark of Ciba-Geigy Corporation (USA) for high-performance epoxy resins used for prepregs and adhesives applications.

Tadanac. Trade name of Cominco Limited (Canada) for a series of bluish-white or silvery-gray lead-base alloys containing 72-89% lead, 0-21% antimony and 0-12% arsenic. They have a melting range of 320-410°C (608-770°F).

tadzhikite. A brown mineral composed of calcium rare-earth titanium borate silicate, $Ca_3Ln_2(Ti,Al,Fe)B_4Si_4O_{22}$. Crystal system, monoclinic. Density, 3.73 g/cm³. Occurrence: Turkestan.

taeniolite. A colorless to brown mineral of the mica group composed of potassium lithium magnesium fluoride silicate, $K_{0.6}(Mg,Li)_3Si_4O_{10}F_2$. Crystal system, monoclinic. Density, 2.80-2.83 g/cm³; refractive index, 1.534.

taenite. (1) A silver white to grayish white mineral of the gold group composed of iron nickel, γ-(Fe,Ni). Crystal system, cubic. Density, 7.80 g/cm³.

(2) A face-centered-cubic (fcc) phase found in meteorites, and composed of about 25-50% nickel, and the balance iron. It is thought to be analogous to austenite in iron-carbon alloys. See also kamacite; plessite.

Tafaloy. Trade name of TAFA Inc. (USA) for zinc, tin-zinc, zinc-aluminum-copper and nickel-chromium alloys used for thermal-spray coatings.

Taffen. Trade name of Exxon-Mobil Chemical Company (USA) for a series of thermoplastic composites. *Taffen STC* is a thermoplas-tic composite consisting of a glass-reinforced polyolefin-based resin. *Taffen PP* refers to polypropylene sheeting. Used for structural and load-bearing applications.

taffeta. A stiff, medium-weight fabric of silk, linen or synthetics (e.g., rayon) in a plain weave. It has a smooth, glossy surface on both sides, and is used especially for dresses, ribbons, and drapery.

taffetized fabrics. Cotton fabrics finished to produce a partially glazed surface resembling *taffeta.*

Tafmer. Trademark of Mitsui Chemicals, Inc. (Japan) for a series of colorless, transparent, gelatin-free alpha-olefin copolymers. *Tafmer A* and *Tafmer P* are ethylene/alpha olefins and *Tafmer XR* is a propylene/alpha-olefin copolymer.

TAF steels. A class of heat-resistant alloy steels developed at the University of Tokyo, Japan, and composed of 0.18% carbon, 0.5% silicon, 1.0% manganese, 10.5% chromium, 1.5% molybdenum, 0.2% vanadium, 0.15% niobium, less than 0.1% nickel, 0.04% boron, 0.02% nitrogen, and the balance iron. Their optimum long-term creep-rupture strength is developed by a heat treatment involving oil quenching from 1150°C (2100°F) to 700°C (1290°F) and subsequent air cooling to room temperature. Used chiefly for ultra-super-critical power plant applications, e.g., turbine rotors.

Taftilon. Trademark of INSA ASA (Turkey) for nylon 6 fibers and filament yarns.

taggers tin. Light-gage *tinplate* with a thickness ranging between 0.15 and 0.30 mm (0.006 and 0.012 in.).

Tagnite. Trademark of Technology Application Group Inc. (USA) for abrasion- and corrosion-resistant coatings used for magnesium and magnesium alloys.

Tagoto. Trade name of Asahi Glass Company Limited (Japan) for patterned glass.

tag paper. A strong, rigid, durable paperboard that often contains jute fibers and is used for shipping tags.

Taiclear. Trade name of Tosoh Company (Taiwan) for clear acrylic plastics based on polymethyl methacrylate (PMMA).

Taifun. Trade name of Dörrenberg Edelstahl GmbH (Germany) for a tough, oil-hardening cold-work steel containing 1.05% carbon, 0.9% manganese, 1% chromium, 1.15% tungsten, and the balance iron. Used for tools and dies.

Tail Shaft Bronze. Trade name of Jenney Cylinder Company (USA) for a leaded bronze containing 87% copper, 6% tin, 3% lead and 4% zinc. Used for pump liners, bearings, and centrifugal castings.

Taimelan. Trade name of Denki Kagaku Kogyo Kabushiki Kaisha (Japan) for a polymer alloy of acrylonitrile-butadiene-styrene (ABS) and polyvinyl chloride (PVC).

Taipolene. Trade name of Vulcan Chemical Corporation (USA) for a series of polypropylenes.

Tairilac. Trademark of Formosa Plastics Corporation (Taiwan) for acrylonitrile-butadiene-styrene (ABS) resins, and several polymer alloys of ABS and polycarbonate (PC), styrene-acrylonitrile (SAN) or polymethyl methacrylate (PMMA), respectively.

Tairilin. (1) Trademark of Nan Ya Plastics Corporation America (USA) for polyester staple fibers and filament yarns.

(2) Trade name of Formosa Plastics Corporation (Taiwan) for a polyethylene terephthalate (PET) film.

Tairipro. Trademark of Formosa Plastics Corporation (Taiwan) for polypropylene plastics and products.

Tairirex. Trademark of Formosa Plastics Corporation (Taiwan) for polystyrene plastics and products.

Tairisan. Trademark of Formosa Plastics Corporation (Taiwan) for styrene-acrylonitrile (SAN) plastics.

Tairiyon. Trademark of Formosa Chemicals & Fiber Corporation (Taiwan) for viscose rayon staple fibers and filament yarns.

Tairylan. Trademark of Formosa Plastics Corporation (Taiwan) for polyacrylonitrile (PAN) fibers.

Taitalac. Trademark of Taita Plastics for acrylonitrile-butadiene-styrene (ABS) resins and products.

takanelite. A gray-black mineral composed of calcium manganese oxide trihydrate, $(Mn,Ca)Mn_4O_9 \cdot 3H_2O$. Crystal system, hexagonal. Density, 3.41 g/cm³. Occurrence: Japan.

Take. Trade name of Nippon Sheet Glass Company Limited (Japan) for a patterned glass.

Take 1. Trademark of Coltene-Whaledent (USA) for an addition-polymerizing silicone-elastomer-based dental impression material.

takeuchiite. A black mineral composed of magnesium iron manganese titanium borate, $(Mg,Mg)_2(Mn,Fe,Ti)BO_5$. Crystal system, orthorhombic. Density, 3.93 g/cm³. Occurrence: Sweden.

takovite. A blue-green mineral of the sjogrenite group composed of nickel aluminum oxide carbonate hydroxide tetrahydrate, $Ni_6Al_2(OH)_{16}(CO_3,OH) \cdot 18H_2O$. Crystal system, rhombohedral (hexagonal). Density, 2.70 g/cm³. Occurrence: Serbia.

Taktene. (1) Trademark of Bayer Corporation (USA) for light-colored high-purity butadiene rubber made by solution polymerization using a lithium or cobalt catalyst. Available in several grades, it is used as a modifier in high-impact polystyrene and acrylonitrile-butadiene copolymers.

(2) Trademark of Bayer Corporation (USA) for abrasion-resistant, resilient butadiene rubber with good aging and flex-cracking resistance and good low-temperature flexibility. Available in standard and masterbatch grades, it is used for tires, conveyor belting, footwear soles, V-belts, seals, profiles, and injection-molded goods.

Talabot. Trade name of Société Nouvelle du Saut-du-Tarn (France) for low- and medium-carbon low-alloy structural steels, and numerous hot-work and cold-work tool steels.

Talandra. Trade name of Fibreglass Limited (UK) for reinforced glass-fiber materials.

talc. (1) A white, apple-green or gray mineral with pearly or greasy luster composed of magnesium silicate hydroxide, $Mg_3Si_4O_{10}(OH)_2$, and having a layered structure. Crystal system, monoclinic. Density, 2.58-2.78 g/cm³; melting point, above 1400°C (2550°F); hardness, 1.0-1.5 Mohs; refractive index, 1.5915. It has high thermal resistance, low thermal and electrical conductivity, good resistance to acids and alkalies, and good fire resistance. Occurrence: China, Russia, USA (California, Georgia, Maryland, Montana, Nevada, New York, North Carolina, Texas, Vermont, Virginia, Washington). Used for ceramics, electroceramics, dinnerware, refractories, gas-burner tips, electrical insulation and wall tile, as a filler in paints, rubber, putty, plaster and oilcloth, as a dusting agent and lubricant, and in paper, slate pencils and crayons. Also known as *talcum*.

(2) A massive, impure form of the mineral talc with pearly, greasy feel and containing clay, feldspar and alkaline-earth oxides. Density, 2.7-2.8 g/cm³; hardness, 1.0-1.5 Mohs; good dielectric properties. Used for electrical insulators, in the manufacture of electroceramics, and as a raw material for *cordierite* electroceramics and refractories. Also known as *massive talc; soaprock; soapstone; steatite.*

Talcoprene. Trademark of The Polymer Group (USA) for talc-filled polypropylene plastics.

talcum powder. See French chalk.

talcum. See talc (1).

Talide. Trade name of Metal Carbides Corporation (USA) for a series of cemented carbides containing varying amounts of tungsten, tantalum, titanium, carbon and cobalt. Used for cutting tools, dies, gages, guides, and wear plates.

Talladium No Bel-T. Trade name of Talladium Inc. (USA) for a beryllium-free nickel-chromium dental bonding alloy.

Talladium Premium V. Trade name of Talladium Inc. (USA) for a corrosion-resistant nickel-chromium dental bonding alloy.

T-Alloy. Trade name of CCS Braeburn Alloy Steel (USA) for a series of tungsten-type hot-work steels (AISI types H21, H22, H24 and H25) containing 0.25-0.5% carbon, 3-4% chromium,

9-15% tungsten, 0.4-0.6% vanadium, and the balance iron. Used for hot-work tools, forging dies, and die casting and molding dies for brass.

talmessite. A colorless mineral of the fairfieldite group composed of calcium magnesium arsenate dihydrate, $Ca_2Mg(AsO_4)_2 \cdot 2H_2O$. Crystal system, triclinic. Density, 3.14 g/cm³; refractive index, 1.685. Occurrence: Iran.

talmi gold. (1) See Abyssinian gold (1).

(2) A copper alloy containing 8-12% zinc and 0.5-1% gold. It has a gold facing applied by welding, and is used for cheap jewelry, ornaments, and hardware.

(3) A yellowish or gold-colored copper-base alloy containing up to 10% aluminum.

talnakhite. A chalcopyrite-like mineral of the tetrahydrite group composed of copper iron sulfide, $Cu_9(Fe,Ni)_8S_{16}$. Crystal system, cubic. Density, 4.29 g/cm³; hardness, 144-178 Vickers. Occurrence: Siberia, Russia. See also chalcopyrite.

Talnex. Trade name of MRC Polymers (USA) for a series of unreinforced and glass fiber-reinforced polyacetal (POM) resins.

Talpa. Trade name of Mitsui Chemicals, Inc. (Japan) for a series of polyether sulfone (PES) and polyether ketone (PEK) resins.

Tam. Trade name of NL Industries (USA) for a series of master alloys used in the manufacture of ferrous and nonferrous alloys. The master alloys for nonferrous-alloy melts include cuprotitanium, magnesium-zirconium, nickel-titanium and molybdenum-titanium, and the foundry alloys for iron and steel include various grades of ferrotitanium, ferrocarbon-titanium, manganese-titanium and silicon-titanium.

tamarack. See American larch.

tamarugite. A colorless mineral composed of sodium aluminum sulfate hexahydrate, $NaAl(SO_4)_2 \cdot 6H_2O$. Crystal system, monoclinic. Density, 2.06 g/cm³; refractive index, 1.487. Occurrence: Chile.

Tamashima. Japanese trade name for rayon fibers and yarns used for textile fabrics.

Tamax. Trademark of Charles Taylor & Sons Company (USA) for a high-grade *mullite* made from calcined Indian *cyanite* or similar high-purity aluminum silicate. It contains about 68% Al_2O_3 and 32% SiO_2. Commercially available in the form of bricks and special shapes, it is used in the manufacture of glass-melting furnaces, refractories for ferrous and nonferrous melting furnaces, and furnace and kiln linings.

Tamcast. Trademark of NL Industries (USA) for a series of mold flux powders for continuous-casting applications in the iron and steel industry.

tampico. A stiff, hard, flexible fiber obtained from a sisal plant (*Agave rigida*) used in the manufacture of power brushes, and for buffing and polishing wheels. See also istle.

Tamtrol. Trademark of NL Industries (USA) for a series of ladle flux systems for continuous-casting applications in the iron and steel industry.

Tamul. Trademark of Charles Taylor & Sons Company (USA) for synthetic *mullite* obtained from selected grades of *bauxite* by sintering at prescribed temperatures. Used in the manufacture of refractory products, such as cements, ramming mixes, etc., and in the construction of metallurgical furnaces and kilns.

Tanagin. Trade name of Takayasu Company Limited (Japan) for a series of polyamide 6 and 6,6 (nylon 6 and 6,6) resins.

tancoite. A colorless to pale pink mineral composed of hydrogen lithium sodium aluminum phosphate hydroxide, $HLiNa_2Al(PO_4)_2(OH)$. Crystal system, orthorhombic. Density, 2.75 g/cm³; refractive index, 1.563. Occurrence: Canada (Manitoba).

Tandem. (1) Trade name of Eyre Smelting Company (UK) for tin- and lead-base bearing metals. They possess excellent anti-friction properties and are used for high-load bearings and, depending on the particular composition, may be suitable for applications involving high shock loads.

(2) Trade name of S. Fry & Company Limited (UK) for zinc castings used for press tools.

Tandem Foam. Trademark of Ludlow Composites Corporation (USA) for a firm composite consisting of two closed-cell polyvinyl chloride latex foams cast together. Used for cushioning and shock absorption applications.

Tanegum. Trade name of Cross Polimeri for a polyvinyl chloride modified with polyurethane and/or polyethylene terephthalate.

taneyamalite. A greenish gray to yellow mineral composed of sodium magnesium manganese iron silicate, $Na(Mn,Mg,Fe)_{12}Si_{12}(O,OH)_{44}$. Crystal system, triclinic. Density, 3.30 g/cm³; refractive index, 1.664. Occurrence: Japan.

tangare. See andiroba.

tangile. The wood of the Philippine mahogany tree *Shorea polysperma*. It is similar in appearance to *American mahogany*, but coarser-textured and less decay-resistant. Tangile has a dark reddish-brown heartwood, a thick light-red sapwood, an open grain, moderate hardness, and works fairly well. Average weight, 592 kg/m³ (37 lb/ft³). Source: Philippine Islands, Malaya. Used for interior trim, paneling, plywood, cabinets, furniture, siding, fixtures and in boat construction. Also known as *Bataan mahogany*. Also spelled tanguile.

tangle sheet. A sheet of mica that has split well in some sections, but has torn in others.

Tango. See Meryl Tango.

tanguile. See tangile.

Tanikalon. Trademark of Taniyama Chemical Industry Company (Japan) for polyethylene monofilaments.

tanimbuca. See verdolago.

tank glass. Glass either melted or suitable for melting in a tank-type glass furnace.

Tannenbaum. Trade name of Krupp Stahl AG (Germany) for a water-hardening steel containing 0.55% carbon, 0.1-0.4% silicon, 0.3-0.7% manganese, and the balance iron. Used for machine-tool parts, gears, bolts, shafts, and mandrels.

tanoak. The hard, heavy, strong, abrasion-resistant wood of the broad-leaved tree *Lithocapus densiflorus*. The sapwood is light reddish-brown turning dark reddish-brown with age. Aged sapwood and aged heartwood are nearly indistinguishable. Source: USA (Southwestern Oregon; Southern California). Used for flooring, and in the manufacture of tannin for leather tanning.

tantalaeschynite. A brownish black to black mineral of the columbite group composed of calcium cerium yttrium titanium niobium tantalum oxide, $(Y,Ce,Ca)(Ta,Ti,Nb)_2O_6$. Crystal system, orthorhombic. Density, 5.75 g/cm³. Occurrence: Brazil.

tantalic acid anhydride. See tantalum pentoxide.

tantalic bromide. See tantalum pentabromide.

tantalic chloride. See tantalum pentachloride.

tantalic fluoride. See tantalum pentafluoride.

tantalic iodide. See tantalum pentaiodide.

tantalite. An iron-black to brownish black mineral with a dark red to black streak and a submetallic luster. It is composed of tantalum niobium iron manganese oxide, $(Fe,Mn)(Ta,Nb)_2O_6$, with the content of tantalum oxide (Ta_2O_5) exceeding that of niobium oxide (Nb_2O_5). Crystal system, orthorhombic. Density, 7.95 g/cm³; hardness, 6 Mohs. Occurrence: Africa (Congo, Nigeria), Brazil, Canada, Malaysia, USA (Idaho, South Dakota). Used as an ore of tantalum and niobium, and in lamp filaments. See also columbite; niobite.

Tantaloy. (1) Trade name Fansteel Metals (USA) for a powder-metallurgy material containing 92.5% tantalum and 7.5% tungsten. It has a density of 16.8 g/cm^3 (0.61 lb/in.^{3}), high strength, excellent corrosion resistance even in severe environments, and is used for springs and other elastic parts.

(2) Trade name of Fansteel Metals (USA) for a superhard tool material composed of tantalum carbide (TaC) with a cobalt binder. Used for cutting tools for stainless steel.

tantalum. A grayish to black, ductile metallic element belonging to group VB (Group 5) of the Periodic Table. It is available in the form of powder, pellets, sheet, foil, microfoil, ingots, bars, rods, tubing, wire and single crystals. The single crystals are usually grown by the float-zone technique. Tantalum occurs naturally in *tantalite* and *niobite* and in various rare minerals. Crystal system, cubic; crystal structure, body-centered cubic. Density, 16.6 g/cm^3; melting point, 2996°C (5425°F); boiling point, 5425°C (9797°F); hardness, 90-200 Vickers; refractive index, 2.05; superconductivity critical temperature, 4.47K; atomic number, 73; atomic weight, 180.948; divalent, trivalent, tetravalent, pentavalent. It has very high tensile strength, a low expansion coefficient, fair electrical conductivity (13.9% IACS), excellent corrosion resistance, and excellent resistance to many acids, alkalies and seawater. Used in high-melting alloys, as an alloying addition to stainless steels to improve corrosion resistance, as a stabilizer in steels (usually with niobium), as a carbide (TaC) in cemented carbide, as a filament for lamps and electronic tubes, for chemical process equipment, laboratory apparatus and high-speed cutting tools, in the manufacture of electronic equipment, electrolytic capacitors, rectifiers, electronic circuitry and thin-film components, in lightning arrestors and surge suppressors, for weights, in dental and surgical instruments, sutures and implants, for rocket nozzles, aircraft and missile components, and for heat exchangers in nuclear reactors. Tantalum powder is also used as an X-ray opacifier. Symbol: Ta.

tantalum alloys. A group of tantalum-based refractory-metal alloys with alloying additions, such as tungsten (W), hafnium (Hf), and/or niobium (Nb). Examples include *90Ta-10W* with high strength at high temperatures and used mainly for aerospace applications, *97.5Ta-2.5W* with high formability and used in welded tubing and heat exchangers and *60Ta-40Nb* with high strength and corrosion resistance, and used for capacitors, cemented carbides, heat exchangers, condensers, lined vessels, thermowells, heating elements, heat shields, spinnerets, and for specialized nuclear and aerospace applications.

tantalum aluminide. A compound of tantalum and aluminum supplied in the form of a fine powder for use in ceramics and high-temperature applications. Formula: TaAl.

tantalum beryllide. Any of the following compounds of tantalum and beryllium used in ceramics: (i) *Tantalum diberyllide.* Density, 8.95 g/cm^3. Formula: TaBe$_2$; (ii) *Tantalum triberyllide.* Density, 8.18-8.23 g/cm^3. Formula: TaBe$_3$; and (iii) *Tantalum dodecaberyllide.* Density, 4.18 g/cm^3; melting point, 1850°C (3362°F). Formula: TaBe$_{12}$.

tantalum boride. Any of the following compounds of tantalum and boron, most of which are used in ceramics: (i) *Tantalum monoboride.* Refractory crystals. Crystal system, orthorhombic. Density, 14.2-14.3 g/cm^3; melting point, 2040°C (3704°F). Formula: TaB; (ii) *Tantalum diboride.* Black crystals. Crystal system, hexagonal. Density, 11.2 g/cm^3; melting point, 3140-

3200°C (5684-5792°F); low thermal expansion. Formula: TaB$_2$; (iii) *Ditantalum boride.* Melting point, 1900°C (3452°F). Formula: Ta$_2$B; (iv) *Tritantalum diboride.* Melting point, 2038°C (3700°F). Formula: Ta$_3$B$_2$; and (v) *Tritantalum tetraboride.* A fine powder. Density, 13.6 g/cm^3; melting point, 2650°C (4802°F). Used as an alloying addition in superalloys. Formula: Ta$_3$B$_4$.

tantalum bromide. See tantalum pentabromide.

tantalum carbide. Any of the following compounds of tantalum and carbon: (i) *Tantalum monocarbide.* A golden-brown to brownish refractory crystalline solid, or fine powder made by heating tantalum oxide and carbon at high temperatures. Crystal system, cubic. Crystal structure, halite. Density, 13.9-14.5 g/cm^3; melting point, 3880-3915°C (7016-7079°F); boiling point, 5500°C (9932°F); hardness, 1790 Vickers (above 9 Mohs); excellent chemical resistance at ordinary and elevated temperatures. Used for cemented carbides, cutting tools, dies, abrasives and high-temperature coatings. Formula: TaC; and (ii) *Ditantalum carbide.* A refractory, crystalline compound of tantalum and carbon. Crystal system, hexagonal. Density, 15.1 g/cm^3; melting point, 3327°C (6021°F). Used for cutting tools and dies. Formula: Ta$_2$C.

tantalum carbide coatings. Impermeable, uniform, electrically conductive high-temperature protective coatings based on tantalum monocarbide (TaC) with excellent resistance to wear and chemical attack by corrosive liquids and gases. They have a density of 15.0 g/cm^3 (0.54 lb/in.^{3}), a melting point of about 3880°C (7016°F), an upper service temperature of about 700°C (1290°F), and are used on graphite, e.g., in epitaxial processes for the manufacture of compound semiconductors by metal-organic chemical vapor deposition (MOCVD) and liquid-phase epitaxy (LPE).

tantalum chloride. See tantalum pentachloride.

tantalum-copper-niobium composite. A superconducting composite in wire form composed of a tantalum-coated copper core surrounded by niobium filaments. The average filament diameter is 5 µm (200 µin.).

tantalum diberyllide. See tantalum beryllide (i).

tantalum diboride. See tantalum boride (ii).

tantalum dioxide. A brown, crystalline powder. Crystal system, tetragonal. Density, 10.0 g/cm^3. Used in ceramics, chemistry and mate-rials research. Formula: TaO$_2$. Also known as *tantalum oxide.*

tantalum diselenide. A crystalline compound of tantalum and selenium. Crystal system, hexagonal. Density, 6.7 g/cm^3. Used as a dry lubricant, in ceramics, and in materials research. Formula: TaSe$_2$. Also known as *tantalum selenide.*

tantalum disilicide. See tantalum silicide (i).

tantalum disulfide. See tantalum sulfide (ii).

tantalum ditelluride. A crystalline compound of tantalum and selenium. Crystal system, monoclinic. Density, 9.4 g/cm^3. Used as a dry lubricant, in ceramics and in materials research. Formula: TaTe$_2$. Also known as *tantalum telluride.*

tantalum dodecaberyllide. See tantalum beryllide (iii).

tantalum fluoride. See tantalum pentafluoride.

tantalum-hafnium carbide. A compound of tantalum, hafnium and carbon available as a fine powder. Melting point, 3942°C (7128°F). Used for refractory components. Formula: Ta$_4$HfC$_5$.

tantalum iodide. See tantalum pentaiodide.

tantalum monoboride. See tantalum boride (i).

tantalum monocarbide. See tantalum carbide (i).

tantalum mononitride. See tantalum nitride (i).

tantalum monosulfide. See tantalum sulfide (i).

tantalum nitride. Any of the following compounds of tantalum and nitrogen: (i) *Tantalum mononitride.* Brown, bronze or black crystals, or fine powder. Crystal system, hexagonal. Density, 13.7-16.3 g/cm³; melting point, 3310-3410°C (5990-6170°F); very high hardness. Used for abrasives, cutting tools, ceramics, and resistors. Formula: TaN; (ii) *Ditantalum nitride.* Crystals or powder. Melting point, loses nitrogen at 1900°C (3450°F). Used for ceramics, and cutting tools. Formula: Ta₂N; and (iii) *Tritantalum nitride.* A fine powder. Density, 14.1 g/cm³; melting point, 3400°C (6152°F). Used in ceramics. Formula: Ti₃N.

tantalum oxide. See tantalum dioxide; tantalum pentoxide.

tantalum pentabromide. Yellow, moisture-sensitive, crystalline powder (98% pure). Density, 4.67-4.99 g/cm³; melting point, 240°C (464°F) [also reported as 265°C or 509°F]; boiling point, 349°C (660°F). Used in the production of tantalum metal. Formula: TaBr₅. Also known as *tantalic bromide; tantalum bromide.*

tantalum pentachloride. Yellow, hygroscopic crystals or yellowish white, moisture-sensitive crystalline powder (99+% pure). Crystal system, monoclinic. Density, 3.68 g/cm³; melting point, 216-220°C (421-428°F); boiling point, 239-242°C (462-468°F). Used in the production of tantalum metal. Formula: TaCl₅. Also known as *tantalic chloride; tantalum chloride.*

tantalum pentafluoride. White, hygroscopic crystals or off-white, moisture-sensitive crystalline powder (99+% pure). Crystal system, monoclinic. Density, 4.74-5.0 g/cm³; melting point, 95-97°C (203-207°F); boiling point, 229°C (444°F). Used in the production of tantalum metal. Formula: TaF₅. Also known as *tantalic fluoride; tantalum fluoride.*

tantalum pentaiodide. A black, moisture-sensitive crystalline powder (99+% pure). Density, 5.80 g/cm³; melting point, 496°C (924°F); boiling point, 543°C (1009°F). Used in the production of tantalum metal. Formula: TaI₅. Also known as *tantalic iodide; tantalum iodide.*

tantalum pentoxide. Colorless crystals or white powder (99+% pure). Crystal system, rhombohedral. Density, 7.6-8.2 g/cm³; melting point, 1785°C (3245°F). Used in the production of tantalum metal, in the preparation of tantalum carbide, in optical glass and ferroelectric components, in camera lenses, for piezoelectric, maser and laser applications, and in dielectric layers for electronics. Formula: Ta₂O₅. Also known as *tantalic acid anhydride; tantalum oxide.*

tantalum phosphide. A compound of tantalum and phosphor. Density, 11.1 g/cm³; melting point, 1660°C (3020°F). Used in ceramics and electronics. Formula: TaP.

tantalum-potassium fluoride. White, silky needles used as intermediate in the production of pure tantalum metal. Formula: K₂TaF₇. Also known as *potassium fluotantalate; potassium-tantalum fluoride.*

tantalum powder. Tantalum (purity, 99.8+%) in the form of a powder with a particle size ranging from less than 10 to over 350 μm (0.0004 to 0.0138 in.). Used as a catalyst, as an X-ray opacifier, and in powder metallurgy, ceramics, and refractories.

tantalum powder-metallurgy product. A product made from tantalum powder that is either prealloyed or an elemental blend of tantalum and a master alloy by pressing and subsequent sintering. Used in high-strength tantalum-tungsten and tantalum-niobium parts.

tantalum selenide. See tantalum diselenide.

tantalum silicide. Any of the following compounds of tantalum and silicon used in ceramics, electronics and materials research: (i) *Tantalum disilicide.* A transition-metal disilicide available as a fine, gray powder (99.5+% pure) and as thin films. Crystal system, hexagonal. Density, 9.14 g/cm³; melting point, above 2200°C (3990°F); hardness, 1560 kgf/mm². The thin-film form is used extensively in the synthesis of submicron complementary metal-oxide semiconductor (CMOS) gates in the semiconductor industry. Formula: TaSi₂; (ii) *Ditantalum silicide.* Density, 13.54 g/cm³; melting point, 2500°C (4532°F). Formula: Ta₂Si; (iii) *Pentatantalum silicide.* Density, 12.86 g/cm³; melting point, 2510°C (4550°F). Formula: Ta₅Si; and (iv) *Pentatantalum trisilicide.* Density, 13.06 g/cm³; melting point, about 2500°C (4530°F). Formula: Ta₅Si₃.

tantalum sulfide. Any of the following compounds of titanium and sulfur: (i) *Tantalum monosulfide.* A grayish powder. Density, 9.20 g/cm³. Used in ceramics and electronics. Formula: TaS; and (ii) *Tantalum disulfide.* Black crystals, or fine black powder (99+% pure). Crystal system, hexagonal. Density, 6.86 g/cm³. Melting point, above 3000°C (5432°F). Used in electronics, and as a dry lubricant. Formula: TaS₂.

tantalum telluride. See tantalum ditelluride.

tantalum triberyllide. See tantalum beryllide (ii).

tantalum-tungsten alloys. A group of refractory powder-metallurgy alloys containing tantalum and tungsten as the two principal elements, and sometimes alloying additions of niobium, hafnium, and/or carbon. They have high hardness and strength, high melting points (e.g., the 90Ta-10W alloy melts at 3080°C and 5576°F). Used for electronic tube parts, chemical applications and rocket motor parts.

tanteuxenite. A black mineral of the columbite group composed of uranium titanium oxide, (U,Fe,V)(Ti,Sn)₂O₆. Density, 4.70 g/cm³. Occurrence: Italy.

tantiron. An acid- and corrosion-resistant alloy containing 1% carbon, 0.4% manganese, 13.5-15% silicon, 0.18% phosphorus, 0.05% sulfur, and the balance iron. Used for chemical equipment, machine parts, vessels, kettles, and pipes.

Tantung. (1) Trade name of Fansteel Metals (USA) for a group of cast alloys of tantalum and tungsten with excellent shock and impact resistance and high hardness used as tool materials for general-purpose machining at speeds above those recommended for high-speed steels, but below those normally used for carbides. It retains its cutting edge hardness at temperatures up to 816°C (1500°F), and regains its original hardness upon cooling down.

(2) Trade name of Fansteel Metals (USA) for a series of wear- and corrosion-resistant cast cobalt-base alloys containing 43-50% cobalt, 25-32% chromium, 14-21% tungsten, 2-4.5% tantalum, 0-4.5% niobium (columbium), 0-5% iron, 1-3% manganese and 2-4% carbon. They have good toughness, excellent abrasion resistance, high red-hardness, and are used for cutting tools, knives, blades, dies, etc.

Tapdie. Trade name of Columbia Tool Steel Company (USA) for oil- or water-hardening cold-work steels (AISI type O7) containing 1.25% carbon, 1.4% tungsten, 0.45% chromium, 0.2% vanadium, and the balance iron. They have great depths of hardening, good toughness and machinability, and good to moderate wear resistance. Used for a variety of cutting tools.

tape. (1) A colored, invisible or transparent tape, usually polymer-based, having an adhesive substance applied to one or both surface. Example include packaging, masking and sealing tape. Also known as *adhesive tape.*

(2) A ribbon composed of continuous or discontinuous fi-

bers oriented in direction of the tape axis and parallel to each other, and bonded together by a matrix resin forming the continuous phase. Also known as *composite tape*.

(3) A prepreg made by arranging alternate plies of unidirectional tape at different directions with respect to the longitudinal fiber axis. Also known as *multidirectional tape prepreg; multioriented prepreg*.

(4) A series of collimated untwisted bundles of continuous fibers impregnated with an uncured matrix resin, in which the fibers are anisotropic, i.e., reinforce primarily in one direction (the "fiber direction"). Typical tape sizes range from 25 to over 1500 mm (1 to over 59 in.) in width and from 0.08 to 0.25 mm (0.003 to 0.01 in.) in thickness. Also known as *unidirectional tape prepreg*.

(5) A composite superconductor in ribbon or strip form.

(6) A long, flat, thin structure usually consisting of paper or a thermoplastic and having textile-like properties.

(7) A thin fabric usually woven in a plain weave and used as a reinforcement for fabrics and for nonstructural applications.

tape-cast ceramics. Thin, flexible ceramic sheets produced by first casting thin layers of slurries, prepared from ceramic powders, organic binders and plasticizers, onto impervious polymer films or glass plates, and allowing them to dry by evaporation, followed by firing or sintering at high-temperatures.

tape prepreg. (1) See tape (3).

(2) See tape (4).

tape sealant. See sealing tape (2).

Tapestry. Trade name of PPG Industries Inc. (USA) for a translucent, semi-opaque, rough-figured glass.

tapestry brick. A brick with a rough, textured surface used for decorative applications.

tape yarn. A yarn consisting of a paper, polyethylene, polypropylene or other tape with a large aspect ratio (i.e., width-to-thickness ratio).

Tapilon. Trademark of Toray Industries, Inc. (Japan) for nylon 6 and nylon 6,6 fibers and filament yarns.

Tapioca. Trade name of SA Glaverbel (Belgium) for patterned glass.

tapioca paste. An inexpensive vegetable glue made from tapioca, i.e., the starch obtained from the root of the tropical American cassava plant (genus *Manihot)*, and used for bonding cardboard boxes and plywood, and for envelopes, labels, postage stamps, etc.

tapiolite. An orange mineral of the rutile group composed of iron tantalum oxide, $FeTa_2O_6$. It can also be made synthetically from ferric oxide (FeO) and tantalum penoxide (Ta_2O_5) under prescribed conditions. Crystal system, tetragonal. Density, 7.82 g/cm³. Occurrence: France.

tar. A thick, black or dark brown, viscous liquid or semisolid resulting from the destructive distillation of organic materials, such as wood, coal, petroleum, oil-shale, or peat. Used as a chemical raw material (e.g., for aniline dyes, machine oils, pharmaceuticals, etc.), in roadbuilding, and in roofing materials. See also coal tar; pine tar; wood tar.

taramellite. A bronze-purple mineral of the axinite group composed of barium iron magnesium titanium chloride borate silicate, $Ba_4(Fe,Ti,Mg,V)_4B_2Si_8O_{29}Cl$. Crystal system, orthorhombic. Density, 3.92 g/cm³. Occurrence: Italy.

taramite. A blue-green mineral of the amphibole group composed of sodium calcium iron aluminum silicate hydroxide, $Na_2CaFe_5Al_2Si_6O_{22}(OH)_2$. Crystal system, monoclinic. Density,

3.50 g/cm³. Occurrence: Tanzania.

taranakite. A white-gray mineral composed of potassium aluminium hydrogen phosphate hydrate, $H_6K_3Al_5(PO_4)_8 \cdot 18H_2O$. Crystal system, rhombohedral (hexagonal). Density, 2.09 g/cm³; refractive index, 1.507. Occurrence: Italy.

tarapacaite. A bright canary yellow mineral composed potassium chromium oxide, α-K_2CrO_4. Crystal system, orthorhombic. Density, 2.74 g/cm³; refractive index, 1.722.

Taramet Sterling. Trade name of Taracorp IMACO, Inc. (USA) for lead-free solders.

tarasovite. A mineral of the smectite group composed of potassium sodium aluminum silicate dihydrate, $NaKAl_8(Si,Al)_{16}O_{40}(OH)_8 \cdot 2H_2O$. Crystal system, monoclinic. Density, 2.36 g/cm³; refractive index, 1.578. Occurrence: Russian Federation.

tar-bearing basic ramming mix. A refractory *ramming mix* composed of basic grains to which tar has been added, and used in furnace linings.

tar-bearing ramming mix. A refractory *ramming mix* to which tar has been added, used to form monolithic furnace linings.

Tarbicol. Trade name of Ato Findley (France) for vinyl copolymers and adhesives used for wood-mosaic floors.

tar felt. A sheet- or felt-like substance saturated with liquid tar and used for roofing and waterproofing applications.

Tarflen. Trade name of Zaklady Azotowe (Russia) for polytetrafluoroethylenes.

Tarflon. Trade name of Idemitsu Petrochemicals (Japan) for polycarbonate and polystyrene resins.

Target. (1) Trademark of Target Concrete Products Limited (Canada) for prepackaged concrete, mortars, grouts, and other products.

(2) Trademark of Anchor Packing Division, Robco Inc. (Canada) for asbestos gasket sheets.

Targis. Trademark of Ivoclar Vivadent AG (Liechtenstein) for a dental ceramer composite that combines the esthetic properties of ceramics with the ease of handling of polymeric veneering materials. It is similar in translucency to ceramics and nearly has the wear of natural dental enamel. Used for crowns and bridges. Also available is *Targis Ceramer*, a dental composite resin for veneering applications.

TargisGold. Trademark of Ivoclar Vivadent AG (Liechtenstein) for a palladium-free, extra-hard dental casting alloy (ADA type IV) containing 70.0% gold, 13.7% silver, 10.0% copper, 3.59% platinum, and less than 1.0% tin, zinc and iridium respectively. It has a yellow color, a density of 15.6 g/cm³ (0.56 lb/in.³), a melting range of 860-925°C (1580-1697°F), an as-cast hardness of 220 Vickers, low elongation, and excellent biocompatibility. Used for crowns, onlays, posts and bridges.

Targranit. British trademark for granite chippings precoated with tar. Used in for civil engineering applications.

Tar-Heel. Trade name of Tar Heel Mica Company (USA) for mica and mica products.

Tarmac. Trade name of Tarmac America (USA) for interlocking paving stones for commercial applications.

tar macadam. A material composed of crushed stone or gravel bonded with tar or a tar-bitumen mix, and used for roadbuilding. See also macadam.

Tarnac. German trade name for a *Manganin*-type electrical resistance alloy composed of copper, manganese and nickel. It has a low temperature coefficient of resistance, high electrical resistivity, good strength, and is used for electrical instruments.

Tarnamid. Trade name of Zaklady Azotowe (Russia) for polyamide 6 (nylon 6) plastics and products.

Tarniban. Trade name of Technic, Inc. (USA) for anti-tarnish products for copper and silver.

Tarnoform. Trade name of Zaklady Azotowe (Russia) for acetal (polyoxymethylene) resins.

Taroblend. Trade name of Bakelite AG (Germany) for polymer alloys of polycarbonate (PC) and acrylonitrile-butadiene-styrene (ABS).

Tarodur. Trade name of Bakelite AG (Germany) for acrylonitrile-butadiene-styrene (ABS) resins.

Taroform. Trade name of Bakelite AG (Germany) for acetal (polyoxymethylene) resins.

tar oil. See creosote.

Tarolon. Trade name of Bakelite AG (Germany) for polycarbonate resins.

Tarolox. Trade name of Bakelite AG (Germany) for polybutylene terephthalate (PBT) and polyethylene terephthalate (PET) resins.

Taroloy. Trade name of Bakelite AG (Germany) for polymer blends of polcarbonate (PC) and polybutylene terephthalate (PBT) resins.

Taromid. Trade name of Bakelite AG (Germany) for a polyamide 6 and 6,6 (nylon 6 and 6,6) resins.

tar paper. A heavy construction paper saturated or coated with coal tar, and available in rolls, typically 915 mm (36 in.) wide and up to 44 m (144 ft.) long. Used for roofing and general construction applications.

tarpaulin. (1) A sheet of coarse, strong, waterproofed fabric, such as canvas or duck, used to protect vehicles, equipment, construction materials and other goods against the weather.

(2) A textile fabric made of jute and used for packaging and lining applications.

Tarpaving. British trademark for *tar macadam* used for paving playgrounds and sidewalks.

Tar-Pon. Trade name of Sigma Coatings, Inc. (USA) for coal-tar epoxy coatings.

Tarpoon. Trade name for a water-repellent, closely woven poplin-type cotton fabric used for jackets and playwear.

tarred rope. An abrasion- and water-resistant rope produced from yarns or fibers that have been dipped into and saturated with tar.

Tarsiyon. Trademark of Formosa Chemicals & Fiber Corporation (Taiwan) for viscose rayon staple fibers and filament yarns.

Tartan. Trademark of MCP, Industries, Inc. (USA) for polyurethane-resin coatings used as covering materials for running and racing tracks, etc.

Tarvinyl I-S. Trade name of Zaklady Azotowe (Russia) for polyvinyl chloride plastics.

Tasil. Trademark of Charles Taylor & Sons Company (USA) for a high-grade *mullite* made from calcined Indian *cyanite* (Al_2SiO_5). It contains about 59% alumina (Al_2O_3), and has a very low content of impurities, such as iron, titania and alkalies. *Tasil* is commercially available in the form of bricks, special shapes, refractory cements and ramming mixes, and has excellent resistance to very high temperatures. Used in the manufacture of glass-melting furnaces, refractories for metallurgical furnaces, and furnace and kiln linings for very high temperatures.

Tasinlon. Trademark of Chung Shing Textile Company (Taiwan) for nylon 6 fibers and filament yarns, and polyester staple fibers and filament yarns.

Taslon. Trademark of Acelon Chemicals & Fiber Corporation (Taiwan) for nylon 6 fibers and filament yarns.

Tasmanian blackwood. See Australian blackwood.

Tasmanian oak. The strong, tough wood of any of several trees of the genus *Eucalyptus* including *E. obliqua*, *E. regnans* and *E. gigantea*. It is not a true oak, but resembles plain oak in appearance. Average weight, 610 kg/m³ (38 lb/ft³). Source: Australia, Tasmania. Used for flooring, and as a replacement for oak.

Tata. (1) Trade name of Tata Iron & Steel Company (India) for a series of plain-carbon, high-speed, hot-work tool steels, stainless steels, low-carbon magnetic steels, low-carbon structural steels, austenitic manganese steels, and a wide range of alloy steels (e.g., chromium, chromium-molybdenum, chromium-molybdenum-vanadium, nickel, nickel-chromium and nickel-chromium-molybdenum grades).

(2) Trade name of Tata Refractories (India) for silica refractory brick used for coke oven and regenerator walls, glass tank roofs, and soaking pits.

tatarskite. A colorless mineral composed of calcium magnesium chloride carbonate sulfate hydroxide heptahydrate, $Ca_6Mg_2(SO_4)_2(CO_3)_2Cl_4(OH)_4 \cdot 7H_2O$. Density, 2.34 g/cm³; refractive index, 1.654. Occurrence: Russian Federation, Ukraine.

Ta-Tenz. Trade name of Techni-Cast Corporation (USA) for an alloy of 91.4% aluminum, 0.7% copper, 0.4% magnesium and 7.5% zinc. Used for aircraft components, automotive components, and machinery parts.

Tateho. Trademark of Tateho Chemical Industry Company, Limited (USA) for discontinuous silicon carbide and silicon nitride reinforcing whiskers used for engineering composites.

Tatmo. Trade name of Latrobe Steel Company (USA) for a series of molybdenum-type high-speed steels (AISI types M1-M7) containing 0.8-1.1% carbon, 3.7-4% chromium, 1.5-1.75% tungsten, 1-2% vanadium, 8-8.75% molybdenum, 0-8.25% cobalt, and the balance iron. They have great depths of hardening, high hot hardness, excellent wear resistance, and are used for dies and cutting tools. Formerly known as *Electrite Tatmo*.

Tatren. Trade name of Slovnaft (Slovakia) for polyphenylene sulfide (PPS) resins.

Tauril. Trademark of Anchor Packing Division, Robco Inc. (Canada) for asbestos gasket sheets.

Taurus. Trade name of David Brown Foundries Company (UK) for an extensive series of copper-base casting alloys including several red brasses and leaded red brasses, yellow brasses and leaded yellow brasses, manganese and leaded manganese bronzes, tin bronzes and lead tin bronzes, and various nickel silvers, and high-nickel copper alloys. Usually supplied in the centrifugally-cast and/or sand-cast condition.

tausonite. A mineral of the perovskite group composed of strontium titanium oxide, $SrTiO_3$. It can also be made synthetically. Crystal system, cubic. Density, 5.12 g/cm³. Occurrence: Russian Federation.

tavorite. A mineral composed of lithium iron phosphate hydroxide, $LiFe(PO_4)(OH)$. Density, 3.95 g/cm³. Occurrence: Brazil.

tawhai. See New Zealand beech.

Tawilco. Trade name of Taylor-Wilson Manufacturing Company (USA) for a heat- and corrosion-resistant cast iron used for gears, machinery parts, and housings.

Taycor. Trademark of Charles Taylor & Sons Company (USA) for a corundum-based refractory prepared by sintering high-purity alumina (Al_2O_3) at a temperature above 1980°C (3595°F) at which the transformation to tabular alumina (α-alumina) takes place. It contains up to 90% Al_2O_3, and is commercially available in the form of bricks, special shapes, cements, and ramming mixes. *Taycor* has exceptional abrasion resistance, excellent load-bearing properties, and good resistance to slags at high temperatures. Used in the construction of metallurgical furnaces, kiln linings and furniture, etc.

Taylor Ball. Trade name of Charles Taylor & Sons Company (USA) for a ball clay.

Taylor process wire. See Taylor wire.

Taylor-White. British trade name for a high-speed steel containing about 0.75-1.15% carbon, 0.2% manganese, 0.25% silicon, 8-8.5% tungsten, 1.8-3% chromium, and the balance iron. Used for tools, cutters, and dies.

Taylor wire. An extremely fine metal or metal-alloy wire of relatively short length made by inserting a wire of larger diameter into a glass or quartz tube and stretching the two together at high temperature. Such a wire can also be produced by first drawing the larger-diameter wire through a bead of molten glass and then through a die. Also known as *Taylor process wire.*

Taylor Zircon. Trade name of Charles Taylor & Sons Company (USA) for refractories made from high-grade refined zirconium silicate ($ZrSiO_4$). Commercially available in the form of bricks, special shapes, cements, and ramming mixes. Used in refractories for melting furnaces and kiln furniture.

tazheranite. An orange mineral of the fluorite group composed of calcium titanium zirconium oxide, $(Zr,Ca,Ti)O_2$. Crystal system, cubic. Density, 5.01 g/cm³; refractive index, 2.35. Occurrence: Russian Federation.

T-beam. A hot-rolled structural steel shape consisting of a web and a flange arranged in the shape of the letter "T." Also known as *tee bar.*

TB-tex. Trade name of Vetreria Italiana Balzaretti Modigliani SpA (Italy) for glass-fiber staple yarn and sliver.

T-Cell. Trade name of Sentinel Polyolefins LLC (USA) for polyethylene roll foam supplied in tape grades.

TD nickel. See thoria-dispersed nickel.

TD NiCr. See thoria-dispersed nickel-chromium.

TDN Nylon. Trademark of Teijin DuPont Nylon Limited (Japan) for nylon 6 and nylon 6,6 fibers and filament yarns.

teak. The hard, durable wood of any of several trees of the genera *Baikiaea, Chlorophora, Flindersia, Hymenea* and *Tectona.* Also known as *teakwood.* For properties and applications, see African teak, Australian teak, East Indian teak, Rhodesian teak and Surinam teak.

teakwood. See teak.

tea lead. British alloy composed of 98% lead and 2% tin, and used in lead foil for packaging tea.

teallite. A black mineral of the herzenbergite group composed of lead tin sulfide, $PbSnS_2$. It can also be made synthetically, and may contain some zinc. Crystal system, orthorhombic. Density, 6.36 g/cm³. Occurrence: Bolivia.

TEC. Trade name of Tennessee East man Corporation (USA) for cellulose acetate plastics.

Teca. Trade name for cellulose acetate fibers used for wearing apparel and as industrial fabrics.

Tecacryl. Trade name Ensinger GmbH (Germany) for acrylic resins.

Tecadur. Trade name of Ensinger GmbH (Germany) for polyesters based on polybutylene terephthalate (PBT) or polyethylene terephthalate (PET) resins.

Tecafine. Trade name of Ensinger GmbH (Germany) for a series polyolefin resins including high-molecular-weight polyethylene (HMWPE), ultrahigh-molecular-weight polyethylene (UHMWPE), polypropylene (PP) and polymethylpentene (PMP) resins.

Tecaflon. Trade name of Ensinger GmbH (Germany) for a series of polymers including polytetrafluoroethylene (PTFE), polyvinylidene fluoride (PVDF) and various fluoropolymers.

Tecaform. Trade name of Ensinger GmbH (Germany) for acetal (polyoxymethylene) resins.

Tecaglide. Trade name of Ensinger GmbH (Germany) for polyamide 6 (nylon 6) resins with low friction and good lubricity, used for bearings.

Tecalan. Trademark for oil and gasoline-resistant polyamide (nylon) tubing for high temperatures and pressures.

Tecalor. Trade name of Ensinger GmbH (Germany) for a thermoplastic polyimide (TPI) resin.

Tecamid. Trade name of Ensinger GmbH (Germany) for a series of polyamides (nylons) including nylon 4,6, nylon 6, nylon 6,6, nylon 6/6T and nylon 11.

Tecanat. Trade name of Ensinger GmbH (Germany) for polycarbonate (PC) resins.

Tecanyl. Trade name of Ensinger GmbH (Germany) for polyphenylene ether (PPE) resins.

Tecapeek. Trade name of Ensinger GmbH (Germany) for polyetheretherketone (PEEK) resins.

Tecapek. Trade name of Ensinger GmbH (Germany) for polyetherketone (PEK) resins.

Tecapei. Trade name of Ensinger GmbH (Germany) for polyetherimide (PEI) resins.

Tecapro. Trade name of Ensinger GmbH (Germany) for a polypropylene alloy.

Tecaran. Trade name of Ensinger GmbH (Germany) for acrylonitrile-butadiene-styrene (ABS) plastics.

Tecarim. Trade name of Ensinger GmbH (Germany) for polyamide 6 (nylon 6) resins.

Tecason. Trade name of Ensinger GmbH (Germany) for a series of polymers based on polysulfone, polyether sulfone or polyphenylene sulfone.

Tecast. Trade name of Ensinger GmbH (Germany) for a series of cast polyamide (nylon) products.

Tecate Gold. Trade name of Aardvark Clay & Supplies (USA) for a red-brown clay (cone 10) with medium texture.

Tecatron. Trade name of Ensinger GmbH (Germany) for polyphenylene sulfide (PPS) resins.

Tecavinyl. Trade name of Ensinger GmbH (Germany) for polyvinyl chloride (PVC) plastics.

Tecdur. Trade name of LEIS Polytechnik GmbH (Germany) for polybutylene terephthalate (PBT) resins.

Techalloy. Trade name of Techalloy Company Inc. (USA) for a series of austenitic, ferritic and martensitic stainless steels, nickel-chromium and nickel-chromium-iron electrical resistance wire alloys, and several iron-nickel glass-sealing alloys.

Techceram. Trade name for a flame-sprayed alumina core ceramic used for dental applications.

Tech-gard. Trademark of Pratt & Lambert, Inc. (USA) for a series of primers and maintenance coatings including red iron oxide-alkyd primers for the corrosion protection of structural steel, high-performance modified soya-alkyd primers for the protection of steel in industrial atmospheres, and corrosion-resistant, high-gloss alkyd-type maintenance coatings for the protection of steel, aluminum, wood and masonry.

Techlon. Trade name of Vamptech (USA) for industrial-grade polyamides (nylons) including nylon 6 and nylon 6,6.

Techmore. Trademark of Mitsui Chemicals, Inc. (Japan) for epoxy resins with high heat resistance used in the manufacture of electrical, semiconductor and printed circuits.

technetium. A silver gray metallic element of Group VIIB (Group 7) of the Periodic Table. First made by the bombardment of molybdenum with deuterons, it is now usually prepared from the reactor fission products of uranium and plutonium. Crystal system, hexagonal. Density, 11.5 g/cm^3; melting point, 2130°C (3866°F); atomic number, 43; atomic weight, 98.906; tetravalent, pentavalent, hexavalent, heptavalent; half-life, 2.6×10^6 years. It has a high superconductivity transition temperature, good corrosion-inhibiting properties, and is used as a radioactive tracer in metallurgy, as a radiation source, in superconductors, and in the generation of high-strength magnetic fields. Symbol: Tc.

Techniace. Trade name of Nippon A&L (Japan) for acrylonitrile-butadiene-styrene (ABS) and polycarbonate (PC) resins and ABS/PC alloys.

technical ceramics. A term referring to ceramics used in the manufacture of technical or industrial products in contrast to the more traditional ceramics (e.g., pottery, earthenware and stoneware) used chiefly in arts and crafts. *Technical ceramics* are based on raw materials, such as silicon dioxide, aluminum or beryllium oxide, boron and silicon carbides, aluminum, boron and silicon nitrides, lanthanum or thorium oxide, titanium diboride, zirconium oxide or silicate, or mixtures thereof. Examples of technical ceramics include glass-ceramics, structural clay products, whitewares, refractories, cermets, cement and concrete, lime, plaster, abrasives, engineering ceramics, etc.

technical fabrics. See industrial fabrics.

technical glass. Commercial glass made by supercooling of a soda-lime-silicate melt without crystallization. It consists essentially of 72-76% silica (SiO_2), 14-16% soda (Na_2O), 5-10% lime (CaO), 0-4% magnesia (MgO) and 1-2% alumina (Al_2O_3). *Technical glass* has a relatively low melting temperature, low electrical and thermal conductivity, low thermal expansion, high light transmissivity, high durability, good workability, good resistance to air, water and most chemicals (except hydrofluoric acid), and high hardness. Used for window and plate glass, containers, light bulbs, industrial products, chemical, electrical and electronic applications, etc. See also soda-lime glass.

Technical Iron. Russian trade name for a commercially pure iron containing 0-0.025% carbon, 0-0.03% manganese and 0-0.03% silicon. It has high magnetic permeability, low coercive force, and is used for magnetic circuits and deflectors.

technical leather. A tough, durable leather obtained from cattlehide by vegetable or chrome tanning, and used for technical or industrial applications, such as torque transmission or conveyor belting, friction disks, engine gaskets, rollers for textile machinery, automobile interiors, and polishing and buffing wheels.

technical papers. A generic term for high-quality papers chiefly used for technical and industrial applications, and usually excluding writing, printing, drawing, packaging, wrapping and sanitary papers. Examples of technical papers include photographic and blueprint papers, insulating and capacitor papers, filter papers, backing papers for coated abrasives, and laminated papers.

Techniclad. Trade name of MetalBond Technologies LLC (USA) for chemical-resistant and wear-resistant electrophoretic coatings, supplied in clear, colored and opaque finishes for application to metal substrates.

Technicoll. Trade name of Beiersdorf AG (Germany) for an extensive range of industrial adhesives.

Technigalva. Trade name of Pasminco Europe (Mazak) Limited (UK) for a galvanizing alloy containing zinc and nickel.

Techni-Gold. Trade name of Technic Inc. (USA) for corrosion-resistant, non-cyanide alkaline gold electrodeposits.

Techni-Silver. Trade name of Technic Inc. (USA) for cyanide and non-cyanide silver electrodeposits.

Techno. (1) Trade name of Techno Polymer Company (USA) for transparent, shock-resistant acrylonitrile-butadiene-styrene resins with a density of 1.07 g/cm^3 (0.039 lb/in.^{3}), a processing temperature of 230°C (446°F), and excellent processibility and dimensional stability. Used for the manufacture of machinery parts, computer and office equipment, appliance parts, tool handles, etc.

(2) See Meryl Techno.

Technoflon. Trade name of Solvay Solexis (USA) for a group of fluorocarbon copolymers and terpolymers and various other fluoroelastomers.

Technoform. Trademark of Technoform Caprano + Brunnhofer KG (Germany) for thermoplastic extrusion products.

Technoloy. Trade name of Traghetto (Italy) for polymer alloys of polyamide (nylon) and polypropylene.

Technoprene. Trade name of Enichem SpA (Italy) for glass fiber-reinforced polypropylene.

Technoprofil. Trade name of Technoprofil Breidenbach & Blau GmbH (Germany) for plastic profiles made of acrylics, polyamides (nylons), polyethylenes, polypropylenes or polyvinyl chlorides.

Technora. Trademark of Teijin Fibers Limited, Industrial Fibers Group (Japan) for a high-performance para-aramid (p-aramid) fiber made from poly-paraphenylene terephthalamide (PPTA). It has a density of 1.39 g/cm^3 (0.050 lb/in.^{3}), a decomposition temperature of 500°C (930°F), high thermal stability, high ultimate tensile strength (3.04 GPa) and elastic modulus (70 GPa), good creep and fatigue resistance, good toughness, good electrical properties (insulator), and good chemical resistance including acids and alkalies. Used as reinforcement fibers for advanced engineering composites including cement and polymers, for electrical and thermal insulation applications, as a rubber reinforcement, for ropes, and for protective products.

Technyl. Trade name of Rhodia Engineering Plastics (France) for a series of polyamide (nylon) resins including: (i) *Technyl A* polyamide 6,6 (nylon 6,6) resins supplied in various grades including unmodified, glass or carbon fiber-reinforced, glass bead- or mineral-filled, silicone-, molybdenum disulfide- or polytetrafluoroethylene-lubricated, fire-retardant, high-impact, supertough and UV-stabilized, and used for connectors, fasteners, cable ties, etc.; (ii) *Technyl B* flame-retardant alloys of polyamide 6 (nylon 6) and polyamide 6,6 (nylon 6,6); (iii) *Technyl C* polyamide 6 (nylon 6) resins supplied in various grades including unmodified, glass or carbon fiber-reinforced, glass bead- or mineral-filled, silicone-, molybdenum disulfide- or polytetrafluoroethylene-lubricated, casting, elastomer-copolymer, stampable sheet, fire-retardant, high-impact, supertough and UV-stabilized; and (iv) *Technyl D* polyamide 6,10 (nylon 6,10) resins supplied in various grades including unmodified, glass or carbon fiber-reinforced, silicone- or polytetrafluoroethylene-lubricated and fire-retardant.

Technystar. Trade name of Rhodia Engineering Plastics (France) for a series of polyamide (nylon) resins.

Techster. Trade name of Rhodia Engineering Plastics (France) for a series of polyesters including *Techster E* amorphous and supertough glass-reinforced polyethylene terephthalates, and *Techster T* polybutylene terepthalates supplied in several grades including unmodified, glass fiber-reinforced, mineral-filled, fire-retardant and UV-stabilized.

Techton 1200. Trademark of Roxul Inc. (Canada) for mandrel-wound, preformed mineral wool thermal insulation supplied in the form of rigid, strong, water-repellent pipe sections with excellent thermal resistance. Used for protecting steam and process piping operating in the temperature range of -50 to +650°C (-120 to +1200°F).

Techtron. Trademark of Quadrant Polymer Corporation (USA) for polyphenylene sulfides (PPS) with good high-temperature properties to 220°C (425°F) and good corrosion resistance, supplied in the form of extruded and compression-molded bars, blocks, rods, sheets and tubes.

Tech-Tronic. Trade name of Techalloy Company Inc. (USA) for a series of iron-base alloys containing 0.50-13.5% nickel, 0.06-0.15% carbon, 4-14% manganese, 0.03% sulfur, 1% silicon, 16.5-23.5% chromium, 0-3% molybdenum, 0.04-0.06% phosphorus, 0.20-0.45% nitrogen, 0-0.3% niobium and 0.1-0.3% vanadium. Commercially available in the form of wire and rods, they are used for cold-headed parts, fasteners, marine hardware, springs, screens, wire racks, and cages.

Tecnisilk. Trademark of Polisilk SA (Spain) for polypropylene fibers and filament yarns.

Tecnoflon. Trademark of Ausimont USA, Inc. for an extensive series of copolymer, terpolymer and peroxide fluoroelastomers available in cure-incorporated and non-curative grades.

Tecnoperl. Trademark of Sovitec SA (France) for glass microbeads used for hardening cold acrylic-based coatings for retroreflective road-marking applications.

Tecnoprene. Trade name of The Polymer Group (USA) for glass-reinforced polypropylenes.

Teco. Trade name of Tungsten Electric Corporation (USA) for a series of sintered carbides consisting of the carbides of tungsten, vanadium, chromium, thorium or molybdenum with cobalt or nickel binders. Used for cutting tools, hardfacing applications, bearings, and wear parts.

Tecoflex. Trademark of Thermedics Inc. (USA) for a series of thermoplastic polyurethanes supplied in medical, food and drug, radiopaque and high-modulus. Many are melt-processible and the medical and drug grades are suitable for catheters, tubing and other medical applications.

Tecoplast. Trademark of Thermedics Inc. (USA) for polyurethane plastics for medical applications.

Teco-Sil. Trademark of C-E Minerals (USA) for electrically fused high-purity silica (99.7%) supplied as grains and powders in various grain sizes, and used for foundry, refractory and investment casting applications.

Tecothane. Trademark of Thermedics Inc. (USA) for a medical-grade polyurethane.

Tectron. Trade name for nylon fibers and products.

Tectig. Trade name of Libbey-Owens-Ford Company (USA) for transparent, electrically conductive, thermal-insulating glass.

Tectophane. Trade name of British Cellophane (UK) for cellophane-type regenerated cellulose.

tectosilicates. Silica materials, such as quartz, orthoclase feldspars and nepheline, in which all four oxygen atoms of the SiO_4 tetrahedra are shared with adjacent tetrahedra resulting in a three-dimensional framework structure. Also known as *framework silicates*.

Tecto-Tel. Trade name of Glasfaser Gesellschaft mbH (Germany) for glass-fiber boards used for insulating flat roofs.

Tedistac. Trade name of Enichem SpA (Italy) for a series of polyurethanes.

Tedlar. Trademark of E.I. DuPont de Nemours & Company (USA) for polyvinyl fluoride (PVF) products including oriented and cast films, resins and adhesives. They have a density of 1.38 g/cm³ (0.050 lb/in.³), outstanding mechanical properties, excellent resistance to weathering and chemicals including solvents and staining agents, excellent UV resistance, low permeability to gases, and good electrical properties. The film grades are used for the encapsulation of fiberglass insulation and ceiling tiles, as protective laminates for fiber-reinforced automotive panels, in exterior glazing applications, for release films in transfer printing and printed circuit boards, for electrical applications, e.g., motor winding, cable wrapping, tapes, etc. The resins adhere well to many substrates and are used as coatings for electronic and electrical applications, and in corrosion-resistant finishes for chemical applications.

Tedur. Trademark of Albis/Bayer Corporation (USA) for heat-resistant polyphenylene sulfide (PPS) resins. Available in glass fiber-, glass bead- and/or mineral-reinforced grades, they have good chemical resistance, continuous service temperatures to 240°C (464°F), and are used for electrical and electronic components, chemical processing equipment, and automotive components.

tee bar. See T-beam.

Teenax. Trade name of Wallace Murray Corporation (USA) for an oil-hardening cold-work tool steel (AISI type O1) containing 0.9% carbon, 1.25% manganese, 0.5% chromium, 0.5% tungsten, 0.15% vanadium, and the balance iron. It has a great depth of hardening, good machinability and nondeforming properties, and is used for taps, chisels, punches, and shear blades.

teepleite. A white mineral composed of sodium boron chloride hydroxide, $Na_2B(OH)_4Cl$. Crystal system, tetragonal. Density, 2.08 g/cm³; refractive index, 1.519.

Tefabloc. Trade name of The Tessenderlo Group (USA) for a series of thermoplastic elastomers based on styrene-butylene-styrene (SBS) or styrene-ethylene-butylene-styrene (SEBS) block polymers.

Tefanyl. Trade name of The Tessenderlo Group (USA) for a series of polyvinyl chlorides.

Tefaprene. Trade name of The Tessenderlo Group (USA) for a series of thermoplastic elastomers.

Teflon. Trademark of E.I. Du Pont de Nemours & Company (USA) for polytetrafluoroethylene (PTFE) resins commercially available in the form of rods, tubes, film, sheets, insulated wire, fibers, monofilaments, multifilaments, granules, fine powder, extrusion and molding powders and water-base dispersions, and as glass- or bronze-filled grades. They have a density of 2.2 g/cm³ (0.08 lb/in.³), high toughness, very low coefficients of friction, moderate strength (when unreinforced), excellent dielectric properties, a melting point of 312°C (594°F), relatively high thermal expansivity, very high thermal stability, a service temperature range of -260 to +260°C (-436 to +500°F), good weatherability, excellent chemical resistance in almost all environments, excellent resistance to moisture and ozone, high melt viscosity, excellent resistance to ultraviolet radiation, good resistance to nuclear radiation, excellent resistance to oxidation, good flame retardancy, good resistance to acids, alcohols, alkalies, aromatic hydrocarbons, greases, oils, solvents, halogens, ketones and oxidizing agents. Used for seals, gaskets, piping, tubing, felt for filters, flexible hose, valves, packings, chemical processing equipment, tanks, laboratory ware including beakers, bottles, wash bottles and sample containers, bearings, electronic parts, printed-circuit board laminates, electrical insulation, wire coating and tape, enamels, high-temperature coatings, coatings for metals and fabrics, non-stick coatings on cookware, in grinding wheels, for bonding industrial diamonds to metal, in coating glass fibers for composites, and in monofilaments for filters. *Note:* Fluorocarbons other than PTFE are now also sold under this trademark, e.g., perfluoroalkoxy fluoropolymer (PFA), fluorinated ethylene-propylene (FEP) resins, etc.

Teflon AF. Trademark of E.I. Du Pont de Nemours & Company (USA) for amorphous fluoropolymer thin film materials including *Teflon-AF 1600,* a copolymer of tetrafluoroethylene and 66 mol% 2,2-bis(trifluoromethyl)-4,5-difluoro-1,3-dioxole. *Teflon AF* has a low dielectric constant (approximately 1.9) and zero cold-creep, and is produced by laser ablation, vacuum pyrolysis, or ultraviolet direct liquid-injection/atomization techniques. Used as interlayer dielectrics for ultra-large-scale-integrated circuit applications.

Teflon FEP. Trademark of E.I. Du Pont de Nemours & Company (USA) for melt-processable fluorinated ethylene propylene (FEP) supplied in unmodified and glass-fiber-coupled grades, and as film. The unmodified grade has a density of 2.15 g/cm^3 (0.078 lb/in.^{3}), low water absorption (0.01%), high translucency, good resistance to abrasion, cut-through and impact, good low-temperature impact resistance, good weatherability and radiation resistance, excellent chemical resistance, high flexibility, and a service temperature range of -250 to +200°C (-418 to +392°F). Used for chemical process equipment, coatings and linings, insulation, glazing film, etc.

Teflon PFA. Trademark of E.I. DuPont de Nemours & Company (USA) for a translucent thermoplastic fluoropolymer resin based on perfluoroalkoxy alkane (PFA). Supplied in unreinforced and glass-reinforced grades, it has a density of 2.13-2.16 g/cm^3 (0.077-0.078 lb/in.^{3}), a melting range of 300-310°C (570-590°F), a durometer hardness of 60 Shore D, excellent thermal resistance, negligible water absorption, excellent resistance to acids, bases, ozone and ultraviolet radiation, good abrasion resistance, good dielectric properties, a low coefficient of friction, and a service temperature range of -195 to +260°C (-320 to +500°F). Used for piping, flexible tubing, valves, pumps, and filter cartridges.

Teflon NXT. Trademark of E.I. DuPont de Nemours & Company (USA) for chemically modified polytetrafluoroethylene resins providing improved performance and fabricating capabilities. They can be readily welded and shaped by thermoforming and joined by heating with the addition of adhesives. Typical uses include pipe and vessel linings, electrical connectors, fluid-handling parts, etc.

Teflon TFE. Trademark of E.I. Du Pont de Nemours & Company (USA) for opaque white, essentially chemically inert polytetrafluoroethylenes. They have a density of 2.2 g/cm^3 (0.08 lb/in.^{3}), excellent flame retardancy, very low coefficients of friction, good high-temperature stability, and are used for chemical pipe and valves, seals, O-rings, stopcocks, and separatory funnel plugs.

Tefzel. Trademark of E.I. DuPont de Nemours & Company (USA) for translucent ethylene tetrafluoroethylene (ETFE) resins and films. Supplied in unfilled, glass and/or carbon fiber-filled and coating powder grades. The unmodified grades are translucent white and have a density of 1.7 g/cm^3 (0.06 lb/in.^{3}), excellent resistance to hydrocarbons, alkalies, and weak and dilute acids, a maximum service temperature of 150°C (300°F), poor resistance to gamma radiation, high rigidity, a water absorp-

tion of less than 0.1%, and are used for laboratory ware, e.g., clamps, adapter, fittings, etc. Also included under this trademark are polytetrafluoroethylene (PTFE) fibers and filament yarns.

Tege. Trade name of Saint-Gobain (France) for insulating units consisting of two panes of glass fused together along the edges.

Tegla. Trade name of Technische Glaswerkstätten Werner Schleyer GmbH (Germany) for a specialty glass used for electronic data processing equipment and photocopiers.

Tego. (1) Trademark of Rohm & Haas Company (USA) for thin tissue saturated with phenol-formaldehyde resin and supplied in the form of rolls for use in hot-press bonding of plywood veneers and in the manufacture of furniture and wall paneling.

(2) Trade name of Th. Goldschmidt AG (Germany) for lead-base bearing alloys with excellent antifriction properties containing 15-18% antimony, 1.5-3% tin, 1-2% copper and 0.3-0.8% arsenic. Also included under this trade name are various lead compounds, such as minium, litharge and lead silicates.

Tegocoll. Trademark of Th. Goldschmidt AG (Germany) for liquid synthetic resin adhesives used for bonding metals.

Tegofan. Trademark of Th. Goldschmidt AG (Germany) for cross-linkable acrylic and polyester resin adhesives used for interior and exterior applications.

Tegofilm. Trademark of Th. Goldschmidt AG (Germany) for synthetic resin glues used for bonding wood to wood or metals, and metals to metals.

Tegonav. Trade name of Th. Goldschmidt AG (Germany) for red lead (minium).

Tego-Tex. Trade name of Th. Goldschmidt AG (Germany) for a series thermosetting resins supplied in the form of decorative and/or protective films for laminating wood products, and for overlays and underlays as well as backing films, core films for laminates, and protective films for the manufacture of plastic laminates.

Teichmann's crystal. See hemin.

Teijin. Trademark of Teijin Company Limited (Japan) for nylon 6, rayon and cellulose acetate fibers and yarns.

Teijin-Acetate. Trademark of Teijin Company Limited (Japan) for cellulose acetate fibers and filament yarns.

Teijinconex. Trademark of Teijin Fibers Limited, Industrial Fibers Group (Japan) for a white-colored, meta-linked aromatic polyamide fiber composed of polymetaphenylene isophthalic amide. It has excellent heat, flame and combustion resistance, high dimensional stability even at high temperatures, a decomposition temperature of about 400°C (750°F), a high glass-transition temperature, excellent chemical stability, and is comparable in strength, elongation and density to polyester fibers. Used for clothing and industrial fabrics.

Teijin-Nylon. Trademark of Teijin Company Limited (Japan) for nylon 6 fibers and filament yarns.

Teijin-Tetoron. See Tetoron.

Teijido. Trademark of Teijin Company Limited (Japan) for polyvinylidene chloride (saran) fibers.

teineite. A sky blue mineral composed of copper tellurate dihydrate, $CuTeO_3 \cdot 2H_2O$. Crystal system, orthorhombic. Density, 3.80 g/cm^3; refractive index, 1.782. Occurrence: Japan.

Tekblend. Trade name of GA. GI srl (Italy) for polymer blends of polycarbonate and acrylonitrile-butadiene-styrene resins.

Tekbond. Trade name of Teknor Apex (USA) for styrene-ethylene-butylene-styrene (SEBS) block copolymers.

Teklamid. Trade name of Tekno Polimer AS (Czech Republic) for polyamides (nylons) including nylon 6 and nylon 6,6.

Teklan. Trade name for a soft, nonflammable modacrylic fiber. It has high strength, a low density, and good resistance to bacteria, household chemicals, and sunlight. Used for woven and knitted dresses, children's wear, curtains, and furnishings.

Teklon. Trade name of GA. GI s.r.l. (Italy) for polycarbonate plastics.

Tekmide. Trade name of GA. GI s.r.l. (Italy) for a series of polyamides (nylons) including nylon 6 and nylon 6,6.

Tekmilon. Trademark of Mitsui Petrochemical Company (Japan) for polyethylene fibers and filament yarns.

Teknoplen. Trade name of Tekno Polimer AS (Czech Republic) for polypropylene plastics.

Teknor Apex. Trade name of Teknor Apex (USA) for plasticized polyvinyl chlorides available in three elongation ranges: 0-100%, 100-300% and above 300%.

Teknoster. Trade name of Tekno Polimer AS (Czech Republic) for polybutylene terephthalate (PBT) plastics.

Tekral. Trade name of GA. GI s.r.l. (Italy) for acrylonitrile-butadiene-styrene (ABS) resins.

Tekron. Trade name of Teknor Apex (USA) for a styrene block copolymer.

Tekuform. Trade name of Tekuma Kunststoff (Germany) for polytetrafluoroethylenes (PTFE) and acetal (polyoxymethylene) plastics.

Tekulon. Trade name of Tekuma Kunststoff (Germany) for polycarbonate resins and plastics.

Tekumid. Trade name of Tekuma Kunststoff (Germany) for polyamides (nylons) including nylon 6 and nylon 6,6.

Telar. Trademark of Filament Fiber Technology, Inc. (USA) for a resilient, spun polypropylene filament yarn supplied in 30-300 denier counts and in a wide range of colors. It is available in flat and false-twist textured types, and may contain special additives, such as flame retardants, UV stabilizers, or photochromic or antimicrobial agents. *Telar* provides excellent abrasion, chemical and wrinkle resistance, good wet strength, controlled shrinkage, and is used for activewear, swimwear, hosiery and socks, outdoor and industrial fabrics, and wall coverings.

telargpalite. A light gray mineral composed of palladium silver telluride, $(Pd,Ag)_3Te$. Crystal system, cubic. Occurrence: Russian Federation.

Telcalloy. Trade name of Telcon Metals Limited (UK) for a series of electrical resistance alloys containing 75-95% copper and 5-25% nickel. The electrical conductivity ranges from 5.5% IACS for a 75Cu-25Ni alloy to 18% IACS for a 95Cu-5Ni alloy.

Telcar. Trade name of BF Goodrich Chemical Company (USA) for elastomers based on thermoplastic olefins (including ethylene) and used for auto-motive applications, such as paintable body filler panels and air deflectors, and for insulation applications.

Telcon. (1) Trade name of Telcon Metals Limited (UK) for a series of iron-nickel, copper-nickel, iron-nickel-cobalt and iron-nickel-chromium alloys used for bimetals, electrical resistances, speedometers, tachometers, and transformers.

(2) Trade name of Telcon Metals Limited (UK) for a series of tough, corrosion-resistant beryllium coppers containing about 2% beryllium, 0-0.2% cobalt, and the balance copper. Used for diaphragms, springs, fuse clips, fasteners, bushings, and electrical contacts.

Telconstan. Trademark of Telcon Metals Limited (UK) for an annealed resistance alloy containing 55% copper and 45%

nickel. Commercially available in the form of foil, rod or wire, it has a density of 8.9 g/cm³ (0.32 lb/in.³), a melting point of 1225-1300°C (2237-2372°F), a hardness of 100-300 Brinell, high electrical resistivity, a low temperature coefficient of resistance, a low coefficient of expansion, high heat resistance and a maximum service temperature in air of 500°C (932°F). Used for electrical resistances.

Telcoseal. Trademark of Telcon Metals Limited (UK) for a series of sealing alloys containing either 49-58% iron and 42-51% nickel, 54% iron, 29% nickel and 17% cobalt, or 52-48% iron, 42-47% nickel and 5-6% chromium. Usually supplied in the annealed condition, they have low electrical conductivity (usually below 5% IACS), and are used for sealing glass or ceramics to metals.

Telcothene. Trade name of Telcon Plastics (UK) for polyethylene resins.

Telcovin. Trade name of Telcon Plastics (UK) for polyvinyl chlorides and other vinyl plastics.

Telectal. Trade name of Metallgesellschaft Reuterweg (Germany) for nonhardenable aluminum alloys containing 1.5% silicon and 0.1% lithium, and used for electrical conductors and light-alloy parts.

Telecut. Trade name of Teledyne Vasco (USA) for a high-speed tool steel containing 1.07% carbon, 0.5% silicon, 0.25% manganese, 0-0.03% sulfur, 0-0.03% phosphorus, 1.5% tungsten, 1.15% vanadium, 9.5% molybdenum, 3.75% chromium, and the balance iron.

Teledium. Trade name of Goodlass Wall & Lead Industries Limited (UK) for a corrosion-resistant lead containing 0.02-0.1% tellurium. Used for water supply pipes, chemical equipment, sheathing for telephone and powder cables, and batteries.

Teleglas. Trade name of PPG Industries Inc. (USA) for neutral gray sheet glass used for implosion-proof television screens.

telegraph bronze. A free-cutting bronze containing 80% copper, 7.5% lead, 7.5% zinc and 5% tin. Used for screw-machine products, machine parts, and switches.

Telemet. Trade name of Allegheny Ludlum Steel (USA) for a heat- and corrosion-resistant steel containing up to 0.25% carbon, 16-23% chromium, 0-2% manganese, 0-1% silicon, and the balance iron. Used for television picture tubes and glass-to-metal seals.

Telendra. Trade name of Fibreglass Limited (UK) for glass-fiber reinforcement materials.

Telfos. Trade name of C. Clifford Limited (UK) for a free-machining grade of wrought phosphor bronze containing 95% copper, 5% tin and a trace of phosphorus. Usually supplied in bar form, it is used for machinery parts, fasteners, and hardware.

Telisol. Trade name of Saint-Gobain (France) for glass wool metal mesh blankets used as insulation in industrial equipment.

Telloy. (1) Trademark of R.T. Vanderbilt Company, Inc. (USA) for a finely ground tellurium powder used as a toughener in rubber compounding.

(2) Trade name of Inland Steel Company (USA) for a free-machining alloy steel containing small quantities of tellurium.

tellurantimony. A dark gray, opaque mineral of the tetradymite group composed of antimony telluride, Sb_2Te_3. Crystal system, rhombohedral (hexagonal). Density, 6.51 g/cm³. Occurrence: Canada (Quebec).

Tellurit. Trade name of Foseco Minsep NV (Netherlands) for tellurium-bearing mold dressings used to produce a chilled layer on the surface of gray iron. Selective chilling and uniform depth of chill are thus ensured.

tellurite. A white or yellowish mineral composed of tellurium oxide, TeO_2. Crystal system, orthorhombic. Density, 4.90 g/cm³; refractive index, 2.18. Occurrence: Mexico.

tellurium. A nonmetallic element of Group VIA (Group 16) of the Periodic Table. It is commercially available in the form of ingots, pieces, granules, powder, sticks, slabs, tablets and crystals, rarely found in native form, and usually occurs combined with gold, silver or other metals (e.g., in *calaverite* and *sylvanite*). Density, 6.25 g/cm³ (crystalline) and 6.00 g/cm³ (amorphous); melting point, 450°C (842°F); boiling point, 990°C (1814°F); hardness, 2.3 Mohs; atomic number, 52; atomic weight, 127.60; divalent, tetravalent, hexavalent. It is a p-type semiconductor and has high electrical resistivity. Crystalline tellurium is a silvery-white material with hexagonal structure, and amorphous tellurium a brownish black powder. Two allotropic forms of crystalline tellurium are known: (i) *Alpha tellurium* below 348°C (658°F) and (ii) *Beta tellurium* above 348°C (658°F). *Tellurium* is used in lead alloys to increase strength, in copper alloys to improve free-machining properties, in steel to improve machinability, in cast iron to control depth of chill, in certain cast Alnico alloys to aid formation of columnar grains, in thermoelectric alloys, in electrical resistors and battery plate protectors, as a glass-forming agent, as a yellow, green or blue colorant in ceramics, glass and glazes, as a vulcanizing agent in rubber, as a catalyst, in thermoelectric devices for spacecraft, and in semiconductors and electronic components. Symbol: Te.

tellurium bronze. See tellurium copper.

tellurium cable alloy. See tellurium lead.

tellurium copper. A phosphorus-deoxidized copper containing up to 0.5% tellurium and 0.01% phosphorus, respectively. It has a high electrical conductivity (95% IACS), excellent machinability, good cold and hot workability and forgeability, good corrosion resistance, and is used for screw-machine products, machine elements, bolts, studs, electrical equipment, transformer and circuit-breaker terminals, current-carrying parts, transistors bases, motor and switch parts, electrical connectors, welding torch and soldering iron tips, and plumbing fittings. Also known as *tellurium bronze*.

tellurium dioxide. A heavy, white to off-white, crystalline powder (99+% pure) with tetragonal or orthorhombic crystal structure. The tetragonal form occurs in nature as the mineral *paratellurite* and the orthorhombic form as the mineral *tellurite*. It can also be made synthetically and is commercially available in high-purity grades (99.995-99.9995%). Density, 5.67 g/cm³ (tetragonal), 4.90 g/cm³ (orthorhombic); melting point, 733°C (1351°F), boiling point, 1245°C (2273°F). Used in metallurgy, ceramics, electronics, chemistry and materials research. The crystals are used in electronics for modulator and switch applications. Also known as *tellurium oxide; tellurous acid anhydride*.

tellurium disulfide. A red powder turning dark brown with age and used in ceramics and lubricants. Formula: TeS_2. Also known as *tellurium sulfide*.

tellurium lead. A lead alloy containing 0.18-0.20% arsenic, 0.13-0.14% tin, 0.06-0.08% bismuth, 0.06-0.10% tellurium, 0.06% copper, and traces of antimony and silver. It has good strength, hardness and toughness, good resistance to corrosion by sulfuric acid, and is used in sheathing for telephone and power cables, in pipes with improved resistance to hydraulic pressure and for chemical equipment. Also known as *tellurium cable alloy*.

tellurium oxide. See tellurium dioxide; tellurium trioxide.

tellurium steel. A free-cutting alloy or stainless steel containing small quantities of tellurium (typically 0.01%) to increase machinability. Tellurium is sometimes preferred to other free-cutting additives (such as sulfur) because it does not affect the overall physical and mechanical properties of steels.

tellurium sulfide. See tellurium disulfide.

tellurium trioxide. Yellow-orange crystals with a density of 5.07 g/cm³ and a melting point of 430°C (806°F). Used in chemistry and materials research. Formula: TeO_3. Also known as *tellurium oxide*.

Tellurium Tubes. Trade name of Foseco Minsep NV (Netherlands) for pelleted tellurium metal supplied in copper wrapping. Used as a ladle addition to gray cast iron to control chill depth and increase hardness.

tellurnickel. See melonite.

tellurobismuthite. A tin-white to steel-gray mineral of the tetrahedrite group composed of bismuth telluride, Bi_2Te_3. It can also be made synthetically. Crystal system, rhombohedral (hexagonal). Density, 8.40 g/cm³. Occurrence: Canada (Ontario), Japan.

tellurohauchecornite. A bronze-yellow mineral of the chalcopyrite group composed of nickel bismuth tellurium sulfide, $Ni_9(Bi,Te)_2S_8$. Crystal system, tetragonal. Density, 6.40 g/cm³. Occurrence: Canada (Ontario).

telluropalladinite. A cream-colored mineral with a yellowish tint. It is composed of palladium telluride, Pd_9Te_4, and can also be made synthetically. Crystal system, monoclinic. Density, 10.25 g/cm³. Occurrence: USA (Montana).

tellurous acid anhydride. See tellurium dioxide.

Telmar. Trade name of Telcon Metals Limited (UK) for a series of *maraging steels* containing 0.01% carbon, 18-19% nickel, 7.5-9.0% cobalt, 3.2-4.9% molybdenum, 0.2-0.6% titanium, and the balance iron.

Telnic. Trademark of Chase Brass & Copper, Company, Inc. (USA) for a phosphor bronze containing 1% nickel, 0.5% tellurium, 0.2% phosphorus, and the balance copper. Usually supplied in the form of rods, it has good machinability and workability, and is used for machine parts and hardware.

telomer. A special type of polymer in which the end groups of the macromolecules lack the ability to react with other monomers during synthesis impeding the formation of large, chemically identical molecules.

Telphy. Trade name of Creusot-Loire (France) for nonmagnetic nickel alloys used for motors and electrical equipment.

Telusa. Trade name for rayon fibers and yarns used for textile fabrics.

Telwolle. Trade name of Linzer Glasspinnerei Franz Haider AG (Austria) for glass wool made by a special process known as the "Tel process."

temagamite. A mineral composed of mercury palladium telluride, $HgPd_3Te_3$. Crystal system, orthorhombic. Density, 9.50 g/cm³. Occurrence: USA, Canada (Ontario).

Temboard. Trademark of Armstrong-Newport Company (USA) for fiberboard, composite board and hardboard used for decorative interior paneling applications.

Tem-Bond. Trade name of Kerr Dental (USA) for a zinc oxide/eugenol dental cement used for temporary luting and restoration applications.

TemDent. Trade name of Kerr Dental (USA) for a dental resin used for temporary bridge and crown restorations.

Temedge. Trade name of ASG Industries Inc. (USA) for patterned glass with chip-resistant, fire-polished edges, available in strip widths for jalousies, counter dividers, and shelving.

Tempalex. Trade name of Central Glass Company Limited (Japan) for a toughened glass.

Tempalite. Trade name of Guardian Industries Corporation (USA) for a toughened glass.

Tempalloy. Trade name of ComAlloy International Corporation (USA) for a series of high-temperature polypropylene resins supplied in unreinforced, glass-reinforced and hybrid-reinforced grades for high-strength injection molded products.

Tempaloy. Trade name of Anaconda Company (USA) for corrosion-resistant, wrought copper-base alloys containing 4-5% nickel, 0.8-1% silicon, 0-9.5% aluminum, 0-1% manganese, and 0-2.5% iron. They have high strength, good cold workability and good to excellent corrosion resistance.

Tempalto. Trade name of Bull's Metal & Marine Limited (UK) for an alloy containing 56% copper, 14% nickel, 30% lead, and a trace of phosphorus. Used for high-temperature metallic packing.

Tempalux. Trade name of Westlake Plastics Company (USA) for a series of conductive polyether-imide (PEI) plastics.

Tempar-Glas. Trade name of Virginia Glass Products Corporation (USA) for toughened glass and spandrel glass including toughened glass with ceramic color fused to one side.

Tempax. Trademark of Schott America Glass & Scientific Products Inc. (USA) for a high-temperature borosilicate sheet glass used for sight glasses, and as a heat-resistant substrate for filters, headlight covers, projector covers, etc.

TempBond. Trade name of Kerr Dental (USA) for a resin-based dental cement used as a lute for temporary restorations.

TempBond Clear. Trade name of Kerr Dental (USA) for a fast-setting, dual-cure, eugenol-free, resin-based dental cement that is flexible and translucent when cured. It has good handling and mixing properties and is fluoride-releasing. Used for temporary restorations, e.g., crowns, inlays, onlays and bridges.

Temp D. Trade name of Kerr Dental (USA) for zinc oxide/eugenol dental cement.

Temper. Trade name of Sanderson Kayser Limited (UK) for a series of water-hardening plain-carbon tool steels containing 0.6-1.4% carbon, and the balance iron. Used for punches, shear blades, chisels, stone bits, crowbars, hammers, scissors, knives, woodworking tools, drills, etc.

Tempera. Trade name of Robert-Leyer-Pritzkow & Co. (Germany) for a series of corrosion-, creep- and heat-resistant steels including several austenitic stainless grades.

Temperato. Trade name of Fratelli Ragazzi SpA (Italy) for a toughened glass.

temperature-compensator alloys. See compensator alloys.

temperature-indicating compound. A thermosensitive material supplied in the form of crayons, pellets and liquids, that will melt at a prescribed temperature and is thus used in welding, heat treatment, forging, molding and other operations to indicate when a metal has reached a certain surface temperature.

temperature-indicating lacquer. A lacquer used to determine the temperature of a surface. It is applied to the surface of a material and changes color when the temperature exceeds a given value.

temper carbon. (1) A form of nodular, finely divided *graphite* found in *malleable cast iron*. It is slowly produced by reheating *white cast iron* to about 870°C (1600°F). Also known as *annealing carbon; temper graphite*.

(2) Fine amorphous particles of carbon produced in certain steels during prolonged annealing as a result of *cementite*

decomposition. Also known as *annealing carbon; temper graphite*.

Temperdie. (1) Trade name of Pennsylvania Steel Corporation (USA) for an oil-hardening hot-work steel containing 0.4% carbon, 1% manganese, 1.5% chromium, 0.25% vanadium, 1% molybdenum, and the balance iron. Used for dies and punches.

(2) Trade name of Atlas Steels Limited (Canada) for special-purpose steels. The most common grade, *Temperdie 40*, is a precipitation hardening low-alloy die steel containing 0.24% carbon, 0.6% manganese, 0.3% silicon, 1.2% chromium, 0.26% molybdenum, 3.5% nickel, 1.2% aluminum, and the balance iron. Usually supplied in the quenched and tempered condition, it has good machinability and nondeforming properties, and is used for plastic molds, zinc die-casting cores, and dies.

tempered glass. Glass that has undergone a tempering treatment, i.e., has been heated to a temperature above the glass-transition region, yet below the softening point, and then cooled to room temperature in a jet of air or an oil bath. The differences in the cooling rates for the surface and the interior result in residual stresses. After cooling, the glass sustains compressive stresses on the surfaces and tensile stresses in the interior regions. The external stress necessary to break this glass is thus much higher than that for untempered glass. When broken, it will fracture into granular, not jagged pieces like ordinary glass. Used for doors, automobile windshields and eyeglass lenses. Also known as *tempered safety glass; toughened glass*.

tempered hardboard. Hardboard subjected to a tempering treatment or manufactured with special additives to increase abrasion resistance, rigidity, density, surface hardness and/or water resistance. Available in thicknesses from 2.1 to 9.5 mm (0.083 to 0.375 in.) with a water absorption increasing from 8 to 30% with decreasing thickness. It is usually more durable than nontempered hardboard, and can have one (S1S) or two smooth sides (S2S). See also hardboard.

tempered lead. See alkali lead.

tempered martensite. Martensite that has been heated (tempered) to produce *ferrite* (body-centered cubic iron) and a fine dispersion of *cementite* (iron carbide). The tempered product has significantly enhanced ductility and toughness. Formerly known as *troostite*. See also martensite.

tempered mortar. Mortar softened by mixing with water.

tempered pitch-bonded basic refractories. Pitch-bonded basic refractories subjected to a special heat treatment to significantly reduce bond softening on reheating. See also pitch-bonded basic refractories.

tempered safety glass. See tempered glass.

tempered service hardboard. A type of *hardboard* similar to *tempered hardboard*, but with a lower percentage of additives (e.g. moisture inhibitors). Depending on the type and amount of additives, properties such as abrasion resistance, rigidity, surface hardness or water resistance may vary intermediate between those of nontempered and tempered hardboard. It is available in thicknesses from 3.2 to 9.5 mm (0.125 to 0.375 in.), and can have one (S1S) or two (S2S) smooth sides. The water absorption of the S1S type increases from 14 to 20% with decreasing thickness, while that of the S2S type increases from 18 to 25%. Also known as *service-tempered hardboard*. See also hardboard; tempered hardboard.

tempered steel. A hardened or normalized steel reheated to a temperature below the transformation range for the purpose of decreasing hardness, increasing ductility and toughness and relieving quenching stresses.

Temperex. (1) Trade name of Vidrierias de Llodio SA (Spain) for a toughened sheet glass.

(2) Trademark of Enthone-OMI, Inc. (USA) for hard gold electroplates used on metallic substrates.

temper graphite. See temper carbon.

tempering sand. A *foundry sand* moistened to obtain the dampness necessary for molding.

Temperit. Trade name of Vidrierias de Llodio SA (Spain) for a toughened plate glass.

Temperite. (1) Trade name of General Cable Corporation (USA) for a series of fusible alloys of lead, tin and cadmium. Available in the form of strips, they have a melting range of 150-330°C (300-625°F) and are used for temperature-indicating equipment.

(2) Trade name of Asahi Glass Company Limited (Japan) for a toughened glass.

(3) Trade name of Vidrobrás (Brazil) for a toughened roughcast plate glass.

Temperlite. Trade name of C-E Glass (USA) for toughened glass.

temper-rolled steel. A sheet steel with high yield strength that, after being hot-rolled and pickled, is subjected to a light cold-rolling pass in a temper mill to impart the desired flatness, metallurgical properties and surface finish.

Temper Tough. Trade name of Darwin & Milner Inc. (USA) for an oil-hardening special-purpose tool steel (AISI type L6) containing 0.75% carbon, 0.6% manganese, 1.15% silicon, 0.8% chromium, 0.15% vanadium, 0.3% molybdenum, and the balance iron. It has good deep-hardening properties and toughness, and is used for hand tools, saws, shear blades, dies, and punches.

Temper-Tuf. Trade name of Hamilton of Indiana Inc. (USA) for a toughened glass.

Tempest. Trade name of Osborn Steels Limited (UK) for a corrosion-resistant high-chromium steel containing 0.16% carbon, 17% chromium, 1.8% nickel, and the balance iron. Used especially for glass molds and inserts, and furnace parts.

Temphase. Trade name of SDS/Kerr (USA) for a chemical-cure, two-component bismethylacrylate (BMA) type composite resin 41% filled with submicron fumed silica and silane-treated barium glass. Used in dentistry for temporary crown and bridge restorations.

Tempil. Trade name Tempil Division of Big Three Industries, Inc. (USA) for temperature-indicating pellets made from certain alloys with predetermined melting points. They indicate temperatures in 7°C (12.5°F) steps in the 45-204°C (113-400°F) range, in 28°C (50°F) steps in the 204-1093°C (400-2000°F) range and in 56°C (100°F) steps from 1093-1371°C (2000-2500°C). When applied to a hot metal surface, the melting of a particular pellet indicates that the surface temperature of the metal is greater than the temperature indicated on the pellet. See also temperature-indicating compound.

Tempilstik. Trade name of Tempil Division of Big Three Industries, Inc. (USA) for temperature-indicating crayons made from certain alloys with predetermined melting points. They are supplied for temperature ranges and applications similar to those of *Tempil* pellets. See also temperature-indicating compound.

Tempit Ultra. Trademark of Centrix Corporation (USA) for a biocompatible, durable, wear-resistant, light-activated diurethane dimethacrylate resin filled up to 30% with micron-sized silica particles. Used in dentistry for temporary restorations. Fluoride-releasing *Tempit Ultra-F* is also available.

Template. Trade name of Bull's Metal & Marine Limited (UK)

for a zinc casting alloy.

Templex. Trade name of Templex CA (Venezuela) for toughened glass.

Templin. Trade name for a zinc oxide/eugenol dental cement.

Tempo. (1) Trade name of Pennsylvania Steel Corporation (USA) for an oil-hardening, nondeforming steel containing 0.9% carbon, 1.15% manganese, 0.5% chromium, 0.5% tungsten, and the balance iron. Used for gages, punches, and measuring tools.

(2) Trade name of J.M. Ney Company (USA) for a white-gold dental casting alloy containing 55% palladium and 35% silver. It has a melting range of 1171-1249°C (2140-2280°F), a casting temperature of 1371°C (2500°F), and is used for porcelain-to-metal restorations.

TempoCem. Trademark of DMG Hamburg (Germany) for a zinc oxide/eugenol dental cement that provides good adhesion and very thin film thickness. Supplied in high- and low-strength grades, it is used for temporary luting of crowns and bridges.

Tempofit. Trade name for a dental resin used for temporary crown and bridge restorations.

Tempolite. Trade name of Ato Findley (France) for urea-formaldehyde adhesives used in the woodworking and furniture industries.

temporary coatings. Solvent- or water-based coatings applied to a substrate for temporary protection or decoration, and easily removable by chemical or mechanical means. They are usually of the organic type and based on cellulosics, or acrylic or vinyl resins. Also known as *strippable coatings*.

Tempra. Trade name for rayon fibers and yarns used for textile fabrics.

Tempraver. Trade name of Vetreria di Vernante SpA (Italy) for a toughened glass.

Temprex. Trade name of Temprex Glass Corporation (USA) for a toughened glass.

Temp-R-Glas. Trade name of Connecticut Hard Rubber Company (USA) for *Teflon*-coated glass fibers made into closely woven release sheets for laminate fabrication.

TempRite. Trademark of BF Goodrich Chemical Company (USA) for polyvinyl chlorides and chlorinated polyvinyl chlorides supplied as powders, granules and particulates.

Temp-R-Lite. Trade name of Shatterproof Glass Corporation (USA) for a toughened glass.

Tempron. Trade name of GC America, Inc. (USA) for a dental acrylic resin used for temporary bridges, crowns and dentures.

Temp-R-Plate. Trade name of Ford Motor Company (USA) for a toughened plate glass.

Temp-Seal. Trade name of Lee & Sons Service & Manufacturing Inc. (USA) for an insulating glass.

Tempstik. Trade name of Tempil Division of Big Three Industries, Inc. (USA) for temperature-indicating crayons covering the entire temperature range (up to 800°C or 1472°F) for preheating in welding and certain heat treatments.

Temrex. Trade name of Kerr Dental (USA) for a zinc oxide–eugenol dental cement.

Temwood. Trademark of Tembec Inc. (Canada) for wood-based building materials and special lightweight, grainless fiberboard.

Tenac. Trademark of Asahi Kasei Kogyo Kabushiki Kaisha (Japan) for polyacetal (polyoxymethylene) homopolymers and copolymers supplied in unmodified, UV-stabilized and polytetrafluoroethylene-lubricated grades. They are available in the form of powders and pellets for the manufacture automotive parts, e.g., gears, cams and rollers, electronic equipment (tape recorders and video recorders), electric components, and fasteners.

Tenacin. Trademark of Dentsply Corporation (USA) for zinc oxyphosphate dental luting cements.

Tenacite. Trade name for a phenolic molding compound.

Tenapso. Trade name of Société Nouvelle des Acieries de Pompey (France) for a plain-carbon steel containing 0.07% carbon, 0.2% silicon, 0.35% manganese, 0.35% copper, and the balance iron. It has good resistance to atmospheric corrosion and is used for building and general fabrication.

Tenasco. Trade name for rayon fibers and yarns used for textile fabrics.

Tenax. Trademark of Tenax Fibers GmbH & Co. KG (Germany) for a series of rayon-derived carbon fibers and filaments.

Tenaxas. Trade name of Eyre Smelting Company (UK) for tin-base bearing metals containing varying amounts of lead. Used for reciprocating-engine bearings.

Tenbor. Trade name of British Steel plc (UK) for a heat-treatable steel containing 0.25-0.3% carbon, 0.25% silicon, 1.25% manganese, 0.005% sulfur, 0.025% phosphorus, 0.03% titanium, 0.002% boron, and the balance iron. Used for wear-resistant components.

Tencel. Trademark of Tencel Limited (UK) for fibrillating and non-fibrillating lyocell cellulosic textile staple fibers that can be used unblended or blended with a wide range of other fibers including nylon, rayon and spandex in the manufacture of evening and casual wear, household textiles, etc. See also lyocell fibers.

Tenco. Trade name of Sanderson Kayser Limited (UK) for a high-speed tool steel containing 0.7-0.8% carbon, 4% chromium, 1% molybdenum, 18% tungsten, 9% cobalt, 2% vanadium and the balance iron. Used for tools, dies and cutters.

Tendan. Trade name for rayon fibers and yarns used for textile fabrics.

Tendrelle. Trademark of Courtaulds Limited (UK) for nylon 6,6 staple fibers and filament yarns.

Teneka. Trade name of TENEKA-Folien (Germany) for self-adhesive screen-printing films and foils and various self-adhesive plastic films for covering books and maps.

Tenelon. Trade name of US Steel Corporation for an austenitic manganese steel containing 0.08-0.12% carbon, 17-18.5% chromium, 14.5-16% manganese, 0.3-1% silicon, 0-0.75% nickel, 0.35% nitrogen, and the balance iron. It has excellent corrosion resistance, good abrasion and wear properties, and good characteristics at elevated temperatures. Used for abrasion-, corrosion-, and wear-resistant parts.

Tenem. Trade name of NGK Metals Corporation (USA) for a strong, wear- and corrosion-resistant alloy composed of copper, tin and beryllium.

Tenet. Trade name for a zinc phosphate dental cement.

Tenex. Trademark of Nuodex Inc. (USA) for a light-colored heat-treated wood *rosin* used in electric insulating compounds, solders, noncorrosive soldering fluxes, adhesive tapes, rubber cements, varnishes, and synthetic resins.

tengerite. (1) A white mineral composed of calcium yttrium carbonate hydroxide trihydrate, $CaY_3(OH)_3(CO_3)_4 \cdot 3H_2O$. Crystal system, orthorhombic. Density, 2.77 g/cm³; refractive index, 1.642. Occurrence: Ireland, Norway.

(2) A white mineral composed of rare-earth carbonate monohydrate, $LnCO_3 \cdot H_2O$. It can also be made synthetically. Crystal system, monoclinic. Density, 2.80 g/cm³. Occurrence: Kazakhstan.

Tenit. Trade name of Styria-Stahl Steirische Gusstahlwerke AG

(Austria) for a series of air- or water-hardening hot-work tool steels containing 0.3-0.6% carbon, 0-1% silicon, 0-0.4% manganese, 1-1.1% chromium, 1.8-3.75% tungsten, 0-0.2% vanadium, and the balance iron. Used for hot-work dies, punches, chisels and headers.

Tenite. Trade name of Eastman Chemical Company (USA) for an extensive series of thermoplastic cellulosics and polyolefins. *Tenite Acetate* (also known as *Tenite I* and *Tenite A*) is a cellulose acetate available in transparent, translucent, opaque, colored and colorless forms. It has good strength and toughness, high surface gloss, good chemical resistance, good moldability and machinability, good oil resistance, and a maximum service temperature of 49-93°C (120-200°F). Used for piping, appliance housings and trim, glazing, packaging, eye shades, handles, knobs, etc. *Tenite Butyrate* (also known as *Tenite II* and *Tenite B*) is a cellulose acetate butyrate supplied in the form of film, sheet, tube and pipe. It has a density of about 1.2 g/cm³ (0.04 lb/in.³), high toughness, high transparency, excellent weathering properties, good dimensional stability, good dielectric properties and good resistance to oils and greases. Used for lenses, consumer products, packaging, plastic film and sheeting, piping and tubing, etc. *Tenite Propionate* (also known as *Tenite P*) is a cellulose acetate propionate available as a transparent thermoplastic possessing high strength, toughness and chemical resistance. Used for spectacle frames, and molded and extruded parts. *Tenite PE* is a low-density polyethylene available in film form. It has high oil and grease resistance, good tear strength and high elongation, and is used for packaging applications. *Tenite PET* is a polyethylene terephthalate supplied in various grades including amorphous, crystalline, glass fiber-reinforced, mineral-filled, high-impact, supertough, fire-retardant and UV-stabilized. *Tenite Polyallomer* is a lightweight, solid plastic that is a block copolymer of propylene and ethylene. *Tenite PP* is a polypropylene supplied with 20-40% talc filler, and as a unmodified or UV-stabilized homopolymer. It has high rigidity, good resistance to heat, chemicals, moisture and electricity, good processibility, good tensile and impact strength and a maximum service temperature of 121°C (250°F). Used for piping, tubing, film, fibers, packaging, molded parts, television cabinets, automotive trim, hinges, and wire covering.

Tenkyo. Japanese trade name for rayon fibers and yarns used for textile fabrics.

tennanite. (1) A black mineral of the tetrahedrite group composed of copper iron arsenic sulfide, $(Cu,Fe)_{12}As_4S_{13}$. Crystal system, cubic. Density, 4.61 g/cm³; refractive index, 3.021; hardness, 297-354 Vickers. Occurrence: Switzerland.

(2) A grayish black mineral of the tetrahedrite group composed of copper mercury arsenic sulfide, $(Cu,Hg)_{12}As_4S_{13}$. Crystal system, cubic. Density, 4.91 g/cm³. Occurrence: Russian Federation.

Tenneseal. Trade name of US Steel Corporation for a preformed, zinc-coated low-carbon sheet steel used in building construction especially for siding and roofing applications.

Tennessee Special. Trade name of US Steel Corporation for a water-hardening tool steel containing 0.95% carbon, and the balance iron. Used for general-purpose tools and broaches.

Tenohm. Trade name for a wrought copper-manganese-nickel alloy used for electrical resistance applications.

tenorite. A dull black mineral with a black streak and a dull to metallic luster. It is composed of copper oxide, CuO, and can also be made synthetically. Crystal system, monoclinic. Density, 6.51 g/cm³; hardness, 3 Mohs. Occurrence: Europe; USA (Arizona, Oregon, Tennessee, Utah, Wyoming). Used as an ore of copper. Also known as *black copper.*

Teno Spin. Trademark of Rosenlew Inc. (USA) for a highly elastic polyethylene foil having an adhesive backing on one side. Used for wrapping and packaging applications.

Tensile-Flex. Trade name of International Wire Products (USA) for an oxygen-free high-strength, precipitation-type copper-base alloy wire. It has excellent resistance to hydrogen embrittlement, a long flex life, and high electrical conductivity.

Tensilite. (1) Trade name of American Manganese Bronze Company (USA) for a wear-resistant manganese bronze containing 68.3% copper, 21% zinc, 3.2% manganese, 0.05% silicon, 0.15% lead, 4.8% aluminum and 2.5% iron. Used for gears and worm wheels.

(2) Trade name of American Manganese Bronze Company (USA) for a tough, strong manganese bronze containing 64-67% copper, 24-30% zinc, 2.5-3.8% manganese, 3.1-4.4% aluminum, and 0-1.2% iron. Used for gears and worm wheels.

(3) Trade name for composite materials made by mixing wood pulp with phenolic resin and compressing into sheets and shaped objects for electrical and mechanical applications, and as a paneling material.

Tensiloy. Trade name of British Steel plc (UK) for an electrical steel having high magnetic permeability and flux density.

Tensite. Trade name of Wilbur B. Driver Company (USA) for a heat-resistant nickel alloy containing 2% aluminum. Used for vacuum tube filaments, cathodes and other filaments.

Tensola. Trade name of Gloucester Foundry (UK) for a series of pearlitic cast irons.

Tensolite. Trade name for chlorinated rubber made into fibers.

Ten Star. Trade name of Carpenter Technology Corporation (USA) for a molybdenum-type high-speed tool steel (AISI type M10) containing about 0.95% carbon, 8.1% molybdenum, 4% chromium, 2% vanadium, and the balance iron. It has a great depth of hardening, high hot hardness, excellent wear resistance, and is used for lathe and planer tools, cutters, broaches, and reamers.

Tenu-Mat. Trade name of Manville Corporation (USA) for fiberglass mats used for duct liners.

Tenure. Trademark of Den-Mat Corporation (USA) for self-curing dental bonding agents for bonding dentin, enamel, porcelain and metal. *Tenure A & B* are chemical curing adhesives with high bonding strengths and excellent retention for bonding intraoral surfaces. *Tenure S* is a light-cure hydrophilic agent used in combination with *Tenite A & B* to further enhance the bonding strength. *Tenure Quik* is a fluoride-releasing, low-viscosity, single-component dental bonding resin. It provides high, consistent bond strengths, and bonds well to all intraoral surfaces including dentin, enamel, dental alloys and porcelain ceramics.

Tenzaloy. Trade name of American Smelting & Refining Company (USA) for an aluminum casting alloy containing 8% zinc, 0.8% copper, 0.4% magnesium and 0-0.1% nickel. It has high strength and machinability, and is used for machinery, housings, and light castings.

Teonex. Trade name of DuPont Teijin Films for polyethylene naphthalate (PEN) film products.

Teori. Trade name of Nippon Sheet Glass Company Limited (Japan) for a patterned glass.

Tepaz. Trade name of Haeckerstahl GmbH (Germany) for a series of austenitic, ferritic and martensitic stainless steels.

Tepcon. Trade name of Polyplastics Taiwan Limited for a series of acetal (polyoxymethylene) plastics.

tephroite. A light to brownish gray mineral of the olivine group composed of manganese magnesium silicate, $(Mn,Mg)_2SiO_4$. It can also be made synthetically. Crystal system, orthorhombic. Density, 3.87-4.12 g/cm³; refractive index, 1.766. Occurrence: Japan.

Tepperite. Trade name for a vinyl styrene resin formerly used as a denture base.

Teraglide. Trade name of Morgan Advanced Ceramics (USA) for titanium dioxide ceramics used in textile machinery for thread guides and jets.

Terathane. Trademark of E.I. DuPont de Nemours & Company (USA) for linear-chain polyether glycol polymers based on polytetrahydrofuran, or copolymers and terpolymers of tetrahydrofuran and caprolactone. The polytetrahydrofuran polymers are available in number-average molecular weights from 650 to 2900. Their densities decrease with increasing molecular weight from 0.978 to 0.970 g/cm³ (0.0353 to 0.0350 lb/in.³), while their melting points increase from 11-19°C (52-66°F) to 30-43°C (86-109°F).

terbia. See terbium oxide.

terbium. Silvery-gray, malleable, ductile metallic element of the lanthanide series (rare-earth group) of the Periodic Table. It is commercially available in the form of ingots, lumps, sheets, foils, rods, wire, filings, chips, powder, sponge, and single crystals. It occurs naturally in the same minerals as dysprosium, europium and gadolinium (e.g., *monazite, gadolinite* and *samarskite*). Crystal system, hexagonal. Crystal structure, hexagonal close-packed. Density, 8.272 g/cm³; melting point, 1356°C (2473°F); boiling point, 3123°C (5654°F); hardness, 30-80 Vickers; atomic number, 65; atomic weight, 158.925; trivalent, tetravalent. It has high chemical reactivity and low electrical conductivity (about 1.5% IACS). Used in alloys (e.g., terbiumcobalt and terbium-iron-cobalt alloys), as a phosphor activator, as a dopant in glass, yttria-stabilized zirconia and solid-state electronics, as a catalyst, in high-temperature superconductors, and in fluorescent lamps and X-ray screens. Symbol: Tb.

terbium boride. Any of the following compounds of terbium and boron used in ceramics and materials research: (i) *Terbium tetraboride.* Density. 6.55 g/cm³. Formula: TbB_4; and (ii) *Terbium hexaboride.* Density, 5.39 g/cm³. Formula: TbB_6.

terbium carbide. Any of the following compounds of terbium and carbon used in ceramics and materials research: (i) *Terbium dicarbide.* Density, 7.17 g/cm³. Formula: TbC_2, (ii) *Diterbium carbide.* Density, 8.33 g/cm³. Formula: Tb_2C; and (iii) *Triterbium carbide.* Density, 8.88 g/cm³. Formula: Tb_3C.

terbium chloride. A hygroscopic, white to off-white powder (99.9+% pure) with a density of 4.35 g/cm³, a melting point of 588°C (1090°F). It is also available in the form of white or colorless crystals as the hexahydrate. Used in the preparation of terbium compounds, in electronics, and in materials research. Formula: $TbCl_3$ (anhydrous); $TbCl_3 \cdot 6H_2O$ (hexahydrate).

terbium dicarbide. See terbium carbide (i).

terbium-doped glass. Glass doped with terbium (Tb^{3+}) ions and used for visible and near-infrared isolators.

terbium fluoride. A white powder (99.9+% pure) with a melting point of 1172°C (2141°F) and a boiling point of 2280°C (4136°F). It is also available as the dihydrate with a melting point of 1172°C (2141°F) and a boiling point of 2280°C (4136°F). Used as a source of terbium, and in the preparation of fluoride glasses, electroluminescent thin films and luminescent zinc sulfide. Formula: TbF_3 (anhydrous); $TbF_3 \cdot 2H_2O$ (dihydrate).

terbium-gallium garnet. A synthetic *garnet* of the rare-earth type used for electronic, optoelectronic and microwave applications, and as a superconductor substrate. Abbreviation: TGG.

terbium hexaboride. See terbium boride (ii).

terbium iodide. Hygroscopic, hexagonal crystals with a density of 5.2 g/cm^3 and melting point of 957°C (1755°F). Used in ceramics and materials research. Formula: TbI_3.

terbium nitrate. See terbium nitrate hexahydrate.

terbium nitrate hexahydrate. White, hygroscopic crystals or powder (99.9+% pure) with a density of 4.35 g/cm^3 and melting point of 39.3°C (103°F). Formula: $Tb(NO_3)_3 \cdot 6H_2O$.

terbium nitride. A crystalline compound of terbium and nitrogen used in ceramics and materials research. Crystal system, cubic. Density, 9.55-9.57 g/cm^3. Formula: TbN.

terbium oxide. White, cubic crystals or dark brown-black powder (99.9+% pure) with a melting point of 2387°C (4327°F) [also reported as 2410°C (4370°F)]. Used in electronics, ceramics and materials research. Formula: Tb_2O_3. Also known as *terbia*.

terbium peroxide. A black to dark brown hygroscopic powder used in ceramics and materials research. Formula: Tb_4O_7.

terbium silicide. A crystalline compound of terbium and silicon used in ceramics and materials research. Crystal system, orthorhombic. Density, 6.66 g/cm^3. Formula: $TbSi_2$.

terbium sulfate. A white powder (99.9+% pure) used in chemistry and materials research. Formula: $Tb(SO_4)_3$.

terbium sulfate octahydrate. White or colorless, hygroscopic crystals (99.9+% pure) that are water-soluble and lose $8H_2O$ at 360°C (680°F). Used in chemistry and materials research. Formula: $Tb(SO_4)_3 \cdot 8H_2O$.

terbium sulfide. A red-brown, crystalline compound of terbium and sulfur used in ceramics, electronics and materials research. Crystal system, orthorhombic. Density, 6.35 g/cm^3; band gap, 1.7 eV. Formula: Tb_2S_3.

terbium tetraboride. See terbium boride (i).

Terblend. Trademark of BASF Corporation (USA) for extrusion- and injection-molding grade thermoplastic polymer alloys. *Terblend B* and *Terblend N* are acrylonitrile-butadiene-styrene/polycarbonate (ABS/PC) alloys, and *Terblend S* are synthetic resins composed of alloys of acrylonitrile, styrene, acrylate and polycarbonate (ASA/PC). They have good stiffness, impact strength, heat and chemical resistance. Used for automotive instrument panels and housings, electrical distribution boxes, motor and computer housings, and lawn mower decks.

Terclon. (1) Trademark of Polyfil NV (Belgium) for polypropylene fibers and filament yarns.

(2) Trademark of Tongkook Synthetic Fibers Company (South Korea) for polester fibers and filament yarns.

Terefilm. Trade name of BASF Corporation (USA) for polyester film based on cyclohexylene dimethylene terephthalate. Available in thicknesses as low as 10 μm (394 μin.), it has excellent dielectric properties, high electrical resistivity, good fatigue and tear strength, and a heat distortion temperature of 171°C (340°F). Used for magnetic recording tapes, electrical insulation, and packaging applications.

Terene. Trademark of Terene Fibres India Limited (India) for polyester fibers supplied in wide range of deniers and staple lengths.

Tere-Pak. Trademark of ICI Pakistan Limited (Pakistan) for polester staple fibers.

terephthalic acid. A carboxylic acid obtained by reacting benzene and potassium carbonate over a cadmium catalyst, or by oxidation of *p*-xylene or mixed xylenes and other alkyl aromatics. It is commercially available as white crystals or powder (98+% pure) with a density of 1.51 g/cm^3 and sublimes above 300°C (572°F). Used chiefly in the manufacture of crystalline polyesters in film, fiber or resin form. Abbreviation: TPA. Formula: $C_6H_4(COOH)_2$. Also known as *p-phthalic acid*.

terephthalic polyester resin. A high-quality unsaturated polyester resin prepared from blends of terephthalic acid and fumaric or maleic anhydride. It has good chemical and thermal resistance and good mechanical properties.

Terez. Trade name of Ter Hell Plastic GmbH (Germany) for a series of plastic products based on acrylonitrile-butadiene-styrene, styrene-acrylonitrile, polymethyl methacrylate, polyoxymethylene, polamide, polypropylene or polycarbonate resins.

Terfenol. Trademark of ETREMA Products, Inc. (USA) for a rare-earth–iron alloy composed of (0.3Tb-0.7Dy)Fe_2 and having outstanding magnetostrictive properties. Used in electrical and magnetic equipment, e.g., actuators, sensors and transducers.

Terfling. Trade name of Uddeholm AB (Sweden) for shock-resisting tool steels.

Tergal. Trademark of Rhône-Poulenc–Tergal Fibres SA (France) for polyester fibers and filament yarns used for textile fabrics, blankets, bed sheets, carpets, rugs and embroidery.

Tergal Tech. Trademark of Rhône-Poulenc–Filtec SA (Switzerland) for polyester fibers and filament yarns used for industrial applications.

Teriber. Trade name for polyester fibers and products.

Terigaine. Trade name of Atofina SA (France) for fireproof polybutylene terephthalate monofilaments used in aerospace and electronics.

Terimix. Trademark of Montefibre SpA (Italy) for polyester fibers and filament yarns used for textile fabrics.

Terinda. Trade name for a polyester fiber used for bright and dull knitting yarns.

Terital. Trade name for polyester fibers and yarns used for textile fabrics.

Teristella. Trademark of Montefibre SpA (Italy) for polyester fibers and filament yarns used for textile fabrics.

Terital. Trademark of Montefibre SpA (Italy) for polyester staple fibers and filament yarns used for textile fabrics.

Terital Eco. Trademark of Montefibre SpA (Italy) for polyester staple fibers.

Terlac. Trade name of Silac for styrene-butylene-styrene (SBS) block copolymers.

Terlenka. Trademark of Enka de Colombia SA (Colombia) for polyester staple fibers and filament yarns used for textile fabrics.

terlinguaite. A yellow, brown or green mineral composed of mercury oxide chloride, Hg_2OCl. It can also be made synthetically. Crystal system, monoclinic. Density, 8.70 g/cm^3; refractive index, 2.64.

Terluran. Trademark of BASF Corporation (USA) for a series of acrylonitrile-butadiene-styrene copolymers available in transparent, low-gloss, unreinforced, glass-reinforced, medium- and high-impact, high-heat, fire-retardant, UV-stabilized, structural-foam and plating grades. They have a density of approximately 1.1 g/cm^3 (0.04 $lb/in.^3$), good tensile strength and impact resistance, high scratch and wear resistance, excellent antistatic performance, an upper service temperature of 90-100°C (194-

212°F), excellent resistance to mineral oils and dilute alkalies, good resistance gasoline and dilute acids and poor resistance to trichloroethylene and tetrachlorocarbon. Used for housings, enclosures, shielding, automotive, electrical and electronic applications, hard hats, and covers.

Terlux. Trademark of BASF Corporation (USA) for a methyl methacrylate/acrylonitrile-butadiene-styrene copolymer.

Termafond. Trade name of Montecatini Settore Alluminio (Italy) for a series of wrought aluminum-copper and aluminum-silicon alloys.

Termico Sicilia. Trade name of Agriplast Srl (Italy) for a thermal film composed of synthetic resins with moderate ethylene-vinyl acetate content and selected stabilizers. It provides high mechanical strength, high transparency, good visible light transmission and good resistance to photooxidation, and is antidrop-treated to minimize the risk of fungal diseases. Supplied in two grades, it is used in greenhouses to protect crops and flowers against thermal inversion and unfavorable light conditions.

Termo-Glas. Trademark of HKO GmbH (Germany) for thermal glass-fiber insulation products.

Termolux. Trade name of Cristales y Vidrios SA–Cristavid (Chile) for a composite consisting of two sheets of glass enclosing a layer of glass fibers.

Termovid. Trade name of Sedas de Vidrio SA (Spain) for a double glazing unit with aluminum surround and an external frame of stainless steel.

Termovis. Trade name of Fabbrica Pisana SpA (Italy) for anti-misting rear light glass.

Ternal. Trade name of Aluminium Laufen AG (Switzerland) for an aluminum casting alloy containing 5% silicon. It has good castability, a dense, fine-grained structure, and good machinability, weldability and corrosion resistance. Used for light castings for the aircraft, automotive and marine industries.

Ternalloy. Trademark of Apex Smelting Company (USA) for a series of corrosion-resistant aluminum casting alloys containing 3-4.9% zinc, 0.6-2.4% magnesium, 0.2-0.5% manganese, 0.2-0.4% chromium and 0-0.2% copper. Some are of the natural-aging type, while others respond to heat treatment. They possess good corrosion resistance and excellent machinability. Used for light-alloy castings.

ternary alloy. An alloy that is essentially composed of three chemical elements, e.g., bismuth-lead-tin, copper-nickel-zinc, copper-tin-lead or nickel-chromium-iron alloys.

ternary steel. A steel composed of three chemical elements, namely, iron, carbon and one alloying element, e.g., a silicon steel containing only iron, carbon and silicon. Also known as *simple alloy steel.*

terne. An alloy composed of 80-97% lead and 3-20% tin, and used for hot-dip coatings on iron and steel sheet or plate. See also long terne; short terne.

terne-coated stainless steel. Stainless steel sheet or strip, usually of the austenitic type, coated with *terne,* and used for roofing, flashings, gutters, and architectural brackets.

terne-coated steel. Steel sheet or strip coated with *terne.* It has excellent solderability and paintability, excellent corrosion resistance and formability, and used for gasoline tanks for cars, trucks and tractors, radiator components, air filter containers, radio and television chassis, roofing, flashing, gutters, downspouts, siding, and electrical hardware.

terne coating. A dull, smooth hot-dip coating composed of 80-97% lead and 3 to 20% tin, and used on iron and steel sheet or plate to improve corrosion resistance, solderability, paintability

and formability.

Ternel. Trade name for vinyon (vinyl chloride) fibers with excellent acid, alkali and mildew resistance. Used especially for textile fabrics.

terneplate. Iron or steel plate or sheet covered with an alloy of about 75-80% lead and 15-20% tin. It has a smooth, but dull finish, and is used for gasoline tanks for cars, trucks and tractors, radiator components, air-filter containers, radio and television chassis, roofing, flashing, gutters, downspouts, siding, and electrical hardware. See also terne; long terne; short terne.

Terne Roofing. Trade name of Follansbee Steel Company (USA) for a copper steel coated with a lead-tin alloy (terne). It has improved formability and solderability, and is used for roofing, gutters, flashings, downspouts, drains, and termite shields.

Ternex. Trade name of British Steel plc (UK) for a mild steel.

Terocoating. Trademark of Eutectic Corporation (USA) for a series of hardfacing alloys.

Terocore. Trademark of Henkel Surface Technologies (USA) for a durable, stiff, high-strength, low-density, expandable structural epoxy foam filled with hollow microspheres. Used in structural sections, and autobody cavities.

Terokal. Trademark of Henkel Surface Technologies (USA) for heat-curing, one-component adhesives used for metal-to-metal structural bonding applications.

Terolan. Trade name of Henkel Surface Technologies (USA) for a black, air-setting sealing compound based on synthetic rubber and supplied in cartridge form. It has excellent resistance to aging, heat and cold, and high stability. Used for rubber-to-glass and rubber-to-metal sealing of autombile windows (e.g., windshields and rear and side windows), and door hinges.

Teron. Trade name for polyester fibers and products.

Terophon. Trademark of Henkel Surface Technologies (USA) for rubber-base dispersions in organic solvents used for automotive applications, e.g., as overcoats on exposed underbody assemblies.

Terorehm. Trademark of Henkel Surface Technologies (USA) for urethane epoxy hybrid adhesives used as flexible, paintable autobody sealers.

Teroson. Trademark of Henkel Surface Technologies (USA) for polymer packing, sealing and jointing compounds, rust-preventive compounds for the steelmaking and automotive industries, acoustic insulation materials, waterproofing materials for the building trades, automotive sealants, undercoatings, and car care products.

Terostat. Trademark of Henkel Surface Technologies (USA) for rubber-base adhesives supplied as mastics and tapes for autobody applications.

Terotex. Trademark of Henkel Surface Technologies (USA) for a series of bitumen-rubber base underbody sealants, and wax-base cavity sealants and dip coatings for automotive applications.

Terphane. Trade name of Toray Industries, Inc. (Japan) for polyethylene terephthalate (PET) film.

p-terphenyl. A liquid or crystalline compound supplied in purities of 99+% with a density of 1.23 g/cm^3, a melting point of 212-213°C (414-415°F), a boiling point of 389°C (732°F) and a flash point above 110°C (230°F). The liquid is suitable as a laser dye with a peak lasing wavelength of 345 nm. The single crystals are used in scintillation counters, and can be combined with *pentacene* for laser and other optoelectronic applications. For the manufacture of plastic phosphors, it can also be polymerized with styrene. Formula: $C_6H_5C_6H_4C_6H_5$.

Terpol. Trademark of Sinteticos Slowak SA (Uruguay) for polyester fibers and filament yarns used for textile fabrics.

terpolymer. A polymer consisting of three repeating units or monomers, such as acrylonitrile-butadiene-styrene or ethylenepropylene-diene.

Terra. Trade name of Ludlow Steel Corporation (USA) for an oil-hardening, nondeforming tool steel used for vise jaws, collets, rings, taps, drills, gauges, dies, pistons, and piston rings.

terra alba. (1) A pure, white, finely ground powder composed of uncalcined gypsum (calcium sulfate dihydrate, $CaSO_4 \cdot 2H_2O$). Used as filler and pigment for paper, paints, plastics, etc.

(2) A term sometimes used for any of several white mineral substances, such as blanc fixe, burnt alum, kaolin or magnesia, used as fillers or pigments.

TerraCel. Trade name of Rayonier (France) for a series of cellulose fibers.

Terrachrome. Trade name of Bausch & Lomb Inc. (USA) for brownish-pink spectacle glass.

Terracote. Trademark of Foseco Minsep NV (Netherlands) for an extensive series of powdered core and mold dressings that can be applied as is, or as aqueous slurries.

terra-cotta. A hard, glazed or unglazed, buff, yellow or brownish-red earthenware used for tile, building block, roofing, vases, statuettes, pottery, and decorations on building exteriors. Abbreviation: TC.

terra-cotta clay. A high-quality clay in buff, yellow, red or brownish-red colors that has low shrinkage, strong bonding and dense burning characteristics, and is used in the manufacture of *terracotta*.

terra-cotta block. A building unit of varying size made from burnt clay and having a compressive strength of about 13.8-27.6 MPa (2-4 ksi) on the net section.

terra di siena. An orange to reddish brown earth pigment composed of burnt *sienna* and used as a paint pigment, and as a colorant for slips, bodies and glazes.

Terradust. Trade name of Foseco Minsep NV (Netherlands) for a finely divided sand additive used to provide a smooth highquality finish for nonferrous sand castings, e.g., brasses.

Terrafix. Trademark of Soil Protection Systems Inc. (Canada) for interlocking concrete blocks.

Terrapaint. Trade name of Foseco Minsep NV (Netherlands) for mold and core dressings for steel and gray cast iron. They are available as concentrated pastes, or as aqueous slurries for brush, swab or spray gun application.

terra ponderosa. See artificial barite.

terra rosa. A reddish brown soil that is a variety of the mineral *hematite* and is used as a red colorant for glazes.

terra sigillata. A porous, fine-textured, glossy, embossed red pottery.

TerraScape. Trademark of Tarmac America (USA) for textured, wet-cast patio blocks supplied brick, slate, cobblestone, and exposed aggregate designs.

Terra-Tel. Trade name of Glasfaser Gesellschaft mbH (Germany) for glass-fiber mats used as bases for gravel and soil in roof gardens.

Terratex. (1) Trade name of General Electric Company (USA) for asbestos paper made with a bentonite clay binder. It has excellent dielectric properties, and is used for electrical insulation.

(2) Trademark of Webtec, Inc. (USA) for geotextiles used in building construction and civil engineering.

(3) Trademark of Interface, Inc., Interface Fabrics Group (USA) for a family of ecologically conscious textile fabrics woven from 100% post-industrial recycled polyester. Used for household and industrial applications.

Terra Wave. Trade name of Otsuka Chemical Company (Japan) for a series of polystyrene resins.

terrazzo concrete. A type of concrete made by adding a special aggregate, usually fragments of colored stone, marble or granite, to Portland cement and water. It is either cast-in-place or precast, and polished and smoothed after the concrete has hardened. Used for decorative surfacing applications on mosaictype floors and walls.

Terrific. Trade name of Wardlows Limited (UK) for a high-speed tool steel containing 0.7-0.8% carbon, 4% chromium, 18% tungsten, 1% vanadium, and the balance iron. Used for tools, cutters, and dies.

terry. See terry cloth

terry cloth. (1) Any rough cloth made of uncut looped yarn. Also known as *terry*.

(2) A cotton pile fabric with uncut loops, made in a wide range of patterns. Used for rugs, towels, robes, etc. Also known as *terry*.

terskite. A pale lilac mineral composed of sodium zirconium silicate dihydrate, $Na_4ZrSi_6O_{16} \cdot 2H_2O$. Crystal system, orthorhombic. Density, 2.71 g/cm^3; refractive index, 1.582. Occurrence: Russian Federation.

tertiary calcium phosphate. See calcium phosphate (iii).

teruggite. A colorless mineral composed of calcium magnesium arsenate borate hydrate, $Ca_4MgB_{12}As_2O_{28} \cdot 18H_2O$. Crystal system, monoclinic. Density, 2.15 g/cm^3; refractive index, 1.528. Occurrence: Argentina.

Terulan. Trademark of BASF AG (Germany) for acrylonitrilebutadiene-styrene resins.

Tervex. Trade name for *mullite* made into a lightweight foam with a honeycomb structure and used for structural refractory parts for high temperature applications.

Terylene. Trademark of ICI Chemicals & Polymers Limited (UK) for crease-resistant synthetic polyester staple fibers and filament yarns made from ethylene glycol and terephthalic acid, and frequently mixed with wool, cotton, silk, hemp or other yarns. They have good resistance to weathering, acids, bleaching agents, insects and fungi, and is used for knitted and netted clothing fabrics (e.g., shirts, dresses and suits), upholstery, bedding, table covers, and parachute and balloon cloth.

Tesa. Trademark of Beiersdorf AG (Germany) for a series of adhesive tape products including packaging tape, colored cloth tape, duct tape, freezer tapes, metal tapes, square photo-mounting tape, etc.

Tesafilm. Trademark of Beiersdorf AG (Germany) for adhesive tape used as a masking, sealing or insulating material in the heating, refrigeration and air-conditioning trades, in boiler and piping construction, and for protective applications.

Tesaflex. Trademark of Beiersdorf AG (Germany) for soft, flexible, heat-, cold-, steam- and pressure-resistant adhesive tape with aluminum vapor deposit used for applications similar to those of *Tesafilm*.

Tesa Foam. Trade name of Tape Rite Company, Inc. (USA) for self-adhesive foam tape.

Tesakrepp. Trademark of Beiersdorf AG (Germany) for masking, sealing and packing tape.

Tesametal. Trade name of Beiersdorf AG (Germany) for a series of heat-, cold-, steam- and pressure-resistant, aluminum-base adhesive tape products, with or without release film, for use as a sealing or insulating material in the heating, refrigeration and air-conditioning trades, and in boiler and piping construction.

Tesaplast. Trade name of Beiersdorf AG (Germany) for self-sealing, heat-, cold-, steam- and pressure-resistant bituminous adhesive tape with aluminum coating used for applications similar to those of *Tesametal*.

teschemacherite. A colorless mineral composed of ammonium hydrogen carbonate, $(NH_4)HCO_3$. It can also be made synthetically. Crystal system, orthorhombic. Density, 1.57 g/cm³; refractive index, 1.535. Occurrence: Europe.

Tesil. Trademark of Silon AS (Czech Republic) for polyester staple fibers used for textile fabrics.

testibiopalladite. A steel-gray mineral of the pyrite group composed of palladium bismuth antimony telluride, Pd(Sb,Bi)Te. Crystal system, cubic. Density, 8.93 g/cm³. Occurrence: China.

Testuggine. Trade name of Vetreria di Vernante SpA (Italy) for a sandblasted glass having a tortoise-shell pattern.

Teteron. Trade name of Kwality Textiles Sdn Bhd. (Malaysia) for polyester and viscose rayon fibers and filament yarns used for textile fabrics.

Tetlon. Trade name for polyester fibers and products.

Teton. Trade name of AL Tech Specialty Steel Corporation (USA) for an oil- or water-hardening wear-resistant steel containing 1% carbon, 1.25% chromium, and the balance iron. It has high hardness and good wear resistance, and is used for bearing races and balls, aircraft bearing components, machinery parts, compression dies, and wear-resistant parts.

Tetoron. (1) Trade name of Toray Industries Inc. (Japan) polyester staple fibers and filament yarns.

(2) Trade name of Teijin Company Limited (Japan) for polyester staple fibers and filament yarns used for textile fabrics.

Tetra. Trade name of Westerwald AG (Germany) for glass blocks having a checkerboard pattern.

tetra-auricupride. A synthetic mineral composed of copper gold, AuCu. Crystal system, tetragonal. Density, 15.03 g/cm³.

Tetrabor. (1) Trademark of Elektroschmelzwerk Kempten GmbH (Germany) for boron carbide (B_4C) and sintered boron carbide products supplied as powders, pastes, papers and disks and used for ceramic applications and as lapping and grinding abrasives.

(2) Trademark of Elektroschmelzwerk Kempten GmbH (Germany) for radiation-protection coatings and linings based on boron, boron compounds or boron isotopes, and supplied as foils, plates, pipes, paints, etc. Also included under this trademark are wear- and tear-resistant synthetic additives used for metals, plastics and building materials.

tetraboron carbide. See boron carbide.

tetraboron silicide. See boron silicide (i).

tetrabromosilane. See silicon tetrabromide.

tetrabutylammonium tetrafluoroborate. A hygroscopic, white powder (99% pure) with a melting point of 160-162°C (320-323°F) used as an electrolyte additive in the synthesis of conducting poly(thiophenes). Formula: $[CH_3(CH_2)_3]_4NBF_4$.

tetracalcium aluminoferrate. A deep-brown compound that is an ingredient in Portland cement (typically 8-13 wt%), high-alumina cement and dolomite-silica firebricks, and can also be made synthetically from lime (CaO), alumina (Al_2O_3) and ferric oxide (Fe_2O_3). Crystal system, orthorhombic. Density, 3.72 g/cm³; refractive index, 2.01. Formula: $4CaO·Al_2O_3·Fe_2O_3$. Abbreviation: C_4AF. Formerly also known as *brownmillerite*.

tetracalcium ferrite. See calcium ferrite (iii).

tetracene. See naphthacene.

tetrachloromethane. See carbon tetrachloride.

tetrachlorosilane. See silicon tetrachloride.

tetrachromium boride. See chromium boride (vii).

tetracolumbium silicide. See niobium silicide (iv).

tetracolumbium trinitride. See niobium nitride (iii).

tetracyanoquinodimethane. An organic compound available in the form of orange crystals (98% pure) with a melting point of 287-289°C (548-552°F) used as an organic semiconductor and as an electron-acceptor molecule in the formation of organic charge-transfer superconductors. Formula: $C_{12}H_4N_4$. Abbreviation: TCNQ.

tetradymite. A pale steel-gray mineral with a metallic luster. It is composed of bismuth tellurium sulfide, Bi_2Te_2S, frequently also containing some selenium. Crystal system, rhombohedral (hexagonal). Density, 7.30 g/cm³; melting point, 600°C (1112°F); hardness, 1.5-2 Mohs. Occurrence: Canada, Europe, USA (Arizona, California, Colorado, Montana, New Mexico, Virginia). Used as an ore of bismuth.

tetraethoxysilane. See tetraethyl orthosilicate.

tetraethyl orthosilicate. A silicic acid ester available as a moisture-sensitive, colorless liquid (97+% pure) with a density of 0.934 g/cm³, a boiling point of 168°C (334°F), a flash point of 116°F (46°C) and refractive index of 1.382 used as a binder for sand and refractories, in the preparation of molds in the investment casting process, for weatherproofing and acidproofing stone, brick, concrete, mortar, cement, plaster and refractories, as a hardener for stone, as a source of colloidal silica in heat- and acid-resistant paints, in protective coatings for industrial buildings, in lacquers and moldings, and in organometallic synthesis. Formula: $Si(OC_2H_5)_4$. Abbreviation: TEOS. Also known as *ethyl silicate; ethyl orthosilicate; silicon tetraethoxide; tetraethoxysilane*.

tetraethyl orthotitanate. See tetraethyl titanate.

tetraethyl titanate. A flammable, moisture-sensitive, colorless liquid. The technical grade contains approximately 20% tetravalent titanium and has a density of 1.088 g/cm³, a melting point of 122°C (251°F), a boiling point of 150-152°C (302-306°F)/10 mm, a flash point of 84°F (28°C) and refractive index of 1.5043. Used in hydrolysis to yield narrow distribution titanium oxide particles suitable for sintering. Formula: $(C_2H_5O)_4Ti$. Also known as *tetraethyl orthotitanate; titanium ethylate; titanium ethoxide.*

tetraferroplatinum. A synthetic mineral composed of platinum iron, PtFe. Crystal system, tetragonal. Density, 15.16 g/cm³.

Tetrafil. Trade name of Wilson Fiberfil International (USA) for a series of engineering plastics based on modified polyethylene terephtalate (PET) and supplied in 30, 45 and 55% glass-filled formulations, in mineral-filled compounds and in 30% glass-reinforced compounds with high flame resistance. *Tetrafil* plastics have mold temperatures between 79 and 121°C (175 and 250°F), good mechanical and thermal properties, and good processing characteristics. Used for appliance, automotive and electronic components.

Tetra-Flex. Trademark of National Polychemicals Inc. (USA) for internally plasticized polymethylene polyphenol resins that are permanently flexible, and used for electrical and industrial laminates.

Tetrafluor. Trade name of Tetrafluor (USA) for unfilled and glass fiber-, graphite- or bronze-filled polytetrafluoroethylenes.

tetrafluorosilane. See silicon tetrafluoride.

tetragonal zirconia. A hard, tough, wear-resistant zirconia (ZrO_2)

that experiences a monoclinic-to-tetragonal phase transformation upon heating to temperatures of about 1150°C (2100°F). This transformation results in the formation of cracks that effectively reduce the brittle ceramic to a powder. The addition of about 3-10 wt% calcia (CaO) stabilizes the zirconia and circumvents the formation of cracks. See also fully stabilized zirconia; partially stabilized zirconia; zirconia.

tetragonal zirconia polycrystal. Zirconia in which the addition of approximately 2-3% yttria (Y_2O_3) as a stabilizer has resulted in the formation of a fine-grained tetragonal microstructure at room temperature. The yttria significantly improves the overall strength, fracture toughness and wear resistance of the zirconia. Used for ceramic applications and orthopedic implants, e.g., ball heads of total hip replacements. Abbreviation: TZP.

tetrahedrite. (1) A black, or grayish black mineral with metallic luster composed of copper silver zinc iron antimony sulfide, $(Cu,Ag,Zn,Fe)_{12}Sb_4S_{13}$. Crystal system, cubic. Density, 5.02 g/cm³; hardness, 3.5-4 Mohs. Occurrence: Canada (British Columbia), Sweden. Used as an ore of copper. Also known as *fahlore; gray copper ore.*

(2) A gray mineral composed of copper antimony sulfide, $Cu_{12}Sb_4S_{13}$. It can also be made synthetically. Crystal system, cubic. Density, 5.00 g/cm³; refractive index, 3.021. Occurrence: Algeria.

tetraiodosilane. See silicon tetraiodide.

tetrairidium dodecacarbonyl. See iridium carbonyl.

tetraisopropoxygermanium. See tetraisopropoxygermane.

tetrakalsilite. A mineral of the nepheline group composed of potassium sodium aluminum silicate, $(K,Na)AlSiO_4$. Crystal system, hexagonal. Density, 2.59 g/cm³; refractive index, 1.540. Occurrence: Italy.

tetralithium trizirconium pentasilicate. See lithium zirconium silicate (ii).

tetramagnesium columbate. See magnesium niobate (iv).

tetramagnesium germanate. See magnesium germanate (iii).

tetramagnesium niobate. See magnesium niobate (iv).

tetramagnesium pentaluminate disilicate. See magnesium aluminum silicate (ii).

tetramanganese boride. See manganese boride (v).

tetramer. A polymer molecule consisting of four identical monomer units, e.g, $(C_2H_2)_4$ is a tetramer of C_2H_2.

tetramethylene sulfone. A white, or off-white, crystalline powder (99% pure) with a density of 1.261 g/cm³, a melting point of 27°C (80°F), a boiling point of 285°C (545°F), a flash point of 33°F (165°C) and a refractive index of 1.484 used as a curing agent for epoxies, as a stationary phase in gas liquid chromatography, and in biochemistry and medicine. Formula: $C_4H_8O_2S$. Also known as *dapsone; sulfolane.*

tetramethyltetraselenafulvalene. An organoselenium compound (97% pure) that decomposes at 264°C (507°F) and whose salts and charge-transfer complexes exhibit superconducting properties. Abbreviation: TMTSF.

tetranatrolite. A white mineral of the zeolite group composed of sodium aluminum silicate dihydrate, $Na_2Al_2Si_3O_{10}\cdot 2H_2O$. Crystal system, tetragonal. Density, 2.28 g/cm³; refractive index, 1.481. Occurrence: Canada (Quebec).

tetranickel zirconium. See nickel zirconium (3ii).

tetraniobium silicide. See niobium silicide (iv).

tetraniobium trinitride. See niobium nitride (iii).

tetraphene. An organic compound that contains four benzene rings and is available in the form of yellow crystals (95+% pure) with a melting point of 157-159°C (314-318°F) and a

boiling point of 438°C (820°F), used in organic synthesis. Formula: $C_{18}H_{12}$. Also known as *1,2-benzanthracene.*

1,1,4,4-tetraphenyl-1,3-butadiene. White crystals (99+% pure) with a melting point of 207-209°C (404-408°F) used as primary fluor and wavelength shifter in soluble scintillators. Formula: $(C_6H_5)_2C=CHCH=C(C_6H_5)$. Abbreviation: TPB.

5,10,15,20-tetraphenyl-21H,23H-porphine. A porphine derivative available as purple crystals (95+% pure) that may contain up to 3% chlorin and have a maximum absorption wavelength of 415 nm used in the synthesis of organometallic complexes and as a dyes in chemistry, biochemistry and biotechnology. Formula: $C_{44}H_{30}N_4$. Abbreviation: TPP. Also known as *meso-tetraphenylporphine.*

tetrarhodium dodecacarbonyl. An organometallic compound available in the form of air-, heat- and moisture-sensitive dark red crystals, and used as an organometallic catalyst and for rhodium coatings. Formula: $Rh_6(CO)_{16}$.

tetrataenite. A silver white to grayish white mineral of the gold group composed of iron nickel, γ-(Fe,Ni). Crystal system, cubic. Density, 7.20 g/cm³. Occurrence: USA.

tetrathiafulvalene. Light-sensitive, orange crystals (97% pure) with a melting point of 120-123°C (248-253°F) used as an organic conductor, as an electron donor for supramolecular synthesis, in charge-transfer complex synthesis, for the electron transfer to diazonium salts, and in biochemistry. Formula: $C_6H_4S_4$. Abbreviation: TTF.

tetrawickmanite. A yellow mineral of the sohngeite group composed of manganese tin hydroxide, $MnSn(OH)_6$. Crystal system, tetragonal. Density, 3.65 g/cm³; refractive index, 1.724. Occurrence: USA (North Carolina).

Tetrawire. Trade name of Saint-Gobain (France) for patterned wired glass incorporating an 25 mm (1 in.) square welded wire mesh.

tetrazirconium silicide. See zirconium silicide (v).

tetrazirconium trisilicide. See zirconium silicide (vi).

Tetric. Trademark of Ivoclar Vivadent North America (USA) for a range of light-curing dental composite resins including *Tetric Flow* low-viscosity types and *Tetric Ceram* medium-viscosity types. Used especially for composite, ceramic and porcelain restorations.

Tetrolene. Trade name for polyester fibers and yarns used for textile fabrics.

Tetron. Trade name of DuPont Teijin Films for polyethylene terephthalate (PET) film and film product.

Tetric. Trade name for light-cure hybrid composites for restorative dentistry.

Tetron. Trade name for polyester fibers and products.

Teutolen. Trademark of Teutofaser GmbH (Germany) for polypropylene staple fibers.

Teviron. Trade name of Teijin Fibers Limited (Japan) for vinyon (polyvinyl chloride) fibers and filament yarns with excellent chemical resistance, high flame retardancy, and very good moisture transmission and heat retention. Used for the manufacture of blankets, underwear, and industrial fabrics.

Tevyro. Trademark of Silon AS (Czech Republic) for polyester staple fibers.

Tewesil. Trade name for a silicone elastomer used for dental impressions.

Texalon. Trade name of Texapol Corporation (USA) for a nylon-6,6 engineering resin.

Texama. Trademark of Newlands Textiles Inc. (Canada) for flameproof textile fabrics.

Texapol. Trade name of Texapol Corporation (USA) for a range of engineering polymers.

Texas clays. A group of fast-firing, low-carbon ball clays with excellent firing colors, mined in Texas, USA, and used for floor and wall tile and in the coal-tar industry.

Texas soapstone. A black, platy soapstone composed of magnesium silicate ($MgSiO_3$) commercially available in the raw condition and as blends of 60-70 raw and 30-40% calcined soapstone in air-floated or crushed form. They fire to a white color and provide high thermal-shock resistance, and are used for various ceramic applications.

Texas White. Trade name of Aardvark Clay & Supplies (USA) for medium coarse white stoneware clays (cone 5).

Texfiber. Trademark of Fibertex Corporation (Philippines) for nylon 6 fibers and filament yarns used for textile fabrics.

Texicote. Trademark of Scott Bader & Company, Limited (USA) for a series of acrylic, vinyl and styrene polymer and copolymer emulsions. Available in plasticized and unplasticized form, they are used for adhesives, and as pigment binders for clay colorants and paint vehicles.

Texilac. Trademark of Scott Bader & Company, Inc. (USA) for various acrylic and vinyl polymer and copolymer solutions. Available in plasticized and unplasticized form, they are used in paper coatings, lacquers, nitrocellulose finishes, etc.

Texin. Trademark of Bayer Corporation (USA) for durable, thermoplastic polyester and polyether based polyurethane elastomers and thermoplastic polyurethane/polycarbonate and polyurethane/acrylonitrile-butadiene-styrene alloys. They have good abrasion and wear resistance, toughness and strength, and are used for automotive components, springs, casters, couplings, sleeves, gears, musical instruments (drumsticks), and medical applications.

Texlon. Trademark of Tongkook Synthetic Fibers Company (South Korea) for polyester (polyethylene terephthalate) and spandex fibers and filament yarns used for textile fabrics.

Texolite PF. Trade name of GE Plastics (USA) for a series of phenol-formaldehyde (PF) resins and PF-impregnated cotton and glass fabric, and paper laminates. The phenolic molding materials are supplied in numerous grades including wood or natural fiber-filled general-purpose, mineral-filled high-heat, mica-filled electrical, glass-reinforced high-impact, cotton-filled medium-shock, chopped fabric-filled medium-impact and cellulose-filled shock-resistant foam.

Tex-Opac. Trademark of Sico Inc. (Canada) for exterior acrylic latex paint.

Texover. Trade name of Vidrieria Argentina SA (Argentina) for a soda-free textile glass yarn used as reinforcement and in electrical insulation.

Texpet. Trademark of Tongkook Synthetic Fibers Company (South Korea) for polyethylene terephthalate (PET) resins and products.

Texsico. Trademark of Sico Inc. (Canada) for a latex paint used for thermal insulation applications.

Texti-Glass. Trade name of P. Genin & Compagnie (France) for glass fabric used in fiber-reinforced plastics.

textile clay. A low-silica china clay used as a textile filler.

textile fabrics. See fabrics.

textile fibers. Fibers of natural (e.g., wool or cotton) or synthetic (e.g., polymers, glass or graphite) origin, usually having a length-to-diameter or length-to-width ratio of at least 100:1. They have high tensile strength over a wide temperature range, high elastic moduli and abrasion resistance and good chemical and thermal properties, and are processed into yarns or used in the manufacture of clothing or other textile fabrics.

textile film. A synthetic textile material in film form with a predominantly longitudinal molecular orientation.

textile glass. A generic term for continuous filaments or staple fibers of glass suitable for spinning, weaving, knitting, braiding, or otherwise making into textile fabrics.

textile materials. A generic term for fabrics woven, knitted, braided, spun, felted or twisted from natural or synthetic fibers or yarns, and used for clothing, carpets, furniture fittings, drive belts, etc. The fibers, yarns and yarn intermediates used are also included in this term. Also known as *textiles*.

textiles. See textile materials.

textile webbing. A strong narrow fabric with a minimum weight of approximately 510 g/m or 15 oz/yd.

textile yarns. A class of yarns with an average tenacity of 900 denier or less made for especially apparel and furnishings. See also yarn.

Textilglas. Trade name of Gevetex Textilglas GmbH (Germany) for a textile glass fiber.

Textilion. Trade name for nylon fibers and yarns used for textile fabrics.

Textilmat. Trade name of Fiber Glass Industries Inc. (USA) for glass-fiber reinforcements.

Textilure. Trademark of Engineering Yarns Inc. (USA) for vinyl-coated polyester yarn.

Textilver. Trade name of Compagnie Française des Isolants SA (France) for a glass fabric.

Textite. Trademark of Macnaughton-Brooks Limited (Canada) for textured exterior coatings for concrete walls and floors.

Textolite. Trademark of GE Plastics (USA) for a series of plastic laminates composed of a thermosetting material (e.g., phenolic, melamine, polyester, epoxy resin, or silicone rubber) bonded to asbestos, cotton, linen, paper, nylon, etc. Commercially available in the form of sheets, tubes and rods, they have good heat and electrical resistance, high rigidity, strength, hardness and stability, low moisture absorption, good moldability, castability and processibility, and can be made fairly resistant to chemicals, but are severely attacked by all strong acids and alkalies. Used for insulating materials, printed circuits, shell forms, handles, pulleys, wheels, television and radio cabinets, plugs, fuse blocks, and coil forms.

Textone. Trademark of Canadian Gypsum Company, Limited for gypsum wallboard.

Textra. (1) Trade name of Morton Powder Coatings (USA) for textured powder coatings.

(2) Trade name for rayon fibers and yarns used for textile fabrics.

Textraface. Trade name of PPG Industries Inc. (USA) for a dual-density thermal insulation composed of a core of *Textrafine* textile fibers and a surface layer of *Superfine* glass fibers. Used for air-conditioning and heating ducts.

Textrafine. Trade name of PPG Industries Inc. (USA) for blanket insulation made of reprocessed textile fibers.

Textrafluff. Trade name of PPG Industries Inc. (USA) for processed continuous glass filaments wound on a tube to provide a package with low volume and high density for storage and shipment. Used for removable and reusable insulating blankets.

Textron. (1) Trade name of Honeywell International, Inc. (USA) for silicon monofilaments used as reinforcement in metal-matrix composites.

(2) Trade name for rayon fibers and yarns.

Textura. Trade name for polyester fibers and yarns used for textile fabrics.

Textural. Trademark of Mulco Inc. (USA) for decorative and protective coatings used for masonry surfaces.

textured aramid. A continuous-filament *aramid* yarn processed by subjecting to a jet of high-velocity air to increase its bulkiness and dryness. It is used for protective apparel, and as a replacement for asbestos.

textured board. An exterior or interior plywood panel with a surface texture or pattern produced by molding, machining, embossing, etc. Examples of textured finishes include brushed, channel-groove, kerfed, roughsawn and reverse board & batten. Also known as *textured panel; textured plywood.*

textured brick. A brick whose surface has been, usually intentionally, altered by scratching, scoring, etc.

Textured Colonnade. Trade name of Libbey-Owens-Ford Company (USA) for a patterned glass.

Textured Doric. Trade name of Libbey-Owens-Ford Company (USA) for a patterned glass featuring narrow vertical reeds.

textured glass yarn. A continuous-filament glass yarn processed by subjecting to a jet of high-velocity air to increase its bulkiness and dryness. It is used for protective apparel, and as a replacement for asbestos.

textured metal. A thin metal sheet with surface patterns or designs produced by rolling, engraving, or abrasive blasting to increase the overall rigidity and stiffness. Used for paneling applications. Also known as *rigidized metal.*

textured panel. See textured board.

textured plywood. See textured board.

textured sheet. A metal sheet whose surface has been deliberately textured by rolling, engraving, grit blasting or shot peening to increase strength, rigidity and stiffness, and significantly decrease elongation. Examples include floor plates with a raised pattern, embossed sheets, and rigidized sheets.

textured yarn. (1) A yarn with a soft appearance and feel produced by crimping (wrinkling) the individual filaments, usually at regular intervals. Also known as *bulked yarn; texturized yarn.*

(2) Any continuous-filament textile yarn processed by subjecting to a jet of high-velocity air to increase its bulkiness and dryness. Also known as *bulked yarn; texturized yarn.*

Texturetone. Trade name of Australian Fibre Glass Proprietary Limited (Australia) for acoustical glass-fiber ceiling board.

texturized yarn. See textured yarn.

TFE-Glass. Trade name of Taconic (USA) for polytetrafluoroethylene (*Teflon*) coated glass fabrics used for industrial applications.

T-film. Trademark of ECO Engineering Company (USA) for polytetrafluoroethylene (*Teflon*) film used as a sealant for pipe threads, and as an antiseizing compound.

thadeuite. A yellow to orange mineral composed of calcium iron magnesium manganese fluoride phosphate hydroxide, CaMg-$(Mg,Fe,Mn)_3(PO_4)_2(OH,F)_2$. Crystal system, orthorhombic. Density, 3.25 g/cm³; refractive index, 1.597. Occurrence: Portugal.

Thai-Zex. Trade name of Bangkok Polyethylene Company Limited (Thailand) for high-density polyethylenes.

thalcusite. A mineral of the chalcopyrite group composed of copper thallium iron sulfide, $Cu_3Tl_2FeS_4$. It can also be made synthetically. Crystal system, tetragonal. Density, 6.15 g/cm³. Occurrence: Russian Federation.

thalenite. A colorless, flesh-red or pink mineral composed of yttrium silicate hydroxide, $Y_3Si_3O_{10}(OH)$. Density, 4.40 g/cm³; refractive index, 1.739. Occurrence: USA (Colorado), Sweden.

thalfenisite. A brown mineral composed of thallium copper iron nickel chloride sulfide, $Tl_6(Fe,Ni,Cu)_{25}S_{26}Cl$. Crystal system, cubic. Density, 5.26 g/cm³. Occurrence: Russian Federation.

Thalid. Trade name of Monsanto Chemical Company (USA) for polyester resins.

thallic oxide. See thallium trioxide.

thallic sulfide. See thallium sulfide (i).

thallium. A bluish-white, soft, malleable metallic element of Group IIIA (Group 13) of the Periodic Table. It is commercially available in the form of rods, sticks, ingots, shot, wire, granules and single crystals, and occurs in trace amounts in copper pyrites and iron pyrites. Density, 11.85 g/cm³; melting point, 303.5°C (578°F); boiling point, 1457°C (2655°F); hardness, 2.0 Brinell; atomic number, 81; atomic weight, 204.383; monovalent, trivalent; superconductivity critical temperature, 2.38K; oxidizes in air at room temperature. Two allotropic forms are known: (i) *Alpha thallium* (hexagonal crystal structure) that is stable below 235°C (455°F); and (ii) *Beta thallium* (body-centered cubic crystal structure) that is stable from 235 to 303.5°C (455 to 578°F). *Thallium* is used in alloys (e.g., with mercury and lead), high-index glasses, low-melting glasses and radioactive paints, in infrared detectors and thermometer fillings, for photoelectric applications, in thallium salts, as an activator for scintillators, in the preparation of thallium-substituted superconductors, and in organometallic research. Symbol: Tl.

thallium-activated sodium iodide. See sodium iodide thallide.

thallium amalgam. A liquid alloy of 91.5% mercury and 8.5% thallium. It has a freezing point of -60°C (-76°F) which is lower than that of mercury (-38.87°C or 37.1°F). Used as a replacement for mercury in low-temperature switches, thermometers, etc. Also known as *thallium-mercury alloy.*

thallium antimonide. A compound of thallium and antimony supplied as a fine high-purity powder for use as a semiconductor in electronics. Formula: TlSb.

thallium arsenide. A compound of thallium and arsenic supplied as fine high-purity powder or single crystals for use as a semiconductor in electronics. Formula: Tl_3As.

thallium barium calcium copper oxide. A ceramic compound available as a fine powder (20 μm or 790 μin.), dry processed from high-purity oxides and carbonates, and in the form of a thin film. It has a superconductivity critical temperature of about 125K and is used for high-temperature superconductors, and in superconductivity research. Formula: $Tl_2Ba_2Ca_2Cu_3O_{10}$. Abbreviation: TBCCO.

thallium bromide. See thallous bromide.

thallium carbonate. White moisture-sensitive crystalline powder (99.9+% pure). Crystal system, monoclinic. Density, 7.11 g/cm³; melting point, 272°C (522°F). Used in the preparation of synthetic diamonds and thallium-based superconductors. Formula: Tl_2CO_3. Also known as *thallous carbonate.*

thallium chloride. See thallous chloride.

thallium cyclopendienide. Light brown crystals (95+% pure) with a melting point of 300°C (572°F). It is also available in sublimed form as yellow crystals. Used in organometallic, semiconductor and superconductor research. Formula: TlC_5H_5. Also known as *cyclopentadienylthallium.*

thallium-doped sodium iodide. See sodium iodide thallide.

thallium glass. A type of flint glass in which the lead has been wholly or partly replaced with thallium. It has a high optical density and refracting power.

thallium high-temperature superconductor. Any of several thallium barium calcium copper oxide high-temperature superconductor compounds, such as $Tl_2Ba_2Ca_2Cu_3O_x$ (Tl-2223), $Tl_2Ba_2CaCu_2O_x$ (Tl-2212) or $Tl_4Ba_3Ca_3Cu_4O_x$ (Tl-4334) with superconductivity critical temperatures of well above 100K. They are available as fine powders, usually dry processed from high-purity oxides and carbonates, and as thin films. Abbreviation: Tl-HTSC.

thallium iodide. See thallous iodide.

thallium-lead alloys. A group of corrosion-resistant alloys composed predominantly of thallium and lead. Some of these alloys have superconductive properties. Used for chemical equipment, and as superconductors.

thallium-mercury alloy. See thallium amalgam.

thallium monobromide. See thallium bromide.

thallium monochloride. See thallium chloride.

thallium monoiodide. See thallium iodide.

thallium monoselenide. See thallium selenide.

thallium monoxide. Black, hygroscopic crystals or powder. Crystal system, rhombohedral. Density, 9.52 g/cm³; melting point, 579°C (1074°F); boiling point, 1080°C (1976°F). Used in low-temperature glasses, optical glasses of high refractive index, artificial gems, and as a colorant to impart greenish-yellow shades to lead glass. Formula: Tl_2O. Also known as *thallium oxide; thallous oxide.*

thallium nitrate. See thallous nitrate.

thallium oxide. See thallium monoxide; thallium trioxide.

thallium oxysulfide. A brown powder that is sensitive to visible and infrared radiation, and is used for photosensitive cells and dark signaling applications. Formula: $Tl_2S_2O_6$.

thallium selenide. A compound of thallium and selenium available in the form of gray, high-purity lumps, plates or leaves. Density, 9.05 g/cm³; melting point, 340°C (644°F). Used for semiconductors and in electronics. Formula: Tl_2Se. Also known as *thallium monoselenide; thallous selenide.*

thallium sulfide. Any of the following compounds of thallium and sulfur: (i) *Thallic sulfide.* A black, amorphous powder used in the manufacture of thallium compounds. Formula: Tl_2S_3; and (ii) *Thallous sulfide.* Blue-black, lustrous, microscopic crystals, or amorphous powder. Crystal system, tetragonal. Density, 8.39-8.46 g/cm³; melting point, 448°C (838°F); boiling point, 1367°C (2493°F). Used in infrared-sensitive photocells. Formula: Tl_2S.

thallium telluride. A compound of thallium and tellurium available in the form of high-purity lumps. Used for semiconductors and in electronics. Formula: Tl_2Te.

thallium trioxide. A brown, amorphous powder or black, cubic, crystalline powder (99+% pure). The density of the amorphous form is 9.65 g/cm³ and that of the crystalline form 10.19 g/cm³. The melting point is approximately 717°C (1323°F) [also reported as 834°C (1533°F)]. Used in the preparation of thallium-containing superconductors. Formula: Tl_2O_3. Also known as *thallic oxide; thallium oxide.*

thallous bromide. Yellowish-white, crystalline powder (99+% pure). Crystal system, cubic. Density, 7.557 g/cm³; melting point, 480°C (896°F); boiling point, 815°C (1499°F). It is available in high-purity grades (99.999%). Used as mixed crystals with thallium iodide for infrared radiation transmitters, and in scintillation counters. Formula: TlBr. Also known as *thallium bromide; thallium monobromide.*

thallous carbonate. See thallium carbonate.

thallous chloride. White, crystalline powder (99+% pure) with a density of 7.00 g/cm³, a melting point of 430°C (806°F) and a boiling point of 720°C (1328°F) used as catalyst, chlorinating agent and in sun-tan lamp monitors. Formula: TlCl. Also known as *thallium chloride; thallium monochloride.*

thallous iodide. Yellow, light-sensitive powder (99.9+% pure). Crystal system, cubic. Density, 7.1-7.29 g/cm³; melting point, 440°C (824°F); boiling point, 823°C (1513°F); turns red at 170°C (338°F). Used as mixed crystals with thallium bromide for infrared radiation transmitters. Formula: TlI. Also known as *thallium iodide; thallium monoiodide.*

thallous nitrate. Colorless or white crystals (99.5+% pure). Crystal system, cubic or rhombohedral. Density, 5.5 g/cm³; melting point, 206°C (403°F); boiling point, decomposes at 430°C (806°F). Used in pyrotechnics (green fire) and in superconductivity studies. Formula: $TlNO_3$. Also known as *thallium nitrate.*

thallous oxide. See thallium monoxide.

thallous selenide. See thallium selenide.

thallous sulfide. See thallium sulfide (ii).

thalotide. A compound of thallium, oxygen and sulfur having photoconductive properties. See also thallium oxysulfide.

thaumasite. A colorless mineral of the ettringerite group composed of calcium carbonate silicate sulfate hydroxide dodecahydrate, $Ca_3Si(OH)_6(SO_4)(CO_3)\cdot 12H_2O$. Crystal system, hexagonal. Density, 1.91 g/cm³; hardness, 3.5 Mohs; refractive index, 1.504. Occurrence: Northern Ireland.

The Force. Trade name for a nickel-titanium wire used for dental applications.

theisite. A pale blue-green mineral composed of copper zinc antimony arsenate hydroxide, $Cu_5Zn_5[(As,Sb)O_4)]_2OH_{14}$. Crystal system, orthorhombic. Density, 4.25 g/cm³; refractive index, 1.785. Occurrence: USA (Colorado).

thenardite. A colorless, white or brownish mineral composed of sodium sulfate, Na_2SO_4. It can also be made synthetically. Crystal system, orthorhombic. Density, 2.66 g/cm³; refractive index, 1.476. Occurrence: Europe (Germany, Spain), USA (Arizona, California), Chile. Also known as *verde salt.*

Thenard's blue. See cobalt blue.

theophrastite. A blue- to emerald-green mineral of the brucite group composed of nickel hydroxide, $Ni(OH)_2$. It may also contain magnesium. Crystal system, hexagonal. Density, 3.60-4.10 g/cm³.

Therban. Trademark of Bayer Corporation (USA) for hydrogenated nitrile-butadiene rubber (HNBR) with outstanding heat, oil and gasoline resistance, excellent weathering, ozone and hot air resistance, very good abrasion resistance and compression set properties, and good mechanical properties. Supplied in several grades with acrylonitrile contents ranging from 21 to 43%, it is used in the manufacture of calendered, extruded and compression molded products for the automotive and petroleum industries, seals, hoses, stators, high-quality gaskets, V-belts, timing belts, and for hydraulic hose and couplings, and cable sheathing.

Therfol. Trade name of Verres Industriels SA (Switzerland) for a heating foil embedded in the interlayer of *Therglas.*

Therglas. Trade name of Verres Industriels SA (Switzerland) for electrically conductive laminated glass, having fine wires incorporated in the interlayer.

Therlite. Trade name of Nippon Sheet Glass Company Limited (Japan) for laminated glass having fine electrothermal wires incorporated in the interlayer.

Therlo. (1) Trademark of Driver-Harris Company (USA) for an electrical resistance alloy containing 85% copper, 9.5% manganese and 5.5% aluminum. It has a low temperature coefficient, and is used in instrument shunts.

(2) Trade name of Harrison Alloys Limited (USA) used for an iron-nickel-base superalloy containing 28-30% nickel, 16-18% cobalt, and the balance iron. Used for electrical resistors and glass to metal seals.

Thermabond. Trade name of Dentecon, Inc. (USA) for a nickel-chromium dental bonding alloy.

Therma-cel. Trademark of RBX Corporation (USA) for closed-cell polyethylene foam used for thermal insulation applications. Seam-sealing grades, such as *Therm-cel Seam Seal* and *Therma-cel SSL*, are also available.

Therma-Cube. Trade name of Thermacore International Inc. (USA) for a porous powder-metallurgy material made of bonded copper particles and used in electronics for making heat sinks, e.g., cooling units, fans, etc.

Thermafiber. Trademark of United States Gypsum Company for mineral-fiber insulation wool.

Thermafiber FRF. Trademark of United States Gypsum Company for a mineral fiber designed to enhance the characteristics of plastic products. The spun fiber is supplied in loose form for easy blending and ready dispersion with other ingredients. It has high tensile strength, good heat resistance up to 1093°C (2000°F), and is used as filler or reinforcement for plastics.

Thermaflow. Trademark for a series of thermosetting polyester resins. They have outstanding dielectric properties, low moisture absorption, good heat resistance up to 204°C (400°F), good weatherability, good chemical and flame resistance and good color stability. Used for reinforced structural shapes, boat hulls, autobodies, aircraft glaze, television and radio parts, and radomes.

ThermaGlas. Trade name of ThermaGlas Inc. (USA) for insulating glass.

Thermaglass. Trade name of Thermaglass Units Limited (New Zealand) for a laminate composed of synthetic resin and glass.

Thermalac. Trademark of Sico Inc. (Canada) for clear water-base baking enamels.

Thermalate. Trademark of Haysite Reinforced Plastics (USA) for tough, glass-mat reinforced polyester laminates used for thermal insulation applications, e.g., on press platens. They have high durability and compressive strength, good oil resistance, very low water absorption, and a maximum service temperature of 232°C (450°F).

thermal barrier. Any material that is used to prevent or minimize the transfer of heat or cold from one body or area to another.

thermal barrier coatings. Coatings applied to metal substrates to protect them against oxidation and corrosion at high temperatures, insulate them from thermal shocks and reduces thermal-fatigue effects and substrate wear. They usually consist of metallic bond-coats (often combinations of nickel, cobalt and chromium with additions of aluminum and yttrium) and ceramic topcoats (e.g., alumina, magnesia or zirconia stabilized with yttrium). *Thermal barrier coatings* for commercial superalloy gas-turbine engines often consist of three layers: an aluminum-rich bond-coat, a thermally grown oxide that is grown on the bond coat, and a ceramic topcoat. Thermal barrier coatings are usually applied by thermal spraying (e.g., plasma spraying) to automotive engine components, engines for aircraft, rockets and spacecraft, diesel engines for automobiles, buses, trucks,

power plants, towboats, etc., and heat-treating and brazing fixtures.

thermal black. Carbon black made under controlled conditions by the thermal decomposition of hydrocarbon gases (e.g., methane, natural gas, etc.). It consists primarily of elemental carbon in the form of extremely fine particles of near-spherical shape, and is used in pigments, and as a reinforcing agent in rubber products. Also known as *thermal carbon black*. See also carbon black.

Thermalbond. (1) Trademark of Saint-Gobain Performance Plastics (USA) for self-adhesive polymer sealing and mounting tape.

(2) Trademark of Norton Company (USA) for flexible, cellular plastic foam insulating materials.

(3) Trademark of Williams-Hayward Protective Coatings, Inc. (USA) for decorative and protective coatings with anti-corrosive and anti-weathering properties for use on exterior metal surfaces.

thermal carbon black. See thermal black.

Thermalcore. Trademark of MSC Laminates & Composites Inc. (USA) for composites consisting of a metal layer applied over aluminum foil, or graphite, E-glass or adhesive film, and used for acoustic and thermal containment applications.

thermal fabrics. (1) Textile fabrics, usually with a honeycomb or waffle texture, knitted or woven such as to hold warm air between the threads. Used especially for underwear, blankets, and winter outdoor clothing.

(2) Textile fabrics with wicking properties woven or knitted from soft, spun synthetic fibers or yarns of polyester, polypropylene or polyvinyl chloride, and used for the manufacture of thermal underwear.

thermal glass. A low-expansion-type borosilicate glass having excellent thermal-shock resistance and chemical durability and high toughness. Used for domestic ovenware, other heat-resistant glassware, chemical and laboratory ware, and in glass-fiber manufacture.

ThermalGraph. Trademark of Amoco Performance Products, Inc. (USA) for pitch-based discontinuous graphite fibers, and rigid fiber panels and fabrics. The fibers are used as reinforcements in thermoplastic-resin and metal-matrix composites and the fabrics and panels for thermal insulation applications.

thermal-insulating cement. See insulating cement.

thermal insulators. See heat insulators.

Therma-Lite. Trademark of Robson Thermal Manufacturing, Limited (USA) for formable thermal insulation.

Thermalkyd. Trademark of Sico Limited (Canada) for alkyd baking enamels.

Thermalloy. (1) Trade name of Abex Corporation (USA) for a series of cast austenitic stainless steels, corrosion-resistant cast irons and various nickel-base superalloys.

(2) Trade name of ENCOR Proprietary Limited (UK) for polymer alloys of polycarbonate (PC) and acrylonitrile-butadiene-styrene (ABS).

thermally bonded batting. A textile product made by first incorporating low-melting fibers or polymers into batting materials that are subsequently heated to effect bonding of the batting materials. Used as filler.

thermally bonded nonwoven fabrics. Nonwoven textile fabrics composed of fibrous batts or webs containing heat-sensitive powder, or single- or bicomponent fibers. They are made by first incorporating the heat-sensitive material into the batt or web, and then bonding by the application of heat with or without pressure. Also known as *thermally bonded nonwovens*.

thermally foamed plastics. Expanded plastics produced from pasty polymer compounds by first introducing a gaseous foaming agent, such as carbon dioxide, and then applying heat to decompose or volatilize the latter and effect foam formation.

thermally quenched phosphor. A *phosphor*, usually an organio compound that emits visible radiation when excited by ultraviolet radiation whereby the brightness at room temperature or slightly elevated temperatures varies inversely with temperature. Used in nondestructive inspection to measure temperatures.

Thermal Mount. Trade name of Mathison's (USA) for a heat-activated mounting adhesive consisting of a polyester substrate coated on one side with a pressure-sensitive acrylic adhesive with moisture-resistant, silicone-treated release liner and on the other with thermal copolymer adhesive.

Thermalouver. Trade name of Thermalouver Inc. (USA) for an insulating glass unit having a solar weather control louver screen in the air space.

thermal-spray coatings. A group of coatings produced by heating finely divided metallic or nonmetallic particles to a molten or semi-molten state and propelling onto the substrate. The coatings may be applied by the wire flame spray, powder flame spray, plasma spray, electric-arc spray or high-velocity oxyfuel process. The starting material may be in the form of a powder, wire or rod.

thermal-spray powder. A metal or nonmetal powder or mixture of powders suitable for use with powder flame spray, plasma spray and high-velocity oxyfuel equipment.

Thermal-Z-Coatings. Trade name for lightweight, well-bonded zirconia ceramics applied by arc-plasma spraying directly to automotive brake or exhaust systems or engine components to form a thermal barrier and protect them against high-temperature oxidation and corrosion up to 1400°C (2550°F).

Thermanit. Trade name of Thyssen Edelstahlwerke AG (Germany) for an extensive series of coated austenitic, ferritic and martensitic stainless steel welding electrodes and filler metals.

Therma-Panels. Trade name of Humphrey Products, Inc. (USA) for metal building panels.

Therma Puff. Trade name of Buffalo Batt & Felt Corporation (USA) for polyester batting supplied in plain, stitched and bonded forms.

Thermaqua. Trademark of Sico Inc. (USA) for water-reducible baking enamels.

ThermaStat. Trade name of E.I. DuPont de Nemours & Company (USA) for high-performance polyester fibers and fabrics.

Therma-Tech. Trade name of M.A. Hanna Company (USA) for thermally conductive compounds consisting thermoplastic resin matrices with powder and fiber additions. Used for thermal management applications including appliance heat exchangers and heat sinks for circuit boards.

Thermavac. Trademark of Arlon Inc. (USA) for uncured silicone vacuum bagging materials.

Thermax. (1) Trade name of Celotex Corporation (USA) for insulating board manufactured from shredded wood fibers and a fire-resistant urethane cement.

(2) Trademark of Cancarb Limited (Canada) for high-purity, medium thermal carbon black supplied in several grades. It has excellent thermal insulation properties, low thermal conductivity, low combustibility, low ash and grit levels, low chemical reactivity and excellent high-temperature resistance up to 3500°C (6330°F) in reducing atmospheres. Used for high-temperature insulation in reactors and industrial furnaces, and in

the manufacture of wire and cable jackets, mechanical rubber goods, refractories, plastics and metal carbides.

(3) Trade name of Thyssen Edelstahlwerke AG (Germany) for a series of chromium, chromium-nickel, nickel-chromium, nickel-chromium-cobalt stainless and/or heat-resisting steels.

(4) Trademark of Schott Glas AG (Germany) for a high-temperature sheet glass.

(5) Trade name of E.I. DuPont de Nemours & Company (USA) for a family of high-performance polyester fibers and fabrics.

Thermax Ultra-Pure. Trademark of Cancarb Limited (Canada) for a high-purity thermal carbon black (99.98% pure carbon) available in soft, readily dispersible pellets and in powder form used in the production of chromium, tungsten, vanadium, tantalum and silicon carbides, and as a reducing agent in the production of chromium, beryllium-copper alloys and solar-grade silicon.

Thermazote. Trade name of Expanded Rubber Company (UK) for phenolic and polyethylene resins and plastics.

Thermblack. Trade name for a series of thermal carbon blacks.

Thermeez. Trademark of Cotronics Corporation (USA) for a ceramic putty made of high-purity aluminum oxide formulated with special ceramic binders, which on drying produces a strong ceramic body. Supplied in tube form, it has a melting point above 1760°C (3200°F), good resistance to molten metals and most chemicals and solvents and good thermal and electrical insulation properties. Used for welding, brazing, electrical and thermal insulation, induction heating, liquid metal handling, and insulating pipes. Also included under this trademark are high-temperature pressure-sensitive tapes woven from ceramic fibers for applications requiring high strength, durability and dimensional stability, and good chemical and electrical resistance.

Thermelast. Trade name of Vacuumschmelze GmbH (Germany) for a series of constant-modulus alloys containing varying amounts of iron, nickel, chromium, titanium, molybdenum, beryllium and aluminum. They have high elastic moduli (Young's and shear modulus) and are characterized by a controllable temperature coefficient. Used for balance springs, leaf springs, diaphragms, springs in scales and pressure cells, tuning forks, mechanical filters, oscillators, and delay lines.

Thermenol. Trade name of Colt Industries (UK) for an alloy containing 79-80% iron, 16% aluminum, 3-5% molybdenum, 0-0.3% vanadium and up to 0.05% carbon. It has high oxidation resistance, a coercive force and magnetic permeability that can be varied by heat treatment, and is used for aircraft parts.

Thermex. (1) Trade name of J. Eberspächer (Germany) for a laminated glass with a plastic interlayer containing active chemicals. It is transparent at ordinary temperatures, but becomes reversibly milky under heat or solar radiation.

(2) Trade name of ComAlloy International Corporation (USA) for a series of heat-dissipative polymeric materials.

Thermid. Trademark of National Starch and Chemical Company (USA) for a series of polyimide oligomers with the empirical formula $C_{68}H_{34}N_4O_{12}$, having continuous-service temperatures up to 316°C (600°F). Used as matrix resins for composites and laminates for electronic applications, e.g., multichip modules, and as interlayer dielectrics for thin-film applications.

Thermimphy. Trade name of Creusot-Loire (France) for a steel containing 0.12% carbon, 7% chromium, 0.5% molybdenum, 0.3% vanadium, and the balance iron. Used for high-temperature bolts and parts.

Therminox. Trade name of Société Nouvelle des Acieries de Pompey (France) for a series of austenitic stainless steels typically containing up to 0.2% carbon, 15-25% chromium, 5-20% nickel, and the balance iron. Used for chemical and food-processing equipment, furnace parts, turbine parts, and heat-treating equipment.

Thermisilid. Trade name of Krupp Steel AG (Germany) for a heat- and acid-resistant special cast iron containing 14-16% silicon, used for equipment for the production and processing of acids, and for general chemical equipment.

Thermit. Trade name of Th. Goldschmidt AG (Germany) for lead-base bearing alloys with high, medium, or low tin contents.

thermit. An intimate mixture of aluminum and a metal oxide in fine-powder form. Upon ignition the aluminum combines chemically with the oxygen of the metal oxide producing temperatures up to 2760°C (5000°F) and yields aluminum oxide and a molten metal. Usually, the oxide is either red ferric oxide (Fe_2O_3) or black ferric oxide (Fe_3O_4). Used in welding (e.g., railroad rail joints, gear teeth, fractured crankshafts, rolls, pipe, large steel structures and castings), in the repair of heavy machinery, in the manufacture of metallic chromium, iron, manganese, molybdenum and tungsten, and in incendiary bombs. Also spelled *thermite*.

Thermkon. Trademark of CMW Inc. (USA) for a series of alloys containing tungsten and copper (W-10Cu to W-25Cu) or molybdenum and copper (e.g., Mo-15Cu). They have good electrical conductivity (28-45% IACS), low thermal expansion and high elastic moduli. Used for matching the coefficients of thermal expansion of semiconductor substrates.

Thermo. (1) Trade name of General Glass Corporation (USA) for an insulating glass.

(2) Trademark of Ceradyne, Inc. (USA) for custom-made refractory shapes.

Thermobestos. Trade name of Johns-Manville Company (USA) for high-temperature asbestos cement, blocks and pipe coverings.

Thermobloc. Trade name of NV Durobor (Belgium) for hollow glass blocks.

thermobonding film. Adhesive films made of thermoplastic or thermosetting material and used for bonding plastics, fabrics, metals, etc.

Thermo-Cell. Trade name of Columbia Metal Products Company (USA) for an insulating glass used for windows and doors.

Thermochrom. Trade name of Willworthy Piston Ring Limited (UK) for a cast iron containing 3.2% carbon, 2.2% silicon, 0.8% chromium, 0.2% molybdenum, and the balance iron. Used for piston rings.

thermochromic material. A material, such as a liquid crystal, that changes color when a certain temperature is reached.

thermochromic paint. A paint consisting of a mixture of temperature-indicating materials that change color as a function of temperature. Used in nondestructive inspection. Also known as *thermocolor paint*.

Thermoclad. Trademark of Alcan Rolled Products Company (Canada) for precoated pipe jacketing.

Thermocol. Trademark of Foote Mineral Company (USA) for a ferroniobium (ferrocolumbium) containing about 53% niobium (columbium), up to 0.15% carbon, and the balance iron. It reacts exothermically, and is used to introduce niobium into steel.

thermocolor paint. See thermochromic paint.

Thermocomp. Trademark of LNP Engineering Plastics (USA) for an extensive series of glass or carbon-fiber-reinforced ther-

moplastics. The following compositions are available: *Thermocomp AF* glass fiber-reinforced acrylonitrile-butadiene-styrene; *Thermocomp BF* glass fiber-reinforced styrene-acrylonitrile; *Thermocomp CF* glass fiber-reinforced polystyrene; *Thermocomp DC* 30% carbon fiber-reinforced polycarbonate; *Thermocomp DF* glass fiber-reinforced polycarbonate; *Thermocomp DL* 15% PTFE-lubricated polycarbonate; *Thermocomp EP* high-flow epoxy resin for thin-wall molding); *Thermocomp FP* glass fiber-reinforced FEP and PFA fluoropolymers; *Thermocomp GC* 30% carbon fiber-reinforced polysulfone; *Thermocomp GF* glass fiber-reinforced polysulfone; *Thermocomp GL* (15% PTFE-lubricated polysulfone); *Thermocomp HC* 30% carbon-reinforced nylon 11; *Thermocomp HF* 30% glass-fiber-reinforced nylon 11; *Thermocomp HSG* high-specific-gravity polymer composite consisting of a thermoplastic base resin filled with metallic and/or nonmetallic powders; *Thermocomp IC* 30% carbon fiber-reinforced nylon 6,12; *Thermocomp IF* 10-30% glass-fiber-reinforced nylon 6,12; *Thermocomp IL* silicone- or PTFE-lubricated nylon 6,12; *Thermocomp JC* (30% carbon fiber-reinforced polyether sulfone; *Thermocomp JF* 20-30% glass fiber-reinforced polyether sulfone; *Thermocomp KC* (30% carbon fiber-reinforced acetal resins; *Thermocomp KFX* glass fiber-coupled acetal copolymers; *Thermocomp LF* glass fiber-reinforced fluorinated ethylene propylene; *Thermocomp MF* 20-30% glass fiber-reinforced polycarbonate; *Thermocomp MFX* glass fiber-coupled polypropylene; *Thermocomp OC* 30% carbon fiber-reinforced polyphenylene sulfide; *Thermocomp OF* glass fiber-reinforced polyphenylene sulfide; *Thermocomp OG* glass fiber- or bead-reinforced polyphenylene sulfide; *Thermocomp OL* 20% PTFE-lubricated polyphenylene sulfide; *Thermocomp PB* glass bead-filled nylon 6; *Thermocomp PC* 30% carbon fiber-reinforced nylon 6; *Thermocomp PF* glass fiber- or glass bead-reinforced nylon 6; *Thermocomp PL* silicone- or PTFE-lubricated nylon 6; *Thermocomp QC* 30% carbon fiber-reinforced nylon 6,10; *Thermocomp QF* glass fiber-reinforced and flame-retardant nylon 6,10; *Thermocomp QL* silicone- or PTFE-lubricated nylon 6,10; *Thermocomp RC* carbon fiber-reinforced nylon 6,10; *Thermocomp RF* glass fiber- or glass bead-reinforced nylon 6,6; *Thermocomp RL* 20% PTFE-lubricated nylon 6,6; *Thermocomp SF* glass fiber- or glass bead-reinforced nylon 12; *Thermocomp TF* glass fiber-reinforced polyurethane; *Thermocomp WC* carbon fiber-reinforced poly-butylene terephthalate; *Thermocomp WF* glass fiber-reinforced polybutylene terephthalate; *Thermocomp WL* silicone- or PTFE-lubricated polybutylene terephthalate; and *Thermocomp ZF* glass fiber-reinforced polyphenylene oxide. *Thermocomp* resins are used for automotive and electronic components, appliance housings, etc.

Thermocone. Trade name for a liquid silicone paint compounded with flake graphite. It is resistant to many chemicals and to high temperatures (up to 538°C, or 1000°F), and used as a finish for heat resistance.

Thermo-Cor. Trademark of American Foam Technologies (USA) for rigid phenolic foams used for aerospace, marine, recreational and industrial insulation applications.

Thermo Cote. Trade name of Thermo-Cote, Inc. (USA) for corrosion-resistant, clear, strippable protective coatings based on thermosetting and thermoplastic resins.

thermocouple alloys. A group of alloys suitable for use in thermocouples for the accurate measurement of temperatures over a wide range. Examples include copper-nickel (about 55% copper and 45% nickel), nickel-chromium (about 90% nickel and

10% chromium), nickel-aluminum (about 94-95% nickel, balance aluminum), platinum-rhodium (70-94% platinum and 6-30% rhodium), platinum-iridium (90% platinum and 10% iridium), and tungsten-rhenium (74% tungsten and 26% rhenium).

Thermocules. Trade name of Outlast Technologies (USA) for a phase change material in the form of microthermal spheres directly incorporated (injected) into acrylic, nylon or polyester fibers or yarns, polymeric foams, or a hydrophilic compound and applying as a pattern coating to a base fabric. The microspheres are produced by a proprietary "Adaptive Control" technology and are concentrated at the surface of the fiber, fabric or foam. Apparel containing or treated with this material can adapt to changes in body temperature absorbing excess body heat and releasing it when the body temperature drops below the comfortable level, and thus keep the wearer more comfortable. "Adaptive Control" fabrics are used for outdoor clothing such as ski apparel, ski boot liners and glove liners, for socks, and for comforters, pillows, mattress pads and other bedding products.

Therm-O-Deck. Trade name of Brouk Company (USA) for expanded *perlite* and *vermiculite* roof fill.

Thermodet. Trade name of Flachglas AG (Germany) for sheets and foils of thermoplastic resins used for vacuum-forming applications. Also included under this trade name are thermoplastics supplied in the form of flock, nonwovens, and textile-laminated, foam-pack and mirror foils.

Thermodur. (1) Trade name of Bergische Stahl Industrie (Germany) for an extensive series of cast heat- and corrosion-resistant austenitic, ferritic and martensitic stainless steels and heat-resistant steels.

(2) Trade name of Eisenwerk Würth GmbH & Co. (Germany) for malleable-iron abrasives used for blast cleaning.

(3) Trade name of KS Kolbenschmidt GmbH (Germany) for a high-heat aluminum casting alloy containing 12.5-13.3% silicon, 3.3-3.9% copper, 1.75% nickel, 0.6-1.1% magnesium, 0-0.65% iron, and 0.1-0.35% manganese. They possess high fatigue strength and wear resistance, and are used for automotive engine pistons.

Thermodyne. Trade name of American Smelting & Refining Company (USA) for babbitts containing varying amounts of tin, antimony and lead. Used for bearings.

thermoelastic material. A material, such as a rigid plastic, that exhibits rubber-like elasticity due to an increase of temperature.

Thermo Electric. Trade name of Thermo Electric Company, Inc. (USA) for thermocouple and extension wire.

thermoelectric ceramics. A group of ceramic materials that exhibit thermoelectric properties, i.e., they produce electrical energy when heated. Included in this group are the so-called "mixed-valence" compounds, i.e., mixtures of oxides of transition metals, such as cobalt, nickel and manganese. Used especially in the electrical and power-supply industries.

thermoelectric materials. A generic term for a class of materials used to convert thermal energy directly into electric energy and vice versa, or provide refrigeration by means of electrical energy. It includes various semiconductors (e.g., cadmium and zinc antimonide, cerium, cesium and samarium sulfide, bismuth, gadolinium and lead selenide, bismuth, germanium and lead telluride, etc.), several metals and alloys (e.g., bismuth, cesium, copper, nickel, platinum, rhodium, tellurium, iron-nickel, nickel-copper, etc.), and various ceramic mixtures of transition-metal oxides.

thermoelectric metals. See thermoelectric materials.

Thermofab. Trademark for Albion Industrial Products (Canada) for a family of textured, woven fabrics available as cloth, woven tapes, ropes, sleeves, paper and millboard, gaskets and seals, and used as asbestos substitutes.

Thermofibers. Trade name of Thermofibers SRL (Italy) for a wide range of fibers including polyvinyl alcohol and polypropylene fibers, and natural and mineral fibers used in the cement industries as asbestos replacements. Also included under this trade name are polyvinyl alcohol fibers, and polyvinyl chloride as well as high-tenacity synthetic yarns and fibers for the textile and paper industries.

Thermofil. Trademark of Thermofil, Inc. (USA) for an extensive range of unmodified and modified thermoplastics, reinforced and/or lubricated thermoplastic compounds, pelletized flame-retardant compounds, and electrically conductive thermoplastic composites. The following compositions are available: *Thermofil ABS* acrylonitrile-butadiene-styrene resins supplied in glass fiber-reinforced, fire-retardant, high heat, UV-stabilized, high- and low-impact, low gloss, transparent, plating and structural foam grades; *Thermofil Acetal* acetal polymers, copolymers and supertough acetal/elastomer alloys supplied in unreinforced, glass or carbon fiber-reinforced, silicone-lubricated, supertough and UV-stabilized grades; *Thermofil HIPS* unmodified, UV-stabilized or fire-retardant high-impact polystyrene; *Thermofil Nylon 6* polyamide 6 (nylon 6) resins supplied in standard, casting, elastomer-copolymer, stampable-sheet, glass fiber- or carbon fiber-reinforced, glass bead- or mineral-filled, silicone-, molybdenum disulfide- or polytetrafluoroethylene-lubricated, high-impact, fire-retardant and UV-stabilized grades; *Thermofil Nylon 6,12* polyamide 6,12 (nylon 6,12) resins supplied in standard, glass fiber- or carbon fiber-reinforced, silicone- or polytetrafluoroethylene-lubricated and fire-retardant grades; *Thermofil Nylon 11* polyamide 11 (nylon 11) resins supplied in standard, glass fiber- or carbon fiber-reinforced, flexible, semi-flexible and UV-stabilized grades; *Thermofil Nylon 12* polyamide 12 (nylon 12) resins supplied in standard, glass fiber- or carbon fiber-reinforced, flexible, semi-flexible and UV-stabilized grades; *Thermofil Nylon 66* polyamide 6 (nylon 6) resins supplied in standard, glass fiber- or carbon fiber-reinforced, glass bead- or mineral-filled, molybdenum disulfide- or polytetrafluoroethylene-lubricated, high-impact, fire-retardant, supertough and UV-stabilized grades; *Thermofil PBT* polybutylene terephthalates supplied in standard, glass fiber- or carbon fiber-reinforced, glass bead- or mineral-filled, silicone- or polytetrafluoroethylene-lubricated, fire-retardant, structural-foam and UV-stabilized grades; *Thermofil PC* polycarbonates supplied in standard, glass fiber- or carbon fiber-reinforced, polytetrafluoroethylene-lubricated, fire-retardant, high-flow, structural-foam and UV-stabilized grades; *Thermofil PES* polyether sulfone resins supplied in standard and glass fiber- or carbon fiber-reinforced grades; *Thermofil PET* polyethylene terephthalate resins supplied in amorphous, crystalline, glass-fiber reinforced, mineral-filled, high-impact, supertough and UV-stabilized grades; *Thermofil PI* polyimide resins supplied in standard, glass fiber-reinforced, graphite-, molybdenum disulfide- or polytetrafluoroethylene-lubricated grades; *Thermofil PP* polypropylene resins filled with 20 or 40% calcium carbonate or talc, reinforced with 20% glass fibers or coupled with 30% glass fibers; *Thermofil PPS* polyphenylene sulfide resins reinforced with 40% glass fibers, 30% carbon fibers or glass fibers and glass beads or lubricated with 20% polytetrafluoroethylene; *Thermofil PS* polystyrene resins supplied in standard, glass fiber-reinforced, medium-impact, UV-stabilized, structural-foam, fire-retardant and silicone-lubricated grades; *Thermofil PSUL* polysulfone resins supplied in standard, glass fiber- or carbon fiber-reinforced or polytetrafluoroethylene-lubricated grades; and *Thermofil SAN* styrene-acrylonitriles supplied in standard, glass fiber-reinforced, fire-retardant, high-impact, high-heat and UV-stabilized. *Thermofil* resins and compounds are used as engineering composite matrix resins and in the manufacture of molded products for applications in the aerospace, automotive, chemical, construction, electronic and many other industries.

Thermofine. Trademark of US Granules Corporation (USA) for fine aluminum granules produced from recovered aluminum foil products.

Therm-O-Flake. Trademark of Thermal Ceramics, Inc. (USA) for lightweight, chemically bonded brick of *expanded vermiculite*. It has high toughness, good insulating properties to over 1000°C (1830°F), and is used for thermal insulation applications.

Thermoflux. Trademark of Vacuumschmelze GmbH (Germany) for a soft-magnetic alloy composed of 70% iron and 30% nickel. It has a strongly temperature-dependent flux density, low coercivity, a Curie temperature of 30-120°C (86-248°F), and is used for the temperature compensation of permanent magnets, and temperature-dependent switches.

Thermo-Foil. Trade name of CIE-Nergy, Inc. (Canada) for reflective thermal foil insulation.

thermoformed plastics. Thermoplastic products (e.g., acrylics and polystyrenes) of varying shape and size formed from sheets by first heating the latter to a soft and flowable consistency and then applying air pressure or a vacuum to force the softened sheets onto shaped molds.

Thermoglas. (1) Trade name of Thermoglas Inc. (US) for an insulating glass.

(2) Trade name of RX Industrie-Produkte Graf von Rex GmbH & Co. KG (Germany) for glass-fiber fabrics with good thermal insulating properties.

(3) Trade name of D. Anderson & Son Limited (UK) for glass fiber-based roofing felt.

Therm-O-Glas. Trade name of Therm-O-Glass Company (USA) for insulating glass units with metal edges.

Thermo-12 Gold. Trade name of Johns Manville Company (USA) for high-temperature fiberglass insulation supplied in the form of blocks and pipes.

Thermoguard. Trademark of M&T Chemicals, Inc. (USA) for antimony-based materials (e.g., antimony trioxide) for incorporation into plastics and coatings as flame-retardant pigments and opacifiers.

Thermold. Trade name of Universal Cyclops Corporation (USA) for chromium- and tungsten-type hot-work tool and die steels (AISI types H10 through H26), air-hardening medium-alloy cold-work tool steels (AISI A9). Typical applications for the chromium-type hot-work steels (AISI types H10 through H19) include extrusion, header and trimmer dies and tools, die-casting dies, punches, forging and piercing tools, shear blades and mandrels. The tungsten-type hot-work steels (AISI types H20 through H26) are commonly used for extrusion and die-casting dies, trimming and blanking dies, die and dummy blocks, and punches. The air-hardening medium-alloy cold-work steels (AISI type A9) are used for cold-heading, gripper, coining and forming dies, die inserts, forming rolls, and punches.

Thermo-Lite. Trade name of Thermo-Lite Inc. (USA) for heat-absorbing, heat-reflecting glass.

Thermolite. (1) Trademark of Clayburn Refractories Limited (Canada) for insulating castings and gunning castables.

(2) Trademark of E.I. DuPont de Nemours & Company (USA) for polyester-based, high-performance thermal insulating fibers and fabrics used in apparel, boots, sleeping bags, etc.

(3) Trade name of Oralite Company (UK) for a dough (or bulk) molded acrylic formerly used for denture bases.

Thermolith. Trademark of Dresser Industries, Inc. (USA) for a chromite-based cold-setting refractory mortar used for joining chromite, forsterite and magnesite refractories.

Thermoloft. Trademark of E.I. DuPont de Nemours & Company (USA) for soft, durable, lightweight thermal insulation made from 80% preconsumer recycled polyester. Supplied in several fiber grades, it is used for outerwear.

Therm-O-Lok. Trade name of International Adhesives, Inc. (USA) for hot-melt adhesives.

thermoluminescent coatings. Ultraviolet-sensitive coatings containing a thermoluminescent material (or thermoluminescent phosphor) that emits visible radiation (light) when excited by ultraviolet radiation (black light). Used in nondestructive inspection.

thermoluminescent materials. Materials, such as calcium fluoride, calcium sulfate, lithium borate or lithium fluoride, sometimes activated with metal ions, such as dysprosium or manganese, that are capable of storing ionizing radiation and emitting it in the form of visible radiation when heated. Used in radiation measurement and dosimetry, and for thermoluminescent coatings for nondestructive inspection.

Thermolux. Trade name of Thermolux Glass Company Limited (USA) for a layer of glass tissue sandwiched between two sheets of glass.

Thermo-Mat. Trade name of Fibrous Glass Products Inc. (USA) for glass-fiber insulation coated on one side with a flame-resistant asbestos-fiber mastic.

Thermo-melt. Trademark of Guertin Brothers Coatings & Sealants Limited (Canada) for hot-melt adhesives and sealants.

thermometal. See bimetal.

thermonatrite. A colorless synthetic mineral composed of sodium carbonate monohydrate, $Na_2CO_3 \cdot H_2O$. Crystal system, orthorhombic. Density, 2.26 g/cm³; hardness, 1.5; refractive index, 1.505.

Thermoneal. Trade name of Heppenstall Company (USA) for a series of hot-work tool steels containing 0.3-0.4% carbon, 0.2-0.4% manganese, 0.8-1.2% silicon, 4.25-5.5% chromium, 1-2% molybdenum, 0-1% vanadium, and the balance iron. Used for die-casting, upsetting, press, forging and extrusion dies.

Thermo-Ohio-Plate. Trade name of Ohio Plate Glass Company (USA) for insulating glass.

Thermopane. Trademark of Libbey-Owens-Ford Company (USA) for hermetically sealed window glass consisting of two sheets of glass. A strip of copper or titanium is evaporated along the edges of the glass onto which, after tinning, a strip of lead is soldered. An acoustically and thermally insulating layer of air is left between the two sheets. Used as glazing on buildings and for display cases and windows.

Thermoperm. Trade name of Krupp Stahl AG (Germany) for a temperature-sensitive, magnetic alloy composed of 70% iron and 30% nickel, and used for compensating shunts for electrical equipment.

thermoplastic adhesives. Adhesives based on thermoplastic resins, such as polyesters, polyethylenes, isobutylenes, polyamides, polyimides, or polyvinyl acetates. In general, these adhesives are relatively soft, have poor creep strength and are usually neither suitable for subzero nor high-temperature applications. Used for hot-melt and noncritical applications.

thermoplastic composites. A group of engineering composites having matrices based on high-temperature thermoplastic resins, such as polyamides, polyesters or polysulfones, reinforced with continuous or discontinuous fibers of aramid, carbon or glass. Abbreviation: TPC. Also known as *thermoplastic matrix composites.*

thermoplastic elastomers. A group of polymeric materials that share common properties with both elastomers and thermoplastics, and can be molded without vulcanization. They behave as elastomers at low temperatures, but as thermoplastics at high temperatures. Depending on their application, they may be based on polyetheramides, polyetheresters, nitrile-butadiene rubber, natural rubber, thermoplastic polyolefins, styrene-butadiene-styrene, polyurethane copolymers, melt-processible polymers, etc. Abbreviations: TE; TPE; TPEL; TR. Also known as *elastoplastics.* See also thermoplastic olefins; thermoplastic rubber; thermoplastic urethanes.

thermoplastic engineering plastics. See engineering thermoplastics.

thermoplastic fluoroelastomer. A fluoropolymer having both elastomeric and thermoplastic properties. See also elastomers; fluoroelastomers; thermoplastic elastomers.

thermoplastic fluoropolymers. A group of engineering thermoplastics based on fluorinated polymers and including polytetrafluoroethylene (PTFE), fluorinated ethylene-propylene (FEP), perfluoroalkoxy alkane (PFA), ethylenetetrafluoroethylene (ETFE), polyvinylidene fluoride (PVDF), polychlorotrifluoroethylene (CTFE), ethylene chlorotrifluoroethylene (ECTFE), polyvinyl fluoride (PVF), poly(ethylene-*co*-tetrafluoroethylene) (PE-TFE) and poly(ethylene-*co*-chlorotrifluoroethylene) (PE-CTFE). Abbreviation: TPFP. See also fluoropolymers.

thermoplastic materials. See thermoplastics.

thermoplastic matrix composites. See thermoplastic composites.

thermoplastic molding compound. A mixture of a thermoplastic resin, such as acrylonitrile-butadiene-styrene, polycarbonate, polyester, polyethylene, polyimide, nylon, or polystyrene, usually in the form of pellets or granules, with additives, such as colorants, fillers, flame retardants, plasticizers, pigments, reinforcements and/or stabilizers. Used in molding and extrusion operations.

thermoplastic olefins. A group of thermoplastic polymers made by the polymerization or copolymerization of simple olefins. Examples include polyethylene, polypropylene, polybutylene and polyisoprene. Abbreviation: TPO. Also known as *thermoplastic polyolefins.* See polyolefins; thermoplastic polyethylenes.

thermoplastic polyesters. A group of linear thermoplastic engineering polymers in which the repeating units are joined by ester groups. Examples include polybutylene terephthalate (PBT), polyethylene terephthalate (PET), and polycarbonate (PC). Abbreviation: TPE; TPES.

thermoplastic polyethylenes. Ordinary, non-crosslinked polyethylenes including high-density and ultrahigh-density grades in contrast to crosslinked polyethylene made thermosetting. Abbreviation: TPE. See also polyethylenes; high-density polyethylenes; ultrahigh-density polyethylenes; crosslinked polyethylenes.

ylenes.

thermoplastic polyimides. Linear polyimides made by condensation polymerization of aromatic diamines or aromatic diisocyanates. They have outstanding high-temperature resistance and toughness, an upper temperature capability of 316°C (601°F), low coefficient of thermal expansion, good mechanical properties, low flammability, good dielectric properties, high radiation resistance, and are used for structural parts in the aerospace and automotive industries, as thermal and electrical insulators, and for printed-circuit boards. Abbreviation: TPI.

thermoplastic polymers. Polymeric materials that become reversibly soft, ductile and deformable upon heating or treating with suitable solvents. They are hardened by cooling, but can be repeatedly heated, dissolved, molded, extruded or otherwise shaped. Examples of such polymers include acrylics, acrylonitrile-butadiene-styrenes, cellulosics, fluoroplastics, polyamides, polycarbonates, polyesters, polyethylenes, polypropylenes, polystyrenes, polyvinyl chlorides and several elastomers. Also known as *thermoplastic resins.*

thermoplastic polyolefins. See thermoplastic olefins.

thermoplastic polyurethanes. A group of linear block copolymers consisting of both rigid and flexible blocks. The rigid blocks are made up of repeating groups of a diisocyanate, such as 4,4'-diphenylmethane diisocyanate (MDI), hexamethylene diisocyanate (HDI) or hydrogenated 4,4'-diphenylmethane diisocyanate (HMDI) and a chain extender or short-chain diol, such as 1,4-butanediol, 1,6-hexanediol or ethylene glycol. The flexible blocks consist of repeating groups of a diisocyanate, such as 4,4'-diphenylmethane diisocyanate (MDI) and a polyol or long-chain diol, such as linear polyester or polyether, polypropylene glycol, polytetramethylene glycol, polyethylene adipate, polyhexamethylene adipate or polyhexamethylene carbonate. Commonly processed by blow or injection molding, or extrusion, their strength, flexibility and rigidity can be widely varied by changing the quantity and type of the diisocyanate, polyol and chain extender. They have excellent low-temperature flexibility and heat resistance, excellent resistance to fuel, oil and fungi, excellent hydrolytic stability, and are often sold under trade names and trademarks, such as Estane, Desmopan, Elastogran, Pellethane, or Texin. Used for flexible and rigid foams, blown and extruded film and sheet products, protective coatings, adhesives, fibers, drive belts, hydraulic seals, grease covers and dust shields, tubing and hose covers, protective coverings for wire and cable, industrial rollers and caster wheels, exterior autobody parts, and sport shoes. Abbreviation: TPUR; TPU.

thermoplastic prepregs. A group of prepregs in which the reinforcing sheet material (cloth, mat or paper) is impregnated with a thermoplastic resin. Resins used for this purpose include polyamideimide, polyetheretherketone, polyetherimide, polyphenylene sulfide and polysulfone. Unlike the traditional thermosetting prepregs, they offer several advantages, such as longer shelf life, higher interlaminar fracture toughness, and lower shrinkage.

thermoplastic resins. See thermoplastic polymers.

thermoplastic rubber. A thermoplastic elastomer based on natural (isoprene) or synthetic rubber (e.g., nitrile-butadiene or styrene-butadiene rubber). Abbreviation: TPR; TR. See also thermoplastic elastomers.

thermoplastics. Generic term for a group of linear amorphous or semicrystalline plastics that can be reversibly softened by heating and hardened by cooling. In the softened state they can be shaped by molding or extrusion, and joined by welding. They are composed of long molecules arranged side by side or intertwined. Examples include acrylics, cellulosics, fluorocarbons, polyamides (nylons), polycarbonates, polyethylenes, polymethyl methacrylates, polypropylenes, polyvinyls, polystyrenes and polysulfones. Abbreviation: TP. Also known as *thermoplasts; thermoplastic materials*.

thermoplastic structural foam. A thermoplastic, such as polycarbonate, polyester, polyethylene or polyurethane, having an integral, solid skin and a foamed, cellular core. It has excellent rigidity, and is often used for automotive and aerospace components.

thermoplasts. See thermoplastics.

Thermoplus. Trade name of Flachglas AG (Germany) for thermal insulating glass.

Thermopont. Trade name for a polycarbonate resin formerly used for dentures.

Thermopress. Trade name for an acrylic injection molding resin used for dental applications.

Therm-O-Proof. Trade name of Thermoproof Glass Company (USA) for an insulating glass.

Thermorun. Trade name of Mitsubishi Chemicals (Japan) for thermoplastic elastomers.

Thermo-Seal. Trade name of H.B. Fuller Company (USA) for glass sealants.

Therm-O-Seal. Trademark of Thermal Ceramics, Inc. (USA) for high-temperature insulating and finishing cements.

Thermoseal. Trade name of Thermoseal Glass Corporation (USA) for insulating glass.

Thermoset. Trade name of Thermoset Plastics Inc. (USA) for an extensive series of thermosetting plastics including several epoxy and aluminum- or silver-filled epoxy resins, acrylic and silver-filled acrylic resins, silicone resins, and filled and unfilled polyurethane resins supplied in various grades for a wide range of applications.

thermoset engineering plastics. See engineering thermosets.

thermoset matrix composites. A group of engineering composites having matrices based on thermosetting resins, such as bismaleimides, epoxies, phenolics, polyimides, polyphenylene sulfides or vinyl esters, reinforced with inorganic or organic fibers or fabrics, e.g., aramid, carbon, graphite, glass, etc.

thermoset molding compounds. Mixtures of thermosetting resins, such as phenol-formaldehydes or epoxies, with pigments, reinforcements, plasticizers, fillers (e.g., glass or graphite fibers, wood flour or mineral granules) and other additives. Used in molding and extrusion operations.

thermoset plastics. See thermosets.

thermoset polyurethanes. See thermosetting polyurethanes.

thermosets. Generic term for a group of plastics that cure on heating, on exposure to high-energy radiation (e.g., ultraviolet light), or by addition of catalysts (curing agents), and cannot be softened or remelted. They are composed of long molecular chains between which chemical crosslinks have been set up to produce a three-dimensional structure. They are very resistant to solvents and other chemicals. Examples include alkyds, allyls, amino resins, bismaleimides, cyanate resins, epoxies, phenolics, unsaturated polyesters, silicones and vinyl esters. Abbreviation: TS. Also known as *thermoset plastics; thermosetting plastics.*

thermosetting adhesives. Adhesives based on thermosetting resins, such as phenolics, melamine- and urea-formaldehydes, epoxies, polyvinyl butyrals, cyanoacrylates or unsaturated poly-

esters. In general, they have good bond strength and creep strength, low to moderate peel strength and low impact strength, low- and high-temperature properties vary from good to poor. Used for structural and general-purpose bonding applications.

thermosetting lacquer. A lacquer whose nonvolatile vehicle is a thermosetting resin (e.g., epoxy, phenolic, silicone or urea-formaldehyde) that cures (or solidifies) on heating and cannot be softened or remelted.

thermosetting plastics. See thermosets.

thermosetting polyamides. A group of low-molecular-weight polyamides based on dimerized vegetable oil acids and polyamines (e.g., diethylene triamine or ethylene diamine), and capable of setting up crosslinks. They have higher flexibility and solubility than ordinary (thermoplastic or nylon-type) polyamides, and are used in especially coatings, adhesives, and varnishes.

thermosetting polyesters. See unsaturated polyesters.

thermosetting polymers. Polymeric materials that become hard and rigid upon curing by application of heat or chemical means and cannot be resoftened. Examples of include phenolics, aminos, epoxies, polyesters, polyurethanes, silicones, isoprene and chloroprene rubber, etc. Also known as *thermosetting resins.*

thermosetting polyurethanes. A class of thermosetting (network polymers) that are reaction products of organic diisocyanates (e.g., toluene diisocyanate or polydiphenylmethane diisocyanate) with any of various compounds containing hydroxyl groups (e.g., diols, polyols, polyethers, or polyesters). They available in cellular, flexible, rigid or solid form with typical density ranging from approximately 0.03 g/cm^3 (0.001 lb/in.^{3}) for soft structures to 1.15 g/cm^3 (0.04 lb/in.^{3}) for rigid forms. Used as engineering plastics, and for paints, varnishes, adhesives, fibers, elastomers and foams, in protective coatings, as binders, sizings and sealants, as potting or casting resins, and in biomedicine. Abbreviation: TSUR. Also known as *thermoset polyurethanes*. See also polyurethanes; thermoplastic polyurethanes.

thermosetting resins. See thermosetting polymers.

Thermo-Sil. Trademark of Ceradyne, Inc. (USA) for rebonded, industrial-grade *fused silica* refractories supplied in the form of castables, foams and shapes.

Thermosil. Trade name of Cyprus Foote Mineral Company (USA) for a ferrosilicon composed of approximately 61% silicon and 39% iron, and used as ladle addition and graphitizing inoculant for gray and ductile cast iron.

Thermosol. Trade name of SA Glaverbel (Belgium) for heat-absorbing sheet glass.

Thermo-Span. Trademark of Carpenter Technology Corporation (USA) for an oxidation-resistant controlled-expansion superalloy with a nominal composition of 29.0% cobalt, 24.5% nickel, 5.5% chromium, 4.8% niobium, 0.85% titanium, 0.45% aluminum, 0.35% silicon, 0.004% boron, and the balance iron. The niobium and titanium additions result in improved ductility and strength at elevated temperatures, while the chromium greatly improves the oxidation resistance. The alloy can be solution-treated and aged, and can withstand operating temperatures up to 700°C (1290°F). Other important properties include good environmental resistance and ease of fabrication. Used for seals, casings, rings, and other components in gas-turbine engines.

Thermotec. Trade name of National Gypsum Company (USA) for asbestos cement sheets belonging to the *Gold Bond* line of products.

thermotropic liquid crystal. A liquid crystal whose molecular properties depend upon temperature, i.e., whose liquid crystalline state is obtained by (i) increasing the temperature of a solid and/or decreasing the temperature of a liquid (known as *enantiomorphic liquid crystal*), or (ii) by either raising the temperature of a solid or lowering the temperature of a liquid (known as *monotropic liquid crystal*). See also liquid crystal.

thermotropic liquid crystal polymer. A melt-orienting, thermoplastic variant of a liquid crystal polymer. Most advanced liquid crystal polymers are thermotropic. Abbreviation: TLCP. See also liquid crystal polymers.

Thermotuf. Trademark of LNP Engineering Plastics (USA) for neat and graphite-reinforced nylon 6,6 thermoplastics with outstanding toughness used for safety goggles. *Thermotuf V* refers to a series of toughened, lubricated nylon 6,6 materials available with glass-fiber and/or carbon-fiber reinforcement.

Thermount. Trademark of E.I. DuPont de Nemours & Company (USA) for laminates and prepregs containing DuPont nonwoven 100% aramid reinforcements (Types E-200 or N710). Used for printed wiring boards and semiconductor packaging applications in the aerospace, avionics and telecommunications industries.

Thermovit. Trademark of Saint-Gobain (France) for a heated laminated safety glass.

Thermovitre. Trade name of Splintex Belge SA (Belgium) for double glazing products.

Thermovyl. Trademark of Rhovyl SA (France) for coarse, preshrunk vinyon staple fibers and filament yarns with enhanced thermal stability. Used for textile fabrics.

Thermowear. Trade name of Carpenter Technology Corporation (USA) for a wear-resistant hot-work die steel containing 0.6% carbon, 0.5% manganese, 1% silicon, 4% chromium, 2.5% molybdenum, 1% vanadium, 1.5% niobium (columbium), 0.1% titanium, and the balance iron. It has high red-hardness, good abrasion resistance, fair toughness, and is used for hot-work compression tools, such as dies, inserts, blades, and punches.

Thermsulate. Trademark of ATD Corporation (USA) for a series of thermal and acoustical management materials made of aluminum, aluminum composites, stainless steel or aluminum-stainless steel composites. They can be corrugated, embossed, layered and three-dimensionally formed, and are ideal for applications requiring the simultaneous reflection, insulation and lateral conduction of thermal and acoustic energy. There are several product series including *T3500/3600* laterally conductive aluminum-polyester composites for heat management applications, *T5000* embossed, multi-layered aluminum composites for heat and noise management, *T10000* embossed, multi-layered stainless steel composites for thermal management in engine compartments, and *k5000* corrugated, crush-resistant, directional laterally conductive composites.

Thermur. Trade name of Glas-Fischer (Germany) for high-quality, heat-resistant, multi-pane insulating glass.

Therm X. Trade name of Ideal Tape Company Inc. (USA) for pressure-sensitive tape used for thermal insulation applications.

Thermx. Trade name of Eastman Chemical Company (USA) for copolyesters and *liquid crystal polymers* used for high-heat applications. *Thermx AG2XX* refers to polymers of cyclohexanedimethanol terephthalic acid (PCTA), *Thermx CG* to polycyclohexane (dimethylene) terephthalates (PCTs), *Thermx EG731* to polyethylene terephthalates (PETs) and *Thermx PCT* to 20-40% glass fiber-reinforced polyesters used for electronic compo-

nents, e.g., connectors.

Thermylene. Trade name of Asahi Thermofil (Japan) for polypropylene plastics.

Thermylon. Trade name of Asahi Thermofil (Japan) for polyamide (nylon) plastics including nylon 6 and nylon 6,6.

The Smart Yarns. Trademark of Solutia, Inc. (USA) for soft, colorfast craft yarns made from acrylic fibers (*Acrilan, Bounce-Back* or *Remember)*, and used for hand knitting and crocheting.

Thessco. Trade name of Sheffield Smelting Company Limited (UK) for a series of brazing and soldering materials including several brazing filler metals based either on copper or silver, and usually containing considerable amounts of zinc and/or cadmium and nickel. Also included under this trade name are several silver-based dental amalgams containing varying amounts of tin.

Thessconite. Trade name of Sheffield Smelting Company Limited (UK) for sintered copper-tungsten, silver-tungsten, silver-tungsten carbide, silver-nickel, silver-carbon and silver-cadmia electrical contact alloys.

Thetaloy. Trade name of Pratt & Whitney Cutting Tool & Gage (USA) for a corrosion and heat-resistant nickel alloy containing 25% chromium, 12.5% cobalt, 7% tungsten, 6% iron, 3% molybdenum, 2.5% manganese, and 0.4% carbon. Used for high-temperature applications.

Thiacril. Trademark of Thiokol Chemical Corporation (USA) for a high-elongation polyacrylate rubber used for gaskets, wire insulation, hoses, and tubing.

thick-film material. A material applied to a substrate, such as a capacitor, resistor, circuit component or solid-state device, as a relatively thick film with a typical thickness of 25 µm (1 mil) or more.

thin-film recording media. Ferromagnetic or ferrimagnetic recording media consisting of thin films composed of oriented, magnetically isolated, minute grains with high magnetic anisotropy. They are usually produced by the deposition, evaporation or sputtering of a micron- or submicron thick layer on a metallic or plastic substrate. Examples include electrodeposited NiP, autocatalytically deposited CoP, and electroless CoNiP and CoMoMnP films as well as Co, CoNi and CoNiCr films evaporated on polyester substrates, Co, CoNi, CoRe, CoNiPt and γ-Fe_2O_3 films sputtered on NiP-coated aluminum, Co and/or Ti doped Fe_2O_3 sputtered on anodized aluminum, CoCrTa sputtered on Cr, and CoCr on NiFe. See also magnetic medium; magnetic recording materials; particulate recording media.

thin-film material. A material applied to a substrate, such as a capacitor, resistor, circuit component or solid-state device, as a thin film with a thickness ranging from a few tenth of a nanometer to several micrometers, usually by vapor deposition or diffusion techniques.

thin-film semiconductor. A film-type semiconductor consisting of an electrically insulating substrate, such as silicon dioxide, with a layer of semiconducting amorphous or single-crystalline material, such as gallium arsenide, zinc selenide or cadmium selenide, usually applied by an evaporation or sputtering technique. Abbreviation: TFS.

thin-film superconductor. A superconductor in the form of a thin film usually produced by growing a superconducting element, such as indium or tin, or a superconducting compound, such as yttrium barium copper oxide, on a suitable substrate. Used for cryogenic switch or storage applications, e.g., cryotrons, and for computers and other electronic equipment. Abbreviation:

TFSC. Also known as *superconductor thin film.*

ThinFlo. Trademark of Castolin SA (France) for flux-coated, bare and coated silver brazing alloys for steel, stainless steel, ferrous metals, lead, zinc, nickel and copper and their alloys, and precious metals.

thin layer. A specially prepared mixture of adsorbents (usually silica gel based) spread on a glass slide to a thickness of 0.254 mm (0.01 in.) and used in thin-layer chromatography.

Thinlite. Trade name of Owens-Illinois Inc. (USA) for hollow glass tiles used for complete curtain-wall systems.

thinner. A volatile liquid, usually a solvent based on hydrocarbons (e.g., naphtha) or oleoresins (e.g., turpentine) that dries by evaporation and is used to lower or otherwise modify the consistency and/or tack of a paint, varnish, cement, and adhesives.

thin-shell precast. Precast concrete with thin slabs and web sections.

Thinstrip. Trade name for a commercially pure magnesium (99.8+%) available in the form of thin sheets or strips.

Thinsulate. Trademark of 3M Company (USA) for a thermal insulation material used for clothing, shoes, and boots.

Thinsuliner. Trade name of Combustion Engineering Company (USA) for an insulating cement.

Thin Thrift. Trade name of Crucible Materials Corporation (USA) for a decarb-free cold-work tool steel.

Thiobac. Trademark of TC Thiolon (USA) for a primary backing material for *Thiolon* grass fibers used for synthetic sports surfaces.

thiocarbamide. See thiourea.

thiofuran. See thiophene.

Thiokol. Trademark of Thiokol GmbH (Germany) for a polysulfide rubber made by the condensation of ethylene dichloride and sodium polysulfide. It has outstanding resistance to water, oil, gasoline and solvents, good weatherability, an upper service temperature up to 120°C (250°F), low gas permeability, good aging properties, relatively high flammability, fair to poor mechanical and physical properties, and is used for gaskets, diaphragms, washers, fuel hoses, tank linings, tubing, life jackets, and in liquid form as a sealant.

Thiolan. Trade name for regenerated protein (azlon) fibers.

Thiolon. Trade name of TC Thiolon (USA) for a series of synthetic grass fibers and yarns (including nylons, polyethylenes and polypropylenes) used for the manufacture of sports surfaces for football and soccer fields, tennis courts, golf courses, hockey arenas, etc.

thiole. See thiophene.

thiophene. A colorless, flammable liquid (99+% pure) obtained from coal tar and petroleum and also made synthetically by heating sodium succinate with phosphorus trisulfide. It has a density of 1.064 g/cm³, a melting point of -38°C (-36°F), a boiling point of 84°C (183°F), a flash point of 30°F (-1°C), a refractive index of 1.527, and is used in organic synthesis, in the preparation of conducting polymers (e.g., polythiophene), in the synthesis of thiophene derivatives and as a dye and solvent. Formula: C_4H_4S. Also known as *thiofuran; thiole.*

Thioplast. Trademark of Chemtron Manufacturing Limited (USA) for polysulfide sealants.

thioplastics. A term used in many English-speaking countries outside the USA for plastics based wholly or in part on sulfide-containing monomers. They are usually rubber-like materials obtained by reactions of alkali polysulfides (e.g., sodium polysulfide, Na_2S_x) with organic dihalides (e.g., ethylene dichlo-

ride. Abbreviation: TM.

thiorubber. A term used in many English-speaking countries outside the USA for *polysulfide rubber,* i.e., an elastomer containing combined sulfur. Abbreviation: TR.

thiourea. A compound obtained by reacting cyanamide with hydrogen sulfide and available as white, shiny crystals (99+% pure) with a density of 1.406 g/cm³ and a melting point of 174-178°C (345-352°F). Used as a rubber accelerator, in amino resins, as an intermediate for dyes and pharmaceuticals, as a mold inhibitor, in photographic and photocopying paper, in electroplating, and in biochemistry. Formula: CH_4N_2S. Abbreviation: TU. Also known as *thiocarbamide.*

Thiozell. Trade name for regenerated protein (azlon) fibers.

third-quality fireclay brick. A term usually referring to low-duty fireclay brick, i.e., refractory brick with a pyrometric cone equivalent greater than 15 and less than 29.

13-8 Supertough. Trade name of Teledyne Allvac (USA) for a precipitation-hardened martensitic stainless steel with 13% chromium, 8% nickel, 2% molybdenum, 1% aluminum and controlled levels of nitrogen, sulfur and titanium. Produced by vacuum induction melting/vacuum-arc-remelting process, it has high toughness and strength, exceptional cryogenic properties at temperatures down to -196°C (-321°F) and enhanced stress corrosion cracking resistance. Used for aircraft component forgings, aircraft structural parts, and cryogenic applications.

13-15 compounds. See III-V compounds.

Thirty-Oak. Trade name of Round Oak Steel (UK) for a chromium steel containing 0.22% carbon, 0.2% silicon, 1% manganese, 0.5% copper (optional), and the balance iron. It has good weldability, and is used for structural applications.

Thixalloy. Trade name of Salzburger Aluminum AG (Austria) for a thixotropic aluminum casting alloy containing magnesium, silicon, manganese and chromium. It possesses a refined globular microstructure produced by electromagnetic stirring during continuous casting of the billet.

thixoformed alloys. Alloys that have been shaped in the semisolid state. Such alloys have appreciable melting ranges and, before forming, consist of solid spheroids in a liquid matrix. In this state the alloys are thixotropic; if they are sheared the viscosity falls and they flow like liquids. However, if they are allowed to stand they thicken again. See also Thixomolding alloys.

Thixomolding alloys. Aluminum, magnesium and zinc alloys suitable for use in a high-speed proprietary process known as "Thixomolding", developed by Thixomat Inc., USA, which makes use of the fact these alloys become semi-solid at temperatures between their solidus and liquidus temperatures (i.e., just below their melting temperatures). In the Thixomolding process the alloy first fed into a special injection molding machine, heated to a semi-solid state, and fed via a hopper and connector into a barrel. The alloy is then forced through the barrel while being heated transforming into a thixotropic, semi-solid (viscous) state, and injected at high speed into the cavity of a closed die to form high-quality finished products of designed net or near-net shapes and tight tolerances, such as automotive engine components, camera cases, business machine housings, cell phone keypads, etc. See also thixoformed alloys.

Thixon. Trademark of Dayton Chemical Division of Whittaker Corporation (US) for a series of adhesives and cements supplied in the form of solutions based on natural or synthetic rubbers and other bonding agents dissolved in water, hydrocarbons or other suitable solvents. They have good resistance to heat and many chemicals, and are used for rubber-to-metal bonding, rubber-to-plastic bonding, and for bonding elastomers.

Thixoflex. Trade name of Zhermack (Italy) for silicone elastomers used for dental impressions.

Thixon. Trademark of Whittaker Corporation (USA) for a series of adhesives and cements with good resistance to heat and many chemicals. They are supplied in the form of solutions based on natural or synthetic rubbers and other bonding agents dissolved in water, hydrocarbon or other suitable solvents. Used for bonding elastomers to metals and plastics.

thixotropes. See thixotropic materials.

thixotropic alloys. Alloys, such as those of aluminum and magnesium, which when heated to temperatures between their solidus and liquidus (i.e., just below their melting temperatures) become semi-solid (viscous).

thixotropic clay. A clay that under the influence of mechanical forces (e.g., molding) becomes temporarily fluid, and after cessation of the mechanical disturbance experiences a considerable increase in strength.

thixotropic materials. Semi-solid materials, such as clays or polymer resins, that in the unstressed state maintain their shapes, but upon application of mechanical stress flow like liquids. Such materials have high static, but low dynamic shear strength. Also known as *thixotropes*.

thixotropic paint. A semi-solid or gel-like paint that, under the influence of mechanical forces (e.g., shaking or stirring) becomes temporarily fluid, and after cessation of the mechanical disturbance redevelops its rigidity and stiffness. The thixotropic characteristics are due to the carefully controlled reaction of a relatively small proportion of a polyamide resin with an alkyd resin base vehicle. Also known as *gel paint; jelly paint*.

Thoglas. Trade name of Verres Industriels SA (Switzerland) for a laminated glass consisting of a heated layer sandwiched between two sheets of glass. Similar to *Therglas*.

Thomas. (1) Trade name of Old Hickory Clay Company (USA) for a range of clays and clay products.

(2) Trademark of Thomas Abrasive Products, Inc. (Canada) for greaseless buffing compounds.

Thomas-Gilchrist steel. See Thomas steel.

Thomas slag. A term used in Europe for a finely ground basic slag containing 12% or more phosphorus pentoxide (P_2O_5). It is obtained as a byproduct in the manufacture of Thomas steel (basic Bessemer steel). Used as a fertilizer. Also known as *slag flour. Note:* In North America, this slag is known as *basic phosphate slag*.

Thomas steel. A term used in Europe for steel made in a pear-shaped, basic refractory-lined vessel, known as "Thomas converter" or "basic Bessemer converter," by blowing air through a molten bath of pig iron whereby most of the carbon and impurities are removed by oxidation. Also known as *Thomas-Gilchrist steel. Note:* In North America, this steel is known as *basic Bessemer steel*.

Thomastrip. Trade name of Thomas Steel Strip Company (USA) for a sintered, cold-rolled strip steel used for ball bearings, retainers, gears, and structural applications.

Thompsoglas. Trade name of Hitco Company (USA) for resin-bonded glass-fiber batting used for insulation and vibration damping applications.

thomsenolite. A colorless, white, or light amber mineral composed of sodium calcium aluminum fluoride monohydrate, $NaCaAlF_6 \cdot H_2O$. Crystal system, monoclinic. Density, 2.96 g/cm³;

refractive index, 1.414. Occurrence: Greenland.

thomsonite. A colorless, snow-white or reddish mineral with a vitreous luster. It belongs to the zeolite group and is composed of sodium calcium aluminum silicate hexahydrate, $NaCa_2Al_5Si_5O_{20} \cdot 6H_2O$. Crystal system, orthorhombic. Density, 2.37 g/cm³; hardness, 5.0-5.5 Mohs; refractive index, 1.531. Occurrence: Canada (Nova Scotia), Europe (Iceland, Scotland), USA. Used in ion exchange and water softening.

Thong. Trademark of Thai Rayon Company (Thailand) for viscose rayon staple fibers.

Thoran. German trade name for a cast alloy containing 95.85% tungsten, 3.94% carbon and 0.21% molybdenum. It has a microstructure consisting of ditungsten carbide (W_2C), tungsten carbide (WC) and dimolybdenum carbide (Mo_2C), and is used for hard cutting tools and dies, and tips for high-speed tools.

thorbastneasite. A brown mineral of the bastnaesite group composed of thorium calcium carbonate fluoride hydrate, $Th(Ca,Ce)(CO_3)_2F_2 \cdot xH_2O$. Crystal system, hexagonal. Density, 4.04 g/cm³; refractive index, 1.674. Occurrence: Siberia.

Thordon. Trademark of Thordon Bearings Inc. (USA) for elastomeric materials used in the manufacture of bearings, bushings, wear strips, and guides.

thoreaulite. A brown to yellow mineral composed of tin tantalum oxide, $SnTa_2O_7$. Crystal system, monoclinic. Density, 7.60 g/cm³; refractive index, 2.417. Occurrence: Congo, Zaire.

Thorflex. Trade name of Thomson-Gordon Limited (Canada) for polyurethane elastomer products.

thoria. See thorium dioxide.

thoria-dispersed nickel. A dispersion-hardening nickel alloy produced by the addition of about 3 vol% of thorium dioxide (ThO_2) as finely dispersed particles. It has good high-temperature strength at over 1000°C (1830°F) and is used for afterburners, aircraft gas turbines, high-temperature applications, and fasteners. Abbreviation: TD nickel.

thoria-dispersed nickel-chromium. A dispersion-hardening nickel-chromium alloy produced by the addition of about 3 vol% thorium dioxide (ThO_2) as finely dispersed particles. It has high resistance to oxidation and sulfidation, excellent high-temperature strength, and is used for high-temperature applications, jet engines, aerospace structural components, and furnace hardware. Abbreviation: TD NiCr.

thorianite. Colorless, dark gray, reddish brown or brownish-black, strongly radioactive mineral of the fluorite group that is composed of thorium dioxide, ThO_2, and may also contain up to a total of 8% cerium and lanthanum. It can also be made synthetically. Crystal system, cubic. Density, 9.7 g/cm³; hardness, 7 Mohs; refractive index, 2.2. Occurrence: India, Russia, Europe, USA, Canada. Used as an ore of thorium and as a source of rare-earth metals.

thoria sol. Thorium dioxide (ThO_2) of submicron particle size available as 2% solids solution and used as a contrasting agent for transmission electron microscopy and in combination with organosilanes for the preparation of derivatives.

thoriated tungsten. A group of dispersed second-phase tungsten-base refractory alloys containing 0.5-2% thorium dioxide (ThO_2) and produced by powder-metallurgy methods. The 99.4W-0.6Th alloy has a density of 19.3 g/cm³ (0.70 lb/in.³). *Thoriated tungsten* improves the thermionic electron emission and increases the creep, impact and vibration strength of tungsten wire. Commercially available in the form of bars, rods, plates and wires, it is used for welding electrodes, filaments, electron-discharge tubes, high-temperature structural applications and aerospace parts. Also known as *tungsten-thoria alloys*.

thoria-urania ceramics. A group of high-temperature ceramics composed of thorium dioxide (ThO_2) and uranium dioxide (UO_2) reinforced with niobium or zirconium fibers, and used for fuel elements in nuclear reactors.

Thorite. (1) Trademark of ICI Americas Inc. (USA) for a non-shrinking mortar of troweling or pouring consistency used for patching holes in masonry brickwork and repairing blisters and other discontinuities in reinforced concrete.

(2) Trademark of Neoprene International Corporation (USA) for sound-insulating rustproof coating-like elastomeric sealant.

thorite. A black, brownish yellow, yellow to orange yellow or brownish black to dark brown, radioactive mineral with a vitreous to resinous luster. It belongs to the zircon group and is composed of thorium silicate, $ThSiO_4$. It can also be made synthetically. Crystal system, tetragonal. Density, 4.30-5.30 g/cm^3; hardness, 4.5-5 Mohs. Occurrence: Norway, Sri Lanka. Used as a source of thorium.

thorium. A silvery white, soft, radioactive metallic element of the actinide series of the Periodic Table. It is commercially available in the form of sintered and unsintered crystal bars, sheets, foils, ingots, rods, powder and sintered pellets, and occurs naturally in radioactive minerals, such as *monazite, thorite* and *thorianite*. Density, 11.5 g/cm^3; melting point, 1750°C (3182°F); boiling point, 4790°C (8654°F); hardness, 38-115 Vickers; superconductivity critical temperature, 1.38K; atomic number, 90; atomic weight, 232.0381; divalent, trivalent, tetravalent. It has poor resistance to acids, good resistance to alkalies and water, fair electrical conductivity (about 13% IACS), and good workability (e.g., cold rolling, extrusion and drawing) and weldability. Two allotropic forms are known: (i) *Alpha thorium* (face-centered cubic crystal structure) that is stable below 1400°C (2552°F); and (ii) *Beta thorium* (body-centered cubic crystal structure) that is stable between 1400-1750°C (2552-3182°F). *Thorium* is used as a nuclear reactor fuel, as an alloying element (e.g., in magnesium and tungsten), in nickel alloys (in the form of thoria dispersions), as a deoxidizer for iron and molybdenum, in tungsten filament wire coatings, as a cathode emitter material, in gaslight mantles, crucibles, welding electrodes, sun lamps and photoelectric cells, as a target in X-ray tubes, and in electronics. Symbol: Th.

thorium aluminide. A compound of thorium and aluminum used in ceramics and materials research. Density, 9.67 g/cm^3. Formula: Th_2Al.

thorium anhydride. See thorium dioxide.

thorium beryllide. A compound of thorium and beryllium used in ceramics and materials research. Density, 4.10 g/cm^3; hardness, 1170-1340 Knoop. Formula: $ThBe_{13}$.

thorium bismuthite. A compound of thorium and bismuth used in ceramics and materials research. Melting point, 1800°C (3270°F). Formula: Th_2Bi.

thorium boride. Any of the following compounds of thorium and boron used in ceramics, materials research and nuclear engineering: (i) *Thorium hexaboride*. A deep-red refractory solid. Density, 6.99-7.1 g/cm^3; melting point, 2450°C (4442°F). Formula: ThB_6; and (ii) *Thorium tetraboride*. A gray powder. Density, 8.45 g/cm^3; melting point, above 2500°C (4532°F); low thermal expansion. Formula: ThB_4.

thorium bromide. Colorless or white, water-soluble crystals with a density of 5.67 g/cm^3 and a melting point of 679°C (1254°F), used in chemistry and materials research. Formula: $ThBr_4$. Also

known as *thorium tetrabromide*.

thorium carbide. Any of the following compounds of thorium and carbon used as nuclear fuels: (i) *Thorium monocarbide*. Cubic crystals. Density, 10.65 g/cm^3; melting point, 2625°C (4757°F). Formula: ThC; and (ii) *Thorium dicarbide*. Yellow monoclinic crystals. Density, 9-9.6 g/cm^3; melting point, 2630-2680°C (4766-4856°F); boiling point, 5000°C (9032°F). Formula: ThC_2.

thorium chloride. Colorless or white hygroscopic needles, or white-gray moisture-sensitive powder (99.9+% pure). Crystal system, tetragonal. Density, 4.59 g/cm^3; melting point, 820°C (1508°F) [also reported as 770°C or 1418°F]; decomposes at 928°C (1702°F). Used for incandescent lighting. Formula: $ThCl_4$. Also known as *thorium tetrachloride*.

thorium dicarbide. See thorium carbide (ii).

thorium dioxide. A heavy, very refractory, crystalline, white powder (99.99%) obtained by reduction of thorium nitrate. It occurs in nature as the mineral *thorianite*. Crystal system, cubic. Crystal structure, fluorite. Density, 9.7-10.0 g/cm^3; melting point, 3050°C (5522°F) [also reported as 3390°C or 6134°F]; boiling point, 4400°C (7952°F); hardness, 6.5-7 Mohs. Used for high-temperature crucibles, cermets and high-temperature ceramics, incandescent gas mantles, silica-free optical glass, cathodes and coatings in electron tubes, thoriated tungsten filaments, as a catalyst and nuclear fuel, in flame spraying, and for microscopic stains. Formula: ThO_2. Also known as *thoria; thorium anhydride; thorium oxide*.

thorium diselenide. See thorium selenide (ii).

thorium disilicide. See thorium silicide (ii).

thorium disulfide. See thorium sulfide (ii).

thorium-doped lanthanum aluminate. An oxide of lanthanum and aluminum containing small additions, typically up to 2%, of thorium (Th^{4+}). It is available as a colorless, domain-free single crystalline substrate. Crystal system, cubic. Density, 6.5 g/cm^3; melting point, 2080°C (3776°F); dielectric constant, 24.5. Used as a substrate for deposition of high-temperature superconductors, and for microwave applications. Formula: $Th:LaAlO_3$. Also known as *thorium-doped lanthanum aluminum oxide (Th:LAO)*.

thorium-doped lanthanum aluminum oxide. See thorium-doped lanthanum aluminate.

thorium fluoride. White crystals or powder. Crystal system, monoclinic. Density, 6.1-6.32 g/cm^3; melting point, 1110°C (2030°F); boiling point, 1680°C (3056°F). It reacts with atmospheric moisture above 500°C (932°F) to form thorium oxyfluoride ($ThOF_2$). Used for the production of metallic thorium and magnesium alloys, and in high-temperature ceramics and optical coatings. Formula: ThF_4. Also known as *thorium tetrafluoride*.

thorium heptasulfide. See thorium sulfide (iv).

thorium hexaboride. See thorium boride (i).

thorium iodide. White-yellow crystals. Crystal system, monoclinic. Melting point, 570°C (1058°F); boiling point, 837°C (1539°F). Used in chemistry and materials research. Formula: ThI_4. Also known as *thorium tetraiodide*.

thorium-mischmetal. A mixture of thorium and mischmetal applied as a high-temperature coating on vacuum-tube anodes to serve as a getter. See also getter; thorium; mischmetal.

thorium monocarbide. See thorium carbide (i).

thorium mononitride. See thorium nitride (i).

thorium monophosphide. See thorium phosphide (i).

thorium monoselenide. See thorium selenide (i).

thorium monosilicide. See thorium silicide (i).

thorium monosulfide. See thorium sulfide (i).

thorium nitrate. White, crystalline, hygroscopic, radioactive substance (99.9+% pure). Melting point, decomposes at 500°C (932°F). Used in thoriated tungsten filaments and incandescent gas mantles, and as a reagent in fluorine determination. Formula: $Th(NO_3)_4 \cdot 4H_2O$. Also known as *thorium nitrate tetrahydrate*.

thorium nitrate tetrahydrate. See thorium nitrate.

thorium nitride. Any of the following compounds of thorium and nitrogen used in ceramics, materials research and nuclear engineering: (i) *Thorium mononitride.* Refractory crystals. Crystal system, cubic. Density, 11.6 g/cm^3; melting point, 2820°C (5108°F). Formula: ThN; (ii) *Thorium sesquinitride.* Formula: Th_2N_3; and (iii) *Thorium tetranitride.* Formula: Th_3N_4.

thorium oxalate. White, crystals or powder. Density, 4.64 g/cm^3; melting point, above 300°C (572°F) decomposes to thorium dioxide (ThO_2). Used in materials research, ceramics, and as a source of thorium oxide. Formula: $Th(C_2O_4)_2 \cdot 2H_2O$. Also known as *thorium oxalate dihydrate*.

thorium oxalate dihydrate. See thorium oxalate.

thorium oxide. See thorium dioxide.

thorium phosphide. Any of the following compounds of thorium and phosphorus used in ceramics, materials research and electronics: (i) *Thorium monophosphide.* Density, 7.0 g/cm^3. Formula: ThP; and (ii) *Thorium tetraphosphide.* Density, 8.59 g/cm^3. Formula: Th_3P_4.

thorium powder. Thorium prepared in powder form either by fused salt electrolysis of the double fluoride (ThF_4KF) or by reduction of thoria (ThO_2) with calcium. Finely divided thorium dust or powder is pyrophoric, explosive and can autoignite. Used in nuclear engineering, ceramics, materials research, and electronics.

thorium selenide. Any of the following compound of thorium and selenium used in ceramics, materials research and electronics: (i) *Thorium monoselenide.* Melting point, 1882°C (3420°F). Formula: ThSe; and (ii) *Thorium diselenide.* Orthorhombic crystals. Density, 8.5 g/cm^3. Formula: $ThSe_2$.

thorium sesquinitride. See thorium nitride (ii).

thorium sesquisulfide. See thorium silicide (iii).

thorium silicide. Any of the following compounds of thorium and silicon used in ceramics, materials research and electronics: (i) *Thorium monosilicide.* Density, 8.99-9.03 g/cm^3. Formula: ThSi; (ii) *Thorium disilicide.* Tetragonal crystals. Density, 7.79-8.23 g/cm^3; melting point, 1850°C (3362°F). Formula: $ThSi_2$; and (iii) *Trithorium disilicide.* Density, 9.80-9.81 g/cm^3. Formula: Th_3Si_2.

thorium sulfate. A white, crystalline powder with a density of 2.8 that loses $4H_2O$ at 42°C (107°F) and the remaining $4H_2O$ at 400°C 752°F). Formula: $Th(SO_4)_2 \cdot 8H_2O$. Also known as *thorium sulfate octahydrate*.

thorium sulfate octahydrate. See thorium sulfate.

thorium sulfide. Any of the following compounds of thorium and sulfur: (i) *Thorium monosulfide.* Density, 9.56 g/cm^3; melting point, approximately 2471°C (4480°F). Used in the manufacture of crucibles. Formula: ThS; (ii) *Thorium disulfide.* Dark brown, crystalline compound. Density, 7.3 g/cm^3; melting point, 1905°C (3461°F). Used as a solid lubricant. Formula: ThS_2; (iii) *Thorium sesquisulfide.* Density, 7.87 g/cm^3; melting point, approximately 1949°C (3540°F). Used for manufacture of crucibles. Formula: Th_2S_3; and (iv) *Thorium heptasulfide.* Used for crucibles for molten cerium. Formula: Th_4S_7.

thorium tetraboride. See thorium boride (ii).

thorium tetrabromide. See thorium bromide.

thorium tetrachloride. See thorium chloride.

thorium tetrafluoride. See thorium fluoride.

thorium tetraiodide. See thorium iodide.

thorium tetranitride. See thorium nitride (iii).

thorium tetraphosphide. See thorium phosphide (ii).

Thornel. Trademark of Union Carbide Corporation (USA) for rayon, polyacrylonitrile (PAN), mesophase pitch carbon and graphite fibers, fabrics and mats. They have a density of 1.5 g/cm^3 (0.05 $lb/in.^3$), high-performance properties, high tensile strength and elastic moduli, excellent dimensional stability and stiffness, high thermal stability, an upper service temperature of about 1590°C (2895°F), high uniformity, low water absorption and relatively low thermal conductivity. Used as reinforcements in engineering composites, for missile and rocket nozzles, reentry vehicles, solid-propellant-rocket motor casings, structural parts of aircraft and spacecraft, space panels, space radiators, heat sinks, and deep submergence vessels.

Thorocrete. Trademark of Thoro System Products (USA) for a concrete patching compound with high bonding strength used for driveway, sidewalk, step and patio repairs.

Thorogrip. Trademark of Thoro System Products (USA) for a quick-setting hydraulic anchoring cement used to hold fixtures, such as posts and railings, in concrete or masonry.

thorogummite. A mineral of the zircon group composed of thorium silicate hydroxide, $(Th,U,Ce)(SiO_4)_{1-x}(OH)_{4x}$. The uranium dioxide-bearing form is black and the uranium dioxide-free form is yellowish brown to white in color. Crystal system, tetragonal. Density, 4.49 g/cm^3; refractive index, 1.54-1.64. Occurrence: USA (Texas).

thorosteenstrupine. A dark brown mineral of the amorphous group composed of calcium manganese thorium fluoride silicate hexahydrate, $(Ca,Th,Mn)_3Si_4O_{11}F \cdot 6H_2O$. Density, 3.02 g/cm^3; refractive index, 1.63-1.66. Occurrence: Siberia. Used in ion exchange and water softening.

Thor-Temp. Trademark of Thomson-Gordon Limited (Canada) for high-temperature polymeric bearing materials.

thortveitite. A rare, weakly radioactive, grayish-green mineral composed of scandium silicate, $(Sc,Y)_2Si_2O_7$, and containing 37-42% scandium oxide (Sc_2O_3) and small amounts of yttrium, ytterbium, lutetium, erbium, etc. It can also be made synthetically. Crystal system, monoclinic. Density, 3.58 g/cm^3; refractive index, 1.79. Occurrence: Norway. Used as an ore of scandium.

thorutite. A black mineral of the amorphous group composed of thorium uranium titanium oxide, $(Th,U,Ca)Ti_2O_6$. Density, 5.82 g/cm^3; refractive index, above 2.1.

Thozell. Trade name for rayon fibers and yarns used for textile fabrics.

threadgoldite. A greenish yellow mineral of the autunite group composed of aluminum uranyl phosphate hydroxide octahydrate, $Al(UO_2)_2(PO_4)_2(OH) \cdot 8H_2O$. Crystal system, monoclinic. Density, 3.40 g/cm^3; refractive index, 1.583. Occurrence: Zaire.

Three Dee. Trade name for gutta-percha used in endodontics.

three-dimensional nanostructures. See nanostructured materials.

three-dimensional photonic crystals. See 3D photonic crystals.

III-V compounds. A group of usually semiconducting compounds formed by combining a metallic element of Group IIIA (Group 13), e.g., trivalent aluminum, boron, gallium or indium with a nonmetallic element of Group VA (Group 15), e.g., pentavalent phosphorus, arsenic or antimony of the Periodic Table. Also known as *13-15 compounds*.

Three Gun. Trademark of Chung Shing Textile Company (Taiwan) for nylon 6 fibers and filament yarns used for textile fabrics.

Three-M-ite. Trademark of 3M Company (USA) for coated abrasives composed of resin-bonded, tough alumina (Al_2O_3) mineral on cloth or polyester backing for grinding and finishing applications.

three-quarter-hard wire. An aluminum wire processed to produce a tensile strength intermediate between that of half-hard and hard-drawn wire. See also half-hard wire; hard-drawn wire.

Three Star. Trade name of Midvale-Heppenstall Company (USA) for a high-speed steel containing 0.72% carbon, 18.5% tungsten, 4% chromium, 4% cobalt, 1.2% vanadium, 0.75% molybdenum, and the balance iron. Used for cutting tools.

Three-Twenty. British nonhardenable aluminum alloy containing 20% zinc and 3% copper. Used for light-alloy parts.

Thrift Finish. Trade name of Crucible Materials Corporation (USA) for decarb-free cold-work tool steel.

Thriglas. Trade name of Shepherd Tobias & Company (UK) for burglar-resistant glass consisting of three or more layers of plate glass laminated together.

thrown silk. Silk yarn produced by twisting one or more raw silk threads together. See also raw silk; silk.

Thru-nic. Trade name of Uyemura International Corporation (USA) for electrolytic nickel.

thulia. See thulium oxide.

thulium. A rare, silvery white, malleable, ductile metallic element of the lanthanide series of the Periodic Table. It is commercially available in the form of ingots, lumps, turnings, chips, sheets, rods, foils, wire, sponge and powder. The most important mineral source is *monazite*. Crystal system, hexagonal. Crystal structure, hexagonal close-packed; density, 9.322 g/cm³; melting point, 1545°C (2813°F); boiling point, 1947°C (3537°F); hardness, 53 Brinell; atomic number, 69; atomic weight, 168.934; trivalent. Used for ferrites, ferrite bubble devices, as a catalyst and X-ray source, and in phosphors and superconductors. Symbol: Tm.

thulium-170. A radioactive isotope of thulium with a mass number of 170. It emits beta rays and soft gamma rays, has a half-life of 127 days, and is used in portable X-ray sources. Symbol: ^{170}Tm.

thulium barium copper oxide. A ceramic, superconductive compound available in the form of a fine powder or as a thin film and used as a high-temperature superconductor. Formula: $TmBa_2Cu_3O_x$. Abbreviation: TmBCO.

thulium boride. Any of the following compounds of thulium and boron used in ceramics: (i) *Thulium tetraboride*. Density, 7.09 g/cm³. Formula: TmB_4. and (ii) *Thulium hexaboride*. Density, 5.59 g/cm³. Formula: TmB_6.

thulium bromide. White, hygroscopic crystals or powder (99.9+% pure). Crystal system, cubic. Melting point, 952°C (1510°F). Used in chemistry and materials research. Formula: $TmBr_3$.

thulium carbide. Any of the following compounds of thulium and carbon used in ceramics: (i) *Thulium dicarbide*. Density, 8.17 g/cm³. Formula: TmC_2; (ii) *Thulium sesquicarbide*. Formula: Tm_2C_3; and (iii) *Trithulium carbide*. Density, 9.90 g/cm³. Formula: Tm_3C.

thulium dicarbide. See thulium carbide (i).

thulium chloride. Yellow, hygroscopic crystals or off-white powder (99.9% pure) with a melting point of 821°C (1510°F). Used

in chemistry and materials research. Formula: $TmCl_3$.

thulium chloride heptahydrate. Light green, hygroscopic crystals (99.9% pure) with a melting point of 824°C (1515°F). Used in chemistry and materials research. Formula: $TmCl_3 \cdot 7H_2O$.

thulium fluoride. White crystals or off-white powder (99.9+% pure) with a melting point of 1158°C (2116°F) and a boiling point of 2200°C (3992°F). Used in fluorinated glasses to enhance the ultraviolet and visible upconversion fluorescence. Formula: TmF_3.

thulium hexaboride. See thulium boride (ii).

thulium iodide. Yellow, hygroscopic crystals with a melting point of 1021°C (1870°F) used in chemistry and materials research. Formula: TmI_3.

thulium nitride. A compound of thulium and nitrogen. Density, 10.84 g/cm³. Used in ceramics and materials research. Formula: TmN.

thulium oxalate. Greenish-white precipitate. Melting point, loses $1H_2O$ at 50°C (122°F). Used in the separation of thulium and other rare earths from the common metals. Formula: $Tm_2(C_2O_4)_3 \cdot 6H_2O$.

thulium oxide. Greenish-white crystals or slightly hygroscopic powder (99.9+% pure). Crystal system, cubic. Density, 8.6 g/cm³; melting point, 2425°C (4397°F). Used in the production of thulium metal, as a radiation source in X-ray equipment, after irradiation in nuclear reactors, as a power source for small thermoelectric devices, and in the preparation of high-temperature superconductors. Formula: Tm_2O_3. Also known as *thulia*.

thulium sesquicarbide. See thulium carbide (ii).

thulium silicide. A compound of thulium and silicon used in ceramics and materials research. Formula: $TmSi_2$.

thulium sulfide. A tan-yellow crystalline compound of thulium and sulfur used in ceramics, electronics (e.g., as a semiconductor) and materials research. Crystal system, cubic. Band gap, 2.43 eV. Formula: Tm_2S_3.

thulium tetraboride. See thulium boride (i).

Thumbprint. Trade name of Australian Window Glass Proprietary Limited (Australia) for a patterned glass.

Thurane. Trade name for rigid polyurethane foam available in the form of boards of varying thicknesses and densities.

Thurston brass. A corrosion-resistant brass composed of 55% copper, 44.5% zinc and 0.5% tin, and used for architectural trim.

Thurston button. A free-cutting copper alloy composed of 33-37% zinc, 1.5-6% tin and 12-15% lead, and used for ornamental and architectural parts.

Thyrite. Trademark of General Electric Company (USA) for a molded electrical resistance material composed of silicon carbide (SiC).

Thyssen-Color. Trade name of Thyssen Stahl AG (Germany) for organically coated sheet steel made by coating a cold-rolled, electrolytically or hot-dip galvanized sheet of *Thyssenstahl* with a corrosion-resistant thermosetting or thermoplastic resin.

Thyssenstahl. Trade name of Thyssen Stahl AG (Germany) for alloy steel used for galvannealed, electrolytically galvanized, hot-dip-galvanized or hot-dip aluminized sheet.

Thyssen Welding. Trade name of Bohler Thyssen Welding USA, Inc. for a series of ferrous and nonferrous welding alloys.

TiAlON. Trade name for an advanced ceramic based on titanium aluminum oxynitride (TiAlON), and usually made by sintering titanium nitride, aluminum nitride and aluminum oxide. It has high strength, toughness and wear resistance and good high-temperature properties, and is used for coatings, and refractory and wear parts.

Tiama mahogany. The strong, durable wood of the sapele mahogany *Entandrophragma angolense*. It is a mahogany-like wood, but is not quite as stable. *Tiama mahogany* works and nails well, but has large distortion in use. Average weight, 550 kg/m³ (34 lb/ft³). Source: Tropical Africa especially West Africa (Nigeria, Ivory Coast, etc.). Used in general construction and for window frames, doors, and stairs. Also known as *brown mahogany; edinam; gedu nohor.*

Tiaralon. Trade name for rayon fibers and yarns used for textile fabrics.

Tibase. Trademark of Altair Technologies, Inc. (USA) for titanium dioxide (TiO_2) powders and slurries including nanoparticle *anatase-* and *rutile*-crystal products.

Tibline. Trade name of Tissage de Bourtzwiller SA (France) for glass curtain materials.

Tibor. Trade name of KB Alloys Inc. (USA) for aluminum-titanium-boron alloys with fine as-cast grain size, homogeneous microstructures and good mechanical properties.

Ticat. Trademark of Altair Technologies, Inc. (USA) for titanium dioxide powders and slurries including nanoparticle *anatase-* and *rutile*-crystal products.

ticking. A strong, striped, tightly woven cotton or linen fabric in a twill weave, used for mattress and pillow covers, awnings, tents, and sportswear.

Tico. British trade name for an electrical resistance alloy containing 27.5-30.4% nickel, 1.12% manganese, 1.1% copper, and the balance iron.

Ticomp. Trademark of Tiodize Company, Inc. (USA) for a high-gloss, high-build elastomeric coating that is designed to protect metallic surfaces against abrasive wear and galvanic corrosion.

Ticon. (1) Trademark of National Lead Company (USA) for a series of stannates, titanates and zirconates of barium, bismuth, calcium, cerium, lead, magnesium, etc. Used in the manufacture of ferroelectric and piezoelectric devices.

(2) Trade name for a corrosion-resistant nickel-chromium dental bonding alloy.

Ticonal. Trade name of Philips Electronic & Associated Industries Limited (UK) for a series of sintered or cast permanent magnet alloys of the *Alnico*-type containing varying amounts of iron, nickel, aluminum, cobalt, copper and titanium.

Ticonium. Trade name of Consolidated Car Heating Company (USA) for a corrosion-resistant casting alloy containing 35% nickel, 31% cobalt, 23% chromium, 5-6% iron, 6% molybdenum, and 0.01% carbon. Used as a dental casting alloy and in bone surgery.

Ti Core. (1) Trade name of Vulcanium Anodizing Systems, Division of Industrial Tianium Corporation (USA) for titanium-covered aluminum bar.

(2) Trade name of Essential Dental Solutions (USA) for a titanium-reinforced dental resin used for core build-ups.

Ticuni. Trade name of GTE Products Corporation (USA) for a brazing filler metal containing 70% titanium, 15% copper and 15% nickel. It is commercially available in the form of foils, flexibraze, extrudable paste and preforms. It has a liquidus temperature of 960°C (1760°F), a solidus temperature of 910°C (1670°F), and is used for brazing titanium and stainless steel.

Ticusil. Trade name of GTE Products Corporation (USA) for a silver brazing alloy containing 68.8% silver, 26.7% copper and 4.5% titanium. Commercially available in the form of foils, wire, powder, flexibraze, extrudable paste and preforms. It has

a liquidus temperature of 850°C (1562°F), a solidus temperature of 886°C (1626°F) and is used for brazing titanium and stainless steel.

Tideguard. Trademark of Ameron International Corporation (US) for a series of spray-on cladding materials for use on tidal and splash zone areas of steel offshore structures and pilings.

Tidion FR. Trademark of Montefibre SpA (Italy) for flame-resistant polyester staple fibers used for textile fabrics.

Tidolith. Trade name for a white pigment composed of 85% lithopone and 15% titanium dioxide. See also lithopone.

Tiefgrund. Trade name of Monopol AG (Switzerland) for several primers for emulsion paints.

tiemannite. A lead gray, opaque mineral of the sphalerite group composed of mercury selenide, HgSe. It contains about 71.7% mercury and 28.3% selenium, and can also be made synthetically. Crystal system, cubic. Density, 8.39 g/cm³.

tienshanite. A green mineral composed of sodium barium manganese titanium boron silicate, $Na_2BaMnTiB_2Si_6O_{20}$. Crystal system, hexagonal. Density, 3.29 g/cm³; refractive index, 1.666. Occurrence: Turkestan.

tiers-argent. See drittel silver.

tiff. See artificial barite.

Tiffany. Trade name for a low-gold dental casting alloy.

Tiger. (1) Trade name of Bethlehem Steel Corporation (USA) for a high-speed steel containing 0.7-0.8% carbon, 18% tungsten, 4% chromium, 1% vanadium, and the balance iron. Used for cutting tools, gages and reamers.

(2) Trade name of Matchless Metal Polish Company (USA) for a greaseless buffing compound.

Tiger Bronze. Trade name of Abex Corporation (USA) for a cast bronze containing 75% copper, 16% lead, 6% tin and 2% zinc. The lead particles are evenly dispersed throughout the matrix. It has a low coefficient of friction, and is used for bearings and bushings.

tiger's eye. A fibrous, yellow to brownish-yellow variety of *quartz* (SiO_2) from South Africa formed from *crocidolite* by replacement and displaying a cat's eye-effect. Used as a gemstone.

Tiger Special. Trade name of Bethlehem Steel Corporation (USA) for a tungsten-type high-speed steel containing 0.75% carbon, 4-4.2% chromium, 18% tungsten, 5% cobalt, 0.5% molybdenum, 1-1.1% vanadium, and the balance iron. It has high hot hardness, excellent wear resistance, and is used for lathe and planer tools, milling cutters, broaches, drills, and other tools.

Tigold. (1) Trade name of Tiodize Company, Inc. (USA) for titanium nitride coatings.

(2) Trademark of King Arthur Jewelry, Inc. (USA) for a nonprecious metal alloy resembling 14-karat gold in appearance.

Tijolinho. Trade name of Vidrobrás (Brazil) for a patterned glass.

Tikana. Trademark of Schott DESAG AG (Germany) for a colorless antique glass with slightly uneven and bubble-free surface. Supplied in standard sizes of 2400 × 1600 mm (94.5 × 63.0 in.), it is used in the restoration of historic window glazing.

tikhonenhovite. A colorless to rosy mineral composed of strontium aluminum fluoride hydroxide monohydrate, $SrAl(OH)F_4 \cdot H_2O$. Crystal system, monoclinic. Density, 3.26 g/cm³; refractive index, 1.456. Occurrence: Russian Federation.

Tilan. Trade name for regenerated protein (azlon) fibers.

tilasite. A gray, gray-violet or gray-green mineral composed of calcium magnesium fluoride arsenate, $CaMgAsO_4F$. Crystal system, monoclinic. Density, 3.77 g/cm³; refractive index, 1.660.

Occurrence: India, Sweden.

tile. (1) A relatively thin, flat glazed or unglazed structural product composed of burnt clay, concrete, stone, etc., and used for functional and ornamental applications on walls, floors, roofs, fireplace hearths, and kitchen countertops.

(2) A glazed or unglazed, hollow or concave structural product composed of burnt clay, concrete, stone, etc., and frequently used for drainage applications.

Tile Bond. Trade name of Flexible Products Company (USA) for adhesives used for bonding clay and concrete roof tiles.

Tile-Guard. Trade name of Manville Corporation (USA) for fiberglass mats used for joints of underground drainage tile.

tiling plaster. See Keene's cement.

Tilite. Trade name of Foseco Minsep NV (Netherlands) for self-sinking grain refiner in tablet form used in the manufacture of wrought and cast aluminum and magnesium alloys.

tilleyite. A colorless, or white mineral composed of calcium carbonate silicate, $Ca_5Si_2O_7(CO_3)_2$. Crystal system, monoclinic. Density, 2.84 g/cm³; refractive index, 1.635. Occurrence: USA (California).

Tilon. (1) Trademark of Rex Industrie-Produkte Graf von Rex GmbH & Co. KG (Germany) for asbestos fabrics, rope and yarn.

(2) Trade name for a vinyl acrylate copolymer formerly used for denture bases.

Timang. Trade name of Harsco Corporation (USA) for a series of nonmagnetic, wear-resistant high-manganese steels containing 0.6-0.8% carbon, 12-15% manganese, 3-7% nickel, and the balance iron. Used for welding rods, woven screens, elevator buckets, and earthmoving equipment.

timber. A large, squared piece of wood suitable for building, carpentry or joinery. This term usually refers to sawed lumber 125 mm (5 in.) or larger in both thickness and width.

Timbrelle. Trademark of DuPont (UK) Limited (UK) for nylon 6,6 staple fibers, filament yarns and textured continuous filaments used for textile fabrics.

Time-Graph. Trade name of Time Steel Service Inc. (USA) for an oil-hardening tool steel (AISI type O6) containing 1.45% carbon, 0.8% manganese, 1.2% silicon, 0.2% chromium, 0.25% molybdenum, and the balance iron. Used for dies and tools.

Timeline. Trademark of Dentsply International Inc. (USA) for a light-cure dental resin used for cavity lining.

Timetal. Trademark of Titanium Metals Corporation (USA) for an extensive series of titanium alloys and mill products including sheet, strip, plate, rod, bar, billet, tubing and welding wire. *Timetal 10-2-3* is a high-strength titanium forging alloy, *Timetal 21S* an oxidation-resistant titanium alloy strip, *Timetal 21SRx* an orthopedic grade of Timetal 21S, *Timetal 62S* (Ti-6Al-1.7Fe-0.1Si) a titanium alloy with excellent mechanical properties (i.e., high tensile and yield strength), *Timetal 1100* a creep-resistant titanium alloy containing 6% aluminum, 2.8% tin, 4% zirconium, 0.4% molybdenum and 0.4% silicon, and *Timetal LCB* (Low Cost Beta) a relatively inexpensive high-strength beta-titanium alloy containing 1.5% aluminum, 4.5% iron and 6.8% molybdenum, having low processing temperature (150-205°C or 300-400°F) and used for springs for aerospace applications and suspension systems of advanced automobile racing cars.

Timken. Trade name of Timken Company (USA) for an extensive series of plain-carbon steels and numerous alloy steels including molybdenum, chromium, chromium-molybdenum and tungsten-chromium types. Also included under this trade name are various austenitic, ferritic and martensitic stainless steels.

tin. A ductile, corrosion-resistant metallic element of to Group IVA (Group 14) of the Periodic Table. It is commercially available in the form of bars, rods, ingots, sheets, sticks, foils, microfoils, wire, tape, pipe, shot, granules, beads, powder, single crystals and in mossy form. The single crystals are usually grown by the Bridgeman technique. The only commercial tin ore is *cassiterite*, but it is also found in several other minerals and also occurs native. At 13°C (56°F) pure solid, silver-white tin (beta tin) with body-centered tetragonal crystal structure undergoes a crystallographic transformation ("tin pest") to a gray powder (alpha tin) with face-centered cubic crystal structure. Upon heating above this temperature alpha tin reverts to beta tin without changing its powder form. Density, 7.28 g/cm³; melting point, 232°C (450°F); boiling point, 2270°C (4118°F); hardness, 1.5-1.8 Mohs; atomic number, 50; atomic weight, 118.710; divalent, tetravalent; superconductivity critical temperature, 3.722K; electrical conductivity, about 14.5% IACS. Used in alloys (e.g., bronzes, solders, pewter, babbitts, typemetals and speculum), in low-melting alloys, in superconductors, in electroplating (plating anodes), in corrosion-resistant and hot-dip coatings, for tinplate, terneplate, tin foil, cladding, die casting, opalescent glass, porcelain enamels, coinage, collapsible tubes, bells, tinned wire, and dental amalgams. Symbol: Sn.

tin acetate. See stannous acetate.

tinaksite. A pale yellow mineral composed of potassium sodium calcium titanium silicate hydroxide, $NaK_2Ca_2TiSi_7O_{19}(OH)$. Crystal system, triclinic. Density, 2.82 g/cm³; refractive index, 1.621. Occurrence: Russian Federation.

tin alloys. Alloys containing tin as the principal element. Typical combinations include: (i) *Tin-lead, tin-silver* and *tin-zinc alloys* used in soldering, coating and for tin foil; (ii) *Tin-antimony alloys (or white metals)* used for costume jewelry; (iii) *Tin-antimony-lead alloys* used for typemetals; (iv) *Tin-antimony-copper alloys (or tin babbitts)* used for bearings; (v) *Copper-tin and copper-tin-zinc alloys* used for bronzes and brasses; and (vi) *Dental amalgams* containing silver, tin and copper.

Ti-Namel. Trade name of Inland Steel Company (USA) for a high-strength low-alloy (HSLA) steel containing 0.06% carbon, 0.3% manganese, 0-0.12% copper, 0.05% aluminum, 0.3% titanium, and the balance iron. Used for enameling sheet.

tin anhydride. See stannic oxide.

tin-antimony alloys. A group of alloys containing tin and antimony as the principal elements. Examples include tin-antimony solders containing up to 5% antimony, and white metal alloys (containing up to 15% antimony) used for bearings and in costume jewelry.

tin-antimony-copper alloys. A group of alloys containing tin, antimony and copper as the principal elements in the typical composition range of up to 92% tin, 4-15% antimony and 3-8.5% copper. They have good corrosion resistance, low hardness and fair fatigue resistance. Used for bearings and bushings (tin babbitt), and die castings.

tin-antimony-copper-lead alloys. A group of alloys containing tin, antimony, copper and lead as the principal elements in the typical composition range of up to 76% tin, 12% antimony, 3% copper, and 11% lead. Used for sleeve bearings.

tin-antimony solder. See antimonial tin solder.

tin ash. (1) See stannic oxide.

(2) A mixture of tin oxide (SnO_2) and lead oxide (PbO) used as an opacifier in glazes.

tin babbitts. See tin-base babbitts.

tin-base babbitts. A group of tin-base bearing alloys containing 65-95% tin, 4-12% antimony, 1-8.5% copper, and small additions of zinc, aluminum, arsenic, iron, cadmium and/or bismuth. They have good corrosion resistance, good antiscoring properties, good embeddability and conformability, fair fatigue resistance and a hardness of 17-27 Brinell. Used for bearings employed in heavy-duty service, and for bushings. Also known as *tin babbitts*.

tin-bearing commercial bronze. A corrosion-resistant *commercial bronze* containing 90% copper, 9.5% zinc and 0.5% tin. Used for heavy-duty bushings and bearings.

tin bisulfide. See artificial gold.

tin brass. An alloy of copper and zinc with a small percentage of tin. The typical composition range is 87.5-88.5% copper, 9.5-11.5% zinc and 1-2% tin. It has good corrosion resistance and workability, and is used for weatherstripping, springs, clips, switches, and terminals.

tin bromide. See stannic bromide; stannous bromide.

tin bronze. A wrought or cast alloy of about 70-93% copper and 7-30% tin, often with significant additions of one or more other elements, such as zinc (up to 5%), lead (up to 20%), and/or nickel (up to 5%). Usually deoxidized with phosphorus, it has a face-centered cubic crystal structure, high corrosion, abrasion and wear resistance, good tensile properties, high toughness, good antifriction properties, and good castability. Used for fittings, valves, drain cocks, valve seats, water gages, flow indicators, pump and turbine housings, bells, wheels, gears, helical and worm gears, bearings, bushings, bellows, springs, diaphragms, fasteners, window and door seals, welding rods, spark-resistant tools, wear plates, fourdrinier wire, paint, electrical components, vacuum dryers, blenders, chemical equipment, and in the fine arts.

tincal. See borax.

tincalconite. A colorless, or chalky white mineral composed of sodium borate pentahydrate, $Na_2B_4O_7 \cdot 5H_2O$. It can also be made synthetically. Crystal system, rhombohedral (hexagonal). Density, 1.85 g/cm³; refractive index, 1.461. Occurrence: USA (Mohave Desert, California). Also known as *mohavite*.

tin chloride. See stannic chloride; stannous chloride.

tin chromate. See stannic chromate; stannous chromate.

tin-coated steel. A carbon steel, usually in sheet form, to whose surface a thin coating of tin has been applied by hot dipping, electroplating or immersion. The coating provides a nontoxic, protective, decorative surface, facilitates soldering and/or assists in bonding to another metal. Also known as *tin-plated steel*.

tin coating. A protective or decorative coating of tin applied to the surface of iron or steel sheet by dipping, immersion or electrodeposition.

tin crystals. See stannous chloride.

tin dibromide. See stannous bromide.

tin dichloride. See stannous chloride.

tin dichloride dihydrate. See stannous chloride dihydrate.

tin dichloride pentahydrate. See stannic chloride pentahydrate.

tin die-casting alloys. Casting alloys containing 80-84% tin, 12-14% antimony, 4-6% copper, up to 0.35% lead, 0.08% iron, 0.08% arsenic, 0.01% zinc, and 0.01% aluminum. They have a hardness of 25-35 Brinell, poor to fair strength, and are used in the production of precision die castings with intricate shapes.

tin difluoride. See stannous fluoride.

tin diiodide. See stannous iodide.

tin dioxide. See stannic oxide.

tin disulfide. See artificial gold.

tin-doped indium oxide. See indium tin oxide.

Tinea. Trade name of Compagnie Française de l'Etain (France) for a series of antifriction alloys containing 10-90% tin, 0-42% lead, 6.5-13.5% antimony and 1-6% copper. Used for bearings.

tin enamel. A white porcelain enamel or glaze made with mill additions of tin oxide (SnO_2) to the slip to impart opacity to the fired product.

tin fluoborate. See stannous fluoborate.

tin fluoride. See stannic fluoride; stannous fluoride.

tin fluoroborate. See stannous fluoborate.

tinfoil. (1) Tin (98.8-99.999% pure) rolled into sheets, 0.001-1 mm (0.00004 to 0.04 in.) thick. Microfoil as thin as 0.001 μm (0.04 μin.) on permanent plastic support is also available. Used for packaging, condensers and insulation.
(2) An alloy composed of 92% tin and 8% zinc rolled into foil for food packaging applications.
(3) An alloy composed of 88% tin, 8% lead, 4% copper and 0-0.5% antimony rolled into very thin sheets and used for bearings and packaging foil.

tin glaze. A white glaze that contains additions of tin oxide (SnO_2) to impart opacity to the fired product, usually a pottery or similar earthenware.

tin-glazed ware. Pottery or earthenware, such as delftware or majolica, coated with a white opaque tin glaze.

Tinicosil. Trade name of Cerro Metal Products Company (USA) for a series of corrosion-resistant alloys containing 42-50% copper, 35-50% zinc, 10-16% nickel, 0-2.2% lead, 0-3% manganese and 0-1% iron. Used for hardware, valve parts, instrument parts, screw-machine products, and fishing tackle.

tin-indium alloys. A group of solders composed of 44-52% indium and 48-56% tin. The eutectic temperature for a 52In-48Sn alloy is 118°C (244°F). *Tin-indium alloys* possess low vapor pressure, are capable to wet glass, quartz and ceramics, and are used for glass- and ceramic-to-metal seals.

tin iodide. See stannic iodide; stannous iodide.

Tinite. Trade name of H. Kramer & Company (USA) for a hard, tough alloy of tin, lead and antimony, used for bearings.

tin-lead alloys. A group of alloys in which tin and lead are the principal elements. Examples include: (i) alloys composed of 50-63% tin and 37-50% lead used as soft solders for various applications; (ii) soft, corrosion-resistant alloys composed of 90-95% tin and 5-10% lead used for bearing overlays; and (iii) alloys composed of 70-97% tin and 3-30% lead used for coating metals.

tin-lead-cadmium alloys. A group of alloys of varying amounts of tin, lead and cadmium. Most commercial alloys are close to the eutectic composition of 52% tin, 30% lead and 18% cadmium.

tin-lead coating. A tin-lead alloy coating electroplated from an acid electrolyte (e.g., lead fluoborate, tin fluoborate, fluoboric acid or lead sulfamate) and used to produce corrosion-resistant, protective coatings on steel, or as a soldering promoter for printed circuits and electronic components.

tin-lead solders. A group of tin-base solders containing up to 50% lead. The 62Sn-38Pb eutectic alloy has a melting point of 183°C (361°F). *Tin-lead solder* possess good wetting characteristics, good strength, and are used in electronics, for printed circuits, in tin coating of metals, and in joining copper pipe.

tinman's solder. A soft solder composed of 66.7% tin and 33.3% lead and used in plumbing. It has a solidus temperature of 183°C

(361°F) and a liquidus temperature of 188°C (370°F).

tin monoxide. See stannous oxide.

tin monosulfide. See stannous sulfide.

tinned wire. (1) A carbon or alloy steel wire continuously passed through a molten tin bath and supplied in three tempers: (i) soft tinned, i.e., tinned after being annealed at or near finish size; (ii) medium hard, i.e., produced from heat-treated stock; and (iii) hard tinned, i.e., tinned after being cold drawn to final size.

(2) A copper wire with a relatively thin coating of tin or tin alloy applied by hot dipping, electroplating or cladding. The coating may serve to provide corrosion protection or facilitate soldering to electronic components.

tin-nickel coating. A tin-nickel alloy coating electroplated from an acid electrolyte (e.g., tin fluoborate or fluoboric acid). It produces a silvery-white, corrosion-resistant protective layer, and is used on various metals and alloys. It is also used as an etch resistant, and as a soldering promoter for printed circuits and electronic components.

tinning metal. An alloy of 50% tin and 50% lead used in electrotyping.

Tinol. Trademark of Kueppers Metallwerk GmbH (Germany) for a series of tin solders.

Tinolite. Trade name for a series of carbon blacks.

Tinols. Trade name of MacLee Chemical Company Inc. (USA) for bright tin and tin-lead electrodeposits and deposition processes.

Tinomats. Trade name of MacLee Chemical Company Inc. (USA) for satin tin and tin-lead electrodeposits and deposition processes.

tin oxalate. See stannous oxalate.

tin oxide. See stannic oxide; stannous oxide.

tin perchloride. See stannic chloride.

tin peroxide. See stannic oxide.

tinplate. A thin sheet of iron or carbon steel coated on one or both sides with a layer of tin (typically 0.4 μm or 0.01 mil thick) by electrodepositon or hot dipping. It has good corrosion resistance, good resistance to food acids, good solderability, and is used for tin cans, food containers, containers for fuels and tobacco, and for gaskets, toys, batteries, signs, and filters. See also coke plate; best coke; charcoal plate; electrolytic tinplate.

tin-plated steel. See tin-coated steel.

Tinposit. Trade name of Shipley Ronal (USA) for immersion tin plates.

tin powder. Tin (99.9+% pure) in the form of a fine powder (100-325 mesh) for use in powder metallurgy and ceramics, and for pigments.

tin protochloride. See stannous chloride.

tin protosulfide. See stannous sulfide.

tin protoxide. See stannous oxide.

tin pyrites. See stannite.

tin pyrophosphate. See stannous pyrophosphate.

tin salts. See stannous chloride.

tinsel. (1) An alloy composed of 60% tin and 40% lead and used for decorative and architectural applications.

(2) Very thin platelets of glass used to produce a sparkling effect in glazes and glass.

(3) A thin strip, sheet or thread of paper, plastic or metal used to produce sparkling appearance effects in fabrics, and for decorative applications.

(4) A glittering or sparkling fabric woven with tinsel yarn.

Tinselfil. Trade name for rayon textile fibers.

Tinselon. Trade name for rayon fibers and yarns used for textile fabrics.

tinsel yarn. A yarn or thread that has been covered, coated or otherwise combined with a metallic substance (e.g., aluminum, copper, gold or silver) to add glitter or sparkle to textile fabrics.

tin selenide. See stannic selenide; stannous selenide.

tinsel lead. A lead alloy containing 1.5% antimony and up to 4% tin. Used for decorative applications.

tin-silver solders. A group of solders containing 90-97% tin and 3-10% silver. Depending on the particular composition, they have liquidus temperatures from 221 and 295°C (430 and 563°F), respectively. They have good wetting characteristics and high joint strength, but are relatively expensive. The microstructure of the eutectic composition (96.5Sn-3.5Ag) consists of fine needles of silver stannide (Ag_3Sn) in a tin-rich matrix. *Tin-silver solders* are used for soldering components in electrical and high-temperature service especially in fine instrument work, and for food applications.

tin solders. A large group of solders in which tin is the principal element. Common examples include tin-lead, tin-antimony, tin-silver and tin-zinc solders.

tin stone. See cassiterite.

tin sulfate. See stannous sulfate.

tin sulfide. See stannous sulfide.

Tintac. Trade name of MacDermid Inc. (USA) for a water-soluble protective vinyl film.

Tintasan. Trade name of DESAG AG (Germany) for tinted spectacle glass of the Chance-Crookes type.

Tinted Carnival. Trade name of Chance Brothers Limited (UK) for decorative table glassware made from various rolled, tinted glasses.

tin telluride. See stannous telluride.

tin tetrabromide. See stannic bromide.

tin tetrachloride. See stannic chloride.

tin tetrafluoride. See stannic fluoride.

tin tetraiodide. See stannic iodide.

tinticite. A creamy white mineral composed of iron phosphate hydroxide heptahydrate, $Fe_6(PO_4)_4(OH)_6 \cdot 7H_2O$. Density, 2.82 g/cm³; refractive index, 1.74-1.75. Occurrence: USA (Utah).

tintinaite. A lead gray mineral of the the lillianite group composed of lead antimony sulfide, $Pb_5Sb_8S_{17}$. Crystal system, orthorhombic. Density, 5.52 g/cm³. Occurrence: Canada (Yukon).

tin-titanium. A compound of tin and titanium used in ceramics and materials research. Melting point, 1666°C (3031°F). Formula: $SnTi_3$.

Tintopal. Trade name of James Clark & Eaton Limited (UK) for colored opaque glass with hard, brilliant finish.

tin-vanadium yellow. See vanadium-tin yellow.

tinzenite. A yellow mineral of the axinite group composed of a calcium manganese aluminum boron silicate hydroxide, $(Ca,Mn,Fe)_3Al_2BSi_4O_{15}(OH)$. Crystal system, triclinic. Density, 3.29 g/cm³; refractive index, 1.701. Occurrence: Switzerland.

tin-zinc alloys. A group of alloys containing tin and zinc as the principal constituents. Typical compositions include: (i) alloys of 92% tin and 8% zinc used in foil form for food packaging; and (ii) alloys of about 50-91% tin and 9-50% zinc used as solders for aluminum.

tin-zinc solders. A group of solders containing 50-91% tin and 9-50% zinc. They have a solidus temperature of 199°C (390°F), and liquidus temperatures ranging between 199 and 377°C (390 and 710°F). Tin-zinc solders are primarily used for joining aluminum since they resist galvanic corrosion of the solder joints.

tin-zirconium. Any of the following compounds of tin and zirconium used in ceramics and materials research: (i) $SnZr_4$. Melting point, 1587°C (2889°F); and (ii) Sn_2Zr_3. Melting point, 1932°C (3510°F).

Tiodize. Trade name of Tiodize Company Inc. (USA) for a galling- and wear-resistant electrolytic conversion coating, compatible with other protective coatings, applied to the surface of titanium and many of its alloys at room temperature in an alkaline bath.

Tioga. Trade name of AL Tech Specialty Steel Corporation (USA) for a tough, oil-hardening special-purpose tool steel (AISI type L6) containing 0.67% carbon, 0.65% chromium, 1.4% nickel, 0.2% molybdenum, and the balance iron. It has a great depth of hardening, high toughness, good wear resistance, and is used for lathe and wear-resistant tools.

Tiolon. Trademark of Tiodize Company, Inc. (USA) for a family of polytetrafluoroethylene resin bonded coatings used as lubricants or mold releases. They inhibit friction, wear and corrosion and bond well to metal, rubber, glass, wood and plastic surfaces.

Tiolube. Trademark of Tiodize Company Inc. (USA) for a series of solid dry-film lubricants containing molybdenum disulfide (MoS_2) and used to prevent premature wear and galling and reduce friction coefficients.

Tipersul. Trademark of E.I. DuPont de Nemours & Company (USA) for potassium titanate available as blocks, lumps, and loose fibers with an average fiber diameter of 1 μm (40 μin.). It has a melting point of 1371°C (2500°F) and an upper service temperature of 1204°C (2200°F). Used for acoustical, electrical and thermal insulation, and for filter media.

Tippfil. Trademark of Tiszai Vegyi (Hungary) for polypropylene staple fibers.

tipping solder. A solder containing 70% lead and 30% tin. It has a density of 9.7 g/cm^3 (0.35 lb/in.³), a solidus temperature of 183°C (361°F) and a liquidus temperature of 255°C (491°F). Used in machine and torch soldering.

TI Polymer. Trademark of Toray Industries, Inc. (Japan) for a series of super-engineering resins.

Tiptolene. Trademark of Lankhorst Touwfabrieken BV (Netherlands) for polyethylene fibers and monofilaments.

Ti-Pure. Trademark of E.I. DuPont de Nemours & Company (USA) for a commercially pure *rutile*-based titanium dioxide used as pigment in coatings, paper and plastics.

tiragolloite. An orange mineral composed of manganese arsenate silicate hydroxide, $Mn_4AsSi_3O_{12}(OH)$. Crystal system, monoclinic. Density, 3.84 g/cm^3; refractive index, 1.751. Occurrence: Italy.

tire cord. Fabric casing plies made of twisted or otherwise formed filaments, strands or yarns of high-tenacity rayon, polyamide (nylon 6,6), rayon-polyamide hybrids, polyester, etc., impregnated with rubber and wrapped around steel bead wires in the construction of automobile tires. The fabric plies provide shape and lend strength to the tire.

tire cord wire. Fine, hard-drawn high-carbon steel wire with high tensile strength used in the construction of automobile tire cords, conveyor belts, and as aircraft control cords.

tire textiles. Textiles, such as rayon or nylon tire cord suitable for use in the construction of automotive tires.

tire yarn. A yarn of rayon, nylon, polyester, etc., suitable for use in the construction of rubber-tire carcasses.

tirodite. A mineral of the amphibole group composed of sodium calcium iron magnesium manganese silicate hydroxide, $(Na,Cl)_2(Mg,Mn,Fe)_5Si_8O_{22}(OH)_2$. Crystal system, monoclinic. Density, 3.16 g/cm^3. Occurrence: Italy.

Tisco. Trade name of Harsco Corporation (USA) for an extensive series of stainless heat-resistant and high-manganese steels, several chromium and chromium-nickel cast irons, and several other steels.

Tiscon. Trade name of Tata Iron & Steel Company (India) for a steel containing 0.19-0.23% carbon, 0.75-0.85% manganese, and the balance iron. Used as reinforcing bars.

Tiscral. Trade name of Tata Iron & Steel Company (USA) for a wear- and abrasion-resistant steel containing 0-0.22% carbon, 1.2-1.5% manganese, 0-0.08% phosphorus, 0.5-0.8% chromium, 0.02-0.1% vanadium, 0.005-0.015% titanium, 0.2-0.4% aluminum, and the balance iron. Used for structural applications.

TI-Shield. Trade name for a clad shielding material developed by Texas Industries, Inc. (USA) and composed of a core of ferromagnetic alloy foil metallurgically bonded to two outer layers of high-purity copper foil. It combines the conductive reflection of high-impedance waves of copper with the magnetic reflection of low-impedance waves of ferromagnetic alloys. *TI-Shield* possesses high energy absorption with a minimum field penetration and very good attenuation via high conductivity and magnetic permeability. Used for electromagnetic shielding applications.

tisinalite. A yellow-orange mineral of the combeite group composed of hydrogen sodium calcium iron manganese titanium silicate hydrate, $Na_3H_3(Mn,Ca,Fe)TiSi_6(O,OH)_{18} \cdot 2H_2O$. Crystal system, rhombohedral (hexagonal). Density, 2.66 g/cm^3; refractive index, 1.624. Occurrence: Russian Federation.

Tiska Nirosta. Trade name of Harsco Corporation (USA) for a series of austenitic stainless steels used for applications requiring excellent corrosion resistance and good heat resistance, such as marine parts, fittings, and valves.

Tismo Poticon. Trade name of Otsuka Chemical Company (Japan) for an extensive series of polymers including polyamides (nylons), polybutylene terephthalates, polyphenylene sulfides, polycarbonates and polyoxymethylenes (acetals).

Tissier's metal. An arsenical bronze containing 97% copper, 2% zinc, 0-1% arsenic and 0-0.5% tin. It has good conductivity and is used for hardware and bearings.

tissue. A lightweight, translucent or transparent woven fabric.

tissue faille. A lightweight textile fabric made from a filament yarn in a plain weave with a distinctive narrow widthwise rib. Used especially for blouses and dresses.

tissue paper. A thin, soft, lightweight, relatively crisp paper, either absorbent or smooth, used chiefly for wrapping, cleaning and cleansing applications.

Tissuglas. Trade name of American Machine & Foundry Company (USA) for thin sheets of matted submicron glass fibers made on a specially adapted papermaking machine. They have very high dielectric strength, an upper service temperature of about 650°C (1200°F), and are used as an electrical insulating material, and as a resin-reinforcing medium.

TiSULC steels. See titanium-stabilized ultralow-carbon steels.

Tital. (1) German trademark for aluminum, magnesium and titanium precision investment castings.

(2) Trademark of Unex Corporation (USA) for aluminum alloys used for hydraulic torque wrenches.

(3) Trademark of KB Alloys, Inc. (USA) for titanium-aluminum master alloys used in the manufacture of aluminum al-

loys.

Titalon. Trade name of Titan Verpackungssysteme GmbH (Germany) for woven fabric sheets used for packaging applications.

Titan. (1) Trade name of Safetee Glass Company (USA) for laminated safety glass.

(2) Trade name of Adamas Carbide Corporation (USA) for titanium carbide-base cemented carbides that may also contain additions of molybdenum carbide and nickel. Used for cutting tools.

(3) Trade name of Carpenter Steel Corporation (USA) for a water-hardening tool steel (AISI type W1) containing about 1% carbon, and the balance iron. It has excellent machinability, good to moderate depth of hardening, wear resistance and toughness and low hot hardness. Used for tools and dies.

(4) Trade name of Osborn Steels Limited (UK) for an austenitic manganese steel containing 1.2% carbon, 13% manganese, and the balance iron. Used for chisels and wear-resistant parts.

(5) Trade name of Cerro Metal Products Company (USA) for an extensive series of wrought and cast copper alloys including various leaded and unleaded brasses and bronzes, aluminum bronzes, aluminum-silicon bronzes, manganese bronzes, naval bronzes, silicon bronzes, yellow brasses, cartridge brasses, muntz metals, and nickel silvers.

(6) Trade name of Titan Dental (USA) for a light-cure hybrid composite used in restorative dentistry.

(7) Trade name of E.I. DuPont de Nemours & Company (USA) for a liquid crystal polymer.

Titanaloy. Trade name of Mathieson-Heglar Company (USA) for a wrought zinc alloy containing 1% copper and 0.12% titanium. It has high creep resistance, and is used for roofing, gutters, trim, housings, and fuses.

titanates. (1) Salts of metatitanic acid (H_2TiO_3) or orthotitanic acid (H_4TiO_4). The former salts is known as "metatitanates," and the latter as "orthotitanates." Examples include lead metatitanate ($PbTiO_3$) and magnesium orthotitanate (Mg_2TiO_4).

(2) See titanate ceramics.

titanate ceramics. Electroceramic compositions consisting of titanium oxide (TiO_2) and one or more of the oxides of barium, beryllium, calcium, magnesium, niobium, strontium, tin, zirconium, etc. For example, barium titanate ($BaTiO_3$) is made by a reaction of barium oxide (BaO) and titanium dioxide (TiO_2) while lead zirconate titanate [$Pb(Ti,Zr)O_3$] is a solid solution of lead titanate ($PbTiO_3$) and lead zirconate ($PbZrO_3$). Titanate ceramics have high dielectric constants, high refractive indices, and excellent ferroelectric and piezoelectric properties. Used for capacitors, electrostrictive transducers, electronic and communication equipment, accelerometers, and in ultrasonic cleaning. Also known as *titanates*.

titanated lithopone. White *lithopone* pigments containing considerable amounts of titanium dioxide (TiO_2).

Titanceed. Trade name of Titan Group (Malaysia) for linear-low-density polyethylene.

titanellow. See titanium trioxide.

Titanex. Trade name of Titan Group (Malaysia) for a series of high-density and linear-low-density polyethylenes.

Titania. Trade name of Bridgeport Rolling Mills Company (USA) for a wrought zinc alloy containing 0.8% copper and 0.15% titanium. It has good workability, and is used for architectural and industrial products, and electrical and electronic components.

titania. See titanium dioxide.

titania brick. Titanium oxide (TiO_2), usually in the form of the crystalline polymorph *rutile*, bonded into bricks with lime.

titania ceramics. Ceramics in which titanium oxide (TiO_2) is the principal crystalline constituent. Used in electronics and electrical engineering.

titania-magnesia ceramics. Ceramics in which titania (TiO_2) and magnesia (MgO) are the principal crystalline phases. They are used chiefly as refractories.

titania porcelain. A vitreous, white technical porcelain in which titania (TiO_2) is the principal crystalline constituent. It is used in the manufacture of titania electroceramics and dielectrics (barium titanate, etc.). Also known as *titania whiteware*.

titania porcelain enamel. A white cover-coat enamel for sheet steel made of frits containing 8-20% titanium oxide (TiO_2). Titania-opacified frits containing 8-15% TiO_2 are used for strong to medium strength colors, and opaque frits containing 17-20% TiO_2 find application in pastel colors. TiO_2 significantly improves the acid resistance of the enamel.

titania whiteware. See titania porcelain.

Titanic. Trade name of Osborn Steels Limited (UK) for medium-carbon alloy steels.

titanic acid anhydride. See titanium anhydride.

titanic anhydride. See titanium dioxide.

titanic bromide. See titanium tetrabromide.

titanic chloride. See titanium tetrachloride.

titanic fluoride. See titanium tetrafluoride.

titanic iodide. See titanium tetraiodide.

titanic iron ore. See ilmenite.

titanic oxide. See titanium dioxide.

Titanit. Trade name of Plansee Metallwerk Gesellschaft (Austria) for a series of cemented carbides based on carbides of tungsten, molybdenum, titanium, etc. Used for cutting tools.

Titanite. (1) Trade name of Titanate Alloys Corporation (USA) for an age-hardenable aluminum alloy containing 4% copper and 0.2% titanium. Used for aircraft wheels, fittings, flywheels and axle housings.

(2) Trade name of Lancaster Glass Corporation (USA) for a toughened glass.

titanite. A yellow-brown, gray, green or black mineral with a resinous to adamantine luster. It is composed of calcium titanosilicate, $CaTiSiO_5$. It can also be made synthetically and may contain some yttrium. Crystal system, monoclinic. Density, 3.4-3.6 g/cm^3; melting point, 1386°C (2527°F); hardness, 5.0-5.5 Mohs; refractive index, 1.894. Occurrence: Austria, Norway, Russia, Switzerland, USA (California, New York). Used as an ore of titanium, in colorants, such as chrome-tin pink, and in the production of crystalline effects in glazes. Formula: $CaTiSiO_5$. Also known as *sphene*.

titanium. A silvery, or gray metallic element belonging to Group IVB (Group 4) of Periodic Table. It is commercially available in the form of powder, granules, sheets, foils, microfoils, ingots, bars, slugs, lumps, tubes, rods, wire, sponge and single crystals, and occurs naturally in minerals, such as *ilmenite, rutile* and *titanite*. Density, 4.51 g/cm^3; melting point, 1660°C (3020°F); boiling point, 3287°C (5949°F); hardness, 60-100 Vickers; atomic number, 22; atomic weight, 47.867; divalent, trivalent, tetravalent; superconductivity critical temperature, 0.40K. It has low thermal conductivity and expansion, low electrical conductivity (about 3.5% IACS), high tensile strength and elastic modulus, good high-temperature properties, excellent resistance to atmospheric and seawater corrosion and corrosion by chlorine, chlorinated solvents and sulfur compounds, good resistance to strong alkalies, and poor resistance to concentrated sulfuric acid and hydrochloric acid. Two allotropic forms of *titanium* are known: (i) *Alpha titanium* (hexagonal crystal structure) that is stable below 880°C (1616°F) and (ii) *Beta titanium* (body-centered cubic crystal structure) that is stable above 880°C (1616°F). *Titanium* is used in steels as a strong carbide former, grain refiner and ferrite strengthener, in stainless steels to reduce sensitivity to intergranular corrosion, in master alloys (e.g., ferrotitanium), as an alloying agent in aluminum, molybdenum, manganese, vanadium, zirconium, etc., as a structural material in aircraft engines, missiles and aerospace parts, for marine equipment, shipbuilding, chemical equipment, food-handling equipment, desalination equipment, equipment for pulp and paper industries, textile machinery, heat exchangers, surgical instruments, orthopedic appliances, bone pins, abrasives, cermets, refractories, paint pigments, X-ray tube targets, electrodes in chlorine batteries, in brazing metals to ceramics, and coatings on metals and ceramics. Symbol: Ti.

titanium alloys. A group of extremely strong alloys of titanium that are highly ductile and easily forged and machined, have high strength-to-weight ratios, and possess high corrosion resistance at normal temperature as well as low thermal conductivities and coefficients of expansion. They are available with three different structures: (i) *Alpha alloys,* that have hexagonal-close packed structures and cannot be hardened by heat treatment; (ii) *Beta alloys,* that have body-centered cubic structures and can be age-hardened; and (iii) *Alpha-beta alloys,* that can be hardened by heat treatment. *Titanium alloys* are used as a substitutes for stainless steels in aircraft and aerospace components and equipment, airframe parts, fuel tanks, structural parts, pressure vessels, autoclaves, chemical process equipment, marine parts, rocket parts, compressors, and orthopedics. See also alpha-titanium alloys; alpha-beta titanium alloys; beta-titanium alloys; near-alpha titanium alloys.

titanium aluminide. Any of the following compounds of titanium and aluminum used in ceramics and for high-temperature applications: (i) *Titanium monoaluminide.* Formula: TiAl; (ii) *Titanium dialuminide.* Density, 4.00 g/cm³; melting point, approximately 1640°C (2984°F). Formula: $TiAl_2$; and (iii) *Titanium trialuminide.* Formula: $TiAl_3$. See also gamma titanium aluminide alloys.

titanium aluminum nitride coating. A bronze- or purple-colored high-performance coating of titanium aluminum nitride (TiAlN), usually 1-5 μm (0.04-0.2 mil) thick, produced on aluminum and nickel alloys, cast iron and tool steel by physical vapor deposition (PVD). It has excellent oxidation and wear resistance, a coefficient of friction of 0.7, a hardness of 2800 Vickers and a maximum service temperature in air of 790°C (1450°F).

titanium aluminum oxynitride ceramics. Advanced ceramics based on titanium aluminum oxynitride (TiAlON) and usually made by sintering titanium nitride, aluminum nitride and aluminum oxide. They have high strength, toughness and wear resistance, good high-temperature properties, and are used for coatings, and refractory and wear parts.

titanium-aluminum-vanadium alloys. A group of alloys in which titanium, aluminum and vanadium are the principal constituents, but they may have additions of tin, zirconium and molybdenum. They have a low density, high tensile strength, good weld-ability, and are used for aircraft structural parts, turbine components, airframe components, orthopedic implants, etc. See also alpha-titanium alloys; near-alpha titanium alloys.

Titanium-Bearing Steel. Trade name of Inland Steel Corporation (USA) for a porcelain-enameled heat-resistant, high-strength low-alloy (HSLA) steel containing 0.2% carbon, 0.3% titanium and the balance iron. Used for aircraft engine parts.

titanium beryllide. Any of the the following compounds of titanium and beryllium used in ceramics: (i) *Titanium monoberyllide.* Density, 4.17 g/cm³. Formula: TiBe; (ii) *Titanium diberyllide.* Density, 3.23 g/cm³; melting point, approximately 1427°C (2601°F). Formula: $TiBe_2$; and (iii) *Titanium dodecaberyllide.* Density, 2.29-2.30 g/cm³; melting point, below 1540°C (2804°F). Formula: $TiBe_{12}$.

titanium boride. Any of the following compounds of titanium and boron:(i) *Titanium monoboride.* A crystalline powder. Density, 5.26 g/cm³; melting point, 2060°C (3740°F); hardness, above 9 Mohs. Used as a refractory, for high-temperature electrical conductors, and in cermets. Formula: TiB; (ii) *Titanium diboride.* A gray, crystalline powder or refractory solid. Crystal system, hexagonal. Density, 4.38-4.52 g/cm³; melting point, 3225°C (5837°F); hardness, above 9 Mohs; low electrical resistance. Used as an addition agent in metallurgy, in the manufacture of aluminum, as a refractory for crucibles, in wear-resistant products, for bearings and bearing liners, cutting tools, jet nozzles and venturis, as an ingredient in cermets, as a high-temperature electrical conductor, for arc and electrolytic electrodes, hardfacing and welding rod coatings, resistance elements and contact points, in nuclear steels, in superalloys, and in coatings resistant to attack by molten metals. Formula: TiB_2; and (iii) *Titanium pentaboride.* A crystalline powder. Melting point, 2093°C (3799°F). Used for ceramics. Formula: Ti_2B_5.

titanium borocarbide. A compound of titanium, boron and carbon available as a fine powder and used in ceramics and materials research. Formula: TiB_4C.

titanium borosilicide. A compound of titanium, boron and silicon available as a powder. Used for ceramics and materials research. Formula: TiB_4Si.

titanium bromide. See titanium dibromide; titanium tribromide; titanium tetrabromide.

titanium bronze. An alloy of 25-90% copper and 10-75% titanium. A 90Cu-10Ti alloy melts at approximately 1000°C (1830°F), and a 75Ti-25Cu at approximately 1280°C (2336°F).

titanium carbide. A very hard refractory material available as a fine gray crystalline powder (98+% pure), usually made by reacting titanium dioxide with carbon at high temperatures. Crystal system, cubic. Crystal structure, halite. Density, 4.93-4.94 g/cm³; melting point, 3065°C±15°C (5549°F±27°F); boiling point, 4820°C (8708°F); hardness, above 9 Mohs; excellent heat resistance; high hot hardness; high thermal-shock resistance. Used as an additive (with tungsten carbide) in cutting tools, in bearings, nozzles, heat- and wear-resistant parts, cermets and cemented carbides, arc-melting electrodes, special refractories, high-temperature conductors, in coatings for cemented carbides, for coating metal extrusion dies, and as a particulate reinforcement in metal-matrix composites. Formula: TiC.

titanium carbide cermets. A group of composite materials consisting of titanium carbide (TiC) and a metallic binder, such as nickel, nickel alloy, molybdenum, etc. They are made at high temperatures under controlled atmospheres using powder metallurgy techniques involving pressing and sintering. *Titanium carbide cermets* have high tensile strength and elastic modulus, a hardness of 70-90 Rockwell A, and excellent high-temperature properties. Used for cutting tools, gas-turbine parts, torch tips, hot-mill-roll guides, and valve parts.

titanium carbide coating. A very hard and wear-resistant coating of titanium carbide, typically 5-10 μm (0.2-0.4 mil) thick. Its effectiveness as a coating on metal or carbide tools lies in the decrease of the coefficient of friction between the tool and the work.

titanium carbonitride. A compound of titanium, boron and carbon available as a fine powder, and used for coatings, and in ceramics and materials research. Formula: TiCN.

titanium carbonitride coating. A wear-resistant, blue-gray coating of titanium carbonitride (TiCN), typically 1-10 μm (0.04-0.40 mil) thick, applied by physical vapor deposition (PVD) techniques. It has very high hardness (3000 Vickers), a low coefficient of friction (typically 0.45), a maximum service temperature (in air) of 400°C (750°F), and is often used on high-speed milling and gear-cutting and heavy-duty stamping tools made of abrasive or hard-to-machine metals and alloys, such as cast iron, tool steels, copper, Inconel and copper, aluminum and titanium alloys.

titanium chloride. See titanium dichloride; titanium tetrachloride; titanium trichloride.

titanium-clad steel. A plain-carbon, or low-alloy steel, usually in sheet or plate form, having a layer of titanium firmly bonded to one or both sides by rolling.

titanium cobalt. A compound of titanium and cobalt used as a superconductor. Formula: Ti_2Co.

titanium columbate. See titanium niobate.

titanium copper. (1) A master alloy composed of varying amounts of titanium and copper, and used as a ladle addition for deoxidizing nonferrous metals.

    (2) An intermetallic compound typically containing 96.4-99.5% copper and 0.5-3.4% titanium and supplied in quarter-hard (1/4 H) and extra-hard (EH) tempers in sheep and strip form. It has a density of 8.7 g/cm^3, a melting range of 1040-1070°C (1904-1958°F) and a hardness of 250-300 Vickers (1/4H tempers) and 300-350 Vickers (EH tempers). Used for connectors, relays, switches and other electrical and electronic components. Formula: $TiCuR_1$.

titanium dialuminide. See titanium aluminide (ii).

titanium diberyllide. See titanium beryllide (ii).

titanium diboride. See titanium boride (ii).

titanium dibromide. A flammable, black powder (99.9+% pure) with a density of 4.0 used in chemistry and materials research. Formula: $TiBr_2$. Also known as *titanium bromide*.

titanium dichloride. Black, hygroscopic crystals or flammable, black powder (99.9+% pure). Crystal system, hexagonal. Density, 3.13 g/cm^3; melting point, 1035°C (1895°F); boiling point, 1500°C (2732°F). Used in chemistry and materials research. Formula: $TiCl_2$. Also known as *titanium chloride*.

titanium diiodide. Black, hexagonal crystals with a density of 5.02 g/cm^3, used in chemistry and materials research. Formula: TiI_2. Also known as *titanium iodide*.

titanium dioxide. A compound of titanium and oxygen that is available as a white powder, or sintered lumps (99+% pure) and occurs in three polymorphic forms: *anatase, brookite* and *rutile*. Commercially obtained from the mineral *ilmenite*. Crystal system, tetragonal (anatase and rutile); orthorhombic (brookite). Density, 3.8-4.3 g/cm^3; melting point, above 1560°C (2840°F); boiling point, above 2500°C (4530°F); hardness, 5.5-6.5 Mohs; refractive index, 2.5-2.7; good dielectric properties; low chemical reactivity; good resistance to dilute acids; good resistance to heat and light; greatest hiding powder of all white pigments. Used as a white pigment in paints, paper, rubber and plastics, as an opacifier in porcelain enamels, glazes and glass, as a constituent in ceramic colors, in delustering synthetic fibers, in titania and titanate electroceramics, as a component in various dielectrics, as a constituent in welding rod coatings, in floor coverings, in the radioactive decontamination of skin, in biochemistry and bioengineering, and in high-temperature transducers (single crystals). Formula: TiO_2. Also known as *titania; titanic acid anhydride; titanic anhydride; titanic oxide; titanium oxide; titanium white*.

titanium dioxide pigments. Fine white powders made from *anatase, rutile* or *ilmenite*. They have low chemical reactivity, good resistance to dilute acids, good resistance to heat and light, and the greatest hiding powder of all white pigments. Used in the manufacture of paint, paper, rubber, plastics, ceramics, and fiberglass.

titanium disilicide. See titanium silicide (ii).

titanium disulfide. See titanium sulfide (ii).

titanium ditelluride. A compound of titanium and tellurium used as a solid lubricant. Formula: $TiTe_2$.

titanium dodecaberyllide. See titanium beryllide (iii).

titanium-doped sapphire. Sapphire crystals fully or partially doped with titanium ions. They have high figures of merit, and are used as crystals in tunable lasers. Abbreviation: $Ti:Al_2O_3$.

titanium fluoride. See titanium trifluoride; titanium tetrafluoride.

titanium hydride. A black or gray moisture-sensitive, flammable powder (8% pure). Density, 3.91 g/cm^3; melting point, approximately 660°C (1220°F) (decomposes). Used as a solder in bonding metals to glass, in powder metallurgy, in the manufacture of metal foams, in refractories, as a getter in electronics, as reduc-ing atmospheres for furnaces, and as a source of pure hydrogen. Formula: TiH_2.

titanium iodide. See titanium diiodide; titanium tetraiodide.

titanium manganese bronze. A corrosion-resistant, high-strength manganese bronze containing about 58.5-59.5% copper, 0.5-0.9% iron, 0.9% tin, 0.1-0.3% manganese, 0-1% silicon, 0-0.5% nickel, and the balance titanium. Used for forgings, rods, bolts, nuts, and for welding rods for copper, brass, nickel alloys, steels and cast irons. A free-cutting grade containing about 0.8% lead for screw-machine products is also available.

titanium-matrix composites. High-performance engineering composites comprising titanium-alloy matrices reinforced with continuous or discontinuous fibers of silicon carbide, boron, etc. They have excellent high temperature properties, high tensile strength and elastic modulus, and are used for aerospace applications, e.g., driveshafts, turbine engine parts, fan blades, and fan frame struts. Abbreviation: TMC. Also known as *titanium metal-matrix composites*.

Titanium Memory Wire. Trade name of American Orthodontics (USA) for a wire made of nickel-titanium shape-memory alloy used in dentistry. See also nickel-titanium.

titanium metal-matrix composites. See titanium-matrix composites.

titanium-molybdenum alloys. A group of alloys containing varying amounts of titanium and molybdenum as well as zirconium and tin. They have high tensile strength, good deep hardenability, and good stress-corrosion resistance. Used for aircraft applications, fasteners, etc.

titanium monoaluminide. See titanium aluminide (i).

titanium monoberyllide. See titanium beryllide (i).

titanium monoboride. See titianium boride (i).

titanium mononickelide. See titanium nickelide (i).

titanium monophosphide. See titanium phosphide (i).

titanium monosilicide. See titanium silicide (i).

titanium monosulfide. See titanium sulfide (i).

titanium monoxide. Bronze pellets, cubic crystals, or fine powder (99.9% pure). Density, 4.95 g/cm³; melting point, 1750°C (3180°F); boiling point, above 3000°C (5430°F). Used for optical and high-temperature applications. Formula: TiO.

titanium niobate. A compound of titanium dioxide and niobium pentoxide used in ceramics. Melting point 1483°C (2701°F). Formula: $TiNb_2O_7$ ($TiCb_2O_7$). Also known as *titanium columbate*.

titanium nickelide. Any of the following compounds of titanium and nickel: (i) *Titanium mononickelide*. Used in ceramics and as a particulate in the reinforcement of composites. Formula: TiNi; and (ii) *Dititanium nickelide*. Used in ceramics and materials research. Formula: Ti_2Ni.

titanium nitride. A bronze powder, or golden-brown crystals or brittle plates (99+% pure) that can be produced from titanium tetrachloride ($TiCl_4$) and ammonia (NH_3). Density, 5.22-5.24 g/cm³; melting point, 2930°C (5306°F); hardness, above 9 Mohs; good oxidation resistance; good resistance to hydrochloric, nitric and sulfuric acids. Used for coatings on cemented carbides for cutting tools to impart superior resistance to crater formation and flank wear, as an optical and mechanical coating on metals and carbides, in cermets and special refractories, in semiconductor devices and rectifiers, in alloys, and for high-temperature appli-cations. Formula: TiN.

titanium nitride coating. A dense, nonporous, ultrahard, gold-colored coating of titanium nitride (typically 3-10 μm or 0.12-0.40 mil thick) applied by chemical vapor deposition (CVD) to cemented carbides for cutting tools to impart superior resistance to crater formation and flank wear. It is also applied to gears, press tools and other engineering components to enhance hardness and wear resistance, and as an optical coating on metals and carbides. It has a face-centered cubic structure, a density of about 5.22 g/cm³ (0.189 lb/in.³), a melting point of 2930°C (5306°F), a maximum service temperature (in air) of 600°C (1110°F), a hardness of 85 Rockwell C, and a band gap of 3.4±0.05 eV. Other important properties include high inertness, excellent corrosion and chemical resistance, good elevated-temperature proper-ties, uniform thickness, a low coefficient of friction that prevents galling, and good bonding to most metals and some ceram-ics and plastics which make it suitable for various other applications including surgical devices, food-processing equipment, and electronic applications.

titanium oxalate. Yellow prisms obtained by treating titanous chloride with oxalic acid. Used in the manufacture of titanium metal. Formula: $Ti_2(C_2O_4)_3 \cdot 10H_2O$. Also known as *titanous oxalate*.

titanium oxide. See titanium dioxide; titanium monoxide; titanium sesquioxide; titanium trioxide.

titanium pentaboride. See titanium boride (iii).

titanium peroxide. See titanium trioxide.

titanium phosphide. Any of the following compounds of titanium and phosphorus used in ceramics and materials research: (i) *Titanium monophosphide*. Gray, hexagonal crystals. Density, 4.08-4.27 g/cm³; melting point, 1990°C (3614°F). Formula: TiP; and (ii) *Trititanium phosphide*. Density, 4.64 g/cm³. Formula: Ti_3P.

titanium phthalocyanine dichloride. A *phthalocyanine* derivative containing a tetravalent titanium ion (Ti^{4+}) and two chlorine atoms. It has a dye content of about 95%, a melting point

above 300°C (570°F), and a maximum absorption wavelength of 692 nm. Used as a dye an pigment. Formula: $TiC_{32}H_{16}Cl_2N_8$.

titanium-potassium fluoride. White leaflets or colorless, monoclinic crystals used in the manufacture of titanium metal. Formula: TiK_2F_6. Also known as *potassium-titanium fluoride*.

titanium powder. Titanium (99.5+% pure) in the form of a flammable, moisture-sensitive powder having a typical particle size ranging from 0.5 to 150 μm (0.02 to 6 mils). Used in powder metallurgy and ceramics.

titanium sesquioxide. A titanium oxide compound available in the form of violet-black crystals (99.9% pure). Crystal system, rhombohedral (hexagonal). Crystal system, corundum. Density, 4.49 g/cm³; melting point, 1842°C (3348°F). Used in chemistry and materials research. Formula: Ti_2O_3.

titanium sesquisulfide. See titanium sulfide (iv).

titanium silicide. Any of the following compounds of titanium and silicon: (i) *Titanium monosilicide*. Density, 4.34 g/cm³; melting point, 1760°C (3200°F). Used in ceramics. Formula: TiSi; (ii) *Titanium disilicide*. A black, crystalline powder (99+% pure). Crystal system, orthorhombic. Density, 4.04 g/cm³; melting point, 1500-1540°C (2730-2805°F). Used for high-temperature ceramics, special alloy applications, in the fabrication of transistors, and as a flame or blast impingement-resistant coatings. They are also suitable as reinforcements for structural ceramic composites and as matrix materials for structural silicide compounds. Formula: $TiSi_2$; (iii) *Trititanium silicide*. Used in high-temperature structural ceramics. Formula: Ti_3Si; and (iv) *Pentatitanium trisilicide*. Crystal system, hexagonal. Density, 4.32 g/cm³; melting point, 2130°C (3866°F); good resistance to high temperature oxidation; poor thermal-shock resistance. They are used for high-temperature applications, and in ceramics, but are also suitable as reinforcements for structural ceramic composites and as matrix materials for structural silicide compounds. Formula: Ti_5Si_3.

titanium sponge. See sponge titanium.

titanium-stabilized steel. A premium-grade low-carbon sheet steel containing about 0.05% carbon, 0.30% manganese, 0.5% aluminum, 0.01% phosphorus, 0.02% sulfur, 0.30% titanium, and the balance iron. It has excellent resistance to warpage, and is used for porcelain enameling applications.

titanium-stabilized ultralow-carbon steels. A group of high-strength sheet steels with ultralow carbon contents, typically less than 50 ppm, stabilized with titanium to remove interstitial elements from solid solution. Their low carbon content is obtained by efficient vacuum degassing, and they are usually supplied as continuously annealed (CA TiSULC) or batch annealed (BA TiSULC) grades with high elongation, formability and drawability, and are used in the automotive industries for autobodies, floor pans, oil pans, etc. Abbreviation: TiSULC steels.

titanium steel. (1) A stainless steel with a small percentage of titanium having increased corrosion resistance especially to intergranular corrosion after welding.

(2) A water-hardening steel containing 0.1-0.8% carbon, 0.3-1% titanium, and the balance iron. Used for tools and machinery parts.

titanium sulfide. Any of the following compounds of titanium and sulfur: (i) *Titanium monosulfide*. Brown, hexagonal crystals or brownish powder. Density, 3.85-4.46 g/cm³; melting point, above 1780°C (3236°F). Used in ceramics and materials research. Formula: TiS; (ii) *Titanium disulfide*. Yellow-brown, moisture-sensitive crystals or powder (99.8+% pure). Crystal system, hexagonal. Density, 3.37-4.39 g/cm³; melting point, 1540°C (2804°F). Used for high-temperature ceramics, in materials research, for special alloy applications, as a solid lubricant, and in flame- or blast impingement-resistant coatings. Formula: TiS_2; (iii) *Titanium trisulfide*. Density, 3.25 g/cm³; unstable above 593°C (1100°F). Used in ceramics and materials research. Formula: TiS_3; and (iv) *Titanium sesquisulfide*. Black, hexagonal crystals. Density, 3.56 g/cm³. Used in ceramics and materials research.

titanium tetrabromide. Yellow-orange, moisture-sensitive, cubic crystals (98+% pure) with a density of 2.6-3.37 g/cm³, a melting point of 39°C (102°F) and a boiling point of 230°C (446°F). Used in the manufacture of titanium salts, in pigments, and as a catalyst. Formula: $TiBr_4$. Also known as *titanic bromide; titanium bromide*.

titanium tetrachloride. Colorless, or pale yellow, moisture-sensitive liquid (99.8+% pure). Density, 1.726 g/cm³; melting point, -25°C (-13°F); boiling point, 136°C (277°F). Used in the manufacture of pure titanium, titanium nitride and titanium salts, in the production of iridescence in glasses, in the chemical vapor deposition of titanium carbide and nitride, as a chemical reagent, in titanium pigments, and as a polymerization catalyst. Formula: $TiCl_4$. Abbreviation: tickle. Also known as *titanic chloride; titanium chloride*.

titanium tetrafluoride. A white, moisture-sensitive powder (98+% pure). Density, 2.798 g/cm³; melting point, 284°C (543°F); boiling point, above 400°C (752°F) (sublimes). Used as a flux in the production of rubies and sapphire abrasives. Formula: TiF_4. Also known as *titanic fluoride; titanium fluoride*.

titanium tetraiodide. Red, moisture-sensitive crystals or powder (98+% pure). Crystal system, cubic. Density, 4.30 g/cm³; melting point, 150°C (302°F); boiling point, 377°C (710°F). Used in titanium salts, as a catalyst, and as a chemical reagent. Formula: TiI_4. Also known as *titanic iodide; titanium iodide*.

titanium trialuminide. See titanium aluminide (iii).

titanium tribromide. Blue-black hexagonal crystals used in chemistry and materials research. Formula: $TiBr_3$. Also known as *titanium bromide; titanous bromide*.

titanium trichloride. Purple, flammable, hygroscopic crystals (99+% pure). Crystal system, hexagonal. Density: 2.64 g/cm³; melting point, decomposes above 440°C (824°F). It is also available as a hydrogen-reduced product. Used in organometallic synthesis involving titanium, as a co-catalyst for polyolefin polymerization. Formula: $TiCl_3$. Also known as *titanium chloride; titanous chloride*.

titanium trifluoride. Violet, hexagonal crystals or moisture-sensitive powder with a density of 2.98-3.40 g/cm³, a melting point of 1200°C (2192°F) and a boiling point of 1400°C (2552°F). Used in chemistry and materials research. Formula: TiF_3. Also known as *titanium fluoride; titanous fluoride*.

titanium trioxide. A yellow powder used in the manufacture of yellow tile, and ivory-colored ceramics and dental porcelains and cements. Formula: TiO_3. Also known as *titanellow; titanium peroxide*.

titanium trisulfide. See titanium sulfide (iii).

titanium-vanadium alloys. A group of alloys with varying amounts of titanium and vanadium as the principal elements, along with other elements, such as chromium, aluminum and/or iron. *Titanium-vanadium alloys* have medium to high tensile strength and toughness, and are used for rocket-motor cases, unmanned airborne systems, and airframe springs. See also beta titanium alloys.

titanium white. See titanium dioxide.

titanium yellow. A yellow paint pigment made by calcining antimony, nickel, and titanium oxides at high temperatures. It has good weatherability and lightfastness. Used as a replacement for lemon yellow (barium chromate).

titanium-zinc. An intermetallic compound of titanium and zinc. Density, 6.0 g/cm³; melting point, 1666°C (3031°F). Used in ceramics, materials research and high-temperature applications. Formula: Ti_3Zn.

Titankote +C3. Trade name of Richter Precision Inc. (USA) for a multilayer, chromium carbide-based coating for cutting tools and aluminum and zinc die-casting dies and molds made of tool or die steels. It is applied by physical-vapor deposition (PVD) at 355°C (675°F) to a thickness of up to 3 μm (120 μin).

Titanlene. Trade name of Titan Group (Malaysia) for low-density polyethylene.

Titan Manganese. Trade name of Osborn Steels Limited (UK) for an austenitic manganese steel containing 1.2% carbon, 13% manganese, and the balance iron. Used for chisels and wear-resistant parts.

Titano. Trademark of Altair Technologies, Inc. (USA) for titanium dioxide powders and slurries including nanoparticle *anatase-* and *rutile*-crystal products.

Titanol. Trademark of Forestadent–Bernhard Förster GmbH (Germany) for a wire of nickel-titanium shape-memory alloy used in dentistry. *Titanol Superelastic* contains 53% nickel and 47% titanium.

Titanolith. Trade name for a white *lithopone* pigment containing about 25% zinc sulfide (ZnS), 15% titanium dioxide (TiO_2), and the balance barium sulfate ($BaSO_4$).

titanous bromide. See titanium tribromide.

titanous chloride. See titanium trichloride.

titanous fluoride. See titanium trifluoride.

titanous oxalate. See titanium oxalate.

Titanox. Trademark of Kronos, Inc. (USA) for an extensive series of white pigments containing varying amounts of titanium dioxide (TiO_2) in either the *anatase* or *rutile* mineral form. *Titanox A* is a relatively pure titanium dioxide, *Titanox B* contains 25% titanium dioxide and 75% barium sulfate (blanc fixe), *Titanox C* is a mixture of 30% titanium dioxide and 70% calcium sulfate, and *Titanox L* is a pale-yellow lead titanate. *Titanox* pigments are used in the manufacture of paints, inks, leather, rubber, plastics, paper, textiles, ceramics, roofing granules, welding rod coatings, and floor coverings.

Titanpro. Trade name of Titan Group (Malaysia) for polypropylene plastics.

Titan-Seewasser. German trade name for an aluminum alloy containing 2-4% magnesium, 1.2% manganese and 0-0.2% titanium. It has good corrosion resistance especially to seawater, and is used for marine hardware and chemical plant equipment.

Titanzex. Trade name of Titan Group (Malaysia) for high-density polyethylene.

Titebond. Trademark of Clayburn Refractories Limited (Canada) for firebrick mortar.

TiTech. Trademark of Titech International, Inc. (USA) for a series of titanium casting alloys.

tivanite. A black mineral composed of vanadium titanium oxide hydroxide, $VTiO_3(OH)$. Crystal system, monoclinic. Density, 4.15 g/cm³. Occurrence: Western Australia.

Tivar. Trademark of Poly-Hi Solidur/Menasha Corporation (USA) for corrosion- and wear-resistant ultrahigh-molecular-weight polyethylene with a low-friction, self-lubricating surface, high impact strength, nil water absorption, low weight (0.928-0.940 g/cm³ or 0.033-0.034 lb/in.³) and a service temperature range of -30 to +82°C (-22 to +180°F). Available in the form of rods, tubes, slabs, sheeting and profiles.

Tixogel. Trademark of Süd-Chemie, Inc. (USA) for a range of thixotropic organoclay products.

Tizirbe. Trade name of Wesgo Division of GTE Sylvania Corporation (USA) for a corrosion-resistant alloy of 48% titanium, 48% zirconium and 4% beryllium. Used for brazing titanium and stainless steel.

Tizit. Trade name of Plansee Metallwerk Gesellschaft (Austria) for a series of cemented carbides based on carbides of tungsten, tantalum, titanium, etc. in cobalt binders. Used for cutting tools.

tlalocite. A blue mineral composed of copper zinc chloride tellurate hydroxide hydrate, $(Cu,Zn)_{16}Te_3O_{11}Cl(OH)_{25} \cdot 27H_2O$. Crystal system, monoclinic. Density, 4.55 g/cm³; refractive index, 1.796. Occurrence: Mexico.

tlapallite. A green mineral that is composed of calcium copper tellurium hydrogen oxide sulfate, $H_6Ca_2Cu_3(SO_4)(TeO_3)_4TeO_6$, and may also contain lead. Crystal system, monoclinic. Density, 5.38 g/cm³; refractive index, 2.115. Occurrence: Mexico, USA (Arizona).

T-Lock Amer-Plate. Trademark of Ameron, Inc. (USA) for thick vinyl plastic sheeting containing inert pigments and plasticizers. It is used as an acidproof lining for concrete pipes, ducts, reservoirs, tanks, walls, floors, etc., and for coating or lining steel surfaces.

T-Lux. Trade name for a light-cure dental resin used for trays.

TMA. Trademark of ORMCO Corporation (USA) for an alloy of 78.0% titanium, 11.0% molybdenum, 6.0% zirconium and 4.0% tin, used for orthodontic wire.

T-Metal. Trade name of London Zinc Mills Limited (UK) for a wrought, heat-treated zinc alloy containing 0.8% copper, 0.1% titanium, 0.003% manganese and 0.002% chromium. It has good mechanical properties and creep strength. Used for building applications, and in roofing.

TMM. Trade name of Rogers Corporation (USA) for a temperature-stable microwave material (TMM) composed of a thermosetting plastic filled with ceramic powder, and supplied as a copper-clad laminate. It has good processibility and machinability, a low dissipation factor, and a stable dielectric constant over a wide range of temperatures. Used especially for the manufacture microwave circuits operating in harsh environments.

TNA-100. Trade name of Ruberoid Company (USA) for corrugated and flat asbestos sheeting, paper and rollboard.

Tneme-Crete. Trade name of Tnemec Company, Inc. (USA) for one-coat brick, concrete and masonry paint.

TNT. Trademark of Toray Nylon Thai Company (Thailand) for polyester filament yarns and nylon 6 monofilaments and yarns used for textile fabrics.

Toabo Polypro. Trademark of Toa Wool Spinning & Weaving Company (Japan) for polypropylene staple fibers.

Toba. Trade name of T. Turton & Sons Limited (UK) for an alloy steel containing 0.37% carbon, 1% chromium, 0.2% vanadium, and the balance iron.

tobelite. A mineral of the mica group composed of ammonium aluminum silicate hydroxide, $NH_4Al_2AlSi_3O_{10}(OH)_2$. It can be made synthetically. Crystal system, monoclinic. Density, 2.61 g/cm³.

tobermorite. A pinkish white mineral composed of calcium silicate tetrahydrate, $Ca_5(Si_6O_{18}H_2) \cdot 4H_2O$. Crystal system, ortho-

rhombic. Density, 2.31-2.58 g/cm³; refractive index, 1.571. Tobermo-rite gel is the principal cementing compound in hardened Portland cement. Occurrence: Northern Ireland; USA (California). See also tobermorite gel.

tobermorite gel. A gel-like solid composed of calcium silicate tetrahydrate, $Ca_5(Si_6O_{18}H_2) \cdot 4H_2O$, which, owing to its great surface area, is used as a binder of concrete. See also tobermorite.

Tobin brass. A copper-base alloy containing about 40% zinc, and small amounts of tin. Used for brazing high-copper brasses.

Tobin bronze. A group of corrosion-resistant naval brasses containing 58-62% copper, 0.5-2.3% tin, and the balance zinc. Sometimes small percentages of lead (about 0.1%) are added to improve machinability. They possess excellent resistance to seawater corrosion, good tensile strength, fair to poor machinability and excellent hot forgeability and hot workability. Used for marine and general hardware, fasteners, pump and valve parts, piston rods, propeller shafts, marine equipment and parts, and in welding rods for steel, cast iron and copper alloys.

tochilinite. A mineral of the valleriite group composed of iron magnesium sulfide hydroxide, $4FeS \cdot 3(Mg,Fe)(OH)_2$. Crystal system, rhombohedral (hexagonal). Occurrence: Cyprus, Russian Federation.

tocorenalite. A yellow mineral composed of silver mercury iodide, $(Ag,Hg)I$. Occurrence: Chile, Australia.

todorokite. (1) A brassy mineral of the rutile group available in the form of a dark brown powder composed of manganese oxide hydrate, $(Mn,Ca)Mn_5O_{11} \cdot 4H_2O$. Crystal system, orthorhombic. Density, 3.67 g/cm³; refractive index, above 1.74. Occurrence: Australia.

(2) A mineral, black in bulk and translucent in thin needles, composed of manganese calcium barium oxide hydrate, $(Mn,Ca,Ba)O_x \cdot zH_2O$ (x= 6-7 and z = 1-2). Crystal system, orthorhombic. Density, 3.67 g/cm³; refractive index, above 2.35. Occurrence: Cuba.

toernebohmite. (1) An olive green mineral composed of cerium rare-earth silicate hydroxide, $(Ce,Ln)_3Si_2O_8(OH)$. Crystal system, monoclinic. Density, 4.90 g/cm³; refractive index, 1.852. Occurrence: Sweden.

(2) A green mineral composed of rare-earth aluminum silicate hydroxide, $Ln_2Al(OH)(SiO_4)_2$. Crystal system, monoclinic. Density, 5.12 g/cm³. Occurrence: Sweden.

Togato. Trade name of Nippon Sheet Glass Company Limited (Japan) for plate glass brilliantly cut and acid-etched on the one side, and polished on the other side.

toile. A delicately patterned fabric in a plain weave, usually of fine silk or rayon fibers.

Toile de Jute. Trade name of Boussois Souchon Neuvesel SA (France) for a patterned glass with weave effect.

Toka Black. Trade name for a series of carbon blacks supplied in several grades.

Tokawhisker. Trademark of Tokai Carbon Company Limited (Japan) for discontinuous silicon carbide reinforcing fibers.

Tokiwa. Trade name of Asahi Glass Company Limited (Japan) for a patterned glass with leaf design.

Tokushu. Trade name of Tokusku Seiko Company Limited (Japan) for a series of corrosion-resistant steels containing 0.08-0.35% carbon, 0.3-1.5% molybdenum, 11-15% chromium, 0-1% nickel, 0-0.5% vanadium, 0-1.1% tungsten, and the balance iron. Used for shafts, bearings, valves, knives, tableware, cutlery, surgical instruments, chemical plant equipment, and oil-refinery equipment.

Tokuso Mac-Bond II. Trade name of Tokuyama Corporation (Japan) for a light-curing, self-etching dental bonding system containing bisphenol A–glycidylether dimethacrylate (BisGMA), triethylene glycol dimethacrylate (TEGDMA) and MAC-10.

Tolalan. Trade name for rayon fibers and yarns used for textile fabrics.

tolbachite. A gold-brown to brown mineral composed of copper chloride, $CuCl_2$. Crystal system, monoclinic. Density, 3.40 g/cm^3.

Toledo. Trade name of Darwin Tools Limited (UK) or a series of steels including several plain-carbon steels containing 0.2-0.55% carbon, and several high-carbon steels containing up to 1.5% chromium.

tolovkite. A steel-gray mineral of the pyrite group composed of iridium antimony sulfide, IrSbS. Crystal system, cubic. Density, 10.50 g/cm^3. Occurrence: Russian Federation.

Tolex. Trade name of Textileather Corporation (USA) for polyvinyl chlorides and other vinyl plastics used for the manufacture of textile fabrics and artificial leather.

Toloy. Trade name of Wellman Alloys Limited (UK) for a series of wrought or cast chromium-nickel, nickel-chromium and nickel-chromium-iron alloys. Also included under this trade name are several cast austenitic stainless and heat-resisting steels containing about 0.1-0.5% carbon, 20-27% chromium, 12-37% nickel, and the balance iron. They have good corrosion and heat resistance, and are used for furnace parts and fittings, superheater parts, heat exchangers, and oil refining equipment.

toluelene. See trans-stilbene.

Tomax. Trade name of Shanghai Tops Ondustrial Company Limited (China) for acrylonitrile-butadiene-styrene (ABS) plastics.

tombac. A group of wrought red brasses containing 82.3-98% copper, 6-20% zinc, 0-6% tin and sometimes small amounts of arsenic. They have good to excellent corrosion resistance and good formability, and are used for pipes, tubing, hardware, deep-drawn products, bearings, electrical applications, ornaments, cheap jewelry, and imitation gold. Also known as *tombac metal*.

tombac metal. See tombac.

tombarthite. A brownish black mineral of the monazite group composed of rare-earth silicate hydroxide, $Ln_4(Si,H_4)_4O_{12-x}$-$(OH)_4$. Crystal system, monoclinic. Density, 3.58 g/cm^3; refractive index, 1.639. Occurrence: Norway.

tombasil. A corrosion- and wear-resistant bronze containing 81-83% copper, 13-15% zinc and 4% silicon. It is principally an alloy of *tombac* and silicon, hence the name. *Tombasil* has good strength and medium hardness, and is used for valve stems, pump impellers, brush holders, propellers, shafts, gears, bearings and structural castings.

tomichite. A black mineral composed of iron vanadium titanium arsenic oxide hydroxide, $(V,Fe)_4Ti_3AsO_{13}(OH)$. Crystal system, monoclinic. Density, 4.16 g/cm^3. Occurrence: Western Australia.

Tonac. Trade name of Tonfer Plastics Industrial Company Limited for polymer alloys of acrylonitrile-butadiene-styrene (ABS) and polycarbonate (PC).

Toncan Iron. Trade name of Republic Steel Corporation (USA) for a copper-molybdenum iron containing 0.03% carbon, 0.12% manganese, 0.035% sulfur, 0.01% phosphorus, 0.07% molybdenum, 0.005% silicon, 0.45% copper, and the balance iron. It has good resistance to atmospheric corrosion and good strength, and is used for roofing, siding and culvert pipes.

Tone. Trade name of Union Carbide Corporation for a series of poly(ε-caprolactone) based biopolymers.

Tonen. Trade name of ExxonMobil Chemical Corporation (USA) for an extensive series of synthetic resins.

Tonneau-Tex. Trademark of Bruin Plastics (USA) for a washable, shrink-resistant vinyl/polyester composite fabric with outstanding tensile and tear strength, used for seamless tonneau covers.

Tonox. Trademark of Uniroyal Inc. (USA) for a series of filament-winding epoxy resins composed of varying amounts of *m*-phenylenediamine and 4,4-methylenedianiline.

Tonum. Trade name of Stone Manganese–J. Stone & Company Limited (UK) for a strong, corrosion-resistant copper-base casting alloy containing 35% zinc, 1.7% aluminum, 1.5% manganese and 1% iron. Used for machinery parts, fittings, and marine parts.

Toolcraft. Trade name of E.M.F. Electric Company Limited (Australia) for a wear-resistant steel containing 1% carbon, 1% chromium, and the balance iron. Used for hardfacing electrodes and cutting tools.

Tool Cote. Trade name of Hapco, Inc. (USA) for cast epoxy/urethane composite surface coatings.

tooling resin. A synthetic resin, such as epoxy or silicone, used as tooling or tooling aid, e.g., in the form of foundry cores, core boxes, prototypes, etc.

tool materials. A group of tough, wear-resistant engineering materials used to make tools for cutting, forming or otherwise shaping a material into a part. Included are materials such as tool steels and certain grades of high-alloy, powder-metallurgy and maraging steels, coated and uncoated cemented carbides, cast cobalt-chromium-tungsten-niobium-carbon alloys, oxide ceramics, polycrystalline cubic boron nitride, and diamond.

Tool-N-Die. Trade name of Hobart Welding Products (USA) for a series of tool and die steels supplied in the form of welding electrodes.

Toolrite. Trademark of Alcoa-Aluminum Company of America (USA) for AA 2011 Series wrought aluminum-copper alloys available in various tempers (e.g., T3, T4 and T8). They exhibit good machinability and mechanical properties, excellent surface finish capabilities; fair corrosion resistance and weldability, and have a density of approximately 2.8 g/cm^3 (0.10 lb/in.^{3}), and a hardness of 80 Brinell. Used for adapters, machine parts, spindles, nozzles and hose components, pipe stems and filters, oil line filters, carburetor parts, camera and clock parts, meter shafts, and radio and television components.

tool steels. A group of steels having either high percentages of carbon, high alloy contents, or combinations of both. The typical carbon content is between 0.2-1.6%, and major alloying elements are carbide stabilizers, such as tungsten, molybdenum, chromium, vanadium and manganese. *Tool steels* are usually grouped into seven categories: (i) High-speed steels; (ii) Hot-work steels; (iii) Cold-work steels; (iv) Shock-resisting steels; (v) Mold steels; (vi) Special-purpose steels; and (vii) Water-hardening steels. They possess high toughness and hardness, excellent abrasion and wear resistance, high resistance to softening at elevated temperature, high tempering temperatures, and good to excellent retention of keen cutting edges. Used for hand, cutting, forming and shaping tools, dies, cutlery, and many other products.

Tool Vac. Trade name for a powder-metallurgy steel used for the manufacture of dies and molds. Its porous microstructure allows air and gas to be exhausted through the mold material which results in higher molding temperatures and shorter molding cycles.

Topacal C-5. Trademark of NSI Dental Pty. Limited (Australia) for a thixotropic, creamy tooth surface coating containing Phoscal, a phosphoprotein/calcium phosphate complex.

Topal. Trade name of Allgemeines Deutsches Metallwerk GmbH (Germany) for a series of alloys containing 80-90% copper, the balance being aluminum, iron and manganese. Used for especially shafts, gears, bearings, and dies.

Topalloy. Trade name of ExxonMobil Chemical Corporation (USA) for an extensive series of polymer alloys.

Topas. Trade name of Ticona GmbH (Germany) for a cyclooelefin copolymer (COC).

Topaz. (1) Trade name of J.F. Jelenko & Company (USA) for a hard low-gold dental casting alloy (ADA type III) providing a brilliant gold color, high tarnish resistance, and a high lustrous polish. Used for crowns, fixed bridgework, and hard inlays.

(2) Trade name of Topaz Technologies, Inc. (USA) for a light-cure microfine composite used for dental restorations.

topaz. Colorless, white, pale blue, pale to light yellow, pinkish, bluish, greenish or brownish mineral with a vitreous luster. It is composed of aluminum fluosilicate, $Al_2SiO_4(F,OH)_2$. Crystal system, orthorhombic. Density, 3.4-3.6 g/cm^3; hardness, 8 Mohs; refractive index, 1.612. Occurrence: Brazil, Germany, Norway, Japan, Mexico, Russia, USA (California, Colorado, Maine, Utah). Used as a substitute for, or in combination with *kyanite* in the production of *mullite*-type high-alumina refractories, and also as a gemstone.

Topcast. Trade name of Topcast Srl (Italy) for a low-gold dental casting alloy.

Topcraft. Trade name of Topcraft Corporation (USA) for a silver-free palladium dental bonding alloy.

Topeka. Trademark of Topeka Inc. (USA) for an *asphaltic concrete* containing up to about 25% stone aggregate.

Topel. Trade name for rayon fibers and yarns used for textile fabrics.

Topex. Trade name of Tong Yang Nylon for polybutylene terephthalate plastics.

Top Fix. Trade name of Ato Findley (France) for structural adhesives and sealants used for automotive, electrical, electronic and mechanical applications.

top-grain leather. The flexible, durable high-grade leather obtained from the outside or top layer of cattlehide by removing the hair and associated epidermis only. It does not need embossing, can be used natural, and is highly valued for shoes, clothing and various other applications.

Tophal. Trade name of Gilby-Fodor SA (France) for a series of electrical resistance alloys containing 12-35% chromium, 3-5% aluminum and 60-85% iron. Used for heating elements and electric furnaces.

Tophel. Trademark of Wilbur B. Driver Company (USA) for a thermocouple alloy composed of 90% nickel and 10% chromium. It has a density of 8.63 g/cm^3 (0.31 $lb/in.^3$), a melting point of approximately 1350°C (2460°F), a maximum service temperature of 1260°C (2300°F), high electrical resistivity, good oxidation resistance and stability, and good tensile strength at room temperature. Used as the positive leg of standard type K thermocouples.

Tophet. Trademark of Wilbur B. Driver Company (USA) for a series of electrical resistance and heating alloys containing nickel and chromium, nickel, iron and chromium, or nickel, chromium and aluminum. Also included under this trademark are several austenitic stainless steels, and iron-nickel-chromium superalloys. *Tophet A* is a heat and oxidation-resistant alloy containing 80% nickel and 20% chromium. Supplied in the form of ingots, forgings, strands, strips, ribbons, sheets, foil, wires, powder and tubes, it has high electrical resistivity, good resistance to mine water, seawater and moist sulfurous atmospheres, and good oxidation resistance at red-heat. Used for electric resistances and heating elements.

Topilene. Trade name of Hyosung Corporation for a series of polypropylenes.

Toplex. Trade name of Multibase Inc. (USA) for a series of polycarbonate/acrylonitrile-butadiene-styrene (PC/ABS) polymer alloys with a broad range of physical properties used for automotive components and appliance parts.

Topliner. Trade name of Manville Corporation (USA) for preformed fiberglass ceiling panel for automobiles.

Toplon. Trademark of Tongyang Nylon Company (South Korea) for nylon 6, polyester and spandex fibers and yarns used for elastic textile fabrics.

Top'n Bond. Trade name of King Packaged Products Company (Canada) for a self-bonding cement for concrete repair and resurfacing applications.

Top Notch. Trade name of Jessop Steel Company (USA) for a shock-resisting tool steel (AISI type S1) containing 0.5% carbon, 2.5% tungsten, 1.25% chromium, 0.25% vanadium, and the balance iron. It has a great depth of hardening, high hardness and strength, excellent toughness, good wear and abrasion resistance, and good hot hardness. Used for chisels, tools, and punches.

Toporex. Trade name of Mitsui Chemicals Corporation (Japan) for a series of polystyrenes.

Topstone. Trade name for a die stone used in dentistry.

Toramomen. Trademark of Toray Industries, Inc. (Japan) for rayon fibers and yarns used for textile fabrics.

Toranil. Trademark of Lake States Chemical Division of St. Regis Paper Company (USA) for calcium lignosulfonates made from desugared and desulfurized coniferous wood extracts. Supplied as brown viscous solutions and tan powders, they are used in refractory materials and adhesives, as binders, in boiler water treatment, and in concrete and leather manufacture. See also lignosulfonate.

Toray. Trademark of Toray Industries, Inc. (Japan) for tough, heat-resistant high-strength pitch carbon reinforcing fibers used for epoxy resin matrices. Also included under this trademark are strong polyamide 6 (nylon6) staple fibers, monofilaments and filament yarns, and polyester staple fibers and filament yarns.

Torayca. Trademark of Toray Industries, Inc. (Japan) for a carbon-fiber yarn consisting of thousands of filaments, and used as reinforcements in plastics, and bridges and similar concrete structures.

Toraycon. Trademark of Toray Industries, Inc. (Japan) for a series of engineering plastics based on polybutylene terephthalate (PBT) resins.

Torayfan. Trademark of Toray Industries, Inc. (Japan) for oriented polypropylene (OPP) film products with excellent barrier properties used for food packaging applications.

Toraylon. Trademark of Toray Industries, Inc. (Japan) for acrylic staple fibers used for the manufacture of textile fabrics.

Toray Nylon. Trademark of Toray Industries, Inc. (Japan) for nylon 6 staple fibers, monofilaments and filament yarns used for the manufacture of textile fabrics.

Toraypef. Trademark of Toray Industries, Inc. (Japan) for a series of polyethylene resins and foams.

Toray Tetoron. See Tetoron.

torbernite. A green, strongly radioactive mineral of the autunite group composed of copper uranyl phosphate hydrate, $Cu(UO_2)_2(PO_4)_2 \cdot xH_2O$. It can also be made synthetically. Crystal system, tetragonal. Density, 3.22 g/cm^3; refractive index, 1.592. Occurrence: USA. Also known as *chalcolite; copper uranite.*

torchon. See torchon lace.

torchon lace. (1) A hand-made linen lace with loosely twisted threads in simple open design. Also known as *torchon.*

(2) A strong, low-cost, machine-made cotton or linen lace, frequently with scalloped edges, produced in coarse threads in simple patterns on a mesh ground. Also known as *torchon.*

Tordal. Trade name of VDM Nickel-Technologie AG (Germany) for free-cutting aluminum alloys containing 4% copper, 0.5% lead and 0.5% bismuth. Used for screw-machine products, machinery parts and fasteners.

Torelina. Trademark of Toray Industries, Inc. (Japan) for a series of engineering plastics based on polyphenylene sulfide (PPS) resins.

Toris. Trade name for rayon fibers and yarns used for textile fabrics.

Torkret. Trademark of Torkret GmbH (Germany) for air-blown concrete, mortar and plaster used for finishing walls.

Torlen. Trademark of Chemitex-Elana (Poland) for a polyester filament yarn used for textile fabrics.

Torlon. Trademark of Amoco Performance Products Inc. (USA) for a series of extruded, injection-molded or compression-molded engineering thermoplastics based on polyamide-imide (PAI) and available in unmodified, fiber-reinforced and graphite- or glass-filled grades. They have a density of 1.4 g/cm^3 (0.05 lb/in.^{3}), excellent tensile and flexural strength properties, exceptional dimensional stability, good elevated and low temperature performance (from cryogenic to 260°C or 500°F), a glass-transition temperature of 335°C (638°F), good thermal stability, good machinability, low coefficients of friction, low wear factors, good self-lubricating properties, excellent galling resistance, good resistance to alpha, beta, gamma and ultraviolet radiation, good resistance to acids, alcohols, aromatic hydrocarbons, greases, oils and ketones, good resistance to automotive and aviation fluids, poor resistance to alkalies, inherent flame resistance, and good dielectric properties. Used for machine elements (e.g., thrust washers, ball and roller bearings, bearing balls, balls for check valves, etc.), fasteners, connectors, and housings, for the aerospace, automotive and electronics industries, electronic and electrical components, and as a thermoplastic matrix resin for advanced composites.

Tormol. Trade name of British Steel Corporation (UK) for a steel containing 0.3% carbon, 2.5% nickel, 0.7% chromium, 0.5% molybdenum, and the balance iron. Used for axles, shafts, gears, crankshafts, and connecting rods.

Tornesit. Trade name of Hercules Company (USA) for *chlorinated rubber.*

Toroglas. Trade name of B/L Systems (USA) for chemically strengthened glass.

Torolac. Trademark of Toray Industries, Inc. (Japan) for a series of engineering plastics based on acrylonitrile-butadiene-styrene (ABS) resins.

Toronto. Trade name of Perma Paving Stone Company (Canada) for interlocking paving stone supplied in standard, full-edge and half-edge designs in natural, red, salmon and dark brown colors, and used for walkways and driveways.

Torpedo. (1) Trade name of Lehigh Steel Corporation (USA) for an oil-hardening, nondeforming tool steel containing 0.9% carbon, 0.5% chromium, 1.3% manganese, 0.5% tungsten and the balance iron. Used for punches, rivet sets and dies.

(2) Trade name for lead-base bearing alloys.

torpedo bronze. A British corrosion-resistant silicon bronze containing 59-62% copper, 36.5-40.5% zinc and 0.5-1.5% silicon. Used for torpedo parts.

torque yarn. A yarn that, when hanging freely, tends to rotate or twist. The opposite of a non-torque yarn.

Torradal. Trade name of Vereinigte Metallwerke Ranshofen-Berndorf (Austria) for heat-treatable, free-cutting aluminum alloy composed of 0.6-1.4% magnesium, 0.6-1.6% silicon, 0.6-1.0% manganese, 0-0.3% chromium, and a total of 0.5-2.5% bismuth, cadmium, lead and tin. Used for screw-machine products.

Torradur. Trade name of Vereinigte Metallwerke Ranshofen-Berndorf (Austria) for a heat-treatable, free-cutting aluminum alloy composed of 3.5-5.0% copper, 0.4-1.8% magnesium, 0.5-1% manganese, and a total of 1.0-3.0% antimony, bismuth, cadmium, lead and tin. Used for screw-machine products.

torreyite. A white to colorless mineral composed of manganese arsenate silicate hydroxide octahydrate, $(Mg,Mn)_9Zn_4(SO_4)_2(OH)_{22} \cdot 8H_2O$. Crystal system, monoclinic. Density, 2.66 g/cm^3; refractive index, 1.584. Occurrence: USA (New Jersey).

tortoise shell. A natural plastic obtained from the horny, organic material (carapace) covering the backs of certain species of marine turtles formerly including the now endangered and protected hawksbill. *Tortoise shell* is essentially composed of *keratin,* can be pressed, stamped or otherwise worked or formed in the heated condition, and takes a high polish. Used for decorative and ornamental applications.

tosudite. A dark blue to azure-blue mineral of the mixed-layer group composed of sodium aluminum silicate hydroxide hydrate, $(Na,K)_xAl_6(Si,Al)_8O_{20}(OH)_{10} \cdot xH_2O$. Crystal system, orthorhombic. Occurrence: Japan.

Total. Trade name for a plasticized acrylic resin used for denture lining.

TotalBond. Trademark of Parkell, Inc. (USA) for a light- and self-cure, radiopaque adhesive dental resin cement toughened with glass, silica and a special trimethylolpropane trimethacrylate (TMPT) based organic filler. It provides high wear resistance and bonds well to all fixed posterior and anterior restorations including crowns and bridges.

Total Comfort. See Anso Total Comfort.

Total Vision. Trademark for invisible polymer coatings.

Totarn. Trade name for rayon fibers and yarns used for textile fabrics.

Toucas. Trade name for a corrosion-resistant alloy composed of 36% copper, 29% nickel, 7% zinc, 7% tin, 7% lead, 7% iron and 7% antimony. Used for ornamental white metal parts.

Touch & Bond. Trade name of Parkell, Inc. (USA) for a light-curing, self-etching and self-priming bonding agent consisting of the adhesive monomer 4-methacryloxyethyl trimellitate anhydride (4-META). It forms a strong, durable bond between composite restorations and tooth structure (i.e., dentin and enamel).

Tough. Trade name of Electro-Steel Company (USA) for a water and oil-hardening steel composed of 0.6% carbon, and the balance iron. Used for tools and dies.

tough alumina. A regular aluminum oxide (Al_2O_3) of blocky shape having a purity of 90-96%.

tough cake. See cake copper.

toughened glass. See tempered glass.

toughened nylon. Nylon, usually of the 6,6 type, whose toughness (or impact strength) has been greatly improved by blending with one or more elastomeric modifiers or with large quantities of a tough thermoplastic resins, such as acrylonitrile-butadiene-styrene or polyphenylene ether. Izod impact strengths of more than 900 J/m (16.9 ft·lbf/in.) can be achieved. Used for sporting goods, such as bicycle wheels, ski boots, rackets and supports for ice and roller skates, consumer products, toys, stone shields, trim clips, and electrical appliances. Abbreviation: TN. Also known as *toughened polyamide*. See also nylon; nylon 6,6.

toughened polyamide. See toughened nylon.

toughened polystyrene. See high-impact polystyrene.

ToughMet. Trademark of Brush Wellman Corporation (USA) for a series of high-performance spinodal hardened copper alloys containing 15% nickel and 8% tin made by Brush Wellman's *EquaCast*™ casting process that produces semi-wrought microstructures, uniform composition and high strength. Other important properties include good corrosion resistance, magnetic properties and bearing durability. The properties can be tailored to particular requirements. Used for severe service applications.

tough pitch copper. Commercially pure copper (99.88+%) whose oxygen content has been lowered to about 0.02-0.04%, either by poling or "pitching" a bath of molten anode copper (a traditional refining technique involving the thrusting of green-wood poles into the melt) or by deoxidizing with hydrocarbons. It has high ductility, high electrical conductivity (100% IACS and above), good corrosion resistance, excellent hot and cold workability and good forgeability. Used for electrical conductors, busbars, electronic components, radio parts, roofing, fasteners, architectural applications, etc. See also fire-refined tough pitch copper; electrolytic tough pitch copper.

ToughRock. Trademark of Georgia-Pacific (USA) for strong, tough, bright-colored gypsum board supplied in a wide range of grades for different applications including standard and flexible wallboard, abuse- or moisture-resistant board, sound-deadening board, ceiling, soffit or veneer board, manufactured housing board, paper-faced sheathing or shaftliners, and area separation board.

tourmaline. A mineral with vitreous to resinous luster that is a complex silicate of aluminum and boron, $(Na,Ca)(Al,Fe,Li,Mg)_3Al_6(BO_3)_3Si_6O_{18}(OH)_4$. It is usually black, brown, blue or green in color, but also occurs in red, yellow, colorless or white varieties. Crystal system, rhombohedral (hexagonal). Density, 2.9-3.4 g/cm³; hardness, 7.0-7.5 Mohs; double refraction properties; piezoelectric and pyroelectric properties. Occurrence: Bolivia, Brazil, Europe, Malagasy, Mexico, Russia, Sri Lanka, USA (California, Connecticut, Maine, New York, Pennsylvania). Used for pressure gages, transducers, optical equipment and oscillator plates, in depth measurement (ships and submersibles) and as a source of boric acid. The transparent varieties are used as gemstones.

Tournay metal. A French brass composed of 82.5% copper and 17.5% zinc, and used for cheap jewelry and buttons.

Tourun Leonard's metal. A tough bearing bronze containing 90% tin and 10% copper.

Tovis. Trademark of Toho Rayon Company Limited (Japan) for viscose rayon staple fibers used for textile fabrics.

tow. (1) The coarse, broken waste fibers of flax, hemp, etc., processed for spinning.

(2) A textile fabric spun from these fibers (1).

TowFlex. Trademark of Applied Fiber Systems, Limited (USA) for a series of flexible, high-impact thermoplastic and thermoset prepregs made by the powder fusion-coating process. Thermoplastics suitable for this purpose include polyamides (e.g., nylon 6), polyetherimides, polyolefins, polyphenylene sulfides, polyetherimides, polyetherketones, polyetheretherketones and some polyesters. Suitable thermosets include bismaleimides, epoxies, polyamides and some unsaturated polyesters. The prepregs are available with 5-95% continuous carbon, aramid or glass fiber content as molded slabs and sheets, laminated sheets, drapable woven fabrics, flexible towpreg tape, unidirectional tape, molding compounds and braided sleeving. Used in the manufacture of composites.

TowFlex CCPP. Trademark of Applied Fiber Systems, Limited (USA) for continuous carbon fiber-reinforced thermoplastic propylenes.

Town Flower. Trademark of Tong-Hwa Synthetic Fiber Company (Taiwan) for polyacrylonitrile (PAN) fibers.

towpreg. See prepreg tow.

Toyo. Brand name of a high-purity tin (99.9+%) containing up to 0.04% lead, 0.03% copper and 0.03% arsenic.

Toyobo. Trademark of Toyobo Company Limited (Japan) for rayon fibers and yarns (*Toyobo Rayon*), polyester staple fibers and filament yarns (*Toyobo Ester*), and nylon 6 filament yarns (*Toyobo Nylon*) used for textile fabrics.

Toyobo Ester. Trademark of Toyoba Company Limited (Japan) for rayon fibers and yarns, polyester staple fibers and filament yarns, and nylon 6 filament yarns used for textile fabrics.

Toyobo HEIM. Trademark of Toyobo Company (Japan) for flame-retardant fibers with excellent self-extinguishing properties supplied in several grades including *Toyobo HEIM-C* for curtains and *Toyobo HEIM-H* for curtains, aircraft blankets, automobile upholstery and other textile goods.

Toyobo Nylon. Trademark of Toyoba Company Limited (Japan) for nylon 6 and nylon 6,6 plastics and filament yarns.

Toyoflon. Trademark of Toray Industries, Inc. (Japan) for fluoropolymer film products.

Toyo Flow. Trademark of Toyoba Company Limited (Japan) for polytetrafluoroethylene (PTFE) staple fibers.

Toyolac. Trademark of Toray Industries Inc. (Japan) for acrylonitrile-butadiene-styrene (ABS) terpolymers and polycarbonate/ABS alloys. The terpolymers are supplied in a wide range of grades including transparent, low-gloss, glass-fiber-reinforced, medium- or high-impact, fire-retardant, high-heat, structural-foam and plating.

Toyotenax. Trademark of Toyobo Company Limited (Japan) for rayon fibers and yarns used for textile fabrics.

T-Phos Copper. Trade name of Metals America (USA) for high-phosphorus certified copper used for plating anodes.

TPO-EPDM alloys. Polymer alloys of a semicrystalline thermoplastic olefin (TPO), such as polyethylene or polypropylene, and elastomeric ethylene-propylene diene monomer (EPDM). They are used for applications requiring the combination of the high impact resistance and good elongation and resiliency of EPDM elastomers with the good chemical resistance, strength and structural properties of polyolefins. Used for automotive bumper parts, tool housings, and sporting goods.

TPUR-ABS alloys. Polymer alloys of thermoplastic polyurethane (TPUR) and acrylonitrile-butadiene-styrene (ABS). They combine the high resiliency and impact resistance of elastomeric TPUR with the high rigidity and good high-temperature resistance of amorphous ABS. Used for automotive components, consumer electronics, housings, etc.

TPUR-PA alloys. Polymer alloys of thermoplastic polyurethane (TPUR) and polyamide (nylon). The elastomeric TPUR greatly improves the impact strength (toughness) of the semicrystalline polyamide (nylon). Used for automotive components, sporting goods and consumer products.

TPUR-PC alloys. Polymer alloys of thermoplastic polyurethane (TPUR) and polycarbonate (PC). The elastomeric TPUR adds resiliency and impact resistance, while the amorphous PC improves rigidity and high-temperature resistance. Used for automobile bumpers.

TPUR-PVC alloys. Polymer blends of thermoplastic polyurethane (TPUR) and polyvinyl chloride (PVC). The TPUR greatly improves the overall toughness, flexibility, and low-temperature characteristics of the PVC.

TPX. Trade name of Mitsui Chemicals, Inc. (Japan) for a polymethylpentene copolymer supplied in rod, sheet, film and granular form. It has a density of 0.835 g/cm³ (0.03 lb/in.³), high clarity and gloss, high hardness, good impact strength, fair resistance to ultraviolet light, good dielectric properties, excellent heat resistance, poor flame resistance, a useful service range of -30 to +115°C (-22 to +239°F), good resistance to acids, alcohols, alkalies, greases, oils and ketones, fair resistance to aromatic hydrocarbons and poor resistance to carbon tetrachloride and cyclohexane. Used for electronic equipment, light reflectors, hospital equipment, and laboratory ware and equipment.

Trace. Trade name of American Fibers & Yarns Company (USA) for polyolefin fibers supplied in several grades including *Trace* and *Trace TR*. Used for textile fabrics.

tracing paper. A thin, translucent or highly transparent paper, usually made from *vellum* or *parchment*, and used for tracing, reproducing and preparing drawings in pencil or ink. It is of high quality, available in rolls and sheets of standard size with a rag content as high as 100%, and may be specially treated to prevent cracking, yellowing, feathering, etc.

Tracolen. Trade name of Trading Company Rotterdam (Netherlands) for a series of polyolefin plastics including various grades of polyethylene and polypropylene.

traditional ceramics. A group of ceramic products including brick, china, porcelain, tile, glass and high-temperature ceramics for which clay is the primary raw material. Not included in this group are advanced and technical ceramics for aerospace, communication and electronic applications.

Traffic Control Fiber System. Trademark of E.I. DuPont de Nemours & Company (USA) for wear-resistant nylon 6,6 filaments and fibers used for textile fabrics including carpets.

traffic paint. A wear- and weather-resistant paint, often white or yellow in color and suitable for marking on paved roads. It usually contains brilliant, and sometimes luminous pigments, and is applied by brushing or spraying. Also known as *road-marking paint.*

Traffolyte. Trade name of Thomas De La Rue (UK) for urea-formaldehyde resins.

Trafoperm. Trade name of Vacuumschmelze GmbH (Germany) for a soft-magnetic iron alloy with preferred magnetic orientation containing up to 3% silicon. It has high saturation flux density, low losses, a Curie temperature of 750°C (1380°F), and is used for pole pieces, medium-frequency and pulse transformer, instrument transformers, welding transformers, chokes, relay components and measuring systems.

tragacanth. A mucilaginous gum obtained from certain shrubs (genus *Astragalus*) of the pea family found in Greece, Turkey, Asia Minor, Iran and southwestern Europe, and used as a binder in glaze and porcelain-enamel slips, as an adhesive to bond dry-process enamels to metals, as a mucilage, in leather dressing, and in textile printing and sizing.

Traliglas. Trade name of NV Syncoglas SA (Belgium) for a glass cloth used for the reinforcement of coating products.

Trance. Trade name of Seymour Products Company (USA) for a free-cutting phosphor bronze containing 90% copper, 5% lead, 5% tin, and a trace of phosphorus. Used for hardware.

Tran-Cor. Trade name of Qarmco International (USA) for a series of hot-rolled grain-oriented electrical steels containing 0-0.003% carbon, 0-4.4% silicon, and the balance iron. Usually supplied in strip form, they have high magnetic permeability, and are used for magnetic cores, transformers, magnetos, motors, generators, and armatures

Tranelec. Trade name of Empire Sheet & Tin Plate Company (USA) for a series of high-permeability iron-silicon alloys used for motors.

Tranquilite. Trade name of Safetee Glass Company Inc. (USA) for gray-colored laminated glass available in three different light transmissions.

tranquillityite. A deep red to black mineral composed of iron titanium zirconium silicate, $Fe_8Zr_2Ti_3Si_3O_{24}$. Crystal system, hexagonal. Density, 4.70 g/cm³; refractive index, 2.12. Occurrence: Sea of Tranquillity, Moon (collected during NASA Apollo Missions).

Transage. Trade name for a series of alpha-beta and beta titanium alloys that have been subjected to a three-stage heat-treating cycle involving annealing and/or aging at different temperature ranges. Used for structural applications and aerospace parts.

Trans-A-Therm. Trade name of Goldsmith Engineering & Chemical (USA) for heat-conductive putties.

Transbond. Trademark of 3M Dental (USA) for a bonding resin used in orthodontics.

transfer glass. Optical glass that after melting has been left in the same pot until cooled to room temperature.

transfer-molded plastics. Thermosetting plastics formed by first softening and melting thoroughly mixed resins under pressure in heated transfer chambers, and then injecting the resin mixture by high pressure through suitable sprues, runners and orifices into the cavities of closed molds where they are allowed to cure. See also molded plastics.

Transflow. Trademark of Norton Performance Plastics (USA) for fluoropolymer tubing available in natural and various colors for food-processing applications.

transformation-hardened steels. Steels with about 0.4-1.2% carbon that have been hardened by preferentially heating the surface to above the austenite transformation temperature (727°C or 1341°F) followed by rapid cooling in a suitable medium to form martensite. The heat source used is usually an intensive flame (flame hardening), electromagnetic induction (induction hardening), or a laser beam (laser-beam hardening).

transformation-toughened alumina. Alumina (Al_2O_3) with greatly increased fracture toughness due to the dispersion of small particles of partially stabilized zirconia (ZrO_2). Other ceramic oxides, such as lime (CaO) or magnesia (MgO), may also be used as stabilizers. Abbreviation: TTA. Also known as *dispersion-toughened alumina.* See also partially stabilized zirconia.

transformation-toughened zirconia. Zirconia (ZrO_2) whose fracture toughness has been enhanced by dispersing small particles of partially stabilized zirconia resulting in a mechanism known as *transformation toughening*, which involves the stress-induced phase transformation of the tetragonal zirconia grains to the monoclinic structure. Other ceramic oxides, such as lime (CaO) or magnesia (MgO), may also be used as stabilizers. Toughening can also result from the introduction of ceramic whiskers (e.g., silicon carbide). Abbreviation: TTZ. Also known as *dispersion-toughened zirconia*. See also partially stabilized zirconia; tetragonal zirconia.

transformer steels. Low-carbon steels (less than 0.1%) containing about 1-4% silicon in solid solution. They are hot-rolled at temperatures of about 200-300°C (390-570°F), and are commonly annealed and processed to provide a preferred grain orientation. Commercially available in the form of strip and sheet, they have ferritic microstructures, soft magnetic properties, high resistivity and permeability, and are used for transformer cores and dynamo and motor parts. Also known as *electrical transformer steels*.

Transglace. French trade name for polished plate glass.

Transil. Trade name of British Steel plc (UK) for a series of hot-rolled transformer steels containing 0-0.1% carbon, 3.2-4.2% silicon, and the balance iron. Supplied in sheet and strip form.

Transite. Trade name of BNZ Materials, Inc. (USA) for boards and pipes molded either from asbestos fibers and Portland cement by a high-pressure process, or from non-asbestos calcium silicate. They have high structural strength at elevated temperatures, excellent thermal insulating properties and thermal-shock resistance, good fire resistance to about 1090°C (2000°F), and are used for fire-resistant walls, roofing, and structural insulation.

transition elements. The chemical elements in Groups IB, IIB, IIIB, IVB, VB, VIB, VIIB and VIII (Groups 3 through 12) of the Periodic Table including elements 21 through 30 (scandium through zinc), 39 through 48 (yttrium through cadmium), 57 through 80 (lanthanum through mercury excluding the lanthanides) and all known elements from 89 (actinium) on. They are characterized by partially filled *d* or *f* electron states and, in some cases, one or two electrons in the next higher energy shell. All are physically and chemically similar, i.e., exclusively metallic in character and tend to become positively charged. Also known as *transition metals*. See also non-transition elements.

transition-metal complex. See coordination compound.

transition-metal compound. A chemical compound in which one element is a transition metal and the other either a transition metal, or an element of Groups IA, IIA, IIIA, IVA, VA, VIA, VIIA (Groups 1, 2, and 13 through 17) of the Periodic Table.

transition-metal oxide. A chemical compound in which one element is oxygen and the other a transition metal.

transition metals. See transition elements.

transit-mixed concrete. See truck-mixed concrete.

translucent concrete. A concrete-glass combination used in thin, rectangular precast or prestressed concrete members.

translucent glass. Glass that transmits light with varying degrees of diffusion and impedes or obscures vision such that objects seen through it are not clearly distinguishable. Also known as *obscure glass*.

translucent materials. Materials through which light is transmitted diffusely to the degree that objects viewed through a specimen of the material are not clearly distinguishable.

translucent paper. (1) A semitransparent paper, usually made from highly refined wood pulp, whose transparency must be such that writing on the back of the paper is legible from the front. The smoothness of this paper is especially uniform and lower than that of tracing papers. See also tracing paper.

(2) A semitransparent paper through which light is transmitted diffusely, i.e., an object can only be clearly distinguished through it when in immediate contact.

translucent tracing paper. A collective term for drawing and writing papers that have been made translucent during manufacture, or by subsequent refining procedures (e.g., oiling).

transmutation glaze. A glaze or glass whose color has been changed by the addition of another colorant or impurity into the batch, or by melting in a crucible in which a glaze or glass of a different color previously has been melted.

Transpan. Trade name of Virginia Glass Products Corporation (USA) for toughened polished plate glass having precision patterns of ceramic colors silk-screened to the back.

Transparent. Trade name of Haeckerstahl GmbH (Germany) for a series of stainless steels containing 0.2-0.4% carbon, 0.3% manganese, 0-0.4% silicon, 13% chromium, and the balance iron. Used for cutlery, surgical and dental equipment, turbine blades and valves.

transparent coating. A clear, colorless or tinted porcelain enamel, glaze or other coating through which the substrate can be seen.

transparent conducting oxides. A class of transparent oxide materials that are thermally reflecting and/or non-iridescent, usually prepared by chemical vapor deposition from organometallic precursors in the form of submicron thin films. Examples include certain n- and p-type conductors including silicon dioxide, strontium copper oxide ($SrCu_2O_2$) and fluorine-doped tin oxide. They are suitable for use on glass for the manufacture of energy-efficient, low-emissivity windows. Abbreviation: TCOs.

transparent glass. Glass, such as ordinary window glass, which transmits a high percentage of the incident light with relatively little absorption, reflection and scattering, and through which objects can be readily seen. Also known as *high-transmission glass*.

transparent glaze. A colored or colorless glaze applied to the surface of a ceramic product. It transmits incident light with relatively little absorption, reflection and scattering and thus the ceramic product can be readily seen through the glaze. See also opaque glaze; semitransparent glaze.

transparent materials. Materials that are capable of transmitting light with relatively little absorption and reflection.

transparent nylon. A transparent type of nylon obtained from a mixture of adipic and azelaic acids and 4,4'-diisocyanatodiphenylmethane or methane diphenyl diisocyanate. Used in electrical connectors.

Transparite. Trade name for transparent sheeting produced in a multistage process (viscose process) by treating cellulose with sodium hydroxide and carbon disulfide, and adding plasticizers and sometimes a waterproof coating.

Transparoi. Trade name of Boussois Souchon Neuvesel (France) for translucent walls incorporating glass bricks or figured glass.

Transpex 2. Trade name of ICI Limited (UK) transparent polystyrenes.

transportation fabrics. Abrasion-, light-, soil- and flame-resistant textile fabrics used exclusively for seat covers in airplanes, buses and trains, and other transportation applications.

Transtherm. Trademark of Insulating Materials Inc. (USA) for thermal management products used for electrical insulating applications, and as interfaces between heat sinks and heat generating equipment. *Transtherm T*, supplied in several grades, refers to silicone elastomers available in fiberglass- and aluminum foil-reinforced grades filled with alumina and/or boron nitride, and *Transtherm TP* are fiberglass-reinforced, alumina-filled polyesters. *Transtherm* products have a density of 2.1 g/cm^3 (0.08 lb/in.^{3}), and continuous service temperature of 180°C (356°F) for the silicones and 150°C (300°F) for the polyesters.

Transtinyl. Trade name of Youngstown Alloy Castings Company (USA) for an abrasion-resisting alloy steel containing 0.7% carbon, 1% chromium, and the balance iron. Used for tools.

Transulite. Trade name of Manville Corporation (USA) for fiberglass thermal insulation.

transuranic elements. The group of chemical elements having atomic numbers greater than that of uranium (92) and including neptunium (93), plutonium (94), americium (95), curium (96), berkelium (97), californium (98), einsteinium (99), fermium (100), mendelevium (101), nobelium (102), lawrencium (103), rutherfordium (also known as element 104, kurchatovium or unnilquadium), hahnium (also known as element 105, nielsbohrium and unnilpentium), and elements 106 (also known as unnilhexium or seaborgium), 107 (unnilseptium), 108 (unniloctium), 109 (unnilennium), 110 (unununnillium) and beyond. All known transuranic elements are radioactive, produced artificially by nuclear bombardment, and are members of the actinide series. Also known as *transuranics; transuranium elements.*

transuranic metals. See transuranic elements.

transuranics. See transuranic elements.

transuranium elements. See transuranic elements.

transuranium metals. See transuranic elements.

trap. See traprock.

trap rock. A dense, partially altered, usually fine-grained, dark-colored volcanic rock, such as basalt, diabase, diorite or gabbro. It has an average weight of 2995 kg/m^3 (187 lb/ft^3), typical average ultimate tensile and compressive strengths of 5.5 MPa (0.8 ksi) and 138 MPa (20 ksi), respectively, and is often crushed into pieces for use as railroad ballast, or in road construction. Also known as *trap*.

traskite. A brownish red mineral of the combeite group composed of barium iron titanium silicate hydroxide hexahydrate, $Ba_9Fe_2Ti_2Si_{12}O_{36}(OH)_6·6H_2O$. Crystal system, hexagonal. Density, 3.71-3.76 g/cm^3; refractive index, 1.714. Occurrence: USA (California).

Travema. Trade name for rayon fibers and yarns used for textile fabrics.

travertine. A white, dense, partly crystalline form of *calcite* (calcium carbonate, $CaCO_3$) with a porous, layered structure, deposited by hot and cold spring waters. Occurrence: Italy; USA (Georgia, Montana, California). It is highly valued as a hard building material.

Travis. (1) Trade name for a nytril fiber, i.e., a fiber based essentially on a vinylidene dinitrile polymer. It has good mildew and outdoor weathering resistance, good heat resistance, good to excellent resistance to acids, fair resistance to cold dilute alkalies, and poor resistance to strong alkalies. Used for textile fabrics.

(2) Trade name for rayon fibers and yarns.

Traycoat. Trademark of ZYP Coatings Inc. (USA) for paintable barrier-separation coatings for application to graphite and refractory metal trays employed in powder-metallurgy (P/M) vacuum sintering. They are supplied in several compositions including *Traycoat-A*, an aluminum oxide-based coating suit-

able for use with high-cobalt alloys, and refractory and reactive metals, and *Traycoat-T*, a titanium nitride-based coating designed for use with rare-earth alloys, and refractory and reactive metals.

TrayDoh. Trade name for a pasty thermoplastic resin used in dentistry.

treble cloth. See triple cloth.

treasurite. A mineral of the lillianite group composed of silver lead bismuth sulfide, $Ag_7Pb_6Bi_{15}S_{12}$. Crystal system, monoclinic. Density, 7.06 g/cm^3. Occurrence: USA (Colorado).

treated fabrics. Textile fabrics coated, impregnated or otherwise finished to enhance one or more chemical or physical properties, such as strech, water repellency, crease resistance, and resistance to soil, insects, fungi and rot.

treated iron. A partially or fully inoculated cast iron in the molten condition containing all principal alloying elements and nodulizing alloys.

Treble Extra. Trade name of Firth Brown Limited (UK) for a series of plain-carbon tool steels containing 0.7-1.4% carbon, and the balance iron.

trechmannite. A red mineral composed of silver arsenic sulfide, $AgAsS_2$. Crystal system, rhombohedral (hexagonal). Density, 4.78 g/cm^3; refractive index, 2.58. Occurrence: Switzerland.

Trefsin. Trademark of Advanced Elastomer Systems (USA) for a series of black thermoplastic elastomers with excellent permeation resistance to water, air and other fluids and gases, good heat and weathering resistance, good resistance to many chemicals, a hardness of 65 Shore A, and a processing temperature of 200°C (390°F). They can be processed by blow and injection molding and extrusion, and are used for chemical equipment, linings, and outdoor tanks.

Trellis. Trade name of C-E Glass (USA) for wired glass with narrow rectangular mesh.

Trelon. Trademark for very durable nylon fibers.

Trem Bronze. Trade name of Greenleaf Corporation (USA) for a tough, corrosion-resistant silicon bronze composed of 98% copper and 2% tin, and used for springs, clips, hardware, and fasteners.

Tremclad. Trade name of Tremco Inc. (USA) for a series of rust protection products including various rust paints supplied in a wide range of colors in can and aerosol form, high-head enamel paints in aerosol cans, primers for protecting galvanized metal and zinc-coated steel, and rust-preventive red oxide primers for iron and steel parts.

tremolite. A white, gray or light green mineral with vitreous to silky luster. It belongs to the amphibole asbestos group and is composed of calcium magnesium silicate hydroxide, $Ca_2Mg_5Si_8O_{22}(OH)_2$, usually containing iron, sodium and/or aluminum. It occurs in prismatic crystals, bladed aggregates or silky fibers, and can also be made synthetically. Crystal system, monoclinic. Density, 2.9-3.3 g/cm^3; hardness, 5-6 Mohs; refractive index, 1.613; good resistance to acids and high temperatures. Occurrence: Europe (Switzerland), South Africa, USA (California, Maryland, New York). Used for acid-resisting applications, ceramics and paints.

Trempex. Trade name of Trempex AB (Sweden) for toughened glass.

Trentino. Trademark of Aquafil SpA (Italy) for polypropylene fibers and filament yarns.

Treopax. Trademark of National Lead Company (USA) for a ceramic material containing more than 90% zirconium dioxide and up to 5% silicon dioxide. It has a melting point of 2480°C

(4496°F), and is used for welding-rod coatings, and as a color stabilizer in titanium enamels.

Trespaphan. Trade name of Dor Moplefan (Italy) for an oriented polypropylene film.

Tretex Protektor. Trade name of Solutions Globales (France) for a flame-resistant textile fabric based on *Trevira CS* polyester fibers and also containing antistatic fibers. Used in the automotive coating and painting industries.

Trevalon. Trademark of Dentsply International (USA) for a heat-cure acrylic resin used for dentures. *Trevalon Hi* is a high-impact grade.

Trevira. Trademark of Hoechst Celanese Corporation (USA) for polyester staple fibers. monofilaments and filament yarns based on polyethylene terephthalate (PET) that has been made flame-resistant by permanently incorporating a flame retardant into the polymer structure. Although it will melt in the presence of a flame, it will extinguish itself after the flame is removed. Used for clothing and sewing yarn and thread.

trevorite. (1) A black opaque mineral of the spinel group composed of iron nickel oxide, $(Ni,Fe)Fe_2O_4$. Crystal system, cubic. Density, 5.21 g/cm³. Occurrence: South Africa.

(2) A black opaque mineral of the spinel group composed of nickel iron oxide, $NiFe_2O_4$. It can also be made synthetically. Crystal system, cubic. Density, 5.16 g/cm³; refractive index, 2.3. Occurrence: Southern Africa.

triacetate. See triacetate fiber.

triacetate fiber. A generic name for a strong, flexible synthetic fiber obtained from cellulose acetate (cellulose ethanoate) in which at least 92% of the hydroxyl groups are acetylated (ethanoylated), i.e., have been reacted with a mixture containing acetic anhydride. Also known as *triacetate*. See also acetate fiber.

Triad DuaLine. Trademark of Dentsply International (USA) for a dual-cure acrylic resin used for relining dentures.

Triad VLC. Trade name of Dentsply International (USA) for a visible-light-cure dental acrylic resin used for impression trays.

Triafil. Trademark of Bayer AG (Germany) for cellulose triacetate film used for electric insulation.

Triafol. Trademark of Bayer AG (Germany) for acetone and methylene chloride soluble acetobutyrate film used for replicating rough material surfaces.

Triage. Trade name of GC America Inc. (USA) for a self-bonding, chemical-curing, fluoride-releasing, radiopaque low-viscosity glass-ionomer resin belonging to the *Fuji* line of products. Used as a pit and fissure sealant, and for filling orthodontic access openings.

Trialbene. Trade name for cellulose triacetate fibers used for wearing apparel and industrial fabrics.

Tri-Alloy. Trade name of Ford Motor Company (USA) for a leaded copper containing 55.5-60.5% copper, 35-40% lead and 4.5% silver. Usually cast on a strip steel backing, they are used for heavy-duty high-speed engine and crankshaft bearings.

Trialloy. Trade name of Tri-Clover (USA) for a corrosion-resistant nickel alloy used for sanitary fittings and valves.

trialuminum disilicate. See aluminum silicate (1).

Triana. (1) Trade name of PPG Industries Inc. (USA) for fine-filament glass-fiber yarn.

(2) Trademark of Rhodia SA (France) for polyacrylonitrile (PAN) staple fibers.

Triangle. Trade name of Triangle Conduit & Cable Company (USA) for a series of coppers and copper-base alloys including various cadmium coppers, architectural bronzes, cartridge brasses, nickel silvers, and copper nickels.

triangulite. A bright yellow mineral of the meta-autunite group composed of aluminum uranyl phosphate hydroxide pentahydrate, $Al_3(UO_2)_4(PO_4)_4(OH)_2·5H_2O$. Crystal system, triclinic. Density, 3.70 g/cm³; refractive index, 1.665. Occurrence: Zaire.

Trianti. Trade name for glass fibers and glass-fiber products.

Triax. Trademark of Bayer Corporation (USA) for an extensive series of thermoplastic alloys including acrylonitrile-butadiene-styrene/polyamide (*Triax 1000* Series) and acrylonitrile-styrene/polycarbonate (*Triax 2000* Series) alloys, supplied in injection molding and extrusion grades, unreinforced for general purposes or glass-fiber-reinforced for special applications. They have very high impact strength, outstanding processibility, excellent flow characteristics, good chemical and heat resistance, good abrasion resistance, good surface finish and reduced moisture sensi-tivity. Used for lawn and garden equipment, power-tool housings, sporting goods, and automotive components. Also included under this trademark are polyamide/styrene-acrylonitrile (PA/SAN) alloys.

Triaxis Glass. Trademark of Houston Stained Glass Supply (USA) for a dichroic glass.

triazine resins. See cyanate resins.

Tribaloy. Trademark of Haynes International (USA) for a group of cobalt- and nickel-base alloys that are extremely wear-resistant, and have good resistance to impact, thermal shock, heat, oxidation and corrosion, and high hot-hardness. The cobalt-base alloys contain about 47-62% cobalt, 28-35% molybdenum, 8-18% chromium, 2-10% silicon, 0-3% nickel and iron, respectively, and up to 0.1% carbon, and the nickel-base alloys contain about 50% nickel, 32% molybdenum, 15% chromium, 3% silicon, up to 1% cobalt, up to 1% iron and 0.1% carbon. *Tribaloy* alloys are available as alloyed powders for plasma spraying or powder metallurgy, as hardfacing rods, and as remelt casting stock. Used for pumps, valves, pistons, vanes, ball and roller bearings, sleeve bearings, arbors, seals, components for chemical and marine equipment, and for nuclear plant equipment.

tribarium aluminate. See barium aluminate (i).

tribasic calcium phosphate. See calcium phosphate (iii).

tribasic zinc phosphate. See zinc phosphate.

tribasic zinc phosphate tetrahydrate. See zinc phosphate tetrahydrate.

Tribit. Trademark of Sam Yang Kasei (Japan) for polybutylene terephthalate (PBT) resins supplied in general-purpose, high-flow, fast cycle, flame-retardant and glass-fiber-reinforced grades.

TriBlend. Trademark of US Silica Company (USA) for a silica sand aggregate supplied in several blends and grades for industrial floor topping applications.

Triblue. Trade name of Atofina Chemicals, Inc. (USA) for chromium conversion coatings.

Tri-Blue. Trade name of Luster-On Products, Inc. (USA) for a protective coating used on zinc plates.

Tribocoat. Trademark of Mold-Tech, Division of Roehlen Industries (USA) for a composite nickel-polyfluorotetraethylene electroless coating with good dry lubricity and antifriction properties, good salt-spray corrosion resistance, an operating temperature range from -45 to +350°C (-50 to +670°F), and an as-deposited hardness of 36 Rockwell C.

Tribocor. Trademark of Fansteel Inc. (USA) for surface-nitrided niobium-titanium-tungsten alloys used for applications requiring simultaneous resistance to corrosion and wear.

Tribol. Trade name of Stapleton Technologies (USA) for wear-resistant electroless coatings.

Tribolite. Trade name of Wear Management Services Inc. (USA) for a series of antifriction alloys designed to resist and control wear, seizing, galling and scoring. Used for bearings.

tribological materials. A group of materials used to improve the antifriction, lubrication, antigalling, antiseizing, antiscoring and/or wear properties of a metal, alloy, polymer, ceramic or composite. Included are lubricants, certain polymers, such as polytetrafluoroethylene, antifriction metals and alloys, certain cobalt and nickel-base alloys, and wear-resistant materials and coatings.

triboluminescent material. A material that emits light upon mechanical stimulation, e.g., by friction produced by the material while sliding on or rubbing against another material.

triboron silicide. See boron silicide (iii).

Tribrite. Trade name of Accurate Engineering Laboratories, Division of Rin Inc. (USA) for blue-bright conversion coating for zinc products.

tributyl borate. A colorless liquid obtained from boric acid and *n*-butanol. It has a density of 0.855-0.857 g/cm³, a melting point of -70°C (-94°F), a boiling point of 230-235°C (446-455°F), a flash point of 200°F (93°C) and a refractive index of 1.1410. Used in organic and organometallic synthesis, as a welding flux, as an intermediate in borohydrate synthesis, and a flame retardant for textiles. Formula: $(C_4H_9)_3BO_3$. Also known as *butyl borate.*

tricalcium aluminate. See calcium aluminate (i).

tricalcium columbate. See calcium niobate (i).

tricalcium disilicate. See calcium silicate (iii).

tricalcium magnesium disilicate. See calcium magnesium silicate (iv).

tricalcium niobate. See calcium niobate (i).

tricalcium orthophosphate. See calcium phosphate (iii).

tricalcium pentaaluminate. See calcium aluminate (iv).

tricalcium phosphate. See calcium phosphate (iii).

tricalcium silicate. See calcium silicate (iv).

Tricel. Trademark of Courtaulds Limited (UK) for hard-wearing cellulose triacetate fibers and filament yarns that blend well with other fibers, and resists wrinkling and soiling. Used for woven and knitted fabrics.

Tricelon. Trade name for a yarn blend of *Tricel* and nylon used for knitting or weaving soft, lightweight fabrics with excellent drape, e.g., blouses, dresses and lingerie.

Tricent. Trade name of International Nickel Inc. (USA) for a tough, shock-resistant alloy steel containing 0.43% carbon, 1.6% silicon, 0.8% manganese, 1.8% nickel, 0.85% chromium, 0.38% molybdenum, 0.08% vanadium, and the balance iron. Used for landing gears and aircraft structures.

trichochromes. See pheochromes.

trichosiderins. See pheochromes.

Tri-Chrome. Trade name of Atofina Chemicals, Inc. (USA) for chromium electroplates and plating processes.

Tri-Chrome Plus. Trade name of Atofina Chemicals, Inc. (USA) for trivalent chromium electroplates and plating processes.

trichromium diboride. See chromium boride (v).

trichromium disilicide. See chromium silicide (iv).

trichromium phosphide. See chromium phosphide (iii).

trichromium silicide. See chromium silicide (iii).

Tri-Clover. Trade name of Tri-Clover (USA) for a water-hardening machinery steel containing 0.3-0.5% carbon, and the balance iron.

tricobalt boride. See cobalt boride (iv).

tricobalt tetroxide. See cobaltocobaltic oxide.

tricolumbium aluminide. See niobium aluminide (iii).

tricolumbium diboride. See niobium boride (iv).

tricolumbium digermanide. See niobium germanide (i).

tricolumbium disilicide. See niobium silicide (iii).

tricolumbium germanide. See niobium germanide (iii).

tricolumbium tetraboride. See niobium boride (iii).

tricomponent fiber. (1) A fiber consisting of three chemically and/or physically dissimilar polymers.

(2) A ceramic fiber consisting of three different elements, such as silicon, titanium and carbon, or silicon, zirconium and carbon.

Tri-Core. Trade name of Alpha Metals Inc. (USA) for a self-fluxing soft solder composed of 40-60% lead and the balance tin.

Tricosal. Trade name of Chemische Fabrik Gruenau GmbH (Germany) for protective coatings used on buildings.

tricot. A hand- or machine-knitted fabric of wool, cotton or synthetic (e.g., acetate, nylon, rayon or polyester) fibers. It is warp-knitted, i.e., with crosswise ribs on the back and thin ribs or wales on the face. Used for clothing.

tricotine. A fabric similar to *gabardine*, but woven with a characteristic steep double-twill weave. Used for dresses, trousers, and women's sportswear.

Trident. Trade name of Latrobe Steel Company (USA) for a cold-work tool steel containing 0.57% carbon, 0.65% silicon, 0.3% manganese, 0.6% chromium, 0.35% molybdenum, 0.2% vanadium, and the balance iron. Used for chisels, mandrels, punches, and dies.

Tridescent. Trade name of Luster-On Products Inc. (USA) for clear, trivalent chromium conversion coatings used for zinc and zinc-alloy parts.

Tridip. Trade name of A Brite Company (USA) for trivalent chromate conversion coatings.

Tri Dip Yellow. Trade name of Alchem Corporation (USA) for a corrosion-resistant, iridescent yellow trivalent chromium coating used for zinc and most zinc alloys.

tridymite. A colorless or white mineral with a vitreous luster. It belongs to the nepheline group and is a crystalline polymorph of silica, SiO_2. It is stable only above 870°C (1598°F), and can also be made synthetically from quartz (SiO_2) and sodium tungstate (Na_2WO_4) by heating at 1300°C (2370°F) for about 24 hours. At high temperatures it has a hexagonal crystal structure which changes to orthorhombic upon recrystallization. The synthetic product has a monoclinic crystal structure. Density, 2.24-2.30 g/cm³; hardness, 7 Mohs; refractive index, 1.470. Used in ceramic bodies to improve thermal shock resistance and minimize crazing.

tridysprosium carbide. See dysprosium carbide (i).

trierbium carbide. See erbium carbide (i).

triethylgallium. An organometallic compound supplied in the form of a colorless pyrophoric liquid (99+% pure) and in high-purity (99.999%) electronic grades. It has a density of 1.059 g/cm³, a melting point, -82.3°C (-116.1°F), a boiling point of 143°C (289°F). Used in organometallic research, electronics and plating. Abbreviation: TEGa. Formula: $(C_2H_5)_3Ga$. Also known as *gallium triethyl.*

triethylindium. An organometallic compound supplied in the form of a colorless, air- and moisture sensitive, pyrophoric liquid (99+% pure) and high-purity (99.999%) electronic grades. It has a density of 1.260 g/cm³, a melting point, -32°C (-25.6°F), a boiling point of 184°C (363°F), and is used in organometallic research, electronics and plating. Abbreviation: TEIn. Formula: $(C_2H_5)_3In$. Also known as *indium triethyl*.

trifiber prints. Quick-drying, antistatic textile fabrics made from a blend of 60% triacetate and 40% polyester fibers. They uniquely combine the durability of synthetics with the softness of natural fibers.

Triflex. (1) Trade name of Triplex Safety Glass Company Limited (UK) for a composite windshield for pressurized aircraft that has a clear plastic interlayer extending beyond the edge of the assembly to be retained in the glazing frame.

(2) Trademark of Twinpak Inc. (Canada) for plastic-reinforced packaging material.

Trifoil. Trademark of Tri-Point Industries, Inc. (USA) for an aluminum foil or sheet having an adhesive coating applied to one side and a *Teflon* coating to the other. Used for chemical and food-processing plant, conveyors, etc.

trigadolinium carbide. See gadolinium carbide (i).

triglyceride. An ester of glycerol and fatty acids in which three of the hydroxyl groups of the glycerol have been replaced by identical or different acid radicals, e.g., glyceryl tributyrate, glyceryl trilaurate or glyceryl tripalmitate. It is obtained from animal and vegetable matter and is the major constituent of fats and oils. Used as a source of fatty acid and derivatives, in the manufacture of edible oils and fats and monoglycerides, and in biochemistry, biotechnology and medicine.

trigonite. A yellow or brown to black mineral composed of lead manganese hydrogen arsenite, $Pb_3MnH(AsO_3)_3$. Crystal system, monoclinic. Density, 6.10 g/cm^3; refractive index, 2.10. Occurrence: Sweden.

triholmium carbide. See holmium carbide (iii).

triiron dodecacarbonyl. A flammable, air-sensitive black crystals with a melting point of 165°C (329°F) used to remove sulfur from substituted thiophene-containing complexes yielding ferroles. Formula: $Fe_3(CO)_{12}$. Also known as *dodecacarbonyliron; iron dodecacarbonyl*.

trikalsilite. A colorless mineral of the nepheline group composed of potassium sodium aluminum silicate, $K_{0.7}Na_{0.3}AlSiO_4$. It can also be made synthetically. Crystal system, hexagonal. Density, 2.63 g/cm^3. Occurrence. Central Africa.

Tri-Krome Blue. Trade name of Pavco Inc. (USA) for a bright blue trivalent chromate coating applied by single dipping.

trilam. A laminated fabric in a plain weave made from polyester film-coated polyester *scrim*, and used for sails.

Trilan. Trade name for cellulose triacetate fibers used for wearing apparel and industrial fabrics.

Trilen. Trademark of Sam Yang Kasei (Japan) for polypropylene resins supplied in general-purpose and special grades.

Trilene. (1) Trademark of Berkley & Company (USA) for nylon 6 fibers and monofilaments.

(2) Trade name of Tri Polyta Indonesia Tbk. for polypropylene plastics.

(3) Trade name of Uniroyal Chemical Company (USA) for several low-molecular-weight polyethylenes used for the mabufacture of sealants and elastomers.

Tri-Light. Trade name of Invicta Bridge & Engineering Company Limited (UK) for industrial roofing system comprising *Plyglass* double glazing units.

Trillium. Trademark of World Alloys & Refining, Inc. (USA) for a corrosion-resistant chromium-cobalt dental ceramic alloy with a density of 7.9 g/cm^3 (0.29 $lb/in.^3$), a melting range of 1300-1370°C (2375-2500°F), a hardness of 366 Vickers, very high yield strength, and low elongation. Used for partial restora-

tions. *Trillium II* is a corrosion-resistant nickel-chromium dental ceramic alloy with a density of 8.2 g/cm^3 (0.30 $lb/in.^3$), a melting range of 1260-1316°C (2300-2400°F), a hardness of 350 Vickers, high yield strength, and low elongation. Used for partial restorations.

Tri-Lok. Trade name of Tri-Lok Company (USA) for a water-hardening steel containing 0.4% carbon, and the balance iron. Used for gears and shafts.

Trilon. Trademark of Mazzaferro SA (Brazil) for nylon 6 mono-filaments and filament yarns.

Triloy. Trademark of Sam Yang Kasei (Japan) for a series of polymer alloys including various blends of polycarbonate (PC) and polybutylene terephthalate (PC/PBT), polyethylene terephthalate (PC/PET) and acrylonitrile-butadiene-styrene (PC/ABS).

trilutetium carbide. See lutetium carbide (ii).

Trim. Trademark of H.J. Bosworth Company (USA) for acrylic dental resins used for temporary crown and bridge restorations.

trimagnesium columbate. See magnesium niobate (iii).

trimagnesium niobate. See magnesium niobate (iii).

trimanganese aluminum trisilicate. See manganese aluminum silicate (ii).

trimanganese arsenide. See manganese arsenide (iii).

trimanganese diphosphide. See manganese phosphide (iv).

trimanganese phosphide. See manganese phosphide (iii).

trimanganese silicide. See manganese silicide (iii).

trimanganese tetroxide. See manganese tetroxide.

Trimax. Trademark of A&P Technology (USA) for braided triaxial broadgoods.

Trimay. Trade name of R.J. Cyr Company (Canada) for an abrasion-resistant plate material consisting of a high-chromium-carbide ferrous alloy deposited on high-grade steel plate. Used for metallurgical and industrial equipment, such as blast furnaces, coke ovens, hoppers, vibratory feeders, and fan housings.

Trimco. Trade name of Trimout Manufacturing Company (USA) for an oil-hardening steel containing 0.6% carbon, 1.5% nickel, 0.8% chromium, and the balance iron. Used for pipe wrenches.

trim enamel paint. A special surface coating differing from regular house paint by its shorter drying time, higher gloss and fewer brush marks. Used mainly for trim, shutters, screens, etc.

trimer. A polymer molecule consisting of three identical monomer units, e.g., $(CH_2O)_3$, $(C_3H_6)_3$ and $(C_2H_2)_3$ are the trimers of CH_2O, C_3H_6, and C_2H_2, respectively.

trimerite. A pink mineral composed of beryllium calcium manganese silicate, $CaMn_2(BeSiO_4)_3$. Crystal system, monoclinic. Density, 3.47 g/cm^3; refractive index, 1.7202. Occurrence: Sweden.

Trimet. (1) Trade name of Handy & Harman (USA) for copper-base brazing filler metals.

(2) Trade name GTE Sylvania (USA) for heat-resistant tungsten alloy wire used for glass sealing.

Tri-Metal. Trade name for silver-clad sheet brass, i.e., a composite consisting of a shock-resistant brass core that has a covering of silver rolled to both sides. Used to facilitate silver brazing of carbide tips to steel cutting tool bodies.

trimethylgallium. An organometallic compound supplied in the form of a colorless pyrophoric liquid (99+% pure), and in high-purity (99.999+%) electronic grades. It has a density of 1.151 g/cm^3, a melting point of -15.8°C (3.6°F) and boiling point of 55.7°C (132.2°F). Used in organometallic research, and electronics. Abbreviation: TMGa. Formula: $(CH_3)_3Ga$. Also known as *gallium trimethyl*.

trimethylindium. An organometallic compound supplied in the form of white, air- and moisture-sensitive, pyrophoric crystals (99.9+% pure), and as high-purity (99.999%) electronic grades. It has a density of 1.568 g/cm^3, a melting point of 88°C (190°F) and a boiling point of 136°C (277°F). Used in organometallic research and electronics. Abbreviation: TMIn. Formula: $(CH_3)_3In$. Also known as *indium trimethyl*.

Tri-M-ite. Trademark of 3M Company (USA) for coated abrasives composed of resin-bonded sharp silicon carbide mineral on cloth, paper or polyester backing for wet or dry grinding and finishing applications.

trimmed block. Dressed or crude mica split into prescribed thicknesses, and side-trimmed to eliminate imperfections, irregularities and contaminants.

Trimo. Trade name of Uddeholm Corporation (USA) for an air- or oil-hardening cold-work tool steel (AISI type D2) containing 1.5% carbon, 0.3% manganese, 0.4% silicon, 12% chromium, 0.8-0.9% molybdenum, 0.2-2.0% vanadium, and the balance iron. It has good deep-hardening properties, good wear resistance and high hardness. Used for punches, shear blades, cold forming dies, forming and blanking dies, thread rolling dies, reamers and gages.

trimolybdenum aluminide. See molybdenum aluminide (ii).

trimolybdenum diboride. See molybdenum boride (vi).

trimolybdenum disilicide. See molybdenum silicide (iii).

trimolybdenum disulfide. See molybdenum sulfide (iii).

trimolybdenum nitride. See molybdenum nitride (iii).

trimolybdenum oxide. A crystalline compound of molybdenum and oxygen with cubic crystal structure, used in ceramics and materials research. Formula: Mo_3O.

trimolybdenum phosphide. See molybdenum phosphide (iii).

trimolybdenum silicide. See molybdenum silicide (ii).

trimolybdenum sulfide. See molybdenum sulfide (ii).

TrimRite. Trademark of Carpenter Technology Corporation (USA) for a series of hardenable martensitic stainless steels. They possess good cold formability, excellent heat-treating properties and moderate corrosion resistance. Used for self-drilling fasteners, food-processing equipment, shafting, gages, valve parts, guides, conveyor chains, instruments, and roofing nails.

Trimtec. Trademark of Odermath Stahlwerkstechnik GmbH (Germany) for a cored steel wire used to introduce selected alloying elements and other additives into steel melts.

Tri-Neo. Trade name of Tridus International Inc. (USA) for neodymium iron boron magnets. See also neodymium-iron boron magnet.

Tri-Ni. Trade name of Enthone-OMI Inc. (USA) for high-sulfur nickel strike.

trinickel boride. See nickel boride.

trinickel disulfide. See nickel sulfide (v).

trinickelous orthophosphate. See nickel phosphate.

trinickel phosphide. See nickel phosphide (iii).

trinickel zirconium. See nickel zirconium.

Trinidad asphalt. A natural asphalt from Trinidad composed of about 47% bitumen, 28% clay and 25% water.

triniobium aluminide. See niobium aluminide (iii).

triniobium diboride. See niobium boride (iv).

triniobium disilicide. See niobium silicide (iii).

triniobium digermanide. See niobium germanide (i).

triniobium germanide. See niobium germanide (iii).

triniobium tetraboride. See niobium boride (iii).

Trinova. Trade name of Trinova Corporation (USA) for rayon fibers and yarns used for textile fabrics.

Trinyl. Trademark of Inquitex SA (Spain) for nylon 6 staple fibers and filament yarns.

triosmium dodecacarbonyl. Yellow crystals (98% pure) with a melting point of 224°C (435°F) used as a reagent and catalyst. $Os_3(CO)_{12}$. Also known as *osmium carbonyl*.

tripentaerythritol. A white- to ivory-colored, hygroscopic powder available in technical grades with a melting range of 225-240°C (437-464°F). Used in varnishes, hard resins and quick-drying tall-oil vehicles. Formula: $C_{15}H_{35}O_8$.

tripeptide. A peptide resulting from the combination of three amino acid residues by peptide linkages.

Tripet. Trade name of Sam Yang Kasei (Japan) for polyethylene terephthalate (PET) resins.

triphane. See spodumene.

triphylite. (1) A pale gray mineral of the olivine group composed of lithium iron manganese phosphate, $Li(Fe,Mn)PO_4$. Crystal system, orthorhombic. Density, 3.42 g/cm^3; refractive index, 1.687. Occurrence: Sweden.

(2) A synthetic mineral of the olivine group composed of lithium iron phosphate, $LiFe(PO_4)$, made by solid-state reaction at 800°C (1472°F) in nitrogen. Crystal system, orthorhombic. Density, 3.62 g/cm^3.

triple brick. A ceramic product having a nominal size of 135 × 102 × 305 mm (5.33 × 4 × 12 in.).

triple cloth. A plied fabric, often used for industrial applications, produced with three warps and three weft threads, along with a binder thread. The plies are secured by stiching. Also known as *treble cloth*.

Triple Conqueror. Trade name of Joseph Beardshaw & Son Limited (UK) for a wear-resistant tool steel containing 1.3% carbon, 4% tungsten, 1% chromium, 0.3% molybdenum, and the balance iron. Used for drawing and extrusion dies, broaches, etc.

Triple Crescent. Trade name of Spencer Clark Metal Industries Limited (UK) for a chromium-tungsten tool steel containing 1.25% carbon, 1.15% chromium, 4.25% tungsten, 0.3% vanadium, and the balance iron. Used for cutting tools, reamers, broaches, finishing tools, and gages.

Triple Die. Trade name of Teledyne Firth-Sterling (USA) for high-carbon high-chromium type cold-work tool steels containing 2.2% carbon, 12% chromium, 1% molybdenum, 1% vanadium, and the balance iron. They have excellent abrasion and wear resistance, good compressive strength, good deep-hardening properties, and are used for blanking, trimming and stamping dies.

Triplee. Trade name of Matchless Metal Polish Company (USA) for buffing compounds for nonferrous metals and plastics.

Triple Eclair. Trade name of Creusot-Loire (France) for oil-hardening high-speed steel containing 0.78% carbon, 18-18.5% tungsten, 0.6-1% vanadium, 0.4-0.6% molybdenum, and the balance iron. It has good red hardness, and is used for lathe and planer tools, cutters, drills, taps, reamers, and other tools.

Triple Flygo. Trade name of Turton Brothers & Mathews Limited (UK) for cobalt-tungsten high-speed steels containing 0.73% carbon, 4.7% chromium, 1% molybdenum, 18% tungsten, 1% vanadium, 5% cobalt, and the balance iron. Used for cutting tools.

Triple Griffin. Trade name of Darwins Alloy Castings (UK) for a tool steel containing 1.3% carbon, 0.3% silicon, 0.4% manganese, 0.3% chromium, 2.7% tungsten, and the balance iron. It has a case-hardened surface, a ductile, tough core, and is used for tube drawing dies, and cutters.

Triple Mermaid. Trade name of Spear & Jackson Industries Limited (UK) for a cobalt-tungsten high-speed steel (AISI type T4) containing 0.8% carbon, 4.5% chromium, 18% tungsten, 1.2-1.5% vanadium, 5-5.5% cobalt, and the balance iron. It has a good deep-hardening properties, high hot hardness, and good wear resistance. Used for lathe and planer tools, drills, cutters, and other tools.

Triple Six. Trade name of British Steel Corporation (UK) for a chromium steel containing 0.6% carbon, 0.6% chromium, and the balance iron.

Triple Spur. Trade name of T. Turton & Sons Limited (UK) for a cobalt-tungsten high-speed steel containing 0.82% carbon, 4.5% chromium, 1% molybdenum, 20% tungsten, 1.7% vanadium, 11% cobalt, and the balance iron. It has good nonwarping properties and wear resistance, and is used for cutting tools.

Triple Velos. Trade name of Spencer Clark Metal Industries Limited (UK) for a cobalt-tungsten high-speed steel containing 0.75% carbon, 4.2% chromium, 0.5% molybdenum, 18% tungsten, 1.15% vanadium, 5% cobalt, and the balance iron. It has high hot hardness and good wear resistance, and is used for lathe and planer tools, drills, cutters, and other tools.

triple-weight hydrogen. See tritium.

Triplex. (1) Trade name of Triplex Safety Glass Company Limited (UK) for a toughened and laminated glass.

(2) Trade name of Société Industrielle Triplex SA (France) for a laminated glass.

Triplex Hotline. Trade name of Triplex Safety Glass Company Limited (UK) for a special electrically heated toughened glass made by firing a printed pattern of resistance bands into the surface during the toughening process.

Triplex Sundym. Trade name of Triplex Safety Glass Company Limited (UK) for a laminated and toughened glass incorporating green-tinted, heat-absorbing glass.

triplex steel. Steel made from *duplex steel* by superrefining in an electric furnace.

triplite. A brown, brownish black, flesh red or pink mineral composed of $(Mn,Fe)_2PO_4(F,OH)$. Crystal system, monoclinic. Density, 3.50; refractive index, 1.6737. Occurrence: Mozambique.

triploidite. A yellowish brown mineral of the triplite group composed of iron manganese phosphate hydroxide, $(Mn,Fe)_2PO_4(OH)$. Crystal system, monoclinic. Density, 3.81-3.83 g/cm^3; hardness, 5 Mohs; refractive index, 1.726. Occurrence: USA (Connecticut).

triplutonium disilicide. See plutonium silicide (iii).

tripoli. A soft, decomposed, siliceous limestone that, when finely pulverized, is used as an abrasive. It is available in various grades according to fineness, rose, cream, white, gray or light olive in color, and usually has negligible cutting action, but is used for buffing and polishing aluminum, brass and steel, for sanding or grinding wood and metal, and as a filtering material, and paint and rubber filler. Also known as *rottenstone*.

tripolite. See diatomaceous earth.

trippkeite. A greenish blue mineral composed of copper arsenate, $CuAs_2O_4$. It can also be made synthetically. Density, 4.80 g/cm^3; refractive index, 2.12. Occurrence: Chile.

Tripps. Trade name of Sam Yang Kasei (Japan) for polyphenylene sulfide (PPS) resins.

TRIP steels. A group of highly plastic high-alloy steels that exhibit transformation-induced plasticity (TRIP), i.e., a plastic deformation triggered martensitic transformation. These steels have been previously heat-treated to produce metastable *austenite* or metastable austenite plus *martensite* and, after martensitic transformation, exhibit exceptionally high rates of strain aging resulting in high tensile and yield strength at temperatures between room temperature and 500°C (930°F). The completion of the martensitic transformation may or may not require cooling to -195°C (-319°F). A typical composition (in wt%) of a cold-rolled TRIP steel is 0.15% carbon, 1.5% manganese, 1.2% silicon, 0.01% phosphorus, 0.05% aluminum, 0.005% nitrogen, and the balance iron. Used for automotive applications, machinery, and construction equipment.

Tripton. Trade name for a dental adhesive used for dentin bonding applications.

tripuhyrite. A greenish yellow mineral of the rutile group composed of iron antimony oxide, $FeSb_2O_6$. Crystal system, tetragonal. Density, 5.82 g/cm^3; refractive index, 2.20. Occurrence: Brazil; Mexico.

Trirex. Trademark of Sam Yang Kasei (Japan) for an extensive series of polycarbonate resins supplied in various grades including clear, unreinforced, glass fiber-reinforced, UV-stabilized, FDA-compliant and flame-retardant.

trirhenium boride. See rhenium boride (i).

Triron. Trademark of Sam Yang Company (South Korea) for polyester staple fibers and filament yarns.

triruthenium dodecacarbonyl. Orange crystals (99% pure) with a melting point of 150°C (300°F) (decomposes) used as a carbonyl cluster precursor and hydrogen-transfer catalyst. Formula: $Ru_3(CO)_{12}$. Also known as *ruthenium carbonyl*.

trisaccharide. A sugar compound, such as raffinose, produced when three monosaccharides are linked together by eliminating a water molecule.

trisamarium carbide. See samarium carbide (iii).

tris(3,6-dioxaheptyl)amine. An an organic compound (95% pure) with a density of 1.011 g/cm^3, a boiling point above 339°C (642°F), a flash point above 230°F (110°C) and a refractive index of 1.4486 used as a solid-liquid phase transfer catalyst. Formula: $(CH_3OCH_2CH_2OCH_2CH_2)_3N$. Abbreviation: TDA.

trisilane. A colorless liquid with a density of 0.740 g/cm^3, a melting point of 117°C (178°F) and a boiling point of 53°C (127°F), used in organic synthesis, and in the preparation of silicones. Formula: Si_3H_8.

trisodium edetate. See trisodium EDTA.

trisodium EDTA. The trisodium salt of ethylenediaminetetraacetic acid, usually supplied in hydrated form as a white powder (95+% pure) with a melting point above 300°C (572°F), and used as a chelating agent. Formula: $C_{10}H_{13}N_2Na_3O_8 \cdot xH_2O$. Abbreviation: EDTA Na_3. Also known as *trisodium edetate*.

Trisolen. Trade name of Buna Sow Leuna Olefinverbund GmbH (Germany) for ethylene vinyl acetate copolymers.

Tristar. Trade name for a corrosion-resistant nickel-chromium dental bonding alloy.

Tri-Star. (1) Trademark of Tri-Star Company (USA) for heavy-duty corrugated cardboard.

(2) Trade name of Tri-Star Chemical Company (USA) for white glues and defoaming agents.

Tri-Steel. Trade name of Inland Steel Company (USA) for a high-strength, low-alloy (HSLA) steel containing 0.22% carbon, 1.25% manganese, 0.3% silicon, 0.02% vanadium, and the balance iron. Used for railroad and agricultural equipment, and mining equipment.

Tristelle. Trade name of Haynes International (USA) for a series of wear-resistant, iron-based high-performance alloys containing 30-35% chromium, 12% cobalt, 10% nickel, 5% silicon and 1-3% carbon.

tristramite. A yellow to greenish yellow mineral of the rhabdophane group composed of calcium uranium phosphate dihydrate, $(Ca,U)(PO_4) \cdot 2H_2O$. Crystal system, hexagonal. Density, 3.80 g/cm³; refractive index, 1.644. Occurrence: UK.

Trital. Trade name for viscose rayon fibers and yarns used in the manufacture of textile yarns and fabrics.

tritantalum diboride. See tantalum boride (iv).

tritantalum nitride. See tantalum nitride (iii).

tritantalum tetraboride. See tantalum boride (v).

Tri-Ten. Trade name of US Steel Corporation for a series of structural steels containing up to 0.25% carbon, 0-1.3% manganese, 0.2-0.3% silicon, 0-1% nickel, 0-0.6% copper, 0.2% vanadium, and the balance iron. Used for truck bodies, rail and mine cars, cranes, derricks, and shovels.

triterbium carbide. See terbium carbide (iii).

trithorium disilicide. See thorium silicide (iii).

trithulium carbide. See thulium carbide (iii).

trititanium phosphide. See titanium phosphide (ii).

trititanium silicide. See titanium silicide (iii).

tritium. A radioactive isotope of hydrogen with a mass number of 3. Unlike ordinary hydrogen, which has only one proton in its nucleus, it has two neutrons and one proton. *Tritium* is obtained by bombardment of lithium with low-energy neutrons in a nuclear reactor, and also occurs naturally in hydrogen in very small proportions. It is heavier than heavy hydrogen (*deuterium*), emits beta radiation, and has a half-life of 12.5 years. Used in thickness gages, cold cathode tubes, in controlled fusion experiments, activator in self-luminous phosphors, as a label in tracer experiments, in luminous paints, in luminous instrument dials, as bombarding particles in cyclotrons, and in thermonuclear power research. Symbols: T, t, H^3, or ^3H. Also known as *heavy heavy hydrogen; hydrogen 3; triple-weight hydrogen.*

tritium paint. A luminous paint containing tritium (hydrogen 3, ^3H). Since tritium emits beta radiation, no or only minimum shielding is required.

tritomite. (1) A dark brown mineral of the apatite group composed of calcium lanthanum cerium boron silicate, $(La,Nd)(Ce,Pr,Th)_2Ca_2(Si_2B)O_{13}$. Crystal system, hexagonal. Density, 4.20 g/cm³; refractive index, 1.745. Occurrence: Norway.

(2) A blackish brown mineral of the apatite group composed of calcium rare-earth aluminum borate silicate dihydrate, $(Ca,Ln)_4(AlSi_3)B_2O_{16} \cdot 2H_2O$. Crystal system, hexagonal. Density, 3.39 g/cm³; refractive index, 1.678. Occurrence: Canada (Ontario).

(3) A black mineral of the apatite group composed of yttrium cerium calcium boron silicate, $Y_3(Ce, Pr,Th)Ca(Si_2B)O_{13}$. Crystal system, hexagonal. Density, 3.40 g/cm³; refractive index, 1.670. Occurrence: USA (New Jersey).

Triton. Trade name of CCS Braeburn Alloy Steel (USA) for a tough, shock-resisting tool steels (AISI type S2) containing 0.5% carbon, 0.9-1.1% silicon, 0.5-0.6% molybdenum, and the balance iron. Used for chisels, pneumatic tools, and rivet sets.

triturate. A powder made by rubbing or grinding a solid, commonly with the addition of a suitable liquid.

Triumph. (1) Trade name of Jessop-Saville Limited (UK) for a tungsten-type high-speed steels containing 0.68% carbon, 3.75% chromium, 14% tungsten, 0.6% vanadium, and the balance iron. Used for reamers, slitting saws, hacksaws, lathe centers, and other tools. *Triumph Superb* refers to tungsten and cobalt-tungsten high-speed steels containing about 0.7-0.8% carbon, 4.2-4.7% chromium, 18-19% tungsten, 1.6% vanadium,

0-6% cobalt, and the balance iron. *Triumph Superb Double* are cobalt-tungsten high-speed steels containing 0.8% carbon, 4.7% chromium, 20% tungsten, 1.6% vanadium, 10% cobalt, and the balance iron. *Triumph Superb Extra* refers to a series of tungsten high-speed steels containing 0.8% carbon, 0-0.6% molybdenum, 4.25% chromium, 22% tungsten, 1.4% vanadium, and the balance iron.

(2) Trade name of Aurident, Inc. (USA) for a dental ceramic alloy containing 53.5% palladium and 37.5% silver. It has good workability, is compatible with most porcelains, and is used for porcelain-to-metal restorations.

triuranium disilicide. See uranium silicide (v).

triuranium octoxide. See uranous-uranyl oxide.

triuranium silicide. See uranium silicide (iv).

Tri-Van. Trade name of Uddeholm Corporation (USA) for an oil-hardening high-carbon, high-chromium cold-work tool steel (AISI type D4) containing 2.25% carbon, 12% chromium, and the balance iron. It has a great depth of hardening, good abrasion and wear resistance, relatively high hot hardness, and is used for blanking and drawing dies, lamination and coining dies, thread rolling dies, swaging dies, punches, shears, and slitters gages.

trivanadium diboride. See vanadium boride (iv).

trivanadium silicide. See vanadium silicide (ii).

trivinyl. See trivinyl fiber.

trivinyl fiber. A generic term for a synthetic fiber produced from a terpolymer of acrylonitrile (vinyl cyanide), a vinyl monomer and a chlorinated vinyl monomer in which the weight percentage of each of these monomers does not exceed 50% of the total weight. Also known as *trivinyl.*

Trivex. Trade name of PPG Industries (USA) for allyl diglycol carbonate (ADC) plastics.

triytterbium carbide. See ytterbium carbide (i).

triyttrium carbide. See ytterbium carbide (iii).

triyttrium pentasilicide. See yttrium silicide (ii).

trizinc columbate. See zinc niobate (ii).

trizinc niobate. See zinc niobate (ii).

trizirconium disilicide. See zirconium silicide (iv).

trizirconium germanide. See zirconium germanide (iii).

Trocal. Trade name of HT-Troplast AG (Germany) for polyvinyl chloride (PVC) profiles.

Trocellan. German trademark for high-quality crosslinked polyolefin foam (including polyethylene) made by a process that utilizes an inert gas as blowing agent. Used as a gasketing material for automotive, appliance and industrial applications, as a replacement for foamed-rubber gaskets, and for automobile components, flotation devices, athletic equipment, and shell packaging and protective packaging.

Trodaloy. Trade name for a series of beryllium bronzes with good electrical conductivity containing varying amounts of copper, beryllium and/or chromium or cobalt. Used for resistance welding electrodes, soldering tips, and electrical parts.

troegerite. A lemon yellow, strongly radioactive mineral of the meta-autunite group composed of uranyl arsenate dodecahydrate, $UO_2(UO_2)_2(AsO_4)_2 \cdot 12H_2O$. Crystal system, tetragonal. Density, 3.55 g/cm³; refractive index, 1.612.

Trogamid. Trademark of Degussa-Huels America Inc. (USA) for polyamide (nylon) engineering resins for blow molding, injection molding and extrusion applications. They have high strength, good impact resistance, excellent resistance to gasoline, mineral oils, tetrachlorocarbon, and dilute alkalies, and moderate resistance to trichloroethylene and dilute acids. Used for machine elements, e.g., bolts, screws, bearings, bushings, gears and cams, and for hoses, tubing and molded products. *Trogamid T* refers to amorphous, glass-fiber-reinforced and transparent polyamide (nylon) engineering resins.

trogtalite. (1) A mineral of the pyrite group composed of cobalt copper selenide, $(Co,Cu)Se_2$. Crystal system, cubic. Density, 7.10 g/cm^3. Occurrence: Zaire.

(2) A rose-violet synthetic mineral of the pyrite group composed of cobalt selenide, $CoSe_2$. It can also be made synthetically. Crystal system, cubic. Density, 7.09 g/cm^3.

troilite. A bronze-yellow mineral of the nickeline group composed of iron sulfide, FeS. It can also be made synthetically. Crystal system, hexagonal. Density, 4.67 g/cm^3. Occurrence: USA (California).

Trojan. (1) Trade name of Atlas Specialty Steels (Canada) for a tungsten-type high-speed steel (AISI type T2) containing 0.8% carbon, 0.25% manganese, 0.3% silicon, 4% chromium, 18.5% tungsten, 0.5% molybdenum, 2% vanadium, and the balance iron. It has high wear resistance and hot hardness, good deep-hardening properties, low toughness, and relatively poor machinability. Used for cutting tools and lathe tools.

(2) Trade name of Trojan Board Limited (Canada) for solid wood interlocking paneling.

Trojan Babbitt. Trade name of Hoyt Metal Company (USA) for a babbitt metal containing varying amounts of antimony, tin and lead. It has good antifriction properties, and is used for bearings for steam engines and internal combustion engines.

Trolen. Trademark of Dynamit Nobel AG (Germany) for low-density polyethylene supplied in the form of blocks, strips, foils, plates, profiles, rods, and pipes.

Trolit. Trademark of Dynamit Nobel AG (Germany) for cellulosic plastics.

Trolitan. Trademark of Hüls AG (Germany) for phenolic molding compounds available in various grades: natural, fiber or wood filled general-purpose, cellulose-filled shock-resistant, cotton-filled medium-shock, glass-fiber-reinforced high-impact, chopped-fabric-filled medium-impact, and mica-filled electrical.

Trolitax. Trademark of Hüls AG (Germany) for a laminate composed of phenolic-impregnated paper or cardboard.

Trolitul. Trademark of Hüls AG (Germany) for a thermoplastic homopolymer produced by the polymerization of styrene, and available in transparent and colored grades. Its properties (e.g. brittleness) can be modified by copolymerizing with other monomers. *Trolitul* thermoplastics have a glossy surface, high hardness, good strength, a softening temperature of 80°C (176°F), and excellent electrical properties, good machinability and weldability, and good gluing properties. Used for electrical components, e.g., housings, coils, battery cases and lighting panels, machine and appliance housings, refrigerator doors, air-conditioner cases, containers and molded housewares, toys, and as a packaging material.

trolley wire. A round or shaped, bare, solid conductor, usually composed of high-strength hard-drawn copper, cadmium bronze or a similar high-copper alloy, and used as an overhead wire to supply current to the motors of busses, locomotives or street cars through traveling current collectors.

trolleite. A pale green mineral composed of aluminum phosphate hydroxide, $Al_4(PO_4)_3(OH)_3$. Crystal system, monoclinic. Density, 3.09 g/cm^3. Occurrence: USA (California).

Trollfoil. Trade name of Trollplast Inc. (USA) for a plastic articulating foil used in dentistry for bite registration.

Tromalit. Trade name of Ugine Aciers (France) for a series of permanent magnet alloys containing 0-20% cobalt, 12-27% nickel, 10-13% aluminum, 0-3.5% copper, and the balance iron. They have high magnetic permeability, and are used for electrical and magnetic equipment, and magnets in speedometers and motors.

Trona. Trade name of Kerr-McGee Corporation (USA) for elemental boron and boron trichloride compounds.

trona. See urao.

Tronamang. Trademark of Kerr-McGee Corporation (USA) for high-purity electrolytic manganese metal and powder, and manganese-aluminum briquettes, used as alloying additions in the manufacture of steels and nonferrous alloys.

Tronex. (1) Trade name of Tronex Corporation (USA) for unsaturated polyester resins.

(2) Trademark of Wheaton Company (USA) for electronic glassware including bushing and standoff insulators, glass-tube base housings, tubing, and bulbs.

troostite. (1) A variety of the mineral *willemite* (Zn_2SiO_4) in which manganese replaces some of the zinc. It is found in yellowish to flesh-red crystals with resinous luster, and is not of great commercial importance.

(2) A now obsolete term formerly used for a fine mixture of *ferrite* and *cementite* that is harder than *sorbite* and obtained in a steel either by quenching at a cooling rate that is slower than the critical rate, or by tempering at a relatively low temperature. The former is now known as "fine pearlite," and the latter as "tempered martensite." *Note:* In tool steels, the term *troostite* is synonymous with upper bainite. See also bainite; pearlite; tempered martensite.

Tropic. (1) Trade name of Société Industrielle Triplex SA (France) for laminated glass with tinted interlayer.

(2) Trade name for rayon fibers and yarns used for textile fabrics.

tropical walnut. The wood of any of various tropical walnut trees especially *Juglans neotropica* and *J. olanchana*. The former grows on the eastern slopes of the Andes, while the latter is found in northern Central America. *Tropical walnut* is somewhat darker and coarser than the American black walnut, and is used for furniture and cabinetmaking. Also known as *nogal*. See also American walnut.

Troposphere. Trade name of Schein Rexodent (UK) for a dental amalgam alloy.

Trosiplast. Trade name of Hüls AG (Germany) for polyvinyl chloride (PVC) plastics and PVC alloys. *Trosiplast 1* is an unplasticized polyvinyl chloride, *Trosiplast 3* refers to plasticized polyvinyl chlorides supplied in three elongation ranges (0-100%, 100-300%, and above 300%), and *Trosiplast PVC/ABS* is an acrylonitrile-butadiene-styrene/polyvinyl chloride alloy.

Troughguard. Trade name of Premier Refractories and Chemicals Inc. (USA) for graphitic plastic refractory mixes.

Trovicel. Trade name of Röchling Haren KG (Germany) for rigid foam sheeting based on polyvinyl chloride (PVC).

Trovidur. Trade name of Röchling Haren KG (Germany) for a series of polyvinyl chloride (PVC) plastics available in various grades including standard, unplasticized and impact-resistant. Used for machine parts, and electrical and electronic components.

Trovitex. Trade name of Röchling Haren KG (Germany) for rigid foam sheeting based on polyvinyl chloride (PVC).

Troy. Trade name of Time Steel Service Inc. (USA) for an oil-hardening cold-work tool steel (AISI type O7) containing 1.25%

carbon, 0.3% manganese, 0.35% silicon, 0.4% chromium, 0.2% vanadium, 1.4% tungsten, and the balance iron. It has excellent deep-hardening properties, good toughness and machinability and is used for hand tools, cutting tools, and dies.

Trualoy.   Trade name of True Alloys Inc. (USA) for a series of high-conductivity coppers, aluminum casting alloys, aluminum casting bronzes, and leaded tin bearing bronzes.

Tru-Ballistic.   Trademark of Honeywell Performance Fibers (USA) for readily dyeable nylon-6 fibers and fabrics with excellent toughness and abrasion resistance. Used especially for luggage.

Trubrite.   Trade name of Arthur Lee & Sons Limited (UK) for a series of low-carbon high-chromium steels (typically 12-20% chromium) and austenitic chromium-nickel steels (12-18% chromium, 8-12% nickel). Used for cutlery, household appliances, kitchenware, and food processing equipment.

Tru-Cast.   (1) Trade name of Manco Products Company (USA) for a hardenable beryllium casting bronze used for molds and stamping and casting dies.

   (2) Trademark of Syon Corporation (USA) for a series of room-temperature-curing, thermally conductive, two-part epoxy adhesives, and thermosetting resin potting compounds.

Trucast.   Trade name of Engelhard Industries (USA) for a series of dental casting alloys including *Trucast Hard*, a hard, gold-based dental casting alloy (ADA type III) that contains some platinum, is available in the annealed and age-hardened conditions, and has a melting range of 900-950°C (1652-1742°F), *Trucast Medium*, a medium-hard platinum-free gold-based dental casting alloy (ADA type II) with a melting range of 920-955°C (1688-1751°F), and *Trucast Soft*, a soft, platinum-free gold-based dental casting alloy (ADA type I) that is available in the annealed and age-hardened conditions, and has a melting range of 940-970°C (1724-1778°F).

Tru-Chrome.   Trade name of Rocky Mountain Orthodontics, Inc. (USA) for orthodontic wire.

truck-mixed concrete.   Concrete proportioned and mixed in a truck mixer in transit to a job site. Also known as *transit-mixed concrete*.

Tru-Cor.   (1) Trade name of Kidd Drawn Steel (USA) for a water-hardening tool steel (AISI type W1) containing 0.9-1.05% carbon, 0.3-0.5% manganese, 0.15-0.3% silicon, and the balance iron. Used for boring tools, drills, and machine tool parts.

   (2) Trade name of National Standard Company (USA) for flux and metal-cored welding wire supplied in various grades (e.g., stainless steel) and diameters.

Trudie.   Trade name of Champion Steel Company (USA) for an air-hardening high-chromium cold-work die steel containing 1.5% carbon, 12% chromium, 1% molybdenum, 1% vanadium, and the balance iron.

True Blue.   Trade name of Pavco Inc. (USA) for a trivalent chromate conversion coating applied by single dipping.

true hemp.   See hemp (1).

true porcelain.   See hard paste porcelain.

true silk.   Silk produced by the silkworm, the cultivated larva of the Mulberry spinner (*Bombyx mori*), as in contrast to tussah silk which is produced by the larvae of certain species of wild or semi-cultivated Asian moth of the genus *Antheraea*. True silk is extensively used in textile manufacture. See also silk; tussah.

true verdigris.   A collective term for blue and green verdigris based on *basic copper acetate*. It must not be confused with either *artificial malachite* (basic copper carbonate) or *patina* (basic copper sulfate or basic copper chloride) which are known

as "false verdigris." See also blue verdigris; copper subacetate; false verdigris; green verdigris; verdigris.

TrueVitality.   Trademark of Den-Mat Corporation (USA) for a radiopaque hybrid composite resin that is colored like natural teeth. It has very high wear resistance and a triple-cure polymerization mode–heat-curing, self-curing and light-curing. Used in dentistry for the restoration of indirect inlays, onlays, crowns, bridges and laminates, for direct anterior and posterior restorations, and for high-strength temporary crowns and bridges.

Truewear.   Trade name of Jessop Steel Company (USA) for a cold-work tool steel (AISI type D7) containing 2.45% carbon, 12.25% chromium, 4.25% vanadium, 1.1% molybdenum, and the balance iron. It has excellent deep-hardening properties and wear resistance and high hot hardness. Used for punches, pins, dies, forming rolls, molds, and tools.

Truflex.   (1) Trade name of Chance Brothers Limited (UK) for mirrors for light projection apparatus.

   (2) Trade name of Texas Instruments Inc. (USA) for an extensive series of thermostatic bimetals usually of the nickel-chromium-iron type.

Truform.   (1) Trade name of Jessop Steel Company (USA) for an oil-hardening cold-work tool steel (AISI type O1) containing 0.9% carbon, 1.2% manganese, 0.5% tungsten, 0.5% chromium, and the balance iron. It has a great depth of hardening, good nondeforming properties, good machinability, and is used for gages, cutters, dies, taps, reamers, shear blades, dies, and punches.

   (2) Trademark of PPG Industries, Inc. (USA) for tough, durable polyester coil coatings for aluminum, cold-rolled steel and galvanized steel substrates. They have high stain and abrasion resistance, and excellent surface hardness and flexibility. Used for architectural and automotive applications, such as building products, doors, awnings, recreational vehicles and truck trailer sheets, and office furniture.

Truglide.   Trade name of Jessop Steel Company (USA) for an oil-hardening cold-work tool steel (AISI type O6) containing about 1.45% carbon, 1.00% silicon, 0.25% molybdenum, and the balance iron. It has small particles of graphitic carbon uniformly dispersed throughout the matrix. *Truglide* steel has a great depth of hardening, excellent machinability, good wear and abrasion resistance, good toughness, low galling tendency, and is used for structural parts, such as bushings, arbors, bodies and/or shanks for cutting tools, gages, and dies.

Trulite.   Trade name of Trudent Products Inc. (UK) for dental acrylics.

Truly White.   Trademark for a pure white high-quality interior latex paint which absorbs less than 5% of incident light rays. It has good resistance to scrubbing and staining, and is available in gloss, semi-gloss, low-luster and flat finishes.

Trumpet Brass.   Trade name of Waterbury Rolling Mills Inc. (USA) for a brass containing 83% copper, 15.5% zinc and 1.5% tin.

Truplate.   Trade name of McGean-Rohco Inc. (USA) for mechanically deposited zinc and other metals.

truscottite.   A white mineral of the reyerite group composed of calcium silicate hydroxide dihydrate, $Ca_{14}Si_{24}O_{58}(OH)_8 \cdot 2H_2O$. Crystal system, hexagonal. Density, 2.48 g/cm^3; refractive index, 1.55. Occurrence: Sumatra.

Tru-Site.   Trade name of Dearborn Glass Company (USA) for non-glare picture glass.

trustedtite.   A yellow mineral of the spinel group composed of nickel selenide, Ni_3Se_4. Crystal system, cubic. Density, 6.65 g/cm³. Occurrence: Finland.

Tru-Steel. Trademark of Steel Shot Producers, Inc. (USA) for heat-treated steel shot with improved toughness used for cleaning and shot peening of metals.

Tru-Stone. Trade name for a die stone used in dentistry.

Tru-Temp. Trade name of ASG Industries Inc. (USA) for toughened glass.

Tru Temp. Trademark of Birchwood Laboratories, Inc. (USA) for a black oxide coating for iron and steel parts.

Truwear. Trade name of Jessop Steel Company (USA) for a high-carbon, high-chromium cold-work tool steel (AISI type D7) containing 2.2% carbon, 12.5% chromium, 4% vanadium, 1.1% molybdenum, and the balance iron. It has very great depth of hardening, outstanding wear and abrasion resistance, and relatively high hot hardness. Used for header and forming dies.

Trycite. Trademark of Dow Chemical Company (USA) for an oriented polystyrene film used for packaging applications.

tryptophan black. A black natural pigment based on the amino acid tryptophan ($C_{11}H_{12}O_2N_2$).

tschermigite. A synthetic mineral of the alum group composed of ammonium aluminum sulfate dodecahydrate, $NH_4Al(SO_4)_2 \cdot 12H_2O$. Crystal system, cubic. Density, 1.64 g/cm³.

TSP EP. Trade name of Thermoset Plastics (US) for epoxy resins available in general-purpose, flexible, high-heat, and aluminum-, glass-, mineral- or silica-filled casting grades.

tsumcorite. A red-brown mineral composed of iron lead zinc arsenate hydrate, $FePbZn(AsO_4)_2 \cdot H_2O$. Crystal system, monoclinic. Density, 5.20 g/cm³. Occurrence: Southwest Africa.

tsumebite. A black mineral of the brackebuschite group composed of copper lead phosphate sulfate hydroxide, $CuPb_2(PO_4)(SO_4)(OH)$. Crystal system, monoclinic. Density, 6.01 g/cm³; refractive index, 1.920. Occurrence: Southwest Africa.

tsumoite. A silver-white mineral of the tetradymite group composed of bismuth telluride, $BiTe$. Crystal system, hexagonal. Density, 8.16 g/cm³. Occurrence: Japan.

Tsuta. Trade name of Nippon Sheet Glass Company Limited (Japan) for a patterned glass.

tuballoy. A term sometimes used for depleted uranium (DU), a grade of metallic uranium containing less than 0.720 wt% of the fissile isotope uranium-235 (^{235}U) normally found in natural uranium. Primarily used for its high density (18.9 g/cm³ or 0.68 lb/in.³). See also depleted uranium; uranium.

Tubastra. Trade name for rayon fibers and yarns used for textile fabrics.

tube. A hollow cylinder of round, square, rectangular, octagonal or elliptical cross section for holding and transporting fluids or finely divided solids.

Tube Borium. Trade name of Stoody Company (USA) for an abrasion-resistant material composed of 60% tungsten carbide and 40% steel. Used for welding and hardfacing electrodes.

tube brass. A brass containing 60-70% copper and 30-40% zinc and commonly used for heat exchanger and condenser tubing. A free-cutting grade with improved machinability containing 66-66.5% copper, 32.4-33.5% zinc and 0.25-1.6% lead is also available.

Tubeloy. Trade name of American Smelting & Refining Company (USA) for a corrosion-resistant lead containing 0.02% tin, 0.02% calcium and 0.02% magnesium. Used for water service pipe.

Tube Mandrel. Trade name of A. Milne & Company (USA) for a tough tool steel containing 1.15-1.35% carbon, 0.35% chromi-

um, and the balance iron. Used for cutters, mandrels and tools.

tube steel. See steel tubing.

Tubize. Trade name of Tubize Rayon Corporation (USA) for rayon fibers and yarns used for textile fabrics.

tubular. A textile fabric made on a circular knitting machine. Also known as *tubular fabric.*

tubular fabric. See tubular.

tubular product. A term covering all hollow metallic and non-metallic products including tubes, hollow shapes and semihollow shapes commonly used for transporting fluids, and as structural members.

Tubulitec. Trade name for a synthetic resin used in dentistry for cavity lining applications.

tucekite. A pale brass-yellow mineral of the chalcopyrite group composed of nickel antimony sulfide, $Ni_9Sb_2S_8$. Crystal system, tetragonal. Density, 6.14 g/cm³. Occurrence: South Africa.

Tuc-Tur. British trade name for an alloy composed of 63% copper, 22% zinc and 15% nickel, or 59-61% copper, 21-29% zinc, 13-18% tin and 0.3% iron. Used for hardware, fittings and machine parts.

Tucunsil. Trade name of Esgo Division of GTE Sylvania Corporation (USA) for a series of copper-titanium-silver alloys used for brazing titanium and stainless steel.

Tudenza. Trade name for rayon fibers and yarns used for textile fabrics.

Tudor. (1) Trade name of Tudor Safety Glass Company Limited (USA) for laminated and toughened glass.

(2) Trade name of United Clays, Inc. (USA) for a plastic, chemically-resistant blend of Tennessee ball clays and kaolins with superior extrusion and mold/shaping characteristics and outstanding refractory properties. Used as castable or pressing blends, or for wet extrusion applications.

Tufanhard. Trade name of Hobart Welding Products (USA) for a series of tough, hard wear-resistant carbon, stainless, austenitic manganese and alloy steels. Used as weld metal for overlaying and buildup applications.

Tufbaria. Trade name of Techno Polymer Company Limited (UK) for acrylonitrile-butadiene-styrene plastics.

Tufboy. Trademark of Irving Industries Limited (Canada) for alloy steel castings.

Tufcel. Trademark of Toyobo Company Limited (Japan) for modal staple fibers for textile fabrics and high-tenacity rayon fibers for tire cord and other industrial applications.

Tufchem. Trade name of Atofina Chemicals, Inc. (USA) for a wide range of building products including polymer concretes, silicate cements and concretes, gunites and monolithic toppings.

Tufcor. Trade name of Jonas & Culver Limited (UK) for a hardened and tempered plain-carbon steel, with soft, ductile core, containing 0.15% carbon, 0.7% manganese, and the balance iron.

Tufcord. Trademark of SRF Limited (India) for nylon 6 fibers and filament yarns for tire cords.

Tuf-Cote. Trade name of St. Gobain Industrial Ceramics (France) for abrasion-, corrosion-, erosion- and wear-resistant coatings consisting of nickel-chromium-boron-silicon matrices with dispersed tungsten carbide particles. They are applied with an oxyacetylene torch and used on drilling, mining and dredging machinery and equipment.

Tufcote. (1) Trademark of E.I. DuPont de Nemours & Company (USA) for a high-solids baking enamel.

(2) Trademark of Specialty Composites Corporation (USA)

for laminate sheets composed of a layer of synthetic resin or resin-impregnated fibers and a viscoelastic layer with or without foil covering. Used for noise and vibration reduction applications.

(3) Trademark of Specialty Composites Corporation (USA) for unreinforced and reinforced plastic foam sheets with or without an additional unfoamed laminating sheet.

Tufex. Trade name of Shree Vallabh Glass Works Limited (USA) for a toughened glass.

tuff. A soft, porous highly siliceous sedimentary rock formed from *volcanic ash* and composed mainly of fragments of less than 4 mm (0.16 in.) in diameter. Used in the manufacture of abrasives and cement.

Tuffak. Trademark of Atoglas, Atofina Chemicals Inc. (USA) for polycarbonate sheeting used for building and construction applications.

Tuffalloy. Trade name of Hapco Inc. (USA) for a series of liquid molding compounds with thermoplastic properties including Series 187, 267 and 280 casting and potting compounds for applications in the food and medical industries, Series 4200 low-viscosity liquid molding compounds with excellent physical properties, and high impact resistance and heat distortion, and Series 8100 low-viscosity thermoplastic liquid molding compounds with high impact resistance and good processibility.

Tuff Buff. Trade name of Aardvark Clay & Supplies (USA) for a soft buff clay (cone 10) with subtle spotting.

Tuff-Cast. Trade name of Tuff-Hard Corporation (USA) for a high-speed tool steel containing 0.7% carbon, 18% tungsten, 4% chromium, 1% vanadium, and the balance iron. Used for cutting tools, hobs, and drills.

Tuf-Flex. Trademark of Libbey-Owens-Ford Company (USA) for a tempered plate glass.

Tufflite Profax. Trade name for polyolefin fibers and plastics.

TuffRez. Trademark of PolySpec (US) for an extensive series of epoxy- and polyurethane-based products. The epoxies include various topping binders, clear coatings, flexible coatings, low-temperature coatings, medium-viscosity topcoats, high-viscosity receiving coats, flat- or textured-finish epoxy floor and wall coatings, and the polyurethanes several high-gloss coatings and waterborne floor coatings.

Tuff-Rib. Trademark of Baldwin Steel Company (USA) for galvanized, galvalumed and painted steel supplied in the form of roofing and side panels for the manufacture of agricultural, commercial, industrial and residential buildings.

Tuff Skin. Trade name of Bradford Insulation Holdings (SA) Limited (Australia) for glass wool.

Tuff-Temp. Trade name of Tuff-Temp Corporation (USA) for a dental acrylic used for temporary crown and bridge restorations.

Tufkut. Trade name of Great Western Steel Company (USA) for a tough, shock-resistant tool steel containing 0.5% carbon, 0.7% manganese, 1.5% silicon, 0.35% molybdenum, and the balance iron. Used for chisels, forming tools, punches, shear blades, mandrels, and hand tools.

Tuflin. Trade name of Dow Chemical Company (USA) for polyethylene plastics.

Tuflite. Trade name of Duplate Canada Limited for a toughened sheet glass.

Tufnol. Trade name of Tufnol Limited (UK) for bakelite-type phenol-formaldehyde resins and plastics.

Tufnol 10G/40. Trademark of Tufnol Limited (UK) for epoxy and phenolic laminates including *Tufnol 10G/40* glass fabric and glass prepreg epoxy grades, and *Tufnol Kite Brand* lightweight phenolic laminates filled with glass or cotton fabrics or paper. The latter have good strength and wear resistance and good insulating properties, and are used for bearings and special pump parts.

Tufpet. Trade name of Toyobo Company Limited (Japan) for a series of tough polybutylene and polyethylene terephthalates.

Tufplate. Trade name of Duplate Canada Limited for a toughened plate glass.

Tufram. Trademark of General Magnaplate Corporation (USA) for a synergistic coating produced by a proprietary process that combines the hardness of aluminum oxide and the protection of a fluorocarbon topcoat to impart optimum hardness, corrosion resistance and permanent lubricity on aluminum and aluminum-alloy parts. Other advantages of *Tufram* coatings include high abrasion and wear resistance, high dielectric strength, rapid heat and cold transfer, good appearance and close part tolerance.

Tufrov. Trademark of PPG Industries, Inc. (USA) for glass fiber rovings.

Tuf-Stuf. Trade name of Mueller Brass Company (USA) for a series of aluminum bronzes containing 9.5-11.5% aluminum and 2-4% iron, other small additions, e.g., nickel and manganese, and the balance copper. They have high strength and toughness, good corrosion resistance, and are used for machine parts and valves.

Tuftest. Trade name of Permaglass Limited (Australia) for a toughened glass.

Tuf-Test. Trade name of Medart Engineering & Equipment Company (USA) for an abrasion-resistant cast iron used for grinding disks.

Tufthane. Trademark of Tiodize Company, Inc. (USA) for a high build, elastomeric coating that protects metallic and plastic surfaces against galvanic corrosion in graphite composites.

Tufton. Trade name for rayon and polyolefin fibers and yarns used for textile fabrics and plastic products.

Tuftran. Trademark of Rohm & Haas Company (USA) for a hard infrared material with excellent imaging characteristics consisting of a 1 mm (0.04 in.) thick layer of zinc sulfide (ZnS) chemically vapor deposited over a polished zinc selenide (ZnSe) window.

tugarinovite. A black mineral of the rutile group composed of molybdenum oxide, MoO_2. It can also be made synthetically. Crystal system, monoclinic. Density, 6.46 g/cm^3.

tugtupite. A rose to white mineral composed of sodium aluminum beryllium chloride silicate, $Na_4AlBeSi_4O_{12}(Cl,S)$. Crystal system, tetragonal. Density, 2.36 g/cm^3; refractive index, 1.496. Occurrence: Greenland.

tuhualite. A blue mineral of the osumilite group composed of sodium potassium iron silicate monohydrate, $(Na,K)_2Fe_4Si_{12}O_{13} \cdot H_2O$. Crystal system, orthorhombic. Density, 2.89 g/cm^3; refractive index, 1.612. Occurrence: New Zealand.

tulameenite. A mineral of the gold group composed of copper iron platinum, $CuFePt_2$. It can also be made synthetically. Crystal system, tetragonal. Density, 14.90 g/cm^3. Occurrence: Canada (British Columbia).

Tulip. Brand name for a high-purity tin (99.87+% pure) containing up to 0.086% lead, 0.01% copper, 0.008% arsenic and 0.026% other impurities.

Tulipe. Trade name of Société Nouvelle du Saut-du-Tarn (France) for a series of water-hardening steels containing 0.5-0.9% carbon, 0.5-0.9% manganese, 0-0.5% silicon, and the balance iron.

Used for machinery and machine-tool parts, agricultural equipment, and gears, shafts, bolts, and fasteners.

tulip poplar. See American whitewood.

tulipwood. See American poplar.

Tullanox. Trademark of Tulco Inc. (USA) super-hydrophobic precipitated and fumed silicas, supplied in regular and deammoniated grades, and super-hydrophobic liquid coatings. The silicas are used in silica elastomers, epoxies, polyurethanes, adhesives, sealants, coatings, powder-metallurgy products, textile fabrics, and cosmetic products. The coatings are used on antennas, radomes, satellite dishes, electronic equipment, and on various polymeric substrates.

tulle. A fine, stiff, netlike, machine-made fabric, usually of silk or synthetics (e.g., rayon), usually having a hexagonal mesh. Used for bridal veils, evening wear, and ballet costumes. It is named after Tulle, a city in southwestern France.

Tumblebryte. Trade name of Pax Surface Chemicals, Inc. (USA) for burnishing compounds.

tumbler-galvanized material. See hot-galvanized material.

Tumblers. Trade name of Edgar Allen & Company Limited (UK) for austenitic manganese steel castings containing 1% carbon, 12% manganese, and the balance iron. Used for tumblers and wear-resistant parts.

Tumblex. (1) Trade name of Polyflow, Inc. (USA) for high-density tumbling media used in metal finishing operations.

(2) Trade name of Abrasive Finishing, Inc. (USA) for mass finishing media used in metal finishing operations.

tumbling abrasives. A group of abrasive media suitable for use in tumbling operations and including preformed balls, cubes and cylinders of aluminum oxide or silicon carbide.

Tunco. Trade name of Simonds Worden White Company (USA) for a cobalt-tungsten-type high-speed steel (AISI type T4) containing 0.75% carbon, 4.1% chromium, 18% tungsten, 1% vanadium, 0.6% molybdenum, 5% cobalt, and the balance iron. It has a great depth of hardening, high hot hardness, excellent wear resistance, and good machinability. Used for lathe and planer tools, milling cutters, and broaches.

Tuncro. Trade name of Atlantic Steel Corporation (USA) for a hot-work tool steel containing 0.5% carbon, 2% tungsten, 1.4% chromium, 0.25% vanadium, and the balance iron. Used for chisels, punches, pneumatic tools, and hot work dies.

TundiBoard. Trademark of Hoeganaes AB (Sweden) for an inorganic distribution plate used for lining steel tundishes.

tundrite. (1) A mineral composed of sodium cerium titanium silicate hydroxide tetrahydrate, $NaCe_2TiSiO_7(OH)\cdot4H_2O$. Crystal system, triclinic. Density, 3.96 g/cm^3. Occurrence: Russian Federation.

(2) A brownish to greenish yellow mineral composed of sodium cerium titanium carbonate silicate hydroxide dihydrate, $Na_3(Ce, La)_4(Ti,Nb)_2(SiO_4)_2(CO_3)_3O_4(OH)\cdot2H_2O$. Crystal system, triclinic. Density, 3.70 g/cm^3; refractive index, 1.80. Occurrence: Russian Federation.

tunellite. A colorless mineral with subvitreous to pearly luster composed of strontium borate tetrahydrate, $SrB_6O_{10}\cdot4H_2O$. It can also be made synthetically. Crystal system, monoclinic. Density, 2.40 g/cm^3; refractive index, 1.534. Occurrence: USA (California).

Tung-Alloy. Trade name of Resisto-Loy Company, Inc. (USA) for an abrasion-, heat-, and wear-resistant alloy containing 0.1% carbon, 35% tungsten, 16% molybdenum, 8% cobalt, 0.05% boron, and the balance iron. Used for hardfacing electrodes.

Tungaloy. Trade name of Boyd-Wagner Company (USA) for an

oil- or water-hardening die steel containing 1.2% carbon, 0.6% chromium, 0.3% molybdenum, and the balance iron. Used for blanking and forming dies.

Tungo. Trade name of Teledyne Vasco (USA) for a tool steel containing 0.5% carbon, 2% tungsten, 1.65% chromium, 0.25% chromium, and the balance iron. It possesses good shock resistance, moderate hot hardness, and is used for punches, chisels and shear knives.

Tungsil. Trade name of Engelhard Corporation (USA) for a sintered alloy containing 27-75% silver and 25-73% tungsten. It has good resistance to mechanical wear and electrical erosion, and is used for electrical contacts for circuits breakers.

Tungsit. Trade name of Tungsit Electro-Metals Works Limited (UK) for a wear-resistant material consisting of tungsten carbide (WC) and ditungsten carbide (W_2C). Used for hardfacing electrodes.

Tungsite. Trade name of Pyramid Steel Company (USA) for a tough, oil-hardening steel containing 0.33% carbon, 0.94% chromium, 0.48% molybdenum, 0.5% tungsten, and the balance iron. Used for punches, chisels, gages, shear blades and gears.

tungstate white. See barium tungstate.

Tungsteel. Trade name of St. Lawrence Steel Company (USA) for an oil- or water-hardening, shock-resistant tool steel containing 0.35% carbon, 0.65% manganese, 0.35% silicon, 0.75% chromium, 0.45% tungsten, 0.4% molybdenum, and the balance iron. Used for hand tools, and riveting hammers.

tungsten. A hard, brittle, gray, metallic element of Group VIB (Group 6) of the Periodic Table. It is commercially available in the form of powder, granules, wire, rods, sheets, foils, microfoils, tubing, crucibles and single crystals. The crystals are usually grown by the float-zone technique. The principal tungsten ores are *scheelite* and *wolframite*. Crystal system, cubic. Crystal structure, body-centered cubic. Density, 19.3 g/cm^3; melting point, 3410°C (6170°F); boiling point, 5660°C (10220°F); hardness, 360-500 Vickers; superconductivity critical temperature, 0.0154 K; atomic number, 74; atomic weight, 183.85; divalent, tetravalent, pentavalent, hexavalent. It has good electrical conductivity (31% IACS), high tensile strength, high abrasion resistance and good resistance to electrical spark erosion, poor resistance to seawater corrosion, and oxidizes in air at about 400°C (750°F). Used for high-speed tool steel, as an additive in ferrous and nonferrous alloys, in electrical contact materials, as a carbide in cemented carbides, for filament wires in incandescent light bulbs, in heating elements for electric furnaces, in vacuum-metallizing equipment, welding electrodes, spark-plug electrodes, cutting tools, rocket nozzles and other aerospace applications, as a target in X-ray tubes, as an emitter in vacuum tubes, for chemical apparatus, high-speed rotors, and solar-energy devices. Symbol: W.

tungsten alloys. A group of alloys in which tungsten is the principal constituent. Common alloy combinations include: (i) *Tungsten-thoria alloys* for welding electrodes and electron-discharge tubes; (ii) *Tungsten-molybdenum alloys* for applications requiring improved machinability; (iii) *Tungsten-rhenium alloys* for the improvement of the resistance to cold fracture in lamp filaments and thermocouples; and (iv) *Tungsten-nickel-copper* and *tungsten-nickel iron alloys* for components for watches and instruments, counterbalance weights, tool shanks, and shielding against X-rays and gamma rays.

tungsten-base powder-metallurgy materials. A group of metal alloys including tungsten-thoria, tungsten-nickel, tungsten-nickel-copper, tungsten-silver and tungsten-copper, obtained from powder mixtures by pressing and sintering at prescribed conditions of temperature and pressure. See also powder-metallurgy materials; tungsten alloys.

tungsten boride. Any of the following compounds of tungsten and boron: (i) *Tungsten monoboride*. A silvery powder (99% pure) made by heating tungsten and boron in an electric furnace. Crystal system, tetragonal. Density, 9-10.77 g/cm³; melting point, 2900°C (5252°F). Used for refractories and in materials research. Formula: WB; (ii) *Ditungsten boride*. A refractory, black powder. Density, 16.7 g/cm³; melting point, 2770°C (5018°F). Used in ceramics and materials research. Formula: W_2B; (iii) *Tungsten pentaboride*. Refractory solid. Density, 11.0 g/cm³; melting point, 2365°C (4289°F). Formula: W_2B_5; and (iv) *Tungsten bor-ide/ditungsten boride*. A mixture of tungsten monoboride and ditungsten boride available as a fine, black powder. Used for ceramics, refractories, and in materials research. Formula: $WB+W_2B$.

tungsten borocarbide. A compound of tungsten, boron and carbon available as a fine powder. Density, 10.5 g/cm³; melting point, above 3000°C (5432°F). Used as a refractory and for other high-temperature applications. Formula: WB_4C.

tungsten brass. A corrosion-resistant brass containing 60% copper, 2-4% tungsten, 1-14% nickel, 0-3% aluminum, 0-0.2% tin, and the balance zinc. Used for fittings and hardware.

tungsten bromide. See tungsten dibromide; tungsten tribromide; tungsten tetrabromide; tungsten pentabromide; tungsten hexabromide.

tungsten bronze. A wear-resistant bronze containing 90% copper and 10% tungsten used for electrical contacts. Abbreviation: TB.

tungsten carbide. Any of the following compounds of tungsten and carbon: (i) *Tungsten monocarbide*. An iron-gray, crystalline powder (99+% pure) that can be made by heating tungsten and lampblack at 1500-1600°C (2730-2910°F). Crystal system, hexagonal. Density, 15.6-15.7 g/cm³; melting point, 2780°C (5036°F); boiling point, 6000°C (10830°F); hardness, above 9.5 Mohs (almost as a hard as diamond); oxidizes on heating in air; decomposes at high temperatures into ditungsten carbide (W_2C) and carbon. Used as cemented carbide for dies and cutting tools, wear-resistant parts, heat- and wear-resistant coatings, cermets, electrical resistors, and as an abrasive. Formula: WC; and (ii) *Ditungsten carbide*. A gray crystalline powder. Crystal system, hexagonal. Density, 17.2 g/cm³; melting point, 2850°C (5162°F); boiling point, 6000°C (10830°F); hardness, above 9.5 Mohs. Used for cemented carbides, wear-resistant parts, cermets and as an abrasive. Formula: W_2C.

tungsten-carbide cermet. A composite material consisting of tungsten carbide and a metallic binder, such as cobalt or nickel. It is made at high temperatures under controlled atmospheres using powder-metallurgy techniques involving pressing and sintering. *Tungsten-carbide cermet* has a density range of 11-15 g/cm³ (0.40-0.54 lb/in.³), high tensile and compressive strength, a high elastic modulus, a hardness of 90 Rockwell A, high stiffness and dimensional stability, excellent abrasion and wear resistance, and is used for valve parts, gages and tools. See also cermet.

tungsten carbonyl. See tungsten hexacarbonyl.

tungsten cast iron. A cast iron containing a small percentage of tungsten (about 1%) with improved tensile and transverse strengths.

tungsten chloride. See tungsten dichloride; tungsten trichloride; tungsten tetrachloride; tungsten pentachloride; tungsten hexachloride.

tungsten-chromium steels. A group of alloy steels containing about 0.5-0.7% carbon, 1.5-2.0% tungsten, 0.5-1.0% chromium used for applications requiring, good hardness, and improved corrosion and heat resistance. Usually the term refers to the steels in the AISI-SAE 72xx series.

tungsten coating. A thin, hard, wear-resistant layer of tungsten produced on a metal or nonmetal substrate by chemical vapor deposition (CVD) from tungsten hexachloride (WCl_6), tungsten hexafluoride (WF_6) or tungsten hexacarbonyl [$W(CO)_6$].

tungsten-copper alloys. A group of alloys containing 10-28% copper, and the balance tungsten, usually made by wrought powder-metallurgy methods. They have excellent thermal conductivity, low thermal expansion, good machinability and platability, and are used for electronic components, such as heat sinks, circuit board cores, lids, covers, thermal spreaders, chip mounting, spark erosion electrodes, and contacts.

Tungsten Diamond. Trade name of George Cook & Company Limited (UK) for an oil-hardening, wear-resistant tool steel containing 1.38% carbon, 4.5% tungsten, 0.25% silicon, 0.75% chromium, 0.35% manganese, and the balance iron. Used for various cutting and other tools.

tungsten dibromide. A yellow crystalline solid that decomposes at 400°C (752°F) and is used in chemistry and materials research. Formula: WBr_2. Also known as *tungsten bromide*.

tungsten dichloride. A yellow crystalline solid that decomposes above 500°C (932°F) and is used in chemistry and materials research. Formula: WCl_2. Also known as *tungsten chloride*.

tungsten diiodide. Orange crystals with a density of 6.79 g/cm³ used in chemistry and materials research. Formula: WI_2. Also known as *tungsten iodide*.

tungsten dinitride. See tungsten nitride (ii).

tungsten dioxide. A brown, crystalline powder (99.9+% pure) that may turn bluish to purple over time. Density, 12.11 g/cm³; melting point, 1500-1600°C (2732-2912°F). Formula: WO_2.

tungsten diselenide. See tungsten selenide.

tungsten disilicide. See tungsten silicide (i).

tungsten disulfide. See tungsten sulfide (i).

tungsten ditelluride. See tungsten telluride.

tungsten fibers. Stiff, creep-resistant fibers or wires with a typical diameter of 0.1-1.5 mm (0.004-0.06 in.), made of tungsten or a tungsten alloy, such as thoriated tungsten, tungsten-hafnium-carbon, tungsten-rhenium or tungsten-rhenium-hafnium-carbon. They have a density range of 18.5-19.5 g/cm³ (0.67-0.70 lb/in.³), high tensile strength and stress rupture strength, and good thermal conductivity. Used as reinforcement wires for metal-matrix composites.

tungsten fiber-reinforced superalloys. A group of engineering alloys composed of cobalt-, iron- or nickel-base matrices reinforced with strong, stiff continuous tungsten or tungsten-alloy fibers or wires. The tungsten-based fibers are often coated with titanium carbide or titanium nitride to enhance their compatibility with the matrix materials. *Tungsten-fiber-reinforced superalloys* have excellent oxidation and high-temperature resistance, high tensile strength and creep resistance, and are used for aircraft engines, rocket engine turbo-pumps, and turbine engine blades. Abbreviation: TFRS. See also tungsten fibers.

tungsten filaments. Continuous filaments of tungsten used as reinforcements in ceramic, metal and polymer-matrix composites, and as substrate fibers for continuous boron fibers.

tungsten fluoride. See tungsten tetrafluoride; tungsten pentafluoride; tungsten hexafluoride.

tungsten heavy alloys. A group of refractory powder-metallurgy alloys containing more than 80% tungsten, with the balance being either copper, nickel, iron, lead, copper-nickel, nickel-iron, or molybdenum-nickel-iron. Typical compositions are W-5Ni-2Cu, W-5Ni-2Fe and W-8Mo-8Ni-2Fe. Their tensile and yield strengths can be greatly increased by mechanical working, but at the expense of ductility. Tungsten heavy alloys have a density range of 16-19 g/cm^3 (0.58-0.69 $lb/in.^3$), a hardness range of 15-46 Rockwell C, high heat resistance, high electrical and thermal conductivity and low coefficient of thermal expansion. Used for nuclear radiation shielding, radioactive shields, counterweights, sinker weights, balances, gyroscopes, heat sinks for microelectronics, electrical contacts, boring bars, vibration damping, anti-armor kinetic-energy penetrators, and welding electrodes.

tungsten hexabromide. Blue-black crystals with a melting point of 309°C (588°F) used in chemistry and materials research. Formula: WBr_6. Also known as *tungsten bromide*.

tungsten hexacarbonyl. A white, volatile, highly refractive crystalline compound (97+% pure) obtained by reacting tungsten with carbon monoxide at high pressures, or by reducing tungsten hexachloride with iron alloy powders in a carbon monoxide atmosphere. Density, 2.65 g/cm^3; melting point, decomposes at 150°C (302°F). Used in the deposition of tungsten coatings on base metals by thermal decomposition (e.g., chemical vapor deposition) of the carbonyl compound at about 300-600°C (570-1110°F). Formula: $W(CO)_6$. Also known as *tungsten carbonyl*.

tungsten hexachloride. Dark blue or purple, air- and moisture-sensitive crystals (99.9+% pure) made by heating tungsten with dry chlorine at red heat. Crystal system, hexagonal. Density, 3.52 g/cm^3; melting point, 275°C (527°F); boiling point, 347°C (657°F). Used in the deposition of tungsten coatings by hydrogen reduction of tungsten hexachloride (e.g., chemical vapor deposition) at 700-900°C (1290-1650°F), as an additive to tin oxide to produce electrically conducting coatings on glass, in single crystalline tungsten wire, and as a catalyst for olefin polymers. Formula: WCl_6. Also known as *tungsten chloride*.

tungsten hexafluoride. Colorless gas or light yellow liquid (98+% pure) with a density of 3.44 g/cm^3 (as liquid), or 12.9 g/L (as gas), a melting point of 2.5°C (36.5°F), and a boiling point of 19.5°C (67°F). Used in the chemical vapor deposition (CVD) of tungsten coatings at 482°C (900°F), and as a fluorinating agent. Formula: WF_6. Also known as *tungsten fluoride*.

tungsten high-speed steels. A group of high-speed tool steels (AISI subgroup T) containing about 0.65-1.60% carbon, 0.30% manganese, 0.30% silicon, 0-1.25% molybdenum, 11.75-21.00% tungsten, 3.75-5.00% chromium, 0.80-5.25% vanadium, 0-13.00% cobalt, and the balance iron. They are often sold under trademarks or trade names, such as *Acmite, Clarite, Maxite, Nipigon, Sabre, Star Zenith, Supremus, Tunco* or *Twinvan*, and have excellent hot hardness, good deep-hardening properties, good non-warping properties, good abrasion and wear resistance, poor toughness, and good to fair machinability. Used for cutting tools, tool bits, milling cutters, drills, taps, chasers, reamers, broaches, lathe and form tools, lathe centers, dies, punches, and for high-temperature structural parts, bearings and pump parts. Also known as *tungsten-type high-speed tool steels*.

Tungsten Hot Work. Trade name of A. Milne & Company (USA) for a hot-work tool steel containing 0.43% carbon, 2.5% chromi-

um, 0.1% vanadium, 9% tungsten, and the balance iron. Used for hot punches, hot shears, and hot blanking, trimming and drawing dies.

tungsten hot-work steels. A group of hot-work tool steels (AISI subgroups H20 through H39) containing about 0.25-0.55% carbon, 0.15-0.40% manganese, 0.15-0.50% silicon, 8.50-19.00% tungsten, 1.75-12.75% chromium, 0.25-1.25% vanadium, and the balance iron. They are often sold under trademarks or trade names such as *Forge-Die, Kalkos, Marvel, Mohawk, Seneca* or *T-Alloy*, and possess good deep-hardening properties, good hardenability, good toughness, good hot hardness, good impact strength, and moderate wear resistance. Used for extrusion dies and mandrels for high-temperature applications, hot forging and gripper dies, hot punches, and hot shear blades. Also known as *tungsten-type hot-work tool steels*.

tungsten iodide. See tungsten diiodide; tungsten tetraiodide.

tungstenite. A black opaque mineral of the molybdenite group composed of tungsten sulfide, WS_2. It can also be made synthetically. Crystal system, hexagonal (2H polytype); rhombohedral (3R polytype). Density, 7.45-7.80 g/cm^3. Occurrence: USA (Utah).

tungsten magnet steel. A hard magnetic steel with about 0.7% carbon, 6% tungsten, 0.5% chromium, and the balance iron. It has a density of approximately 8.1 g/cm^3 (0.29 $lb/in.^3$), a Curie temperature of approximately 760°C (1400°F), no magnetic orientation, high co-ercive force, high remanence, large hysteresis loop, and good hot forgeability. Used in the manufacture of permanent magnets. See also magnet steel.

tungsten-molybdenum alloys. Refractory alloys of tungsten and varying amounts of molybdenum (up to 15%) in which the molybdenum forms a continuous solid solution with tungsten. They have good to moderate strength, good machinability, and are used for heat-resistant components, and as filler metals for brazing and welding.

tungsten monoboride. See tungsten boride (i).

tungsten monocarbide. See tungsten carbide (i).

tungsten mononitride. See tungsten nitride (i).

Tungsten Nickel. Trade name of Harrison Alloys Inc. (USA) for an alloy composed of 96% nickel and 4% tungsten, and is used in the form of ribbons and strips for high-strength electron tube cathodes.

tungsten-nickel-copper alloys. A group of machinable powder-metallurgy tungsten alloys containing about 89-90% tungsten, 6-7% nickel and 3-4% copper. Used for instrument and watch components, counterbalance weights, tool shanks and shielding against X-rays and gamma rays.

tungsten-nickel-iron alloys. A group of machinable powder-metallurgy tungsten alloys containing about 89-98% tungsten, 1-7% nickel and 1-5% iron. Used for instrument and watch components, counterbalance weights, tool shanks, and shielding against X-rays and gamma rays.

tungsten nitride. Any of the following compounds of tungsten and nitrogen used in ceramics, electronics and materials research: (i) *Tungsten mononitride*. A gray, crystalline powder. Density, 12.1 g/cm^3; melting point, dissociates at approximately 600°C (1112°F). Formula: WN; (ii) *Tungsten dinitride*. Hexagonal crystals with a density 7.7 g/cm^3 and a melting point of 600°C (1112°F) (decomposes). Formula: WN_2; and (iii) *Ditungsten nitride*. Gray, cubic crystals with a density of 12.2 g/cm^3 that dissociate at approximately 800-870°C (1472-1598°F). Formula: W_2N.

tungsten oxide. See tungsten dioxide; tungstic oxide.

tungsten oxychloride. Dark red, moisture-sensitive crystals made by treating tungsten or tungstic oxide with chlorine at elevated temperatures. Density, 11.92 g/cm^3; melting point, 211°C (412°F); boiling point, 227.5°C (442°F). Used in incandescent lamps. Formula: $WOCl_4$.

tungsten pentaboride. See tungsten boride (iii).

tungsten pentabromide. Brown-black, moisture-sensitive solid with a melting point of 286°C (547°F) and boiling point of 333°C (631°F). Used in chemistry and materials research. Formula: WBr_5. Also known as *tungsten bromide*.

tungsten pentachloride. Black, moisture-sensitive crystals with a density of 3.88 g/cm^3, a melting point of 242°C (468°F) and a boiling point of 286°C (547°F). Used in chemistry and materials research. Formula: WCl_5. Also known as *tungsten chloride*.

tungsten pentafluoride. A yellow solid that decomposes above 80°C (176°F). Used in chemistry and materials research. Formula: WF_5. Also known as *tungsten fluoride*.

tungsten powder. Tungsten in form of a powder with a particle size ranging from less than 1 μm (40 μin.) to over 250 μm (0.01 in.) and a purity of 99.75% to over 99.99+%. Used for powder metallurgy parts and protective coatings.

tungsten-reinforced metal-matrix composites. A group of engineering materials with matrices composed of cobalt, copper, iron, nickel, titanium or other metal alloy reinforced with strong, stiff tungsten or tungsten-alloy fibers or wires. They have excellent oxidation resistance, high-temperature resistance, and high tensile strength and creep resistance. Used for aircraft engines, rocket-engine turbopumps, turbine-engine blades, and thrust chamber liners of rocket engines. See also tungsten fiber-reinforced superalloys.

Tungsten RhC. Trade name of Chase Brass & Copper Company, Inc. (USA) for a tungsten alloy containing 4% rhenium carbide (RhC). It has high tensile strength and hardness, good high-temperature properties, and is used for high-temperature applications.

tungsten-rhenium alloys. A group of refractory alloys of tungsten and up to 26% rhenium, in which the rhenium is dissolved in the tungsten, and sometimes small dispersions of thoria, or hafnium carbide. The *75W-25Re* alloy has a density of 19.7 g/cm^3 (0.71 $lb/in.^3$), a melting point of 3100°C (5610°F); high tensile strength and elastic modulus and good ductility; *W-Re-ThO$_2$* and *W-Re-HfC* alloys are produced by powder-metallurgy processing or vacuum-arc remelting and find application as fiber reinforcements for nickel-base superalloys, and in structural applications. They possess exceptionally high-temperature strength and low-temperature ductility. *Tungsten-rhenium alloys* are used to improve the cold-fracture resistance of lamp filaments, in high-temperature thermocouple wire, torsion bars and tension members, and as filler metals for brazing and welding.

tungsten selenide. Gray crystals or fine, grayish powder. Crystal system, hexagonal. Density, 9.2 g/cm^3; very high- and low temperature stability; high vacuum stability. Used as a solid lubricant. Formula: WSe_2. Also known as *tungsten diselenide*.

tungsten silicide. Any of the following compounds of tungsten and silicon: (i) *Tungsten disilicide*. Blue-gray crystals, powder, lumps or other pieces (99.7+% pure). Crystal system, tetragonal. Density, 9.3-9.86 g/cm^3; melting point, 2160°C (3920°F); very high hardness. Used in ceramics, as oxidation-resistant coatings for refractory metals, other refractory applications, and for electrical resistance applications. They are also suitable as reinforcements for structural ceramic composites, and

as matrix materials for structural silicide compounds. Formula: WSi_2; and (ii) *Pentatungsten trisilicide*. A blue-gray refractory powder. Density, 14.4 g/cm^3; melting point, 2320°C (4178°F). Used in high-temperature ceramics. Formula: W_5Si_3.

tungsten silicocarbide. A compound of tungsten, silicon and carbon available in powder form for use in ceramics and materials research. Formula: WSiC.

tungsten-silver alloys. A group of alloys of tungsten and up to 50% silver made by powder metallurgy techniques either by pressing and sintering mixed powders, or by pressing and sintering tungsten powder and infiltrating the resulting compact with silver. They have high melting points, excellent resistance to mechanical wear and electrical erosion, low corrosion resistance, and high electrical resistance. Used in electrical contacts for switching devices, automotive ignitions, vibrators, horns, magnetos, and electric razors.

Tungsten Special. Trade name of Associated Steel Corporation (USA) for an oil- or water-hardening, shock-resistant steel containing 0.34% carbon, 0.5% molybdenum, 0.5% tungsten, 1% chromium, 0.8% manganese, and the balance iron. Used for shafts, chuck jaws, chisels, punches, and gears.

Tungsten Spur. Trade name of T. Turton & Sons Limited (UK) for a tungsten high-speed steel containing 0.67% carbon, 3.8% chromium, 14% tungsten, 0.7% vanadium, and the balance iron. Used for cutting tools.

tungsten steel. A steel containing tungsten as the principal alloying element. Tungsten greatly increases the hardness, tensile and yield strengths, corrosion resistance, hardening temperature, and heat resistance of steel, and the cutting-edge life of tool steel. Examples include tungsten-type high-speed tool steels (up to 21% tungsten), tungsten-type hot-work tool steels (up to 19%), and tungsten magnet steel (typically 6% tungsten).

tungsten sulfide. Any of the following compounds of tungsten and sulfur: (i) *Tungsten disulfide*. Gray crystals or fine, grayish-black powder (99+% pure). It occurs in nature as the mineral *tungstenite*. Crystal stystem, hexagonal. Crystal system, molybdite. Density, 7.5-7.6 g/cm^3; melting point, decomposes at 1250°C (2282°F); has lubricating properties at temperatures above 1300°C (2370°F). Used as a solid lubricant. Formula: WS_2; and (ii) *Tungsten trisulfide*. A chocolate-brown powder. Formula: WS_3.

tungsten telluride. Gray crystals or powder (99.8+% pure). Crystal system, orthorhombic. Density, 9.43 g/cm^3; melting point, 1020°C (1868°F). Used in ceramics and materials research. Formula: WTe_2. Also known as *tungsten ditelluride*.

tungsten tetrabromide. Black, orthorhombic crystals used in chemistry and materials research. Formula: WBr_4. Also known as *tungsten bromide*.

tungsten tetrachloride. Black crystals or gray, crystalline, moisture-sensitive powder (97+% pure) with a density of 4.624 g/cm^3. Used in chemistry and materials research. Formula: WCl_4. Also known as *tungsten chloride*.

tungsten tetrafluoride. Red-brown crystals that decompose above 800°C (1472°F) and are used in chemistry and materials research. Formula: WF_4. Also known as *tungsten fluoride*.

tungsten tetraiodide. Black crystals or powder with a density of 5.2 g/cm^3 used in chemistry and materials research. Formula: WI_4. Also known as *tungsten iodide*.

tungsten-thoria alloys. See thoriated tungsten.

tungsten titanium carbide. A compound of tungsten, titanium and carbon available in powder form, and used in ceramics and materials research. Formula: $WTiC_2$.

tungsten tribromide. Black, hexagonal crystals that decompose above 80°C (176°F). Used in chemistry and materials research. Formula: WBr$_3$. Also known as *tungsten bromide.*

tungsten trichloride. A red solid that decomposes at 550°C (1022°F), and is used in chemistry and materials research. Formula: WCl$_3$. Also known as *tungsten chloride.*

tungsten trioxide. See tungstic oxide.

tungsten trisulfide. See tungsten sulfide (ii).

tungsten-type high-speed tool steels. See tungsten high-speed steels.

tungsten-type hot-work tool steels. See tungsten hot-work steels.

tungsten whiskers. Very fine and short, axially-oriented single-crystal tungsten fibers used as reinforcement in engineering composites having metal, ceramic, or polymer matrices. See also tungsten fibers; whiskers.

tungsten wire. Tungsten (99.9+% pure) in wire form usually supplied in diameters from 0.004 to 1.0 mm (0.00016 to 0.040 in.) for high-temperature thermocouples, electronic components, and spark plugs.

tungstic acid. A yellow powder obtained by treating sodium tungstate with hot sulfuric acid. Density, 5.5 g/cm^3; decomposes at 100°C (212°F); boiling point, 1473°C (2683°F); refractive index, 2.24. A white form (H$_2$WO$_4$·H$_2$O) is obtained by the acidification of tungsten solutions. Used in the production of tungsten metal, in the manufacture of tungsten wire, in the production of plastics, as a mordant in dyeing, and in microscopy. Formula: H$_2$WO$_4$. Also known as *orthotungstic acid; wolframic acid.*

tungstic acid anhydride. See tungstic oxide.

tungstic anhydride. See tungstic oxide.

tungstic oxide. Yellow to yellow-green, crystalline powder (99+% pure), or yellow green sintered lumps (99.5+% pure). Crystal system, orthorhombic. Density, 7.16 g/cm^3; melting point, 1473°C (2683°F). Used as a yellow colorant in ceramics, in fireproofing fabrics, in tungstate preparations for X-ray screens, in the manufacture of electrochromic displays and devices, in the preparation of tungsten metal by reduction, and in certain alloys. Formula: WO$_3$. Also known as *tungstic acid anhydride; tungstic anhydride; tungsten trioxide.*

Tungstide. Trade name for a series of hard, abrasion-resistant coatings produced from a composition consisting of near-colloidal particles of tungsten powder suspended in a liquid plastic with molybdenum sulfide additive. Used for self-lubricating applications.

tungstite. A yellow or yellowish green mineral composed of tungsten oxide monohydrate, WO$_3$·H$_2$O. It can also be made synthetically. Crystal system, orthorhombic. Density, 5.50 g/cm^3; refractive index, 2.24 g/cm^3. Occurrence: Canada.

Tungstophen. Trademark of Bayer AG (Germany) for modified phenol-formaldehyde plastics.

tungstophosphoric acid. A yellowish-white, water-soluble, crystalline solid (99.9+% pure) with a melting point of approximately 95°C (203°F) used as a catalyst for organic reactions, as a plating additive, in the manufacture of organic pigments, as a water-resistant additive to adhesives, cement and plastics, as an antistatic agent for textiles, as a reagent in analytical chemistry, as a metal stain in biochemistry and biology, and as a fixing agent in photography. Formula: H$_3$[P(W$_3$O$_{10}$)$_4$]·xH$_2$O. Also known as *phosphotungstic acid; phospho-12-tungstic acid; phosphowolframic acid.*

tungstosilicates. A group of complex, high-molecular-weight inorganic compounds with high degree of hydration. *Tungstosilicates* decompose strongly in aqueous solutions yielding highly colored tungstate reaction products. Structurally, they consist of a central silicon atom surrounded by octahedra of tungsten oxide.

tungstosilicic acid. A white, water-soluble, cystalline powder (99.9+% pure) used as a catalyst for organic reactions, as a plating additive, as a precipitant and inorganic ion-exchanger, as a mordant, in minerals separation, as a reagent for alkaloids, and in biological and biochemical microscopy. Formula: H$_4$[Si(W$_3$O$_{10}$)$_4$]·xH$_2$O. Also known as *silicotungstic acid; silico-12-tungstic acid; silicowolframic acid.*

Tungtube. Trade name of Airco Vacuum Metals (USA) for an abrasion-resistant material supplied in the form of a steel tube containing tungsten carbide particles. Used as a filler rod for oxyacetylene hardfacing applications.

Tungum. Trade name of Tungum Hydraulics Limited (UK) for wrought and cast brasses containing about 10% zinc, 0.8-1.4% nickel, 0.7-1.2% aluminum, 0.8-1.4% silicon, 0-0.3% iron, and the balance copper. They possess good corrosion and fatigue resistance, and high strength. Used for hydraulic and cryogenic castings and forgings, marine purposes, hardware and tubing.

tungusite. A green mineral of the pyrosmalite group composed of calcium iron silicate hydroxide, Ca$_4$Fe$_2$Si$_6$O$_{15}$(OH)$_6$. Density, 2.6-3.4 g/cm^3; refractive index, 1.670. Occurrence: USA (New Jersey), Russian Federation.

Tungweld. Trade name of Lincoln Electric Company (USA) for a series of abrasion-resistant tungsten carbide alloys used for hardfacing electrodes.

Tungwin. Trade name of Baldwin Steel Company (USA) for an oil or water-hardening, non-tempering steel containing 0.34% carbon, 0.78% manganese, 0.28% silicon, 0.82% chromium, 0.4% molybdenum, 0.51% tungsten, and the balance iron. Used for spindles, shafts, gears, dies, and arbors.

tunisite. A white mineral composed of sodium calcium aluminum hydrogen carbonate hydroxide, NaHCa$_2$Al$_4$(CO$_3$)$_4$(OH)$_{10}$. Crystal system, tetragonal. Density, 2.51 g/cm^3; refractive index, 1.573. Occurrence: Tunisia.

Tuntex. Trademark of Tuntex Fiber Corporation (Taiwan) for polyester staple fibers and filament yarns.

tupelo. The moderately heavy, strong, stiff wood from any of various trees of the genus *Nyssa* belonging to the sourgum family (Nyssaceae) and including water tupelo *(N. aquatica)*, black tupelo *(N. sylvatica)* and ogeche tupelo *(N. ogeche)*, all of which grow in the southeastern and southern United States. The lumber is valued for furniture, boxes, crates, veneer, and paper pulp. See also black tupelo; ogeche tupelo; water tupelo.

tupelo gum. See water tupelo.

Turan. German trade name for textile glass fibers.

Turbadium Bronze. Trade name of Baldwin-Lima-Hamilton Corporation (USA) for a series of corrosion-resistant, high-strength cast bronzes containing 48-51% copper, 43-46.5% zinc, 0.4-0.5% tin, 0.1-0.3% lead, 0-0.2% aluminum, 1.75-2.2% manganese, 1.75-2% nickel and 0-1.4% iron. They have good to excellent erosion resistance, and are used for propellers, impellers, turbine runners, and marine parts.

Turbaloy. Trade name of General Electric Company (USA) for corrosion- and heat-resistant austenitic stainless steels used for gas-turbine and heat-resisting parts.

Turbex. Trade name of Magnolia Antifriction Metal Company (UK) for lead-free, tin-base bearing metals used for bearings operating at high loads and speeds.

Turbide. Trade name for an oxidation-resistant sintered alloy containing carbide particles (TiC and Cr_3C_2) in a nickel matrix. Used for gas-turbine blades.

turbine brass. A brass containing 22-32% zinc, 0.25% lead, 0.4% iron, and the balance copper. Used for turbine parts.

turbine metal. A corrosion- and erosion-resistant alloy of 55% copper, 35% zinc, 3% nickel, and a total of 7% aluminum, iron and manganese. Used for turbine runners.

Turbiston. British trade name for several manganese bronzes.

Turbiston's bronze. A corrosion-resistant, high-strength bronze containing about 55% copper, 41% zinc, 2% nickel, 1% aluminum, 0.8% iron, and 0.2% manganese. It has excellent resistance to seawater, and is used for pistons, marine parts, hardware, and fittings.

Turbo. (1) Trade name of Allgemeines Deutsches Metallwerk GmbH (Germany) for a series of white alloys containing 77-80% copper, 10-12% aluminum and 1-8% nickel. They have good resistance to corrosion by acids and steam, and are used for worm wheels, worm shafts, spindles, and pumps.

(2) Trade name of Enthone-OMI Inc. (USA) for bright nickel plating processes and the resulting electrodeposits.

Turbofil. Trade name of Vetrotex France SA (France) for glass fiber products.

Turbo Glyco. Trade name of Joseph T. Ryerson & Son Inc. (USA) for a lead babbitt containing varying amounts of tin and antimony, and used for turbine bearings.

TurboKrete. Trademark of Rust-Oleum Corporation (USA) for an abrasion-, impact- and chemical-resistant, 100% solids epoxy coating used for repairing concrete and filling deep holes.

turbostratic carbon. Carbon with a two-dimensional crystal structure lacking ordering between the layer planes. It has a density of 1.4-2.1 g/cm^3 (0.05-0.08 $lb/in.^3$), high toughness and excellent wear resistance, and includes the low-temperature isotropic variety of *pyrolytic carbon*, the ultralow-temperature isotropic form of vapor-deposited carbon, and amorphous (or glassy) carbon. Used in the manufacture of carbon reinforcing fibers, and for the production of bioceramic coatings on metal substrates.

Turbotherm. Trade name of Vereinigte Edelstahlwerke (Austria) for a series of corrosion and heat-resistant stainless steels used for jet-engine parts, jet engine turbochargers, gas-turbine blades, and steam equipment.

Turcite. Trademark of Busak & Shamban/Smiths Group (USA) for durable, self-lubricating thermoplastic polytetrafluoroethylene (PTFE) bearing materials for continuous-use temperatures up to 82°C (225°F). They have excellent wear resistance, low friction, low water absorption (0.2%), and are used for transmissions, caster assemblies, home appliances, food-processing equipment, exercise equipment, bicycles, etc. *Turcite* is supplied in the form of rods in several grades including *Turcite PK* with high fatigue and wear resistance, good hygroscopic properties, good dimensional stability and excellent chemical and corrosion resistance, *Turcite TA*, a turquoise-colored general-purpose grade with excellent durability and chemical resistance, and *Turcite TX*, a red-colored high-speed, medium-load grade with very low coefficient of friction (0.22).

Turcoat. Trade name of Atofina Chemicals, Inc. (USA) for zinc, manganese and zinc-manganese phosphate coatings used for automotive applications.

Turcon. Trademark of Busak & Shamban/Smiths Group (USA) for polytetrafluoroethylene (PTFE) plastics.

Turex. (1) Trade name of Société Nouvelle des Acieries de Pompey (France) for a series of hot-work tool and die steels containing about 0.3% carbon, 1-3% chromium, 0.1-0.2% vanadium, 0-4% tungsten, and the balance iron.

(2) Trade name of Textileather Corporation (USA) for polyvinyl chlorides used for artificial leather and synthetic textile fabrics.

Turkey brown. A brown natural earth consisting chiefly of ferric oxide, manganese oxides and clay, and used as a permanent pigment. Also known as *Turkey umber.*

Turkey red. A red paint pigment based on ferric oxide (Fe_2O_3).

Turkey umber. See Turkey brown.

Turkish boxwood. The very hard, yellow wood of the tree *Buxus sempervirens*. It has a fine texture, and an average weight of 910 kg/m^3 (57 lb/ft^3). Source: Europe including British Isles, Asia Minor. Used for wood engraving, turnery, woodwinds, and mathematical instruments. Also known as *European boxwood*.

Turm. Trade name of Gipswerke Dr. Karl Würth GmbH & Co. (Germany) for gypsum products including plaster of Paris, casting, flooring and molding plaster, wall board, and drywall compounds.

Turnbull's blue. A deep blue, inorganic pigment obtained by reacting a ferrous salt with potassium ferricyanide [$K_3Fe(CN)_6$]. Used for paints and blueprints.

Turner's yellow. Basic lead chloride ($PbCl_2 \cdot 7H_2O$) available in the form of a yellow crystalline powder, and used as a pigment. Also known as *Cassel yellow; Montpelier yellow; patent yellow; Verona yellow.*

turn insulating paper. A paper, usually of kraft type and coated with insulating varnish, used to insulate conductors designed to become coils in inductive equipment (e.g., transformers).

turquoise. A blue, bluish green or greenish gray mineral composed of copper aluminum phosphate hydroxide pentahydrate, $CuAl_6(PO_4)_4(OH)_8 \cdot 5H_2O$. It may contain iron oxides. Crystal system, triclinic. Density, 2.84-2.95 g/cm^3; refractive index, 1.62. Occurrence: USA (Virginia), UK, Iran. Used as a gemstone and in ceramics.

Turrit. Trademark of Turrit (Germany) for a highly microporous lightweight lime concrete.

tuscanite. A colorless mineral composed of potassium sodium calcium aluminum carbonate silicate sulfate hydroxide monohydrate, $K(Ca,Na)_6(Si,Al)_{10}O_{22}(SO_4,CO_3,OH)_2 \cdot H_2O$. Crystal system, monoclinic. Density, 2.83 g/cm^3; refractive index, 1.590. Occurrence: Italy.

Tuscan red. A mixture of red iron oxide (Fe_2O_3), and a red dye (alizarin) used as a red pigment.

tusionite. A honey-yellow to cinnamon-brown mineral of the calcite group composed of manganese tin borate, $MnSn(BO_3)_2$. Crystal system, rhombohedral (hexagonal). Density, 4.73 g/cm^3; refractive index, 1.854. Occurrence: Afghanistan.

tussah. (1) A light brown-colored, washable *wild silk* produced by the larvae of certain species of uncultivated Asian moth of the genus *Antheraea*. It has a strong, coarse filament, that is flat in cross section, and can be spun into textiles. Also known as *tussore; tussah silk.*

(2) A fabric made from tussah silk (1).

tussah silk. See tussah (1).

Tusson. Trade name for cuprammonium rayon fibers.

tussore. See tussah.

Tutania. British trade name for *Britannia metal* containing 0-16% antimony, 0.7-2.7% copper, 0.8% lead, 0-1.3% zinc, and the balance tin. Used for bearings, tableware, kitchenware, household ware, platters and dishes.

tutenag. A white alloy, similar to nickel silver, containing varying amounts of copper, zinc and nickel, and sometimes iron or lead. Used for tableware and ornamental applications.

tuyere brick. Refractory brick containing one or more small openings through which air and gaseous fuel are forced into a (blast) furnace.

tvalchrelidzeite. A lead-gray mineral composed of mercury antimony arsenic sulfide, $Hg_{12}(Sb,As)_8S_{15}$. Density, 7.38 g/cm^3. Occurrence: Georgian Republic.

TVA slag. A lightweight expanded slag with a honeycomb-like structure produced by the Tennessee Valley Authority (USA) by treating molten slag with water, high-pressure steam, compressed air, or a combination of these treatments, and crushing the resulting product into pieces about 9.5-12.7 mm (0.375-0.500 in.) in size. Used in the manufacture of lightweight concrete and concrete products, and for lightweight, heat-insulating blocks.

tveitite. A white to pale yellow mineral composed of calcium rare-earth yttrium fluoride, $Ca_{1-x}(Y,Ln)_xF_{2+x}$. Crystal system, monoclinic. Density, 3.94 g/cm^3; refractive index, 1.479. Occurrence: Norway.

Twaron. Trademark of Teijin Twaron BV (Netherlands) for a series of high-performance aromatic polyamide (polyaramid) fibers. They have a density of 1.44 g/cm^3 (0.05 $lb/in.^3$), high thermal stability, high ultimate tensile strength (3.15 GPa or 457 ksi) and elastic modulus (80 GPa or 11.6×10^3 ksi), good creep and fatigue resistance, good toughness, good electrical properties (insulator), and good chemical resistance (but may be attacked by strong acids and bases). A grade (*Twaron HM*) with even higher elastic modulus (124 GPa or 1.8×10^3 ksi) than standard *Twaron* is also available. Used as reinforcement fibers for advanced engineering composites, and also for electrical and thermal insulation applications. Formerly known as *Arenka*.

tweed. (1) Formerly, a coarse, heavy fabric with rough surface finish hand-spun from Scottish sheep's wool.

(2) Now, a woolen cloth with a rough surface, usually twill-woven using yarns of different colors. Used for clothing especially for suits and coats.

12-15 compounds. See II-V compounds.

12-16 compounds. See II-VI compounds.

Twi-Clad. Trade name of T. & W. Ide Limited (UK) for permanent colored fired glass cladding panels.

Twi-Lite. (i) Trade name of Amerada Glass Company (USA) for a laminated safety glass with a gray plastic interlayer.

(2) Trade name of Shatterproof Glass Corporation (USA) for laminated glass with an amber-colored plastic interlayer.

Twill. Trade name of PPG Industries Inc. (USA) for rough-cast glass.

twill. See twill-weave fabrics.

twill braid. A ribbon or cord formed by joining together three fiber strands using the twill-weaving technique in which two warp (longitudinal) threads are interlaced with one weft (filling) thread in alternate rows producing raised diagonal lines.

twill-weave fabrics. Woven fabrics in which the weft (filling) threads pass alternately over one and then under two or more warp (longitudinal) threads producing raised diagonal lines. Also known as *twill*.

Twindow. Trademark of PPG Industries Inc. (USA) for insulating double glazing units.

TwindoWeld. Trademark of PPG Industries Inc. (USA) for double glazing units having electrically fused edges.

twine. A strong strand, usually 5 mm (0.2 in.) or less in diameter, consisting of fibers or yarns of cotton, flax, jute, sisal, rayon, nylon or other material, compacted into a twisted structure, and used for tying or binding parcels, bundles, bales, newspapers, lumber, etc.

Twinfilm. Trademark of Twinpak Inc. (Canada) for plastic film used for liquid packaging applications.

Twinflex. Trademark of Wheelabrator Inc. (USA) for woven filter fabrics.

Twin Glass. Trade name of Rylock Company, Limited (USA) for sealed insulating glass used for doors and windows.

Twinlook. Trademark of Heraeus Kulzer Inc. (USA) for a dual-cure dental resin cement.

Twin-Mo. Trade name of H. Boker & Company (USA) for a high-speed steel (AISI type M2) containing 0.85% carbon, 5.5% tungsten, 4.5% molybdenum, 4% chromium, 1.5% vanadium, and the balance iron. It has excellent wear resistance, hot hardness, and is used for general-purpose lathe tools, broaches, reamers, milling cutters, and drills.

Twin Mo-Co. Trade name of H. Boker & Company (USA) for a molybdenum-type high-speed steel (AISI type M36) containing 0.8% carbon, 6% tungsten, 5% molybdenum, 4% chromium, 2% vanadium, 8% cobalt, and the balance iron. It has a great depth of hardening, excellent hot hardness and good wear resistance. Used for lathe and planer tools, milling cutters, drills, and other tools.

twinnite. A black mineral composed of lead antimony arsenic sulfide, $Pb(Sb,As)_2S_4$. Crystal system, triclinic. Density, 5.32 g/cm^3. Occurrence: Canada (Ontario).

Twin Pane. Trade name of Twin Pane Corporation (USA) for insulating glass.

Twin-Seal. Trade name of Western Glass Company Limited (Canada) for insulating glass.

Twin Six. Trade name of Allegheny Ludlum Steel (USA) for a tool steel containing 0.7% carbon, 6% tungsten, 6% molybdenum, and the balance iron. Used for tools, drills, and hobs.

Twinsulite. (1) Trademark of Ford Motor Company (USA) for sealed window insulating units.

(2) Trademark of C. Vitrerie Franklain Quebec Inc. (Canada) for double-glazing units.

Twinsulite Plus One. Trademark of Ford Motor Company (USA) for sealed window insulating units.

Twintex. Trademark of St. Gobain Performance Plastics (USA) for thermoplastic composites made of commingled unidirectional glass fibers and filaments, such as polyethylene, polypropylene, polybutylene terephthalate or polyethylene terephthalate. Used as reinforcements in autobody panels, and structural parts.

Twin Van. Trade name of CCS Braeburn Alloy Steel (USA) for a tungsten-type high-speed steel (AISI type T2) containing 0.96% carbon, 18.5% tungsten, 4.25% chromium, 2.1% vanadium, 0.7% molybdenum, and the balance iron. It has a great depth of hardening, excellent machinability and wear resistance, high hot hardness, and good toughness. Used for lathe and planer tools, milling cutters, drills, dies, and punches.

Twinwindow. Trade name of Twinwindow (Manchester) Limited (UK) for spacer-type sealed double glazing units.

Twisloc. Trade name for nylon fibers and yarns.

Twist. Trade name of SA Glasverbel (Belgium) for a glass with a striated, loose basket-weave pattern on a contrasting pebbly textured background.

twisted nematic liquid crystal. Liquid crystal arrangement in which a *nematic liquid crystal* is placed between two grooved alignment plates, one of which is then oriented perpendicularly to the other. The liquid crystal molecules tend to align themselves along these grooves, and thus a twisted structure is obtained with a twist of 90° from the top to the bottom plate which makes it possible to utilize the birefringence of the liquid crystal for optical switching applications and active-matrix liquid-crystal displays (AMLCDs). Abbreviation: TN liquid crystal. See also liquid crystal.

twisted yarn. A yarn consisting of a *core yarn* with a wrapping of another filament, strip or yarn in which both yarns twist together and cannot be separately unwrapped.

twistless yarn. A yarn made without twist to enhance its softness, dyeability or other properties.

two-component adhesive. An adhesive provided in two components or parts that are mixed prior to application and cure at room or elevated temperatures. Examples include epoxies, acrylics, polyimides, silicones, etc. Also known as *two-part adhesive*.

two-dimensional nanostructures. See nanostructured materials.

two-dimensional photonic crystals. See 2D photonic crystals.

two-dimensional polymers. See 2D polymers.

two-directional fabrics. See bidirectional fabrics.

II-V compounds. A group of usually semiconducting compounds formed by combining an element of Group IIB (Group 12) of the Periodic Table, e.g., divalent zinc or cadmium, and an element of Group VA (Group 15), e.g., pentavalent arsenic or antimony. Examples include zinc arsenide and cadmium antimonide. Also known as *12-15 compounds*.

II-IV compounds. A group of usually semiconducting compounds formed by combining an element of Group IIA (Group 2) of the Periodic Table, e.g., divalent magnesium or calcium, and an element of Group IVA (Group 14), e.g., tetravalent germanium, silicon or tin. Examples include magnesium germanide and calcium silicide. Also known as *2-14 compounds*.

2-14 compounds. See II-IV compounds.

two-mica granite. See binary mica.

two-part adhesive. See two-component adhesive.

two-phase alloy. An alloy consisting of two phases, i.e., two chemically and structurally homogeneous microstructural portions. Examples include aluminum-silicon alloys, lead-tin alloys, and silver-copper alloys.

two-phase material. A material consisting of two phases, i.e., two chemically and structurally homogeneous microstructural portions. Examples include two-phase alloys, dispersion-strengthened aluminum (i.e., dispersions of aluminum oxide particles in aluminum metal), cemented carbides (e.g., tungsten carbide in cobalt or nickel binders), carbon-filled rubber and fiber-reinforced plastics (e.g., glass fiber-reinforced epoxies, carbon fiber-reinforced polyester.

Twoscore. Trade name of Dunford Hadfields Limited (UK) for a corrosion- and heat-resistant low-carbon, high-chromium steel containing about 0.1-0.2% carbon, 15-20% chromium, 2% nickel, and the balance iron. Used for propeller shafts, seaplane hardware and machine parts.

II-VI compounds. A group of usually semiconducting compounds formed by combining an element of Group IIB (Group 12) of the Periodic Table, e.g., divalent zinc, cadmium or mercury, and an element of Group VIA (Group 16), e.g., hexavalent sulfur, selenium or tellurium. Examples include cadmium sulfide, mercury selenide and zinc telluride. Also known as *12-16 compounds*.

Two Spur. Trade name of T. Turton & Sons Limited (UK) for a cobalt-tungsten high-speed steel containing 0.78% carbon, 4.5% chromium, 0.7% molybdenum, 21% tungsten, 1.7% vanadium, 6% cobalt, and the balance iron. It has good nonwarping properties and wear resistance, and is used for cutting tools.

two-stage phenolic resins. See novolacs.

two-to-one brass. A ductile brass containing two parts (66.7 wt%) copper, and one part (33.3 wt%) zinc. Used for drawn and spun parts. See also high brass; yellow brass.

Tybrene. Trade name of Dow Chemical Company (USA) for a series of strong, tough, thermally stable acrylonitrile-butadiene-styrene (ABS) plastics used for molded articles.

Tycel. Trademark of Lord Corporation (USA) for laminating adhesives used for structural applications.

Tychem CPS. Trade name of Reichhold Chemicals, Inc. (USA) for styrene-butadiene copolymers used as coal-tar additives, and as workability enhancers in cementitious mixtures.

tychite. A colorless mineral composed of sodium magnesium carbonate sulfate, $Na_6Mg_2(SO_4)(CO_3)_4$. Crystal system, cubic. Density, 2.59 g/cm^3; refractive index, 5.10. Occurrence: USA (California).

Tycon. Trade name of Carpenter Technology Corporation (USA) for an oil-hardening alloy steel containing 0.4% carbon, 0.65% manganese, 0.65% chromium, 1.75% nickel, 0.35% molybdenum, and the balance iron. Used for gears, shafts, and axles.

TyCor. Trade name of of WebCor Technologies Inc. (USA) for high-strength sandwich structural panels composed of fiber-reinforced foam cores infused with synthetic resin and co-cured with skin layers. They have a density ranging from 0.1-0.3 g/cm³ (0.004-0.01 lb/in.³), and good fatigue properties and damage tolerance. Used for aircraft applications.

Tycril. Trade name of Reichhold Chemicals, Inc. (USA) for a series of abrasion-resistant synthetic latexes used for textile fabrics.

Ty EZ. Trade name for polyolefin fibers and yarns.

Tygan. Trade name of Fothergill Limited (UK) for saran (polyvinylidene chloride) fibers with good chemical resistance, good weatherability, and good resistance to mildew and insects. Used for carpets and rugs, draperies, upholstery, and industrial fabrics.

Tygavac. Trade name of Tygavac Advanced Materials Limited (UK) for a series of polymeric films and fabrics, and sealing tapes. The films and fabrics are used as breather cloths and vacuum bagging films in vacuum molding, and the tapes are employed in vacuum-bag molding to seal between the bagging film and the molding tool surface.

Tyglas. Trade name of Fothergill & Harvey Limited (UK) for laminated and toughened glass.

Tygobond. Trademark of US Stoneware Company for vinyl- and rubber-based adhesives used for joining porous and semiporous materials.

Tygon. Trade name of St. Gobain Performance Plastics (USA) for a series of transparent thermoplastic polymers based on vinyls that are ordinarily rigid, but can be made flexible with plasticizers. They are low-cost, general-purpose materials, often copolymerized and susceptible to heat distortion. *Tygon* polymers have a durometer hardness of 55 Shore A, excellent resistance dilute and weak acids and bases, good resistance to ozone, aging and oxidation, poor resistance to ultraviolet radiation and strong and concentrated acids and bases, and a service temperature range of -50 to +74°C (-58 to +165°F). Used for pipe, garden hose, highly-flexible tubing, electrical wire insulation, linings of chemical equipment, coatings, floor coverings, and adhesives.

Tygothane. Trademark of St. Gobain Performance Plastics (USA) for polyurethane tubing.

Tygoweld. Trademark for Norton Performance Plastics (USA) for modified structural adhesives.

Tylac. Trademark of Reichhold Chemicals, Inc. (USA) for a series of synthetic rubber latexes and elastomers including various grades of styrene-butadiene, carboxylated styrene-butadiene and modified styrene-butadiene latexes, and several nitrile rubbers and nitrile-rubber latexes. Used in paints and hydraulic cement mixtures. *Tylac CPS* refers to a series of styrene-butadiene and acrylonitrile-butadiene copolymers used in paints and hydraulic cement mixtures to enhance adhesion, flexibility and water resistance.

Tyler. Trade name of Time Steel Service Inc. (USA) for a water-hardening tool steel (AISI type W5) containing 1.1% carbon, 0.6% chromium, 0.3% manganese, 0.25% silicon, and the balance iron. It has excellent machinability, fair toughness, and moderate shock resistance. Used for punches, mandrels, and dies.

Tylerite. Trade name of W.S. Tyler Inc. (USA) for a nickel cast iron containing 3.2% carbon, 2.2% silicon, 1.5% nickel, and the balance iron. Used for watch plates and molding machine parts.the Periodic Table, e.g., divalent magnesium or calcium, and an element of Group IVA (Group 14), e.g., tetravalent germanium, silicon or tin. Examples include magnesium germanide and calcium silicide. Also known as *2-14 compounds*.

Tylok-Plus. Trade name of Dentsply/LD Caulk (USA) for a zinc polycarboxylate dental luting cement.

Tylon. Trade name of Tyne Plastics (UK) for a series of polyamides (nylons) including nylon 6 and nylon 6,6.

Ty-Loy. Trade name of W.S. Tyler Inc. (USA) for an abrasion-resistant alloy steel containing 0.7% carbon, 1.2% chromium, and the balance iron. Used for wire cloth.

Tynab. Trade name of Tyne Plastics (UK) for a series of acrylonitrile-butadiene-styrene (ABS) resins.

Tyne. Brand name for a commercially-pure antimony (99+%).

Tyneside. Trade name of Tyneside Safety Glass Company Limited (UK) for laminated and toughened glass.

Tynec. Trade name of Tyne Plastics (UK) for a series of polycarbonate resins.

Tynep. Trade name of Tyne Plastics (UK) for a series of polyesters including polybutylene terephthalate (PBT) and polyethylene terephthalate (PET).

Tynex. Trademark of E.I. DuPont de Nemours & Company (USA) for nylon filaments available with tapered and round cross sections. Used for abrasives, brushes and paint brushes.

Ty-Pak. Trade name of Manville Corporation (USA) for a fiberglass roving package assembly.

Typar. Trademark of E.I. DuPont de Nemours & Company (USA) for a tough, high-strength nonwoven composed of thermally bonded polypropylene filaments. Used for carpet backing, roof linings, packaging materials, in geotextiles for the construction of road subsoils, railway tracks, and in building construction as a tear-resistant house wrap.

type metals A group of alloys containing 54-95% lead, 2-28% antimony, 2-20% tin, and sometimes copper. They are readily fusible, expand slightly upon solidification and produce sharp castings with hard surfaces. *Type metals* have good resistance to air, inks, oils, water and alkaline cleaning solutions, and are used for printing type. See also electrotype metal; intertype metal; linotype metal; monotype metal; standard type metal; stereotype metal.

type I cement. See Portland cement.

type I superconductor. See soft superconductor.

type II cement. See moderate sulfate-resistant cement.

type II superconductor. See hard superconductor.

type III cement. See high-early-strength cement.

type IV cement. See low-heat cement.

type V cement. See sulfate-resistant cement.

typewriter metal. A corrosion-resistant alloy of 57% copper, 20% nickel, 20% zinc and 3% aluminum. Used for typewriter parts.

typewriter paper. A high-quality bond paper having a rag content of at least 75%, a high degree of whiteness, good strength and rigidity and high surface finish, and often water-marked for superior performance. It is available in letter size, 215.9 × 279.4 mm (8.5 × 11 in.), and legal size, 215.4 × 355.6 mm (8.5 × 14 in.). Usually sold by weight (in lb.) per ream (500 sheets).

Typlex. Trade name of Ziv Steel & Wire Company (USA) for an air- or oil-hardening, shock-resistant steel containing 0.4% carbon, 4% nickel, 1.5% chromium, 0.8% molybdenum, 0.4% manganese, and the balance iron. Used for hot-forging dies, and gages.

Tyranno. Trade name of Ube Industries (USA) for a series of modified continuous silicon carbide (SiC) fibers consisting of either silicon, titanium and carbon or silicon, zirconium and carbon, and having an average diameter of 10-15 μm (300-590 μin.), high tensile strength and elastic moduli, an upper service temperature of 1300°C (2370°F) and good thermal stability. Used as reinforcement fibers for advanced engineering composites.

Tyrant Extra. Trade name of Vereinigte Edelstahlwerke (Austria) for a medium-carbon tool steel containing 0.45% carbon, 1% silicon, 1.05% chromium, 2% tungsten, and the balance iron.

Tyrenka. Trade name of Enka BV (Netherlands) for high-tenacity rayon fibers used for tire cords and industrial fabrics.

Tyrethane. Trade name of W.S. Tyler, Inc. (USA) for polyurethane screens.

tyretskite. A white or brownish mineral that is composed of calcium borate hydroxide, $Ca_2B_5O_8(OH)_3$, and may also contain strontium. Density, 2.85 g/cm³; refractive index, 1.64. Occurrence: Siberia.

Tyrex. Trademark of E.I. DuPont de Nemours & Company (USA) for nylon 6,12 fibers and monofilaments used for automotive tire cord.

Tyril. Trademark of Dow Chemical Company (USA) for strong, transparent styrene-acrylonitrile (SAN) copolymer resins supplied in unmodified, glass-fiber-reinforced, high-impact, high-heat, fire-retardant and UV-stabilized. They have good chemical and heat resistance, good processibility, high stiffness, and are used for automotive components including instrument panel lenses.

Tyrilfoam. Trademark of Dow Chemical Company (USA) for a styrene-acrylonitrile (SAN) film.

Tyrin. Trademark of DuPont Dow Elastomers (USA) for chlorinated thermoplastic low-density polyethylenes that can be processed by extrusion, calendering and injection molding, and are compatible with other resins, such as acrylonitrile-butadiene-styrene, styrene-acrylonitrile, polyethylene and polyvinyl chloride. They possess excellent chemical and ignition resistance, excellent low-temperature properties, good weathering resistance, inherent flexibility with plasticization, and a processing temperature of 177°C (351°F). Used for tough, weather-resistant products. *Tyrin CPE* refers to a series of chlorinated polyethylenes.

Tyrite. Trademark of Lord Corporation (USA) for a series of urethane adhesives with excellent temperature and impact resistance, and high resistance to water and many chemicals. They are used as structural adhesives for metals, plastics, rubber, and textiles.

Ty-Rod. Trademark of W.S. Tyler, Inc. (USA) for long-slot wire cloth.

Tyrolit. Trade name of Tyrolit Schleifmittelgesellschaft mbH & Co. (Germany) for an extensive range of abrasives supplied as powders, pastes and papers, and corundum, silicon carbide and diamond grinding wheels and disks.

tyrolite. A green mineral composed of calcium copper arsenate carbonate hydroxide hexahydrate, $CaCu_5(AsO_4)_2(CO_3)(OH)_4 \cdot 6H_2O$. Crystal system, orthorhombic. Density, 3.18-3.65 g/cm^3. Occurrence: Spain.

Tyron. Trade name for high-tenacity rayon fibers used for tire cords and industrial fabrics.

tyrosine polycarbonates. A family of fully synthetic, resorbable biopolymers produced by the polymerization of a monomer based on the amino acid tyrosine ($C_9H_{11}NO_3$). They hydrolytically degrade into naturally occurring, non-inflammatory byproducts. Current and future applications include tissue engineering, drug delivery, orthopedics, spinal and craniomaxillofacial surgery, and devices, such as screws, nails, pins and plates for the fixation of fractures and osteotomies. See also tyrosine-derived polymers.

tyrosine-derived polymers. A class of soluble, amorphous pseudo-poly(amino acid) polymers based on the amino acid *tyrosine* ($C_9H_{11}NO_3$). In these polymers the amino acid is linked not only through amine bonds, but also incorporates non-amide bonds, such as ester, urethane or carbonate linkages. Examples of such polymers include tyrosine-derived polyarylates and polycarbonates. Used as a biomaterial in tissue engineering, orthopedics and drug delivery. See also pseudo-poly(amino acid); tyrosine polycarbonates.

tyrrellite. A light bronze mineral of the spinel group composed of cobalt copper nickel selenide, $(Cu,Co,Ni)_3Se_4$. Crystal system, cubic. Density, 6.60 g/cm^3. Occurrence: Canada (Saskatchewan).

Tyseley alloy. British alloy containing 87.5% zinc, 3.5% copper, 8.7% aluminum and 0.3% silicon. Used for castings and ornaments.

Tytin. Trademark of Kerr Dental (USA) for a spherical particle alloy (amalgam) containing 60% silver, 28% tin and 12% copper with an optimum alloy-to-mercury ratio of about 1:0.73. It has high early compressive strength and good tensile strength, and is used as a dental amalgam for restorations and core-build-ups.

Tytite. Trade name for polyolefin fibers and products.

tyuyamunite. A canary-yellow mineral of the carnotite group composed of calcium uranyl vanadium oxide octahydrate, $Ca(UO_2)_2V_2O_8 \cdot 8H_2O$. Crystal system, orthorhombic. Density, 3.60 g/cm^3; refractive index, 1.90. Occurrence: USA (Texas).

Tyvek. Trademark of E.I. DuPont de Nemours & Company (USA) for *spun-bonded olefins* (high-density polyethylene fibers) used for the manufacture of packaging, envelopes, reusable and disposable laboratory coats and clothing, and for house wrap employed in building construction.

Tyweave. Trademark of E.I. DuPont de Nemours & Company (USA) for architectural mesh.

Tyweld. Trade name for rayon fibers and yarns used for textile fabrics.

Tyzor. Trademark of E.I. DuPont de Nemours & Company (USA) for an extensive series of organotitanates and organozirconates, i.e., organic compounds of titanium or zirconium, such as tetrabutyl titanate or zirconium bis(diethyl citrato)dipropoxide.

TZC alloys. TZC is a trade name of Universal Cyclops (USA) for a series of high-temperature molybdenum alloys containing 1.2% titanium, 0.3% zirconium and 0.1% carbon. They are similar to *TZM alloys*, but have greatly improved mechanical properties, recrystallization temperatures of about 1550°C (2820°F), and are age hardenable.

TZM alloys. TZM is a trade name for AMAX Corporation (USA) for a series of high-temperature molybdenum alloys containing 0.40-0.55% titanium, 0.06-0.012% zirconium, and 0-0.04% carbon. Supplied in bar, rod, sheet, foil and powder form, they have a density of 10.22 g/cm^3 (0.369 lb/in.^{3}), a melting range of 2500-2600°C (4530-4710°F), high strength and hardness, good high-temperature properties, good corrosion, creep and heat resistance, and a recrystallization temperature of 1400°C (2550°F). Used for structural applications, heat exchangers, high-temperature furnaces, heat engines, radiation shields, nuclear reactors, extrusion dies, and tooling for hot-die forging.

U

Ube. Trade name of Ube Industries (USA) for a series of polymeric materials. Examples include *Ube Nylon 6*, *Ube Nylon 66* and *Ube Nylon 12* supplied in unreinforced, glass fiber- or carbon fiber-reinforced, glass bead- or mineral-filled, fire-retardant and UV-stabilized grades. *Ube Nylon 6* and *Ube Nylon 66* are also supplied in molybdenum disulfide-, silicone- and polytetrafluoroethylene-lubricated grades, and the latter is also available in high-impact and super-tough grades. *Ube Nylon 12* is also supplied in flexible and semiflexible. Also included under thjis trade name are low-density polyethylenes (*Ube PE*) and polypropylene homopolymers (*Ube PP*), supplied in standard and UV-stabilized grades.

Ube-Nitto Polypro. Trademark of Ube-Nitto Kasei Company (Japan) for polypropylene staple fibers and filament yarns.

U Brand. Trade name of Firth Brown Limited (UK) for a series of plain-carbon tool steels supplied with carbon contents ranging from 0.7 to 1.7%.

Ucar. (1) Trademark of Union Carbide Corporation (USA) for a series of high-density, ultrahard metal and ceramic coatings applied to the surfaces of metallic materials by plasma-torch or detonation-gun processes to provide improved wear resistance, frictional properties, heat resistance, and electrical insulation. The hardness values and moduli of rupture obtained depend on the particular coating composition and can be as high as 1200 Vickers [for a 83(W,Ti)C-17Ni coating] and 965 MPa (140 ksi) [for a 62Co-28Mo-8Cr-2Si coating], respectively. *Ucar* coatings are used for aircraft and helicopter parts, such as turbine-engine components, airframes and rotor components, and for oilfield machinery components, chemical and plastic processing equipment, wear sleeves, pump and compressor seals, and textile machinery components.

(2) Trade name of Union Carbide Corporation (USA) for a series of wrought and cast heat and corrosion-resistant nickel-chromium, nickel-chromium-molybdenum and nickel-chromium-cobalt alloys. Used for turbine parts, combustion equipment, furnace parts, valves, and machinery parts.

(3) Trademark of Union Carbide Corporation (USA) for a series of synthetic latexes and water-soluble polymers.

(4) Trademark of Union Carbide Corporation (USA) for hybrid fabrics woven from conductive carbon yarns and insulating glass yarns.

(5) Trademark of Union Carbide Corporation (USA) for a series of copolymers of vinyl chloride and vinyl acetate, sometimes with hydroxypropyl acrylate, vinyl alcohol or maleic acid.

Ucet. Trade name for a series of carbon blacks.

Uchatin's bronze. A wear-resistant bronze composed of 92%

copper and 8% tin, and used for bearings, gears and worm wheels.

Ucinite. Trade name of Ucinite Company (USA) for phenolic resins and plastics.

Uddco. Trade name of Uddeholm Corporation (USA) for several corrosion-resistant high-chromium steels, and various corrosion- and heat-resisting steels used for petroleum refining equipment, heat exchangers, and pump and valve parts.

Uddeholm. Trade name of Uddeholm Corporation (USA) for an extensive line of austenitic, ferritic and martensitic stainless steels.

Udel. (1) Trademark of BP Amoco (USA) for a clear, thermoplastic polysulfone-based engineering resin that is transparent to both microwaves and light. Available in unfilled, glass fiber- or carbon fiber-reinforced, polytetrafluoroethylene-lubricated and extruded grades, it has high-performance properties, good toughness and rigidity, easy moldability, a service temperature range of -100 to +150°C (-150 to +300°F), good resistance to oxidation and hydrolysis, high resistance to degradation by gamma radiation, good resistance to chemicals, such as mineral acids, alkalies, salt solutions, detergents and oils, and very low resistance to chemicals, such as ketones and aromatic and chlorinated hydrocarbons. Used for casings and housings for electronic components, piping, filtration equipment, chemical processing equipment, and food-processing equipment.

(2) Trademark of BP Amoco (USA) for biaxially oriented polypropylene film used for laminations, box windows, and shrink wraps.

Udi. Trade name NV Syncoglas SA (Belgium) for a series of nonwoven, prestressed, glass-fiber rovings supplied on various substrates, such as asbestos-mat layers (*Udi-Asbestos*), chopped-strand mat layers (*Udi-Komplex*) and chopped-strand mat layers (*Udi-Mat*).

Udiglas. Trade name of NV Syncoglas SA (Belgium) for nonwoven, prestressed, glass-fiber rovings supplied on surfacing tissues.

Udimar. Trade name of Special Metals Corporation (USA) for a series of *maraging steels*.

Udimet. Trade name of Special Metals Corporation (USA) for a series of nickel-base superalloys containing 15-20% chromium, 15-19% cobalt, 3-6% molybdenum, 2-5% aluminum, 3-5% titanium, 2-4% iron, 1-2% tungsten, and traces of silicon, manganese, zirconium, boron and carbon. Available in cast and wrought form, they have excellent heat resistance and are used for turbines, power plant equipment, and equipment for the rubber, plastics and paper industries.

Udipreg. Trade name of NV Syncoglas SA (Belgium) for pre-impregnated, unidirectional, nonwoven glass-fiber rovings.

Ufi Gel C. Trade name of Voco-Chemie GmbH (Germany) for a silicone elastomer for denture lining and relining, and obturation applications.

U Foam. Trade name of ICI limited (UK) for urea-formaldehyde resins.

Uformite. Trademark of Rohm & Haas Company (USA) for an extensive series of amino resins made by the polycondensation of melamine, urea or *s*-triazine with formaldehyde. They are usually supplied in the form of colorless or pale solutions in water or other suitable solvent, and used for alkali-resistant solvent or water-base amino resin-modified alkyd coatings on automobiles, aluminum siding, metal awnings, refrigerators, washing machines, etc., in varnishes, in adhesives for plywood and cardboard boxes, in paper coatings, as sizing for wet-strength paper, in foundry core binders, and in textile pigment binders.

Ugine. Trade name of Pechiney/Ugine Aciers (France) for an extensive series of steels including various austenitic, ferritic and martensitic stainless steels, water-hardening plain-carbon tool steels, hot-work tool steels, molybdenum, tungsten and cobalt-tungsten high-speed steels, case-hardening steels, machinery steels, structural steels and heat-resistant steels.

Uginium. Trade name of Pechiney/Ugine Aciers (France) for a stainless steel containing 0.1% carbon, 12% chromium, 12% nickel, and the balance iron. It has good resistance to tarnishing in air, and is used for tableware, valves, valve parts, trim, and cutlery.

Uginox. Trade name of Pechiney/Ugine Aciers (France) for austenitic, ferritic and martensitic stainless steels and several heat-resistant nickel-chromium and chromium-nickel steels.

Ugiplus. Trade name of Pechiney/Ugine Aciers (France) for a series of free-machining alloy steels containing 0.35-0.36% carbon, 1-3% chromium, 0.2-0.3% molybdenum, 0.07% sulfur, and the balance iron. Used for screw-machine products.

Ugistab. Trademark of Aimants Ugimag (France) for a rare-earth magnet based on a neodymium-iron-boron material. It has good thermal stability and corrosion resistance, and is used for electric motors and computer peripherals. See also rare-earth magnet; neodymium-iron-boron magnet.

U-Glas. Trademark of Saint Gobain (France) for a molded textured glass supplied in the form of U sections.

ugrandite. A green mineral of the garnet group composed of calcium magnesium aluminum iron silicate hydroxide, $Ca_3Fe_2(SiO_4)_{3-x}(OH)_{4-x}$. Crystal system, cubic. Density, 3.45 g/cm^3; refractive index, 1.828. Occurrence: China.

Ugikral. Trademark of Plastimer (France) for acrylonitrile-butadiene-styrenes.

UHU. Trade name of UHU GmbH (Germany) for an extensive series of glues and adhesives including various synthetic resin glues, hot melts, spray adhesives, and all-purpose and specialty adhesives. *UHU-plus* is a two-component epoxy adhesive supplied in heat-curing and room-temperature-curing grades and having relatively high peel/tensile strength up to a temperature of about 82°C (180°F). Used for metal bonding applications. *UHU Stic* is a glue supplied in form of a twist-base stick for bonding cardboard, paper and wood. Other *UHU* products include *UHU Alleskleber* all-purpose adhesives, *UHU Sekundenkleber* instant adhesives, *UHU Montagekleber* assembly adhesives, *UHU Klebefix* paper adhesives, *UHU coll* paper and wood adhesives, *UHU plast* plastic adhesives, and *UHU Kleister* adhesive pastes and sizes.

uintaite. See gilsonite.

Ukarb. Trade name for a series of carbon blacks.

Ukigusa. Trade name of Central Glass Company Limited (Japan) for a patterned glass.

uklonskovite. A colorless mineral composed of sodium magnesium sulfate hydroxide dihydrate, $NaMg(SO_4)(OH)·2H_2O$. Density, 2.50 g/cm^3. Occurrence: Chile, Russian Federation.

Ulbraseal. Trade name of Ulbrich Stainless Steels & Special Metals Inc. (USA) for an iron-nickel alloy with low thermal expansion, used for glass-sealing applications, e.g., for automotive headlights and television picture tubes, and resistors and thermostats.

ULCB steels. See ultralow-carbon bainitic steels.

ULC IF steels. See ultralow-carbon interstitial-free steel.

Ulcometal. Trade name of NL Industries (USA) for a bearing alloy composed of 98-99% lead and a total of 1-2% barium and calcium.

Ulcony. Trade name for an alloy composed of 65% copper and 35% lead, and used for heavy-duty bearings operating under poor lubrication conditions.

ulexite. A colorless mineral composed of sodium calcium borate octahydrate, $NaCaB_5O_9·8H_2O$. Density, 1.96 g/cm^3; refractive index, 1.504. Occurrence: Chile, USA (California, Nevada). Used as a source of boron. Also known as *boronatrocalcite*.

ullmannite. A steel-gray to silver-white mineral of the pyrite group that is composed of nickel antimony sulfide, NiSbS, and can also contain arsenic. Crystal system, cubic. Density, 6.65 g/cm^3. Occurrence: Germany, Italy.

ulmin brown. See Vandyke brown.

U-Loy. Trade name of Republic Steel Corporation (USA) for sheet steel containing 0.06-0.1% carbon, 0.25-0.4% manganese, 0.25% copper, and the balance iron. Used for roofing and culverts.

Ultalite. Trade name of Cyco International Proprietary Limited (Australia) for aluminum metal-matrix composites.

Ultem. Trademark of GE Plastics (USA) for high-performance linear amorphous thermoplastics based on polyetherimide (PEI) resins. Available in regular, modified and 10, 20 or 30% glass fiber-reinforced grades, they have excellent toughness and creep resistance, good strength and stiffness, high dimension stability, good corrosion resistance, electrical properties and processibility, outstanding heat resistance, inherent flame resistance, a service temperature range of -40 C to +180°C (-40 to +355°F) and a glass transition temperature of 215°C (422°F). Used for pump housings, oil chambers, pump parts, aircraft panels and seat component parts, and automotive and electronic components.

Ultima Gold. Trademark of Degussa-Ney Dental (USA) for a silver-free dental ceramic alloy containing 77.1% palladium and 2% gold. It has high yield strength and sag resistance, a density of 10.7 g/cm^3 (0.39 $lb/in.^3$), a thermal expansion coefficient compatible with most porcelains, and a hardness of 245 Vickers. Used for porcelain-to-metal restorations.

Ultima Lite. Trademark of Degussa-Ney Dental (USA) for a dental casting alloy containing 78.4% palladium. It has high yield strength, good marginal integrity, a mid-range thermal expansion coefficient compatible with most porcelains, and a hardness of 300 Vickers. Used for porcelain-to-metal restorations.

Ultimet. Trademark of Haynes International Inc. (USA) for a series of cast and wrought cobalt-base superalloys with exceptional corrosion and wear resistance used for chemical processing equipment, and extrusion dies and molds.

Ultimium. Trade name of Eutectic Corporation (USA) for hardfacing materials composed of tungsten-base carbides and used in the form of electrodes for producing hard, wear-resistant overlays on steel and cast-iron crusher liners, conveyors, and wear plates.

Ultimo. Trademark of Atlas Specialty Steels (Canada) for a series of medium-carbon alloy steels. *Ultimo-4* is a carburizing grade machinery steel containing 0.4% carbon, 0.75% manganese, 0.75% chromium, 1.75% nickel, 0.4% molybdenum, 0.03% phosphorus, 0.03% sulfur, with the balance being iron. It has high tensile, yield and core strength, and is used for shafts, spindles, clutches, liners, springs, piston rods, truck and tractor gears, heavy-duty axles, gears, driveshafts, transmission parts, rolls, cement and crusher hammers, and shear blades. *Ultimo-6* is a shock-resisting tool steel containing 0.55% carbon, 0.55% manganese, 0.8% silicon, 1% chromium, 1.6% nickel, 0.75% molybdenum, and the balance iron. It has high toughness and abrasion and wear resistance and is used for hot punches, dies, shear blades, extrusion tools, and forging-die inserts.

Ultipor. Trade name of Pall Corporation (USA) for resin-bonded glass fibers used for particulate filters.

ultra-accelerator. A strong accelerator, usually containing thiuram sulfides and dithiocarbamates, used to shorten the low-temperature vulcanization time of rubber.

Ultra-Bar. Trademark of Quanex Corporation (USA) for an electric resistance-heated, quenched-and-tempered steel bar that is superior to conventionally-treated steel bars.

ultrabasite. See diaphorite.

Ultrablack. Trademark of Loctite Corporation (USA) for high-temperature vulcanizing (HTV) and room-temperature-vulcanizing (RTV) silicone sealants and gasket materials that is resistant to oils, noncorrosive to aluminum, iron and steel, has low odor and volatility, and an operating temperature range of -60 to +330°C (-76 to +625°F). Used for automotive engine gaskets.

Ultra-Blak. Trademark of Electrochemical Products, Inc. (USA) for black oxide finishes used on ferrous alloys.

Ultrablend. (1) Trade name of BASF Corporation (USA) for several polymer blends including polybutylene terephthalate/acrylonitrile-styrene-acrylate and polycarbonate/polybutylene terephthalate alloys.

(2) Trademark of Ultradent Products, Inc. (USA) for a light-cure urethane dimethacrylate (UDMA) based material used in dentistry as a cavity liner. *Ultrablend Plus* is a light-activated, radiopaque version that contains calcium hydroxide and calcium hydroxyapatite, and is used as a dentin liner and protective base.

Ultra-Blue. Trademark of Loctite Corporation (USA) for room-temperature-vulcanizing (RTV) silicone sealant that provides controlled swell in contact with oil creating an even tighter seal. It is highly flexible, resistant to hardening, will not dry out, crack or crumble, does not corrode aluminum, steel and iron, and has an operating temperature range of -60 to +260°C (-76 to +500°F). Used for automotive sealing applications including oil pans, camshaft covers, timing chain covers, valve covers, differential housings, and as a general industrial sealant against hydrocarbons, fluids and greases.

Ultra-bright. Trade name for nylon fibers and yarns used for textile fabrics.

Ultra-Bond. Trade name of Den-Mat Corporation (USA) for a dental resin cement used for bonding dentin and enamel to metal or nonmetal appliances.

Ultra Brite. Trade name of Alchem Corporation (USA) for a tin electrodeposit and deposition process.

Ultra Capital. Trade name of Darwins Alloy Castings (UK) for tungsten high-speed steels containing 0.76% carbon, 4.2% chromium, 18-22% tungsten, 1.25% vanadium, and the balance iron. Used for reamers, drills, cutters, and other tools.

Ultra Capital Plus One. Trade name of Eagle & Globe Steel Limited (Australia) for a cobalt-tungsten high-speed steel containing 0.85% carbon, 4.25% chromium, 18.5% tungsten, 1.2% vanadium, 5% cobalt, and the balance iron. Used for cutting tools, reamers, and dies.

Ultra Capital Plus Two. Trade name of Darwins Alloy Castings (UK) for a cobalt-tungsten high-speed steel containing 0.76% carbon, 4.25% chromium, 20% tungsten, 1.5% vanadium, 10%

cobalt, and the balance iron. Used for cutting tools for hard materials.

Ultra-Capital Steel. Trade name of Darwins Alloy Castings (UK) for a high-speed steel containing 0.7% carbon, 17% tungsten, 3.4% chromium, 1% vanadium, and the balance iron. Used for turning tools, cutters, hobs, and broaches.

UltraCast. Trade name of Old Hickory Clay Company (USA) for a series of high-purity ball clays and ball clay slurries.

Ultracast. Trademark of Lafarge Réfractaires Monolithiques (France) for ultralow cement castables used for steel ladles and tundishes.

UltraC Diamond. Trade name of Surmet Corporation (USA) for wear-resistant diamond-like carbon (DLC) coatings with unique amorphous structures applied in vacuum by plasma deposition with low stress at low substrate temperatures (100°C or 212°F). They have high hardness (up to 8000 DPH), excellent corrosion resistance, very low coefficient of friction, and adhere well to many substrates including metals, ceramics and polymers, producing dark, mirror-flat surfaces. Available in thicknesses exceeding 15 μm (600 μin.), they are used for wear and antifriction applications, e.g., on industrial products such as mechanical seals, in biomedicine, e.g., on orthopedic implants, and in semiconductor wafer processing.

Ultracheck. Trade name of Certain-teed Products Corporation (USA) for glass-fiber mats used for landscaping applications.

Ultrachem. Trade name of EDRO Engineering (USA) for a precipitation-hardening stainless steel used for the manufacture of tooling for the plastics industry. It requires no heat treatment, has a high-luster, polished surface finish, excellent impact and tensile strengths, and good corrosion resistance.

Ultrachrome. Trade name of Enthone-OMI Inc. (USA) for a hard chromium coating.

Utraclear. Trade name of Hapco Inc. (USA) for a series of crystal clear liquid molding compounds with excellent physical properties, high tensile strength and heat distortion, and excellent impact resistance.

Ultra Cobalt. Trade name of Latrobe Steel Company (USA) for a tungsten-type high-speed steel (AISI type T6) containing 0.8% carbon, 4% chromium, 2% vanadium, 18% tungsten, 12% cobalt, and the balance iron. It has good deep-hardening properties, excellent wear resistance and hot hardness, and is used for cutting tools, and drills. Formerly known as *Electrite Ultra Cobalt*.

Ultracoat. (1) Trademark of Endotec (USA) for a highly adherent ultra-high quality titanium nitride (TiN) ceramic coating (approximately 10 μm or 0.4 mil thick) with near-diamond hardness, produced by any of several proprietary processes. When applied to titanium alloy or polymer substrates and properly polished, it provides a surface with high smoothness and durability, and significantly reduces friction and wear. Used on the femoral heads and acetabular cups of hip prostheses.

(2) Trade name for electroless deposits with co-deposited submicron particles of a diamond-like material. The deposits are based on nickel, nickel-cobalt-phosphorus, or nickel-phosphorus with ceramic particles, and can be applied to many substrates including alloy, carbon and stainless steels, aluminum, copper and titanium. They provide high hardness, excellent corrosion, wear and abrasion resistance, low coefficients of friction and good mold-release properties. Used as chromium replacements for various applications including molds, dies and tools.

(3) Trade name for continuous galvanized steel sheets and

coils.

Ultra Copper. Trademark of Loctite Corporation (USA) for a high-temperature-vulcanized (HTV) silicone sealant that is resistant to oils, noncorrosive to aluminum, iron and steel, and has low odor and volatility, and an operating temperature range of -60 to +370°C (-76 to +700°F). Used for automotive gaskets.

Ultracor. Trademark of YLA Cellular Products Company (USA) for a series of ultralight honeycomb core materials with carbon, quartz or aramid (*Kevlar*) fibers used for weight-sensitive applications, e.g., in the aerospace industry.

Ultracryl. Trade name of Masel Orthodontics (USA) for a self-cure dental acrylic resin available in liquid and powder form for orthodontic applications.

Ultracoustic. Trade name of Certain-teed Products Corporation (USA) for nonfammable glass-fiber ceiling boards with travertine texture.

Ultra-Cut. Trade name of Bliss & Laughlin Steel Company (USA) for a free-machining steel containing 0.08-0.12% carbon, 0.3% sulfur, and the balance iron. Used for screw-machine products.

Ultradent LC. Trademark of Ultradent Products, Inc. (USA) for a blue-pigmented, low-viscosity, light-cure dental resin used for block-out applications.

Ultradie. Trade name of Universal-Cyclops Corporation (USA) for a series of high-carbon, high-chromium cold-work tool steels (AISI types D2 and D3). The AISI D2-types contain about 1.5% carbon, 12% chromium, 0.8% molybdenum, 1% vanadium, and the balance iron. They possess high hardness, abrasion and wear resistance, great depths of hardening and good nondeforming properties. The AISI D3-types contain about 2.25% carbon, 12% chromium, 0.8% molybdenum, 0.2% vanadium, and the balance iron. They have excellent abrasion and wear resistance and non-deforming properties, great depths of hardening, and good compressive strength. Uses of *Ultradie* steels include lamination, trimming, stamping and blanking dies, forming, bending and drawing tools, punches, shear knives, and blades.

Ultradur. Trademark of BASF Corporation (USA) for a series of polybutylene terephthalate (PBT) thermoplastic polyesters available in unreinforced, reinforced, molding, extrusion, impact-modified and flame-retardant grades. They have high strength and stiffness, excellent heat-aging properties, good high-temperature stability, good chemical and weathering resistance, and low water absorption. Used for automotive components, electrical and electronic applications, pump and motor housings, and highly stressed engineering parts. *Ultradur B* refers to a series of PBT thermoplastic polyesters available in unreinforced, 10, 20 and 30% glass fiber-reinforced, mineral- or glass bead-filled, flame retardant and UV-stabilized grades.

Ultra Ethylux. Trade name of Westlake Plastics Company (USA) for extruded and molded high-density polyethylene products.

Ultrafil. (1) Trade name of Coltene-Whaledent (USA) for hot-injection *gutta-percha* used for dental filling applications.

(2) Trademark of Pennine Fibre Industries Limited (UK) for polyester staple fibers.

Ultra-Fine. Trademark of Indium Corporation of America (USA) for fine soldering wire based on tin-lead, tin-silver or gold-tin alloys. It has good tensile strength, and is used for electronic device packaging applications.

Ultrafine. Trade name of Certain-teed Products Corporation (USA) for glass-fiber insulation products.

ultra fine-grained material. A material that has a grain size finer than ASTM 13.5, i.e., more than 5793 grains per square inch at 100 diameters magnification, or 89786 grains per square millimeter at 1 diameter magnification.

ultrafine silica. A white, finely divided silica powder composed of spherical particles that are 4-25 μm (0.0002-0.001 in.) in size. It is obtained from silicon tetrachloride by combustion, and used as a filler, pigment and carbon black substitute in light colored rubber products, as a flatting agent in paint, and as a thickener in greases.

Ultraform. Trademark of BASF Corporation (USA) for a series of acetal copolymer resins based on polyoxymethylene (POM) and supplied in regular and specialty compounded grades, e.g., glass-reinforced, impact-enhanced, silicone-lubricated, antistatic, UV-resistant, and extrusion. A supertough acetal/elastomer alloy grade is also available. *Ultraform* resins have excellent heat stability, high lot-to-lot consistency, high strength, toughness and rigidity, excellent resistance to mineral oils, gasoline, tetrachlorocarbon and dilute alkalies, and fair resistance to trichloroethylene and dilute acids. Used for mechanical components and fasteners for automotive, electrical and plumbing applications including gears and rollers, and for water handling components.

Ultraglaze. Trademark of GE Silicones (USA) for a series of silicone-based construction adhesives and glazing compounds.

Ultra-Gold. Trade name of J.F. Jelenko & Company (USA) for a dental bonding alloy containing 87.5% gold, 10% platinum and 1% palladium. Used for porcelain-to-metal restorations.

Ultra Grey. Trademark of Loctite Corporation (USA) for a grayish high-performance silicone sealant that is resistant to oils, noncorrosive to aluminum, iron and steel, has low odor and volatility, and an operating temperature range of -60 to +330°C (-76 to +625°F). Used for automotive gaskets.

ultrahigh molecular weight polyethylenes. A group of linear, low-pressure polyethylene resins with extremely high molecular weights (3×10^6 to 6×10^6). They have a density range of 0.92-0.94 g/cm^3 (0.033-0.034 lb/in.^{3}), outstanding abrasion resistance and impact strength, high tensile strength, good energy absorption and sound-dampening properties, good stress-cracking resistance, excellent properties at cryogenic temperatures, negligible water absorption, an upper service temperature of 95°C (203°F), low thermal expansivity, fair ultraviolet resistance, good dielectric properties, low coefficients of friction, nonstick, self-lubricating surfaces, good resistance to dilute acids, alcohols, alkalies, greases, oils and ketones, fair resistance to concentrated acids, and poor resistance to aromatic hydrocarbons and halogens. Used for seals, pistons and pumps of cryogenic equipment, liners for grain silos, bearings, bushings, gears, bulk material hoppers, dump trucks, railcars, chutes, conveyors troughs and flights, wear strips, slide plates, textile machinery parts, surgical supports and prosthesis (e.g., acetabular cups), pump components, valve parts, recreational equipment (e.g., golf-ball cores and bowling alleys). Abbreviation: UHMWPE. See also polyethylenes.

Ultra High Pure Silicon. Trade name of Surmet Corporation (USA) for a silicon coating applied to ceramic, metallic and polymer substrates by a conformal plasma process at low temperatures (typically 100°C or 212°F).

ultrahigh-strength steels. A group of structural steels with a minimum yield strength of 1.38 GPa (200 ksi) including several medium-carbon low-alloy steels (AISI/SAE 4130, 4140, 4340, 6150 and 8640), medium-alloy air-hardening steels (AISI types H11 Mod and H13), 9Ni-4Co structural steels and several stainless steels. Used for constructional applications, such as bridges, buildings, ships, tankers, and industrial machinery and equipment.

Ultra Hide. Trade name of Glidden Company, Division of SCM Corporation (USA) for an extensive range of interior paint products including latex primers and sealers, latex eggshell, flat and semi-gloss paints and alkyd semi-gloss paints.

Ultrakem. Trademark of Sherwin-Williams Company (USA) for a high-solids baking enamel used on heavy, off-road construction and farm equipment.

Ultralam. (1) Trademark of Rogers Corporation (USA) for a polytetrafluoroethylene (PTFE) base woven laminate with excellent dimensional stability, very low etch shrinkage, good surface quality, high bond strength, very low moisture absorption, and a highly uniform dielectric constant. Used for high-reliability stripline and microstrip circuit applications.

    (2) Trademark of Veneer Technology, Inc. (USA) for hardwood veneer panels.

    (3) Trademark of Stauffer Chemical Company (USA) for a reinforced vinyl-foil-laminated sheathing material used for construction applications.

Ultralevel. Trade name of Pax Surface Chemicals Inc. (USA) for a bright nickel electroplate and plating process.

UltraLight. Trade name of TP Orthodontics (USA) for a light-cure orthodontic bonding resin for brackets and other appliances.

Ultralight. Trade name of Cospray Products Division of Alcan Aluminum Corporation (USA) for very-low-density aluminum alloys that are produced by spray casting, and can be made into high-strength parts by forging, machining or extrusion processes.

Ultra Light-Weld. Trademark of Dymax Corporation (USA) for ultraviolet/visible light curing structural adhesives.

Ultra-Lite. Trade name of Enthone-OMI Inc. (USA) for a bright nickel electroplate and plating process.

Ultralite. (1) Trade name of Certain-teed Products Corporation (USA) for glass-fiber duct insulation and lining products.

    (2) Trademark of InKan Limited (Canada) for a toughened safety glass.

    (3) Trade name of Ultramatic Equipment Company (USA) for lightweight ceramic mass-finishing media.

Ultralloy. (1) Trade name of Ultraloy Corporation (USA) for an alloy composed of 48.5% zinc, 48.5% tin, 2-2.5% copper, 0.6-0.8% silicon and 0.03% silver. Depending on the particular composition, melting temperatures range from 370 to 400°C (700 to 750°F). Used as a solder for aluminum and aluminum alloys.

    (2) Trade name of Hapco Inc. (USA) for a series of low-viscosity, low-shrinkage, high-performance liquid molding compounds for pressure, open and vacuum casting, and low-pressure injection molding (LPIM) applications. They are particularly suitable for the manufacture of short-run and prototype parts.

UltrAlloy. Trademark of Alcoa-Aluminum Company of America (USA) for several wrought aluminum-magnesium alloys (AA 6000 series) with excellent machinability, weldability, anodizability and corrosion resistance, and good joining characteristics. Used for automotive brake pistons and transmission valves, driveshafts, steering yokes, air-conditioning compressor pistons, tripod fittings, and camera and cable-television components.

Ultralon. Trademark of ICI Americas Inc. (USA) for electrically conductive polytetrafluoroethylene coatings for metal sub-

strates. They are used to minimize static buildup on business machines and electronic components and devices.

ultralow-carbon bainitic steels. A group of high-strength sheet steels with ultralow carbon contents and bainitic microstructures. The typical composition range (in wt%) is up to 0.04% carbon, 0.2-1.8% manganese, 0.02-0.25% silicon, 0.03-0.04% aluminum, 0.04-0.06% niobium, small additions of molybdenum, titanium and boron, and the balance iron. They are often supplied as microalloyed, interstitial-free (IF) grades, and have excellent mechanical properties including high yield and tensile strengths, and good cold formability and fatigue resistance. Their ultralow carbon content is achieved through vacuum degassing, and their bainitic microstructure by thermomechanical treatment. Used chiefly for automotive applications especially autobodies. Abbreviation: ULCB steels.

ultralow-carbon interstitial-free steels. A group of high-strength sheet steels with ultralow carbon contents, typically less than 50 ppm, stabilized with titanium and/or niobium to remove interstitial elements from the solid solution. The low carbon content is obtained by efficient vacuum degassing. Supplied as cold-rolled, galvanized (hot-dip or electrogalvanized) and galvannealed products, they have high elongation, formability and drawability, and are used in the automotive industries for autobodies, floor pans and oil pans. Titanium-stabilized ultralow-carbon (TiSULC) steels are usually continuously annealed (CA TiSULC) or batch annealed (BA TiSULC). Abbreviation: ULC IF steels. See also interstitial-free steels.

ultralow-density polyethylenes. Linear copolymers of ethylene and a high percentage of a higher-molecular-weight alkene, such as hexene. They have a density of less than 0.90 g/cm³ (0.032 lb/in.³). Abbreviation: ULDPE. See also low-density polyethylenes.

Ultraloy. (1) Trade name of Solar Basic Industries (USA) for a heat-resistant alloy composed of 55% iron, 37.5% chromium and 7.5% aluminum. It has a maximum service temperature of 1316°C (2400°F), high electrical resistivity, and is used for furnace parts, resistors for heating furnaces, and heating elements.

    (2) Trade name of Alchem Corporation (USA) for a zinc-cobalt electrodeposit and deposition process.

Ultralume. Trade name of Alchem Corporation (USA) for a nickel electrodeposit and deposition process.

Ultralumin. Trade name of Ultralumin Leichtmetall AG (Germany) for a heat-treatable aluminum alloy containing 4.7% copper, 0.75% manganese, 0.2% nickel and traces of cerium. Used for light-alloy parts and forgings.

ULTRA-MAG. Trademark of Flexmag Industries, Inc. (USA) for a flexible, machinable thermoplastic permanent magnet sheeting material with multimagnetic polarization. Supplied in rolls, it may either have an uncoated surface, or can be laminated with a wide range of matte or high-gloss coatings, or pressure-sensitive adhesives. It has a density of 3.8 g/cm³ (0.14 lb/in.³), a maximum energy product of 0.78 × 10⁶ GOe, and excellent outdoor weatherability. Used for manufacture of various magnetic components for a variety of industrial applications.

Ultramag. Trade name of Ultramag (USA) for a high-density platinum-cobalt permanent magnet material with high coercive force and magnetic energy product, a Curie temperature of approximately 500°C (930°F), a maximum service temperature of 350°C (660°F), and good machinability.

ultramarine. See ultramarine blue.

ultramarine blue. A brilliant blue inorganic pigment with reddish hue, formerly also made from powdered *lapis lazuli,* but now commonly made by heating a mixture of china clay, sodium carbonate, carbon and sulfur, followed by grinding. It has good alkali and heat resistance, low hiding power, poor acid resistance and weatherability, and is used as a colorant for machinery and toy enamels, as a paint pigment, in white baking enamel and rubber products, for whitening paper and textiles, in printing inks, and in textile printing. Also known as *ultramarine*. See also French blue.

ultramarine green. An inorganic pigment with a greenish hue obtained as a byproduct in the preparation of *ultramarine blue* from china clay, sodium carbonate, carbon and sulfur.

ultramarine red. A reddish pigment made by calcining *ultramarine blue* in a nitric acid atmosphere.

ultramarine violet. A permanent, reddish blue paint pigment that is a polysulfide of sodium aluminosilicate, made from *ultramarine blue* by heating with ammonium chloride, or chlorine and hydrochloric acid.

ultramarine yellow. See lemon yellow.

Ultramid. Trademark of BASF Corporation (USA) for a series of thermoplastic polyamides (nylon 6, nylon 6,6, nylon 6,66, nylon 6,9, nylon 6,10, and blends from these) supplied in unreinforced, glass fiber- or mineral-reinforced, molybdenum disulfide-lubricated, UV-stabilized and impact-modified grades. *Ultramid* poly-amides possess high tensile strength and rigidity, high impact strength and moduli of elasticity, good thermal, electrical and chemical resistance, good to very good processibility, outstanding flame retardancy, a maximum service temperature of about 100-150°C (210-300°F), excellent resistance to gasoline, mineral oils, tetrachlorocarbon and dilute alkalies, fair resistance to trichloroeth-ylene and dilute acids, poor resistance to alcohols and glycols, and moderate water absorption. Used for tubing, pipes, hose, insulation, gaskets, bearings, bushings, foils, gears, cams, bolts, screws, sutures, bristles, cloth, carriers for microchips, electrical parts, injection-molded parts, and synthetic fibers. *Ultramid A* refers to strong, stiff nylon 6,6, *Ultramid B* to nylon 6 supplied in the form of film, sheet, rods, powder, granules and monofilaments, *Ultramid C3X* to nylon 6,66 and *Ultramid T* to nylon 6,6/6T.

Ultramirage. Trade name of Solutia Inc. (USA) for nylon 6,6 filaments and fibers used for textile fabrics.

Ultranyl. Trade name of BASF Corporation (USA) for polyphenylene oxide/polyamide (PPO/PA) alloys with good chemical resistance, good to moderate dimensional stability at elevated temperatures and low mold shrinkage. Used for automotive components.

Ultra-Pake. Trade name Ceramco (USA) for a dental porcelain used for metal-to-ceramic restorations.

Ultra-Pane. Trade name of Ultra-Pane Inc. (USA) for an insulating glass.

Ultrapas. Trade name of Huels America Inc. (USA) for a light-colored laminate composed of several sheets of paper or other cellulose material bonded with melamine-formaldehyde resin and compressed at elevated temperatures. Used in building construction.

Ultrapave. Trademark of Textile Rubber & Chemical Company (USA) for a durable, modified asphalt concrete based on a hot-melt blend of specialty latex rubber and asphalt. It has good resistance to abuse and extreme environmental conditions. Used for airport runways.

Ultrapek. Trade name of BASF Corporation (USA) for a series of high-performance polyaryletherketone (PAEK) resins supplied in unreinforced and glass-fiber-reinforced grades. The have

high heat resistance and good weatherability and are used for the manufacture of extruded and injection-molded components and equipment for the aerospace, automotive, electrical and electronic industries.

Ultraperm. Trademark of Vacuumschmelze GmbH (Germany) for a series of soft-magnetic crystalline alloys composed of 72-83% nickel and 17-28% iron. They have exceptionally high magnetic permeability, low saturation flux density, and low coercivity. Used for magnetic shielding, sensitive relays, measurement instruments, magnetic heads, magnetic amplifiers, chokes, circuit breakers, transformers and power electronics.

Ultraphan. Trademark of Lonza-Werke GmbH (Germany) for white and colored cellulose acetate film products.

Ultraphos. Trade name of Chemtech Finishing Systems Inc. (USA) for a phosphate coating and coating process.

Ultra Plank. Trademark of Abitibi Building Products (Canada) for wood-fiber siding planks.

Ultra-Plate. Trade name OM Group (USA) for electroless nickel and copper used for printed circuit boards.

UltraPruf. Trademark of GE Silicones (USA) for silicone-base construction sealants.

UltraSeal. Trademark of Ultradent Products, Inc. (USA) for composite resins used in dentistry as pit and fissure sealants. Both *UltraSeal XT* and *UltraSeal XT plus* are wear-resistant, flowable, self-cure composite resins, but contain different amounts of fillers.

Ultrasil. (1) Trade name of Heraeus Amersil, Inc. (USA) for ultrahigh-purity (99.9998+%) silica glass products.

(2) Trade name of Degussa AG (Germany) for precipitated silica.

(3) Trademark of Spectran Specialty Optics Company (USA) for optical fibers.

(4) Trademark of CR Mineral Corporation (USA) for pumice.

(5) Trademark of Ohio Sealants, Inc. (USA) for silicone caulking compounds.

UltraSoft. Trademark for a viscoelastic polymer gel used, e.g., for thermo-injecting into bicycle comfort pads.

Ultrason. Trademark of BASF Corporation (USA) for a series of polyether sulfones (*Ultrason E*) and polysulfones (*Ultrason S*) supplied in unreinforced and glass fiber-reinforced grades. *Ultrason E* are transparent, amorphous engineering polyether sulfone thermoplastics supplied in various unreinforced and glass and carbon fiber-reinforced grades as sheets, rods, films and granules. They have a density of 1.37 g/cm³ (0.049 lb/in.³), good elevated-temperature properties, good long-term thermal-aging resistance, high toughness and strength, poor resistance to fatigue and environmental stress cracking, good resistance to alkalies, dilute acids, alcohols, halogens, greases and oils, and fair radiation resistance. Used for electrical, electronic and industrial equipment.

Ultrasorb. Trade name of Owens-Illinois Inc. (USA) for an emerald-green glass that virtually eliminates light transmission in the 360-400 nm range.

UltraSpan. Trademark of GE Silicones (USA) for a series of silicones used for weatherstripping.

ultrastable material. (1) A chemical compound having an exceptionally high chemical stability, i.e., resistance to decomposition by and/or reaction with other compounds.

(2) A material having an exceptionally high stability to extreme conditions, such as high or low temperatures, radiation, chemical attack, weathering, and mechanical stress.

(3) A solid material, such as a plastic or composite, with an exceptionally high dimensional stability and shape retention tendency.

Ultra Stainless Steels. Trademark of AMETEK Specialty Metal Products (USA) for austenitic stainless steel powders with excellent corrosion resistance, produced from standard compositions (AISI types 303L, 304L and 316L) by adding a small amount of tin, and subsequent blending with a small amount of copper-nickel-tin alloy. They are supplied in three grades: (i) *Ultra 303L* with 0.02% carbon, 18% chromium, 11% nickel, 2% tin, 0.8% copper, 0.8% silicon, 0.2% manganese, 0.2% sulfur, 0.01% phosphorus, and the balance iron; (ii) *Ultra 304L* with 0.02% carbon, 19% chromium, 11% nickel, 2% tin, 0.8% copper, 0.8% silicon, 0.2% manganese, 0.01% sulfur, 0.01% phosphorus, and the balance iron; and (iii) *Ultra 316L* with 0.02% carbon, 17% chromium, 13% nickel, 2.5% molybdenum, 2% tin, 0.8% copper, 0.8% silicon, 0.2% manganese, 0.01% sulfur, 0.01% phosphorus, and the balance iron. *Ultra Stainless Steel* powders are used make powder-metallurgy austenitic stainless steels parts.

Ultrastrand. Trade name of Certain-teed Products Corporation (USA) for glass-fiber reinforcements used in the plastics industry.

Ultra-Suede. Trade name of Red Spot Paint & Varnish Company, Inc. (USA) for pigmented spray coatings used on plastics and metals.

Ultra Superior. Trade name of Stahlwerk Stahlschmidt GmbH & Co. (Germany) for tungsten-type high-speed steels.

Ultratech. Trade name for a corrosion-resistant dental bonding alloy.

Ultra-tech. Trade name of Lea Manufacturing Company (USA) for advanced phosphate conversion coatings with exceptional corrosion resistance, suitable for application to various metal substrates by immersion techniques.

Ultratek. Trademark of Ultratek Metals (USA) for nickel-chromium dental alloys in ingot form used for porcelain-to-metal fusion of crowns, bridges and other restorations.

UltraTemp. Trademark of Ultradent Products, Inc. (USA) for a water-soluble, eugenol-free polycarboxylate dental cement used for cementing and luting crowns, filling small temporary preparations, sealing endodontic access openings, and other dental work.

Ultra-Temp. Trade name of Aremco Products Inc. (USA) for a series of high-temperature ceramic adhesives (e.g., zirconia-based) containing metal powders.

Ultratex. (1) Trademark of Mantex Corporation (USA) for unreinforced and fiberglass-reinforced polyethylene polyol and polyester foams supplied as composite boards and sheets for marine applications.

(2) Trademark of Dofasco Inc. (Canada) for textured steel bars, rods, sheets and billets.

(3) Trade name of Albany International Corporation (USA) for wires and wire cloth used for Fourdrinier machines.

Ultrathene. Trade name of Quantum Chemical (USA) for a series of ethylene vinyl acetate (EVA) copolymers used as adhesion improvers in hot-melt and pressure-sensitive adhesives, in conversion coatings, and as modifiers in thermoplastics. They are available with vinyl acetate contents of 12, 25 and 33%.

Ultra-Therm. Trade name of Certain-teed Products Corporation (USA) for glass-fiber insulation products.

Ultratherm. (1) Trade name of Arlon Flexible Technologies Division (USA) for a lightweight, flexible composite material composed of a thin, nonflammable, thermally resistant woven fiber-mat with strong, heat reflective, metallized surface layers. Used for stamped parts in the automotive, aerospace and appliance industries.

(2) Trade name of J.C. Soding & Halbach (Germany) for a series of weldable, corrosion- and/or heat-resistant austenitic, ferritic and ferritic-pearlitic steel castings.

Ultra Touch. Trade name of BASF Corporation (USA) for nylon 6 fibers used for textile fabrics including clothing.

Ultratrac. (1) Trademark of Haysite Reinforced Plastics (USA) for glass-reinforced polyester sheeting.

(2) Trademark of Alco Industries, Inc. (USA) for fiberglass reinforced plastics supplied in sheets and pultruded shapes, and used for electrical insulation.

Ultratrim. Trade name for synthetic bonding resins used for orthodontic applications.

Ultravan. Trade name of Latrobe Steel Company (USA) for high-speed tool steels containing 1.5% carbon, 6.3% tungsten, 4.25% chromium, 4.75% vanadium, 5% molybdenum, 5% cobalt, and the balance iron. Used for inserted-blade cutting tools. Formerly known as *Electrite Ultravan*.

Ultra-Vat. Trade name for metallic fibers.

ultraviolet absorber. A material, such as a hydroxybenzophenone or benzotriazole, that absorbs ultraviolet radiation in the range of 300-350 nm and converts it to thermal energy. Used as a compounding ingredient in plastics and rubber to prevent discoloration and retard deterioration caused by ultraviolet light (especially sunlight). Abbreviation: UV absorber. Also known as *ultraviolet stabilizer.*

ultraviolet-absorbing glass. See ultraviolet glass.

ultraviolet-cure adhesives. A group of adhesives based on synthetic resins, such as acrylics, epoxies or polyesters, applied to substrates, such as glass, metals or plastics, in the liquid state and subsequently cured rapidly to a dry film by subjecting to ultraviolet radiation at wavelengths between 200 and 400 nm. They possess good dimensional stability and toughness, good abrasion, chemical and heat resistance, and improved substrate adhesion. Abbreviation: UV-cure adhesives.

ultraviolet-cure coatings. Organic coatings applied to metallic or nonmetallic substrates as thin films (up to 75 μm or 3 mils) and cured at or slightly above room temperature by subjecting to ultraviolet radiation. Abbreviation: UV-cure coatings.

ultraviolet glass. A glass made by introducing appropriate quantities of elements or compounds into the batch that absorb ultraviolet radiation without adversely affecting the transmission of visible radiation (light). Compounds and elements suitable for this application include cerium oxide (CeO_2), chromium, cobalt, copper, iron, lead, manganese, neodymium, nickel, titanium, uranium and vanadium. Abbreviation UV glass. Also known as *ultraviolet-absorbing glass*. See also Crooke's glass; Wood's glass.

ultraviolet-transmitting glass. A glass that is essentially free from iron, sulfur and titanium, and contains low or insufficient quantities of ultraviolet-absorbing elements or compounds to significantly impede or block ultraviolet radiation. Borosilicate and phosphate glass is available in ultraviolet-light-transmitting grades. Used for germicidal, black-light and sunlight lamps, fluorescent tubes, and ultraviolet-transmitting windows.

ultraviolet stabilizer. See ultraviolet absorber.

Ultratex. Trademark of Ciba-Geigy Corporation (USA) for a series of finishes based on silicone elastomers, and used for natural and synthetic textile fabrics.

Ultra Wear. Trade name of DSM Engineering Plastics (USA) for wear-resistant ultrahigh-molecular weight polyethylenes.

Ultrawear UHMWPE. Trade name of The Polymer Corporation (USA) for a ultrahigh-molecular-weight polyethylene used for the manufacture of tubing, sheeting and other products.

Ultrenka. Trade name of Enka BV (Netherlands) for rayon fibers and yarns used for textile fabrics.

Ultrima. Trade name for rayon textile fibers and yarns.

Ultron. (1) Trademark of Solutia Inc. (USA) for polyvinyl chloride film, and nylon 6,6 staple fibers, monofilaments and filament yarns. The nylon fibers are mixtures of different fiber shapes and sizes, have soil-hiding abilities, are supplied in several grades including *Ultron 3D* and *Ultron VIP*, and are used for commercial carpets.

(2) Trade name of Ultron Systems Inc. (USA) for polyvinyl chloride film products.

Ultrox. Trademark of M & T Chemicals Inc. (USA) for ceramic glaze opacifiers based on refined zirconium silicate powders.

Ultryl. Trade name of Ultron Systems Inc. (USA) for polyvinyl chlorides.

ulvospinel. A dark gray mineral composed of iron titanate, Fe_2TiO_4. It can also be made synthetically from stoichiometric amounts of ferric oxide (Fe_2O_3) and titanium dioxide (TiO_2) under prescribed conditions. Crystal system, cubic. Density, 4.78 g/cm^3; refractive index, 1.670. Occurrence: Sweden, USA (New Jersey).

UMA. Trade name of Republic Steel Corporation (USA) for a series of chromium and chromium-manganese steels used for machine parts, e.g., springs, gears, shafts, and bearings.

Uma. Trade name of XIM Products, Inc. (USA) for several urethane-modified acrylics.

umangite. A dark cherry red to bluish green mineral composed of copper selenide, Cu_3Se_2. Crystal system, tetragonal. It can also be made synthetically. Density, 6.44 g/cm^3. Occurrence: Argentina.

umber. A heavy earth that consists chiefly of ferric oxide (Fe_2O_3) with small percentages of silica (SiO_2), alumina (Al_2O_3), manganese oxides (MnO, MnO_2 and Mn_2O_3) and lime (CaO), and is used in its natural state (*raw umber*) as a brown pigment or after calcining (*burnt umber*) as a reddish-brown pigment in paints, ceramic bodies and glazes, lithographic inks and wallpaper. See also Caledonian brown; Cyprus umber.

umbite. A colorless or yellowish mineral composed of potassium titanium zirconium silicate monohydrate, $K_2(Zr_{0.8}Ti_{0.2})Si_3O_9 \cdot H_2O$. Density, 2.79 g/cm^3; refractive index, 1.610. Occurrence: Russian Federation.

umbozerite. A green to brown mineral of the amorphous group composed of sodium strontium thorium silicate hydroxide, $Na_3Sr_4ThSi_8O_{23}(OH)$. Crystal system, tetragonal. Density, 3.60 g/cm^3; refractive index, 1.640. Occurrence: Russian Federation.

Umbral. Trade name of Carl Zeiss GmbH (Germany) for a dark protective glass composed of two sheets of glass fused together. It does not transmit infrared or ultraviolet light, and is used for sunglasses.

umohoite. A radioactive mineral composed of uranyl molybdenum oxide hydrate, $UO_2MoO_2 \cdot xH_2O$, and available in three polytypes: 12A, 14A and 17A. The 12A polytype is blue-black or green, the 14A polytype dark blue, and the 17A polytype bluish black in color. Crystal system, orthorhombic (12A); monoclinic (14A and 17A). Density, 4.61 (12A); 4.55 (14A); 4.79 (17A); refractive index, 1.83. Occurrence: USA (Utah, Wyoming).

Una. Trade name of Una Welding Inc. (USA) for a series of low-carbon and stainless steels, and aluminum and copper alloys used for welding rods.

Unalit. Trade name of Unalit (France) for fiberboard made from 98% wood fibers by a wet process. Supplied in various grades in the form of thin, strong hardboard and light, acoustically and thermally insulating softboard products, it is used for paneling and veneering applications, and for furniture.

Unapeint. Trade name of Unalit (France) for hardboard panels of *Panodur* Quality P. They are first-choice panels with a dark brown core and one smooth lacquered and one rough side, supplied in sizes up to 1.7 × 5.5 m (5.5 × 18 ft) and in thicknesses from 2 to 8 mm (0.08 to 0.31 in.).

unalloyed tin. See commercially pure tin.

unalloyed titanium. See commercially pure titanium.

Unamo. Trade name of U.N. Alloy Steel Corporation (USA) for a series of high-speed steels containing 0.8-1% carbon, up to 0.3% manganese, up to 0.3% silicon, 4% chromium, 1-2% vanadium, 1.5-6% tungsten, 5-9% molybdenum, 0-5% cobalt, and the balance iron. They have high red-hardness, and are used for lathe and planer tools, cutters, drills, taps, and a variety of other tools.

Unarapid. Trade name of U.N. Alloy Steel Corporation (USA) for a series of high-speed steels containing 0.8-1.5% carbon, 4-4.5% chromium, 1-4.75% vanadium, 13.5-20% tungsten, 0-0.5% molybdenum, 0-12% cobalt, and the balance iron. They have high red-hardness, and are used for lathe and planer tools, cutters, drills, taps, and a variety of other tools.

Unarcoboard. Trade name of UNARCO (USA) for asbestos and other building boards.

Unavan. Trade name of U.N. Alloy Steel Corporation (USA) for several molybdenum-type high-speed steels (AISI type M).

unbalanced yarn. A yarn that has been twisted such that it will untwist and then retwist in the opposite direction.

Unbreakable. Trade name of Swift Levick & Sons Limited (UK) for a case-hardening steel containing 0.14% carbon, 0.9% manganese, 0.04% sulfur, 0.04% phosphorus, and the balance iron. Used for machine parts.

Unbreakable Metal. Trade name of American Smelting & Refining Company (USA) for a casting alloy composed of zinc, aluminum and copper, and used for slush and permanent-mold castings, lamp bases, toys, and novelty items.

unburnt brick. Brick, such as *adobe* or chemically bonded refractory brick, that develops mechanical strength without firing or burning. Also known as *unfired brick*.

unburnt refractory. Any refractory product (brick, tile, etc.) shaped by any of several processes including chemical bonding, metal encasement or deaeration and compression, but excluding firing and burning processes. Also known as *unfired refractory*.

uncombed cotton yarn. A cotton yarn made from roughly aligned (or carded) fibers.

Unda. Trade name of former Gerresheimer Glas AG (Germany) for a specially patterned glass blocks.

undercoated hardboard. See primed hardboard.

undercoated particleboard. See primed particleboard.

undercoater. See undercoating (1).

undercoating. (1) A coating material based on asphalt or rubberized asphalt that remains flexible when dry and resists stone chipping. Available in spray bottles, cans or pails for brush application, it is used to protect, insulate and/or soundproof automobile chassis, floor pans, wheel walls, trunks, and interior and exterior panels. Also known as *undercoater.*

(2) A special material, usually a lacquer or varnish, that may be applied to hardboard, particleboard, etc., as a base coat to improve paintability.

underglaze. A design, decoration or ceramic color applied directly to the unfired or once-fired surface of a ceramic article and subsequently coated with a clear glaze followed by final firing. Also known as *underglaze decoration.*

underglaze color. A finely milled ceramic color applied to a ceramic article, such as pottery, tile, or terra cotta, before the glaze is put on.

underglaze decoration. See underglaze.

underlay paper. See concrete subgrade paper.

under-ridge tile. Roofing tile, usually of burnt clay or shale, used directly under the tile forming the ridge in the construction of tile roofs.

Under-Water. Trade name of Stone Manganese-J. Stone & Company Limited (UK) for tin-base bearing metals containing 30% zinc, 1.5% copper, 0.5% lead and 0.5% antimony. Used for marine applications.

underwater concrete. A water-resistant concrete usually made with hydraulic cement that will set under water, and used in the manufacture of underwater structures, tunnels, foundations of oil rigs, docks and harbor facilities, hydroelectric dams, and reservoirs.

undoped cutback asphalt. Cutback asphalt made without an adhesion-promoting agent. See also cutback asphalt.

Unel. Trade name for a *spandex* (segmented polyurethane) fiber used for the manufacture of textile fabrics with excellent stretch and recovery properties.

unfinished paper. Paper that has not been passed through dry-end equipment (calender) and is therefore considerably rough on both sides.

unfired brick. See unburnt brick.

unfired refractory. See unburnt refractory.

ungemachite. A colorless to pale yellow, transparent mineral composed of potassium sodium iron sulfate hydroxide nonahydrate, $K_3Na_9Fe(SO_4)_6(OH)_3 \cdot 9H_2O$. Density, 2.29 g/cm³; refractive index, 1.502. Occurrence: Chile.

unglazed tile. A hard, dense floor or wall tile that has a homogeneous body deriving its color, texture and properties from the raw materials used and the particular manufacturing process and thermal history.

ungraded aggregate. Aggregate that has not been classified according to size and thus contains particles of sizes ranging from coarse to fine. See also aggregate.

uniaxial crystal. See optically uniaxial crystal.

Unibestos. Trade name of Pittsburgh Corning Corporation (USA) for asbestos insulation blocks and pipe insulation products.

Unibond. Trade name of Shofu Dental Corporation (Japan) for a beryllium-free nickel-chromium dental bonding alloy.

Unibraze. Trade name of J.W. Harris Company, Inc. (USA) for a series of aluminum-silicon, copper-zinc, copper-phosphorus, magnesium-base and silver-base brazing filler metals as well as several alloy-steel arc-welding electrode products.

Unibrite. Trade name of Atotech USA Inc. for a bright nickel electrodeposit and deposition process.

Unibroach. Trade name of Cyclops Corporation (USA) for a high-speed steel containing 0.8% carbon, 4% chromium, 6.25% tungsten, 6.25% molybdenum, 2.4% vanadium, and the balance iron. Used for cutters, broaches and other finishing tools.

Unichem. Trade name of Colorite Plastics (USA) for plasticized

polyvinyl chlorides supplied in three elongation ranges: up to 100%, 100-300% and above 300%. *Unichem UPVC* refers to unplasticized polyvinyl chlorides supplied irregular, crosslinked, high-impact, UV-stabilized and structural-foam grades.

Uni-Crom N.P. Trade name for a high-chromium dental solder.

Unichrome. Trade name of Atotech USA Inc. for chromium and copper electrodeposits and conversion coatings.

Unico. Trade name of Hidalgo Steel Company Inc. (USA) for an oil-hardening, shock-resistant tool steel containing 0.7% carbon, 0.4% manganese, 0.4% silicon, 0.7% chromium, 1.4% tungsten, and the balance iron. Used for pneumatic tools and equipment.

Unicol. Trade name of Ato Findley (France) for an alcohol-resin adhesive used for floor-covering applications.

Unicomb. Trade name for glass fibers and glass-fiber products.

Unicrete. Trademark of Unicrete Products Limited (Canada) for concrete roofing tiles.

Unicryl. Trade name for self-cure methacrylate-based dental acrylics.

Unicure. Trade name of Ferro Corporation (USA) for nonwoven glass fabrics.

Unicut. Trade name of Cyclops Corporation (USA) for several molybdenum-type high-speed steels (AISI type M3) containing 1-1.2% carbon, 5-6.25% molybdenum, 6-6.25% tungsten, 4% chromium, 2.4-3% vanadium, and the balance iron. They have good deep-hardening properties, excellent wear resistance, high hot hardness, good toughness and are used for lathe and planer tools, drills, taps, end mills, milling cutters, form tools, drawing dies, routers, and a variety of other tools.

Unidal. Trade name of Aluminium Industrie AG (Switzerland) for heat-treatable aluminum alloy containing 4-6% zinc and 0.5-1.5% magnesium. Used for architectural applications.

Uni-Die. Trade name of Columbia Tool Steel Company (USA) for an air-hardening, medium-alloy cold-work tool steel (AISI type A6) containing 0.71% carbon, 2% manganese, 1.3% molybdenum, 1% chromium, and the balance iron. It has good nondeforming properties and good wear resistance, and is used for gages, punches, dies, and forming tools. *Uni-Die Smoothcut* is a free-machining grade that contains a trace of sulfur, and is used for shear blades, and dies.

unidirectional composite. An engineering composite composed of continuous reinforcing fibers of carbon, boron, glass or graphite in a ceramic, metallic or polymeric matrix in which the fibers are anisotropic, i.e., oriented or aligned in the same direction. The mechanical properties of such a composite in the direction of the reinforcing fibers is exceptional, while those in the transverse direction are relatively poor. Abbreviation: UDC. Also known as *unidirectional fiber-reinforced composite.*

unidirectional fabrics. Fabrics made of fibers, filaments or yarns with a unidirectional (one-directional) weave. Also known as *one-directional fabrics.*

unidirectional fiber-reinforced composite. See unidirectional composite.

unidirectional laminate. A *plastic laminate* in which virtually all fibers are oriented or aligned in the same direction.

unidirectional molding compound. A *molding compound* containing reinforcing fibers oriented or aligned in the same direction.

unidirectional tape prepreg. See tape (4).

Unidur. (1) Trade name of Aluminium Walzwerke Singen GmbH (Germany) for wrought aluminum alloys containing about 4-4.5% zinc, 0.8-1.2% magnesium and 0-0.2% chromium. Used for aircraft parts, and automotive components.

(2) Trade name of Ato Findley (France) for vinyl additives for mortars.

Unifab. Trade name of Ferro Corporation (USA) for fine-weave glass-fiber yarns.

Unifast. Trade name of GC America Inc. (USA) for a self-cure acrylic resin used for denture repairs. *Unifast LC* is a dense, nonporous, light-cure acrylic resin with high stain and wear resistance, and good finishability. Supplied in translucent, crown colors (various shades of ivory) and gingival colors (various shades of pink), it is used in dentistry for the temporary restoration of crowns, bridges, inlays, onlays, splints and orthodontic plates. *Unifast Trad* is a self-curing, quick-setting acrylic resin supplied in crown colors (various shades of ivory) and gingival colors (various shades of pink), and used for temporary crowns, impression trays, denture repair, bite registration, orthodontic plates, and various other dental and orthodontic applications.

Unifil. Trade name for polyolefin fibers and yarns used for textile fabrics.

Unifil Bond. Trade name of GC Dental (Japan) for a light-curing, self-etching dental bonding system containing urethane dimethacrylate (UDMA) and hydroxyethyl methacrylate (HEMA).

Unifilo. (1) Trade name of Vetreria Italiana Balzaretti Modigliani SpA (Italy) for glass-fiber continuous strand mats.

(2) Trademark of Vetetrotex CertainTeed Corporation (USA) for a continuous strand mat composed of continuous-melt E-glass fibers and a polymer binder.

Unifilm. Trademark of Scapa Tapes North America, Inc. (USA) for pressure-sensitive adhesive transfer films based on acrylics, and used for bonding foams and fiberglass composites for acoustic and thermal insulation applications.

Uniflex. Trade name of Universal Glass Company (UK) for a flexible mirrored sheet glass.

Unifont. Trade name of Swiss Aluminium Limited (Switzerland) for various aluminum-silicon alloys and aluminum-zinc-magnesium casting alloys.

Uniform. Trade name of Cold Metal Products Company, Inc. (USA) for a series of cold-rolled low- and high-carbon strip steels used for springs, stampings, gaskets, and magnets.

Uniformat. Trade name of Ferro Corporation (USA) for chopped glass-fiber mats.

Uniform Oil Hardening. Trade name of Jessop Steel Company (USA) for a nonshrinking, oil-hardening tool steel containing 0.9% carbon, 1.2% manganese, 0.6% chromium, 0.5% tungsten, and the balance iron. Used for tools and dies.

uniform sand. Sand containing particles of uniform size. Also known as *closely graded sand*. See also nonuniform sand.

Uniglass. Trade name of Uniglass Industries (USA) for glass-fiber fabrics.

Unigold. Trade name of Engelhard Industries (USA) for a series of dental gold alloys containing varying amounts of palladium and platinum, and supplied in the annealed and age-hardened condition. They have a melting range of 850-890°C (1560-1635°F).

Unigleam. Trade name of Zinex Corporation (USA) for bright, ductile, corrosion-resistant electrodeposits used on nickel, copper, tin, lead, iron, steel and pewter.

Unihyde. Trade name of Duralac Inc. (USA) for a one-coat air-drying porcelain enamel.

Unilastic. Trade name for a silicone elastomer used for dental impressions.

Unilock. Trade name of Unilock Limited (Canada) for architec-

tural stone used for retaining walls.

Unilot. Trade name of Degussa-Ney Dental (USA) for a dental gold solder.

Uniloy. Trade name of Cyclops Corporation (USA) for an extensive series of standard and free-machining stainless steels including several austenitic, ferritic, martensitic, precipitation-hardening and heat-resisting grades (AISI types 201 through 502).

Unimach. Trade name of Cyclops Corporation (USA) for a series of low-alloy, hot-work tool, and structural steels.

Unimar. Trade name of Cyclops Corporation (USA) for a series of maraging steels with high strength, hardness and toughness, used for solid rocket cases, pressure vessels, jet engine and gas turbine parts, fasteners, and aluminum die-casting dies and cores.

Unimax. Trademark of A&P Technology (USA) for unidirectional carbon or fiberglass sleevings. The carbon sleevings are braided with carbon axials with elastic bias having a tensile strength of 3.86 GPa (560 ksi) and the fiberglass sleevings are braided with E-glass roving axials with elastic bias. Used as reinforcements in composites.

Unimet. Trade name of TRW Inc. (USA) for a series of tool materials based on sintered carbides.

Unimetal. (1) Trade name of Unimetal Company (USA) for a white metal containing lead, tin and antimony. Used for welding white metals.

(2) Trade name of Unimetal Company (USA) for nickel-chromium dental alloys.

Union. Trade name of Glacier Metal Company (USA) for a tin babbitt containing antimony and lead. Used for bearings.

Unionaloy. Trade name of Duraloy/Blaw-Knox Corporation (USA) for an abrasion-resistant cast iron containing 3.2% carbon, 2.5% silicon, 0.6% manganese, and the balance iron. Used for tube mill plugs, mill guides, and hopper liners.

union cloth. A fabric whose warp and weft are made with different types of fibers. Also known as *union fabric*.

union-dyed fabrics. Fabrics made by using two or more identically dyed fibers.

union fabric. See union cloth.

Union Hymo. Trade name of Republic Steel Corporation (USA) for a free-machining steel containing 0.3-0.4% carbon, 1.35-1.65% manganese, and the balance iron. Used for gears, shafts, and machinery parts. *Union Hymo Steel* is a free-machining case-hardening steel containing 0.15-0.25% carbon, 1-1.3% manganese, and the balance iron. Used for case hardened parts, shafts, and pinions.

Union Maxcut. Trade name of Republic Steel Corporation (USA) for a case-hardening steel containing 0.08% carbon, and the balance iron. Used for machinery and case-hardened parts.

Union McQuaid-Ehn. Trade name of Republic Steel Corporation (USA) for a carburizing steel containing 0.15-0.25% carbon, 0.3-0.6% manganese, 0.045% phosphorus, 0.055% sulfur, and the balance iron. Used for high-quality carburized parts.

Union Metal. Trade name of Union Metal Manufacturing Company (USA) for a steel containing 0.3% carbon, and the balance iron. Used for seamless tubing.

Union Multicut. Trade name of Republic Steel Corporation (USA) for a case-hardening steel containing 0.1-0.2% carbon, and the balance iron.

union yarn. A yarn consisting of different kinds of fibers twisted together.

Uniplast. Trade name of Universal Plastics Company (USA) for phenol-formaldehyde resins and plastics.

Unipol. Trade name of Shell Chemical Company) for an extensive series of polypropylene resins supplied in several grades including molding resins for the manufacture of sheet and film products, fibers and monofilaments for the textile fabric and cordage industries, high-viscosity resins for the manufacture of packaging materials, blow molding, injection molding, thermoforming and extrusion resins for bottles, containers, electrical components and other industrial and commercial goods, and coating grades for hot sealing applications.

Uni-Pore. Trade name of DeWal Industries Inc. (USA) for porous ultrahigh-molecular-weight polyethylene films with high strength, durability and chemical resistance. Supplied with pore sizes ranging from 2.5-50 μm (0.1-2 mils), they are used as a filtration and venting media for industrial and laboratory applications.

Uni-Post L.F. Trade name for a precious metal solder used for dental posts.

Unipox. Trade name of Ato Findley (France) for a line of acrylic and epoxy adhesives and coatings used for bonding floor tiles.

Unipreg. Trade name of Ferro Corporation (USA) for a glass mat impregnated with polyester resin and pressed in heated steel dies.

Unipress T. Trade name of United Clays, Inc. (USA) for a ball clay from Texas, USA, having a low carbon content and superior fired color. Used for wall tile and various general ceramic applications.

Unique. Trade name of Metalor Technologies SA (Switzerland) for a hard dental bonding and casting alloy containing 72.1% gold and 13.6% platinum.

Unirove. Trade name of Ferro Corporation (USA) for woven glass rovings.

Uniseal. (1) Trade name of Cyclops Corporation (USA) for several iron-nickel, iron-chromium and nickel-iron alloys with controlled thermal expansion used for glass-to-metal and ceramic-to-metal sealing applications.

(2) Trademark Para Paints Canada Inc. (USA) for an interior alkyd wall primer.

Unisil. Trade name of British Steel plc (UK) for a series of cold-reduced transformer steels containing less than 0.1% carbon, 3.1% silicon, and the balance iron.

UniSolder. Trade name of Nobil Metal SpA (USA) for a dental gold solder.

Unison. Trademark of Dentsply International (USA) for a fast-setting spherical dental amalgam alloy supplied in self-activating capsules for lustrous dental restorations.

Unistole. Trademark of Mitsui Chemicals, Inc. (Japan) for functional polymers available in two grades: *Unistole P*, a liquid primer for bonding and coating engineering polymers, such as polypropylene as well as rubber and other materials; and *Unistole R*, a modified polyolefin supplied in a hydrocarbon solvent, and used for coating metals and bonding metals to polyolefins.

Unistrand. Trade name of Ferro Corporation (USA) for continuous glass-fiber strands.

Uni-Syn. Trade name for resilient, chemical-resistant sheet packing made by bonding long-fiber *chrysotile asbestos* fibers with synthetic rubber, usually in the presence of special curing compounds. It is available as squares, typically 152 × 152 cm (60 × 60 in.) or 203 × 203 cm (80 × 80 in.) in size, and has a tensile strength of 24.2 MPa (3.5 ksi). Used for industrial and marine engines, and for general marine applications at temperatures up to 538°C (1000°F) and pressures up to 13.8 MPa (2 ksi).

Unitape. Trade name of Ferro Corporation (USA) for a woven glass-fiber tape for local reinforcements and for wrapping pipes.

Unitbond. Trade name of Ivoclar/Williams (USA) for a corrosion-resistant nickel-chromium dental bonding agent.

United. Trade name for saran (polyvinylidene chloride) fibers with good chemical resistance, good weatherability, and good resistance to mildew and insects. Used for carpets and rugs, draperies, upholstery, clothing, industrial fabrics, etc.

United Magnet. Trade name of Republic Steel Corporation (USA) for a magnet steel containing 0.6-0.9% carbon, 0.25-0.55% manganese, 0.25-2.1% chromium, 0-6% tungsten, and the balance iron. Used for electrical machinery and magnets.

Unitemp. Trade name of Cyclops Corporation (USA) for an extensive series of standard and high-nickel stainless steels, high-strength and heat-resistant steels, austenitic nickel-base alloys, and several cobalt-, nickel- and iron-base superalloys.

Unitika. Trade name of Unitika Limited (Japan) for polymer-based fibers regenerated from *chitin* obtained from beetles. Used for medical sutures.

Unitika Ester. Trade name of Unitika Limited (Japan) for polyester fibers and filament yarns.

Unitika Nylon. Trade name of Unitika Limited (Japan) for nylon 6 staple fibers, monofilaments and filament yarns.

Unito. Trade name of Kidd Drawn Steel/H.K.Porter Company (USA) for drill rod made from high-quality water-hardening tool steel, and used for twist drills, taps, pins, and punches.

Unitope. Trade name for glass fibers and glass-fiber products.

Unival. Trade name of Union Carbide Corporation (USA) for polyethylene resins.

Univan. Trade name of Duraloy/Blaw-Knox (USA) for a cast nickel-vanadium steel containing 0.28-0.32% carbon, 1.5% nickel, 1% manganese, 0.10-0.15% vanadium. It has high strength, excellent resistance to shocks and stresses. Used for locomotive frames, gears, pinions, spindles, and crossheads.

Univerbal. Trade name of SA Glaverbel (Belgium) for sheet glass made by the Pittsburgh sheet-glass process, developed by Pittsburgh Plate Glass Company, now PPG Industries (USA), and used for the vertical drawing of glass from the surface of the melt and through a slot of desired thickness.

Universal. (1) Trade name of Chung Yip Company (Hong Kong) for glass fibers used for thermal insulation applications.

(2) Trade name of Fidenz Vetraria SpA (Italy) for glass roofing tiles.

(3) Trade name of Krupp Stahl AG (Germany) for cold-work die and tool steels.

(4) Trade name of Permacon (Canada) for architectural stone supplied in two styles: smooth and split-face slope blocks. Used for retaining walls.

universal steel. (1) A carbon or alloy steel plate formed in a universal rolling mill utilizing sets of horizontal and vertical rolls.

(2) A group of low-carbon alloy steels, usually of the chromium-nickel type (about 1.5-2% chromium and nickel respectively), suitable for a wide range of applications including structural members and machine parts, e.g., gears, shafts and highly stressed parts.

Unix. Trade name of Stahlwerk Stahlschmidt GmbH & Co. (Germany) for a water- or oil-hardening tool steel containing 1.05% carbon, 1.1% tungsten, 0.9% manganese, and the balance iron. Used for cutters and dies.

unkilled steel. See rimmed steel.

unnilquadium. See rutherfordium.

Unoflex. Trade name of Unionglas AG (Germany) for a patterned glass. The raised part of the pattern is stain-finished.

Unopal. Trade name of Unionglas AG (Germany) for cast glass with ceramic color fired onto the patterned side.

Unosil S. Trade name for a silicone elastomer used for dental impressions.

unplasticized polyvinyl chlorides. Thermoplastic polymers that unlike regular polyvinyl chloride have not been compounded with plasticizers. Commercially available in film, sheet, rod, tubing and powder form, they have a density of 1.4 g/cm³ (0.05 lb.in.³), good strength and hardness, good ultraviolet resistance, low thermal expansivity, a service temperature range of -30 to +75°C (-22 to +167°F), good dielectric properties, good resistance to dilute acids and alkalies, fair resistance to concentrated acids, alcohols, greases, oils and halogens, and poor resistance to aromatic hydrocarbons and ketones. Although not flame resistant, they are self-extinguishing. Used for molded parts, containers, water pipes, and structural parts. Abbreviation: UPVC. See also plasticized polyvinyl chlorides.

unreinforced concrete. See plain concrete.

unsaturated hydrocarbon. Any straight-chain compound of hydrogen and carbon, such as acetylene or ethylene, containing carbon atoms that form one or more double or triple covalent bonds, and thus do not bond to a maximum of four other atoms.

unsaturated polyesters. A group of polyesters made by the condensation polymerization of dibasic acids or anhydrides with dihydric alcohols (e.g., ethylene glycol or propylene glycol) with the acids or anhydrides being partially or entirely composed of unsaturated materials, such as fumaric acid or maleic anhydride, with the balance being saturated materials, such as adipic acid or phthalic anhydride. Owing to their ethylenic unsaturation, they can be crosslinked with an ethylenically unsaturated reactive monomer, such as diallyl phthalate, methyl methacrylate, styrene, or vinyl toluene. They have a density of 1.3 g/cm³ (0.05 lb.in.³) (unreinforced), good elevated-temperature properties, good mechanical properties especially when reinforced, excellent chemical resistance, and moderate to excellent electrical properties. Used for appliance housings, circuit boards, switchgear, computer equipment, chemical tanks and process vessels, wastewater treatment equipment, valves, fittings, pipes, tubing, building panels, swimming pools, floor grating, doors, boats, canoes and kayaks, docks, bathtubs, shower stalls, furniture, adhesives and coatings, and for the manufacture of glass-fiber mat or roving prepregs. Abbreviation: UP. Also known as *thermosetting polyesters*.

unslaked lime. See calcium oxide.

unsymmetric laminate. A laminate in which the sequence of the component layers (or plies) below the midsurface or midplane is not a mirror image of the stacking sequence above it.

UnTil. Trade name of Sci-Pharm, Inc. (USA) for a self-cure, eugenol-free dental resin cement that bonds to dentin and metal, and is used as a luting agent for intermediate and temporary crown and bridge restorations.

Unyte. Trade name for a thermosetting urea formaldehyde resin commercially available in powder form, either colored, translucent or opaque. It has good moldability, fair machining properties, good resistance to alcohols, ketones, esters, hydrocarbons and oils, and is used for electrical parts and insulation.

upalite. An amber-yellow mineral of the phosphuranylite group composed of aluminum uranyl oxide phosphate hydroxide heptahydrate, $Al(UO_2)_3O(OH)(PO_4)_2 \cdot 7H_2O$. Crystal system, monoclinic. Density, 3.90 g/cm³; refractive index, 1.666. Occurrence: Zaire.

upholstery leather. A thin, highly finished leather with a soft, uniform texture for use in furniture, automobile, bus and airplane seats, coverings, etc. It is usually obtained from the grain side (top and subsequent grain cuts) of split cattlehide.

Upilex. Trademark of Ube Industries, Limited (Japan) for several polyimide resins available in film form with a density of 1.42 g/cm³ (0.051 lb.in.³), high toughness, good fatigue properties, excellent abrasion, creep and heat; good resistance to alpha, beta, gamma and X-ray radiation, good resistance to ultraviolet radiation, good resistance to acids, alcohols, aromatic hydrocarbons, greases, oils and ketones, poor resistance to alkalies, good dielectric properties, low thermal expansivity, and a service temperature range of -270 to +320°C (-454 to +608°F). Used for capacitors, printed circuit boards, and electrical insulation.

Upimol. Trade name of Ube Industries, Limited (Japan) for polyimide molding resins providing high toughness, excellent abrasion, creep and heat resistance, good radiation resistance, good resistance to acids, alcohols, aromatic hydrocarbons, greases, oils and ketones, good dielectric properties, high temperature insulation properties, low thermal expansivity, and a service temperature range of -270 to +320°C (-454 to 608°F). Used for engine components, gaskets, seals, diaphragms, belts, bearings, circuit and instrument insulation, consumer electronics, components in instruments and devices, and electronic components.

UpJohn. Trade name of UpJohn Company (USA) for several polyimide resins with a density of 1.42 g/cm³ (0.051 lb.in.³), high thermal stability, excellent high-temperature properties and radiation resistance, high wear resistance, and inherently low flammability and smoke emission. Used for automotive and mechanical parts.

upland cotton. Cotton obtained from a low, multi-branched shrub (*Gossypium hirsutum*) grown as an annual crop in the US and many other countries. The staple length of the fiber typically ranges from about 22-32 mm (0.9-1.3 in.). See also cotton.

Uposil. Trade name for a silicone elastomer used for dental impressions.

upper bainite. See bainite.

upper leather. Leather, usually chrome- or vegetable-tanned, made chiefly from the skins or hides of calves, cattle, goats, pigs, sheep or horses, and used for shoe and boot uppers.

Urac. Trademark of American Cyanamid Corporation (USA) for thermosetting condensation products of urea and formaldehyde, used as bonding agents in the manufacture of particleboard and plywood, and in the assembly or installation of furniture, cabinets, paneling, and countertops.

Urafil. Trademark of Wilson-Fiberfil International (USA) for glass-fiber-reinforced polyurethanes.

Ural. Trade name of Siberian-Ural Aluminum Company (Russia) for a series of aluminum alloys containing 1.5-3.5% zinc, 4-5% copper and 1-2.5% magnesium.

Uralane. Trademark of Ciba-Geigy Corporation (USA) for a series of clear or amber-colored high-strength urethane adhesives that bond well to polycarbonates, polysulfones and acrylics, fiberglass, glass, stainless steel and aluminum. They are UV-stable, cure within two hours at 66°C (150°F), or within two days at room temperature, and have a maximum service temperature of 82°C (180°F). Used in the aircraft and aerospace industries, and for computers, optical devices, electronic components, and signs. Also supplied under this trademark are several urethane casting materials for the manufacture of tooling components.

uralborite. A colorless mineral composed of calcium borate dihydrate, $CaB_2O_4 \cdot 2H_2O$. Density, 2.60 g/cm^3; refractive index, 1.609. Occurrence: Russia.

Uralite. Trademark of Plastimer (France) for modified urea-formaldehyde plastics.

uralkyd. A high-molecular-weight alkyd resin modified with both a drying oil, such as linseed or safflower oil, and a urethane resin. Coatings produced from *uralkyd* have excellent hardness, durability, and chemical and weather resistance. Also known as *urethane alkyd*.

uralolite. A colorless mineral composed of beryllium calcium phosphate hydroxide tetrahydrate, $CaBe_3(OH)_2(PO_4)_2 \cdot 4H_2O$. Density, 2.14 g/cm^3; refractive index, 1.525. Occurrence: Russia.

uramphite. A pale to bottle green mineral of the meta-autunite group composed of ammonium uranyl phosphate trihydrate, $(NH_4)(UO_2)(PO_4) \cdot 3H_2O$. Crystal system, tetragonal. Density, 3.70; refractive index, 1.585. Occurrence: Russia.

Urania. Trade name of Vetreria Tiburtina Valeria (Italy) for a toughened glass.

urania. See uranium dioxide.

urania ceramics. Heat- and corrosion-resistant ceramics containing substantial amounts of uranium dioxide (UO_2) and used for nuclear reactor applications.

urania-thoria. A mixture of urania (UO_2) and thoria (ThO_2) used as a nuclear reactor fuel. See also thoria-urania ceramics.

uranic oxide. See uranium dioxide.

uraninite. A black, steel-gray, brownish-black, grayish or greenish, strongly radioactive mineral of the fluorite group composed of uranium dioxide, UO_2, and usually also containing varying amounts of uranium trioxide, thorium, cerium, lanthanum, yttrium, lead, radium and helium. *Pitchblende* is a massive variety. Crystal structure, cubic. Density, 9-11 g/cm^3; hardness, 5.5-6 Mohs. Occurrence: Canada, UK, USA (North Carolina, Pennsylvania, Washington). Used as a source of uranium and radium.

uranium. A radioactive, silvery-white, ductile, metallic element of the actinide series of the Periodic Table. It is commercially available in the form of ingots, rods, turnings, plate, sheet, foil, powder and wire. *Natural uranium* contains 99.275% of the isotope ^{238}U, 0.720% of the isotope ^{235}U and 0.005% of the isotope ^{234}U. Important uranium ores are *autunite, coffinite, carnotite, davidite, uraninite (pitchblende), uranophane* and *torbernite*. Density, 19.05 g/cm^3; melting point, 1132°C (2070°F); boiling point, 3818°C (6904°F); hardness, 187-250 Vickers; atomic number, 92; atomic weight, 238.029; trivalent, tetravalent, hexavalent. It has good malleability, low electrical conductivity (about 5% IACS), and good superconductive properties. Three allotropic forms are known: (i) *Alpha uranium* (orthorhombic crystal structure) that is stable to 667°C (1233°F); (ii) *Beta uranium* (tetragonal crystal structure) stable from 667 to 775°C (1233 to 1427°F); and (iii) *Gamma uranium* (body-centered cubic crystal structure) that is stable from 775 to 1132°C (1427 to 2070°F). *Uranium* is used as a nuclear fuel, for X-ray targets and gyrocompass, and as a coloring agent in yellow glass and glazes. Symbol: U.

uranium-233. A fissionable isotope of uranium with a mass number of 233 made by neutron bombardment of thorium-232. It has a half-life of 1.62×10^5 years, emits alpha radiation, and is used as a nuclear fuel in molten-salt and breeder reactors, and in thermonuclear reactors. Symbol: ^{233}U; U-233.

uranium-234. A natural isotope of uranium with a mass number of 234 occurring to the extent of 0.005% in natural uranium. It has a half-life of 2.48×10^5 years, and is used in nuclear research, and in fission detectors and neutron counters. Symbol: ^{234}U; U-234, UII. Also known as *uranium II*.

uranium-235. A rather long-lived radioactive isotope of uranium (uranium-235) which gives rise to the actinium decay series. It occurs to the extent of 0.720% in natural uranium. It emits alpha particles, has an atomic number of 92, a half-life of 7.13 $\times 10^8$ years, disintegrates to thorium-231, and is used in nuclear science and engineering (weapons, reactors, etc.). Symbol: ^{235}U; U-235; AcU. Also known as *actinium-urani-um; actinouranium*.

uranium-238. The natural isotope of uranium with a mass number of 238 occurring to the extent of 99.275% in *natural uranium*. Symbol: ^{238}U; U-238; UI. Also known as *uranium I*.

uranium I. See uranium-238.

uranium II. See uranium-234.

uranium diacetate oxide dihydrate. See uranyl acetate dihydrate.

uranium alloys. A group of alloys in which uranium is the principal constituent and including alloys, such as (i) *U-0.75Ti*, a precipitation-hardenable alloy composed of 99.25% depleted uranium and 0.75% titanium, made by vacuum induction melting or casting, and having an outstanding combination of strength and ductility, a density of 18.6 g/cm^3 (0.67 $lb.in.^3$), and a melting point of 1200°C (2190°F); (ii) *U-2Mo*, a high-strength alloy composed of 98% depleted uranium and 2% molybdenum, made by vacuum induction melting and casting, and having a density of 18.5 g/cm^3 (0.66 $lb.in.^3$), and a melting point of 1150°C (2100°F); and (iii) *U-6.0Nb*, a ductile alloy composed of 94% depleted uranium and 6% niobium, made by consumable electrode vacuum-arc melting, and having excellent corrosion resistance and high elevated-temperature strength.

uranium aluminide. A compound of uranium and aluminum with a density of 8.26 g/cm^3 (0.298 $lb.in.^3$) and a melting point of 1587°C (2890°F). Formula: UAl_2.

uranium-ammonium carbonate. See ammonium uranyl-carbonate.

uranium antimonide. Any of the following compounds of uranium and antimony used in ceramics and materials research: (i) *Uranium monoantimonide*. Melting point, 1850°C (3362°F). Formula: USb; (ii) *Uranium tetraantimonide*. Melting point 1693°C (3079°F). Formula: U_3Sb_4; and (iii) *Uranium triantimonide*. Melting point, 1800°C (3272°F). Formula: U_4Sb_3.

uranium-barium oxide. See barium diuranate.

uranium beryllide. Any of the following compounds of uranium and beryllium used in ceramics and materials research: (i) *Uranium tetraberyllide*. Density, 9.38 g/cm^3; melting point, 2480°C (4496°F); hardness, 2500 Vickers. Formula: UBe_4; and (ii) *Uranium dodecaberyllide*. Density, 5.86 g/cm^3; melting point, 2232°C (4050°F). Formula: UBe_{12}.

uranium boride. Any of the following compounds of uranium and boron used in ceramics and materials research: (i) *Uranium diboride*. Density, 12.73 g/cm^3; melting point 2370°C (4298°F); hardness, 1400 Vickers. Formula: UB_2; (ii) *Uranium tetraboride*. Density, 9.32-9.38 g/cm^3; melting point, 2482°C (4500°F); hardness, 2500 Vickers. Formula: UB_4; and (iii) *Uranium dodecaboride*. Density, 5.86 g/cm^3; melting point 2232°C (4050°F). Formula: UB_{12}.

uranium bromide. See uranium tribromide; uranium tetrabromide.

uranium carbide. Any of the following compounds of uranium and carbon: (i) *Uranium monocarbide.* Gray, cubic crystals. Density, 13.63 g/cm³; melting point, 2790°C (5054°F). Formula: UC; (ii) *Uranium dicarbide.* Density, 11.28 g/cm³; melting point, 2350°C (4262°F); boiling point, 4370°C (7898°F). It is available in the form of gray, tetragonal crystals, pellets or microspheres, and used as a nuclear reactor fuel. Formula: UC_2; and (iii) *Uranium sesquicarbide.* Density, 12.7-12.88 g/cm³; melting point, 1777°C (3231°F). Formula: U_2C_3.

uranium chloride. See uranium trichloride; uranium tetrachloride; uranium pentachloride; uranium hexachloride.

uranium diacetate oxide dihydrate. See uranyl acetate dihydrate.

uranium diboride. See uranium boride (i).

uranium dicarbide. See uranium carbide (ii).

uranium dichloride oxide. See uranyl chloride.

uranium dichloride oxide trihydrate. See uranyl chloride.

uranium dinitrate oxide. See uranyl nitrate.

uranium dinitrate oxide hexahydrate. See uranyl nitrate.

uranium dinitride. See uranium nitride (ii).

uranium dioxide. Brown-black crystals, pellets or powder (99.8+% pure). It occurs in nature as the mineral *uraninite.* Crystal system, cubic. Crystal structure, fluorite. Density, 10.96 g/cm³; melting point, 2760°C (5000°F); hardness, 5.5 Mohs; low thermal conductivity. Used in ceramics, in the production of red, yellow and orange pigments for glazes, as a source of uranium for the fluorides used in isotope separation, in the packing of nuclear fuel rods, in thermistors, and as a metal stain in life-science microscopy. Formula: UO_2. Also known as *urania; uranic oxide; uranium oxide.*

uranium dioxide cermets. A group of composite materials made by powder metallurgy and consisting of uranium dioxide and a metallic binder, such as molybdenum, niobium, titanium or zirconium. They possess good thermal properties, and are used in nuclear-reactor ceramics.

uranium disilicide. See uranium silicide (ii).

uranium disulfide. See uranium sulfide (ii).

uranium dodecaberyllide. See uranium beryllide (ii).

uranium dodecaboride. See uranium boride (iii).

uranium fluoride. See uranium tetrafluoride; uranium pentafluoride; uranium hexafluoride; uranium trifluoride.

uranium glass. Yellowish, orange or greenish glass produced by the addition of a colorant, such as uranium (U), uranium trioxide (UO_3) or sodium diuranate ($Na_2U_2O_7 \cdot 6H_2O$).

uranium hexachloride. Green, hexagonal crystals. Density, 3.6 g/cm³; melting point, 177°C (351°F). Formula: UCl_6. Also known as *uranium chloride.*

uranium hexafluoride. Colorless, white or pale yellow, volatile, deliquescent crystals. Crystal system, monoclinic. Density, 4.68-5.09 g/cm³; melting point, 64-65°C (147-149°F). It is a highly corrosive, radioactive compound whose vapor exhibits properties of a nearly perfect gas. Used in the gaseous diffusion process for separating isotopes of uranium. Formula: UF_6. Abbreviation: hex. Also known as *uranium fluoride.*

uranium hydride. Brown or gray-black crystals or powder. Crystal system, cubic. Density, 10.92-11.1 g/cm³; good electrical conductivity. Used in the preparation of powdered uranium metal by decomposition, and in the separation of hydrogen isotopes. Formula: UH_3.

uranium iodide. See uranium tetraiodide; uranium triiodide.

uranium monoaluminide. See uranium antimonide (i).

uranium monoantimonide. See uranium antimonide (i).

uranium monocarbide. See uranium carbide (i).

uranium mononitride. See uranium nitride (i).

uranium monophosphide. See uranium phosphide (i).

uranium monoselenide. See uranium selenide (i).

uranium monosilicide. See uranium silicide (i).

uranium monosulfide. See uranium sulfide (i).

uranium nitride. Any of the following compound of uranium and nitrogen: (i) *Uranium mononitride.* Gray, cubic crystals. Density, 14.32 g/cm³; melting point, 2805°C (5081°F). Used in ceramics and for nuclear applications. Formula: UN; (ii) *Uranium dinitride.* Density, 11.73 g/cm³; melting point, unstable above 704-787°C (1300-1449°F). Used in ceramics and for nuclear applications. Formula: UN_2; (iii) *Uranium sesquinitride.* Cubic crystals. Density, 11.24 g/cm³; decomposes above 704°C (1300°F). Formula: U_2N_3; and (iv) *Uranium tetranitride.* Yellow crystals. Formula: U_3N_4.

uranium ocher. See gummite.

uranium oxide. See uranium dioxide; uranium trioxide; uranium tetroxide; uranous-uranic oxide.

uranium oxychloride. See uranyl chloride.

uranium oxynitrate. See uranyl nitrate.

uranium pentachloride. Dark green, deliquescent needles that appear red in transmitted light. They have a melting point of approximately 120°C (248°F) and are soluble in absolute alcohol. Formula: UCl_5. Also known as *uranium chloride.*

uranium pentafluoride. Pale blue, hygroscopic, tetragonal crystals. Density, 5.81 g/cm³; melting point, 348°C (658°F). Formula: UF_5. Also known as *uranium fluoride.*

uranium pentoxide. Black powder. Formula: U_2O_5.

uranium peroxide. See uranium tetroxide.

uranium phosphate. A compound of uranium dioxide and phosphorus pentoxide with a melting point of 1550°C (2822°F). Formula: UP_2O_7.

uranium phosphide. Any of the following compounds of uranium and phosphorus used in ceramics and for nuclear applications: (i) *Uranium monophosphide.* Density, 9.68 g/cm³. Formula: UP; and (ii) *Uranium tetraphosphide.* Density, 9.83 g/cm³. Formula: U_3P_4.

uranium powder. Uranium metal (purity 99.7%) in the form of a finely divided pyrophoric powder used for nuclear and nonnuclear (e.g., colorant) applications.

uranium selenide. Any of the following compounds of uranium and selenium used in ceramics and materials research: (i) *Uranium monoselenide.* Density, 11.3 g/cm³; melting point, above 1850°C (3362°F). Formula: USe; and (ii) *Uranium sesquiselenide.* Melting point, 1570°C (2858°F). Formula: U_2Se_3.

uranium sesquicarbide. See uranium carbide (iii).

uranium sesquinitride. See uranium nitride (iii).

uranium sesquiselenide. See uranium selenide (ii).

uranium sesquisulfide. See uranium sulfide (iii).

uranium silicide. Any of the following compounds of uranium and silicon used in ceramics and for nuclear applications: (i) *Uranium monosilicide.* Density, 10.4 g/cm³; melting point 1598°C (2908°F); hardness, 745 Knoop. Formula: USi; (ii) *Uranium disilicide.* Density, about 8.5 g/cm³; melting point, above 1150°C (2102°F); hardness, 700 Knoop. Formula: USi_2; (iii) *Uranium trisilicide.* Density, 8.15 g/cm³; melting point, 1650°C (3002°F); hardness, 445 Knoop. Formula: US_3; (iv) *Triuranium silicide.* Density, 15.58 g/cm³; melting point, about 960°C (1760°F). Formula: U_3Si; and (v) *Triuranium disilicide.* Density, 12.2 g/cm³; melting point, 1666°C (3031°F); hardness, 796 Knoop. Formula: U_3Si_2.

uranium-strontium oxide. A yellow pigment used in porcelain and ceramic glazes. Formula: SrU_2O_7. Also known as *strontium diuranate.*

uranium sulfide. Any of the following compounds of uranium and sulfur used in ceramics and for nuclear applications: (i) *Uranium monosulfide.* Density, 10.87 g/cm³; melting point, about 2370°C (4298°F). Formula: US; (ii) *Uranium disulfide.* Gray-black, tetragonal crystals. Density, 7.52 g/cm³; melting point, 1850°C (3362°F). Formula: US_2; and (iii) *Uranium sesquisulfide.* Gray-black needles. Density, 8.78 g/cm³; melting point, 1927°C (3501°F). Formula: U_2S_3.

uranium telluride. A compound of uranium and tellurium used in ceramics and materials research. Density, 8.8 g/cm³; melting point, 1550-1650°C (2822-3002°F). Formula: UTe.

uranium tetraantimonide. See uranium antimonide (ii).

uranium tetraberyllide. See uranium beryllide (i).

uranium tetraboride. See uranium boride (ii).

uranium tetrabromide. Brown, deliquescent crystals or leaves. Density, 4.84 g/cm³; melting point, 519°C (966°F). Formula: UBr_4. Also known as *uranium bromide.*

uranium tetrachloride. A dark green, moisture-sensitive, crystalline powder (99.7+% pure). Crystal system, octahedral. Density, 4.72-4.87 g/cm³; melting point, 590°C (1094°F); boiling point of 792°C (1457°F); soluble in alcohol and water. Formula: UCl_4. Also known as *uranium chloride.*

uranium tetrafluoride. A highly corrosive, radioactive, green, nonvolatile, monoclinic, crystalline powder made by treating uranium dioxide with hydrogen fluoride. Density, 6.70 g/cm³; melting point, 1036°C (1897°F); boiling point, 1417°C (2583°F). Used in the preparation of uranium metal and uranium hexafluoride. Formula: UF_4. Also known as *green salt; uranium fluoride.*

uranium tetraiodide. Black, hygroscopic crystals or needles. Density, 5.1 g/cm³; melting point, 506°C (943°F). Formula: UI_4. Also known as *uranium iodide.*

uranium tetranitride. See uranium nitride (iv).

uranium tetraphosphide. See uranium phosphide (ii).

uranium tetroxide. Pale yellow, hygroscopic crystals. Density, 2.5 g/cm³; melting point, 115°C (239°F) (decomposes). Used as a red, orange, or yellow colorant in ceramics. Formula: $UO_4 \cdot xH_2O$. Also known as *uranium peroxide.*

uranium triantimonide. See uranium antimonide (iii).

uranium tribromide. Red or dark brown, water-soluble crystals or needles with a melting point of 727°C (1341°F). Formula: UBr_3. Also known as *uranium bromide.*

uranium trichloride. Green, hygroscopic crystals. Density, 5.51 g/cm³; melting point, 837°C (1539°F). Formula: UCl_3. Also known as *uranium chloride.*

uranium trifluoride. Black, hexagonal crystals with a density of 8.9 g/cm³. Formula: UF_3. Also known as *uranium fluoride.*

uranium triiodide. Black, hygroscopic crystals with a melting point of 766°C (1411°F). Formula: UI_3. Also known as *uranium iodide.*

uranium trioxide. Red, orange or yellow, radioactive crystals or powder. Density, 7.3-8.34 g/cm³; melting point, decomposes when heated. Used as an orange coloring agent in ceramics and glass. Formula: UO_3. Also known as *uranyl oxide; orange oxide; orange uranium oxide.*

uranium trisilicide. See uranium silicide (iii).

uranium-uranyl oxide. See uranous-uranic oxide.

uranium yellow. See sodium diuranate.

uranium-zinc acetate oxide. See uranyl zinc acetate.

uranocene. An organometallic coordination compound (molecular sandwich) obtained by bonding an uranium atom to two cyclopentadienyl (C_5H_5) rings, one ring located above and the other below the uranium atom plane, and the *p*-molecular orbitals of the rings sharing electrons with the *f*-orbitals of the uranium atom. Used in organic and organometallic research. Formula: $(C_5H_5)_2U$.

uranocircite. A yellow, radioactive mineral of the autunite group composed of barium uranyl phosphate decahydrate, $Ba(UO_2)_2(PO_4)_2 \cdot 10H_2O$. Crystal system, tetragonal. Density, 3.46 g/cm³; refractive index, 1.583. Occurrence: Germany.

uranophane. A pale to lemon-yellow, radioactive mineral of the uranophane group composed of calcium oxonium uranyl silicate trihydrate, $Ca(H_3)_2(UO_2)_2(SiO_4)_2 \cdot 3H_2O$. Crystal system, monoclinic. Density, 3.83 g/cm³; refractive index, 1.666. Occurrence: USA (New Hampshire).

uranopilite. A yellow, radioactive mineral composed of uranyl sulfate hydroxide dodecahydrate, $(UO_2)_6SO_4(OH)_{10} \cdot 12H_2O$. Crystal system, monoclinic. Density, 3.96 g/cm³; refractive index, 1.624. Occurrence: Canada (Saskatchewan), UK.

uranospathite. A yellow, radioactive mineral of the autunite group composed of the hydrogen uranyl aluminum phosphate hydrate, $HAl(UO_2)_4(PO_4)_4 \cdot 40H_2O$. Crystal system, tetragonal. Density, 2.49 g/cm³; refractive index, 1.511. Occurrence: UK.

uranospaerite. A reddish-orange, radioactive mineral composed of bismuth uranium oxide trihydrate, $Bi_2U_2O_9 \cdot 3H_2O$. Crystal system, monoclinic. Density, 6.36 g/cm³; refractive index, 1.981. Occurrence: Germany.

uranospinite. A pale yellow, radioactive mineral of the autunite group composed of calcium uranyl arsenate decahydrate, $Ca(UO_2AsO_4)_2 \cdot 10H_2O$. It occurs naturally, but can also be made synthetically. Crystal system, monoclinic. Density, 3.25 g/cm³; refractive index, 1.619.

uranotantalite. See samarskite.

uranous-uranic oxide. Olive-green to black, radioactive crystals, granules or powder containing 84.8% uranium. It occurs in nature as the mineral *pitchblende*. Density, 8.39 g/cm³; melting point, decomposes to UO_2 on heating to 1300°C (2372°F). Used in nuclear technology, in the preparation of other uranium compounds, and as a pigment. Formula: U_3O_8. Also known as *triuranium octoxide; uranium-uranyl oxide; uranyl uranate.*

uranpyrochlore. A brown to yellow, radioactive mineral of the pyrochlore group composed of calcium lead uranium niobium tantalum oxide, $(U,Ca,Pb)(Nb,Ta)_2O_7$. Crystal system, cubic. Density, 7.27 g/cm³. Occurrence: Madagascar.

Uranus. Trade name of Creusot-Loire (France) for an extensive series of stainless steels including various austenitic and heat-resistant grades used for automotive components, chemical, petrochemical, oil-refinery, heat-treating and nuclear-plant equipment, industrial furnaces, gas-turbine parts, piping, valves, and pumps.

uranyl acetate dihydrate. Yellow, water-soluble crystalline compound. Density, 2.89 g/cm³; loses $2H_2O$ at 110°C (230°F); decomposes at 275°C (527°F). Used as a chemical reagent, in copying inks, as a bacterial oxidation activator, and as a negative stain in electron microscopy. Formula: $UO_2(O_2C_2H_3)_2 \cdot 2H_2O$. Also known as *uranium diacetate oxide dihydrate.*

uranyl-ammonium carbonate. See ammonium uranyl carbonate.

uranyl chloride. A compound of uranium dioxide and chlorine available in anhydrous and hydrated form. The anhydrous compound is available as yellow, hygroscopic crystals and the trihydrate as a yellow, hygroscopic powder. Used in ceramics and chemistry. Formula: UO_2Cl_2 (anhydrous); $UO_2Cl_2 \cdot 3H_2O$ (trihydrate). Also known as *uranium dichloride oxide; uranium oxychloride.*

uranyl nitrate. Yellow, hygroscopic crystals (99+% pure). Crystal system, orthorhombic. Density, 2.807 g/cm³; melting point, 60.2°C (140°F); boiling point, 118°C (244°F); refractive index, 1.497. Used as a source of uranium dioxide, in the introduction of uranium into nonaqueous solvents, in ceramic glazes and porcelain enamels as a red, yellow or orange colorant, in photography, in volumetric analysis, in electrophoresis, and in electron microscopy as a negative stain. Formula: $UO_2(NO_3)_2 \cdot 6H_2O$. Abbreviation: UNH. Also known as *uranium dinitrate oxide; uranium dinitrate oxide hexahydrate; uranium oxynitrate; yellow salt.*

uranyl oxide. See uranium trioxide.

uranyl uranate. See uranous-uranic oxide.

uranyl zinc acetate. A yellow, crystalline compound. Formula: $UO_2Zn(O_2C_2H_3)_3$. Also known as *uranium zinc acetate oxide.*

urao. A white, gray or yellow mineral with a vitreous, sparkling luster composed of sodium hydrogen carbonate hydrate, $Na_3H(CO_3)_2 \cdot 2H_2O$. Crystal system, monoclinic. Density, 2.11-2.15 g/cm³; hardness, 2.5-3.0 Mohs; refractive index, 1.494. Occurrence: Africa, Egypt, Hungary, USA (Wyoming), Venezuela. Used as a source of sodium compounds. Also known as *trona.*

Urbach. Trade name of Carl Urbach & Co. Stahlwerk KG (Germany) for a series of high-speed steels containing 0.76-1.3% carbon, 4-4.75% chromium, 8-18.5% tungsten, 1.1-3.8% vanadium, 0.7-0.85% molybdenum, 0-10% cobalt, and the balance iron. Used for lathe and planer tools, drills, cutters, and dies.

Urbanit. Trademark of Hoechst AG (Germany) for unmodified and modified phenol- and urea-formaldehyde plastics.

Urbical. Trade name of Inter-Africa Dental (Zaire) for a self-curing, radiopaque dental paste of flowable consistency that contains 25% calcium hydroxide, and is used in restorative dentistry for capping, lining and temporary filling applications. *Urbical LC* is a light-curing type that provides high compressive strength and short setting times, and is used as a cavity liner.

urea. (1) A colorless mineral composed of carbamide, $CO(NH_2)_2$. Crystal structure, tetragonal. Density, 1.33 g/cm³; melting point, 133-135°C (271-275°F); refractive index, 1.480. Occurrence: Australia.

(2) A compound that occurs in nature as a mineral (see above), in urine and body fluids, and can also be made synthetically from liquid ammonia and liquid carbon dioxide. It is commercially available in various grades (99+% pure) in the form of white crystals or powder with a density of 1.335 g/cm³, a melting point of 133-135°C (271-275°F) and a refractive index of 1.480. Used in the preparation of synthetic resins, in flameproofing agents, as a viscosity modifier for starch or casein paper coatings, as a stabilizer for explosives, in hydrocarbon separation, as a chemical intermediate, as a fertilizer, in animal feed, in chemical synthesis, and in the biosciences. Formula: $CO(NH_2)_2$. Also known as *carbamide; carbonyl diamide.*

urea-formaldehyde adhesives. (1) Colloidal dispersions of urea-formaldehyde resin in water usually with the addition of modifiers and secondary binders. See also urea-formaldehyde resins.

(2) See urea-formaldehyde resin glues.

urea-formaldehyde foams. Flexible cellular plastics produced from urea-formaldehyde resins by the action of a blowing agent. They have high resiliency and softness, and are used chiefly for furniture, and automobile components. Abbreviation: UFS.

urea-formaldehyde resin glues. Glues based on urea-formaldehyde resin and available in dry powder form mixed with hardening agents and/or catalysts. Prior to application, they are mixed with water to a creamy consistency. They are moisture resistant, dry to a light brown color, harden through chemical action when water is added, and set at room temperature in 4-8 hours. Used chiefly for bonding wood and wood products. Also known as *urea resin glues.*

urea-formaldehyde resins. A group of amino resins obtained as products of condensation reactions between urea [carbamide, $CO(NH_2)_2$] and formaldehyde. Uncured urea-formaldehydes are water-soluble and used in adhesives, coatings, textile finishes, laminates, etc. They are cured (crosslinked) by controlled heating and pressing in the presence of a catalyst (curing agent), and often made into molding compounds by modifying with fillers (wood flour, cellulose, etc.) and injection-molded into strong, hard, durable parts, e.g., wiring devices, handles, knobs, dinnerware, etc. Abbreviation: UF. Also known as *urea resins.*

urea hydrogen peroxide. See urea peroxide.

urea peroxide. An addition compound of urea and hydrogen peroxide available in the form of white crystals or a crystalline powder (97+% pure) with a melting point of 90-93°C (194-199°F). Used as an oxidizer, as a source of water-free hydrogen peroxide for the preparation of solutions, in the modification of starches, and as a polymerization initiator. Formula: $CH_6N_2O_3$. Also known as *carbamide peroxide; percarbamide; urea hydrogen peroxide.*

urea resin glues. See urea-formaldehyde resin glues.

urea resins. See urea-formaldehyde resins.

Urecoll. Trademark of BASF AG (Germany) for an extensive series of urea-formaldehyde plastics supplied in several grades and product forms.

Ureklad. Trade name of Atlas Minerals & Chemicals, Inc. (USA) for polyurethane resin coatings for application to steel or concrete surfaces.

Urelite. Trade name of Hexcel Corporation (USA) for polyurethane elastomers supplied in hard-cast, hard-microcellular, soft-microcellular, reinforced-microcellular, semi-rigid and structural-foam grades.

urena fiber. A strong bast fiber obtained from the stalk of a jute plant (*Urena lobata*) cultivated in Central Africa, Southeast Asia and Central America. It is virtually indistinguishable from commercial *jute* and thus used as a substitute, and in the manufacture of textile fabrics. Also known as *aramina fiber; Congo jute.*

Urepan. Trademark of Bayer Corporation (USA) for high-performance polyurethane-based elastomers used for automotive and machine parts, and electrical equipment.

Uresin. Trademark of Hoechst AG (Germany) for urea-formaldehyde resins and plastics.

urethane acrylic polymers. See acrylamate polymers.

urethane alkyd. See uralkyd.

urethane coatings. See polyurethane coatings.

urethane elastomer. See polyurethane elastomer.

urethane foam. See polyurethane foam.

urethane foam sealants. See polyurethane foam sealants.

urethane hybrids. See acrylamate polymers.

urethane plastics. Plastics based on urethane resins. See also polyurethanes.

urethane resins. See polyurethanes.

urethanes. See polyurethanes.

Urex. French trade name for urea-formaldehyde resins.

uricite. A mineral composed of uric acid, $C_5H_4N_4O_3$. Crystal system, monoclinic. Density, 1.85 g/cm^3.

Urocristal. French trade name for urea-formaldehyde resins.

Uroflex. Trade name of BASF Corporation (USA) for polyurethane elastomers supplied in hard-cast, hard-microcellular, soft-microcellular, reinforced-microcellular, semi-rigid and structural-foam grades.

Urol. Trade name of AFORA (Spain) for a free-machining alloy steel containing 0.05-0.15% carbon, 1-1.5% manganese, up to 0.06% silicon, up to 0.07% phosphorus, 0.3-0.4% sulfur, 0.15-0.3% lead, and the balance iron. Used for screw-machine products, fasteners, etc.

uronic acid. Any of a class of aldose sugar derivatives, such as glucuronic acid and hyaluronic acid, in which the terminal carbon has been oxidized from a primary alcohol to a carbonyl group.

Uropal. (1) Trade name of Carl Zeiss GmbH (Germany) for a nearly clear ophthalmic glass giving good protection against both ultraviolet and infrared radiation.

(2) Trade name of Plastimer (France) for urea-formaldehyde plastics.

Uroplast. Trademark of Chemtron Manufacturing Limited (USA) for polyurethane sealants.

Uropol. Trademark of Atlas Chemical Company (USA) for polyurethane foams and foam products.

uroporphyrins. A group of *porphyrin* derivatives either found naturally in urine or made synthetically. Synthetic porphyrins are usually supplied in powder form and include purple-colored *uroporphyrin I dihydrochloride* ($C_{40}H_{38}N_4O_{16}\cdot2HCl$) and rust-colored *uroporphyrin I octamethyl ester* ($C_{48}H_{54}N_4O_{16}$). All are used as pigments or dyes. The formula of natural uroporphyrin is $C_{40}H_{38}N_4O_{16}$.

Urotuf. Trade name of Reichhold Chemicals, Inc. (USA) for fast drying, abrasion-resistant, oil-modified urethane resins used for paint and coating applications.

Urushi. Trade name of Shin-Etsu Polymer Company Limited (Japan) for laminated plastics including polyvinyl chloride.

urylon. See urylon fiber.

urylon fiber. A synthetic fiber made by polycondensation of nonamethylenediamine and urea (carbamide). It has a density of 1.07 g/cm^3 (0.039 lb.in.³), a softening point of 205°C (401°F), a melting point above 325°C (617°F), and loses its strength when heated at temperatures above 150°C (300°F), or by long-term exposure to visible radiation. Used chiefly for blending with other synthetic fibers. Also known as *urylon*.

ursilite. A lemon yellow mineral of the weeksite group composed of magnesium uranyl silicate nonahydrate, $Mg_2(UO_2)_2Si_5O_{16}\cdot9H_2O$. Density, 3.25 g/cm^3.

urvantsevite. A grayish white mineral of the bismuth lead palladium, $Pd(Bi,Pb)_2$. Crystal system, hexagonal. Density, 9.66 g/cm^3. Occurrence: Russian Federation.

Urtal. Trade name of Montecatini (Italy) for acrylonitrile-butadiene-styrene plastics.

Usalloy. Trade name of United States Pipe & Foundry Company (USA) for a corrosion-resistant steel containing up to 0.2% carbon, 1.25% manganese and 0.05% sulfur, 0.25% or more nickel, 0.20% or more copper, 1.25% or more chromium, and the balance iron. Used for bolts and nuts.

Usamet. Trade name of Elgiloy Limited Partnership (USA) for a stainless steel containing 0.1% carbon, 2% manganese, 1.4% silicon, 17% chromium, 7.2% nickel, and the balance iron. Used for strips and springs.

Usco. Trade name of US Reduction Company (USA) for a series cast aluminum products including various aluminum-copper, aluminum-silicon, aluminum-silicon-magnesium, aluminum-magnesium, aluminum-magnesium-zinc alloys as well as aluminum-copper alloys with silicon, zinc and/or nickel additions.

ushkovite. A pale yellow to orange-yellow mineral of the paravauxite group composed of magnesium iron phosphate hydroxide octahydrate, $MgFe_2(PO_4)_2(OH)_2\cdot8H_2O$. Crystal structure, triclinic. Density, 2.38 g/cm^3; refractive index, 1.637. Occurrence: Russia.

Usidécoupe. Trade name of Sollac-Usinor (France) for a continuously annealed electrical sheet made from continuously cast extra-mild steel containing less than 0.05% carbon, 0.3% manganese, 0.03% phosphorus, 0.02% sulfur, 0.05% silicon, 0.02% or more aluminum, and the balance iron. It has uniform mechanical properties especially good flatness and surface cleanness, good stampability and weldability, relatively good corrosion resistance, tight thickness tolerances, a yield strength of 290-350 MPa (42-51 ksi), an ultimate tensile strength of 310-370 MPa (45-54 ksi), a maximum elongation of 36%, and a hardness of about 60 Rockwell B. Used in the manufacture of automotive alternators, window wiper and winder motors, small hand tools and domestic appliances.

Usinelec. Trade name of Sollac-Usinor (France) for continuously annealed semi-processed electrical sheet made from continuously cast low- or ultra-low carbon steel and supplied in five grades. The optimum magnetic properties can be obtained by a subsequent annealing treatment. *Usinelec 1 & 2* are used for small industrial motors (e.g., for washing machines), transformers for microwave ovens and motors for refrigerator compressors and ventilation systems. *Usinelec 3 & Usinilec HSP* are pre-decarburized grades with low power losses used especially for industrial motors and refrigerator compressors. *Usinelec UV* is a pre-decarburized grade and has small power losses and good magnetic permeability.

Usivolt. Trade name of Sollac-Usinor (France) for fully-processed electrical sheet steels with guaranteed magnetic properties supplied as coils in several conventional quality grades in four thickness ranges from 0.35 to 1.00 mm (0.01 to 0.04 in.). Depending on the particular sheet thickness, their yield and ultimate tensile strengths range from 270 to 500 MPa (39 to 72 ksi) and 400 to 630 MPa (58 to 91 ksi) respectively, their elongations from 15 to 38%, and their hardnesses from 120 to 230 Vickers. Used in the manufacture of magnetic circuits for transformers, motors and other electrical equipment for the electrical, electronic, domestic-appliance, automotive and building industries. *Usivolt HF* refers to fully-processed electrical sheet steels with good magnetic properties supplied as coils in low thickness quality, and used in the manufacture of rotary electrical machinery with operating frequencies of 0.1-5.0 kHz. *Usivolt HLE* are fully-processed electrical sheet steels with high yield strength, good magnetic permeability and low core losses. Supplied in two grades, they are used in the manufacture of rotary electrical machinery with operating frequencies of 0.1-5.0 kHz. *Usivolt HP* are fully-processed electrical sheet steels with high magnetic permeability, excellent thermal conductivity, and low scatter in magnetic and mechanical properties.

Uskol. Trade name of Uniroyal Corporation (USA) for a series of synthetic rubbers.

usovite. A brown mineral composed of barium magnesium aluminum fluoride, $Ba_2Mg(AlF_6)_2$. Crystal system, orthorhombic. Refractive index, 1.442. Occurrence: Russia.

ussingite. A pale to dark violet red mineral composed of sodium aluminum silicate hydroxide, $Na_2AlSi_3O_8(OH)$. Crystal system, triclinic. Density, 2.51 g/cm³; refractive index, 1.508. Occurrence: Russian Federation.

ustarasite. A silvery gray to gray mineral composed of lead bismuth antimony sulfide, $Pb(Bi,Sb)_6S_{10}$. Occurrence: Siberia.

Utex. Trade name of H.D. Symons & Company (USA) for cellulose acetate plastics.

U-Tex. Trademark of Para Paints Canada Inc. for polyurethane varnishes.

Uthane. Trade of Hexcel Corporation (USA) for polyurethane coatings.

Utica. Trade name of AL Tech Specialty Steel Corporation (USA) for an oil-hardening cold-work tool steel (AISI type O7) containing 1.25% carbon, 1.5% tungsten, 0.4% chromium, 0.2% vanadium, and the balance iron. It has good nondeforming properties, good deep-hardening properties, good toughness and machinability, good to moderate wear resistance, and is used for dies, cutting tools, drills, punches and several other tools.

utile. See Assie mahogany.

Utilex. Trade name of Utilex & Company (UK) for cellulose acetate plastics.

Utilitas. Trade name of Jessop Steel Company (USA) for a water-hardening tool steel containing 0.6-1.4% carbon, and the balance iron. Used for general-purpose tools.

Utility. (1) Trade name of Lehigh Steel Corporation (USA) for a water-hardening tool steel containing 0.6-0.9% carbon, 0.25-0.6% manganese, and the balance iron. Used for blacksmith tools, chisels, and drills.

(2) Trade name of Thomas Bolton Limited (UK) for a low-cost copper alloy used for bearings.

(3) Trademark of Inco Limited (Canada) for nickel metal and several foundry additives obtained from nickel oxide by reduction melting.

Utiloy. (1) Trade name of Utility Electric Steel Foundry Company (USA) for a series of steel castings including several corrosion- and heat-resistant types.

(2) Trade name for low-gold dental casting alloys.

Utter. Trade name of Uddeholm Corporation (USA) for an oil-hardening steel containing 0.5% carbon, 0.9% silicon, 0.9% chromium, and the balance iron. Used for springs.

UV absorber. See ultraviolet absorber.

UV-cure adhesives. See ultraviolet-cure adhesives.

UV-cure coatings. See ultraviolet-cure coatings.

UV glass. See ultraviolet glass.

UV stabilizer. See ultraviolet absorber.

Uval. Trade name of Uval Limited (UK) for double glazing units.

uvanite. A brownish yellow, radioactive mineral composed of uranium vanadium oxide hydrate, $U_2V_6O_{21}\cdot15H_2O$. Crystal system, orthorhombic. Refractive index, 1.879. Occurrence: USA (Utah).

uvarovite. Emerald-green to green-yellow mineral of the garnet group composed of calcium chromium silicate, $Ca_3Cr_2(SiO_4)_3$. Crystal system, cubic. Density, 3.42-3.90 g/cm³; hardness, 6.5-7.5 Mohs; refractive index, 1.86. Used as a coloring agent and abrasive.

Uvasin. Trade name of Schott Glas AG (Germany) for a pink eye-protection glass.

Uverite. Trademark of Harshaw Chemical Company (USA) for a special antimony-titanium opacifying agent composed of oxides of calcium, titanium and antimony and some calcium fluoride. The approximate formula is $7CaO\cdot CaF_2\cdot6TiO_2\cdot2Sb_2O_5$. Used for porcelain enamels.

Uvex. Trade name of Eastman Chemical Company (USA) for several cellulose acetate propionate polymers.

Uvilex. Trade name of Schott DESAG AG (Germany) for hand-a blown ultraviolet-absorbing glass with pale golden color.

Uviolglas. Trademark of Schott Glas AG (Germany) for a glass that transmits ultraviolet light, and is used chiefly for special lamp bulbs and tubes.

Uvisol. Trademark of Schott DESAG AG (Germany) for a special filter glass.

uvite. A brownish black mineral of the tourmaline group composed of calcium magnesium aluminum borate silicate hydroxide, $CaMg_3(Al,Mg)_6(BO_3)_3Si_6O_{18}(OH)_4$. Crystal system, rhombohedral (hexagonal). Density, 3.06 g/cm³; refractive index, 1.641. Occurrence: Czech Republic.

UVShield. Trademark of Solutia Inc. (USA) for a clear, untinted automotive window film that eliminates 99.9% of the incident UVA and UVB rays.

uytenbogaardtite. A gray white mineral with brownish tinge composed of gold silver sulfide, Ag_3AuS_2. It can also be made synthetically. Crystal system, tetragonal. Density, 8.30 g/cm³. Occurrence: Indonesia.

UZIN. Trade name of UZIN Dr. Utz Ges. mbH (Austria) for a range of building products including adhesives for bonding PVC, textile and linoleum laminates and corkboard, and several compounds for the insulation of parquet flooring and for various other floor leveling and insulating applications.

V

VA-Alloy. British trade name for a nonhardenable aluminum alloy containing 13.7% zinc, 5.2% copper, 0.7% iron, and 0.2% vanadium. Used for light-alloy parts.

Vac-Arc. Trademark of Timken Latrobe Steel (USA) for vacuum-melted ultra-high-strength steels used for aircraft landing gears, missile cases, and aerospace components, such as fasteners, shafts, and cams. *Vac-Arc Regent* is a vacuum-melted deep-hardening chromium steel (AISI-SAE 52100) containing about 1% carbon, 0.25-0.45% manganese, 0.15-0.35% silicon, 0.025% sulfur, 0.05% phosphorus, 1.5% chromium, and the balance iron. It has good machinability, hardenability and wear resistance, high dimensional stability, and is used for ball-bearing components for the aircraft and aerospace industries, and for wear plates.

Vaccutherm. Trademark of Krupp Stahl AG (Germany) for a series of corrosion-resistant, hardenable steels containing 0.15-0.2% carbon, 11.5-12% chromium, 0.6-1% molybdenum, 0-0.6% nickel, 0-0.5% tungsten, 0-0.3% vanadium, 0-0.2% niobium (columbium), and the balance iron. Used for chemical plant and oil-refinery equipment, flatware, cutlery, surgical instruments, valves, bearings, gears, shafts, pivots, springs, and stainless hardware.

Vac-Melt. Trade name of George W. Prentiss & Company (USA) for a series of resistance alloys containing 60-77.5% nickel, 15-20% chromium, 0.5-16.5% iron, 0-7% molybdenum and 2-4% manganese. They have excellent heat resistance up to approximately 1200°C (2190°F), and are used for electrical equipment.

Vacodil. Trademark of Vacuumschmelze GmbH (Germany) for a series of low-expansion alloys composed of 64-74% iron, 20-46% nickel, and up to 6% manganese. They have very low linear coefficients of thermal expansion at ordinary temperatures, and are used for temperature measuring and control equipment, and for glass-to-metal sealing in the semiconductor industry.

Vacodur. Trademark of Vacuumschmelze GmbH (Germany) for a soft-magnetic alloy composed of aluminum and iron. It has high mechanical hardness and wear resistance, and is used for magnetic recording heads.

Vacodym. Trademark of Vacuumschmelze GmbH (Germany) for a series of high-performance permanent-magnet alloys based on neodymium, iron and boron, and produced by powder-metallurgy techniques. They have high-field strengths with small magnetic volumes, high magnetic energy products, good high-temperature stability, and a maximum service temperature of 220°C (454°F). Used for disk drives, motors and drive units, and automotive applications.

Vacofer. Trademark of Vacuumschmelze GmbH (Germany) for a series of soft-magnetic, high-purity iron alloys (99.98%) produced by powder-metallurgy techniques. They have low coercivity, high saturation flux density, a steep rise of magnetization characteristics, and a Curie temperature of 770°C (1418°F). Used especially for relays, armatures and pole shoes.

Vacoflex. Trademark of Vacuumschmelze GmbH (Germany) for a series of thermostat bimetals that consist of two metals (usually iron and nickel) with low, but different coefficients of thermal expansion bonded together. They are commercially available in various grades in the form of disks, tapes, strips and stamped parts. Used for probes in automotive temperature measurement and control functions, and in thermostats.

Vacoflux. Trademark of Vacuumschmelze GmbH (Germany) for a series of soft-magnetic alloys containing 50% cobalt and 50% iron. They have extremely high saturation flux densities, medium coercivity, and a Curie temperature of 950°C (1730°F). Used for pole shoes, cores, magnetic lenses, telephone diaphragms, relay components, telephone receivers, chokes, converters, modulators, and magnetic amplifier.

Vacomax. Trademark of Vacuumschmelze GmbH (Germany) for a series of anisotropically hard magnetic materials based on samarium cobalt (Sm_2Co_7 or $SmCo_5$) with additions of iron, and produced by powder-metallurgy techniques. They are available in various forms and grades, and have high magnetic energy products and coercivities, low sensitivity to demagnetization fields, high field strengths at small magnet volume, high temperature stability and corrosion resistance, and a maximum service temperature of 350°C (690°F). Used for magnetic bearings and couplings, electric motors and generators, micromotors, electric clocks, magnetic switches, and sensors.

Vacomet. Trade name of Vacuumschmelze GmbH (Germany) for high-purity chromium, cobalt, nickel and iron, and various high-purity cobalt-chromium, cobalt-nickel, nickel-chromium and nickel-iron alloys used as target materials in the manufacture of thin films for resistors and magnetic recording heads.

Vacon. Trademark of Vacuumschmelze GmbH (Germany) for a series of controlled-expansion and glass-sealing alloys containing 28-29% nickel, 17-23% cobalt, and the balance iron. They have low coefficients of thermal expansion that can be adjusted exactly to electrically insulating glasses or ceramics. Used for semiconductor devices, diesel spark plugs, and glass-to-metal and ceramic-to-metal seals.

Vacoperm. Trademark of Vacuumschmelze GmbH (Germany) for a series of soft-magnetic alloys composed of 72-83% nickel, with the balance being iron. They have high magnetic permeabilities, low static coercivities and a round magnetic hysteresis loop. The most versatile alloy of this series is *Vacoperm 70* also known as *Mumetall*. *Vacoperm* alloys are used for circuit breakers, instrument transformers, high- and low-frequency transformers, chokes, medium-frequency transformer, magnetic heads, leakage current protective switches, measurement instruments, and relays.

Vacoplus. Trademark of Vacuumschmelze GmbH (Germany) for a thermocouple alloy composed of 90% nickel and 10% chromium. It has a high electrical resistivity, a melting point of 1350°C (2460°F), and a maximum service temperature of 870-1260°C (1600-2300°F). Used for thermocouples for oxidizing environments.

Vacovit. Trademark of Vacuumschmelze GmbH (Germany) for a series of alloys composed of 47-54% nickel, 46-53% iron, and 0-6% chromium. They possess low coefficients of thermal expansion that can be adjusted exactly to electrically insulating glasses and ceramics, and a Curie temperature range of 360-525°C (680-977°F). Used for switches, bushings, electronic components, and glass-to-metal and ceramic-to-metal seals.

Vacozet. Trademark of Vacuumschmelze GmbH (Germany) for a series of magnetically semihard cobalt-iron-nickel alloys with up to 85% cobalt. They are available in the form of rods and strips, can be fused to glass, and have a density of 8.1 g/cm^3 (0.29 lb/in.^{3}), a hardness of 400-420 Vickers (rolled), a Curie temperature of 800°C (1470°F), a maximum service temperature of 400°C (750°F), and good coercivity and remanence flux density. Used for bistable relays, and reed contacts.

Vacromium. Trademark of Vacuumschmelze GmbH (Germany) for a series of nonferroelectric, heat-resistant nickel-chromium and nickel-chromium-iron alloys. They have exceptionally good physical properties, high heat resistance and good resistance to oxidation and corrosion. Used in the production of electric resistors and all types of industrial and household heating.

Vacryflux. Trademark of Vacuumschmelze GmbH (Germany) for a series of high-field superconductors including *Vacryflux 5001,* composed of niobium-titanium (NbTi) in a copper or copper-nickel matrix, and *Vacryflux NS,* composed of niobium-tin (Nb$_3$Sn) or niobium-tantalum-tin in a copper-tin matrix. The compound superconductors are available as single-core, multi-core and filament grades. Filament diameters range from 250 μm (0.01 in.) for single-core superconductors made of *Vacryflux 5001* to 35 μm (0.0014 in.) for superconductors made of *Vacryflux NS. Vacryflux* can produce high magnetic flux densities, and is used in the construction of solenoids, in installations for elementary particle physics and nuclear fusion plants, in the production of nuclear magnetic resonance (NMR) magnets and for magnets used in ore preparation.

Vacu. Trade name of Metalor Technologies SA (Switzerland) for several palladium-free dental solders including *Vacu PF* containing 75% gold and having a melting range of 750-800°C (1380-1470°F), and *Vacu 2* also containing 75% gold, but having a melting range of only 645-730° (1195-1345°F).

Vacuflex. Trade name of Omniflex (USA) for cast and blown flexible films used for foam-in-place applications.

Vaculite. (1) Trade name of John Healey (London) Limited (UK) for hollow glass blocks.

(2) Trade name for a bright, highly flexible aluminum-coated paper made by vacuum metallizing, and used for packaging applications, and as a barrier paper.

Vacumeltrol. Trademark of Carpenter Technology Corporation (USA) for strong, heat-resistant nickel-chromium-cobalt-molybdenum alloys.

Vacuminus. Trademark of Vacuumschmelze GmbH (Germany) for a thermocouple alloy containing 95% nickel and 5% chromium.

vacuum bag moldings. Plastic moldings produced by first placing a flexible sheet, bleeder cloth and release film over a resin-impregnated reinforcement on a mold, sealing the edges, applying a vacuum between sheet and reinforcement, and then curing after the entrapped air has been removed. See also bleeder cloth; release film.

Vacuum Formable. Trademark of Ludlow Composites Corporation (USA) for a vacuum-formable closed-cell polyvinyl chloride latex foam.

vacuum-formed plastics. Thermoplastic or thermosetting products formed by first putting the starting material in sheet form into the clamping device of a stationary frame, heating, and then drawing it down by vacuum into a shaped mold.

vacuum-carburized steel. Alloy steel gas-carburized at pressures below atmospheric pressure and temperatures of 980-1050°C (1795-1920°F) in a furnace with an atmosphere consisting only of an enriching hydrocarbon gas, such as natural gas, pure methane or propane. The gas is dissociated at the steel surface and the carbon is directly absorbed into the surface with the liberation of hydrogen. The resulting steel has a hard surface (or case) and a tough core.

vacuum coating. A coating with a thickness usually ranging from less than 1 μm (0.04 mil) to several tens of microns (over 0.4 mil), condensed from vapors onto a metallic or nonmetallic substrate, such as a polymer, semiconductor, glass or ceramic. Used on jewelry, electronic components, decorative plastics, and as a functional coating, and as a dielectric coating on glass. Also known as *vacuum deposit.*

vacuum concrete. A quick-setting concrete subjected to a vacuum to remove entrapped air and excess water from the surface. It has high strength, increased surface hardness, and improved durability and crushing resistance.

vacuum-degassed steel. Steel from which dissolved gases, such as hydrogen, oxygen and nitrogen, have been removed by subjecting the molten metal to a vacuum. Various vacuum degassing methods are employed including ladle degassing, stream degassing, DH (Dortmund-Hörder-Hüttenunion) process, RH (Ruhrstahl-Henrichshütte-Heraeus) process and VOD (vacuum oxygen decarburization) process. *Vacuum-degassed steel* has improved properties (e.g., ductility, corrosion resistance, tensile and impact strength, etc.), higher purity, fewer nonmetallic inclusions and less tendency to segregation and center porosity.

vacuumed clay. Clay that has been vacuum-treated to eliminate air bubbles in order to increase its density and promote the green strength of ceramic bodies.

vacuum deposit. See vacuum coating.

vacuum-evaporated coating. See evaporation vacuum coating.

vacuum-injection-molded plastics. Reinforced plastics formed by placing the reinforcing material between two mating molds, injecting a premixed, room-temperature-curing liquid resin into the bottom mold and drawing it up through the reinforcement (which thus becomes saturated with resin) by a vacuum, usually applied through the top mold. Abbreviation: VIM plastics.

vacuum-melted steel. Steel melted in the absence of air, e.g., in a vacuum-arc or vacuum-induction furnace. Vacuum melting greatly reduces the amount of gases (dissolved hydrogen, oxygen and nitrogen) and nonmetallic inclusions (e.g., oxides) in the steel. The absence of gases improves many steel properties including ductility, magnetic properties, impact strength, and fatigue strength.

vacuum-metallized coating. (1) See vacuum coating.

(2) See evaporation vacuum coating.

vacuum-metallized film. See evaporation vacuum coating.

vacuum-oxygen decarburized steel. A highly refined stainless steel produced by first transferring molten electric-furnace steel to a stainless ladle-furnace, heating and stirring the melt by an induced electrical current, and then introducing selected alloying additions through a hopper, and oxygen through a water-cooled lance. Also known as *VOD steel.*

Vacuum-Seal. Trade name of Vacuum-Seal Corporation (USA) for insulating glass units.

vacuum-sintered products. Dense ceramic or powder-metallurgy products consolidated after forming and pressing by sintering in a furnace under vacuum conditions.

vaesite. A gray mineral of the pyrite group composed of nickel sulfide, NiS_2. Crystal system, cubic. Density, 4.45 g/cm³. Occurrence: Central Africa (Congo, Zaire).

Vairin. Trade name of Elastofibre (Italy) for spandex (elastane) fibers and monofilaments with excellent resistance to abrasion, perspiration and body oils, used for the manufacture of textile fabrics with excellent stretch and recovery properties.

Valand. Trade name of Uddeholm Corporation (USA) for an oil-hardening hot-work tool steel containing 0.3% carbon, 2.4-3% chromium, 0-2% cobalt, 0-1.7% nickel, 8.5-9.4% tungsten, up to 0.3% vanadium, and the balance iron. Used for hot-working tools, die-casting dies, extrusion dies, rams and liners, and punches.

Valena. Trademark of Val Lesina SpA (Italy) for polyester filament yarns used for textile fabrics.

Valenciennes lace. A flat *bobbin lace* whose ground and pattern are made with the same fine threads.

valency compounds. Compounds, such as aluminum antimonide (AlSb) or magnesium lead (Mg_2Pb), formed by two chemically dissimilar metals. They show ordinary chemical valence, have ionic, covalent or mixed bonds, are usually nonmetallic in nature. Many of these compounds are brittle and have poor electrical conductivity. See also intermetallic compounds.

Valenite. Trade name of Valenite Corporation (USA) for a series of cemented carbides composed of carbides of tungsten, molybdenum and/or titanium in cobalt or nickel matrices. Depending upon the particular composition, they are used in the manufacture of cutting tools for various hard materials (e.g., steels, cast irons, nonferrous metals and nonmetallic materials), measuring tools, gauges, dies, and punches.

valentinite. Colorless, white or gray mineral with an adamantine to silky luster. It is composed of antimony trioxide, Sb_2O_3, occurs in nature and can also be made synthetically. Crystal system, orthorhombic. Density, 5.57-5.83 g/cm³; hardness, 2-3 Mohs; refractory index, 2.35. Occurrence: Algeria, Balkan peninsula, Germany, Italy. Used as an ore of antimony. See also senarmonite.

Valqua. Trade name of Valspar Corporation (USA) for water-reducible baking enamels used in metal finishing.

Valéron. Trademark of Van Leer-Keyes (USA) for a range of plastic laminates and films for flexible packaging applications including paper/plastic laminates and polyethylene films, containers and bags.

Valiant. (1) Trade name of Osborn Steels Limited (UK) for a water-hardening tool steel containing 1.5% carbon, 4% tungsten, 0.5% chromium, and the balance iron. It keeps a keen cutting edge, and is used for cutting tools, and engraving cutters.

(2) Trade name of Ivoclar Vivadent AG (Liechtenstein) for a dental amalgam alloy.

Valite. (1) Trademark of Plenco–Plastics Engineering Company (USA) for a series of granular injection-, compression- and transfer-molded novolac phenolics. See also novolacs.

(2) Trade name of Valite Corporation (USA) for plastics derived from *bagasse*.

Valiw. Trade name of Acciaierie Valbruna SpA (Italy) for hardenable, corrosion-resistant steel containing 0.2% carbon, 12.5% chromium, 1.5% molybdenum, 1% tungsten, 0.75% nickel, 0.25% vanadium, and the balance iron. Used for surgical instruments, knives, gears, shafts, and hardware.

Vallade OB. Trade name of ICI Limited (UK) for acrylonitrile-butadiene-styrene plastics.

Vallendar clay. An enameling clay formerly mined at Vallendar, Germany.

Valley Stone. Trademark of Cyde Everett Limited (Canada) for concrete building blocks.

valley tile. A curved roofing tile that is usually V-shaped and used in the valley of a roof, i.e., in the intersection at the bottom of two sloping roofs. It is the opposite of a hip tile.

Vallinox. Trade name of Vallourec SA (France) for a series of austenitic, ferritic and martensitic steels.

valleriite. A bronze-colored mineral composed of copper iron magnesium aluminum sulfide hydroxide, $CuFeS_2·1.53[(Mg,Al)(OH)_2]$. Crystal system, rhombohedral (hexagonal). Density, 3.09 g/cm³. Occurrence: South Africa.

Valmag. Trade name of Swift Levick & Sons Limited (UK) for corrosion- and wear-resistant austenitic manganese steels containing considerable amounts of chromium and nickel. Usually supplied in the solution-treated and aged condition, and used for valves and valve parts.

Valox. Trademark of GE Plastics (USA) for an extensive series of engineering thermoplastics including polybutylene terephthalate (PBT) resins and polybutylene terephthalate/polyethylene terephthalate (PBT-PET) blends supplied in various grades including regular, glass fiber- or carbon fiber-reinforced, mineral- and/or glass bead-filled, silicone- or polytetrafluoroethylene-lubricated, fire-retardant, UV-stabilized and structural-foam. They have excellent durability and dimensional stability, exceptional toughness and fatigue endurance, high heat resistance, low thermal expansion, low water absorption, excellent dielectric strength, good chemical creep and impact resistance, inherent low friction, good ultraviolet stability, good self-sealing capabil-ity, low flammability, good processibility, and excellent surface finishes. Used for electrical and electronic devices (e.g., computers, pocket calculators and data terminals), material- and fluid-handling systems, electrical connectors and switches, air valves and pumps, housings, and consoles.

Valqua. Trade name of Valspar Corporation (USA) for a water-reducible baking enamel.

Valray. Trade name of Inco Alloys International Limited (UK) for a heat-resistant alloy composed of about 80% nickel and 20% chromium, sometimes with small amounts of carbon, manganese and silicon. It has a density of 8.4 g/cm³ (0.30 lb/in.³), a melting point of 1400°C (2550°F), high electrical resistivity, relatively high strength, and is used in coatings for automobile exhaust valves, and as a facing alloy.

Valrez. Trademark of Valchem, Chemical Division (USA) for a series of urea derivatives and modified urea-formaldehyde resins used in the manufacture of wash-and-wear clothing.

Valspar. Trademark of Valspar Corporation (USA) for wood protection products including high-gloss spar varnishes for exterior and marine applications, wood stains, liquid polyurethane coatings for furniture and toys, and various paint, varnish and stain removers.

Valtec. Trade name of Himont USA Inc. for standard and impact-modified polypropylene and polypropylene copolymer resins usually supplied in pellet or bead form for the manufacture of blow-molded and thermoformed cookware and industrial equipment.

Valtra. Trade name of Chevron Chemical Company (USA) for a polystyrene resin designed for fast injection molding of components for consumer products and household appliances. They possess high impact strength, good stiffness, high gloss, and balanced flow properties.

Valtrac. Trade name of Davis & Geck, Inc. (USA) for a biodegradable polymer based on poly(glycolic acid) and used for anastomosis rings.

VALUe-Stat. Trademark of Scharr Industries Inc. (USA) for a metallized polyethylene/polyester laminate film with excellent antistatic properties. Used for static shielding applications.

Valutap. Trade name of Teledyne Vasco (USA) for a die steel containing 1.2% carbon, 1.6% tungsten, 0.7% chromium, 0.2% vanadium, and the balance iron. Used for dies, gages and punches.

Valux. Trademark of 3M Company (USA) for a light-cure, small particle hybrid composite resin paste used for anterior dental restorations.

valve alloys. A somewhat imprecise term for a group of different alloys including several bronzes, brasses, nickel and nickel-copper alloys, stainless steels, cast irons and lead alloys used for the manufacture of valves.

Valve Bearings. British trade name for a babbitt composed of 71% tin, 24% antimony and 5% copper, and used for bearings and valve components.

valve bronze. A corrosion-resistant cast bronze containing 80-86% copper, 2-10% tin, 3-15% zinc, and up to 6% lead. It has good machinability and moderate strength, and is used for valves, pipe fittings and parts, and pump fittings.

valve copper. A corrosion-resistant cast copper containing 86-92% copper, 3-7.5% tin, 1-6% zinc, 0-3% nickel and up to 2% lead. It has good machinability and moderate strength, and is used for valves, and pipe and pump parts.

Valve-Loy. Trade name of Resisto-Loy Company, Inc. (USA) for a steel containing 0.6% carbon, 35% chromium, 10% molybdenum, 8% tungsten, 6% cobalt, 0.02% boron, and the balance iron. Used for hardfacing electrodes and valve faces.

valve metal. A copper alloy containing 9% zinc, 7% lead and 3% tin. Used for valves and valve parts.

valve steels. A group of high-quality steels used for valves and valve parts, such as stems, seats, bodies and actuators. For valves operating at room temperature or slightly elevated temperatures (e.g., check and relief valves, shut-off valves, and metering and regulating valves) various austenitic, ferritic and martensitic stainless steels are used including austenitic types AISI 303, 316, 321 and 329, and martensitic types AISI 414, 420, 440B and 440C. For valves operating at temperatures above 760°C (1400°F) (e.g., automobile and aircraft exhaust valves) several heat- and wear-resistant stainless steels with good resistance to scaling and oxidation are used, e.g., martensitic types AISI 416, 410, 422 and 431, ferritic types AISI 409 and 446, austenitic type AISI 309 as well as precipitation-hardening semi-austenitic type 17-7 PH. Valve parts requiring hard, wear-resistant surfaces are usually made of nitrided steels.

Valvex. Trade name of Swift Lewick & Sons Limited (UK) for valve steels containing 0.45% carbon, 3.5% silicon, 8% chromium, and the balance iron. They are usually supplied in the hardened and tempered condition.

Valvic. Trade name of Swift Lewick & Sons Limited (UK) for a series of stainless steels containing about 0.8% carbon, 2.1% silicon, 20% chromium, 1.5% nickel, and the balance iron. Used for valves and valve parts.

Valvit. Trade name for a bearing and valve bronze containing 91% copper and 9% tin.

Valvo. Trade name of Swift Lewick & Sons Limited (UK) for austenitic stainless steels containing about 0.4% carbon, 13-14% chromium, 10-14% nickel, 2.7% tungsten, and the balance iron. Used for valves and valve parts.

Vamac. Trademark of E.I. DuPont de Nemours & Company (USA) for a tough, low-compression-set ethylene/acrylic elastomer. It has excellent high-temperature durability and resistance to water, hot oil, transmission and power-steering fluids, service lubricants, and ozone and weathering, and good mechanical strength, low-temperature flexibility and vibration damping properties. Used for automotive applications including seals, O-rings, gaskets, specialty hose, cable and wire jacketing, and low-smoke flooring.

Vampire. Trade name of Osborn Steels Limited (UK) for a non-shrinking, oil-hardening tool steel containing 0.9% carbon, 1.5% manganese, 0.3% chromium, and the balance iron. Used for dies, press tools, gages, and punches.

Van. Trade name of LTV Steel (USA) for a series of high-strength low-alloy (HSLA) steels with good weldability, formability and impact toughness used for truck frames, crane booms, and automobile bumpers.

Vanachrom. Trade name of Swift Lewick & Sons Limited (UK) for a chromium-vanadium hot-work die steel containing 0.35% carbon, 5.5% chromium, 1.35% molybdenum, 1% vanadium, and die balance iron.

Van-Ad. Trade name for a master alloy composed of 75% vanadium and 25% titanium used as a vanadium addition for titanium alloys.

vanadic acid anhydride. See vanadium pentoxide.

vanadic sulfate. See vanadyl sulfate.

vanadic sulfide. See vanadium sulfide (iii).

vanadinite. A ruby red, orange red, brownish or yellow mineral with resinous to adamantine luster. It belongs to the apatite group and is composed of lead chlorovanadate, $Pb_5Cl(VO_4)_3$. It can also be made synthetically. Crystal system, hexagonal. Density, 6.66-7.10 g/cm³; hardness, 3 Mohs; refractive index, 2.628. Occurrence: Africa, Mexico, Russia, Scotland, Spain, USA (Arizona, New Mexico). Used as an ore of vanadium and lead. Also known as *vanadite*.

Vanadis. Trade name of Uddeholm Corporation (USA) for a high-performance, high-alloy powder-metallurgy cold-work tool steel with a homogeneous, uniform, fine carbide microstructure. Its dimensional stability after heat treatment exceeds that of high-performance AISI cold-work tool steels, and it has excellent wear resistance, strength and toughness and good machinability. Used for powder-metallurgy parts.

vanadite. See vanadinite.

Vanadium. (1) Trade name of Latrobe Steel Company (USA) for a tungsten-type high-speed steel (AISI type T3) containing 1.1% carbon, 4% chromium, 18% tungsten, 3% vanadium, and the balance iron. It has good wear resistance and cutting ability, and is used for hard cutting tools. Formerly known as *Electrite Vanadium*.

(2) Trade name of Teledyne Vasco (USA) for a series of chromium-vanadium-type special-purpose tool steels (AISI type L2) containing 0.45-1.1% carbon, 0.2-0.35% silicon, 0.1-0.3% manganese, 0.7-1.5% chromium, 0.15-0.25% vanadium, and the balance iron. They have excellent resistance to decarburization, excellent toughness, great depths of hardening, high shock resistance, high strength and hardness, good to moderate machinability, and moderate wear resistance. Used for hand tools, woodworking and blacksmithing tools, cutting and finishing tools, caulking tools, stamps, dies and die holders, rock

drills, gages, ball bearings, gears, springs, shafts, cams, joints, pins, and piston rods.

vanadium. A silvery-white ductile metallic element of Group VB (Group 5) of the Periodic Table. It is commercially available in granules, rods, ingots, sheet, foil, microfoil, wire, powder, turnings, lumps and single crystals. The single crystals are usually grown by the float-zone or the radio-frequency technique. The chief vanadium ores are *carnotite, patronite, roscoelite* and *vanadinite*. Crystal system, cubic. Crystal structure, body-centered cubic. Density, 6.1 g/cm³, melting point, 1890°C (3434°F), boiling point, 3380°C (6116°F); hardness, 80-150 Vickers; atomic number, 23; atomic weight, 50.942; divalent, trivalent, tetravalent, pentavalent. It has a superconductivity critical temperature of 5.40K, low electrical conductivity (about 7-8% IACS), good resistance to air, alkalies, seawater and most reducing acids, good corrosion resistance, good structural strength, and low thermal expansion. Used as a carbide stabilizer and grain refiner in steel manufacture, as ferrovanadium in steel alloys (e.g., corrosion- and scaling-resistant steels, high-temperature steels, spring steels and tool steels), as a target material for X-rays, in vanadium compounds, in superconductors, as a catalyst for synthetic rubber, as a bonding agent in cladding titanium to steel, in construction materials, and in jet engines. Symbol: V.

vanadium aluminide. A compound of vanadium and aluminum with a melting point of 1670°C (3038°F), used in ceramics and materials research. Formula: V_5Al_8.

vanadium-aluminum. A master alloy containing 50-86% vanadium, 13-49% aluminum and 0.1-0.2% carbon.

vanadium beryllide. Any of the following compounds of vanadium and beryllium used in ceramics: (i) *Vanadium diberyllide.* Density, 3.80 g/cm³; melting point, above 1650°C (3000°F). Formula: VBe_2; and (ii) *Vanadium dodecaberyllide.* Density, 2.35 g/cm³. Formula: VBe_{12}.

vanadium boride. Any of the following compounds of vanadium and boron used in ceramics: (i) *Vanadium monoboride.* Refractory compound. Density, 5.1 g/cm³; melting point, above 2100°C (3810°F); hardness, 8-9 Mohs; high electrical resistivity. Formula: VB; (ii) *Vanadium diboride.* A refractory solid or fine powder. Crystal system, hexagonal. Density, 5.5 g/cm³; melting point, 2450°C (4440°F); has highly anisotropic thermal expansion. Formula: VB_2; (iii) *Vanadium triboride.* Density, 5.5 g/cm³; melting point, 2300°C (4170°F). Formula: VB_3; (iv) *Trivanadium diboride.* Melting point, 2066°C (3751°F). Formula: V_3B_2; and (v) *Vanadium tetraboride.* Melting point, 2271°C (4120°F). Formula: V_3B_4.

vanadium bromide. See vanadium tribromide.

vanadium borocarbide. A compound of vanadium, boron and carbon that is available in the form of a fine powder, and used in ceramics and materials research. Formula: VB_4C.

vanadium-boron. A master alloy containing 40-50% vanadium, 6-8% boron, 0-5% titanium, 2-3% aluminum, and the balance iron. Used to introduce vanadium into steel.

vanadium brass. A brass composed of 70% copper, 29.5% zinc and 0.5% vanadium, and used in condenser tubes, sheets, and hardware.

vanadium bromide. See vanadium dibromide.

vanadium bronze. A high-strength bronze composed of 61% copper, 38.5% zinc and 0.5% vanadium, and used for pipes.

vanadium carbide. Any of the following compounds of vanadium and carbon: (i) *Vanadium monocarbide.* Black crystals or fine powder. Crystal system, cubic. Crystal structure, halite.

Density, 5.71-5.77 g/cm³; melting point, 2830°C (5126°F); boiling point, 3900°C (7052°F); hardness, 2800-2950 Vickers. Used as a component in ceramic cutting tool compositions, in alloys for cutting tools, and as an addition to steels. Formula: VC; and (ii) *Divanadium carbide.* Hexagonal crystals. Density, 5.75 g/cm³; melting point, approximately 2166°C (3931°F). Used in ceramics and materials research. Formula: V_2C.

vanadium carbonyl. See vanadium hexacarbonyl.

Vanadium Castdie. Trade name of Columbia Tool Steel Company (USA) for an oil-hardening hot-work steel containing 0.35% carbon, 5% chromium, 1.5% molybdenum, 1% vanadium, and the balance iron. Used for tools and dies.

vanadium chloride. See vanadium dichloride.

vanadium diberyllide. See vanadium beryllide (i).

vanadium diboride. See vanadium boride (ii).

vanadium dibromide. Orange-brown crystals. Crystal system, hexagonal. Density, 4.58 g/cm³; boiling point, 800°C (1470°F) (sublimes). Used in chemistry and materials research. Formula: VBr_2. Also known as *vanadous bromide; vanadium bromide.*

vanadium dichloride. Apple-green, hygroscopic, platy crystals (95+% pure). Crystal system, hexagonal. Density, 3.09-3.23 g/cm³; boiling point, 910°C (1670°F) (sublimes). Used in the purification of hydrogen chloride from arsenic, and as a strong reducing agent. Formula: VCl_2. Also known as *vanadous chloride; vanadium chloride.*

vanadium dioxide. A compound of vanadium and oxygen supplied as a blue-black powder (99+% pure) with a density of 4.34 g/cm³ and a melting point of 1967°C (3573°F). It exhibits antiferromagnetic properties, and is used in ceramics, materials research, and electronics. Formula: VO_2. Also known as *vanadium oxide.*

vanadium disilicide. See vanadium silicide (i).

vanadium disulfide. See vanadium sulfide (iv).

vanadium diiodide. Red-violet, hexagonal crystals with a density of 5.44 g/cm³, used in chemistry and materials research. Formula: VI_2. Also known as *vanadous iodide.*

vanadium dodecaberyllide. See vanadium beryllide (ii).

Vanadium Extra. Trade name of Columbia Tool Steel Company (USA) for a water-hardening tool steel (AISI type W2) containing 1.06% carbon, 0.3% manganese, 0.25% silicon, 0.2% vanadium, and the balance iron. It possesses excellent machinability, good to moderate deep-hardening properties, wear resistance and toughness, and low hot hardness. Used for dies, mandrels, and lathe centers.

Vanadium Firedie. Trade name of Columbia Tool Steel Company (USA) for an oil-hardening hot-work steel (AISI type H13) containing 0.35% carbon, 5% chromium, 1.5% molybdenum, 1% vanadium, and the balance iron. Used for aluminum and magnesium casting dies.

vanadium fluoride. See vanadium penta-fluoride.

vanadium gallium. A compound of vanadium and gallium with a superconductivity critical temperature of 16.5K, and a critical magnetic flux density of 22 teslas. Used as a superconductor. Formula: V_3Ga.

vanadium hexacarbonyl. Blue-green, moisture-sensitive, paramagnetic crystals or powder that sublimes at 50°C (122°F)/15 mm, and decomposes at 60-70°C (140-158°F). It is also available as the diglyme-stabilized sodium salt (bisdiglymesodium hexacarbonyl vanadate) in the form of moisture-sensitive, yellow crystals with a melting point of 173-176°C (343-348°F). Used in the manufacture of fuel additives, in the preparation of plating compounds, and as a general chemical intermediate.

Formula: V(CO)$_6$. Also known as *vanadium carbonyl.*

Vanadium Hot Work. Trade name for a hot-work tool steel containing 0.5% carbon, 0.5% chromium, 1% vanadium, and the balance iron.

vanadium hydride. A compound of vanadium and hydrogen that is available in the form of a fine high-purity powder for use in ceramics and materials research. Formula: VH.

vanadium iodide. See vanadium diiodide.

vanadium mica. A bluish green *muscovite* mica mineral in which part of the aluminum has been replaced by vanadium. The general formula is K(Al,V)$_2$(Si,Al)$_4$O$_{10}$(OH)$_2$. The mineral *roscoelite* is a variety. Crystal system, monoclinic. Density 2.80-2.92 g/cm^3; refractive index, 1.608. Occurrence: Australia, USA (Arizona, California, Colorado, Utah). Used as a source of vanadium.

vanadium monoboride. See vanadium boride (i).

vanadium monocarbide. See vanadium carbide (i).

vanadium mononitride. See vanadium nitride (i).

vanadium monoxide. A compound of vanadium and oxygen available in the form of green crystals. Density, 5.758 g/cm^3; melting point, 1790°C (3254°F). Used in ceramics and materials research. Formula: VO. Also known as *vanadyl; vanadium oxide.*

vanadium monosulfide. See vanadium sulfide (i).

vanadium 2,3-naphthalocyanine oxide. See vanadyl 2,3-naphthalocyanine.

vanadium nitride. Any of the following compounds of vanadium and nitrogen: (i) V*anadium mononitride.* A black powder (99+% pure). Crystal system, cubic; density, 6.13 g/cm^3; melting point, 2050°C (3722°F). Used in refractories and ceramics, and in materials research. Formula: VN; and (ii) *Divanadium nitride.* Density, 5.99 g/cm^3. Used in ceramics and materials research. Formula: V$_2$N.

vanadium oxide. See vanadium monoxide; vanadium dioxide; vanadium trioxide; vanadium tetroxide; vanadium pentoxide.

vanadium pentafluoride. A brownish compound with a density of 1.177 g/cm^3, a melting point of 19.5°C (67°F) and a boiling point of 111°C (231°F). Used in the preparation of vanadium and organovanadium compounds. Formula: VF$_5$. Also known as *vanadium fluoride.*

vanadium pentasulfide. See vanadium sulfide (iii).

vanadium pentoxide. An orange-brown crystalline powder or yellow-orange lumps (98+% pure). Crystal system, orthorhombic. Density, 3.36 g/cm^3; melting point, 690°C (1274°F); boiling point, decomposes above 1750°C (3180°F). Used as a red, green, pink, or yellow ceramic colorant, as a colorant flux in glasses, porcelain enamels and glazes, in glass as an ultraviolet transmission inhibitor, as a glass former, in ferrovanadium, as a catalyst and photographic developer, and in vanadium salts. Formula: V$_2$O$_5$. Also known as *vanadic acid anhydride.*

Vanadium Permendur. Trade name of Bell Telephone Laboratories (USA) for a permanent magnet alloy of 49% iron, 49% cobalt and 2% vanadium. It has high permeability at high flux density, and is used for electrical equipment, pole pieces, and diaphragms.

vanadium phosphide. A compound of vanadium and phosphorus used in ceramics and materials research. Density, 5.0 g/cm^3; melting point, 1315°C (2400°F). Formula: VP.

vanadium phthalocyanine oxide. See vanadyl phthalocyanine.

Vanadium Potts Best. Trade name of Horace T. Potts Company (USA) for a tool steel containing 0.9-1.2% carbon, 0.2% vanadium, and the balance iron. Used for cutters and tools.

vanadium powder. Vanadium (99.5+% pure) in ground form with particle sizes ranging from less than 5 to over 800 μm (0.0002 to over 0.03 in.), used in electronics, superconductivity research, and powder metallurgy.

vanadium sesquioxide. See vanadium trioxide.

vanadium silicide. Any of the following compounds of vanadium and silicon: (i) *Vanadium disilicide.* Metallic prisms or a fine powder. Density, 4.42 g/cm^3; melting point, 1700°C (3092°F); exhibits metallic behavior. Used in ceramics, for electrochemically stable cathodes, and for hydrogen storage applications, e.g., fuel cells. Formula: VSi$_2$. (ii) *Trivanadium silicide.* Cubic crystals. Density, 5.74 g/cm^3; melting point, 1732-2049°C (3150-3720°F); superconductivity critical temperature, 17.1K. Used in superconductor research. Formula: V$_3$Si; and (iii) *Pentavanadium trisilicide.* Density, 5.27 g/cm^3; melting point, 2150°C (3902°F). Used in ceramics and materials research. Formula: V$_5$Si$_3$.

Vanadium Standard. Trade name of Columbia Tool Steel Company (USA) for a water-hardening tool steel (AISI type W2) containing 1.06% carbon, 0.3% manganese, 0.25% silicon, 0.2% vanadium, and the balance iron. It has excellent machinability, good to moderate deep-hardening properties, wear resistance and toughness, and low hot hardness. Used for mandrels, cold chisels, and blacksmith tools.

vanadium steel. A low-alloy steel which has 0.1-0.15% vanadium added to increase its tensile strength, hardness, hot hardness, toughness and corrosion and fatigue resistance, reduce grain growth, and remove oxygen and, possibly, nitrogen.

vanadium sulfate oxide. See vanadyl sulfate.

vanadium sulfide. Any of the following compounds of vanadium and sulfur: (i) *Vanadium monosulfide.* Density, 4.89 g/cm^3; melting point, 1900°C (3452°F). Used in ceramics and materials research. Formula: VS; (ii) *Vanadium trisulfide.* A green-black powder (99.5+% pure). Density, 4.70 g/cm^3; melting point, 1927°C (3501°F). Used in ceramics and materials research. Formula: V$_2$S$_3$; and (iii) *Vanadium pentasulfide.* A fine, black or black-green powder. Density, 3.0 g/cm^3; melting point, decomposes on heating. Used for vanadium compounds, and in ceramics and materials research. Formula: V$_2$S$_5$. Also known as *vanadic sulfide; vanadium sulfide;* and (iv) V*anadium disulfide.* Density, 4.20 g/cm^3; melting point, decomposes. Used as a solid lubricant, and as an electrode in lithium-base batteries. Formula: V$_2$S$_2$.

vanadium tetraboride. See vanadium boride (v).

vanadium tetroxide. A bluish black powder (99.9+%) with a density of 4.34 g/cm^3, and a melting point of 1967°C (3573°F) used as a catalyst at high temperatures, in refractory compositions fusing at temperatures above 1540°C (2804°F). It forms a low-porosity body with beryllia (BeO). Refractories containing vanadium tetroxide tend to be unstable in air. Formula: V$_2$O$_4$.

vanadium-tin yellow. An inorganic pigment composed of a mixture of vanadium pentoxide (V$_2$O$_5$) and tin oxide (SnO$_2$). Also known as *tin-vanadium yellow; vanadium yellow.*

Vanadium Tool. British trade name for an oil-hardening tool steel containing 0.9% carbon, 1% chromium, 0.2% vanadium, and the balance iron. Used for general-purpose tools.

vanadium triboride. See vanadium boride (iii).

vanadium trioxide. Black crystals or powder (97+% pure). It occurs in nature as the mineral *karelianite.* Density, 4.87 g/cm^3; melting point, above 1970°C (3578°F). Used as a catalyst for the conversion of ethylene to ethanol, and in ceramics. Formula: V_2O_3. Also known as *vanadium oxide*.

vanadium trisulfide. See vanadium sulfide (ii).

vanadium yellow. See vanadium-tin yellow.

vanadium zirconium. A compound of vanadium and zirconium with a melting point of 1500°C (2732°F) used in ceramics and materials research. Formula: V_2Zr.

vanadium-zirconium blue. See zirconium-vanadium blue.

vanadium-zirconium turquoise. See zirconium-vanadium turquoise.

vanadocene. An organometallic coordination compound (molecular sandwich) consisting of a vanadium atom (V^{2+}) and two five-membered cyclopenediene (C_5H_5) rings, one located above and the other below the vanadium atom plane. *Vanadocene* is supplied in the form of flammable, moisture-sensitive purple crystals (usually in the sublimed state) with a melting point of 165-167°C (329-333°F), and used in organic and organometallic research. Formula: $(C_5H_5)_2V^{2+}$. Also known as *dicyclopentadienylvandium; bis(cyclopentadienyl)vanadium*.

vanadous bromide. See vanadium dibromide.

vanadous chloride. See vanadium dichloride.

vanadyl. See vanadium monoxide.

vanadyl etioporphyrin. A red-purple crystalline compound which is a *porphyrin* derivative containing vanadium oxide (VO). Used as a pigment and dye. Formula: $(C_{32}H_{36}N_4)VO$.

vanadyl 2,3-naphthalocyanine. A *2,3-naphthalocyanine* derivative containing vanadium oxide. It has a melting point above 300°C (572°F) and a maximum absorption wavelength of 817 nm. Used as a dye and pigment. Formula: $(C_{48}H_{24}N_8)VO$. Abbreviation: VONc. Also known as *vanadium 2,3-naphthalocyanine oxide*.

vanadyl octaethylporphyrin. A red-purple crystalline compound which is a *porphyrin* derivative containing vanadium oxide (VO). Used as a pigment and dye. Formula: $(C_{36}H_{44}N_4)VO$.

vanadyl phthalocyanine. A purple, crystalline *phytalocyanine* derivative that contains vanadium oxide, and exhibits good photoconductive properties. It can also be doped with tetrafluorotetracyanoquinodimethane (F_4-TCNQ) to exhibit semiconducting properties. Used as a dye and pigment, e.g., in photovoltaic devices. Formula: $(C_{32}H_{16}N_8)VO$. Abbreviation: VOPc. Also known as *vanadium phthalocyanine oxide*.

vanadyl sulfate. Blue crystals or lumps (99+% pure) with a melting point of 105°C (221°F). Used as a catalyst, as a reducing agent, as a mordant, in the preparation of aniline black, as a green and blue pigment in glass, porcelain enamels and glazes, and in biological and biochemical microscopy for the preparation of vanadomolybdate stains. Formula: $VOSO_4 \cdot 2H_2O$. Also known as *vanadic sulfate; vanadium sulfate oxide*.

vanalite. A bright yellow mineral composed of sodium aluminum vanadium oxide hydrate, $NaAl_8V_{10} \cdot 30H_2O$. Crystal system, monoclinic. Density, 2.30 g/cm³. Occurrence: Kazakhstan.

Vanalium. Trade name for a corrosion- and erosion-resistant aluminum casting alloy containing about 14% zinc, 5% copper, 0.75% iron, and 0.25% vanadium. Used for aircraft and automotive engines and parts.

Van Allen. British trade name for corrosion-resistant dental alloy containing 64% gold, 18.75% silver, 9% copper, 8% palladium, and 0.25% aluminum.

Vanasil. Trade name of Gould Inc. (USA) for a series of aluminum casting alloys containing 21-23% silicon, 0.9-1.5% copper, 2-2.5% nickel, 0.75-1.25% magnesium, 0.1% vanadium, and 0.15% titanium. Used for pistons, cylinder sleeves and liners, and compressor blades.

Van Chip. Trade name of Teledyne Firth-Sterling (USA) for a molybdenum-type high-speed steel (AISI type M3) containing 1.15% carbon, 4.1% chromium, 6% tungsten, 5.75% molybdenum, 3% vanadium, and the balance iron. Used for cutting, finishing and shaping tools.

Vanceva Color. Trademark of Solutia Inc. (USA) for a tough, colored polymeric interlayer for automotive window applications. Laminated between two glass sheets, it provides protection from flying glass fragments in case of accidents, and reduces outside noise and incident ultraviolet rays.

Vanceva Secure. Trademark of Solutia Inc. (USA) for a tough, high-security polymeric interlayer for automotive window applications. Laminated between two glass sheets, it provides all-around intrusion resistance, protection against flying fragments in case of accidents, and reduces outside noise and incident ultraviolet rays.

Vanco. Trade name of Darwins Alloy Castings (UK) for tungsten- and cobalt-tungsten high-speed steels containing 1.5% carbon, 4% chromium, 12% tungsten, 5% vanadium, 5% cobalt, and the balance iron. They have excellent cutting ability and wear resistance, high hot hardness, and are used for cutting tools, tool bits, and milling cutters.

Vancro. (1) Trade name of Hoytland Steel Company (USA) for a tough, water-hardening tool steel containing 0.45% carbon, 0.7% manganese, 1% chromium, 0.15% vanadium, and the balance iron. Used for tools and dies.

(2) Trade name of Acciairie Valbruna SpA (Italy) for air-hardening cold-work steel containing 1% carbon, 5.15% chromium, 1.15% molybdenum, 0.55% manganese, 0.3% silicon, 0.25% vanadium, and the balance iron. Used for dies, mandrels, shear blades, and punches.

Van Cut. Trade name of Teledyne Vasco (USA) for a series of molybdenum-type high-speed steels (AISI types M3-1 and M3-2) containing 1-1.2% carbon, 0.25-0.35% silicon, 0.20-0.30% manganese, 6-6.5% tungsten, 6-6.5% molybdenum, 3.75-4.25% chromium, 2.4-3% vanadium, and the balance iron. They possess good deep-hardening properties, good heat and wear resistance, and high strength and toughness. Used for lathe and planer tools, drills, milling cutters and various other cutting and finishing tools, and for form tools, dies, saws, and woodworking tools.

Vand-Alloy. Trade name of MacDermid Inc. (USA) for electroless nickel coating.

Vand-Aloy. Trade name of Cemco International Inc., Division of Vanguard Holdings Inc. (USA) for an electroless nickel coating.

Vandar. Trademark of Ticona LLC/Hoechst-Celanese Corporation (USA) for a series of thermoplastic polyester alloys supplied in unfilled, filled, glass-reinforced and injection-molding grades. They have excellent chemical resistance, excellent paintability, outstanding stiffness, high impact strength at high and low temperatures, good stiffness, and negligible water absorption. Used for automotive components, electronic and electrical applications, household appliances, and sports and recreational equipment.

vandenbrandeite. A dark green to nearly black, radioactive mineral of the becquerelite group composed of copper uranium oxide dihydrate, $CuUO_4 \cdot 2H_2O$. Crystal system, triclinic. Density, 5.03 g/cm³; refractive index, 1.792. Occurrence: Congo, Zaire.

vandendriesscheite. A yellowish-orange, radioactive mineral of the becquerelite group composed of lead uranium oxide dodecahydrate, $PbU_7O_{22} \cdot 12H_2O$. Crystal system, orthorhombic. Density, 5.45 g/cm^3; refractive index, 1.850. Occurrence: Central Africa.

Vanderloy. Trade name of Van der Horst Corporation (USA) for an electrolytic iron (99.9% pure) used for electroformed molds and dies.

Vandex. Trademark of R.T. Vanderbilt Company, Inc. (USA) for a fine, dark gray metallic selenium powder with a density of 4.8 g/cm^3 (0.17 $lb/in.^3$) and a melting point above 217°C (423°F). Used in rubber vulcanization.

Van Die Car. Trade name of Ziv Steel & Wire Company (USA) for a wear-resistant steel containing 1.1% carbon, 0.3% vanadium, 0.3% chromium, and the balance iron. Used for punches, knives, and dies.

Vandl-Pruf. Trade name of Multiplate Glass Company (USA) for a laminated safety glass with a standard thickness of 9.5 mm (0.375 in.).

Vandura silk. An early man-made protein fiber introduced in 1894.

Vandyke brown. A dark-brown, naturally occurring pigment composed of mixtures of iron oxide and decomposed vegetable matter. It is found in peat and lignite beds, and was originally obtained from lignitic ocher found near Cassel, Germany. Also known as *Cassel brown; Cassel earth; Rubens brown; ulmin brown.*

Vandyke red. (1) A reddish-brown pigment made from copper ferrocyanide $[Cu_2Fe(CN)_6 \cdot xH_2O]$.

(2) A general term sometimes used for red varieties of ferric oxide (Fe_2O_3) pigments.

Vanex. Trademark of Vanguard Piping Systems, Inc. (USA) for crosslinked polyethylene (PEX) used for durable, flexible, chemical and freeze-break resistant tubing employed in hot and cold water distribution and hydronic heating systems.

Vanguard. (1) Trade name of Osborn Steels Limited (UK) for an oil-hardening cold-work tool steel containing 1.3% carbon, 1% tungsten, 1.5% chromium, 0.75% manganese, and the balance iron. Used for heading and forming dies, punches, and gages.

(2) Trade name of Vanguard Inc. (USA) for a thermoplastic sheeting material.

Vania. Trade name of Saint-Gobain (France) for a glass with a cathedral-type pattern on both sides.

Vanick. Trade name of Malleable Iron Fitting Company (USA) for a nickel-vanadium cast iron containing 2.5% carbon, 2.5% silicon, 0.5% nickel, 0.5% manganese, a trace of vanadium, and the balance iron. It has good strength and wear resistance, and is used for general castings.

Vanidur. Trade name for a corrosion- and heat-resistant, stabilized austenitic cast iron containing 0.1% carbon, 18% chromium, 10% nickel, 1% vanadium, 0.6% titanium, and the balance iron. Used for welded structures and tanks, and chemical and dairy equipment.

Vanimoloy. Trade name of Istituto Sperimentali Metalli Leggeri (Italy) for a cast iron containing 3% carbon and varying amounts of nickel, silicon and manganese. Used for valve lifters.

Vanite. Trade name of Columbia Tool Steel Company (USA) for a tungsten-type high-speed steel (AISI type T2) containing 0.82% carbon, 4% chromium, 2% vanadium, 0.6% vanadium, 18% tungsten, and the balance iron. It has good deep-hardening properties, excellent machinability and wear resistance, high hot hardness, good toughness, and is used for tools, cutters, reamers, and broaches.

Vanleda. Trade name of Firth Brown Limited (UK) for a cobalt-tungsten high-speed steel containing 1.5% carbon, 4.5% chromium, 12.2% tungsten, 5% vanadium, 4.7% cobalt, and the balance iron. Used for a variety of cutting tools.

Van-Lom. Trade name of Teledyne Vasco (USA) for a molybdenum-type high-speed steel (AISI type M10) containing 0.85-1.0% carbon, 0.20-0.35% silicon, 0.1-0.3% manganese, 3.8-4.25% chromium, 8-8.5% molybdenum, 1.8-2.0% vanadium, and the balance iron. It has good deep-hardening properties, high hot hardness, high cutting ability, and excellent wear resistance. Used for a variety of cutting and finishing tools (e.g., lathe and planer tools. milling cutters, drills and broaches), and for dies, punches, shear blades, and bearings.

Vanlon. Trademark of Vanlon Fibers Industrial Corporation (Taiwan) for nylon 6 and polyester fibers and filament yarns.

vanmeersscheite. A yellow to pale yellow, radioactive mineral of the phosphuranylite group composed of uranium uranyl phosphate hydroxide tetrahydrate, $U(UO_2)_3(PO_4)_2(OH)_6 \cdot 4H_2O$. Crystal system, orthorhombic;. Density, 4.67 g/cm^3; refractive index, 1.715. Occurrence: Zaire.

Vanquish. Trade name of Osborn Steels Limited (UK) for a water-hardening tool steel (AISI type W2) containing 0.95% carbon, 0.25% vanadium, and the balance iron. Used for various cutting, finishing and shaping tools.

Vanstar. Trade name Acieries Nouvelle de Pompey (France) for a tool steel containing 1% carbon, 5% chromium, and the balance iron.

Vantage. Trade name of Osborn Steels Limited (UK) for a tungsten-type high-speed steel containing 1.25% carbon, 4.5% chromium, 13.5% tungsten, 4% vanadium, and the balance iron. Used for cutting and finishing tools.

vanthoffite. A colorless mineral composed of sodium magnesium sulfate, $Na_6Mg(SO_4)_4$. It can also be made synthetically. Crystal system, monoclinic. Density, 2.69 g/cm^3; refractive index, 1.488. Occurrence: Austria.

vanuralite. A citron-yellow mineral of the carnotite group composed of an aluminum uranyl vanadium oxide hydroxide hydrate, $Al(UO_2)_2V_2O_8(OH) \cdot 11H_2O$. Crystal system, monoclinic. Density, 3.62 g/cm^3; refractive index, 1.85. Occurrence: Gabon.

vanuranylite. A yellow mineral of the carnotite group composed of oxonium uranyl vanadium oxide hydrate, $(H_3O)_2(UO_2)_2V_2O_8 \cdot 3.6H_2O$. Crystal system, monoclinic. Density, 3.64 g/cm^3; refractive index, 1.92-1.95.

Vanylon. Trademark of Vanylon (Taiwan) for nylon 6 monofilaments and filament yarns.

vapor barrier. A material used to protect the structure and insulation of a building from moisture damage. An effective vapor barrier material must be durable and resistant to the flow of water vapor. Examples of such materials include polyethylene sheeting, aluminum foil, exterior-grade plywood, vinyl wall paper and some types of paints and insulation materials. Also known as *vapor barrier material.*

vapor barrier material. See vapor barrier.

vapor-cured rubber. Rubber vulcanized by first subjecting to sulfur monochloride (S_2Cl_2) fumes and then neutralizing with magnesium carbonate $(MgCO_3)$. Used for rubber goods with thin cross sections.

vapor-deposited coatings. See vapor-deposition coatings.

vapor-deposition coatings. Coatings produced by condensing metallic or nonmetallic materials from their vapors onto metallic or nonmetallic substrates in a high-vacuum chamber. The coatings may serve decorative (e.g., costume jewelry or automobile trim), optical or electronic (e.g., miniature electronics), or protective (e.g., autobodies or industrial equipment) purposes. Also known as *vapor-deposited coatings; vapor deposits.*

vapor deposits. See vapor-deposition coatings.

VaporFab. Trademark of Inco Specialty Powder Products (USA) for a series of conductive nickel-coated carbon fibers produced by a proprietary process. The uniform, thin nickel layer provides ease of handling, controllable conductivity, and retained mechanical integrity. Used as reinforcements in metal-matrix composites, and in the electromagnetic and radio-frequency interference (EMI/RFI) shielding of injection molding composites.

vapor glaze. A glaze consisting of varying amounts of lead, sodium and boric oxide that during firing will vaporize from the melt, and upon cooling will condense on a ceramic body.

Varathane. Trademark of Flecto Coatings Limited (Canada) for rust paints, plastic paints for interior decoration applications, and several oil varnishes.

Varcum. Trade name of Reichhold Chemicals, Inc. (USA) for several heat-reactive, single-stage, liquid phenolic resins. They are high-resin solids at low viscosity, and have good drying and curing properties. Used in abrasives, as backing materials, and for the impregnation of certain cotton fibers.

Varglas. Trade name of Suflex Limited (UK) for glass-fiber sleevings impregnated with varnish and used for insulation applications.

Variantex. Trademark of Wilhelmi Werke AG (Germany) for acoustic insulation boards with textured surface patterns, uncoated, nonflammable chipboard for interior building applications, and support boards for veneers and coatings.

variegated copper ore. See bornite.

variegated yarn. A dyed or printed yarn with many different colors.

Vari-form. Trademark of The Carborundum Company (USA) for refractory castable mixes.

VariGlass VLC. Trade name of Dentsply/Caulk (USA) for a light-cure glass-ionomer dental cement.

Variline. Trade name of Burndy Corporation (USA) for nylon fibers and yarns used for textile fabrics.

Varilux. Trademark of Sovitec (France) for a tough, retroreflective mixture composed of 80% glass microbeads and 20% glass aggregate, used to enhance the visibility of road markings.

Varioflex. Trade name of Alfatec (Austria) for a thin, flexible, wear-resistant fiber-reinforced oxide ceramic used to make thermal insulation boards and vacuum-formed components for high-temperature furnaces and aluminum foundries. It has a maximum service temperature of 1200°C (2190°F).

Variolink II. Trade name of Ivoclar Vivadent AG (Liechtenstein) for a composite resin bonding system used for dental applications.

Varioperm. British trade name for a temperature-sensitive, soft-magnetic alloy composed of 70% iron and 30% nickel, and used for compensating shunts and electrical equipment.

Variosol. Trade name of Schott Glas AG (Germany) for photochromic optical glass.

variscite. A green to bluish green mineral composed of aluminum phosphate dihydrate, $AlPO_4 \cdot 2H_2O$. Crystal system, orthorhombic. Density, 2.50-2.61 g/cm^3; refractive index, 1.56-1.59. Occurrence: USA (Utah, Arkansas), Spain. Used as a gemstone.

Vari-Tran. Trade name of Libbey-Owens-Ford Company (USA) for a chrome alloy coating used on glass to reduce the transmission of visible light and solar heat.

varnish. A liquid composition consisting of resinous substances (e.g., rosins or gums) dissolved in an oil (e.g., linseed or tung oil), alcohol, or turpentine. When applied in a thin layer, it gives a smooth, relatively hard, glossy, transparent or translucent appearance to wood, metal and other substrates. Colorants may or may not be added. Varnish films on wood are resistant to alcohol, heat and water.

varnished cambric. A linen or cotton cloth coated on both sides with an insulating varnish or oil. It has good to fair moisture resistance, good resistance to ordinary oils and greases, and is used for electric cable and transformer insulation.

varnished paper. See insulating paper.

varnish paper. See insulating paper.

varnish stain. An alcohol- and water-resistant finish for wood that consists of a *varnish* tinted with pigments or dyes. It is hard, relatively slow drying, and is used mainly for interior applications. See also alcohol stain; oil stain; sealer stain; water stain; wood stain.

Varnon. Trademark of Harbison-Walker Refractories Company (USA) for dense, strong, hard-fired superduty firebrick. It possesses excellent resistance to reducing gases, such as carbon monoxide, good high-temperature volume stability, and high rigidity under soaking heat conditions. Used as checker bricks for glass tank regenerators, and as a lining material for metallurgical furnaces, rotary kilns, shaft kilns, carbon baking furnaces, and incinerators.

Vaross. Trade name of Associated Spring Company (USA) for a nonmagnetic, corrosion-resistant low-expansion alloy composed of 65% iron and 35% nickel, and used for springs.

Varsico. Trademark of Sico Inc. (Canada) for several protective varnishes.

varulite. A yellowish-green mineral composed of sodium manganese phosphate, $(Na,Ca)Mn(Mn,Fe)_2(PO_4)_3$. Crystal system, monoclinic. Density, 3.58 g/cm^3. Occurrence: Sweden.

Vasco. Trade name of Teledyne Vasco (USA) for an extensive series of steels including various austenitic, ferritic, martensitic and precipitation-hardening stainless grades, vacuum-remelted alloy grades, molybdenum-type high-speed steels (AISI types M2, M7, M15, M42, etc.), cobalt-tungsten-type high-speed tool steels (AISI type T15) and several high- and ultra-high strength steels. *Vasco Chromold (CVM)* is a consumable vacuum-melted case-hardening mold steel (AISI type P5) containing 0.06-0.12% carbon, 0.15-0.25% silicon, 0.1-0.4% manganese, 2.15-2.45% chromium, and the balance iron. It has high cleanliness, high core strength, good ductility and fatigue strength, and is used for glass molds, plastic and rubber dies and high-production die cavities. *Vasco Hypercut* is a molybdenum-type high-speed steel (AISI type M42) containing 1.07% carbon, 0.22% silicon, 0.22% manganese, 1.5% tungsten, 3.75% chromium, 1.15% vanadium, 9.5% molybdenum, 8% cobalt, and the balance iron. It has exceptionally high hardness in the heat-treated condition, high impact strength, and is suitable for cutting nickel and cobalt-base superalloys, heat-resistant titanium alloys and ultrahigh-strength steels. Used for cutting, finishing and shaping tools, and for thread-rolling dies. *Vasco Supreme* is a tungsten-type high-speed tool steel (AISI type T15) containing 1.55-1.60% carbon, 0.15-0.35% silicon, 0.15-0.35% manganese, 4.5-5% chromium, 12-13% tungsten, 4.75-5.25% vanadium, 4.75-5.25% cobalt, and the balance iron. It has a high wear resistance and hot hardness, high cutting efficiency, good machinability, high toughness and high hardness in tempered condition. Used for a variety of cutting, finishing and shaping tools, and for dies, and wear parts. *Vasco Supreme A* is a molybdenum-type high-speed steel (AISI type M15) containing 1.55-1.60% carbon, 0.15-0.35% silicon, 0.15-0.35% manganese, 4.5-5% chromium, 6.25-6.75% tungsten, 4.75-5.25% vanadium, 2.75-3.25% molybdenum, 4.75-5.25% cobalt, and the balance iron.

Vascodyne. Trademark of Teledyne Vasco (USA) for a high-speed steel containing 1% carbon, 0.85% silicon, 0.25% manganese, 1.6% tungsten, 4% molybdenum, 3.75% chromium, 1.95% vanadium, up to 0.03% sulfur, upto 0.03% phosphorus, and the balance iron. Used in the metal-cutting industries.

Vascojet. Trademark of Teledyne Vasco (USA) for a series of fatigue-resistant high-strength and ultrahigh strength steels used for structural and machine parts, such as landing gears, airframes, springs, fasteners, axles, and rotors.

VascoMax. Trademark of Teledyne Vasco (USA) for a series of consumable vacuum-melted maraging steels containing up to 0.03% carbon, up to 0.1% silicon, up to 0.1% manganese, 18.0-18.5% nickel, 3.0-4.8% molybdenum, 0-7.8% cobalt, 0.4-0.6% titanium, 0-0.1% aluminum, and the balance iron. Some grades have additions of up to 1.4% titanium. Supplied in the heat-treated and aged condition, they have high tensile strength, exceptional ductility and toughness after heat treatment, good shock resistance, good elevated-temperature strength, and high hardness. Used for aircraft parts, jet-engine shafts, missile parts, rocket cases, recoil springs, actuators, welding electrodes, dies, die holders and die inserts, molds, and various cutting tools and fixtures.

Vascowear. Trademark of Teledyne Vasco (USA) for a tool and die steel containing 1.12% carbon, 1.2% silicon, 0.3% manganese, 7.75% chromium, 2.4% vanadium, 1.6% molybdenum, 1.1% tungsten, and the balance iron. It has high compressive strength, excellent toughness, and is used for finishing and forming rolls, extrusion and other dies, and shear blades.

Vaseline. Trademark of Chesebrough Pond's Inc. (USA) for a petroleum jelly used as a lubricant and ointment. See also petrolatum.

vashegyrite. (1) A chalky white mineral composed of aluminum phosphate hydroxide hydrate, $Al_{11}(PO_4)_9(OH)_6 \cdot 38H_2O$. Crystal system, orthorhombic. Density, 1.93 g/cm^3; refractive index, 1.488. Occurrence: Czech Republic.

(2) A mineral composed of aluminum phosphate hydroxide hydrate, $Al_4(PO_4)_3(OH)_3 \cdot 11H_2O$. Crystal system, orthorhombic. Refractive index, 1.547. Occurrence: Czech Republic.

(3) A pale greenish white mineral composed of aluminum phosphate hydroxide hydrate, $Al_6(PO_4)_5(OH)_3 \cdot 23H_2O$. Crystal system, orthorhombic. Density, 1.477 g/cm^3; refractive index, 1.477. Occurrence: Czech Republic.

Vaska's catalyst. An organometallic compound based on carbonylchlorobis(triphenylphosphine)iridium and available in the form of lemon-yellow crystals (99.99% pure) with a melting point of 215°C (419°F). Used as a catalyst for organic and organometallic reactions. Formula: $[(C_6H_5)_3P]_2Ir(CO)Cl$. Also known as *Vaska's compound*.

Vaska's compound. See Vaska's catalyst.

Vatar. Trademark of Ausimont USA for a series of modified ethylene chlorotrifluoroethylene (ECTFE) resins. The *Vatar 2000 Series* offers fluoropolymers with a good combination of flexibility and electrical properties for use in flexible jacketing for communication cables, and *Vatar XT* is a fluoropolymer modified to dampen the pair-to-pair crosstalk effect and used for extruded electronic and other components.

vaterite. A colorless mineral composed of calcium carbonate, $CaCO_3$. Crystal system, hexagonal. Density, 2.66 g/cm^3; refractive index, 1.559.

Vatool. Trade name of Disston Inc. (USA) for a water-hardening tool steel containing 0.95% carbon, 0.25% vanadium, up to 0.15% nickel and chromium, respectively, and the balance iron. Used for shear blades, dies, rivet sets, and taps.

Vaucher's alloy. A bearing alloy containing 75% zinc, 18% tin, 4.5% lead and 2.5% antimony.

vauquelinite. A mineral of the fornacite group composed of lead copper chromium oxide phosphate hydroxide, $Pb_2Cu(CrO_4)(PO_6)(OH)$. Crystal system, monoclinic. Density, 6.16 g/cm^3. Occurrence: Russia.

Vautid. (1) Trademark of Vautid-Verschleiss-Technik (Germany) for hard, wear-resistant hardfacing materials, e.g., chilled cast irons, ceramics and plastics.

(2) Trademark of Abresist Corporation (USA) for bimetal castings composed of an impact-resistant steel-base alloy and an abrasion-resistant surface alloy, separated by a strong boundary layer which minimizes mixing of the two alloys. Used for hammer and sinter crushers, hammer-mill beaters, clinker crushers, mixers, and conveyors.

vauxite. A sky to venetian blue mineral composed of iron aluminum phosphate hydroxide hydrate, $FeAl_2(PO_4)_2(OH)_2 \cdot 6H_2O$. Crystal system, triclinic; density, 2.39 g/cm^3; refractive index, 1.555. Occurrence: Bolivia.

vayrynenite. A pale pink mineral composed of beryllium manganese phosphate hydroxide, $Be(Mn,Fe)PO_4(OH)$. Crystal system, monoclinic. Density, 3.21 g/cm^3; refractive index, 1.658. Occurrence: Finland.

V-board. Weather-resistant corrugated *fiberboard* made of high-wet-strength *paperboard*.

V-Bond. Trade name of 3M Dental (USA) for a light-cure dental resin cement.

VCA Alloy. Trade name for a high-strength titanium alloy containing 13% vanadium, 11% chromium and 3% aluminum. It has a service temperature range of -54 to +316°C (-65 to +600°F). Used for missile cases, fasteners, and welded pressure vessels.

VCI paper. See volatile corrosion inhibitor paper.

VDM-LC Nickel. Trade name of ThyssenKrupp VDM GmbH (Germany) for a series of nickels (99.0+% pure) containing varying amounts of iron, manganese, copper, magnesium and/or carbon. Used for chemical and other processing equipment and various other applications requiring corrosion resistance.

VDM Nickel. Trade name of ThyssenKrupp VDM GmbH (Germany) for a series of nickels (93.0+% pure) containing varying amounts of titanium, iron, aluminum, manganese, copper, silicon, sulfur and carbon. Used for chemical and other processing equipment, and in the form of a filler metal for arc and electroslag welding applications.

VDM Titan. Trade name of ThyssenKrupp VDM GmbH (Germany) for a series of titaniums (99.0+% pure) containing varying amounts of iron, palladium, carbon, hydrogen, nitrogen, oxygen. Used for chemical, petrochemical, offshore, power plant and cooling equipment.

VDM Zirconium. Trade name of VDM Nickel Technologie AG (Germany) for a series of zirconiums containing 0-5% hafnium, 0-2% tin, a total of 0.2-0.4 iron and chromium, and 0.16% or more oxygen. Used for corrosion-resistant applications, e.g., chemical plant equipment.

veatchite. A colorless mineral composed of strontium borate monohydrate, $Sr_2B_{11}O_{16}(OH)_5 \cdot H_2O$. Crystal system, monoclinic. Density, 2.78 g/cm^3; refractive index, 1.553. Occurrence: USA (California).

Vectolite. Trademark of Crucible Materials Corporation (USA) for a lightweight, brittle permanent magnet material produced by compressing and sintering a mixture of powders composed of 30% ferric oxide (Fe_2O_3), 44% ferroferric oxide (Fe_3O_4) and 26% cobaltic oxide (Co_2O_3). It has a density of 3.2 g/cm^3 (0.12 $lb/in.^3$), high coercive force, and very high electrical resistance.

VEctomer. Trademark of Allied Signal Inc. (USA) for a wide range of vinyl ethers and vinyl ether based compounds.

Vectra. Trademark of Ticona LLC/Hoechst Celanese Corporation (USA) for a series of liquid crystal polymers that are thermoplastic copolyesters or polyesteramides, supplied in various specialty grades including unfilled, mineral- or graphite-filled, and carbon or glass fiber-reinforced. They have a density range of 1.4-1.6 g/cm^3 (0.05-0.06 $lb/in.^3$), high tensile strength, high modulus of elasticity, good dimensional stability and toughness, excellent processibility, moldability and extrudability, negligible mold shrinkage, good chemical and flame resistance, excellent electrical properties, very low water absorption, good resistance to ultraviolet and alpha radiation, high heat distortion temperature (180°C or 356°F), and good retention of strength and toughness even at cryogenic temperatures. Used chiefly in the electrical and electronics industries for connectors, printed-wiring board components, relay and capacitor housings, sockets, brackets, surface-mount parts, pump housings, pump parts, impellers, windshield washer pump gears, automotive fuel rails, cruise control components, coil formers, and switches.

Vectran. (1) Trade name of Hoechst Celanese Corporation (USA) for polyolefin and polyester fibers.

(2) Trade name of Kuraray Company (Japan) for polyacrylate fibers and filament yarns.

Vectris. Trade name of Ivoclar Williams (USA) for a reinforced dental composite used for crown and bridge restorations.

Vedal. British trade name for a nonhardenable aluminum alloy containing silicon and magnesium, and used for light-alloy parts.

Vedas. Trade name of Osborn Steels Limited (UK) for a water-hardening tool steel (AISI type W1) containing 0.8-1.2% carbon, and the balance iron. Supplied in four carbon-content grades they are used for springs, cutters, and drills.

Vedel. Trademark of Winter & Co. GmbH (Germany) for a vinyl-base artificial leather.

Vedoc. Trade name of Ferro Corporation (USA) for epoxy powder coatings used for automotive, appliance and industrial applications.

Vedril. Trade name of Atoglas, Atofina Chemicals, Inc. (USA) and Montecatini SpA (Italy) for acrylic (polymethyl methacrylate) resins and products.

Vee-Glass. Trade name of Vokes Limited (UK) for glass-filament air filter panels.

veenite. A steel gray, or black mineral composed of lead antimony arsenic sulfide, $Pb_2(Sb,As)_2S_5$. Crystal system, orthorhombic. Density, 5.92 g/cm^3. Occurrence: Canada (Ontario).

Vega. (1) Trade name of Saint Gobain (France) for a molded glass slab whose top surface has a broken stick design, and whose bottom surface has a design featuring concentric prismatic rings and tetrahedrons radiating around a central lens.

(2) Trademark of Carpenter Technology Corporation (USA) for an air-hardening medium-alloy cold-work tool steel (AISI

type A6) containing 0.7% carbon, 0.3% silicon, 2% manganese, 1.25-1.35% molybdenum, 1% chromium, and the balance iron. It has good nondeforming and deep-hardening properties, and good wear resistance. Used for forming, blanking and trimming dies, and forming tools.

(3) Trade name for a silica brick with excellent heat resistance to over 1700°C (3090°F) under load, used for refractories.

Vegaboard. Trademark of Indresco Inc. (USA) for nonhydraulic refractory bonding materials.

Vegard PET. Trademark of Kuraray Company Limited (Japan) for glass-reinforced polyethylene terephthalate (PET) supplied in sheet form. It has a high service temperature (250°C or 480°F), high tensile strength, good sound-dampening properties and good stampability and recyclability. Used for automotive parts, such as engine covers, bumper beams, battery trays and exhaust shields, and for beverage bottles.

vegetable black. Carbon obtained by the destructive distillation or incomplete combustion of wood, nutshells, fruit pits, vines, wine lees or other vegetable carbonaceous materials. Used as a fine paint pigment and extender. Also known as *vegetable charcoal*.

vegetable charcoal. See vegetable black.

vegetable fibers. A class of alkali-resistant natural fibers which are predominantly composed of cellulose. The most important fibers in this class can be grouped according to structure into four types: (i) soft seed-hair fibers, e.g., cotton and kapok; (ii) tough bast fibers, e.g., flax, hemp, jute, ramie and sunn; (iii) tough vascular fibers, e.g., henequen, sisal and yucca; and (iv) grass and straw stems, e.g., broom and esparto. Other structural types include fruit-case fibers, e.g., coir and other palm fibers, and leaf-skin strips, e.g., raffia. Used in the manufacture of paper, textile fabrics, and cordage. See also hard vegetable fibers; soft vegetable fibers.

vegetable fiber yarn. A yarn produced from spun vegetable fibers, e,g., cotton or flax.

vegetable glue. A glue based on a vegetable source, such as starch or dextrin, found in potatoes, corn, rice, or other crops, and usually mixed with gums or resins. Examples include tapioca paste, soybean starch cellulosics and rubber latex. Used in particular in the manufacture of postage stamps, envelopes, labels, and inexpensive plywood.

vegetable parchment. Unsized and unfilled paper of high wet strength that resembles natural parchment, but is based on cotton rags or *alpha cellulose* on which a layer of semi-transparent gelatinous amyloid or cellulose hydrate has been produced by short-term immersion into sulfuric acid immediately followed by washing. Used for documents and food packaging applications. See also amyloid; cellulose; parchment.

vegetable pitch. A resinous material obtained by distillation of certain vegetable materials, e.g., the resin obtained from a coniferous tree, and used in the manufacture of varnishes, paper, and electrical insulation.

vegetable silk. The soft, silky fibers obtained from the seed capsules of plants, such as the kapok tree, or the milkweed.

vegetable wax. A wax high in fatty acids and obtained from vegetable sources, i.e., trees, shrubs, or other plants. Examples include candelilla and carnauba waxes. Used in varnishes, waterproofing compositions, leather dressing, furniture and leather polishes, sealing wax, paper sizing, candles, and soaps. See also candelilla wax; carnauba wax; waxes.

Veiligglas. Trade name of NV Veiligglas (Netherlands) for a laminated glass.

Vekton. Trademark of Chemplast Inc. (USA) for extruded and cast nylons (nylon 6 and nylon 6,6) with excellent abrasion and corrosion resistance and good mechanical properties supplied in unfilled, oil-filled, graphite-filled, molybdenum sulfide-filled and high-heat grades in the form of plates, sheets, strips, rods and tubes. Used for machine components, such as disks, gear blocks, threaded rods, and bushing stock.

Velana. Trademark of Silon AS (Czech Republic) for polyester staple fibers.

Velbex. Trade name of BIP Plastics (UK) for unplasticized polyvinyl chlorides and other vinyl plastics.

Velcro. Trademark of Velcro Industries BV (Netherlands) for a fastening system for textiles including clothing. It consists of two strips or patches of nylon, one covered with minute loop-shaped filaments and the other with hooks.

Velglass. Trade name of Vidrierias de Llodio SA (Spain) for a non-glare glass finely etched on both sides, and used for picture framing.

Velicren. Trademark of Montefibre SpA (Italy) for a polyacrylonitrile staple fiber. *Velicren FR* is a modacrylic fiber with excellent moth and mildew resistance, relatively high tenacity and abrasion resistance, good resistance to acids and alkalies, and fair to poor resistance to ketones. Used for clothing, furnishings, upholstery, and industrial fabrics.

Velinvar. Japanese trade name for a low-expansion alloy composed of 55-63% cobalt, 24-38% iron and 7-13% vanadium, and used for instruments and chronometers.

vellum. (1) A fine grade of parchment made from calfskin, and used chiefly as a writing surface, for tracing, and in bookbinding.

(2) A strong, thick, high-grade rag-pulp writing paper made to resemble original vellum (1), and used for tracing and reproducing original drawings in pencil or ink.

(3) A fine, transparent fabric made from cotton, and used as a tracing cloth.

vellum glaze. See satin glaze.

Vel-Mix. Trade name of Kerr Dental (USA) for a dental die stone supplied as a powder for mixing with water.

Velodal. Trade name of Wieland-Werke AG (Germany) for an age-hardenable aluminum alloy containing 4.5% zinc and 1% magnesium. It has good weldability, and is used for vehicles, machinery and chemical process equipment.

Velodur. Trade name of Wieland-Werke AG (Germany) for age-hardenable high-strength aluminum alloy containing 6.7% zinc, 2.3% manganese, 2% copper, 0.3% chromium and 0.1% manganese. Used in transportation and aircraft construction.

Velofa. Trade name of Vestische Lochblechfabrik GmbH (Germany) for perforated steel sheet and plate used in screening machinery for mines, quarries, and sand and gravel pits.

Veloflex. Trade name of Firestone Tire & Rubber Company (USA) for flexible polyvinyl chlorides.

Velon. Trademark of Firestone Tire & Rubber Company (USA) for plastic sheeting, films and fibers based on saran (vinylidene chloride). Used for fabrics, screens, etc. *Velon LP* polyolefin fibers and products are also available.

Velonigrin. Trade name for browning agents for iron and steel products.

Vel-Opac. Trademark of Sico Inc. (Canada) for an interior alkyd semi-gloss enamel paint.

Velos. Trade name of Spencer Clark Metal Industries Limited (UK) for tungsten, tungsten-molybdenum and cobalt-molybde-num high-speed steels used for cutting tools.

Velosiv. Trade name of Societa Italiana Vetro SpA (Italy) for glass-fiber mats used as support for waterproofing, and as anticorrosive wrapping for internal pipes.

Velostat. Trademark of 3M Company (USA) for electrically conductive plastics used for clothing, containers, hoses, and foam products.

velour. (1) In general, any of a number of knitted or woven fabrics resembling *velvet* and having a short nap or pile that makes it soft to the touch. Used for upholstery, draperies and clothing.

(2) Specifically, a fine, soft, closely knitted or woven wool fabric with a short nap and a finished surface, used for luxurious clothing.

(3) A chrome-tanned leather made from the skins of calves, pigs or goats, and having a velvety surface appearance due to grinding. Also known as *velour leather.*

velour leather. See velour (3).

velour paper. Paper whose surface has been flocked with fine pieces of cotton, nylon, rayon or wool, and which may or may not be embossed or textured.

Velours. Trade name of Atotech USA Inc. (USA) for a satin nickel electrodeposit and deposition process.

veloutine. (1) Flannel having a short nap or pile and resembling velvet. See also flannel.

(2) A fine, soft ribbed fabric woven from silk and either cotton or wool fibers.

Velvabar. Trade name of Copperweld Steel Company (USA) for alloy steel bars having a smooth, bright, scale-free surface finish and more precise straightness than conventional alloy steel bars.

Velva Dri. Trade name of Ashland Chemical Company (USA) for refractory coatings used for foundry applications.

Velvet. Trade name of Union Bronze Company (USA) for a lead babbitt containing varying amounts of tin and antimony. It has good antifriction properties, and is used for bearings.

velvet. A closely woven textile fabric, usually of acrylic, cotton, nylon, rayon or silk, having a thick, short-cut nap or pile. It is soft and smooth to the touch, and obtained by first weaving two layers of fabric together and then shearing them apart at the faces. Used for dresses, evening wear, draperies, upholstery, linings, and trim.

velveteen. A fabric usually of cotton, cotton blend or rayon and having a short, soft-cut nap or pile resembling *velvet*, but woven singly, not face to face. Used for apparel, linings, coverings, upholstery, draperies, shoe uppers, etc.

Velvetex. Trade name for a series of carbon blacks.

Velvetouch. Trade name of S.K. Wellman Corporation (USA) for a sintered material composed of copper, lead, tin and graphite, and used for bearings, linings, and clutch and brake disks.

Velvet Satin. Trademark of Tibbetts Paints Limited (Canada) for an interior latex paint with satin finish.

Velvex. Trade name of ASG Industries Inc. (USA) for a slightly diffusing glass with shallow, wavy pattern.

Venango. Trade name of Cytemp Specialty Steel Division (USA) for a tough, shock-resistant tool steel (AISI type S2) containing 0.5% carbon, 0.45% manganese, 1.1% silicon, 0.2% vanadium, 0.5% molybdenum, and the balance iron. Used for pneumatic and hand tools, pipe cutters, chisels, and stamps. *Venango Special* is a shock-resisting tool steel (AISI type S2) containing 0.65% carbon, 0.5% manganese, 1.1% silicon, 0.2% vanadium, 0.5% molybdenum, and the balance iron. It has outstanding toughness, good deep-hardening properties, good strength, hardness, shock resistance, and moderate wear resistance and machinability. Used for pneumatic and hand tools, chisels, and stamps.

veneer. (1) A thin sheet, layer or strip of wood with a thickness of about 0.3-6.3 mm (0.01-0.25 in.). "Sliced" veneer is made by moving a log or flitch against a knife, "sawed" veneer is produced by simple sawing, and "rotary cut" veneer is cut with a special lathe by rotating a log against the edge of a knife. *Veneer* is used for decorative applications, and in the manufacture of particleboard, plywood, and other wood panel materials.

(2) A relatively thin decorative coating produced on ceramic surfaces.

veneer brick. A hard, weather-resistant brick with low water absorption, often made of sandstone or limestone and suitable for use in finishing exterior walls. It is placed on top of foundation walls and over plywood sheathing.

Veneer Cement. Trade name for a light-cure dimethacrylate (DMA) cement used as a veneering material for dental restorations.

veneered particleboard. Particleboard corestock overlaid with wood veneer or similar materials, and used for furniture, cabinetwork, panels, dividers, and wainscots. See also particleboard; particleboard corestock.

veneer fiberboard. Fiberboard made from wood veneer by first cutting into short fibers and then bonding with phenolic resin. See also fiberboard.

Veneer Opal. Trade name of Plyglass Limited (UK) for a flat glass with a veneer of plastic-bonded, pigmented glass gauze.

veneer plaster. A rapid-drying calcined *gypsum plaster* applied in one or two thin coats over specially prepared gypsum wallboard to provide a strong, hard, abrasion-resistant interior wall finish.

veneer plaster base. Gypsum wallboard serving as a base for veneer plaster. See also gypsum wallboard.

Venetian. Trade name of Hartley, Wood Company (UK) for a type of antique glass.

Venetian. See Venetian cloth.

Venetian cloth. A strong, closely woven fabric in a twill weave, usually of cotton or wool, and having a warp face and a smooth, lustrous surface texture. Used for coatings and suitings. Originally from Venice, Italy. Also known as *Venetian.*

Venetian glass. Fine, very delicate and highly decorated glassware originally made in Venice, Italy.

Venetian lace. A heavy *needlepoint lace* with a raised floral design resembling rose point, but with larger designs. Also known as *Venetian rose point.* See also grospoint.

Venetian red. A pure, brick-red pigment made synthetically by calcining *green copperas* ($FeSO_4 \cdot 7H_2O$) and lime (CaO). It also occurs in nature as a variety of *hematite red*, and is composed of 15-40% high-grade ferric oxide (Fe_2O_3) and 60-80% calcium sulfate ($CaSO_4$). Used in ceramics and paints.

Venetian rose point. See Venetian lace.

Venetian white. A white inorganic pigment composed of a mixture of equal parts of *white lead* [$2PbCO_3 \cdot Pb(OH)_2$] and *barite* ($BaSO_4$).

Veneziana. Trade name of Vetreria di Vernante SpA (Italy) for a glass with linear pattern.

Venivici. Trade name of Krupp Stahl AG (Germany) for a high-speed steel containing 0.82% carbon, 4% chromium, 1.6% vanadium, 0.85% molybdenum, 8.7% tungsten, and the balance iron.

Used for lathe and planer tools, and various other cutting and finishing tools.

Ventaglio. Trade name of Fabbrica Pisana SpA (Italy) for a patterned glass with fan design.

Ventflex. Trade name of Diffusion Textiles (UK) for waterproof fabrics with excellent breathability and stretchability. Used for clothing.

venturin. A yellow pigment of varying composition supplied in powder form and used in japanning to produce colors imitating gold.

Venus. (1) Trade name of Saha-Union (Thailand) for rayon fibers and yarns used for textile fabrics.

(2) Trade name of Heraeus Kulzer, Inc. (USA) for a microhybrid dental resin composite based on bisphenol A glycidyl dimethacrylate (BisGMA) and containing up to 78 wt% fillers consisting of barium aluminum boron fluoride silica glass and dispersed silicon dioxide. Supplied in several shades and opacities, it is used for core build-ups, veneering, indirect restorations, and cavity prepatations.

Vera. Trade name of Eisenwerk Würth GmbH & Co. (Germany) for cast steel blasting abrasives.

Vera Bond. Trade name of Aalba Dent Inc. (USA) for a nickel-chromium dental bonding alloy.

Veracetex. Trade name of Glaceries Réunies SA (Belgium) for a laminated sheet glass.

Veralixe. Trade name of Compagnie Française des Isolants SA (France) for glass fabrics.

Veraloy. Trade name of Goldsmith & Revere (USA) for a dispersed phase high-copper dental amalgam alloy supplied in the form of capsules.

Vera PDS/PDI. Trade name of Aaalba Dent Inc. (USA) for a corrosion-resistant cobalt-chromium dental casting alloy.

Veraphon. Trade name of Vetreria di Vernante SpA (Italy) for insulating glass units.

Vercolor. Trade name of Vetreria di Vernante SpA (Italy) for tinted, patterned and wired glass.

Vercor. Trade name of Wattez NV (Netherlands) for enameled, decorative glass composed of a clear sheet of glass with patterns enameled in silk gloss.

verde salt. See thenardite.

verdigris. See basic copper acetate.

verdolago. The medium-textured wood from the tall hardwood tree *Terminalia amazonica* of the Combretaceae family. It has light-yellowish heartwood with dark stripes, good durability, except in contact with soil, low shrinkage and permeability, and good machinability and finishability. Average weight, 660 kg/m³ (41 lb/ft³). Source: Humid tropical and transitional subtropical forests, especially from the Amazon region of South America. Used for carpentry work, flooring, furniture, structural timber, stairways and veneer, and for plywood, tool handles, and turnery. Also known as *nargusta; tanimbuca.*

Verel. Trademark of Eastman Kodak Company (USA) for a modified acrylic (modacrylic) fiber made from acrylonitrile and vinylidene chloride. It has excellent chemical resistance, good dimensional stability, good resistance to weathering in sunlight, mildew and microorganisms, and is used for knitted and woven fabrics, blankets, paint rollers, polishing cloth, draperies, work clothing, fire curtain shields, pile fabrics, and blended with cotton or wool for industrial applications.

Verelite. Trade name of Dow Chemical Company (USA) for a series of polystyrenes.

Verester. Trade name of Holding Textiles ex-Soieries F. Ducharne

SA (France) for a glass cloth used for fiber-reinforced plastics.

Veribest Drill Rod. Trade name of Diehl Steel Company (USA) for oil-hardening steel drill rod containing 0.9% carbon, 1.1% manga-nese, 0.6% chromium, 0.9% tungsten, 0.24% vanadium, and the balance iron. Used for dies, tools, and punches.

veridian. A paint pigment composed of hydrated chromium oxide.

Veri-dull. Trade name for highly delustered rayon fibers and yarns used for textile fabrics.

Verigrane. Trade name of Boussois Souchon Neuvesel SA (France) for a fire-polished decorative glass.

Verilite. Trade name of Verilite Metals Company (USA) for age-hardenable aluminum casting alloys containing 1-2.5% copper, 0-1.5% chromium, 0.7% iron, 0-0.5% manganese, 0.4% silicon and 0.3-1.5% nickel. Used for aircraft cylinder heads.

Veriloy. Trade name of Driver Harris Company (USA) for a heat- and corrosion-resistant cast steel containing 0.35-0.45% carbon, 20-22% nickel, 10-12% chromium, and the balance iron. It has good resistance to 1500°C (2732°F), and is used for furnace parts and chemical engineering equipment.

Verinor. Trade name Degussa Dental (USA) for a medium-gold dental bonding alloys.

Verisol. Trade name of Mathieux & Compagnie Ets. Verisol SARL (France) for industrial glass fabrics and reinforcements used for protective coatings and coverings.

Verisolith. Trade name of Saint-Gobain (France) for toughened glass blocks.

Veritas. Trademark of Degussa-Ney Dental (Germany) for a biocompatible, white dental casting alloy containing 40% gold, 44.7% palladium and 5% silver. It has a light gray oxide layer, good castability and handling properties, high ductility, excellent marginal integrity, a hardness of 232 Vickers, and is used in restorative dentistry for metal-to-ceramic restorations.

vermeil. Silver coated or plated with gold.

vermicular iron. See compacted graphite cast iron.

vermiculite. A group of micaceous minerals, either green in color or yellowish brown with bronze-yellow luster. They are available in the form of small, lightweight granules or flakes in a wide range of particle sizes. The general formula is $(Mg,Fe)_3$-$(Si,Al,Fe)_4O_{10} \cdot 4H_2O$. Crystal system, monoclinic. Density, 2.1-2.3 g/cm^3; hardness, 1.5-3.0 Mohs; refractive index, 1.54. It has a platelet-like crystal structure, exfoliates from 6 to 20 times its original size when heated to above 950°C (1740°F), and has high porosity, and a liquid absorption capacity of 200-500%. Occurrence: India, South Africa, USA (Colorado, Montana, North Carolina, South Carolina, Texas, Wyoming), Transval. Used as aggregate in lightweight concrete and acoustic and fireproof plaster, in acoustic and asbestos tile, in refractory insulators and thermal insulating concrete floors, as a filler in plastics, rubber, paints and caulking compounds, in fireproofing compounds, for wallpaper printing, and for packing applications.

vermiculite concrete. A lightweight concrete made with *exfoliated vermiculite* as the aggregate.

vermiculite plaster. A plaster usually made by mixing gypsum with *vermiculite* and water, and applied to interior walls as a base coat.

vermilion. A bright, somewhat orange-red, permanent pigment usually composed of mercuric sulfide (HgS). It was formerly obtained from the mineral *cinnabar*, but is now made synthetically by heating mercury and sulfur. Used in paints, rubber and plastics.

vermilion red. See English vermilion; vermilion.

Vermonite. Trade name of Vernon-Benshoff Company (USA) for acrylic resins.

Vermont marble. Marble, usually white mottled with gray, but also black, green, red and mottled in color, mined in the State of Vermont, USA.

vernadite. A black mineral of the rutile group composed of manganese hydroxide, $Mn(OH)_4$. Crystal system, tetragonal. Density, 3.00 g/cm^3. Occurrence: Russia.

Vernial. Trade name for Vereinigte Deutsche Nickel-Werke AG (Germany) for wrought aluminum alloys containing 0.6-1.4% magnesium, 0.6-1.6% silicon, 0.6-1% manganese and up to 0.3% chromium. They possess good forming and welding properties, medium strength, good corrosion resistance, and are used for structural parts, gutters, boats, fan blades and window frames.

Vernicon. Trade name of Vereinigte Deutsche Nickel-Werke AG (Germany) for a resistance alloy similar in properties to *constantan* and containing 56% copper and 44% nickel. Used for electrical resistances, and thermocouples.

Vernicorr. Trade name of Vereinigte Deutsche Nickel-Werke AG (Germany) for a non-heat treatable, corrosion-resistant wrought aluminum alloy containing 0.9-1.4% manganese and up to 0.3% magnesium. Used for roofing and trim.

Vernidur. Trade name of Vereinigte Deutsche Nickel-Werke AG (Germany) for age-hardenable aluminum alloys containing 2.5-5% copper, 0.2-1.8% magnesium and 0.3-1.5% manganese. They have high strength, and are used for fasteners, and for structural parts for the automotive and aircraft industries.

Vernikorr. Trade name of Vereinigte Deutsche Nickel-Werke AG (Germany) for an aluminum alloy containing 0.5-1.5% manganese and up to 0.3% chromium. It has good forming and welding properties, and is used for cooking utensils, heat exchangers, tanks, and furniture.

Vernicron. Trade name of Monopol AG (Switzerland) for one-coat stoving enamels supplied in several grades.

Vernisil. Trade name of Vereinigte Deutsche Nickel-Werke AG (Germany) for a series of highly corrosion-resistant nickel silvers used for fasteners, table flatware, cutlery, marine hardware, bathroom fixtures, optical parts, and electrical and electronic parts including wires and bars.

Vernonite. Trade name of Rohm & Haas Company (USA) for an acrylic resin used for dentures.

Verona yellow. See Turner's yellow.

Verondulit. Trade name of Boussois Souchen Neuvesel SA (France) for a corrugated glass.

Verone. Trade name for a silicone elastomer used for dental impressions.

Veronica. Trade name of Kind & Co. Edelstahlwerk (Germany) for an extensive series of high-speed steels containing 0.8-1.45% carbon, 4-11.5% chromium, 1.5-18% tungsten, 0.7-9.5% molybdenum, 1.2-3.5% vanadium, 3-16% cobalt, and the balance iron. Used for lathe, planer and various other cutting and finishing tools, and for blanking and forming dies.

verplanckite. A brownish orange to brownish yellow mineral composed of barium manganese silicate hydroxide trihydrate, $Ba_2(Mn,Fe,Ti)Si_2O_6(O,OH,Cl,F)_2 \cdot 3H_2O$. Crystal system, hexagonal. Density, 3.52 g/cm^3; refractive index, 1.683. Occurrence: USA (California).

Verracier. Trade name of Splintex Belge SA (Belgium) for a toughened glass.

Verranne. Trade name of Société du Verre Textile SA (France) for glass yarn spun from discontinuous fibers.

Verre Cordelé. Trade name of Saint-Gobain (France) for a streaky antique glass.

VersaClad. Trademark of Protective Metal Alloys Inc. (USA) for hardfacing materials composed of nickel and cobalt alloy powders blended with tungsten carbide and fused onto various base metals to provide extremely wear-resistant, low-distortion, crack-free surfaces on parts, such as fan blades, paddles, valve and seal faces, and skid shoes for helicopters.

VersaCore. Trade name of VersaCore International (USA) for a type of particleboard composed of oak corestock overlaid with wood veneer, and used as a building board, and for cabinets, panels, and furniture.

Versadur. Trade name of HPG International Inc. (USA) for extruded thermoplastic sheeting.

Versaflex. Trade name of GLS Corporation (USA) for several thermoplastic elastomers.

Versafloat. Trademark of Versar Inc. (USA) for syntactic plastic foam structures used for energy absorption, core filling, and buoyancy applications.

Versal. Trade name of Versevorder Metallwerk (Germany) for a non-heat treatable alloy composed of aluminum and magnesium. It has excellent resistance to atmospheric and seawater corrosion, good strength, and is used in the aircraft and automotive industries, in the food processing industry, for housings, and panels.

Versalite. (1) Trademark of Versar Inc. (USA) for lightweight syntactic foam structures used for energy absorption, core filling and buoyancy applications.

(2) Trademark of Centrix, Inc. (USA) for a light-cure hybrid composite used for dental restorations.

Versalloy. Trademark of Atlas Specialty Steels (Canada) for an austenitic stainless steel containing 0.11% carbon, 17% chromium, 8% nickel, and the balance iron. Usually supplied in the annealed condition, it is used especially for household appliances, cutlery, chemical equipment, and food-processing equipment.

VERSAlloy. Trademark of Atlas Specialty Steels (Canada) for a series of hard, wear-resistant, low-ductility, sintered cobalt- and nickel-base braze and hardfacing alloys. Supplied in rod and powder form, they are produced by a proprietary powder-metallurgy process. *VERSAlloy Plus* is a hardfacing alloy supplied in the form of rods for gas-tungsten-arc and oxyacetylene welding, and as powder and wear tips. It offers 25-50% improvement in wear resistance over standard *VERSAlloy* products.

Versalon. Trademark of Schering AG (Germany) for a series of thermoplastic polyamide plastics. They possess high tensile and impact strengths, high elastic moduli, high rigidity, good thermal, electrical and chemical resistance, a maximum service temperature of 93-149°C (200-300°F), relatively high water absorption, and poor resistance to glycols and alcohols. Used for gaskets, tubing, insulation, bearings, cams, gears, bristles, and fabrics. A special grade with good adhesion and flexibility used for hot-melt adhesives, heat-seal coatings and electrical encapsulation is also available.

Versamid. Trademark of Schering AG (Germany) for a series of thermoplastic and reactive polyamide resins. The reactive resins copolymerize with epoxies, and are used for hot-melt and structural adhesives, heat-sealing coatings, paints, varnishes and inks.

Versasteel. Trade name of Crucible Materials Corporation (USA) for an air-hardening tool and die steel containing 1% carbon,

0.3% man-ganese, 2% silicon, 4.25% chromium, 2.5% molybdenum, 1.15% vanadium, 0.3% tungsten, and the balance iron. Used for tools and dies.

Versa-Temp. Trademark of Sultan Chemists, Inc. (USA) for a self-curing bismethacrylate-based dental resin supplied in several shades, and used for temporary crown and bridge restorations.

Versathane. Trademark of Air Products and Chemicals Inc. (USA) for toluene-2,4-diisocyanate (TDI) prepolymers.

Versatile. Trade name of Osborn Steels Limited (UK) for an oil-hardening, shock-resisting tool steel containing 0.4% carbon, 1.5% nickel, 1% chromium, and the balance iron. Used for swaging tools, gears, bolts, crankshafts, and axles.

Versatool. Trade name of Crucible Materials Corporation (USA) for a tool steel containing 1% carbon, 2% silicon, 0.3% manganese, 4.25% chromium, 2.5% molybdenum, 1.15% vanadium, 0.3% tungsten. It has good wear resistance, high hardness, good elevated temperature properties, and is used for dies and shears.

Verseyblack. Trade name of Zinex Corporation (USA) for a deep dark black finish applied over brass.

Versiflex. Trade name of Carbide & Carbon Chemicals Corporation (USA) for flexible polyvinyl chlorides.

Versi-Foam. Trade name of RHH Foam Systems, Inc. (USA) for several polyurethane foams.

Versil. Trade name of Versil Limited (UK) for a range of glass-fiber products.

versiliaite. A black mineral composed of iron zinc antimony arsenic oxide sulfide, $(Fe,Zn,Fe)_8(Sb,Fe,As)_{16}O_{32}S$. Crystal system, orthorhombic. Density, 5.12 g/cm^3. Occurrence: Italy.

Versilient. Trade name of Versil Limited (UK) for acoustical glass-fiber insulation.

Versilok. Trademark of Lord Corporation (USA) for a family of acrylic, modified-acrylic and acrylic/epoxy hybrid structural adhesives used for bonding aluminum, steel, ceramics, plastics, and other materials in the aircraft, computer, business-machine and general electronics industries.

Versite. (1) Trademark of Versar, Inc. (USA) for silicon carbide and silicon nitride (Si_3N_4) whiskers available in various grades for use in reinforced composites.

(2) Trademark of Versite, Inc. (USA) for corrosion- and erosion-resistant coatings used on metallic substrates.

(3) Trademark of Versar, Inc. (USA) for metal-matrix composites supplied in wire, bar, rod, sheet and strip form.

Vertane. Trade name of Sovirel (France) for a green-tinted spectacle glass.

Vertex. (1) Trade name of IVG Composites Inc. (USA) for a family of high-strength glass fibers used for filament winding, prepreg molding, resin transfer molding, pultrusion, and other composite processing techniques.

(2) Trade name of Vertex-Dental BV (Netherlands) for an extensive series of dental resins including several cold- and heat-curing resins for denture bases, denture lining and denture repairs, and cold-curing resins for orthodontics and dental trays.

(3) Trade name of Vertex NP (Czech Republic) for an extensive range of glass-fiber products.

vertical-grained lumber. See edge-grained lumber.

Vertidraw. Trade name of PPG Industries Inc. (USA) for a glass with a thickness down to 0.05 in. (1.27 mm) produced by the *Vertidraw* (vertical-drawing) process.

Vertiglas. Trade name of PPG Industries Inc. (USA) for thin glass products made by the *Vertidraw* (vertical-drawing) process.

Verton. Trademark of LNP Engineering Plastics Inc. (USA) for a series of lightweight thermoplastic composites based on polyamide 6,6 (nylon 6,6) or polyphthalamide reinforced with long-strand glass fibers. They have excellent high-temperature and humidity resistance, high impact strength, stiffness and dimensional stability, good low and elevated temperature properties, good chemical resistance, good resistance to automotive and aviation fuels and lubricants, and are used for components for instruments, devices and equipment, automotive components, ring and pinion gears for small motors, rotors and vanes, and for sporting goods. *Verton MFX* refers to UV-resistant long glass fiber-reinforced polypropylene composites used for automotive applications, and tool housings. They have excellent mechanical properties including high impact and tensile strengths, and good surface appearance.

Vertoval. Trade name of Vergo SA (France) for a green-tinted spectacle glass.

vertumnite. A colorless mineral composed of calcium aluminum silicate hydroxide trihydrate, $Ca_4Al_4Si_4O_6(OH)_{24}\cdot 3H_2O$. Crystal system, monoclinic. Density, 2.15 g/cm^3; refractive index, 1.535. Occurrence: Italy.

Very Best. (1) Trade name of Bisset Steel Company (USA) for a water-hardening tool steel containing 0.8-1.2% carbon, 0.2% vanadium, and the balance iron. Used for cutting tools.

(2) Trade name of Boyd-Wagner Company (USA) for a tough, hard, water-hardening tool steel containing 1.05% carbon, 0.3% manganese, 0.5% chromium, 0.1% vanadium, and the balance iron. Used for mandrels, drills, and dies.

very-high-carbon steel. Strong, hard carbon steel containing between 0.90-1.50% carbon. See also high-carbon steel.

very-high impact polystyrenes. Special grades of high-impact polystyrene with extremely high toughness and impact strength and good dimensional stability, used for packaging and disposables, appliances, consumer electronics, toys, and furniture. Abbreviation: VHIPS. See also high-impact polystyrenes.

very-low-density polyethylenes. Branched-chain polyethylenes, quite similar to standard *low-density polyethylenes*, but with a density of less than 0.910 g/cm^3 (0.033 $lb/in.^3$). Abbreviation: VLDPE.

vesignieite. A green mineral composed of barium copper vanadium oxide hydroxide, $BaCu_3(VO_4)_2(OH)_2$. Crystal system, monoclinic. Density, 4.43 g/cm^3; refractive index, 2.129. Occurrence: China.

Vespel. Trademark of E.I. DuPont de Nemours & Company (USA) for an extensive series of high-temperature polyimide resins available in various grades including unfilled, glass-fiber-reinforced, graphite- and/or fluorocarbon (*Teflon*) filled, and molybdenum disulfide-filled. They possess excellent resilience, compressive strength and wear resistance, good dimensional stability and resistance to friction, good impact resistance, high tensile strengths and moduli of elasticity, good retention of properties at temperatures from cryogenic to 288°C (550°F), excellent heat, abrasion, creep and radiation resistance and good insulating properties. Used for brake shoes, glide blocks, rubbing blocks, bearing races and cages, insulators, insulating bushings, motor, wire, circuit and instrument insulation, bushings, couplings, circuit boards, electrical coil bobbins, valves and valve seats, thrust washers, and as binders for diamond abrasives.

Vestamelt. Trademark of Degussa-Huels AG (Germany) for hotmelt adhesives based on polyamide and polyether sulfone copolymers.

Vestamid. Trademark of Degussa-Huels AG (Germany) for an extensive series of engineering thermoplastics based on polyamides (nylon 6,12 and nylon 12) and supplied in several grades. They have high impact strength, high abrasion and corrosion resistance, low moisture sensitivity, an upper continuous-use temperature of 250°C (480°F), and are used for molded parts. *Vestamid D* refers to nylon 6,12 thermoplastics supplied in unreinforced and 30% carbon- or glass fiber-reinforced grades, *Vestamid E* to flexible and semiflexible nylon 12 thermoplastics, and *Vestamid L* to nylon 6,12 thermoplastics supplied in unreinforced, 30% glass fiber-reinforced, 50% glass bead-filled, fire-retardant and UV-stabilized grades.

Vestan. (1) Trade name of Bayer AG (Germany) for a synthetic fiber based on polyethylene terephthalate.

(2) Trade name of Bayer AG (Germany) for a synthetic fiber based on polyvinylidene chloride (saran), and used for industrial and clothing fabrics.

Vesteel. Trade name of British Steel plc (UK) for a mild steel used for porcelain-enameling applications.

Vestoblend. Trademark of Hüls AG (Germany) for unprocessed plastics and plastic mixtures.

Vestodur. Trademark of Hüls AG (Germany) for polybutylene terephthalate (PBT) resins supplied in various grades including neat, glass fiber-filled, glass bead- and/or mineral-filled, fire-retardant and UV-stabilized.

Vestogral. Trademark of Hüls AG (Germany) for synthetic rubber supplied in powder, crumb or bale form with or without reinforcing fillers.

Vestogrip. Trademark of Hüls AG (Germany) for synthetic rubber supplied in powder, crumb or bale form with or without reinforcing fillers.

Vestogum. Trademark of Espe Fabrik Pharmazeutischer Präparate GmbH & Co. KG (Germany) for elastomeric dental modeling materials.

Vestolan. Trademark of Hüls AG (Germany) for a series of polyethylene fibers and monofilaments.

Vestolen. Trademark of Hüls AG (Germany) for a series of polyolefin resins available in various grades, and used for containers, water pipes, tubing, battery cases, blow-molded parts, and packaging foil. *Vestolen A* refers to polyethylenes supplied in standard, UV-stabilized and fiber-reinforced grades. *Vestolen BT* refers to polyisobutylene resins, and *Vestolen P* designates an extensive series of polypropylene homopolymers and copolymers available in various grades including standard, glass-reinforced, talc-filled, UV-stabilized, fire-retardant and structural foam, film and sheeting. They have a density of 0.91 g/cm^3 (0.033 $lb/in.^3$), high transparency, good toughness, a maxi-mum service temperature of 140°C (284°F), good to excellent weldability, good resistance to elevated temperatures and many chemicals, excellent resistance to dilute acids and alkalies, moderate resistance to gasoline, mineral oils and trichloroethylene. Their uses and special properties may vary considerably from grade to grade.

Vestolit. Trademark of Hüls AG (Germany) for plasticized and unplasticized polyvinyl chlorides. The plasticized grades are supplied in three elongation ranges: 0-100%, 100-300% and above 300%, while the unplasticized grades are supplied in standard, high-impact and structural-foam grades.

Vestopal. Trademark of Hüls AG (Germany) for an extensive series of unsaturated thermosetting polyester resins and alloys. *Vestopal W* are room-temperature-polymerizing styrene-polyesters. The polymerized products are fully crosslinked, pale yellow resins suitable for use as embedding media in light and electron microscopy.

Vestoplast. Trademark of Hüls AG (Germany) for amorphous polyolefins.

Vestopreg. Trademark of Hüls AG (Germany) for reinforced plastics supplied as pellets, rods, sheets and tubes.

Vestopren. Trademark of Hüls AG (Germany) for thermoplastic elastomers including *Vestopren TP* thermoplastic olefins.

Vestoran. Trademark of Degussa-Hüls AG (Germany) for styrene-acrylonitrile (SAN) resins with a density of 1.08 g/cm³ (0.039 lb/in.³), good impact resistance, maximum service temperature of 90-100°C (194-212°F), excellent resistance to mineral oils and dilute alkalies, good resistance to gasoline and dilute acids, and poor resistant to tetrachloroethylene and tetrachlorocarbon. Used for molded products, precision parts, battery boxes, etc. Also included under this trademark are polyphenylene ether (PPE) and polyphenylene oxide (PPO) polymers supplied in unreinforced and glass fiber-reinforced grades.

Vestosint. Trademark of Hüls AG (Germany) for polyamide (nylon 12) coating powders.

Vestowax. Trademark of Hüls AG (Germany) for a *polyethylene wax.*

Vesturit. Trademark of Hüls AG (Germany) for thermoplastic polyester resins used as binders for varnishes and lacquers, and for molded parts.

Vestyplex. Trademark of Hüls AG (Germany) for polymeric compositions used in the manufacture of heat-sealable plastic film for sealing and packaging applications.

Vestypor. Trademark of Hüls AG (Germany) for a series of expandable polystyrenes.

Vestyron. Trademark of Hüls AG (Germany) for thermoplastic polystyrenes available in standard and high-impact grades. The standard polystyrenes are available in unstabilized, UV-stabilized and structural foam grades, while the high-impact polystyrenes are supplied in regular, UV-stabilized and fire-retardant grades. *Vestyron* polystyrenes are clear or colored with glossy surfaces, high hardness and brittleness, good impact strength (when modified), and an upper service temperature of 80-95°C (176-203°F), excellent electrical properties, good machining, gluing and welding properties, excellent resistance to dilute alkalies, good resistance to dilute acids, moderate resistance to mineral oils, poor resistance to gasoline, trichloroethylene and tetrachlorocarbon. Used for electrical components, e.g., housings, coils, battery cases and lighting panels, and machine and appliance housings, refrigerator doors, air-conditioner cases, containers and molded household wares, toys, and packaging materials.

vesuvianite. (1) A brown, yellow or green mineral composed of calcium magnesium iron aluminum silicate hydroxide, $Ca_{10}(Mg,Fe)_2Al_4Si_9O_{34}(OH)_4$. Crystal system, tetragonal. Density, 3.40 g/cm³. Occurrence: Italy. See also idocrase; californite.

(2) A mineral composed of calcium magnesium aluminum silicate hydroxide, $Ca_{19}Mg_4Al_{10}Si_{17}O_{68}(OH)_8$, and made synthetically from a mixture of calcium hydroxide, magnesium hydroxide and aluminum silicate gel. Crystal system, tetragonal. Density, 3.27 g/cm³; refractive index, 1.701.

Vesuvius. Trade name of Firth-Vickers Stainless Steels Limited (UK) for a heat- and corrosion-resistant steel containing 0.1% carbon, 30% chromium, 1.7% nickel, and the balance iron. Used for grids, fire bars, and furnace parts.

veszelyite. A green mineral composed of copper zinc phosphate hydroxide dihydrate, $(Cu,Zn)_3(PO_4)(OH)_3·2H_2O$. Crystal system, monoclinic. Density, 3.53 g/cm³; refractive index, 1.658. Occurrence: Rumania.

Vetroflam. Trademark of Saint-Gobain (France) for a non-insulated, fire-resistant glass.

Vetroflex. Trademark of Saint Gobain (France) for glass-fiber wool used for thermal and acoustical insulation applications.

Vetroflocco. Trade name of Vetrocoke SpA (Italy) for glass wool.

Vetroloid. Trade name of British Celanese (UK) for cellulose acetate plastics.

Vetrolon. Trade name of Gevetex-Textilglas GmbH (Germany) for glass yarn used in the manufacture of decorative glass fabrics.

Vetrolux. Trade name of Thermolux AG (Switzerland) for double glazing units.

Vetroplastic. Trade name of Fibres de Verre SA (Switzerland) for plastic-impregnated glass fiber matting attached to bituminized paper and used for building insulation panels.

Vetrotessile. Trade name for high-strength glass fibers and glass-fiber products including reinforcements.

Vetrotex. Trademark of Vetrotex SA (France) for textile glass fibers.

Vexar. Trademark of E.I. DuPont de Nemours & Company (USA) for polypropylenes and low- and high-density polyethylenes used for the manufacture of plastic netting for packaging applications.

V-Gnathos. Trade name of Metalor Technologies SA (Switzerland) for several dental bonding alloys including *V-Gnathos PF*, a hard alloy containing 86.8% gold and 11.7% platinum, and *V-Gnathos Supra*, an extra-hard alloy containing 86.5% gold and 11.5% platinum.

Viaduct. Trade name of Darwins Alloy Castings (UK) for a tough, shock-resistant tool steel containing 0.4% carbon, 1.1% chromium, 1.9% tungsten, 0.3% vanadium, and the balance iron. Used for chisels and shear blades.

Viag. Trade name of VAW Vereinigte Aluminum-Werke AG (Germany) for a heat-treatable aluminum alloy containing 4.5-5% copper. It has good strength, fair to poor corrosion resistance, good machinability, and is used for light-alloy parts.

Vialbra. Trade name for a wrought aluminum brass composed of composed of 76% copper, 22% zinc and 2% aluminum, and used for machine parts, electrical components, and corrosion-resistant parts.

Vialux. Trademark of Sovitec (France) for retroreflective glass beads used to enhance the visibility of road markings at night and in rain.

Viamin. Trademark of Vianova (Italy) for urea-formaldehyde resins.

Viapal. Trademark of Vianova (Italy) for unsaturated polyester resins.

Viaphen. Trademark of Vianova (Italy) for phenol-formaldehyde resins.

Vibond. Trade name of Hicks, Bullick & Company Limited (UK) for bonded nylon sewing thread.

Vibrac. Trade name of British Steel Corporation (UK) for a series of shock-resisting steels containing 0.27-0.45% carbon, 0.6% manganese, 2.3-2.8% nickel, 0.6% chromium, 0.6% molybdenum, and the balance iron. Used for tool holders, die blocks, gears, shafts, crankshafts, and axles.

Vibralloy. Trade name of Allegheny Ludlum Steel Corporation (USA) for an alloy composed of 49-53% iron, 38-42% nickel and 9% molybdenum. It has high magnetic permeability, and is used for vibrating reeds, mechanical filters, diaphragms, and instrumentation.

Vibraloy. Trade name of Audubon Metalwove Belt Corporation (USA) for an abrasion-resistant steel containing 0.9% carbon, 0.8% chromium, and the balance iron. Used for vibrating screens.

vibrated concrete. Concrete compacted during and after pouring so as to eliminate voids including entrapped air, and water.

Vibrathane. Trademark of Uniroyal Chemical Company (USA) for a series of raw materials, such as isocyanates, polyesters, polyester and polyether prepolymers, polyether glycols, liquid casting resins and millable gums, used in the manufacture of polyurethane foams and elastomers.

vibration-damping steel sheets. Laminate products made by bonding two thin outer sheets of steel to an interlayer of a highly viscoelastic synthetic resin (e.g., polyester), either with thermoplastic adhesives, or by rolling. The bonding of rolled vibration-damping steel sheets is ensured by filling the resin with conductive particles, e.g., nickel. These laminate products have excellent vibration and noise damping characteristics, high durability, and good formability and (spot) weldability, and are used for automotive applications, such as passenger compartment partitions, oil pans, and dash panels. Abbreviation: VDSS.

vibration insulators. See isolators.

Vibren. Trade name Nuova Rayon SpA (Italy) for rayon fibers and yarns used for textile fabrics.

Vibresist. Trade name of Atlas Specialty Steels (Canada) for an oil-hardening high-carbon hollow drill steel containing 0.95% carbon, 0.3% manganese, 0.25% silicon, 1% chromium, 0.25% molybdenum, and the balance iron. It has high strength, stiffness and wear resistance, and is used for hollow rock drills, and mining drills.

Vibrin. Trademark of Uniroyal Chemical Company (USA) for a series of resin compositions of polyesters and crosslinking monomers (e.g., triallyl cyanurate) that when catalyzed will polymerize to infusible solids. They have good strength at elevated temperatures, and are used for casting, impregnating, laminating and molding automotive and aircraft structures, wall panels, tabletops and countertops, boat hulls, chemically inert tanks, large-diameter pipe, and for paper coatings.

Vibrin-Mat. Trade name of W.R. Grace & Company (USA) for glass-fiber sheet-molding compound prepregs.

Vibro. Trade name of CCS Braeburn Alloy Steel (USA) for a shock-resisting hot-work tool steel (AISI type S1) containing 0.5% carbon, 1.4% chromium, 1.9-2.1% tungsten, 0.3% vanadium, and the balance iron. It has good deep-hardening properties, high hardness and strength, excellent toughness, good wear and abrasion resistance, and good hot hardness. Used for tools and dies.

vibrocast pipe. A concrete pipe made by pouring concrete into a stationary vertical form or mold and subjecting either to external or internal vibratory forces.

Vibron. Trade name of Vibron Limited (Canada) for thermosetting resins based on allyl polymers and having good electrical and thermal properties. See also allyls.

Vicalloy. Trademark of Wilbur B. Driver Company (USA) for a series of high-permeability magnetic alloys containing 35-62% cobalt, 6-16% vanadium, and the balance iron. *Vicalloy I* is a permanent magnet material composed of 51-52% cobalt, 38-39% iron and 10% vanadium. It has no magnetic orientation, a Curie temperature of 855°C (1570°F), high coercive force and remanent magnetization, a relatively high magnetic energy product, a maximum service temperature of 450°C (840°F). *Vicalloy II* is a permanent magnet material composed of 52% cobalt, 35% iron and 13% vanadium, and having a Curie temperature of 855°C (1570°F), and a higher coercive force, remanent magnetization and maximum energy product than Vicalloy I. Its magnetic orientation is developed by rolling or other mechanical working. *Vicalloy* alloys are used for electrical and magnetic equipment, such as magnets, magnetic memory, recording tapes, recording heads, magnetic clutches, and hysteresis motors.

Vicara. Trademark of Ardil (Italy) for an azlon fiber produced from *zein* (a protein derived from cornmeal) and supplied in staple and tow form. It has good resistance to mildew, moths, acids, dilute alkalies and boiling water, moderate wet strength, low flammability and shrinkage, and a soft hand. *Vicara* can be blended with cotton, rayon, wool and synthetic fibers (e.g., nylon), and is used for dresses, suits, coats, socks, knitwear, and blankets.

Vici. Trade name for a chrome-tanned glazed goatskin.

Viclan. Trademark of ICI Limited (UK) for saran (polyvinylidene chloride) plastics.

Viclon. Trademark of Kureha Chemical Industry Company (Japan) for polyvinyl chloride and polyvinylidene chloride fibers and monofilaments.

Vicoa Vinyl. Trade name of Hunt-Wilde Corporation (USA) for polyvinyl chloride compounds used for injection molding and extrusion applications.

Vicol. (1) Trademark of Khanna Adhesives & Chemicals Limited (Canada) for a superbonding, all-purpose *white glue*.

(2) Trademark of Rhodia Engineering Plastics (France) for vinyl acetate copolymers used for acetate-yarn sizing.

Vi-Comp. Trade name of Austenal Inc. (USA) for a corrosion-resistant cobalt-chromium dental bonding alloy.

ViCorr. Trade name of Fibergrate/Grating Pacific, LLC (USA) for a fiberglass-reinforced resin used for grating applications. It provides high toughness, excellent chemical resistance (including acidic and caustic environments), and very low flammability.

Vicropal. Trade name of Vicropal SAIC (Argentina) for colored opalescent tiles.

Vicryl. Trademark of Ethicon Inc., Division of Johnson & Johnson (USA) for a biodegradable, bioabsorbable copolymer of 90% *glycolide* and 10% *L-lactide* used for tissue engineering applications, e.g., surgical sutures. *Vicryl* mesh is used for dental guided-tissue-regeneration (GTR) membranes.

Victor. (1) Trade name of Crucible Materials Corporation (USA) for a water-hardening tool steel containing 1.05% carbon, and the balance iron. Used as drill rod for the manufacture of drills, punches, dowels, pins, amd shafts.

(2) Trade name of Osborn Steels Limited (UK) for a chromium-molybdenum tool steel containing 1% carbon, 5.25% chromium, 1.1% molybdenum, 0.45% vanadium, and the balance iron. Used for gages and press tools.

(3) Trade name of Dana Corporation (USA) for asbestos sheet gaskets.

Victor bronze. A corrosion-resistant bronze containing 39% zinc, 1.5% aluminum, 1% iron, 0.33% vanadium, and the balance copper. Used for pipes.

Victoria. Trade name of Lehigh Steel Corporation (USA) for a molybdenum-type high-speed steel (AISI type M2) containing 0.8% carbon, 6% tungsten, 5% molybdenum, 4% chromium, 2% vanadium, and the balance iron. Used for drills, taps, hobs, reamers, and broaches.

Victoria aluminum. A British aluminum alloy containing copper, zinc, silicon and iron. Used for light-alloy parts. See also Partinium.

Victoria green. See malachite green.

Victorieux Saint Juery. Trade name of Société Nouvelle du Saut-du-Tarn (France) for a hot-work steel containing 0.7% carbon, 9.5% tungsten, 2.5% chromium, 0.1% vanadium, and the balance iron. Used for dies and tools.

Victor metal. A corrosion-resistant cast nickel silver containing 50% copper, 35% zinc and 15% nickel. It has good corrosion resistance, machinability and castability, and is used for cast fittings, hardware, and valves.

Victory. Trade name of Unitek (USA) for a corrosion-resistant cobalt-chromium dental casting alloy.

Victory Cobalt. Trade name of Teledyne Vasco (USA) for a molybdenum-type high-speed steel (AISI type M36) containing 0.85% carbon, 8.5% cobalt, 6% tungsten, 5% molybdenum, 4% chromium, 2% vanadium, and the balance iron. It has good deep-hardening properties, high hot hardness, and excellent wear resistance. Used for cutting tools, such as lathe and planer tools, milling cutters, and drills.

Victrex. Trademark of Victrex USA Inc. (USA) for a family of engineering thermoplastics based on polyethersulfone (PES) or polyetheretherketone (PEEK), and supplied in various unreinforced and glass fiber- or carbon fiber-reinforced grades, and as powder coatings. They possess good elevated-temperature mechanical properties, and high toughness, strength and long-term load-bearing capabilities. The PES types are transparent and amorphous, while the PEEK types are semicrystalline polymers. Used for aerospace, electrical, electronic and industrial equipment.

Victron. Trade name for polystyrene and vinyl acetal resins.

Viculoy. Trade name of Akron Bronze & Aluminum Inc. (USA) for a series of strong, corrosion-resistant, age-hardenable, cast beryllium coppers containing 1.6-2.2% beryllium. Used for gears, shafts, marine parts, die molds, non-sparking tools, dies, welding electrodes, circuit breakers, and contacts.

vicuna. A very soft, fine, wool-like weaving fiber obtained from the undercoat hair of the vicuna (*Lama vicuna*), an animal of the llama family native to Bolivia, Ecuador and Peru. Used in the manufacture high-grade coats, jackets, shawls, sweaters and other clothing. Also spelled vicuña.

Vidaflex. Trade name of Jones, Stroud & Company Limited (UK) for sleevings, tapes, fabrics and other products made of glass fibers.

Vidar Supreme. Trade name of Uddeholm Corporation (USA) for a premium chromium-type hot-work tool steel (AISI H11). It has good deep-hardening properties, high hardenability, excellent toughness, high ductility, good hot hardness, moderate wear resistance, and is used for dies, punches, and shear blades.

Vidcolor. Trade name of Vidplan SA (Uruguay) for a colored sheet glass.

Vidlon. Trademark of Yambolen (Bulgaria) for nylon 6 fibers and filament yarns used for textile fabrics.

Vidrarte. Trade name of Companhia Nacional de Vidros e Molduras (Brazil) for laminated glass and mirrors.

Vidrofenol. Trade name of Vidrieria Argentina SA (Argentina) for a continuous filament glass tissue that is reinforced with textile glass yarns, and may or may not be saturated with asphalt. Used for anticorrosive pipe coverings.

Vidroflex. Trade name of Vidrieria Argentina SA (Argentina) for a continuous glass-fiber quilt used for vibration absorption appli-

cations and battery separators.

Vidropal. Trade name of Vidrierias de Llodio SA (Spain) for a ceramic-coated, toughened glass.

Vidro Rochedo. Trade name of Companhia Vidreira Nacional Ltda.-Covina (Portugal) for a toughened glass.

Vidrotel. Trade name of Vidrieria Argentina SA (Argentina) for glass-fiber insulation products.

Vidur. Trade name Carlos Tarrida Monge (Spain) for a toughened glass.

Vidurex. Trade name of Santa Lucia Cristal SACIF (Argentina) for a toughened automotive sheet glass.

Vienna lime. A calcined dolomite composed essentially of calcium and magnesium carbonates, and used as an abrasive in certain buffing compounds especially to produce high colors on nickel or copper plated parts, and for secondary coloring of aluminum and copper-zinc alloys.

Vienna white. Pure *white lead* [$2PbCO_3 \cdot Pb(OH)_2$] originally made in Austria, and used as a paint base. Also known as *Kremnitz white; Krems white.*

Viennese Ornaments. Austrian trade name for a corrosion-resistant nickel silver composed of 55% copper, 25% zinc and 20% nickel, and used for ornamental parts.

Viennese Sheet. Austrian trade name a for a corrosion-resistant nickel silver composed of 60% copper, 20% zinc and 20% nickel, and used for ornaments and hardware.

Viennese Tableware. Austrian trade name for a corrosion-resistant nickel silver composed of 50% copper, 25% zinc and 25% nickel, and used for ornaments and cutlery.

Vifi Cord. Trade name of Vitro-Fibras SA (Mexico) for continuous impregnated glass-fiber yarn used as a rubber reinforcement.

Vifiltro. Trade name of Vitro-Fibras SA (Mexico) for air filter media based on glass fibers.

vigezzite. An orange yellow mineral of the columbite group composed of calcium cerium niobium tantalum titanium oxide, (Ca,Ce)(Nb,Ta,Ti)$_2$O$_6$. Crystal system, orthorhombic. Density, 5.54 g/cm^3; refractive index, 2.315. Occurrence: Italy.

Vigilant. Trade name of Joseph Beardshaw & Son Limited (UK) for a water-hardening tool steel containing 0.5-1% carbon, and the balance iron. Used for chisels, hard tools, and springs.

Vigilpane. Trade name of Libbey-Owens-Ford Company (USA) for laminated anti-burglar glass composed of two panes of 3.2 mm (0.125 in.) thick polished plate glass with a penetration-resisting plastic interlayer.

Vigopas. Trademark of Raschig GmbH (Germany) for unsaturated polyester resins.

Vigorol. Trademark of Raschig GmbH (Germany) for a series of phenol- and cresol-formaldehyde resins.

viitaniemiite. A colorless mineral that is composed of sodium calcium aluminum fluoride phosphate hydroxide, NaCaAl(PO$_4$)(F,OH)$_3$, and may also contain some manganese. Crystal system, monoclinic. Density, 3.06 g/cm^3; refractive index, 1.544. Occurrence: Canada (Quebec), Finland.

Viking. (1) Trade name of Uddeholm Corporation (USA) for a high-alloy tool steel containing 0.5% carbon, 1% silicon, 0.5% manganese, 8% chromium, 1.5% molybdenum, 0.5% vanadium, and the balance iron. Used for dies, rolls, and tube-drawing tools.

(2) Trade name of CCS Braeburn Alloy Steel (USA) for an oil-hardening tool steel containing 1.05% carbon, 1.35% chromium, 0.45% molybdenum, and the balance iron. Used for mandrels and rams. *Viking Extra* is an oil-hardening tool steel

containing 1.05-1.15% carbon, 1.25-1.4% chromium, 0.3-0.5% molybdenum, and the balance iron. Used for tools, cutters, dies, and roller bearings.

(3) Trade name of Emmaboda Glasverk AB (Sweden) for a toughened glass.

(4) Trade name of Peterson Window Corporation (USA) for insulating glass.

Vikmanshyttan. Trade name of Vikmanshyttan AB (Sweden) for a series of corrosion- and/or heat-resistant austenitic, ferritic and martensitic stainless steels including various free-machining and welding grades.

Viko. Trade name of Hicks, Bullick & Company Limited (UK) for polyester-cotton core-spun sewing threads.

Vilit. Trademark of Hüls AG (Germany) for several polyvinyl chloride copolymers used as binders for lacquers and varnishes.

villamaninite. An iron-black mineral of the pyrite group composed of copper iron nickel sulfide, $(Cu,Fe,Ni)S_2$. Crystal system, cubic. Density, 4.50 g/cm³. Occurrence: Spain.

villiaumite. A colorless mineral of the halite group composed of sodium fluoride, NaF. It can also be made synthetically. Crystal system, cubic. Density, 2.76 g/cm³; refractive index, 1.3270. Occurrence: Czech Republic, Western Africa.

Viloft. Trademark of Courtaulds Limited (UK) for viscose rayon staple fibers used for lofty clothing.

Vilon. Trademark of Nitivy Company (Japan) for polyvinyl alcohol (vinal) fibers and filament yarns used for textile fabrics.

Vimet. Trademark of Saint-Gobain (France) for an antireflective glass.

Vimetal. British trade name for a cobalt alloy containing 17% nickel, 9% chromium, 5.5% iron, 4% tungsten, 4% molybdenum, 3% niobium, 1.5% titanium and 0.05% carbon. Used for high-temperature applications.

VIM plastics. See vacuum-injection-molded plastics.

vimsite. A colorless mineral composed of calcium borate hydroxide, $CaB_2O_2(OH)_4$. Crystal system, monoclinic. Density, 2.54 g/cm³; refractive index, 1.614. Occurrence: Russia.

Vimlite. Trade name of Celanese Corporation (USA) for cellulose acetate plastics.

Vimur. Trade name of Santa Lucia Cristal SACIF (Argentina) for an enameled toughened sheet glass.

Vinac. Trademark of Air Products & Chemicals, Inc. (USA) for polyvinyl acetate products supplied in the form of beads, powders, emulsions and solutions. The beads are glass-like spheres used for hot-melt adhesives, solvent paints, paper coatings and inks. The powders are white and free-flowing, and used for special adhesives in concrete and joint cements. The emulsions contain up to 57% solids, and are used as binders in adhesive bases, and as concrete adhesives, pigmented paper coatings and textile finishes. The solutions contain 50-51% polyvinyl acetate polymer dissolved in methanol, and are used for heat-seal adhesives.

VinaGARD. Trademark of Vintex Corporation (Canada) for a range of flame-retardant, UV-resistant, heavy-duty, double vinyl-coated protective fabrics with good dimensional stability, good shrink, tear and abrasion resistance, and good rot, fungus and mildew resistance. Used for protective apparel, tents, mine ducting, and case binding.

Vinal. Trade name of PPG Industries (USA) for a strong, water-resistant textile fiber based on polyvinyl alcohol.

vinal. See vinal fiber.

vinal fiber. A generic term for a chemically resistant man-made fiber whose fiber-forming substance is any long-chain synthetic polymer made up of 50 wt% or more vinyl alcohol units ($-CH_2-CHOH-$), and in which the sum of the vinyl alcohol and acetal units is 85 wt% or more of the fiber. They have excellent resistance to water, fungi and mildew, and are used for fishing nets, garments (e.g., gloves, hats and stockings), and rainwear. Also known as *vinal*.

VinaSIGN. Trademark of Vintex Corporation (Canada) for soft, supple, flame-retardant, printable, vinyl extrusion-coated high-strength textiles with good crazing and cracking resistance and enhanced low-temperature pliability, used for banners, billboards, displays, and signs.

VinaSoft. Trade name for a plasticized acrylic resins used for lining dentures.

Vincel. Trademark of Courtaulds Limited (UK) for a viscose rayon staple fiber with good washability, often blended with cotton for the manufacture of rainwear.

Vin-Clad. Trademark of US Stoneware Company (USA) for a series of dry, free-flowing, polyvinyl chloride powders used for decorative and protective coatings.

Vinco. Trade name of CCS Braeburn Alloy Steel (USA) for a tungsten-type high-speed steel (AISI type T1) containing 0.7% carbon, 0.25% manganese, 1% vanadium, 3.75-4.25% chromium, 17.5-18.5% tungsten, and the balance iron. It has high hot hardness, excellent machinability and wear resistance, good toughness, and is used for shear knives, cutters, and dies. *Vinco Hot Work* is a tungsten-type hot-work tool steel (AISI type H26) containing 0.5% carbon, 4% chromium, 18% tungsten, 1% vanadium, and the balance iron. It has excellent wear resistance, high hot hardness, and is used for knives, punches, and dies and die inserts.

Vindure. Trade name for a high-grade, transparent, water-resistant tracing paper with high rag content and good dimensional stability.

vine black. A fine, black pigment made from charcoal obtained by burning grapevine stems, and used for coloring inks. Also known as *Frankfurt black*.

Vinex. Trade name of Vinex Kunststoff GmbH (Germany) for unplasticized polyvinyl chloride supplied as granules and dry blends for the manufacture of injection-molded and extruded products.

Vinidur. Trademark of BASF Corporation (USA) for white, unplasticized polyvinyl chloride available in standard and high-impact grades in the form of pastes, liquids, chips and granules.

Vinivit. Trade name of Bayer Corporation (USA) for polyvinyl chlorides.

Vinnapas. Trademark of Wacker-Chemie GmbH (Germany) for polyvinyl acetate resins.

Vinnol. Trademark of Wacker-Chemie GmbH (Germany) for unplasticized and plasticized polyvinyl chlorides supplied in granule, powder, dispersion and powder mixture forms. They have moderate tensile strength, a maximum service temperature of 60-80°C (140-176°F), good resistance to dilute acids and alkalies, gasoline and mineral oils, and good dielectric properties. Used for tire tubes, hoses, flexible tubing, adhesives, coatings, and paints.

Vinnolit. Trademark of Wacker-Chemie GmbH (Germany) for UV-stabilized polyvinyl chlorides, and standard and high-impact unplasticized polystyrene resins.

Vinoflex. Trademark of BASF Corporation (USA) for a thermoplastic elastomer based on polyvinyl chloride, and containing up 40% plasticizer. It is available in compounded, uncompounded, plasticized (PVC) and unplasticized (UPVC) grades. The plasticized grades have good flexibility, a density of 1.2-1.3 g/cm³ (0.043-0.047 lb/in.³), high toughness, moderate strength, a hardness of 60-86 Shore, an elongation range of 0-100%; a service temperature range of -30 to +80°C (-22 to +176°F), good resistance to dilute acids and alkalies, mineral oils and gasoline, poor resistance to trichloroethylene and tetrachloroethylene, and excellent weathering resistance. Used for building components, interior decoration products, tubing and hose, floor covering, foils, in chemical equipment (e.g., vessels, pipes and exhaust systems), credit cards, electrical cable and plug applications, automotive components, and food packaging.

vinogradovite. A colorless to white mineral composed of aluminum phosphate hydroxide hydrate, $Na_4Ti_4Si_8O_{22} \cdot xH_2O$. Crystal system, monoclinic. Density, 2.88 g/cm³; refractive index, 1.770. Oc-currence: Russian Federation.

Vinsol. Trademark of Hercules, Inc. (USA) for a series of dark red or brown, brittle, thermoplastic resins that are essentially petroleum hydrocarbon distillates of pinewood. Supplied as flakes, fine powders, aqueous dispersions and solids, they have a melting point of 115°C (239°F), and are used for electrical insulating and dark coatings, as air-entraining agents in concrete, and in adhesives, emulsions, inks and thermoplastics.

Vintage Halo. Trade name of 3M Dental (USA) for a dental porcelain.

Vintage Opal. Trade name of 3M Dental (USA) for an opalescent dental porcelain used for metal-to-ceramic restorations.

Vintex. Trademark of Vintex Corporation (Canada) for waterproof polyvinyl-coated industrial polyester fabrics used for tarpaulins, barrier sheets, etc.

Vinybel. Trade name of Caselit for plasticized polyvinyl chlorides.

vinyl acetal resins. See polyvinyl acetals.

vinyl acetate plastics. Plastics based on polyvinyl acetate.

vinyl acetate resins. See polyvinyl acetates.

Vinylaire. Trade name for a series of foamed vinyl plastisols. See also plastisol.

vinylal. See vinylal fiber.

vinylal fiber. A term used by the ISO (International Organization for Standardization) for a man-made fiber whose fiber-forming substance is any long-chain synthetic polymer made up of 85 wt% or more vinyl alcohol units ($-CH_2-CHOH-$) with varying levels of acetalization. Also known as *vinylal*.

Vinylan. Trade name for a vinal fiber with excellent resistance to water, fungi, mildew and most chemicals, used for textile fabrics including garments.

vinylated alkyds. A mixture of alkyds and polymethylstyrene or polystyrene.

vinylbenzene. See styrene.

Vinyl-Bond. Trade name of ICI Paints (US) for vinyl paints.

Vinyl-Brite. Trade name of Bridges Smith & Company (USA) for vinyl latex paint used for interior and exterior applications.

vinyl-capped addition polyimide. An addition polyimide in which the addition reaction is terminated by a vinyl resin. Abbreviation: VCAP. See also addition polyimides.

vinyl cement. An adhesive or glue that is particularly suitable for bonding vinyl plastics.

vinyl chloride plastics. Plastics based on polyvinyl chloride resins. See also polyvinyl chlorides.

vinyl chloride resins. See polyvinyl chlorides.

vinyl chlorides. See polyvinyl chlorides.

vinyl-coated fabrics. Textile fabrics coated with any of a group of polymers derived from vinyl monomers. The coatings increase both toughness and elasticity, enhance strength and reduce the moisture absorption of the textiles.

vinyl-coated glass yarn. Glass yarn composed of continuous filaments coated with plasticized vinyl chloride resin.

vinyl coatings. A group of fast, air-drying coatings prepared from polyvinyl formal, polyvinyl acetal, polyvinyl butyral, polyvinyl chloride, etc. They have good chemical resistance especially to acids and alkalies, good abrasion and impact resistance, good flexibility and formability, low water permeability, high dielectric resistance, and are used for linings for cans, tanks and pipelines, marine superstructures and shore installations, offshore drilling rigs, ship bottoms, locks, chemical-processing equipment, petroleum equipment, metal awnings, railroad hopper cars, and dairy and brewery equipment.

vinyl compounds. (1) A term referring to a group of highly reactive esters, such as vinyl acetate, vinyl alcohol and vinyl chloride, which contain the vinyl ($CH_2=CH-$) group. Used in the manufacture of plastics. Also known as *vinyls*.

(2) A general term for compounds that contain the vinyl ($CH_2=CH-$) group, e.g., acrylonitrile ($CH_2=CHCN$), methyl methacrylate [$CH_2=C(CH_3)COOCH_3$], and styrene ($C_6H_5CH=CH_2$). Also known as *vinyls*.

Vinyl-Cote. Trademark of The Glidden Company (USA) for vinyl chloride coatings used for marine and industrial applications.

vinyl cyanide. See acrylonitrile.

vinyl ester resins. A group of engineering thermosets that are the products of reactions of acrylic or methacrylic acid with epoxy resins, such as bisphenol A or epoxy novolacs, dissolved in a reactive vinyl monomer, such as styrene. They have outstanding chemical properties, good mechanical properties, and are used as matrix resins for polymer-matrix composites, and for electrical equipment, chemical, plating and wastewater treatment equipment, tanks, linings, pumps, and pipes. Also known as *vinyl esters*.

vinyl ether resins. A group of polymers based on vinyl ethyl ether ($CH_2=CHOC_2H_5$), vinyl methyl ether ($CH_2=CHOCH_3$), or vinyl butyl ether ($CH_2=CHOC_4H_9$).

Vinylex. Trademark of Tudor Safety Glass Company Limited (UK) for a laminated safety glass. *Vinylex Diffusa* is a white, translucent laminated glass used for light diffusion applications, and for applications where normally sandblasted or acid-embossed glass would be employed. *Vinylex Shadowlite* is a laminated glass with brownish-gray tint.

Vinyl Flat. Trademark of Porter Paints (USA) for a line of vinyl wall paints used for interior applications.

vinyl foam. A vinyl resin processed into a closed- or open-cell foam by the chemical action of a blowing or foaming agent, or by mechanical action using high pressure and an inert gas. Used for flotation devices and thermal insulation applications.

vinyl formal-phenolic adhesives. Structural adhesives based on blends of thermoplastic polyvinyl formals and thermosetting phenolics, and supplied as dispersions in solution, and in film form.

vinyl halide. A compound, such as vinyl chloride ($CH_2=CHCl$), in which a halogen (X) is directly attached to a doubly bonded carbon ($-C=C-X-$).

vinylidene resins. See polyvinylidenes.

Vinylite. Trademark of VinylWorks, Inc. (USA) for a series of flexible and rigid synthetic thermoplastic resins that are copolymers of vinyl chloride and vinyl acetate and supplied as translucent and opaque grades in various colors. They are stiff at low temperatures, have good abrasion, flame and weather resistance, good processibility, good machinability, good dielectric strength, excellent resistance to oils and aliphatic hydrocarbons, a maximum service temperature of 138°C (280°F), and a refractive index of 1.4665. Used for floor and wall coverings, upholstery, furniture, tubing, insulation, safety-glass interlayers, and formerly for phonographic records. They are also used in solvent-type adhesives for porcelain, metal, mica, stone and glass.

Vinyl Lux. Trade name of Lilly Industries, Inc., Perfection Paint Division (USA) for vinyl wall paints.

Vinyloid. Trade name for a vinyl acetate resin used for molded products.

Vinylon. Trade name used in Japan for a vinal (polyvinyl alcohol) fiber.

vinylphenyl-POSS. A moisture-sensitive, solid polyhedral oligomeric silsesquioxane (POSS) monomer with a melting point of 375°C (707°F) and flash point above 230°F (110°C), used in the preparation of advanced polymeric materials, and as a model for silica surfaces. See also polyhedral oligomeric silsesquioxanes.

Vinyl-Plex. Trade name of Tower Paint Manufacturing (USA) for vinyl paints used for exterior and interior applications.

Vinyl Plus. Trademark of Domco Industries Limited (Canada) for vinyl tiles.

vinyl polymers. See polyvinyls.

vinyl polysiloxanes. Addition-reaction silicones made from polysiloxane and vinyl resin, usually in the presence of a platinum catalyst. They are often used in dentistry for impressions and bite registration purposes. Abbreviation: VPS. See also polysiloxanes.

vinylpyridine copolymer. An elastomer made by copolymerizing vinylpyridine ($C_5H_4NCH=CH_2$) latex with butadiene-styrene. Abbreviation: VPC.

vinyl resins. See polyvinyls.

vinyls. See vinyl compounds; polyvinyls.

Vinylseal. Trade name of Bakelite Corporation (USA) for polyvinyl chlorides and other vinyl plastics.

Vinyl Suede. Trademark of Porter Paints (USA) for premium vinyl wall paints used for interior applications.

Vinylum. Trade name for an aluminum powder dispersed in a vinyl copolymer and used for highly reflective high-temperature coatings.

Vinyon. Trade name of Avisco (USA) for a series of textile fibers based on thermoplastic copolymers of vinyl chloride and vinyl acetate, or acrylonitrile. They possess good strength and excellent resistance to acids, alkalies, water, sunlight, bacteria, moths and mildew, fair to poor resistance to organic solvents and aromatic hydrocarbons, and do not support combustion. Used for carpets, pressed felts, bonded fabrics, rubber-coated elastic fabrics, and heat-sealable paper.

vinyon. See vinyon fiber.

vinyon fiber. A generic term for a self-extinguishing man-made fiber whose fiber-forming substance is any long-chain synthetic polymer made up of 85 wt% or more vinyl chloride units ($-CH_2-CHCl-$). They have excellent resistance to water, sunlight, bacteria and moths, and a rather low softening point. Also known as vinyon.

Viola. Trade name of Ilssa-Viola SpA (Italy) for a series of corrosion- and/or heat-resistant austenitic, ferritic and martensitic

stainless steels including various free-machining and welding grades.

violarite. A light to violet-gray mineral of the spinel group composed of nickel iron sulfide, $(Fe,Ni)_3S_4$. Crystal system, cubic. Density, 4.65 g/cm³. Occurrence: Canada (Ontario), USA (California, Nevada).

Violet Label. Trade name of Ackerlind Steel Company Limited (USA) for a non-deforming tool steel containing 0.9% carbon, 0.5% chromium, 0.2% vanadium, 1.5% tungsten, and the balance iron. Used for cutting and drawing dies, tools, and gages.

violet phosphorus. See black phosphorus.

Vipax. Tade name of Vidrobrás (Brazil) for glass roofing lenses.

Viper. (1) Trade name of Osborn Steel Limited (UK) for medium-carbon tool steels containing 0.6% carbon, 0.6% chromium, and the balance iron. Used for lathe centers and cutting tools.

(2) Trade name of Osborn Steels Limited (UK) for an oil- or water-hardened steel containing 1% carbon, 0.45% manganese, 1% chromium, and the balance iron. Used for ball bearings, liners, sleeves, and gages.

Vipla. Trademark of Montecatini (Italy) for polyvinyl chloride plastics.

Viplast. Trademark of Montecatini (Italy) for plasticized polyvinyl chlorides.

Viplavil. Trademark of Montecatini (Italy) for polyvinyl chloride copolymers.

Vipolit. Trademark of Lonza-Werke GmbH (Germany) for several modified polyvinyl chlorides.

Viracon. Trade name of 3M/Viracon (USA) for a strengthened glass laminated with a proprietary liquid-crystal-polymer privacy film.

virgilite. A colorless mineral of the quartz group composed of lithium aluminum silicate, $Li_xAl_xSi_{3-x}O_6$. Crystal system, hexagonal. Density, 2.46 g/cm³; refractive index, approximately 1.52. Occurrence: Peru.

virgin aluminum. See primary aluminum.

virgin fiber. A new fiber, i.e., a fiber that has never been reclaimed from manufactured or used textile products.

virgin filament. A monofilament as it comes from the extruder, spinneret, or drawing die.

Virginia pine. The moderately heavy, hard, strong, stiff wood from the small- to medium-sized pine tree Pinus virginiana. The heartwood is orange, and the sapwood almost white. It has very high knottiness, low durability, high shrinkage, and high shock resistance. Source: USA (from New Jersey and Virginia to Northern Alabama, throughout the Appalachian Mountains and in the Ohio Valley). Used for lumber, railroad ties, as pulpwood, and as a fuel. Also known as Jersey pine; scrub pine.

Virginia silver. A nickel silver containing varying amounts of copper, nickel and zinc, and used for ornaments.

virgin material. (1) Any material that has never been used or processed.

(2) A plastic material that has never been used or processed. It is usually in the form of granules, pellets, powder, flock, or liquid. Also known as virgin plastic.

virgin metal. See primary metal.

virgin plastic. See virgin material (2).

virgin wool. New and unused wool, or wool spun or woven only once.

Virgo. Trade name of Creusot-Loire (France) for a series of corrosion-resistant austenitic, ferritic and martensitic stainless steels, numerous heat-, creep- and oxidation-resistant steels, and several cobalt- and nickel-base superalloys.

Virillium. British trade name for a corrosion-resistant dental casting alloy containing 67.9% cobalt, 24.1% chromium, 1.4% nickel, 5.3% molybdenum, and 1.3% iron.

Virion. Trade name for rayon fibers and yarns used for textile fabrics.

Viro. Trade name of Zinex Corporation (USA) for several surface finishes and finishing processes including *Viro-Black,* a shiny black corrosion- and wear-resistant conductive surface finish applied over copper, brass, nickel, palladium, lead, tin, or steel, *Viro-Brass,* a bright, fine-grained brass electrodeposit and noncyanide plating process, and *Viro-Gold,* a bright gold electrodeposit and noncyanide plating process.

Vis. Trade name of Fabbrica Pisana SpA (Italy) for a laminated plate glass.

Visarm. Trade name of Fabbrica Pisana SpA (Italy) for a thick, laminated plate glass.

VisarSeal. Trade name for an unfilled light-cure resin used in restorative dentistry for sealing applications.

Visaterm. Trade name of Fabbrica Pisana SpA (Italy) for a green-tinted laminated safety glass similar to *Triplex Sundym.*

Viscacelle. Trade name of British Cellophane (UK) for cellophane-type regenerated cellulose.

Viscalon. Trade name for rayon fibers and yarns used for textile fabrics.

Viscasil. Trademark of GE Silicones (USA) for a series of silicone elastomers.

Viscocel. Trademark of Unnafibras Textil Ltda. (Brazil) for viscose rayon filament yarns used for textile fabrics.

Viscocord. Trademark of Glanzstoff Austria AG (Austria) for high-tenacity viscose rayon fibers and filament yarns used for industrial fabrics including tire cords.

viscoelastic materials. A group of materials, including in particular polymers, that under stress exhibit viscoelastic deformation, i.e., mechanical deformation involving both elastic (solid-like) and viscous (fluid-like) characteristics. The deformation is dependent upon time, temperature and the magnitude of the stress.

Viscofil. Trademark of Glanzstoff Austria AG (Austria) for viscose rayon filament yarns used for textile fabrics.

Viscol. Trade name of Cantanzaro Filati (Italy) for rayon fibers and yarns used for textile fabrics.

Viscoloid. Trade name of E.I. Du Pont de Nemours & Company (USA) for *celluloid*-type cellulose nitrate plastics.

Viscor. Trade name for rayon fibers and yarns.

Viscosa Sicrem. Trademark of SICREMA SpA (Italy) for viscose rayon filament yarns used for textile fabrics.

Viscose. Trade name of Lenzing Fibers Corporation (USA) for viscose rayon fibers.

viscose. (1) An aqueous solution of sodium cellulose xanthate and sodium hydroxide. It is a highly viscous, yellow liquid used for processing into textile fibers, foils, sponges, cellophane, etc.

(2) Cellulose in the form of fibers or sheets obtained by first dissolving a cellulosic substance, such as wood or cotton, and then regenerating it by extrusion, e.g., by spinning through the minute openings of a spinneret as in the viscose process. See also viscose rayon.

viscose fibers. See viscose (2); viscose rayon.

viscose rayon. Filaments of rayon made by spinning or extruding *viscose* (1) through the minute openings of a spinneret into a bath of sulfuric acid, sodium and zinc salts. They have medium to high tenacity, good resistance to dilute alkalies, good absorp-tive properties, good drapability, good resistance to moths, fair resistance to mildew and sunlight, burn quickly without melting, and can be blended with natural and other synthetic fibers. Used in filament or staple form for apparel, carpets, rugs, blankets, upholstery, and industrial fabrics. Also known as *viscose fiber.* See also cuprammonium rayon; nitrocellulose rayon; rayon fiber.

Viscount. Trade name of Latrobe Steel Company (USA) for a series of free-machining chromium-type hot-work tool steels (AISI type H13) containing 0.35-0.4% carbon, 1% silicon, 0.3-0.8% manganese, 5-5.25% chromium, 1% vanadium, 1.2-1.5% molybdenum, varying amounts of sulfur, and the balance iron. They have high hot hardness, high tensile strength, good abrasion resistance, good resistance to heat checking, great depths of hardening, excellent resistance to decarburization, and good machinability and toughness. Used for extrusion, die-casting and other dies, die inserts, shear blades, forging blocks and various other metalworking tools.

viscous materials. Liquids (e.g., tar), semisolids (e.g., wax) or gases that exhibit internal resistance to movement or flow. Deformation in this type of materials is time-dependent and permanent.

viscous coal tar. An amorphous, resinous compound that is the residue formed in the production of gas from coal. Used in roofing compositions, road construction, and wood preservatives. See also coal tar.

viseite. A mineral composed of sodium calcium aluminum phosphate silicate hydroxide hydrate, $NaCa_5Al_{10}(SiO_4)_3(PO_4)_5(OH)_{14} \cdot 16H_2O$. Density, 2.20 g/cm³. Occurrence: Belgium.

vishnevite. (1) A blue, white mineral of the cancrinite group composed of sodium aluminum silicate sulfate trihydrate, $(Na,Ca,K)_8(Al,Si)_{12}O_{24}(SO_3,CO_2) \cdot 3H_2O$. Crystal system, hexagonal. Density, 2.45 g/cm³; refractive index, 1.501. Occurrence: Russia.

(2) A blue or white mineral of the cancrinite group composed of sodium aluminum silicate sulfate trihydrate, $Na_8(Al,Si)_{12}O_{24}SO_4 \cdot 3H_2O$. Crystal system, hexagonal. Density, 2.46 g/cm³; refractive index, 1.501. Occurrence: Russia.

Visi-Guard. Trademark of Bruin Plastics (USA) for fluorescent fabrics supplied in a wide range of colors in solids and mesh for use in safety clothing, signs, barrier tapes, and flags.

Visil. Trade name of Saetari Oy (Finland) for fire-resistant rayon fibers and yarns used for industrial and household fabrics.

Visio. Trademark of 3M ESPE Dental (USA) for dental bonding agents and light-cure dental resins including *Visio-Bond,* a dental bonding agent, *Visio-Dispers,* a microfine composite resin for anterior restorations, *Visio-Fil,* a hybrid composite resin for anterior restorations, *Visio-Gem,* a dental resin used for laboratory applications, *Visio-Molar,* a hybrid composite resins for posterior restorations, and *Visio-Molar,* a dental composite resin used as a fissure sealant.

Vision. (1) Trademark of Schott Glas AG (Germany) for glass-ceramics with excellent thermal-shock resistance and high strength.

(2) Trademark of Aurident, Inc. (USA) for a silver-free dental ceramic alloy containing 80% palladium, 2% gold, and a small addition of iridium for grain refinement. It has high yield strength at elevated temperatures, high margin creep resistance, is compatible with most porcelains, and used for porcelain-to-metal restorations.

Vision Block. Trade name of PPG Industries Inc. (USA) for laminates consisting of multiple pieces of glass applied to interior plies of plastic, and used for the protection of military personnel against enemy fire.

Vision 2. Trade name for a dental adhesive system for dentin/enamel bonding.

Vision-Lite. Trademark of Saint-Gobain (France) for anti-reflective glass.

VisKing. Trademark of Ethyl Corporation (USA) for plastic films, sheeting, tubing, rods and bars used for wrapping or packaging applications.

Viskoflex. Trade name for a highly flexible elastomer used for dental impressions.

vismirnovite. A pale yellow mineral of the sohngeite group composed of tin zinc hydroxide, $ZnSn(OH)_6$. Crystal system, cubic. Density, 4.13 g/cm³; refractive index, 1.735. Occurrence: Czech Republic.

Visqueen. Trade name of British Visqueen (UK) for polyethylene resins.

Vista. (1) Trade name of Owens-Corning (USA) for clear, transparent, hollow glass blocks.

(2) Trademark of Solutia Inc. (USA) for a polymeric window film with neutral appearance, used for commercial and residential applications. It significantly reduces both incident solar and ultraviolet rays, controls excessive heat gain, and minimizes fading of furniture, carpets and wall coverings.

(3) Trade name of Time Steel Service Inc. (USA) for a series of chromium-, molybdenum- and tungsten-type hot-work tool steels.

(4) Trade name of Vista Chemical Company (USA) for vinyl resins.

(5) Trade name for a low-gold dental casting alloy.

Vistacast. Trade name Steele's (Contractors) Limited (UK) for a toughened tinted rough-cast glass.

Vistaclad. Trade name of Steele's (Contractors) Limited (UK) for a toughened and fired glass.

Vistaflex. Trademark of Advanced Elastomer Systems (USA) for a series of translucent thermoplastic polyolefin elastomers that can be processed by injection molding and extrusion techniques. They possess a snappy "rubber-band" performance, a Vicat softening point of 48°C (118°F), a hardness of 60 Shore A, and a processing temperature of 154°C (309°F). The molded or extruded products appear smooth, and are used for automotive components, insulation applications, and as modifiers in polyethylene and polypropylene to enhance flexibility and impact resistance.

Vistalite. Trade name of Steele's (Contractors) Limited (UK) for a toughened, translucent double-sides rough-cast glass.

Vistalon. Trademark of Exxon Mobil Corporation (USA) for an ethylene-propylene rubber with excellent ozone and weathering resistance, good heat resistance, good low-temperature flexibility, and good resistance to acids and detergents. Used for automotive applications, appliances, cable insulation, electrical equipment, and industrial hose and tubing.

Vistanex. Trademark of Exxon Mobil Corporation (USA) for an ozone-resistant polyisobutylene elastomer prepared at low temperatures.

Vista-Safe. Trade name of Laminated Glass Corporation (USA) for a bullet-resistant laminated glass.

Vistatex. Trade name of Steele's (Contractors) Limited (UK) for a toughened rough-cast glass with one smooth and one irregular and finely ribbed surface. The ribbed side has a ceramic color fired on.

Vistel. Trade name of Vista Chemical Company (USA) for several polyvinyl chlorides.

Visto. Trade name of Dörrenberg Edelstahl GmbH (Germany) for a series of hot-work and high-speed tool steels, and several alloy machinery steels.

Vistra. Trademark of Koeln-Rottweiler AG–Premnitz Plant (Germany) for a spun rayon made by the viscose process. *Vistra XTH* is a high-tenacity grade.

Vistralon. Trade name for viscose rayon fibers and yarns.

Vistram. Trade name of Bayer AG (Germany) for polyurethane films and film products.

Vistron. Trade name of Vistron Corporation (USA) for viscose rayon fibers and yarns used for textile fabrics.

visual purple. See rhodopsin.

visual yellow. A substance formed from *rhodopsin* when exposed to strong light. In the dark, part of it is reconverted into rhodopsin and some is converted into *retinol* (vitamin A).

Visulure. Trade name of Glidden Industrial Coatings (USA) for metalescent fluorocarbon coatings.

Visurit. Trade name of Saint-Gobain (France) for a toughened glass having a peripheral circular zone which is heat-treated differently than the rest.

Vita. (1) Trade name of LeVita Metal Alloy Company (USA) for an extensive series of high-speed steels containing 0.7% carbon, 0-8.5% molybdenum, 4-4.5% chromium, 1.5-22% tungsten, 1-2% vanadium, 0-12% cobalt, and the balance iron. Used for lathe and planer tools.

(2) Trade name of Vita Zahnfabrik (Germany) for an extensive series of dental porcelains and acrylic restoratives including *Vita*, a feldspathic porcelain, *Vita Hi-Ceram* a high-ceramic porcelain for dental cores, *Vita In-Ceram*, a high-ceramic alumi-nous porcelain for dental cores, *Vita K+B*, a self-cure acrylic resin for crown and bridge restorations, *Vita-Omega*, a porcelain for metal-to-ceramic restorations and *Vita Spray-On*, a spray-on porcelain for metal-to-ceramic restorations.

Vitadur. Trade name of Vita Zahnfabrik (Germany) for a series of dental porcelains including *Vitadur Alpha*, a feldspathic porcelain for ceramic crown completions and *Vitadur-N*, an alumina-reinforced high-strength porcelain composite veneering applications.

Vitafilm. Trademark of Goodyear Tire & Rubber Company (USA) for plastic packaging film supplied in rolls and sheets.

Vitafoam. Trademark of Vita International Limited (UK) for polyurethane foam products used for stuffing and upholstery, as carpet underlays, and for furniture.

Vitakon. Trade name of Vinatex (USA) for styrene-butadiene rubber.

Vital. Trade name of Metallwerke Schwarzwald GmbH (Germany) for aluminum casting alloys containing 1.2% zinc, 0.9-1% copper and 0.6-0.9% silicon. Used for light-alloy parts.

Vitalescence. Trademark of Ultradent Products, Inc. (USA) for a 45% filled, light-cured wetting resin used with other dental composite materials.

Vital-X. Trade name of Osborn Steels Limited (UK) for a nondeforming, oil-hardening tool steel containing 2% carbon, 13% chromium, and the balance iron. Used for dies, rolls, punches, and shear blades.

Vitallium. Trade name of Howmet Corporation (USA) for a series of corrosion-resistant cast cobalt-base superalloys containing 28-32% chromium, 5-7% molybdenum, and up to 0.75% manganese, 0.8% silicon and 0.5% carbon, respectively. Used for dental and orthopedic applications, e.g., prostheses and hip replacements.

Vitalu. British trade name for an aluminum die-casting alloy containing 5% silicon, and used for light-alloy parts.

vitamins. A group of approximately fifteen organic nutrients, essential for growth and health, that the body cannot synthesize and must thus be obtained from the diet. They are either fat- or water-soluble, and are obtained from natural products, or made synthetically. Examples include vitamin A (retinol) and vitamin C (ascorbic acid). Vitamins can now be incorporated into biopolymers for *in vivo* slow-release applications.

Vitanol. Trade name for a nickel-titanium wire used for dental applications.

Vitapane. Trade name of Arvey Corporation (USA) for cellulose acetate plastics.

Vitapol. Trademark of Vitafoam (USA) for polyvinyl chloride foam and foam products.

Vitaprene. Trademark of Vitafoam (USA) for polyurethane foam and foam products.

Vitapruf. Trade name of British Vita plc (UK) for polymer-coated textile fabrics.

Vitata. Trade name for rayon fibers and yarns used for textile fabrics.

Vitathane. Trademark of Vitafoam (USA) for polyurethane foam and foam products.

Vitawrap. Trademark of Goodyear Tire & Rubber Company (USA) for plastic packaging film products.

Vitax. Trademark of Hitachi Chemical Company, Limited (Japan) for synthetic resins supplied as liquids, pastes, powders and pellets.

Vitec. Trademark for fire-retardant polyurethane foams.

Vitel. Trademark of Goodyear Tire & Rubber Company (USA) for a series of polyester resins supplied in various grades for a wide range of applications including extrusion into high-strength monofilaments and films, spinning into high-tenacity fibers for clothing, industrial fabrics and tire cord, hot-melt for adhesive resin manufacture, and extrusion coating for foils, paper, textiles and plastics.

vitellin. A phosphoprotein in egg yolk. Also known as *ovovitellin*.

Vitem. Trade name of Santa Lucia Cristal SACIF (Argentina) for a toughened sheet glass used for architecture.

VIT-High Carbon Coke. Trademark for a high-carbon, low-sulfur foundry coke.

Vitinox. Trade name of Creusot-Loire (France) for a series of free-machining austenitic stainless steels containing up to 0.12% carbon, 18% chromium, 8-10% nickel, up to 0.6% molybdenum, 0-3.5% copper, a trace of sulfur, and the balance iron. Used for free-machining parts.

Viton. (1) Trademark of DuPont Dow Elastomers (USA) for a series of white transparent, or black opaque fluorocarbon elastomers that are copolymers of vinylidene fluoride and hexafluoropropylene. Supplied in regular and low-temperature grades, they have a density range of 1.7-2.0 g/cm^3 (0.06-0.07 lb/in.^{3}), outstanding oil and chemical resistance, excellent resistance to extreme temperatures and compounded oils, good resistance to acids, greases and halogens, good to fair resistance to alcohols, alkalies and aromatic hydrocarbons, poor resistance to ketones and strong and concentrated alkalies, excellent resistance to ozone and ultraviolet radiation, poor resistance to gamma radiation, good mechanical properties, low thermal expansivity, a Durometer hardness of about A 75, a service temperature range of -50 to +316°C (-58 to +600°F), and good self-extinguishing properties. Used for aircraft and automotive parts, vacuum equipment, seals and gaskets, diaphragms, tub-

ing, industrial equipment, and low-temperature and radiation equipment. *Viton Extreme* is a specialty grade of *Viton* fluorocarbon elastomer with improved resistance to low-molecular-weight esters, ketones, aldehydes, fluids, caustics and amines. It is made with ethylene, tetrafluoroethylene and perfluoromethyl vinyl ether, has a maximum service temperature in air of 204°C (400°F), and is used for gaskets and transfer-compression molded seals.

(2) Trade name for acrylic resins including *Viton H* and *Viton M* used for dentures.

Vitox. Trademark of Morgan Bioceramics (UK) for alumina ceramics used in orthopedics for femoral heads.

Vitrablok. Trade name of Vitrablok a.s. (Czech Republic) for glass blocks.

Vitrabond. Trademark of 3M Company (USA) for light-cured glass-ionomer dental cements.

Vitracast. Trademark of CV Materials, Limited (USA) for porcelain enamel compositions used to produce protective coatings on continuously cast metals.

Vitral. Trade name of Aluminium Laufen AG (Switzerland) for a heat-treatable, corrosion-resistant aluminum alloy containing 0.4-0.8% magnesium and 0.3-0.7% silicon. It has good formability, and is used ornamental and architectural applications.

Vitralite. (1) Trade name of CertainTeed Products Corporation (USA) for glass-fiber insulating blankets.

(2) Trademark of Pratt & Lambert, Inc. (USA) for several exterior and interior enamel paints.

Vitralon. Trade name of Pratt & Lambert, Inc. (USA) for several chemical coatings, zinc-rich powders, and coating primers.

Vitrasilk. Czech trade name for double glazing units with glass-fiber inlays.

Vitrax. Czech trade name for window and figured rolled glass.

Vitrea. Czech trade name for thick sheet glass.

Vitreac. Trade name of Vitrios Planos Lirquen SA (Chile) for a 4.75 mm (0.187 in.) thick patterned glass.

Vitrebond. Trademark of 3M Dental (USA) for a light-cure, fluoride-releasing glass-ionomer cement with high bond strength, used in restorative dentistry for liner/base bonding applications.

Vitredil. Trade name of Montedison SA (Italy) for plexiglas-type acrylics.

Vitrelloy. Trade name of Howmet Corporation (USA) for an amorphous alloy composed of 61-63% zirconium, 11-12% titanium, 12-12.5% copper, 10-11% nickel, and 3-3.5% beryllium. It has high strength and fracture toughness, good resistance to permanent deformation, a relatively low density (about 6.1 g/cm^3 or 0.22 lb/in.^{3}), and a surface hardness of 50 Rockwell C. Used for aerospace parts, and golf-club head inserts.

Vitremer. Trademark of 3M Dental (USA) for a light or self-cure, fluoride-releasing glass-ionomer dental cement with good adhe-sion to tooth structure (dentin and enamel) and excellent physical properties, used for core-build-ups and restorations. *Vitremer Luting* is a resin-modified, fluoride-releasing glass-ionomer dental cement for luting restorations.

Vitrenamel. Trade name of US Steel Corporation (USA) for porcelain-enameling sheet steels used for household appliances.

Vitreosil. Trade name of Saint Gobain (France) for *fused quartz* composed of 99.8% silicon dioxide (SiO2) and supplied in transparent, translucent and opaque grades. It has a density of 2.2 g/cm^3 (0.08 lb/in.^{3}), good heat resistance and strength, low thermal expansivity, good thermal-shock resistance, good ultraviolet transmission, and high electrical resistance. Used for insulation applications, chemical and heat-treating equipment, and furnace parts.

vitreous antimony. See antimonial glass.

vitreous antimony sulfide. See antimonial glass.

vitreous china. Any dense, glazed or unglazed *whiteware* product with low water absorption, used in the manufacture of artware, dinnerware and sanitary ware.

vitreous carbon. See amorphous carbon.

vitreous clay pipe. A clay pipe that is first fired in a kiln to bring on vitrification and then glazed for water tightness. Used for drainage applications.

vitreous copper. See chalcocite.

vitreous enamel. See porcelain enamel.

vitreous material. See amorphous; noncrystalline material.

vitreous sanitary ware. Vitreous *whiteware* products, such as lavatories, sinks, toilet bowls or urinals, used for hygienic and sanitary applications.

vitreous selenium. See amorphous selenium.

vitreous silica. See fused silica.

vitreous silver. See argentite.

vitreous substance. A substance derived from or consisting of glass.

vitreous slip. A *slip coating* composed of a ceramic material or mixture of materials that will produce a glassy or vitrified surface when applied and fired on a ceramic body.

Vitresil. Trademark of American Fused Quartz (USA) for high-purity fused silica usually supplied as a woven fabric containing 99.8+% silica (SiO_2), and used as a filter medium.

Vitresist. Trade name of D. Brown Industries Limited (UK) for an austenitic stainless steel containing up to 0.06% carbon, 2% silicon, 23% chromium, 23% nickel, 5.5% molybdenum, 1.5% copper, 0.4% niobium, and the balance iron. It has high corrosion resistance, and good resistance to hot sulfuric acid. Used for chemical equipment, storage tanks and vessels, containers, paper and pulp handling equipment, and brewing equipment.

Vitrex. (1) Trademark of Atlas Minerals & Chemical, Inc. (USA) for an acidproof, chemical-hardening silicate cement. It has good resistance to most acids up to 1150°C (2100°F).

(2) Trade name of Manifattura Specchi e Vetri Felice Quentin (Italy) for a toughened plate glass and thick sheet glass.

(3) Trade name of Santa Lucia Cristal SACIF (Argentina) for a toughened automotive sheet glass.

Vitric. (1) Trade name for an enameled corrugated sheet steel used for structural applications.

(2) Trademark of Warren Paint & Color Company (USA) for several enamel paints.

vitrifiable color. A ceramic color made by mixing a metallic oxide with a glaze, and used as decoration.

vitrification clay. A clay that when heated to relatively high temperatures will tend to vitrify, but does not undergo deformation until the vitrification temperature is reached.

vitrified brick. A very hard brick that during firing has reached its vitrification temperature. Used for building and paving applications.

vitrified clay pipe. A strong, acid-, chemical- and heat-resistant pipe with a glassy surface finish made from crushed and blended clays by forming into a tubular shape, drying, and firing at suitable temperatures. Used for drainage applications.

vitriol. Any of various sulfates of metals, such as copper, iron or zinc. *Copper vitriol* ($CuSO_4 \cdot 5H_2O$) is also known as blue vitriol, *iron vitriol* ($FeSO_4 \cdot 7H_2O$) as green vitriol, and *zinc vitriol* ($FeSO_4 \cdot 7H_2O$) as white vitriol.

Vitrix. Trade name of Krupp Stahl AG (Germany) for an oil-hardening, shock-resisting steel containing 0.35% carbon, 1.3% chromium, 4.5% nickel, 0.6% manganese, and the balance iron. Used for gears, shafts, crankshafts, dies, and fasteners.

vitroceramics. See glass-ceramics.

Vitroclad. Trade name Plyglass Limited (UK) for colored glass infill panels.

Vitro-Coat. Trade name of Manville Corporation (USA) for fiberglass and resin compounds used for corrosion protection applications.

Vitrocor. Trade name of Vitro-Fibras SA (Mexico) for light, acoustic ceiling panels made of glass fibers.

Vitroducto. Trade name of Vitro-Fibras SA (Mexico) for a glass-fiber board used in the manufacture of air ducts.

Vitrofib. Trade name Fibras Minerales SA (Spain) for glass-fiber insulation products.

Vitro-Fibras. Trade name of Vitro-Fibras SA (Mexico) for textile glass fibers and glass insulating materials.

Vitrofil. Trade name of Vitrofil (Italy) for a continuous, twisted and plied glass-fiber yarn used in the textile industries.

Vitro-Flex. Trade name of Manville Corporation (USA) for fiberglass reinforcing mats used for polymer-matrix composites.

Vitroflex. Trade name of Manville Corporation (USA) for plain or tinted silvered sheet glass supplied in sections on cloth.

Vitroform. Trade name of Vitro-Fibras SA (Mexico) for glass-fiber roof insulation products.

Vitrolac. Trade name of RCA Records (USA) for polyvinyl chloride formerly used for phonographic records.

Vitrolain. Trademark of Star Porcelain Company (USA) for a high-strength molded electrical porcelain with very low porosity.

Vitrolite. Trade name of Pilkington Brothers Limited (UK) for a type of colored opaque glass with a hard, brilliant fire-finished surface.

Vitrolumen. Trade name of Cristaleria Espanola SA (Spain) for layers of glass fibers enclosing double-glazing units.

Vitrolux. Trade name of Libbey-Owens-Ford Company (USA) for heat-strengthened, 6.35 mm (0.25 in.) thick, polished plate or float glass having a vitreous color fired onto its back surface.

Vitron. Trade name of Manville Corporation (USA) for glass fibers, filament yarns and woven rovings.

Vitropanel. Trade name of Vitro-Fibras SA (Mexico) for glass-fiber boards covered with plastics.

Vitroperm. Trademark of Vacuumschmelze GmbH (Germany) for several permanent magnet alloys.

Vitrophon. Trademark of Fr. Xaver Bayer Isolierglasfabrik KG (Germany) for laminated safety glass with good acoustic properties supplied in various compositions.

vitropyr. A term sometimes used for *volcanic glass*.

Vitrosa-Alpha. Trade name of Vetrocoke SpA (Italy) for a high-efficiency glass insulator.

Vitroslab. Trade name of Plyglass Limited (UK) for a cladding panel composed of a double glazing unit having a colored glass fiber in its cavity and a backing of insulating material.

Vitrosmalt. Trade name of Fabbrica Pisana SpA (Italy) for a fire-enameled glass.

Vitrosil. Trade name of Montecatini (Italy) for reinforced phenol-formaldehyde plastics.

Vitro-Strand. Trade name of Manville Corporation (USA) for fiberglass used for glass and resin spray applications.

Vitrotayl. Trade name of Vitro-Fibras SA (Mexico) for glass-fiber acoustic ceiling panels.

Vitrotec. Trade name of Vitro-Fibras SA (Mexico) for glass-fiber roof insulation products.

Vitrotex. Trade name of Fibras Minerales SA (Spain) for glass fiber yarn used for textile fabrics.

Vitrotherm. Trademark of Fr. Xaver Bayer Isolierglasfabrik KG (Germany) for an acoustic, thermal and sun-reflecting insulating glass.

Vitrovac. Trade name of Vacuumschmelze GmbH (Germany) for a series of amorphous soft magnetic, non-magnetostrictive alloys composed chiefly of iron, nickel and cobalt with selected alloying additives, such as boron and silicon. Manufactured by rapid quenching from the molten state into thin strips about 25-50 μm (1-2 mils) thick and 1-50 mm (1-2 in.) wide, they have high permeability, saturation flux density and low core losses, and are used for inductive-component cores, magnetoelastic sensors, anti-theft devices, cable shielding, and magnetic springs and heads.

Vitrovyc. Trade name of Vidrio Plano de Mexico SA (Mexico) for a ceramic-coated cladding glass.

Vitrox. Trademark of W.R. Grace & Company (USA) for an incompletely calcined aluminum silicate (Al_2SiO_5) prepared from the mineral *cyanite* in fine and coarse grades, and used as *grog* in the manufacture of firebrick and pottery.

Vituf. Trademark of Goodyear Tire & Rubber Company (USA) for unsaturated polyester fibers and extrusion resins used for electrical wire jackets, building wire, missile ground cables, and aircraft, automotive and electronic hook-up wire.

vitusite. A white mineral composed of sodium cerium lanthanum neodymium phosphate, $Na_3(Ce,La,Nd)(PO_4)_2$. Crystal system, orthorhombic. Density, 3.70; refractive index, 1.6465. Occurrence: Greenland.

Viva. Trade name of Hicks, Bullick & Company Limited (UK) for high-tenacity, spun polyester staple sewing thread.

Vivaglass. Trademark of Ivoclar Vivadent AG (Liechtenstein) for a series of glass-ionomer dental cements including *Vivaglass Base*, an encapsulated cement used as a base filler, *Vivaglass Cem*, a luting cement, *Vivaglass Fil*, an encapsulated cement used as a filler, and *Vivaglass Liner*, a light-cure cement used as a lining material.

Vivak. Trade name of Sheffield Plastics, Inc. (USA) for extruded glycol-modified polyethylene terephthalate plastic sheeting.

Vival. Trade name of Compagnie Français des Métaux (France) for a wrought aluminum alloy containing 1% magnesium, 0.5% silicon and 0.5% manganese. Used for light-alloy parts.

Vivalloy HR. Trade name for a dental amalgam alloy.

Vivana. Trademark of Badische Corporation (USA) for vinylidene chloride polymers and interpolymers supplied in the form of fibers, filaments and yarns. Also included under this trademark are nylon fibers.

Vive la Crepe. Trade name for nylon fibers and yarns used for textile fabrics, especially clothing.

vivianite. A colorless, green, or grayish purple blue mineral composed of iron phosphate octahydrate, $Fe_3(PO_4)_2 \cdot 8H_2O$. It can also be made synthetically. Crystal system, monoclinic. Density, 2.69 g/cm³; refractive index, 1.63. Also known as *blue iron earth; blue ocher.*

Vivostar. Trade name for a medium-gold dental bonding alloy.

Vixir. Trade name of Resine SA (France) for polyvinyl chlorides.

Vizor. Trade name of Osborn Steels Limited (UK) for a water-hardening tool steel (AISI type W1) containing 0.8-1.2% carbon, and the balance iron. Supplied in four grades with different carbon content, and used for springs, drills, cutters, dies, and punches.

V-Kut. Trade name of Columbia Tool Steel Company (USA) for a water- or oil-hardening tool steel containing 0.71% carbon, 0.25% manganese, 0.28% silicon, 0.8% chromium, 0.2% vanadium, and the balance iron. Used for rock drills, shear blades, and stamps.

vladimirite. A pale rose mineral composed of calcium hydrogen arsenate pentahydrate, $Ca_5H_2(AsO_4)_4 \cdot 5H_2O$. Crystal system, monoclinic. Density, 3.14 g/cm³; refractive index, 1.655. Occurrence: Russia.

vlasorite. A colorless mineral composed of sodium zirconium silicate, $Na_2ZrSi_4O_{11}$. Crystal system, monoclinic. Density, 2.97-3.03 g/cm³; refractive index, 1.623. Occurrence: South Atlantic (Ascension Island).

Vlieseline. Trademark of Carl Freudenberg (Germany) for a fleecy mixture of various fibers with synthetic resins and rubber, used as a substitute for *buckram*, and for inlays in collars and cuffs on shirts, jackets, etc.

Voco. Trade name of Voco-Chemie GmbH (Germany) for a light-cure dental acrylic resin used for impression tray and baseplate applications.

Vodex. Trade name of Murex Limited (UK) for a plain-carbon steel containing 0.06% carbon, 0.4% manganese, and the balance iron. Used for welding electrodes.

VOD steel. See vacuum-oxygen decarburized steel.

Voelklingen. Trade name of Forges et Acieries de Voelklingen (France) for a series of steels including various highly corrosion-resistant austenitic, ferritic and martensitic stainless grades and several nondeforming, hot-work and plain-carbon tool grades. Also spelled "Völklingen."

voglite. An emerald-green to bright grass-green, radioactive mineral composed of calcium copper uranyl carbonate hexahydrate, $Ca_2Cu(UO_2)(CO_3)_4 \cdot 6H_2O$. Crystal system, monoclinic. Density, 3.06 g/cm³; refractive index, 1.547. Occurrence: Czech Republic, USA (Utah).

Vogt. Trade name for polyolefin fibers and products.

voile. A thin, light, sheer or semisheer, slightly crisp fabric, usually made of cotton, cotton-blend, silk, rayon or wool, in a plain weave. Used for blouses, light dresses, curtains, etc.

Voizit. German trade name for powder-metallurgy materials made by blending 96.5% iron powder with 3.5% graphite, pressing and sintering for a short period of time at about 1110-1130°C (2030-2065°F). They have an average porosity of 30-40%, and are used for antifriction metal and bearings.

Vokar. A Russian sintered alloy with high hardness containing 78-86% tungsten, 8-15% carbon, and the balance manganese and iron. Used for cutting tools.

Volara. Trademark of Voltek, Division of Sekisui America Corporation (USA) for a series of fine-celled, fire-retardant cross-linked polyolefin foam sheet materials that can easily be processed by profile cutting, die cutting, vacuum forming, and lamination. They possess low shrinkage, high tensile strength, good cushioning and insulating properties. Used for automotive and appliance gaskets, aircraft seating, air-conditioner insulation, laminating to fabrics or plastic films, and for automotive vinyl/foam composites, swimming pool covers, and double-coated mounting tape.

volatile corrosion inhibitor. A corrosion inhibitor that evaporates over time thus providing only temporary corrosion protection. It may be supplied in the form of a wrapping paper impregnated with a volatile, corrosion-inhibiting chemical, or as a spray or paint. Abbreviation: VCI.

volatile corrosion inhibitor paper. A corrosion inhibitor in the form of a wrapping paper impregnated with a volatile, corrosion-inhibiting chemical. Abbreviation: VCI paper.

volborthite. A green mineral composed of copper vanadium oxide trihydrate, $Cu_3(VO_4)_2 \cdot 3H_2O$. Crystal system, monoclinic. Density, 3.42 g/cm^3; refractive index, 2.02. Occurrence: USA (California).

volcanic ash. Ash and small pieces of lava thrown up by an erupting volcano. It is high in silica (SiO_2), and is used as a cementing ingredient in concrete, and as a fine abrasive and scouring agent. See also pumicite.

volcanic clay. Bentonite clay obtained from chemically altered, devitrified volcanic ash or *tuff*.

volcanic glass. A highly siliceous, natural glass, such as *obsidian*, formed by the rapid cooling of molten lava. Sometimes referred to as "vitropyr."

volcanic sand. Crushed and sized sand obtained from rock formed by the consolidation of molten lava. Used as an aggregate in the building trades.

Volcano. Trade name of Lehigh Steel Corporation (USA) for a high-speed steel containing 0.7% carbon, 4% chromium, 1% vanadium, 18% tungsten, 5% cobalt, and the balance iron. Used for high-temperature die work.

Volclay. Trademark of AMCOL International Corporation (USA) for *bentonite clay* used in the manufacture of water-impermeable barrier sheets, waterproofing membranes, and foundation moisture barrier panels.

Volco. Trade name of Volco Brass & Copper Company (USA) for an extensive series of brasses containing 63-95% copper, 5-35% zinc, and up to 14% nickel. Used for hardware, jewelry, screen wire, plumbing fixtures, fasteners, reflectors, cartridges, clips, flexible hose, stampings, and coins.

Vole. Trade name of Tufnol Limited (UK) for phenolic resin laminates reinforced with fine- or medium-weave cotton fabrics.

Voler. Trade name for special-purpose molybdenum disulfide solid lubricants.

Volextra. Trademark of Voltek, Division of Sekisui America Corporation (USA) for a coated polyolefin foam product.

Volkaril. Trademark of Plastimer (France) for modified polyvinyl chlorides.

Völklingen. See Voelklingen.

volkovskite. A colorless mineral composed of calcium borate dihydrate, $CaB_6O_9(OH)_2 \cdot 2H_2O$. Crystal system, monoclinic. Density, 2.30 g/cm^3; refractive index, 1.539. Occurrence: Russia.

Vollrip. Trademark of Rippenstreckmetall-Gesellschaft mbH (Germany) for black-lacquered or galvanized ribbed expanded metal.

Volomit. Trade name for a cemented carbide composed of tungsten carbide (WC) and molybdenum carbide (MoC), and tungsten alloys containing 4.5% carbon and 2% iron. Used for dies and cutting tools.

Voloy. Trade name of Comalloy International Corporation (USA) for a series of engineering thermoplastics including (i) several unreinforced and glass fiber-reinforced polypropylenes (some with high thermal resistance), (ii) several high-strength high-temperature glass-fiber- and/or mineral-reinforced polybutylene and polyethylene terephthalates, and (ii) various glass-fiber- and/or mineral-reinforced nylon 6 and nylon 6,6 resins. Depending on the particular type of plastic, they are used for automotive, electrical, electronic and/or industrial applications.

Volpex. Trade name for vinyon fibers used for textile fabrics.

voltaite. A greenish-black to black mineral composed of potassium iron sulfate hydrate, $K_2Fe_5(SO_4)_{12} \cdot 18H_2O$. Crystal system,

cubic. Density, 2.70-2.72 g/cm^3; hardness, 3-4 Mohs; refractive index, 1.600. Occurrence: USA (Arizona).

Voltal. Trade name of Busch-Jäger Lüdenscheider Metallwerke (Germany) for an age-hardenable aluminum alloy containing 4.7% copper, 2-4% silicon, 2.5% zinc, and up to 1.1% iron. Used for light-alloy parts.

Voltalef. Trade name of Atofina SA (France) for several polychlorotrifluoroethylene (PCTFE) polymers supplied as greases, oils and powders for electronic applications.

Voltek. Trademark of Voltek, Division of Sekisui Chemicals (Japan) for a lightweight, flexible, fine-celled poly-olefin foam insulating material supplied in the form of sheets, boards and tubes. It has a high mechanical strength, a service temperature range from subzero to 150°C (300°F), and good resistance to chemicals and the environment. Used for water pipe insulation, thermal beverage containers, and heat-reflective camp mats.

Volumit. Trade name for a hard, sintered alloy composed of tungsten carbide (WC) and molybdenum carbide (MoC), and used for cutting tools and dies.

Volunteer. Trade name of Old Hickory Clay Company (USA) for clays and clay products.

Volvic. Trade name of Aubert & Duval (France) for an oil-hardening hot-work tool steel containing 0.25% carbon, 3.1% chromium, 9% tungsten, 0.4% vanadium, and the balance iron. Used for rivet sets, extrusion press parts, and punches.

Volvit. Trade name of KM-Kabelmetall AG (Germany) for a bearing bronze composed of 91% copper and 9% tin.

volynskite. A mineral composed of silver bismuth telluride, AgBiTe$_2$. Crystal system, orthorhombic. Occurrence: Armenia.

Vonnel. Trademark of Mitsubishi Rayon Company Limited (Japan) for high-performance acrylic (polyacrylonitrile) staple fibers used for the manufacture of textile fabrics including clothing, carpets, and interior decorations.

vonsenite. A black mineral of the ludwigite group composed of iron oxide borate, $Fe_3(BO_3)O_2$. Crystal system, orthorhombic. Density, 4.77 g/cm^3. Occurrence: Spain, USA (California).

Voplex. Trademark of Voplex Corporation (USA) for polyolefin and vinyon (vinyl chloride) fibers used for textile fabrics.

Vorane. Trademark of Dow Chemical Company (USA) for a series of polyurethane chemicals and products including various coatings, elastomers and foams.

Voranol. Trademark of Dow Chemical Company (USA) for a series of polyurethanes.

Vos. Trade name of Lonza-Werke GmbH (Germany) for polymethyl methacrylate resins.

Vought. Trade name of Lockheed Martin/Vought Aircraft Industries (USA) for a series of high-temperature oxidation-resistant modified silicide coatings containing silicon, chromium and either boron or aluminum. They are deposited on molybdenum, niobium and their alloys by pack cementation.

vozhminite. A yellowish mineral with brown tint composed of nickel antimony arsenic sulfide, $Ni_4(As,Sb)S_2$. Crystal system, hexagonal. Density, 6.20 g/cm^3. Occurrence: Czech Republic, Russia.

vrbaite. A gray-black mineral composed of thallium mercury antimony arsenic sulfide, $Tl_4Hg_3Sb_2As_8S_{20}$. Crystal system, orthorhombic. Density, 5.30 g/cm^3. Occurrence: Greece.

V-Star 50. Trade name of Bethlehem Steel Corporation (USA) for a high-strength structural steel (ASTM A572 Gr. 50) with controlled carbon content and alloying additions of niobium and vanadium. It is available in various forms including bars, shapes and piling, and has enhanced strength and weldability. Used for a wide range of applications including shipbuilding, railroad cars, offshore drilling rigs, oilfield structures and power transmission towers.

V-Thane. Trade name of Hallam Polymers Engineering Limited (UK) for polyurethane materials and products.

V-Tool. Trade name of Textron Inc. (USA) for a water-hardening tool steel containing 0.9-1% carbon, 0.25-0.4% manganese, 0.2-0.3% vanadium, and the balance iron. Used for various cutting finishing tools including drills, broaches and reamers.

vuagnatite. A colorless to pale brown mineral of the descloizite group composed of calcium aluminum silicate hydroxide, $CaAl(SiO_4)(OH)$. Crystal system, orthorhombic. Density, 3.42 g/cm^3; refractive index, 1.724. Occurrence: USA (California).

Vue. Trade name of Pittsburgh Corning Corporation (USA) for non-light-directing transparent glass blocks with smooth and clear exterior and interior surfaces.

Vuelite. Trade name of Monsanto Chemical Corporation (USA) for a transparent cellulose acetate used for fluorescent light fixtures.

Vuepack. Trade name of Monsanto Chemical Corporation (USA) for cellulose acetate plastics used for lighting and other electrical applications.

Vulcalock. Trademark of B.F. Goodrich (USA) for a rubber-based adhesive used for bonding rubber to metal.

Vulcan. (1) Trade name of Phosphor Bronze Company Limited (UK) for a tin-base bearing metal used for thin walled linings, high-speed bearings, locomotive diesel engines and other high-duty applications.

(2) Trademark of Cabot Corporation (USA) for carbon black.

vulcanite. (1) A light bronze to yellow-bronze mineral composed of copper telluride, CuTe. It can also be made synthetically from the elements by heating at low pressure. Crystal system, orthorhombic. Density, 7.11 g/cm^3; refractive index, 1.488. Occurrence: USA (Colorado).

(2) See ebonite.

vulcanized fiber. A hard, tough, resilient, horn-like cellulosic material, usually gray, black, brown or red in color, usually made by impregnating wood or paper with a gelatinizing agent (e.g., sulfuric acid, cuprammonium, or a zinc chloride solution), compressing two or more layers, rolling to the desired thickness, removing the gelatinizing agent by leaching, and then drying. Vulcanized fiber has a density of 1.1-1.4 g/cm^3 (0.04-0.05 lb/in.^{3}), very high resistance to chemical action, good electrical properties and mechanical strength, good dimensional properties, moderate moisture resistance, and good resistance to oil, gasoline, benzene and alcohols. Used for electrical insulation, gaskets and seals, washers, handles, bushings, brake shoes, and facings for vises, gears, suitcases and knife handles. Abbreviation: VF.

vulcanized oil. A white or brown, soft, mealy substance prepared by reacting sulfur or sulfur chloride with a vegetable oil, such as castor, corn or rapeseed oil. It has a density of 1.04 g/cm^3 (0.038 lb/in.^{3}), and is used directly in the manufacture of erasers and soft rubber goods, and indirectly as a compounding ingredient in rubber, and as a rubber substitute. Also known as *factice*. See also black factice; brown factice; white factice.

vulcanized India rubber. India rubber that has been treated, usually by the introduction of sulfur and resulting polymer-chain crosslinking, to increase its resiliency, durability, and rupture and abrasion resistance. Abbreviation: VIR. See also India rubber.

vulcanized rubber. Polyisoprene rubber that has undergone a nonreversible chemical reaction in which the physical properties were changed either by (i) reacting with sulfur or other crosslinking agents (e.g., peroxides, metallic oxides, chlorinated quinones, or nitrobenzenes) usually with application of heat, or (ii) subjecting to high-energy radiation (e.g., gamma rays, or an electron beam). Crosslinking converts the rubber from a weak thermoplastic to a strong thermoset thereby enhancing its elastic modulus and yield properties. It has high strength, high electrical resistivity, good thermal-shock resistance, a service temperature range of -55 to +80°C (-67 to +176°F), a hardness of 35-65 Shore A, fair to poor heat resistance, good thermal shock resistance, poor abrasion resistance (unless modified with carbon black), excellent resistance to dilute and weak acids and bases, good resistance to strong and concentrated bases, fair ozone resistance, poor resistance to strong and concentrated acids, hydrocarbons, unsaturated fats and oils and ultraviolet radiation, and high gas permeability. Used for automotive tires, flexible tubing, hose, battery boxes, electric insulation, tank linings, conveyor-belt covers, footwear, foam rubber, and specialized mechanical rubber goods. Also known as *crosslinked rubber*. Abbreviation: VR. See also natural rubber; polyisoprene; synthetic rubber.

vulcan metal. A British corrosion-resistant copper alloy containing 11-12% aluminum, 4.4% iron, 1.5% nickel, 1% silicon, 0.4% tin, 0.7% chromium, and 0-0.25% zinc. Used for structural parts.

Vulcon. Trade name of Tecknit Europe Limited (UK) for a series of conductive elastomers.

Vulkaresen. Trade name of Reichhold Chemical Company (USA) for modified phenol-formaldehyde resins, compounds and laminates.

Vulkem. Trademark of Mameco International (USA) for elastic polyurethane compounds. They are tough, waterproof, bond well to many substrates, and are supplied in cartridges, pails and drums for use in sealing, caulking and/or waterproofing cargo container and truck trailer bodies, recreational vehicles, railcars, boats, log homes, cooling towers, and fasteners.

Vulkafix. Trademark of Rudol-Fabrik (Germany) for several rubber-based adhesives.

Vulkide. Trademark of ICI Limited (UK) for a series of polypropylene and acrylonitrile-butadiene-styrene resins.

Vulkollan. Trademark of Bayer Corporation (USA) for a series of hard-cast polyurethane elastomers that can be processed by compression or open-cast molding, spin casting, or by techniques resulting in low-density cellular foams. They have good wear resistance and shock absorption properties, a hardness range of 65-96 Shore, a service temperature range of -30 to +80°C (-22 to +176°F), and are used for diaphragms, seals, wipers, couplings, gears, rolls, and wheels.

Vulta Foam. Trade name of General Latex & Chemical Corporation (USA) for polyurethane foam.

Vultesol. Trademark of General Latex & Chemical Corporation (USA) for vinyl plastisols.

Vultex. Trademark of General Latex & Chemical Corporation (USA) for a natural rubber powder used in paste form for adhesives, cements and coatings.

vuonnemite. A yellow mineral composed of sodium titanium niobium phosphate silicate, $Na_{10}TiNb_2(PO_4)_2Si_4O_{17}$. Crystal system, triclinic. Density, 3.13 g/cm^3; refractive index, 1.651. Occurrence: Russia.

vuorelainenite. A brownish gray mineral of the spinel group composed of iron manganese chromium vanadium oxide, $(Mn,Fe)(V,Cr)_2O_4$. Crystal system, cubic. Density, 4.64 g/cm^3. Occurrence: Sweden.

VW Magnesium. British trade name for a magnesium casting alloy containing 8% aluminum, 0.6% zinc, 0.3% manganese and 0.005% beryllium. Used for light-alloy parts.

Vybak. Trademark of BIP Plastics (UK) for polyvinyl chlorides and other vinyl plastics.

Vybran. Trademark for acrylic fibers used for the manufacture of clothing.

Vycor. Trademark of Corning Inc. (USA) for a type of commercial glass composed of 96% silica (SiO_2) and 4% boric oxide (B_2O_3). It is made from borosilicate glass by a process that involves chemical leaching. *Vycor* has a density of 2.18 g/cm^3 (0.079 lb/in.^{3}), a refractive index of 1.458, exceptionally high thermal-shock resistance, an extremely low coefficient of expansion (approximately 0.8×10^{-6}/°C), a high softening point (1500°C or 2730°F), high transparency, a maximum service temperature of 900°C (1650°F), and good chemical resistance. Used for laboratory and industrial glassware, e.g., beakers, containers, crucibles, cylinders, dishes, flasks, rods and tubes.

Vycra. Trade name of Bayer AG (Germany) for polyurethane fibers and products.

Vycron. Trade name for polyester fibers.

Vydac. Trademark of The Se/pa/ra/tions Group (USA) for silicas used in chromatography and other separation technologies.

Vydont. Trade name for a vinyl ester formerly used for denture bases.

Vydran. Trademark for acrylic fibers used for the manufacture of clothing.

Vydyne. Trade name of Solutia Inc. (USA) for polyamide 6,6 (nylon 6,6) molding resins, available in regular, glass-fiber or carbon-fiber reinforced, mineral-filled, fire-retardant, UV-stabilized and high-impact grades, and as blends of nylon 6,6 and nylon 6. They have excellent corrosion and impact resistance, excellent resistance to chipping and denting, and good load-bearing capacity. Used for electronic components, and automotive and appliance parts.

Vyes. Trademark of Union Carbide Corporation (USA) for a series of solution vinyl resins employed as modifiers to improve the performance properties of alkyd, epoxy, polyester and urethane resins in high-solids coatings.

VY-Flex. Trade name of Anchor Packing Company (USA) for asbestos and non-asbestos spiral and ring packing products.

Vyflex. (1) Trade name of Lavergne Performance Compound Division (USA) for a series of thermoplastic olefins (TPOs) with excellent flexural moduli and impact resistance.

(2) Trade name for rigid unplasticized polyvinyl chloride supplied in sheet form for the manufacture of ducts and chemical tanks.

(3) Trade name of Plastic Coatings Limited (UK) for a series of flexible vinyl coatings.

VyGuard. Trademark of Ameron International Corporation (US) for a series of maintenance coatings.

Vylene. Trade name of Lavergne Performance Compound Division (USA) for a series of filled polypropylenes supplied in injection molding, extrusion and other grades.

Vylon. Trade name of Lavergne Performance Compound Division (USA) for a series of polyamide 6 and 6,6 (nylon 6 and 6,6) resins supplied in a wide range of grades.

Vylor. Trade name for nylon fibers and yarns.

Vynafoam. Trade name for vinyl plastosols used to produce foamed structures by spraying. Used for gaskets, seals, and as thermal insulation for refrigerator doors.

Vyncolite Phenolic. Trade name of Vyncolite (USA) for a series of reinforced or filled phenolics supplied in wood-filled general-purpose, cotton-filled medium-shock, mineral-filled high-heat, mica-filled electrical, and glass fiber-reinforced high-impact grades.

Vynylal. Trade name for a vinal fiber with excellent resistance to water, fungi, mildew and many chemicals, used for textile fabrics including garments.

Vyon. Trade name of Povair (USA) for porous polyethylenes and filtering foams.

Vyoron. Trade name for polyester fibers.

Vypet. Trade name of Lavergne Performance Compound Division (USA) for a series of glass-reinforced polyethylene terephthalate resins and compounds.

Vypro. Trade name of Lavergne Performance Compound Division (USA) for post-consumer polypropylenes.

Vyprene. Trade name of Lavergne Performance Compound Division (USA) for a series of thermoplastic elastomers supplied in a wide range of grades.

Vyram. Trademark of Monsanto Chemical Company (USA) for a thermoplastic elastomer alloy of polypropylene and natural rubber. It has excellent wear resistance, good flexibility, high resistance to compression set, and is used for automotive applications.

Vyrene. Trademark for a *spandex* fiber with excellent resistance to abrasion, perspiration and body oils, used for the manufacture of textile fabrics with excellent stretch and recovery properties.

vysotskite. A mineral composed of nickel palladium sulfide, $(Pd,Ni)S$. It can also be made synthetically. Crystal system, tetragonal. Density, 8.40 g/cm^3. Occurrence: Russian Federation.

Vyteen. Trade name of Lavergne Performance Compound Division (USA) for polycarbonate resins.

vyuntspakhkite. A colorless mineral composed of yttrium aluminum silicate hydroxide, $Y_4Al_3Si_5O_{18}(OH)_5$. Crystal system, monoclinic. Density, 4.02 g/cm^3; refractive index, 1.692. Occurrence: Russian Federation.

W

Wabcoloy. Trade name of Westinghouse Air Brake Company (USA) for a wear-resistant cast iron containing 2.6% carbon, 2.5% silicon, 0.6% manganese, 1.1% nickel, 0.9% molybdenum, and the balance iron. Used for crankshafts, and compressor cylinders.

Wabenmuster. Trade name of former Glas- und Spiegel-Manufaktur AG (Germany) for glass products with honeycomb designs known internationally under the trade name "Honeycomb."

Wabik Metal. Trade name of Empire Sheet & Tin Plate Company (USA) for a magnetically soft alloy composed of 96% iron and 4% silicon. It has high saturation induction, and is used for armatures and motors.

Wabolit. German trade name for a patterned glass with honeycomb design.

Wacker. Trade name of Wacker-Chemie GmbH (Germany) for an extensive series of silicone products including polymers, resins, standard, high-temperature vulcanizing (HTV) and room-temperature vulcanizing (RTV) rubbers, oils, defoamers, parting agents and silanes. The standard silicone rubbers are used for electrical insulation, contacts in electronic devices, and breathing apparatuses, the HTV silicone rubbers for flame-resistant subway tunnel cables and moisture- and heat-resistant automotive ignition cables and spark-plug connecters. The RTV silicone rubbers are used for heat-resistant gaskets and elastic seals, gaskets.

wadeite. A colorless mineral composed of potassium zirconium silicate, $K_2ZrSi_3O_9$, that occurs naturally and can also made synthetically by heating zirconium dioxide (ZrO_2), potassium carbonate (K_2CO_3) and quartz (SiO_2) under prescribed conditions. Crystal system, hexagonal. Density, 3.10 g/cm^3; refractive index, 1.625. Occurrence: Western Australia.

wadding. A soft sheet consisting fibers, especially carded cotton, loosely held together, and used for padding, packing, stuffing, upholstering, etc.

Wafel. Trade name of Glasindustrie Pietermann BV (Netherlands) for a waffle-patterned glass.

wafer. A large, thin, flat, disk-shaped slice of a crystal composed of high-purity semiconductor material (e.g., silicon) on which hundreds of individual circuit chips (integrated circuits) are manufactured. After the circuitry has been applied, the wafer is cut into pieces to make individual chips.

waferboard. A construction panel made by heat compression of long, thin wood chips or flakes (usually from a soft-grain hardwood, such as aspen) of equal or unequal size and thickness, and bonding with a waterproof adhesive, e.g., a phenolic resin. Both sides of the panel have the same textured surface, and the standard panel size is 1.2 × 2.4 m (4 × 8 ft.) and the thickness ranges from 6.4 to 19 mm (0.25 to 0.75 in.). The panels can be stained, painted or finished in natural fashion, and are used as structural panels, as wall panels, and for doors and door paneling, fencing, outdoor sheds and form structures, and decorative inside projects. Also known as *waferwood.*

waferwood. See waferboard.

Waffle. Trade name of Libbey-Owens-Ford Company (USA) for a patterned glass.

waffle. A knit or woven fabric with a texture or pattern resembling a honeycomb.

Waffle-Crete. Trade name of Advance Precast Limited (Canada) for precast concrete wall panels with waffle patterns.

waffle film. A low- or high-density polyethylene film that has a waffle pattern (diamond or lozenge) on one or both surfaces, and is supplied in clear and colored grades in thicknesses from 15 to 250 μm (0.0006 to 0.01 in.) for medical, pharmaceutical and surface-protection applications.

Wagner & Guhr's aluminum. A soft solder composed of 80% tin and 20% zinc, and used for soldering aluminum and its alloys.

wagnerite. A yellow, grayish, red or brick-brown mineral of the triplite group that is composed of magnesium fluoride phosphate, $Mg_2(PO_4)F$, and may also contain iron, manganese and calcium. Crystal system, monoclinic. Density, 3.45-3.47 g/cm^3; refractive index, 1.615. Occurrence: Sweden.

Wagner's alloy. A tin-base bearing alloy containing 10% antimony, 3% zinc, 1% copper and 0.8% bismuth.

Wagner's formula. A corrosion-resistant alloy composed of 50-66% copper, 19-31% zinc and 13-18% nickel, and used as a base for plated tableware.

wahoo. See winged elm.

Wai-Met. Trade name of Wai-Met Alloys Company (USA) for heat-resistant cobalt-chromium-iron alloys used for furnace equipment.

wairakite. A colorless to white mineral of the analcite group composed of calcium aluminum silicate dihydrate, $CaAl_2(SiO_3)_4 \cdot 2H_2O$. It can also be made synthetically. Crystal system, monoclinic. Density, 2.26 g/cm^3. Occurrence: New Zealand. Used as a zeolite.

Wakaba. Trade name of Nippon Sheet Glass Company Limited (Japan) for a patterned glass.

wakabayashilite. A golden-yellow mineral composed of antimony arsenic sulfide, $(As,Sb)_{11}Si_{18}$. Crystal system, hexagonal. Density, 3.98 g/cm^3. Occurrence: USA (Nevada).

Wakefield. Trade name of Wakefield Corporation (USA) for a series of copper- and iron-base powders used in powder-metallurgy.

wakefieldite. (1) An anthracite-black natural mineral composed of cerium lead vanadium oxide, $(Ce,Pb)VO_4$. Crystal system, tetragonal. Density, 5.30 g/cm^3; refractive index, above 2.0. Occurrence: Zaire.

(2) A yellow-white synthetic mineral of the zircon group composed of yttrium vanadium oxide, YVO_4. Crystal system, tetragonal. Density, 4.25 g/cm^3; refractive index, 1.98.

WAKOL. Trade name of Wakol-Chemie GmbH (Austria) for an extensive series of construction and industrial adhesives, several adhesives for the manufacture of paperboard and shoes, and various drywall and spackling compounds.

Wal-Cast. Trade name of Wahl Refractories, Inc. (USA) for castable refractories.

W-Al-Co. Trade name of Welding Alloy Manufacturing Company (USA) for a series of welding alloys composed of 95% aluminum and 5% silicon, and used for arc-welding electrodes and gas-welding rods for aluminum and its alloys.

Walcoloy. Trade name of Wall Colmonoy Corporation (USA) for stainless wire stock used for the production of thermal-spray coatings.

Walco Metal. Trade name of NL Industries (USA) for lead alloys used for tank linings and chromium plating anodes.

Walkerite. Trade name for a phenol-formaldehyde resin formerly used for denture bases.

wallaby leather. The strong, flexible, durable leather obtained from the skin of the Australian wallaby, a small to medium-sized member of the kangaroo family. Used for shoes and gloves.

wallboard. A term referring to asbestos cement, gypsum, laminated plastics, wood pulp or other materials made into large rigid sheets or panels, and used in the finishing of interior walls and ceilings. Examples include plasterboard, particleboard, plywood, hardboard and gypsum board.

Wallcoat. Trade name of Haviland Products Company (USA) for paint-booth coatings.

Wallex. Trade name of Wall Colmonoy Corporation (USA) for a series of cobalt-chromium-tungsten, cobalt-chromium-nickel-tungsten and tungsten-cobalt-chromium-nickel castings, powders and electrodes used for hardfacing applications. Also included under this trade name are high-carbon high-chromium steels containing some molybdenum and manganese. *Wallex* products have good corrosion, impact and heat resistance and low coefficients of friction. The cobalt-bearing grades may have tungsten carbide particles added.

wallisite. A mineral composed of copper thallium lead arsenic sulfide, $CuPbTlAs_2S_5$. Crystal system, triclinic. Density, 5.71 g/cm^3. Occurrence: Switzerland.

wallkilldellite. A dark red mineral composed of calcium manganese arsenate hydroxide hydrate, $Ca_4Mn_6As_4O_{16}(OH)_8 \cdot 18H_2O$. Crystal system, hexagonal. Density, 2.85 g/cm^3; refractive index, 1.728. Occurrence: USA (New Jersey).

Wallkyd. Trade name of Reichhold Chemicals, Inc. (USA) for non-penetration alkyd resins used in flat wall paints.

Wallmate. Trademark of Dow Chemical Company (USA) for moisture-resistant, rigid *Styrofoam* polystyrene insulation that has a high R-value (heat flow resistance) and is supplied in slotted, 0.6 × 2.4 m (2 × 8 ft.) sheets for the insulation of basement walls and crawlspaces.

W-Alloy. Trade name for a hard aluminum alloy containing 12% copper, 4.5% zinc, and 1% tungsten. Used for leakproof, light-alloy castings.

Wallpaper Plus. Trade name of VDM Technology Corporation (USA) for a composite consisting of a sheet of nickel-chromium-molybdenum alloy *C-276* clad to carbon steel plate by a selective explosion bonding process.

Wallpol. Trade name of Reichhold Chemicals, Inc. (USA) for vinyl acetate homopolymer emulsions used in interior wall paints. *Wallpol CPS* refers vinyl acetate and vinyl acrylic copolymers used as binders and film formers in paints and coatings.

wall tile. (1) A thin, flat, glazed tile used predominantly as the exposed surface in the construction of interior building walls.

(2) A tile consisting of a block of fired clay or a hollow shape of concrete, and used in wall construction.

Walmang. Trade name of Wall Colmonoy Corporation (USA) for an austenitic manganese steel containing 0.75% carbon, 14% manganese, 0.6% silicon, 0-1% molybdenum, 0.3.5% nickel, and the balance iron. Used in the form of hardfacing electrodes

for the repair of high-manganese alloy parts.

walnut. The hard, heavy, strong wood from any of a genus *(Juglans)* of hardwood trees and including especially the American walnut *(J. nigra)*, the butternut *(J. cinerea)*, the European walnut *(J. regia)* and the tropical walnut *(J. neotropica* and *J. olanchana)*. It is valued for making furniture, cabinets and veneer. See also American walnut; butternut; European walnut; African walnut; East India walnut; Queensland walnut.

walnut shell media. Brown- to tan-colored crushed shells of walnuts with a density of about 1.3-1.4 g/cm^3 (0.04-0.05 $lb/in.^3$) and a hardness 1.0-4.0 Mohs, available in coarse, medium and fine grades with mesh sizes typically ranging from 60 to 325. They are used for abrasive blast cleaning, removal of rust, scale, paints, lacquers, carbon deposits, etc., and for deflashing molded plastic parts.

walpurgite. A colorless or wax- to straw-yellow, radioactive mineral composed of bismuth uranyl oxide arsenate trihydrate, $Bi_4(UO_2)(AsO_4)_2O_4 \cdot 3H_2O$. Crystal system, triclinic. Density, 5.95 g/cm^3; refractive index, 1.975. Occurrence: Germany.

Walramite. Trade name for a sintered tungsten carbide used for cutting tools.

walrus leather. Tough, tanned, split or unsplit leather with an attractive natural grain obtained from the thick skins of walruses, and used for ornamental applications, buffing wheels, and traveling goods, e.g., bags and suitcases. Also known as *walrus*. See also sealskin leather.

walstromite. A colorless mineral composed of barium calcium silicate, $BaCa_2Si_3O_9$. Crystal system, triclinic. Density, 3.67 g/cm^3; refractive index, 1.684. Occurrence: USA (California).

Wal-Super Bond. Trade name of Wahl Refractories, Inc. (USA) for high-temperature refractory cements.

Walter pine. See spruce pine.

Wamato metal. A tin-base bearing alloy containing 4.5% copper, 4.5% antimony, 1% nickel, 1% lead and 0-1% cadmium. Used for the main bearings of internal combustion engines.

WAMBRAZE. Williams Advanced Materials (US) for an extensive range of precious and nonprecious high-temperature brazing alloys supplied in a wide range of forms and shapes, such as ribbons, rods, wires, preforms and powders. The compositions available include pure copper, nickel, palladium, platinum and silver, gold-copper and copper-gold alloys, gold-copper-tin, gold-palladium, gold-indium, gold-germanium, gold-tin and gold-nickel alloys, various silver- and palladium-base alloys, and many more. Used for general and microelectronic applications.

WAM Solder. Trade name of Williams Advanced Materials (US) for an extensive series of precious and nonprecious solder alloys supplied in a wide range of forms and shapes, such as ribbons, rods, wires and preforms. The compositions available include pure lead, tin and indium, lead-tin and tin-lead alloys, lead-antimony alloys, lead-silver and lead-tin-silver alloys, lead-indium and lead-indium-silver alloys, tin-gold, tin-silver, tin-antimony, gold-germanium and bismuth-lead-tin alloys, and many more. Most of them are suitable for general and microelectronic applications.

Wando. Trade name of Cytemp Specialty Steel (USA) for an oil-hardening cold-work tool steel (AISI type O1) containing 0.95% carbon, 1.2% manganese, 0.5% chromium, 0.5% tungsten, 0.2% vanadium, and the balance iron. It has good deep-hardening and nondeforming properties and machinability, and is used for punches, shear blades, saws, gages, dies, and various other tools.

Wanlin. Trade name of Hanson Desimpel (Belgium) for building bricks used for facing and lining applications.

Waploc. Trade name of VAW-Vereinigte Aluminium-Werke AG (Germany) for *oxide ceramics* of varying porosity exhibiting excellent thermal-shock resistance and good machinability and dimensional stability. They are made in various shapes and sizes, and used for cutting tools and high-temperature applications.

Wapresta. Trade name of Vereinigte Edelstahlwerke (Germany) for an oil-hardening, shock-resistant tool steel containing 0.3% carbon, 3.75% tungsten, 1.1% chromium, 0.18% vanadium, and the balance iron. Used for pneumatic tools, chisels, punches, and dies.

Waraloy. Trade name of Hewitt Metals Corporation (USA) for a series of corrosion-resistant solders composed of lead, tin and silver.

War Babbitt. Trade name of Duquesne Smelting Corporation (USA) for a tin babbitt containing varying amounts of lead and copper, and used for bearings.

Warcosine. Trademark of Warwick Wax Company Inc. (USA) for a white, microcrystalline petroleum wax.

wardite. A colorless, white, yellow, pale green or blue-green mineral composed of sodium aluminum phosphate hydroxide dihydrate, $NaAl_3(PO_4)_2(OH)_4 \cdot 2H_2O$. Crystal system, tetragonal. Density, 2.76-2.81 g/cm^3; refractive index, 1.59. Occurrence: USA (New Hampshire, Utah), Canada (Yukon).

wardsmithite. A mineral composed of calcium magnesium borate hydrate, $Ca_5MgB_{24}O_{42} \cdot 30H_2O$. Crystal system, tetragonal. Density, 1.88 g/cm^3. Occurrence: USA (California).

Warerite. Trade name of Warerite Limited (UK) for phenol-, melamine- and urea-formaldehyde plastics.

warikahnite. A pale yellow to colorless mineral composed of zinc arsenate dihydrate, $Zn_3(AsO_4)_2 \cdot 2H_2O$. Crystal system, triclinic. Density, 4.24 g/cm^3; refractive index, 1.753. Occurrence: Namibia.

Warman. Trade name of Warman Steel Casting Company (USA) for a series of corrosion- and/or heat-resistant low-carbon high-chromium steels.

Warm Bronze. Trade name of Waterbury Rolling Mills Inc. (USA) for a corrosion-resistant bronze containing 88% copper, 10% zinc and 2% tin. Used for hardware, and plumbing applications.

Warmpresstahl. Trade name of Stahlwerke C. Kabel (Germany) for hot-work tool steels used for dies, rams, and liners.

warm-setting adhesive. See intermediate-temperature-setting adhesive.

Warne's metal. A corrosion-resistant white-metal alloy composed of 37% tin, 26% nickel, 26% bismuth and 11% cobalt, and used for jewelry, and as a replacement for silver in ornaments.

warp-faced fabrics. Woven fabrics whose face is formed by warp yarns, and thus have more warp threads (or ends) than weft threads (or picks) on it.

warp-knit fabrics. Flat, smooth, run-resistant fabrics, such as *atlas* or *tricot*, made by interlocking loops in a longitudinal direction. Also known as *warp knits*.

warp knits. See warp-knit fabrics.

Warplis Drill Rod. Trade name of Teledyne Pittsburgh Tool Steel (USA) for an oil-hardening tool steel (AISI type O1) containing 0.9% carbon, 0.5% chromium, 1.1% manganese, 0.5% tungsten, 0.15% vanadium, and the balance iron. Used for tools, dies, jigs, and gages.

warp yarn. A yarn used for the warps (longitudinal threads) of a woven fabric.

warringtonite. See broachantite.

warwickite. (1) A dark brown to dull black mineral of the ludwigite group composed of magnesium iron titanium oxide borate, $(Mg,Fe)_3Ti(BO_3)_2O_2$. Crystal system, orthorhombic. Density, 3.35 g/cm^3; hardness, 3-4 Mohs; refractive index, 1.809. Occurrence: USA (New York).

(2) A yellowish white mineral of the ludwigite group composed of magnesium titanium oxide borate, $Mg_3Ti(BO_3)_2O_2$. Crystal system, orthorhombic. Density, 3.30 g/cm^3. Occurrence: Siberia.

wash. A coating composed of an emulsion or suspension of refractory fillers (e.g., chromite, silica, or zircon flour) and a suspension agent (e.g., bentonite, or sodium alginate) in alcohol or water. It is applied to mold cavities to facilitate the release of a casting or ware after forming.

Washboard. Trade name of SA Glaverbel (Belgium) for a patterned glass with ribbed-type design.

Washconite. Trade name of Washington Iron Works, Inc. (USA) for an alloy cast iron containing 2.9% total carbon, 1.6% silicon, 1.5% nickel, 0.25% chromium, and the balance iron. Used for gas- and diesel-engine cylinders, liners and pistons.

Washcote. Trade name of Harnischfeger Corporation (USA) for a carbon steel containing 0.2% carbon, and the balance iron. Used for general-purpose welding electrodes.

washed clay. Clay with low silica and grit content purified by stirring into water to form a thin slurry and then allowing the impurities to be removed by setting.

washer brass. A type of yellow brass composed of 62% copper and 38% zinc, and used for hardware, fasteners and washers.

Washi. Trade name of Toppan Printing Company Limited (Japan) for a handmade paper.

washing soda. See sodium carbonate decahydrate.

Washington. Trade name of Jessop Steel Company (USA) for a series of water-hardening tool steels (AISI type W1) containing 0.6-1.4% carbon, varying amounts of vanadium, chromium, manganese and silicon, and the balance iron. They have excellent machinability, good to moderate deep-hardening properties, wear resistance and toughness, low hot hardness, and are used for tools, cutters, and dies.

Washington Penn. Trade name of Washington Penn Plastic Company, Inc. (USA) for a series of polypropylene resins available in unfilled and glass fiber-, talc-, mica- or calcium carbonate-filled grades.

Wasit. Trade name of Wanfrieder Schmirgelwerk (Germany) for high-performance abrasive papers and fabrics for hand and machine sanding and grinding applications in the woodworking and metal industries.

Waspaloy. Trademark of United Technologies Corporation (USA) for a series of nickel-base superalloys containing 19-20% chromium, 13-14% cobalt, 4-4.3% molybdenum, 3% titanium, up to 2% iron, up to 0.7% manganese, 1-1.5% aluminum, 0.1% copper, up to 0.08% carbon, 0.05% zirconium and 0.006% boron. Available in cast and wrought form, they have face-centered cubic (fcc) matrix structures strengthened by intermetallic compound precipitations. They have a density of 8.18 g/cm^3 (0.296 lb/in.^{3}) and a melting range of 1340-1390°C (2445-2535°F). Other important properties include excellent heat resistance, high electrical resistivity, low coefficient of thermal expansion, high tensile strength, and high stress-rupture strength up to 760°C (1400°F). Used for rotating gas-turbine engine parts, high-temperature bolts, and welding wire.

waste paper. A recycled or used paper that is re-used as feedstock in the manufacture of paper and paper products.

watch alloy. A group of corrosion-resistant alloys used for watch cases. The following compositions are available: (i) 50% copper, 47.2% nickel and 2.8% cadmium; (ii) 70% palladium, 25% copper, 4% silver and 1% nickel; and (iii) 37.5% gold, 27% copper, 23% silver and 12.5% palladium.

watch case bezel. A corrosion-resistant alloy composed of 60-63% copper, 21-24% zinc and 16% nickel, and used for watch-case bezels.

watch case metal. A corrosion-resistant alloy composed of 55-65% copper, 16-30% zinc and 10-28% nickel, and used for watchcases.

Watchguard. Trade name of PPG Industries, Inc. (USA) for a laminated glass composed of a 1.5 mm (0.06 in.) thick vinyl interlayer between two pieces of 3.2 mm (0.125 in.) thick polished plate glass.

watchmaker's alloy. A free-cutting alloy composed of 59% copper, 40% zinc and 1% lead, and used for watch parts.

Watco. Trade name of Rust-Oleum Corporation (USA) for a series of wood finishes including satin waxes and Danish oils.

water ash. The strong wood from the small tree *Fraxinus caroliniana*. It is similar in properties to *pumpkin ash*. Source: Swamps of the southern United States from Virginia and Carolina to Florida and along the Gulf coast to Texas. Used for furniture, handles, baskets and crates. Also known as *Carolina ash*.

water-base adhesives. A group of adhesives composed of natural (animal or vegetable) materials, such as hide, fish bones, casein or starch, or synthetic materials, such as polyvinyl alcohol or vinyl acetate copolymers, dissolved or dispersed in water. Often supplied on porous, or nonporous substrates, they have low flammability and toxicity, good solvent resistance, long shelf lives, poor water and low-temperature resistance, poor creep resistance and strength under load, fair to poor heat resistance and poor electrical properties. Used for bonding paper, paperboard, wood, glass, porcelain, plastics, metal foil, leather and fabrics, and for packaging applications, e.g., labeling.

water-base paint. An *organic coating* whose thinner is essentially water. It may be of the (i) water-soluble type containing low-molecular-weight resins dissolved in water, (ii) emulsion type made by emulsifying the film former or binder (usually a latex, oil or resin) in water, or (iii) dispersion type composed of very small, finely divided particles of synthetic resin suspended or dispersed in water. Also known as *water-borne paint; water-thinned paint.*

water-borne paint. See water-base paint.

Waterbury. Trade name of Waterbury Rolling Mills Inc. (USA) for a series of nickel silvers and phosphor bronzes. The nickel silvers are composed of about 55-66% copper, 5-18% nickel, 17-31.5% zinc and 0-2% lead, and have good to excellent corrosion resistance; lead-bearing nickel silvers are also free-cutting. They are used for ornaments, cutlery, hardware, and screw-machine products. The phosphor bronzes contain about 3.5-11% tin, 0.03-0.35% phosphorus, and the balance copper. They have good wear resistance, and are used for springs, clips, and diaphragms.

Watercem. Trade name of TISSI Dental (Italy) for a glass-ionomer cement used in restorative dentistry for luting applications.

water cement. A cement, such as *hydraulic cement*, that will set to a hard product by chemical reaction with water. It is also capable of setting and hardening under water.

watercolor paper. A good-grade of drawing paper often with a rag content of 100% and having a surface texture capable of taking watercolors and withstanding severe scraping action.

Watercrat. Trade name of Marshall Steel Company (USA) for a tool steel containing 1.05% carbon, 0.5% chromium, 0.35% manganese, 0.2% silicon, and the balance iron. Supplied in the form of precision-ground flat stock, it is used for tools, dies, punches, and shear blades.

Waterdie. Trade name of Columbia Tool Steel Company (USA) for water-hardening tool steels (AISI type W). *Waterdie Extra* (AISI type W5-2) contains 1% carbon, 0.25% silicon, 0.5% chromium, 0.35% molybdenum, and the balance iron. It has good wear and deep-hardening properties, excellent machinability, fair toughness, and is used for dies, rolls, drill bushings, and lathe centers. *Waterdie Standard* (AISI type W5-3) contains 1% carbon, 0.5% chromium, 0.35% molybdenum, 0.25% silicon, and the balance iron. Its properties are similar to those of Waterdie Extra, but it has somewhat lower toughness, and is used for automotive tools, wear plates, etc.

watered fabrics. See moiré fabrics.

water-extended polyester. A polyester resin that contains a considerable amount of water and is used as a casting formulation. See also polyester resins.

water-filled plastics. Thermoplastics made by mixing a resin powder with water and a suitable catalyst and pouring into a mold. They cure to a hard, rigid, closed-cell material with a water content of about 50-80%. Used in the manufacture of ornamental articles, lamp stands, statuaries, and models.

water glass. See potash water glass; soda water glass.

water-ground mica. A finely ground *mica* powder of a particle size that 90% of the particles pass the 325-mesh (45-µm) US standard screen. Used as rubber and paint filler.

water-hardening tool steels. A group of tool steels (AISI group W) that contain carbon as the principal alloying element (usually 0.60-1.50%) and small amounts of other elements including manganese and silicon (0.10-0.40%), chromium (0.15-0.60%), nickel (0.20%), molybdenum (0.10%), tungsten (0.15%) and vanadium (0.1-0.3%), with the balance being iron. They are often sold under trade names or trademarks, such as Alva Extra, Coldie, Comet, Crow, Pompton, Python, Red Label or Titan, and have excellent resistance to decarburization, high toughness, good machinability, good resistance to softening at elevated temperatures, low to medium wear resistance, and medium resistance to cracking. Used for chisels, screwdriver blades, cold punches, nail sets, vise jaws, anvil faces, chuck jaws, cutting tools, such as milling cutters, reamers, taps, threading dies, wood augers or planer tools, and die parts.

water hickory. The hard, heavy wood of the hardwood tree *Carya aquatica* belonging to the pecan hickory family. Owing to its large shrinkage, it is considered inferior to true hickories whose wood it resembles. The heartwood is reddish brown with some dark streaks. Source: Coastal Plain rivers and swamps of the southern United States from North Carolina to Florida, along the Gulf coast to Texas, and along the Mississippi River valley. Used for tool and implement handles, flooring, pallets, veneer, and furniture. See also hickory; pecan.

water lime. See hydraulic lime.

Waterlite. Trade name of Watertown Manufacturing Company (USA) for polystyrenes.

watermarked paper. A fine writing or printing paper having a faint mark produced by the pressure of a projecting pattern or design during manufacture. The watermark can be clearly seen by holding the paper to the light. Basis weights run from about 98 to 171 g/m² (0.02 to 0.035 lb/ft²).

water paint. A paint whose vehicle is dissolved in water, e.g., a casein paint.

waterproof cement. Cement containing a hydrophobic (water-repelling) admixture, such as calcium stearate, or having an impervious surface coating. Also known as *waterproofed cement*

waterproof concrete. Concrete containing a waterproofing admixture or having an impervious surface coating applied to decrease permeability.

waterproofed cement. See waterproof cement.

waterproof fabrics. Textile fabrics made fully resistant to water penetration by coating, laminating or otherwise treating with an oil, elastomer, synthetic resin or other suitable chemical. See also water-repellent fabrics.

waterproofing agent. A chemical substance used to render a material, such as cement, masonry or textiles, impervious to water penetration. It can be incorporated as an integral ingredient (waterproofing compound), or applied as a surface coating or film.

waterproofing compound. See waterproofing agent.

waterproof paper. Paper made waterproof by treating with a copper-ammonium solution, coating with a rubber latex, or incorporating one or more synthetic resins (e.g., an acrylic, polyvinyl acetate and/or styrene-butadiene) during manufacture. Used as a dampproofing in building and road construction.

water putty. Any of several powder compounds that when mixed with water produce a putty suitable for filling cracks and holes.

water reducer. An admixture, such as a lignosulfonate salt, used in the manufacture of concrete or mortar to reduce the amount of water per batch. It does not affect the slump characteristics, but increases the workability. Also known as *water-reducing agent*.

water-reducing agent. See water reducer.

water repellent. A hydrophobic material, such as wax, soaps or natural or synthetic resins, used to render a surface resistant to wetting by water, but not completely waterproof. Used for treating textiles, leather, paper, and wood.

water-repellent cement. A *hydraulic cement* that has a water-repellent admixture incorporated during manufacture.

water-repellent fabrics. Textile fabrics that have been treated with wax or synthetic resins (e.g., fluorocarbons or silicones) to render them resistant to wetting by water, but usually not completely waterproof. See also waterproof fabrics.

water-repellent leather. Leather treated with a hydrophobic chemical to reduce the tendency of water to spread and wet its surface. In order to permit the leather to "breathe," it is not made completely waterproof.

water-repellent paper. (1) A water-washable paper that contains a special surface coating and is used for wallpaper.

(2) Any paper treated to resist penetration by water.

water-resistant barrier. A substance that is capable of retarding the transmission of liquids, such as water.

water-resistant fabrics. Fabrics that resist surface wetting and penetration, either owing to the nature of the fibers or yarns used in their manufacture, or produced by treating with suitable chemicals, e.g., synthetic resins.

water-resistant leather. (1) Leather stuffed with water-resistant substances, such as greases, waxes or oils, and used in the manufacture of work clothes, heavy-duty work boots, and shoes.

(2) Upper leather treated with hydrophobic chemicals to reduce water absorption or penetration. See also upper leather.

water-resistant paper. Paper that has certain chemicals added to make it more resistant to deterioration by water.

water-resistant porcelain enamel. A ground-coat porcelain enamel made water-resistant by the addition of a high percentage of silicon dioxide (SiO_2) and zirconium oxide (ZrO_2), usually in the form of *zircon*, to the frit. Used as a coating on sheet steel.

water-soluble coating. A coating, usually of the ceramic or polymeric type, whose ingredients are dissolved in water.

water-soluble gum. A gum, such as acacia, karaya or tragacanth, that can be dissolved in water. Used to make mucilages.

water-soluble polymers. High-molecular-weight polymers that are either of natural (e.g., acacia or karaya gum), semisynthetic (e.g., carboxymethylcellulose or modified starch) or synthetic (e.g., polyvinyl alcohol or polyvinyl pyrrolidone) origin, and dissolve or swell in water at normal or slightly elevated temperatures.

water-soluble resin. A semisynthetic, water-soluble polymer, such as methylcellulose, carboxymethylcellulose, or modified starch, that has been chemically treated to increase its water solubility. Used for adhesives, textile sizes, and coatings.

water stain. A wood stain composed of a dye, such as aniline, dissolved in water. It is rather inexpensive and gives an even color, but is slow drying and tends to raise the grain. See also alcohol stain; oil stain; sealer; varnish stain; wood stain.

water-struck brick. A *soft-mud brick* formed in a wetted or dampened mold to prevent sticking during removal.

water-thinned paint. See water-base paint.

Watertite. Trademark of PennKote Limited (Canada) for a series of hot-melt adhesives and joint sealants.

Watertown Arsenal steel. Trade name for a molybdenum-type high-speed steel containing 0.8% carbon, 0.5% silicon, 0.4% manganese, 9.5% molybdenum, 4% chromium, 1-2% tungsten, 0.9-1.5% vanadium, and the balance iron. It has good deep-hardening and non-warping properties, high hot hardness, good wear resistance, and is used for cutting tools.

Watertown Ware. Trade name of Watertown Manufacturing Company (USA) for melamine-formaldehyde plastics.

water tupelo. The moderately heavy, strong and stiff wood from the tree *Nyssa aquatica*. It is similar to *Ogeechee tupelo* and *sourgum*, and has light brownish-gray heartwood and light-colored sapwood. *Water tupelo* has a close grain, a uniform texture, high hardness, and moderate shock resistance. The wood from the enlarged tree base is much lighter in weight than that above. Source: Coastal Plain swamps and rivers of southern United States, from southern Virginia to Georgia, and along the Gulf Coast from northeastern Florida to Texas as well as along the Mississippi River valley. Used as lumber for furniture, baskets, crates, boxes and pallets, and rotary-peeled for plywood and veneer. Also known as *cotton gum; swamp gum; swamp tupelo; tupelo gum*.

water-vapor-resistant barrier. See water-vapor retarder.

water-vapor retarder. A substance that retards or impedes the transmission of water vapor. Also known as *water-vapor-resistant barrier*.

Watsonite. Trade name for a mica paper.

Watts nickel. An electrodeposited or electroformed nickel coating produced from an aqueous plating bath (Watts bath) containing nickel chloride, nickel sulfate and boric acid. See also nickel coating; electroformed nickel.

Waukesha. (1) Trade name of Waukesha Foundry, Inc. (USA) for a series of cast corrosion-resistant nickel-base alloys used for chemical, food-processing and dairy equipment, and for pumps, bearings, and bushings.

(2) Trade name of Waukesha Foundry, Inc. (USA) for a series of corrosion-resistant cast alloys composed mainly of copper (54-67%) and nickel (23-32%), and used for dairy and food-processing equipment.

Wausau. Trade name of Wausau Motor Parts (USA) for an alloy cast iron containing 3.1% total carbon, 0.6% combined carbon, 2.2% silicon, 0.8% manganese, 1.2% molybdenum, 0.5% nickel, 0.15% chromium, and the balance iron. Used for valve seats.

Wave. Trade name of Southern Dental Industries Limited (Australia) for a light-cure, radiopaque, fluoride-releasing, translucent pyrogenic silica urethane dimethacrylate (UDMA) based hybrid composite resin that is 65% filled with strontium glass particles having an average size of 1.5 µm (60 µin.). It has flowable consistency, high flexibility, high compressive and tensile strengths, excellent wear resistance, excellent color stability and brilliant polishability, and a very low modulus of elasticity. Used in dentistry for restorations, enamel and porcelain repairs, small core-build-ups, for veneering, and as a pit and fissure sealant, and liner.

Wavelite. Trade name of Nippon Sheet Glass Company Limited (Japan) for a corrugated glass.

wavellite. A white, yellow, green or black mineral with a vitreous luster. It is composed of aluminum phosphate hydroxide pentahydrate, $Al_3(PO_4)_2(OH)_3 \cdot 5H_2O$. Crystal system, orthorhombic. Density, 2.30-2.33 g/cm³; hardness, 3.5-4 Mohs; refractive index, 1.535. Occurrence: USA (Arkansas, Pennsylvania, Wisconsin).

wavy-grained wood. Wood in which the fibers and other longitudinal elements form a pattern of rather uniform undulations or waves.

wawa. See obeche.

wax calf. A heavy, wax-finished calfskin leather.

wax-coated fabrics. Textile fabrics whose appearance, hand and/or water repellency have been enhanced by coating with a suitable wax, such as paraffin.

waxed paper. Paper coated, impregnated or otherwise treated with a wax, such as paraffin, to render it resistant to air, water and/or grease. It may also be creped or textured, and is used chiefly for wrapping applications. Also known as *wax paper.*

waxes. A generic term for low-melting, easily deformable organic substances of high molecular weight. They may be hydrocarbons, or esters of fatty acids and alcohols. Chemically, they belong to the lipids (i.e., they are fatty, water-insoluble substances found in animal and plant cells). *Waxes* are water repellent, combustible, dielectric, and have smooth textures. They can be subdivided into the following groups: (i) *Natural waxes,* such as vegetable waxes (e.g., carnauba or candelilla), mineral waxes (e.g., ceresin, ozocerite or paraffin) or animal waxes (e.g., beeswax, lanolin, shellac or insect wax); and (ii) *Synthetic* and *semisynthetic waxes* derived from paraffins during crude oil refining, or by synthesis (e.g., Fischer-Tropsch synthesis). Used as lubricants and polishes, in candles, sealants and crayons, as binders in ceramics and investment casting, in waterproofing, electrical insulation, and paper coatings.

Waxpac. Trademark of Dexter Corporation (USA) for an ethylene vinyl acetate *Hysol* hot-melt adhesive used for sealing waxcoated corrugated cartons.

wax paper. See waxed paper.

wax resist. A protective wax coating applied to patterned areas of a surface to prevent or inhibit glazes, coloring inks or etchants

from adhering to them.

Waylite. Trademark of Waylite Corporation (USA) for a lightweight cellular material obtained by treating molten blast-furnace slag with water or high-pressure steam. Used as an aggregate in the manufacture of lightweight concrete and concrete products. See also blast-furnace slag.

W-board. A grade of corrugated *fiberboard* that is similar to *V-board*, but lower in quality, and commonly used for interior applications or intermediate containers.

W-Decarb. Trade name of Kennametal Inc. (USA) for commercially pure tungsten (99.6-99.9%) containing 0.02-0.1% iron, and used as a bond hardener, as a filler for free-cutting applications, and as a wear-rate modifier in hot-pressed bond systems.

wean-galvanized material. See hot-galvanized material.

Wearaloy. Trade name of Adirondack Steel Casting Company (USA) for a wear- and abrasion-resistant steel containing 0.5% carbon, 1.5% nickel, 0.9% chromium, 0.2% molybdenum, and the balance iron. Used especially for construction equipment components.

Wear-Arc. Trade name of Chemetron Corporation (USA) for a series of abrasion- and wear-resistant hardfacing materials including various chromium, chromium-boron, manganese-nickel and corrosion-resistant chromium-nickel steels, commercially available in the form of covered arc-welding electrodes.

Wear-Cote. Trademark of Wear-Cote International Inc. (USA) for electroless nickel with submicron particles used for producing wear-resistant coatings on metallic substrates. It has excellent corrosion resistance and lubricity, and high hardness, and is used for valves, pumps, fluid-power equipment, food-processing and packaging equipment, aerospace and electronic hardware, textile machinery, and plastic molds.

Wear-Dated. Trademark of Solutia Inc. (USA) for wear-resistant nylon 6,6 filaments and fibers available in several grades including *Wear-Dated, Wear-Dated II, Wear-Dated Assurance* and *Wear-Dated Freedom.* Used for residential carpets. Acrylic fibers for upholstery are also included under this trademark.

Wear Devil. Trade name of Champion Rivet Company (USA) for a series of abrasion- and wear-resistant hardfacing materials commercially available in electrode form and including various chromium alloy steels and cobalt-chromium-tungsten and cobalt-chromium-molybdenum-tungsten alloys.

Wearex. Trade name of Darwins Alloy Castings (UK) for a high-speed steel containing 0.7% carbon, 4% chromium, 14.5% tungsten, 1% vanadium, and the balance iron. Used for lathe and planer tools, shaping and finishing tools, milling cutters, saw blades, and woodworking tools.

Weargard. Trade name of Centrifugal Products Inc. (USA) for an abrasion-resisting cast iron containing 3.6% carbon, 3.2% nickel, 1.3% chromium, 0.2% molybdenum, and the balance iron. Used for overlays, liners and sleeves.

Wearite. Trade name of Mueller Brass Company (USA) for a series of tough, extruded aluminum bronzes containing 10-13.6% aluminum, 3-5% iron, 0-0.5% others, and the balance copper. Used for machine-tool beds, slides, cams, bushings, liners, and bearings.

Wearloy. (1) Trade name for a high-strength, wear-resistant cast iron.

(2) Trade name of Frank Foundries Corporation (USA) for a steel containing about 0.7% carbon, 1.5% nickel, 0.6% chromium, and the balance iron. Used for brake drums of automobiles including trucks and buses.

Wearmang. Trade name of Atlas Specialty Steels (Canada) for a wear- and abrasion-resistant steel containing 0.3% carbon, 1.5% manganese, 0.25% manganese, 0.25% silicon, 0.2% molybdenum, and the balance iron. Used for wear plates and mining equipment.

Wear-O-Matic. Trade name of Chemetron Corporation (USA) for a series of abrasion- and wear-resistant hardfacing materials including a wide range of chromium, chromium-molybdenum, nickel-molybdenum, austenitic manganese, corrosion-resistant chromium-nickel and heat-resistant high-chromium steels, commercially available in the form of arc-welding electrodes.

Wearpact. Trade name of American Steel Foundries (USA) for an impact- and abrasion-resistant steel containing 0.23-0.33% carbon, 1.3-1.8% manganese, 0.4% silicon, 0.4-1% chromium, 0.4-0.6% molybdenum, 0.5% cerium, a trace of boron, and the balance iron. Used for crushers.

wear-resistant fabrics. Textile fabrics that resist deterioration due to normal or excessive wear. Industrial fabrics are often of the wear-resistant type.

Weartuf Steel. Trade name of Horace T. Potts Company (USA) for a wear- and abrasion-resistant steel containing 0.9% carbon, 1.2% manganese, 0.6% silicon, and the balance iron. Used for scraper blades, conveyor buckets, liners, mixers, hoppers, and screens.

Wearweld. Trade name of Lincoln Electric Company (USA) for a steel containing 0.37% carbon, 2.2% manganese, 0.15% silicon, 3.3% chromium, and the balance iron. It has good resistance to metal-to-metal wear, and is used for hardfacing and arc-welding electrodes.

Wearwell. Trade name of Stelco Steel (Canada) for an abrasion-resistant steel containing 0.3-0.35% carbon, 1.3-1.65% manganese, 0.15-0.3% silicon, and the balance iron. Used for riveted and bolted structures.

WeatherBloc. Trademark of Sterling Fibers Inc. (USA) for weather-resistant acrylic fibers supplied in solution-dyed (*Weather-Bloc*) and light-stabilized (*WeatherBloc Natural*) form. They have outstanding resistance to ultraviolet light, excellent dimensional stability, good mildew and chlorine resistance, and are used for outdoor furniture, awnings, marine applications, upholstery, etc.

Weathergrained. Trade name of Libbey-Owens-Ford Company (USA) for a patterned glass.

Weatherite. Trademark of Crandon Paper Mills, Inc. (USA) for *kraft paper* treated with a black waterproofing coating, and used in building construction.

Weatherpane. Trademark of PPG Industries Inc. (USA) for insulating glass panes.

weather-resistant fabrics. Textile fabrics, such as canvases and tarpaulins, which are resistant to a wide range of severe weather conditions.

Weather-Tite. (1) Trademark of BDH Two, Inc. (USA) for a water-resistant, stainable high-performance wood glue widely used in carpentry.

(2) Trademark of Emco Limited (Canada) for roofing shingles.

Weath-R-Proof. Trade name of Thermoproof Glass Company (USA) for an insulating glass.

webbing. (1) A long, narrow woven fabric used in upholstery and for belts.

(2) A long, narrow, usually multi-ply fabric woven in a coarse weave.

Webbite. Trade name of NL Industries (USA) for an alloy com-posed of 93-95% aluminum and 5-7% titanium, and used as an addition agent to improve aluminum alloys by adding titanium.

weberite. A mottled gray mineral composed of sodium magnesium aluminum fluoride, Na_2MgAlF_7. Crystal system, orthorhombic. Density, 2.96 g/cm^3; refractive index, 1.348. Occurrence: Greenland.

Webert Alloy. Trade name of Anaconda Company (USA) for a copper casting alloy containing 14% zinc, and 4% silicon and manganese, respectively. Used for pressure-die castings.

Webrax. Trade name of Hermes Abrasives Limited (USA) for a resin- and grain-impregnated nonwoven abrasive web.

Webril. Trademark of Kendall Company (USA) for thin nonwoven fabrics made of natural and/or synthetic fibers and used as backing materials for polymers films, and as replacements for knitted, netted and woven fabrics.

Webusiv. Trade name of Heinz Weissenburger GmbH (Germany) for a coal slag-based sandblasting abrasive.

Weco. Trademark of Weckerle GmbH Lackfabrik (Germany) for various industrial lacquers and varnishes.

weddellite. A colorless, white, yellowish-brown or brown mineral composed of calcium oxalate dihydrate, $CaC_2O_4 \cdot 2H_2O$. It can also be made synthetically. Crystal system, tetragonal. Density, 1.94 g/cm^3; refractive index, 1.523. Occurrence: Antartica (Weddell Sea).

Wedge. Trade name of Pittsburgh Corning Corporation (USA) for a sculptured glass with wedge design.

wedge brick. A brick having its two main faces sloping toward each other at an acute angle.

Wedico. Trade name of Wesenfeld, Dicke & Co. (Germany) for bismuth, several bismuth, lead and tin alloys, and various type metals.

weeksite. A yellow mineral composed of potassium uranyl silicate tetrahydrate, $K_2(UO_2)_2(Si_2O_5)_3 \cdot 4H_2O$. Crystal system, orthorhombic. Density, 4.10 g/cm^3; refractive index, 1.603. Occurrence: USA (Utah).

Wefahütte. Trade name of Westfalenhütte Dortmund AG (Germany) for a series of low- and medium-carbon alloy steels including several case-hardening, heat-treatable and tool-steel types. Used for machinery parts, such as gears, shafts, bolts, and fasteners, for tools, such as punches and upsetters, for a variety of dies, and for machine-tool parts. Also spelled "Wefahutte," or "Wefahuette."

Wefesiv. Trade name of Heinz Weissenburger GmbH (Germany) for an iron-free coal slag-based blasting abrasive used for cleaning high-grade steels.

weft-knit fabrics. Stretchable knit fabrics, such as *jersey*, made by interlocking loops in a lateral (widthwise) direction. Also known as *weft knits*.

weft knits. See weft-knit fabrics.

weft yarn. A yarn used for the wefts (filling threads) of a woven fabric. Also known as *filling; filling yarn*.

Wegner. Trade name for an antifriction alloy of 80% tin and 20% zinc, used for bearings.

Wegold. Trade name of Wegold Edelmetalle AG (Germany) for an extensive series of dental casting alloys including several high-gold types (e.g., *Wegold Biocomp, Wegold i, Wegold JN* and *Wegold M-PF*), gold types (e.g., *Wegold G*), gold-platinum types (e.g., *Wegold Biologic, Wegold GS, Wegold H* and *Wegold U*), gold-palladium types (e.g., *Wegold EC, Wegold D4* and *Wegold SF*) and palladium-base alloys (e.g., *Wegold Ag, Wegold DG, Wegold MT* and *Wegold N2*). Depending on the particular composition, they are used for crowns, bridges, fillings and/or other restorations.

wegscheiderite. A colorless mineral composed of sodium hydrogen carbonate, $Na_5CO_3 \cdot 3NaHCO_3$. Crystal system, triclinic. Density, 2.34 g/cm^3; refractive index, 1.519. Occurrence: USA (Wyoming).

Wehralloy. Trade name of Wehr Steel Corporation (USA) for a series of corrosion- and/or heat-resistant steels including various austenitic, ferritic and martensitic grades.

weibullite. A light gray mineral composed of lead bismuth selenide sulfide, $Pb_5Bi_8Se_7S_{11}$. Crystal system, orthorhombic. Density, 6.97-7.06 g/cm^3. Occurrence: Sweden.

Weiger. Trade name for a corrosion-resistant alloy of 77-80% silver, 18-20% copper and 2-5% platinum, used as a silver solder.

weighting agent. A compound, such as clay, chalk or zinc acetylacetonate, used to reduce the luster and cost of textile fabrics and enhance their feel or hand.

Weights. Trade name for a corrosion-resistant alloy of 90% copper, 8% tin and 2% zinc, used for weights.

weilerite. A yellowish white mineral of the alunite group that is composed of barium aluminum hydrogen arsenate hydroxide, $BaAl_3H(AsO_4)_2(OH)_6$, and may also contain sulfate (SO_4). Crystal system, rhombohedral (hexagonal). Density, 3.92 g/cm^3; refractive index, 1.645. Occurrence: Germany.

weilite. A white mineral composed of calcium hydrogen arsenate, $CaHAsO_4$. Crystal system, triclinic. Density, 3.48 g/cm^3; refractive index, 1.688.

Weingärtner. Trade name of Emil Weingärtner & Co. (Germany) for an extensive series of steel including various case hardening, carburizing, heat-treatable and nitriding types, free-cutting types, structural types, plain-carbon, cold-work, hot-work and high-speed tool types as well as various corrosion-resistant stainless and heat-resistant types. Also included under this trade name are several silver steels, spring steels and high-precision steels. Also spelled "Weingaertner."

Weir. Trade name for carbon, low-alloy, and corrosion-, heat-and/or wear-resistant steel castings.

Weiralead. Trade name of National Steel Corporation (USA) for a hot-dip lead-coated steel used for structural sheets.

Weircoloy. Trade name of National Steel Corporation (USA) for a low-carbon steel composed of 0.2% carbon, 0.5% copper, and the balance iron. Used for galvanizing structures, copper bearings, and steel sheets.

Weirkote. Trademark of Weirton Steel Corporation (USA) for zinc- and zinc alloy-coated ferrous sheet and strip. *Weirkote Plus* is a low-carbon flat-rolled steel in coil or sheet form coated with *Galfan,* a coating alloy of 95% zinc and 5% aluminum/mischmetal. It possesses excellent formability and corrosion resistance, does not exhibit microcracking and flaking, and is used for automotive parts, painted building panels, marine equipment, air-conditioning equipment, household appliances, roofing, garage doors, gutters, downspouts, siding, and window frames.

Weirite. Trade name of National Intergroup Inc. (USA) for electrolytic tinplate used for roofing and containers.

Weirzin. Trade name of National Intetgroup Inc. (USA) for sheet steel coated electrolytically with zinc, and used for structural parts.

weissbergite. A steel-gray mineral composed of thallium antimony sulfide, $TlSbS_2$. It can also be made synthetically by heating thallium, antimony and sulfur, or thallous sulfide and antimony sulfide in stoichiometric proportions under prescribed

conditions. Crystal system, triclinic. Density, 5.79 g/cm^3. Occurrence: USA (Nevada).

Weisspunkt. Trade name of SWB Stahlformguss-Gesellschaft mbH (Germany) for several high-speed tool steels.

weissite. A bluish black mineral composed of copper selenide, Cu_5Te_3. It can also be made synthetically with slightly different formula ($Cu_{2-x}Te$) by heating copper and tellurium under prescribed conditions. Crystal system, hexagonal. Density, 4.47 g/cm^3; refractive index, 1.864. Occurrence: USA (Colorado).

Welch's alloy. A corrosion-resistant alloy of 52% tin and 48% silver, used as a dental alloy and solder.

Welcon. Trade name of Japan Steel Works Limited (Japan) for a series of structural steels including various high-strength and high-strength low-alloy (HSLA) grades. They are supplied in plate form and possess good weldability, formability and atmospheric corrosion resistance. Used for structural members in welded constructions including bridges and buildings, pressure vessels, mining and construction equipment, crane booms, truck frames, bus bodies, oil tankers, and power-plant equipment.

Weldalite. Trademark for lightweight, wrought aluminum-copper-lithium alloys developed by Lockheed Martin Company (USA) and produced by Alcoa–Aluminum Company of America and McCook Metals LLC (USA). They are available in various tempers and have good weldability, an average density of approximately 2.6 g/cm^3 (0.09 $lb/in.^3$), and a hardness of 123 Brinell. Used for structural and aerospace applications.

Weldanka. Trade name of Dunford Hadfields Limited (UK) for wrought and austenitic stainless steels containing 0.08-0.2% carbon, 18-25% chromium, 7-12% nickel, up to 0.8% niobium, up to 0.4% titanium, and the balance iron. They possess high corrosion and heat resistance, and are used for turbine blading, aircraft parts, chemical plant and equipment, and pulp and paper equipment.

Weld-Arc. Trade name of Champion Rivet Company (USA) for a carbon steel containing 0.1% carbon, and the balance iron. Used in the form of arc-welding rods for welding mild steels.

Weldbest Albronze. Trade name of Weldwire Company, Inc. (USA) for an all-purpose aluminum bronze used in the form of welding electrodes for welding copper and copper alloys.

Weld-Cool. Trademark of Aearo Company (USA) for a gold-coated, hardened filter glass used for welding safety glasses.

welded tubing. See seamed tubing.

welded-wire fabric. See mesh (3).

welded-wire fabric reinforcement. See mesh reinforcement.

Weldex. Trade name of Pittsburgh Brass Manufacturing Company (USA) for a malleable cast iron containing 2.2% carbon, 2.4% silicon, 0.03% phosphorus, 0.01% sulfur, 0.25% manganese, and the balance iron.

Weld-Fast. Trade name of Revere Copper Products, Inc. (USA) for a copper-tin alloy used for bronze welding rods.

welding alloys. Ferrous or nonferrous alloys in powder, wire or rod form used as filler metals in both gas and electric-arc welding, and in hardfacing. See also welding powder; welding rod; welding wire.

welding electrode. See arc-welding electrode.

welding fluxes. Fusible mineral materials containing oxides of manganese, silicon, titanium, aluminum, calcium, zirconium and magnesium as well as other compounds, such as calcium fluoride. They are used in welding nonferrous metals and certain steels and cast irons to prevent oxidation, and dissolve, or aid in the removal of oxides and other undesirable contaminations. Used especially in submerged-arc and electroslag welding, and in oxyacetylene braze welding.

welding glass. A special optical glass, usually tempered and tinted green, brown or gold, employed in welding goggles, helmets and face shields to reduce glare and protect the eyes against harmful radiation and flying sparks. See also filter glass.

welding powder. A welding flux in the form of a powder. Also known as *powdered flux.* See also welding fluxes.

welding rod. Filler metal in the form of a rod or wire which does not conduct electricity, and is melted and added to the weld puddle. Used in welding and braze welding. Also known as *filler rod; filler wire; welding wire.* See also filler (1).

welding wire. See welding rod.

Weldmaco. Trade name of Welding Materials Company (USA) for a carbon steel containing 0.2% carbon, and the balance iron. Used for welding rods.

Weld Mesh. Trademark of Bekaert Corporation (USA) for all-welded wire mesh composed of low-carbon steel and supplied in rolls and sheets with wire sizes from 22 to 2 Gauge and mesh sizes from 0.25×0.25 in. $(6.4 \times 6.4$ mm) to 4×4 in. $(101 \times 101$ mm).

Weld-On. Trade name of IPS Corporation (USA) for several plastic adhesives and cements.

Weldon. Trade name of United Clays, Inc. (USA) for a strong, plastic porcelain with moderately high levels of colloidal carbon and good electrical insulating properties, used in the manufacture of electrical porcelains, and for other applications in the porcelain industry.

Weldon/Victoria. Trade name of United Clays, Inc. (USA) for a strong, white-firing clay from Tennessee, USA, which contains organic carbon, and is used for tile, electrical porcelain and other ceramic applications.

Weldural. Trade name of Hoogovens Aluminum Walzprodukte (Germany) for wrought aluminum-copper alloys (AA 2000 series) supplied in the form of sheets and plates with good machinability and weldability, high dimensional stability, fair corrosion resistance, an average density of 2.84 g/cm³ (0.103 lb/in.³), and a hardness of about 130 Brinell. Used for precise mechanical parts, and machine components.

Weldwood. Trademark of D.A. Veneers Limited Division of Weldwood of Canada Limited for several urea-formaldehyde adhesives.

welinite. A reddish brown mineral composed of manganese tungsten silicate, $Mn_3Si_{0.6}W_{0.4}O_7$. Crystal system, hexagonal. Density, 4.47 g/cm³; refractive index, 1.864. Occurrence: Sweden.

Wellamid. Trademark of Wellman Inc. (USA) for an extensive series of polyamide 6, 6,6 and 66,6 (nylon 6, 6,6 and 66,6) engineering resins supplied in various grades including regular, carbon fiber- or glass fiber-reinforced, silicone-, molybdenum disulfide- or polytetrafluoroethylene-lubricated, glass bead- or mineral-filled, fire-retardant, casting, elastomer-copolymer, stampable-sheet, high-impact, supertough and UV reinforced. Used for automotive components, lawn and garden equipment, consumer products, and electrical components. See also Eco-Lon.

Wellcast. Trade name of Wellman Dynamics Corporation (USA) for a series of casting alloys based on aluminum or copper. The aluminum-base alloys include several aluminum-copper, aluminum-magnesium and aluminum-silicon types, and the copper-base alloys various bronzes, brasses, nickel silvers and copper nickels.

well-graded aggregate. An aggregate with a particle size distribution that will produce a concrete with minimum void space and

thus maximum density.

Wellene. Trademark of Wellman Inc. (USA) for polyester staple fibers used for textile fabrics.

Wellit. Trade name of Vereinigte Glaswerke (Germany) for a corrugated glass.

Wellite. Trade name of Welwyn Plastics (UK) for regenerated cellulose.

Wellkey. Trade name of Teijin Fibers Limited (Japan) for polyester filaments and staple fibers used for the manufacture of textile fabrics.

Wellon. Trademark of Wellman Inc. (USA) for nylon 6 and nylon 6,6 staple fibers.

wellsite. A colorless to white mineral of the zeolite group composed of barium aluminum silicate hexahydrate, $(Ba,Ca,K_2)Al_2$-$Si_6O_{16} \cdot 6H_2O$. Density, 2.25 g/cm³; refractive index, 1.500. Occurrence: Czech Republic.

Wellstrand. Trademark of Wellman Inc. (USA) for nylon and polyester fibers used for textile fabrics.

Welmet. Trade name of Welland Electric Steel Foundry (Canada) for a series of corrosion-resistant cast iron-nickel-chromium alloys used for pipes, valves, fittings, and chemical equipment.

Wel-Met Bronze. Trade name of Wel-Met Company (USA) for a powder-metallurgy bronze containing 84-86% copper, 7-8% tin, 5-6% lead and 1-2% graphite. Used for self-lubricating, oil-impregnated bearings.

Wel-Met Steel. Trade name of Wel-Met Company (USA) for a powder-metallurgy alloy containing 89-92% iron, 7-10% copper and 0.5-2.0% graphite. Used for self-lubricating porous bearings.

Welnet. Trademark of Drahtwerke Roesler Soest GmbH & Co. KG (Germany) for spot-welded wire netting.

weloganite. A lemon-yellow mineral of the mckelveyite group composed of sodium strontium zirconium carbonate trihydrate, $Na_2Sr_3Zr(CO_3)_6 \cdot 3H_2O$. Crystal system, triclinic. Density, 3.22 g/cm³; refractive index, 1.646. Occurrence: Canada (Quebec).

Welsbach alloy. A pyrophoric material composed of 60% cerium, 30% iron and 10% rare earths, and formerly used for lighter flints.

welshite. A reddish black mineral of the aenigmatite group composed of beryllium calcium magnesium iron antimony silicate, $Ca_2Mg_4FeSbBe_2Si_4O_{20}$. Crystal system, triclinic. Density, 3.77 g/cm³. Occurrence: Sweden.

Wel-Ten. Trade name of Nippon Steel Inc. and Yawata Iron & Steel Company Limited (Japan) for a series of structural steels including various high-strength, high-strength low-alloy (HSLA) and quenched-and-tempered low-alloy grades. Some grades are suitable for sub-zero applications. Supplied in plate form, they have excellent weldability, good formability, high strength and toughness, and good atmospheric corrosion resistance. Used for structural members in welded constructions including bridges and buildings, and for earth-moving equipment, storage tanks and pressure vessels, penstock, mining and construction equipment, crane booms, and truck frames.

Welvic. Trade name of ICI Limited (UK) for polyvinyl chlorides and other vinyl plastics.

wenkite. A light gray mineral of the cancrinite group composed of barium calcium aluminum silicate sulfate hydroxide, Ca_5Ba_4-$Al_9Si_{11}S_3O_{53}(OH)_4$. Crystal system, hexagonal. Density, 3.13 g/cm³; refractive index, 1.595. Occurrence: Italy.

Wentex. Trade name of Wentus Kunststoff GmbH (Germany) for a low-pressure polyethylene film.

Wentofan. Trade name of Wentus Kunststoff GmbH (Germany) for a polypropylene film.

Wentolen. Trade name of Wentus Kunststoff GmbH (Germany) for a high-pressure polyethylene film supplied in standard and heat-shrinkable grades.

Werhodit. Trade name of Krebsoege Co. (Germany) for a series of ferrous powder-metallurgy materials.

wermlandite. A green-gray mineral composed of calcium magnesium aluminum iron carbonate hydroxide hydrate, $Ca_2Mg_{14}(Al,Fe)CO_3(OH)_{42} \cdot 29H_2O$. Crystal system, hexagonal. Density, 1.93 g/cm³; refractive index, 1.493. Occurrence: Sweden.

wernerite. See scapolite.

Werzalit. Trade name of Werzelit-Werke (Germany) for building wall and facing tiles, plastic-coated window sills, and chipboard for interior and exterior applications.

Wesgo. Trademark of GTE Products Corporation, Wesgo Division (USA) for alumina ceramics.

Wessel's alloy. A corrosion-resistant nickel silver containing about 51-66% copper, 19-32% nickel, 12.5-17% zinc, 0-0.5% iron and 0-2% silver. Used for ornamental parts, architectural trim and silverware. Also known as *Wessel's silver.*

Wessling. Trade name of Gusstahl-Handels GmbH (Germany) for an acid-resistant cast steel containing 0.3% carbon, 17.6% silicon, 0.8% manganese, and the balance iron. Used for pumps, valves, drains, fittings, and a variety of other castings.

Wessonmetal. Trade name of Wesson Company (USA) for a series of tool materials consisting of sintered mixtures of one or more powdered carbides bonded together in a metallic matrix. Included are several straight grades containing only tungsten carbide in a cobalt matrix as well as various complex grades containing mixtures of either tungsten carbide or titanium carbide in cobalt matrices, titanium carbide and molybdenum carbide in nickel matrices, or mixtures of the carbides of tungsten, titanium, tantalum and niobium in cobalt matrices. They have high hardness, high abrasion and wear resistance and high compressive strength, and are used for machining and cutting tools.

West. Trade name of West Steel Casting Company (USA) for an extensive series of cast steels used for tools, dies and machinery parts.

Westa. Trade name of Westa-Westdeutsche Edelstahlhandelsgesellschaft (Germany) for an extensive series of case-hardening alloy steels including chromium, chromium-molybdenum and nickel-chromium grades, several die and tool steels including plain-carbon, cold-work, hot-work and high-speed grades, and various stainless steels.

Westac. Trade name for several adhesives based on polyvinyl acetate.

West African cedar. See sapele.

Westar. Trademark of Westar Timber Limited (Canada) for softwood lumber.

West Coast hemlock. See Alaskan pine.

Western Alloy. Trade name of Olin Corporation (USA) for a series of wrought coppers and copper-base alloys including several silver-bearing and silver-free high-conductivity coppers, leaded and unleaded brasses and bronzes, and nickel silvers.

western ash. The hard, strong, stiff wood of the broad-leaved tree *Fraxinus latifolia.* The heartwood is brown and the sapwood almost white. *Western ash* has a high shock resistance and an average weight of 609 kg/m³ (38 lb/ft³). Source: Pacific Coast of North America. Used mainly as a fuelwood, but also for cooperage and containers. Also known as *Oregon ash.*

western black willow. The tough, fairly soft, brownish-yellow

wood of the broad-leaved tree *Salix lasiandra.* It has good non-shrinking properties and an average weight of 448 kg/m³ (28 lb./ft³). Source: Western United States. Used for furniture, toys, novelty items, cricket bats, artificial limbs, crates, cooperage, and fenceposts.

western hemlock. See Alaskan pine.

western hemlock fir. See Alaskan pine.

western larch. The hard, tough wood of the tall softwood tree *Larix occidentalis.* The heartwood is yellowish-brown, and the sapwood yellowish-white. It has a straight, close grain, high stiffness, moderate strength, good shock resistance, relatively good decay resistance, and good machining and finishing qualities. Average weight, 577 kg/m³ (36 lb/ft³). Source: Northwestern United States and southwestern Canada. Used as dimension lumber for building construction, railroad ties, piles, poles and posts. Some use is made of higher-grade wood for interior finish and flooring. Also known as *hackmatack; western tamarack; Montana larch; mountain larch.*

western red cedar. The reddish-brown, non-resinous wood of the very large arborvitae *Thuja plicata.* It is similar to *redwood* except for its cedar-like odor and the pronounced transition from spring to summer growth. *Western red cedar* has outstanding durability under all conditions, but may promote corrosion of metal in contact with it. Average weight, 370 kg/m³ (23 lb/ft³). Source: Pacific coast of North America, especially Washington State and throughout British Columbia. Used for roofing shingles, siding, glasshouse construction, structural timbers, and utility poles, and for interior joinery work and cabinetmaking. Also known as *giant arborvitae; Pacific red cedar; shinglewood.*

western spruce. See Sitka spruce.

western tamarack. See western larch.

western white pine. The wood of the coniferous tree *Pinus monticola.* The heartwood is creamy to pale reddish-brown turning dark on exposure, and the sapwood is yellowish white. *Western white pine* has a straight grain, an even texture, contains some resin ducts, machines and finishes well, and exhibits high stability after seasoning. Average weight, 432 kg/m³ (27 lb/ft³). Source: Northwestern United States (Idaho, Washington, Montana and Oregon) and southern British Columbia. Used in building construction, for boxes, matches, millwork, sheathing, paneling, subfloor, siding, trim and finish, for carving and woodworking, and for foundry patterns. Also known as *Idaho white pine.*

western yellow pine. See ponderosa pine.

westerveldite. A gray, opaque synthetic mineral composed of iron arsenide, FeAs. Crystal system, orthorhombic. Density, 7.86.

Westerwald clay. A refractory clay obtained from the Westerwald, a mountainous region in Germany northeast of the City of Coblentz. It is usually high in silica (SiO_2), but may contain over 30% alumina (Al_2O_3).

Westfalia. Trade name of W. Ossenberg & Cie. Edelstahlwerke (Germany) for a series of austenitic, ferritic and martensitic stainless steels including various stabilized and free-machining grades.

Westfälische. Trade name of Westfälische Stahlgesellschaft AG (Germany) for a series of machinery and structural steels as well as several tool steels including plain-carbon, hot-work, cold-work and high-speed types.

Westig. Trade name of Westig GmbH (Germany) for an extensive series of plain-carbon, cold-work, hot-work and high-speed tool steels, high-grade alloy steels (e.g., chromium, chromium-molybdenum, chromium-vanadium and nickel-chromium-molybdenum types), free-machining steels, structural steels, case-hardening steels, ball-bearing and spring steels, silver steels, and various austenitic, ferritic and martensitic stainless steels. They are supplied in a wide range of forms including strips, wires, bars and structural shapes.

Westig-Duro-Biflex. Trademark of Westig GmbH (Germany) for electron-beam-welded specialty bimetallic strip steel used for metal saws.

West Indian boxwood. See zapatero.

West Indian ebony. See cocoawood.

West Indian locust. See Surinam teak.

Westroc. Trade name of Westroc Industries Limited (Canada) for gypsum wallboard supplied in standard, fire-resistant, moisture-resistant and lightweight grades. Also included under this trade name are various drywall compounds.

West Virginia spruce. See red spruce.

Weta. Trade name for a refractory composed of silicon carbide (SiC) powder mixed with selected silicates and metals. It is resistant to acids, alkalies and thermal shock.

Wet Bond. Trade name of Pulpdent Corporation (USA) for a dental adhesive used for dentin and enamel bonding applications.

wet-laid composites. Composites, such as fiberboard or certain nonwoven fabrics, produced on conventional papermaking machinery from natural or synthetic fibers (e.g., wood or textile fibers) with or without binders by pressing or interlocking with or without the application of heat.

wet-laid nonwoven fabrics. Loose, porous textile fabrics made on papermaking machinery from relatively short natural or synthetic fibers, yarns or rovings with a suitable binder (e.g., an adhesive) by pressing or interlocking with or without the application of heat. Used for filtration products, tissue, napkins, and drapes. Also known as *wet-laid nonwovens*. See also dry-laid nonwoven fabrics; nonwoven fabrics.

wet lay-up resins. A group of resins, such as epoxies, unsaturated polyesters and vinyl esters, suitable for processing by wet lay-up techniques, i.e., for application of the liquid resin during putting into place (lay-up) of the reinforcement.

wet machine board. A dense, strong, stiff homogeneous board, typically between 1.3 and 25 mm (0.05 and 1.00 in.) in thickness, produced by the gradual formation on a roll of a number of wet sheets of refined cellulose fibers from a continuous web on an one-cylinder wet machine. The wet sheet is then pressed, dried and calendered to the required thickness. Water resistance and formability are enhanced by the incorporation of asphalt or synthetic resins.

wet mix. A mix containing batch ingredients blended in the dry state and then mixed with water prior to processing.

wet-mix shotcrete. Shotcrete containing all ingredients including the necessary mixing water. It can be conveyed by pneumatic means or moved by displacement. See also shotcrete.

Wetordry. Trade name of 3M Company (USA) for waterproof coated abrasives (e.g., sand papers) used in metal finishing.

Wetrelon. Trade name for nylon fibers and products.

Wet Stick. Trade name of Rexnord Inc. (USA) for a roofing cement in cartridge form that bonds even on wet surfaces.

wet-spun filaments. Synthetic or other man-made filaments produced by extruding the dissolved material through spinnerets, either directly or through a small air gap into a coagulating liquid.

wet-spun yarn. A yarn composed of a vegetable fiber, such as flax or hemp, that has been spun from roving soaked in hot water.

wet-strength paper. Paper with enhanced resistance to bursting, rupturing or tearing in the wet condition. This resistance is either due to a special treatment and fiber interlocking technique, a treatment with special chemicals, or impregnation with a synthetic resin (e.g., a melamine or urea resin). Used for official documents, maps, and wrapping purposes. Also known as *wet-strong paper*.

WeveBac. Trade name for polyolefin fibers and yarns used for textile fabrics.

Wewesiv. Trade name of Heinz Weissenburger GmbH (Germany) for a sandblasting material made by the thermoselect process.

Wexco. Trade name of Wexco Corporation (USA) for a series of alloys including *Wexco 666*, a low-friction nickel-boron-base iron alloy with a martensitic structure and a cementite matrix, *Wexco 777*, a corrosion- and wear-resistant alloy containing tungsten carbide in a chromium-boron-nickel matrix, and *Wexco SSS*, an alloy with a corrosion-resistant dual-phase microstructure of high chromium and boron content in cobalt-nickel.

Weyerhaeuser. Trademark of Weyerhaeuser Limited (Canada) for quality lumber and lumber products including oriented strand board, plywood, and panels.

Weymouth pine. See eastern white pine.

wheel brass. A free-cutting brass composed of 68% copper, 30% zinc and 2% lead, and used for wheels and hardware.

Wheelerite. Trade name of Electro-Foundry Company (USA) for a ductile cast iron containing 3-3.5% total carbon, 1.5-2.5% silicon, 0.05-0.15% magnesium, and the balance iron. Used for gears, housings, and machine-tool castings.

Wheeling Metal. Trade name of Continental Foundry & Machine Company (USA) for a water-hardening steel containing 0.5% carbon, and the balance iron. Used for rolls.

Whelco. Trade name of Metalsource Corporation (USA) for a series of steels including various corrosion-resistant low-carbon high-chromium steels, and numerous tool steels including plain-carbon, hot-work, high-speed and low-alloy types.

wherryite. A light green mineral composed of copper lead oxide chloride carbonate sulfate hydroxide, $Pb_4Cu(CO_3)(SO_4)_2(OH,Cl)_2O$. Crystal system, monoclinic. Density, 6.45 g/cm^3; refractive index, 2.01. Occurrence: USA (Arizona).

whetstone. A hard, fine-grained rock that is usually highly siliceous and used for sharpening knives and cutting tools.

whipcord. A strong worsted fabric, usually cotton, in a twill weave and with fine diagonal ridges. Used for clothing, e.g., suits, and for draperies and upholstery. Also known as *artillery twill*.

whewellite. A colorless, yellowish or brownish mineral composed of calcium oxalate monohydrate, $C_2CaO_4 \cdot H_2O$. It can also be made synthetically. Crystal system, monoclinic. Density, 2.23 g/cm^3; refractive index, 1.5513. Occurrence: Germany.

whisker. See whiskers.

Whiskerloy. Trade name for an extensive series of engineering composites consisting of aluminum- or magnesium-alloy matrices reinforced with sapphire, carbon or silicon carbide whiskers.

whisker-reinforced ceramics. A class of ceramic materials that have polycrystalline ceramic matrices composed of toughened alumina, silicon nitride, spinel, mullite, glass, zirconia or cordierite with dispersions of strong microscopic whiskers of silicon carbide or silicon nitride as reinforcement. They possess excellent strengths and elastic moduli, excellent toughness, good creep resistance, good performance under severe stress and temperature conditions, and are used for cutting tools for machining operations, and aerospace applications.

whisker-reinforced metal-matrix composites. A class of advanced high-performance engineering composites that consist of uniform dispersions of needle-like single crystals of silicon carbide or silicon nitride in homogeneous metallic matrices (mainly aluminum, titanium, magnesium or copper and their alloys). They have high tensile strengths and stiffness, good fracture toughness and enhanced elevated-temperature performance, high compressive strengths, good dimensional stability, high fatigue resistance, excellent wear resistance, good corrosion resistance, and are used for aerospace applications, such as missile wings, missile components, floor panels for aircraft, aircraft fasteners, turbine-engine parts, turbine-compressor wheels, pistons, gears, and pulleys.

whiskers. Very fine and short, axially-oriented, single-crystal fibers. They may be composed of a metal (e.g., aluminum, cobalt, iron, nickel, rhenium or tungsten), a ceramic (e.g., alumina, beryllia, boron carbide, boron nitride, sapphire, silicon carbide or silicon nitride), carbon, boron, or other materials. Whiskers range from 1 to 25 μm (40 to 980 μin.) in diameter, and have typical length-to-diameter ratios of 50-15000, exceptionally high tensile strengths and elastic moduli, and good high-temperature properties. Used as reinforcements for metal, ceramic, plastic, glass and graphite matrices, and in ablative materials for aircraft and aerospace vehicles.

Whispertone. Trade name of Manville Corporation (USA) for fire-resistant glass-fiber acoustical panel used for residential applications.

Whitcarb. Trade name of Whittaker Corporation (USA) for a fine, white calcium carbonate powder used as a compounding ingredient in rubber.

Whitcon. Trade name of Whittaker Corporation (USA) for a fine fluorocarbon powder used as a filler in plastics, elastomers and greases, and as a dry lubricant.

white alumina. A white-colored, recrystallized, friable alumina abrasive used especially for toolroom grinding applications.

white arsenic. See arsenic trioxide.

white ash. See American ash.

white basswood. The relatively soft wood of the hardwood tree *Tilia heterophylla*. The yellowish-brown heartwood merges gradually into the white sapwood. *White basswood* has a fine, even texture, a straight grain, low to medium strength and durability, large shrinkage, high resistance to warpage in use, good gluing and nailing qualities and works very well. Average weight, 420 kg/m³ (26 lb/ft³). Source: Eastern United States (Appalachians from Pennsylvania to southern Georgia and into northern Alabama). Used for furniture, cabinetwork, sash and door frames, paneling, Venetian blinds, pianoforte manufacture, drawing boards, woodenware, containers, boxes, cooperage, veneer, pulpwood, and as corestock for plywood and carving. Also known as *white lime*.

white bearing metals. See babbitts.

white beech. (1) The strong, hard, heavy, light-colored wood of any of several species of beech trees (genus *Fagus)* especially the American beech *(F. grandifolia).* Source: North America. Used for tool handles, cheap furniture, veneer, shoe lasts, and as a fuel. See also American beech.

(2) The hard, strong, tough wood of the tree *Carpinus betulus*. Although the white beech (actually a birch) is not related to the more common *red beech* (a true beech), both trees have much in common, e.g., they are shade trees with smooth silvery gray bark. *White beech* has excellent strength proper-

ties (better than oak), high shock resistance, turns well and takes a very fine polish. Average weight, 750 kg/m³ (47 lb/ft³). Source: Central and southern Europe including British Isles; Asia Minor. Used for tool handles, mallets, cogs, shoe lasts, wood screws, other woodenware, for flooring and piano work, and as a fuel. Also known as *European hornbeam.*

White Benedict Metal. Trade name for a leaded nickel silver containing 18% zinc, 16.5% nickel, 4.5% lead and 1% tin, and the balance copper. See also Benedict Metal (2).

white birch. (1) See American white birch.

(2) The wood of any of various birches with white bark, such as the European silver birch *(Betula pendula).* See also European birch; silver birch.

white brass. (1) A brass composed of 65% copper, 32-33% zinc and 2-3% tin, and used for ornamental parts.

(2) An inverse brass composed of 50-66% zinc and 34-50% copper, and used for castings and ornamental parts.

(3) A brass composed of about 55-65% copper, 15-25% zinc and up to 30% nickel, and used for white-colored castings.

(4) A tough, hard bearing alloy composed of 65% tin, 27-31% zinc and 3-6% copper, and used for automotive applications.

(5) A high-zinc brass containing some copper, and used for novelty items and jewelry.

white bronze. (1) A light-colored bronze containing an exceptionally high amount of tin.

(2) A bronze composed of 20% zinc, 20% manganese, 0-2% aluminum, and the balance copper. Used for jewelry, hardware, and decorative castings.

white buffing compound. A very fine, soft white silica powder used in color buffing to produce lustrous, scratch-free surface finishes. See also buffing compound.

white button alloy. A corrosion-resistant alloy composed of 48-53% copper, 23-25% zinc, 22-24.4% nickel and 2-2.4% iron, used for white buttons.

white button metal. See button alloy.

white carbon. See carbon white.

white carbon black. A fine, white *fumed silica* powder obtained from silicon tetrachloride, and used as a pigment, filler and carbon black substitute in light-colored rubber goods.

white cast iron. An extremely hard, brittle cast iron formed when on solidification the carbon dissolved in the molten iron is not precipitated as graphite, but remains chemically combined with iron as hard *cementite* (iron carbide, Fe_3C). The formation of graphite can be prevented by rapid cooling (chilling), by lowering the silicon contents, by increasing the manganese content accordingly, or by heat treating ordinary cast iron. *White cast iron* typically contains between 2.1 and 4% carbon, has a characteristic silvery-white crystalline fracture surface, high compressive strength, good retention of strength and hardness at elevated temperature, excellent abrasion and wear resistance, and fair to poor machinability and hot workability. Used for the manufacture of malleable iron castings, and for strong, tough, abrasion-resistant castings, chilled-iron rolls, and grinding balls. Abbreviation: WCI. Also known as *hard cast iron; hard iron; white iron.* See also chilled cast iron.

white caustic. See sodium hydroxide.

white cedar. See Atlantic white cedar; eastern white cedar.

white cement. A white, finely ground, essentially iron-free *Portland cement* usually obtained from pure *calcite limestone* and containing opacifying fillers, such as chalk and white-burning clay. It may also be obtained from less pure materials by sintering in a reducing atmosphere. Used in building construction, and in dry coloring. Also known as *white Portland cement*.

white clay. See kaolin.

white cobalt. See cobaltite.

white copper. A term referring to a group of silvery-white, corrosion-resistant alloys composed of copper, zinc and nickel. The preferred term for these alloys is *nickel silvers*.

white copperas. See zinc sulfate heptahydrate.

white corundum. A pure, colorless variety of *corundum* (99+% Al_2O_3) that lacks the coloring impurities of *ruby* (i.e., chromic oxide) and *sapphire* (i.e., cobalt, chromium and titanium).

White Crown Clay. Trade name of Southeastern Clay Company (USA) for a kaolin clay.

White Crystal. Trade name of Corning Glass Works (USA) for an infrared-transmitting glass.

white cypress. See baldcypress.

white deal. See Norway spruce.

White Diamond. Trade name of Matchless Metal Polish Company (USA) for buffing compositions used in metal finishing.

white dogwood. See flowering dogwood.

white elm. See American elm.

White Economy. Trade name for a low-gold dental casting alloy.

White End. Trade name of Evans Steel Company (USA) for an oil-hardening steel containing 0.7% carbon, 1.5% chromium, 0.2% vanadium, and the balance iron. Used for rolls.

white factice. A white, soft, mealy substance prepared by a reaction of sulfur chloride with rapeseed oil, and used directly in the manufacture of erasers and soft rubber goods, and indirectly as a compounding ingredient in rubber, and as a rubber substitute. See also factice; vulcanized oil.

white feldspar. See albite.

white fir. See Norway spruce.

white flint glass. A colorless *flint glass* with high light dispersion used for optical instruments. See also optical flint glass.

white flux. A white powder composed of sodium carbonate (Na_2CO_3), sodium nitrate ($NaNO_3$) and sodium nitrite ($NaNO_2$), and used as a powerful oxidizer for welding metals.

white fused alumina. See fused white alumina.

white glass. See milk glass; opal glass.

white glue. A fast-bonding polyvinyl acetate adhesive used for construction purposes, e.g., for bonding porous materials, such as wood, leather, fabrics, etc.

white gold. Any of several white-colored corrosion-resistant alloys containing (i) 90% gold and 10% palladium, (ii) 20-60% nickel and 40-80% gold, (iii) 75-85% gold, 8-10% nickel and 2-9% zinc, (iv) 37-75% gold, 17-28% copper and 12-25% nickel, or (v) 3-17% zinc, 60% gold and 40% platinum. They possess good workability, and are used for jewelry. See also palladium gold; platinum gold.

white gold solders. A group of white, corrosion-resistant gold solders containing 30-82% gold, 10-55% silver, 1-4% copper, 1-17% zinc, 0-12% nickel and 0-2% cadmium. Used in electronics, and for jewelry and ornaments.

white graniteware. A very strong, white *earthenware*. See also ironstone china.

white graphite. A name sometimes used for commercial boron nitride in the form white crystals or powder especially when used as a lubricant.

white ground coat. A white or opaque porcelain enamel that can be applied directly to a sheet steel as a ground coat.

whiteheart malleable iron. White cast iron that has been converted by a prolonged annealing process involving heating followed by slow cooling under carefully controlled conditions of temperature and time with the resulting microstructure consisting of temper carbon in a ferritic matrix. It has a light-colored fracture surface, hence the name "whiteheart." A typical chemical composition is up to 0.8% total carbon, up to 0.4% combined carbon, 0.1-0.4% manganese, 0.5-1.1% silicon, 0.25% sulfur, 0.1% phosphorus, and the balance iron. *Whiteheart malleable iron* has high strength, ductility and toughness, good shock resistance, good machinability, brazability and solderability, and is used for railroad and automotive castings, thin-walled parts, fittings, levers, keys, padlock components, brake drums, and chain links. Also known as *whiteheart malleable cast iron*.

white hickory. See mockernut hickory.

white holly. See American holly.

white iron. See white cast iron.

white iron pyrite. See marcasite.

whiteite. (1) A pale tan-colored mineral of the jahnsite group composed of calcium iron magnesium manganese aluminum phosphate octahydrate, $Ca(Fe,Mn)Mg_2Al_2(PO_4)_4(OH)_2 \cdot 8H_2O$. Crystal system, monoclinic. Density, 2.58 g/cm³; refractive index, 1.585. Occurrence: Brazil, USA (California).

(2) A tan to gray mineral of the jahnsite group composed of calcium iron magnesium aluminum phosphate hydroxide octahydrate, $CaFeMg_2Al_2(PO_4)_4(OH)_2 \cdot 8H_2O$. Crystal system, monoclinic. Density, 2.61 g/cm³; refractive index, 1.585. Occurrence: Canada (Yukon).

White Label. (1) Trade name of A. Milne & Company (USA) for water-hardening tool steels (AISI types W1 and W2) containing 0.7-1.3% carbon.

(2) Trade name of Peninsular Steel Company (USA) for a high-carbon, high-chromium cold-work tool steel (AISI type D2) containing 1.5-1.6% carbon, 11.5-12.5% chromium, 0.7-0.9% molybdenum, 0.8-1% vanadium, and the balance iron. It has good abrasion and wear resistance, good nondeforming properties, and is used for bending and other forming tools, and a variety of dies.

(3) Trade name of Houghton & Richards Inc. (USA) for a tungsten-type high-speed steel containing 0.7% carbon, 18% tungsten, 4% chromium, 1% vanadium, and the balance iron. It has high hot hardness, good wear resistance and toughness, and is used for cutting, machining and finishing tools.

(4) Trade name of F.E. Knight, Inc. (USA) for a molding rubber used for jewelry applications.

white latten. A copper-base alloy containing zinc and tin, and supplied in the form of thin sheets.

white lead. A white, amorphous powder (99+% pure) made from metallic lead. It occurs in nature as the mineral *hydrocerussite*, has a density of 6.14 g/cm³ (0.222 lb/in.³), and decomposes at 400°C (750°F). Used as a fluxing constituent in glazes, porcelain enamels and glass, in organic and organometallic research, and as a pigment for paints and putties. Formula: $2PbCO_3 \cdot Pb(OH)_2$. Also known as *basic lead carbonate; lead flake; lead subcarbonate*.

white lead ore. See cerussite.

white lead putty. A white, dough-like mixture of at least 10% *white lead*, calcium carbonate and linseed oil, used for setting glass in window frames, filling imperfections in wood or metal surfaces, as a filler for patterns, and for general sealing and caulking applications.

white lead silicate. See basic lead silicate.

white lead sulfate. See basic white lead.

Whiteley. Trade name for a corrosion-resistant dental alloy containing 45-55% gold, 30-35% palladium and 15-20% platinum.

Whitelight. Trade name of White Metal Rolling & Stamping Corporation (USA) for a series of wrought aluminum and magnesium alloys usually supplied in the extruded condition.

white lime. See white basswood.

white locust. See black locust.

white maple. See silver maple.

white-metal alloys. See babbitts.

white-metal bearing alloys. See babbitts.

white metals. (1) A group of white-colored, low-melting alloys composed chiefly of lead, tin and antimony. The three most important classes are: (i) *Fusible alloys;* (ii) *Type metals;* and (iii) *Bearing alloys (babbitts).* See also antifriction metals.

(2) A group of white-colored, low-melting metals including antimony, bismuth, cadmium, lead, tin and zinc.

white mica. See muscovite.

white nickel. A casting alloy composed of 55-64% copper, 18% nickel, 0.35% iron, and the balance zinc. Used for trimmings, brackets, levers, fittings, and plumbing applications. Also known as *white nickel brass.*

white nickel alloy. A strong *cupronickel* composed of 65% copper, 32.25% nickel and 2.75% aluminum, and used for heat- and corrosion-resistant parts, bearings, and bushings.

white nickel brass. See white nickel.

white oak. See American white oak.

white olivine. See forsterite.

White Pine. American trade name for *lake copper* (99.9% pure).

white pine. See eastern white pine; western white pine.

white poplar. The soft, tough wood of the large silvery white poplar tree *Populus alba,* native to Europe, but also introduced in the northern United States and southern Canada. It has a fine, straight grain, moderate durability, and works well. Average weight: 450 kg/m³ (28 lb/ft³). Used for paper pulp, construction lumber, woodenware, and packaging. Also known as *abele.*

white Portland cement. See white cement.

white quebracho. The light, hard, durable wood of any of several trees (genus *Schinopsis)* of the cashew family, especially *S. lorentzii* and *S. balansae.* Source: Argentina, Brazil, Paraguay. Used for lumber, and as a source of tannin for leather tanning. See also quebracho; red quebracho.

white sandalwood. The light-colored, fragrant heartwood of the evergreen tree *Santalum album.* It has a close grain and high hardness, and is used for cabinetwork, ornamental carved objects, chests, boxes, and fans.

white schorl. See albite.

white seraya. The wood of the tree *Parashorea plicata.* It is similar to *yellow seraya,* but lighter in weight. Average weight: 530 kg/m³ (33 lb/ft³). Source: Malay Peninsula, Borneo, Sabah. Used for interior construction work, joinery, and ships' decking under cover. Also known as *white meranti.*

white shellac. White flakes of *shellac* made by bleaching ordinary orange or yellow shellac with alkalies..

white spruce. The almost white, straight-grained wood of the tall spruce *Picea glauca* supplied together with black and red spruce. Its heartwood and sapwood are almost indistinguishable. *White spruce* has good stability after drying, moderate strength, moderate durability when used outdoors, no appreciable odor, and is only slightly resinous. It closely resembles *European spruce* (white deal). Average weight, 420 kg/m³ (26 lb/ft³). Source:

Canada; northeastern United States (Lake states and New England). Used for framing lumber, flooring, millwork, boxes, crates, packing cases, piano sounding boards, ladders, and as a pulpwood. Also known as *shingle spruce.*

white tellurium. See sylvanite.

white tin. See beta tin; tin.

white tombac. A copper-zinc alloy (brass) made white by the addition of arsenic. See also tombac.

White Tombasil. Trade name of Illingworth Steel Company (USA) for a silvery, corrosion-resistant, low-melting alloy composed of copper, zinc, manganese, nickel and lead, and used for marine hardware, building hardware, pumps, and swimming-pool fixtures.

white vitriol. See zinc sulfate heptahydrate.

whiteware. A clay-based ceramic product that becomes white or ivory-colored after firing at high temperatures. *Whitewares* include china, earthenware, plumbing sanitary ware, porcelain, semivitreous ware, and tile. Also known as *ceramic whiteware.*

white walnut. See butternut.

whitewash. (1) A dilute suspension of lime hydrate (calcium hydroxide) or calcium carbonate in water. It has the appearance and consistency of milk, and is applied with a brush to walls as a temporary paint-like coating. Also known as *lime wash; lime water; milk of lime.*

(2) A suspension of casein, trisodium phosphate and lime paste used as a paint, and for decorative applications.

white willow. The soft, but tough wood of the tree *Salix alba* native to Europe, but also found in southeastern Canada and the USA (South Dakota to Georgia and Missouri). It is not of major commercial importance, but sometimes used for cooperage and novelty items, and as a fuelwood. See also black willow.

White-Wood. Trade name of Nudo Products, Inc. (USA) for aluminum, vinyl and fiberglass paneling used for ceilings, walls and signs.

whitewood. See American whitewood; Norway spruce.

white zinc oxide. See zinc oxide.

whiting. See chalk (2).

whiting putty. A dough-like mass made by mixing a dry, white pigment powder composed of 95% calcium carbonate and 5% tinting pigment (e.g., *white lead*) with linseed oil. See also putty.

whitlockite. (1) A white mineral composed of calcium magnesium phosphate, $(Ca,Mg)_3(PO_4)_2$. Crystal system, rhombohedral (hexagonal). Density, 3.12 g/cm³; refractive index, 1.629. Occurrence: USA (New Hampshire).

(2) A colorless to white, gray or yellowish synthetic mineral composed of calcium phosphate, β-$Ca_3(PO_4)_2$. Crystal system, rhombohedral (hexagonal). Density, 3.12 g/cm³; refractive index, 1.629.

whitmoreite. A brown to green-brown mineral of the arthurite group composed of iron phosphate hydroxide tetrahydrate, $Fe \cdot Fe_2(PO_4)_2(OH)_2 \cdot 4H_2O$. Crystal system, monoclinic. Density, 2.87 g/cm³; refractive index, 1.725. Occurrence: USA (New Hampshire).

whole wood. A term referring to any commercially used wood material and including lumber, poles, fencing, panel materials and pulpwood.

Wibalin. Trademark of Winter & Co. GmbH (Germany) for an artificial leather imitating suede.

wickenburgite. A colorless mineral composed of calcium lead aluminum silicate hydroxide, $CaPb_3Al_2Si_{10}O_{24}(OH)_6$. Crystal system, hexagonal. Density, 3.85 g/cm³; refractive index,

1.6918. Occurrence: USA (Arizona).

Wickerweave. Trade name of former Glas- und Spiegel-Manufaktur AG (Germany) for a glass with basketweave pattern. Also known in German-speaking countries under the trade name "Korbgeflecht."

WickGard. Trademark of Honeywell Performance Fibers (USA) for low-wick polyester fibers used for the manufacture of tarps, truck covers, outdoor furniture, etc.

wickmanite. A yellow or brownish greenish mineral of the sohngeite group composed of manganese tin hydroxide, $MnSn(OH)_6$. Crystal system, cubic. Density, 3.89; refractive index, 1.705. Occurrence: Sweden.

wicksite. A dark blue to almost black mineral composed of sodium calcium iron magnesium phosphate dihydrate, $NaCa_2Fe_5Mg(PO_4)_6 \cdot 2H_2O$. Crystal system, orthorhombic. Density, 3.54 g/cm³; refractive index, 1.718. Occurrence: Canada (Yukon).

Wicromal. Trade name of J. & A. Erbsloh Aluminium (Germany) for an aluminum alloy containing 0.5-1.5% manganese and up to 0.3% chromium. It has good workability and weldability, and is used for cooking utensils, heat exchangers, tanks, and furniture.

Widalox. Trade name of Krupp Widia GmbH (Germany) for an aluminum oxide (Al_2O_3) based ceramic material containing special additives. It has high hardness, high hot hardness, and is used for cutting tools.

Widarock. Trade name for a hard die stone used for dental laboratory applications.

Widder. German trade name for a series of solders containing 75-90% lead and 10-25% tin.

widenmannite. A yellow mineral composed of lead uranyl carbonate, $Pb_2UO_2(CO_3)_3$. Crystal system, orthorhombic. Density, 6.89 g/cm³; refractive index, 1.905. Occurrence: Germany.

Widia. Trade name of Krupp Widia GmbH (Germany) for an extremely hard and strong composite material composed of sintered tungsten carbide (3-20 wt%) in a cobalt matrix. It may also contain small amounts of titanium carbide and/or tantalum carbide, and has a high melting point, relatively poor shock resistance, and good abrasion and wear resistance. Used for tool bits, drill bits, cutting tools, dies, reamers, and shaper tools.

Widiadur. Trade name of Krupp Widia GmbH (Germany) for several cemented carbides.

Widie. German trade name for a solder composed of 70% lead and 30% zinc, and having a solidus temperature of 183°C (361°F) and a liquidus temperature of 255°C (491°F).

Widoplan. Trade name of Wilhelmi Werke GmbH & Co. KG (Germany) for chipboard.

Widotex. Trade name of Wilhelmi Werke GmbH & Co. KG (Germany) for textured chipboard.

Wiegold. Trade name of Wieland-Werke AG (Germany) for a gold-colored alloy containing 87% copper, 11% tin, 1% nickel, and 1% tin. Used for art objects.

Wieland. (1) Trade name of Wieland-Werke AG (Germany) for an extensive series of cast and wrought copper alloys including several aluminum and nickel bronzes, phosphor bronzes, leaded and unleaded tin bronzes, leaded and unleaded red brasses, and leaded and unleaded coppers.

(2) Trade name of Wieland-Werke AG (Germany) for a series of aluminum products including high-purity aluminum and various alloys of aluminum and magnesium, aluminum, magnesium and silicon, and other aluminum alloys.

Wiesilber. Trade name of Wieland-Werke AG (Germany) for a nickel silver containing 46-67% copper, 7-19% nickel, 0-2%

lead, and the balance zinc. Used for marine hardware.

wightmanite. A colorless mineral composed of magnesium borate dihydrate, $Mg_5(BO_3)O(OH)_5 \cdot 2H_2O$. Crystal system, triclinic. Density, 2.59 g/cm³; refractive index, 1.603. Occurrence: USA (California).

Wikilana. Trade name for rayon fibers, yarns and fabrics.

Wikmans. Trade name of Wikmanshytte Bruks AB (Sweden) for a series of machinery steels, several austenitic, ferritic and martensitic stainless steels, and a wide range of tool steels including water-hardening plain-carbon, high-speed, hot-work, cold-work, shock-resisting, and die and mold types.

Wilco. Trade name of Engelhard Corporation (USA) for a series of sintered electrical contact materials based on gold, osmium, palladium, platinum or silver. Also included under this trade name are several silver-base brazing alloys and various thermometals.

Wilcoloy. Trade name of Engelhard Corporation (USA) for several strong high-conductivity beryllium, cadmium and chromium coppers and copper-beryllium-cobalt alloys used for spring contacts, electrical springs, and switch and relay parts.

wilcoxite. A colorless to white mineral composed of magnesium aluminum fluoride sulfate hydrate, $MgAl(SO_4)_2F \cdot 18H_2O$. Crystal system, triclinic. Density, 1.58 g/cm³; refractive index, 1.436. Occurrence: USA (New Mexico).

wild black cherry. See American cherry.

wild cherry. See American cherry.

wild silk. Silk produced by the larvae of certain species of uncultivated Asian moths of the genus *Antheraea* (tussah silk) and by the larva of the Anaphe moth (anaphe silk). It has a strong, coarse filament and can be spun into textiles. See also anaphe; silk; tussah.

Wilkinite. Trade name for a sodium bentonite supplied in the form of a very fine powder. It has excellent absorption and adsorption capacity, a great affinity for water, and is used in ceramic bodies and refractories, as a suspension agent in porcelain enamels and glazes, and in papermaking.

Wilkinson's catalyst. A transition-metal complex that consists of a central rhodium atom, three triphenylphosphine $[(C_6H_5)_3P]$ ligands bonded to the rhodium atom by three unshared electron pairs, and one chlorine (Cl) ligand. It is supplied in the form of maroon crystals, and used in homogeneous catalysis, e.g., for the hydroboration of olefins and the decarbonylation of aldehydes. It is also available on 2% divinylbenzene (DVB) crosslinked polystyrene support. Formula: $[(C_6H_5)_3P]_3RhCl$.

wilkmanite. A pale grayish yellow mineral of the nickeline group composed of nickel selenide, Ni_3Se_4. Crystal system, monoclinic. Density, 6.88 g/cm³. Occurrence: Finland.

Wilkoro. Trade name of The Wilkinson Company, Inc. (USA) for gold-based dental and medical alloys.

Will-Ceram. Trademark of Ivoclar Vivadent AG (Liechtenstein) for feldspathic dental porcelains and various ceramic dental alloys. *Will-Ceram Lite-Cast* is a beryllium-free ceramic dental alloy containing 68.5% nickel, 15.5% chromium, 14.0% molybdenum, 1.0% aluminum, and less than 1% silicon and manganese, respectively. It has a white color, a density of 8.5 g/cm³ (0.31 lb/in.³), a melting range of 1330-1390°C (2425-2535°F), a hardness of 195 Vickers, excellent biocompatibility, good finishability, and moderate elongation. Used for single crowns, posts, and bridges. *Will-Ceram W* is an extra-hard ceramic dental alloy containing 54.0% gold, 26.4% palladium, 15.5% silver, 2.5% tin, 1.5% indium and less than 1.0% ruthenium, rhenium and lithium, respectively. It has a white color, a density of 13.8 g/cm³ (0.50 lb/in.³), a melting range of 1230-1280°C (2245-2335°F), an as-cast hardness of 240 Vickers, excellent biocompatibility, good castability, and moderate elongation. Used for crowns, onlays, posts, bridges, and model castings. *Will-Ceram Y* is a ceramic dental alloy containing 84.0% gold, 7.1% platinum, 5.7% palladium, 1.5% silver, and less than 1.0% tin, indium, rhenium, iron and lithium, respectively. It has a rich yellow color, a density of 17.4 g/cm³ (0.63 lb/in.³), a melting range of 1170-1210°C (2140-2210°F), an as-cast hardness of 185 Vickers, excellent biocompatibility, good processibility, and low elongation. Used for crowns, onlays, posts, and bridges. *Will-Ceram Y-Lite* is a ceramic dental alloy containing 75.0% gold, 18.8% palladium, 2.0% silver, 2.0% tin, 2.0% indium, and less than 1.0% copper and rhenium, respectively. It has a yellow color, a density of 16.1 g/cm³ (0.58 lb/in.³), a melting range of 1150-1250°C (2100-2280°F), a hardness after ceramic firing of 225 Vickers, low elongation and excellent biocompatibility and resistance to oral conditions. Used for crowns, onlays, posts, and bridges.

willemite. A colorless, white, yellowish-green, red or brown mineral of the phenakite group with a vitreous to resinous luster. It is composed of zinc orthosilicate, Zn_2SiO_4, and contains theoretically 58.5% zinc. The mineral *troostite* is a manganese-bearing variety. It can also be made synthetically. Crystal system, rhombohedral (hexagonal). Density, 3.30-4.25 g/cm³; hardness, 5.5 Mohs; refractive index, 1.691. It shows intense bright-green fluorescence in ultraviolet radiation, and sometimes exhibits a marked phosphorescence. Occurrence: Africa, Greenland, USA (New Jersey, New Mexico). Used as an ore of zinc, as a phosphor, and in the production of crystalline glazes. The yellow and green species are used as gemstones. Also known as *siliceous calamine.*

willemseite. A green mineral of the talc group composed of magnesium nickel silicate hydroxide, $(Ni,Mg)_3Si_4O_{10}(OH)_2$. Crystal system, monoclinic. Density, 3.31 g/cm³; refractive index, 1.652. Oc-currence: South Africa.

willhendersonite. A colorless mineral of the zeolite group composed of potassium calcium aluminum silicate pentahydrate, $KCaAl_3Si_3O_{12} \cdot 5H_2O$. Crystal system, triclinic. Density, 2.18 g/cm³; refractive index, 1.511. Occurrence: Germany, Italy.

Williams. Trade name of Williams Precious Metals (USA) for an extensive series of soldering alloys including various tin-lead, tin-antimony, tin-silver, tin-zinc, lead-silver, lead-silver-tin, indium-base and bismuth-base alloys as well as numerous copper-, gold- and silver-base alloys including various grades of copper-gold, gold-silver, gold-copper, gold-copper-silver, silver-copper, silver-copper-nickel, gold-silicon, gold-germanium, gold-nickel, silver-copper-palladium, silver-copper-titanium, silver silicon, and gold-silver-copper alloys.

Williamson's blue. Any of a group of iron-bearing blue pigments. See also iron blue.

Willi Glas. Trade name for a glass-ceramic used as dental substrate material.

Willmid. Trademark of Illbruck GmbH (Germany) for polymer foams supplied in the form of plates, blocks, strips and shapes, and used for household appliances, heat shielding, automotive parts, and acoustic applications. *Willmid-FM* is a lightweight, fire-resistant modified polyimide foam material with a low thermal conductivity, a continuous-use temperature range of -195 to +300°C (-320 to +570°F), and excellent resistance to hot flames above 815°C (1500°F). Used for insulation applications.

willow. The light, fairly soft, tough wood from any of several broad-leaved trees and shrubs of the genus *Salix* growing in North America, especially the black willow (*S. nigra*) and the western black willow (*S. lasiandra*). A European variety, the white willow (*S. alba*), is also of some commercial importance. *Willow* has good nonshrinking properties, and is used for construction work, furniture, plywood, wall paneling, toys, novelty items, cricket bats, artificial limbs, cart bottoms, crates, cooperage, veneer, exselsior, charcoal, pulpwood, and fenceposts. See also black willow; white willow; European willow.

willow blue. A colorant consisting of *cobalt blue* diluted with white silica powder or a similar white powder.

willow calf. Grained, willow bark-tanned calfskin.

willyamite. A steel-gray to silver-white mineral of the pyrite group composed of cobalt nickel antimony sulfide, $(Co,Ni)SbS$. Crystal system, cubic. Density, 6.76 g/cm³. Occurrence: Australia (New South Wales).

Wilmil. Trade name of William Mills & Company Limited (UK) for cast aluminum-silicon and aluminum-silicon-magnesium alloys.

Wilmott's aluminum. An alloy composed of 86% tin and 14% bismuth, used as an aluminum solder.

Wil-O-dont. Trade name for a light-cure resin used for orthodontic applications.

Wimet. Trade name of Sandvik Hard Materials Limited (UK) for a series of sintered cemented carbides consisting of tungsten carbide particles in cobalt matrix binders, sometimes with small additions of tantalum, titanium, niobium and/or vanadium carbide. Used for cutting tools, metal and rock drills, extrusion and forming tools and dies, woodworking tools, and wear parts. Also included under this trade name are cemented carbides coated with titanium carbide or titanium nitride.

winchite. A blue mineral of the amphibole group composed of sodium calcium magnesium iron manganese aluminum silicate hydroxide, $NaCa(Mg,Fe,Mn,Al)_5Si_8O_{22}(OH)_2$. Crystal system, monoclinic. Density, 2.97 g/cm³; refractive index, 1.646. Occurrence: India.

window glass. Sheet glass used in regular windows and produced by floating, i.e., by continuously flowing melted *soda-lime glass* onto the flat surface of molten tin contained in a vat. As it flows over the tin a ribbon of glass is formed that has smooth, parallel surfaces. Upon cooling the glass forms a rigid continuous sheet that is then annealed, inspected and cut to size. Nominal thicknesses range from about 2.5 to 22.0 mm (0.094 to 0.875 in.). It is inferior to ground and polished plate glass.

Windsor. (1) Trade name of Jessop Steel Company (USA) for an air-hardening medium-alloy cold-work tool steel (AISI type A2) containing about 1% carbon, 0.45% manganese, 0.25% silicon, 5.25% chromium, 0.25% vanadium, 1% molybdenum, and the balance iron. It possesses good deep-hardening and nondeforming properties, and good wear resistance and toughness. Used for various dies.

    (2) Trade name of Aurident, Inc. (USA) for a gold- and silver-free high-strength dental ceramic alloy containing 78% palladium. It has good processibility, excellent mechanical integrity, and is compatible with high-expansion porcelains. Used for porcelain-to-metal restorations.

winged elm. See cork elm.

Wingtack. Trademark of Goodyear Tire & Rubber Company (USA) for light yellow, high-quality, tackifying hydrocarbon resins for hot melts and pressure-sensitive adhesives.

Wink. Trade name for metallic fibers.

Winnofil. Trademark of Solvay Performance Chemicals (USA) for a family of ultrafine precipitated coated calcium carbonates used as fillers.

Winns Bronze. Trade name for a copper-base alloy containing 32% zinc, 2.2% nickel, 0.75% lead and 0.25% iron.

Winns Superheat. Trade name for a copper-base alloy containing 35-36% zinc, 2% tin, 0.4% lead, 0.1% arsenic and 0.2% iron, used for valves and fittings for superheated steam environments.

winstanleyite. A yellow mineral composed of titanium tellurium oxide, $TiTe_3O_8$. Crystal system, cubic. Density, 5.57 g/cm^3; refractive index, 2.34. Occurrence: USA (Arizona).

Winsulite. Trade name of Winsulite Division of Economy Glass Corporation (USA) for insulating glass units.

Wintrel. Trademark of Winter & Co. GmbH (Germany) for an artificial leather.

wiping cloth. A soft, strong, lint-free cotton cloth used for wiping glass and instruments.

wiping solder. A commercial soft solder that is composed of up to 60% lead and 40% tin, and may have up to 2.5% antimony added. It has a solidus temperature of 183°C (361°F), a liquidus temperature of 238°C (460°F), and is used for soldering automobile radiator cores and heating units, and for joining cable sheath and lead pipes. Also known as *common solder; plumber's wiping solder.*

Wipla. German trade name for an austentic stainless steel containing 0.2% carbon, 18% chromium, 8% nickel, and the balance iron. Used for corrosion-resistant machinery parts.

Wipolan. Trade name for natural and synthetic textile fibers including azlon (regenerated protein) fibers and vinal (vinyl alcohol/vinyl acetate) fibers. Used for textile fabrics including clothing.

Wiptam. Trade name of Krupp Stahl AG (Germany) for a wrought cobalt-base superalloy containing 28.3% chromium, 24.4% nickel, 1.1% silicon, 0.7% manganese and 0.1% carbon. It has good corrosion and wear resistance, and is used as a dental alloy, e.g., for dentures and dental wire.

wire. (1) Any metal or alloy, such as aluminum, copper or steel, drawn out into a thin, flexible rod or thread.

(2) A metallic fiber with relative large diameter. Typical metals used include steel, molybdenum and tungsten. Wires are used chiefly as a radial steel reinforcement in automobile tires, in filament-wound rocket casings, and in wire-wound high-pressure hoses.

wire brass. A brass containing 65-72% copper, 27-35% zinc, 0.3% lead and 0.2% tin. It has high ductility and good cold workability, and is used for wire applications.

wire cloth. See wire fabric.

wire-cut brick. Building or refractory brick cut from extruded columns of clay by means of a taut wire.

wired glass. A sheet of glass having a layer of meshed wire incorporated that resists shattering when broken. Used in building construction, e.g., for windows, skylights, etc. Also known as *wired safety glass; wire glass; wire safety glass.*

Wire Drawing Alloy. Trade name of Columbia Tool Steel Company (USA) for an abrasion-resisting tool and die steel containing 2.35% carbon, 10.5% tungsten, 1.9% chromium, 1.65% manganese, 0.55% molybdenum, and the balance iron. Used for tools, and wire-drawing dies.

Wire-N-Cast. Trade name of Wahl Refractories, Inc. (USA) for wire-reinforced castable refractories.

wired safety glass. See wired glass.

wire fabric. A fabric of fine wire, plain-woven or crimped in squares or rectangles, and used for sieves and filters, window protection screens, or as protective guards or strainers. The wire may be of steel, iron, copper, brass, or other metal or alloy, and is available in mesh sizes ranging from No. 3/4 to No. 100 and with a typical diameter of 0.20-3.43 mm (0.008-0.135 in.). Filter cloth is usually finer (typically 100-400 mesh). Also known as *wire cloth.* The term mesh refers to the number of openings per linear inch. See also wire gauze.

wire gauze. A net-like structure whose openings are formed by a series of fine, crossed metal wires. It is usually finer than a *wire fabric.*

wire glass. See wired glass.

wire rod. A round metal rod used as a starting material in wire-drawing.

wire rope. A rope made by twisting strands of iron, phosphor-bronze, traction steel or plow steel wire about a metallic or nonmetallic core. Common core materials include spring steel, cotton, asbestos, and polyvinyl plastics. See also cable.

wire safety glass. See wired glass.

Wirespray. Trade name of Wall Colmonoy Corporation (USA) for wire stock used for the production of thermal-spray coatings. Various materials are available including carbon steels, stainless steels, zinc and zinc alloys, nickel, nickel-chromium-iron, nickel-copper, molybdenum, commercially pure coppers, brasses, bronzes, nickel silvers and copper nickels, lead and tin babbitts, and aluminum and aluminum alloys.

Wire Waybrite. Trade name of Nippon Sheet Glass Company Limited (Japan) for a wired corrugated glass.

Wireweld. Trade name of British Celanese (UK) for cellulose acetate plastics.

Wirilene. Trade name for polyolefin fibers and yarns used for textile fabrics.

Wiron 88. Trade name of BEGO (Germany) for a beryllium-free nickel-chromium dental bonding alloy.

Wironit. Trade name of Thyssen Edelstahlwerke AG (Germany) for an extensive series of austenitic, ferritic, martensitic and precipitation-hardening stainless steels, heat-resisting steels, and nickel-base superalloys. The corrosion-resistant cobalt-chromium grades are used as dental casting alloys.

Wirosil. Trade name of BEGO (Germany) for a silicone elastomer used as a dental duplicating and impression materials.

Wirsbo-Pex. Trade name of Wirsbo-Pex Platzer Schwedenbau GmbH (Germany) for crosslinked polyethylene tubing and piping used for heating, ventilation and air-conditioning applications.

Wiscon-Cast. Trade name of Wisconsin Centrifugal–A Metaltek International Company (USA) for continuously cast bronze products.

wiserite. A reddish to brown mineral composed of manganese borate dihydrate, $Mn_4B_2O_7 \cdot 2H_2O$. Crystal system, monoclinic. Density, 3.42 g/cm^3. Occurrence: Switzerland.

Wisil. Trade name of Krupp Stahl AG (Germany) for a corrosion-resistant cobalt-base dental alloy containing 27% chromium, 4.5% molybdenum, 1% manganese, 0.4% silicon, 0.35% carbon, and 0.55% other additions.

Wissco. Trade name of Wickwire Spencer Steel Company (USA) for a series of chromium and chromium-nickel stainless steels used for wires, springs, and welding rods.

Wissler High Speed. British trade name for a cast, nonferrous tool material containing 15-40% tungsten, 15-35% chromium, 15-50% nickel, or cobalt, 0.75-2.5% carbon, and 0.5-2.5% boron. Used for dies, and high-speed cutting tools.

Wistel. Trademark of Montefibre SpA (Italy) for polyester filament yarns used for textile fabrics.

Witco. Trade name of Witco Chemical Corporation for an extensive series of carbon blacks supplied in a wide range of grades.

witherite. A colorless or yellowish- to grayish-white mineral of the aragonite group with a vitreous to resinous luster. It is composed of barium carbonate, $BaCO_3$, and can also be made synthetically. Crystal system, orthorhombic. Density, 4.27-4.35 g/cm^3; melting point, 1360°C (2480°F); hardness, 3-3.75 Mohs; refractive index, 1.679. Occurrence: UK, USA (California, Kentucky, Michigan, Minnesota, Wisconsin). Used in optical, plate and tableware glass, in pottery bodies, as a flux in glazes and porce-lain enamels, for structural clay bodies to prevent efflorescence, as a pigment (*blanc fixe*) and extender, and as a source of barium.

Witherm. Trade name of Thyssen Edelstahlwerke AG (Germany) for a series of high-temperature steels of the nickel-chromium, chromium-silicon or chromium-nickel-silicon type, and several corrosion-resistant stainless steels of the chromium or chromium-nickel type.

Witness. Trade name of Columbus Dental (USA) for a bondable hydrocolloid used in dentistry.

Witten. Trade name of Thyssen Edelstahlwerke AG (Germany) for an extensive series of case-hardening steels, alloy steels (e.g., chromium, chromium-molybdenum, chromium-molybdenum-vanadium, chromium-nickel and molybdenum-vanadium types), austenitic, ferritic and martensitic stainless steels, and tools steels including several water-hardening plain-carbon, cold-work, hot-work, high-speed and mold types.

wittichenite. A steel-gray to tin-white mineral composed of copper bismuth sulfide, Cu_3BiS_3. Crystal system, orthorhombic. Density, 6.01-6.20 g/cm^3. Occurrence: Germany, USA (Montana).

wittite. A light lead-gray mineral of the lillianite group that is composed of lead bismuth sulfide, $Pb_5Bi_6(S,Se)_{14}$, and may also contain silver, arsenic and/or selenium. Crystal system, monoclinic. Density, 5.5-6 g/cm^3. Occurrence: Sweden, Russian Federation.

Wizard. Trade name of Ziv Steel & Wire Company (USA) for a tough, shock-resisting oil- or water-hardening tool steel containing about 0.45% carbon, 0.3% manganese, 1% chromium, 1% tungsten, 0.2% molybdenum, and the balance iron. Used for pneumatic tools, hand chisels, punches, and swaging dies.

Woco. Trade name of Bethlehem Steel Corporation (USA) for a tough, shock-resisting tool steel containing 0.45% carbon, 1.5% chromium, 2.25% tungsten, 0.25% vanadium, and the balance iron. Used for pneumatic chisels, shear blades, stamps, punches, and dies.

wodginite. (1) A brown to black mineral of the columbite group composed of manganese niobium tantalum tin oxide, $(Ta,Mn,Nb,Sn)O_2$. Crystal system, orthorhombic. Density, 7.32-7.40 g/cm^3. Occurrence: Russian Federation.

(2) A reddish brown-black mineral composed of manganese tin tantalum oxide, $(Ta,Mn,Sn)O_2$. Crystal system, monoclinic. Density, 7.16-7.36 g/cm^3. Occurrence: Australia, Canada (Manitoba).

woehlerite. A light yellow to brown mineral composed of sodium calcium zirconium niobium silicate, $NaCa_2(Zr,Nb)Si_2O_8(O,O,F,F)$. Crystal system, monoclinic. Density, 3.42 g/cm^3; refractive index, 1.716. Occurrence: Norway.

woelsendorfite. A reddish orange mineral of the becquerel group composed of lead calcium uranium oxide dihydrate, $(Pb,Ca)U_2$-$O_7\cdot2H_2O$. Crystal system, orthorhombic. Density, 6.80 g/cm^3; refractive index, 2.05. Occurrence: Germany.

Wogulan. German trade name for several polyamide resins.

Wolcrylon. Trade name of Wolfen Co. for acrylic (polyacrylonitrile) fibers used for the manufacture of clothing.

wolfachite. A silver- to tin-white mineral composed of nickel arsenic antimony sulfide, $Ni(As,Sb)S$. Crystal system, cubic. Density, 6.6-6.8 g/cm^3. Occurrence: Germany.

wolfeite. A pinkish, reddish-brown, yellowish-brown or wine yellow mineral of the triplite group composed of iron manganese phosphate hydroxide, $(Fe,Mn)_2(PO_4)(OH)$. Crystal system, monoclinic. Density, 3.79 g/cm^3; refractive index, 1.742. Occurrence: USA (New Hampshire).

Wolfin. Trademark of Degussa (Germany) for a series of synthetic resins including phenolics, polypropylenes, polvinyl chlorides, polystyrenes and copolyesters.

Wolfram. (1) Czech trade name for a borosilicate glass used for tungsten sealing applications.

(2) Trade name of Crucible Materials Corporation (USA) for a tungsten-type high-speed steel (AISI type T1) containing 0.75% carbon, 4% chromium, 18% tungsten, 1% vanadium, and the balance iron. Used for cutting, machining and finishing tools.

Wolframant. (1) Trade name of Plansee Metallwerk-Gesellschaft (Austria) for sintered carbides composed of powdered tungsten carbide in a cobalt binder, and used for boring-crown inserts, cutting tools, and drills.

(2) Trade name of Thyssen Edelstahlwerke AG (Germany) for a high-speed steel containing 0.7% carbon, 18% tungsten, 4% chromium, 1% vanadium, 5% cobalt, and the balance iron. Used for cutting, machining and finishing tools.

wolfram brass. An alloy composed of 60% copper, 22% zinc, 14% nickel, and 4% tungsten.

wolfram bronze. A sintered alloy composed of 90% copper and 10% tungsten, and used for welding electrodes and contact materials.

Wolfram Cobalt. Trade name of Vulcan Steel & Tool Company, Limited (USA) for a tungsten-type high-speed steel (AISI type T4) containing 0.75% carbon, 4% chromium, 18% tungsten, 1% vanadium, 5% cobalt, and the balance iron. It has high hot hardness, excellent wear resistance, good machinability, and is used for a wide variety of cutting, machining, finishing and shaping tools.

wolframic acid. See tungstic acid.

wolframinium. A lightweight, corrosion-resistant high-aluminum alloy containing 1.4% antimony, 0.4% tungsten, 0.3% copper, 0.2% iron and 0.1% tin. Used for autobody work.

wolframite. (1) A mineral group containing several tungstates including the iron-rich *ferberite* ($FeWO_4$), the manganese-rich *huebnerite* ($MnWO_4$), the zinc-rich *sanmartinite* ($ZnWO_4$) and the mineral *wolframite* itself.

(2) An isomorphous mixture of the minerals *ferberite* and *huebnerite* with the general formula is $(Fe,Mn)WO_4$. It is grayish or brownish-black, and has a submetallic to resinous luster. Crystal system, monoclinic. Density, 7.0-7.5 g/cm^3; hardness, 5-5.5 Mohs; refractive index, 2.22. Occurrence: Australia, Bolivia, Europe, USA (Colorado, Nevada, South Dakota). Used as an important ore of tungsten.

(3) A brown synthetic mineral composed of iron manganese tungstate $FeMn(WO_4)_2$, and made by the fusion of sodium tungstate dihydrate ($Na_2WO_4\cdot2H_2O$), ferrous chloride tetrahydrate ($FeCl_2\cdot4H_2O$) and manganous chloride dihydrate (Mn-

$Cl_2 \cdot 2H_2O$) at 800°C (1472°F). Crystal system, monoclinic. Density, 7.37 g/cm³; hardness, 5-5.5 Mohs; refractive index, 2.22.

wolfram white. See barium tungstate.

Woliplast. Trade name of Woliplast-Folien GmbH (Germany) for flexible polyvinyl chloride films.

Wolkendekor. Trade name of former Gerresheimer Glas AG (Germany) for glass blocks with patterns resembling clouds.

wollastonite. A white, gray, yellow, brown or red mineral of the pyroxenoid group with a vitreous to pearly luster. It is composed of calcium metasilicate, $CaSiO_3$, and contains theoretically 48.3% lime (CaO) and 51.7% silica (SiO_2). Crystal system, triclinic. Density, 2.8-2.9 g/cm³; hardness, 4.5-5.5 Mohs; melting point (incongruent), 1544°C (2811°F); refractive index, 1.62-1.63. Occurrence: Japan, Mexico, Rumania, USA (California, New York). Used in refractories, cements, dielectric bodies and glazes, wallboard, mineral wool, whiteware bodies, as a flux in welding-rod coatings, as mineral filler for ceramics, paints, plastics and rubber, and in silica gels and paper coatings. Also known as *tabular spar.*

Wollastonite N. Trade name of BPI Inc. (USA) for crushed and sized calcium silicate ($CaSiO_3$) used for metallurgical applications.

Wollaston wire. An extremely fine wire (less than 1 μm or 40 μin.) made by first inserting a length of bare drawn wire (e.g., platinum or gold) in a tight-fitting sheath made of a metal, such as silver, drawing the wire and sheath as an individual rod through a series of dies to the desired size, and then dissolving away the outer covering with a suitable acid. Used for microfuses, and as a wire for electroscopes and hot-wire instruments.

Wolverine Alloy. Trade name of Wolverine Tube, Inc. (USA) for a series of commercially pure aluminums and coppers, and various wrought copper-base alloys (e.g., brasses and bronzes, or copper nickels) as well as wrought aluminum-base alloys (e.g., aluminum-manganese or aluminum-magnesium).

Wompco. Trade name of Worthington Pump Inc. (USA) for a cast iron containing 2.9% total carbon, 0.90% combined carbon, 0.74% manganese, 2% silicon, 0.1% sulfur, and the balance iron. Used for castings, cylinders, and gears.

Wonico. Trade name for an alloy composed of 80% tungsten, 15% nickel and 5% cobalt. It has a very low coefficient of thermal expansion, and is used for glass-sealing applications.

wood. A natural polymeric composite that consists of strong, flexible cellulose fibers surrounded and held together by a stiff material known as *lignin*. It has anisotropic properties arising from its microstructure that includes grains, biological cells, microfibrils and polymeric molecules of cellulose. *Wood* has a high strength-to-weight ratio and a calorific value of 7-20 MJ/kg (3000-8600 Btu/lb.). Botanically, it is classified into hardwood and softwood, the former being produced by deciduous or broad-leaved trees (e.g., ashes, birches, elms, hickories, maples, oaks and walnuts) and the latter by coniferous or evergreen trees (e.g., cedars, cypresses, firs, hemlocks, larches, pines, redwoods and spruces). Used for pulp and paper, building construction, plywood and other panel materials, furniture, boxes, crates and packaging materials, cooperage, fuel, charcoal, rayon and cellophane, excelsior, filler and extender, turpentine, rosin, pine oil, etc.

wood aggregate. A *lightweight aggregate*, such as wood chips, fibers, flakes or flour, used with a suitable cementing medium to make concrete, mortar, etc. See also wood-fiber concrete.

Wood Band Saw. Trade name of Colt Industries (UK) for a water- or oil-hardening tool steel containing 0.7% carbon, 2% chromium, and the balance iron. Used for saws, especially wood band saws.

wood cellulose. Chemical cellulose derived from wood. See also chemical cellulose.

wood-cement board. A composite panel made from excelsior (wood shavings or strands) by bonding with inorganic cement.

wood charcoal. A black, brittle, highly porous form of *amorphous carbon* that is the solid residue of the destructive distillation of wood, e.g., by charring in a retort or kiln without supply of air. Usually available in the form of powder, lumps or briquettes, it is used as a fuel for heating and cooking, in arc-light electrodes, as a decolorizing and filtering medium, and in explosives, e.g., black powder.

Wood-Com. Trademark of Natural Fiber Composites, Inc. (USA) for a series of strong, lightweight engineering plastics composed of polypropylene filled with wood fibers or flour derived from waste wood, paper, or newsprint.

wood composite. A structural composite material, such as fiberboard, flakeboard, hardboard or particleboard composed of wood and other materials (e.g., vegetable fibers and/or bonding agents).

Wood Dough. Trade name of DAP, Inc. (USA) for a polymeric filler used to repair cracks, holes, gouges, scratches and similar imperfections in wood.

wood-fiber concrete. A special *lightweight concrete* made with wood fibers as the principal aggregate. See also wood aggregate.

wood-fibered plaster. A calcined gypsum plaster reinforced or strengthened with ground or shredded wood fibers.

wood-fiber-reinforced plastics. Composite materials whose plastic matrices (e.g., polyurethane or phenolic), have been reinforced with wood fibers.

wood fibers. The strong, flexible cellulose fibers of wood used as an aggregate in concrete and mortar, as fillers in wood composites, such as fiberboard, and as reinforcements in wood-reinforced plastics.

Woodfield. Trade name for prefinished hardboard panels supplied with a size of 1.2 × 2.4 m (4 × 8 ft.) and a thickness of 6.4 mm (0.25 in.), and used for interior wall paneling.

wood filler. Any inert material used to fill the holes and irregularities in planed or sanded wood surfaces, and/or decrease the surface porosity before applying finish coatings. Open-grain woods like ash, hickory, mahogany, oak, poplar or walnut require fillers.

wood flour. Very finely ground softwood or hardwood particles (typically 100-425 μm, or 0.004-0.017 in. in size) similar to wheat flour in appearance. Used as filler, extender and/or reinforcing agent in plastics, rubber, linoleum flooring, paperboard and explosives, and in fine-polishing operations, *Sorel cement* manufacture, and sand casting of metals.

Wood Glu. Trade name for an aqueous dispersion of polyvinyl acetate used as an adhesive for wood and paper products.

wood glue. Natural or synthetic glue used in the manufacture of wood products. Natural wood glues are made from substances, such as dextrin, casein (milk curd), animal blood and bones, while synthetic wood glues are derived from polymer resins, such as ureas, melamines, phenolics, resorcinolss or polyvinyls. See also adhesives; glues.

woodhouseite. A colorless to white mineral of the alunite group composed of calcium aluminum phosphate sulfate hydroxide, $CaAl_3(PO_4)(SO_4)(OH)_6$. Crystal system, rhombohedral (hexagonal). Density, 3.00 g/cm³; refractive index, 1.636. Occurrence: USA (California).

Wood-I-Beam. Trademark of Georgia-Pacific Corporation (USA) for engineered lumber supplied in the form of joists and headers.

wood I-joist. A lightweight, engineered wood product with an I-shaped configuration consisting of a flange, usually made of *dimension lumber* or *laminated veneer lumber* (LVL), and a web composed of *plywood* or *oriented strandboard* (OSB). It is supplied in lengths up to 60 ft. (18 m) for structural applications, especially floor joists. Also known as *I-beam*.

wood laminate. An assembly made by gluing together layers of wood (e.g., veneer or lumber) with an adhesive so that the grain of all laminations is essentially parallel. Used for laminated structural members, e.g., decking, beams and arches. Also known as *laminated wood*. See also built-up laminated wood; glue-laminated wood; plywood.

Woodland. Trade name of Solignum Inc. (Canada) for a weatherproof, UV-resistant, clear natural wood finish with semi-gloss sheen.

wood lath. See lath (1).

wood particles. Flakes, chips, splinters or shavings of wood used in the manufacture of *particleboard* by bonding together under pressure with a synthetic resin (e.g., phenol- or urea-formaldehyde), and as fillers in molded melamine, phenolic or urea plastics.

wood-plastic combination. A composite made by impregnating wood with a suitable synthetic resin, such as an acrylic (e.g., methyl methacrylate). The resin is cured by radiation, or by the addition of a suitable catalyst followed by heating. *Wood-plastic combination* has a density of 0.9-1.1 g/cm³ (0.03-0.04 lb/in.³), excellent indentation resistance, high hardness, and is used for flooring (especially parquet), sporting goods, tabletops, and gunstocks. Abbreviation: WPC. See also plastic wood.

wood pulp. A generic term for fibrous cellulosic materials which are held in aqueous suspension, and are derived from wood by removing some or all of the noncellulosic matter (e.g., lignin). Used in the manufacture of paper, paperboard, cardboard, rayon, and other products. Also known as *paper pulp; pulp*. See also chemical conversion pulp; groundwood pulp; mechanical pulp.

wood pulp yarn. A yarn composed essentially of long, thin, narrow wood pulp-derived paper strips spun or twisted together. Used especially for floor coverings including carpets.

wood rosin. An amorphous, hard, amber-colored resin obtained from the stumps or dead wood of pine trees (chiefly longleaf and Caribbean pine) by steam distillation. It is composed principally of isomers of abietic acid. *Wood rosin* has a density of approximately 1.08 g/cm³ (0.039 lb/in.³), a melting point of 100-150°C (210-300°F), and is used for paper sizing, as an ingredient in paints, varnishes, plastics, synthetic rubber, adhesives, mastics and sealants, in soldering and insulating compounds, and in core oils.

woodruffite. A dark brownish gray mineral of the rutile group composed of zinc manganese oxide tetrahydrate, $(Zn,Mn)_2Mn_5O_{12}·4H_2O$. Crystal system, tetragonal. Density, 4.01 g/cm³. Occurrence: India, USA (New Jersey).

Wood's alloy. See Wood's metal.

Wood Saw. Trade name of Disston Inc. (USA) for a tool steel containing 0.75% carbon, 0.7% nickel, 0.2% chromium, and the balance iron. Used for wood saws.

Wood's fusible alloy. See Wood's metal.

Wood's fusible metal. See Wood's metal.

Wood's glass. A special type of glass that has a high transmission factor for ultraviolet radiation and is nearly opaque to visible radiation.

Wood's metal. (1) A term usually referring to a fusible alloy containing 50% bismuth, 25% lead, 12.5% tin and 12.5% cadmium. It has a density of approximately 9.7 g/cm³ (0.35 lb/in.³), a melting point of 71°C (160°F), and is used for solders, fuses, safety plugs, and sprinkler plugs. Also known as *Wood's alloy; Wood's fusible alloy; Wood's fusible metal*.

(2) A term referring to any of a group of fusible alloys containing 40.5-52.5% bismuth, 25-35% lead, 9-22.5% tin, and up to 12.5% cadmium. They have a melting range of approximately 65-105°C (149-221°F), and are used for solders, fuses, and safety, and sprinkler plugs. Also known as *Wood's fusible alloy; Wood's fusible alloy; Wood's fusible metal*.

wood stain. A finish coat for wood containing a dye or a pigment. Transparent wood stains dissolve in the solvent and are carried into fibers of the wood coloring like a dye. Dye stains are available in water-, alcohol- and oil-soluble types. Pigment stains contain finely divided color pigments dissolved in linseed oil, turpentine, varnish, etc. Unlike paints, wood stains preserve and/or enhance the natural grain and texture of wood. Used on furniture, flooring, exterior and interior walls of houses, shakes, decks, patios, fences, etc. See also alcohol stain; oil stain; penetrating stain; sealer stain; varnish stain; water stain.

Wood-Stik. Trademark of National Casein Company (USA) for casein adhesives supplied in stick form for use in bonding wood, wood panels, and furniture.

wood sugar. See xylose.

wood tar. See pine tar.

Wood Tone Putty. Trademark of H.F. Staples & Company, Inc. (USA) for a wood filler.

wood veneer. See veneer (1).

woodwardite. A greenish- to turquoise-blue mineral composed of copper aluminum sulfate hydroxide hydrate, $Cu_4Al_2(SO_4)(OH)_{12}·xH_2O$. Density, 2.38 g/cm³. Occurrence: UK.

Woodweld. Trademark of Georgia-Pacific Resins, Inc. (USA) for spray-dried liquid urea- and phenol-formaldehyde binders supplied in powdered or liquid form, and used in the manufacture of plywood, waferboard, oriented strandboard and other wood products.

wood wool. Short, fine, curled shavings or strands of a soft wood (e.g. aspen, basswood, cottonwood or poplar) used as packing, stuffing and cushioning materials, and for particleboard and other lightweight building boards. Also known as *excelsior.*

wool. (1) The soft, curly hair or fur of sheep, goats, rabbits, camels, alpaca, llamas and vicunas. The staple textile fibers obtained from this hair are typically 50-200 mm (2-8 in.) long and can be readily woven and spun. Wool has a density of 1.3 g/cm³ (0.05 lb/in.³), good thermal insulating properties, good physical and felting qualities, decomposes at 126°C (259°F), and is used for clothing, blankets, upholstery, carpets, felt, and thermal insulation products.

(2) A term used for technical products supplied in the form fleecy masses of fibers resembling wool, e.g., glass wool, slag wool, or steel wool.

wool-blend stretch fabrics. Strechable textile fabrics made from blends of wool and *spandex*, or wool, polyester and spandex. Used especially for clothing.

woolens. (1) Any fabric made of wool.

(2) Fabrics, heavier than *worsteds*, spun from carded, but uncombed wool yarns.

woolen-type fabrics. Fabrics made either entirely of wool yarn, or of wool-yarn weft (filling) threads and cotton-yarn warp threads.

woolen yarn. Yarn spun from carded, but uncombed wool fibers. It is coarser and shorter than *worsted yarn*. Also known as *wool yarn*.

wool felt. A nonwoven textile fabric composed wholly of virgin wool fibers, or virgin wool fibers in combination with re-used wool or synthetic fibers, and interlocked by rolling or pressing, application of moisture and heat, and/or chemical action. Virgin wool felt is used for hats, sportswear and table coverings; other wool felts are supplied in the form of mats or rolls for insulation, padding and lining applications.

Woolon. Trade name for water- and chemical resistant vinal fibers used especially for garments.

wool pitch. The black, amorphous residue obtained during the distillation of wool fat, and used in lubricating greases for the necks of tinplate rolling mills, in papermaking, and for electrical insulation applications.

woolskin. Sheepskin tanned without removing the fleece.

wool yarn. See woolen yarn.

wootz. A good-grade of steel, originally from southern India, made directly from iron ore in a crucible. Also known as *Indian steel; wootz steel.*

wootz steel. See wootz.

workability agent. An admixture, such as a water reducer or wetting agent, used in concrete, mortar and several other plastic mixes to increase their workability (plasticity).

worked lumber. Lumber that has not only been dressed, but also matched, ship-lapped or patterned.

work lead. Pig lead as it comes from the blast furnace. It is impure and must usually be refined and desilvered prior to use.

World. Trade name of World Alloys & Refining, Inc. (USA) for an extensive range of dental alloys including several yellow-colored gold-base ceramic alloys containing 40-83% gold and varying amounts of palladium and/or silver, and various white palladium-silver alloys containing 36-70% silver, 18-35% palladium and 0-30% gold. Used for porcelain-bonding applications, crown and bridge restorations, and as implant materials and/or post and core materials. *World 2000* is a silver-free dental ceramic alloy containing 99% gold and having a rich, yellow 24-karat color. Used for bonding most regular and low-fusing porcelains.

Worry Free. Trade name for nylon fibers and yarns used for clothing, diapers, and in the form of webbing for fire escape ladder.

worsted fabrics. See worsteds.

worsteds. Firm, durable, smooth-finished fabrics made from *worsted yarn*, and used for dresses, coatings, and suitings. Also known as *worsted fabrics.*

worsted-type fabrics. Fabrics made either entirely of *worsted yarn*, or worsted weft (filling) threads and cotton warp threads.

worsted yarn. A smooth, firm, compact yarn or thread with long, parallel fibers made from wool or cotton by removing the short fibers by combing or gilling. It is now also made from synthetic fibers. Used for smooth-finished fabrics, carpets, and knitting.

Worthite. Trade name of Worthington Corporation (USA) for an austenitic stainless steel containing up to 0.07% carbon, 2.5-3.5% silicon, up to 1% manganese, 18-20% chromium, 22-25% nickel, 2.5-3% molybdenum, 1.5-2% copper, and the balance iron. It is available in wrought and cast form, and has excellent corrosion resistance, and resistance to wet, dry, and hot sulfur dioxide, sulfurous acid, and various sulfites. Used for pumps operating in corrosive environments, and for general construction applications.

Wortle Plate. Trade name for a plain-carbon steel containing 2.5% carbon, and the balance iron. Used for the manufacture of wire-drawing dies.

Wotan. Trade name of Hans Kanz Metallwerke (Germany) for a series of tungsten and cobalt-tungsten high-speed steels.

woven fabrics. Fabrics made by interlacing yarns, fibers or filaments at more or less right angles to produce a web. The yarns running lengthwise are known as the *warps* and the crosswise threads as the *woofs* (or wefts, or fillings). Depending on the pattern produced by interlacing, various weaves can be distinguished including plain, harness, leno, satin and twill weave. See also nonwoven fabrics.

woven fabric prepregs. Prepregs made by interlacing yarns, fibers or filaments at more or less right angles and impregnating with synthetic resin. See also woven fabrics; prepregs.

woven fibrous composites. Composite materials composed of metallic or nonmetallic matrices reinforced with fibers (e.g., textile, glass, carbon, graphite or aramid) interlaced in any of various weaves (e.g., plain, harness, leno, satin or twill). See also fiber-reinforced composites.

woven preform. A preshaped reinforcement of continuous aramid, carbon, glass or other fibers woven into a cloth or mat and injected or impregnated with a hot-melt matrix resin (e.g., an epoxy, bismaleimide or polyimide). Such preforms are often multidirectional.

woven roving. A heavy fabric made by weaving aramid or glass-fiber roving or yarn bundles. See also roving.

woven stretch fabrics. Woven fabrics that can be stretched by 20% or more in either lengthwise or crosswise direction, and on removal of the load exhibit an almost complete recovery.

woven-wire fabrics. Cold-drawn steel wires twisted together by mechanical means to form hexagonal openings, and used as prefabricated steel reinforcements in civil engineering and building construction.

wove paper. A soft, smooth, closely finished paper, usually white in color, and with a uniform, unlined surface, free of watermarks. Used for valuable book covers, envelopes, and scratch pads.

Wrap. Trade name of Progress Paint Manufacturing Company (USA) for a water-reducible acrylic polyester coating.

wrapping paper. A strong, tough, coarse paper, usually a brown *kraft paper* made from sulfate or mixed pulps, and used for wrapping parcels. When used for wrapping presents, it may also be a strong, white or colored paper with printed designs or patterns and one coated surface.

Wrap-It. (1) Trade name of Cotronics Corporation (USA) for a lightweight, resilient wet felt made by combining high-purity refractory fibers with special inorganic binders. After cutting to shape and molding and drying to form, it has good resistance to most chemicals and solvents, and good high-temperature resistance up to 1260°C (2300°F). Used for welding supports, as thermal insulation or fireproofing, and for handling of molten metals.

(2) Trademark of Roll-O-Sheets Canada Limited for plastic film used for food wrapping.

(3) Trademark of Myers Industries, Inc. (USA) for rubberized tape used for wrapping and cushioning applications.

wrap-spun yarn. A plied yarn consisting of a twistless core yarn wrapped with a binder yarn.

wrap yarn. A yarn with a wrapping of one or more other yarns, used for interlinings.

Wrightlon. Trademark of Airtech Europe SA (France) for aliphatic hydrocarbon release films for vacuum-bag molding applications. They have a density of 0.84 g/cm^3 (0.030 $lb/in.^3$), an upper service temperature of 204°C (400°F), and are particularly suited for use with epoxies.

wrinkle-resistant fabrics. See crease-resistant fabrics.

writing paper. A white, strong, well-sized, hard-finished, ruled or unruled fine-quality paper containing 50, 75 or 100% cotton or linen rags. Used for writing, printing and typing. See also rag paper.

wroewolfeite. A blue mineral composed of copper sulfate hydroxide hydrate, $Cu_4(SO_4)(OH)_6 \cdot 2H_2O$. Crystal system, monoclinic. Density, 3.27-3.30 g/cm^3; refractive index, 1.682. Occurrence: USA (Massachusetts).

wrought alloys. Metal alloys that are initially cast, but then drawn, extruded, forged or rolled into final, relatively simple shapes. They are relatively ductile and responsive to both cold and hot working.

wrought aluminum. Aluminum in rolled, stamped, drawn, extruded or forged form. It is 99+% pure and usually contains small additions of copper, manganese, silicon, magnesium and/or zinc.

wrought aluminum alloys. A group of *aluminum alloys* worked by hot or cold rolling, forging, swaging, upsetting, drawing, extrusion, pressing, or other processes. Examples include wrought aluminum-copper, aluminum-manganese, aluminum-silicon, aluminum-magnesium and aluminum-zinc alloys.

wrought brass. A *brass* containing 9-44% zinc, 0-2% lead, traces of aluminum, iron, manganese, nickel, antimony and tin, with the balance being copper. It has excellent hot and cold workability and forgeability, and is used for electrical parts, jewelry, flexible hose, condenser and heat-exchanger tubing, deep-drawn parts, tubes, sheets, wires, hardware, hinges, clasps, padlocks, fasteners (e.g., springs, screws, nuts, bolts or washers), fittings, watch and clock parts, structural shapes, and coinage.

wrought bronze. A *bronze* containing 1-9% tin, up to 0.4% phosphorus, traces of iron, nickel, lead and zinc, with the balance being copper. It has excellent cold workability, high elongation, good strength and corrosion resistance, good to moderate wear resistance, and is used for fasteners (e.g., screws, nuts, bolts, cotter pins, washers, or springs), diaphragms, electrical parts, pole-line hardware, door and window seals, wire mesh, flexible hose, tubing, machine parts, bushings, chemical hardware, and welding rod.

wrought carbon steels. See plain-carbon steels.

wrought heat-resistant alloys. A group of heat-resistant alloys available in the form of billets, bars, sheets, tubing and wire. Included are iron-base alloys, such as austenitic, ferritic, martensitic and precipitation-hardening stainless steels as well as nickel-, iron-nickel- and cobalt-base superalloys. See also heat-resistant alloys; stainless steels; superalloys.

wrought iron. A ferrous material that is an intimate mixture of relatively pure iron and slag. It is manufactured from *pig iron* in a reverberatory or puddling furnace, and is soft, ductile, malleable, tough, easily shaped in the hot and cold condition, readily welded and resists rusting. A typical composition is 0.02-0.06% carbon, 0.075-0.15% silicon, 0.03% manganese, 0.10-0.15% phosphorus, 0.006-0.015% sulfur, a total of 0.05% residual elements (e.g., chromium, nickel, cobalt, copper and molybdenum), 1-4% slags, and the balance iron. The overall composi-

tion of the slags used is 50-60% FeO, 5-15% Fe_2O_3, 15-20% SiO_2, 2-8% MnO, 4-7% CaO+MgO, 1-3% Al_2O_3, and 2-5% P_2O_5. Used for ornamental ironwork, smoke stacks, pipelines, chain links, gates, staybolts, railings, and decorative furniture.

wrought magnesium alloys. A group of *magnesium alloys* containing small amounts of aluminum, manganese, rare earths, thorium, zinc, and/or zirconium. They have outstanding workability and machinability, and high strength-to-weight ratios.

wrought metals. Metals that are initially cast, but then drawn, extruded, forged or rolled into final, relatively simple shapes. They are relatively ductile and responsive to both cold and hot working.

wrought red brass. Any of a group of red-colored alloys composed of 80-90% copper, 5-15% zinc, up to 10% lead and up to 5% tin. Commercially available in the form of flat products, tubes, pipes, and wire, they have a density of 8.75 g/cm^3 (0.32 $lb/in.^3$), a melting point of 1024°C (1875°F) for 85Cu-15Zn, good resistance to corrosion (including atmospheric), high ductility and malleability, excellent cold workability, good hot workability, good to fair machinability (improves with lead content), good strength, resistance to dezincification, surface finish and good conductivity (about 35% IACS). Used for weatherstripping, fasteners, fittings, plumbing, conduit, pipe, tubing, hardware, jewelry, pen caps, pencil ferrules, condenser tubes, radiator cores, heat-exchanger tubing, and drawn and stamped parts. Also known as *red brass; rich low brass.*

wrought stainless steels. A group of *stainless steels* containing at least 11% chromium along with varying amounts of nickel, manganese, silicon, phosphorus, sulfur, molybdenum, copper, niobium (columbium), titanium, tantalum, aluminum and nitrogen. Five grades are identified: (i) *Austenitic stainless steel;* (ii) *Ferritic stainless steel;* (iii) *Martensitic stainless steel;* (iv) *Precipitation-hardenable stainless steel;* and (v) *Duplex stainless.* Depending on the particular grade, the carbon content ranges between about 0.01 and 1.2%. In addition to being corrosion-resistant, all are characterized by good strength, ductility and toughness, and are available in the form of plate, sheet, strip, foil, bar, wire, semifinished products, pipes, and tubing.

wrought yellow brass. Brass composed of 63-66% copper and 34-37% zinc, and commercially supplied in the form of flats, rods and wires. It has excellent workability, good corrosion resistance and hot formability, relatively low strength, fair to poor machinability, and is used for springs, pins, rivets, screws, automobile parts, lamp fixtures, fasteners, locks, hinges, hardware, radiator cores, tanks, and flashlight parts. See also yellow casting brass.

W-Tap. Trade name of Latrobe Steel Company (USA) for an oil-hardening cold-work tool steel (AISI type O7) containing 1.23-1.25% carbon, 1.35-1.5% tungsten, 0.45% chromium, 0.3% molybdenum, 0.2% vanadium, and the balance iron. It has good deep-hardening properties, good toughness and machinability, good to moderate wear resistance, and is used for various cutting and finishing tools including taps and drills.

Wuest. Trade name for a solder composed of 50-65% zinc, 20-30% aluminum and 15-20% copper. Used for soldering aluminum and its alloys.

wuestite. A black, opaque synthetic mineral of the halite group that is a nonstoichiometric oxide of iron, $Fe_{1-x}O$ (x = 0.05), behaves as an extrinsic semiconductor, and can occur in steel as a microconstituent. Crystal system, cubic. Density, 5.74 g/cm^3; refractive index, 2.32; hardness, 100-490 Vickers; metastable below 570°C (1058°F). Also spelled wüstite.

Wulcro. Trade name of Vulcan Steel & Tool Company, Limited (UK) for a tool steel containing 1.3% carbon, 1% chromium, 4.5% tungsten, 0.15% vanadium, and the balance iron. Used for extruding dies.

wulfenite. A pale yellow, yellow, orange or brownish mineral of the scheelite group with adamantine to resinous luster. It is composed of lead molybdate, $PbMoO_4$. Crystal system, tetragonal. Density, 6.7-7.0 g/cm^3; hardness, 2.75-3 Mohs; refractive index, 2.362. Occurrence: Australia, Austria, Germany, Hungary, USA (Arizona, Massachusetts, Nevada, New Mexico, New York, Pennsylvania, Utah). Used as an ore of molybdenum. Formula: $PbMoO_4$. Also known as *yellow lead ore*.

wulfingite. A white mineral composed of zinc hydroxide, ε-$Zn(OH)_2$. Crystal system, orthorhombic. Density, 3.03 g/cm^3.

Wundus. Trade name of Sanderson Kayser Limited (UK) for a cobalt-tungsten-type high-speed steel containing 0.7% carbon, 22% tungsten, 6% cobalt, and the balance iron. Used for milling cutters, drills, finishing and shaping tools, and cutting tools for extremely hard metals.

wurtzite. A brownish-black mineral composed of zinc sulfide (ZnS), and sometimes containing small amounts of iron and manganese. It has the same composition as the mineral *sphalerite*, but crystallizes in a different structure. Crystal system, hexagonal (2H, 8H and 10H polytypes); density, 4.04-4.10 g/cm^3; refractive index, 2.356; Occurrence: South America, USA (Missouri, Ohio, Pennsylvania). Used as an ore of zinc.

wurtzite-type boron nitride. Boron nitride having a hexagonal crystal structure resembling that of the mineral *wurtzite*. It is produced by a modification of the crystallization process, or by detonation, and has a light-gray color, excellent cutting properties, and is used as a superabrasive, for shock-resistant cutting inserts, and for steel polishing applications. Abbreviation: WBN; w-BN. See also boron nitride.

Wuss-Guss. Trade name of Wuss-Guss, Metallgiesserei GmbH (Germany) for aluminum, aluminum-alloy, brass and bronze sand castings, and art-metal castings.

wüstite. See wuestite.

wyartite. A blackish mineral composed of calcium uranium carbonate hydroxide tetrahydrate, $Ca_3U_7C_2O_{22}(OH)_{16} \cdot 4H_2O$. Crystal system, orthorhombic. Density, 4.7 g/cm^3. Occurrence: Central Africa.

wych elm. The tough wood of the hardwood tree *Ulmus glabra*. It has a straight, relatively fine grain, and good bending properties. Average weight, 670 kg/m^3 (42 lb/ft^3). Source: Central Europe and British Isles. Used for chairs and other furniture, constructional purposes, boatbuilding, shafts, and agricultural implements. Also known as *mountain elm; Scotch elm*.

Wycliffe. Trade name of Follsain-Wycliffe Foundries, Limited (UK) for blackheart malleable cast iron used for agricultural and automobile parts, and in shipbuilding.

Wyex. Trade name for a series of carbon blacks.

wyllieite. A deep green mineral composed of sodium iron aluminum phosphate, $Na_2Fe_2Al(PO_4)_3$. Crystal system, monoclinic. Density, 3.6-3.7 g/cm^3; refractive index, 1.691. Occurrence: USA (South Dakota).

Wyndaloy. Trade name of Wyndale Manufacturing Company (USA) for nonmagnetic, corrosion-resistant hardenable alloys composed of about 60% copper, 20% nickel and 20% manganese. They have high electrical resistivity and wear resistance, and are used for valves, pistons, hardware, fasteners, pump components, and resistance strip and wire.

Wynene. Trade name for polyethylene fibers used for textile fabrics and other products.

Wynite. Trade name of Follsain-Wycliffe Foundries, Limited (UK) for a high-strength nickel-chromium cast iron used for jigs and pumps.

X

Xaloy. Trade name of Xaloy Inc. (USA) for several hard, wear-resistant white cast irons, corrosion- and wear-resistant nickel-chromium alloys, cobalt-nickel-chromium alloys and various alloys composed of tungsten carbide dispersed in nickel-silicon-boron matrices. Used for injection molding equipment and components, pump components, and for hardfacing applications.

X-Alloy. British trade name for an aluminum casting alloy containing 3.5-3.6% copper, 1.25% iron, 0.6-0.7% nickel, 0.6-0.7% silicon, and 0.6% magnesium. Used for pistons.

Xantal. Trade name of Alluminio SA (Italy) for a series of cast and wrought corrosion-resistant aluminum bronzes containing 81-90.5% copper, 8-11% aluminum, 0.2-4% iron, 0.2-4% nickel, 0.05-1% zinc and 0-0.8% manganese. Used for castings, propellers, and marine hardware.

Xantar. Trademark of DSM Engineered Plastics (USA) for polycarbonate resins with high heat-deflection temperature, and excellent dimensional stability and impact resistance. They are available in unreinforced, glass fiber-reinforced, fire-retardant, high-flow, UV-stabilized and structural-foam grades.

xanthates. A class of compounds prepared by treating an alcohol with carbon disulfide and an aqueous metal hydroxide. They are usually supplied in the form of water-soluble, yellow crystals or pellets. Examples include potassium ethyl xanthate ($KS_2COC_2H_5$) and sodium isopropyl xanthate [$(CH_3)_2CHOC(S)SNa$]. Used chiefly as collector agents in ore flotation, and as reagents in analytical chemistry.

xanthene dye. A class of dyes, such as eosin ($C_{20}H_8Br_4O_5$), that are structurally related to xanthene [$CH_2(C_6H_4)_2O$] containing the aromatic *chromophore* group C_6H_4.

xanthiosite. A golden yellow mineral composed of nickel arsenate, $Ni_3(AsO_4)_2$. Crystal system, monoclinic. Density, 5.37-5.39 g/cm^3. Occurrence: Germany.

xanthochroite. See greenockite.

xanthoconite. A yellowish mineral composed of silver arsenic sulfide, Ag_3AsS_3. Crystal system, monoclinic. Density, 5.54 g/cm^3. Occurrence: Czech Republic, Mexico.

xanthophyll. A yellow, water-insoluble, oxygenated *carotenoid* pigment found in leaves, fruits, flowers and egg yolk. Formula: $C_{40}H_{56}O_2$. Also known as *carotenol*.

Xanthos. Trade name Sovirel (France) for a protective plate glass used in the manufacture of welders' goggles.

xanthoxenite. A pale yellow to brownish yellow mineral of the fairfieldite group composed of calcium iron phosphate hydroxide trihydrate, $Ca_4Fe_2(PO_4)_4(OH)_2 \cdot 3H_2O$. Crystal system, triclinic. Density, 2.97 g/cm^3; refractive index, 1.715. Occurrence:

Czech Republic, USA (New Hampshire).

Xantopren. Trade name of Heraeus Kulzer Inc. (USA) for silicone elastomers and resins used for dental impressions.

X-Cavalloy. Trade name of Ingersoll-Dresser Pump Company (USA) for a stainless-steel casting alloy with excellent cavitation erosion resistance, used for pump impellers.

Xena. Trade name for rayon fibers and yarns used for textile fabrics.

Xenith. Trade name of Taskem Inc. (USA) for zinc and zinc alloy electroplates and plating processes.

Xeno. Trade name of Dentsply/Caulk (USA) for a light-curing, self-etching, fluoride-containing single-step dental resin bonding/adhesive system containing nano-sized fillers.

xenon. A colorless, odorless, heavy, rare gaseous or liquid element of the Group VIIIA (or Group 18) of the Periodic Table (noble gas group). It is present in the atmosphere in the proportion of about 0.000008 vol%. Density of gas, 5.897 g/L; density of liquid at boiling point, 1.987 g/cm^3; boiling point of liquid, -108.1°C (-162.6°F); liquefaction temperature, -106.9°C (-160°F); atomic number, 54; atomic weight, 131.29; divalent, tetavalent, hexavalent, octavalent; chemically inactive, but not inert. Used in fluorescent, sun and projection lamps, flash lamps for photography, in luminescent tubes and vacuum tubes, in electronic flashlights, in paint testers, in fluorimetry, and in lasers. Symbol: Xe.

xenon-135. A radioactive isotope of xenon with a mass number of 135 produced in nuclear reactors. It is a good neutron absorber and has a half-life, 9.2 hours. Used in nuclear engineering. Symbol: ^{135}Xe.

xenon chloride. A compound of xenon and chlorine supplied as white crystals, and used for excimer lasers. Formula: XeCl.

xenon difluoride. A moisture-sensitive, white, crystalline compound of xenon and fluorine with a density of 4.32 g/cm^3 and a melting point of 128-130°C (262-266°F), used as fluorinating and oxidizing agent. Formula: XeF_2.

xenon fluoride. See xenon difluoride; xenon tetrafluoride; xenon hexafluoride.

xenon hexafluoride. A moisture-sensitive, colorless, crystalline compound of xenon and fluorine with a melting point of 50°C (122°F). The resulting yellow liquid has a boiling point of 75°C (167°F). Used as fluorinating agent. Formula: XeF_6.

xenon tetrafluoride. A moisture-sensitive, colorless, crystalline compound of xenon and fluorine that can be prepared by mixing and heating gaseous fluorine and xenon to 400°C (752°) in a nickel vessel. Used as fluorinating agent, and for optoelectronic applications. Formula: XeF_4.

xenon trioxide. A colorless, nonvolatile, solid compound of xenon and oxygen that can be dissolved in water to form xenic acid, a stable weak acid used as a powerful oxidizer. *Xenon trioxide* reacts with metals, such as sodium, in alkaline solution yielding "metal perxenates" (e.g., sodium perxenate, $Na_4XeO_6 \cdot 8H_2O$). Formula: XeO_3.

xenotime. (1) A pale yellow mineral of the zircon group composed of yttrium erbium phosphate, $(Y,Er)PO_4$. Crystal system, tetragonal. Density, 4.68 g/cm^3. Occurrence: USA (North Carolina), New Zealand.

(2) A yellowish brown, reddish brown, pink, yellow, brown or green mineral composed of yttrium phosphate, YPO_4. It can also be made synthetically from yttrium oxide and phosphoric acid. Crystal system, tetragonal. Density, 4.75 g/cm^3; refractive index, 1.816. See also yttrium phosphate.

Xenoy. Trademark of GE Plastics (USA) for an extensive series of lightweight polycarbonates and thermoplastic alloys of polycarbonate and polybutylene terephthalate, supplied in standard and glass-reinforced grades. They have excellent chemical and corrosion resistance, outstanding impact strength, good absorption of shocks and vibrations, excellent creep and heat resistance, excellent lubricity, good electrical properties and ultraviolet-light stability, good dimensional stability and durability, good moldability, low moisture absorption, and upper service temperatures above 95°C (200°F). Used for autobody parts, bumpers, electrical and electronic components, outdoor equipment, lawnmowers, medical equipment, luggage, transport cases for scientific, electronic, medical and camera equipment, liquid-handling systems, handling pallets, fuel tanks, power-tool components, terminal junction boxes, fasteners, and fixtures.

XeraBraze I. Trade name of Metalor Technologies SA (Switzerland) for a nonprecious dental solder with a melting range of 1110-1200°C (2030-2190°F).

XeraFit. See Metalor XeraFit.

xerogel. (1) A polymer that on immersion into a suitable solvent tends to swell yielding particles with a three-dimensional polymer-chain network.

(2) A porous solid prepared by controlled evaporation of the liquid in a *gel*. It can be sintered to a fully dense glass. Also known as *zerogel*.

xilingolite. A galena white synthetic mineral composed of lead bismuth sulfide, $Pb_{1-x}Bi_{0.66x}S$, and containing 18 mol% bismuth sulfide (Bi_2S_3). Crystal system, orthorhombic. Density, 7.13 g/cm³.

xingzhongite. A steel-gray to yellow mineral composed of copper iridium sulfide, (Ir,Cu)S. Crystal system, cubic. Density, 6.64 g/cm³. Occurrence: China.

xitieshanite. A bright green mineral composed of iron sulfate hydroxide heptahydrate, $Fe(SO_4)(OH) \cdot 7H_2O$. Crystal system, monoclinic. Density, 1.99-2.02 g/cm³; refractive index, 1.570. Occurrence: China.

XL Oralite. Trade name of Oralite Company (UK) for a cellulose nitrate formerly used for denture bases.

X-Mat. Trademark of Owens-Corning Fiberglas (USA) for fabrics made by stitchbonding one or more layers of a 45-degree unidirectional glass-fiber fabric to a veil or chopped strand mat. Used as reinforcements in composites for marine and industrial applications.

xocomecatlite. A green mineral composed of copper tellurate hydroxide, $Cu_3TeO_4(OH)_4$. Crystal system, orthorhombic. Density, 4.65 g/cm³. Occurrence: Mexico.

xocotlite. A colorless to pinkish mineral composed of calcium silicate hydroxide, $Ca_6Si_6O_{17}(OH)_2$. It can also be made synthetically from an equimolar mixture of slaked lime and powdered quartz in an autoclave under prescribed conditions. Crystal system, monoclinic. Density, 2.71 g/cm³; refractive index, 1.583. Occurrence: Scotland.

X-Ply. Trademark of Boise Cascade Corporation (USA) for corrugated cardboard whose corrugations or flutes run in different directions. Used for packaging applications.

X-PLOR. Trade name of Solutions Globales (France) for a moisture-, acid-, dust- and stain-resistant textile fabric composed of 80% polyester and 20% cotton fibers. Used for outdoor recreational and work clothes including jackets, parkas, pants and overalls.

XR Bond. Trade name of Kerr Dental (USA) for a dental adhesive used for bonding dentin.

XR Ionomer. Trade name of Kerr Dental (USA) for a glass-ionomer dental cement.

XR-Polymer. Trade name of E.I. DuPont de Nemours & Company (USA) for a polysulfone-based thermoplastic engineering resin.

X-Span. Trademark of Arndt-Palmer Inc. (Canada) for fiberglass body fillers.

X-static. Trademark of Fox River Mills (USA) for silver-coated nylon fibers for gloves and outdoor hosiery.

X-Supermal. Trade name of Dresser Industries (USA) for a wear-resistant steel containing 1-1.6% carbon, 1.1-1.3% silicon, 0.3% manganese, and the balance iron. Used for chain links, conveyor equipment, and buckets.

Xtol. Trade name for cellulose acetate fibers used for the manufacture of lingerie, pyjamas, shirts, ties and swimwear, and in staple form in blends for suitings, sportswear, knitting yarns, household textiles, carpets, and cable insulations.

XT Polymer. Trade name of Cyro Industries (USA) for a series of acrylic-based multipolymer compounds used for injection-molding applications.

XTRA. Trade name for a series of textile fibers including *XTRA-dul* rayon fibers for yarns and fabrics, and *XTRA-tuff* polyester fibers for fabrics, special paper, and gloves.

Xuper. Trade name of Eutectic Corporation (USA) for a series of copper-zinc or silver-based brazing alloys and welding alloys (bare and flux-coated electrodes), and several alloy powders for metal-spraying applications.

Xydar. Trademark of BP Amoco (USA) for high-performance thermoplastic engineering resins based on liquid crystal polymers consisting essentially of *p,p*-bisphenol, *p*-hydroxybenzoic acid and terephthalic acid monomers. Supplied in unreinforced and reinforced grades, they have excellent creep and high-temperature resistance, excellent strength even at elevated temperatures, high durability, and a continuous-use temperature of 240°C (464°F) for unreinforced grades and up to 343°C (650°F) for reinforced grades. Other important properties include high dielectric strength, good resistance to fuels, solvents and aggressive chemicals, inherent flame resistance, good processibility by conventional injection molding, good stain and stick resistance, and good microwave transparency over a wide range of temperatures. Used for automotive parts including engine components, fire-wall insulation, under-hood connectors, lamp receptacles, valve and belt covers, thrust washers, cookware, and components of microwave ovens, clothes dryers, steam irons, kitchen appliances and power tools.

Xylac. Trade name of Whitford Corporation (USA) for several waterborne and high-solids enamels used for decorative applications.

Xylan. (1) Trade name of Whitford Corporation (USA) for low-friction fluoropolymer coatings and dry-film lubricants.

(2) Trade name Bodycote Metallurgical Coatings Limited (UK) for an extensive series of fluoropolymer organic coatings including *Xylan 1070* spray-applied coatings which have thicknesses ranging from 70-100 μm (2.8-3.9 mils) are heat-cured at 200°C (390°F) for approximately 20 minutes, and the *Xylan 5200* series of dip-spin-applied organic coatings which have thicknesses ranging from 10-15 μm (0.4-0.6 mil) and are also heat-cured at 200°C (390°F) for approximately 20 minutes. Used especially on automotive components including fasteners.

xylan. A pentosan polysaccharide that is found along with cellulose in wood and straw, and on complete hydrolysis by acids yields only *xylose*. Chemically, it is a 1,4-β-linked polypyranoside with rigid backbone structure.

Xylee. Trade name of Glanzstoff Austria AG (Austria) for thermoplastic polyester and polyether urethane elastomers.

Xylethon. Trademark of Durawear Corporation (USA) for an industrial plastomeric material having a frictionless, nonsticking surface that, in contrast to many other plastics, has a tendency to polish rather than become serrated and stringy. It is available in the form of sheeting for materials handling applications.

xylodine. A yarn consisting of a textile fiber and a strip of coated paper twisted together with glue.

Xylolin. German trade name for a paper yarn.

xylolith. See magnesium oxychloride cement.

Xylon. (1) Trademark of American Sandpaper Company (USA) for a coated abrasive paper.

(2) Trademark of Rexall Drug & Chemical Company (USA) for thermoplastic molding resin pellets.

(3) Trade name of Fiberfil (USA) for a series of nylon fibers and products.

Xylon/VO. Trade name of DSM Engineering Plastics (USA) for a high-performance nylon 6,6 resin with excellent chemical, heat and flame resistance, and good electrical properties. Used for automotive, electrical and electronic components.

Xylonite. Trademark of Bakelite Limited (UK) for a thermoplastic material composed of nitrocellulose and camphor, resembling *celluloid*, and used in photoelasticity for making models.

xylose. An aldopentose obtained from wood, straw, corncobs or wood pulp wastes on hydrolysis with hot dilute acids. It is commercially available as a white crystalline, dextrorotatory powder with a melting point of 144-153°C (291-307°F), and used as a source of ethanol, as a sweetener, in dyeing and tanning, and in the biosciences, e.g., for the synthesis of biopolymers. Formula: $C_5H_{10}O_5$. Also known as *wood sugar*.

Xypex. Trademark of Xypex Chemical Corporation (Canada) for a waterproofing compound used to seal concrete pores against water seepage by crystallization growth.

Xytrex. Trademark of EGC Corporation (USA) for a family of custom-blend resins, such as polyetheretherketone (PEEK), polyetherimide (PEI), polyether sulfone (PES) and polyphenylene sulfide (PPS). They exhibit good strength retention at high temperatures, good load-bearing capabilities and chemical resistance and good electrical insulation properties. Used for molded and extruded parts, e.g., tubing, bars, rods, disks, sheets and blocks.

Y

Yablonovite. The first photonic crystal developed by E. Yablono-vitch and co-workers at Bell Communications Research, New Jersey, USA in 1991. It exhibited a three-dimensional photonic band gap, did not propagate microwaves, and was produced by drilling 1 mm (0.04 in.) holes into a block of dielectric material with a refractive index of 3.6. See also photonic crystals; 3D photonic crystals.

Yacare. Trade name of Vidrieria Argentina SA (Argentina) for a glass with irregular, mosaic type pattern.

yacca gum. See acaroid resin.

yafsoanite. A brown mineral composed of calcium lead zinc tellurium oxide, $(Zn,Ca,Pb)_3TeO_6$. Crystal system, cubic. Density, 5.55 g/cm^3; refractive index, 1.800. Occurrence: Russian Federation.

YAG. See yttrium-aluminum garnet.

yagiite. A colorless mineral of the osumilite group composed of sodium magnesium aluminum silicate, $(Na_3K)_3Mg_4(Al,Mg)_6$-$(Si,Al)_{24}O_{60}$. Crystal system, hexagonal. Density, 2.70 g/cm^3; refractive index, 1.536.

yak fibers. Natural protein fibers obtained from the wool or hair of the yak (*Bos gruniens*), a large, long-haired central Asian animal related to cattle and the North American buffalo. *Yak fibers* can be spun into rope and woven into cloth.

Yale bronze. A free-cutting bronze containing 7-8% zinc, 0.5-1.5% tin, 0.7-1.5% lead, and the balance copper. Used for screw-machine products, bushings, nuts, and bolts.

Y-Alloy. Trade name of Sterling International Technology Limited (UK) for a *duralumin*-type high-temperature aluminum casting alloy containing 3.5-4.5% copper, 1.2-1.7% magnesium, 1.8-2.3% nickel, up to 0.7% silicon, up to 0.6% iron and 0.2% titanium. Available in sand- and die-cast form, it has good strength and corrosion resistance, and is used for automobile pistons and cylinder heads.

Yalova. Trademark of Yalova Eliat AS (Turkey) for acrylic (poly-acrylonitrile) and modacrylic (modified acrylic) fibers used for textile fabrics.

Yambolen. Trademark of Yambolen (Bulgaria) for polyester staple fibers and filament yarns used for textile fabrics.

yang. The wood of various trees of the species *Dipterocarpus* growing in Thailand. Similar wood from Burma and the Andaman Islands is known as *gurjun* and that from Borneo and the Malay Peninsula as *keruing*. The wood of these three species is interchangeable for most practical purposes. Average weight, 720 kg/m^3 (45 lb/ft^3). Used as constructional timber in railroad construction, in building as a substitute for oak, and for parquet flooring.

yard lumber. See non-stress-graded lumber.

yarn. A twisted bundle of continuous filaments, fibers or strands composed of natural or synthetic material suitable for knitting, spinning, weaving or otherwise converting into textile fabrics for use in making clothing, automotive tires, or as a reinforcement. See also industrial yarns; textile yarns.

yarn-dyed fabrics. Textile fabrics woven or knitted from dyed (colored) yarns. Used for especially for tapestries and striped or plaided clothing.

yaroslvaite. A white mineral composed of calcium aluminum fluoride hydroxide monohydrate, $Ca_3Al_2F_{10}(OH)_2 \cdot H_2O$. Crystal system, orthorhombic. Density, 3.09 g/cm^3.

yavapaiite. A pink mineral composed of potassium iron sulfate, $KFe(SO_4)_2$. It can also be made synthetically. Crystal system, monoclinic. Density, 2.88-2.93 g/cm^3; refractive index, 1.684. Occurrence: USA (Arizona).

Yaw-Ten. Trade name of Yawata Iron & Steel Company Limited (Japan) for a series of low-carbon structural steels typically containing 0.25-0.50% copper. They possess good resistance to atmospheric corrosion, and good weldability and formability. Used for buildings, structures, vehicle bodies, bridges, and mining, construction and other industrial machinery and equipment.

YD-Nicral. Trade name of Special Metals Corporation (USA) for an yttria-dispersed nickel-chromium alloy composed of 16% chromium, 4.8% aluminum, 1.0% yttria (Y_2O_3), and the balance nickel. It has good high-temperature and stress-rupture strength.

yeast. A collective name for unicellular organisms of varying shape or size of the fungi family belonging to the order Saccharomycetales, especially the genus *Saccharomyces cerevisiae*. Used in the fermentation of sugars, beverages, medicine and bread, in the biosynthesis of proteins, in the manufacture of biomaterials, and as a source of enzymes, nucleic acids and vitamins.

yeatmanite. A dark brown mineral composed of manganese zinc antimony silicate, $Mn_7Zn_8Sb_2Si_4O_{28}$. Crystal system, triclinic. Density, 4.91 g/cm^3; refractive index, 1.895. Occurrence: USA (New Jersey).

yedlinite. A red-violet mineral composed of lead chromium oxide chloride hydroxide, $Pb_6CrCl_6(O,OH)_8$. Crystal system, rhombohedral (hexagonal). Density, 5.85 g/cm^3; refractive index, 2.125. Occurrence: USA (Arizona).

Yellow. Trade name of World Alloys & Refining, Inc. (USA) for a white-colored copper-free dental ceramic alloy containing 35% palladium, 30% silver and 2% gold. It has a density of 10.5 g/cm^3 (0.38 $lb/in.^3$), a melting range of 932-960°C (1710-1760°F), a hardness of 180 Vickers, and low elongation. Used for crown and bridge restorations, as an implant material, and as a post and core material.

yellow acaroid. The yellow or orange gum resin derived from the aloe-like trees *Xanthorrhoea preissii* and *X. tateana* growing in Australia. It is an *acaroid resin* available in the form of small pieces and derives its alternate name "black boy gum" from the appearance of the tree. Used in varnishes, inks and wood stains. Also known as *black boy gum; black boy resin*.

yellow angico. Angico (or queenwood) is the very hard, reddish-brown wood of the curupay tree (*Angico rigada*). The light-brown variety of this wood is known as "angico vermelho," or "yellow angico." It has a close grain, and an average weight of 1121 kg/m^3 (70 lb/ft^3). Source: Brazil. Used for furniture and cabinetwork.

yellow arsenious sulfide. See arsenic sulfide (iii).

yellow birch. The heavy, hard, strong wood of the tall birch tree *Betula alleghaniensis*. The heartwood is light reddish-brown and the sapwood white. *Yellow birch* has a fine uniform texture, good shock resistance and considerable shrinkage. Average weight, 690 kg/m³ (43 lb/ft³). Source: Canada, USA (Northeastern states, Lake states, and along Appalachian Mountains to northern Georgia). Used as lumber and veneer for furniture, chairs, crates, boxes, framing for upholstered work, kitchen utensils, tools handles, woodenware, cooperage, doors, flooring and general utility work, and as plywood for cabinets and furniture. Also known as *silver birch*.

yellow brass. See wrought yellow brass; yellow casting brass.

yellow buckeye. The light, soft, tough wood of the large deciduous trees *Aesculus octandra* belonging to the horse-chestnut family. The heartwood is yellowish-white and the sapwood white. *Yellow buckeye* has a uniform texture, a straight grain, low shock resistance and is difficult to machine. Source: USA (Eastern and central states especially along the Appalachian Mountains from Pennsylvania to North Carolina and to northern Alabama). Used for lumber, pulpwood, boxes, planing-mill products, furniture, and artificial limbs. Also known as *sweet buckeye*.

yellow cake. (1) A term usually referring to *uranous-uranic oxide* (triuranium octoxide), U_3O_8, a yellow uranium oxide.

(2) A term referring to a mixture of uranium oxides that may include sodium diuranate ($Na_2U_2O_7$), ammonium diuranate [$(NH_4)_2U_2O_7$] as well as triuranium octoxide (U_3O_8).

yellow casting brass. A brass composed of 58-72% copper, 22-40% zinc, 1-2% tin and 1-4% lead, and commercially available as centrifugal, continuous, permanent-mold, plaster, sand and die castings. It has good machinability and corrosion resistance, low to moderate strength and elongation, relatively low hardness, and is used for machine parts, bushings, hardware, fittings, ornamental castings, ship trimmings, valves and cocks, plumbing fixtures and fittings, and battery clamps. See also wrought yellow brass.

yellow cedar. See Alaska cedar.

yellow copper. See chalcopyrite.

yellow copperas. See copiapite.

yellow cypress. See Alaska cedar.

yellow deal. See northern pine.

yellow fir. See Douglas fir.

yellow glass. (1) A term sometimes used for *soda-lime glass*.

(2) A glass that has been colored yellow by the addition of certain coloring agents, such as uranium or iron compounds.

yellow gold. A corrosion-resistant jewelry alloy composed of 50-53% gold, 25% silver and 22-25% copper.

yellow guayacan. The hard, strong wood from the tree *Tecoma guayacan* growing throughout Central America. Used for construction purposes, furniture, and woodenware.

Yellow H. Trade name of Pavco Inc. (USA) for a chromate conversion coating.

yellow ingot metal. A *brass ingot metal* containing 65% copper, 2% lead, 1% tin, and the balance zinc. Used in the manufacture of ingots for subsequent casting or further working, and for plumber's fitting.

yellow iron oxide. A yellow pigment of very fine particle size made by reacting ferric oxide, calcium sulfate and water. It has high tinctorial strength and outstanding lightfastness and alkali resistance, and is used in plastics, rubber and paints. Formula: $Fe_2O_3 \cdot H_2O$.

Yellow Label. (1) Trade name of T. Turton & Sons Limited (UK)

for a plain-carbon tool steel containing 0.9% carbon, 0.03% sulfur, 0.03% phosphorus, and the balance iron.

(2) Trade name of Peninsular Steel Company (USA) for a nondeforming tool steel containing 0.9% carbon, 1.2% manganese, 0.5% chromium, 0.5% tungsten, 0.2% vanadium, and the balance iron. Used for tools and dies.

(3) Trade name of Wallace Murray Corporation (USA) for a tool steel containing 0.8-1.1% carbon, 0.15-0.25% vanadium, and the balance iron. Used for milling cutters, dies, tools, and shapers.

yellow lauan. The yellow- to tan-colored wood of any of various trees of the genus *Shorea*. It has low to medium strength and relatively large warpage, and is used for construction purposes and woodenware, but is greatly inferior to *red lauan* (Philippine mahogany).

yellow lead ore. See wulfenite.

yellow lead oxide. See lead monoxide.

yellow locust. See black locust.

yellow meranti. See yellow seraya.

yellow mercuric oxide. Yellow crystals or orange-yellow powder (99% pure). Crystal system, orthorhombic. Density, 11.14 g/cm³; melting point, decomposes at 500°C (932°F). Used in ceramics, in the manufacture of mercury compounds, and in biochemistry and medicine. Formula: HgO. Also known as *yellow mercury oxide; yellow precipitate*.

yellow metal. (1) An archaic term for the noble metal "gold."

(2) See Muntz metal.

yellow ocher. An opaque mixture composed chiefly of *limonite*, FeO(OH)· xH₂O, with some clay and silica. It has an earth tone of yellow to orange-yellow hue. Used as a pigment. Also known as *Cassel yellow; mineral yellow; minette*.

yellow oxide. See sodium diuranate.

yellow pine. The relatively hard wood from any of several species of pine especially the longleaf pine (*Pinus palustris*), the shortleaf pine (*P. echinata*) and the ponderosa pine (*P. ponderosa*).

yellow poplar. See American whitewood.

yellow potassium prussiate. See potassium ferrocyanide.

yellow precipitate. See yellow mercuric oxide.

yellow prussiate of potash. See potassium ferrocyanide.

yellow prussiate of soda. See sodium ferrocyanide.

yellow pyrite. See chalcopyrite.

yellow resin. See pine gum.

yellow salt. See uranyl nitrate.

yellow seraya. The Sabah name for the wood from various species of *Shorea*. It is also known by the Malaysian and Sarawak name "yellow meranti." *Yellow seraya* is similar to *red seraya* in general properties, but somewhat paler and usually slightly stronger and denser. Average weight, 660 kg/m³ (41 lb/ft³). Used for superior joinery, and extensively for interior construction work. Also known as *yellow meranti*.

yellow shellac varnish. A yellow solution made by dissolving *shellac* in grain alcohol or sometimes wood alcohol. Used for coating foundry patterns.

Yellow Special. Trade name of Metalor Technologies SA (Switzerland) for an extra-hard dental casting alloy containing 41% gold and 1.7% palladium.

yellow spruce. See Sitka spruce.

yellow tellurium. See sylvanite.

yellow ultramarine. A yellow pigment composed of barium chromate ($BaCrO_4$) and used in paints and glazes.

yellow ware. A yellow- or buff-colored semivitreous ware or earthenware that may be coated with a transparent, colorless glaze.

yew. The reddish-brown, heavy, strong wood from coniferous trees of the genus *Taxus*, especially the Pacific yew (*T. brevifolia*) and the English yew (*T. baccata*). Used for archery bows, fenceposts, furniture and cabinetwork. See also English yew; Pacific yew.

yftisite. A yellowish mineral composed of yttrium rare-earth titanium oxide silicate fluoride hydroxide, $(Y,Dy,Er,Yb)_4TiO(SiO_4)_2$-$(F,OH)_6$. Crystal system, orthorhombic. Density, 3.96 g/cm^3; refractive index, 1.705. Occurrence: Russian Federation.

YIG. See yttrium-iron garnet.

Ykalon. Trademark of Mitsubishi Chemical Company (Japan) for a series of polypropylene and phenol-formaldehyde plastics.

Yocomite. Trade name of James Yocurn & Son Inc. (USA) for a nickel cast iron containing 2.8-3% carbon, 3% nickel, 0.75% chromium, and the balance iron. Used for dies.

yoderite. A purple mineral composed of magnesium aluminum silicate hydroxide, $(Mg,Al)_2Si(O,OH)_5$. Crystal system, monoclinic. Density, 3.39 g/cm^3; refractive index, 1.691. Occurrence: Tanzania.

Yo-Flex. Trade name of Youngstown Steel (USA) for a series of high-strength hot-dipped galvanized sheet steels containing about 0.13% carbon, 0.45% manganese, 0.01% phosphorus, 0.025% sulfur, and the balance iron.

yofortierite. A rose to violet mineral of the sepiolite group composed of manganese silicate hydroxide hydrate, $Mn_5Si_8O_{20}$-$(OH)_2(OH_2)_4 \cdot xH_2O$. Crystal system, monoclinic. Density, 2.18 g/cm^3. Occurrence: Canada (Quebec).

Yo-Lead. Trade name of Youngstown Steel (USA) for a series of free-machining grades of resulfurized and rephosphorized low-carbon steels containing about 0.14% carbon, 1.1% manganese, 0.35% sulfur, 0.06% phosphorus, 0.15-0.35% lead, and the balance iron. Used for screw-machine products, yokes, studs, bolts, nuts, and universal joints.

Yoloy. Trade name of Youngstown Steel (USA) for an extensive series of copper-bearing low-alloy construction steels and high-strength low-alloy steels. They have good weldability and formability, good corrosion resistance, and are used for construction, mining and transportation equipment, industrial machinery, railroad cars, trailers, autobodies, truck frames, buildings, bridges, structures, and pressure vessels.

Yo-Man. Trade name of Youngstown Steel (USA) for a copper-bearing high-strength low-alloy (HSLA) steel with improved resistance to atmospheric corrosion, used for construction equipment and structural applications.

Yo-Namel. Trade name of Youngstown Steel (USA) for porcelain enameled, low-carbon sheet steels.

Yoonsteel. Trade name of Yoonsteel (Malaysia) Sdn. Bhd. for an extensive series of low- and medium-carbon steel castings, wear-resistant manganese steels, low-alloy structural steels, corrosion- and/or heat-resisting stainless steels, gray and ductile cast irons, wear-resistant alloy cast irons, and corrosion- and heat-resistant cast irons.

Yorcalbro. Trade name of IMI Yorkshire Alloys Limited (UK) for an aluminum brass containing 76-78% copper, 19.5-22% zinc, 1.8-2.3% aluminum and 0.02-0.04% arsenic. It has good resistance to erosion and seawater corrosion, and is used for condenser and heat-exchanger tubes, and seawater pipelines.

Yorcaston. Trade name of Yorkshire Imperial Metals Limited (UK) for a wrought bronze containing 88% copper and 12% tin. It has good resistance to seawater corrosion, and is used for con-

denser and heat-exchanger tubing.

Yorcoron. Trade name of Yorkshire Imperial Metals Limited (UK) for a wrought copper nickel containing 65% copper, 30-31% nickel, 2% iron and 2% manganese. It possesses very good abrasion and corrosion resistance, relatively high strength, and good hot and cold formability. These properties vary greatly with heat treatment. Used for plates and tubes.

Yorcunic. Trade name of Yorkshire Imperial Metals Limited (UK) for an alloy of 87% copper, 10% nickel, 2% iron and 1% manganese. It has excellent resistance to seawater corrosion, and good hot and cold workability. Used for evaporator tubes and feedwater heaters.

Yorcunife. Trade name of Yorkshire Imperial Metals Limited (UK) for an alloy of 92.8% copper, 5.5% nickel, 1.2% iron and 0.5% manganese. It has excellent resistance to seawater corrosion, and good formability, bendability and weldability. Used for seawater pipelines, saltwater pipes, and condenser, heat-exchanger and evaporator tubes and components.

York. Trade name of Time Steel Service Inc. (USA) for an air-hardening steel containing 0.75% carbon, 2% manganese, 1.35% molybdenum, 1% chromium, 0.3% silicon, and the balance iron. Used for tools and dies.

Yorkcrete. Trademark of Yorkton Concrete Products Limited (Canada) for concrete blocks.

Yorklite. Trademark of Yorkton Concrete Products Limited (Canada) for lightweight concrete blocks.

Yorkshire. Trade name of Yorkshire Imperial Metals Limited (UK) for a copper nickel containing 30% nickel, 0.8% manganese, 0.7% iron, and the balance copper. It possesses excellent resistance to seawater corrosion, good strength, and is used for heat exchangers, feedwater heaters, and evaporator and condenser tubes.

yoshimuraite. A dark to orange brown mineral composed of barium manganese titanium silicate phosphate hydroxide, Ba_2Mn_2-$(Ti,Fe)O(Si_2O_7)(P,S)O_4(OH)$. Crystal system, triclinic. Density, 4.13 g/cm^3; refractive index, 1.777. Occurrence: Japan.

Youngstown. Trade name of Youngstown Steel (USA) for a series of electrical steels containing up to 4% silicon. Used for armatures, dynamos, and transformers.

Ytong. Trademark of Ytong AG (Germany) for a porous, lightweight concrete made by the addition of foaming agents to ordinary *lime concrete*. It sets under steam and pressure at approximately 180°C (355°F), and is used for masonry, roofing and thermal insulation applications.

ytterbia. See ytterbium oxide.

ytterbium. A soft, ductile, malleable, metallic element with bright silvery luster of the lanthanide series (rare-earth group) of the Periodic Table. It is commercially available in the form of ingots, rods, lumps, chips, turnings, sheet, foil, wire and powder. The chief ores are *bastnaesite* and *monazite*. Density, 6.977 g/cm^3; melting point, 819°C (1506°F); boiling point, 1194°C (2181°F); hardness, 25 Brinell; atomic number, 70; atomic weight, 173.04, divalent, trivalent. Three allotropic forms are known: (i) *Alpha ytterbium* (face-centered cubic crystal structure) that exists at room temperature; (ii) *Beta ytterbium* (body-centered cubic crystal structure) that exists at high temperatures (798-819°C or 1468-1506°F), and at room temperature and at a pressure of more than 40 kbar; and (ii) *Gamma ytterbium* (hexagonal crystal structure) that exists at low temperatures (below 270K or -3.15°C). *Ytterbium* is used for special alloys, in lasers, phosphors and ceramic capacitors, in high-temperature superconductors, ferrite devices and catalysts, as a dopant for garnets, and as a portable x-ray source. Foils are used to measure pressures, and as stress transducers. Symbol: Yb.

ytterbium-169. A radioactive isotope of ytterbium with a mass number of 169 that decays by electron capture, has a half-life of 33 days, and is used in industrial radiography of thin materials and power-plant boiler tubes. Symbol: ^{169}Yb.

ytterbium barium copper oxide. A ceramic compound available in the form of a fine, superconducting powder, and as thin films. It is quite similar to *yttrium barium copper oxide*, but with ytterbium replacing all of the yttrium. Used for high-temperature superconductors. Formula: $YbBa_2Cu_3O_x$. Abbreviation YbBCO.

ytterbium boride. Any of the following compounds of ytterbium and boron used in ceramics and materials research: (i) *Ytterbium triboride*. Density, 6.74 g/cm^3. Formula: YbB_3; (ii) *Ytterbium tetraboride*. Density, 7.31 g/cm^3. Formula: YbB_4; and (iii) *Ytterbium hexaboride*. Density, 5.57 g/cm^3; hardness, 3800 Vickers. Formula: YbB_6.

ytterbium bromide. A white crystalline compound with a melting point above 300°C (570°F). Formula: $YbBr_3$.

ytterbium carbide. Any of the following compounds of ytterbium and carbon used in ceramics and materials research: (i) *Triytterbium carbide*. Density, 10.26 g/cm^3. Formula: Yb_3C; and (ii) *Ytterbium dicarbide*. Density, 8.10 g/cm^3. Formula: YbC_2.

ytterbium chloride. A white, hygroscopic powder (99.9+% pure) with melting point of 875°C (1607°F). Used in ceramics and materials research. Formula: $YbCl_3$.

ytterbium chloride hexahydrate. White, water-soluble, orthorhombic crystals (99.9+% pure) with a density of 2.575 g/cm^3, a melting point of 150-155°C (302-311°F), and a boiling point of 180°C (356°F) (loses $6H_2O$). Used in chemistry, ceramics and materials research. Formula: $YbCl_3 \cdot 6H_2O$.

ytterbium dibromide. Yellow crystals with a melting point of 673°C (1243°F), used in ceramics, electronics and materials research. Formula: $YbBr_2$.

ytterbium dicarbide. See ytterbium carbide (ii).

ytterbium dichloride. Green crystals with a density of 5.27 g/cm^3 and a melting point of 721°C (1330°F), used in ceramics, electronics and materials research. Formula: $YbCl_2$.

ytterbium diiodide. Black crystals with a melting point of 772°C (1422°F), used in ceramics, electronics and materials research. Formula: YbI_2.

ytterbium disilicate. See ytterbium silicate (iii).

ytterbium disilicide. See ytterbium silicide.

ytterbium disulfide. See ytterbium sulfide (ii).

ytterbium erbium sulfide. Orthorhombic crystals with a melting point of 2013°C±30°C (3655°F±54°F), used in ceramics, electronics and materials research. Formula: $YbEr_2S_4$.

ytterbium fluoride. A hygroscopic, white powder (99.9+% pure) with a density of 8.168 g/cm^3, a melting point of 1157°C (2114°F) and a boiling point of 2200°C (3992°F), used in ceramics, superconductor studies, electronics, and in the preparation of fluorinated glasses. Formula: YbF_3.

ytterbium hexaboride. See ytterbium boride (iii).

ytterbium-iron garnet. A synthetic *garnet* of the rare-earth type that has ferrimagnetic properties and corresponds to the general formula $Yb_3Fe_5O_{12}$. Used for magnetic and electronic applications, and in resonators at microwave frequencies. Abbreviation: YbIG.

ytterbium monosulfide. See ytterbium sulfide (i).

ytterbium monosilicate. See ytterbium silicate (i).

ytterbium nitride. A compound of ytterbium and nitrogen used in ceramics and materials research. Density, 11.33 g/cm^3. Formula: YbN.

ytterbium oxide. A rare-earth oxide available in the form of colorless, cubic crystals, or a heavy, white powder (99.9+% pure). Density, 9.1-9.2 g/cm^3; melting point, 2335°C (4235°F). Used for special alloys, carbon rods for industrial lighting, and as a component in electrically conducting ceramics, glass-ceramics, special glasses and refractories, and phosphors. Formula: Yb_2O_3. Also known as *ytterbia*.

ytterbium sesquisulfide. See ytterbium sulfide (iii).

ytterbium selenide. A compound of ytterbium and selenium with a melting point above 1520°C (2768°F), used in semiconductors and ceramics. Formula: YbSe.

ytterbium silicate. Any of the following compounds of ytterbium oxide and silicon dioxide used in ceramics and materials research: (i) *Ytterbium monosilicate*. Melting point, 1979°C (3594°F); hardness, 5-7 Mohs. Formula: Yb_2SiO_5; (ii) *Ytterbium trisilicate*. Melting point, 1949°C (3540°F); hardness, 5-7 Mohs; Formula: $Yb_4Si_3O_{12}$; and (iii) *Ytterbium disilicate*. Occurs in nature as the mineral *keiviite* and can also be made synthetically from ytterbia and silica. Crystal system, monoclinic. Density, 6.15 g/cm^3; melting point, 1777°C (3231°F); hardness, 5-7 Mohs. Formula: $Yb_2Si_2O_5$.

ytterbium silicide. A compound of ytterbium and silicon available in the form of hexagonal crystals with a density of 7.54 g/cm^3. Used in ceramics and materials research. Formula: $YbSi_2$. Also known as *ytterbium disilicide*.

ytterbium sulfide. Any of the following compounds of ytterbium and sulfur used in ceramics, electronics, and materials research: (i) *Ytterbium monosulfide*. Density, 6.75 g/cm^3. Formula: YbS; (ii) *Ytterbium disulfide*. Formula: YbS_2; (iii) *Ytterbium sesquisulfide*. Bright yellow crystals. Crystal system, cubic. Density, 6.04 g/cm^3; band gap, 2.38 eV; semiconductive properties. Formula: Yb_2S_3; and (iv) *Ytterbium tetrasulfide*. Density, 6.74 g/cm^3. Formula: Yb_3S_4.

ytterbium telluride. A compound of ytterbium and tellurium with a melting point of 1738°C (3160°F), used in ceramics, and in electronics and materials research as a semiconductor. Formula: YbTe.

ytterbium tetraboride. See ytterbium boride (ii).

ytterbium tetrasulfide See ytterbium sulfide (iv).

ytterbium triboride. See ytterbium boride (i).

ytterbium trisilicate. See ytterbium silicate (ii).

Yttralox. Trade name for a transparent ceramic made from yttria (Y_2O_3). It melts above 2200°C (4000°F), and is used for lasers, high-intensity lamps, high-temperature lenses, and infrared windows.

yttria. See yttrium oxide.

yttria aluminate. A compound of yttrium, aluminum and oxygen available in the form of single crystals with low dielectric constant for microwave and high-frequency applications, and as a superconductor substrate. Formula: $YAlO_3$. See also yttrium aluminum perovskite.

yttria-dispersed iron alloys. A group of iron-based alloys with about 20% chromium, 5% aluminum and 0.5% titanium that have been hardened and strengthened by a dispersion (typically 0.5 wt%) of small-diameter yttria (Y_2O_3) particles. They have good high-temperature properties including tensile strength and creep resistance, and are used for high-temperature turbine-engine nozzles and seals, thermal protection systems for spacecraft, heat shields, and aircraft and spacecraft control surfaces.

yttria-dispersed nickel alloys. A group of nickel-based alloys with about 20% chromium, 1.0% iron, 0.5% titanium and 0.3% aluminum that have been hardened and strengthened by a dispersion (typically 0.6 wt%) of small-diameter yttria (Y_2O_3) particles. They have good high-temperature properties including tensile strength and creep resistance, and are used for high-temperature turbine-engine nozzles and seals, thermal protection systems for spacecraft, heat shields, and aircraft and spacecraft control surfaces.

yttrialite. (1) A black, brown or greenish mineral of the amorphous group composed of yttrium thorium silicate, $(Y,Th)_2Si_2O_7$. Crystal system, monoclinic. Density, 4.30 g/cm³; refractive index, 1.738. Occurrence: Sweden.

(2) A mineral of the amorphous group composed of yttrium thorium silicate, $Y_{1.95}Th_{0.05}Si_2O_7$. It can also be made synthetically from yttrium silicate ($Y_2Si_2O_7$) and thoria (ThO_2) under prescribed conditions. Crystal system, monoclinic. Density, 4.00 g/cm³. Occurrence: Sweden.

yttria-stabilized zirconia. A white ceramic material containing 92-97% zirconia (ZrO_2) stabilized with 3-8% yttria (Y_2O_3). It is commercially available in the form of sheets, nonwoven fabrics, chopped fibers and powder, and has a cubic crystal structure, a density of 5.80-5.90 g/cm³ (0.210-0.213 lb/in.³), a melting point of 2500°C (4530°F), an upper continuous-use temperature of 2200°C (3990°F), a hardness of about 1250 Vickers, a dielectric constant of about 27, and high compressive strength. Used in the reinforcement of metals and ceramics, in high-temperature applications, and as a bioceramic in medicine, e.g., in orthopedics for hip replacements. Abbreviation: YSZ. Also known as *yttria-toughened zirconia (YTZ); yttria-toughened zirconia polycrystal (YTZP); cubic zirconia.* See also fully stabilized zirconia; lime-stabilized zirconia; magnesia-stabilized zirconia; partially stabilized zirconia; tetragonal zirconia polycrystal.

yttria-toughened zirconia. See yttria-stabilized zirconia.

yttria-toughened zirconia polycrystal. See yttria-stabilized zirconia.

yttric rare earths. See heavy rare earths.

yttrics. See heavy rare earths.

yttrium. A grayish black or dark gray metallic element with silvery luster belonging to Group IIIB (Group 3) of the Periodic Table. It is commercially available in the form of chips, ingots, rods, foil, powder and wire. The chief ores are *bastnaesite, monazite* and various minerals of the amorphous and columbite groups of minerals. Density, 4.478 g/cm³; melting point, 1522°C (2772°F); boiling point, 3338°C (6040°F); 30-140 Brinell; atomic number, 39; atomic weight, 88.906; trivalent; low neutron capture cross section (about 1.3 Barns); low electrical conductivity (about 3% IACS). Two forms are known: (i) *Alpha yttrium* (hexagonal close-packed crystal structure) existing at room temperature; and (ii) *Beta yttrium* (body-centered cubic crystal structure) existing at high temperatures (1460-1522°C or 2660-2772°F). *Yttrium* is used in iron, aluminum and magnesium alloys to increase strength, in oxidation-resistant alloys, as a deoxidizer for vana-dium and other nonferrous metals, as a grain refiner in chromi-um, molybdenum, zirconium and titanium alloys, in nuclear technology, for garnets and microwave ferrites, for electronic components, radar equipment, television screens and camera lenses, in lasers, in refractories, in coatings for high-temperature alloys, as a host material for rare-earth phosphors, in special semiconductors, and in high-temperature superconductors. Symbol: Y.

yttrium acetate hydrate. White, water-soluble, hygroscopic crystals (99.9% pure). Melting point, 285°C (545°F) (decomposes); loses H_2O at 100-110°C (210-230°F). Used in analytical chemistry and superconductivity studies. Formula: $Y(O_2C_2H_3)_3 \cdot xH_2O$.

yttrium acetylacetonate. A metal alkoxide in the form of whitish yellow crystals (99.9+% pure), soluble in acetone and toluene, and having a melting point of 130-133°C (266-271°F). It is also supplied as the trihydrate. Used in the preparation of luminescent films on glass, and as an organometallic precursor for superconductivity research. Formula: $Y(O_2C_5H_7)_3$ (anhydrous); $Y(O_2C_5H_7)_3 \cdot 3H_2O$ (trihydrate). Also known as *yttrium 2,4-pentanedionate.*

yttrium aluminate. Any of the following compounds of yttrium oxide and aluminum oxide used in ceramics, electronics and materials research: (i) *Yttrium monoaluminate.* Density, 5.50 g/cm³. Formula: $Y_2Al_2O_6$; (ii) *Diyttrium aluminate.* Melting point, 2838°C (5140°F). Formula: $Y_4Al_2O_9$; and (iii) *Diyttrium pentaluminate.* Melting point, 1982°C (3600°F). Formula: $Y_6Al_{10}O_{24}$.

yttrium-aluminum garnet. A synthetic garnet of the rare-earth type having ferrimagnetic properties and corresponding to the general formula $Y_3Al_5O_{12}$. Yttrium-aluminum garnet crystals are nonhygroscopic, and can be doped with chromium, cerium, erbium, holmium, neodymium, thorium, thulium or ytterbium ions. Crystal system, cubic. Density, 4.57 g/cm³; melting point, 1970°C (3578°F); hardness, 8.5 Mohs. Used in lasers, scintillators, and for electronic and microwave applications. Abbreviation: YAG. See also garnet; cerium-doped yttrium aluminum garnet; gallium-doped yttrium aluminum garnet.

yttrium aluminum perovskite. A nonhygroscopic compound of yttrium, aluminum and oxygen, supplied in the form of single crystals. Crystal system, rhombohedral. Density, 5.37-6.1 g/cm³; melt-ing point, 1875°C (3407°F); hardness, 8.5+ Mohs; dielectric constant, 16-20. Used for entrance windows, scintillator crystals in gamma and X-ray counters, electron microscopes, electron and X-ray imaging screens, and tomography units. It may also contain dopants, such as cerium, erbium, neodymium and thulium. Formula: $YAlO_3$. Abbreviation: YAP.

yttrium antimonide. A compound of yttrium and antimony supplied in the form of cubic crystals or a high-purity powder with a density of 5.97 g/cm³, a melting point of 2310°C (4190°). Used as a semiconductor. Formula: YSb.

yttrium arsenide. A compound of yttrium and arsenic supplied in the form of cubic crystals, or as a high-purity powder with a density of 5.59 g/cm³, and used as a semiconductor. Formula: YAs.

yttrium barium copper hydroxy carbonate. A fine gray-brown powder containing about 10% yttrium, 22% barium and 33% copper. It requires heat treatment to form the superconducting *yttrium barium copper oxide.* Used for high-temperature superconductors, e.g., in the manufacture of $YBa_2Cu_3O_x$. Abbreviation: YBCHC.

yttrium barium copper iron oxide. A ceramic compound available in the form of a fine, superconducting powder. It is similar to *yttrium barium copper oxide,* but contains iron. Used in the manufacture of high-temperature superconductors. Abbreviation: YBCFO.

yttrium barium copper oxide. A fine, black, superconducting high-purity ceramic powder containing about 13-14% yttrium, 40-42% barium and 27-31% copper. It is also available as a thin film material. It belongs to the group of *1-2-3 supercon-*

ductors. Bulk powder density, 1.0 g/cm^3; sintered density, above 5.7 g/cm^3; superconductivity critical temperature, 88-95K. Used in superconductivity research, and as a raw material for the manufacture of high-temperature superconducting components, e.g., wires, bearings, sputtering targets, thin films, thick films, etc. Formula: YBa$_2$Cu$_3$O$_x$ (x = 6.5-7.0). Abbreviations: YBCO; Y-123.

yttrium barium copper oxide carbonate. A fine, black, high-purity, ceramic powder containing about 12% yttrium, 36-38% barium and 24-25% copper. It requires heat treatment to become superconductive, and is used for high-temperature superconductors. Formula: (Y$_2$O$_3$)$_{0.5}$(BaCO$_3$)$_2$(CuO)$_3$. Abbreviation: YBCOC.

yttrium barium copper zinc oxide. A ceramic compound available in the form of a fine, superconducting powder. It is similar to *yttrium barium copper oxide,* but contains zinc. Used for the manufacture of high-temperature superconductors. Abbreviation: YBCZO.

yttrium barium magnesium oxide. A ceramic compound available in the form of a fine, superconducting powder similar to *yttrium barium copper oxide,* but with magnesium replacing all of the copper. Used in the manufacture of high-temperature superconductors. Abbreviation: YBMO.

yttrium barium nickel oxide. A ceramic compound available in the form of a fine, superconducting powder similar to *yttrium barium copper oxide,* but with nickel replacing all of the copper. Used in the manufacture of high-temperature superconductors. Abbreviation: YBNO.

yttrium barium zinc oxide. A ceramic compound available in the form of a fine, superconducting powder similar to *yttrium barium copper oxide,* but with zinc replacing all of the copper. Used in the manufacture of high-temperature superconductors. Abbreviation: YBZO.

yttrium beryllide. A compound of yttrium and beryllium with a density of 2.56 g/cm^3, used in ceramics and materials research. Formula: YBe$_{13}$.

yttrium boride. Any of the following compounds of yttrium and boron used in ceramics and materials research: (i) *Yttrium diboride.* Density, 2.91 g/cm^3. Formula: YB$_2$; (ii) *Yttrium triboride.* Density, 3.97 g/cm^3. Formula: YB$_3$;(iii) *Yttrium tetraboride.* Density, 4.36 g/cm^3. Formula: YB$_4$; and (iv) *Yttrium hexaboride.* Refractory solid. Density, 3.72 g/cm^3; hardness, 3260 Vickers. Formula: YB$_6$.

yttrium carbide. Any of the following compounds of yttrium and boron used in ceramics and materials research: (i) *Yttrium monocarbide.* Melting point, 1950°C (3542°F). Formula: YC; (ii) *Yttrium dicarbide.* A yellow refractory, microcrystalline compound. Density, 4.13-4.33 g/cm^3; melting point, 2400°C (4352°F); hardness, 700 Vickers. Formula: YC$_2$; (iii) *Triyttrium carbide.* Density, 5.41 g/cm^3. Formula: Y$_3$C; and (iv) *Yttrium sesquicarbide.* Melting point, 1800°C (3272°F); hardness, 900 Vickers. Formula: Y$_2$C$_3$.

yttrium carbonate. A white to reddish, hygroscopic powder (99.9+% pure) used as a phosphor in refractory gas mantles, and in superconductivity studies. Formula: Y$_2$(CO$_3$)$_3$·3H$_2$O. Also known as *yttrium carbonate trihydrate.*

yttrium carbonate trihydrate. See yttrium carbonate.

yttrium chloride. White, hygroscopic crystals, powder or crystalline chunks (99.9+% pure) containing about 45-46% yttrium. Crystal system, monoclinic. Density, 2.61-2.67 g/cm^3; melting point, 721°C (1330°F). Used in superconductivity research, as a nonaqueous synthetic reagent, and as a starting material for

preparing various organometallic yttrium compounds. Formula: YCl$_3$.

yttrium chloride hexahydrate. White, hygroscopic, orthorhombic, high-purity crystals (99.9+% pure) containing about 29-30% yttrium. It has a density of 2.18 g/cm^3, and is used in superconductivity studies. Formula: YCl$_3$·6H$_2$O.

yttrium diboride. See yttrium boride (i).

yttrium dicarbide See yttrium carbide (ii).

yttrium disilicate. See yttrium silicate (ii).

yttrium disilicide. See yttrium silicide (i).

yttrium disulfide. See yttrium sulfide (ii).

yttrium 2-ethylhexanoate. A white powder (99.9+% pure) supplied in standard and superconductor grades (about 15-17.5% yttrium) . Used in superconductivity studies, and in the production of thin-film by ion-beam patterning. Formula: Y[O$_2$C$_2$H-(C$_2$H$_5$)C$_4$H$_9$]$_3$. Also known as *yttrium octoate.*

yttrium ferrite. A compound of yttrium oxide and ferric oxide available as an off-white powder with a density of 5.17 g/cm^3 and a melting point of 1560°C (2840°F). Used in ceramics, electronics and materials research. Formula: Y$_6$Fe$_2$O$_{12}$.

yttrium fluoride. A white, hygroscopic powder, or gray high-purity chunks (99.9+% pure) with a density of 4.01 g/cm^3 and a melting point of 1152°C (2106°F), used for the preparation of fluoride-doped superconductors, in vacuum deposition, in the optical characterization of yttrium fluoride thin films, and for indium fluoride and other fluoride-based glasses. Formula: YF$_3$.

yttrium-gallium iron garnet. A synthetic *garnet* of the rare-earth type with ferrimagnetic properties. It is similar to *yttrium iron garnet,* but contains some gallium. Used in lasers, and for electronic and microwave applications. Abbreviation: YGaIG.

yttrium germanate. See yttrium germanium oxide.

yttrium germanium oxide. A superconducting ceramic compound of yttrium oxide and germanium oxide, used in superconductivity studies, electronics and optoelectronics. Formula: Y$_2$GeO$_5$. Abbreviation: YGO. Also known as *yttrium germanate.*

yttrium hexaboride. See yttrium boride (iv).

yttrium hexafluoroacetylacetonate. White powder or crystals containing about 12% yttrium, and having a melting point of 166-170°C (331-338°F) and a decomposition temperature of 240°C (464°F). A dihydrate is also available. Used in superconductivity research to produce thin films by chemical-vapor deposition. Formula: Y(O$_2$C$_5$HF$_6$)$_3$. Also known as *yttrium hexafluoro-2,4-pentanedionate.*

yttrium heptasulfide. See yttrium sulfide (iv).

yttrium hexafluoro-2,4-pentanedionate. See yttrium hexafluoro-acetylacetonate.

yttrium-iron garnet. A synthetic *garnet* of the rare-earth type with ferrimagnetic properties corresponding to the general formula Y$_3$Fe$_5$O$_{12}$. It may be doped with gallium or other ions. Used in lasers, electronic transmitters, tuned oscillators, and in microwave devices as filters for selecting or tuning, and transmitters and transducers of sound energy. Abbreviation: YIG.

yttrium isopropoxide. A fine, off-white, moisture-sensitive powder (95+% pure) that contains about 33-37% yttrium, and has a melting point of 200-210°C (392-410°F). Used in the preparation of certain superconducting ceramics, in the preparation of thin films, and as an intermediate for fabrication of *yttrium barium copper oxide* superconducting fibers. Formula: Y(OC$_3$H$_7$)$_3$.

yttrium lithium fluoride. A crystalline compound of yttrium fluoride (YF$_3$) and lithium fluoride (LiF) that can be doped with erbium, holmium, neodymium, thulium or ytterbium ions for use as laser crystals. Formula: YLiF$_4$. Abbreviation: YLF.

yttrium monoaluminate. See yttrium aluminate (i).

yttrium monocarbide See yttrium carbide (i).

yttrium monosilicate. See yttrium silicate (i).

yttrium monosulfide. See yttrium sulfide (i).

yttrium naphthenate. A viscous liquid typically containing 7-13% yttrium, and used chiefly as a paint and varnish drier.

yttrium neodecanoate. An off-white powder supplied in superconductor grades containing 11-14% yttrium, and used in organometallic research, and in the preparation of high-temperature superconductors. Formula: $Y[O_2C_2(CH_3)_2(CH_2)_5CH_3]_3$.

yttrium nitrate. See yttrium nitrate hexahydrate; yttrium nitrate tetrahydrate.

yttrium nitrate hexahydrate. Colorless or reddish, hygroscopic crystals with a density of 2.68 g/cm³. Used in superconductivity research. Formula: $Y(NO_3)_3 \cdot 6H_2O$.

yttrium nitrate tetrahydrate. White, water-soluble crystals (99.9+% pure) typically containing 25.0-26.1% yttrium and having a density of 2.68 g/cm³. Used as an oxidizer, and in superconductivity studies. Formula: $Y(NO_3)_3 \cdot 4H_2O$.

yttrium nitride. A compound of yttrium and nitrogen used in ceramics and materials research. It has a density of 5.90 g/cm³ and a melting point of approximately 2670°C (4840°F). Formula: YN.

yttrium octoate. See yttrium 2-ethylhexanoate.

yttrium oxide. A rare-earth oxide available in the form of white crystals, a white or yellowish white, hygroscopic powder, sintered pieces, or gray-white tablets containing about 73-79% yttrium. It is supplied in standard purity (99.8-99.9%) and high-purity electronic grades (99.99+%), and as a 10% solids solution in water. It has a density of 5.01-5.03 g/cm³, a melting point of 2410°C (4370°F) [also reported as 2439°C or 4422°F] and low thermal expansivity. Used in the manufacture of red phosphors for television tubes, in the production of microwave filters (yttrium-iron garnets), with zirconia (ZrO_2) in the manufacture of special high-temperature refractories (e.g., zirconia and silicon nitride ceramics), in optical glass, in arc welding, incandescent gas mantles and microconcrete, and in superconductivity research for the solid-phase preparation of superconductors by vacuum deposition. The 10% solids aqueous solution is used as a particle-size standard in polymer and biomedical research. Formula: Y_2O_3. Also known as *yttria; yttrium sesquioxide*.

yttrium 2,4-pentanedionate. See yttrium acetylacetonate.

yttrium phosphide. A compound of yttrium and phosphorus that is available as cubic crystals or off-white powder with a density of 4.32-4.40 g/cm³, and used in the preparation of high-purity semiconductors. Formula: YP.

yttrium scandium gadolinium garnet. A soft magnetic material made from yttrium oxide (Y_2O_3), scandium oxide (Sc_2O_3) and gadolinium oxide (Gd_2O_3), and used for lasers and electronic applications. It can also be doped with erbium, chromium and neodymium, or chromium and erbium ions. Abbreviation: YSGG.

yttrium sesquicarbide See yttrium carbide (iv).

yttrium sesquisulfide. See yttrium sulfide (iii)

yttrium silicate. Any of the following compounds of yttrium oxide and silicon dioxide used in ceramics and materials research: (i) *Yttrium monosilicate*. Density, 4.49 g/cm³; melting point, 1979°C (3594°F); hardness, 5-7 Mohs. Formula: Y_2SiO_5; (ii) *Yttrium disilicate*. Density, 4.06 g/cm³; melting point, 1777°C (3231°F); hardness, 5-7 Mohs. Formula: $Y_2Si_2O_7$; and (iii) *Yttrium trisilicate*. Density, 4.39 g/cm³; melting point, 1949°C (3540°F); hardness, 5-7 Mohs. Formula: $Y_4Si_3O_{12}$.

yttrium silicide. Any of the following compounds of yttrium and silicon used in ceramics and materials research: (i) *Yttrium disilicide*. Density, 4.35 g/cm³; melting point, 1521°C (2770°F). Formula: YSi_2; and (ii) *Triyttrium pentasilicide*. Formula: Y_3Si_5.

yttrium sulfate. A white, water-soluble, crystalline powder (99.9+% pure) with a density of 2.52 g/cm³ and a melting point of approximately 1000°C (1830°F) (decomposes). Used as a reagent and in ceramics. Formula: $Y_2(SO_4)_3$.

yttrium sulfate octahydrate. Small, reddish-white crystals (99.9+% pure) with a density of 2.558 g/cm³. It loses $8H_2O$ at 120°C (248°F) and decomposes at 700°C (1292°F). Used in ceramics, and as a reagent. Formula: $Y_2(SO_4)_3 \cdot 8H_2O$.

yttrium sulfide. Any of the following compounds of yttrium and sulfur used in ceramics, electronics and materials research: (i) *Yttrium monosulfide*. A yellow-white powder. Density, 4.95 g/cm³; melting point, 2038°C (3700°F). Formula: YS; (ii) *Yttrium disulfide*. A yellowish powder. Density, 4.35 g/cm³; melting point, 1660°C (3020°F). Formula: YS_2; (iii) *Yttrium sesquisulfide*. Light tan crystals, or yellow powder (99.9+% pure). Crystal system, monoclinic. Density, 3.87 g/cm³; melting point, 1925°C (3497°F); band gap, 2.58 eV. Formula: Y_2S_3; and (iv) *Yttrium heptasulfide*. A yellow-brown powder. Density, 4.18 g/cm³; melting point, 1521°C (2770°F). Formula: Y_5S_7.

yttrium telluride. A compound of yttrium and tellurium with a melting point of 1943°C (3529°F), used in ceramics and materials research. Formula: Y_2Te_3.

yttrium tetraboride. See yttrium boride (iii).

yttrium titanate. A white powder composed of yttrium oxide and titanium dioxide, and used in ceramics and materials research. Formula: $Y_2(TiO_3)_3$.

yttrium triboride. See yttrium boride (ii).

yttrium trisilicate. See yttrium silicate (iii)

yttrium vanadate. White crystals usually grown by the Czochralski technique, and used as red phosphors in television tubes (often with small additions of europium vanadate), fiberoptics, and doped with neodymium ions as laser crystals. Crystal system, tetragonal. Density, 4.22 g/cm³; melting point, 1825°C (3317°F); hardness, 5 Mohs; transmission range, 400-4000 nm. Formula: YVO_4. See also neodymium-doped yttrium vanadate.

yttropyrochlore. A black mineral of the columbite group composed of yttrium cerium niobium titanium oxide, $(Y,Ce,Nd,Th)(Nb,Ti,Ta)_2O_6$. Crystal system, orthorhombic. Density, 4.91 g/cm³. Occurrence: Canada (Ontario).

yttrotungstite. A yellow mineral composed of yttrium tungsten oxide hydroxide, $YW_2O_6(OH)_3$. Crystal system, monoclinic. Density, 5.96 g/cm³; refractive index, 1.98. Occurrence: Malaysia.

Yuan Pao. Trademark of Jang Dah Nylon Industrial Company (Taiwan) for nylon 6 filament yarns used for textile fabrics.

Yucatan sisal. See henequen.

yucca fibers. The tough, coarse cellulose fibers obtained from the leaves and stems of yucca plants (genus *Yucca*) belonging to the lily family, and native to the tropical and subtropical America. Used especially for cordage and brooms, and sometimes for the manufacture of clothing and baskets.

yugawaralite. A colorless mineral of the zeolite group composed of calcium aluminum silicate tetrahydrate, $CaAl_2Si_2O_{16} \cdot 4H_2O$. Crystal system, monoclinic. Density, 2.20 g/cm³; refractive index, 1.497. Occurrence: Iceland.

yukonite. A nearly black mineral with brown tinge composed of calcium iron arsenate hydroxide hydrate, $Ca_3Fe_7(AsO_4)_6(OH)_9$.

18H$_2$O. Occurrence: Canada (Yukon).

Yulon. Trademark of Kemicna Tovarna Moste (Slovenia) for nylon 6 filament yarns used for textile fabrics.

Yuva. Trade name for nylon textile fibers and yarns.

Z

ZA alloys. Trade name for a group of zinc-aluminum foundry alloys that are hypereutectic compositions containing about 8-27% aluminum, 0.8-2.5% copper, 0.1% iron, small additions of magnesium, lead, cadmium and tin, with the balance being zinc. Commercially available as sand, permanent-mold, shell-mold and high-pressure die castings, they possess good mechanical properties including high strength and hardness, good castability, excellent bearing properties, wear resistance and machinability and low casting temperatures. Used for gears, racks and housings, and as replacements for other materials, such as cast iron, steel, bronze and aluminum. "ZA" stands for "Zinc Alloy" and is a letter designation usually followed by a numeral, e.g., ZA-8 or ZA-27, in which the numeral refers to the average aluminum content in percent. See also zinc-aluminum alloys.

ZAC alloys. A group of high-temperature creep-resistant magnesium alloys that contain zinc, aluminum and calcium, and are used for thixomolded automotive components.

Zaclon. Trademark of E.I. DuPont de Nemours & Company (USA) for a series of welding and soldering fluxes based on active zinc ammonium chloride combined with additives.

Zadur. Trade name of Düsseldorf-Heerdt GmbH & Co. KG (Germany) for an extensive series of steels including various high-speed steels, hot-work die and tool steels, cold-work die and tool steels, water-hardening tool steels, alloy machinery steels, case-hardening steels, and corrosion- and/or heat-resistant austenitic, ferritic and martensitic stainless steels.

zaffer. Cobalt in impure form used in the preparation of *smalt*.

Zaflex. Trade name of Atlas Powder Company (USA) for cellulose nitrate film materials.

zaherite. A white mineral composed of aluminum sulfate hydroxide hydrate, $Al_{12}(SO_4)_5(OH)_{26}\cdot20H_2O$. Density, 2.01 g/cm^3; refractive index, 1.4981. Occurrence: Pakistan.

zairite. A greenish mineral of the alunite group composed of bismuth aluminum iron phosphate hydroxide, $Bi(Fe,Al)_3(PO_4)_2(OH)_6$. Crystal system, rhombohedral (hexagonal). Density, 4.37-4.32 g/cm^3; refractive index, 1.82-1.83. Occurrence: Zaire.

Zakaf. Trade name of Atlas Powder Company (USA) for cellulose nitrate plastics.

Zama. Italian trade name for a zinc die-casting alloy containing 4% aluminum and small amounts of magnesium. Used for instrument cases.

Zamak. Trade name of Metallgesellschaft AG (Germany) for a series of zinc die-casting alloys containing about 4% aluminum, 0.1% magnesium and 3% copper. Used for household goods, machinery and equipment, chemical equipment, and technical products.

Zamium. Trade name for an alloy containing 60% nickel, 40% chromium, and traces of manganese and tungsten. Used for heat- and corrosion-resistant parts.

Zam Metal. Trade name of Hanson-Van Winkle-Munning Company (USA) for acid-resistant zinc-base alloys containing small additions of aluminum and mercury. Used for zinc plating anodes.

Zani. Trademark of Inco Limited (Canada) for a creep-resistant, hot-chamber die-castable metal composite composed of a zinc-alloy matrix with a fine dispersion of nickel aluminide. Used for applications up to 120°C (248°F) including automotive under-hood components, electric motors, connectors, and threaded parts.

Zanite. (1) Trade name of E.I. DuPont de Nemours & Company (USA) for liquid crystal polymers.

(2) Trademark of ITW Polymer Castings (USA) for a high-precision polymer concrete made by combining finely graded quartz aggregate and selected fillers with an epoxy binder. It has high strength and good dimensional stability and vibration damping properties, and is used for machine bases, and machine and instrument structural components, e.g., as a replacement for steel or cast iron.

Zantrel. Trademark of Rhône-Poulenc SA (France) for high-modulus rayon fibers with enhanced strength, and abrasion and pilling resistance. Used for textile fabrics including clothing. *Zantrel-Meryl* are rayon fibers and yarns used for textile fabrics.

Zanzibar copal. See animi gum.

Zanzibar gum. See animi gum.

zapatalite. A pale blue mineral composed of copper aluminum phosphate hydroxide tetrahydrate, $Cu_3Al_4(PO_4)_3(OH)_9\cdot4H_2O$. Crystal system, tetragonal. Density, 3.02 g/cm^3; refractive index, 1.646. Occurrence: Mexico.

zapatero. The hard, durable, pale yellow wood of the tree *Gossypiospermum praecox*. It has a straight grain, a fine, uniform texture, and is free from knots. Source: Central America, Venezuela, West Indies. Used for tool handles, musical instruments, rulers, mathematical instruments, inlays, and as a replacement for *Turkish boxwood*. Also known as *Maracaibo boxwood; West Indian boxwood.*

Zapizell. Trade name of Wilhelm Köpp Zellkautschuk (Germany) for a cellular rubber.

Zapon. Trademark for a colorless lacquer made from nitrocellulose (*collodion*) or acetylcellulose using alcohols and acetic acid. It dries very rapidly, leaving a thin, transparent, protective film on the substrate. Used for producing protective coatings on metals, paintings and documents.

Zapon Leathercloth. Trade name of former Locomotive & Rubber Waterproofing Company (UK) for cellulose nitrate-based artificial leather.

Zapp. Trade name of Robert Zapp Werkstofftechnik GmbH (Germany) for an extensive series of steels including numerous wrought or cast austenitic, ferritic, martensitic and precipitation-hardening stainless steels, cold-work and hot-work tool steels, water-hardening tool steels, oil-hardening nondeforming die and tool steels, shock-resistant tool steels, tungsten-, cobalt-tungsten- and molybdenum-type high-speed steels, and several alloy machinery and case-hardening steels.

Zappone Copper. Trade name of Zappone Manufacturing Company (USA) for roof shingles with copper accent.

zaratite. A vitreous, emerald-green mineral composed of nickel carbonate hydroxide tetrahydrate, $Na_3(CO_3)(OH)_4\cdot4H_2O$. Density, 2.63 g/cm³; refractive index, 1.602.

Zarsil. Trademark of Ferro Corporation (USA) for zirconium silicate media with a density of 3.8 g/cm³ (0.14 lb/in.³), and a hardness of 7.0 Mohs. Used for high-speed milling in small-media grinding mills, and in grinding and dispersing organic and inorganic pigments, pastes, coatings or chemicals.

zavaritskite. A gray mineral composed of bismuth oxide fluoride, BiOF. It can also be made synthetically. Crystal system, tetragonal. Density, 8.11 g/cm³; refractive index, 2.213. Occurrence: Russian Federation.

ZBLAN. A compound of the heavy metal fluoride glass family composed of zirconium tetrafluoride, barium difluoride, lanthanum trifluoride, aluminum trifluoride and sodium fluoride and having the formula $ZrF_4BaF_2LaF_3AlF_3NaF$. It can be doped with rare-earth metal ions (e.g., Pr^{3+}), and is used in the form of crystals for laser applications. See also heavy metal fluoride glasses.

Z-Bond. Trade name of Pax Surface Chemicals, Inc. (USA) for a zincating compound used for aluminum processing.

Z-Brick. Trademark of Z-Brick Canada, Division of Manwest Brickfacings Limited (Canada) for facing brick.

Z-Brite. Trade name of Enthone-OMI, Inc. (USA) for a bright zinc electroplate and plating process.

zebrano. The hard, heavy wood obtained from the date palm *(Phoenix dactylifera)* and the coconut palm *(Cocos nucifera)*. It is light brown with dark brown parallel stripes. Source: Africa, Asia, Central and South America, Middle East, southern Europe, southern United States. Used for high-quality furniture and paneling. Also known as *zebra wood.*

zebra wood. See zebrano.

Zebrazone. Trade name of Triplex Safety Glass Company Limited (UK) for zone-toughened glass that has the differential toughening in the vision zone arranged in vertical strips. Used for automobile windshields.

Zedabronze. Trade name of Le Bronze Industriel (France) for a series of zinc-manganese bronzes containing 18-25% zinc, 2.5-5.5% manganese, 5-6.5% aluminum, 0-3.5% iron, and the balance copper. Used for gears, bushings and bearings.

Zedply. Trademark of Zeidler Forest Industries Limited (Canada) for spruce plywood.

Zefkrome. Trademark of Dow Chemical Company (USA) for acrylic fibers used especially for carpets.

Zefran. Trademark of BASF Corporation (USA) for an acrylic fiber in white staple form based on acrylonitrile polymerized with pyrrolidone and supplemented with a dye-receptive component. It has good strength and heat resistance up to 254°C (489°F), good resistance to weak alkalies and common solvents, and good to excellent resistance to acids, mildew and weathering in sunlight. Used for pile and industrial fabrics, carpets, blankets, draperies, filter media, and apparel. Polyamide 6 (nylon 6) fibers for similar applications are also included under this trademark.

Zefron. Trademark of BASF Corporation (USA) for flame-retardant polyester fibers used for carpets and upholstery.

Zeftron. Trademark of BASF Corporation (USA) for nylon-6 staple fibers and filament yarns used for upholstery and floor covering applications including carpets.

Zefsport. Trade name of BASF Corporation (USA) for nylon 6 fibers.

Zehla. Trade name for rayon fibers and yarns used for the manufacture of textile fabrics.

zein. A white to slightly yellow resinous protein material that belongs to the prolamine class and is derived from corn. It contains considerable quantities of free amino acids especially cystine, lysine and tryptophan and has a density of about 1.23 g/cm³ (0.044 lb/in.³). Used in microencapsulation, biomedicine, fiber and laminated board manufacture, in paper coatings, and in printing inks. See also zein fibers.

zein fibers. Zein spun into fibers and used in the manufacvture of fiberboard and, blended with cotton, rayon or nylon fibers, for making wooly fabrics. See also zein.

Zeisse's salt. An organometallic compound that is available in the form of yellow, hygroscopic crystals with a density of 2.88 g/cm³ (0.104 lb/in.³) and a melting point of 220°C (428°F). Used as a catalyst. Formula: $K[PtCl_3(C_2H_4)]\cdot xH_2O$ (x ≤ 1).

Zeissig green. A greenish underglaze color made by calcining 43.5% barium chromate, 34.8% whiting and 21.7% boric acid. Used as a pottery decoration.

zektzerite. A colorless to pink mineral of the osumilite group composed of lithium sodium zirconium silicate, $LiNaZrSi_6O_{15}$. Crystal system, orthorhombic. Density, 2.79 g/cm³; refractive index, 1.584. Occurrence: USA (Washington).

Zelco. Trade name of Zelco (USA) for an aluminum solder containing 83% zinc, 15% aluminum and 2% copper.

Zelec. Trademark of E.I. DuPont de Nemours & Company (USA) for electroconductive powders composed of antimony-doped tin oxide (SnO_2) alone, or an antimony-doped tin oxide (SnO_2) outer shell applied to an inert core of mica, silicon dioxide or titanium dioxide. Used as static-dissipative additives in coatings and plastics.

ZelieBloc. Trade name of BNZ Materials, Inc. (USA) for lightweight, high-purity insulating firebrick made from high-purity refractory clay. Available in various standard shapes and sizes, it has high compressive strength, a high insulating value and good dimensional tolerance. Used for refractory linings or insulation in flues, furnaces, kilns, and other high-temperature installations.

Zellamid. Trade name of Zell-Metall (Germany) for polyamides.

zellerite. A lemon-yellow mineral composed of calcium uranyl carbonate pentahydrate, $CaUO_2(CO_3)_2\cdot5H_2O$. Crystal system, orthorhombic. Density, 3.25 g/cm³; refractive index, 1.559. Occurrence: USA (Wyoming).

Zellidur. Trade name of Zell-Metall (Germany) for polyvinyl chlorides.

Zellvag. Trade name for cuprammonium rayon fibers and yarns used for textile fabrics.

Zeloctite. Trade name for a cellulose lacquer.

Zelux. Trademark of Westlake Plastics Company (USA) for an impact-resistant transparent polycarbonate resin used for automotive, electrical and electronic equipment and components, diffusers and jet-pump impellers.

zemannite. A brown mineral composed of sodium iron zinc hydrogen tellurite hydrate, $Na_xH_{2-x}(ZnFe)_2(TeO_3)_3\cdot xH_2O$. Crystal system, hexagonal. Density, 4.05 g/cm³; refractive index, 1.85. Occurrence: Mexico.

Zemid. Trademark of DuPont Canada for a family of reinforced, mineral-filled polyethylene resins exhibiting good retention of toughness and stiffness at temperatures as low as -40°C (-40°F). Used for molded products.

zenella. A lining fabric in a twill weave, usually made with a rayon, silk or worsted weft and a combed cotton warp.

Zendel. Trademark of Union Carbide Corporation (USA) for tough polyethylene film and sheeting supplied in blown and cast forms, and used for packaging, netting for the agricultural and construction industries, and in the manufacture of adhesives.

Zenite. (1) Trade name of Zenith Foundry Company (USA) for a high-strength alloy cast iron containing 2.85-3.15% carbon, 1.6-1.8% silicon, 0.8% manganese, 0.65-0.85% nickel, 0.15-0.35% copper, 0.2-0.4% molybdenum, 0-0.2% chromium, and the balance iron. Used for gears, shafts and housings.

(2) Trademark of E.I. DuPont de Nemours & Company (USA) for a family of glass- or mineral-reinforced liquid crystal polymer (LCP) resins available in black, white and natural colors. They have high heat resistance, excellent dimensional stability, outstanding fatigue and creep resistance, very good resistance to many chemicals including concentrated acids, alkalies and hydrocarbons, high dielectric strength, good melt processability, and low thermal expansion and mold-shrinkage properties. Used in the electrical, electronics, telecommunications, automotive and aerospace industries for precision parts, such as bobbins, connectors, relays, sockets, motor components, sensor devices and ignition systems.

Zenith. Trade name of Zenith Foundry Company (USA) for tungsten-type high-speed tool steels, and high-strength alloy cast irons.

Zenith Alpha Base. Trade name of PermaCem (USA) for a glass-ionomer dental cement used as a base lining material.

Zenith Alpha Silver. Trade name of PermaCem (USA) for a glass-ionomer/cermet dental cement with silver additions.

Zeno. Trade name of Zenith Foundry Company (USA) for a cast iron containing 3.3% carbon, a total of 0.5% manganese and silicon, and the balance iron. Used for machine housings, gears, and shafts.

ZenTron. Trade name of Owens-Corning Fiberglas (USA) for an aluminosilicate glass fiber with excellent adhesion properties and fatigue resistance, and high tensile strength. It has a catenary-free, single-end-roving construction, can be hybridized with aramid and carbon tows, and is used as a reinforcement for pultruded, filament-wound and molded composites.

Zeo. Trademark of J.M. Huber Corporation (USA) for a series of hydrous silicas used as reinforcing agents for rubber compounds, and as flatting agents in lacquers and varnishes.

Zeodent. Trademark of J.M. Huber Corporation (USA) for precipitated synthetic silica.

Zeodur. Trademark of The Permutit Company, Inc. (USA) for processed *glauconite* used as a cation exchanger in water softening.

Zeofree. Trademark of J.M. Huber Corporation (USA) for synthetic precipitated silica.

Zeokarb. Trademark of The Permutit Company, Inc. (USA) for an ion-exchange resin composed of sulfonated coal and supplied in two ionic forms: hydrogen and sodium.

Zeolex. Trademark of J.M. Huber Corporation (USA) for synthetic precipitated sodium aluminosilicate used as a *zeolite* reinforcing agent.

zeolites. (1) A class of colorless, white, yellow or reddish natural or synthetic hydrous tectosilicates with aluminosilicate tetragonal framework having the approximate composition $Na_2O \cdot Al_2O_3 \cdot xSiO_2 \cdot yH_2O$. Used in ion-exchange reactions (e.g., in water softeners), as catalysts, adsorbents and desiccants, and as heating and cooling agents in solar collectors.

(2) A group of hydrous silicate minerals including in particular *analcime, chalbazite, erionite, harmotome, heulandite, mordenite, natrolite* and *stilbite*. Density range, 2-2.5 g/cm³; hardness range, 3.5-5.5 Mohs. Used in ceramics, and for ion-exchange applications.

Zeolon. Trade name of Norton Company (USA) for a natural zeolite mineral of the *mordenite* subgroup used as an ion exchanger.

Zeonex. Trade name of Nippon Zeon Company, Limited (Japan) for synthetic resins including polymethylpentene (PMP) supplied as liquids, granules and powders. Used for ion-exchange applications.

zeophyllite. (1) A milky white mineral composed of calcium fluoride silicate hydroxide hexahydrate, $Ca_{13}Si_{10}[O(OH)]_{28}[F-(OH)]_{10} \cdot 6H_2O$. Crystal system, monoclinic. Density, 4.30 g/cm³; refractive index, 1.738. Occurrence: Sweden.

(2) A colorless to greenish mineral composed of calcium fluoride silicate hydroxide dihydrate, $Ca_4(Si_3O_8)(OH)_2F_2 \cdot 2H_2O$. Crystal system, triclinic. Density, 2.75 g/cm³; refractive index, 1.5685. Occurrence: Czech Republic.

Zeo-Rex. Trade name for a sulfonated phenol-formaldehyde resin used as a cation exchanger.

Zeosyl. Trademark of J.M. Huber Corporation (USA) for synthetic precipitated silica.

Zeothix. Trademark of J.M. Huber Corporation (USA) for a series of synthetic precipitated amorphous silica pigments supplied as micronized powders. *Zeothix 95* is used as a flatting agent in clear, pigmented, water- and solvent-borne paint and coating systems. *Zeothix 177* and *Zeothix 265* are used as rheological agents in solvent-borne adhesive, sealant, coating, ink and gel-coat systems.

Zephor Bronze. Trade name of Bridgeport Rolling Mills Company (USA) for a corrosion-resistant phosphor bronze containing 95% copper, 4.2-5% tin and 0.03-0.35% phosphorus. Used for electrical switches, washers, springs, and fasteners.

Zephran. Trade name for polyester fibers.

Zephyr. Trade name of J.M. Ney Company (USA) for a non-oxidizing precious-metal dental alloy that contains 80% gold, has a fusing temperature of 1093°C (2000°F), and is used for retention loops of acrylic facings.

zephyr. A light, soft, airy fabric made from fine cotton, wool or synthetic yarn in a plain weave. It often has a white ground and colored stripes, and is made by using intermittent coarse threads to produce a ribbed or corded effect. Used especially for blouses, dresses and shirts.

zephyr yarn. A soft, low-twist worsted yarn spun from very fine wool, or a wool-silk or wool-synthetic blend.

Zeprex. Trademark of American Siporex Corporation (USA) for lightweight concrete with good structural, insulating and fire-resistant properties supplied in the form of floor, wall, building and insulating slabs, beams, blocks, roofing and core materials for the building and construction trades.

Zeraloy. Trade name of Swift Lewick & Sons Limited (UK) for a low-expansion alloy composed of 64% iron, 31% nickel and 5% cobalt, and used for instruments.

Zeramic. Trademark of GBC Corporation (USA) for a zero-expansion β-spodumene (β-LiAlSi₂O₆) ceramic with excellent thermal-shock resistance. Used in glassmaking, and for porcelain enamels, glazes and ceramic bodies. See also spodumene.

Zeranin. Trademark of Isabellenhütte Heusler GmbH KG (Germany) for an electrical resistance alloy composed of 88% copper, 6% manganese and 6% germanium. It has an upper service temperature of 140°C (284°F).

Zergal. Italian trade name for aluminum alloys containing about 4-7% zinc, 2-3% magnesium, 0.5-2.2% copper, and small amounts of manganese, titanium and zirconium. They develop high strength by heat treatment, are readily machinable, and are used mainly in the aircraft industry.

Zerlon. Trademark of Dow Chemical Company (USA) for tough methyl methacrylate-styrene copolymers used as molding materials.

zero-carbon steel. See decarburized enameling steel.

ZeroCeram. Trade name for a series of machinable alumina-based ceramics with zero thermal expansion, a maximum operating temperature of about 1200°C (2200°F), a density of 2.1 g/cm³ (0.08 lb/in.³), and a hardness of 5 Mohs. Used for ceramic parts with excellent thermal-shock resistance.

Zerochrome. Trade name of Zinex Corporation (USA) for chromium-free finishes that resemble chromium finishes in durability and appearance. Used as alternative for hexavalent chromium coatings.

Zero Clearance. Trade name of Lydall, Inc. (USA) for a multilayer polymer composite used for acoustical and thermal insulation applications.

zero-dimensional nanostructures. See nanostructured materials.

Zerodur. Trademark of Schott Glas AG (Germany) for a transparent zero-expansion glass-ceramic used in the manufacture of lenses, prisms, mirror blanks and tubes, gage blocks, and heatable chemical reaction vessels. Its coefficient of linear thermal expansion is extremely small (about $0.02-0.10 \times 10^{-6}$/K) and homogeneous in the temperature range from 0-50°C (32-122°F).

Zerofil. Trade name for a substance made by coating rock wool with bituminous material. Used for low-temperature insulation applications.

zerogel. (1) An apparently solid, dried *gel*.
(2) See xerogel (2).

Zerogen. Trademark of J.M. Huber Corporation (USA) for white magnesium hydroxide [$Mg(OH)_2$] powders.

Zerok. Trademark of Atlas Minerals & Chemicals, Inc. (USA) for a series of synthetic vinyl resin coatings with good resistance to alkalies and oxidizing acids and fumes. Used as protection against fumes, splashing and corrosion on steel and concrete tanks, dikes, floors, and structural building members.

Zeron 100. Trade name of Corralloy Inc. (USA) for a readily weldable superduplex stainless steel with outstanding resistance to crevice, pitting, erosion, stress-cracking and seawater corrosion at normal and elevated temperatures, and a wide range of corrosive chemicals. Supplied in the form of bars, plates, billets and finished products (e.g., flanges, fittings, tubes and pipes), it is used for applications such as offshore oil and gas equipment, seawater and firewater systems, chemical and petrochemical process equipment, pulp and paper equipment, power generation systems, pressure vessels, and heat exchangers.

zeron metal. A heat-treated malleable cast iron containing 2-2.5% carbon, 0.8-1.0% manganese, and the balance iron. Used for wrenches, hardware and fittings.

Zerosil. Trade name of Dreve-Dentamid GmbH (Germany) for a soft silicone elastomer used for dental impressions.

zero-twist yarn. A single or plied yarn produced without twist.

Zerox 101. Trade name of Atlas Minerals & Chemicals, Inc. (USA) for self-priming, one-component vinyl resin coatings for brush, roller or spray application to interior and exterior metal, wood and concrete surfaces.

Zeroxy. Trademark of Cheil Synthetics Inc. (South Korea) for polyester staple fibers used for textile fabrics.

Zetafax. Trademark of Dow Chemical Company (USA) for ethylene-acrylic resins with excellent chemical resistance and good adhesion to metals. Used for protective coatings.

Zetalabor. Trade name of ZHERMACK (Italy) for a silicone resin used in dentistry for laboratory duplications.

ZetaMat 3-D. Trademark of PPG Industries Inc. (USA) for glass-fiber mats used as three-dimensional reinforcements in the manufacture of engineering composites. They are supplied in two series: (i) *Series 1000* featuring randomly oriented continuous glass strands and (ii) *Series 2000* consisting of randomly oriented continuous strands combined with unidirectional roving.

Zetaplus. Trade name of ZHERMACK (Italy) for a silicone elastomer used for dental impressions.

Zetek. Trade name for a nytril fiber, i.e., a fiber based essentially on a vinylidene dinitrile polymer. It has good mildew and outdoor weathering resistance, good heat resistance, good to excellent resistance to acids, fair resistance to cold dilute alkalies, and poor resistance to strong alkalies. Used for textile fabrics.

Zetonia. German trade name for a white antifriction alloy containing tin, lead, and antimony. Used for bearings and bushings.

zeunerite. A yellowish green mineral of the autunite group composed of copper uranyl arsenate hydrate, $Cu(UO_2)_2(AsO_4)_2 \cdot xH_2O$ (x = 10-16). It can also be made synthetically. Crystal system, tetragonal. Density, 3.4 g/cm³; refractive index, 1.61. Occurrence: Germany.

Zeus. (1) Trade name of Krupp Stahl AG (Germany) for an alloy steel containing 0.12% carbon, 0.3-0.5% manganese, up to 0.05% silicon, up to 0.4% phosphorus, 0.035% sulfur, 0.05% titanium, and the balance iron. Used for welding electrodes.
(2) German trade name for an alloy containing 80% copper and 20% silver. Used for safety fuses.
(3) Trade name of ZEUS–Tecnobay Spa Divisione Dentale (Italy) for a dental gypsum for implants.

Zeus Magic White. Trade name of ZEUS–Tecnobay Spa Divisione Dentale (Italy) for a white die stone used for dental applications.

Zeusuperock. Trade name of ZEUS–Tecnobay Spa Divisione Dentale (Italy) for an extra-hard die stone used for dental applications.

Zevescal. Trade name of Calumet Steel Castings Corporation (USA) for abrasion- and wear-resistant high-carbon steel castings.

ZHM alloys. A group of high-temperature molybdenum alloys containing 0.5% zirconium, 1.5% hafnium and 0.2% carbon. They are similar to *TZM alloys*, but have greatly improved mechanical properties and recrystallization temperatures of about 1550°C (2820°F). Used for high-temperature applications including aerospace components and industrial equipment.

Z-Guss. Trade name of Aluminiumwerke Maulbronn (Germany) for an age-hardenable magnesium casting alloy containing 6% aluminum and 3% zinc. Used for light-alloy castings.

zibeline. A heavy, often dark-colored and striped fabric, usually made of wool or a wool-acrylic blend and having a long, rough, unidirectional nap. Used for coats and capes.

Zicral. Trade name of Pechiney SA (France) for a wrought aluminum alloy containing 7-8.5% zinc, 1.75-3% magnesium, 1.5% copper, 0.1-0.4% chromium, 0.1-0.6% manganese and 0.7% silicon. It has good corrosion resistance and strength, and is used in aircraft construction.

Ziegler catalyst. See Ziegler-Natta catalyst.

Ziegler-Natta catalyst. A catalyst composed of a transition-metal halide, such as titanium tetrachloride ($TiCl_4$), and a metal alkyl, such as triethylaluminum [$(C_2H_5)_3Al$]. The active catalyst, a titanium complex holding an ethyl group, is formed by a reaction of the salt and the alkyl in a hydrocarbon solvent. Used in the conversion of ethylene to linear polyethylene, and in the stereospecific polymerization of propylene to crystalline polypropylene and isoprene to *cis*-1,4-polyisoprene. Also known as *Ziegler catalyst. Note:* A wide range of stereospecific catalysts containing other transition-metal halides and metal alkyls or hydrides are now also called "Ziegler catalysts."

ziesite. A mineral composed of copper vanadium oxide, β-$Cu_2V_2O_7$. It can also be made synthetically. Crystal system, monoclinic. Density, 3.86 g/cm³. Occurrence: Sweden.

Zigrinato. Trade name of Fabbrica Pisana SpA (Italy) for a patterned glass.

Zilloy. Trademark of New Jersey Zinc Company (USA) for zinc-base alloys including several wrought zinc alloys containing up to 1% copper, 0.8% cadmium and 0.1% magnesium. Used for roofing applications.

Zimal. Trade name of Birmingham Aluminum Casting Company (UK) for a zinc die-casting alloy containing 4.0-4.2% aluminum and 3.0% copper. It has good corrosion resistance, and is used for carburetors and gears.

Zimalium. German trade name for an aluminum alloy containing 2.8-11.5% zinc and 3.7-7.5% magnesium.

Zimo. Trade name of Monopol AG (Switzerland) for zinc dust primers.

zinalsite. A white to pinkish mineral of the kaolinite-serpentine group composed of zinc aluminum silicate pentahydrate, $Zn_3Al_2(SiO_4)_3 \cdot 5H_2O$. Density, 2.48 g/cm³. Occurrence: Kazakhstan.

zinc. A lustrous, bluish-white metallic element of Group IIB (Group 12) of the Periodic Table. It is commercially available in the form of pieces, rods, ingots, wire, shot, granules, powder, dust, mossy, splatters, semicircular bars, sticks, sheet, foil, wire, and single crystals. The single crystals are usually grown by the Bridgeman technique. Important zinc ores include *sphalerite (most important), smithsonite, franklinite, hemimorphite, willemite* and *zincite.* Crystal system, hexagonal. Crystal structure, hexagonal close-packed. Density, 7.14 g/cm³; melting point, 419.5°C (787°F); boiling point, 907°C (1665°F); atomic number, 30; atomic weight, 65.39; divalent. It has low to intermediate hardness (2.5 Mohs), relatively good electrical conductivity (about 29% IACS), a superconductivity critical temperature of 0.85K, moderate strength and toughness, is very little affected by air and moisture at ordinary temperatures, and is brittle at ordinary temperatures and ductile and malleable at 100-150°C (212-302°F). Used in galvanizing iron, steel and other metals, and for corrosion-resistant coatings, electrodeposits, metal spraying materials, soft solders, alloys (particularly brass, bronze, nickel silver, copper nickel, and babbitt), die-casting metals, zinc compounds, roofing and plumbing supplies, gutters and automotive components, as the negative electrode in many batteries, and in paints, rubber pigments, photoengravers' and printing plates, cable wrappings, electrical fuses, and water and gas valves. Symbol: Zn.

zinc-65. A radioactive isotope of zinc with a mass number of 65 obtained by pile irradiation of zinc metal and cyclotron bombardment of copper-65 with deuterons. It has a half-life of 250 days and emits beta and gamma radiation. Used as a tracer nuclide in the study of wear mechanisms in metals and alloys, and the functions of additives in lubricating oils and the nature of activators in phosphors, and in galvanizing processes. Symbol: ^{65}Zn.

zinc-67. A natural (stable) isotope having a mass number of 67,

and occurring to the extent of 4.11% in zinc. Used in high-precision measuring instruments, and atomic clocks. Symbol: ^{67}Zn.

zinc acetate. White, water-soluble crystals or powder (98+% pure) made by treating zinc oxide with acetic acid on zinc oxide. Density, 1.735 g/cm³; loses $2H_2O$ at 100°C (212°F); decomposes at 200°C (392°F). Used in ceramic glazes, as a wood preservative, as a crosslinking agent for polymers, in medicine and biochemistry, and as a mordant. Formula: $Zn(C_2H_3O_2)_2 \cdot 2H_2O$. Also known as *zinc acetate dihydrate.*

zinc acetate dihydrate. See zinc acetate.

zinc alloys. A group of alloys in which zinc is the predominant constituent, typically present to the extent of 70-100%. Included are zinc-base casting alloys especially those of zinc and aluminum, and zinc and tin, used for the manufacture of parts with high dimensional tolerance as well as wrought zinc alloys available in rolled, forged, extruded or drawn form. Common applications of zinc alloys include die-castings used for decorative parts, padlocks, automotive components, levers, pawls, gear, office equipment, and protective coatings. See also wrought zinc alloys; zinc casting alloys.

Zincalium. British trade name for a non-heat-treatable aluminum alloy containing 0.8-8.3% zinc and 0.8-8.3% magnesium. Used for light-alloy parts.

zinc aluminate. A ceramic compound of zinc oxide (ZnO) and aluminum oxide (Al_2O_3) with a spinel-type structure. It is found in nature as the mineral *gahnite.* Crystal system, cubic. Density, 4.48-4.62 g/cm³; melting point, 1950°C (3542°F); hardness, 7.5-8.0 Mohs; refractive index, 1.790. Used for refractories, and in electronics. Formula: $ZnAl_2O_4$.

zinc-aluminum alloys. A group of cast and wrought alloys in which zinc and aluminum are the principal constituents. Included are: (i) *Zinc-aluminum die-casting alloys* which contain on average 3-8% aluminum, and are used for fittings, clock housings, office equipment, household appliances, and toys; (ii) *Zinc-aluminum foundry alloys* (ZA alloys) which contain about 8-27% aluminum; and (iii) *Wrought zinc-aluminum alloys* having up to 22% aluminum and available in rolled, forged, extruded or drawn form. Wrought alloys based on the eutectoid composition of 78% zinc and 22% aluminum exhibit superplastic properties.

zinc-aluminum casting alloys. A group of casting alloys in which zinc and aluminum are the principal constituents. Included are: (i) *Die-casting alloys* with an average aluminum content of 3-8%; and (ii) *Foundry alloys* (ZA alloys) which contain about 8-27% aluminum. See also zinc-aluminum alloys; zinc die castings; ZA alloys.

zinc aluminum selenide. A crystalline compound of zinc, aluminum and selenium. Crystal system, tetragonal. Crystal structure, "defect" chalcopyrite. Density, 4.37 g/cm³. Used as a semiconductor. Formula: $ZnAlSe_4$.

zinc-aluminum solder. A special eutectic solder of 95% zinc and 5% aluminum, having a melting point of 382°C (720°F), and used for soldering aluminum. See also zinc solder.

zinc-ammonium chloride. See ammonium tetrachlorozincate.

zinc antimonide. A compound of zinc and antimony supplied in the form of silvery-white, orthorhombic, high-purity crystals with a density of 6.33 g/cm³ and a melting point of 570°C (1058°F). It becomes unstable at approximately 1000°C (1832°F), and is used as a p-type semiconductor, and in thermoelectric devices. Formula: $ZnSb_2$.

zinc arsenide. A compound of zinc and arsenic supplied in the form of a fine, white powder with a density of 5.53 g/cm³, and used as a semiconductor. Formula: Zn_3As_2.

zincate. A compound formed by the reaction of zinc with ammonia, or an alkali metal, e.g., potassium zincate (K_2ZnO_2). Zincate solutions, composed chiefly of zinc oxide and sodium hydroxide, are used in immersion coating of aluminum.

zinc babbitt. An *antifriction alloy* containing 26% tin, 5% copper, 3% antimony, and the balance zinc. Used for bearings.

zinc baryta white. See lithopone.

zinc bichromate. See zinc dichromate.

zinc blende. See sphalerite.

zinc bloom. See hydrozincite.

zinc borate. A compound of zinc oxide (ZnO) and boric oxide (B_2O_3) supplied in the form of white crystals or amorphous powders having a melting point of 980°C (1796°F). Used as flux in ceramic compositions, in fireproofing compositions for textiles, and as a fungi and mildew inhibitor. Formula: $3ZnO \cdot 2B_2O_3$.

zinc bromide. A white, hygroscopic, crystalline powder (98+% pure). Crystal system, orthorhombic. Density, 4.2-4.5 g/cm³; melting point, 394°C (741°F); boiling point, 650°C (1202°F); refractive index, 1.5452. Used in radiation shields, photographic emulsions, and in rayon manufacture. Formula: $ZnBr_2$.

zinc cadmium selenide. A compound of varying amounts of zinc, cadmium and selenium used for semiconductor applications. Abbreviation: ZnCdSe.

zinc-cadmium solders. A group of alloys of zinc and cadmium used to solder aluminum. They provide good corrosion resistance when used with the proper flux. Examples of commercially used solders include 90Zn-10Cd and 60Zn-40Cd.

zinc cadmium sulfide. A yellowish fluorescent compound of zinc, cadmium and sulfur used as a phosphor and semiconductor. Abbreviation: ZnCdS.

zinc carbonate. A white, crystalline powder that can be made synthetically and also occurs in nature as the mineral *smithsonite*. Crystal system, trigonal. Density, 4.42-4.45 g/cm³; loses CO_2 at 300°C (572°F). Used in ceramics, Bristol and other glazes, as a pigment, as a fireproofing agent, as a filler for rubber and plastic compositions, and in zinc salts. Formula: $ZnCO_3$.

zinc carbonate hydroxide. See zinc subcarbonate.

zinc casting alloys. A group of zinc-base casting alloys that can be subdivided into the following three categories: (i) *Zinc die-casting alloys* made with high-grade zinc and containing about 3.5-4.5% aluminum, 0.25-1% copper, and 0.04% magnesium. They are relatively dense and fine grained, and have good castability and processibility and good mechanical properties and dimensional tolerances; (ii) *Zinc foundry alloys* or *"ZA alloys"* which are hypereutectic compositions containing from 8-27% aluminum, 0.8-2.5% copper, 0.1% iron, and small additions of magnesium, lead, cadmium and tin. They can be produced by various methods including sand, permanent-mold, shell-mold and high-pressure die casting. ZA alloys have good mechanical properties, excellent bearing properties, wear resistance and machinability, good castability, low casting temperature, and are used for gears, racks, shaft and housings; and (iii) *Slush-casting* and *forming-die alloys* which are usually eutectic compositions containing 94.5-95.25% zinc and 4.75-5.5% aluminum. They are used for lighting fixtures, lamp bases, casket hardware and small statues. Zinc forming-die alloys available under the trade name "Kirksite" contain about 4% aluminum and 3% copper, are made by plaster-mold or sand casting, and are used for dies and punches.

zinc cement. Relatively quick-setting cement based on zinc oxide (ZnO) usually supplied as a pasty substance in zinc chloride solution.

zinc chloride. White, granular, hygroscopic crystals or crystalline powder (98+% pure). Crystal system, cubic. Density, 2.91 g/cm³; melting point, 290°C (554°F); boiling point, 732°C (1350°F). Used for special cements, glass-etching compositions, dental cements, adhesives, parchment and pigments, as an ingredient in soldering fluxes, as burnishing and polishing compound for steel and galvanized iron, as a catalyst, dehydrating agent and wood preservative, and in textile processing and electroplating. Formula: $ZnCl_2$.

zinc chromate. Lemon-yellow crystals or powder made from zinc oxide and chromic oxide, and used as a yellow pigment (*zinc yellow*), as a ceramic colorant, in metal primers (e.g., for steel and iron) with good rust-inhibitive properties, in automotive and other paints, in varnishes, and in epoxy laminates. Formula: $ZnCrO_4$.

zinc chrome. See buttercup yellow.

zinc chromite. A compound containing 74% zinc oxide and 22% chromic oxide, with the balance being other oxides and elements. Formula: $ZnCr_2O_4$. Also known as *zinc chromium oxide*.

zinc chromium oxide. See zinc chromite.

zinc-coated iron. See galvanized iron.

zinc-coated sheet. See galvanized sheet.

zinc-coated steel. See galvanized steel.

zinc-coated steel wire. See galvanized steel wire.

zinc coatings. See galvanized coatings; zinc plate.

zinc cobalt selenide. A compound of varying amounts of zinc, cobalt and selenium, used for semiconductor applications. Abbreviation: ZnCoSe.

zinc columbate. See zinc niobate.

zinc crown glass. An optical *crown glass* containing considerable quantities of zinc oxide as an auxiliary flux.

zinc cyanide. A white powder with a density of 1.852 g/cm³ and a melting point of 800°C (1472°F) (decomposes), used in electroplating iron and steel. Formula: $Zn(CN)_2$.

zinc dichromate. Orange-yellow powder used as a pigment. Formula: $ZnCr_2O_7 \cdot 3H_2O$. Also known as *zinc bichromate*.

zinc die-casting alloys. See zinc casting alloys.

zinc diethyl. See diethylzinc.

zinc dimethyl. See dimethylzinc.

zinc dioxide. A white to yellowish white powder usually composed of 45-60% ZnO_2 and 40-55% ZnO. Density, 1.571 g/cm³; melting point, decomposes rapidly above 150°C (302°F); dissolves in acids, alcohol, acetone and water. Used as a curing agent for rubber and elastomers, and in high-temperature oxidation. Formula: ZnO_2. Also known as *zinc peroxide*.

zinc diphenyl. See diphenylzinc.

Zinc Duralumin. British trade name for a strong aluminum alloy containing 20% zinc, 2.5% copper, 0.5% magnesium and 0.5% manganese. It is heat treatable, but not hardenable, and is used for light-alloy parts.

zinc dust. Gray, moisture-sensitive, finely divided zinc metal (97+% pure; particle size less than 325 mesh) condensed from vapor. It is mainly used in zinc-rich paints, zinc diffusion coatings, galvanized coatings, zinc-rich primers for iron and steel, thermal spraying powders, pigments, elastomers, resins, sealants and lubricants, in batteries, in the manufacture of zinc compounds, and as a catalyst and purifier.

zinc ethyl. See diethylzinc.

zinc ferrite. A ferrimagnetic ceramic with a normal cubic spinel crystal structure (*ferrospinel*). It has a density of 5.33 g/cm³, a melting point of about 1590°C (2895°F), excellent magnetic properties at high frequencies, and high resistivity and corrosion resistance. Used for soft magnets in various applications. Formula: $ZnFe_2O_4$.

zinc fluoride. White, hygroscopic needles, crystals or powder (99+% pure). Crystal system, tetragonal. Density, 4.95 g/cm³; melting point, 872°C (1600°F); boiling point, approximately 1500°C (2730°F). Used as an opacifier and flux in porcelain enamels and glazes, in galvanizing and electroplating, in phosphors, and in wood preservatives. Formula: ZnF_2.

zinc fluoride tetrahydrate. White, orthorhombic crystals (98+% pure) with a density of 2.255 g/cm³, used in ceramics, chemistry and electroplating. Formula: $ZnF_2 \cdot 4H_2O$.

zinc fluoborate. A colorless liquid usually supplied as a 40 or 48% solution for use in acid baths for electroplating, in bonderizing of wire and strip, and as a resin curative. Formula: $Zn(BF_4)_2$. Also known as *zinc fluoroborate*.

zinc fluoroborate. See zinc fluoborate.

zinc fluorosilicate. White, moisture-sensitive crystals (99+% pure). Density, 2.104 g/cm³; melting point, decomposes at 100°C (212°F). Used as a concrete hardener, and as a preservative. Formula: $ZnSiF_6 \cdot 6H_2O$. Also known as *zinc hexafluorosilicate; zinc fluorosilicate hexahydrate; zinc silicofluoride; zinc silicofluoride hexahydrate*.

zinc fluorosilicate hexahydrate. See zinc fluorosilicate.

zinc foam. Zinc made into a lightweight, closed-cell foam by introducing an inert gas. Used for structural applications, and for applications requiring absorption of shocks and vibrations. Also known as *foamed zinc*.

zinc foil and sheet. Flat-rolled zinc products usually made in thicknesses ranging from 0.0025-6.0 mm (0.0001-0.236 in.). Typical alloying additions include cadmium, copper, lead and titanium. Zinc foil and sheet products possess good formability, solderability, weldability and corrosion resistance, and are used in building and construction, in electrical contacts and dry-cell batteries, for architectural applications, and for photoengraving plates and coinage.

zinc formate. The zinc salt of formic acid available as white, water-soluble crystals or powder with a density of 2.207 g/cm³. It loses $2H_2O$ at 140°C (284°F), and is used as a waterproofing agent, as a catalyst in the synthesis of methanol, and as an antiseptic. Formula: $Zn(CH_2O)_2 \cdot 2H_2O$. Also known as *zinc formate dihydrate*.

zinc formate dihydrate. See zinc formate.

zinc gallium sulfide. A compound of zinc, gallium and sulfur. Crystal system, tetragonal. Crystal structure, thiogallate. Density, 3.8 g/cm³; hardness, 305-365 Vickers; melting point, 1248°C±30°C (2278°F±54°F). Used as a semiconductor. Formula: $ZnGa_2S_4$.

zinc germanium arsenide. A compound of zinc, germanium, and arsenic. Crystal system, tetragonal. Crystal structure, chalcopyrite. Density, 5.32 g/cm³; hardness, 6800 Knoop; melting point, 877°C (1611°F); band gap, 0.85 eV. Used as a semiconductor. Formula: $ZnGeAs_2$.

zinc germanium diphosphide. See zinc germanium phosphide.

zinc germanium phosphide. A compound of zinc, germanium and phosphorus. Crystal system, tetragonal. Crystal structure, chalcopyrite. Density, 4.17 g/cm³; hardness, 8100 Knoop; transmission (infrared and visible), 65%; melting point, 992°C (1818°F); maximum service temperature in air, 740°C (1364°F); band gap, 2.2 eV. Used as a semiconductor, in electronics, and in nonlinear optics. Formula: $ZnGeP_2$. Also known as *zinc germanium diphosphide*.

zinc germanium phosphide arsenide. A compound of varying amounts of zinc, germanium, phosphorus and arsenic. Crystal system, tetragonal. Crystal structure, chalcopyrite. Hardness, 490-510 Knoop; band gap, 1.76-1.78 eV; maximum service temperature in air, 475°C (887°F). Used as a semiconductor. Formula: $ZnGeP_{2-x}As_x$ (x = 0.2 or 0.4).

zinc glass. Ordinary *soda-lime glass* in which some of the lime (CaO) is replaced by zinc oxide (ZnO).

Zinc Glo. Trade name of Alchem Corporation (USA) for a zinc electrodeposit and deposition process.

zinc green. A group of green pigments composed essentially of mixtures of *buttercup yellow* (zinc chrome) and *Prussian blue* (ferric ferrocyanide). They have high brilliancy, good light fastness and poor resistance to alkalies or water. Used in flat interior wall paints.

Zincgrip. Trade name of Armco International (USA) for a *mild steel* sheet with a nonflaking zinc coating used for deep-drawing applications.

zinc halide. A compound of zinc and a halogen, e.g., zinc bromide, zinc chloride or zinc fluoride.

zinc hexafluorosilicate. See zinc fluorosilicate.

zinc indium selenide. A compound of zinc, indium and selenium. Crystal system, tetragonal. Crystal structure, "defect" chalcopyrite. Density, 5.44 g/cm³; melting point, 977°C (1791°F); band gap, 1.82 eV. Used as a semiconductor. Formula: $ZnIn_2Se_4$.

zinc indium telluride. A compound of zinc, indium and tellurium. Crystal system, tetragonal. Crystal structure, "defect" chalcopyrite. Density, 5.83 g/cm³; melting point, 802°C (1476°F); band gap, 1.2 eV. Used as a semiconductor. Formula: $ZnIn_2Te_4$.

zinc iodide. A white, crystalline, hygroscopic powder (98+% pure). Crystal system, cubic. Density, 4.67-4.74 g/cm³; melting point, 446°C (834°F); boiling point, 625°C (1157°F) (decomposes). Used chiefly as an analytical reagent, and as an antiseptic. Formula: ZnI_2.

zinc iron selenide. A compound of zinc, iron and selenium used for semiconductor applications. Abbreviation: ZnFeSe.

zincite. An orange-yellow to deep red mineral with a subadamantine luster. It belongs to the wurtzite group, is composed of zinc oxide, ZnO, and contains theoretically 80.3% zinc. Crystal system, hexagonal. Density, 5.4-5.7 g/cm³; melting point, 1670°C (3038°F); hardness, 4-4.5 Mohs; refractive index, 2.013. Occurrence: USA (New Jersey). Used as an ore of zinc, and in the production of high-purity zinc oxide. Also known as *red zinc oxide; red oxide of zinc; red zinc ore; spartalite*.

zinc laurate. A white powder with a melting point of 128°C (262°F) used in paints and varnishes, and in rubber compounding. Formula: $Zn(C_{12}H_{23}O_2)_2$.

zinc linoleate. A brown solid compound containing about 8.5-10% zinc, and used as a drier for paints containing cobalt or manganese soaps. Formula: $Zn(C_{18}H_{31}O_2)_2$.

zinc manganese selenide. A compound semiconductor containing varying amounts of zinc, manganese and selenium. Abbreviation: ZnMnSe.

zinc manganese telluride. A compound semiconductor containing varying amounts of zinc, manganese and tellurium. Abbreviation: ZnMnTe.

zinc methyl. See dimethylzinc.

zinc molybdate. An off-white powder (98+% pure) with a melting point of approximately 900°C (1650°F) used as an adherence-promoting agent in white porcelain enamels. Formula:

$ZnMoO_4$.

zinc molybdate dihydrate. A solid substance with a density of 3.3 g/cm^3 and a melting point of 1650°C (3000°F). Used as a starting material for growing single crystals, and as a corrosion-preventive agent. Formula: $ZnMoO_4·2H_2O$.

zinc monocolumbate. See zinc niobate (i).

zinc mononiobate. See zinc niobate (i).

zinc monotitanate. See zinc titanate.

zinc naphthenate. An amber, viscous liquid that contains 8-10% zinc, and is often supplied as a 67 wt% solution in mineral spirits. Used as a drier and wetting agent in paints, varnishes, resins, in insulating materials, as a wood preservative, and in waterproofing textiles. Formula: $Zn(C_7H_5O_2)_2$.

zinc niobate. Any of the following compounds of zinc oxide and niobium pentoxide (columbium pentoxide) used in ceramics and materials research, and also known as *zinc columbate*: (i) *Zinc mononiobate.* Melting point, 1398°C (2548°F). Formula: $ZnNb_2O_6$ ($ZnCb_2O_6$). Also known as *zinc monocolumbate*; and (ii) *Trizinc niobate.* Melting point 1305°C (2381°F). Formula: $Zn_3Nb_2O_8$ ($Zn_3Cb_2O_8$). Also known as *trizinc columbate*.

zinc nitrate. Colorless, water-soluble crystals or lumps (98+% pure). Crystal system, tetragonal. Density, 2.065 g/cm^3; melting point, 36°C (96°F); loses $6H_2O$ between 105 and 131°C (221 and 266°F). Used as a reagent, as an intermediate, as an acidic catalyst, as a mordant, as a latex coagulant, and in biochemistry and medicine. Formula: $Zn(NO_3)_2·6H_2O$. Also known as *zinc nitrate hexahydrate*.

zinc nitrate hexahydrate. See zinc nitrate.

zinc nitride. A compound of zinc and nitrogen available as blue-gray cubic crystals or gray crystalline powder, and used as a semiconductor. Formula: Zn_3N_2.

zincobotryogen. A deep orange mineral composed of zinc iron sulfate hydroxide heptahydrate, $ZnFe(SO_4)_2(OH)·7H_2O$. Crystal system, monoclinic. Density, 2.30 g/cm^3; refractive index, 1.551. Occurrence: China.

Zinco Brite. Trade name of Alchem Corporation (USA) for a zinc-cobalt electrodeposit and deposition process.

zincocopiapite. A yellowish green mineral composed of zinc iron sulfate hydroxide octadecahydrate, $ZnFe_4(SO_4)_6(OH)_2·18H_2O$. Crystal system, triclinic. Density, 2.18 g/cm^3; refractive index, 1.554. Occurrence: China.

zinc oleate. A white to light reddish brown, greasy, granular powder containing about 8.5-10.5% zinc. It has a melting point of 70°C (158°F), and is used as a drier in paints, resins and varnishes. Formula: $Zn(C_{18}H_{33}O_2)_2$.

Zincolith. Trade name for a *lithopone* composed of about 2 parts of barium sulfate ($BaSO_4$) and 1 part of zinc sulfide (ZnS). Used as a white pigment.

Zincon. Trade name of London Zinc Mills Limited (UK) for commercially pure (99.2%) and high-purity (99.99%) zinc, usually supplied in sheet and strip form. The commercially pure grades may contain some lead, and are used for roofing and dry cell cases.

zinc orthophosphate. See zinc phosphate.

zinc orthophosphate tetrahydrate. See zinc phosphate tetrahydrate.

zinc orthosilicate. See zinc silicate (1).

zinc oxide. White, yellowish-white or grayish crystalline powder or lumps (99+% pure) supplied in two crystalline forms: (i) Cubic zinc oxide. Crystal structure, sphalerite (or zinc blende). Density, 5.675 g/cm^3; melting point, 1975°C (3587°F); hardness, 5 Mohs; and (ii) Hexagonal zinc oxide. Crystal structure, wurtzite (or zincite). Density, 5.606 g/cm^3; melting point, 1977°C (3591°F), hardness, 4 Mohs; refractive index, 2.013; band gap, 3.2 eV. The hexagonal form occurs in nature as the mineral *zincite*. Zinc oxide is used as a paint pigment (*zinc white*), as a mold-growth inhibitor in paints, as a ultraviolet light absorber in plastics, in ceramics, as an opacifier and fluxing ingredient in glass, glazes and enamels, as an ingredient in high-grade fluoride opal glass, tank window glass and certain optical glasses, in Bristol glazes, in porcelain enamel for cast iron, in zinc salts, as a catalyst, in semiconductors, special piezo-electric compositions, phosphors, magnetic ferrites and transducers, as a photoconduc-tor in copying machines and color photography, as an accelerator and activator, and in dental cements. Single-crystal substrates are used for the deposition of high-temperature oxide superconductors, and III-V (13-15) semiconductor compounds. Formula: ZnO. Also known as *flowers of zinc, white zinc oxide*.

zinc oxide catalyst. A catalyst having a surface area of about 35 m^2/g and consisting of 85-95% zinc oxide, 3-7% aluminum oxide and 0.5-3% calcium oxide.

zinc oxide–eugenol cement. A quick-setting *dental cement* made by mixing a powder consisting of about 69% zinc oxide, 29.3% white rosin, 1% zinc stearate and 0.7% zinc acetate, with a liquid consisting of 85% *eugenol* and 15% olive oil. It can be reinforced with alumina (Al_2O_3) and ethoxybenzoic acid (EBA), or polymethyl methacrylate, and is used chiefly for the temporary luting of restoration work, the cementation of veneers and composite inlays, for pulp capping and dental impressions, and as a cavity liner and base.

zinc palmitate. White, amorphous powder with a density of 1.121 g/cm^3 and a melting point of approximately 100°C (212°F) used as a flatting agent in lacquers, as a suspending agent for paint pigments, in rubber compounding, and as a lubricant in plastics. Formula: $Zn(C_{16}H_{31}O_2)_2$.

zinc peroxide. See zinc dioxide.

zinc phenyl. See diphenylzinc.

zinc phosphate. A white crystalline powder (98+% pure). Crystal system, orthorhombic. Density, 3.998 g/cm^3 (at 15°C/59°F); melting point, 900°C (1650°F). Used for conversion coating of steel, aluminum and other metals, in phosphors, and in dental cements. Formula: $Zn_3(PO_4)_2$. Also known as *zinc orthophosphate; tribasic zinc phosphate*.

zinc phosphate cement. A *dental cement* made by mixing a powder of about 90% zinc oxide (ZnO) and 8% magnesium oxide (MgO) with phosphoric acid. The set cement consists of unreacted ZnO particles (average diameter about 8-10 μm or 310-390 μin.) embedded in a zinc phosphate matrix. Used particularly as a base cement, for luting applications, and for the cementation of precision castings.

zinc phosphate coatings. Light to dark gray coatings produced from a dilute solution of phosphoric acid and zinc, and applied to the surface of iron, steel and galvanized steel by immersion, spraying or both. Immersion coatings on steel substrates range in weight from 1.61-43 g/m^2 (150-4000 mg/ft^2) and spray coatings from 1.08-10.8 g/m^2 (100-1000 mg/ft^2). *Zinc phosphate coatings* are used for increased corrosion and/or wear resistance, as bases for paints, and as cold-forming aids.

zinc phosphate primer. A zinc phosphate coating applied to the surface of iron and steel to increase corrosion resistance and enhance the adhesion of subsequent paint coats.

zinc phosphate tetrahydrate. A white, crystalline powder. It occurs in nature as the mineral *hopeite*. Crystal system, orthorhombic. Density, 3.04 g/cm³; melting point, above 105°C (221°F). Used in dental cements, and in the production of phosphors. Formula: $Zn_3(PO_4)_2 \cdot 4H_2O$. Also known as *zinc orthophosphate tetrahydrate; tribasic zinc phosphate tetrahydrate*.

zinc phosphide. A compound composed of zinc and phosphorus available in the form of a black powder or dark gray crystals. Crystal system, tetragonal. Density, 4.55 g/cm³; melting point, above 420°C (788°F); boiling point, 1100°C (2012°F) [also reported as 1160°C (2120°F)]. Used for semiconductor applications, and in electronics. Formula: Zn_3P_2.

zinc phthalocyanine. A purple phthalocyanine derivative that contains a central zinc atom. It has a typical dye content of about 97%, a maximum absorption wavelength of 701 nm, and is used as a dye and pigment, e.g., for photovoltaic cells. Formula: $Zn(C_{32}H_{16}N_8)$. Abbreviation: ZnPc.

zinc-plated steel. See electrogalvanized steel.

zinc plate. A layer of zinc produced on a metallic substrate (e.g., iron or steel) by electrodeposition to provide sacrificial protection. The following plating solutions are used in commerce: (i) *Cyanide zinc baths* containing varying amounts of zinc cyanide, sodium cyanide, sodium hydroxide, sodium carbonate, sodium polysulfide and selected brighteners; (ii) *Alkaline noncyanide baths* containing zinc oxide, sodium hydroxide and selected additives, and (iii) *Acid chloride zinc baths* containing varying amounts of zinc chloride, ammonium chloride, potassium chloride, sodium chloride, boric acid and selected brighteners.

zinc polonide. A compound of zinc and polonium. Crystal system, cubic. Crystal structure, sphalerite. Used as a semiconductor. Formula: ZnPo.

zinc polycarboxylate cement. A *dental cement* made by mixing a powder composed of 90% zinc oxide (ZnO) and 10% magnesium oxide (MgO) with a liquid based on polyacrylic acid. The cured cement consists of unreacted ZnO particles embedded in a zinc carboxylate crosslinked polymer matrix. It forms a relatively strong bond between the tooth structure (dentin and enamel) and metal appliances, and is usually radiopaque and water-activated. Used chiefly for luting applications, for the cementation of bridges, crowns and inlays, and as a cavity liner or base. It is is also suitable for use with single units and short-span fixed partial dentures.

zinc potassium chromate. See buttercup yellow.

zinc powder. Atomized molten zinc (99+% pure) in particle sizes ranging from 100-325 mesh used in the production of zinc-rich paints and zinc diffusion coatings on steel (*sherardized coatings*), in metal primers and autobody coatings, in the preparation of *lithopone*, for thermal spraying, coating and galvanizing, in pipe-thread compounds, in metal refining, in dry-cell batteries, in the pulp and paper industries and pyrotechnics, as a reducing agent, catalyst and purifier, in rubber processing, in the manufacture of zinc salts and other zinc compounds, and in the production of decorative effects in resins.

zinc protoporphyrin. A purple, crystalline compound that is a *porphyrin* derivative characterized by a central divalent zinc atom. Used as a pigmemt and dye. Formula: $Zn(C_{34}H_{32}N_4O_4)$. See also protoporphyrin (2).

zinc pyrophospate. A white powder with a density of 3.75 g/cm³, used as a pigment. Formula: $Zn_2P_2O_7$.

zinc resinate. A zinc compound based on zinc abietate [$Zn(C_{20}H_{29}O_2)_2$] and available as a powder, yellowish liquid or clear amber lumps. Used as a wetting agent, hardener, dispersant, paint and varnish drier, and in synthetic resins.

zinc-rich primer. A *primer* for iron and steel that is based on *zinc dust* and commonly applied in 75-µm (3-mils) dry-film thickness to provide galvanic (sacrificial) protection. Abbreviation: ZRP. See also sacrificial coating.

Zincrolyte. Trade name of Enthone-OMI Inc. (USA) for a zinc-alloy electrodeposit and deposition process.

Zincrometal. Trademark of Hoesch Stahl AG (Germany) for a precoated sheet steel whose coating combines a topcoat of zinc-rich epoxy resin with a zinc-chromium undercoating, and is applied prior to mechanical processing. This unique combination results in a product with exceptional corrosion resistance and formability suitable for the manufacture of autobodies.

zinc scandium sulfide. A compound of zinc, scandium and sulfur. Crystal system, cubic. Crystal structure, spinel. Hardness, 324-410 Vickers; melting point, 1750°C±30°C (3182°F±54°F). Used as a semiconductor. Formula: $ZnSc_2S_4$.

zinc selenide. Yellowish to reddish crystals or yellow powder available in standard grades (99.9+% pure) and electronic grades (up to 99.999% pure). Crystal system, cubic. Crystal structure, sphalerite. Density, 5.42 g/cm³; melting point, 1517°C (2763°F); hardness, 1350 Knoop; refractive index, 2.89; dielectric constant, 9.2; band gap, 2.58 eV; infrared transmission, 70%. Used for windows in infrared optical equipment, for semiconductor applications, in phosphors, and as an evaporated interference layer material in metallography. Formula: ZnSe.

zinc selenide telluride. A nonhygroscopic compound of zinc selenide and zinc telluride that is available in the form of high-purity single crystals. It has high radiation and thermal stability, high conversion efficiency, intrinsic luminescence in the red spectral region, and is used in scintillation-silicon photodiodes, radiation monitors, medical X-ray and tomography equipment, nondestructive testing and inspection systems, and spectrometers for soft X-rays and alpha and beta radiation. Formula: ZnSe(Te).

zinc sheet and foil. See zinc foil and sheet.

zinc silicate. (1) A compound of zinc oxide and silicon dioxide, supplied in the form of white crystals. It occurs in nature as the mineral *willemite*. Density, 4.103 g/cm³; melting point, 1509°C (2748°F); hardness, 5-7 Mohs. Used for ceramics, phosphors and primers. Formula: Zn_2SiO_4. Also known as *zinc orthosilicate*.

(2) A compound of zinc oxide and silicon dioxide supplied as white powder, or crystals. Crystal system, hexagonal. Density, 3.52 g/cm³; melting point, 1437°C (2619°F). Formula: $ZnSiO_3$. Used in ceramics and materials research.

zinc silicofluoride. See zinc fluorosilicate.

zinc silicofluoride hexahydrate. See zinc fluorosilicate.

zinc silicon arsenide. A compound of zinc, silicon, and arsenic. Crystal system, tetragonal. Crystal structure, chalcopyrite. Density, 4.7 g/cm³; hardness, 9200 Knoop; melting point, 1038°C (1900°F); band gap, 1.7 eV. Used as a semiconductor. Formula: $ZnSiAs_2$.

zinc silicon phosphide. A compound of zinc, silicon and phosphorus. Crystal system, tetragonal. Crystal structure, chalcopyrite. Density, 3.39 g/cm³; melting point, 1367°C (2493°F); band gap, 2.3 eV; hardness, 1100 Knoop; good transmission in visible and infrared range. Used as a semiconductor. Formula: $ZnSiP_2$.

zinc slush-casting alloys. A group of eutectic zinc-aluminum alloys with or without small additions of copper. Zinc slush-casting alloy A (UNS Z34510) contains 4.75% aluminum and 0.25% copper, has a density of 6.5 g/cm³ (0.23 lb/in.³) and a melting point of 380°C (716°F) and zinc slush-casting alloy B (UNS

Z30500) contains 5.5% aluminum, has a density of 6.45 g/cm³ (0.233 lb/in.³) and a melting point of 380°C (716°F). *Zinc slush-casting alloys* are used for casket hardware, lighting fixtures, lamp bases, and hollow parts.

zinc soap. A water-insoluble soap formed from zinc and an organic acid, e.g., zinc stearate is a *metallic soap* formed by zinc and stearic acid.

zinc solders. A group of solders in which zinc is the predominant constituent, and which are used mainly for soldering aluminum. Included are: (i) *Zinc-tin solder* composed of 70% zinc and 30% tin and having solidus and liquidus temperature of 199°C (390°F) and 375°C (708°F), respectively; (ii) *Zinc-cadmium solders* such as 90Zn-10Cd and 60Zn-40Cd; and (iii) *Zinc-aluminum solder* with 95% zinc and 5% aluminum which is the eutectic composition having a melting point of 382°C (720°F). See also zinc-aluminum solder; zinc-cadmium solder; zinc-tin solder.

zinc spar. See smithsonite.

zinc spinel. See gahnite.

zinc stearate. A metallic soap available in the form of a pure white, hydrophobic powder with a density of 1.095 g/cm³ and a melting point of 130°C (266°F). Used as a paint and lacquer drier, as a lubricant, as a mold-release agent, as a filler, antifoamer and dusting powder, as a heat and light stabilizer in rubber, and in the biosciences. Formula: $Zn(O_2C_{18}H_{35})_2$.

zinc subcarbonate. A compound composed of zinc hydroxide and zinc carbonate and supplied as a white powder (97+% pure) with a density of 4.398 g/cm³. Formula: $3Zn(OH)_2 \cdot 2ZnCO_3$. Also known as *zinc carbonate hydroxide*.

zinc sulfate monohydrate. Colorless or white crystals or free-flowing powder used in electroplating, in rayon manufacture, and in dyestuffs. Formula: $ZnSO_4 \cdot H_2O$.

zinc sulfate heptahydrate. Colorless crystals or granular, crystalline powder. It occurs in nature as the mineral *goslarite*. Crystal system, orthorhombic. Density, 1.96 g/cm³; melting point, 100°C (212°F); boiling point, loses $7H_2O$ at 280°C (536°F). Used in the manufacture of rayon, in ore flotation, and as a wood preservative. Formula: $ZnSO_4 \cdot 7H_2O$. Also known as *white copperas; white vitriol; zinc vitriol*.

zinc sulfide. A compound of zinc and sulfur existing in two allotropic forms: (i) *Alpha zinc sulfide* (α-ZnS): Colorless or white crystals or yellowish-white powder (99.9+% pure). Occurs in nature as the minerals *wurtzite* and *zincite*. Crystal system, hexagonal. Crystal structure, wurtzite (or zincite). Density, 4.1 g/cm³; melting point, 1827°C (3321°F); boiling point, sublimes at 1180°C (2156°F); hardness, 250 Knoop; 3.5-4 Mohs; refractive index, 2.356; band gap, 3.67 eV; infrared transmission, 72%; and (ii) *Beta zinc sulfide* (β-ZnS): Gray-white crystals or powder (99.9+% pure). Occurs in nature as the mineral *sphalerite*. Crystal system, cubic. Crystal structure, sphalerite (or zinc blende). Density, 4.079 g/cm³; transformation temperature (change to alpha form), 1020°C (1868°F); melting point, 1827°C (3321°F); hardness, 1780 Knoop or 3.5-4 Mohs; refractive index, 2.356; dielectric constant, 8.9; band gap, 3.54 eV; high infrared transmission. *Zinc sulfide* is used as a white pigment in paints, in luminous paints, as an ingredient in lithopone pigments, in white and opaque glasses, as a semiconductor, as a constituent in X-ray and television tubes, in phosphors and similar products, and as an ingredient in rubber and plastics. Formula: ZnS.

zinc sulfide monohydrate. Colorless, white or yellowish crystals or powder with a melting point of 1049°C (1920°F) used

as a white pigment for paint and rubber, in white glass and opaque glass, and in paper coatings. Formula: $ZnS \cdot H_2O$.

zinc sulfide white. See lithopone.

zinc telluride. Gray powder or reddish crystals (99.99+% pure) available in two crystalline forms: (i) cubic with sphalerite (zinc blende) structure; and (ii) hexagonal with wurtzite (zincite) structure. Important properties of zinc telluride include: Density, 6.34 g/cm³; hardness, 900 Knoop; refractive index, 3.56; melting point, 1295°C (2363°F); refractive index, 3.56; dielectric constant, 10.4; band gap, 2.26 eV. Used as intrinsic semiconductor (Group II-VI or Group 12-16), and as a phosphor and photoconductor. Formula: ZnTe.

zinc tetroxychromate. A yellowish powder used as a yellow pigment in corrosion- and blister-resistant metal primers. Formula: $ZnCrO_4 \cdot 4Zn(OH)_2$.

zinc tin antimonide. A compound of zinc, tin and antimony. Crystal system, cubic. Crystal structure, sphalerite. Density, 5.67 g/cm³; melting point, 597°C (1107°F); hardness, 2500 Knoop; band gap, 0.4 eV. Used as a semiconductor. Formula: $ZnSnSb_2$.

zinc tin arsenide. A compound of zinc, tin and arsenide available in two forms: (i) Regular zinc tin arsenide. Crystal system, tetragonal. Crystal structure, chalcopyrite. Density, 5.53 g/cm³; melting point, 775°C (1427°F); hardness, 4550 Knoop; band gap, 0.65 eV; and (ii) High-temperature zinc tin arsenide. Crystal system, cubic. Crystal structure, sphalerite. Density, 5.53 g/cm³; melting point, 777°C (1431°F). Used as a semiconductor. Formula: $ZnSnAs_2$.

zinc tin phosphide. A compound of zinc, tin and phosphorus available in two crystalline forms: (i) Cubic zinc tin phosphide. Crystal structure, sphalerite (zinc blende); melting point, 927°C (1701°F); band gap, 2.1 eV; and (ii) Tetragonal zinc tin phosphide. Crystal structure, chalcopyrite; hardness, 6500 Knoop; band gap, 1.45 eV. Used as a semiconductor. Formula: $ZnSnP_2$.

zinc tinsel. An alloy composed of 60% zinc and 40% lead , used for decorative and architectural applications.

zinc-tin solder. A solder of 70% zinc and 30% tin having a solidus and liquidus temperature of 199°C (390°F) and 375°C (708°F), respectively. Used to solder aluminum. See also zinc solder.

zinc titanate. Any of the following compounds used chiefly for dielectric applications: (i) *Zinc monotitanate*. An off-white powder (96+% pure) made from zinc oxide and titanium dioxide. Melting point, above 1500°C (2730°F). Formula: $ZnTiO_3$; and (ii) *Dizinc titanate*. A powder made from titanium trioxide and zinc. Melting point, above 1500°C (2730°F). Formula: Zn_2TiO_3.

zinc vitriol. See zinc sulfate heptahydrate.

zinc white. Zinc oxide (ZnO) used as a commercial pigment and available as an extremely white, permanent powder having exceptionally high ultraviolet light absorption, but only moderate opacity and covering power. In the coating industry the terms "Chinese white" and "zinc white" are used interchangeably. Used in white paints and coatings, and as a filler in lime soap grease. Also known as *Chinese white*.

zinc yellow. See buttercup yellow.

zinc-zippeite. A yellow mineral of the zippeite group composed of zinc uranyl sulfate hydroxide hydrate, $Zn_2(UO_2)_6(SO_4)_3(OH)_{10} \cdot 16H_2O$. It can also be made synthetically. Crystal system, orthorhombic. Density, 3.30 g/cm³; refractive index, 1.77.

zinc-zirconium silicate. A compound of zinc oxide, zirconium dioxide and silicon dioxide, available in the form of a white powder. Density, 4.8 g/cm³; Melting point, 2080°C (3776°F). Used as an opacifier in ceramic glazes. Formula: $ZnZrSiO_5$.

Zinic. Trade name of MacDermid Inc. (USA) for a zinc-nickel electrodeposit and deposition process.

Zinkal. Trade name of Zinkberatungsstelle GmbH (Germany) for a series of structural zinc alloys.

zinkenite. A steel-gray mineral with bright metallic luster composed of lead antimony sulfide, $Pb_2Sb_{14}O_{27}$. Crystal system, hexagonal. Density, 5.22-5.36 g/cm^3; hardness, 3-3.5 Mohs. Occurrence: Germany.

Zinlac. Trade name of William Zinsser Company (USA) for synthetic shellac plastics based on corn *zein* protein.

Zinnal. Trade name of Vereinigte Silberhammerwerke Hetzel (Germany) for a corrosion-resistant aluminum sheet with a tin cladding on both sides. Used for instrument cases.

Zinnbronze. Trade name of VDM Nickel-Technologie AG (Germany) for a series of wrought or cast tin bronzes containing up to 20% tin and up to 0.4% phosphorus with the balance being copper. They possess excellent corrosion resistance, high hardness and strength, and good wear resistance. Used for turbine wheels, pump and turbine housings, springs, diaphragms, fasteners, tubing, fittings, bearings, bushings, bells, and wear plates.

Zinnorite. Trade name for a fused zirconia (99.2+% pure) with a small amount of silica, supplied in the form of a fine powder with a melting point above 2700°C (4890°F). Used for refractory applications.

zinnwaldite. A violet, yellow or gray mineral of the mica group composed of potassium lithium iron aluminum silicate hydroxide, $K(Li,Fe)_2Si_4O_{10}(OH)_2$. Crystal system, monoclinic. Density, 3.00 g/cm^3; refractive index, 1.573. Occurrence: Germany.

Zinol. Trademark of Heatbath Corporation (USA) for a black conversion coating used on zinc plate and zinc die castings.

zinox. A hydrated zinc oxide used in enamels.

Zinroc. Trade name of Accu Bite Dental Supply (USA) for a zinc oxide–eugenol dental cement.

Zinros. Trademark of Nuodex, Inc. (USA) for a pale-colored *zinc resinate* with high melting point used in rubber compounds, adhesives, and printing inks.

Zionomer. Trade name of DenMat Corporation (USA) for a glass-ionomer dental cement.

Zip. Trade name of Ziv Steel & Wire Company (USA) for a tungsten-type high-speed steel containing 0.65-0.75% carbon, 4-4.5% chromium, 0.85-1% molybdenum, 1.65-2.0% vanadium, 17.5-19% tungsten, 7-7.5% cobalt, and the balance iron. Used for cutting, forming and shaping tools.

Zipcoat. Trade name of Corrocoat Limited (UK) for a glass flake-filled coating system used for one-coat or two-coat application to structural steelwork and industrial machinery and equipment.

Zip Grip. Trade name of Devcon Corporation (USA) for a series of fast-curing, solvent-free, one-component cyanoacrylate adhesives that cure almost instantly when exposed to atmospheric moisture, and are used for bonding metals, plastics, rubber and wood.

Zippalloy. Trade name of Seymour Products Company (USA) for a corrosion-resistant *high brass* composed of 87% copper and 13% zinc, and supplied in various tempers including half-hard, hard and spring. Used for electrical components, lamps, hardware, watches, and jewelry.

zippeite. (1) An orange-yellow, radioactive mineral composed of uranyl sulfate hydroxide octahydrate, $(UO_2)_3(SO_4)_2(OH)_2 \cdot 8H_2O$. Crystal system, monoclinic. Density, 3.66 g/cm^3; refractive index, 1.717.

(2) A yellow, radioactive mineral composed of potassium uranyl sulfate hydroxide tetrahydrate, $K_4(UO_2)_6(SO_4)_3(OH)_{10} \cdot 4H_2O$. It can also be made synthetically. Crystal system, orthorhombic. Density, 3.30 g/cm^3; refractive index, 1.716. Occurrence: Czech Republic.

Zirbeads. Trademark of Zircoa, Inc. (USA) for dispersion media supplied in the form of hard, nonporous, nonmagnetic, nonconductive, chemically inert zirconia (ZrO_2) beads of uniform shape and size and off-white to gray color. They have high resistance to corrosion chipping and fracture, and are used as reinforcements for composites.

Zirblast. Trade name of SEPR Ceramic Beads & Powders (USA) for ceramic blasting media.

Zircadyne. Trademark of Wah Chang, Allegheny Technologies Company (USA) for alloyed and unalloyed zirconium. The unalloyed grades are used for corrosion-resistant process equipment in the chemical industry, and the alloys for nuclear reactor and chemical applications requiring good resistance to hydrochloric, nitric, phosphoric and sulfuric acids, organic acids (e.g., formic and acetic acid), strong alkalies, and molten salts.

Zircal. French trade name for an age-hardenable aluminum alloy containing 7-8.5% zinc, 2.5% magnesium, 1.5% copper and 0.25% chromium. Used for light-alloy parts.

Zircaloy. Trademark of Westinghouse Electric Corporation (USA) for a series of zirconium alloys containing about 1.2-1.7% tin, 0-0.15% oxygen, 0.07-0.35% iron, 0.05-0.15% chromium, 0.03-0.08% nickel and up to 0.05% carbon. They have low neutron cross sections, high strength, and good high-temperature and corrosion resistance. Used for pressure tubes in nuclear reactors, structural reactor components, and as cladding materials for nuclear fuel elements.

Zircar. (1) Trademark of Zircar Products, Inc. (USA) for a series of *yttria-stabilized zirconia* ceramics containing 92% zirconia (ZrO_2) and 8% yttria (Y_2O_3). They are commercially available in the form of discontinuous fibers, bulk fibers, mats, blankets and fabrics, and have a density of 5.6-5.9 g/cm^3 (0.20-0.21 lb/in.^{3}), low thermal conductivity, and an upper continuous-use temperature of about 2200°C (3990°F). Used as reinforcements in metals and ceramics, and for thermal insulation and high-temperature filtering applications.

(2) Trademark of Zircar Products, Inc. (USA) for a series of fibrous ceramics including various advanced ceramic materials based on alumina, alumina-silica, zirconia and other refractory oxide compositions.

Zircoa. Trademark of Zircoa Inc. (USA) for a series of advanced zirconia (ZrO_2) ceramics. *Zircoa A* is composed of high-purity monoclinic zirconia, produced from zircon sand, and used as a pigment, refractory and abrasive, as an ingredient in glass, and as an opacifier in ceramic coatings. *Zircoa B* is a stabilized cubic zirconia used in the manufacture of various ceramic products. *Zircoa-Cast* are self-setting castable zirconia refractories that mold-set hydraulically, and are used in the manufacture of strong, heat-resistant shapes.

Zircofrax. Trademark of Harbison-Carborundum Corporation (USA) for a series of superrefractories made from zirconium oxide (ZrO_2) and zirconium silicate ($ZrSiO_4$). They have excellent high-temperature resistance, high strength and thermal conductivity, an average porosity of 25%, low permeability, good resistance to acids and acid slags, and are used as refractory bricks and shapes for chemical and metallurgical furnace applications and ceramic-kiln furniture.

Zircograf. Trade name of Pechiney Electrométallurgie (France) for a cast iron inoculant containing 66% silicon, 6% manganese, 6% zirconium, 1.5% calcium, 1.3% aluminum, and the balance iron.

zircon. A colorless, yellow, gray, green, blue, red or brown to black mineral with an adamantine luster. It is composed of zirconium silicate, $ZrSiO_4$, and may also contain about 1-3% hafnium. Crystal system, tetragonal. Density, 4.60-4.68 g/cm³; softening temperature, 850-950°C (1560-1740°F); melting point, 2250°C (4080°F); hardness, 7.5 Mohs; refractive index, 1.923. Occurrence: Australia, Brazil, Canada (Ontario), India, Sri Lanka, USA (Colorado, Florida, New York, North Carolina), Western Africa. Used as an acid refractory material in molding sand, foundry cores and precision molds for the casting of alloys, in abrasives and grinding wheels, as an important ore of zirconium, as a source of zirconium oxide and hafnium, as an opacifier in porcelain enamels and glazes, in electrical, refractory and technical porcelains and electrically resisting cements, as a catalyst, in silicone rubbers, and as a gemstone. Also known as *zirconite*.

Zirconal. German trade name for a non-heat-treatable aluminum alloy containing 15% copper, 8% manganese and 0.5% silicon. Used for light-alloy parts.

zircon ceramics. A group of ceramic whitewares in which zircon ($ZrSiO_4$) is the primary crystalline phase. They are characterized by high refractoriness and good dielectric properties, and are used in refractory products, refractory porcelain, and electrical and technical porcelains.

zircon grog. A crushed and sized *grog* composed of about 66% zirconia (ZrO_2), 32% silica (SiO_2) and 2% other oxides. Used for refractory applications.

Zircon H-W. Trademark of Harbison-Walker Refractories Company (USA) for a ceramic material made from purified natural zircon ($ZrSiO_4$) by any of several processes including air ramming, impact pressing and slip casting. It has a density of 3.7 g/cm³ (0.13 lb/in.³), low shrinkage under soaking heat in excess of 1590°C (2895°F), good resistance to thermal spalling and fluxing conditions and wetting and penetration by molten glass. Used for lining glass-tank bottoms and floors, in sodium metaphosphate and sodium silicate furnaces, for taphole blocks in the melting of nonferrous alloys, and in the form of nozzles for continuous steel casting.

zirconia. A compound of zirconium and oxygen available as white crystals or powder (98+% pure). It occurs in nature as the mineral *baddeleyite*, and can also be made synthetically by heating zirconium carbonate or zirconium hydroxide. Crystal system, monoclinic. Density, 5.49-5.73 g/cm³; melting point, approximately 2715°C (4920°F); boiling point, 5000°C (9030°F); hardness, 6.5 Mohs; refractive index, 2.13. It has outstanding thermal resistance, good dielectric properties, and can be stabilized with calcia (CaO), yttria (Y_2O_3) or hafnia (ZrO_2). Used as a source of zirconium metal, in metallurgy, as an opacifier in porcelain enamels, glazes and special glasses, as a refractory structural material in nuclear applications, in refractory linings for furnaces, as an abrasive and polishing medium, in cermets, ferrites and titanates, in highly corrosion-resistant ceramics, high-temperature refractories and electrical and thermal insulation, for setter plates in the firing of ceramics, in wind-tunnel liners, and in the production of piezoelectric crystals, high-frequency induction coils and heat-resistant fibers. Formula: ZrO_2. Also known as *zirconic anhydride; zirconium anhydride; zirconium dioxide; zirconium oxide*. See also calcia-stabilized zirconia; hafnia-stabilized zirconia; yttria-stabilized zirconia.

Zirconia Alundum. Trademark of Norton Company (USA) for zirconia and alumina abrasives.

zirconia brick. Refractory brick composed of up to 94% zirconia (ZrO_2) and 6% or more lime (CaO) and silica (SiO_2). It has a continuous-use temperature above 1900°C (3450°F), and is highly resistant to basic slags. Used as a lining material for metallurgical furnaces. Also known as *zirconium oxide brick*.

zirconia ceramic coating. A zirconium oxide (ZrO_2) based ceramic coating with low thermal conductivity, applied to metallic substrates to serve as thermal barrier. Also known as *zirconium oxide coating*.

zirconia ceramics. Refractory materials based on zirconia (ZrO_2) and often stabilized with oxides, such as calcia or yttria (e.g., 10.5%), to avoid cracking and mechanical weakening during cooling and heating. They have excellent high-temperature chemical inertness and corrosion resistance, good electrical conductivity above 800°C (1470°F), and low thermal conductivity. Used as refractory structural materials, in refractory linings for furnaces and furnace parts, as high-temperature refractories, e.g., for heat-element supports, in electrical and thermal insulation, in ferrites and titanates, in oxygen sensors and thermo-couple protection tubes, and as bioceramics in orthopedics for total hip replacements. Other applications can be found under zirconia. Also known as *zirconium oxide ceramics*.

zirconia fabrics. Discontinuous zirconia fibers made into a flexible planar structure by weaving, knitting, felting or pressing, and used for the reinforcement of metals and ceramics, as thermal insulation, and for high-temperature filtering applications. Also known as *zirconium oxide fabrics*. See also zirconia felt.

zirconia felt. A flexible ceramic textile made by mechanical interlocking of zirconia fibers and available in sheets of varying thickness, and in various preformed shapes. It has excellent high temperature properties (up to 2200°C or 3990°F), good resistance to corrosive environments, very low thermal conductivity, and is used as thermal insulation for furnaces and electronic parts, as chemical barrier in powder-metallurgy sintering applications, and as insulation for gas diffusion burners. Also known as *zirconium oxide felt*. See also zirconia fabrics.

zirconia fibers. A group of heat-resistant, polycrystalline fibers consisting of zirconia (ZrO_2) stabilized with about 5 wt% lime (CaO), up to 8% yttria (Y_2O_3) and/or varying amounts of magnesia (MgO). They have an average fiber diameter of 1-10 µm (39-390 µin.), an upper service temperature of 1650-2200°C (3000-3990°F), low thermal conductivity, and are used as reinforcements for metals and ceramics, and in the form of mats, blankets and fabrics for thermal insulation and high-temperature filtering applications. Also known as *zirconium oxide fibers*.

zirconia foam. Refractory bricks and shapes of zirconia (ZrO_2) having a sponge-like, cellular structure due to purposely introduced gaseous cells distributed throughout. Used for thermal insulation applications. Also known as *zirconium oxide foam*.

zirconia porcelain. A ceramic whiteware in which zirconia (ZrO_2) is the essential crystalline phase. It has excellent thermal properties, good dielectric properties, and is used for electrical and thermal-insulating applications. Also known as *zirconia whiteware; zirconium oxide porcelain*.

zirconia powder. Stabilized or unstabilized zirconia (ZrO_2) supplied in the form of a fine, white powder (98-99.99% pure) with a particle size ranging from submicron to over 150 µm (40 to over 5900 µin.). Used in metallurgy, as an opacifier in porcelain enamels, glazes and special glasses, as an abrasive in polishing and grinding compounds, in the manufacture of refractories, and for thermally sprayed coatings. Also known as *zirconium oxide powder*.

zirconia refractory. A refractory material composed principally of crystalline zirconia (ZrO_2) and supplied in the form of bricks, shapes, fabrics, and fibers. It has a low thermal conductivity and high heat resistance. Stabilized zirconia refractories can be used at temperatures above 2200°C (3990°F). Also known as *zirconium oxide refractory*.

zirconia-silica fibers. Continuous ceramic fibers composed of zirconium oxide (ZrO_2) and silicon dioxide (SiO_2), usually in a ratio of about 2:1. Available in the form of continuous fibers, yarn, roving and fabrics, they possess moderate reinforcement properties, excellent mechanical durability and resistance to flame penetration, and are used for firewall construction and thermal insulation applications.

zirconia-toughened alumina. An alumina-matrix composite that has been transformation-toughened by dispersing small particles of T-phase zirconia or partially stabilized zirconia within the matrix. Oxides of calcium, cerium, magnesium or yttrium are often added as stabilizers. Used for applications requiring high-temperature resistance, good wear resistance and extended fracture toughness. Abbreviation: ZTA. See also *alumina-matrix composites*.

zirconia-toughened ceramics. Ceramic-matrix composites that have been transformation-toughened by dispersing small particles of partially stabilized zirconia within the matrix. Typical matrix materials include alumina and zirconia, and calcium, cerium, magnesium and/or yttrium oxides are added as stabilizers. Used in applications requiring high-temperature resistance and extended fracture toughness. Abbreviation: ZTC. See also *ceramic-matrix composites*.

zirconia whiteware. See *zirconia porcelain*.

zirconic anhydride. See *zirconia*.

Zirconite. Trademark of National Lead Company (USA) for a ceramic material composed of 97+% zircon ($ZrSiO_4$) and supplied in the form of flour and sand. It has a melting point of 2220-2260°C (4028-4100°F) and is used for foundry cores, facings and molds.

zirconite. See *zircon*.

zirconite sand. See *zircon sand*.

zirconium. A rare, ductile, grayish, refractory metallic element of Group IVB (Group 4) of the Periodic Table. It is commercially available in the form of rods, bars, powder, sponge, wire, plate, sheet, foil, tubing, single crystals, crystal bars and crystal-bar turnings. The most important zirconium ores are *baddeleyite* and *zircon*. Density, 6.49 g/cm³; melting point, 1852°C (3366°F); boiling point, 4377°C (7911°F); hardness, 85-100 Vickers; atomic number, 40; atomic weight, 91.224; divalent, trivalent, tetravalent. It has a superconductivity critical temperature of 0.61K, good corrosion resistance, low electrical conductivity (about 4% IACS), and a low neutron absorption coefficient (about 0.182 Barns). It exists in two forms: (i) *Alpha zirconium* which is present at room temperature and has a hexagonal close-packed crystal structure; and (ii) *Beta zirconium* which is present at high temperatures (862-1852°C or 1584-3366°F) and has a body-centered cubic crystal structure. *Zirconium* is used as an alloying element in steels (e.g, as a carbide former, deoxidizer, desulfurizer and/or scavenger) and in nickel-chromium and other nonferrous alloys, as a deoxidizer in metal castings, in alloys for wires and lamp filaments, in alloys with

niobium (columbium) and zinc to make low-temperature, superconductive magnets, as a structural material in nuclear reactors, as a cladding material for fuel elements and in alloys (e.g., *Zircaloy*) for nuclear-energy applications. It is also used as a high-intensity arc light, as an efficient getter for electronic tubes, in catalytic converters, in ceramics and refractories, in metal-to-glass and metal-to-ceram-ic seals, for percussion caps, as a constituent in flashlight powder, in special welding fluxes, and for laboratory crucibles and spinnerets. Symbol: Zr.

zirconium-95. A radioactive isotope of zirconium with a mass number of 95 obtained in a mixture with niobium (columbium) from the fission products of nuclear fuels. It has a half-life of 63 days, emits beta and gamma radiation, and is used for tracing the flow of petroleum in pipelines, in the determination of the rate of catalyst circulation in petroleum cracking plants, and in the study of cracking and polymerization of hydrocarbons. Symbol: ^{95}Zr.

zirconium acetylacetonate. A hygroscopic, white, crystalline powder (98% pure) containing about 18.5-18.9% zirconium. It has a density of 1.415 g/cm³, a melting point of 171-173°C (339-343°F) and is soluble in water, benzene, acetone, pyridine, ethanol and toluene. Used in organic and organometallic synthesis, e.g., as a reagent or catalyst, as a crosslinker for polyalkoxy, polyester and polyol resins, and as an additive to greases and lubricants. Formula: $Zr(O_2C_5H_7)_4$. Also known as *zirconium tetraacetylacetonate; zirconium 2,4-pentanedionate*.

zirconium alloys. A group of alloys based on zirconium that can be subdivided into the following two categories: (i) *Alpha stabilizers* which raise the transformation temperature; and (ii) *Beta stabilizers* which lower it. Zirconium alloys with niobium (e.g., Zr-2.5Nb) and zinc are used to make low-temperature, superconductive magnets. The most common zirconium alloys are those supplied under the trademark *Zircaloy*. Used as structural alloys in nuclear reactors, and as cladding materials for nuclear fuel elements.

zirconium aluminate. A compound of zirconium oxide and aluminum oxide used as a component in high-temperature refractories. Formula: $ZrAl_2O_5$.

zirconium aluminide. Any of the following compounds of zirconium and aluminum used in ceramics and materials research: (i) *Zirconium monoaluminide*. Melting point, 1637°C (2979°F). Formula: ZrAl; and (ii) *Zirconium tetraaluminide*. Density, 5.30 g/cm³; melting point, 1532°C (2790°F). Formula: Zr_4Al_3.

zirconium-aluminum. A master alloy containing 55-65% zirconium, 39-45% aluminum and 0.1% carbon.

zirconium aluminum titanate. A ceramic compound with low thermal expansion, good resistance to thermal shock, and good high-temperature properties at temperatures exceeding 1000°C (1830°F). Used for thermal linings, catalyst supports, and for lining of engine exhaust ports. Abbreviation: ZAT.

zirconium anhydride. See *zirconia*.

zirconium antimonide. A compound of zirconium and antimony used in ceramics, materials research and electronics. Melting point, 1900°C (3452°F). Formula: Zr_2Sb.

zirconium-barium-lanthanum-aluminum-sodium glass. A fluorozirconate glass frequently doped with erbium, praseodymium, thulium, or cerium. Used in laser diodes and for optical fibers. Abbreviation: ZBLAN glass. See also *ZBLAN*.

zirconium basic carbonate. A white powder that is composed of zirconium carbonate hydroxide and zirconium dioxide and decomposes at 135°C (275°F). Formula: $Zr(OH)_2CO_3·ZrO_2$. See also *basic zirconium carbonate; basic zirconium carbonate hy-*

drate.

zirconium beryllide. Any of the following compounds of zirconium and beryllium with good strength at elevated temperatures, often used as moderators in nuclear reactors: (i) $ZrBe_{13}$. Density, 2.72 g/cm³; melting point, 1930°C (3506°F); and (ii) Zr_2Be_{17}. Density, 3.08 g/cm³; melting point, 1980°C (3596°F).

zirconium boride. Any of the following compounds of zirconium and boron: (i) *Zirconium monoboride*. Density, 6.7 g/cm³; melting point, 798-1248°C (1468-2278°F). Used in ceramics and materials research. Formula: ZrB; (ii) *Zirconium diboride*. Gray crystals, or microcrystalline powder. Crystal system, hexagonal. Density, 6.08-6.17 g/cm³; melting point, 3245°C (5873°F); hardness, 8 Mohs; excellent thermal shock resistance; poor oxidation resistance above 1100°C (2010°F). Used as a superrefractory for aircraft and rocket applications (e.g., nozzles), in electrodes for metal refining, in crucibles, as a metallurgical additive, as a cathode in high-temperature electrochemical systems, in thermocouple protection tubes, as a high-temperature electrical conductor, in cutting tools, in coatings for refractory metals (e.g., tantalum), and in metal-casting refractory molds, refractory pouring spouts, combustion chamber lines and other high-temperature products. Formula: ZrB_2; and (iii) *Zirconium dodecaboride*. Density, 3.63 g/cm³; melting point, 2566°C (4651°F). Used in ceramics and materials research. Formula: ZrB_{12}.

zirconium-boron composites. A group of high-temperature engineering composites including: (i) ZrB_2·B. Density, 5.2-5.4 g/cm³; melting point, above 2980°C (5396°F); and (ii) ZrB_2·$MoSiO_2$. Density, 4.87-5.50 g/cm³; melting point, above 2370°C (4298°F).

zirconium bromide. See zirconium tetrabromide.

zirconium bronze. See zirconium copper.

zirconium carbide. Gray crystals or fine, black, pyrophoric powder (99+% pure) produced by heating zirconia with carbon at about 2000°C (3630°F). Crystal system, cubic. Density, 6.73 g/cm³; melting point, 3540°C (6400°F); boiling point, 5100°C (9210°F); hardness, 8-9 Mohs. Used for abrasives, as a refractory, in metal cladding, as a coating material for graphite, for incandescent filaments, cutting-tool components, high-temperature electrical conductors, composites and cermet compounds, as a metallurgical additive, and as a source of zirconium and zirconium compounds. Formula: ZrC.

zirconium carbide coatings. Impermeable, uniform high-temperature protective coatings based on zirconium monocarbide (ZrC). They have a density of 6.7 g/cm³ (0.24 lb/in.³), a melting point of about 3540°C (6405°F), an upper service temperature of about 800°C (1470°F), and excellent resistance to wear and chemical attack by corrosive liquids and gases. Used on graphite in epitaxial processes for the manufacture of compound semiconductors by metal-organic chemical vapor deposition (MOCVD) and liquid-phase epitaxy (LPE).

zirconium carbonate. See basic zirconium carbonate.

zirconium chloride. See zirconium tetrachloride.

zirconium cobalt. A fine alloy powder composed of up to 50% zirconium and 50% or more cobalt. Used for powder-metallurgy parts.

zirconium copper. (1) Deoxidized copper containing about 0.15% zirconium. It is commercially available in the form of rods and wire, and has excellent cold and hot workability, good forgeability, age-hardening properties, strength, electrical conductivity and softening resistance. Used for circuit breakers, commutators, switches, rectifiers, soldering and welding tips, and elec-

tronic components and devices. Also known as *zirconium bronze*.

(2) A fine alloy powder composed of 70% zirconium and 30% copper, and used for powder-metallurgy parts.

zirconium diboride. See zirconium boride (ii).

zirconium dichloride oxide. See zirconyl chloride.

zirconium dichloride oxide octahydrate. See zirconyl chloride octahydrate.

zirconium dioxide. See zirconia.

zirconium diphosphide. See zirconium phosphide (ii).

zirconium disilicide. See zirconium silicide (ii).

zirconium disulfide. See zirconium sulfide (ii).

zirconium dodecaboride. See zirconium boride (iii).

zirconium-ferrosilicon. A master alloy available in various grades, commonly containing 40-47% silicon, 40-45% iron, 9-12% zirconium, and a maximum of 0.20% carbon. Used to introduce zirconium into steel.

zirconium fluoride. See zirconium tetrafluoride.

zirconium germanide. Any of the following compounds of zirconium and germanium used in ceramics, materials research and electronics: (i) *Dizirconium germanide*. Melting point, 1910-2277°C (3470-4131°F). Formula: Zr_2Ge; and (ii) *Zirconium sesquigermanide*. Melting point, 1550°C (2822°F). Formula: Zr_2Ge_3; and (iii) *Trizirconium germanide*. Melting point. 1587°C (2889°F). Formula: Zr_3Ge.

zirconium halide. A compound of zirconium and a halogen, e.g., zirconium bromide ($ZrBr_2$), zirconium chloride ($ZrCl_4$) or zirconium fluoride (ZrF_4).

zirconium hydride. A compound of zirconium that is available in the form of a fine, gray-black, high-purity powder containing about 1.7-2.1% combined hydrogen. Crystal system, tetragonal. Density, 5.47-5.60 g/cm³; autoignition temperature, 270°C (518°F). Used as a vacuum-tube getter, in powder metallurgy for making sintered metals, as a source of hydrogen, as a metal-foaming agent, in nuclear reactors as a neutron moderator, as a reducing agent, and as a hydrogenation catalyst. Formula: ZrH_2.

zirconium hydroxide. A white, bulky, amorphous powder with a density of 3.25 g/cm³ and a melting point of 550°C (1022°F) (decomposes to ZrO_2), used as a colorant in glass manufacture, and as a source of zirconium oxide and zirconium sulfate. Formula: $Zr(OH)_4$.

zirconium iodide. See zirconium tetraiodide.

zirconium-iron pink. A colorant for ceramic glazes that produces a pink color, is composed of chromic oxide (Cr_2O_3) and zirconium dioxide (ZrO_2), and has a maturing temperature of 1220-1280°C (2228-2336°F). Also known as *iron-zirconium pink*.

zirconium-lead alloy. A pyrophoric alloy containing 50% zirconium and up to 50% lead, sometimes with additions of tin. Used in cigarette lighters.

zirconium monoaluminide. See zirconium aluminide (i).

zirconium monoboride. See zirconium boride (i).

zirconium monophosphide. See zirconium phosphide (i).

zirconium monosilicide. See zirconium silicide (i).

zirconium monosulfide. See zirconium sulfide (i).

Zirconium Mullite M. Trade name of BPI Inc. (USA) for a high-melting alumina-zirconia products used for refractory applications. A typical composition is 65% alumina (Al_2O_3), 17% zirconia (ZrO_2), 14% silica (SiO_2), 0.1% lime (CaO), 0.1% ferric oxide (Fe_2O_3), a total of 0.5% nickel and chromium, and a trace of cobalt.

zirconium naphthenate. A metallic naphthenate in the form of an amber-colored, transparent, heavy, viscous liquid. It has a density of 1.05 g/cm³ and is used in certain porcelain enamels and glazes, in lubricants, and in paints and varnishes.

zirconium nickel. A fine alloy powder composed of 30-70% zirconium, with the balance being nickel. The melting point of a powder composed of 70% zirconium and 30% nickel is 1150°C (2100°F). Used for powder-metallurgy parts.

zirconium-niobium alloys. Alloys containing about 97.5% zirconium and 2.5% niobium (columbium). The niobium acts as a mild beta-phase stabilizer. *Zirconium-niobium alloys* have good high-temperature resistance, good strength, moderate corrosion resistance, and low neutron cross sections. Used for low-temperature superconductor magnets and in nuclear reactors.

zirconium nitride. Yellow crystals or brass-brown, crystalline powder (98+% pure). Crystal system, cubic. Density, 7.09 g/cm³; melting point, 2980°C (5396°F); hardness, above 8 Mohs. Used for refractories, crucibles, cermets and advanced ceramics. Formula: ZrN.

zirconium-opacified porcelain enamel. A cover-coat enamel for cast irons with an approximate melted-oxide composition of the frit of 28% silica (SiO_2), 18% lead monoxide (PbO), 10% sodium monoxide (Na_2O), 9% boric oxide (B_2O_3), 9% calcium oxide (CaO), 6% zinc oxide (ZnO), 4% aluminum oxide (Al_2O_3), 4% potassium oxide (K_2O), 6% fluorine (F) and 6% zirconium oxide (ZrO_2).

zirconium opacifiers. A collective term for zircon ($ZrSiO_4$) and zirconia (ZrO_2) used in powder form in glasses, glazes and porcelain enamels to impart or increase diffuse reflection, refraction or diffraction, and produce an opaque appearance by reducing the transparency.

zirconium oxide. See zirconia.

zirconium oxide brick. See zirconia brick.

zirconium oxide ceramics. See zirconia ceramics.

zirconium oxide coating. See zirconia ceramic coating.

zirconium oxide fabrics. See zirconia fabrics.

zirconium oxide felt. See zirconia felt.

zirconium oxide fibers. See zirconia fibers.

zirconium oxide foam. See zirconia foam.

zirconium oxide porcelain. See zirconia porcelain.

zirconium oxide powder. See zirconia powder.

zirconium oxide refractory. See zirconia refractory.

zirconium oxychloride. See zirconyl chloride.

zirconium oxychloride octahydrate. See zirconyl chloride octahydrate.

zirconium 2,4-pentanedionate. See zirconium acetylactonate.

zirconium phosphide. Any of the following compounds of zirconium and phosphorus used in ceramics and materials research: (i) *Zirconium monophosphide.* Crystalline compound. Density, 5.43-5.57 g/cm³. Formula: ZrP; and (ii) *Zirconium diphosphide.* Orthorhombic crystals. Density, 5.1 g/cm³. Formula: ZrP_2.

zirconium-potassium chloride. A compound of zirconium tetrachloride and potassium chloride used as a source of zirconium in certain magnesium alloys. Formula: $ZrCl_4$·KCl. Also known as *potassium-zirconium chloride.*

zirconium-potassium fluoride. White crystals with a density of 3.48 g/cm³. Used as a grain refiner in magnesium and aluminum, in welding fluxes, and in optical glass. Formula: K_2ZrF_6. Also known as *potassium fluozirconate; potassium zirconifluoride.*

zirconium powder. A fine, air-sensitive powder containing 99.2+% zirconium with small additions of other elements, such as iron, magnesium and chromium. It is usually supplied wet in particle sizes from 200 to 400 mesh, dampened with 25+% water. Used for powder-metallurgy parts, photographic flashbulbs, fragmentation devices, and pyrotechnic and special ignition compounds.

zirconium pyrophosphate. A white solid with good high-temperature properties to 1550°C (2820°F) and low coefficient of thermal expansion, used for refractories, phosphors and as an olefin polymerization catalyst. Formula: ZrP_2O_7.

zirconium sesquigermanide. See zirconium germanide (ii).

zirconium sesquisulfide. See zirconium sulfide (iv).

zirconium silicate. A compound of zirconium oxide (ZrO_2) and silicon dioxide (SiO_2) that is available in the form of a white powder (98+% pure). It occurs in nature as the mineral *zirconite.* Crystal system, tetragonal. Density, 4.56 g/cm³; melting point, 2550°C (4620°F). It is also supplied as a 10%-solids solution containing submicron zirconium silicate particles. Used as an ingredient in refractories, porcelain enamels, abrasives and grinding wheels, in precision molds for the casting of alloys, in electrically resisting cements, in the production of zirconium boride ceramics, for other electrical and technical ceramics, and as a component of auxiliary electrodes for the electrochemical determination of silicon. The 10% solids solution is used as a standard for polymer and biomedical research. Formula: $ZrSiO_4$.

zirconium silicide. Any of the following compounds of zirconium and silicon: (i) *Zirconium monosilicide.* Density, 5.56 g/cm³; melting point, 2121°C (3850°F); hardness, 1020-1180 Vickers. Used in ceramics and materials research. Formula: ZrSi; (ii) *Zirconium disilicide.* Gray crystalline powder. Crystal system, orthorhombic. Density, 4.88 g/cm³; melting point 1604°C (2919°F); hardness, 830-1060 Vickers. Used in coatings with high resistance to flame or blast impingement, and in special alloys. Formula: $ZrSi_2$; (iii) *Dizirconium silicide.* Density, 5.99 g/cm³; melting point, 2166°C (3931°F); hardness, 1180-1280 Vickers. Used in ceramics and materials research. Formula: Zr_2Si; (iv) *Trizirconium disilicide.* Melting point, 2210°C (4010°F). Used in ceramics and materials research. Formula: Zr_3Si_2; (v) *Tetrazirconium silicide.* Melting point, 1631°C (2968°F). Used in ceramics and materials research. Formula: Zr_4Si; (vi) *Tetrazirconium trisilicide.* Melting point, 2227°C (4041°F). Used in ceramics and materials research. Formula: Zr_4Si_3; (vii) *Pentazirconium trisilicide.* Density, 5.90 g/cm³; melting point, 2250°C (4082°F); hardness, 1280-1390 Vickers. Used in ceramics and materials research. Formula: Zr_5Si_3; and (viii) *Hexazirconium pentasilicide.* Melting point, 2250°C (4082°F). Used in ceramics and materials research. Formula: Zr_6Si_5.

Zirconium Spinel. Trademark of National Lead Company (USA) for a synthetic *spinel* composed of about 40% zirconium oxide (ZrO_2), 21% silicon dioxide (SiO_2), 20% aluminum oxide (Al_2O_3) and 19% zinc oxide (ZnO). It has a melting point of 1710°C (3110°F), and is used as an opacifier in ceramic glazes.

zirconium sponge. See sponge zirconium.

zirconium steel. A fine-grained steel containing about 0.2-0.6% carbon, 0.1-0.6% zirconium, traces of manganese, sulfur, phosphorus, silicon, and the balance iron. It has good fatigue properties and shock resistance, high ductility and enhanced machinability. Used for machinery and equipment parts including shafts.

zirconium sulfide. Any of the following compounds of zirconium and sulfur: (i) *Zirconium monosulfide.* Density, 4.56 g/cm³;

melting point, 2100°C (3810°F). Used in ceramics and materials research. Formula: ZrS; (ii) *Zirconium disulfide.* A red-brown crystalline powder (99+% pure). Density, 3.82-3.87 g/cm³; melting point 1480°C (2695°F). Used in ceramics and materials research and as lubricants. Formula: ZrS_2; (iii) *Zirconium trisulfide.* Density, 3.82 g/cm³; melting point, above 871°C (1600°F). Used in ceramics and materials research. Formula: ZrS_3; and (iv) *Zirconium sesquisulfide.* Density, 4.29 g/cm³. Used in ceramics and materials research. Formula: Zr_2S_3.

zirconium tetraacetylacetonate. See zirconium acetylacetonate.

zirconium tetraaluminide. See zirconium aluminide (ii).

zirconium tetrabromide. White, cubic crystals or off-white, moisture-sensitive, crystalline powder (98+% pure) with a density of 3.98-4.20 g/cm³ and a melting point of 450°C (842°F). Used in zirconium compounds. Formula: $ZrBr_4$. Also known as *zirconium bromide.*

zirconium tetrachloride. White, moisture-sensitive crystals or powder supplied in high-purity (99.9+%) and reactor (99.5+% pure) grades. Crystal system, monoclinic. Density, 2.8 g/cm³; melting point, approximately 320°C (608°F) (sublimes). Used as a source of zirconium metal, in zirconium and organozirconium compounds, in water repellents for textiles, as a tanning agent, and in nuclear reactors. Formula: $ZrCl_4$. Also known as *zirconium chloride.*

zirconium tetrafluoride. A white, crystalline powder. Crystal system, monoclinic. Density, 4.43 g/cm³; melting point, 932°C (1710°F); refractive index, 1.59. Used as a component of molten salts for nuclear reactors, and in *ZBLAN* and other fluorozirconate glasses. Formula: ZrF_4. Also known as *zirconium fluoride.*

zirconium tetraiodide. A reddish-brown, moisture-sensitive, crystalline powder (99+% pure). Crystal system, cubic. Density, 4.85 g/cm³; melting point, 499°C (930°F)/6.3 atm. Used in zirconium compounds. Formula: ZrI_4. Also known as *zirconium iodide.*

zirconium trisulfide. See zirconium sulfide (iii).

zirconium tungsten. Tungsten-base refractory alloys containing small percentages of zirconium, and used in direct-current tungsten electrodes for uniform welds.

zirconium-vanadium blue. A blue pigment composed of about 60-70% zirconium oxide (ZrO_2), 26-36% silicon dioxide (SiO_2), 3-5% vanadium pentoxide (V_2O_5), and up to 5% of an alkaline oxide, such as sodium oxide (Na_2O). Used in ceramic glazes. Also known as *vanadium-zirconium blue; vanadium-zirconium turquoise; zirconium-vanadium turquoise.*

zirconium-vanadium turquoise. See zirconium-vanadium blue.

zirconium-zinc. A compound of zirconium and zinc that at very low temperatures becomes ferromagnetic and exhibits superconductivity. Used for computer memories and solid-state electronics. Formula: $ZrZn_2$.

zircon-magnesite cement. A finely ground refractory cement composed of 50% dead-burnt magnesia, 35% zircon and 15% fused magnesia in a water glass binder. Used for laying and bonding refractory bricks and linings.

zirconolite. See zirkelite.

zircon porcelain. A special high-temperature electrical porcelain made from a mixture of 60-70% zirconium orthosilicate ($ZrSiO_4$), 20-30% alkaline earth-based flux and 10-20% clay. It has high mechanical strength, excellent thermal-shock resistance, an upper service temperature of 1700°C (3090°F), high dielectric strength, and is used for spark plugs, furnace trays, crucibles, combustion boats, and thermocouple tubes.

zircon refractory. A special, high-temperature refractory in which zirconium orthosilicate ($ZrSiO_4$) is the essential crystalline phase.

zircon sand. A natural sand (*beach sand*) containing considerable amounts, on average about 60-75%, of the mineral *zircon* ($ZrSiO_4$) together with titania (TiO_2), and related materials. It is available in particle sizes from 100-200 mesh, and has a melting range of 2010-2120°C (3650-3850°F), high heat resistance, high thermal conductivity, and low thermal expansion. Occurrence: Australia, Brazil, India, Sri Lanka, USA (Florida), West Africa. Used as a source of zirconium and titanium, in dry sand for steel castings, in mold facings, refractory bricks, and as an opacifier in ceramics. Also known as *zirconite sand.*

zircon whiteware. A whiteware product in which zirconium silicate ($ZrSiO_4$) is the essential crystalline phase.

zirconyl chloride. A corrosive, anhydrous compounds that is usually supplied as a 30 wt% solution in hydrochloric acid and has a density of 1.344 g/cm³ and a melting point of -15°C (5°F). Formula: $ZrOCl_2$. Also known as b*asic zirconium chloride; zirconium oxychloride; zirconium dichloride oxide.*

zirconyl chloride octahydrate. White, hygroscopic crystals or powder (98% pure) with a density of 1.910 g/cm³, and losing $6H_2O$ at 150°C (300°F) and $8H_2O$ at 210°C (410°F). It is obtained by treating zirconium dioxide with hydrochloric acid. Used as a chemical reagent, in zirconium salts, as an additive in greases and textiles, in dyes and water repellents, and in biology, biochemistry and medicine. Formula: $ZrOCl_2 \cdot 8H_2O$. Also known as *basic zirconium chloride octahydrate; zirconium oxychloride octahydrate; zirconium dichloride oxide octahydrate.*

Zircopax. Trademark of National Lead Company (USA) for zirconium silicate ($ZrSiO_4$) with a purity of 94-96.5+%. The zirconia-to-silica ratio is about 2:1. It has good color stability and resistance to crazing, and is used as an opacifier for ceramic glazes.

zircophyllite. A dark brown, nearly black mineral of the astrophyllite group composed of sodium potassium iron manganese zirconium silicate hydroxide, $(K,Na,Ca)_3S(Mn,Fe)_7O(Zr,Nb)_7 \cdot Si_8O_{27}(OH,F)_4$. Density, 3.34 g/cm³; refractive index, 1.738. Occurrence: Russian Federation.

zircosulfate. A colorless synthetic mineral composed of zirconium sulfate tetrahydrate, $Zr(SO_4)_2 \cdot 4H_2O$. Crystal system, orthorhombic. Density, 2.85 g/cm³; refractive index, 1.646.

zirkelite. A black mineral of the pyrochlore group composed of calcium zirconium titanium oxide, $CaZrTi_2O_7$, and containing considerable amounts of zirconium oxide, titanium oxide, niobium oxide, calcium oxide, and small percentages of thulium oxide, uranium dioxide, ferrous oxide, ferric oxide, and manganese oxide. Crystal system, monoclinic. Density, 4.44 g/cm³. Occurrence: Germany, Russian Federation. Used as source of zirconium oxide and titanium oxide. Also known as *zirconolite.*

zirkite. A variety of the mineral *baddeleyite* used as a source of zirconia (ZrO_2).

Zirkonal. German trade name for a non-hardenable, wrought aluminum alloy containing 15% copper, 8% manganese, and 0.5% silicon. Used for strong, light-alloy parts.

Zirmonite. Trademark of Ferro Corporation (USA) for transformation-toughened alumina ceramics that contain dispersed, submicron zirconia (ZrO_2) and can be processed by dry pressing, isostatic pressing, injection molding and extrusion. They have high strength, high resistance to wear and chemical attack, and are used for equipment for the pulp and paper, chemical-processing, oil-production and mineral-handling and processing industries.

Zirmul. Trademark of Charles Taylor & Sons Company (USA) for bonded alumina-zirconia-silica refractories used for industrial furnaces and glassmaking equipment.

Ziron Black. Trade name of Pavco Inc. (USA) for a shiny black coating produced on zinc-iron alloy plates by a single dip process.

Zirox B. Trademark of National Lead Company (USA) for zirconia (ZrO_2) used as a polishing compound for marble, granite, precision glass and ophthalmic lenses.

Zirpor. Trade name of Zircar Ceramics Inc. (USA) for a series of lightweight, sintered, microporous ceramics with low thermal conductivity. *Zirpor-1* is a rigid ceramic product composed of 55% silica (SiO_2), 40% zirconium silicate ($ZrSiO_4$) and 5% other additions. It has a maximum service temperature of 950°C (1740°F) and excellent machinability. *Zirpor-2* and *Zirpor-3* are pressed ceramic sheets composed of 80% silica (SiO_2), 15% silicon carbide (SiC) and 5% other additions.

Zirshot. Trade name of SEPR Ceramic Beads & Powders (USA) for ceramic blasting media.

zirsinalite. A colorless mineral of the columbeite group composed of sodium calcium zirconium silicate, $Na_6CaZrSi_6O_{18}$. Crystal system, rhombohedral (hexagonal). Density, 2.88 g/cm³; refractive index, 1.610. Occurrence: Russia.

Zirspra. Trade name of Zircoa Inc. (USA) for yttria-stabilized zirconia powders, used for plasma-sprayed coatings on aircraft gas-turbine components.

Zirten. Trade name of Japan Steel Works Limited (Japan) for a high-strength low-alloy (HSLA) steel containing up to 0.16% carbon, 0.35-0.65% silicon, 0.3-1.2% manganese, up to 0.12% phosphorus, 0.25-0.55% copper, up to 0.5% nickel, 0.4-0.8% chromium, up to 0.15% zirconium, and the balance iron. It possesses excellent resistance to atmospheric corrosion, and is used for bridges, booms, and railroad and mine cars.

Zisium. British trade name for a non-hardenable wrought aluminum alloy containing 15% zinc, 1-3% copper and up to 1% tin. Used for strong, light-alloy parts.

Ziskon. Trade name for a non-hardenable alloy composed of 25-60% aluminum and 40-75% zinc, and used for light-alloy parts and castings.

Zitex. Trademark of Norton Performance Plastics (USA) for chemically resistant porous polytetrafluoroethylene (PTFE) filter materials used for aerospace, automotive, chemical processing, electrical, electronic and medical applications.

Zivan. Trade name of Ziv Steel & Wire Company (USA) for several medium-carbon alloy machinery steels.

Zivco. Trade name of Ziv Steel & Wire Company (USA) for water-hardening tool steels (AISI types W1 and W2).

Zment. Trade name for a *dental cement* used for temporary crown and bridge restorations.

Z-Metal. Trade name of Castings Corporation (USA) for a series of spheroidized and pearlitic malleable cast irons composed of 2-2.6% total carbon, 0.9-1.1% silicon, 0.7-1.25% manganese, and the balance iron. They have good strength, good corrosion and wear resistance, and are used for gears, crankshafts, nozzles, axle housings, wrenches, hammer heads, air drills, pipe clamps, and bolts.

Z-Nickel. See Duranickel.

ZnNi EG. Trade name of Bethlehem Steel Corporation (USA) for paintable zinc-nickel electroplated sheet steels with excellent corrosion resistance, and enhanced formability, weldability

and surface appearance. Used especially for automotive applications.

ZnNi-UC EG. Trade name of Bethlehem Steel Corporation (USA) for paintable zinc-nickel electroplated sheet steels with thin urethane-based organic topcoats. They have excellent corrosion resistance and enhanced formability, weldability and surface appearance, and are used especially for automotive applications.

Zoco. Trademark of Zochem Division of Hudson Bay Mining & Smelting Company (Canada) for a zinc oxide powder.

Zocofax. Trademark of Zochem Division of Hudson Bay Mining & Smelting Company (Canada) for zinc oxide products.

Zodiac. (1) Trade name of Henry Wiggins & Company Limited (USA) for an electrical resistance alloy containing 64% copper, 20% nickel and 16% zinc.

(2) Trade name of Atotech USA Inc. for bright nickel coatings produced by electrodeposition

ZOE Plus. Trade name of Temrex Corporation (USA) for a zinc oxide–eugenol dental cement used for the temporary luting of restorations.

zoisite. A gray, green, brown or rose-red mineral of the epidote group composed of calcium aluminum silicate hydroxide, $Ca_2Al_3Si_3O_{12}(OH)$. It can also be made synthetically. Crystal system, orthorhombic. Density, 3.15-3.33 g/cm³; hardness, 6-6.5; refractive index, 1.699. Occurrence: Austria, Italy, Norway, Switzerland, USA (Massachusetts, Pennsylvania, Tennessee). Used as an ornamental stone and gemstone.

Zo-Mod. Trademark of ZYP Coatings, Inc. (USA) for paintable coatings used to protect ceramic materials against liquid metals, slags, fluxes, or gases. Formulated to provide thermodynamic stability in corrosive environments up to 1500°C (2730°F), they are used on furnace linings, retorts, trays, and fibrous and bulk ceramics.

Zonalin. Trade name of Kemdent (UK) for zinc oxide–eugenol dental cements.

Zonarez. Trademark of Arizona Chemical Company (USA) for a series of linear, thermoplastic polyterpene resins obtained by the polymerization of terpene ($C_{10}H_{16}$) hydrocarbons composed essentially of β-pinene and dipentene. They have low molecular weights, improved tack and adhesion characteristics and excellent aging properties. Used for pressure-sensitive adhesives for labels and tapes, solvent-based and emulsion adhesives, hot-melt adhesives and coatings, paints, in caulking and general sealant compounds, and in can sealants, investment casting waxes, rubber cements, inks, concrete waterproofing agents, and varnishes.

Zonelex. Trade name of Central Glass Company Limited (Japan) for a partially toughened glass.

Zonelite. Trade name of Asahi Glas Company Limited (Japan) for a differentially toughened glass.

zone-refined material. A high-purity material, such as aluminum, germanium, gold, silicon, tantalum or tungsten, refined using a process that involves repeated melting and crystallization. Used for solid-state devices, semiconductors, and in the production of high-purity single crystals.

Zono-Coustic. Trademark of W.R. Grace & Company (USA) for acoustical plasters.

Zonolite. Trademark of W.R. Grace & Company (USA) for an extensive series of asbestos and non-asbestos (e.g., *exfoliated vermiculite*) building products including high-temperature cements, acoustical plasters, fireproofing materials, and sprayable insulation materials. *Zonolite Insulpave* is a lightweight asphaltic insulating fill, and *Zonolite Monokote* are sprayable, cementitious structural fire protection materials.

Zonyl. Trademark of E.I. DuPont de Nemours & Company (USA) for an extensive series of perfluoroalkyl-based products including several fluoromonomers, fluorotelomers and fluorosur-factants.

Zoolite. Italian trade name for casein plastics.

Zopaque. Trademark of SCM Corporation (USA) for pure, white titanium dioxide powder prepared from the mineral *ilmenite* by a special process. It has a controlled crystal growth rate, and is used as an opacifier in ceramics and glass, and in rubber compounding.

Zorbax. Trade name of Agilent Technologies (USA) for a series of silica gels used in high-pressure liquid chromatography.

Zorite. Trade name of Michiana Products Corporation (USA) for a heat- and corrosion-resistant alloy containing 35% nickel, 15-17% chromium, 1.75% manganese, 1% silicon, 0.5% carbon, and the balance iron. It has a maximum service temperature of 1900°C (3450°F), and is used for furnaces, heaters, and electrical resistances.

zorite. A rose-red mineral composed of sodium titanium aluminum silicate trihydrate, $Na_3(Ti,Al)_2[Si_2(O,OH)_7]_2 \cdot 3H_2O$. Crystal system, orthorhombic. Density, 2.36 g/cm³. Occurrence: Russian Federation.

Z-Prime. Trademark of Zehrung Corporation (USA) for paintable coatings used to protect ceramic materials against liquid metals, slags, fluxes or gases, and formulated to provide thermodynamic stability in corrosive environments up to a temperature of 1500°C (2730°F). Used on furnace linings, retorts, trays, and fibrous and bulk ceramics.

Z-Rim. Trade name of Orpac Inc. (USA) for strong, lightweight, non-fibrous heat insulating boards that physically resemble firebrick, but are produced from yttria-stabilized high-purity zirconia. They have good high-temperature properties to 2200°C (3990°F), and are easily cut and machined. See also yttria-stabilized zirconia.

Zunit. Trade name of Robert Zapp Werkstofftechnik GmbH (Germany) for a series of corrosion-resistant stainless steels and heat-resistant steels. *Zunit-Guss* refers a series of cast corrosion- and heat-resistant steels of the iron-chromium or iron-chromium-nickel type, used for industrial furnaces, furnace parts, and heat-treating equipment.

zunyite. A colorless mineral composed of aluminum chloride fluoride silicate hydroxide, $Al_{13}Si_5O_{20}(OH,F)_{18}Cl$. Crystal system, cubic. Density, 2.87 g/cm³. Occurrence: Algeria, USA (Colorado).

zussmanite. A green mineral composed of potassium iron silicate hydroxide, $K(Fe,Mg,Mn)_{13}(Si,Al)_{18}O_{42}(OH)_{14}$. Crystal system, rhombohedral (hexagonal). Density, 3.15 g/cm³; refractive index, 1.643. Occurrence: USA (California).

zvyagintsevite. A gray mineral of the gold group composed of lead palladium, $PbPd_3$. It can also be made synthetically. Crystal system, cubic. Density, 13.32 g/cm³. Occurrence: Siberia.

zwieselite. A dark brown mineral of the triplite group composed of iron manganese fluoride phosphate, $(Fe,Mn,Ca)_2PO_4(F,OH)$. Crystal system, monoclinic. Density, 3.89-3.92 g/cm³; refractive index, 1.69. Occurrence: Czech Republic.

Zyaliate. Trade name of Vesuvius McDanel (USA) for a moldable paste composed of colloidal ceramics and refractory ceramic fibers, and used in the repair of burners, pipe insulation, brazing fixtures, and ceramic insulation components.

Zycon. Trade name for regenerated protein (azlon) fibers used for the manufacture of textile fabrics.

Zycron. (1) Trademark of Zircoa, Inc. (USA) for a series of *transformation-toughened zirconia* (TTZ) materials composed of 97% zirconia (ZrO_2) and 3% magnesia (MgO). They have good impact, wear and corrosion resistance. *Zycron L* has high tensile strength, while *Zycron H* exhibits outstanding high-temperature performance up to 980°C (1795°F).

(2) Trademark of Homeline Corporation (USA) for a scented plastic film supplied in bag form.

Zyex. Trademark of Zyex Limited (UK) for polyetheretherketone (PEEK) monofilaments, staple fibers and filament yarns used for textile fabrics.

zykaite. A grayish white mineral with yellowish green tint composed of iron arsenate sulfate hydroxide hydrate, $Fe_4(AsO_4)_3(SO_4)(OH) \cdot 15H_2O$. Crystal system, orthorhombic. Density, 2.50 g/cm³. Occurrence: Czech Republic.

Zyklon. Trade name of Otto Wolff Handelsgesellschaft (Germany) for hot-work tool steels and shock-resistant machinery steels.

Zylar. Trade name of Novacor Chemicals Inc. (USA) for acrylic resins.

Zylam. Trademark of DuPont Advanced Material Systems (USA) for chemical- and heat-resistant fiber-reinforced polyetherketoneketone (PEKK) thermoplastic composites.

Zylar. Trade name of Novacor Chemicals Inc. (USA) for acrylic resins used in the manufacture of thermoformed, molded and extruded articles, film and sheeting.

Zylon. Trademark of Toyobo Company Limited (Japan) for high-performance poly(*p*-phenylene-2,6-benzobisoxazole) (PBO) fibers and PBO fiber/epoxy-matrix unidirectional composites. *Zylon-HM* fibers have a density of 1.56 g/cm³ (0.056 lb/in.³), an ultimate tensile strength of 5.80 GPa (841 ksi) and an elastic modulus of 269 GPa (39×10^3 ksi). Used as reinforcing fibers for composites.

Zymaxx. Trademark of DuPont Composites (USA) for a composite material composed of a polytetrafluoroethylene (PTFE) matrix reinforced with high-aspect-ratio carbon fibers. Supplied in stock shapes and as custom-machined parts, it possesses excellent chemical resistance, good dimensional stability over a wide temperature range, good resistance to wear and compressive creep, and high toughness. Used for valve seals and seats, pump and compressor components, gaskets, and other parts employed in chemical-, petroleum- and glass-processing equipment, and for fluid-power engineering equipment.

Zyranox. Trademark of Matroc Bioceramics (UK) for zirconium oxide based bioceramics used for bone and joint prostheses and orthopedic implants.

Zytel. Trademark of E.I DuPont de Nemours & Company (USA) for a series of nylon (polyamide) resins including nylon 6, nylon 6,6, nylon 66,6, nylon 6,12, and various nylon alloys. They are supplied in the form of molding and extrusion powders and as soluble resins in various grades including high gloss, silicone-, molybdenum disulfide- or polytetrafluoroethylene-lubricated, fire-retardant, supertough, impact-modified, UV-stabilized, and glass or carbon fiber-reinforced. They have high impact strength, toughness, stiffness and dimensional stability, high heat-deflection temperatures, low thermal conductivity, good dielectric properties and chemical resistance, and are used for fasteners, casters, recreational equipment, automotive parts, consumer products, such as furniture, power tools, lawnmowers and window frames, and for electrical and electronic components.

Appendix

Bibliographic References
Helpful Sources of Information
Abbreviations and Acronyms

I. Databases

1. Metals and Alloys

1.1 *Alloy Finder V2.0*, ASM International, Materials Park, OH, USA.

1.2 *ASM Aluminum Databank*, ASM International, Materials Park, OH, USA.

1.3 *ASM Stainless Steels Databank*, ASM International, Materials Park, OH, USA.

1.4 *ASM Structural Steels Databank*, ASM International, Materials Park, OH, USA

1.5 *Biomaterials Properties Database*, University of Michigan, School of Dentistry, Ann Arbor, MI, USA.

1.6 *CDA Database*, Copper Development Association, Inc., New York, NY, USA.

1.7 *COPPERDATA*, Copper Development Association, Inc., New York, NY, USA.

1.8 *EUROPAGES–The European Business Directory*: http://www.europages.com.

1.9 *Guide to Stainless Steels*, Int. Ed., Nickel Development Institute, Birmingham, UK.

1.10 *High-Temperature Materials Data Bank (HTM-DB)*, Commission of the European Committees, Institute for Advanced Materials, Petten, The Netherlands.

1.11 *Magnesium Data Set*, ASM International, Materials Park, OH, USA.

1.12 *Magnesium Information System*, International Magnesium Association, McLean, VA, USA.

1.13 *Matweb: The Online Materials Information Resource*: http://www.matweb.com.

1.14 *METADEX*, Cambridge Scientific Abstracts, Beachwood, OH, USA: http://www.csa.com.

1.15 *Metals Datafile*, Cambridge Scientific Abstracts, Beachwood, OH, USA: http://www.csa.com.

1.16 *Stainless Steel Data Set*, ASM International, Materials Park, OH, USA.

1.17 *Structural Steel Data Set*, ASM International, Materials Park, OH, USA.

1.18 *Superconducting Materials Database*, National Research Institute for Metals (NRIM), Tsukuba-shi, Ibaraki, Japan: http://www.nrim.go.jp.

1.19 *Thomas Register of American Manufacturers*: http://www3.thomasregister.com.

1.20 *Tin and Its Uses Database*, ITRI Limited, Uxbridge, Middlesex, UK: http://www.itri.co.uk.

1.21 *Titanium Data Disk*, Titanium Information Group, Kidderminster, UK: http://www.titaniuminforgroup.co.uk.

1.22 *Titanium Data Set*, ASM International, Materials Park, OH, USA.

1.23 *Woldman's Engineering Alloys*, 7th ed., ASM International, Materials Park, OH, USA.

2. Plastics and Elastomers

2.1 *CENTOR's Plastics Databank*, CenTOR Corporation, Irvine, CA, USA.

2.2 *EUROPAGES–The European Business Directory*: http://www.europages.com

2.3 *International Plastics Selector*, D.A.T.A. Business Publishing, Englewood, CO, USA.

2.4 *Matweb: The Online Materials Information Resource*: http://www.matweb.com.

2.5 *METADEX*, Cambridge Scientific Abstracts, Beachwood, OH, USA: http://www.csa.com.

2.6 *Nylon Data Set*, ASM International, Materials Park, OH, USA.

2.7 *PLASPEC*, Plastics Technology, New York, NY, USA.

2.8 *Plastics Digest on CD-ROM*, D.A.T.A. Business Publishing, Englewood, CO, USA.

2.9 *Plastics Information Services Database*, Plastics Informations Systems, Columbus, OH, USA.

2.10 *Plastics Technology Materials Selection Database*, D & S Data Resources, Yardley, PA, USA.

2.11 *POLYMAT*, FIZ Chemie Berlin, Germany.

2.12 *POLYMAT light*, FIZ Chemie Berlin, Germany.

2.13 *The ILI Materials Database*, Rapra Technology, Limited, Shawbury, Shropshire, UK: http://www.rapra.net.

2.14 *Thermoplastics Data Set*, ASM International, Materials Park, OH, USA.

2.15 *Thermoset Plastics Data Set*, ASM International, Materials Park, OH, USA.

2.16 *Thomas Register of American Manufacturers*: http://www3.thomasregister.com.

3. Ceramics and Glasses

3.1 *CENTOR's Ceramics Database*, CenTOR Corporation, Irvine, CA, USA.

3.2 *Ceramics Abstracts*, Cambridge Scientific Abstracts, Beachwood, OH, USA: http://www.csa.com.

3.3   *EUROPAGES–The European Business Directory*: http://www.europages.com.

3.4   *MatWeb: The Online Materials Information Resource*: http://www.matweb.com

3.5   *METADEX*, Cambridge Scientific Abstracts, Beachwood, OH, USA: http://www.csa.com.

3.6   *NIST Structural Ceramics Database,* National Institute of Standards and Technology (NIST), Gaithersburg, MD, USA: http://www.ceramics.nist.gov.

3.7   *Thomas Register of American Manufacturers*: http://www3.thomasregister.com.

3.8   *WebSCD (Structural Ceramics Database),* National Institute of Standards and Technology (NIST), Gaithersburg, MD: http://www.ceramics.nist.gov.

4.   Composite Materials

4.1   *Advanced Composite Materials,* Industrial Products Research, Ministry of International Trade and Industry (MITI), Tsukuba Science City, Ibaraki, Japan.

4.2   *Advanced Composites Bulletin,* Elsevier Advanced Technology (EAT), Kidlington, Oxon, UK.

4.3   *Centor's Composites Database*, CenTOR Corporation, Irvine, CA, USA.

4.4   *EUROPAGES–The European Business Directory*: http://www.europages.com.

4.5   *MatWeb: The Online Materials Information Resource*: http://www.matweb.com.

4.6   *METADEX*, Cambridge Scientific Abstracts, Beachwood, OH, USA: http://www.csa.com.

4.7   *Thomas Register of American Manufacturers*: http://www3.thomasregister.com.

5.   Other Engineering/Engineered Materials

5.1   *Adhesives Digest on CD-ROM*, D.A.T.A. Business Publishing, Englewood, CO, USA.

5.2   *Biomaterials Properties Database,* University of Michigan School of Dentistry, Ann Arbor, MI, USA.

5.3   *EUROPAGES–The European Business Directory*: http://www.europages.com.

5.4   *LiqCryst Database (of Thermotropic Liquid Crystals)*, University of Hamburg and LCI Publishers, Germany.

5.5   *MatWeb: The Online Materials Information Resource*: http://www.matweb.com.

5.6   *MEMS (Micro-Electrical Mechanical Systems) Materials Database Reference,* USC Information Sciences Institute, Marina Del Ray, CA, USA.

5.7   *Thomas Register of American Manufacturers*: http://www3.thomasregister.com.

5.8   *Textile Fibers and Fabrics*: http://www.fibre2fashion.com.

5.9   *The Photonics Directory*, Laurin Publishing: http://www.photonics.com.

5.10   *Woods of the World*, ForestWorld.com, Inc., Colchester, VT, USA.

II.   Handbooks, Encyclopedias and Textbooks

1.   Metals and Alloys

1.1   *Active Metals: Preparation, Characterization, Applications*, A. Furstner, Ed., Wiley-VCH, Weinheim, Germany, 1996.

1.2   *Alloys*, F. Habashi, Wiley-VCH, Weinheim, Germany, 1998.

1.3   *ASM Handbook®*, Vol. 1: Properties and Selection: Iron, Steels, and High-Performance Alloys, ASM International, Materials Park, OH, 1990.

1.4   *ASM Handbook®*, Vol. 2: Properties and Selection: Nonferrous Alloys and Special Purpose Materials, ASM International, Materials Park, OH, 1991.

1.5   *ASM Handbook®*, Vol. 5: Surface Engineering, ASM International, Materials Park, OH, 1994.

1.6   *ASM Handbook®*, Vol. 6: Welding, Brazing, and Soldering, ASM International, Materials Park, OH, 1993.

1.7   *ASM Handbook®*, Vol. 7: Powder Metal Technologies and Applications, ASM International, Materials Park, OH, 1998.

1.8   *ASM Handbook®*, Vol. 15: Casting, ASM International, Materials Park, OH, 1988.

1.9   *ASM Handbook®*, Vol. 20: Materials Selection and Design, ASM International, Materials Park, OH, 1997.

1.10   *ASM Handbook®*, Desk Edition, 2nd ed., ASM International, Materials Park, OH, 1995.

1.11   *ASM Specialty Handbooks®*, Stainless Steels, J.R. Davies, Ed., ASM International, Materials Park, OH, 1994.

1.12   *ASM Specialty Handbook®*, Carbon and Alloy Steels, J. R. Davies, Ed., ASM International, Materials Park, OH, 1996.

1.13   *ASM Specialty Handbook®*, Cast Irons, J. R. Davies, Ed., ASM International, Materials Park, OH, 1996.

1.14   *ASM Specialty Handbook®*, Aluminum and Aluminum Alloys, J. R. Davies, Ed., ASM International, Materials Park, OH, 1993.

1.15   *ASM Specialty Handbook®*, Magnesium and Magnesium Alloys, J. R. Davies, Ed., ASM International, Materials Park, OH, 1999.

1.16   *ASM Specialty Handbook®*, Nickel, Cobalt, and Their Alloys, J. R. Davies, Ed., ASM International, Materials Park, OH, 2000.

1.17   *ASM Specialty Handbook®*, Tool Materials, J. R. Davies, Ed., ASM International, Materials Park, OH, 1995.

1.18   *ASM Specialty Handbook®*, Heat-Resistant Materials, J. R. Davies, Ed., ASM International, Materials Park, OH, 1997.

1.19   *Intermetallics*, G. Sauthoff, Wiley-VCH, Weinheim, Germany, 1998.

1.20   *Powder Metallurgy of Iron and Steel*, R.M. German, Wiley-VCH, Weinheim, Germany, 1998.

1.21   *Pure Metals Properties: A Scientific and Technical Handbook*, ASM International and Freund Publishing House, USA, 1999.

1.22   *Stahlschlüssel (Key to Steels)*, 18th ed., C.W. Wegst, Verlag Stahlschlüssel, Düsseldorf, Germany, 1998.

1.23   *Steel Castings Handbook*, 6th ed., ASM International, Materials Park, OH, 1995.

1.24   *Tool Steels*, 5th ed., ASM International, Materials Park, OH, 1998.

1.25   *Woldman's Engineering Alloys*, 9th ed., J. Frick, Ed., ASM International, Materials Park, OH, 2000.

1.26   *Worldwide Guide to Equivalent Irons and Steels*, 4th ed., ASM International, Materials Park, OH, 2000.

1.27   *Worldwide Guide to Equivalent Nonferrous Metals and Alloys*, 3rd ed., ASM International, Materials Park, OH, 1996.

2.   Plastics and Elastomers

2.1   *An Introduction to Plastics*, H.-G. Elias, Wiley-VCH, Weinheim, Germany, 1993.

2.2   *ASM Engineered Materials Handbook®*, Volume 2: Engineering Plastics, ASM International, Materials Park, OH, USA, 1988.

2.3   *Degradable Polymers: Principles and Applications*, G. Scott and D. Gilead, Eds., Chapman & Hall, New York, NY, USA, 1995.

2.4   *Fundamental Principles of Polymeric Materials*, 2nd ed., S.L. Rosen, John Wiley & Sons, New York, NY, USA, 1993.

2.5   *Fundamentals of Polymer Science: An Introductory Text*, P.C. Painter and M.M. Coleman, Technomic Publishing, Lancaster, PA, USA, 1994.

2.6   *Handbook of Organic Conductive Molecules and Polymers*, H.S. Nalwa, Ed., Wiley-VCH, Weinheim, Germany, 1997.

2.7   *Handbook of Plastic and Rubber Additives*, M. Ash and I. Ash, Gower Publishing, Brookfield, VT, USA, 1995.

2.8   *Handbook of Plastic Compounds, Elastomers, and Resins: An International Guide by Category, Tradename, Composition, and Supplier*, M. Ash and I. Ash, Wiley-VCH, Weinheim, Germany, 1992.

2.9   *Handbook of Plastics and Elastomers*, C.A. Harper, Ed., McGraw-Hill, New York, NY, USA, 1975.

2.10   *Handbook of Reinforced Plastics*, Van Nostrand-Reinhold, New York, NY, USA, 1978

2.11   *International Plastics Handbook*, H-J. Saechtling, Hanser Publishers, Munich, Germany, 1983.

2.12   *Interpenetrating Polymer Networks*, D. Klempner, L.H. Sperling and L.A. Utravki, Eds., American Chemical Society, Washington, DC, USA, 1994.

2.13   *Introduction to Ionomers*, A. Eisenberg and J.S. Kim, John Wiley & Sons, New York, NY, USA, 1998.

2.14   *Modern Fluoropolymers*, J. Scheirs, Wiley-VCH, Weinheim, Germany, 1997.

2.15   *Modern Plastics Encyclopedia*, McGraw-Hill, New York, NY, USA, 1988

2.16   *Plastics Materials*, 7th ed., ASM International, Materials Park, OH, 1999.

2.17   *Polymer Handbook*, 4th ed., J. Brandrup, E.H. Immergut and E.A. Grulke, Eds., John Wiley & Sons, New York, NY, USA, 1999.

2.18   *Polymers: An Encyclopedic Sourcebook of Engineering Properties*, J.I. Kroschwitz, Ed., John Wiley & Sons, New York, NY, USA, 1987.

2.19   *Reaction Polymers: Polyurethanes, Epoxies, Unsaturated Polyesters, Phenolics, Special Monomers, and Additives; Chemistry, Technology, Applications, Markets*, W.F. Gum, W. Riese and H. Ulrich, Eds., Hanser-Gardner Publications, Cincinnati, OH, USA, 1992.

2.20   *Semiconducting Polymers*, G. Hadziioannou and P. VanHutten, Wiley-VCH, Weinheim, Germany, 1999.

2.21   *The Plastics Compendium*, RAPRA Technology Limited, Shawbury, UK.

2.22   *Thermoplastics: Materials Engineering*, L. Mascia, Applied Science Publishers, Limited, London, UK, 1982.

3.   Ceramics and Glasses

3.1   *ASM Engineered Materials Handbook®*, Volume 4: Ceramics and Glasses, ASM International, Materials Park, OH, USA, 1991.

3.2   *Dictionary of Ceramic Science and Engineering*, L.S. O'Bannon, Plenum Press, New York, NY, USA.

3.3   *Engineered Materials Handbook®*, Desk Edition, ASM International, Materials Park, OH, USA, 1995.

3.4   *Fundamentals of Ceramics*, M. Barsoum, McGraw-Hill, New York, NY, USA, 1997.

3.5   *Glass Science*, 2nd ed., R.H. Doremus, Wiley-Interscience, New York, NY, USA, 1994.

3.6   *Handbook of Ceramic Hard Materials*, R. Riedel, Wiley-VCH, Weinheim, Germany, 1999.

3.7   *Introduction to Ceramics*, W.D. Kingery, H.K. Bowen and D.R. Uhlmann, John Wiley & Sons, Inc., New York, NY, USA, 1976.

3.8   *Precursor-Derived Ceramics*, F. Aldinger, J. Bill and F. Wakai, Eds, Wiley-VCH, Weinheim, Germany, 1999.

3.9   *Principles of Electronic Ceramics*, L.L. Hench and J.K. Kerst, Wiley-Interscience, New York, NY, USA, 1990.

3.10   *Science of Ceramic Processing*, Wiley-Interscience, New York, NY, USA, 1986.

4.   Composite Materials

4.1   *ASM Engineered Materials Handbook®*, Volume 1: Composites , 1987, ASM International, Materials Park, OH, USA, 1987.

4.2   *ASM Engineered Materials Reference Book*, 2nd ed., ASM International, Materials Park, OH, USA, 1994.

4.3   *Composite Materials Handbook*, M.M. Schwartz, McGraw-Hill, New York, NY, USA.

4.4   *Engineered Materials Handbook®*, Desk Edition, 1995, ASM International, Materials Park, OH, USA, 1995.

4.5   *Handbook of Composite Reinforcements*, S.M. Lee, Ed., Wiley-VCH, Weinheim, Germany, 1993

4.6   *Handbook of Composites*, G. Lubin, Ed., Van Nostrand-Reinhold, New York, NY, USA, 1982

4.7   *International Encyclopedia of Composites*, S.M. Lee, Ed., Wiley-VCH, Weinheim,Germany, 1990

4.8   *Properties of Concrete*, 4th ed., A.M. Neville, John Wiley & Sons, Inc., New York, NY, USA, 1996.

5.   Other Engineering/Engineered Materials

5.1   *Amorphous Silicon*, K. Tanaka, E. Maruyama, T. Shimada and H. Okamoto, Wiley-VCH, Weinheim, Germany, 1999.

5.2   *ASM Engineered Materials Handbook®*, Volume 3: Adhesives and Sealants, ASM International, Materials Park, OH, USA, 1990.

5.3   *Automotive Paints and Coatings*, G. Fettis, Ed., Wiley-VCH, Weinheim, Germany, 1994.

5.4   *Biodegradable Hydrogels for Drug Delivery*, K. Park, W.S.W., Shalaby and H. Park, Technomic, Lancaster, PA, USA, 1993.

5.5   *Biomimetic Materials Chemistry*, S. Mann, Ed., Wiley-VCH, Weinheim, Germany, 1996

5.6   *Chemistry of Advanced Materials: An Overview*, L.V.,Interrante and M.J. Hampden-Smith, Eds., Wiley-VCH, Weinheim, Germany, 1998.

5.7   *CVD of Compound Semiconductors*, A.C. Jones and P. O'Brien, Wiley-VCH, Weinheim, Germany, 1997.

5.8   *CVD of Nonmetals*, W.S., Rees, Jr., Ed., Wiley-VCH, Weinheim, Germany, 1996.

5.9   *Dendritic Molecules, Concepts - Syntheses - Perspectives*, G.R. Newkome and C.N. Moorefield and F.

Vögtle, Wiley-VCH, Weinheim, Germany, 1996.

5.10   *Electronic Properties of Engineering Materials*, J.D. Livingston, Wiley-VCH–MIT, New York, NY, USA, 1999.

5.11   *Electroplating Engineering Handbook*, 3rd ed., A.K. Graham, Ed., Van Nostrand-Reinhold, New York, NY, USA, 1971.

5.12   *Engineered Materials Handbook®*, Desk Edition, ASM International, Materials Park, OH, USA, 1995.

5.13   *Ferrocenes: Homogeneous Catalysis, Organic Synthesis, Materials Science*, A. Togni and T. Hayashi, Wiley-VCH, Weinheim, Germany, 1994.

5.14   *Ferroelectric and Antiferroelectric Liquid Crystals*, S.T. Lagerwall, Wiley-VCH, Weinheim, Germany, 1999.

5.15   *Handbook of Advanced Electronic and Photonic Materials and Devices*, H.S. Nalwa, Ed., Academic Press, Inc., San Diego, CA, USA, 2000.

5.16   *Handbook of Battery Materials*, J.O. Besenhard, Wiley-VCH, Weinheim, Germany, 1998.

5.17   *Handbook of Chemical Vapor Deposition: Principles, Technology, and Applications*, H.O. Pierson, Noyes Publications, Park Ridge, NJ, USA, 1992.

5.18   *Handbook of Fillers for Plastics*, H. Katz., Ed., Van Nostrand-Reinhold, New York, USA, 1987

5.19   *Handbook of Liquid Crystals*, D. Demus, J.W. Goodby, G.W. Gray, H.W. Spiess and V. Vill, Wiley-VCH, Weinheim, Germany, 1998.

5.20   *Handbook of Properties of Optical Materials*, D. Nikogosyan, Wiley-VCH, Weinheim, Germany, 1997.

5.21   *Handbook of Paint and Coating Raw Materials*, M. Ash and I. Ash, Ashgate Publishing, Brookfield, VT, USA, 1996.

5.22   *Inorganic Materials*, 2nd ed., D.W. Bruce and D. O'Hare, Wiley-VCH, Weinheim, Germany, 1996.

5.23   *Industrial Inorganic Pigments*, 2nd ed., G. Buxbaum, Ed., Wiley-VCH, Weinheim, Germany, 1998.

5.24   *Industrial Organic Pigments*, 2nd ed., W. Herbst and K. Hunger, Wiley-VCH, Weinheim, Germany, 1997.

5.25   *Materials and Devices for Electrical Engineers* and Physicists, R. Colclaser and S. Diehl-Nagle, McGraw Hill, New York, NY, USA, 1985.

5.26   *Metallocenes, Synthesis – Reactivity – Applications*, A. Togni and R.L. Halterman, Wiley-VCH, Weinheim, Germany, 1998.

5.27   *Microoptics*, S. Sinzinger and J. Jahns, Wiley-VCH, Weinheim, Germany, 1999.

5.28   *Nanoparticles and Nanostructured Films*, Preparation, Characterization and Applications, J.H. Fendler, Ed., Wiley-VCH, Weinheim, Germany, 1998.

5.29   *Nanotechnology: Molecularly Designed Materials*, G.M. Chow and K.E. Gonsalves, Eds., American Chemical Society, Washington, DC, USA, 1996.

5.30   *Organic Coatings, Science and Technology*, 2nd ed., Z.W. Wicks, F.N. Jones and S.P. Pappas, Wiley-VCH, Weinheim, Germany, 1999.

5.31   *Organic Molecular Solids: Properties and Applications*, W. Jones, Ed., CRC Press, Boca Raton, FL, USA, 1997.

5.32   *Physical Properties of Liquid Crystals*, G.W. Gray, Ed., Wiley-VCH, Weinheim, Germany, 1999.

5.33   *Plasma-Spray Coating: Principles and Applications*, R.B. Heimann, Wiley-VCH, Weinheim, Germany, 1996.

5.34   *Principles of Electrical Engineering Materials: Devices*, S.O. Kasap, McGraw-Hill, New York, NY, USA, 1997.

5.35   *Quantum Dot Heterostructures*, D. Bimberg and M. Grundmann and N.N. Ledentsov, Wiley-VCH, Weinheim, Germany, 1998.

5.36   *Semiconductor Material and Device Characterization*, D.K, Schroder, Wiley-Interscience, New York, NY, USA, 1990.

5.37   *Surface Coatings Science & Technology*, 2nd ed., S. Paul, Wiley-VCH, Weinheim, Germany, 1995.

5.38   *The Chemistry of Metal CVD*, T. Kodas and M. Hampden-Smith, Wiley-VCH, Weinheim, Germany, 1994.

5.39   *Wood Book*, R. Bard, Coles Publishing Company Limited, Toronto, Canada, 1980.

III. Selected Dictionaries, Encyclopedias, Textbooks and Other Sources of Information on Engineering Materials and Related Topics

1. *Academic Press Dictionary of Science and Technology*, Christopher Morris, Ed., Academic Press, Inc., San Diego, CA, USA, 1992.
2. *A Dictionary of Mining, Mineral, and Related Terms*, P.W. Thrush, Ed., US Bureau of Mines, Department of the Interior, Washington, DC, USA, 1968.
3. *Advanced Engineering Materials*, International Journal, Wiley-VCH, Weinheim, Germany, 2000-2003
4. *Advanced Materials & Processes*, International Journal, ASM International, Materials Park, OH, USA, 1990-2003
5. *A Field Guide to Eastern Trees*, G.A. Petrides, Houghton Mifflin Company, Boston, MA, USA, 1988.
6. *A Field Guide to Rocks and Minerals*, F.H. Pough, Houghton Mifflin Company, Boston, MA, USA, 1976.
7. *ASM Materials Engineering Dictionary*, J. R. Davis, Ed., ASM International, Materials Park, OH, USA, 1996.
8. *Biotechnology from A to Z*, William Bains, IRL Press, Oxford, UK, 1993.
9. *Cement and Concrete Terminology*, Publication SP-19, American Concrete Institute, Detroit, MI, USA.
10. *Chemical Synonyms and Trade Names*, W. Gardner, The Technical Press Limited, London, UK, 1968
11. *Chemistry of Advanced Materials*, L.V. Interrante, Ed., Wiley-VCH, Weinheim, Germany, 1998.
12. *Chromatography Today*, C.F. Poole and S.K. Poole, Elsevier Science Publishing, New York, NY, USA, 1991.
13. *Compilation of ASTM Standard Definitions*, 7th ed., ASTM, Philadelphia, PA, USA, 1990.
14. *Concise Encyclopedia of Polymer Science and Engineering*, J. Kroschwitz, John Wiley & Sons, New York, NY, USA, 1990.
15. *CRC Handbook of Chemistry and Physics*, 80th ed., D.R. Lide, Ed., CRC Press, Boca Raton, FL, USA, 1999.
16. *CRC Practical Handbook of Materials Science*, C. Lynch, CRC Press, Boca Raton, FL, USA, 1989.
17. *Dictionary of Chemical Names and Synonyms*, P. Howard and M. Neal, Lewis Publishing, 1992.
18. *Dictionary of Metallurgy*, C.D. Brown, John Wiley & Sons, Inc., New York, NY, USA, 1997.
19. *Elements of Materials Science and Engineering*, 6th ed., L.H. Van Vlack, Addison-Wesley Publishing Company, Reading, MA, USA, 1989.
20. *Foundations of Materials Science and Engineering*, 2nd ed., W.F. Smith, McGraw Hill, New York, NY, USA, 1993.
21. *Hawley's Condensed Chemical Dictionary*, 13th ed., R.J. Lewis, Sr., Van Nostrand Reinhold, New York, NY, USA, 1997.
22. *Hackh's Chemical Dictionary*, 4th ed., J. Grant, Ed., McGraw-Hill, New York, NY, USA, 1969.
23. *Journal of Protective Coatings & Linings (JPCL)*, SSPC–The Society for Protective Coatings & Technology Publishing Company, Pittsburgh, PA, USA, 1995-2003
24. *Machinery's Handbook*, 23rd ed., H.H. Ryffel, Ed., Industrial Press Inc., New York, NY, USA, 1988.
25. *Materials Handbook*, 14th ed., G.S. Brady, H.R. Clauser and J. Vaccari, McGraw-Hill, 1997.
26. *Materials Science for Engineers*, 3rd ed., J. F. Shackelford, Macmillan Publishing Company, New York, NY, USA, 1992.
27. *Materials Science and Engineering*, W.D. Callister, Jr., John Wiley & Sons, New York, NY, USA, 2000.
28. *McGraw-Hill Dictionary of Earth Sciences*, S.P. Parker, Ed., McGraw-Hill, New York, NY, USA, 1984.
29. *McGraw-Hill Dictionary of Scientific and Technical Terms*, 5th ed., McGraw-Hill, New York, NY, USA, 1993.
30. *Mechanics of Materials*, 2nd ed., F. Beer and E. R. Johnston, Jr., McGraw-Hill, New York, NY, USA, 1992.
31. *Metallic Materials Specification Handbook*, 2nd ed., R.B. Ross, E. & F.N. Spon Limited, London, UK, 1972.
32. *Mineral Powder Diffraction File, Data Book*, JCPDS–International Center for Diffraction Data, Swarthmore, PA, USA, 1986.
33. *Modern Carpentry*, W.H. Wagner, The Goodheart-Willcox Company, Inc., South Holland, IL, USA, 1983.
34. *Modern Semiconductor Device Physics*, S.M. Sze, Wiley-Interscience, New York, NY, USA, 1998.
35. *Modern Spectroscopy*, 3rd ed., J.M. Hollas, John Wiley & Sons, New York, NY, USA, 1996.
36. *MRS Bulletin*, Materials Research Society, Warrendale, PA, USA, 1990-2003.
37. *Optoelectronics: An Introduction to Materials and Devices*, J. Singh, McGraw-Hill, New York, NY, USA, 1996.
38. *Plastics Engineering Handbook*, 4th ed., J. Frados, Ed., Van Nostrand-Reinhold, New York, NY, USA, 1976.
39. *Plastics Process Engineering*, J.L. Throne, Marcel Dekker, Inc., New York, NY, USA, 1979.
40. *Plastics Products Design Handbook*, Part A & B, Marcel Dekker, Inc., New York, NY, USA, 1981-83.
41. *Plastics Technology Handbook*, M. Chanda and S.K. Roy, Marcel Dekker, Inc., New York, NY, USA, 1987.

42. *Polymer Science Dictionary*, M.S.M. Alger, Elsevier Science Publishing, New York, NY, USA, 1989.
43. *Polymer Technology Dictionary*, T. Whealen, Chapman & Hall, New York, NY, USA, 1994.
44. *Principles of Materials Science and Engineering*, 3rd ed., W.F. Smith, McGraw Hill, New York, NY, USA, 1996.
45. *Properties of Materials*, M.A. White, Oxford University Press, Inc. New York, NY, USA, 1999.
46. *Smithells Metals Reference Book*, 6th ed., E.A. Brandes, Ed., Butterworths & Co. (Publishers) Ltd., London, UK, 1983.
47. *Engineering Materials and Their Applications*, 4th ed., R.A. Flinn and P.K. Trojan, Houghton Mifflin Company, Boston, MA, USA, 1990.
48. *Scientific and Technical Acronyms, Symbols, and Abbreviations*, U. Erb and H. Keller, John Wiley & Sons, Inc., New York, NY, USA, 2001.
49. *Special Metal Finishing, Guidebook & Directory*, American Electroplaters and Surface Finishers Society, Orlando, FL, USA, 2001.
50. *Special Metal Finishing, Organic Finishing Guidebook & Directory*, American Electroplaters and Surface Finishers Society, Orlando, FL, USA, 2001.
51. *SPI Handbook of Technology and Engineering of Reinforced Plastics*, J.G. Mohr, Ed., Society of the Plastics Industry, New York, NY, USA, 1981.
52. *Structure and Properties of Engineering Materials*, 4th ed., D.P. Henkel, R. Gordon and A. Pense, McGraw-Hill, New York, NY, USA, 1977.
53. *Structure of Metals*, 3rd revised ed., C. Barrett and T.B. Massalski, Pergamon Press Ltd., Oxford, UK, 1980.
54. *The Merck Index*, 12th ed., S. Budavari, Ed., Merck & Co., Rahway, NJ, USA, 1996.
55. *The Science & Design of Engineering Materials with Materials in Focus CD-ROM*, 2nd ed., J. Schaffer, A. Saxena, S. Antolovich, T. Sanders and S. Warner, McGraw-Hill, New York, NY, USA, 1999.
56. *The Science and Engineering of Materials*, 4th ed., D.R. Askeland and P.P. Phulé, Brooks/Cole, Division of Thomson Learning, Inc., Pacific Grove, CA, USA, 2003.
57. *Van Nostrand's Scientific Encyclopedia*, 8th ed., D.M. Considine, Van Nostrand Reinhold, New York, NY, USA, 1994.
58. *Whittington's Dictionary of Plastics*, 2nd ed., Technomic Publishing, Lancaster, PA, USA, 1978.
59. *World Directory of Manufactured Fiber Producers*, Fiber Economics, Inc., Washington, DC, USA, 1997

IV. Selected Websites of Manufacturers and Suppliers of Engineering/Engineered Materials:

1. Metals and Alloys

Alcan Aluminum Company	http://www.alcan.com
Alcoa—Aluminum Company of America	http://www.alcoa.com
Alfa Aesar, A Johnson Matthey Company	http://www.alfa.com
Allegheny Ludlum Corporation	http://www.alleghenyludlum.com
Allied Signal/Honeywell Amorphous Metals	http://www.metglas.com
Allvac, An Allegheny Teledyne Company	http://www.allvac.com
Ametek Specialty Metal Products	http://www.ametek.com
Armco Inc.	http://www.armco.com
Arris International Corporation	http://www.arris-intl.com
Atlas Steels Inc.	http://www.atlassteels.com
Bellman-Melcor Inc.	http://www.bellmanmelcor.com
Belmont Metals Inc.	http://www.belmontmetals.com
Bethlehem Steel Corporation	http://www.bethsteel.com
Boehler Steel	http://www.bohlersteel.com
Brush Wellman	http://www.brushwellman.com
Carpenter Technology Corporation	http://www.cartech.com
CMW Inc.	http://www.cmwinc.com
Crucible Materials Corporation	http://www.crucibleservice.com
CSM Industries, Inc.	http://www.csm-moly.com
Degussa Corporation	http://www.degussa.com

Deloro Stellite Company	http://www.stellite.com
Dura-Bar, Div. Wells Manufacturing Company	http://www.dura-bar.com
Eastern Alloys, Inc.	http://www.eazall.com
Eutectic Corporation	http://www.eutectic-usa.com
G.O. Carlson, Inc.	http://www.gocarlson.com
Goodfellow Corporation	http://www.goodfellow.com
Handy & Harman	http://www.handyharman.com
Hardface Alloys, Inc.	http://www.hardfacealloys.com
Haynes International Inc.	http://www.haynesintl.com
High Performance Alloys, Inc.	http://www.hpalloy.com
Hoeganaes North America Inc.	http://www.hoganas.com
Inco Limited	http://www.incoltd.com
Inco Special Powder Products	http://www.incospp.com
Indium Corporation of America	http://www.indium.com
Ivoclar Vivadent AG (Dental Alloys)	http://www.ivoclarvivadent.com
Kennametal	http://www.kennametal.com
LTV Steel	http://www.ltvsteel.com
Lucas-Milhaupt, Inc.	http://www.preciousmetalsgroup.com
MacSteel	http://www.macsteel.com
Mi-Tech Metals Inc.	http://www.mi-techmetals.com
Oremet-Wah Chang/Allegheny Teledyne	http://www.twc.com
Pechiney (Metals and Alloys)	http://www.pechiney.com
Protective Metal Alloys Inc.	http://www.pmawear.com
Pyron Corporation	http://www.pyroncorp.com
QuesTek Innovations LLC (Steels and Alloys)	http://www.questek.com
Reynolds Metals Company	http://www.rmc.com
Rhenium Alloys, Inc.	http://www.rhenium.com
RMI Titanium Company	http://www.rmititanium.com
Rolled Alloys	http://www.rolledalloys.com
Sandvik Steel Company	http://www.sandvik.com
Specialloy Inc.	http://www.copperalloys.com
Special Metals Corporation	http://www.specialmetals.com
Slater Steels	http://www.slaterstainless.com
Sollac (Steels)	http://www.sollac.fr
Superior Graphite Company	http://www.graphitesgc.com
ThyssenKrupp VDM GmbH, Germany	http://www.thyssenkruppvdm.com
Timken Steel	http://www.timken.com
Timminco Metals	http://www.timminco.com
Titanium Industries, Inc.	http://www.titanium.com
Titanium Metals Corporation	http://www.timet.com
Trans World Alloys	http://www.twalloys.com
Tri-Cast, Inc.	http://www.tri-cast.com
Uddeholm AB	http://www.uddeholmtooling.com
Ulbrich Stainless Steels & Special Metals	http://www.ulbrich.com
Vacuumschmelze GmbH, Germany	http://www.vacuumschmelze.de
Valimet Inc.	http://www.valimet.com
VAW of America Inc.	http://www.vaw.com
Wall Colmonoy Corporation	http://www.wallcolmonoy.com
World Alloys & Refining, Inc.	http://www.worldalloys.com

2. Plastics and Elastomers

Accurate Plastics, Inc.	http://www.acculam.com
Atofina Chemicals Inc.	http://www.atofinachemicals.com
BASF Corporation	http://www.basf.com
Bayer Corporation	http://www.bayer.com

BP Amoco Polymers	http://www.bpamocochemicals.com
Dow Chemical Company	http://www.dow.com
DSM Engineering Plastic Products	http://www.dsmepp.com
Ebbtide Polymers Corporation	http://www.ebbtidepolymers.com
EGC Corporation	http://www.egccorp.com
E.I. DuPont de Nemours & Co.	http://www.dupont.com
Endura Plastics Inc.	http://www.endura.com
EniChem, Italy	http://www.enichemnet.com
Epoxies Etc.	http://www.epoxies.com
Firestone Polymers	http://www.firestone.com
GE Plastics	http://www.geplastics.com
Goodfellow Corporation	http://www.goodfellow.com
Huntsman Chemical Corporation	http://www.huntsman.com
Lati SpA	http://www.lati.com
LNP Engineering Plastics	http://www.lnp.com
Mitsui Chemicals America, Inc.	http://www.mitsuichemicals.com
Plaskolite, Inc.	http://www.plaskolite.com
Polymer Technology Group, Inc.	http://www.polymertech.com
PolyOne Corporation	http://www.polyone.com
Quadrant Engineering Plastic Products	http://www.quadrantepp.com
Reichhold Chemicals Inc.	http://www.reichhold.com
Rohm & Haas Company	http://www.rohmhaas.com
The Goodyear Tire & Rubber Company	http://www.goodyear.com
Wellman Inc.	http://www.wellman.com

3. Ceramics and Glasses

3M Ceramic Textiles and Composites	http://www.mmm.com/ceramics
ART Inc. (Ceramics)	http://www.art-inc.com
BASF Corporation	http://www.basf.com
BNZ Materials, Inc. (Ceramics)	http://www.bnzmaterials.com
Astro-Met Inc. (Ceramics)	http://www.astromet.com
Carbo Ceramics	http://www.carboceramics.com
Carborundum Corporation	http://www.carbobn.com
Ceradyne Inc. (Ceramics)	http://www.ceradyne.com
Goodfellow Corporation	http://www.goodfellow.com
Heany Industries (Ceramics)	http://www.heany.com
Lydall Technical Papers	http://www.lytherm.com
Morgan Advanced Ceramics	http://www.matroc.com
Morton Advanced Materials (Ceramics)	http://www.mortoncvd.com
Norton Pakco Industrial Ceramics	http://www.nortonpakco.com
Owens-Corning, Inc. (Glass)	http://www.owens-corning.com
PPG Industries (Glass and Glass-Ceramics)	http://www.ppg.com
Reade Advanced Materials	http://www.reade.com
Schott AG (Glass and Glass-Ceramics)	http://www.schott.com
Saint-Gobain (Glass and Glass Ceramics)	http://www.saintgobain.com
Thermal Ceramics	http://www.thermalceramics.com
Zircar Products, Inc. (Ceramics and Composites)	http://www.zircar.com

4. Composite Materials

AlliedSignal Inc.	http://www.alliedsignal.com
BASF Corporation	http://www.basf.com
BFD Inc.	http://www.bfd-inc.com
Blagden Chemicals, Ltd.	http://www.cellobond.com
Brush Wellman, Inc.	http://www.brushwellman.com

Grafil, Inc.   http://www.grafil.com
Hexcel Corporation   http://www.hexcel.com
Metal Matrix Cast Components, Inc.   http://www.mmccinc.com
Owens-Corning, Inc.   http://www.owens-corning.com
Poco Graphite   http://www.poco.com
Reade Advanced Materials   http://www.reade.com
Refractory Composites Inc.   http://www.refractorycomposites.com

5. Other Engineering/Engineered Materials

3M Company   http://www.mmm.com
3M-ESPE (Dental Materials)   http://www.mmm.com/dental
A&A Company Inc. (Coatings)   http://www.aacoinc.com
A Brite Company (Coatings)   http://www.abrite.com
Acordis Group (Synthetic Fibers)   http://www.acordis.com
Advanced Materials (CVD)   http://www.cvdmaterials.com
AFG Industries (Low-E Glass Products)   http://www.afg.com
AGS Minéraux SA, France (Minerals)   http://www.ags-mineraux.com
Alchem Corporation   http://www.alchemcorp.com
AlliedSignal (Amorphous Materials)   http://www.alliedsignal.com
Ameripol Synpol Corporation (Chemicals)   http://www.ameripol.com
Ameron International (Coatings and Linings)   http://www.ameroncoatings.com
Arthrotek–Biomet Company (Biomaterials)   http://www.arthrotek.com
Asahi Kasei Fibers Corporation   http://www.asahi-kasei.co.jp
Atofina Chemicals Inc. (Coatings)   http://www.atofinachemicals.com
Atotech USA Inc. (Coatings)   http://www.atotech.com
Aurident, Inc. (Dental Alloys)   http://www.aurident.com
BASF Corporation (Adhesives, Coatings and Fibers) http://www.basf.com
Bayer Corporation (Adhesives and Fibers)   http://www.bayer.com
Benjamin Moore & Company (Paints)   http://www.benjaminmoore.com
Biocomposites Limited, UK   http://www.biocomposites.com
Biomet Merck (Biomaterials)   http://www.biometmerck.com
Bisco Dental Products   http://www.bisco.com
Brycoat Inc. (Coatings)   http://www.brycoat.com
Carboline, StonCor Corp. (Coatings)   http://www.carboline.com
Ceilcote, Div. SGL ACOTEC (Coatings and Linings) http://www.ceilcoteccl.com
Coltene–Whaledent (Dental Materials)   http://www.coltenewhaledent.com
Cotronics Corporation (Adhesives)   http://www.cotronics.com
Courtaulds plc, UK (Plastics and Fibers)   http://www.courtaulds.com
Covion Organic Semiconductors GmbH, Germany   http://www.covion.com
Crystallume (Coatings)   http://www.crystallume.com
Degussa-Ney Dental   http://www.neydental.com
Den-Mat Corporation (Dental Materials)   http://www.denmat.com
Dentsply International (Dental Materials)   http://www.dentsply.com
E.I. DuPont de Nemours & Co. (Synthetic Fibers) http://www.dupont.com
EniChem, Italy (Synthetic Fibers)   http://www.enichemnet.com
Enthone-OMI Inc. (Coatings)   http://www.enthone-omi.com
EpiWorks, Inc. (Electronic Materials)   http://www.epiworks.com
ETEX Corporation (Biomaterials)   http://www.etexcorp.com
GC America, Inc. (Dental Materials)   http://www.gcamerica.com
General Magnaplate Corporation (Coatings)   http://www.magnaplate.com
General Polymers Corporation (Coatings and Linings) http://www.generalpolymers.com
GluGuru Adhesives Tech Center   http://www.gluguru.com
Hayden Corporation (Coatings)   http://www.haydencorp.com
Heatbath Corporation (Coatings)   http://www.heatbath.com

Honeywell Performance Fibers	http://www.honeywell.com
Huber Engineered Materials (Minerals)	http://www.huberemd.com
IBM Corporation (Electronics and Optoelectronics)	http://www.ibm.com
Implex Corporation (Biomaterials)	http://www.implex.com
Kennametal Engineered Products Group	http://www.kennametalepg.com
Kerr Dental	http://www.kerrdental.com
Heraeus-Kulzer, Inc. (Dental Materials)	http://www.kulzer.com
Hoechst Celanese/Ticona (Fibers)	http://www.ticona.com
Integran Technologies Corporation (Nanomaterials)	http://www.integran.com
Ivoclar North America (Dental Materials)	http://www.ivoclarna.com
J.F. Jelenko & Co. (Dental Alloys)	http://www.jelenko.com
Liquid Crystals Group, Uni Hamburg (Germany)	http://www.liqcryst.chemie.uni-hamburg.de
Luster-On Products Inc. (Coatings)	http://www.luster-on.com
MacDermid Inc. (Coatings)	http://www.macdermid.com
McGean (Coatings)	http://www.mcgean.com
Nanodyne Corporation (Nanomaterials)	http://www.nanodyne.com
NanoMaterials Inc.	http://www.nanomaterials.com
Nanophase Technologies Corporation (Nanopowders)	http://www.nanophase.com
National Nonwovens	http://www.nationalnonwovens.com
Nimet Industries, Inc. (Coatings)	http://www.nimet.com
Norton Diamond Film	http://www.nortondiamondfilm.com
Oakite Products Inc. (Coatings)	http://www.oakite.com
OsteoBiologics, Inc. (Biomaterials)	http://www.obi.com
Owens Corning Fiberglass	http://www.owenscorning.com
Pavco Inc. (Coatings)	http://www.pavco.com
Pelseal Technologies LLC (Adhesives)	http://www.pelseal.com
Plasti-Fab/PFB Corporation (Building Materials)	http://www.plastifab.com
PolySpec, Inc. (Coatings and Linings)	http://www.polyspec.com
PPG High Performance Coatings	http://www.ppghpc.com
Reade Advanced Materials (Minerals)	http://www.reade.com
Resil Chemicals Pvt. Ltd., UK (Fabrics and Fibers)	http://www.resil.com
Rhino Linings USA, Inc. (Coatings and Linings)	http://www.rhinolinings.com
Roxul Inc., Canada (Building Materials)	http://www.roxul.com
RTG Thermal Spray Coatings	http://www.rtgcoatings.com
Sermatech (Coatings)	http://www.sermatech.com
Shofu Dental Corporation	http://www.shofu.com
Sigma-Aldrich Company (Chemicals)	http://www.sigma-aldrich.com
Sharp Laboratories of Europe Ltd. (Optoelectronics)	http://www.nanomaterials.com
Solutia, Inc. (Synthetic Fibers)	http://www.solutia.com
Sulzer Metco (Coatings)	http://www.sulzermetco.com
Superior Environmental Products Inc. (Coatings)	http://www.novocoat.com
Swicofil AG (Textile Fabrics and Fibers)	http://www.swicofil.com
Taskem Inc. (Coatings)	http://www.taskem.com
Technic, Inc. (Coatings)	http://www.technic.com
The Arnold Engineering Company	http://www.grouparnold.com
Tiodize Company Inc. (Coatings)	http://www.tiodize.com
Toho Carbon Fibers Inc., Japan	http://www.tohocarbonfibers.com
Toray Industries Inc. (Fibers)	http://www.toray.com
Tygavac Advanced Materials (Industrial Fabrics)	http://www.tygavac.co.uk
Ultradent Products, Inc. (Dental Materials)	http://www.ultradent.com
Ultramet, Inc. (Coatings)	http://www.ultramet.com
USBiomaterials Corporation	http://www.usbiomat.com
Wellman Inc. (Synthetic Fibers)	http://www.wellman.com
Weyerhaeuser Inc. (Building Products)	http://www.weyerhaeuser.com
Whip Mix Corporation (Dental Materials)	http://www.whipmix.com
Zinex Corporation (Coatings)	http://www.zinex.com

V.   Frequently Used Abbreviations and Acronyms

AA	Standard aluminum and aluminium alloy numbering system of the Aluminum Association.
ACI	Alloy casting designation system of the US Alloy Casting Institute.
ADA	Dental casting alloy classification system of the American Dental Association.
AISI	Ferrous metal and alloy classification system of the American Iron and Steel Institute.
AISI-SAE	Joint tool steel classification system of the American Iron and Steel Institute and the Society of Automotive Engineers International.
AMS	Aerospace Material Specifications.
ANSI	Materials-related standards of the American National Standards Institute.
ASTM	Standards of the American Society for Testing and Materials.
AWS	Standards and specifications of the American Welding Society.
EMI/RFI	Electromagnetic Interference/Radio-Frequency Interference.
EPA	US Environmental Protection Agency.
IACS	Electrical conductivity of a material as a percentage of the International Annealed Copper Standard (%IACS) which is equal to 1724.1 divided by the material's electrical resistivity in nano-ohm meters.
ISO	International Organization for Standardization.
MIL-SPEC	Military Specifications of the US Department of Defense.
MIL-STD	Military Standards of the US Department of Defense.
SAE	Standards and specifications of the US Society of Automotive Engineers International.
UNS	Unified Numbering System for metals and alloys.